BFO750 **CAS:7440-**
BERYLLIUM
DOT: UN 1966/UN 1567
mf: Be mw: 9.01

PROP: A silvery-white, relatively soft, lustrous metal, ductile at red heat. Unreactive to H_2O and air; dissolves vigorously in dil acids. Be Reacts with aq alkalis or H_2. Mp: 1287–1292°, bp: 2970°, d: 1.85.

SYNS: BERYLLIUM-9 □ BERYLLIUM COMPOUNDS, n.o.s. (UN 1566) (DOT) □ BERYLLIUM, powder (UN 1567) (DOT) □ GLUCINIUM □ GLUCINUM □ RCRA WASTE NUMBER P015

TOXICITY DATA WITH REFERENCE
dnd-esc 30 μmol/L MUREAV 89,95,81
dni-nml-ivn 30 μmol/kg PHMCAA 12,298,70
dnd-hmn:hla 30 μmol/L MUREAV 89,95,81
dnd-mus:ast 30 μmol/L MUREAV 89,95,81
itr-rat TDLo:13 mg/kg:NEO ENVRAL 21,63,80
ivn-rbt TDLo:20 mg/kg:ETA LANCAO 1,463,50
ihl-hmn TCLo:300 mg/m³:PUL AEHLAU 9,473,64
ivn-rat LD50:496 μg/kg LAINAW 15,176,66

CONSENSUS REPORTS: NTP 7th Annual Report On Carcinogens. IARC Cancer Review: Group 1 IMEMDT 58,41,93; Human Sufficient Evidence IMEMDT 58,41,93; Animal Sufficient Evidence IMEMDT 1,17,72; Animal Sufficient Evidence IMEMDT 23,143,80; Animal Sufficient Evidence IMEMDT 58,41,93. Beryllium and its compounds are on the Community Right-To-Know List. Reported in EPA TSCA Inventory.

OSHA PEL: TWA 0.002 mg(Be)/m³; STEL 0.005 mg(Be)/m³/30M; CL 0.025 mg(Be)/m³
ACGIH TLV: TWA 0.002 mg/m³, Suspected Human Carcinogen.
DFG TRK: Animal Carcinogen, Suspected Human Carcinogen. Grinding of beryllium metal and alloys: 0.005 mg/m³ calculated as beryllium in that portion of dust that can possibly be inhaled; other beryllium compounds: 0.002 mg/m³ calculated as beryllium in that portion of dust that can possibly be inhaled
NIOSH REL: CL not to exceed 0.0005 mg(Be)/m³
DOT CLASSIFICATION: 6.1; *Label:* Poison (UN 1566); DOT Class: 6.1; *Label:* Poison, Flammable Solid (UN 1567)

SAFETY PROFILE: Confirmed carcinogen with experimental carcinogenic, neoplastigenic, and tumorigenic data. A deadly poison by intravenous route. Human systemic effects by inhalation: lung fibrosis, dyspnea, and weight loss. Human mutation data reported. See also BERYLLIUM COMPOUNDS. A moderate fire hazard in the form of dust or powder, or when exposed to flame or by spontaneous chemical reaction. Slight explosion hazard in the form of powder or dust. Incompatible with halocarbons. Reacts incandescently with fluorine or chlorine. Mixtures of the powder with CCl_4 or trichloroethylene will flash or spark on impact. When heated to decomposition in air it emits very toxic fumes of BeO. Reacts with Li and P.

For occupational chemical analysis use NIOSH: Beryllium, 7102; Elements, 7300.

HR: – Hazard Rating indicating relative hazard for toxicity, fire, and reactivity with 3 denoting the worst hazard level.
See Introduction: paragraph 3, p. xiii

CAS: – The American Chemical Society's Chemical Abstracts Service number. A complete CAS number cross-index is located in Section 2.
See Introduction: paragraph 4, p. xiii.

Cited References: – A code which represents the cited reference for the toxicity data. Complete bibliographic citations in order are listed in Section 4.
See Introduction: paragraph 16, p. xxiii

Consensus Reports: – Supply additional information to enable the reader to make knowledgeable evaluations of potential chemical hazards.
See Introduction: paragraph 17, p. xxiii.

Safety Profiles: – These are text summaries of the toxicity, flammability, reactivity, incompatibility, and other dangerous properties of the entry.
See Introduction: paragraph 19, p.xxvi.

Analytical Methods: – References to OSHA and NIOSH occupational analytical methods.
See Introduction: paragraph 19, p.xxvii.

DISCLAIMER

Extreme care has been taken in preparation of this work. However, neither the publisher nor the authors shall be held responsible or liable for any damages resulting in connection with or arising from the use of any of the information in this book.

I(T)P® A division of International Thomson Publishing, Inc.
The ITP logo is a trademark under license

Printed in the United States of America

For more information, contact:

Van Nostrand Reinhold
115 Fifth Avenue
New York, NY 10003

Chapman & Hall GmbH
Pappelallee 3
69469 Weinheim
Germany

Chapman & Hall
2-6 Boundary Row
London SEI 8HN
United Kingdom

International Thomson Publishing Asia
221 Henderson Road #05-10
Henderson Building
Singapore 0315

Thomas Nelson Australia
102 Dodds Street
South Melbourne, 3205
Victoria, Australia

International Thomson Publishing Japan
Hirakawacho Kyowa Building, 3F
2-2-1 Hirakawacho
Chiyoda-ku, 102 Tokyo
Japan

Nelson Canada
1120 Birchmount Road
Scarborough, Ontario
Canada M1K 5G4

International Thomson Editores
Campos Eliseos 385, Piso 7
Col. Polanco
11560 Mexico D.F. Mexico

1 2 3 4 5 6 7 8 9 10 TPP 01 00 99 98 97 96

Library of Congress Cataloging-in-Publication Data

Lewis, Richard J., Sr.
Sax's dangerous properties of industrial materials / Richard J. Lewis, Sr. — 9th ed.
p. cm.
Includes bibliographical references and index.
ISBN SET 0-442-02025-2
ISBN VOLUME I 0-442-02255-7
ISBN VOLUME II 0-442-02256-5
ISBN VOLUME III 0-442-02257-3
1. Hazardous substances—Handbooks, manuals, etc. I. Sax, N. Irving (Newton Irving). Dangerous properties of industrial materials. 9th ed. II. Title.
T55.3.H3L53 1995
604.7—dc20 95-46514
CIP

Sax's Dangerous Properties of Industrial Materials

Ninth Edition

RICHARD J. LEWIS, SR.

VAN NOSTRAND REINHOLD

I(T)P™ A Division of International Thomson Publishing Inc.

New York • Albany • Bonn • Boston • Detroit • London • Madrid • Melbourne
Mexico City • Paris • San Francisco • Singapore • Tokyo • Toronto

Sax's Dangerous Properties of Industrial Materials

Ninth Edition

Dedicated to Grace Ross Lewis
for her invaluable participation
in every aspect of this revision.

Acknowledgments

I extend thanks to Marianne J. Russell, Nancy Olsen, and Bob Esposito for their encouragement. My thanks and best wishes to Renee Guilmette and especially Geraldine Albert for their expert professional advice and assistance in converting the manuscript to this volume.

Contents

Volume I:

Preface *ix*

Key to Abbreviations *xi*

Introduction *xiii*

Section 1: DOT Guide Number Cross-Index *1*

Section 2: CAS Number Cross-Index *5*

Section 3: Synonym Cross-Index *55*

Section 4: References *789*

Volume II: General Chemicals: Entries A—G *1-1758*

Volume III: General Chemicals: Entries H—Z *1759-3435*

Preface

This ninth edition of *Dangerous Properties of Industrial Materials,* a three-volume set, represents a major revision and updating of the eighth edition. The objective of the book, however, remains the same: to promote safety by providing the most up-to-date hazard information available. The growth in the availability of toxicological and hazard control reports continues unabated. This book cannot contain all the published data and continue to provide the accessibility for which it is known. To continue to provide complete hazard assessments for the maximum number of entries, data for each entry has been selectively reduced. In particular, carcinogenic and reproductive data lines above those required to establish the hazard of the entry have been excluded. Complete data for these entries are available in the books *Carcinogenically Active Chemicals* and *Reproductively Active Chemicals,* both available from the publisher, and in the CD-ROM version of this work.

Over two-thirds of the entries have been revised for this edition, and 2,000 new entries have been added. Some less useful entries will appear only in the CD-ROM editions of *Dangerous Properties of Industrial Materials.* Most of the new entries were selected because they are on the EPA TSCA Inventory. These are reported to be used in commerce in the United States. Emphasis was placed on adding and updating physical properties, updating all DOT Classifications, and the addition of references to both OSHA and NIOSH occupational analytical methods.

Numerous synonyms have been added to assist in locating the many materials that are known under a variety of systematic and common names. The synonym cross-index contains the entry name as well as each synonym. This index should be consulted first to locate a material by name. Synonyms are given in English as well as other major languages such as French, German, Dutch, Polish, Japanese, and Italian.

Many additional physical and chemical properties have been added. Whenever available, physical descriptions, formulas, molecular weights, melting points, boiling points, explosion limits, flash points, densities, autoignition temperatures, and the like have been supplied.

A court order has vacated the OSHA Air Standards set in 1989 and contained in 29CFR 1910.1000. OSHA has decided to enforce only pre-1989 air standards. I have elected to include both the Transitional Limits scheduled to be effective on December 31, 1992, and the Final Rule limits, that were scheduled to be effective on September 1, 1989. These represent the current best judgment as to appropriate workplace air levels. While they may not be enforceable by OSHA, they are better guides than the OSHA Air Standards adopted in 1969.

The following classes of data are new or have been updated for all entries for which they apply:

1. For the first time, OSHA and NIOSH occupational analytical methods are referenced by method name or number for over 700 entries.
2. Also new to this addition are over 100 biological entities or their toxic products.
3. ACGIH TLVs and BEIs reflect the latest recommendations and now include "intended changes."
4. German MAK and BAT reflect the latest recommendations.
5. NTP 7th Annual Report On Carcinogens entries are identified.
6. DOT classifications were updated reflecting the HM-181 rulemaking.
7. CAS numbers are provided for additional entries.

Each entry concludes with a safety profile, a textual summary of the hazards presented by the entry. The discussion of human exposures includes target organs and specific effects reported. Carcinogenic and reproductive assessments have been completely revised for this edition.

Fire and explosion hazards are briefly summarized in terms of conditions of flammable or reactive hazard. Where feasible, fire-fighting materials and methods are discussed. Materials that are known to be incompatible with an entry are listed here.

Also included in the safety profile are comments on disaster hazards that serve to alert users of materials to the dangers that may be encountered on entering storage premises during a fire or other emergency. Although the presence of water, steam, acid fumes, or powerful vibrations can cause the decomposition of many materials into dangerous compounds, of particular concern are high temperatures (such as those resulting from a fire) because these can cause many otherwise mild chemicals to emit highly toxic gases or vapors such as NO_x, SO_x, acids, and so forth, or evolve vapors of antimony, arsenic, mercury, and the like.

The book, which consists of three volumes, is divided as follows:

The first volume contains a CAS number cross-index, a synonym cross-index, and the complete citations for bibliographic references given in the data section.

Section 1 contains the DOT Guide number cross-index for the listed materials.

Section 2 contains the CAS number cross-index for CAS numbers for the listed materials.

Section 3 contains the prime name and synonym cross-index for the listed materials.

Section 4 contains the complete bibliographic references.

The main section of the book is contained in Volumes II and III. It lists and describes approximately 20,000 materials in alphabetical order by entry name.

Please refer to the Introduction for an explanation of the sources of data and codes used.

Every effort has been made to include the most current and complete information. The author welcomes comments or corrections to the data presented.

Richard J. Lewis, Sr.

Key to Abbreviations

abs—absolute
ACGIH—American Conference of Governmental Industrial Hygienists
af—atomic formula
alc—alcohol
alk—alkaline
amorph—amorphous
anhyd—anhydrous
approx—approximately
aq—aqueous
atm—atmosphere
autoign—autoignition
aw—atomic weight
BEI—ACGIH Biological Exposure Indexes
bp—boiling point
b range—boiling range
CAS—Chemical Abstracts Service
cc—cubic centimeter
CC—closed cup
CL—ceiling concentration
COC—Cleveland open cup
compd(s)—compound(s)
conc—concentration, concentrated
contg—containing
cryst—crystal(s), crystalline
d—density
D—day(s)
decomp—decomposition
deliq—deliquescent
dil—dilute
DOT—U.S. Department of Transportation
EPA—U.S. Environmental Protection Agency
eth—ether
(F)—Fahrenheit
FCC—Food Chemical Codex
FDA—U.S. Food and Drug Administration
flam—flammable
flash p—flash point
fp—freezing point
g—gram
glac—glacial
gran—granular, granules
H—hour(s)
HR:—hazard rating
htd—heated
htg—heating
hygr—hygroscopic
IARC—International Agency for Research on Cancer
immisc—immiscible
incomp—incompatible
insol—insoluble
IU—International Unit
kg—kilogram (one thousand grams)
L—liter
lel—lower explosive limit
liq—liquid
M—minute(s)
m^3—cubic meter
mf—molecular formula
mg—milligram
misc—miscible
mL—milliliter
mm—millimeter
mod—moderately
mp—melting point
mppcf—million particles per cubic foot
mw—molecular weight
μ—micro
μg—microgram
ng—nanogram
NIOSH—National Institute for Occupational Safety and Health
nonflam—nonflammable
NTP—National Toxicology Program
OBS—obsolete
OC—open cup
org—organic
ORM—other regulated material (DOT)
OSHA—Occupational Safety and Health Administration
Pa—Pascals
PEL—permissible exposure level
pet, petr—petroleum
pg—picogram (one trillionth of a gram)
Pk—peak concentration
pmole—picomole
powd—powder
ppb—parts per billion (v/v)
pph—parts per hundred (v/v)(percent)
ppm—parts per million (v/v)
ppt—parts per trillion (v/v)
prac—practically
prep—preparation
PROP—properties
refr—refractive
rhomb—rhombic
S, sec—second(s)
sl, slt—slight
sltly—slightly

sol—soluble
soln—solution
solv(s)—solvent(s)
spar—sparingly
spont—spontaneous(ly)
STEL—short-term exposure limit
subl—sublimes
TCC—Tag closed cup
tech—technical
temp—temperature
TLV—Threshold Limit Value
TOC—Tag open cup
TWA—time weighted average
uel—upper explosive limit
unk—unknown, unreported
ULC, ulc—Underwriters Laboratory Classification
USDA—U.S. Department of Agriculture
vac—vacuum
vap—vapor
vap d—vapor density
vap press—vapor pressure
visc—viscosity
vol—volume
W—week(s)
Y—year(s)
%—percent(age)
>—greater than
<—less than
<=—less than or equal to
=>—greater than or equal to
°—degrees of temperature in Celsius (centigrade)
°F—temperature in Fahrenheit

Sax's Dangerous Properties of Industrial Materials

Ninth Edition

Volume II
A-G

AAC250 CAS:8021-27-0 HR: 1
ABIES ALBA OIL

PROP: Colorless to pale-yellow oil from the steam distillation of the crushed cones of *Abies Alba Mill* (FCTXAV 12,807,74).

SYNS: OIL of ABIES ALBA □ OIL of FUR □ OIL of SILVER FIR □ OIL of SILVER PINE □ SILVER FIR NEEDLE OIL □ SILVER FIR OIL □ SILVER PINE OIL □ TEMPLIN OIL

TOXICITY DATA WITH REFERENCE
skn-rbt 500 mg/24H MOD FCTXAV 12,807,74
orl-rat LD50:>5 g/kg FCTXAV 12,809,74
skn-rbt LD50:>5 g/kg FCTXAV 12,809,74

CONSENSUS REPORTS: Reported in EPA TSCA Inventory.

SAFETY PROFILE: Low toxicity by ingestion or skin contact. A skin irritant. When heated to decomposition it emits acrid smoke and irritating fumes.

AAC500 CAS:514-10-3 HR: 3
ABIETIC ACID
mf: $C_{20}H_{30}O_2$ mw: 302.50

PROP: Yellow powder. Mp: 172–175°.

SYNS: 13-ISOPROPYLPODOCARPA-7,13-DIEN-15-OIC ACID □ SILVIC ACID

TOXICITY DATA WITH REFERENCE
ivn-mus LD50:180 mg/kg CSLNX* NX#02819

CONSENSUS REPORTS: Reported in EPA TSCA Inventory.

SAFETY PROFILE: Poison by intravenous route. Combustible. Slight explosion hazard as dust. When heated to decomposition it emits acrid smoke and irritating fumes.

AAC875 CAS:55077-30-0 HR: 3
ABOVIS
mf: $C_{10}H_{20}NO_4\cdot 1/2C_{10}H_6O_6S_2$ mw: 361.42

PROP: Crystals. Mp: 189–191°.

SYNS: (2-ACETYLLACTOYLOXYETHYL)TRIMETHYLAMMONIUM HEMI-1,5-NAPHTHALENEDISULFONATE □ (2-ACETYLLACTOYLOXYETHYL)TRIMETHYLAMMONIUM 1,5-NAPHTHALENEDISULFONATE □ 2-(2-(ACETYLOXY)-1-OXOPROPOXY)-N,N,N-TRIMETHYLETHANAMINIUM 1,5-NAPHTHALENEDISULFONATE (2:1) □ ACLATONIUM NAPADISILATE □ CHOLINE 1,5-NAPHTHALENEDISULFONATE (2:1), DILACTATE, DIACETATE □ TM 723

TOXICITY DATA WITH REFERENCE
orl-mus TDLo:600 g/kg (9W male/2W pre-6D preg):TER OYYAA2 18,923,79
orl-rat TDLo:60 g/kg (30D male):REP OYYAA2 18,749,79
orl-rat LDLo:15 g/kg IYKEDH 12,1204,81
scu-rat LD50:986 mg/kg OYYAA2 13,497,77
ivn-rat LD50:46 mg/kg OYYAA2 13,497,77
orl-mus LD50:15 g/kg USXXAM #3903137
scu-mus LD50:826 mg/kg OYYAA2 13,497,77
ivn-mus LD50:41,900 μg/kg IYKEDH 12,30420,81

SAFETY PROFILE: Poison by intravenous route. Moderately toxic by subcutaneous route. An experimental teratogen. Other experimental reproductive effects. When heated to decomposition it emits toxic fumes of NO_x and SO_x. A cholinergic agent. See also SULFONATES.

AAD000 CAS:1393-62-0 HR: 3
ABRIN

PROP: Yellowish-white powder. Sol in solns of sodium chloride, usually with turbidity. Incubation at 60° for 30 M fails to remove toxic effect, but at 80° most of the toxicity is lost.

SYNS: ABRINS □ AGGLUTININ □ CRAB'S EYES □ INDIAN LICORICE SEED □ JUMBLE BEAD □ PRAYER BEAD □ TOXALBUMIN

TOXICITY DATA WITH REFERENCE
dni-mus-ast 50 μg/kg TOXIA6 11,379,73
orl-hmn LDLo:7 μg/kg MEIEDD 10,1,83
orl-rat LDLo:300 mg/kg AMIHAB 12,468,55
orl-mus LD50:6638 mg/kg ARZNAD 21,888,71
ipr-mus LD50:20 μg/kg 85GDA2 8(1),107,82
ivn-mus LD50:20 μg/kg MEIEDD 10,1,83
orl-rbt LDLo:21 mg/kg AMIHAB 12,468,55
orl-gpg LD50:299 mg/kg ARZNAD 21,888,71

SAFETY PROFILE: A deadly poison to humans by ingestion. Poison by ingestion, intravenous, and intraperitoneal routes. Mutation data reported. When heated to decomposition it emits acrid fumes and irritating smoke. See also RICIN. Note: Do not confuse with abrine.

AAD100 HR: 3
ABRUS PRECATORIUS L., seed kernel extract

TOXICITY DATA WITH REFERENCE
orl-rat TDLo:6 g/kg (male 60D pre):REP AEFTAA 18,217,87
ipr-mus LD50:550 ng/kg CTYAD8 18,196,87
scu-mus LD50:200 μg/kg TOXIA6 6,211,69
scu-gpg LDLo:430 μg/kg TOXIA6 7,211,69

SAFETY PROFILE: Poison by intraperitoneal and subcutaneous routes. Experimental reproductive effects. When heated to decomposition it emits acrid smoke and irritating fumes.

AAD250 **CAS:93164-88-6** **HR: 2**
ACACIA (EXTRACT)

PROP: Indian plant belonging to the family *Leguminosae* (IJEBA6 7,250,69).

SYN: BABUL STEM BARK EXTRACT

TOXICITY DATA WITH REFERENCE
ipr-mus LD50:500 mg/kg IJEBA6 7,250,69

SAFETY PROFILE: Moderately toxic by intraperitoneal route. When heated to decomposition it emits acrid smoke and irritating fumes.

AAD500 **HR: 2**
ACACIA FARNESIANA (Linn.) Willd., extract excluding roots

PROP: Indian plant belonging to the family *Mimosaceae* (IJEBA6 22,487,84).

TOXICITY DATA WITH REFERENCE
orl-rat TDLo:150 mg/kg (female 12-14D post):REP IJEBA6 22,487,84
ipr-mus LD50:562 mg/kg IJEBA6 22,487,84

SAFETY PROFILE: Moderately toxic by intraperitoneal route. Experimental reproductive effects. When heated to decomposition it emits acrid smoke and irritating fumes.

AAD750 **HR: 2**
ACACIA VILLOSA

PROP: Aqueous extract from the root of the plant (JNCIAM 52,1579,74).

SYN: WATAPANA SHIMARON

TOXICITY DATA WITH REFERENCE
scu-rat TDLo:198 mg/kg/22W-I:NEO JNCIAM 52,1579,74
imp-ham TDLo:1660 mg/kg:CAR JNCIAM 53,1259,74
scu-rat TD:300 g/kg/60W-I:NEO,REP JNCIAM 52,445,74

SAFETY PROFILE: Experimental reproductive effects. Questionable carcinogen with experimental neoplastigenic and carcinogenic data. When heated to decomposition it emits smoke and acrid fumes.

AAD875 **CAS:5892-41-1** **HR: 3**
ACAMYLOPHENINE DIHYDROCHLORIDE
mf: $C_{19}H_{32}N_2O_2 \cdot 2ClH$ mw: 393.45

SYNS: AVACAN ☐ CAMYLOFINE DIHYDROCHLORIDE ☐ CAMYLOFINE HYDROCHLORIDE ☐ CAMYLOFIN HYDROCHLORIDE

TOXICITY DATA WITH REFERENCE
orl-mus LD50:760 mg/kg MEIEDD 10,239,83
scu-mus LD50:1350 mg/kg MEIEDD 10,239,83
ivn-mus LD50:49,200 μg/kg MEIEDD 10,239,83

SAFETY PROFILE: Poison by intravenous route. Moderately toxic by ingestion and other routes. When heated to decomposition it emits toxic fumes of NO_x and HCl. See also ESTERS.

AAE000 **CAS:3697-25-4** **HR: 2**
4,10-ACE-1,2-BENZANTHRACENE
mf: $C_{20}H_{14}$ mw: 254.34

SYNS: 1,2-DIHYDROBENZ(e)ACEANTHRYLENE ☐ 5,6-DIHYDROBENZENE(e)ACEANTHRYLENE

TOXICITY DATA WITH REFERENCE
scu-mus TDLo:4 mg/kg:ETA AJCAA7 33,499,38

SAFETY PROFILE: Questionable carcinogen with experimental tumorigenic data. When heated to decomposition it emits acrid smoke and fumes.

AAE100 **CAS:37517-30-9** **HR: 3**
ACEBUTOLOL
mf: $C_{18}H_{28}N_2O_4$ mw: 336.48

PROP: Crystals. Mp: 119–123°.

SYNS: (±)-ACEBUTOLOL ☐ dl-ACEBUTOLOL ☐ 1-(2-ACETYL-4-n-BUTYRAMIDOPHENOXY)-2-HYDROXY-3-ISOPROPYLAMINOPROPANE ☐ 3'-ACETYL-4'-(2-HYDROXY-3-(ISOPROPYLAMINO)PROPOXY)BUTYRANILIDE ☐ (±)-N-(3-ACETYL-4-(2-HYDROXY-3-((1-METHYLETHYL)AMINO)PROPOXY)PHENYL)BUTANAMIDE ☐ 5'-BUTYRAMIDO-2'-(2-HYDROXY-3-ISOPROPYLAMINOPROPOXY)ACETOPHENONE ☐ PRENT

TOXICITY DATA WITH REFERENCE
unr-wmn TDLo:1080 mg/kg (1-39W preg):REP BMJOAE 283,1077,81
orl-wmn TDLo:120 mg/kg (34-39W preg):TER DPTHDL 4(Suppl 1),109,82
orl-wmn TDLo:152 mg/kg:CVS,BPR JTCTDW 20,69,83
ivn-dog LD50:4 mg/kg MASODV 16,13,80

SAFETY PROFILE: Moderately toxic by intravenous route. Human systemic effects by ingestion: developmental abnormalities of the cardiovascular and respiratory systems; effects on newborn in biochemical and metabolic abnormalities and reduced growth statistics. A human teratogen. When heated to decomposition it emits toxic fumes of NO_x. A beta-adrenergic blocker.

AAE125 **CAS:34381-68-5** **HR: 3**
ACEBUTOLOL HYDROCHLORIDE
mf: $C_{18}H_{28}N_2O_4 \cdot ClH$ mw: 372.94

SYNS: ACETOBUTOLOL HYDROCHLORIDE ☐ dl-1-(2-ACETYL-4-BUTYRAMIDOPHENOXY)-2-HYDROXY-3-ISOPROPYLAMINOPROPANE HYDROCHLORIDE ☐ 3'-ACETYL-4'-(2-HYDROXY-3-(ISOPROPYLAMINO)PROPOXY)BUTYRANILIDE HYDROCHLORIDE ☐ M&B 17,803A ☐ SECTRAL

TOXICITY DATA WITH REFERENCE
orl-rat TDLo:1100 mg/kg (7-17D preg):REP OYYAA2 15,885,78
orl-rat TDLo:1100 mg/kg (7-17D preg):TER OYYAA2 15,885,78
orl-rat LD50:6620 mg/kg OYYAA2 20,883,80
ipr-rat LD50:222 mg/kg OYYAA2 15,837,78
scu-rat LD50:1310 mg/kg OYYAA2 15,837,78
ivn-rat LD50:103 mg/kg OYYAA2 15,837,78
orl-mus LD50:4050 mg/kg NIIRDN 6,19,82
ipr-mus LD50:185 mg/kg OYYAA2 15,837,78
scu-mus LD50:291 mg/kg OYYAA2 15,837,78
ivn-mus LD50:53 mg/kg NIIRDN 6,19,82

orl-rbt LD50:296 mg/kg OYYAA2 15,837,78
ivn-rbt LD50:41 mg/kg OYYAA2 16,837,78

SAFETY PROFILE: Poison by ingestion, subcutaneous, intravenous, and intraperitoneal routes. An experimental teratogen. Other experimental reproductive effects. When heated to decomposition it emits toxic fumes of NO_x and HCl.

AAE250 CAS:827-61-2 HR: 3
ACECLIDINE
mf: $C_9H_{15}NO_2$ mw: 169.25

SYNS: 3-ACETOXYQUINUCLIDINE GLAUCOSTAT □ 3-QUINUCLIDINOL ACETATE

TOXICITY DATA WITH REFERENCE
scu-rat LD50:225 mg/kg ARZNAD 18,320,68
ivn-rat LD50:45 mg/kg ARZNAD 18,320,68
orl-mus LD50:165 mg/kg ARZNAD 18,320,68
scu-mus LD50:102 mg/kg ARZNAD 18,320,68
ivn-mus LD50:36 mg/kg RPTOAN 35(2),55,72

SAFETY PROFILE: Poison by ingestion, subcutaneous, and intravenous routes. When heated to decomposition it emits toxic fumes of NO_x.

AAE500 CAS:3685-84-5 HR: 3
ACEFEN
mf: $C_{12}H_{16}ClNO_3{\cdot}ClH$ mw: 294.20

SYNS: AMIPOLNE □ 235 ANP HYDROCHLORIDE □ BRENAL □ CELLATIVE □ CENTROPHENOXINE □ CERUTIL □ (p-CHLOROPHENOXY)ACETIC ACID 2-(DIMETHYLAMINO)ETHYL ESTER HYDROCHLORIDE □ CLOCETE □ DIMETHYLAMINOETHYL p-CHLOROPHENOXYACETATE HYDROCHLORIDE □ DIMETHYLAMINOETHYL 4-CHLOROPHENOXYACETATE HYDROCHLORIDE □ DIMETHYLAMINOETHYL ESTER of p-CHLOROPHENOXYACETIC ACID HYDROCHLORIDE □ HELFERGIN □ LUCIDRIL □ LUCIDRYL HYDROCHLORIDE □ MARUCOTOL □ MECLOFENOXATE HYDROCHLORIDE □ METHOXYNAL □ NSC 113619 □ PROSEROUT

TOXICITY DATA WITH REFERENCE
orl-rat LD50:865 mg/kg KSKZAN 16(2),59,78
ipr-mus LD50:660 mg/kg NIIRDN 6,814,82
scu-mus LD50:1560 mg/kg NIIRDN 6,814,82
orl-mus LD50:1750 mg/kg CRSBAW 153,1914,59
ipr-mus LD50:845 mg/kg CRSBAW 153,1914,59
ivn-mus LD50:350 mg/kg CRSBAW 153,1914,59
ivn-rbt LDLo:150 mg/kg CRSBAW 153,1914,59

SAFETY PROFILE: Poison by intravenous route. Moderately toxic by ingestion and intraperitoneal routes. When heated to decomposition it emits very toxic fumes of Cl^-, NO_x, and HCl.

AAE625 CAS:53164-05-9 HR: 3
ACEMETACIN
mf: $C_{21}H_{18}ClNO_6$ mw: 415.83

PROP: Very fine, pale-yellow crystals from petr eth. Mp: 150–153°.

SYNS: ACM □ (1-(p-CHLORBENZOYL)-5-METHOXY-2-METHYLINDOL-3-ACETOXY)ESSIGSAEURE (GERMAN) □ 1-(4-CHLOROBENZOYL)-5-METHOXY-2-METHYL-1H-INDOLE-3-ACETIC ACID CARBOXYMETHYL ESTER □ ((1-(4-CHLOROBENZOYL)-5-METHOXY-2-METHYLINDOLE-3-YL)ACETOXY)ACETIC ACID □ K-708 □RANTUDIL □ TV 1322

TOXICITY DATA WITH REFERENCE
orl-rat TDLo:44 mg/kg (7-17D preg):REP OYYAA2 22,765,81
orl-rat TDLo:44 mg/kg (7-17D preg):TER OYYAA2 22,765,81
orl-rat LD50:24 mg/kg ARZNAD 30,1398,80
ipr-rat LD50:23 mg/kg ARZNAD 30,1398,80
scu-rat LD50:28 mg/kg ARZNAD 30,1398,80
ivn-rat LD50:28 mg/kg ARZNAD 30,1398,80
ims-rat LD50:19 mg/kg ARZNAD 30,1398,80
orl-mus LD50:18 mg/kg ARZNAD 30,1398,80
ipr-mus LD50:23 mg/kg ARZNAD 30,1398,80
scu-mus LD50:23 mg/kg ARZNAD 30,1398,80
ivn-mus LD50:34 mg/kg ARZNAD 30,1398,80

SAFETY PROFILE: Poison by ingestion, subcutaneous, intraperitoneal, intravenous, and intramuscular routes. An experimental teratogen. Other experimental reproductive effects. When heated to decomposition it emits toxic fumes of Cl^- and NO_x. An anti-inflammatory agent.

AAE750 HR: 1
ACENAPHTHALENE
mf: $C_{10}H_6(CH_2)_2$ mw: 154.2

PROP: White, elongated crystals. Mp: 95°, bp: 277.5°, d: 1.024 @ 99°/4°, vap press: 10 mm @ 131.2°, vap d: 5.32. Insol in water, sltly sol in hot alc, eth, and chloroform.

SYN: 1,8-ETHYLENE NAPHTHALENE

TOXICITY DATA WITH REFERENCE
mma-sat 490 µmol/L/2H CNREA8 39,4152,79

SAFETY PROFILE: Mutation data reported. A skin and mucous membrane irritant. May cause acute vomiting if swallowed in large quantities. Combustible. When heated to decomposition it emits acrid smoke and irritating fumes.

AAF000 CAS:5779-79-3 HR: 2
ACENAPHTHANTHRACENE
mf: $C_{20}H_{14}$ mw: 254.34

SYNS: BENZ(k)ACEPHENANTHRENE □ 4,5-DIHYDROBENZ(k)ACEPHENANTHRYLENE □ 3:4-DIMETHYLENE-1:2-BENZANTHRACENE

TOXICITY DATA WITH REFERENCE
skn-mus TDLo:960 mg/kg/40W-I:ETA PRLBA4 129,439,40

SAFETY PROFILE: Questionable carcinogen with experimental tumorigenic data. When heated to decomposition it emits acrid smoke and irritating fumes.

AAF250 CAS:4657-93-6 HR: 3
5-ACENAPHTHENAMINE
mf: $C_{12}H_{11}N$ mw: 169.24

PROP: Colorless needles, sol in ethanol. Mp: 108°.

SYNS: 5-AMINOACENAPHTHENE □ 1,2-DIHYDRO-5-ACENAPHTHYLENAMINE

TOXICITY DATA WITH REFERENCE
ipr-mus TDLo:3744 mg/kg/78W-I:ETA NEZAAQ 24,263,69
imp-mus TDLo:160 mg/kg:CAR NEZAAQ 24,263,69
ivn-mus LD50:56 mg/kg CSLNX* NX#01911

CONSENSUS REPORTS: IARC Cancer Review: Group 3 IMEMDT 7,56,87; Animal Inadequate Evidence IMEMDT 16,243,78.

SAFETY PROFILE: Poison by intravenous route. Questionable carcinogen with experimental carcinogenic and tumorigenic data. When heated to decomposition it emits toxic fumes of NO_x.

AAF275 CAS:83-32-9 HR: 2
ACENAPHTHENE
mf: $C_{12}H_{10}$ mw: 154.22

SYNS: ACENAPHTHYLENE, 1,2-DIHYDRO- □ 1,8-ETHYLENENAPHTHALENE □ NAPHTHYLENEETHYLENE □ PERIETHYLENENAPHTHALENE

TOXICITY DATA WITH REFERENCE
mmo-omi 3 mg MIKBA5 54,360,85
ipr-rat LD50:600 mg/kg GTPZAB 14(6),46,70

CONSENSUS REPORTS: Reported in EPA TSCA Inventory.

SAFETY PROFILE: Moderately toxic by intraperitoneal route. Mutation data reported. When heated to decomposition it emits acrid smoke and irritating vapors.

For occupational chemical analysis use NIOSH: Polynuclear Aromatic Hydrocarbons (HPLC), 5506; (GC), 5515.

AAF300 CAS:82-86-0 HR: 2
ACENAPHTHENEDIONE
mf: $C_{12}H_6O_2$ mw: 182.18

SYN: 1,2-ACENAPHTHYLENEDIONE

TOXICITY DATA WITH REFERENCE
unr-rat LD50:728 mg/kg RPTOAN 41,146,78

CONSENSUS REPORTS: Reported in EPA TSCA Inventory.

SAFETY PROFILE: Moderately toxic by an unspecified route. When heated to decomposition it emits acrid smoke and irritating vapors.

AAF500 CAS:208-96-8 HR: 2
ACENAPHTHYLENE
mf: $C_{12}H_8$ mw: 152.20

SYN: CYCLOPENTA(de)NAPHTHALENE

TOXICITY DATA WITH REFERENCE
mma-sat 1 mmol/L/2H CNREA8 39,4152,79
ipr-rat LD50:1700 mg/kg GTPZAB 14(6),46,70

CONSENSUS REPORTS: Reported in EPA TSCA Inventory.

SAFETY PROFILE: Moderately toxic by intraperitoneal route. Mutation data reported. When heated to decomposition it emits acrid smoke and irritating fumes.

For occupational chemical analysis use NIOSH: Polynuclear Aromatic Hydrocarbons (HPLC), 5506; (GC), 5515.

AAF625 CAS:72064-79-0 HR: 3
ACEPREVAL
mf: $C_{28}H_{38}O_7$ mw: 486.66

SYNS: 21-(ACETYLOXY)-11-β-HYDROXY-17-((1-OXOPENTYL)OXY) PREGNA-1,4-DIENE-3,20-DIONE □ PREDNISOLONE VALERATE ACETATE □ PREDNISOLONE-17-VALERATE-21-ACETATE □ PVA □ 11-β, 17-α,21-TRIHYDROXY-1,4-PREGNADIENE-3,20-DIONE-21-ACETATE-17-VALERATE

TOXICITY DATA WITH REFERENCE
scu-rat TDLo:55 mg/kg (7-17D preg):REP OYYAA2 20,67,80
scu-rat TDLo:110 mg/kg (7-17D preg):TER OYYAA2 20,67,80
ipr-mus LD50:1360 mg/kg OYYAA2 20,195,80
scu-mus LD50:1150 mg/kg OYYAA2 20,195,80
scu-rbt LD50:100 mg/kg OYYAA2 20,195,80

SAFETY PROFILE: Poison by subcutaneous route. Moderately toxic by other routes. An experimental teratogen. Other experimental reproductive effects.

AAF750 CAS:3598-37-6 HR: 3
ACEPROMAZINE MALEATE
mf: $C_{19}H_{22}N_2OS{\cdot}C_4H_4O_4$ mw: 442.57

SYNS: 2-ACETYL-10-(3-(DIMETHYLAMINO)PROPYL)PHENOTHIAZINE, MALEATE □ ACETYLPROMAZINE MALEATE (1:1) □ ATRAVET □ 10-(3-(DIMETHYLAMINO)PROPYL)PHENOTHIAZIN-2-YL METHYL KETONE MALEATE (1:1) □ MALEATE ACIDE de l'ACETYL-3-DIMETHYLAMINO-3-PROPYL-10-PHENOTHIAZINE (FRENCH) □ NOTENSIL □ PREGICIL □ SOPRONTIN

TOXICITY DATA WITH REFERENCE
orl-rat LD50:400 mg/kg AIPTAK 123,78,59
ivn-rat LD50:95 mg/kg MEIEDD 10,5,83
orl-mus LDLo:270 mg/kg AIPTAK 113,53,57
scu-mus LD50:175 mg/kg AIPTAK 113,53,57
ivn-mus LD50:65 mg/kg APTOA6 19,87,62

SAFETY PROFILE: Poison by ingestion, subcutaneous, and intravenous routes. When heated to decomposition it emits highly toxic fumes of NO_x and SO_x. See also KETONES.

AAF800 CAS:13461-01-3 HR: 3
ACEPROMETAZINE
mf: $C_{19}H_{22}N_2OS$ mw: 326.49

PROP: A liquid.

SYNS: ACEPROMETHAZINE □ 1664 CB □ ETHANONE, 1-(10-(2-(DIMETHYLAMINO)PROPYL)-10H-PHENOTHIAZIN-2-YL)-(9CI) □ KETONE, 10-(2-(DIMETHYLAMINO)PROPYL)PHENOTHIAZIN-2-YL METHYL

TOXICITY DATA WITH REFERENCE
orl-mus LD50:517 mg/kg AAREAV 21,543,64
scu-mus LD50:240 mg/kg AAREAV 21,543,64

DOT CLASSIFICATION: 3; *Label:* Flammable Liquid

SAFETY PROFILE: Poison by subcutaneous route. Moderately toxic by ingestion. A flammable liquid. When heated to decomposition it emits toxic vapors of NO_x and SO_x.

AAG000 **CAS:105-57-7** **HR: 3**
ACETAL
DOT: UN 1088
mf: $C_6H_{14}O_2$ mw: 118.20

PROP: Colorless, volatile liquid; agreeable odor, nutty aftertaste. Mp: −100°, bp: 102.7°, flash p: −5°F (CC), lel: 1.65%, uel: 10.4%, d: 0.831, autoign temp: 446°F, vap press: 10 mm @ 8.0°, vap d: 4.08. Sltly sol in water; misc in alc and eth.

SYNS: ACETAAL (DUTCH) ☐ ACETAL DIETHYLIQUE (FRENCH) ☐ ACETALE (ITALIAN) ☐ 1,1-DIAETHOXY-AETHAN (GERMAN) ☐ DIAETHYLACETAL (GERMAN) ☐ 1,1-DIETHOXY-ETHAAN (DUTCH) ☐ 1,1-DIETHOXYETHANE ☐ DIETHYL ACETAL ☐ 1,1-DIETOSSIETANO (ITALIAN) ☐ ETHYLIDENE DIETHYL ETHER ☐ USAF DO-45

TOXICITY DATA WITH REFERENCE
skn-rbt 10 mg/24H MLD JIHTAB 31,60,49
eye-rbt 500 mg JIHTAB 31,60,49
orl-rat LD50:4600 mg/kg MDZEAK 8,244,67
ihl-rat LCLo:4000 ppm/4H JIHTAB 31,343,49
ipr-rat LD50:900 mg/kg 14CYAT 2,1982,63
orl-mus LD50:3500 mg/kg GISAAA (3),12,77
ipr-mus LD50:500 mg/kg NTIS** AD277-689
orl-rbt LD50:3545 mg/kg PSEBAA 29,730,32

CONSENSUS REPORTS: Reported in EPA TSCA Inventory.

DOT CLASSIFICATION: 3; *Label:* Flammable Liquid

SAFETY PROFILE: Moderately toxic by ingestion, inhalation, and intraperitoneal routes. A skin and eye irritant. A narcotic. Dangerous fire hazard when exposed to heat or flame; can react vigorously with oxidizing materials. Forms heat-sensitive explosive peroxides on contact with air. When heated to decomposition it emits acrid smoke and fumes. See also ETHERS and ALDEHYDES.

AAG250 **CAS:75-07-0** **HR: 3**
ACETALDEHYDE
DOT: UN 1089
mf: C_2H_4O mw: 44.06

PROP: Colorless, fuming liquid; pungent, fruity odor. Mp: −123.5°, bp: 20.8°, lel: 4.0%, uel: 57%, flash p: −36°F (CC), d: 0.804 @ 0°/20°, autoign temp: 347°F, vap d: 1.52. Misc in water, alc, and eth.

SYNS: ACETALDEHYD (GERMAN) ☐ ACETIC ALDEHYDE ☐ ALDEHYDE ACETIQUE (FRENCH) ☐ ALDEIDE ACETICA (ITALIAN) ☐ ETHANAL ☐ ETHYL ALDEHYDE ☐ FEMA No. 2003 ☐ NCI-C56326 ☐ OCTOWY ALDEHYD (POLISH) ☐ RCRA WASTE NUMBER U001

TOXICITY DATA WITH REFERENCE
eye-hmn 50 ppm/15M JIHTAB 28,262,46
skn-rbt 500 mg open MLD UCDS** 12/13/63
eye-rbt 40 mg SEV UCDS** 12/13/63
mma-sat 10 µL/plate EVHPAZ 21,79,77
dnr-esc 10 µL/plate EVHPAZ 21,79,77
sce-hmn:lym 20 ppm/48H MUREAV 58,115,78
ipr-rat TDLo:300 mg/kg (female 8–13D post):REP TJADAB 36,31A,87
ipr-rat TDLo:50 mg/kg (12D preg):TER DADEDV 9,339,82
ihl-rat TCLo:735 ppm/6H/2Y-I:CAR TXCYAC 41,213,86
ihl-ham TCLo:2040 ppm/7H/52W-I:ETA EJCAAH 18,13,82
ihl-hmn TCLo:134 ppm/30M:PUL JAMAAP 165,1908,57
orl-rat LD50:661 mg/kg AGACBH 4,125,74
ihl-rat LC50:37 g/m³/30M APTOA6 6,299,50
ipr-rat LDLo:500 mg/kg JBCHA3 152,41,44
ihl-mus LC50:1500 ppm/4H DTLVS* 4,3,80
scu-rat LD50:640 mg/kg APTOA6 6,299,50
scu-mus LD50:560 mg/kg APTOA6 6,299,50
ivn-mus LD50:212 mg/kg JOANAY 128,65,79
skn-rbt LD50:3540 mg/kg UCDS** 12,13,63
ihl-ham LC50:17,000 ppm/4H PEXTAR 24,162,79
itr-ham LD50:96 mg/kg PEXTAR 24,162,79

CONSENSUS REPORTS: NTP 7th Annual Report On Carcinogens. IARC Cancer Review: Group 2B IMEMDT 7,77,87; Animal Sufficient Evidence IMEMDT 36,101,85; Human Inadequate Evidence IMEMDT 36,101,85. On Community Right-To-Know List. Reported in EPA TSCA Inventory. EPA Genetic Toxicology Program.

OSHA PEL: TWA 100 ppm; STEL 150 ppm
ACGIH TLV: TWA 100 ppm; STEL 150 ppm (Proposed: CL 25, Animal Carcinogen)
DFG MAK: 50 ppm (90 mg/m³), Suspected Carcinogen
DOT CLASSIFICATION: 3; *Label:* Flammable Liquid

SAFETY PROFILE: Confirmed carcinogen with experimental carcinogenic and tumorigenic data. Poison by intratracheal and intravenous routes. A human systemic irritant by inhalation. An experimental teratogen. Other experimental reproductive effects. A skin and severe eye irritant. A narcotic. Human mutation data reported. A common air contaminant. Highly flammable liquid. Mixtures of 30–60% of the vapor in air ignite above 100°. It can react violently with acid anhydrides, alcohols, ketones, phenols, NH_3, HCN, H_2S, halogens, P, isocyanates, strong alkalies, and amines. Reactions with cobalt chloride, mercury(II) chlorate, or mercury(II) perchlorate form sensitive, explosive products. Polymerizes violently in the presence of traces of metals or acids. Reaction with oxygen may lead to detonation. When heated to decomposition it emits acrid smoke and fumes.

For occupational chemical analysis use OSHA: #ID-68 or NIOSH: Acetaldehyde GC 2538; HPLC 3507.

AAG500 **CAS:75-39-8** **HR: 2**
ACETALDEHYDE AMMONIA
DOT: UN 1841
mf: $C_2H_4O \cdot H_3N$ mw: 61.10

PROP: White, crystalline solid. Bp: 110°, mp: 97°. Very sol in water, alc; sltly sol in eth.

SYNS: ACETALDEHYDE, AMINE SALT ☐ ALDEHYDE AMMONIA ☐ 1-AMINOETHANOL ☐ α-AMINOETHYL ALCOHOL ☐ ETHANOL, 1-AMINO-(8CI,9CI)

DOT CLASSIFICATION: 9; *Label:* CLASS 9

SAFETY PROFILE: It readily decomposes into acetaldehyde and ammonia when heated, causing the hazards of these substances. Moderate fire and explosion hazard when exposed to heat or flame. Can react with oxidizing materials. When heated to decomposition it emits toxic fumes of NH_3 and NO_x.

AAG850 CAS:105-82-8 HR: 1
ACETALDEHYDE-DI-n-PROPYL ACETAL
mf: $C_8H_{18}O_2$ mw: 146.26

SYNS: ACETALDEHYDE, DIPROPYL ACETAL ☐ 1,1-DIPROPOXYETHANE ☐ DIPROPYL ACETAL ☐ n-PROPYL ACETAL

TOXICITY DATA WITH REFERENCE
skn-rbt 500 mg/24H MOD FCTXAV 17,897,79

CONSENSUS REPORTS: Reported in EPA TSCA Inventory.

SAFETY PROFILE: A skin irritant. When heated to decomposition it emits acrid smoke and irritating fumes.

AAH000 CAS:16568-02-8 HR: 3
ACETALDEHYDE-N-METHYL-N-FORMYLHYDRAZONE
mf: $C_4H_8N_2O$ mw: 100.14

SYNS: ACETALDEHYDE-N-FORMYL-N-METHYLHYDRAZONE ☐ ETHYLIDENE GYROMITRIN ☐ GYROMITRIN ☐ N-METHYL-N-FORMYL HYDRAZONE of ACETALDEHYDE

TOXICITY DATA WITH REFERENCE
scu-mus TDLo:600 mg/kg/12W-I:CAR NEOLA4 28,559,81
orl-mus TD:5200 mg/kg/52W-I:ETA FEPRA7 39(3,Pt.2),884,80
unk-chd LDLo:10 mg/kg MGLHAE 65,453,74
unk-hmn LDLo:20 mg/kg MGLHAE 65,453,74
orl-rat LD50:320 mg/kg FCTXAV 15,575,77
orl-mus LD50:344 mg/kg MUREAV 54,167,78
orl-rbt LD50:50 mg/kg NATWAY 62,395,75

CONSENSUS REPORTS: IARC Cancer Review: Group 3 IMEMDT 7,56,87; Animal Limited Evidence IMEMDT 7,391,87. EPA Genetic Toxicology Program.

SAFETY PROFILE: Poison via ingestion and possibly other routes. Questionable carcinogen with experimental carcinogenic and tumorigenic data. When heated to decomposition it emits toxic fumes of NO_x.

AAH100 CAS:17167-73-6 HR: 3
ACETALDEHYDE METHYLHYDRAZONE
mf: $C_3H_8N_2$ mw: 72.13

SYNS: ACETALDEHYDE, N-METHYLHYDRAZONE ☐ AMFH

TOXICITY DATA WITH REFERENCE
orl-mus TDLo:208 mg/kg/1Y-I:ETA JJIND8 67,881,81
orl-mus LD50:390 mg/kg TXAPA9 45,429,78

SAFETY PROFILE: Poison by ingestion. Questionable carcinogen with experimental tumorigenic data. When heated to decomposition it emits toxic fumes of NO_x.

AAH250 CAS:107-29-9 HR: 3
ACETALDEHYDE OXIME
DOT: UN 2332
mf: C_2H_5NO mw: 59.08

PROP: A water-sol, crystalline material; sol in alc, eth. Mp: (α) 46.5°, mp: (β) 12°, d: 0.966, bp: 114.5°, flash p: ≤ 72°F.

SYNS: ACETALDOXIME ☐ ALDOXIME ☐ ETHANAL OXIME ☐ ETHYLIDENEHYDROXYLAMINE ☐ USAF AM-5

TOXICITY DATA WITH REFERENCE
mma-mus:lyms 230 mg/L MUREAV 204,149,88
msc-mus:lyms 15 g/L MUREAV 204,149,88
ipr-mus LD50:100 mg/kg NTIS** AD277-689
unk-mus LD50:1150 mg/kg PCJOAU 12,227,78

CONSENSUS REPORTS: Reported in EPA TSCA Inventory.

DOT CLASSIFICATION: 3; *Label:* Flammable Liquid

SAFETY PROFILE: Poison via intraperitoneal route. Mutation data reported. A dangerous fire hazard with a flash point at room temperature. When heated to decomposition it emits toxic fumes of NO_x. See also ALDEHYDES.

AAH500 CAS:918-04-7 HR: 2
ACETALDEHYDE SODIUM SULFITE
mf: $C_2H_5O_2SO_2Na \cdot 1/2H_2O$ mw: 166.2

PROP: White crystals decomp by acid; sol in water; insol in alc.

SYNS: ACETALDEHYDE SODIUM BISULFITE ☐ AZETALDEHYD-SCHWEFLIGSAUREN NATRIUMS (GERMAN) ☐ SODIUM-1-HYDROXYETHANESULFONATE

TOXICITY DATA WITH REFERENCE
orl-rbt LDLo:1220 mg(SO_2)/kg AHYGAJ 57,87,06

CONSENSUS REPORTS: Reported in EPA TSCA Inventory.

SAFETY PROFILE: Moderately toxic by ingestion based upon SO_2 content. When heated to decomposition it emits toxic fumes of SO_x and Na_2O. See also ALDEHYDES and SULFITES.

AAH750 CAS:107-89-1 HR: 3
ACETALDOL
DOT: UN 2839
mf: $C_4H_8O_2$ mw: 88.12

PROP: Clear, white-to-yellow syrupy liquid. Bp: 83° @ 20 mm, flash p: 150°F (OC), d: 1.11, autoign temp: 482°F, vap d: 3.04.

SYNS: ALDOL ☐ 3-BUTANOLAL ☐ 3-HYDROXYBUTANAL ☐ β-HYDROXYBUTYRALDEHYDE ☐ 3-HYDROXYBUTYRALDEHYDE ☐ OXYBUTANAL ☐ OXYBUTYRIC ALDEHYDE

TOXICITY DATA WITH REFERENCE
skn-rbt 10 mg/24H MLD JIHTAB 31,60,49
skn-rbt 10 mg/24H open MLD AIHAAP 23,95,62
eye-rbt 100 mg MLD UCDS** 4/21/67
orl-rat LD50:2180 mg/kg JIHTAB 31,60,49

skn-rbt LD50:140 mg/kg UCDS** 4/21/67

CONSENSUS REPORTS: Reported in EPA TSCA Inventory.

DOT CLASSIFICATION: 6.1; *Label:* Poison

SAFETY PROFILE: Poison via skin contact. Moderately toxic by ingestion. A skin and eye irritant. A flammable liquid and fire hazard when exposed to heat or flame; emits crotonaldchydc and water when heated. See CROTONALDEHYDE. Can react with oxidizing materials.

AAI000 CAS:60-35-5 HR: 3
ACETAMIDE
mf: C_2H_5NO mw: 59.08

PROP: Colorless crystals; mousy odor. Mp: 81°, bp: 221.2°, d: 1.159 @ 20°/4°, vap press: 1 mm @ 65°. Decomp in hot water.

SYNS: ACETIC ACID AMIDE □ ACETIMIDIC ACID □ AMID KYSELINY OCTOVE (POLISH) □ ETHANAMIDE □ METHANECARBOXAMIDE □ NCI-C02108

TOXICITY DATA WITH REFERENCE
oms-mus/ast 10 pph IDZAAW 51,53,76
otr-ham:emb 1 mg/L IJCNAW 19,642,77
orl-rbt TDLo:13 g/kg (6-18D post):TER ARZNAD 30,1557,80
orl-rbt TDLo:39 g/kg (6-18D post):REP ARZNAD 30,1557,80
orl-rat TDLo:431 g/kg/1Y-C:CAR JEPTDQ 3(5-6),149,80
orl-mus TDLo:517 g/kg/1Y-C:CAR JEPTDQ 3(5-6),149,80
orl-rat TD:546 g/kg/52W-C:NEO TXAPA9 14,163,69
orl-rat LD50:7000 mg/kg JRPFA4 4,219,62
ipr-rat LD50:10,300 mg/kg ARZNAD 20,1242,70
scu-rat LD50:10 g/kg OYYAA2 4,451,70
ivn-rat LD50:12,500 mg/kg NYKZAU 64(1),42S,68
unr-rat LD50:2300 mg/kg ARZNAD 18,645,68
orl-mus LD50:12,900 mg/kg NYKZAU 64(1),42S,68
ipr-mus LD50:1000 mg/kg JJIND8 62,911,79
scu-mus LD50:8300 mg/kg OYYAA2 4,451,70
ivn-mus LD50:10 g/kg NYKZAU 64(1),42S,68
ivn-rbt LD50:7500 mg/kg NYKZAU 64(1),42S,68
ivn-ckn LDLo:33,410 mg/kg ARZNAD 20,1242,70

CONSENSUS REPORTS: IARC Cancer Review: Group 2B IMEMDT 7,56,87; Animal Sufficient Evidence IMEMDT 7,389,87. On Community Right-To-Know List. Reported in EPA TSCA Inventory.

DFG MAK: Suspected Carcinogen

SAFETY PROFILE: Suspected carcinogen with experimental carcinogenic and neoplastigenic data. Moderately toxic by intraperitoneal and possibly other routes. An experimental teratogen. Other experimental reproductive effects. Mutation data reported. See also AMIDES. When heated to decomposition it emits toxic fumes of NO_x.

AAI100 CAS:103416-59-7 HR: 2
ACETAMIDE, 2-(DIETHYLAMINO)-N-(1,3-DIMETHYL-4-(o-FLUOROBENZOYL)-5-PYRAZOLYL)-, MONOHYDROCHLORIDE
mf: $C_{18}H_{23}FN_4O_2$•ClH mw: 382.91

SYNS: 2-(DIETHYLAMINO)-N-(1,3-DIMETHYL-4-(o-FLUOROBENZOYL)-5-PYRAZOLYL)ACETAMI de HYDROCHLORIDE □ 2-(DIETHYLAMINO)-N-(4-(2-FLUOROBENZOYL)-1,3-DIMETHYL-1H-PYRAZOL-5-YL) ACETAMIDE HYDROCHLORIDE □ PD 109394

TOXICITY DATA WITH REFERENCE
mma-sat 1 µmol/plate CRNGDP 7,2019,86
orl-rat TD:4550 mg/kg/13W-C:ETA AJPAA4 124,392,86

SAFETY PROFILE: Questionable carcinogen with experimental tumorigenic data. Mutation data reported. When heated to decomposition it emits toxic fumes of F^-, NO_x, and HCl.

AAI125 CAS:85723-21-3 HR: 2
ACETAMIDE, N-(4-(2-FLUOROBENZOYL)-1,3-DIMETHYL-1H-PYRAZOL-5-YL)-2-((3-(2-METHYL-1-PIPERIDINYL)PROPYL)AMINO)-, (Z)-2-BUTENEDIOATE (1:2)
mf: $C_{23}H_{32}FN_5O_2$ mw: 429.60

TOXICITY DATA WITH REFERENCE
orl-rat TD:4550 mg/kg/13W-C:ETA AJPAA4 124,392,86

SAFETY PROFILE: Questionable carcinogen with experimental tumorigenic data. When heated to decomposition it emits toxic fumes of NO_x.

AAI250 CAS:59-66-5 HR: 3
5-ACETAMIDE-1,3,4-THIADIAZOLE-2-SULFONAMIDE
mf: $C_4H_6N_4O_3S_2$ mw: 222.26

SYNS: 2-ACETAMIDO-5-SULFONAMIDO-1,3,4-THIADIAZOLE □ ACETAMIDOTHIADIAZOLESULFONAMIDE □ ACETAMOX □ ACETOZALAMIDE □ ACETAZOLAMID □ ACETAZOLAMIDE □ ACETAZOLEAMIDE □ 2-ACETYLAMINO-1,3,4-THIADIAZOLE-5-SULFONAMIDE □ N-(5-(AMINOSULFONYL)-1,3,4-THIADIAZOL-2-YL)ACETAMIDE □ CARBONIC ANHYDRASE INHIBITOR NO. 6063 □ CIDAMEX □ DEFILTRAN □ DEHYDRATIN □ DIACARB □ DIAKARB □ DIAMOX □ DIDOC □ DILURAN □ DIURAMID □ DIURETICUM-HOLZINGER □ DIUTAZOL □ DONMOX □ EDEMOX □ EUMICTON □ FONURIT □ GLAUPAX □ GLUPAX □ MUIRAMID □ NATRIONEX □ NEPHRAMIDE □ PHONURIT □ N-(5-SULFAMOYL-1,3,4-THIADIAZOL-2-YL)ACETAMIDE □ VETAMOX

TOXICITY DATA WITH REFERENCE
orl-mus TDLo:10 g/kg (female 8-12D post):REP TCMUD8 6,361,86
orl-rat TDLo:3300 mg/kg (1-22D preg):TER TJADAB 1,51,68
orl-man TDLo:54 mg/kg/5D-I:PUL AIMDAP 143,1278,83
ipr-rat LD50:2750 mg/kg NYKZAU 56(4),134S,60
orl-mus LD50:4300 mg/kg ABMGAJ 21,193,68
ipr-mus LD50:1175 mg/kg RPTOAN 39,255,76
scu-mus LD50:3 mg/kg DRUGAY 6,15,82
ivn-mus LD50:3 mg/kg DRUGAY 6,15,82

CONSENSUS REPORTS: Reported in EPA TSCA Inventory.

SAFETY PROFILE: Poison by subcutaneous and intravenous routes. Moderately toxic by intraperitoneal route. Human systemic effects by ingestion: dyspnea. An experimental teratogen by many routes. Other experimental reproductive effects. When heated to decomposition it emits very toxic fumes of NO_x and SO_x. A carbonic anhydrase inhibitor and diuretic used to treat glaucoma.

AAI750 CAS:440-58-4 HR: 1
3-ACETAMIDO-5-(ACETAMIDOMETHYL)-2,4,6-TRI-IODOBENZOIC ACID
mf: $C_{12}H_{11}I_3N_2O_4$ mw: 627.95

SYNS: 3-(ACETYLAMINO)-5-((ACETYLAMINO)METHYL)-2,4,6-TRIIODOBENZOIC ACID □ AMET (GERMAN) □ AMETRIODINIC ACID □ B-4130 □ α-5-DIACETAMIDO-2,4,6-TRIIODO-m-TOLUIC ACID □ IODAMIDE □ JODAMID (GERMAN) □ JODOMIRON □ SH 926 □ UROMIRO □ UROMIRON

TOXICITY DATA WITH REFERENCE
ipr-rat LD50:17,900 mg/kg ARZNAD 15,222,65
ivn-rat LD50:11,400 mg/kg ARZNAD 15,222,65
ivn-mus LD50:10,800 mg/kg MEIEDD 10,725,83
ivn-rbt LD50:13,200 mg/kg ARZNAD 15,222,65
ipr-gpg LD50:15 g/kg ARZNAD 15,222,65

SAFETY PROFILE: Mildly toxic by intraperitoneal and intravenous routes. When heated to decomposition it emits very toxic fumes of NO_x and HI.

AAJ125 CAS:1713-07-1 HR: 1
3-ACETAMIDO-5-AMINO-2,4,6-TRIIODOBENZOIC ACID
mf: $C_9H_7I_3N_2O_3$ mw: 571.88

TOXICITY DATA WITH REFERENCE
mma-sat 1 mg/plate PWPSA8 23,249,80
mnt-hmn:lym 40 mg/L RADLAX 129,199,78
cyt-hmn:lym 2000 ppm RADLAX 129,199,78
ivn-mus LD50:7200 mg/kg JPETAB 116,394,56

SAFETY PROFILE: Mildly toxic by intravenous route. Human mutation data reported. When heated to decomposition it emits toxic fumes of I^- and NO_x.

AAJ150 CAS:89-52-1 HR: 2
2-ACETAMIDOBENZOIC ACID
mf: $C_9H_9NO_3$ mw: 179.19

SYNS: o-ACETAMIDOBENZOIC ACID □ o-ACETOAMINOBENZOIC ACID □ N-ACETYLAMINOBENZOIC ACID □ 2-(ACETYLAMINO)BENZOIC ACID □ ACETYLANTHRANILIC ACID □ N-ACETYLANTHRANILIC ACID □ ANTHRANILIC ACID, N-ACETYL- □ BENZOIC ACID, 2-(ACETYLAMINO)-(9CI) □ 2-CARBOXYACETANILIDE

TOXICITY DATA WITH REFERENCE
orl-mus LD50:1114 mg/kg FRPSAX 38,847,83

CONSENSUS REPORTS: Reported in EPA TSCA Inventory.

SAFETY PROFILE: Moderately toxic by ingestion. When heated to decomposition it emits toxic vapors of NO_x.

AAJ250 CAS:63906-75-2 HR: 3
2-ACETAMIDO-4,5-BIS-(ACETOXYMERCURI)THIAZOLE
mf: $C_9H_{10}Hg_2N_2O_5S$ mw: 659.45

TOXICITY DATA WITH REFERENCE
ipr-mus LDLo:15 mg/kg CBCCT* 6,63,54

CONSENSUS REPORTS: Mercury and its compounds are on the Community Right-To-Know List.

OSHA PEL: CL 0.1 mg(Hg)/m^3 (skin)
ACGIH TLV: TWA 0.1 mg(Hg)/m^3 (skin)
NIOSH REL: (Mercury, Aryl And Inorganic): CL 0.1 mg/m^3 (skin)

SAFETY PROFILE: Poison by intraperitoneal route. When heated to decomposition it emits very toxic fumes of Hg, NO_x, and SO_x. See also MERCURY COMPOUNDS.

AAJ350 CAS:3025-96-5 HR: 3
4-ACETAMIDOBUTYRIC ACID
mf: $C_6H_{11}NO_3$ mw: 145.18

SYNS: γ-ACETYLAMINOBUTYRIC ACID □ BUTYRIC ACID, 4-ACETAMIDO- □ DF 469

TOXICITY DATA WITH REFERENCE
scu-rat TDLo:105 mg/kg (male 21D pre):REP ARANDR 10,239,83
ivn-mus LD50:425 mg/kg AIPTAK 145,233,63

SAFETY PROFILE: Poison by intravenous route. Experimental reproductive effects. When heated to decomposition it emits toxic fumes of NO_x

AAK400 CAS:19361-41-2 HR: 2
3-ACETAMIDOFLUORANTHENE
mf: $C_{18}H_{13}NO$ mw: 259.32

SYNS: 3-ACETYLAMINO-FLUORANTHEN □ 3-ACETYLAMINOFLUORANTHENE □ N-FLUORANTHEN-3-YLACETAMIDE □ N-3-FLUORANTHENYLACETAMIDE

TOXICITY DATA WITH REFERENCE
mma-sat 1 µg/plate NTIS** PB86-213733
orl-rat TDLo:4050 mg/kg/16W-C:ETA ONCOAR 8,233,55

SAFETY PROFILE: Questionable carcinogen with experimental tumorigenic data. Mutation data reported. When heated to decomposition it emits toxic fumes of NO_x.

AAK750 CAS:1068-90-2 HR: 1
ACETAMIDOMALONIC ACID DIETHYL ESTER
mf: $C_9H_{15}NO_5$ mw: 217.25

SYN: DIETHYLESTER KYSELINY ACETYLAMINOMALONOVE (CZECH)

TOXICITY DATA WITH REFERENCE
eye-rbt 500 mg/24H MLD 28ZPAK -,130,72

CONSENSUS REPORTS: Reported in EPA TSCA Inventory.

SAFETY PROFILE: An eye irritant. When heated to

decomposition it emits toxic fumes of NO_x. See also ESTERS.

AAL000 CAS:50309-20-1 HR: 2
7-ACETAMIDO-1-METHYL-4-(p-(p-((1- METHYLPYRIDINIUM-4-YL)AMINO)BENZAMIDO) ANILINO)QUINOLINIUM DI-p-TOLUENESULFONATE
mf: $C_{31}H_{30}N_6O_2 \cdot 2C_7H_7O_3S$ mw: 861.07

TOXICITY DATA WITH REFERENCE
dnd-mus:lym 340 nmol/L JMCMAR 22,134,79
ipr-mus LD10:97 mg/kg JMCMAR 22,134,79

SAFETY PROFILE: Moderately toxic by intraperitoneal route. Mutation data reported. When heated to decomposition it emits very toxic fumes of NO_x and SO_x.

AAL300 CAS:55123-66-5 HR: 2
(S)-2-(2-ACETAMIDO-4-METHYLVALERAMIDO)-N-(1-FORMYL-4-GUANIDINOBUTYL)-4-METHYL-VALERAMIDE
mf: $C_{20}H_{38}N_6O_4$ mw: 426.64

SYN: VALERAMIDE, 2-(2-ACETAMIDO-4-METHYLVALERAMIDO)-N-(1-FORMYL-4-GUANIDINOBUTYL)-4-METHYL-(S)-

TOXICITY DATA WITH REFERENCE
orl-mus TDLo:57,600 mg/kg/69W-C:NEO GANNA2 71,913,80

SAFETY PROFILE: Questionable carcinogen with experimental neoplastigenic data. When heated to decomposition it emits toxic fumes of NO_x.

AAL500 CAS:24143-08-6 HR: 2
5-ACETAMIDO-3-(5-NITRO-2-FURYL)-6H-1,2,4-OXADIAZINE
mf: $C_9H_8N_4O_5$ mw: 252.21

SYN: N-(3-(5-NITRO-2-FURYL)-6H-1,2,4-OXADIAZINYL)ACETAMIDE

TOXICITY DATA WITH REFERENCE
mma-sat 1 µg/plate MUREAV 40,9,76
dnr-sat 500 nmol/well CNREA8 34,2266,74
mmo-esc 300 nmol/well CNREA8 34,2266,74
mrc-esc 500 nmol/well CNREA8 34,2266,74
pic-esc 500 µg/L MUREAV 26,3,74
orl-rat TDLo:21 g/kg/28W-C:CAR CNREA8 29,2212,69

SAFETY PROFILE: Questionable carcinogen with experimental carcinogenic data. When heated to decomposition it emits toxic fumes of NO_x.

AAL750 CAS:531-82-8 HR: 3
2-ACETAMIDO-4-(5-NITRO-2-FURYL)THIAZOLE
mf: $C_9H_7N_3O_4S$ mw: 253.25

SYNS: 2-ACETAMINO-4-(5-NITRO-2-FURYL)THIAZOLE ☐ 2-ACETYLAMINO-4-(5-NITRO-2-FURYL)THIAZOLE ☐ N-(4-(5-NITRO-2-FURANYL)-2-THIAZOLYL)ACETAMIDE ☐ N-(4-(5-NITRO-2-FURYL)-2-THIAZOLYL)ACETAMIDE ☐ N-(4-(5-NITRO-2-FURYL)THIAZOL-2-YL) ACETAMIDE

TOXICITY DATA WITH REFERENCE
mma-sat 100 ng/plate MUREAV 40,9,76
dnr-sat 500 nmol/well CNREA8 34,2266,74
mmo-esc 300 nmol/well CNREA8 34,2266,74
mrc-esc 500 nmol/well CNREA8 34,2266,74
orl-rat TDLo:43 g/kg/46W-C:CAR CNREA8 30,936,70
orl-mus TDLo:2400 mg/kg/14W-C:CAR CNREA8 30,2320,70
orl-dog TDLo:27 g/kg/2Y-C:ETA JNCIAM 45,535,70
orl-rat TD:47 g/kg/46W-C:NEO JNCIAM 54,841,75

CONSENSUS REPORTS: IARC Cancer Review: Group 2B IMEMDT 7,56,87; Animal Sufficient Evidence IMEMDT 1,181,72; IMEMDT 7,185,74.

SAFETY PROFILE: Suspected carcinogen with experimental carcinogenic, tumorigenic, and neoplastigenic data. Mutation data reported. When heated to decomposition it emits very toxic fumes of SO_x and NO_x.

AAM250 CAS:4120-77-8 HR: 2
2-ACETAMIDOPHENATHRENE
mf: $C_{16}H_{13}NO$ mw: 235.30

SYNS: 2-ACETAMINOPHENANTHRENE ☐ 2-ACETYLAMINOPHENANTHRENE ☐ 2-PHENANTHRYLACETAMIDE ☐ N-2-PHENANTHRYLACETAMIDE ☐ N-(2-PHENANTHRYL)ACETAMIDE

TOXICITY DATA WITH REFERENCE
dnr-ham:fbr 1 µmol/L JNCIAM 54,1287,75
orl-rat TDLo:900 mg/kg/13W-C:CAR CNREA8 19,210,59
scu-rat TDLo:75 mg/kg/3W-I:ETA CNREA8 26,2239,66

SAFETY PROFILE: Questionable carcinogen with experimental carcinogenic and tumorigenic data. Mutation data reported. When heated to decomposition it emits toxic fumes of NO_x. See also AMIDES.

AAM875 CAS:85-36-9 HR: 1
3-ACETAMIDO-2,4,6-TRIIODOBENZOIC ACID
mf: $C_9H_6I_3NO_3$ mw: 556.86

SYNS: ACETRIZOIC ACID ☐ 3-(ACETYLAMINO)-2,4,6-TRIIODOBENZOIC ACID ☐ ACIDO 3-ACETILAMINO-2,4,6-TRIIODOBENZOICO (ITALIAN)

TOXICITY DATA WITH REFERENCE
unr-rat LD50:9650 mg/kg JAPMA8 42,721,53
orl-mus LD50:20 g/kg FRPSAX 18,33,63
ivn-mus LD50:8000 mg/kg FRPSAX 18,33,63

SAFETY PROFILE: Mildly toxic by ingestion and intravenous routes. When heated to decomposition it emits toxic fumes of I^- and NO_x.

AAN000 CAS:129-63-5 HR: 1
3-ACETAMIDO-2,4,6-TRIIODOBENZOIC ACID SODIUM SALT
mf: $C_9H_5I_3NO_3 \cdot Na$ mw: 578.84

SYNS: ACETIODONE ☐ ACETRIZOATE SODIUM ☐ ACETRIZOIC ACID SODIUM SALT ☐ BRONCHOSELECTAN ☐ CYSTOKON ☐ DIAGINOL ☐ FORTOMBRINE-N ☐ IODOPACT ☐ IODOPAQUE ☐ JODOPAX ☐ MP 1023 ☐ PYELOKON-FR ☐ SALPIX ☐ SODIUM-3-ACETAMIDO-2,4,6-TRIIODOBENZOATE ☐ SODIUM-3-ACETYLAMINO-2,4,6-TRIIODOBENZOATE ☐ SODIUM ACETRIZOATE ☐ THIXOKON ☐ TRI-ABRODIL ☐ TRIIODRAST ☐ TRIIODYL ☐ TRIIOTRAST ☐ 2,4,6-TRIJOD-3-ACETAMINOBENZOESAEURE NATRIUM (GERMAN) ☐ TRIOPAC 200 ☐ TRIOPAS ☐ TRIUMBREN ☐ TRIUROL ☐ TRIUROPAN ☐ UROKON SODIUM ☐ VESAMIN ☐ VISOTRAST ☐ VROKON

TOXICITY DATA WITH REFERENCE
ivn-mus LD50:9956 mg/kg JACSAT 74,4365,52
ims-mus LD50:12,156 mg/kg JPETAB 117,307,56
ivn-rat LDLo:7500 mg/kg CLDND*
ivn-dog LD50:6300 mg/kg JPETAB 116,394,56
ivn-cat LD50:5650 mg/kg CLDND*
ivn-rbt LD50:5200 mg/kg JPETAB 116,394,56

SAFETY PROFILE: Mildly toxic by intravenous route. When heated to decomposition it emits very toxic fumes of NO_x, Na_2O, and HI.

AAN250 CAS:100700-23-0 HR: 2
2-((4-(3-ACETAMIDO-2,4,6-TRIIODOPHENOXY)BUTOXY)METHYL)BUTYRIC ACID SODIUM SALT
mf: $C_{17}H_{22}I_3NO_5 \cdot Na$ mw: 724.09

TOXICITY DATA WITH REFERENCE
orl-mus LD50:2200 mg/kg FRPSAX 31,349,76
ivn-mus LD50:610 mg/kg FRPSAX 31,349,76

SAFETY PROFILE: Moderately toxic by ingestion and intravenous routes. When heated to decomposition it emits very toxic fumes of I^-, Na_2O, and NO_x.

AAN500 CAS:101651-76-7 HR: 2
2-(2-(3-ACETAMIDO-2,4,6-TRIIODOPHENOXY)ETHOXY)ACETIC ACID SODIUM SALT
mf: $C_{12}H_{12}I_3NO_5 \cdot Na$ mw: 653.94

TOXICITY DATA WITH REFERENCE
orl-mus LD50:9800 mg/kg FRPSAX 31,349,76
ivn-mus LD50:1500 mg/kg FRPSAX 31,349,76

SAFETY PROFILE: Moderately toxic by intravenous route. Mildly toxic by ingestion. When heated to decomposition it emits very toxic fumes of I^-, Na_2O, and NO_x.

AAN750 CAS:100700-24-1 HR: 2
2-(2-(3-ACETAMIDO-2,4,6-TRIIODOPHENOXY)ETHOXY)BUTYRIC ACID SODIUM SALT
mf: $C_{14}H_{16}I_3NO_5 \cdot Na$ mw: 682.00

TOXICITY DATA WITH REFERENCE
orl-mus LD50:3900 mg/kg FRPSAX 31,349,76
ivn-mus LD50:1320 mg/kg FRPSAX 31,349,76

SAFETY PROFILE: Moderately toxic by ingestion and intravenous routes. When heated to decomposition it emits very toxic fumes of I^-, Na_2O, and NO_x.

AAO500 CAS:101651-77-8 HR: 2
2-(2-(3-ACETAMIDO-2,4,6-TRIIODOPHENOXY)ETHOXY)-2-(o-TOLYL)ACETIC ACID SODIUM SALT
mf: $C_{19}H_{18}I_3NO_5 \cdot Na$ mw: 744.07

TOXICITY DATA WITH REFERENCE
orl-mus LD50:3550 mg/kg FRPSAX 31,349,76
ivn-mus LD50:585 mg/kg FRPSAX 31,349,76

SAFETY PROFILE: Moderately toxic by ingestion and intravenous routes. When heated to decomposition it emits very toxic fumes of I^-, Na_2O, and NO_x.

AAO750 CAS:101651-78-9 HR: 2
2-(2-(3-ACETAMIDO-2,4,6-TRIIODOPHENOXY)ETHOXY)-2-(p-TOLYL)ACETIC ACID SODIUM SALT
mf: $C_{19}H_{18}I_3NO_5 \cdot Na$ mw: 744.07

TOXICITY DATA WITH REFERENCE
orl-mus LD50:3430 mg/kg FRPSAX 31,349,76
ivn-mus LD50:658 mg/kg FRPSAX 31,349,76

SAFETY PROFILE: Moderately toxic by ingestion and intravenous routes. When heated to decomposition it emits very toxic fumes of I^-, Na_2O, and NO_x.

AAP250 CAS:100700-25-2 HR: 2
2-((3-(3-ACETAMIDO-2,4,6-TRIIODOPHENOXY)PROPOXY)METHYL)BUTYRIC ACID SODIUM SALT
mf: $C_{16}H_{20}I_3NO_5 \cdot Na$ mw: 710.06

TOXICITY DATA WITH REFERENCE
orl-mus LD50:3700 mg/kg FRPSAX 31,349,76
ivn-mus LD50:750 mg/kg FRPSAX 31,349,76

SAFETY PROFILE: Moderately toxic by ingestion and intravenous routes. When heated to decomposition it emits very toxic fumes of I^-, Na_2O, and NO_x.

AAQ250 CAS:2832-40-8 HR: 2
ACETAMINE YELLOW CG
mf: $C_{15}H_{15}N_3O_2$ mw: 269.33

SYNS: ACETAMIDE, N-(4-((2-HYDROXY-5-METHYLPHENYL)AZO)PHENYL)- ☐ 4-ACETAMIDO-2′-HYDROXY-5′-METHYLAZOBENZENE ☐ ACETATE FAST YELLOW G ☐ ACETOQUINONE LIGHT YELLOW ☐ ACETOQUINONE LIGHT YELLOW 4JLZ ☐ ALTCO SPERSE FAST YELLOW GFN NEW ☐ AMACEL YELLOW G ☐ ARTISIL DIRECT YELLOW G ☐ ARTISIL YELLOW G ☐ ARTISIL YELLOW 2GN ☐ CALCOSYN YELLOW GC ☐ CALCOSYN YELLOW GCN ☐ CELLITON DISCHARGE YELLOW GL ☐ CELLITON FAST YELLOW G ☐ CELLITON FAST YELLOW GA ☐ CELLITON FAST YELLOW GA-CF ☐ CELLITON YELLOW G ☐ CELUTATE YELLOW GH ☐ C.I. 11855 ☐ C.I. 3/11855 ☐ CIBACETE YELLOW GBA ☐ CIBACET YELLOW GBA ☐ CIBACET YELLOW 2GC ☐ C.I. DISPERSE YELLOW 3 ☐ CILLA FAST YELLOW G ☐ CI SOLVENT YELLOW 92 ☐ CI SOLVENT YELLOW 99 ☐ DIACELLITON FAST YELLOW G ☐ DISPERSE FAST YELLOW G ☐ DISPERSE YELLOW G ☐ DISPERSE YELLOW 3 ☐ DISPERSIVE YELLOW 3T ☐ DISPERSE YELLOW Z ☐ DISPERSOL FAST YELLOW G ☐ DISPERSOL PRINTING YELLOW G ☐ DISPERSOL YELLOW A-G ☐ DURGACET YELLOW G ☐ DUROSPERSE YELLOW G ☐ EASTONE YELLOW GN ☐ ESTEROQUINONE LIGHT YELLOW 4JL ☐ ESTONE YELLOW GN ☐ FENACET FAST YELLOW G ☐ FENACET YELLOW G ☐ GENACRON YELLOW G ☐ HISPACET FAST YELLOW G ☐ HISPERSE YELLOW G ☐ N-(4-((2-HYDROXY-5-METHYLPHENYL)AZO)PHENYL)ACETAMIDE ☐ 4′-((6-HYDROXY-m-TOLYL)AZO)ACETANILIDE ☐ INTERCHEM ACETATE YELLOW G ☐ INTERCHEM DISPERSE YELLOW GH ☐ INTRAPERSE YELLOW GBA ☐ INTRASPERSE YELLOW GBA EXTRA ☐ KAYALON FAST YELLOW G ☐ KAYASET YELLOW G ☐ KCA ACETATE FAST YELLOW G ☐ MICROSETILE YELLOW GR ☐ MIKETON FAST YELLOW G ☐ NACELAN FAST YELLOW CG ☐ NCI-C53781 ☐ NOVALON YELLOW 2GN ☐ NYLOQUINONE LIGHT YELLOW 4JL ☐ NYLOQUINONE YELLOW 4J ☐ OSTACET YELLOW P2G ☐ PALACET YELLOW GN ☐ PALANIL YELLOW G ☐ PAMACEL YELLOW G-3 ☐ PERLITON YELLOW G ☐ RELITON YELLOW C ☐ RESIREN YELLOW TG ☐ SAFARITONE YELLOW G ☐ SAMARON YELLOW PA3 ☐ SERINYL HOSIERY YELLOW GD ☐ SERIPLAS YELLOW GD ☐ SERISOL FAST YELLOW GD ☐ SETACYL YELLOW G ☐ SETACYL YELLOW 2GN ☐ SETACYL YELLOW P-2GL ☐ SILOTRAS YELLOW TSG ☐ SUPRACET FAST YELLOW G ☐ SYNTEN YELLOW 2G ☐ SYNTON YELLOW 2G ☐ TERASIL YELLOW GBA EXTRA ☐ TERASIL

YELLOW 2GC □ TERTRANESE YELLOW N-2GL □ TULADISPERSE FAST YELLOW 2G □ VONTERYL YELLOW G □ VONTERYL YELLOW R □ YELLOW RELITON G □ YELLOW Z □ ZLUT DISPERZNI 3 □ ZLUT ROZPOUSTEDLOVA 77

TOXICITY DATA WITH REFERENCE
mmo-sat 10 µg/plate SCIEAS 236,933,87
cyt-frg-par 2800 µL/7D CYTBAI 25,175,79
orl-rat TDLo:35 g/kg/14D-C:REP,NEO NTPTR* NTP-TR-222,82
orl-rat TDLo:180 g/kg/2Y-C:CAR NTPTR* NTP-TR-222,82
orl-mus TDLo:433 g/kg/2Y-C:CAR NTPTR* NTP-TR-222,82
orl-mus TD:216 g/kg/2Y-C:ETA NTPTR* NTP-TR-222,82
ipr-rat LD50:8190 mg/kg GISAAA 53(10),92,88
ipr-mus LD50:8080 mg/kg GISAAA 53(10),92,88

CONSENSUS REPORTS: Community Right-To-Know List. Reported in EPA TSCA Inventory. IARC Cancer Review: Group 3 IMEMDT 48,149,90; Animal Inadequate Evidence IMEMDT 8,97,75; NTP Carcinogenesis Bioassay (feed); Clear Evidence: mouse, rat NTPTR* NTP-TR-222,82.

SAFETY PROFILE: Suspected carcinogen with experimental tumorigenic and carcinogenic data. Low toxicity by intraperitoneal route. An allergen. Mutation data reported. When heated to decomposition it emits toxic fumes of NO_x.

AAQ500 CAS:103-84-4 HR: 3
ACETANILIDE
mf: C_8H_9NO mw: 135.18

PROP: White, shining, crystalline scales. Mp: 113.5°, bp: 305°, flash p: 345°F (OC), d: 1.2105 @ 4°/4°, autoign temp: 1004°F, vap press: 1 mm @ 114.0°, vap d: 4.65. Somewhat sol in water, alc, and eth.

SYNS: ACETAMIDE, N-PHENYL- □ ACETAMIDOBENZENE □ ACETANIL □ ACETANILID □ ACETIC ACID ANILIDE □ ACETOANILIDE □ ACETYLAMINOBENZENE □ ACETYLANILINE □ N-ACETYLANILINE □ AN □ ANILINE, N-ACETYL- □ ANTIFEBRIN □PHENAL-GENE □ PHENALGIN □ N-PHENYLACETAMIDE □ USAF EK-3

TOXICITY DATA WITH REFERENCE
mnt-mus-ipr 50 mg/kg JPMSAE 80,761,91
orl-hmn TDLo:14 mg/kg/D:PUL,KID,BLD 34ZIAG -,62,69
orl-man LDLo:56 mg/kg/H-I:CNS,GIT,MET AJMSA9 122,770,01
orl-man TDLo:405 mg/kg:CNS,PUL JAMAAP 12,103,1889
orl-rat LD50:800 mg/kg JPETAB 54,159,35
ipr-rat LD50:540 mg/kg JAPMA8 48,204,59
orl-mus LD50:1210 mg/kg TXAPA9 19,20,71
ipr-mus LD50:500 mg/kg NTIS** AD277-689
orl-cat LDLo:250 mg/kg JPHAA3 28,70,39
orl-cat LDLo:250 mg/kg JAPMA8 28,70,39
ivn-cat LDLo:8500 µg/kg JAPMA8 30,91,41
orl-rbt LDLo:1500 mg/kg JPETAB 29,466,63
orl-gpg LDLo:200 mg/kg HBAMAK 4,1290,35

CONSENSUS REPORTS: Reported in EPA TSCA Inventory. EPA Genetic Toxicology Program.

SAFETY PROFILE: A human poison by an unspecified route. Poison by ingestion and intravenous routes. Moderately toxic by intraperitoneal route. Human systemic effects by ingestion: hallucinations and distorted perceptions, sleepiness, constipation, cyanosis, respiratory stimulation, kidney damage, methemoglobinemia-carboxhemoglobinemia, and decreased body temperature. Mutation data reported. When heated to decomposition it emits toxic fumes of NO_x. Combustible when exposed to heat or flame. See also ANILINE.

AAR500 CAS:3572-06-3 HR: 2
ACETATE of 4-(HYDROXYPHENYL)-2-BUTANONE
mf: $C_{12}H_{14}O_3$ mw: 206.26

SYNS: 4-(p-ACETOXYPHENYL)-2-BUTANONE □ ENT 32,833 □ 4-(p-HYDROXYPHENYL)-2-BUTANONE ACETATE □ p-(3-OXOBUTYL)PHENYL ACETATE

TOXICITY DATA WITH REFERENCE
orl-rat LD50:3038 mg/kg TXAPA9 31,421,75

CONSENSUS REPORTS: Reported in EPA TSCA Inventory.

SAFETY PROFILE: Moderately toxic by ingestion. When heated to decomposition it emits acrid smoke and irritating fumes.

AAR750 CAS:63868-93-9 HR: 3
(ACETATO)BIS(HEPTYLOXY)PHOSPHINYLMERCURY
mf: $C_{16}H_{33}HgO_5P$ mw: 537.05

SYN: (BIS-(HEPTYLOXY)PHOSPHINYL)MERCURY ACETATE

TOXICITY DATA WITH REFERENCE
ipr-mus LDLo:63 mg/kg CBCCT* 8,103,56

CONSENSUS REPORTS: Mercury and its compounds are on the Community Right-To-Know List.

OSHA PEL: CL 0.1 mg(Hg)/m³ (skin)
ACGIH TLV: TWA 0.1 mg(Hg)/m³ (skin)
NIOSH REL: (Organomercury) TWA 0.01 mg(Hg)/m³ (skin)

SAFETY PROFILE: Poison by intraperitoneal route. See also MERCURY COMPOUNDS. When heated to decomposition it emits very toxic fumes of Hg and PO_x.

AAS000 CAS:63868-94-0 HR: 3
(ACETATO)BIS(HEXYLOXY)PHOSPHINYLMERCURY
mf: $C_{14}H_{29}HgO_5P$ mw: 508.99

SYN: (BIS(HEXYLOXY)PHOSPHINYL)MERCURY ACETATE

TOXICITY DATA WITH REFERENCE
ipr-mus LDLo:125 mg/kg CBCCT* 8,103,56

CONSENSUS REPORTS: Mercury and its compounds are on the Community Right-To-Know List.

OSHA PEL: CL 0.1 mg(Hg)/m³ (skin)
ACGIH TLV: TWA 0.1 mg(Hg)/m³ (skin)
NIOSH REL: (Organomercury) TWA 0.01 mg(Hg)/m³ (skin)

SAFETY PROFILE: Poison by intraperitoneal route. See also MERCURY COMPOUNDS. When heated to decomposition it emits very toxic fumes of Hg and PO_x.

AAS250 CAS:5421-48-7 HR: 3
(ACETATO)(DIETHOXYPHOSPHINYL)MERCURY
mf: $C_6H_{13}HgO_5P$ mw: 396.75

SYN: (DIETHOXY-PHOSPHINYL)MERCURY ACETATE

TOXICITY DATA WITH REFERENCE
ipr-mus LDLo:7800 µg/kg CBCCT* 8,103,56

CONSENSUS REPORTS: Mercury and its compounds are on the Community Right-To-Know List.

OSHA PEL: CL 0.1 mg(Hg)/m³ (skin)
ACGIH TLV: TWA 0.1 mg(Hg)/m³ (skin)
NIOSH REL: (Organomercury) TWA 0.01 mg(Hg)/m³

SAFETY PROFILE: Poison by intraperitoneal route. See also MERCURY COMPOUNDS. When heated to decomposition it emits very toxic fumes of Hg and PO_x.

AAS500 CAS:21450-81-7 HR: 3
(ACETATO)(2,3,5,6-TETRAMETHYLPHENYL)MERCURY
mf: $C_{12}H_{16}HgO_2$ mw: 392.87

SYN: (2,3,5,6-TETRAMETHYLPHENYL)MERCURY ACETATE

TOXICITY DATA WITH REFERENCE
ivn-mus LD50:32 mg/kg CSLNX* NX#05139

CONSENSUS REPORTS: Mercury and its compounds are on the Community Right-To-Know List.

OSHA PEL: CL 0.1 mg(Hg)/m³ (skin)
ACGIH TLV: TWA 0.1 mg(Hg)/m³ (skin)
NIOSH REL: (Mercury, Aryl And Inorganic): CL 0.1 mg/m³ (skin)

SAFETY PROFILE: Poison by intravenous route. See also MERCURY COMPOUNDS. When heated to decomposition it emits toxic fumes of Hg.

AAS750 CAS:1424-27-7 HR: 1
ACETAZOLAMIDE SODIUM
mf: $C_4H_5N_4O_3S_2{\cdot}Na$ mw: 244.24

SYNS: ACETAZOLAMIDE SODIUM SALT □ SODIUM ACETAZOLAMIDE

TOXICITY DATA WITH REFERENCE
ipr-mus TDLo:2000 mg/kg (9D preg):TER TJADAB 20,289,79
ipr-mus TDLo:2000 mg/kg (9D preg):REP TJADAB 20,289,79
ivn-mus LD50:6 g/kg YAKUD5 21,775,79

SAFETY PROFILE: An experimental teratogen. Other experimental reproductive effects. When heated to decomposition it emits very toxic fumes of NO_x, Na_2O, and SO_x.

AAT250 CAS:64-19-7 HR: 3
ACETIC ACID
DOT: UN 2789/UN 2790
mf: $C_2H_4O_2$ mw: 60.06

PROP: Clear, colorless liquid; pungent odor. Mp: 16.7°, bp: 118.1°, flash p: 109°F (CC), lel: 5.4%, uel: 16.0% @ 212°F, d: 1.049 @ 20°/4°, autoign temp: 869°F, vap press: 11.4 mm @ 20°, vap d: 2.07. Misc in water, alc, and eth.

SYNS: ACETIC ACID (aqueous solution) (DOT) □ ACETIC ACID, glacial or acetic acid solution, >80% acid, by weight (UN 2790) (DOT) □ ACETIC ACID, GLACIAL □ ACETIC ACID solution, >10% but not >80% acid, by weight (UN 2790) (DOT) □ ACIDE ACETIQUE (FRENCH) □ ACIDO ACETICO (ITALIAN) □ AZIJNZUUR (DUTCH) □ ESSIGSAEURE (GERMAN) □ ETHANOIC ACID □ ETHYLIC ACID □ FEMA No. 2006 □ GLACIAL ACETIC ACID □ METHANECARBOXYLIC ACID □ OCTOWY KWAS (POLISH) □ VINEGAR ACID

TOXICITY DATA WITH REFERENCE
skn-hmn 50 mg/24H MLD TXAPA9 31,481,75
skn-rbt 20 mg/24H MOD 85JCAE-,304,86
skn-rbt 525 mg open SEV UCDS** 8/7/63
skn-rbt 50 mg/24H MLD TXAPA9 31,481,75
eye-rbt 50 µg open SEV AMIHBC 4,119,51
eye-rbt 5 mg/30S RNS MLD TXCYAC 23,281,82
mmo-esc 300 ppm/3H AMNTA4 85,119,51
sln-dmg-ihl 1000 ppm/24H THAGA6 39,330,69
sln-dmg-orl 1000 ppm THAGA6 39,330,69
cyt-grl-par 40 µmol/L NULSAK 9,119,66
orl-rat TDLo:700 mg/kg (18D post):REP NTOTDY 4,105,82
orl-hmn TDLo:1470 µg/kg:GIT AIHAAP 33,624,72
ihl-hmn TCLo:816 ppm/3M:NOSE,EYE,PUL AMIHAB 21,28,60
unk-man LDLo:308 mg/kg 85DCAI 2,73,70
orl-rat LD50:3310 mg/kg JIHTAB 23,78,41
ihl-rat LCLo:16,000 ppm/4H JIHTAB 23,78,41
ihl-mus LC50:5620 ppm/1H MELAAD 48,559,57
ivn-mus LD50:525 mg/kg APTOA6 18,141,61
orl-rbt LDLo:600 mg/kg CRSBAW 83,136,20
skn-rbt LD50:1060 mg/kg UCDS** 8/7/63
scu-rbt LDLo:600 mg/kg CRSBAW 83,136,20
rec-rbt LDLo:600 mg/kg CRSBAW 83,136,20

CONSENSUS REPORTS: Reported in EPA TSCA Inventory.

OSHA PEL: TWA 10 ppm
ACGIH TLV: TWA 10 ppm; STEL 15 ppm
DFG MAK: 10 ppm (25 mg/m³)
DOT CLASSIFICATION: 8; *Label:* Corrosive

SAFETY PROFILE: A human poison by an unspecified route. Moderately toxic by various routes. A severe eye and skin irritant. Can cause burns, lachrymation, and conjunctivitis. Human systemic effects by ingestion: changes in the esophagus, ulceration or bleeding from the small and large intestines. Human systemic irritant effects and mucous membrane irritant. Experimental reproductive effects. Mutation data reported. A common air contaminant. A flammable liquid. A fire and explosion hazard when exposed to heat or flame; can react vigorously with oxidizing materials. To fight fire, use CO_2, dry chemical, alcohol foam, foam and mist. When heated to decomposition it emits irritating fumes.

Potentially explosive reaction with 5-azidotetrazole; bromine pentafluoride; chromium trioxide; hydrogen peroxide; potassium permanganate; sodium peroxide; and phosphorus trichloride. Potentially violent reactions with acetaldehyde and acetic anhydride. Ignites on contact with potassium-tert-butoxide. Incompatible with chromic acid; nitric acid; 2-amino-ethanol; NH_4NO_3;

ClF_3; chlorosulfonic acid; (O_3+ diallyl methyl carbinol); ethylenediamine; ethylene imine; (HNO_3 + acetone); oleum; $HClO_4$; permanganates; $P(OCN)_3$; KOH; NaOH; xylene.

For occupational chemical analysis use OSHA: #ID-118 or NIOSH: Acetic Acid, 1603.

AAT550 CAS:108419-32-5 HR: 1
ACETIC ACID, C_{7-9}-BRANCHED ALKYL ESTERS, C_8-rich

TOXICITY DATA WITH REFERENCE
orl-rat TDLo: 10 g/kg (female 6-15D post): REP FAATDF 13,303,89
orl-rat LD50: 5 g/kg FAATDF 13,303,89

CONSENSUS REPORTS: Reported in EPA TSCA Inventory.

SAFETY PROFILE: Slightly toxic by ingestion. Experimental reproductive effects. When heated to decomposition it emits acrid smoke and irritating vapors.

AAU000 CAS:150-84-5 HR: 1
ACETIC ACID, CITRONELLYL ESTER
mf: $C_{12}H_{22}O_2$ mw: 198.34

PROP: Found in oils of Citronella Ceylon, geranium, and about 20 other oils (FCTXAV 11,1011,73). Colorless liquid; fruity odor. D: 0.883–0.893, refr index: 1.440–1.450, flash p: 212°F. Sol in alc and fixed oils; insol in glycerin, propylene glycol, and water @ 229°.

SYNS: ACETIC ACID-3,7-DIMETHYL-6-OCTEN-1-YL ESTER □ CITRONELLYL ACETATE (FCC) □ 2,6-DIMETHYL-2-OCTEN-8-OL ACETATE □ 3,7-DIMETHYL-6-OCTEN-1-YL ACETATE □ FEMA No. 2311

TOXICITY DATA WITH REFERENCE
skn-hmn 20 mg/48H MLD FCTXAV 11,1011,73
skn-rbt 500 mg/24H FCTXAV 11,1011,73
orl-rat LD50: 6800 mg/kg FCTXAV 11,1011,73

CONSENSUS REPORTS: Reported in EPA TSCA Inventory.

SAFETY PROFILE: Mildly toxic by ingestion. A human skin irritant. See also ESTERS. Combustible liquid. When heated to decomposition it emits acrid smoke and irritating fumes.

AAU250 CAS:18461-55-7 HR: 3
ACETIC ACID-4,6-DINITRO-o-CRESYL ESTER
mf: $C_9H_8N_2O_6$ mw: 240.19

SYNS: 4,6-DINITRO-o-KRESYLESTER KYSELINY OCTOVE (CZECH) □ DNOK-ACETAT (CZECH)

TOXICITY DATA WITH REFERENCE
skn-rbt 500 mg/24H MOD 28ZPAK -,131,72
eye-rbt 100 mg/24H SEV 28ZPAK -,131,72
ipr-mus LDLo: 63 mg/kg CBCCT* 6,146,54
orl-rat LD50: 46 mg/kg 28ZPAK -,131,72

NIOSH REL: (Dinitro ortho-Cresyl) TWA 0.2 mg/m^3

SAFETY PROFILE: Poison by ingestion and intraperitoneal routes. A skin and severe eye irritant. When heated to decomposition it emits toxic fumes of NO_x.

AAU750 CAS:1516-17-2 HR: 2
ACETIC ACID-2,4-HEXADIEN-1-OL ESTER
mf: $C_8H_{12}O_2$ mw: 140.20

SYNS: 2,4-HEXADIEN-1-OL ACETATE □ 2,4-HEXADIENYL ACETATE □ SORBYL ACETATE

TOXICITY DATA WITH REFERENCE
skn-rbt 500 mg/24H MLD 85JCAE-,360,86
eye-rbt 250 µg/24H SEV 85JCAE-,360,86
orl-rat LD50: 4360 µL/kg TXAPA9 28,313,74
skn-rbt LD50: 2520 µL/kg TXAPA9 28,313,74

CONSENSUS REPORTS: Reported in EPA TSCA Inventory.

SAFETY PROFILE: Moderately toxic by ingestion and skin contact. A skin and severe eye irritant. See also ESTERS. When heated to decomposition it emits acrid smoke and irritating fumes.

AAV500 CAS:3610-27-3 HR: 1
ACETIC ACID, 2-(2-(2-METHOXYETHOXY)ETHOXY) ETHYL ESTER
mf: $C_9H_{18}O_5$ mw: 206.27

PROP: Liquid. Bp: 130°, flash p: 260°F (OC), d: 1.094, vap d: 7.11.

SYNS: ETHANOL, 2-(2-(2-METHOXYETHOXY)ETHOXY)-, ACETATE (8CI,9CI) □ 2-(2-(2-METHOXYETHOXY)ETHOXY)ETHANOL ACETATE □ 2-(2-(2-METHOXYETHOXY)ETHOXY)ETHYL ACETATE □ 2 (2-(2-METHOXYETHOXY)ETHOXY)ETHYLESTER KYSELINY OCTOVE □ METHOXYTRIGLYCOL ACETATE □ 3,6,9-TRIOXADECYLESTER KYSELINY OCTOVE

TOXICITY DATA WITH REFERENCE
skn-rbt 500 mg/24H MLD 85JCAE-,714,86
eye-rbt 500 mg AMIHBC 10,61,54
orl-rat LD50: 11 g/kg AMIHBC 10,61,54
skn-rbt LD50: 8000 mg/kg AMIHBC 10,61,54

CONSENSUS REPORTS: Reported in EPA TSCA Inventory.

SAFETY PROFILE: Mildly toxic by ingestion and skin contact. An eye irritant. See also ESTERS. Combustible. To fight fire, use alcohol foam, CO_2, dry chemical. When heated to decomposition it emits acrid smoke and irritating fumes.

AAW000 CAS:56856-83-8 HR: 3
ACETIC ACID METHYLNITROSAMINOMETHYL ESTER
mf: $C_4H_8N_2O_3$ mw: 132.14

SYNS: α-ACETOXY DIMETHYLNITROSAMINE □ ACETOXYMETHYL-METHYL-NITROSAMIN (GERMAN) □ ACETOXYMETHYL METHYLNITROSAMINE □ N-α-ACETOXYMETHYL-N-METHYLNITROSAMINE □ 1-ACETOXY-N-NITROSODIMETHYLAMINE □ AMMN □ ANN (GERMAN) □ DMN-OAC □ MAMN □ METHYL(ACETOXYMETHYL)NITROSAMINE □ N-NITROSO-N-(ACETOXY)METHYL-N-METHYLAMINE □ N-NITROSO-N-METHYL-N-ACETOXYMETHYLAMINE

TOXICITY DATA WITH REFERENCE
slt-dmg-par 100 µmol/L CNREA8 35,3780,75
cyt-dmg-par 100 µmol/L CNREA8 35,3780,75
mmo-esc 25 µmol/plate GANNA2 70,663,79
orl-rat TDLo: 13 mg/kg: CAR JJIND8 63,93,79
ipr-rat TDLo: 13 mg/kg: CAR JJIND8 63,93,79
ivn-rat TDLo: 13 mg/kg: ETA JJIND8 63,93,79
ipr-mus TDLo: 10 mg/kg (11D preg): TER ARTODN 52,45,83
rec-rat TDLo: 12 mg/kg/46W-I: ETA,REP HEGAD4 30,30,83
ipr-rat LD: 13 mg/kg: NEO JJIND8 58,1531,77
ipr-rat LD: 13 mg/kg: NEO,REP VTPHAK 16,574,79
orl-rat LD50: 130 mg/kg ONCOBS 38,18,81
ipr-rat LD50: 25 mg/kg JNCIAM 58,1533,77
scu-rat LD50: 25 mg/kg ZEKBAI 91,217,78
ivn-rat LD50: 25 mg/kg ZEKBAI 91,217,78
rec-rat LD50: 24 mg/kg ZEKBAI 91,217,78

SAFETY PROFILE: Suspected carcinogen with experimental carcinogenic, neoplastigenic, and tumorigenic data. Poison by ingestion, subcutaneous, intravenous, and intraperitoneal routes. Experimental teratogenic data. Human mutation data reported. When heated to decomposition it emits toxic fumes of NO_x. See also NITROSAMINES, N-NITROSO COMPOUNDS, and ESTERS.

AAW250 CAS:10476-95-6 HR: 3
ACETIC ACID-2-METHYL-2-PROPENE-1,1-DIOL DIESTER
mf: $C_8H_{12}O_4$ mw: 172.20

SYNS: DIACETATE ☐ 2-METHYL-2-PROPENE-1,1-DIOL ☐ 2-PROPENE-1,1-DIOL, 2-METHYL-, DIACETATE

TOXICITY DATA WITH REFERENCE
orl-rat LD50: 440 mg/kg AIHAAP 30,470,69
ihl-rat LCLo: 62 ppm/1H AIHAAP 30,470,69
ipr-mus LDLo: 250 mg/kg CBCCT* 5,61,53
skn-rbt LD50: 44 mg/kg AIHAAP 30,470,69

CONSENSUS REPORTS: Reported in EPA TSCA Inventory. EPA Extremely Hazardous Substances List.

SAFETY PROFILE: Poison by inhalation, skin contact, and intraperitoneal routes. Moderately toxic by ingestion. See also ESTERS. When heated to decomposition it emits acrid smoke and irritating fumes.

AAW500 CAS:1118-39-4 HR: 1
ACETIC ACID MYRCENYL ESTER
mf: $C_{12}H_{20}O_2$ mw: 196.32

PROP: Liquid with woody cologne-like odor. Bp: 53° @ 0.5 mm.

SYNS: ACETIC ACID-2-METHYL-6-METHYLENE-7-OCTEN-2-YL ESTER ☐ 3-METHYLENE-7-METHYL-1-OCTEN-7-YL ACETATE ☐ 2-METHYL-6-METHYLENE-7-OCTEN-2-OL ACETATE ☐ 2-METHYL-6-METHYLENE-7-OCTEN-2-YL ACETATE ☐ MYRCENYL ACETATE

TOXICITY DATA WITH REFERENCE
skn-rbt 500 mg/24H MOD FCTXAV 14,601,76
orl-rat LD50: 6300 mg/kg FCTXAV 14(6),601,76

CONSENSUS REPORTS: Reported in EPA TSCA Inventory.

SAFETY PROFILE: Mildly toxic by ingestion. A skin irritant. See also ESTERS. When heated to decomposition it emits acrid smoke and irritating fumes.

AAW750 CAS:117-98-6 HR: 1
ACETIC ACID-VETIVEROL ESTER
mf: $C_{17}H_{27}O_2$ mw: 263.44

SYNS: 1,2,3,3a,4,5,6,8a-OCTAHYDRO-2-ISOPROPYLIDENE-6-AZULENOL-4,8-DIMETHYL ACETATE ☐ VETIVER ACETATE ☐ VETIVEROL ACETATE ☐ VETIVERT ACETATE ☐ VETIVERYL ACETATE

TOXICITY DATA WITH REFERENCE
skn-rbt 500 mg/24H MOD FCTXAV 12,1011,74

CONSENSUS REPORTS: Reported in EPA TSCA Inventory.

SAFETY PROFILE: A skin irritant. See also ESTERS. When heated to decomposition it emits acrid smoke and irritating fumes.

AAX175 CAS:9003-22-9 HR: 1
ACETIC ACID, VINYL ESTER, POLYMER with CHLOROETHYLENE
mf: $(C_4H_6O_2 \cdot C_2H_3Cl)_n$

SYNS: A 15 (polymer) ☐ ACETIC ACID ETHENYL ESTER POLYMER with CHLORETHENE (9CI) ☐ BAKELITE LP 70 ☐ BAKELITE VLFV ☐ BAKELITE VMCC ☐ BAKELITE VYNS ☐ BREON 351 ☐ CHLOROETHYLENEVINYL ACETATE POLYMER ☐ CORVIC 236581 ☐ DENKALAC 61 ☐ DIAMOND SHAMROCK 744 ☐ EXON 450 ☐ EXON 454 ☐ GEON 135 ☐ HOSTAFLEX VP 150 ☐ LEUCOVYL PA 1302 ☐ NORVINYL P 6 ☐ OPALON 400 ☐ PLIOVAC AO ☐ POLYVINYL CHLORIDE–POLYVINYL ACETATE ☐ PVC CORDO ☐ RHODOPAS 6000 ☐ SARPIFAN HP 1 ☐ SCONATEX ☐ SOLVIC 523KC ☐ SUMILIT PCX ☐ TENNUS 0565 ☐ TYGON ☐ VAGD ☐ VINNOL H 10/60 ☐ VINYL ACETATE–VINYL CHLORIDE COPOLYMER ☐ VINYL ACETATE–VINYL CHLORIDE POLYMER ☐ VINYL CHLORIDE–VINYL ACETATE POLYMER ☐ VINYLITE VYDR 21 ☐ VLVF ☐ VMCC ☐ VYNW

TOXICITY DATA WITH REFERENCE
imp-mus TDLo: 1200 mg/kg: ETA JNCIAM 58,1443,77

CONSENSUS REPORTS: IARC Cancer Review: Animal Limited Evidence IMEMDT 19,377,79. Reported in EPA TSCA Inventory.

SAFETY PROFILE: Suspected carcinogen with experimental tumorigenic data. When heated to decomposition it emits toxic fumes of HCl.

AAX250 CAS:9003-20-7 HR: 1
ACETIC ACID VINYL ESTER POLYMERS
mf: $(C_4H_6O_2)_n$

PROP: Clear, water-white solid resin. Sol in benzene, acetone; insol in water.

SYNS: ACETIC ACID ETHENYL ESTER HOMOPOLYMER ☐ ASAHISOL 1527 ☐ ASB 516 ☐ AYAA ☐ AYAF ☐ BAKELITE AYAA ☐ BAKELITE LP 90 ☐ BASCOREZ ☐ BOND CH 18 ☐ BOOKSAVER ☐ BORDEN 2123 ☐ CEVIAN A 678 ☐ D 50 ☐ DANFIRM ☐ DARATAK ☐ DCA 70 ☐ DUVILAX BD 20 ☐ ELMER'S GLUE ALL ☐ EP 1463 ☐ FORMVAR 1285 ☐ GELVA CSV 16 ☐ GOHSENYL E 50 Y ☐ KURARE

OM 100 □ LEMAC 1000 □ MERCKOGEN 6000 □ MOVINYL 114 □ NATIONAL 120-1207 □ POLYVINYL ACETATE (FCC) □ PROTEX (POLYMER) □ RHODOPAS M □ SOVIOL □ SP 60 ESTER □ TOA-BOND 40H □ UCAR 130 □ VA 0112 □ VINAC B 7 □ VINYL ACETATE HOMOPOLYMER □ VINYL ACETATE POLYMER □ VINYL ACETATE RESIN □ VINYL PRODUCTS R 10688 □ WINACET D

TOXICITY DATA WITH REFERENCE
orl-rat LD:>25 g/kg JACTDZ 11,465,92
orl-mus LD:>25 g/kg JACTDZ 11,465,92

CONSENSUS REPORTS: IARC Cancer Review: Animal Inadequate Evidence IMEMDT 19,341,79. Reported in EPA TSCA Inventory.

SAFETY PROFILE: Very low toxicity by ingestion. When heated to decomposition it emits acrid smoke and irritating fumes. See also ESTERS.

AAX500 CAS:108-24-7 HR: 3
ACETIC ANHYDRIDE
DOT: UN 1715
mf: $C_4H_6O_3$ mw: 102.10

PROP: Colorless, very mobile, strongly refractive liquid; very strong, irritating acetic odor. Mp: −73.1°, bp: 139.55°, flash p: 129°F (CC), d: 1.082 @ 20°/4°, lel: 2.9%, uel: 10.3%, autoign temp: 734°F, vap press: 10 mm @ 36.0°, vap d: 3.52. Sltly sol in water; sol in org solvs. Decomp in hot water and hot alc; misc in alc and eth.

SYNS: ACETANHYDRIDE □ ACETIC ACID, ANHYDRIDE (9CI) □ ACETIC OXIDE □ ACETYL ANHYDRIDE □ ACETYL ETHER □ ACETYL OXIDE □ ANHYDRIDE ACETIQUE (FRENCH) □ ANHYDRID KYSELINY OCTOVE □ ANIDRIDE ACETICA (ITALIAN) □ AZIJNZUURANHYDRIDE (DUTCH) □ ESSIGSAEUREANHYDRID (GERMAN) □ ETHANOIC ANHYDRATE □ OCTOWY BEZWODNIK (POLISH)

TOXICITY DATA WITH REFERENCE
skn-rbt 10 mg/24H open MLD AMIHBC 4,119,51
skn-rbt 540 mg open MLD UCDS** 8/7/63
eye-rbt 250 µg open SEV AMIHBC 4,119,51
orl-rat LD50:1780 mg/kg AMIHBC 4,119,51
ihl-rat LC50:1000 ppm/4H 34ZIAG -,607,69
skn-rbt LD50:4000 mg/kg UCDS** 8/7/63

CONSENSUS REPORTS: Reported in EPA TSCA Inventory.

OSHA PEL: CL 5 ppm
ACGIH TLV: CL 5 ppm (Proposed: TWA 5 ppm)
DFG MAK: 5 ppm (20 mg/m³)
NIOSH REL: Acetic Anhydride: CL 5 ppm
DOT CLASSIFICATION: 8; *Label:* Corrosive

SAFETY PROFILE: Moderately toxic by inhalation, ingestion, and skin contact. A skin and severe eye irritant. A flammable liquid. A fire and explosion hazard when exposed to heat or flame. Potentially explosive reactions with barium peroxide, boric acid, chromium trioxide, 1,3-diphenyltriazene, hydrochloric acid + water, hypochlorous acid, nitric acid, perchloric acid + water, peroxyacetic acid, potassium permanganate, tetrafluoroboric acid, 4-toluenesulfonic acid + water, and acetic acid + water. Reactions with ethanol + sodium hydrogen sulfate, and hydrogen peroxide form explosive products. Reactions with ammonium nitrate + hexamethylenetetrammonium acetate + nitric acid form as products the military explosives RDX and HMX. Reacts violently with N-tert-butylphthalimic acid + tetrafluoroboric acid, chromic acid, glycerol + phosphoryl chloride, and metal nitrates (e.g., copper or sodium nitrates). Incompatible with 2-aminoethanol, aniline, chlorosulfonic acid, (CrO_3 + acetic acid), ethylenediamine, ethyleneimine, glycerol, oleum, HF, permanganates, NaOH, Na_2O_2, H_2SO_4, water, N_2O_2, (glycerol + phosphoryl chloride). When heated to decomposition it emits toxic fumes; can react vigorously with oxidizing materials, will react violently on contact with water or steam. Used in production of drugs of abuse. To fight fire, use CO_2, dry chemical, water mist, alcohol foam. See also ANHYDRIDES.

For occupational chemical analysis use NIOSH: Acetic Anhydride, 3506.

AAX750 CAS:93-29-8 HR: 2
ACETISOEUGENOL
mf: $C_{12}H_{14}O_3$ mw: 206.26

PROP: White crystals; clove odor. Flash p: 153°F. Sol in alc, chloroform, eth; insol in water.

SYNS: 4-ACETOXY-3-METHOXY-1-PROPENYLBENZENE □ ACETYLISOEUGENOL □ FEMA No. 2470 □ ISOEUGENOL ACETATE □ ISOEUGENYL ACETATE (FCC) □ 2-METHOXY-4-PROPENYLPHENYL ACETATE

TOXICITY DATA WITH REFERENCE
orl-rat LD50:3450 mg/kg FCTXAV 13,681,75

CONSENSUS REPORTS: Reported in EPA TSCA Inventory.

SAFETY PROFILE: Moderately toxic by ingestion. Combustible liquid. When heated to decomposition it emits acrid smoke and irritating fumes.

AAY000 CAS:102-01-2 HR: 2
ACETOACETANILIDE
mf: $C_{10}H_{11}NO_2$ mw: 177.22

PROP: White, crystalline solid. Mp: 86°, bp: decomp, flash p: 365°F (COC), d: 1.260 @ 20°, vap press: 0.01 mm @ 20°.

SYNS: AAN □ ACETANILIDE, 2-ACETYL- □ ACETOACETAMIDOBENZENE □ ACETOACETANILID □ ACETOACETIC ACID ANILIDE □ ACETOACETIC ANILIDE □ ((ACETOACETYL)AMINO)BENZENE □ ACETOACETYLANILINE □ ACETYLACETANILIDE □ α-ACETYLACETANILIDE □ N-(ACETYLACETYL)ANILINE □ ANILID KYSELINY ACETOCTOVE □ BUTANAMIDE, 3-OXO-N-PHENYL-(9CI) □ β-KETOBUTYRANILIDE □ 3-OXO-N-PHENYLBUTANAMIDE □ N-PHENYLACETOACETAMIDE □ USAF EK-1239

TOXICITY DATA WITH REFERENCE
orl-rat LD50:5400 mg/kg LONZA# 08FEB79
orl-mus LD50:3400 mg/kg GTPZAB 31(1),49,87
ipr-mus LD50:300 mg/kg NTIS** AD277-689
orl-rbt LD50:3925 mg/kg GTPZAB 31(1),49,87

CONSENSUS REPORTS: Reported in EPA TSCA Inventory.

SAFETY PROFILE: Poison by intraperitoneal route. Moderately toxic by ingestion. A weak allergen. See also

ACETANILIDE. Combustible when exposed to heat or flame. See ANILINE and CYANIDE for disaster hazard. When heated to decomposition it emits toxic NO_x fumes. To fight fire, use alcohol foam, water mist, CO_2, dry chemical.

AAY250 **CAS:101-92-8** **HR: 2**
ACETOACET-p-CHLORANILIDE
mf: $NC_{10}H_{10}O_2Cl$ mw: 211.65

PROP: Crystals. Mp: 107°, bp: decomp, flash p: 350°F (COC), d: 1.438 @ 20°, vap press: 0.01 mm @ 20°, vap d: 7.31.

SYNS: ACETOACETANILIDE, p-CHLORO- □ ACETOACET-p-CHLOROANILIDE □ BUTANAMIDE, N-(4-CHLOROPHENYL)-3-OXO-(9CI) □ p-CHLOROACETOACETANILIDE □ 4′-CHLOROACETOACETANILIDE □ N-(4-CHLOROPHENYL)-3-OXOBUTANAMIDE

TOXICITY DATA WITH REFERENCE
ipr-mus LDLo: 500 mg/kg CBCCT* 4,225,52

CONSENSUS REPORTS: Reported in EPA TSCA Inventory.

SAFETY PROFILE: Moderately toxic by intraperitoneal route. See also ACETANILIDE. Combustible when exposed to heat or flame. Dangerous: see ANILINE and CYANIDE. Can react vigorously with oxidizing materials. To fight fire, use water, foam, CO_2, water mist, dry chemical. When heated to decomposition it emits toxic fumes of Cl^-, CN^-, and NO_x.

AAY300 **CAS:20139-55-3** **HR: 2**
ACETOACET-4-CHLORO-2-METHYLANILIDE
mf: $C_{11}H_{12}ClNO_2$ mw: 225.69

SYNS: o-ACETOACETANISIDIDE, 4′-CHLORO- □ BUTANAMIDE, N-(4-CHLORO-2-METHYLPHENYL)-3-OXO- □ BUTYRANILIDE, 4′-CHLORO-2′-METHYL-3-OXO-

TOXICITY DATA WITH REFERENCE
orl-rat LD50:3500 mg/kg LONZA# 22SEP81

CONSENSUS REPORTS: Reported in EPA TSCA Inventory.

SAFETY PROFILE: Moderately toxic by ingestion. When heated to decomposition it emits toxic vapors of NO_x.

AAY600 **CAS:93-70-9** **HR: 1**
o-ACETOACETOCHLORANILIDE
mf: $C_{10}H_{10}ClNO_2$ mw: 211.66

SYNS: AAoC □ ACETOACETANILIDE, o-CHLORO- □ ACETOACETANILIDE, 2′-CHLORO- □ ACETOACET-o-CHLORANILIDE □ ACETOACET-o-CHLOROANILIDE □ ACETOACETYL-2-CHLOROANILIDE □ BUTANEAMIDE, N-(2-CHLOROPHENYL)-3-OXO- □ o-CHLOROACETOACETANILIDE □ 2′-CHLOROACETOACETANILIDE □ N-(2-CHLOROPHENYL)ACETOACETAMIDE □ 3-OXO-N-(2-CHLOROPHENYLBUTANAMIDE)

TOXICITY DATA WITH REFERENCE
orl-rat LD50:11,600 mg/kg LONZA# 10JUL81

CONSENSUS REPORTS: Reported in EPA TSCA Inventory.

SAFETY PROFILE: Slightly toxic by ingestion. When heated to decomposition it emits toxic vapors of NO_x and Cl^-.

AAY750 **CAS:21282-96-2** **HR: 3**
2-ACETOACETOXYETHYL ACRYLATE
mf: $C_9H_{12}O_5$ mw: 200.21

SYNS: ACETOACETIC ACID, 2-HYDROXYETHYL ESTER, ACRYLATE (8CI) □ AKRYLOYLOXYETHYLESTER KYSELINY ACETOCTOVE □ BUTANOIC ACID, 3-OXO-, 2-((1-OXO-2-PROPENYL)OXY)ETHYL ESTER (9CI) □ 2-HYDROXYETHYL ACETOACETATE ACRYLATE

TOXICITY DATA WITH REFERENCE
skn-rbt 100 mg/24H MOD 85JCAE-,730,86
eye-rbt 20 mg/24H MOD 85JCAE-,730,86
orl-rat LD50:1300 mg/kg TXAPA9 28,313,74
skn-rbt LD50:280 mg/kg TXAPA9 28,313,74

SAFETY PROFILE: Poison by skin contact. Moderately toxic by ingestion. A skin and eye irritant. See also ESTERS. When heated to decomposition it emits acrid smoke and irritating fumes.

AAZ000 **CAS:122-82-7** **HR: 3**
ACETOACET-p-PHENETIDIDE
mf: $C_{12}H_{15}NO_3$ mw: 221.28

PROP: Crystals. Mp: 108.5°, bp: decomp, flash p: 325°F (OC), d: 1.220 @ 20°, vap press: 0.02 mm @ 20°, vap d: 7.63.

SYNS: p-ACETOACETOPHENETIDIDE □ 4-ETHOXYACETOACETANILIDE □ 4′-ETHOXYACETOACETANILIDE

TOXICITY DATA WITH REFERENCE
orl-rat LD50:176 mg/kg FRPSAX 19,822,64

CONSENSUS REPORTS: Reported in EPA TSCA Inventory.

SAFETY PROFILE: Poison by ingestion. See also ACETANILIDE. Combustible. To fight fire, use water, foam, CO_2, water spray, mist, dry chemical. When heated to decomposition it emits toxic fumes of NO_x.

ABA000 **CAS:93-68-5** **HR: 2**
ACETOACET-o-TOLUIDIDE
mf: $C_{11}H_{13}NO_2$ mw: 191.25

PROP: Crystals. Mp: 106°, bp: decomp, d: 1.300 @ 20°, vap press: 0.01 mm @ 20°, flash p: 320°F (COC).

SYNS: 2-ACETOACETYLAMINOTOLUENE □ ACETOACETYL-2-METHYLANILIDE □ 2′-METHYLACETOACETANILIDE

TOXICITY DATA WITH REFERENCE
orl-rat LD50:1600 mg/kg KODAK* -,N-229,76
orl-mus LD50:1600 mg/kg KODAK* -,N-229,76

CONSENSUS REPORTS: Reported in EPA TSCA Inventory.

SAFETY PROFILE: Moderately toxic by ingestion. When heated to decomposition it emits toxic fumes of NO_x.

ABA500 CAS:92-15-9 HR: 2
ACETOACETYL-o-ANISIDINE
mf: $C_{11}H_{13}NO_3$ mw: 207.25

PROP: Crystals. Mp: 86.6°, flash p: 325°F (OC), d: 1.132 @ 86.6°/20°, vap d: 7.0.

SYNS: o-ACETOACETANISIDE □ ACETOACET-o-ANISIDIN (CZECH) □ ACETOACETIC ACID-o-ANISIDIDE □ 2-ACETOACETYLAMINOANISOLE □ ACETOACETYL-o-ANISIDE □ ACETOACETYL-o-ANISINE □ o-METHOXYACETOACETANILIDE □ 2-METHOXYACETOACETANILIDE □ 2'-METHOXYACETOACETANILIDE

TOXICITY DATA WITH REFERENCE
skn-rbt 500 mg/24H MLD 28ZPAK -,116,72
eye-rbt 500 mg/24H MOD 28ZPAK -,116,72

CONSENSUS REPORTS: Reported in EPA TSCA Inventory.

SAFETY PROFILE: A skin and eye irritant. When heated to decomposition it emits toxic fumes of NO_x. Combustible when exposed to heat or flame or oxidizing materials. To fight fire, use CO_2, mist, dry chemicals.

ABA750 CAS:1271-55-2 HR: 3
ACETOFERROCENE
mf: $C_{12}H_{12}FeO$ mw: 228.09

PROP: Orange crystals from heptane. Mp: 85–86°.

SYNS: ACETYLFERROCENE □ 1-ACETYLFERROCENE □ FERROCENE, ACETYL- □ MONACETYLFERROCENE

TOXICITY DATA WITH REFERENCE
orl-rat LDLo:5 mg/kg EPASR* 8EHQ-1285-0578
skn-rat LDLo:500 mg/kg EPASR* 8EHQ-1285-0578
ivn-mus LD50:75 mg/kg CSLNX* NX#08812
ocu-rbt LDLo:30 mg/kg TOLED5 38,103,87

CONSENSUS REPORTS: Reported in EPA TSCA Inventory.

DOT CLASSIFICATION: 3; *Label:* Flammable Liquid

SAFETY PROFILE: Poison by ocular and intravenous routes. A flammable liquid. When heated to decomposition it emits acrid smoke and irritating fumes.

ABB000 CAS:968-81-0 HR: 3
ACETOHEXAMIDE
mf: $C_{15}H_{20}N_2O_4S$ mw: 324.43

PROP: Crystals from aq ethanol. Mp: 188–189°.

SYNS: 1-(p-ACETYLBENZENESULFONYL)-3-CYCLOHEXYLUREA □ 4-ACETYL-N-((CYCLOHEXYLAMINO)CARBONYL)-BENZENESULFONAMIDE □ CYCLAMIDE □ DIMELIN □ DIMELOR □ DYMELOR □ NCI-CO03247 □ ORDIMEL □ TSIKLAMID

TOXICITY DATA WITH REFERENCE
unr-wmn TDLo:900 mg/kg (26-39W preg):REP BMJOAE 2,187,64
orl-rat LD50:5000 mg/kg TXAPA9 18,185,71

CONSENSUS REPORTS: NCI Carcinogenesis Bioassay (feed); No Evidence: mouse, rat NCITR* NCI-CG-TR-50,78. Reported in EPA TSCA Inventory.

SAFETY PROFILE: Human reproductive effects by an unspecified route: stillbirth. Mildly toxic by ingestion. When heated to decomposition it emits very toxic fumes of SO_x and NO_x.

ABB250 CAS:546-88-3 HR: 2
ACETOHYDROXAMIC ACID
mf: $C_2H_5NO_2$ mw: 75.08

PROP: Hygroscopic crystals. Mp: 89–92°.

SYNS: ACETHYDROXAMSAEURE (GERMAN) □ ACETIC ACID, OXIME □ ACETOHYDROXIMIC ACID □ ACETYLHYDROXAMIC ACID □ AHA □ METHYLHYDROXAMIC ACID

TOXICITY DATA WITH REFERENCE
mmo-sat 160 μmol/plate JOPHDQ 3,557,80
dns-rat:lvr 5 mmol/L MUREAV 145,201,85
mma-ham:lng 20 mmol/L MUREAV 152,225,85
ipr-rat TDLo:750 mg/kg (12D preg):REP JMXSAE 2,230,74
ipr-rat TDLo:750 mg/kg (12D preg):TER JMXSAE 2,230,74
ipr-mus LD50:1300 mg/kg PSEBAA 92 660,56

SAFETY PROFILE: Moderately toxic by intraperitoneal route. An experimental teratogen. Other experimental reproductive effects. Mutation data reported. When heated to decomposition it emits toxic fumes of NO_x.

ABB500 CAS:513-86-0 HR: 3
ACETOIN
DOT: UN 2621
mf: $C_4H_8O_2$ mw: 88.12

PROP: Sltly yellow liquid or crystalline solid; buttery odor. D: 1.016, bp: 147–148°, refr index: 1.417, mp: 15°, flash p: 106°F. Misc with water, alc, propylene glycol; insol in vegetable oil.

SYNS: ACETYL METHYL CARBINOL □ 2-BUTANOL-3-ONE □ 2,3-BUTANOLONE □ DIMETHYLKETOL □ FEMA No. 2008 □ 3-HYDROXY-2-BUTANONE □ 1-HYDROXYETHYL METHYL KETONE □ γ-HYDROXY-β-OXOBUTANE

TOXICITY DATA WITH REFERENCE
skn-rbt 500 mg/24H MOD CNREA8 33,3069,73
orl-rat TDLo:12,600 mg/kg (42D male):REP FCTXAV 10,131,72
scu-rat LDLo:14 g/kg FCTXAV 17,509,79

CONSENSUS REPORTS: Reported in EPA TSCA Inventory.

DOT CLASSIFICATION: 3; *Label:* Flammable Liquid

SAFETY PROFILE: Experimental reproductive effects. Mildly toxic by subcutaneous route. A moderate skin irritant. Flammable liquid. When heated to decomposition it emits acrid smoke and fumes. See also KETONES.

ABC000 CAS:116-09-6 HR: 2
ACETOL (1)
mf: $C_3H_6O_2$ mw: 74.09

$HOCH_2CO{\cdot}CH_3$

PROP: Colorless liquid. D: 1.084 @ 20°/4°, fp −17°

(approx), mp: −7°, bp: 145–146° decomp. Misc in water, alc, and eth.

SYNS: HYDROXYACETONE □ 1-HYDROXY-2-PROPANONE

TOXICITY DATA WITH REFERENCE
mmo-sat 500 μg/plate ABCHA6 47,2461,83
orl-rat LD50:2200 mg/kg JIHTAB 30,63,48

CONSENSUS REPORTS: Reported in EPA TSCA Inventory.

SAFETY PROFILE: Moderately toxic by ingestion. Mutation data reported. An allergen. Implicated in aplastic anemia. A 10 gram dose may be fatal to an adult. Skin contact, inhalation, or ingestion can cause asthma, sneezing, irritation of eyes and nose, hives, and eczema. Combustible when exposed to heat or flame. When heated to decomposition it emits acrid smoke and fumes.

ABC250 CAS:828-00-2 HR: 2
ACETOMETHOXANE
mf: $C_8H_{14}O_4$ mw: 174.22

PROP: Yellow to amber, clear liquid. D: 1.068–1.075 @ 25°/25°, bp: 66–68° @ 3 mm, fp: <−25°. Sol in water and org solvs. Misc in water.

SYNS: ACETIC ACID-2,6-DIMETHYL-m-DIOXAN-4-YL ESTER □ ACETOMETHOXAN □ 6-ACETOXY-2,4-DIMETHYL-m-DIOXANE □ DDOA □ DIMETHOXANE □ 2,6-DIMETHYL-m-DIOXAN-4-OL ACETATE □ 2,6-DIMETHYL-m-DIOXAN-4-YL ACETATE □ DIOXIN (bactericide) (OBS.) □ G1V GARD DXN □ NCI-C56213

TOXICITY DATA WITH REFERENCE
mma-sat 5500 μg/plate ENMUDM 8(Suppl 7),1,86
sln-dmg-par 1 pph ENMUDM 7,677,85
orl-rat TDLo:948 g/kg/88W-I:CAR JNCIAM 53,791,74
orl-mus TD:25,750 mg/kg/2Y-C:ETA NTPTR* NTP-TR-354,89
orl-rat LD50:1930 mg/kg GCTB** 3/25/77
orl-mus LDLo:2800 mg/kg NTPTR* NTP-TR-354,89

CONSENSUS REPORTS: IARC Cancer Review: Group 3 IMEMDT 7,56,87; Animal Limited Evidence IMEMDT 15,177,77. NTP Carcinogenesis Studies (gavage): Equivocal Evidence: MOUSE NTPTR* NTP-TR-354,89; (gavage): No Evidence: RAT NTPTR* NTP-TR-354,89.

SAFETY PROFILE: Questionable carcinogen with experimental carcinogenic data. Moderately toxic by ingestion. Mutation data reported. See also ESTERS. When heated to decomposition it emits acrid smoke and fumes.

ABC475 CAS:941-98-0 HR: 2
1′-ACETONAPHTHONE
mf: $C_{12}H_{10}O$ mw: 170.22

PROP: Crystals. Mp: 34°, bp: 302°, flash p: >230°F, d: 1.120.

SYNS: 1-ACETONAPHTHALENE □ α-ACETONAPHTHONE □ 1-ACETONAPHTHONE □ 1-ACETYLNAPHTHALENE □ ETHANONE, 1-(1-NAPHTHALENYL)-(9CI) □ METHYL α-NAPHTHYL KETONE □ METHYL 1-NAPHTHYL KETONE □ α-METHYL NAPHTHYL KETONE □ 1-(1-NAPHTHALENYL)ETHANONE □ α-NAPHTHYL METHYL KETONE □ 1-NAPHTHYL METHYL KETONE

TOXICITY DATA WITH REFERENCE
skn-rbt 500 mg/24H MLD FCTOD7 20,755,82
orl-rat LD50:1560 mg/kg FCTOD7 20,755,82

CONSENSUS REPORTS: Reported in EPA TSCA Inventory.

DOT CLASSIFICATION: 3; *Label:* Flammable Liquid

SAFETY PROFILE: Moderately toxic by ingestion. A skin irritant. A combustible liquid. When heated to decomposition it emits acrid smoke and irritating fumes.

ABC500 CAS:93-08-3 HR: 2
2′-ACETONAPHTHONE
mf: $C_{12}H_{10}O$ mw: 170.22

PROP: White needles; orange-blossom odor. Flash p: 264°F. Mp: 56°, bp: 301–303°. Sol in fixed oils; sltly sol in propylene glycol; insol in glycerin.

SYNS: β-ACETONAPHTHALENE □ ACETONAPHTHONE □ β-ACETONAPHTHONE □ 2-ACETONAPHTHONE □ β-ACETYLNAPHTHALENE □ 2-ACETYLNAPHTHALENE □ FEMA No. 2723 □ METHYL-β-NAPHTHYL KETONE (FCC) □ METHYL-2-NAPHTHYL KETONE □ β-METHYL NAPHTHYL KETONE □ 1-(2-NAPHTHALENYL)ETHANONE □ β-NAPHTHYL METHYL KETONE □ 2-NAPHTHYL METHYL KETONE □ ORANGE CRYSTALS

TOXICITY DATA WITH REFERENCE
skn-hmn 100% FCTXAV 13,867,75
orl-mus LD50:599 mg/kg MDZEAK 8,244,67

CONSENSUS REPORTS: Reported in EPA TSCA Inventory.

DOT CLASSIFICATION: 3; *Label:* Flammable Liquid

SAFETY PROFILE: Moderately toxic by ingestion. A human skin irritant. Flammable liquid. When heated to decomposition it emits acrid smoke and fumes.

ABC750 CAS:67-64-1 HR: 3
ACETONE
DOT: UN 1090/UN 1091
mf: C_3H_6O mw: 58.09

PROP: Volatile, colorless liquid; fragrant mint-like odor. Mp: −94.6°, bp: 56.2° @ 20 mm, refr index: 1.356, flash p: 0°F (CC), lel: 2.6%, uel: 12.8%, d: 0.7972 @ 15°, autoign temp: (color) 869°F, vap press: 240 hPa @ 20°, vap d: 2.00. Misc in water, alc, org solvs, and eth.

SYNS: ACETON (GERMAN, DUTCH, POLISH) □ ACETONE OILS (DOT) □ CHEVRON ACETONE □ DIMETHYLFORMALDEHYDE □ DIMETHYLKETAL □ DIMETHYL KETONE □ FEMA No. 3326 □ KETONE, DIMETHYL □ KETONE PROPANE □ β-KETOPROPANE □ METHYL KETONE □ PROPANONE □ 2-PROPANONE □ PYROACETIC ACID □ PYROACETIC ETHER □ RCRA WASTE NUMBER U002

TOXICITY DATA WITH REFERENCE
eye-hmn 500 ppm JIHTAB 25,282,43
skn-rbt 395 mg open MLD UCDS** 5/7/70
skn-rbt 500 mg/24H MLD 28ZPAK -,42,72
eye-rbt 3950 μg SEV AJOPAA 29,1363,46
eye-rbt 20 mg/24H MOD 85JCAE-,280,86

cyt-smc 200 mmol/tube HEREAY 33,457,47
sln-smc 47,600 ppm ANYAA9 407,186,83
ihl-mam TCLo:31,500 µg/m³/24H (1-13D preg):REP GTPZAB 26(6),24,82
orl-man TDLo:2857 mg/kg 34ZIAG -,64,69
orl-man TDLo:2857 mg/kg DIAEAZ 15,810,66
ihl-man TCLo:12,000 ppm/4H:CNS AOHYA3 16,73,73
ihl-man TDLo:440 µg/m³/6M GISAAA 42(8)42,77
ihl-man TDLo:10 mg/m³/6H GISAAA 42(8)42,77
ihl-hmn TCLo:500 ppm:EYE JIHTAB 25,282,43
ihl-man TCLo:12,000 ppm/4H:GIT AOHYA3 16,73,73
ivn-rat LD50:5500 mg/kg NPIRI* 1,1,74
orl-rat LD50:5800 mg/kg JTEHD6 15,609,85
ihl-rat LC50:50,100 mg/m³/8H AIHAAP 20,364,59
ipr-rat LDLo:500 mg/kg JPPMAB 11,150,59
ivn-rat LD50:5500 mg/kg NPIRI* 1,1,74
orl-mus LD50:3000 mg/kg PCJOAU 14,162,80
ihl-mus LCLo:110 g/m³/1H AGGHAR 5,1,33
ipr-mus LD50:1297 mg/kg SCCUR* -,1,61
ivn-mus LDLo:4 g/kg FAONAU 48A,86,70
orl-dog LDLo:8 g/kg FAONAU 48A,86,70
orl-rbt LD50:5340 mg/kg FAONAU 48A,86,70
skn-rbt LD50:20 g/kg UCDS** 5/7/70

CONSENSUS REPORTS: On Community Right-To-Know List. Reported in EPA TSCA Inventory.

OSHA PEL: TWA 750 ppm; STEL 1000 ppm
ACGIH TLV: TWA 750 ppm; STEL 1000 ppm
DFG MAK: 500 ppm (1200 mg/m³)
NIOSH REL: (Ketones) 10H TWA 590 mg/m³
DOT CLASSIFICATION: 3; *Label:* Flammable Liquid

SAFETY PROFILE: Moderately toxic by various routes. A skin and severe eye irritant. Human systemic effects by inhalation: changes in EEG, changes in carbohydrate metabolism, nasal effects, conjunctiva irritation, respiratory system effects, nausea and vomiting, and muscle weakness. Human systemic effects by ingestion: coma, kidney damage, and metabolic changes. Narcotic in high concentration. In industry, no injurious effects have been reported other than skin irritation resulting from its defatting action, or headache from prolonged inhalation. Experimental reproductive effects. A common air contaminant. Highly flammable liquid. Dangerous disaster hazard due to fire and explosion hazard; can react vigorously with oxidizing materials.

Potentially explosive reaction with nitric acid + sulfuric acid, bromine trifluoride, nitrosyl chloride + platinum, nitrosyl perchlorate, chromyl chloride, thiotrithiazyl perchlorate, and (2,4,6-trichloro-1,3,5-triazine + water). Reacts to form explosive peroxide products with 2-methyl-1,3-butadiene, hydrogen peroxide, and peroxomonosulfuric acid. Ignites on contact with activated carbon, chromium trioxide, dioxygen difluoride + carbon dioxide, and potassium-tert-butoxide. Reacts violently with bromoform, chloroform + alkalies, bromine, and sulfur dichloride. Incompatible with CrO, (nitric + acetic acid), NOCl, nitryl perchlorate, permonosulfuric acid, NaOBr, (sulfuric acid + potassium dichromate), (thio-diglycol + hydrogen peroxide), trichloromelamine, air, HNO_3, chloroform, and H_2SO_4. To fight fire, use CO_2, dry chemical, alcohol foam. Used in production of drugs of abuse.

For occupational chemical analysis use OSHA: #ID-69 or NIOSH: KETONES I (desorption in CS_2) 1300.

ABD000 CAS:57-15-8 HR: 3
ACETONE CHLOROFORM
mf: $C_4H_7Cl_3O$ mw: 177.46

PROP: Hydrated crystals, camphor odor. Mp: 97° (78° anhyd), bp: 167°.

SYNS: ANHYDROUS CHLOROBUTANOL ☐ CHLORBUTANOL ☐ CHLORBUTOL ☐ CHLORETONE ☐ CHLOROBUTANOL ☐CLORTRAN ☐ HCP ☐ METHAFORM ☐ SEDAFORM ☐ TRICHLORO-tert-BUTYL ALCOHOL ☐ tert-TRICHLOROBUTYL ALCOHOL ☐ β,β,β-TRICHLORO-tert-BUTYL ALCOHOL ☐ 1,1,1-TRICHLORO-2-METHYL-2-PROPANOL

TOXICITY DATA WITH REFERENCE
skn-rbt 850 µg MLD XEURAQ MDDC-1715
eye-rbt 9180 µg/30S MLD XEURAQ MDDC-1715
mmo-sat 20 µmol/plate MUREAV 90,91,81
cyt-smc 10 mmol/tube HEREAY 33,457,47
orl-dog LDLo:238 mg/kg AIPTAK 8,77,01
orl-rbt LDLo:213 mg/kg AIPTAK 8,77,01
par-frg LDLo:800 mg/kg AIPTAK 8,77,01

CONSENSUS REPORTS: Reported in EPA TSCA Inventory.

SAFETY PROFILE: Poison by ingestion. A narcotic. A skin and eye irritant. Mutation data reported. See also CHLORAL HYDRATE, which acts similarly. Dangerous; can react with oxidizing materials. Combustible when exposed to heat or flame. When heated to decomposition it emits toxic fumes of Cl^-. See also PHOSGENE.

ABD500 CAS:115-24-2 HR: 3
ACETONE DIETHYLSULFONE
mf: $C_7H_{16}O_4S_2$ mw: 228.35

PROP: Crystals. D: 1.183, mp: 124–126°, bp: 300° (sltly decomp). Sol in water, alc, and eth.

SYNS: ACETONE BIS(ETHYL SULFONE) ☐ 2,2-BIS(ETHYLSULFONYL)PROPANE ☐ DIETHYLSULFONDIMETHYLMETHANE ☐ PROPANE DIETHYL SULFONE ☐ SULFONAL ☐ SULFONMETHANE

TOXICITY DATA WITH REFERENCE
unk-man LDLo:147 mg/kg 85DCAI 2,73,70
orl-dog LDLo:900 mg/kg HBAMAK 4,1404,35
orl-rbt LDLo:3000 mg/kg HBAMAK 4,1404,35
orl-gpg LDLo:8500 mg/kg HBAMAK 4,1404,35

SAFETY PROFILE: A human poison by unspecified route. Moderately toxic by ingestion. Mutation data reported. When heated to decomposition it emits toxic fumes of SO_x.

ABE000 HR: 3
ACETONE PEROXIDE

PROP: Liquid or absorbed on cornstarch. The trimeric form is crystalline. Mp: 97°.

SAFETY PROFILE: Severe skin and eye irritant. Flammable by spontaneous chemical reaction, can react vigorously with reducing materials. The trimeric form is shock-sensitive and static-electricity-sensitive and may detonate.

ABE250 CAS:110-20-3 HR: 3
ACETONE SEMICARBAZONE
mf: $C_4H_9N_3O$ mw: 115.16

PROP: Mp: 190–199° (decomp). Sol in cold water; sltly sol in cold alc; insol in eth.

TOXICITY DATA WITH REFERENCE
ivn-mus LD50:90 mg/kg JPETAB 122,110,58

CONSENSUS REPORTS: Reported in EPA TSCA Inventory.

SAFETY PROFILE: Poison by intravenous route. When heated to decomposition it emits toxic fumes of NO_x.

ABE500 CAS:75-05-8 HR: 3
ACETONITRILE
DOT: UN 1648
mf: C_2H_3N mw: 41.06

PROP: Colorless liquid, almond-ethereal, aromatic odor. Mp: −45°, bp: 81.1°, flash p: 42°F (COC), d: 0.7868 @ 20°/20°, vap d: 1.42, vap press: 100 mm @ 27°, lel: 4.4%, uel: 16%, autoign temp: 975°F. Misc in water, alc, and org solvs. Immisc in petr eth.

SYNS: ACETONITRIL (GERMAN, DUTCH) □ CYANOMETHANE □ CYANURE de METHYL (FRENCH) □ ETHANENITRILE □ ETHYL NITRILE □ METHANECARBONITRILE □ METHANE, CYANO- □ METHYL CYANIDE □ METHYLKYANID □ NCI-C60822 □ RCRA WASTE NUMBER U003 □ USAF EK-488

TOXICITY DATA WITH REFERENCE
sln-smc 47,600 ppm MUREAV 149,339,85
skn-rbt 10 mg/24H JIHTAB 30,63,48
skn-rbt 500 mg open MLD UCDS** 3/18/65
eye-rbt 20 mg SEV JIHTAB 30,63,48
orl-ham TDLo:300 mg/kg (8D preg):TER TJADAB 27,313,83
orl-ham TDLo:400 mg/kg (8D preg):REP TJADAB 27,313,83
orl-hmn TDLo:570 mg/kg:CNS APTOA6 41,340,77
ihl-hmn TCLo:160 ppm/4H 34ZIAG -,65,69
orl-rat LD50:2730 mg/kg TXAPA9 19,699,71
ihl-rat LC50:7551 ppm/8H JOCMA7 1,634,59
ipr-rat LD50:850 mg/kg JOCMA7 1,634,59
scu-rat LD50:3500 mg/kg 85GMAT -,16,82
ivn-rat LD50:1680 mg/kg JOCMA7 1,634,59
par-rat LD50:1100 mg/kg 85GMAT -,16,82
orl-mus LD50:269 mg/kg ARTODN 55,47,84
ihl-mus LC50:2693 ppm/1H CTOXAO 18,991,81
ipr-mus LD50:175 mg/kg TXAPA9 59,589,81
scu-mus LD50:4480 mg/kg 85GMAT -,16,82
ihl-dog LCLo:16,000 ppm/4H JOCMA7 1,634,59

CONSENSUS REPORTS: On Community Right-To-Know List. Reported in EPA TSCA Inventory.

OSHA PEL: TWA 40 ppm; STEL 60 ppm
ACGIH TLV: TWA 40 ppm; STEL 60 ppm (skin)
DFG MAK: 40 ppm (70 mg/m³)
NIOSH REL: (Nitriles) TWA 34 mg/m³

DOT CLASSIFICATION: 3; *Label:* Flammable Liquid, Poison

SAFETY PROFILE: Poison by ingestion and intraperitoneal routes. Moderately toxic by several routes. An experimental teratogen. Other experimental reproductive effects. A skin and severe eye irritant. Human systemic effects by ingestion: convulsions, nausea or vomiting, and metabolic acidosis. Human respiratory system effects by inhalation. Mutation data reported. Dangerous fire hazard when exposed to heat, flame, or oxidizers. Explosion Hazard: See also CYANIDE and NITRILES. When heated to decomposition it emits highly toxic fumes of CN^- and NO_x. Potentially explosive reaction with lanthanide perchlorates and nitrogen-fluorine compounds. Exothermic reaction with sulfuric acid at 53°C. Will react with water, steam, acids to produce toxic and flammable vapors. Incompatible with oleum, chlorosulfonic acid, perchlorates, nitrating agents, indium, dinitrogen tetraoxide, N-fluoro compounds (e.g., perfluorourea + acetonitrile), HNO_3, SO_3. To fight fire, use foam, CO_2, dry chemical.

For occupational chemical analysis use NIOSH: Acetonitrile, 1606.

ABF000 CAS:127-06-0 HR: 2
ACETONOXIME
mf: C_3H_7NO mw: 73.11

PROP: Crystals. D: 0.97; mp: 60–61°; bp: 134–135°. Very sol in water, alc, and eth. Sol in ligroin eth.

SYNS: ACETOXIME □ β-ISONITROSOPROPANE □ 2-PROPANONE OXIME

TOXICITY DATA WITH REFERENCE
orl-rat LD:>500 mg/kg NCNSA6 5,26,53
ipr-mus LD50:4000 mg/kg JPETAB 119,522,57

CONSENSUS REPORTS: Reported in EPA TSCA Inventory.

SAFETY PROFILE: Moderately toxic by ingestion and intraperitoneal routes. When heated to decomposition it emits toxic fumes of NO_x.

ABF500 CAS:117-52-2 HR: 3
3-(α-ACETONYLFURFURYL)-4-HYDROXYCOUMARIN
mf: $C_{17}H_{14}O_5$ mw: 298.31

PROP: White powder; practically insol in water, sol in alcs. Mp: 124°.

SYNS: COUMAFURYL □ CUMAFURYL (GERMAN) □ FOUMARIN □ 3-(α-FURYL-β-ACETYLAETHYL)-4-HYDROXYCUMARIN (GERMAN) □ 3-(1-FURYL-3-ACETYLETHYL)-4-HYDROXYCOUMARIN □ KRUMKIL □ RATAFIN □ RAT-A-WAY

TOXICITY DATA WITH REFERENCE
orl-rat LDLo:400 mg/kg 85GYAZ -,115,71
orl-rat LD50:25 mg/kg FMCHA2 -,D146,80
orl-mus LD50:14,700 μg/kg FMCHA2 -,D146,80

SAFETY PROFILE: Poison by ingestion and possibly other routes. See also WARFARIN.

A

ABF750 CAS:152-72-7 HR: 3
3-(α-ACETONYL-p-NITROBENZYL)-4-HYDROXY-COUMARIN
mf: $C_{19}H_{15}NO_6$ mw: 353.35

SYNS: ACENOCOUMARIN □ ACENOCOUMAROL □ ACENOCUMAROL □ ACENOKUMARIN □ 3-(α-ACETONYL-p-NITROBENZYL)-4-HYDROXY-COUMARIN □ ASCUMAR □ 2H-1-BENZOPYRAN-2-ONE, 4-HYDROXY-3-(1-(4-NITROPHENYL)-3-OXOBUTYL)- □ G-23350 □ 4-HYDROXY-3-(1-(4-NITROPHENYL)-3-OXOBUTYL)-2H-1-BENZOPYRAN-2-ONE □ NICOUMALONE □ 3-(α-(p-NITROPHENOL)-β-ACETYLETHYL)-4-HYDROXYCOUMARIN □ NITROPHENYLACETYLETHYL-4-HYDROXYCOUMARINE □ 3-(α-p-NITROPHENYL-β-ACETYLETHYL)-4-HYDROXYCOUMARIN □ 3-(α-(4'-NITROPHENYL)-β-ACETYLETHYL)-4-HYDROXYCOUMARIN □ NITROVARFARIAN □ NITROWARFARIN □ SINCOUMAR □ SINKUMAR □ SINTHROM □ SINTHROME □ SINTROM □ SINTROMA □ SYNCOUMAR □ SYNCUMAR □ SYNTROM □ ZOTIL

TOXICITY DATA WITH REFERENCE
unr-wmn TDLo:24 mg/kg (1-28W preg):TER AFPEAM 36,63,79
orl-rat LD50:513 mg/kg 29ZVAB -,3,69
orl-mus LD50:1470 mg/kg THERAP 11,85,56
ipr-mus LD50:115 mg/kg MEIEDD 11,6,89

SAFETY PROFILE: Poison by intraperitoneal route. Moderately toxic by ingestion. A human teratogen by an unspecified route. When heated to decomposition it emits toxic fumes such as NO_x. See also WARFARIN.

ABG000 CAS:5714-00-1 HR: 3
ACETOPHENAZINE
mf: $C_{23}H_{29}N_3O_2S{\cdot}2C_4H_4O_4$ mw: 643.77

PROP: Solid. Mp: 175–178°.

SYNS: ACETOPHENAZINE MALEATE □ 2-ACETYL-10-(3-(4-(β-HYDROXYETHYL)PIPERAZINYL)PROPYL)PHENOTHIAZINE □ 1-(2-HYDROXYETHYL)-4-(3-(2-ACETYL-10-PHENOTHIAZYL)PROPYL)PIPERAZINE □ 1-(10-(3-(4-(2-HYDROXYETHYL)-1-PIPERAZINYL)PROPYL)-10H-PHENOTHIAZIN-2-YL)ETHANONE □ 10-(3-(4-(2-HYDROXYETHYL)-1-PIPERAZINYL)PROPYL)PHENOTHIAZIN-2-YL METHYL KETONE □ SCH 6673 □ TINDAL

TOXICITY DATA WITH REFERENCE
eye-rbt 112 mg SEV AMIHAB 14,250,56
orl-rat LD50:415 mg/kg 27ZQAG -,11,72
ipr-rat LD50:60 mg/kg 27ZQAG -,11,72
ivn-rat LD50:39 mg/kg 27ZQAG -,11,72
ivn-mus LD50:71 mg/kg CSLNX* NX#01100
orl-bwd LD50:75 mg/kg TXAPA9 21,315,72

SAFETY PROFILE: Poison by ingestion, intraperitoneal, and intravenous routes. Severe eye irritant. See also KETONES.

ABG250 CAS:591-33-3 HR: 2
m-ACETOPHENETIDIDE
mf: $C_{10}H_{13}NO_2$ mw: 179.24

PROP: Plates. Mp: 97–99°.

SYNS: m-ETHOXYACETANILIDE □ 3-ETHOXYACETANILIDE □ 3'-ETHOXYACETANILIDE □ N-(3-ETHOXYPHENYL)ACETAMIDE (9CI)

TOXICITY DATA WITH REFERENCE
orl-mus LD50:1250 mg/kg TXAPA9 19,20,71

CONSENSUS REPORTS: Reported in EPA TSCA Inventory.

SAFETY PROFILE: Moderately toxic by ingestion. See also p-ACETOPHENETIDIDE. When heated to decomposition it emits toxic fumes of NO_x.

ABG350 CAS:581-08-8 HR: 2
o-ACETOPHENETIDIDE
mf: $C_{10}H_{13}NO_2$ mw: 179.24

SYNS: ACETAMIDE, N-(2-ETHOXYPHENYL)-(9CI) □ ACETANILIDE, 2'-ETHOXY- □ 2-ETHOXYACETANILIDE □ 2'-ETHOXYACETANILIDE □ N-(2-ETHOXYPHENYL)ACETAMIDE

TOXICITY DATA WITH REFERENCE
orl-mus LD50:680 mg/kg TXAPA9 19,20,71

CONSENSUS REPORTS: Reported in EPA TSCA Inventory.

SAFETY PROFILE: Moderately toxic by ingestion When heated to decomposition it emits toxic vapors of NO_x.

ABG750 CAS:62-44-2 HR: 3
p-ACETOPHENETIDIDE
mf: $C_{10}H_{13}NO_2$ mw: 179.24

PROP: Solid. Mp: 137–138°, bp: 242–245°.

SYNS: 1-ACETAMIDO-4-ETHOXYBENZENE □ ACETO-p-PHENALIDE □ p-ACETOPHENETIDE □ ACETO-p-PHENETIDIDE □ ACETOPHENETIDIN □ ACETOPHENETIDINE □ ACETO-4-PHENETIDINE □ ACETOPHENETIN □ ACET-p-PHENALIDE □ ACET-p-PHENETIDIN □ ACETPHENETIDIN □ p-ACETPHENETIDIN □ ACETYLPHENETIDIN □ N-ACETYL-p-PHENETIDINE □ ACHROCIDIN □ ANAPAC □ APC □ ASA COMPOUND □ BROMO SELTZER □ BUFF-A-COMP □ CITRAFORT □ CODEMPIRAL □ COMMOTIONAL □ CONTRADOL □ CORICIDIN □ CORIFORTE □ CORYBAN-D □ DAPRISAL □ DARVON COMPOUND □ DASIKON □ EMPIRIN COMPOUND □ p-ETHOXYACETANILIDE □ 4-ETHOXYACETANILIDE □ N-p-ETHOXYPHENYLACETAMIDE □ N-(4-ETHOXYPHENYL)ACETAMIDE □ FENACETINA □ FIORINAL □ MELABON □ PARACETOPHENETIDIN □ PERCOBARB □ PERCODAN □ p-PHENACETIN □ RCRA WASTE NUMBER U187 □ SINUTAB □ TETRACYDIN □ XARIL □ ZACTIRIN COMPOUND

TOXICITY DATA WITH REFERENCE
mma-sat:333 μg/plate IARCCD 27,283,80
sce-mus-ipr 165 mg/kg JTEHD6 16,355,85
orl-rat TDLo:50,336 mg/kg (17W male):REP JCNDBK 11,96,71
orl-rat TDLo:24 g/kg (female 1-20D post):TER KLWOAZ 43,364,65
orl-wmn TDLo:80 g/kg/63Y-I:CAR JOURAA 113,653,75
orl-man TDLo:57 g/kg/47Y I:CAR JOURAA 113,653,75
scu-mus TDLo:19,200 mg/kg/24W-I:ETA VOONAW 32(5),63,86
orl-mus LD:484 g/kg/96W-C:NEO IJCNAW 29,439,82
unr-man LDLo:74 mg/kg 85DCAI 2,73,70
orl-rat LD50:3600 mg/kg ARZNAD 24,600,74
ipr-rat LD50:634 mg/kg NYKZAU 62,11,66
orl-mus LD50:866 mg/kg ARZNAD 28,1644,78
ihl-mus LC50:33,900 mg/m³ GISAAA 34(10),36,69
ipr-mus LD50:540 mg/kg YKKZAJ 81,659,61
scu-mus LD50:1625 mg/kg ARZNAD 8,25,58
ivn-dog LDLo:260 mg/kg NTIS** PB282-666
orl-rbt LD50:2500 mg/kg GTPZAB 21(9),53,77

scu-rbt LD50: 1 g/kg ARZNAD 21,719,71
orl-gpg LD50: 1870 mg/kg TXAPA9 2,23,60
orl-ham LD50: 1690 mg/kg PHARAT 8,572,53

CONSENSUS REPORTS: NTP 7th Annual Report on Carcinogens. IARC Cancer Review: Group 2A IMEMDT 7,310,87; Animal Inadequate Evidence IMEMDT 13,141,77; Human Limited Evidence IMEMDT 13,141,77; IMEMDT 24,135,80; Animal Limited Evidence IMEMDT 24,135,80; IMEMDT 24,135,80. Reported in EPA TSCA Inventory.

SAFETY PROFILE: Confirmed carcinogen producing tumors of the kidney and bladder. A human poison by an unspecified route. Poison by intravenous and possibly other routes. Moderately toxic by several routes. Human systemic effects by ingestion: cyanosis, liver damage, and methemoglobinemia-carboxhemoglobinemia. Experimental teratogenic data. Other experimental reproductive effects. Mutation data reported. Experimental reproductive effects. Chronic effects consist of weight loss, insomnia, shortness of breath, weakness, and often aplastic anemia. When heated to decomposition it emits toxic fumes of NO_x.

ABH000 CAS:98-86-2 HR: 3
ACETOPHENONE
mf: C_8H_8O mw: 120.16

PROP: Colorless liquid or plates; sweet, pungent odor. Mp: 19.7°, bp: 202.3°, d: 1.026 @ 20°/4°, vap d: 4.14, vap press: 1 mm @ 15°, autoign temp: 1060°F. Very sol in propylene glycol and fixed oils; sol in alc, chloroform, and eth; sltly sol in water; insol in glycerin.

SYNS: ACETYLBENZENE □ BENZOYL METHIDE □ DYMEX □ FEMA No. 2009 □ HYPNONE □ KETONE METHYL PHENYL □ METHYL PHENYL KETONE □ 1-PHENYLETHANONE □ PHENYL METHYL KETONE □ USAF EK-496

TOXICITY DATA WITH REFERENCE
skn-rbt 10 mg/24H open JIHTAB 26,269,44
skn-rbt 515 mg open MLD UCDS•• 12/27/71
eye-rbt 771 µg SEV AJOPAA 29,1363,46
cyt-smc 10 mmol/tube HEREAY 33,457,47
orl-rat LD50: 815 mg/kg GTPZAB 26(8),53,82
orl-mus LD50: 740 mg/kg GTPZAB 26(8),53,82
scu-mus LDLo: 330 mg/kg HDTU•• -,-,33
ipr-mus LD50: 200 mg/kg NTIS•• AD277-689

CONSENSUS REPORTS: Reported in EPA TSCA Inventory.

ACGIH TLV: (Proposed: 10 ppm)
DOT CLASSIFICATION: 3; *Label:* Flammable Liquid

SAFETY PROFILE: Poison by intraperitoneal and subcutaneous routes. Moderately toxic by ingestion. A skin and severe eye irritant. Mutation data reported. Narcotic in high concentration. A hypnotic. Flammable liquid. To fight fire use foam, CO_2, dry chemical. When heated to decomposition it emits acrid smoke and fumes. See also KETONES.

ABH150 CAS:613-91-2 HR: 3
ACETOPHENONE, OXIME
mf: C_8H_9NO mw: 135.18

SYN: ETHANONE, 1-PHENYL-, OXIME

TOXICITY DATA WITH REFERENCE
orl-mus LD50: 2 g/kg MEXPAG 11,137,64
unr-mus LD50: 450 mg/kg PCJOAU 12,227,78

CONSENSUS REPORTS: Reported in EPA TSCA Inventory.

SAFETY PROFILE: Poison by an unspecified route. Slightly toxic by ingestion. When heated to decomposition it emits toxic vapors of NO_x.

ABH500 CAS:61-00-7 HR: 3
ACETOPROMAZINE
mf: $C_{19}H_{22}N_2OS$ mw: 326.49

PROP: Orange oil. Bp: 208–210°@ 0.08 mm.

SYNS: ACEPROMAZINA □ ACEPROMAZINE □ ACEPROMIZINA □ ACETAZINE □ ACETHYLPROMAZIN □ 3-ACETYL-10-(3-DIMETHYLAMINOPROPYL)PHENOTHIAZINE □ ACETYLPROMAZINE □ ANATRAN □ ANERGAN □ ATRAVET □ ATSETOZIN □ AY-57,062 □ AZEPROMAZINE □ 1522 CB □ 10-(3-DIMETHYLAMINOPROPYL)PHENOTHIAZINE-3-ETHYLONE □ 1-(10-(3-(DIMETHYLAMINO)PROPYL)-10H-PHENOTHIAZIN-2-YL)ETHANONE □ 10-(3-DIMETHYLAMINOPROPYL) PHENOTHIAZIN-3-YLMETHYL KETONE □ LISERGAN □ NOTENQUIL □ NOTENSIL □ NOTESIL □ PLEGECYL □ PLEGICIN □ PLIVAPHEN □ SOPRINTIN □ SOPRONTIN □ SOPROTIN □ SV-1522 □ VETRANQUIL □ WY-1172

TOXICITY DATA WITH REFERENCE
ipr-rat LD50: 140 mg/kg FATOAO 24,136,61
orl-mus LDLo: 200 mg/kg AIPTAK 113,53,57
ipr-mus LD50: 350 mg/kg RMNIBN 81,105,77
scu-mus LD50: 130 mg/kg AIPTAK 113,53,57
ivn-mus LD50: 59 mg/kg AIPTAK 115,1,58

DOT CLASSIFICATION: 3; *Label:* Flammable Liquid

SAFETY PROFILE: Poison by ingestion, intravenous, and subcutaneous routes. A flammable liquid. When heated to decomposition it emits toxic fumes of SO_x and NO_x. See also KETONES. An animal tranquilizer.

ABH750 CAS:1071-73-4 HR: 2
ACETOPROPYL ALCOHOL
mf: $C_5H_{10}O_2$ mw: 102.15

PROP: Liquid, D: 1.008, bp: 208° @ 730 mm (decomp).

SYN: 3-ACETYLPROPANOL

TOXICITY DATA WITH REFERENCE
orl-rat LD50: 6400 mg/k TNICS• 13,118,73
ihl-rat LC: >2 g/m³/4H TNICS• 13,118,73
orl-mus LD50: 1960 mg/kg GISAAA 43(8),103,78
ihl-mus LC: >2 g/m³/4H TNICS• 13,118,73
orl-rbt LDLo: 3500 mg/kg GISAAA 43(8),103,78
orl-gpg LD50: 2260 mg/kg GISAAA 43(8),103,78

SAFETY PROFILE: Moderately toxic by ingestion and inhalation. When heated to decomposition it emits acrid smoke.

ABI000 **CAS:350-03-8** **HR: 3**
3-ACETOPYRIDINE
mf: C_7H_7NO mw: 121.15

PROP: Liquid. Bp: 220°, bp: 106° @ 12 mm. Sol in water.

SYNS: β-ACETYLPYRIDINE □ 3-ACETYLPYRIDINE □ METHYL PYRIDYL KETONE □ METHYL-β-PYRIDYL KETONE □ METHYL-3-PYRIDYL KETONE □ PYRIDINE, 3-ACETYL-

TOXICITY DATA WITH REFERENCE
sln-smc 5000 ppm MUREAV 163,23,86
orl-rat LD50:46 μ/kg JACTDZ 1,681,92
ipr-mus LD50:182 mg/kg JPMSAE 64,528,75
orl-qal LD50:422 mg/kg AECTCV 12,355,83
orl-bwd LD50:178 mg/kg AECTCV 12,355,83

CONSENSUS REPORTS: Reported in EPA TSCA Inventory.

DOT CLASSIFICATION: 3; *Label:* Flammable Liquid

SAFETY PROFILE: Poison by ingestion. Moderately toxic by intraperitoneal route. Mutation data reported. A flammable liquid. When heated to decomposition emits toxic fumes of NO_x. See also KETONES.

ABI250 **CAS:87-11-6** **HR: 3**
ACETOPYRROTHINE
mf: $C_8H_8N_2O_2S_2$ mw: 228.30

PROP: Brilliant-yellow needles from 1-butanol. Mp: 273–276° (decomp), bp: 200 @ 0.1 (subl). Produced by *Streptomyces albus* (ANTCAO 2,357,52).

SYNS: 6-ACETAMIDO-4-METHYL-1,2-DITHIOLO(4,3-B)PYRROL-5(4H)-ONE □ 3-ACETAMIDO-5-METHYLPYRROLIN-4-ONE(4,3-D)-1,2-DITHIOLE □ 6-(ACETYLAMINO)-4-METHYL-1,2-DITHIOLO(4,3-B)PYRROL-5(4H)-ONE □ N-(4,5-DIHYDRO-4-METHYL-5-OXO-1,2-DITHIOLO(4,3-B)PYRROL-6-YL)ACETAMIDE □ THIOLUTIN

TOXICITY DATA WITH REFERENCE
orl-mus LD50:25 mg/kg MEIEDD 10,1338,83
scu-mus LD50:25 mg/kg ANTCAO 2,357,52

SAFETY PROFILE: Poison by ingestion and subcutaneous routes. When heated to decomposition it emits very toxic fumes of NO_x and SO_x.

ABI500 **CAS:88-15-3** **HR: 3**
2-ACETOTHIENONE
mf: C_6H_6OS mw: 126.18

PROP: Liquid. D: 1.16 @ 24 mm, mp: 9°, bp: 213–214°.

SYNS: 2-ACETOTHIOPHENE □ 2-ACETYLTHIOPHENE

TOXICITY DATA WITH REFERENCE
ipr-mus LD50:40 mg/kg NTIS** AD691-490

CONSENSUS REPORTS: Reported in EPA TSCA Inventory.

DOT CLASSIFICATION: 3; *Label:* Flammable Liquid

SAFETY PROFILE: Poison by intraperitoneal route. A flammable liquid. When heated to decomposition emits toxic fumes of SO_x.

ABI750 **CAS:537-92-8** **HR: 2**
m-ACETOTOLUIDIDE
mf: $C_9H_{11}NO$ mw: 149.21

SYNS: 3-ACETAMIDOTOLUENE □ ACETO-m-AMINOTOLUENE □ ACETOTOLUIDE □ N-ACETYL-m-TOLUIDINE □ m-METHYLACETANILIDE □ 3-METHYLACETANILIDE □ 3′-METHYLACETANILIDE □ m-TOLYLACETAMIDE □ N-m-TOLYLACETAMIDE

TOXICITY DATA WITH REFERENCE
orl-mus LD50:1450 mg/kg TXAPA9 19,20,71

CONSENSUS REPORTS: Reported in EPA TSCA Inventory.

SAFETY PROFILE: Moderately toxic by ingestion. See also p-ACETOTOLUIDIDE. When heated to decomposition it emits toxic fumes of NO_x.

ABJ000 **CAS:120-66-1** **HR: 2**
o-ACETOTOLUIDIDE
mf: $C_9H_{11}NO$ mw: 149.21

PROP: Needles. Mp: 110°, bp: 296°. Sol in C_6H_6, $CHCl_3$.

SYNS: ACETYL-o-TOLUIDINE □ o-ACETOTOLUIDE □ o-METHYLACETANILIDE □ 2-METHYLACETANILIDE □ 2′-METHYLACETANILIDE

TOXICITY DATA WITH REFERENCE
mma-sat 1 mg/plate NTPTB* JAN 82
mma-sat 47 nmol/plate MUREAV 137,39,84
orl-mus LD50:1450 mg/kg TXAPA9 19,20,71

CONSENSUS REPORTS: Reported in EPA TSCA Inventory.

SAFETY PROFILE: Moderately toxic by ingestion. Mutation data reported. See also p-ACETOTOLUIDIDE. When heated to decomposition it emits toxic fumes of NO_x.

ABJ250 **CAS:103-89-9** **HR: 2**
p-ACETOTOLUIDIDE
mf: $C_9H_{11}NO$ mw: 149.21

PROP: Crystals from alc. Bp: 307°, fp: 335°F (CC), mp: 146°, d: 1.212, vap d: 5.14.

SYNS: p-ACETAMIDOTOLUENE □ p-ACETOTOLUIDE □ 4-ACETOTOLUIDE □ 4-(ACETYLAMINO)TOLUENE □ N-ACETYL-p-TOLUIDIDE □ ACETYL-p-TOLUIDINE □ p-METHYLACETANILIDE □ 4-METHYLACETANILIDE □ 4′-METHYLACETANILIDE

TOXICITY DATA WITH REFERENCE
orl-rat LD50:2640 mg/kg MarJV# 29MAR77
orl-mus LD50:980 mg/kg TXAPA9 19,20,71

CONSENSUS REPORTS: Reported in EPA TSCA Inventory.

SAFETY PROFILE: Moderately toxic by ingestion. See also ACETANILIDE. Combustible. When heated to decomposition it emits toxic fumes of NO_x. To fight fire use water, foam, CO_2, dry chemical.

ABJ750 CAS:26541-56-0 HR: 2
N-ACETOXY-4-ACETAMIDOBIPHENYL
mf: $C_{16}H_{15}NO_3$ mw: 269.32

SYNS: ACETAMIDE, N-(ACETYLOXY)-N-(1,1′-BIPHENYL)-4-YL-(9CI) □ ACETANILIDE, 4-PHENYL-, N-ACETATE (ester) □ ACETIC ACID, (N-ACETYL-N-(4-BIPHENYL)AMINO) ESTER □ ACETIC ACID, ESTER with N-4-BIPHENYLYLACETOHYDROXAMIC ACID □ ACETOHYDROXAMIC ACID, N-(4-BIPHENYLYL)-, ACETATE □ N-ACETOXY-4-ACETYLAMINOBIPHENYL □ N-ACETOXY-4-BIPHENYLACETAMIDE □ N-(ACETYLOXY)-N-(1,1′-BIPHENYL)-4-YLACETAMIDE □ N-(4-BIPHENYLYL)ACETOHYDROXAMIC ACETATE □ N-(4-BIPHENYLYL) ACETOHYDROXAMIC ACID ACETATE □ N,O-DIACETYL-N-(4-BIPHENYLYL)HYDROXYLAMINE

TOXICITY DATA WITH REFERENCE
mmo-sat 25 µg/plate CBINA8 26,11,79
mma-sat 1 µg/plate CBINA8 26,11,79
skn-mus TDLo:4309 µg/kg:NEO JNCIAM 54,491,75

SAFETY PROFILE: Questionable carcinogen with experimental neoplastigenic data. Mutation data reported. See also ESTERS. When heated to decomposition it emits toxic fumes of NO_x.

ABK000 CAS:64058-72-6 HR: 2
ACETOXY(2-ACETAMIDO-5-NITROPHENYL)MERCURY
mf: $C_{10}H_{10}HgN_2O_5$ mw: 438.81

SYN: 2′-(ACETOXYMERCURI)-4′-NITROACETANILIDE

TOXICITY DATA WITH REFERENCE
ipr-rat LDLo:500 mg/kg NCNSA6 5,8,53

OSHA PEL: CL 0.1 mg(Hg)/m³ (skin)
ACGIH TLV: TWA 0.1 mg(Hg)/m³ (skin)
NIOSH REL: (Mercury, Aryl And Inorganic): CL 0.1 mg/m³ (skin)

SAFETY PROFILE: Moderately toxic by intraperitoneal route. See also MERCURY COMPOUNDS. When heated to decomposition it emits very toxic fumes of Hg and NO_x.

ABK250 CAS:26541-57-1 HR: 2
N-ACETOXY-2-ACETAMIDOPHENANTHRENE
mf: $C_{18}H_{15}NO_3$ mw: 293.34

SYNS: ACETIC ACID (N-ACETYL-N-(2-PHENANTHRYL)AMINO)ESTER □ ACETIC ACID ESTER with N-(2-PHENANTHRYL)ACETOHYDROXAMIC ACID □ N-ACETOXY-2-ACETYLAMINOPHENANTHRENE □ N-ACETOXY-4-PHENANTHRYLACETAMIDE □ N-(2-PHENANTHRYL) ACETOHYDROXAMIC ACETATE

TOXICITY DATA WITH REFERENCE
mmo-bcs 14 mol CNREA8 30,1473,70
oms-bcs 10 g/L CNREA8 30,1473,70
dns-hmn:fbr 10 mmol/L/5H IJCNAW 16,284,75
mmo-sat 5 µg/plate CBINA8 26,11,79
mma-sat 50 ng/plate CBINA8 26,11,79
dnd-mam:lym 625 mg/L CNREA8 35,1416,75
skn-mus TDLo:584 µg/kg:NEO JNCIAM 54,491,75

SAFETY PROFILE: Questionable carcinogen with experimental neoplastigenic data. Human mutation data reported. See also ESTERS. When heated to decomposition it emits toxic fumes of NO_x.

ABL000 CAS:6098-44-8 HR: 2
N-ACETOXY-N-ACETYL-2-AMINOFLUORENE
mf: $C_{17}H_{15}NO_3$ mw: 281.33

SYNS: ACETIC ACID (N-ACETYL-N-(2-FLUORENYL)AMINO) ESTER □ N-ACETOXY-2-ACETAMIDOFLUORENE □ N-ACETOXY-2-ACETYLAMINOFLUORENE □ N-ACETOXY-2-FLUORENYLACETAMIDE □ N-(FLUOREN-2-YL)ACETOHYDROXAMIC ACETAMIDE

TOXICITY DATA WITH REFERENCE
mma-sat 1500 ng/plate CBINA8 54,71,85
sce-ham:oth 1800 nmol/L CRNGDP 6,1627,85
dns-hmn:lym 10 µmol/L CALEDQ 2,311,77
dns-hmn:leu 10 µmol/L CRNGDP 1,547,80
dni-hmn:hla 100 µmol/L/30M-C JEPTDQ 2(1),65,78
scu-rat TDLo:21 mg/kg:NEO CNREA8 37,1461,77
imp-rat TDLo:28 mg/kg:ETA CNREA8 37,111,77

CONSENSUS REPORTS: EPA Genetic Toxicology Program.

SAFETY PROFILE: Questionable carcinogen with experimental tumorigenic and neoplastigenic data. Human mutation data reported. When heated to decomposition it emits toxic fumes of NO_x.

ABL250 CAS:26488-34-6 HR: 2
trans-N-ACETOXY-4-ACETYL-AMINOSTILBENE
mf: $C_{18}H_{17}NO_3$ mw: 295.36

SYN: trans-N,o-DIACETYL-N-(p-STYRYLPHENYL)HYDROXYLAMINE

TOXICITY DATA WITH REFERENCE
mrc-smc 10 ppm ZEKBAI 74,412,70
orl-rat TDLo:180 mg/kg/20W-I:CAR ZEKBAI 74,200,70

CONSENSUS REPORTS: EPA Genetic Toxicology Program.

SAFETY PROFILE: Questionable carcinogen with experimental carcinogenic data. Mutation data reported. When heated to decomposition it emits toxic fumes of NO_x.

ABL500 CAS:3061-65-2 HR: 3
2-ACETOXYACRYLONITRILE
mf: $C_5H_5NO_2$ mw: 111.11

SYNS: α-ACETOXYACRYLONITRILE □ α-CYANOVINYL ACETATE

TOXICITY DATA WITH REFERENCE
skn-rbt 10 mg/24H open MLD AIHAAP 23,95,62
orl-rat LD50:100 mg/kg AIHAAP 23,95,62
ihl-rat LCLo:125 ppm/4H AIHAAP 23,95,62
skn-rbt LD50:140 mg/kg AIHAAP 23,95,62

CONSENSUS REPORTS: Cyanide and its compounds are on the Community Right-To-Know List.

SAFETY PROFILE: Poison by inhalation, ingestion, and skin contact. A skin irritant. See also NITRILES. When heated to decomposition it emits toxic fumes of NO_x and CN^-.

ABM000 **HR: 2**
3-β-ACETOXY-BIS NOR-Δ^5-CHOLENIC ACID
mf: $C_{24}H_{36}O_4$ mw: 388.60

SYN: 3-β-ACETOXYPREGN-6-ENE-20-CARBOXYLIC ACID

TOXICITY DATA WITH REFERENCE
scu-mus TDLo: 200 mg/kg/90D-I: ETA NATUAS 209,1026,66

SAFETY PROFILE: Questionable carcinogen with experimental tumorigenic data. When heated to decomposition it emits acrid smoke and irritating fumes.

ABM250 **CAS:1515-76-0** **HR: 3**
1-ACETOXY-1,3-BUTADIENE
mf: $C_6H_8O_2$ mw: 112.14

SYN: ACETIC ACID-1,3-BUTADIENYL ESTER

TOXICITY DATA WITH REFERENCE
skn-rbt 100 μg/24H open AIHAAP 23,95,62
orl-rat LDLo: 710 mg/kg AIHAAP 23,95,62
ihl-rat LCLo: 63 ppm/4H AIHAAP 23,95,62
skn-rbt LD50: 420 mg/kg AIHAAP 23,95,62

SAFETY PROFILE: Poison by inhalation. Moderately toxic by other routes. A skin irritant. Mutation data reported. When heated to decomposition it emits acrid smoke.

ABN000 **CAS:2885-39-4** **HR: 3**
ACETOXYCYCLOHEXIMIDE
mf: $C_{17}H_{25}NO_6$ mw: 339.43

PROP: Crystals. Mp: 140°.

SYNS: ACETYLOXYCYCLOHEXIMIDE □ 3-(2-(5-ACETOXY-3,5-DIMETHYL-2-OXOCYCLOHEXYL)-2-HYDROXYETHYL)GLUTARIMIDE □ AXM □ E-73 ACETATE □ NSC 32743 □ STREPTOVITACIN E 73

TOXICITY DATA WITH REFERENCE
oms-hmn: hla 29 μmol/L BCPCA6 14,205,65
dni-hmn: hla 1 μmol/L BCPCA6 14,205,65
orl-rat LD50: 158 μg/kg JPETAB 136,400,62
ipr-rat LD50: 170 μg/kg JPETAB 136,400,62
scu-rat LD50: 190 μg/kg JPETAB 136,400,62
ipr-mus LD50: 19 mg/kg JPETAB 136,400,62
ivn-dog LDLo: 920 μg/kg JPETAB 136,400,62

SAFETY PROFILE: Deadly poison by ingestion, intravenous, intraperitoneal, and subcutaneous routes. Human mutation data reported. When heated to decomposition it emits toxic fumes such as NO_x.

ABN250 **CAS:24684-58-0** **HR: 2**
11-ACETOXY-15-DIHYDROCYCLOPENTA(a)PHENANTHRACEN-17-ONE
mf: $C_{19}H_{14}O_3$ mw: 290.33

SYN: 11-HYDROXY-15,16-DIHYDROCYCLOPENTA(a)PHENANTHRACEN-17-ONE ACETATE (ESTER)

TOXICITY DATA WITH REFERENCE
skn-mus TDLo: 108 mg/kg/1Y-I: ETA PEXTAR 11,69,69

SAFETY PROFILE: Questionable carcinogen with experimental tumorigenic data. See also ESTERS. When heated to decomposition it emits acrid smoke and irritating fumes.

ABN500 **CAS:38539-23-0** **HR: 2**
1-ACETOXY-1,4-DIHYDRO-4-(HYDROXYAMINO)QUINOLINE ACETATE (ESTER)
mf: $C_{13}H_{13}N_2O_4$ mw: 261.28

SYN: O,O′-DIACETYL 4-HYDROXYAMINOQUINOLINE-1-OXIDE

TOXICITY DATA WITH REFERENCE
mmo-smc 50 mg/L IGSBAL 85,127,72
scu-rat TDLo: 5 mg/kg/2W-I: ETA PSEBAA 136,1206,71
scu-mus TDLo: 30 mg/kg/4W-I: NEO PSEBAA 136,1206,71

SAFETY PROFILE: Questionable carcinogen with experimental tumorigenic and neoplastigenic data. Mutation data reported. See also ESTERS. When heated to decomposition it emits toxic fumes of NO_x.

ABN625 **HR: 3**
1-ACETOXYDIMERCURIO-1-PERCHLORATODIMERCURIOPROPEN-2-ONE
mf: $C_4H_3ClHg_4O_7$ mw: 1000.84

$(CH_2COOH)OHgHgC(:C:O)HgHgOClO_3$

SAFETY PROFILE: Dangerously explosive. When heated to decomposition it emits toxic fumes of Hg and Cl^-. See also MERCURY COMPOUNDS; PERCHLORATES; and EXPLOSIVES.

ABN700 **CAS:66827-45-0** **HR: 3**
β-ACETOXY-N,N-DIMETHYLPHENETHYLAMINE
mf: $C_{12}H_{17}NO_2$ mw: 207.30

SYNS: ACETIC ACID α-(DIMETHYLAMINOMETHYL)BENZYL ESTER □ β-ACETYLOXY-β-PHENYLETHYL DIMETHYLAMINE □ N,N-DIMETHYL-β-ACETOXY β-PHENYLETHYLAMINE

TOXICITY DATA WITH REFERENCE
ipr-mus LD50: 310 mg/kg EJMCA5 13,277,78
scu-mus LDLo: 722 mg/kg AIPTAK 47,96,34
ivn-rbt LDLo: 72 mg/kg AIPTAK 47,96,34

SAFETY PROFILE: Poison by intravenous and intraperitoneal routes. Moderately toxic by subcutaneous route. When heated to decomposition it emits toxic fumes of NO_x. See also AMINES and ESTERS.

ABN725 **CAS:61691-82-5** **HR: 2**
1′-ACETOXYESTRAGOLE
mf: $C_{12}H_{14}O_3$ mw: 206.26

SYN: p-METHOXY-α-VINYLBENZYL ALCOHOL ACETATE (ester)

TOXICITY DATA WITH REFERENCE
mmo-sat 200 nmol/plate CRNGDP 7,2089,86
dnd-hmn: fbr 500 μmol/L CRNGDP 3,935,82
ipr-mus TDLo: 20,626 μg/kg: CAR CNREA8 47,2275,87

CONSENSUS REPORTS: EPA Genetic Toxicology Program.

SAFETY PROFILE: Questionable carcinogen with experimental carcinogenic data. Human mutation data reported. When heated to decomposition it emits acrid smoke and fumes. See also ESTERS.

ABO000 CAS:60-31-1 HR: 3
2-ACETOXYETHYLTRIMETHYLAMMONIUM CHLORIDE
mf: $C_7H_{16}NO_2 \cdot Cl$ mw: 181.69

PROP: Deliquescent crystals, powder. Mp: 149–152°.

SYNS: ACECOLINE □ ACETYLCHOLINE CHLORIDE □ ACETYLCHOLINE HYDROCHLORIDE □ ACETYLCHOLINIUM CHLORIDE □ 2-(ACETYLOXY)-N,N,N-TRIMETHYLETHANAMINIUM CHLORIDE □ ACH CHLORIDE □ ARTEROCOLINE □ CHOLINE CHLORIDE ACETATE □ (2-HYDROXYETHYL)TRIMETHYLAMMONIUM CHLORIDE ACETATE □ OVISOT □ TL 1505

TOXICITY DATA WITH REFERENCE
orl-rat LD50:2500 mg/kg JPETAB 58,337,36
scu-rat LD50:250 mg/kg JPETAB 58,337,36
ivn-rat LD50:22 mg/kg JPETAB 58,337,36
orl-mus LD50:3000 mg/kg JPETAB 58,337,36
scu-mus LD50:170 mg/kg JPETAB 58,337,36
ivn-mus LD50:10 mg/kg JPETAB 119,541,57
par-frg LDLo:200 mg/kg AEPPAE 166,437,32

CONSENSUS REPORTS: Reported in EPA TSCA Inventory.

SAFETY PROFILE: Poison by subcutaneous, intravenous, and parenteral routes. Moderately toxic by ingestion. When heated to decomposition it emits very toxic fumes of NO_x and Cl^-. A cholinergic agent. See also CHOLINE ACETATE (ESTER).

ABO250 CAS:38105-27-0 HR: 2
N-ACETOXYFLUORENYLACETAMIDE
mf: $C_{17}H_{15}NO_3$ mw: 281.33

SYNS: ACETIC ACID ESTER with N-(FLUOREN-3-YL)ACETOHYDROXAMIC ACID □ N-ACETOXY-3-FLUORENYLACETAMIDE □ N-(FLUOREN-3-YL)ACETOHYDROXAMIC ACETATE

TOXICITY DATA WITH REFERENCE
ipr-rat TDLo:350 mg/kg/4W-I:NEO CNREA8 35,447,75
ims-rat TDLo:210 mg/kg/8W-I:ETA CNREA8 35,447,75

SAFETY PROFILE: Questionable carcinogen with experimental tumorigenic and neoplastigenic data. See also ESTERS. When heated to decomposition it emits toxic fumes of acetic acid and NO_x.

ABO500 CAS:55080-20-1 HR: 2
N-ACETOXY-4-FLUORENYLACETAMIDE
mf: $C_{17}H_{15}NO_3$ mw: 281.33

SYNS: ACETIC ACID(N-ACETYL-N-(4-FLUORENYL)AMINO)ESTER □ ACETIC ACID, ESTER with N-(FLUOREN-4-YL)ACETOXYHYDROXAMIC ACID □ N-(FLUOREN-4-YL)ACETOHYDROXAMIC ACETATE

TOXICITY DATA WITH REFERENCE
ipr-rat TDLo:350 mg/kg/4W-I:ETA CNREA8 35,447,75

SAFETY PROFILE: Questionable carcinogen with experimental tumorigenic data. See also ESTERS. When heated to decomposition it emits toxic fumes of NO_x.

ABO740 CAS:2050-43-3 HR: 2
2′,4′-ACETOXYLIDIDE
mf: $C_{10}H_{13}NO$ mw: 163.24

PROP: Crystals from aqueous ethanol. Mp: 129–130°, bp: 170° @ 10 mm.

SYNS: 2,4-DIMETHYLACETANILIDE □ 2′,4′-DIMETHYLACETANILIDE

TOXICITY DATA WITH REFERENCE
orl-mus LD50:1300 mg/kg TXAPA9 19,20,71

CONSENSUS REPORTS: Reported in EPA TSCA Inventory.

SAFETY PROFILE: Moderately toxic by ingestion. When heated to decomposition it emits toxic fumes of NO_x.

ABO750 CAS:29968-75-0 HR: 2
N-ACETOXY-2-FLUORENYLBENZAMIDE
mf: $C_{22}H_{17}NO_3$ mw: 343.40

SYN: N-FLUOREN-2-YL BENZOHYDROXAMIC ACID ACETATE

TOXICITY DATA WITH REFERENCE
orl-rat TDLo:5200 mg/kg/21W-I:ETA CNREA8 30,1485,70

SAFETY PROFILE: Questionable carcinogen with experimental tumorigenic data. When heated to decomposition it emits toxic fumes of NO_x.

ABP760 CAS:2198-53-0 HR: 2
2′,6′-ACETOXYLIDIDE
mf: $C_{10}H_{13}NO$ mw: 163.24

PROP: Crystals. Mp: 177°. Sol in C_6H_6.

SYN: 2,6-DIMETHYLACETANILIDE

TOXICITY DATA WITH REFERENCE
orl-mus LD50:620 mg/kg TXAPA9 19,20,71

CONSENSUS REPORTS: Reported in EPA TSCA Inventory.

SAFETY PROFILE: Moderately toxic by ingestion. When heated to decomposition it emits toxic fumes of NO_x.

ABP770 CAS:2198-54-1 HR: 2
3′,4′-ACETOXYLIDIDE
mf: $C_{10}H_{13}NO$ mw: 163.24

PROP: Crystals from aqueous ethanol. Mp: 99°. Sol in C_6H_6.

SYNS: 3,4-DIMETHYLACETANILIDE □ 3′,4′-DIMETHYLACETANILIDE

TOXICITY DATA WITH REFERENCE
orl-rat TDLo:9900 mg/kg/56W-C:ETA CNREA8 16,525,56
orl-mus LD50:1030 mg/kg TXAPA9 19,20,71

SAFETY PROFILE: Moderately toxic by ingestion. Questionable carcinogen with experimental tumorigenic data. When heated to decomposition it emits toxic fumes of NO_x.

ABQ000 CAS:6283-24-5 HR: 3
p-(ACETOXYMERCURI)ANILINE
mf: $C_8H_9HgNO_2$ mw: 351.77

PROP: Colorless prisms from $(CHCl_3)$. Mp: 166–167°. Insol in water, Et_2O. Sltly sol in aqueous ethanol, $CHCl_3$.

SYNS: (ACETATO)(p-AMINOPHENYL)MERCURY □ p-AMINOPHENYLMERCURIC ACETATE

TOXICITY DATA WITH REFERENCE
ipr-mus LDLo:13 mg/kg JPETAB 31,87,27
ivn-mus LD50:18 mg/kg CSLNX* NX#04750

CONSENSUS REPORTS: Reported in EPA TSCA Inventory. On Community Right-To-Know List.

OSHA PEL: CL 0.1 mg(Hg)/m³ (skin)
ACGIH TLV: TWA 0.1 mg(Hg)/m³ (skin)
NIOSH REL: (Mercury, Aryl And Inorganic): CL 0.1 mg/m³ (skin)

SAFETY PROFILE: Poison by intravenous and intraperitoneal routes. See also MERCURY COMPOUNDS, ANILINE. When heated to decomposition it emits very toxic fumes of NO_x and Hg.

ABQ250 CAS:54481-45-7 HR: 3
2-(ACETOXYMERCURI)-4-NITROANILINE
mf: $C_8H_8HgN_2O_4$ mw: 396.77

SYN: ACETATO(2-AMINO-5-NITROPHENYL)MERCURY

TOXICITY DATA WITH REFERENCE
ipr-rat LDLo:250 mg/kg NCNSA6 5,12,53

OSHA PEL: CL 0.1 mg(Hg)/m³ (skin)
ACGIH TLV: TWA 0.1 mg(Hg)/m³ (skin)
NIOSH REL: (Mercury, Aryl And Inorganic): CL 0.1 mg/m³ (skin)

SAFETY PROFILE: Poison by intraperitoneal route. See also MERCURY COMPOUNDS, NITRO COMPOUNDS of AROMATIC HYDROCARBONS. When heated to decomposition it emits very toxic fumes of Hg and NO_x.

ABQ375 HR: 3
1-ACETOXYMERCURIO-1-PERCHLORATOMERCURIOPROPEN-2-ONE
mf: $C_4H_3ClHg_2O_7$ mw: 599.66

SAFETY PROFILE: Dangerously explosive. When heated to decomposition it emits toxic fumes of Cl^- and Hg. See also MERCURY COMPOUNDS, EXPLOSIVES, and PERCHLORATES.

ABQ600 CAS:83876-62-4 HR: 2
4-ACETOXY-7-METHYLBENZ(c)ACRIDINE
mf: $C_{20}H_{15}NO_2$ mw: 301.36

SYN: BENZ(c)ACRIDIN-4-OL, 7-METHYL-, ACETATE (ESTER)

TOXICITY DATA WITH REFERENCE
scu-mus TDLo:72 mg/kg/12W-I:ETA JMCMAR 26,303,83

SAFETY PROFILE: Questionable carcinogen with experimental tumorigenic data. When heated to decomposition it emits toxic fumes of NO_x.

ABR125 CAS:70715-92-3 HR: 2
N-(ACETOXYMETHYL)-N-ISOBUTYLNITROSAMINE
mf: $C_7H_{14}N_2O_3$ mw: 174.23

SYNS: N-ISOBUTYL-N-(ACETOXYMETHYL)NITROSAMINE □ N-NITROSO-N-(ACETOXYMETHYL)-N-ISOBUTYLAMINE

TOXICITY DATA WITH REFERENCE
mmo-sat 5 µmol/plate GANNA2 70,663,79
mmo-esc 25 µmol/plate GANNA2 70,663,79
dnr-bcs 1 µmol/plate GANNA2 70,663,79
scu-rat TDLo:50 mg/kg/10W-I:ETA JCROD7 104,13,82
scu-rat TD:66 mg/kg/10W-I:CAR IAPUDO 41,619,82

SAFETY PROFILE: Questionable carcinogen with experimental carcinogenic and tumorigenic data. Mutation data reported. When heated to decomposition it emits toxic fumes of NO_x. See also NITROSAMINES.

ABR250 CAS:2517-98-8 HR: 2
7-ACETOXYMETHYL-12-METHYLBENZ(a)ANTHRACENE
mf: $C_{22}H_{18}O_2$ mw: 314.40

SYN: 12-METHYLBENZ(a)ANTHRACENE-7-METHANOL ACETATE (ESTER)

TOXICITY DATA WITH REFERENCE
mmo-sat 4 nmol/plate CBINA8 58,253,86
add-uns:lym 50 g/L RCOCB8 22,345,78
orl-rat TDLo:100 mg/kg:ETA JMCMAR 10,932,67
scu-rat TDLo:150 mg/kg/39D-I:NEO CNREA8 31,1951,71

SAFETY PROFILE: Questionable carcinogen with experimental tumorigenic and neoplastigenic data. Mutation data reported. See also ESTERS. When heated to decomposition it emits acrid smoke and fumes.

ABR625 HR: 3
ACETOXYMETHYLPHENYLNITROSAMINE
mf: $C_9H_{11}N_2O_3$ mw: 195.22

TOXICITY DATA WITH REFERENCE
mmo-sat 64,400 pmol/plate CALEDQ 15,289,82
scu-ham TDLo:42 mg/kg/27W-I:ETA CALEDQ 15,289,82
scu-ham LD50:117 mg/kg CALEDQ 15,289,82

SAFETY PROFILE: Poison by subcutaneous route. Questionable carcinogen with experimental tumorigenic data. Mutation data reported. When heated to decomposition it emits toxic fumes of NO_x. See also NITROSAMINES.

ABS750 CAS:830-03-5 HR: 3
p-ACETOXYNITROBENZENE
mf: $C_8H_7NO_4$ mw: 181.16

PROP: Leaflets from aqueous ethanol. Mp: 81–82°.

SYNS: p-NITROPHENOL ACETATE □ p-NITROPHENYL ACETATE □ 4-NITROPHENYL ACETATE

TOXICITY DATA WITH REFERENCE
ivn-mus LD50:180 mg/kg CSLNX* NX#00217

CONSENSUS REPORTS: Reported in EPA TSCA Inventory.

SAFETY PROFILE: Poison by intravenous route. When heated to decomposition it emits toxic fumes of NO_x.

ABT750 CAS:53198-41-7 HR: 2
1-ACETOXY-N-NITROSODIPROPYLAMINE
mf: $C_8H_{16}N_2O_3$ mw: 188.26

SYNS: ACETIC ACID-1-(PROPYLNITROSAMINO)PROPYL ESTER ☐ N-(α-ACETOXY)PROPYL-N-N-PROPYLNITROSAMINE ☐ 1-(PROPYLNITROSAMINO)PROPYL ACETATE

TOXICITY DATA WITH REFERENCE
mmo-sat 100 nmol/plate MUREAV 49,187,78
scu-ham TDLo:410 mg/kg/33W-I:CAR ZKKOBW 90,127,77
scu-ham LD50:500 mg/kg ZKKOBW 90,127,77

SAFETY PROFILE: Moderately toxic by subcutaneous route. Mutation data reported. Questionable carcinogen with experimental carcinogenic data. See also N-NITROSO COMPOUNDS, NITROSAMINES, ESTERS, and AMINES. When heated to decomposition it emits toxic fumes of NO_x.

ABU000 CAS:51-98-9 HR: 3
17-ACETOXY-19-NOR-17-α-PREGN-4-EN-20-YN-3-ONE
mf: $C_{22}H_{28}O_3$ mw: 340.50

PROP: Crystals from Me_2CO/hexane. Mp: 161–163°.

SYNS: 17-β-ACETOXY-19-NOR-17-α-PREGN-4-EN-20-YN-3-ONE ☐ (17-α)-17-(ACETYLOXY)-19-NORPREGN-4-EN-20-YN-3-ONE ☐ 17-ACETYLOXY(17-α)-19-NORPREGN-4-ESTREN-17-β-OL-ACETATE-3-ONE ☐ 17-ENT ☐ 17-α-ETHINYL-19-NORTESTOSTERONE ACETATE ☐ 17-α-ETHINYL-19-NORTESTOSTERONE-17-β-ACETATE ☐ 17-α-ETHYNYL-17-β-ACETOXY-19-NORANDROST-4-EN-3-ONE ☐ 17-α-ETHYNYL-17-HYDROXYESTR-4-EN-3-ONE ACETATE ☐ 17-α-ETHYNYL-19-NORTESTOSTERONE ACETATE ☐ 17-HYDROXY-19-NOR-17-α-PREGN-4-EN-20-YN-3-ONE ACETATE ☐ 17-β-HYDROXY-19-NOR-17-α-PREGN-4-EN-20-YN-3-ONE ACETATE ☐ NORETHINDRONE-17-ACETATE ☐ 19-NORETHISTERONE ACETATE ☐ 19-NORETHYNYLTESTOSTERONE ACETATE ☐ NORETHYSTERONE ACETATE ☐ NORLUTATE ☐ NORLUTINE ACETATE ☐ ORLUTATE

TOXICITY DATA WITH REFERENCE
dlt-mus-orl 1120 mg/kg/4W MUREAV 26,535,74
spm-mus-orl 1120 mg/kg/4W MUREAV 26,535,74
orl-wmn TDLo:2190 μg/kg (52W pre):REP BMJOAE 2,730,69
unr-wmn TDLo:15 mg/kg (female 13-30W post):TER OBGNAS 22,210,63
orl-rat TDLo:303 μg/kg/2Y-C:ETA JTEHD6 6,895,80

CONSENSUS REPORTS: IARC Cancer Review: Animal Limited Evidence IMEMDT 21,441,79; Animal Sufficient Evidence IMEMDT 6,179,74. EPA Genetic Toxicology Program.

SAFETY PROFILE: Suspected carcinogen with experimental tumorigenic data. Human reproductive effects by ingestion and implant routes: menstrual cycle changes, postpartum effects, and changes in fertility. A human teratogen by an unspecified route with developmental abnormalities of the urogenital system. Mutation data reported. When heated to decomposition it emits acrid smoke and irritating fumes. Used in the treatment of menstrual disorders and uterine bleeding.

ABU500 CAS:62-38-4 HR: 3
ACETOXYPHENYLMERCURY
DOT: UN 1674
mf: $C_8H_8HgO_2$ mw: 336.75

PROP: Lustrous crystals. Mp: 149–152°. Sltly sol in water.

SYNS: ACETATE PHENYLMERCURIQUE (FRENCH) ☐ (ACETATO)PHENYLMERCURY ☐ ACETIC ACID, PHENYLMERCURY DERIV. ☐ (ACETOXYMERCURI)BENZENE ☐ AGROSAN ☐ AGROSAND ☐ AGROSAN GN 5 ☐ ALGIMYCIN ☐ ANTIMUCIN WDR ☐ BENZENE, (ACETOXYMERCURI)- ☐ BENZENE, (ACETOXYMERCURIO)- ☐ BUFEN ☐ CEKUSIL ☐ CELMER ☐ CERESAN ☐ CERESAN UNIVERSAL ☐ CERESOL ☐ CONTRA CREME ☐ DYANACIDE ☐ FEMMA ☐ FENYLMERCURIACETAT (CZECH) ☐ FMA ☐ FUNGITOX OR ☐ GALLOTOX ☐ HL-331 ☐ HONG KIEN ☐ HOSTAQUICK ☐ KWIKSAN ☐ LEYTOSAN ☐ LIQUIPHENE ☐ MERCURIPHENYL ACETATE ☐ MERCURY (II) ACETATE, PHENYL- ☐ MERCURY, ACETOXYPHENYL- ☐ MERGAMMA ☐ MERSOLITE ☐ MERSOLITE 8 ☐ METASOL 30 ☐ NORFORMS ☐ NYLMERATE ☐ OCTAN FENYLRTUTNATY (CZECH) ☐ PAMISAN ☐ PHENMAD ☐ PHENOMERCURIC ACETATE ☐ PHENYLMERCURIACETATE ☐ PHENYL MERCURIC ACETATE ☐ PHENYLMERCURY ACETATE ☐ PHENYLQUECKSILBERACETAT (GERMAN) ☐ PHIX ☐ PMA ☐ PMAC ☐ PMACETATE ☐ PMAL ☐ PMAS ☐ PURASAN-SC-10 ☐ PURATURF 10 ☐ QUICKSAN ☐ RCRA WASTE NUMBER P092 ☐ SANITIZED SPG ☐ SC-110 ☐ SCUTL ☐ SEEDTOX ☐ SHIMMEREX ☐ SPOR-KIL ☐ TAG ☐ TAG 331 ☐ TAG HL 331 ☐ TAG FUNGICIDE ☐ TRIGOSAN ☐ ZIARNIK

TOXICITY DATA WITH REFERENCE
dnr-esc 2 mmol/L MJDHDW 28,F39,80
sce-ham:lym 30 mg/L DBABEF 8,105,84
scu-mus TDLo:110 μg/kg (8D preg):TER ARINAU 3,88,56
ipr-uns TDLo:125 μg/kg (female 8D post):REP TXCYAC 6,281,76
orl-rat LD50:41 mg/kg JACTDZ 1,175,92
orl-mus LD50:13,250 μg/kg YAKUD5 22,291,80
ipr-mus LD50:13 mg/kg AMSVAZ 143,365,52
scu-mus LD50:12 mg/kg TOIZAG 9,101,62
ivn-mus LD50:18 mg/kg CSLNX* NX#00921
orl-ckn LD50:60 mg/kg TXAPA9 2,344,60
orl-qal LD50:71 mg/kg AXVMAW 34,383,80
ipr-uns LD50:10 mg/kg TXCYAC 6,281,76

CONSENSUS REPORTS: IARC Cancer Review: Group 2B, Human Inadequate Evidence IMEMDT 58,239,93. EPA Extremely Hazardous Substances List. Reported in EPA TSCA Inventory. EPA Genetic Toxicology Program. Mercury and its compounds are on the Community Right-To-Know List.

OSHA PEL: CL 0.1 mg(Hg)/m^3 (skin)
ACGIH TLV: TWA 0.1 mg(Hg)/m^3 (skin)
NIOSH REL: (Mercury, Aryl And Inorganic): CL 0.1 mg/m^3 (skin)
DOT CLASSIFICATION: 6.1; *Label:* Poison

SAFETY PROFILE: Poison by ingestion, intravenous, intraperitoneal, subcutaneous, and possibly other routes. An experimental teratogen. Other experimental reproductive effects. Mutation data reported. See also

MERCURY COMPOUNDS. When heated to decomposition it emits toxic fumes of Hg.

ABU800 CAS:2114-33-2 HR: 1
2-ACETOXY-1-PHENYLPROPANE
mf: $C_{11}H_{14}O_2$ mw: 178.25

SYNS: ACETIC ACID, α-METHYL-PHENETHYL ESTER □ BENZYLMETHYLCARBINYL ACETATE □ METHYLBENZYLCARBINYL ACETATE □ α-METHYL β-PHENYLETHYL ACETATE □ 1-PHENYL-2-PROPANOL ACETATE

TOXICITY DATA WITH REFERENCE
skn-rbt 500 mg/24H MLD FCTOD7 20,737,82

CONSENSUS REPORTS: Reported in EPA TSCA Inventory.

SAFETY PROFILE: A skin irritant. When heated to decomposition it emits acrid smoke and irritating fumes.

ABV250 CAS:17427-00-8 HR: 3
3-ACETOXYPHENYLTRIMETHYLAMMONIUM IODIDE
mf: $C_{11}H_{16}NO_2 \cdot I$ mw: 321.18

SYN: NU 2017

TOXICITY DATA WITH REFERENCE
orl-mus LD50:800 mg/kg JPETAB 99,16,50
scu-mus LD50:125 mg/kg JPETAB 99,16,50
ivn-mus LD50:3700 μg/kg JPETAB 99,16,50

SAFETY PROFILE: A poison via subcutaneous and intravenous routes. When heated to decomposition it emits very toxic fumes of NO_x and I^-.

ABW500 CAS:26594-44-5 HR: 2
N-ACETOXY-N-(4-STILBENYL) ACETAMIDE
mf: $C_{18}H_{17}NO_3$ mw: 295.36

SYNS: ACETIC ACID-(N-ACETYL-N-(p-STYRYLPHENYL)AMINO) ESTER □ ACETIC ACID-ESTER with N-(p-STYRYLPHENYL)ACETOHYDROXAMIC ACID □ N-ACETOXY-4-ACETAMIDOSTILBENE □ N,O-DIACETYL-N-(p-STYRYLPHENYL)HYDROXYLAMINE □ N-(p-STYRYLPHENYL)ACETOHYDROXAMIC ACETATE □ N-(p-STYRYLPHENYL)ACETOHYDROXAMIC ACID ACETATE

TOXICITY DATA WITH REFERENCE
mmo-sat 5 μg/plate CBINA8 26,11,79
dns-hmn:fbr 10 μg/L/5H IJCNAW 16,284,75
dns-hmn:hlas 100 nmol/L CNREA8 38,2621,78
skn-mus TDLo:2360 μg/kg:NEO JJIND8 54,491,75

CONSENSUS REPORTS: EPA Genetic Toxicology Program.

SAFETY PROFILE: Questionable carcinogen with experimental neoplastigenic data by skin contact. Human mutation data reported. When heated to decomposition it emits highly toxic fumes of NO_x.

ABW550 CAS:2628-16-2 HR: 2
4-ACETOXYSTYRENE
mf: $C_{10}H_{10}O_2$ mw: 162.20

SYNS: p-ACETOXYSTYRENE □ C-908 □ 4-ETHENYLPHENOL ACETATE □ PHENOL, 4-ETHENYL-, ACETATE □ PHENOL, p-VINYL-, ACETATE (6CI,7CI,8CI) □ p-VINYLPHENOL ACETATE □ 4-VINYLPHENYL ACETATE

TOXICITY DATA WITH REFERENCE
eye-rbt 100 mg MLD EPASR* 8EHQ-1190-1082
orl-rat LD50:1503 mg/kg EPASR* 8EHQ-1190-1082
skn-rat LD50:>2 g/kg EPASR* 8EHQ-1190-1082

SAFETY PROFILE: Moderately toxic by ingestion. Slightly toxic by skin contact. An eye irritant. When heated to decomposition it emits acrid smoke and irritating vapors.

ABW600 CAS:13121-71-6 HR: 3
ACETOXYTRICYCLOHEXYLSTANNANE
mf: $C_{20}H_{36}O_2Sn$ mw: 427.25

PROP: Rod-like crystals from aqueous ethanol. Mp: 61–63°.

SYN: STANNANE, ACETOXYTRICYCLOHEXYL-

TOXICITY DATA WITH REFERENCE
orl-rat LD50:178 mg/kg PHARAT 37,801,82

OSHA PEL: TWA 0.1 mg(Sn)/m³
ACGIH TLV: TWA 0.1 mg(Sn)/m³ (skin)

SAFETY PROFILE: Poison by ingestion. When heated to decomposition it emits toxic fumes of Sn.

ABW750 CAS:1907-13-7 HR: 3
ACETOXYTRIETHYLSTANNANE
mf: $C_8H_{18}O_2Sn$ mw: 264.95

SYNS: ACETOXYTRIETHYLTIN □ TRIAETHYLZINNACETAT (GERMAN) □ TRIETHYLTIN ACETATE

TOXICITY DATA WITH REFERENCE
orl-rat LD50:4 mg/kg BJIMAG 15,15,58
ivn-rat LD50:4200 μg/kg BJIMAG 15,15,58
ivn-mus LD50:8 mg/kg CSLNX* NX#02839

OSHA PEL: TWA 0.1 mg(Sn)/m³ (skin)
ACGIH TLV: TWA 0.1 mg(Sn)/m³ (skin) (Proposed: TWA 0.1 mg(Sn)/m³; STEL 0.2 mg(Sn)/m³ (skin))
NIOSH REL: (Organotin Compounds) TWA 0.1 mg(Sn)/m³

SAFETY PROFILE: Poison by ingestion and intravenous routes. See also TIN COMPOUNDS. When heated to decomposition it emits acrid smoke and irritating Sn^+ fumes.

For occupational chemical analysis use NIOSH: organotin compounds 5504.

ABX000 CAS:2897-46-3 HR: 3
ACETOXYTRIHEXYLSTANNANE
mf: $C_{20}H_{42}O_2Sn$ mw: 433.31

SYNS: ACETOXYTRIHEXYLTIN □ TRIHEXYLTIN ACETATE □ TRI-N-HEXYLZINNACETAT (GERMAN)

TOXICITY DATA WITH REFERENCE
orl-rat LD50:1000 mg/kg BJIMAG 15,15,58
skn-rat LD50:500 mg/kg 85JCAE-,1254,86
ivn-rat LDLo:6 mg/kg BJIMAG 15,15,58

OSHA PEL: TWA 0.1 mg(Sn)/m³ (skin)
ACGIH TLV: TWA 0.1 mg(Sn)/m³ (skin) (Proposed: TWA 0.1 mg(Sn)/m³; STEL 0.2 mg(Sn)/m³ (skin))
NIOSH REL: (Organotin Compounds) TWA 0.1 mg(Sn)/m³

SAFETY PROFILE: Poison by skin contact and intravenous routes. Moderately toxic by ingestion. See also TIN COMPOUNDS. When heated to decomposition it emits acrid smoke and Sn^+ fumes.

For occupational chemical analysis use NIOSH: organotin compounds 5504.

ABX125 CAS:5711-19-3 HR: 3
ACETOXYTRIMETHYLPLUMBANE
mf: $C_5H_{12}O_2Pb$ mw: 311.36

PROP: White crystals from aqueous ethanol. Mp: 192–194°.

SYN: ACETATE de TRIMETHYLPLOMB (FRENCH)

TOXICITY DATA WITH REFERENCE
ipr-rat LD50:66 mg/kg APFRAD 24,17,66
orl-mus LD50:82 mg/kg APFRAD 24,17,66
ipr-mus LD50:34 mg/kg APFRAD 24,17,66

SAFETY PROFILE: Poison by ingestion and intraperitoneal routes. When heated to decomposition it emits toxic fumes of Pb.

ABX150 CAS:919-28-8 HR: 2
ACETOXYTRIOCTYLSTANNANE
mf: $C_{26}H_{54}O_2Sn$ mw: 517.49

SYNS: (ACETYLOXY)TRIOCTYLSTANNANE □ STANNANE, ACETOXYTRIOCTYL-

TOXICITY DATA WITH REFERENCE
orl-rat LD50:30 g/kg PHARAT 37,801,82

OSHA PEL: TWA 0.1 mg(Sn)/m³
ACGIH TLV: TWA 0.1 mg(Sn)/m³ (skin)

SAFETY PROFILE: Slightly toxic by ingestion. When heated to decomposition it emits toxic fumes of Sn.

ABX175 CAS:2587-75-9 HR: 2
ACETOXYTRIPENTYLSTANNANE
mf: $C_{17}H_{36}O_2Sn$ mw: 391.22

SYN: STANNANE, ACETOXYTRIPENTYL-

TOXICITY DATA WITH REFERENCE
orl-rat LD50:447 mg/kg PHARAT 37,801,82

OSHA PEL FINAL: TWA 0.1 mg(Sn)/m³ (skin)
ACGIH TLV: TWA 0.1 mg(Sn)/m³ (skin)

SAFETY PROFILE: Moderately toxic by ingestion. When heated to decomposition it emits toxic fumes of Sn.

ABX250 CAS:900-95-8 HR: 3
ACETOXYTRIPHENYLSTANNANE
mf: $C_{20}H_{18}O_2Sn$ mw: 409.07

PROP: White, crystalline solid. Mp: 120–123°. Practically insol.

SYNS: ACETATE de TRIPHENYL-ETAIN (FRENCH) □ ACETATO di STAGNO TRIFENILE (ITALIAN) □ ACETATOTRIPHENYLSTANNANE □ ACETOXY-TRIPHENYL-STANNAN (GERMAN) □ ACETOXY-TRIPHENYL-STANNANE □ ACETOXYTRIPHENYLTIN □ (ACETYLOXY)TRIPHENYL-STANNANE (9CI) □ BATASAN □ BRESTAN □ ENT 25,208 □ FENOLOVO ACETATE □ FENTIN ACETAAT (DUTCH) □ FENTIN ACETAT (GERMAN) □ FENTIN ACETATE □ FENTINE ACETATE (FRENCH) □ FINTIN ACETATO (ITALIAN) □ GC 6936 □ HOE-2824 □LIROMA-TIN □ LIROSTANOL □ PHENTIN ACETATE □ PHENTINOACETATE □ SUZU □ TINESTAN □ TINESTAN 60 WP □ TIN TRIPHENYL ACETATE □ TPTA □ TPZA □ TRIFENYLTINACETAAT (DUTCH) □ TRIPHENYLACETO STANNANE □ TRIPHENYLTIN ACETATE □ TRIPHENYL-ZINNACETAT (GERMAN) □ TUBOTIN □ VP 1940

TOXICITY DATA WITH REFERENCE
uns-ham:ovr 60 µg/L MUREAV 300,5,93
orl-rat TDLo:54 mg/kg (7-15D preg):REP NTOTDY 4,247,82
orl-rat TDLo:50 mg/kg:TER BECTA6 24,936,80
orl-mus TDLo:132 g/kg/78W-I:NEO NTIS** PB223-159
orl-rat LD50:125 mg/kg TIUSAD 43,9,58
skn-rat LD50:450 mg/kg ARZNAD 19,934,69
ipr-rat LD50:8500 µg/kg BJIMAG 23,222,66
ivn-rat LD50:18 mg/kg GUCHAZ 6,281,73
orl-mus LD50:81 mg/kg BJIMAG 23,222,66
ipr-mus LD50:7900 µg/kg BJIMAG 23,222,66
scu-mus LD50:44 mg/kg GUCHAZ 6,281,73
ivn-mus LD50:18 mg/kg CSLNX* NX#00648
orl-rbt LD50:30 mg/kg 85DPAN -,-,71/76
ipr-rbt LD50:10 mg/kg ARZNAD 13,432,63
orl-gpg LD50:21 mg/kg 85GYAZ -,127,71

CONSENSUS REPORTS: EPA Extremely Hazardous Substances List. Reported in EPA TSCA Inventory.

OSHA PEL: TWA 0.1 mg(Sn)/m³ (skin)
ACGIH TLV: TWA 0.1 mg(Sn)/m³ (skin) (Proposed: TWA 0.1 mg(Sn)/m³; STEL 0.2 mg(Sn)/m³ (skin))
NIOSH REL: (Organotin Compounds) TWA 0.1 mg(Sn)/m³

SAFETY PROFILE: Poison by ingestion, intraperitoneal, intravenous, and subcutaneous routes. Moderately toxic by skin contact. Questionable carcinogen with experimental neoplastigenic data. An experimental teratogen. Other experimental reproductive effects. A fungicide and algicide used as a wood preservative. When heated to decomposition it emits acrid smoke and Sn^+ fumes. See also TIN COMPOUNDS.

For occupational chemical analysis use NIOSH: organotin compounds 5504.

ABX325 CAS:13266-07-4 HR: 3
ACETOXYTRIPROPYLPLUMBANE
mf: $C_{11}H_{24}O_2Pb$ mw: 395.54

PROP: Needles from petroleum ether. Mp: 128°.

SYN: ACETATE de TRIPROPYLPLOMB (FRENCH)

TOXICITY DATA WITH REFERENCE
orl-rat LD50:214 mg/kg APFRAD 24,17,66
ipr-rat LD50:17 mg/kg APFRAD 24,17,66
orl-mus LD50:236 mg/kg APFRAD 24,17,66
ipr-mus LD50:24 mg/kg APFRAD 24,17,66

SAFETY PROFILE: Poison by ingestion and intraperitoneal routes. When heated to decomposition it emits toxic fumes of Pb.

ABX500 CAS:97-44-9 HR: 3
ACETPHENARSINE
mf: $C_8H_{10}AsNO_5$ mw: 275.11

PROP: Crystalline material. Mp: 220–221°. Decomp @ 240–250°. Sltly water sol.

SYNS: 3-ACETAMIDO-4-HYDROXY-PHENYLARSONIC ACID □ ACETARSOL □ ACETARSONE □ 3-ACETYLAMINO-4-HYDROXYPHENYLARSONIC ACID □ (3-(ACETYLAMINO)-4-HYDROXYPHENYL)ARSONINE (9CI) □ N-ACETYL-4-HYDROXY-m-ARSANILIC ACID □ AMARSAN □ AMOEBAL □ ARSONIC ACID □ ARSPHEN □ DEVEGAN □ DISPARICIDA □ DYNARSAN □ EHRLICH 594 □ F 190 □ 190 F □ FOURNEAU 190 □ GINARSOL □ GOYL □ GYNOPLIX □ KHAROPHEN □ KUBARSOL □ LIMARSOL MALAGRIDE □ MEXYL □ MONARGAN □ NILACID □ ORALCID □ ORARSAN □ OSARSAL □ OSARSOLE □ OSVARSAN □ PALLICID □ PAROXYL □ SPIROCID □ SPIROZID □ STOVARSAL □ STOVARSOL □ STOVARSOLAN □ SVC □ VAGISEPT □ VAGOFLOR

TOXICITY DATA WITH REFERENCE
dnd-esc 20 μmol/L MUREAV 89,95,81
orl-wmn TDLo:86 mg/kg/8D:RSP,SKN,MET AJMSA9 174,819,27
orl-man TDLo:89 mg/kg/9D:RSP,END,SKN AJMSA9 174,819,27
ivg-wmn LDLo:155 mg/kg/2D-I:CNS,GIT,MET BMJOAE 1,1282,61
ivg-wmn LDLo:1576 mg/kg/2D-I:CNS,GIT,KID BMJOAE 2,242,60
ivn-rat LDLo:300 mg/kg ADSYAF 25,799,32
orl-mus LD50:4 mg/kg CLDND* NX#03309
ivn-mus LD50:180 mg/kg CSLNX* NX#03309
orl-cat LDLo:150 mg/kg PSEBAA 27,267,30
orl-rbt LDLo:125 mg/kg PSEBAA 27,267,30
ivn-rbt LDLo:120 mg/kg ADSYAF 25,799,32

CONSENSUS REPORTS: Arsenic and its compounds are on the Community Right-To-Know List.

OSHA PEL: TWA 500 μg(As)/m³
ACGIH TLV: TWA 0.2 mg(As)/m³

SAFETY PROFILE: Poison by ingestion and intravenous routes. Human systemic effects by ingestion: respiratory system, endocrine system, dermatitis, and fever. Human systemic effects by intravaginal route: hallucinations, distorted perceptions, convulsions, nausea or vomiting, decreased urine volume, and fever. Mutation data reported. See also ARSENIC COMPOUNDS. When heated to decomposition it emits very toxic fumes of NO_x and As.

ABX750 CAS:123-54-6 HR: 3
ACETYL ACETONE
DOT: UN 2310
mf: $C_5H_8O_2$ mw: 100.13

PROP: Colorless to sltly yellow liquid; pleasant odor. Mp: −23.2°, bp: 139° @ 746 mm, flash p: 105°F (OC), d: 0.952–0.962, refr index: 1.402, vap d: 3.45, autoign temp: 644°F. Misc in alc, eth, chloroform, acetone, glacial acetic acid, and propylene glycol; insol in glycerin and water.

SYNS: ACETOACETONE □ DIACETYLMETHANE □ FEMA No. 2841 □ PENTANEDIONE □ 2,4 PENTANEDIONE (FCC)

TOXICITY DATA WITH REFERENCE
skn-rbt 10 mg/24H JIHTAB 26,269,44
skn-rbt 488 mg open MLD UCDS** 7/8/71
eye-rbt 20 mg SEV AJOPAA 29,1363,46
dlt-rat-ihl 694 ppm/6h/5D TXCYAC 5,463,89
ihl-rat TCLo:398 ppm/6H (female 6-15D post):REP TXCYAC 6,461,90
orl-rat LD50:55 mg/kg GISAAA 52(10),88,87
ihl-rat LCLo:1000 ppm/4H JIDHAN 31,343,49
ipr-rat LDLo:400 mg/kg BCPCA6 13,285,64
orl-mus LD50:951 mg/kg 38MKAJ 2C,4773,82
ipr-mus LD50:750 mg/kg NTIS** AD691-490
skn-rbt LD50:810 mg/kg DCTODJ 9,133,86
ihl-rat TCLo:805 ppm/6H/9D I FAATDF 7,329,86
ihl-rat TCLo:650 ppm/6H/14W-I FAATDF 7,329,86

CONSENSUS REPORTS: Reported in EPA TSCA Inventory.

DOT CLASSIFICATION: 3; *Label:* Flammable Liquid

SAFETY PROFILE: Poison by ingestion and intraperitoneal routes. Moderately toxic by inhalation. A skin and severe eye irritant. Experimental reproductive effects. Mutation data reported. Flammable liquid when exposed to heat or flame. Incompatible with oxidizing materials. To fight fire, use alcohol foam, CO_2, dry chemical.

ABX800 CAS:78600-25-6 HR: 3
3-ACETYLACONITINE HYDROBROMIDE
mf: $C_{36}H_{49}NO_{12}{\cdot}BrH$ mw: 768.78

TOXICITY DATA WITH REFERENCE
orl-mus LD50:2500 μg/kg CYLPDN 2(2),82,81
ipr-mus LD50:700 μg/kg CYLPDN 2(2),82,81
scu-mus LD50:1400 μg/kg CYLPDN 2(2),82,81

SAFETY PROFILE: Poison by ingestion, subcutaneous and intraperitoneal routes. When heated to decomposition it emits toxic fumes of NO_x and HBr.

ABY000 CAS:28322-02-3 HR: 3
4-ACETYLAMINOFLUORENE
mf: $C_{15}H_{13}NO$ mw: 223.29

SYNS: 4-ACETYLAMINOFLUOREN (GERMAN) □ N-FLUOREN-4-YLACETAMIDE □ N-4-FLUORENYLACETAMIDE

TOXICITY DATA WITH REFERENCE
mma-sat 50 μg/plate PMRSDJ 1,285,81
otr-ham:kdy 25 mg/L PMRSDJ 1,638,81

orl-rat TDLo:4175 mg/kg/17W-C:ETA ONCOAR 8,233,55
orl-rat TD:5240 mg/kg/57W-C:ETA,REP JNCIAM 24,149,60
ipr-mus LD50:364 mg/kg PMRSDJ 1,682,81

CONSENSUS REPORTS: EPA Genetic Toxicology Program.

SAFETY PROFILE: Poison by intraperitoneal route. Questionable carcinogen with experimental tumorigenic data. Experimental reproductive effects. Mutation data reported. When heated to decomposition it emits toxic fumes of NO_x.

ABY150 CAS:57229-41-1 HR: 2
2-ACETYLAMINO-9-FLUORENOL
mf: $C_{15}H_{13}NO_2$ mw: 239.29

SYNS: 9-HYDROXY-2-FLUORENYLACETAMIDE ☐ N-(9-HYDROXYFLUOREN-2-YL)ACETAMIDE

TOXICITY DATA WITH REFERENCE
orl-rat TDLo:4644 mg/kg/32W-C:ETA CNREA8 15,188,55

SAFETY PROFILE: Questionable carcinogen with experimental tumorigenic data. When heated to decomposition it emits toxic fumes of NO_x.

ABY250 CAS:3096-50-2 HR: 2
2-ACETYLAMINOFLUORENONE
mf: $C_{15}H_{11}NO_2$ mw: 237.27

SYNS: 2-ACETYLAMINO-9-FLUORENONE ☐ 9-OXO-2-FLUORENYLACETAMIDE ☐ N-(9-OXO-2-FLUORENYL)ACETAMIDE

TOXICITY DATA WITH REFERENCE
orl-rat TDLo:4740 mg/kg/32W-C:CAR CNREA8 15,188,55
orl-rat TD:6075 mg/kg/65W-C:CAR JNCIAM 24,149,60

SAFETY PROFILE: Questionable carcinogen with experimental carcinogenic data. When heated to decomposition it emits toxic fumes of NO_x.

ABY900 CAS:140-40-9 HR: 3
2-ACETYLAMINO-5-NITROTHIAZOLE
mf: $C_5H_5N_3O_3S$ mw: 187.19

PROP: Needles from alc, elongated plates from acetic acid. The commercial product may be yellow. Mp: 264–265°. Sol in aq solns of NaOH and NH_3 with deep orange color.

SYNS: ACETAMIDE, N-(5-NITRO-2-THIAZOLYL)- ☐ ACETYL ENHEPTIN ☐ ACINITRAZOL ☐ ACINITRAZOLE ☐ AMETOTERINA ☐ AMINITROZOL ☐ AMINITROZOLE ☐ CYZINE PREMIX ☐ ENHEPTIN A ☐ GYNOFON ☐ LAVOFLAGIN ☐ NITAZOL ☐ NITAZOLE ☐ NITHIAMIDE ☐ 5-NITRO-2-ACETILAMINOTIAZOLO ☐ N-(5-NITRO-2-THIAZOLYL)ACETAMIDE ☐ PLEOCIDE ☐ TRICHLORAD ☐ TRICHOCID ☐ TRICHOMAN ☐ TRICHORAD ☐ TRICHORAL ☐ TRICOGEN ☐ TRICOLAVAL ☐ TRICORAL ☐ TRICOSTERIL ☐ TRIKOLAVAL ☐ TRITHEON

TOXICITY DATA WITH REFERENCE
mmo-sat 500 nmol/L MUREAV 118,153,83
mmo-esc 20 μmol/L MUREAV 118,153,83
mmo-klp 20 μmol/L MUREAV 118,153,83
mrc-smc 200 ppm MUREAV 118,153,83
orl-rat LD50:>400 mg/kg ANTCAO 5,540,55
ipr-rat LD50:>200 mg/kg ANTCAO 5,540,55
scu-rat LD:>3200 mg/kg ANTCAO 5,540,55
orl-mus LD50:1 g/kg FRPSAX 19,301,64
ipr-mus LD50:>300 mg/kg ANTCAO 5,540,55
scu-mus LD:>3200 mg/kg ANTCAO 5,540,55
orl-dog LD50:125 mg/kg ANTCAO 5,540,55
orl-ckn LD50:800 mg/kg ANTCAO 5,540,55

SAFETY PROFILE: Poison by ingestion. Mutation data reported. When heated to decomposition it emits toxic fumes of SO_x and NO_x.

ACA125 HR: 3
5-ACETYLAMINO-2,4,6-TRIIODO ISOPHTHALIC ACID DI-(N-METHYL-2,3-DIHYDROXYPROPYLAMIDE)
mf: $C_{18}H_{24}I_3N_3O_7$ mw: 775.15

TOXICITY DATA WITH REFERENCE
ivn-mus LD50:31,965 mg/kg USXXAM #4001323
ice-mus LD50:1670 mg/kg USXXAM #4001323
par-rbt LD50:165 mg/kg USXXAM #4001323

SAFETY PROFILE: Poison by parenteral route. Moderately toxic by intracerebral route. When heated to decomposition it emits toxic fumes of I^- and NO_x.

ACA750 CAS:73637-16-8 HR: 2
9-ACETYL-1,7,8-ANTHRACENETRIOL
mf: $C_{16}H_{11}O_4$ mw: 267.27

SYNS: 10-ACETYL-1,8,9-ANTHRACENETRIOL ☐ 10-ACETYLANTHRALIN ☐ 1,8-DIHYDROXY-10-ACETYL-9-ANTHRONE

TOXICITY DATA WITH REFERENCE
skn-mus TDLo:32 mg/kg/53W-I:NEO JMCMAR 21,26,78

SAFETY PROFILE: Questionable carcinogen with experimental neoplastigenic data. When heated to decomposition it emits acrid smoke and fumes.

ACB250 CAS:460-07-1 HR: 3
1-ACETYLAZIRIDINE
mf: C_4H_7NO mw: 85.12

SYN: ACETYLETHYLENEIMINE

TOXICITY DATA WITH REFERENCE
scu-rat TDLo:78 mg/kg/26W-I:NEO BJPCAL 9,306,54
scu-rat TD:80 mg/kg/16W-I:ETA BJPCAL 9,306,54
ipr-mus LD50:13 mg/kg NCISA* PH-43-63-1132

CONSENSUS REPORTS: Reported in EPA TSCA Inventory. EPA Genetic Toxicology Program.

SAFETY PROFILE: Poison by intraperitoneal route. Questionable carcinogen with experimental tumorigenic and neoplastigenic data. When heated to decomposition it emits toxic fumes of NO_x.

ACC000 CAS:3366-61-8 HR: 2
N-ACETYLBENZIDINE
mf: $C_{14}H_{14}N_2O$ mw: 226.30

PROP: Needles from aqueous ethanol. Mp: 199°.

SYNS: 4′-ACETAMIDOBENZIDINE □ N-(4′-AMINO(1,1′-BIPHENYL)-4-YL)-ACETAMIDE □ 4′-(p-AMINOPHENYL)ACETANILIDE

TOXICITY DATA WITH REFERENCE
mma-sat 5 μg/plate ENMUDM 6,145,84
dnd-rat-ipr 25 mg/kg CNREA8 42,2678,82
dnd-rat:lvr 100 mg/L CRNGDP 5,407,84
orl-rat LD50:1630 mg/kg 28ZPAK -,131,72

SAFETY PROFILE: Moderately toxic by ingestion. Mutation data reported. When heated to decomposition it emits toxic fumes of NO_x.

ACC100 CAS:1646-26-0 HR: 2
2-ACETYLBENZOFURAN
mf: $C_{10}H_8O_2$ mw: 160.18

SYNS: 2-ACETYLCOUMARONE □ 1-(2-BENZOFURANYL)ETHANONE □ BENZO(b)FURAN-2-YL METHYL KETONE □ 2-BENZOFURANYL METHYL KETONE □ ETHANONE, 1-(2 BENZOFURANYL)-(9CI) □ KETONE, 2-BENZOFURANYL METHYL

TOXICITY DATA WITH REFERENCE
ipr-mus LD50:1200 mg/kg EJMCA5 12,383,77

CONSENSUS REPORTS: Reported in EPA TSCA Inventory.

DOT CLASSIFICATION: 3; *Label:* Flammable Liquid

SAFETY PROFILE: Moderately toxic by intraperitoneal route. A flammable liquid. When heated to decomposition it emits acrid smoke and irritating vapors.

ACC250 CAS:644-31-5 HR: 3
ACETYL BENZOYL PEROXIDE (solid)
mf: $C_9H_8O_4$ mw: 180.17

PROP: White crystals. Mp: 36–37°, bp: 130° @ 19 mm. Sol in oils, alc, eth, and chloroform.

DOT CLASSIFICATION: Forbidden

SAFETY PROFILE: Poison by inhalation and ingestion. Severe irritant. A powerful oxidizing agent that is corrosive to the skin and mucous membranes. See also PEROXIDES, ORGANIC. Dangerous; shock or heat will cause detonation with evolution of toxic fumes; will react with water or steam to produce heat; can react vigorously with reducing materials. Flammable by spontaneous chemical reaction. To fight fire, use CO_2 or dry chemical. When heated to decomposition it emits acrid smoke and fumes.

ACC750 CAS:63018-98-4 HR: 2
2-ACETYL-3:4-BENZPHENANTHRENE
mf: $C_{20}H_{14}O$ mw: 270.34

SYN: 5-ACETYL BENZO(C)PHENANTHRENE

TOXICITY DATA WITH REFERENCE
skn-mus TDLo:720 mg/kg/30W-I:ETA PRLBA4 131,170,42

SAFETY PROFILE: Questionable carcinogen with experimental tumorigenic data. When heated to decomposition it emits acrid smoke and irritating fumes.

ACD000 CAS:4463-22-3 HR: 2
N-ACETYL-4-BIPHENYLHYDROXYLAMINE
mf: $C_{14}H_{13}NO_2$ mw: 227.28

SYNS: 4-BIPHENYLACETHYDROXAMIC ACID □ N-HYDROXY-AABP □ N-HYDROXY-4-ACETAMIDOBIPHENYL □ N-4-(N-HYDROXYACETAMIDO)BIPHENYL □ N-HYDROXY-4-ACETAMIDODIPHENYL □ N-HYDROXY-4-ACETYLAMINOBIPHENYL □ N-HYDROXY-N-4-BIPHENYLACETAMIDE

TOXICITY DATA WITH REFERENCE
mnt-ham:ovr 290 μmol/L MUREAV 88,397,81
dns-hmn:oth 1 μmol/L JJIND8 72,847,84
dnd-rat-ipr 25 mg/kg COINAV 256,115,77
dns-rat:oth 10 μmol/L CNREA8 43,3974,82
dns-mus:oth 10 μmol/L CNREA8 43,3974,82
dns-rbt:oth 10 μmol/L CNREA8 45,221,85
dns-dog:oth 1 μmol/L CNREA8 42,3974,82
sce-ham:ovr 1440 μmol/L MUREAV 88,397,81
orl-rat TDLo:2400 mg/kg/16W-C:ETA CNREA8 21,1465,61
ipr-rat TDLo:91 mg/kg:CAR CNREA8 41,2450,81

SAFETY PROFILE: Questionable carcinogen with experimental carcinogenic and tumorigenic data. Human mutation data reported. When heated to decomposition it emits toxic fumes of NO_x.

ACD250 CAS:3733-45-7 HR: 3
N-(N-ACETYL-3-(p-(BIS(2-CHLOROETHYL)AMINO) PHENYL)ALANYL-3-PHENYLALANINE) ETHYL ESTER

SYN: ETHYL ESTER of N-ACETYL-dl-SARCOLYSYL-l-PHENYLALANINE

TOXICITY DATA WITH REFERENCE
orl-rat LD50:115 mg/kg FATOAO 33,472,70
ims-rat LD50:33 mg/kg FATOAO 33,472,70
rec-rat LD50:64 mg/kg FATOAO 33,472,70

SAFETY PROFILE: Poison by ingestion and intramuscular routes. When heated to decomposition it emits very toxic fumes of Cl^- and NO_x. See also ESTERS.

ACD500 CAS:18869-73-3 HR: 3
1-ACETYL-3,3-BIS(p-HYDROXYPHENYL)OXINDOLE DIACETATE
mf: $C_{26}H_{21}NO_6$ mw: 443.48

PROP: Crystals from aqueous ethanol. Mp: 201–202°.

SYNS: 1-ACETYL-3,3-BIS(4-(ACETYLOXY)PHENYL)-1,3-DIHYDRO-2H-INDOL-2-ONE □ ISATEX □ LAXAGEN □ LAXAGETTEN □ PHENISATIN □ TRIACETYLDIPHENOLISATIN □ TRISATIN □ UNILAX

TOXICITY DATA WITH REFERENCE
orl-rat LD50:500 mg/kg JAPMA8 42,468,53
ipr-rat LD50:350 mg/kg JAPMA8 42,468,53

CONSENSUS REPORTS: Reported in EPA TSCA Inventory.

SAFETY PROFILE: Poison by intraperitoneal route. Moderately toxic by ingestion. When heated to decomposition it emits toxic fumes of NO_x. A cathartic.

ACD750 **CAS:506-96-7** **HR: 3**
ACETYL BROMIDE
DOT: UN 1716
mf: C_2H_3BrO mw: 122.96

PROP: Colorless, fuming liquid; turns yellow in air. Mp: −96.5°, bp: 76.7°, d: 1.52 @ 9.5°/4°. Decomp in water and alc; misc in benzene, ether, and chloroform.

TOXICITY DATA WITH REFERENCE
ipr-mus LD50:250 mg/kg GTPZAB 20(12),52,76
ihl-uns LC50:48 g/m³ GTPZAB 18(4),55,74

CONSENSUS REPORTS: Reported in EPA TSCA Inventory.

DOT CLASSIFICATION: 8; *Label:* Corrosive

SAFETY PROFILE: Poison by ingestion, inhalation, skin contact, and intraperitoneal routes. See also HYDROBROMIC ACID and ACETIC ACID. Violent reaction on contact with water, steam, methanol, or ethanol produces toxic and reactive HBr. When heated to decomposition it emits highly corrosive and toxic fumes of carbonyl bromide and bromine. To fight fire, use dry chemical, CO_2.

ACE000 **CAS:77-66-7** **HR: 2**
1-ACETYL-3-(2-BROMO-2-ETHYLBUTYRYL)UREA
mf: $C_9H_{15}BrN_2O_3$ mw: 279.17

PROP: Mp: 109°. Sltly sol in water. Sol in EtOH, EtOAc.

SYNS: ABASIN □ ABSIN □ ACECARBROMAL □ACETCARBRO-MAL □ ACETKARBROMAL □ ACETYL ADALIN □ N-((ACETYLAMINO)CARBONYL)-2-BROMO-2-ETHYLBUTANAMIDE □ ACETYLBROMODIETHYLACETYLCARBAMIDE □ N-ACETYL-N-BROMODIETHYLACETYLCARBAMIDE □ N-ACETYL-N-BROMODIETHYLACETYLUREA □ N-ACETYL-N′-α-BROMO-α-ETHYLBUTYRYLCARBAMIDE □ 1-ACETYL-3-(α-BROMO-α-ETHYLBUTYRYL)UREA □ACETYLCARBROMAL □ ADITYL □ CARBASED □ DAROLON □ IBATRAN □PAXAREL □ SEDAMYL □ SEDMYNOL □ SEDTRAN

TOXICITY DATA WITH REFERENCE
orl-hmn TDLo:7 mg/kg:PSY 27ZQAG -,423,72
orl-mus LD50:1600 mg/kg CLDND*

CONSENSUS REPORTS: Reported in EPA TSCA Inventory.

SAFETY PROFILE: Moderately toxic by injection. Human systemic effects by ingestion: toxic psychosis. When heated to decomposition it emits very toxic fumes of Br^- and NO_x. A sedative.

ACE500 **CAS:2813-95-8** **HR: 3**
o-ACETYL-2-sec-BUTYL-4,6-DINITROPHENOL
mf: $C_{12}H_{14}N_2O_6$ mw: 282.28

PROP: Oil or crystals. Mp: 26–27°.

SYNS: ACETIC ACID, (2,4-DINITRO-6-s-BUTYLPHENYL) ESTER □ ACETIC ACID, (4,6-DINITRO-2-s-BUTYLPHENYL) ESTER □ O-ACETYL-2-sec-BUTYL-4,6-DINITROPHENOL □ ARETIT □ ARETIT (the phenol) □ 2-sek.BUTYL-4,6-DINITROFENYLESTER KYSELINY OCTOVE (CZECH) □ 2-sec-BUTYL-4,6-DINITROPHENOL ACETATE (ester) □ 2-sec-BUTYL-4,6-DINITROPHENYLACETATE □ 6-sec-BUTYL-2,4-DINITROPHENYLACETATE □ 2,4-DINITRO-6-sec-BUTYLFENYLESTER KYSELINY OCTOVE (CZECH) □ 2,4-DINITRO-6-sek.BUTYL-PHENYLACETAT (GERMAN) □ 4,6-DINITRO-2-s-BUTYLPHENYL ACETATE □ DINOSEB-ACETATE □ DINOSEBE ACETATE □ HOE 2904 □ β-(2-HYDROXY-3,5-DINITROPHENYL)BUTANE ACETATE □ IVOSIT □ 2-(1-METHYLPROPYL)-4,6-DINITROPHENYL ACETATE □ PHENOL, 2-sec-BUTYL-4,6-DINITRO-, ACETATE (ESTER) (8CI) □ PHENOL, 2-(1-METHYLPROPYL)-4,6-DINITRO-, ACETATE (ESTER) (9CI) □ PHENOTAN

TOXICITY DATA WITH REFERENCE
skn-rbt 500 mg/24H MOD 28ZPAK -,131,72
eye-rbt 500 mg/24H MOD 28ZPAK -,131,72
orl-rat LD50:60 mg/kg FMCHA2-,C26,91
ihl-rat LC50:1300 mg/m³/4H 85JFAN A160,83
orl-ckn LD50:40 mg/kg GUCHAZ 6,229,73

SAFETY PROFILE: Poison by ingestion. Moderately toxic by inhalation. A skin and eye irritant. See also ESTERS and NITRO COMPOUNDS of AROMATIC HYDROCARBONS. When heated to decomposition it emits toxic fumes of NO_x. A herbicide.

ACF000 **CAS:36573-63-4** **HR: 3**
3′-o-ACETYLCALOTROPIN
mf: $C_{31}H_{42}O_{10}$ mw: 574.73

PROP: Crystals from MeOH. Mp: 308–309°. A glycoside isolated from *Asclepius cunssuica* (ARZNAD 28,1095,78).

SYN: ASCLEPIN

TOXICITY DATA WITH REFERENCE
ipr-mus LDLo:15 mg/kg ARZNAD 28,1095,78
ivn-pgn LDLo:400 µg/kg ARZNAD 28,1095,78
scu-frg LDLo:5 mg/kg ARZNAD 28,1095,78

SAFETY PROFILE: Poison by intravenous, subcutaneous, and intraperitoneal routes. When heated to decomposition it emits acrid smoke and irritating fumes.

ACF100 **CAS:1888-91-1** **HR: 2**
ACETYLCAPROLACTAM
mf: $C_8H_{13}NO_2$ mw: 155.22

SYNS: N-ACETYLCAPROLACTAM □ ACETYLKAPROLAKTAM □ 2H-AZEPIN-2-ONE, 1-ACETYLHEXAHYDRO-

TOXICITY DATA WITH REFERENCE
orl-uns LD50:1300 mg/kg 85JCAE-,884,86

CONSENSUS REPORTS: Reported in EPA TSCA Inventory.

SAFETY PROFILE: Moderately toxic by ingestion. When heated to decomposition it emits toxic vapors of NO_x.

ACF250 **CAS:80449-58-7** **HR: 1**
ACETYL CEDRENE

PROP: Prepared by acetylation of the hydrocarbon portion of cedarwood oil in the presence of an acid catalyst.

SYN: VERTOFIX COEUR

TOXICITY DATA WITH REFERENCE
skn-rbt 500 mg/24H MOD FCTXAV 16,637,78
orl-rat LD50:5200 mg/kg FCTXAV 16,637,78

SAFETY PROFILE: Mildly toxic by ingestion. A skin irritant. When heated to decomposition it emits acrid smoke and fumes.

ACF750 CAS:75-36-5 HR: 3
ACETYL CHLORIDE
DOT: UN 1717
mf: C_2H_3ClO mw: 78.50

PROP: Colorless, pungent liquid. Fuming in air. Mp: −112°, bp: 51–52°, flash p: 40°F (CC), autoign temp: 734°F, d: 1.1051 @ 20°/4°, vap d: 2.70. lel: 5%. Decomp in water and alc; misc in benzene, ether, and chloroform. Sol in Et_2O, C_6H_6.

SYNS: ACETIC ACID CHLORIDE □ ACETIC CHLORIDE □ ETHANOYL CHLORIDE □ RCRA WASTE NUMBER U006

TOXICITY DATA WITH REFERENCE
ihl-hmn TCLo:2 ppm/1M:IRR TGNCDL 2,28,61
orl-rat LD50:910 mg/kg GTPZAB 32(3),48,88

CONSENSUS REPORTS: Reported in EPA TSCA Inventory.

DOT CLASSIFICATION: 3; *Label:* Flammable Liquid, Corrosive

SAFETY PROFILE: Poison by inhalation. Moderately toxic by ingestion. A human systemic irritant by inhalation. Violent hydrolysis reaction with water or steam produces heat, acetic acid, HCl, and other corrosive chlorides. May decompose during preparation. Dangerous fire hazard when exposed to heat or flame. Explosion hazard by spontaneous chemical reaction with dimethyl sulfoxide or ethanol. Also incompatible with PCl_3. When heated to decomposition it emits highly toxic fumes of phosgene and Cl. To fight fire, use CO_2 or dry chemical. See also CHLORIDES.

ACG125 CAS:39426-77-2 HR: 3
1-(3-(3-ACETYL-4-(p-(CHLOROPHENYL)PIPERIDINO) PROPYL)-4-METHYLPIPERAZINE) TRIHYDROCHLORIDE
mf: $C_{21}H_{32}ClN_3O{\cdot}3ClH$ mw: 487.39

SYNS: 4-ACETYL-4-(3-CHLOROPHENYL)-1-(3-(4-METHYLPIPERAZINO)-PROPYL)PIPERIDINE TRIHYDROCHLORIDE □ TROJCHLOROWODOREK 4-ACETYLO-4-(3-CHLOROFENYLO)-1-(3-(4-METYLOPIPERAZYNO)-PROPYLO)-PIPERYDYNY

TOXICITY DATA WITH REFERENCE
orl-mus LD50:2290 mg/kg APPHAX 37,579,80
scu-mus LD50:2694 mg/kg APPHAX 37,579,80
ivn-mus LD50:148 mg/kg APPHAX 37,579,80

SAFETY PROFILE: Poison by intravenous route. Moderately toxic by ingestion and other routes. When heated to decomposition it emits toxic fumes of NO_x and HCl.

ACG250 CAS:38838-26-5 HR: 3
N-ACETYL COLCHINOL
mf: $C_{20}H_{23}NO_5$ mw: 357.44

SYNS: N-ACETYL-COLCHINOL (GERMAN) □ (S)-N-(3-HYDROXY-9,10,11-TRIMETHOXY-5H-DIBENZO(a,c)CYCLOHEPTEN-5-YL)-ACETAMIDE (9CI)

TOXICITY DATA WITH REFERENCE
oms-mus-ipr 28 mg/kg CANCAR 3,130,50
oms-mus-par 56 mg/kg CANCAR 3,130,50
spm-mus-par 56 mg/kg CANCAR 3,130,50
unk-rat LDLo:200 mg/kg CANCAR 3,124,50
ipr-mus LD50:56 mg/kg CANCAR 3,124,50
unk-cat LDLo:10 mg/kg CANCAR 3,124,50

SAFETY PROFILE: Poison by intraperitoneal and other unspecified routes. Mutation data reported. When heated to decomposition it emits toxic fumes of NO_x.

ACG300 CAS:3179-56-4 HR: 3
ACETYL CYCLOHEXANEPERSULFONATE
mf: $C_8H_{14}O_5S$ mw: 222.28

SYNS: ACETYL CYCLOHEXYLSULFONYL PEROXIDE □ ACETYL CYCLOHEXANESULFONYL PEROXIDE, >82% wetted with <12% water (DOT) □ LUPERSOL 228Z □ PEROXIDE, ACETYL CYCLOHEXYLSULFONYL

CONSENSUS REPORTS: Reported in EPA TSCA Inventory.

DOT CLASSIFICATION: Forbidden

SAFETY PROFILE: A very unstable peroxide. When heated to decomposition it emits toxic vapors of SO_x.

ACH000 CAS:616-91-1 HR: 3
N-ACETYL-l-CYSTEINE
mf: $C_5H_9NO_3S$ mw: 163.21

PROP: Crystals from water. Mp: 109–110°.

SYNS: l-α-ACETAMIDO-β-MERCAPTOPROPIONIC ACID □ACE-TEIN □ ACETYLCYSTEINE □ N-ACETYLCYSTEINE □ N-ACETYL-N-CYSTEINE □ N-ACETYL-l-CYSTEINE (9CI) □ N-ACETYL-3-MERCAPTOALANINE □ AIRBRON □ BRONCHOLYSIN □ FLUIMUCETIN □ FLUIMUCIL □ FLUMICIL □ INSPIR □ MERCAPTURIC ACID □ (R)-MERCAPTURIC ACID □ MUCOLYTICUM □ MUCOLYTICUM LAPPE □ MUCOMYST □ MUCOSOLVIN □ NAC □ NAC-TB □ NSC 111180 □ PARVOLEX □ RESPAIRE

TOXICITY DATA WITH REFERENCE
mma-sat 8 μg/plate CRNGDP 7,431,86
orl-rat LD50:5050 mg/kg TXAPA9 18,185,71
ivn-rat LD50:1140 mg/kg EJRDD2 61(Suppl 111),45,80
orl-mus LD50:7888 mg/kg THEWA6 30,1926,80
ipr-mus LD50:400 mg/kg NTIS** AD691-490
ivn-mus LD50:3800 mg/kg JMCMAR 10,1172,67

CONSENSUS REPORTS: Reported in EPA TSCA Inventory.

SAFETY PROFILE: Poison by intraperitoneal route. Moderately toxic by other routes. Mutation data reported. When heated to decomposition it emits very toxic fumes of NO_x and SO_x.

ACH075 CAS:50722-38-8 HR: 3
3-ACETYLDEOXYNIVALENOL
mf: $C_{17}H_{22}O_7$ mw: 338.39

PROP: Crystals from Et_2O/pentane. Mp: 185.5–186°.

SYNS: DEHYDRONIVALENOL MONOACETATE □ DEOXYNIVALENOL MONOACETATE

TOXICITY DATA WITH REFERENCE
skn-gpg 3384 ng MLD FAATDF 4(2, Pt 2),S124,84
orl-mus LD50:34 mg/kg FAATDF 4(2, Pt 2),S124,84
ipr-mus LD50:47 mg/kg 41KEAL -,108,78

SAFETY PROFILE: Poison by ingestion and intraperitoneal routes.

ACH090 CAS:22439-58-3 HR: 3
2-ACETYLDIBENZOTHIOPHENE

PROP: A liquid.
mf: $C_{14}H_{10}OS$ mw: 226.30

SYNS: KETONE, 2-DIBENZOTHIENYL METHYL □ DIBENZOTHIEN-2-YL METHYL KETONE

TOXICITY DATA WITH REFERENCE
ivn-mus LD50:180 mg/kg CSLNX* NX#01248

DOT CLASSIFICATION: 3; *Label:* Flammable Liquid

SAFETY PROFILE: Poison by intravenous route. A flammable liquid. When heated to decomposition it emits toxic vapors of SO_x.

ACH125 CAS:59183-18-5 HR: 3
ACETYL-1,1-DICHLOROETHYL PEROXIDE
mf: $C_4H_6Cl_2O_3$ mw: 173.00

$CH_3CO•OOCCl_2CH_3$

SAFETY PROFILE: A viscous liquid explosive, sensitive to friction and heat. When heated to decomposition it emits toxic fumes of Cl^-. See also PEROXIDES.

ACH375 CAS:73987-00-5 HR: 3
16-ACETYLDIGITALINUM VERUM

SYN: 16-ACETATE DIGITOXIN

TOXICITY DATA WITH REFERENCE
ivn-cat LDLo:255 μg/kg JMPCAS 5,988,62
orl-frg LD50:36,800 μg/kg JJPAAZ 9,91,60
scu-frg LD50:490 μg/kg JJPAAZ 9,91,60

SAFETY PROFILE: Poison by ingestion, subcutaneous and intravenous routes.

ACH500 CAS:1111-39-3 HR: 3
ACETYLDIGITOXIN-α
mf: $C_{43}H_{66}O_{14}$ mw: 807.09

PROP: Platelets from MeOH. Mp: 217–221°.

SYNS: α-ACETYLDIGITOXIN □ ACYLANID

TOXICITY DATA WITH REFERENCE
orl-cat LD50:250 μg/kg AIPTAK 159,1,66
ivn-cat LD50:514 μg/kg JPETAB 111,365,54

SAFETY PROFILE: Poison by ingestion and intravenous routes. When heated to decomposition it emits acrid smoke and fumes. See also DIGITOXIN.

ACH750 CAS:1264-51-3 HR: 3
ACETYLDIGITOXIN-β
mf: $C_{43}H_{66}O_{14}$ mw: 807.09

SYNS: β-ACETYLDIGOXIN □ DIGITOXIGENIN + 2-DIGITOXOSE + 1-ACETYL-(4)-DIGITOSE (GERMAN) □ DIGITOXIGENIN + 2-DIGITOXOSE + ACETYL-(3)-DIGITOXOSE (GERMAN)

TOXICITY DATA WITH REFERENCE
ivn-cat LD50:476 μg/kg JPETAB 111,365,54
orl-gpg LD50:50 mg/kg AIPTAK 159,1,66
ivn-gpg LDLo:1750 μg/kg ARZNAD 15,481,65

SAFETY PROFILE: Poison by ingestion and intravenous routes. When heated to decomposition it emits acrid smoke and fumes. See also DIGITOXIN.

ACI000 CAS:5511-98-8 HR: 3
ACETYLDIGOXIN-α
mf: $C_{43}H_{66}O_{15}$ mw: 823.09

PROP: Prisms from MeOH/$CHCl_3$.

SYNS: α-ACETYLDIGOXIN □ DIGORID A □ DIGOXIGENIN + ZUCKERKETTE WIE BEI ACETYL-DIGITOXIN A (GERMAN)

TOXICITY DATA WITH REFERENCE
orl-cat LD50:200 μg/kg AIPTAK 159,1,66
ivn-cat LD50:466 μg/kg JPETAB 111,365,54
idu-cat LDLo:494 μg/kg ARZNAD 20,1765,70
orl-gpg LD50:3300 μg/kg AIPTAK 159,1,66
ivn-gpg LDLo:1380 μg/kg ARZNAD 15,483,65

SAFETY PROFILE: Deadly poison by ingestion, intravenous, and intraduodenal routes. When heated to decomposition it emits acrid smoke and fumes. See also DIGITOXIN.

ACI250 CAS:5355-48-6 HR: 3
ACETYLDIGOXIN-β
mf: $C_{43}H_{66}O_{15}$ mw: 823.09

PROP: Needles from EtOH/$CHCl_3$.

SYNS: β-ACETYLDIGOXIN □ DIGORID B □ DIGOXIGENIN + ZUCKERKETTE WIE BEI ACETYL-DIGITOXIN-α (GERMAN) □ HEXAMETHYLENEIMINE-3,5-DINITROBENZOATE

TOXICITY DATA WITH REFERENCE
orl-dog LD50:422 μg/kg ARZNAD 24,1914,74
ivn-cat LD50:430 μg/kg JPETAB 111,365,54
idu-cat LDLo:413 μg/kg ARZNAD 19,687,69
orl-gpg LD50:2400 μg/kg ARZNAD 15,481,65
ivn-gpg LDLo:1500 μg/kg ARZNAD 15,481,65

SAFETY PROFILE: Deadly poison by ingestion, intravenous, and intraduodenal routes. When heated to decomposition it emits acrid smoke and fumes. See also DIGITOXIN.

ACI375 CAS:21380-82-5 HR: 3
ACETYLDIMETHYLARSINE
mf: C_4H_9AsO mw: 148.04

PROP: Colorless liquid. Bp: 60°.

CONSENSUS REPORTS: Arsenic and its compounds are on the Community Right-To-Know List.

SAFETY PROFILE: A poison. Ignites on contact with air. When heated to decomposition it emits toxic fumes of As. See also ARSENIC COMPOUNDS.

ACI500 CAS:2386-25-6 HR: 3
3-ACETYL-2,4-DIMETHYL-PYRROLE
mf: $C_8H_{11}NO$ mw: 137.20

PROP: Solid. Mp: 137°.

SYN: 2,4-DIMETHYLPYRROL-3-YL METHYL KETONE

TOXICITY DATA WITH REFERENCE
ipr-rat LD50:250 mg/kg JMCMAR 11,1251,68
ipr-mus LD50:400 mg/kg JMCMAR 11,1251,68
ivn-mus LD50:71 mg/kg CSLNX* NX#04669

DOT CLASSIFICATION: 3; *Label:* Flammable Liquid

SAFETY PROFILE: Poison by intravenous and intraperitoneal routes. A flammable liquid. When heated to decomposition it emits toxic fumes of NO_x. See also KETONES.

ACI550 CAS:1500-94-3 HR: 3
3-ACETYL-2,5-DIMETHYL-PYRROLE
mf: $C_8H_{11}NO$ mw: 137.20

SYNS: 2,5-DIMETHYLPYRROL-3-YL METHYL KETONE □ KETONE, 2,5-DIMETHYLPYRROL-3-YL METHYL

TOXICITY DATA WITH REFERENCE
ipr-rat LD50:225 mg/kg JMCMAR 11,1251,68
ipr-mus LD50:553 mg/kg JMCMAR 11,1251,68

DOT CLASSIFICATION: 3; *Label:* Flammable Liquid

SAFETY PROFILE: A poison by intraperitoneal route. A flammable liquid. When heated to decomposition it emits toxic vapors of NO_x.

ACI750 CAS:74-86-2 HR: 3
ACETYLENE
DOT: UN 1001
mf: C_2H_2 mw: 26.04

PROP: Colorless gas, garlic-like odor. Flammable. Bp: −84.0° (subl), lel: 2.5%, uel: 82%, mp: −81.8°, flash p: 0°F (CC), d: 1.173 g/L @ 0°, autoign temp: 581°F, vap press: 40 atm @ 16.8°, vap d: 0.91; d: (liquid) 0.613 @ −80°. D: (solid) 0.730 @ −85°. Sltly sol in water; mod sol in ethanol and acetic acid; very sol in Me_2CO; almost misc in ether.

SYNS: ACETYLEN □ ACETYLENE, dissolved (DOT) □ ETHINE □ ETHYNE □ NARCYLEN

TOXICITY DATA WITH REFERENCE
ihl-hmn TCLo:20 pph:CNS,RSP 34ZIAG -,67,69
ihl-hmn LCLo:50 pph/5M TABIA2 3,231,33
ihl-uns LCLo:50 pph/5M AEPPAE 138,65,28

CONSENSUS REPORTS: Reported in EPA TSCA Inventory.

OSHA PEL: CL 2500 ppm
ACGIH TLV: Simple asphyxiant.
NIOSH REL: (Acetylene) 10H TWA no exposure >2500 ppm
DOT CLASSIFICATION: Forbidden; DOT Class 2.1; *Label:* Flammable Gas

SAFETY PROFILE: Mildly toxic by inhalation. Human systemic effects by inhalation: headache and dyspnea. Narcotic in high concentration. In general industrial practice, acetylene does not constitute a serious toxic hazard. It is a very dangerous fire hazard when exposed to heat, flame, or oxidizers. Moderate explosion hazard when exposed to heat or flame or by spontaneous chemical reaction. At high pressures and moderate temperatures, and in the absence of air, acetylene has been known to decompose explosively. Reacts with copper to form the explosive copper acetylide. Incompatible with brass, copper salts, copper carbide, powdered Co, Hg, Hg salts, K, Ag and Ag salts, RbH, CsH, halogens, HNO_3, NaH, oxidants. Acetylene + halide + UV can explode. Molten K ignites in C_2H_2 and then explodes. C_2H_2 reacts vigorously with trifluoromethyl hypofluorite. With O_2, C_2H_2 can detonate very powerfully. See ACETYLIDES. When ignited, it burns with an intensely hot flame; can react vigorously with oxidizing materials.

When mixed with O_2 in proportions of 40% or more, acetylene acts as a narcotic and has been used in anesthesia. Acetylene acts as a simple asphyxiant by diluting the O_2 in the air to a level that will not support life. However, the presence of impurities in commercial acetylene may result in the production of symptoms before an asphyxiant concentration is reached. Thus: 10% in air produces a slight intoxication, 20% produces a staggering gait, 30% produces general incoordination, 33% leads to unconsciousness in 7 minutes, up to 80% produces complete anesthesia, increased blood pressure, narcosis, and stimulated respiration.

Dizziness, headache, mild gastric symptoms, and (in high concentration) semi-asphyxia and brief loss of consciousness have all been reported. See ARGON for a discussion of simple asphyxiants. To fight fire, use CO_2, water spray, or dry chemical. Stop flow of gas.

ACJ000 HR: 3
ACETYLENE CHLORIDE
mf: CHCCl mw: 60.48

PROP: A gas. Bp: −31°, vap d: 2.0, mp: −126°.

SYN: CHLOROETHYNE

SAFETY PROFILE: Dangerous fire hazard by spontaneous chemical reaction. Spontaneously flammable in air. Shock will explode it. When heated to decomposition it emits highly toxic fumes of phosgene; can react vigorously with oxidizing materials. See also ACETYLENE COMPOUNDS and CHLORINATED HYDROCARBONS, ALIPHATIC.

ACJ125 HR: 3
ACETYLENE COMPOUNDS and ALKYNES

SAFETY PROFILE: The carbon–carbon triple bond is explosively unstable in many acetylenic compounds. Both the lower alkynes (i.e., propyne, butadyne, etc.)

and higher compounds may undergo explosive decomposition. The presence of halogens and heavy metal derivatives may increase these explosive tendencies. See also ACETYLENE, ACETYLIDES, and specific compounds.

ACJ250 **CAS:543-21-5** **HR: 3**
ACETYLENEDICARBOXAMIDE
mf: $C_4H_4N_2O_2$ mw: 112.10

PROP: Crystals from MeOH (aqueous). Mp: 216–218°, (decomp). Produced by *Str. reticuli var. Aquamyceticus* and is identical to Cellocidin.

SYNS: ACETYLENEDICARBOXYLIC ACID DIAMIDE □AQUAMY-CIN □ 2-BUTYNEDIAMIDE □ CELLOCIDIN □ LENAMYCIN □ RENAMYCIN

TOXICITY DATA WITH REFERENCE
orl-mus LD50:89,200 μg/kg JPIFAN (1),15,69
skn-mus LD50:667 mg/kg JPIFAN (1),15,69
ivn-mus LD50:11 mg/kg JAJAAA 11,81,58

CONSENSUS REPORTS: Reported in EPA TSCA Inventory.

SAFETY PROFILE: Poison by ingestion and intravenous routes. Moderately toxic by skin contact. When heated to decomposition it emits toxic fumes of NO_x. See also ACETYLENE COMPOUNDS and ALKYNES.

ACJ500 **CAS:928-04-1** **HR: 3**
ACETYLENEDICARBOXYLIC ACID MONOPOTASSIUM SALT
mf: C_4HO_4•K mw: 152.15

SYNS: MONOPOTASSIUM SALT of ACETYLENEDICARBOXYLIC ACID □ U-4783

TOXICITY DATA WITH REFERENCE
orl-mus LD50:63 mg/kg TXAPA9 17,733,70
ipr-mus LD50:32 mg/kg TXAPA9 17,733,70
ivn-mus LD50:89 mg/kg TXAPA9 17,733,70

CONSENSUS REPORTS: Reported in EPA TSCA Inventory.

SAFETY PROFILE: Poison by ingestion, intravenous, and intraperitoneal routes. When heated to decomposition it emits acrid smoke and fumes of KO_x. See also ACETYLENE COMPOUNDS.

ACK000 **CAS:156-60-5** **HR: 2**
trans-ACETYLENE DICHLORIDE
mf: $C_2H_2Cl_2$ mw: 96.94

PROP: Colorless liquid, pleasant odor. Mp: −50°, bp: 48°, flash p: 36°F, autoign temp: 860°F, lel: 9.7%, uel: 12.8%, d: 1.2743 @ 25°/4°, vap press: 400 mm @ 30.8°, vap d: 3.34.

SYNS: trans-DICHLOROETHYLENE □ trans-1,2-DICHLOROETHYLENE (MAK) □ RCRA WASTE NUMBER U079

TOXICITY DATA WITH REFERENCE
skn-rbt 500 mg/24H MOD JACTDZ 1,11,90
eye-rbt 10 mg MOD JACTDZ 1,11,90
mma-smc 80 mmol/L TCMUD8 4,365,84
ihl-rat TCLo:12,000 ppm/6H (female 7-16D post):REP FAATDF 20,225,93
ihl-hmn TCLo:4800 mg/m³/10M:CNS AHBAAM 116,131,36
orl-rat LD50:1235 mg/kg TXCYAC 7,141,77
ipr-rat LD50:7411 mg/kg TXCYAC 7,141,77
orl-mus LD50:2122 mg/kg DCTODJ 8,373,85
ipr-rat LD50:7536 mg/kg TXCYAC 7(2),141,77
ihl-mus LCLo:75,000 mg/m³/2H AHBAAM 116,131,36
ipr-mus LD50:4019 mg/kg TXCYAC 7,141,77
ihl-cat LCLo:43,000 mg/m³/6H AHBAAM 116,131,36

CONSENSUS REPORTS: Reported in EPA TSCA Inventory.

DFG MAK: 200 ppm (790 mg/m³)

SAFETY PROFILE: Moderately toxic by ingestion. Mildly toxic by inhalation. Human systemic effects by inhalation: sleep, hallucinations, and distorted perceptions. Experimental reproductive effects. A skin and eye irritant. Mutation data reported. Exposure to high vapor concentration can cause nausea, vomiting, weakness, tremor, and cramps. Recovery is usually prompt following removal from exposure. Dermatitis may result from defatting action on skin. Dangerous fire hazard when exposed to heat, flame, or oxidizers. Moderate explosion hazard in the form of vapor when exposed to flame. Violent reaction with difluoromethylene dihypofluorite. Forms shock-sensitive explosive mixtures with dinitrogen tetraoxide. Reaction with solid caustic alkalies or their concentrated solutions produces chloracetylene gas, that ignites spontaneously in air. Reacts violently with N_2O_4, KOH, Na, NaOH. Moderate explosion hazard in the form of vapor when exposed to flame. Can react vigorously with oxidizing materials. To fight fire, use water spray, foam, CO_2, dry chemical. When heated to decomposition it emits toxic fumes of Cl^-. See also CHLORIDES; CHLORINATED HYDROCARBONS, ALIPHATIC; and ACETYLENE COMPOUNDS.

ACK250 **CAS:79-27-6** **HR: 3**
ACETYLENE TETRABROMIDE
mf: $C_2H_2Br_4$ mw: 345.68

PROP: Colorless to yellow liquid. Bp: 151° @ 54 mm, fp: −1°, d: 2.9638 @ 20°/4°, mp: 0.1°, autoign temp: 635°F.

SYNS: MUTHMANN'S LIQUID □ TBE □ 1,1,2,2-TETRABROMAETHAN (GERMAN) □ TETRABROMOACETYLENE □ 1,1,2,2-TETRABROMOETANO (ITALIAN) □ S-TETRABROMOETHANE □ 1,1,2,2-TETRABROMOETHANE □ 1,1,2,2-TETRABROOMETHAAN (DUTCH)

TOXICITY DATA WITH REFERENCE
skn-rbt 500 mg/24H MOD AIHAAP 24,28,63
eye-rbt 100 mg MLD AIHAAP 24,28,63
mmo-sat 10 μg/plate TECSDY 15,101,87
dnr-esc 29,640 μg/disc MUREAV 41,61,76
skn-mus TDLo:130 g/kg/74W-I:NEO JJIND8 63,1433,79
orl-rat LD50:1200 mg/kg VRDEA5 (3),80,67
ihl-rat LC50:549 mg/m³/4H 85GMAT -,107,82
skn-rat LD50:5250 mg/kg 85GMAT-,107,82
orl-mus LD50:269 mg/kg 85GMAT -,107,82
ipr-mus LD50:443 mg/kg ABMGAJ 41,945,82
ipr-mus LD50:443 mg/kg ABMGAJ 41,945,82
orl-gpg LD50:400 mg/kg AIHAAP 30,251,69

orl-rbt LD50:400 mg/kg AMIHBC 2,407,50

CONSENSUS REPORTS: Reported in EPA TSCA Inventory. EPA Genetic Toxicology Program.

OSHA PEL: TWA 1 ppm
ACGIH TLV: TWA 1 ppm
DFG MAK: 1 ppm (14 mg/m^3)

SAFETY PROFILE: Poison by inhalation, ingestion, and intraperitoneal routes. An eye and skin irritant and a narcotic. Questionable carcinogen with experimental neoplastigenic data. Mutation data reported. When heated it emits highly toxic fumes of carbonyl bromide and Br^-. See also ACETYLENE COMPOUNDS and BROMIDES.

For occupational chemical analysis use NIOSH: see 1,1,2,2-tetrabromoethane, 2003.

ACL000 CAS:2597-54-8 HR: 2
N-ACETYL ETHYL CARBAMATE
mf: $C_6H_9NO_3$ mw: 131.15

SYN: ACETYLURETHANE

TOXICITY DATA WITH REFERENCE
ipr-mus TD:7300 mg/kg/10W-I:NEO IJCNAW 4,318,69
ipr-mus TD:3650 mg/kg/5W-I:ETA IJCNAW 4,318,69

SAFETY PROFILE: Questionable carcinogen with experimental neoplastigenic and tumorigenic data. See also CARBAMATES. When heated to decomposition it emits toxic fumes of NO_x.

ACL500 CAS:52217-47-7 HR: 2
N′-ACETYL ETHYLNITROSOUREA
mf: $C_5H_9N_3O_3$ mw: 159.17

TOXICITY DATA WITH REFERENCE
orl-rat TDLo:520 mg/kg/52W-I:ETA PPTCBY 2,73,72
orl-rat LD50:550 mg/kg PPTCBY 2,85,72

SAFETY PROFILE: Moderately toxic by ingestion. Questionable carcinogen with experimental tumorigenic data. When heated to decomposition it emits toxic fumes of NO_x.

ACL750 CAS:88-29-9 HR: 3
ACETYL ETHYL TETRAMETHYL TETRALIN
mf: $C_{18}H_{26}O$ mw: 258.44

PROP: White crystals.

SYNS: ACETYLETHYL TETRAMETHYLTETRALIN □ 6-ACETYL-1,1,4,4-TETRAMETHYL-7-ETHYL-1,2,3,4,-TETRALIN □ 7-ACETYL-1,1,4,4-TETRAMETHYL-1,2,3,4-TETRAHYDRONAPHTHALENE □ AETT □ ETHANONE-1-(3-ETHYL-5,6,7,8-TETRAHYDRO-5,5,8,8-TETRAMETHYL-2-NAPHTHALENYL)(9CI) □ 3′-ETHYL-5′,6′,7′,8′-TETRAHYDRO-5′,5′,8′-TETRAMETHYL-2′-ACETONAPHTHONE □ 1-(3-ETHYL-5,6,7,8-TETRAHYDRO-5,5,8,8-TETRAMETHYL-2-NAPHTHALENYL)-ETHANONE □ MUSK 36A □ POLYCYCLIC MUSK □ VERSALIDE

TOXICITY DATA WITH REFERENCE
skn-rbt 500 mg/24H MLD FCTXAV 17,357,79
orl-rat LD50:260 mg/kg FCTXAV 19,753,81
skn-rat LD50:584 mg/kg FCTXAV 17,357,79
orl-mus LDLo:470 mg/kg AECTCV 14,111,85

CONSENSUS REPORTS: Reported in EPA TSCA Inventory.

SAFETY PROFILE: Poison by ingestion. Moderately toxic by skin contact. A skin and eye irritant. Exposure causes blue coloration of internal organs and central nervous system effects, e.g., hyperexcitability, tremors, lack of coordination, hunched back, and loss of weight. It is slowly metabolized and excreted via feces. Symptoms persist for 90 days after exposure. Severity of symptoms seems proportional to length of exposure. It is freely absorbed via human skin. When heated to decomposition it emits acrid smoke and fumes.

ACM000 CAS:557-99-3 HR: 3
ACETYL FLUORIDE
mf: C_2H_3FO mw: 62.05

PROP: Liquid or gas. D: 1.002 @ 15°/4°; mp: −60°, bp: 20.8°. Sltly sol in alc, ether, acetone, and benzene.

SYNS: FLUORID KYSELINY OCTOVE □ METHYLCARBONYL FLUORIDE

TOXICITY DATA WITH REFERENCE
ihl-mus LC50:2500 mg/m^3 85JCAE-,325,86
ihl-dog LCLo:2000 mg/m^3/30M 11FYAN 3,74,63

CONSENSUS REPORTS: Reported in EPA TSCA Inventory.

OSHA PEL: TWA 2.5 mg(F)/m^3
ACGIH TLV: TWA 2.5 mg(F)/m^3

SAFETY PROFILE: Poison by inhalation. See also FLUORIDES. When heated to decomposition it emits toxic fumes of F^-.

ACM250 CAS:7242-07-1 HR: 3
16-ACETYLGITOXIN
mf: $C_{43}H_{66}O_{15}$ mw: 823.09

TOXICITY DATA WITH REFERENCE
ivn-rat LD50:16,500 μg/kg AIPTAK 155,165,65
ipr-mus LD50:6800 μg/kg AIPTAK 155,165,65
ipr-cat LD50:148 μg/kg AIPTAK 155,165,65
ivn-cat LDLo:110 μg/kg AIPTAK 155,165,65
orl-cat LD50:120 μg/kg AIPTAK 159,1,66
orl-gpg LD50:2500 μg/kg AIPTAK 159,1,66

SAFETY PROFILE: Deadly poison by ingestion, intraperitoneal, and intravenous routes. When heated to decomposition it emits acrid smoke and fumes. See also GITOXIN.

ACM750 CAS:1068-57-1 HR: 3
ACETYL HYDRAZIDE
mf: $C_2H_6N_2O$ mw: 74.10

PROP: Needles from ethanol. Mp: 67°, bp: 127° @ 18 mm.

SYNS: ACETHYDRAZIDE □ ACETOHYDRAZIDE □ N-ACETYLHYDRAZINE □ ENT 61,241 □ ETHANEHYDRAZONIC ACID □ MONOACETYLHYDRAZINE

TOXICITY DATA WITH REFERENCE
mmo-sat 500 μg/plate IJEBA6 19,939,81

mmo-omi 70 mg/L MUREAV 173,233,86
mnt-mus-ipr 120 mg/kg CALEDQ 23,235,84
dni-mus-ipr 150 mg/kg IJEBA6 19,939,81
ipr-mus LD50:153 mg/kg JPETAB 122,110,58
scu-rbt LDLo:116 mg/kg JPETAB 30,87,27
orl-bwd LD50:42,200 μg/kg AECTCV 12,355,83

CONSENSUS REPORTS: Reported in EPA TSCA Inventory.

SAFETY PROFILE: Poison by ingestion, subcutaneous, and intraperitoneal routes. Mutation data reported. Exposure can cause hemolysis and liver damage. See also PHENYLHYDRAZINE. When heated to decomposition it emits toxic fumes of NO_x.

ACN250 CAS:534-33-8 HR: 3
N-ACETYL-4-HYDROXYARSANILIC ACID compounded with DIETHYLAMINE (1:1)
mf: $C_8H_{10}AsNO_5 \cdot C_4H_{11}N$ mw: 348.27

SYNS: ACETARSIN ☐ ACETARSONE DIETHYLAMINE SALT ☐ ACETILARSANO ☐ ACETYLARSAN ☐ N-ACETYL-4-HYDROXY-m-ARSANILIC ACID DIETHYLAMINE SALT ☐ 2-AMINOPHENOL-4-ARSONIC ACID DIETHYLAMINE SALT ☐ ARSAPHENAN ☐ DIETHYLAMINE ACETARSONE ☐ DIETHYLAMINE-3-ACETYLAMINO-4-HYDROXYPHENYLARSONATE ☐ GOLARSYL ☐ SYNTHARSOL

TOXICITY DATA WITH REFERENCE
cyt-hmn:leu 1 nmol/L AEMBAP 91,117,78
cyt-hmn:fbr 1 nmol/L AEMBAP 91,117,78

CONSENSUS REPORTS: Arsenic and its compounds are on the Community Right-To-Know List.

OSHA PEL: TWA 0.5 mg(As)/m^3
ACGIH TLV: TWA 0.2 mg(As)/m^3

SAFETY PROFILE: A poison. Human mutation data reported. See also ARSENIC COMPOUNDS. When heated to decomposition it emits very toxic fumes of NO_x and As.

ACN300 CAS:39543-84-5 HR: 3
2-ACETYL-4-(2-HYDROXY-3-tert-BUTYLAMINOPROPOXY)BENZOFURAN
mf: $C_{17}H_{23}NO_4$ mw: 305.41

PROP: A liquid.

SYNS: 1-(4-(3-((1,1-DIMETHYLETHYL)AMINO)-2-HYDROXYPROPOXY)-2-BENZOFURANYL)ETHA NONE ☐ ETHANONE, 1-(4-(3-((1,1-DIMETHYLETHYL)AMINO)-2-HYDROXYPROPOXY)-2-BENZOFURANYL)- ☐ KETONE, 4-(3-(tert-BUTYLAMINO)-2-HYDROXYPROPOXY)-2-BENZOFURANYL METHYL

TOXICITY DATA WITH REFERENCE
ivn-mus LD50:35 mg/kg GWXXBX #2223184

DOT CLASSIFICATION: 3; *Label:* Flammable Liquid

SAFETY PROFILE: A poison by intravenous route. A flammable liquid. When heated to decomposition it emits toxic vapors of NO_x.

ACN310 CAS:39543-94-7 HR: 3
2-ACETYL-7-(2-HYDROXY-3-sec-BUTYLAMINOPROPOXY)BENZOFURAN
mf: $C_{17}H_{23}NO_4$ mw: 305.41

PROP: A liquid.

SYNS: ETHANONE, 1-(7-(2-HYDROXY-3-((1-METHYLPROPYL)AMINO)PROPOXY)-2-BENZOFURANYL)- ☐ KETONE, 7-(3-(sec-BUTYLAMINO)-2-HYDROXYPROPOXY)-2-BENZOFURANYL METHYL

TOXICITY DATA WITH REFERENCE
ivn-mus LD50:75 mg/kg GWXXBX #2223184

DOT CLASSIFICATION: 3; *Label:* Flammable Liquid

SAFETY PROFILE: A poison by intravenous route. A flammable liquid. When heated to decomposition it emits toxic vapors of NO_x.

ACN320 CAS:39543-80-1 HR: 3
2-ACETYL-7-(2-HYDROXY-3-tert-BUTYLAMINOPROPOXY)BENZOFURAN
mf: $C_{17}H_{23}NO_4$ mw: 305.41

PROP: A liquid.

SYNS: KETONE, 7-(3-(tert-BUTYLAMINO)-2-HYDROXYPROPOXY)-2-BENZOFURANYL METHYL ☐ 1-(7-(3-((1,1-DIMETHYLETHYL)AMINO)-2-HYDROXYPROPOXY)-2-BENZOFURANYL)ETHA NONE ☐ ETHANONE, 1-(7-(3-((1,1-DIMETHYLETHYL)AMINO)-2-HYDROXYPROPOXY)-2-BENZOFURANYL)-

TOXICITY DATA WITH REFERENCE
ivn-mus LD50:45 mg/kg GWXXBX #2223184

DOT CLASSIFICATION: 3; *Label:* Flammable Liquid

SAFETY PROFILE: A poison by intravenous route. A flammable liquid. When heated to decomposition it emits toxic vapors of NO_x.

ACN500 CAS:65734-38-5 HR: 2
N-ACETYL-N′-(p-HYDROXYMETHYL)PHENYLHYDRAZINE
mf: $C_9H_{12}N_2O_2$ mw: 180.23

TOXICITY DATA WITH REFERENCE
orl-mus TDLo:74 g/kg/85W-C:NEO CNREA8 38,177,78
scu-mus TDLo:13 g/kg/26W-I:ETA JTEHD6 8,1,81

SAFETY PROFILE: Questionable carcinogen with experimental neoplastigenic and tumorigenic data. When heated to decomposition it emits toxic fumes such as NO_x.

ACN875 CAS:4254-22-2 HR: 3
ACETYL HYPOBROMITE
mf: $C_2H_3BrO_2$ mw: 138.95

PROP: Crystals.

SAFETY PROFILE: A dangerously unstable explosive. When heated to decomposition it emits toxic fumes of Br^-.

ACO000 **HR: 3**
ACETYLIDES

SAFETY PROFILE: Severe explosion hazard when shocked or exposed to heat. Acetylides are very sensitive to shock, friction, and heat. They explode readily and are among the few commercial explosives that contain no oxygen or nitrogen and therefore produce no gas. The explosion simply results from the large amount of heat instantaneously produced. Acetylides are used for detonating compositions, or in combination with lead azide in detonating rivets, where the acetylides reduce the flash point of the more insensitive azides. They are in a class with the fulminates and the azides as primary detonants. Because these materials are so sensitive to shock and temperature, they must be handled with extreme care. They must be kept cool, and should be kept wet if they are to be stored. (See FULMINATES for suggested precautions in storage and handling of acetylides.) Metal powders, such as finely divided Cu or Ag, should not be stored or kept with acetylene or acetylides because it is possible for these substances to react with these metal powders to form very sensitive acetylides, which although they are not dangerous in themselves, can cause enough of a flash to ignite a possibly explosive mixture of gases and thus cause an explosion in a warehouse or storage area. Examples of commercially used acetylides are silver acetylide and copper acetylide. See also ACETYLENE. See also individual compounds.

ACO250 **CAS:2466-76-4** **HR: 3**
N-ACETYLIMIDAZOLE
mf: $C_5H_6N_2O$ mw: 110.13

PROP: Mp: 101.5–102.5°.

SYN: 1-ACETYLIMIDAZOLE

TOXICITY DATA WITH REFERENCE
ipr-mus TDLo:50 mg/kg/I:NEO JNCIAM 54,495,75
ipr-mus LDLo:250 mg/kg StoGD# 27May75

CONSENSUS REPORTS: Reported in EPA TSCA Inventory.

SAFETY PROFILE: Poison by intraperitoneal route. Questionable carcinogen with experimental neoplastigenic data. When heated to decomposition it emits toxic fumes of NO_x.

ACO300 **CAS:53330-94-2** **HR: 3**
5-ACETYLINDOLE
mf: $C_{10}H_9NO$ mw: 159.20

PROP: A liquid.

SYNS: ACETYL-5-INDOLE □ KETONE, INDOL-5-YL METHYL

TOXICITY DATA WITH REFERENCE
ipr-mus LD50:450 mg/kg EJMCA5 9,453,74

DOT CLASSIFICATION: 3; *Label:* Flammable Liquid

SAFETY PROFILE: Moderately toxic by intraperitoneal route. A flammable liquid. When heated to decomposition it emits toxic vapors of NO_x.

ACO320 **CAS:16078-34-5** **HR: 3**
5-ACETYLINDOLINE
mf: $C_{10}H_{11}NO$ mw: 161.22

PROP: A liquid.

SYNS: INDOLINE, 5-ACETYL- □ KETONE, 5-INDOLINYL METHYL

TOXICITY DATA WITH REFERENCE
ivn-mus LD50:320 mg/kg CSLNX* NX#02219

DOT CLASSIFICATION: 3; *Label:* Flammable Liquid

SAFETY PROFILE: A poison by intravenous route. A flammable liquid. When heated to decomposition it emits toxic vapors of NO_x.

ACO500 **CAS:507-02-8** **HR: 3**
ACETYL IODIDE
DOT: UN 1898
mf: C_2H_3IO mw: 169.95

PROP: Brown, transparent, fuming liquid. Bp: 108°, d: 2.067 @ 20°/4°, decomp in water and alc; sol in ether.

CONSENSUS REPORTS: Reported in EPA TSCA Inventory.

DOT CLASSIFICATION: 8; *Label:* Corrosive

SAFETY PROFILE: A toxic, corrosive material. Reacts with water or steam to produce toxic and corrosive fumes. Dangerous to use. When heated to decomposition it emits toxic fumes of I^-. See also IODIDES.

ACO750 **CAS:1078-38-2** **HR: 2**
1-ACETYL-2-ISONICOTINOYLHYDRAZINE
mf: $C_8H_9N_3O_2$ mw: 179.20

SYNS: ACETYL ISONIAZID □ N-ACETYLISONIAZID □ N-ACETYLISONICOTINYLHYDRAZIDE □ 4-PYRIDINECARBOXYLIC ACID-2-ACETYLHYDRAZIDE

TOXICITY DATA WITH REFERENCE
dni-mus-ipr 1 g/kg IJFBA6 19,939,81
orl-mus TDLo:380 g/kg/68W-C:NEO EJCAAH 9,285,73

CONSENSUS REPORTS: EPA Genetic Toxicology Program.

SAFETY PROFILE: Questionable carcinogen with experimental neoplastigenic data. Mutation data reported. When heated to decomposition it emits toxic fumes of NO_x.

ACP000 **CAS:39293-24-8** **HR: 3**
ACETYLKIDAMYCIN
mf: $C_{46}H_{58}N_2O_{13}$ mw: 847.06

TOXICITY DATA WITH REFERENCE
oms-hmn:hla 1 mg/L JANTAJ 29,1334,76
ipr-rat LD50:35 mg/kg 85ERAY 2,1452,78
ivn-rat LD50:140 mg/kg 85ERAY 2,1452,78
orl-mus LD50:600 mg/kg 85ERAY 2,1452,78
ipr-mus LD50:50 mg/kg 85ERAY 2,1452,78
ivn-mus LD50:200 mg/kg 85ERAY 2,1452,78
ivn-rbt LD50:25 mg/kg 85ERAY 2,1452,78

SAFETY PROFILE: Poison by intravenous and intraperi-

toneal routes. Moderately toxic by ingestion. Human mutation data reported. When heated to decomposition it emits toxic fumes of NO_x.

ACP500 **CAS:63938-24-9** **HR: 3**
1-ACETYLLYSERGIC ACID DIETHYLAMIDE BITARTRATE
mf: $C_{22}H_{27}N_3O_2•2C_4H_4O_6$ mw: 661.68

SYN: 1-ACETYL-9,10-DIDEHYDRO-N,N-DIETHYL-6-METHYLERGOLINE-8-β-CARBOXAMIDE BITARTRATE

TOXICITY DATA WITH REFERENCE
orl-hmn TDLo:1500 ng/kg:PSY PSDTAP 8,59,67
ivn-rbt LD50:1600 µg/kg 27ZQAG -,93,72

SAFETY PROFILE: Deadly poison by intravenous route. Human systemic effects by ingestion of very small amounts: EEG changes, hallucinations, distorted perceptions, and changes in psychophysiological test scores. When heated to decomposition it emits toxic fumes of NO_x. See also other lysergic acid derivatives.

ACP750 **CAS:50485-03-5** **HR: 3**
d-1-ACETYL LYSERGIC ACID MONOETHYLAMIDE
mf: $C_{20}H_{23}N_3O_2$ mw: 337.46

SYNS: 1-ACETYL-9,10-DIDEHYDRO-N-ETHYL-6-METHYLERGOLINE-8-β-CARBOXAMIDE □ 1-ACETYLLYSERGIC ACID ETHYLAMIDE

TOXICITY DATA WITH REFERENCE
orl-hmn TDLo:75 µg/kg:PSY PSDTAP 8,59,67
ivn-rbt LD50:5 mg/kg 27ZQAG -,94,72

SAFETY PROFILE: Poison by ingestion and intravenous routes. Ingesting very small amounts produces psychotropic effects in humans. When heated to decomposition it emits toxic fumes of NO_x. See also various lysergic acid entries.

ACQ250 **CAS:1190-93-8** **HR: 3**
ACETYLMERCAPTOACETIC ACID
mf: $C_4H_6O_3S$ mw: 134.16

PROP: Yellow oil. Bp: 158–159° @ 17 mm.

SYNS: ACETIC ACID, (ACETYLTHIO)-(9CI) □ ACETIC ACID, MERCAPTO-, ACETATE (8CI) □ (ACETYLTHIO)ACETIC ACID □ S-ACETYLTHIOGLYCOLIC ACID □ MERCAPTOACETIC ACID ACETATE □ USAF EK-P-5430

TOXICITY DATA WITH REFERENCE
ipr-mus LD50:150 mg/kg NTIS** AD277-689

CONSENSUS REPORTS: Reported in EPA TSCA Inventory.

SAFETY PROFILE: Poison by intraperitoneal route. When heated to decomposition it emits toxic fumes of SO_x.

ACQ270 **CAS:1115-47-5** **HR: 1**
N-ACETYL-dl-METHIONINE
mf: $C_7H_{13}NO_3S$ mw: 191.27

SYNS: ACETYL-dl-METHIONINE □ dl-N-ACETYLMETHIONINE □ METHIONINE, N-ACETYL-, dl- □ dl-METHIONINE, N-ACETYL-(9CI)

TOXICITY DATA WITH REFERENCE
ipr-mus LD50:6700 mg/kg AIPTAK 91,163,52

CONSENSUS REPORTS: Reported in EPA TSCA Inventory.

SAFETY PROFILE: Slightly toxic by intraperitoneal route. When heated to decomposition it emits toxic vapors of NO_x and SO_x.

ACQ275 **CAS:65-82-7** **HR: 3**
N-ACETYL-l-METHIONINE
mf: $C_7H_{13}NO_3S$ mw:191.24

PROP: Colorless or white crystals or powder; odorless. Sol in water, alc, alkali and mineral acids; insol in ether.

SYNS: ACETYLMETHIONINE □ N-ACETYLMETHIONINE □ METHIONAMINE

TOXICITY DATA WITH REFERENCE
ivn-mus LD50:435 mg/kg RPOBAR 2,262,70

CONSENSUS REPORTS: EPA TSCA Chemical Inventory, JUNE 1993

SAFETY PROFILE: Poison by intravenous route. When heated to decomposition emits toxic fumes of NO_x.

ACQ666 **CAS:1477-40-3** **HR: 3**
α-1-ACETYLMETHADOL
mf: $C_{23}H_{31}NO_2$ mw: 353.55

SYNS: l-α-ACETYLMETHADOL □ levo-α-ACETYLMETHADOL □ 3-HEPTANOL, 6-(DIMETHYLAMINO)-4,4-DIPHENYL-, ACETATE (ester), (3S,6S)-(−)- □ LAAM

TOXICITY DATA WITH REFERENCE
mmo-nsc 200 mg/L DCTODJ 4,19,81
cyt-hmn:lym 70 mg/L ENMUDM 1,180,79
mma-mus:lym 25 mg/L DCTODJ 4,19,81
trn-mus-unr 7 mg/kg DCTODJ 4,19,81
orl-man TDLo:8570 µg/kg (4W male):REP CLPTAT 19,371,76
orl-mus LD50:173 mg/kg JPETAB 110,135,54
ipr-mus LD50:56 mg/kg PBBHAU 9,195,78
scu-mus LD50:111 mg/kg ANYAA9 281,321,76

SAFETY PROFILE: Poison by ingestion, subcutaneous, and intraperitoneal routes. Human reproductive effects by ingestion. Experimental reproductive effects. Human mutation data reported. When heated to decomposition it emits acrid smoke and fumes.

ACQ690 **CAS:43033-72-3** **HR: 3**
l-α-ACETYLMETHADOL HYDROCHLORIDE
mf: $C_{23}H_{31}NO_2•ClH$ mw: 390.01

SYNS: (3S,6S)-(−)-6-(DIMETHYLAMINO)-4,4-DIPHENYL-3-HEPTANOL ACETATE (ester) HYDROCHLORIDE □ 3-HEPTANOL, 6-(DIMETHYLAMINO)-4,4-DIPHENYL-, ACETATE (ester), HYDROCHLORIDE, (3S, 6S)-(−)- □ LAAM HYDROCHLORIDE

TOXICITY DATA WITH REFERENCE
orl-rat TDLo:7400 µg/kg (female 15D pre):REP NETOD7 5,479,83
orl-rat TDLo:7400 µg/kg (female 15D pre):TER NETOD7 5,479,83

orl-rat TDLo: 7061 mg/kg/2Y-C: CAR FAATDF 11,626,88
orl-mus LD50: 71 mg/kg FAATDF 11,626,88
par-mus LD50: 11 mg/kg JPETAB 145,11,64

SAFETY PROFILE: Poison by ingestion and parenteral routes. Questionable carcinogen with experimental carcinogenic data. An experimental teratogen. Other experimental reproductive effects. When heated to decomposition it emits toxic fumes of NO_x and HCl.

ACQ700 CAS:57543-56-3 HR: 3
3-ACETYL-6-METHOXY-2H-1-BENZOPYRAN
mf: $C_{12}H_{12}O_3$ mw: 204.24

PROP: A liquid.

SYN: KETONE, 6-METHOXY-2H-1-BENZOPYRAN-3-YL METHYL

TOXICITY DATA WITH REFERENCE
ipr-mus LD50: 1000 mg/kg EJMCA5 11,81,76

DOT CLASSIFICATION: 3; *Label:* Flammable Liquid

SAFETY PROFILE: Moderately toxic by intraperitoneal route. A flammable liquid. When heated to decomposition it emits acrid smoke and irritating vapors.

ACQ730 CAS:57543-55-2 HR: 3
3-ACETYL-7-METHOXY-2H-1-BENZOPYRAN
mf: $C_{12}H_{12}O_3$ mw: 204.24

PROP: A liquid.

SYN: KETONE, 7-METHOXY-2H-1-BENZOPYRAN-3-YL METHYL

TOXICITY DATA WITH REFERENCE
ipr-mus LD50: 1000 mg/kg EJMCA5 11,81,76

DOT CLASSIFICATION: 3; *Label:* Flammable Liquid

SAFETY PROFILE: Moderately toxic by intraperitoneal route. A flammable liquid. When heated to decomposition it emits acrid smoke and irritating vapors.

ACQ760 CAS:57543-54-1 HR: 3
3-ACETYL-8-METHOXY-2H-1-BENZOPYRAN
mf: $C_{12}H_{12}O_3$ mw: 204.24

PROP: A liquid.

SYN: KETONE, 8-METHOXY-2H-1-BENOZPYRAN-3-YL METHYL

TOXICITY DATA WITH REFERENCE
ipr-mus LD50: 1000 mg/kg EJMCA5 11,81,76

DOT CLASSIFICATION: 3; *Label:* Flammable Liquid

SAFETY PROFILE: Moderately toxic by intraperitoneal route. A flammable liquid. When heated to decomposition it emits acrid smoke and irritating vapors.

ACQ790 CAS:77523-56-9 HR: 3
2-ACETYL-7-METHOXYNAPHTHO(2,1-b)FURAN

PROP: A liquid.
mf: $C_{15}H_{12}O_3$ mw: 240.27

SYNS: ETHANONE, 1-(7-METHOXYNAPHTHO(2,1-b)FURAN-2-YL)- □ KETONE, 7-METHOXYNAPHTHO(2,1-b)FURAN-2-YL METHYL □ R 7237

TOXICITY DATA WITH REFERENCE
mmo-sat 5 μmol/plate MUREAV 88,355,81
mma-sat 5 μmol/plate MUREAV 88,355,81
oth-esc 1 nmol/tube MUTAEX 1,217,86

DOT CLASSIFICATION: 3; *Label:* Flammable Liquid

SAFETY PROFILE: Mutation data reported. A flammable liquid. When heated to decomposition it emits acrid smoke and irritating vapors.

ACR000 CAS:62-51-1 HR: 3
o-ACETYL-β-METHYLCHOLINE CHLORIDE
mf: $C_8H_{18}NO_2 \cdot Cl$ mw: 195.72

PROP: Mp: 172–173°. Very sol in water and alc, decomp in alkalies and ether.

SYNS: AMECHOL □ (2-HYDROXYPROPYL)TRIMETHYLAMMONIUMCHLORIDE ACETATE □ MECHOLYL □ METHACHOLINE CHLORIDE □ METHACHOLINIUM CHLORIDE □ METHYLACETYL CHOLINE □ β-METHYLACETYLCHOLINE CHLORIDE □ TRIMETHYL-β-ACETOXYPROPYLAMMONIUM CHLORIDE

TOXICITY DATA WITH REFERENCE
orl-rat LD50: 750 mg/kg JPETAB 58,337,36
scu-rat LD50: 75 mg/kg JPETAB 58,337,36
ivn-rat LD50: 20 mg/kg JPETAB 58,337,36
orl-mus LD50: 740 mg/kg PHTXA6 22,210,59
ipr-mus LD50: 160 mg/kg TXAPA9 28,227,74
scu-mus LD50: 90 mg/kg JPETAB 58,337,36
ivn-mus LD50: 15 mg/kg JPETAB 58,337,36
ivn-gpg LDLo: 3750 μg/kg AIPTAK 106,245,56

CONSENSUS REPORTS: Reported in EPA TSCA Inventory.

SAFETY PROFILE: Poison by subcutaneous, intravenous, and intraperitoneal routes. Moderately toxic by other routes. When heated to decomposition it emits very toxic fumes of Cl^- and NO_x.

ACR050 CAS:33266-07-8 HR: 3
2-ACETYL-2-METHYL-1,3-DITHIOLANE
mf: $C_6H_{10}OS_2$ mw: 162.28

PROP: A liquid.

SYNS: 1,3-DITHIOLANE, 2-ACETYL-2-METHYL- □ KETONE, METHYL 2-METHYL-1,3-DITHIOLAN-2-YL

TOXICITY DATA WITH REFERENCE
ipr-mus LD50: 1500 mg/kg EJMCA5 17,235,82

DOT CLASSIFICATION: 3; *Label:* Flammable Liquid

SAFETY PROFILE: Moderately toxic by intraperitoneal route. A flammable liquid. When heated to decomposition it emits toxic vapors of SO_x.

ACR100 HR: 3
α-ACETYL-6-METHYLERGOLINE-8-β-PROPIONAMIDE
mf: $C_{20}H_{25}N_3O_2$ mw: 339.48

SYN: ERGOLINE-8-β-PROPIONAMIDE, α-ACETYL-6-METHYL-

TOXICITY DATA WITH REFERENCE
orl-rat TDLo:4 mg/kg (female 5D post):REP ARZNAD 33,1094,83
orl-mus LD50:400 mg/kg ARZNAD 33,1094,83

SAFETY PROFILE: Poison by ingestion. Experimental reproductive effects. When heated to decomposition it emits toxic fumes of NO_x.

ACR300 CAS:83-63-6 HR: 3
N-ACETYL-N-(2-METHYL-4-((2-METHYLPHENYL)AZO) PHENYL)ACETAMIDE
mf: $C_{18}H_{19}N_3O_2$ mw: 309.40

PROP: Brick-red needles or stout red prisms. Mp: 65°.

SYNS: DERMAGAN □ DERMAGEN □ DIACETAZOTOL □ DIACETOTOLUIDE □ o-DIACETOTOLUIDIDE, 4″-(o-TOLYLAZO)-(8CI) □ DIACETYLAMINOAZOTOLUENE □ N,N-DIACETYL-o-TOLYLAZO-o-TOLUIDINE □ DIAMAZO □ DIMAZON □ EPIDERMOL □EPITHE-LONE □ GRANULIN □ PELLIDOL □ PELLIDOLE □ PERIPHERMIN □ 4-o-TOLYLAZO-o-DIACETOTOLUIDE □ 4′-(o-TOLYLAZO)-o-DIACETOTOLUIDIDE

CONSENSUS REPORTS: IARC Cancer Review: Group 3 IMEMDT 7,56,87; Animal Inadequate Evidence IMEMDT 8,113,75

SAFETY PROFILE: Questionable carcinogen. When heated to decomposition it emits dangerous and toxic fumes of NO_x.

ACR400 CAS:28895-91-2 HR: 3
ACETYLMETHYLNITROSOUREA
mf: $C_4H_7N_3O_3$ mw: 145.14

SYNS: ACETYL-METHYL-NITROSO-HARNSTOFF (GERMAN) □ N′-ACETYL-METHYLNITROSOUREA □ N-METHYL-N-NITROSO-N′-ACETYLUREA □ 1-METHYL-1-NITROSOACETYLUREA

TOXICITY DATA WITH REFERENCE
cyt-ham:fbr 500 mg/L/20H MUREAV 48,337,77
orl-rat TDLo:468 mg/kg/47W-I:ETA ZEKBAI 74,23,70
orl-rat LD50:200 mg/kg XENOBH 3,271,73

SAFETY PROFILE: Poison by ingestion. Questionable carcinogen with experimental tumorigenic data. Mutation data reported. When heated to decomposition it emits toxic fumes of NO_x.

ACR500 CAS:1053-74-3 HR: 3
3-ACETYL-10-(3′-N-METHYL-PIPERAZINO-N′-PROPYL)PHENOTHIAZIN
mf: $C_{22}H_{27}N_3OS$ mw: 381.58

PROP: A liquid.

SYNS: KETONE, METHYL 10-(3-(4-METHYL-1-PIPERAZINYL)PROPYL)PHENOTHIAZIN-2-YL □ METHYL 10-(3-(4-METHYL-1-PIPERAZINYL)PROPYL)PHENOTHIAZIN-2-YL KETONE

TOXICITY DATA WITH REFERENCE
orl-rat LD50:650 mg/kg AIPTAK 115,1,58
ivn-mus LD50:87,500 µg/kg AIPTAK 115,1,58

DOT CLASSIFICATION: 3; *Label:* Flammable Liquid

SAFETY PROFILE: A poison by intravenous route. Moderately toxic by ingestion. A flammable liquid. When heated to decomposition it emits toxic vapors of SO_x.

ACR750 CAS:1696-20-4 HR: 2
4-ACETYLMORPHOLINE
mf: $C_6H_{11}NO_2$ mw: 129.18

PROP: Liquid. Mp: 14°, bp: decomp, flash p: 235°F (OC), d: 1.1164, vap press: 0.02 mm @ 20°, vap d: 4.46. Sol in water.

SYN: N-ACETYLMORPHOLINE

TOXICITY DATA WITH REFERENCE
skn-rbt 500 mg/24H MLD 85JCAE-,889,86
eye-rbt 500 mg open AMIHBC 10,61,54
orl-rat LD50:6130 mg/kg AMIHBC 10,61,54
skn-rbt LD50:7500 mg/kg AMIHBC 10,61,54
par-mus LDLo:2400 mg/kg CBCCT* 7,691,55

CONSENSUS REPORTS: Reported in EPA TSCA Inventory.

SAFETY PROFILE: Moderately toxic by parenteral route. A skin and eye irritant. See also MORPHOLINE. Combustible when exposed to heat or flame; can react vigorously with oxidizing materials. To fight fire use alcohol foam. When heated to decomposition it emits toxic fumes of NO_x.

ACS000 CAS:63224-44-2 HR: 2
N-ACETYL-N-MYRISTOYLOXY-2-AMINOFLUORENE
mf: $C_{29}H_{39}NO_3$ mw: 449.69

SYNS: N-ACETYL-N-TETRADECANOYLOXY-2-AMINOFLUORENE □ N-(FLUOREN-2-YL)-o-TETRADECANOYLACETOHYDROXAMIC ACID □ N-MYRISTOYLOXY-AAF □ N-MYRISTOYLOXY-N-ACETYL-2-AMINOFLUORENE

TOXICITY DATA WITH REFERENCE
dns-hmn:fbr 10 µmol/L/5H IJCNAW 16,284,75
msc-ham:lng 50 µmol/L/3H CALEDQ 6,67,79
scu-rat TD:115 mg/kg/6W-I:CAR CRNGDP 2,655,81
scu-rat TDLo:114 mg/kg/5W-I:NEO CNREA8 37,1461,77

SAFETY PROFILE: Questionable carcinogen with experimental carcinogenic and neoplastigenic data. Human mutation data reported. When heated to decomposition it emits toxic fumes of NO_x.

ACS375 HR: 2
N-ACETYLNEOMYCIN

TOXICITY DATA WITH REFERENCE
ipr-mus LD50:3250 mg/kg AACHAX -,227,65
scu-mus LD50:9250 mg/kg AACHAX -,227,65
ivn-mus LD50:625 mg/kg AACHAX -,227,65

SAFETY PROFILE: Moderately toxic by several routes.

ACS500 CAS:65041-92-1 HR: 3
8-ACETYLNEOSOLANIOL
mf: $C_{20}H_{28}O_9$ mw: 412.48

SYNS: 3-HYDROXY-4β,8-α-15-TRIACETOXY-12,13-EPOXYTRICHO-

THEC-9-ENE □ NEOSOLANIOL MONOACETATE □ 4-β,8-α-15-TRIACETOXY-3-α-HYDROXY-12,13-EPOXYTRICHOTHEC-9-ENE □ TRICHOTHEC-9-ENE, 12,13-EPOXY-4-β,8-α-15-TRIACETOXY-3-α-HYDROXY-

TOXICITY DATA WITH REFERENCE
skn-rat 80 ng SEV JAFCAU 26,246,78
skn-rbt 160 ng MOD JAFCAU 26,246,78
skn-gpg 80 ng MLD JAFCAU 26,246,78
orl-gpg LD50:500 μg/kg DFSCDX 4,135,83
orl-ckn LD50:789 μg/kg JAFCAU 26,246,78

SAFETY PROFILE: Poison by ingestion. A severe skin irritant. When heated to decomposition it emits acrid smoke and fumes.

ACS750 CAS:591-09-3 HR: 3
ACETYL NITRATE
mf: $C_2H_3NO_4$ mw: 105.06

PROP: Colorless, hygroscopic, fuming, mobile liquid. Bp: 22° @ 70 mm; d: 1.24 @ 15°/4°.

SYN: ACETIC ACID, ANHYDRIDE with NITRIC ACID (1:1)

TOXICITY DATA WITH REFERENCE
eye-hmn 4 ppm/12M IAPWAR 4,79,61

SAFETY PROFILE: Corrosive to the eye. Violently unstable. Reacts explosively with ethyl-3,4-dihydroxybenzenesulfonate + oleum, HgO, and other active oxides. Solutions may explode violently above 60°C and the pure material explodes above 100°C. When heated to decomposition it emits toxic fumes of NO_x and/or explodes. See also NITRATES.

ACT250 CAS:5275-69-4 HR: 3
2-ACETYL-5-NITROFURAN
mf: $C_6H_5NO_4$ mw: 155.12

SYN: (5-NITRO-2-FURYL) METHYL KETONE

TOXICITY DATA WITH REFERENCE
mmo-omi 1000 ppm APMBAY 6,45,58
mmo-sat 8 μg/plate CNREA8 35,3611,75
scu-rat LD50:200 mg/kg SGOBA9 83,73,46
orl-mus LD50:400 mg/kg SGOBA9 83,73,46

CONSENSUS REPORTS: EPA Genetic Toxicology Program.

DOT CLASSIFICATION: 3; *Label:* Flammable Liquid

SAFETY PROFILE: Poison by subcutaneous route. Moderately toxic by ingestion. Mutation data reported. See also KETONES. A flammable liquid. When heated to decomposition it emits toxic fumes of NO_x.

ACT300 CAS:32116-24-8 HR: 3
2-ACETYL-4-NITROPYRROLE
mf: $C_6H_6N_2O_3$ mw: 154.14

PROP: A liquid.

SYNS: KETONE, METHYL (4-NITRO-2-PYRROLYL) □ PYRROLE, 2-ACETYL-4-NITRO-

TOXICITY DATA WITH REFERENCE
mmo-sat 80 μg/plate CNREA8 35,3611,75

DOT CLASSIFICATION: 3; *Label:* Flammable Liquid

SAFETY PROFILE: Mutation data reported. A flammable liquid. When heated to decomposition it emits toxic vapors of NO_x.

ACT330 CAS:32116-25-9 HR: 3
2-ACETYL-5-NITROPYRROLE
mf: $C_6H_6N_2O_3$ mw: 154.14

PROP: A liquid.

SYNS: KETONE, METHYL (5-NITRO-2-PYRROLYL) □ PYRROLE, 2-ACETYL-5-NITRO-

TOXICITY DATA WITH REFERENCE
mmo-sat 20 μg/plate CNREA8 35,3611,75

DOT CLASSIFICATION: 3; *Label:* Flammable Liquid

SAFETY PROFILE: Mutation data reported. A flammable liquid. When heated to decomposition it emits toxic vapors of NO_x.

ACU125 HR: 2
7-ACETYL-5-OXO-5H-(1)BENZOPYRANO(2,3-b)PYRIDINE
mf: $C_{14}H_9NO_3$ mw: 239.24

SYNS: KETONE-METHYL-5 OXO-5H-(1)BENZOPYRANO(2,3-b)PYRIDYL □ Y-9000

TOXICITY DATA WITH REFERENCE
orl-rat LD50:1679 mg/kg NYKZAU 74,179,78
ipr-rat LD50:409 mg/kg NYKZAU 74,179,78
orl-mus LD50:2326 mg/kg NYKZAU 74,179,78
ipr-mus LD50:473 mg/kg NYKZAU 74,179,78

SAFETY PROFILE: Moderately toxic by ingestion and intraperitoneal routes. When heated to decomposition it emits toxic fumes of NO_x. See also KETONES.

ACU500 CAS:42978-43-8 HR: 2
6-ACETYLOXYMETHYLBENZO(a)PYRENE
mf: $C_{23}H_{16}O_2$ mw: 324.39

SYN: 6-ACETOXY METHYL BENZO(a)PYRENE

TOXICITY DATA WITH REFERENCE
mmo-sat 1 nmol/plate PAACA3 24,93,83
dnd-rat:lym 500 mg/L CBINA8 25,35,79
dnd-mam:lym 500 mg/L CBINA8 25,35,79
scu-rat TDLo:584 μg/kg/60D-I:NEO CBINA8 29,159,80
scu-rat TD:100 mg/kg/40D-I:ETA JMCMAR 16,714,73

SAFETY PROFILE: Questionable carcinogen with experimental neoplastigenic and tumorigenic data. Mutation data reported. When heated to decomposition it emits acrid smoke and fumes.

ACV000 CAS:34627-78-6 HR: 2
5-(1-ACETYLOXY-2-PROPENYL)-1,3-BENZODIOXOLE
mf: $C_{12}H_{12}O_4$ mw: 220.24

SYN: 1'-ACETOXYSAFROLE

TOXICITY DATA WITH REFERENCE
mmo-sat 25 μg/plate JJIND8 62,893,79

dnr-esc 25 mg/L JJIND8 62,873,79
dnd-hmn:oth 500 µmol/L CRNGDP 3,935,82
orl-rat TDLo:62 g/kg/36W-C:NEO CNREA8 33,590,73
scu-rat TDLo:529 mg/kg/10W-I:CAR CNREA8 43,1124,83
ipr-mus TDLo:22 mg/kg:CAR CNREA8 47,2275,87

CONSENSUS REPORTS: EPA Genetic Toxicology Program.

SAFETY PROFILE: Questionable carcinogen with experimental carcinogenic and neoplastigenic data. Human mutation data reported. When heated to decomposition it emits acrid smoke and fumes.

ACV500 CAS:110-22-5 HR: 3
ACETYL PEROXIDE
mf: $C_4H_6O_4$ mw: 118.04

$CH_3CO{\cdot}OOCO{\cdot}CH_3$

PROP: Solid or colorless crystals or liquid with very pungent odor. D: 1.18, mp: 30°, bp: 63° @ 21 mm. Sltly sol in cold water, decomp.

SYNS: ACETYL PEROXIDE, not >25% in solution (UN 2084) (DO) □ ACETYL PEROXIDE, solid, or >25% in solution (DOT) □ DIACETONE PEROXIDES, solid, or >25% in solution (DOT) □ DIACETYL PEROXIDE (MAK)

TOXICITY DATA WITH REFERENCE
eye-rbt 60 mg/1M rns SEV ZAARAM 8,25,58
unk-mus TDLo:283 mg/kg:ETA RARSAM 3,193,63

CONSENSUS REPORTS: Reported in EPA TSCA Inventory.

DFG MAK: Strong Skin Effects
DOT CLASSIFICATION: Forbidden

SAFETY PROFILE: Severe skin and eye irritant. Questionable carcinogen with experimental tumorigenic data. Dangerous fire hazard by spontaneous chemical reaction. A powerful oxidizing agent; can cause ignition of organic materials on contact. Severe explosion hazard when shocked or exposed to heat. It may explode spontaneously in storage and should be used as soon as prepared. It will react with water or steam to produce heat; can react vigorously with reducing materials; emits toxic fumes on contact with acid or acid fumes. To fight fire use CO_2, dry chemical.

Storage and Handling: Must be kept below 27° and not warmed over 30°. Do not add to hot materials. Do not add accelerator to this material. Store in original container with vented cap. Avoid bodily contact. This material is nearly always stored and handled as a 25% solution in an inert solvent. See also ACETYL PEROXIDE 25% solution (in dimethyl phthalate); and PEROXIDES, ORGANIC.

ACX500 CAS:13402-08-9 HR: 2
1-ACETYL-3-PHENYLETHYLACETYLUREA
mf: $C_{13}H_{16}N_2O_3$ mw: 248.31

SYNS: N-((ACETYLAMINO)CARBONYL)-α-ETHYLBENZENEACETAMIDE □ ACETYLPHENETURIDE □ CRAMPOL □ CRAMPOLE □ N-α-ETHYLPHENYLACETYL-N'-ACETYL UREA □ P-398

TOXICITY DATA WITH REFERENCE
ipr-rat LD50:543 mg/kg NIIRDN 6,17,82
ipr-mus LD50:560 mg/kg NIIRDN 6,17,82
orl-rat LD50:1174 mg/kg ARZNAD 18,524,68
orl-mus LD50:1165 mg/kg ARZNAD 18,524,68

SAFETY PROFILE: Moderately toxic by ingestion and intraperitoneal routes. May have human reproductive effects. When heated to decomposition it emits toxic fumes of NO_x. An anticonvulsant.

ACX750 CAS:114-83-0 HR: 3
ACETYLPHENYLHYDRAZINE
mf: $C_8H_{10}N_2O$ mw: 150.20

PROP: Prisms. Mp: 130–132°. Sol in hot water and alc; sltly sol in ether.

SYNS: ACETIC ACID PHENYLHYDRAZONE □ β-ACETYLPHENYLHYDRAZINE □ N-ACETYL-N'-PHENYLHYDRAZINE □ 1-ACETYL-2-PHENYLHYDRAZINE □ APH □ FENYLHYDRAZID KYSELINY OCTOVE □ HYDRACETIN □ N'-PHENYLACETHYDRAZIDE □ PYRODIN □ PYRODINE

TOXICITY DATA WITH REFERENCE
mmo-sat 333 µg/plate EMMUEG 11(Suppl 12),1,88
orl-mus TDLo:31 g/kg/79W-I:NEO BJCAAI 39,584,79
orl-mus LD50:270 mg/kg PCJOAU 14,162,80
ipr-mus LDLo:150 mg/kg NTIS** AD691-490

CONSENSUS REPORTS: Reported in EPA TSCA Inventory.

SAFETY PROFILE: Poison by ingestion and intraperitoneal routes. Questionable carcinogen with experimental neoplastigenic data. See also HYDRAZINE. When heated to decomposition it emits toxic fumes of NO_x.

ACY700 CAS:76298-68-5 HR: 2
cis-2-ACETYL-3-PHENYL-5-TOSYL-3,3a,4,5-TETRAHYDROPYRAZOLO(4,3-c)QUINOLINE
mf: $C_{25}H_{23}N_3O_3S$ mw: 445.57

SYN: 2H-PYRAZOLO(4,3-c)QUINOLINE, 3,3a,4,5-TETRAHYDRO-2-ACETYL-5-((4-METHYLPHENYL)SULFONYL)-3-PHENYL-, cis-

TOXICITY DATA WITH REFERENCE
orl-rat TDLo:100 mg/kg (female 1-5D post):REP IJOCAP 19,297,80
ipr-mus LD50:800 mg/kg IJOCAP 19,297,80

SAFETY PROFILE: Moderately toxic by intraperitoneal route. Experimental reproductive effects. When heated to decomposition it emits toxic fumes of NO_x and SO_x.

ACY750 HR: 2
12-O-ACETYL-PHORBOL-13-DECA-(Δ-2)-ENOATE
mf: $C_{32}H_{45}O_8$ mw: 557.77

TOXICITY DATA WITH REFERENCE
skn-mus 50 µg MLD PLMEAA 22,241,72
skn-mus TDLo:107 g/kg/12W-I:NEO PLMEAA 22,241,72

SAFETY PROFILE: Questionable carcinogen with experimental neoplastigenic data. A skin irritant. When heated to decomposition it emits acrid smoke and fumes.

ACZ000 **CAS:20839-15-0** **HR: 2**
12-O-ACETYL-PHORBOL-13-DECANOATE
mf: $C_{32}H_{48}O_8$ mw: 560.80

SYN: PHORBOL ACETATE, CAPRATE

TOXICITY DATA WITH REFERENCE
skn-mus 49 µg MLD PLMEAA 22,241,72
skn-mus TDLo:26 mg/kg/32W-I:NEO NATWAY 54,282,67

SAFETY PROFILE: A skin irritant. Questionable carcinogen with experimental neoplastigenic data. When heated to decomposition it emits acrid smoke and irritating fumes.

ADA000 **CAS:17433-31-7** **HR: 3**
1-ACETYL-2-PICOLINOLHYDRAZINE
mf: $C_8H_{11}N_3O_2$ mw: 179.20

SYNS: N-ACETYL-N'-ISONICOTINYL HYDRAZIDE □ 1-ACETYL-2-PICOLINOYLHYDRAZINE □ AZAPICYL □ NCI-C04739 □ NSC-68626 □ P-2292 □ 2-PYRIDINECARBOXYLIC ACID-2-ACETYLHYDRAZIDE (9CI)

TOXICITY DATA WITH REFERENCE
ipr-rat TDLo:9750 mg/kg/26W-I:NEO RRCRBU 52,1,75
ipr-mus TDLo:9750 mg/kg/26W-I:NEO,REP RRCRBU 52,1,75
orl-rat LD50:673 mg/kg NCIAL* -,169,65
ivn-rat LD50:470 mg/kg NCIAL* -,169,65
orl-mus LD50:410 mg/kg NCIAL* -,169,65
ivn-mus LD50:255 mg/kg NCIAL* -,169,65

CONSENSUS REPORTS: NCI Carcinogenesis Studies (ipr): Clear Evidence: mouse, rat RRCRBU 52,1,75.

SAFETY PROFILE: Poison by ingestion and intravenous routes. Questionable carcinogen with experimental neoplastigenic data. Experimental reproductive effects. When heated to decomposition it emits toxic fumes such as NO_x.

ADA250 **CAS:618-42-8** **HR: 3**
1-ACETYLPIPERIDINE
mf: $C_7H_{13}NO$ mw: 127.21

PROP: Mp: 131–133°, d: 1.011, bp: 226°. Misc in water, sol in alc.

SYN: N-ACETYLPIPERIDIN (GERMAN)

TOXICITY DATA WITH REFERENCE
scu-rbt LDLo:300 mg/kg BDCGAS 34,2408,01

CONSENSUS REPORTS: Reported in EPA TSCA Inventory.

SAFETY PROFILE: Poison by subcutaneous route. When heated to decomposition it emits toxic fumes of NO_x.

ADA350 **CAS:22047-25-2** **HR: 2**
2-ACETYL PYRAZINE
mf: $C_6H_6N_2O$ mw: 122.13

PROP: Colorless to pale-yellow crystals or liquid; sweet popcorn-like odor. Mp: 75–78°, d: 1.100–1.115 @ 20°, refr index: 1.530–1.540 @ 25°. Sol in acids, alc, ether, and water @ 230°.

SYN: FEMA No. 3126

SAFETY PROFILE: A skin and eye irritant. When heated to decomposition emits toxic fumes of NO_x.

ADA365 **CAS:1122-54-9** **HR: 3**
4-ACETYLPYRIDINE
mf: C_7H_7NO mw: 121.15

SYNS: KETONE, METHYL 4-PYRIDYL □ METHYL 4-PYRIDYL KETONE □ PYRIDINE, 4-ACETYL-

TOXICITY DATA WITH REFERENCE
mrc-smc 9900 ppm MUREAV 163,23,86
sln-smc 6200 ppm MUREAV 163,23,86
ipr-mus LD50:1400 mg/kg JMCMAR 14,551,71

CONSENSUS REPORTS: Reported in EPA TSCA Inventory.

DOT CLASSIFICATION: 3; *Label:* Flammable Liquid

SAFETY PROFILE: Moderately toxic by intraperitoneal route. Mutation data reported. A flammable liquid. When heated to decomposition it emits toxic vapors of NO_x.

ADA725 **CAS:50-78-2** **HR: 3**
ACETYLSALICYLIC ACID
mf: $C_9H_8O_4$ mw: 180.17

PROP: Colorless needles, crystals. Mp: 135°, fp: 118°. Very sltly sol in alc, sol in benzene. Solubility in water = 1% @ 37°, in ether = 5% @ 20°.

SYNS: AC 5230 □ ACENTERINE □ ACESAL □ ACETAL □ ACETICYL □ ACETILSALICILICO □ ACETILUM ACIDULATUM □ ACETISAL □ ACETOL □ ACETONYL □ ACETOPHEN □ ACETOSAL □ ACETOSALIC ACID □ ACETOSALIN □ o-ACETOXYBENZOIC ACID □ 2-ACETOXYBENZOIC ACID □ ACETYLIN □ 2-(ACETYLOXY)BENZOIC ACID □ ACETYLSAL □ ACETYLSALICYLSAEURE (GERMAN) □ ACIDE ACETYLSALICYLIQUE (FRENCH) □ ACIDO o-ACETIL-BENZOICO (ITALIAN) □ ACIDO ACETILSALICILICO (ITALIAN) □ ACIDUM ACETYLSALICYLICUM □ ACIMETTEN □ ACISAL □ ACYLPYRIN □ ASA □ A.S.A. □ A.S.A. EMPIRIN □ ASAGRAN □ ASATARD □ ASPALON □ ASPERGUM □ ASPIRDROPS □ ASPIRIN □ ASPIRINE □ ASPRO □ ASTERIC □ BENASPIR □ BIALPIRINIA □ CAPRIN □ o-CARBOXYPHENYL ACETATE □ COLFARIT □ CONTRHEUMA RETARD □ CRYSTAR □ DELGESIC □ DOLEAN pH 8 □ DURAMAX □ ECM □ ECOTRIN □ EMPIRIN □ ENDYDOL □ ENTERICIN □ ENTEROPHEN □ ENTEROSARINE □ ENTROPHEN □ EXTREN □ GLOBOID □ HELICON □ IDRAGIN □ MEASURIN □ NEURONIKA □ NOVID □ POLOPIRYNA □ RHEUMIN TABLETTEN □ RHODINE □ SALACETIN □ SALCETOGEN □ SALETIN □ SOLPYRON □ XAXA

TOXICITY DATA WITH REFERENCE
dni-hmn:lym 100 µmol/L FEPRA7 36,1748,77
cyt-hmn:fbr 100 mg/L ACYTAN 16,41,72
orl-wmn TDLo:17,280 mg/kg (female 1-39W post):REP JOPDAB 92,478,78
orl-wmn TDLo:17,280 mg/kg (female 1-39W post):TER JOPDAB 92,478,78
orl-cld TDLo:10 mg/kg/1D-I:PUL,SYS CTOXAO 18,247,81
orl-man TDLo:857 mg/kg:CNS,PUL HUTODJ 7,161,88
orl-wmn TDLo:525 mg/kg/5D-I:SYS AIMEAS 80,74,74
orl-wmn TDLo:480 mg/kg/5D-I:SYS NEJMAG 296,418,77
orl-man TDLo:1625 mg/kg:SYS CPEDAM 24,678,85
orl-inf TDLo:120 mg/kg:PUL,SYS BMJOAE 1,1081,79

orl-cld LDLo:104 mg/kg:PUL,GIT LANCAO 2,809,52
orl-cld TDLo:39 mg/kg/13D-I:SYS AJDCAI 139,453,85
orl-hmn TDLo:669 mg/kg/11D:SYS AJHPA9 35,330,78
orl-hmn TDLo:2880 mg/kg/8W:EAR ARZNAD 33,631,83
orl-hmn TDLo:480 mg/kg/7D-I:EAR,CNS ARZNAD 25,281,75
unr-man LDLo:294 mg/kg 85DCAI 2,73,70
orl-rat LD50:200 mg/kg 34ZIAG -,67,69
ipr-rat LD50:340 mg/kg NYKZAU 62,11,66
orl-mus LD50:250 mg/kg ARZNAD 5,572,55
ipr-mus LD50:280 mg/kg JPPMAB 4,872,52
scu-mus LD50:1020 mg/kg DRFUD4 9,91,84
orl-dog LD50:700 mg/kg ARZNAD 21,719,71

CONSENSUS REPORTS: EPA Genetic Toxicology Program. Reported in EPA TSCA Inventory.

OSHA PEL: TWA 5 mg/m^3
ACGIH TLV: TWA 5 mg/m^3

SAFETY PROFILE: Poison by ingestion, intraperitoneal, and possibly other routes. Human systemic effects by ingestion: acute pulmonary edema, body temperature increase, changes in kidney tubules, coma, constipation, dehydration, hematuria, hepatitis, nausea or vomiting, respiratory stimulation, somnolence, tinnitus, urine volume decreased. Implicated in aplastic anemia. A 10 gram dose to an adult may be fatal. A human teratogen. Human reproductive effects by ingestion and possibly other routes: menstrual cycle changes, parturition, various effects on newborn including apgar score, developmental abnormalities of the cardiovascular and respiratory systems. Experimental animal reproductive effects. Human mutation data reported. An allergen; skin contact, inhalation, or ingestion can cause asthma, sneezing, irritation of eyes and nose, hives, and eczema. Combustible when exposed to heat or flame. When heated to decomposition it emits acrid smoke and fumes.

ADA750 CAS:493-53-8 HR: 2
o-ACETYLSALICYLIC ACID, SODIUM SALT
mf: $C_9H_7O_4{\cdot}Na$ mw: 202.15

PROP: Crystals from acetone and ether. Mp: 218° (slt decomp). Very sol in water and alc; sltly sol in acetone.

SYNS: ACETYLSALICYLIC ACID SODIUM SALT ☐ ACETYLSALICYLSAEURE NATRIUMSALZ (GERMAN) ☐ ASPIRIN-NATRIUM (GERMAN) ☐ SODIUM ASPIRIN

TOXICITY DATA WITH REFERENCE
ivn-hmn TDLo:306 μg/kg:BLD GWXXBX #2810425
ipr-rat LD50:1450 mg/kg NYKZAU 79,357,82
scu-mus LDLo:700 mg/kg HDTU** -,-,33
ipr-mus LDLo:500 mg/kg JACSAT 63,1437,41
scu-frg LDLo:909 mg/kg HBAMAK 4,1290,35

SAFETY PROFILE: Moderately toxic by intraperitoneal and subcutaneous routes. Human systemic effects by intravenous route: unspecified changes in the blood. When heated to decomposition it emits toxic fumes of Na_2O. See also ACETOL.

ADB250 CAS:58086-32-1 HR: 3
o-ACETYLSTERIGMATOCYSTIN
mf: $C_{20}H_{14}O_7$ mw: 366.34

TOXICITY DATA WITH REFERENCE
dns-rat:lvr 1 μmol/L MUREAV 173,217,86
mmo-sat 100 μg/plate CNREA8 38,536,78
mma-sat 1 μg/plate CNREA8 38,536,78
mrc-bcs 1 μg/disc CNREA8 36,445,76
ipr-rat LD50:11,300 μg/kg 41KEAL -,108,78
ipr-rat TDLo:245 mg/kg/23W-I:CAR GANNA2 69,237,78
ipr-rat LD50:11,300 μg/kg 41KEAL-,108,78

SAFETY PROFILE: Poison by intraperitoneal route. Questionable carcinogen with experimental carcinogenic data. Mutation data reported. When heated to decomposition it emits acrid smoke and fumes.

ADC300 CAS:1866-15-5 HR: 2
ACETYLTHIOCHOLINE IODIDE
mf: $C_7H_{16}NOS{\cdot}I$ mw: 289.20

SYNS: ACETYLTHIOCHOLINE DIIODIDE ☐ S-ACETYLTHIOCHOLINE IODIDE ☐ 2-(ACETYLTHIO)-N,N,N-TRIMETHYLETHANAMINIUM IODIDE ☐ AMMONIUM, (2-MERCAPTOETHYL)TRIMETHYL-, IODIDE ACETATE ☐ CHOLINE, S-ACETYLTHIO-, IODIDE ☐ ETHANAMINIUM, 2-(ACETYLTHIO)-N,N,N-TRIMETHYL-, IODIDE (9CI) ☐ (2-MERCAPTOETHYL)TRIMETHYLAMMONIUM IODIDE ACETATE

TOXICITY DATA WITH REFERENCE
ivn-mus LD50:1800 μg/kg CSLNX* NX#02898

CONSENSUS REPORTS: Reported in EPA TSCA Inventory.

SAFETY PROFILE: Moderately toxic by intraperitoneal route. When heated to decomposition it emits toxic vapors of NO_x, SO_x, and I^-.

ADC750 CAS:584-26-9 HR: 3
1-ACETYL-2-THIOHYDANTOIN
mf: $C_5H_6N_2O_2S$ mw: 158.19

PROP: Plates from ethanol. Mp: 175–176°. Insol in water and ether; sltly sol in alc.

SYNS: 4-IMIDAZOLIDINONE, 1-ACETYL-2-THIOXO- ☐ USAF B-7 ☐ USAF BE-0405

TOXICITY DATA WITH REFERENCE
orl-mus LD50:600 mg/kg KHFZAN 24(4),32,90
ipr-mus LD50:200 mg/kg NTIS** AD277-689
ivn-mus LD50:320 mg/kg CSLNX* NX#00834

CONSENSUS REPORTS: Reported in EPA TSCA Inventory.

SAFETY PROFILE: Poison by intravenous and intraperitoneal routes. Moderately toxic by ingestion. When heated to decomposition it emits very toxic fumes such as SO_x and NO_x.

ADD250 CAS:591-08-2 HR: 3
ACETYL THIOUREA
mf: $C_3H_6N_2OS$ mw: 118.17

PROP: Needles. Mp: 165–167°. Sol in hot water and alc; sltly sol in ether.

SYNS: 1-ACETYL-2-THIOUREA □ RCRA WASTE NUMBER P002 □ USAF EK-4890

TOXICITY DATA WITH REFERENCE
orl-rat LD50:50 mg/kg JPETAB 90,260,47
ipr-rat LDLo:400 mg/kg JPETAB 97,478,49
orl-mus LDLo:94 mg/kg AECTCV 14,111,85
ipr-mus LD50:100 mg/kg NTIS** AD277-689

CONSENSUS REPORTS: Reported in EPA TSCA Inventory.

SAFETY PROFILE: Poison by ingestion and intraperitoneal routes. When heated to decomposition it emits very toxic fumes of NO_x and SO_x. See also SULFIDES.

ADD750 CAS:77-89-4 HR: 2
ACETYL TRIETHYL CITRATE
mf: $C_{14}H_{22}O_8$ mw: 318.36

PROP: Bp: 197° @ 15 mm.

SYNS: ATEC □ CITRIC ACID, ACETYL TRIETHYL ESTER □ CITROFLEX A 2 □ 1,2,3-PROPANETRICARBOXYLIC ACID, 2-(ACETYLOXY)-, TRIETHYL ESTER (9CI) □ TRICARBALLYLIC ACID, β-ACETOXYTRIBUTYL ESTER □ TRIETHYL ACETYLCITRATE □ TRIETHYL CITRATE, ACETATE □ TRIETHYLESTER KYSELINY ACETYLCITRONOVE

TOXICITY DATA WITH REFERENCE
orl-rat LD50:7 g/kg IPSTB3 3,93,76
ipr-mus LD50:1150 mg/kg JPMSAE 53,774,64
orl-cat LDLo:7500 mg/kg TXAPA9 1,283,59

CONSENSUS REPORTS: Reported in EPA TSCA Inventory.

SAFETY PROFILE: Moderately toxic by intraperitoneal route. Mildly toxic by ingestion. See also ESTERS. When heated to decomposition it emits acrid smoke and fumes.

ADD875 CAS:2260-08-4 HR: 2
ACETYLTRIIODOTHYRONINE FORMIC ACID
mf: $C_{15}H_9I_3O_5$ mw: 649.95

PROP: Mp: 238°

SYNS: ACETIROMATE □ 4-(4-(ACETYLOXY)-3-IODOPHENOXY)-3,5-DIIODO-BENZOIC ACID

TOXICITY DATA WITH REFERENCE
orl-rat LD50:4600 mg/kg IYKEDH 5,383,74
ipr-rat LD50:500 mg/kg IYKEDH 5,383,74
scu-rat LD50:520 mg/kg IYKEDH 5,383,74
orl-mus LD50:3700 mg/kg IYKEDH 5,383,74
ipr-mus LD50:1 g/kg IYKEDH 5,383,74
scu-mus LD50:2500 mg/kg IYKEDH 5,383,74

SAFETY PROFILE: Moderately toxic by ingestion and other routes. When heated to decomposition it emits toxic fumes of I^-.

ADE000 CAS:477-27-0 HR: 3
N-ACETYL TRIMETHYLCOLCHICINIC ACID
mf: $C_{21}H_{23}NO_6$ mw: 385.45

PROP: Needles from alcohol. Mp: 175–177°. Sol in $CHCl_3$.

SYNS: 7-ACETAMIDO-6,7-DIHYDRO-10-HYDROXY-1,2,3-TRIMETHOXY-BENZO(a)HEPTALEN-9(5H)-ONE □ 7-ACETAMIDO-10-HYDROXY-1,2,3-TRIMETHOXY-6,7-DIHYDROBENZO(a)HEPTALEN-9(5H)-ONE □ O^{10}-DEMETHYLCOLCHICINE

TOXICITY DATA WITH REFERENCE
oms-mus-ipr 42 mg/kg CANCAR 3,130,50
oms-mus-par 84 mg/kg CANCAR 3,130,50
spm-mus-par 84 mg/kg CANCAR 3,130,50
orl-hmn LDLo:43 μg/kg PCOC** -,250,66
unk-rat LDLo:30 mg/kg CANCAR 3,125,50
ipr-mus LD50:84 mg/kg CANCAR 3,124,50
ivn-mus LD50:1 mg/kg COREAF 241,1889,55

SAFETY PROFILE: A deadly human poison by ingestion. An experimental poison by intravenous, intraperitoneal, and possibly other routes. Mutation data reported. When heated to decomposition it emits toxic fumes of NO_x. See also COLCHICINE.

ADE050 CAS:19005-95-9 HR: 3
3-ACETYL-2,4,5-TRIMETHYL-PYRROLE
mf: $C_9H_{13}NO$ mw: 151.23

PROP: A liquid.

SYNS: KETONE, METHYL 2,4,5-TRIMETHYLPYRROL-3-YL □ METHYL 2,4,5-TRIMETHYLPYRROL-3-YL KETONE

TOXICITY DATA WITH REFERENCE
ipr-mus LD50:233 mg/kg JMCMAR 11,1251,68

DOT CLASSIFICATION: 3; *Label:* Flammable Liquid

SAFETY PROFILE: A poison by intraperitoneal route. A flammable liquid. When heated to decomposition it emits toxic vapors of NO_x.

ADE075 CAS:1218-34-4 HR: 2
ACETYLTRYPTOPHAN
mf: $C_{13}H_{14}N_2O_3$ mw: 246.29

SYNS: ACETYL-l-TRYPTOPHAN □ N-ACETYL-l-TRYPTOPHAN □ N-ACETYLTRYPTOPHAN □ ACETYL-l-TRP □ (S)-N-ACETYLTRYPTOPHAN □ AC-TRY

TOXICITY DATA WITH REFERENCE
ipr-rat TDLo:1650 mg/kg (female 7-17D post):REP IYKEDH 11,690,80
orl-rat TDLo:55 g/kg (7-17D preg):TER IYKEDH 11,690,80
orl-rat LD50:15,000 mg/kg IYKEDH 11,635,80
ipr-rat LD50:3,900 mg/kg IYKEDH 11,635,80
orl-mus LD50:10,800 mg/kg IYKEDH 11,635,80
ipr-mus LD50:3,580 mg/kg IYKEDH 11,635,80

CONSENSUS REPORTS: Reported in EPA TSCA Inventory.

SAFETY PROFILE: Moderately toxic by some routes. An experimental teratogen. Other experimental reproductive effects. When heated to decomposition it emits toxic fumes of NO_x.

ADE500 **CAS:129-17-9** **HR: 3**
ACID BLUE 1
mf: $C_{27}H_{31}N_2O_6S_2 \cdot Na$ mw: 566.71

PROP: Violet powder. Very sol in water; sol in ethanol.

SYNS: ACID BLUE V ☐ ACID BRIGHT AZURE Z ☐ ACID BRILLIANT BLUE VF ☐ ACID BRILLIANT BLUE Z ☐ ACID BRILLIANT SKY BLUE Z ☐ ACID LEATHER BLUE V ☐ AIZEN BRILLIANT ACID PURE BLUE VH ☐ ALPHAZURINE 2G ☐ AMACID BLUE V ☐ ANHYDRO-4,4'-BIS (DIETHYLAMINO)TRIPHENYLMETHANOL-2',4''-DISULPHONIC ACID, MONOSODIUM SALT ☐ BLEU PATENTE V ☐ BLUE 1084 ☐ 1085 BLUE ☐ BLUE URS ☐ BLUE VRS ☐ BRILLIANT ACID BLUE A EXPORT ☐ BRILLIANT ACID BLUE V EXTRA ☐ BRILLIANT ACID BLUE VS ☐ BRILLIANT BLUE GS ☐ BUCACID PATENT BLUE VF ☐ CARMIN BLUE VS ☐ CARMINE BLUE VF ☐ C.I. 712 ☐ C.I. 42045 ☐ C.I. ACID BLUE 1 ☐ C.I. ACID BLUE 3 ☐ C.I. ACID BLUE 1, SODIUM SALT ☐ C.I. FOOD BLUE 3 ☐ COSMETIC GREEN BLUE R25396 ☐ 4,4'-DI(DIETHYLAMINO)-4',6'-DISULPHOTRIPHENYLMETHANOL ANHYDRIDE, SODIUM SALT ☐ DISULFINE BLUE VN ☐ DISULPHINE VN ☐ DISULPHINE BLUE VN 150 ☐ E 131 ☐ EDICOL SUPRA BLUE VR ☐ ERIO BRILLIANT BLUE V ☐ ERIOGLAUCINE ☐ ERIOGLAUCINE SUPRA ☐ FENAZO BLUE XF ☐ FENAZO BLUE XV ☐ FOOD BLUE 3 ☐ HEXACO BLUE VRS ☐ HEXACOL BLUE VRS ☐ HIDACID BLUE V ☐ INTRACID PURE BLUE V ☐ KITON PURE BLUE V ☐ KITON PURE BLUE V.FQ ☐ L-BLAU 3 ☐ LEATHER BLUE G ☐ LISSAMINE TURQUOISE VN ☐ MERANTINE BLUE VF ☐ MODR KYSELA 1 ☐ MODR POTRAVINARSKA 3 ☐ PATENTBLAU V ☐ PATENT BLUE ☐ PATENT BLUE V ☐ PATENT BLUE VF ☐ PATENT BLUE VF-CF ☐ PATENT BLUE VF SPECIAL ☐ PATENT BLUE VS ☐ PONTACYL BRILLIANT BLUE ☐ PONTACYL BRILLIANT BLUE V ☐ SCHULTZ Nr. 826 ☐ SODIUM BLUE VRS ☐ SODIUM PATENT BLUE V ☐ SULFACID BRILLIANT BLUE 6J ☐ SULFAN BLUE ☐ SULPHAN BLUE ☐ SUMITOMO PATENT PURE BLUE VX ☐ TETRACID CARMINE BLUE V ☐ XYLENE BLUE VS

TOXICITY DATA WITH REFERENCE
mma-sat 1 mg/plate ENMUDM 8(Suppl 7),1,86
mma-sat 1 mg/plate ENMUDM 8(Suppl 7),1,86
scu-rat TDLo:3000 mg/kg/33W-I:NEO BJCAAI 27,230,73
ims-rat TDLo:2070 mg/kg/50W-I:CAR JNCIAM 37,845,66
scu-rat TD:4050 mg/kg/45W-I:ETA FCTXAV 9,463,71
ivn-man LDLo:33 µg/kg:ALR 34ZIAG-,611,69
ipr-mus LD50:3000 mg/kg FCTXAV 5,165,67
ivn-mus LD50:1200 mg/kg SCPHA4 47,39,79

CONSENSUS REPORTS: IARC Cancer Review: Group 3 IMEMDT 7,56,87. Reported in EPA TSCA Inventory.

SAFETY PROFILE: Deadly human poison by intravenous route. Human systemic effects by intravenous route: anaphylaxis. Moderately toxic by several routes. Questionable carcinogen with experimental carcinogenic, tumorigenic, and neoplastigenic data. Mutation data reported. When heated to decomposition it emits very toxic fumes of NO_x, NH_3, Na_2O, and SO_x. See also SULFONATES.

ADE750 **CAS:3861-73-2** **HR: 2**
ACID BLUE 92
mf: $C_{26}H_{16}N_3O_{10}S_3 \cdot 3Na$ mw: 695.60

PROP: Blue crystals. Sol in water, 2-ethoxyethanol; sltly sol in ethanol.

SYNS: ACID BLUE A ☐ ACID LEATHER BLUE R ☐ ACID WOOL BLUE RL ☐ ACILAN FAST NAVY BLUE R ☐ AIREDALE BLUE RL ☐ AMACID FAST BLUE R ☐ ANAZOLENE, SODIUM ☐ 4-((4-ANILINO-5-SULFO-1-NAPHTHYL)AZO)-5-HYDROXY-2,7-NAPHTHALENEDIFULFONIC ACID TRISODIUM ☐ BENZYL BLUE R ☐ BENZYL FAST BLUE R ☐ BUCACID FAST WOOL BLUE R ☐ CALCOCID FAST BLUE SR ☐ C.I. 13390 ☐ C.I. ACID BLUE 92 ☐ C.I. ACID BLUE 92, TRISODIUM SALT ☐ CIRENE BRILLIANT BLUE R ☐ COLACID BLUE A ☐ COOMASSIE BLUE ☐ COOMASSIE BLUE MEDICINAL ☐ COOMASSIE BLUE RL ☐ CYANINE ACID BLUE R ☐ CYANINE ACID BLUE R NEW ☐ FAST ACID BLUE RL ☐ FAST WOOL BLUE R ☐ FENAZO BLUE SR ☐ HISPACID FAST BLUE R ☐ MEDIUM BLUE EMBL ☐ PONTACYL FAST BLUE R ☐ SODIUM AMAZOLENE ☐ SODIUM ANAZOLENE ☐ SULFONINE ACID BLUE R ☐ SULPHON ACID BLUE R ☐ SULPHON ACID BLUE RA ☐ TERTRACID FAST BLUE SR ☐ TRISODIUM-4'-ANILINO-8-HYDROXY-1, 1'-AZONAPHTHALENE-3,6,5'-TRISULFONATE ☐ VONDAMOL FAST BLUE R ☐ WOOL BLUE RL ☐ WOOL FAST BLUE R

TOXICITY DATA WITH REFERENCE
dnd-esc 10 µmol/L MUREAV 89,95,81
ivn-mus LDLo:450 mg/kg BHJUAV 21,492,59

CONSENSUS REPORTS: Reported in EPA TSCA Inventory.

SAFETY PROFILE: Moderately toxic by intravenous route. Mutation data reported. When heated to decomposition it emits very toxic fumes of SO_x, NO_x, and Na_2O.

ADF000 **CAS:3087-16-9** **HR: 2**
ACID BRILLIANT GREEN BS
mf: $C_{27}H_{26}N_2O_7S_2 \cdot Na$ mw: 577.66

SYNS: ACID GREEN 50 ☐ ACID LEATHER GREEN S ☐ ACILAN GREEN BS ☐ AMACID WOOL GREEN S ☐ BRILLIANTSAEURE GRUEN BS ☐ BUCACID WOOL GREEN ☐ CALOCID GREEN S ☐ CALOCID GREEN SB ☐ C.I. 44090 ☐ C.I. ACID GREEN 50, MONOSODIUM SALT ☐ C.I. FOOD GREEN 4 ☐ E 142 ☐ EDICOL SUPRA GREEN B ☐ ERIO GREEN S ☐ FOOD GREEN S ☐ GREEN 5 ☐ 12078 GREEN ☐ GREEN BS ☐ GREEN S ☐ HEXACOL GREEN S ☐ HIDACID WOOL GREEN ☐ KITON GREEN S ☐ LISSAMINE GREEN B ☐ LISSAMINE GREEN BN ☐ NAPHTHAZINE GREEN S ☐ PHARMACID GREEN S ☐ SCHULTZ Nr. 836 ☐ SUMITOMO WOOL GREEN S ☐ UNITERTRACID GREEN BS ☐ VERT ACIDE BRILLIANT BS ☐ VONDACID GREEN S ☐ WATER GREEN SX ☐ WOOL GREEN 5 ☐ WOOL GREEN B ☐ WOOL GREEN BS ☐ WOOL GREEN BSNA ☐ WOOL GREEN MS ☐ WOOL GREEN S ☐ WOOL GREEN S (BIOLOGICAL STAIN) ☐ WOOL GREEN SG ☐ ZELEN KYSELA 50 ☐ ZELEN KYSELA BS ☐ ZELEN POTRAVINARSKA 4

TOXICITY DATA WITH REFERENCE
mrc-smc 2840 µmol/L FCTXAV 19,419,81
mma-sat 1 mg/plate MUREAV 89,21,81
mrc-smc 2840 µmol/L FCTXAV 19,419,81
par-rat TDLo:470 mg/kg/2Y-I:ETA FAONAU 46A,57,69
orl-rat LD50:2 g/kg JPPMAB 16,65,64

CONSENSUS REPORTS: Reported in EPA TSCA Inventory. EPA Genetic Toxicology Program.

SAFETY PROFILE: Moderately toxic by ingestion. Questionable carcinogen with experimental tumorigenic data. Experimental reproductive effects. Mutation data reported. When heated to decomposition it emits very toxic fumes of Na_2O, SO_x, and NO_x.

ADF250 **CAS:12788-93-1** **HR: 3**
ACID BUTYL PHOSPHATE
DOT: UN 1718
mf: $C_4H_{10}O_4P$ mw: 153.1

PROP: Water-white liquid; sol in alc, acetone, and

toluene; insol in water, petroleum, and naphtha. D: 1.120–1.125 @ 25°/40°, flash p: 230°F (COC).

SYNS: n-BUTYL ACID PHOSPHATE □ BUTYL PHOSPHORIC ACID

DOT CLASSIFICATION: 8; *Label:* Corrosive

SAFETY PROFILE: Toxic and corrosive. Combustible when exposed to heat or flame. When heated to decomposition it emits highly toxic fumes of PO_x. See also ESTERS and PHOSPHORIC ACID.

ADF500 **HR: 3**
ACID CARBOYS, EMPTY

SAFETY PROFILE: *Warning:* These containers may contain concentrated vapors or even some liquid acid remaining from their original contents. Therefore, they can give rise to all the hazards of their original contents.

ADG250 **CAS:18472-87-2** **HR: 3**
ACID RED 92
mf: $C_{20}H_2Br_4Cl_4O_5 \cdot 2Na$ mw: 829.64

PROP: Orange-red crystals or powder. Sol in water and ethanol.

SYNS: AIZEN ACID PHLOXINE PB □ C.I. 45410 □ C.I. ACID RED 92 □ CYANOSIN □ CYANOSIN (ACID DYE) □ CYANOSINE □ D and C RED NO. 28 □ EOSIN BLUE □ EOSINE BLUE □ EOSINE BLUISH □ FOOD DYE RED No. 104 □ FOOD RED No. 104 □ JAPAN RED 104 □ ORIENT WATER PINK 2 □ PHLOXIN B □ PHLOXINE B □ PHLOXINE P □ RED 104 □ 11969 RED □ RED No. 104 □ 3427 VERI PUR PINK

TOXICITY DATA WITH REFERENCE
mmo-omi 200 mg/L MUREAV 34,187,76
orl-rat TDLo:63 g/kg (1-22D preg):REP SKEZAP 16,34,75
orl-rat TDLo:63 g/kg (1-22D preg):TER SKEZAP 16,34,75
ivn-mus LD50:310 mg/kg TXAPA9 44,225,78

CONSENSUS REPORTS: Reported in EPA TSCA Inventory. EPA Genetic Toxicology Program.

SAFETY PROFILE: Poison by intravenous route. An experimental teratogen. Other experimental reproductive effects. When heated to decomposition it emits very toxic fumes of Br^-, Cl^-, and Na_2O.

ADG400 **HR: 3**
ACKEE

PROP: A 30- to 40-foot-tall tree with 5-part compound leaves and small green-white flowers. A bright-red pod contains 3 shiny black seeds in a white, waxy matrix. It grows in Florida, Hawaii, and the West Indies.

SYNS: AKEE □ AKI □ ARBRE FRICASSE (HAITI) □ BLIGHIA SAPIDA □ SESO VEGETAL (CUBA, PUERTO RICO)

SAFETY PROFILE: The white matrix of the immature fruit and its attachment to the seeds contain the toxic hypoglycin A. In the ripe fruit these parts are edible. Systemic effects by ingestion may include: vomiting, convulsions, coma, hypoglycemia, and death. Symptoms may begin immediately or may appear after a delay of 6 to 10 hours. In Jamaica poisoning is common in the winter and is called "vomiting sickness." See also 2-METHYLENECYCLOPROPANYLALANINE.

ADG425 **CAS:66789-14-8** **HR: 3**
ACLACINOMYCIN Y
mf: $C_{42}H_{51}NO_{15}$ mw: 809.94

PROP: Solid. Mp: 153–155°.

SYNS: ACLACINOMYCIN Y1 □ MA 144 Y

TOXICITY DATA WITH REFERENCE
dni-mus:leu 190 nmol/L JANTAJ 34,1596,81
oms-mus:leu 12 nmol/L JANTAJ 34,1596,81
ipr-mus LD50:40 mg/kg JANTAJ 33,80-64,80

SAFETY PROFILE: Poison by intraperitoneal route. Mutation data reported. When heated to decomposition it emits toxic fumes of NO_x.

ADG500 **CAS:509-20-6** **HR: 3**
ACONINE

PROP: Amorphous shaped solid. Mp: 132°.

TOXICITY DATA WITH REFERENCE
ivn-mus LD50:117 mg/kg YHHPAL 19,641,84
ivn-cat LD50:400 mg/kg ARZNAD 5,324,55
ivn-gpg LD50:275 mg/kg ARZNAD 5,324,55

SAFETY PROFILE: Poison by intravenous route. When heated to decomposition it emits toxic fumes of NO_x. An antipyretic agent.

ADH000 **CAS:499-12-7** **HR: 3**
ACONITIC ACID
mf: $C_6H_6O_6$ mw: 174.12

PROP: White, crystalline powder. Mp: 192° (decomp). Sol in water, alc; very sltly sol in ether.

SYNS: ACHILLEIC ACID □ CITRIDIC ACID □ EQUISETIC ACID □ 1-PROPENE-1,2,3-TRICARBOXYLIC ACID

TOXICITY DATA WITH REFERENCE
ivn-mus LD50:180 mg/kg CSLNX* NX#00189

CONSENSUS REPORTS: Reported in EPA TSCA Inventory.

SAFETY PROFILE: Poison by intravenous route. A synthetic flavoring substance and adjuvant. When heated to decomposition it emits acrid smoke and fumes.

ADH750 **CAS:302-27-2** **HR: 3**
ACONITINE (crystalline)
mf: $C_{34}H_{49}NO_{11}$ mw: 647.76

PROP: White, crystalline alkaloid; feeble bitter taste. Mp: 204°. Very sparingly sol in water.

SYNS: ACETYL BENZOYL ACONINE □ ACONITANE □ ACONITIN CRISTALLISAT (GERMAN)

TOXICITY DATA WITH REFERENCE
ipr-rat LDLo:125 μg/kg PSEBAA 26,221,28
ivn-rat LD50:80 μg/kg ARZNAD 5,324,55
orl-hmn LDLo:28 mg/kg:CNS,GIT 34ZIAG-,72,69

orl-mus LD50:1 mg/kg 85GDA2 8(1),159,82
scu-mus LDLo:100 mg/kg HDTU** -,-,33
ivn-mus LD50:166 μg/kg 85GDA2 8(1),159,82
ipr-mus LD50:2708 μg/kg CYLPDN 2,170,81
ivn-dog LDLo:350 μg/kg HBAMAK 4,1291,35
scu-cat LDLo:400 μg/kg HBAMAK 4,1291,35
ivn-cat LD50:70 μg/kg ARZNAD 5,324,55
scu-rbt LDLo:131 μg/kg HBAMAK 4,1291,35
scu-gpg LDLo:50 μg/kg JPHAA3 12,957,23
ivn-gpg LD50:60 μg/kg ARZNAD 5,324,55
scu-pgn LDLo:66 μg/kg HBAMAK 4,1291,35
scu-frg LDLo:586 μg/kg HBAMAK 4,1291,35

SAFETY PROFILE: Poison by all routes, including absorption through the skin. Human systemic effects by ingestion: excitement, diarrhea and other gastrointestinal effects. Used to produce heart arrhythmia in experimental animals and as an antipyretic agent. When heated to decomposition it emits highly toxic fumes of NO_x.

ADH875 CAS:6055-69-2 HR: 3
ACONITINE HYDROCHLORIDE
mf: $C_{34}H_{47}NO_{11}{\bullet}ClH$ mw: 682.28

TOXICITY DATA WITH REFERENCE
scu-cat LDLo:134 μg/kg FDWU** -,-,31
scu-gpg LDLo:112 μg/kg FDWU** -,-,31
scu-pgn LDLo:45,500 ng/kg FDWU** -,-,31
scu-frg LDLo:586 μg/kg FDWU** -,-,31

SAFETY PROFILE: Poison by subcutaneous route. When heated to decomposition it emits toxic fumes of NO_x and HCl.

ADI250 HR: 3
ACONITUM CARMICHAELI

PROP: Raw tubers which are the source of processed aconite roots used as an oriental medicine in Japan (YKKZAJ 97,359,77).

TOXICITY DATA WITH REFERENCE
orl-mus LD50:5490 mg/kg YKKZAJ 97,359,77
ipr-mus LD50:190 mg/kg YKKZAJ 97,359,77
scu-mus LD50:200 mg/kg YKKZAJ 97,359,77
ivn-mus LD50:490 mg/kg YKKZAJ 97,359,77

SAFETY PROFILE: Poison by intraperitoneal and subcutaneous routes. Moderately toxic by intravenous route. Mildly toxic by ingestion. When heated to decomposition it emits acrid smoke and fumes.

ADI625 HR: 3
ACORN TANNIN

SYN: TANNIN from ACORN

TOXICITY DATA WITH REFERENCE
ipr-mus LD50:100 mg/kg JPPMAB 9,98,57
scu-mus LD50:100 mg/kg JPPMAB 9,98,57
ivn-mus LD50:150 mg/kg JPPMAB 9,98,57
ims-mus LD50:75 mg/kg JPPMAB 9,98,57

SAFETY PROFILE: Poison by subcutaneous, intramuscular, intravenous, and intraperitoneal routes.

ADI775 CAS:92-81-9 HR: 2
ACRIDAN
mf: $C_{13}H_{11}N$ mw: 181.25

SYNS: ACRIDANE □ ACRIDINE, 9,10-DIHYDRO-(9CI) □ CARBAZINE □ 9,10-DIHYDROACRIDINE

TOXICITY DATA WITH REFERENCE
orl-rat LD50:2140 mg/kg JPMSAE 63,1068,74
scu-mus LD50:3630 μg/kg PSEBAA 78,392,51

SAFETY PROFILE: Poison by subcutaneous route. Moderately toxic by ingestion. When heated to decomposition it emits toxic vapors of NO_x.

ADJ375 CAS:581-29-3 HR: 3
3-ACRIDINAMINE (9CI)
mf: $C_{13}H_{10}N_2$ mw: 194.25

PROP: Yellow crystals. Mp: 224° (dried). Sol in aqueous ethanol; spar sol in Me_2CO, C_6H_6.

SYNS: 2-AMINOACRIDINE (EUROPEAN) □ 3-AMINOACRIDINE

TOXICITY DATA WITH REFERENCE
mmo-sat 20 μg/plate JOUOD4 6,257,84
mmo-omi 80 μg/L JMOBAK 3,762,61
scu-mus LD50:170 mg/kg BJEPA5 28,1,47

CONSENSUS REPORTS: EPA Genetic Toxicology Program.

SAFETY PROFILE: Poison by subcutaneous route. Mutation data reported. When heated to decomposition it emits toxic fumes of NO_x.

ADJ500 CAS:260-94-6 HR: 3
ACRIDINE
DOT: UN 2713
mf: $C_{13}H_9N$ mw: 179.23

PROP: Small, colorless needles or prisms. Mp: 110.5°, bp: 346°, d: 1.005 @ 19.7°/4°, vap press: 1 mm @ 129.4°. Sltly sol in hot water; sol in alc, ether, and CS_2.

SYNS: 9-AZAANTHRACENE □ 10-AZAANTHRACENE □ BENZO(b)QUINOLINE □ 2,3-BENZOQUINOLINE □ DIBENZO(b,e)PYRIDINE

TOXICITY DATA WITH REFERENCE
mmo-sat 230 nmol/L ENMUDM 3,11,81
dnd-mam:lym 100 μmol/L JMOBAK 3,18,61
dnd-ckn:leu 100 μmol/L JMOBAK 3,18,61
orl-rat LD50:2 g/kg GTPZAB 14(9),56,70
orl-mus LD50:500 mg/kg GTPZAB 14(9),56,70
scu-mus LD50:400 mg/kg BJEPA5 28,1,47
ivn-rbt LD50:100 mg/kg BJEPA5 28,1,47

CONSENSUS REPORTS: Reported in EPA TSCA Inventory.

OSHA PEL: TWA 0.2 mg/m^3
DOT CLASSIFICATION: 6.1; *Label:* KEEP AWAY FROM FOOD

SAFETY PROFILE: Poison by ingestion, subcutaneous, and intravenous routes. Mutation data reported. A skin, eye, and mucous membrane irritant. When heated to decomposition it emits toxic fumes of NO_x.

ADJ550 CAS:7101-57-7 HR: 3
ACRIDINE-9-CARBOXAMIDE, N,N-DIETHYL-1,2,3,4-TETRAHYDRO-
mf: $C_{18}H_{22}N_2O$ mw: 282.42

PROP: A liquid.

SYNS: ACRIDINE-9-CARBOXAMIDE, 1,2,3,4-TETRAHYDRO-N,N-DIETHYL- □ KETONE, DIETHYLAMINO(1,2,3,4-TETRAHYDRO-9-ACRIDINYL)

TOXICITY DATA WITH REFERENCE
ipr-mus LD50:250 mg/kg JMCMAR 9,483,66

DOT CLASSIFICATION: 3; *Label:* Flammable Liquid

SAFETY PROFILE: A poison by intraperitoneal route. A flammable liquid. When heated to decomposition it emits toxic vapors of NO_x.

ADJ625 CAS:951-80-4 HR: 3
3,9-ACRIDINEDIAMINE (9CI)
mf: $C_{13}H_{11}N_3$ mw: 209.27

PROP: Crystals from ethanol/Et_2O. Mp: 146°.

SYNS: 2,5-DIAMINOACRIDINE (EUROPEAN) □ 3,9-DIAMINOACRIDINE

TOXICITY DATA WITH REFERENCE
mmo-sat 20 µg/plate JOUOD4 6,257,84
mmo-omi 8 mg/L JMOBAK 3,762,61
mmo-omi 19 µmol/L GENTAE 90,1,78
scu-mus LD50:140 mg/kg BJEPA5 28,1,78

SAFETY PROFILE: Poison by subcutaneous route. Mutation data reported. When heated to decomposition it emits toxic fumes of NO_x.

ADJ750 CAS:17784-47-3 HR: 3
ACRIDINE HYDROCHLORIDE
mf: $C_{13}H_9N{\cdot}ClH$ mw: 215.69

SYNS: ACRIDINE MONOHYDROCHLORIDE □ ACRIDINIUM CHLORIDE

TOXICITY DATA WITH REFERENCE
mmo-omi 80 mg/L JMOBAK 3,762,61
dnd-mam:lym 10 pph BIPMAA 11,2537,72
scu-mus LD50:300 mg/kg QJPPAL 10,649,37

SAFETY PROFILE: Poison by subcutaneous route. Mutation data reported. When heated to decomposition it emits very toxic fumes of HCl and NO_x.

ADJ875 CAS:146-59-8 HR: 2
ACRIDINE MUSTARD
mf: $C_{21}H_{25}Cl_2N_3O{\cdot}2ClH$ mw: 479.31

SYNS: (6-CHLORO-9-(3-ETHYL-2-CHLOROETHYL)AMINOPROPYLAMINO)-2-METHOXYACRIDINE DIHYDROCHLORIDE □ 9-(3-(ETHYL (2-CHLOROETHYL)AMINO)PROPYLAMINO)-6-CHLORO-2-METHOXYACRIDINE DIHYDROCHLORIDE □ ICR 170 □ 2-METHOXY-6-CHLORO-9-(3-(ETHYL-2-CHLOROETHYL)AMINOPROPYLAMINO)ACRIDINE DIHYDROCHLORIDE

TOXICITY DATA WITH REFERENCE
mmo-sat 500 ng/plate MUREAV 136,185,84
slt-dmg-orl 20,860 µmol/L ENMUDM 6,153,84
ipr-mus TDLo:4 mg/kg (1D pre):REP MUREAV 13,171,71
ivn-mus TDLo:4800 µg/kg/28D-I:NEO CNREA8 36,2423,76
ipr-mus LD20:2 mg/kg JMCMAR 15,739,72
ivn-mus LDLo:5 mg/kg CNREA8 36,2423,76

CONSENSUS REPORTS: EPA Genetic Toxicology Program.

SAFETY PROFILE: Poison by intravenous and intraperitoneal routes. Questionable carcinogen with experimental neoplastigenic data. Experimental reproductive effects. Human mutation data reported. When heated to decomposition it emits toxic fumes of NO_x and HCl.

ADK000 CAS:2465-29-4 HR: 2
ACRIDINE RED
mf: $C_{15}H_{14}N_2O{\cdot}ClH$ mw: 274.77

PROP: Sltly sol in water; sol in alc; insol in ether.

SYNS: ACRIDINE RED 3B □ ACRIDINE RED, HYDROCHLORIDE □ DIMETHYLDIAMINOXANTHENYL CHLORIDE

TOXICITY DATA WITH REFERENCE
sln-dmg-orl 1000 ppm AMNTA4 87,295,53
scu-rat TDLo:1215 mg/kg/59W I:ETA GANNA2 47,153,56

CONSENSUS REPORTS: Reported in EPA TSCA Inventory.

SAFETY PROFILE: Questionable carcinogen with experimental tumorigenic data. Mutation data reported. When heated to decomposition it emits very toxic fumes of HCl and NO_x.

ADK250 CAS:191-27-5 HR: 2
ACRIDINO(2,1,9,8-klmna)ACRIDINE
mf: $C_{20}H_{10}N_2$ mw: 278.32

SYN: 6,12-DIAZAANTHANTHRENE

TOXICITY DATA WITH REFERENCE
imp-rat TDLo:600 mg/kg:ETA NEOLA4 18,591,71

SAFETY PROFILE: Questionable carcinogen with experimental tumorigenic data. When heated to decomposition it emits toxic fumes of NO_x.

ADL500 CAS:54301-15-4 HR: 3
4'-(9-ACRIDINYLAMINO)METHANESULFON-m-ANISIDE MONOHYDROCHLORIDE
mf: $C_{21}H_{19}N_3O_3S{\cdot}ClH$ mw: 429.95

PROP: Crystals. Mp: 231–232°.

SYNS: m-AMSA HYDROCHLORIDE □ NCI-C03190 □ NSC 141549

TOXICITY DATA WITH REFERENCE
dnd-mus:leu 2500 µg/L CNREA8 38,1329,78
dns-mus-ipr 5 mg/kg CNREA8 38,1329,78
oms-ham:ovr 2 mg/L JNCIAM 60,1147,78
cyt-ham:ovr 2 mg/L JNCIAM 60,1147,78
sce-ham:ovr 50 µg/L JNCIAM 60,1155,78
orl-mus LD50:181 mg/kg NCISP* JAN86
ipr-mus LD50:20,560 µg/kg NCISP* JAN86
scu-mus LD50:110 mg/kg NCISP* JAN86

SAFETY PROFILE: Poison by ingestion, subcutaneous, and intraperitoneal routes. Mutation data reported. See also SULFONATES. When heated to decomposition it emits very toxic fumes of NO_x, SO_x, and HCl.

ADL750 **CAS:51264-14-3** **HR: 3**
4′-(9-ACRIDINYLAMINO)METHANESULPHON-m-ANISIDIDE
mf: $C_{21}H_{19}N_3O_3S$ mw: 393.49

SYNS: 4′-(9-ACRIDINYLAMINO)-3′-METHOXYMETHANESULFONANILIDE □ 4′-(9-ACRIDINYLAMINO)METHYLSULFONYL-m-ANISIDINE □ AMSA □ m-AMSA □ AMSACRINE □ m-AMSA METHANESULFONATE □ AMSIDINE □ AMSINE □ NSC 141549 □ NSC 249992

TOXICITY DATA WITH REFERENCE
msc-mus:lym 1 μg/L ENMUDM 8(Suppl 6),23,86
dnd-ham:lng 500 nmol/L CNREA8 45,3143,85
ipr-rat TDLo:8 mg/kg (female 6-9D post):TER FAATDF 7,214,86
ipr-rat TDLo:8 mg/kg (female 6-9D post):REP FAATDF 7,214,86
ivn-man LDLo:5405 μg/kg/3H-C:BLD AIMDAP 143,165,83
ivn-hmn TDLo:12 mg/kg:GIT CNREA8 38,3712,78
orl-mus LD50:53,420 μg/kg NCISP* JAN86
ipr-mus LD50:15,470 μg/kg NCISP* JAN86
scu-mus LD50:110 mg/kg NCISP* JAN86
orl-dog LD50:50 mg/kg CTRRDO 66,1939,82
ivn-mus LD50:33.7 mg/kg CTRRDO 64,855,80

CONSENSUS REPORTS: EPA Genetic Toxicology Program.

SAFETY PROFILE: Poison by ingestion, intravenous, subcutaneous, and intraperitoneal routes. Human systemic effects by intravenous route: nausea or vomiting, thrombosis distant from injection site, and bone marrow changes. An experimental teratogen. Other experimental reproductive effects. Mutation data reported. When heated to decomposition it emits very toxic fumes of NO_x and SO_x.

ADM000 **HR: 3**
4′-(9-ACRIDINYLAMINO)-2′-METHOXYMETHANESULFONANILIDE
mf: $C_{21}H_{19}N_3O_3S$ mw: 393.49

SYNS: N-(4-(9-ACRIDINYLAMINO)-3-METHOXYPHENYL)METHANESULFONAMIDE □ m-AMSA

TOXICITY DATA WITH REFERENCE
dnd-mus:leu 10 μmol/L BICHAW 20,6553,81
dnd-mus:oth 40 μmol/L ANBCA2 125,91,82
cyt-mus:lym 1 mg/L ENMUDM 8(Suppl 6),23,86
msc-mus:lym 100 μg/L ENMUDM 8(Suppl 6),23,86
dnd-mam:lym 100 mmol/L CBINA8 44,53,83
mmo-sat 162 μmol/L JMCMAR 23,269,80
sce-hmn:lym 50 μg/L MUREAV 68,295,79
ivn-hmn TDLo:34 mg/kg:CVS,BLD CTRRDO 62,1421,78
ipr-mus LD10:110 mg/kg JMCMAR 23,269,80

SAFETY PROFILE: Poison by intraperitoneal route. Human mutation data reported. Human systemic effects by intravenous route: thrombosis distant from injection site, leukopenia, and thrombocytopenia. When heated to decomposition it emits very toxic fumes of NO_x and SO_x.

ADM500 **CAS:59988-01-1** **HR: 3**
N-(4-(ACRIDINYL-9-AMINO)-3-METHOXYPHENYL) ETHANESULFONAMIDE METHANESULFONATE
mf: $C_{22}H_{21}N_3O_3S•CH_5O_3S$ mw: 504.64

TOXICITY DATA WITH REFERENCE
mma-sat 93,200 nmol/L JMCMAR 22,251,79
ipr-mus LD10:10,500 μg/kg JMCMAR 22,251,79

SAFETY PROFILE: Poison by intraperitoneal route. Mutation data reported. See also SULFONATES. When heated to decomposition it emits very toxic fumes of SO_x and NO_x.

ADN250 **CAS:57164-89-3** **HR: 3**
4′-(9 ACRIDINYLAMINO)-3′-METHYLMETHANESULFONANILIDE
mf: $C_{21}H_{19}N_3O_2S$ mw: 377.49

TOXICITY DATA WITH REFERENCE
mmo-sat 132 μmol/L JMCMAR 23,269,80

SAFETY PROFILE: Mutation data reported. When heated to decomposition it emits very toxic fumes of NO_x and SO_x.

ADO250 **CAS:53221-85-5** **HR: 3**
N-(p-(ACRIDIN-9-YLAMINO)PHENYL)BUTANESULFONAMIDE, HYDROCHLORIDE
mf: $C_{23}H_{23}N_3O_2S•ClH$ mw: 442.01

TOXICITY DATA WITH REFERENCE
mma-sat 49,300 nmol/L JMCMAR 22,251,79
ipr-mus LD10:350 mg/kg JMCMAR 22,251,79

SAFETY PROFILE: Poison by intraperitoneal route. Mutation data reported. When heated to decomposition it emits very toxic fumes of HCl, SO_x, and NO_x.

ADO500 **CAS:53221-86-6** **HR: 2**
N-(p-(9-ACRIDINYLAMINO)PHENYL)-1-ETHANESULFONAMIDE
mf: $C_{21}H_{19}N_3O_2S$ mw: 377.49

TOXICITY DATA WITH REFERENCE
mmo-sat 24 μmol/L JMCMAR 23,269,80
ipr-mus LD10:330 mg/kg JMCMAR 21,430,78

SAFETY PROFILE: Moderately toxic by intraperitoneal route. Mutation data reported. See also SULFONATES. When heated to decomposition it emits very toxic fumes of SO_x and NO_x.

ADO750 **CAS:53221-83-3** **HR: 3**
N-(p-(ACRIDIN-9-YLAMINO)PHENYL)-ETHANESULFONAMIDE, HYDROCHLORIDE
mf: $C_{21}H_{19}N_3O_2S•ClH$ mw: 413.95

TOXICITY DATA WITH REFERENCE
ipr-mus LD10:330 mg/kg JMCMAR 22,251,79
mma-sat 81,600 nmol/L JMCMAR 22,251,79

SAFETY PROFILE: Poison by intraperitoneal route. See also SULFONATES. Mutation data reported. When heated to decomposition it emits very toxic fumes of Cl^-, SO_x, and NO_x.

ADP000 CAS:66147-69-1 HR: 2
N-(p-(ACRIDIN-9-YLAMINO)PHENYLHEXANESULFONAMIDE) HYDROCHLORIDE)
mf: $C_{25}H_{27}N_3O_2S{\cdot}ClH$ mw: 470.07

TOXICITY DATA WITH REFERENCE
mma-sat 9400 nmol/L JMCMAR 22,251,79
ipr-mus LD10:120 mg/kg JMCMAR 22,251,79

SAFETY PROFILE: Moderately toxic by intraperitoneal route. Mutation data reported. When heated to decomposition it emits very toxic fumes such as Cl^-, SO_x, and NO_x.

ADP500 CAS:75775-83-6 HR: 3
N-(p-(ACRIDIN-9-YLAMINO)PHENYL)METHANESULFONAMIDE HYDROCHLORIDE
mf: $C_{20}H_{17}N_3O_2S{\cdot}ClH$ mw: 399.92

TOXICITY DATA WITH REFERENCE
mma-sat 110 µmol/L JMCMAR 22,251,79
ipr-mus LD10:66 mg/kg JMCMAR 22,251,79

SAFETY PROFILE: Poison by intraperitoneal route. Mutation data reported. See also SULFONATES. When heated to decomposition it emits very toxic fumes of NO_x, SO_x, and HCl.

ADP750 CAS:66147-68-0 HR: 3
N-(p-(ACRIDIN-9-YLAMINO)PHENYL)PENTANESULFONAMIDE HYDROCHLORIDE
mf: $C_{24}H_{25}N_3O_2S{\cdot}ClH$ mw: 456.04

TOXICITY DATA WITH REFERENCE
mma-sat 15,600 nmol/L JMCMAR 22,251,79
ipr-mus LD10:70 mg/kg JMCMAR 22,251,79

SAFETY PROFILE: Poison by intraperitoneal route. Mutation data reported. See also SULFONATES. When heated to decomposition it emits very toxic fumes of NO_x, SO_x, and HCl.

ADQ000 CAS:53221-88-8 HR: 2
N-(p-(9-ACRIDINYLAMINO)PHENYL)-1-PROPANESULFONAMIDE
mf: $C_{22}H_{21}N_3O_2S$ mw: 391.52

TOXICITY DATA WITH REFERENCE
mmo-sat 24 µmol/L JMCMAR 23,269,80
ipr-mus LD10:350 mg/kg JMCMAR 21,430,78

SAFETY PROFILE: Poison by intraperitoneal route. See also SULFONATES. Mutation data reported. When heated to decomposition it emits very toxic fumes of SO_x and NO_x.

ADQ250 HR: 2
N-(p-(ACRIDIN-9-YLAMINO)PHENYL)PROPANESULFONAMIDE HYDROCHLORIDE
mf: $C_{22}H_{21}N_3O_2S{\cdot}ClH$ mw: 427.98

TOXICITY DATA WITH REFERENCE
mma-sat 50,100 nmol/L JMCMAR 22,251,79
ipr-mus LD10:350 mg/kg JMCMAR 22,251,79

SAFETY PROFILE: Poison by intraperitoneal route. See also SULFONATES. Mutation data reported. When heated to decomposition it emits very toxic fumes of HCl, SO_x, and NO_x.

ADR000 CAS:107-02-8 HR: 3
ACROLEIN
DOT: UN 1092
mf: C_3H_4O mw: 56.07

$H_2C{=}CHCOH$

PROP: Colorless or yellowish liquid; lachrymatory, disagreeable, choking odor. Mp: −87.7°, bp: 52.5°, flash p: <0°F, d: 0.841 @ 20°/4°, autoign temp: unstable (455°F), lel: 2.8%, uel: 31%, vap d: 1.94. Sol in water, alc, and ether.

SYNS: ACQUINITE □ ACRALDEHYDE □ ACROLEINA (ITALIAN) □ ACROLEINE (DUTCH, FRENCH) □ ACRYLALDEHYD (GERMAN) □ ACRYLALDEHYDE □ ACRYLIC ALDEHYDE □ AKROLEIN (CZECH) □ AKROLEINA (POLISH) □ ALDEHYDE ACRYLIQUE (FRENCH) □ ALDEIDE ACRILICA (ITALIAN) □ ALLYL ALDEHYDE □ AQUALINE □ BIOCIDE □ CROLEAN □ ETHYLENE ALDEHYDE □ MAGNACIDE H □ NSC 8819 □ PROPENAL (CZECH) □ 2-PROPENAL □ PROP-2-EN-1-AL □ 2-PROPEN-1-ONE □ PROPYLENE ALDEHYDE □ RCRA WASTE NUMBER P003 □ SLIMICIDE

TOXICITY DATA WITH REFERENCE
eye-hmn 500 ppb/12M IAPWAR 4,79,61
skn-rbt 5 mg open SEV UCDS** 6/18/71
skn-rbt 500 mg/24H SEV 28ZPAK -,41,72
eye-rbt 1 mg SEV UCDS** 6/18/71
eye-rbt 50 µg/24H SEV 28ZPAK -,41,72
sce-ham:ovr 10 µmol/L CGCGBR 26,108,80
mma-sat 50 µg/plate NTPTB* JAN 82
ivn-rbt TDLo:6 mg/kg (9D preg):REP ARZNAD 30,2080,80
ihl-man TCLo:1 ppm:IRR,IMM BMJOAE 2,913,56
ihl-hmn LCLo:5500 ppb 34ZIAG -,73,69
ihl-hmn LCLo:153 ppm/10M NTIS** PB214-270
ihl-chd TCLo:300 ppb/2H:PUL NPMDAD 8,2469,79
idr-man LDLo:250 mg/kg AEXPBL 43,351,1900
orl-rat LD50:46 mg/kg FMCHA2 -,C24,89
ihl-rat LC50:300 mg/m³/30M APTOA6 6,299,50
ipr-rat LD50:4 mg/kg TXAPA9 71,84,83
scu-rat LD50:50 mg/kg APTOA6 6,299,50
orl-mus LD50:40 mg/kg BIJOAK 34,1196,40
ihl-mus LC50:66 ppm/6H IAANBS 26,281,70
ipr-mus LD50:9008 µg/kg NCISP* JAN86
scu-mus LD50:30 mg/kg APTOA6 6,299,50
ihl-cat LCLo:1570 mg/m³/8H APTOA6 6,299,50

CONSENSUS REPORTS: IARC Cancer Review: Group 3 IMEMDT 7,78,87; Animal Inadequate Evidence IMEMDT 36,133,85; IMEMDT 19,479,79; Human Inadequate Evidence IMEMDT 36,133,85. Community

Right-To-Know List. EPA Extremely Hazardous Substances List. Reported in EPA TSCA Inventory.

OSHA PEL: TWA 0.1 ppm; STEL 0.3 ppm
ACGIH TLV: TWA 0.1 ppm; STEL 0.3 ppm
DFG MAK: 0.1 ppm (0.25 mg/m^3)
DOT CLASSIFICATION: 6.1; *Label:* Poison, Flammable Liquid

SAFETY PROFILE: Human poison by inhalation and intradermal routes. Poison experimentally by most routes. Human systemic irritant and pulmonary system effects by inhalation include: lacrimation, delayed hypersensitivity with multiple organ involvement, and respiratory system damage. Severe eye and skin irritant. Experimental reproductive effects. Human mutation data reported. Questionable carcinogen. Dangerous fire hazard when exposed to heat, flame, or oxidizers. An explosion hazard. Incompatible with amines, SO_2, metal salts, oxidants, (light + heat). Violent polymerization reaction on contact with strong acid, strong base, weak acid conditions (e.g., nitrous fumes, sulfur dioxide, carbon dioxide), thiourea, or dimethylamine. When heated to decomposition it emits highly toxic fumes; can react vigorously with oxidizing materials. To fight fire, use CO_2, dry chemical, or alcohol foam.

For occupational chemical analysis use OSHA: #52 or NIOSH: Acrolein, 2501.

ADR250 CAS:869-29-4 HR: 3
ACROLEIN DIACETATE
mf: $C_7H_{10}O_4$ mw: 158.17

PROP: Liquid. Mp: −36.6°, bp: 107° @ 50 mm, flash p: 180°F (OC), d: 1.0749 @ 20°/20°, vap d: 5.46.

SYNS: ALLYLIDENE DIACETATE □ DIACETOXYPROPENE □ 1,1-DIACETOXYPROPENE-2 □ 3,3-DIACETOXYPROPENE □ SD-345 □ SHELL 345 □ SHELL SD 345

TOXICITY DATA WITH REFERENCE
skn-rbt 10 mg/24H JIHTAB 30,63,48
skn-rbt 500 mg open SEV UCDS** 12/27/71
eye-rbt 10 mg SEV UCDS** 12/27/71
orl-rat LD50:35 mg/kg SCCUR*-,1,61
ihl-rat LCLo:8 ppm/4H UCDS** 12/27/71
orl-mus LD50:37,500 µg/kg SCCUR*-,1,61
ihl-mus LCLo:853 ppm/15M SCCUR*-,1,61
skn-rbt LD50:320 µL/kg UCDS** 12/27/71
skn-gpg LDLo:500 mg/kg SCCUR*-,1,61

SAFETY PROFILE: Poison by ingestion and inhalation, and skin contact. A severe skin and eye irritant. Flammable when exposed to heat or flame; can react with oxidizing materials. When heated to decomposition it emits acrid smoke and fumes. To fight fire, water may be used to blanket the fire; also, foam, CO_2, dry chemical.

ADR500 CAS:100-73-2 HR: 2
ACROLEIN DIMER
DOT: UN 2607
mf: $C_6H_8O_2$ mw: 112.14

PROP: Liquid, sol in water. D: 1.0775 (20°), bp: 151.3°, fp: −100°, flash p: 118°F (OC).

SYNS: ACROLEIN DIMER, stabilized (DOT) □ 3,4-DIHYDRO-2H-PYRAN-2-CARBOXALDEHYDE □ 2,3-DIHYDRO-1,4-PYRAN-2-KARBOXALDEHYD □ 2-FORMYL-3,4-DIHYDRO-2H-PYRAN □ 5-HEXENAL, 2,6-EPOXY- □ PYRAN ALDEHYDE

TOXICITY DATA WITH REFERENCE
skn-rbt 500 mg open MLD UCDS** 7/27/65
eye-rbt 750 ug open SEV AMIHBC 10,61,54
skn-rbt 500 mg open MLD UCDS** 7/27/65
eye-rbt 750 µg SEV AMIHBC 10,61,54
orl-rat LD50:4920 mg/kg AMIHBC 10,61,54

DOT CLASSIFICATION: 3; *Label:* Flammable Liquid

SAFETY PROFILE: Mildly toxic by ingestion. A skin and severe eye irritant. A flammable liquid when exposed to heat, flame, or powerful oxidizing agents. To fight fire, use alcohol foam and multipurpose dry chemical. When heated to decomposition it emits acrid smoke and fumes.

ADR750 CAS:7008-42-6 HR: 2
ACRONYCINE
mf: $C_{20}H_{19}NO_3$ mw: 321.40

PROP: Yellow needles from ethanol. Mp: 175–176°.

SYNS: ACROMYCINE □ ACRONINE □ COMPOUND 42339 □ 3,12-DIHYDRO-6-METHOXY-3,3,12-TRIMETHYL-7H-PYRANO(2,3-C)ACRIDIN-7-ONE □ NCI-C01536 □ NSC 403169

TOXICITY DATA WITH REFERENCE
dni-mus:leu 1 µmol/L CNREA8 33,2310,73
orl-mus LD50:522 mg/kg NCISP* JAN 86
ipr-mus LD50:613 mg/kg NCISP* JAN 86
cyt-mus:fbr 10 mg/L/24H ARZNAD 27,1549,77
ipr-rat TDLo:1170 mg/kg/1Y-I:CAR NCITR* NCI-CG-TR-49,78
ipr-rat TD:585 mg/kg/1Y-I:NEO NCITR* NCI-CG-TR-49,78
ipr-rat TD:1800 mg/kg/1Y-I:CAR NCITR* NCI-CG-TR-49,78
ipr-mus TD:530 mg/kg/8W-I:ETA CNREA8 33,3069,73

CONSENSUS REPORTS: NCI Carcinogenesis Bioassay (ipr); Inadequate Studies: mouse NCITR* NCI-CG-TR-49,78; Clear Evidence: rat NCITR* NCI-CG-TR-49,78.

SAFETY PROFILE: Moderately toxic by ingestion and intraperitoneal routes. Questionable carcinogen with experimental carcinogenic, neoplastigenic, and tumorigenic data. Mutation data reported. When heated to decomposition it emits toxic fumes of NO_x.

ADS150 HR: 2
ACROSTICHUM AUREUM Linn., extract

PROP: Indian plant belonging to the family Pteridiaceae (IJEBA6 15,208,77).

TOXICITY DATA WITH REFERENCE
orl-rat TDLo:150 mg/kg (female 12-14D post):REP IJEBA6 15,208,77
ipr-mus LD50:750 mg/kg IJEBA6 15,208,77

SAFETY PROFILE: Moderately toxic by intraperitoneal route. Experimental reproductive effects. When heated to decomposition it emits acrid smoke and irritating fumes.

A

ADS250 CAS:79-06-1 HR: 3
ACRYLAMIDE
DOT: UN 2074
mf: C_3H_5NO mw: 71.09

PROP: White, crystalline solid. Leaflets from (C_6H_6). Mp: 84.5 ± 0.3°, bp: 125° @ 25 mm, d: 1.122 @ 30°, vap press: 1.6 mm @ 84.5°, vap d: 2.45. Very sol in water, alc, and ether.

SYNS: ACRYLIC AMIDE ☐ AKRYLAMID (CZECH) ☐ AMID KYSELINY AKRYLOVE ☐ ETHYLENECARBOXAMIDE ☐ PROPENAMIDE ☐ 2-PROPENAMIDE ☐ RCRA WASTE NUMBER U007 ☐ VINYL AMIDE

TOXICITY DATA WITH REFERENCE
skn-rbt 50 mg/3D MLD TXAPA9 6,172,64
skn-rbt 500 mg/24H MLD 85JCAE-,337,86
eye-rbt 10 mg/30S RNS MLD TXAPA9 6,172,64
eye-rbt 100 mg/24H MOD 28ZPAK-,54,72
sce-rat-orl 600 mg/kg/10D-C ENMUDM 7(Suppl 3),79,85
dlt-mus-ipr 125 mg/kg MUREAV 173,35,86
orl-rat TDLo:200 mg/kg (7-16D preg):REP TOLED5 7,233,81
orl-rat TDLo:1456 mg/kg/2Y-C:CAR TXAPA9 85,154,86
ipr-mus TDLo:24 mg/kg/8W-I:NEO CNREA8 44,107,84
orl-mus TDLo:300 mg/kg/2W-I:CAR CALEDQ 24,209,84
orl-rat LD:1456 mg/kg/2Y-C:CAR,REP TXAPA9 85,154,86
orl-rat LD50:124 mg/kg AMPMAR 36,58,75
skn-rat LD50:400 mg/kg GISAAA 44(10),73,79
ipr-rat LD50:90 mg/kg AMPMAR 36,58,75
orl-mus LD50:107 mg/kg ARTODN 47,179,81
ipr-mus LD50:170 mg/kg TXAPA9 33,142,75
orl-rbt LD50:150 mg/kg TXAPA9 6,172,64
skn-rbt LDLo:1000 mg/kg TXAPA9 6,172,64
skn-rbt LD50:1680 µL/kg JACTDZ 1,115,90
orl-gpg LDLo:252 mg/kg TXAPA9 6,172,64
scu-gpg LD50:170 mg/kg MELAAD 47,192,56

CONSENSUS REPORTS: NTP 7th Annual Report On Carcinogens. IARC Cancer Review: Group 2B IMEMDT 7,56,87; Animal Sufficient Evidence IMEMDT 39,41,86. EPA Extremely Hazardous Substances List. Community Right-To-Know List. Reported in EPA TSCA Inventory.

OSHA PEL: TWA 0.03 mg/m³ (skin)
ACGIH TLV: Suspected Human Carcinogen, TWA 0.03 mg/m³ (skin)
DFG MAK: Animal Carcinogen, Suspected Human Carcinogen
NIOSH REL: TWA 0.3 mg/m³
DOT CLASSIFICATION: 6.1; *Label:* KEEP AWAY FROM FOOD

SAFETY PROFILE: Confirmed carcinogen with experimental carcinogenic and neoplastigenic data. Poison by ingestion, skin contact, and intraperitoneal routes. Experimental reproductive effects. Mutation data reported. A skin and eye irritant. Intoxication from it has caused a peripheral neuropathy, erythema, and peeling of palms. In industry, intoxication is mainly via dermal route, next via inhalation, and last via ingestion. Time of onset varied from 1–24 months to 8 years. Symptoms were, via dermal route, a numbness, tingling, and touch tenderness. In a couple of weeks, coldness of extremities; later, excessive sweating, bluish-red and peeling of palms, marked fatigue and limb weakness. It is dangerous because it can be absorbed through the unbroken skin. From animal experiments it seems to be a central nervous system toxin. Adult rats fed an average of 30 mg/kg for 14 days were all partially paralyzed and had reduced their food consumption by 50 percent. Polymerizes violently at its melting point. When heated to decomposition it emits acrid fumes and NO_x.

For occupational chemical analysis use OSHA: #21.

ADS750 CAS:79-10-7 HR: 3
ACRYLIC ACID
DOT: UN 2218
mf: $C_3H_4O_2$ mw: 72.07

$H_2C{=}CHCO{\cdot}OH$

PROP: Liquid with acrid odor. Misc in water, benzene, alc, chloroform, ether, and acetone. Mp: 13°, bp: 141° (polymerizes), d: 1.062, vap press: 10 mm @ 39.9°, flash p: 130°F (OC), vap d: 2.45.

SYNS: ACROLEIC ACID ☐ ACRYLIC ACID, inhibited (DOT) ☐ ACRYLIC ACID, GLACIAL ☐ ETHYLENECARBOXYLIC ACID ☐ GLACIAL ACRYLIC ACID ☐ KYSELINA AKRYLOVA ☐ PROPENE ACID ☐ PROPENOIC ACID ☐ 2-PROPENOIC ACID (9CI) ☐ RCRA WASTE NUMBER U008 ☐ VINYLFORMIC ACID

TOXICITY DATA WITH REFERENCE
skn-rbt 500 mg open SEV UCDS** 2/2/65
eye-rbt 1 mg SEV UCDS** 2/2/65
eye-rbt 250 µg/24H SEV 85JCAE-,309,86
orl-rat TDLo:100 g/kg (13W pre-3W post):REP DCTODJ 6,1,83
ipr-rat TDLo:7329 µg/kg (5-15D preg):TER JDREAF 51,1632,72
skn-mus TDLo:37,440 mg/kg/78W-I:CAR EPASR* 8EHQ-0586-0592
scu-mus TDLo:2912 mg/kg/52W-I:ETA CBINA8 61,189,87
orl-rat LD50:33,500 µg/kg 85GMAT -,16,82
ihl-rat LCLo:4000 ppm/4H TXAPA9 28,313,74
ipr-rat LD50:22 mg/kg JDREAF 51,1632,72
orl-mus LD50:2400 mg/kg BIJOAK 34,1196,40
ihl-mus LCLo:5300 mg/m³/2H 85GMAT -,16,82
scu-mus LD50:1590 mg/kg JPPMAB 21,85,69
skn-rbt LD50:280 mg/kg TXAPA9 28,313,74

CONSENSUS REPORTS: IARC Cancer Review: Group 3 IMEMDT 7,56,87; Human Inadequate Evidence IMEMDT 19,47,79. Community Right-To-Know List. Reported in EPA TSCA Inventory.

OSHA PEL: TWA 10 ppm (skin)
ACGIH TLV: 2 ppm (skin)
DOT CLASSIFICATION: 8; *Label:* Corrosive

SAFETY PROFILE: Poison by ingestion, skin contact, and intraperitoneal routes. An experimental teratogen. Other experimental reproductive effects. A severe skin and eye irritant. Questionable carcinogen with experimental carcinogenic and tumorigenic data. Corrosive. Flammable liquid. May undergo exothermic polymerization at room temperature. May become explosive if confined. A fire hazard when exposed to heat or flame.

For occupational chemical analysis use OSHA: #28.

ADT000 CAS:2206-89-5 HR: 3
ACRYLIC ACID-β-CHLOROETHYL ESTER
mf: $C_5H_7ClO_2$ mw: 134.57

SYNS: 2-CHLOROETHANOL ACRYLATE □ CHLOROETHYL ACRYLATE □ β-CHLOROETHYL ACRYLATE □ 2-CHLOROETHYL ACRYLATE □ 2-PROPENOIC ACID-2-CHLOROETHYL ESTER

TOXICITY DATA WITH REFERENCE
skn-rbt 10 mg/24H open SEV AMIHBC 4,119,51
eye-rbt 50 μg open SEV AMIHBC 4,119,51
mmo-sat 333 μg/plate ENMUDM 9(Suppl 9),1,87
orl-rat LD50:180 mg/kg AMIHBC 4,119,51
ihl-rat LCLo:250 ppm/4H AMIHBC 4,119,51

SAFETY PROFILE: Poison by inhalation and ingestion. A severe skin and eye irritant. Mutation data reported. See also ESTERS. When heated to decomposition it emits toxic fumes of Cl^-.

ADT050 CAS:17831-71-9 HR: 2
ACRYLIC ACID, DIESTER with TETRAETHYLENE GLYCOL
mf: $C_{14}H_{22}O_7$ mw: 302.36

SYNS: ACRYLIC ACID, OXYBIS(ETHYLENEOXYETHYLENE) ESTER □ 2-PROPENOIC ACID, OXYBIS(2,1-ETHANEDIYLOXY-2,1-ETHANEDIYL)ESTER □ TETRAETHYLENE GLYCOL DIACRYLATE

TOXICITY DATA WITH REFERENCE
skn-rbt 500 mg/24H MOD JTEHD6 19,149,86
eye-rbt 100 mg SEV JTEHD6 19,149,86
mnt-mus:lym 2 mg/L MUTAEX 4,381,89
cyt-mus:lym 2 mg/L MUTAEX 4,381,89
skn-mus TDLo:16 g/kg/80W-I:ETA JTEHD6 19,149,86

CONSENSUS REPORTS: Reported in EPA TSCA Inventory.

SAFETY PROFILE: Moderate skin and severe eye irritant. Questionable carcinogen with experimental tumorigenic data. Mutation data reported. When heated to decomposition it emits acrid smoke and irritating fumes.

ADT111 CAS:106-71-8 HR: 3
ACRYLIC ACID ESTER with HYDRACRYLONITRILE
mf: $C_6H_7NO_2$ mw: 125.14

PROP: Liquid, sol in water, d: 1.069, bp: polymerizes, fp: −16.9°, flash p: 255°F (COC), vap d: 4.3.

SYNS: ACRYLIC ACID-2-CYANOETHYL ESTER □ CYANOETHYL ACRYLATE □ 2-CYANOETHYL ACRYLATE □ 2-CYANOETHYL PROPENOATE □ HYDRACRYLONITRILE ACRYLATE □ 2-PROPENOIC ACID-2-CYANOETHYL ESTER

TOXICITY DATA WITH REFERENCE
skn-rbt 500 mg open MOD UCDS** 9/27/60
skn-rbt 10 mg/24H open SEV AIHAAP 23,95,62
eye-rbt 5 mg MLD UCDS** 9/27/60
orl-rat LD50:180 mg/kg UCDS** 9/27/60
skn-rbt LD50:220 μL/kg AIHAAP 23,95,62

CONSENSUS REPORTS: Reported in EPA TSCA Inventory. Cyanide and its compounds are on the Community Right-To-Know List.

SAFETY PROFILE: Poison by ingestion and skin contact. A skin and eye irritant. See also ESTERS and NITRILES. A fire hazard when exposed to heat or flame. When heated to decomposition it emits toxic fumes of NO_x and CN^-.

For occupational chemical analysis use OSHA: #ID-55.

ADT250 CAS:4074-88-8 HR: 3
ACRYLIC ACID, 2-ETHOXYETHANOL DIESTER
mf: $C_{10}H_{14}O_5$ mw: 214.24

SYNS: ACRYLIC ACID, OXYDIETHYLENE ESTER (8CI) □ DIACRYALTE DIETHYLENE GLYCOL □ DIETHYLENE GLYCOL DIACRYLATE □ OXYDIETHYLENE ACRYLATE □ OXYDIETHYLENE DIACRYLATE □ 2-PROPENOIC ACID, OXYDI-2,1-ETHANEDIYL ESTER (9CI) □ TGA 2

TOXICITY DATA WITH REFERENCE
skn-rbt 500 mg SEV JTEHD6 19,149,86
eye-rbt 100 mg SEV JTEHD6 19,149,86
orl-rat LD50:250 mg/kg GISAAA 55(6),86,90
orl-mus LD50:550 mg/kg GISAAA 55(6),86,90
skn-rbt LD50:180 μL/kg TXAPA9 28,313,74

CONSENSUS REPORTS: Reported in EPA TSCA Inventory.

SAFETY PROFILE: Poison by ingestion and skin contact. A severe skin and eye irritant. See also ESTERS. When heated to decomposition it emits acrid smoke and fumes.

ADT500 CAS:106-74-1 HR: 2
ACRYLIC ACID, 2-ETHOXYETHYL ESTER
mf: $C_7H_{12}O_3$ mw: 144.19

PROP: Liquid. D: 0.982, bp: 22° @ 78 mm.

SYNS: ACRYLIC ACID-2-ETHOXYETHANOL ESTER □ CELLOSOLVE ACRYLATE □ ETHOXYETHYL ACRYLATE □ 2-ETHOXYETHYL ACRYLATE □ 2-ETHOXYETHYL-2-PROPENOATE □ ETHYLENE GLYCOL MONOETHYL ETHER ACRYLATE □ ETHYLENE GLYCOL MONOETHYL ETHER PROPENOATE □ 2-PROPENOIC ACID-2-ETHOXYETHYL ESTER

TOXICITY DATA WITH REFERENCE
skn-rbt 10 mg/24H open MLD AMIHBC 10,61,54
skn-rbt 500 mg open MLD UCDS** 6/6/69
eye-rbt 20 mg open SEV AMIHBC 10,61,54
orl-rat LD50:1070 mg/kg UCDS** 9/15/64
ihl-rat LCLo:500 ppm/4H AMIHBC 10,61,54
skn-rbt LD50:1010 mg/kg AMIHBC 10,61,54

CONSENSUS REPORTS: Reported in EPA TSCA Inventory.

SAFETY PROFILE: Moderately toxic by various routes. A skin and severe eye irritant. See also ESTERS. When heated to decomposition it emits acrid smoke and fumes.

ADU250 CAS:103-11-7 HR: 3
ACRYLIC ACID-2-ETHYLHEXYL ESTER
mf: $C_{11}H_{20}O_2$ mw: 184.31

PROP: A liquid. Fp: −90°, bp: 130° @ 50 mm, flash p:

180°F (OC), d: 0.8869 @ 20°/20°, vap press: 1 mm @ 50°, vap d: 6.35.

SYNS: 2-ETHYLHEXYL ACRYLATE □ 2-ETHYLHEXYL-2-PROPENOATE □ OCTYL ACRYLATE □ 2-PROPENOIC ACID-2-ETHYLHEXYL ESTER

TOXICITY DATA WITH REFERENCE
skn-rbt 20 mg/24H MOD 85JCAE -,372,86
skn-rbt 500 mg open MLD UCDS** 11/3/71
skn-rbt 10 mg/24H open SEV AMIHBC 4,119,51
eye-rbt 5 mg SEV AJOPAA 29,1363,46
eye-rbt 500 mg/24H MLD 85JCAE -,372,86
skn-mus TDLo: 187 g/kg/78W-I:CAR JTEHD6 16,55,85
skn-mus TD: 240 g/kg/2Y-C:NEO EPASR* 8EHQ-1079-0262
orl-rat LD50: 6500 µL/kg UCDS** 11/3/71
ipr-rat LD50: 1670 mg/kg AMPMAR 36,58,75
orl-mus LD50: 4400 mg/kg GTPZAB 26(9),52,82
ihl-mus LCLo: 600 mg/m³ GTPZAB 26(9),52,82
ipr-mus LD50: 1326 mg/kg JDREAF 51,526,72
skn-rbt LD50: 8480 mg/kg AMIHBC 4,119,51

CONSENSUS REPORTS: Reported in EPA TSCA Inventory.

SAFETY PROFILE: Moderately toxic by inhalation and various other routes. A severe skin and eye irritant. Questionable carcinogen with experimental carcinogenic and neoplastigenic data. A flammable liquid. A fire hazard when exposed to heat or flame. To fight fire use alcohol foam, CO_2, dry chemical. When heated to decomposition it emits acrid smoke and irritating fumes. See also ESTERS.

ADU750 CAS:122-93-0 HR: 3
ACRYLIC ACID 2-(5'-ETHYL-2-PYRIDYL)ETHYL ESTER
mf: $C_{12}H_{15}NO_2$ mw: 205.28

PROP: Liquid, very sltly water-sol. D: 1.0458 @ 20°, bp: 181° @ 50 mm, fp: −75°.

SYNS: 2-(5-ETHYL-2-PYRIDYL)ETHYL ACRYLATE □ 2-(5-ETHYL-2-PYRIDYL)ETHYL PROPENOATE

TOXICITY DATA WITH REFERENCE
skn-rbt 10 mg/24H open SEV AIHAAP 23,95,62
orl-rat LD50: 4920 mg/kg AIHAAP 23,95,62
skn-rbt LD50: 2230 mg/kg AIHAAP 23,95,62

SAFETY PROFILE: Moderately toxic by skin contact. Mildly toxic by ingestion. A severe skin irritant. See also ESTERS. Flammable. Store away from heat, sparks, or powerful oxidizers. To fight fire, use foam, CO_2, dry chemicals. When heated to decomposition it emits toxic fumes of NO_x.

ADV000 CAS:2499-95-8 HR: 1
ACRYLIC ACID HEXYL ESTER
mf: $C_9H_{16}O_2$ mw: 156.25

SYNS: AGEFLEX n-HA □ HEXYL ACRYLATE □ N-HEXYL ACRYLATE □ HEXYL-2-PROPENOATE □ 2-PROPENOIC ACID, HEXEL ESTER

TOXICITY DATA WITH REFERENCE
orl-rat LD50: 26 g/kg AIHAAP 30,470,69
skn-rbt LD50: 5660 mg/kg AIHAAP 30,470,69

CONSENSUS REPORTS: Reported in EPA TSCA Inventory.

SAFETY PROFILE: Mildly toxic by skin contact. See also ESTERS. When heated to decomposition it emits acrid smoke and fumes.

ADV250 CAS:818-61-1 HR: 2
ACRYLIC ACID-2-HYDROXYETHYL ESTER
mf: $C_5H_8O_3$ mw: 116.13

PROP: Liquid. D: 1.011 @ 23.4°, bp: 12° @ 90–92 mm.

SYNS: 2-(ACRYLOYLOXY)ETHANOL □ BISOMER 2HEA □ ETHYLENE GLYCOL ACRYLATE □ ETHYLENE GLYCOL MONOACRYLATE □ HYDROXYETHYL ACRYLATE □ β-HYDROXYETHYL ACRYLATE □ 2-HYDROXYETHYL ACRYLATE □ 2-PROPENOIC ACID-2-HYDROXYETHYL ESTER (9CI)

TOXICITY DATA WITH REFERENCE
skn-rbt 500 mg open MOD UCDS** 3/23/73
skn-rbt 10 mg/24H open MLD AMIHBC 4,119,51
eye-rbt 1 mg SEV UCDS** 3/23/73
eye-rbt 20 mg/24H MOD 85JCAE-,666,86
mnt-mus: lym 18 mg/L MUTAEX 4,381,89
cyt-mus: lym 15 mg/L MUTAEX 4,381,89
orl-rat LD50: 650 µL/kg UCDS** 3/23/73
ihl-rat LCLo: 500 ppm/4H AMIHBC 4,119,51
skn-rbt LD50: 1010 mg/kg AMIHBC 4,119,51

CONSENSUS REPORTS: Reported in EPA TSCA Inventory.

SAFETY PROFILE: Moderately toxic by ingestion, inhalation, and skin contact. A moderate skin and severe eye irritant. Mutation data reported. When heated to decomposition it emits acrid smoke and fumes. See also ESTERS.

ADW100 CAS:25916-47-6 HR: 2
ACRYLIC ACID, POLYMER, ZINC SALT
mf: $(C_3H_4O_2)_x \cdot xZn$

SYNS: 2-PROPENOIC ACID, HOMOPOLYMER, ZINC SALT □ ZINC POLYACRYLATE □ ZINC POLYCARBOXYLATE

TOXICITY DATA WITH REFERENCE
imp-mus TDLo: 1600 mg/kg/1Y-C:ETA JBMRBG 9,69,75

CONSENSUS REPORTS: Reported in EPA TSCA Inventory. Zinc and its compounds are on the Community Right-To-Know List.

SAFETY PROFILE: Questionable carcinogen with experimental tumorigenic data. See also ZINC COMPOUNDS. When heated to decomposition it emits toxic fumes of ZnO, acrid fumes, and CO.

ADW200 CAS:9003-01-4 HR: 3
ACRYLIC ACID, POLYMERS
mf: $(C_3H_4O_2)_4$ mw: 168.06

SYNS: ACRYLIC ACID RESIN □ ACRYLIC POLYMER □ ACRYLIC RESIN □ ACRYSOL A 1 □ ACRYSOL A 3 □ ACRYSOL A 5 □ ACRYSOL AC 5 □ ACRYSOL ASE-75 □ ACRYSOL WS-24 □ ALCOGUM □ ANTIPREX A □ ANTIPREX 461 □ AROLON □ ARON □ ARON A 10H □

ATACTIC POLY(ACRYLIC ACID) □ CARBOMER 940 □ CARBOMER 934P □ CARBOPOL 934 □ CARBOPOL 940 □ CARBOPOL 941 □ CARBOPOL 960 □ CARBOPOL 961 □ CARBOPOL 934P □ CARBOSET □ CARBOSET 515 □ CARBOSET RESIN NO. 515 □ CARPOLENE □ DISPEX C40 □ G-CURE □ GOOD-RITE K 37 □ GOOD-RITE K-700 □ GOOD-RITE K 702 □ GOOD-RITE K727 □ GOOD-RITE WS 801 □ HALOFLEX 202 □ HALOFLEX 208 □ JUNLON 110 □ JURIMER AC 10H □ JURIMER AC 10P □ NALFLOC 636 □ NEOCRYL A-1038 □ OLD 01 □ PAA-25 □ PA 11M □ P 11H □ POLYACRYLATE □ POLY(ACRYLIC ACID) □ POLYTEX 973 □ PRIMAL ASE 60 □ 2-PROPENOIC ACID HOMOPOLYMER (9CI) □ R968 □ RACRYL □ 76 RES □ REVACRYL A 191 □ ROHAGIT SD 15 □ SYNTHEMUL 90-588 □ TECPOL □ TEXCRYL □ VERSICOL E 7 □ VERSICOL E9 □ VERSICOL E15 □ VERSICOL S 25 □ VISCALEX HV 30 □ VISCON 103 □ WS 24 □ WS 801 □ XPA □ ZINPOL

TOXICITY DATA WITH REFERENCE
orl-rat LD50:2500 mg/kg ACIEAY 14,94,75
orl-mus LD50:4600 mg/kg FRPPAO 25,721,70
ipr-mus LD50:39 mg/kg JMCMAR 21,652,78
ivn-mus LD50:70 mg/kg ZMEIAV (9),14,79
orl-gpg LD50:2500 mg/kg FRPPAO 25,721,70

CONSENSUS REPORTS: IARC Cancer Review: Group 3 IMEMDT 7,56,87; Human No Adequate Data IMEMDT 19,47,79; Animal No Adequate Data IMEMDT 19,47,79.

SAFETY PROFILE: Poison by intravenous and intraperitoneal routes. Moderately toxic by ingestion. Questionable carcinogen with no adequate data. When heated to decomposition it emits acrid smoke and fumes.

ADW300 CAS:9007-16-3 HR: 2
ACRYLIC ACID, POLYMER with SUCROSEPOLYALLYL ETHER

SYNS: CARBOMER 934 □ CARBOPOL 934 □ SUCROSE, POLYALLYL ETHER, POLYMER with ACRYLIC ACID

TOXICITY DATA WITH REFERENCE
orl-rat LD50:4100 mg/kg JACTDZ 1(2),109,82
orl-mus LD50:4550 mg/kg JACTDZ 1(2),109,82
orl-gpg LD50:2500 mg/kg GRCSB* GC-36,54,60

CONSENSUS REPORTS: Allyl and its compounds are on the Community Right-To-Know List.

SAFETY PROFILE: Moderately toxic by ingestion. See also ETHERS and POLYMERS. When heated to decomposition it emits acrid smoke and fumes.

ADW750 CAS:71073-91-1 HR: 3
ACRYLIC ACID, TELOMER with TRICHLOROACETIC ACID
mf: $C_2HCl_3O_2{\cdot}3/2C_3H_4O_2$ mw: 265.49

TOXICITY DATA WITH REFERENCE
orl-mus LD50:2750 mg/kg EJMCA5 14,119,79
ipr-mus LD50:300 mg/kg EJMCA5 14,119,79

SAFETY PROFILE: Poison by intraperitoneal route. Moderately toxic by ingestion. When heated to decomposition it emits toxic fumes of Cl^-.

ADX000 CAS:3076-04-8 HR: 2
ACRYLIC ACID TRIDECYL ESTER
mf: $C_{16}H_{30}O_2$ mw: 254.46

SYNS: 2-PROPENOIC ACID TRIDECYL ESTER □ TRIDECYL ACRYLATE

TOXICITY DATA WITH REFERENCE
skn-rbt 10 mg/24H open SEV AIHAAP 23,95,62
orl-rat LD50:44,700 mg/kg AIHAAP 23,95,62
skn-rbt LD50:6300 mg/kg AIHAAP 23,95,62

CONSENSUS REPORTS: Reported in EPA TSCA Inventory.

SAFETY PROFILE: Mildly toxic by skin contact and ingestion. A severe skin irritant. See also ESTERS. When heated to decomposition it emits acrid smoke and fumes.

ADX250 HR: 3
ACRYLOAMIDE
mf: $C_3H_7ClN_2$ mw: 106.57

PROP: An antibiotic produced by the strain *Streptomyces sp.* No. D274-2.

TOXICITY DATA WITH REFERENCE
ipr-mus LD50:38 mg/kg 85ERAY 2,1158,78
scu-mus LD50:38 mg/kg 85ERAY 2,1158,78
ivn-mus LD50:44 mg/kg 85ERAY 2,1158,78

SAFETY PROFILE: Poison by intraperitoneal, subcutaneous, and intravenous routes. When heated to decomposition it emits very toxic fumes of Cl^- and NO_x.

ADX500 CAS:107-13-1 HR: 3
ACRYLONITRILE
DOT: UN 1093
mf: C_3H_3N mw: 53.07

PROP: Colorless, mobile liquid; mild odor. Mp: −82°, bp: 77.3°, fp: −83°, flash p: 30°F (TCC), lel: 3.1%, uel: 17%, d: 0.806 @ 20°/4°, autoign temp: 898°F, vap press: 100 mm @ 22.8°, vap d: 1.83, flash p: (of 5% aq soln): <50°F. Sol in water.

SYNS: ACRITET □ ACRYLNITRIL (GERMAN, DUTCH) □ ACRYL-ON □ ACRYLONITRILE, inhibited (DOT) □ ACRYLONITRILE MONOMER □ AKRYLONITRYL (POLISH) □ CARBACRYL □ CIANURO di VINILE (ITALIAN) □ CYANOETHYLENE □ CYANURE de VINYLE (FRENCH) □ ENT 54 □ FUMIGRAIN □ MILLER'S FUMIGRAIN □ NITRILE ACRILICO (ITALIAN) □ NITRILE ACRYLIQUE (FRENCH) □ PROPENENITRILE □ 2-PROPENENITRILE □ RCRA WASTE NUMBER U009 □ TL 314 □ VCN □ VENTOX □ VINYL CYANIDE □ VINYLKYANID

TOXICITY DATA WITH REFERENCE
bfa-rat/sat 30 mg/kg TXCYAC 16,67,80
dns-rat:lvr 1 mmol/L PMRSDJ 5,371,85
slt-dmg-orl 1520 μmol/L PMRSDJ 5,325,85
skn-hmn 500 mg nse INMEAF 17,199,48
skn-rbt 10 mg/24H open JIHTAB 30,63,48
skn-rbt 500 mg MLD SCCUR* -,1,61
eye-rbt 20 mg SEV JIHTAB 30,63,48
ipr-ham TDLo:641 mg/kg (female 8D post):TER TJADAB 23,325,81
orl-rat TDLo:650 mg/kg (female 6-15D post):REP DOWCC* 03NOV76

orl-rat TDLo:18,200 mg/kg/52W-C:CAR FCTOD7 24,129,86
ihl-rat TCLo:5 ppm/52W-I:ETA MELAAD 68,401,77
orl-rat LD:3640 mg/kg/52W-C:NEO DOWCC* MAR77
ihl-hmn TCLo:16 ppm/20M:EYE,PUL INMEAF 17,199,48
ihl-man LCLo:1 g/m³/1H:CNS,GIT ZAARAM 16,1,66
skn-chd LDLo:2015 mg/kg:CNS,RSP,GIT DMWOAX 75,1087,50
orl-rat LD50:78 mg/kg JOHYAY 3,106,59
ihl-rat LC50:425 ppm/4H TXAPA9 29,81,74
skn-rat LD50:148 mg/kg GISAAA 41(10),103,76
ihl-mus LCLo:315 ppm/4H NTIS** PB280-478
ipr-mus LD50:46 mg/kg TXAPA9 59,589,81
orl-mus LD50:27 mg/kg JHEMA2 3,106,59
scu-mus LD50:35 mg/kg JHEMA2 3,106,59
ihl-dog LCLo:110 ppm/4H JIHTAB 24,27,42

CONSENSUS REPORTS: NTP 7th Annual Report On Carcinogens. IARC Cancer Review: Group 2A IMEMDT 7,79,87; Human Limited Evidence IMEMDT 19,73,79; Animal Limited Evidence IMEMDT 19,73,79. Community Right-To-Know List. EPA Extremely Hazardous Substances List. Reported in EPA TSCA Inventory.

OSHA PEL: TWA 2 ppm; CL 10 ppm/15M; Cancer Hazard.
ACGIH TLV: Suspected Human Carcinogen, TWA 2 ppm (skin).
DFG TRK: 3 ppm (7 mg/m³), Animal Carcinogen, Suspected Human Carcinogen
NIOSH REL: TWA 1 ppm; CL 10 ppm/15M
DOT CLASSIFICATION: 3; *Label:* Flammable Liquid, Poison

SAFETY PROFILE: Confirmed human carcinogen with experimental carcinogenic, neoplastigenic, and tumorigenic data. Poison by inhalation, ingestion, skin contact, and other routes. Human systemic effects by inhalation and skin contact: conjunctive irritation, somnolence, general anesthesia, cyanosis, and diarrhea. An experimental teratogen. Other experimental reproductive effects. Human mutation data reported. Dangerous fire hazard when exposed to heat, flame, or oxidizers. Moderate explosion hazard when exposed to flame. Can react vigorously with oxidizing materials (see also CYANIDE).

Acrylonitrile closely resembles hydrocyanic acid in its toxic action. By inhibiting the respiratory enzymes of tissue, it renders the tissue cells incapable of oxygen absorption. Poisoning is acute; there is little evidence of cumulative action on repeated exposure. Exposure to low concentration is followed by flushing of the face and increased salivation; further exposure results in irritation of the eyes and nose, photophobia, deepened respiration. If exposure continues, shallow respiration, nausea, vomiting, weakness, an oppressive feeling in the chest, and occasionally headache and diarrhea are other complaints. Several cases of mild jaundice accompanied by mild anemia and leucocytosis have been reported. Urinalysis is generally negative, except for an increase in bile pigment. Serum and bile thiocyanates are raised. See also HYDROCYANIC ACID. Unstable and easily oxidized. Explosive polymerization may occur on storage with silver nitrate. Potentially explosive reactions with benzyltrimethylammonium hydroxide + pyrrole, tetrahydrocarbazole + benzyltrimethylammonium hydroxide. Violent reactions with strong acids (e.g., nitric or sulfuric), strong bases, azoisobutyronitrile, dibenzoyl peroxide, di-tert-butylperoxide, or bromine. Incompatible with $AgNO_3$ and amines. To fight fire use CO_2, dry chemical, or alcohol foam. When heated to decomposition it emits toxic fumes of NO_x and CN^-. See also NITRILES and CYANIDE.

For occupational chemical analysis use OSHA: #37 or NIOSH: Acrylonitrile, 1604.

ADX750 HR: 2
ACRYLONITRILE POLYMER with 1,3-BUTADIENE, and STYRENE, COMBUSTION PRODUCTS

mf: $(C_3H_3N)_x$

SYNS: ABS (pyrolysis products) □ ACELAN, combustion products □ ACRIBEL, combustion products □ ACRYL, combustion products □ ACRYLONITRILE-BUTADIENE-STYRENE (pyrolysis products) □ AKSA, combustion products □ ANILANA, combustion products □ BI-LOFT, combustion products □ BULANA, combustion products □ CASHMILON, combustion products □ CRUMERON, combustion products □ DOLAN, combustion products □ EXLAN, combustion products □ FINA, combustion products □ MALON, combustion products □ ORLON, combustion products □ POLYACRYLONITRILE □ 2-PROPENENITRILE HOMOPOLYMER (9CI) □ ZEFRAN, combustion products

TOXICITY DATA WITH REFERENCE
ihl-mus LC50:10 g/m³/30M PWPSA8 21,167,78

SAFETY PROFILE: Moderately to highly toxic by inhalation. Upon decomposition it emits toxic fumes of NO_x and CN^-.

ADY250 CAS:9003-00-3 HR: 2
ACRYLONITRILE POLYMER with CHLOROETHYLENE

mf: $(C_3H_3N{\cdot}C_2H_3Cl)_n$

SYNS: ACROPOR □ ACROPOR AN □ ACROPOR AN 200 □ ACROPOR AN 450 □ ACROPORE □ DYNEL □ DYNEL NYGL □ KANEKALON □ 2-PROPENENITRILE, POLYMER with CHLOROETHENE (9CI) □ SKhN6 □ VINYON N

TOXICITY DATA WITH REFERENCE
imp-mus TDLo:18 mg/kg:ETA CNREA8 15,333,55

CONSENSUS REPORTS: Reported in EPA TSCA Inventory. Cyanide and its compounds are on the Community Right-To-Know List.

SAFETY PROFILE: Questionable carcinogen with experimental tumorigenic data. See also NITRILES. When heated to decomposition it emits very toxic fumes of Cl^-, CN^-, and NO_x.

ADY500 CAS:9003-54-7 HR: 3
ACRYLONITRILE POLYMER with STYRENE

mf: $(C_8H_8{\cdot}C_3H_3N)_x$

SYNS: ACRILAFIL □ ACRYLONITRILE-STYRENE COPOLYMER □ ACRYLONITRILE-STYRENE POLYMER □ ACRYLONITRILE-STYRENE RESIN □ ACS □ AS 61CL □ BAKELITE RMD 4511 □ CEVIAN HL □ DIALUX □ ESTYRENE AS □ KOSTIL □ LITAC □ LURAN □ LUSTRAN □ POLYSTYRENE-ACRYLONITRILE □ 2-PROPENENITRILE POLYMER with ETHENYLBENZENE □ REXENE 106 □ SANREX

□ SN 20 □ STYREN-ACRYLONITRILEPOLYMER □ STYRENE-ACRYLONITRILE COPOLYMER □ TERULAN KP 2540 □ TYRIL

TOXICITY DATA WITH REFERENCE
orl-rat LD50:1800 mg/kg CEHYAN 25,22,80
orl-mus LD50:1000 mg/kg CEHYAN 25,22,80

CONSENSUS REPORTS: IARC Cancer Review: Human No Adequate Data IMEMDT 19,73,79; Animal No Adequate Data IMEMDT 19,73,79; Group 3 IMEMDT 7,56,87. Reported in EPA TSCA Inventory. Cyanide and its compounds are on the Community Right-To-Know List.

SAFETY PROFILE: Moderately to highly toxic by ingestion. See also NITRILES. When heated to decomposition it emits toxic fumes of NO_x and CN^-.

ADY750 CAS:4836-08-2 HR: 2
2-ACRYLOXYETHYLDIMETHYLSULFONIUM METHYL SULFATE

TOXICITY DATA WITH REFERENCE
orl-rat LD50:1870 mg/kg AIHAAP 23,95,62
skn-rbt LD50:2000 mg/kg AIHAAP 23,95,62

SAFETY PROFILE: Moderately toxic by ingestion and skin contact. See also SULFATES. When heated to decomposition it emits toxic fumes of SO_x.

ADZ000 CAS:814-68-6 HR: 3
ACRYLOYL CHLORIDE
mf: C_3H_3ClO mw: 90.51

PROP: Bp: 75°

SYNS: ACRYLIC ACID CHLORIDE □ ACRYLYL CHLORIDE □ 2-PROPENOYL CHLORIDE

TOXICITY DATA WITH REFERENCE
ihl-rat LCLo:25 ppm/4H BJIMAG 27,1,70
ihl-mus LC50:92 mg/m³/2H 85GMAT -,17,82
ivn-mus LD50:180 mg/kg CSLNX* NX#03367

CONSENSUS REPORTS: EPA Extremely Hazardous Substances List. Reported in EPA TSCA Inventory.

SAFETY PROFILE: Poison by inhalation and intravenous routes. When heated to decomposition it emits toxic fumes of Cl^-.

AEA000 HR: 3
ACTINIC RADIATION

SAFETY PROFILE: Outdoor workers, such as fishermen, sailors, soldiers, and farmers, show a high incidence of skin cancer. The commonest acute manifestation of actinic radiation effects on skin is sunburn.

AEA109 CAS:24397-89-5 HR: 3
ACTINOBOLIN
mf: $C_{13}H_{20}N_2O_6$ mw: 300.31

PROP: Amorphous, fluffy, very hygroscopic powder. Amphoteric. Freely sol in water or alc; mod sol in methanol and ethanol. Unstable in basic solutions.

SYNS: 4-(2-AMINOPROPIONAMIDO)-3,4,4a,5,6,7-HEXAHYDRO-5,6,8-TRIHYDROXY-3-METHYLISOCOUMARIN □ NSC 31083

TOXICITY DATA WITH REFERENCE
pic-esc 240 mg/L ZAPOAK 8,139,68
ipr-mus LD50:2844 mg/kg NCISP* JAN86
scu-mus LD50:1828 mg/kg NCISP* JAN86
ivn-mus LD50:6250 μg/kg JANTAJ 32,1069,79
unr-mus LDLo:2000 mg/kg 85ERAY 2,1368,78

SAFETY PROFILE: Poison by intravenous route. Moderately toxic by other routes. Mutation data reported. When heated to decomposition it emits toxic fumes of NO_x. An antimicrobial agent and experimental cariostat.

AEA250 CAS:1338-58-5 HR: 3
ACTINOGAN

SYNS: NSC 53396 □ NSC A15920

TOXICITY DATA WITH REFERENCE
ipr-mus LD50:200 mg/kg ARZNAD 17,693,67
ivn-mus LDLo:17 mg/kg ARZNAD 17,693,67

SAFETY PROFILE: Poison by intraperitoneal and intravenous routes.

AEA500 CAS:1402-38-6 HR: 3
ACTINOMYCIN

SYNS: AURANTIN □ ONCOSTATIN

TOXICITY DATA WITH REFERENCE
oms-omi 1250 μg/L SOGEBZ 4,100,68
cyt-hmn:lym 200 μg/L/2H CCPHDZ 3,143,79
ivn-rat TDLo:150 μg/kg (9D preg):TER TJADAB 30(1),32A,84
orl-rat LDLo:1 mg/kg JPETAB 74,25,42
ipr-rat LDLo:1 mg/kg JPETAB 74,25,42
scu-rat LDLo:1 mg/kg JPETAB 74,25,42
ivn-rat LDLo:1 mg/kg JPETAB 74,25,42
orl-mus LDLo:10 mg/kg JPETAB 74,25,42
ipr-mus LDLo:1 mg/kg TDKNAF 14,60,55
scu-mus LDLo:250 μg/kg JPETAB 74,25,42
ivn-mus LDLo:250 μg/kg JPETAB 74,25,42
orl-rbt LDLo:1 mg/kg JPETAB 74,25,42
ipr-rbt LDLo:1 mg/kg JPETAB 74,25,42
scu-rbt LDLo:1 mg/kg JPETAB 74,25,42
ivn-rbt LDLo:1 mg/kg JPETAB 74,25,42

SAFETY PROFILE: Poison by ingestion, intraperitoneal, subcutaneous, and intravenous routes. An experimental teratogen. Other experimental reproductive effects. Human mutation data reported.

AEA625 CAS:85086-83-5 HR: 3
ACTINOMYCIN 23-21

SYN: SINOACTINOMYCIN

TOXICITY DATA WITH REFERENCE
orl-mus LD50:9400 μg/kg YHTPAD 19,283,84
ipr-mus LD50:515 μg/kg YHTPAD 19,283,84
ivn-mus LD50:1070 μg/kg YHTPAD 19,283,84
ivn-dog LD50:500 μg/kg YHTPAD 19,283,84

SAFETY PROFILE: Poison by ingestion, intravenous, and intraperitoneal routes.

AEA750 CAS:8052-16-2 HR: 3
ACTINOMYCIN C
mf: $C_{62}H_{89}N_{11}O_{17}$ mw: 1260.62

PROP: Red solid. Mp: 252° (decomp).

SYNS: ACTINOCHRYSIN □ CACTINOMYCIN □ DACTINOMYCIN (10%), ACTINOMYCIN C2 (45%), and ACTINOMYCIN C3 (45%) mixture □ HBF 386 □ NSC-18268 □ SANDAMYCIN

TOXICITY DATA WITH REFERENCE
pic-esc 120 mg/L ZAPOAK 8,139,68
slt-dmg-orl 5 ppm MUREAV 173,197,86
dni-hmn:oth 475 µg/L 26QZAP 2,395,72
oms-hmn:oth 47 µg/L 26QZAP 2,395,72
dni-hmn:oth 475 µg/L 26QZAP 2,395,72
oms-hmn:47 µg/L 26QZAP 2,395,72
scu-rat TDLo:350 µg/kg (6-10D preg):TER OSDIAF 14,107,65
scu-rat TDLo:350 µg/kg (6-10D preg):REP OSDIAF 14,107,65
ivn-rat LD50:100 mg/kg ARZNAD 20,1461,70
par-rat LD50:100 mg/kg RRCRBU 52,76,75
ipr-mus LD50:1110 µg/kg AEPPAE 230,559,57
ivn-mus LD50:1 mg/kg 85GDA2 4(2),53,80

SAFETY PROFILE: Deadly poison by intravenous, parenteral, intraperitoneal, and possibly other routes. An experimental teratogen. Other experimental reproductive effects. Human mutation data reported. When heated to decomposition it emits toxic fumes of NO_x. An antibiotic.

AEB000 CAS:50-76-0 HR: 3
ACTINOMYCIN D
mf: $C_{62}H_{86}N_{12}O_{16}$ mw: 1255.60

PROP: Red rhomboids + $3H_2O$ (ethanol). Mp: 246–247°.

SYNS: ACT □ ACTINOMYCINDIOIC D ACID, DILACTONE □ ACTINOMYCIN I □ AD □ COSMEGEN □ DACTINOMYCIN □ DILACTONE ACTINOMYCINDIOIC D ACID □ HBF 386 □ LYOVAC COSMEGEN □ MERACTINOMYCIN □ NCI-C04682 □ NSC 3053 □ ONCOSTATIN K

TOXICITY DATA WITH REFERENCE
dnd-hmn:hla 400 µg/L/15M ECREAL 103,175,76
cyt-hmn:lym 200 µg/L/2H CCPHDZ 3,143,79
ipr-rat TDLo:100 µg/kg (female 18-19D post):REP TJADAB 17,44A,78
ipr-mus TDLo:100 µg/kg (female 8D post):TER SEIJBO 11,5,71
ipr-rat TDLo:2600 µg/kg/17W-I:CAR CNREA8 30,2271,70
scu-mus TDLo:280 µg/kg/18W-I:ETA APJAAG 17,495,67
ipr-rat TD:1700 µg/kg/26W-I:NEO RRCRBU 52,1,75
ivn-hmn TDLo:40 µg/kg/4D-I:SKN NEJMAG 281,1094,69
orl-rat LD50:7200 µg/kg ANYAA9 89,348,60
ipr-rat LD50:100 µg/kg AOGLAR 23,219,76
ivn-rat LD50:460 µg/kg ANYAA9 89,348,60
scu-rat LD50:800 µg/kg ANYAA9 89,348,60
ivn-rat LD50:460 µg/kg ANYAA9 89,348,60
orl-mus LD50:13 mg/kg ANYAA9 89,348,60
ipr-mus LD50:750 µg/kg CTRRDO 61,103,77

CONSENSUS REPORTS: IARC Cancer Review: Group 3 IMEMDT 7,80,87; NCI Carcinogenesis Studies (ipr); Clear Evidence: rat RRCRBU 52,1,75; No Evidence: mouse RRCRBU 52,1,75.

SAFETY PROFILE: Poison by ingestion, intravenous, subcutaneous, and intraperitoneal routes. An experimental teratogen. Other experimental reproductive effects. Human systemic effects by intravenous and possibly other routes: dermatitis, bone marrow damage, and gastrointestinal effects. A human systemic skin irritant by intravenous route. Human mutation data reported. Questionable carcinogen with experimental carcinogenic, neoplastigenic, and tumorigenic data. When heated to decomposition it emits toxic fumes of NO_x.

AEB500 CAS:6980-13-8 HR: 3
ACTINOMYCIN K
mf: $C_{11}H_{12}ClN_3O_4$ mw: 285.71

SYNS: 7-CHLORO-3-β-d RIBOFURANOSYL-3H-IMIDAZO(4,5-b)PYRIDINE □ KASUGAMYCIN □ KENGSHENGMYCIN

TOXICITY DATA WITH REFERENCE
ivn-hmn TDLo:80 µg/kg/25D-I:BLD XPHPAW 441,116,74
orl-rat LD50:22 g/kg 28ZEAL 5,136,76
orl-mus LD50:21 g/kg 28ZEAL 5,136,76
ipr-mus LD50:745 µg/kg 85ERAY 2,1268,78

SAFETY PROFILE: Poison by intraperitoneal route. Mildly toxic by ingestion. Human blood effects by intravenous route. When heated to decomposition it emits very toxic fumes of Cl^- and NO_x.

AEB750 CAS:102488-99-3 HR: 3
ACTINOMYCIN L

SYN: ACTINOMYCIN 2104L

TOXICITY DATA WITH REFERENCE
scu-mus TDLo:4725 µg/kg/35W-I:NEO BKNJA5 2,105,59

CONSENSUS REPORTS: IARC Cancer Review: Animal Sufficient Evidence IMEMDT 10,29,76.

SAFETY PROFILE: Confirmed carcinogen with experimental neoplastigenic data.

AEC000 CAS:12623-78-8 HR: 3
ACTINOMYCIN S

SYN: ACTINOMYCIN 1048A

TOXICITY DATA WITH REFERENCE
scu-mus TDLo:240 µg/kg/16W-I:NEO BKNJA5 2,105,59

CONSENSUS REPORTS: IARC Cancer Review: Animal Sufficient Evidence IMEMDT 10,29,76.

SAFETY PROFILE: Confirmed carcinogen with experimental neoplastigenic data.

AEC200 **CAS:1402-61-5** **HR: 3**
ACTINOMYCIN X2
mf: $C_{61}H_{89}N_{12}O_{17}$ mw: 1262.62

PROP: Crystals from petroleum ether. Mp: 249.5–250.5°.

SYNS: ACTINOMYCIN BV □ ACTINOMYCIN DV □ ACTINOMYCIN J1 □ ACTINOMYCIN S3 □ ACTINOMYCIN-V

TOXICITY DATA WITH REFERENCE
ipr-mus LD50:300 μg/kg JANTAJ 38,1625,85
scu-mus LD50:300 μg/kg 38KLAC -,427,77
ivn-mus LD50:1 mg/kg 85GDA2 4(2),48,80

SAFETY PROFILE: Poison by subcutaneous, intravenous, and intraperitoneal routes. Mutation data reported. When heated to decomposition it emits toxic fumes of NO_x.

AEC250 **CAS:59680-34-1** **HR: 3**
ACTINOXANTHIN
mf: $C_{437}H_{667}N_{121}O_{155}S_4$ mw: 10219.664

PROP: An antibiotic produced by the strain *Actinomyces globisporus.*

SYN: ACTINOXANTHINE

TOXICITY DATA WITH REFERENCE
ipr-mus LD50:240 μg/kg 85ERAY 2,1414,78
scu-mus LD50:1800 μg/kg 85ERAY 2,1414,78

SAFETY PROFILE: Poison by intraperitoneal and subcutaneous routes. When heated to decomposition it emits very toxic fumes of SO_x and NO_x.

AEC625 **CAS:58814-86-1** **HR: 3**
ACULEACIN A
mf: $C_{50}H_{81}N_7O_{16}$ mw: 1036.38

PROP: Amorphous powder. Mp: 162–166°.

TOXICITY DATA WITH REFERENCE
dni-omi 20 mg/L JANTAJ 35,210,82
ipr-mus LD50:600 mg/kg 85GDA2 4(1),361,80
ivn-mus LD50:350 mg/kg JANTAJ 30,297,77
ims-mus LD50:600 mg/kg USXXAM #3978210

SAFETY PROFILE: Poison by intravenous route. Moderately toxic by other routes. Mutation data reported. When heated to decomposition it emits toxic fumes of NO_x.

AEC700 **CAS:59277-89-3** **HR: 2**
ACYCLOVIR
mf: $C_8H_{11}N_5O_3$ mw: 225.24

PROP: Crystals from ethanol. Mp: (decomp).

SYNS: ACICLOVIR □ ACYCLOGUANOSINE □ 2-AMINO-1,9-DIHYDRO-9-((2-HYDROXYETHOXY)METHYL)-6H-PURIN-6-ONE □ BW 248U □ 9-(2-HYDROXYTHEOXYMETHYL)GUANINE □ WELLCOME-248U □ ZOVIRAX

TOXICITY DATA WITH REFERENCE
cyt-hmn:lym 250 mg/L/48H FAATDF 3,587,83
msc-mus:lym 400 mg/L/4H FAATDF 3,587,83
scu-rat TDLo:100 mg/kg (female 10D post):REP TJADAB 36,31A,87
scu-rat TDLo:400 mg/kg (female 9-11D post):TER ARTODN 61,468,88
orl-wmn TDLo:100 mg/kg/5D-I:SKN BMJOAE 289,1424,84
orl-wmn TDLo:80 mg/kg/4D-I:CNS AIMEAS 111,187,89
ivn-man TDLo:134 μg/kg/1D-I:CNS LANCAO 2,385,85
ivn-wmn TDLo:101 mg/kg/2D-I:CNS AJMEAZ 94,212,93
ipr-rat LD50:860 mg/kg IYKEDH 16,866,85
scu-rat LD50:620 mg/kg IYKEDH 16,866,85
ivn-rat LD50:910 mg/kg IYKEDH 16,866,85

SAFETY PROFILE: Moderately toxic. Human systemic effects by ingestion or intravenous routes: allergic dermatitis, somnolence, and hallucinations. An experimental teratogen. Other experimental reproductive effects. Human mutation data reported. When heated to decomposition it emits acrid smoke and irritating fumes.

AEC725 **CAS:69657-51-8** **HR: 2**
ACYCLOVIR SODIUM SALT
mf: $C_8H_{11}N_5O_3 \cdot Na$ mw: 248.23

SYNS: ACYCLOGUANOSINE SODIUM (OBS.) □ 2-AMINO-1,9-DIHYDRO-9-((2-HYDROXYETHOXY)METHYL)-6H-PURIN-6-ONE MONOSODIUM SALT □ 1,9-DIHYDRO-2-AMINO-9-((2-HYDROXYETHOXY) METHYL)-6H-PURIN-6-ONE SODIUM SALT □ SODIUM ACYCLOVIR □ ZOVIRAX SODIUM

TOXICITY DATA WITH REFERENCE
scu-rat TDLo:300 mg/kg (female 10D post):REP ARTODN 62,8,88
scu-rat TDLo:300 mg/kg (female 10D post):TER ARTODN 62,8,88
ivn-man TDLo:107 mg/kg/5D-I DICPBB 22,306,88
ivn-cld TDLo:248 mg/kg/80H-I:GIT DICPBB 20,371,86
ipr-rat LD50:1210 mg/kg FAATDF 3,573,83
scu-rat LD50:650 mg/kg FAATDF 3,573,83
ipr-mus LD50:999 mg/kg FAATDF 3,573,83
ivn-mus LD50:405 mg/kg FAATDF 3,573,83

SAFETY PROFILE: Moderately toxic by intraperitoneal and several other routes. Human systemic effects by intravenous route: nausea or vomiting. An experimental teratogen. Other experimental reproductive effects. When heated to decomposition it emits toxic fumes of NO_x and Na_2O.

AEC750 **CAS:21829-25-4** **HR: 3**
ADALAT
mf: $C_{17}H_{18}N_2O_6$ mw: 346.37

PROP: Yellow crystals. Mp: 172–174°. Sol in Me_2CO, $CHCl_3$.

SYNS: BAY 1040 □ BAY A 1040 □ CITILAT □ CORDIPIN □ 1,4-DIHYDRO-2,6-DIMETHYL-4-(2-NITROPHENYL)-3,5-PYRIDINEDICARBOXYLIC ACID DIMETHYL ESTER □ NIFEDIN □ NIFEDIPINE □ NIFELAT □ 4-(2′-NITROPHENYL)-2,6-DIMETHYL-3,5-DICARBOMETHOXY-1,4-DIHYDROPYRIDINE □ OXCORD □ PROCARDIA

TOXICITY DATA WITH REFERENCE
orl-wmn TDLo:200 μg/kg (female 27W post):TER GOBIDS 24,151,87

orl-wmn TDLo: 200 μg/kg (female 27W post): REP GOBIDS 24,151,87
orl-man TDLo: 105 mg/kg/26W-I:PNS LANCAO 339,1382,92
orl-wmn TDLo: 2400 μg/kg/3D-I:CNS BMJOAE 304,1225,92
orl-wmn TDLo: 31 mg/kg/11W-I AHJOA2 108,611,84
orl-wmn TDLo: 800 μg/kg/1D-I: BLD,GIT PGMJAO 62,1029,86
orl-hmn TDLo: 143 μg/kg ARZNAD 35,518,85
orl-wmn TDLo: 600 μg/kg/45M-I: BPR,GIT AIMDAP 147,556,87
orl-man TDLo: 143 μg/kg/1D:BPR ARZNAD 22,380,72
orl-cld TDLo: 70 mg/kg:CVS,GLN PEDIAU 86,91,90
orl-man TDLo: 714 μg/kg:CVS AEMED3 22,196,93
orl-rat LD50: 1022 mg/kg ARZNAD 22,1,72
ipr-rat LD50: 230 mg/kg YKYUA6 28,1451,77
orl-mus LD50: 310 mg/kg JJPAAZ 40,399,86
ipr-mus LD50: 185 mg/kg PCJOAU 16,817,82
orl-rbt LD50: 504 mg/kg ARZNAD 35,915,85

SAFETY PROFILE: Poison by ingestion, intravenous, and intraperitoneal routes. Human systemic effects by ingestion: decreased blood pressure, BP lowering, cardiomyopathy, changes in regional blood flow, hyperglycemia, nausea or vomiting, respiratory depression, toxic psychosis. An experimental teratogen. Experimental reproductive effects. See also ESTERS. When heated to decomposition it emits toxic fumes of NO_x.

AED250 CAS:665-66-7 HR: 3
1-ADAMANTANAMINE HYDROCHLORIDE
mf: $C_{10}H_{17}N•ClH$ mw: 187.74

PROP: Crystals from $EtOH/Et_2O$. Sol in water and ethanol; prac insol in Et_2O.

SYNS: ADAMANTANAMINE HYDROCHLORIDE □ ADAMANTINE HYDROCHLORIDE □ ADAMANTYLAMINE HYDROCHLORIDE □ 1-ADAMANTYLAMINE HYDROCHLORIDE □ AMANTADINE HYDROCHLORIDE □ AMAZOLON □ AMINOADAMANTANE HYDROCHLORIDE □ 1-AMINOADAMANTENE HYDROCHLORIDE □ EXP 105-1 □ MANTADAN □ NSC 83653 □ SYMMETREL □ TRICYCLO(3.3.1.1.$^{(3,7)}$)DECAN-1-AMINE, HYDROCHLORIDE (9CI) □ VIROFRAL

TOXICITY DATA WITH REFERENCE
orl-wmn TDLo: 182 mg/kg (female 1-91D post): TER LANCAO 2,607,75
orl-rat TDLo: 240 mg/kg (female 9-14D post): REP GNRIDX 4,44,70
orl-man TDLo: 24 mg/kg/1D-I:CNS AEMED3 19,668,90
orl-man LDLo: 43 mg/kg CJPSDF 31,757,86
orl-man TDLo: 13 mg/kg/5D-I:CNS AJPSAO 143,1170,85
orl-man LDLo: 286 mg/kg:CNS AJPSAO 145,267,88
orl-rat LD50: 800 mg/kg IYKEDH 19,164,88
ipr-rat LD50: 150 mg/kg TXAPA9 15,642,69
ivn-rat LD50: 90 mg/kg IYKEDH 19,164,88
orl-mus LD50: 700 mg/kg TXAPA9 15,642,69
ipr-mus LD50: 198 mg/kg IYKEDH 19,164,88
scu-mus LD50: 290 mg/kg IYKEDH 19,164,88
orl-gpg LD50: 360 mg/kg TXAPA9 15,642,69

CONSENSUS REPORTS: Reported in EPA TSCA Inventory.

SAFETY PROFILE: Human poison by ingestion. Poison by ingestion, intraperitoneal, and intravenous routes. A human teratogen with developmental abnormalities of the circulatory system. Experimental reproductive effects. Human systemic effects by ingestion: distorted perceptions, euphoria, excitement, hallucinations. When heated to decomposition it emits very toxic fumes of NO_x and HCl.

AED750 CAS:54099-11-5 HR: 3
1-ADAMANTANEACETIC ACID-2-(DIETHYLAMINO) ETHYL ESTER, ETHYL IODIDE
mf: $C_{20}H_{36}NO_2•I$ mw: 449.47

TOXICITY DATA WITH REFERENCE
orl-mus LD50: 600 mg/kg FRPSAX 32,129,77
ipr-mus LD50: 61 mg/kg FRPSAX 32,129,77

SAFETY PROFILE: Poison by intraperitoneal route. Moderately toxic by ingestion. See also ESTERS and IODIDES. When heated to decomposition it emits very toxic fumes of I^- and NO_x.

AEE000 CAS:54099-12-6 HR: 3
1-ADAMANTANEACETIC ACID-3-(DIMETHYLAMINO) PROPYL ESTER, ETHYL IODIDE
mf: $C_{19}H_{36}NO_2•I$ mw: 437.46

TOXICITY DATA WITH REFERENCE
orl-mus LD50: 450 mg/kg FRPSAX 32,129,77
ipr-mus LD50: 30 mg/kg FRPSAX 32,129,77

SAFETY PROFILE: Poison by intraperitoneal route. Moderately toxic by ingestion. See also ESTERS and IODIDES. When heated to decomposition it emits very toxic fumes of I^- and NO_x.

AEE100 CAS:880-52-4 HR: 2
N-(1-ADAMANTYL)ACETAMIDE
mf: $C_{12}H_{19}NO$ mw: 193.32

SYN: ACETAMIDE, N-(1-ADAMANTYL)-

TOXICITY DATA WITH REFERENCE
ipr-mus LD50: 520 mg/kg PCJOAU 14,185,80

CONSENSUS REPORTS: Reported in EPA TSCA Inventory.

SAFETY PROFILE: Moderately toxic by intraperitoneal route. When heated to decomposition it emits toxic vapors of NO_x.

AEF250 CAS:35507-78-9 HR: 3
5-(1-ADAMANTYL)-2,4-DIAMINO-6-METHYLPYRIMIDINE ETHYLSULFONATE
mf: $C_{15}H_{22}N_4•C_2H_6O_3S$ mw: 368.55

SYN: DAMP-ES

TOXICITY DATA WITH REFERENCE
ipr-rat LD50: 23 mg/kg JNCIAM 60,1029,78
ipr-mus LD50: 40 mg/kg JNCIAM 60,1029,78

SAFETY PROFILE. Poison by intraperitoneal route. See also SULFONATES and AMINES. When heated to decomposition it emits very toxic fumes of NO_x and SO_x.

AEF500 **CAS:31635-40-2** **HR: 3**
N-1-ADAMANTYL-N-(2-(DIMETHYLAMINO)ETHOXY) ACETAMIDE HYDROCHLORIDE
mf: $C_{16}H_{28}N_2O_2 \cdot ClH$ mw: 316.92

SYN: 1-(DIMETHYLAMINOETHOXYACETAMIDO)ADAMANTANE HYDROCHLORIDE

TOXICITY DATA WITH REFERENCE
orl-dog LD50:170 mg/kg ARZNAD 23,577,73
orl-rat LD50:630 mg/kg ARZNAD 23,577,73
ivn-mus LD50:71 mg/kg ARZNAD 23,577,73

SAFETY PROFILE: Poison by ingestion and intravenous route. See also AMINES. When heated to decomposition it emits very toxic fumes of NO_x and HCl. An antiviral agent.

AEG000 **CAS:40284-08-0** **HR: 3**
N-(2-ADAMANTYL)-2-MERCAPTOACETAMIDINE HYDROCHLORIDE
mf: $C_{12}H_{20}N_2S \cdot ClH$ mw: 260.86

TOXICITY DATA WITH REFERENCE
orl-mus LD50:35 mg/kg JMCMAR 15,1313,72
ipr-mus LD50:17 mg/kg JMCMAR 15,1313,72

SAFETY PROFILE: Poison by ingestion and intraperitoneal routes. When heated to decomposition it emits very toxic fumes of NO_x, SO_x, and HCl.

AEG129 **CAS:69804-02-0** **HR: 3**
S-(N-(1-ADAMANTYLMETHYLAMIDINO)METHYL) PHOSPHOROTHIOATE MONOSODIUM SALT
mf: $C_{13}H_{22}N_2O_3PS \cdot Na$ mw: 340.22

TOXICITY DATA WITH REFERENCE
ipr-mus LD50:408 mg/kg PCJOAU 13,22,79
par-mus LD50:98 mg/kg PCJOAU 13,22,79

SAFETY PROFILE: Poison by parenteral and intraperitoneal routes. When heated to decomposition it emits very toxic fumes of NO_x, PO_x, Na_2O, and SO_x.

AEG250 **CAS:22545-60-4** **HR: 3**
N-(1-ADAMANTYLMETHYL)-2-MERCAPTOACETAMIDINE HYDROCHLORIDE
mf: $C_{13}H_{22}N_2S \cdot ClH$ mw: 274.89

TOXICITY DATA WITH REFERENCE
orl-mus LD50:65 mg/kg JMCMAR 15,1313,72
ipr-mus LD50:22 mg/kg JMCMAR 15,1313,72

SAFETY PROFILE: Poison by ingestion and intraperitoneal routes. When heated to decomposition it emits very toxic fumes of HCl, SO_x, and NO_x.

AEG500 **CAS:40284-10-4** **HR: 3**
N-(3-(1-ADAMANTYL)PROPYL)-2-MERCAPTOACETAMIDINE HYDROCHLORIDE HYDRATE (10:10:3)
mf: $C_{15}H_{26}N_2S \cdot ClH \cdot 3/10H_2O$ mw: 308.35

TOXICITY DATA WITH REFERENCE
orl-mus LD50:350 mg/kg JMCMAR 15,1313,72
ipr-mus LD50:25 mg/kg JMCMAR 15,1313,72

SAFETY PROFILE: Poison by ingestion and intraperitoneal routes. When heated to decomposition it emits very toxic fumes of NO_x, SO_x, and HCl.

AEG625 **CAS:1225-60-1** **HR: 3**
ADANTON HYDROCHLORIDE
mf: $C_{16}H_{19}N_3S \cdot ClH$ mw: 321.90

PROP: Crystals from MeCN. Mp: 222–223°. Sol in water.

SYNS: ANDANTOL □ D 201 HYDROCHLORIDE □ N-DIMETHYLAMINOISOPROPYLTHIOPHENYLPYRIDYLAMINE HYDROCHLORIDE □ 10-(2-DIMETHYLAMINO-2-METHYLETHYL)-10H-PYRIDO (3,2-b)(1,4)BENZOTHIAZINE HYDROCHLORIDE □ 10-(2-DIMETHYLAMINOPROPYL)-1-AZAPHENOTHIAZINE HYDROCHLORIDE □ 10-(2-DIMETHYLAMINOPROPYL-(1)-4-AZAPHENTHIAZIN HYDROCHLORID (GERMAN) □ 10-(2-DIMETHYLAMINOPROPYL)-9-THIA-1,10-DIAZAANTHRACENE HYDROCHLORIDE □ ISOTHIPENDYL HYDROCHLORIDE □ NILERGEX HYDROCHLORIDE □ THERUHISTIN HYDROCHLORIDE □ UDANTOL HYDROCHLORIDE

TOXICITY DATA WITH REFERENCE
orl-rat LD50:1220 mg/kg NIIRDN 6,72,82
orl-mus LD50:222 mg/kg ARZNAD 8,489,58
ipr-mus LD50:65 mg/kg ARZNAD 18,435,68

SAFETY PROFILE: Poison by ingestion and intraperitoneal routes. When heated to decomposition it emits toxic fumes of NO_x, SO_x, and HCl.

AEG750 **CAS:1229-29-4** **HR: 3**
ADAPIN
mf: $C_{19}H_{21}NO \cdot ClH$ mw: 315.87

PROP: Mp: 188–189°.

SYNS: CIDOXEPIN HYDROCHLORIDE □ CURATIN □ 11-DIMETHYLAMINO PROPYLIDENE-6H-DIBENZ(b,e)OXEPIN □ 11-(3-(DIMETHYLAMINO)PROPYLIDENE)-6,11-DIHYDRODIBENZ(b,e)OXEPIN HYDROCHLORIDE □ N,N-DIMETHYLDIBENZ(b,e)OXEPIN-$\Delta^{11(6H,\gamma)}$-PROPYLAMINE HYDROCHLORIDE □ DOXEPIN HYDROCHLORIDE □ NSC-108160 □ 1-PROPANAMINE, 3-DIBENZ(b,e)OXEPIN-11(6H)-YLIDENE-N,N-DIMETHYL-, HYDROCHLORIDE □ SINEQUAN

TOXICITY DATA WITH REFERENCE
orl-rat TDLo:1620 mg/kg (female 9-14D post):REP OYYAA2 5,913,71
orl-rat TDLo:540 mg/kg (female 9-14D post):TER OYYAA2 5,913,71
orl-wmn TDLo:112 mg/kg/4W-I JCLPDE 44,106,83
orl-hmn LDLo:90 mg/kg JATOD3 2,18,78
orl-hmn TDLo:9300 µg/kg:CNS JAMAAP 237,2632,77
orl-wmn TDLo:141 mg/kg/12W-I:EAR SMJOAV 76,1204,83
orl-rat LD50:147 mg/kg 27ZQAG -,72,72
ipr-rat LD50:84 mg/kg OYYAA2 6,889,72
scu-rat LD50:155 mg/kg OYYAA2 6,889,72
ivn-rat LD50:13 mg/kg 27ZQAG -,72,72
orl-mus LD50:180 mg/kg OYYAA2 6,889,72
ipr-mus LD50:79 mg/kg 27ZQAG -,72,72
scu-mus LD50:160 mg/kg 27ZQAG -,72,72

SAFETY PROFILE: A human poison by ingestion. An experimental poison by ingestion, subcutaneous, intraperitoneal, and intravenous routes. Human systemic effects by ingestion: hallucinations, distorted perceptions, muscle spasms and change in heart rate, and tinnitus. An experimental teratogen. Experimental reproductive effects. When heated to decomposition it emits toxic fumes of NO_x and HCl. A psychotherapeutic agent.

AEG875 **CAS:5118-29-6** **HR: 3**
ADAPTOL
mf: $C_{21}H_{25}N$ mw: 291.47

SYNS: 3-(10,10-DIMETHYL(10H)-ANTHRACENYLIDENE)-N,N-DIMETHYL-1-PROPANAMINE (9CI) □ 9-(3-DIMETILAMINOPROPYLIDEN)-10,10-DIMETIL-9,10-DIIDROANTHRACENE (ITALIAN) □ DIXERAN □ MELITRACEN □ MELITRACENE □ N 7001 □ N,N,10,10-TETRAMETHYL-$\Delta^{9(10),\gamma}$-ANTHRACENEPROPYLAMINE □ THYMEOL □ TRAUSABUM □ TRAUSABUN □ U-24973

TOXICITY DATA WITH REFERENCE
orl-rat LD50:170 mg/kg FRPPAO 25,519,70
ipr-rat LD50:96 mg/kg FRPPAO 25,519,70
orl-mus LD50:315 mg/kg FRPPAO 25,519,70
ipr-mus LD50:131 mg/kg FRPPAO 25,519,70
ivn-mus LD50:52 mg/kg FRPPAO 25,519,70

SAFETY PROFILE: Poison by ingestion, intravenous, and intraperitoneal routes. When heated to decomposition it emits toxic fumes of NO_x.

AEH000 **CAS:73-24-5** **HR: 3**
ADENINE
mf: $C_5H_5N_5$ mw: 135.15

PROP: Needles. Mp: 360–365° (anhyd) decomp.

SYNS: ADENINIMINE □ 6-AMINOPURINE □ 6-AMINO-1H-PURINE □ 6-AMINO-3H-PURINE □ 6-AMINO-9H-PURINE □ 1,6-DIHYDRO-6-IMINOPURINE □ 3,6-DIHYDRO-6-IMINOPURINE □ LEUCO-4 □ 1H-PURIN-6-AMINE □ USAF CB-18 □ VITAMIN B4

TOXICITY DATA WITH REFERENCE
pic-esc 1 g/L ZAPOAK 12,583,72
cyt-mus-ipr 10 mmol/L NULSAK 17,199,74
ipr-mus TDLo:200 mg/kg (female 10D post):TER OFAJAE 49,75,72
ipr-mus TDLo:200 mg/kg (female 10D post):REP OFAJAE 49,75,72
orl-rat LD50:227 mg/kg TXAPA9 47,229,79
ipr-rat LD50:198 mg/kg JPETAB 104,20,52
orl-mus LD50:783 mg/kg DRUGAY 6,19,82
ipr-mus LD50:100 mg/kg NTIS** AD277-689
scu-mus LDLo:1 g/kg ANYAA9 60,251,54

CONSENSUS REPORTS: Reported in EPA TSCA Inventory.

SAFETY PROFILE: Poison by intraperitoneal route. Moderately toxic by ingestion. An experimental teratogen. Experimental reproductive effects. Mutation data reported. When heated to decomposition it emits toxic fumes of NO_x.

AEH100 **CAS:24356-66-9** **HR: 1**
ADENINE ARABINOSIDE
mf: $C_{10}H_{13}N_5O_4 \cdot H_2O$ mw: 285.30

SYNS: ARA-A □ 9-β-d-ARABINOFURANOSYLADENINEMONOHYDRATE □ 9-β-d-ARABINOFURANOSYL-9H-PURINE-6-AMINE MONOHYDRATE □ SPONGOADENOSINE □ VIDARABINE □ VIRA-A

TOXICITY DATA WITH REFERENCE
orl-mus LD50:>7950 mg/kg AACHAX-,180,68
ipr-mus LD50:4677 mg/kg AACHAX-,180,68

SAFETY PROFILE: Mildly toxic by ingestion and intraperitoneal routes. When heated to decomposition it emits toxic fumes of NO_x.

AEH250 **CAS:700-02-7** **HR: 2**
ADENINE-1-N-OXIDE
mf: $C_5H_5N_5O$ mw: 151.15

PROP: Crystals from water. Mp: 297–307° decomp.

TOXICITY DATA WITH REFERENCE
scu-rat TDLo:1300 mg/kg/26W-I:NEO CNREA8 30,184,70

SAFETY PROFILE: Questionable carcinogen with experimental neoplastigenic data. When heated to decomposition it emits toxic fumes of NO_x.

AEH500 **CAS:321-30-2** **HR: 3**
ADENINE SULFATE
mf: $C_5H_5N_5 \cdot 1/2H_2O_4S$ mw: 821.71

SYNS: ADENINSULFAT □ 1H-PURIN-6-AMINE, SULFATE

TOXICITY DATA WITH REFERENCE
ipr-rat LD50:200 mg/kg AIPTAK 232,302,78
ipr-mus LD50:750 mg/kg TXAPA9 47,229,79

CONSENSUS REPORTS: Reported in EPA TSCA Inventory.

SAFETY PROFILE: Poison by intraperitoneal route. When heated to decomposition it emits toxic vapors of NO_x and SO_x.

AEH750 **CAS:58-61-7** **HR: 2**
ADENOSINE
mf: $C_{10}H_{13}N_5O_4$ mw: 267.28

PROP: Solid. Mp: 234–236°.

SYNS: ADENINE RIBOSIDE □ ADENOSIN (GERMAN) □ β-ADENOSINE □ β-d-ADENOSINE □ 6-AMINO-9-β-d-RIBOFURANOSYL-9H-PURINE □ BONITON □ MYOCOL □ NUCLEOCARDYL □ 9-β-d-RIBOFURANOSIDOADENINE □ SANDESIN □ USAF CB-10

TOXICITY DATA WITH REFERENCE
pic-esc 1 g/L ZAPOAK 12,583,72
oms-hmn:oth 100 μmol/L JIDEAE 65,52,75
cyt-mus-ipr 20 mmol/L NULSAK 17,199,74
dnd-mam:lym 60 mmol/L PNASA6 48,686,62
ivn-wmn TDLo:360 μg/kg/1H-I: CVS AJEMEN 10,326,92
ivn-man TDLo:257 μg/kg: CVS,PUL AJEMEN 11,249,93
ivn-man TDLo:171 μg/kg:CNS AJEMEN 11,192,93
ipr-mus LD50:500 mg/kg NTIS** AD277-689

CONSENSUS REPORTS: Reported in EPA TSCA Inventory.

SAFETY PROFILE: Moderately toxic by intraperitoneal route. Human systemic effects by intravenous route: coma, convulsions, cyanosis, fall in BP, pulse rate decrease, pulse rate increase. Human mutation data reported. When heated to decomposition it emits toxic fumes of NO_x.

AEI000 CAS:53-79-2 HR: 3
ADENOSINE-3′-(α-AMINO-p-METHOXYHYDROCINNAMAMIDO)-3′-DEOXY-N,N-DIMETHYL
mf: $C_{22}H_{29}N_7O_5$ mw: 471.58

PROP: Plates from 2-propanol. Mp: 175.5–177.0°.

SYNS: ACHROMYCIN (PURINE DERIVATIVE) □ 3′-(l-α-AMINO-p-METHOXYHYDROCINNAMAMIDO)-3′-DEOXY-N,N-DIMETHYLADENOSINE □ (S)-3′-((2-AMINO-3-(4-METHOXYPHENYL)-1-OXOPROPYL) AMINO)-3′-DEOXY-N,N-DIMETHYLADENOSINE □ CL 13,900 □ 6-DIMETHYLAMINO-9-(3′-(p-METHOXY-l-PHENYLALANYLAMINO)-β-d-RIBOFURANOSYL)-PURINE □ NSC-3055 □ PUROMYCIN □ STYLOMYCIN

TOXICITY DATA WITH REFERENCE
dnr-esc 100 μg/disc CNREA8 34,1658,74
cyt-dmg:oth 100 mg/L CLDFAT 2,97,73
dni-oin:oth 100 mg/L IJEBA6 16,1027,78
dni-mus:lym 100 μmol/L PLMEAA 34,231,78
dni-mus:fbr 420 μmol/L JCLBA3 58,410,73
unr-rat TDLo:100 mg/kg (female 7D post):TER 85DJA5 -,95,71
par-rbt TDLo:5 μg/kg (1D preg):REP ENDOAO 79,858,66
orl-mus LD50:20 mg/kg 85GDA2 5,302,81
ipr-mus LD50:25 mg/kg 85GDA2 5,302,81
ivn-mus LD50:15 mg/kg 85GDA2 5,302,81

SAFETY PROFILE: Poison by ingestion, intravenous, and intraperitoneal routes. An experimental teratogen. Experimental reproductive effects. Human mutation data reported. When heated to decomposition it emits toxic fumes of NO_x. An antibiotic.

AEI250 CAS:35788-21-7 HR: 3
ADENOSINE-5′-CARBOXAMIDE
mf: $C_{10}H_{12}N_6O_4$ mw: 280.28

SYN: β-d-1-(6-AMINO-9H-PURIN-9-YL)-1-DEOXYRIBOFURANURONAMIDE

TOXICITY DATA WITH REFERENCE
orl-mus LD50:50 mg/kg JMCMAR 23,313,80
ipr-mus LD50:5 mg/kg JMCMAR 23,313,80

SAFETY PROFILE: Poison by ingestion and intraperitoneal routes. When heated to decomposition it emits toxic fumes of NO_x.

AEI500 HR: 3
ADENOSINE-5′-(N-CYCLOBUTYL)CARBOXAMIDE
mf: $C_{14}H_{18}N_6O_4$ mw: 334.38

SYN: 1-(6-AMINO-9H-PURIN-9-YL)-N-CYCLOBUTYL-1-DEOXYRIBOFURANURONAMIDE

TOXICITY DATA WITH REFERENCE
orl-mus LD50:5 mg/kg JMCMAR 23,313,80
ipr-mus LD50:2 mg/kg JMCMAR 23,313,80

SAFETY PROFILE: Poison by ingestion and intraperitoneal routes. When heated to decomposition it emits toxic fumes of NO_x.

AEI750 CAS:35920-40-2 HR: 3
ADENOSINE-5′-(N-CYCLOPENTYL)CARBOXAMIDE
mf: $C_{15}H_{20}N_6O_4$ mw: 348.41

TOXICITY DATA WITH REFERENCE
orl-mus LD50:200 mg/kg JMCMAR 23,313,80
ipr-mus LD50:200 mg/kg JMCMAR 23,313,80

SAFETY PROFILE: Poison by ingestion and intraperitoneal routes. When heated to decomposition it emits toxic fumes of NO_x.

AEJ000 CAS:50908-62-8 HR: 3
ADENOSINE-5′-(N-CYCLOPROPYL)CARBOXAMIDE
mf: $C_{13}H_{16}N_6O_4$ mw: 320.35

SYN: 1-(6-AMINO-9H-PURIN-9-YL)-N-CYCLOPROPYL-1-DEOXYRIBOFURANURONAMIDE

TOXICITY DATA WITH REFERENCE
orl-mus LD50:5 mg/kg JMCMAR 23,313,80
ipr-mus LD50:2 mg/kg JMCMAR 23,313,80

SAFETY PROFILE: Poison by ingestion and intraperitoneal routes. When heated to decomposition it emits toxic fumes of NO_x.

AEJ250 CAS:72209-26-8 HR: 3
ADENOSINE-5′-(N-CYCLOPROPYL)CARBOXAMIDE-N′-OXIDE
mf: $C_{13}H_{16}N_6O_5$ mw: 336.35

SYN: 1-(6-AMINO-9H-PURIN-9-YL)-N-CYCLOPROPYL-1-DEOXYRIBOFURANURONAMIDE-N-OXIDE

TOXICITY DATA WITH REFERENCE
orl-mus LD50:5 mg/kg JMCMAR 23,313,80
ipr-mus LD50:5 mg/kg JMCMAR 23,313,80

SAFETY PROFILE: Poison by ingestion and intraperitoneal routes. When heated to decomposition it emits toxic fumes of NO_x.

AEJ500 CAS:58048-25-2 HR: 3
ADENOSINE-5′-(N-CYCLOPROPYLMETHYL)CARBOXAMIDE
mf: $C_{14}H_{18}N_6O_4$ mw: 334.38

SYN: 1-(6-AMINO-9H-PURIN-9-YL)-N-CYCLOPROPYLMETHYL-1-DEOXYRIBOFURANURONAMIDE

TOXICITY DATA WITH REFERENCE
orl-mus LD50:200 mg/kg JMCMAR 23,313,80
ipr-mus LD50:20 mg/kg JMCMAR 23,313,80

SAFETY PROFILE: Poison by ingestion and intraperitoneal routes. When heated to decomposition it emits toxic fumes of NO_x.

AEJ750 CAS:35788-31-9 HR: 3
ADENOSINE-5′-(N-(2-(DIMETHYLAMINO)ETHYL))CARBOXAMIDE
mf: $C_{14}H_{21}N_7O_4$ mw: 351.42

SYN: 1-(6-AMINO-9H-PURIN-9-YL)-N-(2-(DIMETHYLAMINO)-ETHYL-1-DEOXYRIBOFURANURONAMIDE)

TOXICITY DATA WITH REFERENCE
orl-mus LD50:500 mg/kg JMCMAR 23,313,80
ipr-mus LD50:20 mg/kg JMCMAR 23,313,80

SAFETY PROFILE: Poison by intraperitoneal route. Moderately toxic by ingestion. When heated to decomposition it emits toxic fumes of NO_x.

AEK000 CAS:39491-47-9 HR: 3
ADENOSINE-5′-(N,N-DIMETHYL)CARBOXAMIDE HYDRATE
mf: $C_{12}H_{16}N_6O_4 \cdot H_2O$ mw: 326.36

SYN: 1-(6-AMINO-9H-PURIN-9-YL)-1-DEOXY-N,N-DIMETHYLRIBOFURANURONAMIDE HYDRATE

TOXICITY DATA WITH REFERENCE
orl-mus LD50:1000 mg/kg JMCMAR 23,313,80
ipr-mus LD50:50 mg/kg JMCMAR 23,313,80

SAFETY PROFILE: Poison by intraperitoneal route. Moderately toxic by ingestion. When heated to decomposition it emits toxic fumes of NO_x.

AEK100 CAS:58-64-0 HR: 2
ADENOSINE DIPHOSPHATE
mf: $C_{10}H_{15}N_5O_{10}P_2$ mw: 427.24

SYNS: ADENOSINE 5′-DIPHOSPHATE □ ADENOSINE DIPHOSPHORIC ACID □ ADENOSINE 5′-DIPHOSPHORIC ACID □ ADENOSINE PYROPHOSPHATE □ ADENOSINE 5′-PYROPHOSPHATE □ ADENOSINE 5′-PYROPHOSPHORIC ACID □ ADENOSINE, 5′-(TRIHYDROGEN DIPHOSPHATE) (9CI) □ ADENOSINE, 5′-(TRIHYDROGEN PYROPHOSPHATE) □ 5′-ADENYLPHOSPHORIC ACID □ ADP □ 5′-ADP □ ADP (NUCLEOTIDE)

TOXICITY DATA WITH REFERENCE
oth-hmn:oth 100 μmol/L JIDEAE 65,52,75
ipr-mus LD50:3333 mg/kg PCJOAU 20,160,86

CONSENSUS REPORTS: Reported in EPA TSCA Inventory.

SAFETY PROFILE: Moderately toxic by intraperitoneal route. Human mutation data reported. When heated to decomposition it emits toxic vapors of NO_x and PO_x.

AEK250 CAS:35920-39-9 HR: 3
ADENOSINE-5′-(N-ETHYL)CARBOXAMIDE HEMIHYDRATE
mf: $C_{12}H_{16}N_6O_4 \cdot 1/2H_2O$ mw: 317.35

SYN: 1-(6-AMINO-9H-PURIN-9-YL)-1-DEOXY-N-ETHYLRIBOFURANURONAMIDE HEMIHYDRATE

TOXICITY DATA WITH REFERENCE
orl-mus LD50:5 mg/kg JMCMAR 23,313,80
ipr-mus LD50:500 μg/kg JMCMAR 23,313,80

SAFETY PROFILE: Poison by ingestion and intraperitoneal routes. When heated to decomposition it emits toxic fumes of NO_x.

AEK500 CAS:72209-27-9 HR: 3
ADENOSINE-5′-(N-ETHYL)CARBOXAMIDE-N′-OXIDE
mf: $C_{12}H_{16}N_6O_5$ mw: 324.34

SYN: 1-(6-AMINO-9H-PURIN-9-YL)-1-DEOXY-N-ETHYLRIBOFURANURONAMIDE- N-OXIDE

TOXICITY DATA WITH REFERENCE
orl-mus LD50:20 mg/kg JMCMAR 23,313,80
ipr-mus LD50:2 mg/kg JMCMAR 23,313,80

SAFETY PROFILE: Poison by ingestion and intraperitoneal routes. When heated to decomposition it emits toxic fumes of NO_x.

AEK750 CAS:57872-78-3 HR: 3
ADENOSINE-5′-(N-HEXYL)CARBOXAMIDE HEMIHYDRATE
mf: $C_{16}H_{24}N_6O_4 \cdot 1/2H_2O$ mw: 373.47

SYN: 1-(6-AMINO-9H-PURIN-9-YL)-1-DEOXY-N-HEXYLRIBOFURANURONAMIDE HEMIHYDRATE

TOXICITY DATA WITH REFERENCE
orl-mus LD50:500 mg/kg JMCMAR 23,313,80
ipr-mus LD50:200 mg/kg JMCMAR 23,313,80

SAFETY PROFILE: Poison by intraperitoneal route. Moderately toxic by ingestion. When heated to decomposition it emits toxic fumes of NO_x.

AEL000 CAS:35788-28-4 HR: 3
ADENOSINE-5′-(N-(2-HYDROXYETHYL))CARBOXAMIDE
mf: $C_{12}H_{16}N_6O_5$ mw: 324.34

SYN: 1-(6-AMINO-9H-PURIN-9-YL)-1-DEOXY-N-(2-HYDROXYETHYL) RIBOFURANURONAMIDE

TOXICITY DATA WITH REFERENCE
orl-mus LD50:5 mg/kg JMCMAR 23,313,80
ipr-mus LD50:2 mg/kg JMCMAR 23,313,80

SAFETY PROFILE: Poison by ingestion and intraperitoneal routes. When heated to decomposition it emits toxic fumes of NO_x.

AEL250 CAS:35788-29-5 HR: 3
ADENOSINE-5′-(N-ISOPROPYL)CARBOXAMIDE
mf: $C_{13}H_{18}N_6O_4$ mw: 322.37

SYN: 1-(6-AMINO-9H-PURIN-9-YL)-1-DEOXY-N-ISOPROPYLRIBOFURANURONAMIDE

TOXICITY DATA WITH REFERENCE
orl-mus LD50:5 mg/kg JMCMAR 23,313,80
ipr-mus LD50:5 mg/kg JMCMAR 23,313,80

SAFETY PROFILE: Poison by ingestion and intraperitoneal routes. When heated to decomposition it emits toxic fumes of NO_x.

AEL500 **CAS:54925-45-0** **HR: 3**
ADENOSINE-5′-(N-METHOXY)CARBOXAMIDE HYDRATE
mf: $C_{11}H_{14}N_6O_5 \cdot H_2O$ mw: 328.33

SYN: 1-(6-AMINO-9H-PURIN-9-YL)-1-DEOXY-N-METHOXYRIBOFURANURONAMIDE HYDRATE

TOXICITY DATA WITH REFERENCE
orl-mus LD50:50 mg/kg JMCMAR 23,313,80
ipr-mus LD50:20 mg/kg JMCMAR 23,313,80

SAFETY PROFILE: Poison by ingestion and intraperitoneal routes. When heated to decomposition it emits toxic fumes of NO_x.

AEL750 **CAS:35788-27-3** **HR: 3**
ADENOSINE-5′-(N-METHYL)CARBOXAMIDE HEMIHYDRATE
mf: $C_{11}H_{14}N_6O_4 \cdot 1/2H_2O$ mw: 303.32

SYN: 1-(6-AMINO-9H-PURIN-9-YL)-1-DEOXY-N-METHYLRIBOFURANURONAMIDE HEMIHYDRATE

TOXICITY DATA WITH REFERENCE
orl-mus LD50:20 mg/kg JMCMAR 23,313,80
ipr-mus LD50:5 mg/kg JMCMAR 23,313,80

SAFETY PROFILE: Poison by ingestion and intraperitoneal routes. When heated to decomposition it emits toxic fumes of NO_x.

AEM000 **CAS:57872-80-7** **HR: 3**
ADENOSINE-5′-(N-PROPYL)CARBOXAMIDE
mf: $C_{13}H_{18}N_6O_4$ mw: 322.37

SYN: 1-(6-AMINO-9H-PURIN-9-YL)-1-DEOXY-N-PROPYLRIBOFURANURONAMIDE

TOXICITY DATA WITH REFERENCE
orl-mus LD50:200 mg/kg JMCMAR 23,313,80
ipr-mus LD50:5 mg/kg JMCMAR 23,313,80

SAFETY PROFILE: Poison by ingestion and intraperitoneal routes. When heated to decomposition it emits toxic fumes of NO_x.

AEM100 **CAS:987-65-5** **HR: 3**
ADENOSINE 5′-(TETRAHYDROGENTRIPHOSPHATE), DISODIUM SALT
mf: $C_{10}H_{14}N_5O_{13}P_3 \cdot 2Na$ mw: 551.18

SYNS: ADENOSINE TRIPHOSPHATE DISODIUM □ ADETPHOS □ ATP DISODIUM □ ATP DISODIUM SALT □ DISODIUM ADENOSINE TRIPHOSPHATE □ DISODIUM ADENOSINE 5′-TRIPHOSPHATE □ DISODIUM ATP □ DISODIUM DIHYDROGEN ATP □ SODIUM ATP

TOXICITY DATA WITH REFERENCE
orl-rat LD50:>2 g/kg DRUGAY 6,20,82
scu-rat LD50:>2 g/kg DRUGAY 6,20,82
ivn-rat LD50:380 mg/kg DRUGAY 6,20,82
orl-mus LD50:>2 g/kg DRUGAY 6,20,82
scu-mus LD50:>2 g/kg DRUGAY 6,20,82
ivn-mus LD50:266 mg/kg DRUGAY 6,20,82

CONSENSUS REPORTS: Reported in EPA TSCA Inventory.

SAFETY PROFILE: Poison by intravenous route. Slightly toxic by ingestion and subcutaneous route. When heated to decomposition it emits toxic vapors of NO_x and SO_x.

AEM250 **CAS:15237-44-2** **HR: 2**
ADENOSINE-5′-(TETRAHYDROGENTRIPHOSPHATE) SODIUM SALT
mf: $C_{10}H_{16}N_5O_{13}P_3 \cdot 7Na$ mw: 668.15

SYNS: ATP Na SALT □ NaATP □ SODIUM ATP □ SODIUM ADENOSINE TRIPHOSPHATE □ SODIUM ADENOSINE-5′-TRIPHOSPHATE

TOXICITY DATA WITH REFERENCE
ipr-rat LD50:1379 mg/kg OYYAA2 4,689,70
ipr-mus LD50:1000 mg/kg ARZNAD 7,24,57

SAFETY PROFILE: Moderately toxic by intraperitoneal route. See also PHOSPHATES. When heated to decomposition it emits very toxic fumes of PO_x, Na_2O, and NO_x.

AEM500 **CAS:35170-28-6** **HR: 2**
5′-ADENYLIC ACID POTASSIUM SALT
mf: $C_{10}H_{14}N_5O_7P \cdot K$ mw: 386.36

SYNS: ADENOSINE-5′-MONOPHOSPHATE POTASSIUM SALT □ ADENOSINE-5′-MONOPHOSPHORIC ACID POTASSIUM SALT □ ADENOSINE-5′-PHOSPHATE POTASSIUM SALT □ ADENOSINE-5′-PHOSPHORIC ACID POTASSIUM SALT □ 5′-AMP POTASSIUM SALT

TOXICITY DATA WITH REFERENCE
orl-rat LD50:11,250 mg/kg OYYAA2 4,689,70
ipr-rat LD50:1310 mg/kg OYYAA2 4,689,70
scu-rat LD50:1493 mg/kg OYYAA2 4,689,70
orl-mus LD50:13,791 mg/kg OYYAA2 4,689,70
ipr-mus LD50:1955 mg/kg OYYAA2 4,689,70
scu-mus LD50:1937 mg/kg OYYAA2 4,689,70
ivn-mus LD50:536 mg/kg OYYAA2 4,689,70

SAFETY PROFILE: Moderately toxic by intraperitoneal, subcutaneous, and intravenous routes. Mildly toxic by ingestion. See also PHOSPHATES. When heated to decomposition it emits very toxic fumes of NO_x, K_2O, and PO_x.

AEM750 **CAS:13474-03-8** **HR: 2**
5′-ADENYLIC ACID, SODIUM SALT
mf: $C_{10}H_{14}N_5O_7P \cdot 7Na$ mw: 508.19

SYNS: ADENOSINE 5′-(DIHYDROGEN PHOSPHATE), SODIUM SALT □ ADENOSINE 5′-MONOPHOSPHATE SODIUM SALT □ AMP SODIUM SALT □ 5′-AMP SODIUM SALT □ NaAMP □ SODIUM ADENOSINE-5′-MONOPHOSPHATE □ SODIUM AMP

TOXICITY DATA WITH REFERENCE
ipr-rat LD50:2049 mg/kg OYYAA2 4,689,70
ipr-mus LD50:2000 mg/kg ARZNAD 7,24,57

SAFETY PROFILE: Moderately toxic by intraperitoneal route. When heated to decomposition it emits very toxic fumes of NO_x, Na_2O and PO_x.

AEN000 CAS:628-94-4 HR: 2
ADIPAMIDE
mf: $C_6H_{12}N_2O_2$ mw: 144.20

PROP: Crystals. Mp: 220°. Sol in alc.

SYNS: ADIPIC ACID DIAMIDE □ ADIPIC DIAMIDE □ 1,4-BUTANEDICARBOXAMIDE □ HEXANEDIAMIDE (9CI) □ NCI-C02095

TOXICITY DATA WITH REFERENCE
orl-rat TDLo:1270 mg/kg:CAR JEPTDQ 3(5-6),149,80
orl-rat LDLo:500 mg/kg JPETAB 90,260,47
orl-mus LD50:6000 mg/kg BIJOAK 34,1196,40

CONSENSUS REPORTS: Reported in EPA TSCA Inventory.

SAFETY PROFILE: Moderately toxic by ingestion. Questionable carcinogen with experimental carcinogenic data. When heated to decomposition it emits toxic fumes of NO_x.

AEN250 CAS:124-04-9 HR: 3
ADIPIC ACID
mf: $C_6H_{10}O_4$ mw: 146.16

PROP: White monoclinic prisms. Mp: 152°, flash p: 385°F (CC), d: 1.360 @ 25°/4°, vap press: 1 mm @ 159.5°, vap d: 5.04, autoign temp: 788°F, bp: 337.5°. Very sol in alc. Sol in acetone, water = 1.4% @ 15°; 0.6% @ 15° in ether.

SYNS: ACIFLOCTIN □ ACINETTEN □ ADILACTETTEN □ ADIPINIC ACID □ 1,4-BUTANEDICARBOXYLIC ACID □ FEMA No. 2011 □ 1,6-HEXANEDIOIC ACID □ KYSELINA ADIPOVA (CZECH) □ MOLTEN ADIPIC ACID

TOXICITY DATA WITH REFERENCE
eye-rbt 20 mg/24H SEV 28ZPAK -,51,72
orl-rat LD50:>11 g/kg GISAAA 48(9),72,83
ipr-rat LD50:275 mg/kg JAFCAU 5,759,57
orl-mus LD50:1900 mg/kg JAFCAU 5,759,57
ipr-mus LD50:275 mg/kg TXAPA9 32,566,75
ivn-mus LD50:680 mg/kg JAFCAU 5,759,57

CONSENSUS REPORTS: Reported in EPA TSCA Inventory.

ACGIH TLV: (Proposed: TWA 5 mg/³)

SAFETY PROFILE: Poison by intraperitoneal route. Moderately toxic by other routes. A severe eye irritant. Combustible when exposed to heat or flame; can react with oxidizing materials. When heated to decomposition it emits acrid smoke and fumes.

AEN750 CAS:1985-84-8 HR: 1
ADIPIC ACID BIS(3,4-EPOXY-6-METHYLCYCLOHEXYLMETHYL) ESTER
mf: $C_{22}H_{34}O_6$ mw: 394.56

SYNS: BIS(3,4-EPOXY-6-METHYLCYCLOHEXYLMETHYL)ADIPATE □ DI(3,4-EPOXY-6-METHYLCYCLOHEXYLMETHYL)ADIPATE □ HEXANEDIOIC ACID BIS(4-METHYL-7-OXABICYCLO(4.1.0)HEPT-3-YL) METHYL ESTER

TOXICITY DATA WITH REFERENCE
skn-rbt 500 mg open MLD UCDS** 9/23/70
orl-rat LD50:4290 mg/kg AIHAAP 24,305,63

CONSENSUS REPORTS: Reported in EPA TSCA Inventory.

SAFETY PROFILE: Mildly toxic by ingestion. A skin irritant. See also ESTERS. When heated to decomposition it emits acrid smoke and irritating fumes.

AEO000 CAS:103-23-1 HR: 2
ADIPIC ACID BIS(2-ETHYLHEXYL) ESTER
mf: $C_{22}H_{42}O_4$ mw: 370.64

PROP: Liquid. D: 0.927 @ 20°/4°, Bp: 181–185° @ 2 mm.

SYNS: ADIPOL 2EH □ BEHA □ BIS(2-ETHYLHEXYL) ADIPATE □ BISOFLEX DOA □ DEHA □ DI-2-ETHYLHEXYL ADIPATE □ DIOCTYL ADIPATE □ DOA □ EFFEMOLL DOA □ ERGOPLAST AdDO □ FLEXOL A 26 □ HEXANEDIOIC ACID, BIS(2-ETHYLHEXYL) ESTER □ HEXANEDIOIC ACID, DIOCTYL ESTER □ KODAFLEX DOA □ MONOPLEX DOA □ NCI-C54386 □ OCTYL ADIPATE □ PLASTOMOLL DOA □ PX-238 □ REOMOL DOA □ RUCOFLEX PLASTICIZER DOA □ SICOL 250 □ TRUFLEX DOA □ VESTINOL OA □ WICKENOL 158 □ WITAMOL 320

TOXICITY DATA WITH REFERENCE
eye-rbt 500 mg open AMIHBC 4,119,51
skn-rbt 500 mg open MLD UCDS** 1/12/72
pic-esc 25 µg/well MUREAV 260,349,91
dlt-mus-ipr 1000 mg/kg TXAPA9 32,566,75
ipr-rat TDLo:15 g/kg (5-15D preg):TER JPMSAE 62,1596,73
orl-mus TDLo:1038 g/kg/2Y-C:CAR NTPTR* NTP-TR-212,82
orl-mus TD:2163 g/kg/2Y-C:CAR NTPTR* NTP-TR-212,82
orl-rat LD50:9110 mg/kg AMIHBC 4,119,51
ivn-rat LD50:900 mg/kg MRLR** No.256,54
orl-mus LD50:15 g/kg JACTDZ 3(3),101,84
ivn-rbt LD50:540 mg/kg MRLR** No.256,54

CONSENSUS REPORTS: IARC Cancer Review: Group 3 IMEMDT 7,56,87; Animal Limited Evidence IMEMDT 29,257,82. NTP Carcinogenesis Bioassay (feed); Clear Evidence: mouse NTPTR* NTP-TR-212,82; No Evidence: rat NTPTR* NTP-TR-212,82. Community Right-To-Know List. Reported in EPA TSCA Inventory.

SAFETY PROFILE: Moderately toxic by intravenous route. Mildly toxic by ingestion. Experimental reproductive effects. Mutation data reported. An eye and skin irritant. Questionable carcinogen with experimental carcinogenic data. See also ESTERS. When heated to decomposition it emits acrid smoke and irritating fumes.

AEO250 CAS:63905-29-3 HR: 2
ADIPIC ACID-3-CYCLOHEXENYLMETHANOL DIESTER
mf: $C_{20}H_{30}O_4$ mw: 334.50

TOXICITY DATA WITH REFERENCE
orl-rat LD50:3730 mg/kg TXAPA9 28,313,74
skn-rbt LD50:7070 mg/kg TXAPA9 28,313,74

CONSENSUS REPORTS: Reported in EPA TSCA Inventory.

SAFETY PROFILE: Moderately toxic by ingestion. When

heated to decomposition it emits acrid smoke and irritating fumes.

AEO500 CAS:2998-04-1 HR: 3
ADIPIC ACID DIALLYL ESTER
mf: $C_{12}H_{18}O_4$ mw: 226.30

SYNS: ALLYL ADIPATE □ HEXANEDIOIC ACID-DI-2-PROPENYL ESTER

TOXICITY DATA WITH REFERENCE
orl-rat LDLo: 420 mg/kg SCCUR* -,3,61
orl-mus LD50: 180 mg/kg SCCUR* -,3,61
skn-rbt LDLo: 1 g/kg SCCUR* -,3,61

CONSENSUS REPORTS: Reported in EPA TSCA Inventory. Allyl compounds are on the Community Right-To-Know List.

SAFETY PROFILE: Poison by ingestion. Moderately toxic by skin contact. See also ALLYL COMPOUNDS and ESTERS. When heated to decomposition it emits acrid smoke and irritating fumes.

AEO750 CAS:105-99-7 HR: 2
ADIPIC ACID DIBUTYL ESTER
mf: $C_{14}H_{26}O_4$ mw: 258.40

SYNS: BUTYL ADIPATE □ DIBUTYL ADIPATE □ DI-N-BUTYL ADIPATE □ DIBUTYL ADIPINATE □ DIBUTYL HEXANEDIOATE □ EXPERIMENTAL TICK REPELLENT 3 □ HEXANEDIOIC ACID–DIBUTYL ESTER

TOXICITY DATA WITH REFERENCE
skn-rbt 10 mg/24H MLD AMIHBC 4,119,51
eye-rbt 500 mg AMIHBC 4,119,51
ipr-rat TDLo: 1049 mg/kg (5-15D preg): TER JPMSAE 62,1596,73
orl-rat LD50: 12,900 mg/kg 28ZEAL 5,72,76
ipr-rat LD50: 5244 mg/kg JPMSAE 62,1596,73
orl-rat LD50: 12,900 mg/kg 28ZEAL 5,72,76
ihl-rat LC: >17 mg/m³/4H GISAAA 55(6),86,90
ipr-rat LD50: 5244 μL/kg JPMSAE 62,1596,73
orl-mus LD50: 16,890 mg/kg GISAAA 55(6),86,90
skn-rbt LD50: 20,000 mg/kg AMIHBC 4,119,51

CONSENSUS REPORTS: Reported in EPA TSCA Inventory.

SAFETY PROFILE: Mildly toxic by several routes. An experimental teratogen. Skin and eye irritant. See also ESTERS. When heated to decomposition it emits acrid smoke and irritating fumes.

AEP000 HR: 1
ADIPIC ACID DIDECYL ESTER (mixed isomers)
mf: $C_{26}H_{50}O_4$ mw: 426.76

TOXICITY DATA WITH REFERENCE
skn-rbt 10 mg/24H open MLD AIHAAP 23,95,62
orl-rat LD50: 21 g/kg AIHAAP 23,95,62
skn-rbt LD50: 8410 mg/kg AIHAAP 23,95,62

SAFETY PROFILE: Mildly toxic by ingestion and skin contact. A skin irritant. See also ESTERS. When heated to decomposition it emits acrid smoke and irritating fumes.

AEP250 CAS:7790-07-0 HR: 2
ADIPIC ACID-(DI-2-(2-ETHYLBUTOXY)ETHYL) ESTER
mf: $C_{22}H_{42}O_6$ mw: 402.64

SYN: DI-2-(2-ETHYLBUTOXY)ETHYL ADIPATE

TOXICITY DATA WITH REFERENCE
skn-rbt 10 mg/24H open MLD AMIHBC 10,61,54
eye-rbt 500 mg open AMIHBC 10,61,54
orl-rat LD50: 3250 mg/kg AMIHBC 10,61,54
skn-rbt LD50: 4240 mg/kg AMIHBC 10,61,54

SAFETY PROFILE: Moderately toxic by ingestion. Mildly toxic by skin contact. Skin and eye irritant. See also ESTERS. When heated to decomposition it emits acrid smoke and irritating fumes.

AEP500 CAS:10022-60-3 HR: 2
ADIPIC ACID DI(2-ETHYLBUTYL) ESTER
mf: $C_{18}H_{34}O_4$ mw: 314.52

SYN: DI(2-ETHYLBUTYL) ADIPATE

TOXICITY DATA WITH REFERENCE
eye-rbt 500 mg open AMIHBC 10,61,54
orl-rat LD50: 5620 mg/kg AMIHBC 10,61,54
skn-rbt LD50: 17 g/kg AMIHBC 10,61,54

SAFETY PROFILE: Moderately toxic by skin contact. Mildly toxic by ingestion. An eye irritant. See also ESTERS. When heated to decomposition it emits acrid smoke and irritating fumes.

AEP750 CAS:141-28-6 HR: 2
ADIPIC ACID DIETHYL ESTER
mf: $C_{10}H_{18}O_4$ mw: 202.28

SYNS: DIETHYL ADIPATE □ DIETHYL HEXANEDIOATE □ ETHYL ADIPATE □ ETHYL-Δ-CARBOETHOXYVALERATE

TOXICITY DATA WITH REFERENCE
dlt-mus-ipr 1100 mg/kg TXAPA9 32,566,75
ipr-rat TDLo: 837 mg/kg (female 5-15D post): REP JPMSAE 62,1596,73
ipr-mus LD50: 2190 mg/kg TXAPA9 32,566,75

CONSENSUS REPORTS: Reported in EPA TSCA Inventory.

SAFETY PROFILE: Moderately toxic by intraperitoneal route. Experimental reproductive effects. Mutation data reported. See also ESTERS. When heated to decomposition it emits acrid smoke and irritating fumes.

AEQ000 CAS:110-32-7 HR: 1
ADIPIC ACID, DI(2-HEXYLOXYETHYL) ESTER
mf: $C_{22}H_{42}O_6$ mw: 402.64

SYNS: BIS(2-(HEXYLOXY)ETHYL)ADIPATE □ DIHEXYLOXYETHYL ADIPATE □ DI-(2-(2-HEXYLOXY)ETHYL)ESTER KYSELINY ADIPOVE □ HEXANEDIOIC ACID, BIS(2-(HEXYLOXY)ETHYL)ESTER (9CI)

TOXICITY DATA WITH REFERENCE
skn-rbt 10 mg/24H open MLD AMIHBC 10,61,54
skn-rbt 500 mg/24H MLD 85JCAE-,717,86
eye-rbt 500 mg open AMIHBC 10,61,54
eye-rbt 500 mg/24H MLD 85JCAE-,717,86
orl-rat LD50: 4290 mg/kg AMIHBC 10,61,54

skn-rbt LD50:12,310 µL/kg AMIHBC 10,61,54

CONSENSUS REPORTS: Reported in EPA TSCA Inventory.

SAFETY PROFILE: Mildly toxic by intraperitoneal and skin contact routes. A skin and eye irritant. Mutation data reported. When heated to decomposition it emits acrid smoke and irritating fumes.

AEQ250 CAS:1071-93-8 HR: 2
ADIPIC ACID DIHYDRAZIDE
mf: $C_6H_{14}N_4O_2$ mw: 174.24

SYNS: ADIPIC DIHYDRAZIDE □ HEXANEDIOIC ACID DIHYDRAZIDE

TOXICITY DATA WITH REFERENCE
par-mus LDLo:4000 mg/kg CBCCT* 7,685,55

CONSENSUS REPORTS: Reported in EPA TSCA Inventory.

SAFETY PROFILE: Moderately toxic by parenteral route. When heated to decomposition it emits toxic fumes of NO_x.

AEQ500 CAS:6624-70-0 HR: 2
ADIPIC ACID DIISOPENTYL ESTER
mf: $C_{16}H_{30}O_4$ mw: 286.46

SYNS: BIS(3-METHYLBUTYL) ADIPATE □ DIISOAMYL ADIPATE □ DI(3-METHYLBUTYL)ADIPATE □ HEXANEDIOIC ACID, BIS(3-METHYLBUTYL) ESTER

TOXICITY DATA WITH REFERENCE
orl-gpg LD50:25 g/kg GWXXBX #2703360
ivn-rat LD50:640 mg/kg MRLR** No.256,54
ivn-rbt LD50:640 mg/kg MRLR** No.256,54

SAFETY PROFILE: Moderately toxic by intravenous route. See also ESTERS. When heated to decomposition it emits acrid smoke and irritating fumes.

AEQ750 CAS:6900-06-7 HR: 3
ADIPIC ACID DI-2-PROPYNYL ESTER

TOXICITY DATA WITH REFERENCE
orl-rat LD50:200 mg/kg AIHAAP 23,95,62
skn-rbt LD50:440 mg/kg AIHAAP 23,95,62

SAFETY PROFILE: Poison by ingestion. Moderately toxic by skin contact. See also ESTERS. When heated to decomposition it emits acrid smoke and irritating fumes.

AER000 HR: 2
ADIPIC ACID, UREA mixed with CARBOXYMETHYLCELLULOSE ACIDS

PROP: Consists of 97.3% urea, 0.6% adipic acid, and 2.1% carboxymethylcellulose acids (ANYAA9 75,543,59).

TOXICITY DATA WITH REFERENCE
ivg-mus TDLo:91 g/kg/76W-I:ETA ANYAA9 75,543,59

SAFETY PROFILE: Questionable carcinogen with experimental tumorigenic data.

AER250 CAS:111-69-3 HR: 3
ADIPONITRILE
DOT: UN 2205
mf: $C_6H_8N_2$ mw: 108.16

PROP: Water-white liquid, practically odorless. Mp: 2.3°, bp: 295°, flash p: 199.4°F (OC), d: 0.965 @ 20°/4°, vap d: 3.73. Sol in EtOH, $CHCl_3$; insol in H_2O, Et_2O, CS_2.

SYNS: ADIPIC ACID DINITRILE □ ADIPIC ACID NITRILE □ ADIPODINITRILE □ 1,4-DICYANOBUTANE □ HEXANEDINITRILE □ HEXANEDIOIC ACID DINITRILE □ NITRILE ADIPICO (ITALIAN) □ TETRAMETHYLENE CYANIDE

TOXICITY DATA WITH REFERENCE
orl-rat LD50:155 mg/kg GISAAA 49(12),40,84
ihl-rat LC50:1710 mg/m³/4H TOXID9 1,76,81
orl-mus LD50:172 mg/kg ARTODN 57,88,85
ipr-mus LD50:40 mg/kg NTIS** AD691-490
orl-rbt LD50:22 mg/kg GISAAA 49(12),40,84
scu-gpg LD50:50 mg/kg MELAAD 46,221,55

CONSENSUS REPORTS: EPA Extremely Hazardous Substances List. Reported in EPA TSCA Inventory. Cyanide and its compounds are on the Community Right-To-Know List.

ACGIH TLV: TWA 2 ppm (skin)
NIOSH REL: TWA 18 mg/m³
DOT CLASSIFICATION: 6.1; *Label:* KEEP AWAY FROM FOOD

SAFETY PROFILE: Poison by inhalation, ingestion, subcutaneous, and intraperitoneal routes. The nitrile group will behave as a cyanide when ingested or absorbed in the body. It produces disturbances of the respiration and circulation, irritation of the stomach and intestines, and loss of weight. Its low vapor pressure at room temperature makes exposure to harmful concentrations of its vapors unlikely if handled with reasonable care in well-ventilated areas. Flammable when exposed to heat or flame. When heated to decomposition it emits toxic fumes of CN^-. Can react with oxidizing materials. To fight fire, use foam, CO_2, dry chemical. See also HYDROCYANIC ACID and NITRILES.

AER500 CAS:35108-88-4 HR: 3
ADOBIOL
mf: $C_{18}H_{29}NO_4 \cdot ClH$ mw: 359.94

PROP: Mp: 151–154°.

SYNS: BUFETOLOL HYDROCHLORIDE □ 1-(tert-BUTYLAMINO)-3-(o-((TETRAHYDROFURFURYL)OXY)PHENOXY)-2-PROPANOL HYDROCHLORIDE

TOXICITY DATA WITH REFERENCE
orl-rat TDLo:300 mg/kg (9-14D preg):REP OYYAA2 6,1267,72
orl-mus TDLo:60 mg/kg (female 7-12D post):TER OYYAA2 6,1267,72
orl-rat LD50:1088 mg/kg DRUGAY 6,681,82
scu-rat LD50:1814 mg/kg DRUGAY 6,681,82
ivn-rat LD50:59,400 µg/kg DRUGAY 6,681,82

orl-mus LD50:402 mg/kg DRUGAY 6,681,82
scu-mus LD50:501 mg/kg DRUGAY 6,681,82

SAFETY PROFILE: Poison by intravenous route. Moderately toxic by ingestion and subcutaneous routes. An experimental teratogen. Other experimental reproductive effects. When heated to decomposition it emits very toxic fumes of HCl and NO_x. An antiarrhythmic drug.

AER666 CAS:51460-26-5 HR: 1
ADONA TRIHYDRATE
mf: $C_{10}H_{11}N_4O_5S \cdot Na \cdot 3H_2O$ mw: 376.36

PROP: Yellow-orange needles from aq methanol. Decomp 227–228°. Soluble in water.

SYNS: AC-17 TRIHYDRATE □ ADONA TRIHYDRATE □ ADRENOCHROME SULFONATE AC 17 TRIHYDRATE □ CARBAZOCHROME SODIUM SULFONATE TRIHYDRATE □ 1H-INDOLE-2-SULFONIC ACID, 5-((AMINOCARBONYL)HYDRAZONO)-2,3,5,6-TETRAHYDRO-1-METHYL-6-OXO-, MONOSODIUM SALT, TRIHYDRATE □ SODIUM 1-METHYL-5-SEMICARBAZONO-6-OXO-2,3,5,6-TETRAHYDROINDOLE-3-SULFONATE TRIHYDRATE

TOXICITY DATA WITH REFERENCE
ipr-rat TDLo:1280 mg/kg (female 7-14D post):TER OYYAA2 4,39,70
ipr-rat TDLo:1280 mg/kg (female 7-14D post):REP OYYAA2 4,39,70
orl-mus LD50:>10 g/kg DRUGAY 6,183,82
ivn-mus LD50:>600 mg/kg DRUGAY 6,183,82
orl-dog LD50:>50 g/kg DRUGAY 6,183,82
ivn-dog LD50:>600 mg/kg DRUGAY 6,183,82

SAFETY PROFILE: Very low toxicity by ingestion and intravenous routes. An experimental teratogen. Other experimental reproductive effects. When heated to decomposition it emits toxic fumes of SO_x, NO_x, and Na_2O.

AER750 CAS:8002-01-5 HR: 3
ADONIDIN

TOXICITY DATA WITH REFERENCE
ivn-pgn LDLo:2829 µg/kg YHHPAL 10,561,63
ivn-cat LDLo:3 mg/kg 27ZWAY E.1,78
ivn-rbt LDLo:5 mg/kg 27ZWAY E.1,78
scu-frg LDLo:4 mg/kg 27ZWAY E.1,78

SAFETY PROFILE: Poison by intravenous and subcutaneous routes.

AES000 CAS:51-42-3 HR: 3
ADRENALIN BITARTRATE
mf: $C_9H_{13}NO_3 \cdot C_4H_6O_6$ mw: 333.33

SYNS: ADRENALINE ACID TARTRATE □ (−)-ADRENALINE ACID TARTRATE □ ADRENALINE BITARTRATE □ (−)-ADRENALINE BITARTRATE □ 1-ADRENALINE BITARTRATE □ 1-ADRENALINE-d-BITARTRATE □ ADRENALINE HYDROGEN TARTRATE □ 1-ADRENALINE HYDROGEN TARTRATE □ (−)-ADRENALINE HYDROGEN TARTRATE □ ADRENALINE TARTRATE □ (−)-ADRENALINE TARTRATE □ l-ADRENALINE TARTRATE □ ASMATANE MIST □ (−)-3,4-DIHYDROXY-α-((METHYLAMINO)METHYL)BENZYL) ALCOHOL (+)-TARTRATE (1:1) SALT □ EPINEPHRINE BITARTRATE □ (−)-EPINEPHRINE BITARTRATE □ l-EPINEPHRINE BITARTRATE □ EPINEPHRINE-d-BITARTRATE □ 1-EPINEPHRINE-d-BITARTRATE □ EPINEPHRINE HYDROGEN TARTRATE □ 1-EPINEPHRINE TARTRATE □ IOP □ LYOPHRIN □ MEDIHALER-EPI □ SUPRARENIN

TOXICITY DATA WITH REFERENCE
dni-mus:oth 1 µmol/L CNREA8 43,3514,83
scu-mus TDLo:2400 µg/kg (2D male):REP JRPFA4 22,375,70
scu-rat LD50:8300 µg/kg NIIRDN 6,122,82
ivn-rat LD50:82 µg/kg AIPTAK 137,155,62
orl-mus LD50:4 mg/kg APTOA6 31,49,72
ipr-mus LD50:7800 µg/kg APTOA6 31,43,72
scu-mus LD50:11,100 µg/kg APTOA6 55,73,84
ivn-mus LD50:1780 µg/kg JPETAB 81,269,44

SAFETY PROFILE: Poison by ingestion, subcutaneous, intraperitoneal, and intravenous routes. Experimental reproductive effects. Mutation data reported. When heated to decomposition it emits toxic fumes of NO_x. See also VASOTONIN and other adrenalin compounds.

AES250 CAS:150-05-0 HR: 3
d-ADRENALINE
mf: $C_9H_{13}NO_3$ mw: 183.23

PROP: Light brown or nearly white crystals. Mp: 211–212°. Very sltly sol in water, alc, 1:1 chloroform, and ether.

SYNS: 1-(+)-ADRENALINE □ d-EPINEPHRINE

TOXICITY DATA WITH REFERENCE
scu-rat LDLo:80 mg/kg JPHYA7 38,259,09
ivn-rat LD50:800 µg/kg JPETAB 95,502,49
scu-mus LDLo:4 mg/kg HBAMAK 4,1294,35
ivn-mus LD50:38 mg/kg JPETAB 95,502,49
scu-dog LDLo:5 mg/kg HBAMAK 4,1294,35
ivn-dog LDLo:1 mg/kg HBAMAK 4,1294,35
ivn-cat LDLo:500 µg/kg HBAMAK 4,1294,35
scu-rbt LDLo:10 mg/kg HBAMAK 4,1294,35
ivn-rbt LDLo:50 µg/kg HBAMAK 4,1294,35
scu-gpg LDLo:1 mg/kg HBAMAK 4,1294,35
scu-frg LDLo:5000 mg/kg HBAMAK 4,1294,35

SAFETY PROFILE: Poison by subcutaneous and intravenous routes. Can cause contact dermatitis. Usually the symptoms are of short duration and clear up spontaneously. Combustible when heated. Upon decomposition it emits toxic fumes of NO_x.

AES500 CAS:55-31-2 HR: 3
1-ADRENALINE CHLORIDE
mf: $C_9H_{13}NO_3 \cdot ClH$ mw: 219.69

SYNS: ADRENALIN CHLORIDE □ ADRENALIN HYDROCHLORIDE □ (−)-ADRENALINE HYDROCHLORIDE □ 1-ADRENALINE HYDROCHLORIDE □ 1,2-BENZENEDIOL, 4-(1-HYDROXY-2-(METHYLAMINO) ETHYL)-, HYDROCHLORIDE, (R)- (9CI) □ 1-1-(3,4-DIHYDROXYPHENYL)-2-METHYLAMINO-1-ETHANOL HYDROCHLORIDE □ EPINEPHRINE CHLORIDE □ 1-EPINEPHRINE CHLORIDE □ (−)-EPINEPHRINE HYDROCHLORIDE □ 1-EPINEPHRINE HYDROCHLORIDE □ GELATIN-EPINEPHRINE □ 1-METHYLAMINOETHANOLCATHECHOL HYDROCHLORIDE □ NCI-C55663 □ SUPRANEPHRIN SOLUTION □ SUPRARENIN HYDROCHLORIDE

TOXICITY DATA WITH REFERENCE
scu-rat TDLo:3200 μg/kg (13-20D preg):REP CJPPA3 43,473,65
scu-rat TDLo:3200 μg/kg (13-20D preg):TER CJPPA3 43,473,65
orl-rat LD50:24 mg/kg AIPTAK 180,155,69
scu-rat LD50:5 mg/kg AIPTAK 180,155,69
ivn-rat LDLo:50 μg/kg JPETAB 24,101,24
orl-mus LDLo:50 mg/kg ARZNAD 13,51,63
ipr-mus LD50:4664 μg/kg JPETAB 90,110,47
scu-mus LD50:1980 μg/kg JPETAB 87,214,46
ivn-mus LD50:140 μg/kg EJPHAZ 9,289,70

CONSENSUS REPORTS: Reported in NTP Carcinogenesis Studies (Inhalation); Inadequate Study: RAT, MOUSE NTPTR* NTP-TR-380,90.

SAFETY PROFILE: Poison by ingestion, intravenous, subcutaneous, and intraperitoneal routes. An experimental teratogen. Other experimental reproductive effects. When heated to decomposition it emits very toxic fumes of Cl^- and NO_x. See also VASOTONIN and other adrenalin compounds.

AES625 CAS:329-63-5 HR: 3
dl-ADRENALINE HYDROCHLORIDE
mf: $C_9H_{13}NO_3$•ClH mw: 219.69

SYNS: (±)-ADRENALINE HYDROCHLORIDE □ (±)-3,4-DIHYDROXY-α-((METHYLAMINO)METHYL)BENZYL ALCOHOL HYDROCHLORIDE □ (±)-EPINEPHRINE HYDROCHLORIDE □ dl-EPINEPHRINE HYDROCHLORIDE

TOXICITY DATA WITH REFERENCE
ipr-rat LD50:1250 μg/kg YAKUD5 24,1705,82
ivn-rat LD50:70 μg/kg YAKUD5 24,1705,82
orl-mus LD50:90 mg/kg YAKUD5 24,1705,82
ipr-mus LD50:7800 μg/kg JPETAB 92,369,48
ivn-mus LDLo:5 mg/kg JPETAB 92,108,48

CONSENSUS REPORTS: Reported in EPA TSCA Chemical Inventory.

SAFETY PROFILE: Poison by ingestion, intravenous, and intraperitoneal routes. When heated to decomposition it emits toxic fumes of NO_x and HCl.

AES639 CAS:54-06-8 HR: 3
ADRENOCHROME
mf: $C_9H_9NO_3$ mw: 179.19

SYNS: ADRAXONE □ 2,3-DIHYDRO-3-HYDROXY-1-METHYL-1H-INDOLE-5,6-DIONE (9CI) □ 3-HYDROXY-1-METHYL-5,6-INDOLINEDIONE □ USAF UCTL-7

TOXICITY DATA WITH REFERENCE
ipr-rat LD50:150 mg/kg AIPTAK 106,90,56
ipr-mus LD50:100 mg/kg NTIS** AD277-689
ivn-mus LD50:128 mg/kg AIPTAK 106,90,56

SAFETY PROFILE: Poison by intravenous and intraperitoneal routes. When heated to decomposition it emits toxic fumes of NO_x.

AES650 CAS:9002-60-2 HR: 1
ADRENOCORTICOTROPHIC HORMONE

PROP: White powder. Freely sol in water. Appreciably soluble in 60 to 70% alc or acetone. One U.S.P. unit, one international unit, one Armour unit, or one potency unit denotes the same activity.

SYNS: ACETHROPAN □ ACORTAN □ ACORTO □ ACTH □ ACTHAR □ ACTON □ ACTONAR □ ADRENAL CORTEX HORMONE □ ADRENOCORTICOTROPHIN □ ADRENOCORTICOTROPIC HORMONE □ ADRENOCORTICOTROPIN □ ADRENOMONE □ ADRENOTROPHIN □ ALFATROFIN □ CIBACTHEN □ CORSTILINE □ CORTICOTROPHIN □ CORTICOTROPIN □ CORTICOTROPIN-LIKE SUBSTANCES □ CORTIPHYSON □ CORTROPHIN □ CORTROPHYSON □ DYNAMONE □ EXACTHIN □ ISLACTID □ PITUITARY GLAND ADRENO CORTICO-TROPIC HORMONE □ REACTHIN □ SOLACTHYL □ TUBEX

TOXICITY DATA WITH REFERENCE
scu-mus TDLo:3600 μg/kg (female 12-17D post):REP TJADAB 35,229,87
par-rbt TDLo:10,500 μg/kg (female 19-26D post):TER JOENAK 14,284,56
unr-inf TDLo:240 mg/kg/16W-I:SYS LANCAO 1,901,84

SAFETY PROFILE: Human systemic effects: kidney changes. An experimental teratogen. Other experimental reproductive effects.

AES750 CAS:23214-92-8 HR: 3
ADRIAMYCIN
mf: $C_{27}H_{29}NO_{11}$•ClH mw: 543.57

PROP: Isolated from cultures of *Streptomyces peucetius var. Caesius.*

SYNS: ADM □ ADRIAMYCIN-HCl □ ADRIAMYCIN SEMIQUINONE □ ADRIBLASTINA □ DOXORUBICIN □ DX □ FI 106 □ 14-HYDROXYDAUNOMYCIN □ 14'-HYDROXYDAUNOMYCIN □ 14-HYDROXYDAUNORUBICINE □ KW-125 □ NCI-C01514 □ NSC-123127

TOXICITY DATA WITH REFERENCE
mmo-smc 184 μmol/L MGGEAE 174,39,79
cyt-hmn:leu 20 μg/L CNREA8 31,32,71
unr-rat TDLo:4500 μg/kg (female 10-12D post):REP TJADAB 33,39C,86
ivn-mus TDLo:7 mg/kg (female 7-13D post):TER YKRYAH 6,1152,73
ivn-rat TDLo:5 mg/kg:CAR CNREA8 43,5248,83
ivn-rat TD:8 mg/kg:NEO EXPEAM 27,1209,71
ivn-rat TD:10 mg/kg:CAR JJIND8 66,81,81
ivn-rat TD:5 mg/kg:ETA PAACA3 21,309,80
ivn-hmn TDLo:15 mg/kg/D:CVS,GIT,SKN CANCAR 34,518,74
ivn-hmn TDLo:380 mg/kg/31W:CVS,GIT,SKN CANCAR 34,518,74
ipr-rat LD50:16 mg/kg OYYAA2 6,1075,72
orl-mus LD50:570 mg/kg ANTBAL 28,298,83
ipr-mus LD50:10,700 μg/kg HYDXET 20,303,89
scu-mus LD50:15,980 μg/kg KSRNAM 7,1052,73
ivn-mus LD50:8950 μg/kg KSRNAM 7,1052,73
ivn-dog LD50:2400 μg/kg DCTODJ 6,21,83

CONSENSUS REPORTS: NTP 7th Annual Report On Carcinogens. IARC Cancer Review: Group 2A IMEMDT 7,82,87; Animal Sufficient Evidence IMEMDT 7,82,87; Human Inadequate Evidence IMEMDT 7,82,87.

SAFETY PROFILE: Confirmed carcinogen with experimental carcinogenic, neoplastigenic, and tumorigenic data. Poison by intraperitoneal, subcutaneous, parenteral, and intravenous routes. Human systemic effects by intravenous route: cardiac myopathy including infarction, nausea or vomiting, and effects on the hair. An experimental teratogen. Other experimental reproductive effects. Human mutation data reported. When heated to decomposition it emits very toxic fumes of NO_x and HCl.

AET250 CAS:51898-39-6 HR: 3
ADRIAMYCIN-14-OCTANOATEHYDROCHLORIDE
mf: $C_{35}H_{43}NO_{12}$ mw: 669.79

TOXICITY DATA WITH REFERENCE
dnd-mam:lym 3490 nmol/L CBINA8 20,97,78
ivn-mus LD50:19 mg/kg 31TFAO 3,987,74

SAFETY PROFILE: Poison by intravenous route. Mutation data reported. When heated to decomposition it emits toxic fumes of NO_x.

AET500 HR: 3
AEROMONAS HYDROPHILA A_3 ENDOTOXIN

SYN: ENDOTOXIN, AEROMONAS HYDROPHILA A_3

TOXICITY DATA WITH REFERENCE
ipr-mus LD50:42,950 μg/kg CUMIDD 1,175,78

SAFETY PROFILE: A poison by intraperitoneal route. When heated to decomposition it emits acrid smoke and irritating vapors.

AET750 CAS:1402-68-2 HR: 3
AFLATOXIN

TOXICITY DATA WITH REFERENCE
mnt-rat-orl 8500 μg/kg DCTODJ 10,291,87
dlt-mus-ipr 68 mg/kg NATUAS 219,385,68
par-ham TDLo:4 mg/kg (8D preg):TER DABBBA 34,5251,73
orl-rat TDLo:7788 μg/kg/13W-C:ETA NATUAS 202,1016,64
orl-rat TDLo:2250 μg/kg (10-21D preg):ETA,REP CNREA8 33,262,73
orl-hmn LDLo:229 μg/kg/8W LANCAO 1,1061,75
orl-mky LD50:1750 μg/kg FCTXAV 14,227,76
ims-mky LD50:2020 mg/kg FCTXAV 14,227,76
orl-qal LDLo:4 mg/kg BPOSA4 21,29,80

CONSENSUS REPORTS: NTP 7th Annual Report On Carcinogens. IARC Cancer Review: Group 1 IMEMDT 7,83,87; Human Sufficient Evidence IMEMDT 7,83,87; Animal Sufficient Evidence IMEMDT 7,83,87.

SAFETY PROFILE: Confirmed human carcinogen with experimental tumorigenic data. Human poison by ingestion. An experimental teratogen. Other experimental reproductive effects. Mutation data reported. See also various aflatoxins.

AEU250 CAS:1162-65-8 HR: 3
AFLATOXIN B1
mf: $C_{17}H_{12}O_6$ mw: 312.29

PROP: A crystalline material. Mp: 268°.

SYNS: AFBI □ AFLATOXIN B

TOXICITY DATA WITH REFERENCE
pic-sat 1 μg/L ENMUDM 1,121,79
cyt-hmn:lym 19,200 nmol/L TOLED5 7,245,81
sce-hmn:lym 19,200 nmol/L TOLED5 7,245,81
mma-sat 10 ng/plate FCTOD7 22,355,84
scu-rat TDLo:1200 μg/kg (female 11-14D post):REP TOLED5 18(Suppl 1),113,83
ipr-ham TDLo:4 mg/kg (female 8D post):TER NATUAS 215,638,67
orl-rat TDLo:380 μg/kg/68W-C:CAR CNREA8 27,2370,67
ipr-rat TDLo:2 mg/kg (female 18-21D post):NEO,TER JJIND8 64,1349,80
ipr-rat TDLo:6 mg/kg/8W-I:CAR CNREA8 31,1936,71
ivn-rat TDLo:8 mg/kg (female 15D post):NEO,TER IARCCD 4,100,73
orl-mus TDLo:150 mg/kg/6W-I:NEO TXAPA9 82,19,86
scu-mus TDLo:30 mg/kg (female 15-22D post):NEO,TER BEXBAN 82,1687,76
orl-mky TDLo:168 mg/kg/6Y-C:ETA JJIND8 57,67,76
orl-rat LD50:4800 μg/kg CNREA8 27,2370,67
ipr-rat LD50:6 mg/kg TXAPA9 25,458,73
orl-mus LD50:9 mg/kg APPYAG 12,303,74
ipr-mus LD50:9500 μg/kg LSPPAT 13,1143,73
ipr-dog LDLo:1 mg/kg PAVEAC 3,331,66
orl-mky LD50:2200 μg/kg TXAPA9 19,169,71
orl-cat LD50:550 μg/kg CNREA8 29,236,69
orl-pig LD50:620 μg/kg APPYAG 12,303,74
orl-gpg LD50:2 mg/kg TXAPA9 19,169,71
ipr-gpg LD50:1400 μg/kg JPBAA7 91,277,66
orl-ham LD50:10 mg/kg CNREA8 29,236,69
ipr-ham LD50:6 mg/kg ARPAAQ 83,53,67
orl-dck LD50:335 μg/kg PSEBAA 123,151,66
orl-dom LDLo:2 mg/kg NATUAS 225,1062,70

CONSENSUS REPORTS: IARC Cancer Review: Group 1 IMEMDT 7,83,87; Animal Sufficient Evidence IMEMDT 10,51,76; 1,145,72. EPA Genetic Toxicology Program.

SAFETY PROFILE: Confirmed human carcinogen with experimental tumorigenic, neoplastigenic, and carcinogenic data. Acute poison by ingestion, intraperitoneal, and possibly other routes. Experimental teratogenic and reproductive effects. Mutation data reported. When heated to decomposition it emits acrid smoke. See also various aflatoxins.

AEU500 CAS:58209-98-6 HR: 2
AFLATOXIN B1-2,3-DICHLORIDE
mf: $C_{17}H_{12}Cl_2O_6$ mw: 383.19

SYNS: AFLATOXIN B1 DICHLORIDE □ CYCLOPENTA(c)FURO(3′,2′:4,5)FURO(2,3-h)(1)BENZOPYRAN-1,11-DIONE, 2,3,6a,8,9,9a-HEXAHYDRO-8,9-DICHLORO-4-METHOXY-, (6aS-(6a-α-8-β,9-α-9aα-))- □ 2,3-DICHLOROAFLATOXIN B(sub 1)

TOXICITY DATA WITH REFERENCE
dnd-hmn:fbr 4 μmol/L CBINA8 50,59,84
msc-hmn:fbr 4 nmol/L CBINA8 50,59,84
sln-dmg-par 200 nmol/L CNREA8 38,2608,78

scu-rat TDLo: 197 μg/kg/10W-I:NEO CNREA8 35,3811,75

SAFETY PROFILE: Questionable carcinogen with experimental neoplastigenic data. Human mutation data reported. When heated to decomposition it emits toxic fumes of Cl^-. See also various aflatoxins.

AEU750 CAS:7220-81-7 HR: 3
AFLATOXIN B2
mf: $C_{17}H_{14}O_6$ mw: 314.31

PROP: Yellow crystals with blue fluorescence from MeOH.

SYN: DIHYDROAFLATOXIN B1

TOXICITY DATA WITH REFERENCE
mma-sat 370 ng/plate MUREAV 130,79,84
dnd-rat-par 40 μg/kg/2D-C BBRCA9 83,1354,78
sce-ham:lng 3100 μg/L CRNGDP 1,759,80
dnd-mam:lym 50 μmol/L CRNGDP 3,423,82
dns-rat:lvr 10 μmol/L/1H CNREA8 37,1845,77
mma-ham:lng 83 μmol/L MUREAV 46,27,77
ipr-rat TDLo:600 mg/kg/8W-I:ETA CNREA8 31,1936,71
orl-dck LD50:1700 μg/kg NATUAS 200,1101,63

CONSENSUS REPORTS: IARC Cancer Review: Animal Sufficient Evidence IMEMDT 10,51,76; Animal Limited Evidence IMEMDT 1,145,72.

SAFETY PROFILE: Confirmed human carcinogen with experimental tumorigenic data. Poison by ingestion. Mutation data reported. When heated to decomposition it emits acrid smoke and fumes. See also various aflatoxins.

AEV000 CAS:1165-39-5 HR: 3
AFLATOXIN G1
mf: $C_{17}H_{12}O_7$ mw: 328.29

PROP: Needles from MeOH exhibiting green fluoresence. Mp: 247–250°. Metabolite of *Aspergillus flavus link ex fries*.

TOXICITY DATA WITH REFERENCE
mma-sat 31 ng/plate MUREAV 130,79,84
cyt-mky:kdy 2 mg/L/2H-C JNCIAM 48,1647,72
orl-rat TDLo:4100 μg/kg/22W-I:CAR CNREA8 35,2469,75
orl-rat TDLo:5600 μg/kg/2W-I:CAR CNREA8 31,1936,71
scu-rat TDLo:12 mg/kg/30W-I:NEO BJCAAI 19,392,65
ipr-rat LD50:14,900 μg/kg JPTLAS 102,209,70
orl-dck LD50:785 μg/kg PSEBAA 123,151,66

CONSENSUS REPORTS: IARC Cancer Review: Animal Sufficient Evidence IMEMDT 10,51,76.

SAFETY PROFILE: Confirmed human carcinogen with experimental carcinogenic and neoplastigenic data. Poison by ingestion and intraperitoneal routes. Mutation data reported. When heated to decomposition it emits acrid smoke and irritating fumes. See also various aflatoxins.

AEV250 HR: 3
AFLATOXIN G1 mixed with AFLATOXIN B1

PROP: Metabolites of *Aspergillus flavus link ex fries*, Aflatoxin G1, 56.4%; Alfatoxin B1, 37.7%.

TOXICITY DATA WITH REFERENCE
orl-rat TDLo:66 mg/kg/66W-C:ETA BJCAAI 20,134,66
scu-rat TDLo:1760 μg/kg/44W-I:NEO BJCAAI 19,392,65
itr-rat TDLo:72 mg/kg/30W-I:CAR BJCAAI 20,134,66

SAFETY PROFILE: Confirmed human carcinogen with experimental carcinogenic, neoplastigenic, and tumorigenic data. See also various aflatoxins.

AEV500 CAS:7241-98-7 HR: 3
AFLATOXIN G2
mf: $C_{17}H_{14}O_7$ mw: 330.31

PROP: Crystals with green fluoresence from ethanol. Mp: 237–240°.

SYN: DIHYDROAFLATOXIN G1

TOXICITY DATA WITH REFERENCE
dns-rat:lvr 10 μmol/L/1H CNREA8 37,1845,77
sce-ham:lng 3300 μg/L CRNGDP 1,759,80
orl-dck LD50:2450 μg/kg NATUAS 200,1101,63

CONSENSUS REPORTS: IARC Cancer Review: Animal Inadequate Evidence IMEMDT 1,145,72. EPA Genetic Toxicology Program.

SAFETY PROFILE: Suspected carcinogen. Acute poison by ingestion. Mutation data reported. When heated to decomposition it emits acrid smoke and irritating fumes. See also various aflatoxins.

AEW000 CAS:6795-23-9 HR: 3
AFLATOXIN M1
mf: $C_{17}H_{12}O_7$ mw: 328.29

PROP: Crystals from MeOH exhibiting blue-violet fluor esence. Mp: 299° (decomp).

SYN: 4-HYDROXYAFLATOXIN B1

TOXICITY DATA WITH REFERENCE
dnd-rat-orl 3600 ng/kg CBINA8 32,249,80
cyt-rat-orl 1 mg/kg JNCIAM 47,585,71
mma-sat 200 ng/plate JEPTDQ 2,1099,79
dns-rat:lvr 600 ng/plate TOXID9 1,42,81
orl-rat TDLo:8 mg/kg/8W-I:ETA FCTXAV 12,381,74
orl-rat LDLo:1500 μg/kg JNCIAM 47,585,71

CONSENSUS REPORTS: IARC Cancer Review: Group 2B; Animal Sufficient Evidence IMEMDT 10,51,76; Animal Sufficient Evidence IMEMDT 56,245,93; Human Inadequate Evidence IMEMDT 56,245,93. EPA Genetic Toxicology Program.

SAFETY PROFILE: Confirmed carcinogen with experimental tumorigenic data. Poison by ingestion. Mutation data reported. When heated to decomposition it emits acrid smoke and irritating fumes. See also various aflatoxins.

AEW500 **CAS:29611-03-8** **HR: 2**
AFLATOXIN Ro
mf: $C_{17}H_{14}O_6$ mw: 314.31

PROP: Crystals from C_6H_6/hexane. Mp: 224–226°.

SYNS: AFL □ AFLATOXICOL □ AFLATOXICOL NATURAL EPIMER

TOXICITY DATA WITH REFERENCE
mma-sat 25 ng/plate PNASA6 73,2241,76
orl-rat TDLo: 1092 g/kg/1Y-C:CAR JJIND8 66,1159,81

SAFETY PROFILE: Suspected carcinogen with experimental carcinogenic data. Mutation data reported. When heated to decomposition it emits acrid smoke and irritating fumes. See also various aflatoxins.

AEW625 **CAS:56287-74-2** **HR: 3**
AFLOQUALONE
mf: $C_{16}H_{14}FN_3O$ mw: 283.33

PROP: Pale-yellow prisms from 2-propanol. Mp: 195–196°.

SYNS: 6-AMINO-2-(FLUOROMETHYL)-3-(2-METHYLPHENYL)-4(3H)-QUINAZOLINONE (9CI) □ 6-AMINO-2-FLUOROMETHYL-3-(o-TOLYL)-4(3H)-QUINAZOLINONE □ AROFT □ AROFUTO

TOXICITY DATA WITH REFERENCE
orl-rat LD50: 249 mg/kg IYKEDH 14,297,83
ipr-rat LD50: 385 mg/kg KSRNAM 17,991,83
scu-rat LD50: 823 mg/kg KSRNAM 17,991,83
orl-mus LD50: 397 mg/kg IYKEDH 14,297,83
ipr-mus LD50: 272 mg/kg IYKEDH 14,297,83
scu-mus LD50: 591 mg/kg IYKEDH 14,297,83

SAFETY PROFILE: Poison by ingestion and intraperitoneal routes. Moderately toxic by other routes. When heated to decomposition it emits toxic fumes of F^- and NO_x.

AEW750 **CAS:47897-65-4** **HR: 3**
AFRIDOL BLUE
mf: $C_{32}H_{18}Cl_2N_6O_{14}S_4 \cdot 4Na$ mw: 1001.66

TOXICITY DATA WITH REFERENCE
scu-rat TDLo: 50 mg/kg (8D preg):REP BIJOAK 91,14P,64
scu-rat TDLo: 50 mg/kg (8D preg):TER BIJOAK 91,14P,64

SAFETY PROFILE: An experimental teratogen. Other experimental reproductive effects. When heated to decomposition it emits very toxic fumes of SO_x, NO_x, Na_2O, and Cl^-.

AEX000 **CAS:2315-02-8** **HR: 3**
AFRIN HYDROCHLORIDE
mf: $C_{16}H_{24}N_2O \cdot ClH$ mw: 296.88

PROP: Mp: 300–303° (decomp).

SYNS: AFRAZINE □ AFRIN □ 2-(4-tert-BUTYL-2,6-DIMETHYL-3-HYDROXYBENZYL)-2-IMIDAZOLINIUM CHLORIDE □ 6-tert-BUTYL-3-(2-IMIDAZOLIN-2-YLMETHYL)-2,4-DIMETHYLPHENOL HYDROCHLORIDE □ (2,6-DIMETHYL-4-TERTIARYBUTYL-3-HYDROXYPHENYL)METHYLIMIDAZOLINE HYDROCHLORIDE □ DURATION □ H 990 □ ILIADIN □ NAFRINE □ OXYMETAZOLINE CHLORIDE □ OXYMETAZOLINE HYDROCHLORIDE □ SCH 9384

TOXICITY DATA WITH REFERENCE
ipr-rat TDLo: 1800 µg/kg (1-3D preg):REP IJMRAQ 67,478,78
orl-rat LD50: 680 µg/kg ARZNAD 30,1760,80
scu-rat LD50: 1630 µg/kg ARZNAD 11,1016,61
ivn-rat LD50: 1070 µg/kg ARZNAD 11,1016,61
orl-mus LD50: 4700 µg/kg OYYAA2 1,74,67
ipr-mus LD50: 48 mg/kg FRPPAO 21,204,66
scu-mus LD50: 34 mg/kg ARZNAD 11,1016,61
ivn-mus LD50: 2700 µg/kg OYYAA2 1,74,67

SAFETY PROFILE: Poison by ingestion, subcutaneous, intraperitoneal, and intravenous routes. Experimental reproductive effects. When heated to decomposition it emits very toxic fumes of NO_x and HCl. An adrenergic agent.

AEX250 **CAS:9002-18-0** **HR: 1**
AGAR

PROP: Extracted from the red algae *Rhodopyceae*. Unground: in thin, translucent, membranous strips, ground: pale buff powder. Sol in boiling water; insol in cold water and organic solvents.

SYNS: AGAR-AGAR □ AGAR AGAR FLAKE □ AGAR-AGAR GUM □ BENGAL GELATIN □ BENGAL ISINGLASS □ CEYLON ISINGLASS □ CHINESE ISINGLASS □ DIGENEA SIMPLEX MUCILAGE □ GELOSE □ JAPAN AGAR □ JAPAN ISINGLASS □ LAYOR CARANG □ NCI-C50475

TOXICITY DATA WITH REFERENCE
orl-rat LD50: 11 g/kg FDRLI* 124,-,76
orl-mus LD50: 16 g/kg FDRLI* 124,-,76
orl-rbt LD50: 5800 mg/kg FDRLI* 124,-,76
orl-ham LD50: 6100 mg/kg FDRLI* 124,-,76

CONSENSUS REPORTS: NTP Carcinogenesis Bioassay (feed); No Evidence: mouse, rat NTPTR* NTP-TR-230,82. Reported in EPA TSCA Inventory.

SAFETY PROFILE: Mildly toxic by ingestion. When heated to decomposition it emits acrid smoke and fumes.

AEX750 **CAS:39277-47-9** **HR: 3**
AGENT ORANGE
mf: $C_{12}H_{14}Cl_2O_3 \cdot C_{12}H_{13}Cl_3O_3$ mw: 588.76

SYNS: 2,4-d,n-BUTYL ESTER mixed with 2,4,5-T,n-BUTYL ESTER (1:1) □ 2,4,5-T,n-BUTYL ESTER mixed with 2,4-d,n-BUTYL ESTER □ 2,4-DICHLOROPHENOXYACETIC ACID BUTYL ESTER and 2,4,5-TRICHLOROPHENOXYACETIC ACID (45.5%:48.2%)

TOXICITY DATA WITH REFERENCE
orl-mus TDLo: 1180 mg/kg (12-15D preg):TER AECTCV 6,33,77

SAFETY PROFILE: Contains toxic impurities. An experimental teratogen. See also ESTERS. When heated to decomposition it emits toxic fumes of Cl^-.

AEX850 **CAS:644-06-4** **HR: 2**
AGERATOCHROMENE
mf: $C_{13}H_{16}O_3$ mw: 220.29

PROP: Pale-yellow needles from MeOH or oil. Mp: 49–50°, bp: 145–150° at 4 mm.

SYNS: 2H-1-BENZOPYRAN, 6,7-DIMETHOXY-2,2-DIMETHYL- □ 6,7-DIMETHOXY-2,2-DIMETHYL-2H-BENZO(b)PYRAN □ PRECOCENE 2 □ PRECOCENE II

TOXICITY DATA WITH REFERENCE
dnd-rat:lvr 25 µmol/L CALEDQ 26,311,85
dns-rat:lvr 1 µmol/L CALEDQ 26,311,85
ipr-mus TDLo:27,536 µg/kg:CAR CNREA8 47,2275,87

CONSENSUS REPORTS: Reported in EPA TSCA Inventory.

SAFETY PROFILE: Questionable carcinogen with experimental carcinogenic data. Mutation data reported. When heated to decomposition it emits acrid smoke and irritating fumes.

AEY000 CAS:103-16-2 HR: 1
AGERITE

PROP: Solid. Mp: 39–39.5°.
mf: $C_{13}H_{12}O_2$ mw: 200.25

SYNS: AGERITE ALBA □ ALBA-DOME □ BENOQUIN □ BENZOQUIN □ BENZYL HYDROQUINONE □ p-BENZYLOXYPHENOL □ DEPIGMAN □ HYDROQUINONE BENZYL ETHER □ HYDROQUINONE MONOBENZYL ETHER □ p-HYDROXYPHENYL BENZYL ETHER □ MONOBENZONE □ MONOBENZYL ETHER HYDROQUINONE □ MONOBENZYL HYDROQUINONE □ 4-(PHENYLMETHOXY)PHENOL □ PIGMEX

TOXICITY DATA WITH REFERENCE
skn-gpg 5%/48H MLD JSCCA5 28,357,77
orl-mus TDLo:163 g/kg/78W-I:ETA NTIS** PB223-159
scu-mus TDLo:1000 mg/kg:NEO NTIS** PB223-159
ipr-rat LD50:4500 mg/kg MEIEDD 11,983,89

CONSENSUS REPORTS: Reported in EPA TSCA Inventory.

SAFETY PROFILE: Mild acute toxicity by intraperitoneal route. A skin irritant. Questionable carcinogen with experimental neoplastigenic and tumorigenic data. See also ETHERS. When heated to decomposition it emits acrid smoke and irritating fumes.

AEY125 HR: 3
AGKISTRODON CONTORTRIX VENOM

SYN: VENOM, SNAKE, AGKISTRODON CONTORTRIX

TOXICITY DATA WITH REFERENCE
ipr-mus LD50:10,500 µg/kg 14FHAR -,409,63
ivn-mus LD50:10,920 µg/kg 14FHAR -,409,63
ipr-mam LD50:10,500 µg/kg CLPTAT 8,849,67
ivn-mam LD50:10,920 µg/kg CLPTAT 8,849,67

SAFETY PROFILE: Poison by intravenous and intraperitoneal routes.

AEY130 HR: 3
AGKISTRODON PISCIVORUS VENOM

SYN: VENOM, SNAKE, AGKISTRODON PISCIVORUS

TOXICITY DATA WITH REFERENCE
ipr-mus LD50:5110 µg/kg 14FHAR -,409,63
scu-mus LD50:15 mg/kg JOIMA3 67,299,51
ivn-mus LDLo:2250 µg/kg TOXIA6 3,187,66
ivn-dog LDLo:750 µg/kg 19DDA6 1,269,67
ivn-rbt LD50:5 mg/kg PSEBAA 116,696,64
ipr-mam LD50:5110 µg/kg CLPTAT 8,849,67
ivn-mam LD50:4 mg/kg CLPTAT 8,849,67

SAFETY PROFILE: Poison by subcutaneous, intravenous, and intraperitoneal routes.

AEY135 HR: 3
AGKISTRODON RHODOSTOMA VENOM

SYN: VENOM, SNAKE, AGKISTRODON RHODOSTOMA

TOXICITY DATA WITH REFERENCE
ivn-rat LDLo:300 µg/kg JCINAO 45,1202,66
ipr-mus LD50:4977 µg/kg TOXIA6 9,131,71
scu-mus LD50:16,100 µg/kg 19DDA6 1,323,67
ivn-mus LD50:2820 µg/kg TOXIA6 7,239,69
ivn-dog LD50:900 µg/kg 19DDA6 1,323,67
ims-dog LD50:1900 µg/kg 19DDA6 1,323,67
ivn-rbt LDLo:13 µg/kg JCINAO 45,1202,66
ivn-mam LD50:6200 µg/kg CLPTAT 8,849,67

SAFETY PROFILE: Poison by subcutaneous, intramuscular, intravenous, and intraperitoneal routes.

AEY375 CAS:548-42-5 HR: 3
AGROCLAVINE
mf: $C_{16}H_{18}N_2$ mw: 238.36

PROP: Rods from ether; decomp 198–203°. Needles from acetone; decomp @ 205–206°. Mp: 208 209° (decomp). Freely sol in alc, chloroform, pyridine; sol in benzene, ether; very sltly sol in water.

SYN: 8,9-DIDEHYDRO-6,8-DIMETHYLERGOLINE

TOXICITY DATA WITH REFERENCE
mmo-sat 2 µmol/plate CNREA8 47,1811,87
orl-rat TDLo:50 mg/kg (1-5D preg):REP BJPCAL 33,215P,68
ipr-mus LD50:25 mg/kg ARZNAD 35,1760,85
ivn-mus LD50:25,500 µg/kg NYKZAU 58,386,62

SAFETY PROFILE: Poison by intraperitoneal route. Experimental reproductive effects. Mutation data reported. When heated to decomposition it emits toxic fumes of NO_x.

AEY400 CAS:13118-10-0 HR: 3
AHR 376
mf: $C_{18}H_{25}NO_3 \cdot ClH$ mw: 339.90

SYNS: α-CYCLOPENTYLMANDELIC ACID-1-METHYL-3-PYRROLIDINYL ESTER HYDROCHLORIDE □ 1-METHYL-3-PYRROLIDYL-α-PHENYLCYCLOPENTANEGLYCOLATE HYDROCHLORIDE

TOXICITY DATA WITH REFERENCE
orl-hmn TDLo:14 µg/kg/D CPAJAK 11,5141,66
orl-mus LD50:500 mg/kg JMPCAS 2,523,60
ipr-mus LD50:250 mg/kg JMPCAS 2,523,60

SAFETY PROFILE: Poison by intraperitoneal route. Moderately toxic by ingestion. Human systemic effects by ingestion: hallucinations, distorted perceptions, and toxic psychosis. When heated to decomposition it emits toxic fumes of NO_x and HCl.

AFG000 **HR: 3**
AIPYSURUS LAEVIS VENOM (AUSTRALIA)

SYN: VENOM, SEA SNAKE, AIPYSURUS LAEVIS (AUSTRALIA)

TOXICITY DATA WITH REFERENCE
ivn-mus LD50: 250 μg/kg 85EGD4 5,357,78
ims-mus LD50: 130 μg/kg 85EGD4 5,357,78

SAFETY PROFILE: Poison by intravenous and intramuscular routes.

AFG250 **HR: 2**
AIR, refrigerated liquid
DOT: UN 1002/UN 1003

PROP: Bluish, mobile liquid. $O_2 + N_2$. Bp: −189° (liq); flash p: none; autoign temp: none.

SYNS: AIR, compressed (UN 1002) (DOT) □ AIR, refrigerated liquid (cryogenic liquid) (UN 1003) (DOT) □ AIR, refrigerated liquid, (cryogenic liquid) non-pressurized (UN 1003) (DOT)

DOT CLASSIFICATION: 2.2; *Label:* Nonflammable Gas, Oxidizer (UN 1003); DOT Class: 2.2; *Label:* Nonflammable Gas (UN 1002); DOT Class: Nonflammable Gas; *Label:* Nonflammable Gas.

SAFETY PROFILE: Liquid air can cause tissue damage due to low temperature. Personnel exposed to compressed air may develop caisson disease (the bends, the chokes) if decompression is too rapid. Moderate explosion hazard when containers under pressure are shocked or exposed to heat or flame. Flammable materials, e.g., ethyl ether, hydrocarbons, or charcoal, which have been in contact with liquid air may explode very easily. Ordinary oxidation is greatly accelerated in compressed air. Moderately dangerous disaster hazard; can react vigorously with reducing materials.

AFG500 **CAS:569-64-2** **HR: 3**
AIZEN MALACHITE GREEN
mf: $C_{23}H_{25}N_2 \cdot Cl$ mw: 364.95

SYNS: ACRYL BRILLIANT GREEN B □ ADC MALACHITE GREEN CRYSTALS □ ANILINE GREEN □ BASIC GREEN 4 □ BENZALDEHYDE GREEN □ BRONZE GREEN TONER A-8002 □ BURMA GREEN B □ CHINA GREEN (BIOLOGICAL STAIN) □ C.I. 42000 □ C.I. BASIC GREEN 4 □ DIABASIC MALACHITE GREEN □ DIAMOND GREEN B □ FAST GREEN □ HIDACO MALACHITAE GREEN BASE □ LIGHT GREEN N □ NEW VICTORIA GREEN EXTRA I □ SOLID GREEN CRYSTALS O □ TETROPHENE GREEN M □ TETRAMETHYL DIAPARA-AMIDO-TRIPHENYL CARBINOL □ VICTORIA GREEN

TOXICITY DATA WITH REFERENCE
dnd-mam:lym 10 pph BIPMAA 11,2537,72
orl-mus LD50: 80 mg/kg ARZNAD 1,5,51
ipr-mus LD50: 4200 μg/kg ARZNAD 1,5,51

CONSENSUS REPORTS: Reported in EPA TSCA Inventory. Community Right-To-Know List.

SAFETY PROFILE: Poison by ingestion and intraperitoneal routes. Mutation data reported. When heated to decomposition it emits very toxic fumes of NO_x and Cl^-.

AFG625 **HR: 1**
AJAX, LEMON (scouring powder)

SYN: LEMON AJAX

TOXICITY DATA WITH REFERENCE
skn-rbt 500 mg MLD FCTOD7 20,563,82
eye-rbt 100 mg MOD FCTOD7 20,573,82
eye-rbt 100 mg/45 rns MLD FCTOD7 20,573,82

SAFETY PROFILE: A skin and eye irritant.

AFG750 **CAS:483-04-5** **HR: 3**
AJMALICINE
mf: $C_{21}H_{24}N_2O_3$ mw: 352.47

PROP: Prisms from MeOH. Mp: 259° (decomp).

SYNS: ALKALOID C □ ALKALOID II □ 16,17-DIDEHYDRO-19-METHYLOXAYOHIMBAN-16-CARBOXYLIC ACID METHYL ESTER □ HYDROSARPAN □ LAMURAN □ PY-TETRAHYDROSERPENTINE □ RANITOL □ RAUBASINE □ RAUMALINA □ SARPAN □ SUBSTANCE II □ TENSYL □ TETRAHYDROSERPENTINE □ VINCAIN □ VINCEINE □ Δ-YOHIMBINE

TOXICITY DATA WITH REFERENCE
orl-rat LDLo: 750 mg/kg AEPPAE 233,72,58
ivn-gpg LDLo: 20 mg/kg ARZNAD 23,600,73
orl-chd TDLo: 12,500 μg/kg: CNS,PUL CHETBF 76,97,79
ipr-rat LD50: 200 mg/kg 27ZQAG -,117,72
ivn-rat LD50: 24 mg/kg 27ZQAG -,117,72
orl-mus LD50: 400 mg/kg 27ZQAG -,117,72
ipr-mus LD50: 165 mg/kg 27ZQAG -,117-72
ivn-mus LD50: 20 mg/kg AEPPAE 233,72,58
orl-rbt LD50: 500 mg/kg 27ZQAG -,117,72
ivn-rbt LD50: 20 mg/kg 27ZQAG -,117,72

SAFETY PROFILE: Poison by ingestion, intraperitoneal, and intravenous routes. Human systemic effects by ingestion: general anesthesia, convulsions and lung effects. When heated to decomposition it emits highly toxic NO_x. An antihypertensive agent and tranquilizer.

AFH000 **CAS:4373-34-6** **HR: 3**
AJMALICINE HYDROCHLORIDE
mf: $C_{21}H_{24}N_2O_3 \cdot ClH$ mw: 388.93

SYNS: AJMALICINE MONOHYDROCHLORIDE □ RAUBASNE HYDROCHLORIDE □ γ-YOHIMBINE HYDROCHLORIDE

TOXICITY DATA WITH REFERENCE
orl-mus LDLo: 50 mg/kg LDBU** -,3,32
ivn-mus LD50: 56 mg/kg CSLNX* NX#00444

SAFETY PROFILE: Poison by ingestion and intravenous routes. When heated to decomposition it emits very toxic fumes of NO_x and HCl. See also AJMALICINE.

AFH250 **CAS:4360-12-7** **HR: 3**
AJMALINE
mf: $C_{20}H_{26}N_2O_2$ mw: 326.48

PROP: Solid. Mp: 158–160° from MeOH solvate. Mp: 205–207° (anhyd).

SYNS: CARDIORYTHMINE □ GILURYTMAL □ IGNAZIN □ MERABITOL □ RAUGALLINE □ RAUWOLFIN □ RAUWOLFINE □ RHYTMA-

TON □ RITMOS □ SIDDIQUI □ TACHMALIN □ TAJMALIN □ TAKYCOR

TOXICITY DATA WITH REFERENCE
scu-rat LD50:216 mg/kg FRPSAX 19,865,64
ivn-rat LD50:26 mg/kg PHARAT 31,36,76
orl-mus LD50:255 mg/kg FRPSAX 19,865,64
scu-mus LD50:180 mg/kg FRPSAX 19,865,64
ivn-gpg LDLo:28 mg/kg FRPSAX 19,865,64
orl-qal LDLo:316 mg/kg EESADV 6,149,82
orl-bwd LD50:178 mg/kg AECTCV 12,355,83
ipr-mus LD50:75 mg/kg NTIS** AD691-490
ivn-mus LD50:21 mg/kg JETOAS 8(3),188,75

SAFETY PROFILE: Poison by ingestion, subcutaneous, intraperitoneal, and intravenous routes. When heated to decomposition it emits toxic fumes of NO_x. An antihypertensive agent and tranquilizer.

AFH275 CAS:2552-89-8 HR: 3
AJMALINE BIS(CHLOROACETATE) (ester) HYDROCHLORIDE
mf: $C_{24}H_{28}Cl_2N_2O_4{\cdot}ClH$ mw: 515.90

SYNS: DCAA □ DIMONOCLOROACETILAJMALINA CLORIDRATO (ITALIAN)

TOXICITY DATA WITH REFERENCE
scu-rat LD50:389 mg/kg FRPSAX 19,865,64
orl-mus LD50:570 mg/kg FRPSAX 19,865,64
scu-mus LD50:355 mg/kg FRPSAX 19,865,64
ivn-mus LD50:111 mg/kg FRPSAX 19,865,64
ivn-gpg LDLo:149 mg/kg FRPSAX 19,865,64

SAFETY PROFILE: Poison by subcutaneous and intravenous routes. Moderately toxic by ingestion. When heated to decomposition it emits toxic fumes of NO_x and HCl. See also ESTERS, other Ajmalines.

AFH280 CAS:4410-48-4 HR: 3
AJMALINE HYDROCHLORIDE
mf: $C_{20}H_{26}N_2C_2{\cdot}7ClH$ mw: 581.70

PROP: Amber prisms + $2H_2O$. Mp: 133–134°. Mp: 253–255° (anhyd).

SYN: CHLORHYDRATE de RAUGALLINE (FRENCH)

TOXICITY DATA WITH REFERENCE
orl-mus LD50:440 mg/kg AIPTAK 127,163,60
ipr-mus LD50:105 mg/kg AIPTAK 127,163,60
ivn-mus LD50:26 mg/kg AIPTAK 127,163,60
ivn-cat LDLo:2 mg/kg AIPTAK 216,63,75

SAFETY PROFILE: Poison by intravenous and intraperitoneal routes. Moderately toxic by ingestion. When heated to decomposition it emits toxic fumes of NO_x and HCl. See also other Ajmalines.

AFH500 HR: 3
AK PS
mf: $C_7H_6N_2O_2S_2$ mw: 214.27

PROP: Produced by *Streptomyces flavochromogenes Iwayaensis.*

SYN: ANTIBIOTIC AK PS

TOXICITY DATA WITH REFERENCE
ipr-mus LD50:5 mg/kg JANTAJ 33,80-35,80
ivn-mus LD50:5 mg/kg JANTAJ 33,80-35,80

SAFETY PROFILE: Poison by intraperitoneal and intravenous routes. When heated to decomposition it emits very toxic fumes of NO_x and SO_x.

AFH550 CAS:25331-92-4 HR: 3
AL-1612
mf: $C_{19}H_{28}N_2O_3$ mw: 332.49

SYNS: 5-(1,4-DIOXA-8-AZASPIRO(4.5)DEC-8-YLMETHYL)-3-ETHYL-6,7-DIHYDRO-2-METHYL-INDOL-4(5H)-ONE □ 3-ETHYL-5-(4,4-ETHYLENEDIOXYPIPERIDINO-1-METHYL)-6,7-DIHYDRO-2-METHYLINDOL-4(5H)-ONE

TOXICITY DATA WITH REFERENCE
orl-rat LD50:240 mg/kg ARZNAD 23,1314,73
orl-mus LD50:200 mg/kg ARZNAD 23,1314,73
orl-dog LD50:50 mg/kg ARZNAD 23,1314,73
orl-rbt LD50:100 mg/kg ARZNAD 23,1314,73

SAFETY PROFILE: Poison by ingestion. When heated to decomposition it emits toxic fumes of NO_x.

AFH600 CAS:302-72-7 HR: 1
dl-ALANINE
mf: $C_3H_7NO_2$ mw: 89.09

PROP: Needles or prisms, or white crystalline powder; odorless with a sweet taste. Mp: 295° (decomp). Slightly sol in water, insol in Et_2O.

SYNS: (+-)-ALANINE □ dl-α-ALANINE □ (R,S)-ALANINE □ dl-2-AMINOPROPIONIC ACID □ (+-)-2-AMINOPROPIONIC ACID □ dl-α-AMINOPROPIONIC ACID

TOXICITY DATA WITH REFERENCE
orl-rat TDLo:1820 g/kg/26W-C TXAPA9 37,491,76

SAFETY PROFILE: Low toxicity by ingestion. When heated to decomposition emits toxic fumes of NO_x.

AFH750 CAS:5854-93-3 HR: 3
l-ALANOSINE
mf: $C_3H_7N_3O_4$ mw: 149.13

PROP: Crystals. Mp: 190°.

SYNS: ALANOSINE □ l-2-AMINO-3-(HYDROXYNITROSAMINO)PROPIONIC ACID □ l-2-AMINO-3-((N-NITROSO)HYDROXYLAMINO)PROPIONIC ACID □ 3-(HYDROXYNITROSOAMINO)-l-ALANINE (9CI)

TOXICITY DATA WITH REFERENCE
ipr-mus LD50:600 mg/kg USXXAM #3676490
scu-mus LD50:845 mg/kg NCISP* JAN86
ivn-mus LD50:300 mg/kg USXXAM #3676490

SAFETY PROFILE: Poison by intravenous route. Moderately toxic by intraperitoneal and subcutaneous routes. When heated to decomposition it emits toxic fumes of NO_x. An experimental insect reproduction inhibitor.

AFI500 **CAS:38819-28-2** **HR: 3**
4-N-d-ALANYL-2,4-DIAMINO-2,4-DIDEOXY-l-ARABINOSE
mf: $C_8H_{17}N_3O_4$ mw: 219.28

SYN: PRUMYCIN

TOXICITY DATA WITH REFERENCE
dni-hmn:hla 10 mg/L JANTAJ 33,226,80
ipr-rat LD50:70 mg/kg JANTAJ 32,347,79
orl-mus LDLo:750 mg/kg 85ERAY 2,1167,78
ipr-mus LD50:155 mg/kg JANTAJ 33,226,80
ivn-mus LDLo:160 mg/kg 85ERAY 2,1167,78

SAFETY PROFILE: Poison by intraperitoneal and intravenous routes. Moderately toxic by ingestion. Human mutation data reported. When heated to decomposition it emits toxic fumes of NO_x.

AFI625 **CAS:1397-84-8** **HR: 3**
ALAZOPEPTIN
mf: $C_{15}H_{20}N_6O_5$ mw: 364.41

PROP: Crystals.

SYNS: AA223 LEDERLE □ l-ALLYL-(6-DIAZO-5-OXO)-l-NORLEUCYL-(6-DIAZO-5-OXO)-l-NORLEUCINE □ AMBOMYCIN □ LEDERLE AA223

TOXICITY DATA WITH REFERENCE
mmo-omi 1 mg/plate JGAMA9 11,129,65
ipr-rat TDLo:600 µg/kg (7-8D preg):REP PSEBAA 97,888,58
ipr-rat LD50:150 mg/kg PSEBAA 97,888,58

SAFETY PROFILE: Poison by intraperitoneal route. Experimental reproductive effects. Mutation data reported. When heated to decomposition it emits toxic fumes of NO_x.

AFI750 **CAS:39301-00-3** **HR: 3**
ALBITOCIN

TOXICITY DATA WITH REFERENCE
iut-gpg TDLo:250 µg/kg (1D pre):REP JPPMAB 16,369,64
ipr-rat LD50:800 µg/kg JPPMAB 19,760,67
ipr-mus LD50:5900 µg/kg JPPMAB 19,792,67
ivn-mus LD50:6 mg/kg JPPMAB 19,792,67
ivn-mky LD50:2500 µg/kg JPPMAB 19,792,67
ivn-rbt LD50:1800 µg/kg JPPMAB 19,792,67

SAFETY PROFILE: Poison by intravenous and intraperitoneal routes. Experimental reproductive effects. When heated to decomposition it emits acrid smoke and irritating fumes.

AFI850 **CAS:70536-17-3** **HR: 3**
ALBUMIN MACRO AGGREGATES

SYNS: ALBUMIN □ MAA

TOXICITY DATA WITH REFERENCE
ivn-rat LD50:17 mg/kg IJNMCI 5,51,78
ivn-mus LD50:18 mg/kg IJNMCI 5,51,78
inv-gpg LD50:19 mg/kg IJNMCI 5,51,78

SAFETY PROFILE: Poison by intravenous route. When heated to decomposition it emits acrid smoke and irritating fumes.

AFI980 **CAS:66734-13-2** **HR: 2**
ALCLOMETASONE DIPROPIONATE
mf: $C_{28}H_{37}ClO_7$ mw: 521.10

PROP: Crystals from Me_2CO/MeOH/isopropyl ether. Mp: 212–216°.

SYNS: PREGNA-1,4-DIENE-3,20-DIONE, 7-CHLORO-11-HYDROXY-16-METHYL-17,21-BIS(1-OXOPROPOXY)-, (7-α-11-β,16-α)- □ SCH 22219

TOXICITY DATA WITH REFERENCE
scu-rat TDLo:11 mg/kg (female 7-17D post):REP OYYAA2 33,301,87
scu-rat TDLo:1100 µg/kg (female 7-17D post):TER OYYAA2 33,301,87
scu-rat LD50:3593 mg/kg KSRNAM 21,1253,87
scu-mus LD50:2506 mg/kg KSRNAM 21,1253,87

SAFETY PROFILE: Moderately toxic by subcutaneous route. An experimental teratogen. Other experimental reproductive effects. When heated to decomposition it emits toxic fumes of HCl.

AFJ000 **HR: 3**
ALCOHOL, DENATURED
DOT: NA 1986/NA 1987

PROP: Liquid. Composed of alcohol and denaturants.

SYNS: DENATURED ALCOHOL (DOT) □ DENATURED SPIRITS

DOT CLASSIFICATION: 3; *Label:* Flammable Liquid (NA 1987); DOT Class: 3; *Label:* Flammable Liquid, Poison (NA 1986)

SAFETY PROFILE: Potentially poisonous by ingestion. Toxicity depends upon alcohols in question, generally ethanol with methanol as a denaturant. A flammable liquid and dangerous fire hazard; can react vigorously with oxidizing materials. Moderate explosion hazard. See ETHANOL, METHYL ALCOHOL, and n-PROPYL ALCOHOL.

AFJ250 **HR: 3**
ALCOHOLS, N.O.S.

PROP: A generic term applied to a series of compounds, the simplest of which have the general formula $C_nH_{2n+1}OH$. (See also specific compound.)

SAFETY PROFILE: No general statement can be made due to wide variations in toxic effects. Dangerous fire hazard when exposed to heat or flame. Can react violently in contact with ($H_2O + H_2SO_4$), HOCl, Cl_2, isocyanates, $LiAl_4$, N_2O_4, $HClO_4$, H_2SO_5 (Caro's acid), $Ba(ClO_4)_2$, $(CH_2)_2O$, acetaldehyde, diethyl aluminum bromide, hexamethylene diisocyanate, triisobutyl aluminum.

AFJ400 **CAS:357-56-2** **HR: 3**
ALCOID
mf: $C_{26}H_{32}N_2O_2$ mw: 392.59

PROP: Crystals. Mp: 180–184°. Practically insol in water. Sol in 0.1N HCl, ethanol, methanol, acetone, ethyl acetate, benzene, chloroform, ether, and most organic solvents.

SYNS: DAURAN ☐ DEXTROMORAMIDE ☐ DIMORLIN ☐ (+)-2,2-DIPHENYL-3-METHYL-4-MORPHOLINOBUTYRYLPYRROLIDINE ☐ JETRIUM ☐ JETRIUM R ☐ LINFADOL ☐ MCP 875 ☐ MORAMIDE ☐ NARCOLO ☐ PALFADONNA ☐ PALFIUM ☐ PYRROLAMIDOL ☐ PYRROLAMIDOLUM ☐ R 875 ☐ SKF 5137 ☐ TROXILAN ☐ YETRIUM

TOXICITY DATA WITH REFERENCE
scu-wmn TDLo:150 μg/kg:CNS,PUL BMJOAE 1,211,59
ims-hmn TDLo:57 μg/kg:CNS,PUL NYSJAM 61,83,61
ims-wmn TDLo:200 μg/kg:PUL,GIT PRACAK 197,348,66
ivn-rat LD50:13 mg/kg JPPMAB 9,730,57
scu-mus LD50:140 mg/kg BJPCAL 17,433,61
ivn-mus LD50:21 mg/kg JPPMAB 9,730,57

SAFETY PROFILE: Poison by subcutaneous and intravenous routes. Human systemic effects by subcutaneous and intramuscular routes: coma, cyanosis, respiratory depression, and nausea or vomiting. Caution: May be habit forming. This is a controlled substance (opiate) listed in the U.S. Code of Federal Regulations, Title 21 Part 1308.11 (1985). When heated to decomposition it emits toxic fumes of NO_x.

AFJ500 **CAS:52-01-7** **HR: 3**
ALDACTAZIDE
mf: $C_{24}H_{32}O_4S$ mw: 416.62

PROP: Crystals. Mp: 134–135°.

SYNS: 7-α-ACETYLTHIO-3-OXO-17-α-PREGN-4-ENE-21,17-β-CARBOLACTONE ☐ 7-α-ACETYLTHIO-3-OXO-17-β-PREGN-4-ENE-21,17-β-CARBOLACTONE ☐ ALDACTIDE ☐ ALDACTONE ☐ ALDACTONE A ☐ 3-(3-KETO-7-α-ACETYLTHIO-17-β-HYDROXY-4-ANDROSTEN-17-α-YL) PROPIONIC ACID LACTONE ☐ OSIREN ☐ OSYROL ☐ 3′-(3-OXO-7-α-ACETYLTHIO-17-β-HYDROXYANDROST-4-EN-17-β-YL)PROPIONIC ACID LACTONE ☐ 17-α-PREGN-4-ENE-21-CARBOXYLIC ACID, 1-HYDROXY-7-α-MERCAPTO-3-OXO-α-LACTONE ☐ SC 9420 ☐ SC 15983 ☐ SPIRESIS ☐ SPIRIDON ☐ SPIROCTANIE ☐ SPIRO(17H-CYCLOPENTA(a)PHENAUTHRENE-17,2′-(3′H)-FURAN) ☐ SPIRO(17H-CYCLOPENTA(a)PHENANTHRENE-17,2′(5′H)FURAN), PREGN-4-ENE-21-CARBOXYLIC ACID DERIV. ☐ SPIROLACTONE ☐ SPIROLAKTON ☐ SPIROLANG ☐ SPIRONE ☐ SPIRONOLACTONE ☐ SPIRONOLACTONE A ☐ URACTONE ☐ VEROSPIRON ☐ VEROSPIRONE

TOXICITY DATA WITH REFERENCE
unr-wmn TDLo:280 mg/kg (female 35D pre):REP AIMEAS 69,685,68
orl-rat TDLo:360 mg/kg (female 13-21D post):TER ACENA7 95,540,80
orl-wmn TDLo:70 mg/kg/5W-I:BLD BMJOAE 289,731,84
orl-wmn TDLo:122 mg/kg/61D-I NZMJAX 83,147,76
orl-hmn TDLo:5600 mg/kg:SYS JLCMAK 80,224,72
orl-man TDLo:5714 μg/kg/4D-I:PUL,BLD DICPBB 21,974,87
ipr-rat LD50:277 mg/kg DRUGAY 6,381,82
ipr-mus LD50:260 mg/kg DRUGAY 6,381,82
ipr-rbt LD50:866 mg/kg JEPTDQ 1(5),641,78

CONSENSUS REPORTS: IARC Cancer Review: Group 3 IMEMDT 7,344,87; Animal Limited Evidence IMEMDT 24,259,80; Human Inadequate Evidence IMEMDT 24,259,80. Reported in EPA TSCA Inventory.

SAFETY PROFILE: Poison by intraperitoneal route. Human reproductive effects by ingestion and possibly other routes: men, impotence and breast development; women, menstrual cycle changes or disorders, changes in the breasts and lactation. An experimental teratogen. Other experimental reproductive effects. Other human systemic effects by ingestion: agranulocytosis, kidney tubule damage, increased urine volume, and changes in blood sodium and calcium levels. Questionable carcinogen. When heated to decomposition it emits toxic fumes of SO_x. Used to treat hypertension, edema of congestive heart failure, cirrhosis, and kidney failure.

AFJ625 **CAS:5534-09-8** **HR: 2**
ALDECIN
mf: $C_{28}H_{37}ClO_7$ mw: 521.10

PROP: Solid. Mp: 117–120° (decomp).

SYNS: BECLACIN ☐ BECLOFORTE ☐ BECLOMETASONE DIPROPIONATE ☐ BECLOMETASONE-17,21-DIPROPIONATE ☐ BECLOMETHASONE DIPROPIONATE ☐ BECLOVAL ☐ BECLOVENT ☐ BECOTIDE ☐ BENCONASE ☐ BP2 ☐ CLENIL A ☐ DIPROPIONATE BECLOMETHASONE ☐ ENTYDERMA ☐ INALONE O ☐ INALONE R ☐ KORBUTONE ☐ PROPADERM ☐ RINO-CLENIL ☐ SANASTHYMYL ☐ SCH 18020W ☐ VANCENASE ☐ VANCERIL ☐ VIAROX

TOXICITY DATA WITH REFERENCE
ihl-rat TCLo:168 μg/kg/3M (female 8-17D post):REP OYYAA2 13,185,77
ihl-mus TCLo:406 μg/kg/3M (female 7-15D post):TER OYYAA2 13,195,77
orl-rat LD50:>3750 mg/kg OYYAA2 12,863,76
ihl-rat LC50:>51,600 μg/m³/2H JZKEDZ 2,97,76
ipr-rat LD50:>3 g/kg OYYAA2 12,863,76
scu-rat LD50:>1500 mg/kg OYYAA2 12,863,76
orl-mus LD50:>5 g/kg DRUGAY 6,746,82
ihl-mus LC50:>2880 μg/kg DRUGAY 6,746,82
scu-mus LD50:>5 g/kg DRUGAY 6,746,82

SAFETY PROFILE: Moderately toxic by ingestion and other routes. An experimental teratogen. Other experimental reproductive effects. When heated to decomposition it emits toxic fumes of Cl^-.

AFJ700 **CAS:7779-41-1** **HR: 1**
ALDEHYDE C-10 DIMETHYLACETAL
mf: $C_{12}H_{26}O_2$ mw: 202.38

SYNS: DECANAL, DIMETHYLACETAL ☐ DECYLALDEHYDE DMA ☐ 1,1-DIMETHOXYDECANE ☐ 10,10-DIMETHOXYDECANE

TOXICITY DATA WITH REFERENCE
skn-rbt 500 mg/24H MOD FCTXAV 17,759,79

CONSENSUS REPORTS: Reported in EPA TSCA Inventory.

SAFETY PROFILE: A skin irritant. When heated to decomposition it emits acrid smoke and irritating fumes.

AFJ800 **HR: 2**
ALDEHYDES

PROP: A class of chemicals with the general formula R•CHO, and characterized by an unsaturated carbonyl group (C=O).

SAFETY PROFILE: Aldehydes are widely used in many industrial processes. The US production of acetaldehyde in 1982 was 281,000 tons. The world production of acrolein in 1975 was 59,000 tons. Aldehydes occur in nature and are gaseous by-products of incomplete combustion of wood and coal, in exhaust from gasoline and diesel engines, industrial waste gases and fumes, tobacco smoke, and wood fires. Formaldehyde and acetaldehyde are carcinogens. Many of the aldehydes are mutagens. They are reactive compounds participating in oxidation, reduction, addition, and polymerization reactions. All the aldehydes possess anesthetic properties, but this is obscured by their highly irritating action on the eyes and mucous membranes of the respiratory tract. The lower aldehydes, very soluble in water, act chiefly on the eyes and tissues of the upper respiratory tract. The higher aldehydes, less soluble in water, tend to penetrate more deeply into the respiratory system and may affect the lungs. Some higher aldehydes and also the aromatic aldehydes may exhibit much lower toxicity. See also specific compounds.

AFJ850 **CAS:91315-15-0** **HR: 2**
ALDIMORPH

TOXICITY DATA WITH REFERENCE
orl-rat TDLo:1400 mg/kg (female 4-18D post):TER WZERDH 32(1-2),19,83
orl-rat LD50:3500 mg/kg WZERDH 32(1-2),19,83

SAFETY PROFILE: Moderately toxic by ingestion. An experimental teratogen. When heated to decomposition it emits acrid smoke and irritating fumes.

AFK000 **CAS:1646-88-4** **HR: 3**
ALDOXYCARB
mf: $C_7H_{14}N_2O_4S$ mw: 222.29

PROP: Crystals. Mp: 140–142°. Sltly sol in water.

SYNS: ENT AI3-29261 □ 2-METHYL-2-(METHYLSULFONYL)PROPANAL-o-((METHYLAMINO)CARBONYL)OXIME □ 2-METHYL-2-(METHYLSULFONYL)PROPIONALDEHYDE-o-(METHYLCARBAMOYL)OXIME □ STANDAK □ UC-21865

TOXICITY DATA WITH REFERENCE
orl-rat LD50:20 mg/kg TSCAT* FYI-OTS-0885-0443
ihl-rat LC50:140 mg/m³/4H PEMNDP 9,18,91
skn-rat LD50:1 g/kg SPEADM 78-1,61,78
ipr-rat LD50:21 mg/kg TSCAT* FYI-OTS-0885-0443
ivn-rat LD50:14,900 μg/kg TSCAT* FYI-OTS-0885-0443
skn-rbt LD50:200 mg/kg FMCHA2-,C14,89
orl-dck LD50:33,500 μg/kg PEMNDP 9,18,91

SAFETY PROFILE: Poison by ingestion, intraperitoneal, skin contact, and intravenous routes. When heated to decomposition it emits very toxic fumes of NO_x and SO_x. An insecticide.

AFK250 **CAS:309-00-2** **HR: 3**
ALDRIN
DOT: UN 2761/NA 2762
mf: $C_{12}H_8Cl_6$ mw: 364.90

PROP: Crystals. Mp: 104–105°. Sol in Me_2CO, C_6H_6; sltly sol in pet ether; almost insol in H_2O.

SYNS: ALDREX □ ALDREX 30 □ ALDRIN, cast solid (DOT) □ ALDRINE (FRENCH) □ ALDRITE □ ALDROSOL □ ALTOX □ COMPOUND 118 □ DRINOX □ ENT 15,949 □ HEXACHLOROHEXAHYDRO-endo-exo-DIMETHANONAPHTHALENE □ 1,2,3,4,10,10-HEXACHLORO-1,4,4a,5,8,8a-HEXAHYDRO-1,4,5,8-DIMETHANONAPHTHALENE □ 1,2,3,4,10,10-HEXACHLORO-1,4,4a,5,8,8a-HEXAHYDRO-exo-1,4,-endo-5,8-DIMETHANONAPHTHALENE □ 1,2,3,4,10,10-HEXACHLORO-1,4,4a,5,8,8a-HEXAHYDRO-1,4-endo-exo-5,8-DIMETHANONAPHTHALENE □ HHDN □ NCI-C00044 □ OCTALENE □ RCRA WASTE NUMBER P004 □ SEEDRIN

TOXICITY DATA WITH REFERENCE
cyt-hmn:lym 1900 mg/L MUREAV 31,103,75
cyt-hmn:leu 19,125 μg/L PHTHDT 6,147,79
orl-dog TDLo:73 mg/kg (female 44W pre):REP JAVMA4 123,28,53
orl-ham TDLo:50 mg/kg (female 7D post):TER TJADAB 9,11,74
orl-rat TDLo:200 mg/kg/2Y-C:NEO FCTXAV 2,551,64
orl-mus TDLo:270 mg/kg/80W-I:CAR NCITR* NCI-CG-TR-21,78
orl-mus TD:540 mg/kg/80W-I:CAR NCITR* NCI-CG-TR-21,78
orl-rat TD:188 mg/kg/2Y-C:ETA TXAPA9 11,88,67
orl-hmn TDLo:14 mg/kg:CNS 34ZIAG -,83,69
orl-chd LDLo:1250 μg/kg 34ZIAG -,83,69
orl-rat LD50:39 mg/kg SPEADM 74-1,-,74
ihl-rat LCLo:5800 μg/m³/4H 85GMAT -,73,82
skn-rat LD50:98 mg/kg TXAPA9 14,515,69
ipr-rat LD50:150 mg/kg TXAPA9 11,302,67
orl-mus LD50:44 mg/kg SPEADM 74-1,-,74
ipr-mus LDLo:50 mg/kg SOGEBZ 2,80,66
ivn-mus LD50:21 mg/kg 32ZDAL -,52,70
skn-rbt LDLo:15 mg/kg JEENAI 46,702,53

CONSENSUS REPORTS: IARC Cancer Review: Group 3 IMEMDT 7,88,87; Human Inadequate Evidence IMEMDT 5,25,74; Animal Inadequate Evidence IMEMDT 5,25,74; NCI Carcinogenesis Bioassay (feed); Clear Evidence: mouse NCITR* NCI-CG-TR-21,78; Inadequate Studies: rat NCITR* NCI-CG-TR-21,78. EPA Genetic Toxicology Program. EPA Extremely Hazardous Substances List. Community Right-To-Know List.

OSHA PEL: TWA 0.25 mg/m³ (skin)
ACGIH TLV: TWA 0.25 mg/m³
DFG MAK: 0.25 mg/m³
NIOSH REL: (Aldrin) Reduce to lowest detectable level.
DOT CLASSIFICATION: 6.1; *Label:* Poison

SAFETY PROFILE: Poison by ingestion, skin contact, intravenous, and intraperitoneal routes. Human systemic effects by ingestion: excitement, tremors, and nausea or vomiting. An experimental teratogen. Other experimental reproductive effects. Continued acute exposure causes liver damage. Human mutation data reported. Questionable carcinogen with experimental carcinogenic, neoplastigenic, and tumorigenic data. See also

CHLORINATED HYDROCARBONS. When heated to decomposition it emits toxic fumes of Cl^-.

For occupational chemical analysis use NIOSH: Aldrin and Lindane, 5502.

AFK500 CAS:8067-82-1 HR: 3
ALFADIONE
mf: $C_{23}H_{34}O_5 \cdot C_{21}H_{32}O_3$ mw: 723.10

SYNS: ALFATESINE (FRENCH) □ ALPHADIONE □ ALTHESIN □ CT-1341 □ 3-α-HYDROXY-5-α-PREGNANE-11,20-DIONE mixed with 3-α, 21-DIHYDROXY-5-α-PREGNANE-11,20-DIONE 21-ACETATE (3:1) □ SAFFAN

TOXICITY DATA WITH REFERENCE
ivn-wmn LDLo:125 mg/kg PRACAK 222,249,79
ipr-rat LD50:79 mg/kg THERAP 32,375,77
ivn-rat LD50:36 mg/kg JZKEDZ 1,119,75
ipr-mus LD50:140 mg/kg JEMAAJ 62,191,79
ivn-mus LD50:47 mg/kg JEMAAJ 62,191,79
ivn-rbt LD50:12,700 μg/kg PESTD5 16,208,75

SAFETY PROFILE: Poison by intravenous route in humans. Also poison by intraperitoneal route experimentally. When heated to decomposition it emits acrid smoke and fumes.

AFK750 HR: 1
ALFALFA MEAL

SAFETY PROFILE: An allergen. Skin contact may cause dermatitis. Flammable when exposed to heat or flame or by spontaneous chemical reaction. Avoid moisture content extremes. Fires may smolder for 72 hours before becoming noticeable.

AFK875 CAS:23930-19-0 HR: 3
ALFAXALONE
mf: $C_{21}H_{32}O_3$ mw: 332.53

PROP: Columns, colorless prisms from ether. Mp: 172–174°.

SYNS: ALPHAXALONE □ GR 2/234 □ 3-α-HYDROXY-5-α-PREGNANE-11,20-DIONE □ 3-HYDROXYPREGNANE-11,20-DIONE

TOXICITY DATA WITH REFERENCE
orl-rat LD50:297 mg/kg IYKEDH 8,680,77
ipr-rat LD50:116 mg/kg IYKEDH 8,680,77
ivn-rat LD50:19,400 μg/kg YKYUA6 28,1337,77
orl-mus LD50:880 mg/kg YKYUA6 28,1337,77
ipr-mus LD50:430 mg/kg IYKEDH 8,680,77
scu-mus LD50:5220 mg/kg IYKEDH 8,680,77
ivn-mus LD50:36,900 μg/kg JZKEDZ 1,119,75
ivn-rbt LD50:9360 μg/kg IYKEDH 8,680,77

SAFETY PROFILE: Poison by ingestion, intravenous, and intraperitoneal routes. When heated to decomposition it emits acrid smoke and fumes.

AFK900 CAS:53988-42-4 HR: 1
ALFONAL K

TOXICITY DATA WITH REFERENCE
skn-rbt 500 mg/24H MLD 28ZPAK -,265,72
eye-rbt 100 mg/24H MOD 28ZPAK -,265,72
orl-rat LD50:15,800 mg/kg 28ZPAK -,265,72

SAFETY PROFILE: An eye irritant.

AFK950 HR: 1
ALGERIAN IVY

PROP: Commonly cultivated climbing vines that produce black berries. They are used as outdoor wallcover and as house plants. They grow wild in some areas.

SYNS: CANARY IVY □ ENGLISH IVY □ HEDERA CANARIENSIS □ HEDERA HELIX □ MADEIRA IVY □ IVY □ YEDRA (CUBA)

SAFETY PROFILE: The leaves and berries contain hederin, a poisonous saponin. Ingestion of these plant parts may cause a burning pain in the throat, vomiting, and diarrhea. See also SAPONIN.

AFL000 CAS:9005-32-7 HR: 2
ALGINIC ACID

PROP: Extracted from brown seaweeds. White to yellow-white fibrous powder; odorless and tasteless. Sol in alkaline solutions; insol in organic solvents.

SYNS: KELACID □ LANDALGINE □ NORGINE □ PLOYMANNURONIC ACID □ SAZZIO

TOXICITY DATA WITH REFERENCE
ipr-rat LD50:1600 mg/kg AIPTAK 111,167,57
ipr-mus LDLo:1000 mg/kg TXAPA9 23,288,72

CONSENSUS REPORTS: Reported in EPA TSCA Inventory.

SAFETY PROFILE: Moderately toxic by intraperitoneal route. When heated to decomposition it emits acrid smoke and irritating fumes.

AFL500 CAS:84-96-8 HR: 3
ALIMEMAZINE
mf: $C_{18}H_{22}N_2S$ mw: 298.48

SYNS: ALIMEZINE □ BAYER 1219 □ 10-(3-(DIMETHYLAMINO)-2-METHYLPROPYL)PHENOTHIAZINE □ METHYLPROMAZINE □ REPELTIN □ TERALEN □ N,N,β-TRIMETHYL-10H-PHENOTHIAZINE-10 PROPANAMINE □ TRIMEPRAZINE

TOXICITY DATA WITH REFERENCE
orl-rat LD50:210 mg/kg ANPBAZ 61,669,61
ivn-rat LD50:35 mg/kg ANPBAZ 61,669,61
orl-mus LD50:300 mg/kg DNEUD5 7,45,80
ivn-mus LD50:33 mg/kg PSCBAY 2,17,63

SAFETY PROFILE: Poison by ingestion and intravenous routes. When heated to decomposition it emits very toxic fumes of NO_x and SO_x.

AFL750 CAS:3689-50-7 HR: 3
ALIMEMAZINE-S,S-DIOXIDE
mf: $C_{18}H_{22}N_2O_2S$ mw: 330.48

PROP: Crystals from heptane. Mp: 115°.

SYNS: 10-(3-(DIMETHYLAMINO)-2-METHYLPROPYL)PHENOTHIAZINE-5,5-DIOXIDE □ DIOXO-9,9-(DIMETHYLAMINO-3-METHYL-2-

PROPYL)-10-PHENOTHIAZINE (FRENCH) □ DOSEGRAN □ DOXERGAN □ DYSEDON □ IMAKOL □ OXOMEMAZINE □OXYMEMAZINE □ 6847 R.P.

TOXICITY DATA WITH REFERENCE
eye-rbt 100 mg MLD FCTOD7 20,573,82
eye-rbt 100 mg/4S rns MLD FCTOD7 20,573,82
orl-mus LD50:140 mg/kg THERAP 26,1203,71
ipr-mus TDLo:185 mg/kg OCMJAJ 7,33,61
orl-mus LD50:220 mg/kg AIPTAK 135,364,62
scu-mus LD50:260 mg/kg AIPTAK 135,364,62
ivn-mus LD50:35 mg/kg AIPTAK 135,364,62

SAFETY PROFILE: Poison by ingestion, subcutaneous, and intravenous routes. An eye irritant. When heated to decomposition it emits very toxic fumes of NO_x and SO_x.

AFM000 CAS:62851-48-3 HR: 3
ALIOMYCIN

PROP: An antibiotic produced by *Streptomyces Acidomyceticus.*

TOXICITY DATA WITH REFERENCE
orl-mus LD50:2650 mg/kg 85ERAY 2,1001,78
ipr-mus LD50:45 mg/kg 85ERAY 2,1001,78

SAFETY PROFILE: Poison by intraperitoneal route. Moderately toxic by ingestion.

AFM250 HR: 3
ALIPHATIC and AROMATIC EPOXIDES

SAFETY PROFILE: Suspected carcinogen with experimental tumors of the skin, lung, and blood-forming tissues.

AFM375 CAS:8015-55-2 HR: 2
ALIPUR
mf: $C_{11}H_{22}N_2O \cdot C_{11}H_{10}ClNO_2$ mw: 422.02

SYNS: BUTYNYL-3N-3-CHLOROPHENYLCARBAMATE mixed with 3-CYCLOOCTYL-1,1-DIMETHYL UREA □ CHLORBUFAN mixed with CYCEURON □ CYCEURON plus CHLORBUFAN □ 3-CYCLOOCTYL-1,1-DIMETHYL UREA mixed with BUTYNYL-3N-3-CHLOROPHENYLCARBAMATE □ HS 55

TOXICITY DATA WITH REFERENCE
cyt-mus-unr 500 mg/kg TGANAK 14(6),41,80
cyt-mus-orl 500 mg/kg CYGEDX 14(6),38,80
orl-rat LD50:1125 mg/kg 85GMAT-,17,82
orl-mus LD50:696 mg/kg 85GMAT-,17,82

CONSENSUS REPORTS: EPA Genetic Toxicology Program.

SAFETY PROFILE: Moderately toxic by ingestion. Mutation data reported. When heated to decomposition it emits toxic fumes of Cl^- and NO_x.

AFM400 CAS:3952-78-1 HR: 3
ALIZARIN FLUORINE BLUE
mf: $C_{19}H_{15}NO_8$ mw: 385.35

SYNS: ACETIC ACID, (((3,4-DIHYDROXY-2-ANTHRAQUINONYL) METHYL)IMINO)DI- □ ALIZARIN COMPLEXON □ ALIZARIN COMPLEXONE □ ALIZARINE COMPLEXON □ ALIZARINE COMPLEXONE □ ALIZARINE FLUORINE BLUE □ ALIZARINKOMPLEXON □ 3-AMINOMETHYLALIZARIN-N,N-DIACETIC ACID □ 1,2-DIHYDROXY-ANTHRACHINON-3-METHYLEN-IMINODIESSIGSAEURE □ (((3,4-DIHYDROXY-2-ANTHRAQUINONYL)METHYL)IMINO)DIACETIC ACID □ GLYCINE, N-(CARBOXYMETHYL)-N-((9,10-DIHYDRO-3,4-DIHYDROXY-9,10-DIOXO-2-ANTHRACENY L)METHYL)-

TOXICITY DATA WITH REFERENCE
ivn-mus LD50:170 mg/kg EXPEAM 28,180,72

CONSENSUS REPORTS: Reported in EPA TSCA Inventory.

SAFETY PROFILE: A poison by intravenous route. When heated to decomposition it emits toxic vapors of NO_x.

AFM500 HR: 1
ALKALIES

PROP: A term loosely applied to the hydroxides and carbonates of the alkali metals and alkaline earth metals, as well as the bicarbonate and hydroxide of ammonium. They can neutralize acids, change the color of indicators, and impart a soapy taste and feel to aq solns.

SAFETY PROFILE: Variable toxicity. As a group, they constitute the commonest causes of contact dermatitis. Systemically ammonia is most troublesome. See also AMMONIA. See also specific compound.

AFM750 HR: 3
ALKALOID SALTS

SYN: ALKALOIDS

SAFETY PROFILE: Nearly all alkaloid salts are poisonous. Some are also allergens. See specific alkaloid salt. Dangerous; when heated to decomposition they emit highly toxic fumes.

AFN250 HR: 2
ALKANES

PROP: All colorless neutral liquids with light odors. See also individual alkanes as listed (n-pentane, n-hexane, n-heptane, n-octane).

SAFETY PROFILE: Hexane can cause neuropathy with chronic exposure. Other alkanes or mixtures may have the same effect. Many are dangerous fire hazards when exposed to flame, heat, or oxidizers.

AFN500 CAS:72674-05-6 HR: 2
α-ALKENESULFONIC ACID

SYNS: AOS □ α-OLEFIN SULFONATE □ α-OLEFIN SULPHONATE

TOXICITY DATA WITH REFERENCE
eye-rbt 1% SEV YKGKAM 21,334,72
orl-mus TDLo:3 g/kg (female 6-15D post):TER TXCYAC 3,107,75
orl-mus TDLo:3 g/kg (female 6-15D post):REP TXCYAC 3,107,75

SAFETY PROFILE: An experimental teratogen. Other experimental reproductive effects. A severe eye irritant.

When heated to decomposition it emits toxic fumes of SO_x. See also SULFONATES.

AFN750 **HR: 3**
ALKENYL DIMETHYLETHYL AMMONIUM BROMIDE

PROP: Alkenyl indicates a mixture of aliphatic hydrocarbon radicals with approximately 18 unsaturated carbons.

SYN: ONYXIDE (ONYX OIL & CHEM CO)

TOXICITY DATA WITH REFERENCE
orl-rat LD50:500 mg/kg SSCHAH 25,125,49
orl-gpg LD50:158 mg/kg SSCHAH 25,125,49

SAFETY PROFILE: Poison by ingestion. When heated to decomposition it emits very toxic fumes of NH_3, NO_x and Br^-. See also BROMIDES.

AFO200 **HR: 2**
3-(ALKYLAMINO)PROPIONITRILE

PROP: A straight chain with 17 to 20 carbons.

SYN: PROPIONITRILE, 3-(ALKYLAMINO)-

TOXICITY DATA WITH REFERENCE
orl-rat TDLo:1364 mg/kg (female 1-22D post):TER GISAAA 48(1),82,83
orl-rat LD50:6200 mg/kg GISAAA 48(1),82,83
orl-mus LD50:2800 mg/kg GISAAA 48(1),82,83

SAFETY PROFILE: Moderately toxic by ingestion. An experimental teratogen. When heated to decomposition it emits toxic fumes of NO_x.

AFO250 **HR: 2**
ALKYL ARYL SULFONATE

PROP: A synthetic anionic detergent containing a minimum of 40% sodium alkyl aryl sulfonate, approximately 2% moisture, 1% unsulfonated oil, and the balance sodium sulfate.

SYNS: D-40 □ WITCONATE

TOXICITY DATA WITH REFERENCE
orl-rat LD50:2320 mg/kg JAPMA8 42,489,53
orl-mus LD50:2010 mg/kg JAPMA8 42,489,53
orl-rbt LD50:1730 mg/kg JAPMA8 42,489,53
orl-ham LD50:1131 mg/kg JAPMA8 42,489,53

SAFETY PROFILE: Moderately toxic by ingestion. When heated to decomposition it emits toxic fumes of SO_x.

AFO500 **CAS:42615-29-2** **HR: 3**
ALKYLBENZENESULFONATE

SYNS: ABS □ BENZENESULFONIC ACID, ALKYL DERIVATIVES □ LAS □ LINEAR ALKYLBENZENE SULFONATE □ LINEAR ALKYLBENZENE SULPHONATE

TOXICITY DATA WITH REFERENCE
skn-rat 5% MLD FCTXAV 18,55,80
orl-mus TDLo:100 mg/kg (female 6-15D post):TER SKEZAP 17,295,76
skn-mus TDLo:3200 mg/kg (female 1-4D post):REP LIFSAK 26,49,80
orl-rat LD50:437 mg/kg 34ZIAG -,690,69
orl-mus LD50:1407 mg/kg SKEZAP 4,15,63
scu-mus LD50:1989 mg/kg SKEZAP 4,15,63
ivn-mus LD50:157 mg/kg SKEZAP 4,15,63

SAFETY PROFILE: Poison by intravenous route. Moderately toxic by ingestion and subcutaneous routes. An experimental teratogen. Other experimental reproductive effects. A skin irritant. Very reactive with F_2. See also SULFONATES. When heated to decomposition it emits toxic fumes of SO_x.

AFO750 **HR: 2**
p-n-ALKYLBENZENESULFONIC ACID DERIVATIVE, SODIUM SALT

PROP: Alkyl derivative contains from C_{10} to C_{13}.

SYN: p-N-ALKYLBENZENSULFONAN SODNY (CZECH)

TOXICITY DATA WITH REFERENCE
skn-hmn 2500 µg/24H MLD AKEDAX 235,180,69
skn-rat 10 mg/16H MOD JSCCA5 22,411,71
skn-rbt 500 mg/24H MOD 28ZPAK -,195,72
skn-rbt 10 mg MLD JSCCA5 22,411,71
eye-rbt 250 µg/24H SEV 28ZPAK -,195,72
orl-rat LD50:1870 mg/kg 28ZPAK -,195,72

SAFETY PROFILE: Moderately toxic by ingestion. A skin and severe eye irritant. See also SULFONATES. When heated to decomposition it emits toxic fumes of SO_x.

AFP000 **HR: 2**
ALKYLBENZENESULFONIC ACID SODIUM SALT

PROP: Alkyl is C_{12} derived from propylene tetramer and is highly branched.

TOXICITY DATA WITH REFERENCE
skn-hmn 2500 µg/24H MLD AKEDAX 235,180,69
skn-rat 10 mg/16H MLD JSCCA5 22,411,71
skn-rbt 10 mg MLD JSCCA5 22,411,71

SAFETY PROFILE: A human skin irritant. See also SULFONATES. When heated to decomposition it emits toxic fumes of SO_x.

AFP075 **HR: 2**
ALKYL DIMETHYL BENZALKONIUM CHLORIDE

PROP: Alkyl represents a mixture of fatty acid radicals.

SYN: BDM-CHLORIDE (RUSSIAN)

TOXICITY DATA WITH REFERENCE
unr-rat LD50:2020 mg/kg GISAAA 45(11),73,80
ipr-mus LD50:445 mg/kg JSCCA5 28,667,77
unr-mus LD50:1450 mg/kg GISAAA 45(11),73,80
unr-rbt LD50:750 mg/kg GISAAA 45(11),73,80
unr-gpg LD50:725 mg/kg GISAAA 45(11),73,80

SAFETY PROFILE: Moderately toxic by some routes. When heated to decomposition it emits toxic fumes of Cl^-. See also CHLORIDES.

AFP250 CAS:8001-54-5 HR: 3
ALKYL DIMETHYLBENZYL AMMONIUM CHLORIDE

PROP: Yellowish-white amorph powder. Very sol in H_2O, Me_2CO; almost insol in Et_2O. Alkyl group contains from C_8-C_{18}.

SYNS: ALKYLDIMETHYL(PHENYLMETHYL)QUATERNARY AMMONIUM CHLORIDES □ AMMONYX □ ARQUAD DMMCB-75 □ BARQUAT MB-50 □ BAYCLEAN □ BENZALKONIUM CHLORIDE □ BIOQUAT 50-24 □ BTC □ CATAMINE AB □ DRAPOLENE □ GARDIQUAT 1450 □ HYAMINE 3500 □ INTEXAN LB-50 □ KATAMINE AB □ NEO GERM-I-TOL □ ONYX BTC (ONYX OIL & CHEM CO) □ PHENEENE GERMICIDAL SOLUTION and TINCTURE □ QUATERNARY AMMONIUM COMPOUNDS, ALKYLBENZYLDIMETHYL, CHLORIDES □ RODALON □ TRITON K-60 □ VIKROL RQ □ ZEPHIRAN CHLORIDE

TOXICITY DATA WITH REFERENCE
skn-hmn 150 µg/3D-I MLD 85DKA8 -,127,77
eye-hmn 50 µg SEV AJOPAA 27,1118,44
eye-mky 2 mg/24H SEV TXAPA9 6,701,64
skn-rbt 50 mg/24H MOD 33NFA8 -,2,75
eye-rbt 100 ug AROPAW 34,99,45
eye-rbt 1 mg/24H SEV TXAPA9 6,701,64
dnr-bcs 50 µg/L MUREAV 193,21,88
sce-ham-emb 1 mg/L SHIGAZ 74,1365,87
ivg-rat TDLo: 100 mg/kg (female 1D post): TER JJATDK 5,398,85
ivg-rat TDLo: 50 mg/kg (female 1D post): REP JJATDK 5,398,85
orl-wmn TDLo: 266 mg/kg HUTODJ 7,191,88
orl-rat LD50: 240 mg/kg KSRNAM 4,219,70
ipr-rat LD50: 14,500 µg/kg KSRNAM 4,219,70
ivn-rat LD50: 13,900 µg/kg KSRNAM 4,219,70

SAFETY PROFILE: A human poison by ingestion. An experimental poison by ingestion, intraperitoneal, and intravenous routes. An experimental teratogen. Other experimental reproductive effects. A human skin and severe eye irritant. Mutation data reported. When heated to decomposition it emits very toxic fumes of NO_x, NH_3, and Cl^-. See also CHLORIDES. An antimicrobial agent.

AFP100 CAS:63449-41-2 HR: 3
ALKYL(C_{14-16})DIMETHYLBENZYL AMMONIUM CHLORIDES

SYNS: ROCCAL □ TRET-O-LITE XC 511

TOXICITY DATA WITH REFERENCE
skn-rat LD50: 1420 mg/kg PCJOAU 12,1593,78
orl-mus LD50: 150 mg/kg PCJOAU 12,1593,78
ivn-mus LD50: 16 mg/kg JAPMA8 38,428,49

CONSENSUS REPORTS: Reported in EPA TSCA Inventory.

SAFETY PROFILE: A poison by ingestion and intravenous routes. Moderately toxic by skin contact. When heated to decomposition it emits toxic vapors of NH_4^-, NO_x, and Cl^-.

AFP750 CAS:8023-53-8 HR: 3
ALKYL(C_8H_{18})DIMETHYL-3,4-DICHLOROBENZYLAMMONIUM CHLORIDE

SYNS: ALKYL(C_8H_{17} to $C_{18}H_{37}$) DIMETHYL-3,4-DICHLOROBENZYL AMMONIUM CHLORIDE □ DICHLOROBENZALKONIUM CHLORIDE □ TETROSAN

TOXICITY DATA WITH REFERENCE
eye-rbt 1% SEV JAPMA8 38,428,49
orl-rat LD50: 730 mg/kg SSCHAH 25,125,49
orl-mus LD50: 2000 mg/kg JAPMA8 38,428,49
ivn-mus LD50: 50 mg/kg JAPMA8 38,428,49
orl-gpg LD50: 316 µg/kg SSCHAH 25,125,49

SAFETY PROFILE: A deadly poison by ingestion. Poison by intravenous route. A severe eye irritant. Can cause liver and kidney damage. A moderate allergen. Mutation data reported. See also ESTERS and CHLORIDES. When heated to decomposition it emits very toxic fumes of NO_x, NH_3, and Cl^-.

AFQ000 HR: 3
ALKYLNITRILE
mf: $C_9H_{19}CN$ to $C_{17}H_{35}CN$ mw: 151.2 to 265.3

SYN: NITRIL MASTNE KYSELINY S (CZECH)

TOXICITY DATA WITH REFERENCE
eye-rbt 500 mg/24H MLD 28ZPAK -,160,72

CONSENSUS REPORTS: Cyanide and its compounds are on the Community Right-To-Know List.

SAFETY PROFILE: An eye irritant. See also NITRILES. When heated to decomposition it emits toxic fumes of NO_x and CN^-.

AFQ250 HR: 2
ALKYL PHENYL POLYETHYLENE GLYCOL ETHER

PROP: Liquid. Mp: −5°; d: 1.0643 @ 20°/20°; autoign temp: 590°F.

CONSENSUS REPORTS: Glycol ethers are on the Community Right-To-Know List.

SAFETY PROFILE: Moderately toxic by ingestion. Mildly toxic by dermal contact and inhalation. Water solutions of less than 1% have irritating properties comparable to soap. See also GLYCOLS. Combustible when exposed to heat or flame. Incompatible with oxidizing materials. To fight fire, use water, foam, CO_2, dry chemical.

AFQ500 HR: 3
ALKYL PYRIDINES R

TOXICITY DATA WITH REFERENCE
skn-rbt 500 mg open MLD UCDS** 8/5/71
eye-rbt 5 mg SEV UCDS** 8/5/71
orl-rat LD50: 2240 mg/kg UCDS** 9/5/71
skn-rbt LD50: 356 mg/kg UCDS** 8/5/71

SAFETY PROFILE: Poison by skin contact. Moderately toxic by ingestion. A skin and severe eye irritant.

AFQ575 CAS:5977-35-5 HR: 3
ALKYROM
mf: $C_{12}H_{15}Cl_2NO_2$ mw: 276.18

SYNS: 3-BIS(2-CHLOROETHYL)AMINO-4-METHYLBENZOIC ACID

☐ 3-BIS(2-CHLOROETHYL)AMINO-p-TOLUIC ACID ☐ IOB 82 ☐ NSC-1461711

TOXICITY DATA WITH REFERENCE
cyt-hmn:leu 300 µg/L CCROBU 57,29,73
ipr-rat LD50:17 mg/kg CCROBU 54,319,70
unr-rat LD50:17 mg/kg NEOLA4 27,271,80

CONSENSUS REPORTS: EPA Genetic Toxicology Program.

SAFETY PROFILE: Poison by intraperitoneal and possibly other routes. Human mutation data reported. When heated to decomposition it emits toxic fumes of Cl^- and NO_x. See also NITRO COMPOUNDS of AROMATIC HYDROCARBONS and CHLORINATED HYDROCARBONS, AROMATIC.

AFQ625 HR: 1
ALLAMANDA

PROP: An ornamental crawling or climbing shrub with large yellow flowers and 4- to 6-inch, lance-shaped leaves. It is commonly cultivated in Florida, Hawaii, and the West Indies.

SYNS: ALLAMANDA CATHARTICA ☐ CANARIO (PUERTO RICO) ☐ CAUTIVA (PUERTO RICO) ☐ FLOR de BARBERO (CUBA) ☐ LANI-ALI'I (HAWAII) ☐ NANI-ALI'I (HAWAII) ☐ YELLOW ALLAMANDA

SAFETY PROFILE: Most parts of the plant contain an unidentified cathartic toxin. Ingestion of any part of the plant results in mild catharsis.

AFQ750 CAS:2207-75-2 HR: 1
ALLANTOXANIC ACID, POTASSIUM SALT
mf: $C_4H_3N_3O_4$•K mw: 196.2

SYNS: POTASSIUM AZAOROTATE ☐ POTASSIUM OXONATE ☐ POTASSIUM-s-TRIAZINE-2,4-DIONE-6-CARBOXYLATE ☐ 1,4,5,6-TETRAHYDRO-4,6-DIOXO-s-TRIAZINE-2-CARBOXYLIC ACID, POTASSIUM SALT

TOXICITY DATA WITH REFERENCE
orl-mus TDLo:10,800 µg/kg (female 9-11D post):REP TXCYAC 6,289,76
orl-rat TDLo:3 g/kg (female 8-9D post):TER TXCYAC 6,299,76
par-hmn TDLo:20 mg/kg:GIT CLPTAT 6,436,65

SAFETY PROFILE: Human gastrointestinal tract effects by parenteral route. An experimental teratogen. Other experimental reproductive effects. When heated to decomposition it emits toxic fumes of NO_x and K_2O.

AFR000 HR: 3
ALLENE
mf: $H_2C:C:CH_2$ mw: 40.06

PROP: Colorless, unstable, flammable gas; sweet odor. D: 1.787, mp: −146°, bp: −32°, lel: 2.1%.

SAFETY PROFILE: Unknown toxicity. Probably anesthetic. Dangerous fire hazard when exposed to heat, flame or powerful oxidizers. Moderate explosion hazard when exposed to flame, or compressed to >2 atm. To fight fire, stop flow of gas.

AFR250 CAS:584-79-2 HR: 3
ALLETHRIN
mf: $C_{19}H_{26}O_3$ mw: 302.45

PROP: A viscous liquid.

SYNS: (+)-ALLELRETHONYL (+)-cis,trans-CHRYSANTHEMATE ☐ d-ALLETHRIN ☐ d-trans ALLETHRIN ☐ ALLETHRIN I ☐ ALLEVIATE ☐ ALLYL CINERIN ☐ ALLYL HOMOLOG of CINERIN I ☐ d,l-2-ALLYL-4-HYDROXY-3-METHYL-2-CYCLOPENTEN-1-ONE-d,l-CHRYSANTHEMUM MONOCARBOXYLATE ☐ 3-ALLYL-4-KETO-2-METHYLCYCLOPENTENYL CHRYSANTHEMUMMONOCARBOXYLATE ☐ 3-ALLYL-2-METHYL-4-OXO-2-CYCLOPENTEN-1-YL CHRYSANTHEMATE ☐ dl-3-ALLYL-2-METHYL-4-OXOCYCLOPENT-2-ENYL dl-cis trans CHRYSANTHEMATE ☐ ALLYLRETHRONYL dl-cis-trans-CHRYSANTHEMATE ☐ BIOALTRINA ☐ BIOALLETHRIN ☐ CINERIN I ALLYL HOMOLOG ☐DEPALLETHRIN ☐ ENT 17,510 ☐ EXTHRIN ☐ FDA 1446 ☐ FMC 249 ☐ NECARBOXYLIC ACID ☐ NIA 249 ☐ OMS 468 ☐ PALLETHRINE ☐PYNAMIN ☐ PYNAMIN-FORTE ☐ PYRESIN ☐ PYRESYN ☐ SYNTHETIC PYRETHRINS

TOXICITY DATA WITH REFERENCE
mma-sat 500 µg/plate MUREAV 116,185,83
cyt-ham:lng 1900 ng/L/27H MUREAV 66,277,79
orl-rat LD50:685 mg/kg FMCHA2-,C255,91
ihl-rat LCLo:13,800 mg/m³/4H AMIHBC 2,420,50
ivn-rat LDLo:4 mg/kg PCBPBS 2,308,72
unr-rat LD50:680 mg/kg 30ZDA9-,131,71
skn-mus LD50:1200 mg/kg ABCHA6 27,684,63
ipr-mus LD50:38 mg/kg JAFCAU 31,250,83
ice-mus LDLo:4 mg/kg TXAPA9 66,290,82
orl-rbt LD50:4290 mg/kg SPEADM 78-1,7,78
skn-rbt LD50:11,332 mg/kg AMIHBC 2,420,50
ipr-rbt LD50:11,200 mg/kg WRPCA2 3,28,64
orl-qal LD50:2030 mg/kg PEMNDP 9,19,91

CONSENSUS REPORTS: Reported in EPA TSCA Inventory, EPA Genetic Toxicology Program.

SAFETY PROFILE: Poison by intravenous, intracerebral, and intraperitoneal routes. Moderately toxic by ingestion and skin contact. An allergen. An insecticide. It can cause liver and kidney damage by all routes of entry into the body. Lung congestion may occur due to exposure. Local contact may cause contact dermatitis. Inhalation may cause asthma, coughing, wheezing, running nose and eyes. Mutation data reported. See also ALLYL COMPOUNDS and ESTERS. Slight fire hazard. When heated to decomposition it emits acrid fumes.

AFR500 CAS:34624-48-1 HR: 3
(+)-cis-ALLETHRIN
mf: $C_{19}H_{26}O_3$ mw: 302.45

SYN: (+)-(Z)-2,2-DIMETHYL-3-(2-METHYLPROPENYL)-CYCLOPROPANECARBOXYLIC ACID ESTER with 2-ALLYL-4-HYDROXY-3-METHYL-2-CYCLOPENTEN-ONE

TOXICITY DATA WITH REFERENCE
ihl-rat LCLo:260 mg/m³/2H EVHPAZ 14,15,76
orl-mus LD50:210 mg/kg EVHPAZ 14,15,76
ihl-mus LDLo:260 mg/m³/2H EVHPAZ 14,15,76

SAFETY PROFILE: Poison by ingestion and inhalation. See also ALLYL COMPOUNDS and ESTERS. When heated to decomposition it emits acrid smoke and irritating fumes.

AFR750 **CAS:28434-00-6** **HR: 3**
trans-(+)-ALLETHRIN
mf: $C_{19}H_{26}O_3$ mw: 302.45

SYNS: AI 3-29024 □ d-ALLETHROLONE CHRYSANTHEMUMATE □ (+)-ALLETHRONYL (+)-trans-CHRYSANTHEMUMATE □ 2-ALLYL-4-HYDROXY-3-METHYL-2-CYCLOPENTEN-1-ONE □ S-BIOALLETHRIN □ S-trans-BIOALLETHRIN □ (+)-2,2-DIMETHYL-3-(2-METHYLPROPENYL)-CYCLOPROPANECARBOXYLIC ACID-(E)-,ESTER with (+)- □ ESBIOL □ ESBIOL CONCENTRATE 90%

TOXICITY DATA WITH REFERENCE
orl-rat LD50:430 mg/kg EVHPAZ 14,15,76
ihl-rat LC50:1600 mg/m³/3H EVHPAZ 14,15,76
orl-mus LD50:250 mg/kg EVHPAZ 14,15,76
ihl-mus LC50:2720 mg/m³/3H EVHPAZ 14,15,76
scu-frg LD50:1700 µg/kg PCBPBS 20,217,83

SAFETY PROFILE: Deadly poison by subcutaneous route. Poison by ingestion. Moderately toxic by inhalation. See also ALLETHRIN, ALLYL COMPOUNDS, and ESTERS. When heated to decomposition it emits acrid and irritating fumes. An insecticide.

AFS000 **HR: 3**
ALLETHRIN RACEMIC MIXTURE
mf: $C_{19}H_{26}O_3 \cdot 4Cl_9H_{26}O_3$ mw: 1512.25

SYN: 4-HYDROXY-3-METHYL-2-CYCLOPENTEN-1-ONE, cis- mixed with trans-2,2-DIMETHYL-3-(2-METHYL-PROPENYL)CYCLOPROPANE-CARBOXYLIC ACID ESTER with 2-ALLYL-4-HYDROXY-3-METHYL-2-CYCLOPENTEN-1-ONE (1:4)

TOXICITY DATA WITH REFERENCE
orl-rat LD50:720 mg/kg EVHPAZ 14,15,76
ihl-rat LCLo:260 mg/m³/2H EVHPAZ 14,15,76
orl-mus LD50:500 mg/kg EVHPAZ 14,15,76
ihl-mus LCLo:260 mg/m³/2H EVHPAZ 14,15,76

SAFETY PROFILE: Poison by inhalation. Moderately toxic by ingestion. When heated to decomposition it emits toxic fumes of Cl^-. See also ALLYL COMPOUNDS.

AFS250 **HR: 2**
ALLICIN
mf: $C_6H_{10}OS_2$ mw: 162.3

PROP: A colorless, oily liquid; sharp garlic odor, d: 1.112 @ 20°/4°.

SAFETY PROFILE: Moderately toxic irritant by ingestion and inhalation. When heated to decomposition it emits toxic fumes of SO_x.

AFS500 **CAS:52-43-7** **HR: 3**
ALLOBARBITAL
mf: $C_{10}H_{12}N_2O_3$ mw: 208.24

PROP: Plates from water. Mp: 173°.

SYNS: ALLOBARBITONE □ ALLYLBARBITURAL □ ALNOX □ ALOBARBITAL □ BARBALLYL □ BARBIDAL □ CURRAL □ DIADOL □ DIAL □ DIALLYLBARBITAL □ DIALLYLBARBITURIC ACID □ 5,5-DIALLYLBARBITURIC ACID □ DIALLYLMAL □ 5,5-DI-2-PROPENYL-2,4,6 (1H,3H,5H)-PYRIMIDINETRIONE (9CI) □ DORM □ DORMALLYL □ MALIL □ MALILUM □ NOVALLYL □ NSC-9324

TOXICITY DATA WITH REFERENCE
ipr-rat LD50:127 mg/kg APSCAX 24,7,51
scu-rat LD50:110 mg/kg AEPPAE 152,341,30
ipr-mus LD50:85 mg/kg AITDAQ 7,95,59
scu-mus LDLo:200 mg/kg HDTU**-,-,33
ivn-mus LD50:218 mg/kg KSRNAM 16,2161,82
orl-rbt LDLo:50 mg/kg HBAMAK 4,1289,35
ipr-rbt LDLo:100 mg/kg JAPMA8 25,597,36
ivn-rbt LD50:147 mg/kg KSRNAM 13,791,79

SAFETY PROFILE: A poison by ingestion, intraperitoneal, subcutaneous, and intravenous routes. When heated to decomposition it emits toxic fumes of NO_x. A sedative and hypnotic agent. See also BARBITURATES and ALLYL COMPOUNDS.

AFS625 **CAS:5486-77-1** **HR: 3**
ALLOCLAMIDE
mf: $C_{16}H_{23}ClN_2O_2$ mw: 310.86

PROP: Crystals from abs alc + ether. Mp: 125–127°. Sol in abs alc.

SYNS: 2-(ALLYLOXY)-4-CHLORO-N-(2-DIETHYLAMINO)ETHYL) BENZAMIDE □ 4-CHLORO-N-(2-(DIETHYLAMINO)ETHYL)-2-(2-PROPENYLOXY)-BENZAMIDE (9CI) □ (4-CHLORO-N-(2-DIETHYLAMINO) ETHYL)-2-(2-PROPENYLOXY)BENZAMIDE □ 264CE

TOXICITY DATA WITH REFERENCE
scu-mus LD50:155 mg/kg JJPAAZ 20,1,70
ivn-mus LD50:65 mg/kg NIIRDN 6,53,82
ivn-dog LDLo:50 mg/kg JJPAAZ 20,1,70

SAFETY PROFILE: Poison by subcutaneous and intravenous routes. When heated to decomposition it emits toxic fumes of Cl^- and NO_x.

AFS640 **CAS:5107-01-7** **HR: 3**
ALLOCLAMIDE HYDROCHLORIDE
mf: $C_{16}H_{23}ClN_2O_2 \cdot ClH$ mw: 347.32

PROP: Crystals from EtOH and Et_2O. Mp: 125–127°.

SYN: 2-ALLYLOXY-4-CHLORO-N-(2-(DIETHYLAMINO)ETHYL)BENZAMIDE HYDROCHLORIDE

TOXICITY DATA WITH REFERENCE
orl-mus LD50:746 mg/kg IYKEDH 4,90,73
scu-mus LD50:155 mg/kg IYKEDH 4,90,73
ivn-mus LD50:61 mg/kg IYKEDH 4,90,73
ivn-dog LDLo:50 mg/kg JJPAAZ 16,342,66

SAFETY PROFILE: Poison by subcutaneous and intravenous routes. Moderately toxic by ingestion. When heated to decomposition it emits toxic fumes of NO_x and HCl.

AFS750 **HR: 2**
ALLODAN
mf: $C_8H_6Cl_8$ mw: 385.74

TOXICITY DATA WITH REFERENCE
orl-rat LD50:940 mg/kg GTPZAB 8,30,64
skn-rat LD50:1000 mg/kg GTPZAB 8(4),30,64
orl-mus LD50:750 mg/kg GTPZAB 8(4),30,64

SAFETY PROFILE: Moderately toxic by ingestion and

skin contact. When heated to decomposition it emits toxic fumes of Cl^-.

AFT000 CAS:77-02-1 HR: 3
ALLONAL
mf: $C_{10}H_{14}N_2O_3$ mw: 210.26

PROP: Crystals. Mp: 140–141.5°.

SYNS: ALLIONAL □ 5-ALLYL 5-ISOPROPYLBARBITURATE □ ALLYLISOPROPYLBARBITURIC ACID □ 5-ALLYL-5-ISOPROPYLBARBITURIC ACID □ ALLYLISOPROPYLMALONYLUREA □ ALLYLPROPYMAL □ ALURATE □ APROBARBITAL □ APROBARBITONE □ APROZAL □ ISONAL □ ISOPROPYLALLYLBARBITURIC ACID □ NUMAL

TOXICITY DATA WITH REFERENCE
ipr-rat LDLo:100 mg/kg JPETAB 44,325,32
ipr-mus LD50:200 mg/kg ARZNAD 12,389,62
scu-mus LD50:350 mg/kg ARZNAD 8,25,58
orl-rbt LDLo:160 mg/kg JPETAB 44,325,32
ipr-rbt LDLo:90 mg/kg JPETAB 44,325,32

SAFETY PROFILE: Poison by ingestion and intraperitoneal routes. When heated to decomposition it emits toxic fumes of NO_x. See also BARBITURATES and ALLYL COMPOUNDS. A sedative and hypnotic agent.

AFT250 CAS:63732-62-7 HR: 3
ALLOPSEUDOCODEINE HYDROCHLORIDE
mf: $C_{18}H_{21}NO_3 \cdot ClH$ mw: 335.86

TOXICITY DATA WITH REFERENCE
scu-mus LDLo:300 mg/kg JPETAB 51,35,34
scu-rbt LDLo:200 mg/kg JPETAB 51,35,34

SAFETY PROFILE: Poison by subcutaneous route. When heated to decomposition it emits very toxic fumes of NO_x and HCl.

AFT500 CAS:62-67-9 HR: 3
ALLORPHINE
mf: $C_{19}H_{21}NO_3$ mw: 311.41

PROP: Mp: 208–209°. Sol in dil alkalis.

SYNS: N-ALLYL-7,8-DEHYDRO-4,5-EPOXY-3,6-DIHYDROXYMORPHINAN □ N-ALLYL-N-DESMETHYLMORPHINE □ N-ALLYLNORMORPHINE □ ANARCON □ ANTOFIN □ ANTORPHINE □ LETHIDROME □ LETIDRONE □ LITHIDRONE □ NALLINE □ NALORFINA □ NALORPHINE □ NALORPHINIUM □ NANM

TOXICITY DATA WITH REFERENCE
par-hmn TDLo:200 μg/kg:CNS PAREAQ 8,175,56
scu-hmn TDLo:71 μg/kg:PSY FEPRA7 15,442,56
scu-rat LD50:474 mg/kg AIPTAK 165,112,67
ivn-rat LD50:226 mg/kg AIPTAK 165,112,67
orl-mus LD50:1140 mg/kg 27ZQAG -,268,72
ipr-mus LD50:492 mg/kg PAREAQ 8,175,56
scu-mus LD50:500 mg/kg AIPTAK 165,112,67
ivn-mus LD50:127 mg/kg AIPTAK 165,112,67
par-mus LD50:670 mg/kg PAREAQ 8,175,56
scu-mky LDLo:400 mg/kg FEPRA7 13,369,54
ivn-mky LDLo:100 mg/kg FEPRA7 13,369,54
ivn-rbt LDLo:50 mg/kg PAREAQ 8,175,56

SAFETY PROFILE: Poison by intravenous and subcutaneous routes. Moderately toxic by ingestion and other routes. Human systemic effects by parenteral and subcutaneous routes: central nervous system changes, excitement, tremors, hallucinations and distorted perceptions, and antianxiety effects. When heated to decomposition it emits very toxic fumes of NO_x. See also ALLYL COMPOUNDS.

AFT750 CAS:50-71-5 HR: 3
ALLOXAN
mf: $C_4H_2N_2O_4$ mw: 142.08

HNCO•NHCO•CO•CO (ring: first and last groups joined)

PROP: Orthorhombic crystals from AcOH or by subl. Mp: 256° (decomp). D: 1.70. Sol in water, alc, benzene, and acetone.

SYNS: MESOXALYLCARBAMIDE □ MESOXALYLUREA □ 2,4,5,6(1H,3H)-PYRIMIDINETETRONE □ 2,4,5,6-PYRIMIDINTETRON (CZECH) □ 2,4,5,6-TETRAOXOHEXAHYDROPYRIMIDINE

TOXICITY DATA WITH REFERENCE
eye-rbt 500 mg/24H MLD 85JCAE-,862,86
cyt-mus-ipr 50 mg/kg JOHEA8 65,345,74
ipr-rat TDLo:150 mg/kg (MGN):REP METAAJ 20,401,71
ipr-mus TDLo:200 mg/kg (female 8D post):TER JEEMAF 14,63,65
orl-rat LD50:5210 mg/kg 28ZOAH -,150,72
ivn-rat LD50:300 mg/kg CRSBAW 142,1335,48
ipr-mus LDLo:300 mg/kg PSEBAA 67,154,48
scu-mus LDLo:400 mg/kg PSEBAA 67,154,48
ivn-mus LDLo:200 mg/kg PSEBAA 67,154,48
ivn-rbt LDLo:300 mg/kg AJCPAI 16,257,46
rec-rbt LDLo:180 mg/kg AJCPAI 16,257,46
ivn-pgn LDLo:150 mg/kg PSEBAA 58,31,45
ivn-dck LDLo:250 mg/kg HBTXAC 5,6,59
ivn-dom LDLo:200 mg/kg HBTXAC 5,6,59

CONSENSUS REPORTS: Reported in EPA TSCA Inventory.

SAFETY PROFILE: Poison by intraperitoneal, intravenous, subcutaneous, and rectal routes. Moderately toxic by ingestion. An experimental teratogen. Other experimental reproductive effects. Mutation data reported. Produces diabetes in experimental animals. Decomposes in storage to release CO_2. Do not store in sealed container. Explodes when heated above 170°C. When heated to decomposition it emits toxic fumes of NO_x.

AFU000 CAS:87-39-8 HR: 3
ALLOXAN-5-OXIME
mf: $C_4H_3N_3O_4$ mw: 157.10

PROP: Rhombic or orthorhombic crystals. Mp: 240–241° (decomp). Mod sol in water; sol in alc.

SYNS: 5-HYDROXYIMINOBARBITURIC ACID □ 5-ISONITROSOBARBITURIC ACID □ 2,4,5,6(1H,3H)-PYRIMIDINETETRONE 5-OXIME □ VIOLURIC ACID

TOXICITY DATA WITH REFERENCE
ivn-mus LD50:100 mg/kg CSLNX* NX#05202

CONSENSUS REPORTS: Reported in EPA TSCA Inventory.

SAFETY PROFILE: Poison by intravenous route. When heated to decomposition it emits toxic fumes of NO_x.

AFU250 **HR: 2**
ALLOXANTIN
mf: $C_8H_6N_4O_8 \cdot 2H_2O$ mw: 322.19

PROP: Crystalline powder; on exposure to air turns red; yellow @ 225°.

SYN: UROXIN

SAFETY PROFILE: Moderately toxic by ingestion. On a chronic basis caused disturbed carbohydrate metabolism leading to diabetes. Moderately dangerous; when heated to decomposition it emits toxic fumes of NO_x.

AFU750 **CAS:591-87-7** **HR: 3**
ALLYL ACETATE
DOT: UN 2333
mf: $C_5H_8O_2$ mw: 100.13

PROP: Liquid. Vap d: 3.45, bp: 103–104°, d: 0.928. Flash p: 72°F. Insol in water.

SYNS: ACETIC ACID ALLYL ESTER □ ACETIC ACID-2-PROPENYL ESTER □ 3-ACETOXYPROPENE

TOXICITY DATA WITH REFERENCE
skn-rbt 10 mg/24H MLD JIHTAB 31,60,49
eye-rbt 100 mg/24H MOD 85JCAE-,355,86
orl-rat LD50:130 mg/kg JIHTAB 31,60,49
ihl-rat LC50:1000 ppm/1H AMIHAB 21,28,60
orl-mus LD50:170 mg/kg FCTXAV 2,327,64
skn-rbt LD50:1021 mg/kg JIHTAB 31,60,49

CONSENSUS REPORTS: Reported in EPA TSCA Inventory.

DOT CLASSIFICATION: 3; *Label:* Flammable Liquid, Poison

SAFETY PROFILE: Poison by ingestion. Moderately toxic by inhalation and skin contact. A skin and eye irritant. When heated to decomposition it emits acrid smoke and irritating fumes. Dangerous fire hazard. See also ALLYL COMPOUNDS.

AFV500 **CAS:107-18-6** **HR: 3**
ALLYL ALCOHOL
DOT: UN 1098
mf: C_3H_6O mw: 58.09

$H_2C{=}CHCH_2OH$

PROP: Limpid liquid; pungent odor. Mp: −129°, fp: −50°, bp: 96–97°, lel: 2.5%, uel: 18%, flash p: 70°F (CC), d: 0.854 @ 20°/4°, autoign temp: 713°F, vap press: 10 mm @ 10.5°, vap d: 2.00. Misc in water, alc, and ether.

SYNS: ALCOOL ALLILCO (ITALIAN) □ ALCOOL ALLYLIQUE (FRENCH) □ ALLILOWY ALKOHOL (POLISH) □ ALLYL AL □ ALLYLALKOHOL (GERMAN) □ ALLYLIC ALCOHOL □ 3-HYDROXYPROPENE □ ORVINYLCARBINOL □ PROPENOL □ PROPEN-1-OL-3 □ 1-PROPEN-3-OL □ 2-PROPEN-1-OL □ PROPENYL ALCOHOL □ 2-PROPENYL ALCOHOL □ RCRA WASTE NUMBER P005 □ SHELL UNDRAUTED A □ VINYLCARBINOL □ WEED DRENCH

TOXICITY DATA WITH REFERENCE
eye-hmn 25 ppm SEV AMIHAB 18,303,58
skn-rbt 10 mg/24H open JIHTAB 30,63,48
eye-rbt 4270 µg SEV AJOPAA 29,1363,46
mmo-sat 100 µmol/L MUREAV 93,305,82
mma-sat 50 µg/plate TCMUD8 1,259,80
ihl-man LCLo:1000 ppm/1H 34ZIAG -,86,69
orl-rat LD50:64 mg/kg JIHTAB 30,63,48
ihl-rat LC50:165 ppm/4H AMIHAB 18,303,58
ipr-rat LD50:37 mg/kg TXAPA9 83,108,86
orl-mus LD50:96 mg/kg AMIHAB 18,303,58
ihl-mus LC50:500 mg/m³/2H 85GMAT -,17,82
ipr-mus LC50:60 mg/kg AMIHAB 18,303,58
ivn-mus LD50:78 mg/kg AIPTAK 135,330,62
ihl-mky LCLo:1000 ppm/4H CRTXB2 5,189,77
skn-rbt LD50:45 mg/kg JIHTAB 30,63,48

CONSENSUS REPORTS: EPA Extremely Hazardous Substances List. Reported in EPA TSCA Inventory.

OSHA PEL: TWA 2 ppm; STEL 4 ppm (skin)
ACGIH TLV: TWA 2 ppm; STEL 4 ppm (skin)
DFG MAK: 2 ppm (5 mg/m³)
DOT CLASSIFICATION: 6.1; *Label:* Poison, Flammable Liquid

SAFETY PROFILE: Poison by inhalation, ingestion, skin contact, subcutaneous, intraperitoneal, and possibly other routes. A skin, severe eye (human), and systemic irritant. Mutation data reported. Dangerous fire and explosion hazard when exposed to heat, flame, or oxidizers. Explosive or violent reaction with sulfuric acid, alkali + 2,4,6-trichloro-1,3,5-triazine, or 2,4,6-tris(bromoamino)-1,3,5-triazine. Reaction with carbon tetrachloride produces explosively unstable halogenated C_4 epoxides. Incompatible with chlorosulfonic acid, HNO_3, H_2SO_4, oleum, NaOH, diallyl phosphite, PCl_3, and tri-n-bromomelamine. When heated to decomposition it emits acrid smoke and fumes. To fight fire, use CO_2, alcohol foam, dry chemical. See also ALLYL COMPOUNDS.

For occupational chemical analysis use NIOSH: Alcohols III, 1402.

AFV750 **CAS:66941-48-8** **HR: 3**
5-ALLYL-5-(1-(ALLYTHIO)ETHYL)BARBITURIC ACID SODIUM SALT
mf: $C_{12}H_{15}N_2O_3S \cdot Na$ mw: 290.34

TOXICITY DATA WITH REFERENCE
orl-rat LD50:346 mg/kg JAPMA8 35,231,46
ivn-rat LD50:95 mg/kg JPETAB 88,343,46

SAFETY PROFILE: Poison by ingestion and intravenous routes. When heated to decomposition it emits very toxic fumes of SO_x, Na_2O, and NO_x. See also BARBITURATES and ALLYL COMPOUNDS.

AFW000 **CAS:107-11-9** **HR: 3**
ALLYLAMINE
DOT: UN 2334
mf: C_3H_7N mw: 57.11

PROP: Colorless liquid, burning taste, sharp ammonia odor. Bp: 56.5–58°, d: 0.761 @ 20°/4°, flash p: −20°F, autoign temp: 705°F, vap d: 2.00, lel: 2.2%, uel: 22%. Misc in water, alc, and ether.

SYNS: 3-AMINOPROPENE ☐ 3-AMINOPROPYLENE ☐ MONOALLYLAMINE ☐ 2-PROPENAMINE ☐ 2-PROPEN-1-AMINE

TOXICITY DATA WITH REFERENCE
skn-rbt 500 mg/24H SEV AEHLAU 1,343,60
eye-rbt 50 mg/20S rns SEV AEHLAU 1,343,60
cyt-rat-orl 2500 ng/kg GISAAA 48(1),80,83
ihl-man TCLo:2500 ppb/5M:EYE,PUL AEHLAU 1,343,60
ihl-mam LC50:320 mg/M³ TPKVAL 14,80,75
orl-rat LD50:106 mg/kg AEHLAU 1,343,60
ihl-rat LC50:286 ppm/4H AEHLAU 1,343,60
orl-mus LD50:57 mg/kg AEHLAU 1,343,60
ipr-mus LD50:49 mg/kg AEHLAU 1,343,60
skn-rbt LD50:35 mg/kg AEHLAU 1,343,60
unr-mam LD50:783 mg/kg TPKVAL 14,80,75

CONSENSUS REPORTS: EPA Extremely Hazardous Substances List. Reported in EPA TSCA Inventory.

DOT CLASSIFICATION: 6.1; *Label:* Poison, Flammable Liquid

SAFETY PROFILE: Poison by inhalation, ingestion, intraperitoneal, and skin contact. Human systemic effects by inhalation: lacrimation and lung effects. A systemic irritant. Mutation data reported. A severe eye and skin irritant. Extraordinary precautions against fumes are advised. Dangerous fire and explosion hazard when exposed to heat, flame, or oxidizers. Highly reactive. When heated to decomposition it emits toxic fumes of NO_x. To fight fire, use alcohol foam, CO_2, dry chemical. See also ALLYL COMPOUNDS and AMINES.

AFW250 CAS:77966-30-4 HR: 0
2-(ALLYLAMINO)-6′-CHLORO-o-ACETOTOLUIDIDE HYDROCHLORIDE
mf: $C_{12}H_{15}ClN_2O{\cdot}ClH$ mw: 275.20

SYNS: 2-(ALLYLAMINO)-2′-CHLORO-6′-METHYLACETANILIDE HYDROCHLORIDE ☐ C 3124

TOXICITY DATA WITH REFERENCE
ipr-rat LD50:460 mg/kg ARZNAD 8,407,58
ipr-mus LD50:375 mg/kg ARZNAD 8,407,58
scu-mus LD50:1070 mg/kg ARZNAD 8,407,58

SAFETY PROFILE: Poison by intraperitoneal route. Moderately toxic by subcutaneous route. When heated to decomposition it emits very toxic fumes of Cl^-, NO_x, and HCl. See also ALLYL COMPOUNDS and AMINES.

AFW500 CAS:642-44-4 HR: 2
1-ALLYL-6-AMINO-3-ETHYLURACIL
mf: $C_9H_{13}N_3O_2$ mw: 195.25

PROP: Crystals from EtOAc and Et_2O. Mp: 143–144° (anhyd).

SYNS: ALACIL ☐ ALLACYL ☐ 1-ALLYL-6-AMINO-3-ETHYL-2,4(1H,3H) PYRIMIDINEDIONE ☐ 1-ALLYL-3-ETHYL-6-AMINOTETRAHYDROPYRIMIDINEDIONE ☐ 6-AMINO-3-ETHYL-1-(2-PROPENYL)-2,4(1H,3H-)-PYRIMIDINEDIONE ☐ AMINOMETRADINE ☐ AMINOMETRAMIDE ☐ CATAPYRIN ☐ 1-ETHYL-3-ALLYL-6-AMINOURACIL ☐ KATAPYRIN ☐ MICTINE ☐ MINCARD ☐ S.C. 3497

TOXICITY DATA WITH REFERENCE
orl-rat LD50:2300 mg/kg FEPRA7 14,392,55
ipr-rat LD50:500 mg/kg CLDND* 13,125,58
ipr-mus LD50:560 mg/kg CLDND*

SAFETY PROFILE: Moderately toxic by ingestion and intraperitoneal routes. When heated to decomposition it emits toxic fumes of NO_x. See also ALLYL COMPOUNDS. A diuretic agent.

AFW750 CAS:140-67-0 HR: 2
p-ALLYLANISOLE
mf: $C_{10}H_{12}O$ mw: 148.22

PROP: Isolated from rind of *Persea Gratissima Garth*, and from Oil of Estragon; found in oils of Russian Anise, Basil, Fennel, Turpentine, and others (FCTXAV 14,601,76). Colorless to sltly yellow liquid; anise odor. Bp: 102° at 16 mm, d: 0.960–0.968, refr index: 1.519–1.524, flash p: 178°F. Sol in alc; insol in water.

SYNS: 4-ALLYL-1-METHOXYBENZENE ☐ CHAVICOL METHYL ETHER ☐ ESDRAGOL ☐ ISOANETHOLE ☐ p-METHOXYALLYLBENZENE ☐ 1-METHOXY-4-(2-PROPENYL)BENZENE ☐ METHYL CHAVICOL ☐ NCI-C60946 ☐ TARRAGON

TOXICITY DATA WITH REFERENCE
skn-rbt 500 mg/24H MOD FCTXAV 14,601,76
mmo-sat 1 μmol/plate FCTXAV 14,603,76
mma-sat 1 μmol/plate MUREAV 60,143,79
bfa-rat/sat 2500 mg/kg NUCADQ 1,10,79
dnd-mus-ipr 80 mg/kg CRNGDP 5,1613,84
orl-mus TDLo:97 g/kg/1Y-C:NEO CNREA8 43,1124,83
ipr-mus TDLo:111 mg/kg:CAR CNREA8 47,2275,87
scu-mus TDLo:140 mg/kg/22D-I:CAR JNCIAM 57,1323,76
orl-rat LDLo:1230 mg/kg FCTXAV 14(6),601,76
ipr-rat LD50:1030 mg/kg COREAF 246,1465,58
orl-mus LD50:1250 mg/kg FCTXAV 2,327,64
ipr-mus LD50:1260 mg/kg COREAF 246,1465,58

CONSENSUS REPORTS: Reported in EPA TSCA Inventory.

SAFETY PROFILE: Moderate acute toxicity by many routes. A skin irritant. Questionable carcinogen with experimental carcinogenic and neoplastigenic data. Mutation data reported. Combustible liquid. When heated to decomposition it emits acrid smoke and irritating fumes. See also ALLYL COMPOUNDS. A spice used in foods, liqueurs, and perfumes.

AFX000 CAS:300-57-2 HR: 2
ALLYLBENZENE
mf: C_9H_{10} mw: 118.19

PROP: Liquid. D: 0.901, bp: 160–163° @ 748 mm.

TOXICITY DATA WITH REFERENCE
orl-rat LDLo:4620 mg/kg FCTXAV 2,327,64
orl-mus LD50:2900 mg/kg TXAPA9 7,18,65

CONSENSUS REPORTS: Reported in EPA TSCA Inventory.

SAFETY PROFILE: Moderately toxic by ingestion. When heated to decomposition it emits acrid smoke and irritating fumes. See also ALLYL COMPOUNDS.

A

AFX250 **HR: 3**
ALLYL BENZENE SULFONATE
mf: $C_9H_{10}O_3S$ mw: 198.2

SAFETY PROFILE: A highly reactive and flammable compound. Residue from vacuum distillation @ 92–135° @ 2.6 mbar exploded after removal of heat source. When heated to decomposition it emits toxic fumes of SO_x. See also ALLYL COMPOUNDS and SULFONATES.

AFX500 **CAS:64058-13-5** **HR: 3**
5-ALLYL-5-BENZYL-2-THIOBARBITURIC ACID SODIUM SALT
mf: $C_{14}H_{14}N_2O_2S{\cdot}Na$ mw: 297.35

SYN: SODIUM ALLYLBENZYL THIOBARBITURATE

TOXICITY DATA WITH REFERENCE
orl-rat LDLo:40 mg/kg JPETAB 60,125,37
ipr-rat LDLo:20 mg/kg JPETAB 60,125,37

SAFETY PROFILE: Poison by ingestion and intraperitoneal routes. When heated to decomposition it emits very toxic fumes of NO_x, Na_2O, and SO_x. See also BARBITURATES and ALLYL COMPOUNDS.

AFX750 **CAS:63905-38-4** **HR: 3**
ALLYL-BIS(β-CHLOROETHYL)AMINE HYDROCHLORIDE
mf: $C_7H_{13}Cl_2N{\cdot}ClH$ mw: 218.57

TOXICITY DATA WITH REFERENCE
ipr-mus LD50:3 mg/kg CANCAR 2,1055,49
scu-mus LDLo:4 mg/kg JPETAB 91,224,47

SAFETY PROFILE: Poison by intraperitoneal and subcutaneous routes. When heated to decomposition it emits very toxic fumes of Cl^-, NO_x, and HCl. See also ALLYL COMPOUNDS and AMINES.

AFY000 **CAS:106-95-6** **HR: 3**
ALLYL BROMIDE
DOT: UN 1099
mf: C_3H_5Br mw: 120.99

PROP: Colorless liquid, pungent odor. Mp: −119°, bp: 71.3°, flash p: 30°F, d: 1.3980 @ 20°/4°, autoign temp: 563°F, vap d: 4.17, lel: 4.4%, uel: 7.3%. Insol in water.

SYNS: BROMALLYLENE □ 1-BROMO-2-PROPENE □ 3-BROMOPROPENE □ 3-BROMOPROPYLENE

TOXICITY DATA WITH REFERENCE
dns-hmn:hla 500 μmol/L CALEDQ 20,263,83
ihl-mam LC50:4110 mg/m³ GTPZAB 18(4),55,74
ipr-mam LD50:88 mg/kg GTPZAB 18(4),55,74
mmo-sat 1 μmol/plate BCPCA6 29,993,80
ihl-rat LC50:10,000 mg/m³/30M FAVUAI 7,35,75
ipr-mus LD50:108 mg/kg JPCEAO 320(1),133,78
orl-gpg LD50:30 mg/kg WQCHM* 4,-,74

CONSENSUS REPORTS: Reported in EPA TSCA Inventory.

DOT CLASSIFICATION: 3; *Label:* Flammable Liquid, Poison

SAFETY PROFILE: Poison by ingestion and intraperitoneal routes. Mildly toxic by inhalation. Human mutation data reported. See also ALLYL CHLORIDE and ALLYL COMPOUNDS. Dangerous fire and explosion hazard when exposed to heat, flame, or oxidizers. When heated to decomposition it emits toxic fumes of Br^-. To fight fire, use alcohol foam, water spray or mist, CO_2, dry chemical.

AFY250 **CAS:2051-78-7** **HR: 3**
ALLYL BUTANOATE
mf: $C_7H_{12}O_2$ mw: 128.19

SYNS: ALLYL BUTYRATE □ VINYL CARBINYL BUTYRATE

TOXICITY DATA WITH REFERENCE
skn-hmn 20 mg/48H MLD FCTXAV 15,611,77
skn-rbt 500 mg/24H MOD FCTXAV 15,611,77
orl-rat LD50:250 mg/kg TXAPA9 6,378,64
skn-rbt LD50:530 mg/kg FCTXAV 15,611,77

CONSENSUS REPORTS: Reported in EPA TSCA Inventory.

SAFETY PROFILE: Poison by ingestion. Moderately toxic by skin contact. A human skin irritant. See also ESTERS and ALLYL COMPOUNDS. When heated to decomposition it emits acrid smoke and irritating fumes.

AFY300 **CAS:66941-49-9** **HR: 3**
5-ALLYL-5-(2-BUTENYL)-2-THIOBARBITURIC ACID SODIUM SALT
mf: $C_{11}H_{13}N_2O_2S{\cdot}Na$ mw: 260.31

TOXICITY DATA WITH REFERENCE
ipr-rat LD50:172 mg/kg JAPMA8 34,183,45
ivn-rbt LD50:73 mg/kg JAPMA8 34,183,45

SAFETY PROFILE: Poison by intraperitoneal and intravenous routes. When heated to decomposition it emits very toxic fumes of NO_x, SO_x, and Na_2O. See also BARBITURATES and ALLYL COMPOUNDS.

AFY500 **CAS:115-44-6** **HR: 3**
5-ALLYL-5-sec-BUTYLBARBITURIC ACID
mf: $C_{11}H_{16}N_2O_3$ mw: 224.29

SYNS: 5-ALLYL-5-(1-METHYLPROPYL) BARBITURIC ACID □ BUTABITAL □ sec-BUTYL ALLYL BARBITURIC ACID □ LATUSATE □ LOTUSATE □ 5-(1-METHYLPROPYL)-5-(2-PROPENYL)-2,4,6(1H,3H,5H)-PYRIMIDINETRIONE (9CI) □ PROFUNDOL □ TALBUTAL □ WIN 5095

TOXICITY DATA WITH REFERENCE
orl-rat LD50:57,500 μg/kg TXAPA9 21 315,72
ipr-rat LDLo:75 mg/kg JPETAB 44,325,32
ivn-rat LDLo:68 mg/kg CLDND* 44,325,32
ipr-rbt LDLo:55 mg/kg JPETAB 44,325,32
ivn-rbt LDLo:50 mg/kg JACSAT 57,1961,35
orl-pgn LD50:56 mg/kg TXAPA9 21,315,72

SAFETY PROFILE: Poison by ingestion, intraperitoneal, and intravenous routes. Human psychotropic effects by ingestion. When heated to decomposition it emits toxic NO_x. See also BARBITURATES and ALLYL COMPOUNDS. A sedative and hypnotic agent.

AFY750 CAS:2095-58-1 HR: 3
ALLYL-sec-BUTYL THIOBARBITURIC ACID
mf: $C_{11}H_{16}N_2O_2S$ mw: 240.35

SYN: 5-ALLYL-5-sec-BUTYL-2-THIOBARBITURIC ACID

TOXICITY DATA WITH REFERENCE
ivn-mus LD50:150 mg/kg ARZNAD 4,441,54
orl-rbt LDLo:500 mg/kg JPETAB 60,189,37
ivn-rbt LDLo:60 mg/kg JPETAB 60,189,37
rec-rbt LDLo:110 mg/kg JPETAB 60,189,37

SAFETY PROFILE: Poison by intravenous and rectal routes. When heated to decomposition it emits very toxic fumes of SO_x and NO_x. See also BARBITURATES and ALLYL COMPOUNDS.

AGA000 CAS:64058-14-6 HR: 3
5-ALLYL-5-sec-BUTYL-2-THIOBARBITURIC ACID SODIUM SALT
mf: $C_{11}H_{15}N_2O_2S{\bullet}Na$ mw: 262.33

TOXICITY DATA WITH REFERENCE
orl-rat LDLo:125 mg/kg JPETAB 60,125,37
ipr-rat LDLo:100 mg/kg JPETAB 60,125,37
ivn-rat LDLo:120 mg/kg JPETAB 60,125,37
orl-dog LDLo:130 mg/kg JPETAB 60,125,37
ivn-dog LDLo:60 mg/kg JPETAB 60,125,37
ivn-rbt LDLo:40 mg/kg JPETAB 60,125,37

SAFETY PROFILE: Poison by ingestion, intraperitoneal, and intravenous routes. When heated to decomposition it emits very toxic fumes of NO_x, SO_x, and Na_2O. See also BARBITURATES and ALLYL COMPOUNDS.

AGA250 CAS:66941-53-5 HR: 3
5-ALLYL-5-(1-BUTYLTHIO)ETHYL)BARBITURIC ACID SODIUM SALT
mf: $C_{13}H_{19}N_2O_3S{\bullet}Na$ mw: 306.39

SYN: SODIUM-5-ALLYL-5-(1-(BUTYLTHIO)ETHYL) BARBITURATE

TOXICITY DATA WITH REFERENCE
orl-rat LD50:639 mg/kg JAPMA8 35,231,46
ivn-rat LD50:90 mg/kg JPETAB 88,343,46
ivn-rbt LD50:35 mg/kg JAPMA8 35,244,46

SAFETY PROFILE: Poison by intravenous route. Moderately toxic by ingestion. When heated to decomposition it emits very toxic fumes of NO_x, SO_x, and Na_2O. See also BARBITURATES and ALLYL COMPOUNDS.

AGA500 CAS:123-68-2 HR: 3
ALLYL CAPROATE
mf: $C_9H_{16}O_2$ mw: 156.25

PROP: Bp: 186–188°. Insol in water; sol in alc and ether.

SYNS: ALLYL HEXANOATE (FCC) □ FEMA No. 2032 □ 2-PROPENYL-N-HEXANOATE

TOXICITY DATA WITH REFERENCE
skn-hmn 20 mg/48H MLD FCTXAV 11,1079,73
mrc-bcs 18 µg/disc OEKSDJ 9,177,78
orl-rat LD50:218 mg/kg FCTXAV 2,327,64
skn-rbt LD50:300 mg/kg FCTXAV 11,477,73
orl-gpg LD50:280 mg/kg FCTXAV 2,327,64

CONSENSUS REPORTS: Reported in EPA TSCA Inventory.

SAFETY PROFILE: Poison by ingestion and skin contact. Mutation data reported. An irritant to human skin. When heated to decomposition it emits acrid smoke and irritating fumes. See also ALLYL COMPOUNDS and ESTERS.

AGA750 CAS:2114-11-6 HR: 2
ALLYL CARBAMATE
mf: $C_4H_7NO_2$ mw: 101.12

PROP: Bp: 73–75° @ 2 mm.

SYN: CARBAMIC ACID, ALLYL ESTER

TOXICITY DATA WITH REFERENCE
sce-mus-ipr 100 µmol/kg CNREA8 42,2165,82
ipr-mus TDLo:279 mg/kg/4W-I:NEO CNREA8 29,2184,69

CONSENSUS REPORTS: Reported in EPA TSCA Inventory.

SAFETY PROFILE: Questionable carcinogen with experimental neoplastigenic data. Mutation data reported. See also ALLYL COMPOUNDS and CARBAMATES. When heated to decomposition it emits toxic NO_x.

AGB000 CAS:63884-80-0 HR: 3
ALLYLCARBAMIC ESTER of m-OXYPHENYLDIMETHYLAMINE HYDROCHLORIDE
mf: $C_{12}H_{16}N_2O_2{\bullet}ClH$ mw: 256.76

SYNS: AR-19 □ N-ALLYL CARBAMIC ACID-3-DIMETHYLAMINOPHENYL ESTER HYDROCHLORIDE

TOXICITY DATA WITH REFERENCE
orl-mus LDLo:500 mg/kg JPETAB 43,413,31
ivn-mus LD50:150 mg/kg NTIS** PB158-508

SAFETY PROFILE: Poison by intravenous route. Moderately toxic by ingestion. See also ALLYL COMPOUNDS, ESTERS, and CARBAMATES. When heated to decomposition it emits very toxic fumes of NO_x and HCl.

AGB250 CAS:107-05-1 HR: 3
ALLYL CHLORIDE
DOT: UN 1100
mf: C_3H_5Cl mw: 76.53

$H_2C{=}CHCH_2Cl$

PROP: Colorless liquid with pungent odor. Mp: −136.4°, bp: 44.6°, d: 0.938 @ 20°/4°, fp: −134.5°, flash p: −25°F, lel: 2.9%, uel: 11.2%, autoign temp: 905°F, vap d: 2.64. Misc in org sols. Sltly sol in water.

SYNS: ALLILE (CLORURO DI) (ITALIAN) □ ALLYLCHLORID (GERMAN) □ ALLYLE (CHLORURE D') (FRENCH) □ CHLORALLYLENE □ CHLOROALLYLENE □ 3-CHLOROPRENE □ 1-CHLORO PROPENE-2 □ 1-CHLORO-2-PROPENE □ 3-CHLOROPROPENE □ 3-CHLORO-1-PROPENE □ α-CHLOROPROPYLENE □ 3-CHLOROPROPYLENE □ 3-CHLORO-1-PROPYLENE □ 3-CHLORPROPEN (GERMAN) □ NCI-C04615 □ 2-PROPENYL CHLORIDE

TOXICITY DATA WITH REFERENCE
mmo-esc 20 µL/plate MUREAV 153,57,85

mmo-omi 10 μL/plate CBINA8 30,9,80
skn-rbt 10 mg/24H JIHTAB 30,63,48
eye-rbt 469 mg AJOPAA 29,1363,46
eye-gpg 290 ppm/6H JIHTAB 22,79,40
orl-mus TDLo:4 g/kg (female 7-14D post):TER NTIS** PB86-197605
ipr-rat TDLo:1200 mg/kg:REP EPASR* 8EHQ-0381-0386
orl-mus TDLo:50 g/kg/78W-I:ETA NCITR* NCI-CG-TR-73,78
orl-rat LD50:700 mg/kg JIDHAN 30,63,48
ihl-rat LC50:11 g/m³/2H EESADV 6,19,82
orl-mus LD50:425 mg/kg EESADV 6,19,82
ihl-mus LC50:11,500 mg/m³/2H EESADV 6,19,82
ipr-mus LD50:155 mg/kg SCCUR* -,1,61
ivn-dog LD50:7150 μg/kg JPETAB 89,109,47
skn-rbt LD50:2066 mg/kg JIDHAN 30,63,48

CONSENSUS REPORTS: IARC Cancer Review: Group 3 IMEMDT 7,56,87; Animal Inadequate Evidence IMEMDT 36,39,85; NCI Carcinogenesis Bioassay (gavage); No Evidence: rat NCITR* NCI-CG-TR-73,78; Clear Evidence: mouse NCITR* NCI-CG-TR-73,78. Reported in EPA TSCA Inventory. EPA Genetic Toxicology Program. Community Right-To-Know List.

OSHA PEL: TWA 1 ppm; STEL 2 ppm
ACGIH TLV: TWA 1 ppm; STEL 2 ppm
DFG MAK: 1 ppm (3 mg/m³), Suspected Carcinogen.
NIOSH REL: TWA 1 ppm; CL 3 ppm/15M
DOT CLASSIFICATION: 3; *Label:* Flammable Liquid, Poison

SAFETY PROFILE: Suspected carcinogen with experimental tumorigenic data. Poison by intraperitoneal and intravenous routes. Moderately toxic by ingestion, inhalation, and skin contact. Experimental teratogenic and reproductive effects. A skin and eye irritant. Human mutation data reported. Chronic exposure may cause liver and kidney damage. The vapors of allyl chloride are quite irritating to the eyes, nose, and throat. Contact of the liquid with the skin may lead, in addition to local vasoconstriction and numbness, to rapid absorption and distribution through the body. If remedial measures are not taken promptly, such contact may result in burns and internal injuries. Inhalation may cause headache, dizziness, and in high concentration, loss of consciousness; however, even in low concentration, its odor in most cases is irritating enough to give warning of its presence. Concentration of the vapors high enough to cause serious effects, including damage to the lungs, especially on repeated exposure, may not be intolerable. Consequently, the warning characteristics should never be disregarded. In general, precautions should be taken AT ALL TIMES to avoid spillage and accumulation of noticeable concentration of the vapors in the atmosphere. Acute exposure in experimental animals has resulted in marked inflammation of lungs, irritation of skin, and swelling of the kidneys. Chronically exposed animals have shown degenerative changes in the liver and kidneys. Reported human exposures have been principally cases of irritation of the eyes, skin, and respiratory tract, sometimes accompanied by aches and pains in the bones. Liver and kidney injury is possible.

Dangerous fire and explosion hazard when exposed to heat, flame, or oxidizers. Vigorous or explosive reaction above −70°C with alkyl aluminum chlorides (e.g., trichlorotriethyl dialuminum, ethyl aluminum dichloride, or diethyl aluminum chloride) + aromatic hydrocarbons (e.g., benzene or toluene). Violently exothermic polymerization reaction with Lewis acids (e.g., aluminum chloride, boron trifluoride, or sulfuric acid) and metals (e.g., aluminum, magnesium, zinc, or galvanized metals). Incompatible with HNO_3, ethylene imine, ethylenediamine, chlorosulfonic acid, oleum, NaOH. To fight fire, use CO_2, alcohol foam, dry chemical. See also CHLORINATED HYDROCARBONS, ALIPHATIC, ALLYL COMPOUNDS, and CHLORIDES.

Storage and Handling: Keep cool, away from heat sources. Maintain good ventilation. Work in a fume hood or with closed system if possible; otherwise, use adequate ventilation so that the odor of allyl chloride does not persist. If it should be necessary to enter an area in which the odor of allyl chloride is at all noticeable, use a gas mask equipped with an "organic vapor" canister. Do not disregard the warning odor or eye irritation of allyl chloride.

For occupational chemical analysis use NIOSH: Allyl Chloride, 1000.

AGB500 CAS:2937-50-0 HR: 3
ALLYL CHLOROCARBONATE
DOT: UN 1722
mf: $C_4H_5ClO_2$ mw: 120.54

PROP: Liquid. Bp: 106–114°, flash p: 88°F (CC), d: 1.14, vap d: 4.2.

SYNS: ALLYL CHLOROFORMATE □ ALLYL CHLOROFORMATE (DOT) □ ALLYLESTER KYSELINY CHLORMRAVENCI □ CHLOROFORMIC ACID ALLYL ESTER

TOXICITY DATA WITH REFERENCE
orl-rat LD50:244 mg/kg GTPZAB 28(5),51,84
ihl-rat LC50:32,400 μg/m³ GTPZAB 28(5),51,84
orl-mus LD50:210 mg/kg GTPZAB 28(5),51,84
ihl-mus LD50:23,100 μg/m³ GTPZAB 28(5),51,84

CONSENSUS REPORTS: Reported in EPA TSCA Inventory.

DOT CLASSIFICATION: 8; *Label:* Corrosive, Poison

SAFETY PROFILE: Poison by inhalation and ingestion. Corrosive. Dangerous when exposed to heat, open flame (or sparks), or powerful oxidizers. Can react with oxidizing materials. To fight fire, use alcohol foam, spray or mist, dry chemical. When heated to decomposition it emits toxic fumes of Cl^-. See also ALLYL COMPOUNDS and ESTERS.

AGB750 CAS:4638-03-3 HR: 3
ALLYLCHLOROHYDRIN ETHER
mf: $C_6H_{11}ClO_2$ mw: 150.62

SYNS: ALLYL (3-CHLORO-2-HYDROXYPROPYL) ETHER □ 1-ALLYLOXY-3-CHLORO-2-PROPANOL

TOXICITY DATA WITH REFERENCE
skn-rbt 20 mg/24H MOD 85JCAE-,521,86
eye-rbt 750 μg/24H SEV 28ZPAK -,79,72
skn-rbt 20 mg/24H MOD 85JCAE-,521,86
eye-rbt 750 μg/24H SEV 85JCAE-,521,86

orl-mus LD50:240 mg/kg SCCUR*-,1,61

SAFETY PROFILE: Poison by ingestion. A moderate skin and severe eye irritant. See also ETHERS and ALLYL COMPOUNDS. When heated to decomposition it emits toxic fumes of Cl^-.

AGC000 CAS:1866-31-5 HR: 2
ALLYL CINNAMATE
mf: $C_{12}H_{12}O_2$ mw: 188.24

PROP: Colorless to light-yellow liquid; cherry odor. D: 1.052 @ 25°/25°, bp: 150–152° @ 15 mm. Insol in water; sol in alc; very sol in ether.

SYNS: ALLYL-3-PHENYLACRYLATE □ PROPENYL CINNAMATE □ VINYL CARBINYL CINNAMATE

TOXICITY DATA WITH REFERENCE
skn-hmn 20 mg/48H FCTXAV 15,611,77
orl-rat LD50:1520 mg/kg FCTXAV 2,327,64
skn-rbt LD50:>5 g/kg FCTXAV 15,615,77

CONSENSUS REPORTS: Reported in EPA TSCA Inventory.

SAFETY PROFILE: Moderately toxic by ingestion. Human skin irritant. When heated to decomposition it emits acrid smoke and irritating fumes. See also ALLYL COMPOUNDS and ESTERS.

AGC125 HR: 3
ALLYL COMPOUNDS

PROP: Compounds containing the chemical group $H_2C{=}CHCH_2{\cdot}R$

SAFETY PROFILE: Allyl isovalerate is a poison and causes liver damage and may be an experimental carcinogen. Eugenol is a skin sensitizer in humans and may be an experimental carcinogen. Allyl isothiocyanate is a poison, a severe skin and mucous irritant, and may be an experimental carcinogen. Many allyl compounds are dangerous. They are common in the workplace and the environment. There are numerous uses as chemical intermediates in industry. Chronic allyl chloride exposure causes reversible liver kidney and peripheral nerve damage in humans. Some are naturally occurring (e.g., allyl isothiocyanate and eugenol). Some are used as food additives and flavoring agents. Alkenylbenzenes, including methyl eugenol, have been found in the essential oil and juice of oranges treated with harvesting agents. Some alkenylbenzenes have carcinogenic activity. In general, allyl compounds are reactive and some have the ability to alkylate macromolecules either directly or after metabolic activation. Most are probably metabolized to allyl alcohol which is metabolized to acrolein.

Several allyl compounds are highly flammable and reactive. Triflates (trifluoromethanesulfonate esters) of allyl alcohol and its derivatives are very reactive and are storage hazards. They react violently with aprotic solvents (DMF or DMSO). See also individual entries and ACROLEIN.

AGC150 CAS:13361-32-5 HR: 3
ALLYL CYANOACETATE
mf: $C_6H_7NO_2$ mw: 125.14

SYNS: ACETIC ACID, CYANO-, ALLYL ESTER □ ACETIC ACID, CYANO-, 2-PROPENYL ESTER □ CYANOACETIC ACID, ALLYL ESTER

TOXICITY DATA WITH REFERENCE
orl-rat LD50:160 mg/kg LONZA# 13FEB81

CONSENSUS REPORTS: Reported in EPA TSCA Inventory.

SAFETY PROFILE: Poison by ingestion. When heated to decomposition it emits toxic vapors of NO_x.

AGC200 HR: 3
6-ALLYL-α-CYANOERGOLINE-8-PROPIONAMIDE
mf: $C_{21}H_{24}N_4O$ mw: 348.49

SYN: ERGOLINE-8-PROPIONAMIDE, 6-ALLYL-α-CYANO-

TOXICITY DATA WITH REFERENCE
orl-rat TDLo:300 μg/kg (female 5D post):REP ARZNAD 33,1094,83
orl-mus LD50:400 mg/kg ARZNAD 33,1094,83

SAFETY PROFILE: Poison by ingestion. Experimental reproductive effects. When heated to decomposition it emits toxic fumes of NO_x.

AGC250 CAS:4728-82-9 HR: 2
ALLYL CYCLOHEXANEACETATE
mf: $C_{11}H_{18}O_2$ mw: 182.29

SYNS: ALLYL CYCLOHEXYLACETATE □ CYCLOHEXYLACETIC ACID ALLYL ESTER

TOXICITY DATA WITH REFERENCE
skn-hmn 20 mg/48H MLD FCTXAV 15,611,77
skn-rbt 500 mg/24H MOD FCTXAV 15,611,77
orl-rat LD50:900 mg/kg FCTXAV 15,611,77
skn-rbt LD50:1250 mg/kg FCTXAV 15,611,77

CONSENSUS REPORTS: Reported in EPA TSCA Inventory.

SAFETY PROFILE: Moderately toxic by ingestion and skin contact. Irritating to human skin. See also ALLYL COMPOUNDS and ESTERS. When heated to decomposition it emits acrid smoke and irritating fumes.

AGC500 CAS:2705-87-5 HR: 3
ALLYL CYCLOHEXANEPROPIONATE
mf: $C_{12}H_{20}O_2$ mw: 196.32

PROP: Colorless liquid; pineapple odor. D: 0.945–0.950, refr index: 1.457–1.463, flash p: 212°F. Misc in alc, chloroform, ether; insol in glycerin and water.

SYNS: 3-ALLYLCYCLOHEXYL PROPIONATE □ ALLYL HEXAHYDROPHENYLPROPIONATE □ FEMA No. 2026

TOXICITY DATA WITH REFERENCE
orl-rat LD50:585 mg/kg FCTXAV 2,327,64
orl-gpg LD50:380 mg/kg FCTXAV 2,327,64

CONSENSUS REPORTS: Reported in EPA TSCA Inventory.

SAFETY PROFILE: Poison by ingestion. When heated to decomposition it emits acrid smoke and irritating fumes. Combustible liquid. See ALLYL COMPOUNDS and ESTERS.

AGC750 **CAS:66827-50-7** **HR: 3**
2-ALLYL-2-CYCLOHEXYLACETIC ACID-3-(DIETHYL-AMINO)-2,2-DIMETHYLPROPYLESTER HYDRO-CHLORIDE
mf: $C_{20}H_{37}NO_2{\bullet}ClH$ mw: 360.04

SYN: CYCLOHEXYLALLYL-ESSIGSAEUREESTER DES 3-DIAETHYLAMI-NO-2,2-DIMETHYL-1-PROPANOL (GERMAN)

TOXICITY DATA WITH REFERENCE
ivn-mus LDLo:100 mg/kg AEPPAE 173,86,33
par-frg LDLo:500 mg/kg AEPPAE 173,86,33

SAFETY PROFILE: Poison by intravenous route. Moderately toxic by parenteral route. See also ALLYL COMPOUNDS and ESTERS. When heated to decomposition it emits very toxic fumes of HCl and NO_x.

AGD000 **CAS:66941-60-4** **HR: 3**
5-ALLYL-5-(2-CYCLOPENTENYL)-2-THIOBARBITURIC ACID
mf: $C_{12}H_{14}N_2O_2S$ mw: 250.34

TOXICITY DATA WITH REFERENCE
ipr-rat LD50:100 mg/kg JACSAT 65,2091,43
ivn-rat LD50:100 mg/kg JACSAT 65,2091,43

SAFETY PROFILE: Poison by intraperitoneal and intravenous routes. When heated to decomposition it emits very toxic fumes of NO_x and SO_x. See also BARBITURATES and ALLYL COMPOUNDS.

AGD250 **CAS:142-22-3** **HR: 3**
ALLYL DIGLYCOL CARBONATE
mf: $C_{12}H_{18}O_7$ mw: 274.30

PROP: Liquid. Bp: 162°, flash p: 378°F (OC), d: 1.14.

SYNS: CARBONIC ACID, ALLYL ESTER, DIESTER with DIETHYLENE GLYCOL □ CR 39 □ DAGC □ DIALLYL DIGLYCOL CARBONATE □ DIETHYLENE GLYCOL, BIS(ALLYL CARBONATE)- □ 01M □ NOURYSET 200 □ OXYDIETHYLENEDICARBONIC ACID DIALLYL ESTER □ RAV 7 □ 2,5,8,10-TETRAOXATRIDEC-12-ENOIC ACID, 9-OXO-, 2-PROPENYL ESTER (9CI) □ TRANSALLYL CR 39 □ TS 16

TOXICITY DATA WITH REFERENCE
skn-hmn 2%/48H CODEDG 2,183,76
skn-rbt TDLo:14,859 mg/kg (female 6-18D post):REP EPASR* 8EHQ-0787-0666 FLWP
ipr-mus LD50:270 mg/kg JPETAB 90,338,47

CONSENSUS REPORTS: Reported in EPA TSCA Inventory.

SAFETY PROFILE: Poison by intraperitoneal route. A human skin irritant. The allyl compounds are generally toxic. Experimental reproductive effects. Combustible when exposed to heat or flame; can react with oxidizing material. To fight fire, use water mist or spray, foam, CO_2, dry chemical. When heated to decomposition it emits acrid smoke and irritating fumes. See also ALLYL COMPOUNDS and ESTERS.

AGD500 **CAS:130-83-6** **HR: 3**
6-ALLYL-6,7-DIHYDRO-5H-DIBENZ(c,e)AZEPINE PHOSPHATE
mf: $C_{17}H_{17}N{\bullet}H_3O_4P$ mw: 333.35

PROP: Crystals. Mp: 211–215° (decomp).

SYNS: AZAPETINE PHOSPHATE □ AZEPINE PHOSPHATE □ 6,7-DI-HYDRO-6-(2-PROPENYL)-5H-DIBENZ(c,e)AZEPINE PHOSPHATE □ ILIDAR □ ILIDAR PHOSPHATE □ RO 2-3248

TOXICITY DATA WITH REFERENCE
skn-hmn 2%/48H CODEDG 2,183,76
skn-rbt TDLo:14,859 mg/kg (female 6-18D post):TER EPASR* 8EHQ-0787-0666 FLWP
ipr-mus LD50:270 mg/kg JPETAB 90,338,47

CONSENSUS REPORTS: Reported in EPA TSCA Inventory.

SAFETY PROFILE: Poison by intraperitoneal route. See also ALLYL COMPOUNDS and PHOSPHATES. When heated to decomposition it yields highly toxic fumes of PO_x and NO_x. An anti-adrenergic agent.

AGD750 **CAS:63918-66-1** **HR: 3**
6-ALLYL-6,7-DIHYDRO-3,9-DICHLORO-5H-DIBENZ(c,e)AZEPINE
mf: $C_{17}H_{15}Cl_2N$ mw: 304.23

TOXICITY DATA WITH REFERENCE
ipr-mus LD50:316 mg/kg JPETAB 103,10,51
ivn-mus LD50:47 mg/kg JPETAB 103,10,51

SAFETY PROFILE: Poison by intraperitoneal and intravenous routes. When heated to decomposition it yields highly toxic fumes of chlorides and NO_x. See also ALLYL COMPOUNDS.

AGE000 **CAS:63918-56-9** **HR: 3**
6-ALLYL-6,7-DIHYDRO-6-METHYL-5H-DIBENZ(c,e)A-ZEPINIUM IODIDE
mf: $C_{18}H_{20}N{\bullet}I$ mw: 377.29

SYN: RO 2-3742

TOXICITY DATA WITH REFERENCE
ipr-mus LD50:83 mg/kg JPETAB 103,10,51
ivn-mus LD50:6 mg/kg JPETAB 103,10,51

SAFETY PROFILE: Poison by intraperitoneal and intravenous routes. See also ALLYL COMPOUNDS and IODIDES. When heated to decomposition it yields highly toxic fumes of iodides and NO_x.

AGE250 **CAS:93-15-2** **HR: 3**
4-ALLYL-1,2-DIMETHOXYBENZENE
mf: $C_{11}H_{14}O_2$ mw: 178.25

PROP: Colorless to pale-yellow liquid; clove, carnation odor. Bp: 255°, d: 1.032–1.036, refr index: 1.532, fp: −4, flash p: 212°F. Sol in fixed oils; insol in glycerin and propylene glycol.

SYNS: 1-ALLYL-3,4-DIMETHOXYBENZENE □ 4-ALLYLVERATROLE □ 1,2-DIMETHOXY-4-ALLYLBENZENE □ 1-(3,4-DIMETHOXYPHENYL)-2-PROPENE □ ENT 21,040 □ 1,3,4-EUGENOL METHYL ETHER □ EUGENYL METHYL ETHER □ FEMA No. 2475 □ METHYL EUGENOL (FCC) □ VERATROLE METHYL ETHER

TOXICITY DATA WITH REFERENCE
skn-rbt 500 mg/24H FCTXAV 13,681,75
dnd-mus-ipr 80 mg/kg FCTXAV 13,857,75
orl-rat LD50:1179 mg/kg TXAPA9 31,421,75
ipr-mus LD50:540 mg/kg AIPTAK 199,226,72
ivn-mus LD50:112 mg/kg AIPTAK 199,226,72
skn-rbt LD50:>2025 mg/kg TXAPA9 31,421,75

CONSENSUS REPORTS: Reported in EPA TSCA Inventory.

SAFETY PROFILE: Poison by intravenous route. Moderately toxic by ingestion and intraperitoneal routes. A skin irritant. Mutation data reported. Combustible liquid. When heated to decomposition it emits acrid smoke and irritating fumes. Some other alkenylbenzenes have carcinogenic activity. See also EUGENOL, ALLYL COMPOUNDS and ETHERS.

AGE500 CAS:523-80-8 HR: 2
1-ALLYL-2,5-DIMETHOXY-3,4-METHYLENEDIOXY-BENZENE
mf: $C_{12}H_{14}O_4$ mw: 222.26

PROP: Needles. D: 1.015 @ 20°/4°, mp: 30°, bp: 294°. Apiol is the essential oil from *Petroliselium sativum* seeds (BSIBAC 14,291,39).

SYNS: APIOL □ PARSLEY APIOL □ PARSLEY CAMPHOR

TOXICITY DATA WITH REFERENCE
dnd-mus-ipr 400 mg/kg CRNGDP 5,1613,84
scu-mus LDLo:1000 mg/kg BSIBAC 14,291,39
scu-frg LDLo:1515 mg/kg AEXPBL 35,342,1895

CONSENSUS REPORTS: Reported in EPA TSCA Inventory.

SAFETY PROFILE: Moderately toxic by subcutaneous route. Mutation data reported. When heated to decomposition it emits acrid smoke and irritating fumes. See also ALLYL COMPOUNDS.

AGE625 CAS:691-35-0 HR: 3
ALLYLDIMETHYLARSINE
mf: $C_5H_{11}As$ mw: 146.06

$H_2C{=}CHCH_2As(CH_3)_2$

CONSENSUS REPORTS: Arsenic and its compounds are on the Community Right-To-Know List.

SAFETY PROFILE: Ignites in air if a large surface area is exposed (e.g., small particles on filter paper). When heated to decomposition it emits toxic fumes of As. See also ARSENIC COMPOUNDS.

AGE750 CAS:56717-11-4 HR: 3
1-ALLYL-1-(3,7-DIMETHYLOCTYL)PIPERIDINIUM BROMIDE
mf: $C_{18}H_{36}N{\cdot}Br$ mw: 346.46

PROP: Pale-yellow wax. Very sol in water. Mp: 75°.

SYNS: 1-ALLYL-1-(3,7-DIMETHYLOCTYL)-PIPERIDIUMBROMID (GERMAN) □ ALLYL-TETRA-HYDROGERANYL-PIPERIDINIUMBROMID (GERMAN) □ PIPROCTANYLIUMBROMID (GERMAN)

TOXICITY DATA WITH REFERENCE
orl-rat LD50:360 mg/kg 85DPAN -,-,71/76
skn-rat LD50:115 mg/kg 85DPAN -,-,71/76

SAFETY PROFILE: Poison by ingestion and skin contact. See also ALLYL COMPOUNDS and BROMIDES. When heated to decomposition it emits very toxic fumes of HBr and NO_x.

AGF000 CAS:33132-87-5 HR: 3
β-ALLYL-N,N-DIMETHYLPHENETHYLAMINE
mf: $C_{13}H_{19}N$ mw: 189.33

TOXICITY DATA WITH REFERENCE
orl-mus LD50:210 mg/kg CHTPBA 6,453,71
ivn-mus LD50:37 mg/kg CHTPBA 6,453,71

SAFETY PROFILE: Poison by ingestion and intravenous routes. When heated to decomposition it emits toxic fumes of NO_x. See also ALLYL COMPOUNDS.

AGF250 CAS:743-45-3 HR: 2
5-ALLYL-1,3-DIPHENYLBARBITURIC ACID
mf: $C_{19}H_{16}N_2O_3$ mw: 320.37

SYN: 5-ALLYL-1,3-DIPHENYL-2,4,6(1H,3H,5H)-PYRIMIDINETRIONE

TOXICITY DATA WITH REFERENCE
ipr-rat LD50:1000 mg/kg JMCMAR 7,342,64
ipr-mus LD50:533 mg/kg ARZNAD 17,1519,67

SAFETY PROFILE: Moderately toxic by intraperitoneal route. When heated to decomposition it emits toxic fumes of NO_x. See also BARBITURATES and ALLYL COMPOUNDS.

AGF300 CAS:2179-57-9 HR: 3
ALLYL DISULFIDE
mf: $C_6H_{10}S_2$ mw: 146.28

SYNS: ALLYL DISULPHIDE □ DIALLYL DISULFIDE □ DIALLYL DISULPHIDE □ DISULFIDE, DI-2-PROPENYL (9CI) □ 4,5-DITHIA-1,7-OCTADIENE □ 2-PROPENYL DISULPHIDE

TOXICITY DATA WITH REFERENCE
orl-rat LD50:260 mg/kg FCTOD7 26,297,88
skn-rbt LD50:3600 mg/kg FCTOD7 26,297,88

CONSENSUS REPORTS: Reported in EPA TSCA Inventory.

SAFETY PROFILE: Poison by ingestion. Moderately toxic by skin contact. When heated to decomposition it emits toxic vapors of SO_x.

AGF500 CAS:10138-39-3 HR: 2
ALLYL-3,4-EPOXY-6-METHYLCYCLOHEXANECARBOXYLATE
mf: $C_{11}H_{16}O_3$ mw: 196.27

SYNS: 3,4-EPOXY-6-METHYLCYCLOHEXANECARBOXYLIC ACID,ALLYL ESTER □ 4-METHYL-7-OXABICYCLO(4.1.0)HEPTANE-3-CARBOXYLIC ACID, ALLYL ESTER

TOXICITY DATA WITH REFERENCE
skn-rbt 10 mg/24H open MLD AIHAAP 23,95,62
orl-rat LD50:500 mg/kg AIHAAP 23,95,62
skn-rbt LD50:2830 mg/kg AIHAAP 23,95,62

SAFETY PROFILE: Moderately toxic by skin contact and ingestion. A skin irritant. See also ALLYL COMPOUNDS and ESTERS. When heated to decomposition it emits acrid smoke and irritating fumes.

AGG000 HR: 3
ALLYL ETHYL ETHER
mf: $C_5H_{10}O$ mw: 86.1

PROP: Liquid. Flash p: <75°F.

SAFETY PROFILE: Highly flammable and reactive. Forms explosive peroxides in storage. Can explode during distillation probably due to peroxide formation. When heated to decomposition it emits acrid, irritating fumes. See also ALLYL COMPOUNDS and ETHERS.

AGG250 CAS:6654-31-5 HR: 3
2-ALLYL-5-ETHYL-2′-HYDROXY-9-METHYL-6,7-BENZOMORPHAN
mf: $C_{18}H_{25}NO$ mw: 271.44

SYN: 2,6-METHANO-3-BENZAZOCIN-8-OL-3-ALLYL-6-ETHYL-1,2,3,4,5,6-HEXAHYDRO-11-METHYL

TOXICITY DATA WITH REFERENCE
unk-hmn TDLo:70 μg/kg:PSY SCIEAS 137,541,62
scu-rat LD50:174 mg/kg JPETAB 143,141,64
ivn-rat LD50:32 mg/kg JPETAB 143,141,64
scu-mus LD50:182 mg/kg JPETAB 143,141,64
ivn-mus LD50:32 mg/kg 31ZPAG 2,175,66

SAFETY PROFILE: Poison by subcutaneous and intravenous routes. Human psychotropic effects by an unspecified route. When heated to decomposition it emits toxic fumes of NO_x. See also ALLYL COMPOUNDS.

AGG500 HR: 3
ALLYL FLUORIDE
mf: C_3H_5F mw: 60.07

PROP: Colorless gas. Bp: −10°.

SYN: 3-FLUOROPROPENE

SAFETY PROFILE: Poison by inhalation and ingestion. A strong irritant. See also ALLYL COMPOUNDS and FLUORIDES. When heated to decomposition it emits highly toxic fumes of F^-. Incompatible with water or steam to produce toxic and corrosive fumes.

AGG750 CAS:406-23-5 HR: 3
ALLYL FLUOROACETATE
mf: $C_5H_7FO_2$ mw: 118.12

TOXICITY DATA WITH REFERENCE
ihl-mus LCLo:500 mg/m³ NDRC** -,9,44
scu-mus LD50:6 mg/kg JCSOA9 -,916,49

SAFETY PROFILE: Poison by inhalation and subcutaneous routes. See also ALLYL COMPOUNDS and FLUORIDES. When heated to decomposition it emits toxic fumes of F^-.

AGH000 CAS:1838-59-1 HR: 3
ALLYL FORMATE
DOT: UN 2336
mf: $C_4H_6O_2$ mw: 86.10

$H_2C{=}CHCH_2OCO{\cdot}H$

PROP: Liquid, sltly water-sol, sol in organic solvents. D: 0.948 @ 18°/4°, bp: 83°, flash p: <−50°F.

SYNS: ALLYLESTER KYSELINY MRAVENCI □ FORMIC ACID, ALLYL ESTER

TOXICITY DATA WITH REFERENCE
ihl-rat LC50:980 mg/m³ GTPZAB 28(6),55,84
orl-mus LD50:96 mg/kg GTPZAB 28(6),55,84
ihl-mus LC50:610 mg/m³ GTPZAB 28(6),55,84

CONSENSUS REPORTS: Reported in EPA TSCA Inventory.

DOT CLASSIFICATION: 3; *Label:* Flammable Liquid, Poison

SAFETY PROFILE: Poison by ingestion. Moderately toxic by inhalation. Very flammable and reactive. Dangerous fire hazard. See also ALLYL COMPOUNDS and ESTERS. When heated to decomposition it yields irritating smoke and fumes.

AGH150 CAS:106-92-3 HR: 3
ALLYL GLYCIDYL ETHER
DOT: UN 2219
mf: $C_6H_{10}O_2$ mw: 114.16

PROP: Bp: 153.9°, fp: −100° (forms glass), flash p: 135°F (OC), d: 0.9698 @ 20°/4°, vap press: 21.59 mm @ 60°, vap d: 3.94.

SYNS: AGE □ ALLIL-GLICIDIL-ETERE (ITALIAN) □ 1-ALLILOSSI-2,3 EPOSSIPROPANO (ITALIAN) □ ALLYL-2,3-EPOXYPROPYL ETHER □ ALLYLGLYCIDAETHER (GERMAN) □ 1-ALLYLOXY-2,3-EPOXY-PROPAAN (DUTCH) □ 1-ALLYLOXY-2,3-EPOXYPROPAN (GERMAN) □ 1-(ALLYLOXY)-2,3-EPOXYPROPANE □ NCI-C56666 □ OXYDE d'ALLYLE et de GLYCIDYLE (FRENCH) □ ((2-PROPENYLOXY)METHYL)OXIRANE

TOXICITY DATA WITH REFERENCE
skn-rbt 500 mg/24H SEV 28ZPAK -,135,72
eye-rbt 97 mg SEV AMIHAB 14,250,56
eye-rbt 250 μg/24H SEV 28ZPAK -,135,72
mmo-sat 100 μg/plate NTPTR* NTP-TR-376,90
mma-sat 1 mg/plate NTPTR* NTP-TR-376,90
sln-dmg-orl 5500 ppm ENMUDM 7,349,85
orl-rat LD50:922 mg/kg 28ZPAK -,135,72

ihl-rat LCLo:860 ppm/4H 28ZPAK -,135,72
orl-mus LD50:390 mg/kg AMIHAB 14,250,56
ihl-mus LC50:270 ppm/4H AMIHAB 14,250,56
skn-rbt LD50:2550 mg/kg AMIHAB 14,250,56

CONSENSUS REPORTS: Reported in EPA TSCA Inventory.

OSHA PEL: TWA 5 ppm; STEL 10 ppm
ACGIH TLV: TWA 5 ppm; STEL 10 ppm (skin)
DFG MAK: Confirmed Animal Carcinogen, Suspected Human Carcinogen.
NIOSH REL: (Glycidyl Ethers) CL 45 mg/m³/15M
DOT CLASSIFICATION: 3; *Label:* Flammable Liquid, Poison

SAFETY PROFILE: Confirmed animal carcinogen. Poison by ingestion. Moderately toxic by inhalation and skin contact. Mutation data reported. A severe skin and eye irritant. Can cause central nervous system depression and pulmonary edema. Mutation data reported. A flammable liquid when exposed to heat or flame; can react with oxidizing materials. To fight fire, use foam, CO_2, dry chemical. When heated to decomposition it emits acrid smoke and irritating fumes. See also ALLYL COMPOUNDS.

For occupational chemical analysis use NIOSH: Allyl Glycidyl Ether S346.

AGH250 CAS:142-19-8 HR: 2
ALLYL HEPTANOATE
mf: $C_{10}H_{18}O_2$ mw: 170.28

PROP: Colorless to pale yellow liquid; fruity, sweet, pineapple odor. D: 0.880, refr index: 1.426, flash p: 154°F.

SYNS: ALLYL ENANTHATE □ ALLYL HEPTOATE □ ALLYL HEPTYLATE □ FEMA No. 2031 □ 2-PROPENYL HEPTANOATE

TOXICITY DATA WITH REFERENCE
skn-hmn 20 mg/48H MLD FCTXAV 15,611,77
skn-rbt 500 mg/24H MOD FCTXAV 15,611,77
orl-rat LD50:500 mg/kg TXAPA9 6,378,64
orl-mus LD50:630 mg/kg TXAPA9 7,18,65
skn-rbt LD50:810 mg/kg FCTXAV 15,611,77
orl-gpg LD50:444 mg/kg FCTXAV 2,327,64

CONSENSUS REPORTS: Reported in EPA TSCA Inventory.

SAFETY PROFILE: Moderately toxic by ingestion and skin contact. A human skin irritant. See also ALLYL COMPOUNDS and ESTERS. Combustible liquid. When heated to decomposition it emits acrid smoke and irritating fumes.

AGH500 CAS:52207-83-7 HR: 2
ALLYLHYDRAZINE HYDROCHLORIDE
mf: $C_3H_8N_2$•ClH mw: 108.59

TOXICITY DATA WITH REFERENCE
orl-mus TDLo:9800 mg/kg/35W-C:CAR BJCAAI 34,90,76
orl-mus TDLo:25 mg/kg:ETA PAACA3 16,61,75

SAFETY PROFILE: Questionable carcinogen with experimental carcinogenic and tumorigenic data. When heated to decomposition it emits very toxic fumes of HCl and NO_x. See also ALLYL COMPOUNDS.

AGH750 HR: 3
ALLYL HYDROPEROXIDE
mf: $C_3H_6O_2$ mw: 74.1

SAFETY PROFILE: Highly toxic. A potentially explosive liquid. Unstable to heat, light, and solid alkalies. Mixtures with sand are impact sensitive. Upon decomposition it emits acrid smoke and fumes. See also PEROXIDES, ORGANIC; and ALLYL COMPOUNDS.

AGI000 CAS:152-02-3 HR: 3
N-ALLYL-3-HYDROXYMORPHINAN
mf: $C_{19}H_{25}NO$ mw: 283.45

PROP: Mp: 180–182°.

SYNS: (−)-3-HYDROXY-N-ALLYLMORPHINAN □ l-3-HYDROXY-N-ALLYL MORPHINAN □ (−)-LEVALLORPHAN □ LORFAN □ NALOXIPHAN □ RO-1-7700

TOXICITY DATA WITH REFERENCE
orl-rat LD50:949 mg/kg 27ZIAQ -,-,65
ipr-rat LD50:185 mg/kg 27ZIAQ ,140,73
ipr-mus LD50:184 mg/kg 27ZIAQ -,-,65
scu-mus LD50:200 mg/kg ANYAA9 281,321,76

SAFETY PROFILE: Poison experimentally by subcutaneous and intraperitoneal routes. Moderately toxic by ingestion. When heated to decomposition it emits toxic fumes of NO_x. See also ALLYL COMPOUNDS.

AGI250 CAS:556-56-9 HR: 3
ALLYL IODIDE
DOT: UN 1723
mf: C_3H_5I mw: 167.98

PROP: Yellow liquid, pungent odor. Mp: −99°, bp: 103.1°, d: 1.825 @ 20°/4°, vap d: 5.8.

SYNS: 3-IODOPROPENE □ 3-IODO-1-PROPENE □ 3-IODOPROPYLENE □ 1-PROPENE, 3-IODO-(9CI)

TOXICITY DATA WITH REFERENCE
mmo-sat 1 μmol/plate BCPCA6 29,993,80
dns-hmn:hlas 50 μmol/L CALEDQ 20,263,83

CONSENSUS REPORTS: Reported in EPA TSCA Inventory.

DOT CLASSIFICATION: 3; *Label:* Flammable Liquid, Corrosive

SAFETY PROFILE: Poison by inhalation and ingestion. Mutation data reported. A powerful irritant. A flammable liquid. Incompatible with oxidizing materials. To fight fire, use water, foam, CO_2, dry chemical. When heated to decomposition it emits highly toxic fumes of I^-. See also ALLYL COMPOUNDS and IODIDES.

AGI500 CAS:79-78-7 HR: 2
ALLYL α-IONONE
mf: $C_{16}H_{24}O$ mw: 232.40

PROP: Colorless to yellow liquid; fruity, woody odor. D:

0.928–0.935, refr index: 1.503–1.507, flash p: 212°F. Sol in alc; insol in water @ 265°.

SYNS: FEMA No. 2033 ☐ 1-(2,6,6-TRIMETHYL-2-CYCLOHEXEN-1-YL)-1,6-HEPTADIEN-3-ONE

TOXICITY DATA WITH REFERENCE
skn-rbt 500 mg MLD FCTXAV 11,1079,73
skn-rbt LD50:>5 g/kg FCTXAV 11,493,73

CONSENSUS REPORTS: Reported in EPA TSCA Inventory.

SAFETY PROFILE: A skin irritant. Combustible liquid. When heated to decomposition it emits acrid smoke and irritating fumes. See also ALLYL COMPOUNDS.

AGI750 CAS:77-26-9 HR: 3
ALLYLISOBUTYLBARBITURATE
mf: $C_{11}H_{16}N_2O_3$ mw: 224.29

PROP: Prisms. Mp: 138–139°.

SYNS: ALISOBUMAL ☐ ALLYLBARBITAL ☐ ALLYLBARBITONE ☐ ALLYLBARBITURIC ACID ☐ ALLYLISOBUTYLBARBITAL ☐ 5-ALLYL-5-ISOBUTYLBARBITURIC ACID ☐ 5-ALLYL-5-(2'-METHYL-N-PROPYL) BARBITURIC ACID ☐ BUTALBARBITAL ☐ BUTALBITAL ☐ ISO-BUTYLALLYLBARBITURIC ACID ☐ ISOBUTYLALLYLBARTURIC ACID ☐ ITOBARBITAL ☐ 5-(2-METHYLPROPYL)-5-(2-PROPENYL)-2,4,6(1H,3H,5H)-PYRIMIDINETRIONE (9CI) ☐ OPTALIDON ☐ SANDOPTAL ☐ TETRALLOBARBITAL

TOXICITY DATA WITH REFERENCE
orl-wmn TDLo:400 mg/kg:CVS,PUL,GIT SAVEAB 10,181A,39
orl-chd TDLo:10 mg/kg:CNS,KID SAVEAB 10,209A,39
scu-rat LD50:160 mg/kg AEPPAE 152,341,30
orl-pgn LD50:75 mg/kg TXAPA9 21,315,72
orl-dck LD50:237 mg/kg TXAPA9 21,315,72
orl-bsd LD50:75 mg/kg TXAPA9 21,315,72

SAFETY PROFILE: Poison by ingestion and subcutaneous routes. Human systemic effects by ingestion: toxic psychosis, coma, reduced blood pressure, respiratory depression, nausea and vomiting, and kidney effects. When heated to decomposition it emits toxic NO_x. See also ALLYL COMPOUNDS, BARBITURATES, and various barbital compounds. Used as a sleep aid.

AGJ000 CAS:1476-23-9 HR: 3
ALLYL ISOCYANATE
DOT: UN 2206/UN 2207/UN 2478/UN 3080
mf: C_4H_5NO mw: 83.10

PROP: Liquid. Bp: 85°.

SYN: ISOCYANIC ACID, ALLYL ESTER

TOXICITY DATA WITH REFERENCE
ivn-mus LD50:18 mg/kg CSLNX* NX#03769

CONSENSUS REPORTS: Reported in EPA TSCA Inventory.

DOT CLASSIFICATION: 6.1; *Label:* KEEP AWAY FROM FOOD (UN 2207); DOT Class: 6.1; *Label:* Poison (UN 2206); DOT Class: 6.1; *Label:* Poison, Flammable Liquid (UN 3080); DOT Class: 3; *Label:* Flammable Liquid, Poison (UN 2478)

SAFETY PROFILE: A poison. See also ALLYL COMPOUNDS and ESTERS. A flammable liquid. When heated to decomposition it emits toxic fumes of NO_x.

AGJ250 CAS:57-06-7 HR: 3
ALLYL ISOTHIOCYANATE
DOT: UN 1545
mf: C_4H_5NS mw: 99.16

$H_2C{=}CHCH_2N{:}C{:}S$

PROP: Colorless to pale-yellow liquid oil; irritating odor with mustard taste. Mp: −80°, d: 1.015 @ 15°/4°, bp: 150.7°, fp: −80°, flash p: 115°F, d: 1.013–1.016 @ 25°/25°, vap press: 10 mm @ 38.3°, vap d: 3.41, refr index: 1.527–1.531. Misc with alc, carbon disulfide, and ether.

SYNS: AITC ☐ ALLYL ISORHODANIDE ☐ ALLYL ISOSULFOCYANATE ☐ ALLYL ISOTHIOCYANATE, stabilized (DOT) ☐ ALLYL MUSTARD OIL ☐ ALLYLSENFOEL (GERMAN) ☐ ALLYL SEVENOLUM ☐ ALLYL THIOCARBONIMIDE ☐ ARTIFICIAL MUSTARD OIL ☐ CARBOSPOL ☐ FEMA No. 2034 ☐ ISOTHIOCYANATE d'ALLYLE (FRENCH) ☐ 3-ISOTHIOCYANATO-1-PROPENE ☐ MUSTARD OIL ☐ NCI-C50464 ☐ OIL of MUSTARD, artificial ☐ OLEUM SINAPIS VOLATILE ☐ 2-PROPENYL ISOTHIOCYANATE ☐ REDSKIN ☐ SENF OEL (GERMAN) ☐ SYNTHETIC MUSTARD OIL ☐ VOLATILE OIL of MUSTARD

TOXICITY DATA WITH REFERENCE
eye-rbt 2 mg AEPPAE 219,119,53
mmo-sat 100 µg/plate ABCHA6 44,4017,80
mma-sat 1 µmol/plate CBINA8 38,303,82
scu-rat TDLo:100 mg/kg (8-9D preg):TER FCTXAV 18,159,80
scu-rat TDLo:200 mg/kg (8-9D preg):REP FCTXAV 18,159,80
orl-rat TDLo:12,875 mg/kg/2Y-I:NEO NTPTR* NTP-TR-234,82
skn-mus TDLo:12 g/kg/12W-I:ETA AICCA6 11,699,55
orl-rat LD50:112 mg/kg AEPPAE 219,119,53
ipr-rat LDLo:80 mg/kg ARZNAD 16,870,66
scu-rat LD50:92 mg/kg FCTXAV 18,159,80
orl-mus LD50:308 mg/kg AMRL** TR-73-83,73
ipr-mus LDLo:4 mg/kg TXAPA9 23,288,72
scu-mus LD50:80 mg/kg ARZNAD 5,505,55
skn-rbt LD50:88 mg/kg TXAPA9 42,417,77
ivn-rbt LDLo:12 mg/kg BIJOAK 4,107,09

CONSENSUS REPORTS: IARC Cancer Review: Group 3 IMEMDT 7,56,87; Animal Limited Evidence IMEMDT 36,55,85; NTP Carcinogenesis Bioassay (gavage); No Evidence: mouse NTPTR* NTP-TR-234,82; Clear Evidence: rat NTPTR* NTP-TR-234,82. Reported in EPA TSCA Inventory.

DOT CLASSIFICATION: 6.1; *Label:* Poison

SAFETY PROFILE: Suspected carcinogen with experimental neoplastigenic and tumorigenic data. Poison by ingestion, skin contact, intravenous, subcutaneous, and intraperitoneal routes. Experimental teratogenic and reproductive effects. An eye irritant. An allergen. May cause contact dermatitis. Mutation data reported. A flammable liquid. Highly reactive. When heated to decomposition (above 250°) or on contact with acid or acid fumes it emits highly toxic fumes of CN^-, SO_x, and

NO_x. To fight fire, use foam, CO_2, dry chemical. See also ALLYL COMPOUNDS and ESTERS.

AGJ375 **CAS:3052-45-7** **HR: 3**
ALLYLLITHIUM
mf: C_3H_5Li mw: 48.01

$H_2C{=}CHCH_2Li$

PROP: Colorless solid. Sol in ethers; sltly sol in hydrocarbons.

SAFETY PROFILE: Ignites on contact with air. See also LITHIUM COMPOUNDS.

AGJ500 **HR: 3**
ALLYL MERCAPTAN
mf: C_3H_6S mw: 74.15

$H_2C{=}CHCH_2SH$

PROP: Water-white liquid with a strong garlic odor, darkens on standing. D: 0.925 @ 23°/4°, bp: 68°, flash p: 14°F.

SYN: 2-PROPENE-1-THIOL

SAFETY PROFILE: Poison by inhalation and ingestion. Strong irritant to skin and mucous membranes. When heated to decomposition it emits highly toxic fumes of SO_x. Very dangerous fire hazard. To fight fire, use water mist or spray, alcohol foam, CO_2, or dry chemical. See also ALLYL COMPOUNDS and MERCAPTANS.

AGJ750 **CAS:64037-65-0** **HR: 3**
2-ALLYLMERCAPTO-2-ETHYLBUTYRAMIDE
mf: $C_9H_{17}NOS$ mw: 187.33

SYNS: α-ALLYLMERCAPTO-α, α-DIETHYLACETAMIDE □ 2-ALLYLTHIO-2-ETHYLBUTYRAMIDE

TOXICITY DATA WITH REFERENCE
orl-mus LD50:891 mg/kg JMCMAR 6,351,63
ipr-mus LD50:400 mg/kg NTIS** AD691-490

SAFETY PROFILE: Poison by intraperitoneal route. Moderately toxic by ingestion. When heated to decomposition it emits very toxic fumes of SO_x and NO_x. See also ALLYL COMPOUNDS.

AGK000 **CAS:63915-89-9** **HR: 3**
2-ALLYLMERCAPTOISOBUTYRAMIDE
mf: $C_7H_{13}NOS$ mw: 159.27

SYN: α-ALLYLMERCAPTOISOBUTYRAMIDE

TOXICITY DATA WITH REFERENCE
orl-mus LD50:2000 mg/kg JMCMAR 6,351,63
ipr-mus LD50:250 mg/kg NTIS** AD691-490

SAFETY PROFILE: Poison by intraperitoneal route. Moderately toxic by ingestion. When heated to decomposition it emits toxic fumes of NO_x and SO_x. See also ALLYL COMPOUNDS.

AGK250 **CAS:87-09-2** **HR: 3**
ALLYLMERCAPTOMETHYLPENICILLIN
mf: $C_{13}H_{18}N_2O_4S_2$ mw: 330.45

PROP: Crystals from Me_2CO.

SYNS: ALLYLMERCAPTOMETHYLPENICILLINIC ACID □ ALLYLTHIOMETHYLPENICILLIN □ ALMECILLIN □ AT □ CER-o-CILLIN □ PENICILLIN AT □ PENICILLIN O

TOXICITY DATA WITH REFERENCE
mul-wmn LDLo:35,827 μg/kg/6D-I:CVS JOALAS 31,455,60
ice-mus LD50:45 mg/kg JLCMAK 34,126,49
ice-dog LD50:6490 μg/kg JLCMAK 34,126,49
isp-dog LD50:38 mg/kg JLCMAK 34,126,49
ice-rbt LD50:5610 μg/kg JLCMAK 34,126,49

SAFETY PROFILE: Poison by intracerebral and intraspinal routes. Human systemic effects by multiple routes: changes in blood vessels. When heated to decomposition it emits very toxic fumes of NO_x and SO_x. See also PENICILLIN and ALLYL COMPOUNDS.

AGK500 **CAS:96-05-9** **HR: 2**
ALLYL METHACRYLATE
mf: $C_7H_{10}O_2$ mw: 126.17

SYNS: AGEFLEX AMA □ ALLYLESTER KYSELINY METHAKRYLOVE □ METHACRYLIC ACID, ALLYL ESTER

TOXICITY DATA WITH REFERENCE
skn-rbt 20 mg/24H MOD 85JCAE-,371,86
eye-rbt 500 mg/24H MLD 85JCAE-,371,86
orl-rat LD50:430 mg/kg AIHAAP 30,470,69
ihl-rat LCLo:500 ppm AIHAAP 30,470,69
skn-rbt LD50:500 mg/kg AIHAAP 30,470,69

CONSENSUS REPORTS: EPA Genetic Toxicology Program. Reported in EPA TSCA Inventory.

SAFETY PROFILE: Moderately toxic by ingestion, inhalation, and skin contact. A skin and eye irritant. See also ALLYL COMPOUNDS and ESTERS. When heated to decomposition it emits acrid smoke and irritating fumes.

AGK750 **CAS:6728-21-8** **HR: 2**
ALLYL METHANESULFONATE
mf: $C_4H_8O_3S$ mw: 136.18

SYNS: ALLYL MESYLATE □ METHANESULFONIC ACID, 2-PROPENYL ESTER (9CI)

TOXICITY DATA WITH REFERENCE
mmo-sat 1 μmol/plate BCPCA6 29,993,80
dns-hmn:hla 50 μmol/L CALEDQ 20,263,83
skn-mus TDLo:540 mg/kg/10W-I:NEO CNREA8 17,64,57

SAFETY PROFILE: Questionable carcinogen with experimental neoplastigenic data. Human mutation data reported. See also ALLYL COMPOUNDS and SULFONATES. When heated to decomposition it emits toxic fumes of SO_x.

AGL000 CAS:10402-33-2 HR: 2
4-ALLYL-2-METHOXYPHENYLPHENYLACETATE
mf: $C_{18}H_{18}O_3$ mw: 282.36

SYNS: BENZENEACETIC ACID-2-METHOXY-4-(2-PROPENYL)PHENYL ESTER ◻ EUGENOL PHENYLACETATE ◻ EUGENYL PHENYLACETATE

TOXICITY DATA WITH REFERENCE
skn-rbt 500 mg/24H MOD FCTXAV 16,637,78

CONSENSUS REPORTS: Reported in EPA TSCA Inventory.

SAFETY PROFILE: A skin irritant. See also ALLYL COMPOUNDS and ESTERS. When heated to decomposition it emits acrid smoke and irritating fumes.

AGL250 CAS:66941-77-3 HR: 3
5-ALLYL-5-(1-METHYLALLYL)-2-THIOBARBITURIC ACID SODIUM SALT
mf: $C_{11}H_{13}N_2O_2S{\cdot}Na$ mw: 260.31

TOXICITY DATA WITH REFERENCE
ipr-rat LD50:126 mg/kg JAPMA8 34,183,45
ivn-rbt LD50:58 mg/kg JAPMA8 34,183,45

SAFETY PROFILE: Poison by intraperitoneal and intravenous routes. When heated to decomposition it emits very toxic fumes of SO_x, Na_2O, and NO_x. See also BARBITURATES and ALLYL COMPOUNDS.

AGL375 CAS:77-27-0 HR: 3
5-ALLYL-5-(1-METHYLBUTYL)-2-THIOBARBITURIC ACID
mf: $C_{12}H_{18}N_2O_2S$ mw: 254.38

PROP: Crystals from dil ethanol. Mp: 132–133°. Often used as the sodium salt, $C_{12}H_{17}N_2NaO_2S$.

SYNS: DIHYDRO-5-(1-METHYLBUTYL)-5-(2-PROPENYL)-2-THIOXO-4,6(1H,5H)-PYRIMIDINEDIONE (9CI) ◻ SURITAL ◻ THIAMYLAL ◻ THIOSECONAL

TOXICITY DATA WITH REFERENCE
ivn-rat LD50:66 mg/kg PSEBAA 89,292,55
ivn-dog LD50:36,300 μg/kg PSEBAA 89,292,55
ivn-rbt LD50:26 mg/kg PSEBAA 89,292,55

SAFETY PROFILE: Poison by intravenous route. Caution: Abuse may lead to habituation or addiction. When heated to decomposition it emits toxic fumes of SO_x and NO_x.

AGL500 CAS:63937-27-9 HR: 3
2-ALLYL-3-METHYL-4-HYDROXY-2-CYCLOPENTEN-1-ONE DIMETHYLCARBAMATE
mf: $C_{12}H_{17}NO_3$ mw: 223.30

SYNS: ADK (CZECH) ◻ 2-ALLYL-3-METHYL-2-CYCLOPENTEN-1-ON-4-YL-N,N'-DIMETHYL-KARBAMAT (CZECH)

TOXICITY DATA WITH REFERENCE
skn-rbt 500 mg/24H MLD 28ZPAK -,164,72
orl-rat LD50:53,200 μg/kg 28ZPAK -,164,72

SAFETY PROFILE: Poison by ingestion. A skin irritant. See also CARBAMATES and ALLYL COMPOUNDS. When heated to decomposition it emits toxic fumes of NO_x.

AGL750 CAS:55902-04-0 HR: 2
N-ALLYL-3-METHYL-N-α-METHYLPHENETHYL-6-OXO-1(6H)-PYRIDAZINE ACETAMIDE
mf: $C_{19}H_{23}N_3O_2$ mw: 325.45

TOXICITY DATA WITH REFERENCE
orl-rat LD50:1710 mg/kg EJMCA5 9,644,74
ipr-rat LD50:628 mg/kg EJMCA5 9,644,74

SAFETY PROFILE: Moderately toxic by ingestion and intraperitoneal routes. When heated to decomposition it emits toxic fumes of NO_x. See also ALLYL COMPOUNDS.

AGL875 CAS:7651-40-3 HR: 3
(±)-5-ALLYL-5-(1-METHYL-2-PENTYNYL)-2-THIOBARBITURIC ACID
mf: $C_{13}H_{16}N_2O_2S$ mw: 264.37

SYNS: (±)-DIHYDRO-5-(1-METHYL-2-PENTYNYL)-5-(2-PROPENYL)-2-THIOXO-4,6(1H,5H)-PYRIMIDINEDIONE ◻ LILLY 22113 ◻ THIOHEXITAL

TOXICITY DATA WITH REFERENCE
ivn-rat LD50:32 mg/kg JMCMAR 24,1241,81
ipr-mus LD50:64 mg/kg JMCMAR 24,1241,81
ivn-mus LD50:54 mg/kg JMCMAR 24,1241,81

SAFETY PROFILE: Poison by intravenous and intraperitoneal routes. When heated to decomposition it emits toxic fumes of SO_x and NO_x.

AGM000 CAS:66941-81-9 HR: 3
5-ALLYL-5-(1-METHYLPROPENYL)BARBITURIC ACID
mf: $C_{11}H_{14}N_2O_3$ mw: 222.27

TOXICITY DATA WITH REFERENCE
orl-mus LD50:525 mg/kg JACSAT 61,353,39
ipr-mus LD50:380 mg/kg JACSAT 61,353,39

SAFETY PROFILE: Poison by intraperitoneal route. Moderately toxic by ingestion. When heated to decomposition it emits toxic fumes of NO_x. See also BARBITURATES and ALLYL COMPOUNDS.

AGM060 CAS:10045-34-8 HR: 3
1-ALLYL-2-NITROIMIDAZOLE
mf: $C_6H_7N_3O_2$ mw: 153.16

TOXICITY DATA WITH REFERENCE
orl-mus LD50:158 mg/kg AACHAX -,478,65
ipr-mus LD50:144 mg/kg AACHAX -,478,65
scu-mus LD50:144 mg/kg AACHAX -,478,65

SAFETY PROFILE: Poison by ingestion, subcutaneous, and intraperitoneal routes. When heated to decomposition it emits toxic fumes of NO_x. See also ALLYL COMPOUNDS.

AGM125 HR: 2
1-(ALLYLNITROSAMINO)-2-PROPANONE
mf: $C_6H_{10}N_2O_2$ mw: 142.18

SYNS: NAOP ☐ N-NITROSOALLYL-2-OXOPROPYLAMINE

TOXICITY DATA WITH REFERENCE
mma-sat 10 µg/plate TCMUE9 1,13,84
orl-rat TDLo:995 mg/kg/74W-I:ETA IAPUDO 57,617,84

SAFETY PROFILE: Questionable carcinogen with experimental tumorigenic data. Mutation data reported. When heated to decomposition it emits toxic fumes of NO_x. See also NITROSAMINES.

AGM250 CAS:22235-85-4 HR: 3
8-ALLYL-(±)-1-α-H,5-α-H-NORTHROPAN-3-α-OL
mf: $C_{10}H_{17}NO$ mw: 167.28

SYNS: N-ALLYNORATROPINE ☐ NALTROPINE

TOXICITY DATA WITH REFERENCE
ivn-rat LD50:60 mg/kg ARZNAD 13,567,63
ipr-mus LD50:165 mg/kg AIPTAK 154,210,65

SAFETY PROFILE: Poison by intravenous and intraperitoneal routes. When heated to decomposition it emits toxic fumes of NO_x. See also ALLYL COMPOUNDS.

AGM500 CAS:4230-97-1 HR: 2
ALLYL OCTANOATE
mf: $C_{11}H_{20}O_2$ mw: 184.31

PROP: Colorless liquid; fruity odor. D: 0.8550.861, refr index: 1.425, flash p: 151°F. Sol in alc, fixed oils; sltly sol in propylene glycol; insol in glycerin and water @ 260°.

SYNS: ALLYL CAPRYLATE ☐ FEMA No. 2037 ☐ OCTANOIC ACID ALLYL ESTER ☐ OCTANOIC ACID-2-PROPENYL ESTER

TOXICITY DATA WITH REFERENCE
skn-rbt 310 mg/24H MOD FCTXAV 16,637,78
orl-rat LD50:570 mg/kg FCTXAV 16,637,78

CONSENSUS REPORTS: Reported in EPA TSCA Inventory.

SAFETY PROFILE: Moderately toxic by ingestion. A skin irritant. See also ALLYL COMPOUNDS and ESTERS. When heated to decomposition it emits acrid smoke and irritating fumes.

AGM750 CAS:14520-53-7 HR: 3
2-ALLYLOXYBENZAMIDE
mf: $C_{10}H_{11}NO_2$ mw: 177.22

SYN: o-(ALLYLOXY)BENZAMIDE

TOXICITY DATA WITH REFERENCE
cpr-mus LD50:320 mg/kg PMDCAY 5,59,67
orl-rat LD50:780 mg/kg JPETAB 108,450,53
ipr-rat LD50:250 mg/kg JPETAB 108,450,53

SAFETY PROFILE: Poison by intraperitoneal route. When heated to decomposition it emits toxic fumes of NO_x. See also ALLYL COMPOUNDS.

AGN000 CAS:22131-79-9 HR: 2
(4-ALLYLOXY-3-CHLOROPHENYL)ACETIC ACID
mf: $C_{11}H_{11}ClO_3$ mw: 226.67

PROP: Prisms from cyclohexane. Mp: 92–93°.

SYNS: ALCLOPHENAC ☐ ALLOPYDIN ☐ ARGUN ☐ 2-CHLORO-4-(2-PROPENYLOXY)BENZENEACETIC ACID ☐ EPINAL ☐ MEDIFENAC ☐ MERVAN ☐ NEOSTEN ☐ NEOSTON ☐ PRINALGIN ☐ REUFENAC ☐ W 7320 ☐ ZUMARIL

TOXICITY DATA WITH REFERENCE
orl-rat TDLo:240 mg/kg (8-15D preg):TER IYKEDH 5,72,74
orl-rat TDLo:1 g/kg (6-15D preg):REP ARZNAD 20,618,70
orl-rat LD50:1050 mg/kg ARZNAD 20,618,70
ipr-rat LD50:465 mg/kg IYKEDH 5,38,74
orl-mus LD50:1100 mg/kg ARZNAD 20,618,70
ipr-mus LD50:508 mg/kg IYKEDH 5,38,74

SAFETY PROFILE: Moderately toxic by ingestion and intraperitoneal routes. An experimental teratogen. Other experimental reproductive effects. When heated to decomposition it emits toxic fumes of Cl^-. See also ALLYL COMPOUNDS. An analgesic, antipyretic, and anti-inflammatory agent.

AGN250 CAS:24049-18-1 HR: 2
(4-(ALLYLOXY)-3-CHLOROPHENYL)ACETIC ACID SODIUM SALT
mf: $C_{11}H_{10}ClO_3{\cdot}Na$ mw: 248.65

SYN: ALCLOFENAC SODIUM SALT

TOXICITY DATA WITH REFERENCE
orl-rat LD50:1050 mg/kg FRPSAX 32,286,77
ipr-rat LD50:530 mg/kg FRPSAX 32,286,77

SAFETY PROFILE: Moderately toxic by ingestion and intraperitoneal routes. When heated to decomposition it emits toxic fumes of Cl^- and Na_2O. See also ALLYL COMPOUNDS.

AGN500 CAS:63887-51-4 HR: 3
o-ALLYLOXY-N,N-DIETHYLBENZAMIDE
mf: $C_{14}H_{19}NO_2$ mw: 233.34

TOXICITY DATA WITH REFERENCE
orl-rat LD50:325 mg/kg JPETAB 108,450,53
ipr-rat LD50:140 mg/kg JPETAB 108,450,53

SAFETY PROFILE: Poison by ingestion and intraperitoneal routes. When heated to decomposition it emits toxic fumes of NO_x. See also ALLYL COMPOUNDS.

AGN750 CAS:63887-52-5 HR: 3
o-ALLYLOXY-N,N-DIMETHYLBENZAMIDE
mf: $C_{12}H_{15}NO_2$ mw: 205.28

TOXICITY DATA WITH REFERENCE
orl-rat LD50:740 mg/kg JPETAB 108,450,53
ipr-rat LD50:350 mg/kg JPETAB 108,450,53

SAFETY PROFILE: Poison by intraperitoneal route. Moderately toxic by ingestion. When heated to decomposition it emits toxic fumes of NO_x. See also ALLYL COMPOUNDS.

AGO000 CAS:111-45-5 HR: 3
2-ALLYLOXYETHANOL
mf: $C_5H_{10}O_2$ mw: 102.15

SYNS: 2-ALLOXYETHANOL (CZECH) □ USAF DO-47

TOXICITY DATA WITH REFERENCE
skn-rbt 500 mg/24H MOD 28ZPAK -,99,72
eye-rbt 750 µg/24H SEV 28ZPAK -,99,72
ipr-mus LD50:250 mg/kg NTIS** AD277-689
par-mus LDLo:2000 mg/kg CBCCT* 7,687,55

CONSENSUS REPORTS: Reported in EPA TSCA Inventory.

SAFETY PROFILE: Poison by intraperitoneal route. Moderately toxic by parenteral route. A severe eye and moderate skin irritant. When heated to decomposition it emits acrid smoke and irritating fumes. See also ALLYL COMPOUNDS.

AGO250 CAS:63887-17-2 HR: 3
o-ALLYLOXY-N-(β-HYDROXYETHYL)BENZAMIDE
mf: $C_{12}H_{15}NO_3$ mw: 221.28

TOXICITY DATA WITH REFERENCE
orl-rat LD50:1150 mg/kg JPETAB 108,450,53
ipr-rat LD50:350 mg/kg JPETAB 108,450,53

SAFETY PROFILE: Poison by intraperitoneal route. Moderately toxic by ingestion. When heated to decomposition it emits toxic fumes of NO_x. See also ALLYL COMPOUNDS.

AGO500 CAS:63731-92-0 HR: 3
6-ALLYLOXY-2-METHYLAMINO-4-(N-METHYLPIPERAZINO)-5-METHYLTHIOPYRIMIDINE
mf: $C_{14}H_{23}N_5OS$ mw: 309.48

TOXICITY DATA WITH REFERENCE
orl-mus LD50:400 mg/kg JMCMAR 18,553,75
ivn-mus LD50:56 mg/kg JMCMAR 18,553,75

SAFETY PROFILE: Poison by ingestion and intravenous routes. When heated to decomposition it emits very toxic fumes of NO_x and SO_x. See also ALLYL COMPOUNDS.

AGO750 CAS:29181-23-5 HR: 3
(+)-1-(o-ALLYLOXYPHENOXY)-3-ISOPROPYLAMINO-2-PROPANOLHYDROCHLORIDE
mf: $C_{15}H_{23}NO_3$•HCl mw: 301.46

SYN: CIBA 42155-BA

TOXICITY DATA WITH REFERENCE
orl-rat LD50:760 mg/kg ARZNAD 20,1890,70
ivn-dog LD50:20 mg/kg ARZNAD 20,1890,70
ivn-rbt LDLo:20 mg/kg ARZNAD 20,1890,70

SAFETY PROFILE: Poison by intravenous route. Moderately toxic by ingestion. When heated to decomposition it emits very toxic fumes of HCl and NO_x. See also ALLYL COMPOUNDS.

AGP000 CAS:6452-73-9 HR: 3
(−)-1-(o-ALLYLOXYPHENOXY)-3-ISOPROPYLAMINO-2-PROPANOLHYDROCHLORIDE
mf: $C_{15}H_{23}NO_3$•HCl mw: 301.46

PROP: Solid. Mp: 107–109°.

SYN: CIBA 42244-BA

TOXICITY DATA WITH REFERENCE
orl-rat LD50:900 mg/kg ARZNAD 20,1890,70
ivn-dog LD50:20 mg/kg ARZNAD 20,1890,70
ivn-rbt LDLo:20 mg/kg ARZNAD 20,1890,70

SAFETY PROFILE: Poison by intravenous route. Moderately toxic by ingestion. When heated to decomposition it emits very toxic fumes of HCl and NO_x. See also ALLYL COMPOUNDS.

AGP250 CAS:6452-54-6 HR: 2
3-(ALLYLOXYPHENOXY)-1,2-PROPANEDIOL
mf: $C_{12}H_{16}O_4$ mw: 224.28

TOXICITY DATA WITH REFERENCE
ipr-mus LD50:650 mg/kg JPETAB 93,470,48
scu-mus LD50:582 mg/kg JPETAB 93,470,48

SAFETY PROFILE: Moderately toxic by intraperitoneal and subcutaneous routes. When heated to decomposition it emits acrid smoke and irritating fumes. See also ALLYL COMPOUNDS.

AGP500 CAS:123-34-2 HR: 2
3-ALLYLOXY-1,2-PROPANEDIOL
mf: $C_6H_{12}O_3$ mw: 132.18

SYNS: α-ALLYL GLYCEROL ETHER □ 1-ALLYLOXY-2,3-PROPANEDIOL □ ETHER, ALLYL GLYCERYL □ GLYCERIN 1-ALLYL ETHER □ GLYCEROL α-ALLYL ETHER

TOXICITY DATA WITH REFERENCE
skn-rbt 535 mg/24H MLD AMIHBC 2,574,50
eye-rbt 107 mg AMIHBC 2,574,50
eye-rbt 100 mg/24H MOD 85JCAE-,632,86
orl-mus LD50:4200 mg/kg FEPRA7 8,477,49
scu-mus LD50:1135 mg/kg JPETAB 93,470,48

CONSENSUS REPORTS: Reported in EPA TSCA Inventory.

SAFETY PROFILE: Moderately toxic by ingestion and subcutaneous routes. A skin and eye irritant. See also ETHERS and ALLYL COMPOUNDS. When heated to decomposition it emits acrid smoke and irritating fumes.

AGP750 CAS:3088-44-6 HR: 2
3-ALLYLOXYPROPIONITRILE
mf: C_6H_9NO mw: 111.16

SYN: β-ALLYLOXY-PROPIONITRILE

TOXICITY DATA WITH REFERENCE
skn-rbt 500 mg/24H MLD 85JCAE-,917,86
eye-rbt 500 mg/24H MLD 85JCAE-,917,86
orl-rat LD50:1300 µL/kg TXAPA9 28,313,74
ipr-mus LDLo:500 mg/kg CBCCT* 4,380,52

CONSENSUS REPORTS: Cyanide and its compounds are on the Community Right-To-Know List.

SAFETY PROFILE: Moderately toxic by ingestion and intraperitoneal routes. A skin and eye irritant. See also NITRILES and ALLYL COMPOUNDS. When heated to decomposition it emits toxic fumes of NO_x and CN^-.

AGQ000 CAS:12012-95-2 HR: 2
ALLYL PALLADIUM CHLORIDE DIMER

PROP: Pale-yellow crystals from C_6H_6. Mp: 160°. Sol in C_6H_6, $CHCl_3$, Me_2CO, MeOH.

TOXICITY DATA WITH REFERENCE
skn-rbt 100 mg/24H SEV AEHLAU 30,168,75

SAFETY PROFILE: A severe skin irritant. See also PALLADIUM and ALLYL COMPOUNDS. When heated to decomposition it emits toxic fumes of Cl^-.

AGQ250 CAS:4255-24-7 HR: 3
α-ALLYL PHENETHYLAMINEHYDROCHLORIDE
mf: $C_{11}H_{15}N{\cdot}HCl$ mw: 174.29

PROP: Solid. Mp: 159–161°.

SYNS: ALETAMINE HYDROCHLORIDE □ 1-ALLYL-2-PHENYL-ETHYLAMINE HYDROCHLORIDE

TOXICITY DATA WITH REFERENCE
scu-rat LD50:280 mg/kg AIPTAK 159,442,66
orl-mus LD50:380 mg/kg TXAPA9 21,302,72
ivn-mus LD50:30 mg/kg 27ZQAG -,332,72

SAFETY PROFILE: Poison by ingestion, subcutaneous, and intravenous routes. When heated to decomposition it emits toxic fumes of HCl. See also ALLYL COMPOUNDS.

AGQ500 CAS:1745-81-9 HR: 3
o-ALLYL PHENOL
mf: $C_9H_{10}O$ mw: 134.19

PROP: Mp: 10°, bp: 230°, d: 1.033 @ 18°/4°. Sol in water, alc, chloroform, and ether.

SYN: 2-ALLYL PHENOL

TOXICITY DATA WITH REFERENCE
skn-mus TDLo:8400 mg/kg/30W-I:CAR CNREA8 19,413,59
skn-mus TD:3360 mg/kg/12W-I:NEO CNREA8 19,413,59
ipr-mus LDLo:256 mg/kg CBCCT* 1,127,51

CONSENSUS REPORTS: Reported in EPA TSCA Inventory.

SAFETY PROFILE: Poison by intraperitoneal route. Questionable carcinogen with experimental carcinogenic and neoplastigenic data. When heated to decomposition it emits acrid smoke and fumes. See also ALLYL COMPOUNDS and PHENOLS.

AGQ750 CAS:7493-74-5 HR: 2
ALLYL PHENOXYACETATE
mf: $C_{11}H_{12}O_3$ mw: 192.23

PROP: Colorless to light-yellow liquid; heavy fruit odor.

SYN: ACETATE P.A.

TOXICITY DATA WITH REFERENCE
orl-rat LD50:475 mg/kg FCTXAV 13,681,75
skn-rbt LD50:820 mg/kg FCTXAV 13,681,75

CONSENSUS REPORTS: Reported in EPA TSCA Inventory.

SAFETY PROFILE: Moderately toxic by ingestion and skin contact. When heated to decomposition it emits acrid smoke and irritating fumes. See also ALLYL COMPOUNDS.

AGQ775 CAS:21905-27-1 HR: 3
ALLYL PHENYL ARSINIC ACID
mf: $C_9H_{11}AsO_2$ mw: 226.12

SYNS: ALLYLHYDROXYPHENYLARSINE OXIDE □ ARSINE OXIDE, ALLYLHYDROXYPHENYL-

TOXICITY DATA WITH REFERENCE
ivn-mus LD50:100 mg/kg CSLNX* NX#06910

OSHA PEL: TWA 0.5 mg(As)/m^3

SAFETY PROFILE: Poison by intravenous route. When heated to decomposition it emits toxic fumes of As.

AGQ875 CAS:115-43-5 HR: 3
5-ALLYL-5-PHENYLBARBITURIC ACID
mf: $C_{13}H_{12}N_2O_3$ mw: 244.27

PROP: Crystals, bitter taste. Mp: 156–157.5°. Readily sol in alc, chloroform. One gram dissolves in 580 mL water, in 10 mL ether, in 500 mL benzene, in 4000 mL carbon tetrachloride, in 17,500 mL petr ether.

SYNS: ACIDO 5 FENIL 5 ALLILBARBITURICO (ITALIAN) □ ALLOFENYL □ ALLOPHENYLUM □ ALLPHASEM □ ALPHEBA □ ALPHENAL □ ALPHENATE □ FENALLYMAL □ LUBERGAL □ LUXOMNIN □ PHENALLYMAL □ PHENALLYMALUM □ 5-PHENYL-5-ALLYLBARBITURIC ACID □ 5-PHENYL-5-(2-PROPENYL)-2,4,6(1H,3H,5H)-PYRIMIDINETRIONE (9CI) □ PHENYRAL □ PROPHENAL □ TUBERGAL

TOXICITY DATA WITH REFERENCE
ipr-rat LD50:233 mg/kg SFTIAE 56,31,52
orl-mus LD50:280 mg/kg FRPSAX 17,390,62
ipr-mus LD50:265 mg/kg FRPSAX 17,390,62

SAFETY PROFILE: Poison by ingestion and intraperitoneal routes. Caution: Abuse may lead to habituation or addiction. When heated to decomposition it emits toxic fumes of NO_x.

AGR000 CAS:1746-13-0 HR: 3
ALLYL PHENYL ETHER
mf: $C_9H_{10}O$ mw: 134.19

PROP: Liquid with geranium odor. D: 0.986 @ 15°, bp: 192–195°. Insol in water.

SYNS: (2-PROPENYLOXY)BENZENE □ USAF DO-23

TOXICITY DATA WITH REFERENCE
ipr-mus LD50:100 mg/kg NTIS** AD277-689
ivn-mus LD50:63 mg/kg CSLNX* NX#01855

CONSENSUS REPORTS: Reported in EPA TSCA Inventory.

SAFETY PROFILE: Poison by intravenous and intraperitoneal routes. See also ETHERS. When heated to decomposition it emits acrid smoke and irritating fumes. See also ALLYL COMPOUNDS and ETHERS.

AGR125 CAS:2597-09-3 HR: 3
2-ALLYL-2-PHENYL-4-PENTENOIC ACID 2-(DIETHYLAMINO)ETHYL ESTER HYDROCHLORIDE
mf: $C_{20}H_{29}NO_2 \cdot ClH$ mw: 351.96

SYN: CFT 1201

TOXICITY DATA WITH REFERENCE
orl-rat LD50:1500 mg/kg AEPPAE 225,453,55
ipr-rat LD50:170 mg/kg AEPPAE 225,453,55
scu-rat LD50:1500 mg/kg AEPPAE 225,453,55

SAFETY PROFILE: Poison by intraperitoneal route. Moderately toxic by ingestion and subcutaneous routes. When heated to decomposition it emits toxic fumes of NO_x and HCl.

AGR250 CAS:7341-63-1 HR: 3
ALLYL PHENYL THIOUREA
mf: $C_{10}H_{12}N_2S$ mw: 192.30

SYN: 1-ALLYL-3-PHENYL-2-THIOUREA

TOXICITY DATA WITH REFERENCE
orl-rat LD50:750 mg/kg JPETAB 90,260,47
ivn-mus LD50:56 mg/kg CSLNX* NX#01120

SAFETY PROFILE: Poison by intravenous route. Moderately toxic by ingestion. When heated to decomposition it emits very toxic fumes of NO_x and SO_x. See also ALLYL COMPOUNDS.

AGR500 CAS:2179-59-1 HR: 1
ALLYL PROPYL DISULFIDE
mf: $C_6H_{12}S_2$ mw: 148.30

PROP: Liquid with pungent odor. Bp: 66–69° @ 16 mm.

OSHA PEL: TWA 2 ppm; STEL 3 ppm
ACGIH TLV: TWA 2 ppm; STEL 3 ppm
DFG MAK: 2 ppm (12 mg/m³)
NIOSH REL: (Allyl Propyl Disulfide): TWA 2 ppm; STEL 3 ppm

SAFETY PROFILE: A powerful irritant. Moderately flammable by exposure to heat, flame, or oxidizers. When heated to decomposition it emits highly toxic SO_x. To fight fire, use foam, CO_2, dry chemical. See also ALLYL COMPOUNDS.

AGR750 CAS:15151-00-5 HR: 3
1-ALLYLQUINALDINUM BROMIDE
mf: $C_{13}H_{14}N \cdot Br$ mw: 264.19

TOXICITY DATA WITH REFERENCE
ipr-mus LDLo:64 mg/kg CBCCT* 2,190,50

CONSENSUS REPORTS: Reported in EPA TSCA Inventory.

SAFETY PROFILE: Poison by intraperitoneal route. See also ALLYL COMPOUNDS and BROMIDES. When heated to decomposition it emits very toxic fumes of NO_x and Br^-.

AGS000 CAS:7539-12-0 HR: 3
ALLYLSUCCINIC ANHYDRIDE
mf: $C_7H_8O_3$ mw: 140.15

PROP: Liquid. Bp: 133–140° @ 16 mm.

TOXICITY DATA WITH REFERENCE
skn-rbt 10 mg/24H open SEV AIHAAP 23,95,62
orl-rat LD50:1070 mg/kg AIHAAP 23,95,62
skn-rbt LD50:320 mg/kg AIHAAP 23,95,62

SAFETY PROFILE: Poison by skin contact. Moderately toxic by ingestion. A severe skin irritant. See also ANHYDRIDES and ALLYL COMPOUNDS. When heated to decomposition it emits acrid smoke and irritating fumes.

AGS250 CAS:592-88-1 HR: 3
ALLYL SULFIDE
mf: $C_6H_{10}S$ mw: 114.22

$(H_2C{=}CHCH_2)_2S$

PROP: Colorless liquid, garlic odor. Mp: −83°, bp: 139°, d: 0.8881, vap d: 3.90.

SYNS: ALLYL MONOSULFIDE ☐ DIALLYL MONOSULFIDE ☐ DIALLYL SULFIDE ☐ DIALLYL THIOETHER ☐ OIL GARLIC ☐ THIOALLYL ETHER ☐ 3,3-THIOBIS(1-PROPENE)

TOXICITY DATA WITH REFERENCE
orl-rat LD50:2980 mg/kg FCTOD7 26,299,88
skn-rbt LD50:>5 g/kg FCTOD7 26,299,88
ivn-rbt LDLo:330 mg/kg BIJOAK 4,107,09

CONSENSUS REPORTS: Reported in EPA TSCA Inventory.

SAFETY PROFILE: Poison by intravenous route. Moderately toxic by ingestion. An irritant to skin, eyes, and mucous membranes. When heated to decomposition it emits toxic SO_x. Explosive reaction with N-bromosuccinimide. See also SULFIDES and ALLYL COMPOUNDS.

AGS375 HR: 3
3-(ALLYL-(TETRAHYDRONAPHTHYL)AMINO)-N,N-DIETHYL-PROPIONAMIDE
mf: $C_{20}H_{30}N_2O \cdot ClH$ mw: 350.98

SYNS: 3-(ALLYL-(TETRAHYDRONAPHTHYL)AMINO)-N,N-DIETHYLPROPIONAMIDE HYDROCHLORIDE ☐ N-d,l-ac-TETRAHYDRO-β-NAPHTHYL-N-ALLYL-β-ALANINDIAETHYLAMID-HYDROCHLORID (GERMAN)

TOXICITY DATA WITH REFERENCE
orl-rat LD50:127 mg/kg AIPTAK 128,500,60
scu-rat LD50:44,500 μg/kg AIPTAK 128,500,60
ivn-rat LD50:18,900 μg/kg AIPTAK 128,500,60
orl-mus LD50:210 mg/kg AIPTAK 128,500,60
ipr-mus LD50:118 mg/kg AIPTAK 128,500,60
ivn-mus LD50:23,500 μg/kg AIPTAK 128,500,60
ivn-rbt LD50:1900 μg/kg AIPTAK 128,500,60

SAFETY PROFILE: Poison by ingestion, intravenous, and intraperitoneal routes. When heated to decomposition it emits toxic fumes of NO_x and HCl.

AGS500 CAS:2530-99-6 HR: 3
1-ALLYLTHEOBROMINE
mf: $C_{10}H_{12}N_4O_2$ mw: 220.26

SYN: ALLYLTHEOBROMINE

TOXICITY DATA WITH REFERENCE
orl-hmn TDLo:26 mg/kg:CNS,GIT JPETAB 86,113,46
scu-mus LDLo:125 mg/kg AIPTAK 25,361,21
orl-mus LD50:191 mg/kg JPETAB 116,343,56
irp-mus LD50:102 mg/kg JPETAB 116,343,56
ivn-mus LD50:40 mg/kg JPETAB 86,113,46
scu-rbt LDLo:100 mg/kg AIPTAK 25,361,21
ivn-rbt LDLo:50 mg/kg AIPTAK 25,361,21

SAFETY PROFILE: Poison by ingestion, subcutaneous, intraperitoneal and intravenous routes. Human systemic effects by ingestion: changes in motor activity and nausea or vomiting. When heated to decomposition it emits highly toxic NO_x. See also ALLYL COMPOUNDS.

AGS750 CAS:764-49-0 HR: 3
ALLYL THIOCYANATE
mf: C_4H_5NS mw: 99.16

PROP: Colorless, pungent oil. D: 1.056 @ 15 mm, mp: −102.5°, bp: 152–161°. Sol in water, alc, and ether.

SYNS: ALLYLRHODANID (GERMAN) □ ALLYL SULFOCYANIDE □ THIOCYANIC ACID, ALLYL ESTER

TOXICITY DATA WITH REFERENCE
ipr-rat LDLo:100 mg/kg ARZNAD 16,870,66
scu-rbt LDLo:12 mg/kg AEPPAE 150,257,30

SAFETY PROFILE: Poison by intraperitoneal and subcutaneous routes. See also THIOCYANATES and ALLYL COMPOUNDS. When heated to decomposition it emits very toxic NO_x, SO_x, and CN^-.

AGT000 CAS:3766-55-0 HR: 3
4-ALLYLTHIOSEMICARBAZIDE
mf: $C_4H_9N_3S$ mw: 131.22

TOXICITY DATA WITH REFERENCE
par-rat LD50:500 mg/kg ARZNAD 12,260,62
ivn-mus LD50:56 mg/kg CSLNX* NX#00434

SAFETY PROFILE: Poison by intravenous route. Moderately toxic by parenteral route. When heated to decomposition it emits very toxic fumes of NO_x and SO_x. See also ALLYL COMPOUNDS.

AGT250 CAS:3571-74-2 HR: 3
2-(ALLYLTHIO)-2-THIAZOLINE
mf: $C_6H_9NS_2$ mw: 159.28

SYNS: 4,5-DIHYDRO-2-(2-PROPENYLTHIO)THIAZOLE (9CI) □ EN-28,450

TOXICITY DATA WITH REFERENCE
orl-rat LD50:110 mg/kg 28ZEAL 4,40,69
skn-rbt LD50:340 mg/kg 28ZEAL 4,40,69

SAFETY PROFILE: Poison by ingestion and skin contact. When heated to decomposition it emits very toxic fumes of NO_x and SO_x. See also ALLYL COMPOUNDS.

AGT500 CAS:109-57-9 HR: 3
1-ALLYL-2-THIOUREA
mf: $C_4H_8N_2S$ mw: 116.20

PROP: Monoclinic, rhombic or colorless prisms. D: 1.219, mp: 77–78°. Sol in hot water; insol in benzene, alc; very sltly sol in ether.

SYNS: ALLYLTHIOCARBAMIDE □ 1-ALLYLTHIOUREA □ N-ALLYLTHIOUREA □ AMINOSIN □ (2-PROPENYL)THIOUREA □ RHODALLIN □ RHODALLINE □ THIOSINAMIN □ THIOSINAMINE □ THIOCYNAMINE □ U 19571

TOXICITY DATA WITH REFERENCE
mmo-sat 150 μg/plate ABCHA6 44,3017,80
orl-rat LD50:200 mg/kg JPETAB 90,260,47
ipr-rat LD50:500 mg/kg JPETAB 89,186,47
scu-rat LDLo:50 mg/kg HBAMAK 4,1289,35
scu-mus LDLo:700 mg/kg HBAMAK 4,1289,35
ivn-dog LDLo:110 mg/kg HBAMAK 4,1289,35

CONSENSUS REPORTS: Reported in EPA TSCA Inventory

SAFETY PROFILE: Poison by ingestion, subcutaneous, and intravenous routes. Moderately toxic by intraperitoneal route. Mutation data reported. When heated to decomposition it emits very toxic NO_x and SO_x. See also ALLYL COMPOUNDS

AGT750 CAS:1530-48-9 HR: 3
ALLYL TRI-N-BUTYLPHOSPHONIUMCHLORIDE
mf: $C_{15}H_{32}P\cdot Cl$ mw: 278.89

TOXICITY DATA WITH REFERENCE
ivn-mus LD50:18 mg/kg CSLNX* NX#03135

CONSENSUS REPORTS: Reported in EPA TSCA Inventory.

SAFETY PROFILE: Poison by intravenous route. When heated to decomposition it emits very toxic fumes of PO_x and Cl^-. See also ALLYL COMPOUNDS.

AGU250 CAS:107-37-9 HR: 3
ALLYL TRICHLOROSILANE
DOT: UN 1724
mf: $C_3H_5Cl_3Si$ mw: 175.52

PROP: Colorless liquid, pungent, irritating odor. D: 1.222 @ 20°/4°, bp: 116° @ 750 mm, flash p: 95°F (COC).

SYNS: ALLYLTRICHLOROSILANE, stabilized (DOT) □ SILANE, TRICHLOROALLYL- □ TRICHLOROALLYLSILANE

TOXICITY DATA WITH REFERENCE
ivn-mus LD50:56 mg/kg CSLNX* NX#04219

CONSENSUS REPORTS: Reported in EPA TSCA Inventory.

DOT CLASSIFICATION: 8; *Label:* Corrosive, Flammable Liquid

SAFETY PROFILE: Poison by intravenous route. Corrosive. See ALLYL COMPOUNDS. When heated to decomposition it emits toxic Cl^-. A dangerous fire hazard. To fight fire, use foam, mist, spray, dry chemical.

AGU400 CAS:71500-21-5 HR: 2
ALLYL 3,5,5-TRIMETHYLHEXANOATE
mf: $C_{12}H_{22}O_2$ mw: 198.34

SYNS: ALLYL TRIMETHYLHEXANOATE □ HEXANOIC ACID, 3,5,5-TRIMETHYL-, ALLYL ESTER □ 2-PROPENYL 3,5,5-TRIMETHYLHEXANOATE □ 3,5,5-TRIMETHYLHEXANOIC ACID ALLYL ESTER

TOXICITY DATA WITH REFERENCE
skn-rbt 500 mg/24H SEV FCTOD7 20,639,82
orl-rat LD50:1400 mg/kg FCTOD7 20,639,82

SAFETY PROFILE: Moderately toxic by ingestion. A severe skin irritant. When heated to decomposition it emits acrid smoke and irritating fumes.

AGU500 CAS:76-63-1 HR: 3
ALLYLTRIPHENYLTIN
mf: $C_{21}H_{20}Sn$ mw: 391.10

PROP: Needle-like crystals from alc. Mp: 73–74.5°. Sol in most org solvs.

SYNS: ALLYLTRIPHENYL STANNANE □ DOWCO 187 □ ENT 50,909 □ TRIPHENYL-2-PROPENYL-STANNANE (9CI)

TOXICITY DATA WITH REFERENCE
ivn-mus LD50:100 mg/kg CSLNX* NX#02200

OSHA PEL: TWA 0.1 mg(Sn)/m³ (skin)
ACGIH TLV: TWA 0.1 mg(Sn)/m³ (skin) (Proposed: TWA 0.1 mg(Sn)/m³; STEL 0.2 mg(Sn)/m³ (skin))
NIOSH REL: (Organotin Compounds) TWA 0.1 mg(Sn)/m³

SAFETY PROFILE: Poison by intravenous route. See also TIN COMPOUNDS and ALLYL COMPOUNDS. When heated to decomposition it emits toxic smoke and irritating fumes.

For occupational chemical analysis use NIOSH: Organotin Compounds 5504.

AGU750 CAS:37425-13-1 HR: 2
9-ALLYL-2-(4-(2-TRITYLETHYL)-1-PIPERAZINYL)-9H-PURINEDIMETHANESULFONATE
mf: $C_{33}H_{34}N_6 \cdot 2CH_4O_3S$ mw: 706.95

TOXICITY DATA WITH REFERENCE
mmo-nsc 1 g/L GENTAE 48,597,63
orl-mus LDLo:2000 mg/kg CHTPBA 7,192,72

SAFETY PROFILE: Mutation data reported. Moderately toxic by ingestion. See also SULFONATES and ALLYL COMPOUNDS. When heated to decomposition it emits very toxic fumes of SO_x and NO_x.

AGV000 CAS:557-11-9 HR: 3
ALLYLUREA
mf: $C_4H_8N_2O$ mw: 100.14

PROP: Needles from alc. Mp: 85°. Very sol in water and alc; very sltly sol in petroleum ether, toluene.

SYNS: ALLYLCARBAMIDE □ N-ALLYLUREA □ 1-ALLYLUREA □ MONOALLYLUREA □ N-2-PROPENYLUREA □ 2-PROPENYLUREA □ UREA, 2-PROPENYL-(9CI)

TOXICITY DATA WITH REFERENCE
mmo-sat 1 mg/plate EMMUEG 19(Suppl 21),2,92
orl-rat LDLo:250 mg/kg NCNSA6 5,47,53
orl-mus LDLo:1070 mg/kg AECTCV 14,111,85

CONSENSUS REPORTS: Reported in EPA TSCA Inventory.

SAFETY PROFILE: Poison by ingestion. Mutation data reported. When heated to decomposition it emits toxic fumes of NO_x. See also ALLYL COMPOUNDS.

AGV250 CAS:3917-15-5 HR: 3
ALLYL VINYL ETHER
mf: C_5H_8O mw: 84.13

PROP: Very sltly sol in water. D: 0.8, bp: 67°, flash p: <68°F (OC).

TOXICITY DATA WITH REFERENCE
orl-rat LD50:550 mg/kg AIHAAP 23,95,62
ihl-rat LCLo:8000 ppm/4H AIHAAP 23,95,62

SAFETY PROFILE: Moderately toxic by inhalation and ingestion. See also ALLYL COMPOUNDS and ETHERS. Dangerous fire and explosion hazard from heat, sparks,or powerful oxidizers. To fight fire, use alcohol foam, dry chemical, or mist. Water may be ineffective. When heated to decomposition it yields acrid, irritating fumes. Becomes shock and heat sensitive on storage.

AGV875 HR: 2
ALOE

PROP: A perennial succulent cultivated outdoors in the tropics and a popular houseplant elsewhere. The leaves are thick and hard with spines on the edges. The sap of some aloes is used commercially to produce carthartic glycosides. The gel in the leaves is used as a skin moisturizer and palliative.

SYNS: LALOI (HAITI) □ PANINI-'AWA'AWA (HAWAII) □ SABILA (CUBA, PUERTO RICO) □ SEMPERVIVUM (JAMAICA) □ SINKLE BIBLE (JAMAICA) □ STAR CACTUS (HAWAII) □ ZABILA (MEXICO, DOMINICAN REPUBLIC) □ ZAVILA (PUERTO RICO)

SAFETY PROFILE: The sap contains the toxic barbaloin, an anthraquinone glycoside. Ingestion causes a strong purgative action within 12 hours and may color the urine red. Repeated doses may cause kidney damage.

AGV890 **CAS:25384-17-2** **HR: 3**
ALPERIDINE HYDROCHLORIDE
mf: $C_{18}H_{25}NO_2 \cdot ClH$ mw: 323.90

SYNS: dl-1-ALLYL-1-METHYL-4-PHENYL-4-PIPERIDINOL PROPIONATE HYDROCHLORIDE ☐ ALLYLPRODINE HYDROCHLORIDE ☐ dl-3-METHYL-3-ALLYL-4-PROPIONOXYPIPERIDINE HYDROCHLORIDE ☐ NIH 7440 HYDROCHLORIDE ☐ Ro 2-7113

TOXICITY DATA WITH REFERENCE
orl-rat LD50:125 mg/kg AIPTAK 109,171,57
scu-rat LD50:80 mg/kg AIPTAK 109,171,57
ivn-rat LD50:13,400 μg/kg AIPTAK 109,171,57
orl-mus LD50:536 mg/kg AIPTAK 109,171,57
scu-mus LD50:333 mg/kg AIPTAK 109,171,57
ivn-mus LD50:45,800 μg/kg AIPTAK 109,171,57
ivn-dog LD50:25,600 μg/kg AIPTAK 109,171,57
ivn-rbt LD50:7800 μg/kg AIPTAK 109,171,57

SAFETY PROFILE: Poison by ingestion, subcutaneous, and intravenous routes. When heated to decomposition it emits toxic fumes of NO_x and HCl. Caution: May be habit forming. This is a controlled substance (opiate) listed in the U.S. Code of Federal Regulations, Title 21 Part 1308.11 (1985).

AGW000 **CAS:13707-88-5** **HR: 3**
ALPRENOL HYDROCHLORIDE
mf: $C_{15}H_{23}NO_2 \cdot ClH$ mw: 285.85

PROP: Solid. Mp: 108–110°.

SYNS: 1-(o-ALLYLPHENOXY)-3-(ISOPROPYLAMINO)-2-PROPANOL HYDROCHLORIDE ☐ APTIN ☐ BETAPTIN ☐ GUBERNAL ☐ OXPRENOLOL

TOXICITY DATA WITH REFERENCE
orl-man TDLo:571 μg/kg:CVS KIZSB8 6(4),209,75
orl-rat LD50:590 mg/kg KSRNAM 11,1321,77
scu-rat LD50:290 mg/kg KSRNAM 11,1321,77
scu-mus LD50:215 mg/kg KSRNAM 11,1321,77
orl-mus LD50:184 mg/kg AIPTAK 202,79,73
orl-dog LD50:383 mg/kg GNRIDX 3,614,69
scu-dog LD50:92 mg/kg GNRIDX 3,614,69
ivn-dog LD50:18 mg/kg GNRIDX 3,614,69

SAFETY PROFILE: Poison by ingestion, subcutaneous, intravenous and intraperitoneal routes. Human systemic effects by ingestion: change in heart rate. When heated to decomposition it emits very toxic fumes of HCl and NO_x. See also ALLYL COMPOUNDS.

AGW250 **CAS:13655-52-2** **HR: 3**
ALPRENOLOL
mf: $C_{15}H_{23}NO_2$ mw: 249.39

PROP: Solid. Mp: 57–58°.

SYNS: ALFEPROL (RUSSIAN) ☐ 1-(o-ALLYLPHENOXY)-3-(ISOPROPYLAMINO)-2-PROPANOL ☐ H 56/28 ☐ 1-((1-METHYLETHYL)AMINO)-3-(2-(2-PROPENYL)PHENOXY)-2-PROPANOL

TOXICITY DATA WITH REFERENCE
orl-wmn LDLo:210 mg/kg UGLAAD 139,2017,77
ipr-mus LD50:90 mg/kg EJMCA5 18,151,83
ivn-mus LD50:20 mg/kg ARZNAD 27,1022,77
orl-mam LD50:184 mg/kg PCJOAU 8,137,74
ipr-mam LD50:102 mg/kg PCJOAU 8,137,74

SAFETY PROFILE: A human poison by ingestion. Poison experimentally by ingestion, intravenous, and intraperitoneal routes. When heated to decomposition it emits toxic fumes of NO_x. See also ALLYL COMPOUNDS. A beta-adrenergic blocker.

AGW275 **HR: 2**
ALPROSTADIL-α-CYCLODEXTRIN CLATHRATE
mf: $C_{20}H_{34}O_5 \cdot 7C_{36}H_{60}O_{30}$ mw: 7165.26

SYN: 11,15-DIHYDROXY-9-OXO-PROST-13-EN-1-OIC ACID, (11-α, 13E,15S)-, and α-CYCLODEXTRIN

TOXICITY DATA WITH REFERENCE
orl-rat LD50:7600 mg/kg IYKEDH 11,181,80
ipr-rat LD50:830 mg/kg IYKEDH 11,181,80
scu-rat LD50:660 mg/kg IYKEDH 11,181,80
ivn-rat LD50:640 mg/kg IYKEDH 11,181,80
iat-rat LD50:720 mg/kg IYKEDH 11,181,80
orl-mus LD50:6200 mg/kg IYKEDH 11,181,80
ipr-mus LD50:660 mg/kg IYKEDH 11,181,80

SAFETY PROFILE: Moderately toxic by several routes. When heated to decomposition it emits acrid smoke and irritating fumes.

AGW300 **CAS:8001-95-4** **HR: 3**
ALSEROXYLON

TOXICITY DATA WITH REFERENCE
ipr-rat LD50:260 mg/kg NIIRDN 6,43,82
orl-mus LD50:532 mg/kg NIIRDN 6,43,82
ipr-mus LD50:172 mg/kg NIIRDN 6,43,82
ipr-dog LD50:50 mg/kg NIIRDN 6,43,82
ipr-rbt LD50:84 mg/kg NIIRDN 6,43,82

SAFETY PROFILE: Poison by intraperitoneal route. Moderately toxic by ingestion. When heated to decomposition it emits acrid smoke and irritating fumes.

AGW375 **HR: 3**
ALSTONINE HYDROCHLORIDE
mf: $C_{21}H_{20}N_2O_3 \cdot ClH$ mw: 384.86

TOXICITY DATA WITH REFERENCE
ivn-rat LD50:14,400 μg/kg JPETAB 90,57,47
ivn-mus LD50:8800 μg/kg JPETAB 90,57,47
ivn-dog LDLo:10 mg/kg JPETAB 90,57,47

SAFETY PROFILE: Poison by intravenous route. When heated to decomposition it emits toxic fumes of NO_x and HCl.

AGW476 **CAS:641-38-3** **HR: 3**
ALTERNARIOL
mf: $C_{14}H_{10}O_5$ mw: 258.24

PROP: Needles from alc (aq). Mp: 350° (decomp).

SYNS: AOH ☐ 1-METHYL-3,7,9-TRIHYDROXY-6H-DIBENZO(b,d)PYRAN-6-ONE

TOXICITY DATA WITH REFERENCE
scu-mus TDLo:400 mg/kg (9-12D preg):TER EVHPAZ 4,87,73
scu-mus TDLo:400 mg/kg (9-12D preg):REP EVHPAZ 4,87,73
ipr-mus LDLo:100 mg/kg EVHPAZ 4,87,73

SAFETY PROFILE: Poison by intraperitoneal route. An experimental teratogen. Other experimental reproductive effects. When heated to decomposition it emits acrid smoke and fumes.

AGW550 **HR: 3**
ALTERNARIOL MONOMETHYL ETHER and ALTERNARIOL (1:1)

SYNS: ALTERNARIOL and ALTERNARIOL MONOMETHYL ETHER (1:1) □ AME and AOH (1:1) □ AOH and AME (1:1) □ 6H-DIBENZO (b,d)PYRAN-6-ONE, 1-METHYL-3,7,9-TRIHYDROXY-and 3,9-DIHYDROXY-7-METHOXY-1-METHYL-DIBENZO(b,d)PYRAN-6-ONE (1:1)

TOXICITY DATA WITH REFERENCE
scu-mus TDLo:100 mg/kg (female 9-12D post):TER EVHPAZ 4,87,73
scu-mus TDLo:100 mg/kg (female 9-12D post):REP EVHPAZ 4,87,73
ipr-mus LDLo:200 mg/kg EVHPAZ 4,87,73

SAFETY PROFILE: Poison by intraperitoneal route. An experimental teratogen. Other experimental reproductive effects. When heated to decomposition it emits acrid smoke and irritating fumes.

AGW625 **CAS:73309-75-8** **HR: 3**
ALTOSIDE
mf: $C_{30}H_{40}O_{10}$ mw: 560.70

PROP: Crystals from MeOH/Et_2O. Mp: 226.3–228.3°.

SYNS: ALTOSID (GERMAN) □ 3-β-((α-d-GLUCOPYRANOSYL)OXY)-14-HYDROXY-19-OXO-BUFA-4,20,22-TRIENOLIDE

TOXICITY DATA WITH REFERENCE
dlt-dmg-skn 10 pph CYGEDX 13(6),37,79
ivn-cat LD50:80 μg/kg 85ELDJ -,188,63
unr-cat LDLo:77 μg/kg 85ELDJ -,70,63

SAFETY PROFILE: Deadly poison by intravenous and possibly other routes. When heated to decomposition it emits acrid smoke and fumes.

AGW750 **CAS:569-58-4** **HR: 1**
ALUMINON
mf: $C_{22}H_{23}N_3O_9$ mw: 473.48

PROP: Reddish-brown powder. Sol in H_2O; sltly sol in EtOH; insol in non-polar solvs.

SYNS: AMMONIUM AURINTRICARBOXYLATE □ AURINE-TRICARBOXYLATE d'AMMONIUM (FRENCH) □ AURINTRICARBOXYLIC ACID AMMONIUM SALT □ C.I. MORDANT VIOLET 39, TRIAMMONIUM SALT (8CI) □ LYSOFON □ TRIAMMONIUM AURINTRICARBOXYLATE

TOXICITY DATA WITH REFERENCE
dni-mus-ipr 211 μmol/kg VAAZA2 23,137,77
orl-man TDLo:857 μg/kg (60D male):REP CRSBAW 156,1701,62
orl-rat LD50:9 g/kg GTPZAB 32(3),48,88

CONSENSUS REPORTS: Reported in EPA TSCA Inventory.

SAFETY PROFILE: Human reproductive effects by ingestion: changes in male fertility. Experimental reproductive effects. Mutation data reported. When heated to decomposition it emits toxic fumes of NH_3 and NO_x.

AGX000 **CAS:7429-90-5** **HR: 3**
ALUMINUM
DOT: UN 1309/UN 1396/NA 9260
af: Al aw: 26.98

PROP: Hard, strong, silvery-white ductile metal: in bulk form protected from oxidation in air by coherent Al_2O_3 coating. Mp: 660°, bp: 2494° @ 24 mm, d: 2.702, vap press 1 mm @ 1284°. Sol in HCl, H_2SO_4, hot water and alkalies.

SYNS: A 00 □ A 95 □ A 99 □ A 995 □ A 999 □ AA 1099 □ AA1199 □ AD 1 □ AD1M □ ADO □ AE □ ALAUN (GERMAN) □ ALLBRI ALUMINUM PASTE and POWDER □ ALUMINA FIBRE □ ALUMINIUM BRONZE □ ALUMINUM FLAKE □ ALUMINUM 27 □ ALUMINUM A00 □ ALUMINUM DEHYDRATED □ ALUMINUM METAL (OSHA) □ ALUMINUM, molten (NA 9260) (DOT) □ ALUMINUM POWDER □ ALUMINUM POWDER, coated (UN 1309) (DOT) □ ALUMINUM POWDER, uncoated (UN 1396) (DOT) □ ALUMINUM PYRO POWDERS (OSHA) □ ALUMINUM WELDING FUMES (OSHA) □ AO A1 □ AR2 □ AV00 □ AV000 □ C.I. 77000 □ EMANAY ATOMIZED ALUMINUM POWDER □ JISC 3108 □ JISC 3110 □ L16 □ METANA ALUMINUM PASTE □ NORAL ALUMINUM □ NORAL EXTRA FINE LINING GRADE □ NORAL INK GRADE ALUMINUM □ NORAL NON-LEAFING GRADE □ PAP-1

CONSENSUS REPORTS: Community Right-To-Know List (fume or dust). Reported in EPA TSCA Inventory.

OSHA PEL: Total Dust: TWA 15 mg/m³; Respirable Fraction: TWA 5 mg/m³; Pyro Powders and Welding Fumes: 5 mg/m³; Soluble Salts and Alkyls: 2 mg/m³.
ACGIH TLV: Metal and Oxide: TWA 10 mg/m³ (dust); Pyro Powders and Welding Fumes: TWA 5 mg/m³; Soluble Salts and Alkyls: TWA 2 mg/m³
DFG MAK: 6 mg/m³; BAT: 170 μg/L in urine at end of shift
DOT CLASSIFICATION: 9; *Label:* Class 9 (NA 9260); DOT Class: 4.1; *Label:* Flammable Solid (UN 1309); DOT Class: 4.3; *Label:* Dangerous When Wet (UN 1396)

SAFETY PROFILE: Although aluminum is not generally regarded as an industrial poison, inhalation of finely divided powder has been reported to cause pulmonary fibrosis. It is a reactive metal and the greatest industrial hazards are with chemical reactions. As with other metals the powder and dust are the most dangerous forms. Dust is moderately flammable and explosive by heat, flame, or chemical reaction with powerful oxidizers. To fight fire, use special mixtures of dry chemical.

Powdered aluminum undergoes the following dangerous interactions: explosive reaction after a delay period with $KClO_4$ + $Ba(NO_3)_2$ + KNO_3 + H_2O, also with $Ba(NO_3)_2$ + KNO_3 + sulfur + vegetable adhesives + H_2O. Mixtures with powdered AgCl, NH_4NO_3 or NH_4NO_3 + $Ca(NO_3)_2$ + formamide + H_2O are powerful explosives. Mixture with ammonium peroxodisulfate + water is explosive. Violent or explosive 'thermite' reaction when heated with metal oxides, oxosalts (nitrates, sulfates), or sulfides, and with hot copper oxide worked with an iron or steel tool. Potentially explosive reaction with CCl_4 during ball milling operations. Many violent or explosive reactions with the following halocarbons have occurred in industry: bromomethane, bromotrifluoromethane, CCl_4, chlorodifluoromethane, chloroform, chloromethane, chloromethane + 2-methylpropane, dichlorodifluoromethane, 1,2-dichloroethane, dichloromethane, 1,2-dichloropropane, 1,2-difluorotetrafluoroethane, fluorotrichloroethane, hexachloroethane + alcohol, polytrifluoroethylene oils and greases, tetrachloroethylene, tetrafluoromethane, 1,1,1-trichloroethane, trichloroethylene, 1,1,2-trichlorotrifluoroethane, and trichlorotrifluoroethane-dichlorobenzene. Potentially explosive reaction with chloroform amidinium nitrate. Ignites on contact with vapors of $AsCl_3$, SCl_2, Se_2Cl_2, and PCl_5. Reacts violently on heating with Sb or As. Ignites on heating in $SbCl_3$ vapor. Ignites on contact with barium peroxide. Potentially violent reaction with sodium acetylide. Mixture with sodium peroxide may ignite or react violently. Spontaneously ignites in CS_2 vapor. Halogens: ignites in chlorine gas, foil reacts vigorously with liquid Br_2, violent reaction with H_2O + I_2. Violent reaction with hydrochloric acid, hydrofluoric acid, and hydrogen chloride gas. Violent reaction with disulfur dibromide. Violent reaction with the nonmetals phosphorus, sulfur, and selenium. Violent reaction or ignition with the interhalogens: bromine pentafluoride, chlorine fluoride, iodine chloride, iodine pentafluoride, and iodine heptafluoride. Burns when heated in CO_2. Ignites on contact with O_2, and mixtures with O_2 + H_2O ignite and react violently. Mixture with picric acid + water ignites after a delay period. Explosive reaction above 800°C with sodium sulfate. Violent reaction with sulfur when heated. Exothermic reaction with iron powder + water releases explosive hydrogen gas.

Aluminum powder also forms sensitive explosive mixtures with oxidants such as: liquid Cl_2 and other halogens, N_2O_4, tetranitromethane, bromates, iodates, $NaClO_3$, $KClO_3$, and other chlorates, $NaNO_3$, aqueous nitrates, $KClO_4$ and other perchlorate salts, nitryl fluoride, ammonium peroxodisulfate, sodium peroxide, zinc peroxide, and other peroxides, red phosphorus, and powdered polytetrafluoroethylene (PTFE).

Bulk aluminum may undergo the following dangerous interactions: exothermic reaction with butanol, methanol, 2-propanol, or other alcohols, sodium hydroxide to release explosive hydrogen gas. Reaction with diborane forms pyrophoric product. Ignition on contact with niobium oxide + sulfur. Explosive reaction with molten metal oxides, oxosalts (nitrates, sulfates), sulfides, and sodium carbonate. Reaction with arsenic trioxide + sodium arsenate + sodium hydroxide produces the toxic arsine gas. Violent reaction with chlorine trifluoride. Incandescent reaction with formic acid. Potentially violent alloy formation with palladium, platinum at mp of Al, 600°C. Vigorous dissolution reaction in methanol + carbon tetrachloride. Vigorous amalgamation reaction with mercury (II) salts + moisture. Violent reaction with molten silicon steels. Violent exothermic reaction above 600°C with sodium diuranate.

For occupational chemical analysis use OSHA: #ID-125G or NIOSH: Aluminum, 7013; Elements, 7300.

AGX125 **CAS:12607-92-0** **HR: 3**
ALUMINUM ACEGLUTAMIDE
mf: $C_{35}H_{59}Al_3N_{10}O_{24}$ mw: 1084.98

PROP: White powder. Mp: 221° (decomp).

SYNS: ACEGLUTAMIDE ALUMINUM ☐ N-ACETYL-l-GLUTAMINE ALUMINUM SALT ☐ GLUMAL ☐ KW 110 ☐ PENTAKIS(N^2-ACETYL-l-GLUTAMINATO)TETRAHYDROXYTRIALUMINUM

TOXICITY DATA WITH REFERENCE
orl-rbt TDLo:20 g/kg (female 7-16D post):REP KSRNAM 8,959,74
orl-mus TDLo:28 g/kg (female 7-13D post):TER KSRNAM 8,959,74
ipr-rat LD50:4200 mg/kg USXXAM #3787466
ivn-rat LD50:400 mg/kg USXXAM #3787466
orl-mus LD50:13,100 mg/kg NYKZAU 68,602,72
ipr-mus LD50:5 g/kg USXXAM #3787466
ivn-mus LD50:460 mg/kg USXXAM #3787466

SAFETY PROFILE: Poison by intravenous route. Moderately toxic by some other routes. An experimental teratogen. Other experimental reproductive effects. When heated to decomposition it emits toxic fumes of NO_x. See also ALUMINUM COMPOUNDS.

AGX250 **HR: 1**
ALUMINUM AMMONIUM SULFATE
mf: $Al_2(SO_4)_3(NH_4)_2SO_4 \cdot 24H_2O$ mw: 906

PROP: Colorless crystals; odorless with sweet taste. D: 1.645, mp: 94.5°, bp: loses 20 waters @ 120°. Sol in water, glycerin; insol in alc.

SAFETY PROFILE: Irritating if inhaled or ingested. Upon decomposition it emits toxic fumes of NO_x and SO_x.

AGX300 **CAS:39108-14-0** **HR: 3**
ALUMINUM AZIDE
mf: AlN_9 mw: 153.04

PROP: White moisture-sensitive solid. Sol in THF; insol in Et_2O, C_6H_6.

SAFETY PROFILE: Shock sensitive explosive. When heated to decomposition it emits toxic fumes of NO_x. See also AZIDES and ALUMINUM COMPOUNDS.

AGX500 **HR: 3**
ALUMINUM BOROHYDRIDE
mf: AlB_3H_{12} mw: 71.53

PROP: Liquid. Bp: 44.5°, mp: −64.5°, vap press: 400 mm @ 28.1°.

SYN: ALUMINUM TETRAHYDROBORATE

SAFETY PROFILE: Dangerous by spontaneous chemical reaction; ignites spontaneously in air, particularly in moist air. Explodes in O_2 at temperatures as low as 20°. An explosive range of 5 to 90%. Incompatible with water; steam; oxidizing materials; acid; acid fumes; will react with water or steam to produce heat, H_2, or toxic fumes. To fight fire, use CO_2, dry chemical. See HYDRIDES and BORON COMPOUNDS.

AGX750 **CAS:7727-15-3** **HR: 2**
ALUMINUM BROMIDE
DOT: UN 1725/UN 2580
mf: $AlBr_3$ mw: 266.71

PROP: White to yellow-red lumps. Mp: 97.5°, bp: 263.3° @ 748 mm, d: 3.2, vap press: 1 mm @ 81.3°.

SYNS: ALUMINUM BROMIDE, anhydrous (UN 1725) (DOT) □ ALUMINUM BROMIDE, solution (UN 2580) (DOT) □ ALUMINUM TRIBROMIDE □ TRIBROMOALUMINUM

TOXICITY DATA WITH REFERENCE
orl-rat LD50:1598 mg/kg PHTXA6 60,280,87
ipr-rat LD50:815 mg/kg PHTXA6 60,280,87
orl-mus LD50:1623 mg/kg PHTXA6 60,280,87
ipr-mus LD50:1068 mg/kg PHTXA6 60,280,87

CONSENSUS REPORTS: Reported in EPA TSCA Inventory.

ACGIH TLV: TWA 2 mg(Al)/m^3
DOT CLASSIFICATION: 8; *Label:* Corrosive

SAFETY PROFILE: A toxic, corrosive material. See also BROMIDES and ALUMINUM COMPOUNDS. Mixtures with sodium or potassium explode violently upon impact. When heated to decomposition it emits toxic fumes of Br^-. Do not add H_2O to anhydrous material. Hydrolysis can be violent.

AGY000 **CAS:12794-92-2** **HR: 1**
ALUMINUM BROMIDE HYDROXIDE

SYNS: AL BROMOHYDRATE □ ALUMINUM BROMHYDROXIDE □ ALUMINUM BROMOHYDROL □ ALUMINUM HYDROXYBROMIDE

TOXICITY DATA WITH REFERENCE
skn-hmn 90 mg/3D-I MLD 85DKA8 -,127,77

ACGIH TLV: TWA 2 mg(Al)/m^3

SAFETY PROFILE: A human skin irritant. See also BROMIDES. When heated to decomposition it emits toxic fumes of Br^-.

AGY100 **HR: 1**
ALUMINUM CALCIUM SILICATE

SYN: CALCIUM ALUMINUM SILICATE

SAFETY PROFILE: A nuisance dust.

AGY250 **HR: 2**
ALUMINUM CARBIDE
mf: Al_4C_3 mw: 143.91

PROP: Yellow crystals or powder, hygroscopic. Mp: 2100°, bp: decomp @ 2200°, d: 2.36.

SAFETY PROFILE: Decomposed by water. Mixture with lead dioxide or potassium permanganate reacts incandescently when warmed. Dust can cause pulmonary irritation. See also ALUMINUM COMPOUNDS.

AGY500 **HR: 3**
ALUMINUM CHLORATE
mf: $Al(ClO_3)_3$ mw: 277.4

PROP: Colorless, deliquescent crystals. Mp: decomp.

SAFETY PROFILE: Flammable by spontaneous chemical reaction; a powerful oxidizer; may ignite upon contact with combustibles. Moderate explosion hazard when shocked, exposed to heat or by spontaneous chemical reaction with reducing agents. When contaminated, may become sensitized. Dangerous; shock or heat will explode it. Evaporation emits ClO_2. See also CHLORIDES and CHLORATES.

AGY750 **CAS:7446-70-0** **HR: 3**
ALUMINUM CHLORIDE
DOT: UN 1726/UN 2581
mf: $AlCl_3$ mw: 133.33

PROP: White or colorless hexagonal deliquescent crystals or moisture-sensitive plates. D: 2.44, mp: 192° @ 2.5 atm, bp: subl @ 181°, vap press: 1 mm @ 100.0°. Violently sol in water; sol in alc and ether.

SYNS: ALLUMINIO(CLORURO DI) (ITALIAN) □ ALUMINUMCHLORID (GERMAN) □ ALUMINUM CHLORIDE (1:3) □ ALUMINUM CHLORIDE, anhydrous (DOT) □ ALUMINUM CHLORIDE, solution (DOT) □ ALUMINUM TRICHLORIDE □ CHLORURE d'ALUMINUM (FRENCH) □ PEARSALL □ TRICHLOROALUMINUM

TOXICITY DATA WITH REFERENCE
dnd-rat:ast 500 µmol/L JBCHA3 261,3370,86
cyt-mus-ipr 444 mg/kg NULSAK 15,180,72
orl-rat TDLo:11,512 mg/kg (female 8-22D post):REP NETOD7 8,115,86
ipr-rat TDLo:375 mg/kg (female 9-13D post):TER ANANAU 138,365,75
orl-rat LD50:3730 mg/kg EQSSDX 1,1,75
orl-mus LD50:3805 mg/kg BJIMAG 23,305,66

CONSENSUS REPORTS: Reported in EPA TSCA Inventory.

ACGIH TLV: TWA 2 mg(Al)/m^3
DOT CLASSIFICATION: 8; *Label:* Corrosive

SAFETY PROFILE: Moderately toxic by ingestion. Experimental teratogenic and reproductive effects. Mutation data reported. The dust is an irritant by ingestion, inhalation, and skin contact. Highly exothermic polymerization reactions with alkenes. Incompatible with nitrobenzenes or nitrobenzene + phenol. Highly exothermic reaction with water or steam produces toxic fumes of HCl. See also ALUMINUM COMPOUNDS, CHLORIDES, and HYDROCHLORIC ACID.

AGZ000 **CAS:7784-13-6** **HR: 2**
ALUMINUM CHLORIDE HEXAHYDRATE
mf: $AlCl_3 \cdot 6H_2O$ mw: 241.45

PROP: Colorless crystals from water. Mp: 100° (decomp).

SYNS: ALUMINUM(III) CHLORIDE, HEXAHYDRATE ☐ ALUMINUM TRICHLORIDEHEXAHYDRATE

TOXICITY DATA WITH REFERENCE
skn-hmn 7500 µg/3D-I MLD 85DKA8 -,127,77
dnd-mam:lym 40 µmol/L JCHODP 7,411,76
ivn-mus TDLo:483 g/kg (8D preg):TER ENVRAL 33,22,86
orl-rat LD50:3311 mg/kg PHTXA6 60,280,87
ipr-rat LD50:728 mg/kg PHTXA6 60,280,87
orl-mus LD50:1990 mg/kg PHTXA6 60,280,87
ipr-mus LD50:940 mg/kg PHTXA6 60,280,87

ACGIH TLV: TWA 2 mg(Al)/m^3

SAFETY PROFILE: Moderately toxic by ingestion and intraperitoneal routes. An experimental teratogen. Corrosive and irritating to tissue. Mutation data reported. When heated to decomposition it emits toxic fumes of Cl^-. See also ALUMINUM COMPOUNDS.

AHA000 CAS:12042-91-0 HR: 2
ALUMINUM CHLORIDE HYDROXIDE
mf: $Al_2ClH_5O_5$ mw: 174.46

SYNS: ALUMINUM CHLORHYDRATE ☐ ALUMINUM CHLORHYDROL ☐ ALUMINUM CHLORHYDROXIDE ☐ ALUMINUM CHLOROHYDROXIDE ☐ ALUMINUM HYDROXIDE CHLORIDE ☐ ALUMINUM HYDROXYCHLORIDE ☐ ASTRINGEN ☐ BASIC ALUMINUM CHLORATE ☐ CHLORHYDROL ☐ CHLOROPENTAHYDROXYDIALUMINUM ☐ LOCRON EXTRA ☐ MICRO DRY ☐ WICKENOL 324

TOXICITY DATA WITH REFERENCE
skn-hmn 150 mg/3D-I MLD 85DKA8 -,127,77

CONSENSUS REPORTS: Reported in EPA TSCA Inventory

ACGIH TLV: TWA 2 mg(Al)/m^3

SAFETY PROFILE: A mild human skin irritant. See also ALUMINUM COMPOUNDS, CHLORIDES, and SODIUM HYDROXIDE. When heated to decomposition it emits toxic fumes of Cl^-.

AHA125 CAS:3495-54-3 HR: 2
ALUMINUM CHLORIDE NITROMETHANE
mf: $AlCl_3CH_3NO_2$ mw: 194.38

PROP: White moisture-sensitive crystals from C_6H_6/cyclohexane. Mp: 85°.

SAFETY PROFILE: Mixture with alkenes reacts explosively.

AHA130 CAS:13596-11-7 HR: 1
ALUMINUM CHLORIDE OXIDE
mf: AlClO mw: 78.43

PROP: A solid.

SYNS: ALUMINUM OXIDE CHLORIDE ☐ ALUMINUM OXYCHLORIDE

TOXICITY DATA WITH REFERENCE
unr-rat LD50:4320 mg/kg GISAAA 43(4),12,78
unr-mus LD50:5200 mg/kg GISAAA 43(4),12,78
unr-gpg LD50:5430 mg/kg GISAAA 43(4),12,78

CONSENSUS REPORTS: Reported in EPA TSCA Inventory.

SAFETY PROFILE: Mildly toxic by unreported routes. When heated to decomposition it emits toxic vapors of Cl^-.

AHA135 CAS:1317-25-5 HR: 1
ALUMINUM CHLOROHYDROXYALLANTOINATE
mf: $C_4H_9Al_2ClN_4O_7$ mw: 314.58

SYNS: ALCA ☐ ALCLOXA ☐ ALUMINIUM CHLOROHYDROXYALLANTOINATE ☐ ALUMINUM, CHLORO((2,5-DIOXO-4-IMIDAZOLIDINYL)UREATO)TETRAHYDROXYDI- ☐ ALUMINUM, CHLOROTETRAHYDROXY((2-HYDROXY-5-OXO-2-IMIDAZOLIN-4-YL)UREATO)DI- ☐ CHLORHYDROXYALUMINUM ALLANTOINATE ☐ CHLOROTETRAHYDROXY((2-HYDROXY-5-OXO-2-IMIDAZOLIN-4-YL)UREATO)DIALUMINUM

TOXICITY DATA WITH REFERENCE
eye-rbt 24% MLD YKKZAJ 102,89,82
orl-rat LD50:>8 g/kg DRUGAY -,68,90
scu-rat LD50:>8 g/kg DRUGAY -,68,90
orl-mus LD50:>8 g/kg DRUGAY -,68,90
scu-mus LD50:>8 g/kg DRUGAY -,68,90
unr-mus LD50:15,800 mg/kg AMPMAR 40,63,79

CONSENSUS REPORTS: Reported in EPA TSCA Inventory.

ACGIH TLV: TWA 2 mg(Al)/m^3

SAFETY PROFILE: Slightly toxic by several routes. An eye irritant. When heated to decomposition it emits toxic vapors of NO_x and Cl^-.

AHA150 CAS:24818-79-9 HR: 2
ALUMINUM CLOFIBRATE
mf: $C_{20}H_{21}AlCl_2O_7$ mw: 471.29

SYNS: ALFIBRATE ☐ ALUFIBRATE ☐ ATHEROLIP ☐ BIS(2-(p-CHLOROPHENOXY)-2-METHYLPROPIONATO)HYDROXYALUMINUM ☐ HYDROXYBIS(2-(p-CHLOROPHENOXY)ISOBUTYRIC ACID) ALUMINUM

TOXICITY DATA WITH REFERENCE
ipr-rat LD50:3300 mg/kg JJPAAZ 23,281,73
scu-rat LD50:5100 mg/kg NIIRDN 6,237,82
ipr-mus LD50:3300 mg/kg NIIRDN 6,237,82
scu-mus LD50:5850 mg/kg NIIRDN 6,237,82

SAFETY PROFILE: Moderately toxic by some routes. When heated to decomposition it emits toxic fumes of Cl^-. See also ALUMINUM COMPOUNDS.

AHA175 HR: 2
ALUMINUM COMPOUNDS

SAFETY PROFILE: Aluminum compounds have many commercial uses and are commonly found in industry. Many of these materials are active chemically and thus exhibit dangerous toxic and reactive properties. Inhalation of fine aluminum oxide particles is associated with Shaver's disease. The halides are generally irritants. See also ALUMINUM and individual compounds.

AHA250 **CAS:7047-84-9** **HR: 1**
ALUMINUM DEXTRAN
mf: $C_{18}H_{37}AlO_4$ mw: 344.48

PROP: Powder. A complex containing aluminum and dextran, a chain of molecular weight 2500, corresponding to a chain of 15 anhydroglucose units.

SYNS: ALUMINUM MONOSTEARATE □ ALUMINUM STEARATE (ACGIH) □ STEARIC ACID, ALUMINIUM SALT

CONSENSUS REPORTS: EPA TSCA Chemical Inventory.

ACGIH TLV: TWA 10 mg/m^3

SAFETY PROFILE: A nuisance dust. When heated to decomposition it emits acrid smoke and fumes. See also ALUMINUM COMPOUNDS.

AHA275 **CAS:300-92-5** **HR: 1**
ALUMINUM DISTEARATE
mf: $C_{36}H_{71}AlO_5$ mw: 611.05

SYNS: ALUMINUM DISTEARATE (ACGIH) □ ALUMINUM HYDROXIDE DISTEARATE □ ALUMINUM, HYDROXYBIS(OCTADECANOATO-O)-(9CI) □ ALUMINUM, HYDROXYBIS(STEARATO)- □ ALUMINUM HYDROXYDISTEARATE □ SPECIAL M

CONSENSUS REPORTS: Reported in EPA TSCA Inventory.

ACGIH TLV: TWA 10 mg/m^3

SAFETY PROFILE: A nuisance dust. When heated to decomposition it emits acrid smoke and irritating vapors.

AHA750 **HR: 3**
ALUMINUM ETHYLATE
mf: $Al(OC_2H_5)_3$ mw: 162.15

PROP: Liquid. Decomp by H_2O. Bp: 200° @ 6–8 mm; mp: 140°.

SAFETY PROFILE: Strong irritant to skin, eyes, and mucous membranes by inhalation.

AHA875 **CAS:16449-54-0** **HR: 2**
ALUMINUM FLUFENAMATE
mf: $C_{42}H_{27}AlF_9N_3O_6$ mw: 867.70

SYNS: ALFENAMIN □ ALUFENAMINE □ ALUMINUM (TRIS(2-((3-TRIFLUOROMETHYL)PHENYL)AMINO)BENZOATO-N,o)-o-PYRIN □ TRIS(N-(α,α,α-TRIFLUORO-m-TOLYL)ANTHRANILATO)ALUMINUM □ TS 1801

TOXICITY DATA WITH REFERENCE
orl-mus TDLo:3 g/kg (female 7-12D post):REP TOIZAG 17,159,70
orl-rat LD50:550 mg/kg TOIZAG 17,153,70
ipr-rat LD50:420 mg/kg NIIRDN 4,1289,35
scu-rat LD50:725 mg/kg NIIRDN 6,704,82
orl-mus LD50:1460 mg/kg NIIRDN 6,704,82
ipr-mus LD50:1560 mg/kg NIIRDN 6,704,82

SAFETY PROFILE: Moderately toxic by ingestion and other routes. Experimental reproductive effects. When heated to decomposition it emits toxic fumes of F^- and NO_x. See also ALUMINUM COMPOUNDS.

AHB000 **CAS:7784-18-1** **HR: 3**
ALUMINUM FLUORIDE
mf: AlF_3 mw: 83.98

PROP: Solid, colorless crystals. Mp: 1291°, subl. @ 1260°, d: 2.88, vap press: 1 mm @ 1238°, bp: 1537°. Sparingly sol in water; insol in org solvs.

SYNS: ALUMINUM FLUORURE (FRENCH) □ ALUMINUM TRIFLUORIDE □ FLUORID HLINITY (CZECH)

TOXICITY DATA WITH REFERENCE
eye-rbt 500 mg/24H SEV 28ZPAK -,20,72
orl-mus LD50:103 mg/kg GISAAA 30(4),16,65
scu-frg LDLo:1680 mg/kg CRSBAW 124,133,37

CONSENSUS REPORTS: Reported in EPA TSCA Inventory.

OSHA PEL: TWA 2.5 mg(F)/m^3
ACGIH TLV: TWA 2 mg(Al)/m^3; 2.5 mg(F)/m^3
NIOSH REL: (Fluorides, Inorganic) TWA 2.5 mg(F)/m^3

SAFETY PROFILE: A poison by ingestion. Moderately toxic by subcutaneous route. A severe eye irritant. Violently impact-sensitive when in contact with Na and K. When heated to decomposition it emits toxic fumes of F^-. See also FLUORIDES and ALUMINUM COMPOUNDS.

AHB250 **CAS:73680-58-7** **HR: 3**
ALUMINUM FLUOROSULFATE, HYDRATE

TOXICITY DATA WITH REFERENCE
ivn-mus LD50:56 mg/kg CSLNX* NX#00137

OSHA PEL: TWA 2.5 mg(F)/m^3
ACGIH TLV: TWA 2 mg(Al)/m^3; 2.5 mg(F)/m^3
NIOSH REL: (Fluorides, Inorganic) TWA 2.5 mg(F)/m^3

SAFETY PROFILE: Poison by intravenous route. See also ALUMINUM COMPOUNDS, SULFATES, and FLUORIDES. When heated to decomposition it emits very toxic fumes of F^- and SO_x.

AHB375 **CAS:7360-53-4** **HR: 2**
ALUMINUM FORMATE
mf: $C_3H_3AlO_6$ mw: 162.03

PROP: Amorphous white crystals.

SAFETY PROFILE: Aqueous solution explodes when heated in air.

AHB500 **CAS:7784-21-6** **HR: 3**
ALUMINUM HYDRIDE
DOT: UN 2463
mf: AlH_3 mw: 30.01

PROP: Colorless powder or white amorph or microcrystal air- and moisture-sensitive solid. Usually made in THF or Et_2O and used *in situ*. Several solid forms exist. Solids $AlH_3.nEt_2O$, $n = 0.3–3$ obtained from Et_2O. Sol in Et_2O, THF.

SYNS: ALANE ☐ ALUMINUM TRIHYDRIDE ☐ α-ALUMINUM TRIHYDRIDE

CONSENSUS REPORTS: Reported in EPA TSCA Inventory.

ACGIH TLV: TWA 2 mg(Al)/m^3
DOT CLASSIFICATION: 4.3; *Label:* Dangerous When Wet

SAFETY PROFILE: Hydrides of some metals (such as AsH_3) are extremely toxic. Dangerous fire hazard. An unstable material which is spontaneously flammable in air or O_2. Evolves explosive H_2 upon contact with moisture. Severe explosion hazard by chemical reaction wherein H_2 gas is produced, also in contact with methyl ethers contaminated by CO_2. Mixtures with tetrazole derivatives are explosive. Reacts with oxidizing materials. On contact with acid or acid fumes, it can emit toxic fumes. See also HYDRIDES and ALUMINUM COMPOUNDS.

AHB625 CAS:26351-01-9 HR: 3
ALUMINUM HYDRIDE-DIETHYL ETHER
mf: $AlH_3 \cdot C_4H_{10}O$ mw: 104.13

SAFETY PROFILE: Reacts violently with moist air or water. When heated to decomposition it emits acrid smoke and fumes. See also ALUMINUM COMPOUNDS and ETHERS.

AHB750 CAS:17013-07-9 HR: 3
ALUMINUM HYDRIDE-TRIMETHYL AMINE
mf: $AlH_3 \cdot C_3H_9N$ mw: 89.1

SAFETY PROFILE: An unstable, dangerous compound. Explodes on contact with water. Ignites in moist air. When heated to decomposition it emits toxic fumes of NO_x. See also AMINES, HYDRIDES, and ALUMINUM COMPOUNDS.

AHC000 CAS:21645-51-2 HR: 3
ALUMINUM HYDROXIDE
mf: AlH_3O_3 mw: 78.01

PROP: White, crystalline powder, balls, or granules. Solid from water. D: 2.42, mp: loses H_2O @ 300°. Practically insol in water; sol in mineral acids, alkalis, and caustic soda.

SYNS: AF 260 ☐ ALCOA 331 ☐ ALUMIGEL ☐ ALUMINA HYDRATE ☐ ALUMINA HYDRATED ☐ ALUMINA TRIHYDRATE ☐ α-ALUMINA TRIHYDRATE ☐ ALUMINIC ACID ☐ ALUMINUM HYDRATE ☐ ALUMINUM(III) HYDROXIDE ☐ ALUMINUM HYDROXIDE GEL ☐ ALUMINUM OXIDE HYDRATE ☐ ALUMINUM OXIDE TRIHYDRATE ☐ ALUMINUM TRIHYDRAT ☐ ALUMINUM TRIHYDROXIDE ☐ ALUSAL ☐ AMBEROL ST 140F ☐ AMPHOJEL ☐ BACO AF 260 ☐ BRITISH ALUMINUM AF 260 ☐ C.I. 77002 ☐ GHA 331 ☐ H 46 ☐ HIGILITE ☐ HYDRAL 705 ☐ LIQUIGEL ☐ PGA ☐ TRIHYDRATED ALUMINA

TOXICITY DATA WITH REFERENCE
orl-chd TDLo:122 g/kg/4D:GIT,MET JOPDAB 92,592,78
orl-cld TDLo:122 g/kg/4D: GIT JOPDAB 92,592,78
unr-inf TDLo:39 g/kg/24D-I NEJMAG 310,1079,84
ipr-rat LDLo:150 mg/kg LANCAO 1,564,72

CONSENSUS REPORTS: Reported in EPA TSCA Inventory.

ACGIH TLV: TWA 2 mg(Al)/m^3

SAFETY PROFILE: Poison by intraperitoneal route. Human systemic effects by ingestion: fever, osteomalacia, and gastrointestinal effects. When coprecipitated with bismuth hydroxide and reduced by H_2, it is violently flammable in air. Incompatible with chlorinated rubber.

AHC250 CAS:24623-77-6 HR: 3
ALUMINUM HYDROXIDE OXIDE
mf: $AlHO_2$ mw: 59.99

PROP: White solid. Sol in hot NaOH.

SYNS: ALUMINUM METAHYDROXIDE ☐ HYDRATED ALUMINA

TOXICITY DATA WITH REFERENCE
itr-rat LDLo:90 mg/kg JPBAA7 69,81,55

CONSENSUS REPORTS: Reported in EPA TSCA Inventory.

ACGIH TLV: TWA 2 mg(Al)/m^3

SAFETY PROFILE: Poison by intratracheal route. See also ALUMINUM COMPOUNDS.

AHC500 HR: 3
ALUMINUM IODIDE
mf: AlI_3 mw: 407.7

PROP: White leaflets. Mp: 191°, bp: 360°, d: 3.98 @ 25°, vap press: 1 mm @ 178.0° (sublimes).

SAFETY PROFILE: Incompatible with water. See ALUMINUM COMPOUNDS and IODIDES.

AHC600 CAS:555-31-7 HR: 1
ALUMINUM(II) ISOPROPYLATE
mf: $C_9H_{21}O_3 \cdot Al$ mw: 204.28

SYNS: ALUMINUM ISOPROPOXIDE ☐ TRIISOPROPOXYALUMINUM

TOXICITY DATA WITH REFERENCE
orl-rat LD50:11,300 mg/kg AIHAAP 30,470,69

CONSENSUS REPORTS: Reported in EPA TSCA Inventory.

ACGIH TLV: TWA 2 mg(Al)/m^3

SAFETY PROFILE: Slightly toxic by ingestion. When heated to decomposition it emits acrid smoke and irritating vapors.

AHD250 HR: 3
ALUMINUM MAGNESIUM PHOSPHIDE
DOT: UN 1419
mf: Mg_3AlP_3 mw: 192.8

SYN: MAGNESIUM ALUMINUM PHOSPHIDE (DOT)

ACGIH TLV: TWA 2 mg(Al)/m^3
DOT CLASSIFICATION: 4.3; *Label:* Dangerous When Wet, Poison

SAFETY PROFILE: A poison. Dangerous fire hazard. Evolves spontaneously flammable PH_3 in contact with water. See also PHOSPHIDES, PHOSPHINE, ALUMINUM COMPOUNDS, and MAGNESIUM COMPOUNDS.

AHD500 **HR: 3**
ALUMINUM METHYL
mf: $Al(CH_3)_3$ mw: 72.07

PROP: Colorless liquid. Bp: 130°; mp: 0°.

SAFETY PROFILE: Related alkyl aluminum compounds are poisonous and strong irritants. Very flammable by spontaneous chemical reaction with air. Incompatible with water, halogenated hydrocarbons, and oxidizing materials. When heated to decomposition it emits toxic fumes. To fight fire, do not use water, foam, or halogenated extinguishing agents. Use dry chemical.

AHD750 **CAS:13473-90-0** **HR: 3**
ALUMINUM(III) NITRATE (1:3)
DOT: UN 1438
mf: $N_3O_9 \cdot Al$ mw: 213.01

PROP: White crystals or very hygroscopic solid. Bp: 50° @ 0.01 mm.

SYNS: ALUMINUM NITRATE (DOT) ☐ ALUMINUM TRINITRATE ☐ NITRIC ACID, ALUMINUM SALT ☐ NITRIC ACID, ALUMINUM(3+) SALT

TOXICITY DATA WITH REFERENCE
skn-rbt 500 mg MLD FCTOD7 20,563,82
eye-rbt 100 mg SEV FCTOD7 20,573,82
eye-rbt 100 mg/4S rns MLD FCTOD7 20,573,82
orl-rat LD50:4280 mg/kg 85INA8 5,22,86

CONSENSUS REPORTS: Reported in EPA TSCA Inventory.

ACGIH TLV: TWA 2 mg(Al)/m³
DOT CLASSIFICATION: 5.1; *Label:* Oxidizer

SAFETY PROFILE: A poison. A severe eye and mild skin irritant. A powerful oxidizer. When heated to decomposition it emits toxic NO_x. See NITRATES and ALUMINUM COMPOUNDS. A nitrating agent.

AHD900 **CAS:7784-27-2** **HR: 2**
ALUMINUM(III) NITRATE, NONAHYDRATE (1:3:9)
mf: $N_3O_9 \cdot Al \cdot 9H_2O$ mw: 375.19

PROP: Colorless plates. Mp: 73.5°. Sol in H_2O, Me_2CO, alcohols.

SYNS: ALUMINUM NITRATE NONAHYDRATE ☐ ALUMINUM TRINITRATE NONAHYDRATE ☐ NITRIC ACID, ALUMINUM SALT, NONAHYDRATE (8CI,9CI)

TOXICITY DATA WITH REFERENCE
orl-rat TDLo:7740 mg/kg (female 1-22D post):REP RCOCB8 57,129,87
orl-rat TDLo:1620 mg/kg (female 6-14D post):TER TJADAB 38,253,88
orl-rat LD50:3671 mg/kg HYSAAV 31(7-9),204,66
ipr-rat LD50:901 mg/kg PHTXA6 60,280,87
orl-mus LD50:3980 mg/kg PHTXA6 60,280,87
ipr-mus LD50:1587 mg/kg PHTXA6 60,280,87

ACGIH TLV: TWA 2 mg(Al)/m³

SAFETY PROFILE: Moderately toxic by ingestion and intraperitoneal route. An experimental teratogen. Other experimental reproductive effects.

AHE000 **HR: 3**
ALUMINUM NITRIDE
mf: AlN mw: 41

PROP: White or colorless crystals. Mp: 2200°, bp: sublimes @ 2000°, d: 3.26.

SAFETY PROFILE: A poison. Will react with water or steam to produce toxic or corrosive fumes. Incompatible with water or steam. See also NITRIDES, AMMONIA, and ALUMINUM COMPOUNDS.

AHE250 **CAS:1344-28-1** **HR: 2**
ALUMINUM OXIDE (2:3)
mf: Al_2O_3 mw: 101.96

PROP: White powder or solid. Mp: 2050°, bp: 2977°, d: 3.5–4.0, vap press: 1 mm @ 2158°. Sol in hot NaOH.

SYNS: A 1 (sorbent) ☐ A1-0109 P ☐ ABRAREX ☐ ACTIVATED ALUMINUM OXIDE ☐ ALCOA F 1 ☐ ALMITE ☐ ALON ☐ ALUMINA ☐ α-ALUMINA (OSHA) ☐ β-ALUMINA ☐ γ-ALUMINA ☐ ALUMINUM OXIDE ☐ α-ALUMINUM OXIDE ☐ β-ALUMINUM OXIDE ☐ γ-ALUMINUM OXIDE ☐ ALUMINUM SESQUIOXIDE ☐ ALUMITE ☐ ALUNDUM ☐ BROCKMANN, ALUMINUM OXIDE ☐ CAB-O-GRIP ☐ COMPALOX ☐ DIALUMINUM TRIOXIDE ☐ DISPAL ☐ DOTMENT 324 ☐ FASERTON ☐ G 2 (OXIDE) ☐ KHP 2 ☐ LUCALOX ☐ MICROGRIT WCA ☐ PS 1 ☐ RC 172DBM

TOXICITY DATA WITH REFERENCE
ipl-rat TDLo:90 mg/kg:ETA BJCAAI 28,173,73
imp-rat TDLo:200 mg/kg:NEO JJIND8 67,965,81

CONSENSUS REPORTS: Community Right-To-Know List. Reported in EPA TSCA Inventory.

OSHA PEL: Total Dust: TWA 10 mg/m³; Respirable Fraction: TWA 5 mg/m³
ACGIH TLV: TWA (nuisance particulate) 10 mg/m³ of total dust (when toxic impurities are not present, e.g., quartz <1%)
DFG MAK: 6 mg/m³ (fume)

SAFETY PROFILE: Inhalation of finely divided particles may cause lung damage (Shaver's disease). Questionable carcinogen with experimental neoplastigenic and tumorigenic data by implantation. Exothermic reaction above 200°C with halocarbon vapors produces toxic HCl and phosgene. See also ALUMINUM COMPOUNDS.

For occupational chemical analysis use NIOSH: Nuisance Dust, Total, 0500; Nuisance Dust, Respirable, 0600.

AHE750 **CAS:20859-73-8** **HR: 3**
ALUMINUM PHOSPHIDE
DOT: UN 1397
mf: AlP mw: 57.95

PROP: Dark-gray or dark-yellow moisture-sensitive crystals stable up to 10°. D: 2.85 @ 25°/4°. Mp: >1000°.

SYNS: AIP □ AL-PHOS □ ALUMINUM FOSFIDE (DUTCH) □ ALUMINUM MONOPHOSPHIDE □ CELPHIDE □ CELPHOS □ DELICIA □ DETIA GAS EX-B □ FOSFURI di ALLUMINIO (ITALIAN) □ FUMITOXIN □ PHOSPHURES d'ALUMINUM (FRENCH) □ PHOSTOXIN □ QUICKPHOS □ RCRA WASTE NUMBER P006

TOXICITY DATA WITH REFERENCE
orl-hmn LD50:20 mg/kg 85ARAE 3,38,76
ihl-mam LCLo:1 ppm PCOC** -,25,66

CONSENSUS REPORTS: EPA Extremely Hazardous Substances List. Reported in EPA TSCA Inventory.

ACGIH TLV: TWA 2 mg(Al)/m^3
DOT CLASSIFICATION: 4.3; *Label:* Dangerous When Wet, Poison

SAFETY PROFILE: A human poison by inhalation and ingestion. Dangerous; in contact with water, steam, or alkali it slowly yields PH_3, which is spontaneously flammable in air. Explosive reaction on contact with mineral acids produces phosphine. When heated to decomposition it yields toxic PO_x. See also ALUMINUM COMPOUNDS, PHOSPHIDES, and PHOSPHINE.

AHE875 **CAS:24704-64-1** **HR: 3**
ALUMINUM PHOSPHINATE
mf: $AlH_6O_6P_3$ mw: 221.95

$Al[P(H)(O)OH]_3$

SAFETY PROFILE: A poison. When heated to decomposition it emits toxic and spontaneously flammable fumes of phosphine. See also ALUMINUM COMPOUNDS and PHOSPHINE.

AHF000 **HR: 3**
ALUMINUM PICRATE
mf: $Al(C_6H_2O(NO_2)_3)_3$ mw: 711.3

PROP: A solid.

SAFETY PROFILE: A poison. A powerful irritant. Very flammable by reaction with reducing materials. Severe explosion hazard when shocked or exposed to heat. See also EXPLOSIVES, HIGH. When heated to decomposition it emits highly toxic fumes of NO_x and explodes.

AHF100 **CAS:10043-67-1** **HR: 1**
ALUMINUM POTASSIUM SULFATE
mf: O_8S_2•Al•K mw: 258.20

PROP: Transparent crystals, colorless hexagonal plates, or white crystalline powder; odorless with sweet taste. Sol in water, glycerin; insol in alc.

SYNS: ALUMINUM POTASSIUM ALUM □ ALUMINUM POTASSIUM DISULFATE □ ALUMINUM POTASSIUM SULFATE, ALUM □ ALUMINUM POTASSIUM SULFATE, ANHYDROUS □ ALUM POTASSIUM □ BURNT ALUM □ DIALUMINUM DIPOTASSIUM SULFATE □ POTASSIUM ALUM □ POTASSIUM ALUMINUM SULFATE

TOXICITY DATA WITH REFERENCE
mic-mic-uns 50 µmol/L MUREAV 264,135,91
uns-rat-orl 824 mg/kg/7D-C CYTBAI 66,105,91
orl-rat TDLo:4355 mg/kg/26W-I GISAAA 43(5),101,78
orl-rat TDLo:8040 mg/kg/67D-C FAATDF 10,616,88

CONSENSUS REPORTS: EPA TSCA Chemical Inventory, JUNE 1993

SAFETY PROFILE: A nuisance dust. Mutation data reported.

AHF500 **CAS:1302-76-7** **HR: 2**
ALUMINUM(III) SILICATE (2:1)
mf: O_5Si•2Al mw: 162.05

PROP: Usually blue long bladed crystals. Color often varies in single crystals; also white, gray, green, yellow, pink or nearly black.

SYNS: ALUMINUM OXIDE SILICATE □ CERAMIC FIBRE □ CYANITE □ DISTHENE □ KYANITE □ OIL-DRI □ SAFE-N-DRI □ SILICIC ACID ALUMINUM SALT □ SNOW TEX □ VALFOR

TOXICITY DATA WITH REFERENCE
ipl-rat TDLo:90 mg/kg:ETA BJCAAI 28,173,73

ACGIH TLV: TWA 2 mg(Al)/m^3
DOT CLASSIFICATION: 4.3; *Label:* Dangerous When Wet

SAFETY PROFILE: Questionable carcinogen with experimental tumorigenic data by implantation. See also ALUMINUM COMPOUNDS.

AHF750 **HR: 3**
ALUMINUM SODIUM LACTATE

SYN: ALUMINUMNATRIUMLACTAT (GERMAN)

TOXICITY DATA WITH REFERENCE
scu-mus LDLo:200 mg/kg HBAMAK 4,1289,35
scu-dog LDLo:15 mg/kg HBAMAK 4,1289,35
scu-cat LDLo:150 mg/kg HBAMAK 4,1289,35
scu-rbt LDLo:750 mg/kg HBAMAK 4,1289,35
scu-frg LDLo:600 mg/kg HBAMAK 4,1289,35

ACGIH TLV: TWA 2 mg(Al)/m^3

SAFETY PROFILE: Poison by subcutaneous route. See also ALUMINUM COMPOUNDS. When heated to decomposition it emits acrid smoke and irritating fumes.

AHG000 **CAS:11138-49-1** **HR: 2**
ALUMINUM SODIUM OXIDE
DOT: UN 1819/UN 2812
mf: $NaAlO_2$ mw: 82.0

PROP: White, hygroscopic powder. Mp: 1650°.

SYNS: β ALUMINA □ β''-ALUMINA □ J 242 □ NALCO 680 □ SODIUM ALUMINATE, solid (UN 2812) (DOT) □ SODIUM ALUMINATE, solution (UN 1819) (DOT) □ SODIUM ALUMINUM OXIDE □ SODIUM POLYALUMINATE

TOXICITY DATA WITH REFERENCE
ACGIH TLV: TWA 2 mg(Al)/m^3
DOT CLASSIFICATION: 4.3; *Label:* Dangerous When Wet

SAFETY PROFILE: Moderate irritant to skin, eyes, and mucous membranes. A corrosive substance. When heated to decomposition it emits toxic fumes of Na_2O.

AHG500 **HR: 2**
ALUMINUM SODIUM SULFATE
mf: $NaAl(SO_4)_2 \cdot 12H_2O$ mw: 458.29

PROP: Colorless crystals. Mp: 61°; d: 1.675. Anhydrous: sol in alc; sltly sol in water. Dodecahydrate: sol in water and alc.

SYNS: SODA ALUM □ SODIUM ALUMINUM SULFATE

SAFETY PROFILE: A weak sensitizer. A general-purpose food additive. Local contact may cause contact dermatitis. An irritant. See also SULFATES and ALUMINUM COMPOUNDS. When heated to decomposition it emits toxic fumes of SO_x and Na_2O.

AHG750 **CAS:10043-01-3** **HR: 2**
ALUMINUM SULFATE (2:3)
mf: $O_{12}S_3 \cdot 2Al$ mw: 342.14

PROP: White powder; sweet taste. Mp: decomp @ 770°, d: 2.71. Solubility in water 36.4% @ 20°.

SYNS: ALUM □ ALUMINUM TRISULFATE □ CAKE ALUM □ DIALUMINUM SULPHATE □ DIALUMINUM TRISULFATE □ SULFURIC ACID, ALUMINUM SALT (3:2)

TOXICITY DATA WITH REFERENCE
scu-mus TDLo:27,371 µg/kg (30D male):REP JRPFA4 7,21,64
orl-mus LD50:6207 mg/kg BJIMAG 23,305,66
ipr-mus LD50:1735 mg/kg COREAF 256,1043,63

CONSENSUS REPORTS: Reported in EPA TSCA Inventory.

ACGIH TLV: TWA 2 mg(Al)/m^3

SAFETY PROFILE: Moderately toxic by ingestion and intraperitoneal routes. Experimental reproductive effects. Hydrolyzes to form sulfuric acid, which irritates tissue, especially lungs. When heated to decomposition it emits toxic fumes of SO_x.

AHG875 **CAS:16962-07-5** **HR: 3**
ALUMINUM TETRAHYDROBORATE
DOT: UN 2870
mf: $Al \cdot 3BH_4$ mw: 71.53

SYNS: ALUMINUM BOROHYDRIDE (DOT) □ ALUMINUM BOROHYDRIDE in devices (DOT) □ ALUMINUM HYDROBORATE □ BORATE (1-), TETRAHYDRO-, ALUMINUM (3:1) (9CI)

ACGIH TLV: TWA 2 mg(Al)/m^3
DOT CLASSIFICATION: 4.2; *Label:* Spontaneously Combustible, Dangerous When Wet

SAFETY PROFILE: A poison. Spontaneously flammable in air. Explodes in oxygen with traces of water. Incompatible with alkenes and water. See also ALUMINUM COMPOUNDS and BORON COMPOUNDS.

AHH000 **HR: 3**
ALUMINUM THALLIUM SULFATE
mf: $AlTl(SO_4)_2 \cdot 12H_2O$ mw: 639.6

PROP: Cubic, octagonal, colorless crystals. Mp: 91°, d: 2.32 @ 20°/4°.

CONSENSUS REPORTS: Thallium and its compounds are on the Community Right-To-Know List.

SAFETY PROFILE: A poison. See also THALLIUM COMPOUNDS and ALUMINUM COMPOUNDS.

AHH125 **CAS:12003-96-2** **HR: 2**
ALUMINUM-TITANIUM ALLOY (1:1)
mf: AlTi mw: 74.8814

SAFETY PROFILE: Incompatible with chlorine; bromine; iodine; and hydrogen chloride vapors. See also ALUMINUM COMPOUNDS and TITANIUM COMPOUNDS.

AHH750 **HR: 3**
ALUMINUM TRIPROPYL
mf: $Al(C_3H_7)_3$ mw: 156.24

PROP: Liquid.

SYN: TRIPROPYL ALUMINUM

SAFETY PROFILE: Related alkyl aluminum compounds are poisons. Very flammable by spontaneous reaction with air. Incompatible with halogenated hydrocarbons. Hydrolyzes to evolve flammable vapor. To fight fire, do not use water, foam or halogenated extinguishing agents. Use dry chemical or a special powder extinguisher.

AHH825 **CAS:637-12-7** **HR: 1**
ALUMINUM TRISTEARATE
mf: $C_{54}H_{105}O_6 \cdot Al$ mw: 877.57

SYNS: ALUGEL 34TN □ ALUMINIUM STEARATE □ ALUMINUM STEARATE □ METASAP XX □ OCTADECANOIC ACID, ALUMINUM SALT □ ROFOB 3 □ SA 1500 □ STEARIC ACID, ALUMINUM SALT □ TRIBASIC ALUMINUM STEARATE

CONSENSUS REPORTS: Reported in EPA TSCA Inventory.

ACGIH TLV: TWA 10 mg/m^3

SAFETY PROFILE: A nuisance dust. When heated to decomposition it emits acrid smoke and irritating vapors.

AHI250 **CAS:963-07-5** **HR: 3**
ALYPIN
mf: $C_{16}H_{26}N_2O_2$ mw: 278.44

PROP: White, crystalline powder.

SYNS: ALYPINE □ AMYDRICAINE □ BENZOPROPYL □ 1-(DIMETHYLAMINO)-2-((DIMETHYLAMINO)METHYL)-2-BUTANOL BENZO-

ATE,(ESTER) □ DIMETHYLAMINOSTOVAINE □ 2-ETHYL-1,3-BIS(DIMETHYLAMINO)-2-PROPANOL BENZOATE

TOXICITY DATA WITH REFERENCE
scu-rat LDLo: 200 mg/kg PHREA7 12,262,32
ivn-rat LDLo: 10 mg/kg PHREA7 12,262,32
scu-mus LDLo: 260 mg/kg PHREA7 12,262,32
scu-dog LDLo: 70 mg/kg PHREA7 12,262,32
scu-cat LDLo: 60 mg/kg PHREA7 12,262,32
ivn-cat LDLo: 10 mg/kg PHREA7 12,262,32
scu-rbt LDLo: 96 mg/kg PHREA7 12,262,32
ivn-rbt LDLo: 10 mg/kg PHREA7 12,262,32
ipr-gpg LDLo: 100 mg/kg PHREA7 12,262,32
scu-gpg LDLo: 72 mg/kg PHREA7 12,262,32
ivn-gpg LDLo: 15 mg/kg PHREA7 12,262,32
scu-frg LDLo: 200 mg/kg PHREA7 12,262,32

SAFETY PROFILE: Poison by subcutaneous, intravenous, and intraperitoneal routes. An allergen. See also ESTERS. When heated to decomposition it emits toxic fumes of NO_x.

AHI500 CAS:73990-29-1 HR: 3
AMANITA RUBESCENS TOXIN

SYNS: MUSHROOM AMNITA RUBESCENS TOXIN □ RUBESCENSLYSIN

TOXICITY DATA WITH REFERENCE
ivn-rat LD50: 1400 µg/kg TOXIA6 17(Suppl. 1),165,79
ivn-mus LD50: 300 µg/kg TOXIA6 17(Suppl. 1),165,79

SAFETY PROFILE: Deadly poison by intravenous route.

AHI625 CAS:23109-05-9 HR: 3
α-AMANITINE
mf: $C_{39}H_{54}N_{10}O_{14}S$ mw: 919.09

PROP: Needles. Mp: 254–255° (decomp).

SYN: α AMANITIN (8CI, 9CI)

TOXICITY DATA WITH REFERENCE
oms-omi 700 mg/L AMICCW 122,161,79
dni-dmg:oth 20 mg/L IJEBA6 15,973,77
oms-mus:ast 2 mg/L JANTAJ 36,155,83
ipr-mus LD50: 100 µg/kg NEJMAG 269,223,63

CONSENSUS REPORTS: EPA Genetic Toxicology Program.

SAFETY PROFILE: Poison by intraperitoneal route. Mutation data reported. When heated to decomposition it emits toxic fumes of SO_x and NO_x.

AHI630 HR: 2
AMARBEL EXTRACT

PROP: Indian plant belonging to the family *Convolvulaceae* (IJEBA6 7,250,69)

SYNS: ALGUSI, extract □ CUSCUTA REFLEXA Roxb., extract excluding roots

TOXICITY DATA WITH REFERENCE
orl-rat TDLo: 5600 mg/kg (female 1-7D post): REP IJMRAQ 70,517,79
ipr-mus LD50: 750 mg/kg IJEBA6 7,250,69

SAFETY PROFILE: Moderately toxic by intraperitoneal route. Experimental reproductive effects. When heated to decomposition it emits acrid smoke and irritating fumes.

AHI635 HR: 2
AMARYLLIS

PROP: Bulb-producing, flowering plants that are cultivated as ornamentals in gardens and indoors.

SYNS: A. BELLADONNA □ AZUCENA de MEJICO (MEXICO) □ BARBADOS LILY □ BELLADONNA LILY □ CAPE BELLADONNA □ HIPPEASTRUM (VARIOUS SPECIES) □ LIRIO □ NAKED LADY LILY □ TARARACO □ TARARACO DOBLE (CUBA)

SAFETY PROFILE: The bulbs contain the emetic narcissine (lycorine). Ingestion of a large amount of the bulbs may cause nausea, vomiting, and diarrhea.

AHI750 HR: 3
AMATOL

PROP: A high explosive. Composition: NH_4NO_3, 80%; and TNT, 20%; d: 1.47.

SAFETY PROFILE: Moderately toxic by inhalation and ingestion routes. An allergen. May cause contact dermatitis. See also NITRATES. Dangerous fire hazard. An explosive by shock, spontaneous chemical reaction, or exposure to flame. Decomposition emits highly toxic fumes.

AHI875 CAS:539-21-9 HR: 3
AMBAZONE
mf: $C_8H_{11}N_7S$ mw: 237.32

PROP: Brown crystals from water. Mp: 195° (decomp). Insol in H_2O and EtOH.

SYNS: AMBAZON □ 1-AMIDINOHYDRAZONO-4-THIOSEMICARBAZONO-2,5-CYCLOHEXADIENE □ ANGINON □ p-BENZOQUINONE AMIDINOHYDRAZONE THIOSEMICARBAZONE □ BENZOQUINONE GUANYLHYDRAZONE THIOSEMICARBAZONE □ DC 0572 □ FARINGOSEPT □ GUANOTHIAZON □ INVERSAL □ IVERSAL □ IVERTOL □ ((4-OXO-2,5-CYCLOHEXADIEN-1-YLIDENE)AMINO)GUANIDINE THIOSEMICARBAZONE □ PRIMAL □ PROMASSOL

TOXICITY DATA WITH REFERENCE
mmo-sat 100 µg/plate STBIBN 117,99,87
dnr-mic-uns 10 g/L STBIBN 117,99,87
orl-rat LD50: 750 mg/kg NIIRDN 6,57,82
ipr-rat LD50: 200 mg/kg NIIRDN 6,57,82
orl-mus LD50: 1000 mg/kg NIIRDN 6,57,82
orl-gpg LD50: 80 mg/kg NIIRDN 6,57,82

SAFETY PROFILE: Poison by ingestion and intraperitoneal routes. Mutation data reported. When heated to decomposition it emits toxic fumes of SO_x and NO_x.

AHJ000 CAS:9000-02-6 HR: 1
AMBERGRIS TINCTURE

PROP: Concretion from intestine of sperm whale, composed mostly of cholesterol.

SYNS: AMBER □ AMBRA □ GRAY AMBER

TOXICITY DATA WITH REFERENCE
skn-rbt 500 mg/24H MLD FCTXAV 14,659,76

SAFETY PROFILE: A mild skin irritant. When heated to decomposition it emits acrid smoke and irritating fumes.

AHJ250 **CAS:18683-91-5** **HR: 3**
AMBROXOL
mf: $C_{13}H_{18}Br_2N_2O$ mw: 378.15

SYNS: trans-4-((2-AMINO-3,5-DIBROMOBENCIL)AMINO) CICLOHEXANOL (SPANISH) ☐ N-(2-AMINO-3,4-DIBROMOCICLOHEXIL)-trans-4-AMINOCICLOHEXANOL (SPANISH) ☐ N-(2-AMINO-3,4-DIBROMOCYCLOHEXYL)-trans-4-AMINOCYCLOHEXANOL ☐ N-(trans-4-HIDROXICICLOHEXIL)-(2-AMINO-3,5-DIBROMOBENCIL)AMINA (SPANISH) ☐ N-(trans-4-HYDROXYCYCLOHEXYL)-(2-AMINO-3,5-DIBROMOBENZYL)-AMINE ☐ NA-872

TOXICITY DATA WITH REFERENCE
orl-rat LD50:13,400 mg/kg MDACAP 15,523,79
ipr-rat LD50:380 mg/kg MDACAP 15,523,79
orl-mus LD50:2720 mg/kg MDACAP 15,523,79
ipr-mus LD50:268 mg/kg MDACAP 15,523,79
ivn-mus LD50:138 mg/kg MDACAP 15,523,79
orl-gpg LD50:1180 mg/kg MDACAP 15,523,79
ipr-gpg LD50:280 mg/kg MDACAP 15,523,79

SAFETY PROFILE: Poison by intraperitoneal and intravenous routes. Moderately toxic by ingestion. When heated to decomposition it emits very toxic fumes of Br^- and NO_x.

AHJ500 **CAS:23828-92-4** **HR: 3**
AMBROXOL HYDROCHLORIDE
mf: $C_{13}H_{18}Br_2N_2O{\cdot}ClH$ mw: 414.61

SYNS: trans-4-((2-AMINO-3,5-DIBROMOBENZYL)AMINO)CYCLOHEXANOL HYDROCHLORIDE ☐ MUCOSOLVAN ☐ NA 872

TOXICITY DATA WITH REFERENCE
orl-rat TDLo:88 mg/kg (female 7-17D post):REP OYYAA2 21,271,81
orl-rat TDLo:33 g/kg (7-17D preg):TER OYYAA2 21,271,81
orl-rat LD50:4203 mg/kg IYKEDH 12,263,81
ipr-mus LD50:268 mg/kg ARZNAD 28,889,78
scu-rat LD50:1489 mg/kg IYKEDH 12,263,81
ivn-rat LD50:100 mg/kg OYYAA2 21,281,81
orl-mus LD50:2380 mg/kg OYYAA2 21,281,81
scu-mus LD50:1060 mg/kg IYKEDH 12,263,81
ivn-mus LD50:138 mg/kg ARZNAD 28,889,78
orl-dog LD50:500 mg/kg ARZNAD 28,889,78

SAFETY PROFILE: Poison by intraperitoneal and intravenous routes. Moderately toxic by ingestion and subcutaneous routes. An experimental teratogen. Other experimental reproductive effects. When heated to decomposition it emits very toxic fumes of Br^-, NO_x, and HCl.

AHJ750 **CAS:52645-53-1** **HR: 3**
AMBUSH
mf: $C_{21}H_{20}Cl_2O_3$ mw: 391.31

PROP: Crystals or pale-yellow viscous liquid. Bp: 220° @ 0.05 mm.

SYNS: AI3-29158 ☐ BW-21-Z ☐ ECTIBAN ☐ EXMIN ☐ FMC 33297 ☐ FMC 41655 ☐ ICI-PP 557 ☐ KESTREL (Pesticide) ☐ NDRC-143 ☐ NIA 33297 ☐ OUTFLANK ☐ OUTFLANK-STOCKADE ☐ PERMETHRIN (USDA) ☐ PERMETRIN (HUNGARIAN) ☐ PERMETRINA (PORTUGUESE) ☐ 3-PHENOXYBENZYL (±)-3-(2,2-DICHLOROVINYL)-2,2-DIMETHYLCYCLOPROPANECARBOXYLATE ☐ (3-PHENOXYPHENYL)METHYL-3-(2,2-DICHLORETHENYL)-2,2-DIMETHYLCYCLOPROPANECARBOXYLATE ☐ POUNCE ☐ PP 557 ☐ S-3151 ☐ SBP-1513 ☐ TALCORD ☐ WL 43479

TOXICITY DATA WITH REFERENCE
skn-rbt 500 mg/24H MLD NTIS** AD-A047 284
cyt-mus-orl 150 mg/kg PHABDI 21,227,81
orl-rat TDLo:250 mg/kg (6-15D preg):REP BECTA6 29,84,82
orl-rat LD50:410 mg/kg NTIS** AD-A047-284
ihl-rat LC50:685 mg/m³ YKYUA6 30,1635,79
skn-rat LD50:2500 mg/kg YKYUA6 35,1315,84
scu-rat LD50:6600 mg/kg BOCKAE 41,143,76
orl-mus LD50:540 mg/kg BOCKAE 41,143,76
ihl-mus LC50:685 mg/m³ YKYUA6 30 1635,79
ipr-mus LD50:514 mg/kg TXAPA9 66,153,82
ivn-mus LD50:31 mg/kg TXAPA9 66,153,82
ice-mus LDLo:600 µg/kg TXAPA9 66,290,82

SAFETY PROFILE: Poison by inhalation, intravenous, and intracerebral routes. Moderately toxic by ingestion. Experimental reproductive effects. Mutation data reported. A skin irritant. When heated to decomposition it emits toxic fumes of Cl^-. See also ESTERS.

AHJ875 **HR: 2**
AMERICAN BITTERSWEET

PROP: A climbing, vine-like shrub with oval, toothed leaves. The fruit is an orange-yellow capsule about 0.5 inch long with seeds imbedded in a red pulp. It grows in the region bounded by North Carolina, New Mexico, Saskatchewan, and Quebec. It is commonly used in dried floral arrangements and is supplied by florists.

SYNS: BITTERSWEET ☐ BOURREAU DES ARBRES (CANADA) ☐ CELASTRUS SCANDENS ☐ CLIMBING BITTERSWEET ☐ CLIMBING ORANGE ROOT ☐ FALSE BITTERSWEET ☐ FEVER TWIG ☐ RED ROOT ☐ ROXBURY WAXWORK ☐ SHRUBBY BITTERSWEET ☐ STAFF TREE

SAFETY PROFILE: All parts of the plant and particularly the fruit contain unknown toxins which may cause vomiting and diarrhea upon ingestion.

AHK000 **HR: 3**
AMERICIUM
af: Am aw: 243

PROP: A silvery, somewhat malleable radioactive metal. Mp: 994°, bp: 2607°, d: 13.67 @ 20°.

SAFETY PROFILE: A poison. Bone-seeking, long-lived radioactive element. Flammable, see POWDERED METALS. In a disaster, this highly toxic radioactive material can be disseminated over a wide area, causing a long-lived inhalation hazard. Americium is difficult to remove from surfaces or from the body once it enters.

AHK250 **HR: 3**
AMERICIUM TRICHLORIDE
mf: $AmCl_3$ mw: 349.4

SAFETY PROFILE: See AMERICIUM. Due to its alpha particle radioactivity, it can cause radiolysis and build pressure in sealed containers and eventually explode.

AHK500 **CAS:50-07-7** **HR: 3**
AMETYCIN
mf: $C_{15}H_{18}N_4O_5$ mw: 334.37

PROP: Deep blue-violet crystals. Sol in H_2O, MeOH, Me_2CO.

SYNS: 7-AMINO-9 α-METHOXYMITOSANE □ MIT-C □ MITO-C □ MITOCIN-C □ MITOMYCIN □ MITOMYCIN-C □ MITOMYCINUM □ MMC □ MUTAMYCIN □ MUTAMYCIN (MITOMYCIN for INJECTION) □ MYTOMYCIN □ NCI-C04706 □ NSC 26980 □ RCRA WASTE NUMBER U010

TOXICITY DATA WITH REFERENCE
mma-sat 5 μg/plate CNREA8 38,2148,78
cyt-hmn:hla 10 μmol/L TXCYAC 51,181,65
cyt-hmn:fbr 100 μg/L TRBMAV 27,409,69
ivn-mus TDLo:1 mg/kg (female 13D post):TER NATUAS 279,531,79
ipr-mus TDLo:1750 μg/kg (male 1D pre):REP MUREAV 13,433,71
ipr-rat TDLo:3000 μg/kg/26W-I:NEO RRCRBU 52,1,75
ivn-rat TDLo:2600 μg/kg/8W-I:CAR ARZNAD 20,1461,70
scu-mus TDLo:280 μg/kg/18W-I:CAR APHGBP 17,495,67
ivn-wmn TDLo:1800 μg/kg:PUL LANCAO 2,1037,80
unr-wmn TDLo:2100 μg/kg/40W-I:BLD AIMDAP 143,1617,83
unr-man TDLo:1350 μg/kg/21W-I:SYS AIMDAP 143,803,83
orl-rat LD50:30 mg/kg CNREA8 20,1354,60
ipr-rat LD50:2 mg/kg ADTEAS 3,181,68
scu-rat LD50:3250 μg/kg NIIRDN 6,798,82
ivn-rat LD50:3 mg/kg ARZNAD 20,1467,70
orl-mus LD50:23 mg/kg CNREA8 20,1354,60
ipr-mus LD50:4 mg/kg JAJAAA 13,27,60
scu-mus LD50:7800 μg/kg NIIRDN 6,798,82
ivn-mus LD50:4 mg/kg JAJAAA 13,27,60
ivn-dog LD50:1 mg/kg CNREA8 20,1354,60

CONSENSUS REPORTS: IARC Cancer Review: Group 2B IMEMDT 7,56,87; Animal Sufficient Evidence IMEMDT 10,171,76; NCI Carcinogenesis Studies (ipr); Clear Evidence: rat RRCRBU 52,1,75; No Evidence: mouse RRCRBU 52,1,75. EPA Extremely Hazardous Substances List. EPA Genetic Toxicology Program. Reported in EPA TSCA Inventory.

SAFETY PROFILE: Suspected carcinogen with experimental carcinogenic and neoplastigenic data. Poison by ingestion, subcutaneous, intravenous, and intraperitoneal routes. Human systemic effects by intravenous route: dyspnea and lung fibrosis, hemolysis with or without anemia, changes in tubules (including acute renal failure, acute tubular necrosis), normocytic anemia. Experimental teratogenic and reproductive effects. Human mutation data reported. When heated to decomposition it emits toxic fumes of NO_x. See also CARBAMATES and ESTERS.

AHK625 **CAS:61941-56-8** **HR: 3**
AMFENAC SODIUM MONOHYDRATE
mf: $C_{15}H_{13}NO_3{\cdot}Na{\cdot}H_2O$ mw: 296.30

PROP: Yellow crystals. Mp: 254–255.5°.

SYNS: AHR 5850D MONOHYDRATE □ 2-AMINO-3-BENZOYLBENZENEACETIC ACID SODIUM SALT HYDRATE □ SODIUM (2-AMINO-3-BENZOYLPHENYL)ACETATE MONOHYDRATE

TOXICITY DATA WITH REFERENCE
orl-rat TDLo:5 mg/kg (17-21D preg):REP OYYAA2 20,185,85
orl-rat TDLo:1344 mg/kg (9W male/2W pre-7D preg):TER OYYAA2 20,117,85
orl-rat LD50:311 mg/kg AGACBH 7,133,77
ipr-rat LD50:240 mg/kg JTSCDR 9,87,84
scu-rat LD50:240 mg/kg IYKEDH 16,1461,85
ivn-rat LD50:277 mg/kg JTSCDR 9,87,84
ims-rat LD50:277 mg/kg IYKEDH 16,1461,85
orl-mus LD50:615 mg/kg AGACBH 7,133,77
ipr-mus LD50:540 mg/kg IYKEDH 16,1461,85
scu-mus LD50:580 mg/kg JTSCDR 9,87,84
ivn-mus LD50:550 mg/kg JTSCDR 9,87,84
ims-mus LD50:540 mg/kg JTSCDR 9,87,84

SAFETY PROFILE: Poison by subcutaneous, intramuscular, intravenous, and intraperitoneal routes. Moderately toxic by ingestion. An experimental teratogen. Other experimental reproductive effects. When heated to decomposition it emits toxic fumes of NO_x and Na_2O.

AHK750 **CAS:82-02-0** **HR: 3**
AMICARDINE
mf: $C_{14}H_{12}O_5$ mw: 260.26

PROP: Needles from MeOH (aq) or Et_2O. Mp: 150.3°.

SYNS: AMIPTAN □ AMMICARDINE □ AMMI-KHELLIN □ AMMIPURAN □ AMMISPASMIN □ AMMIVIN □ AMMIVISNAGEN □ BENECARDIN □ BI-KELLINA □ CARDIO-KHELLIN □ CHELLIN □ CHELLINA (ITALIAN) □ CHORAFURONE □ CORONIN □ DELTOSIDE □ 5,8-DIMETHOXY-2-METHYL-4′,5′-FURANO-6,7-CHROMONE □ 5,8-DIMETHOXY-2-METHYL-6,7-FURANOCHROMONE □ 4,9-DIMETHOXY-7-METHYL-5H-FURO(3,2-G)(1)BENZOPYRAN-5-ONE □ 5,8-DIMETHOXY-2-METHYL-4′,5′-FURO-6,7-CHROMONE □ 4,9-DIMETHOXY-7-METHYL-5-OXO-1,8-DIOXABENZ-(F)INDENE □ 4,9-DIMETHOXY-7-METHYL-5-OXOFURO(3,2-G)(1)BENZOPYRAN □ 4,9-DIMETHOXY-7-METHYL-5-OXOFURO(3,2-G)-1,2-CHROMENE □ ESKEL □ EUPHORIN □ GYNOKHELLAN □ INTERKELLIN □ KELICORIN □ KELINCOR □ LYNAMINE □ MEFURINA □ METHAFRONE □ NORKEL □ SIMESKELLINA □ VASOKELLINA □ VISNAGALIN

TOXICITY DATA WITH REFERENCE
scu-rat TDLo:18 g/kg (female 9-14D post):REP TOIZAG 18,81,71
scu-mus TDLo:60 g/kg (female 7-12D post):TER TOIZAG 18,81,71
orl-rat LD50:68,800 μg/kg ARZNAD 11,915,61
ipr-rat LD50:70 mg/kg JDGRAX 9(1-2),33,77
ivn-rat LD50:34 mg/kg ARZNAD 11,848,61
orl-mus LD50:50,800 μg/kg ARZNAD 11,915,61

ipr-mus LD50:155 mg/kg FRPSAX 13,561,58
ivn-mus LD50:30,600 µg/kg ARZNAD 11,915,61
ims-mus LD50:83 mg/kg JDGRAX 7(2),1,75

CONSENSUS REPORTS: EPA Genetic Toxicology Program.

SAFETY PROFILE: Poison by ingestion, intraperitoneal, intramuscular, and intravenous routes. An experimental teratogen. Other experimental reproductive effects. When heated to decomposition it emits acrid smoke and irritating fumes. A vasodilator.

AHL000 CAS:17650-86-1 HR: 3
AMICETIN
mf: $C_{29}H_{42}N_6O_9$ mw: 618.77

PROP: Needles from water. Mp: 165–169° (from cold H_2O). Mp: 244–245° (from hot H_2O).

SYNS: ALLOMYCIN □ D-13 □ NSC 5340 □ SACROMYCIN □ U-4761

TOXICITY DATA WITH REFERENCE
mic-pro-eug 1500 mg/L NEOLA4 19,579,72
dni-pro-eug 1500 mg/L NEOLA4 19,579,72
orl-rat LD50:3600 mg/kg UPJOH• 2(6),-,71
orl-mus LD50:2 g/kg 85GDA2 5,214,81
ipr-mus LD50:530 mg/kg 85GDA2 5,214,81
scu-mus LD50:57 mg/kg 85FZAT-,120,67
ivn-mus LD50:6800 µg/kg 85FZAT-,120,67
ims-mus LDLo:300 mg/kg JAJAAA 8,148,55

SAFETY PROFILE: Poison by subcutaneous, intravenous, and intraperitoneal routes. Moderately toxic by ingestion. Mutation data reported. When heated to decomposition it emits toxic fumes of NO_x.

AHL250 HR: 3
AMICETIN CITRATE

TOXICITY DATA WITH REFERENCE
scu-rat LD50:600 mg/kg 85ERAY 1,155,78
ivn-rat LD50:200 mg/kg 85ERAY 1,155,78
scu-mus LD50:600 mg/kg 85ERAY 1,155,78
ivn-mus LD50:90 mg/kg 85ERAY 1,155,78

SAFETY PROFILE: Poison by intravenous route. Moderately toxic by subcutaneous route. When heated to decomposition it emits acrid smoke and irritating fumes.

AHL500 CAS:1421-68-7 HR: 3
AMIDEFRINE MESYLATE
mf: $C_{10}H_{16}N_2O_3S{\bullet}CH_4O_3S$ mw: 340.45

PROP: Crystals from alc. Mp: 207–209°.

SYNS: AMIDEPHRINE MESYLATE □ AMIDEPHRINE MONOMETHANESULFONATE □ DRICOL □ FENTRINOL □ 3'-(1-HYDROXY-2-(METHYLAMINO)ETHYL)METHANESULFONANILIDE METHANESULFONATE □ 3'-(1-HYDROXY-2-(METHYLAMINO)ETHYL)METHANESULFONANILIDE MONOMETHANESULFONATE SALT □ N-(3-(1-HYDROXY-2-(METHYLAMINO)ETHYL)PHENYL)-METHANESULFONAMIDE MONOMETHANESULFONATE SALT □ (2-METHYLAMINE-1-HYDROXYETHYL)METHANESULFONANILIDE METHANESULFONATE □ 3'-(2-(METHYLAMINO)-1-HYDROXYETHYL)METHANESULFONANILIDE METHANESULFONATE □ MJ 5190 □ NALDE □ PRODUCT 5190

TOXICITY DATA WITH REFERENCE
eye-rbt 10 mg MLD TXAPA9 23,589,72
orl-rat LD50:13 mg/kg TXAPA9 23,589,72
ipr-rat LD50:5 mg/kg TXAPA9 23,589,72
orl-mus LD50:2284 mg/kg TXAPA9 23,589,72
ipr-mus LD50:780 mg/kg TXAPA9 23,589,72
scu-mus LD50:1990 mg/kg IJNEAQ 4,219,65
ivn-mus LD50:190 mg/kg IJNEAQ 4,219,65
ipr-dog LD50:4800 µg/kg IJNEAQ 4,219,65
ivn-dog LD50:1400 µg/kg TXAPA9 23,589,72
orl-cat LDLo:2 mg/kg TXAPA9 23,589,72
orl-rbt LD50:12 mg/kg TXAPA9 23,589,72
idr-rbt LD50:7500 µg/kg TXAPA9 23,589,72

SAFETY PROFILE: Poison by ingestion, intraperitoneal, intravenous, and intradermal routes. Moderately toxic by subcutaneous route. An eye irritant. When heated to decomposition it emits very toxic fumes of NO_x and SO_x. A vasoconstrictor and nasal decongestant. See also SULFONATES.

AHL750 HR: D
AMIDES

PROP: Organic compounds containing the structural group $-CONH_2$, and closely related to the organic acids with the grouping $-COOH$. Common examples are: acetamide (CH_3CONH_2) and urea ($CO(NH_2)_2$).

SAFETY PROFILE: Most of the saturated amides have low toxicity, but the unsaturated and N-substituted amides are irritants and may be absorbed via skin contact. Can cause injury to the liver, kidney, and brain.

AHL875 HR: 3
p-AMIDINOBENZOIC ACID BUTYL ESTER
mf: $C_{12}H_{16}N_2O_2$ mw: 220.30

TOXICITY DATA WITH REFERENCE
skn-rbt 1% SEV JAPMA8 41,202,52
skn-rbt 1000 ppm MOD JAPMA8 41,202,52
scu-mus LD50:350 mg/kg JAPMA8 41,202,52
ivn-mus LD50:70 mg/kg JAPMA8 41,202,52

SAFETY PROFILE: Poison by subcutaneous and intravenous routes. A severe skin irritant. When heated to decomposition it emits toxic fumes of NO_x. See also ESTERS.

AHL880 HR: 3
p-AMIDINOBENZOIC ACID HEXYL ESTER
mf: $C_{14}H_{20}N_2O_2$ mw: 248.36

TOXICITY DATA WITH REFERENCE
skn-rbt 1% SEV JAPMA8 41,202,52
skn-rbt 1000 ppm MOD JAPMA8 41,202,52
scu-mus LD50:550 mg/kg JAPMA8 41,202,52
ivn-mus LD50:150 mg/kg JAPMA8 41,202,52

SAFETY PROFILE: Poison by intravenous route. Moderately toxic by subcutaneous route. A severe skin irritant. When heated to decomposition it emits toxic fumes of NO_x. See also ESTERS.

AHL885 **HR: 3**
p-AMIDINOBENZOIC ACID PENTYL ESTER
mf: $C_{13}H_{18}N_2O_2$ mw: 234.33

TOXICITY DATA WITH REFERENCE
skn-rbt 1% SEV JAPMA8 41,202,52
scu-mus LD50:450 mg/kg JAPMA8 41,202,52
ivn-mus LD50:150 mg/kg JAPMA8 41,202,52

SAFETY PROFILE: Poison by intravenous route. Moderately toxic by subcutaneous route. A severe skin irritant. When heated to decomposition it emits toxic fumes of NO_x. See also ESTERS.

AHL890 **HR: 3**
p-AMIDINOBENZOIC ACID PROPYL ESTER
mf: $C_{11}H_{14}N_2O_2$ mw: 206.27

TOXICITY DATA WITH REFERENCE
skn-rbt 1% SEV JAPMA8 41,202,52
scu-mus LD50:550 mg/kg JAPMA8 41,202,52
ivn-mus LD50:65 mg/kg JAPMA8 41,202,52

SAFETY PROFILE: Poison by intravenous route. Moderately toxic by subcutaneous route. A severe skin irritant. See also ESTERS.

AHN000 **CAS:10319-70-7** **HR: 3**
S-(AMIDINOMETHYL) HYDROGEN THIOSULFATE
mf: $C_2H_6N_2O_3S_2$ mw: 170.22

TOXICITY DATA WITH REFERENCE
orl-mus LD50:300 mg/kg JMCMAR 15,1313,72
ipr-mus LD50:87 mg/kg JMCMAR 15,1313,72

SAFETY PROFILE: Poison by ingestion and intraperitoneal routes. See also THIOSULFATES. When heated to decomposition it emits very toxic fumes of NO_x and SO_x.

AHN625 **CAS:53142-01-1** **HR: 3**
AMIDINOMYCIN
mf: $C_9H_{18}N_4O$ mw: 198.31

SYNS: N-(2-AMIDINOETHYL)-3-AMINOCYCLOPENTANOCARBOXAMIDE □ 3-AMINO-N-(3-AMINO-3-IMINOPROPYL)CYCLOPENTANE CARBOXAMIDE □ MYXOVIROMYCIN

TOXICITY DATA WITH REFERENCE
orl-mus LD50:140 mg/kg 85FZAT -,127,67
scu-mus LD50:20,400 μg/kg 85FZAT -,444,67
ivn-mus LD50:18 mg/kg 85ERAY 2,1214,78

SAFETY PROFILE: Poison by ingestion, subcutaneous, and intravenous routes. When heated to decomposition it emits toxic fumes of NO_x.

AHO250 **CAS:57-67-0** **HR: 2**
N[1]-AMIDINOSULFANILAMIDE
mf: $C_7H_{10}N_4O_2S$ mw: 214.27

PROP: Colorless, monoclinic crystals or needles from aqueous solns. Mp: 189–190°. Sltly sol in alc; insol in ether; solubility in water = 0.19% @ 37°.

SYNS: ABIGUANIL □ 4-AMINO-N-(AMINOIMINOMETHYL)BENZENESULFONAMIDE □ p-AMINOBENZENESULFONYLGUANIDINE □ N-p-AMINOBENZENESULPHONYLGUANIDINE MONOHYDRATE □ ATERIAN □ 4-AMINO-N-(DIAMINOMETHYLENE)BENZENESULFONAMIDE □ BENZENESULFONAMIDE, 4-AMINO-N-(DIAMINOMETHYLENE)- □ N_1-(DIAMINOMETHYLENE)SULFANILAMIDE □ GANIDAN □ GUAMIDE □ GUANICIL □ GUANIDAN □ GUANIDINE, SULFANILYL- □ N (sup 1)-GUANYLSULFANILAMIDE □ RESULFON □ RP 2275 □ RUOCID □ SHIGATOX □ SUGANYL □ SULFAGUANIDINE □ SULFAGUINE □ SULFANILGUANIDINE □ SULFANILYLGUANIDINE □ SULFOGUANIDINE □ SULFOGUENIL □ SULFOQUANIDINE □ SULGIN □ SULPHAGUANIDINE

TOXICITY DATA WITH REFERENCE
orl-rat TDLo:22,500 mg/kg (female 1-15D post):TER AFPEAM 7,180,50
orl-rat TDLo:22,500 mg/kg (1-15D preg):REP AFPEAM 7,180,50
ipr-mus LDLo:500 mg/kg JHHBAI 67,163,40

CONSENSUS REPORTS: Reported in EPA TSCA Inventory.

SAFETY PROFILE: Moderately toxic by intraperitoneal route. An experimental teratogen. Other experimental reproductive effects. See also SULFONATES. When heated to decomposition it emits very toxic fumes of SO_x and NO_x.

AHO750 **CAS:919-76-6** **HR: 3**
AMIDITHION
mf: $C_7H_{16}NO_4PS_2$ mw: 273.33

PROP: Solid.

SYNS: 2-MERCAPTO-N-(2-METHYOXYETHYL)-ACETAMIDE S-ESTER with O,O-DIMETHYL PHOSPHORODITHIOATE □ C 2446 □ CIBA 2446 □ CIBA THIOCRON □ O,O-DIMETHYL-S-(2-METHOXYETHYLCARBAMOYLMETHYL)DITHIOPHOSPHATE □ O,O-DIMETHYL-S-(2-METHOXYETHYLCARBAMOYL METHYL)PHOSPHORODITHIOATE □ ENT 27,160 □ S-(2-((2-METHOXYETHYL)AMINO-2-OXOETHYL) O,O-DIMETHYL) PHOSPHORODITHIOATE □ S-(N-2-METHOXYETHYLCARBAMOYLMETHYL)DIMETHYL PHOPHOROTHIOLOTHIONATE □ THIOCRON

TOXICITY DATA WITH REFERENCE
orl-rat LD50:600 mg/kg WRPCA2 9,119,70
skn-rat LD50:1600 mg/kg 28ZEAL 5,9,76
orl-ckn LD50:94 mg/kg TXAPA9 11,49,67

SAFETY PROFILE: Poison by ingestion. Moderately toxic by skin contact. When heated to decomposition it emits very toxic fumes of SO_x, PO_x, and NO_x. See also ESTERS.

AHP000 **CAS:137-09-7** **HR: 3**
AMIDOL
mf: $C_6H_8N_2O•2ClH$ mw: 197.08

PROP: Grayish-white crystals.

SYNS: ACROL □ 2,4-DIAMINOPHENOL HYDROCHLORIDE □ DIANOL □ NCI-C60026

TOXICITY DATA WITH REFERENCE
mma-sat 330 ng/plate ENMUDM 5(Suppl 1),3,83
ipr-mus LDLo:50 mg/kg RBPMAZ 22,1,52

CONSENSUS REPORTS: Reported in EPA TSCA Inventory. EPA Genetic Toxicology Program. NTP Carcinogenesis Studies (gavage): Some Evidence:

Mouse NTPTR* NTP-TR-401,92; (gavage): No Evidence: Rat NTPTR* NTP-TR-401,92.

SAFETY PROFILE: Poison by intraperitoneal route. Mutation data reported. A mild irritant and allergen. When heated to decomposition it emits toxic fumes of NO_x and HCl.

AHP125 **CAS:21590-92-1** **HR: 3**
AMIDOLINE
mf: $C_{23}H_{29}N_3O_2$ mw: 379.55

PROP: Crystals from ligroin. Mp: 106–107°.

SYNS: 2-ETHYL-2,3-DIHYDRO-3-((4-(2-(1-PIPERIDINYL)ETHOXY) PHENYL)AMINO)-1H-ISOINDOL-1-ONE ☐ 2-ETHYL-3-(β-PIPERIDINO-p-PHENETIDINO)PHTHALIMIDINE ☐ ETOMIDOLINE ☐ K 2680 ☐ SMEDOLIN

TOXICITY DATA WITH REFERENCE
ivn-mus TDLo: 7 mg/kg (female 7-13D post): TER KSRNAM 8,1812,74
orl-rat TDLo: 280 mg/kg (8-14D preg): REP KSRNAM 8,1802,74
orl-rat LD50: 695 mg/kg KSRNAM 8,1730,74
scu-rat LD50: 94 mg/kg KSRNAM 8,1730,74
ivn-rat LD50: 42 mg/kg KSRNAM 8,1730,74
orl-mus LD50: 168 mg/kg KSRNAM 8,1730,74
scu-mus LD50: 109 mg/kg KSRNAM 8,1730,74
ivn-mus LD50: 36 mg/kg KSRNAM 8,1730,74

SAFETY PROFILE: Poison by ingestion, subcutaneous, and intravenous routes. An experimental teratogen. Other experimental reproductive effects. When heated to decomposition it emits toxic fumes of NO_x.

AHP250 **HR: 3**
AMIDO SULFURYL AZIDE
mf: $H_2N_4O_2S$ mw: 122.2

SAFETY PROFILE: A dangerously unstable, explosive compound. See also AZIDES and SULFATES. When heated to decomposition it emits highly toxic fumes of NO_x and SO_x. Shock sensitive.

AHP375 **CAS:40709-23-7** **HR: 3**
AMIKHELLIN HYDROCHLORIDE
mf: $C_{18}H_{21}NO_5•ClH$ mw: 367.86

SYNS: CHLORHYDRATE d'AMIKHELLINE (FRENCH) ☐ 2-METHYL-5-HYDROXY-8-(β-DIAETHYLAMINO-AETHOXY)FURANO-6,7:2'3'-CHROMON HYDROCHLORIDE ☐ NOKHEL

TOXICITY DATA WITH REFERENCE
dnd-omi 110 μmol/L BICMBE 55,1415,73
dnd-omi 10 mmol/L BICMBE 55,1415,73
cyt-nml:oth 270 μmol/L/6H-C AAMMAU 61,210,72
dnd-mam:lym 100 μmol/L BICMBE 55,1415,73
scu-mus LD50: 14,380 μg/kg ARZNAD 13,140,63

SAFETY PROFILE: Poison by subcutaneous route. Mutation data reported. When heated to decomposition it emits toxic fumes of NO_x and HCl.

AHP500 **CAS:95-38-5** **HR: 2**
AMINE 220
mf: $C_{22}H_{42}N_2O$ mw: 350.66

PROP: Liquid. Bp: 235° @ 1 mm, flash p: 465°F (OC), d: 0.9300 @ 20°/20°, vap d: 12.1.

SYNS: 2-(8-HEPTADECENYL)-2-IMIDAZOLINE-1-ETHANOL ☐ 1-HYDROXYETHYL-2-HEPTADECENYLGLYOXALIDINE ☐ 1-(2-HYDROXYETHYL)-2-HEPTADECENYLGLYOXALIDINE ☐ 1-(2-HYDROXYETHYL)-2-N-HEPTADECENYL-2-IMIDAZOLINE ☐ 1-(2-HYDROXYETHYL)-2-HEPTADECENYL-2-IMIDAZOLINE ☐ NALCAMINE G-13

TOXICITY DATA WITH REFERENCE
ivn-mus LD50: 88 mg/kg JAPMA8 38,428,49

CONSENSUS REPORTS: Reported in EPA TSCA Inventory.

SAFETY PROFILE: Poison by intravenous route.. Combustible; can react with oxidizing materials. To fight fire, use foam, CO_2, dry chemical. When heated to decomposition it emits toxic fumes of NO_x.

AHP760 **HR: 1**
AMINES, FATTY

PROP: A normal aliphatic amine derived from fats and oils. May be saturated or unsaturated, primary, secondary or tertiary, but the alkyl groups are straight-chain and have an even number of carbons in each. The length varies from 8 to 22 carbon atoms.

SAFETY PROFILE: Generally of mild toxicity. Used as organic bases, soaps, plasticizers, tire cords, fabric softeners, water-resistant asphalt, hair conditioners, cosmetics, and medicinals.

AHQ000 **CAS:102-28-3** **HR: 3**
3'-AMINOACETANILIDE
mf: $C_8H_{10}N_2O$ mw: 150.20

PROP: Mp: 86.5–87.5°, bp: decomp @ 787°. Sol in water, acetone, alc, and ether; sltly sol in benzene; insol in ligroin.

SYNS: m-ACETAMINOANILINE ☐ m-(ACETYLAMINO)ANILINE ☐ 3-ACETYLAMINOANILINE ☐ N-ACETYL-m-FENYLENEDIAMIN (CZECH) ☐ N-ACETYL-m-PHENYLENEDIAMINE ☐ 3-AMINOACETANILID (CZECH) ☐ m-AMINOACETANILIDE

TOXICITY DATA WITH REFERENCE
eye-rbt 500 mg/24H SEV 28ZPAK -,129,72
mmo-sat 333 μg/plate EMMUEG 11(Suppl 12),1,88
ivn-mus LD50: 320 mg/kg CSLNX* NX#02899

CONSENSUS REPORTS: Reported in EPA TSCA Inventory.

SAFETY PROFILE: Poison by intravenous route. Moderately toxic by ingestion. A severe eye irritant. Mutation data reported. When heated to decomposition it emits toxic fumes of NO_x.

AHQ250 **CAS:122-80-5** **HR: 2**
4′-AMINOACETANILIDE
mf: $C_8H_{10}N_2O$ mw: 150.20

PROP: Needles from water. Mp: 164°.

SYNS: p-ACETAMIDOANILINE □ 4-ACETAMIDOANILINE □ p-ACETOAMINOANILINE □ p-(ACETYLAMINO)ANILINE □ 4-(ACETYLAMINO)ANILINE □ N-ACETYL p-FENYLENDIAMIN (CZECH) □ ACETYL-p-PHENYLENEDIAMINE □ 4′-AMINOACETANILID (CZECH) □ p-AMINOACETANILIDE □ 4-AMINOACETANILIDE □ N-(p-AMINOPHENYL) ACETAMIDE □ C.I. 76005 □ C.I. OXIDATION BASE 19 □ FOURRINE 88 □ FOURRINE A

TOXICITY DATA WITH REFERENCE
eye-rbt 100 mg/24H SEV 28ZPAK -,130,72
mmo-sat 100 μg/plate EMMUEG 11(Suppl 12),1,88

CONSENSUS REPORTS: Reported in EPA TSCA Inventory.

SAFETY PROFILE: A severe eye irritant. Mutation data reported. When heated to decomposition it emits toxic fumes of NO_x. See also AMINES and AMIDES.

AHQ300 **CAS:88-64-2** **HR: 3**
3-AMINOACETANILIDE-4-SULFONIC ACID
mf: $C_8H_{10}N_2O_4S$ mw: 230.26

SYNS: ACETANILIDE-4-SULFONIC ACID, 3-AMINO □ BENZENESULFONIC ACID, 4-ACETAMIDO-2-AMINO-

TOXICITY DATA WITH REFERENCE
ivn-mus LD50:320 mg/kg CSLNX* NX#00379

CONSENSUS REPORTS: Reported in EPA TSCA Inventory.

SAFETY PROFILE: Poison by intravenous route. When heated to decomposition it emits toxic vapors of NO_x and SO_x.

AHR000 **CAS:5466-22-8** **HR: 3**
AMINOACETONITRILE SULFATE
mf: $C_4H_8N_4 \cdot H_2O_4S$ mw: 210.24

TOXICITY DATA WITH REFERENCE
scu-rbt TDLo:500 mg/kg (female 12-31D post):TER AROPAW 69,602,63
orl-rat LDLo:100 mg/kg TJADAB 5,33,72

CONSENSUS REPORTS: Cyanide and its compounds are on the Community Right-To-Know List.

SAFETY PROFILE: Poison by ingestion. An experimental teratogen. Other experimental reproductive effects. See also NITRILES and SULFATES. When heated to decomposition it emits very toxic fumes of NO_x, SO_x, and CN^-.

AHR240 **CAS:99-92-3** **HR: 3**
p-AMINOACETOPHENONE
mf: C_8H_9NO mw: 135.18

PROP: Crystalline. Mp: 106°, bp: 293–295°. Sol in C_6H_6, alc, hot acids, and ether. Insol in water.

SYNS: 4-ACETYLANILINE □ 4′-AMINOACETOPHENONE □ p-AMINOACETYLBENZENE □ USAF EK-631

TOXICITY DATA WITH REFERENCE
ipr-mus LD50:381 mg/kg GEPHDP 14,465,83 NTIS** AD277-689
ipr-rat LD50:260 mg/kg JPETAB 80,31,44
orl-mus LD50:596 mg/kg GEPHDP 14,465,83
ipr-mus LD50:300 mg/kg NTIS** AD277-689
orl-bwd LD50:133 mg/kg AECTCV 12,355,83

CONSENSUS REPORTS: Reported in EPA TSCA Inventory.

SAFETY PROFILE: Poison by ingestion and intraperitoneal routes. When heated to decomposition it emits toxic fumes of NO_x. See also AROMATIC AMINES.

AHR250 **CAS:613-89-8** **HR: 2**
2-AMINOACETOPHENONE
mf: C_8H_9NO mw: 135.18

PROP: Yellow, oily liquid. Bp: 251° (slt decomp); insol in water; sol in alc and ether.

SYNS: ω-AMINOACETOPHENONE □ PHENACYLAMINE

TOXICITY DATA WITH REFERENCE
scu-mus TDLo:2000 mg/kg/7W-I:CAR PGPKA8 14,12,69
scu-gpg TDLo:1480 mg/kg/13W-I:NEO,REP BEXBAN 84,1156,77
scu-ham TDLo:1600 mg/kg/10W-I:ETA VOONAW 25(6),81,79

SAFETY PROFILE: Experimental reproductive effects. Questionable carcinogen with experimental carcinogenic, neoplastigenic, and tumorigenic data. When heated to decomposition it emits toxic fumes of NO_x. See also AROMATIC AMINES.

AHR500 **CAS:99-03-6** **HR: 2**
3′-AMINOACETOPHENONE
mf: C_8H_9NO mw: 135.18

PROP: Yellow, oily liquid or pale-yellow plates from alc. Bp: 251° (slt decomp), mp: 97–99°. Insol in water; sol in alc and ether.

SYNS: m-ACETYLANILINE □ 3-ACETYLANILINE □ β-AMINOACETOPHENONE □ m-AMINOACETOPHENONE □ m-AMINOACETYLBENZENE

TOXICITY DATA WITH REFERENCE
skn-rbt 500 mg/24H MLD 85JCAE-,730,86
eye-rbt 500 mg open AMIHBC 10,61,54
mma-sat 10 mg/L ENMUDM 5,803,83
cyt-mus-scu 400 mg/kg TGANAK 2,538,68
orl-rat LD50:1870 mg/kg AMIHBC 10,61,54
skn-rbt LD50:4340 mg/kg AMIHBC 10,61,54

CONSENSUS REPORTS: Reported in EPA TSCA Inventory.

SAFETY PROFILE: Moderately toxic by ingestion. Mildly toxic by skin contact. A skin and eye irritant. Mutation data reported. When heated to decomposition it emits toxic fumes of NO_x. See also AROMATIC AMINES.

AHS000 CAS:581-28-2 HR: 3
2-AMINOACRIDINE
mf: $C_{13}H_{10}N_2$ mw: 194.25

PROP: Yellow solvated crystals from alc. Mp: 213–214°. Sparingly sol in Me_2CO, C_6H_6.

SYNS: 2-ACRIDINAMINE □ 3-AMINOACRIDINE (EUROPEAN)

TOXICITY DATA WITH REFERENCE
mma-sat 600 ng/plate TXCYAC 34,247,85
mmo-omi 16 mg/L JMOBAK 3,762,61
dnd-mam:lym 15 μmol/L JMOBAK 13,138,65
scu-mus LD50:330 mg/kg BJEPA5 28,1,47

SAFETY PROFILE: Poison by subcutaneous route. Mutation data reported. See also AMINES. When heated to decomposition it emits toxic fumes of NO_x.

AHS500 CAS:90-45-9 HR: 3
9-AMINOACRIDINE
mf: $C_{13}H_{10}N_2$ mw: 194.25

PROP: Yellow crystals from EtOH or Me_2CO. Mp: 233° (anhyd).

SYNS: 9AA □ 9-ACRIDINAMINE □ AMINACRINE □ 5-AMINOACRIDINE □ IZOACRIDINA □ MONACRIN

TOXICITY DATA WITH REFERENCE
mmo-sat 3 μg/plate KSRNAM 19,3212,85
slt-dmg-unr 5000 ppm/6H MUREAV 120,233,83
dns-ham:lvr 1 μmol/L ENMUDM 6,1,84
ipr-mus LD50:68 mg/kg IJOCAP 26,318,87
scu-mus LD50:80 mg/kg BJEPA5 28,1,47

CONSENSUS REPORTS: EPA Genetic Toxicology Program.

SAFETY PROFILE: Poison by intraperitoneal and subcutaneous routes. Mutation data reported. When heated to decomposition it emits toxic fumes of NO_x. See also AMINES.

AHS750 CAS:134-50-9 HR: 3
AMINOACRIDINE HYDROCHLORIDE
mf: $C_{13}H_{10}N_2 \cdot ClH$ mw: 230.71

PROP: Crystals.

SYNS: ACRAMINE YELLOW □ 9-ACRIDINAMINE MONOHYDROCHLORIDE □ AMINACRINE HYDROCHLORIDE □ 5-AMINOACRIDINE HYDROCHLORIDE □ 9-AMINOACRIDINE MONOHYDROCHLORIDE □ MONACRIN □ MONACRIN HYDROCHLORIDE □ NSC-7571

TOXICITY DATA WITH REFERENCE
mmo-sat 10 μg/plate KSRNAM 16,6240,82
mma-sat 50 μg/plate KSRNAM 17,70,83
dnr-esc 20 μL/disc MUREAV 97,1,82
dnr-bcs 20 μL/disc MUREAV 97,1,82
cyt-ham:lng 600 μg/L GMCRDC 27,95,81
dnd-man:lym 10 pph BIPMAA 11,2537,72
orl-mus LD50:78 mg/kg 29ZVAB -,7,69

CONSENSUS REPORTS: Reported in EPA TSCA Inventory. EPA Genetic Toxicology Program.

SAFETY PROFILE: Poison by ingestion. Mutation data reported. See also AMINES. When heated to decomposition it emits very toxic fumes of HCl and NO_x.

AHT000 CAS:60566-40-7 HR: 3
9-AMINOACRIDINE PENICILLIN

SYNS: 9AAP □ PENICILLIN, compounded with 9-AMINOACRIDINE

TOXICITY DATA WITH REFERENCE
orl-mus LD50:227 mg/kg JAPMA8 38,498,49
scu-mus LD50:562 mg/kg JAPMA8 38,498,49

SAFETY PROFILE: Poison by ingestion. Moderately toxic by subcutaneous route. See also PENICILLIN. When heated to decomposition it emits toxic fumes of NO_x.

AHT825 HR: 3
4′-(2-AMINO-9-ACRIDINYLAMINO)METHANESULFONANILIDE
mf: $C_{20}H_{18}N_4O_2S$ mw: 378.48

TOXICITY DATA WITH REFERENCE
mmo-sat 166 μmol/L JMCMAR 23,269,80
ipr-mus LD10:35 mg/kg JMCMAR 23,269,80

SAFETY PROFILE: Poison by intraperitoneal route. Mutation data reported. See also SULFONATES. When heated to decomposition it emits very toxic fumes of NO_x and SO_x.

AHT850 CAS:53222-25-6 HR: 3
6-AMINO-4-((3-AMINO-4-(((4-((1-METHYLPYRIDINIUM-4-YL)AMINO)PHENYL)AMINO)CARBONYL)PHENYL)AMINO)-1-METHYLQUINOLINIUM),DIIODIDE
mf: $C_{29}H_{29}N_6O \cdot 2I$ mw: 731.44

TOXICITY DATA WITH REFERENCE
dnd-mus:lym 490 nmol/L JMCMAR 22,134,79
ipr-mus LD10:20 mg/kg JMCMAR 22,134,79

SAFETY PROFILE: Poison by intraperitoneal route. Mutagenic data reported. When heated to decomposition it emits very toxic fumes of NO_x and I^-.

AHT900 CAS:63991-48-0 HR: 2
6-AMINO-2-(3′-AMINOPHENYL)BENZIMIDAZOLE, DIHYDROCHLORIDE
mf: $C_{13}H_{12}N_4 \cdot 2ClH$ mw: 297.21

SYN: 6-AMINO-2-(3′-AMINOFENYL)BENZIMIDAZOL HYDROCHLORID (CZECH)

TOXICITY DATA WITH REFERENCE
skn-rbt 500 mg/24H MOD 28ZPAK -,145,72
eye-rbt 20 mg/24H SEV 28ZPAK -,145,72
orl-rat LD50:2330 mg/kg 28ZPAK -,145,72

SAFETY PROFILE: Moderately toxic by ingestion. A skin and severe eye irritant. When heated to decomposition it emits toxic fumes of HCl and NO_x. See also AROMATIC AMINES.

AHT950 CAS:63991-49-1 HR: 2
6-AMINO-2-(4'-AMINOPHENYL)BENZIMIDAZOLE DIHYDROCHLORIDE
mf: $C_{13}H_{12}N_4 \cdot 2ClH$ mw: 297.21

SYN: 6-AMINO-2-(4'-AMINOFENYL)BENZIMIDAZOL HYDROCHLORID (CZECH)

TOXICITY DATA WITH REFERENCE
skn-rbt 500 mg/24H MOD 28ZPAK -,145,72
eye-rbt 20 mg/24H SEV 28ZPAK -,145,72
orl-rat LD50:2060 mg/kg 28ZPAK -,145,72

SAFETY PROFILE: Moderately toxic by ingestion. Moderate skin and severe eye irritant. When heated to decomposition it emits very toxic fumes of HCl and NO_x. See also AROMATIC AMINES.

AIA250 CAS:2840-26-8 HR: 2
3-AMINO-p-ANISIC ACID
mf: $C_8H_9NO_3$ mw: 167.18

PROP: Needles from water. Mp: 204°.

SYNS: 3-AMINO-4-METHOXYBENZOIC ACID □ KYSELINA-3-AMINO-4-METHOXYBENZOOVA (CZECH)

TOXICITY DATA WITH REFERENCE
eye-rbt 500 mg/24H SEV 28ZPAK -,115,72
orl-rat LD50:8290 mg/kg 28ZPAK -,115,72

SAFETY PROFILE: Mildly toxic by ingestion. Severe eye irritant. When heated to decomposition it emits toxic fumes of NO_x.

AIA500 CAS:13244-33-2 HR: 2
4-AMINOANISOLE-3-SULFONIC ACID
mf: $C_7H_9NO_4S$ mw: 203.23

SYNS: 2-AMINO-5-METHOXY BENZENESULFONIC ACID □ KYSELINA-4-AMINOANISOL-3-SULFONOVA □ 4-METHOXY-2-SULFOANILINE

TOXICITY DATA WITH REFERENCE
skn-rbt 500 mg/24H MLD 28ZPAK -,184,72
eye-rbt 500 mg/24H SEV 28ZPAK -,184,72
orl-rat LD50:10 g/kg 28ZPAK -,184,72

CONSENSUS REPORTS: Reported in EPA TSCA Inventory.

SAFETY PROFILE: Mildly toxic by ingestion. A mild skin and severe eye irritant. See also SULFONATES and AMINES. When heated to decomposition it emits very toxic fumes of SO_x and NO_x.

AIA750 CAS:82-45-1 HR: 2
1-AMINOANTHRAQUINONE
mf: $C_{14}H_9NO_2$ mw: 223.24

PROP: Orange to red needles. Mp: 256°. Bp: sublimes. Insol in water; sol in HCl, alc, benzene, ether, and chloroform.

SYNS: 1-AMINO-9,10-ANTHRACENEDIONE □ 1-AMINOANTHRACHINON (CZECH) □ α-AMINOANTHRAQUINONE □ 1-AMINO-9,10-ANTHRAQUINONE □ α-ANTHRAQUINONYLAMINE □ C.I. 37275 □ DIAZO FAST RED AL

TOXICITY DATA WITH REFERENCE
eye-rbt 100 mg/24H MOD 28ZPAK -,121,72
dnd-mus-ipr 250 mg/kg ATSUDG (5),355,82
orl-rat TDLo:2400 mg/kg/60W-I:ETA TXAPA9 8,346,66
ipr-rat LD50:1500 mg/kg GTPZAB 21(12),27,77
ipr-mus LD50:6026 mg/kg GTPZAB 9(3),20,65

CONSENSUS REPORTS: Reported in EPA TSCA Inventory.

SAFETY PROFILE: Moderately toxic by intraperitoneal route. An eye irritant. Questionable carcinogen with experimental tumorigenic data. Mutation data reported. When heated to decomposition it emits toxic NO_x. See also AMINES.

AIB000 CAS:117-79-3 HR: 3
2-AMINOANTHRAQUINONE
mf: $C_{14}H_9NO_2$ mw: 223.24

PROP: Red needles from alc. Mp: 302–306°. Bp: sublimes. Insol in water and ether; sol in alc and benzene.

SYNS: 2-AMINO-9,10-ANTHRACENEDIONE □ 2-AMINO-9,10-ANTHRAQUINONE □ β-AMINOANTHRAQUINONE □ β-ANTHRAQUINONYLAMINE □ NCI-C01876

TOXICITY DATA WITH REFERENCE
mmo-sat 1 mg/plate ENMUDM 7(Suppl 5),1,85
mma-sat 1 μg/plate ENMUDM 7(Suppl 5),1,85
mma-esc 1 mg/plate ENMUDM 7(Suppl 5),1,85
orl-rat TDLo:115 g/kg/78W-C:CAR NCITR* NCI-CG-TR-144,78
orl-mus TDLo:655 g/kg/78W-C:CAR NCITR* NCI-CG-TR-144,78
orl-rat TD:225 g/kg/78W-C:NEO NCITR* NCI-CG-TR-144,78
orl-mus TD:330 g/kg/78W-C:ETA NCITR* NCI-CG-TR-144,78
ipr-rat LD50:1500 mg/kg GTPZAB 21(12),27,77

CONSENSUS REPORTS: NTP 7th Annual Report on Carcinogens. IARC Cancer Review: Group 3 IMEMDT 7,56,87; Animal Limited Evidence IMEMDT 27,191,82. NCI Carcinogenesis Bioassay (feed); Clear Evidence: mouse, rat NCITR* NCI-CG-TR-144,78. Community Right-To-Know List. Reported in EPA TSCA Inventory.

SAFETY PROFILE: Confirmed carcinogen with experimental carcinogenic, neoplastigenic, and tumorigenic data. Moderately toxic via intraperitoneal route. Mutation data reported. When heated to decomposition it emits toxic NO_x. See also AMINES.

AIB250 CAS:81-46-9 HR: 1
N-(4-AMINOANTHRAQUINONYL)BENZAMIDE
mf: $C_{21}H_{15}N_2O_3$ mw: 342.37

SYNS: 1-AMINO-4-BENZAMIDOANTHRAQUINONE □ 1-AMINO-4-BENZOYLAMINOANTHRACHINON (CZECH) □ 4-AMINO-1-BENZOYLAMINOANTHRAQUINONE □ 1-AMINO-4-(BENZOYLAMINO)ANTHRAQUINONE □ N-(4-AMINO-9,10-DIHYDRO-9,10-DIOXO-1-ANTHRACENTY)-BENZAMIDE □ CORINTH FLOUR

TOXICITY DATA WITH REFERENCE
eye-rbt 500 mg/24H MLD 28ZPAK -,124,72

CONSENSUS REPORTS: Reported in EPA TSCA Inventory.

SAFETY PROFILE: An eye irritant. When heated to decomposition it emits toxic fumes of NO_x. See also AMIDES.

AIB300 **CAS:83-07-8** **HR: 3**
4-AMINOANTIPYRINE
mf: $C_{11}H_{13}N_3O$ mw: 203.27

SYNS: AAP □ 4-AMINOANTIPYRENE □ AMINOANTIPYRIN □ AMINOANTIPYRINE □ AMINOAZOPHENAZONE □ AMINOPHENAZONE □ 4-AMINOPHENAZONE □ 4-AMMINOANTIPIRINA □ AMPYRONE □ 1,5-DIMETHYL-2-PHENYL-4-AMINOPYRAZOLINE □ METAPIRAZONE □ 3-PYRAZOLIN-5-ONE, 4-AMINO-2,3-DIMETHYL-1-PHENYL- □ 3H-PYRAZOL-3-ONE, 4-AMINO-1,2-DIHYDRO-1,5-DIMETHYL-2-PHENYL- □ SOLNAPYRIN-A

TOXICITY DATA WITH REFERENCE
mmo-sat 5 μmol/plate MUREAV 206,317,88
dnr-esc 312 μg/well MUREAV 133,161,84
orl-rat LD50:1700 mg/kg BCFAAI 117,638,78
ipr-rat LD50:1200 mg/kg JPETAB 99,171,50
orl-mus LD50:800 mg/kg CCCCAK 47,636,82
ipr-mus LD50:270 mg/kg ARZNAD 10,820,60

CONSENSUS REPORTS: Reported in EPA TSCA Inventory.

SAFETY PROFILE: Poison by intraperitoneal route. Moderately toxic by ingestion. Mutation data reported. When heated to decomposition it emits acrid smoke and irritating vapors.

AIB340 **CAS:1123-54-2** **HR: 3**
6-AMINO-8-AZAPURINE
mf: $C_4H_4N_6$ mw: 136.14

SYNS: 7-AMINO-1H-v-TRIAZOLO(4,5-d)PYRIMIDINE □ 8-AZAADENINE □ 8-AZAPURINE, 6-AMINO- □ 1H-v-TRIAZOLO(4,5-d)PYRIMIDIN-7-AMINE

TOXICITY DATA WITH REFERENCE
dnd-esc 20 μmol/L MUREAV 89,95,81
ipr-mus LD50:315 mg/kg PCJOAU 11,224,77

CONSENSUS REPORTS: Reported in EPA TSCA Inventory.

SAFETY PROFILE: Poison by intraperitoneal route. Mutation data reported. When heated to decomposition it emits toxic vapors of NO_x.

AIC000 **CAS:3398-09-2** **HR: 2**
p-AMINO-2′:3-AZOTOLUENE
mf: $C_{14}H_{15}N_3$ mw: 225.32

PROP: Golden-yellow needles from alc, yellow-brown from ligroin. Mp: 80°.

SYN: 4′-AMINO-3,2′-AZOTOLUENE

TOXICITY DATA WITH REFERENCE
orl-mus TDLo:30 g/kg/57W-C:ETA BJCAAI 3,387,49

SAFETY PROFILE: Questionable carcinogen with experimental tumorigenic data. When heated to decomposition it emits toxic fumes of NO_x.

AIC250 **CAS:97-56-3** **HR: 3**
2-AMINO-5-AZOTOLUENE
mf: $C_{14}H_{15}N_3$ mw: 225.32

PROP: Yellow leaflets from alc. Mp: 100°.

SYNS: AAT □ o-AAT □ o-AMIDOAZOTOLUOL (GERMAN) □ AMINOAZOTOLUENE (indicator) □ o-AMINOAZOTOLUENE (MAK) □ 4′-AMINO-2,3′-AZOTOLUENE □ 4′-AMINO-2:3′-AZOTOLUENE □ o-AMINOAZOTOLUENO (SPANISH) □ o-AMINOAZOTOLUOL □ 4-AMINO-2′,3-DIMETHYLAZOBENZENE □ 4′-AMINO-2,3′-DIMETHYLAZOBENZENE □ o-AT □ BRASILAZINA OIL YELLOW R □ BUTTER YELLOW □ C.I. 11160 □ C.I. 11160B □ C.I. SOLVENT YELLOW 3 □ 2′,3-DIMETHYL-4-AMINOAZOBENZENE □ FAST GARNET GBC BASE □ FAST OIL YELLOW □ FAST YELLOW AT □ FAST YELLOW B □ HIDACO OIL YELLOW □ 2-METHYL-4-((2-METHYLPHENYL)AZO)BENZENAMINE □ OAAT □ OIL YELLOW □ OIL YELLOW 21 □ OIL YELLOW 2681 □ OIL YELLOW AT □ OIL YELLOW A □ OIL YELLOW C □ OIL YELLOW I □ OIL YELLOW 2R □ OIL YELLOW T □ ORGANOL YELLOW 25 □ SOMALIA YELLOW R □ SUDAN YELLOW RRA □ o-TOLUENEAZO-o-TOLUIDINE □ o-TOLUOL-AZO-o-TOLUIDIN (GERMAN) □ 5-(o-TOLYLAZO)-2-AMINOTOLUENE □ 4-(o-TOLYLAZO)-o-TOLUIDINE □ TULABASE FAST GARNET GB □ TULABASE FAST GARNET GBC □ WAXAKOL YELLOW NL

TOXICITY DATA WITH REFERENCE
mma-sat 25 ng/plate CNREA8 45,6155,85
dnr-esc 25 mg/L JNCIAM 62,873,79
orl-rat TDLo:15 g/kg/57W-C:NEO BJCAAI 3,387,49
orl-mus TDLo:480 mg/kg (16-21D post):NEP,TER BEXBAN 85,201,78
scu-mus TDLo:4000 mg/kg (15-21D post):NEO,TER BEXBAN 78,1402,74
imp-mus TDLo:80 mg/kg:CAR BJCAAI 22,825,68
mul-mus TDLo:400 mg/kg/I:ETA CNREA8 1,397,41
orl-rat LDLo:1500 mg/kg CNREA8 26,619,66
orl-mus LDLo:800 mg/kg JNCIAM 10,927,50
scu-mus LDLo:1200 mg/kg JNCIAM 10,927,50
orl-dog LD50:300 mg/kg 85JCAE-,1315,86

CONSENSUS REPORTS: NTP 7th Annual Report on Carcinogens. IARC Cancer Review: Group 2B IMEMDT 7,56,87; Animal Sufficient Evidence IMEMDT 8,61,75. Community Right-To-Know List. Reported in EPA TSCA Inventory. EPA Genetic Toxicology Program.

DFG MAK: Animal Carcinogen; Suspected Human Carcinogen

SAFETY PROFILE: Confirmed carcinogen with experimental carcinogenic, neoplastigenic, and tumorigenic data. Poison by ingestion. Moderately toxic by subcutaneous route. An experimental teratogen. Human mutation data reported. When heated to decomposition it emits toxic fumes of NO_x. See also AROMATIC AMINES.

AIC500 **CAS:3963-79-9** **HR: 2**
4′-AMINO-4,2′-AZOTOLUENE
mf: $C_{14}H_{15}N_3$ mw: 225.32

PROP: Yellow plates from alc, golden-yellow needles from ligroin. Mp: 127°.

SYN: 4-(p-TOLYLAZO)-m-TOLUIDINE

TOXICITY DATA WITH REFERENCE
orl-mus TDLo:30 g/kg/57W-C:CAR BJCAAI 3,387,49

SAFETY PROFILE: Questionable carcinogen with experimental carcinogenic data. When heated to decomposition it emits toxic fumes of NO_x.

AIC750 CAS:18936-75-9 HR: 2
10-AMINOBENZ(a)ACRIDINE
mf: $C_{17}H_{12}N_2$ mw: 244.31

SYN: BENZ(a)ACRIDIN-10-AMINE

TOXICITY DATA WITH REFERENCE
scu-mus TDLo:72 mg/kg/9W-I:ETA CHDDAT 267,981,68

SAFETY PROFILE: Questionable carcinogen with experimental tumorigenic data. See also AROMATIC AMINES. When heated to decomposition it emits toxic fumes of NO_x.

AIC825 CAS:556-18-3 HR: 2
4-AMINOBENZALDEHYDE
mf: C_7H_7NO mw: 121.15

SYNS: p-AMINOBENZALDEHYDE □ BENZALDEHYDE, 4-AMINO-

TOXICITY DATA WITH REFERENCE
ipr-mus LD50:912 mg/kg FEPRA7 6,348,47

CONSENSUS REPORTS: Reported in EPA TSCA Inventory.

SAFETY PROFILE: Moderately toxic by intraperitoneal route. When heated to decomposition it emits toxic vapors of NO_x.

AID250 CAS:6957-91-1 HR: 3
p-AMINOBENZALDEHYDETHIOSEMICARBAZONE
mf: $C_8H_{10}N_4S$ mw: 194.28

SYN: 4-AMINOBENZALDEHYDE THIOSEMICARBAZONE

TOXICITY DATA WITH REFERENCE
orl-mus LD50:500 mg/kg JPPMAB 2,764,50
ivn-mus LD50:180 mg/kg CSLNX* NX#00784

SAFETY PROFILE: Poison by intravenous route. Moderately toxic by ingestion. See also ALDEHYDES. When heated to decomposition it emits very toxic fumes of NO_x and SO_x.

AID500 CAS:98-16-8 HR: 3
m-AMINOBENZAL FLUORIDE
DOT: UN 2948
mf: $C_7H_6F_3N$ mw: 161.14

PROP: Colorless liquid with aniline-like odor. Mp: 3°, bp: 189°, d: 1.303 @ 15.5°/15.5°, vap d: 5.56.

SYNS: m-AMINOBENZOTRIFLUORIDE □ 3-AMINOBENZOTRIFLUORIDE □ m-(TRIFLUOROMETHYL)ANILINE □ 3-(TRIFLUOROMETHYL)ANILINE □ 3-(TRIFLUOROMETHYL)BENZENAMINE □ USAF MA-4

TOXICITY DATA WITH REFERENCE
orl-rat LD50:480 mg/kg 85GMAT-,20,82
unr-rat LD50:480 g/kg TPKVAL 14,118,75
orl-rbt LD50:615 mg/kg 85GMAT -,20,82
unk-rat LD50:480 mg/kg TPKVAL 14,118,75
ihl-rat LC50:440 mg/m³/4H TPKVAL 10,131,68
orl-mus LD50:220 mg/kg TPKVAL 10,131,68
ihl-mus LC50:690 mg/m³/2H TPKVAL 10,131,68
ipr-mus LD50:50 mg/kg NTIS** AD277-689

CONSENSUS REPORTS: EPA Extremely Hazardous Substances List. Reported in EPA TSCA Inventory.

DOT CLASSIFICATION: 6.1; *Label:* Poison

SAFETY PROFILE: Poison by inhalation, ingestion, and intraperitoneal routes. May be moderately toxic by other routes. See also AMINES and FLUORIDES. When heated to decomposition it emits very toxic fumes of F^- and NO_x.

AID620 CAS:88-68-6 HR: 1
2-AMINOBENZAMIDE
mf: $C_7H_8N_2O$ mw: 136.17

SYNS: o-AMINOBENZAMIDE □ ANTHRANILAMIDE □ ANTHRANILIMIDIC ACID □ BENZAMIDE, o-AMINO- □ BENZAMIDE, 2-AMINO- (9CI) □ 2-CARBAMOYLANILINE

TOXICITY DATA WITH REFERENCE
ipr-rat LD50:>400 mg/kg CHTPBA 2,202,67
orl-brd LD50:1 g/kg AECTCV 12,355,83

CONSENSUS REPORTS: Reported in EPA TSCA Inventory.

SAFETY PROFILE: Slightly toxic by ingestion and intraperitoneal routes. When heated to decomposition it emits toxic vapors of NO_x.

AID625 CAS:3544-24-9 HR: 1
3-AMINOBENZAMIDE
mf: $C_7H_8N_2O$ mw: 136.17

PROP: Crystals from C_6H_6. Mp: 115°.

SYNS: m-AMINOBENZAMIDE □ 3-AMINO-BENZAMIDE (9CI)

TOXICITY DATA WITH REFERENCE
sce-hmn:lym 1 mmol/L MUREAV 122,223,83
dns-ham:ovr 10 mmol/L ECREAL 143,377,83
sce-ham:ovr 1 mmol/L ENMUDM 6,203,84
orl-bwd LD50:1 g/kg AECTCV 12,355,83

SAFETY PROFILE: Low toxicity by ingestion. Human mutation data reported. When heated to decomposition it emits toxic fumes of NO_x.

AID700 CAS:1197-55-3 HR: 2
4-AMINOBENZENEACETIC ACID
mf: $C_8H_9NO_2$ mw: 151.18

SYNS: ACETIC ACID, (p-AMINOPHENYL)- □ (p-AMINOPHENYL) ACETIC ACID □ 4-AMINOPHENYLACETIC ACID □ p-AMINO-α-TOLUIC ACID □ BENZENEACETIC ACID, 4-AMINO-(9CI) □ 4-CARBOXYMETHYLANILINE

TOXICITY DATA WITH REFERENCE
mma-sat 2500 μg/plate DEMAEP 2,163,86
ipr-mus LD50:3500 mg/kg FRPSAX 13,286,58

CONSENSUS REPORTS: Reported in EPA TSCA Inventory.

SAFETY PROFILE: Moderately toxic by intraperitoneal route. Mutation data reported. When heated to decomposition it emits toxic vapors of NO_x.

AID750 **HR: 3**
p-AMINO BENZENE DIAZONIUMPERCHLORATE
mf: $C_6H_6ClN_3O_4$ mw: 219.6

SAFETY PROFILE: Extremely shock sensitive explosive. Very dangerous. When heated to decomposition it explodes and emits very toxic fumes of Cl^- and NO_x. See PERCHLORATES.

AIE000 **CAS:98-44-2** **HR: 1**
1-AMINO-2,5-BENZENEDISULFONIC ACID
mf: $C_6H_7NO_6S_2$ mw: 253.26

SYNS: 2-AMINO-p-BENZENEDISULFONIC ACID ☐ 2-AMINO-1,4-BENZENEDISULFONIC ACID ☐ 2-AMINO-BENZENE-1,4-DISULFONIC ACID ☐ 2-AMINO-1,4-DISULFOBENZENE ☐ ANILINE-2,5-DISULFONIC ACID ☐ 1-ANILINO-2,5-DISULFONIC ACID ☐ 2,5-DISULFO-1-AMINOBENZENE ☐ 2,5-DISULFOANILINE ☐ KYSELINA ANILIN-2,5-DISULFONOVA (CZECH) ☐ 4-SULFOMETANILIC ACID

TOXICITY DATA WITH REFERENCE
skn-rbt 500 mg/24H MLD 28ZPAK -,182,72
eye-rbt 100 mg/24H MOD 28ZPAK -,182,72

CONSENSUS REPORTS: Reported in EPA TSCA Inventory.

SAFETY PROFILE: A skin and eye irritant. See also SULFONATES. When heated to decomposition it emits very toxic fumes of NO_x and SO_x.

AIE500 **CAS:2447-57-6** **HR: 2**
6-(4-AMINOBENZENESULFONAMIDO)-4,5-DIMETHOXYPYRIMIDINE
mf: $C_{12}H_{14}N_4O_4S$ mw: 310.36

PROP: Crystals from alc. Mp: 190–194°.

SYNS: 4-AMINO-N-(5,6-DIMETHOXY-4-PYRIMIDINYL)BENZENESULFONAMIDE ☐ N'-(5,6-DIMETHOXY-4-PYRIMIDYL)SULFANILAMIDE ☐ FANASIL ☐ FANZIL ☐ RO 4-4393 ☐ SULFADOXINE ☐ 4-SULFANILAMIDO-5,6-DIMETHOXYPYRIMIDINE ☐ SULFORTHOMIDINE ☐ SULPHORMETHOXINE

TOXICITY DATA WITH REFERENCE
orl-mus LD50:5200 mg/kg MEIEDD 10,1276,83
ipr-mus LD50:2900 mg/kg MEIEDD 10,1276,83
scu-mus LD50:2900 mg/kg MEIEDD 10,1276,83

SAFETY PROFILE: Moderately toxic by intraperitoneal and subcutaneous routes. Mildly toxic by ingestion. See also SULFONATES. When heated to decomposition it emits very toxic fumes of SO_x and NO_x.

AIE750 **CAS:729-99-7** **HR: 2**
2-(p-AMINOBENZENESULFONAMIDO)-4,5-DIMETHYLOXAZOLE
mf: $C_{11}H_{13}N_3O_3S$ mw: 267.33

PROP: Solid. Mp: 193–194°.

SYNS: p-AMINOBENZENESULFONYL-2-AMINO-4,5-DIMETHYLOXAZOLE ☐ 2-(p-AMINOBENZOLSULFONAMIDO)-4,5-DIMETHYLOXAZOL (GERMAN) ☐ 4-AMINO-N-(4,5-DIMETHYL-2-OXAZOLYL)BENZENESULFONAMIDE ☐ 2-(p-AMINOPHENYLSULFONYLAMINO)-4,5-DIMETHYLOXAZOLE ☐ DEPOMIDE ☐ N[1]-(4,5-DIMETHYL-2-OXAZOLYL)-SULFANILAMIDE ☐ 4,5-DIMETHYL-2-SULFANILAMIDOOXAZOLE ☐ JUSTAMIL ☐ NUPRIN ☐ OXASULFA ☐ SDMO ☐ SULFABUTIN ☐ SULFADIMETHYLOXAZOLE ☐ SULFAMOXOLE ☐ SULFAMOXOLUM ☐ SULFANO ☐ SULFAVIGOR ☐ SULFMIDIL ☐ SULFUNE ☐ SULFUNO ☐ TARDAMID ☐ TARDAMIDE

TOXICITY DATA WITH REFERENCE
ipr-rat LD50:2500 mg/kg ARZNAD 26,634,76
ipr-mus LD50:1800 mg/kg ARZNAD 26,634,76
orl-mus LD50:15,200 mg/kg ARZNAD 10,612,60
scu-mus LD50:1285 mg/kg ARZNAD 10,612,60
ivn-mus LD50:1000 mg/kg ARZNAD 10,612,60

SAFETY PROFILE: Moderately toxic by subcutaneous, intraperitoneal, and intravenous routes. Mildly toxic by ingestion. When heated to decomposition it emits very toxic fumes of SO_x and NO_x.

AIF000 **CAS:526-08-9** **HR: 2**
3-(p-AMINOBENZENESULFONAMIDO)-2-PHENYLPYRAZOLE
mf: $C_{15}H_{14}N_4O_2S$ mw: 314.39

PROP: Crystals from alc. Mp: 179–183°. Spar sol in H_2O, EtOH, Et_2O.

SYNS: 4-AMINO-N-(1-PHENYL-1H-PYRAZOL-5-YL)BENZENESULFONAMIDE ☐ DEPOCID ☐ DEPTOSULFONAMIDE ☐ EFTOLON ☐ FIRMAZOLO ☐ INAMIL ☐ ISAROL ☐ MERIAN ☐ MICROTAN PIRAZOLO ☐ ORISUL ☐ ORISULF ☐ PAIDAZOLO ☐ N'-(1-PHENYLPYRAZOL-5-YL)SULFANILAMIDE ☐ N[1]-(1-PHENYLPYRAZOL-5-YL)-SULFANILAMIDE (8CI) ☐ 1-PHENYL-5-SULFANILAMIDOPYRAZOLE ☐PLISUL-FAN ☐ RAZIOSULFA ☐ SP ☐ SPP ☐ SULFAAFENAZOLO (ITALIAN) ☐ SULFABID ☐ 5-SULFANILAMIDO-1-PHENYLPYRAZOLE ☐ SULFAPHENAZOLE ☐ SULFAPHENAZON ☐ SULFAPHENYLPIPAZOL ☐ SULFAPHENYLPYRAZOLE ☐ SULPHENAZOLE

TOXICITY DATA WITH REFERENCE
scu-rat LD50:900 mg/kg NIIRDN 6 387,82
ivn-rat LD50:525 mg/kg NIIRDN 6,387,82
orl-mus LD50:4507 mg/kg NIIRDN 6,387,82
scu-mus LD50:660 mg/kg NIIRDN 6,387,82
ivn-mus LD50:470 mg/kg NIIRDN 6,387,82
scu-rbt LD50:950 mg/kg NIIRDN 6,387,82
ivn-rbt LD50:440 mg/kg NIIRDN 6,387,72

SAFETY PROFILE: Moderately toxic by subcutaneous and intravenous routes. Mildly toxic by ingestion. See also SULFONATES. When heated to decomposition it emits very toxic fumes of SO_x and NO_x.

AIF250 **CAS:1126-34-7** **HR: 1**
m-AMINOBENZENESULFONIC ACID SODIUM SALT
mf: $C_6H_6NO_3S \cdot Na$ mw: 195.18

SYN: METANILAN SODNY (CZECH)

TOXICITY DATA WITH REFERENCE
eye-rbt 750 μg/24H SEV 28ZPAK -,179,72

CONSENSUS REPORTS: Reported in EPA TSCA Inventory.

SAFETY PROFILE: A severe eye irritant. See also SULFONATES. When heated to decomposition it emits very toxic fumes of SO_x, NO_x, and Na_2O.

AIF500 CAS:137-07-5 HR: 3
2-AMINOBENZENETHIOL
mf: C_6H_7NS mw: 125.20

PROP: Needles or liquid. Mp: 23–26°, bp: 227.2°, flash p: 175°F, d: 1.168, vap d: 4.3.

SYNS: o-AMINOTHIOPHENOL □ 2-AMINOTHIOPHENOL □ o-MERCAPTOANILINE □ USAF EK-4376

TOXICITY DATA WITH REFERENCE
orl-rat LDLo:500 mg/kg 34ZIAG -,90,69
ipr-mus LD50:25 mg/kg NTIS** AD277-689
ivn-mus LD50:100 mg/kg CSLNX* NX#02532

CONSENSUS REPORTS: Reported in EPA TSCA Inventory.

SAFETY PROFILE: Poison by intraperitoneal route. Moderately toxic by ingestion. Moderately flammable. Can react with oxidizing materials. To fight fire, use water, foam, CO_2, mist or spray, dry chemical.

AIF750 CAS:1193-02-8 HR: 3
p-AMINOBENZENETHIOL
mf: C_6H_7NS mw: 125.20

PROP: Granular crystals. Mp: 46°, bp: 142° @ 15 mm.

SYNS: 4-AMINOBENZENETHIOL □ p-AMINOPHENYLMERCAPTAN □ p-AMINOTHIOPHENOL □ 4-AMINOTHIOPHENOL □ p-MERCAPTOANILINE □ 4-MERCAPTOANILINE

TOXICITY DATA WITH REFERENCE
mma-sat 50 μg/plate MUREAV 67,123,79
orl-mus LDLo:320 mg/kg AECTCV 14,111,85
orl-qal LD50:42,200 μg/kg AECTCV 12,355,83
orl-bwd LD50:42,200 μg/kg AECTCV 12,355,83

CONSENSUS REPORTS: EPA Genetic Toxicology Program.

SAFETY PROFILE: Poison by ingestion. Mutation data reported. When heated to decomposition it emits very toxic fumes of NO_x and SO_x.

AIG000 CAS:934-32-7 HR: 3
2-AMINOBENZIMIDAZOLE
mf: $C_7H_7N_3$ mw: 133.17

PROP: Aqueous leaflets or plates from water. Mp: 222–224°. Sol in water, alkalies, alc, acetone; very sltly sol in ether.

SYN: USAF EK-4037

TOXICITY DATA WITH REFERENCE
mma-sat 710 μmol/L ENMUDM 3,11,81
mmo-sat 100 μg/plate MUREAV 15,273,72
orl-rat TDLo:426 mg/kg (8-15D preg):TER THERAP 31,505,76
orl-rat LDLo:500 mg/kg NCNSA6 5,22,53
orl-mus LD40:600 mg/kg JACSAT 67,905,45
ipr-mus LD50:100 mg/kg NTIS** AD277-689
ivn-mus LD50:126 mg/kg 29QHAQ -,246,74

CONSENSUS REPORTS: Reported in EPA TSCA Inventory.

SAFETY PROFILE: Poison by intravenous and intraperitoneal routes. Moderately toxic by ingestion. An experimental teratogen. Mutation data reported. When heated to decomposition it emits toxic NO_x. See also AROMATIC AMINES.

AIH000 CAS:52329-60-9 HR: 3
2-AMINO-6-BENZIMIDAZOLYL PHENYLKETONE
mf: $C_{14}H_{11}N_3O$ mw: 237.28

SYNS: 2-AMINO-5-BENZOYLBENZIMIDAZOLE □ BENZIMIDAZOLE, 2-AMINO-5-BENZOYL- □ G-1029 □ KETONE, 2-AMINO-5-BENZIMIDAZOLYL PHENYL □ METHANONE, (2-AMINO-1H-BENZIMIDAZOL-5-YL) PHENYL- □ R 18986

TOXICITY DATA WITH REFERENCE
orl-rat TDLo:755 mg/kg (8-15D preg):TER THERAP 31,505,76
orl-mus LD50:450 mg/kg MPPBAB 48,29,79
ipr-mus LD50:257 mg/kg MPPBAB 48,29,79

DOT CLASSIFICATION: 3; *Label:* Flammable Liquid

SAFETY PROFILE: Acute poison by intraperitoneal route. Moderately toxic by ingestion. An experimental teratogen. See also KETONES. A flammable liquid. When heated to decomposition it emits very toxic fumes of NO_x.

AIH500 CAS:99-05-8 HR: 2
m-AMINOBENZOIC ACID
mf: $C_7H_7NO_2$ mw: 137.15

PROP: Needles from water. D: 1.511, mp: 173–178°. Sol in water, alc, and ether.

SYNS: 3-AMINOBENZOIC ACID □ ANILINE-3-CARBOXYLIC ACID □ 3-CARBOXYANILINE □ MADA

TOXICITY DATA WITH REFERENCE
orl-mus LD50:6300 mg/kg QJPPAL 19,483,46
ipr-mus LDLo:500 mg/kg CBCCT* 6,53,54
orl-brd LD50:750 mg/kg AECTCV 12,355,83

CONSENSUS REPORTS: Reported in EPA TSCA Inventory.

SAFETY PROFILE: Moderately toxic by ingestion and intraperitoneal route. When heated to decomposition it emits toxic fumes of NO_x.

AIH600 CAS:150-13-0 HR: 2
p-AMINOBENZOIC ACID
mf: $C_7H_7NO_2$ mw: 137.15

PROP: Yellowish to red crystals or prisms. D: 1.374, mp: 187–188.5°. Sol in water, alkalis, EtOH, Et_2O, C_6H_6.

SYNS: ACIDO p-AMINOBENZOICO □ AMBEN □ AMINOBENZOIC ACID □ γ-AMINOBENZOIC ACID □ 4-AMINOBENZOIC ACID □ 1-AMINO-4-CARBOXYBENZENE □ ANTICANITIC VITAMIN □ ANTI-CHROMOTRICHIA FACTOR □ bacL VITAMIN H1 □ BENZOIC ACID, 4-AMINO- □ p-CARBOXYANILINE □ 4-CARBOXYANILINE □ p-CAR-

BOXYPHENYLAMINE □ CHROMOTRICHIA FACTOR □ anti-CHROMOTRICHIA FACTOR □ KYSELINA p-AMINOBENZOOVA □ PABA □ PABANOL □ PARAMINOL □ PARANATE □ SUNBRELLA □ TRICHOCHROMOGENIC FACTOR □ VITAMIN BX □ VITAMIN H

TOXICITY DATA WITH REFERENCE
dnd-mus-ipr 1 g/kg ARTODN 5,355,82
orl-rat TDLo: 2500 mg/kg (female 1-22D post): REP AJANA2 110,29,62
orl-mus LD50: 2850 mg/kg PSEBAA 49,184,42
orl-dog LD50: 1000 mg/kg PSEBAA 49,184,42
orl-rbt LD50: 1830 mg/kg FEPRA7 10,289,51
ivn-rbt LD50: 2000 mg/kg FEPRA7 1,71,42

CONSENSUS REPORTS: IARC Cancer Review: Group 3 IMEMDT 7,56,87; Animal Inadequate Evidence IMEMDT 16,249,78. Reported in EPA TSCA Inventory.

SAFETY PROFILE: Moderately toxic by ingestion and intravenous routes. Ingesting large doses can cause nausea, vomiting, skin rash, methemoglobinemia, and possibly toxic hepatitis. Experimental reproductive effects. Mutation data reported. Combustible. When heated to decomposition it emits toxic fumes of NO_x. A topical sunscreen.

AIK500 CAS:73698-75-6 HR: 3
p-AMINOBENZOIC ACID-2-(2-(2-(2-(2-(DIETHYLAMINO)ETHOXY)ETHOXY)ETHOXY)ETHOXY)ETHYL ESTER, HYDROCHLORIDE
mf: $C_{21}H_{36}N_2O_6 \cdot ClH$ mw: 449.05

TOXICITY DATA WITH REFERENCE
eye-rbt 1% SEV JPETAB 48,371,33
unk-rat LDLo: 60 mg/kg JPETAB 48,371,33

SAFETY PROFILE: A poison by unspecified route. Severe eye irritant. See also ESTERS. When heated to decomposition it emits very toxic fumes of HCl and NO_x.

AIK750 CAS:73698-76-7 HR: 3
p-AMINOBENZOIC ACID-2-(2-(2-(2-(DIETHYLAMINO)ETHOXY)ETHOXY)ETHOXY)ETHYL ESTER, HYDROCHLORIDE
mf: $C_{19}H_{32}N_2O_5 \cdot ClH$ mw: 404.99

TOXICITY DATA WITH REFERENCE
eye-rbt 1% SEV JPETAB 48,371,33
unk-rat LDLo: 30 mg/kg JPETAB 48,371,33

SAFETY PROFILE: A poison by unspecified route. Severe eye irritant. See also ESTERS. When heated to decomposition it emits very toxic fumes of HCl and NO_x.

AIL000 CAS:73698-77-8 HR: 3
p-AMINOBENZOIC ACID-2-(2-(2-(DIETHYLAMINO)ETHOXY)ETHOXY)ETHYL ESTER, HYDROCHLORIDE
mf: $C_{17}H_{28}N_2O_4 \cdot ClH$ mw: 360.93

TOXICITY DATA WITH REFERENCE
eye-rbt 1% SEV JPETAB 48,371,33
unk-rat LDLo: 30 mg/kg JPETAB 48,371,33

SAFETY PROFILE: A poison by unspecified route. Severe eye irritant. See also ESTERS. When heated to decomposition it emits very toxic fumes of HCl and NO_x.

AIL250 CAS:73698-78-9 HR: 3
p-AMINOBENZOIC ACID-2-(2-(DIETHYLAMINO)ETHOXY)ETHYL ESTER, HYDROCHLORIDE
mf: $C_{15}H_{24}N_2O_3 \cdot ClH$ mw: 316.87

TOXICITY DATA WITH REFERENCE
eye-rbt 1% SEV JPETAB 48,371,33
unk-rat LDLo: 60 mg/kg JPETAB 48,371,33

SAFETY PROFILE: A poison by an unspecified route. Severe eye irritant. See also ESTERS. When heated to decomposition it emits very toxic fumes of HCl and NO_x.

AIL500 CAS:63917-76-0 HR: 3
(p-AMINOBENZOIC ACID-3-(β-DIETHYLAMINO)ETHOXY)PROPYL ESTER
mf: $C_{16}H_{26}N_2O_3$ mw: 294.44

TOXICITY DATA WITH REFERENCE
ivn-mus LD50: 55 mg/kg RCPRAN 15,143,54
scu-gpg LD50: 31 mg/kg RCPRAN 15,143,54

SAFETY PROFILE: Poison by intravenous and subcutaneous routes. See also ESTERS. When heated to decomposition it emits toxic fumes of NO_x.

AIL750 CAS:59-46-1 HR: 3
p-AMINOBENZOIC ACID-2-DIETHYLAMINOETHYL ESTER
mf: $C_{13}H_{20}N_2O_2$ mw: 236.35

PROP: Hygroscopic plates from ligroin or Et_2O. Mp: 61°. Sol in H_2O, EtOH, C_6H_6.

SYNS: ALLOCAINE □ 4-AMINOBENZOIC ACID DIETHYLAMINOETHYL ESTER □ p-AMINOBENZOYLDIETHYLAMINOETHANOL □ DIETHYLAMINOETHYL-p-AMINOBENZOATE □ β-DIETHYLAMINOETHYL-4-AMINOBENZOATE □ 2-DIETHYLAMINOETHYL-p-AMINOBENZOATE □ GEROVITAL □ JENACAINE □ NEOCAINE □ NISSOCAINE □ NOROCAINE □ NOVOCAINE □ PROCAINE □ PROCAINE, BASE □ SCUROCAINE □ SPINOCAINE

TOXICITY DATA WITH REFERENCE
orl-man LDLo: 147 mg/kg 85DCAI 2,73,70
ims-wmn TDLo: 1600 μg/kg: CNS JNNPAU 34,20,71
ipr-rat LDLo: 280 mg/kg TXAPA9 1,156,59
scu-rat LD50: 600 mg/kg AIPTAK 104,388,56
ivn-rat LDLo: 45 mg/kg PHREA7 12,190,32
orl-mus LD50: 350 mg/kg ARZNAD 16,1275,66
ipr-mus LD50: 124 mg/kg RPOBAR 2,213,70
scu-mus LD50: 300 mg/kg ARZNAD 5,376,55
ivn-mus LD50: 45 mg/kg JAPMA8 45,382,56

SAFETY PROFILE: Poison by ingestion, intraperitoneal, intravenous, and subcutaneous routes. Moderately toxic by parenteral route. Human systemic effects by intramuscular route: lack of muscular control, rigidity, and possibly catalepsy. See also ESTERS. When heated to decomposition it emits toxic fumes of NO_x. Used as a local anesthetic.

AIM000 CAS:17599-08-5 HR: 3
p-AMINOBENZOIC ACID-3-(DIETHYLAMINO)PROPYL ESTER HYDROCHLORIDE
mf: $C_{14}H_{22}N_2O_2$•ClH mw: 286.84

SYN: p-AMINO BENZOYL DIETHYL AMINO PROPANOL HYDROCHLORIDE

TOXICITY DATA WITH REFERENCE
unk-rat LD50:125 mg/kg ARZNAD 17,1012,67
scu-mus LD50:300 mg/kg JPETAB 24,160,24

SAFETY PROFILE: Poison by subcutaneous and other unspecified routes. See also ESTERS. When heated to decomposition it emits very toxic fumes of HCl and NO_x.

AIM250 CAS:15154-37-7 HR: 3
p-AMINOBENZOIC ACID-3-(DIISOPROPYLAMINO) PROPYL ESTER HYDROCHLORIDE
mf: $C_{16}H_{26}N_2O_2$•ClH mw: 314.90

SYN: p-AMINO BENZOYL DIISOPROPYL AMINO PROPANOL HYDROCHLORIDE

TOXICITY DATA WITH REFERENCE
unk-rat LD50:75 mg/kg ARZNAD 17,1012,67
scu-mus LD50:150 mg/kg JPETAB 24,160,24

SAFETY PROFILE: Poison by subcutaneous and other unspecified routes. See also ESTERS. When heated to decomposition it emits very toxic fumes of HCl and NO_x.

AIN000 CAS:15154-36-6 HR: 3
p-AMINOBENZOIC ACID-2-(DIPROPYLAMINO)ETHYL ESTER HYDROCHLORIDE
mf: $C_{15}H_{24}N_2O_2$•ClH mw: 300.87

SYN: p-AMINO BENZOYL DI-N PROPYL AMINO ETHANOL HYDROCHLORIDE

TOXICITY DATA WITH REFERENCE
unk-rat LD50:100 mg/kg ARZNAD 17,1012,67
scu-mus LD50:550 mg/kg JPETAB 24,160,24

SAFETY PROFILE: A poison by unspecified route. Moderately toxic by subcutaneous route. See also ESTERS and AMINES. When heated to decomposition it emits very toxic fumes of HCl and NO_x.

AIN150 CAS:619-45-4 HR: 3
p-AMINOBENZOIC ACID METHYL ESTER
mf: $C_8H_9NO_2$ mw: 151.18

SYNS: BENZOIC ACID, p-AMINO-, METHYL ESTER □ METHYL p-AMINOBENZOATE

TOXICITY DATA WITH REFERENCE
ipr-mus LD50:237 mg/kg JMCMAR 17,900,74

CONSENSUS REPORTS: Reported in EPA TSCA Inventory.

SAFETY PROFILE: Poison by intraperitoneal route. When heated to decomposition it emits toxic vapors of NO_x.

AIQ875 CAS:72977-18-5 HR: 2
p-AMINOBENZOIC ACID PHOSPHATE
mf: $C_7H_7NO_2$•H_3O_4P mw: 235.15

TOXICITY DATA WITH REFERENCE
orl-rat LD50:8 g/kg GISAAA 49(10),82,84
orl-mus LD50:3100 mg/kg GISAAA 49(10),82,84
orl-rbt LD50:2500 mg/kg GISAAA 49(10),82,84
orl-gpg LD50:2580 mg/kg GISAAA 49(10),82,84

SAFETY PROFILE: Moderately toxic by ingestion. When heated to decomposition it emits toxic fumes of NO_x and PO_x.

AIQ880 CAS:69780-82-1 HR: 3
m-AMINOBENZOIC ACID-2-(2-PIPERIDYL)ETHYL ESTER HYDROCHLORIDE
mf: $C_{14}H_{20}N_2O_2$•ClH mw: 284.82

SYNS: 2-PIPERIDINEETHANOL-m-AMINOBENZOATE (ester) HYDROCHLORIDE □ β-2-PIPERIDYLETHYL-m-AMINOBENZOATE HYDROCHLORIDE

TOXICITY DATA WITH REFERENCE
scu-mus LDLo:87 mg/kg JACSAT 61,1713,39
scu-rbt LDLo:500 mg/kg ANESAV 1,305,40
ivn-rbt LDLo:20 mg/kg ANESAV 1,305,40
isp-rbt LDLo:24,620 µg/kg ANESAV 1,305,40
scu-gpg LDLo:315 mg/kg ANESAV 1,305,40

SAFETY PROFILE: Poison by intravenous and intraspinal routes. Moderately toxic by subcutaneous route. When heated to decomposition it emits toxic fumes of NO_x and HCl. See also ESTERS.

AIQ885 CAS:69780-83-2 HR: 3
o-AMINOBENZOIC ACID 2-(2-PIPERIDYL)ETHYL ESTER HYDROCHLORIDE
mf: $C_{14}H_{20}N_2O_2$•ClH mw: 284.82

SYNS: 2-PIPERIDINEETHANOL-o-AMINOBENZOATE (ester) HYDROCHLORIDE □ β-2-PIPERIDYLETHYL-o-AMINOBENZOATE HYDROCHLORIDE

TOXICITY DATA WITH REFERENCE
scu-mus LDLo:23 mg/kg JACSAT 61,1713,39
scu-rbt LDLo:150 mg/kg ANESAV 1,305,40
ivn-rbt LDLo:15 mg/kg ANESAV 1,305,40
isp-rbt LDLo:29,970 µg/kg ANESAV 1,305,40
scu-gpg LDLo:153 mg/kg ANESAV 1,305,40

SAFETY PROFILE: Poison by subcutaneous, intravenous, and intraspinal routes. When heated to decomposition it emits toxic fumes of NO_x and HCl. See also ESTERS.

AIQ890 CAS:69780-84-3 HR: 3
p-AMINOBENZOIC ACID-2-(2-PIPERIDYL)ETHYL ESTER HYDROCHLORIDE
mf: $C_{14}H_{20}N_2O_2$•ClH mw: 284.82

SYNS: 2-PIPERIDINEETHANOL-p-AMINOBENZOATE (ester) HYDROCHLORIDE □ β-2-PIPERIDYLETHYL-p-AMINOBENZOATE HYDROCHLORIDE

TOXICITY DATA WITH REFERENCE
scu-mus LDLo:20 mg/kg JACSAT 61,1713,39

scu-rbt LDLo: 250 mg/kg ANESAV 1,305,40
ivn-rbt LDLo: 17 mg/kg ANESAV 1,305,40
isp-rbt LDLo: 10,920 μg/kg ANESAV 1,305,40
scu-gpg LDLo: 196 mg/kg ANESAV 1,305,40

SAFETY PROFILE: Poison by subcutaneous, intravenous, and intraspinal routes. When heated to decomposition it emits toxic fumes of NO_x and HCl. See also ESTERS.

AIR125 CAS:2237-30-1 HR: 2
3-AMINOBENZONITRILE
mf: $C_7H_6N_2$ mw: 118.15

PROP: Crystals from CCl_4. Mp: 53–54°, bp: 288–290°.

SYNS: m-AMINOBENZONITRILE □ m-ANTHRANILONITRILE □ BENZONITRILE, 3-AMINO-(9CI) □ m-CYANOANILINE □ 3-CYANOANILINE

TOXICITY DATA WITH REFERENCE
mma-sat 100 mg/L ENMUDM 5,803,83
orl-qal LD50: 562 mg/kg AECTCV 12,355,83
orl-bwd LD50: 562 mg/kg AECTCV 12,355,83

CONSENSUS REPORTS: Cyanide and its compounds are on the Community Right-To-Know List.

SAFETY PROFILE: Moderately toxic by ingestion. Mutation data reported. When heated to decomposition it emits toxic fumes of NO_x and CN^-. See also NITRILES.

AIR250 CAS:1137-41-3 HR: 3
p-AMINOBENZOPHENONE
mf: $C_{13}H_{11}NO$ mw: 197.25

PROP: Leaflets from alc. Mp: 124°, bp: 246 @ 13 mm. Very sltly sol in cold water, very sol in alc.

SYN: USAF A-233

TOXICITY DATA WITH REFERENCE
ipr-mus LD50: 300 mg/kg NTIS**AD277-689
orl-bwd LD50: 562 mg/kg AECTCV 12,355,83

CONSENSUS REPORTS: Reported in EPA TSCA Inventory.

SAFETY PROFILE: Poison by intraperitoneal route. See also KETONES and AMINES. When heated to decomposition it emits toxic fumes of NO_x.

AIS250 CAS:65793-50-2 HR: 3
3-AMINOBENZO-6,7-QUINAZOLINE-4-ONE
mf: $C_{12}H_9N_3O$ mw: 211.24

SYN: 3-AMINOBENZO(g)QUINAZOLIN-4(3H)-ONE

TOXICITY DATA WITH REFERENCE
orl-man TDLo: 1143 μg/kg: CNS IJEBA6 15,1131,77
orl-rat LD50: 540 mg/kg IJEBA6 15,1125,77
ipr-rat LD50: 266 mg/kg IJEBA6 15,1125,77
orl-mus LD50: 525 mg/kg IJEBA6 15,1125,77
ipr-mus LD50: 160 mg/kg IJEBA6 15,1125,77
ivn-mus LD50: 60 mg/kg IJEBA6 15,1125,77
ipr-cat LD50: 54 mg/kg IJEBA6 15,1125,77

SAFETY PROFILE: Poison by intraperitoneal and intravenous route. Moderately toxic by ingestion. Human central nervous system effects by ingestion. See also KETONES. When heated to decomposition it emits toxic fumes of NO_x.

AIS500 CAS:136-95-8 HR: 3
2-AMINOBENZOTHIAZOLE
mf: $C_7H_6N_2S$ mw: 150.21

PROP: Crystals or leaflets from water. Mp: 132°. Sol in Et_2O, EtOH, $CHCl_3$.

SYNS: 2-AMINOBENZTHIAZOLE □ USAF EK-3941 □ USAF XR-27

TOXICITY DATA WITH REFERENCE
mma-sat 2500 μg/plate FCTOD7 23,695,85
orl-mus LD50: >1 g/kg FSTEAI 45,223,90
ipr-mus LD50: 200 mg/kg NTIS** AD277-689
ivn-mus LD50: 126 mg/kg JPETAB 105,486,52

CONSENSUS REPORTS: Reported in EPA TSCA Inventory.

SAFETY PROFILE: Low toxicity by ingestion. Poison by intraperitoneal and intravenous routes. Mutation data reported. Dangerous; when heated to decomposition it emits highly toxic fumes of NO_x and SO_x.

AIS550 CAS:7442-07-1 HR: 3
6-AMINO-2-BENZOTHIAZOLETHIOL
mf: $C_7H_6N_2S_2$ mw: 182.27

SYNS: BENZOTHIAZOLE, 6-AMINO-2-MERCAPTO- □ BENZOTHIAZOLE, 2-MERCAPTO-6-AMINO- □ USAF XR-30

TOXICITY DATA WITH REFERENCE
orl-rat LD: >500 mg/kg NCNSA6 5,12,53
ipr-mus LD50: 150 mg/kg NTIS** AD277-689

CONSENSUS REPORTS: Reported in EPA TSCA Inventory.

SAFETY PROFILE: Poison by intraperitoneal route. When heated to decomposition it emits very toxic fumes of SO_x and NO_x.

AIS600 CAS:4570-41-6 HR: 3
2-AMINOBENZOXAZOLE
mf: $C_7H_6N_2O$ mw: 134.15

PROP: Leaflets from C_6H_6. Mp: 129–130°.

TOXICITY DATA WITH REFERENCE
orl-rat LD50: 600 mg/kg MDCHAG 4(1),336,64
ipr-rat LD50: 275 mg/kg MDCHAG 4(1),336,64
orl-mus LD50: 678 mg/kg MDCHAG 4(1),336,64
ipr-mus LD50: 392 mg/kg MDCHAG 4(1),336,64
ivn-mus LD50: 238 mg/kg JPETAB 105,486,52

SAFETY PROFILE: Poison by intravenous and intraperitoneal routes. Moderately toxic by ingestion. When heated to decomposition it emits toxic fumes of NO_x.

AIS625 HR: 3
p-AMINOBENZOYLAMINOMETHYLHYDROCOTARNINE
mf: $C_{20}H_{23}N_3O_4$ mw: 369.46

TOXICITY DATA WITH REFERENCE
scu-mus LDLo:100 mg/kg IJMRAQ 25,713,38
scu-dog LDLo:25 mg/kg IJMRAQ 25,713,38
scu-cat LDLo:25 mg/kg IJMRAQ 25,713,38
scu-gpg LDLo:80 mg/kg IJMRAQ 25,713,38

SAFETY PROFILE: Poison by subcutaneous route. When heated to decomposition it emits toxic fumes of NO_x.

AIT000 CAS:5892-15-9 HR: 3
AMINOBENZOYLDIBUTYLAMINOPROPANOL HYDROCHLORIDE
mf: $C_{18}H_{30}N_2O_2$•ClH mw: 342.96

PROP: Crystals from alc. Mp: 157–158.5°.

SYN: p-AMINOBENZOIC ACID-3-(DIBUTYLAMINO)PROPYL ESTER, HYDROCHLORIDE

TOXICITY DATA WITH REFERENCE
ipr-rat LDLo:200 mg/kg JPETAB 18,467,22
unk-rat LD50:75 mg/kg ARZNAD 17,1012,67
scu-mus LD50:100 mg/kg JPETAB 24,160,24
ipr-cat LDLo:200 mg/kg JPETAB 18,467,22
ivn-cat LDLo:15 mg/kg JPETAB 18,467,22

SAFETY PROFILE: Poison by subcutaneous, intraperitoneal, and possibly other routes. See also ESTERS and AMINES. When heated to decomposition it emits very toxic fumes of HCl and NO_x.

AIT250 CAS:51-05-8 HR: 3
p-AMINOBENZOYLDIETHYLAMINOETHANOL HYDROCHLORIDE
mf: $C_{13}H_{20}N_2O_2$•ClH mw: 272.81

PROP: Mono- or triclinic, six-sided plates. Mp: 153–156°. Sol in H_2O, EtOH; insol in Et_2O.

SYNS: ALLOCAINE □ p-AMINOBENZOIC ACID-2-DIETHYLAMINOETHYL ESTER, HYDROCHLORIDE □ 4-AMINOBENZOIC ACID 2-(DIETHYLAMINO)ETHYL ESTER, HYDROCHLORIDE □ AMINOCAINE □ ANADOLOR □ ANESTHESOL □ ANESTIL □ ATOXICOCAINE □ BERNOCAINE □ CETAIN □ CHLOROCAINE □ DIETHYLAMINOETHANOL-4-AMINOBENZOATE HYDROCHLORIDE □ 2-DIETHYLAMINOETHYL-p-AMINOBENZOATE HYDROCHLORIDE □ DUGERASE □ ETHOCAINE □ IROCAINE □ ISOCAINE-ASID □ ISOCAINE-HEISLER □ JUVOCAINE □ KEROCAINE □ LACTOCAINE □ NAUCAINE □ NEOCAINE □ NOVOCAIN-CHLORHYDRAT (GERMAN) □ NOVOCAINE HYDROCHLORIDE □ NOVOCAIN HYDROCHLORID (GERMAN) □ PARACAIN □ PLANOCAINE □ PROCAINE HYDROCHLORIDE □ SCUROCAINE □ SEVICAINE □ SYNCAINE □ TOPOKAIN □ WESTOCAINE

TOXICITY DATA WITH REFERENCE
mul-man TDLo:2286 mg/kg/8D-I:SYS AJKDDP 7,502,86
orl-rat LD50:200 mg/kg ZENBAX 6B,183,51
ipr-rat LD50:160 mg/kg KSRNAM 13,791,79
ivn-rat LD50:38 mg/kg ANESAV 3,398,42
orl-mus LD50:175 mg/kg ZENBAX 6B,183,51
orl-rat LD50:200 mg/kg ZNTFA2 6B,183,51
ipr-rat LD50:160 mg/kg KSRNAM 13,791,79
ivn-rat LD50:38 mg/kg ANESAV 3,398,42
unr-rat LD50:150 mg/kg ARZNAD 17,1012,67
orl-mus LD50:175 mg/kg ZNTFA2 6B,183,51
ipr-mus LD50:165 mg/kg ARZNAD 26,793,76
scu-mus LD50:339 mg/kg BCFAAI 107,310,68
ivn-mus LD50:46,200 μg/kg TXAPA9 1,454,56
ims-mus LD50:500 mg/kg THERAP 9,332,54
ivn-dog LD50:63 mg/kg JDREAF 20,425,41

CONSENSUS REPORTS: Reported in EPA TSCA Inventory. EPA Genetic Toxicology Program.

SAFETY PROFILE: Poison by ingestion, subcutaneous, intravenous, and intraperitoneal routes. Human systemic effects: acute renal failure. May have human reproductive effects. See also ESTERS. When heated to decomposition it emits very toxic fumes of HCl and NO_x. Used as a local anesthetic.

AIT500 CAS:5988-31-8 HR: 3
p-AMINO BENZOYL DI-ISO-PROPYL AMINO ETHANOL HYDROCHLORIDE
mf: $C_{15}H_{24}N_2O_2$•ClH mw: 300.87

SYN: p-AMINOBENZOIC ACID-2-(DIISOPROPYLAMINO)ETHYL ESTER, HYDROCHLORIDE

TOXICITY DATA WITH REFERENCE
unr-rat LD50:100 mg/kg ARZNAD 17,1012,67
scu-mus LD50:400 mg/kg JPETAB 24,160,24

SAFETY PROFILE: Poison by subcutaneous and other unspecified routes. See also ESTERS. When heated to decomposition it emits very toxic fumes of HCl and NO_x.

AIT750 CAS:532-62-7 HR: 3
p-AMINOBENZOYLDIMETHYLAMINO-1,2-DIMETHYLPROPANOL HYDROCHLORIDE
mf: $C_{14}H_{22}N_2O_2$•ClH mw: 286.84

SYNS: p-AMINOBENZOIC ACID 3-(DIMETHYLAMINO)-1,2-DIMETHYLPROPYL ESTER, HYDROCHLORIDE □ BUTAMIN □ 3-DIMETHYLAMINO-1,2-DIMETHYLPROPYL p-AMINOBENZOATE HYDROCHLORIDE □ 4-(DIMETHYLAMINO)-3-METHYL-2-BUTANOL 4-AMINOBENZOATE (ester) HYDROCHLORIDE □ 3-DIMETHYL-1,2-DIMETHYLPROPYL p-AMINOBENZOATE HYDROCHLORIDE □ TOTOCAINE HYDROCHLORIDE □ TUTOCAINE HYDROCHLORIDE

TOXICITY DATA WITH REFERENCE
scu-mus LDLo:350 mg/kg PHREA7 12,190,32
ivn-mus LDLo:50 mg/kg PHREA7 12,190,32
ipr-dog LDLo:82 mg/kg HBTXAC 1,308,55
ivn-dog LDLo:15 mg/kg PHREA7 12,190,32
scu-rbt LDLo:200 mg/kg PHREA7 12,190,32
ivn-rbt LDLo:15 mg/kg PHREA7 12,190,32
isp-rbt LDLo:16 mg/kg JPETAB 57,221,36
ipr-gpg LDLo:250 mg/kg PHREA7 12,190,32
scu-gpg LDLo:193 mg/kg PHREA7 12,190,32
ivn-gpg LDLo:30 mg/kg PHREA7 12,190,32

SAFETY PROFILE: Poison by subcutaneous, intravenous, intraperitoneal, and intraspinal routes. When heated to decomposition it emits very toxic fumes of HCl and NO_x. Used as a surface and infiltration anesthetic.

AIU000 **CAS:17599-09-6** **HR: 3**
p-AMINO BENZOYL DI-N-PROPYL AMINOPROPANOL HYDROCHLORIDE
mf: $C_{16}H_{26}N_2O_2{\cdot}ClH$ mw: 314.90

SYN: p-AMINOBENZOIC ACID 3-(DIPROPYLAMINO)PROPYL ESTER, HYDROCHLORIDE

TOXICITY DATA WITH REFERENCE
unr-rat LD50:100 mg/kg ARZNAD 17,1012,67
scu-mus LD50:200 mg/kg JPETAB 24,160,24

SAFETY PROFILE: Poison by subcutaneous and possibly other unspecified routes. See also ESTERS and AMINES. When heated to decomposition it emits very toxic fumes of HCl and NO_x.

AIU250 **CAS:67031-48-5** **HR: 3**
β-4-AMINOBENZOYLOXY-β-PHENYLETHYL DIMETHYLAMINE
mf: $C_{17}H_{20}N_2O_2$ mw: 284.39

SYN: p-AMINOBENZOIC ACID-(2-(DIMETHYLAMINO)-1-PHENYL) ETHYL ESTER

TOXICITY DATA WITH REFERENCE
scu-mus LDLo:57 mg/kg AIPTAK 47,96,34
ivn-rbt LDLo:11 mg/kg AIPTAK 47,96,34

SAFETY PROFILE: Poison by subcutaneous and intravenous routes. See also ESTERS. When heated to decomposition it emits toxic fumes of NO_x.

AIU500 **CAS:51579-82-9** **HR: 3**
2-AMINO-3-BENZOYLPHENYLACETIC ACID
mf: $C_{15}H_{13}NO_3$ mw: 255.29

PROP: Mp: 121–123° (decomp).

SYNS: AMFENACO (SPANISH) □ 2-AMINO-3-BENZOYL BENZENEACETIC ACID

TOXICITY DATA WITH REFERENCE
orl-rat LD50:615 mg/kg DRFUD4 3,340,78
orl-mus LD50:311 mg/kg DRFUD4 3,340,78

SAFETY PROFILE: Poison by ingestion. When heated to decomposition it emits toxic fumes of NO_x.

AIV500 **CAS:69-53-4** **HR: 3**
AMINOBENZYLPENICILLIN
mf: $C_{16}H_{19}N_3O_4S$ mw: 349.44

PROP: Solid. Mp: 199–202° (decomp).

SYNS: ACILLIN □ ADOBACILLIN □ ALPEN □ AMBLOSIN □ AMCILL □ AMFIPEN □ d-(−)-α-AMINOBENZYLPENICILLIN □ d-(−)-α-AMINOPENICILLIN □ 6-(d(−)-α-AMINOPHENYLACETAMIDO)PENICILLANIC ACID □ (AMINOPHENYLMETHYL)-PENICILLIN □ AMIPENIX S □ AMPERIL □ AMPI-BOL □ AMPICILLIN (USDA) □ d-AMPICILLIN □ d-(−)-AMPICILLIN □ AMPICILLIN A □ AMPICILLIN ACID □ AMPICILLIN ANHYDRATE □ AMPICIN □ AMPIKEL □ AMPIMED □ AMPIPENIN □ AMPLISOM □ AMPLITAL □ AMPY-PENYL □ AUSTRAPEN □ AY-6108 □ BINOTAL □ BONAPICILLIN □ BRITACIL □ BRL □ BRL 1341 □ COPHARCILIN □ CYMBI □ DIVERCILLIN □ DOKTACILLIN □ GRAMPENIL □ GUICITRINA □ GUICITRINE □LIFEAMPIL □ MARISILAN □ NSC-528986 □ NUVAPEN □ OMNIPEN □ P-50 □ PENBRISTOL □ PENBRITIN □ PENBRITIN PAEDIATRIC □ PENBRITIN SYRUP □ PENBROCK □ PENICLINE □ PENTREX □ PENTREXL □ PFIZERPEN A □ POLYCILLIN □ PONECIL □ PRINCIPEN □ QIDAMP □ RO-AMPEN □ SEMICILLIN □ SK-AMPICILLIN □ SYNPENIN □ TOKIOCILLIN □ TOLOMOL □ TOTACILLIN □ TOTALCICLINA □ TOTAPEN □ ULTRABION □ ULTRABRON □ VICCILLIN □ VICCILLIN S □ VICILLIN □ WY-5103

TOXICITY DATA WITH REFERENCE
dnr-esc 20 μL/plate MUREAV 97,1,82
pic-esc 10 ng/plate CNREA8 43,2819,83
orl-rat TDLo:2500 mg/kg (female 4-13D post):TER BEXBAN 91,169,81
orl-man TDLo:400 mg/kg/4W-I:BLD,MET AIMEAS 69,91,68
orl-wmn TDLo:160 mg/kg/4D-I:BLD AIMEAS 99,573,83
ipr-rat LD50:4500 mg/kg TXAPA9 18,185,71
ivn-rat LD50:6200 mg/kg DRUGAY-,88,90
ipr-mus LD50:3250 mg/kg EKFMA7 9,83,80
ivn-mus LD50:4600 mg/kg YKKZAJ 97,987,77
ice-mus LD50:380 mg/kg NKRZAZ 26,196,80

CONSENSUS REPORTS: EPA Genetic Toxicology Program. IARC Cancer Review: Group 3 IMEMDT 50,153,90; Animal Limited Evidence IMEMDT 50,153,90; Human Inadequate Evidence IMEMDT 50,153,90.

SAFETY PROFILE: Poison by intracerebral route. Moderately toxic by intraperitoneal route. Human systemic effects by ingestion: fever, angranulocytosis, and other blood effects. An experimental teratogen. Mutation data reported. When heated to decomposition it emits very toxic fumes of NO_x and SO_x.

AIV625 **HR: 3**
4-AMINO-N-(1-BENZYL-4-PIPERIDYL)-5-CHLORO-o-ANISAMIDE HYDROXYSUCCINATE
mf: $C_{20}H_{24}ClN_3O_2{\cdot}C_4H_6O_5$ mw: 507.97

TOXICITY DATA WITH REFERENCE
orl-rat LD50:2780 mg/kg IYKEDH 16,866,85
ipr-rat LD50:159 mg/kg IYKEDH 16,866,85
scu-rat LD50:4850 mg/kg IYKEDH 16,866,85
ivn-rat LD50:39 mg/kg IYKEDH 16,866,85
ims-rat LD50:2080 mg/kg IYKEDH 16,866,85
orl-mus LD50:510 mg/kg IYKEDH 16,866,85
ipr-mus LD50:145 mg/kg IYKEDH 16,866,85
scu-mus LD50:305 mg/kg IYKEDH 16,866,85
ivn-mus LD50:51 mg/kg IYKEDH 16,866,85
ims-mus LD50:290 mg/kg IYKEDH 16,866,85

SAFETY PROFILE: Poison by subcutaneous, intramuscular, intravenous and intraperitoneal routes. Moderately toxic by ingestion. When heated to decomposition it emits toxic fumes of Cl^- and NO_x.

AIV750 **CAS:4363-03-5** **HR: 2**
4-AMINO-3-BIPHENYLOL
mf: $C_{12}H_{11}NO$ mw: 185.24

PROP: Plates from alc (aq). Mp: 182–184°.

SYN: 4-AMINO-3-HYDROXYBIPHENYL

TOXICITY DATA WITH REFERENCE
scu-mus TDLo:216 mg/kg/3D:CAR JNCIAM 41,403,68

SAFETY PROFILE: Questionable carcinogen with experimental carcinogenic data. When heated to decomposition it emits toxic fumes of NO_x.

AIW000 CAS:1204-79-1 HR: 2
4′-AMINO-4-BIPHENYLOL
mf: $C_{12}H_{11}NO$ mw: 185.24

PROP: Plates from alc (aq). Mp: 273°.

SYN: 4-AMINO-4′-HYDROXYBIPHENOL

TOXICITY DATA WITH REFERENCE
scu-mus TDLo:216 mg/kg/3D:CAR JNCIAM 41,403,68

SAFETY PROFILE: Questionable carcinogen with experimental carcinogenic data. When heated to decomposition it emits toxic fumes of NO_x.

AIW250 CAS:1204-59-7 HR: 2
3-AMINO-4-BIPHENYLOL HYDROCHLORIDE
mf: $C_{12}H_{11}NO \cdot ClH$ mw: 221.70

SYNS: 3-AMINO-4-HYDROXYDIPHENYL HYDROCHLORIDE □ 4-HYDROXY-3-AMINODIPHENYL HYDROCHLORIDE

TOXICITY DATA WITH REFERENCE
imp-mus TDLo:50 mg/kg:CAR BJCAAI 12,222,58

SAFETY PROFILE: Questionable carcinogen with experimental carcinogenic data by implantation. When heated to decomposition it emits very toxic fumes of HCl and NO_x.

AIW500 CAS:65146-47-6 HR: 3
2-AMINO-5-BIPHENYLYLIMIDAZOLEHYDROCHLORIDE
mf: $C_{15}H_{13}N_3 \cdot ClH$ mw: 271.77

TOXICITY DATA WITH REFERENCE
orl-mus LD50:720 mg/kg ARZNAD 27,1889,77
ipr-mus LD50:50 mg/kg ARZNAD 27,1889,77

SAFETY PROFILE: Poison by intraperitoneal route. Moderately toxic by ingestion. When heated to decomposition it emits very toxic fumes of HCl and NO_x.

AIW750 CAS:74039-01-3 HR: 1
2-(4′-AMINO-1,1′-BIPHENYL-4-YL)-2H-NAPHTHO(1,2-d)TRIAZOLE-6,8-DISULFONIC ACID, DIPOTASSIUM SALT
mf: $C_{22}H_{14}N_4O_6S_2 \cdot 2K$ mw: 572.72

SYN: 2-(4′-AMINOXENYL)NAFTO-α,β-TRIAZOL-6,8-DISULFONAN DRASELNY (CZECH)

TOXICITY DATA WITH REFERENCE
skn-rbt 500 mg/24H MLD 28ZPAK -,196,72
eye-rbt 500 mg/24H MLD 28ZPAK -,196,72
orl-rat LD50:7210 mg/kg 28ZPAK -,196,72

SAFETY PROFILE: Mildly toxic by ingestion. A skin and eye irritant. See also SULFONATES. When heated to decomposition it emits very toxic fumes of SO_x, NO_x, and K_2O.

AIX000 CAS:1031-47-6 HR: 3
5-AMINO-1-BIS(DIMETHYLAMIDE)PHOSPHORYL-3-PHENYL-1,2,4-TRIAZOLE
mf: $C_{12}H_{19}N_6OP$ mw: 294.34

PROP: Solid from alc (aq). Mp: 167–168°. Sol in most org solvs; spar sol in H_2O.

SYNS: 5-AMINO-1-BIS(DIMETHYLAMIDO)PHOSPHORYL-3-PHENYL-1,2,4-TRIAZOLE □ 5-AMINO-1-(BIS(DIMETHYLAMINO)PHOSPHINYL)-3-PHENYL-1,2,4-TRIAZOLE □ 5-AMINO-3-FENIL-1-BIS(-DIMETILAMINO)-FOSFORIL-1,2,4-TRIAZOLO (ITALIAN) □ 5-AMINO-3-FENYL-1-BIS (DIMETHYL-AMINO)-FOSFORYL-1,2,4-TRIAZOOL (DUTCH) □ 5-AMINO-3-PHENYL-1-BIS(DIMETHYLAMINO)-PHOSPHORYL-1H-1,2,4-TRIAZOL (GERMAN) □ 5-AMINO-3-PHENYL-1-BIS (DIMETHYL-AMINO)-PHOSPHORYLE-1,2,4-TRIAZOLE(FRENCH) □ 5-AMINO-3-PHENYL-1,2,4-TRIAZOLE-1-YL-N,N,N′,N′-TETRAMETHYLPHOSPHODIAMIDE □ 5-AMINO-3-PHENYL-1,2,4-TRIAZOLYL-1-BIS(DIMETHYLAMIDO)PHOSPHATE □ 5-AMINO-3-PHENYL-1,2,4-TRIAZOLYL-N,N,N′N′-TETRAMETHYL-PHOSPHONAMIDE □ p-(5-AMINO-3-PHENYL-1H-1,2,4-TRIAZOL-1-YL)-N,N,N′-TETRAMETHYL PHOSPHONIC DIAMIDE □ BIS (DIMETHYLAMINO)-3-AMINO-5-PHENYLTRIAZOLYL PHOSPHINE OXIDE □ ENT 27,223 □ NIAGARA 5943 □ 3-PHENYL-5-AMINO-1,2,4-TRIAZOLYL-(1)-(N,N′-TETRAMETHYL) DIAMIDOPHOSPHONATE □ TRIAMIFOS (GERMAN, DUTCH, ITALIAN) □ TRIAMIPHOS □ TRIAMPHOS □ WEPSIN □ WEPSYN □ WEPSYN 155 □ WP 155

TOXICITY DATA WITH REFERENCE
mrc-asn 400 ppm ENMUDM 2,359,80
orl-rat TDLo:6575 μg/kg:TER TXCYAC 2,327,74
orl-rat LD50:20 mg/kg FMCHA2 -,D333,80
skn-rat LD50:48 mg/kg WRPCA2 9,119,70
unk-rat LD50:10 mg/kg 30ZDA9 -,427,71
ipr-rat LD50:15 mg/kg EJPHAZ 16,361,71
orl-mus LD50:10 mg/kg ARSIM* 20,27,66
skn-rbt LD50:1500 mg/kg GUCHAZ 6,508,73
skn rbt LD50:1500 mg/kg

CONSENSUS REPORTS: EPA Extremely Hazardous Substances List.

SAFETY PROFILE: Poison by ingestion, skin contact, intraperitoneal, and possibly other routes. An experimental teratogen. Mutation data reported. When heated to decomposition it emits very toxic fumes of PO_x and NO_x.

AIX250 CAS:56-18-8 HR: 3
AMINOBIS(PROPYLAMINE)
DOT: UN 2269
mf: $C_6H_{17}N_3$ mw: 131.26

PROP: Liquid. Bp: 100° @ 2 mm.

SYNS: AMINOBIS(PROPYLAMINE) □ BIS-(3-AMINOPROPYL) AMINE □ 3,3-DIAMINODIPROPYLAMINE □ 3,3′-DIAMINODIPROPYLAMINE □ DIPROPYLENETRIAMINE □ IMINOBIS(PROPYLAMINE) □ 3,3′-IMINOBIS(PROPYLAMINE) □ 1,3-PROPANEDIAMINE, N-(3-AMINOPROPYL)- □ PROPYLAMINE, 3,3′-IMINOBIS-

TOXICITY DATA WITH REFERENCE
skn-rbt 470 mg open MOD UCDS** 6/13/68
eye-rbt 47 mg SEV UCDS** 6/13/68
orl-rat LD50:738 mg/kg ZHYGAM 20,393,74
orl-mus LD50:435 mg/kg ZHYGAM 20,393,74
orl-rbt LD50:210 mg/kg ZHYGAM 20,393,74
skn-rbt LDLo:110 mg/kg AIHAAP 23,95,62

CONSENSUS REPORTS: Reported in EPA TSCA Inventory.

DOT CLASSIFICATION: 8; *Label:* Corrosive

SAFETY PROFILE: Poison by skin contact. Moderately toxic by ingestion. A skin and severe eye irritant. When heated to decomposition it emits toxic fumes of NO_x. An explosive.

AIX500 CAS:64037-07-6 HR: 3
2-AMINO-5-BROMOBENZOXAZOLE
mf: $C_7H_5BrN_2O$ mw: 213.05

TOXICITY DATA WITH REFERENCE
orl-rat LD50:1000 mg/kg MDCHAG 4(1),338,64
ipr-rat LD50:160 mg/kg MDCHAG 4(1),338,64
orl-mus LD50:819 mg/kg MDCHAG 4(1),336,64
ipr-mus LD50:180 mg/kg MDCHAG 4(1),336,64

SAFETY PROFILE: Poison by intraperitoneal route. Moderately toxic by ingestion. When heated to decomposition it emits very toxic fumes of Br^- and NO_x.

AIX750 CAS:52112-66-0 HR: 3
2-AMINO-6-BROMOBENZOXAZOLE
mf: $C_7H_5BrN_2O$ mw: 213.05

TOXICITY DATA WITH REFERENCE
orl-rat LD50:500 mg/kg MDCHAG 4(1),338,64
ipr-rat LD50:200 mg/kg MDCHAG 4(1),338,64
orl-mus LD50:560 mg/kg MDCHAG 4(1),336,64
ipr-mus LD50:294 mg/kg MDCHAG 4(1),336,64

SAFETY PROFILE: Poison by intraperitoneal route. Moderately toxic by ingestion. See also BROMIDES. When heated to decomposition it emits very toxic fumes of Br^- and NO_x.

AIY000 CAS:64037-09-8 HR: 2
2-AMINO-5-BROMO-6-CHLOROBENZOXAZOLE
mf: $C_7H_4BrClN_2O$ mw: 247.49

TOXICITY DATA WITH REFERENCE
orl-mus LD50:2000 mg/kg MDCHAG 4(1),336,64
ipr-mus LD50:450 mg/kg MDCHAG 4(1),336,64

SAFETY PROFILE: Moderately toxic by ingestion and intraperitoneal route. When heated to decomposition it emits very toxic fumes of Br^-, Cl^-, and NO_x.

AIY250 CAS:64037-08-7 HR: 3
2-AMINO-6-BROMO-5-CHLOROBENZOXAZOLE
mf: $C_7H_4BrClN_2O$ mw: 247.49

TOXICITY DATA WITH REFERENCE
orl-rat LD50:140 mg/kg MDCHAG 4(1),338,64
ipr-rat LD50:140 mg/kg MDCHAG 4(1),338,64
orl-mus LD50:658 mg/kg MDCHAG 4(1),336,64
ipr-mus LD50:240 mg/kg MDCHAG 4(1),336,64

SAFETY PROFILE: Poison by ingestion and intraperitoneal routes. Moderately toxic by ingestion. When heated to decomposition it emits very toxic fumes of Cl^-, Br^-, and NO_x.

AIY500 CAS:116-82-5 HR: 1
1-AMINO-2-BROMO-4-HYDROXYANTHRAQUINONE
mf: $C_{14}H_8BrNO_3$ mw: 318.14

SYN: 1-AMINO-2-BROM-4-HYDROXYANTHRACHINON (CZECH)

TOXICITY DATA WITH REFERENCE
eye-rbt 500 mg/24H MLD 28ZPAK -,83,72

CONSENSUS REPORTS: Reported in EPA TSCA Inventory.

SAFETY PROFILE: An eye irritant. When heated to decomposition it emits very toxic fumes of Br^- and NO_x.

AIY750 CAS:73791-29-4 HR: 1
1-AMINO-2-BROMO-4-(2-(2-HYDROXYETHYL)SULFONYL-4-METHYLPHENYLAMINO)ANTHRAQUINONE
mf: $C_{23}H_{20}BrN_3O_5S$ mw: 530.43

SYN: MODR ALIZARINOVA CISTA B (CZECH)

TOXICITY DATA WITH REFERENCE
eye-rbt 500 mg/24H MLD 28ZPAK -,241,72

SAFETY PROFILE: An eye irritant. See also SULFONATES. When heated to decomposition it emits very toxic fumes of Br^-, NO_x, and SO_x.

AIY850 CAS:56741-95-8 HR: 2
2-AMINO-5-BROMO-6-PHENYL-4(1H)-PYRIMIDINONE
mf: $C_{10}H_8BrN_3O$ mw: 266.12

PROP: Crystals from alc (aq). Mp: 268–270°.

SYNS: BROPIRAMINE □ BROPIRIMINE □ 4(1H)-PYRIMIDINONE, 2-AMINO-5-BROMO-6-PHENYL- □ U-54461

TOXICITY DATA WITH REFERENCE
mnt-mus-orl 500 mg/kg MUREAV 252,239,91
cyt-mus:lym 200 mg/L MUREAV 252,221,91
orl-rat TDLo:1450 mg/kg (female 15-22D post):REP TJADAB 35,30A,87
orl-rat TDLo:2900 mg/kg (female 15-22D post):TER TJADAB 35,30A,87
orl-rat LD50:>3200 mg/kg DRFUD4 9,567,84

SAFETY PROFILE: Moderately toxic by ingestion. An experimental teratogen. Other experimental reproductive effects. Mutation data reported. When heated to decomposition it emits toxic fumes of NO_x, Br^-.

AIZ000 CAS:66064-11-7 HR: 3
N-AMINO-2-(m-BROMOPHENYL)SUCCINIMIDE
mf: $C_{10}H_9BrN_2O_2$ mw: 269.12

TOXICITY DATA WITH REFERENCE
orl-mus LD50:3682 mg/kg ARZNAD 29,290,79
ipr-mus LD50:387 mg/kg EJMCA5 13,465,78

SAFETY PROFILE: Poison by intraperitoneal route. Moderately toxic by ingestion. When heated to decomposition it emits very toxic fumes of Br^- and NO_x.

AJA000 **CAS:5003-71-4** **HR: 3**
1-AMINO-3-BROMOPROPANE HYDROBROMIDE
mf: $C_3H_6BrN\cdot BrH$ mw: 216.93

PROP: Mp: 171–172°.

SYNS: 3-BROMO-1-PROPANAMINE HYDROBROMIDE □ 3-BROMO-PROPYLAMINE HYDROBROMIDE

TOXICITY DATA WITH REFERENCE
ipr-mus TDLo: 1150 mg/kg/8W-I: ETA CNREA8 39,391,79
ipr-mus LD50: 109 mg/kg YKKZAJ 97,1117,77

CONSENSUS REPORTS: Reported in EPA TSCA Inventory.

SAFETY PROFILE: Poison by intraperitoneal route. Questionable carcinogen with experimental tumorigenic data. When heated to decomposition it emits very toxic fumes of HBr and NO_x.

AJA250 **CAS:96-20-8** **HR: 3**
2-AMINOBUTAN-1-OL
mf: $C_4H_{11}NO$ mw: 89.16

PROP: Water-white liquid. Mp: −2°, bp: 178°, flash p: 165°F (OC), d: 0.944 @ 20°/20°, vap d: 3.06.

SYNS: 2-AMINO-1-BUTANOL □ 2-AMINO-n-BUTYL ALCOHOL □ BUTANOL-2-AMINE

TOXICITY DATA WITH REFERENCE
orl-mus LD50: 2300 mg/kg 20PKA3 -,-,67
ipr-mus LDLo: 250 mg/kg CBCCT* 5,338,53
ivn-mus LD50: 316 mg/kg CSLNX* NX#00036

CONSENSUS REPORTS: Reported in EPA TSCA Inventory.

SAFETY PROFILE: Poison by intravenous and intraperitoneal routes. Moderately toxic by ingestion. Moderately flammable when exposed to heat, flame, or oxidizing materials. To fight fire, use water spray, alcohol foam, dry chemical. When heated to decomposition it yields NO_x. See also ALCOHOLS and AMINES.

AJA375 **CAS:52712-76-2** **HR: 3**
4-AMINO-2-(4-BUTANOYLHEXAHYDRO-1H-1,4-DIAZEPIN-1-YL)-6,7-DIMETHOXYQUINAZOLINE HYDROCHLORIDE
mf: $C_{19}H_{27}N_5O_3\cdot ClH$ mw: 409.97

PROP: Solid. Mp: 280–282°.

TOXICITY DATA WITH REFERENCE
orl-rat TDLo: 169 mg/kg (female 17-22D post): REP KSRNAM 17,930,83
orl-rat TDLo: 1100 mg/kg (7-17D preg): TER KSRNAM 17,914,83
orl-rat LD50: 980 mg/kg KSRNAM 17,843,83
scu-rat LD50: 365 mg/kg KSRNAM 17,843,83
ivn-rat LD50: 50 mg/kg KSRNAM 17,843,83
ims-rat LD50: 152 mg/kg KSRNAM 17,843,83
orl-mus LD50: 1201 mg/kg KSRNAM 17,843,83

SAFETY PROFILE: Poison by subcutaneous, intravenous, and intramuscular routes. Moderately toxic by ingestion and other routes. An experimental teratogen. Other experimental reproductive effects. When heated to decomposition it emits toxic fumes of NO_x and HCl.

AJA500 **CAS:3624-87-1** **HR: 3**
3-AMINO-2-BUTOXYBENZOIC ACID-2-DIETHYLAMINOETHYL ESTER HYDROCHLORIDE
mf: $C_{17}H_{28}N_2O_3\cdot ClH$ mw: 344.93

SYNS: 2-BUTOXY-3-AMINOBENZOIC ACID β-DIETHYLAMINOETHYL ESTER HYDROCHLORIDE □ 2′-DIETHYLAMINOETHYL-3-AMINO-2-BUTOXYBENZOATE HYDROCHLORIDE □ β DIETHYLAMINOETHYL-2-BUTOXY-3-AMINOBENZOATE HYDROCHLORIDE □ METHAMBUCAINE HYDROCHLORIDE □ METHAMBUTOXYCAINE HYDROCHLORIDE □ PRIMACAINE □ PRIMACAINE HYDROCHLORIDE

TOXICITY DATA WITH REFERENCE
ivn-rat LD50: 11 mg/kg CLDND*
ipr-mus LD50: 192 mg/kg CLDND*
scu-mus LD50: 392 mg/kg CLDND*
ivn-mus LD50: 23 mg/kg CLDND*
ivn-cat LD50: 8 mg/kg CLDND*
ivn-rbt LD50: 17 mg/kg CLDND*
ipr-gpg LD50: 212 mg/kg CLDND*

SAFETY PROFILE: Poison by subcutaneous, intravenous, and intraperitoneal routes. See also ESTERS. When heated to decomposition it emits very toxic fumes of HCl and NO_x.

AJA550 **CAS:104-13-2** **HR: 3**
p-AMINOBUTYLBENZENE
mf: $C_{10}H_{15}N$ mw: 149.26

SYNS: 1-AMINO-4-BUTYLBENZENE □ ANILINE, 4-BUTYL □ BENZENAMINE, 4-BUTYL-(9CI) □ p-n-BUTYLANILINE □ 4-BUTYL-BENZENAMINE

TOXICITY DATA WITH REFERENCE
ipr-mus LD50: 81 mg/kg JMCMAR 17,900,74

CONSENSUS REPORTS: Reported in EPA TSCA Inventory.

SAFETY PROFILE: Poison by intraperitoneal route. When heated to decomposition it emits toxic vapors of NO_x.

AJA750 **CAS:3037-72-7** **HR: 3**
(4-AMINOBUTYL)DIETHOXYMETHYLSILANE
mf: $C_9H_{23}NO_2Si$ mw: 205.42

SYN: Δ-AMINOBUTYLMETHYLDIETHOXYSILANE

TOXICITY DATA WITH REFERENCE
skn-rbt 10 mg/24H open MLD AIHAAP 23,95,62
orl-rat LDLo: 6500 mg/kg AIHAAP 23,95,62
skn-rat LD50: 45 mg/kg JPMSAE 60,1113,71
skn-mus LD50: 45 mg/kg JPMSAE 60,1113,71
skn-rbt LD50: 45 mg/kg AIHAAP 23,95,62

CONSENSUS REPORTS: Reported in EPA TSCA Inventory, EPA Extremely Hazardous Substances List.

SAFETY PROFILE: Poison by skin contact. Moderately toxic by ingestion. A skin irritant. When heated to decomposition it emits toxic fumes of NO_x.

AJB000 **CAS:33132-75-1** **HR: 3**
p-AMINO-β-sec-BUTYL-N,N-DIMETHYLPHENETHYLAMINE
mf: $C_{14}H_{22}N_2$ mw: 218.38

TOXICITY DATA WITH REFERENCE
orl-mus LD50:185 mg/kg CHTPBA 6,453,71
ivn-mus LD50:45 mg/kg CHTPBA 6,453,71

SAFETY PROFILE: Poison by ingestion and intravenous routes. When heated to decomposition it emits toxic fumes of NO_x.

AJB250 **CAS:118-68-3** **HR: 3**
3-(2-AMINOBUTYL)INDOLE ACETATE
mf: $C_{12}H_{16}N_2 \cdot C_2H_4O_2$ mw: 248.36

PROP: Solid. Mp: 165–166°.

SYNS: α-ETHYLTRYPTAMINE ACETATE □ dl-α-ETHYLTRYPTAMINE ACETATE □ ETRYPTAMINE ACETATE □ INDOLE-3-(2-AMINOBUTYL) ACETATE

TOXICITY DATA WITH REFERENCE
orl-rat LD50:49 mg/kg TXAPA9 4,547,62
ipr-mus LD50:72 mg/kg TXAPA9 4,547,62
ivn-mus LD50:45 mg/kg CSLNX* NX#00376

SAFETY PROFILE: Poison by ingestion, intraperitoneal, and intravenous routes. See also ESTERS. When heated to decomposition it emits toxic fumes of NO_x.

AJB500 **CAS:18237-16-6** **HR: 3**
3-(4-AMINOBUTYL)INDOLE HYDROCHLORIDE
mf: $C_{12}H_{16}N_2 \cdot ClH$ mw: 224.76

SYN: Δ-INDOLYBUTYLAMINE HYDROCHLORIDE

TOXICITY DATA WITH REFERENCE
ipr-mus LD50:222 mg/kg RPTOAN 33,180,70
ivn-mus LD50:83 mg/kg RPTOAN 33,180,70

SAFETY PROFILE: Poison by intraperitoneal and intravenous routes. When heated to decomposition it emits very toxic fumes of NO_x and HCl.

AJC000 **CAS:30653-83-9** **HR: 3**
5-AMINO-N-BUTYL-2-PROPARGYLOXYBENZAMIDE
mf: $C_{14}H_{18}N_2O_2$ mw: 246.34

PROP: Crystals from alc. Mp: 85–87°.

SYNS: 5-AMINO-N-BUTYL-2-(2-PROPYNYLOXY)BENZAMIDE □ MY 41-6 □ PARSAL □ PARSALMIDE □ 2-PROPARGILOSSI-5-AMINO-N-(n-BUTIL)-BENZAMIDE (ITALIAN)

TOXICITY DATA WITH REFERENCE
orl-rat LD50:864 mg/kg DRFUD4 2,55,77
orl-mus LD50:428 mg/kg DRFUD4 2,55,77
ivn-mus LD50:148 mg/kg DRFUD4 2,55,77

SAFETY PROFILE: Poison by intravenous route. Moderately toxic by ingestion. When heated to decomposition it emits toxic fumes of NO_x.

AJC250 **CAS:3069-30-5** **HR: 2**
(4-AMINOBUTYL)TRIETHOXYSILANE
mf: $C_{10}H_{25}NO_3Si$ mw: 235.45

SYN: 4-(TRIETHOXYSILYL)BUTYLAMINE

TOXICITY DATA WITH REFERENCE
skn-rbt 10 mg/24H open MLD AIHAAP 23,95,62
orl-rat LD50:1620 mg/kg AIHAAP 23,95,62
skn-rbt LD50:2500 mg/kg AIHAAP 23,95,62

SAFETY PROFILE: Moderately toxic by ingestion and skin contact. A skin irritant. When heated to decomposition it emits toxic fumes of NO_x.

AJC375 **CAS:3251-08-9** **HR: 2**
4-AMINOBUTYRAMIDE
mf: $C_4H_{10}N_2O$ mw: 102.16

SYNS: AMGABA □ AMIDE of GABA

TOXICITY DATA WITH REFERENCE
ipr-rat LD50:900 mg/kg AITEAT 13,70,65
ipr-mus LD50:1080 mg/kg AITEAT 13,70,65
ivn-rbt LDLo:350 mg/kg AITEAT 13,70,65

SAFETY PROFILE: Moderately toxic by some routes. When heated to decomposition it emits toxic fumes of NO_x. See also AMINES and AMIDES.

AJC500 **CAS:34562-99-7** **HR: 3**
γ-AMINOBUTYRIC ACID CETYL ESTER
mf: $C_{20}H_{41}NO_2$ mw: 327.62

SYNS: CETYL-γ-AMINOBUTYRATE □ CETYL GABA

TOXICITY DATA WITH REFERENCE
ipr-mus LD50:155 mg/kg NEPHBW 19,217,80
ivn-mus LD50:22 mg/kg NEPHBW 19,217,80

SAFETY PROFILE: Poison by intravenous and intraperitoneal routes. See also ESTERS. When heated to decomposition it emits toxic fumes of NO_x.

AJC625 **CAS:3251-07-8** **HR: 2**
4-AMINOBUTYRIC ACID METHYL ESTER
mf: $C_5H_{11}NO_2$ mw: 117.17

SYNS: MEGABA □ METHYL ESTER of GABA

TOXICITY DATA WITH REFERENCE
ipr-rat LD50:950 mg/kg AITEAT 13,70,65
ipr-mus LD50:1300 mg/kg AITEAT 13,70,64
ivn-rbt LDLo:300 mg/kg AITEAT 13,70,65

SAFETY PROFILE: Moderately toxic by some routes. When heated to decomposition it emits toxic fumes of NO_x. See also AMINES and ESTERS.

AJC750 **CAS:1688-71-7** **HR: 3**
4′-AMINOBUTYROPHENONE
mf: $C_{10}H_{13}NO$ mw: 163.24

SYNS: p-AMINOBUTYROPHENONE □ 1-(4-AMINOPHENYL)-1-BUTANONE □ 1-BUTANONE, 1-(4-AMINOPHENYL)-(9CI) □ BUTYROPHENONE, 4′-AMINO-

A

TOXICITY DATA WITH REFERENCE
orl-rat LD50:84 mg/kg GEPHDP 14,465,83
orl-mus LD50:133 mg/kg GEPHDP 14,465,83
ipr-mus LD50:183 mg/kg GEPHDP 14,465,83
orl-qal LD50:178 mg/kg AECTCV 12,355,83
orl-brd LD50:42,200 μg/kg AECTCV 12,355,83

CONSENSUS REPORTS: Reported in EPA TSCA Inventory.

SAFETY PROFILE: Poison by ingestion and intraperitoneal routes. When heated to decomposition it emits toxic vapors of NO_x.

AJD000 CAS:60-32-2 HR: 2
6-AMINOCAPROIC ACID
mf: $C_6H_{13}NO_2$ mw: 131.20

PROP: Leaflets. Mp: 204–206°. Sol in water; insol in EtOH.

SYNS: ACEPRAMINE ☐ ACS ☐ AFIBRIN ☐ AMICAR ☐ AMINOCAPROIC ACID ☐ ω-AMINOCAPROIC ACID ☐ ε-AMINOCAPROIC ACID ☐ ω-AMINOHEXANOIC ACID ☐ AMINOKAPRON ☐ CAPRAMOL ☐ CAPRALENSE ☐ CAPROCID ☐ CAPROLISIN ☐ CL 10304 ☐ CY 116 ☐ EACA ☐ EACA KABI ☐ EACS ☐ EPSAMON ☐ EPSICAPRON ☐ HEMOCAPROL ☐ HEMOPAR ☐ HEPIN ☐ IPSILON ☐ 177 J.D. ☐ ε-LEUCINE ☐ ε-NORLEUCINE ☐ NSC-26154 ☐ RESPRAMIN

TOXICITY DATA WITH REFERENCE
eye-rbt 500 mg/24H MLD 28ZPAK -,128,72
orl-rat TDLo:153 g/kg (40D pre/1-21D preg):REP APTOA6 22,340,65
orl-man TDLo:1778 mg/kg/8D-I:SYS AJKDDP 8,441,86
mul-man TDLo:14,400 mg/kg/59D-I BJURAN 60,81,87
ipr-rat LD50:7000 mg/kg PHMCAA 3,62,61
ivn-rat LD50:3300 mg/kg PHMCAA 3,62,61
orl-mus LD50:14,300 mg/kg NIIRDN 6,79,82
ivn-mus LD50:4900 mg/kg AAREAV 22,481,65
ivn-dog LDLo:2150 mg/kg PHMCAA 3,62,61
ivn-gpg LDLo:19,800 mg/kg AAREAV 22,481,65

CONSENSUS REPORTS: Reported in EPA TSCA Inventory.

SAFETY PROFILE: Moderately toxic by intravenous route. Human systemic effects by ingestion: changes in tubules (including acute renal failure, acute tubular necrosis), hematuria, and increased body temperature. Experimental reproductive effects. An eye irritant. When heated to decomposition it emits toxic fumes such as NO_x.

AJD250 CAS:38237-76-2 HR: 3
p-AMINO CAPROPHENONE
mf: $C_{12}H_{17}NO$ mw: 191.30

SYN: p-HEXANOYLANILINE

TOXICITY DATA WITH REFERENCE
orl-rat LD50:216 mg/kg GEPHDP 14,465,83
orl-mus LD50:299 mg/kg GEPHDP 14,465,83
ipr-mus LD50:35 mg/kg JMCMAR 17,900,74

SAFETY PROFILE: Poison by ingestion and intraperitoneal routes. When heated to decomposition it emits toxic fumes such as NO_x. See also AROMATIC AMINES.

AJD500 CAS:64686-82-4 HR: 3
4-AMINO-5-CARBAMYL-3-BENZYLTHIAZOLE-2(3H)-THIONE
mf: $C_{11}H_{11}N_3OS_2$ mw: 265.37

TOXICITY DATA WITH REFERENCE
orl-mus LD50:250 mg/kg ARZNAD 27,1652,77
ipr-mus LD50:100 mg/kg ARZNAD 27,1652,77

SAFETY PROFILE: Poison by ingestion and intraperitoneal routes. When heated to decomposition it emits very toxic fumes of NO_x and SO_x.

AJD750 CAS:26148-68-5 HR: 3
AMINO-α-CARBOLINE
mf: $C_{11}H_9N_3$ mw: 183.2

PROP: Crystals from $CHCl_3$/hexane or EtOH. Mp: 202–203°.

SYNS: 2-AMINO-α-CARBOLINE ☐ 2-AMINO-9H-PYRIDO(2,3-B)INDOLE

TOXICITY DATA WITH REFERENCE
sce-hmn:lym 4000 μg/L MUREAV 77,65,80
mmo-sat 1 μg/plate ABCHA6 43,1155,79
mma-sat 1 μg/plate CALEDQ 10,141,80
dnr-bcs 10 μL/plate ABCHA6 45,2031,81
slt-dmg-orl 400 ng/kg JJCREP 76,468,85
orl-mus TDLo:37,600 mg/kg/98W-C:CAR EVHPAZ 67,129,86
orl-mus TD:50,424 mg/kg/82W-C:CAR CRNGDP 5,815,84

CONSENSUS REPORTS: Cancer Review: Group 2B IMEMDT 7,56,87; Animal Sufficient Evidence IMEMDT 40,245,86.

SAFETY PROFILE: Suspected carcinogen with experimental carcinogenic data. Human mutation data reported. When heated to decomposition it emits toxic fumes of NO_x.

AJE000 CAS:3688-35-5 HR: 3
AMINOCHLORAMBUCIL
mf: $C_{14}H_{20}Cl_2N_2O_2$ mw: 319.26

SYNS: 2-AMINO-4-(p-(BIS(2-CHLOROETHYL)AMINO)PHENYL)BUTYRIC ACID ☐ α-AMINO-γ-(p-DICHLOROETHYLAMINO)-PHENYLBUTYRIC ACID ☐ CB-1385

TOXICITY DATA WITH REFERENCE
sln-dmg-unk 10 mmol/L ANYAA9 160,228,69
sln-dmg-par 3 mmol/L GENRA8 1,173,60
sln-dmg par 3 mmol/L GENRA8 1,173,60
ipr-rat LDLo:25 mg/kg BCPCA6 5,192,60
scu-mus LD10:20 mg/kg EJCAAH 10,667,74

CONSENSUS REPORTS: EPA Genetic Toxicology Program.

SAFETY PROFILE: Poison by subcutaneous and intraperitoneal routes. Mutation data reported. When heated to decomposition it emits very toxic fumes of Cl^- and NO_x.

AJE250 CAS:50416-18-7 HR: 2
2-AMINO-6'-CHLORO-o-ACETOTOLUIDIDE, HYDROCHLORIDE
mf: $C_9H_{11}ClN_2O \cdot ClH$ mw: 235.13

SYNS: 2-AMINO-2'-CHLORO-6'-METHYLACETANILIDE, HYDROCHLORIDE □ C 3104

TOXICITY DATA WITH REFERENCE
eye-rbt 2% MLD ARZNAD 8,407,58
ipr-rat LD50:525 mg/kg ARZNAD 8,407,58
ipr-mus LD50:565 mg/kg ARZNAD 8,407,58
scu-mus LD50:840 mg/kg ARZNAD 8,407,58

SAFETY PROFILE: Moderately toxic by intraperitoneal and subcutaneous routes. An eye irritant. When heated to decomposition it emits very toxic fumes of Cl^-, NO_x, and HCl. See also AROMATIC AMINES.

AJE325 CAS:117-11-3 HR: 1
1-AMINO-5-CHLOROANTHRAQUINONE
mf: $C_{14}H_8ClNO_2$ mw: 257.68

SYNS: 1-CHLOR-5-AMINOANTHRACHINON (CZECH) □ 5-CHLORO-1-AMINOANTHRAQUINONE

TOXICITY DATA WITH REFERENCE
eye-rbt 500 mg/24H MLD 28ZPAK -,87,72

CONSENSUS REPORTS: Reported in EPA TSCA Inventory.

SAFETY PROFILE: An eye irritant. When heated to decomposition it emits very toxic fumes of Cl^- and NO_x.

AJE350 CAS:94110-08-4 HR: 2
3-AMINO-4-CHLOROBENZOIC ACID 2-((DIMETHYLAMINO)ETHYL) ESTER HYDROCHLORIDE
mf: $C_{11}H_{15}ClN_2O_2 \cdot ClH$ mw: 279.19

SYNS: 3-AMINO-4-CLORO-BENZOATO di DIMETILAMINOETILE CLORIDRATO (ITALIAN) □ 4-CLORO-3-AMINOBENZOATO di DIMETILAMINOETILE CLORIDRATO (ITALIAN) □ REC 1-0060

TOXICITY DATA WITH REFERENCE
orl-mus LD50:2480 mg/kg BCFAAI 97,457,58
ipr-mus LD50:880 mg/kg BCFAAI 97,457,58
unr-mam LD50:880 mg/kg FRPSAX 13,574,58

SAFETY PROFILE: Moderately toxic by ingestion and other routes. When heated to decomposition it emits toxic fumes of NO_x and HCl. See also ESTERS.

AJE400 CAS:719-59-5 HR: 2
2-AMINO-5-CHLOROBENZOPHENONE
mf: $C_{13}H_{10}ClNO$ mw: 231.69

SYNS: (2-AMINO-5-CHLOROPHENYL)PHENYLMETHANONE □ METHANONE, (2-AMINO-5-CHLOROPHENYL)PHENYL-

TOXICITY DATA WITH REFERENCE
ipr-mus LD50:681 mg/kg IJSIDW 44,1,82

CONSENSUS REPORTS: Reported in EPA TSCA Inventory.

SAFETY PROFILE: Moderately toxic by intraperitoneal route. When heated to decomposition it emits toxic vapors of NO_x and Cl^-.

AJE500 CAS:19952-47-7 HR: 3
2-AMINO-4-CHLOROBENZOTHIAZOLE
mf: $C_7H_5ClN_2S$ mw: 184.65

TOXICITY DATA WITH REFERENCE
orl-mus LD50:2400 mg/kg JPETAB 105,486,52
ivn-mus LD50:71 mg/kg JPETAB 105,486,52

CONSENSUS REPORTS: Reported in EPA TSCA Inventory.

SAFETY PROFILE: Poison by intravenous route. Moderately toxic by ingestion. When heated to decomposition it emits very toxic fumes of Cl^-, SO_x, and NO_x.

AJE750 CAS:95-24-9 HR: 3
2-AMINO-6-CHLOROBENZOTHIAZOLE
mf: $C_7H_5ClN_2S$ mw: 184.65

TOXICITY DATA WITH REFERENCE
orl-mus LD50:398 mg/kg JPETAB 105,486,52
ivn-mus LD50:76 mg/kg JPETAB 105,486,52

CONSENSUS REPORTS: Reported in EPA TSCA Inventory.

SAFETY PROFILE: Poison by ingestion and intravenous route. When heated to decomposition it emits very toxic fumes of SO_x, NO_x, and Cl^-.

AJF250 CAS:64037-10-1 HR: 3
2-AMINO-4-CHLOROBENZOXAZOLE
mf: $C_7H_5ClN_2O$ mw: 168.59

TOXICITY DATA WITH REFERENCE
orl-rat LD50:500 mg/kg MDCHAG 4(1),338,64
orl-mus LD50:378 mg/kg MDCHAG 4(1),336,64
ipr-mus LD50:54 mg/kg MDCHAG 4(1),336,64

SAFETY PROFILE: Poison by ingestion and intraperitoneal route. When heated to decomposition it emits very toxic fumes of NO_x and Cl^-.

AJF500 CAS:61-80-3 HR: 3
2-AMINO-5-CHLOROBENZOXAZOLE
mf: $C_7H_5ClN_2O$ mw: 168.59

SYNS: 5-CHLORO-2-BENZOXAZOLAMINE □ DEFLEXOL □ FLEXILON □ FLEXIN □ MCN-485 □ USAF MA-12 □ ZOXAMIN □ ZOXAZOLAMINE □ ZOXINE

TOXICITY DATA WITH REFERENCE
orl-hmn TDLo:14 mg/kg/D:CNS JAMAAP 160,745,56
orl-rat LD50:782 mg/kg FEPRA7 16,319,57
ipr-rat LD50:102 mg/kg JPETAB 129,75,60
orl-mus LD50:540 mg/kg AIPTAK 128,112,60
ipr-mus LD50:100 mg/kg NTIS** AD277-689
ivn-mus LD50:376 mg/kg 29QHAQ -,-,74
ivn-dog LD50:117 mg/kg FEPRA7 16,319,57
orl-ham LD50:670 mg/kg JPETAB 129,75,60
ipr-ham LD50:268 mg/kg JPETAB 129,75,60

SAFETY PROFILE: Poison by intraperitoneal and intravenous routes. Moderately toxic by ingestion. Human systemic effects by ingestion: muscle rigidity. When heated to decomposition it yields toxic fumes of Cl^- and NO_x.

AJF750 CAS:52112-68-2 HR: 3
2-AMINO-6-CHLOROBENZOXAZOLE
mf: $C_7H_5ClN_2O$ mw: 168.59

TOXICITY DATA WITH REFERENCE
orl-rat LD50:226 mg/kg MDCHAG 4(1),338,64
ipr-rat LD50:123 mg/kg MDCHAG 4(1),338,64
orl-mus LD50:600 mg/kg MDCHAG 4(1),336,64
ipr-mus LD50:347 mg/kg MDCHAG 4(1),336,64

SAFETY PROFILE: Poison by ingestion and intraperitoneal route. When heated to decomposition it emits very toxic fumes of Cl^- and NO_x.

AJG750 CAS:64037-11-2 HR: 3
2-AMINO-7-CHLOROBENZOXAZOLE
mf: $C_7H_5ClN_2O$ mw: 168.59

TOXICITY DATA WITH REFERENCE
orl-rat LD50:410 mg/kg MDCHAG 4(1),338,64
orl-mus LD50:590 mg/kg MDCHAG 4(1),336,64
ipr-mus LD50:180 mg/kg MDCHAG 4(1),336,64

SAFETY PROFILE: Poison by intraperitoneal route. Moderately toxic by ingestion. When heated to decomposition it emits very toxic fumes of NO_x and Cl^-.

AJH000 CAS:364-62-5 HR: 3
4-AMINO-5-CHLORO-N-(2-(DIETHYLAMINO)ETHYL)-N-ANISAMIDE
mf: $C_{14}H_{22}ClN_3O_2$ mw: 299.84

SYNS: 4-AMINO-5-CHLORO-N-(2-(DIETHYLAMINO)ETHYL)-2-METHOXYBENZAMIDE □ 5-CHLORO-2-METHOXYPROCAINAMIDE □ DEL □ N-(DIETHYLAMINOETHYL)-2-METHOXY-4-AMINO-5-CHLOROBENZAMIDE □ MAXOLON □ METACLOPROMIDE □ METHOCHLOPRAMIDE □ 2-METHOXY-5-CHLOROPROCAINAMIDE □ METOCHLOPRAMIDE □ METOCLOL □ MORIPERAN □ PLASIL □ PRIMPERAN □ RELIVERAN

TOXICITY DATA WITH REFERENCE
dnd-hmn:leu 100 nmol/L CRNGDP 12,1613,91
orl-man TDLo:34 mg/kg (male 60D pre):REP WJMDA2 144,359,86
orl-wmn TDLo:3600 μg/kg/6D-I AIMEAS 97,621,82
orl-cld TDLo:900 μg/kg:EYE ADCHAK 55,310,80
orl-man TDLo:111 mg/kg/37W-I:PNS JJMDAT 23,152,84
ivn-cld TDLo:2 mg/kg/1D-C JOPDAB 104,138,84
ivn-wmn TDLo:2400 μg/kg:BLD NEJMAG 307,1346,82
ivn-man TDLo:14 μg/kg:BPR,CNS AIMEAS 104,125,86
orl-rat LD50:750 mg/kg NIIRDN 6,838,82
ipr-rat LD50:114 mg/kg NIIRDN 6,838,82
scu-rat LD50:340 mg/kg NIIRDN 6,838,82
ivn-rat LD50:50 mg/kg BCFAAI 115,649,76
orl-mus LD50:270 mg/kg NIIRDN 6,838,82
ipr-mus LD50:96 mg/kg NIIRDN 6,838,82
scu-mus LD50:190 mg/kg NIIRDN 6,838,82
ivn-mus LD50:33 mg/kg NIIRDN 6,838,82

SAFETY PROFILE: Poison by ingestion, subcutaneous, and intravenous routes. Human systemic effects by ingestion or intravenous routes: tremors, high blood pressure and abnormal catecholamine levels in the sympathetic nervous system, diplopia. Experimental reproductive effects. Mutation data reported. When heated to decomposition it emits very toxic fumes of NO_x and Cl^-.

AJH125 CAS:27260-19-1 HR: 3
4-AMINO-5-CHLORO-N-(2-(ETHYLAMINOETHYL)-o-ANISAMIDE)
mf: $C_{12}H_{18}ClN_3O_2$ mw: 271.78

SYNS: 4-AMINO-5-CHLORO-N-(2-ETHYLAMINOETHYL)-2-METHOXYBENZAMIDE □ 4-AMINO-5-CHLORO-N-(2-ETILAMINOETIL)-2-METOSSIBENZAMIDE (ITALIAN) □ DEETILATO METOCLOPRAMIDE (ITALIAN) □ DEETILMETOCLOPRAMIDE (ITALIAN) □ DEM

TOXICITY DATA WITH REFERENCE
ipr-rat LD50:242 mg/kg BCFAAI 115,649,76
ivn-rat LD50:120 mg/kg BCFAAI 115,649,76
ipr-mus LD50:245 mg/kg BCFAAI 115,649,76
ivn-mus LD50:41 mg/kg BCFAAI 115,649,76

SAFETY PROFILE: Poison by intravenous and intraperitoneal routes. When heated to decomposition it emits toxic fumes of Cl^- and NO_x.

AJH129 CAS:3443-15-0 HR: 3
6-AMINO-2-(2-CHLOROETHYL)-2,3-DIHYDRO-4H-1,3-BENZOXAZIN-4-ONEHYDROCHLORIDE
mf: $C_{10}H_{11}ClN_2O_2 \cdot ClH$ mw: 263.14

SYNS: A 350 □ AMINOCHLORTHENOXAZIN HYDROCHLORIDE □ 2-(β-CHLOROETHYL)-2,3-DIHYDRO-4-OXO-6-AMINO-1,3-BENZOXAZINE HYDROCHLORIDE □ ICI 350

TOXICITY DATA WITH REFERENCE
orl-rat LD50:619 mg/kg ARZNAD 14,124,64
ipr-rat LD50:607 mg/kg ARZNAD 13,884,63
orl-mus LD50:2250 mg/kg ARZNAD 13,884,63
ivn-mus LD50:293 mg/kg ARZNAD 13,884,63

SAFETY PROFILE: Poison by intravenous route. Moderately toxic by ingestion and intraperitoneal routes. When heated to decomposition it emits very toxic fumes of Cl^- and NO_x. An antipyretic and analgesic agent.

AJH250 CAS:116-84-7 HR: 1
1-AMINO-5-CHLORO-4-HYDROXYANTHRAQUINONE
mf: $C_{14}H_8ClNO_3$ mw: 273.68

SYN: 1-AMINO-4-HYDROXY-5-CHLORANTHRACHINON (CZECH)

TOXICITY DATA WITH REFERENCE
eye-rbt 500 mg/24H MLD 28ZPAK -,83,72

SAFETY PROFILE: An eye irritant. When heated to decomposition it emits very toxic fumes of Cl^- and NO_x.

AJH500 CAS:5857-94-3 HR: 1
3-AMINO-5-CHLORO-4-HYDROXYBENZENESULFONIC ACID
mf: $C_6H_6ClNO_4S$ mw: 223.64

SYNS: 5-CHLORO-4-HYDROXYMETANILIC ACID □ KYSELINA 2-CHLOR-6-AMINOFENOL-4-SULFONOVA (CZECH)

TOXICITY DATA WITH REFERENCE
eye-rbt 100 mg/24H MOD 28ZPAK -,181,72

CONSENSUS REPORTS: Reported in EPA TSCA Inventory.

SAFETY PROFILE: An eye irritant. See also SULFONATES. When heated to decomposition it emits very toxic fumes of Cl^-, SO_x, and NO_x.

AJH750 **CAS:1750-46-5** **HR: 3**
2-AMINO-5-CHLORO-6-HYDROXYBENZOXAZOLE
mf: $C_7H_5ClN_2O_2$ mw: 184.59

TOXICITY DATA WITH REFERENCE
orl-rat LD50:1070 mg/kg MDCHAG 4(1),338,64
orl-mus LD50:960 mg/kg MDCHAG 4(1),336,64
ipr-mus LD50:357 mg/kg MDCHAG 4(1),336,64

SAFETY PROFILE: Poison by intraperitoneal route. Moderately toxic by ingestion. When heated to decomposition it emits very toxic fumes of Cl^- and NO_x.

AJI000 **CAS:2139-00-6** **HR: 2**
2-AMINO-5-CHLORO-6-METHOXYBENZOXAZOLE
mf: $C_8H_7ClN_2O_2$ mw: 198.62

TOXICITY DATA WITH REFERENCE
orl-mus LD50:1600 mg/kg MDCHAG 4(1),336,64
ipr-mus LD50:490 mg/kg MDCHAG 4(1),336,64

SAFETY PROFILE: Moderately toxic by ingestion and intraperitoneal routes. When heated to decomposition it emits very toxic fumes of NO_x and Cl^-.

AJI250 **CAS:2797-51-5** **HR: 3**
2-AMINO-3-CHLORO-1,4-NAPHTHOQUINONE
mf: $C_{10}H_6ClNO_2$ mw: 207.62

PROP: Yellow-brown needles from AcOH. Mp: 196–198°.

SYNS: 2-CHLORO-3-AMINO-1,4-NAPHTHOQUINONE □ 06K □ 06K-QUINONE □ 06K-50W □ MOGETON GRANULE

TOXICITY DATA WITH REFERENCE
orl-rat LD50:1360 mg/kg FMCHA2 -,C161,83
orl-mus LD50:1260 mg/kg FMCHA2 -,C161,83
ipr-mus LD50:800 mg/kg JMCMAR 26,570,83
ivn-mus LD50:320 mg/kg CSLNX* NX#03360

SAFETY PROFILE: Poison by intravenous route. Moderately toxic by ingestion and intraperitoneal routes. When heated to decomposition it emits very toxic fumes of Cl^- and NO_x. See also AROMATIC AMINES.

AJI300 **CAS:91481-02-6** **HR: 3**
1-(4-AMINO-5-(2-CHLOROPHENYL)-2-METHYL-1H-PYRROL-3-YL)ETHANONE
mf: $C_{13}H_{13}ClN_2O$ mw: 248.73

PROP: A liquid.

SYNS: ETHANONE, 1-(4-AMINO-5-(2-CHLOROPHENYL)-2-METHYL-1H-PYRROL-3-YL)- □ KETONE, (4-AMINO-5-(o-CHLOROPHENYL)-2-METHYLPYRROL-3-YL) METHYL

TOXICITY DATA WITH REFERENCE
orl-mus LD50:500 mg/kg FRPSAX 39,538,84

DOT CLASSIFICATION: 3; *Label:* Flammable Liquid

SAFETY PROFILE: Moderately toxic by ingestion. A flammable liquid. When heated to decomposition it emits toxic vapors of NO_x and Cl^-.

AJI330 **CAS:56463-73-1** **HR: 3**
1-(4-AMINO-5-(p-CHLOROPHENYL)-2-METHYL-1H-PYRROL-3-YL)ETHANONE
mf: $C_{13}H_{13}ClN_2O$ mw: 248.73

PROP: A liquid.

SYNS: ETHANONE, 1-(4-AMINO-5-(p-CHLOROPHENYL)-2-METHYL-1H-PYRROL-3-YL)- □ KETONE, (4-AMINO-5-(p-CHLOROPHENYL)-2-METHYLPYRROL-3-YL) METHYL

TOXICITY DATA WITH REFERENCE
orl-mus LD50:1 g/kg FRPSAX 39,538,84

DOT CLASSIFICATION: 3; *Label:* Flammable Liquid

SAFETY PROFILE: Low toxicity by ingestion. A flammable liquid. When heated to decomposition it emits toxic vapors of NO_x.

AJI500 **CAS:50510-11-7** **HR: 3**
2-AMINO-5-((p-CHLOROPHENYL)THIOMETHYL)-2-OXAZOLINE
mf: $C_{10}H_{11}ClN_2OS$ mw: 242.74

SYN: 4,5-DIHYDRO-5-((p-CHLOROPHENYL)THIOMETHYL)OXAZOLAMINE

TOXICITY DATA WITH REFERENCE
orl-mus LD50:383 mg/kg JMCMAR 16,510,73
ipr-mus LD50:215 mg/kg JMCMAR 16,510,73

SAFETY PROFILE: Poison by ingestion and intraperitoneal routes. When heated to decomposition it emits very toxic fumes of Cl^-, NO_x, and SO_x.

AJI600 **CAS:34839-12-8** **HR: 3**
dl-1-AMINO-3-CHLORO-2-PROPANOL HYDROCHLORIDE
mf: $C_3H_8ClNO\cdot ClH$ mw: 146.03

SYN: (±)-1-AMINO-3-CHLORO-2-PROPANOL HYDROCHLORIDE

TOXICITY DATA WITH REFERENCE
orl-rat TDLo:210 mg/kg (14D male):REP CCPTAY 9,451,74
orl-rat LD50:165 mg/kg CCPTAY 9,451,74

CONSENSUS REPORTS: EPA Genetic Toxicology Program.

SAFETY PROFILE: Poison by ingestion. Experimental reproductive effects. When heated to decomposition it emits toxic fumes of Cl^- and NO_x.

AJI650 **CAS:41663-73-4** **HR: 3**
2-AMINO-5-CHLOROTHIAZOLE
mf: $C_3H_3ClN_2S$ mw: 134.59

TOXICITY DATA WITH REFERENCE
mmo-sat 1 mmol/L MUREAV 118,153,83
mmo-klp 500 µmol/L MUREAV 118,153,83

ivn-mus LD50:180 mg/kg CSLNX* NX#02306

SAFETY PROFILE: Poison by intravenous route. Mutation data reported. When heated to decomposition it emits toxic fumes of Cl^-, SO_x, and NO_x.

AJJ250 CAS:88-51-7 HR: 1
6-AMINO-4-CHLORO-m-TOLUENESULFONIC ACID
mf: $C_7H_8ClNO_3S$ mw: 221.67

SYNS: 2B ACID □ BENZENESULFONIC ACID, 2-AMINO-4-CHLORO-4-CHLORO-5-METHYL- □ BRILLIANT TONING RED AMINE □ 2-CHLORO-4-AMINOTOLUENE-5-SULFONIC ACID □ KYSELINA 2-CHLOR-4-TOLUIDIN-5-SULFONOVA (CZECH) □ RED 2B ACID

TOXICITY DATA WITH REFERENCE
eye-rbt 500 mg/24H MOD 28ZPAK -,184,72
orl-rat LD50:12 g/kg 28ZPAK -,184,72

CONSENSUS REPORTS: Reported in EPA TSCA Inventory.

SAFETY PROFILE: Mildly toxic by ingestion. An eye irritant. See also SULFONATES. When heated to decomposition it emits very toxic fumes of Cl^-, NO_x, and SO_x.

AJJ500 CAS:2448-39-7 HR: 3
6-AMINOCOUMARIN COUMARIN-3-CARBOXYLIC ACID SALT
mf: $C_{19}H_{10}NO_6$ mw: 348.30

TOXICITY DATA WITH REFERENCE
orl-mus LD50:103 mg/kg YKKZAJ 83,1124,63
scu-mus LD50:55 mg/kg YKKZAJ 83,1124,63

SAFETY PROFILE: Poison by ingestion and subcutaneous routes. When heated to decomposition it emits toxic fumes of NO_x.

AJJ750 CAS:63989-79-7 HR: 3
6-AMINOCOUMARIN HYDROCHLORIDE
mf: $C_9H_7NO_2 \cdot ClH$ mw: 197.63

TOXICITY DATA WITH REFERENCE
orl-mus LD50:623 mg/kg YKKZAJ 83,1124,63
scu-mus LD50:353 mg/kg YKKZAJ 83,1124,63

SAFETY PROFILE: Poison by subcutaneous route. Moderately toxic by ingestion. When heated to decomposition it emits very toxic fumes such as HCl and NO_x.

AJJ800 CAS:6264-93-3 HR: 3
2-AMINO-2,4,6-CYCLOHEPTATRIEN-1-ONE
mf: C_7H_7NO mw: 121.15

PROP: Yellow prisms from $CHCl_3$/hexane or by subliming. Mp: 106–107°.

SYN: 2-AMINOTROPONE

TOXICITY DATA WITH REFERENCE
ipr-mus LD50:176 mg/kg CPBTAL 20,60,72
scu-mus LD50:175 mg/kg CPBTAL 20,60,72
ivn-mus LD50:333 mg/kg CPBTAL 20,60,72

SAFETY PROFILE: Poison by subcutaneous, intravenous, and intraperitoneal routes. When heated to decomposition it emits toxic fumes of NO_x.

AJJ875 CAS:3485-14-1 HR: 3
(1-AMINOCYCLOHEXYL)PENICILLIN
mf: $C_{15}H_{23}N_3O_4S$ mw: 341.47

PROP: Crystals. Mp: 182–183° (anhydrate), Mp: 156–158° (decomp). Sol in water.

SYNS: AC-PC □ 6-(1-AMINOCYCLOHEXANECARBOXAMIDO) PENICILLANIC ACID □ AMINOCYCLOHEXYLPENICILLIN □ CALTHOR □ CICLACILLIN □ CICLACILLUM □ CITOSARIN □ CYCLACILLIN □ CYCLAPEN □ SYNGACILLIN □ ULTRACILLIN □ VASTCILLIN □ VATRACIN □ VIPICIL □ WY 4508 □ WYVITAL

TOXICITY DATA WITH REFERENCE
orl-hmn TDLo:210 mg/kg/7D-I:GIT,SKN CHTHBK 22,154,76
orl-rat LD50:5010 mg/kg CHTHBK 22,154,76
ipr-rat LD50:5010 mg/kg CHTHBK 22,154,76
scu-rat LD50:6500 mg/kg TAKHAA 29,117,70
orl-mus LD50:5010 mg/kg CHTHBK 22,154,76
ipr-mus LD50:3776 mg/kg CHTHBK 22,154,76
scu-mus LD50:7500 mg/kg TAKHAA 29,117,70
orl-dog LD50:2500 mg/kg CHTHBK 22,154,76

SAFETY PROFILE: Moderately toxic by ingestion and other routes. Human systemic effects by ingestion: dermatitis and diarrhea. Experimental reproductive effects. When heated to decomposition it emits toxic fumes of SO_x and NO_x.

AJK000 HR: 3
1-AMINO-2-(o-CYCLOHEXYLPHENOXY)PROPIONALDOXIME
mf: $C_{15}H_{22}N_2O_2$ mw: 262.39

SYN: MG 18415

TOXICITY DATA WITH REFERENCE
orl-mus LD50:620 mg/kg ARZNAD 29,729,79
ipr-mus LD50:200 mg/kg ARZNAD 29,729,79

SAFETY PROFILE: Poison by intraperitoneal route. Moderately toxic by ingestion. When heated to decomposition it emits toxic fumes of NO_x.

AJK250 CAS:52-52-8 HR: 3
1-AMINOCYCLOPENTANE-1-CARBOXYLIC ACID
mf: $C_6H_{11}NO_2$ mw: 129.18

PROP: Prisms. Mp: 328–329°.

SYNS: ACPC □ 1-AMINO-1-CYCLOPENTANECARBOXYLIC ACID □ CB 1639 □ CYCLOLEUCINE □ NSC 1026 □ WR 14,997 □ X 201

TOXICITY DATA WITH REFERENCE
orl-hmn TDLo:60 mg/kg:CNS,GIT JMPCAS 3,1,61
orl-rat LD50:290 mg/kg JMPCAS 3,1,61
ivn-rat LD50:340 mg/kg JMCMAR 3,1,61
orl-mus LD50:309 mg/kg JMPCAS 3,1,61
ipr-mus LD50:119 mg/kg NCISP* JAN86
scu-mus LD50:375 mg/kg NCISP* JAN86
orl-dog LD50:300 mg/kg JMPCAS 3,1,61
ivn-dog LD50:300 mg/kg JMCMAR 3,1,61
orl-gpg LD50:140 mg/kg JMPCAS 3,1,61

CONSENSUS REPORTS: Reported in EPA TSCA Inventory. EPA Genetic Toxicology Program.

SAFETY PROFILE: Poison by ingestion, subcutaneous, intraperitoneal, and intravenous routes. Human systemic effects by ingestion: anorexia, nausea and vomiting. When heated to decomposition it emits toxic fumes of NO_x.

AJK500 CAS:60676-83-7 HR: 3
4-AMINO-N-CYCLOPROPYL-3,5-DICHLOROBENZAMIDE
mf: $C_{10}H_{10}Cl_2N_2O$ mw: 245.12

SYNS: N-CYCLOPROPYL-4-AMINO-3,5-DICHLOROBENZAMIDE □ N-CYCLOPROPYL-3,5-DICHLORO-4-AMINOBENZAMIDE

TOXICITY DATA WITH REFERENCE
orl-rat LD50:170 mg/kg 27ZQAG -,400,72
ipr-rat LD50:160 mg/kg 27ZQAG -,400,72
orl-mus LD50:195 mg/kg 27ZQAG -,400,72
ipr-mus LD50:265 mg/kg JMCMAR 6,528,63

SAFETY PROFILE: Poison by ingestion and intraperitoneal routes. When heated to decomposition it emits very toxic fumes of Cl^- and NO_x.

AJK750 CAS:1951-25-3 HR: 3
AMINODARONE
mf: $C_{25}H_{29}I_2NO_3$ mw: 645.35

SYNS: AMIODARONE □ 2-BUTYL-3-BENZOFURANYL p-((2-DIETHYLAMINO)ETHOXY)-m,m-DIIODOPHENYL KETONE □ 2-BUTYL-3-(3,5-DIIODO-4-(2-DIETHYLAMINOETHOXY)BENZOYL)BENZOFURAN □ 2-N-BUTYL-3′,5′-DIIODO-4′-N-DIETHYLAMINOETHOXY-3-BENZOYLBENZOFURAN □ L. 3428 □ LABAZ

TOXICITY DATA WITH REFERENCE
orl-man TDLo:133 mg/kg/23D-I:SKN LANCAO 1,51,84
ipr-mus LD50:254 mg/kg EJTXAZ 8,122,75
ivn-mus LD50:178 mg/kg EJTXAZ 8,188,75

DOT CLASSIFICATION: 3; *Label:* Flammable Liquid

SAFETY PROFILE: Poison by intravenous and intraperitoneal routes. Human systemic effects by ingestion: photosensitivity of the skin. A flammable liquid. When heated to decomposition it emits very toxic fumes of I^- and NO_x. A coronary vasodilator.

AJL250 CAS:63041-30-5 HR: 2
9-AMINO-1,2,5,6-DIBENZANTHRACENE
mf: $C_{22}H_{15}N$ mw: 293.38

SYN: 7-AMINODIBENZ(a,h)ANTHRACENE

TOXICITY DATA WITH REFERENCE
skn-mus TDLo:1250 mg/kg/52W-I:ETA PRLBA4 117,318,35

SAFETY PROFILE: Questionable carcinogen with experimental tumorigenic data. When heated to decomposition it emits toxic fumes of NO_x.

AJL500 CAS:81-49-2 HR: 1
1-AMINO-2,4-DIBROMOANTHRAQUINONE
mf: $C_{14}H_7Br_2NO_2$ mw: 381.04

SYNS: 1-AMINO-2,4-DIBROMANTHRACHINON (CZECH) □ 2,4-DIBROMO-1-ANTHRAQUINONYLAMINE □ NCI-C55458

TOXICITY DATA WITH REFERENCE
eye-rbt 500 mg/24H MLD 28ZPAK -,88,72
mmo-sat 333 μg/plate ENMUDM 5(Suppl 1),3,83
mma-sat 333 μg/plate NTPTB* JAN 82

CONSENSUS REPORTS: Reported in EPA TSCA Inventory.

SAFETY PROFILE: An eye irritant. Mutation data reported. When heated to decomposition it emits very toxic fumes of Br^- and NO_x.

AJL750 CAS:52112-67-1 HR: 2
2-AMINO-5,7-DIBROMOBENZOXAZOLE
mf: $C_7H_4Br_2N_2O$ mw: 291.95

TOXICITY DATA WITH REFERENCE
orl-mus LD50:1050 mg/kg MDCHAG 4(1),336,64
ipr-mus LD50:780 mg/kg MDCHAG 4(1),336,64

SAFETY PROFILE: Moderately toxic by ingestion and intraperitoneal route. When heated to decomposition it emits very toxic fumes of Br^- and NO_x.

AJL875 CAS:102207-73-8 HR: 3
2-AMINO-4-DIBUTYLAMINOETHOXYPYRIMIDINE
mf: $C_{14}H_{26}N_4O$ mw: 266.44

SYNS: 2-AMINO-4-(2-DIBUTYLAMINOETHOXY)PYRIMIDINE □ OR-1550

TOXICITY DATA WITH REFERENCE
orl-rat LD50:1000 mg/kg AIPTAK 106,50,56
ipr-rat LD50:75 mg/kg AIPTAK 106,50,56
ipr-mus LD50:157 mg/kg AIPTAK 106,50,56
ivn-mus LD50:44 mg/kg AIPTAK 106,50,56
orl-dog LD50:450 mg/kg AIPTAK 106,50,56
ivn-dog LD50:35 mg/kg AIPTAK 106,50,56
orl-rbt LD50:1260 mg/kg AIPTAK 106,50,56
ivn-rbt LD50:46 mg/kg AIPTAK 106,50,56

SAFETY PROFILE: Poison by intravenous and intraperitoneal routes. Moderately toxic by ingestion. When heated to decomposition it emits toxic fumes of NO_x.

AJM000 CAS:133-90-4 HR: 2
3-AMINO-2,5-DICHLOROBENZOIC ACID
mf: $C_7H_5Cl_2NO_2$ mw: 206.03

PROP: Purplish-white powder. Mp: 200–201°.

SYNS: ACP-M-728 □ AMBIBEN □ AMOBEN □ CHLORAMBEN □ 2,5-DICHLORO-3-AMINOBENZOIC ACID □ NCI-C00055 □ ORNAMENTAL WEED □ VEGABEN

TOXICITY DATA WITH REFERENCE
mmo-sat 10 mg/plate ENMUDM 5(Suppl 1),3,83
mma-sat 1 mg/plate NTPTB* JAN 82
cyt-mus-ipr 58,500 μg/kg CARYAB 33,527,80
cyt-mus-orl 234 mg/kg CARYAB 33,527,80

orl-mus TDLo:672 g/kg/80W-C:CAR NCITR* NCI-CG-TR-25,77
orl-mus TD:1344 g/kg/80W-C:CAR NCITR* NCI-CG-TR-25,77
orl-rat LD50:3500 mg/kg RREVAH 10,97,65
orl-mus LD50:3725 mg/kg GISAAA 45(4),74,80
skn-rbt LD50:3136 mg/kg WRPCA2 7,135,68

CONSENSUS REPORTS: NCI Carcinogenesis Bioassay Completed; Results Positive: mouse NCITR* NCI-CG-TR-25,77; Results Negative: rat NCITR* NCI-CG-TR-25,77. Community Right-To-Know List. Reported in EPA TSCA Inventory.

SAFETY PROFILE: Moderately toxic by ingestion. Questionable carcinogen with experimental carcinogenic data. Mutation data reported. When heated to decomposition it emits highly toxic fumes such as Cl^- and NO_x. See also AROMATIC AMINES.

AJM500 CAS:64037-12-3 HR: 3
2-AMINO-5,6-DICHLOROBENZOXAZOLE
mf: $C_7H_4Cl_2N_2O$ mw: 203.03

TOXICITY DATA WITH REFERENCE
orl-mus LD50:1200 mg/kg MDCHAG 4(1),336,64
ipr-mus LD50:300 mg/kg MDCHAG 4(1),336,64

SAFETY PROFILE: Poison by intraperitoneal route. Moderately toxic by ingestion. When heated to decomposition it emits very toxic fumes of Cl^- and NO_x.

AJM600 CAS:91480-92-1 HR: 3
1-(4-AMINO-5-(3,4-DICHLOROPHENYL)-2-METHYL 1H-PYRROL-3-YL)ETHANONE
mf: $C_{13}H_{12}Cl_2N_2O$ mw: 283.17

PROP: A liquid.

SYNS: ETHANONE, 1-(4-AMINO-5-(3,4-DICHLOROPHENYL)-2-METHYL-1H-PYRROL-3-YL)- □ KETONE, (4-AMINO-5-(3,4-DICHLOROPHENYL)-2-METHYLPYRROL-3-YL) METHYL

TOXICITY DATA WITH REFERENCE
orl-mus LD50:1 g/kg FRPSAX 39,538,84

DOT CLASSIFICATION: 3; *Label:* Flammable Liquid

SAFETY PROFILE: Low toxicity by ingestion. A flammable liquid. When heated to decomposition it emits toxic vapors of NO_x.

AJM750 CAS:50510-12-8 HR: 3
2-AMINO-5-((3,4-DICHLOROPHENYL)THIOMETHYL)-2-OXAZOLINE
mf: $C_{10}H_{10}Cl_2N_2OS$ mw: 277.18

TOXICITY DATA WITH REFERENCE
orl-mus LD50:562 mg/kg JMCMAR 16,510,73
ipr-mus LD50:383 mg/kg JMCMAR 16,510,73

SAFETY PROFILE: Poison by intraperitoneal route. Moderately toxic by ingestion. When heated to decomposition it emits very toxic fumes of Cl^-, NO_x, and SO_x.

AJN250 CAS:2381-85-3 HR: 3
5-AMINO-9-(DIETHYLAMINO)BENZO(a)PHENOXAZIN-7-IUM SULFATE (2:1)
mf: $C_{40}H_{40}N_6O_2{\cdot}O_4S$ mw: 732.92

PROP: Bright blue crystals. Sol in H_2O, EtOH.

SYNS: C.I. 51180 □ C.I. BASIC BLUE 12 □ CRESOL FAST VIOLET □ CRESYL FAST VIOLET □ NILE BLUE □ NILE BLUE A □ NILE BLUE AX □ NILE BLUE BASE □ NILE BLUE CHLORIDE □ NILE BLUE HYDROCHLORIDE

TOXICITY DATA WITH REFERENCE
mic-mic-uns 1 ppm POASAD 34,114,53
ivn-mus LDLo:65 mg/kg TXAPA9 44,225,78

CONSENSUS REPORTS: Reported in EPA TSCA Inventory.

SAFETY PROFILE: Poison by intravenous route. Mutation data reported. See also SULFATES. When heated to decomposition it emits very toxic fumes of NO_x and SO_x.

AJN375 CAS:102207-75-0 HR: 3
2-AMINO-4-DIETHYLAMINOETHOXYPYRIMIDINE
mf: $C_{10}H_{18}N_4O$ mw: 210.32

SYNS: 2-AMINO-4-(2-DIETHYLAMINOETHOXY)PYRIMIDINE □ OR-1556

TOXICITY DATA WITH REFERENCE
orl-rat LD50:2000 mg/kg AIPTAK 106,50,56
ipr-rat LD50:75 mg/kg AIPTAK 106,50,56
ipr-mus LD50:252 mg/kg AIPTAK 106,50,56
ivn-mus LD50:174 mg/kg AIPTAK 106,50,56
orl-dog LD50:2750 mg/kg AIPTAK 106,50,56
ivn-dog LD50:165 mg/kg AIPTAK 106,50,56
orl-rbt LD50:875 mg/kg AIPTAK 106,50,56
ivn-rbt LD50:174 mg/kg AIPTAK 106,50,56

SAFETY PROFILE: Poison by intravenous and intraperitoneal routes. Moderately toxic by ingestion. When heated to decomposition it emits toxic fumes of NO_x.

AJN500 CAS:51-06-9 HR: 3
p-AMINO-N-(2-DIETHYLAMINOETHYL)BENZAMIDE
mf: $C_{13}H_{21}N_3O$ mw: 235.37

PROP: Solid. Mp: 46–48°, bp: 210–215° @ 2 mm.

SYNS: p-AMINOBENZOIC DIETHYLAMINOETHYLAMIDE □ 4-AMINO-N-(2-(DIETHYLAMINO)ETHYL)-BENZAMIDE (9CI) □ NOVOCAINAMIDE □ NOVOCAINE AMIDE □ NOVOCAMID □ PROCAINAMIDE □ PROCAINE AMIDE □ PROCAMIDE □ PRONESTYL

TOXICITY DATA WITH REFERENCE
orl-man TDLo:8579 mg/kg/43W-I:PUL AJMEAZ 76,146,84
orl-man TDLo:29 mg/kg:CNS AJCDAG 57,340,86
orl-wmn TDLo:1826 mg/kg/13W-I:CVS AHJOA2 83,798,72
orl-hmn TDLo:2280 mg/kg/22W:MSK BHJUAV 34,284,72
ivn-man TDLo:583 mg/kg/12D-C:CVS AHJOA2 109,375,85
ivn-rat LD50:110 mg/kg RPTOAN 33,292,70
orl-mus LD50:525 mg/kg CCCCAK 42,3628,77
orl-mus LD50:525 mg/kg CCCCAK 42,3628,77
ipr-mus LD50:178 mg/kg PJPPAA 37,551,85
ivn-mus LD50:49 mg/kg PJPPAA 37,551,85

orl-dog LDLo: 2210 mg/kg TXAPA9 21,253,72
ivn-rbt LD50: 125 mg/kg PJPPAA 32,833,80
ivn-gpg LD50: 280 mg/kg FRPSAX 12,77,57

SAFETY PROFILE: Poison by intravenous and intraperitoneal routes. Moderately toxic by ingestion. Human systemic effects by ingestion: cardiac abnormalities, joint effects, cough, tremors, dyspnea, and other lung effects. When heated to decomposition it emits toxic fumes of NO_x.

AJN750 CAS:63887-34-3 HR: 3
p-AMINO-N-(2-DIETHYLAMINOETHYL)BENZAMIDE SULFATE
mf: $C_{13}H_{21}N_3O{\cdot}H_2O_4S$ mw: 333.45

SYNS: PROCAINAMIDE SULFATE □ PROCAINE AMIDE SULFATE □ SUPICAINE AMIDE SULFATE

TOXICITY DATA WITH REFERENCE
ivn-rat LD50: 165 mg/kg RPOBAR 2,318,70
ivn-mus LD50: 146 mg/kg RPOBAR 2,318,70

SAFETY PROFILE: Poison by intravenous route. See also SULFATES. When heated to decomposition it emits very toxic fumes of NO_x and SO_x.

AJO000 CAS:61827-74-5 HR: 3
N-(2-AMINO-5-DIETHYLAMINOPHENETHYL)METHANE SULFONAMIDEHYDROCHLORIDE
mf: $C_{13}H_{23}N_3O_2S{\cdot}HCl$ mw: 321.91

TOXICITY DATA WITH REFERENCE
orl-rat LDLo: 400 mg/kg KODAK* -,-,71
ipr-rat LD50: 50 mg/kg KODAK* -,-,71

SAFETY PROFILE: Poison by ingestion and intraperitoneal routes. See also SULFONATES. When heated to decomposition it emits very toxic fumes of NO_x, SO_x, and HCl.

AJO250 CAS:2198-58-5 HR: 3
p-AMINO DIETHYLANILINE HYDROCHLORIDE
mf: $C_{10}H_{16}N_2{\cdot}ClH$ mw: 200.74

PROP: Bp: 217.5°. Sltly sol in water, sol in ether.

SYN: N,N-DIETHYL-p-PHENYLENEDIAMINE HYDROCHLORIDE

TOXICITY DATA WITH REFERENCE
orl-rat LDLo: 200 mg/kg KODAK* -,-,71
ipr-rat LDLo: 25 mg/kg KODAK* -,-,71
ivn-mus LD50: 24 mg/kg CSLNX* NX#07893

CONSENSUS REPORTS: Reported in EPA TSCA Inventory.

SAFETY PROFILE: Poison by ingestion, intraperitoneal, and intravenous routes. When heated to decomposition it emits very toxic fumes of HCl and NO_x.

AJO500 CAS:134-58-7 HR: 3
5-AMINO-1,6-DIHYDRO-7H-v-TRIAZOLO(4,5-d)PYRIMIDIN-7-ONE
mf: $C_4H_4N_6O$ mw: 152.14

PROP: Crystals from water. Mp: 305° (decomp).

SYNS: 8 AG □ 5-AMINO-1,4-DIHYDRO-7H-1,2,3-TRIAZOLO(4,5-d) PYRIMIDIN-7-ONE (9CI) □ 5-AMINO-7-HYDROXY-1H-v-TRIAZOLO(d) PYRIMIDINE □ 5-AMINO-1H-v-TRIAZOLO(d)PYRIMIDIN-7-OL □ 5-AMINO-v-TRIAZOLO(4,5-d)PYRIMIDIN-7-OL □ AZAGUANINE □ AZA-GUANINE-8 □ 8-AZAGUANINE □ AZAN □ AZG □ B-28 □ GUANAZOL □ GUANAZOLO □ NSC-749 □ PATHOCIDIN □ PATHOCIDINE □ SF-337 □ SK 1150 □ TRIAZOLOGUANINE

TOXICITY DATA WITH REFERENCE
dni-mus:lym 66 µmol/L CJBBDU 62,280,84
oms-mus:leu 1 µmol/L AEZRA2 20,351,82
ipr-mus TDLo: 80 mg/kg (female 7D post): REP JEEMAF 6,593,58
ipr-mus TDLo: 80 mg/kg (female 8D post): TER JEEMAF 6,593,58
ipr-rat LD50: 1000 mg/kg ADTEAS 3,181,68
orl-mus LD50: 1500 mg/kg OSDIAF 17,491,68
ipr-mus LD50: 100 mg/kg 85GDA2 5,193,81

CONSENSUS REPORTS: EPA Genetic Toxicology Program.

SAFETY PROFILE: Poison by intraperitoneal route. Moderately toxic by ingestion. Mutation data reported. An experimental teratogen. Other experimental reproductive effects. When heated to decomposition it emits toxic fumes of NO_x. Inhibits protein synthesis.

AJO625 CAS:102207-76-1 HR: 3
2-AMINO-4-DI-ISOBUTYLAMINOETHOXYPYRIMIDINE
mf: $C_{14}H_{26}N_4O$ mw: 266.44

SYNS: 2-AMINO-4-(2-DIISOBUTYLAMINOETHOXY)PYRIMIDINE □ OR-1578

TOXICITY DATA WITH REFERENCE
ipr-rat LD50: 800 mg/kg AIPTAK 106,50,56
ipr-mus LD50: 762 mg/kg AIPTAK 106,50,56
ivn-mus LD50: 72 mg/kg AIPTAK 106,50,56
orl-dog LD50: 2500 mg/kg AIPTAK 106,50,56
ivn-dog LD50: 75 mg/kg AIPTAK 106,50,56
orl-rbt LD50: 2000 mg/kg AIPTAK 106,50,56
ivn-rbt LD50: 16 mg/kg AIPTAK 106,50,56

SAFETY PROFILE: Poison by intravenous route. Moderately toxic by ingestion and other routes. When heated to decomposition it emits toxic fumes of NO_x.

AJO750 CAS:73747-29-2 HR: 3
3-AMINO-2-(2-(DIISOPROPYLAMINO)ETHOXY)BUTYROPHENONEDIHYDROCHLORIDE
mf: $C_{18}H_{30}N_2O_2{\cdot}2ClH$ mw: 379.42

SYN: REC 7-0591

TOXICITY DATA WITH REFERENCE
orl-rat LD50: 238 mg/kg ARZNAD 16,1275,66
scu-rat LD50: 111 mg/kg ARZNAD 16,1275,66
ivn-rat LD50: 3 mg/kg ARZNAD 16,1275,66
orl-mus LD50: 42 mg/kg ARZNAD 16,1275,66
ipr-mus LD50: 38 mg/kg ARZNAD 16,1275,66
scu-mus LD50: 42 mg/kg ARZNAD 16,1275,66
ivn-mus LD50: 7800 µg/kg ARZNAD 16,1275,66
ivn-cat LD50: 5 mg/kg ARZNAD 16,1275,66

SAFETY PROFILE: Poison by ingestion, subcutaneous, intravenous, and intraperitoneal routes. See also KE-

TONES. When heated to decomposition it emits very toxic fumes of NO_x and HCl.

AJO800 CAS:91480-90-9 HR: 3
1-(4-AMINO-5-(3,4-DIMETHOXYPHENYL)-2-METHYL-1H-PYRROL-3-YL)ETHANONE
mf: $C_{15}H_{18}N_2O_3$ mw: 274.35

PROP: A liquid.

SYNS: ETHANONE, 1-(4-AMINO-5-(3,4-DIMETHOXYPHENYL)-2-METHYL-1H-PYRROL-3-YL)- ☐ KETONE, (4-AMINO-5-(3,4-DIMETHOXYPHENYL)-2-METHYLPYRROL-3-YL) METHYL

TOXICITY DATA WITH REFERENCE
orl-mus LD50:1 g/kg FRPSAX 39,538,84

DOT CLASSIFICATION: 3; *Label:* Flammable Liquid

SAFETY PROFILE: Low toxicity by ingestion. A flammable liquid. When heated to decomposition it emits toxic vapors of NO_x.

AJP000 CAS:19216-56-9 HR: 3
1-(4-AMINO-6,7-DIMETHOXY-2-QUINAZOLINYL-4-(2-FURANYLCARBONYL)) PIPERAZINE
mf: $C_{19}H_{21}N_5O_4$ mw: 383.45

PROP: Solid. Mp: 278–280°.

SYNS: FURAZOSIN ☐ 2-(4-(2-FUROYL)PIPERAZIN-1-YL)-4-AMINO-6,7-DIMETHOXYQUINAZOLINE ☐ PRAZOSIN

TOXICITY DATA WITH REFERENCE
orl-man TDLo:2571 µg/kg/60D-I:REP AMSVAZ 213,319,83
orl-wmn TDLo:20 µg/kg:BPR,GIT AIMEAS 97,455,82
orl-man TDLo:1143 µg/kg:BPR AMSVAZ 213,157,83
orl-hmn TDLo:280 µg/kg:CNS,CVS BMJOAE 2,508,76
orl-hmn TDLo:1260 µg/kg:CNS,KID BMJOAE 1,622,78

SAFETY PROFILE: Human systemic effects by ingestion of very small amounts: somnolence, hallucinations, distorted perceptions, changes in motor activity, decreased blood pressure, nausea or vomiting, and kidney effects. Experimental reproductive effects. When heated to decomposition it emits toxic fumes of NO_x.

AJP125 CAS:102207-77-2 HR: 3
2-AMINO-4-DIMETHYLAMINOETHOXYPYRIMIDINE
mf: $C_8H_{14}N_4O$ mw: 182.26

SYNS: 2-AMINO-4-(2-DIMETHYLAMINOETHOXY)PYRIMIDINE ☐ OR-1549

TOXICITY DATA WITH REFERENCE
orl-rat LD50:1500 mg/kg AIPTAK 106,50,56
ipr-rat LD50:75 mg/kg AIPTAK 106,50,56
ipr-mus LD50:478 mg/kg AIPTAK 106,50,56
ivn-mus LD50:252 mg/kg AIPTAK 106,50,56
orl-dog LD50:1500 mg/kg AIPTAK 106,50,56
ivn-dog LD50:450 mg/kg AIPTAK 106,50,56
orl-rbt LD50:3750 mg/kg AIPTAK 106,50,56
ivn-rbt LD50:440 mg/kg AIPTAK 106,50,56

SAFETY PROFILE: Poison by intravenous and intraperitoneal routes. Moderately toxic by ingestion. When heated to decomposition it emits toxic fumes of NO_x.

AJP250 CAS:92-31-9 HR: 3
3-AMINO-7-DIMETHYLAMINO-2-METHYLPHENAZATHIONIUM CHLORIDE
mf: $C_{15}H_{16}N_3S{\cdot}Cl$ mw: 305.85

PROP: Dark green powder with bronze lustre. Sol in water giving a blue to violet soln.

SYNS: BLUTENE ☐ BLUTENE CHLORIDE ☐ C.I. 925 ☐ C.I. 52040 ☐ C.I. BASIC BLUE 17 ☐ DIMETHYLTOLUTHIONINE CHLORIDE ☐ F KLOT ☐ KLOT ☐ SCHULTZ No. 1041 ☐ TOLAZUL ☐ TOLONIUM CHLORIDE ☐ TOLUIDINE BLUE ☐ TOLUIDINE BLUE O ☐ TOLUIDENE BLUE O CHLORIDE

TOXICITY DATA WITH REFERENCE
unr-man TDLo:43 mg/kg/6D:GIT,BLD 34ZIAG -,597,69
ipr-rat LD50:215 mg/kg AEPPAE 204,288,47
ivn-rat LD50:28,930 µg/kg SMBUA9 9,96,51
ivn-mus LD50:27,560 µg/kg SMBUA9 9,96,51

CONSENSUS REPORTS: Reported in EPA TSCA Inventory. EPA Genetic Toxicology Program.

SAFETY PROFILE: Poison by intravenous and intraperitoneal routes. Human systemic effects by an unspecified route: nausea or vomiting and blood effects. When heated to decomposition it emits very toxic fumes of Cl^-, SO_x, and NO_x.

AJQ000 CAS:63731-93-1 HR: 3
2-AMINO-4-γ-DIETHYLAMINOPROPYLAMINO-5,6-DIMETHYLPYRIMIDINE
mf: $C_{13}H_{25}N_5$ mw: 251.43

TOXICITY DATA WITH REFERENCE
orl-mus LDLo:250 mg/kg JCSOA9 -,357,46

SAFETY PROFILE: Poison by ingestion. When heated to decomposition it emits toxic fumes of NO_x.

AJQ100 CAS:109-55-7 HR: 3
1-AMINO-3-DIMETHYLAMINOPROPANE
mf: $C_5H_{14}N_2$ mw: 102.21

$(CH_3)_2N(CH_2)_3NH_2$

PROP: Colorless liquid. Mp: <−70°, bp: 132–135°, flash p: 100°F (OC), d: 0.8100 @ 30°, vap press: 10 mm @ 30°, vap d: 3.52.

SYNS: N,N-DIMETHYL-N-(3-AMINOPROPYL)AMINE ☐ 3-(DIMETHYLAMINO)PROPYLAMINE ☐ N,N-DIMETHYL-1,3-DIAMINOPROPANE ☐ N,N-DIMETHYL-1,3-PROPANEDIAMINE ☐ N,N-DIMETHYL-1,3-PROPYLENEDIAMINE

TOXICITY DATA WITH REFERENCE
skn-rbt 100 µg/24H open AIHAAP 23,95,62
eye-rbt 5 mg MOD UCDS** 12/15/71
orl-rat LDLo:1870 mg/kg AIHAAP 23,95,62
skn-rbt LD50:600 µL/kg UCDS** 12/15/71

CONSENSUS REPORTS: Reported in EPA TSCA Inventory.

SAFETY PROFILE: Moderately toxic by ingestion and skin contact. A skin and eye irritant. Very flammable when exposed to heat, flame, or oxidizers. Reaction with 1,2-dichloroethane produces explosive acetylene gas. This and other amines ignite on contact with cellulose nitrate of high surface area. To fight fire, use alcohol foam, CO_2, dry chemical. When heated to decomposition it emits toxic fumes of NO_x. See also AMINES.

AJQ250 **CAS:553-24-2** **HR: 3**
AMINODIMETHYLAMINOTOLUAMINOZINE HYDROCHLORIDE
mf: $C_{15}H_{16}N_4 \cdot ClH$ mw: 288.81

PROP: Black or very dark green powder. Mp: 290° (decomp). Sol in H_2O and EtOH.

SYNS: 3-AMINO-7-DIMETHYLAMINO-2-METHYLPHENAZINE HYDROCHLORIDE □ 3-AMINO-7-(DIMETHYLAMINO)-2-METHYL-PHENAZINE MONOHYDROCHLORIDE □ C.I. 50040 □ C.I. BASIC RED 5 □ C.I. BASIC RED 5, MONOHYDROCHLORIDE □ KERNECHTROT □ MICHROME No. 226 □ NEUTRAL RED □ NEUTRAL RED CHLORIDE □ NEUTRAL RED W □ NUCLEAR FAST RED □ TOLUYLENE RED □ $N^{(8)}$, $N^{(8)}$,3-TRIMETHYL-2,8-PHENAZINEDIAMINE MONOHYDROCHLORIDE

TOXICITY DATA WITH REFERENCE
cyt-ckn-par 85 μg/kg 47JMAE -,137,82
mma-sat 10 μg/plate MUREAV 48,109,77
mmo-esc 2500 ppt/3H AMNTA4 85,119,51
ivn-rat LD50:112 mg/kg FEPRA7 10,337,51
ivn-mus LD50:142 mg/kg FEPRA7 10,337,51
ivn-rbt LD50:96,600 μg/kg SMBUA9 9,96,51

CONSENSUS REPORTS: Reported in EPA TSCA Inventory. EPA Genetic Toxicology Program.

SAFETY PROFILE: Poison by intravenous route. Mutation data reported. When heated to decomposition it emits very toxic fumes of HCl and NO_x.

AJQ500 **CAS:21554-20-1** **HR: 3**
4-AMINO-3′,5′-DIMETHYL-4′-HYDROXYAZOBENZENE
mf: $C_{14}H_{15}N_3O$ mw: 241.32

TOXICITY DATA WITH REFERENCE
orl-rat TDLo:40 g/kg/2Y-C:ETA AABIAV 52,33,63
ipr-rat LD50:350 mg/kg AABIAV 52,33,63

SAFETY PROFILE: Poison by intraperitoneal route. Questionable carcinogen with experimental tumorigenic data. When heated to decomposition it emits toxic fumes of NO_x.

AJQ600 **CAS:77094-11-2** **HR: 3**
2-AMINO-3,4-DIMETHYLIMIDAZO(4,5-f)QUINOLINE
mf: $C_{12}H_{12}N_4$ mw: 212.28

PROP: Solid. Mp: 296–298° (sealed tube).

SYNS: 3,4-DIMETHYL-3H-IMIDAZO(4,5-f)QUINOLIN-2-AMINE □ MeIQ

TOXICITY DATA WITH REFERENCE
mma-sat 100 ng/plate CRNGDP 7,273,86
slt-dmg-orl 100 ng/kg JJCREP 76,468,85
dns-rat:lng 3 μmol/L ENMUDM 7,245,85
dnd-mus:leu 100 μmol/L MUREAV 144,57,85
dns-ham:lng 3 μmol/L ENMUDM 7,245,85
msc-ham:lng 25 mg/L MUREAV 118,91,83
dns-gpg:lng 10 μmol/L ENMUDM 7,245,85
orl-rat TDLo:5040 mg/kg/40W-C:CAR CRNGDP 10,610,89
orl-mus TDLo:7476 mg/kg/89W-C:CAR CRNGDP 7,1889,86

CONSENSUS REPORTS: IARC Cancer Review: Group 2B IMEMDT 56,197,93; Animal Sufficient Evidence IMEMDT 56,197,93; Animal Inadequate Evidence IMEMDT 40,275,86; Human No Adequate Data IMEMDT 40,275,86; Human Inadequate Evidence IMEMDT 56,197,93.

SAFETY PROFILE: Confirmed carcinogen with experimental carcinogenic data. Mutation data reported. When heated to decomposition it emits toxic fumes of NO_x.

AJQ675 **CAS:77500-04-0** **HR: 3**
2-AMINO-3,8-DIMETHYLIMIDAZO(4,5-f)QUINOXALINE
mf: $C_{11}H_{11}N_5$ mw: 213.27

PROP: Crystals. Mp: 295–300° (sealed tube).

SYNS: 2-AMINO-3,8-DIMETHYL-3H-IMIDAZO(4,5-f)QUINOXALINE □ 3,8-DIMETHYL-3H-IMIDAZO(4,5-f)QUINOXALIN-2-AMINE

TOXICITY DATA WITH REFERENCE
mma-sat 5 ng/plate MUREAV 144,131,85
slt-dmg-orl 100 ng/kg JJCREP 76,468,85
msc-ham:ovr 300 mg/L MUTAEX 2,483,87
orl-rat TDLo:8580 mg/kg/61W-C:CAR CRNGDP 9,71,88
orl-mus TDLo:42,336 mg/kg/84W-C:CAR CRNGDP 8,665,87

CONSENSUS REPORTS: IARC Cancer Review: Group 2B IMEMDT 56,211,93; Animal Sufficient Evidence IMEMDT 56,211,93; Animal Inadequate Evidence IMEMDT 40,283,86; Human No Adequate Data IMEMDT 40,283,86; Human Inadequate Evidence IMEMDT 56,211,93.

SAFETY PROFILE: Confirmed carcinogen with experimental carcinogenic data. Mutation data reported. When heated to decomposition it emits toxic fumes of NO_x.

AJQ750 **HR: 3**
2-AMINO-6-DIMETHYL-4-(p-(p-((p-((1-METHYLPYRIDINIUM-3-YL)CARBAMOYL)PHENYL)CARBABENZAMIDO)ANILINO)PYRIMIDIMIUM), DIIODIDE
mf: $C_{33}H_{32}N_8O_3 \cdot 2I$ mw: 842.53

TOXICITY DATA WITH REFERENCE
dnd-mus:lym 840 nmol/L JMCMAR 22,134,79
ipr-mus LD10:20 mg/kg JMCMAR 22,134,79

SAFETY PROFILE: Poison by intraperitoneal route. Mutation data reported. When heated to decomposition it emits very toxic fumes of I^- and NO_x.

AJR000 **CAS:4302-87-8** **HR: 3**
p-AMINO-N,α-DIMETHYLPHENETHYLAMINE
mf: $C_{10}H_{16}N_2$ mw: 164.28

SYNS: 1-(p-AMINOPHENYL)-2-METHYLAMINOPROPAN (GERMAN) □ α-(4-AMINOPHENYL)-β-METHYLAMINO-PROPANE □ 1-(p-AMINOPHENYL)-2-METHYLAMINOPROPANE

TOXICITY DATA WITH REFERENCE
orl-rat LD50:300 mg/kg AEPPAE 195,647,40
ipr-rat LDLo:85 mg/kg AEPPAE 195,647,40

scu-rat LD50: 280 mg/kg AIPTAK 159,442,66

SAFETY PROFILE: Poison by ingestion, intraperitoneal, and subcutaneous routes. See also AMINES. When heated to decomposition it emits toxic fumes of NO_x.

AJR100 CAS:56464-19-8 HR: 3
1-(4-AMINO-1,2-DIMETHYL-5-PHENYL-1H-PYRROL-3-YL)ETHANONE
mf: $C_{14}H_{16}N_2O$ mw: 228.32

PROP: A liquid.

SYNS: 4-AMINO-1,2-DIMETHYL-5-PHENYLPYRROL-3-YLETHANONE □ ETHANONE, 1-(4-AMINO-1,2-DIMETHYL-5-PHENYL-1H-PYRROL-3-YL)- □ KETONE, (4-AMINO-1,2-DIMETHYL-5-PHENYLPYRROL-3-YL) METHYL

TOXICITY DATA WITH REFERENCE
orl-mus LD50: 1 g/kg FRPSAX 39,538,84

DOT CLASSIFICATION: 3; *Label:* Flammable Liquid

SAFETY PROFILE: Low toxicity by ingestion. A flammable liquid. When heated to decomposition it emits toxic vapors of NO_x.

AJR400 CAS:31272-21-6 HR: 3
5-AMINO-1,3-DIMETHYL-4-PYRAZOLYL o-FLUOROPHENYL KETONE
mf: $C_{12}H_{12}FN_3O$ mw: 233.27

SYNS: (5-AMINO-1,3-DIMETHYL-1H-PYRAZOL-4-YL)(2-FLUOROPHENYL)METHANONE □ KETONE, 5-AMINO-1,3-DIMETHYLPYRAZOL-4-YL o-FLUOROPHENYL □ METHANONE, (5-AMINO-1,3-DIMETHYL-1H-PYRAZOL-4-YL)(2-FLUOROPHENYL)- □ PD 71627

TOXICITY DATA WITH REFERENCE
mma-sat 10 nmol/plate CRNGDP 7,2019,86
orl-rat TD: 2366 mg/kg/13W-C:ETA AJPAA4 124,392,86

DOT CLASSIFICATION: 3; *Label:* Flammable Liquid

SAFETY PROFILE: Questionable carcinogen with experimental tumorigenic data. Mutation data reported. A flammable liquid. When heated to decomposition it emits toxic fumes of F^- and NO_x.

AJR500 CAS:68808-54-8 HR: 3
3-AMINO-1,4-DIMETHYL-5H-PYRIDO(4,3-b)INDOLE ACETATE
mf: $C_{13}H_{13}N_3 \cdot C_2H_4O_2$ mw: 271.35

PROP: Pale-brown needles or small prisms from EtOAc. Mp: 252–262°.

SYNS: 1,4-DIMETHYL-5H-PYRIDO(4,3-b)INDOL-3-AMINE ACETATE □ 1,4-DIMETHYL-5H-PYRIDO(4,3-b)INDOL-3-AMINE MONOACETATE □ TRP-P-1 (ACETATE)

TOXICITY DATA WITH REFERENCE
slt-dmg-orl 200 ppm MUREAV 122,315,83
mma-sat 1 µg/plate CPBTAL 26,611,78
orl-rat TDLo: 1539 mg/kg/29W-C:CAR JJCREP 76,815,85
orl-mus TDLo: 11 g/kg/89W-C:CAR SCIEAS 213,346,81

SAFETY PROFILE: Suspected carcinogen with experimental carcinogenic data. Mutation data reported. When heated to decomposition it emits toxic fumes of NO_x.

AJR750 CAS:35572-78-2 HR: 2
2-AMINO-4,6-DINITROTOLUENE
mf: $C_7H_7N_3O_4$ mw: 197.17

PROP: Yellow crystals from water.

SYNS: 3,5-DINITRO-o-TOLUIDINE □ 2-METHYL-3,5-DINITROBENZENAMINE

TOXICITY DATA WITH REFERENCE
mmo-sat 10 µg/plate ENMUDM 4,163,82
orl-rat LD50: 1394 mg/kg NTIS** AD-A080-146
orl-mus LD50: 1522 mg/kg NTIS** AD-A080-146

SAFETY PROFILE: Moderately toxic by ingestion. Mutation data reported. See also AMINES and NITRO COMPOUNDS of AROMATIC HYDROCARBONS. When heated to decomposition it emits toxic fumes of NO_x.

AJS100 CAS:92-67-1 HR: 3
4-AMINODIPHENYL
mf: $C_{12}H_{11}N$ mw: 169.24

PROP: Leaflets or colorless crystals from alc (aq). Mp: 53°, bp: 302°, d: 1.160 @ 20°/20°, autoign temp: 842°F.

SYNS: p-AMINOBIPHENYL □ 4-AMINOBIPHENYL □ 4-AMINODIFENIL (SPANISH) □ p-AMINODIPHENYL □ BIPHENYLAMINE □ (1,1'-BIPHENYL)-4-AMINE □ p-BIPHENYLAMINE □ 4-BIPHENYLAMINE □ PARAAMINODIPHENYL □ p-PHENYLANILINE □ XENYLAMIN (CZECH) □ XENYLAMINE

TOXICITY DATA WITH REFERENCE
mma-sat 2 µg/plate ENMUDM 5(Suppl 1),3,83
dnd-esc 30 µmol/L MUREAV 89,95,81
msc-hmn:fbr 60 mg/L MUREAV 121,71,83
dnd-rat:lvr 30 µmol/L SinJF# 26OCT82
dns-mus-orl 200 mg/kg MUREAV 125,291,84
orl-rat TDLo: 4524 mg/kg/40W-C:ETA ARZNAD 12,270,62
scu-mus TDLo: 216 mg/kg/3D-I:CAR JNCIAM 41,403,68
orl-mus TD: 5460 µg/kg:CAR EJCAAH 21,865,85
orl-mus TD: 5460 µg/kg EJCODS 21,865,85
orl-rat LD50: 500 mg/kg JIHTAB 29,1,47
orl-mus LD50: 205 mg/kg EJCODS 21,865,85
ipr-mus LDLo: 250 mg/kg CBCCT* 6,54,54
orl-dog LDLo: 25 mg/kg SCIEAS 167,992,70
orl-rbt LD50: 690 mg/kg JIHTAB 29,1,47

CONSENSUS REPORTS: NTP 7th Annual Report on Carcinogens. IARC Cancer Review: Group 1 IMEMDT 7,91,87; Human Limited Evidence IMEMDT 1,74,72; Animal Sufficient Evidence IMEMDT 1,74,72; Human Sufficient Evidence IMEMDT 28,151,82. Reported in EPA TSCA Inventory. EPA Genetic Toxicology Program. Community Right-To-Know List.

OSHA PEL: Cancer Suspect Agent
ACGIH TLV: Confirmed Human Carcinogen
DFG MAK: Human Carcinogen

SAFETY PROFILE: Confirmed human carcinogen with experimental carcinogenic and tumorigenic data. Poison by ingestion and intraperitoneal routes. Human mutation data reported. An irritant. Effects resemble those of benzidine. See also BENZIDINE. Slight to moderate fire hazard when exposed to heat, flames (sparks), or powerful oxidizers. To fight fire, use water spray, mist, dry chemical. When heated to decomposition it emits toxic fumes of NO_x. See also AROMATIC AMINES.

AJS225 **HR: 2**
2-AMINODIPYRIDO(1,2-a:3′,2′-d)IMIDAZOLE HYDROCHLORIDE
mf: $C_{10}H_8N_4 \cdot ClH$ mw: 220.68

SYN: DIPYRIDO(1,2-a:3′,2′-d)IMIDAZOLE, 2-AMINO-, HYDROCHLORIDE

TOXICITY DATA WITH REFERENCE
slt-dmg-orl 100 ng/kg JJCREP 76,468,85
orl-rat TDLo:4116 mg/kg/24W-C:CAR GANNA2 75,207,84

SAFETY PROFILE: Questionable carcinogen with experimental carcinogenic data. Mutation data reported. When heated to decomposition it emits toxic fumes of NO_x and HCl.

AJS250 **CAS:16268-87-4** **HR: 3**
2-AMINO-4,6-DIPYRROLIDINOTRIAZINE
mf: $C_{11}H_{18}N_6$ mw: 234.35

TOXICITY DATA WITH REFERENCE
orl-mus LD50:600 mg/kg JMCMAR 13,1081,70
ivn-mus LD50:56 mg/kg CSLNX* NX#03988

SAFETY PROFILE: Poison by intravenous route. Moderately toxic by ingestion. See also AMINES. When heated to decomposition it emits toxic fumes of NO_x.

AJS500 **CAS:101-50-8** **HR: 1**
4-AMINO-3,4′-DISULFOAZOBENZENE
mf: $C_{12}H_{11}N_3O_6S_2$ mw: 357.38

SYNS: 4-AMINOAZOBENZENE-3,4′-DISULFONIC ACID □ 6-AMINO-3,4′-AZODI-BENZENESULFONIC ACID □ 2-AMINO-5-((4-SULFOPHENYL)AZO)-BENZENESULFONIC ACID □ 4-(4-AMINO-3-SULFOPHENYLAZO)BENZENESULFONIC ACID □ KYSELINA 4-AMINOAZOBENZEN-3,4′-DISULFONOVA (CZECH)

TOXICITY DATA WITH REFERENCE
eye-rbt 500 mg/24H SEV 28ZPAK -,192,72
orl-rat LD50:14,800 mg/kg 28ZPAK -,192,7

CONSENSUS REPORTS: Reported in EPA TSCA Inventory.

SAFETY PROFILE: Mildly toxic by ingestion. A severe eye irritant. When heated to decomposition it emits very toxic fumes of SO_x and NO_x. See also SULFONATES.

AJS750 **CAS:146-37-2** **HR: 3**
4-AMINO-1-DODECYLQUINALDINIUM ACETATE
mf: $C_{22}H_{35}N_2 \cdot C_2H_3O_2$ mw: 386.64

PROP: Crystals from Me_2CO. Mp: 170–171°.

SYNS: 1-DODECYL-4-AMINOQUINALDINIUM ACETATE □ N-DODECYL-4-AMINOQUINALDINIUM ACETATE □ LAURODIN □ LAUROLINIUM ACETATE

TOXICITY DATA WITH REFERENCE
orl-mus LD50:132 mg/kg JPPMAB 15,129,63
ipr-mus LD50:2 mg/kg JPPMAB 15,129,63
scu-mus LD50:30 mg/kg JPPMAB 15,129,63
ivn-mus LD50:6 mg/kg JPPMAB 15,129,63

SAFETY PROFILE: Poison by ingestion, intraperitoneal, subcutaneous, and intravenous routes. When heated to decomposition it emits toxic fumes of NO_x.

AJS900 **CAS:2697-65-6** **HR: 3**
2-AMINO-ETHANESELENOL HYDROCHLORIDE
mf: $C_2H_7NSe \cdot ClH$ mw: 160.52

SYN: ETHANESELENOL, 2-AMINO-, HYDROCHLORIDE

TOXICITY DATA WITH REFERENCE
ipr-mus LD50:10 mg/kg JMCMAR 12,510,69

OSHA PEL: TWA 0.2 mg(Se)/m^3
ACGIH TLV: TWA 0.2 mg(Se)/m^3

SAFETY PROFILE: Poison by ingestion. When heated to decomposition it emits toxic fumes of NO_x, Se, and HCl.

AJS950 **CAS:2697-60-1** **HR: 3**
2-AMINOETHANESELENOSULFURIC ACID
mf: $C_2H_7NO_3SSe$ mw: 204.12

SYN: SELENOSULFURIC ACID, 2-AMINOETHYL ESTER

TOXICITY DATA WITH REFERENCE
ipr-mus LD50:18 mg/kg JMCMAR 12,510,69

OSHA PEL: TWA 0.2 mg(Se)/m^3
ACGIH TLV: TWA 0.2 mg(Se)/m^3

SAFETY PROFILE: Poison by intraperitoneal route. When heated to decomposition it emits toxic fumes of NO_x, SO_x, and Se.

AJT250 **CAS:60-23-1** **HR: 3**
2-AMINOETHANETHIOL
mf: C_2H_7NS mw: 77.16

PROP: Crystals from alc. Mp: 99–100°. Sol in MeOH, EtOH; spar sol in water.

SYNS: 2-AMINOETHYL MERCAPTAN □ BECAPTAN □ CISTEAMINA (ITALIAN) □ CYCTEINAMINE □ CYSTEAMIDE □ CYSTEAMINE □ DECARBOXYCYSTEINE □ LAMBRATEN □ MEA □ MECRAMINE □ MERCAMINE □ MERCAPTAMINE □ β-MERCAPTOETHYLAMINE □ (2-MERCAPTOETHYL)AMINE □ THIOETHANOLAMINE

TOXICITY DATA WITH REFERENCE
pic-esc 50 mg/L APMBAY 12,234,64
cyt-ham:ovr 1 mmol/L CALEDQ 5,199,78
dns-ham:fbr 1 mmol/L CALEDQ 5,199,78
sce-ham:ovr 100 μmol/L MUREAV 68,351,79
orl-rat TDLo:42 g/kg (70D pre-21D post):REP TXAPA9 11,523,67
ipr-rat LD50:232 mg/kg ARZNAD 5,421,55
scu-rat LD50:84 mg/kg OSDIAF 5,128,56
orl-mus LD50:625 mg/kg JMCMAR 18,798,75
ipr-mus LD50:250 mg/kg JMCMAR 12,510,69
scu-mus LD50:84 mg/kg OSDIAF 5,128,56
ivn-mus LD50:190 mg/kg CHDDAT 262,206,66
ivn-rbt LD50:150 mg/kg ARZNAD 5,421,55

CONSENSUS REPORTS: EPA Genetic Toxicology Program.

SAFETY PROFILE: Poison by intravenous, subcutaneous, and intraperitoneal routes. Moderately toxic by ingestion. Experimental reproductive effects. Mutation data reported. When heated to decomposition it emits very toxic fumes of SO_x and NO_x.

AJT500 CAS:2937-53-3 HR: 3
2-AMINOETHANETHIOSULFURIC ACID
mf: $C_2H_6NO_4S_2$ mw: 172.21

SYN: USAF EK-8413

TOXICITY DATA WITH REFERENCE
ipr-mus LD50:400 mg/kg NTIS** AD691-490

CONSENSUS REPORTS: Reported in EPA TSCA Inventory.

SAFETY PROFILE: Poison by intraperitoneal route. See also AMINES and SULFATES. When heated to decomposition it emits very toxic fumes of SO_x and NO_x.

AJT750 CAS:17026-81-2 HR: 2
3-AMINO-4-ETHOXYACETANILIDE
mf: $C_{10}H_{14}N_2O_2$ mw: 194.26

SYNS: 2-AMINO-4-ACETAMINIFENETOL (CZECH) □ NCI-C01887

TOXICITY DATA WITH REFERENCE
eye-rbt 500 mg/24H MLD 28ZPAK -,115,72
mmo-sat 1 mg/plate ENMUDM 7(Suppl 5),1,85
mma-sat 33,300 ng/plate ENMUDM 7(Suppl 5),1,85
orl-rat TDLo:130 g/kg/78W-C:ETA NCITR* NCI-CG-TR-112,78
orl-mus TDLo:524 mg/kg/78W-C:CAR NCITR* NCI-CG-TR-112,78
orl-rat LD50:631 mg/kg NCIMR* NIH-71-E-2144

CONSENSUS REPORTS: NTP Carcinogenesis Bioassay (feed): Clear Evidence: mouse NCITR* NCI-TR-112,78; Inadequate Studies: rat NCITR* NCI-TR-112,78. Reported in EPA TSCA Inventory.

SAFETY PROFILE: Moderately toxic by ingestion. An eye irritant. Questionable carcinogen with experimental carcinogenic and tumorigenic data. Mutation data reported. When heated to decomposition it emits toxic fumes of NO_x.

AJU000 HR: 2
2-AMINO-3-ETHOXYCARBONYL-5-BENZYL-4,5,6,7-TETRAHYDROTHIENO (2,3-c)PYRIDINE HYDROCHLORIDE
mf: $C_{17}H_{20}NO_2S \cdot ClH$ mw: 338.90

SYNS: ETHYL-2-AMINO-6-BENZYL-3-THIENO(2,3-c)PYRIDINECARBOXYLATE HYDROCHLORIDE □ Y-3642-HCl

TOXICITY DATA WITH REFERENCE
orl-rat LD50:4750 mg/kg YKKZAJ 90(11),1439,70
ipr-rat LD50:1520 mg/kg YKKZAJ 90(11),1439,70
orl-mus LD50:2050 mg/kg YKKZAJ 90(11),1439,70
ipr-mus LD50:620 mg/kg YKKZAJ 90(11),1439,70

SAFETY PROFILE: Moderately toxic by ingestion and intraperitoneal routes. When heated to decomposition it emits toxic fumes of NO_x, SO_x, and HCl.

AJU250 CAS:929-06-6 HR: 2
2-AMINOETHOXYETHANOL
DOT: UN 3055
mf: $C_4H_{11}NO_2$ mw: 105.16

SYNS: 2-(2-AMINOETHOXY)ETHANOL □ DIGLYCOLAMINE

TOXICITY DATA WITH REFERENCE
skn-rbt 10 mg/24H open SEV AMIHBC 4,119,51
eye-rbt 50 µg/24H SEV 85JCAE-,624,86
orl-rat LD50:5660 mg/kg AMIHBC 4,119,51
skn-rbt LD50:1190 mg/kg AMIHBC 4,119,51

CONSENSUS REPORTS: Reported in EPA TSCA Inventory.

DOT CLASSIFICATION: 8; *Label:* Corrosive

SAFETY PROFILE: Moderately toxic by skin contact. Mildly toxic by ingestion. Severe eye and skin irritant. Corrosive and a powerful irritant. When heated to decomposition it emits toxic fumes of NO_x.

AJU500 CAS:118-28-5 HR: 1
5-AMINO-6-ETHOXY-2-NAPHTHALENESULFONIC ACID
mf: $C_{12}H_{13}NO_4S$ mw: 267.32

SYNS: C.I. 38480 □ ETHOXY CLEVE'S ACID □ KYSELINA 1-AMINO-2-ETHOXYNAFTALEN-6-SULFONOVA (CZECH) □ KYSELINA ETHOXY-CLEVE-1,6 (CZECH)

TOXICITY DATA WITH REFERENCE
eye-rbt 100 mg/24H MOD 28ZPAK -,191,72
orl-rat LD50:12 g/kg 28ZPAK -,191,72

CONSENSUS REPORTS: Reported in EPA TSCA Inventory.

SAFETY PROFILE: Mildly toxic by ingestion. An eye irritant. When heated to decomposition it emits very toxic fumes of NO_x and SO_x. See also SULFONATES.

AJU625 CAS:1501-84-4 HR: 3
1-(1-AMINOETHYL)ADAMANTANE HYDROCHLORIDE
mf: $C_{12}H_{21}N \cdot ClH$ mw: 215.80

PROP: Solid. Mp: 373–375° (sealed tube).

SYNS: EXP 126 □ JP 61 □ MERADAN □ MERADANE □ α-METHYL-1-ADAMANTANEMETHYLAMINE HYDROCHLORIDE □ α-METHYLTRICYCLO(3.3.1.1^{3,7})DECANE-1-METHANAMINE HYDROCHLORIDE □ REMANTADIN □ RIMANTADINE HYDROCHLORIDE

TOXICITY DATA WITH REFERENCE
orl-rat LD50:640 mg/kg VOONAW 28(9),23,82
ipr-rat LD50:135 mg/kg KHFZAN 11(6),73,77
ipr-mus LD50:135 mg/kg PCJOAU 11,798,77

SAFETY PROFILE: Poison by intraperitoneal route. Moderately toxic by ingestion. When heated to decomposition it emits toxic fumes of NO_x and HCl.

AJU875 CAS:25682-07-9 HR: 3
2-AMINOETHYLAMMONIUM PERCHLORATE
mf: $C_2H_9ClN_2O_4$ mw: 160.56

SAFETY PROFILE: Explodes upon heating. When heated to decomposition it emits toxic fumes of Cl^-, NH_3, and NO_x. See also PERCHLORATES.

AJV000 CAS:132-32-1 HR: 3
3-AMINO-9-ETHYLCARBAZOLE
mf: $C_{14}H_{14}N_2$ mw: 210.30

PROP: Solid. Mp: 98–100°. In cancer bioassay both free amine and hydrochloride salt were used NCITR* NCI-CG-TR-93,78.

SYN: 3-AMINO-N-ETHYLCARBAZOLE

TOXICITY DATA WITH REFERENCE
orl-rat TDLo:33 g/kg/78W-C:CAR NCITR* NCI-CG-TR-93,78
orl-mus TDLo:87 g/kg/78W-C:CAR NCITR* NCI-CG-TR-93,78
ipr-mus LD50:150 mg/kg NTIS** AD691-490

CONSENSUS REPORTS: Reported in EPA TSCA Inventory.

DFG MAK: Suspected Carcinogen

SAFETY PROFILE: Suspected carcinogen with experimental carcinogenic data. Poison by ingestion and intraperitoneal routes. When heated to decomposition it emits toxic fumes of NO_x.

AJV250 CAS:6109-97-3 HR: 3
3-AMINO-9-ETHYLCARBAZOLEHYDROCHLORIDE
mf: $C_{14}H_{14}N_2 \cdot ClH$ mw: 246.76

PROP: In cancer bioassay both free amine and hydrochloride salt used NCITR* NCI-CG-TR-93,78.

SYN: NCI-C03043

TOXICITY DATA WITH REFERENCE
mma-sat 1 µg/plate ENMUDM 5(Suppl 1),3,83
orl-rat TDLo:33 g/kg/78W-C:CAR NCITR* NCI-CG-TR-93,78
orl-mus TDLo:87 g/kg/78W-C:CAR NCITR* NCI-CG-TR-93,78
orl-rat LD50:234 mg/kg JPETAB 99,450,50

CONSENSUS REPORTS: NCI Carcinogenesis Bioassay Completed; Results Positive: mouse, rat NCITR* NCI-CG-TR-93,78.

SAFETY PROFILE: Suspected carcinogen with experimental carcinogenic data. Poison by ingestion. Mutation data reported. When heated to decomposition it emits very toxic fumes of NO_x and HCl.

AJV500 CAS:1197-18-8 HR: 2
trans-4-AMINOETHYLCYCLOHEXANE-1-CARBOXYLIC ACID
mf: $C_8H_{15}NO_2$ mw: 157.24

PROP: Crystals from Me_2CO/EtOH (aq). Mp: 386–392° (decomp). Very spar sol in EtOH, and Et_2O.

SYNS: AMCHA ☐ trans-AMCHA ☐ AMIKAPRON ☐ trans-p-(AMINOMETHYL)CYCLOHEXANECARBOXYLICACID ☐ trans-1-AMINOMETHYLCYCLOHEXANE-4-CARBOXYLIC ACID ☐ trans-4-AMINOMETHYL-1-CYCLOHEXANECARBOXYLIC ACID ☐ AMSTAT ☐ ANVITOFF ☐ BAY 3517 ☐ CL 65336 ☐ CYCLOCAPRON ☐ DV-79 ☐ EMORHALT ☐ EXACYL ☐ FRENOLYSE ☐ HEXAPROMIN ☐ HEXATRON ☐ RIKAVARIN ☐ RP 18,429 ☐ SPIRAMIN ☐ TAMCHA ☐ TRANEX ☐ TRANEXAMIC ACID ☐ TRANHEXAMIC ACID ☐ TRANSAMLON ☐ UGUROL

TOXICITY DATA WITH REFERENCE
orl-mus TDLo:9 mg/kg (female 7-12D post):REP OYYAA2 5,415,71
orl-rat TDLo:9 mg/kg (female 9-14D post):TER OYYAA2 5,415,71
orl-rat LD50:3000 mg/kg APTOA6 22,340,65
ipr-rat LD50:4200 mg/kg MEIEDD 10,1269,83
scu-rat LD50:4620 mg/kg NIIRDN 6,512,82
ivn-rat LD50:1200 mg/kg APTOA6 22,340,65
scu-mus LD50:5310 mg/kg NIIRDN 6,512,82
ivn-mus LD50:1350 mg/kg NIIRDN 6,512,82
ivn-dog LD50:1110 mg/kg NIIRDN 6,512,82

SAFETY PROFILE: Moderately toxic by ingestion and intravenous routes. An experimental teratogen. Other experimental reproductive effects. When heated to decomposition it emits toxic fumes such as NO_x. A hemostatic agent.

AJV850 CAS:63991-14-0 HR: 3
α-(1-AMINOETHYL)-2,4-DIMETHOXYBENZYL ALCOHOL HYDROCHLORIDE
mf: $C_{11}H_{17}NO_3 \cdot ClH$ mw: 247.75

SYN: BENZYL ALCOHOL-α-(1-AMINOETHYL)-2,4-DIMETHOXY HYDROCHLORIDE

TOXICITY DATA WITH REFERENCE
scu-rat LDLo:320 mg/kg JPETAB 71,62,41
ivn-rbt LDLo:21 mg/kg JACSAT 53,4149,31

SAFETY PROFILE: Poison by subcutaneous and intravenous routes. When heated to decomposition it emits very toxic fumes of NO_x and Cl^-.

AJW000 CAS:111-41-1 HR: 2
N-AMINOETHYLETHANOLAMINE
mf: $C_4H_{12}N_2O$ mw: 104.18

$HOC_2H_4NHC_2H_4NH_2$

PROP: Colorless liquid. Bp: 243.7°, flash p: 216°F, d: 1.0304 @ 20°/20°, autoign temp: 695°F, vap press: <0.01 mm @ 20°, vap d: 3.59. Misc in H_2O, EtOH; spar sol in Et_2O.

SYNS: AMINOETHYL ETHANOLAMINE ☐ ETHANOLETHYLENE DIAMINE ☐ N-HYDROXYETHYL-1,2-ETHANEDIAMINE ☐ N-(β-HYDROXYETHYL)ETHYLENEDIAMINE ☐ N-(2-HYDROXYETHYL)ETHYLENEDIAMINE ☐ MONOETHANOLETHYLENEDIAMINE

TOXICITY DATA WITH REFERENCE
skn-rbt 10 mg/24H open JIHTAB 26,269,44
skn-rbt 445 mg open MLD UCDS** 11/29/63
eye-rbt 50 mg SEV UCDS** 7/19/65
mmo-sat 2800 µg/plate ENMUDM 9(Suppl 9),1,87
orl-rat LD50:3000 mg/kg UCDS** 7/19/65
skn-rat LD50:2250 mg/kg 85GMAT -,64,82
mmo-sat 2800 µg/plate ENMUDM 9(Suppl 9),1,87
orl-rat LD50:3 g/kg UCDS** 7/19/65
skn-rat LD50:2250 mg/kg 85GMAT-,64,82
ipr-rat LD50:120 mg/kg EVSSAV 2,289,68

ivn-rat LD50:417 mg/kg 85GMAT -,64,82
ims-rat LD50:2 g/kg 85GMAT -,64,82
orl-mus LD50:3550 mg/kg 85GMAT -,64,82
orl-rbt LD50: 2 g/kg 85GMAT -,64,82
orl-rat LD50:3 g/kg UCDS** 7/19/65
skn-rat LD50:2250 mg/kg 85GMAT-,64,82
ipr-rat LD50:120 mg/kg EVSSAV 2,289,68
scu-rat LD50:2250 mg/kg EVSSAV 2,289,68
ivn-rat LD50:417 mg/kg EVSSAV 2,289,68
ims-rat LD50:2 g/kg EVSSAV 2,289,68
orl-mus LD50:3550 mg/kg EVSSAV 2,289,68
orl-rbt LD50:2 g/kg EVSSAV 2,289,68
skn-rbt LD50:3560 μL/kg UCDS** 7/19/65
orl-gpg LD50:1500 mg/kg 85GMAT -,64,82
skn-gpg LD50:1800 mg/kg JIHTAB 26,269,44

CONSENSUS REPORTS: Reported in EPA TSCA Inventory.

SAFETY PROFILE: Moderately toxic by ingestion, skin contact and several other routes. A severe eye irritant and moderate skin irritant. Mutation data reported. Combustible. To fight fire use alcohol foam, mist, dry chemical. As with other amines it ignites on contact with cellulose nitrate of high surface area. When heated to decomposition it emits toxic fumes of NO_x.

AJW250 **HR: 3**
6-AMINO-1-ETHYL-4-p-((p-((1-ETHYLPYRIDINIUM-4-YL)AMINO)2-AMINOPHENYL)CARBAMOYL)ANILINO)QUINOLINIUM DIIODIDE
mf: $C_{31}H_{33}N_7O \cdot 2I$ mw: 773.51

TOXICITY DATA WITH REFERENCE
dnd-mus:lym 710 nmol/L JMCMAR 22,134,79
ipr-mus LD10:6500 μg/kg JMCMAR 22,134,79

SAFETY PROFILE: Poison by intraperitoneal route. Mutation data reported. When heated to decomposition it emits very toxic fumes of NO_x and I^-.

AJW500 **CAS:50309-16-5** **HR: 3**
6-AMINO-1-ETHYL-4-(p-(p-((1-ETHYLPYRIDINIUM-4-YL)AMINO)BENZAMIDO)ANILINO)QUINOLINIUM DIIODIDE
mf: $C_{31}H_{32}N_6O \cdot 2I$ mw: 758.49

TOXICITY DATA WITH REFERENCE
dnd-mus:lym 600 nmol/L JMCMAR 22,134,79
ipr-mus LD10:9 mg/kg JMCMAR 22,134,79

SAFETY PROFILE: Poison by intraperitoneal route. Mutation data reported. When heated to decomposition it emits very toxic fumes of I^- and NO_x.

AJW750 **CAS:42013-69-4** **HR: 3**
6-AMINO-1-ETHYL-4-(p-((p-((1-ETHYLPYRIDINIUM-4-YL)AMINO)PHENYL)CARBAMOYL)ANILINOQUINOLINIUM) DIBROMIDE
mf: $C_{31}H_{32}N_6O \cdot 2Br$ mw: 664.51

TOXICITY DATA WITH REFERENCE
dnd-mus:lym 690 nmol/L JMCMAR 22,134,79
ipr-mus LD10:10 mg/kg JMCMAR 22,134,79

SAFETY PROFILE: Poison by intraperitoneal route. Mutation data reported. When heated to decomposition it emits very toxic fumes of Br^- and NO_x.

AJX000 **CAS:61-54-1** **HR: 3**
3-(2-AMINOETHYL)INDOLE
mf: $C_{10}H_{12}N_2$ mw: 160.24

PROP: Needles from pet ether. Mp: 118°, bp: 137° @ 0.15 mm. Very spar sol in Et_2O, $CHCl_3$, and C_6H_6.

SYNS: (AMINO-2 ETHYL)-3-INDOLE (FRENCH) □ 1H-INDOLE-3-ETHANAMINE □ INDOL-3-ETHYLAMINE □ 2-(3-INDOLYL)ETHYLAMINE □ TRYPTAMINE

TOXICITY DATA WITH REFERENCE
ipr-rat LD50:223,200 μg/kg JPMSAE 66(12),1962,77
ipr-mus LD50:100 mg/kg EJMCA5 9,453,74
scu-mus LD50:500 mg/kg DPHFAK 22,313,70

SAFETY PROFILE: Poison by intraperitoneal route. Moderately toxic by subcutaneous route. When heated to decomposition it emits toxic fumes of NO_x.

AJX250 **CAS:343-94-2** **HR: 3**
3-(2-AMINOETHYL)INDOLE HYDROCHLORIDE
mf: $C_{10}H_{12}N_2 \cdot ClH$ mw: 196.70

PROP: Needles from EtOH/EtOAc. Mp: 248°.

SYNS: β-INDOLAETHYLAMIN-CHLORHYDRAT (GERMAN) □ β-3-INDOLYLETHYLAMINE HYDROCHLORIDE □ INDOLE-3-ETHYLAMINE HYDROCHLORIDE □ β-INDOLE-ETHYLAMINE HYDROCHLORIDE □ TRYPTAMINE HYDROCHLORIDE

TOXICITY DATA WITH REFERENCE
scu-rat LDLo:1300 mg/kg JPMRAB 2,77,27
ipr-mus LD50:197 mg/kg YKKZAJ 91,1620,71
scu-mus LD50:504 mg/kg RPTOAN 33,180,70
ivn-mus LD50:109 mg/kg BJPCAL 23,43,64
scu-rbt LDLo:1000 mg/kg JPMRAB 2,77,27

CONSENSUS REPORTS: Reported in EPA TSCA Inventory.

SAFETY PROFILE: Poison by intravenous and intraperitoneal routes. Moderately toxic by subcutaneous route. When heated to decomposition it emits very toxic NO_x and HCl.

AJX500 **CAS:50-67-9** **HR: 3**
3-(2-AMINOETHYL)INDOL-5-OL
mf: $C_{10}H_{12}N_2O$ mw: 176.24

SYNS: 3-(β-AMINOETHYL)-5-HYDROXYINDOLE □ ANTEMOQUA □ ANTEMOVIS □ DS SUBSTANDE □ ENTERAMINE □ HIPPOPHAIN □ 5-HT □ 5-HTA □ 5-HYDROXY-3-(β-AMINOETHYL)INDOLE □ 5-HYDROXYTRYPTAMINE □ SEROTONIN □ SUBSTANCE DS □ SUBSTANZ DS □ THROMBOCYTIN □ THROMBOTONIN

TOXICITY DATA WITH REFERENCE
scu-rat TDLo:110 mg/kg (female 1-22D post):REP JOPDAB 63,394,63
ipr-rat TDLo:10 mg/kg (18D preg):TER ARPAAQ 81,257,66
scu-rat LD50:285 mg/kg PSCBAY 2,17,63
ivn-rat LD50:30 mg/kg PSCBAY 2,17,63

orl-mus LD50:60 mg/kg MZUZA8 (3),61,85
ipr-mus LD50:160 mg/kg IJPPAZ 17,31,73
scu-mus LD50:601 mg/kg FEPRA7 23,T125,64
ivn-mus LD50:81 mg/kg FATOAO 26,10,63
ims-mus LD50:750 mg/kg AIPTAK 112,319,57

SAFETY PROFILE: Poison by ingestion, intravenous, and intraperitoneal routes. An experimental teratogen. Other experimental reproductive effects. When heated to decomposition it emits toxic fumes of NO_x. A neurotransmitter.

AJX750 CAS:971-74-4 HR: 3
3-(2-AMINOETHYL)INDOL-5-OL CREATININE SULFATE
mf: $C_{10}H_{12}N_2O{\cdot}C_4H_7N_3O{\cdot}H_2O_4S{\cdot}H_2O$ mw: 405.48

PROP: Plates. Mp: 214–216° (decomp).

SYNS: CREATININE SULFATE compounded with 3-(2-AMINOETHYL) INDOLE-5-OL (1:1:1), MONOHYDRATE □ CREATININE SULFATE compounded with 3-(2-AMINOETHYL)INDOL-5-OL (1:1:1) □ 5-HYDROXYTRYPTAMINE CREATININE SULFATE □ 5-HYDROXYTRYPTAMINE CREATININE SULFATE MONOHYDRATE □ SEROTIN CREATININE SULFATE □ SEROTONIN CREATININE SULFATE MONOHYDRATE

TOXICITY DATA WITH REFERENCE
ivn-rat TDLo:10 mg/kg (female 20D post):REP AJPHAP 169,537,52
scu-mus TDLo:10 mg/kg (female 15D post):TER AJOGAH 99,250,67
scu-rat LD50:257 mg/kg JPETAB 105,80,52
ivn-rat LD50:66 mg/kg JPETAB 105,80,52
ipr-mus LD50:405 mg/kg YKKZAJ 94,1620,74
ivn-mus LD50:352 mg/kg JPETAB 105,80,52
ivn-gpg LD50:28 mg/kg JPMSAE 57,1543,68

CONSENSUS REPORTS: EPA Genetic Toxicology Program.

SAFETY PROFILE: Poison by subcutaneous and intravenous routes. Moderately toxic by intraperitoneal route. An experimental teratogen. Other experimental reproductive effects. See also SULFATES. When heated to decomposition it emits very toxic fumes of SO_x and NO_x.

AJY000 CAS:1704-04-7 HR: 3
AMINOETHYLISOSELENOURONIUM BROMIDE HYDROCHLORIDE

SYNS: 2-AMINOAETHYLISOSELENOURONIUMBROMID-HYDROBROMID (GERMAN) □ MONOETHYLISOSELENOURONIUMBROMIDE-HYDROBROMIDE

TOXICITY DATA WITH REFERENCE
ipr-mus LD50:51 mg/kg STRAAA 151,78,76
scu-mus LD50:50 mg/kg STRAAA 151,78,76
ivn-mus LD50:44 mg/kg STRAAA 151,78,76

CONSENSUS REPORTS: On Community Right-To-Know List.

SAFETY PROFILE: Poison by intraperitoneal, subcutaneous, and intravenous routes. When heated to decomposition it emits very toxic fumes of HCl, Br^- and NO_x.

AJY250 CAS:56-10-0 HR: 3
2-β-AMINOETHYLISOTHIOUREA
mf: $C_3H_9N_3S{\cdot}2BrH$ mw: 281.05

PROP: Solid. Mp: 194–195°.

SYNS: AET □ AET BROMIDE □ AET DIHYDROBROMIDE □ AET-2HBR □ β-AMINOAETHYL-ISOTHIURONIUM DIHYDROBROMID(GERMAN) □ 2-AMINOETHYL ESTER CARBAMIMIDOTHIOIC ACID DIHYDROBROMIDE □ 2-(β-AMINOETHYL)ISOTHIOURONIUM BROMIDE HYDROBROMIDE □ S-(β-AMINOETHYL)ISOTHIURONIUM BROMIDE HYDROBROMIDE □ S-(2-AMINOETHYL)ISOTHIURONIUM BROMIDE HYDROBROMIDE □ β-AMINOETHYLISOTHIURONIUM BROMIDE HYDROBROMIDE □ 2-AMINOETHYLISOTHIURONIUM BROMIDE HYDROBROMIDE □ 2-AMINOETHYLISOTHIOURONIUM DIBROMIDE □ 2-AMINOETHYLISOTHIURONIUM DIHYDROBROMIDE □ 2-(2-AMINOETHYL)-2-THIOPSEUDOUREA HYDROBROMIDE □ ANTIRAD □ ANTIRADON □ SURRECTAN □ USAF XR-31

TOXICITY DATA WITH REFERENCE
ipr-rat LD50:288 mg/kg AIPTAK 142,198,63
ivn-rat LD50:85 mg/kg CLCEAL 105,1165,66
ipr-mus LD50:400 mg/kg NTIS** AD277-289
ivn-mus LD50:96 mg/kg CLCEAL 105,1165,66
orl-dog LD50:177 mg/kg AIPTAK 142,510,63
ipr-dog LD50:113 mg/kg AIPTAK 142,510,63
ipr-rbt LD50:236 mg/kg AIPTAK 142,510,63

CONSENSUS REPORTS: Reported in EPA TSCA Inventory.

SAFETY PROFILE: Poison by ingestion, intraperitoneal, subcutaneous, and intravenous routes. When heated to decomposition it emits very toxic NO_x, SO_x, and HBr.

AJY500 CAS:871-25-0 HR: 3
2-AMINOETHYLISOTHIOURONIUMDICHLORIDE
mf: $C_3H_9N_3S{\cdot}2ClH$ mw: 192.13

SYNS: AET DICHLORIDE □ β-AMINOAETHYLISOTHIURONIUM-CHLORID-HYDROCHLORID (GERMAN) □ S-β-AMINOETHYLISOTHIOURONIC DIHYDROCHLORIDE □ 2-AMINOETHYL-2-THIOPSEUDOUREA DICHLORIDE □ 2-(2-AMINOETHYL)-2-THIOPSEUDOUREA DIHYDROCHLORIDE □ USAF XR-32

TOXICITY DATA WITH REFERENCE
par-rat LD50:325 mg/kg TXAPA9 1,8,59
ipr-mus LD50:250 mg/kg NTIS** AD277-689
par-mus LD50:400 mg/kg TXAPA9 1,8,59
par-dog LD50:110 mg/kg TXAPA9 1,8,59
scu-mus LD50:266 mg/kg ARZNAD 8,72,58

SAFETY PROFILE: Poison by subcutaneous, parenteral, and intraperitoneal routes. When heated to decomposition it emits very toxic fumes of HCl, SO_x, and NO_x.

AJY750 CAS:92-09-1 HR: 3
4-AMINO-N-ETHYL-m-(β-METHANESULFONAMIDOETHYL)-m-TOLUIDINE

TOXICITY DATA WITH REFERENCE
orl-rat LDLo:400 mg/kg KODAK* -,-,71
ipr-rat LDLo:10 mg/kg KODAK* -,-,71

CONSENSUS REPORTS: Reported in EPA TSCA Inventory.

SAFETY PROFILE: Poison by ingestion and intraperito-

A

neal routes. When heated to decomposition it emits very toxic fumes of SO_x and NO_x.

AJZ000 **CAS:22137-01-5** **HR: 3**
3-(2-AMINOETHYL)-5-METHOXYBENZOFURAN HYDROCHLORIDE
mf: $C_{11}H_{13}NO_2 \cdot ClH$ mw: 227.71

TOXICITY DATA WITH REFERENCE
ivn-rat LDLo:6 mg/kg RPTOAN 33,246,70
ivn-mus LD50:55 mg/kg RPTOAN 33,246,70

SAFETY PROFILE: Poison by intravenous route. When heated to decomposition it emits very toxic fumes of HCl and NO_x.

AKA000 **CAS:28089-06-7** **HR: 3**
6-(β-AMINOETHYL)-5-METHOXYBENZOFURANHYDROCHLORIDE
mf: $C_{11}H_{12}NO_2 \cdot ClH$ mw: 226.70

SYN: 6-(2-AMINOETHYL)-5-METHOXYBENZOFURAN HYDROCHLORIDE

TOXICITY DATA WITH REFERENCE
ivn-rat LDLo:48 mg/kg RPTOAN 33,246,70
ivn-mus LD50:60 mg/kg RPTOAN 33,246,70

SAFETY PROFILE: Poison by intravenous route. When heated to decomposition it emits very toxic fumes of HCl and NO_x.

AKA250 **CAS:63991-23-1** **HR: 3**
α-(1-AMINOETHYL)-4-METHOXYBENZYLALCOHOL HYDROCHLORIDE
mf: $C_{10}H_{15}NO_2 \cdot ClH$ mw: 217.72

SYN: α-(1-AMINOETHYL)-4-METHOXYBENZYL ALCOHOL HYDROCHLORIDE

TOXICITY DATA WITH REFERENCE
scu-rat LDLo:160 mg/kg JPETAB 71,62,41
ivn-rbt LDLo:35 mg/kg JACSAT 53,4149,31

SAFETY PROFILE: Poison by subcutaneous and intravenous routes. When heated to decomposition it emits very toxic fumes of Cl^- and NO_x.

AKA500 **CAS:52479-18-2** **HR: 3**
2-(AMINOETHYL)-2-METHYL-1,3-BENZODIOXOLE HYDROCHLORIDE
mf: $C_{10}H_{13}NO_2 \cdot ClH$ mw: 215.70

SYN: 2-AMINOETHYL-2-METHYL-1,3-BENZODIOXOLE HYDROCHLORIDE

TOXICITY DATA WITH REFERENCE
ivn-rat LD50:33 mg/kg EJMCA5 12,413,77
ipr-mus LD50:100 mg/kg EJMCA5 12,413,77

SAFETY PROFILE: Poison by intravenous and intraperitoneal routes. When heated to decomposition it emits very toxic fumes of HCl and NO_x.

AKA600 **CAS:56464-20-1** **HR: 3**
1-(4-AMINO-1-ETHYL-2-METHYL-5-PHENYL-1H-PYRROL-3-YL)ETHANONE
mf: $C_{15}H_{18}N_2O$ mw: 242.35

PROP: A liquid.

SYNS: ETHANONE, 1-(4-AMINO-1-ETHYL-2-METHYL-5-PHENYL-1H-PYRROL-3-YL)- □ KETONE, (4-AMINO-1-ETHYL-2-METHYL-5-PHENYL-PYRROL-3-YL) METHYL

TOXICITY DATA WITH REFERENCE
orl-mus LD50:500 mg/kg FRPSAX 39,538,84

DOT CLASSIFICATION: 3; *Label:* Flammable Liquid

SAFETY PROFILE: Moderately toxic by ingestion. A flammable liquid. When heated to decomposition it emits toxic vapors of NO_x.

AKA750 **CAS:2038-03-1** **HR: 3**
N-AMINOETHYLMORPHOLINE
mf: $C_6H_{14}N_2O$ mw: 130.22

PROP: Liquid. Mp: 25.6°, bp: 204.2°, flash p: 347°F (OC), d: 0.9915 @ 20°/20°, vap d: 4.49.

SYNS: β-AMINOAETHYL-MORPHOLIN (GERMAN) □ 4-MORPHOLINEETHANAMINE

TOXICITY DATA WITH REFERENCE
skn-rbt 10 mg/24H open JIHTAB 26,269,44
eye-rbt 50 μg/24H SEV 85JCAE-,888,86
orl-rat LD50:3000 mg/kg JIHTAB 26,269,44
scu-mus LD50:2145 mg/kg JIHTAB 26,269,44
skn-gpg LD50:300 mg/kg JIHTAB 26,269,44

CONSENSUS REPORTS: Reported in EPA TSCA Inventory.

SAFETY PROFILE: Poison by skin contact. Moderately toxic by ingestion and subcutaneous routes. A skin and severe eye irritant. Moderately flammable when exposed to heat, flame, or oxidizing materials. To fight fire, use alcohol foam, dry chemical. When heated to decomposition it emits toxic fumes of NO_x.

AKB000 **CAS:140-31-8** **HR: 3**
N-AMINOETHYLPIPERAZINE
DOT: UN 2815
mf: $C_6H_{15}N_3$ mw: 129.24

PROP: Light-colored liquid. D: 0.9852 @ 20°/20°, mp: −19°, bp: 220.4°, flash p: 200°F (OC), vap d: 4.4.

SYNS: AMINOETHYLPIPERAZINE □ N-(β-AMINOETHYL)PIPERAZINE □ N-(2-AMINOETHYL)PIPERAZINE □ 1-(2-AMINOETHYL)PIPERAZINE □ USAF DO-46

TOXICITY DATA WITH REFERENCE
skn-rbt 100 μg/24H open AIHAAP 23,95,62
skn-rbt 5 mg/24H SEV 85JCAE -,864,86
eye-rbt 20 mg/24H MOD 85JCAE -,864,86
sce-ham:ovr 125 μg/L MUREAV 320,31,94
msc-ham:ovr 500 μg/L MUREAV 320,31,94
otr-mus:lym 1 μL/L ENMUDM 4,390,82
orl-rat TDLo:1680 mg/kg (male 28D pre):REP GISAAA 51(10),66,86
orl-rat LD50:2140 mg/kg AIHAAP 23,95,62

ipr-mus LD50:250 mg/kg NTIS** AD277-689
skn-rbt LD50:880 mg/kg UCDS** 6/13/69

CONSENSUS REPORTS: Reported in EPA TSCA Inventory.

DOT CLASSIFICATION: 8; *Label:* Corrosive

SAFETY PROFILE: Poison by intraperitoneal routes. Moderately toxic by ingestion and skin contact. Experimental reproductive effects. A skin and eye irritant. Mutation data reported. See also AMINES. Moderately flammable when exposed to heat, flame, sparks, or powerful oxidizers. To fight fire, use alcohol foam. When heated to decomposition it emits toxic fumes of NO_x.

AKB125 HR: 2
5-AMINO-2-ETHYLTETRAZOL
mf: $C_3H_7N_5$ mw: 113.12

SAFETY PROFILE: Forms an explosive complex with aluminum hydride. When heated to decomposition it emits toxic fumes of NO_x.

AKB250 CAS:13073-35-3 HR: 2
2-AMINO-4-(ETHYLTHIO)BUTYRIC ACID
mf: $C_6H_{13}NO_2S$ mw: 163.26

PROP: Solid. Mp: 272–274° (decomp).

SYNS: l-2-AMINO-4-(ETHYLTHIO)BUTYRIC ACID □ ETHIONINE □ l-ETHIONINE □ S-ETHYL-l-HOMOCYSTEINE

TOXICITY DATA WITH REFERENCE
dni-esc 2 g/L CYTOAN 50,387,85
dni-hmn:lym 2 mmol/L BBACAQ 520,139,79
otr-ham:emb 5 mg/L CRNGDP 4,291,83
orl-mus TDLo:44,100 mg/kg/2Y-C:CAR CRNGDP 7,1143,86

CONSENSUS REPORTS: Reported in EPA TSCA Inventory. EPA Genetic Toxicology Program.

SAFETY PROFILE: Questionable carcinogen with experimental carcinogenic data. Human mutation data reported. When heated to decomposition it emits very toxic fumes of NO_x and SO_x.

AKB500 CAS:3724-89-8 HR: 3
S-(2-AMINOETHYL)THIOPHOSPHATEMONOSODIUM SALT
mf: $C_2H_7NO_3PS{\cdot}Na$ mw: 179.12

SYNS: 2-AMINO-ETHANETHIOL DIHYDROGEN PHOSPHATE(ester), MONOSODIUM SALT □ CISTAPHOS □ CYSTAPHOS □ CYSTAPHOS SODIUM SALT □ MONOSODIUM-β-AMINOETHYL THIOPHOSPHATE □ SODIUM HYDROGEN-S-(2-AMINOETHYL)PHOSPHOROTHIOATE □ SODIUM HYDROGEN-S-(2-AMINOETHYL)PHOSPHOROTHIOIC ACID □ WR 638

TOXICITY DATA WITH REFERENCE
ipr-rat LD50:555 mg/kg RADOA8 16,249,76
ims-rat LD50:505 mg/kg RADOA8 16,249,76
orl-mus LD50:1433 mg/kg RADOA8 16,249,76
ipr-mus LD50:806 mg/kg RADOA8 16,249,76
ims-mus LD50:1003 mg/kg RADOA8 16,249,76
ipr-gpg LD50:358 mg/kg RADOA8 16,249,76

SAFETY PROFILE: Poison by intraperitoneal route. Moderately toxic by ingestion and intramuscular routes. See also PHOSPHATES and ESTERS. When heated to decomposition it emits very toxic fumes of PO_x, SO_x, NO_x, and Na_2O.

AKC000 CAS:63019-67-0 HR: 2
2-AMINO-N-FLUOREN-2-YLACETAMIDE
mf: $C_{15}H_{14}N_2O$ mw: 238.31

SYN: 2-GLYCYLAMINOFLUORENE

TOXICITY DATA WITH REFERENCE
orl-rat TDLo:1200 mg/kg/20W-I:ETA NATUAS 184,2018,59

SAFETY PROFILE: Questionable carcinogen with experimental tumorigenic data. When heated to decomposition it emits toxic fumes of NO_x.

AKC250 CAS:1682-39-9 HR: 2
2-AMINO-5-FLUOROBENZOXAZOLE
mf: $C_7H_5FN_2O$ mw: 152.14

TOXICITY DATA WITH REFERENCE
orl-rat LD50:1000 mg/kg MDCHAG 4(1),338,64
orl-mus LD50:700 mg/kg MDCHAG 4(1),336,64
ipr-mus LD50:450 mg/kg MDCHAG 4(1),336,64

SAFETY PROFILE: Moderately toxic by ingestion and intraperitoneal routes. When heated to decomposition it emits very toxic fumes of F^- and NO_x.

AKC500 CAS:324-93-6 HR: 3
4-AMINO-4′-FLUORODIPHENYL
mf: $C_{12}H_{10}FN$ mw: 187.23

PROP: Leaflets from alc. Mp: 121°.

SYNS: 4′-FLUORO-4-AMINODIPHENYL □ 4′-FLUORO-4-BIPHENYLAMINE

TOXICITY DATA WITH REFERENCE
mor-mus:emb 500 µg/L JJIND8 52,1167,74
orl-rat TDLo:300 mg/kg:CAR CNREA8 26,619,66
orl-mus TDLo:520 mg/kg/26W-I:NEO BJCAAI 19,297,65
scu-rat TD:2000 mg/kg/W-I:ETA BMBUAQ 14,141,58
orl-rat LDLo:300 mg/kg CNREA8 26,619,66

CONSENSUS REPORTS: EPA Genetic Toxicology Program.

SAFETY PROFILE: Poison by ingestion. Questionable carcinogen with experimental carcinogenic, neoplastigenic, and tumorigenic data. Mutation data reported. When heated to decomposition it emits very toxic fumes of F^- and NO_x.

AKC550 CAS:91480-89-6 HR: 3
1-(4-AMINO-5-(3-FLUOROPHENYL)-2-METHYL-1H-PYRROL-3-YL)ETHANONE
mf: $C_{13}H_{13}FN_2O$ mw: 232.28

PROP: A liquid.

SYNS: ETHANONE, 1-(4-AMINO-5-(3-FLUOROPHENYL)-2-METHYL-

1H-PYRROL-3-YL)- □ KETONE, (4-AMINO-5-(m-FLUOROPHENYL)-2-METHYLPYRROL-3-YL) METHYL

TOXICITY DATA WITH REFERENCE
orl-mus LD50:500 mg/kg FRPSAX 39,538,84

DOT CLASSIFICATION: 3; *Label:* Flammable Liquid

SAFETY PROFILE: Moderately toxic by ingestion. A flammable liquid. When heated to decomposition it emits toxic vapors of NO_x and F^-.

AKC560 CAS:56463-65-1 HR: 3
1-(4-AMINO-5-(4-FLUOROPHENYL)-2-METHYL-1H-PYRROL-3-YL)ETHANONE
mf: $C_{13}H_{13}FN_2O$ mw: 232.28

PROP: A liquid.

SYNS: ETHANONE, 1-(4-AMINO-5-(4-FLUOROPHENYL)-2-METHYL-1H-PYRROL-3-YL)- □ KETONE, (4-AMINO-5-(p-FLUOROPHENYL)-2-METHYLPYRROL-3-YL) METHYL

TOXICITY DATA WITH REFERENCE
orl-mus LD50:1 g/kg FRPSAX 39,538,84

DOT CLASSIFICATION: 3; *Label:* Flammable Liquid

SAFETY PROFILE: Low toxicity by ingestion. A flammable liquid. When heated to decomposition it emits toxic vapors of NO_x and F^-.

AKC600 CAS:125-84-8 HR: 2
AMINOGLUTETHIMIDE
mf: $C_{13}H_{16}N_2O_2$ mw: 232.31

SYNS: p-AMINOGLUTETHIMIDE □ 2-(p-AMINOPHENYL)-2-ETHYLGLUTARIMIDE □ Ba-16038 □ CYTADREN □ ELIPTEN □ 3-ETHYL-3-(p-AMINOPHENYL)-2,6-DIOXOPIPERIDINE □ GLUTARIMIDE, 2-(p-AMINOPHENYL)-2-ETHYL- □ ORIMETEN □ 2,6-PIPERIDINEDIONE, 3-(4-AMINOPHENYL)-3-ETHYL-

TOXICITY DATA WITH REFERENCE
orl-wmn TDLo:3630 mg/kg (female 1-35W post):TER AJDCAI 124,421,72
orl-wmn TDLo:20,500 mg/kg/94W-I:BLD BMJOAE 291,970,85
orl-man LDLo:21 mg/kg/3D-I:PUL AIMEAS 105,633,86
ipr-mus LD50:625 mg/kg JMCMAR 18,736,75

SAFETY PROFILE: Moderately toxic by intraperitoneal route. Human systemic effects by ingestion: agranulocytosis, dyspnea. Human teratogenic effects by ingestion: urogenital developmental abnormalities. When heated to decomposition it emits toxic fumes of NO_x.

AKC625 CAS:23734-88-5 HR: 2
AMINOGLUTETHIMIDE PHOSPHATE
mf: $C_{13}H_{16}N_2O_2 \cdot H_3O_4P$ mw: 330.31

SYNS: AGP □ α-(p-AMINOPHENYL)-α-ETHYLGLUTARIMIDE PHOSPHATE □ 2-(p-AMINOPHENYL)-2-ETHYLGLUTARIMIDE PHOSPHATE

TOXICITY DATA WITH REFERENCE
scu-rbt TDLo:360 mg/kg (female 25-26D post):REP JANSAG 42,131,76
orl-mus LD50:1800 mg/kg PSEBAA 139,100,72

SAFETY PROFILE: Moderately toxic by ingestion. Experimental reproductive effects. When heated to decomposition it emits toxic fumes of PO_x and NO_x. An antisteroidogenic drug. See also PHOSPHATES.

AKC750 CAS:79-17-4 HR: 2
AMINOGUANIDINE
mf: CH_6N_4 mw: 74.11

PROP: Crystalline. Mp: decomp. Sol in H_2O, and EtOH; insol in Et_2O.

SYNS: AMINATE BASE □ GUANYL HYDRAZINE □ HYDRAZINECARBOXIMIDAMIDE

TOXICITY DATA WITH REFERENCE
scu-rat LD50:1258 mg/kg JPETAB 119,444,57
scu-mus LD50:963 mg/kg JPETAB 119,444,57

SAFETY PROFILE: Moderately toxic by subcutaneous route. See also AMINES. All of the oxoacid salts are potentially explosive. When heated to decomposition it emits toxic fumes of NO_x.

AKC800 CAS:1937-19-5 HR: 2
AMINOGUANIDINE HYDROCHLORIDE
mf: $CH_6N_4 \cdot ClH$ mw: 110.57

SYNS: GUANIDINE, AMINO-, HYDROCHLORIDE □ GUANYLHYDRAZINE HYDROCHLORIDE □ HYDRAZINECARBOXIMIDAMIDE HYDROCHLORIDE

TOXICITY DATA WITH REFERENCE
scu-rat LDLo:2984 mg/kg JPETAB 28,251,26

CONSENSUS REPORTS: Reported in EPA TSCA Inventory

SAFETY PROFILE: Moderately toxic by subcutaneous route. When heated to decomposition it emits toxic vapors of NO_x, HCl, and Cl^-.

AKD250 CAS:2834-84-6 HR: 2
AMINOGUANIDINE SULFATE
mf: $CH_6N_4 \cdot H_2O_4S$ mw: 172.19

SYN: AMINOGUANIDINE SULPHATE

TOXICITY DATA WITH REFERENCE
scu-mus TDLo:14,400 mg/kg (1-6D preg):REP JRPFA4 6,179,63
orl-rat LD50:500 mg/kg JPETAB 90,260,47

SAFETY PROFILE: Moderately toxic by ingestion. Experimental reproductive effects. When heated to decomposition it emits toxic fumes of SO_x and NO_x.

AKD375 CAS:10308-82-4 HR: 3
AMINO GUANIDINIUM NITRATE
mf: $CH_7N_5O_3$ mw: 137.1

SAFETY PROFILE: An unstable compound and powerful oxidizer. Aqueous solutions may explode violently when heated to evaporation. When heated to decomposition it emits toxic fumes of NO_x. See also NITRATES.

AKD500 CAS:543-38-4 HR: 2
I,2-AMINO-4-(GUANIDINOOXY)BUTYRIC ACID
mf: $C_5H_{12}N_4O_3$ mw: 176.21

PROP: Crystals from alc. Mp: 184°.

SYNS: 2-AMINO-4-(GUANIDINOOXY)-l-BUTYRIC ACID □ o-((AMINOIMINOMETHYL)AMINO)-l-HOMOSERINE □ CANAVANIN □ l-CANAVANINE

TOXICITY DATA WITH REFERENCE
mmo-omi 10 mg/L MUREAV 12,349,71
dnd-hmn:hla 200 µmol/L ECREAL 107,191,77
dni-ham:oth 2200 µmol/L JCLLAX 75,129,70
ipr-rat LDLo:7 g/kg TXAPA9 91,395,87
scu-rat LD50:5900 mg/kg TXAPA9 91,395,87

SAFETY PROFILE: Moderately toxic by intraperitoneal route. Human mutation data reported. When heated to decomposition it emits toxic fumes of NO_x.

AKD600 CAS:6411-75-2 HR: 3
2-AMINOHEPTANE SULFATE
mf: $C_{14}H_{34}N_2 \cdot H_2O_4S$ mw: 328.58

PROP: Solid. Mp: 230–240°.

SYNS: 2-HEPTANAMINE SULFATE (2:1) □ 2-HEPTYLAMINE SULFATE □ 1-METHYLHEXYLAMINE SULFATE □ TUAMINE SULFATE □ TUAMINOHEPTANE SULFATE

TOXICITY DATA WITH REFERENCE
ipr-rat LD50:60 mg/kg JPETAB 81,235,44
ivn-rat LD50:47,300 µg/kg JAPMA8 42,107,53
ipr-mus LD50:163 mg/kg JPETAB 98,300,50
scu-mus LD50:100 mg/kg JAPMA8 39,12,50
ivn-mus LD50:16,300 µg/kg JAPMA8 42,107,53

SAFETY PROFILE: Poison by subcutaneous, intravenous, and intraperitoneal routes. When heated to decomposition it emits toxic fumes of SO_x and NO_x. See also AMINES and SULFATES.

AKD625 CAS:7790-12-7 HR: 2
7-AMINOHEPTANOIC ACID, ISOPROPYL ESTER
mf: $C_{10}H_{21}NO_2$ mw: 187.32

TOXICITY DATA WITH REFERENCE
skn-rbt 100 µg/24H open AIHAAP 23,95,62
orl-rat LD50:4000 mg/kg AIHAAP 23,95,62
skn-rbt LD50:890 mg/kg AIHAAP 23,95,62

SAFETY PROFILE: Moderately toxic by ingestion and skin contact. A skin irritant. When heated to decomposition it emits toxic fumes of NO_x. See also ESTERS.

AKD750 CAS:2009-03-2 HR: 3
3-(7-AMINOHEPTYL)INDOLE ADIPATE

TOXICITY DATA WITH REFERENCE
ipr-mus LD50:285 mg/kg RPTOAN 33,180,70
ivn-mus LD50:79 mg/kg RPTOAN 33,180,70

SAFETY PROFILE: Poison by intraperitoneal and intravenous routes. When heated to decomposition it emits toxic fumes of NO_x.

AKD775 CAS:90043-86-0 HR: 3
9-AMINO-2,3,5,6,7,8-HEXAHYDRO-1H-CYCLOPENTA (b)QUINOLINE HYDROCHLORIDE HYDRATE
mf: $C_{12}H_{16}N_2 \cdot ClH \cdot H_20$ mw: 242.78

SYN: 2,3,5,6,7,8-HEXAHYDRO-9-AMINO-1H-CYCLOPENTA(b)QUINOLINE HYDROCHLORIDE HYDRATE

TOXICITY DATA WITH REFERENCE
scu-rat LD50:60 mg/kg BAXXDU #2125696
orl-mus LD50:68 mg/kg BAXXDU #2125696
ipr-mus LD50:44 mg/kg BAXXDU #2125696
scu-mus LD50:52 mg/kg BAXXDU #2125696

SAFETY PROFILE: Poison by ingestion, subcutaneous, and intraperitoneal routes. When heated to decomposition it emits toxic fumes of NO_x and HCl. See also AMINES.

AKD875 CAS:60145-64-4 HR: 3
3-AMINO-4-HOMOISOTWISTANE
mf: $C_{11}H_{19}N$ mw: 165.31

SYN: OCTAHYDRO-1,6-METHANONAPHTHALEN-1(2H)-AMINE

TOXICITY DATA WITH REFERENCE
orl-mus LD50:550 mg/kg JKXXAF #78-50338
ipr-mus LD50:102 mg/kg JKXXAF #78-50338
ivn-mus LD50:40 mg/kg JKXXAF #78-50338

SAFETY PROFILE: Poison by intravenous and intraperitoneal routes. Moderately toxic by ingestion. When heated to decomposition it emits toxic fumes of NO_x. See also AMINES.

AKD925 CAS:102-56-7 HR: 3
AMINOHYDROQUINONE DIMETHYL ETHER
mf: $C_8H_{11}NO_2$ mw: 153.20

SYNS: ANILINE, 2,5-DIMETHOXY- □ BENZENAMINE, 2,5-DIMETHOXY-(9CI) □ C.I. 35811 □ 2,5-DIMETHOXYANILINE □ 2,5-DIMETHOXYBENZENAMINE

TOXICITY DATA WITH REFERENCE
orl-mus LD50:120 mg/kg GTPZAB 4(2),30,60
orl-brd LD50:100 mg/kg TXAPA9 21,315,72

CONSENSUS REPORTS: Reported in EPA TSCA Inventory.

SAFETY PROFILE: Poison by ingestion. When heated to decomposition it emits toxic vapors of NO_x.

AKE000 CAS:4502-10-7 HR: 3
2-AMINO-3-HYDROXYACETOPHENONE
mf: $C_8H_9NO_2$ mw: 151.18

SYN: 2-AMINO-3-HYDROXYPHENYL METHYL KETONE

TOXICITY DATA WITH REFERENCE
imp-mus TDLo:80 mg/kg:NEO BJCAAI 11,212,57

DOT CLASSIFICATION: 3; *Label:* Flammable Liquid

SAFETY PROFILE: Questionable carcinogen with experimental neoplastigenic data. See also KETONES. Flammable liquid. When heated to decomposition it emits toxic fumes of NO_x.

AKE250 CAS:116-85-8 HR: 3
1-AMINO-4-HYDROXYANTHRAQUINONE
mf: $C_{14}H_9NO_3$ mw: 239.24

PROP: Red-violet powder or pink plates from (C_6H_6), violet needles (C_6H_6> from pet ether. Mp: 207–208°. Sol in water, HCl, alc, ether, and benzene.

SYNS: 1A-4OA □ ACETATE FAST RED 2B □ ACETOQUINONE LIGHT GOOSEBERRY RL □ ACETYLON FAST PINK B □ AMACEL PINK B □ 1-AMINO-4-HYDROXY-9,10-ANTHRACENEDIONE □ 1-AMINO-4-OXYANTHRAQUINONE □ 9,10-ANTHRACENEDIONE, 1-AMINO-4-HYDROXY-(9CI) □ ANTHRAQUINONE, 1-AMINO-4-HYDROXY- □ ARTISIL DIRECT RED 3BP □ ARTISIL RED 3BP □ CALCOSYN PINK B □ CELANTHRENE RED 3BN □ CELLITON FAST PINK BA-CF □ CELLITON FAST PINK BN □ CELUTATE PINK B □ CELUTATE PINK BN □ CELUTATE PINK BY □ CERVEN DISPERZNI 15 □ CIBACETE RED 3B □ CIBACET RED 3B □ CIBACET RED E3B □ CILLA FAST PINK BN □ C.I. 60710 □ C.I. DISPERSE RED 15 □ C.I. SOLVENT RED 53 □ DIACELLITON FAST PINK B □ DISPERSE FAST PINK B □ DISPERSE RED 15 □ DISPERSE RED 25 □ DISPERSOL ORANGE D-G □ DURANOL RED 2B □ FENACET FAST PINK B □ 1-HYDROXY-4-AMINOANTHRAQUINONE □ 4-HYDROXY-1-ANTHRAQUINONYLAMINE □ INTERCHEM ACETATE PINK BLF □ INTERCHEM HISPERSE PINK BH □ MICROSETILE PINK BN □ NACELAN PINK B □ NEOSETILE PINK BN □ ORACET RED 3B □ PARA M □ PERLITON PINK 3B □ SERISOL FAST RED 2B □ SETACYL PINK 3B □ SUPRACET BRILLIANT RED 2B

TOXICITY DATA WITH REFERENCE
mmo-sat 100 µg/plate MUREAV 40,203,76
mma-sat 100 µg/plate MUREAV 40,203,76
ipr-rat LD50:2700 mg/kg GTPZAB 21(12),27,77
ivn-mus LD50:56 mg/kg CSLNX* NX#00428

CONSENSUS REPORTS: Reported in EPA TSCA Inventory.

SAFETY PROFILE: Poison by intravenous route. Moderately toxic by intraperitoneal route. Mutation data reported. When heated to decomposition it emits toxic fumes of NO_x.

AKE500 CAS:103-18-4 HR: 3
4-AMINO-4′-HYDROXYAZOBENZENE
mf: $C_{12}H_{11}N_3O$ mw: 213.26

TOXICITY DATA WITH REFERENCE
orl-rat TDLo:28 g/kg/2Y-C:ETA AABIAV 52,33,63
orl-rat LD50:1950 mg/kg AABIAV 52,33,63
ipr-rat LD50:300 mg/kg AABIAV 52,33,63

CONSENSUS REPORTS: Reported in EPA TSCA Inventory.

SAFETY PROFILE: Poison by intraperitoneal route. Moderately toxic by ingestion. Questionable carcinogen with experimental tumorigenic data. When heated to decomposition it emits toxic fumes of NO_x.

AKE750 CAS:548-93-6 HR: 2
2-AMINO-3-HYDROXYBENZOIC ACID
mf: $C_7H_7NO_3$ mw: 153.15

PROP: Leaflets from water. Mp: 164°.

SYNS: 3-HYDROXYANTHRANILIC ACID □ 3-HYDROXY-ANTHRANILSAEURE (GERMAN) □ 3-OHAA □ 3-OXYANTHRANILIC ACID

TOXICITY DATA WITH REFERENCE
cyt-hmn:emb 30 mg/L BEXBAN 67,200,69
cyt-hmn:leu 100 mg/L TSITAQ 15,1505,73
scu-mus TDLo:1600 mg/kg/8W-I:CAR VOONAW 22(6),47,76
scu-mus TDLo:185 mg/kg (13-17D post):TER JCREA8 96,163,80
scu-dog TDLo:500 mg/kg/20W-I:ETA VOONAW 26(3),93,80
imp-mus TD:80 mg/kg:NEO BJCAAI 11,212,57
scu-mus TD:2000 mg/kg/7W-I:CAR PGPKA8 14,12,69
imp-mus TDLo:160 mg/kg ANYAA9 108,924,63

SAFETY PROFILE: Questionable carcinogen with experimental carcinogenic, neoplastigenic, and tumorigenic data. An experimental teratogen. Human mutation data reported. When heated to decomposition it emits toxic fumes such as NO_x.

AKF000 CAS:536-25-4 HR: 3
3-AMINO-4-HYDROXYBENZOIC ACID METHYL ESTER
mf: $C_8H_9NO_3$ mw: 167.18

PROP: Light-brown needles from C_6H_6, AcOH, $CHCl_3$ or EtOH (aq). Mp: 142°.

SYNS: AMINOBENZ □ ORTHOCAINE □ ORTHODERM □ ORTHOFORM

TOXICITY DATA WITH REFERENCE
orl-dog LDLo:1 g/kg HBAMAK 4,1289,35
ipr-dog LDLo:250 mg/kg HBAMAK 4,1289,35

CONSENSUS REPORTS: Reported in EPA TSCA Inventory.

SAFETY PROFILE: Poison by intraperitoneal route. Moderately toxic by ingestion. See also ESTERS. When heated to decomposition it emits toxic fumes of NO_x.

AKF250 CAS:73728-82-2 HR: 2
4-AMINO-3-HYDROXYBIPHENYL SULFATE
mf: $C_{12}H_{11}NO{\cdot}H_2O_4S$ mw: 283.32

SYNS: 4-AMINO-3-BIPHENYLOL HYDROGEN SULFATE □ 3-HYDROXY-4-AMINODIPHENYL SULPHATE

TOXICITY DATA WITH REFERENCE
imp-mus TDLo:100 mg/kg:CAR BJCAAI 10,539,56

SAFETY PROFILE: Questionable carcinogen with experimental carcinogenic data. See also SULFATES. When heated to decomposition it emits very toxic fumes of SO_x and NO_x.

AKF375 CAS:352-21-6 HR: 1
4-AMINO-3-HYDROXYBUTYRIC ACID
mf: $C_4H_8NO_3$ mw: 118.13

PROP: dl-Form: Crystals from dil alc. Decomp 218°. Sol in water; very sparingly sol in methanol, alc, ether, chloroform, ethyl acetate. d(+)-Form: Crystals from water. Decomp 214°. l(−)-Form: Crystals from water or water + ethanol. Decomp 212°.

SYNS: γ-AMINO-β-HYDROXYBUTYRIC ACID □ BUKSAMIN □ GA-

BOB □ GABOMADE □ GAMIBETAL □ β-HYDROXY-α-AMINOBUTYRIC ACID □ β-OXY-GABA

TOXICITY DATA WITH REFERENCE
ipr-rat LD50:7000 mg/kg NIIRDN 6,195,82
ipr-mus LD50:7000 mg/kg NIIRDN 6,195,82
scu-mus LD50:7000 mg/kg NIIRDN 6,195,82
ivn-mus LD50:7000 mg/kg NIIRDN 6,195,82
unr-mus LD50:7080 mg/kg BTMNA7 25,297,62

SAFETY PROFILE: Mildly toxic by several routes. When heated to decomposition it emits toxic fumes of NO_x. See also AMINES.

AKF500 CAS:64058-65-7 HR: 2
3-AMINO-4-(2-HYDROXY)ETHOXYBENZENARSONIC ACID
mf: $C_8H_{12}AsNO_5$ mw: 277.13

TOXICITY DATA WITH REFERENCE
ivn-rat LDLo:1700 mg/kg JPETAB 63,122,38
ims-rat LDLo:2000 mg/kg JPETAB 63,122,38

CONSENSUS REPORTS: Arsenic compounds are on the Community Right-To-Know List.

OSHA PEL: TWA 0.5 mg(As)/m^3

SAFETY PROFILE: Moderately toxic by intravenous and intramuscular routes. See also ARSENIC COMPOUNDS. When heated to decomposition it emits very toxic fumes of NO_x and As.

AKF750 CAS:64048-94-8 HR: 3
(3-AMINO-4-(2-HYDROXYETHOXY)PHENYL)ARSINE OXIDE
mf: $C_8H_{12}AsNO_3$ mw: 245.13

TOXICITY DATA WITH REFERENCE
orl-rat LDLo:200 mg/kg JPETAB 63,122,38
ivn-rat LDLo:15 mg/kg JPETAB 63,122,38
ims-rat LDLo:16 mg/kg JPETAB 63,122,38

CONSENSUS REPORTS: Arsenic compounds are on the Community Right-To-Know List.

OSHA PEL: TWA 0.5 mg(As)/m^3

SAFETY PROFILE: Poison by ingestion, intravenous, and intramuscular routes. See also ARSENIC COMPOUNDS. When heated to decomposition it emits very toxic fumes of NO_x and As.

AKG000 CAS:69226-39-7 HR: 1
4-AMINO-N-(2-HYDROXYETHYL)-o-TOLUENESULFONAMIDE
mf: $C_9H_{14}N_2O_3S$ mw: 230.31

SYN: N-HYDROXYETHYLAMID KYSELINY 4-AMINOTOLUEN-2-SULFONOVE (CZECH)

TOXICITY DATA WITH REFERENCE
skn-rbt 500 mg/24H MLD 28ZPAK -,200,72
eye-rbt 100 mg/24H SEV 28ZPAK -,200,72
orl-rat LD50:8900 mg/kg 28ZPAK -,200,72

SAFETY PROFILE: Mildly toxic by ingestion. A skin and severe eye irritant. When heated to decomposition it emits very toxic fumes of NO_x and SO_x.

AKG250 CAS:60573-88-8 HR: 3
α-AMINO-3-HYDROXY-5-ISOXAZOLEACETIC ACID HYDRATE
mf: $C_5H_6N_2O_4{\cdot}H_2O$ mw: 176.15

SYNS: α-AMINO-2,3-DIHYDRO-3-OXO-5-ISOXAZOLEACETIC ACID □ α-AMINO-3-HYDROXY-5-ISOXAZOLESSIGSAURE HYDRAT (GERMAN) □ AMINO-(3-HYDROXY-5-ISOXAZOLYL)ACETIC ACID □ IBOTENIC ACID □ IBOTENSAURE (GERMAN) □ ISOTENIC ACID □ PRAMUSCIMOL

TOXICITY DATA WITH REFERENCE
orl-rat LD50:129 mg/kg ARZNAD 18,311,68
ivn-rat LD50:42 mg/kg ARZNAD 18,311,68
orl-mus LD50:38 mg/kg ARZNAD 18,311,68
ivn-mus LD50:15 mg/kg ARZNAD 18,311,68
scu-rbt LDLo:45 mg/kg AIPTAK 5,161,1899
scu-frg LDLo:4000 mg/kg AIPTAK 5,161,1899

SAFETY PROFILE: Poison by ingestion, subcutaneous, and intravenous routes. When heated to decomposition it emits toxic fumes of NO_x.

AKG500 CAS:4439-84-3 HR: 1
2-AMINO-5-HYDROXYLEVULINIC ACID
mf: $C_5H_9NO_4$ mw: 147.15

SYNS: 2-AMINO-5-HYDROXY-4-OXOPENTANOIC ACID □ H-899 □ HON □ Δ-HYDROXY-γ-OXO-l-NORVALINE □ 5-HYDROXY-4-OXONORVALINE

TOXICITY DATA WITH REFERENCE
orl-mus LD50:7600 mg/kg JAJAAA 14,39,61
scu-mus LD50:8000 g/kg JAJAAA 14,39,61
ivn-mus LD50:5200 mg/kg JAJAAA 14,39,61

SAFETY PROFILE: Mildly toxic by ingestion, subcutaneous, and intravenous routes. When heated to decomposition it emits toxic fumes of NO_x.

AKH000 CAS:90-20-0 HR: 1
4-AMINO-5-HYDROXY-2,7-NAPHTHALENEDISULFONIC ACID
mf: $C_{10}H_9NO_7S_2$ mw: 319.32

SYNS: C.I. 35570 □ H ACID □ KYSELINA 1-AMINO-8-NAFTOL-3,6-DISULFONOVA (CZECH) □ KYSELINA H (CZECH)

TOXICITY DATA WITH REFERENCE
eye-rbt 500 mg/24H MLD 28ZPAK -,189,72

CONSENSUS REPORTS: Reported in EPA TSCA Inventory.

SAFETY PROFILE: An eye irritant. See also SULFONATES. When heated to decomposition it emits very toxic fumes of NO_x and SO_x.

AKH500 CAS:6837-93-0 HR: 1
4-AMINO-5-HYDROXY-2,7-NAPHTHALENEDISULFONIC ACID-p-TOLUENESULFONATE (ESTER)
mf: $C_{17}H_{15}NO_9S_3$ mw: 473.51

SYNS: 1-AMINO-3,6-DISULFO-8-NAFTYLESTER KYSELINA p-TOLUENSULFONOVE (CZECH) □ KYSELINA o-TOSYL-H (CZECH)

TOXICITY DATA WITH REFERENCE
eye-rbt 100 mg/24H SEV 28ZPAK -,194,72
orl-rat LD50:11,500 mg/kg 28ZPAK -,194,72

SAFETY PROFILE: Mildly toxic by ingestion. A severe eye irritant. See also ESTERS and SULFONATES. When heated to decomposition it emits very toxic fumes of NO_x and SO_x.

AKH750 CAS:83-64-7 HR: 1
4-AMINO-5-HYDROXY-1-NAPHTHALENESULFONIC ACID
mf: $C_{10}H_9NO_4S$ mw: 239.26

PROP: Needles. Sol in water.

SYNS: AMINONAPHTHOL SULFONIC ACID S □ CHICAGO ACID S □ KYSELINA 1-AMINO-8-NAFTOL-4-SULFONOVA (CZECH)

TOXICITY DATA WITH REFERENCE
eye-rbt 500 mg/24H MLD 28ZPAK -,188,72
orl-rat LD50:6210 mg/kg 28ZPAK -,188,72

SAFETY PROFILE: Mildly toxic by ingestion. An eye irritant. See also SULFONATES. When heated to decomposition it emits very toxic fumes of NO_x and SO_x.

AKI000 CAS:87-02-5 HR: 1
7-AMINO-4-HYDROXY-2-NAPHTHALENESULFONIC ACID
mf: $C_{10}H_9NO_4S$ mw: 239.26

SYNS: AMINONAPHTHOL SULFONIC ACID J □ I ACID □ ISOGAMMA ACID □ KYSELINA 2-AMINO-5-NAFTOL-7-SULFONOVA (CZECH)

TOXICITY DATA WITH REFERENCE
eye-rbt 500 mg/24H MLD 28ZPAK -,188,72
orl-rat LD50:11,500 mg/kg 28ZPAK -,188,72

CONSENSUS REPORTS: Reported in EPA TSCA Inventory.

SAFETY PROFILE: Mildly toxic by ingestion. An eye irritant. See also SULFONATES. When heated to decomposition it emits very toxic fumes of NO_x and SO_x.

AKI250 CAS:96-93-5 HR: 1
3-AMINO-4-HYDROXY-5-NITROBENZENESULFONIC ACID
mf: $C_6H_6N_2O_6S$ mw: 234.20

PROP: Prisms from water.

SYN: KYSELINA 6-NITRO-2-AMINOFENOL-4-SULFONOVA (CZECH)

TOXICITY DATA WITH REFERENCE
skn-rbt 500 mg/24H MLD 28ZPAK -,182,72
eye-rbt 20 mg/24H MOD 28ZPAK -,182,72
orl-rat LDLo:5360 mg/kg 28ZPAK -,182,72

SAFETY PROFILE: Mildly toxic by ingestion. A skin and eye irritant. See also SULFONATES. When heated to decomposition it emits very toxic fumes of SO_x and NO_x.

AKI500 CAS:63019-81-8 HR: 2
4-AMINO-3-HYDROXY-4′-NITRODIPHENYLHYDROCHLORIDE
mf: $C_{12}H_{10}N_2O_3 \cdot ClH$ mw: 266.70

SYNS: 4-AMINO-4′-NITRO-3-BIPHENYLOL HYDROCHLORIDE □ 4′-NITRO-4-AMINO-3-HYDROXYDIPHENYL HYDROCHLORIDE □ 4′-NITRO-4-AMINO-3-HYDROXYDIPHENYL HYDROGEN CHLORIDE

TOXICITY DATA WITH REFERENCE
imp-mus TDLo:100 mg/kg:ETA BJCAAI 10,539,56

SAFETY PROFILE: Questionable carcinogen with experimental tumorigenic data. When heated to decomposition it emits very toxic fumes of HCl and NO_x.

AKI750 CAS:17418-58-5 HR: 1
1-AMINO-4-HYDROXY-2-PHENOXYANTHRAQUINONE
mf: $C_{20}H_{13}NO_4$ mw: 331.34

SYNS: CERVEN BRILANTNI OSTACETOVA F-LB (CZECH) □ CERVEN DISPERZNI 60 □ C.I. DISPERSE RED 60 (8CI) □ C.I. DISPERSE RED 71 □ C.I. DISPERSE RED 83 □ DISPERSE POLYESTER PINK 2S □ DISPERSE RED 60 □ DISPERSOL RED B 2B □ DURANOL BRILLIANT RED T 2B □ FORON BRILLIANT RED E 2BL □ HOSTATHERM PINK FBL □ LATYL CERISE N □ MIKETON POLYESTER RED FB □ OSTACET BRILLIANT RED E-LB □ PALANIL RED BF □ RESIREN RED TB □ RESOLIN RED FB □ RESOLIN RED FBE □ RESORIN RED FBE □ SAMARON PINK FBL □ SERILENE BRILLIANT RED 2BL □ SERILENE RED 2BL □ SUMIKARON RED E-FBL □ TERAPRINT □ TERSETILE RUBINE FL □ TRANSETILE RUBINE P-FL

TOXICITY DATA WITH REFERENCE
eye-rbt 500 mg/24H MLD 28ZPAK -,239,72

CONSENSUS REPORTS: Reported in EPA TSCA Inventory.

SAFETY PROFILE: An eye irritant. When heated to decomposition it emits toxic fumes of NO_x.

AKI900 CAS:102516-61-0 HR: 3
3-(((3-AMINO-4-HYDROXYPHENYL)PHENYLARSINO)THIO)ALANINE
mf: $C_{15}H_{17}AsN_2O_3S$ mw: 380.32

SYN: ALANINE, 3-(((3-AMINO-4-HYDROXYPHENYL)PHENYLARSINO)THIO)-

TOXICITY DATA WITH REFERENCE
ivn-mus LDLo:40 mg/kg PHBUA9 2,19,54

OSHA PEL: TWA 0.5 mg(As)/m^3

SAFETY PROFILE: Poison by intravenous route. When heated to decomposition it emits toxic fumes of NO_x, SO_x, and AS.

AKJ000 CAS:58152-03-7 HR: 3
1-N-(S-3-AMINO-2-HYDROXYPROPIONYL)BETAMYCIN
mf: $C_{22}H_{43}N_5O_{12}$ mw: 569.70

PROP: Powder.

SYN: 1-N-(S-3-AMINO-2-HYDROXYPROPIONYL) GENTAMYCIN B

TOXICITY DATA WITH REFERENCE
ipr-mus LD50:5000 mg/kg DRFUD4 4,525,79
ivn-mus LD50:330 mg/kg DRFUD4 4,525,79

SAFETY PROFILE: Poison by intravenous route. Mildly toxic by intraperitoneal route. When heated to decomposition it emits toxic fumes of NO_x.

AKJ250 CAS:5423-12-1 HR: 2
3-AMINO-4(1-(2-HYDROXY)PROPOXY)BENZENEARSONIC ACID
mf: $C_9H_{14}AsNO_5$ mw: 291.16

TOXICITY DATA WITH REFERENCE
ivn-rat LDLo:2500 mg/kg JPETAB 63,122,38
ims-rat LDLo:3000 mg/kg JPETAB 63,122,38

CONSENSUS REPORTS: Arsenic compounds are on the Community Right-To-Know List.

OSHA PEL: TWA 0.5 mg(As)/m^3

SAFETY PROFILE: Moderately toxic by intravenous and intramuscular routes. See also ARSENIC COMPOUNDS. When heated to decomposition it emits very toxic fumes of NO_x and As.

AKJ500 CAS:63717-25-9 HR: 2
S-3-AMINO-2-HYDROXYPROPYL SODIUMHYDROGEN PHOSPHOROTHIOATETETRAHYDRATE
mf: $C_3H_9NO_4PS{\cdot}Na{\cdot}4H_2O$ mw: 281.23

TOXICITY DATA WITH REFERENCE
orl-mus LD50:2500 mg/kg JMCMAR 18,803,75
ipr-mus LD50:2200 mg/kg JMCMAR 18,803,75

SAFETY PROFILE: Moderately toxic by ingestion and intraperitoneal routes. When heated to decomposition it emits very toxic fumes of NO_x, PO_x, Na_2O, and SO_x.

AKJ750 CAS:2835-95-2 HR: 2
4-AMINO-2-HYDROXYTOLUENE
mf: C_7H_9NO mw: 123.17

PROP: Plates from water. Mp: 161°.

SYNS: 5-AMINO-o-CRESOL □ 5-AMINO-2-METHYLPHENOL

TOXICITY DATA WITH REFERENCE
mmo-sat 333 µg/plate EMMUEG 11(Suppl 12),1,88
orl-rat LD50:3600 mg/kg FCTXAV 15,607,77
orl-qal LD50:750 mg/kg AECTCV 12,355,83
orl-rat LD50:3600 mg/kg FCTXAV 15,607,77

CONSENSUS REPORTS: Reported in EPA TSCA Inventory. EPA Genetic Toxicology Program.

SAFETY PROFILE: Moderately toxic by ingestion. Mutation data reported. When heated to decomposition it emits toxic fumes of NO_x. See also PHENOLS.

AKK000 CAS:21644-95-1 HR: 3
4-AMINO-4′-HYDROXY-2,3′,5′-TRIMETHYLAZOBENZENE
mf: $C_{15}H_{17}N_3O$ mw: 255.35

TOXICITY DATA WITH REFERENCE
orl-rat LDLo:600 mg/kg AABIAV 52,33,63
ipr-rat LD50:142 mg/kg AABIAV 52,33,63

SAFETY PROFILE: Poison by intraperitoneal route. Moderately toxic by ingestion. When heated to decomposition it emits toxic fumes of NO_x.

AKK250 CAS:360-97-4 HR: 2
5-AMINOIMIDAZOLE-4-CARBOXAMIDE
mf: $C_4H_6N_4O$ mw: 126.14

PROP: Crystals from alc. Mp: 170–171°.

SYNS: AIC □ AICA □ Ba 2756 □ DIAZOL-C □ 5-IMIDAZOLECARBOXAMIDE, 4-AMINO-

TOXICITY DATA WITH REFERENCE
mmo-sat 10 µg/plate JTEHD6 2,1095,77
mmo-esc 400 mg/L MUREAV 190,89,87
orl-rat TDLo:6390 mg/kg/21W-C:ETA JJIND8 54,951,75
ipr-mus LD50:2 g/kg GANNA2 59,207,68

CONSENSUS REPORTS: Reported in EPA TSCA Inventory.

SAFETY PROFILE: Low toxicity by ingestion. Questionable carcinogen with experimental tumorigenic data. Mutation data reported. When heated to decomposition it emits toxic fumes of NO_x.

AKK625 CAS:34879-34-0 HR: 2
5-AMINO-4-IMIDAZOLECARBOXAMIDE UREIDOSUCCINATE
mf: $C_5H_8N_2O_5{\cdot}CaH_6N_4O$ mw: 302.29

SYN: CARBAICA

TOXICITY DATA WITH REFERENCE
ivn-rat LD50:1190 mg/kg DECRDP 6,471,80
ivn-mus LD50:640 mg/kg DECRDP 6,471,80
ivn-gpg LD50:440 mg/kg DECRDP 6,471,80

SAFETY PROFILE: Moderately toxic by intravenous route. When heated to decomposition it emits toxic fumes of NO_x.

AKK750 CAS:581-64-6 HR: 3
7-AMINO-3-IMINO-3H-PHENOTHIAZINEMONOHYDROCHLORIDE
mf: $C_{12}H_9N_3S{\cdot}ClH$ mw: 263.76

PROP: Blackish-green needles from HCl (aq). Sol in hot H_2O, and H_2SO_4; sltly sol in cold H_2O.

SYNS: 3,7-DIAMINOPHENOTHIAZIN-5-IUM CHLORIDE □ KATALYSIN □ LAUTHSCHES VIOLETT (GERMAN) □ THIONIN □ THIONINE

TOXICITY DATA WITH REFERENCE
dnr-bcs 2 mg/disc TRENAF 27,153,76
ipr-rat LD50:215 mg/kg AEPPAE 204,288,47
ivn-rat LD50:7400 µg/kg SMBUA9 9,96,51
ipr-mus LD50:400 mg/kg NTIS** AD691-490

ivn-mus LD50:8030 μg/kg SMBUA9 9,96,51

CONSENSUS REPORTS: Reported in EPA TSCA Inventory. EPA Genetic Toxicology Program.

SAFETY PROFILE: Poison by intraperitoneal route. Mutation data reported. When heated to decomposition it emits very toxic fumes of NO_x, SO_x, and HCl.

AKL000 CAS:2338-18-3 HR: 3
2-AMINOINDANE HYDROCHLORIDE
mf: $C_9H_{11}N{\cdot}ClH$ mw: 169.67

PROP: Crystals from HCl or EtOH. Mp: 234–239°.

SYNS: 2-AMINOINDAN HYDROCHLORIDE □ 2,3-DIHYDRO-1H-INDEN-2-AMINE HYDROCHLORIDE (9CI) □ 2-INDANYLAMINE HYDROCHLORIDE □ SU 8629 HYDROCHLORIDE

TOXICITY DATA WITH REFERENCE
orl-mus LD50:500 mg/kg JPETAB 133,400,61
ipr-mus LD50:170 mg/kg JPETAB 133,400,61
scu-mus LD50:158 mg/kg JPETAB 133,400,61
orl-dog LDLo:50 mg/kg JPETAB 133,400,61

SAFETY PROFILE: Poison by ingestion, subcutaneous, and intraperitoneal routes. When heated to decomposition it emits toxic fumes of NO_x and HCl. See also AMINES.

AKL100 CAS:13935-78-9 HR: 3
AMINOINDANOL HYDROCHLORIDE
mf: $C_9H_{11}NO{\cdot}ClH$ mw: 185.67

SYN: 2-AMINO-1-HYDROXYHYDROINDENE HYDROCHLORIDE

TOXICITY DATA WITH REFERENCE
scu-rat LD50:350 mg/kg AEPPAE 169,114,33
scu-mus LD50:800 mg/kg AEPPAE 169,114,33
scu-gpg LD50:400 mg/kg AEPPAE 169,114,33

SAFETY PROFILE: Poison by subcutaneous route. When heated to decomposition it emits very toxic fumes of NO_x and HCl.

AKL500 CAS:64037-13-4 HR: 3
2-AMINO-5-IODOBENZOXAZOLE
mf: $C_7H_5IN_2O$ mw: 260.04

TOXICITY DATA WITH REFERENCE
orl-rat LD50:1000 mg/kg MDCHAG 4(1),338,64
orl-mus LD50:800 mg/kg MDCHAG 4(1),336,64
ipr-mus LD50:150 mg/kg MDCHAG 4(1),336,64

SAFETY PROFILE: Poison by intraperitoneal route. Moderately toxic by ingestion. See also IODIDES. When heated to decomposition it emits very toxic fumes of I^- and NO_x.

AKL625 CAS:550-28-7 HR: 2
AMINOISOMETRADIN
mf: $C_9H_{13}N_3O_2$ mw: 195.25

PROP: Crystals from water. Mp: 175°. Sol in EtOH, Me_2CO, and water (25°). Freely sol in alc and acetone; insol in ether.

SYNS: AMINOISOMETRADINE □ 6-AMINO-1-METALLYL-3-METHYLPYRIMIDINE-2,4-DIONE □ 6-AMINO-3-METHYL-1-(2-METHYLALLYL)-2,4(1H,3H)-PYRIMIDINEDIONE □ 6-AMINO-3-METHYL-1-(2-METHYLALLYL)URACIL □ 6-AMINO-3-METHYL-1-(2-METHYL-2-PROPENYL)-2,4(1H,3H)-PYRIMIDINEDIONE □ AMISOMETRADIN □ AMISOMETRADINE □ 1-METHALLYL-3-METHYL-6-AMINOTETRAHYDROPYRIMIDINEDIONE □ ROLICTON

TOXICITY DATA WITH REFERENCE
orl-rat LD50:1560 mg/kg AIPTAK 126,400,60
ipr-rat LD50:515 mg/kg AIPTAK 126,400,60
orl-mus LD50:610 mg/kg MEIEDD 10,73,83
ipr-mus LD50:415 mg/kg MEIEDD 10,73,83

SAFETY PROFILE: Moderately toxic by ingestion and other routes. When heated to decomposition it emits toxic fumes of NO_x.

AKL750 CAS:55217-61-3 HR: 3
trans-1-AMINO-2-MERCAPTOMETHYLCYCLOBUTANEHYDROCHLORIDE
mf: $C_5H_{11}NS{\cdot}ClH$ mw: 153.69

SYN: (trans)-2-MERCAPTOMETHYLCYCLOBUTYLAMINE HYDROCHLORIDE

TOXICITY DATA WITH REFERENCE
orl-mus LD50:500 mg/kg JMCMAR 18,323,75
ipr-mus LD50:250 mg/kg JMCMAR 18,323,75

SAFETY PROFILE: Poison by intraperitoneal route. Moderately toxic by ingestion. When heated to decomposition it emits very toxic fumes of HCl, SO_x, and NO_x.

AKM000 CAS:2349-67-9 HR: 3
2-AMINO-5-MERCAPTO-1,3,4-THIADIAZOLE
mf: $C_2H_3N_3S_2$ mw: 133.20

PROP: Crystals from water. Mp: 233–234° (decomp).

SYNS: 5-AMINO-2-MERCAPTO-1,3,4-THIADIAZOLE □ 5-AMINO-1,3,4-THIADIAZOLE-2-THIOL □ 2-AMINO-1,3,4-THIADIAZOLE-5-THIOL □ 2-AMINO-Δ^2-1,3,4-THIADIAZOLINE-5-THIONE □ 5-AMINO-1,3,4-THIADIAZOLINE-2-THIONE □ 2-MERCAPTO-5-AMINO-1,3,4-THIADIAZOLE □ NSC 21402 □ USAF PD-25

TOXICITY DATA WITH REFERENCE
ipr-mus LD50:250 mg/kg NTIS** AD691-490

CONSENSUS REPORTS: Reported in EPA TSCA Inventory.

SAFETY PROFILE: Poison by intraperitoneal route. When heated to decomposition it emits very toxic fumes of NO_x and SO_x.

AKM125 CAS:963-34-8 HR: 3
6-AMINOMETHAQUALONE
mf: $C_{16}H_{15}N_3O$ mw: 265.34

SYNS: 6-AMINO-2-METHYL-3-(o-TOLYL)-1(3H)-QUINAZOLINONE □ HB-218 □ 2-METHYL-3-o-TOLYL-6-AMINO-CHINAZOLINON-4 (GERMAN)

TOXICITY DATA WITH REFERENCE
orl-rat LD50:60 mg/kg ARZNAD 21,362,71
ipr-rat LD50:64 mg/kg ARZNAD 21,362,71
orl-mus LD50:125 mg/kg ARZNAD 21,362,71

ipr-mus LD50:91 mg/kg ARZNAD 21,362,71

SAFETY PROFILE: Poison by ingestion and intraperitoneal routes. When heated to decomposition it emits toxic fumes of NO_x.

AKM250 CAS:10165-33-0 HR: 3
1-AMINO-2-METHOXYANTHRAQUINONE
mf: $C_{15}H_{11}NO_3$ mw: 253.27

PROP: Red crystals. Mp: 221–222°.

SYN: 1-A-2-MA (RUSSIAN)

TOXICITY DATA WITH REFERENCE
mma-mus:lym 50 mg/L/4H NTIS** AD-A064-953
msc-mus:lym 200 mg/L/4H NTIS** AD-A064-953
ipr-rat LD50:300 mg/kg GTPZAB 21(12),27,77

SAFETY PROFILE: Poison by intraperitoneal route. Mutation data reported. When heated to decomposition it emits toxic fumes of NO_x.

AKM500 CAS:120-35-4 HR: 3
3-AMINO-4-METHOXY BENZANILIDE
mf: $C_{14}H_{14}N_2O_2$ mw: 242.30

SYN: BENZANILIDE, 3-AMINO-4-METHOXY-

TOXICITY DATA WITH REFERENCE
ivn-mus LD50:320 mg/kg CSLNX* NX#01183

CONSENSUS REPORTS: Reported in EPA TSCA Inventory.

SAFETY PROFILE: Poison by intravenous route. When heated to decomposition it emits toxic vapors of NO_x.

AKM750 CAS:5464-79-9 HR: 3
2-AMINO-4-METHOXYBENZOTHIAZOLE
mf: $C_8H_8N_2OS$ mw: 180.24

SYN: 4-METHOXY-2-AMINOBENZOTHIAZOLE

TOXICITY DATA WITH REFERENCE
mma-sat 10 µg/L CNREA8 39,682,79
orl-mus LD50:562 mg/kg JPETAB 105,486,52
ivn-mus LD50:46 mg/kg JPETAB 105,486,52

SAFETY PROFILE: Poison by intravenous route. Moderately toxic by ingestion. Mutation data reported. When heated to decomposition it emits very toxic fumes as NO_x and SO_x.

AKN000 CAS:64037-14-5 HR: 3
2-AMINO-5-METHOXYBENZOXAZOLE
mf: $C_8H_8N_2O_2$ mw: 164.18

TOXICITY DATA WITH REFERENCE
orl-rat LD50:1000 mg/kg MDCHAG 4(1),338,64
ipr-rat LD50:268 mg/kg MDCHAG 4(1),338,64
orl-mus LD50:1090 mg/kg MDCHAG 4(1),336,64
ipr-mus LD50:432 mg/kg MDCHAG 4(1),336,64

SAFETY PROFILE: Poison by intraperitoneal route. Moderately toxic by ingestion. When heated to decomposition it emits toxic fumes as NO_x.

AKN250 CAS:63040-25-5 HR: 2
4-AMINO-4′-METHOXY-3-BIPHENYLOLHYDROCHLORIDE
mf: $C_{13}H_{13}NO_2 \cdot ClH$ mw: 251.73

SYN: 3-HYDROXY-4′-METHOXY-4-AMINODIPHENYL HYDROCHLORIDE

TOXICITY DATA WITH REFERENCE
imp-mus TDLo:80 mg/kg:ETA BJCAAI 17,127,63

SAFETY PROFILE: Questionable carcinogen with experimental tumorigenic data. When heated to decomposition it emits very toxic fumes of HCl and NO_x.

AKN500 CAS:951-39-3 HR: 2
2-AMINO-3-METHOXYDIPHENYLENE OXIDE
mf: $C_{13}H_{11}NO_2$ mw: 213.25

SYN: 3-METHOXY-2-AMINODIPHENYLENE OXIDE

TOXICITY DATA WITH REFERENCE
orl-rat TDLo:15 g/kg/70W-I:CAR ZEKBAI 61,45,56
orl-rat TD:21 g/kg/76W-C:ETA JNCIAM 39,1069,67

SAFETY PROFILE: Questionable carcinogen with experimental carcinogenic and tumorigenic data. When heated to decomposition it emits toxic fumes of NO_x.

AKN750 CAS:6504-77-4 HR: 3
4-AMINO-N-(2-METHOXYETHYL)-7-((2-METHOXYETHYL)AMINO-2-PHENYL)-6-PTERIDINECARBOXAMIDE
mf: $C_{19}H_{23}N_7O_3$ mw: 397.49

SYN: WY 5256

TOXICITY DATA WITH REFERENCE
orl-rat LD50:300 mg/kg TXAPA9 18,185,71
orl-mus LD50:250 mg/kg TXAPA9 18,185,71

SAFETY PROFILE: Poison by ingestion. When heated to decomposition it emits toxic fumes of NO_x.

AKO000 CAS:68772-17-8 HR: 2
6-AMINO-8-METHOXY-1-METHYL-4-(p-(p-((1-METHYLPYRIDINIUM-4-YL)AMINO)BENZAMIDO)ANILINOQUINOLINIUM) DI-p-TOLUENESULFONATE
mf: $C_{30}H_{30}N_6O_2 \cdot 2C_7H_7O_3S$ mw: 849.06

TOXICITY DATA WITH REFERENCE
dnd-mus:lym 310 nmol/L JMCMAR 22,134,79
ipr-mus LD10:56 mg/kg JMCMAR 22,134,79

SAFETY PROFILE: Moderately toxic by intraperitoneal route. Mutation data reported. See also SULFONATES. When heated to decomposition it emits very toxic fumes of NO_x and SO_x.

AKO100 CAS:91480-88-5 HR: 3
1-(4-AMINO-5-(4-METHOXY-3-METHYLPHENYL)-2-METHYL-1H-PYRROL-3-YL)ETHANONE
mf: $C_{15}H_{18}N_2O_2$ mw: 258.35

PROP: A liquid.

SYNS: ETHANONE, 1-(4-AMINO-5-(4-METHOXY-3-METHYLPHENYL)-

2-METHYL-1H-PYRROL-3-YL)- □ KETONE, (4-AMINO-5-(4-METHOXY-m-TOLYL)-2-METHYLPYRROL-3-YL) METHYL

TOXICITY DATA WITH REFERENCE
orl-mus LD50:1 g/kg FRPSAX 39,538,84

DOT CLASSIFICATION: 3; *Label:* Flammable Liquid

SAFETY PROFILE: Low toxicity by ingestion. A flammable liquid. When heated to decomposition it emits toxic vapors of NO_x.

AKO250 CAS:68772-43-0 HR: 3
7-AMINO-4-(2-METHOXY-p-(p-((1-METHYLPYRIDINIUM-4-YL)AMINO)BENZAMIDO)ANILINO)-1-METHYLQUINOLINIUM)) DIBROMIDE
mf: $C_{30}H_{30}N_6O_2$•2Br mw: 666.48

TOXICITY DATA WITH REFERENCE
dnd-mus:lym 690 nmol/L JMCMAR 22,134,79
ipr-mus LD10:14 mg/kg JMCMAR 22,134,79

SAFETY PROFILE: Poison by intraperitoneal route. Mutation data reported. See also BROMIDES. When heated to decomposition it emits very toxic fumes of NO_x and Br^-.

AKO350 CAS:2379-90-0 HR: 1
1-AMINO-2-METHOXY-4-OXYANTHRAQUINONE
mf: $C_{15}H_{11}NO_4$ mw: 269.27

SYNS: ACETOQUINONE LIGHT PINK RLZ □ 1-AMINO-4-HYDROXY-2-METHOXY-9,10-ANTHRACENEDIONE □ 1-AMINO-4-HYDROXY-2-METHOXYANTHRAQUINONE □ 1A-2MO-4OA □ 9,10-ANTHRACENEDIONE, 1-AMINO-4-HYDROXY-2-METHOXY-(9CI) □ ANTHRAQUINONE, 1-AMINO-4-HYDROXY-2-METHOXY- □ ARTISIL BRILLIANT PINK RFS □ CELLITON FAST PINK RF □ CELLITON FAST PINK RFA-CF □ CERVEN DISPERZNI 4 □ C.I. 60755 □ C.I. DISPERSE RED 4 □ CILLA FAST PINK RF □ DIANIX FAST PINK R □ DISPERSE PINK Zh □ DISPERSE RED-4 □ DISPERSE ROSE Zh □ ESTEROQUINONE LIGHT PINK RLL □ FENACET FAST PINK RF □ INTERCHEM ACETATE FAST PINK DNA □ MIKETON FAST PINK RL □ MIKETON POLYESTER PINK RL □ NYLOQUINONE PINK B □ PALANIL PINK RF □ PERILITON BRILLIANT PINK R □ SAMARON PINK RFL □ SUPRACET FAST PINK 2R

TOXICITY DATA WITH REFERENCE
orl-rat LD50:5 g/kg 85JCAE-,1329,86
ipr-rat LD50:1 g/kg GTPZAB 21(12),27,77

CONSENSUS REPORTS: Reported in EPA TSCA Inventory.

SAFETY PROFILE: Low toxicity by ingestion and intraperitoneal routes. When heated to decomposition it emits toxic vapors of NO_x.

AKO400 CAS:91480-86-3 HR: 3
1-(4-AMINO-5-(3-METHOXYPHENYL)-2-METHYL-1H-PYRROL-3-YL)ETHANONE
mf: $C_{14}H_{16}N_2O_2$ mw: 244.32

PROP: A liquid.

SYNS: ETHANONE, 1-(4-AMINO-5-(3-METHOXYPHENYL)-2-METHYL-1H-PYRROL-3-YL)- □ KETONE, (4-AMINO-5-(m-METHOXYPHENYL)-2-METHYLPYRROL-3-YL) METHYL

TOXICITY DATA WITH REFERENCE
orl-mus LD50:1 g/kg FRPSAX 39,538,84

DOT CLASSIFICATION: 3; *Label:* Flammable Liquid

SAFETY PROFILE: Low toxicity by ingestion. A flammable liquid. When heated to decomposition it emits toxic vapors of NO_x.

AKO430 CAS:56463-62-8 HR: 3
1-(4-AMINO-5-(4-METHOXYPHENYL)-2-METHYL-1H-PYRROL-3-YL)ETHANONE

PROP: A liquid.
mf: $C_{14}H_{16}N_2O_2$ mw: 244.32

SYNS: ETHANONE, 1-(4-AMINO-5-(4-METHOXYPHENYL)-2-METHYL-1H-PYRROL-3-YL)- □ KETONE, (4-AMINO-5-(p-METHOXYPHENYL)-2-METHYLPYRROL-3-YL) METHYL

TOXICITY DATA WITH REFERENCE
orl-mus LD50:1 g/kg FRPSAX 39,538,84

DOT CLASSIFICATION: 3; *Label:* Flammable Liquid

SAFETY PROFILE: Low toxicity by ingestion. A flammable liquid. When heated to decomposition it emits toxic vapors of NO_x.

AKO450 CAS:91481-03-7 HR: 3
1-(4-AMINO-5-(o-METHOXYPHENYL)-2-METHYL-1H-PYRROL-3-YL)ETHANONE
mf: $C_{14}H_{16}N_2O_2$ mw: 244.32

PROP: A liquid.

SYNS: ETHANONE, 1-(4-AMINO-5-(2-METHOXYPHENYL)-2-METHYL-1H-PYRROL-3-YL)- □ KETONE, (4-AMINO-5-(2-METHOXYPHENYL)-2-METHYLPYRROL-3-YL) METHYL

TOXICITY DATA WITH REFERENCE
orl-mus LD50:1 g/kg FRPSAX 39,538,84

DOT CLASSIFICATION: 3; *Label:* Flammable Liquid

SAFETY PROFILE: Low toxicity by ingestion. A flammable liquid. When heated to decomposition it emits toxic vapors of NO_x.

AKO500 CAS:80-35-3 HR: 2
4-AMINO-N-(6-METHOXY-3-PYRIDAZINYL)-BENZENESULFONAMIDE
mf: $C_{11}H_{12}N_4O_3S$ mw: 280.33

PROP: Yellowish-white powder. Mp: 182–183°. Insol in Et_2O; spar sol in H_2O.

SYNS: ALTEZOL □ 3-(p-AMINOBENZENESULFAMIDO)-6-METHOXYPYRIDAZINE □ 3-p-AMINOBENZENESULPHONAMIDO-7-METHOXYPYRIDAZINE □ CL 13494 □ DAVOSIN □ DEPOVERNIL □ DUROX □ KINEKS □ KINEX □ KYNEX □ LEDERKYN □ LENTAC □ LISULFEN □ LONGIN □ MEDICEL □ N[1]-(6-METHOXY-3-PYRIDAZINYL) SULFANILAMIDE □ 6-METHOXY-3-SULFANILAMIDOPYRIDAZINE □ MIDICEL □ MIDIKEL □ MYASUL □ MYLOSUL □ OPINSUL □ PARAMID □ PARAMID SUPRA □ PETRISUL □ PIRIDOLO □ QUINOSEPTYL □ RETAMID □ RETASULFIN □ RP 7522 □ SLOSUL □ SMOP □ SMP □ SPOFADAZINE □ SULFALEX □ 3-SULFA-6-METHOXYPYRIDAZINE □ SULFAMETOXIPIRIDAZINE □ 3-SULFANILAMIDE-6-METHOXYPYRIDAZINE □ 3-SULFANILAMIDO-6-METHOXYPYRIDAZINE □ 6-

SULFANILAMIDO-3-METHOXYPYRIDAZINE □ SULFAPYRIDAZINE □ SULFDURAZIN □ SULFMETHOXIPIRIDAZINE □ SULFOZONA □ SULPHAMETHOXYPYRIDAZINE □ SULTIRENE □ SURIRENE □ VINCES

TOXICITY DATA WITH REFERENCE
orl-rat TDLo:1200 mg/kg (9-14D preg):TER SEIJBO 13,7,73
orl-rat TDLo:8400 mg/kg (male 6W pre):REP JRPFA4 81,259,87
orl-rat LD50:2739 mg/kg ARZNAD 11,459,61
orl-mus LD50:1750 mg/kg ARZNAD 15,1441,65
ipr-mus LD50:1200 mg/kg RPTOAN 37,223,74
scu-mus LD50:4500 mg/kg ARZNAD 10,440,60

SAFETY PROFILE: Moderately toxic by ingestion and intraperitoneal routes. An experimental teratogen. Other experimental reproductive effects. See also SULFONATES. When heated to decomposition it emits very toxic fumes of NO_x and SO_x.

AKO750 CAS:3690-12-8 HR: 3
4-AMINO-2-METHOXY-5-PYRIMIDINEMETHANOL
mf: $C_6H_9N_3O_2$ mw: 155.1

SYNS: 4-AMINO-5-HYDROXYMETHYL-2-METHOXYPYRIMIDINE □ 4-AMINO-2-METHOXY-5-PYRIMIDINEMETHANOL □ BACIMETHRIN □ BACIMETHRINE □ BACIMETRIN □ 2-METHOXY-4-AMINO-5-HYDROXYMETHYLPYRIMIDINE

TOXICITY DATA WITH REFERENCE
ipr-mus LD50:300 mg/kg 85ERAY 3,1597,78
ivn-mus LD50:300 mg/kg 85ERAY 3,1597,78

SAFETY PROFILE: Poison by intraperitoneal and intravenous routes. When heated to decomposition it emits toxic fumes of NO_x.

AKP250 CAS:1220-94-6 HR: 2
4-AMINO-1-METHYLAMINOANTHRAQUINONE
mf: $C_{15}H_{12}N_2O_2$ mw: 252.29

SYNS: ACETOQUINONE LIGHT VIOLET N □ AMACEL VIOLET 6B □ 1-AMINO-4-(METHYLAMINO)-9,10-ANTHRACENEDIONE □ 9,10-ANTHRACENEDIONE, 1-AMINO-4-(METHYLAMINO)-(9CI) □ CELLITON FAST VIOLET 6B □ CELLITON FAST VIOLET 6BA-CF □ C.I. 61105 □ CILLA FAST VIOLET 6B □ C.I. DISPERSE VIOLET 4 □ C.I. SOLVENT VIOLET 12 □ DIACELLITON FAST VIOLET BF □ DISPERSE FAST VIOLET B □ DISPERSE VIOLET 4S □ DISPERSOL VIOLET B □ DURANOL BRILLIANT VIOLET B □ FENACET FAST VIOLET 6B □ INTERCHEM ACETATE VIOLET 6B □ KAYALON FAST VIOLET BB □ 1-MA-4-AA □ MICROSETILE VIOLET B □ NACELAN VIOLET 4B □ ORACET VIOLET B □ ORACET VIOLET BN □ SERISOL FAST VIOLET 6B □ SUPRACET VIOLET 2B □ VIOLET DISPERZNI 4 □ VIOLET ROZPOUSTEDLOVA 12

TOXICITY DATA WITH REFERENCE
ipr-rat LD50:1000 mg/kg GTPZAB 21(12),27,77

CONSENSUS REPORTS: Reported in EPA TSCA Inventory.

SAFETY PROFILE: Moderately toxic by intraperitoneal route. When heated to decomposition it emits toxic fumes of NO_x.

AKP500 CAS:17463-44-4 HR: 2
dl-α-AMINO-β-METHYLAMINOPROPIONIC ACID
mf: $C_4H_{10}N_2O_2$ mw: 118.16

PROP: First isolated from seeds of *Cycas circinalis* (FEPRA7 31,1473,72).

TOXICITY DATA WITH REFERENCE
ipr-rat LDLo:840 mg/kg FEPRA7 31,1473,72
ipr-mus LDLo:1680 mg/kg FEPRA7 31,1473,72
ipr-ckn LDLo:400 mg/kg FEPRA7 31,1473,72

SAFETY PROFILE: Moderately toxic by intraperitoneal route. See also AMINES. When heated to decomposition it emits toxic fumes of NO_x.

AKP750 CAS:82-28-0 HR: 3
1-AMINO-2-METHYLANTHRAQUINONE
mf: $C_{15}H_{11}NO_2$ mw: 237.27

SYNS: ACETATE FAST ORANGE R □ ACETOQUINONE LIGHT ORANGE JL □ 1-AMINO-2-METHYL-9,10-ANTHRACENEDIONE □ ARTISIL ORANGE 3RP □ CELLITON ORANGE R □ C.I. 60700 □ C.I. DISPERSE ORANGE 11 □ CILLA ORANGE R □ DISPERSE ORANGE □ DURANOL ORANGE G □ 2-METHYL-1-ANTHRAQUINONYLAMINE □ MICROSETILE ORANGE RA □ NCI-C01901 □ NYLOQUINONE ORANGE JR □ PERLITON ORANGE 3R □ SERISOL ORANGE YL □ SUPRACET ORANGE R

TOXICITY DATA WITH REFERENCE
mma-sat 33 μg/plate EMMUEG 11(Suppl 12),1,88
orl-rat TDLo:30 g/kg/78W-C:CAR NCITR* NCI-CG-TR-111,78
orl-mus TDLo:37 g/kg/73W-C:CAR NCITR* NCI-CG-TR-111,78
orl-rat TD:39 g/kg/77W-C:NEO TOLED5 4,71,79
orl-mus TD:307 g/kg/73W-C:ETA IARC** 27,199,82

CONSENSUS REPORTS: NTP 7th Annual Report on Carcinogens. IARC Cancer Review: Group 3 IMEMDT 7,56,87; Animal Limited Evidence IMEMDT 27,199,82. NCI Carcinogenesis Bioassay (feed); Clear Evidence: mouse, rat NCITR* NCI-CG-TR-111,78. Community Right-To-Know List. Reported in EPA TSCA Inventory.

SAFETY PROFILE: Confirmed carcinogen with experimental carcinogenic, neoplastigenic, and tumorigenic data. Mutation data reported. When heated to decomposition it emits toxic fumes of NO_x.

AKQ000 CAS:88-44-8 HR: 1
2-AMINO-5-METHYLBENZENESULFONIC ACID
mf: $C_7H_9NO_3S$ mw: 187.23

PROP: Needles.

SYNS: 4-AMINOTOLUENE-3-SULFONIC ACID □ 6-AMINO-m-TOLUENESULFONIC ACID □ KYSELINA-4-TOLUIDIN-3-SULFONOVA (CZECH) □ PTMS □ PTMSA □ RED 4B ACID □ p-TOLUIDINE-m-SULFONIC ACID

TOXICITY DATA WITH REFERENCE
orl-rat LD50:11,700 mg/kg 28ZPAK -,183,72
eye-rbt 500 mg/24H MOD 28ZPAK -,183,72

CONSENSUS REPORTS: Reported in EPA TSCA Inventory.

SAFETY PROFILE: Mildly toxic by ingestion. An eye

irritant. See also SULFONATES. When heated to decomposition it emits very toxic fumes of NO_x and SO_x.

AKQ500 CAS:1477-42-5 HR: 3
2-AMINO-4-METHYLBENZOTHIAZOLE
mf: $C_8H_8N_2S$ mw: 164.24

SYN: 4-METHYL-2-AMINOBENZOTHIAZOLE

TOXICITY DATA WITH REFERENCE
orl-mus LD50:697 mg/kg JPETAB 105,486,52
ivn-mus LD50:54 mg/kg JPETAB 105,486,52

CONSENSUS REPORTS: Reported in EPA TSCA Inventory.

SAFETY PROFILE: Poison by intravenous route. Moderately toxic by ingestion. When heated to decomposition it emits very toxic fumes of SO_x and NO_x.

AKQ750 CAS:64037-15-6 HR: 3
2-AMINO-5-METHYLBENZOXAZOLE
mf: $C_8H_8N_2O$ mw: 148.18

TOXICITY DATA WITH REFERENCE
orl-mus LD50:640 mg/kg MDCHAG 4(1),336,64
ipr-mus LD50:360 mg/kg MDCHAG 4(1),336,64

SAFETY PROFILE: Poison by intraperitoneal route. Moderately toxic by ingestion. When heated to decomposition it emits toxic fumes of NO_x.

AKR000 CAS:2454-37-7 HR: 2
3-AMINO-α-METHYLBENZYL ALCOHOL
mf: $C_8H_{11}NO$ mw: 137.20

SYN: m-AMINO-α-METHYLBENZYL ALCOHOL

TOXICITY DATA WITH REFERENCE
skn-rbt 500 mg/24H MLD 85JCAE-,691,86
eye-rbt 500 mg open AMIHBC 10,61,54
orl-rat LD50:3100 mg/kg AMIHBC 10,61,54
skn-rbt LD50:3560 µL/kg AMIHBC 10,61,54

SAFETY PROFILE: Moderately toxic by ingestion and skin contact. A skin and eye irritant. When heated to decomposition it emits toxic fumes of NO_x.

AKR250 HR: 3
8-((4-AMINO-1-METHYLBUTYL)AMINO)-6-METHOXYQUINOLINE DIPHOSPHATE
mf: $C_{15}H_{21}N_3O \cdot 2H_3O_4P$ mw: 455.39

SYN: PRIMACHIN (GERMAN)

TOXICITY DATA WITH REFERENCE
ipr-mus LD50:60 mg/kg ARZNAD 20,1775,70
ivn-brd LD50:11 mg/kg ARZNAD 20,1775,70

SAFETY PROFILE: Poison by intraperitoneal and intravenous routes. See also PHOSPHATES. When heated to decomposition it emits very toxic fumes of NO_x and PO_x.

AKR500 CAS:21452-14-2 HR: 3
2-AMINO-4-METHYL-5-CARBOXANILIDOTHIAZOLE
mf: $C_{11}H_{11}N_2OS$ mw: 219.30

SYNS: ALF □ F 849 □ SEEDVAX □ SIDVAX □ UNIROYAL F849

TOXICITY DATA WITH REFERENCE
orl-rat LD50:1410 mg/kg FMCHA2 -,C211,83
unr-mam LD50:141 mg/kg 30ZDA9 -,419,71

SAFETY PROFILE: Poison by an unspecified route. Moderately toxic by ingestion. When heated to decomposition it emits very toxic fumes of NO_x and SO_x.

AKR750 CAS:2051-79-8 HR: 3
4-AMINO-3-METHYL-N,N-DIETHYLANILINEHYDROCHLORIDE
mf: $C_{11}H_{18}N_2 \cdot ClH$ mw: 214.77

TOXICITY DATA WITH REFERENCE
orl-rat LDLo:200 mg/kg KODAK* -,-,71
ipr-rat LD50:25 mg/kg KODAK* -,-,71

CONSENSUS REPORTS: Reported in EPA TSCA Inventory.

SAFETY PROFILE: Poison by ingestion and intraperitoneal routes. When heated to decomposition it emits very toxic fumes of NO_x and HCl.

AKS000 CAS:4781-76-4 HR: 3
2-AMINOMETHYL-2,3-DIHYDRO-4H-PYRAN
mf: $C_6H_{11}NO$ mw: 113.18

SYN: 2-AMINOMETHYL-3,4-DIHYDRO-2H-PYRAN

TOXICITY DATA WITH REFERENCE
orl-rat LD50:1000 mg/kg AIHAAP 30,470,69
ihl-rat LCLo:100 ppm BJIMAG 27,1,70
skn-rbt LDLo:180 mg/kg AIHAAP 30,470,69

SAFETY PROFILE: Poison by skin contact and inhalation. Moderately toxic by ingestion. When heated to decomposition it emits toxic fumes of NO_x.

AKS250 CAS:67730-11-4 HR: 3
2-AMINO-6-METHYLDIPYRIDO(1,2-a:3′,2′-d)IMIDAZOLE
mf: $C_{11}H_{10}N_4$ mw: 198.25

PROP: Yellow prisms from MeOH/EtOAc.

SYNS: GLU-P-1 □ 6-ME-GLU-P-2 □ 6-METHYL DIPYRIDO(1,2-a:3′,2′-d)IMIDAZOL-2-AMINE

TOXICITY DATA WITH REFERENCE
mma-sat 250 ng/plate JJCREP 76,835,85
sce-hmn:lym 1000 µg/L MUREAV 77,65,80
dnd-mus-ipr 10 mg/kg JJCREP 76,835,85
orl-rat TDLo:9100 mg/kg/68W-C:CAR EVHPAZ 67,129,86
orl-mus TDLo:15,200 mg/kg/58W-C:CAR EVHPAZ 67,129,86

CONSENSUS REPORTS: IARC Cancer Review: Group 2B IMEMDT 7,56,87; Animal Sufficient Evidence IMEMDT 40,223,86.

SAFETY PROFILE: Confirmed carcinogen with experimental carcinogenic data. Human mutation data reported. When heated to decomposition it emits toxic fumes of NO_x.

AKS275 **HR: 2**
2-AMINO-6-METHYLDIPYRIDO(1,2-a:3′,2′-d)IMIDAZOLE HYDROCHLORIDE
mf: $C_{11}H_{10}N_4•ClH$ mw: 234.71

SYN: DIPYRIDO(1,2-a:3′,2′-d)IMIDAZOLE, 2-AMINO-6-METHYL-, HYDROCHLORIDE

TOXICITY DATA WITH REFERENCE
slt-dmg-orl 100 ng/kg JJCREP 76,468,85
orl-rat TDLo:3612 mg/kg/24W-C:CAR GANNA2 75,207,84

SAFETY PROFILE: Questionable carcinogen with experimental carcinogenic data. Mutation data reported. When heated to decomposition it emits toxic fumes of NO_x and HCl.

AKS500 **CAS:31416-87-2** **HR: 3**
α-AMINOMETHYL-3-FLUOROBENZYLALCOHOL HYDROBROMIDE
mf: $C_8H_{10}FNO•BrH$ mw: 236.11

SYN: 2-AMINO-1-(3-FLUOROPHENYL)ETHANOL HYDROBROMIDE

TOXICITY DATA WITH REFERENCE
ipr-mus LD50:600 mg/kg JPETAB 106,440,52
ivn-mus LD50:180 mg/kg JPETAB 106,440,52

SAFETY PROFILE: Poison by intravenous route. Moderately toxic by intraperitoneal route. When heated to decomposition it emits very toxic fumes of F^-, Br^-, and NO_x.

AKS750 **CAS:63765-80-0** **HR: 3**
4-AMINO-2-METHYL-3-HEXANOL
mf: $C_7H_{17}NO$ mw: 131.25

SYN: USAF CS-4

TOXICITY DATA WITH REFERENCE
ipr-hmn LDLo:25 mg/kg AMRL** -,5,62
ipr-mus LD50:25 mg/kg NTIS** AD277-689

SAFETY PROFILE: A human poison by intraperitoneal route. When heated to decomposition it emits toxic fumes such as NO_x.

AKT000 **CAS:536-21-0** **HR: 3**
α-(AMINOMETHYL)-m-HYDROXYBENZYL ALCOHOL
mf: $C_8H_{11}NO_2$ mw: 153.20

SYNS: 1-(m-HYDROXYPHENYL)-2-AMINOETHANOL □ 1-(3′-HYDROXYPHENYL)-2-AMINOETHANOL □ m-HYDROXYPHENYLETHANOLAMINE □ 1-(3-HYDROXYPHENYL)-1-HYDROXY-2-AMINOETHANE □ METACARDIOL □ NORENOL □ NORMETOL □ NORPHENYLEPHRINE □ NORSYNEPHRINE □ NOVADRAL □ m-OCTOPAMINE

TOXICITY DATA WITH REFERENCE
orl-rat LD50:390 mg/kg OYYAA2 2,217,68
ipr-rat LD50:32 mg/kg OYYAA2 2,217,68
scu-rat LD50:28,100 μg/kg OYYAA2 2,60,68
ivn-rat LD50:17,400 μg/kg OYYAA2 2,217,68
orl-mus LD50:263 mg/kg RPOBAR 2,295,70
ipr-mus LD50:198 mg/kg RPOBAR 2,295,70
scu-mus LD50:459 mg/kg RPOBAR 2,295,70
ivn-mus LD50:4900 μg/kg RPOBAR 2,295,70

SAFETY PROFILE: Poison by ingestion, subcutaneous, intravenous, and intraperitoneal routes. When heated to decomposition it emits toxic fumes of NO_x.

AKT250 **CAS:104-14-3** **HR: 3**
α-(AMINOMETHYL)-p-HYDROXYBENZYL ALCOHOL
mf: $C_8H_{11}NO_2$ mw: 153.20

SYNS: 1-(p-HYDROXYPHENYL)-2-AMINOETHANOL □ p-HYDROXYPHENYLETHANOLAMINE □ NORDEN □ NORPHEN □ NORSYMPATHOL □ NORSYNEPHRINE □ OCTOPAMINE □ WIN 5512

TOXICITY DATA WITH REFERENCE
ipr-mus LD50:600 mg/kg JPETAB 106,341,52
scu-mus LDLo:1050 mg/kg AIPTAK 101,81,55
ivn-mus LD50:75 mg/kg JPETAB 106,341,52
ivn-gpg LDLo:200 mg/kg AIPTAK 101,81,55

SAFETY PROFILE: Poison by intravenous route. Moderately toxic by intraperitoneal and subcutaneous route. When heated to decomposition it emits toxic fumes of NO_x.

AKT500 **CAS:4779-94-6** **HR: 3**
α-AMINOMETHYL-3-HYDROXYBENZYLALCOHOL HYDROCHLORIDE
mf: $C_8H_{11}NO_2•ClH$ mw: 189.66

SYN: WIN 5501

TOXICITY DATA WITH REFERENCE
orl-rat LD50:390 mg/kg OYYAA2 4,561,70
ipr-rat LD50:32 mg/kg OYYAA2 4,561,70
scu-rat LD50:28 mg/kg OYYAA2 4,561,70
orl-mus LD50:3300 mg/kg OYYAA2 4,561,70
ipr-mus LD50:370 mg/kg JPETAB 106,440,52
ivn-mus LD50:113 mg/kg JPETAB 106,440,52

SAFETY PROFILE: Poison by ingestion, intraperitoneal, subcutaneous, and intravenous routes. When heated to decomposition it emits very toxic fumes of HCl and NO_x.

AKT600 **CAS:76180-96-6** **HR: 3**
2-AMINO-3-METHYLIMIDAZO(4,5-f)QUINOLINE
mf: $C_{11}H_{10}N_4$ mw: 198.25

PROP: Crystals from MeOH (aq).

TOXICITY DATA WITH REFERENCE
sln-dmg-orl 1 mmol/L MUREAV 156,93,85
dnd-mus:lvr 100 μmol/L JJCREP 76,835,85
orl-rat TDLo:4300 mg/kg/56W-C:CAR EVHPAZ 67,129,86
orl-mus TDLo:20,800 mg/kg/97W-C:CAR EVHPAZ 67,129,86
orl-rat TD:3600 mg/kg/43W-C:ETA GANNA2 75,467,84

CONSENSUS REPORTS: IARC Cancer Review: Group 2A IMEMDT 56,165,93; Animal Sufficient Evidence IMEMDT 40,261,86; Animal Sufficient Evidence IMEMDT 56,165,93; Human No Adequate Data IMEMDT 40,261,86; Human Inadequate Evidence IMEMDT 56,165,93.

SAFETY PROFILE: Confirmed carcinogen with experimental carcinogenic and tumorigenic data. Mutation data reported. When heated to decomposition it emits toxic fumes of NO_x.

AKT620 HR: 2
2-AMINO-3-METHYLIMIDAZO(4,5-f)QUINOLINE DIHYDROCHLORIDE
mf: $C_{11}H_{10}N_4 \cdot 2ClH$ mw: 271.17

SYN: IQ DIHYDROCHLORIDE

TOXICITY DATA WITH REFERENCE
orl-rat TDLo: 4081 mg/kg/31W-I: CAR JJCREP 76,570,85

SAFETY PROFILE: Questionable carcinogen with experimental carcinogenic data. When heated to decomposition it emits toxic fumes of NO_x.

AKT750 CAS:2763-96-4 HR: 3
5-AMINOMETHYL-3-ISOXYZOLE
mf: $C_4H_6N_2O_2$ mw: 114.12

PROP: Crystals from EtOH. Mp: 174–176° (decomp).

SYNS: AGARIN □ 5-AMINOMETHYL-3-HYDROXYISOXAZOLE □ 5-(AMINOMETHYL)-3-ISOXAZOLOL □ 5-(AMINOMETHYL)-3(2H)-ISOXAZOLONE □ 3-HYDROXY-5-AMINOMETHYLISOXAZOLE □ 3-HYDROXY-5-AMINOMETHYLISOXAZOLE-AGARIN □ MUSCIMOL □ RCRA WASTE NUMBER P007

TOXICITY DATA WITH REFERENCE
unk-hmn TDLo: 109 μg/kg: CNS,GIT ARZNAD 18,311,68
orl-rat LD50: 45 mg/kg ARZNAD 18,311,68
ivn-rat LD50: 4500 μg/kg ARZNAD 18,311,68
ipr-mus LD50: 2500 μg/kg ARZNAD 18,311,68
scu-mus LD50: 3800 μg/kg ARZNAD 18,311,68
ivn-mus LD50: 5620 μg/kg CSLNX* NX#11824
ivn-rbt LDLo: 10 mg/kg ARZNAD 18,311,68

CONSENSUS REPORTS: Reported in EPA TSCA Inventory. EPA Extremely Hazardous Substances List.

SAFETY PROFILE: Poison by ingestion, subcutaneous, intravenous, and intraperitoneal routes. Human systemic effects by an unspecified route: sleep, nausea or vomiting, hallucinations and distorted perceptions. When heated to decomposition it emits toxic fumes of NO_x.

AKT800 CAS:56463-76-4 HR: 3
1-(4-AMINO-2-METHYL-5-(2-METHYLPHENYL)-1H-PYRROL-3-YL)ETHANONE
mf: $C_{14}H_{16}N_2O$ mw: 228.32

PROP: A liquid.

SYNS: ETHANONE, 1-(4-AMINO-2-METHYL-5-(2-METHYLPHENYL)-1H-PYRROL-3-YL)- □ KETONE, (4-AMINO-2-METHYL-5-(o-TOLYL)PYRROL-3-YL) METHYL

TOXICITY DATA WITH REFERENCE
orl-mus LD50: 1 g/kg FRPSAX 39,538,84

DOT CLASSIFICATION: 3; *Label:* Flammable Liquid

SAFETY PROFILE: Low toxicity by ingestion. A flammable liquid. When heated to decomposition it emits toxic vapors of NO_x.

AKT830 CAS:56463-70-8 HR: 3
1-(4-AMINO-2-METHYL-5-(3-METHYLPHENYL)-1H-PYRROL-3-YL)ETHANONE
mf: $C_{14}H_{16}N_2O$ mw: 228.32

PROP: A liquid.

SYNS: ETHANONE, 1-(4-AMINO-2-METHYL-5-(3-METHYLPHENYL)-1H-PYRROL-3-YL)- □ KETONE, (4-AMINO-2-METHYL-5-(m-TOLYL)PYRROL-3-YL) METHYL

TOXICITY DATA WITH REFERENCE
orl-mus LD50: 1 g/kg FRPSAX 39,538,84

DOT CLASSIFICATION: 3; *Label:* Flammable Liquid

SAFETY PROFILE: Low toxicity by ingestion. A flammable liquid. When heated to decomposition it emits toxic vapors of NO_x.

AKT850 CAS:56463-61-7 HR: 3
1-(4-AMINO-2-METHYL-5-(4-METHYLPHENYL)-1H-PYRROL-3-YL)ETHANONE
mf: $C_{14}H_{16}N_2O$ mw: 228.32

PROP: A liquid.

SYNS: ETHANONE, 1-(4-AMINO-2-METHYL-5-(4-METHYLPHENYL)-1H-PYRROL-3-YL)- □ KETONE, (4-AMINO-2-METHYL-5-(p-TOLYL)PYRROL-3-YL) METHYL

TOXICITY DATA WITH REFERENCE
orl-mus LD50: 1 g/kg FRPSAX 39,538,84

DOT CLASSIFICATION: 3; *Label:* Flammable Liquid

SAFETY PROFILE: Low toxicity by ingestion. A flammable liquid. When heated to decomposition it emits toxic vapors of NO_x.

AKX500 CAS:83-70-5 HR: 3
4-AMINO-2-METHYL-1-NAPHTHOL
mf: $C_{11}H_{11}NO$ mw: 173.23

SYNS: 4-AMINO-2-METHYL-1-NAPHTHALENOL □ 1-HYDROXY-2-METHYL-4-AMINONAPHTHALENE □ KAYVISYN □ 2-METHYL-4-AMINO-1-HYDROXYNAPHTHALENE □ 2-METHYL-4-AMINO-1-NAPHTHOL □ 3-METHYL-4-HYDROXY-1-NAPHTHYLAMINE □ SYNKAMIN □ SYNKAMIN BASE □ VITAMIN K5

TOXICITY DATA WITH REFERENCE
ipr-mus LD50: 250 mg/kg ARZNAD 17,1339,67

CONSENSUS REPORTS: Reported in EPA TSCA Inventory.

SAFETY PROFILE: Poison by intraperitoneal route. When heated to decomposition it emits toxic NO_x.

AKY000 CAS:10187-86-7 HR: 2
3-AMINO-4-METHYL-5-(5-NITRO-2-FURYL)-s-TRIAZOLE

SYN: 4-METHYL-5-(5-NITRO-2-4H-1,2,4-TRIAZOL-3-AMINE)

TOXICITY DATA WITH REFERENCE
orl-man TDLo: 126 mg/kg: GIT JMCMAR 16,312,73
orl-mus LD50: 1460 mg/kg JMCMAR 16,312,73
ipr-mus LD50: 730 mg/kg JMCMAR 16,312,73

SAFETY PROFILE: Moderately toxic by ingestion and intraperitoneal routes. Human gastrointestinal tract effects by ingestion. When heated to decomposition it emits toxic fumes of NO_x.

AKY250 CAS:5581-52-2 HR: 3
2-AMINO-6-(1′-METHYL-4′-NITRO-5′-IMIDAZOLYL) MERCAPTOPURINE
mf: $C_9H_8N_8O_2S$ mw: 292.31

SYNS: 2-AMINO-6-(1-METHYL-4-NITRO-5-IMIDAZOLYL)MERCAPTOPURINE □ 6-BENZYLAMINOPURINE □ BW 57-323 □ BW 57-323H □ GUANERAN □ IRG □ 1-METHYL-4-NITRO-5-(2′-AMINO-6′-PURINYL) MERCAPTOIMIDAZIDE □ NSC-38887 □ 1H-PURINE-2-AMINE, 6-((1-METHYL-4-NITRO-1H-IMIDAZOL-5-YL)-THIO)- □ THIAMIPRINE □ TIAMIPRINE

TOXICITY DATA WITH REFERENCE
ipr-rbt TDLo: 7500 µg/kg (female 6-8D post): TER TNEOAO 18,57,64
ipr-rat TDLo: 5 mg/kg (female 7D post): REP JRPFA4 4,291,62
orl-mus LD50: 450 mg/kg RPTOAN 34,284,71
ipr-mus LD50: 136 mg/kg RPTOAN 34(6),284,71

SAFETY PROFILE: Poison by intraperitoneal route. Moderately toxic by ingestion. An experimental teratogen. Other experimental reproductive effects. When heated to decomposition it emits very toxic SO_x and NO_x.

AKY750 CAS:14370-50-4 HR: 2
2-(AMINOMETHYL)NORBORNANE
mf: $C_8H_{15}N$ mw: 125.24

SYN: (2,5-ENDOMETHYLENECYCLOHEXYLMETHYL)AMINE

TOXICITY DATA WITH REFERENCE
skn-rbt 100 µg/24H open AIHAAP 23,95,62
orl-rat LD50: 1410 mg/kg AIHAAP 23,95,62
skn-rbt LD50: 520 mg/kg AIHAAP 23,95,62

SAFETY PROFILE: Moderately toxic by ingestion and skin contact. A skin irritant. When heated to decomposition it emits toxic fumes of NO_x.

AKY875 CAS:35629-70-0 HR: 2
2-AMINO-4-METHYLOXAZOLE
mf: $C_4H_6N_2O$ mw: 98.10

SAFETY PROFILE: Potentially explosive reaction with hydrogen peroxide and iron(II) catalysts. When heated to decomposition it emits toxic fumes of NO_x.

AKZ000 CAS:2835-99-6 HR: 2
4-AMINO-3-METHYLPHENOL
mf: C_7H_9NO mw: 123.17

PROP: Prisms from EtOH (aq). Mp: 179°.

SYNS: 4-AMINO-m-CRESOL □ 3-METHYL-4-AMINOPHENOL □ PHENOL, 4-AMINO-3-METHYL-

TOXICITY DATA WITH REFERENCE
ipr-mus LD50: 680 mg/kg NNGADV 3,35,78

CONSENSUS REPORTS: Reported in EPA TSCA Inventory.

SAFETY PROFILE: Moderately toxic by intraperitoneal route. When heated to decomposition it emits toxic fumes of NO_x.

ALA000 CAS:50901-84-3 HR: 3
cis-2-AMINO-5-METHYL-4-PHENYL-1-PYRROLINE
mf: $C_{11}H_{14}N_2$ mw: 174.27

TOXICITY DATA WITH REFERENCE
orl-rat LD50: 420 mg/kg EJMCA5 13,161,78
orl-mus LD50: 276 mg/kg EJMCA5 13,161,78
ivn-mus LD50: 14 mg/kg EJMCA5 13,161,78

SAFETY PROFILE: Poison by ingestion and intravenous routes. When heated to decomposition it emits toxic fumes of NO_x.

ALA250 CAS:50901-87-6 HR: 3
trans-2-AMINO-5-METHYL-4-PHENYL-1-PYRROLINE
mf: $C_{11}H_{14}N_2$ mw: 174.27

TOXICITY DATA WITH REFERENCE
orl-mus LD50: 80 mg/kg EJMCA5 13,161,78
ivn-mus LD50: 26 mg/kg EJMCA5 13,161,78

SAFETY PROFILE: Poison by ingestion and intravenous routes. When heated to decomposition it emits toxic fumes of NO_x.

ALA300 CAS:59133-39-0 HR: 3
1-(4-AMINO-2-METHYL-5-PHENYL-1H-PYRROL-3-YL) ETHANONE HYDROCHLORIDE
mf: $C_{13}H_{14}N_2O \cdot ClH$ mw: 250.75

PROP: A liquid.

SYNS: ETHANONE, 1-(4-AMINO-2-METHYL-5-PHENYL-1H-PYRROL-3-YL)-, MONOHYDROCHLORIDE □ KETONE, (4-AMINO-2-METHYL-5-PHENYLPYRROL-3-YL) METHYL, MONOHYDROCHLORIDE

TOXICITY DATA WITH REFERENCE
orl-mus LD50: 1 g/kg FRPSAX 39,538,84

DOT CLASSIFICATION: 3; *Label:* Flammable Liquid

SAFETY PROFILE: Low toxicity by ingestion. A flammable liquid. When heated to decomposition it emits toxic vapors of NO_x, HCl, and Cl^-.

ALA500 CAS:41394-05-2 HR: 2
4-AMINO-3-METHYL-6-PHENYL-1,2,4-TRIAZIN-5(4H)-ONE
mf: $C_{10}H_{10}N_4O$ mw: 202.24

PROP: Crystals or solid. Mp: 166–167°. Sltly sol in water.

SYNS: BAY-DRW 1139 □ DRW 1139 □ GOLTIX □ METAMITON □ METAMITRON (GERMAN) □ 3-METHYL-4-AMINO-6-PHENYL-1,2,4-TRIAZIN(4H)-ON (GERMAN)

TOXICITY DATA WITH REFERENCE
mmo-nsc 1 mg/L ENMUDM 7(Suppl 3),11,85

orl-rat LD50:1447 mg/kg 85ARAE 2,133,77
orl-mus LD50:1450 mg/kg 85DPAN -,-,71/76

SAFETY PROFILE: Moderately toxic by ingestion. Mutation data reported. Moderately toxic by ingestion. When heated to decomposition it emits toxic fumes of NO_x.

ALA750 CAS:55921-66-9 HR: 3
2-AMINO-4-(N-METHYLPIPERAZINO)-5-METHYLTHIO-6-CHLOROPYRIMIDINE
mf: $C_{10}H_{16}ClN_5S$ mw: 273.82

TOXICITY DATA WITH REFERENCE
orl-mus LD50:225 mg/kg JMCMAR 18,553,75
ivn-mus LD50:33 mg/kg JMCMAR 18,553,75

SAFETY PROFILE: Poison by ingestion and intravenous routes. When heated to decomposition it emits very toxic fumes of SO_x, NO_x, and Cl^-.

ALB000 CAS:115-69-5 HR: 3
2-AMINO-2-METHYL-1,3-PROPANEDIOL
mf: $C_4H_{11}NO_2$ mw: 105.16

PROP: A clear liquid. Mp: 110°, bp: 151° @ 10 mm, vap d: 3.63.

SYNS: AMINOGLYCOL □ AMPD □ GENTIMON □ ISOBUTANDIOL-2-AMINE □ PENTAERYTHRITOL DICHLOROHYDRIN

TOXICITY DATA WITH REFERENCE
orl-rat LD50:17 g/kg JACTDZ 9(2),203,90
orl-mus LDLo:140 mg/kg AECTCV 14,111,85
orl-rbt LDLo:1500 mg/kg JIHTAB 22,315,40

CONSENSUS REPORTS: Reported in EPA TSCA Inventory.

SAFETY PROFILE: Poison by ingestion. Combustible. Can react with oxidizing materials. When heated to decomposition it emits toxic fumes of NO_x.

ALB250 CAS:2854-16-2 HR: 2
1-AMINO-2-METHYL-2-PROPANOL
mf: $C_4H_{11}NO$ mw: 89.16

PROP: D: 0.929 @ 20°/20°, bp: 151°.

TOXICITY DATA WITH REFERENCE
orl-rat LDLo:3000 mg/kg SCCUR* -,1,61
orl-mus LD50:2450 mg/kg SCCUR* -,1,61
ihl-mus LCLo:1095 ppm/18H SCCUR* -,1,61
skn-rbt LDLo:1960 mg/kg SCCUR* -,1,61

SAFETY PROFILE: Moderately toxic by ingestion, inhalation, and skin contact. When heated to decomposition it emits toxic fumes of NO_x.

ALB500 CAS:7447-44-1 HR: 2
S-2-AMINO-2-METHYLPROPYL DIHYDROGEN PHOSPHOROTHIOATE
mf: $C_4H_{12}NO_3PS$ mw: 185.20

SYN: S-(2-AMINO-2-METHYLPROPYL)PHOSPHOROTHIOATE

TOXICITY DATA WITH REFERENCE
orl-mus LD50:2800 mg/kg JMCMAR 18,803,75
ipr-mus LD50:750 mg/kg JMCMAR 18,803,75
unr-mus LD50:750 mg/kg JMCMAR 9,911,66

SAFETY PROFILE: Moderately toxic by ingestion, intraperitoneal, and other unspecified routes. See also PHOSPHATES. When heated to decomposition it emits very toxic fumes of SO_x, PO_x, and NO_x.

ALB625 CAS:18591-81-6 HR: 2
3-AMINO-6-METHYL-4-PYRIDAZINETHIOL
mf: $C_5H_7N_3S$ mw: 141.21

SYNS: 3-AMINO-4-MERCAPTO-6-METHYLPYRIDAZIN (GERMAN) □ 3-AMINO-4-MERCAPTO-6-METHYLPYRIDAZINE

TOXICITY DATA WITH REFERENCE
orl-rat LD50:1975 mg/kg PHARAT 37,285,82
orl-mus LD50:1637 mg/kg PHARAT 37,136,82
scu-mus LD50:1637 mg/kg PHARAT 36,698,81

SAFETY PROFILE: Moderately toxic by ingestion and other routes. When heated to decomposition it emits toxic fumes of NO_x and SO_x.

ALB750 CAS:3731-51-9 HR: 3
2-AMINOMETHYLPYRIDINE
mf: $C_6H_8N_2$ mw: 108.16

PROP: Bp: 91° @ 15 mm. Sol in water.

SYNS: 2-PICOLINAMINE □ 2-PICOLYLAMINE □ 2-PYRIDINEMETHYLAMINE □ (2-PYRIDYLMETHYL)AMINE

TOXICITY DATA WITH REFERENCE
ivn-mus LD50:340 mg/kg APFRAD 26,345,68
orl-qal LD50:750 mg/kg AECTCV 12,355,83
orl-bwd LD50:562 mg/kg AECTCV 12,355,83
ivn-mus LD50:340 mg/kg APFRAD 26,345,68

CONSENSUS REPORTS: Reported in EPA TSCA Inventory.

SAFETY PROFILE: Poison by intravenous route. Moderately toxic by ingestion. When heated to decomposition it emits toxic fumes of NO_x.

ALC000 CAS:1603-40-3 HR: 3
2-AMINO-3-METHYLPYRIDINE
mf: $C_6H_8N_2$ mw: 108.16

PROP: Solid. Mp: 26–26.4°, bp: 221.2°, vap d: 3.73. Sol in water.

TOXICITY DATA WITH REFERENCE
orl-rat LD50:100 mg/kg 85JCAE-,841,86
ihl-rat LCLo:650 ppm/6H 85JCAE-,841,86
scu-mus LD50:36 mg/kg AJEBAK 36,491,58
ivn-mus LD50:10 mg/kg CSLNX* NX#01585
skn-gpg LD50:200 mg/kg 85JCAE-,841,86

CONSENSUS REPORTS: Reported in EPA TSCA Inventory.

SAFETY PROFILE: Poison by ingestion, skin contact, subcutaneous, and intravenous routes. Moderately toxic by inhalation. Combustible. When heated to decomposition it emits toxic fumes of NO_x.

ALC250 **CAS:695-34-1** **HR: 3**
2-AMINO-4-METHYLPYRIDINE
mf: $C_6H_8N_2$ mw: 108.16

PROP: Crystals or leaflets (ligroin). Mp: 99°, bp: 230.9°, vap d: 3.73.

SYNS: α-AMINO-γ-PICOLINE □ 2-AMINO-4-PICOLINE □ASCEN-SIL □ 4M2AP □ 4-METHYL-2-AMINOPYRIDINE □ METHYL-4-AMINO-2-PYRIDINE □ 4-PICOLYLAMINE □ RA 1226 □ W 45 □ W 45 RASCHIG

TOXICITY DATA WITH REFERENCE
orl-rat LD50:200 mg/kg 85JCAE-,841,86
scu-rat LD50:160 mg/kg AEPPAE 227,234,55
scu-mus LD50:64 mg/kg NYKZAU 53,227S,57
ivn-mus LD50:39 mg/kg APFRAD 26,345,68
skn-gpg LD50:500 mg/kg 85JCAE-,841,86

CONSENSUS REPORTS: Reported in EPA TSCA Inventory.

SAFETY PROFILE: Poison by ingestion, skin contact, subcutaneous, and intravenous routes. Combustible. When heated to decomposition it emits toxic fumes of NO_x. An analgesic and cardiac stimulant.

ALC500 **CAS:1824-81-3** **HR: 3**
2-AMINO-6-METHYLPYRIDINE
mf: $C_6H_8N_2$ mw: 108.16

PROP: Solid. Mp: 41°, bp: 214.4°, vap d: 3.73. Very sol in water. Insol in ligroin, sol in most org solvs.

TOXICITY DATA WITH REFERENCE
orl-rat LD50:100 mg/kg 85JCAE-,841,86
scu-mus LD50:52 mg/kg AJEBAK 36,491,58
ivn-mus LD50:18 mg/kg CSLNX* NX#00148
skn-gpg LD50:200 mg/kg 85JCAE-,841,86

CONSENSUS REPORTS: Reported in EPA TSCA Inventory.

SAFETY PROFILE: Poison by ingestion, skin contact, subcutaneous, and intravenous routes. When heated to decomposition it emits toxic fumes of NO_x.

ALC750 **CAS:53222-52-9** **HR: 3**
4-((3-AMINO-4-((4-((1-METHYLPYRIDINIUM-4-YL)AMINO)BENZOYL)AMINO)PHENYL)AMINO)-1-METHYLQUINOLINIUM)DIBROMIDE
mf: $C_{29}H_{28}N_6O \cdot 2Br$ mw: 636.45

TOXICITY DATA WITH REFERENCE
dnd-mus:lym 1870 nmol/L JMCMAR 22,134,79
ipr-mus LD10:7 mg/kg JMCMAR 22,134,79

SAFETY PROFILE: Poison by intraperitoneal route. Mutation data reported. See also BROMIDES. When heated to decomposition it emits very toxic fumes of Br^- and NO_x.

ALD750 **CAS:68006-83-7** **HR: 3**
2-AMINO-3-METHYL-9H-PYRIDO(2,3-b)INDOLE
mf: $C_{12}H_{11}N_3$ mw: 197.2

PROP: Crystals from $CHCl_3$/hexane. Mp: 215–218°,

SYN: 2-AMINO-3-METHYL-α-CARBOLINE

TOXICITY DATA WITH REFERENCE
mmo-sat 1 µg/plate ABCHA6 43,1155,79
dnr-bcs 10 µL/plate ABCHA6 45,2031,81
slt-dmg-orl 400 ng/kg JJCREP 76,468,85
mma-sat 10 ng/plate CALEDQ 10,141,80
orl-mus TDLo:32,800 mg/kg/85W-C:CAR EVHPAZ 67,129,86
orl-mus TD:37,380 mg/kg/64W-C:CAR CRNGDP 5,815,84

SAFETY PROFILE: Confirmed carcinogen with experimental carcinogenic data. Mutation data reported. When heated to decomposition it emits toxic fumes of NO_x.

ALD500 **CAS:62450-07-1** **HR: 3**
3-AMINO-1-METHYL-5H-PYRIDO(4,3-b)INDOLE
mf: $C_{12}H_{11}N_3$ mw: 197.26

SYNS: 3-AMINO-1-METHYL-γ-CARBOLINE □ 1-METHYL-3-AMINO-5H-PYRIDO(4,3-b)INDOLE □ TRP-P-2 □ TRYPTOPHAN P2

TOXICITY DATA WITH REFERENCE
mmo-sat 50 ng/plate CRNGDP 5,505,84
mma-sat 50 ng/plate CRNGDP 5,505,84
dnd-mus:lvr 50 µmol/L JJCREP 76,835,85
orl-rat TDLo:4350 mg/kg/2Y-C:NEO CALEDQ 13,23,81
orl-mus TDLo:9648 mg/kg/57W-C:CAR EVHPAZ 67,129,86
scu-mus TDLo:12,500 µg/kg:CAR PPTCBY 37,193,85

CONSENSUS REPORTS: IARC Cancer Review: Group 2B IMEMDT 7,56,87; Animal Sufficient Evidence IMEMDT 31,255,83. EPA Genetic Toxicology Program.

SAFETY PROFILE: Confirmed carcinogen with experimental carcinogenic and neoplastigenic data. Mutation data reported. When heated to decomposition it emits toxic fumes of NO_x.

ALE750 **CAS:72254-58-1** **HR: 2**
3-AMINO-1-METHYL-5H-PYRIDO(4,3-b)INDOLE ACETATE
mf: $C_{12}H_{11}N_3 \cdot C_2H_4O_2$ mw: 257.32

PROP: Prisms. Mp: 242–247°.

SYNS: 5H-PYRIDO(4,3-b)INDOL-3-AMINE, 1-METHYL-1, MONOACETATE □ TRP-P-2(ACETATE)

TOXICITY DATA WITH REFERENCE
slt-dmg-orl 400 ppm MUREAV 122,315,83
mma-sat 1 µg/plate CPBTAL 26,611,78
orl-mus TDLo:14 g/kg/89W-C:CAR SCIEAS 213,346,81

SAFETY PROFILE: Questionable carcinogen with experimental carcinogenic data. Mutation data reported. When heated to decomposition it emits toxic fumes of NO_x.

ALF250 **CAS:127-79-7** **HR: 2**
4-AMINO-N-(4-METHYL-2-PYRIMIDINYL)-BENZENESULFONAMIDE
mf: $C_{11}H_{12}N_4O_2S$ mw: 264.33

PROP: Cream-colored powder, darkens on exposure to light. Mp: 237° (decomp). Spar sol in water.

SYNS: A-310 □ (p-AMINOBENZOLSULFONYL)-2-AMINO-4-METHYL-

PYRIMIDIN (GERMAN) □ CREMOMERAZINE □ DEBENAL-M □ KELAMERAZINE □ MEBACID □ MESULFA □ METHYLPYRIMAL □ N[1]-(4-METHYL-2-PYRIMIDINYL)SULFANILAMIDE □ METHYLSULFAZIN □ PERCOCCIDE □ PIRIMAL-M □ PYRALCID □ PYRIMAL M □ ROMEZIN □ RP 2632 □ 2643-RP □ SEPTACIL □ SULFAMERADINE □ SULFAMERAZIN □ SULFAMETHYLDIAZINE □ SULPHAMERAZINE □ SUMEDINE □ VETA-MERAZINE

TOXICITY DATA WITH REFERENCE
orl-mus TDLo:6 g/kg (female 6-10D post):REP JPETAB 101,362,51
scu-rat LD50:1890 mg/kg ABMGAJ 27,141,71
ivn-rat LD50:1100 mg/kg AEPPAE 211,367,50
ipr-mus LD50:1400 mg/kg ARZNAD 5,213,55
scu-mus LD50:1190 mg/kg ABMGAJ 27,141,71

CONSENSUS REPORTS: Reported in EPA TSCA Inventory.

SAFETY PROFILE: Moderately toxic by intravenous and subcutaneous routes. Experimental reproductive effects. When heated to decomposition it emits very toxic fumes of NO_x and SO_x.

ALF500 CAS:55661-38-6 HR: 3
1-(4-AMINO-2-METHYLPYRIMIDIN-5-YL)METHYL-3-(2-CHLOROETHYL)-3-NITROSOUREA
mf: $C_9H_{13}ClN_6O_2{\cdot}ClH$ mw: 309.19

SYNS: ACNU □ N'-((4-AMINO-2-METHYL-5-PYRIMIDINYL)METHYL)-N-(2-CHLOROETHYL)-N-NITROSOUREA HCl □ 3-((4-AMINO-2-METHYL-5-PYRIMIDINYL)METHYL)-1-(2-CHLOROETHYL)-1-NITROSOUREA HYDROCHLORIDE □ CS-439 □ NIDRAN □ NIMUSTINE HYDROCHLORIDE □ NSC-245382

TOXICITY DATA WITH REFERENCE
mmo-sat 50 μg/plate CNREA8 38,2148,78
mma-sat 50 μg/plate CNREA8 38,2148,78
ipr-rat TDLo:10 mg/kg (female 8D post):TER TJADAB 38,553,88
ipr-rat TDLo:10 mg/kg (female 8D post):REP TJADAB 38,553,88
orl-rat LD50:113 mg/kg 37XLA2 2,1233,78
ipr-rat LD50:52,700 μg/kg IYKEDH 10,884,79
scu-rat LD50:60,800 μg/kg IYKEDH 10,884,79
ivn-rat LD50:15 mg/kg MDACAP 16,424,80
orl-mus LD50:83 mg/kg 37XLA2 2,1233,78
ipr-mus LD50:49,300 μg/kg IYKEDH 10,884,79
scu-mus LD50:72,800 μg/kg IYKEDH 10,884,79

SAFETY PROFILE: Poison by ingestion, intraperitoneal, and intravenous routes. An experimental teratogen. Other experimental reproductive effects. Mutation data reported. When heated to decomposition it emits very toxic fumes of Cl^- and NO_x.

ALF600 CAS:454-41-1 HR: 2
2-AMINO-4-(METHYLSULFINYL)BUTYRIC ACID
mf: $C_5H_{11}NO_3S$ mw: 165.23

SYNS: BUTANOIC ACID, 2-AMINO-4-(METHYLSULFINYL)-(9CI) □ BUTYRIC ACID, 2-AMINO-4-(METHYLSULFINYL)- □ METHIONINE SULFOXIDE □ dl-METHIONINE SULFOXIDE

TOXICITY DATA WITH REFERENCE
ipr-mus TDLo:3500 mg/kg (female 9-15D post):TER ASMUAA 38,193,62
ipr-mus LD50:4000 mg/kg IJRBA3 3,41,61

SAFETY PROFILE: Moderately toxic by intraperitoneal route. An experimental teratogen. When heated to decomposition it emits toxic fumes of NO_x and SO_x.

ALF750 CAS:6628-83-7 HR: 2
2-AMINOMETHYLTETRAHYDROPYRAN
mf: $C_6H_{11}NO$ mw: 113.18

TOXICITY DATA WITH REFERENCE
orl-rat LD50:710 mg/kg AIHAAP 30,470,69
skn-rbt LD50:710 mg/kg AIHAAP 30,470,69

SAFETY PROFILE: Moderately toxic by ingestion and skin contact. When heated to decomposition it emits toxic fumes of NO_x.

ALG250 CAS:73696-62-5 HR: 2
2-AMINO-N-(3-METHYL-2-THIAZOLIDINYLIDENE) ACETAMIDE
mf: $C_6H_{11}N_3OS$ mw: 173.26

TOXICITY DATA WITH REFERENCE
orl-mus LD50:4666 mg/kg JMCMAR 23,773,80
ivn-mus LD50:549 mg/kg JMCMAR 23,773,80

SAFETY PROFILE: Moderately toxic by intravenous route. Mildly toxic by ingestion. When heated to decomposition it emits very toxic fumes of NO_x and SO_x.

ALG375 CAS:55864-39-6 HR: 2
5-AMINO-3-METHYLTHIO-1,2,4-OXADIAZOLE
mf: $C_3H_5N_3OS$ mw: 131.15

SAFETY PROFILE: Decomposes violently at its mp of 97-99°C. Upon decomposition it emits toxic fumes of SO_x and NO_x.

ALG500 CAS:21172-28-1 HR: 3
α-AMINOMETHYL-m-TRIFLUOROMETHYLBENZYL ALCOHOL
mf: $C_9H_{10}F_3NO$ mw: 205.20

TOXICITY DATA WITH REFERENCE
orl-mus LD50:700 mg/kg ARZNAD 27,116,77
ipr-mus LD50:226 mg/kg ISYAM* -,21,70

SAFETY PROFILE: Poison by intraperitoneal route. Moderately toxic by ingestion. When heated to decomposition it emits very toxic fumes of F^- and NO_x.

ALH000 CAS:117-62-4 HR: 1
2-AMINO-1,5-NAPHTHALENEDISULFONIC ACID
mf: $C_{10}H_9NO_6S_2$ mw: 303.32

SYNS: KYSELINA 2-NAFTYLAMIN-1,5-DISULFONOVA (CZECH) □ KYSELINA SULFO-TOBIAOVA (CZECH) □ 5-SULFO-TOBIAS ACID

TOXICITY DATA WITH REFERENCE
skn-rbt 500 mg/24H MLD 28ZPAK -,188,72
eye-rbt 20 mg/24H MOD 28ZPAK -,188,72
orl-rat LD50:5430 mg/kg 28ZPAK -,188,72

CONSENSUS REPORTS: Reported in EPA TSCA Inventory. EPA Genetic Toxicology Program.

SAFETY PROFILE: Mildly toxic by ingestion. A skin and eye irritant. See also SULFONATES. When heated to decomposition it emits very toxic fumes of NO_x and SO_x.

ALH250 CAS:131-27-1 HR: 1
3-AMINO-1,5-NAPHTHALENEDISULFONIC ACID
mf: $C_{10}H_9NO_6S_2$ mw: 303.32

PROP: Prisms.

SYNS: ACID IV □ 2-AMINO-4,8-NAPHTHALENEDISULFONIC ACID □ 7-AMINO-1,5-NAPHTHALENEDISULFONIC ACID □ C ACID □ 4,8-DISULFO-2-NAPHTHALAMINE □ KYSELINA C (CZECH) □ KYSELINA-2-NAFTYLAMIN-4,8-DISULFONOVA (CZECH) □ β-NAPHTHYLAMINEDISULFONIC ACID □ β-NAPHTHYLAMINE-4,8-DISULFONIC ACID □ 2-NAPHTHYLAMINE-4,8-DISULFONIC ACID

TOXICITY DATA WITH REFERENCE
eye-rbt 100 mg/24H MOD 28ZPAK -,189,72
orl-rat LD50:11,400 mg/kg 28ZPAK -,189,72

CONSENSUS REPORTS: Reported in EPA TSCA Inventory.

SAFETY PROFILE: An eye irritant. See also SULFONATES. When heated to decomposition it emits very toxic fumes of NO_x and SO_x.

ALH500 CAS:118-33-2 HR: 2
6-AMINO-NAPHTHALENE-1,3-DISULFONIC ACID
mf: $C_{10}H_9NO_6S_2 \cdot Na$ mw: 326.31

PROP: Rhombic needles. Sol in water.

SYN: 2-NAFTYLAMIN-5,7-DISULFONAN SODNY (CZECH)

TOXICITY DATA WITH REFERENCE
skn-rbt 2 mg/24H SEV 85JCAE-,1059,86
eye-rbt 750 μg/24H SEV 28ZPAK -,189,72
orl-rat LD50:2000 mg/kg 28ZPAK -,189,72

CONSENSUS REPORTS: Reported in EPA TSCA Inventory.

SAFETY PROFILE: Moderately toxic by ingestion. A severe skin and eye irritant. When heated to decomposition it emits very toxic fumes of NO_x and SO_x.

ALH750 CAS:81-16-3 HR: 1
2-AMINO-1-NAPHTHALENESULFONIC ACID
mf: $C_{10}H_9NO_3S$ mw: 223.26

PROP: Anhydrous flakes (hot H_2O) or hydrated needles (cold H_2O).

SYNS: KYSELINA-2-NAFTYLAMIN-1-SULFONOVA (CZECH) □ KYSELINA TOBIASOVA (CZECH) □ 2-NAPHTHYLAMINE-1-SULFONIC ACID □ TOBIAS ACID

TOXICITY DATA WITH REFERENCE
eye-rbt 500 mg/24H MLD 28ZPAK -,187,72
orl-rat LD50:19,400 mg/kg 28ZPAK -,187,72

CONSENSUS REPORTS: Reported in EPA TSCA Inventory.

SAFETY PROFILE: Mildly toxic by ingestion. An eye irritant. When heated to decomposition it emits very toxic fumes of NO_x and SO_x. See also SULFONATES.

ALI000 CAS:84-86-6 HR: 3
4-AMINO-1-NAPHTHALENESULFONIC ACID
mf: $C_{10}H_9NO_3S$ mw: 223.26

PROP: Needles. Spar sol in water with blue fluorescence. Sol in MeOH.

SYNS: 1-AMINONAPHTHALENE-4-SULFONIC ACID □ 1-AMINO-4-SULFONAPHTHALENE □ NAPHTHIONIC ACID □ 1,4-NAPHTHIONIC ACID □ α-NAPHTHYLAMINE-p-SULFONIC ACID □ 1-NAPHTHYLAMINE-4-SULFONIC ACID □ PIRIA'S ACID □ USAF M-5

TOXICITY DATA WITH REFERENCE
ipr-mus LD50:300 mg/kg NTIS** AD277-689

CONSENSUS REPORTS: Reported in EPA TSCA Inventory. EPA Genetic Toxicology Program.

SAFETY PROFILE: Poison by intraperitoneal route. See also SULFONATES. When heated to decomposition it emits very toxic fumes of NO_x and SO_x.

ALI240 CAS:84-89-9 HR: 2
5-AMINO-1-NAPHTHALENESULFONIC ACID
mf: $C_{10}H_9NO_3S$ mw: 223.26

SYN: 1-NAPHTHALENESULFONIC ACID, 5-AMINO-

TOXICITY DATA WITH REFERENCE
orl-rat LD50:>5 g/kg GISAAA 51(1),87,86
ipr-rat LD50:2880 mg/kg GISAAA 51(1),87,86
orl-mus LD50:>5 g/kg GISAAA 51(1),87,86
ipr-mus LD50:2990 mg/kg GISAAA 51(1),87,86

CONSENSUS REPORTS: Reported in EPA TSCA Inventory.

SAFETY PROFILE: Moderately toxic by intraperitoneal route. Low toxicity by ingestion. When heated to decomposition it emits toxic vapors of NO_x and SO_x.

ALI250 CAS:119-79-9 HR: 1
5-AMINO-2-NAPHTHALENESULFONIC ACID
mf: $C_{10}H_9NO_3S$ mw: 223.26

PROP: Needles. Spar sol in water.

SYNS: 1-AMINO-6-NAPHTHALENESULFONIC ACID □ 1-AMINO-6-SULFONAPHTHALENE □ CLEVE'S ACID-1,6 □ CLEVE'S BETA-ACID □ KYSELINA CLEVE (CZECH) □ KYSELINA-1-NAFTYLAMIN-6-SULFONOVA (CZECH) □ 1-NAPHTHYLAMINE-6-SULFONIC ACID □ 5-NAPHTHYLAMINE-2-SULFONIC ACID

TOXICITY DATA WITH REFERENCE
eye-rbt 500 mg/24H MLD 28ZPAK -,187,72
orl-rat LD50:14,200 mg/kg 28ZPAK -,187,72

CONSENSUS REPORTS: Reported in EPA TSCA Inventory.

SAFETY PROFILE: Mildly toxic by ingestion. An eye irritant. See also SULFONATES. When heated to decomposition it emits very toxic fumes of NO_x and SO_x.

ALI300 **CAS:86-60-2** **HR: 2**
7-AMINO-1-NAPHTHALENESULFONIC ACID
mf: $C_{10}H_9NO_3S$ mw: 223.26

PROP: Prisms or needles from water. Spar sol in cold water; mod sol in hot water.

SYNS: BADEN ACID □ BADISCHE ACID □ 2-NAPHTHYLAMINE-8-SULFONIC ACID

TOXICITY DATA WITH REFERENCE
ipr-mus TDLo: 18 g/kg/8W-I: NEO JJIND8 67,1299,81

SAFETY PROFILE: Questionable carcinogen with experimental neoplastigenic data. When heated to decomposition it emits toxic fumes of NO_x and SO_x.

ALI750 **CAS:118-03-6** **HR: 1**
7-AMINO-1,3,6-NAPHTHALENETRISULFONIC ACID
mf: $C_{10}H_9NO_9S_3$ mw: 383.38

SYNS: KYSELINA KOCHOVA (CZECH) □ KYSELINA 2-NAFTYLAMIN-3,6,8-TRISULFONOVA (CZECH)

TOXICITY DATA WITH REFERENCE
skn-rbt 500 mg/24H MLD 28ZPAK -,190,72
eye-rbt 500 mg/24H MLD 28ZPAK -,190,72
orl-rat LD50: 13 g/kg 28ZPAK -,190,72

CONSENSUS REPORTS: Reported in EPA TSCA Inventory.

SAFETY PROFILE: Mildly toxic by ingestion. A skin and severe eye irritant. See also SULFONATES. When heated to decomposition it emits very toxic fumes of NO_x and SO_x.

ALJ000 **CAS:5959-52-4** **HR: 2**
3-AMINO-2-NAPHTHOIC ACID
mf: $C_{11}H_9NO_2$ mw: 187.21

PROP: Yellow scales or leaflets from dilute alcohol or ether. Mp: 219–220°. Sltly sol in hot water.

SYNS: 3-AMINOISONAPHTHOIC ACID □ 3-AMINO-2-NAPHTHALENECARBOXYLIC ACID

TOXICITY DATA WITH REFERENCE
orl-mus LD50: 1600 mg/kg 14CYAT 2,1840,63

CONSENSUS REPORTS: Reported in EPA TSCA Inventory.

SAFETY PROFILE: Moderately toxic by ingestion. When heated to decomposition it emits toxic fumes of NO_x. See also AROMATIC AMINES.

ALJ250 **CAS:42884-33-3** **HR: 1**
2-AMINO-1-NAPHTHOL
mf: $C_{10}H_9NO$ mw: 159.20

PROP: Mp: 255° (decomp). Sol in alc.

SYN: AMINONAPHTHALENOL

TOXICITY DATA WITH REFERENCE
ipr-rat TDLo: 39 mg/kg/13W-I: ETA CNREA8 28,535,68
imp-mus TDLo: 56 mg/kg: CAR SAIGBL 24,186,82

SAFETY PROFILE: Questionable carcinogen with experimental carcinogenic and tumorigenic data. When heated to decomposition it emits toxic fumes of NO_x.

ALJ500 **CAS:86-97-5** **HR: 3**
5-AMINO-2-NAPHTHOL
mf: $C_{10}H_9NO$ mw: 159.20

SYN: 2-NAPHTHOL, 5-AMINO-

TOXICITY DATA WITH REFERENCE
ivn-mus LD50: 180 mg/kg CSLNX* NX#04025

CONSENSUS REPORTS: Reported in EPA TSCA Inventory.

SAFETY PROFILE: Poison by intravenous route. When heated to decomposition it emits toxic vapors of NO_x.

ALJ750 **CAS:118-46-7** **HR: 3**
8-AMINO-2-NAPHTHOL
mf: $C_{10}H_9NO$ mw: 159.20

PROP: Crystals from benzene or ligroin. Mp: 95–97° (decomp). Sol in hot water, alkali, and HCl.

TOXICITY DATA WITH REFERENCE
ivn-mus LD50: 180 mg/kg CSLNX* NX#04024

CONSENSUS REPORTS: Reported in EPA TSCA Inventory.

SAFETY PROFILE: Poison by intravenous route. When heated to decomposition it emits toxic NO_x.

ALK000 **CAS:1198-27-2** **HR: 1**
1-AMINO-2-NAPHTHOL HYDROCHLORIDE
mf: $C_{10}H_9NO \cdot ClH$ mw: 195.66

PROP: Needles from alc. Mp: 201°; sltly sol in water; sol in alc and ether.

SYN: 2-HYDROXY-1-NAPHTHYLAMINE HYDROCHLORIDE

TOXICITY DATA WITH REFERENCE
dnr-esc 500 μg/well/16H CBINA8 15,219,76
imp-mus TDLo: 100 mg/kg: CAR BJCAAI 10,539,56

CONSENSUS REPORTS: Reported in EPA TSCA Inventory. EPA Genetic Toxicology Program.

SAFETY PROFILE: Questionable carcinogen with experimental carcinogenic data. Mutation data reported. When heated to decomposition it emits very toxic fumes of HCl and NO_x.

ALK250 **CAS:41772-23-0** **HR: 2**
2-AMINO-1-NAPHTHOL HYDROCHLORIDE
mf: $C_{10}H_9NO \cdot ClH$ mw: 195.66

PROP: Needles. Mp: 255° (decomp); sol in alc.

SYN: 1-HYDROXY-2-NAPHTHYLAMINE HYDROCHLORIDE

TOXICITY DATA WITH REFERENCE
scu-rat TDLo: 1100 mg/kg/45W-I: ETA BJCAAI 6,412,52
imp-mus TDLo: 80 mg/kg: CAR BJCAAI 17,127,63
imp-mus TD: 100 mg/kg: CAR BJCAAI 10,539,56

CONSENSUS REPORTS: Reported in EPA TSCA Inventory.

SAFETY PROFILE: Questionable carcinogen with experimental carcinogenic and tumorigenic data. When heated to decomposition it emits very toxic fumes of NO_x and HCl.

ALK500 CAS:5959-56-8 HR: 1
4-AMINO-1-NAPHTHOL HYDROCHLORIDE
mf: $C_{10}H_9NO•ClH$ mw: 195.66

SYN: 1-AMINO-4-NAPHTHOL HYDROCHLORIDE

TOXICITY DATA WITH REFERENCE
imp-mus TDLo:50 mg/kg:NEO BJCAAI 12,222,58

CONSENSUS REPORTS: Reported in EPA TSCA Inventory.

SAFETY PROFILE: Questionable carcinogen with experimental neoplastigenic data. When heated to decomposition it emits very toxic fumes of NO_x and HCl.

ALK625 CAS:5438-85-7 HR: 3
2-AMINO-1,4-NAPHTHOQUINONE IMINE HYDROCHLORIDE
mf: $C_{10}H_8N_2O•ClH$ mw: 208.66

SYNS: 2-AMINO-4-IMINO-1(4H)-NAPHTHALENONE HYDROCHLORIDE □ ANQI

TOXICITY DATA WITH REFERENCE
dni-mus:ast 20 µmol/L CPBTAL 17,105,69
oms-mus:ast 20 µmol/L CPBTAL 17,105,69
dnd-mam:lym 100 µmol/L CPBTAL 17,113,69
ipr-mus LD50:5450 µg/kg CPBTAL 17,1432,69

SAFETY PROFILE: Poison by intraperitoneal route. Mutation data reported. When heated to decomposition it emits toxic fumes of NO_x and HCl.

ALK750 CAS:605-92-5 HR: 2
2-AMINO-1-NAPHTHYL ESTER SULFURIC ACID
mf: $C_{10}H_9NO_4S$ mw: 239.26

PROP: Sltly sol in water.

SYNS: 2-AMINO-1-NAPHTHYL HYDROGEN SULFATE □ 2-AMINO-1-NAPHTHYL HYDROGEN SULPHATE

TOXICITY DATA WITH REFERENCE
imp-mus TDLo:50 mg/kg:NEO BJCAAI 12,222,58

SAFETY PROFILE: Questionable carcinogen with experimental neoplastigenic data. See also ESTERS and SULFATES. When heated to decomposition it emits very toxic fumes of NO_x and SO_x.

ALL000 CAS:63976-07-8 HR: 2
2-AMINO-1-NAPHTHYLGLUCOSIDURONIC ACID
mf: $C_{16}H_{17}NO_7$ mw: 335.34

SYN: 2-NAPHTHYLAMINE-1-d-GLUCOSIDURONIC ACID

TOXICITY DATA WITH REFERENCE
imp-mus TDLo:80 mg/kg:NEO BJCAAI 11,212,57

SAFETY PROFILE: Questionable carcinogen with experimental neoplastigenic data. When heated to decomposition it emits toxic fumes of NO_x.

ALL250 CAS:329-89-5 HR: 3
6-AMINONICOTINAMIDE
mf: $C_6H_7N_3O$ mw: 137.16

PROP: Crystals. Mp: 200°.

SYNS: AMINONICOTINAMIDE □ 6-AMINONIKOTINSAEUREAMID (GERMAN) □ 6-AMINONICOTINIC ACID AMIDE □ 6-AMINO-NICOTINSAEUREAMID (GERMAN) □ 6-AN □ 6-ANA □ FDA 0121 □ NSC 21206 □ U-8774

TOXICITY DATA WITH REFERENCE
dlt-rat-ipr 1 mg/kg TXAPA9 19,371,71
orl-mus TDLo:20 mg/kg (female 8-12D post):REP TCMUD8 6,361,86
ipr-rbt TDLo:5 mg/kg (female 12D post):TER TJADAB 16,297,77
ipr-rat LD50:11 mg/kg CAXXA4 #1089763
orl-mus LDLo:320 mg/kg AECTCV 14,111,85
ipr-gpg LD50:10 mg/kg TXAPA9 33,320,75

SAFETY PROFILE: Poison by ingestion and intraperitoneal routes. An experimental teratogen. Other experimental reproductive effects. Mutation data reported. When heated to decomposition it emits toxic fumes of NO_x. A central nervous system depressant.

ALL500 CAS:99-56-9 HR: 2
2-AMINO-4-NITROANILINE
mf: $C_6H_7N_3O_2$ mw: 153.16

PROP: Dark-red needles from water. Mp: 201°. Sol in EtOH, Me_2CO, C_6H_6, $CHCl_3$; mod sol in dil acids.

SYNS: C.I. 76020 □ 1,2-DIAMINO-4-NITROBENZENE □ NCI-C03941 □ 4NDB □ 4-NITRO-1,2-BENZENEDIAMINE □ 4-NITRO-1,2-DIAMINOBENZENE □ p-NITRO-o-PHENYLENEDIAMINE □ 4-NITRO-o-PHENYLENE-DIAMINE □ 4-NITRO-1,2-PHENYLENEDIAMINE □ 4-NOPD

TOXICITY DATA WITH REFERENCE
mmo-sat 1 µg/plate ENMUDM 8(Suppl 7),1,86
dnr-esc 10 mg/L CRNGDP 2,189,81
mmo-asn 200 mg/L MUREAV 97,293,82
ipr-rat TDLo:3 g/kg (30D male):TER SheCW# 25MAR77
scu-mus TDLo:7680 mg/kg (6-15D preg):REP TJADAB 24,253,81
orl-rat LD50:681 mg/kg NCILB* NIH-NCI-E-C-72-3252
orl-mus LD50:681 mg/kg NCILB* NIH-NCI-E-C-72-3252

CONSENSUS REPORTS: IARC Cancer Review: Group 3 IMEMDT 7,56,87; Animal Inadequate Evidence IMEMDT 16,63,78. NCI Carcinogenesis Bioassay (feed); No Evidence: mouse, rat NCITR* NCI-CG-TR-180,79. Reported in EPA TSCA Inventory. EPA Genetic Toxicology Program.

SAFETY PROFILE: Moderately toxic by ingestion. An experimental teratogen. Other experimental reproductive effects. Mutation data reported. When heated to decomposition it emits toxic fumes of NO_x.

ALL750 CAS:5307-14-2 HR: 3
4-AMINO-2-NITROANILINE
mf: $C_6H_7N_3O_2$ mw: 153.16

PROP: Black needles with strong green reflection from water. Mp: 137°.

SYNS: C.I. 76070 ☐ C.I. OXIDATION BASE 22 ☐ 1,4-DIAMINO-2-NITROBENZENE ☐ DURAFUR BROWN ☐ DURAFUR BROWN 2R ☐ DYE GS ☐ FOURAMIEN 2R ☐ FOURRINE 36 ☐ FOURRINE BROWN 2R ☐ NCI-C02222 ☐ 2NDB ☐ 2-NITRO-1,4-BENZENEDIAMINE ☐ 2-NITRO-1,4-DIAMINOBENZENE ☐ NITRO-p-PHENYLENEDIAMINE ☐ 2-NITRO-1,4-PHENYLENEDIAMINE ☐ o-NITRO-p-PHENYLENEDIAMINE (MAK) ☐ 2-NITRO-p-PHENYLENEDIAMINE ☐ 2-NP ☐ 2-N-p-PDA ☐ 2-NPPD ☐ OXIDATION BASE 22 ☐ URSOL BROWN RR ☐ ZOBA BROWN RR

TOXICITY DATA WITH REFERENCE
mmo-sat 5 μg/plate NATUAS 255,506,75
dns-rat:lvr 100 mg/L MUREAV 97,359,82
otr-ham:emb 500 μg/L NCIMAV 58,243,81
sce-ham-orl 125 mg/kg BLFSBY 29B,613,83
cyt-hmn:lym 50 mg/L/24H NATUAS 255,506,75
scu-mus TDLo:1600 mg/kg (female 6-15D post):TER TJADAB 19,37A,79
scu-mus TDLo:2240 mg/kg (female 6-15D post):REP TJADAB 24,253,81
orl-mus TDLo:288 g/kg/78W-C:NEO NCITR* NCI-CG-TR-169,79
orl-mus TD:144 g/kg/78W-C:ETA NCITR* NCI-CG-TR-169,79
orl-rat LD50:2100 mg/kg JSCCA5 23,259,72
ipr-rat LD50:348 mg/kg JTEHD6 2,657,77

CONSENSUS REPORTS: IARC Cancer Review: Group 3 IMEMDT 7,56,87; Animal Inadequate Evidence IMEMDT 16,73,78. NCI Carcinogenesis Bioassay (feed); No Evidence: rat NCITR* NCI-CG-TR-169,79; Clear Evidence: mouse NCITR* NCI-CG-TR-169,79. Reported in EPA TSCA Inventory. EPA Genetic Toxicology Program.

DFG MAK: Suspected Carcinogen

SAFETY PROFILE: Suspected carcinogen with experimental carcinogenic and neoplastigenic data. Poison by intraperitoneal route. Moderately toxic by ingestion. An experimental teratogen. Other experimental reproductive effects. Mutation data reported. When heated to decomposition it emits toxic fumes of NO_x.

ALL800 CAS:4346-51-4 HR: 2
2-AMINO-5-NITROBENZENESULFONIC ACID AMMONIUM SALT
mf: $C_6H_6N_2O_5S{\cdot}H_3N$ mw: 235.24

SYN: BENZENESULFONIC ACID, 2-AMINO-5-NITRO-, AMMONIUM SALT

TOXICITY DATA WITH REFERENCE
orl-rat LD:>10 g/kg GTPZAB 32(4),55,88
ipr-rat LD50:2280 mg/kg GTPZAB 32(4),55,88
orl-mus LD:>10 g/kg GTPZAB 32(4),55,88

CONSENSUS REPORTS: Reported in EPA TSCA Inventory.

SAFETY PROFILE: Moderately toxic by intraperitoneal route. Low toxicity by ingestion. When heated to decomposition it emits toxic vapors of NO_x and NH_3.

ALM000 CAS:1211-40-1 HR: 1
4-AMINO-4′-NITROBIPHENYL
mf: $C_{12}H_{10}N_2O_2$ mw: 214.24

PROP: Red needles from EtOH. Mp: 203–204°.

SYN: 4′-NITRO-4-BIPHENYLAMINE

TOXICITY DATA WITH REFERENCE
mmo-sat 100 μg/plate MUREAV 149,9,85
uns-mus-ipr 25 mg/kg MUREAV 268,255,92
orl-rat TDLo:1440 mg/kg/73W-I:ETA TXAPA9 14,661,69

SAFETY PROFILE: Questionable carcinogen with experimental tumorigenic data. Mutation data reported. When heated to decomposition it emits toxic fumes of NO_x.

ALM250 CAS:3775-55-1 HR: 2
2-AMINO-5-(5-NITRO-2-FURYL)-1,3,4-OXADIAZOLE
mf: $C_6H_4N_4O_4$ mw: 223.19

TOXICITY DATA WITH REFERENCE
orl-rat TDLo:20 g/kg/46W-C:CAR JNCIAM 54,841,75

SAFETY PROFILE: Questionable carcinogen with experimental carcinogenic data. When heated to decomposition it emits toxic fumes of NO_x.

ALM500 CAS:38514-71-5 HR: 1
2-AMINO-4-(5-NITRO-2-FURYL)THIAZOLE
mf: $C_7H_5N_3O_3S$ mw: 211.21

SYN: ANFT

TOXICITY DATA WITH REFERENCE
mma-sat 100 ng/plate MUREAV 40,9,76
mmo-esc 300 nmol/well CNREA8 34,2266,74
dnd-esc 10 μmol/L CBINA8 31,133,80
dnd-mam:lym 50 μmol/L CRNGDP 3,1339,82
orl-rat TDLo:30,212 mg/kg/Y-C:CAR CRNGDP 3,275,82
orl-mus TDLo:68 g/kg/46W-C:NEO CNREA8 33,1593,73
orl-rat TD:40,894 mg/kg/46W-C:ETA PAACA3 21,75,80

CONSENSUS REPORTS: EPA Genetic Toxicology Program.

SAFETY PROFILE: Questionable carcinogen with experimental carcinogenic, neoplastigenic, and tumorigenic data. Mutation data reported. When heated to decomposition it emits very toxic fumes of NO_x and SO_x.

ALM750 CAS:7532-52-7 HR: 2
5-AMINO-3-(5-NITRO-2-FURYL)-s-TRIAZOLE
mf: $C_6H_5N_5O_3$ mw: 195.16

SYN: 3-(5-NITRO-2-FURYL)-1H-1,2,4-TRIAZOL-5-AMINE

TOXICITY DATA WITH REFERENCE
orl-mus LD50:1800 mg/kg JMCMAR 16,312,73
ipr-mus LD50:1460 mg/kg JMCMAR 16,312,73

SAFETY PROFILE: Moderately toxic by intraperitoneal

route. Mildly toxic by ingestion. When heated to decomposition it emits toxic fumes of NO_x.

ALN750 CAS:18264-75-0 HR: 2
1-AMINO-3-NITRO GUANIDINE
mf: $CH_5N_5O_2$ mw: 119.2

SAFETY PROFILE: Very unstable, explosive compound. See also NITRO COMPOUNDS of AROMATIC HYDROCARBONS. When heated to decomposition it emits toxic fumes of NO_x. Detonates @ 190° (mp).

ALO000 CAS:121-88-0 HR: 2
2-AMINO-5-NITROPHENOL
mf: $C_6H_6N_2O_3$ mw: 154.14

PROP: Orange needles from water. Mp: 207–208°. $H_2N(NO_2)C_6H_3OH$

SYNS: C.I. 76535 □ 3-HYDROXY-4-AMINONITROBENZENE □ 2-HYDROXY-4-NITROANILINE □ NCI C55970 □ 3-NITRO-6-AMINOPHENOL □ 5-NITRO-2-AMINOPHENOL □ RODOL YBA □ URSOL YELLOW BROWN A

TOXICITY DATA WITH REFERENCE
mmo-sat 20 µg/plate PNASA6 72,2423,75
mma-sat 1 µmol/plate MUREAV 58,11,78
cyt-ham:lng 1 mg/L ATSUDG (4),41,80
orl-rat TDLo:51,500 mg/kg/2Y-C:ETA NTPTR* NTP-TR-334,88
orl-rat TDLo:3756 mg/kg/16D-I:ETA,REP
orl-rat LD50:>4 g/kg JTEHD6 2,657,77
ipr-rat LD50:>800 mg/kg JTEHD6 2,657,77 NTPTR* NTP-TR-334,88
orl-rat TDLo:52 g/kg/13W-I NTPTR* NTP-TR-334,88
orl-mus TDLo:104 g/kg/13W-I NTPTR* NTP-TR-334,88

CONSENSUS REPORTS: IARC Cancer Review: Group 3 IMEMDT 57,177,93; Animal Limited Evidence IMEMDT 57,177,93; Human Inadequate Evidence IMEMDT 57,177,93. NTP Carcinogenesis Studies (gavage): Some Evidence: rat NTPTR* NTP-TR-334,88. Reported in EPA TSCA Inventory. EPA Genetic Toxicology Program.

SAFETY PROFILE: Questionable carcinogen with experimental tumorigenic data. Experimental reproductive effects. Mutation data reported. Potentially explosive reaction with nitrous acid. When heated to decomposition it emits toxic fumes of NO_x.

ALO500 CAS:61702-43-0 HR: 3
2-AMINO-4-NITROPHENOL SODIUM SALT
mf: $C_6H_5N_2O_3$•Na mw: 176.12

SYN: l'ORTHO, p-AMINONITROPHENOL, SEL SODIQUE (FRENCH)

TOXICITY DATA WITH REFERENCE
ipr-dog LDLo:500 mg/kg AIPTAK 50,20,35
ipr-pgn LDLo:95 mg/kg AIPTAK 50,20,35

CONSENSUS REPORTS: Reported in EPA TSCA Inventory.

SAFETY PROFILE: Poison by intraperitoneal route. Moderately toxic by intraperitoneal route. When heated to decomposition it emits toxic fumes of NO_x and Na_2O.

ALO750 CAS:2871-01-4 HR: 2
2-((4-AMINO-2-NITROPHENYL)AMINO)ETHANOL
mf: $C_8H_{11}N_3O_3$ mw: 197.22

SYNS: ETHANOL, 2-((4-AMINO-2-NITROPHENYL)AMINO)- □ HC RED NO. 3 □ 4-(2-HYDROXYETHYL)AMINO-3-NITROANILINE □ N (sup 1)-(2-HYDROXYETHYL)-2-NITRO-p-PHENYLENEDIAMINE □ NCI-C54922

TOXICITY DATA WITH REFERENCE
mmo-sat 100 µg/plate NTPTR* NTP-TR-281,86
orl-mus TDLo:182 g/kg/2Y-C:CAR NTPTR* NTP-TR-281,86

CONSENSUS REPORTS: IARC Cancer Review: Group 3 IMEMDT 57,153,93; Human Inadequate Evidence IMEMDT 57,153,93; Animal Inadequate Evidence IMEMDT 57,153,93. NTP Carcinogenesis Studies (gavage); Equivocal Evidence: mouse NTPTR* NTP-TR-281,86; No Evidence: rat NTPTR* NTP-TR-281,86. Reported in EPA TSCA Inventory.

SAFETY PROFILE: Questionable carcinogen with experimental carcinogenic data. Mutation data. When heated to decomposition it emits toxic fumes of NO_x.

ALP000 CAS:2104-09-8 HR: 2
2-AMINO-4-(p-NITROPHENYL)THIAZOLE
mf: $C_9H_7N_3O_2S$ mw: 221.25

TOXICITY DATA WITH REFERENCE
orl-rat TDLo:2150 mg/kg/13W-C:CAR JNCIAM 54,841,75
orl-mus TDLo:9600 mg/kg/46W-C:ETA CNREA8 33,1593,73

SAFETY PROFILE: Questionable carcinogen with experimental carcinogenic data. When heated to decomposition it emits toxic NO_x and SO_x.

ALP750 CAS:119-72-2 HR: 1
4-AMINO-4′-NITRO-2,2′-STILBENEDISULFONIC ACID
mf: $C_{14}H_{12}N_2O_8S_2$ mw: 400.40

SYN: KYSELINA 4-NITRO-4′-AMINOSTILBEN-2,2′-DISULFONOVA (CZECH)

TOXICITY DATA WITH REFERENCE
eye-rbt 500 mg/24H MLD 28ZPAK -,194,72
orl-rat LD50:14,200 mg/kg 28ZPAK -,194,72

CONSENSUS REPORTS: Reported in EPA TSCA Inventory.

SAFETY PROFILE: Mildly toxic by ingestion. An eye irritant. When heated to decomposition it emits very toxic fumes of NO_x and SO_x. See also SULFONATES.

ALQ000 CAS:121-66-4 HR: 3
2-AMINO-5-NITROTHIAZOLE
mf: $C_3H_3N_3O_2S$ mw: 145.15

PROP: Solid. Mp: 195–196°.
$SC(NH_2){=}NCH{=}CNO_2$ (ring closure S…C)

SYNS: AMINONITROTHIAZOLE □ AMINONITROTHIAZOLUM □ AMINZOL SOLUBLE □ ENHEPTIN □ ENTRAMIN □ NCI-C03065 □ NITRAMIN □ NITRAMINE □ 5-NITRO-2-AMINOTHIAZOLE □ NITROMIN IDO □ 5-NITRO-2-THIAZOLYLAMINE □ USAF EK-6561

TOXICITY DATA WITH REFERENCE
mmo-sat 500 μg/plate WTMOA3 69,19,82
mma-sat 666 μg/plate ENMUDM 7(Suppl 5),1,85
mmo-esc 50 μmol/L MUREAV 118,153,83
mma-esc 800 μg/plate ENMUDM 7(Suppl 5),1,85
mmo-klp 200 μmol/L MUREAV 118,153,83
orl-rat TDLo:700 mg/kg (14D male):REP TXAPA9 2,418,60
orl-rat TDLo:28 g/kg/2Y-C:CAR NCITR* NCI-CG-TR-53,78
orl-rat TD:23 g/kg/46W-C:NEO JNCIAM 54,841,75
orl-rat TD:12 g/kg/2Y-C:ETA NCITR* NCI-CG-TR-53,78
ipr-mus LD50:200 mg/kg NTIS** AD277-689

CONSENSUS REPORTS: IARC Cancer Review: Group 3 IMEMDT 7,56,87; Animal Limited Evidence IMEMDT 31,71,83. NCI Carcinogenesis Bioassay (feed); No Evidence: mouse NCITR* NCI-CG-TR-53,78; Clear Evidence: rat NCITR* NCI-CG-TR-53,78. Reported in EPA TSCA Inventory.

SAFETY PROFILE: Poison by intraperitoneal route. Experimental reproductive effects. Questionable carcinogen with experimental carcinogenic, tumorigenic, and neoplastigenic data. Mutation data reported. When heated to decomposition it emits very toxic fumes of NO_x and SO_x. Incompatible with HNO_3 and H_2SO_4. An antiprotozoal agent.

ALQ650 CAS:645-88-5 HR: 3
AMINOOXYACETIC ACID
mf: $C_2H_5NO_3$ mw: 91.08

SYNS: AOAA □ (CARBOXYMETHOXY)AMINE □ (o-CARBOXYMETHYL)HYDROXYLAMINE □ U 7524

TOXICITY DATA WITH REFERENCE
mmo-bcs 1 mol/L MUREAV 4,517,67
ipr-mus LD50:40 mg/kg BCPCA6 28,1397,79
scu-mus LD50:40 mg/kg BCPCA6 27,103,78

SAFETY PROFILE: Poison by subcutaneous and intraperitoneal routes. Mutation data reported. When heated to decomposition it emits toxic fumes of NO_x.

ALQ750 CAS:64046-62-4 HR: 3
2-AMINOOXYACETIC ACID BUTYL ESTER, HYDROCHLORIDE
mf: $C_6H_{13}NO_3$•ClH mw: 183.66

TOXICITY DATA WITH REFERENCE
ivn-rat LDLo:50 mg/kg JMPCAS 5,464,62
ipr-mus LD50:69 mg/kg JMPCAS 5,464,62

SAFETY PROFILE: Poison by intravenous and intraperitoneal routes. See also ESTERS. When heated to decomposition it emits very toxic fumes of NO_x and HCl.

ALR250 CAS:6191-22-6 HR: 3
dl-AMINOPENTAMIDE HYDROCHLORIDE
mf: $C_{19}H_{24}N_2O$•ClH mw: 332.91

SYN: α,α-DIPHENYL-γ-DIMETHYLAMINOVALERAMIDE HYDROCHLORIDE

TOXICITY DATA WITH REFERENCE
orl-mus LD50:396 mg/kg JPETAB 100,325,50
ivn-mus LD50:35 mg/kg JPETAB 100,325,50

SAFETY PROFILE: Poison by ingestion and intravenous route. When heated to decomposition it emits very toxic fumes of NO_x and HCl.

ALR500 CAS:31699-72-6 HR: 3
3-(5-AMINOPENTYL)INDOLE ADIPATE
mf: $C_{13}H_{18}N_2$•$C_6H_{10}O_4$ mw: 348.49

SYN: ω-3-INDOLYLAMYLAMINE ADIPINATE

TOXICITY DATA WITH REFERENCE
ipr-mus LD50:215 mg/kg RPTOAN 33,180,70
ivn-mus LD50:79 mg/kg RPTOAN 33,180,70

SAFETY PROFILE: Poison by intraperitoneal and intravenous routes. When heated to decomposition it emits toxic fumes of NO_x.

ALR750 CAS:28832-64-6 HR: 3
AMINOPERIMIDINE
mf: $C_{11}H_9N_3$ mw: 183.23

PROP: Crystals. Mp: 239°. Sol in acids; insol in water.

TOXICITY DATA WITH REFERENCE
unk-mus LDLo:50 mg/kg ATMPA2 32,177,38

CONSENSUS REPORTS: Reported in EPA TSCA Inventory.

SAFETY PROFILE: Poison by unspecified route. When heated to decomposition it emits toxic fumes of NO_x.

ALS000 CAS:4176-53-8 HR: 1
1-AMINOPHENANTHRENE
mf: $C_{14}H_{11}N$ mw: 193.26

PROP: Needles from ligroin. Mp: 145–146°.

TOXICITY DATA WITH REFERENCE
mma-sat ng/plate ENMUDM 6,497,84
orl-rat TDLo:250 mg/kg:ETA ZEKBAI 72,321,69

SAFETY PROFILE: Questionable carcinogen with experimental tumorigenic data. Mutation data reported. When heated to decomposition it emits toxic fumes of NO_x.

ALS250 CAS:63307-29-9 HR: 3
17-(p-AMINOPHENETHYL)-MORPHINAN-3-OL (—)-
mf: $C_{24}H_{30}N_2O$ mw: 362.56

TOXICITY DATA WITH REFERENCE
orl-mus LD50:70 mg/kg 31ZPAG 2,85,66
ivn-mus LD50:12 mg/kg 31ZPAG 2,85,66

SAFETY PROFILE: Poison by ingestion and intravenous routes. When heated to decomposition it emits toxic fumes of NO_x.

ALS500 CAS:63732-42-3 HR: 3
(−)-17-(m-AMINOPHENETHYL)-MORPHINAN-3-OL, HYDROCHLORIDE
mf: $C_{24}H_{30}N_2O{\cdot}ClH$ mw: 399.02

TOXICITY DATA WITH REFERENCE
scu-mus LD50:160 mg/kg 31ZPAG 2,85,66
ivn-mus LD50:8 mg/kg 31ZPAG 2,85,66

SAFETY PROFILE: Poison by subcutaneous and intravenous routes. When heated to decomposition it emits very toxic fumes of NO_x and HCl.

ALS750 CAS:63732-43-4 HR: 3
(±)-17-(p-AMINOPHENETHYL)MORPHINAN-3-OL, HYDROCHLORIDE
mf: $C_{24}H_{30}N_2O{\cdot}ClH$ mw: 399.02

TOXICITY DATA WITH REFERENCE
scu-mus LD50:207 mg/kg 31ZPAG 2,85,66
ivn-mus LD50:41 mg/kg 31ZPAG 2,85,66

SAFETY PROFILE: Poison by subcutaneous and intravenous routes. When heated to decomposition it emits very toxic fumes of NO_x and HCl.

ALT500 CAS:591-27-5 HR: 3
m-AMINOPHENOL
DOT: UN 2946
mf: C_6H_7NO mw: 109.14

PROP: Prisms from toluene. Mp: 123°. Sol in water and alc; sltly sol in ether.

SYNS: m-AMINOFENOL (CZECH) □ 3-AMINO-1-HYDROXYBENZENE □ m-AMINOPHENOL (DOT) □ 3-AMINOPHENOL □ BASF URSOL EG □ C.I. 76545 □ C.I. OXIDATION BASE 7 □ FOURAMINE EG □ FOURRINE 65 □ FOURRINE EG □ FURRO EG □ FUTRAMINE EG □ 3-HYDROXYANILINE □ NAKO TEG □ PELAGOL EG □ RENAL EG □ TERTRAL EG □ URSOL EG □ ZOBA EG

TOXICITY DATA WITH REFERENCE
skn-rbt 12,500 μg/24H MLD FCTXAV 15,607,77
eye-rbt 100 mg/24H MOD 28ZPAK -,109,72
sln-nsc 220 mg/L MUREAV 167,35,86
unr-rat TDLo:1245 mg/kg (1-20D preg):TER GISAAA 48(9),19,83
unr-rat TDLo:1245 mg/kg (1-20D preg):REP GISAAA 48(9),19,83
orl-rat LD50:924 mg/kg GTPZAB 32(1),49,88
ihl-rat LC50:1162 mg/m³ GTPZAB 32(1),49,88
ipr-rat LDLo:1 g/kg AIPTAK 131,151,61
orl-mus LD50:401 mg/kg GTPZAB 32(1),49,88
ipr-mus LD50:150 mg/kg NTIS** AD691-490
scu-cat LDLo:70 mg/kg AEXPBL 72,241,13
orl-qal LD50:750 mg/kg AECTCV 12,355,83

CONSENSUS REPORTS: Reported in EPA TSCA Inventory. EPA Genetic Toxicology Program.

DOT CLASSIFICATION: 6.1; *Label:* KEEP AWAY FROM FOOD

SAFETY PROFILE: Poison by ingestion, subcutaneous, and intraperitoneal routes. An experimental teratogen. Other experimental reproductive effects. Mutation data reported. A skin and eye irritant. When heated to decomposition it emits toxic fumes of NO_x.

ALT000 CAS:95-55-6 HR: 3
2-AMINOPHENOL
DOT: UN 2512
mf: C_7H_7NO mw: 109.14

PROP: Colorless needles. Mp: 173°; bp: subl. Sol in water and alc; very sol in ether.

SYNS: 2-AMINO-1-HYDROXYBENZENE □ o-AMINOPHENOL □ BASF URSOL 3GA □ BENZOFUR GG □ C.I. 76520 □ C.I. OXIDATION BASE 17 □ FOURAMINE OP □ o-HYDROXYANILINE □ 2-HYDROXYANILINE □ NAKO YELLOW EGA □ PARADONE OLIVE GREEN B □ PELAGOL 3GA □ PELAGOL GREY GG □ ZOBA 3GA

TOXICITY DATA WITH REFERENCE
eye-rbt 100 mg MLD FCTOD7 20,573,82
mma-sat 100 μg/plate ENMUDM 5(Suppl 1),3,83
ipr-ham TDLo:150 mg/kg (8D preg):REP TXAPA9 63,264,82
ipr-ham TDLo:150 mg/kg (8D preg):TER TXAPA9 63,264,82
orl-rat LD50:1300 mg/kg RPTOAN 34,307,71
ipr-rat LDLo:300 mg/kg AIPTAK 131,151,61
scu-rat LD50:37 mg/kg YKYUA6 32,1093,81
orl-mus LD50:1250 mg/kg GTPZAB 25(8),50,81
ipr-mus LD50:200 mg/kg NTIS** AD691-490
scu-cat LDLo:37 mg/kg AEXPBL 72,241,13

CONSENSUS REPORTS: Reported in EPA TSCA Inventory.

DOT CLASSIFICATION: 6.1; *Label:* KEEP AWAY FROM FOOD

SAFETY PROFILE: Poison by intraperitoneal and subcutaneous routes. Moderately toxic by ingestion route. An experimental teratogen. Other experimental reproductive effects. An eye irritant. Mutation data reported. When heated to decomposition it emits toxic NO_x. See also AROMATIC AMINES.

ALT250 CAS:123-30-8 HR: 3
4-AMINOPHENOL
DOT: UN 2512
mf: C_6H_7NO mw: 109.14

PROP: Colorless crystals or plates from water; sltly sol in water, alc, and ether; insol in chloroform. Mp: 189.6–190.2°, bp: 284° (decomp).

SYNS: ACTIVOL □ p-AMINOFENOL (CZECH) □ 4-AMINO-1-HYDROXYBENZENE □ p-AMINOPHENOL □ p-AMINOPHENOL (DOT) □ BASF URSOL P BASE □ BENZOFUR P □ CERTINAL □ C.I. OXIDATION BASE 6A □ CITOL □ DURAFUR BROWN RB □ FOURAMINE P □ FOURRINE 84 □ FOURRINE P BASE □ FURRO P BASE □ p-HYDROXYANILINE □ 4-HYDROXYANILINE □ NAKO BROWN R □ PAP □ PARANOL □ PELAGOL GREY P BASE □ PELAGOL P BASE □ RENAL AC □ RODINAL □ TERTRAL P BASE □ URSOL P □ URSOL P BASE □ ZOBA BROWN P BASE

TOXICITY DATA WITH REFERENCE
skn-rbt 12,500 μg/24H MLD FCTXAV 15,607,77
eye-rbt 100 mg MLD BIOFX* 29-4/73
spm-mus-ipr 500 mg/kg/5D MUREAV 69,149,80

mmo-ome 5 mg/L MUREAV 173,233,86
unr-rat TDLo: 563 mg/kg (female 1-20D post): TER GISAAA 48(9),19,83
ipr-ham TDLo: 200 mg/kg (female 8D post): REP TXAPA9 63,264,82
orl-rat LD50: 375 mg/kg BIOFX* 29-4/73
unr-rat LD50: 675 mg/kg GISAAA 50(3),4,85
orl-mus LD50: 420 mg/kg GISAAA 35,28,70
ipr-mus LDLo: 100 mg/kg RBPMAZ 22,1,52
scu-mus LDLo: 470 mg/kg AEPPAE 188,130,38
scu-rat LDLo: 37 mg/kg AEXPBL 72,241,13
orl-bwd LD50: 56,200 µg/kg AECTCV 12,355,83

CONSENSUS REPORTS: Reported in EPA TSCA Inventory. EPA Genetic Toxicology Program.

DOT CLASSIFICATION: 6.1; *Label:* KEEP AWAY FROM FOOD

SAFETY PROFILE: Poison by ingestion, subcutaneous, and intraperitoneal routes. An experimental teratogen. Other experimental reproductive effects. An allergen and skin and eye irritant. Mutation data reported. Can cause contact dermatitis, bronchial asthma, and methemoglobinemia with cyanosis. When heated to decomposition it emits toxic fumes of NO_x.

ALT750 CAS:63957-37-9 HR: 3
m-AMINOPHENOL ANTIMONYL TARTRATE
mf: $C_6H_8NO \cdot C_4H_4O_7Sb$ mw: 395.98

TOXICITY DATA WITH REFERENCE
ipr-mus LD50: 55 mg/kg AJTMAQ 25,263,45

CONSENSUS REPORTS: Antimony compounds are on the Community Right-To-Know List.

OSHA PEL: TWA 0.5 mg(Sb)/m³
ACGIH TLV: TWA 0.5 mg(Sb)/m³
NIOSH REL: (Antimony) TWA 0.5 mg(Sb)/m³

SAFETY PROFILE: Poison by intraperitoneal route. When heated to decomposition it emits very toxic fumes of NO_x and Sb. See also ANTIMONY COMPOUNDS.

ALU000 CAS:63957-38-0 HR: 3
o-AMINOPHENOL ANTIMONYL TARTRATE
mf: $C_6H_8NO \cdot C_4H_4O_7Sb$ mw: 395.98

SYN: o-AMINOPHENOL-OXO(TARTRATO)ANTIMONATE(1-)-

TOXICITY DATA WITH REFERENCE
ipr-mus LD50: 63 mg/kg AJTMAQ 25,263,45

CONSENSUS REPORTS: Antimony compounds are on the Community Right-To-Know List.

OSHA PEL: TWA 0.5 mg(Sb)/m³
ACGIH TLV: TWA 0.5 mg(Sb)/m³
NIOSH REL: (Antimony) TWA 0.5 mg(Sb)/m³

SAFETY PROFILE: Poison by intraperitoneal route. See also ANTIMONY COMPOUNDS. When heated to decomposition it emits very toxic fumes of NO_x and Sb.

ALU250 CAS:63957-39-1 HR: 3
p-AMINOPHENOL ANTIMONYL TARTRATE
mf: $C_6H_8NO \cdot C_4H_4O_7Sb$ mw: 395.98

SYN: p-AMINOPHENOL-OXO(TARTRATO)ANTIMONATE(1-)-

TOXICITY DATA WITH REFERENCE
ipr-mus LD50: 50 mg/kg AJTMAQ 25,263,45

CONSENSUS REPORTS: Antimony compounds are on the Community Right-To-Know List.

OSHA PEL: TWA 0.5 mg(Sb)/m³
ACGIH TLV: TWA 0.5 mg(Sb)/m³
NIOSH REL: (Antimony) TWA 0.5 mg(Sb)/m³

SAFETY PROFILE: Poison by intraperitoneal route. See also ANTIMONY COMPOUNDS. When heated to decomposition it emits very toxic fumes of NO_x and Sb.

ALU500 CAS:51-78-5 HR: 2
p-AMINOPHENOL HYDROCHLORIDE
mf: $C_6H_7NO \cdot ClH$ mw: 145.60

PROP: Colorless prisms. Mp: 306° (decomp). Sol in water and alc.

SYN: 4-AMINOPHENOL HYDROCHLORIDE

TOXICITY DATA WITH REFERENCE
slt-dmg-orl 20 mmol/L MUREAV 240,87,90
ipr-mus LD50: 750 mg/kg NTIS** AD691-490

CONSENSUS REPORTS: Reported in EPA TSCA Inventory.

SAFETY PROFILE: Moderately toxic by intraperitoneal route. Mutation data reported. When heated to decomposition it emits very toxic fumes of HCl and NO_x.

ALU750 CAS:69782-45-2 HR: 2
p-AMINOPHENOL TARTRATE
mf: $C_6H_7NO \cdot C_4H_6O_6$ mw: 259.24

TOXICITY DATA WITH REFERENCE
ivn-rbt LDLo: 1 g/kg AEXPBL 33,216,1894
scu-gpg LDLo: 2 g/kg AEXPBL 33,216,1894
scu-frg LDLo: 1515 mg/kg AEXPBL 33,216,1894

SAFETY PROFILE: Moderately toxic by subcutaneous route. Mildly toxic by intravenous routes. When heated to decomposition it emits toxic fumes of NO_x.

ALU875 CAS:76487-32-6 HR: 3
1-(3-(p-AMINOPHENOXY)PROPYL)-4-(o-METHOXYPHENYL)PIPERAZINE DIHYDROCHLORIDE
mf: $C_{20}H_{27}N_3O_2 \cdot 2ClH$ mw: 414.42

SYNS: 1-(p-AMINOPHENOXY)-3-(N¹)-(o-METHOXYPHENYL)-N⁴-PIPERAZINYL)PROPANE 2HCl □ COMPOUND 74-637

TOXICITY DATA WITH REFERENCE
orl-rat LD50: 416 mg/kg DRFUD4 6,346,81
orl-mus LD50: 319 mg/kg DRFUD4 6,346,81
ipr-mus LD50: 158 mg/kg DRFUD4 6,346,81

SAFETY PROFILE: Poison by ingestion and intraperitoneal routes. When heated to decomposition it emits toxic fumes of NO_x and HCl.

ALV000 CAS:15686-71-2 HR: 3
7-(d-α-AMINOPHENYLACETAMIDO)DESACETOXYCEPHALOSPORANIC ACID
mf: $C_{16}H_{17}N_3O_4S$ mw: 347.42

SYNS: 7-(d-2-AMINO-2-PHENYLACETAMIDO)-3-METHYL-Δ(SUP 3)-CEPHEM-4-CARBOXYLIC ACID □ CEFA-ISKIA □ CEFALOTO □ CEPHALEXIN □ CEPOREX □ CEPOREXIN □ CEPOREXINE □ CEX □ KEFLEX □ KEFORAL □ LARIXIN □ LEXIBIOTICO □ MEDLEXIN □ NEOLEXINA □ ORACEF □ OROXIN □ ORTISPORINA □ S 6437 □ SARTOSONA □ SENCEPHALIN □ SYNCL

TOXICITY DATA WITH REFERENCE
orl-rat TDLo:52 g/kg (female 17-22D post):REP NKRZAZ 27(Suppl 7),865,79
orl-rat TDLo:4 g/kg (female 9-14D post):TER OYYAA2 3,249,69
orl-hmn TDLo:14 mg/kg/D:GIT AACHAX -,361,68
ipr-rat LD50:4 g/kg KSRNAM 3,390,69
scu-rat LD50:6100 mg/kg KSRNAM 3,390,69
orl-mus LD50:1495 mg/kg NKRZAZ 27(Suppl 7),765,79
ipr-mus LD50:400 mg/kg AACHAX-,489,68
scu-mus LD50:1150 mg/kg OYYAA2 3,227,69

SAFETY PROFILE: Poison by intraperitoneal route. Moderately toxic by ingestion and other routes. An experimental teratogen. Other experimental reproductive effects. Human systemic effects by ingestion: nausea, vomiting, and diarrhea. When heated to decomposition it emits very toxic fumes of NO_x and SO_x.

ALV050 CAS:7621-86-5 HR: 2
2-(4-AMINOPHENYL)-5-AMINOBENZIMIDAZOLE
mf: $C_{13}H_{12}N_4$ mw: 224.29

SYNS: 2-(4-AMINOPHENYL)-1H-BENZIMIDAZOL-5-AMINE □ 1H-BENZIMIDAZOL-5-AMINE, 2-(4-AMINOPHENYL)- □ BENZIMIDAZOLE, 5-AMINO-2-(p-AMINOPHENYL)-

TOXICITY DATA WITH REFERENCE
orl-rat LD50:5 g/kg GISAAA 43(9),23,78
orl-mus LD50:5500 mg/kg GISAAA 43(9),23,78

CONSENSUS REPORTS: Reported in EPA TSCA Inventory.

SAFETY PROFILE: Low toxicity by ingestion. When heated to decomposition it emits toxic vapors of NO_x.

ALV100 CAS:73791-39-6 HR: 3
p-AMINOPHENYLARSINE OXIDE DIHYDRATE
mf: $C_6H_6AsNO \cdot 2H_2O$ mw: 219.09

SYNS: ANILINE, p-ARSENOSO-, DIHYDRATE □ 4-ARSENOSOANILINE, DIHYDRATE □ ARSINE, (p-AMINOPHENYL)OXO-, DIHYDRATE

TOXICITY DATA WITH REFERENCE
ipr-mus LD50:4430 μg/kg JPETAB 70,211,40
ivn-mus LD50:100 mg/kg CSLNX* NX#06293

OSHA PEL: TWA 0.5 mg(As)/m^3
ACGIH TLV: TWA 0.2 mg(As)/m^3

SAFETY PROFILE: Poison by intraperitoneal and intravenous routes. When heated to decomposition it emits toxic fumes of NO_x and As.

ALV500 CAS:43087-91-8 HR: 2
5-AMINO-2-PHENYLBENZOTHIAZOLE
mf: $C_{13}H_{10}N_2S$ mw: 226.31

SYNS: FABT (CZECH) □ 2-FENYL-5-AMINOBENZTHIAZOL (CZECH)

TOXICITY DATA WITH REFERENCE
eye-rbt 100 mg/24H SEV 28ZPAK -,203,72
orl-rat LD50:2940 mg/kg 28ZPAK -,203,72

SAFETY PROFILE: Moderately toxic by ingestion. Severe eye irritant. When heated to decomposition it emits very toxic fumes of NO_x and SO_x.

ALV750 CAS:20123-68-6 HR: 3
1-m-AMINOPHENYL-2-CYCLOPROPYLAMINOETHANOLDIHYDROCHLORIDE
mf: $C_{11}H_{16}N_2O \cdot 2ClH$ mw: 265.21

SYN: AB-15

TOXICITY DATA WITH REFERENCE
orl-rat LD50:3250 mg/kg BCPCA6 18,2293,69
ipr-rat LD50:710 mg/kg BCPCA6 18,2293,69
ivn-rat LD50:390 mg/kg BCPCA6 18,2293,69
orl-mus LD50:1060 mg/kg BCPCA6 18,2293,69
ipr-mus LD50:470 mg/kg BCPCA6 18,2293,69
ivn-mus LD50:260 mg/kg BCPCA6 18,2293,69

SAFETY PROFILE: Poison by intravenous route. Moderately toxic by ingestion and intraperitoneal routes. When heated to decomposition it emits very toxic fumes of Cl^- and NO_x.

ALW000 CAS:63979-26-0 HR: 3
1-(4-AMINOPHENYL)-4-(DIETHYLCARBOXAMIDE)-5-METHYL-1,2,3-TRIAZOLE HYDROCHLORIDE

SYN: SKF-183A

TOXICITY DATA WITH REFERENCE
orl-rat LD50:494 mg/kg TXAPA9 1,150,59
ipr-mus LD50:260 mg/kg TXAPA9 1,150,59

SAFETY PROFILE: Poison by intraperitoneal route. Moderately toxic by ingestion. When heated to decomposition it emits very toxic fumes of HCl and NO_x.

ALW250 CAS:98-84-0 HR: 2
1-AMINO-1-PHENYLETHANE
mf: $C_8H_{11}N$ mw: 121.20

SYNS: α-METHYLBENZYLAMINE □ α-PHENYLETHYLAMINE □ 1-PHENYLETHYLAMINE

TOXICITY DATA WITH REFERENCE
skn-rbt 10 mg/24H SEV AMIHBC 4,119,51
eye-rbt 250 μg SEV AMIHBC 4,119,51
orl-rat LD50:940 mg/kg AMIHBC 4,119,51
skn-rbt LD50:780 mg/kg AMIHBC 4,119,51

SAFETY PROFILE: Moderately toxic by ingestion and skin contact. A skin and severe eye irritant. See also AMINES. When heated to decomposition it emits toxic fumes of NO_x.

ALW500 CAS:64038-09-1 HR: 3
5-(p-AMINOPHENYL)-5-ETHYL-1-METHYLBARBITURIC ACID
mf: $C_{13}H_{15}N_3O_3$ mw: 261.31

SYN: PAM

TOXICITY DATA WITH REFERENCE
ims-mus LD50:210 mg/kg TXAPA9 47,305,79
ipr-rat LD50:780 mg/kg PHMCAA 5,237,63
ivn-mus LD50:160 mg/kg ARZNAD 11,809,61

SAFETY PROFILE: Poison by intravenous and intramuscular routes. Moderately toxic by intraperitoneal route. When heated to decomposition it emits toxic fumes of NO_x.

ALW750 CAS:144-14-9 HR: 3
N-β-(p-AMINOPHENYL)ETHYLNORMEPERIDINE
mf: $C_{22}H_{28}N_2O_2$ mw: 352.52

PROP: Solid. Mp: 83°.

SYNS: 1-(p-AMINOPHENETHYL)-4-PHENYLISONIPECOTIC ACID, ETHYL ESTER □ 1-(p-AMINOPHENETHYL)-4-PHENYLPIPERIDINE-4-CARBOXYLIC ACID ETHYL ESTER □ N-(β-(p-AMINOPHENYL)ETHYL)-4-PHENYL-4-CARBETHOXYPIPERIDINE □ ETHYL-1-(p-AMINOPHENETHYL)-4-PHENYLISONIPECOTATE

TOXICITY DATA WITH REFERENCE
orl-rat LD50:175 mg/kg 27ZIAQ -,-,65
ipr-rat LD50:45 mg/kg 27ZIAQ -,44,73
scu-rat LD50:163 mg/kg 27ZIAQ -,-,65
orl-mus LD50:128 mg/kg 27ZIAQ -,44,73
ipr-mus LD50:53 mg/kg 27ZIAQ -,-,65
scu-mus LD50:100 mg/kg 27ZIAQ -,44,73
ivn-mus LD50:25 mg/kg 27ZIAQ -,44,73

SAFETY PROFILE: Poison by ingestion, subcutaneous, intravenous, and intraperitoneal routes. When heated to decomposition it emits toxic fumes of NO_x.

ALW900 CAS:66471-17-8 HR: 2
2-(m-AMINOPHENYL)-3-INDOLECARBOXALDEHYDE, 4-(m-TOLYL)-3-THIOSEMICARBAZONE
mf: $C_{23}H_{21}N_5S$ mw: 399.55

SYN: INDOLE-3-CARBOXALDEHYDE, 2-(m-AMINOPHENYL)-, 4-(m-TOLYL)-3-THIOSEMICARBAZONE

TOXICITY DATA WITH REFERENCE
orl-rbt TDLo:80 mg/kg (female 1D pre):REP IJMRAQ 66,983,77
orl-rbt LD50:800 mg/kg IJMRAQ 66,983,77

SAFETY PROFILE: Moderately toxic by ingestion. Experimental reproductive effects. When heated to decomposition it emits toxic fumes of NO_x and SO_x.

ALX000 CAS:130-17-6 HR: 3
2-(p-AMINOPHENYL)-6-METHYLBENZOTHIAZOLYL-7-SULFONIC ACID
mf: $C_{14}H_{12}N_2O_3S_2$ mw: 320.40

SYNS: 7-BENZOTHIAZOLESULFONIC ACID □ 6-METHYL-2-(p-AMINO PHENYL)

TOXICITY DATA WITH REFERENCE
ivn-mus LD50:178 mg/kg CSLNX* NX#00718

CONSENSUS REPORTS: Reported in EPA TSCA Inventory.

SAFETY PROFILE: Poison by intravenous route. See also SULFONATES. When heated to decomposition it emits very toxic fumes of SO_x and NO_x.

ALX100 CAS:1783-81-9 HR: 2
m-AMINOPHENYL METHYL SULFIDE
mf: C_7H_9NS mw: 139.23

SYNS: m-AMINOTHIOANISOLE □ 3-AMINOTHIOANISOLE □ ANILINE, m-(METHYLTHIO)- □ BENZENAMINE, 3-(METHYLTHIO)-(9CI) □ 3-METHYLMERCAPTOANILINE □ m-(METHYLTHIO)ANILINE □ 3-(METHYLTHIO)BENZENAMINE

TOXICITY DATA WITH REFERENCE
orl-qal LD50:750 mg/kg AECTCV 12,355,83
orl-brd LD50:750 mg/kg AECTCV 12,355,83

CONSENSUS REPORTS: Reported in EPA TSCA Inventory.

SAFETY PROFILE: Moderately toxic by ingestion. When heated to decomposition it emits toxic vapors of NO_x and SO_x.

ALX250 CAS:13425-22-4 HR: 3
2-AMINO-5-PHENYL-OXAZOLINE FORMATE
mf: $C_9H_{10}N_2O•C_4H_4O_4$ mw: 278.29

SYNS: AMINOREXFUMARATE □ MENOCIL

TOXICITY DATA WITH REFERENCE
orl-chd TDLo:1 mg/kg:CNS ATXKA8 26,117,70
orl-hmn TDLo:3 mg/kg:CNS ATXKA8 26,117,70
orl-rat LD50:25 mg/kg ATXKA8 26,117,70

SAFETY PROFILE: Poison by ingestion. Human central nervous system effects by ingestion. When heated to decomposition it emits toxic fumes of NO_x.

ALX500 CAS:61706-44-3 HR: 3
2-(p-AMINOPHENYL)-2-PHENYLPROPIONAMIDE
mf: $C_{15}H_{16}N_2O$ mw: 240.33

SYNS: 2-FENIL-2-(p-AMINOFENIL)PROPIONAMMIDE (ITALIAN) □ 2-PHENYL-2-(p-AMINOPHENYL)PROPIONAMIDE

TOXICITY DATA WITH REFERENCE
orl-rat LD50:1600 mg/kg FRPSAX 31,671,76
orl-mus LD50:260 mg/kg FRPSAX 31,671,76

SAFETY PROFILE: Poison by ingestion. When heated to decomposition it emits toxic fumes of NO_x.

ALX750 CAS:3314-35-0 HR: 3
3-AMINO-1-PHENYL-2-PYRAZOLINE
mf: $C_9H_{11}N_3$ mw: 161.23

SYN: 1-FENYL-3-AMINOPYRAZOLIN (CZECH)

TOXICITY DATA WITH REFERENCE
skn-rbt 500 mg/24H MOD 28ZPAK -,144,72
eye-rbt 100 mg/24H SEV 28ZPAK -,144,72

orl-rat LD50:78 mg/kg 28ZPAK -,144,72

SAFETY PROFILE: Poison by ingestion. Moderately toxic skin irritant. Severe eye irritant. When heated to decomposition it emits toxic fumes of NO_x.

ALX879 HR: 1
4-AMINO-5-PHENYL-3-PYRAZOLYL METHYL KETONE
mf: $C_{11}H_{11}N_3O$ mw: 201.25

SYN: 1-(4-AMINO-5-PHENYL(1H)-PYRAZOL-3-YL)ETHANONE

TOXICITY DATA WITH REFERENCE
orl-mus LD50:1 g/kg FRPSAX 39,618,84

SAFETY PROFILE: Moderately toxic by ingestion. When heated to decomposition it emits toxic fumes of NO_x. See also KETONES.

ALY250 CAS:134-37-2 HR: 2
1-(3-AMINOPHENYL)-2-PYRIDONE
mf: $C_{11}H_{10}N_2O$ mw: 186.23

PROP: Crystals. Mp: 182.5–184.5°.

SYNS: AMINOPHENYLPYRIDONE □ 1-(3-AMINOPHENYL)-2-(1H)-PYRIDINONE □ 1-(m-AMINOPHENYL)-2(1H)-PYRIDONE □ 1-m-AMINOPHENYL-2-PYRIDONE □ AMPHENIDONE □ DORNWAL □ DORNWALL

TOXICITY DATA WITH REFERENCE
orl-rat LD50:2300 mg/kg FEPRA7 19,390,60
orl-mus LD50:1300 mg/kg FEPRA7 19,390,60
ivn-mus LD50:660 mg/kg 27ZQAG -,201,72

SAFETY PROFILE: Moderately toxic by ingestion and intravenous routes. When heated to decomposition it emits toxic fumes of NO_x.

ALY500 CAS:41136-03-2 HR: 3
2-AMINO-5-PHENYLTHIOMETHYL-2-OXAZOLINE
mf: $C_{10}H_{12}N_2OS$ mw: 208.30

TOXICITY DATA WITH REFERENCE
orl-mus LD50:147 mg/kg JMCMAR 16,510,73
ipr-mus LD50:178 mg/kg JMCMAR 16,510,73

SAFETY PROFILE: Poison by ingestion and intraperitoneal routes. When heated to decomposition it emits very toxic fumes of SO_x and NO_x.

ALY675 CAS:4922-98-9 HR: 3
3-AMINO-3-PHENYL-1,2,4-TRIAZOLE
mf: $C_8H_8N_4$ mw: 160.18

PROP: Crystals from water. Mp: 186–187°.

SAFETY PROFILE: Reaction with nitrous acid gives a touch sensitive explosive product. Upon decomposition it emits toxic fumes of NO_x.

ALY750 CAS:59690-88-9 HR: 2
1-(m-AMINOPHENYL)UREA HYDROCHLORIDE
mf: $C_7H_9N_3O{\cdot}ClH$ mw: 187.65

SYN: m-AMINOFENYLMOCOVINA HYDROCHLORID (CZECH)

TOXICITY DATA WITH REFERENCE
eye-rbt 100 mg/24H SEV 28ZPAK -,165,72

SAFETY PROFILE: A severe eye irritant. When heated to decomposition it emits very toxic fumes of NO_x and HCl.

ALZ000 CAS:51249-05-9 HR: 2
AMINOPHON
mf: $C_{18}H_{37}NO_3P$ mw: 346.53

SYNS: 1-(BUTYLAMINO)CYCLOHEXYLPHOSPHONIC ACID DIBUTYL ESTER □ O,O-DIBUTYL-1-BUTYLAMINO-CYCLOHEXYLPHOSPHONATE

TOXICITY DATA WITH REFERENCE
orl-rat LD50:3000 mg/kg EQSFAP 3,686,75
skn-rat LD50:1200 mg/kg EQSFAP 3,686,75
ipr-rat LD50:1385 mg/kg EQSFAP 3,686,75
orl-mus LD50:3475 mg/kg EQSFAP 3,686,75
skn-rbt LD50:500 mg/kg EQSFAP 3,686,75
orl-ham LD50:10,000 mg/kg EQSFAP 3,686,75

SAFETY PROFILE: Moderately toxic by several routes. When heated to decomposition it emits very toxic fumes of PO_x and NO_x.

AMA000 CAS:1990-90-5 HR: 2
4-AMINO-3-PICOLINE
mf: $C_6H_8N_2$ mw: 108.16

PROP: Crystals from C_6H_6/pet ether. Mp: 108–109°.

SYN: PHILLIPS 1908

TOXICITY DATA WITH REFERENCE
orl-rat LD50:446 mg/kg TXAPA9 21,315,72
orl-brd LD50:2400 μg/kg TXAPA9 21,315,72

SAFETY PROFILE: Moderately toxic by ingestion. When heated to decomposition it emits toxic fumes of NO_x.

AMA010 CAS:1603-41-4 HR: 3
6-AMINO-3-PICOLINE
mf: $C_6H_8N_2$ mw: 108.16

SYNS: 2-AMINO-5-METHYLPYRIDINE □ 3-PICOLINE, 6-AMINO- □ 2-PYRIDINAMINE, 5-METHYL-

TOXICITY DATA WITH REFERENCE
orl-rat LD50:200 mg/kg 85JCAE-,841,86
scu-mus LD50:110 mg/kg AJEBAK 36,491,58
skn-gpg LD50:400 mg/kg 85JCAE-,841,86

CONSENSUS REPORTS: Reported in EPA TSCA Inventory.

SAFETY PROFILE: Poison by ingestion, subcutaneous, and skin contact routes. When heated to decomposition it emits toxic vapors of NO_x.

AMA100 CAS:26844-49-5 HR: 3
(4-AMINOPIPERIDINO)METHYL INDOL-3-YL KETONE
mf: $C_{15}H_{19}N_3O$ mw: 257.37

PROP: A liquid.

SYN: KETONE, (4-AMINOPIPERIDINO)METHYL INDOL-3-YL

TOXICITY DATA WITH REFERENCE
ipr-mus LD50:>400 mg/kg JMCMAR 14,1054,71

DOT CLASSIFICATION: 3; *Label:* Flammable Liquid

SAFETY PROFILE: Moderately toxic by intraperitoneal route. A flammable liquid. When heated to decomposition it emits toxic vapors of NO_x.

AMA250 CAS:616-30-8 HR: 3
3-AMINO-1,3-PROPANEDIOL
mf: $C_3H_9NO_2$ mw: 91.13

SYNS: 1-AMINOGLYCEROL □ 2,3-DIHYDROXYPROPYLAMINE

TOXICITY DATA WITH REFERENCE
orl-rat LD50:7500 mg/kg SCCUR* -,1,61
orl-mus LD50:2460 mg/kg SCCUR* -,1,61
ipr-mus LD50:246 mg/kg SCCUR* -,1,61
ipr-rbt LD50:198 mg/kg SCCUR* -,1,61

CONSENSUS REPORTS: Reported in EPA TSCA Inventory.

SAFETY PROFILE: Poison by intraperitoneal route. Moderately toxic by ingestion. When heated to decomposition it emits toxic fumes of NO_x.

AMA500 CAS:78-96-6 HR: 3
1-AMINOPROPAN-2-OL
mf: C_3H_9NO mw: 75.13

$H_2NCH_2CHOHCH_3$

PROP: Liquid, slt ammonia odor, sol in water. D: 0.969, mp: 1.4°, flash p: 171°F, vap d: 2.6.

SYNS: α-AMINOISOPROPYL ALCOHOL □ 1 AMINO-2-PROPANOL □ 2-HYDROXYPROPYLAMINE □ ISOPROPANOLAMINE □ MONO-ISO-PROPANOLAMINE □ THREAMINE

TOXICITY DATA WITH REFERENCE
skn-rbt 485 mg open MOD UCDS** 5/21/71
eye-rbt 970 µg SEV UCDS** 5/21/71
orl-rat LD50:1715 mg/kg GTPZAB 30(7),46,86
ipr-mus LDLo:250 mg/kg CBCCT* 4,232,52
skn-rbt LD50:1640 mg/kg UCDS** 5/21/71

CONSENSUS REPORTS: Reported in EPA TSCA Inventory.

SAFETY PROFILE: Poison by intraperitoneal route. Moderately toxic by ingestion and skin contact. A skin and severe eye irritant. Moderately flammable in presence of heat, flame, sparks, powerful oxidizers. Ignites on contact with cellulose nitrate of high surface area. Catalyzes the explosive polymerization of 2,4-hexadienal. To fight fire, use alcohol foam. When heated to decomposition it emits toxic fumes of NO_x.

AMB000 CAS:138-61-4 HR: 3
AMINOPROPANOL PYROCATECHOLHYDROCHLORIDE
mf: $C_9H_{13}NO_3 \cdot ClH$ mw: 219.69

SYNS: 3,4-DIHYDROXYNOREPHEDRINE HYDROCHLORIDE □ 3,4-DIHYDROXYPHENYLAMINOPROPANOL HYDROCHLORIDE □ 3,4-DIHYDROXYPHENYLPROPANOLAMINE HYDROCHLORIDE □ ISOADRENALINE HYDROCHLORIDE □ α-METHYLNORADRENALINE HYDROCHLORIDE □ NORHOMOEPINEPHRINE HYDROCHLORIDE

TOXICITY DATA WITH REFERENCE
scu-rat LDLo:3 mg/kg JPETAB 71,62,41
ivn-rbt LDLo:11 mg/kg JACSAT 53,4149,31

SAFETY PROFILE: Poison by subcutaneous and intravenous routes. When heated to decomposition it emits very toxic fumes of NO_x and Cl^-.

AMB250 HR: 3
2-AMINO PROPIONITRILE
mf: $C_3H_6N_2$ mw: 70.1

CONSENSUS REPORTS: Cyanide and its compounds are on the Community Right-To-Know List.

SAFETY PROFILE: A poison and dangerous fire hazard. Can explode in storage. See also NITRILES. Upon decomposition it emits toxic fumes of CN^- and NO_x.

AMB500 CAS:151-18-8 HR: 2
3-AMINOPROPIONITRILE
mf: $C_3H_6N_2$ mw: 70.11

$H_2NC_2H_4C{\equiv}N$

PROP: Liquid, amine odor. Bp: 185°.

SYNS: β-AMINOPROPIONITRILE □ BAPN □ β-CYANOETHYLAMINE

TOXICITY DATA WITH REFERENCE
sce-mus:emb 1250 mg/kg ARTODN 47,305,81
ipr-mus TDLo:1250 mg/kg (female 10D post):TER ARTODN 47,305,81
orl-rat TDLo:5 g/kg (female 14-15D post):REP TJADAB 4,227,71
skn-mus LDLo:12,800 mg/kg EMPSAL 39,154,83
ipr-mus LD50:1152 mg/kg EMPSAL 39,154,83

CONSENSUS REPORTS: EPA Genetic Toxicology Program. Reported in EPA TSCA Inventory. Cyanide and its compounds are on the Community Right-To-Know List.

SAFETY PROFILE: Moderately toxic by intraperitoneal route. An experimental teratogen. Other experimental reproductive effects. Mutation data reported. Nitriles usually have cyanide-like effects. See also CYANIDE. Easily oxidized and unstable. A storage hazard; it polymerizes to an explosive yellow solid. When heated to decomposition it emits toxic fumes of CN^- and NO_x. For fire and explosion hazards see CYANIDE.

AMB750 CAS:2079-89-2 HR: 2
β-AMINOPROPIONITRILE FUMARATE
mf: $C_3H_6N_2 \cdot 2C_4H_4O_4$ mw: 302.27

SYNS: β-APN □ BAPN FUMARATE □ DI-β-AMINOPROPIONITRILE FUMARATE □ DI-BAPN FUMARATE

TOXICITY DATA WITH REFERENCE
orl-rat TDLo:1200 mg/kg (female 15-20D post):TER ARPAAQ 81,60,66
orl-rat LDLo:800 mg/kg TJADAB 5,33,72
ipr-mus LD50:5362 mg/kg EMPSAL 39,154,83

orl-ham LD50:5000 mg/kg TJADAB 14,43,76

CONSENSUS REPORTS: Reported in EPA TSCA Inventory. Cyanide and its compounds are on the Community Right-To-Know List.

SAFETY PROFILE: Moderately toxic by ingestion and intraperitoneal routes. An experimental teratogen. When heated to decomposition it emits toxic fumes of NO_x and CN^-. See also NITRILES.

AMC000 CAS:70-69-9 HR: 3
p-AMINOPROPIOPHENONE
mf: $C_9H_{11}NO$ mw: 149.21

PROP: Needles from water. Mp: 140°.

SYNS: 1-(4-AMINOPHENYL)-1-PROPANONE □ ETHYL-p-AMINOPHENYL KETONE □ PAPP □ PARAMINOPROPIOPHENONE □ USAF UCTL-1856

TOXICITY DATA WITH REFERENCE
orl-rat LD50:177 mg/kg BECTA6 30,122,83
ipr-rat LDLo:525 mg/kg CJBPAZ 38,667,60
orl-mus LD50:168 mg/kg GEPHDP 14,465,83
ipr-mus LD50:80 mg/kg NTIS** AD277-689
orl-cat LD50:5600 μg/kg BECTA6 30,122,83
orl-mam LD50:5600 μg/kg BECTA6 30,122,83

CONSENSUS REPORTS: Reported in EPA TSCA Inventory. EPA Extremely Hazardous Substances List.

DOT CLASSIFICATION: 3; *Label:* Flammable Liquid

SAFETY PROFILE: Poison by ingestion and intraperitoneal routes. Ingestion of large doses can cause cyanosis. A flammable liquid. When heated to decomposition it emits toxic fumes of NO_x.

AMC250 CAS:112-33-4 HR: 2
3-AMINOPROPOXY-2-ETHOXY ETHANOL
mf: $C_7H_{17}NO_3$ mw: 163.25

SYN: POLYGLYCOLAMINE H-163

TOXICITY DATA WITH REFERENCE
skn-rbt 500 mg open MLD UCDS** 1/20/72
eye-rbt 2 mg/24H SEV 85JCAE-,633,86
orl-rat LD50:6500 mg/kg AIHAAP 30,470,69
ihl-rat LCLo:20,000 ppm/30M AIHAAP 30,470,69
skn-rbt LD50:5990 mg/kg AIHAAP 30,470,69

CONSENSUS REPORTS: Reported in EPA TSCA Inventory.

SAFETY PROFILE: Mildly toxic by ingestion, inhalation and skin contact. A skin and severe eye irritant. When heated to decomposition it emits toxic fumes of NO_x.

AMC500 CAS:31897-98-0 HR: 3
1-(3-AMINOPROPYL)ADAMANTANEHYDROCHLORIDE
mf: $C_{13}H_{23}N\cdot ClH$ mw: 229.83

SYN: 3-(1-ADAMANTYL)PROPYLAMINE HYDROCHLORIDE

TOXICITY DATA WITH REFERENCE
orl-mus LD50:600 mg/kg JMCMAR 17,602,74
ipr-mus LD50:150 mg/kg JMCMAR 17,602,74

SAFETY PROFILE: Poison by intraperitoneal route. Moderately toxic by ingestion. When heated to decomposition it emits very toxic fumes of HCl and NO_x.

AMC750 CAS:63717-27-1 HR: 3
2-((3-AMINOPROPYL)AMINO) ETHANETHIOL, DIHYDROGEN PHOSPHATE (ester-HYDRATE)
mf: $C_5H_{15}N_2O_3PS\cdot H_2O$ mw: 232.27

SYN: PHOSPHOROTHIOIC ACID, S-ESTER with 2-((3-AMINOPROPYL)AMINO)ETHANETHIOL, HYDRATE

TOXICITY DATA WITH REFERENCE
ipr-mus LD50:700 mg/kg JMCMAR 12,236,69
unk-mus LD50:375 mg/kg JMCMAR 9,911,66

SAFETY PROFILE: Poison by unspecified route. Moderately toxic by intraperitoneal route. When heated to decomposition it emits very toxic fumes of NO_x, PO_x, and SO_x. See also ESTERS.

AMD000 CAS:20537-88-6 HR: 3
AMINOPROPYL AMINOETHYLTHIOPHOSPHATE
mf: $C_5H_{15}N_2O_3PS$ mw: 214.25

PROP: Solid. Mp: 160–161°.

SYNS: AMINOFOSTINE □ 2-((3-AMINOPROPYL)AMINO)-ETHANETHIOL, DIHYDROGEN PHOSPHATE ESTER (9CI) □ S-Ω-(3-AMINOPROPYLAMINO)ETHYL DIHYDROGEN PHOSPHOROTHIOATE □ S-(2-(3-AMINOPROPYLAMINO)ETHYL) PHOSPHOROTHIOATE □ S,2-(3-AMINOPROPYLAMINO)ETHYL-PHOSPHOROTHIOIC ACID □ 2-(3-AMINOPROPYLAMINO)ETHYL THIOPHOSPHATE □ APAETP □ AU-95722 □ ETHIOFOS □ GAMMAPHOS □ NSC-296961 □ SAPEP □ WR 2721 □ YM-08310

TOXICITY DATA WITH REFERENCE
cyt-mus-ipr 300 mg/kg CUSCAM 54,1080,85
ivn-rat TDLo:1300 mg/kg (female 17-22D post):REP OYYAA2 27,875,84
ipr-rat TDLo:300 mg/kg (female 9D post):TER RAREAE 107,49,86
ipr-rat LD50:418 mg/kg RADOA8 16,249,76
ims-rat LD50:396 mg/kg RADOA8 16,249,76
orl-mus LD50:842 mg/kg RADOA8 20,746,80
ipr-mus LD50:321 mg/kg NCISP* JAN86
ivn-mus LD50:557 mg/kg NTIS** PB81-199580
ims-mus LD50:514 mg/kg RADOA8 16,249,76
ivn-dog LDLo:279 mg/kg TOPADD 13,58,85
ipr-gpg LD50:407 mg/kg RADOA8 16,249,76

SAFETY PROFILE: Poison by intravenous, intramuscular, and intraperitoneal routes. Moderately toxic by ingestion. An experimental teratogen. Other experimental reproductive effects. Mutation data reported. See also ESTERS and PHOSPHATES. When heated to decomposition it emits very toxic fumes of SO_x, PO_x, and NO_x.

AMD250 CAS:56643-49-3 HR: 2
S-3-(ω-AMINOPROPYLAMINO)-2-HYDROXYPROPYL DIHYDROGENPHOSPHOROTHIOATE
mf: $C_6H_{17}N_2O_4PS$ mw: 244.28

TOXICITY DATA WITH REFERENCE
orl-mus LD50:1200 mg/kg JMCMAR 18,803,75

A

ipr-mus LD50:875 mg/kg JMCMAR 18,803,75

SAFETY PROFILE: Moderately toxic by ingestion and intraperitoneal routes. See also PHOSPHATES. When heated to decomposition it emits very toxic fumes of PO_x, NO_x, and SO_x.

AMD500 CAS:1945-32-0 HR: 3
N-(3-AMINOPROPYL)-1,4-BUTANEDIAMINE, PHOSPHATE
mf: $C_7H_{19}N_3 \cdot 7H_3O_4P$ mw: 831.29

SYNS: SPD PHOSPHATE □ SPERMIDINE PHOSPHATE

TOXICITY DATA WITH REFERENCE
ipr-mus LD50:468 mg/kg LIFSAK 23,2137,78
ivn-mus LD50:92 mg/kg LIFSAK 23,2137,78

SAFETY PROFILE: Poison by intravenous route. Moderately toxic by intraperitoneal route. See also PHOSPHATES. When heated to decomposition it emits very toxic fumes of PO_x and NO_x.

AME000 CAS:3179-76-8 HR: 3
(3-AMINOPROPYL)DIETHOXYMETHYLSILANE
mf: $C_8H_{21}NO_2Si$ mw: 191.39

TOXICITY DATA WITH REFERENCE
skn-rbt 10 mg/24H open MLD AIHAAP 23,95,62
orl-rat LD50:4760 mg/kg AIHAAP 23,95,62
ipr-mus LD50:40 mg/kg RCRVAB 38(12),975,69
skn-rbt LD50:2520 mg/kg AIHAAP 23,95,62

CONSENSUS REPORTS: Reported in EPA TSCA Inventory.

SAFETY PROFILE: Poison by intraperitoneal route. Moderately toxic by ingestion and skin contact. A skin irritant. When heated to decomposition it emits toxic fumes of NO_x.

AME500 CAS:299-26-3 HR: 3
3-(2-AMINOPROPYL)INDOLE
mf: $C_{11}H_{14}N_2$ mw: 174.27

SYNS: INDOPAN □ α METHYL-β-INDOLAETHYLAMINE (GERMAN) □ α-METHYL-β-INDOLEETHYLAMINE □ α-METHYLTRYPTAMINE

TOXICITY DATA WITH REFERENCE
ipr-rat TDLo:10 mg/kg (female 4-5D post):REP FATOAO 29,224,66
ipr-rat TDLo:20 mg/kg (female 4D post):TER FATOAO 29,224,66
orl-hmn TDLo:384 μg/kg:PSY JNMDAN 131,428,60
orl-rat LD50:22 mg/kg TXAPA9 4,547,62
scu-rat LD50:50 mg/kg FATOAO 29,224,66
ivn-rat LD50:75 mg/kg FATOAO 29,224,66
ipr-mus LD50:20 mg/kg PSYPAG 16,385,70
scu-mus LDLo:500 mg/kg JPMRAB 3,235,29
ivn-mus LDLo:120 mg/kg JPMRAB 3,235,29
scu-rbt LDLo:500 mg/kg JPMRAB 3,235,29
ivn-rbt LDLo:90 mg/kg JPMRAB 3,235,29

SAFETY PROFILE: Poison by ingestion and intraperitoneal routes. Moderately toxic by subcutaneous route. Human psychotropic effects by ingestion. An experimental teratogen. Other experimental reproductive effects. When heated to decomposition it emits toxic fumes of NO_x.

AME750 CAS:18237-15-5 HR: 3
3-(γ-AMINOPROPYL)-INDOLEHYDROCHLORIDE
mf: $C_{11}H_{14}N_2 \cdot ClH$ mw: 210.73

SYNS: HOMOTRYPTAMINE HYDROCHLORIDE □ INDOLE-3-PROPYLAMINE HYDROCHLORIDE □ γ-3-INDOLYLPROPYLAMINE HYDROCHLORIDE

TOXICITY DATA WITH REFERENCE
ipr-mus LD50:235 mg/kg RPTOAN 33,180,70
ivn-mus LD50:98 mg/kg RPTOAN 33,180,70

SAFETY PROFILE: Poison by intravenous and intraperitoneal routes. When heated to decomposition it emits very toxic fumes of HCl and NO_x.

AMF250 CAS:123-00-2 HR: 3
4-AMINOPROPYLMORPHOLINE
mf: $C_7H_{16}N_2O$ mw: 144.25

PROP: Liquid. Mp: −15°, bp: 224.7°, flash p: 220°F (OC), d: 0.9872 @ 20°/20°, vap press: 0.06 mm @ 20°, vap d: 4.97.

SYNS: N-(3-AMINOPROPYL)MORFOLIN □ N-(3-AMINOPROPYL)MORPHOLINE □ MORPHOLINE, N-AMINOPROPYL-

TOXICITY DATA WITH REFERENCE
skn-rbt 10 mg/24H SEV AMIHBC 4,119,51
skn-rbt 500 mg open SEV UCDS** 3/25/70
eye-rbt 1 mg UCDS** 3/25/70
eye-rbt 250 μg/24H SEV 85JCAE-,890,86
orl-rat LD50:3560 mg/kg UCDS** 3/25/70
skn-rbt LD50:1230 mg/kg AMIHBC 4,119,51

CONSENSUS REPORTS: Reported in EPA TSCA Inventory.

SAFETY PROFILE: A corrosive material. Moderately toxic by several routes. A severe skin and eye irritant. Combustible. Can react with oxidizing materials. To fight fire, use alcohol foam, dry chemical. When heated to decomposition it emits toxic fumes of NO_x.

AMF375 CAS:3690-04-8 HR: 2
AMINOPROPYLON
mf: $C_{16}H_{22}N_4O_2$ mw: 302.42

PROP: Prisms from benzene. Mp: 181°. Very sol in water.

SYNS: AMINOPROPYLONE □ AMIPYLO □ N-ANTIPYRINYL-2-(DIMETHYLAMINO)PROPIONAMIDE □ N-(2,3-DIHYDRO-1,5-DIMETHYL-3-OXO-2-PHENYL-1H-PYRAZOL-4-YL)-2-(DIMETHYLAMINO)PROPANAMIDE □ 4-(2-(DIMETHYLAMINO)PROPIONAMIDO)ANTIPYRINE

TOXICITY DATA WITH REFERENCE
orl-mus LD50:2950 mg/kg OYYAA2 13,109,77
ipr-mus LD50:820 mg/kg OYYAA2 13,109,77
ims-mus LD50:2120 mg/kg OYYAA2 13,109,77

SAFETY PROFILE: Moderately toxic by ingestion and other routes. When heated to decomposition it emits toxic fumes of NO_x.

AMF500 **CAS:1075-61-2** **HR: 3**
m-(2-AMINOPROPYL)PHENOL
mf: $C_9H_{13}NO$ mw: 151.23

SYN: α-METHYL-m-TYRAMINE

TOXICITY DATA WITH REFERENCE
scu-mus LD50:17 mg/kg ARZNAD 15,219,65
ivn-mus LD50:82 mg/kg ARZNAD 15,219,65

SAFETY PROFILE: Poison by subcutaneous and intravenous routes. When heated to decomposition it emits toxic fumes of NO_x.

AMF750 **HR: 3**
6-AMINO-1-PROPYL-4-(p-((p-((1-PROPYLPYRIDINIUM-4-YL)AMINO)-2-AMINOPHENYL)CARBAMOYL)ANILINO)QUINOLINIUM) DIIODIDE)
mf: $C_{33}H_{37}N_7O \cdot 2I$ mw: 801.57

TOXICITY DATA WITH REFERENCE
dnd-mus:lym 710 nmol/L JMCMAR 22,134,79
ipr-mus LD10:10 mg/kg JMCMAR 22,134,79

SAFETY PROFILE: Poison by intraperitoneal route. Mutation data reported. See also IODIDES. When heated to decomposition it emits very toxic fumes of NO_x and I^-.

AMG000 **CAS:68772-13-4** **HR: 3**
6-AMINO-1-PROPYL-4-(p-((p-((1-PROPYLPYRIDINIUM-4-YL)AMINO)PHENYL)CARBAMOYL)ANILINO)QUINOLINIUM)DIBROMIDE)
mf: $C_{33}H_{36}N_6O \cdot 2Br$ mw: 692.57

TOXICITY DATA WITH REFERENCE
dnd-mus:lym 680 nmol/L JMCMAR 22,134,79
ipr-mus LD10:10 mg/kg JMCMAR 22,134,79

SAFETY PROFILE: Poison by intraperitoneal route. See also BROMIDES. Mutation data reported. When heated to decomposition it emits very toxic fumes of NO_x and Br.

AMG500 **CAS:17869-27-1** **HR: 2**
1-(3-AMINOPROPYL)-2,8,9-TRIOXA-5-AZA-1-SILABICYCLO(3.3.3) UNDECANE
mf: $C_9H_{20}O_3Si$ mw: 204.38

SYN: 3-AMINOPROPYLSILATRAN (CZECH)

TOXICITY DATA WITH REFERENCE
skn-rbt 500 mg/24H SEV 28ZPAK -,220,72
eye-rbt 20 mg/24H SEV 28ZPAK -,220,72
orl-rat LD50:5800 mg/kg 28ZPAK -,220,72

SAFETY PROFILE: Mildly toxic by ingestion. A severe skin and eye irritant. When heated to decomposition it emits smoke and acrid fumes.

AMG750 **CAS:54-62-6** **HR: 3**
AMINOPTERIDINE
mf: $C_{19}H_{20}N_8O_5$ mw: 440.47

PROP: Yellow needles, sol in sodium hydroxide soln.

SYNS: 4-AMINO-4-DEOXYPTEROYLGLUTAMATE □ 4-AMINO-PGA □ AMINOPTERIN □ 4-AMINOPTEROYLGLUTAMIC ACID □ APGA □ ENT 26,079 □ FOLIC ACID, 4-AMINO- □ NSC 739

TOXICITY DATA WITH REFERENCE
spm-mus-ipr 2 mg/kg/5D PNASA6 78,4425,78
mnt-mus-ipr 5 mg/kg/5D-C CNJGA8 21,319,79
orl-wmn TDLo:2880 μg/kg (female 56-67D post):REP AMDCA5 97,274,59
orl-wmn TDLo:580 μg/kg (female 6-8W post):TER AJOGAH 84,356,62
ims-rat TDLo:1200 μg/kg/17W-C:ETA AMUK** 38,248,62
orl-wmn TDLo:120 μg/kg:GIT AJOGAH 63,1298,52
orl-rat LDLo:2500 μg/kg JPETAB 95,303,49
ipr-rat LD50:3 mg/kg CANCAR 9,955,56
orl-mus LD50:3 mg/kg CKFRAY 28,159,79
ipr-mus LD50:1900 μg/kg JPETAB 95,303,49

CONSENSUS REPORTS: EPA Extremely Hazardous Substances List.

SAFETY PROFILE: Poison by ingestion and intraperitoneal routes. Human and experimental teratogenic data. Other experimental reproductive effects. Mutation data reported. Human systemic effects by ingestion: gastrointestinal. Questionable carcinogen with experimental tumorigenic data. When heated to decomposition it emits toxic fumes of NO_x.

AMH250 **CAS:154-42-7** **HR: 3**
2-AMINOPURINE-6-THIOL
mf: $C_5H_5N_5S$ mw: 167.21

PROP: Needles from water.

SYNS: 2-AMINO-6-MERCAPTOPURINE □ 2-AMINO-6-MP □ 2-AMINO-6-PURINETHIOL □ 2-AMINOPURINE-6(1H)-THIONE □ BW 5071 □ LANVIS □ 6-MERCAPTO-2-AMINOPURINE □ 6-MERCAPTOGUANINE □ NSC-752 □ 6H-PURINE-6-THIONE, 2-AMINO-1,7-DIHYDRO-(9CI) □ TABLOID □ TG □ ThG □ THIOGUANINE □ 6-THIOGUANINE □ TIOGUANIN □ TIOGUANINE □ WELLCOME U3B

TOXICITY DATA WITH REFERENCE
skn-hmn 29 mg/3W CTRRDO 63,619,79
cyt-hmn:fbr 73 mg/L MUREAV 4,353,67
sce-rat:oth 200 nmol/L BCPCA6 34,515,85
dnd-ham:ovr 50 μg/L PAACA3 24,295,83
oms-ham:ovr 50 μg/L PAACA3 24,295,83
cyt-ham:lng 50 μg/L MUREAV 139,149,84
sce-ham:lng 15 μg/L MUREAV 139,149,84
ipr-rat TDLo:25 mg/kg (12D preg):TER PSEBAA 116,685,64
ipr-rat TDLo:10 mg/kg (7D preg):REP JRPFA4 4,291,62
ipr-rat LD50:300 mg/kg JRPFA4 291,62
orl-mus LD50:160 mg/kg EKFMA7 9,56,80
ipr-mus LD50:54 mg/kg EKFMA7 9,56,80

SAFETY PROFILE: Poison by ingestion and intraperitoneal routes. Human mutation data reported. An experimental teratogen. Other reproductive effects. A human skin irritant. When heated to decomposition it emits very toxic fumes of SO_x and NO_x.

AMH500 CAS:58048-24-1 HR: 3
1-(6-AMINO-9H-PURIN-9-YL)-N-CYCLOPROPYL-1-DEOXY-2,3-DIHYDROXYRIBOFURANURONAMIDE DIACETATE
mf: $C_{17}H_{20}N_6O_6$ mw: 404.43

TOXICITY DATA WITH REFERENCE
orl-mus LD50:20 mg/kg JMCMAR 23,313,80
ipr-mus LD50:5 mg/kg JMCMAR 23,313,80

SAFETY PROFILE: Poison by ingestion and intraperitoneal routes. When heated to decomposition it emits toxic fumes of NO_x.

AMH750 CAS:58048-26-3 HR: 3
1-(6-AMINO-9H-PURIN-9-YL)-1-DEOXY-2,3-DIHYDROXY-N-ETHYLRIBOFURANURONAMIDE DIACETATE
mf: $C_{16}H_{20}N_6O_6$ mw: 392.42

TOXICITY DATA WITH REFERENCE
orl-mus LD50:2 mg/kg JMCMAR 23,313,80
ipr-mus LD50:2 mg/kg JMCMAR 23,313,80

SAFETY PROFILE: Poison by ingestion and intraperitoneal routes. When heated to decomposition it emits toxic fumes of NO_x.

AMI000 CAS:504-29-0 HR: 3
2-AMINOPYRIDINE
DOT: UN 2671
mf: $C_5H_6N_2$ mw: 94.13

PROP: White powder or crystals from ligroin. Mp: 58.1, bp: 210.6°. Sol in water and ether; very sol in alc; sltly sol in ligroin.

SYNS: o-AMINOPYRIDINE □ α-AMINOPYRIDINE □ AMINO-2-PYRIDINE □ α-PYRIDINAMINE □ α-PYRIDYLAMINE

TOXICITY DATA WITH REFERENCE
ihl-hmn TCLo:5 ppm/5H:CNS IMSUAI 19,317,50
orl-rat LD50:200 mg/kg 85JCAE-,838,86
ipr-mus LD50:35 mg/kg JMCMAR 8,296,65
scu-mus LD50:70 mg/kg AEPPAE 226,163,55
ivn-mus LD50:23 mg/kg APFRAD 26,345,68
orl-qal LD50:133 mg/kg AECTCV 12,355,83
orl-bwd LD50:31,600 μg/kg AECTCV 12,355,83

CONSENSUS REPORTS: Reported in EPA TSCA Inventory.

OSHA PEL: TWA 0.5 ppm
ACGIH TLV: TWA 0.5 ppm
DFG MAK: 0.5 ppm (2 mg/m³)
DOT CLASSIFICATION: 6.1; *Label:* Poison

SAFETY PROFILE: Poison by ingestion, inhalation, subcutaneous, intravenous, and intraperitoneal routes. Toxic effects resemble strychnine poisoning. Human systemic effects by inhalation: somnolence, convulsions, and antipsychotic effects. Human central nervous system effects by inhalation. When heated to decomposition it emits highly toxic fumes of NO_x.

For occupational chemical analysis use NIOSH: 2-Aminopyridine S158.

AMI250 CAS:462-08-8 HR: 3
3-AMINOPYRIDINE
DOT: UN 2671
mf: $C_5H_6N_2$ mw: 94.13

PROP: Leaflets or crystals from benzene or ligroin. Mp: 64°; bp: 251°. Very sol in water, alc, ether; insol in ligroin.

SYNS: m-AMINOPYRIDINE (DOT) □ AMINO-3-PYRIDINE □ 3-PYRIDINAMINE □ 3-PYRIDYLAMINE

TOXICITY DATA WITH REFERENCE
ipr-mus LD50:28 mg/kg JMCMAR 8,296,65
scu-mus LD50:30 mg/kg AEPPAE 226,163,55
ivn-mus LD50:24 mg/kg APFRAD 26,345,68
orl-qal LD50:178 mg/kg AECTCV 12,355,83
orl-bwd LD50:13,300 μg/kg AECTCV 12,355,83

CONSENSUS REPORTS: Reported in EPA TSCA Inventory.

DOT CLASSIFICATION: 6.1; *Label:* Poison

SAFETY PROFILE: Poison by ingestion, intraperitoneal, subcutaneous, and intravenous routes. When heated to decomposition it emits toxic fumes of NO_x.

AMI500 CAS:504-24-5 HR: 3
4-AMINOPYRIDINE
DOT: UN 2671
mf: $C_5H_6N_2$ mw: 94.13

PROP: Needles or crystals from benzene. Mp: 158°; sol in water; sltly sol in benzene and ether.

SYNS: AMINO-4-PYRIDINE □ γ-AMINOPYRIDINE □ p-AMINOPYRIDINE □ 4-AP □ AVITROL □ 4-PYRIDINAMINE □ 4-PYRIDYLAMINE □ RCRA WASTE NUMBER P008 □ VMI 10-3

TOXICITY DATA WITH REFERENCE
orl-man LDLo:590 μg/kg CTOXAO 16,487,80
orl-rat LD50:21 mg/kg JTCEEM 6(3),175,86
ipr-rat LD50:6500 μg/kg TXAPA9 26,532,73
orl-mus LDLo:42 mg/kg AECTCV 14,111,85
ipr-mus LD50:10 mg/kg JMCMAR 8,296,65
scu-mus LD50:5 mg/kg APFRAD 26,345,68
orl-pgn LD50:7500 μg/kg ASTTA8 (680),157,79
orl-qal LD50:7650 μg/kg ASTTA8 (680),157,79
orl-dck LD50:4200 mg/kg TXAPA9 21,315,72

CONSENSUS REPORTS: Reported in EPA TSCA Inventory. EPA Extremely Hazardous Substances List.

DOT CLASSIFICATION: 6.1; *Label:* Poison

SAFETY PROFILE: Poison by ingestion, subcutaneous, intravenous and intraperitoneal routes. Human systemic effects by ingestion: hallucinations and distorted perceptions, dyspnea, nausea or vomiting. When heated to decomposition it emits toxic fumes of NO_x.

AMI750 CAS:73074-20-1 HR: 3
3-AMINOPYRIDINE HYDROCHLORIDE
mf: $C_5H_6N_2 \cdot ClH$ mw: 130.59

SYN: 3-PYRIDINAMINE HYDROCHLORIDE

TOXICITY DATA WITH REFERENCE
orl-rat LDLo:79 mg/kg 34ZIAG -,93,69
orl-dog LDLo:23 mg/kg 34ZIAG -,93,69
skn-rbt LDLo:327 mg/kg 34ZIAG -,93,69

SAFETY PROFILE: Poison by ingestion and skin contact. When heated to decomposition it emits very toxic fumes of HCl and NO_x.

AMJ000 CAS:1003-40-3 HR: 3
4-AMINOPYRIDINE HYDROCHLORIDE
mf: $C_5H_6N_2$•ClH mw: 130.59

SYN: 4-PYRIDINAMIDE HYDROCHLORIDE

TOXICITY DATA WITH REFERENCE
orl-rat LDLo:28 mg/kg 34ZIAG -,93,69
scu-rat LD50:10,130 μg/kg EKMMA8 18,98,79
scu-mus LD50:11,900 μg/kg EKMMA8 18,98,79
orl-dog LDLo:4 mg/kg 34ZIAG -,93,69
skn-rbt LDLo:327 mg/kg 34ZIAG -,93,69
orl-pig LDLo:18 mg/kg 34ZIAG -,93,69

SAFETY PROFILE: Poison by ingestion, skin contact, and subcutaneous routes. An eye irritant. When heated to decomposition it emits very toxic fumes of HCl and NO_x.

AMJ250 CAS:3535-75-9 HR: 3
4-AMINOPYRIDINE-1-OXIDE
mf: $C_5H_6N_2O$ mw: 110.13

SYNS: 4-AMINO-PYRIDINEN-OXIDE □ PHILLIPS 1863

TOXICITY DATA WITH REFERENCE
orl-rat LD50:75 mg/kg TXAPA9 21,315,72
orl-bwd LD50:85 mg/kg TXAPA9 21,315,72

SAFETY PROFILE: Poison by ingestion. When heated to decomposition it emits toxic fumes of NO_x.

AMJ500 CAS:30194-63-9 HR: 3
4-AMINO-N-(2-(4-(2-PYRIDINYL)-1-PIPERAZINYL)ETHYL)BENZAMIDE
mf: $C_{18}H_{23}N_5O$ mw: 325.46

SYN: S 1688

TOXICITY DATA WITH REFERENCE
orl-mus LD50:441 mg/kg ARZNAD 24,1970,74
ipr-mus LD50:250 mg/kg ARZNAD 24,1964,74
ivn-mus LD50:162 mg/kg ARZNAD 24,1970,74

SAFETY PROFILE: Poison by intravenous and intraperitoneal routes. Moderately toxic by ingestion. When heated to decomposition it emits toxic fumes of NO_x.

AMJ625 CAS:70145-80-1 HR: 3
2-AMINO-5-(4-PYRIDYL)-1,3,4-THIADIAZOLEHYDROCHLORIDE
mf: $C_7H_6N_4S$•ClH mw: 214.69

TOXICITY DATA WITH REFERENCE
orl-mus LD50:500 mg/kg JMCMAR 8,676,65
scu-mus LD50:200 mg/kg JMCMAR 8,676,65

SAFETY PROFILE: Poison by subcutaneous route. Moderately toxic by ingestion. When heated to decomposition it emits very toxic fumes of NO_x, SO_x, and HCl.

AMJ750 CAS:59985-27-2 HR: 2
2-(2-AMINO-4-PYRIMIDINYLVINYL)QUINOXALINE-N,N′-DIOXIDE
mf: $C_{14}H_{11}N_5O_2$ mw: 281.30

SYN: 2-AMINO-4-((2-QUINOXALINYL-N,N′-DIOXIDE)VINYL)PYRIMIDINES

TOXICITY DATA WITH REFERENCE
ipr-rat LDLo:500 mg/kg RVFTBB 7,117,76
orl-mus LDLo:4000 mg/kg RVFTBB 7,117,76
ipr-mus LDLo:1000 mg/kg RVFTBB 7,117,76
orl-rbt LDLo:2000 mg/kg RVFTBB 7,117,76
orl-gpg LDLo:1000 mg/kg RVFTBB 7,117,76

SAFETY PROFILE: Moderately toxic by ingestion and intraperitoneal routes. When heated to decomposition it emits toxic fumes of NO_x.

AMK250 CAS:8015-18-7 HR: 2
AMINOPYRINE-BARBITAL
mf: $C_{13}H_{17}N_3O$•$C_8H_{12}N_2O_3$ mw: 415.55

PROP: Silky needles. Mp: 113–115°. Freely sol in water.

SYNS: BARBIMON □ COTALMON □ GRELAN □ MATANOL □ PFETFFER'S SUBSTANCE □ PYRABITAL □ PYRAMON □ SEDALON □ VERAMID □ VERAMON

TOXICITY DATA WITH REFERENCE
scu-mus TDLo:900 mg/kg (9-11D preg):TER TXCYAC 29,281,84
ipr-rat TDLo:400 mg/kg (1D male):REP KSRNAM 14,723,80
ipr-rat LD50:476 mg/kg OYYAA2 16,229,78
orl-mus LD50:1466 mg/kg OYYAA2 8,453,74
ipr-mus LD50:450 mg/kg OYYAA2 16,229,78
orl-mky LDLo:1 g/kg HBAMAK 4,1289,35
orl-rbt LDLo:1200 mg/kg HBAMAK 4,1289,35

SAFETY PROFILE: Moderately toxic by ingestion and other routes. An experimental teratogen. Other experimental reproductive effects. When heated to decomposition it emits toxic fumes of NO_x.

AMK500 CAS:68-89-3 HR: 3
AMINOPYRINE SODIUM SULFONATE
mf: $C_{13}H_{17}N_3O_4S$•Na mw: 334.38

PROP: Minute crystals. Sol in water.

SYNS: (ANTIPYRINYLMETHYLAMINO)METHANESULFONIC ACID SODIUM SALT □ METHYLAMINOANTIPYRINE SODIUM METHANESULFONATE □ 4-METHYLAMINO-1,5-DIMETHYL-2-PHENYL-3-PYRAZOLONE SODIUM METHANESULFONATE □ METHYLAMINOPHENYLDIMETHYLPYRAZOLONE METHANESULFONATE SODIUM □ 1-PHENYL-2,3-DIMETHYL-5-PYRAZOLONE-4-METHYLAMINOMETHANESULFONATESODIUM □ 1-PHENYL-2,3-DIMETHYLPYRAZOLONE-(5)-4-METHYLAMINOMETHANESULFONICACID SODIUM □ PHENYL DIMETHYL PYRAZOLON METHYL AMINOMETHANE SODIUM SULFONATE □ 4-SODIUM METHANESULFONATE METHYLAMINE-ANTIPYRINE □ SODIUM METHYLAMINOANTIPYRINE METHANESULFONATE □ SODIUM-4-METHYLAMINO-1,5-DIMETHYL-2-PHENYL-3-PYRAZOLONE 4-METHANESULFONATE □ SODIUM NORAMIDOPYRINE METH-

ANESULFONATE □ SODIUM-1-PHENYL-2,3-DIMETHYL-4-METHYLAMINOPYRAZOLON-N-METHANESULFONATE □ SODIUM-1-PHENYL-2,3-DIMETHYL-5-PYRAZOLONE-4-METHYLAMINO METHANESULFONATE □ SODIUM PHENYLDIMETHYLPYRAZOLONMETHYLAMINOMETHANE SULFONATE

TOXICITY DATA WITH REFERENCE
mma-sat: 1 mg/plate AMONDS 3,253,80
cyt-hmn: lym 250 mg/L SOGEBZ 11,528,75
ipr-mus TDLo: 1 g/kg (female 9D post): REP JMTHBU 53,550,73
ipr-mus TDLo: 750 mg/kg (female 9D post): TER JMTHBU 53,550,73
orl-mus TDLo: 536 mg/kg/78W-C: NEO JJIND8 71,1295,83
orl-rat LD50: 3 g/kg ARZNAD 21,719,71
scu-rat LD50: 2117 mg/kg ARZNAD 24,600,74
ivn-rat LD50: 2182 mg/kg ARZNAD 26,703,76
ims-rat LD50: 1625 mg/kg RPTOAN 51,183,88
orl-mus LD50: 2891 mg/kg ARZNAD 24,600,74
ipr-mus LD50: 250 mg/kg AIPTAK 107,322,56
scu-mus LD50: 69 mg/kg RPTOAN 31,53,68
orl-rbt LD50: 2150 mg/kg GTPZAB 23(12),47,79

SAFETY PROFILE: Poison by subcutaneous route. Moderately toxic by several other routes. An experimental teratogen. Other experimental reproductive effects. Human mutation data reported. Questionable carcinogen with experimental neoplastigenic data. See also SULFONATES. When heated to decomposition it emits very toxic fumes of NO_x, Na_2O, and SO_x.

AMK725 CAS:580-17-6 HR: 3
3-AMINOQUINOLINE
mf: $C_9H_8N_2$ mw: 144.19

SYNS: 3-QUINOLINEAMINE □ QUINOLINE, 3-AMINO-

TOXICITY DATA WITH REFERENCE
mma-sat 5 µmol/plate MUREAV 187,191,87
ipr-mus LD50: 150 mg/kg FATOAO 41,708,78
ivn-mus LD50: 180 mg/kg CSLNX* NX#03890

CONSENSUS REPORTS: Reported in EPA TSCA Inventory.

SAFETY PROFILE: Poison by intraperitoneal and intravenous route. Mutation data reported. When heated to decomposition it emits toxic vapors of NO_x.

AML600 CAS:634-60-6 HR: 3
2-AMINORESORCINOL HYDROCHLORIDE
mf: $C_6H_7NO_2 \cdot ClH$ mw: 161.60

SYN: RESORCINOL, 2-AMINO-, HYDROCHLORIDE

TOXICITY DATA WITH REFERENCE
orl-rat LDLo: 500 mg/kg JPETAB 90,260,47
ipr-rat LD50: 30 mg/kg PHBUA9 3,337,55

CONSENSUS REPORTS: Reported in EPA TSCA Inventory.

SAFETY PROFILE: Poison by intraperitoneal route. Moderately toxic by ingestion. When heated to decomposition it emits toxic vapors of NO_x.

AMM000 CAS:3131-60-0 HR: 1
5-AMINO-2-β-d-RIBOFURANOSYL-as-TRIAZIN-3(2H)-ONE
mf: $C_8H_{12}N_4O_5$ mw: 244.24

SYN: 6-AZACYTIDINE

TOXICITY DATA WITH REFERENCE
sln-dmg-par 21 mmol/L BCPCA6 15,299,66
ivn-mus TDLo: 200 mg/kg (5D preg): REP BCPCA6 14,1549,65
ipr-mus TDLo: 2200 mg/kg (MGN): TER BCPCA6 14,1549,65
ipr-rat LD50: 9200 mg/kg RPTOAN 50,50,87
ipr-mus LD50: 14 g/kg BCPCA6 14,1517,65

SAFETY PROFILE: Mildly toxic by intraperitoneal route. An experimental teratogen. Other experimental reproductive effects. Mutation data reported. When heated to decomposition it emits toxic fumes of NO_x.

AMM125 CAS:117-55-5 HR: 1
AMINO-S ACID
mf: $C_{10}H_9NO_6S_2$ mw: 303.32

TOXICITY DATA WITH REFERENCE
orl-rat LD50: 56 g/kg GISAAA 45(3),73,80
orl-mus LD50: 56 g/kg GISAAA 45(3),73,80
orl-rbt LD50: 24 g/kg GISAAA 45(3),73,80
orl-gpg LD50: 24 g/kg GISAAA 45(3),73,80

SAFETY PROFILE: When heated to decomposition it emits toxic fumes of SO_x and NO_x. See also SULFONATES.

AMM250 CAS:65-49-6 HR: 2
4-AMINOSALICYLIC ACID
mf: $C_7H_7NO_3$ mw: 153.14

PROP: Needles, plates, or crystals from EtOH/Et_2O. Mp: 147°. Sol in dil acid or base. Very sol in water and alc; sltly sol in ether.

SYNS: 4-AMINO-2-HYDROXYBENZOIC ACID □ AMINOPAR □ AMINOSALICYLIC ACID □ p-AMINOSALICYLIC ACID □ AMINOX □ APACIL □ APAS □ DEAPASIL □ ENTEPAS □ GABBROPAS □ HELLIPIDYL □ 2-HYDROXY-4-AMINOBENZOIC ACID □ 3-HYDROXY-4-CARBOXYANILINE □ KYSELINA-p-AMINOSALICYLOVA (CZECH) □ NSC 2083 □ OSACYL □ PAMACYL □ PAMISYL □ PARAMYCIN □ PARA-PAS □ PARASAL □ PARASALICIL □ PARASALINDON □ PAS □ PASA □ PASALON □ PASARA □ PAS-C □ PASCORBIC □ PASEM □ PASK □ PASMED □ PASNODIA □ PASOLAC □ PROPASA □ REZI-PAS □ SANIPRIOL-4

TOXICITY DATA WITH REFERENCE
eye-rbt 100 mg/24H MOD 85JCAE-,659,86
cyt-mus: mmr 2 mmol/L/24H-C JTSCDR 5,141,80
eye-rbt 100 mg/24H MOD 28ZPAK -,106,72
cyt-mus-orl 50 mg/kg NULSAK 22,96,79
orl-mus LD50: 4 g/kg JPPMAB 2,764,50
ipr-mus LD50: 4250 mg/kg ZENBAX 6B,183,51
scu-mus LD50: 4 g/kg JPPMAB 2,764,50
ivn-mus LD50: 3898 mg/kg ANTBAL 18,249,73
orl-rbt LD50: 3650 mg/kg FEPRA7 10,289,51

CONSENSUS REPORTS: Reported in EPA TSCA Inventory. EPA Genetic Toxicology Program.

SAFETY PROFILE: Moderately toxic ingestion and other routes. An eye irritant. Mutation data reported. When heated to decomposition it emits toxic fumes of NO_x.

AMM500 **CAS:89-57-6** **HR: 3**
5-AMINOSALICYLIC ACID
mf: $C_7H_7NO_3$ mw: 153.14

PROP: Needles from $NaHSO_3$. Mp: decomp @ 260–283°. Sol in HCl and CS_2; sltly sol in hot water; insol in alc.

SYNS: 5-AMINO-2-HYDROXYBENZOIC ACID □ m-AMINOSALICYLIC ACID □ p-AMINOSALICYLSAEURE (GERMAN)

TOXICITY DATA WITH REFERENCE
orl-man TDLo: 51 mg/kg/5D-I: GIT,SKN,CNS LANCAO 1,917,84
orl-wmn TDLo: 8 mg/kg: GIT,SKN,CNS LANCAO 1,917,84
orl-mus LD50: 5 g/kg ZENBAX 6B,183,51
ipr-mus LDLo: 313 mg/kg CBCCT* 2,58,50

CONSENSUS REPORTS: Reported in EPA TSCA Inventory.

SAFETY PROFILE: Poison by intraperitoneal route. Moderately toxic by ingestion. Human systemic effects by ingestion: hypermotility, diarrhea, dermatitis, increased body temperature. When heated to decomposition it emits toxic fumes of NO_x.

AMM750 **CAS:551-36-0** **HR: 3**
p-AMINOSALICYLIC ACID, 2-(DIETHYLAMINO)ETHYL ESTER, HYDROCHLORIDE
mf: $C_{12}H_{18}N_2O_3$•ClH mw: 274.78

PROP: Prisms from EtOH. Mp: 154°.

SYNS: 4-AMINO-2-HYDROXYBENZOIC ACID, 2-(DIETHYLAMINO) ETHYL ESTER, HYDROCHLORIDE (9CI) □ 4-AMINOSALICYLIC ACID-2-(DIETHYLAMINO)ETHYL ESTER HYDROCHLORIDE □ p-AMINOSALICYLSAEUREDIAETHYLAMINOAETHYLESTER-CHLORHYDRAT (GERMAN) □ C 4201 □ HCl SALZ des p-AMINO-SALICYLSAEURE-DIAETHYLAMINOAETHYLESTER (GERMAN) □ SALICYLIC ACID, 4-AMINO-, 2-(DIETHYLAMINO)ETHYL ESTER, HYDROCHLORIDE □ VERBINDUNG S 557 HCl □ WIN 2022

TOXICITY DATA WITH REFERENCE
orl-rat LD50: 130 mg/kg ZENBAX 6B,183,51
ipr-rat LDLo: 190 mg/kg ARZNAD 1,154,51
ivn-rat LDLo: 36 mg/kg ARZNAD 1,154,51
orl-mus LD50: 135 mg/kg ZENBAX 6B,183,51
ipr-mus LD50: 115 mg/kg ARZNAD 1,154,51
scu-mus LD50: 282 mg/kg ARZNAD 8,708,58
ivn-mus LD50: 37 mg/kg JPETAB 104,40,52

SAFETY PROFILE: Poison by ingestion, subcutaneous, intravenous, and intraperitoneal routes. When heated to decomposition it emits very toxic fumes of NO_x and HCl. See also ESTERS.

AMN000 **CAS:78280-31-6** **HR: 3**
p-AMINOSALICYLIC ACID, 2-(DIMETHYLAMINO)ETHYL ESTER HYDROCHLORIDE
mf: $C_{11}H_{16}N_2O_3$•ClH mw: 260.75

SYNS: 4-AMINO-2-HYDROXYBENZOIC ACID, 2-(DIMETHYLAMINO) ETHYL ESTER, HYDROCHLORIDE □ C 4200 □ 2-(DIMETHYLAMINO) ETHYL-p-AMINOSALICYLATE □ HCl SALZ des p-AMINO-SALICYLSAEURE-DIMETHYLAMINOAETHYL-ESTER (GERMAN)

TOXICITY DATA WITH REFERENCE
ipr-rat LDLo: 290 mg/kg ARZNAD 1,154,51
ivn-rat LDLo: 63 mg/kg ARZNAD 1,154,51
scu-mus LD50: 560 mg/kg ARZNAD 8,708,58
ivn-mus LD50: 67 mg/kg JPETAB 123,269,58

SAFETY PROFILE: Poison by intraperitoneal and intravenous routes. Moderately toxic by subcutaneous route. See also ESTERS. When heated to decomposition it emits toxic fumes of NO_x and HCl.

AMN250 **CAS:6946-29-8** **HR: 3**
p-AMINOSALICYLIC ACID HYDRAZIDE
mf: $C_7H_9N_3O_2$ mw: 167.19

TOXICITY DATA WITH REFERENCE
orl-rbt LDLo: 250 mg/kg CLDND*
unk-dog LDLo: 30 mg/kg CLDND*

SAFETY PROFILE: Poison by ingestion and other unspecified routes. When heated to decomposition it emits toxic fumes of NO_x.

AMN300 **CAS:7722-06-7** **HR: 3**
4-AMINO-1,2,5-SELENADIAZOLE-3-CARBOXAMIDE
mf: $C_3H_4N_4OSe$ mw: 191.07

SYNS: NSC 84963 □ 1,2,5-SELENADIAZOLE-3-CARBOXAMIDE, 4-AMINO-

TOXICITY DATA WITH REFERENCE
ipr-mus LDLo: 4 mg/kg AACHAX-,551,66

OSHA PEL: TWA 0.2 mg(Se)/m^3
ACGIH TLV: TWA 0.2 mg(Se)/m^3

SAFETY PROFILE: Poison by intraperitoneal route. When heated to decomposition it emits toxic fumes of NO_x and Se.

AMN500 **CAS:15267-04-6** **HR: 3**
2-AMINOSELENOAZOLINE
mf: $C_3H_6N_2Se$ mw: 149.07

SYNS: 2-AMINOSELENOAZOLIN (GERMAN) □ 2-ASe

TOXICITY DATA WITH REFERENCE
ipr-mus LD50: 160 mg/kg STRAAA 151,78,76
scu-mus LD50: 177 mg/kg STRAAA 151,78,76
ivn-mus LD50: 151 mg/kg STRAAA 151,78,76

CONSENSUS REPORTS: Selenium compounds are on the Community Right-To-Know List.

OSHA PEL: TWA 0.2 mg(Se)/m^3
ACGIH TLV: TWA 0.2 mg(Se)/m^3
DFG MAK: 0.1 mg(Se)/m^3

SAFETY PROFILE: Poison by intraperitoneal, subcutaneous, and intravenous routes. When heated to decomposition it emits very toxic fumes of NO_x and Se. See also SELENIUM COMPOUNDS.

AMO000 CAS:4309-66-4 HR: 2
trans-4-AMINOSTILBENE
mf: $C_{14}H_{13}N$ mw: 195.28

PROP: Light-yellow needles from EtOH. Mp: 151°.

SYNS: 4-(2-PHENYLETHENYL)BENZENAMINE,(E) □ trans-4-STILBENE □ trans-4-N-STILBENAMINE

TOXICITY DATA WITH REFERENCE
mma-sat 10 μg/plate PNASA6 70,2281,73
orl-rat TDLo:200 mg/kg/13W-C:CAR CNREA8 24,128,64
scu-rat TDLo:26 mg/kg/4W-I:CAR,REP CNREA8 24,128,64
scu-rat TD:63 mg/kg/6W-I:ETA PTRMAD 241,147,48

SAFETY PROFILE: Questionable carcinogen with experimental carcinogenic and tumorigenic data. Experimental reproductive effects. Mutation data reported. When heated to decomposition it emits toxic fumes of NO_x.

AMO250 CAS:3432-10-8 HR: 1
2-(p-AMINOSTYRYL)-6-(p-ACETYLAMINOBENZOYLAMINO)QUINOLINE METHOACETATE
mf: $C_{27}H_{25}N_4O_2 \cdot C_2H_3O_2$ mw: 496.61

SYN: STYRYL 430

TOXICITY DATA WITH REFERENCE
mmo-esc 1 pph CRSBAW 142,453,48
scu-mus TDLo:268 mg/kg:ETA JPBAA7 42,155,36

SAFETY PROFILE: Questionable carcinogen with experimental tumorigenic data. Mutation data reported. When heated to decomposition it emits toxic fumes of NO_x.

AMO750 CAS:74039-02-4 HR: 1
2,(4'-AMINO-3'-SULFO-1,1'-BIPHENYL-4-YL)-2H-NAPHTHO(1,2-4)TRIAZOLE-6,8-DISULFONIC ACID, TRIPOTASSIUM SALT
mf: $C_{22}H_{13}N_4O_9S_3 \cdot 3K$ mw: 690.87

SYN: 2-(4'-AMINOXENYL)NAFTO-α,β-TRIAZOL-6,8,3'TRISULFONAN DRASELNY (CZECH)

TOXICITY DATA WITH REFERENCE
eye-rbt 500 mg/24H SEV 28ZPAK -,196,72
orl-rat LD50:8330 mg/kg 28ZPAK -,196,72

SAFETY PROFILE: Mildly toxic by ingestion. A severe eye irritant. See also SULFONATES. When heated to decomposition it emits very toxic fumes of NO_x, SO_x, and K_2O.

AMP000 CAS:29727-70-6 HR: 1
6-AMINO-5-SULFOMETHYL-2-NAPHTHALENESULFONIC ACID
mf: $C_{11}H_{11}NO_6S_2$ mw: 317.35

SYN: KYSELINA 1-SULFOMETHYL-2-NAFTYLAMIN-6-SULFONOVA (CZECH)

TOXICITY DATA WITH REFERENCE
eye-rbt 500 mg/24H SEV 28ZPAK -,190,72
orl-rat LD50:9200 mg/kg 28ZPAK -,190,72

SAFETY PROFILE: Mildly toxic by ingestion. A severe eye irritant. See also SULFONATES. When heated to decomposition it emits very toxic fumes of NO_x and SO_x.

AMP500 CAS:16760-18-2 HR: 3
3,3'-(2-AMINOTEREPHTHALOYLBIS(IMINO(3-AMINO-p-PHENYLENE)CARBONYLIMINO))BIS(1-ETHYLPYRIDINIUM), DI-p-TOLUENESULFONATE
mf: $C_{36}H_{36}N_9O_4 \cdot 2C_7H_7O_3S$ mw: 1001.21

TOXICITY DATA WITH REFERENCE
dnd-mus:lym 260 nmol/L JMCMAR 22,134,79
ipr-mus LD10:14 mg/kg JMCMAR 22,134,79

SAFETY PROFILE: Poison by intraperitoneal route. See also SULFONATES. Mutation data reported. When heated to decomposition it emits very toxic fumes of SO_x and NO_x.

AMP750 CAS:16802-49-6 HR: 3
3,3'-(2-AMINOTEREPHTHALOYLBIS(IMINO(3-AMINO-p-PHENYLENE)CARBONYLIMINO))BIS(1-PROPYLPYRIDINIUM), DI-p-TOLUENESULFONATE
mf: $C_{38}H_{41}N_9O_4 \cdot 2C_7H_7O_3S$ mw: 1030.28

TOXICITY DATA WITH REFERENCE
dnd-mus:lym 260 nmol/L JMCMAR 22,134,79
ipr-mus LD10:10 mg/kg JMCMAR 22,134,79

SAFETY PROFILE: Poison by intraperitoneal route. Mutation data reported. When heated to decomposition it emits very toxic fumes of NO_x and SO_x.

AMQ000 CAS:16760-14-8 HR: 3
3,3'-(2-AMINOTEREPHTHALOYBIS(IMINO-p-PHENYLENECARBONYLIMINO))BIS(1-ETHYLPYRIDINIUM), DI-p-TOLUENESULFONATE
mf: $C_{36}H_{35}N_7O_4 \cdot 2C_7H_7O_3S$ mw: 972.18

TOXICITY DATA WITH REFERENCE
dnd-mus:lym 530 nmol/L JMCMAR 22,134,79
ipr-mus LD10:14 mg/kg JMCMAR 22,134,79

SAFETY PROFILE: Poison by intraperitoneal route. Mutation data reported. When heated to decomposition it emits very toxic fumes of SO_x and NO_x.

AMQ250 CAS:16760-13-7 HR: 3
3,3'-(2-AMINOTEREPHTHALOYLBIS(IMINO-p-PHENYLENECARBONYLIMINO))BIS(1-METHYLPYRIDINIUM), DI-p-TOLUENESULFONATE
mf: $C_{34}H_{31}N_7O_4 \cdot 2C_7H_7O_3S$ mw: 944.12

TOXICITY DATA WITH REFERENCE
dnd-mus:lym 530 nmol/L JMCMAR 22,134,79
ipr-mus LD10:20 mg/kg JMCMAR 22,134,79

SAFETY PROFILE: Poison by intraperitoneal route. Mutation data reported. See also SULFONATES. When heated to decomposition it emits very toxic fumes of SO_x and NO_x.

AMQ500 **CAS:23757-42-8** **HR: 3**
4-AMINO-2,2,5,5-TETRAKIS(TRIFLUOROMETHYL)-3-IMIDAZOLINE
mf: $C_7H_3F_{12}N_3$ mw: 357.13

PROP: Solid. Mp: 159.7–160.4°.

SYNS: 5-AMINO-2,2,4,4-TETRAKIS(TRIFLUOROMETHYL)IMIDAZOLIDINE □ EXP 338

TOXICITY DATA WITH REFERENCE
ipr-rat TDLo:6500 μg/kg (female 1D pre):REP TXAPA9 18,917,71
orl-rbt TDLo:26 mg/kg (female 6-18D post):TER TXAPA9 18,917,71
orl-rat LD50:19 mg/kg 27ZQAG -,265,72
ipr-rat LD50:12 mg/kg 27ZQAG -,265,72
orl-mus LD50:262 mg/kg 27ZQAG -,265,72
ipr-mus LD50:189 mg/kg 27ZQAG -,265,72
ivn-mus LD50:231 mg/kg 27ZQAG -,265,72
orl-dog LD50:150 mg/kg 27ZQAG -,265,72
orl-gpg LD50:11 mg/kg 27ZQAG -,265,72

SAFETY PROFILE: Poison by ingestion, intraperitoneal, and intravenous routes. An experimental teratogen. Other experimental reproductive effects. When heated to decomposition it emits very toxic fumes of F^- and NO_x.

AMQ750 **CAS:6130-92-3** **HR: 3**
1-AMINO-2,2,6,6-TETRAMETHYLPIPERIDINE
mf: $C_9H_{20}N_2$ mw: 156.31

PROP: Liquid. Bp: 80–83° @ 20 mm.

TOXICITY DATA WITH REFERENCE
orl-mus LD50:261 mg/kg MDCHAG 7,312,67
ivn-mus LD50:44 mg/kg NATUAS 184,1707,59

SAFETY PROFILE: Poison by ingestion and intravenous routes. When heated to decomposition it emits toxic fumes of NO_x.

AMR000 **CAS:4418-61-5** **HR: 3**
AMINOTETRAZOLE
mf: CH_3N_5 mw: 85.09

PROP: Crystals. Mp: 206° (decomp).

SYNS: 5-AMINOTETRAZOLE □ 5-AMINO-1H-TETRAZOLE □ 1H-TETRAZOL-5-AMINE

TOXICITY DATA WITH REFERENCE
ipr-mus LD50:2500 mg/kg RPTOAN 41,249,78

CONSENSUS REPORTS: Reported in EPA TSCA Inventory.

SAFETY PROFILE: Moderately toxic by intraperitoneal route. An unstable material; explodes with KOH. When heated to decomposition it emits toxic fumes of NO_x.

AMR250 **CAS:4005-51-0** **HR: 3**
2-AMINO-1,3,4-THIADIAZOLE
mf: $C_2H_3N_3S$ mw: 101.14

PROP: Pale-yellow crystals from water or alc. Mp: 190–191°.

SYNS: AMINOTHIADIAZOLE □ ATDA □ 1,3,4-THIADIAZOL-2-AMINE

TOXICITY DATA WITH REFERENCE
ipr-rat TDLo:100 mg/kg (female 16D post):REP TJADAB 27,29A,83
orl-rat TDLo:20 mg/kg (female 10D post):TER CPHPA5 12,212,81
ipr-rat LD50:200 mg/kg TAKHAA 35,68,76
scu-rat LD50:200 mg/kg JHMJAX 130,95,72
ipr-mus LD50:6500 mg/kg TAKHAA 35,68,76

CONSENSUS REPORTS: EPA Genetic Toxicology Program.

SAFETY PROFILE: Poison by subcutaneous and intraperitoneal routes. An experimental teratogen. Other experimental reproductive effects. When heated to decomposition it emits very toxic fumes of NO_x and SO_x.

AMR500 **CAS:26861-87-0** **HR: 3**
2-AMINO-1,3,4-THIADIAZOLEHYDROCHLORIDE
mf: $C_2H_3N_3S \cdot ClH$ mw: 137.60

SYNS: 2-AMINO-1,3,4-THIADIAZOLE, MONOHYDROCHLORIDE □ ATDA HYDROCHLORIDE

TOXICITY DATA WITH REFERENCE
ipr-rat TDLo:100 mg/kg (female 11D post):TER TJADAB 21,381,80
ipr-rat TDLo:100 mg/kg (female 10D post):REP TJADAB 7,65,73
ipr-rat LDLo:100 mg/kg TJADAB 7,65,73
ipr-mus LD50:250 mg/kg NTIS** AD691-490

CONSENSUS REPORTS: Reported in EPA TSCA Inventory.

SAFETY PROFILE: Poison by intraperitoneal route. An experimental teratogen. Other experimental reproductive effects. When heated to decomposition it emits very toxic fumes of HCl, SO_x, and NO_x.

AMS000 **CAS:6630-99-5** **HR: 3**
5-AMINO-1,2,3,4-THIATRIAZOLE
mf: CH_2N_4S mw: 102.11

PROP: Crystals. Mp: 128–130°.

SAFETY PROFILE: Very unstable. Explodes weakly at 130°C. Upon decomposition it emits toxic fumes of SO_x and NO_x.

AMS250 **CAS:96-50-4** **HR: 3**
2-AMINOTHIAZOLE
mf: $C_3H_4N_2S$ mw: 100.15

SCH=CHN=CNH$_2$ (ring closure from S to C)

PROP: Light brown or yellow crystals from alc. Mp: 90°, bp: 140° @ 11 mm (decomp). Sltly sol in water, alc, ether; sol in hot alc.

SYNS: ABADOL □ ABADOLE □ AMINOTHIAZOLE □ BASEDOL □ 2-THIAZOLAMINE □ 2-THIAZOLYLAMINE □ 2-THIAZYLAMINE □ USAF EK-P-5501

TOXICITY DATA WITH REFERENCE
mma-sat 3333 µg/plate MUREAV 155,17,85
mmo-klp 1 mmol/L MUREAV 118,153,83
mma-mus:lym 1214 mg/L MUREAV 155,17,85
msc-mus:lym 557 mg/L MUREAV 155,17,85
orl-rat LD50:480 mg/kg JIHTAB 30,71,48
ivn-rat LD50:570 mg/kg AEPPAE 211,367,50
ipr-mus LD50:200 mg/kg NTIS** AD277-689
orl-cat LDLo:120 mg/kg JIHTAB 30,71,48
orl-rbt LD50:370 mg/kg JIHTAB 30,71,48
orl-gpg LDLo:120 mg/kg JIHTAB 30,71,48

CONSENSUS REPORTS: Reported in EPA TSCA Inventory.

SAFETY PROFILE: Poison by ingestion and intraperitoneal routes. Mutation data reported. Spontaneous ignition occurs at 100°. Mixtures with nitric acid or nitric acid + sulfuric acid explode on heating. Incompatible with HNO_3 and H_2SO_4. When heated to decomposition it emits very toxic SO_x and NO_x fumes.

AMS675 CAS:104-96-1 HR: 2
4-AMINOTHIOANISOLE
mf: C_7H_9NS mw: 139.23

SYNS: p-AMINOPHENYL METHYL SULFIDE ☐ p-AMINOTHIOANISOLE ☐ ANILINE, p-(METHYLTHIO)- ☐ BENZENAMINE, 4-(METHYLTHIO)-(9CI) ☐ 4-(METHYLTHIO)ANILINE ☐ 4-(METHYLTHIO) BENZENAMINE ☐ p-THIOANISIDINE ☐ p-THIOMETHOXYANILINE

TOXICITY DATA WITH REFERENCE
orl-qal LD50:562 mg/kg AECTCV 12,355,83

CONSENSUS REPORTS: Reported in EPA TSCA Inventory.

SAFETY PROFILE: Moderately toxic by ingestion. When heated to decomposition it emits toxic vapors of NO_x and SO_x.

AMS750 CAS:1004-40-6 HR: 3
6-AMINO-2-THIOURACIL
mf: $C_4H_5N_3OS$ mw: 143.18

SYN: URACIL, 6-AMINO-2-THIO-

TOXICITY DATA WITH REFERENCE
ipr-mus LD50:370 mg/kg ARZNAD 31,1713,81

CONSENSUS REPORTS: Reported in EPA TSCA Inventory.

SAFETY PROFILE: Poison by intraperitoneal route. When heated to decomposition it emits toxic vapors of NO_x and SO_x.

AMT000 CAS:88-62-0 HR: 2
2-AMINO-p-TOLUENESULFONIC ACID
mf: $C_7H_9NO_3S$ mw: 187.23

PROP: Needles. Sol in water.

SYN: KYSELINA 2-TOLUIDIN-4-SULFONOVA (CZECH)

TOXICITY DATA WITH REFERENCE
eye-rbt 100 mg/24H SEV 28ZPAK -,183,72
orl-rat LD50:8480 mg/kg 28ZPAK -,183,72

CONSENSUS REPORTS: Reported in EPA TSCA Inventory.

SAFETY PROFILE: Mildly toxic by ingestion. A severe eye irritant. See also SULFONATES. When heated to decomposition it emits very toxic fumes of NO_x and SO_x.

AMT250 CAS:133-78-8 HR: 2
4-AMINO-o-TOLUENESULFONIC ACID
mf: $C_7H_9NO_3S$ mw: 187.23

PROP: Needles or plates. Sol in water.

SYN: KYSELINA-3-TOLUIDIN-6-SULFONOVA (CZECH)

TOXICITY DATA WITH REFERENCE
eye-rbt 500 mg/24H SEV 28ZPAK -,183,72

SAFETY PROFILE: A severe eye irritant. See also SULFONATES. When heated to decomposition it emits very toxic fumes of NO_x and SO_x.

AMT500 CAS:139-13-9 HR: 3
AMINOTRIACETIC ACID
mf: $C_6H_9NO_6$ mw: 191.16

PROP: Prismatic crystals from water. Mp: 242° (decomp), bp: 167° @ 13 mm. Sltly sol in water.

SYNS: N,N-BIS(CARBOXYMETHYL)GLYCINE ☐ CHEL 300 ☐ COMPLEXON I ☐ GLYCINE, N,N-BIS(CARBOXYMETHYL)-(9CI) ☐ HAMPSHIRE NTA ACID ☐ KOMPLEXON I ☐ KYSELINA NITRILOTRIOCTOVA ☐ NCI-C02766 ☐ NITRILOTRIACETIC ACID ☐ NTA ☐ TITRIPLEX I ☐ TRIGLYCINE ☐ TRIGLYCOLLAMIC ACID ☐ TRILON A ☐ α-α',α''-TRIMETHYLAMINETRICARBOXYLIC ACID ☐ VERSENE NTA ACID

TOXICITY DATA WITH REFERENCE
orl-rat TDLo:430 g/kg/75W-C:CAR NCITR* NCI-CG-TR-6,77
orl-mus TDLo:832 g/kg/66W-C:NEO NCITR* NCI-CG-TR-6,77
orl-rat LD50:1100 mg/kg ACIEAY 14,94,75
orl-mus LD50:3160 mg/kg NCILB* NIH-NCI-E-C-72-3252
ipr-mus LDLo:125 mg/kg TXAPA9 23,288,72

CONSENSUS REPORTS: NTP 7th Annual Report On Carcinogens. IARC Cancer Review: Group 2B IMEMDT 48,181,90; Animal Sufficient Evidence IMEMDT 48,181,90; Human No Adequate Data IMEMDT 48,181,90. NCI Carcinogenesis Bioassay (feed); Clear Evidence: mouse, rat NCITR* NCI-CG-TR-6,77. Reported in EPA TSCA Inventory. Community Right-To-Know List.

SAFETY PROFILE: Confirmed carcinogen with experimental carcinogenic and neoplastigenic data. Poison by intraperitoneal route. Moderately toxic by ingestion. When heated to decomposition it emits toxic fumes of NO_x.

AMU000 CAS:35695-72-8 HR: 3
3-AMINO-1-TRICHLORO-2-PENTANOL
mf: $C_5H_{10}Cl_3NO$ mw: 206.51

SYN: TCA-PE

TOXICITY DATA WITH REFERENCE
scu-mus LD50:1364 mg/kg SKIZAB 28,231,72

ivn-mus LD50:284 mg/kg SKIZAB 28,231,72

SAFETY PROFILE: Poison by intravenous route. Moderately toxic by subcutaneous route. When heated to decomposition it emits very toxic fumes of $Cl^-$and NO_x.

AMU125 CAS:53516-81-7 HR: 3
((3-AMINO-2,4,6-TRICHLOROPHENYL)METHYLENE) HYDRAZIDE BENZENESULFONIC ACID
mf: $C_{13}H_{10}Cl_3N_3O_2S$ mw: 378.67

SYNS: DENVER RESEARCH CENTER No. DRC-4575 □ DRC-4575

TOXICITY DATA WITH REFERENCE
orl-mus TDLo:62 mg/kg (8-13D preg):REP JTEHD6 3,407,77
orl-mus LD50:33 mg/kg JTEHD6 3,407,77

SAFETY PROFILE: Poison by ingestion. Experimental reproductive effects. When heated to decomposition it emits toxic fumes of Cl^-, SO_x, and NO_x. See also SULFONATES.

AMU500 CAS:35695-70-6 HR: 3
3-AMINO-1-TRICHLORO-2-PROPANOL
mf: $C_3H_6Cl_3NO$ mw: 178.45

SYN: TCA-PR

TOXICITY DATA WITH REFERENCE
scu-mus LD50:1556 mg/kg SKIZAB 28,231,72
ivn-mus LD50:301 mg/kg SKIZAB 28,231,72

SAFETY PROFILE: Poison by intravenous route. Moderately toxic by subcutaneous route. When heated to decomposition it emits very toxic fumes of $Cl^-$and NO_x.

AMU550 CAS:344-72-9 HR: 3
2-AMINO-4-(TRIFLUOROMETHYL)-5-THIAZOLECARBOXYLIC ACID ETHYL ESTER
mf: $C_7H_7F_3N_2O_2S$ mw: 240.22

SYN: 5-THIAZOLECARBOXYLIC ACID, 2-AMINO-4-(TRIFLUOROMETHYL)-, ETHYL ESTER

TOXICITY DATA WITH REFERENCE
ivn-mus LDLo:75 mg/kg CBCCT* 6,142,54

CONSENSUS REPORTS: Reported in EPA TSCA Inventory.

SAFETY PROFILE: Poison by intravenous route. When heated to decomposition it emits toxic vapors of NO_x and SO_x.

AMU625 CAS:3119-15-1 HR: 2
3-AMINO-2,4,6-TRIIODO-BENZOIC ACID
mf: $C_7H_4I_3NO_2$ mw: 514.82

PROP: Crystals. Mp: 196.5–197.5°.

SYN: ACIDO-3-AMINO-2,4,6-TRIIODOBENZOICO (ITALIAN)

TOXICITY DATA WITH REFERENCE
unr-rat LD50:1450 mg/kg JAPMA8 42,721,53
orl-mus LD50:600 mg/kg QJPPAL 19,483,46
ivn-mus LD50:800 mg/kg FRPSAX 18,33,63

SAFETY PROFILE: Moderately toxic by ingestion and other routes. When heated to decomposition it emits toxic fumes of I^- and NO_x.

AMU750 CAS:3115-05-7 HR: 2
N-(3-AMINO-2,4,6-TRIIODOBENZOYL)-N-(2-CARBOXYETHYL)ANILINE
mf: $C_{16}H_{13}I_3N_2O_3$ mw: 662.01

PROP: White or pale-yellow powder. Mp: 133–134.5°. Sol in dioxan, DMF.

SYNS: 3-((3-AMINO-2,4,6-TRIIODOBENZOYL)PHENYLAMINO)PROPIONIC ACID □ N-(3-AMINO-2,4,6-TRIJODBENZOYL)-N-PHENYL-β-AMINOPROPIONSAEURE (GERMAN) □ ISOBENZAMIC ACID □ ST 5066/S (GERMAN)

TOXICITY DATA WITH REFERENCE
orl-rat LD50:2800 mg/kg TXAPA9 14,232,69
ivn-rat LD50:500 mg/kg TXAPA9 14,232,69
orl-mus LD50:2870 mg/kg ARZNAD 11,384,61
ivn-mus LD50:530 mg/kg JMCMAR 13,997,70

SAFETY PROFILE: Moderately toxic by ingestion and intravenous routes. When heated to decomposition it emits very toxic fumes of I^- and NO_x.

AMV375 CAS:1634-73-7 HR: 3
4-((3-AMINO-2,4,6-TRIIODOPHENYL)ETHYLAMINO)-4-OXO-BUTANOIC ACID
mf: $C_{12}H_{13}I_3N_2O_3$ mw: 613.97

SYNS: N-AETHYL-N-(2,4,6-TRIJOD-3-AMINOPHENYL)-SUCCINAMIDSAEURE (GERMAN) □ 3′-AMINO-N-ETHYL-2′,4′,6′-TRIIODOSUCCINANILIC ACID □ RG 235 □ SH 771

TOXICITY DATA WITH REFERENCE
ivn-rat LD50:370 mg/kg PHARAT 27,411,72
orl-mus LD50:650 mg/kg PHARAT 27,391,72
ivn-mus LD50:288 mg/kg PHARAT 27,411,72

SAFETY PROFILE: Poison by intravenous route. Moderately toxic by ingestion. When heated to decomposition it emits toxic fumes of I^- and NO_x.

AMV750 CAS:23217-86-9 HR: 3
2-(3-AMINO-2,4,6-TRIIODOPHENYL)VALERIC ACID
mf: $C_{11}H_{12}I_3NO_2$ mw: 570.94

TOXICITY DATA WITH REFERENCE
orl-mus LD50:2100 mg/kg JMCMAR 13,559,70
ivn-mus LD50:170 mg/kg JMCMAR 13,559,70

SAFETY PROFILE: Poison by intravenous route. Moderately toxic by ingestion. See also IODIDES. When heated to decomposition it emits very toxic fumes of I^- and NO_x.

AMV790 CAS:698-49-7 HR: 3
4-AMINOTROPOLONE
mf: $C_7H_7NO_2$ mw: 137.15

PROP: Needles from MeOH. Mp: 187–188°.

SYN: 4-AMINO-2-HYDROXY-2,4,6-CYCLOPHEPTATRIEN-1-ONE

TOXICITY DATA WITH REFERENCE
ipr-mus LD50:265 mg/kg CPBTAL 20,60,72
scu-mus LD50:400 mg/kg CPBTAL 20,60,72
ivn-mus LD50:177 mg/kg YKKZAJ 92,19,72

SAFETY PROFILE: Poison by subcutaneous, intravenous, and intraperitoneal routes. When heated to decomposition it emits toxic fumes of NO_x.

AMV800 CAS:7021-46-7 HR: 3
5-AMINOTROPOLONE
mf: $C_7H_7NO_2$ mw: 137.15

PROP: Yellow, scaly crystals. Mp: 177–177.5°.

SYN: 5-AMINO-2-HYDROXY-2,4,6-CYCLOHEPTATRIEN-1-ONE

TOXICITY DATA WITH REFERENCE
ipr-mus LD50:230 mg/kg YKKZAJ 91,550,71
scu-mus LD50:521 mg/kg YKKZAJ 91,550,71
ivn-mus LD50:175 mg/kg YKKZAJ 92,19,72

SAFETY PROFILE: Poison by intravenous and intraperitoneal routes. Moderately toxic by subcutaneous route. When heated to decomposition it emits toxic fumes of NO_x.

AMV875 HR: 1
AMINO-TS-ACID
mf: $C_{10}H_9NO_6S_2$ mw: 303.32

TOXICITY DATA WITH REFERENCE
orl-rat LD50:29 g/kg GISAAA 45(3),73,80
orl-mus LD50:29 g/kg GISAAA 45(3),73,80
orl-rbt LD50:18 g/kg GISAAA 45(3),73,80
orl-gpg LD50:42 g/kg GISAAA 45(3),73,80

SAFETY PROFILE: Low order of toxicity. When heated to decomposition it emits toxic fumes of SO_x and NO_x. See also SULFONATES.

AMW000 CAS:2432-99-7 HR: 3
11-AMINOUNDECANOIC ACID
mf: $C_{11}H_{23}NO_2$ mw: 201.35

PROP: Solid. Mp: 190–192°.

SYNS: AMINOUNDECANOIC ACID □ 11-AMINOUNDECYLIC ACID □ NCI-C50613

TOXICITY DATA WITH REFERENCE
otr-ham:emb 2500 mmol/L ENMUDM 8,515,86
sce-ham:ovr 500 mg/L EMMUEG 10(Suppl 10),1,87
orl-rat TDLo:655 g/kg/2Y-C:CAR NTPTR* NTP-TR-216,82
orl-rat TD:328 g/kg/2Y-C:NEO NTPTR* NTP-TR-216,82
orl-rat LDLo:14,700 μg/kg NTPTR* NTP-TR-216,82

CONSENSUS REPORTS: IARC Cancer Review: Group 3 IMEMDT 7,56,87; Animal Limited Evidence IMEMDT 39,239,86. NTP Carcinogenesis Bioassay (feed): Clear Evidence: mouse, rat NTPTR* NTP-TR-216,82. Reported in EPA TSCA Inventory.

SAFETY PROFILE: Poison by ingestion. Questionable carcinogen with experimental carcinogenic and neoplastigenic data. Mutation data reported. When heated to decomposition it emits toxic fumes of NO_x.

AMW250 CAS:873-83-6 HR: 2
6-AMINOURACIL
mf: $C_4H_5N_3O_2$ mw: 127.12

PROP: Needles from water. Mp: decomp. Sol in water, alkalies, NH_4OH, and acids.

TOXICITY DATA WITH REFERENCE
par-mus LDLo:2400 mg/kg CBCCT* 7,696,55

CONSENSUS REPORTS: Reported in EPA TSCA Inventory.

SAFETY PROFILE: Moderately toxic by parenteral route. When heated to decomposition it emits toxic fumes of NO_x.

AMW500 CAS:38237-74-0 HR: 3
p-AMINO VALEROPHENONE
mf: $C_{11}H_{15}NO$ mw: 177.27

TOXICITY DATA WITH REFERENCE
orl-rat LD50:84 mg/kg GEPHDP 14,465,83
orl-mus LD50:94 mg/kg GEPHDP 14,465,83
ipr-mus LD50:120 mg/kg FEPRA7 6,348,47

CONSENSUS REPORTS: Reported in EPA TSCA Inventory.

SAFETY PROFILE: Poison by ingestion and intraperitoneal route. See also AMINES. When heated to decomposition it emits toxic fumes of NO_x.

AMW750 CAS:6623-41-2 HR: 1
2-AMINO-4,5-XYLENOL
mf: $C_8H_{11}NO$ mw: 137.20

PROP: Crystals from toluene. Mp: 173–175°.

SYN: 2-AMINO-4,5-DIMETHYLPHENOL

TOXICITY DATA WITH REFERENCE
imp-mus TDLo:80 mg/kg:CAR BJCAAI 11,212,57

CONSENSUS REPORTS: Reported in EPA TSCA Inventory.

SAFETY PROFILE: Questionable carcinogen with experimental carcinogenic data. See also AMINES. When heated to decomposition it emits toxic fumes of NO_x.

AMX000 CAS:5369-84-6 HR: 3
3-AMINO-4-(2-(2,6-XYLYLOXY)ETHYL)-4H-1,2,4-TRIAZOLE
mf: $C_{12}H_{16}N_4O$ mw: 232.32

SYN: 3,5-BIS(2-FURYL)-1H-1,2,4-TRIAZOLE

TOXICITY DATA WITH REFERENCE
ipr-rat LD50:235 mg/kg JMCMAR 9,42,66
ipr-mus LD50:2000 mg/kg JMCMAR 9,22,66

SAFETY PROFILE: Poison by intraperitoneal route. When heated to decomposition it emits toxic fumes of NO_x.

AMX250 **HR: 2**
AMIODOXYL BENZOATE
mf: $C_7H_8INO_4$ mw: 297.06

PROP: White, odorless, sltly bitter, crystalline powder.

SYN: ARTHRYTIN OXOATE

SAFETY PROFILE: Moderately toxic by ingestion. It is a non-selective, systemic herbicide. Dangerous; when heated to decomposition it emits toxic fumes of NO_x and I^-.

AMX500 **HR: 3**
AMIPURIMYCIN HYDRATE
mf: $C_{30}H_{20}N_7O_8•H_2O$ mw: 624.59

PROP: Isolated from culture filtrate of *Streptomyces novoguineesis* T-36496.

TOXICITY DATA WITH REFERENCE
skn-rbt 200 ppm/10D SEV JANTAJ 30,1,77
orl-rat LD50:20 mg/kg JANTAJ 30,1,77
ivn-rat LD50:1 mg/kg JANTAJ 30,1,77
orl-mus LD50:10 mg/kg JANTAJ 30,1,77
ivn-mus LD50:1 mg/kg JANTAJ 30,1,77

SAFETY PROFILE: Poison by ingestion and intravenous routes. A severe skin irritant. When heated to decomposition it emits toxic fumes of NO_x.

AMX750 **CAS:57-43-2** **HR: 3**
AMITAL
mf: $C_{11}H_{18}N_2O_3$ mw: 226.31

PROP: Slightly bitter crystals or leaflets from water or alc. Mp: 156–158°.

SYNS: AMAL □ AMASUST □ AMOBARBITAL □ AMOBARBITONE □ AMOSPAN □ AMYBAL □ AMYLBARBITONE □ AMYLOBARBITAL □ AMYLOBARBITONE □ AMYTAL □ BARBAMIL □ BARBAMYL □ BARBAMYL ACID □ BINOCTAL □ DORLOTYN □ DORMYTAL □ 5-ETHYL-5-ISOAMYLBARBITURIC ACID □ 5-ETHYL-5-ISOAMYLMALONYL UREA □ ETHYLISOPENTYLBARBITURIC ACID □ 5-ETHYL-5-ISOPENTYLBARBITURIC ACID □ 5-ETHYL-5-(3-METHYLBUTYL)BARBITURIC ACID □ EUNOCTAL □ ISOAMYLETHYLBARBITURIC ACID □ 5-ISOAMYL-5-ETHYLBARBITURIC ACID □ ISOMYL □ ISOMYTAL □ MYLODORM □ NSC 10815 □ PENTYMAL □ PENTYMALUM □ 2,4,6(1H,3H,5H)-PYRIMIDINETRIONE, 5-ETHYL-5-(3-METHYLBUTYL)-(9CI) □ ROBARB □ SCHIWANOX □ SEDNOTIC □ SOMNAL □ STADADORM □ SUMITAL □ TALAMO

TOXICITY DATA WITH REFERENCE
orl-rat LD50:250 mg/kg ARZNAD 21,719,71
ipr-rat LD50:115 mg/kg ARZNAD 21,719,71
scu-rat LD50:190 mg/kg AEPPAE 152,341,30
orl-mus LD50:345 mg/kg JACSAT 61,96,39
ipr-mus LD50:175 mg/kg JMCMAR 10,1078,67
scu-mus LD50:212 mg/kg ARZNAD 15,688,65
orl-dog LDLo:250 mg/kg JPETAB 26,371,25
ivn-dog LD50:58 mg/kg DRUGAY-,63,90
orl-cat LDLo:100 mg/kg PHREA7 19,472,39
ipr-rbt LDLo:90 mg/kg JPETAB 41,465,31
scu-rbt LDLo:170 mg/kg JACSAT 45,243,23
ivn-rbt LD50:49 mg/kg JPETAB 96,209,49

SAFETY PROFILE: A poison by ingestion, intravenous, intraperitoneal, and subcutaneous routes. See also BARBITURATES. When heated to decomposition it emits toxic fumes of NO_x.

AMX825 **CAS:3734-97-2** **HR: 3**
AMITON OXALATE
mf: $C_{10}H_{24}NO_3PS•C_2H_2O_4$ mw: 359.42

PROP: Crystals from Me_2CO. Mp: 98–99°.

SYNS: ACID OXALATE □ CHIPMAN 6199 □ CHIPMAN R-6, 199 □ CITRAM □ (2-(2-DIETHYLAMINO)ETHYL)-O,O-DIETHYL ESTER, OXALATE (1:1) □ S-(2-DIETHYLAMINOETHYL)-O,O-DIETHYLPHOSPHOROTHIOATE HYDROGEN OXALATE □ O,O-DIETHYL-S-(2-DIETHYLAMINO) ETHYLPHOSPHOROTHIOATE HYDROGEN OXALATE □ O,O-DIETHYL-S-(β-DIETHYLAMINO)ETHYL PHOSPHOROTHIOLATE HYDROGEN OXALATE □ O,O-DIETHYL-S-(2-ETHYL-N,N-DIETHYLAMINO) PHOSPHOROTHIOATE HYDROGEN OXALATE □ ENT 20,993 □ HYDROGEN OXALATE of AMITON □ PHOSPHOROTHIOIC ACID □ TETRAM □ TETRAM MONOOXALATE

TOXICITY DATA WITH REFERENCE
ipr-mus LD50:500 µg/kg PAREAQ 11,636,59
orl-rat LD50:3 mg/kg 28ZEAL 4,162,69

CONSENSUS REPORTS: EPA Extremely Hazardous Substances List.

SAFETY PROFILE: A deadly poison by ingestion and intraperitoneal routes. Human systemic effects may include: headache, giddiness, nervousness, impaired vision, weakness, nausea, cramps, diarrhea, muscular weakness and loss of control, convulsions and coma. Flammable. To extinguish fire, use dry chemical, carbon dioxide, water spray, fog, or foam. When heated to decomposition it emits toxic fumes of SO_x, PO_x, and NO_x. A cholinesterase inhibitor used as an insecticide.

AMY000 **CAS:4317-14-0** **HR: 3**
AMITRIPTYLINE-N-OXIDE
mf: $C_{20}H_{23}NO$ mw: 293.44

PROP: Solid. Mp: 228–230°.

SYNS: AMITRIPTYLINOXIDE □ 1-PROPANAMINE, 3-(10,11-DIHYDRO-5H-DIBENZO(A,D)CYCLOHEPTEN-5-YLIDENE)-N,N-DI-METHYL-N-OXIDE

TOXICITY DATA WITH REFERENCE
orl-rat TDLo:2500 mg/kg (female 6-15D post):TER ARZNAD 28,1898,78
orl-rat TDLo:2500 mg/kg (female 6-15D post):REP ARZNAD 28,1898,78
orl-rat LD50:1800 mg/kg ARZNAD 28,1898,78
ipr-rat LD50:110 mg/kg ARZNAD 28,1898,78
ivn-rat LD50:25 mg/kg ARZNAD 28,1898,78
orl-mus LD50:330 mg/kg ARZNAD 28,1898,78
ipr-mus LD50:320 mg/kg ARZNAD 28,1898,78
ivn-mus LD50:87 mg/kg ARZNAD 28,1898,78
orl-dog LD50:330 mg/kg ARZNAD 28,1898,78
orl-rbt LD50:330 mg/kg ARZNAD 28,1898,78
orl-gpg LD50:330 mg/kg ARZNAD 28,1898,78

SAFETY PROFILE: Poison by ingestion, intraperitoneal, and intravenous routes. An experimental teratogen. Other experimental reproductive effects. When heated to decomposition it emits toxic fumes of NO_x.

AMY050 **CAS:61-82-5** **HR: 3**
AMITROLE
mf: $C_2H_4N_4$ mw: 84.10

PROP: Crystals from H_2O, EtOH, or EtOAc. Mp: 159°.

SYNS: AMEROL ☐ AMINOTRIAZOLE ☐ 2-AMINOTRIAZOLE ☐ 3-AMINOTRIAZOLE ☐ 3-AMINO-s-TRIAZOLE ☐ 3-AMINO-1,2,4-TRIAZOLE ☐ 2-AMINO-1,3,4-TRIAZOLE ☐ 3-AMINO-1H-1,2,4-TRIAZOLE ☐ AMINOTRIAZOLE (PLANT REGULATOR) ☐ 3-AMINO-1,2,4-TRIAZOLE (ACGIH) ☐ AMINO TRIAZOLE WEEDKILLER 90 ☐ AMINOTRIAZOL-SPRITZPULVER ☐ AMITOL ☐ AMITRIL ☐ AMITRIL T.L. ☐ AMITROL ☐ AMITROL 90 ☐ AMITROL-T ☐ AMIZOL ☐ AMIZOL D ☐ AMIZOL DP NAU ☐ AMIZOL F ☐ AT ☐ ATA ☐ 3,A-T ☐ AT-90 ☐ AT Liquid ☐ AZAPLANT ☐ AZAPLANT KOMBI ☐ AZOLAN ☐ AZOLE ☐ CAMPAPRIM A 1544 ☐ CYTROL ☐ CYTROL AMITROLE-T ☐ CYTROLE ☐ DIUROL ☐ DIUROL 5030 ☐ DOMATOL ☐ DOMATOL 88 ☐ ELMASIL ☐ EMISOL ☐ EMISOL 50 ☐ EMISOL F ☐ ENT 25445 ☐ FENAMINE ☐ FENAVAR ☐ HERBIDAL TOTAL ☐ HERBIZOLE ☐ KLEER-LOT ☐ ORGA-414 ☐ RADOXONE TL ☐ RAMIZOL ☐ RCRA WASTE NUMBER U011 ☐ SIMAZOL ☐ SOLUTION CONCENTREE T271 ☐ TRIAZOLAMINE ☐ 1H-1,2,4-TRIAZOL-3-AMINE ☐ USAF XR-22 ☐ VOROX ☐ VOROX AA ☐ VOROX AS ☐ WEEDAR ADS ☐ WEEDAR AT ☐ WEEDAZIN ☐ WEEDAZIN ARGINIT ☐ WEEDAZOL ☐ WEEDAZOL GP2 ☐ WEEDAZOL SUPER ☐ WEEDAZOL T ☐ WEEDAZOL TL ☐ WEEDEX GRANULAT ☐ WEEDOCLOR ☐ X-ALL Liquid

TOXICITY DATA WITH REFERENCE
mma-sat 50 μg/plate PMRSDJ 1,351,81
mrc-asn 600 μg/L MUREAV 147,288,85
sln-asn 600 μg/L MUREAV 147,288,85
hma-mus/sat 12 mg/kg JNCIAM 62,911,79
msc-ham:emb 1 mg/L MUREAV 140,205,84
orl-mus TDLo:1935 mg/kg (female 6-14D post):TER NTIS** PB223-160
scu-mus TDLo:4176 mg/kg (female 6-14D post):REP NTIS** PB223-160
orl-rat TDLo:4595 mg/kg/2.5Y-C:CAR TXAPA9 69,161,83
orl-mus TDLo:113 g/kg/3W-I:CAR JNCIAM 42,1101,69
orl-rat TD:3670 mg/kg/2Y-C:NEO SCIEAS 132,296,60
orl-mus TD:366 g/kg/26W-C:ETA TOLED5 29,145,85
orl-rat LD50:1100 mg/kg RREVAH 10,97,65
orl-mus LD50:14,700 mg/kg PCOC** -,33,66
ipr-mus LD50:200 mg/kg NTIS** AD277-689

CONSENSUS REPORTS: NTP 7th Annual Report On Carcinogens. IARC Cancer Review: Group 2B IMEMDT 7,92,87; Human Inadequate Evidence IMEMDT 41,293,86; IMEMDT 7,31,74; Animal Sufficient Evidence IMEMDT 7,31,74; IMEMDT 41,293,86. Reported in EPA TSCA Inventory. EPA Genetic Toxicology Program.

OSHA PEL: TWA 0.2 mg/m³
ACGIH TLV: TWA 0.2 mg/m³
DFG MAK: 0.2 mg/m³ (as total dust)

SAFETY PROFILE: Confirmed carcinogen with experimental carcinogenic, tumorigenic, and neoplastigenic data. Low oral toxicity. Poison by intraperitoneal route. Moderately toxic by ingestion. An experimental teratogen. Other experimental reproductive effects. Mutation data reported. When heated to decomposition it emits toxic fumes of NO_x. An herbicide and plant growth regulator.

AMY250 **HR: 3**
AMMINE PENTAHYDROXO PLATINUM
mf: H_8NO_5Pt mw: 297.2

SAFETY PROFILE: A poison. An explosively unstable compound. Explodes @ >250°. Upon decomposition it emits toxic fumes of NO_x. See also PLATINUM COMPOUNDS.

AMY500 **CAS:7664-41-7** **HR: 3**
AMMONIA
DOT: UN 1005
mf: H_3N mw: 17.04

PROP: Colorless, alkaline, nonflammable gas with extremely pungent odor; liquefied by compression. Mp: −77.7°, bp: −33.35°, lel: 16%, uel: 25%, d: 0.771 g/L @ 0°, 0.817 g/L @ −79°, autoign temp: 1204°F, vap press: 10 atm @ 25.7°, vap d: 0.6. Very sol in water; moderately sol in alc.

SYNS: AM-FOL ☐ AMMONIA ANHYDROUS ☐ AMMONIA, anhydrous, liquefied (DOT) ☐ AMMONIAC (FRENCH) ☐ AMMONIACA (ITALIAN) ☐ AMMONIA Gas ☐ AMMONIAK (GERMAN) ☐ AMMONIA SOLUTIONS, relative density <0.880 at 15 degrees C in water, with >50% ammonia (DOT) ☐ AMONIAK (POLISH) ☐ ANHYDROUS AMMONIA ☐ NITRO-SIL ☐ R 717 ☐ SPIRIT of HARTSHORN

TOXICITY DATA WITH REFERENCE
mmo-esc 1500 ppm/3H AMNTA4 85,119,51
cyt-rat-ihl 19,800 μg/m³/16W BZARAZ 27,102,74
ihl-hmn LCLo:30,000 ppm/5M TJSGA8 45,458,67
ihl-hmn TCLo:20 ppm:IRR AGGHAR 13,528,55
unk-man LDLo:132 mg/kg 85DCAI 2,73,70
ihl-rat LCLo:2000 ppm/4H JIHTAB 31,343,49
ihl-mus LD50:4837 ppm/1H NTIS** PB214-270
ihl-cat LCLo:7000 ppm/1H JIHTAB 26,29,44
ihl-cat TCLo:1000 ppm/10M AEHLAU 35,6,80
ihl-rbt LCLo:7000 ppm/1H JIHTAB 26,29,44
ihl-mam LCLo:5000 ppm/5M AEPPAE 138,65,28

CONSENSUS REPORTS: EPA Extremely Hazardous Substances List. Community Right-To-Know List. Reported in EPA TSCA Inventory.

OSHA PEL: TWA 35 ppm
ACGIH TLV: TWA 25 ppm; STEL 35 ppm
DFG MAK: 50 ppm (35 mg/m³)
NIOSH REL: CL 50 ppm
DOT CLASSIFICATION: 2.3; *Label:* Poison Gas; DOT Class: 2.2; *Label:* Nonflammable Gas

SAFETY PROFILE: A human poison by an unspecified route. Poison experimentally by inhalation. An eye, mucous membrane, and systemic irritant by inhalation. Mutation data reported. A common air contaminant. Difficult to ignite. Explosion hazard when exposed to flame or in a fire. NH_3 + air in a fire can detonate. Potentially violent or explosive reactions on contact with interhalogens (e.g., bromine pentafluoride, chlorine trifluoride), 1,2-dichloroethane (with liquid NH_3), boron halides, chloroformamideium nitrate, ethylene oxide (polymerization reaction), magnesium perchlorate, nitrogen trichloride, oxygen + platinum, or strong oxidants (e.g., potassium chlorate, nitryl chloride, chromyl chloride, dichlorine oxide, chromium trioxide, trioxygen difluoride, nitric acid, hydrogen peroxide, tetramethylammonium amide, thiocarbonyl azide thiocyanate, sulfinyl chloride, thiotriazyl chloride, ammonium peroxodisulfate, fluorine, nitrogen oxide, dinitrogen tetraoxide, and liquid oxygen). Forms sensitive explosive mixtures with air + hydrocarbons, 1-chloro-2,4-dinitrobenzene, 2-, or 4-chloronitrobenzene (above 160°C/30 bar), ethanol + silver nitrate, germanium derivatives, stibine, and chlorine. Reaction with silver chloride, silver nitrate, silver azide, and silver oxide form the explosive silver nitride. Reactions with chlorine azide, bromine, iodine, iodine + potassium, heavy metals and their compounds (e.g. gold(III) chloride, mercury, and potassium thallium amide ammoniate), tellurium halides (e.g. tellurium tetrabromide, and tellurium tetrachloride) and pentaborane(9) give explosive products. Incompatible in contact with Ag, acetaldehyde, acrolein, B, BI_3, halogens, $HClO_3$, ClO, chlorites, chlorosilane, (ethylene dichloride + liquid ammonia), Au, hexachloromelamine, (hydrazine + alkali metals), HBr, HOCl, $Mg(ClO_4)_2$, N_2O_4, NCl_3, NF_3, OF_2, P_2O_5, P_2O_3, picric acid, (K + AsH_3), (K + PH_3), (K + $NaNO_2$), potassium ferricyanide, potassium mercuric cyanide, (Na + CO), Sb, S, SCl_2, tellurium hydropentachloride, trichloromelamine, NO_2Cl, SbH_3, tetramethylammonium amide, $SOCl_2$, and thiotrithiazylchloride. Incandescent reaction when heated with calcium. Emits toxic fumes of NH_3 and NO_x when exposed to heat. To fight fire stop flow of gas.

For occupational chemical analysis use OSHA: #ID-164 or NIOSH: Ammonia, 3505.

AMZ125 CAS:57530-25-3 HR: 2
2-AMMONIOTHIAZOLE NITRATE
mf: $C_3H_5N_3O_3S$ mw: 163.15

SAFETY PROFILE: Explosive decomposition at 142°C. Upon decomposition it emits toxic fumes of SO_x and NO_x.

ANA000 CAS:631-61-8 HR: 3
AMMONIUM ACETATE
mf: $C_2H_4O_2 \cdot H_3N$ mw: 77.10

PROP: Crystals. Mp: 114°, d: 1.07.

SYN: ACETIC ACID, AMMONIUM SALT

TOXICITY DATA WITH REFERENCE
ipr-rat LD50:632 mg/kg ABBIA4 64,342,56
ivn-mus LD50:386 mg/kg MEIEDD 10,74,83
ipr-ckn LDLo:1735 mg/kg BIJOAK 106,699,68

CONSENSUS REPORTS: Reported in EPA TSCA Inventory.

SAFETY PROFILE: Poison by intravenous route. Moderately toxic by intraperitoneal routes. When heated to decomposition it emits toxic fumes of NO_x and NH_3.

ANA500 HR: 3
AMMONIUM (AMINYLENIUM BIS [TRIHYDROBORATE])
mf: $B_2H_{12}N_2$ mw: 61.72

SAFETY PROFILE: A highly reactive hydride. Self ignites in air when heated. When heated to decomposition it emits toxic fumes of NO_x and NH_3. Explodes on heating in air. See also BORON COMPOUNDS and HYDRIDES.

ANA750 CAS:12164-94-2 HR: 3
AMMONIUM AZIDE
mf: H_4N_4 mw: 60.08

PROP: Colorless plates. Mp: 160°, bp: explodes, d: 1.346, vap press: 1 mm @ 59.2° (sublimes).

DOT CLASSIFICATION: Forbidden

SAFETY PROFILE: Poison by inhalation and ingestion. See also AZIDES. Moderately flammable. Unstable. Explosion hazard upon rapid heating.

ANB000 CAS:5251-79-6 HR: 2
AMMONIUM BENZAMIDOOXYACETATE
mf: $C_9H_9NO_4 \cdot H_3N$ mw: 212.23

SYNS: AMMONIUM-2-(BENZAMIDOOXY)ACETATE ☐ BENZADOX ☐ BENZAMIDOOXY ACETIC ACID, AMMONIUM SALT ☐ TOPCIDE

TOXICITY DATA WITH REFERENCE
orl-rat LD50:2500 mg/kg 28ZEAL 5,23,76
skn-rbt LD50:450 mg/kg 28ZEAL 5,23,76

SAFETY PROFILE: Moderately toxic by ingestion and skin contact. When heated to decomposition it emits toxic fumes of NO_x and NH_3.

ANB100 CAS:1863-63-4 HR: 3
AMMONIUM BENZOATE
mf: $C_7H_5O_2 \cdot H_4N$ mw: 139.17

SYNS: BENZOIC ACID, AMMONIUM SALT ☐ VULNOC AB

TOXICITY DATA WITH REFERENCE
orl-rat LD50:825 mg/kg GISAAA 51(1),75,86
orl-mus LD50:235 mg/kg GISAAA 51(1),75,86
ivn-rbt LDLo:400 mg/kg JPETAB 44,81,32

CONSENSUS REPORTS: Reported in EPA TSCA Inventory.

SAFETY PROFILE: Poison by ingestion and intravenous routes. When heated to decomposition it emits toxic vapors of NO_x and NH_3.

ANB250 CAS:1066-33-7 HR: 3
AMMONIUM BICARBONATE (1:1)
mf: $HCO_3 \cdot H_4N$ mw: 79.1

PROP: Hard, colorless to white crystals or solid; faint ammonia odor, stable at room temp, volatile. Decomp below mp. Mp: 107.5° (rapid heating). D: 1.586. Sol in water; insol in alc.

SYNS: ACID AMMONIUM CARBONATE ☐ AMMONIUM CARBONATE ☐ AMMONIUM HYDROGEN CARBONATE ☐ CARBONIC ACID, MONOAMMONIUM SALT ☐ MONOAMMONIUM CARBONATE

TOXICITY DATA WITH REFERENCE
ivn-mus LD50:245 mg/kg AJVRAH 29,897,68

CONSENSUS REPORTS: Reported in EPA TSCA Inventory.

SAFETY PROFILE: Poison by intravenous route. When heated to decomposition it emits toxic fumes of NO_x and NH_3.

ANB500 **CAS:7789-09-5** **HR: 3**
AMMONIUM BICHROMATE
DOT: UN 1439
mf: $Cr_2H_8N_2O_7$ mw: 252.10

PROP: Bright red-orange crystals; air-stable monoclinic crystals. Mp: decomp, d: 2.936. Sol in water and alc.

SYNS: AMMONIO (DICROMATO DI) (ITALIAN) □ AMMONIUMBICHROMAAT (DUTCH) □ AMMONIUMDICHROMAAT (DUTCH) □ AMMONIUMDICHROMAT (GERMAN) □ AMMONIUM DICHROMATE □ AMMONIUM DICHROMATE(VI) □ BICHROMATE d'AMMONIUM (FRENCH)

TOXICITY DATA WITH REFERENCE
scu-gpg LDLo:25 mg/kg EQSSDX 1,1,75

CONSENSUS REPORTS: Reported in EPA TSCA Inventory. Chromium and its compounds are on the Community Right-To-Know List.

OSHA PEL: CL 0.1 $mg(CrO_3)/m^3$
ACGIH TLV: TWA 0.05 $mg(Cr)/m^3$; Confirmed Human Carcinogen
NIOSH REL: (Chromium(VI)) TWA 25 $\mu g(Cr(VI))/m^3$; CL 50 $\mu g/m^3$/15M
DOT CLASSIFICATION: 5.1; *Label:* Oxidizer

SAFETY PROFILE: Poison by inhalation, ingestion, skin contact, and subcutaneous routes. See also CHROMIUM COMPOUNDS. An unstable oxidizer. Moderately flammable; reacts with reducing agents.

For occupational chemical analysis use NIOSH: Chromium Hexavalent 7024.

ANB600 **CAS:10192-30-0** **HR: 2**
AMMONIUM BISULFITE
mf: $H_3N \cdot H_2O_3S$ mw: 99.12

PROP: A solid.

SYNS: AMMONIUM HYDROGEN SULFITE □ AMMONIUM MONOSULFITE □ MONOAMMONIUM SULFITE □ SULFUROUS ACID, MONOAMMONIUM SALT

CONSENSUS REPORTS: Reported in EPA TSCA Inventory.

DOT CLASSIFICATION: 8; *Label:* Corrosive

SAFETY PROFILE: A corrosive solid. When heated to decomposition it emits toxic vapors of NH_4^-.

ANC000 **CAS:13843-59-9** **HR: 3**
AMMONIUM BROMATE
mf: NH_4BrO_3 mw: 145.96

PROP: Colorless crystals. Mp: explodes. Very sol in water.

SYNS: BROMIC ACID, AMMONIUM SALT □ AMMONIUM BROMATE (DOT)

DOT CLASSIFICATION: Forbidden

SAFETY PROFILE: An unstable, explosive oxidizing material. See also BROMATES. Severe explosion hazard.

ANC250 **CAS:12124-97-9** **HR: 2**
AMMONIUM BROMIDE
mf: BrH_4N mw: 97.96

PROP: Colorless or white cubic, sltly hygroscopic crystals. Mp: sublimes @ 4°, bp: 235° in vac, d: 2.429, vap press: 1 mm @ 198.3°. Very sol in water.

SYN: HYDROBROMIC ACID MONOAMMONIATE

TOXICITY DATA WITH REFERENCE
orl-rat LD50:2700 mg/kg GTPZAB 33(10),57,89
orl-mus LD50:2860 mg/kg GTPZAB 33(10),57,89
ipr-mus LD50:559 mg/kg GTPZAB 33(10),57,89
ipr-gpg LD50:535 mg/kg GTPZAB 33(10),57,89

CONSENSUS REPORTS: Reported in EPA TSCA Inventory.

SAFETY PROFILE: Moderately toxic by ingestion and intraperitoneal routes. See also BROMIDES. When heated to decomposition it emits very toxic fumes of NO_x, Br^-, and NH_3. Incompatible with BrF_3; IF_7; K.

ANC750 **HR: 3**
AMMONIUM BROMO SELENATE
mf: $(NH_4)SeBr_6$ mw: 594.5

PROP: Red octagonal crystals. D: 3.326, decomp in water; sltly sol in ether.

CONSENSUS REPORTS: Selenium and its compounds are on the Community Right-To-Know List.

SAFETY PROFILE: Poison and dangerous hazard. See also SELENIUM COMPOUNDS and BROMIDES.

AND250 **HR: 3**
AMMONIUM CADMIUM CHLORIDE
mf: $4NH_4Cl \cdot CdCl_2$ mw: 397.3

PROP: Colorless, rhombic crystals. D: 2.01; sol in water.

CONSENSUS REPORTS: Cadmium and its compounds are on the Community Right-To-Know List.

OSHA PEL: TWA 5 $\mu g(Cd)/m^3$
ACGIH TLV: TWA 0.05 $mg(Cd)/m^3$ (Proposed: TWA 0.01 $mg(Cd)/m^3$ (dust), Suspected Human Carcinogen; 0.002 $mg(Cd)/m^3$ (respirable dust), Suspected Human Carcinogen); BEI: 10 $\mu g/g$ creatinine in urine; 10 $\mu g/L$ in blood
DFG BAT: **Blood:** 1.5 $\mu g/dL$; Urine: 15 $\mu g/dL$; Suspected Carcinogen
NIOSH REL: (Cadmium) Reduce to lowest feasible level.

SAFETY PROFILE: Confirmed human carcinogen. A poison. When heated to decomposition it emits toxic fumes of NH_3, NO_x, and Cl^-. See also CADMIUM COMPOUNDS.

AND500 **HR: 3**
AMMONIUM CALCIUM ARSENATE
mf: $NH_4CaAsO_4 \cdot 6H_2O$ mw: 305.1

PROP: Colorless crystals. Mp: 140° (decomp), d: 1.905 @ 15°. Sltly sol in cold water; sol in hot water; sol in NH_4Cl and NH_4OH.

CONSENSUS REPORTS: Arsenic and its compounds are on the Community Right-To-Know List.

SAFETY PROFILE: A poison. See also ARSENIC COMPOUNDS.

AND750 **CAS:1111-78-0** **HR: 3**
AMMONIUM CARBAMATE
mf: $CH_3NO_2 \cdot H_3N$ mw: 78.09

PROP: White, crystalline, rhombic powder; sol in water and alc; ammonia odor. Sublimes at 60°.

SYN: AMMONIUM AMINOFORMATE

TOXICITY DATA WITH REFERENCE
ivn-rat LD50:39 mg/kg AJVRAH 29,897,68
ivn-mus LD50:77 mg/kg AJVRAH 29,897,68

CONSENSUS REPORTS: Reported in EPA TSCA Inventory.

SAFETY PROFILE: Poison by intravenous route. See also CARBAMATES.

ANE000 **CAS:506-87-6** **HR: 3**
AMMONIUM CARBONATE
mf: $(NH_4)_2CO_3$ mw: 96.11

PROP: Colorless crystals; strong odor of NH_3; sharp taste. Decomposes on standing to ammonium bicarbonate. Mp: 58°. Sltly sol in water.

SYNS: AMMONIUMCARBONAT (GERMAN) ☐ CARBONIC ACID, AMMONIUM SALT ☐ CARBONIC ACID, DIAMMONIUM SALT ☐ DIAMMONIUM CARBONATE

TOXICITY DATA WITH REFERENCE
ivn-mus LD50:96 mg/kg AJVRAH 29,897,68
ivn-dog LDLo:200 mg/kg HBAMAK 4,1289,35
scu-frg LDLo:250 mg/kg HBAMAK 4,1289,35

CONSENSUS REPORTS: Reported in EPA TSCA Inventory.

SAFETY PROFILE: Poison by subcutaneous and intravenous routes. When heated to decomposition it emits toxic fumes of NO_x and NH_3.

ANE250 **CAS:10192-29-7** **HR: 3**
AMMONIUM CHLORATE
mf: ClH_3NO_3 mw: 100.49

PROP: White, unstable, colorless, crystals or needles. Very soluble in water.

SYN: CHLORIC ACID, AMMONIUM SALT

DOT CLASSIFICATION: Forbidden

SAFETY PROFILE: A powerful oxidizer. Moderately flammable due to spontaneous chemical reaction. Explosion hazard due to shock, chemical reaction, or exposure to heat. A storage hazard; it may explode at room temperature. Explodes when heated to 100°C. When contaminated it is very sensitive. Solution in water may explode if heated or dried. When heated to decomposition it emits highly toxic fumes of Cl^- and NO_x. Incompatible with reducing materials; BrF_3; BrF_5.

ANE500 **CAS:12125-02-9** **HR: 3**
AMMONIUM CHLORIDE
mf: $H_4N \cdot Cl$ mw: 53.50

PROP: White, hygroscopic, solid or crystals; salty taste. Bp: 520°, mp: 337.8°, d: 1.520, vap press: 1 mm @ 160.4° (sublimes). Sol in water, alc, and glycerin.

SYNS: AMCHLOR ☐ AMMONERIC ☐ AMMONIUMCHLORID (GERMAN) ☐ AMMONIUM MURIATE ☐ CHLORID AMONNY (CZECH) ☐ DARAMMON ☐ SAL AMMONIA ☐ SAL AMMONIAC ☐ SALAMMONITE ☐ SALMIAC

TOXICITY DATA WITH REFERENCE
eye-rbt 500 mg/24H SEV 28ZPAK -,15,72
eye-rbt 100 mg SEV LPPTAK 24,598,76
cyt-ham:fbr 400 mg/L FCTOD7 22,623,84
orl-rat LD50:1650 mg/kg 28ZPAK -,15,72
ims-rat LD50:30 mg/kg EMSUA8 4,223,46
orl-mus LD50:1300 mg/kg IYKEDH 21,257,90
ipr-mus LD50:1439 mg/kg COREAF 256,1043,63
scu-mus LDLo:500 mg/kg 27ZIAQ -,39,73
orl-dog LDLo:600 mg/kg HBAMAK 4,1289,35
orl-rbt LDLo:1000 mg/kg HBAMAK 4,1289,35
ivn-rbt LDLo:78 mg/kg HBAMAK 4,1289,35
scu-gpg LDLo:72 mg/kg HBAMAK 4,1289,35
ivn-gpg LDLo:220 mg/kg 27ZWAY 1,470,23

CONSENSUS REPORTS: Reported in EPA TSCA Inventory.

OSHA PEL: (Fume) TWA 10 mg/m³; STEL 20 mg/m³
ACGIH TLV: TWA 10 mg/m³; STEL 20 mg/m³

SAFETY PROFILE: Poison by subcutaneous, intravenous, and intramuscular routes. Moderately toxic by other routes. A severe eye irritant. Mutation data reported. Explosive reaction with potassium chlorate or bromine trifluoride. Violent reaction (ignition) with bromine pentafluoride, NH_4, NO_3, and IF_7. Reaction with hydrogen cyanide may give the explosive nitrogen trichloride. When heated to decomposition it emits very toxic fumes of NO_x, Cl^-, and NH_3.

ANE750 **CAS:13820-40-1** **HR: 2**
AMMONIUM CHLOROPALLADATE(II)
mf: $Cl_4H_8N_2Pd$ mw: 284.30

PROP: Olive-green crystals. D: 2.17, mp: decomp.

SYNS: AMMONIUM TETRACHLOROPALLADATE ☐ DIAMMONIUM TETRACHLOROPALLADATE ☐ DIAMMONIUM TETRACHLOROPALLADATE(2-) ☐ PALLADATE(2-), TETRACHLORO-, DIAMMONIUM (8CI) ☐ PALLADATE(2-), TETRACHLORO-, DIAMMONIUM, (SP-4-1)-(9CI)

TOXICITY DATA WITH REFERENCE
skn-rbt 100 mg/24H SEV AEHLAU 30,168,75

CONSENSUS REPORTS: Reported in EPA TSCA Inventory.

SAFETY PROFILE: A severe skin irritant. See also PALLADIUM. When heated to decomposition it emits very toxic fumes of Cl^-, NO_x, and NH_3.

ANF000 CAS:19168-23-1 HR: 3
AMMONIUM CHLOROPALLADATE(IV)
mf: $Cl_6H_8N_2Pd$ mw: 355.20

PROP: Deep red-brown crystals. D: 2.418, mp: decomp.

SYNS: AMMONIUM HEXACHLOROPALLADATE ☐ DIAMMONIUM HEXACHLOROPALLADATE

TOXICITY DATA WITH REFERENCE
skn-rbt 100 mg/24H SEV AEHLAU 30,168,75

CONSENSUS REPORTS: Reported in EPA TSCA Inventory.

SAFETY PROFILE: A poison skin irritant. When heated to decomposition it emits very toxic fumes of NO_x, Cl^-, and NH_3.

ANF250 CAS:16919-58-7 HR: 3
AMMONIUM CHLOROPLATINATE
mf: $Cl_6Pt{\cdot}2H_4N$ mw: 443.89

PROP: Cubic, yellow, crystals or solid. D: 3.065, mp: decomp. Aq solns slowly photoreduce with substitution. Sol in water.

SYNS: AMMONIUM HEXACHLOROPLATINATE(IV) ☐ AMMONIUM PLATINIC CHLORIDE ☐ DIAMMONIUM HEXACHLOROPLATINATE (2-) ☐ PLATINIC AMMONIUM CHLORIDE

TOXICITY DATA WITH REFERENCE
ihl-hmn TCLo:0.9 μg/m³:PUL BJIMAG 2,92,45
orl-rat LD50:195 mg/kg GTPZAB 21(7),55,77

CONSENSUS REPORTS: Reported in EPA TSCA Inventory.

OSHA PEL: TWA 0.002 mg(Pt)/m³
ACGIH TLV: TWA 0.002 mg(Pt)/m³

SAFETY PROFILE: Poison by inhalation and ingestion. Human pulmonary system effects by inhalation. See also PLATINUM COMPOUNDS. An explosively unstable compound. Incompatible with KOH (boiling with alkali yields a product which, after drying, will explode @ 205° or if mixed with combustibles). When heated to decomposition it emits very toxic fumes of Cl^-, NO_x, and NH_3.

ANF500 CAS:7788-98-9 HR: 3
AMMONIUM CHROMATE
mf: $(NH_4)_2CrO_4$ mw: 152.10

PROP: Yellow, crystalline material. Mp: decomp @ 180°, d: 1.91 @ 12°. Sol in cold water.

SYNS: AMMONIUM CHROMATE(VI) ☐ CHROMIC ACID, DIAMMONIUM SALT ☐ DIAMMONIUM CHROMATE ☐ NEUTRAL AMMONIUM CHROMATE

TOXICITY DATA WITH REFERENCE
mmo-sat 35 μg/plate CRNGDP 9,611,88
dnr-esc 25 μg/well MUREAV 133,161,84

CONSENSUS REPORTS: Chromium and its compounds are on the Community Right-To-Know List.

OSHA PEL: CL 0.1 mg(CrO_3)/m³
ACGIH TLV: TWA 0.05 mg(Cr)/m³; Confirmed Human Carcinogen
NIOSH REL: (Chromium(VI)) TWA 25 μg(Cr(VI))/m³; CL 50 μg/m³/15M

SAFETY PROFILE: A poison. Mutation data reported. See also CHROMIUM COMPOUNDS. A powerful oxidizer. An explosion hazard when shocked or heated. When heated to decomposition it emits toxic fumes of NH_3 and NO_x. Incompatible with reducing agents.

For occupational chemical analysis use NIOSH: Chromium Hexavalent 7024.

ANF625 HR: 3
AMMONIUM CHROME ALUMS
mf: $H_8N_2O_4S{\cdot}Cr_2{\cdot}O_{12}S_3{\cdot}24H_2O$ mw: 956.82

SYNS: AMMONIUM SULFATE, and CHROMIC SULFATE, TETRACOSAHYDRATE ☐ CHROMIC AMMONIUM SULFATE

TOXICITY DATA WITH REFERENCE
orl-rat LD50:720 mg/kg 85GMAT ,38,82
skn-rat LDLo:2 g/kg 85GMAT -,38,82
ihl-mus LC50:51 mg/m³/2H 85GMAT -,38,82
skn-mus LD50:110 mg/kg 85GMAT -,38,82
ims-mus LD50:115 mg/kg 85GMAT ,38,82
skn-rbt LDLo:1 g/kg 85GMAT -,38,82

CONSENSUS REPORTS: Chromium and its compounds are on the Community Right-To-Know List.

ACGIH TLV: TWA 0.5 mg(Cr)/m³; Not Classifiable as a Carcinogen

SAFETY PROFILE: Poison by inhalation, skin contact, and intramuscular routes. Moderately toxic by ingestion. When heated to decomposition it emits toxic fumes of SO_x, NO_x, and NH_3. See also AMMONIUM SULFATE and CHROMIUM COMPOUNDS.

ANF750 HR: 3
AMMONIUM CHROMIC SULFATE
mf: $NH_4Cr(SO_4)_2{\cdot}12H_2O$ mw: 478.4

PROP: Green or violet crystals. Mp: 94° (−9H_2O @ 94°), d: 1.720, water sol.

CONSENSUS REPORTS: Chromium and its compounds are on the Community Right-To-Know List.

OSHA PEL: CL 0.1 mg(CrO_3)/m³
ACGIH TLV: TWA 0.05 mg(Cr)/m³; Confirmed Human Carcinogen
NIOSH REL: (Chromium(VI)) TWA 25 μg(Cr(VI))/m³; CL 50 μg/m³/15M

SAFETY PROFILE: A confirmed carcinogen. Poison. See

also CHROMIUM COMPOUNDS and SULFATES. When heated to decomposition it emits toxic fumes of NH_3, NO_x, and SO_x.

ANF800 **CAS:7632-50-0** **HR: 2**
AMMONIUM CITRATE
mf: $C_6H_8O_7{\cdot}xH_3N$ mw: 311.42

PROP: Granules or crystals. D: 1.48. Sol in water; sltly sol in alc.

SYNS: AMMONIUM CITRATE, DIBASIC (DOT) □ CITRIC ACID, AMMONIUM SALT □ DIAMMONIUM CITRATE

TOXICITY DATA WITH REFERENCE
ivn-mus LD50:331 mg/kg JCINAO 37,497,58

CONSENSUS REPORTS: Reported in EPA TSCA Inventory.

SAFETY PROFILE: Experimental poison by intravenous route. A skin and eye irritant. When heated to decomposition it emits acrid smoke and irritating fumes.

ANG000 **HR: 3**
AMMONIUM CYANIDE
mf: NH_4CN mw: 44.1

PROP: Solid, white powder or crystals. Mp: 36° (decomp), bp: subl @ 40°, d: 1.002 @ 100°, vap press: 400 ppm @ 20.5°. Very sol in water and alc; decomp in hot water.

CONSENSUS REPORTS: Cyanide and its compounds are on the Community Right-To-Know List.

SAFETY PROFILE: A poison. See also CYANIDE. When heated to decomposition it emits toxic CN^-, NH_3, NO_x.

ANG125 **CAS:12008-61-6** **HR: 3**
AMMONIUM DECAHYDRODECABORATE (2−)
mf: $B_{10}H_{18}N_2$ mw: 154.26

SAFETY PROFILE: A poison. Product of the reaction with nitrous acid explodes when dry. When heated to decomposition it emits toxic fumes of NO_x and NH_3. See also BORON COMPOUNDS.

ANG250 **HR: 3**
AMMONIUM DIFLUORIDE mixed with HYDROCHLORIC ACID

SYN: WHITE ACID (DOT)

SAFETY PROFILE: A corrosive. Poison by inhalation, ingestion, and skin contact. When heated to decomposition it emits very toxic fumes of F^-, HF, and HCl.

ANG500 **CAS:3226-36-6** **HR: 2**
AMMONIUM DIMETHYL DITHIOCARBAMATE
mf: $C_3H_7NS_2{\cdot}H_3N$ mw: 138.27

SYN: DIRAM A

TOXICITY DATA WITH REFERENCE
orl-rat LD50:1458 mg/kg HYSAAV 32,169,67
orl-mus LD50:592 mg/kg HYSAAV 32,169,67
orl-rbt LD50:450 mg/kg HYSAAV 32,169,67
orl-gpg LD50:1680 mg/kg HYSAAV 32,169,67

SAFETY PROFILE: Moderately toxic by ingestion. See also CARBAMATES. When heated to decomposition it emits very toxic fumes of NO_x, SO_x, and NH_3.

ANG625 **CAS:76556-13-3** **HR: 3**
AMMONIUM-3,5-DINITRO-1,2,4-TRIAZOLIDE
mf: $C_2H_4N_6O_4$ mw: 176.09

SAFETY PROFILE: An explosive. Upon decomposition it emits toxic fumes of NO_x and NH_3. See also EXPLOSIVES.

ANG750 **CAS:25954-13-6** **HR: 1**
AMMONIUM ETHYL CARBAMOYLPHOSPHONATE
mf: $H_4N{\cdot}C_3H_7NO_4P$ mw: 170.13

PROP: Crystals. Very sol in H_2O, MeOH; sltly sol in EtOH; very sltly sol in Me_2CO, C_6H_6.

SYNS: AMMONIUM-AETHYL-CARBAMOYL-PHOSPHONAT (GERMAN) □ AMMONIUM ETHYL CARBAMOYLPHOSPHONATE solution □ DPX 1108 □ FOSAMINE AMMONIUM □ KRENITE □ KRENITE BRUSH CONTROL AGENT

TOXICITY DATA WITH REFERENCE
orl-rat LD50:11 g/kg 85JFAN A218,84
ihl-rat LC50:>57 g/m³/1H 85JFAN A218,84
skn-rbt LD50:>1660 mg/kg PEMNDP 9,442,91
orl-gpg LD50:7380 mg/kg PEMNDP 9,442,91
orl-qal LD50:10,000 mg/kg 85DPAN-,-,71/76
orl-dck LD50:10 g/kg 85JFAN A218,84

SAFETY PROFILE: Mildly toxic by ingestion and inhalation. See also PHOSPHATES. When heated to decomposition it emits very toxic fumes of NO_x, PO_x, and NH_3.

ANG925 **CAS:14221-47-7** **HR: 3**
AMMONIUM FERRIC OXALATE
mf: $C_6FeO_{12}{\cdot}3H_4N$ mw: 374.06

PROP: Green solid.

SYNS: AMMONIUM FERRIOXALATE □ AMMONIUM TRIOXALATOFERRATE(III) □ FERRIC AMMONIUM OXALATE □ FERRIC AMMONIUM OXALATE (DOT) □ TRIAMMONIUM TRIS-(ETHANEDIOATO (2-)-O,O')FERRATE(3-1)

TOXICITY DATA WITH REFERENCE
skn-rbt 500 mg MOD FCTOD7 20,563,82
eye-rbt 100 mg MOD FCTOD7 20,573,82
eye-rbt 100 mg/4S rns MLD FCTOD7 20,573,82

CONSENSUS REPORTS: Reported in EPA TSCA Inventory.

SAFETY PROFILE: An eye and skin irritant. When heated to decomposition it emits toxic fumes of NH_3. See also OXALATES.

ANH000 **CAS:13826-83-0** **HR: 3**
AMMONIUM FLUOBORATE
mf: NH_4BF_4 mw: 104.86

PROP: White, colorless, rhombic crystals. D: 1.871 @ 15°, mp: sublimes. Sol in NH_4OH and water.

SYNS: AMMONIUM BOROFLUORIDE □ AMMONIUM FLUOROBORATE □ AMMONIUM TETRAFLUOROBORATE □ AMMONIUM TETRAFLUOROBORATE(1-)

CONSENSUS REPORTS: Reported in EPA TSCA Inventory.

OSHA PEL: TWA 2.5 $mg(F)/m^3$
NIOSH REL: (Fluorides, Inorganic) TWA 2.5 $mg(F)/m^3$

SAFETY PROFILE: A poison and strong irritant. See also FLUORIDES and BORON COMPOUNDS. When heated to decomposition it emits very toxic fumes of F^-, NO_x, and NH_3.

ANH250 **CAS:12125-01-8** **HR: 3**
AMMONIUM FLUORIDE
DOT: UN 2505
mf: $H_4N{\cdot}F$ mw: 37.05

PROP: White, colorless, deliquescent crystals. Mp: sublimes; d: 1.009 @ 25°. Very sol in water; sltly sol in alc.

SYNS: AMMONIUM FLUORURE (FRENCH) □ NEUTRAL AMMONIUM FLUORIDE

TOXICITY DATA WITH REFERENCE
ipr-rat LD50:32 mg/kg XEURAQ UR-154,1951
scu-frg LDLo:280 mg/kg CRSBAW 124,133,37

CONSENSUS REPORTS: Reported in EPA TSCA Inventory.

OSHA PEL: TWA 2.5 $mg(F)/m^3$
NIOSH REL: TWA 2.5 $mg(F)/m^3$
DOT CLASSIFICATION: 6.1; *Label:* KEEP AWAY FROM FOOD

SAFETY PROFILE: Poison by subcutaneous and intraperitoneal routes. See also FLUORIDES. When heated to decomposition it emits very toxic fumes of F^-, NO_x, and NH_3. Incompatible with ClF_3.

ANH500 **CAS:540-69-2** **HR: 2**
AMMONIUM FORMATE
mf: $CH_2O_2{\cdot}H_3N$ mw: 63.07

PROP: White, deliquescent crystals. Mp: 116°, bp: decomp @ 180°, d: 1.280.

SYN: FORMIC ACID AMMONIUM SALT

TOXICITY DATA WITH REFERENCE
orl-mus LD50:2250 mg/kg ZERNAL 9,332,69
ivn-mus LD50:410 mg/kg ZERNAL 9,332,69

CONSENSUS REPORTS: Reported in EPA TSCA Inventory.

SAFETY PROFILE: Moderately toxic by ingestion and intravenous routes. When heated to decomposition it emits toxic fumes of NO_x and NH_3.

ANH875 **CAS:14481-29-9** **HR: 3**
AMMONIUM HEXACYANOFERRATE(II)
mf: $C_6H_{16}FeN_{10}$ mw: 284.11

CONSENSUS REPORTS: Cyanide and its compounds are on the Community Right-To-Know List.

SAFETY PROFILE: A poison. Reacts explosively with metal nitrates when heated, e.g., cobalt(II) nitrate at 220°C and copper(II) nitrate at 220°C. When heated to decomposition it emits toxic fumes of CN^- and NH_3. See also CYANIDE.

ANI000 **CAS:13815-28-6** **HR: 3**
AMMONIUM HEXAFLUOROFERRATE
mf: $F_6FeH_{12}N_3$ mw: 224.00

SYN: HEXAFLUORO FERRATE (3-) TRIAMMONIUM SALT

TOXICITY DATA WITH REFERENCE
ivn-mus LD50:56 mg/kg CSLNX* NX#04248

OSHA PEL: TWA 2.5 $mg(F)/m^3$
ACGIH TLV: TWA 1 $mg(Fe)/m^3$
NIOSH REL: (Fluorides, Inorganic) TWA 2.5 $mg(F)/m^3$

SAFETY PROFILE: Poison by intravenous route. See also FLUORIDES and IRON COMPOUNDS. When heated to decomposition it emits very toxic fumes of F^-, NO_x, and NH_3.

ANI250 **CAS:16962-40-6** **HR: 3**
AMMONIUM HEXAFLUOROTITANATE
mf: $F_6Ti{\cdot}H_4N_2$ mw: 193.96

PROP: White solid.

TOXICITY DATA WITH REFERENCE
ivn-mus LD50:56 mg/kg CSLNX* NX#00134

CONSENSUS REPORTS: Reported in EPA TSCA Inventory.

OSHA PEL: TWA 2.5 $mg(F)/m^3$
NIOSH REL: (Fluorides, Inorganic) TWA 2.5 $mg(F)/m^3$

SAFETY PROFILE: Poison by intravenous route. See also FLUORIDES, AMMONIA, and TITANIUM COMPOUNDS. When heated to decomposition it emits very toxic fumes of F^- and NO_x.

ANI500 **CAS:13815-31-1** **HR: 3**
AMMONIUM HEXAFLUOROVANADATE
mf: $F_6H_{12}N_3V$ mw: 219.09

SYN: HEXAFLUORO VANADATE (3-) TRIAMMONIUM SALT

TOXICITY DATA WITH REFERENCE
ivn-mus LD50:10 mg/kg CSLNX* NX#04249

OSHA PEL: TWA 2.5 $mg(F)/m^3$
ACGIH TLV: TWA 0.05 $mg(V_2O_5)/m^3$
NIOSH REL: (Vanadium Compounds) CL 0.05 $mg(V)/m^3$/15M

SAFETY PROFILE: Poison by intravenous route. See also FLUORIDES and VANADIUM COMPOUNDS. When heated to decomposition it emits very toxic NH_3, NO_x, VO_x, and fluorides.

ANI750 **HR: 3**
AMMONIUM HEXANITRO COBALTATE
mf: $CoH_{12}N_9O_{12}$ mw: 389.1

CONSENSUS REPORTS: Cobalt and its compounds are on the Community Right-To-Know List.

SAFETY PROFILE: Explodes @ 230°. Also is impact sensitive. Upon decomposition it emits toxic fumes of NO_x. See also COBALT COMPOUNDS and NITRATES.

ANJ000 **CAS:1341-49-7** **HR: 3**
AMMONIUM HYDROGEN FLUORIDE
DOT: UN 1727/UN 2817
mf: F_2H_5N mw: 57.06

PROP: White, colorless crystals. D: 1.51, mp: 126°, bp: 239°. Will etch glass. Very sol in water; sltly sol in alc.

SYNS: ACID AMMONIUM FLUORIDE □ AMMONIUM BIFLUORIDE □ AMMONIUM DIFLUORIDE □ AMMONIUM FLUORIDE comp. with HYDROGEN FLUORIDE (1:1) □ AMMONIUM HYDROFLUORIDE □ AMMONIUM HYDROGEN BIFLUORIDE □ AMMONIUM HYDROGEN DIFLUORIDE □ AMMONIUM HYDROGEN FLUORIDE, solid (UN 1727) (DOT) □ AMMONIUM HYDROGEN FLUORIDE, solution (UN 2817) (DOT)

CONSENSUS REPORTS: Reported in EPA TSCA Inventory.

OSHA PEL: TWA 2.5 $mg(F)/m^3$
ACGIH TLV: TWA 2.5 $mg(F)/m^3$
NIOSH REL: (Fluorides, Inorganic) TWA 2.5 $mg(F)/m^3$
DOT CLASSIFICATION: 8; *Label:* Corrosive (UN 1727); DOT Class: 8; *Label:* Corrosive, Poison (UN 2817)

SAFETY PROFILE: Caustic poison and strong irritant by all routes. See also HYDROFLUORIC ACID. When heated to decomposition it emits very toxic fumes of F^-, NO_x, and NH_3.

ANJ500 **CAS:7803-63-6** **HR: 2**
AMMONIUM HYDROGEN SULFATE
DOT: UN 2506
mf: NH_4HSO_4 mw: 115.11

PROP: White rhombic crystals; sol in water; insol in acetone. Mp: 146.9°, d: 1.78.

SYNS: ACID AMMONIUM SULFATE □ AMMONIUM ACID SULFATE □ AMMONIUM BISULFATE □ AMMONIUM MONOHYDROGEN SULFATE □ MONOAMMONIUM SULFATE □ SULFURIC ACID, MONOAMMONIUM SALT

CONSENSUS REPORTS: Reported in EPA TSCA Inventory.

DOT CLASSIFICATION: 8; *Label:* Corrosive

SAFETY PROFILE: Moderately toxic by ingestion. A corrosive. See also SULFATES. Dangerous; when heated to decomposition it emits highly toxic fumes of sulfuric acid and SO_x, NH_3, and NO_x.

ANJ750 **CAS:12124-99-1** **HR: 3**
AMMONIUM HYDROSULFIDE
mf: NH_4HS mw: 51.11

PROP: Powder or crystals. Mp: 118° (150 atm), d: 1.17, vap press: 400 mm @ 21.8°.

SYNS: AMMONIUM BISULFIDE □ AMMONIUM HYDROGEN SULFIDE □ AMMONIUM HYDROSULFIDE, solution (DOT) □ AMMONIUM MERCAPTAN □ AMMONIUM SULFHYDRATE □ MONOAMMONIUM SULFIDE □ SIRNIK AMONNY □ TRUE AMMONIUM SULFIDE

TOXICITY DATA WITH REFERENCE
orl-rat LD50:168 mg/kg NTIS** AD-A062-138
orl-mus LDLo:80 mg/kg JPETAB 76,179,42
skn-mus LDLo:2457 mg/kg JPETAB 76,179,42
ipr-mus LDLo:10 mg/kg JPETAB 76,179,42
scu-mus LD50:132 mg/kg 28ZPAK-,18,72
ivn-mus LDLo:2 mg/kg JPETAB 76,179,42
ivn-dog LDLo:2 mg/kg JPETAB 76,179,42
skn-rbt LD50:1682 mg/kg JACTDZ 1,712,92
scu-rbt LDLo:7500 μg/kg JPETAB 76,179,42
ivn-rbt LDLo:1500 μg/kg JPETAB 76,179,42
idr-rbt LDLo:30 mg/kg JPETAB 76,179,42
skn-gpg LDLo:692 mg/kg JPETAB 76,179,42
par-gpg LDLo:143 mg/kg JPETAB 76,179,42

CONSENSUS REPORTS: Reported in EPA TSCA Inventory.

SAFETY PROFILE: Poison by ingestion, subcutaneous, and intravenous routes. Moderately toxic by skin contact. Pyroforic in air. See also SULFIDES. When heated to decomposition it emits very toxic fumes of SO_x, NO_x, and NH_3. Incompatible with zinc.

ANK250 **CAS:1336-21-6** **HR: 3**
AMMONIUM HYDROXIDE
DOT: NA 2672
mf: $H_4N{\cdot}HO$ mw: 35.06

PROP: Clear, colorless liquid solution of ammonia; very pungent odor. D: 0.90, mp: −77°. Sol in water. Soln contains not more than 44% ammonia.

SYNS: AMMONIA AQUEOUS □ AMMONIA WATER 29% □ AMMONIA SOLUTIONS, with >10% but not >35% ammonia (UN 2672) (DOT) □ AMMONIA SOLUTIONS, with >35% but not >50% ammonia (UN 2073) (DOT) □ AQUA AMMONIA

TOXICITY DATA WITH REFERENCE
eye-rbt 1 mg/30S RNS SEV TXCYAC 23,281,82
eye-rbt 750 μg SEV AJOPAA 29,1363,46
mmo-sat 10 μL/plate ANYAA9 76,475,58
mmo-esc 10 μL/disc ANYAA9 76,475,58
orl-hmn LDLo:43 mg/kg 34ZIAG -,95,69
ihl-hmn LCLo:5000 ppm 34ZIAG -,95,69
ihl-hmn TCLo:700 ppm:EYE JISMAB 61,271,71
ihl-hmn TCLo:408 ppm:IRR JISMAB 61,271,71
orl-rat LD50:350 mg/kg JIHTAB 23,259,41
orl-cat LDLo:750 mg/kg HBAMAK 4,1289,35
ivn-rbt LDLo:10 mg/kg HBAMAK 4,1289,35

CONSENSUS REPORTS: Reported in EPA TSCA Inventory.

NIOSH REL: (Ammonia) CL 50 ppm

DOT CLASSIFICATION: 8; *Label:* Corrosive (UN 2672); DOT Class: 2.2; *Label:* Nonflammable Gas (UN 2073)

SAFETY PROFILE: A human poison by ingestion. An experimental poison by inhalation and ingestion. A severe eye irritant. Human systemic irritant effects by ocular and inhalation routes. Mutation data reported. Incompatible with acrolein, nitromethane, acrylic acid, chlorosulfonic acid, dimethyl sulfate, halogens, (Au + aqua regia), HCl, HF, HNO_3, oleum, β-propiolactone, propylene oxide, $AgNO_3$, Ag_2O, (Ag_2O + C_2H_5OH), $AgMnO_4$, H_2SO_4. Dangerous; liquid can inflict burns. Use with adequate ventilation. When heated to decomposition it emits NH_3 and NO_x.

ANK500 **HR: 3**
AMMONIUM HYPOPHOSPHITE
mf: H_6NO_2P mw: 83.03

PROP: White granules or rhombic crystals. D: 1.634, mp: 200°, bp: decomp @ 240°. Sol in water, alc, NH_3; insol in acetone.

SAFETY PROFILE: When heated it can liberate highly toxic and flammable PH_3. See also PHOSPHINE. When heated to decomposition it can emit highly toxic fumes of PH_3, PO_x, NH_3, and NO_x.

ANK600 **CAS:7783-18-8** **HR: 2**
AMMONIUM HYPOSULFITE
mf: $O_3S_2{\cdot}2H_4N$ mw: 148.22

PROP: A solid.

SYNS: AMMONIUM THIOSULFATE □ DIAMMONIUM THIOSULFATE □ THIOSULFURIC ACID, DIAMMONIUM SALT

TOXICITY DATA WITH REFERENCE
orl-rat LD50:2890 mg/kg GTPZAB 26(6),54,82
orl-mus LD:>3 g/kg GTPZAB 26(6),54,82
orl-gpg LD50:1098 mg/kg GTPZAB 26(6),54,82

CONSENSUS REPORTS: Reported in EPA TSCA Inventory.

SAFETY PROFILE: Moderately toxic by ingestion. When heated to decomposition it emits toxic vapors of NH_4^- and SO_x.

ANK650 **CAS:27441-86-7** **HR: 2**
AMMONIUM IMIDODISULFONATE
mf: $H_3NO_6S_2{\cdot}xH_3N$ mw: 296.44

SYNS: AMMONIUM IMIDOBISSULFATE □ AMMONIUM IMIDOSULFONATE □ IMIDODISULFURIC ACID, AMMONIUM SALT

TOXICITY DATA WITH REFERENCE
orl-rat LD50:3300 mg/kg GISAAA 52(10),88,87
orl-mus LD50:2700 mg/kg GISAAA 52(10),88,87
orl-gpg LD50:2250 mg/kg GISAAA 52(10),88,87

CONSENSUS REPORTS: Reported in EPA TSCA Inventory.

SAFETY PROFILE: Moderately toxic by ingestion. When heated to decomposition it emits toxic vapors of NH_4^- and SO_x.

ANK750 **HR: 3**
AMMONIUM IODATE
mf: H_4INO_3 mw: 192.94

PROP: Colorless crystals. D: 3.309 @ 21°, mp: 150° (decomp). Sltly sol in cold water; insol in hot water.

SAFETY PROFILE: A powerful, unstable oxidizer. When heated to decomposition it emits very toxic fumes of I^- and NO_x. Has detonated upon contact with a scoop, possibly due to contamination by ammonium periodate. See also IODATES.

ANL000 **HR: 2**
AMMONIUM IODIDE
mf: NH_4I mw: 145

PROP: Colorless, hygroscopic crystals. Mp: subl @ 551°, bp: 220° (vac), d: 2.514 @ 25°, vap press: 1 mm @ 210.9°.

SAFETY PROFILE: Moderately toxic. See also IODIDES. Incompatible with BrF_3; IF_7; K. When heated to decomposition it emits toxic fumes of I^-, NH_3, and NO_x.

ANL100 **CAS:57267-78-4** **HR: 1**
AMMONIUM ISETHIONATE
mf: $C_2H_5O_4S{\cdot}H_3N$ mw: 142.17

SYNS: ETHANESULFONIC ACID, 2-HYDROXY-, AMMONIUM SALT □ 2-HYDROXYETHANESULFONIC ACID AMMONIUM SALT

TOXICITY DATA WITH REFERENCE
eye-rbt 100 mg MLD FCTOD7 20,573,82

CONSENSUS REPORTS: Reported in EPA TSCA Inventory.

SAFETY PROFILE: An eye irritant. When heated to decomposition it emits toxic fumes of SO_x, NH_3, and NO_x.

ANL500 **CAS:10169-00-3** **HR: 2**
AMMONIUM LANTHANUM NITRATE
mf: $H_4N{\cdot}La{\cdot}7NO_3$ mw: 591.03

SYNS: LANTHANUM AMMONIUM NITRATE □ NITRIC ACID, AMMONIUM LANTHANUM SALT □ NITRIC ACID, LANTHANUM AMMONIUM SALT

TOXICITY DATA WITH REFERENCE
orl-rat LD50:3400 mg/kg AIHOAX 1,637,50
ipr-rat LD50:625 mg/kg AIHOAX 1,637,50

SAFETY PROFILE: Moderately toxic by ingestion and intraperitoneal routes. See also NITRATES. When heated to decomposition it emits very toxic fumes of NH_3 and NO_x.

ANL750 **HR: 3**
AMMONIUM MAGNESIUM ARSENATE
mf: $NH_4MgAsO_4{\cdot}6H_2O$ mw: 289.4

PROP: Colorless crystals. Mp: decomp, d: 1.932 @ 15°. Very sltly water-sol.

CONSENSUS REPORTS: Arsenic and its compounds are on the Community Right-To-Know List.

SAFETY PROFILE: When heated to decomposition it emits very toxic fumes of As, NH_3, and NO_x. See ARSENIC COMPOUNDS and MAGNESIUM COMPOUNDS.

ANM000 **HR: 3**
AMMONIUM MAGNESIUM CHROMATE
mf: $(NH_4)_2CrO_4{\cdot}MgCrO_4{\cdot}6H_2O$ mw: 400.5

PROP: Yellow crystals. Mp: decomp, d: 1.84. Very water-sol.

CONSENSUS REPORTS: Chromium and its compounds are on the Community Right-To-Know List.

OSHA PEL: CL 0.1 mg(CrO_3)/m³
ACGIH TLV: TWA 0.05 mg(Cr)/m³; Confirmed Human Carcinogen
NIOSH REL: (Chromium(VI)) TWA 25 µg(Cr(VI))/m³; CL 50 µg/m³/15M

SAFETY PROFILE: A confirmed carcinogen. A poison. See also CHROMIUM COMPOUNDS and MAGNESIUM COMPOUNDS. Moderately flammable; can explode. Incompatible with reducing agents. When heated to decomposition it can emit toxic fumes of NH_3 and NO_x.

ANM250 **CAS:530-31-4** **HR: 1**
AMMONIUM MANDELATE
mf: $C_8H_7O_3{\cdot}H_4N$ mw: 169.20

TOXICITY DATA WITH REFERENCE
orl-rat LDLo:5000 mg/kg AIPTAK 64,79,40
orl-rbt LDLo:5000 mg/kg AIPTAK 64,79,40

SAFETY PROFILE: Mildly toxic by ingestion. When heated to decomposition it emits toxic fumes of NO_x and NH_3.

ANM500 **CAS:5421-46-5** **HR: 3**
AMMONIUM MERCAPTOACETATE
mf: $C_2H_3O_2S{\cdot}H_3N$ mw: 108.15

PROP: Colorless liquid; strong skunk-like odor.

SYNS: AMMONIUM THIOGLYCOLATE ☐ AMMONIUM THIOGLYCOLLATE ☐ THIOGLYCOLLIC ACID, AMMONIUM SALT ☐ USAF MO-2

TOXICITY DATA WITH REFERENCE
ipr-rat LD50:165 mg/kg JPETAB 97,349,49
ipr-mus LD50:100 mg/kg NTIS** AD277-689
ivn-cat LD50:175 mg/kg JPETAB 97,349,49
ivn-rbt LD50:100 mg/kg JPETAB 97,349,49

CONSENSUS REPORTS: Reported in EPA TSCA Inventory.

SAFETY PROFILE: Poison by intravenous and intraperitoneal routes. An allergen; can cause contact dermatitis. Emits hydrogen sulfide. See also SULFIDES. When heated to decomposition it emits very toxic NO_x, SO_x, and NH_3.

ANM625 **CAS:58696-86-9** **HR: 3**
AMMONIUM-3-METHYL-2,4,6-TRINITROPHENOXIDE
mf: $C_7H_8N_4O_7$ mw: 260.16

SAFETY PROFILE: May explode spontaneously in storage. When heated to decomposition or on explosion it emits toxic fumes of NO_x and NH_3.

ANM750 **CAS:13106-76-8** **HR: 3**
AMMONIUM MOLYBDATE
mf: $MoO_4{\cdot}2H_4N$ mw: 196.04

PROP: White solid. Sol in water.

SYNS: AMMONIUM PARAMOLYBDATE ☐ DIAMMONIUM MOLYBDATE ☐ MOLYBDIC ACID DIAMMONIUM SALT

TOXICITY DATA WITH REFERENCE
orl-rat LD50:333 mg/kg 28ZLA8 -,214,61
ipr-rat LDLo:203 mg/kg EQSSDX 1,1,75
orl-cat LDLo:1600 mg/kg EQSSDX 1,1,75
orl-rbt LDLo:1870 mg/kg EQSSDX 1,1,75
scu-rbt LDLo:1600 mg/kg EQSSDX 1,1,75
orl-gpg LDLo:2200 mg/kg EQSSDX 1,1,75
ipr-gpg LDLo:800 mg/kg EQSSDX 1,1,75
scu-gpg LDLo:1380 mg/kg EQSSDX 1,1,75

CONSENSUS REPORTS: Reported in EPA TSCA Inventory.

OSHA PEL: TWA 5 mg(Mo)/m³
ACGIH TLV: TWA 5 mg(Mo)/m³

SAFETY PROFILE: Poison by ingestion and intraperitoneal route. Moderately toxic by other routes. An irritant. See also MOLYBDENUM COMPOUNDS. When heated to decomposition it emits toxic fumes of NH_3 and NO_x.

ANN000 **CAS:6484-52-2** **HR: 3**
AMMONIUM(I) NITRATE(1:1)
DOT: UN 0222/UN 1942/UN 2426
mf: $HNO_3{\cdot}H_3N$ mw: 80.06

PROP: Colorless crystals. Mp: 169.6°, d: 1.725 @ 25°, bp: decomp >210°. Solubility: 192/100 @ 20°.

SYNS: AMMONIUM NITRATE ☐ AMMONIUM NITRATE, liquid (hot concentrated solution) (UN 2426) (DOT) ☐ AMMONIUM NITRATE, with >0.2% combustible substances (UN 0222) (DOT) ☐ AMMONIUM NITRATE, with not >0.2% of combustible substances (UN 1942) (DOT) ☐ AMMONIUM SALTPETER ☐ HERCO PRILLS ☐ NITRIC ACID, AMMONIUM SALT ☐ VARIOFORM I

CONSENSUS REPORTS: Community Right-To-Know List. Reported in EPA TSCA Inventory.

DOT CLASSIFICATION: 5.1; *Label:* Oxidizer (UN 2426); DOT Class: EXPLOSIVE 1.1D; *Label:* EXPLOSIVE 1.1D (UN 0222); DOT Class: 5.1; *Label:* Oxidizer (UN 1942)

SAFETY PROFILE: A powerful oxidizer and an allergen. See also NITRATES. A relatively stable explosive that has, however, caused many industrial explosions. Violent or explosive spontaneous reactions with acetic anhydride + nitric acid, ammonium sulfate + potassium, copper iron(II) sulfide, sawdust, urea, barium nitrate, hot water, and ammonium chloride + water + zinc.

Forms heat- or shock-sensitive explosive mixtures with acetic acid, aluminum + calcium nitrate + formamide (a blasting explosive), ammonia, charcoal + metal oxides (e.g., rust, copper oxide, zinc oxide above 80°C), chloride salts (e.g., ammonium chloride, calcium chloride, iron(III) chloride, and aluminum chloride), cyanoguanidine, fertilizers (e.g., super phosphate + organic materials above 90°C), hydrocarbon oils, powdered metals (e.g., aluminum, antimony, bismuth, cadmium, chromium, cobalt, copper, iron, lead, magnesium, manganese, nickel, tin, zinc, brass, stainless steel, titanium, and potassium), nonmetals (e.g., charcoal, and phosphorus), organic fuels (e.g., wax, oils, and stearates), potassium permanganate, sugar, sulfur, and trinitroanisole. Reaction with alkali metals (e.g., sodium) forms an explosive product. Ignites on contact with ammonium dichromate, potassium dichromate, potassium chromate, barium chloride, sodium chloride, potassium nitrate, and chromium(VI) salts. Can ignite when mixed with acetic acid. Use water in large amounts to fight fire. It is important that the mass of materials be kept cool and that burning be extinguished promptly. Ventilate well. May explode under confinement and high temperatures. When heated to decomposition it emits highly toxic fumes of NO_x. Can react vigorously with reducing materials. Incompatible with, (NH_4Cl + heat), (C + heat), organic matter, P, NaOCl, $NaClO_4$. Occasional explosions in presence of oil, $(NH_4)_2SO_4$ with K or Na.

ANO250 CAS:13446-48-5 HR: 3
AMMONIUM NITRITE
mf: NH_4NO_2 mw: 64.04

PROP: White to yellow crystals. Mp: explodes @ 60–70°, bp: subl @ 30° in vacuo, d: 1.69; very sol in water, dil alk.

SYN: NITROUS ACID, AMMONIUM SALT

CONSENSUS REPORTS: Reported in EPA TSCA Inventory.

DOT CLASSIFICATION: Forbidden

SAFETY PROFILE: Powerful oxidizer. See also NITRITES. Severe explosion hazard when shocked or exposed to heat (60-70°). When heated to decomposition it emits toxic fumes of NO_x and NH_3.

ANO400 HR: 3
AMMONIUM aci-NITROMETHANE
mf: $CH_7O_2N_2$ mw: 79.1

SAFETY PROFILE: A friction-sensitive explosive. See also NITRATES.

ANO500 CAS:135-20-6 HR: 3
AMMONIUM-N-NITROSOPHENYLHYDROXYLAMINE
mf: $C_6H_6N_2O_2{\cdot}H_4N$ mw: 156.19

PROP: Needles from water. Mp: 163–164°. Sol in water and alc; insol in Et_2O.

SYNS: CUPFERRON □ N-HYDROXY-N-NITROSO-BENZENAMINE, AMMONIUM SALT □ KUPFERRON (CZECH) □ NCI-C03258 □ N-NITROSOFENYLHYDROXYLAMIN AMONNY (CZECH) □ N-NITROSOPHENYLHYDROXYLAMIN AMMONIUM SALZ (GERMAN) □ N-NITROSO-PHENYLHYDROXYLAMINE AMMONIUM SALT

TOXICITY DATA WITH REFERENCE
mmo-sat 100 µg/plate ENMUDM 7(Suppl 5),1,85
cyt-grh-orl 1 ppm JCGEDO 1,75,66
eye-rbt 20 mg/24H MOD 85JCAE-,510,86
orl-rat TDLo:123 g/kg/78W-C:CAR NCITR* NCI-CG-TR-100,78
orl-mus TDLo:437 g/kg/78W-C:CAR NCITR* NCI-CG-TR-100,78
orl-rat TD:9040 mg/kg/65W-C:ETA ZEKBAI 69,103,67
orl-rat LD50:199 mg/kg GTPZAB 32(3),48,88
ipr-rat LDLo:50 mg/kg KODAK* -,-,71
ivn-mus LD50:180 mg/kg CSLNX* NX#04968

CONSENSUS REPORTS: NTP 7th Annual Report On Carcinogens. NCI Carcinogenesis Bioassay (feed); Clear Evidence: mouse, rat NCITR* NCI-CG-TR-100,78. Reported in EPA TSCA Inventory. Community Right-To-Know List.

SAFETY PROFILE: Confirmed carcinogen with experimental carcinogenic and tumorigenic data. Poison by intravenous route. An eye irritant. Solutions with thorium salts are unstable explosives above 15°C. Solutions with titanium or zirconium salts are unstable explosives above 40°C. When heated to decomposition it emits very toxic NH_3 and NO_x. See also N-NITROSO COMPOUNDS and AMINES.

ANO750 CAS:1113-38-8 HR: 3
AMMONIUM OXALATE
mf: $C_2H_2O_4{\cdot}2H_3N$ mw: 124.12

PROP: Colorless crystals. Mp: decomp, d: 1.50. Sltly sol in water.

SYNS: ETHANEDIOIC ACID DIAMMONIUM SALT □ OXALIC ACID, DIAMMONIUM SALT

DOT CLASSIFICATION: 6.1; *Label:* KEEP AWAY FROM FOOD

CONSENSUS REPORTS: Reported in EPA TSCA Inventory.

SAFETY PROFILE: A poison. Can react violently with (NaOCl + ammonium acetate). When heated to decomposition it can emit toxic fumes of NH_3 and NO_x. See also OXALATES.

ANO875 HR: 3
AMMONIUM OXOFLUOROMOLYBDATE
mf: $F_4MoO_2{\cdot}2H_4N$ mw: 240.04

TOXICITY DATA WITH REFERENCE
orl-rat TDLo:1400 mg/kg (28D male):REP SHKKAN 23,859,81
orl-rat LD50:242 mg/kg SHKKAN 23,237,81
ipr-rat LD50:58 mg/kg SHKKAN 23,859,81
scu-rat LD50:152 mg/kg SHKKAN 23,859,81
orl-mus LD50:250 mg/kg SHKKAN 23,859,81
ipr-mus LD50:51 mg/kg SHKKAN 23,859,81
scu-mus LD50:72 mg/kg SHKKAN 23,859,81

SAFETY PROFILE: Poison by ingestion, subcutaneous,

and intraperitoneal routes. Experimental reproductive effects. When heated to decomposition it emits toxic fumes of F^-, NO_x, and NH_3. See also MOLYBDENUM COMPOUNDS.

ANO900 CAS:12208-54-7 HR: 1
AMMONIUM PARATUNGSTATE HEXAHYDRATE
mf: $H_{24}N_6O_{24}W_7{\cdot}6H_2O$ mw: 1887.37

TOXICITY DATA WITH REFERENCE
orl-rat LD50:11,300 mg/kg AIHAAP 30,470,69

ACGIH TLV: TWA 1 mg(W)/m³; STEL 3 mg(W)/m³
NIOSH REL: 10H TWA 1 mg(W)/m³

SAFETY PROFILE: Mildly toxic by ingestion. When heated to decomposition it emits toxic fumes of NO_x and W.

ANP000 HR: 3
AMMONIUM PENTA PEROXODICHROMATE
mf: $Cr_2H_8N_2O_{12}$ mw: 332.2

CONSENSUS REPORTS: Chromium and its compounds are on the Community Right-To-Know List.

OSHA PEL: CL 0.1 mg(CrO_3)/m³
ACGIH TLV: TWA 0.05 mg(Cr)/m³0.05 mg(Cr)/m³0.05 mg(Cr)/m³; Confirmed Human Carcinogen
NIOSH REL: (Chromium(VI)) TWA 25 µg(Cr(VI))/m³; CL 50 µg/m³/15M

SAFETY PROFILE: A confirmed carcinogen. An unstable compound. Detonation can be initiated by heat, friction or impact. See also CHROMIUM COMPOUNDS. Explodes @ 50°. When heated to decomposition it emits toxic fumes of NO_x.

ANP250 HR: 3
AMMONIUM PERCHLORATE
mf: NH_4ClO_4 mw: 117.50

PROP: White crystals. Mp: decomp, d: 1.95.

SAFETY PROFILE: Easily ignited by friction. Can explode when mixed with sugar, charcoal, or on contact with hot copper pipes. Can be sensitized by nitryl perchlorate, KIO_4, $KMnO_4$, metals (as co-crystallized impurities). It becomes impact-sensitive when contaminated by powdered carbon, ferrocene, sulfur, organic matter, or powdered metals. When heated to decomposition it emits toxic fumes of NH_3, Cl^-, and NO_x. See PERCHLORATES.

ANP500 HR: 3
AMMONIUM PERCHLORYL AMIDE
mf: $H_5N_2O_3Cl$ mw: 116.6

PROP: Mp: 80°.

SAFETY PROFILE: A shock-sensitive explosive. May detonate @ 80°. When heated to decomposition it emits very toxic fumes of NH_3, NO_x, and Cl^-.

ANP625 CAS:3825-26-1 HR: 3
AMMONIUM PERFLUOROOCTANOATE
mf: $C_8F_{15}O_2{\cdot}H_4N$ mw: 431.13

PROP: Solid.

SYNS: AMMONIUM PENTADECAFLUOROOCTANATE □ AMMONIUM PERFLUOROCAPRILATE □ AMMONIUM PERFLUOROCAPRYLATE □ APFO □ FC-143 □ PERFLUOROAMMONIUM OCTANOATE

TOXICITY DATA WITH REFERENCE
skn-rbt 500 mg MOD TXAPA9 81,348,85
eye-rbt 500 mg/24H MOD AIHAAP 41,576,80
ihl-rat TCLo:25 mg/m³/6H (6-15D preg):REP FAATDF 4,429,84
orl-rat LD50:430 mg/kg AIHAAP 41,576,80
ihl-rat LC50:980 mg/m³/4H FCTOD7 24,1325,86
skn-rat LD50:7 g/kg TXAPA9 81,348,85
skn-rbt LD50:4300 mg/kg TXAPA9 81,348,85

CONSENSUS REPORTS: Reported in EPA TSCA Inventory.

ACGIH TLV: 0.01 mg/m³; Animal Carcinogen

SAFETY PROFILE: Confirmed carcinogen. Poison by inhalation. Moderately toxic by ingestion. An eye and skin irritant. Experimental reproductive effects. When heated to decomposition it emits toxic fumes of F^- and NH_3.

ANP750 HR: 3
AMMONIUM-m-PERIODATE
mf: NH_4IO_4 mw: 209

PROP: Colorless crystals. Mp: explodes, d: 3.056.

SAFETY PROFILE: A contact explosive. See also IODATES and IODIDES. Heat, impact, and touch as from a scoop or an abrasive impact may cause explosion. When heated to decomposition it can emit toxic fumes of NH_3, NO_x, and I^-.

ANQ250 HR: 3
AMMONIUM PEROXO BORATE
mf: $BH_4NO_3{\cdot}1/2H_2O$ mw: 85.86

PROP: White crystals. Mp: decomp; sltly sol in water.

SAFETY PROFILE: Potentially explosive by heat, friction, or impact. See also BORON COMPOUNDS. When heated to decomposition it emits toxic fumes of NO_x and NH_3.

ANQ500 HR: 2
AMMONIUM PEROXO DISULFATE
mf: $H_8N_2O_8S_2$ mw: 228.2

SAFETY PROFILE: See also SULFATES. An unstable compound. Detonated via heat, friction, or impact. A powerful oxidizer. Incompatible with Al, H_2O, powdered metal, Ag salts, Fe, Na_2O_2, zinc, NH_3. When heated to decomposition it emits very toxic fumes of SO_x, NO_x, and NH_3.

ANQ750 **HR: 3**
AMMONIUM PEROXYCHROMATE
mf: $(NH_4)_3CrO_2$ mw: 234.1

PROP: Red-brown crystals. Mp: decomp @ 40°, bp: explodes @ 50°.

CONSENSUS REPORTS: Chromium and its compounds are on the Community Right-To-Know List.

OSHA PEL: CL 0.1 $mg(CrO_3)/m^3$
ACGIH TLV: TWA 0.05 $mg(Cr)/m^3$; Confirmed Human Carcinogen
NIOSH REL: (Chromium(VI)) TWA 25 $\mu g(Cr(VI))/m^3$; CL 50 $\mu g/m^3$/15M

SAFETY PROFILE: A confirmed carcinogen. A poison. See also CHROMIUM COMPOUNDS. Moderately flammable by chemical reaction with reducing agents. A powerful oxidizer. Moderately explosive when heated. When heated to decomposition it emits toxic fumes of NO_x, and NH_3.

ANR000 **CAS:7727-54-0** **HR: 3**
AMMONIUM PERSULFATE
DOT: UN 1444
mf: $O_8S_2 \cdot 2H_4N$ mw: 228.22

$H_4NOSO_2OOSO_2ONH_4$

PROP: Colorless, white, monoclinic crystals. Mp: decomp @ 120°, d: 1.982. Stable as dry solid; decomposes in H_2O forming O_2.

SYNS: AMMONIUM PEROXYDISULFATE □ PERSULFATE d'AMMONIUM (FRENCH)

TOXICITY DATA WITH REFERENCE
ipr-rat LD50:226 mg/kg DTLVS* 4,327,80
ivn-rbt LD50:178 mg/kg DTLVS* 4,327,80
orl-rat LD50:689 mg/kg 85INA8 5,468,86

CONSENSUS REPORTS: Reported in EPA TSCA Inventory.

DOT CLASSIFICATION: 5.1; *Label:* Oxidizer

SAFETY PROFILE: Poison by intravenous and intraperitoneal routes. Moderately toxic by ingestion. A powerful oxidizer that can react vigorously with reducing agents. Releases oxygen when heated. Mixtures with sodium peroxide are explosives sensitive to friction, heating above 75°C, or contact with CO_2 or water. Mixtures with (powdered aluminum + water) or (zinc + ammonia) are explosive. Violent reaction with iron or solutions of ammonia + silver salts. Solution with sulfuric acid is a strong oxidizing cleaning solution. When heated to decomposition it emits toxic fumes of SO_x, NH_3, and NO_x.

ANR250 **CAS:1074-52-8** **HR: 3**
AMMONIUM PHENYLDITHIOCARBAMATE
mf: $C_7H_6NS_2 \cdot H_4N$ mw: 186.31

PROP: Yellow prisms. Mp: 141–143° decomp. Very sol in water.

SYN: PHENYLDITHIOCARBAMIC ACID, AMMONIUM SALT

TOXICITY DATA WITH REFERENCE
ipr-mus LDLo:100 mg/kg JMPCAS 5,846,62

CONSENSUS REPORTS: Reported in EPA TSCA Inventory.

SAFETY PROFILE: Poison by intraperitoneal route. See also CARBAMATES. When heated to decomposition it emits very toxic fumes of NO_x, SO_x and NH_3.

ANR500 **CAS:7783-28-0** **HR: 2**
AMMONIUM PHOSPHATE, DIBASIC
mf: $H_6N_2 \cdot H_3O_4P$ mw: 132.08

PROP: White crystals or powder; salty taste. D: 1.619, mp: 185° (decomp). Sol in water; insol in alc.

SYNS: AMMONIUM PHOSPHATE □ DIAMMONIUM HYDROGEN PHOSPHATE □ DIBASIC AMMONIUM PHOSPHATE □ SECONDARY AMMONIUM PHOSPHATE

CONSENSUS REPORTS: Reported in EPA TSCA Inventory.

SAFETY PROFILE: Low to moderate toxicity. See also PHOSPHATES. When heated to decomposition it emits very toxic fumes of PO_x, NO_x, and NH_3.

ANR750 **CAS:7772-76-1** **HR: 2**
AMMONIUM PHOSPHATE, MONOBASIC
mf: $NH_4H_2PO_4$ mw: 115

PROP: Brilliant-white crystals or powder. D: 1.803 @ 19°, mp: 190°. Sol in water.

SAFETY PROFILE: Incompatible with NaOCl.

ANS000 **HR: 3**
AMMONIUM PHOSPHIDE
mf: $P(NH_4)_3$ mw: 85.07

SAFETY PROFILE: Poison by inhalation and ingestion. See also PHOSPHINE. When heated to decomposition it emits toxic fumes of PO_x, NO_x, and NH_3.

ANS250 **CAS:51503-61-8** **HR: 3**
AMMONIUM PHOSPHITE
mf: H_6NO_3P mw: 99.04

PROP: Needles from water by slow evap.

SYN: AMMONIUM ORTHOPHOSPHITE

TOXICITY DATA WITH REFERENCE
ihl-rat LCLo:580 ppm/1H ZGSHAM 25,279,33
ihl-gpg LCLo:288 ppm/2H ZGSHAM 25,279,33

SAFETY PROFILE: Poison by inhalation. When heated to decomposition it emits very toxic fumes of NO_x, NH_3, and PO_x.

ANS500 **CAS:131-74-8** **HR: 3**
AMMONIUM PICRATE
DOT: UN 0004/UN 1310
mf: $C_6H_3N_3O_7 \cdot H_3N$ mw: 246.16

PROP: Red or yellow, rhombic crystals. D: 1.719, mp: decomp, bp: expl @ 423°. Solubility: 1.1/100 @ 20°.

SYNS: AMMONIUM CARBAZOATE □ AMMONIUM PICRATE, dry or wetted with <10% water, by weight (UN 0004) (DOT) □ AMMONIUM PICRATE, wetted with not <10% water, by weight (UN 1310) (DOT) □ AMMONIUM PICRONITRATE □ EXPLOSIVE D □ OBELINE PICRATE □ PHENOL, 2,4,6-TRINITRO-, AMMONIUM SALT (9CI) □ PICRATOL □ PICRIC ACID, AMMONIUM SALT □ RCRA WASTE NUMBER P009 □ 2,4,6-TRINITROPHENOL AMMONIUM SALT

CONSENSUS REPORTS: Reported in EPA TSCA Inventory.

DOT CLASSIFICATION: EXPLOSIVE 1.1D; *Label:* EXPLOSIVE 1.1D (UN 0004)

SAFETY PROFILE: An allergen. Moderately irritating to skin, eyes, and mucous membranes. Moderately flammable by spontaneous chemical reaction. A powerful oxidizer that reacts vigorously with reducing materials. Dangerous explosive when shocked or heated. The presence of trace metals increases its heat sensitivity. See PICRIC ACID, NITRATES, and EXPLOSIVES, HIGH. When heated to decomposition it emits highly toxic fumes of NO_x.

ANT000 **CAS:12259-92-6** **HR: 3**
AMMONIUM POLYSULFIDE (solution)
DOT: UN 2818

SYNS: AMMONIUM POLYSULFIDE, solution (DOT) □ AMMONIUM SULFIDE (POLY-) □ AMMONIUM SULFIDE, solution, red □ AMMONIUM TRISULFIDE □ AP-S □ DIAMMONIUM TRISULFIDE

CONSENSUS REPORTS: Reported in EPA TSCA Inventory.

DOT CLASSIFICATION: 8; *Label:* Corrosive, Poison

SAFETY PROFILE: Poison due to presence of sulfides and H_2S. See also AMMONIUM HYDROSULFIDE. When heated to decomposition it emits very toxic fumes of NO_x, SO_x, and H_2S.

ANT250 **CAS:64046-00-0** **HR: 2**
AMMONIUM POTASSIUM SELENIDE mixed with AMMONIUM POTASSIUM SULFIDE
mf: $H_4KNSe + NH_4KS$ mw: 136.11 + 89.21 = 225.32

SYN: AMMONIUM POTASSIUM SULFIDE mixed with AMMONIUM POTASSIUM SELENIDE

TOXICITY DATA WITH REFERENCE
orl-rat TDLo:450 mg/kg/2Y-C:ETA CNREA8 3,230,43

CONSENSUS REPORTS: Selenium and its compounds are on the Community Right-To-Know List.

OSHA PEL: TWA 0.2 mg(Se)/m^3
ACGIH TLV: TWA 0.2 mg(Se)/m^3
DFG MAK: 0.1 mg(Se)/m^3

SAFETY PROFILE: Questionable carcinogen with experimental tumorigenic data. See also SULFIDES. When heated to decomposition it emits very toxic fumes of NO_x, NH_3, SO_x, and Se.

ANT300 **CAS:19441-09-9** **HR: 3**
AMMONIUM REINECKATE HYDRATE
mf: $C_4H_{10}N_7S_4 \cdot Cr \cdot H_2O$ mw: 354.47

SYN: CHROMATE(1-), DIAMMINETETRAKIS(ISOTHIOCYANATO)-, AMMONIUM, HYDRATE

TOXICITY DATA WITH REFERENCE
scu-mus LD50:110 mg/kg ABMGAJ 3,28,59
ivn-mus LD50:180 mg/kg CSLNX* NX#01547

OSHA PEL: CL 0.1 mg(CrO_3)/m^3

SAFETY PROFILE: Poison by subcutaneous and intravenous routes. When heated to decomposition it emits toxic fumes of NO_x, SO_x, and CR.

ANT500 **CAS:6381-61-9** **HR: 2**
AMMONIUM SACCHARIN
mf: $C_7H_8N_2O_3S$ mw: 200.23

PROP: White crystals or crystalline powder; intense sweet taste. Sol in water.

SYNS: 1,2-BENZISOTHIAZOLIN-3-ONE 1,1-DIOXIDE AMMONIUM SALT □ DARAMIN □ SACCHARIN AMMONIUM □ SACCHARINATE AMMONIUM

TOXICITY DATA WITH REFERENCE
eye-rbt 100 mg SEV LPPTAK 24,598,76

SAFETY PROFILE: A severe eye irritant. When heated to decomposition emits toxic fumes of NO_x.

ANT600 **CAS:528-94-9** **HR: 3**
AMMONIUM SALICYLATE
mf: $C_7H_5O_3 \cdot H_4N$ mw: 155.17

SYNS: 2-HYDROXYBENZOIC ACID MONOAMMONIUM SALT □ SALICYLIC ACID, MONOAMMONIUM SALT □ SALICYL-VASOGEN

TOXICITY DATA WITH REFERENCE
orl-hmn TDLo:57 mg/kg:GIT JPETAB 36,319,29
par-rat LDLo:600 mg/kg JPETAB 36,319,29
par-mus LDLo:550 mg/kg JPETAB 36,319,29
ivn-dog LDLo:467 mg/kg AIPTAK 51,398,35

CONSENSUS REPORTS: Reported in EPA TSCA Inventory.

SAFETY PROFILE: Poison by intravenous route. Moderately toxic by parenteral route. Human systemic effects by ingestion: nausea or vomiting. When heated to decomposition it emits toxic vapors of NH_3.

ANU000 **HR: 2**
AMMONIUM SALTS of PHOSPHATIDIC ACIDS

TOXICITY DATA WITH REFERENCE
orl-rat LD50:5000 mg/kg FAONAU 53A,215,74
ivn-rat LD50:2000 mg/kg FAONAU 53A,215,74
orl-dog LD50:2000 mg/kg FAONAU 53A,215,74

orl-rbt LD50:5000 mg/kg FAONAU 53A,215,74
ivn-gpg LD50:2000 mg/kg FAONAU 53A,215,74

SAFETY PROFILE: Moderately toxic by ingestion and intravenous routes. When heated to decomposition it emits very toxic fumes of NH_3, PO_x, and NO_x.

ANU200 CAS:1002-89-7 HR: 1
AMMONIUM STEARATE
mf: $C_{18}H_{35}O_2 \cdot H_4N$ mw: 301.58

SYNS: AMMONIUM STEARATE ☐ OCTADECANOIC ACID, AMMONIUM SALT ☐ STEARIC ACID, AMMONIUM SALT

CONSENSUS REPORTS: Reported in EPA TSCA Inventory.

ACGIH TLV: TWA 10 mg/m^3

SAFETY PROFILE: A nuisance dust. When heated to decomposition it emits toxic vapors of NH_3.

ANU650 CAS:7773-06-0 HR: 2
AMMONIUM SULFAMATE
mf: $H_2NO_3S \cdot H_4N$ mw: 114.14

PROP: Deliquescent, hygroscopic, crystalline material (white crystalline solid). Bp: 160° (decomp), mp: 131°. Sol in water, liq NH_3, formamide, and glycerol.

SYNS: AMCIDE ☐ AMICIDE ☐ AMMAT ☐ AMMATE ☐ AMMONIUM AMIDOSULFONATE ☐ AMMONIUM AMIDOSULPHATE ☐ AMMONIUMSALZ der AMIDOSULFONSAEURE (GERMAN) ☐ AMMONIUM SULPHAMATE ☐ AMS ☐ IKURIN ☐ MONOAMMONIUM SULFAMATE ☐ SULFAMATE ☐ SULFAMIC ACID, MONOAMMONIUM SALT ☐ SULFAMINSAEURE (GERMAN)

TOXICITY DATA WITH REFERENCE
orl-rat LD50:2 g/kg AMIHAB 14,178,56
ipr-rat LDLo:800 mg/kg JIHTAB 25,26,43
orl-mus LD50:3100 mg/kg GTPZAB 7(5),56,63

CONSENSUS REPORTS: Reported in EPA TSCA Inventory.

OSHA PEL: TWA 10 mg/m^3; Respirable Fraction: 5 mg/m^3
ACGIH TLV: TWA 10 mg/m^3
DFG MAK: 15 mg/m^3

SAFETY PROFILE: Moderately toxic by ingestion and intraperitoneal routes. Somewhat explosive when heated or by spontaneous chemical reaction in a hot acid solution. A powerful oxidizer. When heated to decomposition it emits very toxic fumes of NH_3, NO_x, and SO_x. See also SULFONATES and SULFAMIC ACID.

ANU750 CAS:7783-20-2 HR: 2
AMMONIUM SULFATE (2:1)
mf: $H_8N_2O_4S$ mw: 132.16

$SO_4 \cdot (NH_4)_2$

PROP: White rhombic crystals. Mp: >280° (decomp), d: 1.77. Sol in water; insol in alc.

SYNS: AMMONIUM SULPHATE ☐ DIAMMONIUM SULFATE ☐ SULFURIC ACID, DIAMMONIUM SALT

TOXICITY DATA WITH REFERENCE
orl-man TDLo:1500 mg/kg:GIT GISAAA 42(2),100,77
orl-rat LD50:3000 mg/kg CNJMAQ 12,216,48
ipr-mus LD50:610 mg/kg UCPHAQ 2,1,41
orl-dom LDLo:3500 mg/kg AJVRAH 32,1229,71

CONSENSUS REPORTS: Community Right-To-Know List. Reported in EPA TSCA Inventory.

SAFETY PROFILE: Moderately toxic by several routes. Human systemic effects by ingestion: hypermotility, diarrhea, nausea or vomiting. See also SULFATES. Incandescent reaction on heating with potassium chlorate. Reaction with sodium hypochlorite gives the unstable explosive nitrogen trichloride. Incompatible with (K + NH_4NO_3), KNO_2, (NaK + NH_4NO_3). When heated to decomposition it emits very toxic fumes of NO_x, NH_3, and SO_x.

ANV800 CAS:13820-41-2 HR: 3
AMMONIUM TETRACHLOROPLATINATE

PROP: Red-brown solid or crystals. Mp: 140–150° (decomp), d: 2.936. Sol in water.
mf: $Cl_4Pt \cdot 2H_4N$ mw: 372.99

SYNS: PLATINATE(2-), TETRACHLORO-, DIAMMONIUM ☐ TETRAMINE PLATINUM(II) CHLORIDE

TOXICITY DATA WITH REFERENCE
ipr-mus LD50:60 mg/kg TXAPA9 49,41,79

CONSENSUS REPORTS: Reported in EPA TSCA Inventory.

OSHA PEL: TWA 0.002 mg(Pt)/m^3
ACGIH TLV: TWA 0.002 mg(Pt)/m^3

SAFETY PROFILE: Poison by intraperitoneal route. When heated to decomposition it emits toxic fumes of NH_3, Cl^-, and Pt.

ANW250 HR: 3
AMMONIUM TETRANITROPLATINATE(II)
mf: $H_8N_6O_8Pt$ mw: 415.3

SAFETY PROFILE: An explosively unstable compound. Sensitive to heat. See also NITRATES and PLATINUM COMPOUNDS.

ANW500 HR: 3
AMMONIUM TETRAPEROXO CHROMATE
mf: $CrH_{12}N_3O_8$ mw: 234.2

CONSENSUS REPORTS: Chromium and its compounds are on the Community Right-To-Know List.

OSHA PEL: CL 0.1 mg(CrO_3)/m^3
ACGIH TLV: TWA 0.05 mg(Cr)/m^3; Confirmed Human Carcinogen
NIOSH REL: (Chromium(VI)) TWA 25 µg(Cr(VI))/m^3; CL 50 µg/m^3/15M

SAFETY PROFILE: A confirmed carcinogen. A poison. Impact explodes @ 50° or in contact with H_2SO_4. See also CHROMIUM COMPOUNDS. Incompatible with

H_2SO_4. When heated to decomposition it emits toxic fumes of NO_x.

ANW750 **CAS:1762-95-4** **HR: 3**
AMMONIUM THIOCYANATE
mf: $CNS \cdot H_4N$ mw: 76.13

PROP: Colorless solid or deliquescent crystals. Mp: 149.6°, bp: decomp @ 170°, d: 1.305. Very sol in H_2O, EtOH; sol in MeOH, Me_2CO; prac insol $CHCl_3$, and EtOAc.

SYNS: AMMONIUM RHODANATE ☐ AMMONIUM RHODANIDE ☐ AMMONIUM SULFOCYANATE ☐ AMMONIUM SULFOCYANIDE ☐ AMTHIO ☐ RHODANID ☐ RHODANIDE ☐ TRANS-AID ☐ USAF EK-P-433 ☐ WEEDAZOL TL

TOXICITY DATA WITH REFERENCE
orl-hmn TDLo:430 mg/kg:GIT,CNS DAKMAJ 102,606,11
orl-rat LD50:750 mg/kg GTPZAB 30(10),51,86
orl-mus LD50:500 mg/kg GTPZAB 30(10),51,86
ipr-mus LDLo:500 mg/kg NTIS** AD277-689
orl-gpg LD50:500 mg/kg GTPZAB 30(10),51,86

CONSENSUS REPORTS: Reported in EPA TSCA Inventory. EPA Genetic Toxicology Program.

SAFETY PROFILE: Poison by ingestion and intraperitoneal routes. Human systemic effects by ingestion: hallucinations and distorted perceptions, nausea or vomiting, and other gastrointestinal effects. See also THIOCYANATES. When heated to decomposition it emits toxic fumes of NH_3, NO_x, SO_x, and CN^-. Incompatible with $KClO_3$ and mixtures with $Pb(NO_3)_2$.

ANX750 **HR: 3**
AMMONIUM TRICHLOROACETATE
mf: $NH_4O_2CCCl_3$ mw: 180.6

SAFETY PROFILE: Poison by inhalation and ingestion. A powerful irritant. When heated to decomposition or on contact with acid or acid fumes it emits toxic fumes of Cl^-, NH_3, and NO_x. Incompatible with water or steam.

ANX800 **CAS:15660-29-4** **HR: 3**
AMMONIUM TRIFLUOROSTANNITE
mf: $F_3Sn \cdot H_4N$ mw: 193.74

TOXICITY DATA WITH REFERENCE
ivn-mus LD50:18 mg/kg CSLNX* NX#00130

OSHA PEL: TWA 2 mg(Sn)/m³; TWA 2.5 mg(F)/m³
ACGIH TLV: TWA 2 mg(Sn)/m³
NIOSH REL: (Fluorides, Inorganic) TWA 2.5 mg(F)/m³

SAFETY PROFILE: Poison by intravenous route. See also FLUORIDES and TIN COMPOUNDS. When heated to decomposition it emits very toxic fumes of NH_3, NO_x, and fluorides.

ANX875 **CAS:63839-60-1** **HR: 3**
AMMONIUM-2,4,5-TRINITROIMIDAZOLIDE
mf: $C_3H_4N_6O_6$ mw: 220.10

SAFETY PROFILE: An explosive comparable in power to RDX, but more thermally stable. Upon decomposition it emits toxic fumes of NO_x and NH_3. See also EXPLOSIVES.

ANY250 **CAS:7803-55-6** **HR: 3**
AMMONIUM VANADATE
DOT: UN 2859
mf: $O_3V \cdot H_4N$ mw: 116.99

PROP: Colorless to yellow crystals or solid. Mp: 200° (decomp), d: 2.326.

SYNS: AMMONIUM METAVANADATE (DOT) ☐ RCRA WASTE NUMBER P119 ☐ VANADIC ACID, AMMONIUM SALT

TOXICITY DATA WITH REFERENCE
mrc-bcs 300 mmol/L MUREAV 77,109,80
ipr-ham TDLo:22,500 μg/kg (5-10D preg):REP ENVRAL 29,256,82
ipr-ham TDLo:11,280 μg/kg (5-10D preg):TER ENVRAL 29,256,82
orl-rat LD50:58,100 μg/kg GISAAA 57(7-8),26,92
ihl-rat LC50:7800 μg/m³/4H GISAAA 57(7-8),26,92
skn-rat LD50:2102 mg/kg GISAAA 57(7-8),26,92
ipr-rat LD50:18 mg/kg ATXKA8 16,182,56
scu-rat LD50:23 mg/kg ATXKA8 16,182,56
itr-rat LDLo:8 mg/kg ATXKA8 16,182,56
scu-mus LDLo:16 mg/kg AJSNAO 1,347,17
ivn-rbt LDLo:1 mg/kg AJSNAO 1,347,17
scu-gpg LDLo:643 μg/kg AJSNAO 1,347,17

CONSENSUS REPORTS: Reported in EPA TSCA Inventory. EPA Genetic Toxicology Program.

ACGIH TLV: TWA 0.05 mg(V_2O_5)/m³
NIOSH REL: (Vanadium Compounds) CL 0.05 mg(V)/m³/15M
DOT CLASSIFICATION: 6.1; *Label:* Poison

SAFETY PROFILE: Poison by ingestion, subcutaneous, intravenous, intratracheal, and intraperitoneal routes. Moderately toxic by skin contact. An experimental teratogen. Other experimental reproductive effects. Mutation data reported. See also VANADIUM COMPOUNDS. When heated to decomposition it emits toxic fumes of NH_3, VO_x, and NO_x.

ANY500 **CAS:69782-62-3** **HR: 3**
AMMONIUM VANADI-ARSENATE
mf: $H_{16}N_4O_2 \cdot As_2O_5V_2$ mw: 515.92

TOXICITY DATA WITH REFERENCE
scu-rat LDLo:34 mg/kg AJSNAO 1,347,17
ivn-rbt LDLo:6 mg/kg AJSNAO 1,347,17

CONSENSUS REPORTS: Arsenic and its compounds are on the Community Right-To-Know List.

NIOSH REL: (Vanadium Compounds) CL 0.05 mg(V)/m³/15M; CL 2 μg(As)/m³/15M

SAFETY PROFILE: Poison by subcutaneous and intravenous routes. See ARSENIC and VANADIUM COMPOUNDS. When heated to decomposition it emits very toxic fumes of NO_x and As.

A

ANY750 **HR: 3**
AMMONIUM VANADO-ARSENATE
mf: $H_{40}N_{10}O_5 \cdot 3As_2O_5 \cdot 4O_4V_2$ mw: 1228.78

TOXICITY DATA WITH REFERENCE
scu-rat LDLo: 246 mg/kg AJSNAO 1,347,17
ivn-rbt LDLo: 75 mg/kg AJSNAO 1,347,17

CONSENSUS REPORTS: Arsenic and its compounds are on the Community Right-To-Know List.

ACGIH TLV: TWA 0.2 mg(As)/m^3 (Proposed: 0.01 mg(As)/m^3; Human Carcinogen)
NIOSH REL: (Vanadium Compounds) CL 0.05 mg(V)/m^3/15M; (Arsenic, Inorganic) CL 2 µg(As)/m^3/15M

SAFETY PROFILE: Poison by subcutaneous and intravenous routes. See ARSENIC and VANADIUM COMPOUNDS. When heated to decomposition it emits very toxic NO_x, NH_3, and As.

ANZ000 **CAS:3566-10-7** **HR: 3**
AMOBAM

PROP: Used to react with zinc sulfate to form Zineb. (28ZEAL 5,11,76)

SYNS: AMBAM □ DITHANE STAINLESS

TOXICITY DATA WITH REFERENCE
orl-rat LD50: 395 mg/kg 28ZEAL 5,11,76

CONSENSUS REPORTS: Reported in EPA TSCA Inventory.

SAFETY PROFILE: Poison by ingestion. When heated to decomposition it emits very toxic fumes of SO_x and NO_x.

AOA075 **HR: 3**
AMOSULALOL HYDROCHLORIDE
mf: $C_{18}H_{24}N_2O_5S \cdot ClH$ mw: 416.96

SYN: YM-09538

TOXICITY DATA WITH REFERENCE
orl-rat TDLo: 3300 mg/kg (7-17D preg): TER KSRNAM 19,6105,85
orl-rat TDLo: 1100 mg/kg (7-17D preg): REP KSRNAM 19,6105,85
scu-rat LD50: 541 mg/kg KSRNAM 19,6121,85
ivn-rat LD50: 105 mg/kg KSRNAM 19,6121,85
orl-mus LD50: 5740 mg/kg KSRNAM 19,6121,85
scu-mus LD50: 394 mg/kg KSRNAM 19,6121,85
ivn-mus LD50: 104 mg/kg KSRNAM 19,6121,85

SAFETY PROFILE: Poison by subcutaneous and intravenous routes. An experimental teratogen. Experimental reproductive effects. When heated to decomposition it emits toxic fumes of SO_x, NO_x, and HCl.

AOA095 **CAS:14028-44-5** **HR: 3**
AMOXAPINE
mf: $C_{17}H_{16}ClN_3O$ mw: 313.79

PROP: Crystals from benzene/petr ether. Mp: 178–180°.

SYNS: AMOXEPINE □ ASENDIN □ 2-CHLORO-11-(1-PIPERAZINYL) DIBENZ(b,f)(1,4)OXAZEPINE □ CL 67772 □ DEMOLOX □ MOXADIL

TOXICITY DATA WITH REFERENCE
orl-wmn LDLo: 40 mg/kg: CVS AEMED3 17,274,88
orl-man TDLo: 70 mg/kg/3W-I: CVS JCLPDE 45,358,84
orl-cld LDLo: 25 mg/kg: EYE,CNS SMJOAV 76,543,83
orl-cld TDLo: 10 mg/kg: CVS JAMAAP 250,1069,83
orl-man TDLo: 4286 µg/kg: MUS SMJOAV 77,94,84
orl-wmn LDLo: 40 mg/kg: CNS,CVS,BPR JAMAAP 250,1069,83
orl-wmn TDLo: 5 mg/kg: BAH JAMAAP 250,1069,83
orl-wmn TDLo: 17 mg/kg: BAH JTCTDW 20,101,83
orl-wmn TDLo: 15 mg/kg/5D-I: GLN JCLPDE 44,347,83
orl-man TDLo: 14 mg/kg: SYS JAMAAP 248,3141,82
orl-man TDLo: 4285 µg/kg/2D-I: BAH AJPSAO 140,115,83
unr-man TDLo: 5714 µg/kg/2D-I: PNS SMJOAV 76,1077,83
orl-rat LD50: 313 mg/kg IYKEDH 11,811,80
ipr-rat LD50: 201 mg/kg AIPTAK 233,107,78
scu-rat LD50: 4500 mg/kg KSRNAM 5,1852,71
orl-mus LD50: 122 mg/kg AIPTAK 233,107,78

SAFETY PROFILE: Poison by ingestion and intraperitoneal routes. Human systemic effects by ingestion: acute renal failure, acute tubular necrosis, BP lowering, coma, convulsions, decreased body temperature, EKG changes, excitement, fasciculations, heart rate changes, hyperglycemia, increased body temperature, miosis, muscle contraction or spasticity, pulse rate increase, rigidity, somnolence. When heated to decomposition it emits toxic fumes of NO_x and Cl^-.

AOA100 **CAS:61336-70-7** **HR: 2**
AMOXICILLIN TRIHYDRATE
mf: $C_{16}H_{19}N_3O_5S \cdot 3H_2O$ mw: 419.50

SYNS: α-AMINO-p-HYDROXYBENZYLPENICILLIN TRIHYDRATE □ (2S-(2-α,5-α,6-β(S*)))-6-((AMINO(4-HYDROXYPHENYL)ACETYL)AMINO)-3,3-DIMETHYL-7-OXO-4-THIA-1-AZABICYCLO(3.2.0)HEPTANE-2-CARBOXYLIC ACID TRIHYDRATE □ BRL 2333 TRIHYDRATE

TOXICITY DATA WITH REFERENCE
orl-mus TDLo: 9100 mg/kg (7-13D preg): TER KSRNAM 7,3113,73
orl-rat TDLo: 162 g/kg (25W male): REP KSRNAM 7,3074,73
ipr-rat LD50: 2870 mg/kg KSRNAM 7,3040,73
ipr-mus LD50: 3590 mg/kg KSRNAM 7,3040,73

SAFETY PROFILE: Moderately toxic. An experimental teratogen. Other experimental reproductive effects. When heated to decomposition it emits toxic fumes of SO_x and NO_x.

AOA125 **CAS:61-19-8** **HR: 1**
AMP
mf: $C_{10}H_{14}N_5O_7P$ mw: 347.26

PROP: Solid. Mp: 196–200°.

SYNS: ADENOSINE-5′-MONOPHOSPHATE □ ADENOSINE-5-MONOPHOSPHORIC ACID □ ADENOSINE-5′-MONOPHOSPHORIC ACID □ ADENOSINE PHOSPHATE □ ADENOSINE-5′-PHOSPHATE □ ADENOSINE-5′-PHOSPHORIC ACID □ ADENOVITE □ ADENYL □ ADENYLIC ACID □ tert-ADENYLIC ACID □ A5MP □ 5-AMP □ 5′-AMP □ AMP (nucleotide) □ CARDIOMONE □ ERGADENYLIC ACID □ LYCEDAN

☐ MUSCLE ADENYLIC ACID ☐ MY-B-DEN ☐ MYOSTON ☐ NSC-20264 ☐ PHOSADEN ☐ PHOSPHADEN ☐ PHOSPHENTASIDE

TOXICITY DATA WITH REFERENCE
oms-hmn:oth 100 µmol/L JIDEAE 65,52,75
oms-mus:oth 50 µmol/L JIDEAE 66,313,76
ipr-rat TDLo:2800 mg/kg (7-13D preg):REP OYYAA2 4,625,70
ipr-mus LD50:4 g/kg PCJOAU 20,160,86

CONSENSUS REPORTS: Reported in EPA TSCA Inventory.

SAFETY PROFILE: Slightly toxic by intraperitoneal route. Experimental reproductive effects. Human mutation data reported. When heated to decomposition it emits toxic fumes of PO_x and NO_x.

AOA250 CAS:60-15-1 HR: 3
AMPHETAMINE
mf: $C_9H_{13}N$ mw: 135.23

SYNS: β-AMINOPROPYLBENZENE ☐ DESOXYNOREPHEDRINE ☐ ELASTONON ☐ FENOPROMIN ☐ α-METHYLPHENETHYLAMINE ☐ MYDRIAL ☐ 1-PHENYL-2-AMINO-PROPAN (GERMAN) ☐ 1-PHENYL-2-AMINOPROPANE ☐ β-PHENYLISOPROPYLAMIN (GERMAN) ☐ (PHENYLISOPROPYL)AMINE ☐ β-PHENYLISOPROPYLAMINE ☐ PROTIOAMPHETAMINE

TOXICITY DATA WITH REFERENCE
orl-inf TDLo:7500 mg/kg:CNS,CVS,SKN AJDCAI 130,507,76
orl-rat LDLo:50 mg/kg AEPPAE 195,647,40
ipr-rat LD50:125 mg/kg JPETAB 132,97,61
scu-rat LD50:39 mg/kg JPETAB 86,280,46
orl-mus LD50:22 mg/kg ARZNAD 32,604,82
ipr-mus LD50:16 mg/kg PSCHDL 51,209,77
scu-mus LD50:2800 µg/kg AEPPAE 233,72,58
ivn-mus LD50:18 mg/kg APTOA6 38,474,76
scu-rbt LDLo:20 mg/kg AEPPAE 192,331,39

SAFETY PROFILE: Poison by ingestion, subcutaneous, intravenous, and intraperitoneal routes. Human systemic effects by ingestion: excitement, changes in heart rate, and sweating. When heated to decomposition it emits very toxic fumes of NO_x. See other amphetamine entries.

AOA500 CAS:51-64-9 HR: 3
d-AMPHETAMINE
mf: $C_9H_{13}N$ mw: 135.23

PROP: Oil. Bp: 102° @ 16 mm.

SYNS: d-2-AMINO-1-PHENYLPROPANE ☐ (+)-AMPHETAMINE ☐ AMSUSTAIN ☐ DEPHADREN ☐ DEXAMPHETAMINE ☐ DEXEDRINE ☐ α-METHYLPHENETHYLAMINE, d-FORM ☐ d-1-PHENYL-2-AMINO-PROPAN (GERMAN) ☐ d-1-PHENYL-2-AMINOPROPANE

TOXICITY DATA WITH REFERENCE
scu-rat TDLo:25 mg/kg (female 5-9D post):REP DABBBA 31,6304,71
orl-man TDLo:42 mg/kg/25W-I BIPCBF 20,1332,85
orl-cld TDLo:3600 µg/kg/10D-I AJPSAO 143,1176,85
orl-rat LD50:38 mg/kg TXAPA9 18,185,71
ipr-rat LDLo:20 mg/kg AEPPAE 195,647,40
scu-rat LD50:200 mg/kg 27ZIAQ -,84,73
orl-mus LD50:40 mg/kg TXAPA9 41,329,77
ipr-mus LD50:4400 µg/kg AIPTAK 161,206,66
scu-mus LD50:20 mg/kg AIPTAK 146,392,63
ivn-mus LD50:25 mg/kg JMCMAR 15,410,72
orl-dog LDLo:6400 µg/kg 27ZIAQ -,84,73
orl-mky LDLo:32 mg/kg 27ZIAQ -,84,73
ipr-grb LD50:17,600 µg/kg GERNDJ 23,165,77
orl-mam LD50:375 mg/kg JMCMAR 8,836,65

CONSENSUS REPORTS: Reported in EPA TSCA Inventory.

SAFETY PROFILE: Poison by ingestion, subcutaneous, intravenous, and intraperitoneal routes. Experimental reproductive effects. Chronic exposure causes central nervous system damage and blood-pressure effects. When heated to decomposition it emits toxic NO_x. See other amphetamine entries.

AOA750 CAS:2706-50-5 HR: 3
AMPHETAMINE HYDROCHLORIDE
mf: $C_9H_{13}N \cdot ClH$ mw: 171.69

SYNS: dl-α-METHYL-PHENETHYLAMINE HYDROCHLORIDE ☐ dl-β-PHENYLISOPROPYLAMINE HYDROCHLORIDE

TOXICITY DATA WITH REFERENCE
ipr-mus LD50:40 mg/kg JMCMAR 8,100,65
ivn-rbt LD50:15 mg/kg JPETAB 79,187,43
scu-gpg LDLo:52 mg/kg JPETAB 47,339,33

SAFETY PROFILE: Poison by subcutaneous, intravenous, and intraperitoneal routes. When heated to decomposition it emits very toxic fumes of HCl and NO_x. See other amphetamine entries.

AOB000 HR: 3
dl-AMPHETAMINE SALT with FINE RESIN

PROP: Amberlite XE-69 is a sulfonic acid cation exchange resin. Mesh size exceeds 200 mesh. (TXAPA9 1,42,59)

SYN: α-METHYL-PHENETHYLAMINE compounded with AMBERLITE XE-69

TOXICITY DATA WITH REFERENCE
orl-rat LD50:195 mg/kg TXAPA9 1,42,59
orl-mus LD50:200 mg/kg TXAPA9 1,42,59

SAFETY PROFILE: Poison by ingestion. When heated to decomposition it emits toxic fumes of NO_x and SO_x. See other amphetamine entries.

AOB250 CAS:60-13-9 HR: 3
dl-AMPHETAMINE SULFATE
mf: $C_{18}H_{26}N_2 \cdot H_2O_4S$ mw: 368.54

SYNS: ACETDRON ☐ ADIPAN ☐ ADIPARTHROL ☐ AKETDRIN ☐ AKTEDRIN ☐ ALENTOL ☐ AMFETAMINA ☐ AMFETAMINE ☐ (±)-2-AMINO-1-PHENYLPROPANE SULFATE ☐ (±)-AMPHETAMINE SULFATE ☐ ANFETAMINA ☐ BENNIE ☐ BENZAMPHETAMINE ☐BENZEDRYNA ☐ BENZIES ☐ BETAFEN ☐ CARTWHEELS ☐ DEOXYNOREPHEDRINE ☐ DESOXYNOREPHEDRINE ☐ HEARTS ☐ IBIOZEDRINE ☐ LINAMPHETA ☐ (±)-α-METHYLPHENETHYLAMINE SULFATE ☐ NCI-C55710 ☐ NOREPHEDRANE ☐ PEACHES ☐ PHARMEDRINE ☐ PHENAMINE ☐ PHENEDRINE ☐ (±)-PHENISOPROPYLAMINE SULFATE

A

□ β-PHENYL ISOPROPYLAMINE SULFATE □ PSYCHEDRINUM □ PSYCHEDRYNA □ RACEPHEN □ ROSES □ STIMULAN

TOXICITY DATA WITH REFERENCE
sln-dmg-unk 1500 g/L CTOXAO 5,395,72
scu-rat TDLo:11 mg/kg (1-22D preg):REP PSCHDL 58,171,78
orl-mus TDLo:900 mg/kg (1-18D preg):TER OFAJAE 41,227,65
orl-hmn TDLo:41 mg/kg:CNS,CVS KLWOAZ 17,1580,38
orl-rat LD50:55 mg/kg ARZNAD 13,711,63
ipr-rat LD50:125 mg/kg JPETAB 132,97,61
scu-rat LD50:160 mg/kg AIMEAS 10,1874,37
orl-mus LD50:24 mg/kg ARZNAD 13,711,63
ipr-mus LD50:13 mg/kg RPTOAN 48,26,85
ipr-mus LD50:13 mg/kg FATOAO 48(1),15,85
scu-mus LD50:7 mg/kg AIPTAK 170,428,67
ivn-mus LD50:68 mg/kg JPETAB 84,12,45
orl-dog LD50:23 mg/kg PSEBAA 118,557,65
ivn-dog LD50:6 mg/kg PSEBAA 118,557,65
ivn-rbt LDLo:22 mg/kg JOPHAN 37,597,39
scu-gpg LD50:105 mg/kg AIPTAK 137,375,62

CONSENSUS REPORTS: EPA Genetic Toxicology Program.

SAFETY PROFILE: Poison by ingestion, subcutaneous, intravenous, and intraperitoneal routes. Human systemic effects by ingestion: altered sleep time, anorexia and change in heart rate. A central nervous system stimulant. An experimental teratogen. Other experimental reproductive effects. Mutation data reported. When heated to decomposition it emits very toxic NO_x and SO_x. See other amphetamine entries.

AOB500 CAS:139-10-6 HR: 3
AMPHETANE PHOSPHATE
mf: $C_9H_{13}N{\cdot}H_3O_4P$ mw: 233.23

SYNS: ACETMIN □ ACTEMIN □ AKTEDRON □ AMPHATE □ AMPHETAMINE PHOSPHATE □ dl-AMPHETAMINE PHOSPHATE □ AMPHOS □ BAR-DEX □ DEPUALONE □ DIETAMINE □ DYNAPHENIL □ dl-α-METHYL-PHENETHYLAMINE PHOSPHATE □ α-METHYLPHENETHYLAMINE PHOSPHATE, dl-MIXTURE □ MONOBASIC racemic AMPHETAMINE PHOSPHATE □ MONOBASIC dl-α-METHYLPHENETHYLAMINE PHOSPHATE □ MONOPHOR □ MONOPHOS □ OBESITABS □ 1-PHENYL-2-AMINOPROPANE MONOPHOSPHATE □PROFETAMINE □ PROFETAMINE PHOSPHATE □ RACEPHEN □ RAPHETAMINE PHOSPHATE

TOXICITY DATA WITH REFERENCE
orl-rat LD50:175 mg/kg TXAPA9 1,42,59
orl-mus LD50:154 mg/kg TXAPA9 1,42,59
ipr-mus LDLo:52 mg/kg JPETAB 127,55,59

SAFETY PROFILE: Poison by ingestion and intraperitoneal routes. When heated to decomposition it emits very toxic fumes of PO_x and NO_x. See other amphetamine entries.

AOB875 CAS:1402-82-0 HR: 3
AMPHOMYCIN
mf: $C_{58}H_{91}N_{13}O_{20}$ mw: 1290.46

PROP: Crystals. Acidic, surface-active polypeptide. Sol in water and the lower alcs; insol in nonpolar solvents.

SYNS: AMFOMYCIN □ GLUMAMYCIN □ U-6658

TOXICITY DATA WITH REFERENCE
orl-mus LD50:500 mg/kg 85GDA2 4(1),317,80
ipr-mus LD50:233 mg/kg CNCRA6 30,9,63
ivn-mus LD50:178 mg/kg 85FZAT -,131,67

SAFETY PROFILE: Poison by intravenous and intraperitoneal routes. Moderately toxic by ingestion. Induces hemolysis. Active against gram-positive bacteria. Suggested as a topical agent for animal and plant infections. When heated to decomposition it emits acrid smoke and irritating fumes.

AOC250 HR: 1
AMPHOTERIC-2

SYNS: AMPHOTERGE K-2 □ 1-CARBOXYMETHYL-1-CARBOXYETHOXYETHYL-2-COCO-IMIDAZOLINIUM BETAINE □ MIRANOL C2M-SF CONC

TOXICITY DATA WITH REFERENCE
skn-hmn 60 mg/3D-I MLD 85DKA8 -,127,77
skn-rbt 10 mg/24H DCTODJ 1,305,78
eye rbt 2 mg DCTODJ 1,305,78

SAFETY PROFILE: An eye irritant. A human skin irritant. When heated to decomposition it emits toxic fumes of NO_x.

AOC275 CAS:71463-34-8 HR: 1
AMPHOTERIC-17
mf: $C_{18}H_{35}N_2O_3{\cdot}HO_4S{\cdot}Na$ mw: 447.61

SYNS: 1H-IMIDAZOLIUM, 4,5-DIHYDRO-1-(CARBOXYMETHYL)-1-(2-HYDROXYETHYL)-2-UNDECYL-, HYDROGEN SULFATE (salt), MONOSODIUM SALT □ MIRANOL MHT

TOXICITY DATA WITH REFERENCE
skn-rbt 10 mg/24H DCTODJ 1,305,78
eye-rbt 2 mg DCTODJ 1,305,78

SAFETY PROFILE: A skin and eye irritant. When heated to decomposition it emits toxic fumes of NO_x, SO_x, and H_2S.

AOC500 CAS:1397-89-3 HR: 3
AMPHOTERICIN B
mf: $C_{47}H_{73}NO_{17}$ mw: 924.21

PROP: Deep-yellow prisms from DMF. Insol in H_2O.

SYNS: AMB □ AMPHOMORONAL □ AMPHOTERICIN beta □ AMPHOTERICINE B □ AMPHOZONE □ FUNGILIN □ FUNGISONE □ FUNGIZONE □ IAB □ IODOACETAMIDE □ MYSTECLIN-F □ NSC 527017 □ TEGOPEN

TOXICITY DATA WITH REFERENCE
spm-rbt-ivn 20 mg/kg/11D JRPFA4 7,13,64
orl-dog TDLo:600 mg/kg (male 30D pre):REP INURAQ 7,90,69
ivn-wmn LDLo:22 mg/kg/4D-I:BLD SMJOAV 76,409,83
ivn-wmn TDLo:20 μg/kg:PUL NEJMAG 315,836,86
ivn-man LDLo:164 μg/kg/5H-I:CVS DICPBB 17,547,83
ivn-rat LD50:11,300 μg/kg DRUGAY 6,36,82
ipr-mus LD50:27,740 μg/kg NCISP* JAN86
ivn-mus LD50:1200 μg/kg PHINDQ 6,164,85

ivn-dog LD50:6 mg/kg BIORAK 43,2043,78

SAFETY PROFILE: Poison by intravenous and intraperitoneal routes. Human systemic effects by intravenous route: leukopenia, lung changes, and cardiac changes. Experimental reproductive effects. Mutation data reported. When heated to decomposition it emits toxic fumes of NO_x.

AOC750 CAS:35375-29-2 HR: 3
AMPHOTERICIN B, METHYL ESTER HYDROCHLORIDE
mf: $C_{47}H_{75}NO_{20}•ClH$ mw: 1010.69

TOXICITY DATA WITH REFERENCE
ipr-mus LD50:1320 mg/kg SCIEAS 179,584,73
ivn-mus LD50:75 mg/kg 85ERAY 2,1019,78
ivn-dog LD50:48 mg/kg SCIEAS 179,584,73

SAFETY PROFILE: Poison by intravenous route. See also ESTERS. Moderately toxic by intraperitoneal route. When heated to decomposition it emits very toxic fumes of NO_x and HCl.

AOC875 CAS:8067-85-4 HR: 2
AMPICILLIN-OXACILLIN MIXTURE
mf: $C_{19}H_{19}N_3O_5S•C_{16}H_{19}N_3O_4S$ mw: 750.91

SYNS: (2S-(2-α,5-α,6-β(S*)))-6-((AMINOPHENYLACETYL)AMINO)-3,3-DIMETHYL-7-OXO-4-THIA-1-AZABICYCLO(3.2.0)HEPTANE-2-CARBOXYLIC ACID mixt. with (2S-(2-α,5-α,6-β))-3,3-DIMETHYL-6-(((5-METHYL-3-PHENYL-4-ISOXAZOLYL)CARBONYL)AMINO)-7-OXO-4-THIA-1-AZABICYCLO(3.2.0)HEPTANE-2-CARBOXYLIC ACID □ OXACILLIN-AMPICILLIN MIXTURE

TOXICITY DATA WITH REFERENCE
ivn-rat LD50:3710 mg/kg NIIRDN 6,58,82
ipr-mus LD50:4700 mg/kg NIIRDN 6,58,82
scu-mus LD50:4940 mg/kg NIIRDN 6,58,82
ivn-mus LD50:3250 mg/kg NIIRDN 6,58,82

SAFETY PROFILE: Moderately toxic. When heated to decomposition it emits toxic fumes of Cl^-, SO_x, and NO_x.

AOD000 CAS:26309-95-5 HR: 2
AMPICILLIN PIVALOYLOXYMETHYL ESTER HYDROCHLORIDE
mf: $C_{22}H_{29}N_3O_6S•ClH$ mw: 500.06

SYNS: ALPHACILINA □ ALPHACILLIN □ 6-(d-α-AMINO PHENYL ACETAMIDO) PENICILLANIC ACID PIVALOYL OXY METHYL ESTER HYDROCHLORIDE □ BEROCILLIN □ CENTURINA □ DEVONIUM □ DIANCINA □ INACILIN □ MAXIFEN □ PIVALOYLOXYMETHYL d-α-AMINOBENZYLPENICILLINATE HYDROCHLORIDE □ PIVAMPICILLIN HYDROCHLORIDE □ PIVATIL □ PONDOCIL □ PONDOCILLIN □ SANGUICILLIN

TOXICITY DATA WITH REFERENCE
orl-rat LD50:5 g/kg AACHAX -,341,70
scu-rat LD50:4500 mg/kg AACHAX -,341,70
orl-mus LD50:3340 mg/kg AACHAX -,341,70
scu-mus LD50:3600 mg/kg AACHAX -,341,70

SAFETY PROFILE: Moderately toxic by ingestion and subcutaneous routes. See also ESTERS. When heated to decomposition it emits very toxic fumes of NO_x, SO_x, and HCl.

AOD125 CAS:7177-48-2 HR: 1
AMPICILLIN TRIHYDRATE
mf: $C_{16}H_{19}N_3O_4S•3H_20$ mw: 403.50

SYNS: AMCAP □ AMCILL □ AMINOBENZYLPENICILLIN TRIHYDRATE □ α-AMINOBENZYLPENICILLIN TRIHYDRATE □ AMPERIL □ AMPICHEL □ AMPIKEL □ AMPINOVA □ AMPLIN □ ANCILLIN □ CYMBI □ DIVERCILLIN □ LIFEAMPIL □ MOREPEN □ NCI-C56086 □ PEN A □ PENSYN □ POLYCILLIN □ PRINCILLIN □ RO-AMPEN □ TRAFARBIOT □ UKOPEN □ VIDOPEN

TOXICITY DATA WITH REFERENCE
orl-rat TDLo:1500 mg/kg (6-11D preg):TER ANTBAL 18,815,73
orl-mus TDLo:9100 mg/kg (female 7-13D post):REP KSRNAM 7,3113,73
orl-rat LD50:10 g/kg ANTBAL 20,653,75
orl-mus LD50:15,200 mg/kg ANTBAL 20,653,75

SAFETY PROFILE: Mildly toxic by ingestion. An experimental teratogen. Other experimental reproductive effects. When heated to decomposition it emits toxic fumes of SO_x and NO_x.

AOD250 CAS:134-53-2 HR: 3
AMPROTROPINE PHOSPHATE
mf: $C_{18}H_{29}NO_3•H_3O_4P$ mw: 405.48

PROP: Bitter crystals. Mp: 142–145°.

SYNS: AP 407 □ 3-DIETHYLAMINO-2,2-DIMETHYLPROPYL TROPATE PHOSPHATE □ 1-PROPANOL, 3-(DIETHYLAMINO)-2,2-DIMETHYL-, TROPATE, PHOSPHATE □ SYNTROPAN □ dl-TROPASAEUREESTER DES 3-DIAETHYLAMINO-2,2-DIMETHYL-1-PROPANOL PHOSPHAT (GERMAN)

TOXICITY DATA WITH REFERENCE
ivn-rat LD50:43 mg/kg JLCMAK 30,700,45
orl-mus LDLo:570 mg/kg JPETAB 60,1,37
scu-mus LDLo:1250 mg/kg JPETAB 60,1,37
ivn-mus LD50:51 mg/kg JLCMAK 30,700,45
scu-cat LDLo:200 mg/kg JPETAB 60,1,37
scu-rbt LDLo:500 mg/kg JPETAB 60,1,37
ivn-rbt LD50:25 mg/kg SMWOAS 76,1282,46
scu-frg LDLo:1500 mg/kg JPETAB 60,1,37
par-frg LDLo:1000 mg/kg AEPPAE 173,86,33

SAFETY PROFILE: Poison by ingestion, subcutaneous, and intravenous routes. Moderately toxic by parenteral route. When heated to decomposition it emits very toxic fumes of PO_x and NO_x. An anticholinergic agent.

AOD375 CAS:60719-84-8 HR: 3
AMRINONE
mf: $C_{10}H_9N_3O$ mw: 187.22

PROP: Crystals from DMF. Mp: 294–297° (decomp).

SYNS: 5-AMINO(3,4'-BIPYRIDIN)-6-(1H)-ONE □ 5-AMINO-5-(4-PYRIDINYL)-2(1H)-PYRIDINONE □ INOCOR □ WIN 40680 □ WINCORAM

TOXICITY DATA WITH REFERENCE
orl-rat TDLo:1 g/kg (female 6-15D post):TER PHARAT 41,214,86
orl-rat TDLo:1 g/kg (female 6-15D post):REP PHARAT 41,214,86

orl-man LDLo: 1429 μg/kg:CVS,SYS,BLD AIMDAP 145,825,85
orl-rat LD50: 102 mg/kg NDADD8 1,259,83
ivn-rat LD50: 75 mg/kg PHARAT 41,209,86
orl-mus LD50: 288 mg/kg TPHSDY 1,143,80
ivn-mus LD50: 150 mg/kg TPHSDY 1,143,80

SAFETY PROFILE: Poison by ingestion and intravenous routes. Human systemic effects by ingestion: cardiac arrhythmias, liver function, thrombocytopenia. An experimental teratogen. Other experimental reproductive effects. When heated to decomposition it emits toxic fumes of NO_x. A cardiotonic agent.

AOD500 CAS:29883-15-6 HR: 3
AMYGDALIN
mf: $C_{20}H_{27}NO_{11}$ mw: 457.48

PROP: Trihydrate. Mp: 214°.

SYNS: d(−)-MANDELONITRILE-β-d-GENTIOBIOSIDE □ d-MANDELONITRILE-β-d-GLUCOSIDO-6-β-d-GLUCOSIDE □ NSC-15780

TOXICITY DATA WITH REFERENCE
hma-mus/sat 250 mg/kg SCIEAS 198,625,77
orl-ham TDLo: 300 mg/kg (female 8D post):TER SCIEAS 215,1513,82
orl-inf LDLo: 50 mg/kg JAMAAP 238,482,77
orl-rat LD50: 522 mg/kg WJMDA2 134,97,81
orl-mus LD50: 443 mg/kg CTOXAO 17,85,80

SAFETY PROFILE: Human poison by ingestion (infant data). Poison experimentally by ingestion. An experimental teratogen. Mutation data reported. When heated to decomposition it emits toxic fumes of NO_x.

AOD725 CAS:628-63-7 HR: 3
n-AMYL ACETATE
DOT: UN 1104
mf: $C_7H_{14}O_2$ mw: 130.21

PROP: Colorless liquid; pear or banana-like odor. Mp: −78.5°, bp: 148° @ 737 mm, ULC: 55–60, lel: 1.1%, uel: 7.5%, flash p: 77°F (CC), d: 0.879 @ 20°/20°, autoign temp: 714°F, vap d: 4.5. Very sltly sol in water; misc in alc and ether.

SYNS: ACETATE d'AMYLE (FRENCH) □ ACETIC ACID, AMYL ESTER □ AMYL ACETATE (DOT) □ AMYL ACETIC ESTER □ AMYLAZETAT (GERMAN) □ AMYLESTER KYSELINY OCTOVE □ BIRNENOEL □ OCTAN AMYLU (POLISH) □ PEAR OIL □ PENT-ACETATE □ 1-PENTANOL ACETATE □ PENTYL ACETATE □ n-PENTYL ACETATE □ 1-PENTYL ACETATE □ PRIMARY AMYL ACETATE

TOXICITY DATA WITH REFERENCE
eye-hmn 300 ppm JIHTAB 25,282,43
ihl-hmn TCLo: 5000 mg/m³/30M:CNS,EYE,PUL AHYGAJ 78,260,13
ihl-hmn TCLo: 200 ppm:CNS NPIRI* 1,3,74
orl-rat LD50: 6500 mg/kg NPIRI* 1,3,74
orl-rbt LD50: 7400 mg/kg 85JCAE-,357,86
ihl-rat LCLo: 5200 ppm/8H DTLVS* 3,12,71
ipr-gpg LDLo: 1500 mg/kg AIHAAP 35,21,74

CONSENSUS REPORTS: Reported in EPA TSCA Inventory.

OSHA PEL: TWA 100 ppm
ACGIH TLV: TWA 100 ppm
DOT CLASSIFICATION: 3; *Label:* Flammable Liquid

SAFETY PROFILE: Moderately toxic by intraperitoneal route. Human systemic effects by inhalation: conjunctiva irritation, headache, and somnolence. A human eye irritant. Apparently more toxic than butyl acetate. Chronic toxicity is of a low order. Dangerous fire hazard when exposed to heat or flame; can react with oxidizing materials. Moderately explosive in the form of vapor when exposed to flame. To fight fire, use alcohol foam, dry chemical. When heated to decomposition it emits acrid smoke and irritating fumes. See also ESTERS, AMYL ALCOHOL, and ACETIC ACID.

For occupational chemical analysis use NIOSH: Esters I, 1450.

AOD735 CAS:626-38-0 HR: 3
sec-AMYL ACETATE
DOT: UN 1104
mf: $C_7H_{14}O_2$ mw: 130.21

PROP: Colorless liquid. Bp: 120°, flash p: 73.4°F (CC), d: 0.862–0.866 @ 20°/20°, vap d: 4.48, lel: 1.1%, uel: 7.5%. Sltly sol in water; misc in alc and ether.

SYNS: 2-ACETOXYPENTANE □ 1-METHYLBUTYL ACETATE □ sek.AMYLESTER KYSELINY OCTOVE □ 2-AMYLESTER KYSELINY OCTOVE □ 1-METHYLBUTYL ACETATE □ 2-PENTANOL, ACETATE □ 2-PENTYL ACETATE

TOXICITY DATA WITH REFERENCE
ihl-hmn TCLo: 200 ppm:EYE JIHTAB 25,282,43
ihl-gpg LCLo: 10,000 ppm/5H PHRPA6 51,811,36

CONSENSUS REPORTS: Reported in EPA TSCA Inventory.

OSHA PEL: TWA 125 ppm
ACGIH TLV: TWA 125 ppm
DOT CLASSIFICATION: 3; *Label:* Flammable Liquid

SAFETY PROFILE: Mildly toxic by inhalation. Human systemic effects by inhalation: conjunctiva irritation. Dangerous fire hazard when exposed to heat or flame; can react with oxidizing materials. Moderately explosive in the form of vapor when exposed to heat or flame. To fight fire, use alcohol foam, dry chemical. When heated to decomposition it emits acrid smoke and irritating fumes.

For occupational chemical analysis use NIOSH: Esters I, 1450.

AOD750 HR: 2
AMYL ACETATE (mixed isomers)
mf: $C_7H_{14}O_2$ mw: 130.21

PROP: Colorless liquid, pear-like odor. Mp: −78.5°, bp: 148° @ 737 mm, ULC: 55–60, lel: 1.1%, uel: 7.5%, flash p: 77°F (CC), d: 0.879 @ 20°/20°, autoign temp: 714°F, vap d: 4.5.

SYN: ACETIC ACID, AMYL ESTER

TOXICITY DATA WITH REFERENCE
skn-rbt 10 mg/24H open MLD AIHAAP 23,95,62
orl-rat LD50:4950 mg/kg AIHAAP 23,95,62

DFG MAK: 100 ppm (525 mg/m^3)

SAFETY PROFILE: A skin irritant. Mildly toxic by ingestion. Dangerous fire hazard; can react with oxidizing materials. Moderately explosive in the form of vapor when exposed to flame. To fight fire, use alcohol foam, dry chemical. When heated to decomposition it emits acrid smoke and irritating fumes.

AOE000 **HR: 3**
AMYL ALCOHOL
mf: $C_5H_{12}O$ mw: 88.1

PROP: Clear liquid. Mp: −79°, bp: 137.8°, flash p: 91°F (CC), d: 0.8168 @ 20°/20°, ULC: 40, lel: 1.2%, uel: 10% @ 212°F, vap press: 1 mm @ 13.6°, 10 mm @ 44.9°, vap d: 3.04. Sol in water; misc in alc and ether.

SYNS: ALCOOL AMYLIQUE (FRENCH) □ N-AMYL ALCOHOL □ AMYL ALCOHOL, NORMAL □ N-AMYLALKOHOL (CZECH) □ N-BUTYLCARBINOL □ N-PENTANOL □ PENTANOL-1 □ PENTAN-1-OL □ PENTASOL □ PENTYL ALCOHOL □ PRIMARY AMYL ALCOHOL

TOXICITY DATA WITH REFERENCE
ipr-mus TDLo:1200 mg/kg/8W-I:ETA CNREA8 33,3069,73
skn-rbt 3200 mg/kg/24H SEV AIHAAP 34,493,73
skn-rbt 500 mg/24H MOD 28ZPAK -,35,72
eye-rbt 81 mg SEV AIHAAP 34,493,73
eye-rbt 20 mg/24H SEV 28ZPAK -,35,72
orl-rat LD50:3030 mg/kg FCTXAV 2,327,64
ihl-rat LCLo:14,000 mg/m^3/6H AIHAAP 34,493,73
ipr-rat LDLo:490 mg/kg AEPPAE 132,214,28
orl-mus LD50:200 mg/kg GISAAA 35(9),88,70
ihl-mus LCLo:14,000 mg/m^3/6H AIHAAP 34,493,73
ivn-cat LDLo:15 mg/kg JPETAB 16,1,20
skn-rbt LD50:4490 mg/kg 31ZTAS -,76,68

CONSENSUS REPORTS: Reported in EPA TSCA Inventory.

SAFETY PROFILE: Moderately toxic by ingestion and skin contact. An eye and upper respiratory irritant by inhalation. A severe skin and eye irritant. Ingestion can cause headache, nausea, vomiting, delirium, and methemoglobin formation. Questionable carcinogen with experimental tumorigenic data. Extremely flammable if exposed to heat, flame, or powerful oxidizers. Moderately explosive when exposed to flame. Incompatible with oxidizing materials, hydrogen trisulfide. To fight fire, use alcohol foam, dry chemical.

AOE200 **CAS:598-74-3** **HR: 3**
iso-AMYLAMINE
DOT: UN 2733/UN 2734
mf: $C_5H_{13}N$ mw: 87.19

SYNS: 2-BUTANAMINE, 3-METHYL-(9CI) □ 1,2-DIMETHYLPROPANAMINE □ 1,2-DIMETHYLPROPYLAMINE □ 3-METHYL-2-BUTANAMINE □ PROPYLAMINE, 1,2-DIMETHYL-

TOXICITY DATA WITH REFERENCE
ipr-mus LD50:279 mg/kg JJPAAZ 17,475,67

CONSENSUS REPORTS: Reported in EPA TSCA Inventory.

DOT CLASSIFICATION: 8; *Label:* Corrosive, Flammable Liquid (UN 2734); DOT Class: 3; *Label:* Flammable Liquid, Corrosive (UN 2733)

SAFETY PROFILE: Poison by intraperitoneal route. A flammable liquid. When heated to decomposition it emits toxic vapors of NO_x.

AOE500 **HR: 2**
AMYL AZIDE
mf: $C_5H_{11}N_3$ mw: 113

SAFETY PROFILE: Moderately toxic irritant and toxic by ingestion and inhalation. Narcotic in high concentration. Can cause a fall in blood pressure. See also AZIDES. An unstable material.

AOE750 **CAS:63018-99-5** **HR: 2**
5-n-AMYL-1:2-BENZANTHRACENE
mf: $C_{23}H_{22}$ mw: 298.45

SYN: 8-PENTYLBENZ(a)ANTHRACENE

TOXICITY DATA WITH REFERENCE
skn-mus TDLo:790 mg/kg/33W-I:ETA PRLBA4 129,439,40

SAFETY PROFILE: Questionable carcinogen with experimental tumorigenic data. When heated to decomposition it emits acrid smoke and irritating fumes.

AOF000 **CAS:2049-95-8** **HR: 1**
tert-AMYLBENZENE
mf: $C_{11}H_{16}$ mw: 148.27

PROP: Liquid. D: 0.874, bp: 189–191°. Insol in water; misc in alc and ether.

SYN: tert-PENTYLBENZENE

TOXICITY DATA WITH REFERENCE
orl-rat LDLo:10 mL/kg AMIHAB 19,403,59

CONSENSUS REPORTS: Reported in EPA TSCA Inventory.

SAFETY PROFILE: Mildly toxic by ingestion. When heated to decomposition it yields irritating fumes and smoke.

AOF250 **CAS:63905-98-6** **HR: 3**
4-AMYL-N-BENZOHYDRYLPYRIDINIUM BROMIDE
mf: $C_{22}H_{26}N{\cdot}Br$ mw: 384.40

SYN: B-45

TOXICITY DATA WITH REFERENCE
ipr-rat LD50:8 mg/kg FEPRA7 9,280,50
scu-rat LD50:4 mg/kg FEPRA7 9,280,50
orl-mus LD50:35 mg/kg FEPRA7 9,280,50
ipr-mus LD50:4500 µg/kg FEPRA7 9,280,50
scu-mus LD50:1600 µg/kg FEPRA7 9,280,50
ivn-mus LD50:1300 µg/kg FEPRA7 9,280,50
ivn-dog LD50:30 mg/kg FEPRA7 9,280,50
scu-rbt LD50:35 mg/kg FEPRA7 9,280,50

ivn-rbt LD50:10 mg/kg FEPRA7 9,280,50
scu-gpg LD50:15 mg/kg FEPRA7 9,280,50

SAFETY PROFILE: Poison by ingestion, intraperitoneal, subcutaneous, and intravenous routes. See also BROMIDES. When heated to decomposition it emits very toxic fumes of NO_x and Br^-.

AOF500 CAS:63990-96-5 HR: 1
AMYL BIPHENYL
mf: $C_{17}H_{20}$ mw: 224.37

PROP: Liquid. Mp: −60°, bp: 305–337°, flash p: 300°F, d: 0.958 @ 20°/20°, vap d: 7.73.

SYN: PENTYLBIPHENYL

TOXICITY DATA WITH REFERENCE
orl-rat LDLo:5000 mg/kg AMIHAB 19,403,59

SAFETY PROFILE: Mildly toxic by ingestion and inhalation routes. Combustible when exposed to heat or flame. Moderately dangerous; when heated to decomposition it emits irritating fumes and smoke. Incompatible with oxidizing materials. To fight fire, use foam, CO_2, dry chemical.

AOF750 HR: 3
d-AMYL BROMIDE
mf: $CH_3(CH_2)_4Br$ mw: 151.1

PROP: Colorless liquid. Bp: 120°, flash p: 90°F, fp: <−30°, d: 1.211 @ 25°/25°.

SAFETY PROFILE: Poison by intraperitoneal route. It can cause liver damage, is narcotic in high concentrations, and is a local irritant. See also BROMIDES. Extremely flammable. To fight fire, use alcohol foam, water mist or spray, dry chemical. When heated to decomposition it emits very toxic bromides. Incompatible with oxidizing materials.

AOF800 CAS:110-53-2 HR: 2
n-AMYL BROMIDE
mf: $C_5H_{11}Br$ mw: 151.07

SYNS: AMYL BROMIDE □ 1-BROMOPENTANE □ PENTANE, 1-BROMO- □ PENTYL BROMIDE □ n-PENTYL BROMIDE □ 1-PENTYL BROMIDE

TOXICITY DATA WITH REFERENCE
ipr-mus LD50:1250 mg/kg GTPZAB 20(12),52,76
ihl-uns LC50:26,800 mg/m³ GTPZAB 18(4),55,74

CONSENSUS REPORTS: Reported in EPA TSCA Inventory.

SAFETY PROFILE: Moderately toxic by intraperitoneal route. Slightly toxic by inhalation. When heated to decomposition it emits toxic vapors of Br^-.

AOG000 CAS:540-18-1 HR: 2
n-AMYL BUTYRATE
mf: $C_9H_{18}O_2$ mw: 158.27

PROP: Colorless liquid. D: 0.871, mp: −73.2°, bp: 186.4°. Sol in water, miscible with alc and ether.

SYNS: AMYL BUTYRATE □ BUTANOIC ACID PENTYL ESTER □ PENTYL BUTYRATE

TOXICITY DATA WITH REFERENCE
orl-rat LD50:12,210 mg/kg FCTXAV 2,327,64
orl-gpg LD50:11,950 mg/kg FCTXAV 2,327,64

CONSENSUS REPORTS: Reported in EPA TSCA Inventory.

SAFETY PROFILE: Mildly toxic by ingestion. When heated to decomposition it emits acrid smoke and irritating fumes.

AOG500 CAS:122-40-7 HR: 2
α-AMYL CINNAMALDEHYDE
mf: $C_{14}H_{18}O$ mw: 202.32

PROP: Pale, yellow oil or liquid; floral jasmine odor. D: 0.963, refr index: 1.554, bp: 174–175° @ 20 mm. Sol in fixed oils; insol in glycerin and propylene glycol

SYNS: α-AMYL CINNAMIC ALDEHYDE □ α-AMYL-β-PHENYLACROLEIN □ FEMA No. 2061 □ JASMINALDEHYDE □ α-PENTYLCINNAMALDEHYDE

TOXICITY DATA WITH REFERENCE
skn-rbt 100 mg/24H SEV CTOIDG 94(8),41,79
skn-gpg 5%/2W MLD ADVEA4 58,121,78
skn-gpg 100 mg/24H MOD CTOIDG 94(8),41,79
orl-rat LD50:3730 mg/kg FCTXAV 2,327,64

CONSENSUS REPORTS: Reported in EPA TSCA Inventory.

SAFETY PROFILE: Moderately toxic by ingestion. A severe skin irritant. See also ALDEHYDES. When heated to decomposition it emits acrid smoke and irritating fumes.

AOG600 HR: 1
AMYL CINNAMATE
mf: $C_{14}H_{19}O_2$ mw: 218.28

PROP: Colorless to pale-yellow liquid; slt cocoa odor. D: 0.992–0.997, refr index: 1.535, flash p: 212°F. Sol in fixed oils; sltly sol in propylene glycol; insol in glycerin @ 310°.

SYNS: FEMA No. 2063 □ ISOAMYL CINNAMATE □ ISOAMYL 3-PENTYL PROPENATE

SAFETY PROFILE: Combustible liquid. When heated to decomposition it emits acrid smoke and irritating fumes.

AOG750 CAS:7493-78-9 HR: 2
AMYL CINNAMIC ACETATE
mf: $C_{16}H_{22}O_2$ mw: 246.38

SYNS: α-N-AMYL-β-PHENYLACRYL ACETATE □ α-PENTYL CINNAMYL ACETATE

TOXICITY DATA WITH REFERENCE
skn-rbt 500 mg/24H FCTXAV 14,659,76

SAFETY PROFILE: Moderately toxic skin irritant. When heated to decomposition it emits acrid smoke and irritating fumes.

AOH000 **CAS:101-85-9** **HR: 2**
α-AMYLCINNAMIC ALCOHOL
mf: $C_{14}H_{20}O$ mw: 204.34

SYNS: α-AMYLCINNAMYL ALCOHOL □ 2-AMYL-3-PHENYL-2-PROPEN-1-OL □ 2-BENZYLIDENE-1-HEPTANOL

TOXICITY DATA WITH REFERENCE
orl-rat LD50:4000 mg/kg FCTXAV 12,807,74

CONSENSUS REPORTS: Reported in EPA TSCA Inventory.

SAFETY PROFILE: Moderately toxic by ingestion. See also ALCOHOLS. When heated to decomposition it emits acrid smoke and irritating fumes.

AOH100 **CAS:68527-78-6** **HR: 1**
AMYL CINNAMYLIDENE METHYL ANTHRANILATE
mf: $C_{22}H_{25}NO_2$ mw: 335.48

SYNS: ANTHRANILIC ACID, N-(2-BENZYLIDENEHEPTYLIDENE)-, METHYL ESTER □ METHYL N-(β-PENTYLCINNAMYLIDENE)ANTHRANILATE

TOXICITY DATA WITH REFERENCE
skn-rbt 500 mg/24H MOD FCTXAV 16,645,78

CONSENSUS REPORTS: Reported in EPA TSCA Inventory.

SAFETY PROFILE: A skin irritant. When heated to decomposition it emits toxic fumes of NO_x.

AOH250 **CAS:53043-14-4** **HR: 2**
6-n-AMYL-m-CRESOL
mf: $C_5H_{11}C_6H_3OHCH_3$ mw: 178.3

PROP: Bp: 258°, flash p: 240°F, d: 0.97.

TOXICITY DATA WITH REFERENCE
orl-rat LD50:1500 mg/kg PSEBAA 32,592,35

SAFETY PROFILE: Moderately toxic by ingestion. Combustible liquid when exposed to heat or flame. Dangerous; when heated to decomposition it emits irritating fumes. Incompatible with oxidizing materials.

AOH750 **CAS:16587-71-6** **HR: 1**
4-tert-AMYLCYCLOHEXANONE
mf: $C_{11}H_{20}O$ mw: 168.31

PROP: Solid. Mp: 96°, bp: 110–113° @ 12 mm.

SYNS: 4-(1,1-DIMETHYLPROPYL)CYCLOHEXANONE □ 4-tert-PENTYLCYCLOHEXANONE

TOXICITY DATA WITH REFERENCE
skn-rbt 500 mg/24H FCTXAV 12,807,74
orl-rat LD50:4700 mg/kg FCTXAV 12,807,74
skn-rbt LD50:4700 mg/kg FCTXAV 12,819,74

CONSENSUS REPORTS: Reported in EPA TSCA Inventory.

SAFETY PROFILE: Mildly toxic by ingestion. A skin irritant. When heated to decomposition it emits smoke and acrid fumes.

AOI000 **HR: 2**
AMYLCYCLOHEXYL ACETATE (mixed isomers)
mf: $C_{13}H_{23}O_2$ mw: 211.2

SYN: PENTYLCYCLOHEXANOL ACETATE

TOXICITY DATA WITH REFERENCE
skn-rbt 500 mg/24H MOD FCTXAV 14,659,76

SAFETY PROFILE: Moderately toxic skin irritant. When heated to decomposition it emits acrid smoke and acrid fumes.

AOI200 **CAS:692-95-5** **HR: 3**
AMYLDICHLORARSINE
mf: $C_5H_{11}AsCl_2$ mw: 216.98

SYNS: N-AMYLDICHLORARSINE □ ARSINE, AMYLDICHLORO- □ ARSINE, DICHLOROPENTYL- □ DICHLOROPENTYLARSINE □ PENTYLDICHLOROARSINE

TOXICITY DATA WITH REFERENCE
ihl-mus LC50:1400 mg/m³/10M NTIS** PB158-508
skn-mus LDLo:4 mg/kg NTIS** PB158-508

OSHA PEL: TWA 0.5 mg(As)/m³

SAFETY PROFILE: Poison by skin contact. Moderately toxic by inhalation. When heated to decomposition it emits toxic fumes of As and Cl^-.

AOI250 **CAS:14779-78-3** **HR: 1**
AMYL-p-DIMETHYLAMINOBENZOATE
mf: $C_{14}H_{21}NO_2$ mw: 235.36

SYN: AMYL DIMETHYL PABA

TOXICITY DATA WITH REFERENCE
skn-hmn 15 mg/3D-I MLD 85DKA8 -,127,77

CONSENSUS REPORTS: Reported in EPA TSCA Inventory.

SAFETY PROFILE: A mild human skin irritant. When heated to decomposition it emits toxic fumes of NO_x.

AOI500 **CAS:58817-05-3** **HR: 1**
AMYLDIMETHYL-p-AMINO BENZOIC ACID
mf: $C_{17}H_{27}NO_2$ mw: 277.45

SYNS: p-DIMETHYLAMINOBENZOIC ACID, PENTYL ESTER □ p-DIMETHYLAMINOBENZOIC ACID, OCTYL ESTER □ OCTYL-DIMETHYL-p-AMINOBENZOIC ACID

TOXICITY DATA WITH REFERENCE
skn-hmn 15 mg/3D-I MLD 85DKA8 -,127,77

SAFETY PROFILE: A mild human skin irritant. See also ESTERS. When heated to decomposition it emits toxic fumes of NO_x.

AOI800 **CAS:25377-72-4** **HR: 3**
n-AMYLENE PENTENE
DOT: UN 1108
mf: C_5H_{10} mw: 70.15

SYNS: AMYLENE □ PENTYLENE

CONSENSUS REPORTS: Reported in EPA TSCA Inventory.

DOT CLASSIFICATION: 3; *Label:* Flammable Liquid

SAFETY PROFILE: Moderately toxic. Very flammable; reacts with heat, flame and oxidizing materials. To fight fire, use foam, CO_2, dry chemical.

AOJ000 HR: 3
AMYLENES, MIXED
DOT: UN 1106
mf: C_5H_{10} mw: 70.58

PROP: Water-white liquid. Bp: 32.2°, flash p: 0°F, d: 0.66 @ 20°.

CONSENSUS REPORTS: Reported in EPA TSCA Inventory.

DOT CLASSIFICATION: 3; *Label:* Flammable Liquid

SAFETY PROFILE: Moderately toxic. Very flammable; reacts with heat, flame, and oxidizing materials. To fight fire, use foam, CO_2, dry chemical.

AOJ500 CAS:638-49-3 HR: 3
n-AMYL FORMATE
mf: $C_6H_{12}O_2$ mw: 116.18

PROP: Clear liquid. D: 0.902, 0.893 @ 15°/4°, mp: −73.5°, bp: 130.4°, flash p: 80°F. Very sltly sol in water; misc in alc and ether.

SYNS: AMYL FORMATE □ PENTYL FORMATE □ n-PENTYL FORMATE

TOXICITY DATA WITH REFERENCE
skn-rbt 500 mg/24H MLD FCTXAV 18,649,80
orl-rat LD50:>5 g/kg FCTXAV 18,649,80
skn-rbt LD50:>5 g/kg FCTXAV 18,649,80
orl-uns LD50:6300 mg/kg GTPZAB 32(10),25,88
ihl-uns LC50:14 g/m^3 GISAAA 51(5),61,86

CONSENSUS REPORTS: Reported in EPA TSCA Inventory.

SAFETY PROFILE: Very low toxicity by several routes. A skin irritant. See also ESTERS. Dangerously flammable; reacts vigorously with heat, flame, oxidizing materials. To fight fire, use foam, CO_2, dry chemical.

AOJ750 CAS:63885-68-7 HR: 3
o-n-AMYL HARMOL HYDROCHLORIDE
mf: $C_{17}H_{20}N_2O \cdot ClH$ mw: 304.85

SYN: AMYL HARMOL HYDROCHLORIDE

TOXICITY DATA WITH REFERENCE
ipr-mus LDLo:200 mg/kg QJPPAL 5,56,32
scu-gpg LDLo:400 mg/kg QJPPAL 5,56,32
scu-frg LDLo:200 mg/kg QJPPAL 5,37,32

SAFETY PROFILE: Poison by intraperitoneal and subcutaneous routes. When heated to decomposition it emits very toxic fumes of HCl and NO_x.

AOK000 CAS:10484-36-3 HR: 1
AMYLISOEUGENOL
mf: $C_{15}H_{22}O_2$ mw: 234.37

SYNS: AMYLOXYISOEUGENOL □ ISOEUGENOL AMYL ETHER □ 2-METHOXY-1-(PENTYLOXY)-4-(1-PROPENYL)-BENZENE □ 1-PENTOXY-2-METHOXY-4-PROPENYLBENZENE

TOXICITY DATA WITH REFERENCE
skn-rbt 500 mg/24H MOD FCTXAV 17,509,79
orl-rat LD50:>5 g/kg FCTXAV 17,513,79
skn-rbt LD50:>5 g/kg FCTXAV 17,513,79

CONSENSUS REPORTS: Reported in EPA TSCA Inventory.

SAFETY PROFILE: Very low toxicity by ingestion and skin contact. A skin irritant. See also ETHERS. When heated to decomposition it emits acrid smoke and irritating fumes.

AOK250 HR: 1
AMYL LACTATE
mf: $C_8H_{16}O_3$ mw: 160.2

PROP: Colorless liquid. Bp: 210°; flash p: 175°F; d: 0.960 @ 20°.

SAFETY PROFILE: An irritant by inhalation and ingestion. See also ESTERS. Moderately flammable. Incompatible with heat, flame, oxidizing materials. To fight fire, use foam, CO_2, dry chemical.

AOK500 HR: 2
AMYL LAURATE
mf: $C_5H_{11}O_2C(CH_2)_{10}CH_3$ mw: 270.44

PROP: Bp: 290°, flash p: 300°F, d: 0.86.

SAFETY PROFILE: It may defat skin and cause contact dermatitis. Combustible. Incompatible with oxidizing materials. To fight fire, use CO_2, dry chemical.

AOK750 CAS:105-30-6 HR: 2
AMYL METHYL ALCOHOL
mf: $C_6H_{14}O$ mw: 102.20

PROP: Liquid. Bp: 130°, flash p: 114°F (CC), d: 0.804, vap d: 3.52.

SYNS: 1,3-DIMETHYL BUTANOL □ ISOHEXYL ALCOHOL □ ISOPROPYL DIMETHYL CARBINOL □ METHYLAMYL ALCOHOL □ METHYL ISOBUTYL CARBINOL □ 2-METHYLPENTANOL-1 □ 2-METHYL-2-PROPYLETHANOL

TOXICITY DATA WITH REFERENCE
skn-rbt 10 mg/24H open MLD AIHAAP 23,95,62
skn-rbt 500 mg/24H MLD 85JCAE -,198,86
eye-rbt 750 μg/24H SEV 85JCAE -,198,86
ihl-hmn TCLo:50 ppm:IRR JIHTAB 28,262,46
orl-rat LD50:1410 mg/kg AMIHBC 10,61,54
skn-rbt LD50:3560 mg/kg AMIHBC 10,61,54

CONSENSUS REPORTS: Reported in EPA TSCA Inventory.

SAFETY PROFILE: Moderately toxic by ingestion and skin contact. A skin and severe eye irritant. Human systemic irritant by inhalation. A flammable liquid; can react with oxidizing materials. To fight fire, use CO_2, dry chemical. When heated to decomposition it emits smoke and acrid fumes.

For occupational chemical analysis use NIOSH: Alcohols III, 1402.

AOL000 CAS:13256-07-0 HR: 3
n-AMYL-N-METHYLNITROSAMINE
mf: $C_6H_{14}N_2O$ mw: 130.22

SYNS: AMN ☐ METHYLAMYLNITROSAMIN (GERMAN) ☐ METHYLAMYLNITROSAMINE ☐ METHYL-N-AMYLNITROSAMINE ☐ N-METHYL-N-NITROSOPENTYLAMINE ☐ METHYL-N-PENTYLNITROSAMINE ☐ N-NITROSO-N-METHYL-N-AMYLAMINE ☐ NITROSOMETHYL-N-PENTYLAMINE

TOXICITY DATA WITH REFERENCE
mma-sat 10 µg/plate TCMUE9 1,13,84
mma-esc 1 µmol/plate GANNA2 75,8,84
dnr-esc 25 µL/well CBINA8 15,219,76
orl-rat TDLo:168 mg/kg/8W-C:CAR NIPAA4 78,1889,81
ipr-rat TDLo:50 mg/kg:NEO CNREA8 39,3644,79
scu-rat TDLo:240 mg/kg/40W-I:CAR CCLCDY 2,263,80
orl-rat TD:330 mg/kg/31W-C:ETA ARZNAD 19,1077,69
orl-rat LD50:120 mg/kg ZEKBAI 69,103,67
ipr-rat LD50:85 mg/kg CNREA8 39,3644,79
scu-rat LD50:120 mg/kg ZEKBAI 69,103,67

CONSENSUS REPORTS: EPA Genetic Toxicology Program.

SAFETY PROFILE: Poison by ingestion, subcutaneous, and intraperitoneal routes. Suspected carcinogen with experimental carcinogenic, neoplastigenic, and tumorigenic data. Mutation data reported. When heated to decomposition it emits toxic NO_x. See also NITROSAMINES and N-NITROSO COMPOUNDS.

AOL250 CAS:1002-16-0 HR: 3
AMYL NITRATE
DOT: UN 1112
mf: $C_5H_{11}NO_3$ mw: 133.17

PROP: Liquid. Bp: 145°, flash p: 125°F (OC), d: 0.99.

SYNS: AMYLESTER KYSELINY DUSICNE ☐ NITRATE d'AMYLE (FRENCH)

TOXICITY DATA WITH REFERENCE
ihl-rat LCLo:3593 ppm AMIHAB 11,290,55
ihl-mus LCLo:1374 ppm AMIHAB 11,290,55
ihl-rbt LCLo:1703 ppm AMIHAB 11,290,55
ihl-gpg LCLo:1703 ppm AMIHAB 11,290,55

DOT CLASSIFICATION: 3; *Label:* Flammable Liquid

SAFETY PROFILE: Moderately toxic by inhalation. A flammable liquid. An oxidizing agent. When heated to decomposition it emits toxic fumes of NO_x.

AOL500 CAS:463-04-7 HR: 3
n-AMYL NITRITE
mf: $C_5H_{11}NO_2$ mw: 117.17

PROP: Clear, yellowish liquid; peculiar, ethereal, fruity odor and pungent, aromatic taste. Bp: 104°, d: 0.853 @ 20°/4°, autoign temp: 408°F, vap d: 4.0.

SYNS: AMYL NITRITE (DOT) ☐ 1-NITROPENTANE ☐ NITROUS ACID, PENTYL ESTER ☐ PENTYL NITRITE

TOXICITY DATA WITH REFERENCE
mmo-sat 1 mg/plate BSIBAC 56,816,80
scu-mus LDLo:30 g/kg HDTU**-,-,33 AIPTAK 49,272,35

CONSENSUS REPORTS: Reported in EPA TSCA Inventory.

SAFETY PROFILE: Moderately toxic by inhalation and ingestion. Causes flushing of skin, rapid pulse, headache, and fall in blood pressure. Mutation data reported. See also NITRITES and ESTERS. Flammable when exposed to heat or flame or by spontaneous chemical reaction. To fight fire, use alcohol foam. An oxidizing material. Vapors explode when heated. It will react with oxidizing or reducing materials. When heated to decomposition it emits toxic fumes of NO_x.

AOL750 CAS:64005-62-5 HR: 2
n-AMYL-N-NITROSOURETHANE
mf: $C_8H_{16}N_2O_3$ mw: 188.26

SYN: N-NITROSO-N-PENTYLCARBAMIC ACID-ETHYL ESTER

TOXICITY DATA WITH REFERENCE
cyt-ham:fbr 63 mg/L/48H MUREAV 48,337,77
orl-rat TDLo:2625 mg/kg/50W-C:CAR GANNA2 73,48,82
orl-rat TD:7350 mg/kg/35W-C:CAR GANNA2 73,48,82
orl-rat TDLo:5880 mg/kg/24W-C:ETA GANNA2 70,653,79

SAFETY PROFILE: Questionable carcinogen with experimental carcinogenic and tumorigenic data. Mutation data reported. See also ESTERS. When heated to decomposition it emits toxic fumes of NO_x. See also N-NITROSO COMPOUNDS.

AOM000 CAS:644-26-8 HR: 3
AMYLOCAINE
mf: $C_{14}H_{21}NO_2$ mw: 235.36

SYNS: AMYLEINE ☐ 1-(DIMETHYLAMINO)-2-METHYL-2-BUTANOL BENZOATE (ESTER) ☐ STOVAINE

TOXICITY DATA WITH REFERENCE
ivn-rat LDLo:25 mg/kg PHREA7 12,262,32
ipr-mus LDLo:170 mg/kg HBAMAK 4,1289,35
scu-mus LDLo:170 mg/kg HBAMAK 4,1289,35
ipr-dog LDLo:100 mg/kg HBAMAK 4,1289,35
scu-dog LDLo:100 mg/kg HBAMAK 4,1289,35

SAFETY PROFILE: Poison by intravenous, subcutaneous, and intraperitoneal routes. See also ESTERS. When heated to decomposition it emits toxic fumes of NO_x.

AOM150 CAS:9047-13-6 HR: 2
AMYLOPECTINE SULPHATE

SYNS: AMYLOPECTIN, HYDROGEN SULFATE ☐ AMYLOPECTIN SULFATE ☐ AMYLOPECTIN SULFATE (SN-263) ☐ SULFATED AMYLOPECTIN

TOXICITY DATA WITH REFERENCE
orl-rat TDLo:621 g/kg/24W-C:CAR JUIZAG 32,479,86
ipr-rat LD50:30 mg/kg TOIZAG 17,111,70
scu-rat LD50:1051 mg/kg TOIZAG 17,111,70
ipr-mus LD50:133 mg/kg TOIZAG 17,111,70
scu-mus LD50:935 mg/kg TOIZAG 17,111,70

SAFETY PROFILE: Poison by subcutaneous and intraperitoneal route. Questionable carcinogen with experimental carcinogenic data. When heated to decomposition it emits toxic fumes of NO_x and SO_x.

AOM250 CAS:14938-35-3 HR: 2
4-n-AMYLPHENOL
mf: $C_{11}H_{16}O$ mw: 164.27

PROP: A liquid. Bp: 342°, vap d: 5.66, flash p: 219°F (OC), d: 0.966.

SYN: p-PENTYLPHENOL

TOXICITY DATA WITH REFERENCE
skn-mus TDLo:4100 mg/kg/12W-I:NEO CNREA8 19,413,59

CONSENSUS REPORTS: Reported in EPA TSCA Inventory.

DOT CLASSIFICATION: 6.1; *Label:* KEEP AWAY FROM FOOD

SAFETY PROFILE: Questionable carcinogen with experimental neoplastigenic data. Moderately flammable. To fight fire, use foam, CO_2, dry chemical. When heated to decomposition it emits acrid smoke and irritating fumes.

AOM325 CAS:136-81-2 HR: 2
o-AMYLPHENOL
mf: $C_{11}H_{16}O$ mw: 164.27

SYNS: o-PENTYLPHENOL □ 2-PENTYLPHENOL □ PHENOL, o-PENTYL- □ PHENOL, 2-PENTYL-(9CI)

TOXICITY DATA WITH REFERENCE
orl-rat LD50:700 mg/kg JPETAB 53,218,35

CONSENSUS REPORTS: Reported in EPA TSCA Inventory.

DOT CLASSIFICATION: 6.1; *Label:* KEEP AWAY FROM FOOD

SAFETY PROFILE: Moderately toxic by ingestion. When heated to decomposition it emits acrid smoke and irritating vapors.

AOM500 HR: 3
2-sec-AMYLPHENOL
mf: $C_{11}H_{16}O$ mw: 164.27

PROP: Clear, straw-colored liquid. D: 0.955–0.971 @ 30°/30°, bp: 235–250°, flash p: 200°F. Very sltly sol in water; sol in oils and organic solvents.

SYN: o-(sec-PENTYL) PHENOL

TOXICITY DATA WITH REFERENCE
skn-mus TDLo:4100 mg/kg/12W-I:NEO CNREA8 19,413,59
ivn-mus LD50:100 mg/kg JMCMAR 23,1350,80

SAFETY PROFILE: Poison by intravenous route. Questionable carcinogen with experimental neoplastigenic data by skin contact. Moderately flammable when exposed to heat or flame. To fight fire, use foam, fog, dry chemical, water mist or spray, multipurpose dry chemical. When heated to decomposition it emits acrid smoke and irritating fumes.

AOM750 CAS:25735-67-5 HR: 2
4-sec-AMYLPHENOL
mf: $C_{11}H_{16}O$ mw: 164.27

PROP: D: <1.0, bp: 482–516°F, flash p: 270°F.

SYN: p-(sec-PENTYL) PHENOL

TOXICITY DATA WITH REFERENCE
skn-mus TDLo:4080 mg/kg/12W-I:ETA CNREA8 19,413,59

SAFETY PROFILE: Questionable carcinogen with experimental tumorigenic data. Combustible when exposed to heat or flame. To fight fire, use dry chemical, water mist, CO_2. When heated to decomposition it emits acrid smoke and fumes.

AON000 CAS:80-46-6 HR: 2
4-tert-AMYLPHENOL
mf: $C_{11}H_{16}O$ mw: 164.27

PROP: Colorless needles. Bp: 250°, mp: 92–93°, flash p: 232°F (OC).

SYNS: AMILPHENOL □ AMYL PHENOL 4T □ p-tert-AMYLPHENOL □ p-(α,α-DIMETHYLPROPYL)PHENOL □ p-(1,1-DIMETHYLPROPYL) PHENOL □ 2-METHYL-2-p-HYDROXYPHENYLBUTANE □ PENTA-PHEN □ p-tert-PENTYLPHENOL □ PTAP □ UCAR AMYL PHENOL 4T

TOXICITY DATA WITH REFERENCE
skn-rbt 100 µg/24H open AIHAAP 23,95,62
eye-rbt 1% SEV UCDS** 8/13/64
eye-rbt 500 mg SEV IHFCAY 6,1,67
orl-rat LD50:1830 mg/kg IHFCAY 6,1,67
skn-rbt LD50:2000 mg/kg UCDS** 8/13/64

CONSENSUS REPORTS: Reported in EPA TSCA Inventory.

SAFETY PROFILE: Moderately toxic by ingestion and skin contact. A skin and severe eye irritant. Combustible. When heated to decomposition it emits toxic fumes. To fight fire, use dry chemical, water mist, CO_2. Incompatible with oxidizing materials.

AON250 CAS:2282-34-0 HR: 3
3-sec-AMYLPHENYL-N-METHYLCARBAMATE
mf: $C_{13}H_{19}NO_2$ mw: 221.33

SYNS: ENT 27,127 □ m-(1-METHYLBUTYL)PHENYL METHYLCARBAMATE

TOXICITY DATA WITH REFERENCE
orl-rat LD50:87 mg/kg 28ZEAL 5,31,76

skn-rbt LD50:680 mg/kg 28ZEAL 5,31,76
orl-ckn LD50:44 mg/kg TXAPA9 11,49,67

SAFETY PROFILE: Poison by ingestion. Moderately toxic by skin contact. See also CARBAMATES. When heated to decomposition it emits toxic fumes of NO_x.

AON350 **HR: 3**
AMYL PROPIONATE
mf: $C_8H_{16}O_2$ mw: 144.21

PROP: Colorless liquid; fruity, apricot-pineapple odor. D: 0.866, refr index: 1.405–1.409, flash p: 106°F. Sol in alc, fixed oils; insol in glycerin, propylene glycol, water @ 160°.

SYNS: FEMA No. 2082 □ ISOAMYL PROPIONATE

SAFETY PROFILE: A flammable liquid. When heated to decomposition it emits acrid smoke and irritating fumes.

AON500 **CAS:32446-40-5** **HR: 3**
n-AMYL THIOCYANATE
mf: $C_6H_{11}NS$ mw: 129.24

PROP: Pale yellow oil. D: 0.905, bp: 197°. Insol in water; sol in alc and ether.

SYN: THIOCYANIC ACID, AMYL ESTER

TOXICITY DATA WITH REFERENCE
ipr-mus LD50:75 mg/kg JACSAT 78,3843,56
scu-mus LD50:75 mg/kg CLDND*

SAFETY PROFILE: Poison by subcutaneous and intraperitoneal routes. See also THIOCYANATES, ESTERS. When heated to decomposition it emits toxic fumes of NO_x and SO_x.

AON750 **CAS:64-43-7** **HR: 3**
AMYTAL SODIUM
mf: $C_{11}H_{17}N_2O_3$•Na mw: 248.29

SYNS: 5-ETHYL-5-ISOPENTYLBARBITURIC ACID SODIUM SALT □ 5-ETHYL-5-(3-METHYLBUTYL)BARBITURIC ACID, SODIUM DERIVATIVE □ 5-ISOAMYL-5-ETHYLBARBITURIC ACID, SODIUM DERIVATIVE □ SODIUM AMYLOBARBITONE □ SODIUM ETHYLISOAMYLBARBITURATE □ SODIUM ISOAMYLETHYL BARBITURATE

TOXICITY DATA WITH REFERENCE
orl-rat LD50:275 mg/kg JPETAB 68,22,40
scu-rat LDLo:90 mg/kg JPETAB 31,1,27
ivn-rat LD50:128 mg/kg JAPMA8 44,152,55
orl-mus LD50:505 mg/kg FRPSAX 14,845,59
ipr-mus LDLo:200 mg/kg JPETAB 31,455,27
scu-mus LDLo:280 mg/kg JPHAA3 26,1248,37
ivn-mus LDLo:200 mg/kg JPHAA3 26,1248,37
orl-dog LD50:99 mg/kg JPETAB 68,22,40

SAFETY PROFILE: Poison by ingestion, subcutaneous, intravenous, and intraperitoneal routes. When heated to decomposition it emits toxic NO_x and Na_2O.

AON825 **HR: 3**
ANABAENA FLOS-AQUAE TOXIN

SYNS: A. FLOS-AQUAE TOXIN □ TOXIN, ANABAENA FLOS-AQUAE NRC-44-1

TOXICITY DATA WITH REFERENCE
orl-rat LDLo:7500 μg/kg SCIEAS 187,542,75
ipr-mus LDLo:300 μg/kg SCIEAS 187,542,75
orl-dck LDLo:1880 μg/kg SCIEAS 187,542,75
orl-ctl LDLo:1800 μg/kg SCIEAS 187,542,75

SAFETY PROFILE: Poison by ingestion and intraperitoneal routes.

AON875 **CAS:494-52-0** **HR: 3**
ANABASINE
mf: $C_{10}H_{14}N_2$ mw: 162.26

PROP: Liquid. Bp: 270–272°, fp 9°, d: 1.0455. Sol in water and in most organic solvents.

SYNS: ANABASIN □ (−)-ANABASIN □ ANABAZIN □ NEONICOTINE □ NEONIKOTIN □ 1-3-(2′-PIPERIDYL)PYRIDINE □ 3-(2-PIPERIDINYL)PYRIDINE □ 3-(2-PIPERIDYL)-PYRIDINE □ 2-(3-PYRIDYL)-PIPERIDINE □ 2-(3′-PYRIDYL) PIPERIDINE □ (−)-2-(3′-PYRIDYL) PIPERIDINE

TOXICITY DATA WITH REFERENCE
orl-pig TDLo:20,800 μg/kg (30-37D preg):TER TJADAB 30,61,84
orl-dog LDLo:50 mg/kg JPETAB 48,95,33
ivn-dog LDLo:3 mg/kg JPETAB 48,95,33
ivn-rbt LDLo:1 mg/kg JPETAB 48,95,33
skn-gpg LDLo:100 mg/kg JPETAB 48,95,33

SAFETY PROFILE: Poison by ingestion, subcutaneous, and intravenous routes. Moderately toxic by skin contact. An experimental teratogen. Insecticide. Acute and subacute toxicity: increased salivation, vertigo, confusion, disturbed vision and hearing, photophobia, cold extremities, nausea, vomiting, diarrhea, syncope, colonic spasms. When heated to decomposition it emits toxic fumes of NO_x.

AOO000 **HR: 2**
ANAGESTONE ACETATE mixed with MESTRANOL (10:1)
mf: $C_{24}H_{36}O_3$ mw: 372.60

SYNS: ANATROPIN mixed with MESTRANOL (10:1) □ MESTRANOL mixed with ANAGESTONE ACETATE (1:10)

TOXICITY DATA WITH REFERENCE
orl-dog TDLo:259 mg/kg/2Y-I:ETA JJIND8 65,137,80

SAFETY PROFILE: Questionable carcinogen with experimental tumorigenic data. When heated to decomposition it emits acrid smoke and irritating fumes.

AOO120 **CAS:64285-06-9** **HR: 3**
ANATOXIN I
mf: $C_{10}H_{15}NO$ mw: 165.26

PROP: Oil.

SYNS: ANATOXIN-a □ ANTX-a □ ETHANONE, 1-(9-AZABICYCLO (4.2.1)NON-2-EN-2-YL)-, (1R)-

TOXICITY DATA WITH REFERENCE
ipr-ham TDLo: 1125 μg/kg (female 12-14D post): TER TOXIA6 18,684,80
ipr-mus LDLo: 250 μg/kg TOXIA6 27,79,89

SAFETY PROFILE: Poison by intraperitoneal route. An experimental teratogen. When heated to decomposition it emits toxic fumes of NO_x.

AOO125 CAS:53-39-4 HR: 2
ANAVAR
mf: $C_{19}H_{30}O_3$ mw: 306.49

PROP: Crystals from 2-propanol. Mp: 235–238°.

SYNS: LONAVAR □ OXANDROLONE □ PROTIVAR □ PROVITAR □ VASOROME

TOXICITY DATA WITH REFERENCE
orl-rat TDLo: 1800 mg/kg (2-19D preg): REP FESTAS 22,735,71
ipr-rat LD50: 4893 mg/kg NYKZAU 65,418,69
orl-mus LD50: 1832 mg/kg NYKZAU 65,418,69
ipr-mus LD50: 922 mg/kg NYKZAU 65,418,69

SAFETY PROFILE: Moderately toxic by ingestion and intraperitoneal routes. Experimental reproductive effects. When heated to decomposition it emits acrid smoke and fumes.

AOO135 HR: 3
ANCISTRODON PISCIVORUS VENOM

SYN: VENOM, SNAKE, ANCISTRODON PISCIVORUS

TOXICITY DATA WITH REFERENCE
ipr-mus LD50: 6200 μg/kg ANREAK 139,305,61
ivn-mus LDLo: 7500 μg/kg 14FHAR -,373,63
ipr-frg LD50: 10 mg/kg ANREAK 139,305,61

SAFETY PROFILE: Poison by intravenous and intraperitoneal routes.

AOO250 HR: 3
ANDROCTONUS AMOREUXI VENOM

SYNS: A. AMOREUXI VENOM □ VENOM, SCORPION, ANDROCTONUS AMOREUXI

TOXICITY DATA WITH REFERENCE
ipr-rat TDLo: 1 mg/kg (8-12D preg): TER TOXIA6 21,177,83
ipr-rat TDLo: 1 mg/kg (8-12D preg): REP TOXIA6 21,177,83
ims-mus LD50: 880 μg/kg TOXIA6 13,253,75
unr-mus LD50: 600 μg/kg TOXIA6 9,1,71

SAFETY PROFILE: Deadly poison by intramuscular and unspecified routes. An experimental teratogen. Other experimental reproductive effects. When heated to decomposition it emits acrid smoke and irritating fumes.

AOO265 HR: 3
ANDROCTONUS AUSTRALIS HECTOR VENOM

SYNS: A. AUSTRALIS HECTOR VENOM □ VENOM, SCORPION, ANDROCTONUS AUSTRALIS HECTOR

TOXICITY DATA WITH REFERENCE
ipr-mus LD50: 97 μg/kg TOXIA6 22,308,84
scu-mus LD50: 420 μg/kg EJBCAI 16,514,70
ice-mus LD50: 700 ng/kg TOXIA6 22,308,84
unr-mus LD50: 9 μg/kg TOXIA6 20,9,82

SAFETY PROFILE: Deadly poison by subcutaneous, intraperitoneal, intracerebral, and possibly other routes.

AOO275 CAS:76-43-7 HR: 3
ANDROFLUORENE
mf: $C_{20}H_{29}FO_3$ mw: 336.49

PROP: Crystals. Decomp @ 270°. Sol in pyridine; sltly sol in acetone, chloroform; sparingly sol in methanol; practically insol in water, ether, benzene, and hexanes.

SYNS: ANDROFLUORONE □ ANDROSTEROLO □ 11-β,17-β-DIHYDROXY-9-α-FLUORO-17-α-METHYL-4-ANDROSTER-3-ONE □ FLUORO-9-α DIHYDROXY-11-β,17-β METHYL-17-α ANDROSTENE-4 ONE-3 (FRENCH) □ 9-FLUORO-11-β-,17-β-DIHYDROXY-17-METHYLANDROST-4-EN-3-ONE □ 9-α-FLUORO-11-β,17-β-DIHYDROXY-17-α-METHYL-4-ANDROSTENE-3-ONE □ 9-α-FLUORO-11-β-HYDROXY-17-METHYLTESTOSTERONE □ 9-α-FLUORO-17-α-METHYL-11-β,17-DIHYDROXY-4-ANDROSTEN-3-ONE □ FLUOTESTIN □ FLUOXIMESTERONE □ FLUOXYMESTERONE □ FLUOXYMESTRONE □ FLUSTERON □ FLUTESTOS □ HALOTESTIN □ 17-α-METHYL-9-α-FLUORO-11-β-HYDROXYTESTERONE □ NEO-ORMONAL □ NSC-12165 □ ORALSTERONE □ ORATESTIN □ ORA-TESTRYL □ TESTORAL □ U 6040 □ ULTANDREN □ ULTANDRENE

TOXICITY DATA WITH REFERENCE
orl-man TDLo: 12 mg/kg (84D male): REP JCEMAZ 44,121,77
scu-rat TDLo: 6 mg/kg (female 17-20D post): TER AVBIB9 13,71,74
orl-hmn TDLo: 400 μg/kg: BIO,SKN,PUL CANCAR 41,758,78
ipr-mus LD50: 2350 mg/kg OYYAA2 14,623,77
orl-rat TDLo: 437,500 μg/kg/35D-C OYYAA2 16,779,78

CONSENSUS REPORTS: EPA Genetic Toxicology Program.

SAFETY PROFILE: Poison by ingestion. Moderately toxic by intraperitoneal route. Human systemic effects by ingestion: dermatitis, changes in respiratory system and transaminase activity. Human reproductive effects by ingestion: spermatogenesis. An experimental teratogen. Other experimental reproductive effects. When heated to decomposition it emits toxic fumes of F^-.

AOO300 CAS:1239-29-8 HR: 2
ANDROFURAZANOL
mf: $C_{20}H_{30}N_2O_2$ mw: 330.52

PROP: Needles from methanol. Mp: 152–153°.

SYNS: DH 245 □ FRAZALON □ FURAZABOL □ 17-β-HDYROXY-17-α-METHYL-5-α-ANDROSTANO(2,3-c)FURAZAN □ 17-METHYL-5-α-ANDROSTANO(2,3-c)(1,2,5)OXADIAZOL-17-β-OL □ 17-α-METHYL-5-α-AN-

DROSTANO(2,3-c)(1,2,5)OXADIAZOL-17-β-OL □ MIOTOLON □ MYOTOLON

TOXICITY DATA WITH REFERENCE
scu-mus TDLo:2100 mg/kg (female 7-12D post):REP KSRNAM 4,2088,70
orl-rat TDLo:6 g/kg (13-18D preg):TER KSRNAM 4,2088,70
orl-mus LD50:1731 mg/kg OYYAA2 3,187,69
ipr-mus LD50:494 mg/kg CPBTAL 14,285,66

SAFETY PROFILE: Moderately toxic by ingestion and other routes. An experimental teratogen. Other experimental reproductive effects. When heated to decomposition it emits toxic fumes of NO_x.

AOO375 CAS:4720-09-6 HR: 3
ANDROMEDOTOXIN
mf: $C_{22}H_{36}O_7$ mw: 412.58

PROP: Crystals from EtOAc/pentane. Mp: 267–270°.

SYNS: ACETYLLANDROMEDOL □ ASEBOTOXIN □ G-I □ GRAYANOTOXANE-3,5,6,10,14,16-HEXOL 14 ACETATE □ GRAYANOTOXIN I □ RHODOTOXIN

TOXICITY DATA WITH REFERENCE
ipr-mus LD50:1310 μg/kg TXAPA9 35,303,76
scu-mus LD50:148 μg/kg JJPAAZ 6,46,56
ivn-cat LDLo:400 μg/kg JJPAAZ 6,46,56
ivn-rbt LDLo:270 μg/kg JJPAAZ 6,46,56
ivn-gpg LD50:1300 μg/kg ARTODN 44,259,80
par-frg LD50:3899 μg/kg JJPAAZ 6,46,56

SAFETY PROFILE: Poison by subcutaneous, parenteral, intravenous and intraperitoneal routes. When heated to decomposition it emits acrid smoke and fumes.

AOO400 CAS:302-96-5 HR: 1
ANDROSTANAZOL
mf: $C_{21}H_{32}N_2O$ mw: 328.55

SYNS: ANDROSTANAZOLE □ 17-β-HYDROXY-17-α-METHYLANDROSTANO(3,2-c)PYRAZOLE □ STANOZOLOL □ WIN 14833 □ WINSTROL □ WINSTROL V

TOXICITY DATA WITH REFERENCE
orl-rat TDLo:3 g/kg (13-18D preg):TER KSRNAM 4,2088,70
scu-rat TDLo:14 mg/kg (14D pre):REP CCPTAY 5,489,72
orl-wmn TDLo:24 mg/kg/17W-I:SYS BMJOAE 294,612,87
orl-man TDLo:4285 mg/kg/30D-I:SYS BMJOAE 294,612,87

SAFETY PROFILE: Human systemic effects by ingestion: jaundice. An experimental teratogen. Other experimental reproductive effects. When heated to decomposition it emits toxic fumes of NO_x.

AOO410 CAS:2297-30-5 HR: 2
ANDROSTENEDIOL DIPROPIONATE
mf: $C_{25}H_{38}O_4$ mw: 402.63

PROP: Solid. Mp: 115–116°.

SYNS: ANDROST-5-ENE-3-β,17-β-DIOL, DIPROPIONATE □ ANDROST-5-ENE-3,17-DIOL, DIPROPANOATE, (3-β,17-β)- (9CI) □ BISEXOVIS □ BISEXOVISTER □ GINANDRIN □ STENANDIOL

TOXICITY DATA WITH REFERENCE
scu-rat TDLo:125 mg/kg (female 16D post):TER ANENAG 16,283,55
scu-rat TDLo:125 mg/kg (female 16D post):REP ANENAG 16,283,55
orl-mus LD50:1185 mg/kg PCJOAU 17,30,83

SAFETY PROFILE: Moderately toxic by ingestion. An experimental teratogen. Other experimental reproductive effects. When heated to decomposition it emits acrid smoke and irritating fumes.

AOO425 CAS:63-05-8 HR: 2
ANDROSTENEDIONE
mf: $C_{19}H_{26}O_2$ mw: 286.45

PROP: Dimorphous: Needles from acetone, or crystals from hexane. Mp: 173–174°.

SYNS: Δ^4-ANDROSTEN-3,17-DIONE □ Δ^4-ANDROSTENE-3,17-DIONE □ Δ-4-ANDROSTENEDIONE □ 4-ANDROSTENE-3,17-DIONE □ ANDROTEX □ SKF 2170

TOXICITY DATA WITH REFERENCE
ims-rat TDLo:80 mg/kg (female 14-21D post):REP JCPPAV 92,13,78
ims-rat TDLo:80 mg/kg (female 14-21D post):TER JCPPAV 92,13,78
scu-mus TDLo:600 mg/kg/72W-I:ETA JNCIAM 19,977,57

SAFETY PROFILE: An experimental teratogen. Other experimental reproductive effects. Questionable carcinogen with experimental tumorigenic data. When heated to decomposition it emits acrid smoke and irritating fumes.

AOO450 CAS:53-43-0 HR: 2
ANDROSTENOLONE
mf: $C_{19}H_{28}O_2$ mw: 288.47

PROP: Dimorphous. Needles: mp: 140–141°. Leaflets: mp: 152–153°. Sol in benzene, alc, and ether; sparingly sol in chloroform and petr ether.

SYNS: 17-CHETOVIS □ trans-DEHYDROANDROSTERONE □ DEHYDROEPIANDROSTERONE □ 5-DEHYDROEPIANDROSTERONE □ DEHYDROISOANDROSTERONE □ 5,6-DEHYDROISOANDROSTERONE □ DHA □ DIANDRON □ DIANDRONE □ 5,6-DIDEHYDROISOANDROSTERONE □ 17-HORMOFORIN □ 3-β-HYDROXY-5-ANDROSTEN-17-ONE □ PRASTERONE □ PSICOSTERONE

TOXICITY DATA WITH REFERENCE
scu-rat TDLo:300 mg/kg (female 15-20D post):TER ANENAG 17,118,56
orl-rat TDLo:1800 mg/kg (2-19D preg):REP FESTAS 22,735,71
orl-mus TDLo:25 g/kg/52D-C:NEO CNREA8 48,2788,88

SAFETY PROFILE: An experimental teratogen. Experimental reproductive effects. Questionable carcinogen with experimental neoplastigenic data. When heated to decomposition it emits acrid smoke and irritating fumes.

AOO475 CAS:521-10-8 HR: 3
ANDROSTESTONE-M
mf: $C_{20}H_{32}O_2$ mw: 304.52

PROP: Crystals from ethyl acetate. Mp: 205.5–206.5°. Insol in water. Sltly sol in some organic solvents.

SYNS: ANDRODIOL □ ANDROSTESTON-M □ CRESTABOLIC □ DIOLANDRONE □ DIOLOSTENE □ ESJAYDIOL □ MAD □ MADIOL □ MASDIOL □ MEGABION (JAPANESE) □ MESTENEDIOL □ METANDIOL □ METANDRIOL □ METENDIOL □ METHANABOL □ METHANDIOL □ METHANDRIOL □ METHANDROLAN □ METHOSTAN □ METHYLANDROSTENDIOL □ METIDIONE □ METILDIOLO □ METOCRYST □ NABADIAL □ NEOSTENE □ NEOSTERON □ NEUTRORMONE □ NEUTROSTERON □ NOTANDRON □ NOTANDRON-DEPOT □ PROTANDREN □ STENEDIOL □ STENIBELL □ STENOSTERONE □ TESTODIOL □ TROFORMONE

TOXICITY DATA WITH REFERENCE
unr-wmn TDLo:64 mg/kg (female 26-39W post):TER GEFRA2 13,216,53

SAFETY PROFILE: Human reproductive effects by an unspecified route: developmental abnormalities of the urogenital system. A human and experimental teratogen. When heated to decomposition it emits acrid smoke and irritating fumes.

AOO490 CAS:532-11-6 HR: 3
ANETHOLE TRITHIONE
mf: $C_{10}H_8OS_3$ mw: 240.36

PROP: Orange-colored prisms from butyl acetate, very bitter taste. Mp: 111°. Practically insol in water. Sol in pyridine, chloroform, benzene, dioxane, and carbon disulfide; sltly sol in ether, acetone, ethyl acetate, acetic acid, alc, cyclohexane, and petr ether.

SYNS: ANETHOLTRITHION □ FELVITEN □ HEPORAL □ 5-(p-METHOXYPHENYL)-1,2-DITHIOCYCLOPENTEN-3-THIONE □ 5-(p-METHOXYPHENYL)-3H-1,2-DITHIOLE-3-THIONE □ 5-(4-METHOXYPHENYL)-3H-1,2-DITHIOLE-3-THIONE (9CI) □ 5-(p-METHOXYPHENYL) TRITHIONE □ MUCINOL □ SKF 1717 □ SUFRALEM □ SULFARLEM □ SULFOGAL □ SULFRALEM □ TIOPROPEN □ TIOTRIFAR □ TRITHIO □ TRITHIOANETHOLE □ TRITHIO-(p-METHOXYPHENYL) PROPENE

TOXICITY DATA WITH REFERENCE
ipr-rat TDLo:20,160 mg/kg (24W pre):REP OYYAA2 12,79,76
ims-rat LD50:35 mg/kg AEPPAE 222,244,54
orl-mus LD50:3850 mg/kg NIIRDN 6,25,82
ipr-mus LD50:1780 mg/kg NIIRDN 6,25,82
orl-gpg LD50:6000 mg/kg NIIRDN 6,25,82

SAFETY PROFILE: Poison by intramuscular route. Moderately toxic by ingestion and intraperitoneal routes. Experimental reproductive effects. When heated to decomposition it emits toxic fumes of SO_x.

AOO500 CAS:956-90-1 HR: 3
ANGEL DUST
mf: $C_{17}H_{25}N \cdot ClH$ mw: 279.89

PROP: Crystals. Mp: 46–46.5°, bp: 135–137°.

SYNS: CI395 □ CN-25,253-2 □ DOA □ ELEPHANT TRANQUILIZER □ ELYSION □ GP-121 □ HOG □ NSC-40902 □ PCP HYDROCHLORIDE □ PEACE PILL □ PHENCYCLIDINE HYDROCHLORIDE □ 1-(1-PHENYLCYCLOHEXYL)PIPERIDINE HYDROCHLORIDE □SERNYL □ SERNYLAN □ SERNYL HYDROCHLORIDE □ TRANK

TOXICITY DATA WITH REFERENCE
scu-mus TDLo:50 mg/kg (female 6-15D post):REP FEPRA7 42,1157,83
ipr-rat TDLo:240 mg/kg (female 15-20D post):TER PBBHAU 11(Suppl),39,79
orl-hmn TDLo:71 µg/kg:CNS JAMAAP 238,515,77
orl-hmn TDLo:71 µg/kg:BAH JAMAAP 238,515,77
orl-hmn LDLo:14 mg/kg JAMAAP 238,515,77
ivn-hmn TDLo:10 µg/kg:CNS,PNS CPAJAK 6,150,61
orl-rat LD50:135 mg/kg NETOD7 3,11,81
orl-mus LD50:77 mg/kg JPPMAB 28,713,76
ipr-mus LD50:59,558 µg/kg SAAMDZ 2,143,81
scu-mus LD50:43 mg/kg JMCMAR 24,496,81
ivn-mus LD50:16 mg/kg LIFSAK 31,803,82
ivn-dog LDLo:50 mg/kg TXCYAC 19,11,81
orl-pgn LD50:237 mg/kg TXAPA9 21,315,72
orl-dck LD50:75 mg/kg TXAPA9 21,315,72
orl-bwd LD50:5600 µg/kg TXAPA9 21,315,72

SAFETY PROFILE: Poison by ingestion, subcutaneous, intravenous, and intraperitoneal routes. Human systemic effects by ingestion and intravenous routes: distorted perceptions, euphoria, excitement, hallucinations, and paresthesia. An experimental teratogen. Other experimental reproductive effects. Often mixed with other drugs of abuse yielding totally unpredictable effects. A controlled substance. When heated to decomposition it emits very toxic fumes of HCl and NO_x.

AOO750 CAS:591-12-8 HR: 2
α-ANGELICA LACTONE
mf: $C_5H_6O_2$ mw: 98.11

SYN: 4-HYDROXYPENT-3-ENOIC ACID LACTONE

TOXICITY DATA WITH REFERENCE
orl-mus LD50:2800 mg/kg DCTODJ 3,249,80
ipr-mus LD50:3000 mg/kg APTOA6 2,109,46

CONSENSUS REPORTS: Reported in EPA TSCA Inventory.

SAFETY PROFILE: Moderately toxic by ingestion and intraperitoneal routes. When heated to decomposition it emits acrid smoke and irritating fumes.

AOO760 CAS:8015-64-3 HR: 2
ANGELICA OIL, root

PROP: Extracted from roots of *Angelica archangelica L.* A pale-yellow to amber liquid; pungent odor with bittersweet taste. Sol in fixed oils; sltly sol in mineral oil; insol in glycerin, propylene glycol.

SYNS: ANGELICA ROOT OIL □ ANGELIKA OEL □ OILS, ANGELICA ROOT

TOXICITY DATA WITH REFERENCE
orl-rat LD50:11,000 mg/kg FCTXAV 13,713,75
orl-mus LD50:2200 mg/kg FCTXAV 13,713,75

CONSENSUS REPORTS: Reported in EPA TSCA Inventory.

SAFETY PROFILE: Moderately toxic by ingestion. When heated to decomposition it emits acrid smoke and irritating fumes.

AOO800 CAS:554-18-7 HR: 3
ANGELI'S SULFONE
mf: $C_{24}H_{34}N_2O_{18}S_3 \cdot 2Na$ mw: 780.76

PROP: White, amorphous powder; sweet-tasting solid. Sol in water; sltly sol in alc; insol in ether, benzene, methanol, ethyl acetate, and pyridine.

SYNS: ACEPROSOL □ ANGELI SULFONE □ p,p'-DIAMINODIPHENYLSULFONE-N,N'-DI(DEXTROSE SODIUM SULFONATE) □ DISODIUM p,p'-DIAMINODIPHENYLSULFONE-N,N'-DIGLUCOSE SULFONATE □ d-GLUCITOL, 1,1'-(SULFONYLBIS(4,1-PHENYLENEIMINO))BIS(1-DEOXY-1-SULFO)-, DISODIUM SALT (9CI) □ GLUCOSULFONE □ GLUCOSULFONE SODIUM □ 501 P □ PROMANIDE □ PROMIN □ PROMIN SODIUM □ PROMOTIN □ PROTOMIN □ 501 SIEGFRIED □ S. N. 166 □ SODIUM GLUCOSULFONE □ SULFONA P □ SOLFONE □ 1, 1'-(SULFONYLBIS(4,1-PHENYLENEIMINO))BIS(1-DEOXY-1-SULFO-d-GLUCITOL) DISODIUM SALT □ p,p'-SULFONYLDIANILINE N,N'-DI-GLUCOSIDE DISODIUM DISULFONALTE □ p,p'-SULFONYLDIANILINE-N,N'-DI-d-GLUCOSE SODIUM BISULFITE □ TASMIN

TOXICITY DATA WITH REFERENCE
orl-mus LD50:3930 μg/kg NIIRDN 6,225,82
scu-mus LD50:6500 μg/kg NIIRDN 6,225,82
ivn-mus LD50:5250 μg/kg NIIRDN 6,225,82

SAFETY PROFILE: Poison by ingestion, subcutaneous, and intravenous routes. When heated to decomposition it emits toxic fumes of SO_x, NO_x, and Na_2O.

AOO825 HR: 2
ANGEL'S TRUMPET

PROP: A small tree or large shrub that may grow to 20 feet. The large flowers are funnel shaped, grow to 10 or 12 inches long and may be white, yellow-white or pink.

SYNS: BELLADONA (HAWAII) □ BRUGMANSIA ARBOREA □ BRUGMANSIA X CANDIDA □ BRUGMANSIA SANGUINEA □ BRUGMANSIA SUAVEOLENS □ CAMPANA (CUBA, PUERTO RICO) □ CORNUCOPIA □ FLORIPONDIO (PUERTO RICO) □ NANA-HONUA (HAWAII)

SAFETY PROFILE: The whole plant contains poisonous belladonna alkaloids. The seeds and dried leaves are used as hallucinogens. Ingestion may cause fever, increased heart rate, dilated pupils, delirium, and high blood pressure. See also BELLADONNA.

AOO875 CAS:131-49-7 HR: 3
ANGIGRAFIN
mf: $C_{11}H_9I_3N_2O_4 \cdot C_7H_{17}NO_5$ mw: 809.17

PROP: Rhombic needles, sltly sweet taste. Mp: 189–193° (decomp). Solubility in water at 20°: 89 g/100 mL.

SYNS: AMIDOTRIZOATE MEGLUMINE □ ANGIOGRAFIN □ BENZOIC ACID, 3,5-DIACETAMIDO-2,4,6-TRIIODO-, compd. with 1-DEOXY-1-(METHYLAMINO)-d-GLUCITOL □ CARDIOGRAFIN □ CYSTOGRAFIN □ DIATRIZOATE MEGLUMINE □ DIATRIZOATE METHYLGLUCAMINE □ DITRIZOATE METHYLGLUCAMINE □ GASTROGRAFIN □ d-GLUCITOL, 1-DEOXY-1-(METHYLAMINO)-, 3,5-BIS(ACETYLAMINO)-2,4,6-TRIIODOBENZOATE (SALT) □ HYPAQUE 13.4 □ HYPAQUE 60 □ HYPAQUE CYSTO □ HYPAQUE M 30 □ HYPAQUE MEGLUMINE □ MEGLUMINE AMIDOTRIZOATE □ MEGLUMINE DIATRIZOATE □ METHYLGLUCAMINE DIATRIXOATE □ RENOGRAFFIN M-76 □ RENOGRAFIN □ RENO M □ RENO M 60 □ RENO-M-DIP □ RENURIX □ UNIPAQUE □ UROVIST

TOXICITY DATA WITH REFERENCE
scu-man TDLo:214 mg/kg 34ZIAG -,392,69
ivn-rat LD50:15,300 mg/kg YACHDS 12(Suppl 1),11,84
ivn-mus LD50:21,200 mg/kg NIIRDN 6,32,82
ice-mus LD50:80 mg/kg THERAP 26,595,71

SAFETY PROFILE: Poison by intracerebral route. Human systemic effects by subcutaneous route: kidney damage and reduced urine volume. When heated to decomposition it emits toxic fumes of I^- and NO_x.

AOO900 CAS:1407-47-2 HR: 3
ANGIOTONIN

PROP: Hydrolyzed by strong acids and bases and above pH 9.5. Sol in organic solvents, in aq solns pH 5–8.

SYNS: ANGIOTENSIN □ HYPERTENSIN

TOXICITY DATA WITH REFERENCE
scu-ham TDLo:200 μg/kg (female 8D post):TER LIFSAK 8,525,69
scu-ham TDLo:20 μg/kg (female 8D post):REP LIFSAK 8,525,69
ivn-rat LDLo:8 mg/kg 27ZIAQ -,43,73

SAFETY PROFILE: Poison by intravenous route. An experimental teratogen. Other experimental reproductive effects. When heated to decomposition it emits acrid smoke and irritating fumes.

AOO925 CAS:1402-83-1 HR: 3
ANGOLAMYCIN
mf: $C_{46}H_{77}NO_{17}$ mw: 916.24

PROP: Crystals from Et_2O. Mp: 133–136°.

TOXICITY DATA WITH REFERENCE
ipr-mus LDLo:280 mg/kg 85FZAT -,135,67
scu-mus LD50:500 mg/kg 85FZAT -,135,67
unr-mus LDLo:1000 mg/kg 85ERAY 1,74,78

SAFETY PROFILE: Poison by several routes. When heated to decomposition it emits toxic fumes of NO_x.

AOP250 CAS:2270-40-8 HR: 3
ANGUIDIN
mf: $C_{18}H_{26}O_7$ mw: 354.44

PROP: Crystals from EtOAc. Mp: 161–162°.

SYNS: ANG 66 □ ANGUIDINE □ DAS □ 4-β,15-DIACETOXY-3-α-HYDROXY-12,13-EPOXYTRICHOTHEC-9-ENE □ DIACETOXYSCIRPENOL □ 4,15-DIACETOXYSCIRPEN-3-OL □ DIAZETOXYSKIRPENOL (GERMAN) □ 12,13-EPOXY-4-β,15-DIAZETOXY-3-α-HYDROXY-TRICHOTHEC-9-ENE □ (3-α,4-β)-12,13-EPOXY-4,15-DIACETATE-TRICHOTHEC-9-ENE-3,4,15-TRIOL □ MM 4462 □ NSC-141537

TOXICITY DATA WITH REFERENCE
skn-gpg 284 ng MLD FAATDF 4,S124,84
dns-rat-orl 3 mg/kg CALEDQ 38,199,87
dnd-mus-ivn 5600 μg/kg PAACA3 19,65,78

ipr-mus TDLo:1 mg/kg (9D post):TER TXCYAC 45,245,87
ipr-mus TDLo:1 mg/kg (9D post):REP TXCYAC 45,245,87
ipr-wmn TDLo:12 mg/kg/5D:GIT CTRRDO 63,789,79
orl-rat LD50:7 mg/kg VHTODE 25,335,83
ipr-rat LD50:750 μg/kg DFSCDX 4,135,83
ivn-rat LD50:1300 μg/kg ARZNAD 18,989,68
orl-mus LD50:7300 μg/kg BIBIAU 10,445,68
ihl-mus LD50:11,300 μg/kg TOXID9 4,12,84
ipr-mus LD50:7839 μg/kg NCISP* JAN86

CONSENSUS REPORTS: EPA Genetic Toxicology Program.

SAFETY PROFILE: A deadly poison by ingestion, inhalation, intravenous, intraperitoneal, and subcutaneous routes. Human systemic effects by intraperitoneal route: muscle weakness, nausea or vomiting, and fever. An experimental teratogen. Other experimental reproductive effects. Mutation data reported. A skin irritant. When heated to decomposition it emits acrid smoke and fumes.

AOP500 **HR: 2**
ANHYDRIDES

PROP: Chemical compounds derived from acids by elimination of a molecule of water. Thus, sulfur trioxide (SO_3) is the anhydride of sulfuric acid (H_2SO_4); carbon dioxide (CO_2) is the anhydride of carbonic acid (H_2CO_3); phthalic acid ($C_6H_4(CO_2H)_2$) minus water gives phthalic anhydride ($C_6H_4(CO_2)O$). This term should not be confused with anhydrous, meaning without water.

SAFETY PROFILE: Anhydrides are acidic and react with bases in tissue. Thus, they tend to attack and irritate tissue.

AOP750 **CAS:35891-69-1** **HR: 3**
ANHYDROMYRIOCIN
mf: $C_{21}H_{37}NO_5$ mw: 383.4

TOXICITY DATA WITH REFERENCE
ipr-rat LD50:37 mg/kg 85ERAY 3,2067,78
orl-mus LD50:100 mg/kg 85ERAY 3,2067,78
ipr-mus LD50:75 mg/kg 85ERAY 3,2067,78

SAFETY PROFILE: Poison by ingestion and intraperitoneal routes. When heated to decomposition it emits toxic fumes such as NO_x.

AOQ000 **CAS:62-53-3** **HR: 3**
ANILINE
DOT: UN 1547
mf: C_6H_7N mw: 93.14

PROP: Colorless, oily liquid which darkens on exposure to light; characteristic odor. Mp: −6°, bp: 184.4°, lel: 1.3%, ULC: 20–25, flash p: 158°F (CC), fp: −6.2°, d: 1.02 @ 20°/4°, autoign temp: 1139°F, vap press: 1 mm @ 34.8°, vap d: 3.22.

SYNS: AMINOBENZENE □ AMINOPHEN □ ANILIN (CZECH) □ ANILINA (ITALIAN, POLISH) □ ANILINE OIL □ BENZENAMINE □ BLUE OIL □ C.I. 76000 □ HUILE d'ANILINE (FRENCH) □ NCI-C03736 □ PHENYLAMINE

TOXICITY DATA WITH REFERENCE
skn-rbt 500 mg/24H MOD 28ZPAK -,65,72
eye-rbt 102 mg SEV BIOFX* 1-5/69
mma-sat 100 μg/plate PJABDW 53,34,77
dnr-esc 25 μL/well/16H CBINA8 15,219,76
bfa-rat/sat 300 mg/kg MUREAV 79,173,80
orl-mus TDLo:4480 mg/kg (female 6-13D post):REP TCMUD8 7,29,87
orl-rat TDLo:11 g/kg/29W-C:NEO APMIAL 26,473,49
unk-hmn LDLo:357 mg/kg JIDHAN 13,87,31
unk-man LDLo:150 mg/kg 85DCAI 2,73,70
orl-rat LD50:250 mg/kg JPETAB 90,260,47
ihl-rat LCLo:250 ppm/4H JIHTAB 31,343,49
skn-rat LD50:1400 mg/kg AGGHAR 15,447,57
ipr-rat LD50:420 mg/kg AGGHAR 15,447,57
ihl-mus LC50:175 ppm/7H NTIS** PB214-270
ipr-mus LD50:492 mg/kg IZSBAI 3,91,65
scu-mus LD50:200 mg/kg ARZNAD 8,107,58
orl-dog LD50:195 mg/kg NTIS** PB214-270
skn-dog LDLo:1540 mg/kg NTIS** PB214-270
ihl-cat LCLo:180 ppm/8H XPHBAO 271,4,41

CONSENSUS REPORTS: IARC Cancer Review: Group 3 IMEMDT 7,99,87; Animal Inadequate Evidence IMEMDT 4,27,74; Human No Evidence IMEMDT 4,27,74. EPA Extremely Hazardous Substances List. Community Right-To-Know List. Reported in EPA TSCA Inventory.

OSHA PEL: TWA 2 ppm (skin)
ACGIH TLV: TWA 2 ppm (skin); BEI: 50 mg/L total p-aminophenol in urine at end of shift.
DFG MAK: 2 ppm (8 mg/m³), Suspected Carcinogen; BAT: 1 mg/L in urine at end of shift.
DOT CLASSIFICATION: 6.1; *Label:* Poison

SAFETY PROFILE: Suspected carcinogen with experimental neoplastigenic data. A human poison by an unspecified route. Poison experimentally by most routes including inhalation and ingestion. Experimental reproductive effects. A skin and severe eye irritant, and a mild sensitizer. In the body, aniline causes formation of methemoglobin, resulting in prolonged anoxemia and depression of the central nervous system; less acute exposure causes hemolysis of the red blood cells, followed by stimulation of the bone marrow. The liver may be affected with resulting jaundice. Long-term exposure to aniline dye manufacture has been associated with malignant bladder growths. A common air contaminant. A combustible liquid when exposed to heat or flame. To fight fire, use alcohol foam, CO_2, dry chemical. It can react vigorously with oxidizing materials. When heated to decomposition it emits highly toxic fumes of NO_x. Spontaneously explosive reactions occur with benzenediazonium-2-carboxylate, dibenzoyl peroxide, fluorine nitrate, nitrosyl perchlorate, red fuming nitric acid, peroxodisulfuric acid, and tetranitromethane. Violent reactions with boron trichloride, peroxyformic acid, diisopropyl peroxydicarbonate, fluorine, trichloronitromethane (145°C), acetic anhydride, chlorosulfonic acid, hexachloromelamine, ($HNO_3 + N_2O_4 + H_2SO_4$), (nitrobenzene + glycerin), oleum, ($HCHO + HClO_4$), perchromates, K_2O_2, β-propiolactone, $AgClO_4$,

Na_2O_2, H_2SO_4, trichloromelamine, acids, peroxydisulfuric acid, FO_3Cl, diisopropyl peroxy-dicarbonate, n-haloimides, and trichloronitromethane. Ignites on contact with sodium peroxide + water. Forms heat- or shock-sensitive explosive mixtures with anilinium chloride (detonates at 240°C/7.6 bar), nitromethane, hydrogen peroxide, 1-chloro-2,3-epoxypropane, and peroxomonosulfuric acid. Reactions with perchloryl fluoride, perchloric acid, and ozone form explosive products.

For occupational chemical analysis use NIOSH: Amines, Aromatic, 2002.

AOQ250 CAS:1300-14-7 HR: 3
ANILINE ANTIMONYL TARTRATE
mf: $C_6H_8N{\bullet}C_4H_4O_7Sb$ mw: 379.98

PROP: White crystals.

SYN: ANTIMONYL ANILINE TARTRATE

TOXICITY DATA WITH REFERENCE
ipr-mus LD50:81 mg/kg AJTMAQ 25,263,45

CONSENSUS REPORTS: Antimony and its compounds are on the Community Right-To-Know List.

OSHA PEL: TWA 0.5 mg(Sb)/m³
ACGIH TLV: TWA 0.5 mg(Sb)/m³
NIOSH REL: (Antimony) TWA 0.5 mg(Sb)/m³

SAFETY PROFILE: Poison by intraperitoneal route. See also ANTIMONY COMPOUNDS and ANILINE. When heated to decomposition it emits very toxic fumes of Sb and NO_x.

AOQ500 HR: 2
ANILINE DYES

SAFETY PROFILE: The finished dyes are generally very much less toxic than many of the intermediates occurring or used in the manufacture of the dyes. Some of the aniline dyes cause local irritating effects to the eyes, mucous membranes, and skin; the basic dyes are believed to be more irritating than the acid dyes. Allergic responses to aniline dyes have been known to occur. See also specific compounds. When heated to decomposition they emit toxic fumes of NO_x and possibly SO_x.

AOQ875 CAS:553-27-5 HR: 3
ANILINE MUSTARD
mf: $C_{10}H_{13}Cl_2N$ mw: 218.14

PROP: Stout prisms from methanol. Mp: 45°, bp: 164°. Sol in hot methanol and ethanol; very sltly sol in ether.

SYNS: N,N-BIS(2-CHLOROETHYL)ANILINE □ N,N-BIS(2-CHLOROETHYL)BENZENAMINE □ β,β′-DICHLORODIETHYLANILINE □ N,N-DI(2-CHLOROETHYL)ANILINE □ LYMPHCHIN □ LYMPHOCIN □ LYMPHOQUIN □ NSC-18429 □ PHENYLBIS(2-CHLOROETHYLAMINE) □ TL 476

TOXICITY DATA WITH REFERENCE
dnd-mus:lym 30 μmol/L CNREA8 44,78,84
orl-rat LD50:239 mg/kg NCIMR* -,469,69
ipr-rat LD50:141 mg/kg BCPCA6 13,969,64
orl-mus LD50:123 mg/kg NCIMR* -,469,69
ihl-mus LCLo:500 mg/m³/10M NDRC** NDCRC-132,Dec,42
ipr-mus LD50:52 mg/kg NCIMR* -,469,69

CONSENSUS REPORTS: EPA Genetic Toxicology Program.

SAFETY PROFILE: Poison by inhalation, ingestion, and intraperitoneal routes. Mutation data reported. When heated to decomposition it emits toxic fumes of Cl^- and NO_x. See also ANILINE DYES.

AOR000 HR: 3
ANILINE OIL DRUMS, EMPTY

SAFETY PROFILE: Combustible if full of vapors, such drums may ignite under the proper conditions. A dangerous disaster hazard if many drums are involved. They emit highly toxic fumes of aniline. See ANILINE.

AOR250 HR: 3
ANILINE VANADATE, DIHYDRATE
mf: $C_6H_7N_2O_5V_2{\bullet}2H_2O$ mw: 325.07

TOXICITY DATA WITH REFERENCE
scu-rat LDLo:68 mg/kg AJSNAO 1,347,17
scu-mus LDLo:128 mg/kg AJSNAO 1,347,17
scu-gpg LDLo:2 mg/kg AJSNAO 1,347,17

ACGIH TLV: TWA 0.05 mg(V_2O_5)/m³
NIOSH REL: (Vanadium Compounds) CL 0.05 mg(V)/m³/15M

SAFETY PROFILE: Poison by subcutaneous route. See also VANADIUM COMPOUNDS. When heated to decomposition it emits toxic fumes of NO_x and VO_x.

AOR500 CAS:548-62-9 HR: 3
ANILINE VIOLET
mf: $C_{25}H_{30}N_3{\bullet}Cl$ mw: 408.03

PROP: Dark green powder or bright blue-violet crystals. Mp: 215° (decomp). Sol in H_2O, EtOH, $CHCl_3$.

SYNS: ADERGON □ AIZEN CRYSTAL VIOLET □ AIZEN CRYSTAL VIOLET EXTRA PURE □ ANILINE VIOLET PYOKTANINE □ ATMONIL □ AVERMIN □ AXURIS □ BADIL □ BASIC VIOLET 3 □ BASIC VIOLET BN □ BISMUTH VIOLET □ BLAUES PYOKTANIN □ BRILLIANT VIOLET 5B □ CALCOZINE VIOLET C □ CALCOZINE VIOLET 6BN □ C.I. 42555 □ C.I. BASIC VIOLET 3 □ CRYSTAL VIOLET □ CRYSTAL VIOLET 6B □ CRYSTAL VIOLET O □ CRYSTAL VIOLET 5BO □ CRYSTAL VIOLET 6BO □ CRYSTAL VIOLET 10B □ CRYSTAL VIOLET AO □ CRYSTAL VIOLET AON □ CRYSTAL VIOLET BASE □ CRYSTAL VIOLET BP □ CRYSTAL VIOLET BPC □ CRYSTAL VIOLET CHLORIDE □ CRYSTAL VIOLET EXTRA PURE □ CRYSTAL VIOLET EXTRA PURE APN □ CRYSTAL VIOLET EXTRA PURE APNX □ CRYSTAL VIOLET FN □ CRYSTAL VIOLET HL2 □ CRYSTAL VIOLET PURE DSC □ CRYSTAL VIOLET PURE DSC BRILLIANT □ CRYSTAL VIOLET SS □ CRYSTAL VIOLET TECHNICAL □ CRYSTAL VIOLET USP □ GENTERSAL □ GENTIAN VIOLET □ GENTIANAVIOLETT □ GENTIAVERM □ GENTICID □ GENTIOLETTEN □ HECTOGRAPH VIOLET SR □ HECTO VIOLET R □ HEXAMETHYLPARAOSANILINE CHLORIDE □ HEXAMETHYL-p-ROSANILINE CHLORIDE □ HEXAMETHYL p-ROSANILINE HYDROCHLORIDE □ HEXAMETHYL VIOLET □ HIDACO BRILLIANT CRYSTAL VIOLET □ HIDACO CRYSTAL VIOLET □ KRISTALL-VIOLETT □ MEROXYL □ MEROXYLAN □ MEROXYLAN-WANDER □ MEROXYLWANDER □ METHYLROSANILINCHLORID □ METHYLROSANILINE

CHLORIDE □ METHYLROSANILINUM CHLORATUM □ METHYL VIOLET 5BNO □ METHYL VIOLET 5BO □ METHYL VIOLET 10B □ METHYL VIOLET 10BD □ METHYL VIOLET 10BK □ METHYL VIOLET 10BN □ METHYL VIOLET 10BNS □ METHYL VIOLET 10BO □ METHYLVIOLETT □ MITSUI CRYSTAL VIOLET □ NCI-C55969 □ OXIURAN □ OXYCOLOR □ OXYOZYL □ PAPER BLUE R □ PARAROSANILINE, N,N,N′,N′,N″,N″-HEXAMETHYL-, CHLORIDE □ PLASTORESIN VIOLET 5BO □ PYOKTANIN □ PYOVERM □ VERMICID □ VIANIN □ VIOCID □ 12416 VIOLET □ VIOLET 6BN □ VIOLET 5BO □ VIOLET CP □ VIOLET GENCIANOVA □ VIOLET KRYSTALOVA □ VIOLET XXIII □ VIOLET ZASADITA 3

TOXICITY DATA WITH REFERENCE
skn-hmn 3 mg/3D-I MLD 85DKA8 -,127,77
skn-hmn 2 mg/2D-I MLD ADVEA4 52,55,72
skn-gpg 6 mg/3D-I ADVEA4 52,55,72
mmo-sat 100 ng/plate MUREAV 89,21,81
cyt-hmn:hla 500 μg/L MUREAV 58,269,78
cyt-hmn:lym 500 μg/L MUREAV 58,269,78
orl-rbt TDLo:7 mg/kg (6-19D post):TER NTIS** PB83-182519
orl-rat TDLo:40 mg/kg (6-15D post):REP NTIS** PB83-155754
orl-rat TDLo:10,950 mg/kg/2Y-C:CAR FCTOD7 27,239,89
orl-mus TDLo:25,750 mg/kg/2Y-C:CAR FAATDF 5,902,85
orl-rat LD50:420 mg/kg ARZNAD 1,5,51
ipr-rat LD50:8900 μg/kg ARZNAD 1,5,51
orl-mus LD50:96 mg/kg ARZNAD 1,5,51
ipr-mus LD50:5100 μg/kg ARZNAD 1,5,51
ivn-mus LDLo:20 mg/kg PSEBAA 31,825,34

CONSENSUS REPORTS: Reported in EPA TSCA Inventory.

SAFETY PROFILE: Poison by ingestion, intravenous, and intraperitoneal routes. An experimental teratogen. Other experimental reproductive effects. A human skin irritant. Human mutation data reported. Questionable carcinogen with experimental carcinogenic data. When heated to decomposition it emits very toxic fumes of NO_x and Cl^-.

AOR625 **HR: 3**
ANILINIUM NITRATE
mf: $C_6H_8N_2O_3$ mw: 156.14

SAFETY PROFILE: Potentially hypergolic reaction with concentrated nitric acid is promoted by: ammonium or sodium vanadates; copper(I) chloride; potassium permanganate; sodium pentacyanonitrosylferrate; and vanadium (V) oxide. When heated to decomposition it emits toxic fumes of NO_x. See also ANILINE and NITRATES.

AOR630 **HR: 3**
ANILINIUM PERCHLORATE
mf: $C_6H_8ClNO_4$ mw: 193.59

SAFETY PROFILE: Mixtures with metal oxides (e.g., manganese dioxide, copper oxide, and nickel oxide) are heat-sensitive explosives. When heated to decomposition it emits toxic fumes of Cl^- and NO_x. See also ANILINE and PERCHLORATES.

AOR635 **CAS:14293-15-3** **HR: 3**
2-ANILINO-4′-(BENZYLOXY)-2-PHENYLACETOPHENONE
mf: $C_{27}H_{23}NO_2$ mw: 393.51

PROP: A liquid.

SYNS: ACETOPHENONE, 2-ANILINO-4′-(BENZYLOXY)-2-PHENYL- □ 4′-(α-ANILINOBENZYLOXY)-2-PHENYLACETOPHENONE □ α-ANILINO-p-BENZYLOXYPHENYL BENZYL KETONE

TOXICITY DATA WITH REFERENCE
orl-rat TDLo:500 mg/kg (female 1-5D post):REP IJEBA6 4,244,66

DOT CLASSIFICATION: 3; *Label:* Flammable Liquid

SAFETY PROFILE: Experimental reproductive effects. A flammable liquid. When heated to decomposition it emits toxic vapors of NO_x.

AOR640 **CAS:5410-78-6** **HR: 3**
4-ANILINODICHLOROARSINE, HYDROCHLORIDE
mf: $C_6H_6AsCl_2N \cdot ClH$ mw: 274.41

SYNS: ANILINE, p-DICHLOROARSINO-, HYDROCHLORIDE □ ARSINE, (p-AMINOPHENYL)DICHLORO-, HYDROCHLORIDE □ p-DICHLOROARSINOANILINE HYDROCHLORIDE

TOXICITY DATA WITH REFERENCE
ivn-mus LD50:6300 μg/kg CSLNX* NX#05113

OSHA PEL: TWA 0.5 mg(As)/m^3

SAFETY PROFILE: Poison by intravenous route. When heated to decomposition it emits toxic fumes of NO_x, As, Cl^-, and HCl.

AOR700 **CAS:14406-57-6** **HR: 3**
2-ANILINO-4′-(2-(DIETHYLAMINO)ETHOXY)-2-PHENYLACETOPHENONE HYDROCHLORIDE
mf: $C_{26}H_{29}N_2O_2 \cdot ClH$ mw: 438.03

PROP: A liquid.

SYNS: ACETOPHENONE, 2-ANILINO-4′-(2-(DIETHYLAMINO)ETHOXY)-2-PHENYL-, HYDROCHLORIDE □ α-ANILINO-p-DIETHYLAMINOETHOXYPHENYL BENZYL KETONE HYDROCHLORIDE

TOXICITY DATA WITH REFERENCE
orl-rat TDLo:500 mg/kg (female 1-5D post):REP IJEBA6 4,244,66

DOT CLASSIFICATION: 3; *Label:* Flammable Liquid

SAFETY PROFILE: Experimental reproductive effects. A flammable liquid. When heated to decomposition it emits toxic vapors of NO_x and HCl.

AOR750 **CAS:122-98-5** **HR: 3**
2-ANILINOETHANOL
mf: $C_8H_{11}NO$ mw: 137.20

PROP: D: 1.1, bp: 268°, flash p: 305°F (OC). Sltly sol in water.

SYNS: N-(2-HYDROXYETHYL)PHENYLAMINE □ 2-(PHENYLAMINO)ETHANOL □ PHENYL ETHANOLAMINE □ N-PHENYLETHANOLAMINE

TOXICITY DATA WITH REFERENCE
skn-rbt 545 mg open MLD UCDS** 8/21/61
eye-rbt 5 mg SEV AJOPAA 29,1363,46
eye-rbt 20 mg SEV AJOPAA 29,1363,46
eye-rbt 20 mg/24H MOD 85JCAE-,691,86
orl-rat LD50:2230 mg/kg UCDS** 8/21/61
ipr-mus LDLo:176 mg/kg JAMAAP 123,761,43
scu-dog LDLo:220 mg/kg JAMAAP 123,761,43
ivn-dog LDLo:165 mg/kg JAMAAP 123,761,43
skn-rbt LD50:63 mg/kg AIHAAP 23,95,62
ivn-rbt LDLo:44 mg/kg JAMAAP 123,761,43

CONSENSUS REPORTS: Reported in EPA TSCA Inventory.

SAFETY PROFILE: Poison by skin contact, intraperitoneal, and intravenous routes. Moderately toxic by ingestion. A skin and severe eye irritant. Combustible when exposed to heat or flame. To fight fire, use dry chemical, water mist. When heated to decomposition it emits toxic fumes of NO_x.

AOS000 CAS:67227-20-7 HR: 3
(2-ANILINOETHYL)HYDRAZONE DIHYDROCHLORIDE
mf: $C_8H_{13}N_3 \cdot 2ClH$ mw: 224.16

TOXICITY DATA WITH REFERENCE
orl-mus LD50:200 mg/kg JMCMAR 6,63,63
ipr-mus LD50:175 mg/kg JMCMAR 6,63,63

SAFETY PROFILE: Poison by ingestion and intraperitoneal routes. When heated to decomposition it emits very toxic fumes of NO_x and HCl.

AOS500 CAS:88-35-7 HR: 2
2-ANILINO-5-NITROBENZENESULFONIC ACID
mf: $C_{12}H_9N_2O_5S \cdot Na$ mw: 316.28

SYN: 4-NITRODIFENYLAMIN-2-SULFONAN SODYN (CZECH)

TOXICITY DATA WITH REFERENCE
skn-rbt 500 mg/24H MLD 28ZPAK -,191,72
eye-rbt 250 µg/24H SEV 28ZPAK -,191,72
orl-rat LD50:2200 mg/kg 28ZPAK -,191,72

SAFETY PROFILE: Moderately toxic by ingestion. A mild skin and severe eye irritant. When heated to decomposition it emits very toxic fumes of SO_x, Na_2O, and NO_x.

AOS750 CAS:101-59-7 HR: 3
ANILINO (p-NITROPHENYL) SULFIDE
mf: $C_{12}H_{10}N_2O_2S$ mw: 246.30

PROP: Orange prisms with blue reflex from C_6H_6, cryst from EtOH. Mp: 146–147°.

SYN: 4-AMINO-4′-NITRODIPHENYL SULFIDE

TOXICITY DATA WITH REFERENCE
mmo-sat 25 µg/plate MUREAV 67,123,79
ivn-mus LD50:180 mg/kg CSLNX* NX#00435

SAFETY PROFILE: Poison by intravenous route. Mutation data reported. See also SULFIDES and NITRO COMPOUNDS of AROMATIC HYDROCARBONS. When heated to decomposition it emits very toxic fumes of NO_x and SO_x.

AOT000 CAS:122-37-2 HR: 3
p-ANILINOPHENOL
mf: $C_{12}H_{11}NO$ mw: 185.24

PROP: Gray, solid leaflets. Mp: 73°, bp: 215–216° @ 12 mm.

SYNS: 4-ANILINOPHENOL □ p-HYDROXYDIFENYLAMIN (CZECH) □ p-HYDROXYDIPHENYLAMINE □ 4-HYDROXYDIPHENYLAMINE □ p-OXYDIPHENYLAMINE □ 4-(PHENYLAMINO)-PHENOL □ N-PHENYL-p-AMINOPHENOL □ PHENYL-p-AMINOPHENOL □ VTI 1

TOXICITY DATA WITH REFERENCE
eye-rbt 250 µg/24H SEV 28ZPAK -,111,72
orl-rat LD50:1220 mg/kg 28ZPAK -,111,72
orl-mus LD50:2120 mg/kg ARZNAD 12,1123,62
ivn-mus LD50:60 mg/kg ARZNAD 12,1123,62

CONSENSUS REPORTS: Reported in EPA TSCA Inventory.

SAFETY PROFILE: Poison by intravenous route. Moderately toxic by ingestion. A severe eye irritant. See also AROMATIC AMINES. When heated to decomposition it emits toxic fumes of NO_x.

AOT100 CAS:1075-76-9 HR: 2
3-ANILINOPROPIONITRILE
mf: $C_9H_{10}N_2$ mw: 146.21

SYNS: β-ANILINOPROPIONITRILE □ N-(CYANOETHYL)ANILINE □ N-(β-CYANOETHYL)ANILINE □ N-(2-CYANOETHYL)ANILINE □ 2-PHENYLAMINOPROPIONITRILE □ PROPANENITRILE, 3-ANILINO- □ PROPANENITRILE, 3-(PHENYLAMINO)- □ PROPIONITRILE, 3-ANILINO-

TOXICITY DATA WITH REFERENCE
orl-rat LD50:4700 mg/kg GTPZAB 32(9),50,88
orl-mus LD50:2560 mg/kg GTPZAB 32(9),32,88

CONSENSUS REPORTS: Reported in EPA TSCA Inventory.

SAFETY PROFILE: Moderately toxic by ingestion. When heated to decomposition it emits toxic vapors of NO_x.

AOT125 CAS:17615-73-5 HR: 2
6-(p-ANILINOSULFONYL)METANILAMIDE
mf: $C_{12}H_{13}N_3O_4S_2$ mw: 327.40

SYNS: SDDS □ 2-SOLFAMONYL-4,4′-DIAMINOPHENYLSULFONE

TOXICITY DATA WITH REFERENCE
orl-rat TDLo:24 g/kg (9-14D preg):TER OYYAA2 9,695,75
orl-rat TDLo:1800 mg/kg (9-14D preg):REP OYYAA2 9,695,75
ipr-mus LD50:5150 mg/kg OYYAA2 2,184,68

SAFETY PROFILE: Moderately toxic by intraperitoneal route. An experimental teratogen. Other experimental reproductive effects. When heated to decomposition it emits toxic fumes of SO_x and NO_x.

AOT250 **HR: 3**
ANILITE

SAFETY PROFILE: A highly explosive mixture composed of liquid NO_2 and carbon disulfide or gasoline. Extremely sensitive to shock.

AOT525 **CAS:135-02-4** **HR: 2**
o-ANISALDEHYDE
mf: $C_8H_8O_2$ mw: 136.16

PROP: Crystals. Mp: 37–39°, bp: 238°, d: 1.127, flash p: 244°F. Sol in water.

SYNS: 2-ANISALDEHYDE □ BENZALDEHYDE, 2-METHOXY-(9CI) □ o-METHOXYBENZALDEHYDE □ 2-METHOXYBENZALDEHYDE □ 6-METHOXYBENZALDEHYDE □ 2-METHOXYBENZENECARBOXALDEHYDE □ SALICYLALDEHYDE METHYL ETHER

TOXICITY DATA WITH REFERENCE
skn-rbt 500 mg/24H MOD FCTXAV 17,855,79
sce-hmn:lyms 125 μmol/L MUREAV 206,17,88
orl-rat LD50:2500 mg/kg FCTXAV 17,855,79

CONSENSUS REPORTS: Reported in EPA TSCA Inventory.

SAFETY PROFILE: Moderately toxic by ingestion. A skin irritant. Mutation data reported. Combustible liquid. When heated to decomposition it emits acrid smoke and irritating fumes.

AOT530 **CAS:123-11-5** **HR: 2**
p-ANISALDEHYDE
mf: $C_8H_8O_2$ mw: 136.15

PROP: Colorless oil; hawthorn odor. D: 1.123 @ 20°/4, refr index: 1.571–1.574, mp: 0°, bp: 247–248°, flash p: 250°F. Misc in alc, ether, fixed oils; sol in propylene glycol; insol in glycerin and water.

SYNS: ANISIC ALDEHYDE □ FEMA No. 2670 □ 4-METHOXYBENZALDEHYDE □ p-METHOXYBENZALDEHYDE (FCC)

TOXICITY DATA WITH REFERENCE
skn-rbt 500 mg/24H MOD FCTXAV 12,807,74
mmo-sat 400 μL/plate BECTA6 24,590,80
orl-rat LD50:1510 mg/kg FCTXAV 2,327,64
orl-gpg LD50:1260 mg/kg FCTXAV 2,327,64

CONSENSUS REPORTS: Reported in EPA TSCA Inventory.

SAFETY PROFILE: Moderately toxic by ingestion. A skin irritant. Mutation data reported. Combustible liquid. When heated to decomposition it emits acrid smoke and irritating fumes.

AOT750 **CAS:2439-77-2** **HR: 2**
o-ANISAMIDE
mf: $C_8H_9NO_2$ mw: 151.18

PROP: Plates from water. Mp: 129°.

SYNS: o-METHOXYBENZAMIDE □ 2-METHOXYBENZAMIDE

TOXICITY DATA WITH REFERENCE
ipr-rat LD50:450 mg/kg JPETAB 108,450,53
orl-mus LD50:1200 mg/kg JPPMAB 4,872,52
ipr-mus LD50:900 mg/kg JPPMAB 4,872,52

SAFETY PROFILE: Moderately toxic by ingestion and intraperitoneal routes. When heated to decomposition it emits toxic fumes of NO_x.

AOU250 **CAS:8007-70-3** **HR: 2**
ANISE OIL

PROP: Consists of (80–90%) of anethole and small quantities of methyl chavicol, p-methoxyacetophenone, and other materials. Found in the dried ripe fruit of *Impinella anisum L.* (FCTXAV 11,855,73). D: 0.978–0.988 @ 25°/25°.

SYNS: ANISEED OIL □ ANIS OEL (GERMAN) □ OIL of ANISE □ STAR ANISE OIL

TOXICITY DATA WITH REFERENCE
mmo-sat 500 μg/plate JAFCAU 30,563,82
dnr-bcs 30 μL/disc TOFOD5 8,91,85
orl-rat LD50:2250 mg/kg FCTXAV 11,855,73

CONSENSUS REPORTS: Reported in EPA TSCA Inventory.

SAFETY PROFILE: Moderately toxic by ingestion. A weak sensitizer. May cause contact dermatitis. Mutation data reported. Combustible liquid. When heated to decomposition it emits acrid smoke and irritating fumes.

AOU500 **CAS:586-38-9** **HR: 3**
m-ANISIC ACID
mf: $C_8H_8O_3$ mw: 152.15

PROP: Needles from water. Mp: 110.5°, bp: 170–172° @ 10 mm. Sol in hot water, alc, and ether.

SYNS: m-METHOXYBENZOIC ACID □ 3-METHOXYBENZOIC ACID

TOXICITY DATA WITH REFERENCE
ipr-mus LDLo:250 mg/kg CBCCT* 6,144,54

CONSENSUS REPORTS: Reported in EPA TSCA Inventory.

SAFETY PROFILE: Poison by intraperitoneal route. When heated to decomposition it emits acrid smoke and irritating fumes.

AOU600 **CAS:100-09-4** **HR: 3**
p-ANISIC ACID
mf: $C_8H_8O_3$ mw: 152.16

SYNS: 4-ANISIC ACID □ DRACONIC ACID □ KYSELINA 4-METHOXYBENZOOVA □ p-METHOXYBENZOIC ACID □ 4-METHOXYBENZOIC ACID

TOXICITY DATA WITH REFERENCE
scu-mus LD50:400 mg/kg CKFRAY 31,236,82

CONSENSUS REPORTS: Reported in EPA TSCA Inventory.

SAFETY PROFILE: Poison by subcutaneous route. When heated to decomposition it emits acrid smoke and irritating vapors.

AOV000 **CAS:94-30-4** **HR: 2**
p-ANISIC ACID, ETHYL ESTER
mf: $C_{10}H_{12}O_3$ mw: 180.21

PROP: Colorless liquid; fruity, anise odor. D: 1.103 @ 25/25, refr index: 1.522–1.526, mp: 7–8°, bp: 263°, flash p: 212°F. Sol in alc and ether; sltly sol in water.

SYNS: ETHYL ANISATE ☐ ETHYL-p-ANISATE (FCC) ☐ ETHYL-p-METHOXYBENZOATE ☐ ETHYL-4-METHOXYBENZOATE ☐ FEMA No. 2420

TOXICITY DATA WITH REFERENCE
orl-rat LD50:2040 mg/kg FCTXAV 14,659,76

CONSENSUS REPORTS: Reported in EPA TSCA Inventory.

SAFETY PROFILE: Moderately toxic by ingestion. See also ESTERS. Combustible liquid. When heated to decomposition it emits acrid smoke and irritating fumes.

AOV250 **CAS:7466-54-8** **HR: 3**
o-ANISIC ACID, HYDRAZIDE
mf: $C_8H_{10}N_2O_2$ mw: 166.20

SYNS: o-METHOXYBENZOHYDRAZIDE ☐ o-METHOXYBENZOIC ACID HYDRAZIDE ☐ 2-METHOXYBENZOIC ACID HYDRAZIDE ☐ o-METHOXYBENZOYLHYDRAZIDE ☐ 2-METHOXYBENZOYL HYDRAZIDE ☐ 2-METHOXYBENZOYLHYDRAZINE

TOXICITY DATA WITH REFERENCE
orl-mus TDLo:6000 mg/kg/22W-I:NEO 34ZRA9 -,869,66

SAFETY PROFILE: Questionable carcinogen with experimental neoplastigenic data. When heated to decomposition it emits toxic fumes of NO_x.

AOV500 **CAS:3290-99-1** **HR: 3**
p-ANISIC ACID, HYDRAZIDE
mf: $C_8H_{10}N_2O_2$ mw: 166.20

SYNS: ANISIC ACID HYDRAZIDE ☐ ANISIC HYDRAZIDE ☐ ANISOYLHYDRAZINE ☐ p-ANISOYLHYDRAZINE ☐ p-METHOXYBENZOIC ACID HYDRAZIDE ☐ 4-METHOXYBENZOIC ACID HYDRAZIDE ☐ p-METHOXYBENZOIC HYDRAZIDE ☐ (p-METHOXYBENZOYL)HYDRAZINE ☐ 4-METHOXYBENZOYL HYDRAZIDE ☐ 4-METHOXYBENZOYLHYDRAZINE

TOXICITY DATA WITH REFERENCE
orl-mus TDLo:12 g/kg/22W-I:NEO 34ZRA9 -,869,66
ivn-mus LD50:178 mg/kg CSLNX* NX#00894

SAFETY PROFILE: Poison by intravenous route. Questionable carcinogen with experimental neoplastigenic data. When heated to decomposition it emits toxic fumes of NO_x.

AOV750 **CAS:121-98-2** **HR: 1**
p-ANISIC ACID, METHYL ESTER
mf: $C_9H_{10}O_3$ mw: 166.19

PROP: Plates from alcohol. Flakes from EtOH or Et_2O. Mp: 48–49°, bp: 255–256°. Insol in water; sol in alc and ether.

SYNS: METHYL-p-ANISATE ☐ METHYL-p-METHOXYBENZOATE

TOXICITY DATA WITH REFERENCE
skn-rbt 500 mg/24H MLD FCTXAV 14,443,76
orl-rat LD50:>5 g/kg FCTXAV 14,481,76
skn-rbt LD50:>5 g/kg FCTXAV 14,481,76

CONSENSUS REPORTS: Reported in EPA TSCA Inventory.

SAFETY PROFILE: Low toxicity by ingestion and skin contact. A skin irritant. See also ESTERS. When heated to decomposition it emits acrid smoke and irritating fumes.

AOV875 **CAS:122-84-9** **HR: 3**
ANISIC KETONE
mf: $C_{10}H_{12}O_2$ mw: 164.22

PROP: Oil or crystals. Mp: 46°, bp: 261–265°.

SYNS: p-ACETONYLANISOLE ☐ ANISKETONE ☐ ANISYL METHYL KETONE ☐ p-METHOXYBENZYL METHYL KETONE ☐ 4-METHOXYBENZYL METHYL KETONE ☐ p-METHOXYPHENYLACETONE ☐ 1-(p-METHOXYPHENYL)-2-PROPANONE

TOXICITY DATA WITH REFERENCE
skn-rbt 500 mg/24H MLD FCTXAV 17,857,79
orl-rat LD50:3330 mg/kg FCTXAV 17,857,79
ipr-mus LD50:560 mg/kg FCTXAV 17,857,79

CONSENSUS REPORTS: Reported in EPA TSCA Inventory.

DOT CLASSIFICATION: 3; *Label:* Flammable Liquid

SAFETY PROFILE: Moderately toxic by ingestion and other routes. A skin irritant. A flammable liquid. See also KETONES.

AOV890 **CAS:536-90-3** **HR: 2**
m-ANISIDINE
mf: C_7H_9NO mw: 123.17

SYNS: m-AMINOANISOLE ☐ 3-AMINOANISOLE ☐ m-ANISYLAMINE ☐ BENZENAMINE, 3-METHOXY-(9CI) ☐ 3-METHOXYANILINE ☐ 3-METHOXYBENZENAMINE

TOXICITY DATA WITH REFERENCE
cyt-ham:ovr 160 mg/L EMMUEG 10(Suppl 10),1,87
sce-ham:ovr 50 mg/L EMMUEG 10(Suppl 10),1,87
orl-qal LD50:562 mg/kg AECTCV 12,355,83
orl-brd LD50:562 mg/kg AECTCV 12,355,83

CONSENSUS REPORTS: Reported in EPA TSCA Inventory.

SAFETY PROFILE: Moderately toxic by ingestion. Mutation data reported. When heated to decomposition it emits toxic vapors of NO_x.

AOV900 **CAS:90-04-0** **HR: 3**
o-ANISIDINE
DOT: UN 2431
mf: C_7H_9NO mw: 123.17

PROP: Yellowish liquid. Mp: 5°, bp: 225°. Sol in acids; insol in H_2O; misc in EtOH, Et_2O, C_6H_6.

SYNS: o-AMINOANISOLE ☐ 2-AMINOANISOLE ☐ 1-AMINO-2-METHOXYBENZENE ☐ 2-ANISIDINE ☐ o-ANISYLAMINE ☐ BENZENA-

MINE, 2-METHOXY-(9CI) □ 2-METHOXY-1-AMINOBENZENE □ o-METHOXYANILINE □ 2-METHOXYANILINE □ 2-METHOXYBENZENAMINE □ o-METHOXYPHENYLAMINE

TOXICITY DATA WITH REFERENCE
mma-sat 333 μg/plate IMEMDT 27,63,82
dni-mus-orl 200 mg/kg MUREAV 46,305,77
orl-rat LD50:2000 mg/kg IMEMDT 27,63,82
orl-mus LD50:1400 mg/kg IMEMDT 27,63,82
orl-rbt LD50:870 mg/kg IMEMDT 27,63,82
orl-bwd LD50:422 mg/kg AECTCV 12,355,83

CONSENSUS REPORTS: IARC Cancer Review: Group 2B IMEMDT 7,56,87; Human Limited Evidence IMEMDT 27,63,82. EPA Genetic Toxicology Program. Reported in EPA TSCA Inventory. Community Right-To-Know List.

OSHA PEL: TWA 0.5 mg/m^3
ACGIH TLV: TWA 0.5 mg/m^3 (skin)
DFG MAK: 0.1 ppm (0.5 mg/m^3)

SAFETY PROFILE: Confirmed carcinogen. Moderately toxic by ingestion. Mutation data reported. When heated to decomposition it emits toxic fumes of NO_x.

For occupational chemical analysis use NIOSH: Anisidine, 2514.

AOW000 CAS:104-94-9 HR: 1
p-ANISIDINE
mf: C_7H_9NO mw: 123.16

PROP: Crystals, plates from aq soln. D: 1.089 @ 55°/55°, mp: 57°, bp: 246°, vap d: 4.28. Sol in hot water, alc, and ether; insol in water.

SYNS: p-AMINOANISOLE □ 4-AMINOANISOLE □ 1-AMINO-4-METHOXYBENZENE □ 4-ANISIDINE □ p-ANISYLAMINE □ p-METHOXYANILINE □ 4-METHOXYANILINE □ 4-METHOXYBENZENAMINE □ 4-METHOXYBENZENEAMINE □ p-METHOXYPHENYLAMINE

TOXICITY DATA WITH REFERENCE
mmo-sat 1 mg/plate ENMUDM 5(Suppl 1),3,83
dnr-esc 39 mmol/L MUREAV 272,145,92
orl-rat LD50:1400 mg/kg AGGHAR 15,447,57
skn-rat LD50:3200 mg/kg AGGHAR 15,447,57
ipr-rat LD50:1400 mg/kg AGGHAR 15,447,57
orl-mus LD50:1410 mg/kg JACTDZ 1,184,92

CONSENSUS REPORTS: IARC Cancer Review: Group 3 IMEMDT 7,56,87; Human Inadequate Evidence IMEMDT 27,63,82. Community Right-To-Know List. Reported in EPA TSCA Inventory.

OSHA PEL: TWA 0.5 mg/m^3
ACGIH TLV: TWA 0.5 mg/m^3 (skin)
DFG MAK: 0.1 ppm (0.5 mg/m^3)

SAFETY PROFILE: Moderately toxic by several routes. A mild sensitizer. May cause a contact dermatitis. Mutation data reported. See also ANILINE. When heated to decomposition it evolves toxic fumes of NO_x.

For occupational chemical analysis use NIOSH: Anisidine, 2514.

AOW500 CAS:64090-82-0 HR: 3
m-ANISIDINE ANTIMONYL TARTRATE

TOXICITY DATA WITH REFERENCE
ipr-mus LD50:28 mg(Sb)/kg AJTMAQ 25,263,45

CONSENSUS REPORTS: Antimony and its compounds are on the Community Right-To-Know List.

OSHA PEL: TWA 0.5 mg(Sb)/m^3
ACGIH TLV: TWA 0.5 mg(Sb)/m^3
NIOSH REL: (Antimony) TWA 0.5 mg(Sb)/m^3

SAFETY PROFILE: Poison by intraperitoneal route. See also ANTIMONY COMPOUNDS. When heated to decomposition it emits very toxic fumes of NO_x and Sb.

AOW750 CAS:64070-14-0 HR: 3
o-ANISIDINE ANTIMONYL TARTRATE

TOXICITY DATA WITH REFERENCE
ipr-mus LD50:29 mg(Sb)/kg AJTMAQ 25,263,45

CONSENSUS REPORTS: Antimony and its compounds are on the Community Right-To-Know List.

OSHA PEL: TWA 0.5 mg(Sb)/m^3
ACGIH TLV: TWA 0.5 mg(Sb)/m^3
NIOSH REL: (Antimony) TWA 0.5 mg(Sb)/m^3

SAFETY PROFILE: Poison by intraperitoneal route. See also ANTIMONY COMPOUNDS. When heated to decomposition it emits very toxic fumes of Sb and NO_x.

AOX000 CAS:64070-15-1 HR: 3
p-ANISIDINE ANTIMONYL TARTRATE

TOXICITY DATA WITH REFERENCE
ipr-mus LD50:28 mg(Sb)/kg AJTMAQ 25,263,45

CONSENSUS REPORTS: Antimony and its compounds are on the Community Right-To-Know List.

OSHA PEL: TWA 0.5 mg(Sb)/m^3
ACGIH TLV: TWA 0.5 mg(Sb)/m^3
NIOSH REL: (Antimony) TWA 0.5 mg(Sb)/m^3

SAFETY PROFILE: Poison by intraperitoneal route. See also ANTIMONY COMPOUNDS. When heated to decomposition it emits very toxic fumes of NO_x and Sb.

AOX250 CAS:134-29-2 HR: 3
o-ANISIDINE HYDROCHLORIDE
mf: $C_7H_9NO \cdot ClH$ mw: 159.63

SYNS: o-AMINOANISOLE HYDROCHLORIDE □ 2-AMINOANISOLE HYDROCHLORIDE □ o-ANISYLAMINE HYDROCHLORIDE □ BENZENAMINE, 2-METHOXY-, HYDROCHLORIDE (9CI) □ C.I. 37115 □ FAST RED BB BASE □ 2-METHOXY-1-AMINOBENZENE HYDROCHLORIDE □ o-METHOXYANILINE HYDROCHLORIDE □ 2-METHOXYANILINE HYDROCHLORIDE □ 2-METHOXYBENZENAMINE HYDROCHLORIDE □ 2-METHOXYBENZENEAMINE HYDROCHLORIDE □ o-METHOXYPHENYLAMINE HYDROCHLORIDE □ NCI-C03747

TOXICITY DATA WITH REFERENCE
mma-sat 10 μg/plate ENMUDM 7(Suppl 5),1,85
mma-esc 1 mg/plate ENMUDM 7(Suppl 5),1,85
orl-rat TDLo:180 g/kg/2Y-C:CAR NCITR* NCI-TR-89,78
orl-mus TDLo:721 g/kg/2Y-C:CAR NCITR* NCI-TR-89,78

orl-mus TD:216 g/kg/78W-C:ETA NCITR* NCI-CG-TR-89,78
orl-mus TD:1803 g/kg/1Y-C:NEO IMEMDT 27,63,82

CONSENSUS REPORTS: NTP 7th Annual Report On Carcinogens. IARC Cancer Review: Group 2B IMEMDT 7,56,87; Animal Sufficient Evidence IMEMDT 27,63,82 Human No Adequate Data IMEMDT 27,63,82. NCI Carcinogenesis Bioassay (feed); Clear Evidence: mouse, rat NCITR* NCI-CG-TR-89,78. Community Right-To-Know List.

SAFETY PROFILE: Confirmed carcinogen with experimental carcinogenic, neoplastigenic, and tumorigenic data. Mutation data reported. When heated to decomposition it emits very toxic fumes of NO_x and HCl.

AOX500 CAS:20265-97-8 HR: 1
p-ANISIDINE HYDROCHLORIDE
mf: $C_7H_9NO{\cdot}ClH$ mw: 159.63

SYN: NCI-C03758

TOXICITY DATA WITH REFERENCE
mmo-sat 33 μg/plate ENMUDM 7(Suppl 5),1,85
mmo-esc 1 mg/plate ENMUDM 7(Suppl 5),1,85
orl-rat TDLo:2163 g/kg/1Y-C:CAR IMEMDT 27,63,82
orl-rat TDLo:116 g/kg/92W-C:ETA NCITR* NCI-CG-TR-116,78

CONSENSUS REPORTS: IARC Cancer Review: Group 3 IMEMDT 7,56,87; Animal Inadequate Evidence IMEMDT 27,63,82; NCI Carcinogenesis Bioassay (feed); No Evidence: mouse NCITR* NCI-CG-TR-116,78; Inadequate Studies: rat NCITR* NCI-CG-TR-116,78. Reported in EPA TSCA Inventory.

SAFETY PROFILE: Questionable carcinogen with experimental carcinogenic and tumorigenic data. Mutation data reported. When heated to decomposition it emits very toxic fumes of NO_x and HCl.

AOX750 CAS:100-66-3 HR: 3
ANISOLE
DOT: UN 2222
mf: C_7H_8O mw: 108.15

PROP: Mobile liquid, clear straw color; phenol, anise odor. Vapor d: 3.72, mp: −37.3°, bp: 153.8°, flash p: 125°F (COC), d: 0.983–0.988, refr index: 1.513–1.518, vap press: 10 mm @ 42.2°, autoign temp: 887°F. Insol in water; sol in alc and ether.

SYNS: BENZENE, METHOXY □ ETHER, METHYL PHENYL □ FEMA No. 2097 □ METHOXYBENZENE □ METHYL PHENYL ETHER □ PHENYL METHYL ETHER

TOXICITY DATA WITH REFERENCE
skn-rbt 500 mg/24H MOD FCTXAV 17,241,79
ihl-mus LC50:3021 mg/m³/2H GTPZAB 28(6),43,84
orl-rat LD50:3700 mg/kg TXAPA9 6,378,64
orl-mus LD50:2800 mg/kg JPETAB 88,400,46

CONSENSUS REPORTS: Reported in EPA TSCA Inventory.

DOT CLASSIFICATION: 3; *Label:* Flammable Liquid

SAFETY PROFILE: Moderately toxic by ingestion and inhalation. A skin irritant. A flammable liquid. To fight fire, use foam, CO_2, dry chemical. When heated to decomposition it emits acrid fumes.

AOY000 CAS:22862-76-6 HR: 3
ANISOMYCIN
mf: $C_{14}H_{19}NO_4$ mw: 265.34

SYNS: ANTIBIOTIC PA-106 □ FLAGECIDIN

TOXICITY DATA WITH REFERENCE
orl-rat LD50:72 mg/kg ANTCAO 5,490,55
ipr-rat LD50:345 mg/kg ANTCAO 5,490,55
scu-rat LD50:230 mg/kg ANTCAO 5,490,55
ivn-rat LD50:167 mg/kg ANTCAO 5,490,55
orl-mus LD50:148 mg/kg ANTCAO 5,490,55
ipr-mus LD50:400 mg/kg ANTCAO 5,490,55
scu-mus LD50:600 mg/kg ANTCAO 5,490,55
ivn-mus LD50:140 mg/kg ANTCAO 5,490,55
orl-gpg LDLo:300 mg/kg ANTCAO 5,490,55

SAFETY PROFILE: Poison by ingestion, intraperitoneal, subcutaneous, and intravenous routes. When heated to decomposition it emits toxic fumes of NO_x.

AOY250 CAS:100-07-2 HR: 3
ANISOYL CHLORIDE
DOT: UN 1729
mf: $C_8H_7ClO_2$ mw: 170.60

$CH_3OC_6H_4CO{\cdot}Cl$

PROP: Needle-like crystals. Mp: 22°, bp: 160–164° @ 35 mm. Insol in water; sol in ether and acetone.

SYNS: p-ANISYOL CHLORIDE □ BENZOYL CHLORIDE, METHOXY-(9CI) □ METHOXYBENZOYL CHLORIDE

CONSENSUS REPORTS: Reported in EPA TSCA Inventory.

DOT CLASSIFICATION: 8; *Label:* Corrosive

SAFETY PROFILE: Corrosive to skin, eyes, mucous membranes, and other tissue. Evolves HCl by hydrolysis. A storage hazard; can explode spontaneously at room temperature. When heated to decomposition it emits toxic fumes of Cl^- and may explode.

AOY300 CAS:73343-67-6 HR: 3
3-ANISOYL-2-MESITYLBENZOFURAN
mf: $C_{25}H_{22}O_3$ mw: 370.47

PROP: A liquid.

SYNS: BENZOFURAN, 2-MESITYL-3-(p-ANISOYL)- □ BENZOFURAN, 2-MESITYL-3-(p-METHOXYBENZOYL)- □ KETONE, 2-MESITYL-3-BENZOFURANYL p-METHOXYPHENYL □ 2-MESITYLBENZOFURAN-3-YL (p-METHOXYPHENYL) KETONE □ METHANONE, (4-METHOXYPHENYL)(2-(2,4,6-TRIMETHYLPHENYL)-3-BENZOFURANYL)- □ (4-METHOXYPHENYL)(2-(2,4,6-TRIMETHYLPHENYL)-3-BENZOFURANYL)METHANONE

TOXICITY DATA WITH REFERENCE
ipr-mus LDLo:2 g/kg EJMCA5 14,517,79

DOT CLASSIFICATION: 3; *Label:* Flammable Liquid

SAFETY PROFILE: Low toxicity by intraperitoneal route. A flammable liquid. When heated to decomposition it emits acrid smoke and irritating vapors.

AOY400 **HR: 1**
ANISYL ACETATE
mf: $C_{10}H_{12}O_3$ mw: 180.20

PROP: Colorless to sltly yellow liquid; fruity, balsamic odor. D: 1.104, refr index: 1.511–1.516, flash p: 210°F. Sol in alc and most oils; insol in glycerin and propylene glycol.

SYNS: FEMA No. 2098 □ p-METHOXYBENZYL ACETATE

SAFETY PROFILE: Combustible liquid. When heated to decomposition it emits acrid smoke and irritating fumes.

AOY450 **CAS:104-92-7** **HR: 2**
ANISYL BROMIDE
mf: C_7H_7BrO mw: 187.05

SYNS: ANISOLE, p-BROMO- □ BENZENE, 1-BROMO-4-METHOXY- (9CI) □ p-BROMANISOLE □ p-BROMOANISOLE □ 4-BROMOANISOLE □ 1-BROMO-4-METHOXYBENZENE □ p-BROMOPHENYL METHYL ETHER □ p-METHOXYBROMOBENZENE □ 4-METHOXYBROMOBENZENE □ p-METHOXYPHENYL BROMIDE □ 4-METHOXYPHENYL BROMIDE

TOXICITY DATA WITH REFERENCE
orl-mus LD50:2200 mg/kg GISAAA 44(12),19,79
ipr-mus LD50:1186 mg/kg GISAAA 44(12),19,79

CONSENSUS REPORTS: Reported in EPA TSCA Inventory.

SAFETY PROFILE: Moderately toxic by ingestion and intraperitoneal routes. When heated to decomposition it emits toxic vapors of Br^-.

APA000 **CAS:27471-67-6** **HR: 3**
N-(o-ANISYL)-2-(p-BUTOXYPHENOXY)-N-(2(DIETHYLAMINO)ETHYL)ACETAMIDE HYDROCHLORIDE
mf: $C_{25}H_{36}N_2O_4 \cdot ClH$ mw: 465.09

TOXICITY DATA WITH REFERENCE
orl-mus LD50:350 mg/kg EJMCA5 10,286,75
ivn-mus LD50:22 mg/kg EJMCA5 10,286,75

SAFETY PROFILE: Poison by ingestion and intravenous routes. When heated to decomposition it emits very toxic fumes of NO_x and HCl.

APE000 **CAS:102-17-0** **HR: 1**
ANISYL PHENYLACETATE
mf: $C_{16}H_{16}O_3$ mw: 256.3

SYNS: p-METHOXYBENZYL PHENYLACETATE □ PHENYLACETIC ACID, p-METHOXYBENZYL ESTER

TOXICITY DATA WITH REFERENCE
skn-rbt 500 mg/24H MLD FCTXAV 10,649,80
orl-rat LD50:>5 g/kg FCTXAV 18,651,80
skn-rbt LD50:>5 g/kg FCTXAV 18,651,80

SAFETY PROFILE: Low toxicity by ingestion and skin contact. A skin irritant. See also ESTERS. When heated to decomposition it emits acrid smoke and irritating fumes.

APE100 **HR: 2**
ANNATTO EXTRACT

PROP: From solvent extraction of *Bixa orellana* L. seeds (JAPMA8 49,218,60). Yellow-red solutions or powder.

SYNS: ACHIOTE □ BIXA ORELLANA

TOXICITY DATA WITH REFERENCE
orl-hmn TDLo:357 mg/kg:SKN ARTODN (Suppl 1),141,78
ipr-mus LD50:700 mg/kg JAPMA8 49,218,60

SAFETY PROFILE: Moderately toxic by intraperitoneal route. Human systemic effects by skin contact. When heated to decomposition it emits acrid smoke and irritating fumes.

APE529 **CAS:66547-10-2** **HR: 3**
ANSAMITOCIN P-4
mf: $C_{33}H_{45}N_2O_9Cl$ mw: 649.25

PROP: Crystals from EtOAc. Mp: 177–180° (decomp).

SYNS: 3-DE(2-(ACETYLMETHYLAMINO)PROPIONYLOXY)-3-HYDROXYMAYTANSINE ISOVALERATE (ESTER) □ MAYTANSINOL ISOVALERATE

TOXICITY DATA WITH REFERENCE
oms-mus/ast 1 μg/kg CNREA8 40,1707,80
cyt-mus/ast 4 μg/kg CNREA8 40,1707,80
ipr-mus LDLo:625 μg/kg JANTAJ 31,78-143,78

SAFETY PROFILE: Poison by intraperitoneal route. Mutation data reported. When heated to decomposition it emits toxic fumes of Cl^- and NO_x. See also ESTERS.

APE625 **CAS:5968-79-6** **HR: 3**
ANTHALLAN HYDROCHLORIDE
mf: $C_{17}H_{25}NO_5 \cdot ClH$ mw: 359.89

SYN: 3-((DIBUTYLAMINO)METHYL)-4,5,6-TRIHYDROXYPHTHALIDE HYDROCHLORIDE

TOXICITY DATA WITH REFERENCE
ipr-mus LD50:300 mg/kg JAPMA8 38,433,49
scu-mus LD50:1080 mg/kg JAPMA8 38,433,49
ivn-mus LD50:80 mg/kg JAPMA8 38,433,49

SAFETY PROFILE: Poison by intravenous and intraperitoneal routes. Moderately toxic by other routes. When heated to decomposition it emits toxic fumes of HCl and NO_x.

APE750 **CAS:191-26-4** **HR: 1**
ANTHANTHRENE
mf: $C_{22}H_{12}$ mw: 276.34

PROP: Golden-yellow plates from C_6H_6. Mp: 261° (decomp).

SYNS: ANTHANTHREN (GERMAN) □ ANTHRANTHRENE □ DIBENZO-(drf,mno)CHRYSENE □ DIBENZO(cd,mk)PYRENE

TOXICITY DATA WITH REFERENCE
mma-sat 1 µg/plate MUREAV 51,311,78
add-uns:lym 30 µmol/L CBINA8 47,87,83
imp-rat TDLo:4150 µg/kg:CAR JJIND8 71,539,83
skn-mus TDLo:263 mg/kg/30W-I:CAR ZKKOBW 89,113,77
skn-mus TD:2100 mg/kg/88W-I:ETA PRLBA4 129,439,40

CONSENSUS REPORTS: IARC Cancer Review: Group 3 IMEMDT 7,56,87; Animal Limited Evidence IMEMDT 32,95,83

SAFETY PROFILE: Questionable carcinogen with experimental carcinogenic and tumorigenic data. Mutation data reported. A polycyclic hydrocarbon found in polluted air. When heated to decomposition it emits acrid fumes.

APF000 CAS:12706-94-4 HR: 3
ANTHELMYCIN
mf: $C_{21}H_{37}N_5O_{14}$ mw: 583.57

PROP: Crystals.

SYNS: ANTIBIOTIC 33876 □ HIKIZIMYCIN □ 2(1H)-PYRIMIDINONE-4-AMINO-1-(4-AMINO-6-0-(3-AMINO-3-DEOXY-β-d-GLUCOPYRANOSYL)-4-DEOXY-d-GLYCERO-d-GALACTO-β-d-GLUCO-UNDECAPYRANOSYL)

TOXICITY DATA WITH REFERENCE
orl-mus LD50:125 mg/kg 85GDA2 5,227,81
ivn-mus LD50:5 mg/kg 85ERAY 2,1107,78

SAFETY PROFILE: Poison by ingestion and intravenous routes. When heated to decomposition it emits toxic fumes of NO_x.

APF750 CAS:189-58-2 HR: 2
ANTHRA(9,1,2-cde)BENZO(h)CINNOLINE
mf: $C_{22}H_{12}N_2$ mw: 304.36

SYN: 1,2-DIAZA-3,4:9,10-DIBENZPYRENE

TOXICITY DATA WITH REFERENCE
scu-mus TDLo:80 mg/kg:ETA IJMRAQ 53,638,65

SAFETY PROFILE: Questionable carcinogen with experimental tumorigenic data. When heated to decomposition it emits toxic fumes of NO_x.

APG050 CAS:610-49-1 HR: 2
1-ANTHRACENAMINE
mf: $C_{14}H_{11}N$ mw: 193.26

PROP: Yellow needles from alcohol. Mp: 130°. Insol in HCl; sol in alc.

SYNS: α-AMINOANTHRACENE □ 1-AMINOANTHRACENE □ 1-ANTHRACYLAMINE □ 1-ANTHRAMINE

TOXICITY DATA WITH REFERENCE
mma-sat 20 µg/plate PNASA6 72,5135,75
dnr-esc 100 mg/L JNCIAM 62,873,79
mrc-smc 5 pph JNCIAM 62,901,79
orl-rat TDLo:7200 mg/kg/27D-I:ETA CNREA8 28,924,68
ipr-mus LD50:1250 mg/kg JJIND8 62,911,79

SAFETY PROFILE: Moderately toxic by intraperitoneal route. Questionable carcinogen with experimental tumorigenic data. Mutation data reported. When heated to decomposition it emits toxic fumes of NO_x.

APG100 CAS:613-13-8 HR: 2
2-ANTHRACENAMINE
mf: $C_{14}H_{11}N$ mw: 193.26

PROP: Yellow leaflets from alcohol. Mp: 238°, bp: subl @ 93° @ 9 mm. Insol in water; sltly sol in alc and ether.

SYNS: β-AMINOANTHRACENE □ 2-AMINOANTHRACENE □ 2-ANTHRACYLAMINE □ 2-ANTHRAMINE □ 2-ANTHRYLAMINE

TOXICITY DATA WITH REFERENCE
mmo-sat 6 nmol/plate BBRCA9 89,259,79
mma-sat 2 µg/plate PNASA6 72,5135,75
dnr-esc 100 mg/L JNCIAM 62,873,79
hma-mus/sat 125 mg/kg JNCIAM 62,911,79
orl-mus TDLo:210 mg/kg (female 1-21D post):TER TJADAB 25(2),61A,82
orl-rat TDLo:45 mg/kg/30D-I:CAR CNREA8 28,924,68
skn-rat TDLo:260 µg/kg/33W-I:ETA BJCAAI 9,631,55
skn-mus TDLo:62 mg/kg/2Y-I:CAR NTIS** CONF-801143
ipr-mus LD50:1500 mg/kg JJIND8 62,911,79

SAFETY PROFILE: Moderately toxic by intraperitoneal route. Suspected carcinogen with experimental carcinogenic and tumorigenic data. An experimental teratogen. Mutation data reported. See also AMINES. When heated to decomposition it emits toxic fumes of NO_x.

APG500 CAS:120-12-7 HR: 2
ANTHRACENE
mf: $C_{14}H_{10}$ mw: 178.24

$C_6H_4:(CH)_2:C_6H_4$

PROP: Colorless crystals, monoclinic plates from EtOH, violet fluorescence when pure. Mp: 217°, lel: 0.6%, flash p: 250°F (CC), d: 1.24 @ 27°/4°, autoign temp: 1004°F, vap press: 1 mm @ 145.0° (subl), vap d: 6.15, bp: 339.9°. Insol in water. Solubility in alc @ 1.9/100 @ 20°; in ether 12.2/100 @ 20°.

SYNS: ANTHRACEN (GERMAN) □ ANTHRACIN □ GREEN OIL □ PARANAPHTHALENE □ TETRA OLIVE N2G

TOXICITY DATA WITH REFERENCE
skn-mus 118 µg MLD CALEDQ 4,333,78
mma-sat 100 µg/plate ABCHA6 43,1433,79
dns-hmn:fbr 10 mg/L CNREA8 38,2091,78
hma-mus/sat 125 mg/kg JNCIAM 62,911,79
dnd-mam:lym 100 µmol BIPMAA 9,689,70
orl-rat TDLo:20 g/kg/79W-I:ETA ZEKBAI 60,697,55
scu-rat TDLo:3300 mg/kg/33W-I:NEO NATWAY 42,159,55
orl-mus LD:>17 g/kg GTPZAB 13(5),59,69
ipr-mus LD50:430 mg/kg PMRSDJ 1,682,81

CONSENSUS REPORTS: IARC Cancer Review: Group 3 IMEMDT 7,56,87; Animal Inadequate Evidence IMEMDT 32,105,83; Human No Adequate Data IMEMDT 32,105,83. Reported in EPA TSCA Inventory. Community Right-To-Know List.

OSHA PEL: TWA 0.2 mg/m³

SAFETY PROFILE: Moderately toxic by intraperitoneal

route. A skin irritant and allergen. Questionable carcinogen with experimental neoplastigenic and tumorigenic data. Mutation data reported. Combustible when exposed to heat, flame, or oxidizing materials. Moderately explosive when exposed to flame, $Ca(OCl)_2$, chromic acid. To fight fire, use water, foam, CO_2, water spray or mist, dry chemical. Explodes on contact with fluorine.

For occupational chemical analysis use OSHA: #ID-58 or NIOSH: Polynuclear Aromatic Hydrocarbons (HPLC), 5506; (GC), 5515.

APG550 CAS:723-62-6 HR: 2
ANTHRACENE-9-CARBOXYLIC ACID
mf: $C_{15}H_{10}O_2$ mw: 222.25

SYNS: 9-ANTHROIC ACID □ 9-CARBOXYANTHRACENE

TOXICITY DATA WITH REFERENCE
pic-esc 127 µg/well MUREAV 260,349,91
ipr-mus LD50:750 mg/kg JMCMAR 11,1020,68

CONSENSUS REPORTS: Reported in EPA TSCA Inventory.

SAFETY PROFILE: Moderately toxic by intraperitoneal route. Mutation data reported. When heated to decomposition it emits acrid smoke and irritating vapors.

APH250 CAS:480-22-8 HR: 2
1,8,9-ANTHRACENETRIOL
mf: $C_{14}H_{10}O_3$ mw: 226.24

PROP: Yellow powder. Mp: 178–180°. Insol in water; sol in fat, hot alc, benzene, and dilute alkalies.

SYNS: ANTHRALIN □ 1,8,9-ANTHRATRIOL □ DIHYDROXYANTHRANOL □ 1,8-DIHYDROXYANTHRANOL □ 1,8-DIHYDROXY-9-ANTHRANOL □ 1,8-DIHYDROXY-9-ANTHRONE □ DIOXYANTHRANOL □ 1,8,9-TRIHYDROXYANTHRACENE

TOXICITY DATA WITH REFERENCE
mmo-sat 100 µg/plate BCSTB5 5,1489,77
mma-sat 100 µg/plate BCSTB5 5,1489,77
dnr-esc 250 µg/plate JNCIAM 62,873,79
mmo-smc 165 nmol/L ADVEA4 51,45,71
skn-mus TDLo:509 mg/kg/53W-I:NEO JMCMAR 21,26,78
skn-mus TD:73 mg/kg/11W-I:ETA GANNA2 59,187,68

CONSENSUS REPORTS: IARC Cancer Review: Group 3 IMEMDT 7,56,87; Animal Limited Evidence IMEMDT 13,75,77.

SAFETY PROFILE: Questionable carcinogen with experimental neoplastigenic and tumorigenic data. Mutation data reported. Skin contact can cause folliculitis. Absorption can cause kidney damage and intestinal disturbances. Combustible when heated. When heated to decomposition it emits acrid smoke and irritating fumes.

APH500 CAS:16203-97-7 HR: 2
1,8,9-ANTHRACENETRIOL TRIACETATE
mf: $C_{20}H_{16}O_6$ mw: 352.36

PROP: Yellow needles. Mp: 209–210°.

SYNS: EXOLAN □ 1,8,9-TRIACETOXYANTHRACENE

TOXICITY DATA WITH REFERENCE
eye-rbt 330 µg SEV BJOPAL 53,819,69

SAFETY PROFILE: A severe eye irritant. When heated to decomposition it emits acrid smoke and irritating fumes.

API000 CAS:4803-27-4 HR: 3
ANTHRAMYCIN
mf: $C_{16}H_{17}N_3O_4$ mw: 315.4

PROP: Small yellow prisms from Me_2CO (aq). Mp: 188–194°.

SYNS: ANTRAMYCIN □ (E)-1H-PYRROLO(2,1-C)(1,4)BENZODIAZEPINE-2-ACRYLAMIDE, 5,10,11,11a-TETRAHYDRO-9,11-DIHYDROXY-8-METHYL-5-OXO

TOXICITY DATA WITH REFERENCE
pic-esc 60 ng/plate CNREA8 43,2819,83
dnd-mam:lym 400 µmol/L BBACAQ 475,521,77
mmo-eug 60 mg/L NEOLA4 19,579,72
dns-hmn:fbr 200 µmol/L JBCHA3 254,605,79
dnd-hmn:fbr 10 µmol/L JBCHA3 254,605,79
ipr-mus LD50:650 µg/kg TOLED5 18,337,83

CONSENSUS REPORTS: EPA Genetic Toxicology Program.

SAFETY PROFILE: Deadly poison by intraperitoneal route. Human mutation data reported. When heated to decomposition it emits toxic fumes of NO_x.

API125 CAS:5544-25-2 HR: 3
ANTHRAMYCIN METHYL ETHER
mf: $C_{17}H_{19}N_3O_4 \cdot H_2O$ mw: 347.41

SYNS: ANTHRAMYCIN-11-METHYL ETHER □ ANTIBIOTIC A □ ROCHE 5-9000 □ (E)-5,10,11,11a-TETRAHYDRO-9-HYDROXY-11-METHOXY-8-METHYL-5-OXO-1H-PYRROLO(2,1-c)(1,4)BENZODIAZEPINE-2-ACRYLAMIDE MONOHYDRATE

TOXICITY DATA WITH REFERENCE
orl-mus LD50:3400 µg/kg 15MBAH -,303,64
ipr-mus LD50:1200 µg/kg 15MBAH -,303,64
scu-mus LD50:2900 µg/kg 15MBAH -,303,64

SAFETY PROFILE: Poison by ingestion, subcutaneous, and intraperitoneal routes. When heated to decomposition it emits toxic fumes of NO_x.

API500 CAS:118-92-3 HR: 2
ANTHRANILIC ACID
mf: $C_7H_7NO_2$ mw: 137.15

PROP: Needle-like crystals or leaflets. Mp: 144–148°, bp: subl, d: 1.412 @ 20°. Solubility: in water: 0.35/100 @ 14°, in 90% alc: 10.7/100 @ 10°, in ether: 16/100 @ 70°.

SYNS: o-AMIDOBENZOIC ACID □ o-AMINOBENZOIC ACID □ 2-AMINOBENZOIC ACID □ 1-AMINO-2-CARBOXYBENZENE □ CARBOXYANILINE □ o-CARBOXYANILINE □ 2-CARBOXYANILINE □ NCI-C01730 □ VITAMIN L

TOXICITY DATA WITH REFERENCE
msc-hmn:lyms 1667 mg/L MUREAV 196,61,88
mma-mus:lyms 250 mg/L EMMUEG 12(Suppl 13),37,88

orl-mus TDLo:34,800 mg/kg (male 8D pre):REP MPHEAE 15,7,66
orl-rat TDLo:16 g/kg/25W-C:ETA APMIAL 26,447,49
scu-mus TDLo:2040 mg/kg (13-17D preg):ETA,REP JCROD7 96,163,80
orl-mus LD50:1400 mg/kg QJPPAL 19,483,46
ipr-mus LD50:2500 mg/kg RPTOAN 37,105,74

CONSENSUS REPORTS: IARC Cancer Review: Group 3 IMEMDT 7,56,87; Animal Inadequate Evidence IMEMDT 16,265,78. NTP Carcinogenesis Bioassay (feed): No Evidence: mouse, rat NCITR* NCI-TR-36,78. Reported in EPA TSCA Inventory.

SAFETY PROFILE: Moderately toxic by ingestion and intraperitoneal route. Experimental reproductive effects. Human mutation data reported. Questionable carcinogen with experimental tumorigenic data. Combustible. When heated to decomposition it emits toxic fumes of NO_x.

API750 CAS:87-29-6 HR: 2
ANTHRANILIC ACID, CINNAMYL ESTER
mf: $C_{16}H_{15}NO_2$ mw: 253.32

PROP: Reddish-yellow powder; balsamic odor. Mp: 60°, flash p: 212°F. Sol in alc, chloroform, ether; insol in water.

SYNS: 2-AMINOBENZOIC ACID-3-PHENYL-2-PROPENYL ESTER □ CINNAMYL ALCOHOL ANTHRANILATE □ CINNAMYL-o-AMINOBENZOATE □ CINNAMYL-2-AMINOBENZOATE □ CINNAMYL ANTHRANILATE (FCC) □ FEMA No. 2295 □ NCI-C03510 □ 3-PHENYL-2-PROPENYLANTHRANILATE □ 3-PHENYL-2-PROPEN-1-YL ANTHRANILATE

TOXICITY DATA WITH REFERENCE
mmo-esc 10 µg/plate ENMUDM 7(Suppl 5),1,85
mic-mus:lym 10 mg/L SCIEAS 236,933,87
orl-rat TDLo:546 g/kg/2Y-C:CAR NCITR* NCI-TR-196,80
orl-mus TDLo:1310 g/kg/2Y-C:CAR NCITR* NCI-TR-196,80
ipr-mus TDLo:12 g/kg/8W-I:NEO CNREA8 33,3069,73
orl-rat LD50:5000 mg/kg FCTXAV 13,681,75
skn-rbt LD50:5000 mg/kg FCTXAV 13,681,75

CONSENSUS REPORTS: IARC Cancer Review: Group 3 IMEMDT 7,56,87; Animal Limited Evidence IMEMDT 31,133,83; Animal Inadequate Evidence IMEMDT 16,287,78; NCI Carcinogenesis Bioassay (feed); Clear Evidence: mouse, rat NCITR* NCI-CG-TR-196,80. Reported in EPA TSCA Inventory.

SAFETY PROFILE: Suspected carcinogen with experimental carcinogenic and neoplastigenic data. Mutation data reported. See also ESTERS. Combustible liquid. When heated to decomposition it emits toxic fumes of NO_x.

API800 CAS:609-86-9 HR: 3
ANTHRANILIC ACID, 3,5-DIIODO-
mf: $C_7H_5I_2NO_2$ mw: 388.93

SYNS: 2-AMINO-3,5-DIIODOBENZOIC ACID □ BENZOIC ACID, 2-AMINO-3,5-DIIODO-(9CI) □ 3,5-DIIODOANTHRANILIC ACID

TOXICITY DATA WITH REFERENCE
unr-rat LD50:180 mg/kg JAPMA8 42,721,53
orl-mus LD50:900 mg/kg QJPPAL 19,483,46

CONSENSUS REPORTS: Reported in EPA TSCA Inventory.

SAFETY PROFILE: Poison by an unspecified route. Moderately toxic by ingestion. When heated to decomposition it emits toxic vapors of NO_x.

APJ000 CAS:7149-26-0 HR: 1
ANTHRANILIC ACID, LINALYL ESTER
mf: $C_{17}H_{23}NO_2$ mw: 273.41

SYNS: 3,7-DIMETHYL-1,6-OCTADIEN-3-YL-o-AMINOBENZOATE □ 1,5-DIMETHYL-1-VINYL-4-HEXEN-1-YL-o-AMINOBENZOATE □ LINALYL-o-AMINOBENZOATE □ LINALYL ANTHRANILATE

TOXICITY DATA WITH REFERENCE
orl-rat LD50:4250 mg/kg FCTXAV 14(5),443,76

CONSENSUS REPORTS: Reported in EPA TSCA Inventory.

SAFETY PROFILE: Mildly toxic by ingestion. See also ESTERS. When heated to decomposition it emits toxic fumes of NO_x.

APJ250 CAS:134-20-3 HR: 2
ANTHRANILIC ACID, METHYL ESTER
mf: $C_8H_9NO_2$ mw: 151.18

PROP: Crystals or plates from alc or colorless liquid; grape odor. D: 1.161–1.169, mp: 24–25°, bp: 225–230° @ 15 mm, flash p: 219°F. Very sol in water, propylene glycol, hot abs alc (23/100); insol in ether, chloroform, and glycerin.

SYNS: o-AMINOBENZOIC ACID METHYL ESTER □ 2-AMINOBENZOIC ACID METHYL ESTER □ o-CARBOMETHOXYANILINE □ 2-CARBOMETHOXYANILINE □ FEMA No. 2682 □ 2-(METHOXYCARBONYL) ANILINE □ METHYL o-AMINOBENZOATE □ METHYL 2-AMINOBENZOATE □ METHYL ANTHRANILATE (FCC) □ METHYLESTER KYSELINY ANTHRANILOVE □ NEROLI OIL, ARTIFICAL

TOXICITY DATA WITH REFERENCE
skn-rbt 500 mg/24H MOD FCTXAV 12,807,74
dnr-bcs 23 mg/disc OIGZDE 34,267,85
orl-mus TDLo:34,800 mg/kg (8D male pre)(21D pre):REP MPHEAE 15,7,66
orl-rat LD50:2910 mg/kg FCTXAV 2,327,64
orl-mus LD50:3900 mg/kg FCTXAV 2,327,64
orl-gpg LD50:2780 mg/kg FCTXAV 2,327,64

CONSENSUS REPORTS: Reported in EPA TSCA Inventory.

SAFETY PROFILE: Moderately toxic by ingestion. Experimental reproductive effects. A skin irritant. See also ESTERS. Combustible liquid. When heated to decomposition it emits toxic fumes of NO_x.

APJ500 CAS:133-18-6 HR: 1
ANTHRANILIC ACID, PHENETHYL ESTER
mf: $C_{15}H_{15}NO_2$ mw: 241.31

PROP: White to yellow crystals; grape odor.

SYNS: BENZOIC ACID, 2-AMINO-, 2-PHENYLETHYL ESTER □ BEN-

ZYLCARBINYL ANTHRANILATE □ β-PHENETHYL-o-AMINOBENZOATE □ PHENETHYL ANTHRANILATE □ 2-PHENYLETHYL-o-AMINOBENZOATE □ PHENYLETHYL ANTHRANILATE □ 2-PHENYLETHYL ANTHRANILATE

TOXICITY DATA WITH REFERENCE
skn-rbt 500 mg/24H MOD FCTXAV 14,659,76

CONSENSUS REPORTS: Reported in EPA TSCA Inventory.

SAFETY PROFILE: A skin irritant. See also ESTERS. When heated to decomposition it emits toxic fumes of NO_x.

APJ750 CAS:1885-29-6 HR: 3
ANTHRANILONITRILE
mf: $C_7H_6N_2$ mw: 118.15

PROP: Needles from CS_2. Mp: 51°, bp: 262–263°. Very sltly sol in water; sol in alc and ether.

SYNS: o-AMINOBENZONITRILE □ 2-AMINOBENZONITRILE □ o-CYANOANILINE □ 2-CYANOANILINE

TOXICITY DATA WITH REFERENCE
ivn-mus LD50:180 mg/kg CSLNX* NX#00381

CONSENSUS REPORTS: Reported in EPA TSCA Inventory. Cyanide and its compounds are on the Community Right-To-Know List.

SAFETY PROFILE: Poison by intravenous route. See also NITRILES. When heated to decomposition it emits toxic fumes of NO_x and CN^-.

APK000 CAS:129-56-6 HR: 3
ANTHRA(1,9-cd)PYRAZOL-6(2H)-ONE
mf: $C_{14}H_8N_2O$ mw: 220.24

SYNS: C.I. 70300 □ PYRAZOLANTHRONE □PYRAZOLEANTHRONE □ 1,9-PYRAZOLOANTHRONE

TOXICITY DATA WITH REFERENCE
ivn-mus LD50:178 mg/kg CSLNX* NX#00640

CONSENSUS REPORTS: Reported in EPA TSCA Inventory.

SAFETY PROFILE: Poison by intravenous route. When heated to decomposition it emits toxic fumes of NO_x.

APK250 CAS:84-65-1 HR: 2
ANTHRAQUINONE
mf: $C_{14}H_8O_2$ mw: 208.22

PROP: Yellow rhombic crystals from $PhNO_2$ or AcOH. Mp: 286°, bp: 376.9°, flash p: 365°F (CC), d: 1.438, vap press: 1 mm @ 190.0°, vap d: 7.16. Sol in hot C_6H_6, toluene, $PhNO_2$; mod sol in EtOH; sltly sol in Et_2O. Insol in water.

SYNS: 9,10-ANTHRACENEDIONE □ ANTHRADIONE □ 9,10-ANTHRAQUINONE □ 9,10-DIOXOANTHRACENE

TOXICITY DATA WITH REFERENCE
mmo-sat 2 μg/plate AEMIDF 43,1354,82
mmo-sat 333 μg/plate EMMUEG 11(Suppl 12),1,88
orl-rat LDLo:15 g/kg 85GMAT-,22,82
ihl-rat LC50:>1300 mg/m³/4H PEMNDP 9,37,91
skn-rat LD50:>1 g/kg 85JFAN A019,83
ipr-rat LD50:3500 mg/kg GTPZAB 21(12),27,77
orl-mus LD50:>5 g/kg PEMNDP 9,37,91

CONSENSUS REPORTS: Reported in EPA TSCA Inventory.

SAFETY PROFILE: Moderately toxic by intraperitoneal route. A mild allergen. Mutation data reported. Combustible when exposed to heat or flame. To fight fire, use water, foam, CO_2, water spray or mist, dry chemical. When heated to decomposition it emits acrid smoke and irritating fumes.

APK500 CAS:25704-81-8 HR: 1
ANTHRAQUINONE BRILLIANT GREEN CONCENTRATE ZH
mf: $C_{36}H_{18}Br_2O_4$ mw: 674.36

SYNS: CALEDON JADE GREEN 2G □ C.I. 59830 □ C.I. VAT GREEN 2

TOXICITY DATA WITH REFERENCE
eye-rbt 500 mg/24H MOD 28ZPAK -,249,72

CONSENSUS REPORTS: Reported in EPA TSCA Inventory.

SAFETY PROFILE: An eye irritant. See also BROMIDES. When heated to decomposition it emits very toxic fumes of Br^-.

APK625 CAS:117-14-6 HR: 2
1,5-ANTHRAQUINONEDISULFONIC ACID
mf: $C_{14}H_8O_8S_2$ mw: 368.34

PROP: Yellowish needles. Mp: 310–311°. Very sol in water.

SYN: 1,5-DISULFOANTHRAQUINONE

TOXICITY DATA WITH REFERENCE
orl-rat LD50:2357 mg/kg GISAAA 45(3),73,80
orl-mus LD50:2357 mg/kg GISAAA 45(3),73,80
orl-rbt LD50:5500 mg/kg GISAAA 45(3),73,80
orl-gpg LD50:2357 mg/kg GISAAA 45(3),73,80

CONSENSUS REPORTS: Reported in EPA TSCA Inventory.

SAFETY PROFILE: Moderately toxic by ingestion. When heated to decomposition it emits toxic fumes of SO_x.

APK635 CAS:82-48-4 HR: 2
1,8-ANTHRAQUINONEDISULFINIC ACID
mf: $C_{14}H_8O_8S_2$ mw: 368.34

PROP: Yellow needles. Mp: 293–294°. Very sol in H_2O, sol in EtOH.

SYN: 1,8-DISULFOANTHRAQUINONE

TOXICITY DATA WITH REFERENCE
orl-rat LD50:1870 mg/kg GISAAA 45(3),73,80
orl-mus LD50:2800 mg/kg GISAAA 45(3),73,80
orl-rbt LD50:4200 mg/kg GISAAA 45(3),73,80
orl-gpg LD50:1870 mg/kg GISAAA 45(3),73,80

CONSENSUS REPORTS: Reported in EPA TSCA Inventory.

SAFETY PROFILE: Moderately toxic by ingestion. When heated to decomposition it emits toxic fumes of SO_x.

APK750 CAS:62399-48-8 HR: 2
((N-ANTHRAQUINON-2-YL)AMINOMETHYLENE)DIMETHYLAMMONIUM CHLORIDE
mf: $C_{17}H_{15}N_2O_2$•Cl mw: 314.79

SYN: N,N-DIMETHYL-N′-(1-ANTHRACHINONYL)FORMAMIDINIUMCHLORID (GERMAN)

TOXICITY DATA WITH REFERENCE
skn-rbt 500 mg/24H MLD 28ZPAK -,124,72
eye-rbt 5 mg/24H SEV 28ZPAK -,124,72
orl-rat LD50:3190 mg/kg 28ZPAK -,124,72

SAFETY PROFILE: Moderately toxic by ingestion. Skin and severe eye irritant. When heated to decomposition it emits very toxic fumes of NO_x, Cl^-, and NH_3.

APK850 CAS:131-14-6 HR: 1
2,6-ANTHRAQUINONYLDIAMINE
mf: $C_{14}H_{10}N_2O_2$ mw: 238.26

PROP: Reddish-brown prisms. Mp: 310–320°. Insol in $CHCl_3$.

SYNS: 9,10-ANTHRACENEDIONE, 2,6-DIAMINO- □ ANTHRAQUINONE, 2,6-DIAMINO- □ 2,6-DIAMINOANTHRACHINON □ 2,6-DIAMINOANTHRAQUINONE □ 2,6-DIAMINO-9,10-ANTHRAQUINONE

TOXICITY DATA WITH REFERENCE
eye-rbt 500 mg/24H MLD 28ZPAK-,122,72
mma-sat 100 µg/plate MUREAV 40,203,76

CONSENSUS REPORTS: Reported in EPA TSCA Inventory.

SAFETY PROFILE: An eye irritant. Mutation data reported. When heated to decomposition it emits toxic fumes of NO_x.

APL250 CAS:61907-23-1 HR: 2
N,N‴-(2,6-ANTHRAQUINONYLENE)BIS(N,N-DIETHYLACETAMIDE)
mf: $C_{26}H_{32}N_4O_2$ mw: 432.62

SYNS: BISAMIDINE □ N′,N‴-(9,10-DIHYDRO-9,10-DI-OXO-2,6-ANTHRACENEDIYL)BIS(N,N-DIETHYLETHANIMIDAMIDE)

TOXICITY DATA WITH REFERENCE
orl-rat LD50:416 mg/kg DRFUD4 4,705,79
orl-ham LD50:850 mg/kg DRFUD4 4,705,79

SAFETY PROFILE: Moderately toxic by ingestion. When heated to decomposition it emits toxic fumes of NO_x.

APL500 CAS:4403-90-1 HR: 2
2,2′-(1,4-ANTHRAQUINONYLENEDIIMINO)BIS(5-METHYLBENZENESULFONIC ACID) DISODIUM SALT
mf: $C_{28}H_{20}N_2O_8S_2$•2Na mw: 622.60

SYNS: C.I. 61 570 (CZECH) □ ZELEN ALIZARINOVA BRILANTNI G-EXTRA (CZECH)

TOXICITY DATA WITH REFERENCE
skn-rbt 500 mg/24H MLD 28ZPAK -,247,72
eye-rbt 500 mg/24H SEV 28ZPAK -,247,72
orl-rat LD50:>10 g/kg GTPZAB 28(7),53,84
orl-mus LD50:6700 mg/kg GTPZAB 28(7),53,84

CONSENSUS REPORTS: Reported in EPA TSCA Inventory.

SAFETY PROFILE: Moderately toxic by ingestion. A skin and severe eye irritant. See also SULFONATES. When heated to decomposition it emits very toxic fumes of NO_x, SO_x, and Na_2O.

APL750 CAS:116-76-7 HR: 1
1,1′-(ANTHRAQUINON-1,4-YLENEDIIMINO)DIANTHRAQUINONE
mf: $C_{42}H_{22}N_2O_6$ mw: 650.66

SYNS: 1,4-BIS-1′-ANTHRACHINONYLAMINO-ANTHRACHINON (CZECH) □ 1,4-TRIANTHRIMID (CZECH)

TOXICITY DATA WITH REFERENCE
eye-rbt 500 mg/24H SEV 28ZPAK -,127,72

CONSENSUS REPORTS: Reported in EPA TSCA Inventory.

SAFETY PROFILE: A severe eye irritant. When heated to decomposition it emits toxic fumes of NO_x.

APM000 CAS:117-03-3 HR: 2
1,1′-(ANTHRAQUINON-1,5-YLENEDIIMINO)DIANTHRAQUINONE
mf: $C_{42}H_{22}N_2O_6$ mw: 650.66

SYNS: 1,5-BIS-1′-ANTHRACHINONYLAMINO-ANTHRACHINON (CZECH) □ 1,5-TRIANTHRIMID (CZECH)

TOXICITY DATA WITH REFERENCE
eye-rbt 500 mg/24H SEV 28ZPAK -,128,72

CONSENSUS REPORTS: Reported in EPA TSCA Inventory.

SAFETY PROFILE: A severe eye irritant. When heated to decomposition it emits toxic fumes of NO_x.

APM250 CAS:73688-63-8 HR: 1
4,4′-(1,4-ANTHRAQUINONYLENEDIIMINODIPHENYL-1,4-ENEDIOXO)BENZENESULFONIC ACID
mf: $C_{38}H_{26}N_2O_{10}S_2$ mw: 734.78

SYN: ZELEN MIDLONOVA BLS (CZECH)

TOXICITY DATA WITH REFERENCE
eye-rbt 20 mg/24H SEV 28ZPAK -,247,72
orl-rat LD50:8550 mg/kg 28ZPAK -,247,72

SAFETY PROFILE: Mildly toxic by ingestion. A severe eye irritant. See also SULFONATES. When heated to decomposition it emits very toxic fumes of SO_x and NO_x.

APM750 **CAS:1715-81-7** **HR: 2**
9-ANTHRONOL
mf: $C_{14}H_{10}O_2$ mw: 210.24

TOXICITY DATA WITH REFERENCE
skn-mus TDLo: 700 mg/kg/73W-I:NEO JMCMAR 21,26,78

SAFETY PROFILE: Questionable carcinogen with experimental neoplastigenic data. When heated to decomposition it emits acrid smoke and irritating fumes.

APM875 **HR: 2**
ANTHURIUM

PROP: A berry-producing plant with thick, heart-shaped, dark green leaves. The "flower" consists of a leaf that has turned a bright color with a spike emerging from it. Brightly colored berries eventually grow from the spike. Native to the tropical areas of America, they are grown in gardens in southern Florida and Hawaii, and as house plants elsewhere.

SYNS: ANTURIO □ FLAMINGO FLOWER □ FLAMINGO LILY □ FLOR de CULEBRA (PUERTO RICO) □ GUINDA (PUERTO RICO) □ HOJA GRANDE (CUBA) □ LENGUNA de VACA (CUBA, PUERTO RICO) □ LOMBRICERO (CUBA) □ PIGTAIL PLANT □ TAIL FLOWER

SAFETY PROFILE: The leaves and stems contain toxic calcium oxalate raphides. Chewing these plant parts results in burning pain in the lips, mouth, and throat, possibly followed by inflammation and blistering. Systemic effects are usually not seen because of the insolubility of calcium oxalate. See also OXALATES.

APN000 **CAS:23605-05-2** **HR: 3**
α-ANTIARBIN
mf: $C_{29}H_{42}O_{11}$ mw: 566.71

PROP: Crystals from H_2O. Mp: 238–240°.

TOXICITY DATA WITH REFERENCE
ivn-cat LD50: 90 µg/kg 85ELDJ -,188,63
ivn-rbt LDLo: 1 mg/kg 27ZWAY E.1,78,-

SAFETY PROFILE: Poison by intravenous route. When heated to decomposition it emits acrid smoke and irritating fumes.

APP500 **CAS:1263-89-4** **HR: 3**
ANTIBIOTIC 1600
mf: $C_{23}H_{45}N_5O_{14}{\cdot}H_2O_4S$ mw: 713.81

SYNS: AMINOSIDINE SULFATE □ AMINOSIDINE SULPHATE □ AMINOSIDIN SULFATE □ FARMIGLUCIN □ FARMINOSIDIN □ Fi 5853 □ GABBROMICINA □ GABBROMYCIN □ GABBRORAL □ GABBROROL □ HUMATIN □ HUMYCIN SULFATE □ PARAMICINA □ PARICINA □ PAROMOMYCIN SULFATE

TOXICITY DATA WITH REFERENCE
dnd-esc 30 µmol/L MUREAV 89,95,81
orl-rat LD50: 21,620 mg/kg NIIRDN 6,595,82
scu-rat LD50: 870 mg/kg NKRZAZ 16,114,68
ivn-rat LD50: 181 mg/kg ANTCAO 9,730,59
ims-rat LD50: 1200 mg/kg NIIRDN 6,595,82
orl-mus LD50: 23,500 mg/kg JJANAX 36,644,83
scu-mus LD50: 123 mg/kg THERAP 16,184,61
ivn-mus LD50: 98 mg/kg JJANAX 36,644,83
ims-mus LD50: 438 mg/kg JJANAX 36,644,83
ivn-mky LDLo: 93 mg/kg ANTCAO 9,730,59

SAFETY PROFILE: Poison by intravenous, subcutaneous, and intramuscular routes. Mildly toxic by ingestion. Mutation data reported. See also SULFATES. When heated to decomposition it emits very toxic fumes of SO_x and NO_x.

APS750 **CAS:37517-28-5** **HR: 3**
ANTIBIOTIC BB-K 8
mf: $C_{22}H_{43}N_5O_{13}$ mw: 585.70

PROP: Crystals. Mp: 203–204° (decomp).

SYN: AMIKACIN

TOXICITY DATA WITH REFERENCE
ims-rbt TDLo: 90 mg/kg (8-16D preg):TER JJANAX 28,366,75
orl-mus LD50: >6 g/kg JJANAX 28,415,75
ipr-mus LD50: 750 mg/kg AMACCQ 13,41,78
scu-mus LD50: 6200 mg/kg JJANAX 28,415,75
ivn-mus LD50: 280 mg/kg JANTAJ 43,858,90
ims-mus LD50: 6200 mg/kg JJANAX 28,415,75

CONSENSUS REPORTS: EPA Genetic Toxicology Program.

SAFETY PROFILE: Poison by intravenous, intraperitoneal, and intramuscular routes. Moderately toxic by intra peritoneal route. An experimental teratogen. When heated to decomposition it emits toxic fumes of NO_x.

APT000 **CAS:39831-55-5** **HR: 3**
ANTIBIOTIC BB-K8 SULFATE
mf: $C_{22}H_{43}N_5O_{13}{\cdot}2H_2O_4S$ mw: 781.86

PROP: Amorphous. Mp: 220–230° (decomp).

SYNS: AMIKACIN SULFATE □ AMIKIN □ AMIKLIN □ BB-K8 □ BIKLIN □ FABIANOL □ NOVAMIN

TOXICITY DATA WITH REFERENCE
scu-rat TDLo: 175 mg/kg (female 8-14D post):REP JJANAX 28,372,75
scu-rat TDLo: 175 mg/kg (female 8-14D post):TER JJANAX 28,372,75
orl-rat LD50: >4 g/kg JJANAX 28,415,75
ipr-rat LD50: 3500 mg/kg JJANAX 28,415,75
scu-rat LD50: 3604 mg/kg PBPSDY 1,125,77
ivn-rat LD50: 234 mg/kg PBPSDY 1,125,77
ims-rat LD50: 2244 mg/kg JJANAX 39,3164,86
ipr-mus LD50: 2930 mg/kg JJANAX 28,415,75
scu-mus LD50: 2470 mg/kg PBPSDY 1,125,77
ivn-mus LD50: 181 mg/kg JJANAX 39,3164,86

SAFETY PROFILE: Poison by intravenous route. Moderately toxic by intraperitoneal and subcutaneous routes. An experimental teratogen. Other experimental reproductive effects. When heated to decomposition it emits very toxic fumes of NO_x and SO_x. See also SULFATES.

APT250 **CAS:51627-14-6** **HR: 2**
ANTIBIOTIC BL-640
mf: $C_{18}H_{18}N_6O_5S_2$ mw: 462.54

SYNS: ANTIBIOTIC BL-S 640 □ BLS 640 □ CEFATRIZINE □ SKF 60771 □ S 640P

TOXICITY DATA WITH REFERENCE
orl-rat TDLo:175 mg/kg (8-14D preg):REP JJANAX 29,129,76
orl-rat TDLo:5600 mg/kg (female 8-14D post):TER JJANAX 29,129,76
orl-rat LD:>6 g/kg JJANAX 29,612,76
ipr-rat LD50:4325 mg/kg JJANAX 29,612,76
ipr-mus LD50:6410 mg/kg JJANAX 29,612,76
ipr-rbt LD50:1500 mg/kg JJANAX 29,612,76
scu-rbt LD50:3000 mg/kg JJANAX 29,612,76

SAFETY PROFILE: Moderately toxic by intraperitoneal and subcutaneous routes. An experimental teratogen. Other experimental reproductive effects. When heated to decomposition it emits very highly toxic fumes of NO_x and SO_x.

APT375 **CAS:69866-21-3** **HR: 3**
ANTIBIOTIC CC 1065
mf: $C_{37}H_{32}N_7O_8$ mw: 702.76

PROP: Clustered needles or amorphous amber-colored foam.

SYNS: CC-1065 □ NSC 298223

TOXICITY DATA WITH REFERENCE
dni-mus:leu 4 µg/L CNREA8 42,999,82
oms-mus:leu 45 µg/L CNREA8 42,999,82
dni-ham:ovr 150 ng/L CNREA8 42,3532,82
oms-ham:ovr 5 µg/L CNREA8 42,3532,82
dnd-mam:lym 7400 nmol/L CNREA8 42,999,82
ipr-mus LD50:6900 ng/kg JANTAJ 37,63,84
ivn-mus LD50:9 µg/kg JANTAJ 37,63,84
ivn-rbt LDLo:1 µg/kg JANTAJ 37,63,84

SAFETY PROFILE: Deadly poison by intravenous and intraperitoneal routes. Mutation data reported. When heated to decomposition it emits toxic fumes of NO_x.

APT750 **CAS:39391-39-4** **HR: 2**
ANTIBIOTIC FR 1923
mf: $C_{23}H_{24}N_4O_9$ mw: 500.51

PROP: Needles. Mp: 214–216° (decomp).

SYN: NOCARDICIN A

TOXICITY DATA WITH REFERENCE
ipr-rat LD50:2600 mg/kg 85ERAY 1,897,78
scu-rat LD50:3100 mg/kg 85ERAY 1,897,78
ipr-mus LD50:2500 mg/kg 85ERAY 1,897,78
scu-mus LD50:2900 mg/kg 85ERAY 1,897,78
ivn-mus LD50:2100 mg/kg 85ERAY 1,897,78

SAFETY PROFILE: Moderately toxic by intraperitoneal, subcutaneous, and intravenous routes. When heated to decomposition it emits toxic fumes of NO_x.

APU000 **HR: 3**
ANTIBIOTIC G-52 SULFATE

PROP: Aminoglycoside produced by a species of the genus *Micromonospora* (JANTAJ 29,483,76).

TOXICITY DATA WITH REFERENCE
ipr-mus LD50:200 mg/kg JANTAJ 29,483,76
scu-mus LD50:400 mg/kg JANTAJ 29,483,76
ivn-mus LD50:50 mg/kg JANTAJ 29,483,76

SAFETY PROFILE: Poison by intraperitoneal, subcutaneous, and intravenous routes. See also SULFATES. When heated to decomposition it emits very toxic fumes of SO_x and NO_x.

APU500 **CAS:57576-44-0** **HR: 3**
ANTIBIOTIC MA 144A1
mf: $C_{42}H_{53}NO_{15}$ mw: 811.96

PROP: Yellow powder. Mp: 129–135° (decomp).

SYNS: ACLACINOMYCIN A □ ANTIBIOTIC MA 144A

TOXICITY DATA WITH REFERENCE
eye-rbt 100 µg MLD JJANAX 33,453,80
dnd-rat:lvr 6300 nmol/L MOPMA3 14,290,78
cyt-rat-ipr 15 mg/kg MUREAV 260,215,91
ipr-rat TDLo:27 mg/kg (17-22D preg/21D post):REP OYYAA2 19,855,80
ipr-rat TDLo:2200 µg/kg (female 7-17D post):TER OYYAA2 19,783,80
orl-rat LD50:58,560 µg/kg JJANAX 33,138,80
ipr-rat LD50:17,930 µg/kg JJANAX 33,138,80
scu-rat LD50:20 mg/kg JJANAX 33,138,80
ivn-rat LD50:25,710 µg/kg JJANAX 33,138,80
orl-mus LD50:30,100 µg/kg ANTBAL 30,918,85
ipr-mus LD50:16,100 µg/kg ANTBAL 30,918,85
scu-mus LD50:22,100 µg/kg ANTBAL 30,918,85
ivn-mus LD50:16,500 µg/kg ANTBAL 30,918,85

SAFETY PROFILE: Poison by ingestion, intraperitoneal, subcutaneous, and intravenous routes. An experimental teratogen. Other experimental reproductive effects. Mutation data reported. An eye and subcutaneous irritant. When heated to decomposition it emits toxic fumes of NO_x.

APV000 **CAS:63710-09-8** **HR: 3**
ANTIBIOTIC MA 144S2
mf: $C_{36}H_{45}NO_{14}$ mw: 715.82

PROP: Dark-red plates. Mp: 162–163°.

SYNS: 1-HYDROXY MA144 S1 □ MA144 S2 □ MUSETTAMYCIN

TOXICITY DATA WITH REFERENCE
dnd-rat:lvr 10 µmol/L MOPMA3 14,290,78
dni-mus:leu 310 nmol/L JANTAJ 34,1596,81
oms-mus:leu 59 nmol/L JANTAJ 34,1596,81
ipr-mus LD50:12,500 µg/kg JKXXAF #78-44555

SAFETY PROFILE: Poison by intraperitoneal route. Mutation data reported. When heated to decomposition it emits toxic fumes of NO_x.

APV750 **CAS:3734-60-9** **HR: 3**
ANTIBIOTIC PA147
mf: $C_6H_6O_3$ mw: 126.12

PROP: Pale-yellow oil. Created from the Streptomycet-en strain, A-415-Z3 with qualities similar to the strains *Str. filipinensis* and *Str. roseochromogenes W.* and *H.* (ARZNAD 17,693,67).

SYNS: ANTIBIOTICUM PA147 (GERMAN) □ 3-CARBOXY-2,4-PENTADIENALLACTOL

TOXICITY DATA WITH REFERENCE
ivn-mus LDLo:250 mg/kg ARZNAD 17,693,67
unk-mus LD50:20 mg/kg ARZNAD 17,693,67

SAFETY PROFILE: Poison by intravenous and other unspecified routes. When heated to decomposition it emits acrid smoke and irritating fumes.

APY500 **CAS:53179-09-2** **HR: 3**
ANTIBIOTIC 66-40 SULFATE
mf: $C_{19}H_{37}N_5O_7 \cdot 5/2H_2O_4S$ mw: 692.81

SYNS: EXTRAMYCIN □ MENSISO □ PATHOMYCIN □ RICKAMICIN SULFATE □ SISOMICIN SULFATE □ SISOMIN

TOXICITY DATA WITH REFERENCE
scu-rat LD50:500 mg/kg JZKEDZ 4,107,78
ivn-rat LD50:49 mg/kg JZKEDZ 4,107,78
ims-rat LD50:404 mg/kg JZKEDZ 4,107,78
ipr-mus LD50:221 mg/kg JANTAJ 23,551,70
scu-mus LD50:272 mg/kg JZKEDZ 4,107,78
ivn-mus LD50:34 mg/kg JANTAJ 23,551,70
ims-mus LD50:280 mg/kg JZKEDZ 4,107,78

SAFETY PROFILE: Poison by intravenous, intramuscular, intraperitoneal, and subcutaneous routes. See also SULFATES. When heated to decomposition it emits very toxic fumes of SO_x and NO_x.

AQB000 **CAS:31282-04-9** **HR: 3**
ANTIHELMYCIN
mf: $C_{20}H_{37}N_3O_{13}$ mw: 527.60

PROP: Amorphous. Mp: 160–180° (decomp).

SYNS: HYGROMIX-8 □ HYGROMYCIN B (USDA)

TOXICITY DATA WITH REFERENCE
mmo-esc 250 mg/L AACHAX 18,798,80
mic-mic-uns 3 g/L ANTBAL 25,822,80
ipr-rat LD50:63 mg/kg GISAAA 38,11,73
ipr-gpg LD50:13 mg/kg GISAAA 38,11,73

SAFETY PROFILE: Poison by intraperitoneal route. Mutation data reported. When heated to decomposition it emits toxic fumes of NO_x.

AQB750 **CAS:7440-36-0** **HR: 3**
ANTIMONY
DOT: UN 2871
af: Sb aw: 121.75

PROP: Silvery or gray, lustrous metalloid. Mp: 630°, bp: 1635°, d: 6.684 @ 25°, vap press: 1 mm @ 886°. Insol in water; sol in hot concentrated H_2SO_4.

SYNS: ANTIMONY BLACK □ ANTIMONY POWDER (DOT) □ ANTIMONY REGULUS □ ANTYMON (POLISH) □ C.I. 77050 □ STIBIUM

TOXICITY DATA WITH REFERENCE
ihl-rat TCLo:50 mg/m³/7H/52W-I:CAR JTEHD6 18,607,86
orl-rat LD50:7 g/kg EQSFAP 1,1,75
ipr-rat LD50:100 mg/kg 85GMAT -,22,82
ipr-mus LD50:90 mg/kg 85GMAT -,22,82
ipr-gpg LD50:150 mg/kg EQSFAP 1,1,75

CONSENSUS REPORTS: Antimony and its compounds are on the Community Right-To-Know List. Reported in EPA TSCA Inventory.

OSHA PEL: TWA 0.5 mg(Sb)/m³
ACGIH TLV: TWA 0.5 mg(Sb)/m³
DFG MAK: 0.5 mg(Sb)/m³
NIOSH REL: TWA 0.5 mg(Sb)/m³
DOT CLASSIFICATION: 6.1; *Label:* KEEP AWAY FROM FOOD

SAFETY PROFILE: An experimental poison by intraperitoneal route. Questionable carcinogen with experimental carcinogenic data. Moderate fire and explosion hazard in the forms of dust and vapor, when exposed to heat or flame. See also POWDERED METALS. When heated or on contact with acid it emits toxic fumes of SbH_3. Electrolysis of acid sulfides and stirred Sb halide yields explosive Sb. It can react violently with NH_4NO_3, halogens, BrN_3, BrF_3, $HClO_3$, ClO, ClF_3, HNO_3, KNO_3, $KMnO_4$, K_2O_2, $NaNO_3$, oxidants.

For occupational chemical analysis use OSHA: #ID-125G or NIOSH: Elements in Blood or Tissue 8005.

AQC000 **CAS:72017-60-8** **HR: 3**
ANTIMONY AMMONIA TRIACETIC ACID
mf: $C_{12}H_{14}N_2O_{12}Sb \cdot 2H_2O$ mw: 536.07

SYNS: ATA-Sb □ Sb-71

CONSENSUS REPORTS: Antimony and its compounds are on the Community Right-To-Know List.

OSHA PEL: TWA 0.5 mg(Sb)/m³
ACGIH TLV: TWA 0.5 mg(Sb)/m³
NIOSH REL: (Antimony) TWA 0.5 mg(Sb)/m³

SAFETY PROFILE: Probably a poison. See also ANTIMONY COMPOUNDS. When heated to decomposition it emits toxic fumes of NO_x, Sb, and NH_3.

AQC250 **CAS:64046-93-1** **HR: 3**
ANTIMONY, BIS(TRICHLORO) compounded with 1 mole of OCTAMETHYL PYROPHOSPHORAMIDE
mf: $Cl_3Sb \cdot 1/2(C_8H_{24}N_4O_3P_2)$ mw: 371.25

TOXICITY DATA WITH REFERENCE
ipr-mus LD50:35 mg/kg JAFCAU 14,512,66

CONSENSUS REPORTS: Antimony and its compounds are on the Community Right-To-Know List.

OSHA PEL: TWA 0.5 mg(Sb)/m³
ACGIH TLV: TWA 0.5 mg(Sb)/m³
NIOSH REL: (Antimony) TWA 0.5 mg(Sb)/m³

SAFETY PROFILE: Poison by intraperitoneal route. See also ANTIMONY COMPOUNDS and PHOSPHATES. When heated to decomposition it emits very toxic fumes of Cl^-, Sb, PO_x, and NO_x.

AQC500 **CAS:10025-91-9** **HR: 3**
ANTIMONY(III) CHLORIDE
DOT: UN 1733
mf: Cl_3Sb mw: 228.10

PROP: Colorless, rhombic, deliq, hygroscopic crystals which fume in the air. D: 3.06, mp: 73.4°, bp: 220°, vap press: 1 mm @ 49.2° (subl). Sol in cold EtOH, CS_2, Et_2O, CCl_4 and H_2O (small amounts); insol in quinoline, other org bases.

SYNS: ANTIMOINE (TRICHLORURE d') □ ANTIMONIO (TRICLORURO di) □ ANTIMONOUS CHLORIDE □ ANTIMONOUS CHLORIDE (DOT) □ ANTIMONTRICHLORID □ ANTIMONY BUTTER □ ANTIMONY CHLORIDE □ ANTIMONY CHLORIDE (DOT) □ ANTIMONY TRICHLORIDE □ ANTIMONY TRICHLORIDE, liquid (DOT) □ ANTIMONY TRICHLORIDE, solid (DOT) □ ANTIMONY TRICHLORIDE, solution (DOT) □ ANTIMOONTRICHLORIDE □ BUTTER of ANTIMONY □ CHLORID ANTIMONITY □ CHLORURE ANTIMONIEUX □ C.I. 77056 □ STIBINE, TRICHLORO- □ TRICHLOROSTIBINE □ TRICHLORURE d'ANTIMOINE

TOXICITY DATA WITH REFERENCE
dnr-bcs 10 mmol/L MUREAV 77,109,80
orl-rat TDLo:4400 μg/kg (female 1-22D post):REP TCMUD8 7,491,87
ihl-hmn TDLo:73 mg/kg:PUL BJIMAG 23,318,66
orl-rat LD50:525 mg/kg MARJV# 29MAR77
ipr-mus LD50:13 mg/kg 85GMAT-,23,82
orl-gpg LD50:574 mg/kg HYSAAV 29(12),16,64

CONSENSUS REPORTS: Reported in EPA TSCA Inventory. Antimony and its compounds are on the Community Right-To-Know List.

OSHA PEL: TWA 500 μg(Sb)/m^3
ACGIH TLV: TWA 0.5 mg(Sb)/m^3
NIOSH REL: (Antimony) TWA 0.5 mg(Sb)/m^3
DOT CLASSIFICATION: 8; *Label:* Corrosive

SAFETY PROFILE: Moderately toxic by ingestion. Human pulmonary system effects by inhalation. Corrosive by vigorous reaction with moisture, generating heat and hydrogen chloride gas (a strong irritant), which can cause pulmonary edema when inhaled. Systemic effects can be caused by the antimony. See also ANTIMONY COMPOUNDS. Experimental reproductive effects. Mutation data reported. When heated to decomposition it emits very toxic fumes of chlorine and antimony. It can react violently with aluminum, potassium, sodium.

AQD000 **CAS:7647-18-9** **HR: 3**
ANTIMONY(V) CHLORIDE
DOT: UN 1730/UN 1731
mf: Cl_5Sb mw: 299.01

PROP: Colorless or red-yellow oil or liquid; offensive odor. Mp: 4°, bp: 140°, d: 2.336, vap press: 1 mm @ 22.7°. Decomp in water; sol in HCl, HBr, CS_2, CCl_4, and $CHCl_3$.

SYNS: ANTIMONIC CHLORIDE □ ANTIMONIO (PENTACLORURO DI) (ITALIAN) □ ANTIMONPENTACHLORID (GERMAN) □ ANTIMONY PENTACHLORIDE □ ANTIMONY PENTACHLORIDE (DOT) □ ANTIMONY PERCHLORIDE □ ANTIMOONPENTACHLORIDE (DUTCH) □ BUTTER of ANTIMONY □ PENTACHLOROANTIMONY □ PENTACHLORURE d'ANTIMOINE (FRENCH) □ PERCHLORURE d'ANTIMOINE (FRENCH)

TOXICITY DATA WITH REFERENCE
mrc-bcs 30 μL/disc MUREAV 77,109,80
orl-rat LD50:1115 mg/kg HYSAAV 29(12),16,64
orl-gpg LD50:900 mg/kg HYSAAV 29(12),16,64

CONSENSUS REPORTS: Reported in EPA TSCA Inventory. Antimony and its compounds are on the Community Right-To-Know List.

OSHA PEL: TWA 500 μg(Sb)/m^3
ACGIH TLV: TWA 0.5 mg(Sb)/m^3
NIOSH REL: (Antimony) TWA 0.5 mg(Sb)/m^3
DOT CLASSIFICATION: 8; *Label:* Corrosive

SAFETY PROFILE: Poison by ingestion. Corrosive. Mutation data reported. See ANTIMONY COMPOUNDS and ANTIMONY(III) CHLORIDE. When heated to decomposition it emits very toxic fumes of Cl^- and Sb.

AQD500 **HR: 3**
ANTIMONY COMPOUNDS

CONSENSUS REPORTS: On Community Right-To-Know List.

SAFETY PROFILE: Most antimony compounds are poisons by ingestion, inhalation, and intraperitoneal routes. See also ANTIMONY. Locally antimony compounds irritate the skin and mucous membranes. Sb^{+++} and hot $HClO_3$ can form an explosive mixture.

AQD750 **CAS:3064-61-7** **HR: 3**
ANTIMONY DIMERCAPTOSUCCINATE
mf: $C_{12}H_6O_{12}S_6Sb_2 \cdot 6Na$ mw: 915.98

PROP: Sol in water.

SYNS: ANTIMONY DIMERCAPTOSUCCINATE(IV) □ SODIUM ANTIMONY-2,3-meso-DIMERCAPTOSUCCINATE □ STIBOCAPTATE

TOXICITY DATA WITH REFERENCE
ivn-rat LD50:195 mg/kg CHTHBK 13,339,68
ipr-mus LD50:2500 mg/kg FAZMAE 17,108,73
scu-mus LD50:2 g/kg BWHOA6 45,371,71

CONSENSUS REPORTS: Antimony and its compounds are on the Community Right-To-Know List.

OSHA PEL: TWA 0.5 mg(Sb)/m^3
ACGIH TLV: TWA 0.5 mg(Sb)/m^3
NIOSH REL: (Antimony) TWA 0.5 mg(Sb)/m^3

SAFETY PROFILE: Poison by intravenous route. Moderately toxic by intraperitoneal and subcutaneous routes. See also ANTIMONY COMPOUNDS and SULFIDES. When heated to decomposition it emits toxic fumes of SO_x, Na_2O, and Sb.

AQE000 CAS:7783-56-4 HR: 3
ANTIMONY(III) FLUORIDE (1:3)
DOT: NA 1549
mf: F_3Sb mw: 178.75

PROP: Colorless, rhombic, very deliq crystals. Mp: 292°, bp: 376° (subl), d: 4.379 @ 20.9°. Readily sol in water with part hydrolysis, sol in polar org solvs. Insol in C_6H_6, chlorobenzene, pet ether.

SYNS: ANTIMOINE FLUORURE (FRENCH) ☐ ANTIMONOUS FLUORIDE ☐ ANTIMONY TRIFLUORIDE ☐ ANTIMONY TRIFLUORIDE, solid or solution (DOT) ☐ STIBINE, TRIFLUORO-(9CI) ☐ TRIFLUOROANTIMONY ☐ TRIFLUOROSTIBINE

TOXICITY DATA WITH REFERENCE
scu-frg LDLo:224 mg/kg CRSBAW 124,133,37

CONSENSUS REPORTS: Reported in EPA TSCA Inventory. Antimony and its compounds are on the Community Right-To-Know List.

OSHA PEL: TWA 0.5 mg(Sb)/m³; TWA 2.5 mg(F)/m³
ACGIH TLV: TWA 2.5 mg(F)/m³; TWA 0.5 mg(Sb)/m³
NIOSH REL: TWA 0.5 mg(Sb)/m³
DOT CLASSIFICATION: 8; *Label:* Corrosive

SAFETY PROFILE: Poison by subcutaneous route. Corrosive to skin and eyes. See also FLUORIDES and ANTIMONY COMPOUNDS. When heated to decomposition it emits very toxic fumes of F^- and Sb.

AQE250 CAS:58164-88-8 HR: 3
ANTIMONY LACTATE
DOT: UN 1550
mf: $C_9H_{15}O_9$•Sb mw: 388.99

PROP: Tan mass, water-sol.

SYNS: ANTIMONY LACTATE, solid (DOT) ☐ LACTIC ACID, ANTIMONY SALT ☐ PROPANOIC ACID, 2-HYDROXY-, TRIANHYDRIDE with ANTIMONIC ACID (H_3SbO_3) (9CI)

CONSENSUS REPORTS: Reported in EPA TSCA Inventory. Antimony and its compounds are on the Community Right-To-Know List.

OSHA PEL: TWA 0.5 mg(Sb)/m³
ACGIH TLV: TWA 0.5 mg(Sb)/m³
NIOSH REL: (Antimony) TWA 0.5 mg(Sb)/m³
DOT CLASSIFICATION: 6.1; *Label:* KEEP AWAY FROM FOOD

SAFETY PROFILE: A poison. See also ANTIMONY COMPOUNDS. When heated to decomposition it emits toxic fumes of Sb.

AQE300 CAS:77824-42-1 HR: 2
ANTIMONYL-2,4-DIHYDROXY-5-HYDROXYMETHYL PYRIMIDINE
mf: $C_{10}H_{10}N_4O_8Sb_2$ mw: 557.74

SYN: 2,4,10,12-TETRAOXA-6,16,17,18-TETRAAZA-3,11-DISTIBATRICYCLO(11.3.1.1^5,9)OCTADECA-1(17),5,7,9(18),13,15-HEXAENE-8,14-DIMETHANOL, 3,11-DIHYDROXY-

TOXICITY DATA WITH REFERENCE
ipr-mus LD50:660 mg/kg JEMAAJ 62,1,79

OSHA PEL: TWA 0.5 mg(Sb)/m³
ACGIH TLV: TWA 0.5 mg(Sb)/m³
NIOSH REL: 10H TWA 0.5 mg(Sb)/m³

SAFETY PROFILE: Moderately toxic by intraperitoneal route. When heated to decomposition it emits toxic fumes of NO_x and Sb.

AQE305 CAS:77824-44-3 HR: 3
ANTIMONYL-2,4-DIHYDROXY PYRIMIDINE
mf: $C_8H_6N_4O_6Sb_2$ mw: 497.68

SYN: 2,4,10,12-TETRAOXA-6,16,17,18-TETRAAZA-3,11-DISTIBATRICYCLO(11.3.1.1^5,9)OCTADECA-1(17),5,7,9(18),13,15-HEXAENE, 3,11-DIHYDROXY-

TOXICITY DATA WITH REFERENCE
ipr-mus LD50:300 mg/kg JEMAAJ 62,1,79

OSHA PEL: TWA 0.5 mg(Sb)/m³
ACGIH TLV: TWA 0.5 mg(Sb)/m³
NIOSH REL: 10H TWA 0.5 mg(Sb)/m³

SAFETY PROFILE: Poison by intraperitoneal route. When heated to decomposition it emits toxic fumes of NO_x and Sb.

AQE320 CAS:77824-43-2 HR: 3
ANTIMONYL-7-FORMYL-8-HYDROXYQUINOLINE-5-SULPHONATE
mf: $C_{20}H_{11}N_2O_{11}S_2Sb$•2Na mw: 687.18

SYN: 5-QUINOLINESULFONIC ACID, 8,8'-((HYDROXYSTIBYLENE)BIS(OXY))BIS(7-FORMYL)-, DISODIUM SALT

TOXICITY DATA WITH REFERENCE
ipr-mus LD50:75 mg/kg JEMAAJ 62,1,79

OSHA PEL: TWA 0.5 mg(Sb)/m³
ACGIH TLV: TWA 0.5 mg(Sb)/m³
NIOSH REL: 10H TWA 0.5 mg(Sb)/m³

SAFETY PROFILE: Poison by intraperitoneal route. When heated to decomposition it emits toxic fumes of NO_x, SO_x, and Sb.

AQE500 CAS:6169-12-6 HR: 1
ANTIMONY LITHIUM THIOMALATENONAHYDRATE
mf: $C_{12}H_9O_{12}S_3$•$9H_2O$•6Li•Sb mw: 766.96

SYNS: ANTHIOMALINE NONAHYDRATE ☐ LITHIUM ANTIMONY THIOMALATE NONAHYDRATE ☐ THIOANTIMONIC(III) ACID, TRIESTER with MERCAPTO SUCCINIC ACID DILITHIUM SALT, NONAHYDRATE

TOXICITY DATA WITH REFERENCE
ims-chd TDLo:73 mg/kg:GIT JAMAAP 125,952,44
ims-hmn TDLo:11 mg/kg:GIT JAMAAP 125,952,44

CONSENSUS REPORTS: Antimony and its compounds are on the Community Right-To-Know List.

OSHA PEL: TWA 0.5 mg(Sb)/m³
ACGIH TLV: TWA 0.5 mg(Sb)/m³
NIOSH REL: (Antimony) TWA 0.5 mg(Sb)/m³

SAFETY PROFILE: Human gastrointestinal tract effects by intramuscular route. See also LITHIUM; ANTIMONY

COMPOUNDS; and ESTERS. When heated to decomposition it emits very toxic fumes of SO_x and Sb.

AQE750 **HR: 3**
ANTIMONY NITRIDE
mf: NSb mw: 135.76

CONSENSUS REPORTS: Antimony and its compounds are on the Community Right-To-Know List.

SAFETY PROFILE: See ANTIMONY COMPOUNDS and NITRIDES. Explosively decomposes upon warming in a vacuum. When heated to decomposition it emits very toxic fumes of Sb, NO_x, and NH_3.

AQF000 **CAS:1309-64-4** **HR: 3**
ANTIMONY OXIDE
mf: O_3Sb_2 mw: 291.50

PROP: White cubes. D: 5.2, mp: 650°, bp: 1550° subl. Very sltly sol in water; sol in KOH and HCl.

SYNS: A 1530 □ A 1582 □ A 1588LP □ AMSPEC-KR □ ANTIMONIOUS OXIDE □ ANTIMONY(3+) OXIDE □ ANTIMONY PEROXIDE □ ANTIMONY SESQUIOXIDE □ ANTIMONY TRIOXIDE □ ANTIMONY WHITE □ ANTOX □ ANZON-TMS □ AP 50 □ BLUE STAR □ CHEMETRON FIRE SHIELD □ C.I. 77052 □ C.I. PIGMENT WHITE 11 □ DECHLORANE A-O □ DIANTIMONY TRIOXIDE □ EXITELITE □ EXTREMA □ FLOWERS of ANTIMONY □ NCI-C55152 □ NYACOL A 1530 □ SENARMONTITE □ THERMOGUARD B □ THERMOGUARD S □ TIMONOX □ TWINKLING STAR □ VALENTINITE □ WEISSPIESSGLANZ □ WHITE STAR

TOXICITY DATA WITH REFERENCE
mrc-bcs 50 mmol/L MUREAV 77,109,80
sce-ham:lng 90 μg/L MUREAV 264,163,91
ihl-rat TCLo:270 μg/m³ (1-21D post):TER GISAAA 52(10),85,87
ihl-rat TCLo:270 μg/m³ (1-21D post):REP GISAAA 52(10),85,87
ihl-rat TCLo:4200 μg/m³/52W-I:CAR AIHAM* 20,1,80
ihl-rat TC:4 mg/m³/1Y-I:ETA PESTC* 8,16,80
ihl-rat TC:1600 μg/m³/52W-I:NEO AIHAM* 20,1,80
ihl-rat TC:50 mg/m³/7H/52W-I:CAR JTEHD6 18,607,86
orl-rat LD50:>20 g/kg JIDHAN 30,63,48
ipr-rat LD50:3250 mg/kg EQSSDX 1,1,75
ipr-mus LD50:172 mg/kg 85GMAT-,23,82
ivn-dog LDLo:3 mg/kg HBAMAK 4,1289,35
scu-rbt LDLo:2500 μg/kg HBAMAK 4,1289,35

CONSENSUS REPORTS: Reported in EPA TSCA Inventory. Antimony and its compounds are on the Community Right-To-Know List.

OSHA PEL: TWA 0.5 mg(Sb)/m³
ACGIH TLV: TWA 0.5 mg(Sb)/m³; Suspected Carcinogen
DFG MAK: Animal Carcinogen, Suspected Human Carcinogen
NIOSH REL: TWA 0.5 mg(Sb)/m³

SAFETY PROFILE: Confirmed carcinogen with experimental carcinogenic and neoplastigenic data. Poison by intravenous and subcutaneous routes. Moderately toxic by other routes. An experimental teratogen. Other experimental reproductive effects. Mutation data reported. See also ANTIMONY COMPOUNDS. When heated to decomposition it emits toxic Sb fumes. Incompatible with chlorinated rubber and heat of 216° and with BrF_3.

AQF250 **CAS:7783-70-2** **HR: 3**
ANTIMONY(V) PENTAFLUORIDE
DOT: UN 1732
mf: F_5Sb mw: 216.75

PROP: Oily, colorless liquid. Very reactive. Mp: 7.0°, bp: 149.5°, d: (liq) 2.99 @ 23°. Sol in water and KF.

SYNS: ANTIMONY FLUORIDE □ ANTIMONY(V) FLUORIDE □ ANTIMONY PENTAFLUORIDE (DOT) □ PENTAFLUOROANTIMONY

TOXICITY DATA WITH REFERENCE
ihl-mus LC50:270 mg/m³ GISAAA 35(7),25,70
ihl-rat TCLo:15 mg/m³/2H/15W-I GISAAA 35(7),25,70

CONSENSUS REPORTS: Reported in EPA TSCA Inventory. Antimony and its compounds are on the Community Right-To-Know List. EPA Extremely Hazardous Substances List.

OSHA PEL: TWA 0.5 mg(Sb)/m³
ACGIH TLV: TWA 0.5 mg(Sb)/m³
NIOSH REL: (Antimony) TWA 0.5 mg(Sb)/m³
DOT CLASSIFICATION: 8; *Label:* Corrosive, Poison

SAFETY PROFILE: A poison by inhalation. A very reactive, corrosive liquid to skin, eyes, mucous membranes. See also FLUORIDES and ANTIMONY COMPOUNDS. Violent reaction with phosphates. When heated to decomposition it emits very toxic fumes of F^- and Sb.

AQF500 **CAS:1315-04-4** **HR: 3**
ANTIMONY PENTASULFIDE
mf: S_5Sb_2 mw: 403.80

PROP: Dark orange-yellow powder or solid. Mp: 75° (decomp), d: 4.120. Insol dilute aqueous acids; sol conc HCl.

SYNS: ANTIMONIAL SAFFRON □ ANTIMONIC SULFIDE □ ANTIMONY RED □ ANTIMONY SULFIDE □ C.I. 77061 □ GOLDEN ANTIMONY SULFIDE

TOXICITY DATA WITH REFERENCE
orl-rat LD:>10 g/kg GISAAA 54(4),68,89
ipr-rat LD50:1500 mg/kg EQSFAP 1,1,75
ipr-mus LD50:458 mg/kg 85GMAT-,23,82

CONSENSUS REPORTS: Reported in EPA TSCA Inventory. Antimony and its compounds are on the Community Right-To-Know List.

OSHA PEL: TWA 0.5 mg(Sb)/m³
ACGIH TLV: TWA 0.5 mg(Sb)/m³
NIOSH REL: (Antimony) TWA 0.5 mg(Sb)/m³

SAFETY PROFILE: Moderately toxic by intraperitoneal route. See also ANTIMONY COMPOUNDS and SULFIDES. Flammable when exposed to heat or by chemical reaction with powerful oxidizers. Use water to fight fire. Moderately explosive when shocked or by spontaneous chemical reaction in contact with powerful oxidizers. When heated to decomposition or on contact with acid or acid fumes it emits highly toxic fumes of oxides of sulfur and antimony. Incompatible with water or steam to produce toxic and flammable vapors and with oxidizers, e.g., $Ag(ClO_3)_2$, $HClO_3$, ClO_2, $Mg(ClO_3)_2$, TlO, $Zn(ClO_3)_2$.

AQF750 CAS:1314-60-9 HR: 2
ANTIMONY PENTOXIDE
mf: O_5Sb_2 mw: 323.50

PROP: Yellowish-white powder or deep yellow crystals. D: 3.78, mp: decomp @ 380°. Insol in water; sltly sol in warm KOH soln.

SYNS: ANTIMONIC "ACID" □ ANTIMONIC OXIDE □ ANTIMONY PENTAOXIDE □ DIANTIMONY PENTOXIDE □ STIBIC ANHYDRIDE

TOXICITY DATA WITH REFERENCE
ipr-rat LD50:4 g/kg EQSFAP 1,1,75
ipr-mus LD50:978 mg/kg 85GMAT-,23,82

CONSENSUS REPORTS: Reported in EPA TSCA Inventory. Antimony and its compounds are on the Community Right-To-Know List.

OSHA PEL: TWA 500 μg(Sb)/m³
ACGIH TLV: TWA 0.5 mg(Sb)/m³
NIOSH REL: (Antimony) TWA 0.5 mg(Sb)/m³

SAFETY PROFILE: Moderately toxic by intraperitoneal route. See also ANTIMONY COMPOUNDS.

AQG000 CAS:35743-94-3 HR: 3
ANTIMONY POTASSIUM DIMETHYLCYSTEINOTARTRATE

PROP: A dimethyl cysteine chelate of sodium antimonyl tartrate with 14.7% Sb content (PSEBAA 129,284,68).

TOXICITY DATA WITH REFERENCE
scu-mus LD50:350 mg/kg PSEBAA 129,284,68
ivn-mus LD50:450 mg/kg PSEBAA 129,204,68

CONSENSUS REPORTS: Antimony and its compounds are on the Community Right-To-Know List.

OSHA PEL: TWA 500 μg(Sb)/m³
NIOSH REL: TWA 0.5 mg(Sb)/m³

SAFETY PROFILE: Poison by subcutaneous route. Moderately toxic by intravenous route. See also ANTIMONY COMPOUNDS. When heated to decomposition it emits very toxic fumes of NO_x and Sb.

AQG250 CAS:28300-74-5 HR: 3
ANTIMONY POTASSIUM TARTRATE
DOT: UN 1551
mf: $C_8H_4O_{12}Sb_2 \cdot 3H_2O \cdot 2K$ mw: 635.88

PROP: Colorless crystals to white powder. D: 2.607, mp: loses H_2O @ 100°.

SYNS: ANTIMONYL POTASSIUM TARTRATE □ EMETIQUE (FRENCH) □ ENT 50,434 □ POTASSIUM ANTIMONYL TARTRATE □ POTASSIUM ANTIMONYL-d-TARTRATE □ POTASSIUM ANTIMONY TARTRATE □ TARTAR EMETIC □ TARTARIZED ANTIMONY □ TARTRATE ANTIMONIO-POTASSIQUE (FRENCH) □ TARTRATED ANTIMONY

TOXICITY DATA WITH REFERENCE
dni-esc 19 μmol/L BCPCA6 23,1451,74
cyt-hmn:fbr 100 μmol/L JDGRAX 7(3),27,75
orl-hmn LDLo:2 mg/kg PCOC** -,1097,66
ivn-hmn TDLo:1392 μg/kg:EYE,PUL LANCAO 210,227,26
ivn-man LDLo:12 mg/kg/1W-I:LIV,KID LANCAO 210,227,26
ivn-man LD50:249 mg/kg/9D-I JTMHA9 21,38,18
orl-rat LD50:115 mg/kg ARSIM* 20,24,66
ipr-rat LD50:11 mg/kg EQSSDX 1,1,75
ims-rat LDLo:33 mg/kg EQSSDX 1,1,75
orl-mus LDLo:600 mg/kg EQSFAP 1,1,75
ipr-mus LD50:33 mg/kg BWHOA6 53,379,76
scu-mus LD50:55 mg/kg PSEBAA 129,284,68
ivn-mus LD50:45 mg/kg FAZMAE 17,108,73

CONSENSUS REPORTS: Antimony and its compounds are on the Community Right-To-Know List.

OSHA PEL: TWA 0.5 mg(Sb)/m³
ACGIH TLV: TWA 0.5 mg(Sb)/m³
NIOSH REL: (Antimony) TWA 0.5 mg(Sb)/m³
DOT CLASSIFICATION: 6.1; *Label:* KEEP AWAY FROM FOOD

SAFETY PROFILE: Human poison by intravenous route, producing liver and kidney changes, somnolence, dyspnea, and pupillary dilation. Poison by ingestion, subcutaneous, intravenous, intramuscular, and intraperitoneal routes. Large doses cause severe liver damage. Human mutation data reported. Used medicinally, the therapeutic dose is close to the toxic dose. Upon decomposition it emits toxic fumes of K_2O and Sb.

AQG500 CAS:64070-11-7 HR: 3
d-ANTIMONY POTASSIUM TARTRATE
mf: $C_4H_4O_7Sb \cdot K$ mw: 324.93

PROP: White crystals.

SYN: POTASSIUM ANTIMONYL-d-TARTRATE

TOXICITY DATA WITH REFERENCE
ipr-mus LD50:45,637 μg/kg AJTMAQ 30,591,50

CONSENSUS REPORTS: Antimony and its compounds are on the Community Right-To-Know List.

OSHA PEL: TWA 0.5 mg(Sb)/m³
ACGIH TLV: TWA 0.5 mg(Sb)/m³
NIOSH REL: (Antimony) TWA 0.5 mg(Sb)/m³

SAFETY PROFILE: Poison by intraperitoneal route. See also ANTIMONY COMPOUNDS. When heated to decomposition it emits toxic fumes of Sb and K_2O.

AQG750 CAS:64070-12-8 HR: 3
dl-ANTIMONY POTASSIUM TARTRATE
mf: $C_4H_4O_7Sb \cdot K$ mw: 324.93

SYNS: POTASSIUM ANTIMONYL-d,l-TARTRATE □ dl-TARTARIC ACID, ANTIMONY POTASSIUM SALT

TOXICITY DATA WITH REFERENCE
ipr-mus LD50:45,637 μg/kg AJTMAQ 30,591,50

CONSENSUS REPORTS: Antimony and its compounds are on the Community Right-To-Know List.

OSHA PEL: TWA 0.5 mg(Sb)/m^3
ACGIH TLV: TWA 0.5 mg(Sb)/m^3
NIOSH REL: (Antimony) TWA 0.5 mg(Sb)/m^3

SAFETY PROFILE: Poison by intraperitoneal route. See also ANTIMONY COMPOUNDS. When heated to decomposition it emits toxic fumes of Sb and K_2O.

AQH000 CAS:11071-15-1 HR: 3
l-ANTIMONY POTASSIUM TARTRATE
mf: $C_4H_5O_7Sb{\cdot}K$ mw: 325.94

SYNS: POTASSIUM ANTIMONYL-l-TARTRATE □ l-TARTARIC ACID, ANTIMONY POTASSIUM SALT

TOXICITY DATA WITH REFERENCE
ipr-mus LD50:31,600 µg/kg YFZADL 2,278,82

CONSENSUS REPORTS: Reported in EPA TSCA Inventory. Antimony and its compounds are on the Community Right-To-Know List.

OSHA PEL: TWA 0.5 mg(Sb)/m^3
ACGIH TLV: TWA 0.5 mg(Sb)/m^3
NIOSH REL: (Antimony) TWA 0.5 mg(Sb)/m^3

SAFETY PROFILE: Poison by intraperitoneal route. See also ANTIMONY COMPOUNDS. When heated to decomposition it emits toxic fumes of Sb and K_2O.

AQH250 CAS:64070-10-6 HR: 3
meso-ANTIMONY POTASSIUM TARTRATE
mf: $C_4H_4O_7Sb{\cdot}K$ mw: 324.93

SYN: POTASSIUM ANTIMONYL-meso-TARTRATE

TOXICITY DATA WITH REFERENCE
ipr-mus LD50:49,640 µg/kg AJTMAQ 30,591,50

CONSENSUS REPORTS: Antimony and its compounds are on the Community Right-To-Know List.

OSHA PEL: TWA 0.5 mg(Sb)/m^3
ACGIH TLV: TWA 0.5 mg(Sb)/m^3
NIOSH REL: (Antimony) TWA 0.5 mg(Sb)/m^3

SAFETY PROFILE: Poison by intraperitoneal route. See also ANTIMONY COMPOUNDS. When heated to decomposition it emits toxic fumes of Sb and K_2O.

AQH500 CAS:15489-16-4 HR: 3
ANTIMONY PYROCATECHOL SODIUM DISULFONATE
mf: $C_{12}H_4O_{16}S_4Sb{\cdot}7H_2O{\cdot}5Na$ mw: 799.24

SYNS: ANTIMONYLBRENZEATECHINDISULFOSAURES NATRIUM (GERMAN) □ ANTIMOSAN □ CORYSTIBIN □ FOUADIN □ NEOANTIMOSAN □ PYROSTIB □ REPRODAL □ SODIUM ANTIMONY BIS (PYROCATECHOL-2,4-DISULFONATE) □ SODIUM ANTIMONY (III) BIS-PYROCATECHOL-3,5-DISULFONATE HEPTAHYDRATE □ SODIUM ANTIMOSAN □ STIBOPHEN □ TRIMON

TOXICITY DATA WITH REFERENCE
ipr-mus LD50:260 mg/kg AJTMAQ 25,263,45
scu-mus LD50:670 mg/kg BWHOA6 45,371,71
ivn-mus LD50:1050 mg/kg FAZMAE 17,108,73
ivn-rbt LD50:90 mg/kg JPETAB 89,196,47
ims-rbt LD50:91 mg/kg JPETAB 87,119,46

CONSENSUS REPORTS: Antimony and its compounds are on the Community Right-To-Know List.

OSHA PEL: TWA 0.5 mg(Sb)/m^3
ACGIH TLV: TWA 0.5 mg(Sb)/m^3
NIOSH REL: (Antimony) TWA 0.5 mg(Sb)/m^3

SAFETY PROFILE: Poison by intraperitoneal, intramuscular, and intravenous routes. Moderately toxic by intraperitoneal route. See also ANTIMONY COMPOUNDS and SULFONATES. When heated to decomposition it emits toxic fumes of Sb and Na_2O.

AQH750 HR: 3
ANTIMONY SODIUM DIMETHYL CYSTEINO TARTRATE

PROP: Made up of 5.8 parts of sodium antimony tartrate and 10 parts of dimethyl cysteine (FAZMAE 17,108,73).

SYN: SODIUM ANTIMONYL DIMETHYLCYSTEINE TARTRATE

TOXICITY DATA WITH REFERENCE
ipr-mus LD50:450 mg/kg PSEBAA 129,284,68
scu-mus LD50:385 mg/kg PSEBAA 129,284,68
ivn-mus LD50:435 mg/kg PSEBAA 129,284,68
ims-mus LD50:325 mg/kg PSEBAA 129,284,68

CONSENSUS REPORTS: Antimony and its compounds are on the Community Right-To-Know List.

OSHA PEL: TWA 500 µg(Sb)/m^3
NIOSH REL: (Antimony) TWA 10H 0.5 mg(Sb)/m^3

SAFETY PROFILE: Poison by subcutaneous and intramuscular routes. Moderately toxic by other routes. See also ANTIMONY COMPOUNDS. When heated to decomposition it emits very toxic Sb fumes and Na_2O.

AQH800 CAS:16037-91-5 HR: 3
ANTIMONY SODIUM GLUCONATE
mf: $C_{12}H_{20}O_{17}Sb_2{\cdot}3Na{\cdot}9H_2O$ mw: 1048.91

SYNS: ESTIBOGLUCONATO SODICO □ d-GLUCONIC ACID, CYCLIC ESTER with ANTIMONIC ACID ($H_8Sb_2O_9$) (2:1),TRISODIUM SALT, NONAHYDRATE □ d-GLUCONIC ACID, 2,4:2′,4′-O-(OXYDISTIBYLIDYNE)BIS-, Sb,Sb′-DIOXIDE, TRISODIUM SALT, NONAHYDRATE □ MYOSTIBIN □ PENTOSTAM □ SODIUM STIBOGLUCONATE □ SOLUSTIBOSAN □ SOLUSTIN □ SOLUSURMIN □ SOLYUSURMIN □ STIBANATE □ STIBANOSE □ STIBATIN □ STIBINOL

TOXICITY DATA WITH REFERENCE
ipr-mus LD50:33 mg/kg CLDND* 11,155,49

OSHA PEL: TWA 0.5 mg(Sb)/m^3
ACGIH TLV: TWA 0.5 mg(Sb)/m^3
NIOSH REL: (Antimony) 10H TWA 0.5 mg(Sb)/m^3

SAFETY PROFILE: Poison by intraperitoneal route. When heated to decomposition it emits toxic fumes of Sb.

AQI000 CAS:12550-17-3 HR: 3
ANTIMONY(III) SODIUM GLUCONATE
mf: $C_6H_8O_7Sb{\cdot}Na$ mw: 336.88

SYNS: SODIUM ANTIMONY GLUCONATE □ SODIUM ANTIMONY (III) GLUCONATE □ TRIVALENT SODIUM ANTIMONYL GLUCONATE

TOXICITY DATA WITH REFERENCE
ipr-mus LD50:3440 µg/kg CLDND* 81,224,44

CONSENSUS REPORTS: Antimony and its compounds are on the Community Right-To-Know List.

OSHA PEL: TWA 0.5 mg(Sb)/m³
ACGIH TLV: TWA 0.5 mg(Sb)/m³
NIOSH REL: TWA 0.5 mg(Sb)/m³

SAFETY PROFILE: Poison by intraperitoneal route. See also ANTIMONY COMPOUNDS. When heated to decomposition it emits toxic fumes of Sb and Na_2O.

AQI250 CAS:16037-91-5 HR: 3
ANTIMONY(V) SODIUM GLUCONATE
mf: $C_{12}H_{20}O_{17}Sb_2$•3Na•9H_2O mw: 1048.91

SYNS: SODIUM ANTIMONY(V) GLUCONATE □ ESTIBOGLUCONATO SODICO □ d-GLUCONIC ACID, CYCLIC ESTER with ANTIMONIC ACID (H8Sb2O9) (2:1),TRISODIUM SALT, NONAHYDRATE □ MYOSTIBIN □ PENTOSTAM □ SODIUM STIBOGLUCONATE □ SOLUSTIBOSAN □ SOLUSTIN □ SOLUSURMIN □ SOLYUSURMIN □ STIBANATE □ STIBANOSE □ STIBATIN □ STIBINOL

TOXICITY DATA WITH REFERENCE
ipr-mus LD50:33 mg/kg CLDND* 11,155,49

CONSENSUS REPORTS: Antimony and its compounds are on the Community Right-To-Know List.

OSHA PEL: TWA 0.5 mg(Sb)/m³
ACGIH TLV: TWA 0.5 mg(Sb)/m³
NIOSH REL: TWA 0.5 mg(Sb)/m³

SAFETY PROFILE: Poison by intraperitoneal route. See also ANTIMONY COMPOUNDS. When heated to decomposition it emits toxic fumes of Sb.

AQI500 CAS:66922-79-0 HR: 3
ANTIMONY SODIUM PROPYLENE DIAMINE TETRAACETIC ACID DIHYDRATE
mf: $C_{11}H_{14}N_2O_8$Sb•Na•2H_2O mw: 483.05

SYNS: PDTA-Sb □ Sb-57

TOXICITY DATA WITH REFERENCE
ipr-mus LD50:131 mg/kg SSINAV 13,789,64

CONSENSUS REPORTS: Antimony and its compounds are on the Community Right-To-Know List.

OSHA PEL: TWA 0.5 mg(Sb)/m³
ACGIH TLV: TWA 0.5 mg(Sb)/m³
NIOSH REL: (Antimony) TWA 0.5 mg(Sb)/m³

SAFETY PROFILE: Poison by intraperitoneal route. See also ANTIMONY COMPOUNDS. When heated to decomposition it emits toxic fumes of NO_x and Na_2O.

AQI750 CAS:34521-09-0 HR: 3
ANTIMONY SODIUM TARTRATE
mf: $C_8H_4O_{12}Sb_2$•2Na mw: 581.60

SYNS: ANTIMONY SODIUM OXIDE-l-(+)-TARTRATE □ NATRIUM-ANTIMONYLTARTRAT (GERMAN) □ SODIUM ANTIMONYL TARTRATE □ SODIUM ANTIMONY TARTRATE □ STIBNAL □ STIBUNAL

TOXICITY DATA WITH REFERENCE
cyt-hmn:leu 2300 pmol/L/48H MUREAV 16,332,72
ivn-hmn TDLo:79 mg/kg/3D-I TEARAI 34(4),62,62
ipr-mus LD50:60 mg/kg PSEBAA 129,284,68
scu-mus LD50:48 mg/kg PSEBAA 129,284,68
ivn-mus LD50:25 mg/kg MEIEDD 10,103,83

CONSENSUS REPORTS: Antimony and its compounds are on the Community Right-To-Know List.

OSHA PEL: TWA 0.5 mg(Sb)/m³
ACGIH TLV: TWA 0.5 mg(Sb)/m³
NIOSH REL: (Antimony) TWA 0.5 mg(Sb)/m³

SAFETY PROFILE: Poison by subcutaneous, intravenous, and intraperitoneal routes. Human toxic effects by intravenous route. Human mutation data reported. See also ANTIMONY COMPOUNDS. When heated to decomposition it emits toxic fumes of Sb.

AQJ250 CAS:7446-32-4 HR: 2
ANTIMONY(III) SULFATE (2:3)
mf: $O_{12}S_3Sb_2$ mw: 531.68

PROP: White powder or colorless deliquescent crystals. Mp: decomp, d: 3.625 @ 4°. Decomposes to basic sulfates by water.

SYNS: ANTIMONOUS SULFATE □ ANTIMONY TRISULFATE □ DIANTIMONY TRISULFATE

CONSENSUS REPORTS: Antimony and its compounds are on the Community Right-To-Know List.

OSHA PEL: TWA 500 µg(Sb)/m³
ACGIH TLV: TWA 0.5 mg(Sb)/m³
NIOSH REL: (Antimony) TWA 0.5 mg(Sb)/m³

SAFETY PROFILE: See ANTIMONY COMPOUNDS and SULFATES. When heated to decomposition it emits very toxic fumes of Sb and SO_x.

AQJ500 HR: 3
ANTIMONY TARTRATE
mf: $C_{12}H_{12}O_{18}$•2Sb mw: 687.74

PROP: White crystals.

SYN: BRECHWEINSTEIN

TOXICITY DATA WITH REFERENCE
orl-rbt LDLo:115 mg/kg UDHU** -,-,37
ims-rbt LD50:90 mg/kg JPETAB 87,119,46

CONSENSUS REPORTS: Antimony and its compounds are on the Community Right-To-Know List.

OSHA PEL: TWA 0.5 mg(Sb)/m³
ACGIH TLV: TWA 0.5 mg(Sb)/m³
NIOSH REL: (Antimony) TWA 0.5 mg(Sb)/m³

SAFETY PROFILE: Poison by intramuscular route. See also ANTIMONY COMPOUNDS. When heated to decomposition it emits toxic fumes of Sb.

AQJ750 **CAS:6923-52-0** **HR: 1**
ANTIMONY TRIACETATE
mf: $C_6H_9O_6$•Sb mw: 298.90

SYNS: ACETIC ACID, TRIANHYDRIDE with ANTIMONIC ACID □ ANTIMONY(III) ACETATE □ OCTAN ANTIMONITY (CZECH)

TOXICITY DATA WITH REFERENCE
skn-rbt 500 mg/24H SEV 28ZPAK -,17,72
eye-rbt 20 mg/24H MOD 28ZPAK-,17,72
mor-ham:emb 3 µmol/L CNREA8 39,193,79
orl-rat LD50:4480 mg/kg 28ZPAK -,17,72

CONSENSUS REPORTS: Reported in EPA TSCA Inventory. Antimony and its compounds are on the Community Right-To-Know List.

OSHA PEL: TWA 0.5 mg(Sb)/m^3
ACGIH TLV: TWA 0.5 mg(Sb)/m^3
NIOSH REL: (Antimony) TWA 0.5 mg(Sb)/m^3

SAFETY PROFILE: Mildly toxic by ingestion. A skin and eye irritant. Mutation data reported. See also ANTIMONY COMPOUNDS. When heated to decomposition it emits acrid smoke and irritating fumes.

AQK000 **CAS:7789-61-9** **HR: 3**
ANTIMONY TRIBROMIDE
DOT: NA 1549
mf: Br_3Sb mw: 361.48

PROP: Yellow or white, deliquescent, crystalline mass becoming amber yellow when fused. Hygroscopic. Decomp by water. Mp: 96°, bp: 280°, d: 4.145, vap press: 1 mm @ 93.9°.

SYNS: ANTIMONY TRIBROMIDE □ ANTIMONY TRIBROMIDE, solid or solution (DOT) □ TRIBROMOSTIBINE

CONSENSUS REPORTS: Antimony and its compounds are on the Community Right-To-Know List. Reported in EPA TSCA Inventory.

OSHA PEL: TWA 0.5 mg(Sb)/m^3
ACGIH TLV: TWA 0.5 mg(Sb)/m^3
DOT CLASSIFICATION: 8; *Label:* Corrosive

SAFETY PROFILE: A poison. Corrosive to skin, eyes, and mucous membranes. Reaction with water liberates HBr and antimony trioxide. Can cause severe burns. See also ANTIMONY COMPOUNDS.

AQK250 **HR: 3**
ANTIMONY TRICHLORIDE OXIDE
mf: Cl_3OSb mw: 244.1

CONSENSUS REPORTS: Antimony and its compounds are on the Community Right-To-Know List.

SAFETY PROFILE: Often a component of violent hazardous materials reactions. See also ANTIMONY COMPOUNDS. When heated to decomposition it emits very toxic fumes of Cl^- and Sb. Incompatible with BF_3.

AQK500 **HR: 3**
ANTIMONY TRIETHYL
mf: $Sb(C_2H_5)_3$ mw: 209.0

PROP: Liquid, water-insol. D: 1.324 @ 16°, mp: −29°, bp: 159.5°.

CONSENSUS REPORTS: Antimony and its compounds are on the Community Right-To-Know List.

SAFETY PROFILE: Alkyl metal compounds are often highly toxic. See also ANTIMONY COMPOUNDS. Dangerous fire hazard by spontaneous chemical reaction. Explodes in air, water, carbon tetrachloride, other halogenated hydrocarbons, dimethyl formamide, and triethyl borine. When heated to decomposition it emits highly toxic fumes of Sb.

AQK750 **HR: 3**
ANTIMONY TRIIODIDE
mf: SbI_3 mw: 502.5

PROP: Red-to-yellow crystals. Mp: 170°, bp: 401°, d: 4.768 @ 22°, vap press: 1 mm @ 163.6°.

CONSENSUS REPORTS: Antimony and its compounds are on the Community Right-To-Know List.

SAFETY PROFILE: Poison by ingestion. See also IODIDES and ANTIMONY COMPOUNDS. Incompatible with sodium, potassium. When heated to decomposition it emits highly toxic Sb fumes and I^-.

AQL000 **HR: 3**
ANTIMONY TRIMETHYL
mf: $Sb(CH_3)_3$ mw: 166.9

PROP: Liquid, sltly sol in water. Bp: 80.6°, d: 1.523 @ 15°

SYN: TRIMETHYL STIBINE

CONSENSUS REPORTS: Antimony and its compounds are on the Community Right-To-Know List.

SAFETY PROFILE: Toxic. See also ANTIMONY TRIETHYL and ANTIMONY COMPOUNDS. Dangerous fire hazard by spontaneous reaction in air. Explodes in water. When heated to decomposition it emits highly toxic fumes of antimony. Incompatible with oxidizing materials, halogenated hydrocarbons.

AQL250 **HR: 3**
ANTIMONY TRIPHENYL
mf: $C_{18}H_{15}Sb$ mw: 353.11

CONSENSUS REPORTS: Antimony and its compounds are on the Community Right-To-Know List.

SAFETY PROFILE: Poison by intraperitoneal and ingestion routes. Upon heating it burns in air. See also ANTIMONY COMPOUNDS. Incompatible with BrF_3.

AQL500 **CAS:1345-04-6** **HR: 3**
ANTIMONY TRISULFIDE
mf: S_3Sb_2 mw: 339.68

PROP: Red-to-black crystals. Mp: 563°, d: 4.64. Sol in H_2SO_4, solubility in water: 0.002/100 @ 20° (decomp).

SYNS: ANTIMONOUS SULFIDE □ ANTIMONY GLANCE □ ANTIMONY ORANGE □ ANTIMONY SESQUISULFIDE □ ANTIMONY SULFIDE □ ANTIMONY TRISULFIDE COLLOID □ ANTIMONY VERMILION □ BLACK ANTIMONY □ C.I. 77060 □ C.I. PIGMENT RED 107 □ CRIMSON ANTIMONY □ DIANTIMONY TRISULFIDE □ LYMPHOSCAN □ NEEDLE ANTIMONY

TOXICITY DATA WITH REFERENCE
ihl-hmn TCLo: 580 μg/m³/35W: BLD,GIT IMSUAI 23,521,54
ipr-rat LDLo: 1390 mg/kg INMEAF 10,15,41
ipr-mus LD50: 209 mg/kg 85GMAT-,23,82

CONSENSUS REPORTS: IARC Cancer Review: Group 3 IMEMDT 47,291,89; Animal Limited Evidence IMEMDT 47,291,89; Human Inadequate Evidence IMEMDT 47,291,89. Reported in EPA TSCA Inventory. Antimony and its compounds are on the Community Right-To-Know List.

OSHA PEL: TWA 500 μg(Sb)/m³
ACGIH TLV: TWA 0.5 mg(Sb)/m³
NIOSH REL: (Antimony) TWA 0.5 mg(Sb)/m³

SAFETY PROFILE: Poison by intraperitoneal route. Human blood and gastrointestinal system effects by inhalation. Questionable carcinogen. See also ANTIMONY COMPOUNDS and SULFIDES. Spontaneously flammable when exposed to strong oxidizers. Flammable when exposed to heat or flame. Moderately explosive by spontaneous reaction with chlorates, perchlorates, ClO, thallic oxide. When heated to decomposition or on contact with acid or acid fumes it emits highly toxic fumes of oxides of sulfur and antimony. Will react with water or steam to produce toxic and flammable vapors.

AQL750 **HR: 3**
ANTIMONY TRITELLURIDE
mf: Sb_2Te_3 mw: 626.4

PROP: Gray powder. Mp: 629°; d: 6.50 @ 13°.

SYN: ANTIMONY TELLURIDE

CONSENSUS REPORTS: Antimony and its compounds are on the Community Right-To-Know List.

SAFETY PROFILE: Probably a poison. See also ANTIMONY COMPOUNDS and TELLURIUM COMPOUNDS. Flammable by spontaneous reaction with strong oxidizers. Moderately explosive by chemical reaction in contact with chlorates and perchlorates. When heated to decomposition or on contact with acid or acid fumes it emits highly toxic fumes of Sb and tellurium. Incompatible with water or steam and oxidizing materials.

AQM000 **CAS:11118-72-2** **HR: 3**
ANTIMYCIN

SYN: FINTROL

TOXICITY DATA WITH REFERENCE
orl-rat LD50: 28 mg/kg TAFSAI 96,320,67
ipr-rat LD50: 1600 μg/kg TAFSAI 96,320,67
orl-mus LD50: 55 mg/kg TAFSAI 96,320,67
ipr-mus LD50: 1700 μg/kg TAFSAI 96,320,67
orl-rbt LD50: 10 mg/kg TAFSAI 96,320,67
orl-gpg LD50: 1800 μg/kg TAFSAI 96,320,67
orl-qal LD50: 39 mg/kg

SAFETY PROFILE: Poison by ingestion and intraperitoneal routes.

AQM250 **CAS:1397-94-0** **HR: 3**
ANTIMYCIN A
mf: $C_{28}H_4N_2O_9$ mw: 512.34

SYNS: ANTIPIRICULLIN □ VIROSIN

TOXICITY DATA WITH REFERENCE
orl-rat LDLo: 30 mg/kg 85ERAY 2,1078,78
ipr-rat LD50: 800 μg/kg CNREA8 13,49,53
scu-rat LD50: 25 mg/kg 85ERAY 2,1078,78
ipr-mus LD50: 820 μg/kg TDKNAF 14,60,55
scu-mus LD50: 21 mg/kg JAJAAA 9,63,56
ivn-mus LD50: 893 μg/kg JAJAAA 9,63,56

CONSENSUS REPORTS: EPA Extremely Hazardous Substances List.

SAFETY PROFILE: Poison by ingestion, intraperitoneal, subcutaneous, and intravenous routes. When heated to decomposition it emits toxic fumes of NO_x.

AQM260 **CAS:27220-59-3** **HR: 3**
ANTIMYCIN A4
mf: $C_{25}H_{34}N_2O_9$ mw: 506.61

TOXICITY DATA WITH REFERENCE
ipr-mus LD50: 7600 μg/kg 85FZAT -,146,67
scu-mus LD50: 25 mg/kg 85FZAT -,146,67
ivn-mus LD50: 900 μg/kg 85FZAT -,146,67

SAFETY PROFILE: Poison by subcutaneous, intravenous, and intraperitoneal routes. When heated to decomposition it emits toxic fumes of NO_x.

AQM500 **HR: 3**
ANTIMYCOIN

PROP: An antifungal agent produced by the strain *Streptomyces aureus* 3569 (85ERAY 2,959,78).

TOXICITY DATA WITH REFERENCE
ipr-mus LD50: 204 mg/kg 85ERAY 2,959,78
scu-mus LD50: 532 mg/kg 85ERAY 2,959,78

SAFETY PROFILE: Poison by intraperitoneal route. Moderately toxic by subcutaneous route.

AQN000 **CAS:60-80-0** **HR: 3**
ANTIPYRINE
mf: $C_{11}H_{12}N_2O$ mw: 188.23

PROP: Fine, white crystals, leaflets, or scales. Mp: 113°, bp: 319° @ 174 mm, d: 1.19. Very sol in water and alc; sltly sol in ether.

SYNS: DIMETHYLOXYQUINAZINE □ 2,3-DIMETHYL-1-PHENYL-3-PYRAZOLIN-5-ONE □ 2,3-DIMETHYL-1-PHENYL-5-PYRAZOLONE □ OXYDIMETHYLQUINAZINE □ PHENAZONE (pharmaceutical) □ 1-PHENYL-2,3-DIMETHYLPYRAZOLE-5-ONE □ 1-PHENYL-2,3-DIMETHYL-5-PYRAZOLONE

TOXICITY DATA WITH REFERENCE
dni-hmn:hlas 80 mmol/L CRNGDP 13,2389,92
orl-rat TDLo:361 g/kg/92W-C:ETA IJCNAW 27,521,81
unk-man LDLo:74 mg/kg 85DCAI 2,73,70
orl-rat LD50:1705 mg/kg ARZNAD 9,401,59
scu-rat LDLo:1570 mg/kg AEPPAE 186,195,37
orl-mus LD50:1310 mg/kg JAPMA8 45,137,56
ipr-mus LD50:750 mg/kg AIPTAK 135,376,62
scu-mus LD50:1000 mg/kg ARZNAD 17,214,67
ivn-mus LD50:500 mg/kg ARZNAD 10,686,60
orl-dog LDLo:500 mg/kg HBAMAK 4,1304,35

CONSENSUS REPORTS: Reported in EPA TSCA Inventory.

SAFETY PROFILE: A human poison by an unspecified route. Moderately toxic via ingestion, subcutaneous, and intravenous routes. Questionable carcinogen with experimental tumorigenic data. Mutation data reported. When heated to decomposition it emits toxic fumes of NO_x.

AQN250 CAS:520-07-0 HR: 2
ANTIPYRINE SALICYLATE
mf: $C_{11}H_{12}N_2O \cdot C_7H_6O_3$ mw: 326.38

PROP: Solid. Mp: 91–92°.

SYNS: ANSAL □ SALAZOLON □ SALIPHENAZON □ SALIPYRAZOLAN □ SALIPYRINE

TOXICITY DATA WITH REFERENCE
unk-mus LDLo:1200 mg/kg HBAMAK 4,1289,35
unk-gpg LDLo:1600 mg/kg HBAMAK 4,1289,35

SAFETY PROFILE: Moderately toxic by unspecified routes. When heated to decomposition it emits toxic fumes of NO_x. An analgesic and antipyretic.

AQN500 CAS:15387-10-7 HR: 3
N-((ANTIPYRINYLISOPROPYLAMINO)METHYL)NICOTINAMIDE
mf: $C_{21}H_{25}N_5O_2$ mw: 379.51

PROP: Solid. Mp: 165–166°.

SYN: NICOTINAMIDOMETHYLAMINOPYRAZOLONE

TOXICITY DATA WITH REFERENCE
ipr-rat LD50:400 mg/kg ARZNAD 20,1024,70
orl-mus LD50:1460 mg/kg ARZNAD 20,1024,70
ipr-mus LD50:1060 mg/kg ARZNAD 20,1024,70
ipr-ham LD50:853 mg/kg ARZNAD 20,1024,70

SAFETY PROFILE: Poison by intraperitoneal route. Moderately toxic by ingestion and intraperitoneal routes. When heated to decomposition it emits toxic fumes of NO_x.

AQN625 CAS:3270-78-8 HR: 3
ANTRYCIDE METHYL SULFATE
mf: $C_{17}H_{21}N_6 \cdot CH_4O_4S \cdot CH_3O_4S$ mw: 532.65

PROP: Creamy-white crystals from MeOH (aq). Mp: 265–266°. Freely sol in water.

SYNS: 4-AMINO-6-((2-AMINO-1,6-DIMETHYLPYRIMIDINIUM-4-YL)AMINO)-1-METHYL-QUINALDINIUM BIS(METHYL SULFATE) □ ANTRYCIDE □ QUINAPYRAMINE □ QUINAPYRAMINE METHYL SUFLATE

TOXICITY DATA WITH REFERENCE
dns-omi 50 μmol/L CNREA8 45,112,85
scu-rat LD50:18 mg/kg BJPCAL 5,25,50
ipr-mus LD50:15 mg/kg BJPCAL 5,25,50
scu-mus LD50:20 mg/kg BJPCAL 5,25,50
ivn-mus LD50:10 mg/kg MEIEDD 10,1162,83
scu-rbt LD50:15 mg/kg BJPCAL 5,25,50
ivn-rbt LD50:5 mg/kg BJPCAL 5,25,50

SAFETY PROFILE: Poison by subcutaneous, intravenous, and intraperitoneal routes. Mutagenic data. When heated to decomposition it emits toxic fumes of NO_x and SO_x.

AQN635 CAS:86-88-4 HR: 3
ANTU
DOT: UN 1651
mf: $C_{11}H_{10}N_2S$ mw: 202.29

PROP: Crystals or prisms with bitter taste. Mp: 198°. Sltly sol in H_2O, sol in Me_2CO.

SYNS: ALPHANAPHTHYL THIOUREA □ ALPHANAPHTYL THIOUREE (FRENCH) □ ALRATO □ ANTURAT □ CHEMICAL 109 □ DIRAX □ KILL KANTZ □ KRYSID □ 1-NAFTIL-TIOUREA (ITALIAN) □ 1-NAFTYLTHIOUREUM (DUTCH) □ 1-NAPHTHALENYLTHIOUREA □ α-NAPHTHALTHIOHARNSTOFF (GERMAN) □ α-NAPHTHOTHIOUREA □ α-NAPHTHYLTHIOCARBAMIDE □ 1-NAPHTHYL-THIOHARNSTOFF (GERMAN) □ α-NAPHTHYLTHIOUREA □ 1-NAPHTHYL THIOUREA (MAK) □ 1-(1-NAPHTHYL)-2-THIOUREA □ N-(1-NAPHTHYL)-2-THIOUREA □ α-NAPHTHYLTHIOUREA (DOT) □ 1-NAPHTHYL-THIOUREE (FRENCH) □ NAPHTOX □ RATTRACK □ RCRA WASTE NUMBER P072 □ SMEESANA □ U-5227 □ USAF EK-P-5976

TOXICITY DATA WITH REFERENCE
mma-sat 500 μmol/L ENMUDM 3,11,81
otr-ham:emb 1600 μg/L NCIMAV 58,243,81
scu-mus TDLo:5 mg/kg:ETA NTIS** PB223-159
unr-man LDLo:588 mg/kg 85DCAI 2,73,70
orl-rat LD50:6 mg/kg AFDOAQ 16,47,52
ipr-rat LD50:2470 μg/kg JPETAB 97,432,49
ipr-mus LD50:10 mg/kg NTIS** AD 277-689
orl-dog LD50:380 μg/kg PCOC** -,57,66
ipr-dog LD50:16 mg/kg PSEBAA 62,22,46
orl-mky LD50:4250 mg/kg 85DPAN -,-,71/76

CONSENSUS REPORTS: IARC Cancer Review: Animal Inadequate Evidence IMEMDT 30,347,83. Reported in EPA TSCA Inventory. EPA Extremely Hazardous Substances List. EPA Genetic Toxicology Program.

OSHA PEL: TWA 0.3 mg/m³
ACGIH TLV: TWA 0.3 mg/m³
DFG MAK: 0.3 mg/m³
DOT CLASSIFICATION: 6.1; *Label:* Poison

SAFETY PROFILE: Poison by ingestion and intraperito-

neal routes. Moderately toxic to humans by an unspecified route. Questionable carcinogen with experimental tumorigenic data. Mutagenic data. A rodenticide used extensively. Death is caused by pulmonary edema. Chronic toxicity has been known to cause dermatitis and a decrease in the white blood cells. When heated to decomposition it emits toxic fumes of NO_x and SO_x.

AQN650 **CAS:24345-16-2** **HR: 3**
APAMINE
mf: $C_{79}H_{131}O_{24}S_4$ mw: 2027.65

PROP: Highly basic compd.

SYN: APAMIN

TOXICITY DATA WITH REFERENCE
ipr-mus LD50:3800 µg/kg TOXIA6 22,308,84
ivn-mus LD50:4 mg/kg NSAPCC 300,189,77
ice-mus LD50:1800 ng/kg TOXIA6 22,308,84
par-mus LD50:600 mg/kg TOXIA6 20,157,82

SAFETY PROFILE: Poison by intravenous, parenteral, intracerebral, and intraperitoneal routes. When heated to decomposition it emits toxic fumes of SO_x and NO_x.

AQN750 **CAS:13539-59-8** **HR: 2**
APAZONE
mf: $C_{16}H_{20}N_4O_2$ mw: 300.40

PROP: Solid. Mp: 228°.

SYNS: AZAPROPAZON (GERMAN) □ AZAPROPAZONE (anhydrous) □ CINNAMIN □ CINNOPROPAZONE □ 1,2-DIHYDRO-3-DIMETHYLAMINO-7-METHYL-1,2-(PROPYLMALONYL)-1,2,4-BENZOTRIAZINE □ 3-DIMETHYLAMINO-7-METHYL-1,2-(n-PROPYLMALONYL)-1,2-DIHYDRO-1,2,4-BENZOTRIAZINE □ 5-DIMETHYLAMINO-9-METHYL-2-PROPYL-1H PYRAZOLO(1,2-a)(1,2,4)BENZOTRIAZINE-1,3(2H)-DIONE □ MI 85 □ MSC-102824 □ PROLIXAN □ RHEUMOX □ SINNAMIN

TOXICITY DATA WITH REFERENCE
orl-rat LD50:1800 mg/kg OYYAA2 15,41,78
ipr-rat LD50:650 mg/kg CMROCX 4,17,76
ivn-rat LD50:660 mg/kg CMROCX 4,17,76
orl-mus LD50:1080 mg/kg CMROCX 4,17,76
ipr-mus LD50:920 mg/kg CMROCX 4,17,76
ivn-mus LD50:680 mg/kg CMROCX 4,17,76
ivn-cat LD50:500 mg/kg CMROCX 4,17,76

SAFETY PROFILE: Moderately toxic by ingestion, intraperitoneal and intravenous routes. When heated to decomposition it emits toxic fumes of NO_x.

AQO000 **CAS:52-46-0** **HR: 3**
APHOLATE
mf: $C_{12}H_{24}N_9P_3$ mw: 387.36

PROP: Orthorhombic needles from *m*-xylene; monoclinic crystals from CS_2. Mp: 150°. Very sol in water.

SYNS: APN □ AZIRIDINE-1,3,5,2,4,6-TRIAZATRIPHOSPHORINE DERIVATIVE □ 1-AZIRIDINYLPHOSPHONITRILE TRIMER □ ENT 26,316 □ HEXA(1-AZIRIDINYL)TRIPHOSPHOTRIAZINE □ 2,2,4,4,6,6-HEXAHYDRO-2,2,4,4,6,6-HEXAKIS(1-AZIRIDINYL)-1,3,5,2, 4,6-TRIAZATRIPHOSPHORINE □ 2,2,4,4,6,6-HEXAKIS(1-AZIRIDINYL)CYCLOTRIPHOSPHAZA-1,3,5-TRIENE □ 2,2,4,4,6,6-HEXAKIS(1-AZIRIDINYL)-2,2,4,4,6,6-HEXAHYDRO-1,3,5,2,4,6-TRIAZATRIPHOSPHORINE □ HEXAKIS-(1-AZIRIDINYL)PHOSPHONITRILE □ HEXAKIS(AZIRIDINYL)PHOSPHOTRIAZINE □ NSC-26812 □ OLIN MO. 2174 □ PHOLATE □ PN6 □ SQ 8388

TOXICITY DATA WITH REFERENCE
cyt-mus-ipr 15 mg/kg PISCAD 59(Pt.3),417,72
cyt-hmn:leu 100 µmol/L CHROAU 24,314,68
ipr-rat TDLo:10 mg/kg (11D preg):TER AEHLAU 16,805,68
ipr-mus TDLo:5 mg/kg (male 5D pre):REP EXPEAM 24,924,68
orl-rat LD50:98 mg/kg TXAPA9 14,515,69
orl-mus LD50:110 mg/kg JAFCAU 14,301,66
ipr-mus LD50:50 mg/kg JMCMAR 29,1341,86

CONSENSUS REPORTS: IARC Cancer Review: Group 3 IMEMDT 7,56,87; Animal Inadequate Evidence IMEMDT 9,31,75. EPA Genetic Toxicology Program.

SAFETY PROFILE: Poison by ingestion and intramuscular routes. An experimental teratogen. Other experimental reproductive effects. Human mutagenic data. When heated to decomposition it emits very toxic fumes of NO_x and PO_x.

AQO250 **CAS:500-55-0** **HR: 3**
APOATROPINE
mf: $C_{17}H_{21}NO_2$ mw: 271.39

PROP: Crystals from Et_2O. Mp: 60–62°.

SYNS: APOATROPIN □ ATROPAMIN □ ATROPAMINE □ ATROPYLTROPEINE □ endo-α-METHYLENEBENZENEACETIC ACID 8-METHYL-8-AZABICYCLO(3.2.1)OCT-3-YL ESTER □ 1-α-H,5-α-H-TROPAN-3-α-OL, ATROPATE (ESTER) □ TROPIC ACID, 3-α-TROPANYL ESTER □ TROPINE, ATROPATE (ESTER)

TOXICITY DATA WITH REFERENCE
orl-mus LD50:160 mg/kg MEIEDD 11,117,89
ipr-mus LD50:10,400 µg/kg NTIS** PB85-203544

SAFETY PROFILE: Poison by ingestion and intraperitoneal routes. See also ESTERS. When heated to decomposition it emits toxic fumes of NO_x. An antispasmodic agent.

AQO500 **CAS:641-81-6** **HR: 2**
APOCHOLIC ACID
mf: $C_{24}H_{38}O_4$ mw: 390.62

PROP: Crystals. Mp: 176–177°.

SYN: 3-α,12-α-DIHYDROXY-5-β-CHOL-8(14)-EN-24-OIC ACID

TOXICITY DATA WITH REFERENCE
scu-mus TDLo:200 mg/kg/13W-I:ETA NATUAS 190,1007,61

SAFETY PROFILE: Questionable carcinogen with experimental tumorigenic data. When heated to decomposition it emits acrid smoke and irritating fumes.

AQO750 **HR: 3**
APOCODEINE
mf: $C_{18}H_{19}NO_2$ mw: 281.34

PROP: White, crystalline solid. Mp: 124°.

SAFETY PROFILE: Poison by inhalation and ingestion. A weak sensitizer and may cause contact dermatitis. See also CODEINE. When heated to decomposition it emits highly toxic fumes of NO_x.

AQP000 **CAS:1937-37-7** **HR: 3**
APOMINE BLACK GX
mf: $C_{34}H_{25}N_9O_7S_2 \cdot 2Na$ mw: 781.78

SYNS: AHCO DIRECT BLACK GX □ AIREDALE BLACK ED □ AIZEN DIRECT DEEP BLACK EH □ AIZEN DIRECT DEEP BLACK GH □ AIZEN DIRECT DEEP BLACK RH □ AMANIL BLACK GL □ AMANIL BLACK WD □ ATLANTIC BLACK BD □ ATLANTIC BLACK C □ ATLANTIC BLACK E □ ATLANTIC BLACK EA □ ATLANTIC BLACK GAC □ ATLANTIC BLACK GG □ ATLANTIC BLACK GXCW □ ATLANTIC BLACK GXOO □ ATLANTIC BLACK SD □ ATUL DIRECT BLACK E □ AZINE DEEP BLACK EW □ AZOCARD BLACK EW □ AZOMINE BLACK EWO □ BELAMINE BLACK GX □ BENCIDAL BLACK E □ BENZAMIL BLACK E □ BENZO DEEP BLACK E □ BENZO LEATHER BLACK E □ BENZOFORM BLACK BCN-CF □ BLACK 2EMBL □ BLACK 4EMBL □ BRASILAMINA BLACK GN □ BRILLIANT CHROME LEATHER BLACK H □ CALCOMINE BLACK □ CALCOMINE BLACK EXL □ CARBIDE BLACK E □ CERN PRIMA 38 □ CHLORAMINE BLACK C □ CHLORAMINE BLACK EC □ CHLORAMINE BLACK ERT □ CHLORAMINE BLACK EX □ CHLORAMINE BLACK EXR □ CHLORAMINE BLACK XO □ CHLORAMINE CARBON BLACK S □ CHLORAMINE CARBON BLACK SJ □ CHLORAMINE CARBON BLACK SN □ CHLORAZOL BLACK E □ CHLORAZOL BLACK E (BIOLOGICAL STAIN) □ CHLORAZOL BLACK EA □ CHLORAZOL BLACK EN □ CHLORAZOL BURL BLACK E □ CHLORAZOL LEATHER BLACK ENP □ CHLORAZOL SILK BLACK G □ CHROME LEATHER BLACK E □ CHROME LEATHER BLACK EC □ CHROME LEATHER BLACK EM □ CHROME LEATHER BLACK G □ CHROME LEATHER BRILLIANT BLACK ER □ C.I. 30235 □ C.I. DIRECT BLACK 38 □ C.I. DIRECT BLACK 38, DISODIUM SALT □ COIR DEEP BLACK C □ COLUMBIA BLACK EP □ DIACOTTON DEEP BLACK □ DIACOTTON DEEP BLACK RX □ DIAMINE DEEP BLACK EC □ DIAMINE DIRECT BLACK E □ DIAPHTAMINE BLACK V □ DIAZINE BLACK E □ DIAZINE DIRECT BLACK E □ DIAZINE DIRECT BLACK G □ DIAZOL BLACK 2V □ DIPHENYL DEEP BLACK G □ DIRECT BLACK A □ DIRECT BLACK BRN □ DIRECT BLACK CX □ DIRECT BLACK CXR □ DIRECT BLACK E □ DIRECT BLACK EW □ DIRECT BLACK EX □ DIRECT BLACK FR □ DIRECT BLACK GAC □ DIRECT BLACK GW □ DIRECT BLACK GX □ DIRECT BLACK GXR □ DIRECT BLACK JET □ DIRECT BLACK META □ DIRECT BLACK METHYL □ DIRECT BLACK N □ DIRECT BLACK RX □ DIRECT BLACK SD □ DIRECT BLACK WS □ DIRECT BLACK Z □ DIRECT BLACK 3 □ DIRECT BLACK 38 □ DIRECT DEEP BLACK E □ DIRECT DEEP BLACK E EXTRA □ DIRECT DEEP BLACK EA-CF □ DIRECT DEEP BLACK EAC □ DIRECT DEEP BLACK EW □ DIRECT DEEP BLACK EX □ ENIANIL BLACK CN □ ERIE BLACK B □ ERIE BLACK BF □ ERIE BLACK GAC □ ERIE BLACK GXOO □ ERIE BLACK JET □ ERIE BLACK NUG □ ERIE BLACK RXOO □ ERIE BRILLIANT BLACK S □ ERIE FIBRE BLACK VP □ FENAMIN BLACK E □ FIBRE BLACK VF □ FIXANOL BLACK E □ FORMALINE BLACK C □ FORMIC BLACK C □ FORMIC BLACK CW □ FORMIC BLACK BA □ FORMIC BLACK MTG □ FORMIC BLACK TG □ HISPAMIN BLACK EF □ INTERCHEM DIRECT BLACK Z □ KAYAKU DIRECT DEEP BLACK EX □ KAYAKU DIRECT DEEP BLACK GX □ KAYAKU DIRECT DEEP BLACK S □ KAYAKU DIRECT LEATHER BLACK EX □ KAYAKU DIRECT SPECIAL BLACK AAX □ LURAZOL BLACK BA □ META BLACK □ MITSUI DIRECT BLACK EX □ MITSUI DIRECT BLACK GX □ NCI-C54557 □ NIPPON DEEP BLACK □ NIPPON DEEP BLACK GX □ PAPER BLACK BA □ PAPER BLACK T □ PAPER DEEP BLACK C □ PARAMINE BLACK B □ PARAMINE BLACK E □ PEERAMINE BLACK E □ PEERAMINE BLACK GXOO □ PHENAMINE BLACK BCN-CF □ PHENAMINE BLACK CL □ PHENAMINE BLACK E □ PHENAMINE BLACK E 200 □ PHENO BLACK EP □ PHENO BLACK SGN □ PONTAMINE BLACK E □ PONTAMINE BLACK EBN □ SANDOPEL BLACK EX □ SERISTAN BLACK B □ TELON FAST BLACK E □ TETRAZO DEEP BLACK G □ TERTRODIRECT BLACK E □ TETRODIRECT BLACK EFD □ UNION BLACK EM □ VONDACEL BLACK N

TOXICITY DATA WITH REFERENCE
eye-rbt 100 mg MOD TSCAT* OTS 215154
mmo-sat 10 μg/plate TOLED5 4,519,79
bfa-rat:sat 300 mg/kg IAEHDW 49,177,81
orl-mus TDLo:5 g/kg (female 8-12D post):REP FAATDF 20,177,93
orl-rat TDLo:6825 mg/kg/13W-C:CAR NCITR* NCI-TR-108,78
orl-mus TDLo:34 mg/kg/60W-C:ETA TJIDAH 88,467,73
orl-rat LD:13,286 mg/kg/13W-C:CAR TXAPA9 54,431,80
orl-rat LD50:7600 mg/kg TSCAT* OTS 215154
ihl-rat LCLo:180 g/m³/1H TSCAT* OTS 215154
orl-rbt LDLo:1262 mg/kg TSCAT* OTS 215154

CONSENSUS REPORTS: NTP 7th Annual Report On Carcinogens. IARC Cancer Review: Animal Sufficient Evidence IMEMDT 29,295,82, Human Limited Evidence IMEMDT 29,295,82. Reported in EPA TSCA Inventory. NTP Carcinogenesis Bioassay (feed): Clear Evidence: rat NCICTR* NCI-TR-108,78; No Evidence: mouse NCICTR NCI-TR-108,78. On Community-Right-To-Know List.

SAFETY PROFILE: Confirmed carcinogen with carcinogenic and tumorigenic data. Moderately toxic by ingestion and inhalation. An eye irritant. Mutation data reported. When heated to decomposition it emits very toxic fumes of NO_x, Na_2O, and SO_2.

For occupational chemical analysis use NIOSH: Dyes, 5013.

AQP250 **CAS:58-00-4** **HR: 3**
APORMORPHINE
mf: $C_{17}H_{17}NO_2$ mw: 267.35

PROP: White, crystalline alkaloid. Mp: 195° (decomp). Sltly sol in water.

SYNS: APOMORFIN □ APOMORPHINE □ 6A-β-APORPHINE-10,11-DIOL

TOXICITY DATA WITH REFERENCE
scu-rat TDLo:7 mg/kg (female 13-19D post):REP TJADAB 33,100C,86
ivn-rat LDLo:40 mg/kg ARZNAD 10,1003,60
orl-mus LD50:300 mg/kg FRPSAX 35,951,80
ipr-mus LD50:160 mg/kg JMCMAR 15,348,72
scu-mus LDLo:13 mg/kg HBAMAK 4,1289,35
ivn-mus LD50:56 mg/kg CSLNX* NX#03170

SAFETY PROFILE: Poison by ingestion, subcutaneous, intravenous, and intraperitoneal routes. Experimental reproductive effects. Central nervous system effects. A powerful emetic. A weak sensitizer and may cause contact dermatitis. When heated to decomposition it emits highly toxic fumes of NO_x.

AQP500 **CAS:314-19-2** **HR: 3**
APORMORPHINE CHLORIDE
mf: $C_{17}H_{17}NO_2 \cdot ClH$ mw: 303.81

SYNS: 6A-β-APORMPHINE-10,11-DIOL HYDROCHLORIDE □ N-METHYLNORAPORMORPHINE HYDROCHLORIDE

TOXICITY DATA WITH REFERENCE
mmo-sat 20 μg/plate MUREAV 137,17,84
ipr-mus LD50:145 μg/kg JMCMAR 18,1194,75
ivn-mus LD50:38 mg/kg TXAPA9 6,334,64

CONSENSUS REPORTS: Reported in EPA TSCA Inventory.

SAFETY PROFILE: Poison by intravenous and intraperitoneal routes. Mutation data reported. See also APORMORPHINE. When heated to decomposition it emits very toxic fumes of NO_x and HCl.

AQP750 **CAS:4361-80-2** **HR: 3**
APOTHESINE
mf: $C_{16}H_{23}NO_2$ mw: 261.40

SYN: CINNAMIC ACID 3-(DIETHYLAMINO) PROPYL ESTER

TOXICITY DATA WITH REFERENCE
ivn-rat LDLo:20 mg/kg AJPHAP 68,120,24
ipr-mus LDLo:700 mg/kg JLCMAK 11,1082,26
scu-mus LDLo:700 mg/kg JLCMAK 11,1082,26
ivn-cat LDLo:20 mg/kg AJPHAP 68,120,24
scu-gpg LDLo:250 mg/kg JLCMAK 11,1082,26

SAFETY PROFILE: Poison by intravenous and subcutaneous routes. Moderately toxic by intraperitoneal route. When heated to decomposition it emits toxic fumes of NO_x.

AQP800 **HR: 3**
APPLE of SODOM (extract)

SYNS: BEC 001 □ SOLANUM SODOMEUM, extract

TOXICITY DATA WITH REFERENCE
unr-rat LD50:41 mg/kg EPXXDW #20029
orl-mus LD50:550 mg/kg EPXXDW #20029
ipr-mus LD50:30 mg/kg EPXXDW #20029

SAFETY PROFILE: Poison by intraperitoneal and possibly other routes. Moderately toxic by ingestion.

AQP875 **HR: 3**
APPLE SEEDS

PROP: Seeds of a deciduous tree widely cultivated in many temperate regions. The fruit is commonly available.

SYNS: MALUS (VARIOUS SPECIES) □ MANZANA (SPANISH) □ POMMIER (FRENCH)

SAFETY PROFILE: The seeds contain a cyanogenetic glycoside. Ingestion of small quantities of the seeds is harmless. After a delay period systemic effects from ingestion may include abdominal pain, vomiting, lethargy, and coma. An adult died from chewing and swallowing a cup of seeds. See also CYANIDE.

AQP890 **HR: 3**
APRICOT PITS

PROP: Fruit bearing trees and shrubs widely cultivated in temperate regions. Various varieties of fruit are commonly available.

SYNS: ALBARICOQUE (SPANISH) □ CEREZA (SPANISH) □ CHERRY □ CHOKE CHERRY □ MELOCOTON (SPANISH) □ PEACH □ PLUM □ PRUNUS (VARIOUS SPECIES) □ SLOE

SAFETY PROFILE: The seed kernel contains cyanogenetic glycosides. After a delay period systemic effects from ingestion may include abdominal pain, vomiting, lethargy, and coma. Most fatalities result from ingestion of apricot pits or their products. See also CYANIDE.

AQQ000 **CAS:50650-74-3** **HR: 3**
APTROL SULFATE
mf: $C_{10}H_{15}N \cdot 1/2H_2O_4S$ mw: 198.30

SYNS: 4-METHYLPHENISOPROPYLAMINE SULFATE □ 1-(4-METHYLPHENYL)-2-PROPYLAMINE SULFATE □ β-p-TOLYL-ISOPROPYLAMINE SULFATE □ p-XYLYLMETHYLCARBINAMINE SULFATE

TOXICITY DATA WITH REFERENCE
orl-man TDLo:1500 μg/kg:GIT JPETAB 100,298,50
ipr-mus LD50:136 mg/kg JPETAB 100,298,50

SAFETY PROFILE: Poison by intraperitoneal route. Human gastrointestinal tract effects by ingestion. See also SULFATES. When heated to decomposition it emits very toxic fumes of SO_x and NO_x.

AQQ100 **CAS:17168-82-0** **HR: 3**
AQUA-1,2-DIAMINOETHANE DIPEROXO CHROMIUM (IV)
mf: $C_2H_{10}CrN_2O_5$ mw: 194.11

$Cr(C_2H_{10}N_2)(O_2)_2 \cdot H_2O$

CONSENSUS REPORTS: Chromium and its compounds are on the Community Right-To-Know List.

SAFETY PROFILE: A poison. A light-sensitive explosive which may explode when heated above 96°C. When heated to decomposition it emits toxic fumes of NO_x. See also CHROMIUM COMPOUNDS and PEROXIDES.

AQQ125 **CAS:17185-68-1** **HR: 3**
AQUA-1,2-DIAMINOPROPANEDIPEROXOCHROMIUM (IV) DIHYDRATE
mf: $C_3H_{12}CrN_2O_5 \cdot 2H_2O$ mw: 244.16

CONSENSUS REPORTS: Chromium and its compounds are on the Community Right-To-Know List.

OSHA PEL: CL 0.1 $mg(CrO_3)/m^3$
ACGIH TLV: TWA 0.05 $mg(Cr)/m^3$; Confirmed Human Carcinogen.
NIOSH REL: (Chromium(VI)) TWA 25 $\mu g(Cr(VI))/m^3$; CL 50 $\mu g/m^3$/15M

SAFETY PROFILE: A confirmed carcinogen. May explode spontaneously at room temperature. Upon decomposition it emits toxic fumes of NO_x. See also CHROMIUM COMPOUNDS.

AQQ250 **CAS:6091-11-8** **HR: 3**
AR-45
mf: $C_{15}H_{21}N_3O_2 \cdot CH_3I$ mw: 417.33

TOXICITY DATA WITH REFERENCE
orl-mus LDLo: 250 mg/kg JPETAB 43,413,31
ivn-mus LDLo: 750 μg/kg JPETAB 43,413,31

SAFETY PROFILE: Poison by ingestion and intravenous routes. When heated to decomposition it emits very toxic fumes of NO_x and I^-.

AQQ500 **CAS:9000-01-5** **HR: 2**
ARABIC GUM
mw: 240,000

PROP: Yellowish-amber lumps. A gum from the stems and branches of *Acacia senegal (L.) Willd.* or of *Acacia* (Fam. Leguminosae). Sol in water; insol in alc.

SYNS: ACACIA ☐ ACACIA DEALBATA GUM ☐ ACACIA GUM ☐ ACACIA SENEGAL ☐ ACACIA SYRUP ☐ AUSTRALIAN GUM ☐ GUM ARABIC ☐ GUM OVALINE ☐ GUM SENEGAL ☐ INDIAN GUM ☐ NCI-C50748 ☐ SENEGAL GUM ☐ STARSOL No. 1 ☐ WATTLE GUM

TOXICITY DATA WITH REFERENCE
eye-rbt 36 mg/5H SEV AROPAW 78,384,67
dlt-rat-orl 54,600 mg/kg/10W-C ENMUDM 8,357,86
orl-rat TDLo: 350 g/kg (male 10W pre): REP ENMUDM 8,357,86
orl-rat LD50: >16 g/kg FDRLI* 124,-,76
orl-mus LD50: >16 g/kg FDRLI* 124,-,76
orl-rbt LD50: 8 g/kg FDRLI* 124,-,76
orl-ham LD50: >18 g/kg FDRLI* 124,-,76

CONSENSUS REPORTS: NTP Carcinogenesis Bioassay (feed); No Evidence: mouse, rat NTPTR* NTP-TR-227,82. Reported in EPA TSCA Inventory.

SAFETY PROFILE: Very low toxicity by ingestion. Inhalation or ingestion has produced hives, eczema, and angiodema. Experimental reproductive effects. A severe eye irritant. A weak allergen. Mutation data reported. Combustible. When heated to decomposition it emits acrid smoke.

AQQ750 **CAS:147-94-4** **HR: 2**
ARABINOCYTIDINE
mf: $C_9H_{13}N_3O_5$ mw: 243.25

PROP: Prisms from EtOH (aq). Mp: 212–213°.

SYNS: AC-1075 ☐ ALEXAN ☐ 4-AMINO-1-ARABINOFURANOSYL-2-OXO-1,2-DIHYDROPYRIMIDIN ☐ 4-AMINO-1-ARABINOFURANOSYL-2-OXO-1,2-DIHYDROPYRIMIDINE ☐ ARABINOCYTIDINE ☐ 4-AMINO-1-β-D-ARABINOFURANOSYL-2(1H)-PYRIMIDINON ☐ 4-AMINO-1-β-D-ARABINOFURANOSYL-2(1H)-PYRIMIDINONE ☐ 1-β-D-ARABINOFURANOSYL-4-AMINO-2(1H)PYRIMIDINONE ☐ 1-ARABINOFURANOSYLCYTOSINE ☐ 1-β-ARABINOFURANOSYLCYTOSINE ☐ 1-(β-D-ARABINOFURANOSYL)CYTOSINE ☐ β-D-ARABINOSYLCYTOSINE ☐ ARABITIN ☐ ARA-C ☐ ARACTIDINE ☐ ARA-CYTIDINE ☐ ARACYTIN ☐ CITARABINA ☐ CYLOCIDE ☐ CYTARABIN ☐ CYTARABINA ☐ CYTARABINE ☐ CYTARABINOSIDE ☐ CYTOSAR ☐ CYTOSAR-U ☐ CYTOSINEARABINOSIDE ☐ CYTOSINE-β-ARABINOSIDE ☐ CYTOSINE β-D-ARABINOSIDE ☐ CYTOSINE, 1-β-D-ARABINOSYL- ☐ IRETIN ☐ NCI-C04728 ☐ NSC 63878 ☐ 2(1H)-PYRIMIDINONE, 4-AMINO-1-β-D-ARABINOFURANOSYL-(9CI) ☐ SPONGOCYTIDINE ☐ U-19,920 ☐ U-19920 A ☐ UDICIL

TOXICITY DATA WITH REFERENCE
skn-hmn 45 mg/3W CTRRDO 63,619,79
eye-hmn 105 mg/7D-I AROPAW 72,535,64
cyt-hmn: leu 50 μmol/L/6H ECREAL 46,276,67
cyt-hmn: lym 3 mg/L/4H SOGEBZ 12,1552,76
mnt-ham-ipr 2 mg/kg/24H MUREAV 40,325,76
ipr-mus TDLo: 50 mg/kg (female 12D post): REP TCMUD8 7,7,87
ipr-rat TDLo: 100 mg/kg (female 11D post): TER DEBIAO 45,103,75
ipr-rat TDLo: 2500 mg/kg/7W-I: ETA CANCAR 40,1935,77
scu-man TDLo: 60 mg/kg/90W-I: EAR,BLD DICPBB 21,798,87
scu-wmn TDLo: 6480 μg/kg/12D-I: BLD NEJMAG 310,1328,84
ivn-chd TDLo: 33,200 μg/kg/240D-I: CNS CANCAR 42,53,78
ivn-hmn TDLo: 17,241 mg/kg/6D-I: SKN AIMEAS 102,556,85
ivn-wmn TDLo: 720 mg/kg/3D-I: CNS NEURAI 35,1475,85
ivn-man TDLo: 649 mg/kg/4D-I: PNS DICPBB 21,177,87
ivn-man LDLo: 1536 mg/kg/43W-I: PNS DICPBB 21,177,87
ivn-man TDLo: 23,500 μg/kg/7D-C: EYE AJOPAA 113,587,92
orl-rat LD50: >5 g/kg DRUGAY 6,321,82
ipr-rat LD50: >5 g/kg DRUGAY 6,321,82
scu-rat LD50: >5 g/kg DRUGAY 6,321,82
ivn-rat LD50: >5 g/kg DRUGAY 6,321,82
orl-mus LD50: 3150 mg/kg DRUGAY 6,321,82
ipr-mus LD50: 3779 mg/kg CNREA8 39,2204,79
scu-mus LD50: >10 g/kg DRUGAY 6,321,82
ivn-mus LD50: >7 g/kg DRUGAY 6,321,82

CONSENSUS REPORTS: NCI Carcinogenesis Studies (ipr); No Evidence: mouse, rat CANCAR 40,1935,77

SAFETY PROFILE: Moderate to low toxicity by ingestion. Human systemic effects: allergic dermatitis, ataxia, blood changes, central nervous system effects conjunctive irritation, degenerative brain changes, hearing acuity change, lacrimation, peripheral nerve fasciculations, spleen changes. An experimental teratogen. Other experimental reproductive effects. A human skin and eye irritant. Questionable carcinogen with experimental tumorigenic data. Human mutagenic data. When heated to decomposition it emits toxic fumes of NO_x.

AQQ900 **CAS:5536-17-4** **HR: 3**
9-β-d-ARABINO FURANOSYL ADENINE
mf: $C_{10}H_{13}N_5O_4$ mw: 267.28

PROP: Needles. Mp: 257–257.5°.

SYNS: ADENINE ARABINOSIDE ☐ ARABINOSYLADENINE ☐ β-d-ARABINOSYLADENINE ☐ 9-ARABINOSYLADENINE ☐ VIDARABIN ☐ VIDARABINE

TOXICITY DATA WITH REFERENCE
skn-rbt 3% MLD IYKEDH 23,360,92
eye-rbt 1% IYKEDH 13,561,82
skn-gpg 1% MLD IYKEDH 13,561,82
dni-hmn: hlas 40 mg/L JANTAJ 35,119,82
dni-hmn: lyms 400 nmol/L CNREA8 40,1405,80
ims-rat TDLo: 2 g/kg (female 6-15D post): TER TJADAB 15,231,77

ivg-rat TDLo:1316 mg/kg (female 15-21D post):REP TJADAB 15,231,77
ivn-wmn LDLo:105 mg/kg/1W-I CMAJAX 132,392,85
ivn-hmn TDLo:300 μg/kg:SYS JIDIAQ 133,A192,76
ivn-hmn TDLo:2 mg/kg:BLD JIDIAQ 134,75,76
ipr-rat LD50:1476 mg/kg IYKEDH 15,688,84
scu-rat LD50:8914 mg/kg IYKEDH 15,688,84
ivn-rat LD50:302 mg/kg IYKEDH 15,688,84
orl-mus LD50:7800 μg/kg 37ASAA 2,962,78
ipr-mus LD50:3057 mg/kg IYKEDH 15,688,84
scu-mus LD50:5086 mg/kg IYKEDH 15,688,84
ivn-mus LD50:442 mg/kg IYKEDH 15,688,84

CONSENSUS REPORTS: Reported in EPA TSCA Inventory.

SAFETY PROFILE: Poison by ingestion and intravenous routes. Moderately toxic by intraperitoneal route. An experimental teratogen. Other experimental reproductive effects. Human systemic effects by intravenous route: central nervous system, blood, and other effects. A skin and eye irritant. Human mutation data reported. When heated to decomposition it emits toxic fumes of NO_x.

AQQ905 CAS:29984-33-6 HR: 2
9-(β-d-ARABINOFURANOSYL)ADENINE-5′-(DIHYDROGEN PHOSPHATE)
mf: $C_{10}H_{14}N_5O_7P$ mw: 347.26

SYNS: ADENINE ARABINOSIDE MONOPHOSPHATE ☐ ARABINOSYLADENINE MONOPHOSPHATE ☐ ADENINE ARABINOSIDE 5′-MONOPHOSPHATE ☐ 5′-ARABINOSYLADENINE MONOPHOSPHATE ☐ 9-(5-o-PHOSPHONO-β-d-ARABINOFURANOSYL)-9H-PURIN-6-AMINE

TOXICITY DATA WITH REFERENCE
ipr-rat LD50:1700 mg/kg DRFUD4 4,547,79
ipr-mus LD50:1200 mg/kg DRFUD4 4,547,79

SAFETY PROFILE: Moderately toxic by intraperitoneal route. When heated to decomposition it emits very toxic fumes of PO_x and NO_x.

AQR000 CAS:69-74-9 HR: 3
1-β-d-ARABINOFURANOSYLCYTOSINE HYDROCHLORIDE
mf: $C_9H_{13}N_3O_5 \cdot ClH$ mw: 279.71

PROP: Crystals from EtOH (aq). Mp: 188–193°.

SYNS: ARABINOSYLCYTOSINE HYDROCHLORIDE ☐ CYLOCIDE ☐ CYTARABINE HYDROCHLORIDE ☐ CYTOSAR HYDROCHLORIDE ☐ CYTOSINE ARABINOSIDE HYDROCHLORIDE ☐ IRETIN ☐ NSC 63878 ☐ SPONGOCYTIDINE HYDROCHLORIDE

TOXICITY DATA WITH REFERENCE
eye-hmn 21 mg/7D-I MLD AJOPAA 60,1074,65
eye-mky 35 mg/15D-I MOD AJOPAA 60,1074,65
eye-rbt 42 mg/14D-I MOD AJOPAA 60,1074,65
dni-mus:lym 10 mg/L EJCAAH 6,379,70
msc-mus:leu 100 μmol/L CNREA8 29,1881,69
ipr-rat TDLo:280 mg/kg (female 15D post):REP JONRA9 34,950,80
ipr-rat LD50:5500 mg/kg OYYAA2 6,1255,72
orl-mus LD50:826 mg/kg NCISP* JAN86
ipr-mus LD50:825 mg/kg OYYAA2 6,1255,72
scu-mus LD50:2262 mg/kg NCISP* JAN86
ivn-dog LD50:172 mg/kg OYYAA2 8,353,74
ivn-mky LD50:396 mg/kg OYYAA2 8,353,74

SAFETY PROFILE: Poison by intravenous route. Moderately toxic by intraperitoneal and subcutaneous routes. Experimental reproductive effects. A human eye irritant. Mutation data reported. When heated to decomposition it emits very toxic fumes of NO_x and HCl.

AQR500 CAS:6742-07-0 HR: 2
1-β-d-ARABINOFURANOSYL-2′,3′,5′-TRIACETATE
mf: $C_{15}H_{19}N_3O_8$ mw: 369.37

SYN: NSC-93150

TOXICITY DATA WITH REFERENCE
ipr-mus LD50:4400 mg/kg NCIHL* -,353,67
ivn-mus LD50:680 mg/kg NCIHL* -,353,67

SAFETY PROFILE: Moderately toxic by intravenous route. Mildly toxic by intraperitoneal route. When heated to decomposition it emits toxic fumes of NO_x.

AQS750 CAS:506-32-1 HR: 3
ARACHIDONIC ACID
mf: $C_{20}H_{32}O_2$ mw: 304.52

PROP: Liquid. Mp: −49.5°, bp: 163° @ 1 mm.

SYNS: (ALL-Z)-5,8,11,14-EICOSATETRAENOIC ACID ☐ ARCHIDONATE

TOXICITY DATA WITH REFERENCE
dns-mus:mmr 10 mg/L CRNGDP 5,1123,84
sce-ham:ovr 320 μmol/L PAACA3 27,95,86
scu-mus TDLo:80 μg/kg (female 1D pre):REP FESTAS 25,636,74
ivn-rat LDLo:100 mg/kg THBRAA 9,67,76
ivn-mus LD50:33 mg/kg JPETAB 224,369,83
ivn-rbt LDLo:1 mg/kg THBRAA 9,67,76

SAFETY PROFILE: Poison by intravenous route. Experimental reproductive effects. Mutation data reported. When heated to decomposition it emits acrid smoke and irritating fumes.

AQS875 CAS:41948-17-8 HR: 3
ARA-C PALMITATE
mf: $C_{25}H_{43}N_3O_6$ mw: 481.71

SYNS: 1-β-d-ARABINOFURANOSYLCYTOSINE-5′-PALMITATE ☐ 1-β-d-ARABINOFURANOSYLCYTOSINE-5′-PALMITOYL ESTER ☐ ARABINOSYLCYTOSINE PALMITATE ☐ ARA-CP ☐ ARACYTIDINE-5′-PALMITATE ☐ CYTOSINE ARABINOSIDE PALMITATE ☐ PALMO-ARA-C

TOXICITY DATA WITH REFERENCE
dni-rat-ipr 200 mg/kg TJADAB 7,219,73
ipr-rat TDLo:200 mg/kg (12D preg):TER TJADAB 7,219,73
ipr-mus LD50:155 mg/kg NCISP* JAN86

SAFETY PROFILE: Poison by intraperitoneal route. An experimental teratogen. Mutation data reported. When heated to decomposition it emits toxic fumes of NO_x.

AQT250 **CAS:1446-17-9** **HR: 3**
ARATEN PHOSPHATE
mf: $C_{18}H_{26}ClN_3 \cdot H_3O_4P$ mw: 417.92

SYNS: 7-CHLORO-4((4-(DIETHYLAMINO)-1-METHYLBUTYL)AMINO)-QUINOLINE PHOSPHATE (1:1) □ CHLOROQUINE PHOSPHATE

TOXICITY DATA WITH REFERENCE
orl-wmn TDLo:2660 mg/kg (2-39W preg):REP AROTAA 80,407,64
orl-wmn TDLo:2660 mg/kg (2-39W preg):TER AROTAA 80,407,64
orl-hmn LDLo:43 mg/kg JFSCAS 5,201,65
orl-wmn TDLo:2740 mg/kg/39W:CNS ARPAAQ 93,209,72
orl-chd LDLo:38 mg/kg PEDIAU 27,95,61

SAFETY PROFILE: Human poison by ingestion. Human systemic effects by ingestion: muscle weakness. A human teratogen. Human reproductive effects: termination of pregnancy; developmental abnormalities of the eye, ear, and musculoskeletal system; and effects on newborn and postnatal development. Moderately toxic to humans by ingestion with musculoskeletal effects. When heated to decomposition it emits very toxic fumes of Cl^-, NO_x, and PO_x.

AQT500 **CAS:39300-45-3** **HR: 3**
ARATHANE
mf: $C_{18}H_{24}N_2O_6$ mw: 364.44

PROP: Liquid.

SYNS: CAPRYLDINITROPHENYL CROTONATE □ 2-CAPRYL-4,6-DINITROPHENYL CROTONATE □ CROTONATE de 2,4-DINITRO 6-(1-METHYL-HEPTYL)-PHENYLE (FRENCH) □ 4,6-DINITRO-2-CAPRYLPHENYL CROTONATE □ 4,6-DINITRO-2-(2-CAPRYL)PHENYL CROTONATE □ DINITRO(1-METHYLHEPTYL)PHENYL CROTONATE □ 2,4-DINITRO-6-(1-METHYLHEPTYL)PHENYL CROTONATE □ 2,4-DINITRO-6-(2-OCTYL)PHENYL CROTONATE □ ENT 24,727 □ (6-(1-METHYL-HEPTYL)-2,4-DINITRO-FENYL)-CROTONAAT (DUTCH) □ (6-(1-METHYL-HEPTYL)-2,3-DINITRO-PHENYL)-CROTONAT (GERMAN) □ 2-(1-METHYLHEPTYL)-4,6-DINITROPHENYL CROTONATE □ (6-(1-METIL-EPITL)-2,4-DINITRO-FENIL)-CROTONATO (ITALIAN)

TOXICITY DATA WITH REFERENCE
mmo-sat 500 μg/plate MUREAV 116,185,83
mmo-smc 5 ppm RSTUDV 6,161,76
orl-mus TDLo:120 mg/kg (female 7-16D post):REP TCMUD8 6,33,86
orl-mus TDLo:50 mg/kg (female 7-16D post):TER TCMUD8 6,375,86
scu-mus TDLo:10 mg/kg:NEO NTIS** PB223-159
orl-rat LD50:1102 mg/kg BCTKAG 8,373,75
ivn-rat LD50:23 mg/kg AIPTAK 119,31,59
orl-mus LD50:49,500 μg/kg BCTKAG 8,373,75
orl-dog LD50:100 mg/kg SPEADM 78-1,25,78
ihl-rat LC50:360 mg/m³/4H PEMNDP 9,303,91
ivn-rat LD50:23 mg/kg AIPTAK 119,31,59
orl-rbt LD50:3 g/kg BCTKAG 8,373,75
skn-rbt LD50:9400 mg/kg SPEADM 74-1,-,74

SAFETY PROFILE: Poison by ingestion and intravenous routes. An experimental teratogen. Other experimental reproductive effects. Mutation data reported. Questionable carcinogen with experimental neoplastigenic data. See NITRATES. When heated to decomposition it emits toxic fumes of NO_x.

AQT575 **CAS:55028-70-1** **HR: 2**
ARBAPROSTIL
mf: $C_{21}H_{34}O_5$ mw: 366.55

SYNS: 15(R)-METHYLPROSTAGLANDIN E2 □ 15(R)-15-METHYL-PROSTAGLANDIN E2

TOXICITY DATA WITH REFERENCE
ims-wmn TDLo:600 ng/kg (12W preg):REP CCPTAY 9,523,74

SAFETY PROFILE: Human reproductive effects by intramuscular route: terminates pregnancy. Other experimental reproductive effects.

AQT650 **HR: 3**
ARECA NUT

PROP: From the Areca palm tree, a native to South Asia. Orange-yellow in color when ripe. The seed, the size of a small egg, is separated from the fibrous pericarp and used fresh, after sun drying, or curing. It is chewed either alone or as a component of mixtures including Betel leaf and/or tobacco. Also known as Betel nut and supari. The nut contains several alkaloids, primarily arecoline, arecaidine, arecolidine, guvacoline, and guvacine.

SYNS: BETEL NUT □ SUPARI (INDIA)

SAFETY PROFILE: 3-(methylnitrosamino)propionaldehyde is an experimental carcinogen. Arecoline is one of the agents responsible for betel quid addiction. It mimics the action of acetylcholine and acts as a stimulant. It is a poison by intraperitoneal route. Reactions in the mouth and during processing can produce from these alkaloids several nitrosamines: N-nitrosoguvacoline, 3-(methylnitrosamino)propionitrile, 3-(methylnitrosamino)propionaldehyde, and N-nitrosoguvacine. N-nitrosoguvacoline and N-nitrosoguvacine have been found in the mouths of betel quid users. Areca nut extracts are experimental carcinogens and mutagens. See also BETEL QUID and SMOKELESS TOBACCO.

AQT750 **CAS:63-75-2** **HR: 3**
ARECOLINE
mf: $C_8H_{13}NO_2$ mw: 155.22

PROP: Oily liquid. Bp: 94° @ 17 mm.

SYNS: ARECAIDINE METHYL ESTER □ ARECOLINE BASE □ METHYL-1,2,5,6-TETRAHYDRO-1-METHYLNICOTINATE □ N-METHYL-Δ-TETRAHYDRONICOTINIC ACID METHYL ESTER □ N-METHYLTETRAHYDROPYRIDINE-β-CARBOXYLIC ACID METHYL ESTER □ 1,2,5,6-TETRAHYDRO-1-METHYLNICOTINIC ACID, METHYL ESTER

TOXICITY DATA WITH REFERENCE
dnd-hmn:Cells-uns 20 μg/tube MUREAV 278,271,92
dni-hmn:Cells-uns 339 μmol/L IJCNAW 47,396,91
skn-ham TDLo:2698 mg/kg/65W-I:NEO JNCIAM 53,1259,74
orl-rat LD50:2500 mg/kg EJMCA5 26,853,91
ipr-rat LD50:40 mg/kg BIJOAK 113,123,69
orl-mus LD50:550 mg/kg EJMCA5 26,853,91
ipr-mus LD50:190 mg/kg AIPTAK 192,88,71
ivn-mus LD50:36 mg/kg FATOAO 28,33,65
unk-mus LDLo:100 mg/kg HBAMAK 4,1289,35

scu-dog LD50:5 mg/kg FAZMAE 17,108,73
unk-dog LDLo:5 mg/kg HBAMAK 4,1289,35

CONSENSUS REPORTS: IARC Cancer Review: Animal Inadequate Evidence IMEMDT 37,141,85.

SAFETY PROFILE: Poison by subcutaneous and intraperitoneal routes. Moderately toxic by ingestion. Questionable carcinogen with experimental neoplastigenic data. It mimics the action of acetylcholine, a neurotransmitter, and is a parasympathetic nervous system stimulant. Its action on the central nervous system can cause tremors. Human mutation data reported. It is easily nitrosated to several nitrosamines. See also ESTERS and NITROSAMINES. It is the major alkaloid found in betel quid. Combustible, can react with oxidizing materials. When heated to decomposition it emits highly toxic fumes of NO_x.

AQU000 CAS:300-08-3 HR: 3
ARECOLINE BROMIDE
mf: $C_8H_{13}NO_2$•BrH mw: 236.14

PROP: Solid Mp: 172°.

SYNS: ARECOLINE HYDROBROMIDE □ METHYL-1,2,5,6-TETRAHYDRO-1-METHYLNICOTINATE, HYDROBROMIDE □ 1,2,5,6-TETRAHYDRO-1-METHYLNICOTINIC ACID, METHYL ESTER, HYDROBROMIDE

TOXICITY DATA WITH REFERENCE
par-rat LD50:270 mg/kg ABMGAJ 28,681,72
scu-mus LDLo:65 mg/kg JPETAB 35,75,29
ivn-mus LD50:18 mg/kg CSLNX* NX#11778

SAFETY PROFILE: Poison by parenteral, subcutaneous, and intravenous routes. When heated to decomposition it emits very toxic fumes of HBr and NO_x.

AQU250 CAS:61-94-9 HR: 3
ARECOLINE HYDROCHLORIDE
mf: $C_8H_{13}NO_2$•ClH mw: 191.68

PROP: Solid. Mp: 157–158°.

SYNS: NICOTINIC ACID-1,2,5,6-TETRAHYDRO-1-METHYL-, METHYL ESTER, HYDROCHLORIDE □ 3-PYRIDINECARBOXYLIC ACID-1,2,5,6-TETRAHYDRO-1-METHYL ESTER, HYDROCHLORIDE

TOXICITY DATA WITH REFERENCE
dni-mus-ipr 60 mg/kg IJEBA6 17,1141,79
oth-mus-ipr 60 mg/kg IJEBA6 17,1141,79
orl-mus TDLo:10,400 mg/kg/1Y-I:CAR JCREA8 107,169,84
ipr-mus LD50:154 mg/kg TXAPA9 28,227,74
ivn-mus LD50:32 mg/kg CSLNX* NX#12238

SAFETY PROFILE: Poison by intraperitoneal and intravenous routes. Questionable carcinogen with experimental carcinogenic data. See also ESTERS. When heated to decomposition it emits very toxic fumes of NO_x and HCl.

AQU500 CAS:30233-80-8 HR: 3
ARESKAP 100

PROP: Monobutylphenyl-phenol sodium monosulfonate. (JAPMA8 38,428,49)

TOXICITY DATA WITH REFERENCE
eye-rbt 1% MLD JAPMA8 38,428,49
orl-mus LD50:3800 mg/kg JAPMA8 38,428,49
ivn-mus LD50:180 mg/kg JAPMA8 38,428,49

SAFETY PROFILE: Poison by intravenous route. Mildly toxic by ingestion. An eye irritant. See also SULFONATES. When heated to decomposition it emits toxic fumes of SO_x.

AQU750 CAS:30233-81-9 HR: 3
ARESKET 300

SYN: MONOBUTYL DIPHENYL SODIUM MONOSULFONATE

TOXICITY DATA WITH REFERENCE
eye-rbt 1% SEV JAPMA8 38,428,49
orl-mus LD50:3500 mg/kg JAPMA8 38,428,49
ivn-mus LD50:250 mg/kg JAPMA8 38,428,49

SAFETY PROFILE: Poison by intravenous route. Moderately toxic by ingestion. A severe eye irritant. See also SULFONATES. When heated to decomposition it emits toxic fumes of SO_x.

AQV000 HR: 3
ARESKLENE 400

SYN: DIBUTYLPHENYL-PHENOL SODIUM DISULFONATE

TOXICITY DATA WITH REFERENCE
eye-rbt 1% SEV JAPMA8 38,428,49
orl-mus LD50:2200 mg/kg JAPMA8 38,428,49
ivn-mus LD50:200 mg/kg JAPMA8 38,428,49

SAFETY PROFILE: Poison by intravenous route. Moderately toxic by ingestion. A severe eye irritant. See also SULFONATES. When heated to decomposition it emits toxic fumes of SO_x.

AQV500 CAS:627-75-8 HR: 2
d-ARGININE HYDROCHLORIDE
mf: $C_6H_{14}N_4O_2$•ClH mw: 210.70

SYNS: ARGININE, MONOHYDROCHLORIDE, d- □ d-ARGININE, MONOHYDROCHLORIDE (9CI)

TOXICITY DATA WITH REFERENCE
ipr-rat LD50:3582 mg/kg ABBIA4 64,319,56

CONSENSUS REPORTS: Reported in EPA TSCA Inventory.

SAFETY PROFILE: Moderately toxic by intraperitoneal route. When heated to decomposition it emits toxic vapors of NO_x, HCl, and Cl^-.

AQW000 CAS:1119-34-2 HR: 2
l-ARGININE MONOHYDROCHLORIDE
mf: $C_6H_{14}N_4O_2$•ClH mw: 210.70

PROP: White crystalline powder; odorless. Mp: 222–235° (decomp). Very sol in water; sltly sol in alc.

SYNS: ARGAMINE □ ARGININE HYDROCHLORIDE □ l-ARGININE HYDROCHLORIDE □ ARGININE MONOHYDROCHLORIDE □ ARGI-

VENE □ DETOXARGIN □ l-HYDROCHLORIDE ARGININE □ LEVARGIN □ MINOPHAGEN A □ R-GENE

TOXICITY DATA WITH REFERENCE
ipr-rat TDLo:90 mg/kg (1-6D preg):TER AJEBAK 51,553,73
orl-rat LD50:12 g/kg JPMSAE 62,49,73
ipr-rat LD50:3793 mg/kg ABBIA4 58,253,55

CONSENSUS REPORTS: Reported in EPA TSCA Inventory.

SAFETY PROFILE: Moderately toxic by intraperitoneal route. Mildly toxic by ingestion. An experimental teratogen. When heated to decomposition it emits very toxic fumes of NO_x and HCl.

AQW250 CAS:7440-37-1 HR: 1
ARGON
af: Ar aw: 39.94

DOT: UN 1006/UN 1951

PROP: Colorless, inert, odorless, tasteless, monatomic gas. Forms no true chemical compds. Forms clathrates with H_2O and hydroquinone. Forms complex with HBr. Mp: −189.2°, bp: −185.7°, d: 1.784 g/L @ 0°, 1.40 @ −186°, 1.65 @ −233°. Solubility in water 3.36 mL/100 g @ 20°.

CONSENSUS REPORTS: Reported in EPA TSCA Inventory.

DOT CLASSIFICATION: 2.2; *Label:* Nonflammable Gas

SAFETY PROFILE: A simple asphyxiant gas. As an inert gas, it has no specific inherent dangerous properties. Gases of this type have no specific toxicity effect, but they act by excluding O_2 from the lungs. The effect of simple asphyxiant gases is proportional to the extent to which they diminish the amount (partial pressure) of O_2 in the air that is breathed. The oxygen may be diminished to 75% of its normal percentage in air before appreciable symptoms develop, and this in turn requires the presence of a simple asphyxiant in a concentration of 33% in the mixture of air and gas. When the simple asphyxiant reaches a concentration of 50%, marked symptoms can be produced. A concentration of 75% is fatal in a matter of minutes. The first symptoms produced by simple asphyxiant gases such as argon are rapid respirations and air hunger. Mental alertness is diminished and muscular coordination is impaired. Later, judgment becomes faulty and all sensations are depressed. Emotional instability often results and fatigue occurs rapidly. As the asphyxia progresses, there may be nausea and vomiting, prostration, and loss of consciousness, and finally, convulsions, deep coma, and death.

AQX250 CAS:124-94-7 HR: 3
ARISTOCORT
mf: $C_{21}H_{27}FO_6$ mw: 394.48

PROP: Crystals. Mp: 269–271°.

SYNS: 9-α-FLUORO-16-α-HYDROXYPREDNISOLONE □ 9-α-FLUORO-11-β,16-α,17,21-TETRAHYDROXYPREGNA-1,4-DIENE-3,20-DIONE □ 9-α-FLUORO-11-β,16-α,17,21-TETRAHYDROXY-1,4-PREGNADIENE-3,20-DIONE □ 9-α-FLUORO-11-β,16-α,17-α,21-TETRAHYDROXYPREGNA-1,4-DIENE-3,20-DIONE □ FLUOXYPREDNISOLONE □ KENACORT □ PREGNA-1,4-DIENE-3,20-DIONE,9-FLUORO-11,16,17,21-TETRAHYDROXY-,(11-β,16-α) □ RODINOLONE □ SK-TRIAMCINOLONE □ 11-β,16-α,17-α,21-TETRAHYDROXY-9-α-FLUORO-1,4-PREGNADIENE-3,20-DIONE □ TRIAMCINOLONE

TOXICITY DATA WITH REFERENCE
oms-hmn-orl 428 μg/kg ARDEAC 103,39,71
unr-wmn TDLo:12,600 μg/kg (female 1-39W post):TER AJOGAH 96,985,66
orl-pig TDLo:10,500 μg/kg (female 15-16W post):REP THGNBO 30,137,88
scu-rat LD50:99 mg/kg TXAPA9 8,250,66

SAFETY PROFILE: Poison by subcutaneous route. An experimental teratogen. Other experimental reproductive effects. Human mutation data reported. When heated to decomposition it emits toxic fumes of F^-. An anti-inflammatory and antiallergic agent.

AQX500 CAS:76-25-5 HR: 3
ARISTOCORT ACETONIDE
mf: $C_{24}H_{31}O_6F$ mw: 434.55

SYNS: ACETOSPAN □ ARISTODERM □ ARISTOGEL □ 9-α-FLUORO-11-β,21-DIHYDROXY-16-α-ISOPROYLIDENEDIOXY-1,4-PREGNADIENE, 3,20-DIONE □ 9-α-FLUORO-16-HYDROXYPREDNISOLONE ACETONIDE □ 9-α-FLUORO-16-α-17-α-ISOPROPYLEDENE DIOXY PREDNISOLONE □ 9-α-FLUORO-16-α-17-α-ISOPROPYLIDENEDIOXY-Δ-1-HYDROCORTISONE □ FLUTONE □ KENACORT-A □ KENALOG □ TRAMACIN □ TRIAMCINCOLONE ACETONIDE □ TRIAMCINOLONE ACETONIDE □ TRIAMCINOLONE-16,17-ACETONIDE □ VETALOG

TOXICITY DATA WITH REFERENCE
dns-hmn:oth 1 nmol/L CNREA8 43,2664,83
oms-hmn-skn 5000 ppm ARDEAC 103,39,71
ims-ham TDLo:100 μg/kg (female 11D post):REP ANREAK 199,135A,81
ims-mus TDLo:10 mg/kg (female 11D post):TER PNASA6 67,779,70
scu-rat LD50:13,100 μg/kg DRFUD4 6,44,81
orl-mus LD50:5 g/kg YAKUD5 21,2117,79
ipr-mus LD50:105 mg/kg DRUGAY 6,516,82
scu-mus LD50:132 mg/kg DRUGAY 6,516,82

CONSENSUS REPORTS: Reported in EPA TSCA Inventory.

SAFETY PROFILE: Poison by subcutaneous and intraperitoneal routes. An experimental teratogen. Other experimental reproductive effects. Human mutation data reported. When heated to decomposition it emits acrid smoke and toxic fumes of F^-.

AQY000 HR: 3
ARISTOLICHIA INDICA L., ALCOHOLIC EXTRACT

PROP: Obtained from extracts of the bitter roots of the Indian shrub *Aristolochia indica L.* (IJEBA6 15,428,77).

TOXICITY DATA WITH REFERENCE
orl-mus TDLo:15 mg/kg (female 6D post):REP IJMRAQ 66,991,77
orl-mus LDLo:100 mg/kg IJMRAQ 66,991,77

SAFETY PROFILE: Poison by ingestion. Experimental reproductive effects. See also ARISTOLOCHINE.

AQY125 **HR: 2**
ARISTOLOCHIC ACID SODIUM SALT
mf: $C1_7H_{10}NO_7{\cdot}Na$ mw: 363.27

SYN: 8-METHOXY-6-NITROPHENANTHRO(3,4-d)-1,3-DIOXOLE-5-CARBOXYLIC ACID SODIUM SALT

TOXICITY DATA WITH REFERENCE
orl-rat TDLo:37 mg/kg/1Y-C:CAR ARTODN 51,107,82
orl-rat TD:90 mg/kg/13W-C:CAR ARTODN 51,107,82

SAFETY PROFILE: Questionable carcinogen with experimental carcinogenic data. When heated to decomposition it emits toxic fumes of NO_x and Na_2O.

AQY250 **CAS:313-67-7** **HR: 3**
ARISTOLOCHINE
mf: $C_{17}H_{11}NO_7$ mw: 341.29

PROP: Crystals from DMF/EtOH, EtOH or MeOH/Et_2O. Mp: 281–286° (decomp). From alcoholic extract of *Aristolochia indico* (CNCRA6 42,35,64).

SYNS: ARISTOLOCHIC ACID □ BIRTHWORT □ 8-METHOXY-6-NITROPHENANTHOL-(3,4-d)-1,3-DIOXOLE-5-CARBOXYLIC ACID □ NSC-50413

TOXICITY DATA WITH REFERENCE
mmo-sat 200 µg/plate MUREAV 113,259,83
msc-rat-orl 45 mg/kg MUREAV 143,143,85
orl-mus TDLo:105 mg/kg/3W-C:ETA ARTODN 61,504,88
ivn-man LDLo:3 mg/kg/2D-I CNCRA6 42,35,64
orl-rat LD50:184 mg/kg ARTODN 59,328,87
ivn-rat LD50:74 mg/kg ARTODN 59,328,87
orl-mus LD50:55,900 µg/kg ARTODN 59,328,87
ipr-mus LD50:14,320 µg/kg KIHSDM (6),2,81
ivn-mus LD50:38,400 µg/kg ARTODN 59,328,87

SAFETY PROFILE: Confirmed carcinogen. Poison by ingestion, intraperitoneal, and intravenous routes. Mutation data reported. When heated to decomposition it emits toxic fumes of NO_x.

From 'International Register of Potentially Toxic Chemicals: April 1982.' Vol 5 No. 1: The Ministry of Health of the Federal Republic of Germany has withdrawn from the national market drugs containing aristolochic acid. The decision resulted from the demonstration of a carcinogenic potential in a three-month ingestion toxicity study undertaken in rats. Aristolochic acid is claimed to promote phagocytosis and to have immunostimulant activity. A growth-inhibiting effect on experimentally induced tumors has been described, but this effect has not been shown to have any clinical relevance. Extracts of species of *Aristolochiacea* have traditionally been used as a bitter, and a broad range of therapeutic effects has been claimed.

AQY385 **CAS:68991-20-8** **HR: 3**
ARMOISE OIL

TOXICITY DATA WITH REFERENCE
orl-mus LD50:370 mg/kg FCTXAV 13,719,75

CONSENSUS REPORTS: Reported in EPA TSCA Inventory.

SAFETY PROFILE: A poison by ingestion. When heated to decomposition it emits acrid smoke and irritating vapors.

AQY400 **CAS:81-04-9** **HR: 1**
ARMSTRONG'S S ACID
mf: $C_{10}H_8O_6S_2$ mw: 288.30

PROP: Crystals or plates. Mp: 240–245° (anhyd). Sol in water and alc; practically insol in ether.

SYNS: ARMSTRONG'S ACID □ 1,5-NAPHTHYLENE DISULFONIC ACID

TOXICITY DATA WITH REFERENCE
orl-rat LD50:30 g/kg GISAAA 45(3),73,80
orl-mus LD50:47 g/kg GISAAA 45(3),73,80
orl-rbt LD50:30 g/kg GISAAA 45(3),73,80
orl-gpg LD50:47 g/kg GISAAA 45(3),73,80

CONSENSUS REPORTS: Reported in EPA TSCA Inventory.

SAFETY PROFILE: Very low oral toxicity. When heated to decomposition it emits toxic fumes of SO_x. See also SULFONATES.

AQY500 **HR: 3**
ARNICA

PROP: An alcoholic infusion

SYNS: MOUNTAIN TOBACCO □ WOLFSBANE

SAFETY PROFILE: Poison by inhalation and ingestion. A moderate irritant and allergen. It can cause gastroenteritis, nervous disturbances, and collapse. May cause contact dermatitis. Combustible when exposed to heat or flame. Incompatible with oxidizing materials.

AQY750 **HR: 3**
AROMATIC AMINES

PROP: Amines that contain one or more rings of unsaturated or cyclic HC, such as benzene. There are vast numbers of such amines. The term is largely due to the characteristic odor.

SAFETY PROFILE: Many of these aromatic amines are recognized as carcinogenic to the human bladder, ureter, and renal pelvis, intestines, lung, liver, and prostate. See also AMINES.

AQZ000 **HR: 2**
AROMATIC SPIRITS of AMMONIA

PROP: Colorless liquid, suffocating odor of ammonia. Composition: 10% by weight of NH_3 in alcohol.

SAFETY PROFILE: See AMMONIA. A dangerous fire hazard due to its alcohol content. Moderately explosive. When heated, it emits toxic fumes of ammonia. Incompatible with oxidizing materials.

AQZ900 **CAS:585-54-6** **HR: 3**
ARSACETIN SODIUM SALT
mf: $C_8H_9AsNO_4 \cdot Na$ mw: 281.09

SYNS: ARSACETIN □ ARSANILIC ACID, N-ACETYL-, SODIUM SALT □ ARSONIC ACID, (4-(ACETYLAMINO)PHENYL)-, MONOSODIUM SALT (9CI) □ SODIUM ACETYLARSANILATE

TOXICITY DATA WITH REFERENCE
scu-rat LDLo:550 mg/kg BIZEA2 184,360,27
ivn-rbt LDLo:300 mg/kg JPETAB 23,107,24

OSHA PEL: TWA 0.5 mg(As)/m^3
ACGIH TLV: TWA 0.2 mg(As)/m^3

SAFETY PROFILE: Poison by intravenous route. Moderately toxic by subcutaneous route. When heated to decomposition it emits toxic fumes of NO_x and As.

ARA000 **CAS:6018-32-2** **HR: 2**
ARSACETIN TETRAHYDRATE
mf: $C_8H_9AsNO_4 \cdot Na \cdot 4H_2O$ mw: 353.17

PROP: White, crystalline powder; odorless and tasteless.

SYNS: p-ACETAMIDOBENZENEARSONIC ACID, SODIUM SALT, TETRAHYDRATE □ N-ACETYL-p-AMINOBENZENEARSONIC ACID, SODIUM SALT, TETRAHYDRATE □ N-ACETYLARSANILIC ACID, SODIUM SALT, TETRAHYDRATE □ ARSACETIN SODIUM SALT, TETRAHYDRATE □ SODIUM ACETYL ARSANILATE

TOXICITY DATA WITH REFERENCE
ivn-rbt LD50:550 mg/kg MEIEDD 11,125,89

CONSENSUS REPORTS: Arsenic and its compounds are on the Community Right-To-Know List.

OSHA PEL: TWA 0.5 mg(As)/m^3
ACGIH TLV: TWA 0.2 mg(As)/m^3

SAFETY PROFILE: Moderately toxic by intravenous route. See also ARSENIC COMPOUNDS. When heated to decomposition it emits very toxic fumes of As and NO_x.

ARA250 **CAS:98-50-0** **HR: 3**
ARSANILIC ACID
mf: $C_6H_8AsNO_3$ mw: 217.06

PROP: Needles from aq solns. Mp: 232°, bp: decomp, $-H_2O$ @ 15°. Insol in Me_2CO, $CHCl_3$, C_6H_6; sltly sol in EtOH; sol in Et_2O, H_2O, conc acids, alkalis.

SYNS: p-AMINOBENZENEARSONIC ACID □ 4-AMINOBENZENEARSONIC ACID □ AMINOPHENYLARSINE ACID □ p-AMINOPHENYLARSINE ACID □ p-AMINOPHENYLARSINIC ACID □ 4-AMINOPHENYLARSONIC ACID □ p-ANILINEARSONIC ACID □ ANTOXYLIC ACID □ p-ARSANILIC ACID □ 4-ARSANILIC ACID □ ATOXYLIC ACID

TOXICITY DATA WITH REFERENCE
orl-rat LD50:>1000 mg/kg TXAPA9 18,185,71
ipr-rat LDLo:400 mg/kg JPETAB 80,393,44
ipr-mus LD50:248 mg/kg APFRAD 37,483,79
ivn-mus LD50:100 mg/kg CSLNX* NX#06774

CONSENSUS REPORTS: IARC Cancer Review: Animal Inadequate Evidence IMEMDT 23,39,80. Reported in EPA TSCA Inventory. Arsenic and its compounds are on the Community Right-To-Know List.

OSHA PEL: TWA 0.5 mg(As)/m^3
ACGIH TLV: TWA 0.2 mg(As)/m^3

SAFETY PROFILE: Poison by intravenous and intraperitoneal routes. Moderately toxic by ingestion. Flammable, decomposes with heat to yield flammable vapors. When heated to decomposition or on contact with acid or acid fumes it emits highly toxic fumes of As and NO_x. See also ARSENIC COMPOUNDS and ANILINE.

For occupational chemical analysis use NIOSH: Arsenic, Organo-, 5022.

ARA500 **CAS:127-85-5** **HR: 3**
ARSANILIC ACID, MONOSODIUM SALT
DOT: UN 2473
mf: $C_6H_7AsNO_3 \cdot Na$ mw: 239.05

PROP: Tetrahydrate: white, odorless, crystalline powder; faint salty taste. Sol in water; somewhat sol in alc.

SYNS: (4-AMINOPHENYL)ARSONIC ACID SODIUM SALT □ ANHYDROUS SODIUM ARSANILATE □ ARSAMIN □ ARSANILIC ACID SODIUM SALT □ ARSINOSOLVIN □ ARSONIC ACID, (4-AMINOPHENYL)-, MONOSODIUM SALT (9CI) □ ATOXYL □ MONOSODIUM (4-AMINOPHENYL)ARSONATE □ NCI-C61176 □ NUARSOL □ PIGLET PRO-GEN V □ PRO-GEN SODIUM □ PROTOXYL □ SOAMIN □ SODIUM AMINARSONATE □ SODIUM p-AMINOBENZENEARSONATE □ SODIUM AMINOPHENOL ARSONATE □ SODIUM p-AMINOPHENYLARSONATE □ SODIUM-ANALINE ARSONATE □ SODIUM ANILARSONATE □ SODIUM ARSANILATE □ SODIUM ARSANILATE (DOT) □ SODIUM p-ARSANILATE □ SODIUM ARSONILATE □ SONATE □ TRYPOXYL

TOXICITY DATA WITH REFERENCE
scu-rat LD50:75 mg/kg BIZEA2 184,360,27
scu-mus LD50:400 mg/kg MEIEDD 10,1230,83
scu-dog LDLo:5 mg/kg HBAMAK 4,1289,35
scu-rbt LDLo:200 mg/kg HBAMAK 4,1289,35

CONSENSUS REPORTS: Arsenic and its compounds are on the Community Right-To-Know List.

OSHA PEL: 8H TWA 0.5 mg(As)/m^3
ACGIH TLV: TWA 0.2 mg(As)/m^3
DOT CLASSIFICATION: 6.1; *Label:* KEEP AWAY FROM FOOD

SAFETY PROFILE: Poison by subcutaneous route. Can cause blindness. When heated to decomposition it emits very toxic fumes of As and NO_x.

ARA750 **CAS:7440-38-2** **HR: 3**
ARSENIC
DOT: UN 1558
af: As aw: 74.92

PROP: Silvery to black, brittle, crystalline, or amorphous metalloid. Mp: 814° @ 36 atm, bp: subl @ 612°, d: black crystals 5.724 @ 14°; black amorphous 4.7, vap press: 1 mm @ 372° (subl). Insol in water; sol in HNO_3.

SYNS: ARSENICALS □ ARSEN (GERMAN, POLISH) □ ARSENIC, metallic (DOT) □ ARSENIC BLACK □ ARSENIC-75 □ COLLOIDAL ARSENIC □ GREY ARSENIC □ METALLIC ARSENIC

TOXICITY DATA WITH REFERENCE
cyt-mus-ipr 4 mg/kg/48H-I EXPEAM 37,129,81
cyt-mus-orl 280 mg/kg/8W MUREAV 113,293,83

orl-rat TDLo:605 μg/kg (35 W preg):REP GISAAA (8)30,77
orl-rat TDLo:580 μg/kg (female 30W pre):TER FATOAO 41,620,78
orl-man TDLo:76 mg/kg/12Y-I:CAR RMCHAW 99,664,71
imp-rbt TDLo:75 mg/kg:ETA ZEKBAI 52,425,42
orl-man TDLo:7857 mg/kg/55Y:SKN CMAJAX 120,168,79
orl-man TDLo:7857 mg/kg/55Y:GIT CMAJAX 120,168,79
orl-rat LD50:763 mg/kg GTPZAB 31(12),53,87
ipr-rat LD50:13,390 μg/kg TXCYAC 64,191,90
orl-mus LD50:145 mg/kg GTPZAB 31(12),53,87
ipr-mus LD50:46,200 μg/kg GTPZAB 31(12),53,87
scu-rbt LDLo:300 mg/kg ASBIAL 24,442,38
scu-gpg LDLo:300 mg/kg ASBIAL 24,442,38

CONSENSUS REPORTS: NTP 7th Annual Report on Carcinogens. IARC Cancer Review: Group 1 IMEMDT 7,100,87; Human Sufficient Evidence IMEMDT 23,39,80; Human Inadequate Evidence IMEMDT 2,48,73. Reported in EPA TSCA Inventory. Arsenic and its compounds are on the Community Right-To-Know List.

OSHA PEL: TWA 0.01 mg(As)/m^3; Cancer Hazard
ACGIH TLV: TWA 0.2 mg(As)/m^3 (Proposed: 0.01 mg(As)/m^3; Human Carcinogen)
DFG TRK:0.2 mg/m^3 calculated as arsenic in that portion of dust that can possibly be inhaled
NIOSH REL: CL 2 μg(As)/m^3
DOT CLASSIFICATION: 6.1; *Label:* Poison

SAFETY PROFILE: Confirmed human carcinogen producing liver tumors. Poison by subcutaneous, intramuscular, and intraperitoneal routes. Human systemic skin and gastrointestinal effects by ingestion. An experimental teratogen. Other experimental reproductive effects. Mutation data reported. Flammable in the form of dust when exposed to heat or flame or by chemical reaction with powerful oxidizers such as bromates, chlorates, iodates, peroxides, lithium, NCl_3, KNO_3, $KMnO_4$, Rb_2C_2, $AgNO_4$, NOCl, IF_5, CrO_3, ClF_3, ClO, BrF_3, BrF_5, BrN_3, RbC_3BCH, CsC_3BCH. Slightly explosive in the form of dust when exposed to flame. When heated or on contact with acid or acid fumes, it emits highly toxic fumes; can react vigorously on contact with oxidizing materials. Incompatible with bromine azide, dirubidium acetylide, halogens, palladium, zinc, platinum, NCl_3, $AgNO_3$, CrO_3, Na_2O_2, hexafluoroisopropylideneamino lithium.

For occupational chemical analysis use OSHA: #ID-105 or NIOSH: Arsenic (Hydride AAS) 7900.

ARB000 CAS:10102-53-1 HR: 3
m-ARSENIC ACID
mf: $AsHO_3$ mw: 123.93

SYN: METAARSENIC ACID

CONSENSUS REPORTS: Reported in EPA TSCA Inventory. Arsenic and its compounds are on the Community Right-To-Know List.

OSHA PEL: TWA 0.01 mg(As)/m^3; Cancer Hazard
ACGIH TLV: TWA 0.2 mg(As)/m^3 (Proposed: 0.01 mg (As)/m^3; Human Carcinogen)
DFG MAK: Human Carcinogen
NIOSH REL: (Arsenic, Inorganic) CL 2 μg(As)/m^3/15M

SAFETY PROFILE: Confirmed human carcinogen. When heated to decomposition it emits toxic fumes of arsenic. See also ARSENIC COMPOUNDS.

ARB250 CAS:7778-39-4 HR: 3
o-ARSENIC ACID
DOT: UN 1553/UN 1554
mf: AsH_3O_4 mw: 141.95

SYNS: ACIDE ARSENIQUE LIQUIDE (FRENCH) □ ARSENATE □ ARSENIC ACID, liquid (DOT) □ ARSENIC ACID, solid (DOT) □ DESICCANT L-10 □ HI-YIELD DESICCANT H-10 □ ORTHOARSENIC ACID □ RCRA WASTE NUMBER P010 □ ZOTOX □ ZOTOX CRAB GRASS KILLER

TOXICITY DATA WITH REFERENCE
cyt-hmn:leu 7200 nmol/L MUREAV 88,73,81
cyt-hmn:fbr 100 ppb MUREAV 88,73,81
ipr-rat TDLo:30 mg/kg (9D preg):TER JTSCDR 4,405,79
orl-rat LD50:48 mg/kg FMCHA2 -,C27,91
orl-dog LDLo:10 mg/kg FDWU** -,-,31
orl-rbt LDLo:5 mg/kg FDWU** -,-,31
orl-pgn LDLo:100 mg/kg FDWU** -,-,31
orl-ckn LDLo:125 mg/kg FDWU** -,-,31

CONSENSUS REPORTS: Reported in EPA TSCA Inventory. Arsenic and its compounds are on the Community Right-To-Know List.

OSHA PEL: TWA 0.01 mg(As)/m^3; Cancer Hazard
ACGIH TLV: TWA 0.2 mg(As)/m^3 (Proposed: 0.01 mg(As)/m^3; Human Carcinogen)
DFG MAK: Human Carcinogen
NIOSH REL: (Arsenic, Inorganic) CL 2 μg(As)/m^3/15M
DOT CLASSIFICATION: 6.1; *Label:* Poison

SAFETY PROFILE: Confirmed human carcinogen. Poison by ingestion. An experimental teratogen. Human mutation data reported. When heated to decomposition it emits toxic fumes of arsenic. See also ARSENIC COMPOUNDS.

ARB750 CAS:7778-44-1 HR: 3
ARSENIC ACID, CALCIUM SALT (2:3)
DOT: UN 1573
mf: $As_2O_8{\cdot}3Ca$ mw: 398.08

PROP: Colorless, amorphous powder. D: 3.620. Solubility in water: 0.013/100 @ 25°.

SYNS: CALCIUM ARSENATE (DOT) □ CALCIUM ORTHOARSENATE □ CALCIUMARSENAT □ CHIP-CAL □ CHIP-CAL GRANULAR □ CUCUMBER DUST □ FENCAL □ FLAC □ KALO □ KALZIUMARSENIAT (GERMAN) □ KILMAG □ PENCAL □ SECURITY □ SPRACAL □ TRICALCIUM ARSENATE □ TRICALCIUMARSENAT (GERMAN) □ TURF-CAL

TOXICITY DATA WITH REFERENCE
itr-rat TDLo:1600 μg/kg:ETA IJCNAW 24,786,79
itr-ham TDLo:120 mg/kg/15W-C:NEO CALEDQ 27,99,85
orl-rat LD50:20 mg/kg AFDOAQ 15,122,51
skn-rat LD50:2400 mg/kg 28ZEAL 5,35,76
orl-mus LD50:794 mg/kg AMRL** TR-72-62,72
orl-dog LD50:38 mg/kg 85DPAN -,-,71/76
orl-rbt LDLo:50 mg/kg JPETAB 39,246,30
orl-mam LD50:35 mg/kg PCOC** -,170,66

CONSENSUS REPORTS: NTP 7th Annual Report On Carcinogens. IARC Cancer Review: Group 1 IMEMDT 7,100,87; Human Sufficient Evidence IMEMDT 23,39,80; Animal No Evidence IMEMDT 2,48,73; Animal Inadequate Evidence IMEMDT 23,39,80. Reported in EPA TSCA Inventory. Arsenic and its compounds are on the Community Right-To-Know List. EPA Extremely Hazardous Substances List.

OSHA PEL: TWA 0.01 mg(As)/m³; Cancer Hazard
ACGIH TLV: TWA 0.2 mg(As)/m³ (Proposed: 0.01 mg(As)/m³; Human Carcinogen)
DFG MAK: Human Carcinogen
NIOSH REL: CL 2 µg(As)/m³/15M
DOT CLASSIFICATION: 6.1; *Label:* Poison

SAFETY PROFILE: Confirmed human carcinogen. Poison by ingestion. Moderately toxic by skin contact. When heated to decomposition it emits toxic fumes of arsenic.

ARC000 CAS:7778-43-0 HR: 3
ARSENIC ACID, DISODIUM SALT
mf: $Na_2HAsO_4 \cdot 7H_2O$ mw: 312.01

PROP: Colorless white powder or solid, effloresces. D: 1.88, mp: $-7H_2O$ @ 130°, bp: decomp @ 150°. Solubility in water: 61/100 @ 15°, sol in glycerin.

SYNS: DISODIUM ARSENATE ☐ DISODIUM ARSENIC ACID ☐ DISODIUM HYDROGEN ARSENATE ☐ DISODIUM HYDROGEN ORTHOARSENATE ☐ DISODIUM MONOHYDROGEN ARSENATE ☐ SODIUM ACID ARSENATE ☐ SODIUM ARSENATE ☐ SODIUM ARSENATE DIBASIC, anhydrous

TOXICITY DATA WITH REFERENCE
cyt-hmn:leu 7200 µmol/L MUREAV 88,73,81
mrc-bcs 100 mmol/L MUREAV 77,109,80
ipr-rat LDLo:34,720 µg/kg JPETAB 58,454,36

CONSENSUS REPORTS: Reported in EPA TSCA Inventory. Arsenic and its compounds are on the Community Right-To-Know List.

OSHA PEL: TWA 0.5 mg(As)/m³: Cancer Hazard
ACGIH TLV: TWA 0.2 mg(As)/m³ (Proposed: 0.01 mg(As)/m³; Human Carcinogen)
NIOSH REL: (Arsenic, Inorganic) CL 2 µg(As)/m³/15M
DFG MAK: Human Carcinogen

SAFETY PROFILE: Confirmed human carcinogen. Poison by intraperitoneal route. Human mutation data reported. When heated to decomposition it emits toxic fumes of arsenic. See ARSENIC COMPOUNDS.

ARC250 CAS:10048-95-0 HR: 3
ARSENIC ACID, DISODIUM SALT, HEPTAHYDRATE
mf: $AsHO_4 \cdot 2Na \cdot 7H_2O$ mw: 427.05

PROP: Prisms. Mp: 40°. Very sol in water; sltly sol in EtOH.

SYNS: DISODIUM ARSENATE, HEPTAHYDRATE ☐ SODIUM ACID ARSENATE, HEPTAHYDRATE ☐ SODIUM ARSENATE, DIBASIC, HEPTAHYDRATE ☐ SODIUM ARSENATE HEPTAHYDRATE

TOXICITY DATA WITH REFERENCE
cyt-dmg-orl 454 ppm SOGEBZ 10,608,74
otr-hmn:fbr 2500 nmol/L CNREA8 47,3815,87
cyt-hmn:leu 7200 nmol/L MUREAV 88,73,81
orl-mus TDLo:120 mg/kg (female 10D post):TER EVHPAZ 19,219,77
ipr-mus TDLo:45 mg/kg (female 9D post):REP AEHLAU 24,62,72
ims-mus LD50:87,360 µg/kg EXMDA4 (440),312,78
scu-gpg LDLo:50 mg/kg BMJOAE 2,217,13

CONSENSUS REPORTS: NTP 7th Annual Report On Carcinogens. Arsenic and its compounds are on the Community Right-To-Know List.

OSHA PEL: TWA 0.01 mg(As)/m³; Cancer Hazard
ACGIH TLV: TWA 0.2 mg(As)/m³ (Proposed: 0.01 mg(As)/m³; Human Carcinogen)
NIOSH REL: (Arsenic, Inorganic) CL 2 µg(As)/m³/15M
DFG MAK: Human Carcinogen

SAFETY PROFILE: Confirmed human carcinogen. Poison by subcutaneous route. An experimental teratogen. Other experimental reproductive effects. Human mutation data reported. See also ARSENIC COMPOUNDS. When heated to decomposition it emits toxic fumes of arsenic.

ARC500 CAS:7774-41-6 HR: 3
ARSENIC ACID, HEMIHYDRATE
mf: $AsH_3O_4 \cdot 1/2H_2O$ mw: 150.96

PROP: White, translucent crystals. Mp: 35.5°, bp: $-H_2O$ @ 160°, d: 2.0–2.5.

SYNS: ARSENIC ACID, solid (DOT) ☐ ORTHOARSENIC ACID HEMIHYDRATE

TOXICITY DATA WITH REFERENCE
ivn-rbt LD50:6 mg/kg MEIEDD 11,126,89

CONSENSUS REPORTS: Arsenic and its compounds are on the Community Right-To-Know List.

OSHA PEL: TWA 0.01 mg(As); Cancer Hazard
ACGIH TLV: TWA 0.2 mg(As)/m³ (Proposed: 0.01 mg(As)/m³; Human Carcinogen)
NIOSH REL: (Arsenic, Inorganic) CL 2 µg(As)/m³/15M
DFG MAK: Human Carcinogen.

SAFETY PROFILE: Confirmed human carcinogen. Poison by intravenous route. When heated to decomposition it emits toxic fumes of arsenic. See also ARSENIC COMPOUNDS.

ARC750 CAS:7645-25-2 HR: 3
ARSENIC ACID, LEAD SALT
DOT: UN 1617
mf: $AsH_3O_4 \cdot 7Pb$ mw: 1592.28

SYNS: ARSENIATE de PLOMB (FRENCH) ☐ LEAD ARSENATE

TOXICITY DATA WITH REFERENCE
orl-rat LD50:100 mg/kg AFDOAQ 15,122,51
orl-mus LD50:1000 mg/kg JPETAB 93,407,48
orl-rbt LD50:125 mg/kg JAPMA8 37,122,48
orl-ckn LD50:450 mg/kg JAPMA8 37,122,48

CONSENSUS REPORTS: Arsenic compounds and lead compounds are on the Community Right-To-Know List.

OSHA PEL: TWA 0.01 mg(As)/m³; Cancer Hazard
ACGIH TLV: TWA 0.15 mg(Pb)/m³
NIOSH REL: (Lead, Inorganic): 10H TWA 0.10 mg(Pb)/m³; (Arsenic, Inorganic): CL 0.002 mg(As)/m³/15M
DFG MAK: Human Carcinogen.
DOT CLASSIFICATION: 6.1; *Label:* Poison

SAFETY PROFILE: Confirmed human carcinogen. Poison by ingestion. See also LEAD COMPOUNDS and ARSENIC COMPOUNDS. When heated to decomposition it emits very toxic fumes of lead and arsenic.

ARD000 CAS:10103-50-1 HR: 3
ARSENIC ACID, MAGNESIUM SALT
DOT: UN 1622
mf: $AsH_3O_4 \cdot 7Mg$ mw: 312.12

PROP: Monoclinic, white crystals. D: 2.60–2.61.

SYNS: ARSENIATE de MAGNESIUM (FRENCH) □ MAGNESIUM ARSENATE □ MAGNESIUM ARSENATE PHOSPHOR

TOXICITY DATA WITH REFERENCE
orl-rat LDLo:280 mg/kg TXAPA9 1,156,59
orl-mus LD50:315 mg/kg IRGGAJ 20,21,63
orl-rbt LDLo:80 mg/kg AIHAAP 19,504,58

CONSENSUS REPORTS: Reported in EPA TSCA Inventory. Arsenic and its compounds are on the Community Right-To-Know List.

OSHA PEL: TWA 0.01 mg(As)/m³; Cancer Hazard
ACGIH TLV: TWA 0.2 mg(As)/m³ (Proposed: 0.01 mg(As)/m³; Human Carcinogen)
DFG MAK: Human Carcinogen
NIOSH REL: (Arsenic, Inorganic) CL 2 µg(As)/m³/15M
DOT CLASSIFICATION: 6.1; *Label:* Poison

SAFETY PROFILE: Confirmed human carcinogen. Poison by ingestion. When heated to decomposition it emits toxic fumes of arsenic. See also ARSENIC COMPOUNDS.

ARD250 CAS:7784-41-0 HR: 3
ARSENIC ACID, MONOPOTASSIUM SALT
DOT: UN 1677
mf: $AsH_2O_4 \cdot K$ mw: 180.04

SYNS: MACQUER'S SALT □ MONOPOTASSIUM ARSENATE □ MONOPOTASSIUM DIHYDROGEN ARSENATE □ POTASSIUM ACID ARSENATE □ POTASSIUM ARSENATE □ POTASSIUM ARSENATE, MONOBASIC □ POTASSIUM DIHYDROGEN ARSENATE □ POTASSIUM HYDROGEN ARSENATE

CONSENSUS REPORTS: NTP 7th Annual Report On Carcinogens. IARC Cancer Review: Human Sufficient Evidence IMEMDT 23,39,80. Reported in EPA TSCA Inventory. Arsenic and its compounds are on the Community Right-To-Know List.

OSHA PEL: TWA 0.01 mg(As)/m³, Cancer Hazard
ACGIH TLV: TWA 0.2 mg(As)/m³ (Proposed: 0.01 mg (As)/m³; Human Carcinogen)
NIOSH REL: (Arsenic, Inorganic) CL 2 µg(As)/m³/15M
DOT CLASSIFICATION: 6.1; *Label:* Poison

SAFETY PROFILE: Confirmed human carcinogen. Mutation data reported. When heated to decomposition it emits toxic fumes of arsenic. See also ARSENIC COMPOUNDS.

ARD500 CAS:15120-17-9 HR: 3
ARSENIC ACID, MONOSODIUM SALT
mf: $AsO_3 \cdot Na$ mw: 145.91

PROP: Needle-like fibrous hygroscopic crystals. Sol in water.

SYNS: ARSENIC ACID, SODIUM SALT (9CI) □ SODIUM ARSENATE □ SODIUM METAARSENATE □ SODIUM MONOHYDROGEN ARSENATE

TOXICITY DATA WITH REFERENCE
sln-dmg-orl 2 µmol/L CNJGA8 11,677,69
slt-dmg-orl 100 µmol CNJGA8 17,55,75

CONSENSUS REPORTS: Arsenic and its compounds are on the Community Right-To-Know List.

OSHA PEL: TWA 0.5 mg(As)/m³: Cancer Hazard
ACGIH TLV: TWA 0.2 mg(As)/m³ (Proposed: 0.01 mg(As)/m³; Human Carcinogen)
DFG MAK: Human Carcinogen
NIOSH REL: (Arsenic, Inorganic) CL 2 µg(As)/m³/15M

SAFETY PROFILE: Confirmed human carcinogen. A poison. Mutation data reported. See also ARSENIC COMPOUNDS. When heated to decomposition it emits toxic fumes of arsenic.

ARD600 CAS:10103-60-3 HR: 3
ARSENIC ACID, MONOSODIUM SALT
mf: $AsH_2O_4 \cdot Na$ mw: 163.93

PROP: Solid. Mp: 118°.

SYNS: MONOSODIUM ARSENATE □ SODIUM ARSENATE □ SODIUM DIHYDROGEN ARSENATE □ SODIUM DIHYDROGEN ORTHOARSENATE

TOXICITY DATA WITH REFERENCE
ivn-rbt LDLo:45 mg/kg JPETAB 23,107,24

CONSENSUS REPORTS: Arsenic and its compounds are on the Community Right-To-Know List.

OSHA PEL: TWA 0.5 mg(As)/m³: Cancer Hazard
ACGIH TLV: TWA 0.2 mg(As)/m³ (Proposed: 0.01 mg(As)/m³; Human Carcinogen)
NIOSH REL: CL 2 µg(As)/m³/15M
DFG MAK: Human Carcinogen

SAFETY PROFILE: Confirmed human carcinogen. Poison by intravenous route. When heated to decomposition it emits toxic fumes of arsenic.

ARD750 CAS:7631-89-2 HR: 3
ARSENIC ACID, SODIUM SALT
DOT: UN 1685
mf: $AsH_3O_4 \cdot 7Na$ mw: 202.94

SYNS: FATSCO ANT POISON □ SODIUM ARSENATE (DOT) □ SODIUM METAARSENATE □ SODIUM ORTHOARSENATE □ SWEENEY'S ANT-GO

TOXICITY DATA WITH REFERENCE
cyt-hmn:lym 2 µmol/L ADREDL 267,91,80
orl-mus TDLo:475 mg/kg (female 8–12D post):REP TCMUD8 6,361,86
orl-mus TDLo:120 mg/kg (female 11D post):TER JEPTDQ 1(6),857,78
scu-mus TDLo:10 mg(As)/kg/20D-C:ETA VDGPAN 55,289,71
ipr-rat LDLo:49 mg/kg JPETAB 58,454,36
ivn-rat LDLo:85 mg/kg JPETAB 33,270,28
orl-rbt LDLo:51 mg/kg JPETAB 33,270,28
ivn-rbt LDLo:28 mg/kg JPETAB 33,270,28

CONSENSUS REPORTS: NTP 7th Annual Report on Carcinogens. IARC Cancer Review: Human Sufficient Evidence IMEMDT 23,39,80; Animal Inadequate Evidence IMEMDT 2,48,73; IMEMDT 23,39,80. Reported in EPA TSCA Inventory. Arsenic and its compounds are on the Community Right-To-Know List.

OSHA PEL: TWA 0.01 mg(As)/m^3; Cancer Hazard
ACGIH TLV: TWA 0.2 mg(As)/m^3 (Proposed: 0.01 mg(As)/m^3; Human Carcinogen)
NIOSH REL: CL 2 µg(As)/m^3/15M
DOT CLASSIFICATION: 6.1; *Label:* Poison

SAFETY PROFILE: Confirmed human carcinogen with experimental tumorigenic data. Poison by ingestion, intravenous, and intraperitoneal routes. An experimental teratogen. Other experimental reproductive effects. Mutation data reported. When heated to decomposition it emits toxic fumes of As and Na_2O. See also ARSENIC COMPOUNDS.

ARE000 CAS:64070-83-3 HR: 3
ARSENIC(V) ACID, TRISODIUM SALT, HEPTAHYDRATE (1:3:7)
mf: $AsO_4{\cdot}3Na{\cdot}7H_2O$ mw: 334.03

SYN: TRISODIUM ARSENATE, HEPTAHYDRATE

TOXICITY DATA WITH REFERENCE
ipr-mus LD50:9 mg/kg COREAF 257,791,63

CONSENSUS REPORTS: Arsenic and its compounds are on the Community Right-To-Know List.

OSHA PEL: TWA 0.01 mg(As)/m^3; Cancer Hazard
ACGIH TLV: TWA 0.2 mg(As)/m^3
NIOSH REL: (Arsenic, Inorganic) CL 2 µg(As)/m^3/15M
DFG MAK: Human Carcinogen

SAFETY PROFILE: Confirmed human carcinogen. Poison by intraperitoneal route. See also ARSENIC COMPOUNDS. When heated to decomposition it emits toxic fumes of arsenic.

ARE250 CAS:8028-75-9 HR: 3
ARSENICAL DIP

SYNS: ARSENICAL DIP, liquid (DOT) □ SHEEP DIP

CONSENSUS REPORTS: Arsenic and its compounds are on the Community Right-To-Know List.

OSHA PEL: TWA 0.5 mg(As)/m^3

SAFETY PROFILE: A poison. See also ARSENIC COMPOUNDS.

ARE500 CAS:8028-73-7 HR: 2
ARSENICAL DUST
DOT: UN 1562

SYNS: ARSENICAL FLUE DUST □ FLUE DUST, ARSENIC CONTAINING

TOXICITY DATA WITH REFERENCE
itr-rat TDLo:120 mg/kg/15W-I:ETA EVHPAZ 19,191,77

CONSENSUS REPORTS: Reported in EPA TSCA Inventory. Arsenic and its compounds are on the Community Right-To-Know List.

OSHA PEL: TWA 0.5 mg(As)/m^3
ACGIH TLV: TWA 0.2 mg(As)/m^3 (Proposed: 0.01 mg(As)/m^3; Human Carcinogen)
NIOSH REL: CL 2 µg(As)/m^3/15M
DOT CLASSIFICATION: 6.1; *Label:* Poison

SAFETY PROFILE: A poison. Questionable carcinogen with experimental tumorigenic data. See also ARSENIC COMPOUNDS.

ARE750 CAS:8028-73-7 HR: 3
ARSENICAL FLUE DUST
DOT: UN 1562

TOXICITY DATA WITH REFERENCE
itr-rat TDLo:120 mg/kg/15W-I:ETA EVHPAZ 19,191,77

CONSENSUS REPORTS: Arsenic and its compounds are on the Community Right-To-Know List. Reported in EPA TSCA Inventory.

OSHA PEL: TWA 0.5 mg(As)/m^3
DFG MAK: Human Carcinogen.
DOT CLASSIFICATION: 6.1; *Label:* Poison

SAFETY PROFILE: Confirmed human carcinogen with experimental tumorigenic data. Poison by inhalation and ingestion. See also ARSENIC COMPOUNDS.

ARF000 HR: 3
ARSENIC BISULFIDE
mf: As_2S_2 mw: 214

PROP: Red-brown crystals. Bp: 565°, mp: (β) 307°, d: (α) 3.506 @ 19°, (β) 3.254 @ 19°.

SYN: REALGAR

CONSENSUS REPORTS: Arsenic and its compounds are on the Community Right-To-Know List.

SAFETY PROFILE: A poison. See also ARSENIC COMPOUNDS and SULFIDES. Flammable in the form of dust when exposed to heat or flame. Explosion hazard when intimately mixed with powerful oxidizers such as Cl_2; KNO_3; chlorates. It will react with water or steam to produce toxic and flammable vapors.

ARF250 **CAS:7784-33-0** **HR: 3**
ARSENIC(III) BROMIDE
DOT: UN 1555
mf: $AsBr_3$ mw: 314.65

PROP: Colorless, deliquescent, rhombic crystals. Mp: 32.8°, bp: 220.0°, vap press: 1 mm @ 41.8°, d: 3.3972 @ 25°, (liq) 3.3282.

SYNS: ARSENIC TRIBROMIDE □ ARSENOUS BROMIDE □ ARSENOUS TRIBROMIDE □ TRIBROMOARSINE

CONSENSUS REPORTS: Reported in EPA TSCA Inventory. Arsenic and its compounds are on the Community Right-To-Know List.

OSHA PEL: TWA 0.01 mg(As)/m³; Cancer Hazard
ACGIH TLV: TWA 0.2 mg(As)/m³ (Proposed: 0.01 mg (As)/m³; Human Carcinogen)
NIOSH REL: (Arsenic, Inorganic) CL 2 μg(As)/m³/15M
DOT CLASSIFICATION: 6.1; *Label:* Poison

SAFETY PROFILE: Confirmed carcinogen. A poison. See also ARSENIC COMPOUNDS and BROMIDES. When heated to decomposition it emits very toxic fumes of As and Br^-.

ARF500 **CAS:7784-34-1** **HR: 3**
ARSENIC CHLORIDE
DOT: UN 1560
mf: $AsCl_3$ mw: 181.27

PROP: Colorless, oily liquid. Freezing to colorless crystals with pearly sheen. Fumes in air. D: 2.15 @ 25°, mp: −16°, bp: 130°. Decomp in water and by UV light; misc in chloroform, CCl_4, ether, iodine, P, S, alkali iodides, oils and fats. Vap d: 6.25, vap press: 10 mm @ 23.5°.

SYNS: ARSENIC BUTTER □ ARSENIC(III) CHLORIDE □ ARSENIOUS CHLORIDE □ ARSENOUS CHLORIDE □ ARSENOUS TRICHLORIDE (9CI) □ CHLORURE d'ARSENIC (FRENCH) □ CHLORURE ARSENIEUX (FRENCH) □ FUMING LIQUID ARSENIC □ TRICHLOROARSINE □ TRICHLORURE d'ARSENIC (FRENCH)

TOXICITY DATA WITH REFERENCE
cyt-hmn:leu 600 nmol/L MUREAV 88,73,81
mrc-bcs 30 μL/disc MUREAV 77,109,80
otr-ham:emb 3 μmol/L CNREA8 39,193,79
ihl-mus LCLo:338 ppm/10M HBTXAC 1,324,56
ihl-cat LCLo:100 mg/m³/1H ZGEMAZ 13,523,21

CONSENSUS REPORTS: Reported in EPA TSCA Inventory. Arsenic and its compounds are on the Community Right-To-Know List. EPA Extremely Hazardous Substances List.

OSHA: Cancer Hazard
ACGIH TLV: TWA 0.2 mg(As)/m³ (Proposed: 0.01 mg(As)/m³; Human Carcinogen)
NIOSH REL: (Arsenic, Inorganic) CL 2 μg(As)/m³/15M
DOT CLASSIFICATION: 6.1; *Label:* Poison

SAFETY PROFILE: A poison via inhalation. See also ARSENIC COMPOUNDS and CHLORIDES. Very poisonous; fumes in air. Mutation data reported. When heated to decomposition it emits very toxic fumes of As and Cl^-. Highly reactive. Explodes with Na, K, and Al on impact.

ARF750 **HR: 3**
ARSENIC COMPOUNDS

SYN: ARSENICALS

CONSENSUS REPORTS: Arsenic and its compounds are on the Community Right-To-Know List.

OSHA PEL: Inorganic: TWA 0.01 mg(As)/m³; Cancer Hazard; Organic: TWA 0.5 mg(As)/m³
ACGIH TLV: TWA 0.2 mg(As)/m³ (Proposed: (inorganic compounds) 0.01 mg(As)/m³; Human Carcinogen)
NIOSH REL: CL 2 μg(As)/m³/15M

SAFETY PROFILE: Inorganic compounds are confirmed human carcinogens producing tumors of the mouth, esophagus, larynx, bladder, and paranasal sinus. Recognized carcinogens of the skin, lungs, and liver. Used as insecticides, herbicides, silvicides, defoliants, desiccants, and rodenticides. Poisoning from arsenic compounds may be acute or chronic. Acute poisoning usually results from swallowing arsenic compounds; chronic poisoning from either swallowing or inhaling. Acute allergic reactions to arsenic compounds used in medical therapy have been fairly common, the type and severity of reaction depending upon the compound. Inorganic arsenicals are more toxic than organics. Trivalent is more toxic than pentavalent. Acute arsenic poisoning (from ingestion) results in marked irritation of the stomach and intestines with nausea, vomiting, and diarrhea. In severe cases, the vomitus and stools are bloody and the patient goes into collapse and shock with weak, rapid pulse, cold sweats, coma, and death. Chronic arsenic poisoning, whether through ingestion or inhalation, may manifest itself in many different ways. There may be disturbances of the digestive system such as loss of appetite, cramps, nausea, constipation, or diarrhea. Liver damage may occur, resulting in jaundice. Disturbances of the blood, kidneys, and nervous system are not infrequent. Arsenic can cause a variety of skin abnormalities including itching, pigmentation, and even cancerous changes. A characteristic of arsenic poisoning is the great variety of symptoms that can be produced. Dangerous; when heated to decomposition, or when metallic arsenic contacts acids or acid fumes, or when water solutions of arsenicals are in contact with active metals such as Fe, Al, or Zn, highly toxic fumes of arsenic are emitted.

ARG000 **HR: 3**
ARSENIC DIETHYL
mf: $As(C_2H_5)_2$ mw: 266.2

PROP: Liquid or oil. Bp: 185–190°, d: about 1.

CONSENSUS REPORTS: Arsenic and its compounds are on the Community Right-To-Know List.

SAFETY PROFILE: A poison. A dangerous fire hazard by spontaneous chemical reaction. Dangerous when heated. Incompatible with oxidizing materials. See also ARSENIC COMPOUNDS.

ARG250 **HR: 3**
ARSENIC DIMETHYL
mf: $As(CH_3)_2$ mw: 210.0

PROP: Colorless to yellow oily liquid. Mp: −6°; bp: 186°; d: 1.15.

CONSENSUS REPORTS: Arsenic and its compounds are on the Community Right-To-Know List.

SAFETY PROFILE: Poison by inhalation and ingestion. See also ARSENIC COMPOUNDS. Flammable. Evolves dangerous fumes of arsenic when heated.

ARG500 **HR: 3**
ARSENIC HEMISELENIDE
mf: As_2Se mw: 228.78

CONSENSUS REPORTS: Arsenic compounds and its compounds as well as selenium and its compounds are on the Community Right-To-Know List.

OSHA PEL: TWA 0.01 mg(As)/m³; Cancer Hazard; TWA 0.2 mg(Se)/m³
ACGIH TLV: TWA 0.2 mg(As)/m³; TWA 0.2 mg(Se)/m³
DFG TRK: 0.2 mg/m³ calculated as arsenic in that portion of dust that can possibly be inhaled; 0.1 mg(Se)/m³
NIOSH REL: CL 2 μg(As)/m³

SAFETY PROFILE: When heated to decomposition it emits fumes of As and Se. Incompatible with oxidizing materials. When heated to decomposition it emits highly toxic fumes of Se and arsenic. See ARSENIC COMPOUNDS and SELENIUM COMPOUNDS.

ARG750 **CAS:7784-45-4** **HR: 3**
ARSENIC IODIDE
mf: AsI_3 mw: 455.62

PROP: Lustrous, orange-red, hexagonal crystals, leaves or platelets. Mp: 146°; bp: 403°; d: 4.38 @ 13°. Sol in H_2O, Et_2O, CS_2, xylene, dioxan; sltly sol in conc HCl.

SYNS: ARSENIC TRIIODIDE ☐ ARSENOUS IODIDE ☐ ARSENOUS TRIIODIDE (9CI) ☐ TRIIODOARSINE

CONSENSUS REPORTS: Reported in EPA TSCA Inventory. Arsenic and its compounds are on the Community Right-To-Know List.

OSHA PEL: TWA 0.01 mg(As)/m³; Cancer Hazard
ACGIH TLV: TWA 0.2 mg(As)/m³ (Proposed: 0.01 mg (As)/m³; Human Carcinogen)
NIOSH REL: (Arsenic, Inorganic) CL 2 μg(As)/m³/15M

SAFETY PROFILE: A poison. See also ARSENIC COMPOUNDS and IODIDES. Can form a shock-sensitive compound with sodium or potassium. When heated to decomposition it emits very toxic fumes of I^- and arsenic.

ARH250 **HR: 3**
ARSENIC PENTASULFIDE
mf: As_2S_5 mw: 310.2

PROP: Brownish-yellow, glassy, amorphous, highly refractive mass. Mp: 500° (subl).

CONSENSUS REPORTS: Arsenic and its compounds are on the Community Right-To-Know List.

OSHA PEL: TWA 0.01 mg(As)/m³
ACGIH TLV: TWA 0.2 mg(As)/m³ (Proposed: 0.01 mg (As)/m³; Human Carcinogen)
NIOSH REL: CL 2 μg(As)/m³/15M

SAFETY PROFILE: See also ARSENIC COMPOUNDS and SULFIDES. Flammable in the form of dust when exposed to heat or flame. Explosive when intimately mixed with powerful oxidizers, such as Cl_2, KNO_3, or chlorates. Will react with water and steam to produce toxic and flammable vapors. Incompatible with water, steam, and strong oxidizers.

ARH500 **CAS:1303-28-2** **HR: 3**
ARSENIC PENTOXIDE
DOT: UN 1559
mf: As_2O_5 mw: 229.84

PROP: White, amorphous, deliquescent solid. Mp: decomp @ 800°, d: 4.32. Sol in alc. Very sol in H_2O.

SYNS: ANHYDRIDE ARSENIQUE (FRENCH) ☐ ARSENIC ACID ☐ ARSENIC ACID ANHYDRIDE ☐ ARSENIC ANHYDRIDE ☐ ARSENIC OXIDE ☐ ARSENIC(V) OXIDE ☐ DIARSENIC PENTOXIDE ☐ RCRA WASTE NUMBER P011 ☐ ZOTOX

TOXICITY DATA WITH REFERENCE
cyt-hmn:leu 1200 nmol/L MUREAV 88,73,81
mrc-bcs 50 mmol/L MUREAV 77,109,80
itt-rat TDLo:4597 μg/kg (male 1D pre):REP JRPFA4 7,21,64
orl-rat LD50:8 mg/kg 28ZEAL 4,50,69
orl-mus LD50:55 mg/kg IRGGAJ 20,21,63
ivn-rbt LDLo:6 mg/kg NTIS** PB214-270

CONSENSUS REPORTS: NTP 7th Annual Report on Carcinogens. IARC Cancer Review: Human Sufficient Evidence IMEMDT 23,39,80. Reported in EPA TSCA Inventory. Arsenic and its compounds are on the Community Right-To-Know List. EPA Extremely Hazardous Substances List.

OSHA PEL: Cancer Hazard
ACGIH TLV: TWA 0.2 mg(As)/m³ (Proposed: 0.01 mg(As)/m³; Human Carcinogen)
DFG MAK: Human Carcinogen
NIOSH REL: CL 2 μg(As)/m³/15M
DOT CLASSIFICATION: 6.1; *Label:* Poison

SAFETY PROFILE: Confirmed human carcinogen. Poison by ingestion and intravenous routes. Experimental reproductive effects. Mutation data reported. Reacts vigorously with Rb_2C_2. When heated to decomposition it emits toxic fumes of arsenic. See also ARSENIC COMPOUNDS.

ARH750 **HR: 3**
ARSENIC PHOSPHIDE
mf: AsP mw: 105.9

PROP: Brown to red powder. Mp: sublimes with decomp.

CONSENSUS REPORTS: Arsenic and its compounds are on the Community Right-To-Know List.

SAFETY PROFILE: Flammable by spontaneous chemical reaction. Phosphine is liberated upon contact with moisture. Dangerous when heated. Incompatible with water or steam; oxidizing materials. See ARSENIC COMPOUNDS and PHOSPHINE.

ARI000 **CAS:1303-33-9** **HR: 3**
ARSENIC SULFIDE
DOT: NA 1557
mf: As_2S_3 mw: 246.04

PROP: Red needles or yellow in polycrystal form. Bp: 707°, d: 3.43, mp: 327°. Insol in water; sol in alkalies.

SYNS: ARSENIC SESQUISULFIDE ☐ ARSENIC SULFIDE YELLOW ☐ ARSENIC SULPHIDE ☐ ARSENIC SULFIDE YELLOW ☐ ARSENIC SULPHIDE ☐ ARSENIC TERSULPHIDE ☐ ARSENIC TRISULFIDE ☐ ARSENIC TRISULFIDE (DOT) ☐ ARSENIC YELLOW ☐ ARSENIOUS SULPHIDE ☐ ARSENOUS SULFIDE ☐ AURIPIGMENT ☐ C.I. 77086 ☐ C.I. PIGMENT YELLOW ☐ DIARSENIC TRISULFIDE ☐ DIARSENIC TRISULPHIDE ☐ KING'S GOLD ☐ KING'S YELLOW ☐ ORPIMENT

TOXICITY DATA WITH REFERENCE
scu-rat TDLo:125 mg/kg:ETA BJCAAI 20,190,66

CONSENSUS REPORTS: IARC Cancer Review: Human Sufficient Evidence IMEMDT 23,39,80. Reported in EPA TSCA Inventory. Arsenic and its compounds are on the Community Right-To-Know List.

OSHA PEL: Cancer Hazard
ACGIH TLV: TWA 0.2 mg(As)/m³ (Proposed: 0.01 mg(As)/m³; Human Carcinogen)
NIOSH REL: (Arsenic, Inorganic) CL 2 µg(As)/m³/15M
DOT CLASSIFICATION: 6.1; *Label:* Poison

SAFETY PROFILE: Confirmed human carcinogen with experimental tumorigenic data. A poison. Reacts violently with H_2O_2, (KNO_3+ S). When heated to decomposition or on contact with acid or acid fumes it emits highly toxic fumes of SO_2, H_2S, and As. Reacts with water or steam to emit toxic and flammable vapors.

ARI250 **CAS:7784-35-2** **HR: 3**
ARSENIC TRIFLUORIDE
mf: AsF_3 mw: 131.92

PROP: Very mobile, colorless liquid which fumes in air and readily hydrolyzed D: 3.01, mp: −5.95, bp: 63°, vap press: 100 mm @ 13.2°, 400 mm @ 41.5°. Insol in water; sol in alc, benzene, and mercury.

SYNS: ARSENIC FLUORIDE ☐ ARSENOUS FLUORIDE ☐ TRIFLUOROARSINE

TOXICITY DATA WITH REFERENCE
ihl-mus LCLo:2000 mg/m³/10M NDRC** NDCrc-132,Aug,42

CONSENSUS REPORTS: Reported in EPA TSCA Inventory. Arsenic and its compounds are on the Community Right-To-Know List.

OSHA PEL: TWA 0.01 mg(As)/m³; Cancer Hazard
ACGIH TLV: TWA 0.2 mg(As)/m³ (Proposed: 0.01 mg(As)/m³; Human Carcinogen)
NIOSH REL: (Arsenic, Inorganic) CL 2 µg(As)/m³/15M

SAFETY PROFILE: Confirmed human carcinogen. A poison by inhalation. Strong reaction with P_2O_3. When heated to decomposition it emits very toxic fumes of As and F^-. See also FLUORIDES and ARSENIC COMPOUNDS.

ARI500 **CAS:8012-54-2** **HR: 3**
ARSENIC TRIIODIDE mixed with MERCURIC IODIDE

SYNS: ARSENIOUS and MERCURIC IODIDE, solution (DOT) ☐ DONOVAN'S SOLUTION

CONSENSUS REPORTS: Arsenic compounds and Mercury compounds are on the Community Right-To-Know List.

ACGIH TLV: TWA 0.1 mg(Hg)/m³ (skin)
NIOSH REL: (Arsenic, Inorganic): CL 0.002 mg(As)/m³/15M; (Mercury, Aryl And Inorganic): CL 0.1 mg/m³ (skin)

SAFETY PROFILE: A poison. See also ARSENIC COMPOUNDS, MERCURY COMPOUNDS, and IODIDES. When heated to decomposition it emits very toxic fumes of Hg, As, and I^-.

ARI750 **CAS:1327-53-3** **HR: 3**
ARSENIC TRIOXIDE
DOT: UN 1561
mf: As_2O_3 mw: 197.84

PROP: Colorless, rhombic crystals (dimer, claudetite), or white powder. D: 4.15, mp: 312°, bp: 460°. Solubility in water: 1.82/100 @ 20°; sol in alc. Cubes: Colorless. D: 3.865, mp: 309°. Solubility in water: 1.2/100 @ 20°.

SYNS: ACIDE ARSENIEUX ☐ ANHYDRIDE ARSENIEUX ☐ ARSENIC BLANC ☐ ARSENIC OXIDE ☐ ARSENIC(III) OXIDE ☐ ARSENIC SESQUIOXIDE ☐ ARSENICUM ALBUM ☐ ARSENIGEN SAURE ☐ ARSENIOUS ACID ☐ ARSENIOUS OXIDE ☐ ARSENIOUS TRIOXIDE ☐ ARSENITE ☐ ARSENOLITE ☐ ARSENOUS ACID ☐ ARSENOUS ACID ANHYDRIDE ☐ ARSENOUS ANHYDRIDE ☐ ARSENOUS OXIDE ☐ ARSENOUS OXIDE ANHYDRIDE ☐ ARSENTRIOXIDE ☐ ARSODENT ☐ CLAUDELITE ☐ CLAUDETITE ☐ CRUDE ARSENIC ☐ DIARSENIC TRIOXIDE ☐ RCRA WASTE NUMBER P012 ☐ WHITE ARSENIC

TOXICITY DATA WITH REFERENCE
mrc-bcs 50 mmol/L MUREAV 77,109,80
orl-wmn TDLo:600 mg/kg (female 30W post):REP AJDCAI 117,328,69
ihl-mus TCLo:28,500 µg/m³/4H (female 9-12D post):TER JJATDK 5,61,85
itr-rat TDLo:16 mg/kg/15W-I:ETA EVHPAZ 19,191,77
itr-ham TDLo:45 mg/kg/15W-I:NEO ARTODN 7,403,84
orl-man LDLo:29 mg/kg:CNS,GIT AEMED3 16,702,87
orl-man LDLo:206 mg/kg:CVS,SYS JTCTDW 29,45,91
orl-hmn LDLo:1429 µg/kg YKYUA6 31,1247,80
orl-man LDLo:2857 mg/kg:SYS,CNS JTCTDW 29,131,91
unr-man LDLo:2941 µg/kg 85DCAI 2,73,70

orl-rat LD50:14,600 µg/kg GISAAA 52(1),21,87
ipr-rat LD50:871 mg/kg GTPZAB 19(3),30,75
scu-rat LDLo:8 mg/kg JPETAB 19,337,22
orl-mus LD50:31,500 µg/kg CHYCDW 14,86,80
scu-mus LD50:9800 µg/kg PSEBAA 78,392,51
ivn-mus LD50:10,700 µg/kg PSEBAA 78,392,51
orl-dog LDLo:10 mg/kg HBAMAK 4,1306,35
orl-rbt LDLo:4 mg/kg NTIS** PB214-270
ivn-rbt LDLo:10,560 µg/kg BIZEA2 70,144,15

CONSENSUS REPORTS: NTP 7th Annual Report On Carcinogens. IARC Cancer Review: Group 1 IMEMDT 7,100,87; Human Limited Evidence IMEMDT 2,48,73; Human Sufficient Evidence IMEMDT 23,39,80; Animal Inadequate Evidence IMEMDT 2,48,73; IMEMDT 23,39,80. Reported in EPA TSCA Inventory. Arsenic and its compounds are on the Community Right-To-Know List. EPA Extremely Hazardous Substances List.

OSHA PEL: TWA 0.01 mg(As)/m^3: Cancer Hazard
ACGIH TLV: Production: Suspected Human Carcinogen (Proposed: 0.01 mg(As)/m^3; Human Carcinogen)
DFG MAK: Human Carcinogen.
NIOSH REL: CL 2 µg(As)/m^3/15M
DOT CLASSIFICATION: 6.1; *Label:* Poison

SAFETY PROFILE: Confirmed human carcinogen with experimental neoplastigenic and tumorigenic data. Poison by ingestion, subcutaneous, and intravenous routes. Human systemic effects by ingestion: sleep changes, muscle weakness, hypermotility, diarrhea, cardiac arrhythmias, coma, fatty degeneration of the liver, depressed renal function tests. An experimental teratogen. Other experimental reproductive effects. Mutation data reported. Reacts vigorously with Rb_2C_2, ClF_3, F_2, Hg, OF_2, $NaClO_3$. See also ARSENIC COMPOUNDS.

For occupational chemical analysis use NIOSH: Arsenic Trioxide, 7901.

ARJ000 **HR: 3**
ARSENIC TRIOXIDE mixed with SELENIUM DIOXIDE (1:1)
mf: AsO_3•O_2Se mw: 233.88

SYN: SELENIUM DIOXIDE mixed with ARSENIC TRIOXIDE (1:1)

TOXICITY DATA WITH REFERENCE
orl-mus TDLo:16 mg/kg/35W-C:ETA BICHBX 9(3),245,78

CONSENSUS REPORTS: Arsenic and its compounds, as well as selenium and its compounds, are on the Community Right-To-Know List.

OSHA PEL: TWA 0.01 mg(As)/m^3; Cancer Hazard
ACGIH TLV: TWA 0.2 mg(As)/m^3; Suspected Carcinogen; 0.2 mg(Se)/m^3
DFG MAK: 0.1 mg(Se)/m^3
NIOSH REL: (Arsenic, Inorganic): CL 0.002 mg(As)/m^3/15M

SAFETY PROFILE: Confirmed human carcinogen with experimental tumorigenic data. See also ARSENIC COMPOUNDS and SELENIUM COMPOUNDS. When heated to decomposition it emits very toxic fumes of As and Se.

ARJ100 **CAS:56320-22-0** **HR: 3**
ARSENIC TRISULFIDE
DOT: NA 1557
mf: AsS_2 mw: 139.04

PROP: A solid.

SYNS: ARSENIC DISULFIDE ☐ ARSENIC SULFIDE (DOT)

OSHA: Cancer Hazard
DOT CLASSIFICATION: 6.1; *Label:* Poison

SAFETY PROFILE: Confirmed human carcinogen. A poison. When heated to decomposition it emits toxic vapors of As and SO_x.

ARJ250 **HR: 3**
ARSENIDES

CONSENSUS REPORTS: Arsenic and its compounds are on the Community Right-To-Know List.

SAFETY PROFILE: Compounds of arsenic and hydrogen or metals (i.e., transitional, alkaline earth, or rare-earth). These materials are dangerous because they readily emit very toxic arsine and arsenic fumes when exposed to heat, moisture, acids, and acid fumes.

ARJ500 **CAS:14060-38-9** **HR: 3**
ARSENIOUS ACID, SODIUM SALT
mf: AsH_3O_3•7Na mw: 286.88

PROP: Colorless or grayish-white powder. D: 1.87.

SYNS: ARSONIC ACID, SODIUM SALT (9CI) ☐ ARSENIOUS ACID, SODIUM SALT POLYMERS ☐ NATRIUMARSENIT (GERMAN) ☐ SODIUM ORTHOARSENITE

TOXICITY DATA WITH REFERENCE
ipr-rat LDLo:9 mg/kg JPETAB 58,454,36
orl-frg LDLo:600 mg/kg HBAMAK 4,1289,35
scu-frg LDLo:200 mg/kg HBAMAK 4,1289,35

CONSENSUS REPORTS: Arsenic and its compounds are on the Community Right-To-Know List.

OSHA PEL: TWA 0.01 mg(As)/m^3; Cancer Hazard
ACGIH TLV: TWA 0.2 mg(As)/m^3 (Proposed: 0.01 mg(As)/m^3; Human Carcinogen)
NIOSH REL: (Arsenic, Inorganic) CL 2 µg(As)/m^3/15M

SAFETY PROFILE: Confirmed human carcinogen. Poison by intraperitoneal and subcutaneous routes. Moderately toxic by ingestion. When heated to decomposition it emits toxic fumes of arsenic.

ARJ750 **CAS:1303-18-0** **HR: 3**
ARSENOPYRITE
mf: AsFeS mw: 162.83

SYNS: ARSENOMARCASITE ☐ MISPICKEL

TOXICITY DATA WITH REFERENCE
ipr-rat LD:>10 g/kg GTPZAB 28(7),53,84
ipr-mus LD:>10 g/kg GTPZAB 28(7),53,84
ivn-mus LDLo:200 mg/kg JNCIAM 1,241,40

CONSENSUS REPORTS: Arsenic and its compounds are on the Community Right-To-Know List.

OSHA PEL: TWA 0.01 mg(As)/m³
ACGIH TLV: TWA 0.2 mg(As)/m³ (Proposed: 0.01 mg(As)/m³; Human Carcinogen)
NIOSH REL: (Arsenic, Inorganic) CL 2 µg(As)/m³/15M

SAFETY PROFILE: Poison by intravenous route. When heated to decomposition it emits very toxic fumes of As and SO_x.

ARJ755 CAS:1122-90-3 HR: 3
p-ARSENOSOANILINE
mf: C_6H_6AsNO mw: 183.05

SYN: ANILINE, p-ARSENOSO-

TOXICITY DATA WITH REFERENCE
ipr-mus LD50:4 mg/kg JMCMAR 9,221,66

OSHA PEL: TWA 0.5 mg(As)/m³

SAFETY PROFILE: Poison by intraperitoneal route. When heated to decomposition it emits toxic fumes of NO_x and As.

ARJ760 CAS:4164-07-2 HR: 3
p-ARSENOSO-N,N-BIS(2-CHLOROETHYL)ANILINE
mf: $C_{10}H_{12}AsCl_2NO$ mw: 308.05

SYN: ANILINE, p-ARSENOSO-N,N-BIS(2-CHLOROETHYL)-

TOXICITY DATA WITH REFERENCE
ipr-mus LD50:5545 µg/kg JMCMAR 9,221,66

OSHA PEL: TWA 0.5 mg(As)/m³

SAFETY PROFILE: Poison by intraperitoneal route. When heated to decomposition it emits toxic fumes of NO_x, As, and Cl^-.

ARJ770 CAS:5185-80-8 HR: 3
p-ARSENOSO-N,N-BIS(2-HYDROXYETHYL)ANILINE
mf: $C_{10}H_{14}AsNO_3$ mw: 271.17

SYN: ANILINE, p-ARSENOSO-N,N-BIS(2-HYDROXYETHYL)-

TOXICITY DATA WITH REFERENCE
ipr-mus LD50:7593 µg/kg JMCMAR 9,221,66

OSHA PEL: TWA 0.5 mg(As)/m³

SAFETY PROFILE: Poison by intraperitoneal route. When heated to decomposition it emits toxic fumes of NO_x and As.

ARJ800 CAS:4164-06-1 HR: 3
p-ARSENOSO-N,N-DIETHYLANILINE
mf: $C_{10}H_{14}AsNO$ mw: 239.17

SYN: ANILINE, p-ARSENOSO-N,N-DIETHYL-

TOXICITY DATA WITH REFERENCE
ipr-mus LD50:2809 µg/kg JMCMAR 9,221,66

OSHA PEL: TWA 0.5 mg(As)/m³

SAFETY PROFILE: Poison by intraperitoneal route. When heated to decomposition it emits toxic fumes of NO_x and As.

ARJ900 CAS:63951-03-1 HR: 3
ARSENOXIDE SODIUM
mf: $C_6H_5AsNO_2\cdot Na$ mw: 221.03

SYN: PHENOL, 2-AMINO-4-ARSENOSO-, SODIUM SALT

TOXICITY DATA WITH REFERENCE
ivn-rat LDLo:20 mg/kg MADCAJ 6,195,37

OSHA PEL: TWA 0.5 mg(As)/m³
ACGIH TLV: TWA 0.2 mg(As)/m³ (Proposed: 0.01 mg(As)/m³; Human Carcinogen)

SAFETY PROFILE: Poison by intravenous route. When heated to decomposition it emits toxic fumes of NO_x and As.

ARK250 CAS:7784-42-1 HR: 3
ARSINE
DOT: UN 2188
mf: AsH_3 mw: 77.95

PROP: Thermally unstable, colorless, gas with mild garlic odor. D: 2.695 g/L; bp: −62.5°; vap d: 2.66; mp: −116°. Readily oxidized to As_2O_3. Very little tendency to protonate. Solubility in water: 28 mg/100 @ 20°. Sol in benzene and chloroform.

SYNS: ARSENIC HYDRID ☐ ARSENIC HYDRIDE ☐ ARSENIC TRIHYDRIDE ☐ ARSENIURETTED HYDROGEN ☐ ARSENOUS HYDRIDE ☐ ARSENOWODOR (POLISH) ☐ ARSENWASSERSTOFF (GERMAN) ☐ HYDROGEN ARSENIDE

TOXICITY DATA WITH REFERENCE
ihl-hmn TCLo:3 ppm:RBC AMIHAB 21,132,60
ihl-hmn LCLo:25 ppm/30M AEHLAU 19,133,69
ihl-man TDLo:338 ppt:GIT AEHLAU 34,224,79
ihl-man TDLo:338 ppt:SYS AEHLAU 34,224,79
ihl-man TDLo:338 ppt:CNS AEHLAU 34,224,79
ihl-rat LCLo:300 mg/m³/15M FATOAO 30(2),226,67
ihl-mus LCLo:70 mg/m³/3H AEXPBL 80,288,17
ihl-dog LCLo:400 mg/m³/15M FATOAO 30(2),226,67
ihl-mky LCLo:70 mg/m³/15M HBAMAK 4,1289,35
ihl-cat LCLo:150 mg/m³/20M HBAMAK 4,1289,35
ihl-rbt LCLo:500 mg/m³/15M FATOAO 30(2),226,67
ihl-frg LCLo:4500 mg/m³/3H AEXPBL 80,288,17

CONSENSUS REPORTS: IARC Cancer Review: Human Sufficient Evidence IMEMDT 23,39,80. Reported in EPA TSCA Inventory. Arsenic and its compounds are on the Community Right-To-Know List. EPA Extremely Hazardous Substances List.

OSHA PEL: TWA 0.05 ppm
ACGIH TLV: TWA 0.05 ppm
DFG MAK: 0.05 ppm (0.2 mg/m³)
NIOSH REL: (Arsine) CL 2 µg(As)/m³/15M
DOT CLASSIFICATION: 2.3; *Label:* Poison Gas, Flammable Gas

SAFETY PROFILE: Confirmed human carcinogen. Poison by inhalation. Human red blood cell, gastrointestinal system, central nervous system, and other systemic effects by inhalation. Flammable when exposed to flame. Moderately explosive when exposed to Cl_2, HNO_3, (K + NH_3), open flame, or powerful shock. Dangerous, more toxic than its oxidation product. When heated to decomposition it emits highly toxic fumes of arsenic. See also ARSENIC, ARSENIC COMPOUNDS, and HYDRIDES.

For occupational chemical analysis use OSHA: #ID-125G or NIOSH: Arsine, 6001.

ARK500 **HR: 3**
ARSINE BORON TRIBROMIDE
mf: $AsH_3 \cdot BBr_3$ mw: 328.6

CONSENSUS REPORTS: Arsenic and its compounds are on the Community Right-To-Know List.

SAFETY PROFILE: A poison. See BROMIDES, ARSENIC COMPOUNDS, and BORON COMPOUNDS. A highly unstable compound. Ignites in air. When heated to decomposition it emits very toxic fumes of As and Br^-.

ARK750 **CAS:67360-94-5** **HR: 3**
ARSINE-TRI-1-PIPERIDINIUM CHLORIDE
mf: $C_{15}H_{33}AsN_3 \cdot 3Cl$ mw: 436.78

SYN: ARSINOTRIS PIPERIDINIUM TRICHLORIDE

TOXICITY DATA WITH REFERENCE
ivn-rbt LDLo:200 mg/kg JPETAB 28,233,26
ims-rbt LDLo:500 mg/kg JPETAB 28,233,26

CONSENSUS REPORTS: Arsenic and its compounds are on the Community Right-To-Know List.

OSHA PEL: TWA 0.5 mg(As)/m³
ACGIH TLV: TWA 0.2 mg(As)/m³
NIOSH REL: CL 2 µg(As)/m³/15M

SAFETY PROFILE: Poison by intravenous route. Moderately toxic by intramuscular route. When heated to decomposition it emits very toxic fumes of As, NO_x and Cl^-. See also ARSENIC COMPOUNDS.

ARK780 **CAS:52740-16-6** **HR: 3**
ARSONIC ACID, CALCIUM SALT (1:1)
mf: $AsH_2O_3 \cdot Ca$ mw: 165.02

CONSENSUS REPORTS: NTP 7th Annual Report on Carcinogens.

SAFETY PROFILE: Confirmed carcinogen. When heated to decomposition it emits toxic fumes of As.

ARK800 **HR: 3**
(3-(p-ARSONOPHENYL)UREIDO)DITHIOBENZOIC ACID
mf: $C_{14}H_{13}AsN_2O_4S_2$ mw: 412.33

SYN: DITHIOCARBOXYPHENYL-p-CARBAMIDOPHENYLARSENOUS OXIDE

TOXICITY DATA WITH REFERENCE
orl-rat LD50:850 mg/kg FEPRA7 6,306,47
ipr-rat LD50:76 mg/kg FEPRA7 6,306,47
ipr-mus LD50:265 mg/kg FEPRA7 6,306,47

CONSENSUS REPORTS: Arsenic and its compounds are on the Community Right-To-Know List.

SAFETY PROFILE: Poison by intraperitoneal route. Moderately toxic by ingestion and other routes. When heated to decomposition it emits toxic fumes of SO_x, NO_x, and arsenic. See also ARSENIC COMPOUNDS.

ARL000 **CAS:538-03-4** **HR: 3**
ARSPHENOXIDE
mf: $C_6H_6AsNO_2 \cdot ClH$ mw: 235.51

SYNS: 2-AMINO-4-ARSENOSOPHENOL HYDROCHLORIDE □ 3-AMINO-4-HYDROXY-PHENARSINE HYDROCHLORIDE □ 3-AMINO-4-HYDROXYPHENYLARSINE OXIDE HYDROCHLORIDE □ 3-AMINO-4-HYDROXYPHENYL ARSINOXIDE HYDROCHLORIDE □ ARSENO 39 □ ARSENOSAN □ ARSENOXIDE □ EHRLICH 5 □ FONTARSAN □ MAPHARSAL □ MAPHARSEN □ MAPHARSIDE □ OXIARSOLAN □ OXOPHENARSINE HYDROCHLORIDE

TOXICITY DATA WITH REFERENCE
ivn-hmn TDLo:400 µg/kg:CNS,BRN AJSGA3 28,218,44
orl-rat LDLo:500 mg/kg NCNSA6 5,12,53
ivn-rat LD50:18,500 µg/kg JACSAT 70,1762,48
orl-mus LD50:110 µg/kg CLDND* 81,284,44
scu-mus LD50:25 mg/kg JPETAB 81,284,44
ivn-mus LD50:20 mg/kg JPETAB 81,284,44
ivn-dog LDLo:12 mg/kg JPETAB 50,198,34

CONSENSUS REPORTS: Arsenic and its compounds are on the Community Right-To-Know List.

OSHA PEL: TWA 0.5 mg(As)/m³
ACGIH TLV: TWA 0.2 mg(As)/m³

SAFETY PROFILE: Poison by ingestion, intravenous, and intraperitoneal routes. Human systemic effects by ingestion: stroke, convulsions, and coma. See also ARSENIC COMPOUNDS. When heated to decomposition it emits very toxic fumes of As, NO_x, and HCl. An antirickettsial and antitrypanosomal agent.

ARL250 **CAS:8022-37-5** **HR: 2**
ARTEMISIA OIL

PROP: Chief constituent is Thujone, and found in the plant *Artemisia absinthium* L. (FCTXAV 13,681,75).

SYNS: ABSINTHIUM □ ARTEMISIA OIL (WORMWOOD) □ OIL, ARTEMISIA

TOXICITY DATA WITH REFERENCE
orl-rat LD50:960 mg/kg FCTXAV 13,681,75

CONSENSUS REPORTS: Reported in EPA TSCA Inventory.

SAFETY PROFILE: Moderately toxic by ingestion. An allergen. Habitual users develop "absinthism" with tremors, vertigo, vomiting, and hallucinations. May cause a contact dermatitis. When heated to decomposition it emits acrid smoke and irritating fumes.

ARL375 **CAS:63968-64-9** **HR: 2**
ARTEMISINE
mf: $C_{15}H_{22}O_5$ mw: 282.37

PROP: Needles. Mp: 156–157°. Sol in most aprotic solvents. Sltly sol in oil.

SYNS: ARTEANNUIN □ ARTEMISININ □ OCTAHYDRO-3,6,9-TRIMETHYL-3,12-EPOXY-12H-PYRANO(4,3-j)-1,2-BENZODIOXEPIN-10(3H)-ONE □ QINGHAOSU (CHINESE) □ QING HAU SAU (CHINESE)

TOXICITY DATA WITH REFERENCE
orl-mus LD50:5105 mg/kg CMJODS 92,811,79
ipr-mus LD50:1558 mg/kg CMJODS 92,811,79
ims-mus LD50:2800 mg/kg CMJODS 92,811,79

SAFETY PROFILE: Moderately toxic by ingestion, intramuscular, and intraperitoneal routes. When heated to decomposition it emits acrid smoke and fumes.

ARL425 CAS:71963-77-4 HR: 3
ARTEMISININELACTOL METHYL ETHER
mf: $C_{16}H_{26}O_5$ mw: 298.42

SYNS: ARTEMETHER □ 3,12-EPOXY-12H-PYRANO(4,3-j)-1,2-BENZODIOXEPIN, DECAHYDRO-10-METHOXY-3,6,9-TRIMETHYL-, (3-α-5a-β,6-β,8a-β,9-α-12-β,12aR)-, (+)- □ METHYL-DIHYDROARTEMISININE

TOXICITY DATA WITH REFERENCE
orl-rat TDLo:200 mg/kg (female 6-15D post):REP CYLPDN 5,118,84
ims-mus LD50:263 mg/kg CYLPDN 2,138,81

SAFETY PROFILE: Poison by intramuscular route. Experimental reproductive effects. When heated to decomposition it emits acrid smoke and irritating fumes.

ARL500 CAS:149-95-1 HR: 3
d-ARTERENOL
mf: $C_8H_{11}NO_3$ mw: 169.20

PROP: Solid. Mp: 215–217° (decomp).

SYN: α (AMINOMETHYL)-3,4-DIHYDROXYBENZYL ALCOHOL

TOXICITY DATA WITH REFERENCE
ivn-rat LD50:1400 μg/kg JPETAB 95,502,49
ivn-mus LD50:18 mg/kg JPETAB 95,502,49

SAFETY PROFILE: Poison by intravenous route. When heated to decomposition it emits toxic fumes of NO_x.

ARL750 CAS:138-65-8 HR: 3
dl-ARTERENOL
mf: $C_8H_{11}NO_3$ mw: 169.20

PROP: Solid. Mp: 191° (decomp). Sltly sol in water.

SYN: α-(AMINOMETHYL)-3,4-DIHYDROXYBENZYL ALCOHOL

TOXICITY DATA WITH REFERENCE
ivn-rat LD50:130 μg/kg JPETAB 95,502,49
ivn-mus LD50:4700 μg/kg JPETAB 95,502,49

SAFETY PROFILE: Poison by intravenous route. When heated to decomposition it emits toxic fumes of NO_x.

ARL875 HR: 2
ARYL ALKYL POLYETHER ALCOHOL

SYNS: ALKYL ARYL POLYETHER ALCOHOLS □ ALKYLARYLPOLYGLYKOLAETHER (GERMAN) □ ALKYL PHENOL POLYGLYCOL ETHERS □ ALKYL PHENOXY POLYETHOXY ETHANOLS

TOXICITY DATA WITH REFERENCE
orl-rat LD50:3200 mg/kg 85GYAZ -,144,71
ipr-mus LD50:3100 mg/kg PSTGAW 3,1,45
orl-rbt LD50:3 g/kg 85GYAZ -,144,71

SAFETY PROFILE: Moderately toxic by ingestion and other routes.

ARM000 CAS:13425-94-0 HR: 3
ASALIN
mf: $C_{22}H_{33}Cl_2N_3O_4$ mw: 474.48

SYNS: N-ACETYL-SARCOLYSIL VALINE ETHYL ETHER □ ASALINE □ AZALINE □ ETHYL ESTER of N-ACETYL-dl-SARCOSYLYL-dl-VALINE

TOXICITY DATA WITH REFERENCE
ipr-rat TDLo:50 mg/kg (female 6D post):TER VOONAW 13(11),79,67
orl-rat LD50:59 mg/kg FATOAO 33,472,70
ipr-rat LD50:187 mg/kg PCJOAU 12,25,78
ims-rat LD50:17 mg/kg FATOAO 33,472,70
rec-rat LD50:40 mg/kg FATOAO 33,472,70

SAFETY PROFILE: Poison by ingestion, intramuscular, rectal, and intraperitoneal routes. An experimental teratogen. See also ESTERS.

ARM250 CAS:1332-21-4 HR: 3
ASBESTOS
DOT: NA 2212

SYNS: AMIANTHUS □ AMOSITE (OBS.) □ AMPHIBOLE □ ASBEST (GERMAN) □ ASBESTOS FIBER □ FIBROUS GRUNERITE □ NCI-C08991 □ SERPENTINE

TOXICITY DATA WITH REFERENCE
mmo-esc 10 mg/plate PSEBAA 177,343,84
mma-esc 10 mg/plate PSEBAA 177,343,84 VOONAW 20(4),47,74
imp-rat TDLo:750 mg/kg:ETA ZEKBAI 62,561,58
ihl-hmn TCLo:1.2 fibers/cc/19Y-C:PUL ARDSBL 104,576,71

CONSENSUS REPORTS: NTP 7th Annual Report On Carcinogens. IARC Cancer Review: Group 1 IMEMDT 7,106,87; Human Sufficient Evidence IMEMDT 2,17,73; IMEMDT 14,11,77; Animal Sufficient Evidence IMEMDT 2,17,73; IMEMDT 14,11,77 Reported in EPA TSCA Inventory. On Community Right-To-Know List. EPA Genetic Toxicology Program.

OSHA PEL: TWA 2 million fibers/m^3; CL 10 million fibers/m^3; Cancer and Lung Disease Hazard
ACGIH TLV: TWA 2 fibers/cc; Confirmed Human Carcinogen; (Proposed: TWA 0.2 fibers/cc)
DFG TRK:(Fine dust particles that are able to reach the alveolar area of the lung) crocidolite: 0.05×10^6 fibers/m^3 (0.025 mg/m^3) (definition of fiber: length greater than 5 μm; diameter less than 3 μm; length/diameter greater than 3:1, equivalent to 1 fiber/cc); chrysotile, amosite, anthophyllite, tremolite, actinolite: 1×10^6 fibers/m^3 (0.05 mg/m^3), applicable when there is more than 2.5% asbestos in the dust; 2.0 mg/m^3, applicable when there is less than or equal to 2.5 weight percent asbestos in fine dust
NIOSH REL: TWA 100,000 fibers/m^3 over 5 μm in length
DOT CLASSIFICATION: 9; *Label:* CLASS 9

SAFETY PROFILE: Confirmed human carcinogen producing lung tumors. Experimental neoplastigenic and tumorigenic data. Human pulmonary system effects by inhalation. Usually at least 4 to 7 years of exposure are required before serious lung damage (fibrosis) results. Mutation data reported. A common air contaminant.

For occupational chemical analysis use NIOSH: Fibers, 7400.

ARM260 CAS:77536-66-4 HR: 3
ASBESTOS, ACTINOLITE
DOT: NA 2212

SYNS: ACTINOLITE ASBESTOS □ ASBESTOS (ACGIH)

TOXICITY DATA WITH REFERENCE
ipr-rat TDLo:50 mg/kg:ETA STRHAV 39,386,79

CONSENSUS REPORTS: IARC Cancer Review: Group 1 IMEMDT 7,106,87; Animal Sufficient Evidence IMEMDT 14,11,77.

OSHA PEL: TWA 2 million fibers/m^3; CL 10 million fibers/m^3; Cancer Hazard
ACGIH TLV: TWA 2 fibers/cc; Confirmed Human Carcinogen; (Proposed: TWA 0.2 fibers/cc)
DFG TRK:(Fine dust particles that are able to reach the alveolar area of the lung) 1×10^6 fibers/m^3 (0.05 mg/m^3), applicable when there is more than 2.5% asbestos in the dust
NIOSH REL: TWA 100,000 fibers/m^3 over 5 µm in length
DOT CLASSIFICATION: 9; *Label:* CLASS 9

SAFETY PROFILE: Confirmed human carcinogen. See also other asbestos entries.

ARM262 CAS:12172-73-5 HR: 3
ASBESTOS, AMOSITE

SYNS: AMOSITE ASBESTOS □ ASBESTOS (ACGIH) □ MYSORITE □ NCI-C60253A

TOXICITY DATA WITH REFERENCE
cyt-ham:ovr 10 mg/L CSHCAL 4,941,77
sce-ham:ovr 10 mg/L JEPTDQ 4(2-3),373,80
msc-ham:lng 10 mg/L MUREAV 68,265,79
ihl-rat TCLo:11 mg/m^3/2Y-I:CAR BJCAAI 29,252,74
ipl-rat TDLo:80 mg/kg:CAR TOLED5 13,143,82
itr-rat TDLo:12 mg/kg/12W-I:ETA TOLED5 13,143,82
imp-rat TDLo:200 mg/kg:NEO JJIND8 67,965,81

CONSENSUS REPORTS: NTP 7th Annual Report on Carcinogens. IARC Cancer Review: Group 1 IMEMDT 7,106,87; Animal Sufficient Evidence IMEMDT 2,17,73; IMEMDT 14,11,77; Human Sufficient Evidence IMEMDT 2,17,73; IMEMDT 14,11,77. NTP Carcinogenesis Studies (feed); No Evidence: hamster NTPTR* NTP-TR-249,83. EPA Genetic Toxicology Program.

OSHA PEL: TWA 2 million fibers/m^3; CL 10 million fibers/m^3; Cancer Hazard
ACGIH TLV: TWA 0.5 fibers/cc; Confirmed Human Carcinogen; (Proposed: TWA 0.2 fibers/cc)
DFG TRK:(Fine dust particles that are able to reach the alveolar area of the lung) 1×10^6 fibers/m^3 (0.05 mg/m^3) applicable when there is more than 2.5% asbestos in the dust
NIOSH REL: TWA 100,000 fibers/m^3 over 5 µm in length

SAFETY PROFILE: Confirmed human carcinogen with experimental carcinogenic, neoplastigenic, and tumorigenic data. Mutation data reported.

For occupational chemical analysis use NIOSH: Fibers, 7400; Asbestos Fibers, 7402.

ARM264 CAS:77536-67-5 HR: 3
ASBESTOS, ANTHOPHYLITE

SYNS: ANTHOPHYLITE □ ASBESTOS (ACGIH) □ AZBOLEN ASBESTOS □ FERROANTHOPHYLLITE

TOXICITY DATA WITH REFERENCE
cyt-ham:ovr 10 mg/L CSHCAL 4,941,77
ihl-rat TCLo:11 mg/m^3/1Y-I:CAR BJCAAI 29,252,74
ipl-rat TDLo:200 mg/kg:NEO BJCAAI 28,173,73
ipl-rat TD:2400 mg/kg/34W-I:ETA IAPUDO 30,343,80

CONSENSUS REPORTS: NTP 7th Annual Report on Carcinogens. IARC Cancer Review: Animal Sufficient Evidence IMEMDT 2,17,73; IMEMDT 14,11,77; Human Sufficient Evidence IMEMDT 14,11,77. EPA Genetic Toxicology Program.

OSHA PEL: TWA 2 million fibers/m^3; CL 10 million fibers/m^3; Cancer Hazard
ACGIH TLV: TWA 2 fibers/cc; Confirmed Human Carcinogen; (Proposed: TWA 0.2 fibers/cc)
DFG TRK:(Fine dust particles that are able to reach the alveolar area of the lung) 1×10^6 fibers/m^3 (0.05 mg/m^3), applicable when there is more than 2.5% asbestos in the dust
NIOSH REL: TWA 100,000 fibers/m^3 over 5 µm in length

SAFETY PROFILE: Confirmed human carcinogen with experimental carcinogenic, neoplastigenic, and tumorigenic data. Mutation data reported.

ARM266 CAS:17068-78-9 HR: 3
ASBESTOS, ANTHOPHYLLITE

SYNS: AZBOLEN ASBESTOS □ 16 F

TOXICITY DATA WITH REFERENCE
otr-ham:emb 3500 µg/m^3 CRNGDP 9,891,88
ipr-rat TDLo:250 mg/kg:ETA ZHYGAM 32,89,86

CONSENSUS REPORTS: NTP 7th Annual Report on Carcinogens.

ACGIH TLV: TWA 2 fibers/cc; Confirmed Human Carcinogen; (Proposed: TWA 0.2 fibers/cc)

SAFETY PROFILE: Confirmed carcinogen with experimental tumorigenic data. Mutation data reported.

For occupational chemical analysis use NIOSH: Fibers, 7400; Asbestos Fibers, 7402.

ARM268 CAS:12001-29-5 HR: 3
ASBESTOS, CHRYSOTILE
DOT: NA 2212

PROP: Silky white to green to brownish fibers.

SYNS: 7-45 ASBESTOS □ ASBESTOS (ACGIH) □ AVIBEST C □ CALIDRIA RG 100 □ CALIDRIA RG 144 □ CALIDRIA RG 600 □ CASSIAR AK □ CHRYSOTILE ASBESTOS □ HOOKER NO. 1 CHRYSOTILE

ASBESTOS □ K6-30 □ METAXITE □ NCI C61223A □ PLASTIBEST 20 □ 5R04 □ RG 600 □ SERPENTINE □ SERPENTINE CHRYSOTILE □ SYLODEX □ WHITE ASBESTOS □ WHITE ASBESTOS, (chrysotile, actinolite, anthophyllite, tremolite) (DOT)

TOXICITY DATA WITH REFERENCE
oms-hmn:fbr 10 mg/L MUREAV 116,369,83
oms-ham:ovr 10 mg/L MUREAV 116,369,83
ihl-man TCLo:400 mppcf/1Y-C:CAR,PUL AEHLAU 28,61,74
orl-rat TDLo:7100 mg/kg/39W-C:CAR ARGEAR 46,437,76
ihl-rat TCLo:11 mg/m^3/26W-I:CAR BJCAAI 29,252,74
itr-rat TDLo:13 mg/kg:ETA ENVRAL 21,63,80
scu-mus TDLo:2400 mg/kg/13W-I:NEO FCTXAV 6,566,68
ihl-hmn TCLo:2.8 fibers/cc/5Y:PUL ENVRAL 23,292,80
ipr-rat LDLo:300 mg/kg AJPAA4 70,291,73

CONSENSUS REPORTS: NTP 7th Annual Report on Carcinogens. IARC Cancer Review: Human Sufficient Evidence IMEMDT 2,17,73; Animal Sufficient Evidence IMEMDT 2,17,73. NTP Carcinogenesis Studies (feed); Some Evidence: rat NTPTR* NTP-TR-295,85. EPA Genetic Toxicology Program.

OSHA PEL: TWA 2 million fibers/m^3; CL 10 million fibers/m^3; Cancer Hazard
ACGIH TLV: TWA 2 fibers/cc; Confirmed Human Carcinogen; (Proposed: TWA 0.2 fibers/cc)
DFG TRK:(Fine dust particles that are able to reach the alveolar area of the lung) 1 × 10^6 fibers/m^3 (0.05 mg/m^3), applicable when there is more than 2.5% asbestos in the dust
NIOSH REL: TWA 100,000 fibers/m^3 over 5 μm in length
DOT CLASSIFICATION: 9; *Label:* CLASS 9

SAFETY PROFILE: Confirmed human carcinogen producing tumors of the lung. Human mutation data reported. Poison by intraperitoneal route. Human systemic effects by inhalation: lung fibrosis, dyspnea, and cough.

For occupational chemical analysis use NIOSH: Fibers, 7400; Asbestos Fibers, 7402.

ARM275 CAS:12001-28-4 HR: 3
ASBESTOS, CROCIDOLITE
DOT: NA 2212
mf: $ONa_2Fe_2O_{33}FeO_8SiO_2H_2O$ mw: 765.98

SYNS: AMORPHOUS CROCIDOLITE ASBESTOS □ ASBESTOS (ACGIH) □ BLUE ASBESTOS (DOT) □ BROWN ASBESTOS (DOT) □ CROCIDOLITE ASBESTOS □ CROCIDOLITE (DOT) □ FIBROUS CROCIDOLITE ASBESTOS □ KROKYDOLITH (GERMAN) □ NCI-C09007

TOXICITY DATA WITH REFERENCE
oms-hmn:fbr 10 mg/L MUREAV 116,369,83
dns-ham:oth 1280 ng/cm^2 CNREA8 42,3669,82
ihl-rat TCLo:11 mg/m^3/1Y-I:CAR BJCAAI 29,252,74
ipr-rat TDLo:100 mg/kg:CAR PBPHAW 14,47,78
imp-rat TDLo:200 mg/kg:NEO JJIND8 67,965,81
ipl-mus TDLo:200 mg/kg:ETA 31BYAP ,92,74

CONSENSUS REPORTS: NTP 7th Annual Report on Carcinogens. IARC Cancer Review: Animal Sufficient Evidence IMEMDT 14,11,77, IMEMDT 2,17,73; Human Sufficient Evidence IMEMDT 14,11,77. EPA Genetic Toxicology Program.

OSHA PEL: TWA 2 million fibers/m^3; CL 10 million fibers/m^3; Cancer Hazard
ACGIH TLV: TWA 0.2 fibers/cc; Confirmed Human Carcinogen
DFG TRK:(Fine dust particles that are able to reach the alveolar area of the lung) crocidolite: 0.05 × 10^6 fibers/m^3 (0.025 mg/m^3) (definition of fiber: length greater than 5 μm; diameter less than 3 μm; length/diameter greater than 3:1, equivalent to 1 fiber/cc)
NIOSH REL: TWA 100,000 fibers/m^3 over 5 μm in length
DOT CLASSIFICATION: 9; *Label:* CLASS 9

SAFETY PROFILE: Confirmed human carcinogen with experimental carcinogenic, neoplastigenic, and tumorigenic data by inhalation. Human mutation data reported.

For occupational chemical analysis use NIOSH: Fibers, 7400; Asbestos Fibers, 7402.

ARM280 CAS:77536-68-6 HR: 3
ASBESTOS, TREMOLITE
DOT: NA 2212

SYNS: ASBESTOS (ACGIH) □ FIBROUS TREMOLITE □ NCI-C08991 □ TREMOLITE ASBESTOS

TOXICITY DATA WITH REFERENCE
ipl-rat TDLo:100 mg/kg:NEO BJCAAI 45,352,82
ipl-ham TDLo:80 mg/kg:ETA 43GRAK-,335,79

CONSENSUS REPORTS: IARC Cancer Review: Human Sufficient Evidence IMEMDT 14,11,77; Animal Sufficient Evidence IMEMDT 14,11,77.

OSHA PEL: TWA 2 million fibers/m^3; CL 10 million fibers/m^3; Cancer Hazard
ACGIH TLV: TWA 2 fibers/cc; Confirmed Human Carcinogen; (Proposed: TWA 0.2 fibers/cc)
DFG TRK:(Fine dust particles that are able to reach the alveolar area of the lung) 1 × 10^6 fibers/m^3 (0.05 mg/m^3), applicable when there is more than 2.5% asbestos in the dust.
NIOSH REL: TWA 100,000 fibers/m^3 over 5 μm in length
DOT CLASSIFICATION: 9; *Label:* CLASS 9

SAFETY PROFILE: Confirmed human carcinogen with experimental tumorigenic and neoplastigenic data.

For occupational chemical analysis use NIOSH: Fibers, 7400; Asbestos Fibers, 7402.

ARM500 CAS:512-85-6 HR: 3
ASCARIDOLE
mf: $C_{10}H_{16}O_2$ mw: 168.26

PROP: Colorless unstable liquid. Mp: 3.3°, bp: 40° @ 2 mm; 115° @ 15 mm, d: 1.010 @ 20°/4°.

SYNS: ASCARIDOL □ ASCARIDOLE (organic peroxide) (DOT) □ ASCARISIN □ 2,3-DIOXABICYCLO(2.2.2)OCT-5-ENE, 1-ISOPROPYL-4-METHYL- □ 1-METHYL-4-(1-METHYLETHYL)-2,3-DIOXABICYCLO (2.2.2)OCT-5-ENE □ 1,4-PEROXIDO-p-MENTHENE-2

TOXICITY DATA WITH REFERENCE
skn-rbt 500 mg MLD SCCUR*-,1,61
skn-mus TDLo:25 g/kg/42W-I:NEO JNCIAM 35,707,65
skn-mus TD:38 g/kg/63W-I:ETA 14JTAF -,275,64
orl-rat LD50:200 mg/kg FEPRA7 7,252,48
orl-mus LD50:400 mg/kg FEPRA7 7,252,48
orl-dog LDLo:250 mg/kg JPETAB 24,359,25

DOT CLASSIFICATION: Forbidden

SAFETY PROFILE: Poison by ingestion. Questionable carcinogen with experimental neoplastigenic and tumorigenic data. Flammable by spontaneous chemical reaction. An oxidizer. Explodes when heated >130° or when exposed to organic acids. Dangerous; heating emits toxic fumes and may explode; reacts with reducing materials. See also CHENOPODIUM OIL, and PEROXIDES, ORGANIC.

ARM750 CAS:38462-04-3 HR: 2
ASCOFURANONE

PROP: Needles from Me_2CO/hexane. Mp: 84°.

SYN: (S-(e,e))-3-CHLORO-4,6-DIHYDROXY-2-METHYL-5-(3-METHYL-7-(TETRAHYDRO-5,5-DIMETHYL-4-OXO-2-FURANYL)-2,6-OCTADIENYL)-BENZALDEHYDE

TOXICITY DATA WITH REFERENCE
ipr-rat LD50:1350 mg/kg JANTAJ 26,681,73
ipr-mus LD50:2220 mg/kg JANTAJ 26,681,73

SAFETY PROFILE: Moderately toxic by intraperitoneal route. See also ALDEHYDES. When heated to decomposition it emits toxic fumes of Cl^-.

ARN000 CAS:50-81-7 HR: 2
l-ASCORBIC ACID
mf: $C_6H_8O_6$ mw: 176.14

PROP: White crystals. Mp: 192°. Sol in water; sltly sol in alc; insol in ether, chloroform, benzene, petroleum ether, fixed oils, and fats.

SYNS: ASCORBIC ACID □ l(+)-ASCORBIC ACID □ASCORBUTI-NA □ CEVITAMIC ACID □ CEVITAMIN □ FEMA No. 2109 □ 3-KETO-l-GULOFURANOLACTONE □ l-3-KETOTHREOHEXURONIC ACID LACTONE □ NATRASCORB INJECTABLE □ NCI-C54808 □ 3-OXO-l-GULOFURANOLACTONE □ VITACIN □ VITAMIN C □ VITAMISIN □ VITASCORBOL □ XITIX □ l-XYLOASCORBIC ACID

TOXICITY DATA WITH REFERENCE
mmg-sat:500 µg/plate ABCHA6 45,327,81
mmo-nsc 2 mmol/L MAGDA3 10,249,79
orl-gpg TDLo:19,500 mg/kg (female 30-58D post):REP ANYAA9 258,401,75
ivn-mus TDLo:800 mg/kg (female 8D post):TER TOIZAG 8,175,61
ivn-man TDLo:2300 mg/kg/2D:BLD AIMEAS 82,810,75
ivn-wmn LDLo:900 mg/kg:SYS AIMDAP 145,950,85
orl-rat LD50:11,900 mg/kg OYYAA2 19,323,80
orl-mus LD50:3367 mg/kg NCISP* JAN86
ivn-mus LD50:518 mg/kg RPOBAR 2,269,70

CONSENSUS REPORTS: NTP Carcinogenesis Bioassay (feed); No Evidence: mouse, rat NTPTR* NTP-TR-247,83; NTPTR* NTP-TR-214,82. Reported in EPA TSCA Inventory.

SAFETY PROFILE: Moderately toxic by ingestion and intravenous routes. Human systemic effects by intravenous route: blood, changes in tubules (including acute renal failure, acute tubular necrosis). An experimental teratogen. Other experimental reproductive effects. Mutation data reported. When heated to decomposition it emits acrid smoke and irritating fumes.

ARN250 CAS:1402-88-6 HR: 3
ASCOSIN

PROP: Produced by *Streptomycete canescus* (ANTCAO 2,472,52)

TOXICITY DATA WITH REFERENCE
orl-mus LD50:500 mg/kg 85ERAY 2,1035,78
ipr-mus LD50:9 mg/kg ANTCAO 2,472,52
scu-mus LD50:72 mg/kg 85ERAY 2,1035,78
ivn-mus LD50:13 mg/kg ANTCAO 2,472,52

SAFETY PROFILE: Poison by intravenous, subcutaneous, and intraperitoneal routes. Moderately toxic by ingestion.

ARN500 CAS:16830-15-2 HR: 2
ASIATICOSIDE

PROP: Crystals. Mp: 235–238°. A glycoside terpene from the plant *Centella asiatica* (CNREA8 32,1463,72).

SYNS: BLASTOESTIMULINA □ CENTELASE □ DERMATOLOGICO □ MADECASSOL

TOXICITY DATA WITH REFERENCE
skn-mus TDLo:400 mg/kg/52W-I:ETA CNREA8 32,1463,72

SAFETY PROFILE: Questionable carcinogen with experimental tumorigenic data. When heated to decomposition it emits acrid smoke and fumes. Promotes healing of wounds.

ARN700 CAS:35844-94-1 HR: 3
ASPAMINOL HYDROCHLORIDE
mf: $C_{20}H_{25}NO{\cdot}ClH$ mw: 331.92

SYNS: 1,1-DIPHENYL-3-N-PIPERIDINOBUTANOL-1 HYDROCHLORIDE □ (±)-γ-METHYL-α,α-DIPHENYL-1-PYRROLIDINEPROPANOL HYDROCHLORIDE

TOXICITY DATA WITH REFERENCE
orl-mus LD50:642 mg/kg NIIRDN 6,333,82
scu-mus LD50:354 mg/kg NIIRDN 6,333,82
ivn-mus LD50:42,700 µg/kg NIIRDN 6,333,82

SAFETY PROFILE: Poison by subcutaneous and intravenous routes. Moderately toxic by ingestion. When heated to decomposition it emits toxic fumes of NO_x and HCl.

ARN800 CAS:9015-68-3 HR: 2
l-ASPARAGINASE

PROP: Crystals or powder. Sol in water.

SYNS: ASPARAGINASE □ l-ASPARAGINASE X □ l-ASPARAGINASI (ITALIAN) □ l-ASPARAGINE AMIDOHYDROLASE □ LEUCOGEN □ NSC-109229

TOXICITY DATA WITH REFERENCE
ivn-mus TDLo:31,815 μg/kg (female 7-13D post):TER YIKUAO 18,271,69
ivn-rat TDLo:3000 iu/kg (female 6-15D post):REP RRCRBU 33,174,70
ipr-mus TDLo:10 iu/kg/4D:NEO BSIBAC 47,418,71
ims-chd TDLo:8145 iu/kg/1W:SYS CANCAR 34,780,74
ipr-rat LD50:8204 mg/kg YIKUAO 18,271,69
scu-rat LD50:8204 mg/kg YIKUAO 18,271,69
ivn-rat LD50:7568 mg/kg YIKUAO 18,271,69
ivn-dog LD50 50,000 iu/kg RRCRBU 33,174,70
ivn-cat LD50 50,000 iu/kg RRCRBU 33,174,70
ivn-rbt LDLo 500 iu/kg RRCRBU 33,174,70

SAFETY PROFILE: Human (child) systemic effects by intramuscular route. An experimental teratogen. Other experimental reproductive effects. Questionable carcinogen with experimental neoplastigenic data.

ARN825 **CAS:22839-47-0** **HR: 1**
ASPARTAME
mf: $C_{14}H_{18}N_2O_5$ mw: 294.34

PROP: White crystalline powder from water or alc; odorless with a sweet taste. Mp: 190°. Sltly sol in water, alc.

SYNS: 3-AMINO-N-(α-CARBOXYPHENETHYL)SUCCINAMIC ACID N-METHYL ESTER, stereoisomer □ ASPARTYLPHENYLALANINE METHYL ESTER □ N-l-α-ASPARTYL-l-PHENYLALANINE 1-METHYL ESTER (9CI) □ CANDEREL □ DIPEPTIDE SWEETENER □ EQUAL □ METHYL ASPARTYLPHENYLALANATE □ 1-METHYL N-l-α-ASPARTYL-l-PHENYLALANINE □ NUTRASWEET □ SWEET DIPEPTIDE

TOXICITY DATA WITH REFERENCE
orl-mus TDLo:4 g/kg (15-18D preg):REP RCPBDC 9,385,85
orl-wmn TDLo:3710 μg/kg:SKN AIMEAS 104,207,86

SAFETY PROFILE: Human systemic effects by ingestion: allergic dermatitis. Experimental reproductive effects. When heated to decomposition it emits toxic fumes of NO_x.

For occupational chemical analysis use NIOSH: Aspartame 5031.

ARN830 **CAS:617-45-8** **HR: 1**
dl-ASPARTIC ACID
mf: $C_4H_7NO_4$ mw: 133.10

PROP: Colorless to white monoclinic crystals; acid taste. Mp: 280° (decomp). Sltly sol in water; insol in alc, ether.

SYN: dl-AMINOSUCCINIC ACID

TOXICITY DATA WITH REFERENCE
ipr-mus LD50:>6 g/kg KHFZAN 24(5),17,90

CONSENSUS REPORTS: Reported in EPA TSCA Inventory.

SAFETY PROFILE: Very low toxicity by intraperitoneal route. When heated to decomposition emits toxic fumes of NO_x.

ARN850 **CAS:56-84-8** **HR: 1**
l-ASPARTIC ACID
mf: $C_4H_7NO_4$ mw: 133.10

PROP: Colorless to white crystals or leaflets; acid taste. Mp: 270°. Sltly sol in water; insol in alc, ether.

SYNS: (S)-AMINOBUTANEDIOIC ACID □ l-AMINOSUCCINIC ACID □ ASPARAGIC ACID □ l-ASPARAGIC ACID □ ASPARAGINIC ACID □ l-ASPARAGINIC ACID □ ASPARTIC ACID □ (l)-ASPARTIC ACID □ l-(+)-ASPARTIC ACID □ (S)-ASPARTIC ACID

TOXICITY DATA WITH REFERENCE
ipr-mus LD50:6 g/kg PCJOAU 25,569,91

CONSENSUS REPORTS: Reported in EPA TSCA Inventory.

SAFETY PROFILE: Low toxicity by intraperitoneal route. When heated to decomposition emits toxic fumes of NO_x.

ARN875 **HR: 3**
ASPERASE

TOXICITY DATA WITH REFERENCE
ipr-rat LD50:65,300 μg/kg FATOAO 45(6),78,82
ivn-rat LD50:52,500 μg/kg FATOAO 45(6),78,82
ipr-mus LD50:167 mg/kg FATOAO 45(6),78,82
ivn-mus LD50:107 mg/kg FATOAO 45(6),78,82
ivn-gpg LD50:8200 μg/kg FATOAO 45(6),78,82

SAFETY PROFILE: Poison by intravenous and intraperitoneal routes.

ARO000 **CAS:490-02-8** **HR: 3**
ASPERGILLIC ACID
mf: $C_{12}H_{20}N_2O_2$ mw: 224.34

SYNS: 2-HYDROXY-3-ISOBUTYL-6-(1-METHYLPROPYL)PYRAZINE 1-OXIDE □ 3-ISOBUTYL-6-sec-BUTYL-2-HYDROXYPYRAZINE-1-OXIDE

TOXICITY DATA WITH REFERENCE
orl-mus LDLo:200 mg/kg JOBAAY 45,433,43
ipr-mus LDLo:150 mg/kg JOBAAY 45,433,43

SAFETY PROFILE: Poison by ingestion and intraperitoneal routes. When heated to decomposition it emits toxic fumes of NO_x.

ARO250 **CAS:67-99-2** **HR: 3**
ASPERGILLIN
mf: $C_{13}H_{14}N_2O_4S_2$ mw: 326.41

PROP: Monoclinic crystals from MeOH. Mp: 221° (decomp).

SYN: GLIOTOXIN

TOXICITY DATA WITH REFERENCE
ipr-mus LDLo:45 mg/kg 85ERAY 3,1919,78
ivn-mus LDLo:45 mg/kg 85ERAY 3,1919,78
ivn-rbt LDLo:45 mg/kg JACSAT 65,2005,43

SAFETY PROFILE: Poison by intraperitoneal and intravenous routes. When heated to decomposition it emits very toxic fumes such as SO_x and NO_x.

ARO500 **CAS:8052-42-4** **HR: 3**
ASPHALT
DOT: NA 1999

PROP: Black or dark-brown mass. Bp: <470°, flash p: 400+°F (CC), d: 0.95–1.1, autoign temp: 905°F.

SYNS: ASPHALT, at or above its Fp (DOT) □ ASPHALT FUMES (ACGIH) □ ASPHALT, PETROLEUM □ ASPHALTUM □ BITUMEN (MAK) □ JUDEAN PITCH □ MINERAL PITCH □ PETROLEUM ASPHALT □ PETROLEUM BITUMEN □ PETROLEUM PITCH □ PETROLEUM ROOFING TAR □ ROAD ASPHALT (DOT) □ ROAD TAR (DOT)

TOXICITY DATA WITH REFERENCE
skn-mus TDLo:130 g/kg/81W-I:CAR HYSAAV 33(4-6),180,68
skn-mus TD:69 g/kg/43W-I:ETA HYSAAV 33(4-6),180,68

CONSENSUS REPORTS: IARC Cancer Review: Group 3 IMEMDT 7,133,87; Human Inadequate Evidence IMEMDT 35,39,85. Reported in EPA TSCA Inventory.

ACGIH TLV: TWA 5 mg/m³
DFG MAK: Suspected Carcinogen
NIOSH REL: (Asphalt Fumes) CL 5 mg/m³/15M
DOT CLASSIFICATION: 3; *Label:* Flammable Liquid

SAFETY PROFILE: Suspected carcinogen with experimental carcinogenic and tumorigenic data. A moderate irritant. May contain carcinogenic components. Combustible when exposed to heat or flame. To fight fire, use foam, CO_2, or dry chemical.

ARO750 **CAS:8052-42-4** **HR: 3**
ASPHALT (CUT BACK)

PROP: A liquid petroleum product, solubility of residue from distillation in carbon tetrachloride = 99.5%. Flash p: <50°F.

SYNS: ROAD ASPHALT (DOT) □ ROAD TAR, liquid (DOT)

SAFETY PROFILE: Contains carcinogenic components. A dangerous fire hazard when exposed to heat or flame. To fight fire, use dry chemical, water mist, fog. When heated to decomposition it emits smoke and irritating acrid fumes.

ARP000 **CAS:2096-42-6** **HR: 3**
ASPICULAMYCIN
mf: $C_{16}H_{25}N_7O_8$ mw: 443.48

PROP: Needles from MeOH. Mp: 211–217° (decomp).

SYNS: ASTEROMYCIN □ 1-(4-DEOXY-4-(SARCOSYL-d-SERYL)AMINO-β-d-GLUCOPYRANURONAMIDE) CYTOSINE □ GOUGEROTIN □ QUINGFENGMYCIN

TOXICITY DATA WITH REFERENCE
ipr-mus LD50:57 mg/kg 85FZAT-,313,67
ivn-mus LD50:57 mg/kg JAJAAA 15,93,62

SAFETY PROFILE: Poison by intraperitoneal and intravenous routes. When heated to decomposition it emits toxic fumes of NO_x.

ARP125 **HR: 2**
ASPIRIN-dl-LYSINE
mf: $C_{15}H_{22}N_2O_6$ mw: 326.39

SYNS: dl-LYSINE ACETYLSALICYLATE □ dl-LYSINE ACETYLSALICYLIC ACID SALT □ dl-LYSINE MONO(2-(ACETYLOXY)BENZOATE

TOXICITY DATA WITH REFERENCE
orl-rat LD50:4350 mg/kg IYKEDH 13,1128,82
scu-rat LD50:1860 mg/kg IYKEDH 13,1128,82
ivn-rat LD50:1525 mg/kg IYKEDH 13,1128,82
orl-mus LD50:3270 mg/kg IYKEDH 13,1128,82
scu-mus LD50:2100 mg/kg IYKEDH 13,1128,82
ivn-mus LD50:950 mg/kg IYKEDH 13,1128,82

SAFETY PROFILE: Moderately toxic by ingestion and other routes. When heated to decomposition it emits toxic fumes of NO_x.

ARP250 **CAS:8003-03-0** **HR: 2**
ASPIRIN, PHENACETIN, and CAFFEINE
mf: $C_{10}H_{13}NO_2 \cdot C_9H_8O_4 \cdot C_8H_{10}N_4O_2$ mw: 553.63

PROP: Composed of 50% aspirin, 46% phenacetin, and 4% caffeine (NCIMR* NIH-71-E-2144).

SYNS: 2-(ACETYLOXY)BENZOIC ACID, mixed with 3,7-DIHYDRO-1,3,7-TRIMETHYL-1H-PURINE-2,6-DIONE and N-(4-ETHOXYPHENYL) ACETAMIDE □ APC (pharmaceutical) □ ASCOPHEN □ CITRAMON □ EMPIRIN COMPOUND □ NCI-C02697 □ OSCOPHEN □ THOMAPYRIN

TOXICITY DATA WITH REFERENCE
ipr-mus LD50:44 mg/kg PCJOAU 15,139,81
orl-rat TDLo:382 g/kg/78W-C:ETA NCITR* NCI-CG-TR-67,78
orl-rat LD50:1420 mg/kg NCIMR* NIH-71-E-2144

CONSENSUS REPORTS: NCI Carcinogenesis Bioassay (feed); Inadequate Studies: mouse, rat NCITR* NCI-CG-TR-67,78.

SAFETY PROFILE: Moderately toxic by ingestion. Questionable carcinogen with experimental tumorigenic data. See also CAFFEINE, p-ACETOPHENETIDIDE. When heated to decomposition it emits toxic fumes of NO_x.

ARP500 **HR: 2**
ASSAM TEA

PROP: Tannin containing fraction of leaf used (JNCIAM 57,207,76).

SYN: CAMELLIA SINENSIS

TOXICITY DATA WITH REFERENCE
scu-rat TDLo:1850 mg/kg/58W-I:NEO JNCIAM 57,207,76

SAFETY PROFILE: Questionable carcinogen with experimental neoplastigenic data. When heated to decomposition it emits acrid smoke and irritating fumes.

ARP625 **CAS:88746-71-8** **HR: 2**
ASTA Z 7557
mf: $C_9H_{18}Cl_2N_2O_5PS_2 \cdot C_6H_{13}N$ mw: 499.48

SYN: Z 7557

TOXICITY DATA WITH REFERENCE
dnd-hmn:lym 50 mg/L INNDDK 2,161,84
dnd-mus-ipr 250 µmol/kg INNDDK 2,181,84
dnd-mus:lym 12,500 µg/kg INNDDK 2,161,84
orl-rat LD50:1 g/kg INNDDK 2,201,84
ivn-rat LD50:250 mg/kg INNDDK 2,201,84
orl-mus LD50:2310 mg/kg INNDDK 2,201,84
ipr-mus LD50:315 mg/kg INNDDK 2,253,84
ivn-mus LD50:500 mg/kg INNDDK 2,201,84

SAFETY PROFILE: Poison by intravenous and intraperitoneal routes. Moderately toxic by ingestion. Human mutation data reported. When heated to decomposition it emits toxic fumes of Cl^-, PO_x, SO_x, and NO_x. See also ESTERS and SULFONATES.

ARP675 CAS:68844-77-9 HR: 3
ASTEMIZOLE
mf: $C_{28}H_{31}FN_4O$ mw: 458.63

PROP: Crystals. Mp: 149.1°.

SYNS: ASTEMIZOL (GERMAN) □ 1-(p-FLUOROBENZYL)-2-((1-(2-(p-METHOXYPHENYL)ETHYL)PIPERID-4-YL)AMINO)BENZIMIDAZOLE □ HISMANAL □ HISTAMINOS □ PARALERGIN

TOXICITY DATA WITH REFERENCE
orl-wmn TDLo:4 mg/kg:BAH HUTODJ 5,43,86
orl-wmn TDLo:4 mg/kg:BAH,CVS,GIT BMJOAE 292,660,86
orl-wmn TDLo:4 mg/kg:CVS JTCTDW 31,121,93
orl-rat LD50:>2560 mg/kg AIPTAK 251,39,81
scu-rat LD50:355 mg/kg ARZNAD 33,381,83
ivn-rat LD50:28 mg/kg ARZNAD 33,381,83
orl-mus LD50:2560 mg/kg AIPTAK 251,39,81
ivn-dog LD50:22 mg/kg ARZNAD 33,381,83
orl-gpg LD50:933 mg/kg ARZNAD 33,381,83

SAFETY PROFILE: Poison by subcutaneous and intravenous routes. Moderately toxic by ingestion. Human systemic effects by ingestion: arrhythmias, coma, nausea or vomiting, somnolence. When heated to decomposition it emits toxic fumes of F^- and NO_x.

ARP875 CAS:14698-07-8 HR: 3
ASVERINE CITRATE
mf: $C_{15}H_{17}NS_2 \cdot C_6H_8O_7$ mw: 467.59

SYNS: ASVERIN-C □ ASVERIN CITRATE □ AT 327 CITRATE □ BITHIODINE □ 3-(DI-2-THIENYLMETHYLENE)-1-METHYLPIPERIDINE CITRATE □ 1-METHYL-3-(DI-2-THIENYLMETHYLENE)PIPERIDINE CITRATE

TOXICITY DATA WITH REFERENCE
orl-mus LD50:867 mg/kg KSRNAM 7,3279,73
scu-mus LD50:376 mg/kg CPBTAL 7,372,59
ivn-dog LD50:75 mg/kg CPBTAL 7,372,59

SAFETY PROFILE: Poison by subcutaneous and intravenous routes. Moderately toxic by ingestion. When heated to decomposition it emits toxic fumes of NO_x and SO_x.

ARQ000 CAS:2185-98-0 HR: 3
AT-581
mf: $C_{14}H_{20}Cl_2N_2O_2 \cdot 2ClH$ mw: 392.18

SYNS: 3-(o-((BIS(2-CHLOROETHYL)AMINO)METHYL)PHENYL)ALANINE DIHYDROCHLORIDE □ o-BIS(2-CHLOROETHYL)AMINOMETHYLPHENYLALANINE HYDROCHLORIDE □ 2-((BIS(2-CHLOROETHYL)AMINO)METHYL)-PHENYLALANINE DIHYDROCHLORIDE (9CI)

TOXICITY DATA WITH REFERENCE
ivn-hmn TDLo:2 mg/kg/10D-I:BLD XPHPAW 441,186,74
ipr-mus LD50:5900 µg/kg SSINAV 13,789,64

SAFETY PROFILE: Poison by intraperitoneal route. Human blood effects by intravenous route. When heated to decomposition it emits very toxic fumes of HCl, NO_x, and Cl^-.

ARQ250 CAS:83-89-6 HR: 3
ATABRINE
mf: $C_{23}H_{30}ClN_3O$ mw: 400.01

PROP: Bright yellow crystals. Mp: decomp @ 248°.

SYNS: ACRICHINE □ ACRINAMINE □ ACRIQUINE □ AKRICHIN □ ANTIMALARINA □ 6-CHLORO-9-((4-(DIETHYL AMINO)-1-METHYL BUTYL)AMINO)-2-METHOXYACRIDINE □ 3-CHLORO-7-METHOXY-9-(1-METHYL-4-DIETHYLAMINOBUTYLAMINO)ACRIDINE □ ERION □ HAFFKININE □ MEPACRINE □ 2-METHOXY-6-CHLORO-9-DIETHYLAMINOPENTYLAMINOACRIDINE □ QUINACRINE

TOXICITY DATA WITH REFERENCE
mma-sat 500 µg/plate TXAPA9 52,237,80
mmo-omi 50 mg/L GENTAE 90,1,78
mmo-sat 200 mg/L JOBAAY 122,549,75
mnt-mus-ipr 280 µmol/kg MUREAV 26,553,74
dns-mam:lym 1 µmol/L CBINA8 8,113,74
iut-wmn TDLo:13,600 µg/kg (female 1D pre):REP CCPTAY 7,333,73
orl-mus LD50:1320 mg/kg FAZMAE 17,108,73
scu-mus LD50:239 mg/kg JPETAB 119,444,57
ivn-mus LD50:50 mg/kg THERAP 25,823,70
orl-ckn LD50:714 mg/kg AEPPAE 201,402,43

SAFETY PROFILE: Poison by intravenous and subcutaneous routes. Moderately toxic by ingestion. Mutation data reported. Experimental reproductive effects. Has been implicated in aplastic anemia. When heated to decomposition, it emits very toxic fumes of Cl^- and NO_x.

ARQ325 CAS:5140-35-2 HR: 3
ATHERILINE
mf: $C_{19}H_{15}NO_5$ mw: 337.35

PROP: Mp: 250–260° (decomp).

SYNS: ATHEROLINE □ 4,5,6,6a-TETRADEHYDRO-9-HYDROXY-1,2,10-TRIMETHOXYNORAPORPHIN-7-ONE

TOXICITY DATA WITH REFERENCE
orl-mus LD50:450 mg/kg APFRAD 38,537,80
ipr-mus LD50:170 mg/kg APFRAD 38,537,80
ivn-mus LD50:90 mg/kg APFRAD 38,537,80

SAFETY PROFILE: Poison by intravenous and intraperitoneal routes. Moderately toxic by ingestion. When heated to decomposition it emits toxic fumes of NO_x.

ARQ500 CAS:56-65-5 **HR: 3**
ATP
mf: $C_{10}H_{16}N_5O_{13}P_3$ mw: 507.22

PROP: Solid. Mp: 143–145° (decomp). Freely sol in water. The anhydrous barium salt is stable, but the hydrated salt slowly decomp forming 5′-adenylic acid and barium pyrophosphate.

SYNS: ADENOSINE TRIPHOSPHATE □ ADENOSINE-5′-TRIPHOSPHATE □ ADENOSINE-5′-TRIPHOSPHORIC ACID □ ADENYLPYROPHOSPHORIC ACID □ ADEPHOS □ ADETOL □ ADYNOL □ ARA-ATP □ 9-β-d-ARABINOFURANOSYLADENINE 5′-TRIPHOSPHATE □ ATIPI □ 5′-ATP □ ATP (nucleotide) □ ATRIPHOS □ GLUCOBASIN □ MYOTRIPHOS □ STRIADYNE □ TRIADENYL □ TRIPHOSADEN □ TRIPHOSPHADEN □ TRIPHOSPHORIC ACID ADENOSINE ESTER

TOXICITY DATA WITH REFERENCE
dni-hmn:lym 10 μmol/L CNREA8 42,2092,82
oms-hmn:oth 10 μmol/L JIDEAE 65,52,75
dni-mus:leu 2 μmol/L CHTHBK 27,61,81
ipr-rat LD50:200 mg/kg BJANAD 53,305,81
ipr-mus LD50:2780 mg/kg PCJOAU 20,160,86

CONSENSUS REPORTS: EPA Genetic Toxicology Program. Reported in EPA TSCA Inventory.

SAFETY PROFILE: Poison by intraperitoneal route. Human mutation data reported. When heated to decomposition it emits toxic fumes of PO_x and NO_x.

ARQ700 **HR: 2**
ATRATOL 80W

PROP: Contains 75% atrazine and 5% prometon. (FMCHA2 -,D24,80)

TOXICITY DATA WITH REFERENCE
orl-rat LD50:2436 mg/kg FMCHA2 -,D24,80
ihl-rat LC50:3190 mg/kg/4H FMCHA2-,D24,80

ACGIH TLV: TWA 5 mg/m³

SAFETY PROFILE: Moderately toxic by ingestion and inhalation. When heated to decomposition it emits acrid smoke and irritating fumes.

ARQ725 CAS:1912-24-9 **HR: 3**
ATRAZINE
mf: $C_8H_{14}ClN_5$ mw: 215.72

PROP: Crystals. Mp: 175—177°. Solubility at 25°: in water: 70 ppm; ether: 12,000 ppm; chloroform: 52,000 ppm; methanol: 18,000 ppm.

SYNS: A 361 □ AATREX □ AATREX 4L □ AATREX NINE-O □ AATREX 80W □ 2-AETHYLAMINO-4-CHLOR-6-ISOPROPYLAMINO-1,3,5-TRIAZIN (GERMAN) □ 2-AETHYLAMINO-4-ISOPROPYLAMINO-6-CHLOR-1,3,5-TRIAZIN (GERMAN) □ AKTIKON □ AKTIKON PK □ AKTINIT A □ AKTINIT PK □ ARGEZIN □ ATAZINAX □ ATRANEX □ ATRASINE □ ATRATOL A □ ATRAZIN □ ATRED □ ATREX □ CANDEX □ CEKUZINA-T □ 2-CHLORO-4-ETHYLAMINEISOPROPYLAMINE-s-TRIAZINE □ 1-CHLORO-3-ETHYLAMINO-5-ISOPROPYLAMINO-s-TRIAZINE □ 1-CHLORO-3-ETHYLAMINO-5-ISOPROPYLAMINO-2,4,6-TRIAZINE □ 2-CHLORO-4-ETHYLAMINO-6-ISOPROPYLAMINO-s-TRIAZINE □ 2-CHLORO-4-ETHYLAMINO-6-ISOPROPYLAMINO-1,3,5-TRIAZINE □ 6-CHLORO-N-ETHYL-N′-(1-METHYLETHYL)-1,3,5-TRIAZINE-2,4-DIAMINE (9CI) □ 2-CHLORO-4-(2-PROPYLAMINO)-6-ETHYLAMINO-s-TRIAZINE □ CRISATRINA □ CRISAZINE □ CYAZIN □ FARMCO ATRAZINE □ FENAMIN □ FENAMINE □ FENATROL □ G 30027 □ GEIGY 30,027 □ GESAPRIM □ GESOPRIM □ GRIFFEX □ HUNGAZIN □ HUNGAZIN PK □ INAKOR □ OLEOGESAPRIM □ PRIMATOL □ PRIMAZE □ RADAZIN □ RADIZINE □ SHELL ATRAZINE HERBICIDE □ STRAZINE □ TRIAZINE A 1294 □ VECTAL □ VECTAL SC □ WEEDEX A □ WONUK □ ZEAZIN □ ZEAZINE

TOXICITY DATA WITH REFERENCE
skn-rbt 38 mg open MLD CIGET* -,-,77
eye-rbt 6320 μg SEV CIGET* -,-,77
skn-mam 500 mg MLD VRDEA5 (5),133,77
eye-mam 100 mg SEV VRDEA5 (5),133,77
sln-nsc 10 mg/L MUREAV 167,35,86
dns-hmn:fbr 3 mmol/L MUREAV 74,77,80
hma-rat/esc 100 mg/kg CECED9 6388,328,80
mma-ham:lng 3 mmol/L MUREAV 74,77,80
orl-rbt TDLo:975 mg/kg (female 7-19D post):TER JTEHD6 24,307,88
scu-mus TDLo:418 mg/kg (female 6-14D post):REP NTIS** PB223-160
orl-rat TDLo:33,775 mg/kg/2Y-C:CAR NEOLA4 37,533,90
orl-mus TDLo:9000 mg/kg/78W-I:ETA NTIS** PB223-159
orl-rat LD50:672 mg/kg FAATDF 7,299,86
ihl-rat LC50:5200 mg/m³/4H FMCHA2 -,C3,83
ipr-rat LD50:235 mg/kg PESTD5 17,351,76
orl-mus LD50:850 mg/kg 85GMAT -,36,82
ipr-mus LD50:626 mg/kg PESTD5 17,351,76
orl-rbt LD50:750 mg/kg 85DPAN -,-,71/76
skn-rbt LD50:7500 mg/kg 28ZEAL 5,15,76
orl-ham LD50:1000 mg/kg TXAPA9 48,A192,79

CONSENSUS REPORTS: EPA Genetic Toxicology Program. Reported in EPA TSCA Inventory.

OSHA PEL: TWA 5 mg/m³
ACGIH TLV: TWA 5 mg/m³
DFG MAK: 2 mg/m³

SAFETY PROFILE: Poison by intraperitoneal route. Moderately toxic by ingestion. Mildly toxic by inhalation and skin contact. An experimental teratogen. Other experimental reproductive effects. Human mutation data reported. A skin and severe eye irritant. Questionable carcinogen with experimental tumorigenic data. When heated to decomposition it emits toxic fumes of Cl^- and NO_x.

ARQ750 CAS:637-07-0 **HR: 3**
ATROMID S
mf: $C_{12}H_{15}ClO_3$ mw: 242.72

SYNS: AMOTRIL □ ANGIOKAPSUL □ ANPARTON □ ANTILIPID □ APOLAN □ ARTERIOFLEXIN □ ARTEROSOL □ ARTES □ARTEVIL □ ATECULON □ ATERIOSAN □ ATHEBRATE □ ATHEROMIDE □ ATHEROPRONT □ ATHRANID-WIRKSTOFF □ ATROLEN □ATROMID □ ATROMIDIN □ ATROVIS □ AY 61123 □ AZIONYL □ BIOSCLERAN □ BRESIT □ CARTAGYL □ α-p-CHLOROPHENOXYISOBUTYRYL ETHYL ESTER □ 2-(4-CHLOROPHENOXY)-2-METHYLPROPANOIC ACID ETHYL ESTER □ 2-(p-CHLOROPHENOXY)-2-METHYLPROPIONIC ACID ETHYL ESTER □ CINNARIZIN □ CITIFLUS □ CLARIPEX □ CLOBERAT □ CLOBRAT □ CLOBREN-SF □ CLOFAR □ CLOFIBRAM □ CLOFIBRAT □ CLOFIBRATO (SPANISH) □ CLOFINIT □ CLOFIPRONT □ CPIB □ DELIVA □ DURA CLOFIBRAT □ ELPI □ EPIB □ ETHYL CHLOROPHENOXYISOBUTYRATE □ ETHYL-p-CHLOROPHENOXYISOBUTYRATE □ ETHYL-α-p-CHLOROPHENOXYISOBUTYRATE □ ETHYL-α-(4-CHLOROPHENOXY)ISOBUTYRATE □ ETHYL-2-(p-CHLOROPHENOXY)ISOBUTYRATE □ ETHYL-α-(p-CHLOROPHE-

NOXY)-α-METHYLPROPIONATE □ ETHYL-α-(4-CHLOROPHENOXY)-α-METHYLPROPIONATE □ ETHYL 2-(p-CHLOROPHENOXY)-2-METHYLPROPIONATE □ ETHYL 2-(4-CHLOROPHENOXY)-2-METHYLPROPIONATE □ ETHYL CLOFIBRATE □ FIBRALEM □ GERASTOP □ HYCLORATE □ ICI 28257 □ KLOFIRAN □ LEVATROM □ LIPAMID □ LIPAVIL □ LIPAVLON □ LIPIDE 500 □ LIPIDSENKER □ LIPOFACTON □ LIPOMID □ LIPONORM □ LIPOREDUCT □ LIPORIL □ LIPOSID □ LIPRIN □ LIPRINAL □ LOBETRIN □ MISCLERON □ NEGALIP □ NEO-ATOMID □ NORMALIP □ NORMAT □ NORMOLIPOL □ NSC-79389 □ OXAN 600 □ PERSANTINAT □ RECOLIP □ REGARDIN □ REGELAN □ ROBIGRAM □ SCROBIN □ SEROFINEX □ SEROTINEX □ SKEROLIP □ SKLEROMEX □ SKLEROMEXE □ SKLEROTABLINEN □ SKLERO-TABULS □ TICLOBRAN □ VINCAMIN COMPOSITUM □ XYDURIL □ YOCLO

TOXICITY DATA WITH REFERENCE
dns-rat:lvr 100 pmol/L CRNGDP 5,1547,84
sce-ham:ovr 100 μmol/L/1H CRNGDP 5,703,84
scu-mus TDLo:960 mg/kg (female 17D post):REP HCMYAL 63,7,79
orl-rbt TDLo:1300 mg/kg (female 6-18D post):TER ARZNAD 31,1831,81
orl-rat TDLo:100 g/kg/72W-C:ETA CNREA8 39,3419,79
orl-man TDLo:1071 μg/kg NEURAI 37,881,87
orl-wmn TDLo:80 mg/kg/2D-I:MET NEJMAG 301,1345,79
orl-man TDLo:171 mg/kg/6D-I IJMDAI 20,1082,84
orl-rat LD50:940 mg/kg AFTOD7 6,255,80
ipr-rat LD50:910 mg/kg JETOAS 5,239,72
orl-mus LD50:1220 mg/kg CPBTAL 32,1568,84
ipr-mus LD50:540 mg/kg ARZNAD 31,1816,81
scu-mus LD50:2000 mg/kg NIIRDN 6,367,82
ivn-mus LD50:30 mg/kg NIIRDN 6,367,82
ipr-dog LD50:500 mg/kg NIIRDN 6,367,82
scu-dog LD50:500 mg/kg NIIRDN 6,367,82

CONSENSUS REPORTS: IARC Cancer Review: Group 3 IMEMDT 7,171,87; Animal Limited Evidence IMEMDT 24,39,80; Human Inadequate Evidence IMEMDT 24,39,80

SAFETY PROFILE: Poison by intravenous route. Moderately toxic by ingestion and other routes. An experimental teratogen. Other experimental reproductive effects. Reduces plasma lipid levels. Human systemic effects by ingestion: muscle weakness, muscle spasms, and fever. Questionable carcinogen with experimental tumorigenic data. When heated to decomposition it emits toxic fumes of Cl^-.

ARR000 CAS:51-55-8 HR: 3
ATROPINE
mf: $C_{17}H_{23}NO_3$ mw: 289.41

PROP: Solid. Mp: 116–117°. Colorless crystalline alkaloid.

SYNS: ATROPIN (GERMAN) □ EYEULES □ dl-HYOSCYAMINE □ 2-PHENYLHYDRACRYLIC ACID-3-α-TROPANYL ESTER □ β-PHENYL-γ-OXYPROPIONSAEURE-TROPYL-ESTER (GERMAN) □ 1-α-H,5-α-H-TROPAN-3-α-OL (±)-TROPATE (ESTER) □ dl-TROPANYL-2-HYDROXY-1-PHENYLPROPIONATE □ TROPIC ACID, ESTER with TROPINE □ TROPIC ACID-3-α-TROPANYL ESTER □ TROPINE TROPATE □ dl-TROPYLTROPATE □ (±)-TROPYL TROPATE

TOXICITY DATA WITH REFERENCE
ivn-wmn TDLo:20 μg/kg (female 26-39W post):TER AJOGAH 82,1055,61
unr-mus TDLo:1600 μg/kg (female 1D pre):REP FESTAS 12,346,61
orl-hmn TDLo:33 μg/kg:EYE JTCTDW 22,581,84/85
ivn-man TDLo:14 μg/kg AIMEAS 101,720,84
ims-man TDLo:175 μg/kg FEPRA7 32,250,73
ims-hmn TDLo:1 μg/kg:EYE 85IVAW 1,L1,82
orl-rat LD50:500 mg/kg AIPTAK 155,393,65
ipr-rat LD50:280 mg/kg JPETAB 105,166,52
scu-rat LD50:250 mg/kg AIPTAK 68,339,42
ivn-rat LD50:73 mg/kg SMWOAS 76,1282,46
ivn-mus LD50:44 mg/kg PCJOAU 11,905,77
ims-rat LD50:920 mg/kg DCTODJ 1,355,78
orl-mus LD50:75 mg/kg AIPTAK 59,149,38
ipr-mus LD50:30 mg/kg JMCMAR 31,683,88
scu-mus LD50:510 mg/kg NYKZAU 53,84S,57
ivn-mus LD50:30 mg/kg JMCMAR 28,1760,85

CONSENSUS REPORTS: Reported in EPA TSCA Inventory.

SAFETY PROFILE: Poison by ingestion, subcutaneous, intravenous, and intraperitoneal routes. Human systemic effects by ingestion and intramuscular routes: visual field changes, mydriasis (pupillary dilation), and muscle weakness. An experimental teratogen. Other experimental reproductive effects. An alkaloid. When heated to decomposition it emits toxic fumes of NO_x.

ARR250 CAS:2472-17-5 HR: 3
ATROPINE SULFATE (1:1)
mf: $C_{17}H_{23}NO_3 \cdot H_2O_4S$ mw: 387.49

TOXICITY DATA WITH REFERENCE
dnd-rat-ipr 200 mg/kg/25D EJPHAZ 7,73,69
ivn-mus LD50:78 mg/kg JLCMAK 30,700,45

SAFETY PROFILE: Poison by intravenous route. Mutation data reported. See also ATROPINE and SULFATES. When heated to decomposition it emits very toxic fumes of NO_x and SO_x.

ARR500 CAS:55-48-1 HR: 3
ATROPINE SULFATE (2:1)
mf: $C_{34}H_{46}N_2O_6 \cdot H_2O_4S$ mw: 676.90

PROP: Crystals. Mp: 190–194°.

SYNS: ATROPIN SIRAN (CZECH) □ ATROPINSULFAT (GERMAN) □ SULFATE d'ATROPINE (FRENCH) □ 1-α-H,5-α-H-TROPAN-3-α-OL (±)-TROPATE (ESTER), SULFATE (2:1) SALT □ dl-TROPANYL-2-HYDROXY-1-PHENYLPROPIONATE SULFATE □ TROPINTRAN

TOXICITY DATA WITH REFERENCE
ipr-rat TDLo:40 mg/kg (male 1D pre):REP ANENAG 24(Suppl 3),1,63
par-mus TDLo:50 mg/kg (female 8D post):TER JPMSAE 62,1626,73
orl-cld TDLo:20 μg/kg ADCHAK 54,222,79
ivn-man TDLo:28 μg/kg/11H-I:SYS AJDCAI 137,291,83
ims-hmn TDLo:28 μg/kg JAPYAA 8,635,56
ocu-wmn TDLo:20 μg/kg/4H-I:CVS AIMDAP 146,45,86
mul-wmn TDLo:44 μg/kg/1D-I AIMDAP 146,45,86
orl-rat LD50:600 mg/kg AIPTAK 155,393,65
ipr-rat LD50:215 mg/kg TXAPA9 11,511,67
scu-rat LD50:540 mg/kg AIPTAK 155,393,65
orl-mus LD50:468 mg/kg AIPTAK 156,467,65

ipr-mus LD50:180 mg/kg JPMSAE 55,849,66
scu-mus LD50:400 mg/kg THERAP 12,412,57
ivn-mus LD50:31 mg/kg ARZNAD 16,637,66

CONSENSUS REPORTS: Reported in EPA TSCA Inventory.

SAFETY PROFILE: Poison by subcutaneous, intravenous, and intraperitoneal routes. Moderately toxic by ingestion. Human (child) pulmonary system effects by ingestion. Human systemic effects: decreased body temperature, cardiac arrhythmias. An experimental teratogen. Other experimental reproductive effects. See also ATROPINE. When heated to decomposition it emits very toxic fumes of NO_x and SO_x.

ARR750 CAS:67-92-5 HR: 3
ATUMIN
mf: $C_{19}H_{35}NO_2{\cdot}ClH$ mw: 346.01

PROP: Crystals from butanone. Sol in water. Mp: 164–166°.

SYNS: BIS(CYCLOHEXYL)CARBOXYLIC ACID DIETHYLAMINOETHYL ESTER HYDROCHLORIDE □ DICYCLOMINE HYDROCHLORIDE □ DIETHYLAMINOCARBETHOXYBICYCLOHEXYL HYDROCHLORIDE □ β-DIETHYLAMINOETHYL-1-CYCLOHEXYLCYCLOHEXANECARBOXYLATE HYDROCHLORIDE □ β-DIETHYLAMINOETHYL-1-CYCLOHEXYLHEXAHYDROBENZOATE HYDROCHLORIDE

TOXICITY DATA WITH REFERENCE
orl-inf TDLo:1 g/kg:BAH,PUL BMJOAE 288,901,84
orl-rat LD50:1290 mg/kg DRUGAY 6,317,82
orl-mus LD50:625 mg/kg JAPMA8 39,305,50
ivn-rbt LD50:35 mg/kg JAPMA8 39,305,50

CONSENSUS REPORTS: Reported in EPA TSCA Inventory.

SAFETY PROFILE: Poison by intravenous route. Moderately toxic by ingestion. Human systemic effects by ingestion: rigidity, dyspnea, cyanosis. When heated to decomposition it emits very toxic fumes of HCl and NO_x.

ARR875 CAS:1674-96-0 HR: 2
ATURBANE HYDROCHLORIDE
mf: $C_{17}H_{24}N_2O_2{\cdot}ClH$ mw: 324.89

PROP: Solid. Mp: 176–177°.

SYN: CHLORHYDRATE de α-PHENYL-α-(β′-DIETHYLAMINOETHYL) GLUTARIMIDE (FRENCH)

TOXICITY DATA WITH REFERENCE
scu-rat TDLo:480 mg/kg (1-12D preg):REP COREAF 258,2666,64
orl-mus LD50:1200 mg/kg ARZNAD 15,534,65

SAFETY PROFILE: Moderately toxic by ingestion. Experimental reproductive effects. When heated to decomposition it emits toxic fumes of NO_x and HCl. An anticholinergic and antispasmodic.

ARS000 CAS:60748-45-0 HR: 3
ATX II

PROP: A polypeptide isolated from the sea anemone, *Anemonia sulcata.* (TOXIA6 16,561,78)

SYN: SEA ANEMONE TOXIN II

TOXICITY DATA WITH REFERENCE
ivn-mus LD50:310 μg/kg TOXIA6 16,561,78
par-mus LD50:1 μg/kg NSAPCC 309,165,79

SAFETY PROFILE: A deadly poison by intravenous and parenteral routes.

ARS125 CAS:74469-00-4 HR: 2
AUGMENTIN
mf: $C_{16}H_{19}N_3O_5S{\cdot}C_8H_9NO_5{\cdot}K$ mw: 603.72

SYNS: AMOXICILLIN mixed with POTASSIUM CLAVULANATE (2:1) □ AUGMENTIN (antibiotic) □ BRL 25000 □ POTASSIUM CLAVULANATE mixed with AMOXICILLIN (1:2)

TOXICITY DATA WITH REFERENCE
orl-rat TDLo:1500 mg/kg (6-15D preg):REP NKRZAZ 31(Suppl 2),238,83
ipr-rat LD50:2774 mg/kg NKRZAZ 31(Suppl 2),113,83
scu-rat LD50:3487 mg/kg NKRZAZ 31(Suppl 2),113,83
ipr-mus LD50:3925 mg/kg NKRZAZ 31(Suppl 2),113,83
scu-mus LD50:6433 mg/kg NKRZAZ 31(Suppl 2),113,83

SAFETY PROFILE: Moderately toxic by some routes. Experimental reproductive effects. When heated to decomposition it emits toxic fumes of SO_x, NO_x, and K_2O.

ARS130 CAS:78173-92-9 HR: 3
AURAMYCIN A
mf: $C_{41}H_{51}NO_{15}$ mw: 797.93

PROP: Yellow powder. Mp: 141°.

TOXICITY DATA WITH REFERENCE
dni-mus:leu 630 nmol/L JANTAJ 34,1596,81
oms-mus:leu 120 nmol/L JANTAJ 34,1596,81
ipr-mus LD50:100 mg/kg JANTAJ 35,82-59,82

SAFETY PROFILE: Poison by intraperitoneal route. Mutation data reported. When heated to decomposition it emits toxic fumes of NO_x.

ARS135 CAS:78173-91-8 HR: 3
AURAMYCIN B
mf: $C_{41}H_{49}NO_{15}$ mw: 795.91

PROP: Yellow powder. Mp: 161°.

TOXICITY DATA WITH REFERENCE
dni-mus:leu 750 nmol/L JANTAJ 34,1596,81
oms-mus:leu 120 nmol/L JANTAJ 34,1596,81
ipr-mus LD50:100 mg/kg JANTAJ 35,82-60,82

SAFETY PROFILE: Poison by intraperitoneal route. Mutation data reported. When heated to decomposition it emits toxic fumes of NO_x.

ARS150 **CAS:34031-32-8** **HR: 3**
AURANOFIN
mf: $C_{20}H_{34}AuO_9PS$ mw: 678.54

PROP: Crystals. Mp: 110–111°.

SYNS: RIDAURA □ SK&F 39162 □ 2,3,4,6-TETRA-o-ACETYL-1-THIO-β-d-GLUCOPYRANOSATO-S-(TRIETHYLPHOSPHINE)GOLD □ TRIETHYLPHOSPHINE GOLD

TOXICITY DATA WITH REFERENCE
dni-hmn:oth 41 μmol/L BCPCA6 34,3243,85
orl-rat TDLo:100 mg/kg (female 6-15D post):TER VTPHAK 15(Suppl 5),89,78
orl-rat TDLo:100 mg/kg (female 6-15D post):REP VTPHAK 15(Suppl 5),89,78
orl-man TDLo:1200 μg/kg/2W-I:GIT JRHUA9 13,228,86
orl-wmn TDLo:5400 μg/kg/10D-I ARHEAW 27,1316,84
orl-rat LD50:265 mg/kg VTPHAK 15(Suppl.5),1,78
ipr-rat LD50:25,500 μg/kg IYKEDH 17,1106,86
scu-rat LD50:235 mg/kg IYKEDH 17,1106,86
ivn-rat LD50:39 mg/kg IYKEDH 17,1106,86
orl-mus LD50:310 mg/kg VTPHAK 15(Suppl.5),1,78

SAFETY PROFILE: Poison by ingestion, intraperitoneal, and intravenous routes. Human systemic effects by ingestion: ulceration or bleeding from stomach. An experimental teratogen. Other experimental reproductive effects. Human mutation data reported. When heated to decomposition it emits very toxic fumes of SO_x and PO_x. See also GOLD COMPOUNDS.

ARS500 **CAS:130-01-8** **HR: 3**
AUREINE
mf: $C_{18}H_{25}NO_5$ mw: 335.44

PROP: Solid. Mp: 232–233°.

SYNS: 12-HYDROXYSENECIONAN-11,16-DIONE □ SENECIONINE

TOXICITY DATA WITH REFERENCE
sln-dmg-par 20 μmol/L ZEVBA5 91,74,60
dns-rat:lvr 700 nmol/L CNREA8 45,3125,85
dnd-mus-ipr 90 mg/kg TOXID9 1,42,81
orl-rat TDLo:140 mg/kg (female 4-10D post):TER JPMSAE 77,461,88
orl-rat TDLo:140 mg/kg (female 4-10D post):REP JPMSAE 77,461,88
ipr-rat LDLo:33 mg/kg CBINA8 12,299,76
ivn-rat LD50:41,200 μg/kg JPETAB 87,382,46
unr-rat LD50:40 mg/kg CNREA8 28,2237,68
ivn-mus LD50:64 mg/kg JPETAB 75,69,42
ivn-ham LD50:61 mg/kg RETOAE 5,53,49

CONSENSUS REPORTS: EPA Genetic Toxicology Program.

SAFETY PROFILE: Poison by intravenous, intraperitoneal, and possibly other routes. An experimental teratogen. Other experimental reproductive effects. Mutation data reported. When heated to decomposition it emits toxic fumes of NO_x.

ARS750 **HR: 3**
AUREMETINE

PROP: Percentage composition: 28% emetine, 16% auramine, and 56% iodine. (AJTMAQ 10,249,30)

TOXICITY DATA WITH REFERENCE
orl-cat LDLo:20 mg/kg AJTMAQ 10,249,30
orl-rbt LDLo:75 mg/kg AJTMAQ 10,249,30

SAFETY PROFILE: Poison by ingestion. When heated to decomposition it emits very toxic fumes of I^- and NO_x.

ART000 **CAS:58194-38-0** **HR: 3**
AUREOFUSCIN
mf: $C_{25}H_{37}NO_{10}$ mw: 511.63

PROP: Solid. Mp: 170° (decomp).

TOXICITY DATA WITH REFERENCE
ipr-mus LD50:25 mg/kg JANTAJ 30,77-12,77
ivn-mus LD50:28 mg/kg JANTAJ 30,77-12,77

SAFETY PROFILE: Poison by intraperitoneal and intravenous routes. When heated to decomposition it emits toxic fumes of NO_x.

ART125 **CAS:70213-45-5** **HR: 3**
AUROMOMYCIN

PROP: Yellow plates.

TOXICITY DATA WITH REFERENCE
dnd-omi 100 mg/L CNREA8 39,2787,79
dnd-hmn:lym 500 μg/L CNREA8 44,3202,84
dnd-mus:lym 60 ng/L/10M CNREA8 39,2787,79
oms-mus:lym 60 μg/L CNREA8 39,2787,79
ivn-mus LD50:3 mg/kg JANTAJ 32,330,79

SAFETY PROFILE: Poison by intravenous route. Human mutation data reported.When heated to decomposition it emits acrid smoke and irritating fumes.

ART250 **CAS:12192-57-3** **HR: 3**
1-AUROTHIO-d-GLUCOPYRANOSE
mf: $C_6H_{11}O_5S{\cdot}Au$ mw: 392.20

PROP: Yellow crystals from EtOH (aq). Sol in water, insol org solvs.

SYNS: AUREOTAN □ AUROMYOSE □ AUROTAN □ AUROTHIOGLUCOSE □ AURUMINE □ AUTHRON □ BRENOL □ (d-GLUCOPYRANOSYLTHIO)GOLD □ (1-d-GLUCOSYLTHIO)GOLD □ GLYSANOL B □ GOLD THIOGLUCOSE □ GTG □ ORONOL □ ROMOSOL □ SOLGANAL □ SOLGANAL B □ (1-THIO-d-GLUCOPYRANOSATO)GOLD □ 1-THIO-GLUCOPYRANOSE, MONOGOLD(1+) SALT □ THIOGLUCOSE d'OR (FRENCH)

TOXICITY DATA WITH REFERENCE
ipr-mus TDLo:400 mg/kg (1D pre):TER PSEBAA 124,1190,67
par-mus TDLo:800 mg/kg (1D pre):REP ENDOAO 78,845,66
par-mus TDLo:400 mg/kg:CAR PAACA3 3,37,59
par-mus TD:750 mg/kg:NEO RFECAC 13,40,68
par-mus TD:750 mg/kg:CAR RFECAC 11,828,66
ims-wmn TDLo:2600 μg/kg/15D-I:GIT ARHEAW 27,230,84

ims-man TDLo: 3357 μg/kg/4W-I:EYE JRHUA9 12,619,85
par-wmn TDLo: 2700 μg/kg/4W-I:SYS JRHUA9 11,843,84
unr-man LDLo: 3 mg/kg SAVEAB 10,101A,39
scu-mus LDLo: 1650 mg/kg EMSUA8 3,146,45
ivn-ckn LD50: 1000 mg/kg POSCAL 52,926,73
ims-ckn LDLo: 300 mg/kg TXAPA9 35,223,76

CONSENSUS REPORTS: IARC Cancer Review: Group 1 IMEMDT 7,56,87; Animal Limited Evidence IMEMDT 13,39,77

SAFETY PROFILE: Confirmed carcinogen with experimental carcinogenic and neoplastigenic data. A deadly human poison by an unspecified route. An experimental poison by intramuscular route. Moderately toxic by subcutaneous and intravenous routes. Human systemic effects: nausea or vomiting, cholestatic jaundice, and eye effects. An experimental teratogen. Other experimental reproductive effects. See also GOLD COMPOUNDS. When heated to decomposition it emits very toxic fumes of SO_x. Used to treat rheumatoid arthritis.

ART500 **CAS:11002-90-7** **HR: 3**
AUROVERTIN
mf: $C_{26}H_{34}O_9$ mw: 490.3

TOXICITY DATA WITH REFERENCE
ivn-mus LD50: 1650 μg/kg 85ERAY 3,2003,78
ivn-dog LDLo: 1 mg/kg 85ERAY 3,2003,78
ivn-rbt LDLo: 1 mg/kg 85ERAY 3,2003,78

SAFETY PROFILE: Poison by intravenous route. When heated to decomposition it emits acrid smoke and irritating fumes.

ARU250 **CAS:55256-53-6** **HR: 3**
AUSTOCYSTIN D
mf: $C_{22}H_{22}O_8$ mw: 414.44

PROP: Crystals from C_6H_6. Mp: 114–116°.

TOXICITY DATA WITH REFERENCE
mma-sat 500 ng/plate MUREAV 58,193,78
orl-mus LD50: 300 mg/kg OYYAA2 3,187,69

SAFETY PROFILE: Poison by ingestion. Mutation data reported. When heated to decomposition it emits acrid smoke and irritating fumes.

ARU750 **HR: 3**
AUSTRALIAN COPPERHEAD SNAKE VENOM

SYNS: A. SUPERBA (AUSTRALIA) VENOM □ AUSTRELAPS SUPERBA (AUSTRALIA) VENOM □ VENOM, AUSTRALIAN ELAPIDAE SNAKE, AUSTRELAPS SUPERBA

TOXICITY DATA WITH REFERENCE
ipr-mus LD50: 280 μg/kg 85EGD4 5,368,78
scu-mus LD50: 500 μg/kg TOXIA6 17,661,79

SAFETY PROFILE: Poison by intraperitoneal and subcutaneous routes.

ARU875 **HR: 3**
AUSTRALIAN DEATH ADDER SNAKE VENOM

SYNS: ACANTHOPHIA ANTARCTICUS VENOM □ VENOM, AUSTRALIAN ELAPIDAE SNAKE, ACANTHOPHIS ANTARCTICUS

TOXICITY DATA WITH REFERENCE
ipr-mus LD50: 160 μg/kg TOXIA6 17,609,79
scu-mus LD50: 338 μg/kg TOXIA6 17,661,79
ims-mus LD50: 80 μg/kg BIJOAK 199,211,81
scu-gpg LDLo: 130 μg/kg MJAUAJ 2,801,71

SAFETY PROFILE: Deadly poison by subcutaneous, intramuscular, and intraperitoneal routes.

ARV000 **HR: 3**
AUSTRALIAN KING BROWN SNAKE VENOM

SYNS: AUSTRALIS □ P. AUSTRALIS VENOM □ PSEUDECHIS AUSTRALIS VENOM □ VENOM, AUSTRALIAN SNAKE, PSEUDECHIS

TOXICITY DATA WITH REFERENCE
ipr-mus LD50: 520 μg/kg TOXIA6 17(Suppl 1),121,79
scu-mus LD50: 2380 μg/kg TOXIA6 17,661,79
ivn-mus LD50: 230 μg/kg TOXIA6 23,73,85

SAFETY PROFILE: Deadly poison by intravenous, intraperitoneal, and subcutaneous routes.

ARV125 **HR: 3**
AUSTRALIAN KING COBRA SNAKE VENOM

SYNS: OPHIOPHAGUS HANNAH VENOM □ VENOM, AUSTRALIAN SNAKE, OPHIOPHAGUS HANNAH

TOXICITY DATA WITH REFERENCE
ipr-mus LD50: 355 μg/kg YHHPAL 19,721,84
scu-mus LD50: 1091 μg/kg TOXIA6 17,661,79
ivn-mus LD50: 125 μg/kg BICMBE 61,791,79

SAFETY PROFILE: Poison by subcutaneous, intravenous, and intraperitoneal routes.

ARV250 **HR: 3**
AUSTRALIAN RED-BELLIED BLACK SNAKE VENOM

SYNS: P. PORPHYRIACUS (AUSTRALIA) VENOM □ PSEUDECHIS PORPHYRIACUS (AUSTRALIA) VENOM □ PSEUDECHIS PORPHYRIACUS VENOM □ VENOM, AUSTRALIAN ELAPIDAE SNAKE, PSEUDECHIS PORPHYRIACUS

TOXICITY DATA WITH REFERENCE
ipr-mus LD50: 700 μg/kg 85EGD4 5,368,78
scu-mus LD50: 2530 μg/kg TOXIA6 17,661,79
ims-frg LDLo: 500 ug/kg TOXIA6 19,749,81

SAFETY PROFILE: Deadly poison by intramuscular, subcutaneous, and intraperitoneal routes.

ARV375 **HR: 3**
AUSTRALIAN ROUGH SCALED SNAKE VENOM

SYNS: TROPIDECHIS CARINATUS VENOM □ VENOM, AUSTRALIAN SNAKE, TROPIDECHIS CARINATUS

TOXICITY DATA WITH REFERENCE
ipr-mus LD50: 125 μg/kg TOXIA6 20,1085,82
scu-mus LD50: 1090 μg/kg TOXIA6 17,661,79

scu-dog LDLo:500 μg/kg MJAUAJ 2,801,71
scu-gpg LDLo:75 μg/kg MJAUAJ 2,801,71
ivn-dom LDLo:21,800 μg/kg MJAUAJ 2,801,71

SAFETY PROFILE: Poison by subcutaneous, intravenous, and intraperitoneal routes.

ARV500 **HR: 3**
AUSTRALIAN TAIPAN SNAKE VENOM

SYNS: O. SCUTELLATUS (AUSTRALIA) VENOM □ OXYURANUS SCUTELLATUS (AUSTRALIA) VENOM □ VENOM, AUSTRALIAN ELAPIDAE SNAKE, OXYURANUS SCUTELLATUS

TOXICITY DATA WITH REFERENCE
ipr-mus LD50:9 μg/kg 85EGD4 5,372,78
scu-mus LD50:64 μg/kg TOXIA6 17,661,79
scu-gpg LDLo:25 ug/kg MJAUAJ 2,801,71

SAFETY PROFILE: Deadly poison by intraperitoneal and subcutaneous routes.

ARV550 **HR: 3**
AUSTRALIAN TIGER SNAKE VENOM

SYNS: NOTECHIS SCUTATUS VENOM □ VENOM, AUSTRALIAN SNAKE, NOTECHIS SCUTATUS

TOXICITY DATA WITH REFERENCE
ipr-mus LD50:50 μg/kg AJTHAB 9,284,60
scu-mus LD50:118 μg/kg TOXIA6 17(Suppl 1),121,79
ivn-rbt LDLo:1 μg/kg PSEBAA 116,696,64
scu-gpg LDLo:20 μg/kg MJAUAJ 2,801,71
ims-frg LDLo:500 μg/kg TOXIA6 19,749,81
ipr-mam LD50:40 μg/kg CLPTAT 8,849,67

SAFETY PROFILE: Deadly poison by subcutaneous, intramuscular, intravenous, and intraperitoneal routes.

ARV625 **HR: 3**
AUSTRELAPS SUPERBA VENOM

SYNS: A. SUPERBA VENOM □ AUSTRELAPS SUPERBUS VENOM □ VENOM, AUSTRALIAN SNAKE, AUSTRELAPS SUPERBA

TOXICITY DATA WITH REFERENCE
ipr-mus LD50:280 μg/kg 85EGD4 -,368,78
scu-mus LD50:6200 μg/kg TOXIA6 18,443,80
ims-frg LDLo:400 μg/kg TOXIA6 19,749,81

SAFETY PROFILE: Poison by subcutaneous, intramuscular, and intraperitoneal routes.

ARV750 **CAS:7437-53-8** **HR: 3**
AVACAN HYDROCHLORIDE
mf: $C_{19}H_{32}N_2O_2 \cdot ClH$ mw: 356.99

TOXICITY DATA WITH REFERENCE
ivn-hmn TDLo:357 μg/kg:EYE DMWOAX 76,479,51
ims-hmn TDLo:357 μg/kg:EYE DMWOAX 76,479,51

SAFETY PROFILE: A human poison by intravenous and intramuscular routes. Very irritating to experimental animals and humans. When heated to decomposition it emits toxic fumes such as Cl^- and NO_x.

ARW000 **CAS:43222-48-6** **HR: 3**
AVENGE
mf: $C_{16}H_{17}N_2 \cdot CH_3O_4S$ mw: 348.45

PROP: Solid. Mp: 155–157°. Very sol in water; sltly sol in alcohols; insol in pet ether.

SYNS: AC 84777 □ DIFENZOQUAT METHYL SULFATE □ 1,2-DIMETHYL-3,5-DIPHENYL-1-H-PYRAZOLIUM METHYL SULFATE □ FINAVEN □ MATAVEN □ YEH-YAN-KU

TOXICITY DATA WITH REFERENCE
orl-rat LD50:270 mg/kg FMCHA2 -,C20,83
skn-rbt LD50:3540 mg/kg FMCHA2 -,C20,83

SAFETY PROFILE: Poison by ingestion. Moderately toxic by skin contact. See also SULFATES. When heated to decomposition it emits very toxic fumes of NO_x and SO_x.

ARW250 **CAS:75-80-9** **HR: 3**
AVERTIN
mf: $C_2H_3Br_3O$ mw: 282.78

PROP: Crystals, needles, or prisms with ethereal odor and aromatic taste. Mp: 80°, bp: 92–93° @ 10 mm. Sltly water-sol; sol in alc and organic solvents.

SYNS: BROMETHOL □ ETHOBROM □ NARCOLAN □ NARKOLAN □ RENARCOL □ TRIBROMETHANOL □ 2,2,2-TRIBROMOETHANOL □ TRIBROMOETHYL ALCOHOL □ 2,2,2-TRIBROMOETHYL ALCOHOL

TOXICITY DATA WITH REFERENCE
orl-rat LDLo:1 g/kg JPETAB 63,183,38
scu-rat LDLo:530 mg/kg AEPPAE 182,348,36
orl-mus LD50:930 mg/kg JPMSAE 56,920,67
orl-rbt LDLo:1100 mg/kg AEPPAE 132,214,28
ipr-rbt LDLo:150 mg/kg AEPPAE 132,214,28
ivn-rbt LDLo:120 mg/kg 27ZWAY E.2,130,-

CONSENSUS REPORTS: Reported in EPA TSCA Inventory.

SAFETY PROFILE: Poison by intravenous and intraperitoneal routes. Moderately toxic by ingestion and other routes. Dangerous when heated; see also BROMIDES.

ARW750 **CAS:151-06-4** **HR: 3**
AVICOL
mf: $C_{10}H_{14}ClN \cdot ClH$ mw: 220.16

PROP: Solid. Mp: 234°.

SYNS: p-CHLORO-α,α-DIMETHYLPHENETHYLAMINE HYDROCHLORIDE □ 4-CHLORO-α,α-DIMETHYLPHENETHYLAMINE HYDROCHLORIDE □ CHLOROPHENTERMINE HYDROCHLORIDE □ 1-(p-CHLOROPHENYL)-2-METHYL-2-AMINOPROPANE HYDROCHLORIDE □ CHLORPHENTERMINE HYDROCHLORIDE □ α,α-DIMETHYL-p-CHLOROPHENETHYLAMINE HYDROCHLORIDE □ LUCOFEN □LUCOFENE □ NSC-76098 □ PRE-SATE □ PRESATE HYDROCHLORIDE □ S-62 □ S 62-2 □ W 2426

TOXICITY DATA WITH REFERENCE
scu-rat TDLo:150 mg/kg (16-20D preg):REP TOXID9 1,31,81
orl-rat LD50:150 mg/kg CHTPBA 5,247,70
ipr-rat LD50:144 mg/kg APTOA6 17,121,60
orl-mus LD50:225 mg/kg ARZNAD 13,711,63
ipr-mus LD50:88 mg/kg CHTPBA 5,247,70

ivn-mus LD50:55 mg/kg JPETAB 137,365,62

SAFETY PROFILE: Poison by ingestion, subcutaneous, intraperitoneal, and intravenous routes. Experimental reproductive effects. When heated to decomposition it emits very toxic fumes of HCl and NO_x. An anorexic agent.

ARW800 **CAS:8024-32-6** **HR: 1**
AVOCADO OIL

SYNS: ALLIGATOR PEAR OIL □ LIPOVAL A

TOXICITY DATA WITH REFERENCE
skn-rbt 500 mg/24H SEV JEPTDQ 4(4),93,80
eye-rbt 100 mg JEPTDQ 4(4),93,80

CONSENSUS REPORTS: Reported in EPA TSCA Inventory.

SAFETY PROFILE: A severe skin and eye irritant. When heated to decomposition it emits acrid smoke and irritating vapors.

ARX125 **HR: 3**
AYUSH-47

PROP: An Indian indigenous preparation containing equal parts of the bark of *Saraca indica, Areca catechu, Coccus lacca,* gold and sugar (IJMRAQ 70,504,79).

TOXICITY DATA WITH REFERENCE
orl-rat TDLo:25 mg/kg (1-5D preg):TER IJMRAQ 70,504,79
orl-rat TDLo:12,500 μg/kg (1-5D preg):REP IJMRAQ 70,504,79
orl-mus LDLo:200 mg/kg IJMRAQ 70,504,79

SAFETY PROFILE: Poison by ingestion. Experimental reproductive effects. When heated to decomposition it emits acrid smoke and irritating fumes.

ARX150 **CAS:21650-02-2** **HR: 3**
AZABICYCLANE CITRATE
mf: $C_{16}H_{23}NO•C_6H_8O_7$ mw: 437.54

PROP: Mp: 207°.

SYNS: 4-β-METHOXY-1-METHYL-4-α-PHENYL-3-α,5-α-PROPANOPIPERIDINE HYDROGEN CITRATE □ 9-β-METHOXY-9-α-PHENYL-3-METHYL-3-AZABICYCLO(3.3.1)NONANE CITRATE

TOXICITY DATA WITH REFERENCE
ipr-mus TDLo:24 mg/kg (7-12D preg):TER IYKEDH 3,195,72
scu-rat LD50:130 mg/kg TXAPA9 17,344,70
orl-mus LD50:300 mg/kg TXAPA9 17,344,70
scu-mus LD50:200 mg/kg TXAPA9 17,344,70
ivn-mus LD50:56 mg/kg TXAPA9 17,344,70

SAFETY PROFILE: Poison by ingestion, subcutaneous, and intravenous routes. An experimental teratogen. When heated to decomposition it emits toxic fumes of NO_x.

ARX300 **CAS:283-24-9** **HR: 3**
3-AZABICYCLO(3.2.2)NONANE
mf: $C_8H_{15}N$ mw: 125.24

TOXICITY DATA WITH REFERENCE
ivn-mus LD50:56 mg/kg CSLNX* NX#01357

CONSENSUS REPORTS: Reported in EPA TSCA Inventory.

SAFETY PROFILE: Poison by intravenous route. When heated to decomposition it emits toxic vapors of NO_x.

ARX500 **CAS:69766-49-0** **HR: 3**
1-AZABICYCLO(3.2.1)OCTAN-6-OL DIPHENYLACETATE HYDROCHLORIDE
mf: $C_{21}H_{23}NO_2•ClH$ mw: 357.91

SYN: 6-DIPHENYLACETOXY-1-AZABICYCLO(3.2.1)OCTANE HYDROCHLORIDE

TOXICITY DATA WITH REFERENCE
ipr-mus LD50:105 mg/kg JPETAB 104,284,52
ivn-mus LD50:33 mg/kg JPETAB 104,284,52
ivn-dog LD50:30 mg/kg JPETAB 104,284,52

SAFETY PROFILE: Poison by intraperitoneal and intravenous routes. When heated to decomposition it emits very toxic fumes of HCl and NO_x.

ARX750 **CAS:69766-48-9** **HR: 3**
1-AZABICYCLO(3.2.1)OCTAN-6-OL-9-FLUORENECARBOXYLATE HYDROCHLORIDE
mf: $C_{21}H_{21}NO_2•ClH$ mw: 355.89

SYN: RO 2-3245

TOXICITY DATA WITH REFERENCE
ipr-mus LD50:137 mg/kg JPETAB 104,284,52
ivn-mus LD50:23 mg/kg JPETAB 104,284,52

SAFETY PROFILE: Poison by intraperitoneal and intravenous routes. When heated to decomposition it emits very toxic fumes of HCl and NO_x.

ARX770 **CAS:69766-47-8** **HR: 3**
AZABICYCLOOCTANOL METHYL BROMIDE DIPHENYLACETATE
mf: $C_{22}H_{26}NO_2•Br$ mw: 416.40

SYN: RO 2-3951

TOXICITY DATA WITH REFERENCE
ipr-mus LD50:46 mg/kg JPETAB 104,284,52
ivn-mus LD50:4 mg/kg JPETAB 104,284,52

SAFETY PROFILE: Poison by intraperitoneal and intravenous routes. See also BROMIDES. When heated to decomposition it emits toxic fumes of Br^- and NO_x.

For occupational chemical analysis use NIOSH: Anisidine 2514.

ARX800 **CAS:1249-84-9** **HR: 3**
AZACOSTEROL DIHYDROCHLORIDE
mf: $C_{25}H_{44}N_2O•2ClH$ mw: 461.63

SYNS: AZACOSTEROL HYDROCHLORIDE □ AZASTEROL □ DIAZA-

COSTEROL HYDROCHLORIDE □ 20,25-DIAZOCHOLESTEROL DIHYDROCHLORIDE □ 17-β-((3-(DIMETHYLAMINO)-PROPYL)METHYLAMINO)ANDROST-5-EN-3-β-OL DIHYDROCHLORIDE □ IMD 760 □ ORNITROL □ SC 12937

TOXICITY DATA WITH REFERENCE
oms-pgn-orl 600 mg/kg/10D-I JRPFA4 15,145,68
spm-pgn-orl 600 mg/kg/10D-I JRPFA4 15,145,68
ipr-rat TDLo:100 mg/kg (male 10D pre):REP JOAND3 7,277,86
orl-rat LD50:470 mg/kg 85ARAE 3,100,76/77
ipr-rat LD50:60 mg/kg FMCHA2 -,C174,83
orl-mus LD50:380 mg/kg FMCHA2 -,C174,83
ipr-mus LD50:92 mg/kg FMCHA2 -,C174,83

SAFETY PROFILE: Poison by ingestion and intraperitoneal routes. Experimental reproductive effects. Mutation data reported. When heated to decomposition it emits toxic fumes of NO_x and HCl.

ARX875 CAS:78110-38-0 HR: 2
AZACTAM
mf: $C_{13}H_{17}N_5O_8S_2$ mw: 435.47

PROP: The first totally synthetic monocyclic β-lactam (monobactam) antibiotic.

SYNS: 2-(((1-(2-AMINO-4-THIAZOLYL)-2-((2-METHYL-4-OXO-1-SULFO-3-AZETIDINYL)AMINO)-2-OXOETHYLIDENE)AMINO)OXY)-2-METHYLPROPANOIC ACID, (2S-(2-α,3β(Z)))- □ AZTHREONAM □ AZTREONAM □ PRIMBACTAM □ SQ 26,776

TOXICITY DATA WITH REFERENCE
ivn-rat TDLo:2970 mg/kg (7-17D preg):REP NKRZAZ 33(Suppl 1),203,85
ivn-rat TDLo:1100 mg/kg (7-17D preg):TER NKRZAZ 33(Suppl 1),203,85
ipr-rat LD50:2549 mg/kg NKRZAZ 33(Suppl 1),143,85
scu-rat LD50:3154 mg/kg NKRZAZ 33(Suppl 1),143,85
ivn-rat LD50:2001 mg/kg KSRNAM 19,468,85
ipr-mus LD50:2897 mg/kg NKRZAZ 33(Suppl 1),143,85
scu-mus LD50:3906 mg/kg NKRZAZ 33(Suppl 1),143,85
ivn-mus LD50:1963 mg/kg NKRZAZ 33(Suppl 1),143,85

SAFETY PROFILE: Moderately toxic by several routes. An experimental teratogen. Other experimental reproductive effects. When heated to decomposition it emits toxic fumes of NO_x and SO_x.

ARY000 CAS:320-67-2 HR: 3
AZACYTIDINE
mf: $C_8H_{12}N_4O_5$ mw: 244.24

PROP: Solid. Mp: 232–234° (decomp).

SYNS: 5-AC □ 5-ACZ □ 4-AMINO-1-β-d-RIBOFURANOSYL-d-TRIAZIN-2(1H)-ONE □ 4-AMINO-1-β-d-RIBOFURANOSYL-1,3,5-TRIAZIN-2(1H)-ONE □ ANTIBIOTIC U 18496 □ AZACITIDINE □ 5-AZACYTIDINE □ 5′-AZACYTIDINE □ LADAKAMYCIN □ MYLOSAR □ NCI-C01569 □ NSC-102816 □ U 18496

TOXICITY DATA WITH REFERENCE
dnd-hmn:fbr 1 μmol/L PNASA6 79,2352,82
dni-hmn:leu 3 μmol/L CNREA8 43,763,83
ipr-rat TDLo:4 mg/kg (female 17D post):REP FCTOD7 22,963,84
ipr-mus TDLo:1 mg/kg (female 7D post):TER TJADAB 30(1),9A,84
ipr-rat TDLo:190 mg/kg/38W-I:ETA CRNGDP 5,1583,84
ipr-mus TDLo:100 mg/kg/50W-I:CAR CALEDQ 37,51,87
unr-mus TDLo:1 mg/kg (16D post):CAR TJADAB 32,33A,85
ipr-mus TD:284 mg/kg/43W-I:NEO NCITR* NCI-TR-42,78
ipr-mus LD:1 mg/kg (female 12D post):NEO,TER CALEDQ 27,81,85
ivn-wmn TDLo:6 mg/kg/10D-I:BLD CCROBU 56,413,72
ivn-wmn TDLo:500 μg/kg:GIT CCROBU 56,413,72
orl-mus LD50:572 mg/kg TXAPA9 19,382,71
ipr-mus LD50:68 mg/kg EXPEAM 22,53,66
ivn-mus LD50:229 mg/kg NTIS** PB84-211432
ivn-dog LD50:7200 μg/kg AVPCAQ 14,285,77
orl-bwd LD50:100 mg/kg AECTCV 12,355,83

CONSENSUS REPORTS: IARC Cancer Review: Group 3 IMEMDT 7,56,87; Animal Limited Evidence IMEMDT 26,37,81; NCI Carcinogenesis Bioassay (ipr); Inadequate Studies: rat NCITR* NCI-CG-TR-42,78; Clear Evidence: mouse NCITR* NCI-CG-TR-42,78. EPA Genetic Toxicology Program.

SAFETY PROFILE: Poison by ingestion, intravenous, and intraperitoneal routes. Human systemic effects by intravenous route: nausea, vomiting and diarrhea, reduction in white cell count (luekopenia and agranulocytosis). An experimental teratogen. Other experimental reproductive effects. Questionable carcinogen with experimental carcinogenic, neoplastigenic, tumorigenic data. Human mutation data reported. A skin irritant. When heated to decomposition it emits toxic fumes of NO_x.

ARY125 CAS:2353-33-5 HR: 3
5-AZADEOXYCYTIDINE
mf: $C_8H_{12}N_4O_4$ mw: 228.24

PROP: Crystals from MeOH. Mp: 191° (decomp).

SYNS: 4-AMINO-1-(2-DEOXY-β-d-erythro-PENTOFURANOSYL)-s-TRIAZIN-2(1H)-ONE □ 5-AZA-2′-DEOXYCYTIDINE

TOXICITY DATA WITH REFERENCE
dns-hmn:leu 5 μmol/L/1H CNREA8 42,519,82
dns-mus:leu 5 μmol/L/1H CNREA8 42,519,82
dni-mus-ipr 4 mg/kg BCPCA6 29,2929,80
dni-ham:ovr 1 μmol/L BBACAQ 697,286,82
ivn-mus LD50:22 mg/kg DCTODJ 4,373,81

SAFETY PROFILE: Poison by intravenous route. Human mutation data reported. When heated to decomposition it emits toxic fumes of NO_x.

ARY500 CAS:63907-29-9 HR: 2
2-AZAHYPOXANTHINE
mf: $C_4H_3N_5O$ mw: 137.12

SYN: 4-OXO-4H-IMIDAZO(4,5-D)-v-TRIAZINE

TOXICITY DATA WITH REFERENCE
orl-rat TDLo:3005 mg/kg/21W-C:ETA JNCIAM 54,951,75

SAFETY PROFILE: Questionable carcinogen with experimental tumorigenic data. When heated to decomposition it emits toxic fumes of NO_x.

ARY750 **CAS:11003-24-0** **HR: 3**
AZALOMYCIN F

PROP: Needles. Mp: 125–127° (decomp). An antibiotic produced by *Streptomyces hygroscopicus var. azalomycetic* first isolated from soil sample collected from around root of azaleas (ARZNAD 18,1396,68).

TOXICITY DATA WITH REFERENCE
orl-mus LD50:580 mg/kg ARZNAD 18,1396,68
ipr-mus LD50:25,900 µg/kg ARZNAD 18,1396,68
scu-mus LD50:162 mg/kg ARZNAD 18,1396,68
ivn-mus LD50:12,500 µg/kg ARZNAD 18,1396,68

SAFETY PROFILE: Poison by intraperitoneal, subcutaneous, and intravenous routes. Moderately toxic by ingestion.

ARZ000 **CAS:146-36-1** **HR: 3**
AZAPETINE
mf: $C_{17}H_{17}N$ mw: 235.35

PROP: Oil. Bp: 176–179° @ 12 mm. Insol in water.

SYNS: 6-ALLYL-6,7-DIHYDRO-5H-DIBENZ(c,e)AZEPINE □ ILIDAR □ ILIDAR BASE

TOXICITY DATA WITH REFERENCE
orl-mus LD50:460 mg/kg BCFAAI 98,702,59
ipr-mus LD50:210 mg/kg CLDND*
scu-mus LD50:725 mg/kg CLDND*
ivn-mus LD50:27 mg/kg CLDND*
ims-mus LD50:600 mg/kg CLDND*
ivn-dog LD50:50 mg/kg CLDND*
ivn-rbt LD50:26 mg/kg CLDND*

SAFETY PROFILE: Poison by intraperitoneal and intravenous routes. Moderately toxic by ingestion, subcutaneous, and intramuscular routes. When heated to decomposition it emits toxic fumes of NO_x.

ASA000 **CAS:22304-30-9** **HR: 2**
AZAPROPAZONE
mf: $C_{16}H_{20}N_4O_2 \cdot 2H_2O$ mw: 336.44

SYNS: APAZONE DIHYDRATE □ AZAPROPAZON DIHYDRAT (GERMAN) □ 1,2-DIHYDRO-3-DIMETHYLAMINO-7-METHYL-1,2-(PROPYLMALONYL)-1,2,4-BENZOTRIAZINE DIHYDRATE □ 3-DIMETHYLAMINO-7-METHYL-1,2-(N-PROPYLMALONYL)-1,2-DIHYDRO-1,2,4-BENZOTRIAZINE DIHYDRATE □ MI 85 DI □ PROLIXAN

TOXICITY DATA WITH REFERENCE
eye-rbt 100 mg MLD CMROCX 4,17,76
orl-rat LD50:1950 mg/kg ARZNAD 23,1215,73
ipr-rat LD50:1101 mg/kg TOIZAG 19,242,72
scu-rat LD50:1101 mg/kg TOIZAG 19,242,72
ivn-rat LD50:710 mg/kg ARZNAD 19,36,69
orl-mus LD50:1080 mg/kg ARZNAD 23,1215,73
ipr-mus LD50:1107 mg/kg TOIZAG 19,242,72
scu-mus LD50:1190 mg/kg TOIZAG 19,242,72
ivn-mus LD50:750 mg/kg ARZNAD 19,36,69

SAFETY PROFILE: Moderately toxic by ingestion, intraperitoneal, subcutaneous and intravenous routes. An eye irritant. When heated to decomposition it emits toxic fumes of NO_x. An anti-inflammatory and analgesic agent.

ASA250 **CAS:74037-31-3** **HR: 2**
AZAPROPAZONE SODIUM
mf: $C_{15}H_{20}N_4O_2 \cdot Na$ mw: 322.39

SYNS: AZAPROPAZON NATRIUMSALZ (GERMAN) □ 1,2-DIHYDRO-3-DIMETHYLAMINO-7-METHYL-1,2-(PROPYLMALONYL)-1,2,4-BENZOTRIAZINE SODIUM SALT □ 3-DIMETHYLAMINO-7-METHYL-1,2-(PROPYLMALONYL)-1,2-DIHYDRO-1,2,4-BENZOTRIAZINE SODIUM SALT □ SODIUM AZAPROPAZONE

TOXICITY DATA WITH REFERENCE
orl-rat LD50:1900 mg/kg CMROCX 4,17,76
ivn-rat LD50:810 mg/kg CMROCX 4,17,76
orl-mus LD50:1950 mg/kg CMROCX 4,17,76
ivn-mus LD50:960 mg/kg CMROCX 4,17,76

SAFETY PROFILE: Moderately toxic by ingestion and intravenous routes. When heated to decomposition it emits toxic fumes of NO_x.

ASA500 **CAS:115-02-6** **HR: 3**
AZASERINE
mf: $C_5H_7N_3O_4$ mw: 173.15

PROP: Light-yellow needles from EtOH (aq). Mp: 146–162° (decomp). Produced by the strain *Streptomyces fragilis* (85ERAY 2,1249,78)

SYNS: AZASERIN □ l-AZASERINE □ AZS □ CI-337 □ CL 337 □ CN-15,757 □ DIAZOACETATE (ESTER)-l-SERINE □ l-DIAZOACETATE (ESTER) SERINE □ DIAZO-ACETIC ACID ESTER with SERINE □ o-DIAZOACETYL-l-SERINE □ NSC-742 □ P-165 □ RCRA WASTE NUMBER U015 □ l-SERINE DIAZOACETATE □ l-SERINE DIAZOACETATE (ester)

TOXICITY DATA WITH REFERENCE
mma-sat 200 ng/plate PNASA6 72,5135,75
mmo-bcs 500 µmol/L EXPEAM 39,530,83
ipr-ham TDLo:2 mg/kg (female 8D post):TER EXPEAM 39,324,83
ipr-rat TDLo:10 mg/kg (female 7-8D post):REP PSEBAA 94,27,57
orl-rat TDLo:150 mg/kg/5W-C:NEO CANCAR 47,1562,81
ipr-rat TDLo:30 mg/kg:CAR CRNGDP 10,311,89
ipr-rat TD:440 mg/kg/13W-I:ETA CNREA8 40,592,80
ipr-rat LD:260 mg/kg/26W-I:CAR CNREA8 35,2249,75
orl-rat LD50:170 mg/kg CANCAR 10,889,57
ipr-rat LD50:70 mg/kg PSEBAA 94,27,57
orl-mus LD50:150 mg/kg CANCAR 10,889,57
ipr-mus LD50:100 mg/kg CANCAR 10,889,57
scu-mus LD50:50 mg/kg 85GDA2 4(1),432,80
ivn-mus LD50:62 mg/kg 85ERAY 2,1249,78
ivn-dog LDLo:30 mg/kg TXAPA9 22,595,72

CONSENSUS REPORTS: IARC Cancer Review: Group 2B IMEMDT 7,56,87; Animal Limited Evidence IMEMDT 10,73,76. EPA Genetic Toxicology Program.

SAFETY PROFILE: Suspected carcinogen with experimental carcinogenic, neoplastigenic, and tumorigenic data. Poison by ingestion, intraperitoneal, and subcutaneous routes. An experimental teratogen. Other experimental reproductive effects. Human mutation data reported. When heated to decomposition it emits toxic fumes of NO_x.

ASA750 CAS:1497-16-1 HR: 2
6-AZASPIRO(3,4)OCTANE-5,7-DIONE
mf: $C_7H_9NO_2$ mw: 139.17

SYN: AZA-6-SPIRO(3,4)OCTANE-DIONE-5,7 (FRENCH)

TOXICITY DATA WITH REFERENCE
orl-mus LD50:1750 mg/kg BSCFAS 3,1119,66
ipr-mus LD50:1450 mg/kg BSCFAS 3,1119,66

SAFETY PROFILE: Moderately toxic by ingestion and intraperitoneal routes. When heated to decomposition it emits toxic fumes of NO_x.

ASA875 CAS:64-60-8 HR: 3
4-(3-AZASPIRO(5.5)UNDEC-3-YL)-4′-FLUORO-BUTYROPHENONE HYDROCHLORIDE
mf: $C_{20}H_{28}FNO \cdot ClH$ mw: 353.95

TOXICITY DATA WITH REFERENCE
orl-rat LD50:200 mg/kg JMCMAR 8,62,65
ipr-rat LD50:50 mg/kg JMCMAR 8,62,65
orl-mus LD50:120 mg/kg JMCMAR 8,62,65
ipr-mus LD50:45 mg/kg JMCMAR 8,62,65

SAFETY PROFILE: Poison by ingestion and intraperitoneal routes. When heated to decomposition it emits toxic fumes of F^-, NO_x, and HCl.

ASB250 CAS:446-86-6 HR: 3
AZATHIOPRINE
mf: $C_9H_7N_7O_2S$ mw: 277.29

PROP: Pale-yellow crystals from Me_2CO (aq). Mp: 243–244° (decomp).

SYNS: AZANIN ☐ AZATIOPRIN ☐ AZOTHIOPRINE ☐ BW 57-322 ☐ CCUCOL ☐ IMURAN ☐ IMUREK ☐ IMUREL ☐ METHYLNITROIMIDAZOLYLMERCAPTOPURINE ☐ 6-(1′-METHYL-4′-NITRO-5′-IMIDAZOLYL)-MERCAPTOPURINE ☐ 6-(METHYL-p-NITRO-5-IMIDAZOLYL)-THIOPURINE ☐ 6-(1-METHYL-p-NITRO-5-IMIDAZOLYL)-THIOPURINE ☐ 6-((1-METHYL-4-NITROIMIDAZOL-5-YL)THIO)PURINE ☐ 6-(1-METHYL-4-NITROIMIDAZOL-5-YLTHIO)PURINE ☐ 6-((1-METHYL-4-NITRO-1H-IMIDAZOL-5-YL)THIO)-1H-PURINE ☐ NCI-C03474 ☐ NSC-39084 ☐ RORASUL

TOXICITY DATA WITH REFERENCE
mma-sat 300 µmol/L EXPEAM 40,370,84
cyt-hmn-unr 1074 mg/kg/4Y MUREAV 94,501,82
orl-rat TDLo:240 mg/kg (female 10-15D post):REP OYYAA2 2,401,68
orl-mus TDLo:50 mg/kg (female 10D post):TER NSAPCC 298,93,77
orl-wmn TDLo:273 mg/kg/13W-C:CAR,BLD BMJOAE 4,235,72
orl-rat TDLo:1932 mg/kg/46W-C:ETA ESKHA5 (102),66,84
orl-mus TDLo:1596 mg/kg/95W-C:NEO JJCREP 80,419,89
scu-mus TDLo:1200 mg/kg/30W-I:CAR ARZNAD 29,662,79
ims-mus TDLo:3500 mg/kg/26W-I:CAR BLOOAW 31,396,68
orl-wmn TD:3 g/kg/3.5Y-C:CAR,KID JAMAAP 237,152,77
orl-wmn TD:5460 mg/kg/6Y-I:CAR,BLD AMSVAZ 207,315,80
orl-man TD:728 mg/kg/43W-C:CAR,BLD AJMSA9 273,335,77
orl-man TD:1565 mg/kg/4Y-C:CAR,BLD AJMEAZ 57,885,74
orl-man TD:3266 mg/kg/3Y-C:CAR,BLD AJMEAZ 57,885,74
orl-wmn TD:2 g/kg/3Y-C:CAR,BLD PGMJAO 53,173,77
orl-man LDLo:395 mg/kg/56W-I:SYS GASTAB 90,446,86
orl-wmn TDLo:500 µg/kg:GIT,SYS JRHUA9 13,1117,86
orl-man TDLo:7500 µg/kg/1W-I:BPR,SYS ARHEAW 24,1453,81
orl-man TDLo:243 mg/kg/1Y-I:SYS GASTAB 90,446,86
unr-cld TDLo:2500 µg/kg/D-I:BLD AJDCAI 134,377,80
orl-rat LD50:535 mg/kg NIIRDN 6,3,82
ipr-rat LD50:300 mg/kg JRPFA4 4,297,62
idu-rat LD50:630 mg/kg RPTOAN 31,223,68
orl-mus LD50:1389 mg/kg NIIRDN 6,3,82
ipr-mus LD50:273 mg/kg RPTOAN 34,284,71
scu-mus LD50:350 mg/kg JMCMAR 18,320,75
idu-mus LD50:2437 mg/kg RPTOAN 31,223,68
idu-rbt LDLo:100 mg/kg RPTOAN 31,223,68

CONSENSUS REPORTS: NTP 7th Annual Report On Carcinogens. IARC Cancer Review: Group 1 IMEMDT 7,119,87 Human Sufficient Evidence IMEMDT 26,47,81; Animal Limited Evidence IMEMDT 26,47,81. NCI Carcinogenesis Studies (ipr); No Evidence: rat CANCAR 40,1935,77; Clear Evidence: mouse CANCAR 40,1935,77. EPA Genetic Toxicology Program.

SAFETY PROFILE: Confirmed human carcinogen producing bladder tumors and leukemia. Poison by subcutaneous, intradermal, and intraperitoneal routes. Moderately toxic by ingestion. Human systemic effects: liver changes, hypermotility, diarrhea, nausea or vomiting, increased body temperature, BP lowering, decreased urine volume or anuria, normocytic anemia, bone marrow changes. An experimental teratogen. Other experimental reproductive effects. Human mutation data reported. When heated to decomposition it emits very toxic fumes of NO_x and SO_x. An immunosuppressant.

ASB750 CAS:123-99-9 HR: 1
AZELAIC ACID
mf: $C_9H_{16}O_4$ mw: 188.23

PROP: Leaflets or needles. Mp: 106.5°, bp: 226° @ 10 mm, d: 1.029 @ 20°/4°, vap press: 1 mm @ 178.3°. Solubility: in water = 0.2/100, very sol in alc, in ether = 2.7/100 @ 15°.

SYNS: ANCHOIC ACID ☐ HEPTANEDICARBOXYLIC ACID ☐ 1,7-HEPTANEDICARBOXYLIC ACID ☐ LEPARGYLIC ACID ☐ NONANEDIOIC ACID

TOXICITY DATA WITH REFERENCE
skn-rbt 500 mg/24H MLD EMERY* S3B,-,64
eye-rbt 3 mg MLD EMERY* S3B,-,64
orl-rat LD50:>5 g/kg NTIS** AD-A067-313

CONSENSUS REPORTS: Reported in EPA TSCA Inventory.

SAFETY PROFILE: Low toxicity by ingestion. A skin and eye irritant. Closely related to glutaric acid and adipic acid. Combustible when exposed to heat or flame; can react with oxidizing materials.

For occupational chemical analysis use NIOSH: Azelaic Acid, 5019.

ASC000 **CAS:109-31-9** **HR: 2**
AZELAIC ACID DIHEXYL ESTER
mf: $C_{21}H_{40}O_4$ mw: 356.61

SYNS: DI-N-HEXYL AZELATE □ NONANEDIOTIC ACID, DIHEXYL ESTER

TOXICITY DATA WITH REFERENCE
orl-rat LD50:16 g/kg AIHAAP 30,470,69
orl-mus LDLo:15,000 mg/kg 34ZIAG -,226,69
orl-rbt LDLo:1000 mg/kg 34ZIAG -,226,69
orl-gpg LDLo:6000 mg/kg TXAPA9 4,247,62

CONSENSUS REPORTS: Reported in EPA TSCA Inventory.

SAFETY PROFILE: Moderately toxic by ingestion. See also ESTERS. When heated to decomposition it emits acrid smoke and irritating fumes.

ASC125 **CAS:58581-89-8** **HR: 3**
AZELASTINE
mf: $C_{22}H_{24}ClN_3O$ mw: 381.94

SYN: 4-(p-CHLOROBENZYL)-2-(HEXAHYDRO-1-METHYL-1H-AZEPIN-4-YL)-1-(2H)-PHTHALAZINONE

TOXICITY DATA WITH REFERENCE
orl-rat LD50:310 mg/kg ARZNAD 31,1184,81
ipr-rat LD50:43,200 μg/kg ARZNAD 31,1184,81
scu-rat LD50:59,600 μg/kg ARZNAD 31,1184,81
ivn-rat LD50:26,900 μg/kg ARZNAD 31,1184,81
orl-mus LD50:124 mg/kg ARZNAD 31,1184,81
ipr-mus LD50:42,800 μg/kg ARZNAD 31,1184,81
scu-mus LD50:54,200 μg/kg ARZNAD 31,1184,81
ivn-mus LD50:35,500 μg/kg ARZNAD 31,1184,81
orl-dog LD50:51,300 μg/kg ARZNAD 31,1184,81
ivn-dog LD50:13,700 μg/kg ARZNAD 31,1184,81

SAFETY PROFILE: Poison by ingestion, subcutaneous, intravenous, and intraperitoneal routes. Experimental reproductive effects. When heated to decomposition it emits toxic fumes of NO_x and Cl^-.

ASC130 **CAS:79307-93-0** **HR: 3**
AZELASTINE HYDROCHLORIDE
mf: $C_{22}H_{24}ClN_3O\cdot ClH$ mw: 418.40

PROP: Solid. Mp: 225–229°.

SYNS: AZELASTIN □ 4-(p-CHLOROBENZYL)-2-(HEXAHYDRO-1-METHYL-1H-AZEPIN-4-YL)-1-(2H)-PHTHALAZINONE HCl

TOXICITY DATA WITH REFERENCE
orl-rat TDLo:755 mg/kg (7-17D preg):TER ARZNAD 31,1225,81
orl-rbt TDLo:390 mg/kg (6-18D preg):REP ARZNAD 31,1225,81
orl-rat LD50:580 mg/kg KSRNAM 20,5231,86
ivn-rat LD50:24,600 μg/kg KSRNAM 20,5231,86
orl-mus LD50:143 mg/kg KSRNAM 20,5231,86
ivn-mus LD50:25,400 μg/kg KSRNAM 20,5231,86
orl-dog LD50:107 mg/kg KSRNAM 20,5235,86

SAFETY PROFILE: Poison by ingestion and intravenous routes. An experimental teratogen. Other experimental reproductive effects. When heated to decomposition it emits toxic fumes of NO_x and HCl.

ASC250 **CAS:24853-80-3** **HR: 3**
AZEPHEN
mf: $C_{16}H_{19}N_4O\cdot ClH$ mw: 333.86

PROP: Solid. Mp: 315–317°.

SYNS: 5-METHYL-3-(4-METHYL-1-PIPERAZINYL)-5H-PYRIDAZINO(3,4-b)(1,4)BENZOXAZINE HYDROCHLORIDE □ 2-(4-METHYL-1-PIPERAZINYL)-10-METHYL-3,4-DIAZAPHENOXAZINDIHYDROCHLORID (GERMAN)

TOXICITY DATA WITH REFERENCE
orl-rat LD50:1 g/kg ZPPLBF 114,787,75
scu-rat LD50:490 mg/kg ZPPLBF 114,787,75
ivn-rat LD50:48 mg/kg ZPPLBF 114,787,75
orl-mus LD50:700 mg/kg PCJOAU 4,118,70
scu-mus LD50:330 mg/kg ZPPLBF 114,787,75
ivn-mus LD50:63 mg/kg RPTOAN 37,2,74
orl-cat LD50:200 mg/kg ZPPLBF 114,787,75

SAFETY PROFILE: Poison by ingestion, intravenous, and subcutaneous routes. When heated to decomposition it emits very toxic fumes of HCl and NO_x. An antidepressant.

ASC500 **CAS:2133-34-8** **HR: 2**
l-2-AZETIDINECARBOXYLIC ACID
mf: $C_4H_7NO_2$ mw: 101.12

PROP: Crystals from MeOH. Sol in water.

TOXICITY DATA WITH REFERENCE
ipr-mus TDLo:500 mg/kg (female 11D post):TER TJADAB 16,123,77
ipr-rat TDLo:300 mg/kg (female 8D post):REP SEIJBO 16,263,76
scu-mus LD50:1000 mg/kg 85GDA2 8(1),101,82

SAFETY PROFILE: Moderately toxic by subcutaneous route. An experimental teratogen. Other experimental reproductive effects. When heated to decomposition it emits toxic fumes of NO_x.

ASC750 **HR: 3**
AZIDES

SAFETY PROFILE: Variable toxicity. Many azides are poisonous, and cause a fall in blood pressure, and some inhibit enzyme action, thus resembling nitrites and cyanides. An azide is a compound of hydrogen or a metal ion and the monovalent $-N_3$ radical. All azide salts and the acid are unstable and some decompose explosively, although lead azide, which is one of the most important azides, is not very sensitive. Dangerous; shock and heat will explode them. When heated to decomposition they emit highly toxic fumes. If exposed to CS_2, they forms violently explosive salts. Organic azides are sensitized by metal salts or traces of strong acid. (See also specific compound.)

ASD000 CAS:78-57-9 HR: 2
AZIDITHION
mf: $C_6H_{12}N_5O_2PS_2$ mw: 281.32

PROP: Crystals from MeOH. Sltly sol in water, most org solvs. Sol in THF, 2-methoxyethanol.

SYNS: 2,4-DIAMINO-6-DIMETHOXYPHOSPHINOTHIONYLTHIOMETHYL-s-TRIAZINE □ 4,6-DIAMINO-s-TRIAZINE-2-METHANETHIOL S-ESTER with O,O-DIMETHYLPHOSPHORODITHIOATE □ S-((4,6-DIAMINO-1,3,5-TRIAZIN-2-YL)-METHYL)-O,O-DIMETHYL-DITHIOFOSFAAT (DUTCH) □ S-((4,6-DIAMINO-1,3,5-TRIAZIN-2-YL)-METHYL)-O,O-DIMETHYL-DITHIOPHOSPHAT (GERMAN) □ 4,6-DIAMINO-1,3,5-TRIAZIN-2-YLMETHYL-O,O-DIMETHYL PHOSPHORODITHIOATE □ S-((4,6-DIAMINO-s-TRIAZIN-2-YL)METHYL)-O,O-DIMETHYL PHOSPHORODITHIOATE □ S-(4,6-DIAMINO-1,3,5-TRIAZIN-2-YLMETHYL)-O,O-DIMETHYL PHOSPHORODITHIOATE □ S-(4,6-DIAMINO-1,3,5-TRIAZIN-2-YLMETHYL) DIMETHYL PHOSPHOROTHIOLOTHIONATE □ 2-DIMETHOXYPHOSPHINOTHIOYLTHIOMETHYL-4,6-DIAMINO-s-TRIAZINE □ O,O-DIMETHYL-S-(4,6-DIAMINO-1,3,5-TRIAZINYL-2-METHYL) DITHIOPHOSPHATE □ O,O-DIMETHYL- S-(4,6-DIAMINO-s-TRIAZIN-2-YLMETHYL)PHOSPHORODITHIOATE □ O,O-DIMETHYL-S-(4,6-DIAMINO-1,3,5-TRIAZIN-2-YL) METHYL PHOSPHORODITHIOATE □ O,O-DIMETHYL-S-(4,6-DIAMINO-1,3,5-TRIAZIN-2-YL)METHYL PHOSPHOROTHIOLOTHIONATE □ DITHIOPHOSPHATE de O,O-DIMETHYLE et de S-((4,6-DIAMINO-1,3,5-TRIAZINE-2-YL)-METHYLE) (FRENCH) □ ENT 25,760 □ MENAZON □ PP175 □ R 15,175 □ SAIPHOS □ SAPHICOL □ SAPHIZON □ SAPHIZON-DP □ SAPHOS □ SAYFOR □ SAYFOS □ SAYPHOS □ SYPHOS

TOXICITY DATA WITH REFERENCE
mmo-smc 5 ppm RSTUDV 6,161,76
orl-rat LD50:890 mg/kg AR3IM* 20,14,66
orl-mus LD50:427 mg/kg SPEADM 78-1,48,78
orl-ckn LD50:487 mg/kg TXAPA9 7,606,65
unr-mam LD50:900 mg/kg 30ZDA9 -,373,71

CONSENSUS REPORTS: EPA Genetic Toxicology Program.

SAFETY PROFILE: Moderately toxic by ingestion and possibly other routes. Mutation data reported. When heated to decomposition it emits very toxic fumes of NO_x, PO_x, and SO_x.

ASD375 CAS:18523-48-3 HR: 3
AZIDOACETIC ACID
mf: $C_2H_3N_3O_2$ mw: 101.06

PROP: Crystals or liquid. Bp: 116° @ 12 mm, d: 1.354 @ 33 mm. Sol in water.

SAFETY PROFILE: The acid in contact with iron or iron salts undergoes rapid exothermic decomposition at 25°C and explodes at 90°C. Upon decomposition it emits toxic fumes of NO_x. See also AZIDES.

ASD500 CAS:4504-27-2 HR: 3
AZIDOACETONE
mf: $C_3H_5N_3O$ mw: 99.10

PROP: Bp: 53–55° @ 2.5 mm.

SAFETY PROFILE: A dangerous fire and storage hazard. It can explode in the dark or when heated. When heated to decomposition it emits toxic fumes of NO_x. See also AZIDES.

ASE000 CAS:57707-64-9 HR: 3
AZIDOACETO NITRILE
mf: $C_2H_2N_4$ mw: 82.10

CONSENSUS REPORTS: Cyanide and its compounds are on the Community Right-To-Know List.

SAFETY PROFILE: An unstable explosive sensitive to impact and heat. When heated to decomposition it emits toxic fumes of NO_x and CN^-. See also NITRILES and AZIDES.

ASE250 HR: 3
N-AZIDO CARBONYL AZEPINE
mf: $C_7H_6N_4O$ mw: 162.20

SAFETY PROFILE: Highly unstable, explosive compound. See also azides. Can explode on distillation. When heated to decomposition it emits toxic fumes of NO_x.

ASE500 CAS:54567-24-7 HR: 3
AZIDOCARBONYL GUANIDINE
mf: $C_2H_4N_6O$ mw: 128.10

SAFETY PROFILE: Violently explosive on rapid heating. When heated to decomposition it emits toxic fumes of NO_x. See also AZIDES.

ASE875 CAS:22952-08-3 HR: 3
AZIDOCODEINE
mf: $C_{18}H_{22}N_4O_2$ mw: 326.44

SYN: 6-DEOXY-6-AZIDODIHYDROISOCODEINE

TOXICITY DATA WITH REFERENCE
orl-rat LD50:120 mg/kg JPPMAB 25,929,73
scu-rat LD50:125 mg/kg JPPMAB 25,929,73
ivn-rat LD50:52 mg/kg JPPMAB 25,929,73

SAFETY PROFILE: Poison by ingestion, subcutaneous, and intravenous routes. When heated to decomposition it emits toxic fumes of NO_x. See also CODEINE and AZIDES.

ASF500 HR: 3
AZIDODIMETHYL BORANE
mf: $C_2H_6BN_3$ mw: 82.9

SAFETY PROFILE: Very unstable. Explodes when heated. When heated to decomposition it emits toxic fumes of NO_x. See also BORANES, BORON COMPOUNDS, and AZIDES.

ASF625 CAS:70664-49-2 HR: 3
2-AZIDO-3,5-DINITROFURAN
mf: $C_4HN_5O_5$ mw: 199.08

SAFETY PROFILE: A heat- and impact-sensitive explosive. Upon decomposition it emits toxic fumes of NO_x. See also AZIDES.

ASF750 **CAS:4472-06-4** **HR: 3**
AZIDODITHIOFORMIC ACID
mf: CHN_3S_2 mw: 119.20

SYNS: AZIDODITHIOCARBONIC ACID (DOT) □ AZIDOTHIOCARBONIC ACID □ CARBONAZIDODITHIOIC ACID □ FORMIC ACID, AZIDODITHIO-

DOT CLASSIFICATION: Forbidden

SAFETY PROFILE: Very unstable. The acid and its salts are shock- and heat-sensitive explosives. Upon decomposition it emits very toxic fumes of NO_x and SO_x. See also AZIDES.

ASF800 **CAS:53422-49-4** **HR: 3**
2-AZIDOETHANOL NITRATE
mf: $C_2H_4NO_3$ mw: 90.07

PROP: A solid.

SYNS: 2-AZIDOETHYL NITRATE □ AZIDOETHYL NITRATE (DOT) □ ETHANOL, 2-AZIDO-, NITRATE (ester)

DOT CLASSIFICATION: Forbidden

SAFETY PROFILE: An explosion hazard.

ASG250 **CAS:4658-28-0** **HR: 2**
2-AZIDO-4-ISOPROPYLAMINO-6-METHYLTHIO-s-TRIAZINE
mf: $C_7H_{11}N_7S$ mw: 225.31

PROP: Crystals or powder. Mp: 95°. Very sltly sol in water.

SYNS: 2-AZIDO-4-ISOPROPYLAMINO-6-METHYLTHIO-1,3,5-TRIAZINE □ 4-AZIDO-N-(1-METHYLETHYL)-6-(METHYLTHIO)-1,3,5-TRIAZIN-2-AMINE □ 4-AZIDO-N-(1-METHYLETHYL)-6-(METHYLTHIO)-1,3,5-TRIAZIN-2-AMINI □ AZIPROTRYN □ AZIPROTRYNE □ AZIRPOTRYNE □ BRASORAN □ C 7019 □ CIBA C 7019 □ ISOPROPYLAMINO-4-AZIDO-6-METHYLTHIO-1,3,5-TRIAZIN (GERMAN) □MESORANIL □ MEZARONIL □ MEZURON

TOXICITY DATA WITH REFERENCE
orl-rat LD50:3600 mg/kg FMCHA2 -,C150,830
orl-rbt LD50:1800 mg/kg GUCHAZ 6,26,73

SAFETY PROFILE: Moderately toxic by ingestion. When heated to decomposition it emits very toxic fumes of NO_x and SO_x.

ASG500 **HR: 3**
N-AZIDO METHYL AMINE
mf: $C_2H_6N_4$ mw: 86.1

SAFETY PROFILE: Very unstable, explosive compound. When heated to decomposition it emits toxic fumes of NO_x. See also AZIDES, AMINES, and EXPLOSIVES.

ASG625 **CAS:59327-98-9** **HR: 3**
2-AZIDOMETHYLBENZENEDIAZONIUM TETRAFLUOROBORATE
mf: $C_7H_6BF_4N_5$ mw: 246.96

SAFETY PROFILE: Explosive reaction with trichloroacetonitrile. When heated to decomposition it emits toxic fumes of F^- and NO_x. See also AZIDES and BORON COMPOUNDS.

ASG675 **CAS:22952-87-0** **HR: 3**
AZIDOMORPHINE
mf: $C_{17}H_{20}N_4O_2$ mw: 312.41

SYN: 6-DEOXY-6-AZIDODIHYDROISOMORPHINE

TOXICITY DATA WITH REFERENCE
orl-rat LD50:62 mg/kg JPPMAB 25,929,73
scu-rat LD50:13 mg/kg JPPMAB 25,929,73
ivn-rat LD50:8100 μg/kg JPPMAB 25,929,73
orl-mus LD50:58 mg/kg JPPMAB 27,99,75
scu-mus LD50:16 mg/kg JPPMAB 27,99,75
ivn-mus LD50:13 mg/kg JPPMAB 27,99,75

SAFETY PROFILE: Poison by ingestion, subcutaneous, and intravenous routes. When heated to decomposition it emits toxic fumes of NO_x. See also AZIDES and MORPHINE.

ASH000 **CAS:35038-46-1** **HR: 3**
5-AZIDOTETRAZOLE
mf: CHN_7 mw: 111.10

SAFETY PROFILE: An unstable explosive. Explodes on contact with acetic acid. Its sodium, potassium, and silver salts are impact- and friction-sensitive explosives. The ammonium salt is a heat-sensitive explosive. When heated to decomposition it emits toxic fumes of NO_x. See also AZIDES.

ASH250 **HR: 3**
3-AZIDO-1,2,4-TRIAZOLE
mf: $C_2H_2N_6$ mw: 110.1

SAFETY PROFILE: Very unstable, explosive compound. Very sensitive to heat. Samples have exploded during analytical combustion. When heated to decomposition it emits toxic fumes of NO_x. See AZIDES.

ASH425 **HR: 3**
AZINOTHRICIN

SYN: α-ETHYL-6-(3-ETHYL-1,5-DIMETHYL-4-OXO-1,5-HEPTADIENYL)-N-(1,8,14,15,18,21,27-HEPTAAZA-21-HYDROXY-7-(1-HYDROXYETHYL-2,6,9.16,19,22-HEXAOXO-4-ISOPROPYL-20-(METHOXYMETHYL)-17,18-D IMETHYL-5-OXATRICYCLO(21.4.0.0(sup 10,15)HEPTACOSAN-3-6L)TETRAHYDRO-α-2-DIHYDROXY-5-METHYL-2H-PYRAN-2-ACETAMIDE

TOXICITY DATA WITH REFERENCE
ipr-mus LD50:3200 μg/kg JANTAJ 39,17,86
scu-mus LD50:420 mg/kg JANTAJ 39,17,86
ivn-mus LD50:10 mg/kg JANTAJ 39,17,86

SAFETY PROFILE: Poison by intravenous and intraperi-

toneal routes. Moderately toxic by other routes. When heated to decomposition it emits toxic fumes of NO_x.

ASH500 **CAS:86-50-0** **HR: 3**
AZINPHOS METHYL
mf: $C_{10}H_{12}N_3O_3PS_2$ mw: 317.34

PROP: Crystals or brown, waxy solid. D: 1.44, mp: 74°. Very sltly sol in water, very sol in $CHCl_3$, toluene.

SYNS: AZINFOS-METHYL (DUTCH) □ AZINPHOS METHYL, liquid (DOT) □ AZINPHOS-METILE (ITALIAN) □ BAY 9027 □ BAYER 17147 □ BENZOTRIAZINE derivative of a METHYL DITHIOPHOSPHATE □ BENZOTRIAZINEDITHIOPHOSPHORIC ACID DIMETHOXY ESTER □ CARFENE □ COTNION METHYL □ CRYSTHION 2L □ CRYSTHYON □ DBD □ S-(3,4-DIHYDRO-4-OXO-BENZO(α)(1,2,3)TRIAZIN-3-YLMETHYL)-O,O-DIMETHYL PHOSPHORODITHIOATE □ S-(3,4-DIHYDRO-4-OXO-1,2,3-BENZOTRIAZIN-3-YLMETHYL)-O,O-DIMETHYL PHOSPHORODITHIOATE □ O,O-DIMETHYL-S-(BENZAZIMINOMETHYL) DITHIOPHOSPHATE □ O,O-DIMETHYL-S-(1,2,3-BENZOTRIAZINYL-4-KETO)METHYL PHOSPHORODITHIOATE □ O,O-DIMETHYL-S-(3,4-DIHYDRO-4-KETO-1,2,3-BENZOTRIAZINYL-3-METHYL) DITHIOPHOSPHATE □ DIMETHYLDITHIOPHOSPHORIC ACID N-METHYLBENZAZIMIDE ESTER □ O,O-DIMETHYL-S-(4-OXO-3H-1,2,3-BENZOTRIZIANE-3-METHYL)PHOSPHORODITHIOATE □ O,O-DIMETHYL-S-(4-OXOBENZOTRIAZINO-3-METHYL)PHOSPHORODITHIOATE □ O,O-DIMETHYL-S-(4-OXO-1,2,3-BENZOTRIAZINO(3)-METHYL) THIOTHIONOPHOSPHATE □ O,O-DIMETHYL-S-((4-OXO-3H-1,2,3-BENZOTRIAZIN-3-YL)-METHYL)-DITHIOFOSFAAT (DUTCH) □ O,O-DIMETHYL-S-((4-OXO-3H-1,2,3-BENZOTRIAZIN-3-YL)-METHYL)-DITHIOPHOSPHAT (GERMAN) □ O,O-DIMETHYL-S-4-OXO-1,2,3-BENZOTRIAZIN-3(4H)-YLMETHYL PHOSPHORODITHIOATE □ O,O-DIMETIL-S-((4-OXO-3H-1,2,3-BENZOTRIAZIN-3-IL)-METIL)-DITIOFOSFATO (ITALIAN) □ ENT 23,233 □ GOTHNION □ GUSATHION □ GUTHION (DOT) □ 3-(MERCAPTOMETHYL)-1,2,3-BENZOTRIAZIN-4(3H)-ONE-O,O-DIMETHYL PHOSPHORODITHIOATE □ 3-(MERCAPTOMETHYL)-1,2,3-BENZOTRIAZIN-4(3H)-ONE-O,O-DIMETHYL PHOSPHORODITHIOATE-S-ESTER □ METHYLAZINPHOS □ N-METHYLBENZAZIMIDE, DIMETHYLDITHIOPHOSPHORIC ACID ESTER □ METHYL GUTHION □ METILTRIAZOTION □ NCI-C00066

TOXICITY DATA WITH REFERENCE
mmo-ssp 25 mmol/L MUREAV 117,139,83
mma-ssp 25 mmol/L MUREAV 117,139,83
cyt-hmn:lng 120 mg/L CNJGA8 17,455,75
cyt-hmn:oth 120 mg/L CNJGA8 17,455,75
orl-rat TDLo:190 mg/kg (6-22D preg/21D post):REP ARTODN 43,177,80
orl-mus TDLo:20 mg/kg (female 8D post):TER TCMUD8 5,3,85
orl-rat TDLo:5110 mg/kg/78W-C:ETA NCITR* NCI-CG-TR-69,78
orl-rat LD50:7 mg/kg JPPMAB 13,435,61
ihl-rat LC50:69 mg/m³/1H NTIS** PB277-077
skn-rat LD50:88 mg/kg 85JCAE -,1182,86
ipr-rat LD50:4900 μg/kg PSEBAA 114,509,63
ivn-rat LD50:7500 μg/kg NTIS** PB277-077
orl-mus LD50:15 mg/kg JPFCD2 15,867,80
skn-mus LD50:65 mg/kg KUMJAX 12,313,59

CONSENSUS REPORTS: NCI Carcinogenesis Bioassay (feed); Inadequate Studies: rat NCITR* NCI-CG-TR-69,78; No Evidence: mouse NCITR* NCI-CG-TR-69,78. EPA Genetic Toxicology Program. EPA Extremely Hazardous Substances List.

OSHA PEL: TWA 0.2 mg/m³ (skin)
ACGIH TLV: TWA 0.2 mg/m³ (skin)
DFG MAK: 0.2 mg/m³

SAFETY PROFILE: Poison by inhalation, ingestion, skin contact, intravenous, and intraperitoneal routes. An experimental teratogen. Other experimental reproductive effects. Human mutation data reported. Questionable carcinogen with experimental tumorigenic data. See also PARATHION and ESTERS. When heated to decomposition it emits very toxic fumes of PO_x, SO_x, and NO_x.

ASH750 **CAS:671-51-2** **HR: 3**
AZIRIDINE CARBOXYLIC ACID ETHYL ESTER
mf: $C_5H_9NO_2$ mw: 115.15

SYNS: N-CARBETHOXYETHYLENIMINE □ N-(ETHOXYCARBONYL)AZIRIDINE □ N-ETHOXYCARBONYLETHYLENEIMINE □ ETHOXYCARBONYL-1-ETHYLENIMINE □ ETHYL AZIRIDINECARBOXYLATE □ ETHYL-1-AZIRIDINECARBOXYLATE □ ETHYL AZIRIDINOCARBOXYLATE □ ETHYL-1-AZIRIDINYLCARBOXYLATE □ ETHYL AZIRIDINYLFORMATE

TOXICITY DATA WITH REFERENCE
cyt-rat-ipr 40 mg/kg BJPCAL 6,357,51
ipr-rat LD50:8600 μg/kg BJPCAL 21,581,63
ivn-mus LD50:180 mg/kg CSLNX* NX#04621

SAFETY PROFILE: Poison by intravenous and intraperitoneal routes. Mutation data reported. See also ESTERS. When heated to decomposition it emits toxic fumes of NO_x.

ASI000 **CAS:1072-52-2** **HR: 3**
1-AZIRIDINE ETHANOL
mf: C_4H_9NO mw: 87.14

PROP: Liquid. Bp: 154–156°.

SYNS: 2-(1-AZIRIDINYL)ETHANOL □ β-HYDROXY-1-ETHYLAZIRIDINE □ N-(β-HYDROXYETHYL)AZIRIDINE □ 2-HYDROXY-1-ETHYLAZIRIDINE □ N-(2-HYDROXYETHYL)AZIRIDINE □ N-HYDROXYETHYL ETHYLENE IMINE □ N-(2-HYDROXYETHYL)ETHYLENIMINE □ 1-(2-HYDROXYETHYL)ETHYLENIMINE

TOXICITY DATA WITH REFERENCE
skn-rbt 545 mg open MOD UCDS** 11/17/64
eye-rbt 1090 mg SEV UCDS** 11/17/64
mmo-sat 1 μg/plate ENMUDM 5(Suppl 1),3,83
sln-oin-dmg-orl 50 ppm ENMUDM 7,325,85
scu-mus TDLo:900 mg/kg/75W-I:NEO JNCIAM 46,143,71
orl-rat LD50:74 mg/kg JIHTAB 31,60,49
ivn-mus LD50:56 mg/kg CSLNX* NX#03613
skn-rbt LD50:71 μL/kg UCDS** 11/17/64

CONSENSUS REPORTS: IARC Cancer Review: Group 3 IMEMDT 7,56,87; Animal Limited Evidence IMEMDT 9,47,75. Reported in EPA TSCA Inventory.

SAFETY PROFILE: Poison by ingestion, skin contact, and intravenous routes. A skin and eye irritant. Questionable carcinogen with experimental neoplastigenic data. Mutation data reported. When heated to decomposition it emits toxic fumes of NO_x.

ASI250 CAS:1072-66-8 HR: 3
1-AZIRIDINE PROPIONITRILE
mf: $C_5H_8N_2$ mw: 96.15

PROP: Bp: 80–84° @ 10 mm.

SYNS: 1-AZIRIDINEPROPANENITRILE □ N-(2-CYANOETHYL)AZIRIDINE □ 1-(2-CYANOETHYL)AZIRIDINE □ N-(β-CYANOETHYL)ETHYLENIMINE □ 1-(2-CYANOETHYL)ETHYLENIMINE

TOXICITY DATA WITH REFERENCE
cyt-rat-ipr 20 mg/kg BJPCAL 6,357,51
orl-rat LDLo: 100 mg/kg NCNSA6 5,22,53

CONSENSUS REPORTS: Cyanide and its compounds are on the Community Right-To-Know List.

SAFETY PROFILE: Poison by ingestion. Mutation data reported. See also NITRILES. When heated to decomposition it emits toxic fumes of NO_x.

ASI300 CAS:4638-44-2 HR: 3
1-AZIRIDINYL m-(BIS(2-CHLOROETHYL)AMINO)PHENYL KETONE
mf: $C_{13}H_{16}Cl_2N_2O$ mw: 287.21

PROP: A liquid.

TOXICITY DATA WITH REFERENCE
unr-rat LD50: 69 mg/kg NEOLA4 27,261,80

DOT CLASSIFICATION: 3; *Label:* Flammable Liquid

SAFETY PROFILE: A poison by an unspecified route. A flammable liquid. When heated to decomposition it emits toxic vapors of NO_x and Cl^-.

ASK875 CAS:57998-68-2 HR: 3
AZIRIDINYLQUINONE
mf: $C_{16}H_{20}N_4O_6$ mw: 364.40

PROP: Orange needles from EtOH. Mp: 230°.

SYNS: AZQ □ 2,5-BIS(1-AZIRIDINYL)-3,6-DIOXO-1,4-CYCLOHEXADIENE-1,4-DICARBAMIC ACID DIETHYL ESTER □ DIAZIQUONE □ 3,6-DIAZIRIDINYL-2,5-BIS(CARBOETHOXYAMINO)-1,4-BENZOQUINONE □ NSC 182986 □ USAN-DIAZIQUONE

TOXICITY DATA WITH REFERENCE
mmo-sat 10 μg/plate TCMUD8 5,319,85
dnd-omi 1 μmol/L PAACA3 24,322,83
dnd-hmn: emb 50 μmol/L PAACA3 24,246,83
dnd-hmn: oth 75 μmol/L CNREA8 44,4447,84
dnd-mus: leu 57 μmol/L PAACA3 24,246,83
dnd-ham: ovr 50 μmol/L CNREA8 44,5634,84
dns-ham: ovr 300 nmol/L PAACA3 24,322,83
ipr-mus LD50: 11,290 μg/kg NCISP* JAN86
ivn-mus LD50: 10,300 μg/kg NTIS** PB80-177934

SAFETY PROFILE: Poison by intravenous and intraperitoneal routes. Human mutation data reported. When heated to decomposition it emits toxic fumes of NO_x. See also CARBAMATES and ESTERS.

ASK925 CAS:538-41-0 HR: 3
p-AZOANILINE
mf: $C_{12}H_{12}N_4$ mw: 212.28

SYNS: ANILINE, 4,4'-AZODI- □ 4,4'-AZOBISBENZENAMINE □ 4,4'-AZODIANILINE □ BENZENAMINE, 4,4'-AZOBIS-(9CI) □ p-DIAMINOAZOBENZENE □ 4,4'-DIAMINOAZOBENZENE

TOXICITY DATA WITH REFERENCE
mma-sat 100 μg/plate MUTAEX 4,115,89
ivn-mus LD50: 180 mg/kg CSLNX* NX#01451

CONSENSUS REPORTS: Reported in EPA TSCA Inventory.

SAFETY PROFILE: Poison by intravenous route. Mutation data reported. When heated to decomposition it emits toxic vapors of NO_x.

ASL250 CAS:103-33-3 HR: 2
AZOBENZENE
mf: $C_{12}H_{10}N_2$ mw: 182.23

PROP: Orange, monoclinic crystals. Mp: 68°, bp: 297°, d: 1.203 @ 20°/4°, vap press: 1 mm @ 103.5°. Insol in water. Solubility in alc = 4.2/100 @ 20° in ether (ligroin) = 12/100 @ 20°.

SYNS: AZOBENZEEN (DUTCH) □ AZOBENZIDE □ AZOBENZOL □ AZOBISBENZENE □ AZODIBENZENE □ AZODIBENZENEAZOFUME □ BENZENEAZOBENZENE □ DIAZOBENZENE □ DIPHENYLDIAZENE □ 1,2-DIPHENYLDIAZENE □ DIPHENYLDIIMIDE □ ENT 14,611 □ NCI-C02926 □ USAF EK-704

TOXICITY DATA WITH REFERENCE
mma-sat 40 μg/plate PNASA6 72,5135,75
orl-rat TDLo: 7350 mg/kg/2Y-C: CAR NCITR* NCI-CG-TR-154,79
scu-rat TDLo: 17 g/kg/2Y-I: ETA CANCAR 3,789,50
orl-mus TDLo: 300 mg/kg/8W-I: NEO TXAPA9 82,19,86
orl-rat TD: 15 g/kg/2Y-C: CAR NCITR* NCI-CG-TR-154,79
orl-rat LD50: 1000 mg/kg ARSIM* 20,2,66
ipr-mus LD50: 500 mg/kg NTIS** AD277-689
unk-mam LDLo: 50 mg/kg BESAAT 12,117,66

CONSENSUS REPORTS: IARC Cancer Review: Group 3 IMEMDT 7,56,87; Animal Limited Evidence IMEMDT 8,75,75; NCI Carcinogenesis Bioassay (feed); Clear Evidence: rat NCITR* NCI-CG-TR-154,79; No Evidence: mouse NCITR* NCI-CG-TR-154,79. Reported in EPA TSCA Inventory.

SAFETY PROFILE: Moderately toxic by ingestion and possibly other routes. Questionable carcinogen with experimental carcinogenic, neoplastigenic, and tumorigenic data. When heated to decomposition it emits toxic fumes of NO_x.

ASL500 CAS:2638-94-0 HR: 2
4,4'-AZOBIS(4-CYANOPENTANOIC ACID)
mf: $C_{12}H_{16}N_4O_4$ mw: 280.32

SYNS: AZOBIS(CYANOVALERIC ACID) □ 4,4'-AZOBIS(4-CYANOVALERIC ACID) □ KYSELINA 4,4'-AZO-BIS-(4-KYANVALEROVA) □ PENTANOIC ACID, 4,4'-AZOBIS(4-CYANO)- (9CI) □ VALERIC ACID, 4,4'-AZOBIS(4-CYANO)-

TOXICITY DATA WITH REFERENCE
ipr-mus LD50:666 mg/kg 85JCAE-,922,86

CONSENSUS REPORTS: Reported in EPA TSCA Inventory.

SAFETY PROFILE: Moderately toxic by intraperitoneal route. When heated to decomposition it emits toxic vapors of NO_x.

ASL750 CAS:78-67-1 HR: 3
AZOBISISOBUTYLONITRILE
DOT: UN 2952
mf: $C_8H_{12}N_4$ mw: 164.24

PROP: Crystals from EtOH. Mp: 107° (decomp).

SYNS: ACETO AZIB □ AIBN □ α,α′-AZOBISISOBUTYLONITRILE □ AZOBISISOBUTYRONITRILE □ 2,2′-AZOBIS(ISOBUTYRONITRILE) □ 2,2′-AZOBIS(2-METHYLPROPIONITRILE) □ AZODIISOBUTYRONITRILE □ α,α′-AZODIISOBUTYRONITRILE □ 2,2′-AZODIISOBUTYRONITRILE □ AZODIISOBUTYRONITRILE (DOT) □ 2,2′-DICYANO-2,2′-AZOPROPANE □ POLY-ZOLE AZDN □ POROFOR 57 □ VAZO 64

TOXICITY DATA WITH REFERENCE
orl-rat LDLo:670 mg/kg 34ZIAG 0,117,69
orl-mus LD50:700 mg/kg MEIEDD 10,132,83
ipr-mus LD50:25 mg/kg NTIS** AD691-490

CONSENSUS REPORTS: Cyanide and its compounds are on the Community Right-To-Know List. Reported in EPA TSCA Inventory.

DOT CLASSIFICATION: 4.1; *Label:* Flammable Solid, Explosive

SAFETY PROFILE: Poison by intraperitoneal route. Moderately toxic by ingestion. Easily oxidized, unstable. Violent exothermic decomposition when heated. Solution in acetone may decompose explosively. Explodes when heated with heptane. When heated to decomposition it emits toxic fumes of NO_x and CN^-. See also NITRILES. A free-radical generator.

ASM000 CAS:64037-73-6 HR: 3
AZOBIS ISOBUTYRAMIDE HYDROCHLORIDE
mf: $C_{10}H_{22}N_6$•ClH mw: 262.84

TOXICITY DATA WITH REFERENCE
orl-rat LDLo:400 mg/kg KODAK* -,-,71
ipr-rat LD50:200 mg/kg KODAK* -,-,71

SAFETY PROFILE: Poison by ingestion and intraperitoneal routes. When heated to decomposition it emits very toxic fumes of NO_x and HCl.

ASM050 CAS:2997-92-4 HR: 3
2,2′-AZOBIS(2-METHYLPROPANIMIDAMIDE) DIHYDROCHLORIDE
mf: $C_8H_{18}N_6$•2ClH mw: 271.24

SYNS: 2,2′-AZOBIS(2-METHYLPROPIONAMIDINE) DIHYDROCHLORIDE □ MS 1 □ MS 1 (CATALYST) □ PROPANIMIDAMIDE, 2,2′-AZOBIS(2-METHYL-), DIHYDROCHLORIDE (9CI) □ PROPIONAMIDINE, 2,2′-AZOBIS(2-METHYL-), DIHYDROCHLORIDE □ V 50

TOXICITY DATA WITH REFERENCE
skn-rbt 500 mg MLD EPASR* 8EHQ-0282-0427S
orl-rat LD50:410 mg/kg EPASR* 8EHQ-0282-0427S
skn-rat LD50:>5900 mg/kg EPASR* 8EHQ-0282-0427S

CONSENSUS REPORTS: Reported in EPA TSCA Inventory.

SAFETY PROFILE: Poison by ingestion. A skin irritant. When heated to decomposition it emits toxic vapors of NO_x and HCl.

ASM250 HR: 3
AZOCHLORAMIDE
mf: $C_2H_4Cl_2N_6$ mw: 183.0

PROP: Bright yellow crystals. Mp: explodes @ 155°.

SYN: CHLOROZODIN

SAFETY PROFILE: Mild allergen. Severe explosion hazard when shocked or exposed to heat; explodes at 155° or more, particularly in the presence of metals. Incompatible with metals. Moderately dangerous; when heated to decomposition it emits toxic fumes of NO_x, Cl^-, and may explode.

ASM300 CAS:123-77-3 HR: 3
AZODICARBONAMIDE
mf: $C_2H_4N_4O_2$ mw: 116.08

PROP: Yellow to orange-red crystalline powder. Mp: above 180° (decomp). Sltly sol in dimethyl sulfoxide; insol in water, organic solvents.

SAFETY PROFILE: Flammable solid. When heated to decomposition emits toxic fumes of NO_x.

ASN000 HR: 3
"AZODRIN"
mf: $C_6H_{14}O_5NP$ mw: 211.2

PROP: Reddish-brown solid, mild ester odor. Bp: 125°.

SYN: MONOCROTOPHOS

SAFETY PROFILE: Poison by ingestion and skin contact. See also ESTERS. A dangerous fire hazard. When heated to decomposition it evolves highly toxic fumes of NO_x and PO_x.

ASN250 CAS:821-14-7 HR: 3
AZO ETHANE
mf: $C_4H_{10}N_2$ mw: 86.16

PROP: Liquid. Bp: 58–59°.

SYN: AZOAETHAN (GERMAN)

TOXICITY DATA WITH REFERENCE
ihl-rat TCLo:1250 mg/kg/1H (female 15D post):TER IARCCD 4,45,73
ihl-rat TDLo:37 mg/kg (22D preg):NEO,TER XENOBH 3,271,73
ihl-rat TCLo:37 mg/kg/1H (22D post):CAR IARCCD 4,45,73
scu-rat TDLo:1250 mg/kg/26W-I:ETA ZEKBAI 67,31,65
ihl-rat TC:4000 ppm/1H (15D post):NEO,TER FCTXAV 6,584,68

ihl-rat TC:300 mg/kg/1H (15D post):NEO,TER EXPEAM 24,561,68
scu-rat LDLo:2200 mg/kg ZEKBAI 67,31,65

SAFETY PROFILE: Moderate acute toxicity. An experimental teratogen. Questionable carcinogen with experimental carcinogenic and tumorigenic data. When heated to decomposition it emits toxic fumes of NO_x. An unstable, dangerously explosive material in concentrated state.

ASN375 **HR: 3**
AZOFORMALDOXIME
mf: $C_2H_2N_4O_2$ mw: 114.06

PROP: Mp: 140°C.

SAFETY PROFILE: Explodes at its melting point. Upon decomposition it emits toxic fumes of NO_x.

ASN400 **CAS:503-28-6** **HR: 3**
AZOMETHANE
mf: $C_2H_6N_2$ mw: 58.08

PROP: Yellow liquid. Fp −78°, bp: 2°. Sol in water, most org solvs.

SAFETY PROFILE: A heat-sensitive explosive. When heated to decomposition it emits toxic fumes of NO_x.

ASN500 **CAS:487-10-5** **HR: 2**
1,1′-AZONAPHTHALENE
mf: $C_{20}H_{14}N_2$ mw: 282.35

PROP: Red needles from acetic acid. Mp: 190°, bp: subl >190°. Insol in water; sol in acetic acid; very sltly sol in alc; very sol in benzene.

TOXICITY DATA WITH REFERENCE
scu-mus TDLo:6300 mg/kg/63W-I:ETA AJCAA7 40,62,40

SAFETY PROFILE: Questionable carcinogen with experimental tumorigenic data. When heated to decomposition it emits toxic fumes of NO_x.

ASN750 **CAS:582-08-1** **HR: 2**
2,2′-AZONAPHTHALENE
mf: $C_{20}H_{14}N_2$ mw: 282.35

PROP: Red prisms from chloroform or tan plates from toluene. Mp: 205°, bp: subl @ 210°. Insol in water; sltly sol in alc; sol in benzene.

SYN: DI-β-NAPHTHYLDIIMIDE

TOXICITY DATA WITH REFERENCE
orl-mus TDLo:8400 mg/kg/42W-I:ETA AJCAA7 40,62,40

SAFETY PROFILE: Questionable carcinogen with experimental tumorigenic data. When heated to decomposition it emits toxic fumes of NO_x.

ASO375 **CAS:27589-33-9** **HR: 3**
AZOSEMIDE
mf: $C_{12}H_{11}ClN_6O_2S_2$ mw: 370.86

PROP: Crystals. Mp: 218–221°.

SYNS: 2-CHLORO-5-(1H-TETRAZOL-5-YL)-N(sup 4)-2-THENYLSULFANILAMIDE □ DIURAPID □ PLE 1053

TOXICITY DATA WITH REFERENCE
orl-mus TDLo:12,500 mg/kg (6-15D preg):TER SEIJBO 24,111,84
orl-rat LD50:2545 mg/kg IYKEDH 18,666,87
ipr-rat LD50:287 mg/kg IYKEDH 18,666,87
ivn-rat LD50:252 mg/kg IYKEDH 18,666,87
orl-mus LD50:6350 mg/kg IYKEDH 18,666,87
scu-mus LD50:762 mg/kg IYKEDH 18,666,87

SAFETY PROFILE: Poison by intraperitoneal and intravenous routes. Moderately toxic by ingestion and subcutaneous routes. An experimental teratogen. When heated to decomposition it emits toxic fumes of Cl^-, SO_x, and NO_x.

ASO501 **CAS:7644-67-9** **HR: 3**
AZOTOMYCIN
mf: $C_{17}H_{23}N_7O_8$ mw: 453.47

SYNS: ANTIBIOTIC 1719 □ DUAZOMYCIN B

TOXICITY DATA WITH REFERENCE
ivn-hmn TDLo:192 mg/kg/4D:GIT CTRRDO 61,1719,77
orl-mus LDLo:105 mg/kg ANTBAL 18,332,73
ipr-mus LDLo:27 mg/kg ANTBAL 18,332,73
scu-mus LDLo:31 mg/kg ANTBAL 18,332,73
ivn-mus LDLo:99 mg/kg ANTBAL 18,332,73

SAFETY PROFILE: Poison by ingestion, intraperitoneal, subcutaneous, and intravenous routes. Human gastrointestinal tract effects by intravenous route. When heated to decomposition it emits toxic fumes of NO_x.

ASO510 **HR: 3**
AZOTOMYCIN SODIUM
mf: $C_{17}H_{23}N_7O_8 \cdot Na$ mw: 475.47

SYNS: ANTIBIOTIC 1719 SODIUM SALT □ AZOTOMYCIN □ AZOTOMYCIN SODIUM SALT □ N-(1-((1-CARBOXY-5-DIAZO-4-OXOPENTYL)CARBAMOYL)-5-DIAZO-4-OXOPENTYL)-GLUTAMINE SODIUM SALT □ 6-DIAZO-2-(2-(4-AMINO-4-CARBOXYBUTYRAMIDO)-6-DIAZO-5-OXOHEXANAMIDO)-HEXANOIC ACID SODIUM □ DIAZOMYCIN B □ DUAZOMYCIN B □ NSC 56654 □ 1719 SODIUM □ SODIUM AZOTOMYCIN

TOXICITY DATA WITH REFERENCE
orl-mus LD50:388 mg/kg NCISP* JAN86
ipr-mus LD50:471 mg/kg NCISP* JAN86
scu-mus LD50:271 mg/kg NCISP* JAN86

SAFETY PROFILE: Poison by ingestion and subcutaneous routes. Moderately toxic by intraperitoneal route. When heated to decomposition it emits toxic fumes of NO_x and Na_2O.

ASO750 CAS:495-48-7 HR: 3
AZOXYBENZENE
mf: $C_{12}H_{10}N_2O$ mw: 198.23

PROP: Yellow, rhombic crystals. D: 1.248 @ 20°/20°; mp: 36°; bp: decomp. Insol in water; solubility in alc = 11.4/100 @ 15°, solubility in ether (ligroin) = 43.5/100 @ 15°.

SYNS: AZOBENZENE OXIDE □ AZOSSIBENZENE (ITALIAN) □ AZOXYBENZEEN (DUTCH) □ AZOXYBENZIDE □ AZOXYBENZOL (GERMAN) □ AZOXYDIBENZENE □ ORDINARY AZOXYBENZENE

TOXICITY DATA WITH REFERENCE
skn-rbt 10 mg/24H MLD AMIHBC 10,61,54
eye-rbt 500 mg AMIHBC 10,61,54
mmo-sat 10 μg/plate PMRSDJ 1,271,81
mmo-esc 50 μg/plate PMRSDJ 1,351,81
orl-rat LD50:620 mg/kg AMIHBC 10,61,54
ipr-rat LD50:940 mg/kg 85JFAN A195,83
orl-mus LD50:515 mg/kg KYDKAJ 21,25,76
ipr-mus LDLo:500 mg/kg CBCCT* 7,389,55
ipr-rat LD50:940 mg/kg 85JFAN A195,83
ipr-mus LDLo:500 mg/kg CBCCT* 7,389,55
skn-rbt LD50:1350 mg/kg 85JFAN A195,83
scu-rbt LDLo:250 mg/kg AEXPBL 35,401,1895

CONSENSUS REPORTS: Reported in EPA TSCA Inventory.

SAFETY PROFILE: Poison by subcutaneous route. Moderately toxic by ingestion, skin contact, and other routes. A skin and eye irritant. Mutation data reported. Combustible. When heated to decomposition it emits toxic fumes of NO_x.

ASP000 CAS:16301-26-1 HR: 3
AZOXYETHANE
mf: $C_4H_{10}N_2O$ mw: 102.16

PROP: Bp: 46°.

SYNS: AZOXYAETHAN (GERMAN) □ DIETHYLDIAZENE-1-OXIDE

TOXICITY DATA WITH REFERENCE
ivn-rat TDLo:25 mg/kg (10D preg):TER XENOBH 3,271,73
orl-rat TDLo:500 mg/kg/20W-I:ETA CNREA8 47,3968,87
scu-rat TDLo:30 mg/kg (11D post):NEO,TER NATWAY 60,555,73
ivn-rat TDLo:50 mg/kg (15D post):NEO,TER IARCCD 4,45,73
ivn-rat TD:50 mg/kg (15D post):NEO,TER XENOBH 3,271,73
scu-rat LD50:240 mg/kg XENOBH 3,271,73
ivn-rat LD50:210 mg/kg IARCCD 4,45,73

SAFETY PROFILE: Poison by subcutaneous and intravenous routes. An experimental teratogen. Questionable carcinogen with experimental carcinogenic and tumorigenic data. When heated to decomposition it emits toxic fumes of NO_x.

ASP250 CAS:25843-45-2 HR: 3
AZOXYMETHANE
mf: $C_2H_6N_2O$ mw: 74.10

PROP: Oil. Bp: 98°.

SYN: AOM

TOXICITY DATA WITH REFERENCE
mma-sat 13,600 μmol/L/20M CNREA8 38,4585,78
sln-dmg-unk 1 mmol/L/3D-C DRISAA 50,138,73
dnd-rat-scu 30 mg/kg CNREA8 38,1589,78
ivn-rat TDLo:30 mg/kg (15D post):TER IARCCD 4,45,73
orl-rat TDLo:20 mg/kg/(22D):ETA XENOBH 3,271,73
scu-rat TDLo:3200 μg/kg:CAR VTPHAK 12,165,75
ivn-rat TDLo:20 mg/kg (22D post):CAR IARCCD 4,45,73
scu-rat LD50:27 mg/kg XENOBH 3,271,73

SAFETY PROFILE: Suspected carcinogen with experimental carcinogenic and tumorigenic data. Poison by subcutaneous route. An experimental teratogen. Mutation data reported. When heated to decomposition it emits toxic fumes of NO_x.

ASP500 CAS:17697-55-1 HR: 2
1-AZOXYPROPANE
mf: $C_6H_{14}N_2O$ mw: 130.22

SYNS: 1,1'-AZOXYPROPANE □ DIPROPYLDIAZENE 1-OXIDE

TOXICITY DATA WITH REFERENCE
orl-rat TDLo:773 mg/kg/26W-I:CAR CRNGDP 8,1947,87

SAFETY PROFILE: Questionable carcinogen with experimental carcinogenic data. When heated to decomposition it emits toxic fumes of NO_x.

ASP510 HR: 2
2-AZOXYPROPANE
mf: $C_6H_{14}N_2O$ mw: 130.22

SYNS: AZOXYISOPROPANE □ BIS(1-METHYLETHYL)DIAZENE 1-OXIDE

TOXICITY DATA WITH REFERENCE
orl-rat TDLo:773 mg/kg/26W-I:CAR CRNGDP 8,1947,87

SAFETY PROFILE: Questionable carcinogen with experimental carcinogenic data. When heated to decomposition it emits toxic fumes of NO_x.

ASP600 CAS:275-51-4 HR: 3
AZULENE
mf: $C_{10}H_8$ mw: 128.18

SYNS: BICYCLO(5.3.0)DECAPENTAENE □ BICYCLO(0.3.5)DECA-1,3,5,7,9-PENTAENE □ BICYCLO(5.3.0)-DECA-2,4,6,8,10-PENTAENE □ CYCLOPENTACYCLOHEPTENE

TOXICITY DATA WITH REFERENCE
orl-rat LD50:>4 g/kg DRUGAY 6,13,82
ipr-rat LD50:180 mg/kg DRUGAY 6,13,82
scu-rat LD50:520 mg/kg DRUGAY 6,13,82
orl-mus LD50:>3 g/kg DRUGAY 6,13,82
ipr-mus LD50:108 mg/kg DRUGAY 6,13,82
scu-mus LD50:145 mg/kg DRUGAY 6,13,82
ivn-mus LD50:56 mg/kg CSLNX* NX#07952

CONSENSUS REPORTS: Reported in EPA TSCA Inventory.

SAFETY PROFILE: Poison by intraperitoneal, intravenous, and subcutaneous routes. When heated to decomposition it emits acrid smoke and irritating vapors.

ASP750 **CAS:6580-41-2** **HR: 2**
AZULENO(5,6,7-cd)PHENALENE
mf: $C_{20}H_{12}$ mw: 252.32

PROP: Moss-green crystals. Mp: 184°.

TOXICITY DATA WITH REFERENCE
scu-mus TDLo:80 mg/kg/4W-I:ETA PAACA3 10,12,69

SAFETY PROFILE: Questionable carcinogen with experimental tumorigenic data. When heated to decomposition it emits acrid smoke and irritating fumes.

B

B

BAB250 **CAS:37661-08-8** **HR: 3**
BACAMPICILLIN HYDROCHLORIDE
mf: $C_{21}H_{27}N_3O_7S \cdot ClH$ mw: 502.03

PROP: Crystals from Me_2CO pet ether. Mp: 171–176° (decomp).

SYNS: AMBACAMP □ BACACIL □ BAPC □ BECAMPICILLIN □ CAMBAXIN □ PENGLOBE □ SPECTROBID

TOXICITY DATA WITH REFERENCE
ivn-rat LD50:176 mg/kg NIIRDN 6,575,82
orl-mus LD50:8529 mg/kg NKRZAZ 27(Suppl 4),17,79
ipr-mus LD50:176 mg/kg NKRZAZ 27(Suppl 4),17,79
scu-mus LD50:9475 mg/kg NKRZAZ 27(Suppl 4),17,79
ivn-mus LD50:184 mg/kg NKRZAZ 27(Suppl 4),17,79

SAFETY PROFILE: Poison by intraperitoneal and intravenous routes. Mildly toxic by ingestion and subcutaneous routes. When heated to decomposition it emits very toxic fumes of NO_x, SO_x, and HCl.

BAB625 **CAS:70458-96-7** **HR: 3**
BACCIDAL
mf: $C_{16}H_{18}FN_3O_3$ mw: 319.37

PROP: Crystals from methylene chloride/methanol. Mp: 227–228°. Hygroscopic in air, forms a hemihydrate.

SYNS: AM-715 □ BARAZAN □ 1,4-DIHYDRO-1-ETHYL-6-FLUORO-4-OXO-7-(1-PIPERAZINYL)-3-QUINOLINECARBOXYLIC ACID □ 1-ETHYL-6-FLUORO-1,4-DIHYDRO-4-OXO-7-(1-PIPERAZINYL)-3-QUINOLINECARBOXYLIC ACID □ MK-366 □ NORFLOXACIN

TOXICITY DATA WITH REFERENCE
dnr-bcs 62,500 µg/L NKRZAZ 29(Suppl 4),938,81
orl-mus TDLo:1250 mg/kg (female 6-15D post):REP NKRZAZ 29(Suppl 4),895,81
orl-rat TDLo:2750 mg/kg (female 7-17D post):REP KSRNAM 15,5251,81
orl-rbt TDLo:1300 mg/kg (female 6-18D post):TER FAATDF 7,272,86
orl-mus TDLo:10 g/kg (male 61D pre):TER NKRZAZ 29(Suppl 4),886,81
orl-mus TDLo:20 g/kg (male 61D pre):TER NKRZAZ 29(Suppl 4),886,81
orl-rbt TDLo:1300 mg/kg (female 6-18D post):REP KSRNAM 16,667,82
orl-rbt TDLo:1625 mg/kg (female 13D pre):REP KSRNAM 16,667,82
orl-rat TDLo:1375 mg/kg (7-17D preg):TER KSRNAM 15,5251,81
orl-man TDLo:94 mg/kg/13D-I:MSK NZMJAX 96,590,83
ivn-rat LD50:245 mg/kg NKRZAZ 29(Suppl 4),766,81
orl-mus LD50:4 g/kg JMCMAR 30,2163,87
ivn-mus LD50:220 mg/kg NKRZAZ 29(Suppl 4),766,81
ims-mus LD50:470 mg/kg NKRZAZ 29(Suppl 4),766,81

SAFETY PROFILE: Poison by intravenous route. Moderately toxic by other routes. Human systemic effects by ingestion: musculoskeletal changes. An experimental teratogen. Other experimental reproductive effects. Mutation data reported. When heated to decomposition it emits toxic fumes of F^- and NO_x.

BAB650 **HR: 3**
BACILLUS CEREUS exo-ENTEROTOXIN

TOXICITY DATA WITH REFERENCE
ivn-mus LDLo:15 mg/kg APMBAY 29,201,75
par-mus LDLo:14 mg/kg BIORAK 38,113,73
ivn-rbt LDLo:3 mg/kg BEXBAN 73,78,72

SAFETY PROFILE: Poison by intravenous and parenteral routes.

BAB700 **HR: 2**
BACILLUS Sp. No. 21 POLYSACCHARIDE

TOXICITY DATA WITH REFERENCE
ims-mus LD50:500 mg/kg ABCHA6 38,1407,74

SAFETY PROFILE: Moderately toxic by intramuscular route. When heated to decomposition it emits acrid smoke and irritating vapors.

BAB750 **CAS:1395-21-7** **HR: 3**
BACILLUS SUBTILIS BPN

PROP: A commercial raw proteolytic enzyme used in laundry detergents (FCTXAV 7,581,69).

SYNS: BACILLOMYCIN (8CI, 9CI) □ BACILLOMYCIN R □ FUNGOCIN □ SUBTILISINS (ACGIH) □ SUBTILISINS BPN

TOXICITY DATA WITH REFERENCE
eye-rbt 3 mg SEV FCTXAV 7,581,69
ipr-mus LD50:75 mg/kg 85ERAY 3,1606,78

OSHA PEL: CL 0.00006 mg/m³
ACGIH TLV: CL 0.00006 mg/m³

SAFETY PROFILE: A poison via intraperitoneal route. A severe eye irritant. When heated to decomposition it emits toxic fumes of NO_x.

BAC000 **CAS:9014-01-1** **HR: 3**
BACILLUS SUBTILIS CARLSBERG

PROP: A commercial raw proteolytic enzyme used in laundry detergents (FCTXAV 7,581,69).

SYNS: ALCALASE □ ALK-ENZYME □ BACILLOPEPTIDASE A □ BACILLOPEPTIDASE B □ BIOPRASE □ COLISTINASE □ E.C. 3.4.4.16 □ E.C. 3.4.21.14 □ MAXATASE □ NAGARSE □ SUBTILISIN (9CI, ACGIH) □ SUBTILISIN CARLSBURG □ SUBTILISIN NOVO □ SUBTILOPEPTI-

DASE A □ SUBTILOPEPTIDASE B □ SUBTILOPEPTIDASE BPN' □ SUBTILOPEPTIDASE C □ THERMOASE PC-10

TOXICITY DATA WITH REFERENCE
eye-rbt 3 mg MOD FCTXAV 7,581,69
orl-rat LD50:3700 mg/kg FCTXAV 7,581,69

CONSENSUS REPORTS: Reported in EPA TSCA Inventory.

ACGIH TLV: CL 0.00006 mg/m³

SAFETY PROFILE: Moderately toxic by ingestion. An eye irritant. When heated to decomposition it emits toxic fumes of NO_x.

BAC020 CAS:9001-92-7 HR: 3
BACILLUS SUBTILIS NEUTRAL PROTEASE

SYNS: A.S. 1.398 □ PROTEASE, BACILLUS SUBTILIS NEUTRAL

TOXICITY DATA WITH REFERENCE
ipr-mus LD50:45 mg/kg CYLPDN 4,214,83

CONSENSUS REPORTS: Reported in EPA TSCA Inventory.

SAFETY PROFILE: A poison by intraperitoneal route. When heated to decomposition it emits acrid smoke and irritating vapors.

BAC040 CAS:68038-71-1 HR: 1
BACILLUS THURINGIENSIS

SYNS: AGRITOL □ BACILLUS THURINGENSIS □ BACILLUS THURINGIENSIS BERLINER □ BACTOSPEIN □ BACTUCIDE □ BACTUR □ BAKTHANE □ BERLINER □ BIOTROL □ BITOKSYBACILLIN □ BTB □ BTB 202 □ CAJRAB □ DIPEL □ GOMELIN □ LARVATROL □ MEGA BT □ SAN 239 □ THURICIDE □ THURINGIN

TOXICITY DATA WITH REFERENCE
orl-rat LD50:>20 g/kg SPEADM 78-1,11,78
skn-rat LD50:>5 g/kg PEMNDP 9,49,91
skn-rbt LD50:>20 g/kg SPEADM 78-1,11,78 FEREAC 54,7740,89

CONSENSUS REPORTS: Reported in EPA TSCA Inventory.

SAFETY PROFILE: Low toxicity by ingestion and skin contact. When heated to decomposition it emits acrid smoke and irritating vapors.

BAC125 CAS:23526-02-5 HR: 3
BACILLUS THURINGIENSIS EXOTOXIN

SYNS: EXOTOXIN □ β-EXOTOXIN (BACILLUS THURINGIENSIS) □ THURINGIENSIN □ THURINGIENSIN A □ THURINTOX □ TURINGIN-1

TOXICITY DATA WITH REFERENCE
sln-dmg-orl 1 pph HEREAY 85,113,77
cyt-hmn:leu 20 pph HEREAY 85,105,77
cyt-rat-orl 4500 g/kg/90D-C HEREAY 85,105,77
ivn-rat LD50:>300 mg/kg NYKZAU 77(3),1P-36P,81
unr-rat LD50:390 mg/kg GISAAA 55(6),86,90
unr-mus LD50:672 mg/kg GISAAA 55(6),86,90
unr-rbt LD50:300 mg/kg GISAAA 55(6),86,90
unr-gpg LD50:175 mg/kg GISAAA 55(6),86,90

SAFETY PROFILE: Poison by unreported routes. Human mutation data reported.

BAC130 HR: 1
BACILLUS THURINGIENSIS var. ISRAELENSIS

SYN: GNATROL

TOXICITY DATA WITH REFERENCE
orl-rat LD50:>5 g/kg FMCHA2-,C155,91
orl-rbt LD50:>2 g/kg FMCHA2-,C155,91

SAFETY PROFILE: Low toxicity by ingestion. When heated to decomposition it emits acrid smoke and irritating vapors.

BAC135 HR: 3
BACILLUS THURINGIENSIS subsp. ISRAELENSIS POLYPEPTIDE crystal

SYN: POLYPEPTIDE, BACILLUS THURINGIENSIS subsp. ISRAELENSIS, crystal preparation

TOXICITY DATA WITH REFERENCE
ipr-mus LD50:770 μg/kg FAATDF 13,310,89

SAFETY PROFILE: A poison by intraperitoneal route. When heated to decomposition it emits acrid smoke and irritating vapors.

BAC140 HR: 3
BACILLUS THURINGIENSIS MORRISONI EXOTOXIN

SYN: EXOTOXIN, BACILLUS THURINGIENSIS MORRISONI

TOXICITY DATA WITH REFERENCE
scu-mus LD50:136 mg/kg JEENAI 78,613,85

SAFETY PROFILE: A poison by subcutaneous route. When heated to decomposition it emits acrid smoke and irritating vapors.

BAC175 CAS:29393-20-2 HR: 2
BACILYSIN
mf: $C_{12}H_{18}N_2O_5$ mw: 270.32

PROP: Amorphous solid.

SYNS: N-l-ALANYL-3-(5-OXO-7-OXABICYCLO(4.1.0)HEPT-2-YL)-l-ALANINE □ α-((2-AMINO-1-OXOPROPYL)AMINO)-5-OXO-7-OXABICYCLO(4.1.0)HEPTANE-2-PROPANOIC ACID □ α-(2-AMINOPROPIONAMIDO)-5-OXO-7-OXABICYCLO(4.1.0)HEPTANE-2-PROPIONIC ACID □ ANTIBIOTIC KM 208 □ BACILLIN □ KM-208 □ TETAINE

TOXICITY DATA WITH REFERENCE
dni-hmn:hla 320 μmol/L BBACAQ 825,199,85
oms-hmn:hla 150 μmol/L BBACAQ 825,199,85
ivn-mus LD50:450 mg/kg 85GDA2 4(1),221,80

SAFETY PROFILE: Moderately toxic by intravenous route. Human mutation data reported. When heated to decomposition it emits toxic fumes of NO_x. An antibiotic.

B

BAC250 **CAS:1405-87-4** **HR: 3**
BACITRACIN

PROP: White to pale-buff, hygroscopic powder; odorless or slt odor. Freely sol in water, alc, methanol, and glacial acetic acid; insol in acetone, chloroform, and ether. When heated to decomposition it emits acrid smoke and irritating fumes.

SYNS: AYFIVIN □ BACIGUENT □ BACI-JEL □ BACILIQUIN □ BACITEK OINTMENT □ FORTRACIN □ PARENTRACIN □ PENITRACIN □ TOPITRACIN □ USAF CB-7 □ ZUTRACIN

TOXICITY DATA WITH REFERENCE
dnd-esc 5 μmol/L MUREAV 89,95,81
ipr-rat LD50:190 mg/kg PSEBAA 64,503,47
ipr-mus LD50:300 mg/kg NTIS** AD277-689
scu-mus LDLo:1300 mg/kg PSEBAA 64,503,47
ivn-mus LD50:360 mg/kg PSEBAA 64,503,47
orl-gpg LD50:2 g/kg ANTCAO 4,304,54

CONSENSUS REPORTS: Reported in EPA TSCA Inventory.

SAFETY PROFILE: A poison by intraperitoneal and intravenous routes. Moderately toxic by ingestion and subcutaneous routes. Mutation data reported.

BAC260 **CAS:55852-84-1** **HR: 1**
BACITRACIN METHYLENE DISALICYLATE

PROP: White to brownish-gray powder. Disagreeable odor. Sol in water, pyridine, ethanol; less sol in acetone, ether, chloroform, pentane, benzene.

SYNS: BMD □ FORTRACIN (BACITRACIN-MD) □ KEMITRACIN 10 □ MD BACITRACIN

TOXICITY DATA WITH REFERENCE
orl-rat LD50:>10 g/kg ANTCAO 4,304,54
orl-gpg LD50:2 g/kg ANTCAO 4,304,54

CONSENSUS REPORTS: Reported in EPA TSCA Inventory.

SAFETY PROFILE: Low oral toxicity. When heated to decomposition it emits acrid smoke and irritating fumes.

BAC275 **CAS:1134-47-0** **HR: 3**
BACLOFEN
mf: $C_{10}H_{12}ClNO_2$ mw: 213.68

PROP: Crystals from water. Mp: 206–208°.

SYNS: β-(AMINOMETHYL)-4-CHLOROBENZENEPROPANOIC ACID □ β-(AMINOMETHYL)-p-CHLOROHYDROCINNAMIC ACID □ γ-AMINO-β-(p-CHLOROPHENYL)BUTYRIC ACID □ Ba 34647 □ BACLON □ C 34647Ba □ β-(p-CHLOROPHENYL)-γ-AMINOBUTYRIC ACID □ β-(4-CHLOROPHENYL)GABA □ CIBA 34,647-Ba □ LIORESAL

TOXICITY DATA WITH REFERENCE
orl-wmn TDLo:18 mg/kg:CNS,PUL JTCTDW 22,11,84
orl-man TDLo:14 mg/kg:BAH JTCTDW 20,59,83
orl-man TDLo:4286 μg/kg:BAH,CVS AJEMEN 4,552,86
orl-rat LD50:145 mg/kg NIIRDN 6,576,82
scu-rat LD50:115 mg/kg IYKEDH 11,181,80
ivn-rat LD50:78 mg/kg IYKEDH 11,181,80
orl-mus LD50:200 mg/kg NIIRDN 6,576,82
scu-mus LD50:103 mg/kg IYKEDH 11,181,80
ivn-mus LD50:31 mg/kg YKYUA6 31,871,80

SAFETY PROFILE: Poison by ingestion, subcutaneous, and intravenous routes. Human systemic effects by ingestion: blood pressure lowering, coma, muscle weakness, pulse rate decrease, respiratory depression. When heated to decomposition it emits toxic fumes of Cl^- and NO_x. A muscle relaxant.

BAC390 **HR: D**
BACTEROIDES FRAGILIS ENDOTOXIN

SYN: ENDOTOXIN, BACTEROIDES FRAGILIS

TOXICITY DATA WITH REFERENCE
ims-gpg TDLo:2500 mg/kg (female 30-68D post):REP AJOGAH 168,714,93

SAFETY PROFILE: Experimental reproductive effects. When heated to decomposition it emits acrid smoke and irritating vapors.

BAC400 **HR: 3**
BACTERIODES FRAGILIS ENDOTOXIN, EXTRACTS

PROP: Phenol-water extracts from BACTERIODES FRAGILIS 62/73 strain. BAPBAN 26,19,78

TOXICITY DATA WITH REFERENCE
ivn-mus LD50:38 mg/kg BAPBAN 26(1),19,78
ivn-rbt LDLo:2 mg/kg BAPBAN 26(1),19,78
idr-rbt LDLo:500 μg/kg BAPBAN 26(1),19,78

SAFETY PROFILE: A poison by intravenous and intradermal routes. When heated to decomposition it emits acrid smoke and irritating vapors.

BAD000 **CAS:64550-80-7** **HR: 2**
BA-10,11-DIOL-8,9-EPOXIDE-1
mf: $C_{18}H_{14}O_3$ mw: 278.32

SYN: 8,9,10,11-TETRAHYDRO-10,11-DIHYDROXY-8-α,9-α-EPOXYBENZ(a)ANTHRACENE

TOXICITY DATA WITH REFERENCE
skn-mus TDLo:22 mg/kg:ETA CNREA8 38,1699,78

SAFETY PROFILE: Questionable carcinogen with experimental tumorigenic data. When heated to decomposition it emits acrid smoke and irritating fumes.

BAD250 **HR: 1**
BAGASSE DUST

SAFETY PROFILE: A nuisance dust from the fibrous residue of cane sugar manufacture. Inhalation can cause bronchial asthma, sneezing, rhinorrhea, pneumonitis, etc. See also COTTON DUST. Fire and explosion hazard when exposed to heat, flame, or oxidizers.

BAD625 **CAS:10309-37-2** **HR: 3**
BAKUCHIOL
mf: $C_{18}H_{24}O$ mw: 256.42

PROP: Oil. Bp: 145–147° @ 0.7 mm.

TOXICITY DATA WITH REFERENCE
orl-mus LD50:2560 mg/kg MZHUDX 42,646,80
ipr-mus LD50:94 mg/kg MZHUDX 42,646,80
ivn-mus LD50:31 mg/kg MZHUDX 42,646,80

SAFETY PROFILE: Poison by intravenous and intraperitoneal routes. Moderately toxic by ingestion. When heated to decomposition it emits acrid smoke and fumes.

BAD750 **CAS:59-52-9** **HR: 3**
BAL
mf: $C_3H_8OS_2$ mw: 124.23

PROP: Viscous, oily liquid; pungent odor, bp: 140° @ 40 mm, vap d: 4.3, d: 1.2385 @ 25°/4°.

SYNS: BRITISH ANTILEWISITE ☐ DICAPTOL ☐ DIMERCAPROL PROPANOL ☐ DIMERCAPTOL ☐ 2,3-DIMERCAPTOL-1-PROPANOL ☐ DIMERCAPTOPROPANOL ☐ 2,3-DIMERCAPTOPROPANOL ☐ 2,3-DIMERCAPTOPROPAN-1-OL ☐ DITHIOGLYCEROL ☐ 1,2-DITHIOGLYCEROL ☐ 2,3-DITHIOPROPANOL ☐ SULFACTIN ☐ USAF ME-1

TOXICITY DATA WITH REFERENCE
scu-mus TDLo:100 mg/kg (11D preg):TER ANREAK 135,261,59
scu-mus TDLo:200 mg/kg (female 12-13D post):TER ANREAK 135,261,59
ims-hmn TDLo:3 mg/kg:BLD,SKN SCIEAS 102,601,45
ipr-rat LD50:105 mg/kg APFRAD 5,172,47
scu-rat LD50:2 g/kg APFRAD 5,172,47
ims-rat LD50:87 mg/kg TXAPA9 36,297,76
orl-mus LD50:217 mg/kg QJPPAL 21,364,48
ipr-mus LD50:25 mg/kg NTIS** AD277-689
ivn-mus LD50:56 mg/kg CSLNX* NX#04985
ims-mus LD50:113 mg/kg AEPPAE 223,408,54
ivn-rbt LD50:50 mg/kg BIJOAK 41,325,47
ims-rbt LD50:50 mg/kg BIJOAK 41,325,47
par-rbt LD50:40 mg/kg APFRAD 5,172,47

CONSENSUS REPORTS: EPA Genetic Toxicology Program. Reported in EPA TSCA Inventory.

SAFETY PROFILE: Poison via ingestion, intramuscular, parenteral, intraperitoneal, and intravenous routes. Experimental teratogenic effects. Human systemic effects by intramuscular route: hemorrhage and dermatitis. Human blood and systemic skin effects by intramuscular route. It causes redness and swelling when applied locally to the skin, but does not produce blisters or ulcers. Intensely irritating to eyes and mucous membranes. Systemic symptoms are caused by injection. When heated to decomposition, it emits toxic fumes of SO_x. Used as an antidote to arsenic, gold, and mercury poisoning.

BAE100 **CAS:8008-88-6** **HR: 1**
BALDRIAN OIL

SYN: BALDRIAN OEL

TOXICITY DATA WITH REFERENCE
orl-rat LD50:15 g/kg PHARAT 14,435,59

CONSENSUS REPORTS: Reported in EPA TSCA Inventory.

SAFETY PROFILE: Low toxicity by ingestion. When heated to decomposition it emits acrid smoke and irritating vapors.

BAE325 **HR: 1**
BALSAM APPLE

PROP: A tree that may grow to 20 or 30 feet on rocks or other trees. The oval, leathery leaves are 3 to 8 inches across. It produces white flowers with pink edges and a gold center, and a golf-ball sized fruit that turns brown and opens when ripe. The trees grow wild in Hawaii, southern Florida, and the West Indies.

SYNS: CLUSIA ROSEA ☐ COPEY ☐ CUPEY ☐ FIGUIER MAUDIT MARRON (HAITI) ☐ PITCH APPLE ☐ SCOTCH ATTORNEY ☐ WILD MAMEE

SAFETY PROFILE: The fruit and sap contain an unidentified poison that causes profuse diarrhea after ingestion.

BAE750 **HR: 1**
BALSAM of PERU

PROP: Dark-brown, viscid liquid; vanilla odor. Sol in fixed oils; sltly sol in propylene glycol; insol in glycerin. Extracted from *Myroxylon pereirae Klotzsch.*

SYNS: BALSAM PERU OIL (FCC) ☐ PERUVIAN BALSAM

SAFETY PROFILE: A mild allergen. Combustible when heated. When heated to decomposition it emits acrid smoke and irritating fumes.

BAF000 **CAS:9000-64-0** **HR: 1**
BALSAM TOLU

PROP: Resin derived from *Toluifera balsamam* (FCTXAV 14,659,76).

SYNS: BALSAMS, TOLU ☐ OPOBALSAM ☐ RESIN TOLU ☐ THOMAS BALSAM ☐ TOLU ☐ TOLU BALSAM GUM ☐ TOLU BALSAM TINCTURE ☐ TOLU RESIN

TOXICITY DATA WITH REFERENCE
skn-rbt 500 mg/24H MLD FCTXAV 14,689,76

CONSENSUS REPORTS: Reported in EPA TSCA Inventory.

SAFETY PROFILE: A mild skin irritant. When heated to decomposition it yields toxic and irritating fumes and smoke.

BAF250 **CAS:8029-29-6** **HR: 2**
BANDANE

PROP: A mixture of isomers containing 60–62% chlorine used as preemergent herbicide (27ZTAP 3,20,69).

SYNS: HALTS ☐ POLYCHLORODICYCLOPENTADIENE ☐ POLYCHLORODICYCLOPENTADIENE ISOMERS

TOXICITY DATA WITH REFERENCE
orl-rat LD50:504 mg/kg WRPCA2 9,119,70
skn-rat LD50:12 g/kg 27ZTAP 3,20,69

SAFETY PROFILE: Moderately toxic by ingestion. Mildly toxic by skin contact. When heated to decomposition it emits toxic fumes of Cl^-. See also CHLORINATED HYDROCARBONS, ALIPHATIC.

BAF325 **HR: 3**
BANEBERRY

PROP: Perennial herbs 1 to 2 feet tall with large compound leaves. It grows small white flowers in the spring and berries in the summer. The color of the berries depends on the species: *A. pachypoda*, white; *A. rubra*, red; *A. spicata*, purple-black. Various species are found in the temperate zones of North America from Canada to Georgia and New Mexico. Some are cultivated.

SYNS: ACTAEA (VARIOUS SPECIES) □ A. PACHYPODA □ A. RUBRA □ A. SPICATA □ COHOSH □ DOLLS EYES □ HERB-CHRISTOPHER □ NECKLACEWEED □ PAIN de COULEUVRE (CANADA) □ POISON de COULEUVRE (CANADA) □ SNAKEBERRY

SAFETY PROFILE: The toxin, whose identity is not known, is found only in the berries and roots. Liquid from these is a strong irritant and forms blisters on the skin and mucous membranes. Ingestion causes pain and inflammation of the lips, mouth, and throat, vomiting and diarrhea with blood, abdominal cramps, kidney damage, and central nervous system effects including dizziness, confusion, fainting, and convulsions.

BAF825 **CAS:1415-73-2** **HR: 3**
BARBALOIN
mf: $C_{21}H_{22}O_9$ mw: 418.43

PROP: Yellow needles from EtOH. Mp: 148°. Sol in water.

SYNS: 10-(1',5'-ANHYDROGLUCOSYL)ALOE-EMODIN-9-ANTHRONE □ 1,8-DIHYDROXY-3-HYDROXYMETHYL-10-(6-HYDROXYMETHYL-3,4,5-TRIHYDROXY-2-PYRANYL)ANTHRONE □ 10-GLUCOPYRANOSYL-1,8-DIHYDROXY-3-(HYDROXYMETHYL)-9(10H)-ANTHRACENONE

TOXICITY DATA WITH REFERENCE
ivn-mus LD50:200 mg/kg 85GDA2 8(2),314,82
orl-cat LDLo:500 mg/kg HBAMAK 4,1298,35
scu-rbt LDLo:200 mg/kg HBAMAK 4,1298,35
scu-pgn LDLo:200 mg/kg HBAMAK 4,1298,35

SAFETY PROFILE: Poison by subcutaneous and intravenous routes. Moderately toxic by ingestion. When heated to decomposition it emits acrid smoke and fumes. A carthartic and purgative.

BAG000 **CAS:57-44-3** **HR: 3**
BARBITAL
mf: $C_8H_{12}N_2O_3$ mw: 184.22

PROP: Faintly bitter crystals from H_2O; polymorphic forms; triagonal crystals, monoclinic prisms, monoclinic needles, and triclinic cryst. Mp: 190°.

SYNS: BARBITONE □ DEBA □ DIEMAL □ DIETHYLBARBITONE □ DIETHYL-BARBITURIC ACID □ 5,5-DIETHYLBARBITURIC ACID □ DIETHYLMALONYLUREA □ 5,5-DIETHYL-2,4,6(1H,3H,5H)-PYRIMIDINETRIONE □ DORMONAL □ ETHYLBARBITAL □ HYPNOGENE □ MALONAL □ SEDEVAL □ URONAL □ VEROLETTIN □ VERONAL □ VESPERAL

TOXICITY DATA WITH REFERENCE
cyt-mus-ipr 33 g/kg IJMRAQ 61,1568,73
cyt-ham:lng 1 g/L ATSUDG (4),41,80
ipr-rat LDLo:300 mg/kg JPETAB 44,325,32
scu-rat LD50:450 mg/kg AEPPAE 152,341,30
orl-mus LD50:600 mg/kg NIIRDN 6,590,82
ipr-mus LD50:178 mg/kg FRPSAX 14,269,59
scu-mus LD50:630 mg/kg YKKZAJ 74,122,54
orl-cat LDLo:280 mg/kg PHREA7 19,472,39
scu-rbt LDLo:250 mg/kg JACSAT 45,243,23

CONSENSUS REPORTS: EPA Genetic Toxicology Program.

SAFETY PROFILE: Poison by ingestion, intravenous, intraperitoneal, and subcutaneous routes. Ingestion causes psychological effects in humans. Mutation data reported. When heated to decomposition it emits toxic fumes of NO_x. See also BARBITURATES. A hypnotic and sedative.

BAG250 **CAS:144-02-5** **HR: 3**
BARBITAL SODIUM
mf: $C_8H_{12}N_2O_3 \cdot Na$ mw: 207.21

PROP: Bitter crystals or powder.

SYNS: BARBITAL Na □ BARBITAL SOLUBLE □ BARBITONE SODIUM □ DIETHYLBARBITURATE MONOSODIUM □ 5,5-DIETHYLBARBITURIC ACID SODIUM deriv. □ DIETHYLMALONYLUREA SODIUM □ EMBINAL □ MEDINAL □ NATRINAL □ NATRIUMBARBITALS (GERMAN) □ NERVOSETON □ 2,4,6(1H,3H,5H)-PYRIMIDINETRIONE, 5,5-DIETHYL-, MONOSODIUM SALT (9CI) □ SODIUM BARBITAL □ SODIUM BARBITONE □ SODIUM DIETHYLBARBITURATE □ SODIUM-5,5-DIETHYLBARBITURATE □ SODIUM ETHYLBARBITAL □ SODIUM MALONYLUREA □ SODIUM VERONAL □ SOLUBLE BARBITAL □ SOPRINAL □ THYALONE □ VERONAL SODIUM

TOXICITY DATA WITH REFERENCE
sce-ham:emb 100 µg/L IJCNAW 20,768,77
unr-rat TDLo:125 mg/kg (female 19D post):REP JCPPAV 45,146,52
scu-rat TDLo:2400 mg/kg (8-19D preg):TER CUSCAM 36,3,67
scu-rat TDLo:2400 mg/kg (8-19D preg):REP CUSCAM 36,3,67
ipr-rat TDLo:200 mg/kg (1D pre):REP SJDBA9 10,574,79
ipr-mus TDLo:390 mg/kg (female 1-6D post):REP ARZNAD 19,1309,69
scu-rat TDLo:2600 mg/kg (female 7-19D post):REP IJEBA6 15,346,77
scu-rat TDLo:6 g/kg (male 20D pre):REP CUSCAM 34,584,65
ipr-mus TDLo:390 mg/kg (female 1-6D post):TER ARZNAD 19,1309,69
ipr-mus TDLo:690 mg/kg (female 1-6D post):TER ARZNAD 19,1309,69
orl-rat TDLo:121 g/kg/72W-C:NEO CRNGDP 11,2149,90
orl-rat LD :30 mg/kg/78W-C:ETA CRNGDP 10,183,89
orl-rat LD50:600 mg/kg ARTODN 54,275,83
scu-rat LDLo:300 mg/kg JPETAB 31,1,27
ivn-rat LD50:280 mg/kg JPETAB 135,213,62
orl-mus LD50:800 mg/kg TXAPA9 27,70,74
ipr-mus LD50:620 mg/kg JPETAB 87,265,46

scu-mus LD50:700 mg/kg JPETAB 109,268,53
ivn-mus LD50:830 mg/kg TXAPA9 27,70,74
orl-dog LDLo:350 mg/kg 27ZWAY 2,-,36
ivn-dog LDLo:300 mg/kg JPETAB 60,125,37
orl-cat LDLo:275 mg/kg 27ZWAY 2,-,36

SAFETY PROFILE: Poison by ingestion, subcutaneous, intravenous, and intraperitoneal routes. Large doses cause marked depression (sometimes preceded by excitation), prolonged coma, and death. Experimental teratogenic and reproductive effects. Allergic skin reactions may occur on contact. Implicated in development of aplastic anemia. Questionable carcinogin with experimental tumorigenic and neoplastigenic data. A truly habit-forming drug. Other experimental reproductive effects. Mutation data reported. Combustible. When heated to decomposition it emits toxic fumes of NO_x and Na_2O. See also BARBITURATES.

BAG500 **HR: 3**
BARBITURATES

SYNS: BARBITAL □ BARBITAL SODIUM □ BARBITONE

SAFETY PROFILE: Salts or derivatives of barbituric acid are central nervous system depressants, and are used as hypnotics, sedatives, and anesthetics. Usually administered orally. They are strongly habit forming. Several compounds including amo-, seco-, and pentabarbital are restricted chemicals. Their use can cause a reaction called barbiturism, which is marked by chills, headache, fever, and cutaneous eruptions. See BARBITAL SODIUM.

BAH250 **CAS:7440-39-3** **HR: 3**
BARIUM
DOT: UN 1400
af: Ba aw: 137.36

PROP: Silver-white, sltly lustrous, somewhat malleable metal. Mp: 727°, bp: 1640°, d: 3.5 @ 20°, vap press: 10 mm @ 1049°. Dissolves in H_2O forming $Ba(OH)_2$ solns. Sol. $NH_3(l)$ blue-black soln.

CONSENSUS REPORTS: Reported in EPA TSCA Inventory. Community Right-To-Know List.

OSHA PEL: TWA 0.5 mg(Ba)/m^3
ACGIH TLV: TWA 0.5 mg(Ba)/m^3
DFG MAK: 0.5 mg(Ba)/m^3
DOT CLASSIFICATION: 4.3; *Label:* Dangerous When Wet

SAFETY PROFILE: Water and stomach acids solubilize barium salts and can cause poisoning. Symptoms are vomiting, colic, diarrhea, slow irregular pulse, transient hypertension, and convulsive tremors and muscular paralysis. Death may occur in a few hours to a few days. Half-life of barium in bone has been estimated at 50 days. Dust is dangerous and explosive when exposed to heat, flame, or chemical reaction. Violent or explosive reaction with water, CCl_4, fluorotrichloromethane, trichloroethylene, and C_2Cl_4. Incompatible with acids, $C_2Cl_3F_3$, $C_2H_2FCl_3$, C_2HCl_3 and water, 1,1,2-trichlorotrifluoroethane, and fluorotrichloroethane. The powder may ignite or explode in air or other oxidizing gases. See also BARIUM COMPOUNDS.

For occupational chemical analysis use NIOSH: Barium, Soluble Compounds, 7056.

BAH500 **CAS:543-80-6** **HR: 3**
BARIUM ACETATE
mf: $C_4H_6O_4 \cdot Ba$ mw: 255.44

PROP: White or colorless crystals. Decomp on heating with $BaCO_3$ formation. Very sol in H_2O.

SYNS: ACETIC ACID, BARIUM SALT □ BARIUM DIACETATE □ OCTAN BARNATY (CZECH)

TOXICITY DATA WITH REFERENCE
orl-rat LD50:921 mg/kg MarJV# 29MAR77
ivn-mus LD50:21 mg/kg TXAPA9 22,150,72
orl-rbt LDLo:236 mg/kg EQSSDX 1,1,75
scu-rbt LDLo:96 mg/kg EQSSDX 1,1,75
ivn-rbt LDLo:12 mg/kg EQSSDX 1,1,75

CONSENSUS REPORTS: Reported in EPA TSCA Inventory. Barium and its compounds are on the Community Right-To-Know List.

OSHA PEL: TWA 0.5 mg(Ba)/m^3
ACGIH TLV: TWA 0.5 mg(Ba)/m^3
DFG MAK: 0.5 mg(Ba)/m^3

SAFETY PROFILE: Poison via ingestion, intravenous, and subcutaneous routes. When heated to decomposition it emits acrid smoke and fumes. See also BARIUM COMPOUNDS.

BAH750 **CAS:12070-27-8** **HR: 3**
BARIUM ACETYLIDE
mf: C_2Ba mw: 161.35

CONSENSUS REPORTS: Barium and its compounds are on the Community Right-To-Know List.

SAFETY PROFILE: Ignites on contact with vapors of water or ethanol in air. Incandescent reaction when heated with: hydrogen @ 150°C; chlorine @ 140°C; bromine @ 130°C; iodine @ 122°C; and selenium @ 150°C. See also BARIUM COMPOUNDS and ACETYLIDES.

BAI000 **CAS:18810-58-7** **HR: 3**
BARIUM AZIDE
DOT: UN 0224/UN 1571
mf: BaN_6 mw: 221.40

PROP: Monoclinic prisms or crystals, decomp on heating with loss of N_2 at about 12°. Mp: evolves N_2 at about 120°, bp: explodes, d: 2.936. Very sol in H_2O; sltly sol in EtOH; insol in Et_2O.

SYNS: BARIUM AZIDE, dry or wetted with <50% water, by weight (UN 0224) (DOT) □ BARIUM AZIDE, wetted with not <50% water, by weight (UN 1571) (DOT)

CONSENSUS REPORTS: Reported in EPA TSCA Inventory. Barium and its compounds are on the Community Right-To-Know List.

OSHA PEL: TWA 0.5 mg(Ba)/m^3
ACGIH TLV: TWA 0.5 mg(Ba)/m^3
DFG MAK: 0.5 mg(Ba)/m^3
DOT CLASSIFICATION: EXPLOSIVE 1.1A; *Label:* EXPLOSIVE 1.1A, Poison (UN 0224); DOT Class: 4.1; *Label:* Flammable Solid, Poison (UN 1571)

SAFETY PROFILE: A poison. Moderate explosion hazard when shocked or heated to 275°. Spontaneously flammable in air. Very unstable. When heated to decomposition it emits toxic fumes of NO_x. See also BARIUM COMPOUNDS (soluble) and AZIDES.

BAI500 **HR: 3**
BARIUM BENZOATE
mf: $Ba(C_7H_5O_2)_2 \cdot 2H_2O$ mw: 415.61

PROP: White, nacreous leaflets. Mp: loses $2H_2O$ @ 100°.

CONSENSUS REPORTS: Barium and its compounds are on the Community Right-To-Know List.

SAFETY PROFILE: Deadly poison. See also BARIUM COMPOUNDS (soluble).

BAI750 **CAS:13967-90-3** **HR: 3**
BARIUM BROMATE
DOT: UN 2719
mf: $Ba(BrO_3)_2 \cdot H_2O$ mw: 411.21

PROP: White or colorless crystals or crystalline powder. Decomp on heating with O_2 evolution and $BaBr_2$ formation. Mp: decomp @ 260°, d: 3.99 @ 18°.

CONSENSUS REPORTS: Barium and its compounds are on the Community Right-To-Know List. EPA TSCA Chemical Inventory.

OSHA PEL: TWA 0.5 mg(Ba)/m^3
ACGIH TLV: TWA 0.5 mg(Ba)/m^3
DFG MAK: 0.5 mg(Ba)/m^3
DOT CLASSIFICATION: 5.1; *Label:* Oxidizer, Poison

SAFETY PROFILE: Very toxic. Fire hazard by chemical reaction with easily oxidized materials. Explodes at 300°. Mixtures with sulfur are unstable storage hazards; igniting immediately at 91°C and after a 2–11 day delay period at room temperature. Incompatible with Al, As, C, Cu, metal sulfides, organic matter, P, and reducing materials. When heated to decomposition, it emits toxic fumes of Br^-. See also BARIUM COMPOUNDS (soluble) and BROMINE.

BAI770 **CAS:15337-60-7** **HR: 2**
BARIUM CADMIUM LAURATE

SYN: LAURIC ACID, BARIUM CADMIUM SALT

TOXICITY DATA WITH REFERENCE
orl-rat LD50:1696 mg/kg GTPZAB 18(3),50,74
orl-mus LD50:516 mg/kg GISAAA 46(4),18,81

CONSENSUS REPORTS: Reported in EPA TSCA Inventory.

OSHA PEL: TWA 0.5 mg(Ba)/m^3
ACGIH TLV: TWA 0.5 mg(Ba)/m^3; 0.05 mg(Cd)/m^3
DFG MAK: 0.5 mg(Ba)/m^3
NIOSH REL: (Cadmium, dust and fume): lowest feasible conc.

SAFETY PROFILE: Moderately toxic by ingestion. When heated to decomposition it emits toxic vapors of Ba^+ and Cd^-.

BAI800 **CAS:1191-79-3** **HR: 3**
BARIUM CADMIUM STEARATE
mf: $C_{72}H_{140}O_8 \cdot Ba \cdot Cd$ mw: 1383.86

SYNS: CADMIUM BARIUM STEARATE □ OCTADECANOIC ACID, BARIUM CADMIUM SALT (4:1:1) (9CI) □ STEARIC ACID, BARIUM CADMIUM SALT (4:1:1)

TOXICITY DATA WITH REFERENCE
orl-rat LD50:1980 mg/kg GISAAA 40(2),102,75
orl-mus LD50:1381 mg/kg 41HTAH -,14,78

CONSENSUS REPORTS: Reported in EPA TSCA Inventory.

OSHA PEL: TWA 5 μg(Cd)/m^3
ACGIH TLV: TWA 0.01 mg(Cd)/m^3; Suspected Carcinogen
NIOSH REL: TWA reduce to lowest feasible level

SAFETY PROFILE: Confirmed human carcinogen. Moderately toxic by ingestion. When heated to decomposition it emits toxic fumes of Ba and Cd.

BAI825 **CAS:4696-54-2** **HR: 2**
BARIUM CAPRYLATE
mf: $C_{16}H_{32}O_4 \cdot Ba$ mw: 425.82

SYNS: BARIUM OCTANOATE □ BARIUM OCTOATE

TOXICITY DATA WITH REFERENCE
orl-rat LD50:1 g/kg GISAAA 39(11),91,74
orl-mus LD50:1100 mg/kg GISAAA 39(11),91,74
orl-gpg LD50:1250 mg/kg GISAAA 39(11),91,74

CONSENSUS REPORTS: Barium and its compounds are on the Community Right-To-Know List.

SAFETY PROFILE: Moderately toxic by ingestion.

BAJ000 **HR: 3**
BARIUM CARBIDE
mf: BaC_2 mw: 161.4

PROP: Gray crystals. D: 3.75.

CONSENSUS REPORTS: Barium and its compounds are on the Community Right-To-Know List.

SAFETY PROFILE: A poison. A fire and explosion hazard by chemical reaction with moisture to form acetylene. Incompatible with Se; S; H_2O. To fight fire, use CO_2, dry chemical. See also BARIUM COMPOUNDS (soluble).

BAJ250 **CAS:513-77-9** **HR: 3**
BARIUM CARBONATE (1:1)
mf: $CO_3 \cdot Ba$ mw: 197.35

PROP: White orthorhombic powder or crystals, becomes hexagonal at 8° and cubic at 976°. Decomp on heating with CO_2 loss. Mp: 1740 @ 90 atm, bp: decomp; d: 4.43. Dissolves in acids to form corresponding Ba salts. Prac insol in H_2O; insol in alc EtOH.

SYNS: BARIUM CARBONATE □ CARBONIC ACID, BARIUM SALT (1:1) □ C.I. 77099 □ C.I. PIGMENT WHITE 10

TOXICITY DATA WITH REFERENCE
ihl-rat TCLo: 3130 μg/m³/24H (female 16W pre): REP GTPZAB 20(7),33,76
ihl-rat TCLo: 1150 μg/m³/24H (16W male): REP GTPZAB 20(7),33,76
orl-man LDLo: 800 mg/kg YKYUA6 28,329,77
orl-wmn TDLo: 800 mg/kg: GIT BMJOAE 289,882,84
orl-hmn TDLo: 11 mg/kg: GIT YKYUA6 31,1247,80
orl-hmn LDLo: 17 mg/kg YKYUA6 28,329,77
orl-hmn TDLo: 29 mg/kg: PNS IJMDAI 3,565,67
orl-rat LD50: 418 mg/kg 85GMAT -,23,82
ivn-rat LDLo: 20 mg/kg EQSSDX 1,1,75
orl-mus LD50: 200 mg/kg 85GMAT -,23,82
ipr-mus LD50: 50 mg/kg 85GMAT -,23,82
orl-dog LDLo: 400 mg/kg PCOC** -,95,66

CONSENSUS REPORTS: Reported in EPA TSCA Inventory. Barium and its compounds are on the Community Right-To-Know List.

OSHA PEL: TWA 0.5 mg(Ba)/m³
ACGIH TLV: TWA 0.5 mg(Ba)/m³
DFG MAK: 0.5 mg(Ba)/m³

SAFETY PROFILE: Poison by ingestion, intravenous, and intraperitoneal routes. Human systemic effects by ingestion: stomach ulcers, muscle weakness, paresthesias and paralysis, hypermotility, diarrhea, nausea or vomiting, lung changes. Experimental reproductive effects. Incompatible with BrF_3 and 2-furanpercarboxylic acid. See also BARIUM COMPOUNDS (soluble).

BAJ500 **CAS:13477-00-4** **HR: 3**
BARIUM CHLORATE
DOT: UN 1445
mf: Cl_2O_6•Ba mw: 304.24

PROP: Colorless prisms or white powder. Mp: loses H_2O @ 414°, d: 3.18.

SYN: CHLORIC ACID, BARIUM SALT

CONSENSUS REPORTS: Reported in EPA TSCA Inventory. Barium and its compounds are on the Community Right-To-Know List.

OSHA PEL: TWA 0.5 mg(Ba)/m³
ACGIH TLV: TWA 0.5 mg(Ba)/m³
DFG MAK: 0.5 mg(Ba)/m³
DOT CLASSIFICATION: 5.1; *Label:* Oxidizer, Poison

SAFETY PROFILE: A poison. For fire and explosion hazards, see CHLORATES. Incompatible with Al, As, C, charcoal, Cu, MnO_2, metal sulfides, S_4N_4, organic matter, P, S. See also BARIUM COMPOUNDS (soluble).

BAK000 **CAS:10361-37-2** **HR: 3**
BARIUM CHLORIDE
mf: $BaCl_2$ mw: 208.24

PROP: Colorless, deliquescent, orthorhombic, flat, crystals. Undergoes orthorhombic to cubic phase transition at 9°. Mp: transition @ 925° to cubic crystals. Bp: 1560°, d: 3.856 @ 24°. Very sol in H_2O; prac insol in EtOH.

SYNS: BARIUM DICHLORIDE □ NCI-C61074 □ SBa 0108E

TOXICITY DATA WITH REFERENCE
mrc-smc 14 mmol/L MUTAEX 1,21,86
itt-rat TDLo: 16,659 μg/kg (1D male): REP JRPFA4 7,21,64
orl-rat LD50: 118 mg/kg FOREAE 7,313,42
scu-rat LD50: 178 mg/kg 27ZIAQ -,53,73
ivn-rat LDLo: 20 mg/kg JLCMAK 15,35,29
orl-mus LDLo: 70 mg/kg EQSSDX 1,1,75
ipr-mus LD50: 184 mg/kg GTPZAB 28(6),45,84
scu-mus LDLo: 10 mg/kg NTIS** AEC-TR-6710
orl-dog LDLo: 90 mg/kg 27ZIAQ -,53,73
scu-dog LDLo: 10 mg/kg EQSSDX 1,1,75

CONSENSUS REPORTS: Reported in EPA TSCA Inventory. Barium and its compounds are on the Community Right-To-Know List. EPA Genetic Toxicology Program.

OSHA PEL: TWA 0.5 mg(Ba)/m³
ACGIH TLV: TWA 0.5 mg(Ba)/m³
DFG MAK: 0.5 mg(Ba)/m³

SAFETY PROFILE: A poison by ingestion, subcutaneous, intravenous, and intraperitoneal routes. Inhalation absorption of barium chloride equals 60–80%; oral absorption equals 10–30%. Experimental reproductive effects. Mutation data reported. See also BARIUM COMPOUNDS (soluble). When heated to decomposition it emits toxic fumes of Cl^-.

BAK125 **CAS:14674-74-9** **HR: 3**
BARIUM CHLORITE
mf: $BaCl_2O_4$ mw: 172.23

CONSENSUS REPORTS: Barium and its compounds are on the Community Right-To-Know List.

SAFETY PROFILE: A poison. Decomposes explosively at 190°C. Ignites on contact with dimethyl sulfate. When heated to decomposition it emits toxic fumes of Cl^-. See also BARIUM COMPOUNDS and CHLORITES.

BAK250 **CAS:10294-40-3** **HR: 3**
BARIUM CHROMATE(VI)
mf: Ba•CrO_4 mw: 255.36

PROP: Heavy, pale yellow, crystalline powder; darkens on heating. D: 4.498 @ 15°. Sol in strong acids; insol in org solvents.

SYNS: BARIUM CHROMATE (1:1) □ BARIUM CHROMATE OXIDE □ BARYTA YELLOW □ CHROMIC ACID, BARIUM SALT (1:1) □ C.I. 77103 □ C.I. PIGMENT YELLOW 31 □ LEMON CHROME □ LEMON YELLOW □ PERMANENT YELLOW □ STEINBUHL YELLOW □ ULTRAMARINE YELLOW

TOXICITY DATA WITH REFERENCE
sce-ham:ovr 100 μg/L MUREAV 156,219,85

CONSENSUS REPORTS: NTP 7th Annual Report On Carcinogens. IARC Cancer Review: Group 1 IMEMDT 7,165,87; Animal Inadequate Evidence IMEMDT 2,100,73; Human Sufficient Evidence IMEMDT 23,205,80. Reported in EPA TSCA Inventory. Barium and its compounds are on the Community Right-To-Know List.

OSHA PEL: TWA 0.1 mg $(C_3O_3)m^3$; 0.5 mg(Ba)/m³
ACGIH TLV: TWA 0.5 mg(Ba)/m³; 0.05 mg(Cr)/m³; Confirmed Human Carcinogen
DFG MAK: 0.5 mg(Ba)/m³
NIOSH REL: TWA 0.001 mg(Cr(VI))/m³

SAFETY PROFILE: Confirmed human carcinogen. A poison. Mutation data reported. Reacts vigorously with reducing materials. See also BARIUM COMPOUNDS (soluble) and CHROMIUM COMPOUNDS. Used in pyrotechnics and as an explosive initiator.

For occupational chemical analysis use NIOSH: Chromium Hexavalent 7024.

BAK500 **HR: 3**
BARIUM COMPOUNDS (soluble)

CONSENSUS REPORTS: Barium and its compounds are on the Community Right-To-Know List.

OSHA PEL: Soluble Compounds: TWA 0.5 mg(Ba)/m³
ACGIH TLV: Soluble Compounds:TWA 0.5 mg/m³
DFG MAK: Soluble Compounds: 0.5 mg/m³
DOT CLASSIFICATION: 6.1; *Label:* Poison

SAFETY PROFILE: The chromate is a human carcinogen. The soluble barium salts, such as the chloride and sulfide, are poisonous when ingested. The insoluble sulfate used in radiography is not acutely toxic. See also BARIUM SULFATE. Few cases of industrial systemic poisoning have been reported, but one investigator describes a fatal case of poisoning attributed to barium oxide, the symptoms being severe abdominal pain with vomiting, dyspnea, rapid pulse, paralysis of the arm and leg, and eventually cyanosis and death. The same investigator produced paralysis in animals with barium oxide and carbonate. The usual result of exposure to the sulfide, oxide, and carbonate is irritation of the eyes, nose, and throat, and of the skin, producing dermatitis. The salts mentioned are somewhat caustic.

BAK750 **CAS:542-62-1** **HR: 3**
BARIUM CYANIDE
DOT: UN 1565
mf: C_2BaN_2 mw: 189.38

PROP: White, crystalline powder.

SYNS: BARIUM CYANIDE, solid (DOT) □ BARIUM DICYANIDE □ RCRA WASTE NUMBER P013

CONSENSUS REPORTS: Reported in EPA TSCA Inventory. Cyanide and its compounds, as well as barium and its compounds, are on the Community Right-To-Know List.

OSHA PEL: TWA 0.5 mg(Ba)/m³
ACGIH TLV: TWA 0.5 mg(Ba)/m³
DFG MAK: 0.5 mg(Ba)/m³
DOT CLASSIFICATION: 6.1; *Label:* Poison

SAFETY PROFILE: A deadly poison. See also CYANIDE and BARIUM COMPOUNDS (soluble). When heated to decomposition it emits toxic fumes of CN^-.

BAL000 **HR: 3**
BARIUM CYANOPLATINITE
mf: $BaPt(CN)_4•4H_2O$ mw: 508.6

PROP: (a) Monoclinic, yellow crystals; (b) rhombic crystals. Mp: loses $2H_2O$ @ 100°; d: (a) 2.076, (b) 2.085.

CONSENSUS REPORTS: Cyanide and its compounds, as well as barium and its compounds, are on the Community Right-To-Know List.

OSHA PEL: TWA 5 mg(CN)/m³
ACGIH TLV: CL 5 mg(CN)/m³ (skin)
DFG MAK: 5 mg/m³
NIOSH REL: (Cyanide) CL 5 mg(CN)/m³/10M

SAFETY PROFILE: A poison. See also BARIUM COMPOUNDS (soluble), CYANIDE, and PLATINUM COMPOUNDS. When heated to decomposition it emits highly toxic fumes of CN^- and NO_x.

BAL250 **HR: 3**
BARIUM DIAZIDE
mf: BaN_6 mw: 221.38

CONSENSUS REPORTS: Barium and its compounds are on the Community Right-To-Know List.

SAFETY PROFILE: A poison. Impact-sensitive when dry; avoid contact with Pb, acids. See also BARIUM COMPOUNDS and AZIDES.

BAL275 **CAS:6332-68-9** **HR: 3**
BARIUM DIBENZYLPHOSPHATE
mf: $C_{28}H_{28}BaO_8P_2$ mw: 691.84

TOXICITY DATA WITH REFERENCE
ivn-mus LD50:56 mg/kg CSLNX* NX#04099

OSHA PEL: TWA 0.5 mg(Ba)/m³
ACGIH TLV: TWA 0.5 mg(Ba)/m³
DFG MAK: 0.5 mg(Ba)/m³

SAFETY PROFILE: Poison by intravenous route. When heated to decomposition it emits toxic fumes of PO_x and Ba.

BAL500 **HR: 3**
BARIUM DICHROMATE
mf: $BaCr_2O_7$ mw: 353.38

PROP: Brownish-red, crystalline masses.

SYN: BARIUM BICHROMATE

CONSENSUS REPORTS: Barium and its compounds, as well as chromium and its compounds, are on the Community Right-To-Know List.

SAFETY PROFILE: A poison. Some chromates are carcinogenic. A moderate fire hazard by chemical reaction with easily oxidized materials. A powerful oxidizer. Incompatible with reducing materials. See also BARIUM COMPOUNDS and CHROMIUM COMPOUNDS.

BAL750 CAS:13862-62-9 HR: 3
BARIUM FLUOBORATE
mf: $B_2F_8{\cdot}Ba$ mw: 310.96

SYNS: BARIUM BIS(TETRAFLUOROBORATE) ☐ BARIUM TETRAFLUOROBORATE

TOXICITY DATA WITH REFERENCE
orl-rat LDLo:250 mg/kg NCNSA6 5,27,53

CONSENSUS REPORTS: Barium and its compounds are on the Community Right-To-Know List.

OSHA PEL: TWA 0.5 mg(Ba)/m³; 2.5 mg(F)/m³
ACGIH TLV: TWA 0.5 mg(Ba)/m³
DFG MAK: 0.5 mg(Ba)/m³
NIOSH REL: (Fluorides, Inorganic) TWA 2.5 mg(F)/m³

SAFETY PROFILE: Poison by ingestion. See also BARIUM COMPOUNDS, BORON COMPOUNDS, and FLUORIDES. When heated to decomposition it emits toxic fumes of F^-.

BAM000 CAS:7787-32-8 HR: 3
BARIUM FLUORIDE
mf: BaF_2 mw: 175.34

PROP: White, colorless powder or cubic crystals. Mp: 1368°, bp: 2137°, d: 4.89. Sltly sol in H_2O.

SYN: BARYUM FLUORURE (FRENCH)

TOXICITY DATA WITH REFERENCE
ipr-mus TDLo:656 mg/kg (female 1-21D post):TER DZZEA7 34,484,79
ipr-mus TDLo:525 mg/kg (1-21D preg):TER DZZEA7 34,484,79
orl-rat LD50:250 mg/kg VAMNAQ (2),28,77
ipr-mus LD50:29,910 μg/kg DZZEA7 34,484,79
scu-frg LDLo:1540 mg/kg CRSBAW 124,133,37

CONSENSUS REPORTS: Reported in EPA TSCA Inventory. Barium and its compounds are on the Community Right-To-Know List.

OSHA PEL: TWA 0.5 mg(Ba)/m³; 2.5 mg(F)/m³
ACGIH TLV: TWA 0.5 mg(Ba)/m³; 2.5 mg(F)/m³
DFG MAK: 0.5 mg(Ba)/m³
NIOSH REL: (Fluorides, Inorganic) TWA 2.5 mg(F)/m³

SAFETY PROFILE: A poison by ingestion and intraperitoneal routes. Moderately toxic by subcutaneous route. An experimental teratogen. See also FLUORIDES and BARIUM COMPOUNDS (SOLUBLE). When heated to decomposition it emits toxic fumes of F^-.

BAM250 CAS:13477-09-3 HR: 3
BARIUM HYDRIDE
mf: BaH_2 mw: 139.38

PROP: Gray, orthorhombic crystals or lumps. Mp: decomp @ 675°, bp: 1400°, d: 4.21 @ 0°.

CONSENSUS REPORTS: Barium and its compounds are on the Community Right-To-Know List.

SAFETY PROFILE: A poison. Rapidly decomposed by water and acids. In powder form, it ignites spontaneously in air and reacts vigorously with water. Coarser material ignites when heated in oxygen. Moisture-sensitive, reacts with H_2O with formation of $Ba(OH)_2$ and H_2. Decomp on heating with evolution of H_2 gas and formation of Sr. It is incompatible with water; acids; and metal halogenates. A dangerous fire hazard because moisture may cause it to ignite. To fight fire, use dry chemical, graphite, CO_2. See also BARIUM COMPOUNDS (soluble) and HYDRIDES.

BAM500 HR: 3
BARIUM HYDROXIDE
mf: $Ba(OH)_2$ mw: 171.35

CONSENSUS REPORTS: Barium and its compounds are on the Community Right-To-Know List.

SAFETY PROFILE: A poison. See also BARIUM COMPOUNDS (soluble). Incompatible with chlorinated rubber.

BAM750 HR: 3
BARIUM HYPOPHOSPHITE
mf: $Ba(H_2PO_2)_2{\cdot}H_2O$ mw: 285.38

PROP: Crystalline powder. Mp: decomp, d: 2.90 @ 17°.

CONSENSUS REPORTS: Barium and its compounds are on the Community Right-To-Know List.

SAFETY PROFILE: A poison. When heated to decomposition it emits highly toxic fumes of PO_x. Incompatible with $KClO_3$. When heated to decomposition it emits toxic fumes of PO_x. See also BARIUM COMPOUNDS (soluble).

BAN000 HR: 3
BARIUM IODATE
mf: $Ba(IO_3)_2$ mw: 487.20

PROP: White, crystalline powder. Mp: decomp, d: 4.998.

CONSENSUS REPORTS: Barium and its compounds are on the Community Right-To-Know List.

SAFETY PROFILE: A poison. A powerful oxidizer. Incompatible with Al; As; C; Cu; metal sulfides; organic matter. When heated to decomposition it emits toxic fumes of I^-. See also BARIUM COMPOUNDS (soluble) and IODATES.

BAN250 CAS:10022-31-8 HR: 3
BARIUM(II) NITRATE (1:2)
DOT: UN 1446
mf: $N_2O_6{\cdot}Ba$ mw: 261.36

PROP: Lustrous, colorless, cubic crystals. Mp: 592°, bp: decomp, d: 3.24 @ 23°. Decomp on heating with evolution of NO_2 and O_2 and formation of BaO. Insol in EtOH.

SYNS: BARIUM DINITRATE ☐ BARIUM NITRATE (DOT) ☐ DUSICNAN BARNATY (CZECH) ☐ NITRATE de BARYUM (FRENCH) ☐ NITRIC ACID, BARIUM SALT

TOXICITY DATA WITH REFERENCE
skn-rbt 500 mg/24H MLD 28ZPAK -,10,72
eye-rbt 100 mg/24H MOD 28ZPAK -,10,72
orl-rat LD50:355 mg/kg 28ZPAK -,10,72
scu-mus LDLo:10 mg/kg NTIS** AEC-TR-6710
ivn-mus LD50:8500 μg/kg TXAPA9 22,150,72
orl-dog LDLo:800 mg/kg YKYUA6 31,1247,80
orl-rbt LDLo:150 mg/kg YKYUA6 31,1247,80
par-rbt LDLo:30 mg/kg MELAAD 30,44,39

CONSENSUS REPORTS: Reported in EPA TSCA Inventory. Barium and its compounds are on the Community Right-To-Know List.

OSHA PEL: TWA 0.5 mg(Ba)/m^3
ACGIH TLV: TWA 0.5 mg(Ba)/m^3
DFG MAK: 0.5 mg(Ba)/m^3
DOT CLASSIFICATION: 5.1; *Label:* Oxidizer, Poison

SAFETY PROFILE: A poison via ingestion, subcutaneous, parenteral, and intravenous routes. An irritant to skin and eyes. When heated to decomposition it emits very toxic fumes of NO_x. An oxidizer. Mixtures with finely divided aluminum magnesium alloys are easily ignitable and extremely sensitive to friction or impact. Such mixtures are used in chemical photoflash applications. Incompatible with (Mg + BaO_2 + Zn), Al, and Mg alloys. When heated to decomposition it emits toxic fumes of NO_x. See also BARIUM COMPOUNDS (soluble) and NITRATES.

BAN500 CAS:12047-79-9 HR: 3
BARIUM NITRIDE
mf: Ba_3N_2 mw: 440.10

PROP: Colorless crystals or deep purple powder. Extremely moisture-sensitive. Reacts with H_2O with formation of $Ba(OH)_2$ and evolution of NH_3. Decomp on heating to form Ba_2N and N_2. Bp: 1000° (vac), d: 4.783 @ 25°/4°.

TOXICITY DATA WITH REFERENCE
orl-mus LD50:46,100 μg/kg TOIZAG 22,119,75

CONSENSUS REPORTS: Barium and its compounds are on the Community Right-To-Know List.

SAFETY PROFILE: A poison by ingestion. Flammable by spontaneous chemical reaction with water to liberate explosive ammonia gas. Dangerous; explodes upon heating and by spontaneous chemical reaction, liberating NH_3 vapor which can form explosive mixtures with air. Violent reaction with air or moisture. See also BARIUM COMPOUNDS (soluble) and AMMONIA.

BAO000 CAS:1304-28-5 HR: 3
BARIUM OXIDE
DOT: UN 1884
mf: BaO mw: 153.34

PROP: White to yellowish-white powder or cubic crystals; moisture-sensitive. Mp: 1913°, bp: 2000° (approx), d: 5.72. Mod sol in EtOH; insol in Me_2CO.

SYNS: BARIUM MONOXIDE ☐ BARIUM PROTOXIDE ☐ BARYTA ☐ CALCINED BARYTA ☐ OXYDE de BARYUM (FRENCH)

TOXICITY DATA WITH REFERENCE
scu-mus LD50:50 mg/kg ZVKOA6 19,186,74

CONSENSUS REPORTS: Reported in EPA TSCA Inventory. Barium and its compounds are on the Community Right-To-Know List.

OSHA PEL: TWA 0.5 mg(Ba)/m^3
ACGIH TLV: TWA 0.5 mg(Ba)/m^3
DFG MAK: 0.5 mg(Ba)/m^3
DOT CLASSIFICATION: 6.1; *Label:* KEEP AWAY FROM FOOD

SAFETY PROFILE: A poison via subcutaneous route. See also BARIUM COMPOUNDS (soluble). Combustible by spontaneous chemical reaction; produces heat on contact with water or steam. Reacts with H_2O, $Ba(OH)_2$. Incompatible with H_2S, hydroxylamine, N_2O_4, triuranium octaoxide, SO_3.

BAO250 CAS:1304-29-6 HR: 3
BARIUM PEROXIDE
DOT: UN 1449
mf: BaO_2 mw: 169.34

PROP: Pale, grayish-white powder. Mp: 450°, bp: loses O_2 @ 800°, d: 4.96. Decomp on heating to BaO and O_2. Dissolves in water with formation of H_2O_2.

SYNS: BARIO (PEROSSIDO di) (ITALIAN) ☐ BARIUM BINOXIDE ☐ BARIUM DIOXIDE ☐ BARIUMPEROXID (GERMAN) ☐ BARIUMPEROXYDE (DUTCH) ☐ BARIUM SUPEROXIDE ☐ DIOXYDE de BARYUM (FRENCH) ☐ PEROXYDE de BARYUM (FRENCH)

TOXICITY DATA WITH REFERENCE
scu-mus LD50:50 mg/kg ZVKOA6 19,186,74

CONSENSUS REPORTS: Reported in EPA TSCA Inventory. Barium and its compounds are on the Community Right-To-Know List.

OSHA PEL: TWA 0.5 mg(Ba)/m^3
ACGIH TLV: TWA 0.5 mg(Ba)/m^3
DFG MAK: 0.5 mg(Ba)/m^3
DOT CLASSIFICATION: 5.1; *Label:* Oxidizer, Poison

SAFETY PROFILE: A poison via subcutaneous route. A powerful oxidizer. Explodes on contact with acetic anhydride. Ignites when mixed with calcium-silicon alloys, powdered aluminum, powdered magnesium, water + organic compounds. Mixtures with propane react violently when heated. The powder ignites when heated to 265°C with selenium. Wood ignites with friction from the peroxide. Incompatible with H_2S, water, peroxyformic acid, hydroxylamine solution, mixture of (Mg + Zn + $Ba(NO_3)_2$), and organic matter. See also BARIUM COMPOUNDS (soluble) and PEROXIDES, INORGANIC.

BAO300 CAS:50864-67-0 HR: 3
BARIUM POLYSULFIDE

SYNS: BARIUMPOLYSULFID ☐ BARIUM SULFIDE ☐ SOLABAR ☐ SOLBAR

TOXICITY DATA WITH REFERENCE
orl-man TDLo: 226 mg/kg AIMDAP 132,891,73
orl-rat LD50: 375 mg/kg FMCHA2-,C34,89

OSHA PEL: TWA 0.5 mg(Ba)/m³
ACGIH TLV: TWA 0.5 mg(Ba)/m³
DFG MAK: 0.5 mg(Ba)/m³

SAFETY PROFILE: Poison by ingestion. Human systemic effects by ingestion: flaccid paralysis without anesthesia, muscle weakness, and dyspnea. When heated to decomposition it emits toxic fumes of SO_x and Ba.

BAO500 HR: 3
BARIUM RHODANIDE
mf: $BaC_6H_4O_2N_2S_4$ mw: 401.6

CONSENSUS REPORTS: Barium and its compounds are on the Community Right-To-Know List.

SAFETY PROFILE: A poison. Explosive. When heated to decomposition it can emit highly toxic fumes of SO_x and NO_x. See also BARIUM COMPOUNDS.

BAO750 CAS:17125-80-3 HR: 3
BARIUM SILICOFLUORIDE
mf: $F_6Si•Ba$ mw: 279.43

PROP: White or colorless rhombohedral crystalline powder. D: 4.29 @ 21°/4°, mp: 300° (decomp). Decomp on heating to form SiF_4 and BaF_2. Sltly sol in H_2O. Insol in EtOH.

SYNS: BARIUM FLUOROSILICATE ☐ BARIUM FLUOSILICATE ☐ BARIUM HEXAFLUOROSILICATE ☐ BARIUM HEXAFLUOROSILICATE (2-) ☐ BARIUM SILICON FLUORIDE ☐ BARIUMSILICOFLUORID ☐ BARIUM SILICON FLUORIDE ☐ SILICATE(2-), HEXAFLUORO-, BARIUM ☐ SILICATE(2-), HEXAFLUORO-, BARIUM (1:1) (9CI) ☐ SILICON FLUORIDE BARIUM SALT

TOXICITY DATA WITH REFERENCE
orl-rat LD50: 175 mg/kg AFDOAQ 15,122,51
orl-rbt LDLo: 175 mg/kg JPETAB 39,246,30

CONSENSUS REPORTS: Reported in EPA TSCA Inventory. Barium and its compounds are on the Community Right-To-Know List.

OSHA PEL: 8H TWA 0.5 mg(Ba)/m³; TWA 2.5 mg(F)/m³
ACGIH TLV: TWA 0.5 mg(Ba)/m³
DFG MAK: 0.5 mg(Ba)/m³
NIOSH REL: (Fluorides, Inorganic) TWA 2.5 mg(F)/m³
DOT CLASSIFICATION: 6.1; *Label:* KEEP AWAY FROM FOOD

SAFETY PROFILE: A poison by ingestion. When heated to decomposition it emits toxic fumes of F^-. See also BARIUM COMPOUNDS (soluble).

BAO825 CAS:6865-35-6 HR: 2
BARIUM STEARATE
mf: $C_{36}H_{72}O_4•Ba$ mw: 706.42

SYNS: BARIUM DISTEARATE ☐ OCTADECANOIC ACID, BARIUM SALT (9CI) ☐ STAVINOR 40 ☐ STEARIC ACID, BARIUM SALT

TOXICITY DATA WITH REFERENCE
orl-rat LD50: 4 g/kg GISAAA 39(11),91,74
orl-mus LD50: 3500 mg/kg GISAAA 39(11),91,74
orl-gpg LD50: 3600 mg/kg GISAAA 39(11),91,74

CONSENSUS REPORTS: Barium and its compounds are on the Community Right-To-Know List.

SAFETY PROFILE: Moderately toxic by ingestion. When heated to decomposition it emits acrid smoke and fumes. See also BARIUM COMPOUNDS.

BAO900 CAS:20236-55-9 HR: 3
BARIUM STYPHNATE
DOT: NA 0473

SYNS: 1,3-BENZENEDIOL, 2,4,6-TRINITRO-, BARIUM SALT, HYDRATE (2:1:1) ☐ RESORCINOL, 2,4,6-TRINITRO-, BARIUM SALT, HYDRATE, (2:1:1)

DOT CLASSIFICATION: Explosive 1.1A; *Label:* Explosive 1.1A

SAFETY PROFILE: An explosive. When heated to decomposition it emits toxic vapors of NO_x and fumes of Ba.

BAP000 CAS:7727-43-7 HR: 2
BARIUM SULFATE
mf: $O_4S•Ba$ mw: 233.40

PROP: White, heavy, orthorhombic, odorless, powder or crystals. Undergoes orthorhombic to monoclinic phase transition at 11°. D: 4.50 @ 15°, mp: 1580°. Sltly sol in H_2O. Insol in water or dilute acids.

SYNS: ACTYBARYTE ☐ ARTIFICIAL BARITE ☐ ARTIFICIAL HEAVY SPAR ☐ BAKONTAL ☐ BARIDOL ☐ BARITE ☐ BARITOP ☐ BAROSPERSE ☐ BAROTRAST ☐ BARYTA WHITE ☐ BARYTES ☐ BAYRITES ☐ BLANC FIXE ☐ C.I. 77120 ☐ C.I. PIGMENT WHITE 21 ☐ CITOBARYUM ☐ COLONATRAST ☐ ENAMEL WHITE ☐ ESOPHOTRAST ☐ EWEISS ☐ E-Z-PAQUE ☐ FINEMEAL ☐ LACTOBARYT ☐ LIQUIBARINE ☐ MACROPAQUE ☐ NEOBAR ☐ ORATRAST ☐ PERMANENT WHITE ☐ PRECIPITATED BARIUM SULPHATE ☐ RAYBAR ☐ REDI-FLOW ☐ SOLBAR ☐ SULFURIC ACID, BARIUM SALT (1:1) ☐ SUPRAMIKE ☐ TRAVAD ☐ UNIBARYT

TOXICITY DATA WITH REFERENCE
mnt-mus-ipr 12,500 μg/kg GWZHEW 12,77,86
ipl-rat TDLo: 200 mg/kg: ETA BJCAAI 28,173,73

CONSENSUS REPORTS: Reported in EPA TSCA Inventory. Barium and its compounds are on the Community Right-To-Know List.

OSHA PEL: Total Dust: TWA 10 mg/m³; Respirable Fraction: 5 mg/m³
ACGIH TLV: TWA (nuisance particulate) 10 mg/m³ of total dust (when toxic impurities are not present, e.g., quartz <1%).

SAFETY PROFILE: Questionable carcinogen with ex-

perimental tumorigenic data. Mutation data reported. A relatively insoluble salt used as an opaque medium in radiography. Soluble impurities can lead to toxic reactions. Heating with aluminum can produce an explosion. Incompatible with aluminum and potassium. When heated to decomposition it emits toxic fumes of SO_x.

BAP250 **CAS:21109-95-5** **HR: 3**
BARIUM SULFIDE
mf: BaS mw: 169.4

PROP: Cubic, colorless crystals. Moisture-sensitive. D: 4.25 @ 15°, mp: 1200°.

CONSENSUS REPORTS: Barium and its compounds are on the Community Right-To-Know List.

SAFETY PROFILE: A poison. Flammable by spontaneous chemical reaction; air, moisture, or acid fumes may cause it to ignite. For explosion and disaster hazards, see SULFIDES. To fight fire, use CO_2, dry chemical. Reacts violently with phosphorus(V) oxide. Mixtures with lead dioxide, potassium chlorate, or potassium nitrite explode when heated. Incompatible with Cl_2O, $Ca(NO_3)_2$, $Sr(NO_3)_2$, $Ca(ClO_3)_2$, $Sr(ClO_3)_2$, $(ClO_3)_2$. See also BARIUM COMPOUNDS (soluble) and SULFIDES.

BAP500 **HR: 3**
BARIUM THIOCYANATE
mf: $C_2BaN_2S_2$ mw: 253.52

CONSENSUS REPORTS: Barium and its compounds are on the Community Right-To-Know List.

SAFETY PROFILE: A deadly poison. Incompatible with potassium chlorate, sodium nitrate. When heated to decomposition it emits toxic fumes of SO_x and NO_x. See also BARIUM COMPOUNDS and THIOCYANATES.

BAP750 **CAS:12009-21-1** **HR: 2**
BARIUM ZIRCONIUM(IV) OXIDE
mf: $O_4Zr_4{\cdot}Ba$ mw: 566.22

PROP: Light gray-buff powder or white powder. D: 5.52, mp: 2510°. Insol in water and alkalies; sltly sol in acid.

SYNS: BARIUM ZIRCONATE □ BARIUM ZIRCONIUM OXIDE □ BARIUM ZIRCONIUM TRIOXIDE □ ZIRCONATE, BARIUM (1:1)

TOXICITY DATA WITH REFERENCE
orl-rat LD50:1980 mg/kg AIHAAP 24,131,63
ipr-rat LD50:420 mg/kg AIHAAP 24,131,63

CONSENSUS REPORTS: Reported in EPA TSCA Inventory. Barium and its compounds are on the Community Right-To-Know List.

OSHA PEL: TWA 5 mg(Zr)/m³; STEL 10 mg(Zr)/m³
ACGIH TLV: TWA 5 mg(Zr)/m³; STEL 10 mg(Zr)/m³
DFG MAK: 5 mg(Zr)/m³

SAFETY PROFILE: Moderately toxic by ingestion and intraperitoneal routes. Inhalation produces interstitial pneumonitis. See also ZIRCONIUM COMPOUNDS and BARIUM COMPOUNDS.

BAQ250 **CAS:65-61-2** **HR: 3**
BASIC ORANGE 3RN
mf: $C_{17}H_{19}N_3{\cdot}ClHZnCl_2$ mw: 438.12

PROP: Mp: 182°.

SYNS: ACRIDINE ORANGE □ ACRIDINE ORANGE NO □ ACRIDINE ORANGE R □ C.I. 46005 □ C.I. BASIC ORANGE 14 □ RHODULINE ORANGE NO □ N,N,N′,N′-TETRAMETHYL-3,6-ACRIDINEDIAMINE MONOHYDROCHLORIDE (9CI)

TOXICITY DATA WITH REFERENCE
dnd-mus:ast 20 µmol/L BBACAQ 374,96,74
dnd-mam:lym 10 pph BIPMAA 11,2537,72
dnd-sal:spr 40 µmol/L BBRCA9 40,1239,70

CONSENSUS REPORTS: EPA Genetic Toxicology Program. Reported in EPA TSCA Inventory. Zinc and its compounds are on the Community Right-To-Know List.

SAFETY PROFILE: Mutation data reported. When heated to decomposition it emits very toxic fumes of HCl, Cl^-, and NO_x. See also ZINC COMPOUNDS.

BAR250 **CAS:8015-73-4** **HR: 2**
BASIL OIL

PROP: Contains about 55% methyl chavicol and 35% of alcohols calculated as lenatoal and other compounds found in the leaves of *Ocimum resilium L.* (FCTXAR 11,855,73). A pale yellow liquid; floral, spicy odor. Sol in fixed oils and propylene glycol; insol in glycerin.

SYNS: BASIL OIL, EUROPEAN TYPE (FCC) □ BASIL OIL, SWEET □ OCIMUM BASILICUM OIL □ OIL OF BASIL □ OILS, BASIL

TOXICITY DATA WITH REFERENCE
skn-mus 100% MLD FCTXAV 11,867,73
orl-rat LD50:1400 mg/kg FCTXAV 11,855,73

CONSENSUS REPORTS: On EPA Extremely Hazardous Substances List by error. Reported in EPA TSCA Inventory.

SAFETY PROFILE: Moderately toxic by ingestion. A skin and eye irritant. When heated to decomposition it emits acrid smoke and irritating fumes.

BAR325 **HR: 2**
BASKET FLOWER

PROP: Bulb-producing ornamental plants. The long, thin leaves emerge from the ground not from a stem. The flowers are white or yellow and grow from a leafless stem. The seeds are carried in a capsule. They are native to the southeastern United States and tropical areas of the Americas, and are commonly cultivated.

SYNS: ALLIGATOR LILY □ CROWN BEAUTY □ HYMENOCALLIS (VARIOUS SPECIES) □ LIRIO (SPANISH) □ SEA DAFFODIL □ SPIDER LILY □ TARARACO BLANCO (CUBA)

SAFETY PROFILE: The bulb contains the poison lycorine and similar alkaloids. Ingestion of large amounts may cause nausea, vomiting, and diarrhea.

BAR500 **HR: 2**
BASORA CORRA

PROP: Aqueous extract from the root of the plant (JNCIAM 52,445,74).

SYN: MELOCHIA TOMENTOSA

TOXICITY DATA WITH REFERENCE
scu-rat TDLo:300 g/kg/60W-I:NEO JNCIAM 52,445,74

SAFETY PROFILE: Questionable carcinogen with experimental neoplastigenic data.

BAR750 **CAS:23509-16-2** **HR: 3**
BATRACHOTOXIN
mf: $C_{31}H_{42}N_2O_6$ mw: 538.75

PROP: Noncrystal. Active principle from the skin of the Columbian arrow poison frog.

SYNS: 20-(2,4-DIMETHYL-1H-PYRROLE-3-CARBOXYLATE) BATRACHOTOXININ A □ 20-α-(2,4-DIMETHYL-1H-PYRROLE-3-CARBOXYLATE) BETRACHOTOXININ A

TOXICITY DATA WITH REFERENCE
ipr-mus LD50:2 μg/kg TOXIA6 7,315,69
scu-mus LD50:2 μg/kg CTOXAO 4,331,71
ivn-mus LDLo:2700 ng/kg TOXIA6 8,85,70

SAFETY PROFILE: A deadly poison by intraperitoneal, intravenous, and subcutaneous routes. When heated to decomposition it emits toxic fumes of NO_x.

BAR800 **CAS:41621-49-2** **HR: 3**
BATRAFEN
mf: $C_{12}H_{17}NO_2 \cdot C_2H_7NO$ mw: 268.40

SYNS: 2-AMINOETHANOL compounded with 6-CYCLOHEXYL-1-HYDROXY-4-METHYL-2(1H)-PYRIDINONE (1:1) □ CIC □ CICLOPIROX ETHANOLAMINE SALT (1:1) □ CICLOPIROXOLAMIN □ CICLOPIROXOLAMINE □ 6-CYCLOHEXYL-1-HYDROXY-4-METHYL-2(1H)-PYRIDINONE compounded with 2-AMINOETHANOL (1:1) □ 6-CYCLOHEXYL-1-HYDROXY-4-METHYL-2(1H)-PYRIDON, 2-AMINOETHANOL-SALZ (GERMAN) □ 6-CYCLOHEXYL-1-HYDROXY-4-METHYL-2(1H)-PYRIDONE, 2-AMINOETHANOL-SALT □ 6-CYCLOHEXYL-1-HYDROXY-4-METHYL-2(1H)-PYRIDONE ETHANOLAMINE SALT □ HOE 296 □ LORPOX □ TERIT

TOXICITY DATA WITH REFERENCE
orl-rat TDLo:8400 mg/kg (28D pre):REP OYYAA2 9,67,75
orl-rat LD50:2350 mg/kg IYKEDH 8,107,77
ipr-rat LD50:146 mg/kg OYYAA2 9,57,75
scu-rat LD50:9800 mg/kg YKYUA6 28,115,77
ivn-rat LD50:72 mg/kg IYKEDH 8,107,77
orl-mus LD50:1740 mg/kg IYKEDH 8,107,77
ipr-mus LD50:83 mg/kg OYYAA2 9,57,75
scu-mus LD50:1730 mg/kg IYKEDH 8,107,77
ivn-mus LD50:71 mg/kg OYYAA2 9,57,75

SAFETY PROFILE: Poison by intravenous and intraperitoneal routes. Moderately toxic by ingestion and subcutaneous routes. Experimental reproductive effects. When heated to decomposition it emits toxic fumes of NO_x.

BAR825 **CAS:64314-28-9** **HR: 3**
BAUMYCIN A1
mf: $C_{34}H_{43}NO_{13}$ mw: 673.78

PROP: Orange-red crystals. Mp: 182–185°.

TOXICITY DATA WITH REFERENCE
pic-esc 50 ng/plate CNREA8 43,2819,83
dni-mus:leu 1700 nmol/L JANTAJ 34,1596,81
oms-mus:leu 560 nmol/L JANTAJ 34,1596,81
ipr-mus LD50:1500 μg/kg JANTAJ 31,78-67,78

SAFETY PROFILE: Poison by intraperitoneal route. Mutation data reported. When heated to decomposition it emits toxic fumes of NO_x.

BAR830 **CAS:64253-71-0** **HR: 3**
BAUMYCIN A2
mf: $C_{34}H_{43}NO_{13}$ mw: 673.78

PROP: Orange-red crystals. Mp: 185–189°.

TOXICITY DATA WITH REFERENCE
pic-esc 50 ng/plate CNREA8 43,2819,83
dni-mus:leu 1900 nmol/L JANTAJ 34,1596,81
oms-mus:leu 710 nmol/L JANTAJ 34,1596,81
ipr-mus LD50:15 mg/kg JANTAJ 31,78-68,78

SAFETY PROFILE: Poison by intraperitoneal route. Mutation data reported. When heated to decomposition it emits toxic fumes of NO_x.

BAS000 **CAS:7682-90-8** **HR: 3**
BAY 75546
mf: $C_{12}H_{17}BrN_3O_3PS$ mw: 394.26

SYN: 3-BROMO-5,7-DIMETHYL PYRAZOLYL-2-PYRIMIDINEPHOSPHOROTHIOIC ACID-O,O-DIETHYL ESTER

TOXICITY DATA WITH REFERENCE
orl-rat LD50:2000 mg/kg TXAPA9 21,315,72
orl-bwd LD50:2400 μg/kg TXAPA9 21,315,72

SAFETY PROFILE: Poison by ingestion. When heated to decomposition it emits very toxic fumes of Br^-, NO_x, PO_x, and SO_x. See also ESTERS.

BAT000 **CAS:145-63-1** **HR: 2**
BAYER 205
mf: $C_{51}H_{40}N_6O_{23}S_6$ mw: 1297.33

PROP: Pinkish-white, hygroscopic powder. Very sol in H_2O; insol in EtOH.

SYNS: ANTRYPOL □ BELGANYL □ CARBANILIDE, 3,3′-BIS((5-((4,6,8-TRISULFO-1-NAPHTHYL)CARBAMOYL)-o-TOLYL)CARBAMOYL)- □ FARMA □ FARMA 939 □ FOURNEAU □ FOURNEAU 309 □ GERMANIN □ NAGANOL □ NAPHURIDE □ SURAMIN □ SURAMINE

TOXICITY DATA WITH REFERENCE
unr-mus TDLo:195 mg/kg (female 9-11D post):REP CRSBAW 167,1717,73
scu-rat TDLo:250 mg/kg (female 9D post):TER CBINA8 58,149,86
ipr-rat TDLo:300 mg/kg (female 7-9D post):REP CRSBAW 167,1518,73
ipr-mus TDLo:195 mg/kg (female 9-11D post):REP CRSBAW 167,1518,73

ipr-mus TDLo: 150 mg/kg (female 9-11D post): TER CRSBAW 167,1518,73
ivn-man TDLo: 46 mg/kg/5W-I: EYE NEJMAG 314,1455,86
ivn-mus LD50: 620 mg/kg ADVPA3 15,289,78

SAFETY PROFILE: Moderately toxic by intravenous route. Human systemic effects by intravenous route: eye effects. An experimental teratogen. Other experimental reproductive effects. When heated to decomposition it emits very toxic fumes of SO_x and NO_x.

BAT500 HR: 2
BAY OIL

PROP: Consists mainly of eugenol and chavicol (55–65%), major portion of balance consists of terpenes (alpha-pinene, myrcene, and dipentene), small quantities of citrol, nerol, cineol, and other terpenoids have also been found (FCTXAV 11,855,73). Yellow or brown liquid; aromatic odor, pungent, spicy taste. Sol in alc and glacial acetic acid.

SYNS: BAY LEAF OIL □ BOIS d'INDE □ LAUREL LEAF OIL □ MYRCIA OIL □ MYRICIA OIL □ OIL of BAY □ OIL of MYRCIA

TOXICITY DATA WITH REFERENCE
orl-rat LD50: 1800 mg/kg FCTXAV 11,855,73

CONSENSUS REPORTS: Reported in EPA TSCA Inventory.

SAFETY PROFILE: Moderately toxic by ingestion. When heated to decomposition it emits acrid smoke.

BAT750 CAS:14816-18-3 HR: 3
BAYTHION
mf: $C_{12}H_{15}N_2O_3PS$ mw: 298.32

PROP: Liquid. D: 1.176° @ 20 mm, fp: 5–6°, bp: 102° @ 0.01 mm (decomp).

SYNS: B 77488 □ BAY 5621 □ BAY 77488 □ BAYRE 77488 □ BENZOYL CYANIDE-o-(DIETHOXYPHOSPHINOTHIOYL)OXIME □ O,O-DIAETHYL-o-(α-CYANBENZYLIDEN-AMINO)-THIONPHOSPHAT (GERMAN) □ O,O-DIAETHYL-o-(α-CYANO-BENZYLIDENAMINO)-MONOTHIOPHOSPHAT (GERMAN) □ α-(((DIETHOXYPHOSPHINOTHIOYL)OXY)IMINO)BENZENEACETONITRILE □ (DIETHOXY-THIOPHOSPHORYLOXYIMINO)-PHENYL ACETONITRILE □ O,O-DIETHYL PHOSPHOROTHIOATE, o-ESTER with PHENYLGLYOXYLONITRILE OXIME □ ENT 27,488 □ 4-ETHOXY-7-PHENYL-3,5-DIOXA-6-AZA-4-PHOSPHAOCT-6-ENE-8-NITRILE-4-SULFIDE □ PHENYLGLYOXYLONITRILE OXIME-O,O-DIETHYL PHOSPHOROTHIOATE □ PHOXIME □ PHOXIN □ SEBACIL □ VALEXONE □ VOLATON

TOXICITY DATA WITH REFERENCE
orl-rat TDLo: 49 mg/kg (74D male): TER GISAAA 45(7),77,80
orl-rat LD50: 300 mg/kg FAATDF 7,299,86
skn-rat LD50: 1000 mg/kg 28ZEAL 5,181,76
orl-mus LD50: 1050 mg/kg 52OLAC -,230,83
orl-dog LD50: 250 mg/kg 28ZEAL 5,181,76
orl-cat LD50: 250 mg/kg 28ZEAL 5,181,76
orl-rbt LD50: 250 mg/kg 85DPAN -,-,71/76

CONSENSUS REPORTS: Cyanide and its compounds are on the Community Right-To-Know List.

SAFETY PROFILE: Poison by ingestion. An experimental teratogen. When heated to decomposition it emits very toxic fumes of CN^-, NO_x, PO_x, and SO_x. See also NITRILES.

BAT830 CAS:63428-82-0 HR: 3
BEAUVERIN

SYNS: BEAUVERIA BASSIANA □ BOVERIN □ BOVERINE

TOXICITY DATA WITH REFERENCE
ipr-mus LD50: 128 mg/kg CYLPDN 6,213,85

CONSENSUS REPORTS: Reported in EPA TSCA Inventory.

SAFETY PROFILE: A poison by intraperitoneal route. When heated to decomposition it emits acrid smoke and irritating vapors.

BAT850 CAS:8021-39-4 HR: 3
BEECHWOOD CRESOATE

SYNS: CRESOATE, WOOD □ RCRA WASTE NUMBER U051

TOXICITY DATA WITH REFERENCE
orl-rat TDLo: 52,416 mg/kg/91D-C OYYAA2 21,899,81
orl-rat TDLo: 210 g/kg/96W-C OYYAA2 28,925,84
orl-mus TDLo: 197 g/kg/91D-C OYYAA2 21,899,81

CONSENSUS REPORTS: NTP 7th Annual Report On Carcinogens. Reported in EPA TSCA Inventory.

SAFETY PROFILE: Confirmed carcinogen. When heated to decomposition it emits acrid smoke and irritating fumes.

BAU000 CAS:8012-89-3 HR: 1
BEESWAX

PROP: Yellow to brownish-yellow, soft to brittle wax. Mp: 62–65°, d: 0.95–0.96. Sol in chloroform, ether, fixed oils; sltly sol in alc.

SYNS: BEESWAX, WHITE □ BEESWAX, YELLOW

SAFETY PROFILE: A mild allergen. Combustible when heated.

BAU255 CAS:39543-79-8 HR: 3
BEFUNOLOL HYDROCHLORIDE
mf: $C_{16}H_{21}NO_4{\cdot}ClH$ mw: 327.84

SYNS: 2-ACETYL-7-((2-HYDROXY-3-ISOPROPYLAMINO)PROPOXY)BENZOFURAN HYDROCHLORIDE □ BENTOX □ BFE 60 □ 1-(7-(2-HYDROXY-3-((1-METHYLETHYL)AMINO)PROPOXY)-2-BENZOFURANYL)ETHANONE HYDROCHLORIDE

TOXICITY DATA WITH REFERENCE
orl-mus TDLo: 1300 mg/kg (female 15-21D post): REP KSRNAM 13,3740,79
orl-rat TDLo: 1400 mg/kg (female 9-15D post): TER KSRNAM 13,3678,79
orl-rbt TDLo: 1300 mg/kg (female 6-18D post): REP KSRNAM 13,3715,79
orl-rat TDLo: 3 g/kg (30D male): REP KSRNAM 13,3232,79
orl-rat LD50: 922 mg/kg KSRNAM 13,4138,79

B

ipr-rat LD50:182 mg/kg IYKEDH 14,484,83
scu-rat LD50:498 mg/kg IYKEDH 14,484,83
orl-mus LD50:950 mg/kg IYKEDH 14,484,83
ipr-mus LD50:184 mg/kg IYKEDH 14,484,83
scu-mus LD50:434 mg/kg IYKEDH 14,484,83
ivn-mus LD50:65 mg/kg IYKEDH 14,484,83

SAFETY PROFILE: Poison by intravenous and intraperitoneal routes. Moderately toxic by ingestion and subcutaneous routes. An experimental teratogen. Other experimental reproductive effects. When heated to decomposition it emits toxic fumes of NO_x and HCl. A beta-adrenergic blocker.

BAU270 **CAS:4696-76-8** **HR: 3**
BEKANAMYCIN
mf: $C_{18}H_{37}N_5O_{10}$ mw: 483.60

PROP: Crystals. Mp: 178–182° (decomp).

SYNS: AMINODEOXYKANAMYCIN □ 2′-AMINO-2′-DEOXYKANAMYCIN □ KANAMYCIN B □ KANENDOMYCIN □ KDM □ NEBRAMYCIN FACTOR 5 □ NEBRAMYCIN V □ NK 1006 □ o-3-AMINO-3-DEOXY-α-d-GLUCOPYRANOSYL-(1-4)-o-(2,6-DIAMINO-2,6-DIDEOXY)-α-d-GLUCOPYRANOSYL-(1-6)-2-DEOXY-d-STREPTAMINE

TOXICITY DATA WITH REFERENCE
ipr-mus LD50:800 mg/kg 85GDA2 1,159,80
scu-mus LD50:750 mg/kg 85GDA2 1,159,80
ivn-mus LD50:132 mg/kg JANTAJ 27,677,74

CONSENSUS REPORTS: EPA Genetic Toxicology Program.

SAFETY PROFILE: Poison by intravenous route. Moderately toxic by intraperitoneal and subcutaneous routes. When heated to decomposition it emits toxic fumes of NO_x.

BAU325 **HR: 2**
BELGENINE

PROP: Extracted from *Mallotus japonicus merel arg* (NIIRDN 6,768,82).

TOXICITY DATA WITH REFERENCE
ipr-rat LD50:3040 mg/kg NIIRDN 6,768,82
ivn-rat LD50:2800 mg/kg NIIRDN 6,768,82
ipr-mus LD50:6410 mg/kg NIIRDN 6,768,82
ivn-mus LD50:5400 mg/kg NIIRDN 6,768,82

SAFETY PROFILE: Moderately toxic by intraperitoneal and intravenous routes.

BAU500 **HR: 3**
BELLADONNA

PROP: An extract from the deadly nightshade plant. The alkaloids atropine and belladonnine are derivatives.

SYN: DEADLY NIGHTSHADE

SAFETY PROFILE: A deadly poison. See also HYOSCYAMINE and ATROPINE. Local contact may cause a contact dermatitis. A poisonous constituent of some berries and plants, and of some folk remedies.

BAU750 **CAS:147-24-0** **HR: 3**
BENADRYL HYDROCHLORIDE
mf: $C_{17}H_{21}NO{\cdot}ClH$ mw: 291.85

PROP: Crystals from $EtOH/Et_2O$. Mp: 161–162°. Sol in H_2O.

SYNS: AMBENYL □ BAX □ BENA □ BENADRYL □ BENDYLATE □ BENOCTEN □ BENZEHIST □ BENZHYDRAMINE HYDROCHLORIDE □ 2-(BENZHYDRYLOXY)-N,N-DIMETHYLETHYLAMINE-HYDROCHLORIDE □ DABYLEN □ DIFENHYDRAMINE HYDROCHLORIDE □ DIMETHYLAMINE BENZHYDRYL ESTER HYDROCHLORIDE □ β-DIMETHYLAMINOETHYL BENZHYDRYL ESTER HYDROCHLORIDE □ DIPHENYLHYDRAMINE HYDROCHLORIDE □ 2-(DIPHENYLMETHOXY)-N,N-DIMETHYL-ETHANAMINE HYDROCHLORIDE □ 2-DIPHENYLMETHOXY-N,N-DIMETHYLETHYLAMINE HYDROCHLORIDE □ DOLESTAN □ ELDADRYL □ FELBEN □ FENYLHIST □ HALBMOND □ α-HYDROXYDIPHENYLMETHANE-β-DIMETHYLAMINOETHYL ETHER HYDROCHLORIDE □ NCI-C56075 □ ROHYDRA □ SK-DIPHENHYDRAMINE □ VALDRENE □ WEHYDRYL

TOXICITY DATA WITH REFERENCE
orl-mus TDLo:800 mg/kg (female 8-12D post):REP TCMUD8 6,361,86
orl-mus TDLo:420 mg/kg (female 1-21D post):REP ARZNAD 18,188,68
par-rat TDLo:48 mg/kg (female 13-16D post):REP AFTOD7 3,157,77
orl-rat TDLo:1 g/kg (6-15D preg):TER NTIS** PB83-180612
orl-rat TDLo:39 mg/kg (female 6-15D post):REP TXAPA9 18,971,71
par-rat TDLo:12 mg/kg (female 10D post):TER AFTOD7 3,157,77
orl-rat TDLo:27,037 mg/kg/2Y-C:ETA NTPTR* NTP-TR-355,89
orl-cld TDLo:12,500 μg/kg:BAH,CVS JOPDAB 90,1017,77
orl-man TDLo:10,714 μg/kg:CNS,BAH,BPR AJEMEN 4,369,86
skn-cld TDLo:60 mg/kg/6H-I:EYE,PSY CPEDAM 25,163,86
orl-rat LD50:500 mg/kg NIIRDN 6,334,82
ipr-rat LD50:82 mg/kg JPETAB 102,250,51
scu-rat LD50:201 mg/kg YKKZAJ 81,261,61
ivn-rat LD50:35 mg/kg YACHDS 12,2769,84
orl-mus LD50:114 mg/kg JPETAB 113,72,55
ipr-mus LD50:56 mg/kg JPETAB 112,318,54
scu-mus LD50:99,200 μg/kg NYKZAU 54,33,58
ivn-mus LD50:20 mg/kg ARZNAD 5,72,55
ivn-dog LD50:24 mg/kg JPETAB 89,227,47

CONSENSUS REPORTS: Reported in NTP Carcinogenesis Studies (Feed); Equivocal Evidence: Rat NTPTR* NTP-TR-355,89; (Feed); No Evidence: Mouse NTPTR* NTP-TR-355,89.

SAFETY PROFILE: Poison by ingestion, subcutaneous, intravenous, and intraperitoneal routes. Human systemic effects by ingestion or skin contact: arrhythmias, ataxia, blood pressure elevation, convulsions, distorted perceptions, eye effects, and hallucinations. Experimental teratogenic and reproductive effects. Questionable carcinogen with experimental tumorigenic data. When heated to decomposition it emits very toxic fumes of NO_x and HCl. See also ESTERS and ETHERS.

BAV000 CAS:3813-05-6 HR: 2
BENAZOLIN
mf: $C_9H_6O_3NClS$ mw: 243.6

PROP: White, crystalline solid. Mp: 193°. Sltly sol in H_2O.

SYNS: BEN-30 □ BENAZALOX □ BEN-CORNOX □ BENOPAN □ BENSECAL □ BENZAR □ 4-CHLORO-2-OXO-3(2H)-BENZOTHIAZOLEACETIC ACID □ 4-CHLORO-2-OXOBENZOTHIAZOLIN-3-YL ACETIC ACID □ CORNOX CWK □ CRESOPUR □ EUNASIN □ EX10781 □ GALIPAN □ GERBITOX □ GRASSLAND WEEDKILLER □ HERBAZOLIN □ KEROPUR □ LEGUMEX EXTRA □ LEY-CORNOX □ LEYMIN □ METIZOLIN □ RD7693 □ TRI-CORNOX SPECIAL

TOXICITY DATA WITH REFERENCE
orl-rat LD50:3000 mg/kg 85ARAE 2,26,77

SAFETY PROFILE: Moderately toxic by ingestion. An herbicide. When heated to decomposition it emits toxic fumes of SO_x, Cl^-, and NO_x. See also CHLORIDES.

BAV250 CAS:14286-84-1 HR: 3
BENCYCLANE FUMARATE
mf: $C_{19}H_{31}NO \cdot C_4H_2O_4$ mw: 403.57

PROP: Crystals from EtOH (aq). Mp: 131–133°.

SYNS: BENCICLANE □ BENCYCLANE □ 3-((1-BENZYLCYCLOHEPTYL)OXY)-N,N-DIMETHYLPROPYLAMINE FUMARATE □ N-(3-(1-BENZYL-CYCLOHEPTYLOXY)-PROPYL)-N,N-DIMETHYL-AMMONIUM-HYDROGENFUMARAT (GERMAN) □ EGYT 201 □ FLUDILAT □ HALIDO

TOXICITY DATA WITH REFERENCE
orl-rat LD50:414 mg/kg 27ZQAG -,383,72
ipr-rat LD50:86 mg/kg 27ZQAG -,383,72
scu-rat LD50:257 mg/kg 27ZQAG -,383,72
ivn-rat LD50:41 mg/kg 27ZQAG -,383,72
orl-mus LD50:446 mg/kg 27ZQAG -,383,72
ipr-mus LD50:132 mg/kg 27ZQAG -,383,72
scu-mus LD50:203 mg/kg 27ZQAG -,383,72
ivn-mus LD50:45 mg/kg AITEAT 15,415,67
ims-mus LD50:150 mg/kg AITEAT 15,415,67
orl-dog LDLo:300 mg/kg ARZNAD 20,1385,70

SAFETY PROFILE: Poison by ingestion, intramuscular, intraperitoneal, subcutaneous, and intravenous routes. When heated to decomposition it emits toxic fumes of NO_x.

BAV325 CAS:20187-55-7 HR: 3
BENDAZOLIC ACID
mf: $C_{16}H_{14}N_2O_3$ mw: 282.32

PROP: Crystals from ethanol. Mp: 160°. Practically insol in water; sol in chloroform, acetone.

SYNS: AF 983 □ BENDAZAC □ ((1-BENZYL-1H-INDAZOL-3-YL)OXY)ACETIC ACID □ BINDAZAC □ ((1-(PHENYLMETHYL)-1H-INDAZOL-3-YL)OXY)-ACETIC ACID (9CI) □ VERSUS □ ZILDASAC

TOXICITY DATA WITH REFERENCE
orl-rat LD50:1200 mg/kg MEIEDD 10,146,83
ipr-rat LD50:319 mg/kg IYKEDH 10,884,79
scu-rat LD50:714 mg/kg IYKEDH 10,884,79
ivn-rat LD50:304 mg/kg MEIEDD 10,146,83
orl-mus LD50:1105 mg/kg MEIEDD 10,146,83
ipr-mus LD50:339 mg/kg IYKEDH 10,884,79
scu-mus LD50:406 mg/kg IYKEDH 10,884,79
ivn-mus LD50:380 mg/kg MEIEDD 10,146,83

SAFETY PROFILE: Poison by intravenous and intraperitoneal routes. Moderately toxic by ingestion and subcutaneous routes. When heated to decomposition it emits toxic fumes of NO_x.

BAV350 CAS:8064-77-5 HR: 3
BENDECTIN
mf: $C_{19}H_{35}NO_2 \cdot C_{17}H_{22}N_2O \cdot C_8H_{11}NO_3 \cdot C_4H_6O_4 \cdot 2ClH$
mw: 940.18

SYNS: DEBENDOX □ LENOTAN

TOXICITY DATA WITH REFERENCE
orl-wmn TDLo:75,600 µg/kg (female 6-11W post):REP BMJOAE 1,691,78
orl-wmn TDLo:648 mg/kg (female 1-39W post):REP BMJOAE 1,691,78
orl-wmn TDLo:60 mg/kg (female 4-18W post):TER BMJOAE 1,691,78
orl-wmn TDLo:60 mg/kg (female 4-18W post):REP BMJOAE 1,691,78
orl-wmn TDLo:75,600 µg/kg (female 6-11W post):TER BMJOAE 1,691,78
orl-wmn TDLo:67 mg/kg (female 4-20W post):TER BMJOAE 1,925,78
orl-wmn TDLo:648 mg/kg (female 1-39W post):TER BMJOAE 1,691,78
orl-rat TDLo:8 g/kg (female 6-15D post):TER TJADAB 29(2),52A,84
orl-rat TDLo:5 g/kg (female 6-15D post):TER TJADAB 29(2),52A,84
orl-mky TDLo:38 mg/kg (female 22-50D post):REP TJADAB 32,191,85
orl-rat TDLo:2 g/kg (female 6-15D post):TER TJADAB 29(2),52A,84

SAFETY PROFILE: Human reproductive effects by ingestion: developmental abnormalities of the gastrointestinal system. Human and experimental teratogenic and reproductive effects. When heated to decomposition it emits toxic fumes of NO_x and HCl.

BAV500 HR: 2
BENLATE and SODIUM NITRITE

SYNS: 1-(BUTYLCARBAMOYL)-2-BENZIMIDAZOLECARBAMIC ACID METHYL ESTER and SODIUM NITRITE (1:6) □ SODIUM NITRITE and BENLATE

TOXICITY DATA WITH REFERENCE
orl-mus TDLo:31 g/kg/26W-I:CAR NEOLA4 24,119,77

SAFETY PROFILE: Questionable carcinogen with experimental carcinogenic data. When heated to decomposition it emits toxic fumes of Na_2O and NO_x. See also CARBAMATES, ESTERS, and NITRITES.

BAV575 CAS:17804-35-2 HR: 3
BENOMYL
mf: $C_{14}H_{18}N_4O_3$ mw: 290.36

PROP: Very sltly sol in H_2O; sol in $CHCl_3$; less sol in other org solvents.

SYNS: ARILATE ☐ BBC ☐ BENLATE 50 ☐ BENOMYL 50W ☐ BNM ☐ 1-(BUTYLCARBAMOYL)-2-BENZIMIDAZOLECARBAMIC ACID, METHYL ESTER ☐ 1-(BUTYLCARBAMOYL)-2-BENZIMIDAZOL-METHYL-CARBAMAT (GERMAN) ☐ 1-(N-BUTYLCARBAMOYL)-2-(METHOXY-CARBOXAMIDO)-BENZIMIDAZOL (GERMAN) ☐ DU PONT 1991 ☐ FUNDASOL ☐ FUNGICIDE 1991 ☐ MBC ☐ METHYL-1-(BUTYLCARBAMOYL)-2-BENZIMIDAZOLYLCARBAMATE ☐ TERSAN 1991

TOXICITY DATA WITH REFERENCE
skn-man 0.1% MLD LANCAO 2,1252,80
sln-smc 123 ppm ANYAA9 407,186,83
sln-hmn:lym 10 mg/L MUREAV 121,139,83
mmo-asn 250 µg/L MUREAV 91,115,81
orl-rat TDLo:936 mg/kg (female 7-22D post):REP TXAPA9 62,44,82
orl-mus TDLo:1 g/kg (female 8-12D post):REP JTEHD6 10,541,82
orl-rat TDLo:85 g/kg (multi):REP ARTODN 4,459,80
orl-mus TDLo:2200 mg/kg (female 7-17D post):TER TXAPA9 62,44,82
orl-rat TDLo:625 mg/kg (female 7-17D post):TER TXAPA9 62,44,82
orl-rat TDLo:3500 µg/kg (male 70D pre):REP TXCYAC 28,103,83
orl-rat TDLo:250 mg/kg (female 12D post):REP BEXBAN 83,247,77
orl-rat TDLo:1250 mg/kg (male 5D pre):REP JTEHD6 13,53,84
orl-mus TDLo:1100 mg/kg (female 7-17D post):TER TXAPA9 62,44,82
orl-rat TDLo:156 mg/kg (female 7-16D post):TER TXAPA9 62,44,82
orl-rat TDLo:250 mg/kg (female 12D post):TER BEXBAN 83,247,77
orl-rat LD50:10 g/kg JHEMA2 24,295,80
ihl-rat LC50:>2 g/m³/4H PEMNDP 9,59,91
skn-rat LD50:>1 g/kg WRPCA2 9,119,70
orl-mus LD50:5600 mg/kg 17QLAD 12,85,77
orl-bwd LD50:100 mg/kg TXAPA9 21,315,72

CONSENSUS REPORTS: Reported in EPA TSCA Inventory. EPA Genetic Toxicology Program.

OSHA PEL: Total Dust: TWA 10 mg/m³; Respirable Fraction: 5 mg/m³
ACGIH TLV: TWA 10 mg/m³

SAFETY PROFILE: Poison by ingestion. Mildly toxic by inhalation. Experimental teratogenic and reproductive effects. Human mutation data reported. A human skin irritant. When heated to decomposition it emits toxic fumes of NO_x. See also CARBAMATES.

BAV625 CAS:29462-18-8 HR: 2
BENTAZEPAM
mf: $C_{17}H_{16}N_2OS$ mw: 296.41

PROP: Mp: 249–250°.

SYNS: 1,3,6,7,8,9-HEXAHYDRO-5-PHENYL-2H-(1)BENZOTHIENO(2,3-e)-1,4-DIAZEPIN-2-ONE ☐ QM-6008 ☐ 6,7-TETRAMETHYLENE-5-PHENYL-1,2-DIHYDRO-3H-THIENO(2,3-e)(1,4)DIAZEPIN-2-ONE ☐ THIADIPONE ☐ TIADIPONE

TOXICITY DATA WITH REFERENCE
orl-rat LD50:2 g/kg ARZNAD 26,926,75
orl-mus LD50:980 mg/kg ARZNAD 25,926,75
ipr-mus LD50:630 mg/kg ARZNAD 25,926,75

SAFETY PROFILE: Moderately toxic by ingestion and intraperitoneal routes. When heated to decomposition it emits toxic fumes of SO_x and NO_x.

BAV750 CAS:1302-78-9 HR: 1
BENTONITE

PROP: A clay containing appreciable amounts of the clay mineral montmorillonite; light yellow or green, cream, pink, gray to black solid. Insol in water and common organic solvents.

SYNS: ALBAGEL PREMIUM USP 4444 ☐ BENTONITE 2073 ☐ BENTONITE MAGMA ☐ HI-JEL ☐ IMVITE I.G.B.A. ☐ MAGBOND ☐ MONTMORILLONITE ☐ PANTHER CREEK BENTONITE ☐ SOUTHERN BENTONITE ☐ TIXOTON ☐ VOLCLAY ☐ VOLCLAY BENTONITE BC ☐ WILKINITE

TOXICITY DATA WITH REFERENCE
orl-mus TDLo:12,000 g/kg/28W-C:ETA ANYAA9 57,678,54
ivn-rat LD50:35 mg/kg BSIBAC 44,1685,68

CONSENSUS REPORTS: Reported in EPA TSCA Inventory.

SAFETY PROFILE: Poison by intravenous route causing blood clotting. Questionable carcinogen with experimental tumorigenic data.

BAW000 CAS:7093-10-9 HR: 2
BENZ(1)ACEANTHRENE
mf: $C_{20}H_{14}$ mw: 254.34

PROP: Pale-yellow plates from C_6H_6/Et_2O. Mp: 176.5–177°.

SYNS: 8:9-ACE-1:2-BENZANTHRACENE ☐ 1,2-DIHYDROBENZ(1)ACEANTHRYLENE ☐ 8:9-DIMETHYLENE-1:2-BENZANTHRACENE

TOXICITY DATA WITH REFERENCE
scu-mus TDLo:800 mg/kg/13W-I:ETA AJCAA7 28,334,36

SAFETY PROFILE: Questionable carcinogen with experimental tumorigenic data. When heated to decomposition it emits acrid smoke and fumes.

BAW250 CAS:205-99-2 HR: 3
BENZ(e)ACEPHENANTHRYLENE
mf: $C_{20}H_{12}$ mw: 252.32

PROP: Needles from C_6H_6 or EtOH. Mp: 168°.

SYNS: 3,4-BENZ(e)ACEPHENANTHRYLENE ☐ 2,3-BENZFLUORANTHENE ☐ 3,4-BENZFLUORANTHENE ☐ BENZO(b)FLUORANTHENE ☐ BENZO(e)FLUORANTHENE ☐ 2,3-BENZOFLUORANTHENE ☐ 3,4-BENZOFLUORANTHENE ☐ 2,3-BENZOFLUORANTHRENE ☐ B(b)F

TOXICITY DATA WITH REFERENCE
mma-sat 31 nmol/plate CRNGDP 6,1023,85
otr-ham:lng 100 µg/L TXCYAC 17,149,80
sce-ham-ipr 900 mg/kg/24H MUREAV 66,65,79
imp-rat TDLo:5 mg/kg:ETA JJIND8 71,539,83
skn-mus TDLo:88 ng/kg/120W-I:CAR ARGEAR 50,266,80
ipr-mus TDLo:5046 µg/kg/15D-I:NEO CALEDQ 34,15,87
scu-mus TDLo:72 mg/kg/9W-I:ETA AICCA6 19,490,63

skn-mus TD:72 mg/kg/60W-I:ETA CANCAR 12,1194,59
imp-rat TD:5 mg/kg:ETA 50NNAZ 7,571,83
skn-mus TD:4037 μg/kg/20D-I:ETA CRNGDP 6,1023,85

CONSENSUS REPORTS: NTP 7th Annual Report On Carcinogens. IARC Cancer Review: Group 2B IMEMDT 7,56,87; Animal Sufficient Evidence IMEMDT 32,147,83; IMEMDT 3,69,73. EPA Genetic Toxicology Program.

ACGIH TLV: Suspected Carcinogen

SAFETY PROFILE: Confirmed carcinogen with experimental carcinogenic and tumorigenic data. Mutation data reported. When heated to decomposition it emits acrid smoke and irritating fumes.

For occupational chemical analysis use NIOSH: Polynuclear Aromatic Hydrocarbons (HPLC), 5506; (GC), 5515.

BAW500 CAS:71-79-4 HR: 3
BENZACINE HYDROCHLORIDE
mf: $C_{18}H_{21}O_3ClH$ mw: 403.28

PROP: Mp: 186–188°.

SYNS: BENZACIN □ BENZACINE □ BENZACIN HYDROCHLORIDE □ DIMETHYLAMINOETHYL BENZILATE, HYDROCHLORIDE □ β-DIMETHYLAMINOETHYL BENZILATE HYDROCHLORIDE □ 2-(DIMETHYLAMINO)ETHYL BENZILATE HYDROCHLORIDE □ DIMETHYLAMINOETHYL BENZYLATE HYDROCHLORIDE □ DIMETHYLAMINOETHYL DIPHENYLHYDROXYACETATE HYDROCHLORIDE □ HK-141

TOXICITY DATA WITH REFERENCE
orl-rat LD50:1035 mg/kg JLCMAK 30,700,45
ivn-rat LD50:30 mg/kg JLCMAK 30,700,45
orl-mus LD50:281 mg/kg JLCMAK 30,700,45
ipr-mus LD50:137 mg/kg PCJOAU 2,201,68
ivn-mus LD50:40 mg/kg JLCMAK 30,700,45

SAFETY PROFILE: Poison by ingestion, intravenous, and intraperitoneal routes. When heated to decomposition it emits toxic fumes of HCl.

BAW750 CAS:225-51-4 HR: 2
BENZ(c)ACRIDINE
mf: $C_{17}H_{11}N$ mw: 229.29

PROP: Brilliant-yellow needles from C_6H_6/pet ether. Mp: 108°.

SYNS: 12-AZABENZ(a)ANTHRACENE □ B(c)AC □ 3,4-BENZACRIDINE □ 7,8-BENZACRIDINE (FRENCH) □ 3,4-BENZOACRIDINE □ α-CHRYSIDINE □ α-NAPHTHACRIDINE □ RCRA WASTE NUMBER U016

TOXICITY DATA WITH REFERENCE
mma-sat 1 nmol/plate GANNA2 70,749,79
sce-ham:ovr 10 μmol/L MUREAV 118,103,83
sce-ham:lng 1 μmol/L MUREAV 118,103,83
skn-mus TDLo:2400 mg/kg/67W-I:ETA IJCAAR 5,183,68
ipr-mus TDLo:9630 mg/kg/3D-I:NEO CNREA8 44,5161,84

CONSENSUS REPORTS: IARC Cancer Review: Group 3 IMEMDT 7,56,87; Animal Sufficient Evidence IMEMDT 3,241,73; Animal Limited Evidence IMEMDT 32,129,83.

SAFETY PROFILE: Questionable carcinogen with experimental neoplastigenic and tumorigenic data. Mutation data reported. When heated to decomposition it emits toxic fumes of NO_x.

BAX000 CAS:3123-27-1 HR: 2
BENZ(c)ACRIDINE-7-CARBONITRILE
mf: $C_{18}H_{10}N_2$ mw: 254.30

SYNS: 7-CYANOBENZ(c)ACRIDINE □ 7-CYANOBENZO(c)ACRIDINE

TOXICITY DATA WITH REFERENCE
scu-mus TDLo:120 mg/kg/9W-I:ETA CHDDAT 267,981,68

CONSENSUS REPORTS: Cyanide and its compounds are on the Community Right-To-Know List.

SAFETY PROFILE: Questionable carcinogen with experimental tumorigenic data. See also NITRILES. When heated to decomposition it emits toxic fumes of NO_x and CN^-.

BAX250 CAS:3301-75-5 HR: 2
BENZ(c)ACRIDINE-7-CARBOXALDEHYDE
mf: $C_{18}H_{11}NO$ mw: 257.30

SYNS: 3,4-BENZACRIDINE-9-ALDEHYDE □ 7-FORMYLBENZ(c)ACRIDINE □ 7-FORMYLBENZO(c)ACRIDINE

TOXICITY DATA WITH REFERENCE
mma-sat 10 μg/plate CRNGDP 7,23,86
scu-mus TDLo:200 mg/kg:ETA VOONAW 1,52,55

SAFETY PROFILE: Questionable carcinogen with experimental tumorigenic data. Mutation data reported. When heated to decomposition it emits toxic fumes of NO_x. See also ALDEHYDES.

BAY250 CAS:63019-50-1 HR: 2
α-(BENZ(c)ACRIDIN-7-YL)-N-(p-(DIMETHYLAMINO) PHENYL)NITRONE
mf: $C_{26}H_{21}N_3O$ mw: 391.50

SYN: α-(9-(3,4-BENZACRIDYL))-N-(p-DIMETHYLAMINO-PHENYL)-NITRONE

TOXICITY DATA WITH REFERENCE
scu-mus TDLo:200 mg/kg:ETA VOONAW 1,52,55

SAFETY PROFILE: Questionable carcinogen with experimental tumorigenic data. When heated to decomposition it emits toxic fumes of NO_x.

BAY275 CAS:1896-62-4 HR: 3
trans-BENZALACETONE
mf: $C_{10}H_{10}O$ mw: 146.20

PROP: A liquid.

SYNS: trans-BENZYLIDENACETONE □ trans-BENZYLIDENEACETONE □ 3-BUTEN-2-ONE, 4-PHENYL-, (E)- □ METHYL trans-STYRYL KETONE □ trans-4-PHENYL-3-BUTENE-2-ONE □ TPBO

TOXICITY DATA WITH REFERENCE
mma-sat 300 μg/plate FCTOD7 20,427,82

DOT CLASSIFICATION: 3; *Label:* Flammable Liquid

SAFETY PROFILE: Mutation data reported. A flammable

liquid. When heated to decomposition it emits acrid smoke and irritating vapors.

BAY300 CAS:98-87-3 HR: 3
BENZAL CHLORIDE
DOT: UN 1886
mf: $C_7H_6Cl_2$ mw: 161.03

PROP: Very refractive liquid. Mp: −16°, bp: 214°, d: 1.29.

SYNS: BENZYL DICHLORIDE □ BENZYLENE CHLORIDE □ BENZYLIDENE CHLORIDE □ BENZYLIDENE CHLORIDE (DOT) □ CHLOROBENZAL □ CHLORURE DE BENZYLIDENE □ (DICHLOROMETHYL)BENZENE □ α-α-DICHLOROTOLUENE □ RCRA WASTE NUMBER U017 □ TOLUENE, α-α-DICHLORO-

TOXICITY DATA WITH REFERENCE
mma-sat 600 nmol/plate/20M MUREAV 54,143,78
mma-esc 600 nmol/plate/20M MUREAV 54,143,78
mrc-bcs 31 μmol/disc MUREAV 54,143,78
skn-mus TDLo:9200 mg/kg/50W-I:CAR GANNA2 72,655,81
skn-mus TD:35,200 mg/kg/42W-I:NEO GANNA2 72,655,81
orl-rat LD50:3249 mg/kg NTIS** PB214-270
ihl-rat LD50:61 ppm/2H IARC** 29,65,82
orl-mus LD50:2462 mg/kg AMRL** TR-72-62/72
ihl-mus LD50:32 ppm/2H IARC** 29,65,82

CONSENSUS REPORTS: IARC Cancer Review: Human Inadequate Evidence IMEMDT 29,65,82; Animal Limited Evidence IMEMDT 29,65,82. Reported in EPA TSCA Inventory. EPA Genetic Toxicology Program. EPA Extremely Hazardous Substances List. Community Right-To-Know List.

DFG MAK: Confirmed Human Carcinogen.
DOT CLASSIFICATION: 6.1; *Label:* Poison

SAFETY PROFILE: Confirmed carcinogen with experimental carcinogenic and neoplastigenic data. Poison by inhalation. Moderately toxic by ingestion. A strong irritant and lachrymator. Causes central nervous system depression. Mutation data reported. When heated to decomposition it emits toxic fumes of Cl^-. See also CHLORINATED HYDROCARBONS, AROMATIC.

BAY500 CAS:100-52-7 HR: 3
BENZALDEHYDE
mf: C_7H_6O mw: 106.13

PROP: Colorless liquid; burning taste with bitter almond odor. Mp: −26°, bp: 179°, fp: −56.9° (to −55°), flash p: 148°F, d: 1.041, autoign temp: 377°F, vap press: 1 mm @ 26.2°, vap d: 3.65, refr index: 1.544. Sltly sol in water; misc in alc, ether, oils.

SYNS: ALMOND ARTIFICIAL ESSENTIAL OIL □ ARTIFICIAL ALMOND OIL □ BENZENECARBALDEHYDE □ BENZENECARBONAL □ BENZOIC ALDEHYDE □ FEMA No. 2127 □ NCI-C56133

TOXICITY DATA WITH REFERENCE
skn-rbt 500 mg/24H MOD FCTXAV 14,659,76
sce-hmn:lym 1 mmol/L MUREAV 206,17,88
slt-mus:lym 400 mg/L EMMUEG 17,196,91
orl-mus TDLo:154 g/kg/2Y-C:NEO NTPTR* NTP-TR-378,90
orl-rat LD50:1300 mg/kg FCTXAV 2,327,64
scu-rat LDLo:5000 mg/kg AIPTAK 27,163,22
orl-mus LD50:28 mg/kg EJTXAZ 9,99,76
ipr-mus LD50:9 mg/kg EJTXAZ 9,99,76
scu-rbt LD50:5000 mg/kg FCTXAV 14,693,76
orl-gpg LD50:1000 mg/kg FCTXAV 2,327,64

CONSENSUS REPORTS: NTP Carcinogenesis Studies (gavage): Some Evidence: mouse; NTP-TR-378,90; No Evidence: rat NTP-TR-378,90. EPA Genetic Toxicology Program. Reported in EPA TSCA Inventory.

SAFETY PROFILE: Poison by ingestion and intraperitoneal routes. Moderately toxic by subcutaneous route. An allergen. Acts as a feeble local anesthetic. Local contact may cause contact dermatitis. Causes central nervous system depression in small doses and convulsions in larger doses. A skin irritant. Questionable carcinogen with experimental tumorigenic data. Mutation data reported. Combustible liquid. To fight fire, use water (may be used as a blanket), alcohol, foam, dry chemical. A strong reducing agent. Reacts violently with peroxyformic acid and other oxidizers. See also ALDEHYDES.

BAY750 CAS:633-03-4 HR: 3
BENZALDEHYDE GREEN
mf: $C_{27}H_{33}N_2 \cdot HO_4S$ mw: 482.69

PROP: Bright green crystals or powder. Mp: 210° (decomp). Sol in H_2O, EtOH, and $CHCl_3$.

SYNS: ADC BRILLIANT GREEN CRYSTALS □ AIZEN DIAMOND GREEN GH □ ANILINE GREEN □ ASTRA DIAMOND GREEN GX □ AVON GREEN A-4379 □ BASIC BRIGHT GREEN □ BRILLIANT GREEN SULFATE □ CALCOZINE BRILLIANT GREEN G □ C.I. 42040 □ C.I. BASIC GREEN 1, SULFATE (1:1) □ DEORLENE GREEN JJO □ DIAMOND GREEN G □ EMERALD GREEN □ ETHYL GREEN □ FAST GREEN JJO □ HIDACO BRILLIANT GREEN □ MALACHITE GREEN G □ MITSUI BRILLIANT GREEN G □ TERTROPHENE BRILLIANT GREEN G □ TOKYO ANILINE BRILLIANT GREEN

TOXICITY DATA WITH REFERENCE
skn-hmn 2 mg/2D-I MLD ADVEA4 52,55,72
skn-gpg 6 mg/3D-I ADVEA4 52,55,72
mmo-smc 100 μg/L VINIT* #542-84
orl-rat LDLo:10 mg/kg GTPZAB 7(2),54,63
ipr-rat LDLo:8 mg/kg PSEBAA 31,825,34
ipr-mus LDLo:5 mg/kg PSEBAA 31,825,34
ivn-mus LDLo:3 mg/kg PSEBAA 31,825,34
ipr-gpg LDLo:3 mg/kg PSEBAA 31,825,34

CONSENSUS REPORTS: Reported in EPA TSCA Inventory.

SAFETY PROFILE: Poison by ingestion, intraperitoneal, and intravenous routes. A mild human skin irritant. Mutation data reported. See also ALDEHYDES and SULFATES. When heated to decomposition it emits very toxic fumes of NO_x, NH_3, and SO_x.

BBA000 CAS:1708-39-0 HR: 2
BENZAL GLYCERYL ACETAL
mf: $C_{10}H_{12}O_3$ mw: 180.22

PROP: Colorless to pale-yellow liquid; mild almond odor. D: 1.183–1.193, refr index: 1.535–1.541, flash p: 165°F.

SYNS: BENZALDEHYDE GLYCERYL ACETAL (FCC) □ BENZYLIDENE GLYCEROL □ BUTYL PHENYL ACETATE □ FEMA No. 2209 □ 2-PHENYL-m-DIOXAN-5-OL

TOXICITY DATA WITH REFERENCE
orl-rat LD50:3150 mg/kg FCTXAV 14,699,76
ipr-mus LD50:1296 mg/kg AIPTAK 85,474,51
skn-rbt LD50:5000 mg/kg FCTXAV 14,699,76

SAFETY PROFILE: Moderately toxic by ingestion and intraperitoneal routes. Mildly toxic by skin contact. Combustible liquid. When heated to decomposition it emits acrid smoke and irritating fumes.

BBA500 HR: 3
BENZALKONIUM CHLORIDE

PROP: White or yellowish-white powder, aromatic odor, very bitter taste.

SYNS: ALKYLDIMETHYLETHYLBENZYL AMMONIUM CHLORIDE □ ALKYL((ETHYLPHENYL)METHYL)DIMETHYL QUATERNARY AMMONIUM CHLORIDES □ BENIROL □ BTC 471 □ CEQUARTYL □ DRAPOLEX □ ENUCLEN □ GERMINOL □ GERMITOL □ OCTYL-OCTADECYL DIMETHYL ETHYLBENZYL AMMONIUM CHLORIDES □ PARALKAN □ ROCCAL □ RODALON □ ZEPHIRAN CHLORIDE □ ZEPHIROL

TOXICITY DATA WITH REFERENCE
eye-rat 2 mg FCTXAV 15,131,77
eye-mus 2 mg SEV FCTXAV 15,131,77
eye-dog 2 mg FCTXAV 15,131,77
eye-rbt mg SEV FCTXAV 15,131,77
eye-rbt 8 µg SEV AJOPAA 78,98,74
eye-rbt 10 mg MLD TXAPA9 55,501,80
eye-gpg 2 mg FCTXAV 15,131,77
eye-ham 2 mg SEV FCTXAV 15,131,77
orl-rat LD50:300 mg/kg 28ZEAL 4,38,69
skn-rat LD50:1420 mg/kg PCJOAU 12,1593,78
orl-mus LD50:150 mg/kg PCJOAU 12,1593,78

SAFETY PROFILE: Poison by ingestion. Moderately toxic by skin contact. A severe eye irritant. A bactericide and fungicide. Dangerous; when heated to decomposition it emits toxic fumes of Cl^- and NO_x. See also CHLORIDES.

BBA625 CAS:39387-42-3 HR: 3
BENZALKONIUM SACCHARINATE

SYNS: AKYL DIMETHYL BENZYL AMMONIUM SACCHARINATE □ ALKYL DIMETHYL BENZALKONIUM SACCHARINATE □ HOLLICHEM HQ 3300 □ ONYXIDE 3300

TOXICITY DATA WITH REFERENCE
orl-rat LD50:990 mg/kg KSRNAM 4,219,70
ipr-rat LD50:37 mg/kg KSRNAM 4,219,70
scu-rat LD50:720 mg/kg KSRNAM 4,219,70
ivn-rat LD50:14,500 µg/kg KSRNAM 4,219,70
orl-mus LD50:920 mg/kg KSRNAM 4,219,70
ipr-mus LD50:33 mg/kg KSRNAM 4,219,70
scu-mus LD50:790 mg/kg KSRNAM 4,219,70
ivn-mus LD50:23 mg/kg KSRNAM 4,219,70

SAFETY PROFILE: Poison by intravenous and intraperitoneal routes. Moderately toxic by ingestion and subcutaneous routes. When heated to decomposition it emits toxic fumes of NO_x.

BBA750 HR: 3
BENZALMALONONITRILE
mf: $C_6H_5CH_2CH(CN)_2$ mw: 156.2

CONSENSUS REPORTS: Cyanide and its compounds are on the Community Right-To-Know List.

SAFETY PROFILE: Poison. See also NITRILES and CYANIDE.

BBB000 CAS:55-21-0 HR: 2
BENZAMIDE
mf: C_7H_7NO mw: 121.15

PROP: Plates from H_2O. Mp: 130°. Bp: 288° (sltly decomp). Sltly sol in H_2O, Et_2O; sol in EtOH.

SYNS: BENZOIC ACID AMIDE □ BENZOYLAMIDE □ PHENYLCARBOXYAMIDE

TOXICITY DATA WITH REFERENCE
sce-hmn:lym 1 mmol/L MUREAV 122,223,83
sce-ham:ovr 1 mmol/L MUREAV 123,63,84
mnt-mam:kdy 1 g/L IJEBA6 18,329,80
cyt-mam:kdy 1 g/L IJEBA6 18,329,80
ipr-rat LD50:781 mg/kg APFRAD 48,23,90
orl-mus LD50:1160 mg/kg TXAPA9 19,20,71.

CONSENSUS REPORTS: Reported in EPA TSCA Inventory. Community Right-To-Know List. Human mutation data reported.

SAFETY PROFILE: Moderately toxic by ingestion and intraperitoneal route. When heated to decomposition it emits toxic fumes of NO_x. See also AMIDES.

BBB500 CAS:63018-69-9 HR: 2
BENZ(a)ANTHRACEN-7-ACETONITRILE
mf: $C_{20}H_{13}N$ mw: 267.34

SYN: 10-CYANOMETHYL-1,2-BENZANTHRACENE

TOXICITY DATA WITH REFERENCE
scu-mus TDLo:600 mg/kg:ETA JNCIAM 1,303,40

CONSENSUS REPORTS: Cyanide and its compounds are on the Community Right-To-Know List.

SAFETY PROFILE: Questionable carcinogen with experimental tumorigenic data. See also NITRILES. When heated to decomposition it emits toxic fumes of NO_x and CN^-.

BBB750 CAS:2381-18-2 HR: 2
BENZ(a)ANTHRACEN-7-AMINE
mf: $C_{18}H_{13}N$ mw: 243.32

PROP: Yellow leaflets or needles. Mp: 174.5–175.5°.

SYN: 10-AMINO-1,2-BENZANTHRACENE

TOXICITY DATA WITH REFERENCE
scu-mus TDLo:1500 mg/kg/23W-I:ETA PRLBA4 129,439,40

SAFETY PROFILE: Questionable carcinogen with experimental tumorigenic data. When heated to decomposition it emits toxic fumes of NO_x. See also AROMATIC AMINES.

BBC000 CAS:56961-60-5 HR: 2
BENZ(a)ANTHRACEN-8-AMINE
mf: $C_{18}H_{13}N$ mw: 243.32

SYN: 5-AMINO-1:2-BENZANTHRACENE

TOXICITY DATA WITH REFERENCE
scu-mus TDLo:400 mg/kg/3W-I:ETA PRLBA4 131,170,42

SAFETY PROFILE: Questionable carcinogen with experimental tumorigenic data. When heated to decomposition it emits toxic fumes of NO_x. See also AROMATIC AMINES.

BBC250 CAS:56-55-3 HR: 3
BENZ(a)ANTHRACENE
mf: $C_{18}H_{12}$ mw: 228.30

PROP: Colorless leaflets or plates from EtOH/AcOH. Mp: 160°, bp: 400°.

SYNS: BA □ BENZANTHRACENE □ 1,2-BENZANTHRACENE □ 1,2-BENZ(a)ANTHRACENE □ 1,2-BENZANTHRAZEN (GERMAN) □ BENZANTHRENE □ 1,2-BENZANTHRENE □ BENZOANTHRACENE □ BENZO(a)ANTHRACENE □ 1,2-BENZOANTHRACENE □ BENZO(a)PHENANTHRENE □ BENZO(b)PHENANTHRENE □ 2,3-BENZOPHENANTHRENE □ 2,3-BENZPHENANTHRENE □ NAPHTHANTHRACENE □ RCRA WASTE NUMBER U018 □ TETRAPHENE

TOXICITY DATA WITH REFERENCE
mma-sat 4 μg/plate CRNGDP 5,747,84
msc-hmn:lym 9 μmol/L DTESD7 10,277,82
dni-hmn:oth 10 μmol/L CNREA8 42,3676,82
dnd-mus-skn 192 μmol/kg CRNGDP 5,231,84
skn-mus TDLo:18 mg/kg:NEO CNREA8 38,1699,78
scu-mus TDLo:2 mg/kg:ETA CNREA8 15,632,55
imp-mus TDLo:80 mg/kg:CAR BJCAAI 22,825,68
skn-mus TD:18 mg/kg:ETA CNREA8 38,1705,78
skn-mus TD:360 mg/kg/56W-I:ETA CNREA8 11,892,51
skn-mus TD:240 mg/kg/1W-I:NEO BJCAAI 9,177,55
ivn-mus LDLo:10 mg/kg JNCIAM 1,225,40

CONSENSUS REPORTS: NTP 7th Annual Report On Carcinogens. IARC Cancer Review: Group 2A IMEMDT 7,56,87; Animal Sufficient Evidence IMEMDT 32,135,83; IMEMDT 3,45,73. EPA Genetic Toxicology Program. Reported in EPA TSCA Inventory.

ACGIH TLV: (Proposed: Suspected Human Carcinogen)

SAFETY PROFILE: Confirmed carcinogen with experimental carcinogenic, neoplastigenic, tumorigenic data by skin contact and other routes. Poison by intravenous route. Human mutation data reported. It is found in oils, waxes, smoke, food, drugs. When heated to decomposition it emits acrid smoke and irritating fumes.

For occupational chemical analysis use NIOSH: Polynuclear Aromatic Hydrocarbons (HPLC), 5506; (GC), 5515.

BBC500 CAS:63018-40-6 HR: 2
1,2-BENZANTHRACENE-10-ACETIC ACID, METHYL ESTER
mf: $C_{21}H_{16}O_2$ mw: 300.37

SYN: BENZ(a)ANTHRACEN-7-ACETIC ACID, METHYL ESTER

TOXICITY DATA WITH REFERENCE
scu-mus TDLo:600 mg/kg:ETA JNCIAM 1,303,40

SAFETY PROFILE: Questionable carcinogen with experimental tumorigenic data. When heated to decomposition it emits acrid smoke and fumes. See also ESTERS.

BBC750 CAS:7505-62-6 HR: 1
BENZ(a)ANTHRACENE-7-CARBOXALDEHYDE
mf: $C_{19}H_{12}O$ mw: 256.31

SYN: 1,2-BENZANTHRACENE-10-ALDEHYDE

TOXICITY DATA WITH REFERENCE
mmo-sat 1 nmol/plate DCTODJ 2,383,79
scu-mus TDLo:280 mg/kg:ETA JNCIAM 1,303,40

SAFETY PROFILE: Questionable carcinogen with experimental tumorigenic data. Mutation data reported. When heated to decomposition it emits acrid smoke and fumes. See also ALDEHYDES.

BBD000 CAS:19926-22-8 HR: 2
BENZ(a)ANTHRACENE-7,12-DICARBOXALDEHYDE
mf: $C_{20}H_{12}O_2$ mw: 284.32

SYN: 7,12-DIFORMYLBENZ(a)ANTHRACENE

TOXICITY DATA WITH REFERENCE
dnd-omi 2 mg/L PNASA6 74,1378,77
skn-mus TDLo:8000 mg/kg:NEO JJIND8 61,135,78

SAFETY PROFILE: Questionable carcinogen with experimental neoplastigenic data by skin contact. Mutation data reported. When heated to decomposition it emits acrid smoke and irritating fumes. See also ALDEHYDES.

BBD250 CAS:60967-88-6 HR: 2
BENZ(a)ANTHRACENE-1,2-DIHYDRODIOL
mf: $C_{18}H_{14}O_2$ mw: 262.32

SYNS: BA-1,2-DIHYDRODIOL □ trans-1,2-DIHYDROXY-1,2-DIHYDROBENZ(a)ANTHRACENE

TOXICITY DATA WITH REFERENCE
mma-sat 10 μmol/L CNREA8 42,1620,82
msc-ham:lng 1200 μg/L/3H BJCAAI 39,540,79
skn-mus TDLo:2100 μg/kg:ETA CNREA8 38,1699,78
skn-mus TD:10 mg/kg:ETA PNASA6 74,3176,77

CONSENSUS REPORTS: EPA Genetic Toxicology Program.

SAFETY PROFILE: Questionable carcinogen with experimental tumorigenic data by skin contact. Mutation data reported. When heated to decomposition it emits acrid smoke and irritating fumes.

BBD500 **CAS:60967-89-7** **HR: 2**
BENZ(a)ANTHRACENE-3,4-DIHYDRODIOL
mf: $C_{18}H_{14}O_2$ mw: 262.32

SYNS: BA-3,4-DIHYDRODIOL □ trans-3,4-DIHYDRO-3,4-DIHYDROXYBENZO(a)ANTHRACENE □ trans-3,4-DIHYDROXY-3,4-DIHYDROBENZ(a)ANTHRACENE

TOXICITY DATA WITH REFERENCE
mma-sat 25 µmol/L BBRCA9 72,680,76
msc-ham:lng 2500 µg/L/3H BJCAAI 39,540,79
skn-mus TDLo:2100 µg/kg:NEO CNREA8 38,1699,79
skn-mus TD:4200 µg/kg:NEO PNASA6 74,3176,77
skn-mus TD:4200 µg/kg:NEO CNREA8 38,1705,78

CONSENSUS REPORTS: EPA Genetic Toxicology Program.

SAFETY PROFILE: Questionable carcinogen with experimental neoplastigenic data by skin contact. Mutation data reported. When heated to decomposition it emits acrid smoke and irritating fumes.

BBD750 **CAS:67335-43-7** **HR: 2**
(+)-(3S,4S)trans-BENZ(a)ANTHRACENE-3,4-DIHYDRODIOL
mf: $C_{18}H_{10}O_3$ mw: 274.28

SYNS: (+)-(3S,4S)-trans-3,4-DIHYDRO-3,4-DIHYDROXYBENZ(a)ANTHRACENE □ (+)-(3S,4S)-trans-3,4-DIHYDRO-3,4-DIHYDROXYBENZO(a)ANTHRACENE

TOXICITY DATA WITH REFERENCE
skn-mus TDLo:4390 µg/kg:ETA CNREA8 38,1705,78

SAFETY PROFILE: Questionable carcinogen with experimental tumorigenic data by skin contact. When heated to decomposition it emits acrid smoke and irritating fumes.

BBE250 **CAS:3719-37-7** **HR: 2**
BENZ(a)ANTHRACENE-5,6-DIHYDRODIOL
mf: $C_{18}H_{14}O_2$ mw: 262.32

SYNS: BA-5,6-DIHYDRODIOL □ BA-5,6-trans-DIHYDRODIOL □ BENZ(a)ANTHRACENE-5,6-trans-DIHYDRODIOL □ trans-5,6-DIHYDROXY-5,6-DIHYDROBENZ(a)ANTHRACENE

TOXICITY DATA WITH REFERENCE
otr-ham:emb 4 mg/L CNREA8 32,1391,72
skn-mus TDLo:21 mg/kg:NEO PNASA6 74,3176,77
skn-mus TD:2100 µg/kg:ETA CNREA8 38,1699,78

CONSENSUS REPORTS: EPA Genetic Toxicology Program.

SAFETY PROFILE: Questionable carcinogen with experimental tumorigenic and neoplastigenic data by skin contact. Mutation data reported. When heated to decomposition it emits acrid smoke and irritating fumes.

BBE750 **CAS:34501-24-1** **HR: 2**
trans-BENZ(a)ANTHRACENE-8,9-DIHYDRODIOL
mf: $C_{18}H_{14}O_2$ mw: 262.32

SYNS: BA-8,9-DIHYDRODIOL □ trans-8,9-DIHYDROXY-8,9-DIHYDROBENZ(a)ANTHRACENE

TOXICITY DATA WITH REFERENCE
mma-sat 25 µmol/L BBRCA9 72,680,76
skn-mus TDLo:2100 µg/kg:NEO CNREA8 38,1699,78
skn-mus TD:4200 µg/kg:ETA PNASA6 74,3176,77

CONSENSUS REPORTS: EPA Genetic Toxicology Program.

SAFETY PROFILE: Questionable carcinogen with experimental tumorigenic and neoplastigenic data by skin contact. Mutation data reported. When heated to decomposition it emits acrid smoke and irritating fumes.

BBF000 **CAS:60967-90-0** **HR: 2**
BENZ(a)ANTHRACENE-10,11-DIHYDRODIOL
mf: $C_{18}H_{14}O_2$ mw: 262.32

SYNS: BA-10,11-DIHYDRODIOL □ trans-10,11-DIHYDROXY-10,11-DIHYDROBENZ(a)ANTHRACENE

TOXICITY DATA WITH REFERENCE
mma-sat 100 µmol/L CNREA8 42,1620,82
skn-mus TDLo:2100 µg/kg:ETA PNASA6 74,3176,77
skn-mus TD:4200 µg/kg:ETA CNREA8 38,1699,78

CONSENSUS REPORTS: EPA Genetic Toxicology Program.

SAFETY PROFILE: Questionable carcinogen with experimental tumorigenic data by skin contact. When heated to decomposition it emits acrid smoke and irritating fumes.

BBF500 **CAS:2564-65-0** **HR: 2**
BENZ(a)ANTHRACENE-7,12-DIMETHANOL
mf: $C_{20}H_{16}O_2$ mw: 288.36

SYNS: 9:10-BISHYDROXYMETHYL-1:2-BENZANTHRACENE □ 7:12-DIHYDROXYMETHYLBENZ(a)ANTHRACENE

TOXICITY DATA WITH REFERENCE
mma-sat 20 nmol/plate 46OJAN -,675,81
mmo-esc 1 g/L/2H GENTAE 39,141,54
scu-mus TDLo:2600 mg/kg/40W-I:ETA PRLBA4 129,439,40

CONSENSUS REPORTS: EPA Genetic Toxicology Program.

SAFETY PROFILE: Questionable carcinogen with experimental tumorigenic data. Mutation data reported. When heated to decomposition it emits acrid smoke and irritating fumes.

BBF750 **CAS:63018-62-2** **HR: 2**
BENZ(a)ANTHRACENE-7,12-DIMETHANOLDIACETATE
mf: $C_{24}H_{20}O_4$ mw: 372.44

SYNS: ACETIC ACID, BENZ(a)ANTHRACENE-7,12-DIMETHANOL DIESTER □ 9,10-BISACETOXYMETHYL-1,2-BENZANTHRACENE

TOXICITY DATA WITH REFERENCE
skn-mus TDLo:1700 mg/kg/71W-I:ETA PRLBA4 129,439,40
scu-mus TDLo:3000 mg/kg/45W-I:ETA PRLBA4 129,439,40

SAFETY PROFILE: Questionable carcinogen with experimental tumorigenic data. See also ESTERS. When heated to decomposition it emits acrid smoke and irritating fumes.

BBG000 CAS:67335-42-6 HR: 2
(−)(3R,4R)-trans-BENZ(a)ANTHRACENE-3,4-DIOL
mf: $C_{18}H_{10}O_3$ mw: 274.28

SYNS: (−)(3R,4R)-trans-3,4-DIHYDRO-3,4-DIHYDROXYBENZ(a)ANTHRACENE □ (−)(3R,4R)trans-3,4-DIHYDRO-3,4-DIHYDROXYBENZO(a)ANTHRACENE

TOXICITY DATA WITH REFERENCE
skn-mus TDLo: 1100 μg/kg: NEO CNREA8 38,1705,78

SAFETY PROFILE: Questionable carcinogen with experimental neoplastigenic data by skin contact. When heated to decomposition it emits acrid smoke and irritating fumes.

BBG500 CAS:63020-45-1 HR: 2
BENZ(a)ANTHRACENE-7-ETHANOL
mf: $C_{20}H_{16}O$ mw: 272.36

SYN: 10-β-HYDROXYETHYL-1: 2-BENZANTHRACENE

TOXICITY DATA WITH REFERENCE
skn-mus TDLo: 1220 mg/kg/51W-I: ETA PRLBA4 131,170,42
scu-mus TDLo: 1700 mg/kg/26W-I: ETA PRLBA4 129,439,40

SAFETY PROFILE: Questionable carcinogen with experimental tumorigenic data. When heated to decomposition it emits acrid smoke and irritating fumes.

BBG750 CAS:17012-91-8 HR: 2
BENZ(a)ANTHRACENE-7-METHANEDIOLDIACETATE (ester)
mf: $C_{23}H_{18}O_4$ mw: 358.41

SYN: 7-DIACETOXYMETHYLBENZ(a)ANTHRACENE

TOXICITY DATA WITH REFERENCE
scu-mus TDLo: 120 mg/kg/6W-I: CAR IJCNAW 2,500,67

SAFETY PROFILE: Questionable carcinogen with experimental carcinogenic data. See also ESTERS. When heated to decomposition it emits acrid smoke and irritating fumes.

BBH000 CAS:63018-59-7 HR: 2
BENZ(a)ANTHRACENE-7-METHANETHIOL
mf: $C_{19}H_{14}S$ mw: 274.39

SYN: 1,2-BENZANTHRYL-10-METHYLMERCAPTAN

TOXICITY DATA WITH REFERENCE
scu-mus TDLo: 80 mg/kg: ETA CNREA8 6,454,46

SAFETY PROFILE: Questionable carcinogen with experimental tumorigenic data. When heated to decomposition it emits toxic fumes of SO_x. See also MERCAPTANS.

BBH250 CAS:16110-13-7 HR: 2
BENZ(a)ANTHRACENE-7-METHANOL
mf: $C_{19}H_{14}O$ mw: 258.33

SYNS: 7-HMBA □ 7-HYDROXYMETHYLBENZ(a)ANTHRACENE □ 10-HYDROXYMETHYL-1,2-BENZANTHRACENE

TOXICITY DATA WITH REFERENCE
dnd-omi 30 μmol/L CBINA8 31,51,80
otr-mus: oth 100 μg/L IJCNAW 13,304,74
dnd-mus: emb 800 μg/L CNREA8 33,2386,73
dnd-mam: lym 30 μmol/L CBINA8 31,51,80
skn-mus TDLo: 56 mg/kg/60W-I: ETA CNREA8 43,2034,83
scu-mus TDLo: 1000 mg/kg/16W-I: ETA PRLBA4 129,439,40
skn-mus TDLo: 700 mg/kg/29W-I: ETA PRLBA4 129,439,40

CONSENSUS REPORTS: EPA Genetic Toxicology Program.

SAFETY PROFILE: Questionable carcinogen with experimental tumorigenic data. Mutation data reported. When heated to decomposition it emits acrid smoke and irritating fumes.

BBH500 CAS:17526-24-8 HR: 2
BENZ(a)ANTHRACENE-7-METHANOL ACETATE
mf: $C_{21}H_{16}O_2$ mw: 300.37

SYNS: ACETIC ACID, BENZ(a)ANTHRACENE-7-METHANOL ESTER □ 10-ACETOXYMETHYL-1,2-BENZANTHRACENE

TOXICITY DATA WITH REFERENCE
mmo-sat 1 nmol/plate DCTODJ 2,383,79
skn-mus TDLo: 16 mg/kg/17W-I: ETA VOONAW 21(10),50,75
scu-mus TDLo: 1200 mg/kg/17W-I: ETA PRLBA4 129,439,40
skn-mus TD: 744 mg/kg/31W-I: ETA PRLBA4 129,439,40

SAFETY PROFILE: Questionable carcinogen with experimental tumorigenic data. Mutation data reported. See also ESTERS. When heated to decomposition it emits acrid smoke and irritating fumes.

BBH750 CAS:63018-57-5 HR: 2
BENZ(a)ANTHRACENE-7-THIOL
mf: $C_{18}H_{12}S$ mw: 260.36

SYNS: 1,2-BENZANTHRYL-10-MERCAPTAN □ 7-MERCAPTOBENZ(a)ANTHRACENE

TOXICITY DATA WITH REFERENCE
scu-mus TDLo: 80 mg/kg: ETA CNREA8 6,454,46

SAFETY PROFILE: Questionable carcinogen with experimental tumorigenic data. When heated to decomposition it emits toxic fumes of SO_x. See also MERCAPTANS.

BBI000 CAS:960-92-9 HR: 2
BENZ(a)ANTHRACEN-5-OL
mf: $C_{18}H_{12}O$ mw: 244.30

PROP: Golden crystals from toluene. Mp: 202–204° (decomp).

SYNS: 3-HYDROXY-1,2-BENZANTHRACENE □ 5-HYDROXYBENZ(a) ANTHRACENE

TOXICITY DATA WITH REFERENCE
dnd-ham:kdy 5 mg/L BCPCA6 20,1297,71
dnd-ham:lng 1 mg/L CBINA8 4,389,71/72
scu-mus TDLo:1240 mg/kg:ETA JNCIAM 1,303,40

CONSENSUS REPORTS: EPA Genetic Toxicology Program.

SAFETY PROFILE: Questionable carcinogen with experimental tumorigenic data. Mutation data reported. When heated to decomposition it emits acrid smoke and irritating fumes.

BBI250 CAS:82-05-3 HR: 3
7H-BENZ(de)ANTHRACEN-7-ONE
mf: $C_{17}H_{10}O$ mw: 230.27

PROP: Pale yellow needles from xylene or EtOH. Mp: 174°, vap press: 1 mm @ 225.0°.

SYNS: 7H-BENZ(de)ANTHRACENE-7-ONE □ BENZANTHRENONE □ BENZANTHRONE □ 7H-BENZO(de)ANTHRACEN-7-ONE □ BENZOANTHRONE □ MS-BENZANTHRONE □ NAPHTHANTHRONE □ 7-OXOBENZ(de)ANTHRACENE

TOXICITY DATA WITH REFERENCE
skn-rbt 500 mg/24H MLD 28ZPAK -,60,72
eye-rbt 100 mg/24H MOD 28ZPAK -,60,72
ipr-rat LD50:1500 mg/kg RPTOAN 40,137,77
ipr-mus LD50:290 mg/kg RPTOAN 40,137,77

CONSENSUS REPORTS: Reported in EPA TSCA Inventory.

SAFETY PROFILE: Poison by intraperitoneal route. Skin and eye irritant. Combustible when heated. Incompatible with nitrobenzene and potassium hydroxide. When heated to decomposition it emits acrid smoke and irritating fumes.

BBI750 CAS:63018-49-5 HR: 2
1,2-BENZANTHRYL-3-CARBAMIDOACETIC ACID
mf: $C_{21}H_{16}N_2O_3$ mw: 344.39

SYN: N-(BENZ(a)ANTHRACEN-5-YLCARBAMOYL)GLYCINE

TOXICITY DATA WITH REFERENCE
scu-mus TDLo:120 mg/kg:ETA CNREA8 6,454,46

SAFETY PROFILE: Questionable carcinogen with experimental tumorigenic data. When heated to decomposition it emits toxic fumes such as NO_x.

BBJ000 CAS:63018-50-8 HR: 2
1,2-BENZANTHRYL-10-CARBAMIDOACETIC ACID
mf: $C_{21}H_{16}N_2O_3$ mw: 344.39

SYN: N-(BENZ(a)ANTHRACEN-7-YLCARBAMOYL)GLYCINE

TOXICITY DATA WITH REFERENCE
scu-mus TDLo:160 mg/kg:ETA CNREA8 6,454,46

SAFETY PROFILE: Questionable carcinogen with experimental tumorigenic data. When heated to decomposition it emits toxic fumes of NO_x.

BBJ250 CAS:63018-56-4 HR: 2
1,2-BENZANTHRYL-10-ISOCYANATE
mf: $C_{19}H_{11}NO$ mw: 269.31

SYN: ISOCYANIC ACID, BENZ(a)ANTHRACEN-7-YL ESTER

TOXICITY DATA WITH REFERENCE
scu-mus TDLo:40 mg/kg:ETA CNREA8 6,454,46

SAFETY PROFILE: Questionable carcinogen with experimental tumorigenic data. See also ESTERS. When heated to decomposition it emits toxic fumes of NO_x.

BBJ500 CAS:1477-19-6 HR: 3
BENZARONE
mf: $C_{17}H_{14}O_3$ mw: 266.31

PROP: Solid. Mp: 126–127°.

SYNS: BENZOFURAN, (2-ETHYL-3-(4'-HYDROXYBENZOYL)) □ 2-ETHYL-3-BENZOFURANYL p-HYDROXYPHENYL KETONE □ 2-ETHYL-3-(p-HYDROXYBENZOYL)BENZOFURAN □ 2-ETHYL-4'-HYDROXY-3-BENZOYLBENZOFURAN □ ETHYL-2 (HYDROXY-4 BENZOYL)-3 BENZOFURANNE □ FRAGIVIX

TOXICITY DATA WITH REFERENCE
orl-mus TDLo:600 mg/kg (7-12D preg):REP KSRNAM 3,961,69
orl-rat TDLo:6 g/kg (female 9-14D post):TER KSRNAM 3,961,69
orl-rat TDLo:1020 mg/kg (9-14D preg):REP KSRNAM 3,961,69
orl-rat TDLo:1020 mg/kg (9-14D preg):TER KSRNAM 3,961,69
ipr-mus LD50:200 mg/kg AIPTAK 154,94,65

CONSENSUS REPORTS: Reported in EPA TSCA Inventory.

DOT CLASSIFICATION: 3; *Label:* Flammable Liquid

SAFETY PROFILE: Poison by intraperitoneal route. An experimental teratogen. Other experimental reproductive effects. A flammable liquid. When heated to decomposition it emits acrid and irritating smoke and fumes. See also KETONES.

BBJ750 CAS:59-97-2 HR: 3
BENZAZOLINE HYDROCHLORIDE
mf: $C_{10}H_{12}N_2 \cdot ClH$ mw: 196.70

PROP: Solid. Mp: 171–172°.

SYNS: ARTERODY □ BENZYLIMIDAZOLINE HYDROCHLORIDE □ 2-BENZYL-2-IMIDAZOLINE MONOHYDROCHLORIDE □ IMIDALINE HYDROCHLORIDE □ PRISCOL □ PRISCOLINE HYDROCHLORIDE □ TOLAVAD □ TOLAZOLINE CHLORIDE □ TOLAZOLINE HYDROCHLORIDE □ TOLPAL

TOXICITY DATA WITH REFERENCE
ivn-inf TDLo:48 mg/kg/47H-C:GIT AUPJB7 22,221,86
ivn-hmn TDLo:150 µg/kg:CVS,SKN FOMDAK 27,729,41
orl-rat LD50:1200 mg/kg NIIRDN 6,511,82
ipr-rat LD50:100 mg/kg NIIRDN 6,511,82
ivn-rat LD50:85 mg/kg NIIRDN 6,511,82
orl-mus LD50:400 mg/kg ARZNAD 21,1992,71
ipr-mus LD50:130 mg/kg ARZNAD 21,1992,71
ivn-mus LD50:60 mg/kg CLDND*

B

CONSENSUS REPORTS: Reported in EPA TSCA Inventory.

SAFETY PROFILE: Poison by ingestion, intravenous, and intraperitoneal routes. Human systemic effects by intravenous route: change in heart rate, sweating, ulceration or bleeding from duodeum, ulceration or bleeding from small intestine, unspecified vascular effects. When heated to decomposition it emits very toxic fumes of NO_x and HCl.

BBK000 **CAS:300-62-9** **HR: 3**
BENZEDRINE
mf: $C_9H_{13}N$ mw: 135.23

PROP: Liquid or oil. Bp: 203°, flash p: <212°F (OC), d: 0.931, vap d: 4.65. Sltly sol in H_2O.

SYNS: ACTEDRON □ ADIPAN □ ALLODENE □ dl-AMPHETAMINE □ ANOREXIDE □ (±)-BENZEDRINE □ dl-BENZEDRINE □ DEOXYNOREPHEDRINE □ (±)-DESOXYNOREPHEDRINE □ racemic-DESOXYNOREPHEDRINE □ ELASTONON □ ISOAMYCIN □ ISOMYN □ MECODRIN □ α-METHYLBENZENEETHANEAMINE □ dl-α-METHYLPHENETHYLAMINE □ (±)-α-METHYLPHENETHYLAMINE □ NOREPHEDRANE □ NOVYDRINE □ ORTEDRINE □ PHENEDRINE □ dl-1-PHENYL-2-AMINOPROPANE □ PROFAMINA □ PROPISAMINE □ PSYCHEDRINE □ RAPHETAMINE □ SIMPATEDRIN □ SYMPAMINE □ SYMPATEDRINE □ WECKAMINE

TOXICITY DATA WITH REFERENCE
dnd-esc 40 μmol/L MUREAV 89,95,81
scu-rat TDLo:11 mg/kg (1-22D preg):REP PSYPAG 40,25,74
unr-man LDLo:2206 μg/kg 85DCAI 2,73,70
orl-rat LD50:30 mg/kg ARZNAD 23,810,73
scu-rat LD50:180 mg/kg JPETAB 85,119,45
orl-mus LD50:21 mg/kg ARZNAD 23,810,73
ipr-mus LD50:5500 μg/kg AIPTAK 161,206,66
scu-mus LD50:15 mg/kg FEPRA7 4,139,45
ivn-mus LD50:15 mg/kg AIPTAK 145,392,63

CONSENSUS REPORTS: Reported in EPA TSCA Inventory. EPA Extremely Hazardous Substances List.

SAFETY PROFILE: A deadly human poison by an unspecified route. An experimental poison by ingestion, subcutaneous, intraperitoneal, and intravenous routes. Experimental reproductive effects. Mutation data reported. A central nervous system stimulant. Overdoses cause hyperactivity, restlessness, insomnia, rapid pulse, rise in blood pressure, dilated pupils, dryness of the throat. Combustible when exposed to heat, flame, or oxidizers. When heated to decomposition it emits toxic fumes of NO_x. To fight fire, use CO_2, dry chemical, alcohol foam, water mist, fog. See other benzedrine entries.

BBK250 **CAS:156-31-0** **HR: 3**
BENZEDRINE SULFATE
mf: $C_{18}H_{26}N_2 \cdot H_2O_4S$ mw: 368.54

SYNS: AMITRENE □ AMPHOIDS S □ AMPHORDS S □ BAR-TIME □ DIAMPHETAMINE SULFATE □ KLINE □ dl-α-METHYLPHENETHYLAMINE SULFATE □ PHENETHYLAMINE, α-METHYL-, SULFATE (2:1) □ 1-PHENYL-2-AMINOPROPANE SULFATE

TOXICITY DATA WITH REFERENCE
ipr-rat LDLo:25 mg/kg JPETAB 100,267,50
scu-rat LDLo:10 mg/kg JPETAB 71,62,41
ipr-mus LD50:75 mg/kg JPETAB 93,114,48
scu-mus LD50:14 mg/kg JPETAB 87,214,46
orl-dog LDLo:20 mg/kg AJMSA9 198,785,39

SAFETY PROFILE: A poison via ingestion, intraperitoneal, and subcutaneous routes. When heated to decomposition it emits very toxic fumes of SO_x and NO_x. See also BENZEDRINE and SULFATES.

BBK500 **CAS:51-63-8** **HR: 3**
d-BENZEDRINE SULFATE
mf: $C_{18}H_{26}N_2 \cdot H_2O_4S$ mw: 368.54

PROP: Plates.

SYNS: ACEDRON □ ADJUDETS □ ADRIXINE □ AFATIN □ ALBEMAP □ AMDEX □ d-AMFETASUL □ AMITRENE □ AMPHAETEX □ AMPHEDRINE □ AMPHEREX □ (+)-AMPHETAMINE SULFATE □ d-AMPHETAMINE SULFATE □ AMSUSTAIN □ APETAIN □ ARDEX □ BETAFEDRINA □ BETAFEDRINE □ d-BETAPHEDRINE □ CARRTIME □ CRADEX □ DADEX □ DADOX d-CITRAMINE □ DELLIPSOIDS □ DEPHADREN □ DESOXYN □ DEXAIME □ DEXALINE □ DEXALME □ DEXAMED □ DEXAMINE □ DEXAMPHAMINE □ DEXAMPHETAMINE □ DEXAMPHETAMINE SULFATE □ DEXAMYL □ DEXEDRINA □ DEXEDRINE SULFATE □ DEXIES □ DEXTROAMPHETAMINE SULFATE □ DEXTRO-α-METHYLPHENETHYLAMINE SULFATE □ DEXTRO-1-PHENYL-2-AMINOPROPANE SULFATE □ DEXTRO-β-PHENYLISOPROPYLAMINE SULFATE □ FASTBALLS □ HEARTS □ (S)-α-METHYL-BENZENEETHANAMINE SULFATE (2:1) □ d-α-METHYLPHENETHYLAMINE SULFATE □ ORANGES □ PELLCAFS □ PELLCAP □ PELLCAPS □ PERKE □ PHENOPROMIN □ d-1-PHENYL-2-AMINOPROPANE SULFATE □ d-β-PHENYLISOPROPYLAMINE SULFATE □ PHETADEX □ PSYCHODRINE □ REVIDEX □ SIMPAMINA-D □ SYMPAMINA-D □ TEMPODEX □ TUPHETAMINE □ TYDEX □ ZAMINE

TOXICITY DATA WITH REFERENCE
orl-wmn TDLo:96 mg/kg (5-39W preg):TER INTSAO 50,79,68
orl-rat TDLo:146 mg/kg (female 30D pre-21D post):REP DABBBA 39,5121,79
scu-rat TDLo:108 mg/kg (female 5-22D post):REP LIFSAK 18,605,76
orl-rat TDLo:440 mg/kg (1-22D preg):REP DABBBA 37,1576,76
scu-rat TDLo:500 μg/kg (male 1D pre):REP PSYPAG 10,44,66
ipr-rat TDLo:70 mg/kg (female 9-10D post):REP ANREAK 172,338,72
ipr-mus TDLo:500 μg/kg (female 8D post):TER LANCAO 2,1021,65
ipr-mus TDLo:100 mg/kg (female 10D post):TER TJADAB 35,27,87
orl-rat LD50:32 mg/kg ARZNAD 33,1411,83
ipr-rat LD50:43,200 μg/kg TXAPA9 29,397,74
ivn-rat LD50:30 mg/kg JPETAB 110,180,54
orl-mus LD50:10 mg/kg JMCMAR 18,71,75
ipr-mus LD50:9700 μg/kg JPETAB 135,240,62
scu-mus LD50:16 mg/kg AIPTAK 184,34,70
ivn-mus LD50:30 mg/kg JPETAB 137,365,62
orl-dog LD50:10 mg/kg PSEBAA 118,557,65
ivn-dog LD50:3 mg/kg PSEBAA 118,557,65
ivn-rbt LD50:10 mg/kg JPETAB 110,180,54
orl-bwd LD50:56,200 μg/kg AECTCV 12M355,83

SAFETY PROFILE: Poison by ingestion, intraperitoneal, subcutaneous, and intravenous routes. A human teratogen that causes developmental abnormalities of the central nervous system. Experimental reproductive effects including other teratogenic effects. A habit-forming stimulant. When heated to decomposition it emits very toxic fumes of SO_x and NO_x. See also other benzedrine compounds and SULFATES.

BBK750 **CAS:51-62-7** **HR: 3**
l-BENZEDRINE SULFATE
mf: $C_{18}H_{26}N_2{\cdot}H_2O_4S$ mw: 368.54

SYNS: (−)-AMPHETAMINE SULFATE □ l-AMPHETAMINE SULFATE □ LEVEDRINE □ l-1-PHENYL-2-AMINOPROPANE SULFATE

TOXICITY DATA WITH REFERENCE
scu-rat LDLo:160 mg/kg JPETAB 71,62,41
ipr-mus LD50:232 mg/kg JPETAB 158,135,67

SAFETY PROFILE: A poison via subcutaneous and intraperitoneal routes. See also SULFATES. When heated to decomposition it emits very toxic fumes of SO_x and NO_x.

BBL000 **CAS:142-04-1** **HR: 3**
BENZENAMINE HYDROCHLORIDE
DOT: UN 1548
mf: $C_6H_7N{\cdot}ClH$ mw: 129.60

PROP: Crystals. Vap d: 4.46, d: 1.22, mp: 198°, bp: 245°, flash p: 380°F (OC).

SYNS: ANILINE CHLORIDE □ ANILINE HYDROCHLORIDE (DOT) □ "ANILINE SALT" □ ANILINIUM CHLORIDE □ CHLORHYDRATE d'ANILINE (FRENCH) □ CHLORID ANILINU (CZECH) □ NCI-C03736 □ PHENYLAMINE HYDROCHLORIDE □ SUL ANILINOVA (CZECH) □ USAF EK-442

TOXICITY DATA WITH REFERENCE
skn-rbt 500 mg/24H MOD 28ZPAK -,65,72
eye-rbt 20 mg/24H MOD 28ZPAK -,65,72
sce-hmn:lym 50 µmol/L BLFSBY 29b,561,84
otr-rat:emb 79,500 ng/plate JJATDK 1,190,81
sce-ham:fbr 10 µmol/L JNCIAM 58,1635,77
orl-rat TDLo:1400 mg/kg (7-20D preg):TER TXAPA9 77,465,85
orl-rat TDLo:130 g/kg/2Y-C:CAR NCITR* NCI-CG-TR-130,78
orl-rat TD:238 g/kg/2Y-C:CAR NCITR* NCI-CG-TR-130,78
orl-rat TD:137 g/kg/60W-C:ETA IARC** 27,39,82
orl-rat TD:2163 g/kg/2Y-C:CAR IARC** 27,39,82
orl-rat TD:4326 g/kg/2Y-C:CAR IARC** 27,39,82
orl-rat LD50:840 mg/kg TXAPA9 42,417,77
ipr-rat LDLo:500 mg/kg NCNSA6 5,11,53
orl-mus LD50:841 mg/kg NTIS** PB214-270
ipr-mus LD50:300 mg/kg NTIS** AD277-689

CONSENSUS REPORTS: IARC Cancer Review: Animal Limited Evidence IMEMDT 27,39,82. NCI Carcinogenesis Bioassay Completed; Results Positive: rat NCITR* NCI-CG-TR-130,78; Results Negative: mouse NCITR* NCI-CG-TR-130,78. Reported in EPA TSCA Inventory. EPA Genetic Toxicology Program.

DOT CLASSIFICATION: 6.1; *Label:* KEEP AWAY FROM FOOD

SAFETY PROFILE: Suspected carcinogen with experimental carcinogenic and tumorigenic data. Poison by intraperitoneal route. Moderately toxic by ingestion. Experimental teratogenic effects. Human mutation data reported. A skin and eye irritant. Combustible when exposed to heat or flame. When heated to decomposition or on contact with acid or acid fumes, it emits highly toxic fumes of aniline and chlorine compounds. Reacts explosively with aniline at 240°C/7.6 bar. Can react vigorously with oxidizing materials. To fight fire, use water, CO_2, water mist or spray, dry chemical. See also ANILINE.

BBL250 **CAS:71-43-2** **HR: 3**
BENZENE
DOT: UN 1114
mf: C_6H_6 mw: 78.12

PROP: Clear, colorless liquid. Mp: 5.51°, bp: 80.093–80.094°, flash p: 12°F (CC), d: 0.8794 @ 20°, autoign temp: 1044°F, lel: 1.4%, uel: 8.0%, vap press: 100 mm @ 26.1°, vap d: 2.77, ULC: 95–100. Very sltly sol in H_2O, misc in most org solvs.

SYNS: (6)ANNULENE □ BENZEEN (DUTCH) □ BENZEN (POLISH) □ BENZIN (OBS.) □ BENZINE (OBS.) □ BENZOL (DOT) □ BENZOLE □ BENZOLENE □ BENZOLO (ITALIAN) □ BICARBURET of HYDROGEN □ CARBON OIL □ COAL NAPHTHA □CYCLOHEXATRIENE □ FENZEN (CZECH) □ MINERAL NAPHTHA □ MOTOR BENZOL □ NCI-C55276 □ NITRATION BENZENE □ PHENE □ PHENYL HYDRIDE □ PYROBENZOL □ PYROBENZOLE □ RCRA WASTE NUMBER U019

TOXICITY DATA WITH REFERENCE
skn-rbt 15 mg/24H open MLD AIHAAP 23,95,62
skn-rbt 20 mg/24H MOD 85JCAE- 25,86
eye-rbt 88 mg MOD AMIHAB 14,387,56
eye-rbt 2 mg/24H SEV 28ZPAK -,23,72
oms-hmn:lym 5 µmol/L CNREA8 45,2471,85
mma-mus:emb 2500 mg/L PMRSDJ 5,639,85
orl-mus TDLo:6500 mg/kg (female 8-12D post):REP TCMUD8 6,361,86
par-mus TDLo:4 g/kg (female 12D post):REP NEZAAQ 25,438,70
ihl-mus TCLo:5 ppm (female 6-15D post):TER TXCYAC 42,171,86
ihl-rbt TCLo:1 g/m³/24H (female 7-20D post):TER ARTODN 8,425,85
ihl-rat TCLo:50 ppm/24H (female 7-14D post):TER JOHYAY 24,363,80
scu-mus TDLo:1100 mg/kg (female 12D post):TER TOXID9 1,125,81
ihl-rbt TCLo:1 g/m³/24H (female 7-20D post):REP ARTODN 8,425,85
ihl-rat TCLo:670 mg/m³/24H (15D pre/1-22D preg):REP HYSAAV 33,327,68
ihl-rat TCLo:150 ppm/24H (female 7-14D post):REP JOHYAY 24,363,80
ipr-mus TDLo:5 mg/kg (male 1D pre):REP TPKVAL 15,30,79
ihl-man TCLo:200 mg/m³/78W-I:CAR,BLD EJCAAH 7,83,71
ihl-hmn TCLo:10 ppm/8H/10Y-I:CAR,BLD TRBMAV 37,153,78
orl-rat TDLo:52 g/kg/52W-I:CAR MELAAD 70,352,79

ihl-rat TCLo:1200 ppm/6H/10W-I:ETA PAACA3 25,75,84
orl-mus TDLo:18,250 mg/kg/2Y-C:CAR NTPTR* NTP-TR-289,86
ihl-mus TCLo:300 ppm/6H/16W-I:ETA TXAPA9 75,358,84
skn-mus TDLo:1200 g/kg/49W-I:NEO BJCAAI 16,275,62
ipr-mus TDLo:1200 mg/kg/8W-I:NEO TXAPA9 82,19,86
scu-mus TDLo:600 mg/kg/17W-I:ETA KRANAW 9,403,32
par-mus TDLo:670 mg/kg/19W-I:ETA KLWOAZ 12,109,33
ihl-hmn TC:150 ppm/15M/8Y-I:CAR,BLD BLOOAW 52,285,78
orl-rat TD:52 g/kg/1Y-I:CAR AJIMD8 4,589,83
orl-rat TD:10 g/kg/52W-I:CAR MELAAD 70,352,79
ihl-man TC:600 mg/m^3/4Y-I:CAR,BLD NEJMAG 271,872,64
ihl-man TC:150 ppm/11Y-I:CAR,BLD BLUTA9 28,293,74
ihl-mus TC:1200 ppm/6H/10W-I:ETA PAACA3 25,75,84
orl-mus TD:2400 mg/kg/8W-I:NEO TXAPA9 82,19,86
ihl-hmn TC:8 ppb/4W-I:CAR,BLD NEJMAG 316,1044,87
ihl-hmn TC:10 mg/m^3/11Y-I:CAR,BLD BJIMAG 44,124,87
ihl-mus TC:300 ppm/6H/16W-I:CAR IMMUAM (3),156,84
ihl-hmn LCLo:2 pph/5M TABIA2 3,231,33
orl-man LDLo:50 mg/kg YAKUD5 22,883,80
ihl-hmn LCLo:20,000 ppm/5M 29ZUA8 -,-,53
ihl-man TCLo:150 ppm/1Y-I:BLD BLUTA9 28,293,74
ihl-hmn TCLo:100 ppm INMEAF 17,199,48
ihl-hmn LCLo:65 mg/m^3/5Y:BLD ARGEAR 44,145,74
orl-rat LD50:3306 mg/kg TXAPA9 19,699,71
ihl-rat LC50:10,000 ppm/7H 28ZRAQ -,113,60
ipr-rat LD50:2890 μg/kg 36YFAG -,302,77
orl-mus LD50:4700 mg/kg HYSAAV 32,349,67
ihl-mus LC50:9980 ppm JIHTAB 25,366,43
ipr-mus LD50:340 mg/kg ANYAA9 243,104,75
orl-dog LDLo:2000 mg/kg HBAMAK 4,1313,35
ihl-dog LCLo:146,000 mg/m^3 HBTXAC 1,324,56
ihl-cat LCLo:170,000 mg/m^3 HBTXAC 1,324,56
ivn-rbt LDLo:88 mg/kg JTEHD6 -(Suppl2),45,77

CONSENSUS REPORTS: NTP 7th Annual Report on Carcinogens. IARC Cancer Review: Group 1 IMEMDT 7,120,87; Human Limited Evidence IMEMDT 7,203,74; Animal Inadequate Evidence IMEMDT 7,203,74; IARC Cancer Review: Animal Limited Evidence IMEMDT 29,93,82; Human Sufficient Evidence IMEMDT 29,93,82. NTP Carcinogenesis Studies (gavage); Clear Evidence: mouse, rat NTPTR* NTP-TR-289,86. EPA Genetic Toxicology Program. Reported in EPA TSCA Inventory. On Community Right-To-Know List.

OSHA PEL: TWA 1 ppm; STEL 5 ppm; Pk 5 ppm/15M/8H; Cancer Hazard
ACGIH TLV: TWA 10 ppm; Suspected Human Carcinogen (Proposed: TWA 0.3 ppm; Confirmed Human Carcinogen); BEI: 50 mg(total phenol)/L in urine at end of shift recommended as a mean value
DFG TRK:5 ppm (16 mg/m^3) Human Carcinogen
NIOSH REL: TWA 0.32 mg/m^3; CL 3.2 mg/m^3/15M
DOT CLASSIFICATION: 3; *Label:* Flammable Liquid

SAFETY PROFILE: Confirmed human carcinogen producing myeloid leukemia, Hodgkin's disease, and lymphomas by inhalation. Experimental carcinogenic, neoplastigenic, and tumorigenic data. A human poison by inhalation. An experimental poison by skin contact, intraperitoneal, intravenous, and possibly other routes. Moderately toxic by ingestion and subcutaneous routes. A severe eye and moderate skin irritant. Human systemic effects by inhalation and ingestion: blood changes, increased body temperature. Experimental teratogenic and reproductive effects. Human mutation data reported. A narcotic. In industry, inhalation is the primary route of chronic benzene poisoning. Poisoning by skin contact has been reported. Recent (1987) research indicates that effects are seen at less than 1 ppm. Exposures needed to be reduced to 0.1 ppm before no toxic effects were observed. Elimination is chiefly through the lungs. A common air contaminant.

A dangerous fire hazard when exposed to heat or flame. Explodes on contact with diborane, bromine pentafluoride, permanganic acid, peroxomonosulfuric acid, and peroxodisulfuric acid. Forms sensitive, explosive mixtures with iodine pentafluoride, silver perchlorate, nitryl perchlorate, nitric acid, liquid oxygen, ozone, arsenic pentafluoride + potassium methoxide (explodes above 30°C). Ignites on contact with sodium peroxide + water, dioxygenyl tetrafluoroborate, iodine heptafluoride, and dioxygen difluoride. Vigorous or incandescent reaction with hydrogen + Raney nickel (above 210°C), uranium hexafluoride, and bromine trifluoride. Can react vigorously with oxidizing materials, such as Cl_2, CrO_3, O_2, $NClO_4$, O_3, perchlorates, ($AlCl_3$ + $FClO_4$), (H_2SO_4 + permanganates), K_2O_2, ($AgClO_4$ + acetic acid), Na_2O_2. Moderate explosion hazard when exposed to heat or flame. Use with adequate ventilation. To fight fire, use foam, CO_2, dry chemical.

Poisoning occurs most commonly via inhalation of the vapor, although benzene can penetrate the skin and cause poisoning. Locally, benzene has a comparatively strong irritating effect, producing erythema and burning, and, in more severe cases, edema and even blistering. Exposure to high concentrations of the vapor (3000 ppm or higher) may result from failure of equipment or spillage. Such exposure, while rare in industry, may cause acute poisoning, characterized by the narcotic action of benzene on the central nervous system. The anesthetic action of benzene is similar to that of other anesthetic gases, consisting of a preliminary stage of excitation followed by depression and, if exposure is continued, death through respiratory failure. The chronic, rather than the acute form, of benzene poisoning is important in industry. It is a recognized leukemogen. There is no specific blood picture occurring in cases of chronic benzol poisoning. The bone marrow may be hypoplastic, normal, or hyperplastic, the changes reflected in the peripheral blood. Anemia, leucopenia, macrocytosis, reticulocytosis, thrombocytopenia, high color index, and prolonged bleeding time may be present. Cases of myeloid leukemia have been reported. For the worker, repeated blood examinations are necessary, including hemoglobin determinations, white and red cell counts, and differential smears. Where a worker shows a progressive drop in either red or white cells, or where the white count remains below <5,000/mm$_3$ or the red count remains below 4.0 million/mm$_3$, on two successive monthly examinations, the worker should be immediately removed from benzene exposure. Elimination is chiefly through the lungs, when fresh air is breathed. The portion that is absorbed is oxidized, and the oxidation products are combined with sulfuric and glycuronic acids and eliminated in the urine. This may be used as a diagnostic sign. Benzene has a definite cumulative action, and exposure to a relatively high concentration is not serious from the point of view of causing damage to the blood-forming system, provided the exposure is not repeated. In acute poisoning, the worker becomes confused and dizzy, complains of tightening of the leg muscles and of pressure over the forehead, then passes into a stage of excitement. If allowed to remain exposed, he quickly becomes stupefied and lapses into coma. In nonfatal cases, recovery is usually complete with no permanent disability. In chronic poisoning the onset is slow, with the symptoms vague; fatigue, headache, dizziness, nausea and loss of appetite, loss of weight, and weakness are common complaints in early cases. Later, pallor, nosebleeds, bleeding gums, menorrhagia, petechiae, and purpura may develop. There is great individual variation in the signs and symptoms of chronic benzene poisoning.

For occupational chemical analysis use OSHA: #12 or NIOSH: Hydrocarbons, Aromatic, 1501; Hydrocarbons, BP 36-126 C, 1500.

BBL500 CAS:122-78-1 HR: 2
BENZENEACETALDEHYDE
mf: C_8H_8O mw: 120.16

PROP: Oily, colorless liquid that polymerizes and grows more viscous on standing; odor similar to lilac and hyacinth. Has been crystallized, mp: 33–34°, d:(25/25) 1.023–1.030, refr index: 1.525–1.545, bp: (10) 78°, n (20/D) 1.524–1.528, flash p: 154°F. Sltly sol in water; sol in alc, ether, and propylene glycol. One part is sol in two parts of 80% alc forming a clear solution.

SYNS: FEMA No. 2874 □ HYACINTHIN □ PAA □ PHENYLACETALDEHYDE (FCC) □ PHENYLACETIC ALDEHYDE □ PHENYLETHANAL □ α-TOLUALDEHYDE □ α-TOLUIC ALDEHYDE

TOXICITY DATA WITH REFERENCE
skn-hmn 2%/48H FCTXAV 17,377,79
orl-rat LD50:1550 mg/kg FCTXAV 17,377,79
orl-mus LD50:3890 mg/kg FCTXAV 17,377,79
orl-gpg LD50:3890 mg/kg FCTXAV 17,377,79

CONSENSUS REPORTS: Reported in EPA TSCA Inventory.

SAFETY PROFILE: Moderately toxic by ingestion. Human skin irritant. Combustible liquid. When heated to decomposition it emits acrid smoke and irritating fumes. See also ALDEHYDES.

BBL750 CAS:98-05-5 HR: 3
BENZENEARSONIC ACID
mf: $C_6H_7AsO_3$ mw: 202.05

PROP: Colorless crystals from water. D: 1.760, mp: 160° decomp. Sol in water.

SYNS: PHENYL ARSENIC ACID □ PHENYLARSONIC ACID

TOXICITY DATA WITH REFERENCE
orl-rat LDLo:50 mg/kg JPETAB 93,287,48
orl-mus LD50:270 μg/kg CLDND* 80,93,44
ivn-rbt LD50:16 mg/kg JPETAB 80,93,44

CONSENSUS REPORTS: Reported in EPA TSCA Inventory. EPA Extremely Hazardous Substances List. Arsenic and its compounds are on the Community Right-To-Know List.

OSHA PEL: TWA 0.5 mg/(As)m³
ACGIH TLV: TWA 0.2 mg(As)/m³

SAFETY PROFILE: A deadly poison by ingestion and intravenous routes. See also ARSENIC COMPOUNDS. When heated to decomposition it emits toxic fumes of As.

BBL825 CAS:4547-69-7 HR: 3
BENZENE-1,3-BIS(SULFONYL AZIDE)
mf: $C_6H_4N_6O_4S_2$ mw: 288.26

$C_6H_4(SO_2N_3)_2$

SAFETY PROFILE: An explosive. Upon decomposition it emits toxic fumes of SO_x and NO_x. See also EXPLOSIVES and AZIDES.

BBM000 CAS:98-80-6 HR: 3
BENZENEBORONIC ACID
mf: $C_6H_7BO_2$ mw: 121.94

PROP: Needles from H_2O. Mp: 216°. Sol in MeOH, EtOH; sltly sol in H_2O and Et_2O.

SYNS: ACIDE PHENYLBORIQUE (FRENCH) □ BOROPHENYLIC ACID □ PHENYLBORIC ACID □ USAF BO-2

TOXICITY DATA WITH REFERENCE
orl-rat LD50:740 mg/kg 14KTAK -,693,64
ipr-mus LD50:500 mg/kg NTIS** AD277-689
ivn-mus LD50:320 mg/kg CSLNX* NX#02033
ivn-dog LDLo:450 mg/kg BANMAC 135,314,51
orl-rbt LDLo:600 mg/kg 14KTAK -,693,64
skn-rbt LDLo:4500 mg/kg 14KTAK -,693,64
ipr-gpg LD50:284 mg/kg BANMAC 135,314,51

CONSENSUS REPORTS: Reported in EPA TSCA Inventory.

SAFETY PROFILE: Poison by intravenous and intraperitoneal routes. Moderately toxic by ingestion. Mildly toxic by skin contact. See also BORON COMPOUNDS. When heated to decomposition it emits acrid smoke and irritating fumes.

BBM250 CAS:2227-79-4 HR: 3
BENZENECARBOTHIOAMIDE
mf: C_7H_7NS mw: 137.21

SYNS: BENZOTHIAMIDE □ BENZOTHIOAMIDE □ THIOBENZAMIDE □ TIOBENZAMIDE (ITALIAN)

TOXICITY DATA WITH REFERENCE
mnt-mus-orl 180 μmol/kg MUREAV 192,141,87
orl-rat TDLo:6300 mg/kg/15W-C:ETA BSIBAC 54,1027,78
orl-rat TD:13 g/kg/38W-C:ETA ARTODN 55,34,84
orl-rat TD:13,300 mg/kg/38W-C:ETA ARTODN 55,34,84
orl-mus LD50:95 mg/kg THERAP 8,237,53
ipr-mus LD50:500 mg/kg PCJOAU 11,1383,77

CONSENSUS REPORTS: Reported in EPA TSCA Inventory.

SAFETY PROFILE: Poison by ingestion. Moderately toxic by intraperitoneal route. Questionable carcinogen with experimental tumorigenic data. Mutation data reported. When heated to decomposition it emits very toxic fumes of NO_x and SO_x.

BBM500 **CAS:63021-32-9** **HR: 2**
BENZENECARBOXALDEHYDE
mf: $C_{19}H_{15}N$ mw: 257.35

SYNS: BENZALDEHYDE FFC ☐ 7-ETHYLBENZ(c)ACRIDINE ☐ 9-ETHYL-3,4-BENZACRIDINE ☐ PHENYLMETHANAL

TOXICITY DATA WITH REFERENCE
scu-mus TDLo:200 mg/kg:ETA VOONAW 1,52,55

SAFETY PROFILE: Questionable carcinogen with experimental tumorigenic data. See also ALDEHYDES. When heated to decomposition it emits toxic fumes of NO_x.

BBM750 **CAS:1670-14-0** **HR: 2**
BENZENECARBOXIMIDAMIDE HYDROCHLORIDE
mf: $C_7H_8N_2 \cdot ClH$ mw: 156.63

SYN: BENZAMIDINE, HYDROCHLORIDE

TOXICITY DATA WITH REFERENCE
ipr-mus LD50:580 mg/kg BIREBV 20,1045,79

CONSENSUS REPORTS: Reported in EPA TSCA Inventory.

SAFETY PROFILE: Moderately toxic by intraperitoneal route. When heated to decomposition it emits toxic vapors of NO_x, HCl, and Cl^-.

BBN250 **CAS:17333-86-7** **HR: 3**
BENZENE DIAZONIUM-2-CARBOXYLATE
mf: $C_7H_4N_2O_2$ mw: 148.12

SAFETY PROFILE: A heat- and shock-sensitive explosive. Explosive or violent reaction with aniline, arylisocyanides, and 1-pyrrolidinylcyclohexene. When heated to decomposition it emits toxic fumes of NO_x. See also EXPLOSIVES.

BBN500 **CAS:100-34-5** **HR: 3**
BENZENE DIAZONIUM CHLORIDE
mf: $C_6H_5ClN_2$ mw: 140.58

PROP: Crystals.

SYN: BENZENE DIAZONIUM CHLORIDE (dry) (DOT)

DOT CLASSIFICATION: Forbidden

SAFETY PROFILE: Potentially explosive when dry. Potentially explosive reaction with potassium o-methyldithiocarbonate. When heated to decomposition it emits toxic fumes of Cl^- and NO_x.

BBN650 **CAS:36211-73-1** **HR: 3**
BENZENEDIAZONIUM HYDROGEN SULFATE
mf: $C_6H_6N_2O_4S$ mw: 202.18

PROP: Prisms.

SAFETY PROFILE: Explodes at 100°C. When heated to decomposition it emits toxic fumes of SO_x and NO_x.

BBN750 **CAS:619-97-6** **HR: 3**
BENZENE DIAZONIUM NITRATE
mf: $C_6H_5N_3O_3$ mw: 167.12

SYN: BENZENE DIAZONIUM NITRATE (dry) (DOT)

DOT CLASSIFICATION: Forbidden

SAFETY PROFILE: An explosive sensitive to friction, impact, and heating to 90°. Upon decomposition it emits toxic fumes of NO_x. See also EXPLOSIVES and NITRATES.

BBN850 **CAS:6925-01-5** **HR: 3**
BENZENEDIAZONIUM-4-OXIDE
mf: $C_6H_4N_2O$ mw: 120.11

SAFETY PROFILE: Decomposes violently at 75°C. When heated to decomposition it emits toxic fumes of NO_x.

BBO125 **CAS:612-31-7** **HR: 3**
BENZENEDIAZONIUM-2-SULFONATE
mf: $C_6H_4N_2O_3S$ mw: 184.17

PROP: Crystals. Mp: 106° (decomp).

SAFETY PROFILE: Explodes on contact with flame or on impact. Upon decomposition it emits toxic fumes of SO_x and NO_x.

BBO250 **CAS:305-80-6** **HR: 3**
BENZENE DIAZONIUM-4-SULFONATE
mf: $C_6H_4N_2O_3S$ mw: 184.17

PROP: Needles from water. Sol in water.

SAFETY PROFILE: An unstable explosive which may explode when touched. Incompatible with metals. Store in small quantities under refrigeration in loosely plugged containers. Upon decomposition it emits toxic fumes of NO_x and SO_x.

BBO325 **CAS:369-57-3** **HR: 3**
BENZENEDIAZONIUM TETRAFLUOROBORATE
mf: $C_6H_5N_2 \cdot BF_4$ mw: 191.94

SYNS: BENZENEDIAZONIUM FLUOBORATE ☐ BENZENEDIAZONIUM FLUOROBORATE ☐ PHENYLDIAZONIUM FLUOROBORATE (SALT) ☐ PHENYLDIAZONIUM TETRAFLUOROBORATE

TOXICITY DATA WITH REFERENCE
mmo-sat 10 μmol/L CNREA8 42,1446,82
scu-ham TDLo:85 mg/kg/71W-I:ETA CALEDQ 15,289,82
scu-ham TD:100 mg/kg/84W-I:ETA CALEDQ 15,289,82
scu-ham TD:119 mg/kg/50W-I:ETA CALEDQ 15,289,82
scu-ham TD:157 mg/kg/66W-I:ETA CALEDQ 15,289,82

scu-ham TD:200 mg/kg/84W-I:ETA CALEDQ 15,289,82
orl-ham LD50:354 mg/kg CALEDQ 15,289,82
scu-ham LD50:166 mg/kg CALEDQ 15,289,82

SAFETY PROFILE: Poison by ingestion and subcutaneous routes. Questionable carcinogen with experimental tumorigenic data. Mutation data reported. When heated to decomposition it emits toxic fumes of NO_x and F^-. See also BORON COMPOUNDS.

BBO400 CAS:19521-84-7 HR: 3
BENZENEDIAZONIUM TRIBROMIDE
mf: $C_6H_5Br_3N_2$ mw: 344.83

SAFETY PROFILE: A sensitive explosive. Upon decomposition it emits toxic fumes of Br^- and NO_x. See also EXPLOSIVES.

BBO500 CAS:88-96-0 HR: 1
1,2-BENZENEDICARBOXAMIDE
mf: $C_8H_8N_2O_2$ mw: 164.18

SYNS: NCI-C03612 □ P-D □ PHTHALAMIDE □ o-PHTHALIC ACID DIAMIDE

TOXICITY DATA WITH REFERENCE
ipr-rat LD50:4004 mg/kg APFRAD 48,23,90
ipr-mus LD50:4104 mg/kg APFRAD 48,23,90

CONSENSUS REPORTS: Reported in EPA TSCA Inventory. NTP Carcinogenesis Bioassay (feed): No Evidence: mouse, rat NCITR* NCI-TR-161,79.

SAFETY PROFILE: Mildly toxic by intraperitoneal route. When heated to decomposition it emits toxic vapors of NO_x.

BBP000 CAS:123-61-5 HR: 3
BENZENE-1,3-DIISOCYANATE
mf: $C_8H_4N_2O_2$ mw: 160.14

PROP: Crystals. Mp: 51–55°, bp: 102–104° @ 8 mm.

SYNS: BENZENE-1,3-DIISOCYANATE □ BENZENE, 1,3-DIISOCYANATO- □ 1,3-DIISOCYANATOBENZENE □ NACCONATE 400 □ m-PHENYLENE DIISOCYANATE □ m-PHENYLENE ISOCYANATE

TOXICITY DATA WITH REFERENCE
ivn-mus LD50:5600 µg/kg CSLNX* NX#07804

CONSENSUS REPORTS: Reported in EPA TSCA Inventory. Cyanide and its compounds are on the Community Right-To-Know List.

NIOSH REL: TWA (Diisocyanates) 0.005 ppm; CL 0.02 ppm/10M

SAFETY PROFILE: A sensitizer at very low concentrations. Deadly poison by intravenous route. When heated to decomposition it emits toxic fumes of NO_x and CN^-. See also ESTERS.

BBP250 CAS:623-26-7 HR: 2
p BENZENEDINITRILE
mf: $C_8H_4N_2$ mw: 128.14

PROP: Crystals, Mp: 222°, vap d: 4.42.

SYNS: 4-CYANOBENZONITRILE □ p-DICYANOBENZENE □ 1,4-DICYANOBENZENE □ NITRIL KYSELINY TEREFTALOVE (CZECH) □ p-PDN □ p-PHTHALODINITRILE □ TEREFTALODINITRIL (CZECH) □ TEREPHTHALONITRILE

TOXICITY DATA WITH REFERENCE
eye-rbt 500 mg/24H MLD 28ZPAK -,159,72
orl-rat LD50:>6400 mg/kg ZAARAM 19,225,69
ipr-rat LD50:4004 mg/kg APFRAD 48,23,90
orl-mus LD50:>300 mg/kg JMCMAR 21,906,78
ipr-mus LD50:699 mg/kg INHEAO 4,11,66

CONSENSUS REPORTS: Reported in EPA TSCA Inventory. Cyanide and its compounds are on the Community Right-To-Know List.

SAFETY PROFILE: Moderately toxic by ingestion and intraperitoneal routes. An eye irritant. When heated to decomposition it emits toxic fumes of CN^- and NO_x. See also NITRILES.

BBP750 CAS:608-73-1 HR: 3
BENZENE HEXACHLORIDE
mf: $C_6H_6Cl_6$ mw: 290.82

$ClCH(CHCl)_4CHCl$ (ring)

PROP: Technical grade contains 68.7% α-BHC, 6.5% β-BHC, and 13.5% γ BHC (JPFCD2 14,305,79). White, crystalline powder. Mp: 113°, vap press: 0.0317 mm @ 20°.

SYNS: BHC (USDA) □ COMPOUND-666 □ DBH □ ENT 8,601 □ GAMMEXANE □ HCCH □ HEXA □ HEXACHLOR □ HEXACHLO-RAN □ HEXACHLOROCYCLOHEXANE □ 1,2,3,4,5,6-HEXACHLOROCYCLOHEXANE □ HEXYLAN □ JACUTIN □ LATKA 666

TOXICITY DATA WITH REFERENCE
mmo-omi 100 mg/L MILEDM 5,103,77
otr-rat-orl 875 mg/kg/7W-I CRNGDP 5,479,84
orl-mus TDLo:9120 mg/kg (22W male):REP BECTA6 26,508,81
orl-rat TDLo:8100 mg/kg (90D male):REP APTOA6 52,12,83
orl-mus TDLo:6720 mg/kg/80W-C:CAR JPFCD2 14(3),305,79
skn-mus TDLo:1600 mg/kg/80W-I:ETA JPFCD2 14(3),305,79
orl-mus TD:800 mg/kg/80W-I:ETA JPFCD2 14(3),305,79
orl-mus TD:12,600 mg/kg/30W-C:CAR JCROD7 99,143,81
orl-mus TD:12,960 mg/kg/26W-C:CAR CMSHAF 1,279,72
orl-mus TD:21,600 mg/kg/52W-C:CAR TUMOAB 69,383,83
orl-mus TD:5400 mg/kg/13W-C:NEO TUMOAB 69,383,83
orl-mus TD:7200 mg/kg/17W-C:ETA IJEBA6 19,1159,81
orl-mus TD:9 g/kg/21W-C:ETA AEHLAU 37,156,82
orl-mus TD:10,800 mg/kg/26W-C:ETA AEHLAU 37,156,82
orl-mus TD:7200 mg/kg/17W-C:ETA CMBID4 27,231,81
orl-mus TD:10,800 mg/kg/26W-C:ETA CMBID4 27,231,81
ihl-man TCLo:400 µg/kg/3D:CNS,GIT,MET GISAAA 49(10),26,84
orl-rat LD50:100 mg/kg ATXKA8 22,115,66
skn-rat LD50:0.9 mg/kg 85DPAN -,-,71/76
orl-mus LD50:59 mg/kg PEMNDP 8,443,87
scu-rbt LD50:75 mg/kg XPHPAW 414,273,55
orl-gpg LDLo:1400 mg/kg MEMOAQ 4,25,50

orl-ckn LD50:597 mg/kg POSCAL 60,2599,81
orl-brd LD50:56 mg/kg TXAPA9 21,315,72

CONSENSUS REPORTS: NTP 7th Annual Report On Carcinogens. IARC Cancer Review: Animal Sufficient Evidence IMEMDT 5,47,74.

SAFETY PROFILE: Confirmed carcinogen with experimental carcinogenic, neoplastigenic, and tumorigenic data by ingestion and skin contact. Poison by ingestion, skin contact, and subcutaneous routes. Human systemic effects by inhalation: headache, nausea or vomiting, and fever. Implicated in aplastic anemia. Experimental reproductive effects. Mutation data reported. Lindane is more toxic than DDT or dieldrin. When heated to decomposition it emits highly toxic fumes of phosgene, HCl, and Cl^-. Potentially violent reaction with dimethylformamide + iron. When heated to decomposition it emits highly toxic fumes of phosgene, HCl, and Cl^-. See other benzenehexachloride entries.

A toxic organochlorine that is persistent in the environment and accumulates in mammalian tissue. For cattle, the oral LD50 <= 100 mg/kg. The various isomers have different actions; the γ (lindane) and α isomers are central nervous system stimulants, the principal symptom being convulsions. The β and Δ isomers are central nervous system depressants. The use of thermal vaporizers with lindane has caused acute poisoning by inhalation.

The dangerous acute dose of the technical mixture has been estimated at about 30 g and the dangerous dose of lindane at about 7 to 15 g. However, as already mentioned, a single dose of 45 mg (or approximately 0.65 mg/kg) of lindane caused convulsions. Lindane shows a marked difference in toxicity to different species. Its toxic effect on laboratory animals compares favorably with that of DDT, but for several domestic animals, notably calves, lindane is more toxic than DDT or dieldrin. On a chronic systemic basis the α, β, and γ isomers are experimental carcinogens. Has been implicated in aplastic anemia.

Dermatitis and perhaps other manifestations based on sensitivity represent a sort of chronic, though probably not systemic intoxication, which has been observed in humans.

The signs and symptoms of confirmed acute poisoning in humans have paralleled those in experimental animals. These signs and symptoms are: excitation, hyperirritability, loss of equilibrium, clonic-tonic convulsions, and later depression.

There is some evidence that the pulmonary edema and vascular collapse may be of neurogenic origin also. The symptoms in animals systemically poisoned by the γ-isomer alone are essentially similar to those caused by mixtures, although the onset may be earlier. Workers acutely exposed to high air concentrations of lindane and its decomposition products show headache, nausea, and irritation of eyes, nose, and throat.

In rare instances, urticaria has followed exposure to lindane vapor. Unlike the signs and symptoms already mentioned, this allergic manifestation occurs only in susceptible individuals, and usually only after a period of sensitization.

BBQ000 CAS:319-84-6 HR: 3
BENZENE HEXACHLORIDE-α-isomer
mf: $C_6H_6Cl_6$ mw: 290.82

PROP: Solid. Mp: 158°.

SYNS: α-BENZENEHEXACHLORIDE □ α-BHC □ ENT 9,232 □ α-HCH □ α-HEXACHLORANE □ HEXACHLORCYCLOHEXAN (GERMAN) □ α-HEXACHLOROCYCLOHEXANE □ α-1,2,3,4,5,6-HEXACHLOROCYCLOHEXANE (MAK) □ 1-α,2-α,3-β,4-α,5-β,6-β-HEXACHLOROCYCLOHEXANE □ α-LINDANE

TOXICITY DATA WITH REFERENCE
dns-rat:lvr 1 μmol/L CNREA8 42,3010,82
cyt-rat-orl 756 mg/kg/3W JNCIAM 54,1245,75
orl-rat TDLo:20 g/kg/48W-C:NEO JNCIAM 54,801,75
orl-mus TDLo:5 g/kg/24W-C:CAR JNCIAM 51,1637,73
orl-mus TD:10 g/kg/24W-C:CAR NAIZAM 25,635,74
orl-rat TD:11,040 mg/kg/86W-C:ETA CNREA8 41,4140,81
orl-rat TD:13,020 mg/kg/92W-I:ETA CNREA8 41,4140,81
orl-mus TD:12,960 mg/kg/26W-C:CAR CMSHAF 1,279,72
orl-mus TD:8820 mg/kg/21W-C:ETA JJIND8 71,1307,83
orl-mus TD:5040 mg/kg/24W-C:ETA SAIGBL 17,54,75
orl-hmn LDLo:14 g/kg 85GYAZ -,54,71
orl-rat LD50:177 mg/kg FATOAO 39,455,76

CONSENSUS REPORTS: NTP 7th Annual Report On Carcinogens. IARC Cancer Review: Animal Sufficient Evidence IMEMDT 20,195,79; IMEMDT 5,47,74. EPA Genetic Toxicology Program. Reported in EPA TSCA Inventory.

DFG MAK: 0.5 mg/m^3

SAFETY PROFILE: Confirmed carcinogen with experimental carcinogenic, tumorigenic, and neoplastigenic data. Poison by ingestion. Mutation data reported. When heated to decomposition it emits toxic fumes of Cl^-. See also BENZENE HEXACHLORIDE and other benzenehexachloride entries.

BBQ500 CAS:58-89-9 HR: 3
BENZENE HEXACHLORIDE-γ-isomer
mf: $C_6H_6Cl_6$ mw: 290.82

PROP: Solid. Mp: 112.5°.

SYNS: AALINDAN □ AFICIDE □ AGRISOL G-20 □ AGROCIDE □ AGRONEXIT □ AMEISENATOD □ AMEISENMITTEL MERCK □ APARSIN □ APHTIRIA □ APLIDAL □ ARBITEX □ BBH □ BEN-HEX □ BENTOX 10 □ γ-BENZENE HEXACHLORIDE □ BEXOL □ BHC □ γ-BHC □ CELANEX □ CHLORESENE □ CODECHINE □ DBH □ DETMOL-EXTRAKT □ DETOX 25 □ DEVORAN □ DOL GRANULE □ DRILL TOX-SPEZIAL AGLUKON □ ENT 7,796 □ ENTOMOXAN □ EXAGAMA □ FORLIN □ GALLOGAMA □ GAMACID □ GAMAPHEX □ GAMENE □ GAMISO □ GAMMA-COL □ GAMMAHEXA □ GAMMAHEXANE □ GAMMALIN □ GAMMOPAZ □ HCCH □ HCH □ γ-HCH □ HECLOTOX □ HEXACHLORAN □ γ-HEXACHLORAN □ γ-HEXACHLORANE □ γ-HEXACHLOROBENZENE □ 1-α,2-α,3-β,4-α,5-α,6-β-HEXACHLOROCYCLOHEXANE □ γ-HEXACHLOROCYCLOHEXANE (MAK) □ 1,2,3,4,5,6-HEXACHLOROCYCLOHEXANE, γ-ISOMER □ HEXATOX □ HEXICIDE □ HGI □ INEXIT □ ISOTOX □ JACUTIN □ KOKOTINE □ KWELL □ LENDINE □ LENTOX □ LIDENAL □ LINDAGRAIN □ LINDANE (ACGIH, DOT, USDA) □ LINTOX □ MILBOL 49 □ MSZYCOL □ NCI-C00204 □ NEO-SCABICIDOL □ NEXIT □ NOVIGAM □ OVADZIAK □ PEDRACZAK □ QUELLADA □ RCRA WASTE NUMBER U129 □ SANG gamma □ STREUNEX □ TAP 85 □ VITON

TOXICITY DATA WITH REFERENCE
dns-ofs:lvr 45 μmol/L HKXUDL 4,268,84
msc-ham:lng 200 mg/L GISAAA 49(5),82,84
orl-rbt TDLo:60 mg/kg (female 9D post):REP BVIPA7 21,85,77
orl-dog TDLo:473 mg/kg (female 1-63D post):REP 32OAAP -,253,73
orl-rat TDLo:100 mg/kg (female 9D post):TER BVIPA7 21,85,77
orl-rbt TDLo:260 mg/kg (female 6-18D post):REP TXCYAC 9,239,78
itt-rat TDLo:10 mg/kg (male 10D pre):REP APTOA6 31,1,72
orl-rat TDLo:200 mg/kg (6-15D preg):TER TXCYAC 9,239,78
orl-mus TDLo:14 g/kg/2Y-C:CAR CRNGDP 8,1889,87
orl-mus TD:25 g/kg/73W-C:NEO FCTXAV 11,433,73
orl-chd LDLo:180 mg/kg:CNS,PUL CMEP** -,1,56
orl-chd TDLo:111 mg/kg:CNS AEHLAU 25,374,72
skn-man TDLo:20 mg/kg/6W I:EYE,CNS AJDCAI 141,125,87
orl-rat LD50:76 mg/kg SPEADM 74-1,-,74
skn-rat LD50:500 mg/kg WRPCA2 9,119,70
ipr-rat LDLo:35 mg/kg AEPPAE 212,463,51
orl-mus LD50:44 mg/kg JEENAI 65,632,72
ipr-mus LD50:125 mg/kg SOGEBZ 2(1),80,66
orl-dog LD50:40 mg/kg SPEADM 74-1,-,74
ivn-dog LDLo:8 mg/kg TIEUA7 5,61,50
orl-rbt LD50:60 mg/kg JHEMA2 22,115,78
skn-rbt LD50:50 mg/kg AFDOAQ 16,3,52
ivn-rbt LDLo:4500 μg/kg JPETAB 92,140,48
orl-gpg LD50:127 mg/kg FEPRA7 6,386,47
orl-ham LD50:360 mg/kg JETOAS 7,159,74
ipr-ham LD50:640 mg/kg ARTODN 58,152,85
ims-bwd LDLo:26 mg/kg TIEUA7 5,61,50

CONSENSUS REPORTS: NTP 7th Annual Report On Carcinogens. IARC Cancer Review: Animal Sufficient Evidence IMEMDT 5,47,74; IMEMDT 20,195,79. NCI Carcinogenesis Bioassay (feed); No Evidence: mouse, rat NCITR* NCI-CG-TR-14,77. EPA Extremely Hazardous Substances List. EPA Genetic Toxicology Program. Community Right-To-Know List. Reported in EPA TSCA Inventory.

OSHA PEL: TWA 0.5 mg/m³ (skin)
ACGIH TLV: TWA 0.5 mg/m³ (skin)
DFG MAK: 0.5 mg/m³

SAFETY PROFILE: Confirmed carcinogen with experimental carcinogenic neoplastigenic data. A human systemic poison by ingestion. Also a poison by ingestion, skin contact, intraperitoneal, intravenous, and intramuscular routes. Human systemic effects by ingestion: convulsions, dyspnea, and cyanosis. Experimental teratogenic and reproductive effects. Mutation data reported. See also BENZENE HEXACHLORIDE and other benzene hexachloride entries. When heated to decomposition it emits toxic fumes of Cl^-, HCl, and phosgene.

For occupational chemical analysis use NIOSH: Aldrin and Lindane, 5502.

BBQ750 **HR: 3**
BENZENEHEXACHLORIDE (mixed isomers)
mf: $C_6H_6Cl_6$ mw: 290.82

PROP: Technical BHC contains about 64% α, 10% β, 13% γ, 9% Δ and 1% ε isomers of 1,2,3,4,5,6-hexachlorocyclohexane (IARC** 5,47,74).

SYNS: BENZAHEX □ BENZEX □ DOL □ DOLMIX □ FBHC □ FHCH □ 1,2,3,4,5,6-HEXACHLOROCYCLOHEXANE (mixture of isomers) □ HEXYCLAN □ KOTOL □ SOPROCIDE □ TECHNICAL BHC □ TECHNICAL HCH

TOXICITY DATA WITH REFERENCE
orl-mus TD:11 g/kg/26W-C:ETA TXCYAC 19,31,81
orl-mus TDLo:13 g/kg/24W-C:NEO GANNA2 62,431,71
unr-man TDLo:643 μg/kg:CNS CMEP** -,1,56
orl-rat LD50:400 mg/kg 85GMAT -,73,82
orl-mus LD50:500 mg/kg 85GMAT -,73,82
orl-cat LDLo:300 mg/kg 85GMAT -,73,82
ihl-cat LCLo:20 mg/m³/6H 85GMAT -,73,82

CONSENSUS REPORTS: IARC Cancer Review: Animal Sufficient Evidence IMEMDT 5,47,74; IMEMDT 20,195,79.

SAFETY PROFILE: Confirmed carcinogen with experimental tumorigenic and neoplastigenic data. Poison by inhalation and ingestion. Human systemic effects by an unspecified route: convulsions. Potentially dangerous reaction with DMF in presence of Fe, also CCl_4. When heated to decomposition it emits highly toxic fumes of Cl^-, HCl, and phosgene. See also BENZENE HEXACHLORIDE and other benzenehexachloride entries.

BBR000 **CAS:319-85-7** **HR: 3**
***trans*-α-BENZENEHEXACHLORIDE**
mf: $C_6H_6Cl_6$ mw: 290.82

PROP: Solid. Mp: 297°.

SYNS: β-BENZENEHEXACHLORIDE □ β-BHC □ ENT 9,233 □ β-HCH □ β-HEXACHLOROBENZENE □ β-HEXACHLOROCYCLOHEXANE □ 1-α,2-β,3-α,4-β,5-α,6-β-HEXACHLOROCYCLOHEXANE □ β-1,2,3,4,5,6-HEXACHLOROCYCLOHEXANE (MAK) □ β-ISOMER □ β-LINDANE

TOXICITY DATA WITH REFERENCE
orl-mus TDLo:18 g/kg/2Y-C:NEO FCTXAV 11,433,73
orl-rat LD50:6 g/kg ALLVAR 43,-,55
orl-mus LDLo:1500 mg/kg PHTXA6 22,273,59

CONSENSUS REPORTS: NTP 7th Annual Report On Carcinogens. IARC Cancer Review: Animal Sufficient Evidence IMEMDT 5,47,74; Animal Limited Evidence IMEMDT 20,195,79. Reported in EPA TSCA Inventory.

DFG MAK: 0.5 mg/m³

SAFETY PROFILE: Confirmed carcinogen with experimental neoplastigenic data. Mildly toxic by ingestion. When heated to decomposition it emits very toxic fumes of Cl^-, HCl, and phosgene. See also BENZENE HEXACHLORIDE and other benzenehexachloride entries.

For occupational chemical analysis use NIOSH: aldrin and lindane, 5502.

BBR325 **CAS:6996-92-5** **HR: 3**
BENZENESELENIC ACID
mf: $C_6H_6O_2Se$ mw: 189.07

PROP: Plates from water. Mp: 124–125°.

CONSENSUS REPORTS: Selenium and its compounds are on the Community Right-To-Know List.

OSHA PEL: TWA 0.2 mg(Se)/m³
ACGIH TLV: TWA 0.2 mg(Se)/m³
DFG MAK: 0.1 mg(Se)/m³

SAFETY PROFILE: Reacts violently with hydrazine derivatives (e.g., benzohydrazide). When heated to decomposition it emits toxic fumes of Se. See also SELENIUM COMPOUNDS.

BBR380 **CAS:21230-20-6** **HR: 3**
BENZENESULFINYL AZIDE
mf: $C_6H_5N_3OS$ mw: 167.18

SAFETY PROFILE: Explodes at room temperature. Upon decomposition it emits toxic fumes of SO_x and NO_x. See also AZIDES.

BBR390 **CAS:4972-29-6** **HR: 3**
BENZENE SULFINYL CHLORIDE
mf: C_6H_5ClOS mw: 160.56

PROP: Plates. Mp: 38°.

SAFETY PROFILE: May explode if stored in a sealed container. When heated to decomposition it emits toxic fumes of SO_x and Cl^-.

BBR500 **CAS:98-10-2** **HR: 2**
BENZENESULFONAMIDE
mf: $C_6H_7NO_2S$ mw: 157.20

PROP: Solid. Mp: 156°,

SYNS: BENZENESULPHONAMIDE □ BENZOSULFONAMIDE □ BSA

TOXICITY DATA WITH REFERENCE
orl-rat LD50:991 mg/kg MarJV# 29MAR77
orl-mus LD50:740 mg/kg GTPZAB 23(12),47,79
ipr-mus LD50:1000 mg/kg JMCMAR 8,548,65

CONSENSUS REPORTS: Reported in EPA TSCA Inventory.

SAFETY PROFILE: Moderately toxic by ingestion and intraperitoneal routes. When heated to decomposition it emits very toxic fumes of SO_x and NO_x.

BBR750 **CAS:1678-25-7** **HR: 2**
BENZENESULFONANILIDE
mf: $C_{12}H_{11}NO_2S$ mw: 233.30

PROP: Solid. Mp: 110°.

SYN: BENZENESULFANILIDE

TOXICITY DATA WITH REFERENCE
ipr-rat TDLo:593 mg/kg/4W-I:ETA CNREA8 30,1485,70

SAFETY PROFILE: Questionable carcinogen with experimental tumorigenic data. When heated to decomposition it emits very toxic fumes of SO_x and NO_x.

BBS250 **CAS:98-11-3** **HR: 3**
BENZENESULFONIC ACID
mf: $C_6H_6O_3S$ mw: 158.18

PROP: Deliquescent plates or tablets. Mp: 43–44°.

SYN: PHENYLSULFONIC ACID

TOXICITY DATA WITH REFERENCE
skn-rbt 100 μg/24H open AIHAAP 23,95,62
skn-rbt 2 mg/24H SEV 85JCAE-,1053,86
eye-rbt 250 μg/24H SEV 85JCAE-,1053,86
orl-rat LD50:890 mg/kg AIHAAP 23,95,62
orl-bwd LD50:75 mg/kg TXAPA9 21,315,72
skn-cat LDLo:10 g/kg JPETAB 84,358,45

CONSENSUS REPORTS: Reported in EPA TSCA Inventory.

SAFETY PROFILE: Poison by ingestion, skin contact, and probably inhalation. A severe skin and eye irritant. See also SULFATES and SULFONATES.

BBS300 **CAS:80-17-1** **HR: 3**
BENZENESULFONIC HYDRAZIDE
DOT: UN 2970
mf: $C_6H_8N_2O_2S$ mw: 172.22

SYNS: BENZENESULFOHYDRAZIDE □ BENZENESULFONIC ACID, HYDRAZIDE □ BENZENESULFONOHYDRAZIDE □ BENZENESULFONYL HYDRAZIDE □ BENZENESULFONYL HYDRAZINE □ BENZENE SULPHONOHYDRAZIDE □ CELOGEN BSH □ ChKhZ 9 □ GENITRON BSH □ HYDRAZIDE BSG □ NITROPORE OBSH □ PHENYLSULFOHYDRAZIDE □ PHENYLSULFONYL HYDRAZIDE □ PHENYLSULFONYLHYDRAZINE □ POROFOR BSH □ POROFOR-BSH-PULVER □ POROFOR ChKhZ 9

TOXICITY DATA WITH REFERENCE
orl-rat LDLo:50 mg/kg IPSTB3 3,93,76

CONSENSUS REPORTS: Reported in EPA TSCA Inventory.

DOT CLASSIFICATION: 4.1; *Label:* Flammable Solid

SAFETY PROFILE: Poison by ingestion. A flammable solid. When heated to decomposition it emits toxic vapors of NO_x and SO_x

BBS750 **CAS:98-09-9** **HR: 3**
BENZENESULFONYL CHLORIDE
DOT: UN 2225
mf: $C_6H_5ClO_2S$ mw: 176.62

PROP: Liquid. D: 1.384 @ 15°/15°, mp: 14.5°, bp: 251–252°,

SYNS: BENZENE SULFONCHLORIDE □ BENZENESULFONIC (ACID) CHLORIDE □ BENZENE SULPHONYL CHLORIDE (DOT) □ BENZENOSULFOCHLOREK (POLISH) □BENZENOSULPHOCHLORIDE □ BSC-REFINE D □ PHENYLSULFONYL CHLORIDE □ RCRA WASTE NUMBER U020

TOXICITY DATA WITH REFERENCE
orl-rat LD50:1960 mg/kg MEPAAX 20,513,69

ipr-rat LD50:76 mg/kg MEPAAX 20,513,69

CONSENSUS REPORTS: Reported in EPA TSCA Inventory.

DOT CLASSIFICATION: 8; *Label:* Corrosive

SAFETY PROFILE: Poison by intraperitoneal route. A dangerous storage hazard. It may explode in a sealed bottle. Explosive reaction with dimethyl sulfoxide. Reacts vigorously with methyl formamide. When heated to decomposition it emits toxic fumes of Cl^- and SO_x. See also SULFONATES.

BBT000 CAS:20611-21-6 HR: 1
2-(BENZENESULFONYL)ETHANOL
mf: $C_8H_{10}O_3S$ mw: 186.24

SYNS: FENYL-β-HYDROXYETHYLSULFON (CZECH) □ 2-(PHENYLSULFONYL)ETHANOL

TOXICITY DATA WITH REFERENCE
skn-rbt 500 mg/24H MLD 28ZPAK -,200,72
eye-rbt 20 mg/24H MOD 28ZPAK -,200,72
orl-rat LD50:5830 mg/kg 28ZPAK -,200,72

SAFETY PROFILE: Mildly toxic by ingestion. A skin and eye irritant. When heated to decomposition it emits toxic fumes of SO_x.

BBT250 CAS:368-43-4 HR: 3
BENZENESULPHONYL FLUORIDE
mf: $C_6H_5FO_2S$ mw: 160.17

PROP: Clear liquid. Bp: 209°, fp: −5°, flash p: 196°F, d: 1.329, vap press: 8 mm @ 80°, vap d: 5.52.

TOXICITY DATA WITH REFERENCE
ipr-rat LD50:100 mg/kg NATUAS 173,33,54

CONSENSUS REPORTS: Reported in EPA TSCA Inventory.

SAFETY PROFILE: A poison by intraperitoneal routes. Slightly irritating to skin. Flammable when exposed to heat or flame. It can react vigorously with oxidizing materials. To fight fire, use water, foam, CO_2, water spray or mist, dry chemical. When heated to decomposition it emits toxic fumes of F^- and SO_x. See also FLUORIDES and SULFATES.

BBU125 CAS:3470-17-5 HR: 2
BENZENETRIFUROXAN
mf: $C_6N_6O_6$ mw: 252.12

PROP: Prisms from AcOH (aq) or EtOH/EtOAc. Mp: 194–195° (slight decomp).

SYNS: BENZOTRIFUROXAN □ BENZOTRIS(c)FURAZAN-2-OXIDE □ BTF

TOXICITY DATA WITH REFERENCE
skn-rbt 500 mg/24H MLD NTIS** DE83013231
eye-rbt 100 mg/24H SEV NTIS** DE83013231
eye-rbt 100 mg/30S rns SEV NTIS** DE83013231
orl-rat LD50:2884 mg/kg NTIS** DE83013231

CONSENSUS REPORTS: Reported in EPA TSCA Inventory.

SAFETY PROFILE: Moderately toxic by ingestion. A mild skin and severe eye irritant. When heated to decomposition it emits acrid smoke and fumes.

BBU250 CAS:533-73-3 HR: 3
1,2,4-BENZENETRIOL
mf: $C_6H_6O_3$ mw: 126.12

PROP: Plates from Et_2O. Mp: 140.5° (subl). Sol in water.

SYNS: HYDROXYHYDROQUINONE □ HYDROXYQUINOL □ OXYHYDROCHINON (GERMAN) □ OXYHYDROQUINONE □ 1,2,4-TRIHYDROXYBENZENE

TOXICITY DATA WITH REFERENCE
oms-hmn:lym 50 μmol/L CNREA8 45,2471,85
sce-hmn:lym 5 μmol/L CNREA8 45,2471,85
scu-mus LD50:120 mg/kg INHEAO 5,143,67
ipr-mus LDLo:125 mg/kg CBCCT* 6,145,54
scu-mus LD50:122 mg/kg INHEAO 5,143,67

CONSENSUS REPORTS: EPA Genetic Toxicology Program. Reported in EPA TSCA Inventory.

SAFETY PROFILE: Poison by subcutaneous and intraperitoneal routes. Human mutation data reported. When heated to decomposition it emits acrid smoke and irritating fumes.

BBU500 HR: 3
BENZENE TRIOZONIDE
mf: $C_6H_6O_9$ mw: 222.11

SAFETY PROFILE: An unstable explosive, sensitive to the slightest touch. Upon decomposition it emits acrid smoke and fumes.

BBU625 CAS:3691-78-9 HR: 3
BENZETHIDIN
mf: $C_{23}H_{29}NO_3$ mw: 367.53

PROP: Liquid. Bp: 220° @ 0.5 mm.

SYNS: BENZETHIDINE □ ETHYL-1-(2-BENZYLOXYETHYL)-4-PHENYLPIPERIDINE-4-CARBOXYLATE □ NIH 7574 □ 4-PHENYL-1-(2-(PHENYLMETHXY)ETHYL)-4-PIPERIDINECARBOXYLIC ACID ETHYL ESTER

TOXICITY DATA WITH REFERENCE
orl-rat LD50:284 mg/kg BJPCAL 15,254,60
scu-rat LD50:600 mg/kg BJPCAL 15,254,60
ivn-mus LD50:10,900 μg/kg BJPCAL 15,254,60

SAFETY PROFILE: Poison by ingestion and intravenous routes. Moderately toxic by subcutaneous route. When heated to decomposition it emits toxic fumes of NO_x. See also ESTERS.

BBU750 CAS:5929-09-9 HR: 3
BENZETHONIUM CHLORIDE MONOHYDRATE
mf: $C_{27}H_{42}NO_2{\cdot}Cl{\cdot}H_2O$ mw: 466.17

SYNS: p-DIISOBUTYLPHENOXYETHOXYETHYLDIMETHYLBEN-

ZYLAMMONIUM CHLORIDE MONOHYDRATE □ HYAMINE 1622 □ PHEMEROL CHLORIDE MONOHYDRATE

TOXICITY DATA WITH REFERENCE
orl-rat LD50:420 mg/kg PCOC** -,121,66
ipr-rat LD50:33 mg/kg PCOC** -,121,66
ivn-rat LD50:19 mg/kg PCOC** -,121,66
ivn-mus LD50:32 mg/kg CSLNX* NX#00430

SAFETY PROFILE: Poison by intraperitoneal and intravenous routes. Moderately toxic by ingestion. When heated to decomposition it emits toxic fumes of NO_x.

BBU800 CAS:5633-14-7 HR: 3
BENZETIMIDE
mf: $C_{23}H_{26}N_2O_2 \cdot ClH$ mw: 398.97

PROP: Solid. Mp: 270–275°.

SYNS: dl-1-BENZYL-4-(2,6-DIOXO-3-PHENYL-3-PIPERIDYL)PIPERIDINE HYDROCHLORIDE □ BZ □ DIOXATRINE □ R 4929 □ SPASMENTRAL

TOXICITY DATA WITH REFERENCE
ivn-rat LD50:37,600 μg/kg ARZNAD 21,1365,71
orl-mus LD50:680 mg/kg OYYAA2 3,283,69
ivn-mus LD50:46 mg/kg ARZNAD 21,1365,71

SAFETY PROFILE: Poison by intravenous route. Moderately toxic by ingestion. When heated to decomposition it emits toxic fumes of NO_x and HCl. An anticholinergic.

BBV000 CAS:52-49-3 HR: 3
BENZHEXOL HYDROCHLORIDE
mf: $C_{20}H_{31}NO \cdot ClH$ mw: 337.98

PROP: Mp: 258.5° (decomp). Sltly sol in Et_2O, C_6H_6.

SYNS: APARKAN □ ARTANE □ ARTANE HYDROCHLORIDE □ ARTANE TRIHEXYPHENIDYL □ BENZHEXOL CHLORIDE □CYCLO-DOL □ α-CYCLOHEXYL-α-PHENYL-1-PIPERIDINEPROPANOL HYDROCHLORIDE □ PACITANE □ PARALEST □ PARGITAN □ PARKINSAN □ PARKOPAN □ PERAGIT □ 1-PHENYL-1-CYCLOHEXYL-3-PIPERIDYL-1-PROPANOL HYDROCHLORIDE □ PIPANOL □ 3-(1-PIPERIDYL)-1-CYCLOHEXYL-1-PHENYL-1-PROPANOL HYDROCHLORIDE □ ROMPARKIN □ SEDRENA □ TREMIN □ TRIESIFENIDILE □TRIEXIFENIDILA □ TRIHEXYLPHENIDYL HYDROCHLORIDE □ TRIPHEDINON □ TRIPHENIDYL □ TSIKLODOL

TOXICITY DATA WITH REFERENCE
orl-wmn TDLo:800 μg/kg:BAH BJPYAJ 145,300,84
orl-man TDLo:400 μg/kg/1W-I:EYE NEURAI 37,832,87
ipr-rat LD50:195 mg/kg 27ZQAG -,311,72
ivn-rat LD50:30 mg/kg 27ZQAG -,311,72
orl-mus LD50:217 mg/kg NIIRDN 6,525,82
ipr-mus LD50:150 mg/kg PHARAT 37,483,82
scu-mus LD50:152 mg/kg 27ZQAG -,311,72
ivn-mus LD50:39 mg/kg 27ZQAG -,311,72
scu-gpg LD50:320 mg/kg AIPTAK 137,375,62

SAFETY PROFILE: Poison by ingestion, intraperitoneal, intravenous, and subcutaneous routes. An anticholinergic agent which causes human psychotropic effects. Human systemic effects by ingestion: distorted perceptions, eye effects, hallucinations, toxic psychosis. When heated to decomposition it emits very toxic fumes of NO_x and HCl.

BBV250 CAS:613-94-5 HR: 3
BENZHYDRAZIDE
mf: $C_7H_8N_2O$ mw: 136.17

PROP: Crystals from water. Mp: 112.5°. Sol in water, acids, EtOH, C_6H_6, and Me_2CO.

SYNS: BENZOHYDRAZIDE □ BENZOHYDRAZINE □ BENZOIC HYDRAZIDE □ BENZOYL HYDRAZIDE

TOXICITY DATA WITH REFERENCE
orl-mus TDLo:15 g/kg/77W-C:CAR EJCAAH 8,341,72
orl-mus TD:13 g/kg/30W-I:NEO 34ZRA9 -,869,65
scu-mus LD50:122 mg/kg JPETAB 122,110,58
scu-rbt LDLo:102 mg/kg JPETAB 30,87,27

CONSENSUS REPORTS: Reported in EPA TSCA Inventory.

SAFETY PROFILE: Poison by subcutaneous route. Questionable carcinogen with experimental carcinogenic and neoplastigenic data. Violent reaction with benzeneseleninic acid. When heated to decomposition it emits toxic fumes of NO_x.

BBV500 CAS:58-73-1 HR: 3
BENZHYDRYL
mf: $C_{17}H_{21}NO$ mw: 255.39

PROP: Oil. Bp: 163–167° @ 3 mm.

SYNS: ALERYL □ ALLEDRYL □ ALLERGAN B □ ALLERGEVAL □ ALLERGICAL □ ALLERGIN □ ALLERGINA □ ALLERGIVAL □ AMIDRYL □ ANTISTOMINUM □ ANTOMIN □ AUTOMIN □BAGAODRYL □ BARAMINE □ BENA □ BENACHLOR □ BENADON □ BENADRIN □ BENADRYL □ BEN-ALLERGIN □ BENAPON □ BENODIN □ BENODINE □ BENYLAN □ BENZANTINE □ BENZHYDRAMINE □ BENZHYDRAMINUM □ BENZHYDRIL □ o-BENZHY DRYLDIMETHYLAMINOETHANOL □ 2-(BENZHYDRYLOXY)-N,N-DIMETHYLETHYLAMINE □ 2-(BENZOHYDRYLOXY)-N,N-DIMETHYLETHYLAMINE □ BETRAMIN □ DABYLEN □ DEBENDRIN □ DERMISTINE □ DERMODRIN □ DESENTOL □ DIABENYL □ DIABYLEN □ DIBONDRIN □ DIFEDRYL □ DIFENHYDRAMIN □ DIFENIDRAMINA (ITALIAN) □ DIHIDRAL □ DIMEDROL □ DIMEDRYL □ β-DIMETHYLAMINO-AETHYL-BENZHYDRYL-AETHER (GERMAN) □ β-DIMETHYLAMINOETHANOL DIPHENYLMETHYL ETHER □ α-(2-DIMETHYLAMINOETHOXY)DIPHENYLMETHANE □ β-DIMETHYL AMINOETHYLBENZHYDRYLETHER □ DIPHANTINE □ DIPHENYLHYDRAMINE □ 2-(DIPHENYLMETHOXY)-N,N-DIMETHYLETHYLAMINE □ DRYISTAN □ DRYLISTAN □ DYLAMON □ ETANAUTINE □ HISTAXIN □ HYADRINE □ IBIODRAL □ MEDIDRYL □ MEPHADRYL □ NAUSEN □ PROBEDRYL □ RESTAMIN □ RESTAMINE □ RIGIDIL □ RIGIDYL □ S51 □ SYNTEDRIL □ SYNTODRIL □ VENA

TOXICITY DATA WITH REFERENCE
dnd-esc 1 mg/L KHFZAN 16(10),11,82
dni-hmn:fbr 12,500 μg/L DNSYAG 29,829,68
oms-hmn:fbr 12,500 μg/L DNSYAG 29,829,68
cyt-hmn:fbr 100 mg/L ACYTAN 16,41,72
scu-mus TDLo:240 mg/kg (6D pre):REP PSEBAA 89,629,55
unr-mus TDLo:20 mg/kg (1D pre):REP FESTAS 12,346,61
orl-hmn TDLo:714 μg/kg:BAH DDREDK 27,33,92
unr-man LDLo:7353 μg/kg 85DCAI 2,73,70
orl-rat LD50:390 mg/kg RPTOAN 40,42,77
ipr-rat LD50:280 mg/kg IJMRAQ 59,614,71
orl-mus LD50:160 mg/kg CHTPBA 7,224,72

B

ipr-mus LD50:56 mg/kg YKYUA6 34,27,83
scu-mus LD50:50 mg/kg BCFAAI 111,293,72
ivn-mus LD50:29 mg/kg RPTOAN 40,42,77
ipr-gpg LD50:75 mg/kg THERAP 28,767,73
scu-gpg LD50:56 mg/kg ARZNAD 4,189,54

CONSENSUS REPORTS: Reported in EPA TSCA Inventory.

SAFETY PROFILE: Deadly human poison by an unspecified route. Poison by ingestion, intravenous, intraperitoneal, and subcutaneous routes. Experimental reproductive effects. Human systemic effects by ingestion: somnolence, alteration of operant conditioning, changes in psychophysiological tests. Human mutation data reported. When heated to decomposition it emits toxic fumes of NO_x. See also ETHERS.

BBV750 CAS:3733-63-9 HR: 3
1-BENZHYDRYL-4-(2-(2-HYDROXYETHOXY)ETHYL)PIPERAZINE
mf: $C_{21}H_{28}N_2O_2$ mw: 340.51

PROP: Bp: 185° @ 0.005 mm.

SYNS: DECLOXIZINE □ 1-(DIPHENYLMETHYL)-4-(2-(2-HYDROXYETHOXY)ETHYL)PIPERAZINE □ 2-(2-((4-DIPHENYLMETHYL)-1-PIPERAZINYL)ETHOXY)ETHANOL HYDROXYDIETHYLPHENAMINE □ UCB 1402

TOXICITY DATA WITH REFERENCE
orl-rat TDLo:1120 mg/kg (female 8-15D post):TER PSDTAP 9,134,68
orl-rat TDLo:840 mg/kg (female 8-15D post):REP PSDTAP 9,134,68
orl-rat LD50:840 mg/kg ARZNAD 18,1002,68
ipr-rat LD50:103 mg/kg ARZNAD 18,1002,68
ivn-rat LD50:47 mg/kg ARZNAD 18,1002,68
orl-mus LD50:470 mg/kg ARZNAD 18,1002,68
ipr-mus LD50:135 mg/kg ARZNAD 18,1002,68
ivn-mus LD50:45 mg/kg ARZNAD 18,1002,68

SAFETY PROFILE: Poison by intraperitoneal and intravenous routes. Moderately toxic by ingestion. An experimental teratogen. Other experimental reproductive effects. When heated to decomposition it emits toxic fumes of NO_x.

BBW500 CAS:132-69-4 HR: 3
BENZIDAMINE HYDROCHLORIDE
mf: $C_{19}H_{23}N_3O•ClH$ mw: 345.91

PROP: Mp: 160°. Very sol in H_2O.

SYNS: AF 864 □ BENALGIN □ BENZINDAMINE HYDROCHLORIDE □ BENZYDAMINE HYDROCHLORIDE □ 1-BENZYL-3-γ-DIMETHYLAMINOPROPOXY-1H-INDAZOLE HYDROCHLORIDE □ 1-BENZYL-3-(3-(DIMETHYLAMINO)PROPOXY)-1H-INDAZOLE HYDROCHLORIDE □ BENZYRIN □ DIFFLAM □ N,N-DIMETHYL-3((1-PHENYLMETHYL)-1H-INDAZOL-3-YL)OXY-1-PROPANAMINE HYDROCHLORIDE □ DORINAMIN □ ENZAMIN □ EPIROTIN □ IMOTRYL □ INDOLIN □ RIRILIM □ RIRIPEN □ SALYZORON □ TAMAS □ TANTUM □ VERAX

TOXICITY DATA WITH REFERENCE
eye-rbt 200 mg rns MOD ARZNAD 22,724,72
orl-mus TDLo:240 mg/kg (7-12D preg):REP SKNEA7 22,109,72
orl-rat TDLo:5600 mg/kg (15-22D preg/21D post):REP EESADV 5,307,81
orl-rat TDLo:1200 mg/kg (10-15D preg):TER OYYAA2 3,271,69
scu-rat TDLo:180 mg/kg (female 10-15D post):TER OYYAA2 3,271,69
orl-rat TDLo:60 mg/kg (female 9-14D post):TER SKNEA7 22,109,72
orl-cld TDLo:50 mg/kg ATXKA8 23,215,68
orl-rat LD50:740 mg/kg YKKZAJ 99,240,79
scu-rat LD50:720 mg/kg OYYAA2 2,70,68
ipr-rat LD50:100 mg/kg TXAPA9 10,148,67
ivn-rat LD50:43,500 μg/kg OYYAA2 2,70,68
orl-mus LD50:440 mg/kg OYYAA2 16,1011,78
ipr-mus LD50:110 mg/kg TXAPA9 10,148,67
scu-mus LD50:218 mg/kg TXAPA9 10,148,67
ivn-mus LD50:33 mg/kg TXAPA9 10,148,67

SAFETY PROFILE: Poison by intraperitoneal, subcutaneous, and intravenous routes. Moderately toxic by ingestion. An experimental teratogen. Other experimental animal reproductive effects. An eye irritant. A nonsteroidal anti-inflammatory analgesic. When heated to decomposition it emits very toxic fumes of HCl and NO_x.

BBW750 CAS:59-98-3 HR: 3
BENZIDAZOL
mf: $C_{10}H_{12}N_2$ mw: 160.24

PROP: Crystals from pet ether. Mp: 66–68°.

SYNS: ARTONIL □ BENZAZOLINE □ 2-BENZYL-2-IMIDAZOLINE □ 2-BENZYL-4,5-IMIDAZOLINE □ 2-BENZYL-4,5-IMIDAZOLINE HYDROCHLORIDE □ CLORIDRATO DI-2-BENZIL-4,5-IMIDAZOLINA (ITALIAN) □ DIVASCOL □ IMIDALIN □ KASIMID □ LAMBRIL □ OLITENSOL □ PERIPHERINE □ PHENYLMETHYLIMIDAZOLINE □ PREFAXIL □ PRISCOL □ PRISCOLINE □ TOLAZOLINE □ VASIMID □ VASODIL □ VASODILATAN

TOXICITY DATA WITH REFERENCE
cyt-ham:lng 62,500 μg/L GMCRDC 27,95,81
orl-mus LD50:350 mg/kg CPBTAL 22,514,74
ipr-mus LD50:160 mg/kg PBPHAW 1,542,65
ivn-mus LD50:40 mg/kg RPTOAN 37,198,74

SAFETY PROFILE: Poison by ingestion, intraperitoneal, and intravenous routes. Mutation data reported. When heated to decomposition it emits toxic fumes of NO_x.

BBX000 CAS:92-87-5 HR: 3
BENZIDINE
DOT: UN 1885
mf: $C_{12}H_{12}N_2$ mw: 184.26

PROP: Grayish-yellow, crystalline powder; white or sltly reddish crystals, powder, or leaf from water or alc. Mp: 127.5–128.7° @ 740 mm, bp: 401.7°, d: 1.250 @ 20°/4°.

SYNS: BENZIDIN (CZECH) □ BENZIDINA (ITALIAN) □ BENZYDYNA (POLISH) □ p,p-BIANILINE □ 4,4'-BIANILINE □ (1,1'-BIPHENYL)-4,4'-DIAMINE (9CI) □ 4,4'-BIPHENYLDIAMINE □ 4,4'-BIPHENYLENEDIAMINE □ C.I. 37225 □ C.I. AZOIC DIAZO COMPONENT 112 □ p,p'-DIAMINOBIPHENYL □ 4,4'-DIAMINOBIPHENYL □ 4,4'-DIAMI-

NO-1,1′-BIPHENYL ☐ p-DIAMINODIPHENYL ☐ 4,4′-DIAMINODIPHENYL ☐ p,p′-DIANILINE ☐ 4,4′-DIPHENYLENEDIAMINE ☐ FAST CORINTH BASE B ☐ NCI-C03361 ☐ RCRA WASTE NUMBER U021

TOXICITY DATA WITH REFERENCE
dnd-hmn:fbr 3 mmol/L ENMUDM 7,267,85
dnd-rat-ipr 63 mg/kg CRNGDP 6,1285,85
msc-mus:lym 500 μg/L MUREAV 125,291,84
dns-ham:lvr 20 nmol/L MUREAV 136,255,84
oms-dog:oth 100 μmol/L CNREA8 44,1893,84
ihl-man TCLo:17,600 μg/m³/14Y-C:CAR,KID AEHLAU 27,1,73
orl-rat TDLo:108 mg/kg/27D-I:CAR CNREA8 28,924,68
ihl-rat TCLo:10 mg/m³/56W-I:ETA BEXBAN 69,68,70
scu-rat TDLo:2025 mg/kg/27W-I:CAR CANCAR 3,789,50
itr-rat TDLo:315 mg/kg/34W:ETA BEXBAN 69,68,70
scu-mus TDLo:8400 mg/kg/35W-I:ETA VOONAW 17(5),61,71
orl-ham TDLo:75 mg/kg/3Y-C:ETA 85DAAC 5,129,66
scu-mus TD:1620 g/kg/45W-I:ETA AICCA6 7,46,50
scu-rat TD:850 mg/kg/32W-I:ETA VOONAW 20(2),53,74
scu-rat TD:800 mg/kg/60W-I:ETA VOONAW 20(8),69,74
orl-rat LD50:309 mg/kg NTIS** PB214-270
orl-mus LD50:214 mg/kg NTIS** PB214-270
ipr-mus LD50:110 mg/kg PMRSDJ 1,682,81
orl-dog LDLo:200 mg/kg AEXPBL 58,167,1907
orl-rbt LDLo:200 mg/kg AEXPBL 58,167,1907

CONSENSUS REPORTS: NTP 7th Annual Report On Carcinogens. IARC Cancer Review: Human Limited Evidence IMEMDT 1,80,72; Human Sufficient Evidence IMEMDT 29,149,82; Animal Sufficient Evidence IMEMDT 1,80,72; IMEMDT 29,149,82. EPA Genetic Toxicology Program. Community Right-To-Know List. Reported in EPA TSCA Inventory.

OSHA: Cancer Suspect Agent
ACGIH TLV: Confirmed Human Carcinogen
DFG MAK: Human Carcinogen.
DOT CLASSIFICATION: 6.1; *Label:* Poison

SAFETY PROFILE: Confirmed human carcinogen producing bladder tumors. Experimental carcinogenic and tumorigenic data. Poison by ingestion and intraperitoneal routes. Human mutation data reported. Can cause damage to blood, including hemolysis and bone marrow depression. On ingestion causes nausea and vomiting, which may be followed by liver and kidney damage. Any exposure is considered extremely hazardous. When heated to decomposition it emits highly toxic fumes of NO_x. See also AROMATIC AMINES.

For occupational chemical analysis use OSHA: #ID-65 or NIOSH: Benzidine in Urine (TLC), 8304; Benzidine in Urine (GC), 8306.

BBX250 CAS:16993-94-5 HR: 2
3,3′-BENZIDINE DICARBOXYLIC ACID, DISODIUM SALT
mf: $C_{14}H_{10}N_2O_4{\cdot}2Na$ mw: 316.24

SYN: 4,4′-DIAMINO-3,3′-BIPHENYLDICARBOXYLIC ACID DISODIUM SALT

TOXICITY DATA WITH REFERENCE
scu-rat TDLo:7 g/kg/77W-I:ETA VOONAW 15(5),60,69
scu-mus TDLo:9 g/kg/43W-I:ETA VOONAW 15(5),60,69

SAFETY PROFILE: Questionable carcinogen with experimental tumorigenic data. When heated to decomposition it emits toxic fumes of NO_x and Na_2O.

BBX500 CAS:117-61-3 HR: 2
2,2′-BENZIDINEDISULFONIC ACID
mf: $C_{12}H_{12}N_2O_6S_2$ mw: 344.38

PROP: Prisms.

SYNS: 6,6′-BIMETANILIC ACID ☐ 4,4′-DIAMINOBIPHENYL-2,2′-DISULFONIC ACID ☐ 4,4′-DIAMINO-2,2′-BIPHENYLDISULFONIC ACID ☐ 4,4′-DIAMINODIPHENYL-2,2′-DISULFONIC ACID ☐ 2,2′-DISULFOBENZIDINE ☐ KYSELINA BENZIDIN-2,2′-DISULFONOVA (CZECH)

TOXICITY DATA WITH REFERENCE
eye-rbt 500 mg/24H SEV 28ZPAK -,191,72

CONSENSUS REPORTS: Reported in EPA TSCA Inventory.

SAFETY PROFILE: A severe eye irritant. See also SULFONATES. When heated to decomposition it emits very toxic fumes of SO_x and NO_x.

BBX750 CAS:531-85-1 HR: 2
BENZIDINE HYDROCHLORIDE
mf: $C_{12}H_{12}N_2{\cdot}2ClH$ mw: 257.18

PROP: Leaflets. Sol in H_2O.

SYNS: (1,1′-BIPHENYL)-4,4′-DIAMINE, DIHYDROCHLORIDE ☐ DIHIDROCLORURO de BENZIDINA (SPANISH)

TOXICITY DATA WITH REFERENCE
mmo-sat 100 nmol/plate MUREAV 136,33,84
sce-ham-ipr 12,500 μg/kg MUREAV 113,33,83
ipr-rat TDLo:62 mg/kg/4W-I:CAR CRNGDP 2,747,81
orl-mus TDLo:3360 mg/kg/80W-C:CAR TXAPA9 64,171,82
orl-mus TD:1600 mg/kg/84W-C:ETA CNREA8 35,2814,75
orl-mus TD:5040 mg/kg/60W-C:CAR TXAPA9 64,171,82
orl-ham TDLo:75 mg/kg/3Y-C:ETA 85DAAC 5,129,66
orl-mus TD:8064 mg/kg/80W-C:CAR EJCAAH 16,1205,80
orl-mus TD:1680 mg/kg/40W-C:ETA TOPADD 9,1,81
orl-mus TD:13,440 mg/kg/1Y-C:ETA FAATDF 4,69,84

CONSENSUS REPORTS: Reported in EPA TSCA Inventory. EPA Genetic Toxicology Program.

SAFETY PROFILE: Suspected carcinogen with experimental carcinogenic and tumorigenic data. Human mutation data reported. When heated to decomposition it emits very toxic fumes of HCl and NO_x.

BBY000 CAS:531-86-2 HR: 3
BENZIDINE SULFATE
mf: $C_{12}H_{12}N_2{\cdot}H_2O_4S$ mw: 282.34

SYN: (1,1′-BIPHENYL)-4,4′-DIAMINE SULFATE (1:1)

TOXICITY DATA WITH REFERENCE
scu-rat TDLo:2475 mg/kg/33W-I:CAR CANCAR 3,789,50
scu-rat TD:3900 mg/kg/52W-I:CAR GTPZAB 19(6),28,75

OSHA: Carcinogen

SAFETY PROFILE: Confirmed human carcinogen with

experimental carcinogenic data. See also BENZIDINE and SULFATES. When heated to decomposition it emits toxic fumes of SO_x and NO_x.

BBY250 **CAS:2051-89-0** **HR: 2**
BENZIDINE-3-SULFURIC ACID
mf: $C_{12}H_{12}N_2O_3S$ mw: 264.32

PROP: Crystals.

SYNS: BENZIDINE-3-SULPHURIC ACID □ 4,4'-DIAMINO-3-BIPHENYL-3-SULFONIC ACID □ 4:4'-DIAMINO-3-DIPHENYLYL HYDROGEN SULFATE □ 3-SULFOBENZIDINE

TOXICITY DATA WITH REFERENCE
imp-mus TDLo:80 mg/kg:ETA BJCAAI 17,127,63

CONSENSUS REPORTS: Reported in EPA TSCA Inventory.

SAFETY PROFILE: Questionable carcinogen with experimental tumorigenic data. See also SULFATES. When heated to decomposition it emits very toxic fumes of NO_x and SO_x.

BBY300 **HR: 2**
BENZIDINE SULPHATE and HYDRAZINE-BENZENE
mf: $C_6H_8N_2 \cdot C_{12}H_{12}N_2 \cdot H_2O_4S$ mw: 390.50

SYN: HYDRAZINE-BENZENE and BENZIDINE SULFATE

TOXICITY DATA WITH REFERENCE
scu-rat TDLo:9100 mg/kg/52W-I:CAR GTPZAB 19(6),28,75

SAFETY PROFILE: Suspected carcinogen with experimental carcinogenic data. When heated to decomposition it emits toxic fumes of NO_x and SO_x.

BBY500 **CAS:3365-94-4** **HR: 2**
BENZIDIN-3-YL ESTER SULFURIC ACID
mf: $C_{12}H_{12}N_2O_4S$ mw: 280.32

SYNS: BENZIDIN-3-YL HYDROGEN SULFATE □ 4,4'-DIAMINO-3-DIPHENYLYL HYDROGEN SULFATE

TOXICITY DATA WITH REFERENCE
imp-mus TDLo:80 mg/kg:ETA BJCAAI 17,127,63

SAFETY PROFILE: Questionable carcinogen with experimental tumorigenic data. See also ESTERS and SULFURIC ACID. When heated to decomposition it emits very toxic fumes of NO_x and SO_x.

BBY750 **CAS:134-81-6** **HR: 2**
BENZIL
mf: $C_{14}H_{10}O_2$ mw: 210.24

PROP: Yellow crystals from alc. Mp: 95°, bp: 346–348°, d: 1.23 @ 15°/4°, vap press: 1 mm @ 128.4°.

SYNS: DIBENZOYL □ DIPHENYL-α,β-DIKETONE □ 1,2-DIPHENYLETHANEDIONE □ DIPHENYLGLYOXAL

TOXICITY DATA WITH REFERENCE
eye-rbt 100 mg/24H SEV 28ZPAK -,43,72
orl-mus LD50:>3 g/kg IYKEDH 15,359,84

CONSENSUS REPORTS: Reported in EPA TSCA Inventory.

SAFETY PROFILE: Low toxicity by ingestion. An eye irritant. Combustible. When heated to decomposition it emits acrid smoke and irritating fumes. See also KETONES.

BBY990 **CAS:76-93-7** **HR: 2**
BENZILIC ACID
mf: $C_{14}H_{12}O_3$ mw: 228.26

SYNS: ACIDE DIPHENYLHYDROXYACETIQUE □ BENZENEACETIC ACID, α-HYDROXY-α-PHENYL-(9CI) □ DIPHENYLGLYCOLIC ACID □ α-α-DIPHENYLGLYCOLIC ACID □ DIPHENYLHYDROXYACETIC ACID □ HYDROXYDIPHENYLACETIC ACID □ α-HYDROXY-α-PHENYLBENZENEACETIC ACID

TOXICITY DATA WITH REFERENCE
orl-mus LD50:2 g/kg AIPTAK 116,154,58
scu-mus LD50:1300 mg/kg AIPTAK 116,154,58

CONSENSUS REPORTS: Reported in EPA TSCA Inventory.

SAFETY PROFILE: Moderately toxic by subcutaneous route. Slightly toxic by ingestion. When heated to decomposition it emits acrid smoke and irritating vapors.

BCA000 **CAS:57-37-4** **HR: 3**
BENZILIC ACID-β-DIETHYLAMINOETHYL ESTER HYDROCHLORIDE
mf: $C_{20}H_{25}NO_3 \cdot ClH$ mw: 363.92

PROP: Crystals from Me_2CO. Mp: 177–178°. Sol in H_2O, insol in Et_2O.

SYNS: ACTOZINE □ AMIOYL □ AMISYL □ AMITAKON □ AMIZIL HYDROCHLORIDE □ ARCADINE □ AY-5406 □ BENACTIZINE HYDROCHLORIDE □ BENACTYZIN (CZECH) □ BENACTYZINE CHLORIDE □ BENACTYZINE HYDROCHLORIDE □ BENAKTIN □ BENZILATE DU DIETHYLAMINO-ETHANOL CHLORHYDRATE (FRENCH) □ CAFRON □ CEDAD □ CEVANOL □ DESTENDO □ β-DIETHYLAMINOETHYL BENZILATE HYDROCHLORIDE □ 2-DIETHYLAMINOETHYL BENZILATE HYDROCHLORIDE □ 2-DIETHYLAMINOETHYL DIPHENYLGLYCOLATE HYDROCHLORIDE □ 2-(DIFENYL-HYDROXYACETOXY)ETHYL-DIETHYLAMMONIUMCHLORID (CZECH) □ DIPHENYLGLYCOLLIC ACID-2-(DIETHYLAMINO)ETHYL ESTER HYDROCHLORIDE □ FOBEX □ IBIOTYZIL □ KATRON □ LEUCIDIL □ NERVACTON □ NERVATIL □ NEURAKTIL □ NEUROBENZIL □ NEUROLEPTONE □ NUTINAL □ PARASAN □ PARPON □ PHOBEX □ PROCALM □ STOIKON □ SUAVITIL □ TRANQUILLIN □ VALLADAN □ WIN 5606

TOXICITY DATA WITH REFERENCE
scu-rat TDLo:500 μg/kg (ID male):REP PSYPAG 10,44,66
orl-hmn TDLo:14 μg/kg:CNS 27ZQAG -,363,72
orl-rat LD50:184 mg/kg TXAPA9 1,42,59
ipr-rat LD50:100 mg/kg APTOA6 11,405,55
orl-mus LD50:160 mg/kg 27ZQAG -,363,72
ipr-mus LD50:76 mg/kg JPETAB 74,274,42
scu-mus LD50:250 mg/kg 27ZQAG -,363,72
ivn-mus LD50:14,300 μg/kg 28ZPAK -,253,72
idr-mus LD50:350 mg/kg AIPTAK 59,149,38
ipr-rbt LD50:100 mg/kg APTOA6 11,405,55
ivn-rbt LD50:15 mg/kg 27ZQAG -,363,72

ipr-gpg LD50: 100 mg/kg APTOA6 11,405,55

CONSENSUS REPORTS: Reported in EPA TSCA Inventory.

SAFETY PROFILE: Poison by ingestion, intraperitoneal, subcutaneous, intradermal, and intravenous routes. Human systemic effects by ingestion of very small amounts: toxic psychosis. Experimental reproductive effects. When heated to decomposition it emits very toxic fumes of NO_x and HCl.

BCA375 CAS:73954-17-3 HR: 3
8-BENZILOYLOXY-6,10-ETHANO-5-AZONIASPIRO (4.5)DECDANE CHLORIDE
mf: $C_{25}N_{30}NO_3 \cdot Cl$ mw: 428.01

SYNS: 6,10-ETHANO-5-AZONIASPIRO(4.5)DECAN-8-OL CHLORIDE BENZILATE □ 3-HYDROXY-SPIRO(8-AZONIABICYCLO(3.2.1)OCTANE-8,1'-PYRROLIDINIUM CHLORIDE) BENZILATE

TOXICITY DATA WITH REFERENCE
orl-rat LD50: 1501 mg/kg IYKEDH 4,90,73
ipr-rat LD50: 103 mg/kg ARZNAD 16,1581,66
scu-rat LD50: 707 mg/kg IYKEDH 4,90,73
ivn-rat LD50: 15,500 μg/kg IYKEDH 4,90,73
orl-mus LD50: 750 mg/kg ARZNAD 16,1581,66
ipr-mus LD50: 50 mg/kg IYKEDH 4,90,73
scu-mus LD50: 203 mg/kg IYKEDH 4,90,73
ivn-mus LD50: 11,200 μg/kg IYKEDH 4,90,73
ims-mus LD50: 89 mg/kg ARZNAD 16,1581,66

SAFETY PROFILE: Poison by subcutaneous, intramuscular, intravenous, and intraperitoneal routes. Moderately toxic by ingestion. When heated to decomposition it emits toxic fumes of Cl^- and NO_x.

BCB000 CAS:67360-95-6 HR: 3
4-BENZILOYLOXY-1,1,2,2,6-PENTAMETHYLPIPERIDINIUM CHLORIDE (β FORM)
mf: $C_{24}H_{32}NO_3 \cdot Cl$ mw: 418.02

SYNS: 4-BENZILYLOXY-1,2,2,6-TETRAMETHYLPIPERIDINE METHOCHLORIDE (β FORM) □ 4-((HYDROXYDIPHENYLACETYL)OXY)-1,1,2,2,6-PENTAMETHYLPIPERIDINIUM CHLORIDE (β FORM)

TOXICITY DATA WITH REFERENCE
orl-mus LD50: 1000 mg/kg JPETAB 85,85,45
ipr-mus LD50: 75 mg/kg JPETAB 85,85,45
scu-mus LD50: 325 mg/kg JPETAB 85,85,45

SAFETY PROFILE: Poison by intraperitoneal and subcutaneous routes. Moderately toxic by ingestion. When heated to decomposition it emits very toxic fumes of Cl^- and NO_x.

BCB250 CAS:67360-95-6 HR: 3
4-BENZILOYLOXY-1,1,2,6,6-PENTAMETHYLPIPERIDINIUM CHLORIDE (α FORM)
mf: $C_{24}H_{32}NO_3 \cdot Cl$ mw: 418.02

SYNS: 4-BENZILYLOXY-1,2,2,6-TETRAMETHYLPIPERIDINE METHOCHLORIDE (α FORM) □ 4-((HYDROXYDIPHENYLACETYL)OXY)-1,1,2,2,6-PENTAMETHYLPIPERIDINIUM CHLORIDE (α FORM)

TOXICITY DATA WITH REFERENCE
ipr-mus LD50: 800 mg/kg JPETAB 85,85,45
scu-mus LD50: 375 mg/kg JPETAB 85,85,45

SAFETY PROFILE: Poison by subcutaneous route. Moderately toxic by intraperitoneal route. When heated to decomposition it emits very toxic fumes of Cl^- and NO_x.

BCB750 CAS:51-17-2 HR: 3
BENZIMIDAZOLE
mf: $C_7H_6N_2$ mw: 118.15

PROP: Tabular crystals or plates. Mp: 170.5°, bp: >360°. Sol in alc; sparingly sol in water.

SYNS: 3-AZAINDOLE □ AZINDOLE □ o-BENZIMIDAZOLE □ 1H-BENZIMIDAZOLE (9CI) □ BENZIMINAZOLE □ 1,3-BENZODIAZOLE □ BENZOIMIDAZOLE □ BZI □ 1,3-DIAZAINDENE □ N,N'-METHENYL-o-PHENYLENEDIAMINE □ NSC 759

TOXICITY DATA WITH REFERENCE
mmo-sat 250 μg/plate CHIMAD 27,68,73
mmo-esc 1 mg/disc APMBAY 6,23,58
dnd-esc 15 mmol/L/48H ANBCA2 75,45,76
pic-esc 1 g/L ZAPOAK 12,583,72
orl-rat LDLo:500 mg/kg NCNSA6 5,22,53
ipr-rat LD50: 385 mg/kg AIPTAK 95,123,53
orl-mus LD50: 2910 mg/kg JPETAB 105,486,52
ivn-mus LD50: 280 mg/kg 29QHAQ -,246,74

CONSENSUS REPORTS: Reported in EPA TSCA Inventory.

SAFETY PROFILE: Poison by intravenous and intraperitoneal routes. Moderately toxic by ingestion. Mutation data reported. When heated to decomposition it emits highly toxic fumes of NO_x.

BCC000 CAS:4414-88-4 HR: 3
2-BENZIMIDAZOLEACETONITRILE
mf: $C_9H_7N_3$ mw: 157.19

SYNS: 2-BENZIMIDAZOLYLACETONITRILE □ 2-KYANMETHYLBENZIMIDAZOL (CZECH)

TOXICITY DATA WITH REFERENCE
ivn-mus LD50: 56 mg/kg CSLNX* NX#04148

CONSENSUS REPORTS: Reported in EPA TSCA Inventory. Cyanide and its compounds are on the Community Right-To-Know List.

SAFETY PROFILE: Poison by intravenous route. See also NITRILES. When heated to decomposition it emits toxic fumes of NO_x.

BCC250 CAS:6898-43-7 HR: 2
BENZIMIDAZOLE METHYLENE MUSTARD
mf: $C_{14}H_{19}Cl_2N_3 \cdot ClH$ mw: 336.72

SYNS: BENZIMIDAZOLE MUSTARD □ 2-(BIS(2-CHLOROETHYL) AMINOMETHYL)-5,5-DIMETHYLBENZIMIDAZOLE HYDROCHLORIDE □ 2-(DI-2-CHLOROETHYL)AMINOMETHYL-5,6-DIMETHYLBENZIMIDAZOLE □ NSC-23892

TOXICITY DATA WITH REFERENCE
ipr-mus TDLo: 12 mg/kg/4W: CAR JNCIAM 36,915,66

SAFETY PROFILE: Questionable carcinogen with ex-

perimental carcinogenic data. When heated to decomposition it emits very toxic fumes of HCl and NO_x.

BCC500 **CAS:583-39-1** **HR: 3**
2-BENZIMIDAZOLETHIOL
mf: $C_7H_6N_2S$ mw: 150.21

PROP: Plates from alc (aq). Mp: 298°. Sol in EtOH; sltly sol in H_2O.

SYNS: ANTIEGENE MB □ ANTIOXIDANT MB (CZECH) □ AOMB □ ASM MB □ 2-MERCAPTOBENZIMIDAZOLE □ MERCAPTOBENZOIMIDAZOLE □ 2-MERCAPTOBENZOIMIDAZOLE □ MERKAPTOBENZIMIDAZOL (CZECH) □ NCI-C60980 □ o-PHENYLENETHIOUREA □ USAF EK-6540 □ USAF XF-21

TOXICITY DATA WITH REFERENCE
skn-rbt 500 mg/24H MLD 28ZPAK -,168,72
eye-rbt 500 mg/24H MLD 28ZPAK -,168,72
orl-mus LD50:750 mg/kg FRZKAP 17(1),36,62
ipr-mus LD50:200 mg/kg NTIS** AD277-689
ivn-mus LD50:180 mg/kg CSLNX* NX#04376

CONSENSUS REPORTS: Reported in EPA TSCA Inventory.

SAFETY PROFILE: Poison by intraperitoneal and intravenous routes. Moderately toxic by ingestion. Skin and eye irritant. When heated to decomposition it emits toxic fumes of SO_x and NO_x. See also MERCAPTANS.

BCD125 **CAS:52096-22-7** **HR: 3**
BENZIMIDAZOLIUM-1-NITROIMIDATE
mf: $C_7H_5N_4O_2$ mw: 177.14

SAFETY PROFILE. Explodes at its mp: 169°C. Upon decomposition it emits toxic fumes of NO_x.

BCD325 **CAS:21035-25-6** **HR: 3**
1-(2-BENZIMIDAZOLYL)-3-METHYLUREA
mf: $C_9H_{10}N_4O$ mw: 190.23

SYN: BCM (NH)

TOXICITY DATA WITH REFERENCE
oms-hmn:leu 1 mg/L THERAP 31,505,76
oms-hmn:oth 2 mg/L THERAP 31,505,76
orl-rat TDLo:152 mg/kg (8-15D preg):TER THERAP 31,505,76
ivn-mus LD50:56 mg/kg CSLNX* NX#03246

SAFETY PROFILE: Poison by intravenous route. Human mutation data reported. An experimental teratogen. When heated to decomposition it emits toxic fumes of NO_x.

BCD750 **CAS:642-72-8** **HR: 3**
BENZINDAMINE
mf: $C_{19}H_{23}N_3O$ mw: 309.45

PROP: Bp: 160° @ 0.05 mm.

SYNS: BENZYDAMINE □ 1-BENZYL-3-(3-(DIMETHYLAMINO)PROPOXY)-1H-INDAZOLE

TOXICITY DATA WITH REFERENCE
orl-rat LD50:9500 mg/kg ARZNAD 22,711,72
orl-mus LD50:460 mg/kg JMCMAR 15,923,72
ipr-mus LD50:109 mg/kg JMCMAR 15,471,72
scu-mus LD50:445 ng/kg OYYAA2 6,1285,72
ivn-mus LD50:25 mg/kg OYYAA2 6,1285,72

SAFETY PROFILE: Poison by intraperitoneal and intravenous routes. Moderately toxic by ingestion and subcutaneous routes. When heated to decomposition it emits toxic fumes of NO_x.

BCE000 **CAS:208-07-1** **HR: 3**
BENZ(e)INDENO(1,2-b)INDOLE
mf: $C_{19}H_{11}N$ mw: 253.31

SYN: 4,5-BENZO-2,3-1′,2′-INDENOINDOLE (FRENCH)

TOXICITY DATA WITH REFERENCE
scu-mus TDLo:120 mg/kg/9W-I:ETA BAFEAG 42,3,55
scu-mus LDLo:40 mg/kg BAFEAG 42,3,55

SAFETY PROFILE: Poison by subcutaneous route. Questionable carcinogen with experimental tumorigenic data. When heated to decomposition it emits toxic fumes of NO_x.

BCE250 **CAS:5585-71-7** **HR: 3**
BENZINDOPYRINE HYDROCHLORIDE
mf: $C_{22}H_{20}N_2$•ClH mw: 348.90

PROP: Solid. Mp: 199–200°.

SYNS: 4-(1-BENZYL-3-INDOLETHYL)PYRIDINE HYDROCHLORIDE □ 1-BENZYL-3-(2-(4-PYRIDYL)ETHYL)INDOLE HYDROCHLORIDE

TOXICITY DATA WITH REFERENCE
orl-hmn TDLo:714 µg/kg/D:PSY AMCTAH 6,521,59
ipr-mus LD50:520 mg/kg JPETAB 125,122,59
ivn-mus LD50:98 mg/kg JPETAB 125,122,59

SAFETY PROFILE: Poison by intravenous route. Moderately toxic by intraperitoneal route. Human psychotropic effects via ingestion. When heated to decomposition it emits very toxic fumes of NO_x and HCl.

BCE475 **CAS:2634-33-5** **HR: 2**
1,2-BENZISOTHIAZOL-3(2H)-ONE
mf: C_7H_5NOS mw: 151.19

SYNS: 1,2-BENZISOTHIAZOLIN-3-ONE □ PROXEL PL

TOXICITY DATA WITH REFERENCE
orl-rat LD50:1020 mg/kg PLRCAT 3,385,71
orl-mus LD50:1150 mg/kg PLRCAT 3,385,71

CONSENSUS REPORTS: Reported in EPA TSCA Inventory.

SAFETY PROFILE: Moderately toxic by ingestion. When heated to decomposition it emits toxic vapors of NO_x and SO_x.

BCE500 **CAS:81-07-2** **HR: 3**
1,2-BENZISOTHIAZOL-3(2H)-ONE-1,1-DIOXIDE
mf: $C_7H_5NO_3S$ mw: 183.19

PROP: White crystals or powder from water; odorless

with sweet taste. Mp: 224° (decomp), bp: subl. Sol in water, alc, chloroform, and ether.

SYNS: ANHYDRO-o-SULFAMINEBENZOIC ACID □ 3-BENZISOTHIAZOLINONE-1,1-DIOXIDE □ o-BENZOIC SULPHIMIDE □ o-BENZOSULFIMIDE □ BENZOSULPHIMIDE □ BENZO-2-SULPHIMIDE □ o-BENZOYL SULFIMIDE □ o-BENZOYL SULPHIMIDE □ 1,2-DIHYDRO-2-KETOBENZISOSULFONAZOLE □ 1,2-DIHYDRO-2-KETOBENZISOSULPHONAZOLE □ 2,3-DIHYDRO-3-OXOBENZISOSULFONAZOLE □ 2,3-DIHYDRO-3-OXOBENZISOSULPHONAZOLE □ GARANTOSE □ GLUCID □ GLUSIDE □ HERMESETAS □ 3-HYDROXYBENZISOTHIAZOL-S,S-DIOXIDE □ INSOLUBLE SACCHARINE □ KANDISET □NA-TREEN □ RCRA WASTE NUMBER U202 □ SACARINA □ SACCAHARIMIDE □ SACCHARINA □ SACCHARIN ACID □ SACCHARINE □SACCHARINOL □ SACCHARINOSE □ SACCHAROL □ SAXIN □ SUCRE EDULCOR □ SUÇRETTE □ o-SULFOBENZIMIDE □ o-SULFOBENZOIC ACID IMIDE □ 2-SULPHOBENZOIC IMIDE □ SYKOSE □ SYNCAL □ ZAHARINA

TOXICITY DATA WITH REFERENCE
cyt-smc 200 mg/L NATUAS 294,263,81
dnd-rat:lvr 3 mmol/L SinJF# 26OCT82
dns-rat:lvr 100 pmol/L CRNGDP 5,1547,84
dnd-mus-ipr 100 mg/kg ATSUDG (5),355,82
sce-ham:lng 100 mg/L BJCAAI 45,769,82
orl-mus TDLo:101 g/kg (female 1-21D post):REP DBTEAD 17,103,69
orl-mus TDLo:155 mg/kg (female 7D post):TER IIZAAX 16,330,64
orl-mus TDLo:16,800 mg/kg (MGN):REP TXCYAC 8,285,77
orl-rat TDLo:2008 g/kg/2Y-C:ETA JAPMA8 40,583,51
orl-mus TDLo:548 g/kg/1Y-C:ETA IJEBA6 24,197,86
skn-mus TDLo:9600 mg/kg/10W-I:ETA BJCAAI 10,363,56
imp-mus TDLo:80 mg/kg:NEO BJCAAI 11,212,57
orl-mus LD50:17 g/kg EXPEAM 35,1364,79

CONSENSUS REPORTS: NTP 7th Annual Report On Carcinogens. IARC Cancer Review: Group 2B IMEMDT 7,334,87, Human Inadequate Evidence IMEMDT 22,111,80; Animal Sufficient Evidence IMEMDT 22,111,80. EPA Genetic Toxicology Program. Reported in EPA TSCA Inventory. Community Right-To-Know List.

SAFETY PROFILE: Confirmed carcinogen with experimental neoplastigenic and tumorigenic data. Mild acute toxicity by ingestion. Experimental teratogenic and reproductive effects. Mutation data reported. When heated to decomposition it emits toxic NO_x and SO_x.

BCE750 CAS:68291-97-4 HR: 2
1,2-BENZISOXAZOLE-3-METHANESULFONAMIDE
mf: $C_8H_8N_2O_3S$ mw: 212.24

PROP: Crystals from EtOAc. Mp: 160–163°.

SYNS: AD-810 □ 3-SULFAMOYLMETHYL-1,2-BENZISOXAZOLE

TOXICITY DATA WITH REFERENCE
orl-rat TDLo:220 mg/kg (female 7-17D post):REP YACHDS 15,4399,87
orl-dog TDLo:1320 mg/kg (female 14-35D post):TER YACHDS 15,4435,87
orl-rat TDLo:660 mg/kg (female 7-17D post):TER YACHDS 15,4399,87
orl-rat TDLo:5160 mg/kg (male 9W pre):REP YACHDS 15,4387,87
orl-rat TDLo:13 mg/kg (male 64D pre):REP YACHDS 15,4387,87
orl-mus TDLo:5 g/kg (female 6-15D post):TER YACHDS 15,4435,87
orl-rat LD50:1992 mg/kg YACHDS 15,4337,87
ipr-rat LD50:733 mg/kg ARZNAD 30,477,80
scu-rat LD50:925 mg/kg YACHDS 15,4337,87
ivn-rat LD50:672 mg/kg YACHDS 15,4337,87
orl-mus LD50:1892 mg/kg ARZNAD 30,477,80
ipr-mus LD50:699 mg/kg ARZNAD 30,477,80
scu-mus LD50:1009 mg/kg YACHDS 15,4337,87
orl-dog LD50:1 g/kg YACHDS 15,4337,87

SAFETY PROFILE: Moderately toxic by ingestion, intraperitoneal, subcutaneous, and intravenous routes. An experimental teratogen. Other experimental reproductive effects. When heated to decomposition it emits very toxic fumes of SO_x and NO_x. An anticonvulsant.

BCE825 CAS:15301-48-1 HR: 3
BENZITRAMIDE
mf: $C_{31}H_{32}N_4O_2$ mw: 492.67

PROP: White, crystalline powder. Mp: 145–149°. Also reported as pale yellow amorphous powder, mp: 124.5–126°. Solubility above 1 g/100 mL in ethyl acetate, acetone, benzene, chloroform. Almost insol in water and dilute acids.

SYNS: BEZITRAMIDE □ BURGODIN □ 1-(3-CYANO-3,3-DIPHENYLPROPYL)-4-(2-OXO-3-PROPIONYL-1-BENZIMIDAZOLINYL)PIPERIDINE □ 1-(1-(3-CYANO-3,3-DIPHENYLPROPYL)-4-PIPERIDYL)-3-PROPIONYL-2-BENZIMIDAZOLINONE □ R-4845

TOXICITY DATA WITH REFERENCE
orl-rat LD50:141 mg/kg MEIEDD 10,170,83
orl-mus LD50:2101 mg/kg MEIEDD 10,170,83
orl-dog LD50:80 mg/kg ARZNAD 21,862,71
orl-gpg LD50:60,400 μg/kg ARZNAD 21,862,71

CONSENSUS REPORTS: Cyanide and its compounds are on the Community Right-To-Know List.

SAFETY PROFILE: Poison by ingestion. Caution: May be habit forming. This is a controlled substance (opiate) listed in the U.S. Code of Federal Regulations, Title 21 Part 1308.12 (1985). When heated to decomposition it emits toxic fumes of NO_x and CN^-.

BCF500 CAS:1491-10-7 HR: 2
BENZO(f)(1)BENZOTHIENO(3,2-b)QUINOLINE
mf: $C_{19}H_{11}NS$ mw: 285.37

SYN: NAPHTHO(1,2-e)THIANAPHTHENO(3,2-b)PYRIDINE

TOXICITY DATA WITH REFERENCE
scu-mus TDLo:72 mg/kg/9W-I:ETA EJCAAH 4,123,68

SAFETY PROFILE: Questionable carcinogen with experimental tumorigenic data. When heated to decomposition it emits very toxic fumes of SO_x and NO_x.

BCF750 **CAS:1491-09-4** **HR: 2**
BENZO(h)(1)BENZOTHIENO(3,2-b)QUINOLINE
mf: $C_{19}H_{11}NS$ mw: 285.37

SYN: NAPHTHO(2,1-e)THIANAPHTHENO(3,2-b)PYRIDINE

TOXICITY DATA WITH REFERENCE
scu-mus TDLo:72 mg/kg/9W-I:ETA EJCAAH 4,123,68

SAFETY PROFILE: Questionable carcinogen with experimental tumorigenic data. When heated to decomposition it emits very toxic fumes of SO_x and NO_x.

BCG000 **CAS:846-35-5** **HR: 2**
BENZO(e)(1)BENZOTHIOPYRANO(4,3-b)INDOLE
mf: $C_{19}H_{11}NS$ mw: 285.37

TOXICITY DATA WITH REFERENCE
scu-mus TDLo:360 mg/kg/25W-I:NEO JNCIAM 46,1257,71

SAFETY PROFILE: Questionable carcinogen with experimental neoplastigenic data. When heated to decomposition it emits very toxic fumes of SO_x and NO_x.

BCG250 **CAS:239-01-0** **HR: 2**
11H-BENZO(a)CARBAZOLE
mf: $C_{16}H_{11}N$ mw: 217.28

PROP: Plates from EtOH. Mp: 226°.

SYN: 1,2-BENZCARBAZOLE

TOXICITY DATA WITH REFERENCE
skn-mus TDLo:840 mg/kg/21W-I:ETA HSZPAZ 236,79,35

SAFETY PROFILE: Questionable carcinogen with experimental tumorigenic data. When heated to decomposition it emits toxic fumes such as NO_x.

BCG500 **CAS:214-17-5** **HR: 2**
BENZO(b)CHRYSENE
mf: $C_{22}H_{14}$ mw: 278.36

PROP: Pale-yellow leaflets from C_6H_6. Mp: 292–294°.

SYNS: 2,3-BENZOCHRYSENE ◻ 3,4-BENZOTETRACENE ◻ BENZO(c)TETRAPHENE ◻ DIBENZO-2,3,7,8-PHENANTHRENE ◻ 3,4-BENZOTETRAPHENE ◻ 1,2:6,7-DIBENZOPHENANTHRENE ◻ 2,3:7,8-DIBENZOPHENANTHRENE

TOXICITY DATA WITH REFERENCE
mma-sat 50 µg/plate MUREAV 174,247,86
skn-mus TDLo:28 mg/kg:NEO JNCIAM 50,1717,73

SAFETY PROFILE: Questionable carcinogen with experimental neoplastigenic data by skin contact. Mutation data reported. When heated to decomposition it emits acrid smoke and irritating fumes.

BCG750 **CAS:194-69-4** **HR: 2**
BENZO(c)CHRYSENE
mf: $C_{22}H_{14}$ mw: 278.36

PROP: Needles from AcOH. Mp: 126–127°.

SYN: 1,2,5,6-DIBENZPHENANTHRENE

TOXICITY DATA WITH REFERENCE
skn-mus TDLo:1630 mg/kg/68W-I:ETA PRLBA4 129,439,40
scu-mus TDLo:2400 mg/kg/36W-I:ETA PRLBA4 129,439,40

SAFETY PROFILE: Questionable carcinogen with experimental tumorigenic data. When heated to decomposition it emits acrid smoke and irritating fumes.

BCH000 **CAS:196-78-1** **HR: 2**
BENZO(g)CHRYSENE
mf: $C_{22}H_{14}$ mw: 278.36

PROP: Needles from AcOH. Mp: 114–115°.

SYNS: 1,2,3,4-DIBENZOPHENANTHRENE ◻ 1,2,3,4-DIBENZPHENANTHRENE

TOXICITY DATA WITH REFERENCE
orl-mus TDLo:15 g/kg/74W-I:ETA PRLBA4 129,439,40
skn-mus TDLo:720 mg/kg/30W-I:ETA PRLBA4 129,439,40
scu-mus TDLo:1400 mg/kg/21W-I:ETA PRLBA4 129,439,40

SAFETY PROFILE: Questionable carcinogen with experimental tumorigenic data. When heated to decomposition it emits acrid smoke and irritating fumes.

BCH250 **CAS:5096-19-5** **HR: 2**
N-6-(3,4-BENZOCOUMARINYL)ACETAMIDE
mf: $C_{15}H_{10}NO_3$ mw: 252.26

SYN: N-(6-OXO-6H-DIBENZO(b,d)PYRAN-1-YL)ACETAMIDE

TOXICITY DATA WITH REFERENCE
orl-rat TDLo:5000 mg/kg:CAR CNREA8 26,619,66
orl-rat TD:9000 mg/kg/27D-I:ETA CNREA8 28,924,68

SAFETY PROFILE: Questionable carcinogen with experimental carcinogenic and tumorigenic data. When heated to decomposition it emits very toxic fumes of NO_x.

BCH750 **CAS:10085-81-1** **HR: 3**
BENZOCTAMINE HYDROCHLORIDE
mf: $C_{18}H_{19}N \cdot ClH$ mw: 285.84

PROP: Solid. Mp: 320–322°.

SYNS: BA 30,803 ◻ 1-METHYLAMINOMETHYLDIBENZO(b,c)BICYCLO(2,2,2)OCTADIENE HYDROCHLORIDE ◻ N-METHYLETHANOANTHRACENE-9-(10H)-METHYLAMINE HYDROCHLORIDE ◻ TACITIN

TOXICITY DATA WITH REFERENCE
ims-rat TDLo:51 mg/kg (2-18D preg):TER BSIBAC 49,1309,73
orl-rat LDLo:700 mg/kg TXAPA9 18,185,71
ivn-rat LD50:26 mg/kg 27ZQAG -,336,72

SAFETY PROFILE: Poison by intravenous route. Moderately toxic by ingestion. Experimental teratogenic effects. A sedative and muscle relaxant. When heated to decomposition it emits very toxic fumes of NO_x and HCl^-.

BCH800 **CAS:6809-93-4** **HR: 3**
1-BENZOCYCLOBUTENYL n-BUTYL KETONE
mf: $C_{13}H_{16}O$ mw: 188.29

SYNS: BICYCLO(4.2.0)OCTA-1,3,5-TRIENE, 7-VALERYL- □ 1-BICYCLO(4.2.0)OCTA-1,3,5-TRIEN-7-YL-1-PENTANONE □ KETONE, BICYCLO(4.2.0)OCTA-1,3,5-TRIEN-7-YL BUTYL □ PENTANONE, 1-BENZOCYCLOBUTYL- □ 1-PENTANONE, 1-BICYCLO(4.2.0)OCTA-1,3,5-TRIEN-7-YL-

TOXICITY DATA WITH REFERENCE
ipr-mus LD50:550 mg/kg JMCMAR 9,656,66

DOT CLASSIFICATION: 3; *Label:* Flammable Liquid

SAFETY PROFILE: Moderately toxic by intraperitoneal route. A flammable liquid. When heated to decomposition it emits acrid smoke and irritating vapors.

BCI000 **CAS:198-46-9** **HR: 2**
BENZO(de)CYCLOPENT(a)ANTHRACENE
mf: $C_{20}H_{12}$ mw: 252.32

SYN: Δ^3-DEHYDRO-3,4-TRIMETHYLENE-ISOBENZANTHRENE-2

TOXICITY DATA WITH REFERENCE
imp-mus TDLo:600 mg/kg/40W-I:ETA JNCIAM 2,241,41

SAFETY PROFILE: Questionable carcinogen with experimental tumorigenic data. When heated to decomposition it emits acrid smoke and irritating fumes.

BCI250 **CAS:240-44-8** **HR: 2**
1H-BENZO(a)CYCLOPENT(b)ANTHRACENE
mf: $C_{21}H_{16}$ mw: 268.37

SYN: 6,7-CYCLOPENTENO-1,2-BENZANTHRACENE

TOXICITY DATA WITH REFERENCE
skn-mus TDLo:820 mg/kg/34W-I:ETA PRLBA4 117,318,35

SAFETY PROFILE: Questionable carcinogen with experimental tumorigenic data. When heated to decomposition it emits acrid smoke and irritating fumes.

BCI500 **CAS:135-87-5** **HR: 3**
BENZODIOXANE HYDROCHLORIDE
mf: $C_{14}H_{19}NO_2 \cdot ClH$ mw: 269.80

PROP: Solid. Mp: 232–236°.

SYNS: BENODAINE HYDROCHLORIDE □ 1-(1,4-BENZODIOXAN-2-YLMETHYL)PIPERIDINEHYDROCHLORIDE □ F 933 □ FOURNEAU 933 □ 2-PIPERIDINOMETHYL-1,4-BENZODIOXAN HYDROCHLORIDE □ 2-(1-PIPERIDYLMETHYL)-1,4-BENZODIOXAN HYDROCHLORIDE □ PIPEROXANE HYDROCHLORIDE

TOXICITY DATA WITH REFERENCE
orl-rat TDLo:735 mg/kg (3-9D preg):REP PSEBAA 100,555,59
orl-mus LD50:502 mg/kg JAPMA8 48,409,59
ipr-mus LD50:175 mg/kg JAPMA8 48,409,59
scu-mus LD50:500 mg/kg THERAP 13,17,58
ivn-mus LD50:26 mg/kg AIPTAK 105,221,56

SAFETY PROFILE: Poison by intraperitoneal and intravenous routes. Moderately toxic by ingestion and subcutaneous routes. Experimental reproductive effects. When heated to decomposition it emits very toxic fumes of NO_x and HCl.

BCJ000 **CAS:5208-87-7** **HR: 3**
1,3-BENZODIOXOLE-5-(2-PROPEN-1-OL)
mf: $C_{10}H_{10}O_3$ mw: 178.20

SYNS: 1'-HYDROXYSAFROLE □ 1,2-METHYLENEDIOXY-4-(1-HYDROXYALLYL)BENZENE □ α-VINYLPIPERONYL ALCOHOL

TOXICITY DATA WITH REFERENCE
mma-sat 1 µmol/plate MUREAV 60,143,79
dnd-rat-ipr 100 mg/kg CNREA8 36,1686,76
oms-mus-ipr 400 µmol/kg CNREA8 41,2664,81
orl-rat TDLo:77 g/kg/73W-C:CAR CNREA8 37,1883,77
scu-rat TDLo:1426 mg/kg/10W-I:CAR CNREA8 43,1124,83
orl-mus TDLo:206 g/kg/56W-C:CAR CNREA8 33,590,73
ipr-mus TDLo:17,820 µg/kg:CAR CNREA8 47,2275,87
orl-mus TD:230 g/kg/52W-C:CAR CNREA8 37,1883,77
orl-rat TD:79 g/kg/34W-C:CAR CNREA8 33,590,73
orl-mus TD:117 g/kg/1Y-C:ETA PAACA3 24,79,83
orl-mus TD:61 g/kg/52W-C:NEO CNREA8 43,5163,83
orl-mus TD:117 g/kg/52W-C:NEO CNREA8 43,5163,83
scu-rat TD:265 mg/kg/10W-I:ETA CNREA8 33,590,73

SAFETY PROFILE: Suspected carcinogen with experimental carcinogenic, neoplastigenic, and tumorigenic data. Human mutation data reported. When heated to decomposition it emits acrid smoke and irritating fumes.

BCJ125 **CAS:32283-21-9** **HR: 3**
1,3-BENZODITHIOLIUM PERCHLORATE
mf: $C_7H_5ClO_4S_2$ mw: 252.69

SAFETY PROFILE: A friction and heat-sensitive explosive. Upon decomposition it emits toxic fumes of Cl^- and SO_x. See also PERCHLORATES.

BCJ150 **CAS:54531-52-1** **HR: 3**
BENZODOL
mf: $(C_6H_7AsO_4 \cdot CH_2O)_n$

PROP: Sol in water, alc, and NaOH.

SYNS: ARSONIC ACID, (4-HYDROXYPHENYL)-, polymer with FORMALDEHYDE □ (4-HYDROXYPHENYL)ARSONIC ACID polymer with FORMALDEHYDE □ POLYBENZARSOL

TOXICITY DATA WITH REFERENCE
orl-mus LD50:>4 g/kg ANTCAO 8,400,58
ipr-mus LD50:235 mg/kg MEIEDD 11,1203,89

OSHA PEL: TWA 0.5 mg(As)/m^3
ACGIH TLV: TWA 0.2 mg(As)/m^3

SAFETY PROFILE: Poison by intraperitoneal route. Low toxicity by ingestion. When heated to decomposition it emits toxic fumes of As.

BCJ500 **CAS:205-82-3** **HR: 3**
BENZO(j)FLUORANTHENE
mf: $C_{20}H_{12}$ mw: 252.32

PROP: Yellow crystals from EtOH. Mp: 165°, bp: 240–260° @ 2 mm.

SYNS: BENZ(j)FLUOROANTHRENE □ 10,11-BENZFLUORANTHENE □ BENZO(1)FLUORANTHENE □ 7,8-BENZOFLUORANTHENE □ B(j) F □ DIBENZO(a,jk)FLUORENE

TOXICITY DATA WITH REFERENCE
mma-sat 10 µg/plate CNREA8 40,4258,80
dnd-mus-skn 3760 nmol/kg PAACA3 25,121,84
imp-rat TDLo:25 mg/kg:CAR JJIND8 71,539,83
skn-mus TDLo:312 mg/kg/26W-I:ETA CANCAR 12,1194,59
ipr-mus TDLo:11,102 µg/kg/15D-I:NEO CALEDQ 34,15,87
imp-rat TD:5 mg/kg:ETA 50NNAZ 7,571,83

CONSENSUS REPORTS: NTP 7th Annual Report On Carcinogens. IARC Cancer Review: Group 2B IMEMDT 7,56,87; Animal Limited Evidence IMEMDT 3,82,73; Animal Sufficient Evidence IMEMDT 32,155,83.

SAFETY PROFILE: Confirmed carcinogen with experimental carcinogenic, neoplastigenic, and tumorigenic data. Mutation data reported. When heated to decomposition it emits acrid smoke and irritating fumes.

BCJ750 CAS:207-08-9 HR: 3
BENZO(k)FLUORANTHENE
mf: $C_{20}H_{12}$ mw: 252.32

PROP: Yellow prisms from C_6H_6 or AcOH. Mp: 217°, bp: 480°.

SYNS: 8,9-BENZOFLUORANTHENE □ 11,12-BENZOFLUORANTHENE □ 11,12-BENZO(k)FLUORANTHENE □ 2,3,1′,8′-BINAPHTHYLENE □ DIBENZO(b,jk)FLUORENE

TOXICITY DATA WITH REFERENCE
mma-sat 10 µg/plate CNREA8 40,4528,80
imp-rat TDLo:5 mg/kg:ETA 50NNAZ 7,571,83
skn-mus TDLo:2820 mg/kg/47W-I:ETA CANCAR 12,1194,59
scu-mus TDLo:72 mg/kg/9W-I:ETA AICCA6 19,490,63

CONSENSUS REPORTS: NTP 7th Annual Report On Carcinogens. IARC Cancer Review: Animal Sufficient Evidence IMEMDT 32,163,83.

SAFETY PROFILE: Confirmed carcinogen with experimental tumorigenic data. Mutation data reported. When heated to decomposition it emits acrid smoke and irritating fumes.

For occupational chemical analysis use NIOSH: Polynuclear Aromatic Hydrocarbons (HPLC), 5506; (GC), 5515.

BCK250 CAS:271-89-6 HR: 2
BENZOFURAN
mf: C_8H_6O mw: 118.14

PROP: Liquid. D: 1.078° @ 15/15, bp: 166.5–168° @ 735 mm.

SYNS: BENZO(b)FURAN □ 2,3-BENZOFURAN □BENZOFURFURAN □ COUMARONE □ NCI-C56166 □ 1-OXINDENE

TOXICITY DATA WITH REFERENCE
msc-mus:lym 100 mg/L EMMUEG 11,91,88
sce-ham:ovr 199 mg/L NTPTR* NTP-TR-370,89
orl-rat TDLo:16,250 mg/kg/13W-I
orl-rat TDLo:7 g/kg/14D-I:REP,ETA NTPTR* NTP-TR-370,89
orl-rat TDLo:61,800 mg/kg/2Y-C:CAR NTPTR* NTP-TR-370,89
orl-mus TDLo:30,900 mg/kg/2Y-C:CAR NTPTR* NTP-TR-370,89
ipr-mus LD50:500 mg/kg EJMCA5 12,383,77

CONSENSUS REPORTS: Reported in EPA TSCA Inventory. NTP Carcinogenesis Studies (gavage): Clear Evidence: Mouse NTPTR* NTP-TR-370,89; (gavage): Some Evidence: Rat NTPTR* NTP-TR-370,89. EPA TSCA Chemical Inventory, JUNE 1993.

SAFETY PROFILE: Suspected carcinogen with experimental carcinogenic data reported. Moderately toxic by intraperitoneal route. Mutation data reported. When heated to decomposition it emits acrid smoke and fumes.

BCL100 CAS:42242-58-0 HR: 2
p-(7-BENZOFURYLAZO)-N,N-DIMETHYLANILINE
mf: $C_{16}H_{15}N_3O$ mw: 265.34

SYNS: ANILINE, p-(7-BENZOFURYLAZO)-N,N-DIMETHYL- □ N,N-DIMETHYL-p-(7-BENZOFURYLAZO)ANILINE

TOXICITY DATA WITH REFERENCE
orl-rat TDLo:1620 mg/kg/13W-C:ETA JMCMAR 16,717,73

SAFETY PROFILE: Questionable carcinogen with experimental tumorigenic data. When heated to decomposition it emits toxic fumes of NO_x.

BCL250 CAS:23844-24-8 HR: 3
BENZOGUANAMINE
mf: $C_{22}H_{32}N_2O_5$ mw: 404.56

PROP: Crystals. Mp: 227°, d: 1.4.

SYNS: 2-ACETOXY-3-DIETHYLCARBAMYL-9,10-DIMETHOXY-1,2,3,4,6,7-HEXAHYDRO-11B-BENZO(a)QUINOLIZINE □BENZOCHINAM-IDE □ BENZOQUINAMIDE □ BENZQUINAMIDE □ BENZQUINAMIDU (POLISH) □ BZQ □ P 2647 □ QUANTRIL □ QUANTRYL

TOXICITY DATA WITH REFERENCE
orl-rat LD50:1050 mg/kg 27ZQAG -,208,72
ivn-rat LD50:100 µg/kg 27ZQAG -,208,72
orl-mus LD50:580 mg/kg 27ZQAG -,208,72
ipr-mus LD50:321 mg/kg DIPHAH 17,145,65
ivn-mus LD50:100 mg/kg 27ZQAG -,208,72
orl-bwd LD50:100 mg/kg TXAPA9 21,315,72

SAFETY PROFILE: Poison by ingestion, intraperitoneal, and intravenous routes. When heated to decomposition it emits toxic fumes of NO_x.

BCL500 CAS:495-18-1 HR: 2
BENZOHYDROXAMIC ACID
mf: $C_7H_7NO_2$ mw: 137.15

PROP: Rhombic tablets. Mp: 128°.

SYNS: BENZOHYDROXAMATE □ BENZOYLHYDROXAMIC ACID □ N-HYDROXYBENZAMIDE □ PHENYLHYDROXAMIC ACID

TOXICITY DATA WITH REFERENCE
mmo-sat 2500 nmol/plate MUREAV 135,139,84
mma-sat 1 µmol/plate MUREAV 56,7,77
orl-rat LD:>500 mg/kg NCNSA6 5,26,53

CONSENSUS REPORTS: Reported in EPA TSCA Inventory. EPA Genetic Toxicology Program.

SAFETY PROFILE: Moderately toxic by ingestion. Mutation data reported. When heated to decomposition it emits toxic fumes of NO_x.

BCL750 CAS:65-85-0 HR: 2
BENZOIC ACID
mf: $C_7H_6O_2$ mw: 122.13

PROP: White crystalline powder, leaflets, or needles from water. Mp: 122°, bp: 249°, flash p: 250°F (CC), d: 1.316, autoign temp: 1060°F, vap press: 1 mm @ 96.0° (sublimes), vap d: 4.21. Very sltly sol in water; sol in alc, ether, chloroform, and fixed oils.

SYNS: ACIDE BENZOIQUE (FRENCH) □ BENZENECARBOXYLIC ACID □ BENZENEFORMIC ACID □ BENZENEMETHANOIC ACID □ BENZOATE □ BENZOESAEURE (GERMAN) □ BENZOIC ACID (DOT) □ CARBOXYBENZENE □ DRACYLIC ACID □ KYSELINA BENZOOVA (CZECH) □ PHENYL CARBOXYLIC ACID □ PHENYLFORMIC ACID □ RETARDER BA □ RETARDEX □ SALVO LIQUID □ SALVO POWDER □ TENN-PLAS

TOXICITY DATA WITH REFERENCE
skn-hmn 22 mg/3D-I MOD 85DKA8 -,127,77
skn-rbt 500 mg/24H MLD BIOFX* 28-4/73
eye-rbt 100 mg SEV BIOFX* 28-4/73
mmo-esc 10 mmol/L ZBPIA9 112,226,59
dni-hmn:lym 5 mmol/L PNASA6 79,1171,82
orl-man LDLo:500 mg/kg FCTXAV 17,715,79
orl-rat LD50:1700 mg/kg IPSTB3 3,93,76
orl-mus LD50:1940 mg/kg IYKEDH 15,359,84
ipr-mus LD50:1460 mg/kg CRSBAW 160,1097,66
orl-dog LD50:2000 mg/kg 27ZTAP 3,22,69
orl-cat LD50:2000 mg/kg 27ZTAP 3,22,69
orl-rbt LDLo:2000 mg/kg HBTXAC 5,23,59
scu-rbt LDLo:2000 mg/kg HBTXAC 5,23,59
orl-gpg LDLo:2 g/kg MMWOAU 77,13,30
ipr-gpg LDLo:1400 mg/kg HBTXAC 5,23,59
scu-frg LDLo:100 mg/kg HBTXAC 5,23,59

CONSENSUS REPORTS: Reported in EPA TSCA Inventory. EPA Genetic Toxicology Program.

SAFETY PROFILE: Moderately toxic by ingestion, subcutaneous, and intraperitoneal routes. A severe eye irritant. A human skin and severe eye irritant. Mutation data reported. Combustible when exposed to heat or flame; can react with oxidizing materials. The powder burns rapidly in oxygen. To fight fire, use water, CO_2, water spray or mist, dry chemical. When heated to decomposition it emits acrid smoke and irritating fumes.

BCM000 CAS:120-51-4 HR: 2
BENZOIC ACID, BENZYL ESTER
mf: $C_{14}H_{12}O_2$ mw: 212.26

PROP: Leaflets found in Peru and tolu balsams, in ylang-ylang and in about 20 other essential oils (FCTXAV 11,1011,73). Colorless oily liquid; slt aromatic odor. Mp. 21°, bp: 324°, flash p: 298°F (CC), d: 1.116, refr index: 1.568, vap d: 7.3, autoign temp: 898°F. Misc with alc, chloroform, ether; insol in glycerin, water.

SYNS: ASCABIN □ ASCABIOL □ BENYLATE □ BENZOIC ACID, PHENYLMETHYL ESTER □ BENZYL ALCOHOL BENZOIC ESTER □ BENZYL BENZENECARBOXYLATE □ BENZYL BENZOATE (FCC) □ BENZYLETS □ BENZYL PHENYLFORMATE □ COLEBENZ □ FEMA No. 2138 □ NOVOSCABIN □ PERUSCABIN □ SCABANCA □ VANZOATE □ VENZONATE

TOXICITY DATA WITH REFERENCE
orl-rat LD50:1700 mg/kg JPETAB 93,26,48
skn-rat LD50:4 g/kg JPETAB 93,26,48
orl-mus LD50:1400 mg/kg JPETAB 93,26,48
orl-cat LD50:2240 mg/kg JPETAB 84,358,45
orl-rbt LD50:1680 mg/kg FCTXAV 11,1015,73
skn-rbt LD50:4000 mg/kg FCTXAV 11,1015,73
orl-gpg LD50:1000 mg/kg JPETAB 93,26,48

CONSENSUS REPORTS: Reported in EPA TSCA Inventory.

SAFETY PROFILE: Moderately toxic by ingestion and skin contact. Combustible liquid. Can react with oxidizing materials. To fight fire, use CO_2, water spray or mist, dry chemical. When heated to decomposition it emits acrid and irritating fumes and smoke. See also ESTERS.

BCM250 CAS:1696-17-9 HR: 2
BENZOIC ACID-N,N-DIETHYLAMIDE
mf: $C_{11}H_{15}NO$ mw: 177.27

PROP: Bp: 280–282°.

SYNS: BENZOIC ACID DIETHYLAMIDE □ BENZOYLDIETHYLAMINE □ N,N-DIETHYLBENZAMIDE □ R 2 □ REBEMID □ REP

TOXICITY DATA WITH REFERENCE
orl-rat LD50:2000 mg/kg FMCHA2 -,D219,80
orl-mus LD50:780 mg/kg MPPBAB 47,77,78
ihl-mus LC50:142 g/m³ MPPBAB 47,77,78
skn-mus LD50:1700 mg/kg MPPBAB 47,77,78

CONSENSUS REPORTS: Reported in EPA TSCA Inventory.

SAFETY PROFILE: Moderately toxic by ingestion and skin contact. When heated to decomposition it emits toxic fumes of NO_x.

BCP000 CAS:67011-39-6 HR: 1
BENZOIC-3-CHLORO-N-ETHOXY-2,6-DIMETHOXY-BENZIMIDIC ANHYDRIDE
mf: $C_{18}H_{18}ClNO_5$ mw: 363.82

SYNS: BENZOMATE □ BENZOXAMATE □ CITRAZON □ ETHYL-o-BENZOYL-3-CHLORO-2,6-DIMETHOXY-BENZOHYDROXIMATE □ NA-53

TOXICITY DATA WITH REFERENCE
orl-rat LD50:15,000 mg/kg 85ARAE 1,92,77
ipr-rat LD50:4217 mg/kg NYKGA7 3,123,76
orl-mus LD50:12 g/kg SPEADM 78-1,23,78
ipr-mus LD50:4264 mg/kg NYKGA7 3,123,76

SAFETY PROFILE: Mildly toxic by ingestion and intraperitoneal routes. When heated to decomposition it emits very toxic fumes of Cl^- and NO_x.

BCP250 **CAS:119-53-9** **HR: 3**
BENZOIN
mf: $C_{14}H_{12}O_2$ mw: 212.26

SYNS: ACETOPHENONE, 2-HYDROXY-2-PHENYL- ☐ BENZOYLPHENYLCARBINOL ☐ BITTER ALMOND OIL CAMPHOR ☐ ETHANONE, 2-HYDROXY-1,2-DIPHENYL- ☐ FENYL-α-HYDROXYBENZYLKETON ☐ α-HYDROXYBENZYL PHENYL KETONE ☐ α-HYDROXY-α-PHENYLACETOPHENONE ☐ 2-HYDROXY-2-PHENYLACETOPHENONE ☐ KETONE, α-HYDROXYBENZYL PHENYL ☐ NCI-C50011 ☐ WY-42956

TOXICITY DATA WITH REFERENCE
mmo-sat 750 μg/plate PMRSDJ 5,187,85
mma-smc 25 mg/L PMRSDJ 5,247,85
dns-rat:lvr 1 mmol/L PMRSDJ 5,371,85
msc-mus:lym 62,500 μg/L PMRSDJ 5,587,85
orl-rat LD50:10 g/kg FCTXAV 11,871,73
skn-rbt LD50:8870 mg/kg FCTXAV 11,871,73

CONSENSUS REPORTS: NCI Carcinogenesis Bioassay (feed); No Evidence: mouse, rat NCITR* NCI-CG-TR-204,80. Reported in EPA TSCA Inventory.

DOT CLASSIFICATION: 3; *Label:* Flammable Liquid

SAFETY PROFILE: Slightly toxic by ingestion and skin contact. Mutation data reported. A flammable liquid. When heated to decomposition it emits acrid smoke and irritating fumes. See also KETONES.

BCP500 **CAS:441-38-3** **HR: 3**
α-BENZOIN OXIME
mf: $C_{14}H_{13}NO_2$ mw: 227.28

SYNS: BENZOINOXIM (CZECH) ☐ CUPRON (CZECH) ☐ CUPRONE ☐ α-OXIME BENZOIN ☐ USAF FA-5

TOXICITY DATA WITH REFERENCE
eye-rbt 500 mg/24H MLD 28ZPAK -,111,72
orl-rat LD:>500 mg/kg NCNSA6 5,9,53
ipr-mus LD50:150 mg/kg NTIS** AD691-490

CONSENSUS REPORTS: Reported in EPA TSCA Inventory.

SAFETY PROFILE: Poison by intraperitoneal route. Mildly toxic by ingestion. An eye irritant. When heated to decomposition it emits toxic fumes of NO_x.

BCP650 **CAS:53-89-4** **HR: 3**
BENZOMETAN
mf: $C_{22}H_{25}N_3O$ mw: 347.50

PROP: Crystals from ethanol. Mp: 181–183° (decomp).

SYNS: BENZOPIPERILONE (ITALIAN) ☐ BENZPIPERILONE ☐ BENZPIPERYLON ☐ 4-BENZYL-1-(1-METHYL-4-PIPERIDYL)-3-PHENYL-3-PYRAZOLIN-5-ONE ☐ 1,2-DIHYDRO-2-(1-METHYL-4-PIPERIDINYL)-5-PHENYL-4-(PHENYLMETHYL)-3H-PYRAZOL-3-ONE (9CI) ☐ HUMEDIL ☐ KB 95 ☐ 1 (N METIL-PIPERIDIL-4')-3-FENIL-4-BENZIL-PIRAZOLONE-5 (ITALIAN) ☐ PPBP ☐ REUDLONIL ☐ TELON

TOXICITY DATA WITH REFERENCE
orl-rat LD50:2700 mg/kg BCFAAI 102,602,63
ivn-rat LD50:160 mg/kg BCFAAI 102,602,63
orl-mus LD50:1880 mg/kg BCFAAI 102,602,63
scu-mus LD50:615 mg/kg BCFAAI 102,602,63
ivn-mus LD50:160 mg/kg BCFAAI 102,602,63
orl-rbt LD50:1700 mg/kg BCFAAI 102,602,63
ivn-rbt LD50:83 mg/kg BCFAAI 102,602,63

SAFETY PROFILE: Poison by intravenous route. Moderately toxic by ingestion and subcutaneous routes. When heated to decomposition it emits toxic fumes of NO_x.

BCP685 **CAS:3811-10-7** **HR: 3**
BENZOMETHAMINE BROMIDE
mf: $C_{22}H_{31}N_2O_2$•Br mw: 435.46

SYNS: N,N-DIETHYL-2-((HYDROXYDIPHENYLACETYL)METHYLAMINO)-N-METHYL-ETHANAMINIUM BROMIDE (9CI) ☐ DIETHYLMETHYL(2-(N-METHYLBENZILAMIDO)ETHYL)AMMONIUM BROMIDE ☐ MC 3199

TOXICITY DATA WITH REFERENCE
orl-mus LD50:2700 mg/kg JPETAB 114,54,55
ipr-mus LD50:136 mg/kg JPETAB 114,54,55
ivn-mus LD50:31,800 μg/kg JPETAB 114,54,55

SAFETY PROFILE: Poison by intravenous and intraperitoneal routes. Moderately toxic by ingestion. When heated to decomposition it emits toxic fumes of Br^- and NO_x.

BCP750 **CAS:192-70-1** **HR: 2**
BENZO(a)NAPHTHO(8,1,2-cde)NAPHTHACENE
mf: $C_{28}H_{16}$ mw: 352.44

PROP: Yellow needles from xylene. Mp: 262–263°.

SYN: NAPHTO(1,2-c-d-e)NAPHTACENE (FRENCH)

TOXICITY DATA WITH REFERENCE
scu-mus TDLo:72 mg/kg/9W-I:ETA CHDDAT 266,301,68

SAFETY PROFILE: Questionable carcinogen with experimental tumorigenic data. When heated to decomposition it emits acrid smoke and irritating fumes.

BCQ000 **CAS:196-79-2** **HR: 2**
BENZO(h)NAPHTHO(1,2-f,s-3)QUINOLINE
mf: $C_{21}H_{13}N$ mw: 279.35

PROP: Crystals from toluene. Mp: 127–128°.

SYN: PYRIDO(3',2':5,6)CHRYSENE

TOXICITY DATA WITH REFERENCE
scu-mus TDLo:72 mg/kg/9W-I:ETA COREAF 252,1711,61

SAFETY PROFILE: Questionable carcinogen with experimental tumorigenic data. When heated to decomposition it emits toxic fumes such as NO_x.

BCQ250 **CAS:100-47-0** **HR: 3**
BENZONITRILE
DOT: UN 2224
mf: C_7H_5N mw: 103.13

PROP: Transparent, colorless oil; almond-like odor. D: 1.246 @ 20°/4°, bp: 191°, mp: −12.8°.

SYNS: BENZENENITRILE ☐ BENZOIC ACID NITRILE ☐ BENZONI-

TRILE (DOT) □ CYANOBENZENE □ FENYLKYANID □ PHENYL CYANIDE

TOXICITY DATA WITH REFERENCE
skn-rbt 500 mg/24H MOD FCTXAV 17(Suppl),695,79
orl-rat LDLo: 720 mg/kg AMRL** TR-74-78,74
ihl-rat LCLo: 950 ppm/8H AMRL** TR-74-78,74
skn-rat LD50: 1200 mg/kg AMRL** TR-74-78,74
orl-mus LD50: 971 mg/kg NEZAAQ 39,423,84
ihl-mus LC50: 6000 mg/m^3 AZMZA6 52(11),60,75
scu-mus LD50: 180 mg/kg MEIEDD 10,156,83
scu-rbt LDLo: 200 mg/kg AIPTAK 5,161,1899
scu-frg LDLo: 1700 mg/kg AIPTAK 5,161,1899
ipr-rat LD50: 740 mg/kg APFRAD 48,23,90
orl-mus LD50: 971 mg/kg NEZAAQ 39,423,84
ipr-mus LD50: 400 mg/kg FCTXAV 17,723,79

CONSENSUS REPORTS: Reported in EPA TSCA Inventory. Cyanide and its compounds are on the Community Right-To-Know List.

DOT CLASSIFICATION: 6.1; *Label:* Poison

SAFETY PROFILE: Poison by intraperitoneal and subcutaneous routes. Moderately toxic by ingestion, inhalation, and skin contact. See also NITRILES. A skin irritant. Combustible liquid. When heated to decomposition it emits toxic fumes of CN^- and NO_x.

BCQ500 CAS:189-55-9 HR: 3
BENZO(rst)PENTAPHENE
mf: $C_{24}H_{14}$ mw: 302.38

PROP: Green-yellow needles from toluene. Mp: 280–282°, bp: 275° @ 0.05 mm (subl).

SYNS: DB(a,i)P □ DIBENZO(a,i)PYRENE □ DIBENZO(b,h)PYRENE □ 1,2,7,8-DIBENZOPYRENE □ 3,4:9,10-DIBENZOPYRENE □ DIBENZ(a,i)PYRENE □ 1,2:7,8-DIBENZPYRENE □ 3,4:9,10-DIBENZPYRENE □ RCRA WASTE NUMBER U064

TOXICITY DATA WITH REFERENCE
mma-sat 20 µg/plate PNASA6 72,5135,75
mrc-esc 600 µg/well MUREAV 46,53,77
dnd-esc 10 µmol/L MUREAV 89,95,81
otr-ham:kdy 80 µg/L BJCAAI 37,873,78
msc-ham:lng 30 µg/L CNREA8 42,1646,82
skn-mus TDLo: 47 mg/kg/39W-I:ETA CANCAR 12,1079,59
scu-mus TDLo: 1000 µg/kg:ETA PEXTAR 11,384,69
scu-ham TDLo: 2 mg/kg:NEO NATUAS 203,308,64
itr-ham TDLo: 33 g/kg/8W-I:ETA ZKKOBW 82,175,74
scu-mus TD: 72 mg/kg/9W-I:ETA COREAF 246,1477,58
skn-mus TD: 141 mg/kg/47W-I:ETA IJMRAQ 53,638,65
scu-ham TD: 16 mg/kg:ETA PEXTAR 11,384,69
scu-mus TD: 80 mg/kg:ETA IJMRAQ 53,638,65
scu-mus TD: 4 mg/kg:ETA PAACA3 13,37,72
scu-mus TD: 4 mg/kg:ETA NATUAS 192,286,61

CONSENSUS REPORTS: NTP 7th Annual Report On Carcinogens. IARC Cancer Review: Group 2B IMEMDT 7,56,87; Animal Sufficient Evidence IMEMDT 3,215,73; IMEMDT 32,337,83. EPA Genetic Toxicology Program.

SAFETY PROFILE: Confirmed carcinogen with experimental neoplastigenic and tumorigenic data. Mutation data reported. When heated to decomposition it emits acrid smoke and irritating fumes.

BCQ750 CAS:63040-53-9 HR: 2
BENZO(rst)PENTAPHENE-5-CARBOXALDEHYDE
mf: $C_{25}H_{14}O$ mw: 330.39

SYN: 5-FORMYL-3,4:9,10-DIBENZOPYRENE

TOXICITY DATA WITH REFERENCE
scu-mus TDLo: 72 mg/kg/9W-I:ETA COREAF 252,1236,61

SAFETY PROFILE: Questionable carcinogen with experimental tumorigenic data. When heated to decomposition it emits acrid smoke and irritating fumes.

BCR000 CAS:191-24-2 HR: 2
BENZO(ghi)PERYLENE
mf: $C_{22}H_{12}$ mw: 276.34

PROP: Yellowish-green fluorescent leaflets from C_6H_6. Mp: 272–273°.

SYNS: 1,12-BENZPERYLENE □ 1,12-BENZOPERYLENE

TOXICITY DATA WITH REFERENCE
mma-sat 2 µg/plate/48H FCTXAV 17,141,79

CONSENSUS REPORTS: IARC Cancer Review: Group 3 IMEMDT 7,56,87, Animal Inadequate Evidence IMEMDT 32,195,83. EPA Genetic Toxicology Program.

SAFETY PROFILE: Questionable carcinogen. Mutation data reported. When heated to decomposition it emits acrid smoke and irritating fumes.

For occupational chemical analysis use NIOSH: Polynuclear Aromatic Hydrocarbons (HPLC), 5506; (GC), 5515.

BCR250 CAS:190-07-8 HR: 2
BENZO(a)PHENALENO(1,9-hi)ACRIDINE
mf: $C_{27}H_{15}N$ mw: 353.43

SYN: BENZO(c)PHENALENO(1,9-I,j)ACRIDINE

TOXICITY DATA WITH REFERENCE
scu-mus TDLo: 72 mg/kg/9W-I:ETA BAFEAG 52,49,65

SAFETY PROFILE: Questionable carcinogen with experimental tumorigenic data. When heated to decomposition it emits toxic fumes such as NO_x.

BCR500 CAS:190-03-4 HR: 2
BENZO(a)PHENALENO(1,9-i,j)ACRIDINE
mf: $C_{27}H_{15}N$ mw: 353.43

SYN: BENZO(h)PHENALENO(1,9-bc)ACRIDINE

TOXICITY DATA WITH REFERENCE
scu-mus TDLo: 72 mg/kg/9W-I:ETA BAFEAG 52,49,65

SAFETY PROFILE: Questionable carcinogen with experimental tumorigenic data. When heated to decomposition it emits toxic fumes of NO_x.

BCR750 CAS:195-19-7 HR: 2
BENZO(c)PHENANTHRENE
mf: $C_{18}H_{12}$ mw: 228.30

PROP: Needles from EtOH or pet ether. Mp: 68°.

SYNS: 3,4-BENZOPHENANTHRENE ☐ 3,4-BENZPHENANTHRENE ☐ TETRAHELICENE

TOXICITY DATA WITH REFERENCE
mma-sat 25 nmol/plate CNREA8 40,2876,80
skn-mus TDLo: 940 mg/kg/39W-I: ETA PRLBA4 117,318,35
skn-mus TD: 1220 mg/kg/51W-I: ETA PRLBA4 129,439,40
skn-mus TD: 1270 mg/kg/53W-I: ETA PRLBA4 123,343,37

CONSENSUS REPORTS: IARC Cancer Review: Group 3 IMEMDT 7,56,87, Animal Inadequate Evidence IMEMDT 32,205,83.

SAFETY PROFILE: Questionable carcinogen with experimental tumorigenic data. Mutation data reported. When heated to decomposition it emits acrid and irritating fumes.

BCS000 CAS:4466-76-6 HR: 2
BENZO(c)PHENATHRENE-8-CARBOXALDEHYDE
mf: $C_{19}H_{12}O$ mw: 256.31

SYN: 2-FORMYL-3:4-BENZPHENANTHRENE

TOXICITY DATA WITH REFERENCE
scu-mus TDLo: 5200 mg/kg/52W-I: ETA PRLBA4 131,170,42

SAFETY PROFILE: Questionable carcinogen with tumorigenic data. When heated to decomposition it emits acrid smoke and irritating fumes.

BCS100 HR: 2
(±)-BENZO(c)PHENANTHRENE-3,4-DIHYDRODIOL
mf: $C_{18}H_{14}O_2$ mw: 262.32

TOXICITY DATA WITH REFERENCE
ipr-mus TDLo: 11 mg/kg/15D-I: NEO CNREA8 46,2257,86

SAFETY PROFILE: Questionable carcinogen with experimental neoplastigenic data. When heated to decomposition it emits toxic and irritating fumes.

BCS103 HR: 2
(+)-BENZO(c)PHENANTHRENE-3,4-DIOL-1,2-EPOXIDE-1
mf: $C_{18}H_{14}O_3$ mw: 278.32

SYNS: BENZO(c)PHENANTHRENE-3,4-DIOL, 1,2,3,4-TETRAHYDRO-1,2-EPOXY-, (Z)-(+)-(1R,2S,3R,4S)- ☐ cis-1-β,2-β-EPOXY-1,2,3,4-TETRAHYDROBENZO(c)PHENANTHRENE-3-α-4-β-DIOL

TOXICITY DATA WITH REFERENCE
skn-mus TDLo: 111 μg/kg: NEO CNREA8 46,2257,86
ipr-mus TDLo: 12 mg/kg/15D-I: NEO CNREA8 46,2257,86

SAFETY PROFILE: Questionable carcinogen with experimental neoplastigenic data. When heated to decomposition it emits toxic and irritating fumes.

BCS105 HR: 2
(+)-BENZO(c)PHENANTHRENE-3,4 DIOL-1,2-EPOXIDE-2
mf: $C_{18}H_{14}O_3$ mw: 278.32

SYNS: BENZO(c)PHENANTHRENE-3,4-DIOL, 1,2,3,4-TETRAHYDRO-1,2-EPOXY-1, (E)-(+)-(1S,2R,3R,4S)- ☐ trans-1-α-2-α-EPOXY-1,2,3,4-TETRAHYDROBENZO(c)PHENANTHRENE-3-α,4-β-DIOL

TOXICITY DATA WITH REFERENCE
skn-mus TDLo: 278 μg/kg: NEO CNREA8 46,2257,86
ipr-mus TDLo: 12 mg/kg/15D-I: NEO CNREA8 46,2257,86

SAFETY PROFILE: Questionable carcinogen with experimental neoplastigenic data. When heated to decomposition it emits toxic and irritating fumes.

BCS110 HR: 2
(−)-BENZO(c)PHENANTHRENE-3,4-DIOL-1,2-EPOXIDE-2
mf: $C_{18}H_{14}O_3$ mw: 278.32

SYNS: BENZO(c)PHENANTHRENE-3,4-DIOL, 1,2,3,4-TETRAHYDRO-1,2-EPOXY-, (E)-(−)-(1R,2S,3S,4R)- ☐ trans-1-β,2-β-EPOXY-1,2,3,4-TETRAHYDROBENZO(c)PHENANTHRENE-3-β, 4-α-DIOL

TOXICITY DATA WITH REFERENCE
skn-mus TDLo: 111 μg/kg: NEO CNREA8 46,2257,86
ipr-mus TDLo: 12 mg/kg/15D-I: NEO CNREA8 46,2257,86

SAFETY PROFILE: Questionable carcinogen with experimental neoplastigenic data. When heated to decomposition it emits toxic and irritating fumes.

BCS250 CAS:119-61-9 HR: 3
BENZOPHENONE
mf: $C_{13}H_{10}O$ mw: 182.23

PROP: Rhombic prisms (stable form), monoclinic prisms (labile form), white crystals; persistent rose-like odor. Mp (α): 49°, mp (β): 26°, mp (γ): 47°, bp: 305.4°, d (α): 1.0976 @ 50°/50°, d (β): 1.108 @ 23°/40°, vap press: 1 mm @ 108.2. Sol in fixed oils; sltly sol in propylene glycol; insol in glycerin.

SYNS: BENZOYLBENZENE ☐ DIPHENYL KETONE ☐ DIPHENYLMETHANONE ☐ FEMA No. 2134 ☐ α-OXODIPHENYLMETHANE ☐ PHENYL KETONE

TOXICITY DATA WITH REFERENCE
orl-mus LD50: 2895 mg/kg JETOAS 9,99,76
ipr-mus LD50: 727 mg/kg JETOAS 9,99,76

CONSENSUS REPORTS: Reported in EPA TSCA Inventory.

DOT CLASSIFICATION: 3; *Label:* Flammable Liquid

SAFETY PROFILE: Moderately toxic by ingestion and intraperitoneal routes. Combustible when heated. Incompatible with oxidizers. When heated to decomposition it emits acrid and irritating fumes. See also KETONES.

BCS325 CAS:131-55-5 HR: 2
BENZOPHENONE-2
mf: $C_{13}H_{10}O_5$ mw: 246.23

PROP: Needles from H_2O. Mp: 196–198°.

SYNS: 2,2′,4,4′-TETRAHYDROXY BENZOPHENONE ☐ 2,4,2′,4′-TETRAHYDROXYBENZOPHENONE ☐ THBP ☐ UVINOL D-50 ☐ UVINUL D-50

TOXICITY DATA WITH REFERENCE
eye-rbt 100 mg MOD JACTDZ 2(5),35,83
mma-sat 100 μg/plate FCTOD7 20,427,82
cyt-mus:lym 200 μg/plate JACTDZ 2(5),35,83
orl-rat LD50:1220 mg/kg JACTDZ 2(5),35,83

CONSENSUS REPORTS: Reported in EPA TSCA Inventory.

SAFETY PROFILE: Moderately toxic by ingestion. An eye irritant. Mutation data reported. When heated to decomposition it emits toxic fumes of NO_x.

BCS400 CAS:574-66-3 HR: 2
BENZOPHENONE, OXIME
mf: $C_{13}H_{11}NO$ mw: 197.25

SYNS: BENZOPHENOXIME □ DIPHENYL KETOXIME □ DIPHENYLMETHANONE OXIME □ (DIPHENYLMETHYLENE)HYDROXYLAMINE □ METHANONE, DIPHENYL-, OXIME (9CI)

TOXICITY DATA WITH REFERENCE
unr-mus LD50:560 mg/kg PCJOAU 12,227,78

CONSENSUS REPORTS: Reported in EPA TSCA Inventory.

SAFETY PROFILE: Moderately toxic by unspecified route. When heated to decomposition it emits toxic vapors of NO_x.

BCS500 CAS:51593-70-5 HR: 3
1-(2H-1-BENZOPYRAN-3-YL)ETHANONE
mf: $C_{11}H_{10}O_2$ mw: 174.21

SYNS: 2H-1-BENZOPYRAN, 3-ACETYL- □ 2H-1-BENZOPYRAN-3-YL METHYL KETONE □ ETHANONE, 1-(2H-1-BENZOPYRAN-3-YL)-(9CI) □ KETONE, 2H-1-BENZOPYRAN-3-YL METHYL

TOXICITY DATA WITH REFERENCE
ipr-mus LD50:1000 mg/kg EJMCA5 11,81,76

DOT CLASSIFICATION: 3; *Label:* Flammable Liquid

SAFETY PROFILE: Moderately toxic by intraperitoneal route. A flammable liquid. When heated to decomposition it emits acrid smoke and irritating vapors.

BCS750 CAS:50-32-8 HR: 3
BENZO(a)PYRENE
mf: $C_{20}H_{12}$ mw: 252.32

PROP: Pale-yellow crystals. Mp: 177°, bp: 312° @ 10 mm. Insol in water; sol in benzene, toluene, and xylene.

SYNS: BENZO(d,e,f)CHRYSENE □ 3,4-BENZOPIRENE (ITALIAN) □ 3,4-BENZOPYRENE □ 6,7-BENZOPYRENE □ BENZ(a)PYRENE □ 3,4-BENZPYREN (GERMAN) □ 3,4-BENZ(a)PYRENE □ 3,4-BENZYPYRENE □ B(a)P □ RCRA WASTE NUMBER U022

TOXICITY DATA WITH REFERENCE
skn-mus 14 μg MLD CALEDQ 4,333,78
dnd-sal:spr 3 g/L BIPMAA 5,477,67
dnd-hmn:oth 1500 nmol/L TCMUD8 1,3,80
msc-hmn:oth 100 nmol/L CRNGDP 1,765,80
ipr-rat TDLo:60 mg/kg (female 16-18D post):REP BNEOBV 38,291,80
orl-mus TDLo:100 mg/kg (female 7-16D post):REP TJADAB 19,37A,79
orl-mus TDLo:1280 mg/kg (female 16D pre-5D post):TER DOESD6 54,410,81
orl-rat TDLo:2 g/kg (female 28D pre):REP EXPEAM 20,224,64
orl-rat TDLo:1344 mg/kg (female 15D pre-5D post):REP DOESD6 54,410,81
scu-rat TDLo:150 mg/kg (female 6-8D post):TER TXCYAC 42,195,86
orl-rat TDLo:40 mg/kg (female 14D post):TER NSAPCC 272,89,72
orl-mus TDLo:100 mg/kg (multi):REP BIREBV 24,183,81
scu-rat TDLo:150 mg/kg (female 6-8D post):REP TXCYAC 42,195,86
orl-mus TDLo:100 mg/kg (female 16D pre-5D post):REP DOESD6 54,410,81
ipr-mus TDLo:1500 mg/kg (male 5D pre):REP PMRSDJ 1,712,81
ipr-ham TDLo:10 mg/kg (male 5D pre):REP JTEHD6 8,929,81
ipr-mus TDLo:200 mg/kg (female 7D post):TER TJADAB 20,365,79
ipr-mus TDLo:300 mg/kg (female 16-18D post):TER JTEHD6 6,569,80
orl-rat TDLo:15 mg/kg:CAR EXPTAX 18,288,80
ipr-rat TDLo:16 mg/kg:ETA BJCAAI 12,65,58
scu-rat TDLo:455 μg/kg/60D-I:NEO CBINA8 29,159,80
ivn-rat TDLo:39 mg/kg/6D-I:ETA CNREA8 29,506,69
ims-rat TDLo:2400 μg/kg:CAR NTIS** DOE/EV/03140-5
ice-rat TDLo:22 mg/kg:ETA CNREA8 29,1927,69
itr-rat TDLo:68 mg/kg/15W-I:CAR 85AGAF-,480,76
imp-rat TDLo:150 μg/kg:CAR JJIND8 72,733,84
orl-mus TDLo:700 mg/kg/75W-I:CAR GISAAA 45(12),14,80
ihl-mus TCLo:200 ng/m³/6H/13W-I:ETA GISAAA 47(7),23,82
skn-mus TDLo:120 mg/k (multi):CAR BEXBAN 71,677,71
skn-mus TDLo:28,500 μg/kg/19W-I:CAR FAATDF 9,297,87
skn-mus TDLo:25 ng/kg/110W-I:CAR ARGEAR 50,266,80
ipr-mus TDLo:10 mg/kg:NEO ARTODN 4,74,80
ipr-mus TDLo:300 mg/kg (16-18D post):CAR JTEHD6 6,569,80
scu-mus TDLo:9 mg/kg:CAR JJIND8 71,309,83
scu-mus TDLo:480 mg/kg (11-15D post):CAR PSEBAA 135,84,70
ivn-mus TDLo:10 mg/kg:ETA JNCIAM 1,225,40
itr-mus TDLo:200 mg/kg/10W-I:NEO PWPSA8 22,269,79
imp-mus TDLo:200 mg/kg:CAR BJCAAI 39,761,79
unr-mus TDLo:80 mg/kg/8D-I:ETA BEBMAE 88(11),592,79
rec-mus TDLo:200 mg/kg:CAR ONCOBS 37,77,80
par-dog TDLo:819 mg/kg/26W-I:ETA JJIND8 65,921,80
imp-dog TDLo:651 mg/kg/21W-C:ETA JJIND8 65,921,80
scu-mky TDLo:40 mg/kg:ETA PSEBAA 127,594,68
skn-rbt TDLo:17 mg/kg/57W-I:ETA HSZPAZ 236,79,35
ivn-rbt TDLo:30 mg/kg (25D post):NEO BEXBAN 85,369,78
itr-rbt TDLo:145 mg/kg/2Y-I:ETA GANNA2 71,197,80
orl-ham TDLo:420 mg/kg/21W-I:ETA ZEKBAI 65,56,62
ihl-ham TCLo:9500 μg/m³/4H/96W-I:ETA JJIND8 66,575,81
scu-ham TDLo:4000 μg/kg:ETA CNREA8 32,360,72

itr-ham TDLo:64 mg/kg:CAR CALEDQ 3,231,77
itr-ham TDLo:120 mg/kg/17W-I:NEO CALEDQ 25,271,85
imp-frg TDLo:45 mg/kg:ETA EXPEAM 20,143,64
imp-rat TD:500 µg/kg:CAR CALEDQ 20,97,83
skn-mus TD:12 mg/kg/20D-I:CAR CRNGDP 6,1483,85
itr-rat TD:200 mg/kg/15W-I:CAR 31BYAP-,199,74
skn-mus TD:26 mg/kg/65W-I:CAR AJPAA4 102,381,81
rec-mus TD:560 mg/kg/14W-I:CAR CALEDQ 20,117,83
scu-mus TD:8 mg/kg:CAR CNREA8 12,657,52
itr-ham TD:360 mg/kg/36W-I:CAR CNREA8 32,28,72
ims-rat TD:3150 µg/kg:CAR PAACA3 21,72,80
skn-mus TD:18 mg/kg/73W-I:CAR EVHPAZ 38,149,81
scu-mus TD:12 mg/kg:CAR GANNA2 62,309,71
scu-rat LD50:50 mg/kg ZEKBAI 69,103,67
ipr-mus LDLo:500 mg/kg TXAPA9 23,288,72
irn-frg LDLo:9 mg/kg CNREA8 24,1969,64

CONSENSUS REPORTS: NTP 7th Annual Report On Carcinogens. IARC Cancer Review: Group 2A IMEMDT 7,56,87; Animal Sufficient Evidence IMEMDT 32,211,83; IMEMDT 3,91,73. Reported in EPA TSCA Inventory.

OSHA PEL: TWA 0.2 mg/m^3

SAFETY PROFILE: Confirmed carcinogen with experimental carcinogenic, neoplastigenic, and tumorigenic data. A poison via subcutaneous, intraperitoneal, and intrarenal routes. Experimental teratogenic and reproductive effects. Human mutation data reported. A skin irritant. A common air contaminant of water, food, and smoke. When heated to decomposition it emits acrid smoke and fumes. See other benzopyrenes.

For occupational chemical analysis use OSHA: #ID-58 or NIOSH: Polynuclear Aromatic Hydrocarbons (HPLC), 5506; (GC), 5515.

BCT000 CAS:192-97-2 HR: 2
BENZO(e)PYRENE
mf: $C_{20}H_{12}$ mw: 252.32

PROP: Prisms from C_6H_6. Mp: 178–179°, bp: 250° @ 3–4 mm (subl).

SYNS: 1,2-BENZOPYRENE □ 4,5-BENZOPYRENE □ 1,2-BENZPYRENE □ B(e)P

TOXICITY DATA WITH REFERENCE
mmo-sat 1 nmol/plate CNREA8 40,1985,80
mma-sat 1 µg/plate ENMUDM 6(Suppl 2),1,84
msc-hmn:oth 12 µmol/L MUREAV 130,127,84
dns-rat:lvr 79 nmol/L CNREA8 42,3010,82
dnd-mus-skn 192 µmol/kg CRNGDP 5,231,84
otr-ham:kdy 25 µg/L TOLED5 7,143,80
ipr-mus TDLo:150 mg/kg (8D preg):TER SEIJBO 21,97,81
ipr-mus TDLo:300 mg/kg (8D preg):REP SEIJBO 21,97,81
ipr-mus TDLo:300 mg/kg (8D preg):TER SEIJBO 21,97,81
orl-mus TDLo:360 mg/kg/43W-I:ETA VRRAAT 20,276,38
skn-mus TDLo:240 mg/kg/30W-I:ETA CALEDQ 7,51,79
scu-mus TDLo:160 mg/kg:ETA AJCAA7 27,474,36
scu-gpg TDLo:140 mg/kg/37W-I:ETA AJCAA7 27,474,36
skn-mus TD:516 mg/kg/43W-I:ETA CANCAR 12,1079,59

CONSENSUS REPORTS: IARC Cancer Review: Group 3 IMEMDT 7,56,87; Animal Inadequate Evidence IMEMDT 32,225,83; Animal Limited Evidence IMEMDT 3,137,73. EPA Genetic Toxicology Program.

SAFETY PROFILE: Questionable carcinogen with experimental tumorigenic data. Experimental teratogenic and reproductive effects. Human mutation data reported. When heated to decomposition it emits acrid smoke and irritating fumes.

For occupational chemical analysis use NIOSH: Polynuclear Aromatic Hydrocarbons (HPLC), 5506; (GC), 5515.

BCT250 CAS:13312-42-0 HR: 2
BENZO(a)PYRENE-6-CARBOXYALDEHYDE
mf: $C_{21}H_{12}O$ mw: 280.33

SYNS: 3,4-BENZPYRENE-5-ALDEHYDE □ 6-FORMYLBENZO(a)PYRENE

TOXICITY DATA WITH REFERENCE
scu-rat TDLo:17 mg/kg/60D-I:NEO CBINA8 29,159,80
skn-mus TDLo:180 mg/kg/20W-I:NEO CBINA8 22(1),53,78
scu-mus TDLo:40 mg/kg:NEO BJCAAI 26,506,72
unk-mus TDLo:80 mg/kg/8D-I:ETA BEBMAE 88(11),592,79
scu-mus TD:200 mg/kg:ETA COREAF 245,876,57
skn-mus TD:32 mg/kg/20W-I:ETA BEXBAN 87,474,79
skn-mus TD:179 mg/kg/20W-I:ETA PAACA3 18,59,77
scu-rat TD:100 mg/kg/40D-I:ETA JMCMAR 16,714,73

SAFETY PROFILE: Questionable carcinogen with experimental tumorigenic and neoplastigenic data. When heated to decomposition it emits acrid smoke and fumes. See also ALDEHYDES.

BCT500 CAS:64048-70-0 HR: 2
BENZO(a)PYRENE-6-CARBOXALDEHYDE THIOSEMICARBAZONE
mf: $C_{22}H_{15}N_3S$ mw: 353.46

SYN: 3,4-BENZPYRENE-5-ALDEHYDE THIOSEMICARBAZONE

TOXICITY DATA WITH REFERENCE
scu-mus TDLo:200 mg/kg:ETA COREAF 245,876,57

SAFETY PROFILE: Questionable carcinogen with experimental tumorigenic data. When heated to decomposition it emits very toxic fumes of SO_x and NO_x. See also ALDEHYDES.

BCT750 CAS:13345-25-0 HR: 2
BENZO(a)PYRENE-7,8-DIHYDRODIOL
mf: $O_2C_{20}H_{14}$ mw: 286.34

SYN: BP-7,8-DIHYDRODIOL

TOXICITY DATA WITH REFERENCE
mma-sat 8 µmol/L CALEDQ 24,281,84
dnd-hmn:fbr 30 µmol/L CBINA8 41,155,82
skn-mus TDLo:4580 µg/kg:NEO CCSUDL 3,371,78
par-mus TDLo:4 mg/kg:ETA BJCAAI 37,657,78
skn-mus TD:2290 µg/kg:ETA CALEDQ 2,115,76

CONSENSUS REPORTS: EPA Genetic Toxicology Program.

SAFETY PROFILE: Questionable carcinogen with experimental neoplastigenic and tumorigenic data. Human mutation data reported. When heated to decomposition it emits acrid smoke and irritating fumes.

BCU000 CAS:60268-85-1 HR: 2
anti-BENZO(a)PYRENE-7,8-DIHYDRODIOL-9,10-OXIDE
mf: $C_{20}H_{14}O_3$ mw: 302.34

SYNS: BENZO(a)PYRENE-7,8-DIHYDRODIOL-9,10-EPOXIDE (anti) □ BP-7,8-DIHYDRODIOL-9,10-EPOXIDE (anti) □ anti-BP-7,8-DIHYDRODIOL-9,10-OXIDE

TOXICITY DATA WITH REFERENCE
dnd-hmn:lym 800 µg/L CRNGDP 3,1107,82
msc-ham:lng 100 µg/L IJCNAW 24,203,79
skn-mus TDLo:2400 µg/kg:ETA CALEDQ 2,115,76
skn-mus TD:4840 µg/kg:ETA CALEDQ 3,23,77

SAFETY PROFILE: Questionable carcinogen with experimental tumorigenic data by skin contact. Human mutation data reported. When heated to decomposition it emits acrid smoke and irritating fumes.

BCU250 CAS:58917-67-2 HR: 1
BENZO(a)PYRENE DIOL EPOXIDE ANTI

SYNS: anti(±)BENZO(a)PYRENE-DIOL-EPOXIDE □ anti-BPDE □ anti-r-7,trans-8-DIHYDROXY-trans-9,10-OXY-7,8,9,10-TETRAHYDROBENZO(a)PYRENE □ BPDE □ BP DIOL EPOXIDE ANTI □ trans-7,8-DIHYDROXY-9,10-OXY-7,8,9,10-TETRAHYDROBENZO(a)PYRENE

TOXICITY DATA WITH REFERENCE
mmo-sat 200 pmol/plate MUREAV 125,95,84
dnd-hmn:fbr 1 µmol/L ENMUDM 7,267,85
dns-hmn:fbr 1500 nmol/L BBACAQ 824,146,85
msc-hmn:fbr 200 nmol/L MUREAV 125,95,84
dnd-ham:ovr 1 mg/L MUREAV 129,365,84
skn-mus TDLo:2400 µg/kg:ETA CNREA8 39,67,79

CONSENSUS REPORTS: EPA Genetic Toxicology Program.

SAFETY PROFILE: Questionable carcinogen with experimental tumorigenic data reported. Human mutation data reported. When heated to decomposition it emits acrid smoke and irritating fumes.

BCU500 CAS:3067-13-8 HR: 2
BENZO(a)PYRENE-1,6-DIONE
mf: $C_{20}H_{10}O_2$ mw: 282.30

PROP: Yellow needles from AcOH. Mp: 295° (2° decomp).

SYNS: 1,6-BENZO(a)PYRENEDIONE □ BENZO(a)PYRENE-1,6-QUINONE □ PB-1,6-QUINONE

TOXICITY DATA WITH REFERENCE
msc-ham:lng 2 mg/L CNREA8 36,3350,76
skn-mus TDLo:4520 µg/kg:NEO CCSUDL 3,371,78

CONSENSUS REPORTS: EPA Genetic Toxicology Program.

SAFETY PROFILE: Questionable carcinogen with experimental neoplastigenic data by skin contact. Mutation data reported. When heated to decomposition it emits acrid smoke and irritating fumes.

BCU750 CAS:3067-14-9 HR: 2
BENZO(a)PYRENE-3,6-DIONE
mf: $C_{20}H_{10}O_2$ mw: 282.30

PROP: Red needles from AcOH. Mp: 291° (decomp).

SYNS: 3,6-BENZO(a)PYRENEDIONE □ BENZO(a)PYRENE-3,6-QUINONE □ BP-3,6-QUINONE

TOXICITY DATA WITH REFERENCE
mma-sat 5 µg/plate ENMUDM 7,839,85
dnd-hmn:fbr 1 µmol/L TOLED5 28,37,85
msc-ham:lng 2 mg/L CNREA8 36,3350,76
skn-mus TDLo:4520 µg/kg:NEO CCSUDL 3,371,78

CONSENSUS REPORTS: EPA Genetic Toxicology Program.

SAFETY PROFILE: Questionable carcinogen with experimental neoplastigenic data by skin contact. Human mutation data reported. When heated to decomposition it emits acrid smoke and irritating fumes.

BCV000 CAS:3067-12-7 HR: 2
BENZO(a)PYRENE-6,12-DIONE
mf: $C_{20}H_{10}O_2$ mw: 282.30

PROP: Solid. Mp: 320–322°.

SYNS: 6,12-BENZO(a)PYRENEDIONE □ BENZO(a)PYRENE-6,12-QUINONE □ 6,12-BENZOPYRENE QUINONE □ BP-6,12-QUINONE

TOXICITY DATA WITH REFERENCE
msc-ham:lng 4 mg/L CNREA8 36,3350,76
skn-mus TDLo:4520 µg/kg:NEO CCSUDL 3,371,78

CONSENSUS REPORTS: EPA Genetic Toxicology Program.

SAFETY PROFILE: Questionable carcinogen with experimental neoplastigenic data. Mutation data reported. When heated to decomposition it emits acrid smoke and irritating fumes.

BCV250 CAS:21247-98-3 HR: 3
BENZO(a)PYRENE-6-METHANOL
mf: $C_{21}H_{14}O$ mw: 282.35

PROP: Pale-yellow crystals from C_6H_6. Mp: 270–271°.

SYN: 6-HYDROXYMETHYLBENZO(a)PYRENE

TOXICITY DATA WITH REFERENCE
dnd-omi 30 µmol/L CBINA8 31,51,80
dnd-mam:lym 500 mg/L CBINA8 25,35,79
bfa-rat/sat 1 mg/kg MUREAV 173,251,86
scu-rat TDLo:100 mg/kg/40D-I:CAR JMCMAR 16,714,73
skn-mus TDLo:180 mg/kg/20W-I:CAR CBINA8 22(1),53,78
scu-mus TDLo:20 mg/kg:NEO BJCAAI 26,506,72
skn-mus TD:4510 µg/kg:NEO CCSUDL 3,371,78
skn-mus TD:180 mg/kg/20W-I:ETA PAACA3 18,59,77
scu-rat TD:508 µg/kg/60D-I:NEO CBINA8 29,159,80

CONSENSUS REPORTS: EPA Genetic Toxicology Program.

SAFETY PROFILE: Suspected carcinogen with experimental carcinogenic, neoplastigenic, and tumorigenic data. Mutation data reported. When heated to decomposition it emits acrid smoke and fumes.

BCV500 CAS:37574-47-3 HR: 2
BENZO(a)PYRENE-4,5-OXIDE
mf: $C_{20}H_{12}O$ mw: 268.32

SYNS: BENZO(1,2)PYRENO(4,5-b)OXIRENE-3b,4b-DIHYDRO □ BENZO(a)PYRENE-4,5-EPOXIDE □ BENZ(a)PYRENE 4,5-OXIDE □ BP-4,5-EPOXIDE □ BP 4,5-OXIDE

TOXICITY DATA WITH REFERENCE
mmo-sat 250 ng/plate ENMUDM 7,839,85
mma-sat 1 μg/plate ENMUDM 7,839,85
mmo-esc 1 μg/plate TCMUD8 5,339,85
dnr-bcs 1- μg/plate CNREA8 45,2600,85
dnd-mam:lym 800 nmol CRNGDP 3,267,82
scu-rat TDLo:40 mg/kg/50D-I:ETA IJCNAW 18,351,76
skn mus TDLo:2144 μg/kg:NEO CCSUDL 3,371,78
skn-mus TD:2160 μg/kg:ETA CNREA8 37,4130,77
skn-mus TD:32 mg/kg/60W-I:ETA PNASA6 73,243,76
skn-mus TD:2160 μg/kg:ETA CALEDQ 2,115,76

CONSENSUS REPORTS: EPA Genetic Toxicology Program.

SAFETY PROFILE: Questionable carcinogen with experimental tumorigenic and neoplastigenic data. Mutation data reported. When heated to decomposition it emits acrid and irritating fumes.

BCV750 CAS:36504-65-1 HR: 2
BENZO(a)PYRENE-7,8-OXIDE
mf: $C_{20}H_{12}O$ mw: 268.32

PROP: Pale-yellow prisms.

SYNS: BENZO(10,11)CHRYSENO(1,2-b)OXIRENE-6-β,7-α-DIHYDRO □ BENZO(a)PYRENE-7,8-DIHYDRO-7,8-EPOXY □ BENZO(a)PYRENE-7,8-EPOXIDE □ 6-β,7 α-DIHYDROBENZO(10,11)CHRYSENO(1,2-b)OXIRENE □ BP 7,8-EPOXIDE □ BP 7,8-OXIDE □ 7,8-EPOXY-7,8-DIHYDROBENZO(a)PYRENE

TOXICITY DATA WITH REFERENCE
mmo-sat 250 ng/plate CNREA8 36,3350,76
mma-sat 25 μmol/L JBCHA3 251,4882,76
skn-mus TDLo:32 mg/kg/60W-I:CAR PNASA6 73,243,76
scu-mus TDLo:10 mg/kg:ETA JJIND8 64,617,80
skn-mus TD:2160 μg/kg:ETA CNREA8 37,4130,77
skn-mus TD:2144 μg/kg:NEO CCSUDL 3,371,78
skn-mus TD:2160 μg/kg:ETA CALEDQ 2,115,76
skn-mus TD:68 mg/kg/42W-I:ETA PNASA6 73,3867,76
skn-mus TD:99 mg/kg:NEO PNASA6 73,243,76

CONSENSUS REPORTS: EPA Genetic Toxicology Program.

SAFETY PROFILE: Questionable carcinogen with experimental carcinogenic, neoplastigenic, and tumorigenic data. Mutation data reported. When heated to decomposition it emits irritating fumes.

BCW000 CAS:36504-66-2 HR: 2
BENZO(a)PYRENE-9,10-OXIDE
mf: $C_{20}H_{12}O$ mw: 268.32

PROP: Pale-yellow prisms.

SYN: BP-9,10-OXIDE

TOXICITY DATA WITH REFERENCE
mmo-sat 250 ng/plate CNREA8 36,3350,76
skn-mus TDLo:2140 μg/kg:ETA CCSUDL 3,371,78
skn-mus TD:2160 μg/kg:ETA CNREA8 37,4130,77

CONSENSUS REPORTS: EPA Genetic Toxicology Program.

SAFETY PROFILE: Questionable carcinogen with experimental tumorigenic data by skin contact. Mutation data reported. When heated to decomposition it emits acrid smoke and irritating fumes.

BCW250 CAS:60448-19-3 HR: 2
BENZO(a)PYRENE-11,12-OXIDE
mf: $C_{20}H_{12}O$ mw: 268.32

SYN: BP-11,12-OXIDE

TOXICITY DATA WITH REFERENCE
mmo-sat 1 μg/plate CNREA8 36,3350,76
msc-ham:lng 5 mg/L CNREA8 36,3350,76
skn-mus TDLo:2140 μg/kg:NEO CCSUDL 3,371,78
skn-mus TD:2160 μg/kg:ETA CNREA8 37,4130,77

CONSENSUS REPORTS: EPA Genetic Toxicology Program.

SAFETY PROFILE: Questionable carcinogen with experimental tumorigenic and neoplastigenic data by skin contact. Mutation data reported. When heated to decomposition it emits acrid smoke and irritating fumes.

BCX000 CAS:56892-30-9 HR: 2
BENZO(a)PYREN-2-OL
mf: $C_{20}H_{12}O$ mw: 268.32

PROP: Crystals from C_6H_6. Mp: 227–228°.

SYN: 2-HYDROXYBENZO(a)PYRENE

TOXICITY DATA WITH REFERENCE
mmo-sat 8500 pmol/L RRBCAD 18,291,81
dnd-hmn:fbr 30 μmol/L CBINA8 41,155,82
mma-sat 2 nmol/plate CNREA8 39,2660,79
mma-ham:lng 25 nmol/plate CNREA8 39,2660,79
msc-ham:lng 10 mg/L CNREA8 36,3350,76
skn-mus TDLo:69 mg/kg/32W-I:NEO CNREA8 37,2608,77
scu-mus TDLo:10 mg/kg:ETA JJIND8 64,617,80

CONSENSUS REPORTS: EPA Genetic Toxicology Program.

SAFETY PROFILE: Questionable carcinogen with experimental tumorigenic and neoplastigenic data. Human mutation data reported. When heated to decomposition it emits acrid smoke and irritating fumes.

BCX250 **CAS:13345-21-6** **HR: 2**
BENZO(a)PYREN-3-OL
mf: $C_{20}H_{12}O$ mw: 268.32

PROP: Yellow crystals from C_6H_6/pet ether. Mp: 226–227° (decomp).

SYNS: BP-3-HYDROXY □ 3-HYDROXYBENZO(a)PYRENE □ 8-HYDROXY-3,4-BENZPYRENE

TOXICITY DATA WITH REFERENCE
dnd-hmn:fbr 30 µmol/L CBINA8 41,155,82
dnr-esc 250 mg/L JNCIAM 62,873,79
msc-ham:lng 12 µmol/L PNASA6 73,607,76
skn-mus TDLo:117 µg/kg:NEO CNREA8 38,678,78
scu-mus TDLo:160 mg/kg/21W-I:ETA BJCAAI 6,400,52
skn-mus TD:4300 µg/kg:ETA CALEDQ 3,23,77
skn-mus TD:2760 mg/kg/69W-I:ETA BJCAAI 6,400,52
skn-mus TD:4280 µg/kg:NEO CCSUDL 3,371,78

CONSENSUS REPORTS: EPA Genetic Toxicology Program.

SAFETY PROFILE: Questionable carcinogen with experimental tumorigenic and neoplastigenic data by skin contact. Human mutation data reported. When heated to decomposition it emits acrid smoke and irritating fumes.

BCX500 **CAS:24027-84-7** **HR: 2**
BENZO(a)PYREN-5-OL
mf: $C_{20}H_{12}O$ mw: 268.32

PROP: Yellow needles from toluene or by sublimation. Mp: 195–196° (decomp).

SYN: 5-HYDROXYBENZO(a)PYRENE

TOXICITY DATA WITH REFERENCE
skn-mus TDLo:4280 µg/kg:ETA CCSUDL 3,371,78
skn-mus TD:4290 µg/kg:ETA CNREA8 38,678,78

CONSENSUS REPORTS: EPA Genetic Toxicology Program.

SAFETY PROFILE: Questionable carcinogen with experimental tumorigenic data by skin contact. When heated to decomposition it emits acrid smoke and irritating fumes.

BCX750 **CAS:33953-73-0** **HR: 2**
BENZO(a)PYREN-6-OL
mf: $C_{20}H_{12}O$ mw: 268.32

PROP: Needles from Et_2O/pet ether. Mp: 207–209°.

SYN: 6-HYDROXYBENZO(a)PYRENE

TOXICITY DATA WITH REFERENCE
mma-sat 25 µmol/L JBCHA3 251,4882,76
mma-ham:lng 3700 nmol/L PNASA6 73,607,76
mmo-sat 7 µg/plate ENMUDM 7,839,85
msc-ham:lng 5 mg/L CNREA8 36,3350,76
skn-mus TDLo:4280 µg/kg:NEO CCSUDL 3,371,78
scu-mus TDLo:12 mg/kg:NEO GANNA2 62,419,71
idr-mus TDLo:2400 µg/kg:ETA GANNA2 62,419,71
skn-mus TD:4290 µg/kg:NEO CNREA8 38,678,78

CONSENSUS REPORTS: EPA Genetic Toxicology Program.

SAFETY PROFILE: Questionable carcinogen with experimental neoplastigenic and tumorigenic data. Mutation data reported. When heated to decomposition it emits acrid smoke and fumes.

BCY000 **CAS:37994-82-4** **HR: 2**
BENZO(a)PYREN-7-OL
mf: $C_{20}H_{12}O$ mw: 268.32

PROP: Yellow plates from C_6H_6/pet ether. Mp: 218–219°.

SYN: 7-HYDROXYBENZO(a)PYRENE

TOXICITY DATA WITH REFERENCE
mmo-sat 16 µg/plate MUREAV 36,379,76
mma-sat 7 µg/plate ENMUDM 7,839,85
dni-omi 200 µg/L PNASA6 74,1378,77
dnd-hmn:fbr 30 µmol/L CBINA8 41,155,82
msc-ham:lng 12 µmol/L PNASA6 73,607,76
skn-mus TDLo:4280 µg/kg:NEO CCSUDL 3,371,78
skn-mus TD:4290 µg/kg:ETA CNREA8 38,678,78

CONSENSUS REPORTS: EPA Genetic Toxicology Program.

SAFETY PROFILE: Questionable carcinogen with experimental tumorigenic and neoplastigenic data by skin contact. Human mutation data reported. When heated to decomposition it emits acrid smoke and fumes.

BCY250 **CAS:17573-21-6** **HR: 2**
BENZO(a)PYREN-9-OL
mf: $C_{20}H_{12}O$ mw: 268.32

PROP: Yellow needles from xylene. Mp: 196°.

SYN: 9-HYDROXYBENZO(a)PYRENE

TOXICITY DATA WITH REFERENCE
mma-sat 7 µg/plate ENMUDM 7,839,85
dnd-hmn:fbr 30 µmol/L CBINA8 41,155,82
dnd-mam:lym 447 nmol CRNGDP 3,267,82
skn-mus TDLo:4280 µg/kg:NEO CCSUDL 3,371,78
skn-mus TD:4290 µg/kg:NEO CNREA8 38,678,78

CONSENSUS REPORTS: EPA Genetic Toxicology Program.

SAFETY PROFILE: Questionable carcinogen with experimental neoplastigenic data by skin contact. Human mutation data reported. When heated to decomposition it emits acrid smoke and irritating fumes.

BCY500 **CAS:56892-31-0** **HR: 2**
BENZO(a)PYREN-10-OL
mf: $C_{20}H_{12}O$ mw: 268.32

PROP: Solid. Mp: 200–201°.

SYN: 10-HYDROXYBENZO(a)PYRENE

TOXICITY DATA WITH REFERENCE
mmo-sat 18,600 pmol/L RRBCAD 18,291,81
skn-mus TDLo:4280 µg/kg:ETA CCSUDL 3,371,78

skn-mus TD:4290 µg/kg:ETA CNREA8 38,678,78

CONSENSUS REPORTS: EPA Genetic Toxicology Program.

SAFETY PROFILE: Questionable carcinogen with experimental tumorigenic data by skin contact. Mutation data reported. When heated to decomposition it emits acrid smoke and irritating fumes.

BCY750 CAS:56892-32-1 HR: 2
BENZO(a)PYREN-11-OL
mf: $C_{20}H_{12}O$ mw: 268.32

PROP: Yellow leaflets from C_6H_6. Mp: 220° (decomp).

SYN: 11-HYDROXYBENZO(a)PYRENE

TOXICITY DATA WITH REFERENCE
dnd-hmn:fbr 30 µmol/L CBINA8 41,155,82
skn-mus TDLo:82 mg/kg/38W-I:ETA CNREA8 37,2608,77

CONSENSUS REPORTS: EPA Genetic Toxicology Program.

SAFETY PROFILE: Questionable carcinogen with experimental tumorigenic data by skin contact. Human mutation data reported. When heated to decomposition it emits acrid smoke and irritating fumes.

BCZ000 CAS:56892-33-2 HR: 2
BENZO(a)PYREN-12-OL
mf: $C_{20}H_{12}O$ mw: 268.32

SYN: 12-HYDROXYBENZO(a)PYRENE

TOXICITY DATA WITH REFERENCE
mmo-sat 1 µg/plate CNREA8 36,3350,76
mma-sat 7 µg/plate ENMUDM 7,839,85
msc-ham:lng 15 mg/L CNREA8 36,3350,76
skn-mus TDLo:4280 µg/kg:NEO CCSUDL 3,371,78
skn-mus TD:4290 µg/kg:NEO CNREA8 38,678,78

CONSENSUS REPORTS: EPA Genetic Toxicology Program.

SAFETY PROFILE: Questionable carcinogen with experimental neoplastigenic data by skin contact. Mutation data reported. When heated to decomposition it emits acrid smoke and irritating fumes.

BDA000 CAS:207-89-6 HR: 2
7H-BENZO(a)PYRIDO(3,2-g)CARBAZOLE
mf: $C_{19}H_{12}N_2$ mw: 268.33

SYN: 1,2-BENZOPYRIDO(3′,2′:5,6)CARBAZOLE

TOXICITY DATA WITH REFERENCE
scu-mus TDLo:72 mg/kg/9W-I:ETA NATUAS 191,1005,61

SAFETY PROFILE: Questionable carcinogen with experimental tumorigenic data. When heated to decomposition it emits toxic fumes such as NO_x.

BDA250 CAS:194-62-7 HR: 2
7H-BENZO(c)PYRIDO(2,3-g)CARBAZOLE
mf: $C_{19}H_{12}N_2$ mw: 268.33

SYN: 5,6-BENZOPYRIDO(3′,2′:3,4)CARBAZOLE

TOXICITY DATA WITH REFERENCE
scu-mus TDLo:72 mg/kg/9W-I:ETA COREAF 257,818,63

SAFETY PROFILE: Questionable carcinogen with experimental tumorigenic data. When heated to decomposition it emits toxic fumes such as NO_x.

BDA500 CAS:194-60-5 HR: 2
7H-BENZO(c)PYRIDO(3,2-g)CARBAZOLE
mf: $C_{19}H_{12}N_2$ mw: 268.33

SYN: 3,4-BENZOPYRIDO(3′,2′:5,6)CARBAZOLE

TOXICITY DATA WITH REFERENCE
scu-mus TDLo:72 mg/kg/9W-I:ETA NATUAS 191,1005,61

SAFETY PROFILE: Questionable carcinogen with experimental tumorigenic data. When heated to decomposition it emits toxic fumes such as NO_x.

BDA750 CAS:239-67-8 HR: 2
13H-BENZO(a)PYRIDO(3,2-i)CARBAZOLE
mf: $C_{19}H_{12}N_2$ mw: 268.33

SYN: 7,8-BENZOPYRIDO(2′,3′:1,2)CARBAZOLE

TOXICITY DATA WITH REFERENCE
scu-mus TDLo:72 mg/kg/9W-I:ETA COREAF 257,818,63

SAFETY PROFILE: Questionable carcinogen with experimental tumorigenic data. When heated to decomposition it emits toxic fumes such as NO_x.

BDB000 CAS:207-88-5 HR: 2
13H-BENZO(g)PYRIDO(2,3-a)CARBAZOLE
mf: $C_{19}H_{12}N_2$ mw: 268.33

SYN: 5,6-BENZOPYRIDO(2′,3′:1,2)CARBAZOLE

TOXICITY DATA WITH REFERENCE
orl-mus TDLo:2880 mg/kg/24W-I:ETA COREAF 257,818,63
scu-mus TDLo:72 mg/kg/9W-I:ETA NATUAS 191,1005,61

SAFETY PROFILE: Questionable carcinogen with experimental tumorigenic data. When heated to decomposition it emits toxic fumes such as NO_x.

BDB250 CAS:207-85-2 HR: 2
13H-BENZO(g)PYRIDO(3,2-a)CARBAZOLE
mf: $C_{19}H_{12}N_2$ mw: 268.33

PROP: Straw-colored needles from EtOH. Mp: 368°.

SYN: 5,6-BENZOPYRIDO(3′,2′:1,2)CARBAZOLE

TOXICITY DATA WITH REFERENCE
orl-mus TDLo:3720 mg/kg/31W-I:ETA COREAF 257,818,63
scu-mus TDLo:72 mg/kg/9W-I:ETA NATUAS 191,1005,61

SAFETY PROFILE: Questionable carcinogen with ex-

perimental tumorigenic data. When heated to decomposition it emits toxic fumes such as NO_x.

BDB500 **CAS:318-03-6** **HR: 2**
11H-BENZO(g)PYRIDO(4,3-b)INDOLE
mf: $C_{15}H_{10}N_2$ mw: 218.27

SYN: 8,9-BENZO-γ-CARBOLINE

TOXICITY DATA WITH REFERENCE
scu-mus TDLo:72 mg/kg/9W-I:NEO CHDDAT 271,1474,70

SAFETY PROFILE: Questionable carcinogen with experimental neoplastigenic data. When heated to decomposition it emits toxic fumes such as NO_x.

BDC250 **CAS:583-63-1** **HR: 3**
o-BENZOQUINONE
mf: $C_6H_4O_2$ mw: 108.10

PROP: Solid. Mp: 60–70° (decomp).

SYNS: 1,2-BENZOQUINONE □ BENZOQUINONE (DOT) □ 3,5-CYCLOHEXADIENE-1,2-DIONE □ o-QUINONE

TOXICITY DATA WITH REFERENCE
mmo-sat 100 ng/plate BECTA6 24,590,80

SAFETY PROFILE: A poison. Mutation data reported. When heated to decomposition it emits acrid smoke and irritating fumes.

BDC750 **CAS:800-24-8** **HR: 3**
BENZOQUINONE AZIRIDINE
mf: $C_{16}H_{22}N_2O_6$ mw: 338.40

SYNS: A-139 □ AZIRIDYL BENZOQUINONE □ BAYER A 139 □ BAYER R39 SOLUBLE □ 2,5-BIS(1-AZIRIDINYL)-3,6-BIS(2-METHOXYETHOXY)-p-BENZOQUINONE □ 2,5-BIS(1-AZIRIDINYL)-3,6-BIS(2-METHOXYETHOXY)-2,5-CYCLOHEXADIENE-1,4-DIONE □ 2,5-BISMETHOXYETHOXY-3,6-BISETHYLENEIMINO-1,4-BENZOQUINONE □ 3,6-BIS(β-METHOXYETHOXY)-2,5-BIS(ETHYLENEIMINO)-p-BENZOQUINONE □ 3,6-BIS(β-METHOXYETHOXY)-2,5-BIS(ETHYLENIMINO)-p-BENZOQUINONE □ E 39 SOLUBLE □ NSC-17262

TOXICITY DATA WITH REFERENCE
dlt-dmg-orl 1 μmol/L MUREAV 14,250,72
ipr-mus TDLo:4 mg/kg/4W:CAR JNCIAM 36,915,66
ivn-dog LDLo:250 μg/kg CCSUBJ 2,203,65
ivn-mky LDLo:500 μg/kg CCSUBJ 2,203,65

CONSENSUS REPORTS: IARC Cancer Review: Group 3 IMEMDT 7,56,87; Animal Limited Evidence IMEMDT 9,51,75. EPA Genetic Toxicology Program.

SAFETY PROFILE: Deadly poison by intravenous route. Questionable carcinogen with experimental carcinogenic data. Mutation data reported. When heated to decomposition it emits toxic fumes of NO_x.

BDD000 **CAS:495-73-8** **HR: 3**
1,4-BENZOQUINONE-N′-BENZOYLHYDRAZONE OXIME
mf: $C_{13}H_{11}N_3O_2$ mw: 241.27

SYNS: BAYER 15080 □ BENCHINOX □ BENGUINOX □ BENQUINOX □ BENZOIC ACID(4-(HYDROXYIMINO)-2,5-CYCLOHEXADIEN-1-YLIDENE) HYDRAZIDE □ p-BENZOQUINONE OXIME BENZOYLHYDRAZONE □ CEREDON □ CERELINE □ CERENOX □ CHINONOXIM-BENZOYLHYDRAZON (GERMAN) □ CHINONOXIME-BENZOYLHYDRAZONE □ COBH □ GBH □ LERENOX □ QGH □ QUINONE OXIME BENZOYLHYDRAZONE □ TILLANTOX □ TSERENOX

TOXICITY DATA WITH REFERENCE
orl-rat LD50:100 mg/kg FMCHA2 -,C48,83
orl-mus LD50:100 mg/kg GUCHAZ 6,34,73

SAFETY PROFILE: Poison by ingestion. When heated to decomposition it emits toxic NO_x.

BDD125 **HR: 3**
BENZOQUINONE-1,4-BIS(CHLOROIMINE)(1,4-BIS(CHLORIMIDO)-2,5-CYCLOHEXADIENE)
mf: $C_6H_4Cl_2N_2$ mw: 175.02

CH=CHC(:NCl)CH=CH C:NCl

SAFETY PROFILE: Explodes on heating. When heated to decomposition it emits toxic fumes of Cl^- and NO_x.

BDD200 **CAS:4377-73-5** **HR: 3**
1,4-BENZOQUINONE DIIMINE
mf: $C_6H_6N_2$ mw: 106.13

HN:C_6H_4:NH

SYN: 1,4-DIIMIDO-2,5-CYCLOHEXADIENE

SAFETY PROFILE: Explosive decomposition on contact with concentrated acids (e.g., sulfuric or nitric acid). Upon decomposition it emits toxic fumes of NO_x.

BDD500 **CAS:3009-34-5** **HR: 3**
p-BENZOQUINONE MONOIMINE
mf: C_6H_5NO mw: 107.11

SYNS: p-BENZOQUINONE IMINE □ p-BENZOQUINONIMINE □ 2,5-CYCLOHEXADIEN-1-ONE, 4-IMINO- □ 4-IMINO-2,5-CYCLOHEXADIEN-1-ONE □ PBQI □ p-QUINONIMINE

TOXICITY DATA WITH REFERENCE
ipr-ham TDLo:200 mg/kg (female 8D post):REP TXAPA9 63,264,82
ipr-ham TDLo:200 mg/kg (female 8D post):TER TXAPA9 63,264,82

SAFETY PROFILE: Experimental reproductive effects. The solid decomposes violently (nearly explosive).When heated to decomposition it emits toxic fumes of NO_x.

BDE000 **CAS:37150-27-9** **HR: 3**
BENZO-1,2,3-THIADIAZOLE-1,1-DIOXIDE
mf: $C_6H_4N_2O_2S$ mw: 168.17

$C_6H_4SO_2N$= N

PROP: Yellow-brown needles.

SAFETY PROFILE: The solid may explode spontaneously or on impact, friction, or heating to 60°C. Upon decomposition it emits toxic fumes of NO_x and SO_x.

BDE250 CAS:91-33-8 HR: 3
BENZOTHIAZIDE
mf: $C_{15}H_{14}ClN_3O_4S_3$ mw: 431.95

SYNS: AQUATAG □ 3-((BENZYLTHIO)METHYL)-6-CHLORO-1,2,4-BENZOTHIADIAZINE-7-SULFONAMIDE-1,1-DIOXIDE □ 3-BENZYLTHIOMETHYL-6-CHLORO-2H-1,2,4-BENZOTHIADIAZINE-7-SULFONAMIDE-1,1-DIOXIDE □ 3-BENZYLTHIOMETHYL-6-CHLORO-7-SULFAMOYL-1,2,4-BENZOTHIADIAZINE-1,1-DIOXIDE □ 3-BENZYLTHIO METHYL-6-CHLORO-7-SULFAMYL-1,2,4-BENZOTHIADIAZINE-1,1-DIOXIDE □ 3-BENZYLTHIOMETHYL-6-CHLORO-7-SULFAMYL-2H-1,2,4-BENZOTHIADIAZINE-1,1-DIOXIDE □ 6-CHLORO-3-(((PHENYLMETHYL)THIO)METHYL)-2H-1,2,4-BENZOTHIADIAZINE-7-SULFONAMIDE DIOXIDE □ EDEMEX □ EXNA □ EXOSALT □ FOVANE □ FREEURIL □ NACLEX □ P 1393 □ PFIZER 1393 □ URESE

TOXICITY DATA WITH REFERENCE
ivn-rat LD50:422 mg/kg JPETAB 128,122,60
ivn-mus LD50:410 mg/kg JPETAB 128,122,60
ivn-dog LDLo:200 mg/kg JPETAB 128,122,60

SAFETY PROFILE: Poison by intravenous route. A diuretic and antihypertensive agent. When heated to decomposition it emits very toxic fumes of SO_x, NO_x, and Cl^-.

BDE500 CAS:95-16-9 HR: 3
BENZOTHIAZOLE
mf: C_7H_5NS mw: 135.19

PROP: Liquid, odor of quinoline, sltly water-sol. D: 1.246 @ 20°/4°, bp: 223–225°.

SYNS: BENZOSULFONAZOLE □ O-2857 □ 1-THIA-3-AZAINDENE □ USAF EK-4812

TOXICITY DATA WITH REFERENCE
orl-rat LD50:466 mg/kg NTIS** AD-A172-647
ihl-rat LC:>1400 mg/m³/6H EPASR* 8EHQ-1190-0987S
ipr-rat LDLo:1 g/kg JPETAB 45,189,32
ivn-rat LDLo:200 mg/kg JPETAB 45,189,32
orl-mus LD50:900 mg/kg DCTODJ 3,249,80
ipr-mus LD50:100 mg/kg NTIS** AD277-689
ivn-mus LD50:95 mg/kg JPETAB 105,486,52
unr-mus LD50:310 mg/kg KHFZAN 9(12),11,75

CONSENSUS REPORTS: Reported in EPA TSCA Inventory.

SAFETY PROFILE: Poison by ingestion, intraperitoneal, intravenous, and possibly other routes. When heated to decomposition it emits very toxic fumes of SO_x, CN^-, and NO_x.

BDE750 CAS:120-78-5 HR: 3
BENZOTHIAZOLE DISULFIDE
mf: $C_{14}H_8N_2S_4$ mw: 332.48

PROP: Cream to pale-yellow powder. Mp: 186°, d: 1.5.

SYNS: ALTAX □ BENZOTHIAZOLYL DISULFIDE □ 2-BENZOTHIAZOLYL DISULFIDE □ BIS(BENZOTHIAZOLYL)DISULFIDE □ BIS(2-BENZOTHIAZYL) DISULFIDE □ DI-2-BENZOTHIAZOLYLDISULFIDE □ DIBENZOTHIAZYL DISULFIDE □ 2,2′-DIBENZOTHIAZYLDISULFIDE □ DIBENZOYLTHIAZYL DISULFIDE □ DIBENZTHIAZYL DISULFIDE □ 2,2′-DITHIOBIS(BENZOTHIAZOLE) □ DWUSIARCZEK DWUBENZOTIAZYLU (POLISH) □ MBTS □ MBTS RUBBER ACCELERATOR □ 2-MERCAPTOBENZOTHIAZOLEDISULFIDE □ 2-MERCAPTOBENZOTHIAZYLDISULFIDE □ ROYAL MBTS □ THIOFIDE □ USAF B-33 □ USAF CY-5 □ USAF EK-5432 □ VULKACIT DM □ VULKACIT DM/MGC

TOXICITY DATA WITH REFERENCE
mma-mus:lym 15 mg/L ENMUDM 5,193,83
par-rat TDLo:400 mg/kg (female 4-11D post):TER BEXBAN 93,107,82
par-rat TDLo:400 mg/kg (4-11D preg):REP BEXBAN 93,107,82
orl-mus TDLo:172 g/kg/78W-I:ETA NTIS** PB223-159
ipr-rat LD50:2600 mg/kg IPSTB3 3,93,76
orl-mus LD50:7 g/kg IPSTB3 3,93,76
ipr-mus LD50:100 mg/kg NTIS** AD277-689
ivn-mus LD50:180 mg/kg CSLNX* NX#02251

CONSENSUS REPORTS: Reported in EPA TSCA Inventory.

SAFETY PROFILE: Poison by intravenous and intraperitoneal routes. Slightly toxic by ingestion. Experimental teratogenic and reproductive effects. Questionable carcinogen with experimental tumorigenic data. Mutation data reported. When heated to decomposition it emits very toxic fumes of SO_x and NO_x. See also SULFIDES.

BDF000 CAS:149-30-4 HR: 3
2-BENZOTHIAZOLETHIOL
mf: $C_7H_5NS_2$ mw: 167.25

PROP: Light-yellow powder or needles from MeOH (aq). Mp: 177–179°. Sltly sol in EtOH, Et_2O, and AcOH; insol in H_2O; sol in alkalis.

SYNS: BENZOTHIAZOLE-2-THIONE □ 2(3H)-BENZOTHIAZOLETHIONE □ 2-BENZOTHIAZOLYL MERCAPTAN □ CAPTAX □ KAPTAX □ MBT □ MERCAPTOBENZOTHIAZOLE □ 2 MERCAPTOBENZOTHIAZOLE □ 2-MERKAPTOBENZOTIAZOL □ 2-MERKAPTOBENZTHIAZOL □ NCI-C56519 □ PENNAC MBT POWDER □ ROKON □ ROTAX □ SULFADENE □ USAF GY-3 □ USAF XR-29 □ VULKACIT MERCAPTO

TOXICITY DATA WITH REFERENCE
sce-ham:ovr 351 mg/L NTPTR* NTP-TR-332,88
par-rat TDLo:400 mg/kg (female 4-11D post):TER BEXBAN 93,107,82
par-rat TDLo:800 mg/kg (2D male/2D pre):TER BEXBAN 93,107,82
par-rat TDLo:800 mg/kg (2D male/2D pre):REP BEXBAN 93,107,82
scu-mus TDLo:4176 mg/kg (female 6-14D post):TER NTIS** PB223-160
orl-rat TDLo:195 g/kg/2Y-I:CAR NTPTR* NTP-TR-332,88
orl-mus TDLo:35 g/kg/78W-I:ETA NTIS** PB223-159
scu-mus TDLo:215 mg/kg:CAR NTIS** PB223-159
orl-mus TD:195 g/kg/2Y-I:ETA NTPTR* NTP-TR-332,88
orl-rat LD50:100 mg/kg IPSTB3 3,93,76
ipr-rat LD50:300 mg/kg MEPAAX 16,35,65
orl-mus LD50:1851 mg/kg VCTDC* 10/12/82
ipr-mus LD50:100 mg/kg NTIS** AD277-689

CONSENSUS REPORTS: NTP Carcinogenesis Studies (gavage); Some Evidence: rat NTPTR* NTP-TR-332,88: (gavage); Equivocal Evidence: mouse NTPTR* NTP-TR-332,88. Reported in EPA TSCA Inventory.

SAFETY PROFILE: Suspected carcinogen with experimental carcinogenic tumorigenic data. Poison by ingestion and intraperitoneal routes. Experimental teratogenic and reproductive effects. Mutation data reported. Incompatible with oxidizers. When heated to decomposition or on contact with acids or acid fumes it emits toxic SO_x and NO_x. See also MERCAPTANS.

BDF250 CAS:95-30-7 HR: 2
2-BENZOTHIAZOLYL-N,N-DIETHYLTHIOCARBAMYL SULFIDE
mf: $C_{12}H_{14}N_2S_3$ mw: 282.46

SYNS: 2-(N,N-DIETHYLDITHIOCARBAMYL)BENZOATHIAZOLE □ ETHYLAC

TOXICITY DATA WITH REFERENCE
orl-rat LD50:6 g/kg IPSTB3 3,93,76
orl-rbt LD50:2700 mg/kg RCTEA4 44,512,71

CONSENSUS REPORTS: Reported in EPA TSCA Inventory.

SAFETY PROFILE: Moderately toxic by ingestion. See also CARBAMATES. When heated to decomposition it emits very toxic fumes of NO_x and SO_x.

BDF750 CAS:95-32-9 HR: 1
2-BENZOTHIAZOLYL MORPHOLINODISULFIDE
mf: $C_{11}H_{12}N_2OS_3$ mw: 284.43

PROP: Solid. Mp: 135–136°.

SYNS: MORFAX □ MORPHOLINO-2-BENZOTHIAZOLYL DISULFIDE □ 2-(MORPHOLINODITHIO)BENZOTHIAZOLE □ N-MORPHOLINYL-2-BENZOTHIAZOLYL DISULFIDE □ 4-MORPHOLINYL-2-BENZOTHIAZYL DISULFIDE □ N-OXYDIETHYL-2-BENZOTHIAZOLSULFENA MID (CZECH) □ SULFENAX MOB (CZECH) □ VULCUREN 2

TOXICITY DATA WITH REFERENCE
orl-mus LD50:3 g/kg SCIEAS 36(1-4),10,89

CONSENSUS REPORTS: Reported in EPA TSCA Inventory.

SAFETY PROFILE: Mildly toxic by ingestion. When heated to decomposition it emits very toxic fumes of NO_x and SO_x. See also SULFIDES.

BDG000 CAS:102-77-2 HR: 3
2-BENZOTHIAZOLYL-N-MORPHOLINOSULFIDE
mf: $C_{11}H_{12}N_2OS_2$ mw: 252.37

SYNS: AMAX /S 2-BENZOTHIAZOLYLSULFENYL MORPHOLINE □ 4-(2-BENZOTHIAZOLYLTHIO)MORPHOLINE □ 2-(MORPHOLINOTHIO) BENZOTHIAZOLE □ MORPHOLINYLMERCAPTOBENZOTHIAZOLE □ 2-(4-MORPHOLINYLTHIO)BENZOTHIAZOLE □ N-(OXYDIETHYLENE) BENZOTHIAZOLE-2-SULFENAMIDE □ SANTOCURE MOR □ SULFENAMIDE M □ USAF CY-7 □ VULCAFOR BSM

TOXICITY DATA WITH REFERENCE
eye-rbt 100 mg/24H MOD 85JCAE-,1099,86
dnr-esc 10 µg/tube ENMUDM 5,193,83
mma-mus:lym 15 mg/L ENMUDM 5,193,83
otr-mus:emb 200 µg/L ENMUDM 5,193,83
msc-mus:lym 10 mg/L ENMUDM 5,193,83
par-rat TDLo:400 mg/kg (female 4-11D post):TER BEXBAN 93,107,82
orl-rat TDLo:8096 mg/kg (1-20D preg):TER AOISDR 43,90,81
scu-mus TDLo:464 mg/kg:NEO NTIS** PB223-159
orl-rat LD50:1980 mg/kg IPSTB3 3,93,76
orl-mus LD50:1870 mg/kg 20ZJAG -,64,68 NTIS** AD277-689

CONSENSUS REPORTS: Reported in EPA TSCA Inventory.

SAFETY PROFILE: Poison by intraperitoneal route. Moderately toxic by ingestion. Questionable carcinogen with experimental neoplastigenic data. Experimental teratogenic effects. An eye irritant. Mutation data reported. See also MERCAPTANS and SULFIDES. When heated to decomposition it emits very toxic fumes of NO_x and SO_x.

BDG250 CAS:1079-33-0 HR: 3
BENZO(b)THIEN-4-YL METHYLCARBAMATE
mf: $C_{10}H_9NO_2S$ mw: 207.26

SYNS: 4-BENZOTHIENYL METHYLCARBAMATE □ BENZO(b)THIOPHENE-4-OL METHYLCARBAMATE □ ENT 27,041 □ MCA-600 □ MOBAM □ MOBAM PHENOL □ MOBIL MC-A-600 □ MOS-708 □ OMS-708

TOXICITY DATA WITH REFERENCE
orl-rat LD50:70 mg/kg TXAPA9 11,546,67
ipr-rat LD50:40,800 µg/kg BWHOA6 44(1-3),241,71
ivn-rat LD50:24,800 µg/kg BWHOA6 44(1-3),241,71
orl-gpg LDLo:50 mg/kg JEENAI 60,733,67
scu-gpg LDLo:25 mg/kg JEENAI 60,733,67
orl-pgn LD50:52,600 µg/kg ASTTA8 (680),157,79
orl-bwd LD50:17,800 µg/kg ASTTA8 (680),157,79

SAFETY PROFILE: Poison by ingestion, intraperitoneal, intravenous, subcutaneous, and possibly other routes. See also CARBAMATES. When heated to decomposition it emits very toxic fumes of NO_x and SO_x.

BDG325 CAS:724-34-5 HR: 3
6-BENZOTHIOPURINE
mf: $C_{12}H_{10}N_4S$ mw: 242.32

SYNS: 6-BENZYLMERCAPTOPURINE □ 6-BENZYL-MP □ 6-(BENZYLTHIO)PURINE □ NSC 29421 □ 6-((PHENYLMETHYL)THIO)-1H-PURINE (9CI)

TOXICITY DATA WITH REFERENCE
orl-rat TDLo:100 mg/kg (7D preg):REP JRPFA4 4,291,62
ipr-mus LD50:501 mg/kg NCISP* JAN86
par-mus LD50:180 mg/kg JPMSAE 71,618,82

SAFETY PROFILE: Poison by parenteral route. Moderately toxic by other routes. Experimental reproductive effects. When heated to decomposition it emits toxic fumes of NO_x and SO_x.

BDH000 CAS:90-16-4 HR: 3
1,2,3-BENZOTRIAZIN-4(1H)-ONE
mf: $C_7H_5N_3O$ mw: 147.15

PROP: Needles from cyclohexane.

SYNS: BENZAZIMIDE □ BENZAZIMIDONE □ BENZOKETCTRIAZ-

INE □ 3H-1,2,3-BENZOTRIAZIN-4-ONE □ 4-KETOBENZOTRIAZINE □ USAF MA-2

TOXICITY DATA WITH REFERENCE
orl-mus TDLo:3480 mg/kg (8D male/21D pre):REP MPHEAE 15,7,66
orl-mus TDLo:2400 mg/kg (15D male):REP MPHEAE 15,7,66
ipr-mus LD50:50 mg/kg NTIS** AD277-689

CONSENSUS REPORTS: Reported in EPA TSCA Inventory.

SAFETY PROFILE: Poison by intraperitoneal route. Experimental reproductive effects. When heated to decomposition it emits toxic fumes of NO_x.

BDH250 CAS:95-14-7 HR: 3
1H-BENZOTRIAZOLE
mf: $C_6H_5N_3$ mw: 119.14

$C_6H_4NHN=N$ (ring)

PROP: Needles from C_6H_6. Mp: 100°. Sol in C_6H_6.

SYNS: 1,2-AMINOZOPHENYLENE □ AZIMIDOBENZENE □ AZIMINOBENZENE □ BENZENE AZIMIDE □ BENZISOTRIAZOLE □ 1,2,3-BENZOTRIAZOLE □ COBRATEC #99 □ 2,3-DIAZAINDOLE □ NCI-C03521 □ NSC-3058 □ 1,2,3-TRIAZAINDENE □ U-6233

TOXICITY DATA WITH REFERENCE
mma-sat 100 µg/plate IARCCD 27,283,80
mmo-esc 333 µg/plate ENMUDM 7(Suppl 5),1,85
mma-esc 33,300 ng/plate ENMUDM 7(Suppl 5),1,85
otr-rat:emb 94 µg/plate JJATDK 1,190,81
orl-rat TDLo:220 g/kg/78W-I:ETA NCITR* NCI-CG-TR-88,78
orl-mus TDLo:770 g/kg/78W-I:ETA NCITR* NCI-CG-TR-88,78
orl-rat LD50:600 mg/kg GISAAA 46(11),70,81
orl-mus LD50:615 mg/kg NTIS** AD-A067-313
ipr-mus LD50:400 mg/kg FATOAO 41,708,78
ipr-mus LD50:1000 mg/kg CNCRA6 30,9,63
ivn-mus LD50:238 mg/kg JPETAB 105,486,52

CONSENSUS REPORTS: NCI Carcinogenesis Bioassay (feed); Inadequate Studies: mouse, rat NCITR* NCI-CG-TR-88,78. Reported in EPA TSCA Inventory.

SAFETY PROFILE: Poison by intravenous route. Moderately toxic by ingestion and intraperitoneal routes. Questionable carcinogen with experimental tumorigenic data. Mutation data reported. May detonate at 220°C or during vacuum distillation. When heated to decomposition it emits toxic fumes of NO_x.

BDH500 CAS:98-08-8 HR: 3
BENZOTRIFLUORIDE
DOT: UN 2338
mf: $C_7H_5F_3$ mw: 146.12

PROP: Water-white liquid, aromatic odor. Mp: −29.1°, bp: 98–99° @ 725 mm, flash p: 54°F (CC), d: 1.197 @ 15.5°/15.5°, vap d: 5.04, vap press: 11 mm @ 0°.

SYNS: BENZENYL FLUORIDE □ BENZYLIDYNE FLUORIDE □ PHENYLFLUOROFORM □ (TRIFLUOROMETHYL)BENZENE □ α,α,α-TRIFLUOROTOLUENE □ ω-TRIFLUOROTOLUENE □ USAF MA-16

TOXICITY DATA WITH REFERENCE
orl-rat LD50:15,000 mg/kg TPKVAL 10,131,68
ihl-rat LC50:70,810 mg/m³/4H 85GMAT -,25,82
orl-mus LD50:10,000 mg/kg TPKVAL 10,131,68
ihl-mus LC50:92,240 mg/m³/2H 85GMAT -,25,82
ipr-mus LD50:100 mg/kg NTIS** AD277-689
scu-frg LDLo:870 mg/kg AEPPAE 130,250,28

CONSENSUS REPORTS: Reported in EPA TSCA Inventory.

DOT CLASSIFICATION: 3; *Label:* Flammable Liquid

SAFETY PROFILE: Poison by intraperitoneal route. Moderately toxic by subcutaneous route. See also FLUORIDES. Dangerous fire hazard. To fight fire, use water, foam, CO_2, spray mist, dry chemical. When heated to decomposition it emits toxic fumes of F^-. Incompatible with oxidizing materials.

BDH750 CAS:215-58-7 HR: 2
BENZO(b)TRIPHENYLENE
mf: $C_{22}H_{14}$ mw: 278.36

PROP: Clear plates, leaflets, or needles from EtOH or AcOH. Mp: 205°.

SYNS: DB(a,c)A □ DIBENZ(a,c)ANTHRACENE □ 1,2:3,4-DIBENZANTHRACENE □ DIBENZO(a,c)ANTHRACENE □ 1,2:3,4-DIBENZOANTHRACENE

TOXICITY DATA WITH REFERENCE
mma-sat 10 µg/plate PNASA6 72,5135,75
dnd-hmn:emb 360 nmol/L CBINA8 22,257,78
dns-hmn:hla 100 nmol/L CNREA8 38,2625,78
msc-ham:lng 1 mg/L PNASA6 73,188,76
skn-mus TDLo:440 mg/kg/65W-I:ETA JNCIAM 44,641,70
scu-mus TDLo:12 mg/kg:ETA IJCNAW 32,765,83
skn-mus TD:40 mg/kg:ETA JNCIAM 44,1167,70
skn-mus TD:22 mg/kg:ETA CNREA8 40,1981,80

CONSENSUS REPORTS: EPA Genetic Toxicology Program. IARC Cancer Review: Group 3 IMEMDT 7,56,87; Animal Limited Evidence IMEMDT 32,289,83.

SAFETY PROFILE: Questionable carcinogen with experimental tumorigenic data. Human mutation data reported. When heated to decomposition it emits acrid smoke and irritating fumes.

BDI000 CAS:86-13-5 HR: 3
BENZOTROPINE
mf: $C_{21}H_{25}NO$ mw: 307.47

SYNS: BENZTROPINE □ 3-α-(DIPHENYLMETHOXY)-1-α-H,5-α-H-TROPANE

TOXICITY DATA WITH REFERENCE
scu-mus LD50:60 mg/kg JMPCAS 4,215,61
ivn-mus LD50:25 mg/kg JMPCAS 4,215,61

SAFETY PROFILE: Poison by subcutaneous and intravenous routes. When heated to decomposition it emits toxic fumes of NO_x.

BDI500 CAS:273-53-0 HR: 3
BENZOXAZOLE
mf: C_7H_5NO mw: 119.13

PROP: Solid. Mp: 31°, bp: 183°. Insol in water.

SYNS: 1-OXA-3-AZAINDENE □ USAF EK-5017

TOXICITY DATA WITH REFERENCE
orl-mus LD50:750 mg/kg MDCHAG 4(1),336,64
ipr-mus LD50:250 mg/kg MDCHAG 4(1),336,64
ivn-mus LD50:179 mg/kg JPETAB 105,486,52

CONSENSUS REPORTS: Reported in EPA TSCA Inventory.

SAFETY PROFILE: Poison by intraperitoneal and intravenous routes. Moderately toxic by ingestion. When heated to decomposition it emits toxic fumes such as NO_x.

BDJ000 CAS:59-49-4 HR: 3
2-BENZOXAZOLINONE
mf: $C_7H_5NO_2$ mw: 135.13

PROP: Solid. Mp: 141–142°.

SYNS: 2-BENZOXAXOLOL □ BENZOXAZOLINONE □ BENZOXAZOLONE □ 2(3H)-BENZOXAZOLONE □ 2-HYDROXYBENZOXAZOLE □ USAF EK-5429

TOXICITY DATA WITH REFERENCE
orl-rat LD50:700 mg/kg MDCHAG 4(1),308,64
orl-mus LD50:554 mg/kg GTPZAB 31(8),36,87
ipr-mus LD50:400 mg/kg NTIS** AD277-689

CONSENSUS REPORTS: Reported in EPA TSCA Inventory.

SAFETY PROFILE: Poison by intraperitoneal route. Moderately toxic by ingestion. When heated to decomposition it emits toxic fumes of NO_x.

BDJ250 CAS:2310-17-0 HR: 3
S-((3-BENZOXAZOLINYL-6-CHLORO-2-OXO)METHYL) O,O-DIETHYLPHOSPHORODITHIOATE
mf: $C_{12}H_{15}ClNO_4PS_2$ mw: 367.82

PROP: Crystals with garlic odor. Mp: 47.5–48.0°. Insol in H_2O and hydrocarbons.

SYNS: AZOFENE □ BENZOPHOSPHATE □ BENZPHOS □ CHIPMAN 11974 □ S-(6-CHLORO-3-(MERCAPTOMETHYL)-2-BENZOXAZOLINONE)-O,O-DIETHYL PHOSPHORODITHIOATE □ 3-(6-CHLORO-2-OXOBENZOXAZOLIN-3-YL)METHYL-O,O-DIETHYL PHOSPHOROTHIOLOTHIONATE □ O,O-DIAETHYL-S-(6-CHLOR-2-OXO-BEN(b)-1,3-OXALIN-3-YL)-METHYL-DIT HIOPHOSPHAT (GERMAN) □ O,O-DIETHYL-S-((6-CHLOOR-2-OXO-BENZOXAZOLIN-3-YL)-METHYL)-DITHIO FOSFAAT (DUTCH) □ O,O-DIETHYL-S-(6-CHLOROBENZOXAZOLINYL-3-METHYL)DITHIOPHOSPHATE □ O,O-DIETHYL-S-((6-CHLORO-2-OXOBENZOXAZOLIN-3-YL)METHYL) PHOSPHORODITHIOATE □ O,O-DIETHYL-S-(6-CHLORO-2-OXO-BENZOXAZOLIN-3-YL)METHYL-PHOSPHORO THIOLOTHIONATE □ 3-DIETHYLDITHIOPHOSPHORYLMETHYL-6-CHLOROBENZOXAZOLONE-2 □ O,O-DIETIL-S-((6-CLORO-2-OXO-BENZOSSAZOLIN-3-IL)-METIL)-DITIOFOSFATO (ITALIAN) □ ENT 27,163 □ FOZALON □ NIA-9241 □ NIAGARA 9241 □ NPH-1091 □ PHASOLON □ PHOSALON □ PHOSALONE □ PHOZALON □ RHODIA RP 11974 □ RUBITOX □ ZOLON □ ZOLONE □ ZOLONE PM □ ZOOLON

TOXICITY DATA WITH REFERENCE
orl-rat LD50:85 mg/kg KSKZAN 16(2),59,78
skn-rat LD50:390 mg/kg WRPCA2 9,119,70
unr-rat LD50:135 mg/kg 30ZDA9 -,371,71
orl-mus LD50:73 mg/kg GTPZAB 19(9),55,75
skn-rbt LD50:1000 mg/kg 85DPAN -,-,71/76
orl-gpg LD50:150 mg/kg GUCHAZ 6,408,73
orl-ckn LD50:661 mg/kg VETNAL 54(11),75,78

CONSENSUS REPORTS: EPA: Farm Worker Field Reentry FEREAC 39,16888,74.

SAFETY PROFILE: Poison by ingestion, skin contact, and possibly other routes. A cholinesterase inhibitor. See also PARATHION. When heated to decomposition it emits very toxic fumes of Cl^-, NO_x, PO_x, and SO_x.

BDJ600 CAS:19379-90-9 HR: 2
BENZOXONIUM CHLORIDE
mf: $C_{23}H_{42}NO_2{\cdot}Cl$ mw: 400.11

SYNS: ABSONAL □ ABSONAL V □ AMMONIUM, BENZYLBIS(2-HYDROXYETHYL)DODECYL-, CHLORIDE □ BACTOFEN □ BELORAN □ BENZENEMETHANAMINIUM, N-DODECYL-N,N-BIS(2-HYDROXYETHYL)-, CHLORIDE (9CI) □ BENZYLDODECYLBIS(2-HYDROXYETHYL)AMMONIUM CHLORIDE (6CI,7CI) □ BIALCOL □ BRADOPHEN □ COHORTAN □ D301 □ DI(2-HYDROXYETHYL)BENZYLDODECYLAMMONIUM CHLORIDE □ N-DODECYL-N,N-BIS(2-HYDROXYETHYL) BENZENEMETHANAMINIUM CHLORIDE □ DODECYL-DI(β-OXYAETHYL)-BENZYL-AMMONIUMCHLORID □ KATANOL C12 □ LOMADES □ OROFAR □ ZY 15021

TOXICITY DATA WITH REFERENCE
orl-rat LD50:750 mg/kg EGESAQ 32,395,88
ipr-mus LD50:1584 mg/kg PCJOAU 10,55,76

CONSENSUS REPORTS: Reported in EPA TSCA Inventory.

SAFETY PROFILE: Moderately toxic by ingestion and intraperitoneal routes. When heated to decomposition it emits toxic vapors of NH_4^+ and Cl^-.

BDJ800 CAS:93-91-4 HR: 2
BENZOYLACETONE
mf: $C_{10}H_{10}O_2$ mw: 162.20

SYNS: ACETOACETOPHENONE □ α-ACETYLACETOPHENONE □ 2-ACETYLACETOPHENONE □ ACETYLBENZOYLMETHANE □ BENZOYLACETON □ 1,3-BUTANEDIONE, 1-PHENYL-

TOXICITY DATA WITH REFERENCE
orl-rat LD:>500 mg/kg NCNSA6 5,28,53
unr-rat LDLo:600 mg/kg BCPCA6 14,1325,65

CONSENSUS REPORTS: Reported in EPA TSCA Inventory.

SAFETY PROFILE: Moderately toxic by ingestion. When heated to decomposition it emits acrid smoke and irritating vapors.

BDK750 CAS:117-05-5 HR: 2
5-BENZOYLAMINO-1-CHLOROANTHRAQUINONE
mf: $C_{21}H_{12}ClNO_3$ mw: 361.79

SYNS: 1-BENZAMIDO-5-CHLORO-ANTHRAQUINONE □ 1-CHLOR-5-BENZOYLAMINOANTHRACHINON (CZECH) □ 1-CHLORO-5-BENZA-

MIDO-ANTHRAQUINONE □ N-(5-CHLORO-9,10-DIHYDRO-9,10-DI-OXO-1-ANTHRACENYL)-BENZAMIDE □ 1-X-5-BAA (RUSSIAN)

TOXICITY DATA WITH REFERENCE
eye-rbt 500 mg/24H MLD 28ZPAK -,89,72
mmo-sat 50 μg/plate MUREAV 40,203,76
ipr-rat LD50:3000 mg/kg GTPZAB 21(12),27,77

CONSENSUS REPORTS: Reported in EPA TSCA Inventory.

SAFETY PROFILE: Moderately toxic by intraperitoneal route. An eye irritant. Mutation data reported. When heated to decomposition it emits very toxic fumes of NO_x and Cl^-.

BDK800 **CAS:135-57-9** **HR: 1**
o-(BENZOYLAMINO)PHENYL DISULFIDE
mf: $C_{26}H_{20}N_2O_2S_2$ mw: 456.60

SYNS: BENZANILIDE, 2′,2‴-DITHIOBIS- □ BENZAMIDE, N,N′-(DI-THIODI-2,1-PHENYLENE)BIS- □ BIS(o-BENZAMIDOPHENYL) DISULFIDE □ BIS(2-BENZAMIDOPHENYL) DISULFIDE □ BIS-o-BENZOYLAMINOFENYL-DISULFID □ o,o′-DIBENZAMIDODIPHENYL DISULFIDE □ DI-o-BENZAMIDOPHENYL DISULPHIDE □ 2,2′-DIBENZOYLAMINODIPHENYL DISULFIDE □ 2′,2‴-DITHIOBISBENZANILIDE □ 2′,2‴-DITHIODIBENZANILIDE □ N,N′-(DITHIODI-2,1-PHENYLENE)BISBENZAMIDE □ PEPTAZIN BAFD □ PEPTISANT 10 □ PEPTON 22

TOXICITY DATA WITH REFERENCE
eye-rbt 500 mg/24H MLD 85JCAE-,1007,86

CONSENSUS REPORTS: Reported in EPA TSCA Inventory.

SAFETY PROFILE: An eye irritant. When heated to decomposition it emits toxic vapors of NO_x and SO_x.

BDL750 **CAS:582-61-6** **HR: 3**
BENZOYL AZIDE
mf: $C_7H_5N_3O$ mw: 147.14

SYNS: BENZAZIDE □ BENZOIC ACID AZIDE

DOT CLASSIFICATION: Forbidden

SAFETY PROFILE: May explode when heated above 120°C. See also AZIDES.

BDL850 **CAS:85-52-9** **HR: 2**
2-BENZOYLBENZOIC ACID
mf: $C_{14}H_{10}O_3$ mw: 226.24

SYNS: BENZOIC ACID, o-BENZOYL- □ BENZOIC ACID, 2-BENZOYL- □ BENZOPHENONE-2-CARBOXYLIC ACID

TOXICITY DATA WITH REFERENCE
orl-rat LD50:4600 mg/kg GTPZAB 15(11),52,71
orl-mus LD50:880 mg/kg GTPZAB 15(11),52,71

CONSENSUS REPORTS: Reported in EPA TSCA Inventory.

SAFETY PROFILE: Moderately toxic by ingestion. When heated to decomposition it emits acrid smoke and irritating vapors.

BDM500 **CAS:98-88-4** **HR: 3**
BENZOYL CHLORIDE
DOT: UN 1736
mf: C_7H_5ClO mw: 140.57

PROP: Colorless, fuming, pungent liquid; decomposes in water. Fp −1°, mp: −0.5°, bp: 197°, flash p: 162°F (CC), d: 1.22 @ 15°/15°, vap press: 1 mm @ 32.1°, vap d: 4.88.

SYNS: BENZENECARBONYL CHLORIDE □ BENZOIC ACID, CHLORIDE □ BENZOYL CHLORIDE (DOT) □ α-CHLOROBENZALDEHYDE

TOXICITY DATA WITH REFERENCE
mmo-sat 1 μmol/plate MUREAV 58,11,78
skn-mus TDLo:9200 mg/kg/50W-I:ETA GANNA2 72,655,81
skn-mus TD:17,600 mg/kg/42W-I:ETA GANNA2 72,655,81
skn-mus TD:35,200 mg/kg/42W-I:ETA GANNA2 72,655,81
ihl-hmn TCLo:2 ppm/1M:NOSE,PUL TGNCDL 2,31,61
orl-rat LDLo:1900 mg/kg 85GMAT-,25,82
ihl-rat LC50:1870 mg/m³/2H 85GMAT-,25,82

CONSENSUS REPORTS: IARC Cancer Review: Group 3 IMEMDT 7,56,87; Human Inadequate Evidence IMEMDT 29,83,82; Animal Inadequate Evidence IMEMDT 29,83,82. Community Right-To-Know List. Reported in EPA TSCA Inventory. EPA Genetic Toxicology Program.

DFG MAK: Confirmed Human Carcinogen
DOT CLASSIFICATION: 8; *Label:* Corrosive

SAFETY PROFILE: Confirmed carcinogen with experimental tumorigenic data by skin contact. Human systemic effects by inhalation: unspecified effects on olfaction and respiratory systems. Corrosive effects on the skin, eyes, and mucous membranes by inhalation. Flammable when exposed to heat or flame. Will react with water or steam to produce heat and toxic and corrosive fumes. Violent or explosive reaction with dimethyl sulfoxide, and aluminum chloride + naphthalene. To fight fire, use alcohol foam, CO_2, dry chemical. Incompatible with dimethyl sulfoxide, (NaN_3 + KOH), water, steam, and oxidizers. When heated to decomposition it emits toxic fumes of Cl^-. See also CHLORIDES and ALDEHYDES.

BDN125 **CAS:62303-19-9** **HR: 3**
2-BENZOYLHYDRAZONO-1,3-DITHIOLANE
mf: $C_{10}H_{10}N_2OS_2$ mw: 238.34

SYNS: BHD □ YU 7802

TOXICITY DATA WITH REFERENCE
sce-hmn:lym 200 μg/L CIYPDA 15,318,84
orl-rat LD50:72,800 μg/kg CIYPDA 15,318,84
orl-mus LD50:111 mg/kg CIYPDA 15,318,84
orl-gpg LD50:383 mg/kg CIYPDA 15,318,84

SAFETY PROFILE: Poison by ingestion. Human mutation data reported. When heated to decomposition it emits toxic fumes of SO_x and NO_x.

BDN200 **CAS:52222-87-4** **HR: 3**
6-BENZOYL-2-NAPHTHOL
mf: $C_{17}H_{12}O_2$ mw: 248.29

SYN: KETONE, 6-HYDROXY-2-NAPHTHYL PHENYL

TOXICITY DATA WITH REFERENCE
otr-ham:kdy 80 μg/L BJCAAI 37,873,78

DOT CLASSIFICATION: 3; *Label:* Flammable Liquid

SAFETY PROFILE: Mutation data reported. A flammable liquid. When heated to decomposition it emits acrid smoke and irritating vapors.

BDO199 **CAS:55398-24-8** **HR: 2**
N-BENZOYLOXY-N-ETHYL-4-AMINOAZOBENZENE
mf: $C_{21}H_{19}N_3O_2$ mw: 345.43

TOXICITY DATA WITH REFERENCE
scu-rat TDLo:262 mg/kg/8W-I:NEO CNREA8 35,880,75

SAFETY PROFILE: Questionable carcinogen with experimental neoplastigenic data. See also AZIDES. When heated to decomposition it emits toxic fumes of NO_x.

BDO500 **CAS:55398-26-0** **HR: 2**
N-BENZOYLOXY-4′-ETHYL-N-METHYL-4-AMINOAZOBENZENE
mf: $C_{22}H_{21}N_3O_2$ mw: 359.46

TOXICITY DATA WITH REFERENCE
scu-rat TDLo:272 mg/kg/8W-I:NEO CNREA8 35,880,75

SAFETY PROFILE: Questionable carcinogen with experimental neoplastigenic data. When heated to decomposition it emits toxic fumes such as NO_x.

BDP000 **CAS:6098-46-0** **HR: 2**
N-BENZOYLOXY-N-METHYL-4-AMINOAZOBENZENE
mf: $C_{20}H_{17}N_3O_2$ mw: 331.40

SYNS: o-BENZOYL-N-METHYL-N-(p-(PHENYLAZO)PHENYL)HYDROXYLAMINE □ N-(BENZOYLOXY)-N-METHYL-4-(PHENYLAZO)-BENZENAMINE

TOXICITY DATA WITH REFERENCE
mmo-sat 100 nmol/plate CALEDQ 1,91,75
dnd-esc 60 mmol/L CNREA8 40,2493,80
dnd-rat:lvr 50 mmol/L CBINA8 31,1,80
scu-rat TD:375 mg/kg/12W-I:NEO CNREA8 27,1600,67
scu-rat TD:318 mg/kg/12W-I:NEO CNREA8 39,3441,79

CONSENSUS REPORTS: EPA Genetic Toxicology Program.

SAFETY PROFILE: Questionable carcinogen with experimental neoplastigenic data. Mutation data reported. See also AZIDES. When heated to decomposition it emits toxic fumes of NO_x.

BDP500 **CAS:42978-42-7** **HR: 2**
6-BENZOYLOXYMETHYLBENZO(a)PYRENE
mf: $C_{28}H_{18}O_2$ mw: 386.46

TOXICITY DATA WITH REFERENCE
scu-rat TDLo:100 mg/kg/40D-I:CAR JMCMAR 16,714,73
scu-rat TD:2898 μg/kg/60D-I:NEO CBINA8 29,159,80

SAFETY PROFILE: Questionable carcinogen with experimental carcinogenic and neoplastigenic data. When heated to decomposition it emits acrid smoke and irritating fumes.

BDQ000 **CAS:55398-25-9** **HR: 2**
N-BENZOYLOXY-4′-METHYL-N-METHYL-4-AMINOAZOBENZENE
mf: $C_{21}H_{19}N_3O_2$ mw: 345.43

TOXICITY DATA WITH REFERENCE
scu-rat TDLo:262 mg/kg/8W-I:NEO CNREA8 35,880,75

SAFETY PROFILE: Questionable carcinogen with experimental neoplastigenic data. See also AZIDES. When heated to decomposition it emits toxic fumes of NO_x.

BDQ250 **CAS:31012-29-0** **HR: 2**
7-BENZOYLOXYMETHYL-12-METHYLBENZ(a)ANTHRACENE
mf: $C_{27}H_{20}O_2$ mw: 376.47

SYN: 12-METHYLBENZ(a)ANTHRACENE-7-METHANOL BENZOATE (ESTER)

TOXICITY DATA WITH REFERENCE
scu-rat TDLo:20 mg/kg/39D-I:NEO CNREA8 31,1951,71

SAFETY PROFILE: Questionable carcinogen with experimental neoplastigenic data. See also ESTERS. When heated to decomposition it emits very acrid smoke and irritating fumes.

BDR750 **CAS:4342-36-3** **HR: 3**
BENZOYLOXYTRIBUTYLSTANNANE
mf: $C_{19}H_{32}O_2Sn$ mw: 411.20

SYNS: TRIBUTYLTIN BENZOATE □ TRI-N-BUTYL-ZINN BENZOATE (GERMAN)

TOXICITY DATA WITH REFERENCE
orl-rat LD50:132 mg/kg ARZNAD 19,934,69
scu-rat LD50:505 mg/kg TRIPA7 -,1,73
orl-mus LD50:108 mg/kg ATXKA8 23,283,68
ivn-mus LD50:178 mg/kg CSLNX* NX#00090

CONSENSUS REPORTS: Reported in EPA TSCA Inventory.

OSHA PEL: TWA 0.1 mg(Sn)/m³ (skin)
ACGIH TLV: TWA 0.1 mg(Sn)/m³ (skin) (Proposed: TWA 0.1 mg(Sn)/m³; STEL 0.2 mg(Sn)/m³ (skin))
DFG MAK: 0.002 ppm (0.05 mg/m³)
NIOSH REL: (Organotin Compounds) TWA 0.1 mg(Sn)/m³

SAFETY PROFILE: Poison by ingestion and intravenous routes. Moderately toxic by subcutaneous route. See also TIN COMPOUNDS. When heated to decomposition it emits acrid smoke and irritating fumes.

For occupational chemical analysis use NIOSH: Organotin Compounds 5504.

BDS000 CAS:94-36-0 HR: 3
BENZOYL PEROXIDE
mf: $C_{14}H_{10}O_4$ mw: 242.24

PROP: White, granular, tasteless, odorless powder or prisms. Mp: 106–108.6° (decomp), bp: decomposes explosively, autoign temp: 176°F. Sol in benzene, acetone, chloroform; sltly sol in alc; insol in water.

SYNS: ACETOXYL □ ACNEGEL □ AZTEC BPO □ BENOXYL □ BENZAC □ BENZAKNEW □ BENZOIC ACID, PEROXIDE □ BENZOPEROXIDE □ BENZOYL □ BENZOYLPEROXID (GERMAN) □ BENZOYLPEROXYDE (DUTCH) □ BENZOYL SUPEROXIDE □ BZF-60 □ CADET □ CADOX □ CLEARASIL BENZOYL PEROXIDE LOTION □ CLEARASIL BP ACNE TREATMENT □ CUTICURA ACNE CREAM □ DEBROXIDE □ DIBENZOYLPEROXID (GERMAN) □ DIBENZOYL PEROXIDE (MAK) □ DIBENZOYLPEROXYDE (DUTCH) □ DIPHENYLGLYOXAL PEROXIDE □ DRY AND CLEAR □ EPI-CLEAR □ FOSTEX □ GAROX □ INCIDOL □ LOROXIDE □ LUCIDOL □ LUPERCO □ LUPEROX FL □ NAYPER B and BO □ NOROX BZP-250 □ NOVADELOX □ OXY-5 □ OXY-10 □ OXYLITE □ OXY WASH □ PANOXYL □ PEROSSIDO di BENZOILE (ITALIAN) □ PEROXYDE de BENZOYLE (FRENCH) □ PERSADOX □ QUINOLOR COMPOUND □ SULFOXYL □ SUPEROX □ THERADERM □ TOPEX □ VANOXIDE □ XERAC

TOXICITY DATA WITH REFERENCE
eye-rbt 500 mg/24H MLD 28ZPAK -,52,72
dnd-hmn:oth 100 µmol/L CNREA8 45,2522,85
dni-ham:oth 56 µmol/L CNREA8 45,2522,85
dns-rat:lvr 100 pmol/L CRNGDP 5,1547,84
skn-mus TDLo:24 g/kg/30W-I:ETA SCIEAS 213,1023,81
orl-rat LD50:7710 mg/kg 28ZPAK -,52,72
orl-mus LD50:5700 mg/kg GISAAA 32(3),31,67
ipr-mus LDLo:250 mg/kg YKYUA6 31,855,80

CONSENSUS REPORTS: IARC Cancer Review: Group 3 IMEMDT 7,56,87, Animal Inadequate Evidence IMEMDT 36,267,85, Human Inadequate Evidence IMEMDT 36,267,85. Reported in EPA TSCA Inventory. EPA Genetic Toxicology Program. Community Right-To-Know List.

OSHA PEL: TWA 5 mg/m^3
ACGIH TLV: TWA 5 mg/m^3
DFG MAK: Weak allergin and skin irritant
NIOSH REL: (Benzoyl Peroxide) TWA 5 mg/m^3

SAFETY PROFILE: Poison by intraperitoneal routes. Can cause dermatitis, asthmatic effects, testicular atrophy, and vasodilation. An allergen and eye irritant. Human mutation data reported. Questionable carcinogen with experimental tumorigenic data. Moderate fire hazard by spontaneous chemical reaction in contact with reducing agents. It ignites readily and burns rapidly. A powerful oxidizer. Dangerous explosion hazard; may explode spontaneously, when heated to above melting point, or when overheated under confinement. It is moderately sensitive to heat, shock, friction, or contact with combustible materials. Explosive decomposition above the mp (103°) forms flammable products.

Explosive or violent reaction on contact with N,N-dimethylaniline, aniline, dimethyl sulfide, lithium tetrahydroaluminate, and N bromosuccinimide + 4-toluic acid. Mixture with carbon tetrachloride + ethylene explodes at elevated temperatures and pressures. Reacts violently in contact with various organic or inorganic acids, alcohols, amines, metallic naphthenates, as well as with polymerization accelerators, e.g., dimethylaniline, and (CCl_4 + C_2H_4). Violent reaction with charcoal when heated above 50°. Decomposition produces dense white smoke of benzoic acid, phenyl benzoate, terphenyls, biphenyls, benzene and carbon dioxide. Vigorous reaction leading to ignition with methylmethacrylate, and vinyl acetate + ethyl acetate. To fight fire, use water spray, foam. All precautions must be taken to guard against fire and explosion hazards. Keep in a cool place, out of the direct rays of the sun, away from sparks, open flames, and other sources of heat, avoid shock, rough handling, friction from grinding, etc. Isolated storage is required; keep away from possible contact with acids, alcohols, ethers, or other reducing agents or polymerization catalysts such as dimethylaniline. Complete instructions on storage and handling available from manufacturer. See also PEROXIDES.

For occupational chemical analysis use NIOSH: Benzoyl Peroxide, 5009.

BDS250 HR: 2
BENZOYL PEROXIDE, WET

PROP: A paste or wetted granular material containing at least 30% water. Autoign temp 176°F.

SAFETY PROFILE: Moderate fire hazard by chemical reaction with reducing agents; a powerful oxidizer. Mixed with a large surplus of water (i.e., 30%), this material is relatively safe. It is most dangerous when it contains very little water (1% or less). To fight fire, use water, foam or spray. Care must be taken to prevent drying out of wet material. See BENZOYL PEROXIDE.

BDS300 CAS:744-80-9 HR: 2
BENZOYLPHENOBARBITAL
mf: $C_{19}H_{16}N_2O_4$ mw: 336.37

PROP: Solid. Mp: 134–135°.

SYNS: BENZOBARBITAL □ BENZONAL □ 1-BENZOYL-5-ETHYL-5-PHENYLBARBITURIC ACID □ 1-BENZOYL-5-ETHYL-5-PHENYL-2,4,6-TRIOXOHEXAHYDROPYRIMIDINE □ BENZOYLLUMINAL □ BENZOY LUMINAL □ 2,4,6(1H,3H,5H)-PYRIMIDINETRIONE, 1 BENZOYL-5-ETHYL-5-PHENYL-(9CI)

TOXICITY DATA WITH REFERENCE
dlt-mus-orl 20 mg/kg CYGEDX 10(2),1,76
orl-mus TDLo:20 mg/kg (male 1D pre):REP CYGEDX 10(2),1,76
orl-mus LD50:982 mg/kg APSXAS 6,177,69

SAFETY PROFILE: Moderately toxic by ingestion. Experimental reproductive effects. Mutation data reported. When heated to decomposition it emits toxic fumes of NO_x.

BDS500 CAS:23107-96-2 HR: 3
***o*-BENZOYL PHENYLACETIC ACID**
mf: $C_{15}H_{12}O_3$ mw: 240.27

SYNS: ACIDE BENZOYL-2-PHENYLACETIQUE (FRENCH) □ 2-BENZOYLPHENYLACETIC ACID

B

TOXICITY DATA WITH REFERENCE
orl-mus LD50:2700 mg/kg EJMCA5 9,397,74
ipr-mus LDLo:300 mg/kg EJMCA5 11,7,76

SAFETY PROFILE: Poison by intraperitoneal route. Moderately toxic by ingestion. When heated to decomposition it emits acrid smoke and irritating fumes.

BDU500 CAS:22071-15-4 HR: 3
2-(m-BENZOYLPHENYL)PROPIONIC ACID
mf: $C_{16}H_{14}O_3$ mw: 254.30

SYNS: ALRHEUMAT □ ALRHEUMUM □ m-BENZOYLHYDRATROPIC ACID □ 3-BENZOYLHYDRATROPIC ACID □ 2-(3-BENZOYLPHENYL) PROPIONIC ACID □ CAPISTEN □ FASTUM □ ISO-K □ KEFENID □ KETOPROFEN □ KETOPRON □ LERTUS □ MEPROFEN □ ORUDIS □ ORUVAIL □ PROFENID □ 19583 RP

TOXICITY DATA WITH REFERENCE
orl-rat TDLo:720 mg/kg (female 30D pre):REP YACHDS 12,1015,84
orl-rat TDLo:720 mg/kg (30D pre):REP YACHDS 12,1015,84
orl-rat TDLo:1 mg/kg (female 21D post):TER OYYAA2 27,117,84
orl-hmn TDLo:714 µg/kg:GIT JCPCBR 24,486,84
orl-wmn TDLo:80 mg/kg/10D-I:SYS BMJOAE 292,97,86
unr-chd TDLo:300 mg/kg/15D-I:BRN,CNS,GIT NEJMAG 300,796,79
orl-rat LD50:62,400 µg/kg ARZNAD 34,280,84
ipr-rat LD50:80 mg/kg NIIRDN 6,265,82
scu-rat LD50:100 mg/kg JNPHAG 2,259,71
ivn-rat LD50:350 mg/kg IYKEDH 9,222,78
rec-rat LD50:84 mg/kg JTSCDR 6,209,81
orl-mus LD50:360 mg/kg PJPPAA 38,107,86
ipr-mus LD50:300 mg/kg EJMCA5 11,7,76
scu-mus LD50:550 mg/kg JNPHAG 2,259,71
ivn-mus LD50:500 mg/kg JNPHAG 2,259,71

CONSENSUS REPORTS: Reported in EPA TSCA Inventory.

SAFETY PROFILE: Poison by ingestion, subcutaneous, intravenous, rectal, and intraperitoneal routes. Human systemic effects by an unspecified route: headache, nausea or vomiting, and degenerative changes in the brain, changes in kidney tubules. An experimental teratogen. Other experimental reproductive effects. When heated to decomposition it emits acrid smoke and irritating fumes. An anti-inflammatory and analgesic agent.

BDV250 CAS:63989-75-3 HR: 3
N-BENZOYL TRIMETHYL COLCHICINIC ACID METHYL ETHER
mf: $C_{27}H_{27}NO_6$ mw: 461.55

SYNS: N-BENZOYL-N-DEACETYL COLCHICINE □ N-BENZOYL TMCA METHYL ETHER

TOXICITY DATA WITH REFERENCE
oms-mus-ipr 8 mg/kg CANCAR 3,130,50
oms-mus-par 32 mg/kg CANCAR 3,134,50
spm-mus-par 32 mg/kg CANCAR 3,134,50
ipr-mus LD50:32 mg/kg MDREP* No.204,49
ims-mus LD50:27,924 mg/kg JMCMAR 26,1365,83
scu-cat LDLo:12,500 µg/kg AEXPBL 72,228,13

SAFETY PROFILE: Poison by intraperitoneal and subcutaneous routes. Mutation data reported. See also COLCHICINE and ETHERS. When heated to decomposition it emits toxic fumes of NO_x.

BDV750 CAS:5929-01-1 HR: 2
1:2-BENZPYRENE PICRATE
mf: $C_{20}H_{12}{\bullet}C_6H_3N_3O_7$ mw: 481.44

SYN: BENZO(a)PYRENE MONOPICRATE

TOXICITY DATA WITH REFERENCE
skn-mus TDLo:1200 mg/kg/50W-I:ETA PRLBA4 117,318,35

SAFETY PROFILE: Questionable carcinogen with experimental tumorigenic data by skin contact. See also NITRATES. When heated to decomposition it emits toxic fumes of NO_x.

BDW000 CAS:113-69-9 HR: 3
BENZQUINAMIDE HYDROCHLORIDE
mf: $C_{22}H_{32}N_2O_5{\bullet}ClH$ mw: 441.02

SYNS: EMETE-CON □ EMETICON □ NSC 64375

TOXICITY DATA WITH REFERENCE
orl-rat LD50:990 mg/kg TXAPA9 18,185,71
orl-mus LD50:580 mg/kg MDACAP 11,9,75
ipr-mus LD50:376 mg/kg TXAPA9 18,185,71

SAFETY PROFILE: Poison by intraperitoneal route. Moderately toxic by ingestion. When heated to decomposition it emits very toxic fumes of Cl^- and NO_x. A tranquilizer and antiemetic.

BDW100 CAS:599-71-3 HR: 1
BENZSULFOHYDROXAMIC ACID
mf: $C_6H_7NO_3S$ mw: 173.20

SYN: HYDROXAMIC ACID, BENZSULFO-

TOXICITY DATA WITH REFERENCE
orl-rat LD:>500 mg/kg NCNSA6 5,42,53
scu-mus LDLo:1 g/kg AIPTAK 12,447,04

CONSENSUS REPORTS: Reported in EPA TSCA Inventory.

SAFETY PROFILE: Slightly toxic by subcutaneous route. When heated to decomposition it emits toxic vapors of NO_x and SO_x.

BDW650 HR: 3
BENZVALENE
mf: C_6H_6 mw: 78.11

SAFETY PROFILE: This strained ring compound is a friction-sensitive explosive. It may be handled safely in an ether solution. Upon decomposition it emits acrid smoke and fumes.

BDX000 **CAS:140-11-4** **HR: 3**
BENZYL ACETATE
mf: $C_9H_{10}O_2$ mw: 150.19

PROP: Colorless liquid; sweet, floral fruity odor. Mp: −51.5°, bp: 134° @ 102 mm, flash p: 216°F (CC), d: 1.06, autoign temp: 862°F, vap press: 1 mm @ 45°, vap d: 5.1, refr index: 1.501. Sol in alc, most fixed oils, propylene glycol; insol in glycerin and water @ 214°.

SYNS: ACETIC ACID BENZYL ESTER □ ACETIC ACID PHENYLMETHYL ESTER □ α-ACETOXYTOLUENE □ BENZYL ETHANOATE □ FEMA No. 2135 □ NCI-C06508

TOXICITY DATA WITH REFERENCE
skn-rbt 100 mg/24H MOD CTOIDG 94(8),41,79
dnr-bcs 21 mg/disc OIGZDE 34,267,85
mma-hmn:lyms 1500 mg/L MUREAV 196,61,88
mma-mus:lyms 500 mg/L MUREAV 196,61,88
msc-mus:lyms 700 mg/L SCIEAS 236,933,87
orl-rat TDLo:258 g/kg/2Y-I:NEO NTPTR* NTP-TR-250,86
orl-mus TDLo:258 g/kg/2Y-I:NEO NTPTR* NTP-TR-250,86
ihl-hmn TCLo:50 ppm:PSY,PUL,GLN TGNCDL 2,31,61
orl-rat LD50:2490 mg/kg FCTXAV 2,327,64
orl-mus LD50:830 mg/kg GISAAA 50(7),17,85
ihl-mus LCLo:1300 mg/m³/22H AGGHAR 5,1,33
ihl-cat LC50:245 ppm/8H AMIHAB 21,28,60
skn-cat LDLo:10 g/kg JPETAB 84,358,45
orl-rbt LD50:2200 mg/kg GISAAA 50(7),17,85
scu-rbt LDLo:3000 mg/kg AGGHAR 5,1,33
orl-gpg LD50:2200 mg/kg GISAAA 50(7),17,85
scu-gpg LDLo:3000 mg/kg AGGHAR 5,1,33

CONSENSUS REPORTS: IARC Cancer Review: Group 3 IMEMDT 7,56,87; Animal Limited Evidence IMEMDT 40,109,86. NTP Carcinogenesis Studies (gavage); Some Evidence: mouse, rat NTPTR* NTP-TR-250,86. Reported in EPA TSCA Inventory.

ACGIH TLV: (Proposed: TWA 10 ppm, Animal Carcinogen)

SAFETY PROFILE: A poison by inhalation. Moderately toxic by ingestion and subcutaneous routes. Human systemic effects by inhalation: an antipsychotic, unspecified respiratory and urinary system effects. Questionable carcinogen with experimental tumorigenic data. Combustible liquid. To fight fire, use alcohol foam, CO_2. When heated to decomposition it emits irritating fumes. See also ESTERS.

BDX090 **CAS:1214-39-7** **HR: 2**
6-BENZYLADENINE
mf: $C_{12}H_{11}N_5$ mw: 225.28

SYNS: ABG 3034 □ ADENINE, N-BENZYL- □ BA □ 6-BA □ BA (GROWTH STIMULANT) □ BAP □ 6-BAP □ BAP (GROWTH STIMULANT) □ BENZYLADENINE □ N-BENZYLADENINE □ N^6-BENZYLADENINE □ BENZYLAMINOPURINE □ N^6-(BENZYLAMINO)PURINE □ 6-(BENZYLAMINO)PURINE □ 6-(N-BENZYLAMINO)PURINE □ N-(PHENYLMETHYL)-1H-PURIN-6-AMINE □ 1H-PURIN-6-AMINE, N-(PHENYLMETHYL)-(9CI) □ SD 4901 □ SQ 4609

TOXICITY DATA WITH REFERENCE
oth-hmn:leu 100 nmol/L EXPEAM 32,29,76
oth-hmn:leu 10 µmol/L EXPEAM 32,29,76
orl-rat LD50:2125 mg/kg TOIZAG 19,336,72
orl-mus LD50:1300 mg/kg TOIZAG 19,336,72
skn-mus LD50:>5 g/kg TOIZAG 19,336,72
scu-mus LD50:>2300 mg/kg TOIZAG 19,336,72

CONSENSUS REPORTS: Reported in EPA TSCA Inventory.

SAFETY PROFILE: Moderately toxic by ingestion and skin contact. Human mutation data reported. When heated to decomposition it emits toxic vapors of NO_x.

BDX500 **CAS:100-51-6** **HR: 3**
BENZYL ALCOHOL
mf: C_7H_8O mw: 108.15

PROP: Found in jasmine, hyacinth, ylang-ylang oils, and at least two dozen other essential oils (FCTXAV 11,1011,73). Water-white liquid; faint, aromatic odor, sharp burning taste. Mp: −15.3°, bp: 205.3°, flash p: 213°F (CC), d: 1.050, autoign temp: 817°F, vap press: 1 mm @ 58.0°, vap d: 3.72, refr index: 1.540. Misc with alc, chloroform, ether, and water @ 206°(decomp). Moderately sol in water.

SYNS: BENZAL ALCOHOL □ BENZENECARBINOL □ BENZENEMETHANOL □ BENZOYL ALCOHOL □ FEMA No. 2137 □ HYDROXYTOLUENE □ α-HYDROXYTOLUENE □ NCI-C06111 □ PHENOLCARBINOL □ PHENYLCARBINOL □ PHENYLMETHANOL □ PHENYLMETHYL ALCOHOL □ α-TOLUENOL

TOXICITY DATA WITH REFERENCE
skn-man 16 mg/48H MLD CTOIDG 94(8),41,79
skn-rbt 10 mg/24H open MLD AMIHBC 4,119,51
eye-rbt 750 µg open SEV AMIHBC 4,119,51
skn-pig 100% MOD FCTXAV 11,1011,73
dnr-bcs 21 mg/disc OIGZSE 34,267,85
orl-mus TDLo:6 g/kg (female 6-13D post):REP TCMUD8 7,29,87
orl-rat LD50:1230 mg/kg FCTXAV 2,327,64
ihl-rat LCLo:2000 ppm/4H JIDHAN 31,343,49
ipr-rat LD50:400 mg/kg NPIRI* 1,6,74
scu-rat LDLo:1700 mg/kg RMSRA6 15,561,1895
ivn-rat LD50:53 mg/kg TXAPA9 18,60,71
orl-mus LD50:1360 mg/kg GISAAA 50(7),81,85
ivn-mus LD50:324 mg/kg AIPTAK 135,330,62
ivn-dog LDLo:50 mg/kg TXAPA9 18,60,71
par-dog LDLo:9 mg/kg TXAPA9 25,153,73
skn-cat LDLo:10 g/kg JPETAB 84,358,45

CONSENSUS REPORTS: EPA Genetic Toxicology Program. Reported in EPA TSCA Inventory.

SAFETY PROFILE: Poison by ingestion, intraperitoneal, intravenous, parenteral routes. Moderately toxic by inhalation, skin contact, and subcutaneous routes. A moderate skin and severe eye irritant. Mutation data reported. Combustible liquid. Mixtures with sulfuric acid decompose explosively at 180°. Exothermic polymerization is catalyzed by HBr + iron when heated above 100°. To fight fire, use alcohol foam, CO_2, dry chemical. When heated to decomposition it emits acrid smoke and fumes. See also ALCOHOLS.

BDY000 CAS:3287-99-8 HR: 3
BENZYLAMINE HYDROCHLORIDE
mf: $C_7H_9N \cdot ClH$ mw: 143.63

PROP: Solid. Mp: 255–257°.

SYNS: BENZYENEMETHAMAMINE HYDROCHLORIDE □ BENZYLAMMONIUM CHLORIDE □ USAF EL-82

TOXICITY DATA WITH REFERENCE
ipr-mus LD50:500 mg/kg NTIS** AD277-689
ivn-mus LD50:220 mg/kg APFRAD 9,390,51

CONSENSUS REPORTS: Reported in EPA TSCA Inventory.

SAFETY PROFILE: Poison by intravenous route. Moderately toxic by intraperitoneal route. When heated to decomposition it emits very toxic fumes of HCl, NH_3, and NO_x. See also AROMATIC AMINES.

BDY250 CAS:77966-31-5 HR: 3
2-(BENZYLAMINO)-6′-CHLORO-o-ACETOTOLUIDIDE HYDROCHLORIDE
mf: $C_{16}H_{17}ClN_2O \cdot ClH$ mw: 325.26

SYN: C 3117

TOXICITY DATA WITH REFERENCE
eye-rbt 2% MLD ARZNAD 8,407,58
ipr-rat LD50:280 mg/kg ARZNAD 8,407,58
scu-mus LD50:1175 mg/kg ARZNAD 8,407,58

SAFETY PROFILE: Poison by intraperitoneal route. Moderately toxic by subcutaneous route. An eye irritant. When heated to decomposition it emits very toxic fumes of Cl^-, NO_x, and HCl.

BDY500 CAS:52400-76-7 HR: 3
2-(2-(BENZYLAMINO)ETHYL)-2-METHYL-1,3-BENZODIOXOLE HYDROCHLORIDE
mf: $C_{17}H_{19}NO_2 \cdot ClH$ mw: 305.83

TOXICITY DATA WITH REFERENCE
ivn-rat LD50:15 mg/kg EJMCA5 12,413,77
ipr-mus LD50:110 mg/kg EJMCA5 12,413,77

SAFETY PROFILE: Poison by intravenous and intraperitoneal routes. When heated to decomposition it emits very toxic fumes of HCl and NO_x.

BDY669 CAS:61-33-6 HR: 3
BENZYL-6-AMINOPENICILLINIC ACID
mf: $C_{16}H_{18}N_2O_4S$ mw: 334.42

PROP: Crystals.

SYNS: ABBOCILLIN □ (5R,6R)-BENXYLPENICILLIN □ BENZOPENICILLIN □ BENZYLPENICILLIN □ BENZYLPENICILLIN G □ BENZYLPENICILLINIC ACID □ CILLORAL □ CILOPEN □ COMPOCILLIN G □ COSMOPEN □ DROPCILLIN □ FREE BENZYLPENICILLIN □ GALOFAK □ GELACILLIN □ LIQUACILLIN □ PENICILLIN G □ PHENYLACETAMIDOPENICILLANIC ACID □ (PHENYLMETHYL) PENICILLINIC ACID □ PRADUPEN □ SPECILLINE G

TOXICITY DATA WITH REFERENCE
dnr-esc 20 µL/disc MUREAV 97,1,82
dnr-bcs 100 µL/plate MUREAV 97,1,82
mmo-omi 12 µg/L ARMKA7 81,1,72
oms-omi 20 µg/L AMACCQ 17,572,80
scu-rat TDLo:2600 mg/kg/65W-I:ETA LANCAO 1,394,86
par-chd TDLo:15,000 units/kg:NOSE,CNS,PUL BJCAAI 17,100,63
orl-rat LD50:8 g/kg ANTCAO 12,249,62
unk-rat LD50:9 g/kg ANTBAL 23, 317,78
orl-mus LD50:>5 g/kg AACHAX-,619,67
ipr-mus LD50:3500 mg/kg AACHAX-,619,67
ivn-mus LD50:329 mg/kg BCPCA6 16,1365,67
ice-mus LD50:5700 µg/kg JLCMAK 34,126,49
unk-mus LD50:7800 mg/kg ANTBAL 23,317,78
ice-rbt LD50:1118 µg/kg JLCMAK 34,126,49
isp-dog LD50:4940 µg/kg JLCMAK 34,126,49
ice-rbt LD50:653 µg/kg JLCMAK 34,126,49
orl-ham LD50:24 mg/kg TXAPA9 14,510,69
scu-ham LD50:96 mg/kg TXAPA9 14,510,69

CONSENSUS REPORTS: EPA Genetic Toxicology Program.

SAFETY PROFILE: Poison by ingestion, intravenous, intracerebral, intraspinal, subcutaneous, and possibly other routes. Human (child) systemic effects by parenteral route: changes in cochlear (inner ear) structure or function, convulsions, and dyspnea. Questionable carcinogen with experimental tumorigenic data. Mutation data reported. When heated to decomposition it emits very toxic fumes of NO_x and SO_x. See other penicillin entries.

BDY750 CAS:103-14-0 HR: 2
4-(BENZYLAMINO)PHENOL
mf: $C_{13}H_{13}NO$ mw: 199.27

SYN: PHENOL, p-(BENZYLAMINO)-

TOXICITY DATA WITH REFERENCE
orl-rat LDLo:500 mg/kg JPETAB 90,260,47

CONSENSUS REPORTS: Reported in EPA TSCA Inventory.

SAFETY PROFILE: Moderately toxic by ingestion. When heated to decomposition it emits toxic vapors of NO_x.

BEA000 CAS:67465-04-7 HR: 3
2-BENZYLAMINOPYRIDINE HYDROCHLORIDE
mf: $C_{12}H_{12}N_2 \cdot ClH$ mw: 220.70

SYNS: 2-BAP HYDROCHLORIDE □ N-(2-PYRIDYL)BENZYLAMINE HYDROCHLORIDE

TOXICITY DATA WITH REFERENCE
orl-mus LD50:1187 mg/kg TXAPA9 37,165,76
ipr-mus LD50:220 mg/kg TXAPA9 37,165,76
ivn-mus LD50:90 mg/kg JPETAB 84,16,45

SAFETY PROFILE: Poison by intravenous and intraperitoneal routes. Moderately toxic by ingestion. When heated to decomposition it emits very toxic fumes of NO_x and HCl. See also AROMATIC AMINES.

BEA275 CAS:101997-51-7 HR: 3
1-(2-(N-BENZYLANILINO)ETHYL)PIPERIDINE HYDROCHLORIDE
mf: $C_{20}H_{26}N_2 \cdot ClH$ mw: 330.94

SYNS: N-β-(BENZILFENILAMINO)ETILPIPERIDINA CLORIDRATO (ITALIAN) □ N-β-(BENZYL-PHENYLAMINO)ETHYLPIPERIDINE HYDROCHLORIDE

TOXICITY DATA WITH REFERENCE
ims-rat LDLo:380 mg/kg FRPSAX 15,562,60
orl-mus LD50:1500 mg/kg FRPSAX 13,3,58
ipr-mus LD50:180 mg/kg FRPSAX 15,562,60
ivn-mus LD50:25 mg/kg FRPSAX 13,3,58
ipr-gpg LD50:110 mg/kg BSCIA3 31,520,49

SAFETY PROFILE: Poison by intravenous, intramuscular, and intraperitoneal routes. Moderately toxic by ingestion. When heated to decomposition it emits toxic fumes of NO_x and HCl.

BEA325 CAS:622-79-7 HR: 3
BENZYL AZIDE
mf: $C_7N_7N_3$ mw: 133.15

PROP: Liquid. D: 1.0655 @ 25 mm, bp: 108° @ 23 mm. Insol in water.

SAFETY PROFILE: A heat-sensitive explosive. Explosive reaction with bis(trifluoromethyl)nitroxide. Upon decomposition it emits toxic fumes of NO_x. See also AZIDES.

BEA500 CAS:36226-64-9 HR: 3
BENZYLBARBITAL
mf: $C_{13}H_{14}N_2O_3$ mw: 246.29

SYNS: 5-BENZYL-5-ETHYLBARBITURIC ACID □ ETHYLBENZYLBARBITURIC ACID □ 5-ETHYL-5-(PHENYLMETHYL)-2,4,6(1H,3H,5H)-PYRIMIDINETRIONE (9CI)

TOXICITY DATA WITH REFERENCE
ipr-mus LD50:73 mg/kg JPETAB 89,356,47
orl-cat LDLo:400 mg/kg JPETAB 26,371,25
scu-rbt LDLo:60 mg/kg JACSAT 45,243,23

SAFETY PROFILE: Poison by ingestion, intraperitoneal, and subcutaneous routes. When heated to decomposition it emits toxic fumes of NO_x. An hypnotic agent. See also BARBITURATES.

BEA825 CAS:621-72-7 HR: 3
2-BENZYLBENZIMIDAZOLE
mf: $C_{14}H_{12}N_2$ mw: 208.28

PROP: Crystals or needles from benzene. Mp: 187°. Practically insol in water; freely sol in glacial acetic acid; sol in alc, hot benzene, and propylene glycol.

SYNS: BENDAZOL □ BENDAZOLE □ 2-BENZYLBENZIMINAZOLE □ DIBASOL □ DIBAZOL □ DIBAZOLE □ 2-(PHENYLMETHYL)-1H-BENZIMIDAZOLE □ TROMASEDAN

TOXICITY DATA WITH REFERENCE
unr-rat TDLo:2 mg/kg (9D preg):REP AKGIAO 43(12),10,67
unr-rat TDLo:2 mg/kg (9D preg):TER AKGIAO 43(12),10,67
orl-mus LD50:100 mg/kg FRZKAP (1),44,83
ipr-mus LD50:240 mg/kg PCJOAU 19,544,85
scu-mus LDLo:504 mg/kg PCJOAU 13,829,79

SAFETY PROFILE: Poison by ingestion and intraperitoneal routes. Moderately toxic by subcutaneous route. Experimental reproductive effects. When heated to decomposition it emits toxic fumes of NO_x.

BEA850 CAS:1421-23-4 HR: 3
N-BENZYLBIGUANIDE HYDROCHLORIDE
mf: $C_9H_{13}N_5 \cdot ClH$ mw: 227.73

SYNS: 1-BENZILBIGUANIDE CLORIDRATO (ITALIAN) □ BENZYLBIGUANIDE HYDROCHLORIDE □ 1-BENZYLBIGUANIDE HYDROCHLORIDE

TOXICITY DATA WITH REFERENCE
orl-rat LD50:481 mg/kg FRPSAX 15,521,60
ipr-rat LD50:108 mg/kg FRPSAX 15,521,60
ipr-mus LD50:195 mg/kg JAJAAA 18,196,65

SAFETY PROFILE: Poison by intraperitoneal route. Moderately toxic by ingestion. When heated to decomposition it emits toxic fumes of NO_x and HCl.

BEB750 CAS:101834-51-9 HR: 3
5-BENZYL-2,2-BIS(TRIFLUOROMETHYL)-4-METHYLOXAZOLIDINE HYDRATE
mf: $C_{12}H_{11}F_6NO \cdot H_2O$ mw: 317.26

TOXICITY DATA WITH REFERENCE
orl-mus LD50:200 mg/kg JMCMAR 13,1215,70
ipr-mus LD50:300 mg/kg JMCMAR 13,1215,70

SAFETY PROFILE: Poison by ingestion and intraperitoneal routes. See also FLUORIDES. When heated to decomposition it emits very toxic fumes of F^- and NO_x.

BEC000 CAS:100-39-0 HR: 2
BENZYL BROMIDE
DOT: UN 1737
mf: C_7H_7Br mw: 171.05

PROP: Clear, refractive liquid; pleasant odor, lachrymator, insol in water. Mp: −4.0°, bp: 198°, d: 1.438 @ 22°/0°, vap d: 5.8.

SYNS: (BROMOMETHYL)BENZENE □ p-(BROMOMETHYL)NITROBENZENE □ BROMOPHENYLMETHANE □ ω-BROMOTOLUENE □ α-BROMOTOLUENE (DOT)

TOXICITY DATA WITH REFERENCE
dns-esc 1300 μmol/L ZKKOBW 92,177,78

CONSENSUS REPORTS: Reported in EPA TSCA Inventory.

DOT CLASSIFICATION: 6.1; *Label:* Poison, Corrosive

SAFETY PROFILE: Intensely irritating and corrosive to skin, eyes, and mucous membranes. Large doses cause central nervous system depression. Mutation data reported. Reaction with molecular sieve produces toxic hydrogen bromide gas. See also BROMIDES.

BEC250 **CAS:103-05-9** **HR: 2**
BENZYL-tert-BUTANOL
mf: $C_{11}H_{16}O$ mw: 164.27

PROP: Bp: 128° @ 17 mm.

SYNS: DIMETHYLPHENYLETHYL CARBINOL □ 1,1-DIMETHYL-3-PHENYLPROPANOL □ 1,1-DIMETHYL-3-PHENYL-1-PROPANOL □ α,α-DIMETHYL-Δ-PHENYLPROPYL ALCOHOL □ 2-METHYL-4-PHENYL-2-BUTANOL □ PHENYLETHYL DIMETHYL CARBINOL

TOXICITY DATA WITH REFERENCE
orl-rat LD50:2200 mg/kg FCTXAV 12,517,74
skn-rbt LD50:3500 mg/kg FCTXAV 12,517,74

CONSENSUS REPORTS: Reported in EPA TSCA Inventory.

SAFETY PROFILE: Moderately toxic by ingestion and skin contact. When heated to decomposition it emits acrid smoke and irritating fumes. See also ALCOHOLS.

BEC500 **CAS:85-68-7** **HR: 2**
BENZYL BUTYL PHTHALATE
mf: $C_{19}H_{20}O_4$ mw: 312.39

PROP: Clear, oily liquid. Mp: <−35°, bp: 370°, flash p: 390°F, d: 1.116 @ 25°/25°, vap d: 10.8.

SYNS: BBP □ 1,2-BENZENEDICARBOXYLIC ACID, BUTYL PHENYLMETHYL ESTER □ BUTYL BENZYL PHTHALATE □ n-BUTYL BENZYL PHTHALATE □ NCI-C54375 □ PALATINOL BB □ SANTICIZER 160 □ SICOL 160 □ UNIMOLL BB

TOXICITY DATA WITH REFERENCE
orl-rat TDLo:21 g/kg (14D male):REP TOXID9 4,136,84
orl-rat TDLo:433 g/kg/2Y-C:CAR NTPTR* NTP-TR-213,82
orl-rat TD:437 g/kg/2Y-C:CAR EVHPAZ 65,271,86
orl-rat LD50:2330 mg/kg IARC** 29,193,82
skn-rat LD50:6700 mg/kg GISAAA 39(6),25,74
orl-mus LD50:4170 mg/kg IARC** 29,193,82
skn-mus LD50:6700 mg/kg GISAAA 39(6),25,74
ipr-mus LD50:3160 mg/kg EVHPAZ 4,3,73
orl-gpg LD50:13,750 mg/kg GTPZAB 24(3),25,80

CONSENSUS REPORTS: IARC Cancer Review: Group 3 IMEMDT 7,56,87, Animal Inadequate Evidence IMEMDT 29,193,82; NTP Carcinogenesis Bioassay (feed); No Evidence: mouse NTPTR* NTP-TR-213,82; Clear Evidence: rat NTPTR* NTP-TR-213,82. Reported in EPA TSCA Inventory. Community Right-To-Know List.

SAFETY PROFILE: Questionable carcinogen with experimental carcinogenic data. Moderately toxic by ingestion, skin contact, and intraperitoneal routes. Experimental reproductive effects. See also ESTERS. Combustible when exposed to heat or flame; can react with oxidizers. To fight fire, use spray or mist, CO_2, dry chemical. When heated to decomposition it emits acrid smoke and irritating fumes.

BED000 **CAS:103-37-7** **HR: 2**
BENZYL n-BUTYRATE
mf: $C_{11}H_{14}O_2$ mw: 178.25

PROP: Colorless liquid; floral plum-like odor. D: 1.006, refr index: 1.492, flash p: 212°F. Sol in fixed oils; insol in glycerin, propylene glycol, water @ 239°.

SYNS: BENZYL n-BUTANOATE □ FEMA No. 2140

TOXICITY DATA WITH REFERENCE
orl-rat LD50:2330 mg/kg FCTXAV 2,327,64

CONSENSUS REPORTS: Reported in EPA TSCA Inventory.

SAFETY PROFILE: Moderately toxic by ingestion. See also ESTERS. Combustible liquid. When heated to decomposition it emits acrid smoke and irritating fumes.

BED250 **CAS:63884-81-1** **HR: 3**
BENZYLCARBAMIC ESTER of 3-OXYPHENYLDIMETHYLAMINE HYDROCHLORIDE
mf: $C_{16}H_{18}N_2O_2{\cdot}ClH$ mw: 306.82

SYNS: AR-22 □ 3-(N′-BENZYLCARBAMOYLOXY)-N,N-DIMETHYLANILINE HYDROCHLORIDE

TOXICITY DATA WITH REFERENCE
orl-mus LDLo:500 mg/kg JPETAB 43,413,31
ivn-mus LDLo:50 mg/kg JPETAB 43,413,31

SAFETY PROFILE: Poison by intravenous route. Moderately toxic by ingestion. See also ESTERS and CARBAMATES. When heated to decomposition it emits very toxic fumes of NO_x and HCl.

BED500 **CAS:64051-16-7** **HR: 3**
BENZYLCARBAMIC ESTER of 3-OXYPHENYLTRIMETHYLAMMONIUM METHYLSULFATE
mf: $C_{17}H_{21}N_2O_2{\cdot}CH_3O_4S$ mw: 396.50

SYNS: AMMONIUM ((3-N-BENZYLCARBAMOYLOXY)PHENYL)TRIMETHYL METHYLSULFATE □ AR-23 □ N-BENZYL-CARBAMIC ACID-3-(TRIMETHYLAMMONIO)PHENYL ESTER, METHYLSULFATE □ (m-HYDROXYPHENYL)TRIMETHYLAMMONIUM METHYLSULFATE BENZYLCARBAMATE

TOXICITY DATA WITH REFERENCE
orl-mus LDLo:33 mg/kg JPETAB 43,413,31
ivn-mus LDLo:100 μg/kg NTIS** PB158-508

SAFETY PROFILE: Poison by ingestion and intravenous routes. See also CARBAMATES; ESTERS; and SULFATES. When heated to decomposition it emits very toxic fumes of SO_x, NH_3, and NO_x.

BED750 **CAS:14504-15-5** **HR: 2**
3-BENZYL-4-CARBAMOYLMETHYLSYDNONE
mf: $C_{11}H_{11}N_3O_3$ mw: 233.25

SYN: 3-BENZYLSYDNONE-4-ACETAMIDE

TOXICITY DATA WITH REFERENCE
orl-rat TDLo:23 mg/kg/13W-I:NEO GANNA2 65,273,74
orl-rat LD50:4450 mg/kg GANNA2 65,273,74

SAFETY PROFILE: Mildly toxic by ingestion. Questionable carcinogen with experimental neoplastigenic data. When heated to decomposition it emits toxic fumes of NO_x.

BEE250 **CAS:103-53-7** **HR: 1**
BENZYLCARBINYL CINNAMATE
mf: $C_{17}H_{16}O_2$ mw: 252.33

SYNS: PHENETHYL CINNAMATE □ β-PHENETHYL CINNAMATE □ PHENYLETHYL CINNAMATE □ β-PHENYLETHYL CINNAMATE

TOXICITY DATA WITH REFERENCE
skn-rbt 500 mg/24H MLD FCTXAV 16,637,78
orl-rat LD50:5 g/kg FCTXAV 16,637,78
orl-mus LD50:4500 mg/kg FCTXAV 16,637,78
orl-gpg LD50:4500 mg/kg FCTXAV 16,637,78

CONSENSUS REPORTS: Reported in EPA TSCA Inventory.

SAFETY PROFILE: Mildly toxic by ingestion. A skin irritant. See also ESTERS. When heated to decomposition it emits acrid smoke and irritating fumes.

BEE375 **CAS:100-44-7** **HR: 3**
BENZYL CHLORIDE
DOT: UN 1738
mf: C_7H_7Cl mw: 126.59

PROP: Colorless liquid, very refractive; irritating, unpleasant odor. Mp: −48°, bp: 99° @ 62 mm, lel: 1.1%, flash p: 153°F, d: 1.11 @ 4°/4°, autoign temp: 1085°F, vap d: 4.36.

SYNS: BENZILE (CLORURO di) (ITALIAN) □ BENZYLCHLORID (GERMAN) □ BENZYLE (CHLORURE de) (FRENCH) □ CHLOROMETHYLBENZENE □ CHLOROPHENYLMETHANE □ α-CHLOROTOLUENE □ ω-CHLOROTOLUENE □ α-CHLORTOLUOL (GERMAN) □ CHLORURE de BENZYLE (FRENCH) □ NCI-C06360 □ RCRA WASTE NUMBER P028 □ TOLYL CHLORIDE

TOXICITY DATA WITH REFERENCE
dnd-hmn:fbr 1 mmol/L MUREAV 145,209,85
dnd-hmn:oth 1 mmol/L MUREAV 145,209,85
otr-ham:emb 1600 μg/L CRNGDP 1,323,80
orl-rat TDLo:1 g/kg (female 6-15 post):TER JTEHD6 17,51,86
ipr-mus TDLo:250 mg/kg (male 5D pre):REP MUREAV 100,345,82
scu-rat TDLo:50 mg/kg:ETA FCTXAV 6,576,68
orl-mus TDLo:31 g/kg/2Y-I:CAR JJIND8 76,1231,86
skn-mus TDLo:9200 mg/kg/50W-I:ETA GANNA2 72,655,81
scu-rat TD:2100 mg/kg/53W-I:ETA ZEKBAI 74,241,70
orl-rat LD50:1231 mg/kg NTIS** PB214-270
ihl-rat LC50:150 ppm/2H IARC** 11,217,76
scu-rat LD50:1 g/kg ZEKBAI 74,241,70
orl-mus LD50:1500 mg/kg 85GMAT -,25,82
ihl-mus LC50:80 ppm/2H IARC** 11,217,76

CONSENSUS REPORTS: IARC Cancer Review: Animal Limited Evidence IMEMDT 29,49,82; Animal Sufficient Evidence IMEMDT 11,217,76; Human Inadequate Evidence IMEMDT 29,49,82. EPA Genetic Toxicology Program. Community Right-To-Know List. Reported in EPA TSCA Inventory. EPA Extremely Hazardous Substances List.

OSHA PEL: TWA 1 ppm
ACGIH TLV: TWA 1 ppm
DFG MAK: Confirmed Carcinogen
NIOSH REL: (Benzyl Chloride) CL 5 mg/m³/15M

DOT CLASSIFICATION: 6.1; *Label:* Poison, Corrosive

SAFETY PROFILE: Confirmed carcinogen with experimental carcinogenic and tumorigenic data. Poison by inhalation. Moderately toxic by ingestion and subcutaneous routes. Experimental reproductive effects. Human mutation data reported. A corrosive irritant to skin, eyes, and mucous membranes. Flammable and moderately explosive when exposed to heat or flame. Can react vigorously with oxidizing materials. May explode during distillation. The decomposition rate can reach explosive violence in presence of metals such as iron. Catalytic impurities (e.g., aluminum, iron, rust) or sodium acetate + pyridine + iron (at 115°C) may cause violent polymerization reactions. Will react with water or steam to produce toxic and corrosive fumes. Incompatible with dimethyl sulfoxide. Used in production of drugs of abuse. When heated to decomposition it emits toxic fumes of Cl^-. See also CHLORINATED HYDROCARBONS, AROMATIC.

For occupational chemical analysis use NIOSH: Hydrocarbons, halogenated, 1003.

BEE500 **CAS:140-18-1** **HR: 2**
BENZYL CHLOROACETATE
mf: $C_9H_9ClO_2$ mw: 184.63

PROP: Oil. Bp: 79–81° @ 0.65 mm.

SYNS: BENZYL-α-CHLOROACETATE □ BENZYL MONOCHLORACETATE □ CHLOROACETIC ACID BENZYL ESTER

TOXICITY DATA WITH REFERENCE
ipr-mus LDLo:500 mg/kg CBCCT* 8,99,56

CONSENSUS REPORTS: Reported in EPA TSCA Inventory.

SAFETY PROFILE: Moderately toxic by intraperitoneal route. See also ESTERS. When heated to decomposition it emits toxic fumes of Cl^-.

BEE750 **CAS:77966-32-6** **HR: 3**
N-BENZYL-6′-CHLORO-2-(DIETHYLAMINO)-o-ACETOTOLUIDIDE HYDROCHLORIDE
mf: $C_{20}H_{25}ClN_2O{\cdot}ClH$ mw: 381.38

SYN: C 3136

TOXICITY DATA WITH REFERENCE
eye-rbt 2% MOD ARZNAD 8,609,58
ipr-rat LD50:96 mg/kg ARZNAD 8,609,58
scu-mus LD50:360 mg/kg ARZNAD 8,609,58

SAFETY PROFILE: Poison by intraperitoneal and subcutaneous routes. An eye irritant. When heated to decomposition it emits very toxic fumes of NO_x, HCl, and Cl^-.

BEE800 **CAS:1833-31-4** **HR: 3**
BENZYLCHLORODIMETHYLSILANE
mf: $C_9H_{13}ClSi$ mw: 184.76

SYN: SILANE, BENZYLCHLORODIMETHYL-

TOXICITY DATA WITH REFERENCE
ivn-mus LD50:56 mg/kg CSLNX* NX#04165

CONSENSUS REPORTS: Reported in EPA TSCA Inventory.

SAFETY PROFILE: Poison by intravenous route. When heated to decomposition it emits toxic vapors of Cl^-.

BEF500 **CAS:501-53-1** **HR: 3**
BENZYL CHLOROFORMATE
DOT: UN 1739
mf: $C_8H_7ClO_2$ mw: 170.60

PROP: Colorless to pale-yellow liquid or oil; odor of phosgene. Mp: 0°, bp: 103° @ 20 mm.

SYNS: BENZYLCARBONYL CHLORIDE □ BENZYL CHLOROCARBONATE (DOT) □ BENZYL CHLOROFORMATE (DOT) □ BENZYLOXYCARBONYL CHLORIDE □ BZCF □ CARBOBENZOXY CHLORIDE □ CARBOBENZYLOXY CHLORIDE □ CHLOROFORMIC ACID BENZYL ESTER

CONSENSUS REPORTS: Reported in EPA TSCA Inventory.

DOT CLASSIFICATION: 8; *Label:* Corrosive

SAFETY PROFILE: Poison by ingestion and inhalation routes. A powerful corrosive irritant. Thermally unstable. Will react with water or steam to produce toxic and corrosive fumes and heat. Iron salts catalyze the explosive decomposition of the ester. When heated to decomposition it emits toxic fumes of Cl^- and phosgene. See also PHOSGENE, ESTERS, and CHLORIDES.

BEF750 **CAS:1322-48-1** **HR: 3**
2-BENZYL-4-CHLOROPHENOL
mf: $C_{13}H_{11}ClO$ mw: 218.69

PROP: Nearly colorless flakes. Mp: 49°, bp: 175° @ 5 mm, d: 1.2 @ 55°/25°.

SYNS: BENZYLCHLOROPHENOL □ 4-CHLORO-α-PHENYLCRESOL

TOXICITY DATA WITH REFERENCE
orl-rat LD50:1700 mg/kg JPMSAE 63,1068,74

CONSENSUS REPORTS: Chlorophenols are on the Community Right-To-Know List.

SAFETY PROFILE: Moderately toxic by ingestion. When heated to decomposition it emits toxic fumes of Cl^-. See also CHLOROPHENOLS.

BEG000 **CAS:501-68-8** **HR: 2**
N-BENZYL-β-CHLOROPROPANAMIDE
mf: $C_{10}H_{12}ClNO$ mw: 197.68

PROP: Large crystals from MeOH. Mp: 94°.

SYNS: BECLAMID □ BECLAMIDE □ BEKLAMID □ BENZCHLOROPROPAMIDE □ BENZCHLORPROPAMID □ BENZOCHLORPROPAMID □ BENZYLAMIDE □ N-BENZYL-β-CHLOROPROPIONAMIDE □ N-BENZYL-3-CHLOROPROPIONAMIDE □ CHLORACON □ CHLORAKON □ CHLOROETHYLPHENAMIDE □ 3-CHLORO-N-(PHENYLMETHYL)PROPANAMIDE □ N-(3-CHLOROPROPIONYL)BENZYLAMINE □HIBI-CON □ KHLORAKON □ NEURACEN □ NIDRANE □ NYDRAN □ NYDRANE □ POSEDRAN □ POSEDRINE □ SECLAR

TOXICITY DATA WITH REFERENCE
orl-rat LD50:3200 mg/kg JPETAB 107,403,53
ipr-rat LD50:770 mg/kg JPETAB 107,403,53
ivn-rat LD50:770 mg/kg 27ZQAG -,384,72
orl-mus LD50:1000 mg/kg PCJOAU 14,99,80
ipr-mus LD50:650 mg/kg 27ZQAG -,384,72

SAFETY PROFILE: Moderately toxic by ingestion, intraperitoneal, and intravenous routes. When heated to decomposition it emits very toxic fumes of Cl^- and NO_x. An anticonvulsant.

BEG300 **CAS:72850-64-7** **HR: 1**
BENZYL 2-CHLORO-4-(TRIFLUOROMETHYL)-5-THIAZOLECARBOXYLATE
mf: $C_{12}H_7ClF_3NO_2S$ mw: 321.71

SYNS: FLURAZOLE □ PHENYLMETHYL 2-CHLORO-4-(TRIFLUOROMETHYL)-5-THIAZOLECARBOXYLATE □ SCREEN □ 5-THIAZOLECARBOXYLIC ACID, 2-CHLORO-4-(TRIFLUOROMETHYL)-, PHENYLMETHYL ESTER

TOXICITY DATA WITH REFERENCE
orl-rat LD50:5010 mg/kg 85AREA 3,123,86

CONSENSUS REPORTS: Reported in EPA TSCA Inventory.

SAFETY PROFILE: Low toxicity by ingestion. When heated to decomposition it emits toxic vapors of SO_x, NO_x, F^-, and Cl^-.

BEG750 **CAS:103-41-3** **HR: 2**
BENZYL CINNAMATE
mf: $C_{16}H_{14}O_2$ mw: 238.30

PROP: Found in balsams of Peru, tolu, styrax, copaiba, and others (FCTXAV 11,1011,73). White crystals; aromatic odor. Mp: 39°, bp: 350.0°, vap press: 1 mm @ 173.8°, flash p: 212°F. Sol in fixed oils; insol in glycerin and propylene glycol.

SYNS: BENZYL ALCOHOL CINNAMIC ESTER □ BENZYL γ-PHENYLACRYLATE □ CINNAMEIN □ trans-CINNAMIC ACID BENZYL ESTER □ FEMA No. 2142 □ 3-PHENYL-2-PROPENOIC ACID PHENYLMETHYL ESTER (9CI)

TOXICITY DATA WITH REFERENCE
skn-rbt 500 mg MLD FCTXAV 11,1011,73
orl-rat LDLo:5530 mg/kg FCTXAV 2,327,64
orl-gpg LD50:3760 mg/kg FCTXAV 2,327,64

CONSENSUS REPORTS: Reported in EPA TSCA Inventory.

SAFETY PROFILE: Moderately toxic by ingestion. A mild allergen and skin irritant. Combustible liquid. See also ESTERS. When heated to decomposition it emits acrid smoke and irritating fumes.

BEH000 **CAS:363-13-3** **HR: 3**
1-BENZYL-2(1H)-CYCLOHEPTIMIDAZOLONE
mf: $C_{15}H_{12}N_2O$ mw: 236.29

PROP: Pale yellow crystals from MeOH. Mp: 181°.

SYNS: BENZYLCYCLOHEPTIMIDAZOL-2(1H)-ONE □ 1-(PHENYLMETHYL)-2(1H)-CYCLOHEPTIMIDAZOLONE

B

TOXICITY DATA WITH REFERENCE
orl-mus LD50:358 mg/kg ARZNAD 18,939,68
ipr-mus LD50:119 mg/kg ARZNAD 18,939,68

SAFETY PROFILE: Poison by ingestion and intraperitoneal routes. When heated to decomposition it emits toxic fumes of NO_x.

BEI250 CAS:101651-55-2 HR: 3
2-(BENZYL(2-(DIETHYLAMINO)ETHYL)AMINO)ACETANILIDE DIHYDROCHLORIDE
mf: $C_{21}H_{29}N_3O \cdot 2ClH$ mw: 412.45

SYN: C 5348

TOXICITY DATA WITH REFERENCE
eye-rbt 2% MLD ARZNAD 9,167,59
scu-mus LD50:375 mg/kg ARZNAD 9,167,59

SAFETY PROFILE: Poison by subcutaneous route. An eye irritant. When heated to decomposition it emits very toxic fumes of HCl and NO_x.

BEI500 CAS:102489-46-3 HR: 3
2-(BENZYL(2-(DIETHYLAMINO)ETHYL)AMINO)-o-ACETOTOLUIDIDE DIHYDROCHLORIDE
mf: $C_{22}H_{31}N_3O \cdot 2ClH$ mw: 426.48

SYN: C 5351

TOXICITY DATA WITH REFERENCE
eye-rbt 2% MLD ARZNAD 9,167,59
scu-mus LD50:160 mg/kg ARZNAD 9,167,59

SAFETY PROFILE: Poison by subcutaneous route. An eye irritant. When heated to decomposition it emits very toxic fumes of NO_x and HCl. See also AMINES.

BEI750 CAS:102489-45-2 HR: 3
2-(BENZYL(2-(DIETHYLAMINO)ETHYL)AMINO)-6'-CHLORO-o-ACETOTOLUIDIDE DIHYDROCHLORIDE
mf: $C_{22}H_{30}ClN_3O \cdot 2ClH$ mw: 460.92

SYN: C 5296

TOXICITY DATA WITH REFERENCE
eye-rbt 2% MLD ARZNAD 9,167,59
scu-mus LD50:32 mg/kg ARZNAD 9,167,59

SAFETY PROFILE: Poison by subcutaneous route. An eye irritant. When heated to decomposition it emits very toxic fumes of NO_x, HCl, and Cl^-.

BEJ250 CAS:77966-34-8 HR: 3
(2-(BENZYL(3-DIETHYLAMINO)PROPYL)AMINO)-o-ACETOTOLUIDIDE DIHYDROCHLORIDE
mf: $C_{23}H_{33}N_3O \cdot 2ClH$ mw: 440.51

SYN: C 5353

TOXICITY DATA WITH REFERENCE
eye-rbt 2% MLD ARZNAD 9,167,59
scu-mus LD50:260 mg/kg ARZNAD 9,167,59

SAFETY PROFILE: Poison by subcutaneous route. An eye irritant. When heated to decomposition it emits very toxic fumes of NO_x and HCl.

BEJ500 CAS:77966-75-7 HR: 3
2-(BENZYL(3-(DIETHYLAMINO)PROPYL)AMINO)-2',6'-ACETOXYLIDIDE DIHYDROCHLORIDE
mf: $C_{24}H_{35}N_3O \cdot 2ClH$ mw: 454.54

SYN: C 5354

TOXICITY DATA WITH REFERENCE
eye-rbt 2% MLD ARZNAD 9,167,59
scu-mus LD50:35 mg/kg ARZNAD 9,167,59

SAFETY PROFILE: Poison by subcutaneous route. An eye irritant. When heated to decomposition it emits very toxic fumes of NO_x and HCl.

BEJ825 HR: 3
N-BENZYL-N',N'-DIETHYL-N-1-NAPHTHYLETHYLENEDIAMINE
mf: $C_{23}H_{28}N_2$ mw: 332.53

TOXICITY DATA WITH REFERENCE
ipr-rat LDLo:45 mg/kg BJPCAL 11,1,56
ipr-mus LD50:63 mg/kg BJPCAL 11,1,56
scu-mus LD50:300 mg/kg BJPCAL 11,1,56

SAFETY PROFILE: Poison by subcutaneous and intraperitoneal routes. When heated to decomposition it emits toxic fumes of NO_x. See also AMINES.

BEJ830 HR: 3
N-BENZYL-N',N'-DIETHYL-N-2-NAPHTHYLETHYLENEDIAMINE
mf: $C_{23}H_{28}N_2$ mw: 332.53

TOXICITY DATA WITH REFERENCE
ipr-rat LDLo:90 mg/kg BJPCAL 11,1,56
ipr-mus LD50:128 mg/kg BJPCAL 11,1,56
scu-mus LD50:459 mg/kg BJPCAL 11,1,56

SAFETY PROFILE: Poison by intraperitoneal route. Moderately toxic by subcutaneous route. When heated to decomposition it emits toxic fumes of NO_x. See also AMINES.

BEL525 CAS:7083-24-1 HR: 3
BENZYL 1-(2-(DIMETHYLAMINO)PROPYL)PYRROL-2-YL, CITRATE KETONE
mf: $C_{17}H_{22}N_2O \cdot C_6H_8O_7$ mw: 462.55

SYN: 1-(2-(DIMETHYLAMINO)PROPYL)-2-(PHENYLACETYL)PYRROLE CITRATE

TOXICITY DATA WITH REFERENCE
ivn-mus LD50:40 mg/kg CHTPBA 1,127,66

DOT CLASSIFICATION: 3; *Label:* Flammable Liquid

SAFETY PROFILE: A poison by intravenous route. A flammable liquid. When heated to decomposition it emits acrid smoke and irritating vapors.

BEL550 **CAS:73747-22-5** **HR: 3**
BENZYLDIMETHYLAMMONIUM HEXAFLUOROARSENATE
mf: $C_9H_{13}N•AsF_6H$ mw: 325.16

SYNS: BENZYLAMINE, N,N-DIMETHYL-, HEXAFLUOROARSENATE (1-) □ N,N-DIMETHYLBENZYLAMINE HEXAFLUOROARSENATE

TOXICITY DATA WITH REFERENCE
ivn-mus LD50:180 mg/kg CSLNX* NX#04251

OSHA PEL: TWA 0.5 mg(As)/m^3

SAFETY PROFILE: Poison by intravenous route. When heated to decomposition it emits toxic fumes of NO_x, F^-, and As.

BEL750 **CAS:151-05-3** **HR: 2**
BENZYLDIMETHYL CARBINYL ACETATE
mf: $C_{12}H_{16}O_2$ mw: 192.28

PROP: Bp: 102–103° @ 10 mm.

SYNS: DIMETHYLBENZYL CARBINOLACETATE □ α,α-DIMETHYLPHENETHYL ACETATE □ α,α-DIMETHYLPHENETHYL ALCOHOL ACETATE □ DMBCA

TOXICITY DATA WITH REFERENCE
skn-rbt 500 mg/24H MOD FCTXAV 12,533,74
orl-rat LD50:3300 mg/kg FCTXAV 12,533,74

CONSENSUS REPORTS: Reported in EPA TSCA Inventory.

SAFETY PROFILE: Moderately toxic by ingestion. A skin irritant. When heated to decomposition it emits acrid smoke and irritating fumes. See also ALCOHOLS.

BEL850 **CAS:10094-34-5** **HR: 1**
BENZYL DIMETHYLCARBINYL n-BUTYRATE
mf: $C_{14}H_{20}O_2$ mw: 220.34

SYNS: BENZYL DIMETHYLCARBINYL BUTYRATE □ BUTYRIC ACID, α-α-DIMETHYLPHENETHYL ESTER □ DIMETHYLBENZYLCARBINYL BUTYRATE □ α-α-DIMETHYLPHENETHYL BUTYRATE

TOXICITY DATA WITH REFERENCE
skn-rbt 500 mg/24H MOD FCTXAV 18,667,80
orl-rat LD50:>5 g/kg FCTXAV 18,667,80
skn-rbt LD50:>5 g/kg FCTXAV 18,667,80

CONSENSUS REPORTS: Reported in EPA TSCA Inventory.

SAFETY PROFILE: Low oral toxicity. A skin irritant. When heated to decomposition it emits acrid smoke and irritating fumes.

BEL900 **CAS:122-18-9** **HR: 2**
BENZYLDIMETHYLCETYLAMMONIUM CHLORIDE
mf: $C_{25}H_{46}N•Cl$ mw: 396.17

SYNS: ACINOL □ AMMONIUM, BENZYLHEXADECYLDIMETHYL-, CHLORIDE □ AMMONYX G □ AMMONYX T □ BAKTONIUM □ BANICOL □ BENZALETAS □ BENZENEMETHANAMINIUM, N-HEXADECYL-N,N-DIMETHYL-, CHLORIDE □ BENZYLDIMETHYLHEXADECYLAMMONIUM CHLORIDE □ BICETONIUM □ BONJELA □ CDBAC □ CETALKONIUM CHLORIDE □ CETOL □ CETYLON □ CETYL ZEPHIRAN □ DEHYQUART CBB □ DEHYQUART CDB □ DMCBAC □ PHARYCIDIN CONCENTRATE □ RODALON □ SPILAN □ TETRASEPTAN □ WIN 357 □ WINZER SOLUTION □ ZETTYN CHLORIDE

TOXICITY DATA WITH REFERENCE
eye-rbt 150 mg MLD ARZNAD 9,349,59
eye-gpg 500 mg MOD ARZNAD 9,349,59
orl-rat LD:>500 mg/kg NCNSA6 5,39,53

CONSENSUS REPORTS: Reported in EPA TSCA Inventory.

SAFETY PROFILE: Moderately toxic by ingestion. An eye irritant. When heated to decomposition it emits toxic vapors of NO_x and Cl^-.

BEM000 **CAS:139-07-1** **HR: 3**
BENZYLDIMETHYLDODECYLAMMONIUM CHLORIDE
mf: $C_{21}H_{38}N•Cl$ mw: 340.05

PROP: Solid. Mp: 31–32°.

SYN: DODECYL DIMETHYL BENZYLAMMONIUM CHLORIDE

TOXICITY DATA WITH REFERENCE
skn-rbt 1 mg/24H OYYAA2 6,329,72
eye-rbt 1 mg OYYAA2 6,329,72
orl-rat LD50:400 mg/kg 85JCAE-,490,86
ipr-rat LD50:100 mg/kg 85JCAE-,490,86

SAFETY PROFILE: Poison by ingestion and intraperitoneal routes. A skin and eye irritant. When heated to decomposition it emits very toxic fumes of NO_x, NH_3, and Cl^-.

BEM250 **CAS:37557-89-4** **HR: 1**
BENZYLDIMETHYLEICOSANYLAMMONIUM CHLORIDE
mf: $C_{29}H_{54}N•Cl$ mw: 452.29

SYN: EICOSANYL DIMETHYL BENZYLAMMONIUM CHLORIDE

TOXICITY DATA WITH REFERENCE
skn-rbt 1 mg/24H OYYAA2 6(2),329,72
eye-rbt 1 mg OYYAA2 6(2),329,72

SAFETY PROFILE: A skin and eye irritant. When heated to decomposition it emits very toxic fumes of NH_3, NO_x, and Cl^-.

BEM325 **HR: 3**
N-BENZYL-N′,N′-DIMETHYL-N-1-NAPHTHYLETHYLENEDIAMINE
mf: $C_{21}H_{24}N_2$ mw: 304.47

TOXICITY DATA WITH REFERENCE
ipr-rat LDLo:90 mg/kg BJPCAL 11,1,56
ipr-mus LD50:135 mg/kg BJPCAL 11,1,56
scu-mus LD50:463 mg/kg BJPCAL 11,1,56

SAFETY PROFILE: Poison by intraperitoneal route. Moderately toxic by subcutaneous route. When heated to decomposition it emits toxic fumes of NO_x. See also AMINES.

BEM330 **HR: 3**
N-BENZYL-N′,N′-DIMETHYL-N-2-NAPHTHYLETHYLENEDIAMINE
mf: $C_{21}H_{24}N_2$ mw: 304.47

TOXICITY DATA WITH REFERENCE
ipr-rat LDLo:135 mg/kg BJPCAL 11,1,56
ipr-mus LD50:265 mg/kg BJPCAL 11,1,56
scu-mus LD50:740 mg/kg BJPCAL 11,1,56

SAFETY PROFILE: Poison by intraperitoneal route. Moderately toxic by subcutaneous route. When heated to decomposition it emits toxic fumes of NO_x. See also AMINES.

BEM500 **CAS:961-71-7** **HR: 3**
N-BENZYL-N′,N′-DIMETHYL-N-PHENYLETHYLENEDIAMINE
mf: $C_{17}H_{22}N_2$ mw: 254.41

PROP: Pale-yellow oil. Bp: 179–180° @ 7 mm.

SYNS: ANTERGAN □ BRIDAL □ DIMETINA □ N,N-DIMETHYL-N′-PHENYL-N′-(PHENYLMETHYL)-1,2-ETHANEDIAMINE (9CI) □ LERGITIN □ NCI-C60719 □ PHENBENZAMINE □ PM245 □ 2339 RP

TOXICITY DATA WITH REFERENCE
ipr-rat LDLo:120 mg/kg BJPCAL 11,1,56
ims-rat LDLo:350 mg/kg FRPSAX 13,3,58
ipr-mus LD50:170 mg/kg FRPSAX 13,3,58
scu-mus LD50:400 mg/kg BJPCAL 11,1,56

SAFETY PROFILE: Poison by subcutaneous, intraperitoneal, and intramuscular routes. When heated to decomposition it emits toxic fumes of NO_x.

BEM750 **CAS:525-02-0** **HR: 3**
1-BENZYL-2,5-DIMETHYL SEROTONIN HYDROCHLORIDE
mf: $C_{19}H_{22}N_2O{\bullet}ClH$ mw: 330.89

PROP: Solid. Mp: 230–231°.

SYNS: 3-(2-AMINOETHYL)-1-BENZYL-5-METHOXY-2-METHYLINDOLE HYDROCHLORIDE □ BAS □ BENANSERIN HYDROCHLORIDE □ BENZYL ANTISEROTONIN □ 1-BENZYL-2-METHYL-3-(2-AMINOETHYL)-5-METHOXYINDOLE HYDROCHLORIDE □ 1-BENZYL-2-METHYL 5 METHOXYTRYPTAMINE HYDROCHLORIDE □ SEROTONIN BENZYL ANALOG □ WOOLLEY'S ANTISEROTONIN

TOXICITY DATA WITH REFERENCE
ipr-rat TDLo:70 mg/kg (8-14D preg):REP PEDIAU 27,318,61
ipr-mus LD50:250 mg/kg JMCMAR 9,819,66

SAFETY PROFILE: Poison by intraperitoneal route. A serotonin antagonist that causes psychotropic effects in humans. Experimental reproductive effects. When heated to decomposition it emits very toxic fumes of HCl and NO_x.

BEN000 **CAS:121-54-0** **HR: 3**
BENZYLDIMETHYL(2-(2-(p-(1,1,3,3-TETRAMETHYLBUTYL)PHENOXY)ETHOXY)ETHYL) AMMONIUM CHLORIDE
mf: $C_{27}H_{42}NO_2{\bullet}Cl$ mw: 448.15

PROP: Colorless crystals or plates. Mp: 164–166°. Very sol in H_2O; sol in Me_2CO, EtOH, and $CHCl_3$.

SYNS: ANTI-GERM 77 □ ANTISEPTOL □ BENZETHONIUM CHLORIDE □ BENZETONIUM CHLORIDE □ BENZYLDIMETHYL-p-(1,1,3,3-TETRAMETHYLBUTYL)PHENOXYETHOXY-ETHYLAMMONIUM CHLORIDE □ BZT □ DIAPP □ DIISOBUTYLPHENOXYETHOXYETHYLDIMETHYL BENZYL AMMONIUM CHLORIDE □ DISILYN □ HYAMINE □ HYAMINE 1622 □ NCI-C61494 □ p-tert-OCTYLPHEN OXYETHOXYETHYLDIMETHYLBENZYL AMMONIUM CHLORIDE □ PHEMERIDE □ PHEMEROL CHLORIDE □ PHEMITHYN □ POLYMINE D □ QUATRACHLOR □ SOLAMINE

TOXICITY DATA WITH REFERENCE
eye-rbt 30 µg SEV PSTGAW 20,16,53
dnr-esc 1500 ng/well MUREAV 133,161,84
sce-ham:emb 1 mg/L SHIGAZ 74,1365,87
scu-rat TDLo:104 mg/kg/1Y-I:NEO CTOXAO 4,185,71
orl-rat LD50:368 mg/kg PSEBAA 120,511,65
ipr-rat LD50:16,500 µg/kg FSDZD4 9,729,83
scu-rat LD50:119 mg/kg NTIS** PB195-158
ivn-rat LD50:19 mg/kg SCHSAV 30,147,54
unr-rat LD50:420 mg/kg MEIEDD 10,152,83
orl-mus LD50:338 mg/kg PSEBAA 120,511,65
ipr-mus LD50:15,500 µg/kg FSDZD4 9,729,83
ivn-mus LD50:30 mg/kg JAPMA8 40,267,51

CONSENSUS REPORTS: Reported in EPA TSCA Inventory.

SAFETY PROFILE: Poison by ingestion, subcutaneous, intraperitoneal, and intravenous routes. A severe eye irritant. Questionable carcinogen with experimental neoplastigenic data. Mutation data reported. When heated to decomposition it emits very toxic fumes of Cl^-, NH_3, and NO_x. A topical anti-infective agent.

BEN250 **CAS:101-49-5** **HR: 2**
2-BENZYLDIOXOLAN
mf: $C_{10}H_{12}O_2$ mw: 164.22

SYN: PHENYLACETALDEHYDE ETHYLENEGLYCOL ACETAL

TOXICITY DATA WITH REFERENCE
orl-rat LD50:2200 mg/kg FCTXAV 14,827,76
skn-rbt LD50:2600 mg/kg FCTXAV 14,827,76

CONSENSUS REPORTS: Reported in EPA TSCA Inventory.

SAFETY PROFILE: Moderately toxic by ingestion and skin contact. When heated to decomposition it emits acrid smoke and irritating fumes. See also ALDEHYDES.

BEN800 **CAS:7492-37-7** **HR: 3**
1 BENZYL DIPROPYL KETONE
mf: $C_{14}H_{20}O$ mw: 204.34

SYNS: 3-BENZYL-4-HEPTANONE □ 4-HEPTANONE, 3-BENZYL-(8CI) □ 4-HEPTANONE, 3-(PHENYLMETHYL)-

TOXICITY DATA WITH REFERENCE
orl-rat LD50:4400 mg/kg JACTDZ 1,2,90

DOT CLASSIFICATION: 3; *Label:* Flammable Liquid

SAFETY PROFILE: Low toxicity by ingestion. A flammable liquid. When heated to decomposition it emits acrid smoke and irritating vapors.

BEO000 CAS:7281-04-1 HR: 3
BENZYLDODECYLDIMETHYL AMMONIUM BROMIDE
mf: $C_{21}H_{38}N{\cdot}Br$ mw: 384.51

PROP: Crystals. Mp: 47°. Sol in H_2O.

SYNS: AMMONYL BR 1244 □ BACFOR BL □ BENZALKONIUM BROMIDE □ BENZENEMETHANAMINIUM, N-DODECYL-N,N-DIMETHYL-, BROMIDE (9CI) □ BENZODODECINIUM BROMIDE □ BENZYLDIMETHYLDODECYLAMMONIUM BROMIDE □ BROMEK DWUMETYLOLAURYLOBENZYLOAMONIOWY □ DIMETHYL LAURYLBENZENE AMMONIUM BROMIDE □ N-DODECYL-N,N-DIMETHYLBENZENEMETHANAMINIUM BROMIDE □ SINNOQUAT BL 80 □ SINNOQUAT BL 95 □ STERINOL □ STERINOLU (POLISH)

TOXICITY DATA WITH REFERENCE
orl-rat LD50:250 mg/kg RPZHAW 17,543,66
ipr-rat LD50:90 mg/kg BCTKAG 7,161,74

CONSENSUS REPORTS: Reported in EPA TSCA Inventory.

SAFETY PROFILE: Poison by ingestion and intraperitoneal routes. See also BROMIDES. When heated to decomposition it emits very toxic fumes of NH_3, NO_x, and Br^-.

BEO250 CAS:103-50-4 HR: 2
BENZYL ETHER
mf: $C_{14}H_{14}O$ mw: 198.28

PROP: Colorless to pale-yellow liquid. Mp: 5°, bp: 182–183° @ 22 mm, flash p: 275°F (CC), d: 1.056, vap d: 6.84, refr index: 1.557.

SYNS: BENZYL OXIDE (CZECH) □ DIBENZYLETHER (CZECH) □ FEMA No. 2371

TOXICITY DATA WITH REFERENCE
skn-rbt 500 mg/24H MLD 28ZPAK -,38,72
eye-rbt 500 mg/24H MLD 28ZPAK -,38,72
orl-rat LD50:2500 mg/kg FCTXAV 16,637,78

CONSENSUS REPORTS: Reported in EPA TSCA Inventory.

SAFETY PROFILE: Moderately toxic by ingestion. Vapors are probably narcotic in high concentration. A skin and eye irritant. Combustible when exposed to heat or flame; can react with oxidizing materials. Moderate explosion hazard by spontaneous chemical reaction. To fight fire, use CO_2, dry chemical. See also ETHERS.

BEP250 CAS:104-57-4 HR: 2
BENZYL FORMATE
mf: $C_8H_8O_2$ mw: 136.16

SYNS: BENZYL ALCOHOL FORMATE □ BENZYL METHANOATE

TOXICITY DATA WITH REFERENCE
dnr-bcs 22 mg/disc OIGZSE 34,267,85
orl-rat LD50:1400 mg/kg FCTXAV 11,1019,73
skn-rbt LD50:2000 mg/kg FCTXAV 11,1019,73

CONSENSUS REPORTS: Reported in EPA TSCA Inventory.

SAFETY PROFILE: Moderately toxic by ingestion and skin contact. Mutation data reported. Probably narcotic in high concentrations. See also ESTERS. When heated to decomposition it emits acrid, irritating fumes.

BEP500 CAS:10453-86-8 HR: 3
5-BENZYL-3-FURYL METHYL(±)-cis,trans-CHRYSANTHEMATE
mf: $C_{22}H_{26}O_3$ mw: 338.48

SYNS: BENZOFUROLINE □ BENZYLFUROLINE □ (5-BENZYL-3-FURYL) METHYL-2,2-DIMETHYL-3-(2-METHYLPROPENYL)-CYCLOPROPANECARBOXYLATE □ CHRYSON □ CHRYSRON □ DIMETHYL-3-(2-METHYL-1-PROPENYL)CYCLOPROPANECARBOXYLATE □ ENT 27,474 □ FMC 17370 □ FOR-SYN □ NIA 17170 □ NRDC 104 □ NSC 195022 □ OMS-1206 □ PREMGARD □ PYNOSECT □ PYRETHERM □ RESMETHRIN □ RESMETRINA (PORTUGUESE) □ SBP-1382 □ S.B. PENICK 1382 □ SYNTHRIN

TOXICITY DATA WITH REFERENCE
orl-rat LD50:1244 mg/kg FAATDF 7,299,86
ihl-rat LC:>420 mg/m³/4H NTIS** AD747-345
skn-rat LD50:4200 mg/kg 85JFAN A362,83
ipr-rat LDLo:19,200 mg/kg NTIS** AD747-345
scu-rat LD50:>5 g/kg BOCKAE 34,157,69
ivn-rat LDLo:160 mg/kg BIOGAL 41(10),283,75
orl-mus LD50:300 mg/kg ABCHA6 37,2681,73
skn-mus LD50:>5 g/kg BOCKAE 34,157,69
ipr-mus LD50:>1 g/kg OYYAA2 3,325,69
scu-mus LD50:>2 g/kg BOCKAE 34,157,69
ihl-dog LC:>420 mg/m³/4H NTIS** AD747-345
ivn-dog LDLo:250 mg/kg NTIS** AD747-345
skn-rbt LD50:2500 mg/kg SPEADM 78-1,9,78

CONSENSUS REPORTS: EPA Genetic Toxicology Program.

SAFETY PROFILE: Poison by ingestion and intravenous routes. Moderately toxic by inhalation and skin contact. When heated to decomposition it emits acrid and irritating fumes. See also ESTERS.

BEP750 CAS:28434-01-7 HR: 3
5-BENZYL-3-FURYLMETHYL(+)-trans-CHRYSANTHEMATE
mf: $C_{22}H_{26}O_3$ mw: 338.48

PROP: Bp: 174° @ 0.0008 mm.

SYNS: BIORESMETHRIN □ BIORESMETHRINE □ BIORESMETRINA (PORTUGUESE) □ NIA-18739 □ NRDC 107 □ (+)-trans-RESMETHRIN □ d-trans-RESMETHRIN □ RU-11484 □ SBP-1390

TOXICITY DATA WITH REFERENCE
ivn-rat LD50:340 mg/kg BIOGAL 41,283,75
orl-mus LD50:590 mg/kg EVHPAZ 14,15,76

SAFETY PROFILE: Poison by intravenous route. Moderately toxic by ingestion. When heated to decomposition it emits acrid smoke and irritating fumes. A pesticide. See also ESTERS.

BEQ000 **CAS:555-96-4** **HR: 3**
BENZYLHYDRAZINE
mf: $C_7H_{10}N_2$ mw: 122.19

PROP: Liquid. Bp: 107° @ 8 mm.

TOXICITY DATA WITH REFERENCE
oms-bcs 10 mmol/L MUREAV 5,343,68
ipr-mus LD50:75 mg/kg THERAP 22,367,67
scu-mus LD50:68 mg/kg ANYAA9 80,568,59

SAFETY PROFILE: Poison by intraperitoneal and subcutaneous routes. Mutation data reported. When heated to decomposition it emits toxic fumes such as NO_x.

BEQ250 **CAS:20570-96-1** **HR: 2**
BENZYLHYDRAZINE DIHYDROCHLORIDE
mf: $C_7H_{10}N_2 \cdot 2ClH$ mw: 195.11

PROP: Solid. Mp: 145° (decomp).

TOXICITY DATA WITH REFERENCE
orl-mus TDLo:10 g/kg/29W-C:NEO ZEKBAI 87,267,76
ipr-mus LD50:11 mg/kg JMCMAR 18,20,75

SAFETY PROFILE: Poison by intraperitoneal route. Questionable carcinogen with experimental neoplastigenic data. When heated to decomposition it emits very toxic fumes of HCl and NO_x.

BEQ500 **CAS:1073-62-7** **HR: 3**
BENZYLHYDRAZINE HYDROCHLORIDE
mf: $C_7H_{10}N_2 \cdot ClH$ mw: 158.65

PROP: Leaflets from EtOH. Mp: 111°.

SYNS: P 1297 □ USAF EL-54 □ Z 102

TOXICITY DATA WITH REFERENCE
orl-mus LD50:90 mg/kg JMPCAS 5,221,62
ipr-mus LD50:50 mg/kg NTIS** AD277-689

SAFETY PROFILE: Poison by ingestion and intraperitoneal routes. When heated to decomposition it emits very toxic fumes of NO_x and Cl^-.

BEQ625 **CAS:73-48-3** **HR: 3**
BENZYLHYDROFLUMETHIAZIDE
mf: $C_{15}H_{14}F_3N_3O_4S_2$ mw: 421.44

PROP: Crystals from MeOH/$CHCl_3$. Mp: 228°. Insol in water, chloroform, benzene, and ether; sol in acetone and alc.

SYNS: APRINOX □ Be 724-A □ BENDROFLUAZIDE □ BENDROFLUMETHIAZIDE □ BENTRIDE □ BENURON □ BENZYDROFLUMETHIAZIDE □ 3-BENZYL-3,4-DIHYDRO-6-(TRIFLUOROMETHYL)-2H-1,2,4-BENZOTHIADIAZINE-7-SULFONAMIDE 1,1-DIOXIDE □ BENZYLRODIURAN □ 3-BENZYL-6-TRIFLUOROMETHYL-7-SULFAMOYL-3,4-DIHYDRO-1,2,4-BENZOTHIADIAZINE-1,1 DIOXIDE □ BERKOZIDE □ DIIFT □ DL 11368 □ BRISTURIC □ BRISTURON □ CENTYL □ FLUMESIL □ FT 8 □ INTOLEX □ NATERETIN □ NATURETIN □ NATURINE □ NEO-NACLEX □ NEO-RONTYL □ NIAGARIL □ NIKION □ ORSILE □ PLURYL □ PLURYLE □ PLUSURIL □ POLIURON □ RELAN BETA □ REPICIN □ SALURAL □ SALURES □ SINESALIN □ SODIURETIC □ THIAZIPIDICO □ 6-TRIFLUOROMETHYL-3-BENZYL-7-SULFAMYL-3,4-DIHYDRO-1,2,4-BENZOTHIADIAZINE-1,1-DIOXIDE □ URLEA

TOXICITY DATA WITH REFERENCE
cyt-ham:lng 200 mg/L GMCRDC 27,95,81
orl-wmn TDLo:3 mg/kg:CNS LANCAO 1,564,82
ipr-mus LD50:4800 mg/kg AEPPAE 238,435,60
ivn-mus LD50:395 mg/kg JPETAB 134,273,61

SAFETY PROFILE: Poison by intravenous route. Human systemic effects by ingestion: convulsions and somnolence. Mutation data reported. When heated to decomposition it emits toxic fumes of F^-, SO_x, and NO_x.

BER500 **CAS:3268-19-7** **HR: 3**
4,6-o-BENZYLIDENE-β-d-GLUCOPYRANOSIDE PODOPHYLLOTOXIN
mf: $C_{35}H_{36}O_{13}$ mw: 664.71

SYNS: NSC 42076 □ PODOPHYLLOTOXIN-BENZILIDEN-GLUCOSID (GERMAN) □ PODOPHYLLOTOXIN-o-BENXYLIDENE-β-d-GLUCOPYRANOSIDE □ PRORESIDOR □ SP G □ SPG 827

TOXICITY DATA WITH REFERENCE
orl-mus LD50:280 mg/kg ARZNAD 11,549,61
ipr-mus LD50:280 mg/kg ARZNAD 11,549,61

SAFETY PROFILE: Poison by ingestion and intraperitoneal routes. When heated to decomposition it emits acrid smoke and irritating fumes.

BES250 **CAS:2782-70-9** **HR: 3**
BENZYLIDENEMETHYLPHOSPHORODITHIOATE
mf: $C_{11}H_{18}O_4P_2S_4$ mw: 404.47

SYNS: S,S'-BENZYLIDENE BIS(O,O-DIMETHYL PHOSPHORODITHIOATE) □ ENT 25,739 □ SD 7438 □ SHELL SD 7,438 □ TOLUENE-α,α-DITHIOL BIS(O,O-DIMETHYL PHOSPHORODITHIOATE)

TOXICITY DATA WITH REFERENCE
orl-rat LD50:280 mg/kg 28ZEAL 4,371,69
orl-mus LD50:176 mg/kg ARSIM* 20,19,66
skn-rbt LD50:2500 mg/kg BESAAT 12,161,66
orl-ckn LD50:5096 mg/kg TXAPA9 11,49,67

SAFETY PROFILE: Poison by ingestion. Moderately toxic by skin contact. When heated to decomposition it emits very toxic fumes of PO_x and SO_x. A pesticide. See also ESTERS.

BES300 **CAS:52098-16-5** **HR: 3**
1-BENZYL-2-INDOLYL HYDROXYMETHYL KETONE
mf: $C_{17}H_{15}NO_2$ mw: 265.33

SYNS: 1-BENZYL-2-(HYDROXYACETYL)INDOLE □ KETONE, 1-BENZYL-2-INDOLYL HYDROXYMETHYL-

TOXICITY DATA WITH REFERENCE
ipr-mus LD50:350 mg/kg PCJOAU 8,74,74

DOT CLASSIFICATION: 3; *Label:* Flammable Liquid

SAFETY PROFILE: A poison by intraperitoneal route. A flammable liquid. When heated to decomposition it emits toxic vapors of NO_x.

BES500 **CAS:122-73-6** **HR: 1**
BENZYL ISOAMYL ETHER
mf: $C_{12}H_{18}O$ mw: 178.30

PROP: Liquid or oil. Bp: 235°, d: 0.965 @ 15.5°/15.5°.

SYNS: BENZYL ISOPENTYL ETHER □ ISOAMYL BENZYL ETHER

TOXICITY DATA WITH REFERENCE
skn-rbt 500 mg/24H MLD FCTXAV 16,647,78

CONSENSUS REPORTS: Reported in EPA TSCA Inventory.

SAFETY PROFILE: A skin irritant. See also ETHERS. Flammable when exposed to heat or flame; can react with oxidizing materials. To fight fire, use foam, CO_2, dry chemical.

BES750 **CAS:120-11-6** **HR: 1**
BENZYL ISOEUGENOL ETHER
mf: $C_{17}H_{18}O_2$ mw: 254.35

SYNS: BENZYL ALCOHOL ETHER with ISOEUGENOL □ BENZYL ISOEUGENOL □ BENZYL-2-METHOXY-4-PROPENYLPHENYL ETHER

TOXICITY DATA WITH REFERENCE
skn-rbt 500 mg/24H MLD FCTXAV 11,1025,73
orl-rat LD50:4900 mg/kg FCTXAV 11,1025,73

CONSENSUS REPORTS: Reported in EPA TSCA Inventory.

SAFETY PROFILE: Mildly toxic by ingestion. A skin irritant. See also ETHERS. When heated to decomposition it emits acrid smoke and irritating fumes.

BET000 **CAS:51-12-7** **HR: 3**
N-BENZYL-β-(ISONICOTINYLHYDRAZINO)PROPIONAMIDE
mf: $C_{16}H_{18}N_4O_2$ mw: 298.38

SYNS: N^1-β-BENZYLCARBAMOYLETHYL-N^2-ISONICOTINOYLHYDRAZINE □ 1-(2-(BENZYLCARBAMOYL)ETHYL)-2-ISONICOTINOYLHYDRAZINE □ (2-(2-BENZYLCARBAMYL)ETHYL)-HYDRAZIDE ISONICOTINIC ACID □ N-BENZYL-β-(ISONICOTINOYLHYDRAZINE) PROPIONAMIDE □ DELMONEURINA □ ESPRIL □ ISALIZINA □ N-I-SONICOTINOYL-N′(β-N-BENZYLCARBOXAMIDOETHYL)HYDRAZINE □ MYGAL □ NIALAMIDE □ NIAMID □ NIAMIDAL □ NIAQUITIL □ NUREDAL □ NYAZIN □ P 1133 □ PSICODISTEN □ 4-PYRIDINECARBOXYLIC ACID 2-(3-OXO-3-((PHENYLMETHYL)AMINO)PROPYL)HYDRAZIDE □ SURGEX

TOXICITY DATA WITH REFERENCE
mma-sat 10 µg/plate PLRCAT 12,423,80
mmo-sat 10 µg/plate MUREAV 40,305,76
dnr-esc 27 µmol/plate JTEHD6 9,287,82
oms-bcs 10 mmol/L MUREAV 5,343,68
dnd-mus-ipr 2450 µmol/kg CNREA8 41,1469,81
sce-mus-ipr 435 mg/kg JTEHD6 9,287,82
scu-mus TDLo:24 mg/kg (1-6D preg):REP JOENAK 27,147,63
orl-rat LD50:1700 mg/kg TXAPA9 1,524,59
ipr-rat LD50:760 mg/kg TXAPA9 1,524,59
orl-mus LD50:590 mg/kg 27ZQAG -,269,72
ipr-mus LD50:200 mg/kg MPHEAE 16,267,67
ivn-mus LD50:120 mg/kg 27ZQAG -,269,72

CONSENSUS REPORTS: EPA Genetic Toxicology Program.

SAFETY PROFILE: Poison by intravenous and intraperitoneal routes. Moderately toxic by ingestion. Mutation data reported. Experimental reproductive effects. An antidepressant. When heated to decomposition it emits toxic fumes of NO_x.

BEU250 **CAS:622-78-6** **HR: 3**
BENZYL-ISOTHIOCYANATE
mf: C_8H_7NS mw: 149.22

PROP: Orange-red, crystalline solid. Mp: 41°, bp: 230°, d: 1.125

SYNS: BENZYL MUSTARD OIL □ BENZYLSENFOEL (GERMAN) □ ISOTHIOCYANIC ACID BENZYL ESTER

TOXICITY DATA WITH REFERENCE
mmo-sat 150 µg/plate ABCHA6 44,3017,80
ipr-rat LDLo:100 mg/kg ARZNAD 16,870,66
ipr-mus LDLo:100 mg/kg ARZNAD 21,121,71
scu-mus LD50:150 mg/kg ARZNAD 5,505,55

CONSENSUS REPORTS: Reported in EPA TSCA Inventory.

SAFETY PROFILE: Poison by intraperitoneal and subcutaneous routes. Intensely irritating. Mutation data reported. Moderate fire hazard via heat, flame, and oxidizers. To fight fire, use water, spray, foam, dry chemical. When heated to decomposition it emits very toxic NO_x and SO_x. See also ESTERS and THIOCYANATES.

BEU500 **CAS:538-28-3** **HR: 3**
BENZYLISOTHIOUREA HYDROCHLORIDE
mf: $C_8H_{10}N_2S•ClH$ mw: 202.72

PROP: Dimorphic. Mp: 146–148°.

SYNS: BENZYLISOTHIOURONIUM CHLORIDE □ 2-BENZYLISOTHIOURONIUM CHLORIDE □ BENZYL THIOPSEUDOUREA HYDROCHLORIDE □ 2-BENZYL-2-THIO-PSEUDOUREA HYDROCHLORIDE □ BENZYLTHIURONIUM CHLORIDE □ S-BENZYLTHIURONIUM CHLORIDE □ BTKH □ ISOTHIOURONIUM CHLORIDE, BENZYL □ 2-THIO-2-BENZYL-PSEUDOUREA HYDROCHLORIDE □ TL 944 □ USAF EK-2124

TOXICITY DATA WITH REFERENCE
orl-rat LD50:150 mg/kg JPETAB 90,260,47
ipr-mus LD50:50 mg/kg NTIS** AD277-689
scu-mus LDLo:80 mg/kg NDRC** No.9-4-1-9,43
ivn-mus LD50:32 mg/kg CSLNX* NX#00167

CONSENSUS REPORTS: Reported in EPA TSCA Inventory.

SAFETY PROFILE: Poison by ingestion, intraperitoneal, subcutaneous, and intravenous routes. When heated to decomposition it emits very toxic fumes of HCl, SO_x, and NO_x.

BEU750 CAS:140-25-0 HR: 1
BENZYL LAURATE
mf: $C_{19}H_{30}O_2$ mw: 290.49

SYNS: BENZYL DODECANOATE □ DODECANOIC ACID BENZYL ESTER

TOXICITY DATA WITH REFERENCE
skn-rbt 500 mg/kg/24H MOD FCTXAV 16,649,78

CONSENSUS REPORTS: Reported in EPA TSCA Inventory.

SAFETY PROFILE: A skin irritant. See also ESTERS. When heated to decomposition it emits acrid smoke and irritating fumes.

BEU800 CAS:35133-55-2 HR: 3
4-BENZYL-α-(4-METHOXYPHENYL)-β-METHYL-1-PIPERIDINEETHANOL
mf: $C_{22}H_{29}NO_2$ mw: 339.52

SYN: RC 61-96

TOXICITY DATA WITH REFERENCE
orl-mus LD50:120 mg/kg ARZNAD 21,1992,71
ipr-mus LD50:45 mg/kg ARZNAD 21,1992,71
ivn-mus LD50:13 mg/kg ARZNAD 21,1992,71

SAFETY PROFILE: Poison by ingestion, intravenous, and intraperitoneal routes. When heated to decomposition it emits toxic fumes of NO_x.

BEX500 CAS:306-07-0 HR: 3
BENZYLMETHYLPROPYNYLAMINE HYDROCHLORIDE
mf: $C_{11}H_{13}N{\cdot}ClH$ mw: 195.71

PROP: Crystals from $EtOH/Et_2O$. Mp: 154–155°. Sol in H_2O: (sol unstable).

SYNS: A 19120 □ N BENZYL N METHYL 2 PROPYNYLAMINE HYDROCHLORIDE □ EUDATINE □ N-METHYL-N-(2-PROPYNYL)BENZYLAMINE HYDROCHLORIDE □ PARGYLINE HYDROCHLORIDE □ USAF A-19120

TOXICITY DATA WITH REFERENCE
scu-mus TDLo:12 mg/kg (1-6D preg):REP JOENAK 27,146,63
ipr-rat TDLo:150 mg/kg (10D male):REP FESTAS 26,232,75
orl-man TDLo:108 mg/kg/26W-I JCLPDE 44,25,83
orl-wmn TDLo:1500 μg/kg/D:CNS,PSY AJPSAO 118,255,61
orl-rat LD50:250 mg/kg 27ZQAG -,401,72
ipr-rat LD50:142 mg/kg ANYAA9 107,1068,63
ivn-rat LD50:175 mg/kg 27ZQAG -,401,72
orl-mus LD50:680 mg/kg ANYAA9 107,1068,63
ipr-mus LD50:300 mg/kg NTIS** AD277-689
orl-dog LD50:175 mg/kg ANYAA9 107,1068,63

SAFETY PROFILE: Poison by ingestion, intraperitoneal, and intravenous routes. Human systemic effects by ingestion: effects on fluid intake, psychological effects. Experimental reproductive effects. When heated to decomposition it emits very toxic fumes of HCl and NO_x.

BEX750 CAS:7368-12-9 HR: 3
(1-BENZYL-3-METHYL-5-PYRAZOLYLOXYETHYL)TRIMETHYLAMMONIUM IODIDE
mf: $C_{16}H_{24}N_3O{\cdot}I$ mw: 401.33

SYN: B-325

TOXICITY DATA WITH REFERENCE
orl-mus LD50:4013 mg/kg ARZNAD 17,214,67
scu-mus LD50:181 mg/kg ARZNAD 17,214,67

SAFETY PROFILE: Poison by subcutaneous route. Moderately toxic by ingestion. See also IODIDES. When heated to decomposition it emits very toxic fumes of NH_3, NO_x, and I^-.

BFA250 CAS:15285-42-4 HR: 3
BENZYL NITRATE
mf: $C_7H_7NO_3$ mw: 153.14

PROP: Oil. Bp: 90–92° @ 10 mm.

SAFETY PROFILE: Explodes above 180°C. Violent reaction with Lewis acids (e.g., sulfuric acid, tin(IV) chloride, boron trifluoride) results in gas evolution. When heated to decomposition it emits toxic fumes of NO_x. See also NITRATES.

BFC200 CAS:101670-78-4 HR: 3
5-BENZYLOXY-3-ISONIPECOTOYLINDOLE
mf: $C_{21}H_{21}N_2O_2$ mw: 333.44

SYNS: INDOLE, 5-BENZYLOXY-3-ISONIPECOTOYL- □ KETONE, 5-BENZYLOXY-3-INDOLYL 4-PIPERIDYL

TOXICITY DATA WITH REFERENCE
ivn-mus LD50:28 mg/kg CSLNX* NX#12896

DOT CLASSIFICATION: 3; *Label:* Flammable Liquid

SAFETY PROFILE: A poison by intravenous route. A flammable liquid. When heated to decomposition it emits toxic vapors of NO_x.

BFC750 CAS:1538-09-6 HR: 2
BENZYLPENCILLINDIBENZYLETHYLENEDIAMINE SALT
mf: $C_{32}H_{36}N_4O_8S_2{\cdot}C_{16}H_{20}N_2$ mw: 909.22

PROP: Crystals. Mp: 123–124°.

SYNS: BEACILLIN □ BEN-P □ BENZACILLIN □ BENZATHINE BENZYLPENICILLIN □ BENZATHINE PENICILLIN □ BENZATHINE PENICILLIN G □ BENZETHACIL □ BENZYLPENICILLIN BENZATHINE □ BICA-PENICILLIN □ BICILLIN □ CEPACILINA □ CEPACILLINA □ CILLENTA □ DBED DIPENCILLIN G □ DBED PENICILLIN □ DEBECILLIN □ DEBECYLINA □ DIAMINE DIPENICILLIN G □ DIAMINOCILLIAN □ DIBENCIL □ DIBENCILLIN □ N,N′-DIBENZYLETHYLENEDIAMINE BIS(BENZYL PENICILLIN) □ DIBENZYLETHYLENEDIAMINE-DI-PENICILLIN G □ N,N′-DIBENZYLETHYLENEDIAMINE, compounded with PENICILLIN G (1:2) □ DIPO-SAFT □ DURABIOTIC □ DURA-PENITA □ DUROPENIN □ EXTENCILLINE □ EXTENICILLINE □ LENTOCILLIN □ LENTOPENIL □ LEOMYPEN □ LONGACILIAN □ LONGICIL □ LPG □ MEGACILLIN SUSPENSION □ MOLDAMIN □ NCI-C56100 □ NEOLIN □ PENADUR □ PENADUR L-A □ PENDEPON □ PEN-DI-BEN □ PENDITAN □ PENDURAN □ PENICILLIN G, compounded with N,N′-DIBENZYLETHYLENEDIAMINE (2:1) □ PENICILLIN G SALT of N,N′-DIBENZYLETHYLENEDIAMINE □ PENI-

DURAL □ PENIDURE □ PENILENTE □ PERMAPEN □ RETARPEN □ TARDOCILLIN □ VETARCILLIN □ VICIN □ WYCILLINA

TOXICITY DATA WITH REFERENCE
par-rat TDLo:22,905 µg/kg (4D preg):REP BEXBAN 82,1076,76
orl-mus LD50:2000 mg/kg NIIRDN 6,774,82
ipr-mus LDLo:460 mg/kg NIIRDN 6,774,82

CONSENSUS REPORTS: Reported in EPA TSCA Inventory.

SAFETY PROFILE: Moderately toxic by ingestion and intraperitoneal routes. Experimental reproductive effects. When heated to decomposition it emits very toxic fumes of NO_x and SO_x. See other penicillin entries.

BFD000 CAS:113-98-4 HR: 3
BENZYLPENICILLINIC ACID POTASSIUM SALT
mf: $C_{16}H_{17}N_2O_4S \cdot K$ mw: 372.51

PROP: Needles from butanol (aq). Mp: 214–217° (decomp). Sol in H_2O.

SYNS: BENZYLPENICILLIN POTASSIUM □ BENZYLPENICILLIN POTASSIUM SALT □ CILLORAL □ COSMOPEN □ CRISTAPEN □ CRYSTAPEN □ ESKACILLIN □ FALAPEN □ FORPEN □ HIPERCILINA □ HYASORB □ HYLENTA □ MEGACILLIN TABLETS □ MONOPEN □ NOTARAL □ PENALEV □ PENICILLIN G POTASSIUM □ PENICILLIN G POTASSIUM SALT □ PENISEM □ PENTID □ PENTIDS □ PFIZERPEN □ POTASSIUM BENZYLPENICILLIN □ POTASSIUM BENZYLPENICILLINATE □ POTASSIUM BENZYLPENICILLIN G □ POTASSIUM PENICILLIN G □ POTASSIUM SALT of BENZYLPENICILLIN □ QIDPEN G □ SCOTCIL □ SK-PENICILLIN G □ SUGRACILLIN □ TABILIN □ TU CILLIN

TOXICITY DATA WITH REFERENCE
spm-rat-unr 200 mg/kg/8D JOURAA 112,348,74
orl-rat LD50:6700 mg/kg AIPTAK 123,295,60
scu-rat LD50:11,250 mg/kg TXAPA9 9,445,66
ivn-rat LD50:243 mg/kg ABANAE 3,534,55/56
orl-mus LD50:6257 mg/kg AIPTAK 125,83,60
ivn-mus LD50:240 mg/kg ABANAE 3,534,55/56
ice-mus LDLo:2 mg/kg PLMEAA 49,103,83
orl-rbt LD50:5848 mg/kg ANTCAO 10,376,60
orl-gpg LDLo:1 g/kg ANTCAO 5,463,55
ipr-gpg LDLo:500 mg/kg ANTCAO 5,463,55
ivn-gpg LD50:303 mg/kg RPOBAR 2,306,70

CONSENSUS REPORTS: EPA Genetic Toxicology Program. Reported in EPA TSCA Inventory.

SAFETY PROFILE: Poison by intracerebral and intravenous routes. Moderately toxic by intraperitoneal route. Mutation data reported. See other penicillin entries. When heated to decomposition it emits toxic fumes of NO_x and SO_x.

BFD250 CAS:69-57-8 HR: 3
BENZYL PENICILLINIC ACID SODIUM SALT
mf: $C_{16}H_{17}N_2O_4S \cdot Na$ mw: 356.40

PROP: Needles from butanol (aq). Mp: 215° (decomp). Sol in H_2O and MeOH.

SYNS: AMERICAN PENICILLIN □ BENZYLPENICILLIN SODIUM □ CRYSTAPEN □ MYCOFARM □ NOVOCILLIN □ PEN-A-BRASIVE □ PENICILLIN-G, MONOSODIUM SALT □ PENICILLIN G, SODIUM □ PENICILLIN G, SODIUM SALT □ PENILARYN □ PENZYLPENICILLIN SODIUM SALT □ SODIUM BENZYLPENICILLIN □ SODIUM BENZYLPENICILLIN G □ SODIUM BENZYLPENICILLINATE □ SODIUM PENICILLIN □ SODIUM PENICILLIN G □ SODIUM PENICILLIN II □ VETICILLIN

TOXICITY DATA WITH REFERENCE
scu-mus TDLo:30 mg/kg (female 14D post):REP CRSBAW 158,528,64
orl-rat TDLo:1500 mg/kg (6-11D preg):TER ANTBAL 18,815,73
scu-rat TDLo:1840 mg/kg/46W-I:ETA BJCAAI 15,85,61
orl-rat LD50:6916 mg/kg AMPMAR 39,259,78
par-rat LD50:2900 µg/kg AACHAX -,863,65
scu-mus LD50:4750 mg/kg NYKZAU 55,23,59
ivn-mus LD50:1500 mg/kg ARZNAD 9,31,59
ims-mus LD50:2800 mg/kg ARZNAD 9,31,59
ice-mus LD50:3800 µg/kg NYKZAU 55,23,59
ims-gpg LDLo:60 mg/kg LBASAE 30,524,80

CONSENSUS REPORTS: EPA Genetic Toxicology Program.

SAFETY PROFILE: Poison by intracerebral, parenteral, and intramuscular routes. Moderately toxic via intravenous route. Mildly toxic by ingestion. Experimental teratogenic and reproductive effects. Questionable carcinogen with experimental tumorigenic data. When heated to decomposition it emits very toxic fumes of NO_x, Na_2O, and SO_x. An antibiotic. See other penicillin entries.

BFD400 HR: 1
BENZYL PHENYLACETATE
mf: $C_{15}H_{14}O_2$ mw: 226.27

PROP: Colorless liquid; sweet, floral odor with honey undertone. D: 1.095–1.099, refr index: 1.553–1.558, flash p: 212°F. Sol in alc, chloroform, ether.

SYN: FEMA No. 2149

SAFETY PROFILE: Combustible liquid. When heated to decomposition it emits acrid smoke and irritating fumes.

BFE770 CAS:25174-65-6 HR: 3
4-BENZYLPIPERAZINYL β-(p-CHLOROPHENYL)PHENETHYL KETONE
mf: $C_{26}H_{27}ClN_2O$ mw: 419.00

SYNS: 1-BENZYL-4-(3-(p-CHLOROPHENYL)-3-PHENYLPROPIONYL) PIPERAZINE □ KETONE, 4-BENZYLPIPERAZINYL β-(p-CHLOROPHENYL)PHENETHYL □ PIPERAZINE, 1-BENZYL-4-(p-CHLORO-β-PHENYLHYDROCINNAMOYL)- □ PIPERAZINE, 1-BENZYL-4-(3-(p-CHLOROPHENYL)-3-PHENYLPROPIONYL)-

TOXICITY DATA WITH REFERENCE
ipr-mus LD50:800 mg/kg JMCMAR 12,860,69

DOT CLASSIFICATION: 3; *Label:* Flammable Liquid

SAFETY PROFILE: Moderately toxic by intraperitoneal route. A flammable liquid. When heated to decomposition it emits toxic vapors of NO_x and Cl^-.

BFG600 **CAS:101-82-6** **HR: 2**
2-BENZYLPYRIDINE
mf: $C_{12}H_{11}N$ mw: 169.24

SYNS: 2-(PHENYLMETHYL)PYRIDINE □ PYRIDINE, 2-BENZYL- □ PYRIDINE, 2-(PHENYLMETHYL)-(9CI)

TOXICITY DATA WITH REFERENCE
scu-mus LD50:1500 mg/kg AEPPAE 227,129,55

CONSENSUS REPORTS: Reported in EPA TSCA Inventory.

SAFETY PROFILE: Moderately toxic by subcutaneous route. When heated to decomposition it emits toxic vapors of NO_x.

BFG750 **CAS:2116-65-6** **HR: 3**
4-BENZYLPYRIDINE
mf: $C_{12}H_{11}N$ mw: 169.24

PROP: Liquid. D: 1.076 @ 0°/0°, bp: 287° @ 742 mm.

TOXICITY DATA WITH REFERENCE
ivn-mus LD50:25 mg/kg CSLNX* NX#12240
orl-bwd LD50:18 mg/kg TXAPA9 21,315,72

CONSENSUS REPORTS: Reported in EPA TSCA Inventory.

SAFETY PROFILE: Poison by ingestion and intravenous routes. A flammable material. Incompatible with oxidizers. When heated to decomposition it emits toxic fumes of NO_x.

BFH100 **CAS:3670-09-5** **HR: 3**
BENZYL 4-PYRIDYL KETONE THIOSEMICARBAZONE

SYN: KETONE, BENZYL(4-PYRIDYL), THIOSEMICARBAZONE

TOXICITY DATA WITH REFERENCE
orl-mus LD50:450 mg/kg JMCMAR 8,676,65

DOT CLASSIFICATION: 3; *Label:* Flammable Liquid

SAFETY PROFILE: Moderately toxic by ingestion. A flammable liquid. When heated to decomposition it emits toxic vapors of SO_x.

BFI400 **CAS:2284-30-2** **HR: 3**
4-BENZYL RESORCINOL
mf: $C_{13}H_{12}O_2$ mw: 200.25

SYN: RESORCINOL, 4-BENZYL-

TOXICITY DATA WITH REFERENCE
ivn-mus LD50:73 mg/kg BJPCAL 22,221,64

CONSENSUS REPORTS: Reported in EPA TSCA Inventory.

SAFETY PROFILE: Poison by intravenous route. When heated to decomposition it emits acrid smoke and irritating vapors.

BFJ750 **CAS:118-58-1** **HR: 2**
BENZYL SALICYLATE
mf: $C_{14}H_{12}O_3$ mw: 228.26

PROP: Thick colorless liquid, pleasant odor. Bp: 208° @ 26 mm, d: 1.175 @ 20°, refr index: 1.579. Sol in fixed oils; insol in glycerin and propylene glycol.

SYNS: BENZYL-o-HYDROXYBENZOATE □ FEMA No. 2151

TOXICITY DATA WITH REFERENCE
orl-rat LD50:2227 mg/kg FCTXAV 11,1029,73

CONSENSUS REPORTS: Reported in EPA TSCA Inventory.

SAFETY PROFILE: Moderately toxic by ingestion. See also BENZYL ALCOHOL, SALICYLIC ACID, and ESTERS. Combustible when exposed to heat or flame. When heated to decomposition it emits acrid smoke and irritating fumes. Incompatible with oxidizing materials.

BFJ825 **CAS:766-06-3** **HR: 3**
BENZYL SILANE
mf: $C_7H_{10}Si$ mw: 122.24

PROP: Liquid. Bp: 149°.

SAFETY PROFILE: Ignites spontaneously in air. Upon decomposition it emits acrid smoke and fumes.

BFJ850 **CAS:1121-53-5** **HR: 3**
BENZYL SODIUM
mf: C_7H_7Na mw: 114.12

PROP: Red crystals. Decomp below mp.

SAFETY PROFILE: Ignites spontaneously in air. Upon decomposition it emits toxic fumes of Na_2O.

BFK000 **CAS:35506-85-5** **HR: 3**
BENZYL SULFITE
mf: $C_{14}H_{14}O_3S$ mw: 262.34

PROP: Liquid. Bp: 193–199° @ 15 mm (part decomp).

SYN: SULFUROUS ACID, DIBENZYL ESTER

TOXICITY DATA WITH REFERENCE
ivn-mus LD50:178 mg/kg CSLNX* NX#02156

SAFETY PROFILE: Poison by intravenous route. See also SULFITES. When heated to decomposition it emits toxic fumes of SO_x.

BFK325 **CAS:1090-53-5** **HR: 3**
1-BENZYL-2-(3-(4,5,6,7-TETRAHYDROBENZISOXAZOYLYL)CARBONYL)HYDRAZINE HYDROCHLORIDE
mf: $C_{15}H_{17}N_3O_4 \cdot ClH$ mw: 307.81

TOXICITY DATA WITH REFERENCE
orl-mus LD50:2082 mg/kg SKNEA7 14,58,64
ipr-mus LD50:723 mg/kg SKNEA7 14,58,64
orl-cat LD50:118 mg/kg SKNEA7 14,58,64

SAFETY PROFILE: Poison by ingestion. Moderately

toxic by intraperitoneal route. When heated to decomposition it emits toxic fumes of NO_x and HCl.

BFK750 **CAS:13402-51-2** **HR: 3**
S-BENZYL THIOBENZOATE
mf: $C_{14}H_{12}OS$ mw: 228.32

PROP: Crystals from EtOH. Mp: 39.5°.

SYN: TIBENZATE

TOXICITY DATA WITH REFERENCE
orl-mus LD50:1550 mg/kg YKKZAJ 89,1179,69
ivn-mus LD50:180 mg/kg CSLNX* NX#02522

SAFETY PROFILE: Poison by intravenous route. Moderately toxic by ingestion. When heated to decomposition it emits toxic fumes of SO_x. See also ESTERS.

BFL000 **CAS:3012-37-1** **HR: 3**
BENZYL THIOCYANATE
mf: C_8H_7NS mw: 149.22

PROP: Orange-red, crystals or solid. Mp: 41–42°, bp: 230°, d: 1.125.

SYNS: BENZYL MUSTARD OIL □ PHENYLMETHYL ESTER THIOCYANIC ACID (9CI) □ SOLVAT 14 □ α-THIOCYANATOTOLUENE □ TROPEOLIN

TOXICITY DATA WITH REFERENCE
ipr-rat LDLo:40 mg/kg ARZNAD 16,870,66
ipr-mus LD50:17 mg/kg PCBPBS 2,95,72
scu-mus LD50:100 mg/kg JJPAAZ 3,99,54

CONSENSUS REPORTS: Reported in EPA TSCA Inventory.

SAFETY PROFILE: Poison by subcutaneous and intraperitoneal routes. See also THIOCYANATES. When heated to decomposition it emits very toxic fumes of NO_x, SO_x, and CN^-.

BFL125 **CAS:1874-58-4** **HR: 3**
BENZYLTHIOGUANINE
mf: $C_{12}N_{11}N_5S$ mw: 257.34

SYNS: 2-AMINO-6-BENZYLMERCAPTOPURINE □ 2-AMINO-6-BENZYL-MP □ 2-AMINO-6-(BENZYLTHIO)PURINE □ 6-BENZYLTHIOGUANINE □ NSC 15747 □ 6-((PHENYLMETHYL)THIO)-1H-PURIN-2-AMINE (9CI) □ SRI 702

TOXICITY DATA WITH REFERENCE
orl-rat TDLo:100 mg/kg (7D preg):REP JRPFA4 4,291,62
ipr-mus LD50:222 mg/kg NCISP* JAN86

SAFETY PROFILE: Poison by intraperitoneal route. Experimental reproductive effects. When heated to decomposition it emits toxic fumes of NO_x and SO_x.

BFL250 **CAS:98-07-7** **HR: 3**
BENZYL TRICHLORIDE
DOT: UN 2226
mf: $C_7H_5Cl_3$ mw: 195.47

PROP: Clear, colorless to yellowish liquid; penetrating odor. Mp: −5°, bp: 221°, d: 1.38 @ 15.5°/15.5°, vap d: 6.77.

SYNS: BENZENYL CHLORIDE □ BENZENYL TRICHLORIDE □ BENZOIC TRICHLORIDE □ BENZOTRICHLORIDE (DOT, MAK) □ BENZYLIDYNE CHLORIDE □ CHLORURE de BENZENYLE (FRENCH) □ PHENYL CHLOROFORM □ PHENYLTRICHLOROMETHANE □ RCRA WASTE NUMBER U023 □ TOLUENE TRICHLORIDE □ TRICHLOORMETHYLBENZEEN (DUTCH) □ TRICHLORMETHYLBENZOL (GERMAN) □ TRICHLOROMETHYLBENZENE □ 1-(TRICHLOROMETHYL) BENZENE □ TRICLOROMETILBENZENE (ITALIAN) □ TRICHLOROPHENYLMETHANE □ α,α,α-TRICHLOROTOLUENE □ ω,ω,ω-TRICHLOROTOLUENE □ TRICLOROTOLUENE (ITALIAN)

TOXICITY DATA WITH REFERENCE
skn-rbt 20 mg/24H MOD 85JCAE-,157,86
eye-rbt 50 μg/24H SEV 85JCAE-,157,86
mma-esc 500 nmol/plate/20M MUREAV 54,143,78
mrc-bcs 2600 nmol/disc MUREAV 54,143,78
ihl-mus TCLo:1620 ppb/30M/22W-I:NEO SAIGBL 28,352,86
skn-mus TDLo:9200 mg/kg/50W-I:CAR GANNA2 130,250,28
ihl-uns LC50:60 mg/m³ GTPZAB 30(3),6,86

CONSENSUS REPORTS: NTP 7th Annual Report On Carcinogens. IARC Cancer Review: Human Limited Evidence IMEMDT 29,73,82; Animal Sufficient Evidence IMEMDT 29,73,82. EPA Genetic Toxicology Program. EPA Extremely Hazardous Substances List. Reported in EPA TSCA Inventory.

DFG MAK: Confirmed Human Carcinogen
DOT CLASSIFICATION: 8; *Label:* Corrosive

SAFETY PROFILE: Confirmed carcinogen with experimental carcinogenic data by skin contact and neoplastigenic data by inhalation. Experimental poison by inhalation. Corrosive to the skin, eyes, and mucous membranes. Large doses can cause central nervous system depression. Mutation data reported. When heated to decomposition it emits toxic fumes of Cl^-. See also CHLORINATED HYDROCARBONS, AROMATIC.

BFL300 **CAS:56-37-1** **HR: 3**
BENZYLTRIETHYLAMMONIUM CHLORIDE
mf: $C_{13}H_{22}N{\cdot}Cl$ mw: 227.81

SYNS: AMMONIUM, BENZYLTRIETHYL-, CHLORIDE □ BENZENEMETHANAMINIUM, N,N,N-TRIETHYL-, CHLORIDE (9CI) □ TEBAC □ N,N,N-TRIETHYLBENZENEMETHANAMINIUM CHLORIDE □ TRIETHYLBENZYLAMMONIUM CHLORIDE

TOXICITY DATA WITH REFERENCE
ivn-mus LD50:18 mg/kg CSLNX* NX#01867

CONSENSUS REPORTS: Reported in EPA TSCA Inventory.

SAFETY PROFILE: Poison by intravenous route. When heated to decomposition it emits toxic vapors of NO_x and Cl^-.

BFM250 **CAS:56-93-9** **HR: 3**
BENZYLTRIMETHYLAMMONIUM CHLORIDE
mf: $C_{10}H_{16}N{\bullet}Cl$ mw: 185.72

PROP: Bp: >135° (some decomp), fp: <−50° (for 61% sol), d: 1.07 @ 20°/20° (61% sol).

SYNS: BENZENEMETHANAMINIUM, N,N,N-TRIMETHYL-, CHLORIDE (9CI) □ BTM □ TMBAC □ N,N,N-TRIMETHYLBENZENEMETHANAMINIUM CHLORIDE □ TRIMETHYLBENZYLAMMONIUM CHLORIDE

TOXICITY DATA WITH REFERENCE
orl-rat LDLo:250 mg/kg NCNSA6 5,39,53
orl-mus LDLo:1600 mg/kg JPMSAE 69,327,80

CONSENSUS REPORTS: Reported in EPA TSCA Inventory.

SAFETY PROFILE: Poison by ingestion. Combustible. When heated to decomposition it emits very toxic fumes of NH_3, NO_x, and Cl^-.

BFM500 **CAS:100-85-6** **HR: 3**
BENZYLTRIMETHYLAMMONIUM HYDROXIDE
mf: $C_{10}H_{16}N{\bullet}HO$ mw: 167.28

PROP: Solid. Fp: 15°. Sol in a variety of solvents.

SYN: TRIMETHYLBENZYLAMMONIUM HYDROXIDE

TOXICITY DATA WITH REFERENCE
scu-mus LDLo:35 mg/kg JPETAB 28,367,26

CONSENSUS REPORTS: Reported in EPA TSCA Inventory.

SAFETY PROFILE: Poison by subcutaneous route. A strong base. When heated to decomposition it emits toxic fumes of NH_3 and NO_x. See also ALKALIES.

BFM750 **CAS:4525-46-6** **HR: 3**
BENZYL TRIMETHYL AMMONIUM IODIDE
mf: $C_{10}H_{16}N{\bullet}I$ mw: 277.17

PROP: Solid. Mp: 181–182°.

SYNS: BENZYLDIMETHYLAMINE METHIODIDE □ PHENMETHYL TRIMETHYLAMMONIUM IODIDE

TOXICITY DATA WITH REFERENCE
ipr-mus LD50:41 mg/kg UCPHAQ 2,161,44
ivn-mus LD50:5600 µg/kg CSLNX* NX#00844

CONSENSUS REPORTS: Reported in EPA TSCA Inventory.

SAFETY PROFILE: Poison by intraperitoneal and intravenous routes. See also IODIDES. When heated to decomposition it emits very toxic fumes of NO_x, NH_3, and I^-.

BFN125 **CAS:538-32-9** **HR: 2**
BENZYLUREA
mf: $C_8H_{10}N_2O$ mw: 150.20

PROP: Crystals. Mp: 147–148°, decomp at 200°. One gram dissolves in 60 mL warm water, 33 mL acetone; sltly sol in benzene, ether.

SYNS: BENZYLCARBAMIDE □ N-BENZYLUREA □ 1-BENZYLUREA □ PHENYLMETHYLUREA

TOXICITY DATA WITH REFERENCE
orl-rat LD50:4410 mg/kg GISAAA 44(3),68,79
orl-mus LD50:570 mg/kg JMCMAR 11,814,68
orl-rat LD50:2700 mg/kg GISAAA 44(3),68,79

SAFETY PROFILE: Moderately toxic by ingestion. When heated to decomposition it emits toxic fumes of NO_x.

BFN500 **CAS:2086-83-1** **HR: 3**
BERBERINE
mf: $C_{20}H_{18}NO_4$ mw: 336.39

PROP: White to yellow crystals. Mp (anhyd): 145°.

SYNS: BERBERIN □ 9,10-DIMETHOXY-2,3-(METHYLENEDIOXY)-7,8,13,13A-TETRAHYDROBERBINIUM

TOXICITY DATA WITH REFERENCE
mmo-sat 100 µmol/L AMACCQ 9,77,76
dnd-esc 10 µmol/L MUREAV 89,95,81
orl-mus LD50:329 mg/kg YKKZAJ 82,726,62
scu-mus LD50:18 mg/kg RPTOAN 31,129,68
scu-rbt LDLo:100 mg/kg HBAMAK 4,1289,35

SAFETY PROFILE: An alkaloid poison by ingestion and subcutaneous routes. In humans, toxic doses lower the body temperature, increase peristalsis, and cause death by central paralysis. Mutation data reported. Should carry a poison label. Should never be ingested without the advice of a physician. Should not be handled excessively since it may be absorbed through the skin and have a toxic effect upon the body. An antimalarial agent. When heated to decomposition it emits highly toxic fumes of NO_x.

BFN550 **HR: 3**
BERBERINE CHLORIDE DIHYDRATE
mf: $C_{20}H_{18}NO_4{\bullet}Cl{\bullet}2H_2O$ mw: 407.88

SYNS: BERBERINE HYDROCHLORIDE BIHYDRATE □ 5,6-DIHYDRO-9,10-DIMETHOXYBENZO(g)-1,3-BENZODIOXOLO(5,6-a)QUINOLIZINIUM CHLORIDE DIHYDRATE

TOXICITY DATA WITH REFERENCE
ipr-rat LD50:138 mg/kg KSRNAM 8,654,74
scu-rat LD50:7970 mg/kg KSRNAM 8,654,74
ivn-rat LD50:46,200 µg/kg KSRNAM 8,654,74
ipr-mus LD50:30 mg/kg KSRNAM 8,654,74
scu-mus LD50:13,900 µg/kg KSRNAM 8,654,74
ivn-mus LD50:7600 µg/kg KSRNAM 8,654,74

SAFETY PROFILE: Poison by subcutaneous, intravenous, and intraperitoneal routes. When heated to decomposition it emits toxic fumes of Cl^- and NO_x. See also CHLORIDES.

BFN600 **CAS:633-65-8** **HR: 3**
BERBERINE HYDROCHLORIDE
mf: $C_{20}H_{18}NO_4{\bullet}Cl$ mw: 371.84

SYNS: BERBERINE CHLORIDE □ BERBINIUM, 7,8,13,13a-TETRADEHYDRO-9,10-DIMETHOXY-2,3-(METHYLENEDIOXY)-, CHLORIDE □ BENZO(g)(1,3)BENZODIOXOLO(5,6-a)QUINOLIZINIUM,5,6-DIHY-

DRO-9,10-DIMETHOXY-, CHLORIDE (9CI) □ BERBERINIUM CHLORIDE

TOXICITY DATA WITH REFERENCE
dnd-uns:lyms 22 μmol/L IJBBBQ 18,245,81
orl-rat LD50:>15 g/kg KSRNAM 8,654,74
orl-mus LD50:>29,586 mg/kg KSRNAM 8,654,74
ipr-mus LD50:37 mg/kg JPETAB 104,253,52

CONSENSUS REPORTS: Reported in EPA TSCA Inventory.

SAFETY PROFILE: Poison by intraperitoneal route. Slightly toxic by ingestion. Mutation data reported. When heated to decomposition it emits toxic vapors of NO_x and Cl^-.

BFN625 CAS:316-41-6 HR: 3
BERBERINE SULFATE
mf: $C_{40}H_{36}N_2O_8 \cdot O_4S$ mw: 768.84

SYNS: BERBERINE SULFATE (2:1) □ BERBERIN SULFATE □ 5,6-DIHYDRO-9,10-DIMETHOXY-BENZO(g)-1,3-BENZODIOXOLO(5,6-a)QUINOLIZINIUM SULFATE (2:1) □ NEUTRAL BERBERINE SULFATE

TOXICITY DATA WITH REFERENCE
ipr-mus LD50:26,400 μg/kg NIIRDN 6,770,82
scu-mus LD50:13,200 μg/kg NIIRDN 6,770,82
ivn-mus LD50:8200 μg/kg NIIRDN 6,770,82

SAFETY PROFILE: Poison by subcutaneous, intravenous, and intraperitoneal routes. When heated to decomposition it emits toxic fumes of NO_x and SO_x. See also SULFATES.

BFN750 CAS:6190-33-6 HR: 3
BERBERINE SULFATE TRIHYDRATE
mf: $C_{40}H_{36}N_2O_8 \cdot O_4S \cdot 3H_2O$ mw: 822.90

SYNS: 5,6-DIHYDRO-9,10-DIMETHOXYBENZO(g)-1,3-BENZODIOXOLO(5,6-a)QUINOLIZINIUM SULFATE TRIHYDRATE □ 7,8,13,13A-TETRADEHYDRO-9,10-DIMETHOXY-2,3-(METHYLENEDIOXY)BERBINIUM SULFATE TRIHYDRATE □ UMBELLATINE SULFATE TRIHYDRATE

TOXICITY DATA WITH REFERENCE
scu-frg LDLo:20 mg/kg HBAMAK 4,1289,35

SAFETY PROFILE: Poison by subcutaneous route. See also BERBERINE and SULFATES. When heated to decomposition it emits very toxic SO_x and NO_x.

BFN990 CAS:68917-15-7 HR: 1
BERGAMOT MINT OIL

SYNS: OILS, MINT, MENTHA CITRATA □ MENTHA CITRATA OIL

TOXICITY DATA WITH REFERENCE
orl-rat LD50:5 g/kg FCTOD7 30,73S,92
skn-gpg LD50:>5 g/kg FCTOD7 30,73S,92

CONSENSUS REPORTS: Reported in EPA TSCA Inventory.

SAFETY PROFILE: Low toxicity by ingestion and skin contact. When heated to decomposition it emits acrid smoke and irritating vapors.

BFO000 CAS:8007-75-8 HR: 1
BERGAMOT OIL rectified

PROP: Yellow-green liquid; agreeable odor. *Composition:* 1-linalyl acetate, 1-linalool, d-limonene, dipentene, bergaptene. By rectification of bergamot oil expressed, under vacuum, to remove completely the furocoumarins and other related nonvolatile residues; found in the fruit of citrus *Bergamia risso et poiteau* (Fam. Rutaceae) (FCTXAV 11,1011,73). D: 0.875–0.880 @ 25°/25°. Misc with alc, glacial acetic acid; sol in fixed oils; insol in glycerin, propylene glycol.

SYNS: BERGAMOTTE OEL (GERMAN) □ OIL of BERGAMOT, cold-pressed □ OIL of BERGAMOT, rectified

TOXICITY DATA WITH REFERENCE
skn-rbt 500 mg/24H MLD FCTXAV 11,1035,73
orl-rat LD50:11,520 mg/kg PHARAT 14,435,59

CONSENSUS REPORTS: Reported in EPA TSCA Inventory.

SAFETY PROFILE: Mildly toxic by ingestion. A mild skin irritant and allergen. Combustible. When heated to decomposition it emits acrid smoke and irritating fumes.

BFO100 CAS:5956-63-8 HR: 2
BERGENIN HYDRATE
mf: $C_{14}H_{16}O_9 \cdot 7H_2O$ mw: 454.44

SYN: 2,4,4a,10b-TETRAHYDRO-3,4,8,10-TETRAHYDROXY-2-(HYDROXYMETHYL)-9-METHOXY-PYRANO(3,2-o)(2)BENZOPYRAN-6(2H)-ONE HYDRATE

TOXICITY DATA WITH REFERENCE
ipr-rat TDLo:2688 mg/kg (84D pre):REP KSRNAM 9,1198,75
ipr-rat LD50:3040 mg/kg KSRNAM 9,1198,75
ivn-rat LD50:2800 mg/kg KSRNAM 9,1198,75
ipr-mus LD50:6410 mg/kg KSRNAM 9,1198,75
ivn-mus LD50:5400 mg/kg KSRNAM 9,1198,75

SAFETY PROFILE: Moderately toxic by several routes. Experimental reproductive effects. When heated to decomposition it emits acrid smoke and fumes.

BFO125 HR: 3
BERSAMA ABYSSINICA Fres. ssp. ABYSSINICA, leaf extract

PROP: African plant belonging to the family Melianthaceae (JPPMAB 14,496,62).

TOXICITY DATA WITH REFERENCE
orl-mus LD50:840 μg/kg JPPMAB 14,496,62
ipr-mus LD50:510 μg/kg JPPMAB 14,496,62
ivn-mky LD50:90 μg/kg JPPMAB 14,496,62
ivn-cat LD50:119 μg/kg JPPMAB 14,496,62

SAFETY PROFILE: Deadly poison by ingestion, intravenous, and intraperitoneal routes.

BFO250 **CAS:12161-82-9** **HR: 3**
BERTRANDITE
mf: $H_{10}O_9Si_2{\cdot}H_2O{\cdot}Be_4$ mw: 264.34

PROP: Colorless, pale yellow, orthorhombic crystals.

SYN: BERYLLIUM SILICATE HYDRATE

CONSENSUS REPORTS: IARC Cancer Review: Group 1 IMEMDT 58,41,93; Human Sufficient Evidence IMEMDT 58,41,93; Animal Sufficient Evidence IMEMDT 1,17,72; Animal Sufficient Evidence IMEMDT 23,143,80; Animal Sufficient Evidence IMEMDT 58,41, 93. Reported in EPA TSCA Inventory. Beryllium and its compounds are on the Community Right-To-Know List.

OSHA PEL: TWA 0.002 mg(Be)/m^3; STEL 0.005 mg(Be) /m^3/30M; CL 0.025 mg(Be)/m^3
ACGIH TLV: TWA 0.002 mg(Be)/m^3, Suspected Human Carcinogen.
NIOSH REL: (Beryllium) CL not to exceed 0.0005 mg (Be)/m^3

SAFETY PROFILE: Confirmed carcinogen. See also BERYLLIUM and BERYLLIUM COMPOUNDS. When heated to decomposition it emits very toxic fumes of BeO.

BFO500 **CAS:1302-52-9** **HR: 3**
BERYL
mf: $Al_2O_{18}Si_6{\cdot}3Be$ mw: 537.53

PROP: Colorless, white, blue-green, green-yellow, yellow, or blue crystals. D: 2.63–2.91.

SYNS: BERYLLIUM ALUMINOSILICATE □ BERYLLIUM ALUMINUM SILICATE □ BERYL ORE

TOXICITY DATA WITH REFERENCE
ihl-rat TCLo: 15 mg/m^3/74W-I: NEO TXAPA9 15,10,69

CONSENSUS REPORTS: NTP 7th Annual Report On Carcinogens. IARC Cancer Review: Group 1 IMEMDT 58,41,93; Human Sufficient Evidence IMEMDT 58,41,93; Animal Sufficient Evidence IMEMDT 1,17,72; Animal Sufficient Evidence IMEMDT 23,143,80; Animal Sufficient Evidence IMEMDT 58,41,93. Reported in EPA TSCA Inventory. Beryllium and its compounds are on the Community Right-To-Know List.

OSHA PEL: TWA 0.002 mg(Be)/m^3; STEL 0.005 mg(Be)/m^3/30M; CL 0.025 mg(Be)/m^3
ACGIH TLV: TWA 0.002 mg(Be)/m^3, Suspected Human Carcinogen.
NIOSH REL: (Beryllium) CL not to exceed 0.0005 mg(Be)/m^3

SAFETY PROFILE: Confirmed carcinogen with experimental carcinogenic, neoplastigenic, and tumorigenic data. See also BERYLLIUM COMPOUNDS and SILICATES. When heated to decomposition it emits toxic fumes of BeO.

BFO750 **CAS:7440-41-7** **HR: 3**
BERYLLIUM
DOT: UN 1966/UN 1567
af: Be aw: 9.01

PROP: A silvery-white, relatively soft, lustrous metal, ductile at red heat. Unreactive to H_2O and air; dissolves vigorously in dil acids. Be Reacts with aq alkalis or H_2. Mp: 1287–1292°, bp: 2970°, d: 1.85.

SYNS: BERYLLIUM-9 □ BERYLLIUM COMPOUNDS, n.o.s. (UN 1566) (DOT) □ BERYLLIUM, powder (UN 1567) (DOT) □ GLUCINIUM □ GLUCINUM □ RCRA WASTE NUMBER P015

TOXICITY DATA WITH REFERENCE
dnd-esc 30 µmol/L MUREAV 89,95,81
dni-nml-ivn 30 µmol/kg PHMCAA 12,298,70
dnd-hmn:hla 30 µmol/L MUREAV 89,95,81
dnd-mus:ast 30 µmol/L MUREAV 89,95,81
itr-rat TDLo: 13 mg/kg: NEO ENVRAL 21,63,80
ivn-rbt TDLo: 20 mg/kg: ETA LANCAO 1,463,50
ihl-hmn TCLo: 300 mg/m^3: PUL AEHLAU 9,473,64
ivn-rat LD50: 496 µg/kg LAINAW 15,176,66

CONSENSUS REPORTS: NTP 7th Annual Report On Carcinogens. IARC Cancer Review: Group 1 IMEMDT 58,41,93; Human Sufficient Evidence IMEMDT 58,41,93; Animal Sufficient Evidence IMEMDT 1,17,72; Animal Sufficient Evidence IMEMDT 23,143,80; Animal Sufficient Evidence IMEMDT 58,41,93. Beryllium and its compounds are on the Community Right-To-Know List. Reported in EPA TSCA Inventory.

OSHA PEL: TWA 0.002 mg(Be)/m^3; STEL 0.005 mg(Be)/m^3/30M; CL 0.025 mg(Be)/m^3
ACGIH TLV: TWA 0.002 mg/m^3, Suspected Human Carcinogen.
DFG TRK: Animal Carcinogen, Suspected Human Carcinogen. Grinding of beryllium metal and alloys: 0.005 mg/m^3 calculated as beryllium in that portion of dust that can possibly be inhaled; other beryllium compounds: 0.002 mg/m^3 calculated as beryllium in that portion of dust that can possibly be inhaled
NIOSH REL: CL not to exceed 0.0005 mg(Be)/m^3
DOT CLASSIFICATION: 6.1; *Label:* Poison (UN 1566); DOT Class: 6.1; *Label:* Poison, Flammable Solid (UN 1567)

SAFETY PROFILE: Confirmed carcinogen with experimental carcinogenic, neoplastigenic, and tumorigenic data. A deadly poison by intravenous route. Human systemic effects by inhalation: lung fibrosis, dyspnea, and weight loss. Human mutation data reported. See also BERYLLIUM COMPOUNDS. A moderate fire hazard in the form of dust or powder, or when exposed to flame or by spontaneous chemical reaction. Slight explosion hazard in the form of powder or dust. Incompatible with halocarbons. Reacts incandescently with fluorine or chlorine. Mixtures of the powder with CCl_4 or trichloroethylene will flash or spark on impact. When heated to decomposition in air it emits very toxic fumes of BeO. Reacts with Li and P.

For occupational chemical analysis use NIOSH: Beryllium, 7102; Elements, 7300.

BFP000 **CAS:543-81-7** **HR: 3**
BERYLLIUM ACETATE
mf: $C_4H_6O_4 \cdot Be$ mw: 127.11

PROP: Plates. Mp: decomp @ 300°.

SYN: BERYLLIUM ACETATE, NORMAL

TOXICITY DATA WITH REFERENCE
ipr-rat LD50:317 mg/kg XEURAQ UR-70,1949

CONSENSUS REPORTS: IARC Cancer Review: Group 1 IMEMDT 58,41,93; Human Sufficient Evidence IMEMDT 58,41,93; Animal Sufficient Evidence IMEMDT 1,17,72; Animal Sufficient Evidence IMEMDT 23,143,80; Animal Sufficient Evidence IMEMDT 58,41,93. Beryllium and its compounds are on the Community Right-To-Know List.

OSHA PEL: TWA 0.002 mg(Be)/m³; STEL 0.005 mg(Be)/m³/30M; CL 0.025 mg(Be)/m³
ACGIH TLV: TWA 0.002 mg(Be)/m³, Suspected Human Carcinogen.
DFG MAK: Animal Carcinogen, Suspected Human Carcinogen.
NIOSH REL: (Beryllium) CL not to exceed 0.0005 mg(Be)/m³

SAFETY PROFILE: Confirmed carcinogen. Poison by intraperitoneal route. See also BERYLLIUM COMPOUNDS. When heated to decomposition it emits toxic fumes of BeO.

BFP250 **CAS:12770-50-2** **HR: 3**
BERYLLIUM ALUMINUM ALLOY

PROP: Alloy is 62% beryllium and 38% aluminum (ENVRAL 21,63,80).

SYNS: ALUMINUM ALLOY, Al,Be ☐ ALUMINUM BERYLLIUM ALLOY

TOXICITY DATA WITH REFERENCE
itr-rat TDLo:13 mg/kg:ETA ENVRAL 21,63,80

CONSENSUS REPORTS: NTP 7th Annual Report On Carcinogens. IARC Cancer Review: Group 1 IMEMDT 58,41,93; Human Sufficient Evidence IMEMDT 58,41,93; Animal Sufficient Evidence IMEMDT 1,17,72; Animal Sufficient Evidence IMEMDT 23,143,80; Animal Sufficient Evidence IMEMDT 58,41,93. Beryllium and its compounds are on the Community Right-To-Know List.

OSHA PEL: TWA 0.002 mg(Be)/m³; STEL 0.005 mg(Be)/m³/30M; CL 0.025 mg(Be)/m³
ACGIH TLV: TWA 0.002 mg(Be)/m³, Suspected Human Carcinogen
DFG MAK: Animal Carcinogen, Suspected Human Carcinogen
NIOSH REL: (Beryllium) CL not to exceed 0.0005 mg(Be)/m³

SAFETY PROFILE: Confirmed carcinogen with experimental carcinogenic and tumorigenic data. See also BERYLLIUM COMPOUNDS. When heated to decomposition it emits very toxic BeO.

BFP500 **CAS:66104-24-3** **HR: 3**
BERYLLIUM CARBONATE
mf: $C_2H_2Be_3O_8$ mw: 181.07

SYNS: BERYLLIUM CARBONATE, BASIC ☐ BERYLLIUMOXIDE CARBONATE ☐ BIS(CARBONATO(2-))DIHYDROXYTRIBERYLLIUM

CONSENSUS REPORTS: NTP 7th Annual Report On Carcinogens. IARC Cancer Review: Group 1 IMEMDT 58,41,93; Human Sufficient Evidence IMEMDT 58,41, 93; Animal Sufficient Evidence IMEMDT 1,17,72; Animal Sufficient Evidence IMEMDT 23,143,80; Animal Sufficient Evidence IMEMDT 58,41,93. Reported in EPA TSCA Inventory. Beryllium and its compounds are on the Community Right-To-Know List.

OSHA PEL: TWA 0.002 mg(Be)/m³; STEL 0.005 mg(Be)/m³/30M; CL 0.025 mg(Be)/m³
ACGIH TLV: TWA 0.002 mg(Be)/m³, Suspected Human Carcinogen
DFG MAK: Animal Carcinogen, Suspected Human Carcinogen
NIOSH REL: CL not to exceed 0.0005 mg(Be)/m³

SAFETY PROFILE: Confirmed carcinogen. See also BERYLLIUM COMPOUNDS. When heated to decomposition it emits toxic BeO dust.

BFP750 **CAS:13106-47-3** **HR: 3**
BERYLLIUM CARBONATE (1:1)
mf: $CO_3 \cdot Be$ mw: 69.02

SYN: CARBONIC ACID BERYLLIUM SALT (1:1)

TOXICITY DATA WITH REFERENCE
ipr-gpg LDLo:300 mg/kg NIHBAZ 181,20,43

CONSENSUS REPORTS: IARC Cancer Review: Group 1 IMEMDT 58,41,93; Human Sufficient Evidence IMEMDT 58,41,93; Animal Sufficient Evidence IMEMDT 1,17,72; Animal Sufficient Evidence IMEMDT 23,143,80; Animal Sufficient Evidence IMEMDT 58,41,93. Reported in EPA TSCA Inventory. Beryllium and its compounds are on the Community Right-To-Know List.

OSHA PEL: TWA 0.002 mg(Be)/m³; STEL 0.005 mg(Be)/m³/30M; CL 0.025 mg(Be)/m³
ACGIH TLV: TWA 0.002 mg(Be)/m³, Suspected Human Carcinogen
DFG MAK: 50 ppm (90 mg/m³)
NIOSH REL: (Beryllium) CL not to exceed 0.0005 mg(Be)/m³

SAFETY PROFILE: Confirmed carcinogen. Poison by intraperitoneal route. See also BERYLLIUM COMPOUNDS. When heated to decomposition it emits highly toxic fumes of BeO.

BFQ000 **CAS:7787-47-5** **HR: 3**
BERYLLIUM CHLORIDE
mf: $BeCl_2$ mw: 79.91

PROP: Colorless, deliquescent needles, or orthorhombic crystals. Undergoes transition to high temp orthorhombic polymorph at 4°. Mp: 415°, bp: 520°, d: 1.899 @ 25°, vap press: 1 mm @ 291° (subl). Very sol in H_2O,

EtOH, Et_2O, or Py; sltly sol in C_6H_6, $CHCl_3$, CS_2; insol. in NH_3, or Me_2CO.

SYN: BERYLLIUM DICHLORIDE

TOXICITY DATA WITH REFERENCE
mmo-esc 10 μmol/L MUREAV 126,9,84
msc-ham:lng 2 mmol/L MUREAV 68,259,79
itr-rat TDLo:1685 μg/kg (3D post):TER GISAAA 51(8),44,86
itr-rat TDLo:1685 μg/kg (5D post):REP GISAAA 51(8),44,86
ihl-rat TCLo:20 μg/m³/1H/17W-I:ETA GTPZAB 19(7),34,75
orl-rat LD50:86 mg/kg HYSAAV 30,169,65
ipr-rat LD50:44 mg/kg EQSSDX 1,1,75
orl-mus LD50:92 mg/kg HYSAAV 30,169,65
ipr-mus LD50:106 mg/kg COREAF 256,1043,63
ipr-gpg LD50:50 mg/kg EQSSDX 1,1,75

CONSENSUS REPORTS: NTP 7th Annual Report On Carcinogens. IARC Cancer Review: Group 1 IMEMDT 58,41,93; Human Sufficient Evidence IMEMDT 58,41,93; Animal Sufficient Evidence IMEMDT 1,17,72; Animal Sufficient Evidence IMEMDT 23,143,80; Animal Sufficient Evidence IMEMDT 58,41,93. EPA Genetic Toxicology Program. Reported in EPA TSCA Inventory. Beryllium and its compounds are on the Community Right-To-Know List.

OSHA PEL: TWA 0.002 mg(Be)/m³; STEL 0.005 mg(Be)/m³/30M; CL 0.025 mg(Be)/m³
ACGIH TLV: TWA 0.002 mg(Be)/m³, Suspected Human Carcinogen
DFG MAK: Animal Carcinogen, Suspected Human Carcinogen
NIOSH REL: (Beryllium) CL not to exceed 0.0005 mg(Be)/m³

SAFETY PROFILE: Confirmed carcinogen with experimental tumorigenic data. Poison by ingestion and intraperitoneal routes. An experimental teratogen. Other experimental reproductive effects. Mutation data reported. When heated to decomposition it emits very toxic fumes of BeO and Cl^-. See also BERYLLIUM COMPOUNDS and CHLORIDES.

BFQ500 HR: 3
BERYLLIUM COMPOUNDS

CONSENSUS REPORTS: IARC Cancer Review: Group 1 IMEMDT 58,41,93; Human Sufficient Evidence IMEMDT 58,41,93; Animal Sufficient Evidence IMEMDT 1,17,72; Animal Sufficient Evidence IMEMDT 23,143,80; Animal Sufficient Evidence IMEMDT 58,41,93. Beryllium and its compounds are on the Community Right-To-Know List.

OSHA PEL: TWA 0.002 mg(Be)/m³; STEL 0.005 mg(Be)/m³/30M; CL 0.025 mg(Be)/m³
ACGIH TLV: TWA 0.002 mg/m³, Suspected Human Carcinogen
DFG TRK: Animal Carcinogen, Suspected Human Carcinogen. Grinding of beryllium metal and alloys: 0.005 mg/m³ calculated as beryllium in that portion of dust that can possibly be inhaled; other beryllium compounds: 0.002 mg/m³ calculated as beryllium in that portion of dust that can possibly be inhaled

SAFETY PROFILE: Confirmed carcinogens. Beryllium compounds can enter the body through inhalation of dusts and fumes, and may act locally on the skin. Even alloys of low beryllium content have been shown to be dangerous. In industry, inhalation of the dust can cause severe lung damage with symptoms appearing within months. Effects have been reported in persons living near processing plants and in families of beryllium workers. The fluoride, ammonium fluoride, sulfate, oxide, and hydroxide occur during extraction from beryllium ore. Exposure to the oxide may occur in processing of beryllium alloys and beryllium ceramics.

The extraction of Be from its ore is attended by exposure to acid salts of the metal, particularly the fluoride (BeF_2), the ammonium fluoride and sulfate ($BeSO_4$), and also to beryllium oxide (BeO), and hydroxide [$Be(OH)_2$]. Exposure to the oxide also occurs in the casting of beryllium alloys and in operations with beryllia ceramics. In the manufacture of fluorescent powders, lamps, and sign tubes there may be exposure to beryllium carbonate and to more complex salts, such as ZnMnBe silicate. Exposure to beryllium compounds encountered in the extraction of the metal or its oxide from the ore, particularly the halide salts, has been attended, in certain individuals, by the development of dermatitis of an edematous and papulovesicular type, chronic skin ulcers, rhinitis, nasopharyngitis, epistaxis, bronchitis, and in severe cases, by the development of an acute pneumonitis, with cough, scanty sputum, low-grade fever, rales, dyspnea, and substernal pain. Radiographs show diffuse haziness throughout both lungs, followed by the appearance of soft, ill-defined opacities. The condition occurs while the worker is exposed, sometimes within 1 or 2 months of starting work, and recovery occurs within 2 months, as a rule, though radiographic changes sometimes persist for longer periods. Occasionally, recovery may not occur and lung fibrosis results. In severe cases of pneumonitis the patient may die. Necropsies have revealed diffuse pulmonary edema, hemorrhagic extravasation, large numbers of plasma cells, and a relative absence of polymorphonuclear infiltration. On the basis of experimental work with animals, certain investigators are of the opinion that the acute upper and lower respiratory effects are due chiefly to the acid radical present in the dust or fume, but this view has little support. A delayed form of lung disease, characterized by the occurrence of granulomatous areas in the lung tissue, has been reported in workers manufacturing fluorescent powders, lamps, and sign tubes, casting beryllium master alloys, and producing beryllium from beryl ore. Symptoms can start during exposure, but they may be delayed up to 5 years or more after the last exposure. The commonest symptoms are coughing, shortness of breath, loss of appetite, loss of weight, and fatigue. Rales are usually present in the bases and axillae, and the red cell count is frequently elevated. Cyanosis is common and the pulse and respiratory rates are often increased. Radiographically, three stages of the disease are described: (1) a diffuse, uniform granular shadowing extending throughout both lung fields; (2) a diffuse reticular pattern on the granular background; (3) the appearance of distinct

B

nodules scattered through the lungs, with some enlargement and blurring of the hilar shadows. The intensity of the shadowing is usually greater in the middle third of the lung fields. The prognosis is poor. Clinical improvement may occur gradually over a period of several years, but there appears to be little tendency for the radiographic shadowing to clear. In certain cases, the disease has progressed gradually for some months or years, with death resulting from respiratory and cardiac failure. In several instances necropsies have shown the presence of a diffuse fibrosis with coarse strands of hyalinized collagen between the alveoli and, in some places, replacing them. The hyalinized areas contained granulomatous foci, the alveolar walls are thickened and fibrosed, the blood vessels being engorged and dilated. In some cases the hilar lymph nodes show granulomatous change and fibrosis. Granulomatous change has also been noted in the liver and hyaline fibrosis in the spleen. Two cases of delayed lung disease coming to autopsy have presented papular lesions on the dorsum of the hands; on the biopsy these showed "sarcoid-like" lesions with central necrosis.

Several cases have been reported in which localized granulomatous lesions developed following penetrating wounds caused by splinters of glass from broken fluorescent light tubes. Several weeks or months following the accident, swellings were noted in the injured areas and excision revealed granulomatous tumors, which in one case was shown to contain beryllium.

There is no specific treatment, but temporary remissions have been produced by ACTH and cortisone.

BFQ750 CAS:12010-12-7 HR: 3
BERYLLIUM COMPOUND with NIOBIUM (12:1)
mf: $Be_{12}Nb$ mw: 201.03

TOXICITY DATA WITH REFERENCE
itr-rat TDLo:2500 µg/kg:ETA ENVRAL 21,63,80

CONSENSUS REPORTS: IARC Cancer Review: Group 1 IMEMDT 58,41,93; Human Sufficient Evidence IMEMDT 58,41,93; Animal Sufficient Evidence IMEMDT 1,17,72; Animal Sufficient Evidence IMEMDT 23,143,80; Animal Sufficient Evidence IMEMDT 58,41,93. Beryllium and its compounds are on the Community Right-To-Know List.

OSHA PEL: TWA 0.002 mg(Be)/m^3; STEL 0.005 mg(Be)/m^3/30M; CL 0.025 mg(Be)/m^3
ACGIH TLV: TWA 0.002 mg(Be)/m^3, Suspected Human Carcinogen
NIOSH REL: (Beryllium) CL not to exceed 0.005 mg(Be)/m^3

SAFETY PROFILE: Confirmed carcinogen with experimental tumorigenic data. When heated to decomposition in air it emits very toxic fumes of BeO. See also BERYLLIUM COMPOUNDS and NIOBIUM.

BFR000 CAS:12232-67-6 HR: 3
BERYLLIUM COMPOUND with TITANIUM (12:1)
mf: $Be_{12}Ti$ mw: 156.02

SYN: TITANIUM compounded with BERYLLIUM (1:12)

TOXICITY DATA WITH REFERENCE
itr-rat TDLo:2500 µg/kg:ETA ENVRAL 21,63,80

CONSENSUS REPORTS: IARC Cancer Review: Group 1 IMEMDT 58,41,93; Human Sufficient Evidence IMEMDT 58,41,93; Animal Sufficient Evidence IMEMDT 1,17,72; Animal Sufficient Evidence IMEMDT 23,143,80; Animal Sufficient Evidence IMEMDT 58,41,93. Beryllium and its compounds are on the Community Right-To-Know List.

OSHA PEL: TWA 0.002 mg(Be)/m^3; STEL 0.005 mg(Be)/m^3/30M; CL 0.025 mg(Be)/m^3
ACGIH TLV: TWA 0.002 mg(Be)/m^3, Suspected Human Carcinogen
NIOSH REL: (Beryllium) CL not to exceed 0.0005 mg(Be)/m^3

SAFETY PROFILE: Confirmed carcinogen with experimental tumorigenic data. See also BERYLLIUM COMPOUNDS and TITANIUM COMPOUNDS. When heated to decomposition it emits very toxic fumes of BeO.

BFR250 CAS:12400-16-7 HR: 3
BERYLLIUM COMPOUND with VANADIUM (12:1)
mf: $Be_{12}V$ mw: 159.06

SYN: TITANIUM compounded with BERYLLIUM (1:12)

TOXICITY DATA WITH REFERENCE
itr-rat TDLo:2500 µg/kg:NEO ENVRAL 21,63,80

CONSENSUS REPORTS: IARC Cancer Review: Group 1 IMEMDT 58,41,93; Human Sufficient Evidence IMEMDT 58,41,93; Animal Sufficient Evidence IMEMDT 1,17,72; Animal Sufficient Evidence IMEMDT 23,143,80; Animal Sufficient Evidence IMEMDT 58,41,93. Beryllium and its compounds are on the Community Right-To-Know List.

OSHA PEL: TWA 0.002 mg(Be)/m^3; STEL 0.005 mg(Be)/m^3/30M; CL 0.025 mg(Be)/m^3
ACGIH TLV: TWA 0.002 mg(Be)/m^3, Suspected Human Carcinogen
NIOSH REL: (Beryllium) CL not to exceed 0.0005 mg(Be)/m^3; (REL to Vanadium) 1.0 mg(V)/m^3

SAFETY PROFILE: Confirmed carcinogen with experimental tumorigenic data. See also BERYLLIUM COMPOUNDS and VANADIUM COMPOUNDS. When heated to decomposition it emits very toxic fumes of BeO and VO_x.

BFR500 CAS:7787-49-7 HR: 3
BERYLLIUM FLUORIDE
mf: BeF_2 mw: 47.01

PROP: Amorphous, colorless, hexagonal crystals. Undergoes transition from low temp quartz to high temp quartz structure types at 2°. Readily forms glass. Mp: 552°, d: 1.986 @ 25°. Subl @ 8°. Very sol in H_2O; sltly sol in EtOH.

SYN: BERYLLIUM DIFLUORIDE

TOXICITY DATA WITH REFERENCE
ihl-rat TCLo:20 µg/m^3/1H/17W-I:ETA GTPZAB 19(7),34,75

ihl-rat TCLo: 49 μg/m³/26W: ETA PEXTAR 2,203,61
orl-rat LD50: 98 mg/kg XEURAQ UR-154,1951
orl-mus LD50: 100 mg/kg XPHPAW 2173,23,72
scu-mus LD50: 20 mg/kg XPHPAW 2173,23,72
ivn-mus LD50: 1800 μg/kg XPHPAW 2173,23,72
ipr-ham LD50: 21 mg/kg XEURAQ UR-154,1951

CONSENSUS REPORTS: NTP 7th Annual Report On Carcinogens. IARC Cancer Review: Group 1 IMEMDT 58,41,93; Human Sufficient Evidence IMEMDT 58,41,93; Animal Sufficient Evidence IMEMDT 1,17,72; Animal Sufficient Evidence IMEMDT 23,143,80; Animal Sufficient Evidence IMEMDT 58,41,93. Beryllium and its compounds are on the Community Right-To-Know List. Reported in EPA TSCA Inventory.

OSHA PEL: TWA 0.002 mg(Be)/m³; STEL 0.005 mg(Be)/m³/30M; CL 0.025 mg(Be)/m³
ACGIH TLV: TWA 0.002 mg(Be)/m³, Suspected Human Carcinogen; 2.5 mg(F)/m³
NIOSH REL: (Beryllium) CL not to exceed 0.0005 mg(Be)/m³

SAFETY PROFILE: Confirmed carcinogen with experimental carcinogenic and tumorigenic data by inhalation. Poison by ingestion, subcutaneous, intravenous, and intraperitoneal routes. See also BERYLLIUM COMPOUNDS and FLUORIDES. Incompatible with Mg. When heated to decomposition, it emits very toxic fumes of BeO and F^-.

BFR750 CAS:7787-52-2 HR: 3
BERYLLIUM HYDRIDE
mf: BeH_2 mw: 11.03

PROP: White solid.

CONSENSUS REPORTS: IARC Cancer Review: Group 1 IMEMDT 58,41,93; Human Sufficient Evidence IMEMDT 58,41,93; Animal Sufficient Evidence IMEMDT 1,17,72; Animal Sufficient Evidence IMEMDT 23,143,80; Animal Sufficient Evidence IMEMDT 58,41,93. Beryllium and its compounds are on the Community Right-To-Know List.

SAFETY PROFILE: Confirmed carcinogen. A dangerous fire hazard. When heated to 220°C it liberates explosive hydrogen gas. Reacts violently with methanol, water, and dilute acids. When heated to decomposition it emits toxic fumes of BeO. See BERYLLIUM COMPOUNDS and HYDRIDES.

BFS000 CAS:13598-15-7 HR: 3
BERYLLIUM HYDROGEN PHOSPHATE (1:1)
mf: $BeHO_4P$ mw: 104.99

SYNS: BERYLLIUM PHOSPHATE □ PHOSPHORIC ACID, BERYLLIUM SALT (1:1) □ PHOSPHOROUS ACID, BERYLLIUM SALT

TOXICITY DATA WITH REFERENCE
ihl-rat TCLo: 3571 μg/m³/17W: ETA PEXTAR 2,203,61
ihl-mky TDLo: 900 μg/kg/17W: ETA PEXTAR 2,203,61
ihl-mky TC: 97 mg/m³/8D-I: ETA IMSUAI 3,1,64
ivn-mus LD50: 16 mg/kg TXAPA9 24,497,73

CONSENSUS REPORTS: NTP 7th Annual Report On Carcinogens. IARC Cancer Review: Group 1 IMEMDT 58,41,93; Human Sufficient Evidence IMEMDT 58,41,93; Animal Sufficient Evidence IMEMDT 1,17,72; Animal Sufficient Evidence IMEMDT 23,143,80; Animal Sufficient Evidence IMEMDT 58,41,93. Beryllium and its compounds are on the Community Right-To-Know List.

OSHA PEL: TWA 0.002 mg(Be)/m³; STEL 0.005 mg(Be)/m³/30M; CL 0.025 mg(Be)/m³
ACGIH TLV: TWA 0.002 mg(Be)/m³, Suspected Human Carcinogen
NIOSH REL: (Beryllium) CL not to exceed 0.0005 mg(Be)/m³

SAFETY PROFILE: Confirmed carcinogen with experimental carcinogenic and tumorigenic data. Poison by intravenous route. See also BERYLLIUM COMPOUNDS and PHOSPHATES. When heated to decomposition it emits very toxic fumes of BeO and PO_x.

BFS250 CAS:13327-32-7 HR: 3
BERYLLIUM HYDROXIDE
mf: H_2O_2•Be mw: 43.03

PROP: Colorless, orthorhomic, amorphous powder or crystals. Decomp on heating with H_2O loss forming BeO. Mp: decomp @ 138°. Practically insol in H_2O.

SYNS: BERYLLIUM DIHYDROXIDE □ BERYLLIUM HYDRATE

TOXICITY DATA WITH REFERENCE
itr-rat TDLo: 1125 μg/kg: ETA TXAPA9 17,299,70
itr-rat TD: 1785 μg/kg/43W-I: ETA ENVRAL 21,63,80
ivn-rat LDLo: 3821 μg/kg XEURAQ UR-70,1949

CONSENSUS REPORTS: NTP 7th Annual Report On Carcinogens. IARC Cancer Review: Group 1 IMEMDT 58,41,93; Human Sufficient Evidence IMEMDT 58,41,93; Animal Sufficient Evidence IMEMDT 1,17,72; Animal Sufficient Evidence IMEMDT 23,143,80; Animal Sufficient Evidence IMEMDT 58,41,93. Beryllium and its compounds are on the Community Right-To-Know List. Reported in EPA TSCA Inventory.

OSHA PEL: TWA 0.002 mg(Be)/m³; STEL 0.005 mg(Be)/m³/30M; CL 0.025 mg(Be)/m³
ACGIH TLV: TWA 0.002 mg(Be)/m³, Suspected Human Carcinogen.
NIOSH REL: (Beryllium) CL not to exceed 0.0005 mg(Be)/m³

SAFETY PROFILE: Confirmed carcinogen with experimental carcinogenic and tumorigenic data. Poison by intravenous route. See also BERYLLIUM COMPOUNDS. When heated to decomposition it emits very toxic fumes of BeO.

BFS750 HR: 3
BERYLLIUM MANGANESE ZINC SILICATE
mf: $BeMnO_4SiZn$ mw: 221.41

SYNS: MANGANESE ZINC BERYLLIUM SILICATE □ ZINC MANGANESE BERYLLIUM SILICATE

TOXICITY DATA WITH REFERENCE
ihl-rat TCLo: 20 mg/m³/4W: ETA PEXTAR 2,203,61
ivn-rbt TDLo: 500 mg/kg/40W-I: ETA PEXTAR 2,203,61

CONSENSUS REPORTS: IARC Cancer Review: Group 1 IMEMDT 58,41,93; Human Sufficient Evidence IMEMDT 58,41,93; Animal Sufficient Evidence IMEMDT 1,17,72; Animal Sufficient Evidence IMEMDT 23,143,80; Animal Sufficient Evidence IMEMDT 58,41,93. Beryllium, manganese, zinc, and their compounds are on the Community Right-To-Know List.

OSHA PEL: TWA 0.002 mg(Be)/m³; STEL 0.005 mg(Be)/m³/30M; CL 0.025 mg(Be)/m³
ACGIH TLV: TWA 0.002 mg(Be)/m³, Suspected Human Carcinogen; TWA 5 mg(Mn)/m³
NIOSH REL: (Beryllium) CL Not to exceed 0.0005 mg(Be)/m³

SAFETY PROFILE: Confirmed carcinogen with experimental tumorigenic data. When heated to decomposition it emits very toxic fumes of BeO and ZnO. See also BERYLLIUM COMPOUNDS, MANGANESE COMPOUNDS, and ZINC COMPOUNDS.

BFT000 CAS:13597-99-4 HR: 3
BERYLLIUM NITRATE
DOT: UN 2464
mf: BeN_2O_6 mw: 133.03

PROP: Deliquescent white amorphous solid or white-yellowish crystals; deliquescent. Mp: 60°, bp: decomp @ 100–200°.

SYNS: BERYLLIUM DINITRATE □ NITRIC ACID, BERYLLIUM SALT

TOXICITY DATA WITH REFERENCE
itt-rat TDLo:10,803 µg/kg (1D male):REP JRPFA4 7,21,64
ivn-mus LD50:3160 µg/kg CUSCAM 55,899,86
ipr-mus LD50:500 µg/kg EQSSDX 1,1,75
scu-mus LDLo:50 mg/kg RDWU** -,-,30
ipr-gpg LDLo:100 mg/kg EQSSDX 1,1,75
scu-frg LDLo:1041 mg/kg RDWU** -,-,30

CONSENSUS REPORTS: IARC Cancer Review: Group 1 IMEMDT 58,41,93; Human Sufficient Evidence IMEMDT 58,41,93; Animal Sufficient Evidence IMEMDT 1,17,72; Animal Sufficient Evidence IMEMDT 23,143,80; Animal Sufficient Evidence IMEMDT 58,41,93. Beryllium and its compounds are on the Community Right-To-Know List. Reported in EPA TSCA Inventory.

OSHA PEL: TWA 0.002 mg(Be)/m³; STEL 0.005 mg(Be)/m³/30M; CL 0.025 mg(Be)/m³
ACGIH TLV: TWA 0.002 mg(Be)/m³, Suspected Human Carcinogen
NIOSH REL: CL not to exceed 0.0005 mg(Be)/m³
DOT CLASSIFICATION: 5.1; *Label:* Oxidizer, Poison

SAFETY PROFILE: Confirmed carcinogen. Poison by intraperitoneal, intravenous, and subcutaneous routes. Experimental reproductive effects. When heated to decomposition it emits very toxic fumes of BeO and NO_x. See also BERYLLIUM COMPOUNDS and NITRATES.

BFT250 CAS:1304-56-9 HR: 3
BERYLLIUM OXIDE
mf: BeO mw: 25.01

PROP: White, amorph powder or white, hexagonal crystals; piezoelectric and pyroelectric. Undergoes hexagonal to tetragonal transition at 21°. Mp: 2507°, bp: 3900° (approx), d: 3.025. Dissolves in conc H_2SO_4 and in fused KOH. Sltly sol in H_2O.

SYNS: BERYLLIA □ BERYLLIUM MONOXIDE □ THERMALOX

TOXICITY DATA WITH REFERENCE
itr-rat TDLo:139 mg/kg (3D post):TER GISAAA 51(8),44,86
itr-rat TDLo:139 mg/kg (3D post):REP GISAAA 51(8),44,86
ihl-rat TCLo:28 mg/m³/17W-C:ETA IMSUAI 40,23,71
itr-rat TDLo:75 mg/kg/15W-I:ETA FKIZA4 71,19,80
ihl-rbt TCLo:17 mg/m³/5H/D/48W-I:ETA AMPLAO 51,473,51
ivn-rbt TDLo:500 mg/kg/6W-I:ETA FEPRA7 5,221,46
imp-rbt TDLo:10 mg/kg:ETA ARPAAQ 88,89,69
ivn-rbt TD:500 mg/kg/48W-I:ETA PEXTAR 2,203,61
itr-rat TD:169 mg/kg/2W-I:ETA AMIHAB 19,190,59
ivn-rbt TD:625 mg/kg/25W-I:ETA AMSHAR 25,99,77
itr-rat TD:79 mg/kg/15W-I:ETA SAIGBL 20,230,78

CONSENSUS REPORTS: NTP 7th Annual Report On Carcinogens. IARC Cancer Review: Group 1 IMEMDT 58,41,93; Human Sufficient Evidence IMEMDT 58,41,93; Animal Sufficient Evidence IMEMDT 1,17,72; Animal Sufficient Evidence IMEMDT 23,143,80; Animal Sufficient Evidence IMEMDT 58,41,93. Beryllium and its compounds are on the Community Right-To-Know List. Reported in EPA TSCA Inventory.

OSHA PEL: TWA 0.002 mg(Be)/m³; STEL 0.005 mg(Be)/m³/30M; CL 0.025 mg(Be)/m³
ACGIH TLV: TWA 0.002 mg(Be)/m³, Suspected Human Carcinogen
NIOSH REL: (Beryllium) CL not to exceed 0.0005 mg(Be)/m³

SAFETY PROFILE: Confirmed carcinogen with experimental tumorigenic data. Experimental teratogenic data. Other experimental reproductive effects. See also BERYLLIUM COMPOUNDS. Incompatible with (Mg + heat). When heated to decomposition it emits very toxic fumes of BeO.

BFT500 CAS:19049-40-2 HR: 3
BERYLLIUM OXYACETATE
mf: $C_{12}H_{18}Be_4O_{13}$ mw: 406.34

PROP: Colorless cubic crystals from $CHCl_3$. Undergoes cubic to orthorhombic transition at 1°. Mp: 284°, bp: 331°. Sol in $CHCl_3$, AcOH; sltly sol in EtOH and Et_2O.

SYNS: BERYLLIUM ACETATE, BASIC □ BERYLLIUM OXIDE ACETATE □ HEXAKIS(µ-ACETATO-O:O')-µ⁴-OXOTETRABERYLLIUM □ HEXAKIS(µ-ACETATO)-µ⁴-OXOTETRABERYLLIUM

CONSENSUS REPORTS: IARC Cancer Review: Group 1 IMEMDT 58,41,93; Human Sufficient Evidence IMEMDT 58,41,93; Animal Sufficient Evidence IMEMDT 1,17,72; Animal Sufficient Evidence IMEMDT 23,143,80; Animal Sufficient Evidence IMEMDT 58,41,

93. Beryllium and its compounds are on the Community Right-To-Know List.

OSHA PEL: TWA 0.002 mg(Be)/m³; STEL 0.005 mg(Be)/m³/30M; CL 0.025 mg(Be)/m³
ACGIH TLV: TWA 0.002 mg(Be)/m³, Suspected Human Carcinogen
NIOSH REL: (Beryllium) CL not to exceed 0.0005 mg (Be)/m³

SAFETY PROFILE: Confirmed carcinogen. See BERYLLIUM COMPOUNDS. When heated to decomposition it emits toxic fumes of BeO.

BFT750 CAS:63990-88-5 HR: 3
BERYLLIUM OXYFLUORIDE
mf: BeF_2O_2 mw: 79.01

TOXICITY DATA WITH REFERENCE
orl-rat LD50:146 mg/kg XEURAQ UR 154,1951
scu-mus LDLo:5 mg/kg BJEPA5 30,375,49
ivn-mus LDLo:3500 μg/kg BJEPA5 30,375,49
ipr-gpg LDLo:10 mg/kg NIHBAZ 181,20,43

CONSENSUS REPORTS: IARC Cancer Review: Group 1 IMEMDT 58,41,93; Human Sufficient Evidence IMEMDT 58,41,93; Animal Sufficient Evidence IMEMDT 1,17,72; Animal Sufficient Evidence IMEMDT 23,143,80; Animal Sufficient Evidence IMEMDT 58,41,93. Beryllium and its compounds are on the Community Right-To-Know List.

OSHA PEL: TWA 0.002 mg(Be)/m³; STEL 0.005 mg(Be)/m³/30M; CL 0.025 mg(Be)/m³
ACGIH TLV: TWA 0.002 mg(Be)/m³, Suspected Human Carcinogen; 2.5 mg(F)/m³
NIOSH REL: (Beryllium) CL not to exceed 0.0005 mg(Be)/m³

SAFETY PROFILE: Confirmed carcinogen. Poison by ingestion, subcutaneous, intravenous, and intraperitoneal routes. See also BERYLLIUM COMPOUNDS and FLUORIDES. When heated to decomposition it emits very toxic fumes of BeO and F^-.

BFU000 CAS:13597-95-0 HR: 3
BERYLLIUM PERCHLORATE
mf: $Be(ClO_4)_2$ mw: 207.91

PROP: Very hygroscopic crystals, sol in water: 148.6 g/100 mL.

CONSENSUS REPORTS: IARC Cancer Review: Group 1 IMEMDT 58,41,93; Human Sufficient Evidence IMEMDT 58,41,93; Animal Sufficient Evidence IMEMDT 1,17,72; Animal Sufficient Evidence IMEMDT 23,143,80; Animal Sufficient Evidence IMEMDT 58,41,93.

OSHA PEL: TWA 0.002 mg(Be)/m³; STEL 0.005 mg(Be)/m³/30M; CL 0.025 mg(Be)/m³
ACGIH TLV: TWA 0.002 mg(Be)/m³, Suspected Human Carcinogen.
NIOSH REL: CL not to exceed 0.0005 mg(Be)/m³

SAFETY PROFILE: Confirmed carcinogen. A powerful oxidant used in propellant and igniter systems. When heated to decomposition it emits toxic fumes of Cl^- and BeO. See also BERYLLIUM COMPOUNDS and PERCHLORATES.

BFU250 CAS:13510-49-1 HR: 3
BERYLLIUM SULFATE (1:1)
mf: $O_4S{\cdot}Be$ mw: 105.07

PROP: Colorless, tetragonal, hygroscopic crystals. Undergoes polymorphic transitions at 5° and 6°. On further heating dissoc without melting to BeO, SO_3, SO_2 and O_2 crystals. Mp: 550–600° (decomp), d: 2.443. Insol in H_2O.

SYN: SULFURIC ACID, BERYLLIUM SALT (1:1)

TOXICITY DATA WITH REFERENCE
mrc-bcs 10 mmol/L MUREAV 77,109,80
otr-mus:fbr 200 μg/L JJIND8 67,1303,81
ihl-rat TCLo:432 μg/m³/26W:ETA PEXTAR 2,203,61
itr-rat TDLo:17 mg/kg/2W-I:ETA AMIHAB 19,190,59
ihl-rat TC:643 μg/m³/39W-I:ETA AMIHAB 19,190,59
orl-rat LD50:82 mg/kg HYSAAV 30,169,65
ihl-rat LCLo:10 mg/m³ EQSSDX 1,1,75
ipr-rat LDLo:18 mg/kg XPHPAW 2173,23,72
scu-rat LD50:1500 μg/kg XPHPAW 2173,23,72
ivn-rat LD50:7200 μg/kg XPHPAW 2173,23,72
itr-rat LDLo:10 mg/kg XPHPAW 2173,23,72
orl-mus LD50:80 mg/kg HYSAAV 30,169,65
ihl-mus LCLo:47 mg/m³ EQSSDX 1,1,75
ipr-mus LD50:1200 mg/kg JJIND8 62,911,79
scu-mus LD50:1500 μg/kg XPHPAW 2173,23,72
ivn-mus LD50:500 μg/kg XPHPAW 2173,23,72
ivn-dog LDLo:600 μg/kg XPHPAW 2173,23,72

CONSENSUS REPORTS: NTP 7th Annual Report On Carcinogens. IARC Cancer Review: Group 1 IMEMDT 58,41,93; Human Sufficient Evidence IMEMDT 58,41,93; Animal Sufficient Evidence IMEMDT 1,17,72; Animal Sufficient Evidence IMEMDT 23,143,80; Animal Sufficient Evidence IMEMDT 58,41,93. Beryllium and its compounds are on the Community Right-To-Know List. Reported in EPA TSCA Inventory.

OSHA PEL: TWA 0.002 mg(Be)/m³; STEL 0.005 mg(Be)/m³/30M; CL 0.025 mg(Be)/m³
ACGIH TLV: TWA 0.002 mg(Be)/m³, Suspected Human Carcinogen
NIOSH REL: (Beryllium) CL not to exceed 0.0005 mg(Be)/m³

SAFETY PROFILE: Confirmed carcinogen with experimental tumorigenic data. Acute poison by inhalation, ingestion, intraperitoneal, subcutaneous, intravenous, and intratracheal routes. See also BERYLLIUM COMPOUNDS and SULFATES. Mutation data reported. When heated to decomposition it emits very toxic fumes of SO_x and BeO.

BFU500 CAS:7787-56-6 HR: 3
BERYLLIUM SULFATE TETRAHYDRATE (1:1:4)
mf: $O_4S{\cdot}Be{\cdot}4H_2O$ mw: 177.15

PROP: Colorless, tetragonal crystals. Decomp on heating with H_2O loss. Very sol in H_2O.

SYNS: BERYLLIUM SULPHATE TETRAHYDRATE ☐ SULFURIC ACID, BERYLLIUM SALT (1:1), TETRAHYDRATE

TOXICITY DATA WITH REFERENCE
mma-sat 3300 ng/plate ENMUDM 6(Suppl 2),1,84
sce-hmn:lym 1 mg/L ENMUDM 3,597,81
ihl-rat TCLo:668 μg/m³/40W-C:CAR CNREA8 27,439,67
ivn-mus LD50:4971 μg/kg TXAPA9 24,497,73

CONSENSUS REPORTS: NTP 7th Annual Report On Carcinogens. IARC Cancer Review: Group 1 IMEMDT 58,41,93; Human Sufficient Evidence IMEMDT 58,41,93; Animal Sufficient Evidence IMEMDT 1,17,72; Animal Sufficient Evidence IMEMDT 23,143,80; Animal Sufficient Evidence IMEMDT 58,41,93. Beryllium and its compounds are on the Community Right-To-Know List.

OSHA PEL: TWA 0.002 mg(Be)/m³; STEL 0.005 mg(Be)/m³/30M; CL 0.025 mg(Be)/m³
ACGIH TLV: TWA 0.002 mg(Be)/m³, Suspected Human Carcinogen
NIOSH REL: CL not to exceed 0.0005 mg(Be)/m³

SAFETY PROFILE: Confirmed carcinogen with experimental carcinogenic data by inhalation. Deadly poison by subcutaneous and intravenous routes. Human mutation data reported. See also BERYLLIUM COMPOUNDS and SULFATES. When heated to decomposition it emits very toxic fumes of BeO and SO_x.

BFU750 **HR: 3**
BERYLLIUM TETRAHYDROBORATE
mf: B_2BeH_8 mw: 38.70

$Be(BH_4)_2$

CONSENSUS REPORTS: IARC Cancer Review: Group 1 IMEMDT 58,41,93; Human Sufficient Evidence IMEMDT 58,41,93; Animal Sufficient Evidence IMEMDT 1,17,72; Animal Sufficient Evidence IMEMDT 23,143,80; Animal Sufficient Evidence IMEMDT 58,41,93. Beryllium and its compounds are on the Community Right-To-Know List.

OSHA PEL: TWA 0.002 mg(Be)/m³; STEL 0.005 mg(Be)/m³/30M; CL 0.025 mg(Be)/m³
ACGIH TLV: TWA 0.002 mg(Be)/m³, Suspected Human Carcinogen
NIOSH REL: CL not to exceed 0.0005 mg(Be)/m³

SAFETY PROFILE: Confirmed carcinogen. Ignites and then explodes in air or on contact with water. Upon decomposition it emits toxic fumes of BeO and BO_x. See also BERYLLIUM COMPOUNDS and BORON COMPOUNDS.

BFV000 **HR: 3**
BERYLLIUM TETRAHYDROBORATETRIMETHYL AMINE
mf: $C_3H_{17}B_2BeN$ mw: 97.78

$Be(BH_4(2 \cdot N(CH_3)_3$

CONSENSUS REPORTS: IARC Cancer Review: Group 1 IMEMDT 58,41,93; Human Sufficient Evidence IMEMDT 58,41,93; Animal Sufficient Evidence IMEMDT 1,17,72; Animal Sufficient Evidence IMEMDT 23,143,80; Animal Sufficient Evidence IMEMDT 58,41,93. Beryllium and its compounds are on the Community Right-To-Know List.

OSHA PEL: TWA 0.002 mg(Be)/m³; STEL 0.005 mg(Be)/m³/30M; CL 0.025 mg(Be)/m³
ACGIH TLV: TWA 0.002 mg(Be)/m³, Suspected Human Carcinogen
NIOSH REL: CL not to exceed 0.0005 mg(Be)/m³

SAFETY PROFILE: Confirmed carcinogen. It will ignite in contact with air or water. When heated to decomposition it emits toxic fumes of BeO, BO_x, and NO_x. See also BERYLLIUM COMPOUNDS and BORON COMPOUNDS.

BFV250 **CAS:39413-47-3** **HR: 3**
BERYLLIUM ZINC SILICATE
mf: $O_2Si \cdot Zn \cdot Be$ mw: 134.47

SYN: ZINC BERYLLIUM SILICATE

TOXICITY DATA WITH REFERENCE
ivn-rbt TDLo:100 mg/kg/2W-I:ETA LANCAO 1,519,50
ims-rbt TDLo:20 mg/kg:ETA BJCAAI 20,778,66
imp-rbt TDLo:10 mg/kg:ETA ARPAAQ 88,89,69
ivn-rbt TDLo:500 mg/kg/10W-I:ETA JBJSA3 38A,543,54
ivn-rbt TD:500 mg/kg/6W-I:ETA FEPRA7 5,221,46

CONSENSUS REPORTS: NTP 7th Annual Report On Carcinogens. IARC Cancer Review: Group 1 IMEMDT 58,41,93; Human Sufficient Evidence IMEMDT 58,41,93; Animal Sufficient Evidence IMEMDT 1,17,72; Animal Sufficient Evidence IMEMDT 23,143,80; Animal Sufficient Evidence IMEMDT 58,41,93. Beryllium and its compounds, as well as zinc and its compounds, are on the Community Right-To-Know List.

OSHA PEL: TWA 0.002 mg(Be)/m³; STEL 0.005 mg(Be)/m³/30M; CL 0.025 mg(Be)/m³
ACGIH TLV: TWA 0.002 mg(Be)/m³, Suspected Human Carcinogen
NIOSH REL: (Beryllium) CL not to exceed 0.0005 mg(Be)/m³

SAFETY PROFILE: Confirmed carcinogen with experimental tumorigenic data. When heated to decomposition it emits toxic fumes of BeO and ZnO. See also BERYLLIUM COMPOUNDS, ZINC COMPOUNDS, and SILICATES.

BFV300 **CAS:58970-76-6** **HR: 3**
BESTATIN
mf: $C_{16}H_{24}N_2O_4$ mw: 308.42

PROP: Needles. Mp: 233–236°.

SYNS: 3-(R)-AMINO-2-(s)-HYDROXY-4-PHENYLBUTANOYL-(2)-LEUCINE ☐ (S-(4*,s*))-N-(3-AMINO-2-HYDROXY-1-OXO-4-PHENYLBUTYL)-l-LEUCINE ☐ NK 421

TOXICITY DATA WITH REFERENCE
orl-rat TDLo:31,850 mg/kg (91D pre):REP OYYAA2 27,401,84
orl-rat TDLo:31,850 mg/kg (91D pre):REP OYYAA2 27,401,84

orl-rat TDLo:31,850 mg/kg (91D pre):REP OYYAA2 27,401,84
ipr-rat LD50:780 mg/kg JJANAX 36,2971,83
scu-rat LD50:1900 mg/kg JJANAX 36,2971,83
ipr-mus LD50:190 mg/kg JJANAX 36,2971,83
scu-mus LD50:1300 mg/kg JJANAX 36,2971,83

SAFETY PROFILE: Poison by intraperitoneal route. Moderately toxic by subcutaneous route. Experimental reproductive effects. When heated to decomposition it emits toxic fumes of NO_x.

BFV350 CAS:54856-23-4 HR: 3
BETAHISTINE MESYLATE
mf: $C_8H_{12}N_2 \cdot 2CH_4O_3S$ mw: 328.44

SYNS: BETAHISTINE MESILATE □ N-METHYL-2-PYRIDINEETHANAMINE DIMETHANESULFONATE

TOXICITY DATA WITH REFERENCE
orl-rat LD50:3030 mg/kg NIIRDN 6,750,82
scu-rat LD50:940 mg/kg NIIRDN 6,750,82
ivn-rat LD50:604 mg/kg NIIRDN 6,750,82
orl-mus LD50:500 mg/kg NIIRDN 6,750,82
scu-mus LD50:1630 mg/kg NIIRDN 6,750,82
ivn-mus LD50:505 mg/kg NIIRDN 6,750,82
orl-gpg LD50:1400 mg/kg NIIRDN 6,750,82
scu-gpg LD50:120 mg/kg NIIRDN 6,750,82
ivn-gpg LD50:22,900 μg/kg NIIRDN 6,750,82

SAFETY PROFILE: Poison by subcutaneous route. Moderately toxic by ingestion and other routes. When heated to decomposition it emits toxic fumes of NO_x and SO_x.

BFV750 CAS:378-44-9 HR: 1
BETAMETHASONE
mf: $C_{22}H_{29}FO_5$ mw: 392.51

SYNS: BETNELAN □ BETSOLAN □ CELESTONE □ 9-α-FLUORO-16-β-METHYLPREDNISOLONE □ 9-α-FLUORO-16-β-METHYL- 1,4-PREGNADIENE-11-β,17-α,21-TRIOL-3,20-DIONE □ 9-FLUORO-11-β,17,21-TRIHYDROXY-16-β-METHYLPREGNA-1,4-DIENE-3,20-DIONE □ 9-α-FLUORO-11-β,17,21-TRIHYDROXY-16-β-METHYLPREGNA-1,4-DIENE- 3,20-DIONE □ 16-β-METHYL-1,4-PREGNADIENE-9-α-FLUORO-11-β,17-α,21-TRIOL- 3,20-DIONE □ NSC-39470 □ Sch 4831

TOXICITY DATA WITH REFERENCE
scu-uns TDLo:138 mg/kg (female 8-18D post):TER JIDOAA 35,387,86
ims-mky TDLo:600 μg/kg (20W preg):TER AJOGAH 127,261,77
scu-uns TDLo:69 mg/kg (female 8-18D post):REP JIDOAA 35,387,86
scu-uns TDLo:138 mg/kg (female 8-18D post):REP JIDOAA 35,387,86
ims-mky TDLo:1600 μg/kg (18W preg):TER ARDSBL 117,377,78
orl-mus LD50:>4500 mg/kg YAKUD5 21,2117,79

SAFETY PROFILE: Low toxicity by ingestion. An experimental teratogen. Other experimental reproductive effects. When heated to decomposition it emits toxic fumes of F^-.

BFV760 CAS:22298-29-9 HR: 3
BETAMETHASONE BENZOATE
mf: $C_{29}H_{33}FO_6$ mw: 496.62

SYNS: BETAMETHASONE 17-BENZOATE □ BETHAMETHASONE 17-BENZOATE □ MS-1112

TOXICITY DATA WITH REFERENCE
scu-rat TDLo:80 μg/kg (female 9–18D post):REP OYYAA2 10,661,75
skn-rbt TDLo:7500 μg/kg (female 7–18D post):REP OYYAA2 10,685,75
skn-rbt TDLo:7500 μg/kg (female 7–18D post):TER OYYAA2 10,685,75
scu-rat TDLo:10 mg/kg (female 9–18D post):TER OYYAA2 10,661,75
scu-rat TDLo:1 mg/kg (female 9–18D post):TER OYYAA2 10,661,75
scu-rat TDLo:80 μg/kg (1–8D preg):REP OYYAA2 10,661,75
scu-rat LD50:194 mg/kg TXAPA9 8,250,66

SAFETY PROFILE: Poison by subcutaneous route. An experimental teratogen. Other experimental reproductive effects. When heated to decomposition it emits toxic fumes of F^-.

BFV765 CAS:5593-20-4 HR: 3
BETAMETHASONE DIPROPIONATE
mf: $C_{28}H_{37}FO_7$ mw: 504.65

PROP: Powder. Mp: 170–179° (decomp).

SYNS: BETAMETHASONE 17,21-DIPROPIONATE □ DIPROSONE □ 9-FLUORO-11-β,17,21-TRIHYDROXY-16-β-METHYLPREGNA-1,4-DIENE-3,20-DIONE, 17,21-DIPROPIONATE □ S-3440

TOXICITY DATA WITH REFERENCE
scu-rat TDLo:7500 μg/kg (female 9-14D post):REP OYYAA2 8,705,74
scu-mus TDLo:936 μg/kg (female 7-12D post):REP OYYAA2 8,705,74
scu-rat TDLo:30 mg/kg (female 9-14D post):REP OYYAA2 8,705,74
scu-mus TDLo:3750 μg/kg (female 7-12D post):TER OYYAA2 8,705,74
scu-rat TDLo:120 mg/kg (female 9-14D post):TER OYYAA2 8,705,74
scu-rbt TDLo:6 μg/kg (female 15D post):REP OYYAA2 21,645,81
scu-mus TDLo:3750 μg/kg (female 7-12D post):REP OYYAA2 8,705,74
scu-rat TDLo:1820 μg/kg (female 26W pre):REP OYYAA2 18,507,79
scu-rat TDLo:910 μg/kg (26W male):REP OYYAA2 18,507,79
scu-rbt TDLo:130 μg/kg (female 6-18D post):TER KSRNAM 11,1672,77
scu-mus TDLo:640 μg/kg (female 12D post):TER OYYAA2 21,645,81
scu-rat TDLo:7500 μg/kg (female 9-14D post):TER OYYAA2 8,705,74
scu-rbt TDLo:32,500 ng/kg (female 6-18D post):TER KSRNAM 11,1672,77
scu-rat TDLo:480 mg/kg (female 9-14D post):TER OYYAA2 8,705,74

ipr-mus LD50:103 mg/kg NIIRDN 6,753,82
scu-mus LD50:78,100 μg/kg NIIRDN 6,753,82

SAFETY PROFILE: Poison by subcutaneous and intraperitoneal routes. An experimental teratogen. Other experimental reproductive effects. When heated to decomposition it emits toxic fumes of F^-.

BFV770 CAS:151-73-5 HR: 2
BETAMETHASONE DISODIUM PHOSPHATE
mf: $C_{22}H_{30}FO_8P{\cdot}2Na$ mw: 518.47

SYNS: BETAMETHASONE-21-DISODIUM PHOSPHATE □ BETAMETHASONE SODIUM PHOSPHATE □ BETNESOL □ 9-FLUORO-11-β,17,21-TRIHYDROXY-16-β-METHYLPREGNA-1,4-DIENE-3,20-DIONE, 21-(DIHYDROGEN PHOSPHATE), DISODIUM SALT

TOXICITY DATA WITH REFERENCE
scu-rat TDLo:468 μg/kg (9-14D preg):REP OYYAA2 8,705,75
scu-rat TDLo:1878 μg/kg (female 9-14D post):REP OYYAA2 8,705,74
scu-mus TDLo:60 mg/kg (female 7-12D post):TER OYYAA2 8,705,74
scu-rat TDLo:468 μg/kg (9-14D preg):TER OYYAA2 8,705,75
scu-rat TDLo:1878 μg/kg (female 9-14D post):REP OYYAA2 8,705,74
scu-rat TDLo:7500 μg/kg (female 9-14D post):TER OYYAA2 8,705,74
scu-mus TDLo:15 mg/kg (female 7-12D post):TER OYYAA2 8,705,74
orl-rat LD50:1877 mg/kg DRUGAY-,1075,90
ipr-rat LD50:1179 mg/kg DRUGAY-,1075,90
ivn-rat LD50:1276 mg/kg YAKUD5 21,2117,79
orl-mus LD50:1607 mg/kg SKIZAB 29,153,73
ipr-mus LD50:1166 mg/kg SKIZAB 29,153,73
scu-mus LD50:1363 mg/kg SKIZAB 29,153,73

SAFETY PROFILE: Moderately toxic by ingestion and other routes. An experimental teratogen. Other experimental reproductive effects. When heated to decomposition it emits toxic fumes of F^-, PO_x and Na_2O.

BFV900 CAS:37717-82-1 HR: 3
BETANIDINE SULFATE
mf: $C_{18}H_{14}N_2O_8{\cdot}7H_2O_4S$ mw: 1072.90

SYNS: BETANIDIN SULFATE □ 2-CARBOXY-1-((2,6-DICARBOXY-2,3-DIHYDRO-4(1H)-PYRIDINYLIDENE)ETHYLIDENE)5,6-DIHYDROXY-1H-INDOLIUM, HYDROXIDE, INNER SALT, SULFATE (SALT)

TOXICITY DATA WITH REFERENCE
orl-rat LD50:3142 mg/kg NIIRDN 6,750,82
ipr-rat LD50:135 mg/kg NIIRDN 6,750,82
scu-rat LD50:681 mg/kg NIIRDN 6,750,82
orl-mus LD50:1059 mg/kg NIIRDN 6,750,82
ipr-mus LD50:170 mg/kg NIIRDN 6,750,82
scu-mus LD50:398 mg/kg NIIRDN 6,750,82

SAFETY PROFILE: Poison by subcutaneous and intraperitoneal routes. Moderately toxic by ingestion. When heated to decomposition it emits toxic fumes of SO_x and NO_x. See also SULFATES.

BFV975 HR: 2
BETEL LEAVES

TOXICITY DATA WITH REFERENCE
orl-rat TDLo:3000 g/kg/43W-C:ETA EXPEAM 35,384,79

CONSENSUS REPORTS: IARC Cancer Review: Animal Inadequate Evidence IMEMDT 37,141,85

SAFETY PROFILE: Questionable carcinogen with experimental tumorigenic data. When heated to decomposition it emits toxic and irritating fumes.

BFW000 CAS:39323-48-3 HR: 3
BETEL NUT

PROP: Mottled brown with fawn color. Extract of 50 gs sun-dried betel nut in 100 mL boiling water (IJCNAW 17,469,76).

SYNS: ARECA CATECHU □ ARECA CATECHU Linn., fruit extract □ ARECA CATECHU Linn., nut extract □ BN □ PINANG □ POOGIPHALAM, nut extract □ SUPARI, nut extract

TOXICITY DATA WITH REFERENCE
mnt-mus-ipr 1600 mg/kg CRNGDP 5,501,84
sce-mus-ipr 62,500 μg/kg/5D-C CRNGDP 7,37,86
msc-ham:lng 5 mg/L CRNGDP 5,501,84
orl-mus TDLo:400 mg/kg (6-15D preg):TER TXCYAC 37,315,85
orl-rat TDLo:1750 mg/kg (female 1-7D post):REP PLMEAA 26,391,74
orl-rat TDLo:3500 mg/kg (1-7D preg):REP IJMRAQ 59,302,71
scu-rat TDLo:2016 mg/kg/42W-I:NEO JNCIAM 60,683,78
orl-mus TDLo:340 g/kg/17W-I:CAR BJCAAI 40,922,79
scu-mus TDLo:1728 mg/kg/13W-I:CAR IJEBA6 18,1159,80
scu-mus TD:48 g/kg/5W-I:NEO IJCNAW 17,469,76
ipr-mus LD50:681 mg/kg IJEBA6 18,594,80

CONSENSUS REPORTS: IARC Cancer Review: Animal Limited Evidence IMEMDT 37,141,85.

SAFETY PROFILE: Suspected carcinogen with experimental carcinogenic and neoplastigenic data. Moderately toxic by intraperitoneal route. Experimental teratogenic and reproductive effects. When heated to decomposition it emits toxic fumes of NO_x. See also ARECA NUT, other betel entries, and SMOKELESS TOBACCO.

BFW010 CAS:89957-52-8 HR: 2
BETEL NUT, polyphenol fraction

SYN: POLYPHENOL FRACTION OF BETEL NUT

TOXICITY DATA WITH REFERENCE
orl-mus TDLo:380 mg/kg/W-C:ETA BJCAAI 40,922,79
scu-mus TDLo:988 mg/kg/13W-I:CAR IJEBA6 18,1159,80

SAFETY PROFILE: Questionable carcinogen with experimental carcinogenic and tumorigenic data. When heated to decomposition it emits acrid smoke and irritating fumes.

BFW050 **HR: 2**
BETEL NUT TANNIN

SYN: TANNIN from BETEL NUT

TOXICITY DATA WITH REFERENCE
sce-mus-ipr 1500 mg/kg/15D C CRNGDP 7,37,86
orl-mus TDLo:27,740 mg/kg/1Y-I:ETA IJEBA6 24,229,86

SAFETY PROFILE: Questionable carcinogen with experimental tumorigenic data. Mutation data reported. When heated to decomposition it emits toxic and irritating fumes.

BFW120 **HR: 3**
BETEL QUID

PROP: Composed of Areca nut, lime, catechu, and possibly tobacco and spices wrapped in a betel leaf. Used throughout the Orient and many Pacific islands, it is chewed in a manner similar to chewing tobacco. The major ingredients are:

AREGA NUT (betel nut, supari in India) is the fruit of the areca palm (*Areca catechu L.*). The ripe, orange-yellow nut is separated from its fibrous pericarp and may be cured in boiling water and then dried.

BETEL LEAF comes from the Betel vine (*Piper betle L.*).

LIME (chuna or chunam in India) is prepared from seashells or quarried stone and mixed with water (slaked) to release calcium hydroxide.

CATECHU (kattha in India) is a resinous extract from the heartwood of the Acacia tree (*A. catechu* or *A. suma*).

TOBACCO is the leaf from the tobacco plant (*N. rustica* or *N. Tabacum*).

SPICES and flavorings, such as cardamom, cloves, grated fresh coconut, and sugar, are sometimes added.

SYN: PAN

SAFETY PROFILE: The areca nut contains several alkaloids that are the primary cause of habituation. The most abundant alkaloid is arecoline that mimics the action of acetylcholine and acts as a stimulant.

Betel quid toxicity is due to the presence of areca nut alkaloids, nitrosamines derived from these compounds, polyphenols, and, when used, tobacco specific nitrosamines. Several nitrosamines derived from areca nut alkaloids have been found in the saliva of quid chewers: N-nitrosoguvacoline, N-nitrosoproline, N′-nitrosonornicotine, N′-nitrosoanatabine, and 4-(methylnitrosamino) 1-(3-pyridyl)-1-butanone. Users of betel quid with tobacco also have nitrosamines derived from tobacco alkaloids in their saliva. Catechu has a high percentage of polyphenols such as kaempferol, dihydroxykaempferol, taxifolin, isorhamnetin, (+)afzelchin, dimeric procyanidin and (−)eipcatechin. Many nitrosamines have been shown to be experimental carcinogens.

There is sufficient evidence that betel quid with tobacco is carcinogenic to humans. There is inadequate evidence that betel quid without tobacco is carcinogenic to humans. There is limited evidence that extracts of betel quid with and without tobacco are carcinogenic to experimental animals. There is limited evidence that areca nut with and without tobacco is carcinogenic to experimental animals. Extracts from the areca nut are mutatgenic. Chewing the quid may cause mouth ulcerations and periodontal disease. There is a high incidence of oral leukoplakia (a precancerous lesion) in betel quid users.

See also specific compounds; ARECA NUT, other betel entries, SMOKELESS TOBACCO, N-NITROSO COMPOUNDS, and NITROSAMINES.

BFW125 **HR: 2**
BETEL QUID EXTRACT

TOXICITY DATA WITH REFERENCE
dni-hmn:lym 25,000 ppm CNREA8 39,4802,79
dni-rat:mmr 25,000 ppm CNREA8 39,4802,79
dni-mus:fbr 25,000 ppm CNREA8 39,4802,79
skn-mus TDLo:720 g/kg/26W-C:ETA BJCAAI 14,597,60
scu-mus TDLo:3376 mg/kg/13W-I:CAR IJEBA6 18,1159,80
skn-mus TD:2920 g/kg/2Y-C:ETA BJCAAI 14,597,60

CONSENSUS REPORTS: IARC Cancer Review: Human Inadequate Evidence IMEMDT 37,141,85; Animal Limited Evidence IMEMDT 37,141,85.

SAFETY PROFILE: Suspected carcinogen with experimental carcinogenic and tumorigenic data by skin contact. Human mutation data reported. See other betel entries.

BFW135 **HR: 3**
BETEL TOBACCO EXTRACT

SYN: JAFFNA TOBACCO

TOXICITY DATA WITH REFERENCE
sce-hmn:lym 10 mg/L TOLED5 8,17,81
mnt-mus:ipr 24 mg/kg CRNGDP 5,501,84
otr-ham:emb 50 mg/L TOLED5 8,17,81
msc-ham:lng 5 mg/L CRNGDP 5,501,84

CONSENSUS REPORTS: IARC Cancer Review: Human Sufficient Evidence IMEMDT 37,141,85; Animal Limited Evidence IMEMDT 37,141,85

SAFETY PROFILE: Confirmed human carcinogen. Human mutation data reported. See also SMOKELESS TOBACCO and other betel entries.

BFW250 **CAS:114-85-2** **HR: 3**
BETHANIDINE SULFATE
mf: $C_{20}H_{30}N_6 \cdot H_2O_4S$ mw: 452.64

PROP: Solid. Mp: 289–290° (decomp).

SYNS: BATEL □ BENTANIDOL □ BENZAIDIN □ BENZANIDINE □ BENZOXINE □ 1-BENZYL-2,3-DIMETHYL-GUANIDINE SULFATE (1:1/2) □ N-BENZYL-N′,N″-DIMETHYLGUANIDINE SULFATE □ BETALING □ BETANIDOLE □ BETHANID □ BETHANIDINE, HEMISULFATE □ BW 467-C-60 □ 467-C-60 □ N,N-DIMETHYL″-(PHENYLMETHYL)GUANIDINE SULPHATE (2:1) □ ESBATAL □ ESTABAL □ EUSMANID □ HYPERSIN □ NSC-106563 □ REGULIN □ TENATHAN

TOXICITY DATA WITH REFERENCE
ivn-rat LD50:20 mg/kg NYKZAU 72,837,76
orl-mus LD50:520 mg/kg BJPCAL 20,36,63
ipr-mus LD50:150 mg/kg BJPCAL 20,36,63

scu-mus LD50:260 mg/kg BJPCAL 20,36,63
ivn-mus LD50:12 mg/kg BJPCAL 20,36,63
ivn-ckn LDLo:100 mg/kg BJPCAL 20,36,63

SAFETY PROFILE: Poison by intraperitoneal, subcutaneous, and intravenous routes. Moderately toxic by ingestion. See also SULFATES. When heated to decomposition it emits toxic fumes of NO_x and SO_x.

BFW500 CAS:319-86-8 HR: 2
Δ-BHC
mf: $C_6H_6Cl_6$ mw: 290.82

PROP: Solid. Mp: 129–132°.

SYNS: Δ-BENZENEHEXACHLORIDE □ ENT 9,234 □ 1-α,2-α,3-α,4-β,5-α,6-β-HEXACHLOROCYCLOHEXANE □ Δ-HEXACHLOROCYCLOHEXANE □ Δ-1,2,3,4,5,6-HEXACHLOROCYCLOHEXANE □ Δ-LINDANE

TOXICITY DATA WITH REFERENCE
orl-rat LD50:1000 mg/kg ARSIM* 20,5,66

CONSENSUS REPORTS: Reported in EPA TSCA Inventory.

SAFETY PROFILE: Moderately toxic by ingestion. When heated to decomposition it emits toxic fumes of Cl^-. See also CHLORINATED HYDROCARBONS, ALIPHATIC.

BFW750 CAS:128-37-0 HR: 2
BHT (food grade)
mf: $C_{15}H_{24}O$ mw: 220.39

PROP: White, crystalline solid; faint characteristic odor. Bp: 265°, fp: 68°, flash p: 260°F (TOC), d: 1.048 @ 20°/4°, vap d: 7.6, mp: 71°. Sol in alc; insol in water and propylene glycol.

SYNS: ADVASTAB 401 □ AGIDOL □ ANTIOXIDANT DBPC □ ANTIOXIDANT 29 □ AO 29 □ AO 4K □ 2,6-BIS(1,1-DIMETHYLETHYL)-4-METHYLPHENOL □ BUKS □ BUTYLATED HYDROXYTOLUENE □ BUTYLHYDROXYTOLUENE □ CAO 1 □ CAO 3 □ CATALIN CAO-3 □ CHEMANOX 11 □ DBMP □ DBPC (technical grade) □ DIBUTYLATED HYDROXYTOLUENE □ 2,6-DI-tert-BUTYL-p-CRESOL (OSHA, ACGIH) □ 2,6-DI-tert-BUTYL-1-HYDROXY-4-METHYLBENZENE □ 3,5-DI-tert-BUTYL-4-HYDROXYTOLUENE □ 2,6-DI-terc. BUTYL-p-KRESOL (CZECH) □ 2,6-DI-tert-BUTYL-p-METHYLPHENOL □ 2,6-DI-tert-BUTYL-4-METHYLPHENOL □ FEMA No. 2184 □ 4-HYDROXY-3,5-DI-tert-BUTYLTOLUENE □ IMPRUVOL □ IONOL □ IONOL (antioxidant) □ 4-METHYL-2,6-DI-terc. BUTYLFENOL (CZECH) □ METHYL DI-tert-BUTYLPHENOL □ 4-METHYL-2,6-DI-tert-BUTYLPHENOL □ NCI-C03598 □ NONOX TBC □ PARABAR 441 □ SUSTANE □ TENOX BHT □ TOPANOL □ VANLUBE PCX

TOXICITY DATA WITH REFERENCE
skn-hmn 500 mg/48H MLD AMIHBC 5,311,52
skn-rbt 500 mg/48H MOD AMIHBC 5,311,52
eye-rbt 100 mg/24H MOD 28ZPAK -,57,72
dni-hmn:lym 20 µmol/L BBRCA9 80,963,78
dns-rat:lvr 100 pmol/L CRNGDP 5,1547,84
spm-mus-ipr 350 mg/kg/5D-I CMMUAO 5,257,78
orl-mus TDLo:12,600 mg/kg (female 1-21D post):REP FEPRA7 31,596,72
orl-rat TDLo:6 g/kg (13W male/13W pre-3W post):REP FCTOD7 24,1,86
orl-rat TDLo:9 g/kg (male 2W pre):REP FCTXAV 19,153,81
orl-rat TDLo:35 g/kg (male 10W pre):REP ENMUDM 8,357,86
orl-mus TDLo:43,800 mg/kg (female 52D pre):REP FCTXAV 3,371,65
orl-rat TDLo:1302 mg/kg (male 1D pre):REP TRENAF 27,93,76
orl-mus TDLo:1200 mg/kg (female 9D post):TER TRENAF 28(2),45,77
orl-rat TDLo:134 g/kg/32W-C:CAR CRNGDP 4,895,83
orl-mus TDLo:435 mg/kg/69W-C:CAR FCTXAV 12,367,74
orl-rat TD:247 g/kg/3Y-C:CAR,REP FCTOD7 24,1,86
orl-rat TD:247 g/kg/3Y-C:NEO,REP FCTOD7 24,1,86
orl-mus TD:1423 mg/kg/43W-C:NEO TXCYAC 38,151,86
orl-rat TD:247 g/kg:CAR FCTOD7 24,1121,86
orl-rat TD:963 g/kg:CAR FCTOD7 24,1071,86
orl-wmn TDLo:80 mg/kg:PSY,GIT NEJMAG 314,648,86
orl-rat LD50:890 mg/kg NEOLA4 24,253,77
orl-mus LD50:650 mg/kg SCIEAS 36(1-4),10,89
ipr-mus LD50:138 mg/kg TXAPA9 61,475,81
ivn-mus LD50:180 mg/kg JMCMAR 23,1350,80
orl-cat LDLo:940 mg/kg AMIHAB 11,93,55
orl-rbt LDLo:2100 mg/kg AMIHAB 11,93,55
orl-gpg LD50:10,700 mg/kg AMIHAB 11,93,55

CONSENSUS REPORTS: IARC Cancer Review: Group 3 IMEMDT 7,56,87; Animal Limited Evidence IMEDT 40,161,86. NCI Carcinogenesis Bioassay Completed; (feed): No Evidence: mouse, rat NCITR* NCI-CG-TR-150,79. Reported in EPA TSCA Inventory. EPA Genetic Toxicology Program.

OSHA PEL: TLV 10 mg/m³
ACGIH TLV: TLV 10 mg/m³

SAFETY PROFILE: Poison by intraperitoneal and intravenous routes. Moderately toxic by ingestion. An experimental teratogen. Other experimental reproductive effects. A human skin irritant. A skin and eye irritant. Questionable carcinogen with experimental carcinogenic and neoplastigenic data. Combustible when exposed to heat or flame. It can react with oxidizing materials. To fight fire, use CO_2, dry chemical. When heated to decomposition it emits acrid smoke and fumes.

BFX000 CAS:613-35-4 HR: 3
4′,4‴-BIACETANILIDE
mf: $C_{16}H_{16}N_2O_2$ mw: 268.34

PROP: Solid. Mp: 317°.

SYNS: N,N′-(1,1′-BIPHENYL)-4,4′-DIYLBIS-ACETAMIDE 4′,4‴-BIACETANILIDE □ N,N′-4,4′-BIPHENYLYLENEBISACETAMIDE □ 4,4′-DIACETYLAMINOBIPHENYL □ N,N′-DIACETYL BENZIDINE □ 4,4′-DIACETYLBENZIDINE

TOXICITY DATA WITH REFERENCE
mma-sat 5 µg/plate ENMUDM 6,145,84
dnd-rat:lvr 100 mg/L CRNGDP 5,407,84
orl-rat TDLo:6300 mg/kg/35W-C:ETA CNREA8 16,525,56
ipr-rat TDLo:64 mg/kg/4W-I:CAR CRNGDP 2,747,81
scu-rat TDLo:900 mg/kg:NEO ARPAAQ 81,146,66
scu-rat TD:4350 mg/kg/39W-I:ETA VOONAW 8(11),11,62
ipr-rat TD:900 mg/kg:NEO ARPAAQ 81,146,66

CONSENSUS REPORTS: IARC Cancer Review: Group

2B IMEMDT 7,56,87, Animal Sufficient Evidence IMEMDT 16,293,78. Reported in EPA TSCA Inventory.

SAFETY PROFILE: Suspected carcinogen with experimental carcinogenic, neoplastigenic, and tumorigenic data. Mutation data reported. When heated to decomposition it emits toxic fumes of NO_x.

BFX125 CAS:3624-96-2 HR: 3
BIALLYLAMICOL DIHYDROCHLORIDE
mf: $C_{28}H_{40}N_2O_2$•2ClH mw: 509.62

PROP: Solid. Mp: 209–210°.

SYNS: BIALAMICOL HYDROCHLORIDE □ BIALLYLAMICOL HYDROCHLORIDE □ α,α′-BIS(DIETHYLAMINO)-5,5′-DIALLYL-m,m′-BITOLYL-4,4′-DIOL DIHYDROCHLORIDE □ CAMOFORM HYDROCHLORIDE □ PAA-701 DIHYDROCHLORIDE □ SN 6771 DIHYDROCHLORIDE

TOXICITY DATA WITH REFERENCE
orl-hmn TDLo:243 mg/kg/9D:MET,GIT 85GLAQ 1,304,46
orl-hmn TDLo:43 mg/kg:GIT 85GLAQ 1,304,46
orl-rat LD50:1649 mg/kg ANTCAO 7,113,57
orl-mus LD50:3950 mg/kg ANTCAO 7,113,57

SAFETY PROFILE: Poison to humans by ingestion with systemic effects: fever and nausea or vomiting. When heated to decomposition it emits toxic fumes of NO_x and HCl. See also ALLYL COMPOUNDS.

BFX250 CAS:2130-56-5 HR: 2
5,5′-BIANTHRANILIC ACID
mf: $C_{14}H_{12}N_2O_4$ mw: 272.28

PROP: Needles. Mp: 300°.

SYNS: 3,3′-BENZIDINEDICARBOXYLIC ACID □ 4,4′-DIAMINOBIPHENYL-3,3′-DICARBOXYLIC ACID □ 4,4′-DIAMINO-3,3′-BIPHENYLDICARBOXYLIC ACID □ 3,3′-DICARBOXYBENZIDINE □ KWAS BENZYDYNODWUKAROKSYLOWY (POLISH)

TOXICITY DATA WITH REFERENCE
pic-esc 100 mmol/L MDMIAZ 31,11,79
scu-rat TDLo:7 g/kg/77W-I:ETA VOONAW 15(5),60,69
orl-mus TDLo:22 g/kg/45W-I:ETA VOONAW 15(5),60,69
scu-mus TDLo:12 g/kg/60W-I:ETA VOONAW 15(5),60,69

CONSENSUS REPORTS: Reported in EPA TSCA Inventory.

SAFETY PROFILE: Questionable carcinogen with experimental tumorigenic data. Mutation data reported. When heated to decomposition it emits toxic fumes of NO_x.

BFX325 CAS:4388-03-8 HR: 3
1,1′-BIAZIRIDINYL
mf: $C_4H_8N_2$ mw: 84.12

$(CH_2CH_2N\text{-})_2$

PROP: Liquid. Fp: −11, bp: 83–83.2° @ 750 mm. Sol in H_2O and EtOH.

SAFETY PROFILE: When heated it reacts violently with oxygen. When heated to decomposition it emits toxic fumes of NO_x.

BFX500 CAS:103-29-7 HR: 3
BIBENZYL
mf: $C_{14}H_{14}$ mw: 182.28

PROP: Flash p: 264°F, autoign temp: 896°F, d: 1.0, vap d: 6.29, bp: 284°, mp: 52°.

SYNS: DIBENZYL □ 1,2-DIPHENYLETHANE

TOXICITY DATA WITH REFERENCE
ipr-mus LD50:2500 mg/kg ARZNAD 19,617,69
ivn-mus LD50:78 mg/kg ARZNAD 19,617,69

CONSENSUS REPORTS: Reported in EPA TSCA Inventory.

SAFETY PROFILE: Poison by intravenous route. Moderately toxic by intraperitoneal route. Combustible. To fight fire, use water, spray, mist, alcohol foam, dry chemical. When heated to decomposition it emits acrid smoke and fumes.

BFY000 CAS:826-62-0 HR: 1
BICYCLO(2.2.1)-HEPT-5-ENE-2,3-DICARBOXYLIC ANHYDRIDE
mf: $C_9H_8O_3$ mw: 164.17

SYN: ANHYDRID KYSELINY 3,6-ENDOMETHYLEN-Δ(SUP 4)-TETRAHYDROFTALOVE (CZECH)

TOXICITY DATA WITH REFERENCE
skn-rbt 500 mg/24H MLD 28ZPAK -,140,72
eye-rbt 5 mg/24H SEV 28ZPAK -,140,72

CONSENSUS REPORTS: Reported in EPA TSCA Inventory.

SAFETY PROFILE: Mild skin and severe eye irritant. When heated to decomposition it emits acrid smoke and irritating fumes. See also ANHYDRIDES.

BFY250 CAS:95-39-6 HR: 2
BICYCLO(2.2.1)HEPT-5-ENE-2-METHYLOL ACRYLATE
mf: $C_{11}H_{14}O_2$ mw: 178.25

SYNS: ACRYLIC ACID-5-NORBORNEN-2-METHYL ESTER □ ACRYLIC ACID-5-NORBORNEN-2-YLMETHYL ESTER □ CYCLOL ACRYLATE □ 2,5-endo-METHYLENE-Δ^3-TETRAHYDROBENZYL ACRYLATE □ 5-NORBORNENE-2-METHANOL ACRYLATE □ 5-NORBORNENE-2-METHYLOLACRYLATE □ 2-PROPENOIC ACID BICYCLO(2,2,1)HEPT-5-EN-2-YLMETHYL ESTER

TOXICITY DATA WITH REFERENCE
orl-rat LD50:1410 mg/kg TXAPA9 28,313,74
skn-rbt LD50:2830 mg/kg TXAPA9 28,313,74

CONSENSUS REPORTS: Reported in EPA TSCA Inventory.

SAFETY PROFILE: Moderately toxic by ingestion and skin contact. See also ESTERS. When heated to decomposition it emits acrid smoke and irritating fumes.

BFY750 **CAS:2886-89-7** **HR: 2**
BICYCLONONADIENE DIEPOXIDE
mf: $C_9H_{12}O_2$ mw: 152.21

SYNS: 1,2:5,6-DIEPOXYHEXAHYDROINDAN □ 4,9-DIOXATETRACYCLO($5.4.0.0^{3,5}.0^{8,10}$)UNDECANE □ 4,10-DIOXATETRACYCLO($5.4.0^{3,5}.0^{1,7}.0^{9,11}$)UNDECANE □ OCTAHYDRO-2H-BISOXIRENO(a,f)INDENE

TOXICITY DATA WITH REFERENCE
orl-rat LD50:2140 mg/kg AIHAAP 30,470,69
skn-rbt LDLo:1770 mg/kg AIHAAP 30,470,69

SAFETY PROFILE: Moderately toxic by ingestion and skin contact. When heated to decomposition it emits acrid smoke and irritating fumes.

BFZ100 **CAS:6809-95-6** **HR: 3**
BICYCLO(4.2.0)OCTA-1,3,5-TRIEN-7-YL BENZYL KETONE
mf: $C_{16}H_{14}O$ mw: 222.30

SYNS: BENZYL BICYCLO(4.2.0)OCTA-1,3,5-TRIEN-7-YL KETONE □ KETONE, BICYCLO(4.2.0)OCTA-1,3,5-TRIEN-7-YL BENZYL

TOXICITY DATA WITH REFERENCE
ipr-mus LD50:1750 mg/kg JMCMAR 9,656,66

DOT CLASSIFICATION: 3; *Label:* Flammable Liquid

SAFETY PROFILE: Moderately toxic by intraperitoneal route. A flammable liquid. When heated to decomposition it emits acrid smoke and irritating vapors.

BFZ110 **CAS:6813-90-7** **HR: 3**
BICYCLO(4.2.0)OCTA-1,3,5-TRIEN-7-YL BENZYL KETONE OXIME
mf: $C_{16}H_{15}NO$ mw: 237.32

SYNS: BENZYL BICYCLO(4.2.0)OCTA-1,3,5-TRIEN-7-YL KETONE OXIME □ KETONE, BICYCLO(4.2.0)OCTA-1,3,5-TRIEN-7-YL BENZYL, OXIME

TOXICITY DATA WITH REFERENCE
ipr-mus LD50:550 mg/kg JMCMAR 9,656,66

DOT CLASSIFICATION: 3; *Label:* Flammable Liquid

SAFETY PROFILE: Moderately toxic by intraperitoneal route. A flammable liquid. When heated to decomposition it emits toxic vapors of NO_x.

BFZ120 **CAS:1075-30-5** **HR: 3**
BICYCLO(4.2.0)OCTA-1,3,5-TRIEN-7-YL METHYL KETONE
mf: $C_{10}H_{10}O$ mw: 146.20

SYNS: BICYCLO(4.2.0)OCTA-1,3,5-TRIENE, 7-ACETYL- □ KETONE, BICYCLO(4.2.0)OCTA-1,3,5-TRIEN-7-YL METHYL

TOXICITY DATA WITH REFERENCE
ipr-mus LD50:550 mg/kg JMCMAR 9,656,66

DOT CLASSIFICATION: 3; *Label:* Flammable Liquid

SAFETY PROFILE: Moderately toxic by intraperitoneal route. A flammable liquid. When heated to decomposition it emits acrid smoke and irritating vapors.

BFZ130 **CAS:6813-93-0** **HR: 3**
BICYCLO(4.2.0)OCTA-1,3,5-TRIEN-7-YL METHYL KETONE O-ACETYLOXIME
mf: $C_{12}H_{13}NO_2$ mw: 203.26

SYNS: O-ACETYL-1-ACETYLBENZOCYCLOBUTENE OXIME □ KETONE, BICYCLO(4.2.0)OCTA-1,3,5-TRIEN-7-YL METHYL, O-ACETYLOXIME

TOXICITY DATA WITH REFERENCE
ipr-mus LD50:550 mg/kg JMCMAR 9,656,66

DOT CLASSIFICATION: 3; *Label:* Flammable Liquid

SAFETY PROFILE: Moderately toxic by intraperitoneal route. A flammable liquid. When heated to decomposition it emits toxic vapors of NO_x.

BFZ140 **CAS:6813-95-2** **HR: 3**
BICYCLO(4.2.0)OCTA-1,3,5-TRIEN-7-YL METHYL KETONE O-ALLYLOXIME
mf: $C_{13}H_{15}NO$ mw: 201.29

SYN: KETONE, BICYCLO(4.2.0)OCTA-1,3,5-TRIEN-7-YL METHYL, O-ALLYLOXIME

TOXICITY DATA WITH REFERENCE
ipr-mus LD50:1320 mg/kg JMCMAR 9,656,66

DOT CLASSIFICATION: 3; *Label:* Flammable Liquid

SAFETY PROFILE: Moderately toxic by intraperitoneal route. A flammable liquid. When heated to decomposition it emits toxic vapors of NO_x.

BFZ150 **CAS:7315-27-7** **HR: 3**
BICYCLO(4.2.0)OCTA-1,3,5-TRIEN-7-YL METHYL KETONE O-BUTYLOXIME
mf: $C_{14}H_{19}NO$ mw: 217.34

SYN: KETONE, BICYCLO(4.2.0)OCTA-1,3,5-TRIEN-7-YL METHYL, O-BUTYLOXIME

TOXICITY DATA WITH REFERENCE
ipr-mus LD50:1750 mg/kg JMCMAR 9,656,66

DOT CLASSIFICATION: 3; *Label:* Flammable Liquid

SAFETY PROFILE: Moderately toxic by intraperitoneal route. A flammable liquid. When heated to decomposition it emits toxic vapors of NO_x.

BFZ160 **CAS:3264-31-1** **HR: 3**
BICYCLO(4.2.0)OCTA-1,3,5-TRIEN-7-YL METHYL KETONE OXIME
mf: $C_{10}H_{11}NO$ mw: 161.22

SYNS: 1-ACETYLBENZOCYCLOBUTENE OXIME □ KETONE, BICYCLO(4.2.0)OCTA-1,3,5-TRIEN-7-YL METHYL, OXIME

TOXICITY DATA WITH REFERENCE
ipr-mus LD50:550 mg/kg JMCMAR 9,656,66

DOT CLASSIFICATION: 3; *Label:* Flammable Liquid

SAFETY PROFILE: Moderately toxic by intraperitoneal route. A flammable liquid. When heated to decomposition it emits toxic vapors of NO_x.

B

BFZ170 **CAS:73747-51-0** **HR: 3**
BICYCLO(4.2.0)OCTA-1,3,5-TRIEN-7-YL PENTYL KETONE OXIME
mf: $C_{13}H_{17}NO$ mw: 203.31

SYNS: 1-BICYCLO(4.2.0)OCTA-1,3,5-TRIEN-7-YL-1-PENTANONE OXIME □ KETONE, BICYCLO(4.2.0)OCTA-1,3,5-TRIEN-7-YL PENTYL, OXIME □ PENTANONE, 1-BENZOCYCLOBUTYL-, OXIME □ 1-PENTANONE, 1-BICYCLO(4.2.0)OCTA-1,3,5-TRIEN-7-YL-, OXIME

TOXICITY DATA WITH REFERENCE
ipr-mus LD50:550 mg/kg JMCMAR 9,656,66

DOT CLASSIFICATION: 3; *Label:* Flammable Liquid

SAFETY PROFILE: Moderately toxic by intraperitoneal route. A flammable liquid. When heated to decomposition it emits toxic vapors of NO_x.

BFZ180 **CAS:6809-94-5** **HR: 3**
BICYCLO(4.2.0)OCTA-1,3,5-TRIEN-7-YL PHENYL KETONE
mf: $C_{15}H_{12}O$ mw: 208.27

SYNS: BICYCLO(4.2.0)OCTA-1,3,5-TRIENE, 7-BENZOYL- □ KETONE, BICYCLO(4.2.0)OCTA-1,3,5-TRIEN-7-YL PHENYL

TOXICITY DATA WITH REFERENCE
ipr-mus LD50:1250 mg/kg JMCMAR 9,656,66

DOT CLASSIFICATION: 3; *Label:* Flammable Liquid

SAFETY PROFILE: Moderately toxic by intraperitoneal route. A flammable liquid. When heated to decomposition it emits acrid smoke and irritating vapors.

BGA250 **CAS:81-21-0** **HR: 0**
BICYCLOPENTADIENE DIOXIDE
mf: $C_{10}H_{12}O_2$ mw: 164.22

SYNS: DICYCLOPENTADIENE DIEPOXIDE □ DICYCLOPENTADIENE DIOXIDE □ 1,2:5,6-DIEPOXYHEXAHYDRO-4,7-METHANOINDAN □ 1,2:5,6-DIEPOXY-3a,4,5,6,7,7a-HEXAHYDRO-4,7-METHANOINDAN □ UNOX 207 □ UNOX EPOXIDE 207 □ UNOX 207X

TOXICITY DATA WITH REFERENCE
skn-rbt 500 mg open MLD UCDS** 10/5/61
orl-rat LD50:210 mg/kg UCDS** 10/5/61
ivn-mus LD50:56 mg/kg CSLNX* NX#04159
skn-rbt LD50:8000 mg/kg UCDS** 10/5/61

CONSENSUS REPORTS: Reported in EPA TSCA Inventory.

SAFETY PROFILE: Poison by ingestion and intravenous routes. Mildly toxic by skin contact. A skin irritant. When heated to decomposition it emits acrid smoke and irritating fumes.

BGA650 **CAS:5164-35-2** **HR: 3**
BICYCLO(2.1.0)PENT-2-ENE
mf: C_5H_6 mw: 66.10

PROP: Unstable liquid.

SAFETY PROFILE: This strained ring compound may explode spontaneously. When heated to decomposition it emits acrid smoke and fumes.

BGA750 **CAS:1464-53-5** **HR: 3**
1,1′-BI(ETHYLENE OXIDE)
mf: $C_4H_6O_2$ mw: 86.10

PROP: Colorless liquid. Bp: 142°, mp: 19°, d: 1.113 @ 18°/4°.

SYNS: BIOXIRANE □ 2,2′-BIOXIRANE □ BUTADIENDIOXYD (GERMAN) □ BUTADIENE DIEPOXIDE □ 1,3-BUTADIENE DIEPOXIDE □ BUTADIENE DIOXIDE □ BUTANE DIEPOXIDE □ DEB □ DIEPOXYBUTANE □ 2,4-DIEPOXYBUTANE □ 1,2:3,4-DIEPOXYBUTANE □ DIOXYBUTADIENE □ ENT 26,592 □ ERYTHRITOL ANHYDRIDE □ RCRA WASTE NUMBER U085

TOXICITY DATA WITH REFERENCE
skn-rbt 10 mg/24H open SEV AMIHBC 10,61,54
skn-rbt 50 mg open SEV UCDS** 4/25/58
eye-rbt 250 μg open SEV AMIHBC 10,61,54
eye-rbt 3 ppm/3D JPCAAC 10,17,60
mmo-asn 20 μmol/L MUREAV 132,161,84
cyt-hmn:bmr 100 μg/L CGCYDF 9,51,83
sce-mus-ivn 193 μmol/kg MUREAV 108,251,83
sce-ham:lng 1 mg/L CNREA8 44,3270,84
ipr-rat TDLo:380 mg/kg/13W-I:ETA BJPCAL 6,235,51
skn-mus TDLo:95 g/kg/78W-I:ETA AIHAAP 24,305,63
unk-mus TDLo:3400 mg/kg:ETA RARSAM 3,193,63
orl-rat LD50:78 mg/kg AMIHBC 10,61,54
ihl-rat LC50:90 ppm/4H SCCUR* -,2,61
orl-mus LD50:72 mg/kg SCCUR* -,2,61
ipr-mus LD50:31 mg/kg AEPPAE 230,559,57
skn-rbt LD50:80 mg/kg AMIHBC 10,61,54

CONSENSUS REPORTS: NTP 7th Annual Report On Carcinogens. EPA Extremely Hazardous Substances List. EPA Genetic Toxicology Program. Community Right-To-Know List. Reported in EPA TSCA Inventory.

SAFETY PROFILE: Confirmed carcinogen with experimental tumorigenic data. Poison by ingestion, inhalation, skin contact, and intraperitoneal routes. Human mutation data reported. A severe skin and eye irritant. When heated to decomposition it emits acrid smoke and irritating fumes.

BGA825 **HR: 3**
BIFONAZOLE
mf: $C_{22}H_{18}N_2$ mw: 310.42

SYNS: BAY H 4502 □ BIFONAZOL □ 1-(α-(4-BIPHENYLYL)BENZYL)IMIDAZOLE □ 1-((4-BIPHENYLYL)PHENYLMETHYL)-1H-IMIDAZOLE □ MYCOSPOR

TOXICITY DATA WITH REFERENCE
orl-rat TDLo:1160 mg/kg (female 16-22D post):REP ARZNAD 33,739,83
orl-mus TDLo:1200 mg/kg (male 1D pre):REP ARZNAD 33,739,83
orl-rbt TDLo:390 mg/kg (female 6-18D post):REP ARZNAD 33,739,83
scu-rat TDLo:1050 mg/kg (female 35D pre):REP OYYAA2 28,23,84
scu-rat TDLo:10,500 mg/kg (female 35D pre):REP OYYAA2 28,23,84
scu-rat TDLo:10,500 mg/kg (male 35D pre):REP OYYAA2 28,23,84
scu-rat TDLo:1050 mg/kg (male 35D pre):REP OYYAA2 28,23,84

orl-rat LD50:1463 mg/kg OYYAA2 28,23,84
ivn-rat LD50:63 mg/kg OYYAA2 28,23,84
orl-mus LD50:2629 mg/kg
ivn-mus LD50:57 mg/kg OYYAA2 28,23,84
orl-rbt LD50:4000 mg/kg ARZNAD 33,739,83

SAFETY PROFILE: Poison by intravenous route. Moderately toxic by ingestion. Experimental reproductive effects. When heated to decomposition it emits toxic fumes of NO_x.

BGB250 **CAS:37247-90-8** **HR: 3**
BIHOROMYCIN (crystalline)
mf: $C_{41}H_{76}O_{13}$ mw: 777.17

PROP: Crystals from hexane. Mp: 130°.

TOXICITY DATA WITH REFERENCE
orl-mus LD50:7 mg/kg 85ERAY 2,1077,78
ipr-mus LD50:19 mg/kg 85ERAY 2,1077,78

SAFETY PROFILE: Poison by ingestion and intraperitoneal routes. When heated to decomposition it emits acrid smoke and irritating fumes.

BGB315 **CAS:3521-84-4** **HR: 2**
BILIGRAFIN FORTE
mf: $C_{20}H_{14}I_6N_2O_6•2C_7H_{17}NO_5$ mw: 1530.26

SYNS: ADIPIODONE MEGLUMINE □ CAVUMBREN □ CHOLOGRAF-N-METHYLGLUCAMINE □ ENDOCISTOBIL □ ENDOGRAFIN □ ENDOGRAPHIN □ INTRABLIX □ IODIPAMIDE MEGLUMINE □ IODIPAMIDE MEGLUMINE SALT □ IODIPAMIDE METHYLGLUCAMINE SALT □ MEGLUMINE IODIPAMIDE □ METHYL GLUCAMINE BILIGRAFIN □ METHYL GLUCAMINE IODIPAMIDE □ ULTRABIL

TOXICITY DATA WITH REFERENCE
ivn-rat LD50:5000 mg/kg NIIRDN 6,5,82
par-rat LD50:1921 mg/kg FRPSAX 28,1011,73
ivn-mus LD50:3195 mg/kg INVRAV 15(Suppl),142,80
ivn-dog LD50:1200 mg/kg FRPSAX 28,996,73
par-rbt LD50:1446 mg/kg FRPSAX 28,1011,73

SAFETY PROFILE: Moderately toxic by several routes. When heated to decomposition it emits toxic fumes of I^- and NO_x.

BGB325 **HR: 2**
BILIGRAFIN SODIUM
mf: $C_{20}H_{12}I_6N_2O_6•2Na$ mw: 1183.72

SYNS: ADIPIC ACID DI-(3-CARBOXY-2,4,6-TRIIODOANILIDE) DISODIUM □ ADIPINSAEURE-DI-(3-CARBOXY-2,4,6-TRIJOD-ANILID) DINATRIUM (GERMAN) □ BILIGRAFIN NATRIUM (GERMAN)

TOXICITY DATA WITH REFERENCE
ivn-rat LD50:3800 mg/kg ARZNAD 14,451,64
ivn-mus LD50:3200 mg/kg ARZNAD 14,451,64
ivn-dog LD50:1900 mg/kg ARZNAD 14,451,64

SAFETY PROFILE: Moderately toxic by intravenous route. When heated to decomposition it emits toxic fumes of I^-, NO_x, and Na_2O.

BGB350 **HR: 1**
BILIVISTAN SODIUM
mf: $C_{18}H_8I_6N_2O_7•2Na$ mw: 1171.66

SYNS: BILIVISTAN NATRIUM (GERMAN) □ DIGLYCOLIC ACID DI-(3-CARBOXY-2,4,6-TRIIODOANILIDE) DISODIUM □ DIGLYCOLSAEURE-DI-(3-CARBOXY-2,4,6-TRIJOD-ANILID) DINATRIUM (GERMAN) □ 3,3'-(OXYDIMETHYLENEBIS(CARBONYLIMINO))BIS(2,4,6-TRIIODOBENZOIC ACID DISODIUM SALT)

TOXICITY DATA WITH REFERENCE
ivn-rat LD50:6000 mg/kg ARZNAD 14,451,64
ivn-mus LD50:5300 mg/kg ARZNAD 14,451,64
ivn-dog LD50:5500 mg/kg ARZNAD 14,451,64

SAFETY PROFILE: Mildly toxic by intravenous route. When heated to decomposition it emits toxic fumes of I^-, NO_x, and Na_2O.

BGB400 **CAS:55268-74-1** **HR: 3**
BILTRICIDE
mf: $C_{19}H_{24}N_2O_2$ mw: 312.45

PROP: Crystals from EtOAc/hexane. Mp: 136–138°. Solubility (g/100 mL): ethanol 9.7; chloroform 56.7; water 0.04.

SYNS: CESOL □ 2-CYCLOHEXYLCARBONYL-1,2,3,6,7,11b-HEXAHYDRO-4H-PYRAZINO(2,1-a)ISOQUINOLIN-4-ONE □ DRONCIT □ EMBAY 8440 □ PRAZIQUANTEL □ PYQUITON

TOXICITY DATA WITH REFERENCE
mmo-sat 150 mg/L ENMUDM 2,234,80
dni-hmn:hla 5 mmol/L MUREAV 93,447,82
bfa-mus/sat 1200 mg/kg CNREA8 38,4478,78
msc-ham:lng 10 mg/L CNREA8 42,2692,82
orl-rat LD50:2840 mg/kg ARZNAD 31,555,81
ipr-rat LD50:586 mg/kg IYKEDH 20,228,89
orl-mus LD50:2454 mg/kg ARZNAD 31,555,81
ipr-mus LD50:376 mg/kg IYKEDH 20,228,89
scu-mus LD50:7172 mg/kg ARZNAD 31,555,81
orl-rbt LD50:1050 mg/kg ARZNAD 31,555,81

CONSENSUS REPORTS: EPA Genetic Toxicology Program.

SAFETY PROFILE: Poison by intraperitoneal route. Moderately toxic by ingestion and other routes. Human mutation data reported. When heated to decomposition it emits toxic fumes of NO_x.

BGB500 **CAS:485-31-4** **HR: 3**
BINAPACRYL
mf: $C_{15}H_{18}N_2O_6$ mw: 322.35

PROP: Crystals or powder. Mp: 65–69°. Very sltly sol in H_2O; sol in most org solvs.

SYNS: ACRICID □ AMBOX □ 2-sec-BUTYL-4,6-DINITROPHENYL-3,3-DIMETHYLACRYLATE □ 2-sec-BUTYL-4,6-DINITROPHENYL-3-METHYL-2-BUTENOATE □ 2-sec-BUTYL-4,6-DINITROPHENYL-3-METHYLCROTONATE □ 2-sec-BUTYL-4,5-DINITROPHENYL SENECIOATE □ DAPACRYL □ 3,3-DIMETHYL-ACRYLATE de 2,4-DINITRO-6-(1-METHYLPROPYLE) PHENYLE (FRENCH) □ 3,3-DIMETHYLACRYLIC ACID 2-sec-BUTYL-4,5-DINITROPHENYL ESTER □ DINAPACRYL □ 4,6-DINITRO-2-sec-BUTYLPHENYL β,β-DIMETHYLACRYLATE □ 2,4-DINITRO-6-sec-BUTYLPHENYL-2-METHYLCROTONATE □ 4,6-DINITROPHENYL-2-sec-BUTYL-3-METHYL-2-BUTENONATE □ DINOSEB METHACRYLATE □

ENDOSAN □ ENT 25,793 □ FMC 9044 □ HOE 2784 □ 3-METHYLCROTONIC ACID 2-sec-BUTYL-4,6-DINITROPHENYL ESTER □ (6-(1-METHYL-PROPYL)-2,4-DINITRO-FENYL)-3,3-DIMETHYL ACRYLAAT (DUTCH) □ (6-(1-METHYL-PROPYL)-2,4-DINITRO-PHENYL)-3,3-DIMETHYL ACRYLAT (GERMAN) □ 2-(1-METHYLPROPYL)-4,6-DINITROPHENYL-β,β-DIMETHACRYLATE □ (6-(1-METIL-PROPIL)-2,4-DINITROFENIL)-3,3-DIMETIL-ACRILATO (ITALIAN) □ MOROCIDE □ MORROCID □ NIA 9044 □ NIAGARA 9044

TOXICITY DATA WITH REFERENCE
mmo-sat 5 mg/plate MUREAV 116,185,83
orl-rat LD50:58 mg/kg TXAPA9 14,515,69
skn-rat LD50:720 mg/kg WRPCA2 9,119,70
orl-mus LD50:1600 mg/kg TXAPA9 7,353,65
skn-mus LD50:750 mg/kg PEMNDP 8,73,87
orl-dog LD50:50 mg/kg GUCHAZ 6,42,73
skn-rbt LD50:750 mg/kg GUCHAZ 6,42,73
orl-gpg LD50:200 mg/kg TXAPA9 7,353,65

SAFETY PROFILE: Poison by ingestion. Moderately toxic by skin contact. Mutation data reported. A cholinesterase inhibitor. When heated to decomposition it emits fumes of NO_x. See also PARATHION and PHOSPHORUS COMPOUNDS.

BGB750 CAS:4488-22-6 HR: 2
(1,1′-BINAPHTHALENE)-2,2′-DIAMINE
mf: $C_{20}H_{16}N_2$ mw: 284.38

PROP: Silvery plates from EtOH. Mp: 191°.

SYN: 2,2′-DIAMINO-1,1′-DINAPHTHYL

TOXICITY DATA WITH REFERENCE
skn-mus TDLo:590 mg/kg/25W-I:ETA AJCAA7 40,62,40
scu-mus TDLo:2700 mg/kg/27W-I:ETA AJCAA7 40,62,40

SAFETY PROFILE: Questionable carcinogen with experimental tumorigenic data. When heated to decomposition it emits toxic fumes of NO_x.

BGC000 CAS:795-95-9 HR: 2
(1,2′-BINAPHTHALENE)-1,2′-DIAMINE
mf: $C_{20}H_{16}N_2$ mw: 284.38

SYN: 1:2′-DIAMINO-1′:2-DINAPHTHYL

TOXICITY DATA WITH REFERENCE
skn-mus TDLo:2880 mg/kg/60W-I:ETA PRLBA4 131,170,72

SAFETY PROFILE: Questionable carcinogen with experimental tumorigenic data by skin contact. When heated to decomposition it emits toxic fumes of NO_x.

BGC100 CAS:602-09-5 HR: 3
(1,1′-BINAPHTHALENE)-2,2′-DIOL
mf: $C_{20}H_{14}O_2$ mw: 286.34

SYNS: β-BINAPHTHOL □ 1,1′-BI-2-NAPHTHOL □ BIS-β-NAPHTHOL □ 2,2′-DIHYDROXYBINAPHTHALENE □ 2,2′-DIHYDROXYDINAPHTHYL □ 2,2′-DINAPHTHOL

TOXICITY DATA WITH REFERENCE
orl-mus LDLo:42 mg/kg AECTCV 14,111,85

CONSENSUS REPORTS: Reported in EPA TSCA Inventory.

SAFETY PROFILE: Poison by ingestion. When heated to decomposition it emits acrid smoke and irritating vapors.

BGC250 CAS:69382-20-3 HR: 3
BINDON ETHYL ETHER
mf: $C_{20}H_{14}O_3$ mw: 302.34

SYNS: BINDON ATHYLATHER □ 2-(3-ETHOXY-1-INDANYLIDENE)-1,3-DINDANDIONE

TOXICITY DATA WITH REFERENCE
ipr-mus TDLo:2500 μg/kg (9D-preg):REP ARTODN 33,191,75
ipr-mus TDLo:5 mg/kg (9D preg):REP ARTODN 33,191,75
ipr-mus TDLo:5 mg/kg (9D preg):TER ARTODN 33,191,75
ipr-mus LDLo:40 mg/kg ARTODN 33,191,75

SAFETY PROFILE: Poison by intraperitoneal route. An experimental teratogen. Other experimental reproductive effects. When heated to decomposition it emits acrid smoke and irritating fumes. See also ETHERS.

BGC500 CAS:57647-35-5 HR: 3
BINODALINE HYDROCHLORIDE
mf: $C_{19}H_{23}N_3 \cdot ClH$ mw: 329.87

PROP: Solid. Mp: 188–190°.

SYNS: BINODALIN HYDROCHLORID (GERMAN) □ 1 (DIMETHYLAMINOETHYL-METHYL)AMINO 3-PHENYLINDOLE HYDROCHLORIDE □ 1-(ω-DIMETHYLAMINOETHYLMETHYL)AMINO-3-PHENYLINDOLE HYDROCHLORIDE □ SGD-SCHA 1059 □ N,N,N′-TRIMETHYL-N′-(3-PHENYL-1H-INDOL-1-YL)-1,2-ETHANEDIAMINE MONOHYDROCHLORIDE

TOXICITY DATA WITH REFERENCE
orl-rat LD50:1160 mg/kg ARZNAD 33,726,83
ivn-rat LD50:26 mg/kg ARZNAD 22,726,83
orl-mus LD50:760 mg/kg ARZNAD 33,726,83
ivn-mus LD50:54 mg/kg ARZNAD 33,726,83
ivn-cat LDLo:50,800 μg/kg ARZNAD 33,726,83

SAFETY PROFILE: Poison by intravenous route. Moderately toxic by ingestion. When heated to decomposition it emits very toxic fumes of HCl and NO_x.

BGC625 CAS:6620-60-6 HR: 1
BINOSIDE
mf: $C_{18}H_{26}N_2O_4$ mw: 334.46

PROP: Crystals. Mp: 142–145°.

SYNS: dl-4-BENZAMIDO-N,N-DIPROPYLGLUTARAMIC ACID □ (±)-4-(BENZOYLAMINO)-5-(DIPROPYLAMINO)-5-OXO-PENTANOIC ACID □ CR 242 □ 242 DL □ GASTRIDENE □ GASTROTOPIC □ MIDELID □ MILID □ MILIDE □ NULSA □ PROGLUMIDE □ PROMIDE (parasympatholytic) □ ULCUTIN □ W 5219 □ XYDE □ XYLAMIDE □ XYLAMIDE (gastroprotective agent)

TOXICITY DATA WITH REFERENCE
orl-rat TDLo:1350 mg/kg (9-14D preg):TER OYYAA2 5,225,71
orl-rat LD50:20 g/kg NIIRDN 6,722,82
ipr-rat LD50:1420 mg/kg NIIRDN 6,722,82

orl-mus LD50:8070 mg/kg MIMEAO 58,3653,67
ipr-mus LD50:1480 mg/kg NIIRDN 6,722,82
ivn-mus LD50:2250 mg/kg MIMEAO 58,3653,67

SAFETY PROFILE: Moderately toxic by several routes. An experimental teratogen. When heated to decomposition it emits toxic fumes of NO_x.

BGC750 **HR: 3**
BIOALLETHRIN
mf: $C_{19}H_{26}O_3$ mw: 302.45

SYNS: d-trans ALLETHRIN □ ALLYL HONOLOG of CINERIN I □ BIOALETRINA (PORTUGUESE) □ (+)-trans-CHRYSANTHEMUMIC ACID ESTER of (+ −)-ALLETHROLONE □ ENT 16,275

TOXICITY DATA WITH REFERENCE
orl-rat LD50:425 mg/kg SPEADM 78-1,7,78
ivn-rat LDLo:4 mg/kg BIOGAL 41(10),283,75
orl-mus LD50:330 mg/kg EVHPAZ 14,15,76

CONSENSUS REPORTS: EPA Genetic Toxicology Program. Reported in EPA TSCA Inventory.

SAFETY PROFILE: Poison by ingestion and intravenous routes. When heated to decomposition it emits acrid and irritating fumes. An insecticide. See other allethrin entries and ALLYL COMPOUNDS.

BGC825 **HR: 2**
BIODIASTASE 1000

PROP: Extracted from *Aspergillus* (KSRNAM 8,2378,74).

TOXICITY DATA WITH REFERENCE
orl-rat TDLo:600 g/kg (30D pre):REP KSRNAM 8,2378,74
orl-rat TDLo:1820 g/kg (26W male):REP KSRNAM 8,2378,74
ipr-rat LD50:678 mg/kg KSRNAM 8,2378,74
scu-rat LD50:7320 mg/kg KSRNAM 8,2378,74
ipr-mus LD50:445 mg/kg KSRNAM 8,2378,74
scu-mus LD50:2220 mg/kg KSRNAM 8,2378,74

SAFETY PROFILE: Moderately toxic by intraperitoneal and subcutaneous routes. Experimental reproductive effects. When heated to decomposition it emits acrid smoke and irritating fumes.

BGD000 **CAS:5697-56-3** **HR: 3**
BIOGASTRONE
mf: $C_{34}H_{50}O_7$ mw: 570.84

PROP: Cream crystals. Mp: 291–294°.

SYNS: BIORAL □ CARBENOXOLONE □ 3-β-(3-CARBOXYPROPIONYLOXY)-11-OXO-OLEAN-12-EN-30-OIC ACID □ 3-β-HYDROXY-11-OXO-OLEAN-12-EN-30-OIC ACID, HYDROGEN SUCCINATE

TOXICITY DATA WITH REFERENCE
orl-man TDLo:120 mg/kg/56D-I BMJOAE 2,150,76
orl-rat LD50:2450 mg/kg OYYAA2 19,323,80
ipr-rat LD50:128 mg/kg OYYAA2 19,323,80
scu-rat LD50:1720 mg/kg OYYAA2 19,323,80
scu-dog LD50:1060 mg/kg OYYAA2 19,323,80
ivn-dog LD50:371 mg/kg OYYAA2 19,323,80

CONSENSUS REPORTS: Reported in EPA TSCA Inventory.

SAFETY PROFILE: Poison by intravenous and intraperitoneal routes. Moderately toxic by ingestion and subcutaneous routes. Human systemic effects by ingestion: muscle weakness and flaccid paralysis. When heated to decomposition it emits acrid smoke and fumes.

BGD250 **CAS:20354-26-1** **HR: 2**
BIOXONE
mf: $C_9H_6Cl_2N_2O_3$ mw: 261.07

PROP: Light-tan solid. Mp: 123–124°. Very sltly sol in H_2O.

SYNS: 2-(3,4-DICHLOROPHENYL)-4-METHYL-1,2,4-OXADIAZOLIDINE-3,5-DIONE □ METHAZOLE □ OXYDIAZOL □ PAXILON □ PROBE □ TUNIC □ VCS 438

TOXICITY DATA WITH REFERENCE
orl-rat LD50:2501 mg/kg FMCHA2 -,C195,83

SAFETY PROFILE: Moderately toxic by ingestion. When heated to decomposition it emits very toxic fumes of Cl^- and NO_x.

BGD500 **CAS:514-65-8** **HR: 3**
BIPERIDEN
mf: $C_{21}H_{29}NO$ mw: 311.51

SYNS: AKINETON □ AKINOPHYL □ BEPERIDEN □ α-(BICYCLO(2.2.1)HEPT-5-EN-2-YL)-α-PHENYL-1-PIPERIDINO PROPANOL □ 1-BICYCLOHEPTENYL-1-PHENYL-3-PIPERIDINO-PROPANOL-1 □ KL 373 □ 3-PIPERIDINO-1-PHENYL-1-BICYCLOHEPTENYL-1-PROPANOL □ 3-PIPERIDINO-1-PHENYL-1-BICYCLO(2.2.1)HEPTEN-(5)-YL-PROPANOL-(1)(GERMAN)

TOXICITY DATA WITH REFERENCE
orl-rat LD50:750 mg/kg NIIRDN 6,636,82
orl-mus LD50:530 mg/kg NIIRDN 6,636,82
ipr-mus LD50:161 mg/kg NIIRDN 6,636,82
orl-dog LD50:340 mg/kg NIIRDN 6,636,82
scu-mus LD50:195 mg/kg AIPTAK 128,204,60
ivn-mus LD50:56 mg/kg AIPTAK 128,204,60

SAFETY PROFILE: Poison by ingestion, subcutaneous, intraperitoneal, and intravenous routes. When heated to decomposition, it emits toxic fumes of NO_x.

BGD750 **CAS:1235-82-1** **HR: 3**
BIPERIDINE HYDROCHLORIDE
mf: $C_{21}H_{29}NO•ClH$ mw: 347.97

SYNS: AKINETON HYDROCHLORIDE □ AKINOPHYL □ α-BICYCLO(2.2.1)HEPT-5-EN-1-YL-α-PHENYL-PIPERIDINEPROPANOL HYDROCHLORIDE □ α-(BICYCLO(2.2.1)HEPT-5-EN-2-YL)-α-PHENYL-1-PIPERIDINEPROPANOL HYDROCHLORIDE □ 1-BICYCLOHEPTENYL-1-PHENYL-3-PIPERIDINOPROPANOL-1 HYDROCHLORIDE □ BIPERIDEN HYDROCHLORIDE □ α-5-NORBORNEN-2-YL-α-PHENYL-PIPERIDINE PROPANOL HYDROCHLORIDE

TOXICITY DATA WITH REFERENCE
orl-rat LD50:750 mg/kg TXAPA9 2,379,60
orl-mus LD50:545 mg/kg 29ZVAB -,17,69
ivn-mus LD50:56 mg/kg MEIEDD 10,175,83
orl-dog LD50:340 mg/kg TXAPA9 2,379,60

SAFETY PROFILE: Poison by ingestion and intravenous routes. When heated to decomposition it emits very toxic fumes of NO_x and HCl. See also BIPERIDIN.

BGE000 CAS:92-52-4 HR: 3
BIPHENYL
mf: $C_{12}H_{10}$ mw: 154.22

PROP: Monoclinic, white scales, with a pleasant odor. Mp: 71°, bp: 255°, flash p: 235°F (CC), d: 0.991 @ 75°/4°, autoign temp: 1004°F, vap d: 5.31, lel: 0.6% @ 232°, uel: 5.8% @ 331°F.

SYNS: BIBENZENE □ 1,1′-BIPHENYL □ DIPHENYL (OSHA) □ LEMONENE □ PHENADOR-X □ PHENYLBENZENE □ PHPH □ XENENE

TOXICITY DATA WITH REFERENCE
eye-rbt 100 mg MLD MONS** 99,37,89
sce-ham:fbr 100 μmol/L JNCIAM 58,1635,77
orl-mus TDLo:56 g/kg:ETA NTIS** PB223-159
scu-mus TDLo:46 mg/kg:NEO NTIS** PB223-159
ihl-hmn TCLo:4400 μg/m³:IRR AEHLAU 26,70,73
orl-rat LD50:2400 mg/kg MONS** 24,268,83
ivn-mus LD50:56 mg/kg CSLNX* NX#00198
orl-rbt LD50:2400 mg/kg NASDA6 28,983,77

CONSENSUS REPORTS: EPA Genetic Toxicology Program. Reported in EPA TSCA Inventory. Community Right-To-Know List.

OSHA PEL: TWA 0.2 ppm
ACGIH TLV: TWA 0.2 ppm
DFG MAK: 0.2 ppm (1 mg/m³)

SAFETY PROFILE: Poison by intravenous route. Moderately toxic by ingestion. A powerful irritant by inhalation in humans. Human systemic effects by inhalation of very small amounts: flaccid paralysis, nausea or vomiting, and other unspecified gastrointestinal effects. Questionable carcinogen with experimental tumorigenic and neoplastigenic data. Mutation data reported. Combustible when exposed to heat or flame; can react with oxidizing materials. To fight fire, use CO_2, dry chemical, water spray, mist, fog. When heated to decomposition it emits acrid smoke and fumes.

For occupational chemical analysis use NIOSH: Biphenyl, 2530.

BGE125 CAS:5728-52-9 HR: 3
4-BIPHENYLACETIC ACID
mf: $C_{14}H_{12}O_2$ mw: 212.25

PROP: Crystals from AcOH. Mp: 164–165°.

SYNS: (4-BIPHENYL) ACETIC ACID □ (1,1′-BIPHENYL)-4-ACETIC ACID □ p-BIPHENYLACETIC ACID □ 4-BIPHENYLACETIC ACID □ 4-CARBOXYMETHYLBIPHENYL □ LY 61017

TOXICITY DATA WITH REFERENCE
scu-rat TDLo:540 mg/kg (17-22D preg/21D post):REP KSRNAM 20,2153,86
scu-rat TDLo:360 mg/kg (female 6-17D post):TER KSRNAM 20,2142,86
scu-rat TDLo:240 mg/kg (6-17D preg):TER KSRNAM 20,2142,86
orl-rat LD50:410 mg/kg KSRNAM 20,2107,86
ipr-rat LD50:495 mg/kg KSRNAM 20,2107,86
scu-rat LD50:148 mg/kg KSRNAM 20,2107,86
orl-mus LD50:675 mg/kg KSRNAM 20,2107,86
ipr-mus LD50:508 mg/kg KSRNAM 20,2107,86
scu-mus LD50:730 mg/kg KSRNAM 20,2107,86
scu-dog LD50:320 mg/kg KSRNAM 20,2107,86
scu-rbt LD50:1280 mg/kg KSRNAM 20,2107,86

SAFETY PROFILE: Poison by subcutaneous route. Moderately toxic by ingestion and intraperitoneal routes. An experimental teratogen. Other experimental reproductive effects. When heated to decomposition it emits acrid smoke and fumes.

BGE250 CAS:90-41-5 HR: 2
2-BIPHENYLAMINE
mf: $C_{12}H_{11}N$ mw: 169.24

PROP: Crystals from EtOH (aq). Mp: 49–50°, bp: 299°.

SYNS: o-AMINOBIPHENYL □ 2-AMINOBIPHENYL □ o-AMINODIPHENYL □ 2-AMINODIPHENYL □ o-BIPHENYLAMINE □ (1,1′-BIPHENYL)-2-AMINE (9CI) □ o-PHENYLANILINE □ 2-PHENYLANILINE

TOXICITY DATA WITH REFERENCE
mmo-sat 33 μg/plate ENMUDM 5(Suppl 1),3,83
pic-esc 250 mg/L CNREA8 41,532,81
orl-rat LD50:2340 mg/kg JIHTAB 29,1,47
orl rbt LD50:1020 mg/kg JIHTAB 29,1,47

CONSENSUS REPORTS: EPA Genetic Toxicology Program. Reported in EPA TSCA Inventory.

SAFETY PROFILE: Moderately toxic by ingestion. Mutation data reported. When heated to decomposition, it emits toxic fumes of NO_x. See also AROMATIC AMINES.

BGE300 HR: 2
4-BIPHENYLAMINE, DIHYDROCHLORIDE
mf: $C_{12}H_{11}N \cdot 2ClH$ mw: 242.16

SYN: 4-AMINOBIPHENYL DIHYDROCHLORIDE

TOXICITY DATA WITH REFERENCE
orl-rat TDLo:2238 mg/kg/31W-I:CAR JJCREP 76,570,85

SAFETY PROFILE: Questionable carcinogen with experimental carcinogenic data. When heated to decomposition it emits toxic fumes of NO_x and HCl.

BGE325 CAS:2185-92-4 HR: 2
2-BIPHENYLAMINE, HYDROCHLORIDE
mf: $C_{12}H_{11}N \cdot ClH$ mw: 205.70

SYN: NCI-C50282

TOXICITY DATA WITH REFERENCE
mma-sat 10 μg/plate SCIEAS 236,933,87
mma-mus:lyms 110 mg/L EMMUEG 12,85,88
cyt-ham:ovr 200 mg/L SCIEAS 236,933,87
orl-mus TDLo:260 g/kg/2Y-C:CAR NTPTR* NTP-TR-233,82
orl-mus TD:262 g/kg/2Y-C:CAR FAATDF 2,201,82

CONSENSUS REPORTS: NCI Carcinogenesis Studies (feed): Clear Evidence: mouse NTPTR* NTP-TR-233,82; No Evidence: rat NTPTR* NTP-TR-233,82.

SAFETY PROFILE: Questionable carcinogen with experimental carcinogenic data. Mutation data reported. When heated to decomposition it emits toxic fumes of NO_x and HCl.

BGF000 **CAS:20743-57-1** **HR: 2**
N-4-BIPHENYLBENZAMIDE
mf: $C_{19}H_{15}NO$ mw: 273.35

SYNS: N-4-BIPHENYLYLBENZAMIDE □ 4′-PHENYLBENZANILIDE

TOXICITY DATA WITH REFERENCE
ipr-rat TDLo:508 mg/kg/4W-I:ETA CNREA8 30,1485,70

SAFETY PROFILE: Questionable carcinogen with experimental tumorigenic data. When heated to decomposition it emits toxic fumes of NO_x.

BGF109 **CAS:492-17-1** **HR: 3**
2,4′-BIPHENYLDIAMINE
mf: $C_{12}H_{12}N_2$ mw: 184.26

PROP: Needles from alc (aq), very sltly sol in alc and ether. Mp: 56°, bp: 363°.

SYNS: o,p′-BIANILINE □ (1,1′-BIPHENYL)-2,4′-DIAMINE □ o,p′-DIAMINOBIPHENYL □ 2,4′-DIAMINODIPHENYL □ o,p′-DIANILINE □ DIFENYLIN □ 2,4′-DIPHENYLDIAMINE □ DIPHENYLINE

TOXICITY DATA WITH REFERENCE
mma-sat 100 μg/plate MUREAV 149,9,85
orl-dog TDLo:7020 mg/kg/5Y-I:ETA NEOLA4 15,3,68
orl-rat LD50:311 mg/kg NEOLA4 15,3,68

CONSENSUS REPORTS: IARC Cancer Review: Group 3 IMEMDT 7,56,87; Animal Inadequate Evidence IMEMDT 16,313,78.

SAFETY PROFILE: A poison by ingestion. Questionable carcinogen with experimental tumorigenic data. Mutation data reported. When heated to decomposition it emits toxic fumes of NO_x. See also AROMATIC AMINES.

BGF250 **CAS:1591-30-6** **HR: 3**
4,4′-BIPHENYLDICARBONITRILE
mf: $C_{14}H_8N_2$ mw: 204.24

PROP: Solid. Mp: 234°.

SYN: NCR DR DCN

TOXICITY DATA WITH REFERENCE
ipr-mus LD50:75 mg/kg NTIS** AD691-490

CONSENSUS REPORTS: Reported in EPA TSCA Inventory. Cyanide and its compounds are on the Community Right-To-Know List.

SAFETY PROFILE: Poison by intraperitoneal route. See also NITRILES. When heated to decomposition it emits toxic fumes of NO_x and CN^-.

BGF899 **CAS:1137-79-7** **HR: 2**
4-BIPHENYLDIMETHYLAMINE
mf: $C_{14}H_{15}N$ mw: 197.30

SYN: 4-DIMETHYLAMINOBIPHENYL

TOXICITY DATA WITH REFERENCE
orl-rat TDLo:19 g/kg/43W-C:CAR CNREA8 16,525,56

SAFETY PROFILE: Questionable carcinogen with experimental carcinogenic data. When heated to decomposition it emits toxic fumes of NO_x. See also AROMATIC AMINES.

BGG000 **CAS:1806-29-7** **HR: 3**
2,2′-BIPHENYLDIOL
mf: $C_{12}H_{10}O_2$ mw: 186.22

PROP: Prisms from toluene; hydrated crystals from H_2O. Mp: 109° (anhyd), mp: 73–75° (hydrate), bp: 325–326°.

SYNS: o,o′-BIPHENOL □ 2,2′-BIPHENOL □ 2,2′-DIHYDROXYBIPHENYL

TOXICITY DATA WITH REFERENCE
oms-hmn:lym 5 μmol/L CNREA8 45,2471,85
sce-hmn:lym 300 μmol/L CNREA8 45,2471,85
ipr-mus LD50:150 mg/kg NTIS** AD691-490
ivn-mus LD50:56 mg/kg CSLNX* NX#07870

CONSENSUS REPORTS: Reported in EPA TSCA Inventory.

SAFETY PROFILE: Poison by intraperitoneal and intravenous routes. Human mutation data reported. When heated to decomposition it emits acrid smoke and irritating fumes.

BGG250 **CAS:1079-21-6** **HR: 3**
2,5-BIPHENYLDIOL
mf: $C_{12}H_{10}O_2$ mw: 186.22

PROP: Needles from EtOH (aq). Mp: 96–98°.

SYNS: (1,1′-BIPHENYL)-2,5-DIOL □ 2,5-DIHYDROXYBIPHENYL □ HYDROQUINONE, PHENYL- □ PHENYLHYDROQUINONE □ o-PHENYLHYDROQUINONE □ 2-PHENYLHYDROQUINONE

TOXICITY DATA WITH REFERENCE
dnd-esc 10 μmol/L CBINA8 76,163,90
add-hmn:leu 100 μmol/L CRNGDP 13,1937,92
ipr-mus LDLo:250 mg/kg CBCCT* 6,222,54
ivn-mus LD50:22 mg/kg BJPCAL 22,221,64

CONSENSUS REPORTS: Reported in EPA TSCA Inventory.

SAFETY PROFILE: Poison by intraperitoneal and intravenous routes. Mutation data reported. When heated to decomposition it emits acrid smoke and irritating fumes.

BGG500 **CAS:92-88-6** **HR: 3**
4,4′-BIPHENYLDIOL
mf: $C_{12}H_{10}O_2$ mw: 186.22

PROP: Needles or plates from EtOH. Mp: 286°.

SYNS: p,p′-BIPHENOL □ USAF DO-30

TOXICITY DATA WITH REFERENCE
oms-hmn:lym 100 nmol/L CNREA8 45,2471,85
sce-hmn:lym 5 μmol/L CNREA8 45,2471,85
orl-rat LD50:9850 mg/kg TXAPA9 28,313,74

ipr-mus LD50:100 mg/kg NTIS** AD277-689
skn-rbt LD50:1780 mg/kg TXAPA9 28,313,74

CONSENSUS REPORTS: Reported in EPA TSCA Inventory.

SAFETY PROFILE: Poison by intraperitoneal route. Moderately toxic by skin contact. Mildly toxic by ingestion. Human mutation data reported. When heated to decomposition it emits acrid smoke and irritating fumes.

BGH000 CAS:20275-19-8 HR: 3
1,1′-(p′p′-BIPHENYLENEBIS(CARBONYLMETHYL))DI-2-PICOLINIUM DIBROMIDE
mf: $C_{28}H_{26}N_2O_2$•2Br mw: 582.38

SYN: (4,4′-BIPHENYLYLENEBIS(2-OXOETHYLENE))-2-PICOLINIUM DIBROMIDE

TOXICITY DATA WITH REFERENCE
ipr-mus LD50:1500 µg/kg JAPMA8 43,79,54
ivn-mus LD50:2900 µg/kg TXAPA9 27,666,74

SAFETY PROFILE: Deadly poison by intraperitoneal and intravenous routes. See also BROMIDES. When heated to decomposition it emits very toxic fumes of NO_x and Br^-.

BGH250 CAS:73-51-8 HR: 3
4,4′-BIPHENYLENEBIS(2-OXOETHYLENE)BIS(DIMETHYL(2-HYDROXYETHYL)AMMONIUM) DIBROMIDE
mf: $C_{24}H_{34}N_2O_4$•2Br mw: 574.42

TOXICITY DATA WITH REFERENCE
ipr-rat LD50:45 µg/kg JPETAB 115,127,55
ipr-mus LD50:20 µg/kg JPETAB 115,127,55
ivn-dog LDLo:75 µg/kg JPETAB 115,127,55
ivn-rbt LDLo:50 µg/kg JPETAB 115,127,55
ipr-gpg LDLo:30 µg/kg JPETAB 115,127,55

SAFETY PROFILE: Deadly poison by intraperitoneal and intravenous routes. See also BROMIDES. When heated to decomposition it emits very toxic fumes of NO_x, NH_3, and Br^-.

BGH500 CAS:77967-05-6 HR: 3
4,4′-BIPHENYLENEBIS(3-OXOPROPYLENE)BIS(DIMETHYL(2-HYDROXYETHYL)AMMONIUM)DIBROMIDE
mf: $C_{26}H_{38}N_2O_4$•2Br mw: 602.48

TOXICITY DATA WITH REFERENCE
ipr-mus LD50:10 mg/kg JPETAB 115,127,55
ipr-rbt LD50:20 mg/kg JPETAB 115,127,55

SAFETY PROFILE: Poison by intraperitoneal route. See also BROMIDES. When heated to decomposition it emits very toxic fumes of NO_x, NH_3, and Br^-.

BGI250 CAS:6810-26-0 HR: 2
4-BIPHENYLHYDROXYLAMINE
mf: $C_{12}H_{11}NO$ mw: 185.24

SYNS: (1,1′-BIPHENYL)-4-AMINE, N-HYDROXY- □ N-4-BIPHENYLYLHYDROXYLAMINE □ N-HYDROXY-4-AMINOBIPHENYL □ 4-HYDROXYAMINOBIPHENYL □ 4-HYDROXYLAMINOBIPHENYL

TOXICITY DATA WITH REFERENCE
mmo-sat 5 µg/plate MUREAV 151,201,85
mmo-esc 2500 nmol/L MUREAV 151,201,85
dns-hmn:oth 1 µmol/L JJIND8 72,847,84
dns-rat:lvr 5 µmol/L ENMUDM 3,11,81
dns-rbt:oth 10 µmol/L CNREA8 45,221,85
scu-mus TDLo:216 mg/kg/3D:CAR JNCIAM 41,403,68

CONSENSUS REPORTS: EPA Genetic Toxicology Program.

SAFETY PROFILE: Questionable carcinogen with experimental carcinogenic data. Human mutation data reported. When heated to decomposition it emits highly toxic fumes of NO_x. See also AROMATIC AMINES.

BGJ250 CAS:90-43-7 HR: 3
2-BIPHENYLOL
mf: $C_{12}H_{10}O$ mw: 170.22

PROP: Needles from pet ether. Mp: 56°, bp: 275°.

SYNS: o-BIPHENYLOL □ (1,1′-BIPHENYL)-2-OL □ o-DIPHENYLOL □ DOWCIDE 1 □ DOWCIDE 1 ANTIMICROBIAL □ 2-HYDROXYBIFENYL (CZECH) □ o-HYDROXYBIPHENYL □ 2-HYDROXYBIPHENYL □ o-HYDROXYDIPHENYL □ 2-HYDROXYDIPHENYL □ KIWI LUSTR 277 □ NCI-C50351 □ OPP □ ORTHOHYDROXYDIPHENYL □ ORTHOPHENYLPHENOL □ ORTHOXENOL □ o-PHENYLPHENOL □ 2-PHENYLPHENOL □ PREVENTOL O EXTRA □ REMOL TRF □ TETROSIN OE □ TORSITE □ TUMESCAL OPE □ USAF EK-2219 □ o-XENOL

TOXICITY DATA WITH REFERENCE
skn-rbt 250 mg MccSB# 15JUN84
skn-rbt 20 mg/24H MOD 85JCAE-,228,86
eye-rbt 50 µg/24H SEV 85JCAE-,228,86
mmo-sat 60 µg/plate ENMUDM 5(Suppl 1),3,83
cyt-hmn:fbr 200 µg/L MUREAV 54,255,78
msc-hmn:emb 20 mg/L MUREAV 156,123,85
msc-hmn:oth 15 mg/L TRENAF 35,399,84
cyt-ham:ovr 100 mg/L MUREAV 141,95,84
orl-rat TDLo:6 g/kg (female 6-15D post):TER NNGADV 3,365,78
orl-rat TDLo:52,168 mg/kg (male 13W pre):REP TRENAF 32-2,33,81
orl-mus TDLo:13,050 mg/kg (female 7-15D post):TER TRENAF 29,89,78
orl-rat TDLo:478 g/kg/91W-C:CAR FCTOD7 22,865,84
orl-rat LD:135 g/kg/26W-C:NEO FCTOD7 25,359,87
orl-rat LD50:2000 mg/kg NNGADV 3,365,78
unr-rat LD50:2700 mg/kg TRENAF 29,89,78
orl-mus LD50:1050 mg/kg NAIZAM 32,425,81
ipr-mus LD50:50 mg/kg NTIS** AD277-689

CONSENSUS REPORTS: IARC Cancer Review: Group 3 IMEMDT 7,56,87, Animal Inadequate Evidence IMEMDT 30,329,83; NTP Carcinogenesis Studies (dermal); No Evidence: mouse NTPTR* NTP-TR-301,86. Reported in EPA TSCA Inventory. On Community Right-To-Know List.

SAFETY PROFILE: A poison by intraperitoneal route. Moderately toxic by ingestion and possibly other routes. An experimental teratogen. Other experimental reproductive effects. Human mutation data reported. Severe eye and moderate skin irritant. Questionable carcinogen with experimental carcinogenic data. When heated to decomposition it emits acrid smoke and irritating fumes.

BGJ500 **CAS:92-69-3** **HR: 3**
4-BIPHENYLOL
mf: $C_{12}H_{10}O$ mw: 170.22

PROP: Needles or plates from EtOH (aq). Mp: 164–165°, bp: 305–308°.

SYNS: p-HYDROXYBIPHENYL □ 4-HYDROXYBIPHENYL □ p-HYDROXYDIPHENYL □ 4-HYDROXYDIPHENYL □ PARAXENOL □ p-PHENYLPHENOL □ 4-PHENYLPHENOL

TOXICITY DATA WITH REFERENCE
orl-mus TDLo:153 g/kg/78W-I:ETA NTIS** PB223-159
scu-mus TDLo:1000 mg/kg:CAR NTIS** PB223-159
ipr-mus LD50:150 mg/kg NTIS** AD691-490

CONSENSUS REPORTS: Reported in EPA TSCA Inventory.

SAFETY PROFILE: Acute poison by intraperitoneal route. Questionable carcinogen with experimental carcinogenic and tumorigenic data. When heated to decomposition it emits acrid, irritating fumes.

BGJ750 **CAS:132-27-4** **HR: 3**
2-BIPHENYLOL, SODIUM SALT
mf: $C_{12}H_9O{\cdot}Na$ mw: 192.20

SYNS: BACTROL □ (1,1′-BIPHENYL)-2-OL, SODIUM SALT □ D.C.S. □ DORVICIDE A □ DOWICIDE □ DOWICIDE A □ DOWICIDE A & A FLAKES □ DOWIZID A □ 2-HYDROXYBIPHENYL SODIUM SALT □ 2-HYDROXYDIPHENYL SODIUM □ 2-HYDROXYDIPHENYL, SODIUM SALT □ MIL-DU-RID □ MYSTOX WFA □ NATRIPHENE □ OPP-Na □ OPP-SODIUM □ ORPHENOL □ PHENOL, o-PHENYL-, SODIUM deriv. □ o-PHENYLPHENOL, SODIUM SALT □ 2-PHENYLPHENOL SODIUM SALT □ PREVENTOL-ON □ PREVENTOL ON & ON EXTRA □ SODIUM 2-BIPHENYLOLATE □ SODIUM (1,1′-BIPHENYL)-2-OLATE □ SODIUM, (2-BIPHENYLYLOXY)- □ SODIUM 2-HYDROXYDIPHENYL □ SODIUM ORTHO PHENYLPHENATE □ SODIUM o-PHENYLPHENATE □ SODIUM 2-PHENYLPHENATE □ SODIUM o-PHENYLPHENOL □ SODIUM o-PHENYLPHENOLATE □ SODIUM o-PHENYLPHENOXIDE □ SOPP □ STOPMOLD B □ TOPANE

TOXICITY DATA WITH REFERENCE
skn-hmn 1 mg MccSB# 15JUN84
skn-rbt 50 mg/24H SEV MccSB# 15JUN84
mmo-asn 16 μmol/L PHYTAJ 66,217,76
sln-asn 52 μmol/L EVHPAZ 31,81,79
orl-mus TDLo:144 g/kg (male 60D pre):TER TRENAF 29,99,78
orl-mus TDLo:900 mg/kg (7-15D preg):TER TRENAF 29,89,78
orl-mus TDLo:72 g/kg (60D male):REP TRENAF 29,99,78
skn-mus TDLo:18,800 mg/kg/47W-I:CAR CRNGDP 10,1163,89
orl-rat TD:269 g/kg/32W-C:ETA GANNA2 74,625,83
orl-rat TD:126 g/kg/13W-C:CAR FCTXAV 19,303,81
orl-rat TD:486 g/kg/2Y-C:ETA NAIZAM 37,270,86
orl-rat TD:223 g/kg/26W-C:NEO FCTOD7 25,359,87
orl-rat LD50:656 mg/kg TRENAF 30(2),57,79
orl-mus LD50:683 mg/kg TRENAF 30,54,79
orl-cat LD50:500 mg/kg FAONAU 38A,47,65
orl-cat LD50:500 mg/kg FAONAU 38A,47,65

CONSENSUS REPORTS: IARC Cancer Review: Group 2B IMEMDT 7,56,87, Animal Limited Evidence IMEMDT 30,329,83. Reported in EPA TSCA Inventory.

SAFETY PROFILE: Suspected carcinogen with experimental carcinogenic, neoplastigenic, and tumorigenic data. Moderately toxic by ingestion. Experimental teratogenic and reproductive effects. A human skin irritant. A severe skin irritant to experimental animals. When heated to decomposition it emits toxic fumes of Na_2O. See also 2-BIPHENYLOL.

BGK000 **CAS:3644-37-9** **HR: 3**
(2-BIPHENYLOXY)TRIBUTYLTIN
mf: $C_{24}H_{37}OSn$ mw: 460.30

SYNS: ((2-BIPHENYLYLOXY)TRIBUTYL)STANNANE □ ((1,1′-BIPHENYL)-2-YLOXY)TRIBUTYL-(9CI) STANNANE □ TRIBUTYL-o-PHENYLPHENOXYTIN □ TRIBUTYLTIN-o-PHENYLPHENOXIDE

TOXICITY DATA WITH REFERENCE
ivn-mus LD50:100 mg/kg CSLNX* NX#01826

OSHA PEL: TWA 0.1 mg(Sn)/m³ (skin)
ACGIH TLV: TWA 0.1 mg(Sn)/m³ (skin) (Proposed: TWA 0.1 mg(Sn)/m³; STEL 0.2 mg(Sn)/m³ (skin))
NIOSH REL: (Organotin Compounds) TWA 0.1 mg(Sn)/m³

SAFETY PROFILE: Poison by intravenous route. See also TIN COMPOUNDS. When heated to decomposition it emits acrid smoke and irritating fumes.

For occupational chemical analysis use NIOSH: Organotin Compounds 5504.

BGK500 **CAS:91-95-2** **HR: 2**
3,3′,4,4′-BIPHENYLTETRAMINE
mf: $C_{12}H_{14}N_4$ mw: 214.30

PROP: Crystals from MeOH. Mp: 178–179°.

SYNS: 3,3′-DIAMINOBENZIDENE □ 3,3′,4,4′-DIPHENYLTETRAMINE □ 3,3′,4,4′-TETRAAMINOBIPHENYL

TOXICITY DATA WITH REFERENCE
mma-sat 100 μg/plate BJCAAI 37,873,78
dnd-esc 20 μmol/L MUREAV 89,95,81
mmo-smc 140 μmol/L MGGEAE 174,39,79
dns-rat:lvr 500 μmol/L ENMUDM 3,11,81
orl-rat TDLo:9000 mg/kg/27D-I:ETA CNREA8 28,924,68
orl-rat LDLo:3000 mg/kg CNREA8 26,619,66
orl-mus LD50:1834 mg/kg GISAAA 46(1),94,81

CONSENSUS REPORTS: Reported in EPA TSCA Inventory.

SAFETY PROFILE: Questionable carcinogen with experimental tumorigenic data. Moderately toxic by ingestion. Mutation data reported. When heated to decomposition it emits toxic fumes of NO_x. See also AROMATIC AMINES.

BGK750 **CAS:7411-49-6** **HR: 3**
3,3′,4,4′-BIPHENYLTETRAMINE TETRAHYDROCHLORIDE
mf: $C_{12}H_{14}N_4•4ClH$ mw: 360.14

PROP: Crystals. Sol in acids.

SYNS: 3,3′-DIAMINOBENZIDINE TETRAHYDROCHLORIDE □ 3,3′,4,4′-TETRAAMINOBIPHENYL TETRAHYDROCHLORIDE

TOXICITY DATA WITH REFERENCE
orl-rat TDLo:260 g/kg/78W-C:ETA JEPTDQ 2,325,78
orl-mus TDLo:260 g/kg/78W-C:NEO JEPTDQ 2,325,78
orl-mus TD:520 g/kg/78W-C:NEO JEPTDQ 2,325,78
ipr-mus LD50:330 mg/kg NCIBR* NIH-NCI-E-68-1311,10,73

CONSENSUS REPORTS: Reported in EPA TSCA Inventory.

SAFETY PROFILE: Poison by intraperitoneal route. Questionable carcinogen with experimental neoplastigenic and tumorigenic data. When heated to decomposition it emits very toxic fumes of HCl and NO_x. See also AROMATIC AMINES.

BGL000 **CAS:13607-48-2** **HR: 2**
N-4-BIPHENYLYLBENZENESULFONAMIDE
mf: $C_{18}H_{15}NO_2S$ mw: 309.40

SYN: N-4-BIPHENYLYL BENZENESULFONAMIDE

TOXICITY DATA WITH REFERENCE
ipr-rat TDLo:634 mg/kg/4W-I:ETA CNREA8 30,1485,70

SAFETY PROFILE: Questionable carcinogen with experimental tumorigenic data. See also SULFONATES. When heated to decomposition it emits very toxic fumes of SO_x and NO_x.

BGL250 **CAS:36330-85-5** **HR: 3**
3-(4-BIPHENYLYLCARBONYL)PROPIONIC ACID
mf: $C_{16}H_{14}O_3$ mw: 254.30

PROP: Solid. Mp: 185–187°.

SYNS: 4-(4-BIPHENYLYL)-4-OXOBUTYRIC ACID □ BUFEMID □ CINOPAL □ CINOPOP □ CL82204 □ DIPHENYL-4-γ-OXO-γ-BUTRIC ACID □ FENBUFEN □ LEDERFEN □ γ-OXO(1,1-BIPHENYL)-4-BUTANOIC ACID □ β,p-PHENYLBENZOYLPROPIONIC ACID

TOXICITY DATA WITH REFERENCE
orl-rat TDLo:1120 mg/kg (female 16-22D post):REP ARZNAD 30,725,80
orl-rat TDLo:280 mg/kg (female 16-22D post):REP ARZNAD 30,725,80
orl-rat TDLo:280 mg/kg (female 16-22D post):TER ARZNAD 30,725,80
orl-rat TDLo:3780 mg/kg (male 63D pre):REP ARZNAD 30,725,80
orl-rat TDLo:140 mg/kg (female 16-22D post):REP ARZNAD 30,725,80
orl-rat TDLo:540 mg/kg (female 30D pre):REP YACHDS 5,3511,77
orl-wmn TDLo:12 mg/kg/1D-I BMJOAE 290,822,85
orl-man TDLo:90 mg/kg/1W-I:PUL HUTODJ 7,35,88
orl-rat LD50:200 mg/kg ARZNAD 30,725,80
ipr-rat LD50:265 mg/kg ARZNAD 30,721,80
scu-rat LD50:247 mg/kg ARZNAD 30,721,80
orl-mus LD50:795 mg/kg ARZNAD 30,721,80
ipr-mus LD50:482 mg/kg PCIPDV 15,132,83
scu-mus LD50:1189 mg/kg IYKEDH 10,884,79

SAFETY PROFILE: Poison by ingestion, intraperitoneal, and subcutaneous routes. Human systemic effects by ingestion: cough, sweating, body temperature. An experimental teratogen. Other experimental reproductive effects. An anti-inflammatory agent. When heated to decomposition it emits acrid smoke and irritating fumes.

BGL400 **CAS:27695-61-0** **HR: 3**
1-(1,1′-BIPHENYL)-4-YL-2-((4-(DICHLOROACETYL)PHENYL)AMINO)-2-HYDROXYETHANONE
mf: $C_{22}H_{17}Cl_2NO_3$ mw: 414.30

SYNS: ETHANONE, 1-(1,1′-BIPHENYL)-4-YL-2-((4-(DICHLOROACETYL)PHENYL)AMINO)-2-HYDROXY- □ KETONE, 1-(1,1′-BIPHENYL)-4-YL-2-((4-(DICHLOROACETYL)PHENYL)AMINO)-2-HYDROXY-

TOXICITY DATA WITH REFERENCE
ipr-mus LD50:3 g/kg ARZNAD 23,573,73

DOT CLASSIFICATION: 3; *Label.* Flammable Liquid

SAFETY PROFILE: Low toxicity by intraperitoneal route. A flammable liquid. When heated to decomposition it emits toxic vapors of NO_x and Cl^-.

BGL500 **CAS:7203-95-4** **HR: 3**
1-BIPHENYLYL-3,3-DIMETHYLTRIAZENE
mf: $C_{14}H_{15}N_3$ mw: 225.32

SYNS: 3,3-DIMETHYL-1-XENYL-TRIAZENE □ 1-XENYL-3,3-DIMETHYLTRIAZIN (CZECH)

TOXICITY DATA WITH REFERENCE
mmo-sat 21 nmol/L JMCMAR 22,473,79
orl-rat LD50:347 mg/kg 28ZPAK -,77,72
ipr-mus LD50:344 mg/kg JMCMAR 19,1299,76

SAFETY PROFILE: Poison by ingestion and intraperitoneal routes. Mutation data reported. When heated to decomposition it emits toxic fumes of NO_x.

BGM000 **CAS:18355-50-5** **HR: 3**
7,7′-(p,p′-BIPHENYLYLENEBIS(CARBONYLIMINO))BIS(2-ETHYLQUINOLINIUM) DITOSYLATE
mf: $C_{36}H_{32}N_4O_2•2C_7H_7O_3S$ mw: 895.12

SYN: 7,7′-(4,4′-BIPHENYLYLENEBIS(CARBONYLIMINO)) BIS(1-ETHYLQUINOLINIUM)DI-p-TOLUENESULFONATE

TOXICITY DATA WITH REFERENCE
add-mus:lym 300 nmol/L JMCMAR 22,134,79
ipr-mus LD10:31 mg/kg JMCMAR 22,134,79

SAFETY PROFILE: Poison by intraperitoneal route. See also SULFONATES. Mutation data reported. When heated to decomposition it emits very toxic fumes of NO_x and SO_x.

BGM100 **CAS:37940-57-1** **HR: 3**
4-BIPHENYLYL ETHYLKETONE
mf: $C_{15}H_{14}O$ mw: 210.29

SYNS: KETONE, 4-BIPHENYL ETHYL □ 4-PHENYLPROPIOPHENONE □ PROPIOPHENONE, 4′-PHENYL-

TOXICITY DATA WITH REFERENCE
ivn-mus LD50:180 mg/kg CSLNX* NX#04519

DOT CLASSIFICATION: 3; *Label:* Flammable Liquid

SAFETY PROFILE: A poison by intravenous route. A flammable liquid. When heated to decomposition it emits acrid smoke and irritating vapors.

BGN000 **CAS:29968-68-1** **HR: 2**
N-4-BIPHENYLYL-N-HYDROXYBENZENESULFONAMIDE
mf: $C_{18}H_{15}NO_3S$ mw: 325.40

SYN: HYDROXY-4-BIPHENYLYLBENZENESULFONAMIDE

TOXICITY DATA WITH REFERENCE
ipr-rat TDLo:618 mg/kg/4W-I:NEO CNREA8 30,1485,70

SAFETY PROFILE: Questionable carcinogen with experimental neoplastigenic data. See also SULFONATES. When heated to decomposition it emits very toxic fumes of SO_x and NO_x.

BGO000 **CAS:1734-91-4** **HR: 3**
2-(2-BIPHENYLYLOXY)TRIETHYLAMINE HYDROCHLORIDE
mf: $C_{18}H_{23}NO{\cdot}ClH$ mw: 305.88

SYNS: DACORENE HYDROCHLORIDE □ 2-(DIETHYLAMINOETHOXY)DIPHENYL HCl □ 1262 F □ F 1262

TOXICITY DATA WITH REFERENCE
ipr-mus LD50:125 mg/kg BJPCAL 1,90,46
scu-mus LDLo:125 mg/kg APFRAD 5,7,47
ivn-mus LD50:27 mg/kg BJPCAL 1,90,46

SAFETY PROFILE: Poison by subcutaneous, intraperitoneal, and intravenous routes. When heated to decomposition it emits very toxic fumes of NO_x and HCl.

BGO500 **CAS:366-18-7** **HR: 3**
2,2′-BIPYRIDINE
mf: $C_{10}H_8N_2$ mw: 156.20

PROP: White crystals or prisms from pet ether. Mp: 69.7°, bp: 272–273°. Sol in H_2O, EtOH, Et_2O, C_6H_6, $CHCl_3$, and dilute acids.

SYNS: BIPYRIDINE □ α,α′-BIPYRIDINE □ α,α′-BIPYRIDYL □ 2,2′-BIPYRIDYL □ 2,2′-BYPYRIDIN □ CI-588 □ α,α′-DIPYRIDYL □ 2,2′-DIPYRIDYL

TOXICITY DATA WITH REFERENCE
mmo-sat 20 μg/plate ABCHA6 45,327,81
mma-sat 20 μg/plate ABCHA6 45,327,81
ipr-rat TDLo:60 mg/kg (12D preg):TER TJADAB 18,63,78
scu-mus TDLo:8000 mg/kg/40W-I:ETA JNCIAM 24,109,60
orl-rat LD50:100 mg/kg JTEHD6 10,363,82
ipr-rat LD50:150 mg/kg PJPPAA 27,619,75
scu-rat LD50:131 mg/kg JPETAB 135,317,62
ipr-mus LD50:200 mg/kg JPETAB 196,478,76

CONSENSUS REPORTS: Reported in EPA TSCA Inventory.

SAFETY PROFILE: Poison by ingestion, subcutaneous, and intraperitoneal routes. Experimental teratogenic data. Questionable carcinogen with experimental tumorigenic data. Mutation data reported. When heated to decomposition it emits toxic fumes of NO_x.

BGO600 **CAS:553-26-4** **HR: 3**
4,4′-BIPYRIDINE
mf: $C_{10}H_8N_2$ mw: 156.20

SYNS: γ,γ′-BIPYRIDYL □ 4,4-BIPYRIDYL □ 4,4′-BIPYRIDYL □ 4,4′-DIPYRIDINE □ γ,γ′-DIPYRIDYL □ 4,4-DIPYRIDYL □ 4,4′-DIPYRIDYL □ 4-(4-PYRIDYL)PYRIDINE

TOXICITY DATA WITH REFERENCE
orl-rat LD50:172 mg/kg JTEHD6 10,363,82

CONSENSUS REPORTS: Reported in EPA TSCA Inventory.

SAFETY PROFILE: Poison by ingestion. When heated to decomposition it emits toxic vapors of NO_x.

BGO750 **CAS:8001-88-5** **HR: 2**
BIRCH TAR OIL

PROP: Brown liquid; leather-like odor. D: 0.886–0.950. Found in the tar of the bark and wood of *Betula pendula Roth* (Fam. Betulaceae) and prepared by steam distillation of the tar obtained by dry distillation of the bark and wood (FCTXAV 11,1011,73). Sol in fixed oils; insol in glycerin, mineral oil, and propylene glycol.

SYN: BIRCH TAR OIL, RECTIFIED (FCC)

TOXICITY DATA WITH REFERENCE
skn-rbt 500 mg/24H FCTXAV 11,1037,73

CONSENSUS REPORTS: Reported in EPA TSCA Inventory.

SAFETY PROFILE: A skin irritant. Moderately irritating to eyes and mucous membranes. A mild allergen. Combustible when exposed to heat or flame; can react with oxidizing materials.

BGO775 **CAS:515-69-5** **HR: 2**
BISABOLOL
mf: $C_{15}H_{26}O$ mw: 222.41

SYNS: (−)-α-BISABOLOL □ α-4-DIMETHYL-α-(4-METHYL-3-PENTENYL)-3-CYCLOHEXENE-1-METHANOL □ 5-HEPTEN-2-OL, 6-METHYL-2-(4-METHYL-3-CYCLOHEXEN-1-YL)- □ 6-METHYL-2-(4-METHYL-3-CYCLOHEXEN-1-YL)-5-HEPTEN-2-OL

TOXICITY DATA WITH REFERENCE
orl-rat LD50:14,850 mg/kg ARZNAD 19,615,69
orl-mus LD50:11,350 mg/kg ARZNAD 19,615,69

CONSENSUS REPORTS: Reported in EPA TSCA Inventory.

SAFETY PROFILE: Moderately toxic by ingestion. When

heated to decomposition it emits acrid smoke and irritating vapors.

BGP250 **CAS:304-28-9** **HR: 3**
2,7-BIS(ACETAMIDO)FLUORENE
mf: $C_{17}H_{16}N_2O_2$ mw: 280.35

SYNS: 2,7-DIACETAMIDOFLUORENE □ 2,7-DIACETYLAMINOFLUORENE □ 2,7-FAA □ N,N'-FLUOREN-2,7-YLBISACETAMIDE □ 2,7-FLUORENYLBISACETAMIDE □ N,N'-FLUOREN-2,7-YLENEBISACETAMIDE □ N,N'-2,7-FLUORENYLENEBISACETAMIDE □ N,N'-(FLUOREN-2,7-YLENE)BIS(ACETYLAMINE) □ N,N'-2,7-FLUORENYLENEDIACETAMIDE

TOXICITY DATA WITH REFERENCE
mmo-sat 100 µg/plate PNASA6 69,3128,72
mma-sat 10 µg/plate PNASA6 72,5135,75
dns-rat:lvr 500 nmol/L ENMUDM 3,11,81
cyt-rat-orl 315 mg/kg/3W JNCIAM 54,1245,75
orl-rat TDLo:4830 mg/kg/46W-C:CAR GANNA2 60,211,69
ipr-rat TDLo:2044 mg/kg/29W-I:CAR JNCIAM 29,977,62
orl-mus TDLo:2100 mg/kg/13W-C:CAR JNCIAM 40,629,68
orl-uns TDLo:1200 mg/kg/34W-C:ETA JJIND8 74,909,85
orl-rat TD:1830 mg/kg/32W-C:ETA NCIMAV 5,85,61
orl-rat TD:2060 mg/kg/34W-C:ETA JNCIAM 29,977,62
orl-rat TD:2250 mg/kg/21W-C:ETA GANNA2 62,471,71
orl-mus TD:6480 mg/kg/26W-C:NEO NAIZAM 33,545,83
ipr-rat TD:8 mg/kg:ETA JJIND8 71,211,83
orl-mus TD:7668 mg/kg/30W-C:CAR ONCOBS 41,101,84
orl-rat TD:2800 mg/kg/32W-C:ETA APHGBP 26,341,76
orl-rat TD:228 g/kg/26W-C:ETA APHGBP 22,477,72
orl-rat TD:2035 mg/kg/34W-C:ETA NCIMAV 5,1,61

CONSENSUS REPORTS: EPA Genetic Toxicology Program.

SAFETY PROFILE: Suspected carcinogen with experimental carcinogenic, neoplastigenic, and tumorigenic data. Mutation data reported. When heated to decomposition it emits toxic fumes of NO_x.

BGP500 **CAS:63981-20-4** **HR: 2**
BIS-4-ACETAMINO PHENYL SELENIUMDIHYDROXIDE
mf: $C_8H_{11}NO_3$•Se mw: 248.16

TOXICITY DATA WITH REFERENCE
orl-rat TDLo:2890 mg/kg/15W-C:ETA SCIEAS 103,762,46

CONSENSUS REPORTS: Selenium and its compounds are on the Community Right-To-Know List.

OSHA PEL: TWA 0.2 mg(Se)/m³
ACGIH TLV: TWA 0.2 mg(Se)/m³
DFG MAK: 0.1 mg(Se)/m³

SAFETY PROFILE: Questionable carcinogen with experimental tumorigenic data. When heated to decomposition it emits very toxic fumes of NO_x and Se. See also SELENIUM COMPOUNDS.

BGP750 **CAS:15172-86-8** **HR: 3**
4,4'-BISACETOPHENONE-α,α'-DI(3-METHYLPYRIDINIUM) DIBROMIDE
mf: $C_{28}H_{26}N_2O_2$•2Br mw: 582.38

SYN: (4,4'-BIPHENYLYLENEBIS(2-OXOETHYLENE))-3-PICOLINIUM DIBROMIDE

TOXICITY DATA WITH REFERENCE
ipr-mus LD50:50 µg/kg JAPMA8 43,79,54
ivn-mus LD50:69 µg/kg TXAPA9 27,666,74

SAFETY PROFILE: Deadly poison by intraperitoneal and intravenous routes. See also BROMIDES. When heated to decomposition it emits very toxic fumes of Br^- and NO_x.

BGQ000 **CAS:5967-09-9** **HR: 3**
BIS(ACETOXYDIBUTYLSTANNANE) OXIDE
mf: $C_{20}H_{42}O_5Sn_2$ mw: 600.00

SYNS: BIS(DIBUTYLACETOXYTIN)OXIDE □ DIACETOXYTETRABUTYLDISTANNOXANE

TOXICITY DATA WITH REFERENCE
ivn-mus LD50:320 mg/kg CSLNX* NX#02081

CONSENSUS REPORTS: Reported in EPA TSCA Inventory.

OSHA PEL: TWA 0.1 mg(Sn)/m³ (skin)
ACGIH TLV: TWA 0.1 mg(Sn)/m³ (skin) (Proposed: TWA 0.1 mg(Sn)/m³; STEL 0.2 mg(Sn)/m³ (skin))
NIOSH REL: (Organotin Compounds) TWA 0.1 mg(Sn)/m³

SAFETY PROFILE: Poison by intravenous route. See also TIN COMPOUNDS. When heated to decomposition it emits acrid smoke and irritating fumes.

For occupational chemical analysis use NIOSH: Organotin Compounds 5504.

BGQ250 **CAS:64058-74-8** **HR: 3**
2,6-BIS(ACETOXYMERCURI)-4-NITROACETANILIDE
mf: $C_{12}H_{12}Hg_2N_2O_7$ mw: 697.44

SYN: BIS(ACETATO-O)(ω-(2-(ACETYLAMINO)-5-NITRO-1,3-PHENYLENE)DI)-MERCURY

TOXICITY DATA WITH REFERENCE
ipr-rat LDLo:500 mg/kg NCNSA6 5,10,53

CONSENSUS REPORTS: Mercury and its compounds are on the Community Right-To-Know List.

OSHA PEL: CL 0.1 mg(Hg)/m³ (skin)
ACGIH TLV: TWA 0.1 mg(Hg)/m³ (skin)
NIOSH REL: (Mercury, Aryl And Inorganic): CL 0.1 mg/m³ (skin)

SAFETY PROFILE: Moderately toxic by intraperitoneal route. See also MERCURY COMPOUNDS. When heated to decomposition it emits very toxic fumes of NO_x and Hg.

BGQ750 **CAS:14024-64-7** **HR: 2**
BIS(ACETYLACETONATO) TITANIUM OXIDE
mf: $C_{10}H_{14}O_5Ti$ mw: 262.14

PROP: Sol in C_6H_6. Insol in pet ether.

SYNS: BIS(2,4-PENTANEDIONATO)TITANIUM OXIDE □ TITANIUM ACETONYL ACETONATE □ TITANIUM OXIDE BIS(ACETYLACETONATE) □ TITANIUM, OXOBIS(2,4-PENTANEDIONATO-O,O') □ TITANYL BIS(ACETYLACETONATE)

TOXICITY DATA WITH REFERENCE
ims-rat TDLo:360 mg/kg/69W-I:ETA NCIUS* PH 43-64-886,AUG,69
ims-rat TD:2025 mg/kg/34W-I:ETA NCIUS* PH 43-64-886,DEC,68
ipr-rat LD50:650 mg/kg NCIUS* PH 43-64-886,JUL,68

SAFETY PROFILE: Moderately toxic by intraperitoneal route. Questionable carcinogen with experimental tumorigenic data. When heated to decomposition it emits acrid smoke and irritating fumes. See also TITANIUM COMPOUNDS.

BGR000 **CAS:13395-16-9** **HR: 3**
BIS(ACETYL ACETONE)COPPER
mf: $C_{10}H_{14}O_4 \cdot Cu$ mw: 261.78

SYNS: BIS(2,4-PENTANEDIONATO)COPPER □ COPPER(II) ACETYLACETONATE □ COPPER BIS(ACETYLACETONATE) □ COPPER BIS (ACETYLACETONE) □ COPPER BIS(2,4-PENTANEDIONATE) □ COPPER DIACETYLACETONATE □ CUPRIC ACETYLACETONATE

TOXICITY DATA WITH REFERENCE
ipr-mus LD50:19 mg/kg CHTHBK 16,371,71
ivn-mus LD50:10 mg/kg CSLNX* NX#00604

CONSENSUS REPORTS: Reported in EPA TSCA Inventory. Copper and its compounds are on the Community Right-To-Know List.

SAFETY PROFILE: Poison by intravenous and intraperitoneal routes. See also COPPER COMPOUNDS. When heated to decomposition it emits acrid smoke and fumes of Cu.

BGR250 **CAS:22750-65-8** **HR: 2**
2,5-BIS(ACETYLAMINO)FLUORENE
mf: $C_{17}H_{16}N_2O_2$ mw: 280.35

SYNS: N,N'-FLUOREN-2,5-YLENEBISACETAMIDE □ 2,5-FLUORENYLENEBISACETAMIDE

TOXICITY DATA WITH REFERENCE
orl-rat TDLo:4550 mg/kg/26W-C:CAR CNREA8 22,1002,62

SAFETY PROFILE: Questionable carcinogen with experimental carcinogenic data. When heated to decomposition it emits toxic fumes of NO_x.

BGR325 **CAS:50588-13-1** **HR: 3**
1,1'-((2-β,3-α,5-α,16-β,17-β)-3,17-BIS(ACETYLOXY) ANDROSTANE-2,16-DIYL)BIS(1-METHYLPIPERIDINIUM) DIBROMIDE
mf: $C_{35}H_{60}N_2O_4 \cdot 2Br$ mw: 732.79

TOXICITY DATA WITH REFERENCE
orl-rat LD50:202 mg/kg IYKEDH 4,90,73
ipr-rat LD50:479 μg/kg IYKEDH 4,90,73
scu-rat LD50:436 μg/kg IYKEDH 4,90,73
ivn-rat LD50:129 μg/kg IYKEDH 4,90,73
orl-mus LD50:21,200 μg/kg IYKEDH 4,90,73
ipr-mus LD50:116 μg/kg IYKEDH 4,90,73
scu-mus LD50:168 μg/kg IYKEDH 4,90,73
ivn-mus LD50:36 μg/kg IYKEDH 4,90,73

SAFETY PROFILE: Poison by ingestion, subcutaneous, intravenous, and intraperitoneal routes. When heated to decomposition it emits toxic fumes of NO_x and Br^-. See also BROMIDES.

BGR500 **CAS:12266-58-9** **HR: 3**
BIS(ACRYLONITRILE) NICKEL (O)
mf: $C_6H_6N_2Ni$ mw: 164.84

PROP: Red crystals.

CONSENSUS REPORTS: Cyanide and its compounds, as well as nickel and its compounds, are on the Community Right-To-Know List.

SAFETY PROFILE: Ignites spontaneously in air. When heated to decomposition it emits toxic fumes of CN^- and NO_x. See also NITRILES and NICKEL COMPOUNDS.

BGR750 **CAS:63906-14-9** **HR: 3**
1,4-BIS(4-ALDOXIMINOPYRIDINIUM)BUTANEDIOL-2,3-BIBROMIDE
mf: $C_{16}H_{20}N_4O_4 \cdot 2Br$ mw: 492.22

SYNS: 1,4-BIS(4-HYDROXYIMINOMETHYL-PYRIDINIUM-(1))-BUTANEDIOL-2,3 DIBROMID (GERMAN) □ R 21

TOXICITY DATA WITH REFERENCE
ipr-mus LD50:130 mg/kg ARZNAD 14,870,64
ivn-mus LD50:64 mg/kg ARZNAD 14,870,64
ims-mus LD50:148 mg/kg ARZNAD 14,870,64

SAFETY PROFILE: Poison by intraperitoneal, intravenous, and intramuscular routes. See also BROMIDES. When heated to decomposition it emits very toxic fumes of Br^- and NO_x.

BGS250 **CAS:114-90-9** **HR: 3**
1,3-BIS(4-ALDOXIMINOPYRIDINIUM) DIMETHYL ETHER BICHLORIDE
mf: $C_{14}H_{16}N_4O_3 \cdot Cl_2$ mw: 359.24

PROP: Solid. Mp: 225° (decomp).

SYNS: BH 6 □ 1,3-BIS(4-HYDROXYIMINOMETHYL-1-PYRIDINIO)-2-OXAPROPANE DICHLORIDE □ BIS(4-HYDROXYIMINOMETHYLPYRIDINIUM-1-METHYL)ETHER DICHLORIDE □ BIS(ISONICOTINALDOXIME 1-METHYL) ETHER DICHLORIDE □ BU-6 □ N,N-DIMETHYLENEOXIDEBIS(PYRIDINIUM-4-ALDOXIME) DICHLORIDE □ N,N-DIMETHYLENOXID-BIS-(PYRIDINIUM-4-ALDOXIM)-DICHLORID (GERMAN) □ ETHER BIS-14-HYDROXY-IMINOMETHYLOPYRIDINE-(1)-METYLODICHLORIDE (POLISH) □ LUEH 6 □ LUH6 □ LUH[6] □ LUH[6]-Cl2 □ LUH6-CHLORIDE □ OBIDOXIME CHLORIDE □ OBIDOXIME DICHLORIDE □ OBIDOXIME HYDROCHLORIDE □ 1,1'-(OXYBIS (METHYLENE))BIS(4-(HYDROXYIMINO)METHYL)PYRIDINIUM DICHLORIDE □ 1,1'-(OXYDIMETHYLENE)BIS(4-FORMYLPYRIDINIUM) DICHLORIDE DIOXIME □ 1,1'-(OXYDIMETHYLENE)BIS(4-FORMYLPYRIDINIUM) DIOXIME DICHLORIDE □ TOKSOBIDIN □ TOXOBI-

DIN □ TOXOGONIN □ TOXOGONIN DICHLORIDE □ TOXOGONINE

TOXICITY DATA WITH REFERENCE
ipr-rat LD50:189 mg/kg RPTOAN 38,168,75
ivn-rat LD50:133 mg/kg ARZNAD 14,5,64
ims-rat LD50:205 mg/kg RPTOAN 38,168,75
orl-mus LD50:2240 mg/kg 28ZEAL 5,168,76
ipr-mus LD50:111 mg/kg FAATDF 3,533,83
scu-mus LD50:183 mg/kg RPTOAN 38,168,75
ivn-mus LD50:70 mg/kg ARZNAD 14,870,64
ims-mus LD50:172 mg/kg ARZNAD 14,870,64

SAFETY PROFILE: Poison by intraperitoneal, intravenous, intramuscular, and subcutaneous routes. Moderately toxic by ingestion. When heated to decomposition it emits toxic fumes of Cl^- and NO_x. See also BROMIDES.

BGS500 CAS:27469-53-0 HR: 3
1-(4,6-BISALLYLAMINO-s-TRIAZINYL)-4-(p,p′-DIFLUOROBENZHYDRYL)-PIPERAZINE
mf: $C_{26}H_{29}F_2N_7$ mw: 477.62

PROP: Solid. Mp: 175–180°.

SYNS: ALMITRINA (SPANISH) □ 2,4-BIS(ALLYLAMINO) 6 (4-(BIS-(p-FLUOROPHENYL)METHYL)-1-PIPERAZINYL)-s-TRIAZINE

TOXICITY DATA WITH REFERENCE
ipr-mus LD50:390 mg/kg DRFUD4 3,717,78
ivn-mus LD50:210 mg/kg DRFUD4 3,717,78

SAFETY PROFILE: Poison by intraperitoneal and intravenous routes. When heated to decomposition it emits very toxic fumes of F^- and NO_x. See also ALLYL COMPOUNDS.

BGS750 CAS:5975-73-5 HR: 1
BIS(3-ALLYLOXY-2-HYDROXYPROPYL) FUMARATE
mf: $C_{16}H_{24}O_8$ mw: 344.40

SYN: BIS-3-ALLOXY-2-HYDROXYPROPYL-1-ESTER KYSELINY FUMAROVE (CZECH)

TOXICITY DATA WITH REFERENCE
skn-rbt 500 mg/24H MLD 28ZPAK -,100,72
eye-rbt 100 mg/24H MOD 28ZPAK -,100,72
orl-rat LD50:9710 mg/kg 28ZPAK -,100,72

SAFETY PROFILE: Mildly toxic by ingestion. A skin and eye irritant. When heated to decomposition it emits acrid smoke and irritating fumes. See also ESTERS and ALLYL COMPOUNDS.

BGS825 CAS:90566-09-9 HR: 2
4,5-BIS(ALLYLOXY)-2-IMIDAZOLINDINONE
mf: $C_9H_{14}N_2O_3$ mw: 198.25

TOXICITY DATA WITH REFERENCE
orl-mus LD50:1850 mg/kg CPBTAL 12,843,64
ipr-mus LD50:1600 mg/kg CPBTAL 12,843,64
scu-mus LD50:1650 mg/kg CPBTAL 12,843,64

SAFETY PROFILE: Moderately toxic by ingestion, subcutaneous, and intraperitoneal routes. When heated to decomposition it emits toxic fumes of NO_x. See also ALLYL COMPOUNDS.

BGT000 CAS:28434-86-8 HR: 3
BIS(4-AMINO-3-CHLOROPHENYL) ETHER
mf: $C_{12}H_{10}Cl_2N_2O$ mw: 269.14

SYNS: 3,3′-DICHLOR-4,4′-DIAMINO-DIPHENYLAETHER (GERMAN) □ 3,3′-DICHLORO-4,4′-DIAMINODIPHENYL ETHER □ 4,4′-OXYBIS(2-CHLOROANILINE) □ 4,4′-OXYBIS(2-CHLORO-BENZENAMINE)

TOXICITY DATA WITH REFERENCE
mmo-sat 100 µg/plate SAIGBL 24,498,82
dns-rat:lvr 1 µmol/L MUREAV 204,683,88
scu-rat TDLo:11 g/kg/27W-I:CAR NATWAY 57,676,70
scu-rat TD:14 g/kg/96W-I:CAR NATWAY 64,394,77

CONSENSUS REPORTS: IARC Cancer Review: Group 2B IMEMDT 7,56,87; Animal Sufficient Evidence IMEMDT 16,309,78.

SAFETY PROFILE: Suspected carcinogen with experimental carcinogenic data. Mutation data reported. When heated to decomposition it emits toxic fumes of Cl^- and NO_x. See also ETHERS.

BGT125 CAS:26493-63-0 HR: 3
BIS(2-AMINOETHYL)AMINE COBALT(III) AZIDE
mf: $C_4H_{13}CoN_{12}$ mw: 288.16

$(HN(C_2H_4NH_2)_2Co)(N_3)_3$

CONSENSUS REPORTS: Cobalt and its compounds are on the Community Right-To-Know List.

SAFETY PROFILE: A dangerous shock-sensitive explosive. Upon decomposition it emits toxic fumes of NO_x. See COBALT COMPOUNDS and AZIDES.

BGT150 CAS:59419-71-5 HR: 3
BIS(2-AMINOETHYL)AMINEDIPEROXOCHROMIUM(IV)
mf: $C_4H_{13}CrN_3O_4$ mw: 219.16

$HN(C_2H_4NH_2)_2Cr(O_2)_2$

CONSENSUS REPORTS: Chromium and its compounds are on the Community Right-To-Know List.

SAFETY PROFILE: Decomposes explosively when heated to 110°C. Upon decomposition it emits toxic fumes of NO_x. See also CHROMIUM COMPOUNDS and PEROXIDES.

BGT250 CAS:314-13-6 HR: 3
4,4′-BIS(1-AMINO-8-HYDROXY-2,4-DISULFO-7-NAPHTHYLAZO)-3,3′-BITOLYL, TETRASODIUM SALT
mf: $C_{34}H_{24}N_6O_{14}S_4 \cdot 4Na$ mw: 960.84

PROP: Blue crystals with brown/green lustre. Sol in H_2O, EtOH, acid, and alkalis.

SYNS: 4,4′ BIS(7-(1-AMINO-8-HYDROXY-2,4-DISULFO)NAPHTHYLAZO)-3,3′ -BITOLYL, TETRASODIUM SALT □ 4,4′ BIS(1-AMINO-8-HYDROXY-2,4-DISULPHO-7-NAPHTHYLAZO)-3,3′ -BITOLYL, TETRASODIUM SALT □ BLEKIT EVANSA (POLISH) □ CHLORAZOL SKY BLUE FF □ C.I. 23860 □ C.I. DIRECT BLUE 53 □ DIAMINE SKY BLUE FF □

DIAZOBLEU □ DIAZOL PURE BLUE FF □ DYE EVANS BLUE □ EB □ EVABLIN □ EVANS BLUE DYE □ GEIGY-BLAU 536 □ T 1824

TOXICITY DATA WITH REFERENCE
mma-sat 33 μg/plate CRNGDP 3,21,82
dns-rat:lvr 100 μmol/L MUREAV 136,255,84
dnd-mus-skn 192 μmol/kg CRNGDP 5,231,84
ipr-rat TDLo:70 mg/kg (female 8D post):TER PSEBAA 127,215,68
ipr-rat TDLo:140 mg/kg (female 8D post):TER TJADAB 2,85,69
ipr-rat TDLo:70 mg/kg (8D preg):REP PSEBAA 127,215,68
ipr-rat TDLo:850 mg/kg/34W-I:ETA APHGBP 33,1,53
ivn-rat LDLo:5 g/kg ARSUAX 48,17,44
ipr-mus LDLo:200 mg/kg BHJUAV 21,492,59
ivn-dog LDLo:3 g/kg ARSUAX 48,17,44
ivn-cat LDLo:1 g/kg ARSUAX 48,17,44
ivn-rbt LDLo:1 g/kg ARSUAX 48,17,44

CONSENSUS REPORTS: IARC Cancer Review: Group 3 IMEMDT 7,56,87, Animal Limited Evidence IMEMDT 8,151,75. Reported in EPA TSCA Inventory. EPA Genetic Toxicology Program.

SAFETY PROFILE: Poison by intraperitoneal route. Moderately toxic by intravenous route. An experimental teratogen. Other experimental reproductive effects. Questionable carcinogen with experimental tumorigenic data. Mutation data reported. When heated to decomposition it emits very toxic fumes of SO_x, Na_2O, and NO_x.

BGT500 CAS:2579-20-6 HR: 3
1,3-BIS(AMINOMETHYL)CYCLOHEXANE

SYNS: 1,3-CYCLOHEXANEBIS(METHYLAMINE) (8CI) □ CYCLOHEXANEDIMETHANAMINE (9CI) □ 1,3-DI(AMINOMETHYL)CYCLOHEXANE □ KODAK SILVER HALIDE SOLVENT HS-103

TOXICITY DATA WITH REFERENCE
orl-rat LD50:880 mg/kg HURC** -,-,73
skn-rat LDLo:100 mg/kg KODAK* -,-,71
ipr-rat LDLo:25 mg/kg KODAK* -,-,71

CONSENSUS REPORTS: Reported in EPA TSCA Inventory.

SAFETY PROFILE: Poison by skin contact and intraperitoneal routes. Moderately toxic by ingestion. When heated to decomposition it emits toxic fumes of NO_x.

BGT750 CAS:2549-93-1 HR: 2
1,4-BIS(AMINOMETHYL)CYCLOHEXANE
mf: $C_8H_{18}N_2$ mw: 142.28

SYNS: BAMCH □ SILVER HALIDE SOLVENT (HS103)

TOXICITY DATA WITH REFERENCE
skn-rbt 500 mg SEV SUNCO* 10/78
orl-rat LD50:530 mg/kg SUNCO* 10/78
skn-rbt LD50:420 mg/kg SUNCO* 10/78

CONSENSUS REPORTS: Reported in EPA TSCA Inventory.

SAFETY PROFILE: Moderately toxic by ingestion and skin contact. A severe skin irritant. When heated to decomposition it emits toxic fumes of NO_x.

BGU000 CAS:63077-09-8 HR: 2
BIS(2-AMINO-1-NAPHTHYL)SODIUM PHOSPHATE
mf: $C_{20}H_{17}N_2O_4P{\cdot}Na$ mw: 403.35

SYN: 2-AMINO-1-NAPHTHOL PHOSPHATE (ESTER) SODIUM SALT

TOXICITY DATA WITH REFERENCE
imp-mus TDLo:80 mg/kg:CAR BJCAAI 17,127,63

SAFETY PROFILE: Questionable carcinogen with experimental carcinogenic data. When heated to decomposition it emits very toxic fumes of PO_x, NO_x, and Na_2O. See also PHOSPHATES and ESTERS.

BGU500 CAS:4485-25-0 HR: 3
2,2-BIS(p-AMINOPHENYL)-1,1,1-TRICHLOROETHANE
mf: $C_{14}H_{13}Cl_3N_2$ mw: 315.64

SYNS: 2,2,-BIS(p-ANILINE)-1,1,1-TRICHLOROETHANE □ p,p'-DIAMINODIPHENYLTRICHLOROETHANE

TOXICITY DATA WITH REFERENCE
orl-rat LDLo:1000 mg/kg JAPMA8 37,461,48
orl-mus LDLo:250 mg/kg JAPMA8 37,461,48

SAFETY PROFILE: Poison by ingestion. When heated to decomposition it emits very toxic fumes such as Cl^- and NO_x.

BGU750 CAS:105-83-9 HR: 3
BIS(γ-AMINOPROPYL)METHYLAMINE
mf: $C_7H_{19}N_3$ mw: 145.29

PROP: Liquid, completely miscible in water. D: 0.9307 @ 20°/20°, bp: 240.6°, fp: −29.6°, flash p: 220°F.

SYNS: BIS(ω-AMINOPROPYL)METHYLAMINE □ BIS(3-AMINOPROPYL)METHYLAMINE □ N,N-BIS(γ-AMINOPROPYL)METHYLAMINE □ N,N-BIS(3-AMINOPROPYL)METHYLAMINE □ 3,7'-DIAMINO-N-METHYLDIPROPYLAMINE □ METHYLBIS(3-AMINOPROPYL)AMINE

TOXICITY DATA WITH REFERENCE
skn-rbt 100 μg/24H open AIHAAP 23,95,62
eye-rbt 5 mg SEV UCDS** 2/28/67
orl-rat LD50:1540 mg/kg UCDS** 2/28/67
ihl-rat LCLo:333 ppm/1H AIHAAP 23,95,62
skn-rbt LDLo:140 mg/kg AIHAAP 23,95,62

CONSENSUS REPORTS: Reported in EPA TSCA Inventory.

SAFETY PROFILE: Poison by inhalation and skin contact. Moderately toxic by ingestion. A skin and severe eye irritant. See also AMINES. Combustible when exposed to heat or flame. To fight fire, use foam, fog, dry chemical. When heated to decomposition it emits toxic fumes of NO_x.

BGV000 CAS:7209-38-3 HR: 3
1,4-BIS(AMINOPROPYL)PIPERAZINE
mf: $C_{10}H_{24}N_4$ mw: 200.38

SYN: BIS(AMINOPROPYL)PIPERAZINE (DOT)

TOXICITY DATA WITH REFERENCE
ivn-mus LD50:3500 μg/kg CPBTAL 20,2459,72

CONSENSUS REPORTS: Reported in EPA TSCA Inventory.

SAFETY PROFILE: Poison by intravenous route. A corrosive material and a powerful irritant to skin, eyes, and mucous membranes. When heated to decomposition it emits toxic fumes of NO_x.

BGV500 CAS:14650-81-8 HR: 2
BIS(2-AMINOTHIOPHENOL), ZINC SALT
mf: $C_{12}H_{12}N_2S_2Zn$ mw: 313.75

PROP: White powder. Sltly sol in DMF, DMSO, Py.

SYNS: o-AMINOTHIOFENOLAT ZINECNATY (CZECH) □ BIS(2-AMINOPHENYLTHIO)ZINC

TOXICITY DATA WITH REFERENCE
skn-rbt 500 mg/24H MOD 28ZPAK -,11,72
eye-rbt 2 mg/24H SEV 28ZPAK -,11,72

CONSENSUS REPORTS: Zinc and its compounds are on the Community Right-To-Know List.

SAFETY PROFILE: A skin and severe eye irritant. See also ZINC COMPOUNDS. When heated to decomposition it emits very toxic fumes of ZnO, NO_x, and SO_x.

BGW000 CAS:4193-55-9 HR: 3
4,4′-BIS((4-ANILINO-6-BIS(2-HYDROXYETHYL)AMINO-w-TRIAZIN-2-YL)AMINO)-2,2′-STILBENEDISULFONIC ACID DISODIUM SALT
mf: $C_{40}H_{40}N_{12}O_{10}S_2$•2Na mw: 959.02

TOXICITY DATA WITH REFERENCE
eye-rbt 35 mg MOD MVCRB3 2,193,73
orl-rat LD50:14,530 mg/kg MVCRB3 2,193,73
ipr-rat LD50:350 mg/kg MVCRB3 2,193,73
scu-mus LD50:1000 mg/kg MVCRB3 2,193,73
orl-gpg LD50:250 mg/kg MVCRB3 2,193,73

SAFETY PROFILE: Poison by ingestion and intraperitoneal routes. Moderately toxic by subcutaneous route. An eye irritant. See also SULFONATES. When heated to decomposition it emits very toxic fumes of NO_x, Na_2O, and SO_x.

BGW100 CAS:3426-43-5 HR: 3
4,4′-BIS((4-ANILINO-6-METHOXY-s-TRIAZIN-2-YL)AMINO)-2,2′-STILBENEDISULFONIC ACID) DISODIUM SALT
mf: $C_{34}H_{28}N_{10}O_8S_2$•2Na mw: 814.82

SYN: DISODIUM-4,4′-BIS((4-ANILINO-6-METHOXY-s-TRIAZIN-2-YL)AMINO)STILBENE-2,2′-DISULFONATE

TOXICITY DATA WITH REFERENCE
eye-rbt 100 mg MOD MVCRB3 2,193,73
ipr-rat LD50:330 mg/kg GISAAA 51(1),87,86

CONSENSUS REPORTS: Reported in EPA TSCA Inventory.

SAFETY PROFILE: Poison by intraperitoneal route. An eye irritant. When heated to decomposition it emits toxic fumes of SO_x and Cl^-.

BGW650 CAS:68979-48-6 HR: 3
1,2-BIS(AZIDOCARBONYL)CYCLOPROPANE
mf: $C_5H_4N_6O_2$ mw: 180.13

$N_3CO•CHCH_2CHCO•N_3$

SAFETY PROFILE: Spontaneously explosive. When heated to decomposition it emits toxic fumes of NO_x. See also AZIDES.

BGW700 HR: 3
BIS(2-AZIDOETHOXYMETHYL)NITRAMINE
mf: $C_6H_{12}N_8O_4$ mw: 260.21

$(N_3C_2H_4OCH_2)_2NNO_2$

SAFETY PROFILE: An impact-sensitive explosive. Upon decomposition it emits toxic fumes of NO_x. See also AZIDES.

BGW710 CAS:17607-20-4 HR: 3
3,3-BIS(AZIDOMETHYL)OXETANE
mf: $C_5H_8N_6O$ mw: 168.16

$CH_2OCH_2C(CH_2N_3)_2$

SAFETY PROFILE: A sensitive explosive. Upon decomposition it emits toxic fumes of NO_x. See also AZIDES.

BGW720 CAS:5284-80-0 HR: 1
1,5-BIS(p-AZIDOPHENYL)-1,4-PENTADIEN-3-ONE
mf: $C_{17}H_{12}N_6O$ mw: 316.35

SYNS: 1,5-BIS-(4-AZIDOFENYL)-1,4-PENTADIEN-3-ON □ DIAZIDODIBENZALACETON □ 1,4-PENTADIEN-3-ONE, 1,5-BIS(4-AZIDOPHENYL)-(9CI)

TOXICITY DATA WITH REFERENCE
skn-rbt 500 mg/24H MLD 85JCAE-,733,86
eye-rbt 500 mg/24H MLD 85JCAE-,733,86

SAFETY PROFILE: A skin and eye irritant. When heated to decomposition it emits acrid smoke and irritating fumes.

BGW750 CAS:526-62-5 HR: 3
2,5-BIS(AZIRIDINO)BENZOQUINONE
mf: $C_{10}H_{10}N_2O_2$ mw: 190.22

SYNS: BAYER G4073 □ 2,5-BIS-ATHYLENIMINOBENZOCHINON-1,4 (GERMAN) □ 2,5-BIS(1-AZIRIDYNYL)BENZOQUINONE □ 2,5-BIS-ETHYLENIMINOBENZOQUINONE □ CHINON I (GERMAN) □ QUINON I

TOXICITY DATA WITH REFERENCE
mmo-sat 10 μL/plate ANYAA9 76,475,58
mma-sat 200 μg/plate SYSWAE 12,41,79
mmo-esc 50 μg/disc APMBAY 6,23,58
sln-dmg-orl 50,000 ppm MUREAV 2,29,65
cyt-hmn:leu 200 μg/L/4H CHROAU 26,475,69
ipr-mus LD50:29,500 μg/kg AEPPAE 230,559,57

CONSENSUS REPORTS: EPA Genetic Toxicology Program.

SAFETY PROFILE: Poison by intraperitoneal route.

Human mutation data reported. When heated to decomposition it emits toxic fumes of NO_x.

BGX500 CAS:1553-36-2 HR: 3
N,N′-BIS(AZIRIDINYLACETYL)-1,8-OCTAMETHYLENE DIAMINE
mf: $C_{16}H_{30}N_4O_2$ mw: 310.50

SYN: N,N′-BIS(AZIRIDINEACETYL)-1,8-OCTAMETHYLENEDIAMINE

TOXICITY DATA WITH REFERENCE
mmo-sat 6410 mg/L MUREAV 31,115,75
cyt-rat-orl 200 µg/kg MUREAV 31,115,75
dlt-rat-orl 100 mg/kg MUREAV 31,115,75
ipr-mus TDLo: 100 mg/kg (5D male): REP EXPEAM 24,924,68
orl-rat LD50: 225 mg/kg MUREAV 31,115,75
orl-mus LD50: 1070 mg/kg EXPEAM 24,924,68
ipr-mus LD50: 88 mg/kg EXPEAM 24,924,68

SAFETY PROFILE: Poison by ingestion and intraperitoneal routes. Experimental reproductive effects. Mutation data reported. When heated to decomposition it emits toxic fumes of NO_x.

BGX750 CAS:24279-91-2 HR: 3
2,5-BIS(1-AZIRIDINYL)-3-(2-CARBAMOYLOXY-1-METHOXYETHYL)-6-METHYL-1,4-BENZOQUINONE
mf: $C_{15}H_{19}N_3O_5$ mw: 321.37

PROP: Red to reddish-brown crystals. Mp: 202° (decomp). Sltly sol in chloroform, acetone, and abs alc. Practically insol in water.

SYNS: 2,5-BIS(1-AZIRIDINYL)-3-(2-HYDROXY-1-METHOXYETHYL)-6-METHYL-p-BENZOQUINONE CARBAMATE (ESTER) ☐ CARBAZILQUINONE ☐ CARBOQUONE ☐ ESQUINON

TOXICITY DATA WITH REFERENCE
mmo-sat 2500 ng/plate TAKHAA 44,96,85
mma-sat 100 µg/plate CNREA8 38,2148,78
mmo-esc 2500 ng/plate TAKHAA 44,96,85
dnr-bcs 4 µg/plate TAKHAA 44,96,85
sce-ham:lng 10 µg/L CNREA8 44,3270,84
orl-rat LD50: 27,300 µg/kg IYKEDH 6,119,75
ipr-rat LD50: 3070 µg/kg IYKEDH 6,119,75
scu-rat LD50: 3990 µg/kg OYYAA2 8,501,74
ivn-rat LD50: 3620 µg/kg IYKEDH 6,119,75
orl-mus LD50: 28,600 µg/kg IYKEDH 6,119,75
ipr-mus LD50: 3440 µg/kg IYKEDH 6,119,75
scu-mus LD50: 4900 µg/kg OYYAA2 8,501,74
ivn-mus LD50: 5430 µg/kg OYYAA2 6,119,75

SAFETY PROFILE: A poison via ingestion, intraperitoneal, subcutaneous, and intravenous routes. Mutation data reported. When heated to decomposition it emits toxic NO_x. See also CARBAMATES.

BGX775 CAS:302-48-7 HR: 3
P,P-BIS(1-AZIRIDINYL)-N-ETHYLPHOSPHINIC AMIDE
mf: $C_6H_{14}N_3OP$ mw: 175.20

PROP: Solid. Mp: 57–61°, bp: 144° @ 5 mm.

SYNS: P,P-BIS(1-AZIRIDINYL)-N-ETHYLAMINOPHOSPHINE OXIDE ☐ ENT 50787 ☐ PHOSPHINIC AMIDE, P,P-BIS(1-AZIRIDINYL)-N-ETHYL-

TOXICITY DATA WITH REFERENCE
pic-esc 11,500 µmol/L HEREAY 68,245,71
mmo-ssp 70 mmol/L HEREAY 68,245,71
ipr-mus TDLo: 500 µg/kg (male 1D pre): REP FOBLAN 20,1,74
ipr-mus LDLo: 41 mg/kg FATOAO 28,70,65

SAFETY PROFILE: Poison by intraperitoneal route. Experimental reproductive effects. Mutation data reported. When heated to decomposition it emits toxic fumes of NO_x and PO_x.

BGY000 CAS:1078-79-1 HR: 3
BIS(1-AZIRIDINYL)(2-METHYL-3-THIAZOLIDINYL) PHOSPHINE OXIDE
mf: $C_8H_{16}N_3OPS$ mw: 233.30

SYNS: IMIPHOS ☐ MARCOPHANE ☐ MARKOFANE

TOXICITY DATA WITH REFERENCE
ipr-rat LD50: 50 mg/kg 21ACAB -,129,68
orl-mus LD50: 225 mg/kg 21ACAB -,129,68
ipr-mus LD50: 142 mg/kg 21ACAB -,129,68

SAFETY PROFILE: Poison by ingestion and intraperitoneal routes. When heated to decomposition it emits very toxic fumes of SO_x, PO_x, and NO_x.

BGY125 CAS:27807-69-8 HR: 3
N-(BIS(1-AZIRIDINYL)PHOSPHINYL)-p-CHLOROBENZAMIDE
mf: $C_{11}H_{13}ClN_3O_2P$ mw: 285.69

TOXICITY DATA WITH REFERENCE
unr-rat LD50: 26 mg/kg PCJOAU 16,626,82
scu-mus LD50: 50 mg/kg 85GDA2 1,263,80
unr-rbt LDLo: 20 mg/kg PCJOAU 6,475,72

SAFETY PROFILE: Poison by subcutaneous and possibly other routes. When heated to decomposition it emits toxic fumes of Cl^-, NO_x, and PO_x.

BGY140 CAS:27807-51-8 HR: 3
N-(BIS(1-AZIRIDINYL)PHOSPHINYL)-p-IODOBENZAMIDE
mf: $C_{11}H_{13}IN_3O_2P$ mw: 377.14

SYNS: A-19 ☐ N-p-IODOBENZOYL-N′,N′,N′,N′-DIETHYLENETRIAMIDE of PHOSPHORIC ACID

TOXICITY DATA WITH REFERENCE
unr-rat LD50: 50 mg/kg PCJOAU 16,626,82
ivn-mus LD50: 572 mg/kg IJEBA6 21,31,83
unr-rbt LDLo: 35 mg/kg PCJOAU 6,475,72

SAFETY PROFILE: Poison by unspecified routes. When heated to decomposition it emits toxic fumes of I^-, NO_x, and PO_x.

BGY500 CAS:2275-81-2 HR: 3
p,p-BIS(1-AZIRIDINYL)-N-PROPYLPHOSPHINIC AMIDE
mf: $C_7H_{16}N_3OP$ mw: 189.23

SYNS: p,p-BIS(1-AZIRIDINYL)-N-PROPYLAMINOPHOSPHINE OXIDE □ ENT 51253 □ PHOSPHINIC AMIDE, p,p-BIS(1-AZIRIDINYL)-N-PROPYL- □ PROPYLAMINO-BIS(1-AZIRIDINYL)PHOSPHINE OXIDE

TOXICITY DATA WITH REFERENCE
pic-esc 2300 µmol/L HEREAY 68,245,71
mmo-ssp 70 mmol/L HEREAY 68,245,71
ipr-mus TDLo:500 µg/kg (male 1D pre):REP FOBLAN 20,1,74
ipr-mus LDLo:25 mg/kg FATOAO 28,70,65

SAFETY PROFILE: Poison by intraperitoneal route. Experimental reproductive effects. Mutation data reported. When heated to decomposition it emits toxic fumes of NO_x and PO_x.

BGY700 CAS:1271-54-1 HR: 3
BIS-BENZENE CHROMIUM
mf: $C_{12}H_{12}Cr$ mw: 208.24

PROP: Air-sensitive brown-black crystals. Mp: 284–285°. Sol in C_6H_6; sltly sol in Et_2O.

SYNS: CHROMIUM, BIS(BENZENE)-(8CI) □ CHROMIUM, BIS(eta^6)-BENZENE)-(9CI) □ CHROMIUM(II), DIPHENYL- □ DIBENZENE-CHROMIUM □ DIPHENYLCHROMIUM

TOXICITY DATA WITH REFERENCE
ivn-mus LD50:17,800 µg/kg CSLNX* NX#02380

OSHA PEL: TWA 0.5 mg(Cr)/m^3
ACGIH TLV: TWA 0.5 mg(Cr)/m^3; Not Classifiable as a Carcinogen

SAFETY PROFILE: Poison by intravenous route. When heated to decomposition it emits toxic fumes of Cr.

BGY720 CAS:12089-29-1 HR: 3
BIS(BENZENE)CHROMIUM IODIDE
mf: $C_{12}H_{12}Cr•I$ mw: 335.14

PROP: Light-sensitive, air-stable yellow solid. Sol in H_2O and EtOH.

SYNS: BIS(BENZENE)CHROMIUM(1+)IODIDE □ CHROMIUM(1+), BIS(BENZENE)-, IODIDE (8CI) □ CHROMIUM(1+), BIS(eta^6)-BENZENE)-, IODIDE (9CI) □ CHROMIUM, BIS(BENZENE)IODO- □ CHROMIUM(III), DIPHENYL-, IODIDE □ DIBENZENECHROMIUM IODIDE □ DIPHENYLCHROMIUM(III) IODIDE

TOXICITY DATA WITH REFERENCE
ivn-mus LD50:18 mg/kg CSLNX* NX#02011

OSHA PEL: TWA 0.5 mg(Cr)/m^3; Not Classifiable as a Carcinogen
ACGIH TLV: TWA 0.5 mg(Cr)/m^3; Not Classifiable as a Carcinogen

SAFETY PROFILE: Poison by intravenous route. When heated to decomposition it emits toxic fumes of Cr and I^-.

BHA000 CAS:63950-89-0 HR: 2
BIS(BENZOATO)DIOXOCHROMIUM TRIHYDRATE
mf: $C_{14}H_{10}CrO_6•3H_2O$ mw: 380.30

SYN: KYSELINA CEROMSALICYLOVA (CZECH)

TOXICITY DATA WITH REFERENCE
eye-rbt 100 mg/24H MOD 28ZPAK -,19,72
orl-rat LD50:4810 mg/kg 28ZPAK -,19,72

CONSENSUS REPORTS: Chromium and its compounds are on the Community Right-To-Know List.

SAFETY PROFILE: Mildly toxic by ingestion. An eye irritant. See also CHROMIUM COMPOUNDS. When heated to decomposition it emits acrid smoke and irritating fumes.

BHA500 CAS:95-35-2 HR: 1
N,N′-BIS(2-BENZOTHIAZOLYLTHIOMETHYLENE) UREA
mf: $C_{17}H_{14}N_4OS_4$ mw: 418.59

SYN: 1,3-BIS((2-BENZOTHIAZOLYLTHIO)METHYL)UREA

TOXICITY DATA WITH REFERENCE
skn-hmn 500 mg/48H MLD AMIHBC 5,311,52
skn-rbt 500 mg MOD AMIHBC 5,311,52
orl-rat LD50:6000 mg/kg AMIHBC 5,311,52

CONSENSUS REPORTS: Reported in EPA TSCA Inventory.

SAFETY PROFILE: Mildly toxic by ingestion. A human skin irritant. When heated to decomposition it emits very toxic fumes of NO_x and SO_x.

BHA750 CAS:155-04-4 HR: 3
BIS(2-BENZOTHIAZOLYLTHIO)ZINC
mf: $C_{14}H_8N_2S_4•Zn$ mw: 397.85

SYNS: 2-BENZOTHIAZOLETHIOL, ZINC SALT (2:1) □ BIS(MERCAPTOBENZOTHIAZOLATO)ZINC □ HERMAT Zn-MBT □ 2-MERCAPTO-BENZOTHIAZOLE ZINC SALT □ OXAF □ PENNAC ZT □ TISPERSE MB-58 □ USAF GY-7 □ VULKACIT ZM □ ZENITE □ ZENITE SPECIAL □ ZETAX □ ZINC-2-BENZOTHIAZOLETHIOLATE □ ZINC BENZOTHIAZOLYL MERCAPTIDE □ ZINC BENZOTHIAZOL-2-YLTHIOLATE □ ZINC BENZOTHIAZYL-2-MERCAPTIDE □ ZINC MERCAPTOBENZOTHIAZOLATE □ ZINC-2-MERCAPTOBENZOTHIAZOLE □ ZINC MERCAPTOBENZOTHIAZOLE SALT □ ZMBT □ ZnMB

TOXICITY DATA WITH REFERENCE
scu-mus TDLo:1000 mg/kg:CAR NTIS** PB223-159
orl-rat LD50:540 mg/kg VCTDC* 12/9/76
ipr-mus LD50:200 mg/kg NTIS** AD277-689

CONSENSUS REPORTS: Reported in EPA TSCA Inventory. Zinc compounds are on the Community Right-To-Know List.

SAFETY PROFILE: Poison by intraperitoneal route. Moderately toxic by ingestion and subcutaneous routes. Questionable carcinogen with experimental carcinogenic data. When heated to decomposition it emits very toxic fumes of SO_x, NO_x, and ZnO. See also ZINC COMPOUNDS and MERCAPTANS.

BHB000 CAS:64092-23-5 HR: 3
BIS(2-BENZOYLBENZOATO)BIS(3-(1-METHYL-2-PYRROLIDINYL)PYRIDINE) NICKEL TRIHYDRATE
mf: $C_{48}H_{46}N_4NiO_6 \cdot 3H_2O$ mw: 887.75

SYN: NICOTINE, COMPOUND, with NICKEL(II)-o-BENZOYL BENZOATE TRIHYDRATE (2:1)

TOXICITY DATA WITH REFERENCE
orl-rat LDLo:150 mg/kg NCNSA6 5,22,53
ipr-rat LDLo:75 mg/kg NCNSA6 5,22,53

CONSENSUS REPORTS: Nickel and its compounds are on the Community Right-To-Know List.

OSHA PEL: TWA 0.1 mg(Ni)/m³
ACGIH TLV: TWA 0.1 mg(Ni)/m³; (Proposed: TWA 0.05 mg(Ni)/m³; Human Carcinogen)
NIOSH REL: (Inorganic Nickel) TWA 0.015 mg(Ni)/m³

SAFETY PROFILE: Suspected carcinogen. Poison by ingestion and intraperitoneal routes. See also NICKEL COMPOUNDS and NICOTINE. When heated to decomposition it emits toxic fumes of NO_x.

BHB100 CAS:94-01-9 HR: 1
1,3-BIS(BENZOYLOXY)BENZENE
mf: $C_{20}H_{14}O_4$ mw: 318.34

SYNS: 1,3-BENZENEDIOL, DIBENZOATE □ RESORCINOL, DIBENZOATE

TOXICITY DATA WITH REFERENCE
ipr-mus LD50:8000 mg/kg JAPMA8 46,185,57

CONSENSUS REPORTS: Reported in EPA TSCA Inventory.

SAFETY PROFILE: Slightly toxic by intraperitoneal route. When heated to decomposition it emits acrid smoke and irritating vapors.

BHB300 CAS:140-28-3 HR: 3
1,2-BIS(BENZYLAMINO)ETHANE
mf: $C_{16}H_{20}N_2$ mw: 240.38

SYNS: BENZATHINE □ BENZATIN □ DBED □ N,N′-DIBENZYLETHYLENEDIAMINE □ ETHYLENEDIAMINE, N,N′-DIBENZYL- □ USAF DO-53

TOXICITY DATA WITH REFERENCE
orl-mus LD50:388 mg/kg CNCRA6 52,579,68
ipr-mus LD50:50 mg/kg NTIS** AD277-689
par-mus LD50:80 mg/kg ANTCAO 4,633,54

CONSENSUS REPORTS: Reported in EPA TSCA Inventory.

SAFETY PROFILE: Poison by ingestion, intraperitoneal, and parenteral routes. When heated to decomposition it emits toxic vapors of NO_x.

BHB500 CAS:74037-60-8 HR: 1
(4,6-BIS(BIS(BUTOXYMETHYL)AMINO)-s-TRIAZIN-2-YLIMINO)DIMETHANOL
mf: $C_{25}H_{50}N_6O_6$ mw: 530.81

SYN: DIMETHYLOL-TETRAKIS-BUTOXYMETHYLMELAMIN (CZECH)

TOXICITY DATA WITH REFERENCE
skn-rbt 500 mg/24H MLD 28ZPAK -,157,72
eye-rbt 500 mg/24H MLD 28ZPAK -,157,72

SAFETY PROFILE: A skin and eye irritant. When heated to decomposition it emits toxic fumes of NO_x.

BHB750 CAS:4420-79-5 HR: 3
2,5-BIS(BIS-(2-CHLOROETHYL)AMINOMETHYL)HYDROQUINONE
mf: $C_{16}H_{24}Cl_4N_2O_2$ mw: 418.22

SYNS: HYDROQUINONE MUSTARD □ NSC 18321 □ WEATHERBEE MUSTARD

TOXICITY DATA WITH REFERENCE
ipr-mus TDLo:28 mg/kg/4W:CAR JNCIAM 36,915,66
ipr-rat LD10:4700 μg/kg CNCRA6 17,1,62
ivn-dog LDLo:900 μg/kg CCSUBJ 2,201,65
ivn-mky LDLo:1800 μg/kg CCSUBJ 2,201,65

SAFETY PROFILE: Deadly poison by intravenous and intraperitoneal routes. A powerful irritant. Questionable carcinogen with experimental carcinogenic data. When heated to decomposition it emits highly toxic fumes of NO_x and Cl^-.

BHB950 CAS:4028-32-4 HR: 1
4,4′-BIS((4-BIS((2-HYDROXYETHYL)AMINO)-6-CHLORO-s-TRIAZIN-2-YL)AMINO)-2,2′-STILBENEDISULFONIC ACID, DISODIUM SALT
mf: $C_{28}H_{30}Cl_2N_{10}O_{10}S_2 \cdot 2Na$ mw: 847.68

TOXICITY DATA WITH REFERENCE
eye-rbt 100 mg MOD MVCRB3 2,193,73

CONSENSUS REPORTS: Reported in EPA TSCA Inventory.

SAFETY PROFILE: An eye irritant. When heated to decomposition it emits toxic fumes of NO_x, SO_x, and Cl^-.

BHC500 CAS:4470-72-8 HR: 2
4,4′-BIS((4-BIS(2-HYDROXYETHYL)AMINO-6-METHOXY-s-TRIAZIN-2-YL)AMINO)-2,2′-STILBENEDISULFONIC ACID DISODIUM SALT
mf: $C_{30}H_{36}N_{10}O_{12}S_2 \cdot 2Na$ mw: 838.86

TOXICITY DATA WITH REFERENCE
eye-rbt 100 mg SEV MVCRB3 2,193,73

SAFETY PROFILE: A severe eye irritant. See also SULFONATES. When heated to decomposition it emits very toxic fumes of NO_x, Na_2O, and SO_x.

BHC750 CAS:12224-02-1 HR: 2
4,4′-BIS((4-BIS((2-HYDROXYETHYL)AMINO)-6-(m-SULFOANILINO)-s-TRIAZIN-2-YL)AMINO)-2,2′-STILBENEDISULFONIC ACID TETRASODIUM SALT
mf: $C_{40}H_{40}N_{12}O_{16}S_4 \cdot 4Na$ mw: 1165.12

TOXICITY DATA WITH REFERENCE
skn-rbt 500 mg/24H MLD MVCRB3 2,193,73
eye-rbt 100 mg MOD MVCRB3 2,193,73
orl-rat LD50:1960 mg/kg GISAAA 51(1),87,86
ipr-rat LD50:1750 mg/kg MVCRB3 2,193,73

orl-mus LD50:1620 mg/kg GISAAA 51(1),87,86
scu-mus LD50:1500 mg/kg MVCRB3 2,193,73
ivn-mus LD50:900 mg/kg MVCRB3 2,193,73

CONSENSUS REPORTS: Reported in EPA TSCA Inventory.

SAFETY PROFILE: Moderately toxic by ingestion, intraperitoneal, subcutaneous, and intravenous routes. A skin and eye irritant. See also SULFONATES. When heated to decomposition it emits very toxic fumes of SO_x, Na_2O, and NO_x.

BHD000 CAS:64036-79-9 HR: 3
BIS(BIS(β-HYDROXYETHYL)SULFONIUMETHYL)SULFIDE DICHLORIDE
mf: $C_{12}H_{28}O_4S_3{\bullet}2Cl$ mw: 403.48

SYN: (THIOETHYLENE)BIS(BIS(2-HYDROXYETHYL)SULFONIUM) DICHLORIDE

TOXICITY DATA WITH REFERENCE
orl-rat LDLo:250 mg/kg NCNSA6 5,9,53
scu-mus LD50:200 mg/kg NTIS** PB158-507

SAFETY PROFILE: Poison by ingestion and subcutaneous routes. When heated to decomposition it emits very toxic fumes of SO_x and Cl^-. See also SULFIDES.

BHD250 CAS:3785-34-0 HR: 3
1,2-BIS(BROMOACETOXY)ETHANE
mf: $C_6H_8Br_2O_4$ mw: 303.96

SYNS: BROMOACETIC ACID ETHYLENE ESTER □ ETHYLENE BIS (BROMOACETATE) □ ETHYLENE BROMOACETATE □ ETHYLENE GLYCOL BIS(BROMOACETATE) □ PANDUROL □ S 13

TOXICITY DATA WITH REFERENCE
ipr-mus LD50:39 mg/kg JNCIAM 31,297,63
ivn-mus LD50:56 mg/kg CSLNX* NX#03918
ivn-dog LD50:15 mg/kg JNCIAM 31,297,63

SAFETY PROFILE: Poison by intraperitoneal and intravenous routes. When heated to decomposition it emits toxic fumes of Br^-. See also BROMIDES and ESTERS.

BHJ000 CAS:2050-47-7 HR: 3
BIS(p-BROMOPHENYL) ETHER
mf: $C_{12}H_8Br_2O$ mw: 328.02

PROP: Crystals from EtOH. Mp: 58.5°, bp: 338–340°.

SYN: USAF DO-61

TOXICITY DATA WITH REFERENCE
ipr-mus LD50:125 mg/kg NTIS** AD277-689

CONSENSUS REPORTS: Reported in EPA TSCA Inventory.

SAFETY PROFILE: Poison by intraperitoneal route. See also ETHERS and BROMIDES. When heated to decomposition it emits toxic fumes of Br^-.

BHJ250 CAS:54-91-1 HR: 3
1,4-BIS(3-BROMOPROPIONYL)-PIPERAZINE
mf: $C_{10}H_{16}Br_2N_2O_2$ mw: 356.10

PROP: Crystals from H_2O. Mp: 106–107°.

SYNS: A 1803 □ A-8103 □ AMEDEL □ NSC-25154 □ PIPOBROMAN □ VERCYTE

TOXICITY DATA WITH REFERENCE
mmo-sat 1 mg/plate CNREA8 38,2148,78
mmo-esc 400 μg/plate TAKHAA 44,96,85
pic-esc 500 mg/L APMBAY 12,234,64
sce-hmn:lum 1 μmol/L CTRRDO 69,505,85
cyt-hmn:leu 1 μmol/L CNREA8 25,275,65
orl-rat TDLo:20 mg/kg (female 7-10D post):TER BTDCAV 13,103,72
orl-mus TDLo:160 mg/kg (female 6-9D post):REP BTDCAV 13,103,72
orl-rat TDLo:80 mg/kg (female 11-14D post):REP BTDCAV 13,103,72
orl-rat TDLo:120 mg/kg (female 11-14D post):TER BTDCAV 13,103,72
orl-mus TDLo:240 mg/kg (female 10-13D post):TER KAIZAN 46,19,71
orl-rat LD50:220 mg/kg IYKEDH 4,467,73
ipr-rat LD50:140 mg/kg IYKEDH 4,467,73
scu-rat LD50:139 mg/kg NIIRDN 6,638,82
orl-mus LD50:382 mg/kg IYKEDH 4,467,73
ipr-mus LD50:285 mg/kg APPHAX 37,249,80
scu-mus LD50:353 mg/kg IYKEDH 4,467,73

SAFETY PROFILE: Poison by ingestion, subcutaneous and intraperitoneal routes. An experimental teratogen. Other experimental reproductive effects. Human mutation data reported. When heated to decomposition it emits very toxic fumes of Br^- and NO_x. See also BROMIDES.

BHJ500 CAS:126-15-8 HR: 2
BISBUTENYLENETETRAHYDROFURFURAL
mf: $C_{13}H_{16}O_2$ mw: 204.29

PROP: Pale-yellow liquid. D: 1.120 @ 20°/20°, bp: 307°.

SYNS: AC-R-11 □ BUTADIEN-FURFURAL COPOLYMER □ 2,3:4,5-BIS(2-BUTYLENE)TETRAHYDRO-2-FURFURAL □ 2,3,4,5-BIS(Δ^2-BUTENYLENE)TETRAHYDROFURFURAL □ 2,3,4,5-BIS(2-BUTENYLENE)TETRAHYDROFURFURAL □ 2,3,4,5-BIS(2-BUTYLENE)TETRAHYDRO-2-FURALDEHYDE □ BIS-Δ^2-BUTYLENETETRAHYDROFURFURAL □ 2,3,4,5-BIS(Δ^2-BUTYLENE)TETRAHYDROFURFURAL □ 2,3:4,5-DI(2-BUTENYL) TETRAHYDROFURFURAL □ ENT 17,596 □ 4A-FORMYL-1,4,4A,5A,6,9,9A,9B-OCTAHYDRODIBENZOFURAN □ 2-FURALDEHYDE, 2,3:4,5-BIS (2-BUTENYLENE)TETRAHYDRO- □ 1,5A,6,9,9A,9B-HEXAHYDRO-4A (4H)-DIBENZOFURANCARBOXALDEHYDE □ MGK 11 □ MGK REPELLENT 11 □ PHILLIPS R-11 □ R-11

TOXICITY DATA WITH REFERENCE
orl-rat LD50:2500 mg/kg MEIEDD 10,1170,83
ivn-rat LD50:2 g/kg YKYUA6 32,605,81

CONSENSUS REPORTS: Reported in EPA TSCA Inventory.

SAFETY PROFILE: Moderately toxic by ingestion. An insect repellant. When heated to decomposition it emits acrid smoke and fumes. See also ALDEHYDES.

BHJ625 **CAS:91216-69-2** **HR: 2**
4,5-BIS(2-BUTENYLOXY)-2-IMIDAZOLIDINONE
mf: $C_{11}H_{18}N_2O_3$ mw: 226.31

SYN: SRC-15

TOXICITY DATA WITH REFERENCE
orl-mus LD50:1700 mg/kg CPBTAL 12,843,64
ipr-mus LD50:770 mg/kg CPBTAL 12,843,64
scu-mus LD50:1750 mg/kg CPBTAL 12,843,64

SAFETY PROFILE: Moderately toxic by ingestion, subcutaneous, and intraperitoneal routes. When heated to decomposition it emits toxic fumes of NO_x.

BHK000 **CAS:117-83-9** **HR: 1**
BIS(2-BUTOXYETHYL)PHTHALATE
mf: $C_{20}H_{30}O_6$ mw: 366.50

SYNS: 2-BUTOXYETHANOL PHTHALATE (2:1) □ β-BUTOXYETHYL PHTHALATE □ BUTYL "CELLOSOLVE" PHTHALATE □ BUTYL GLYCOL PHTHALATE □ DI(BUTOXYETHYL)PHTHALATE □ DIBUTYL CELLOSOLVE PHTHALATE □ DIBUTYLGLYCOL PHTHALATE □ KESSCOFLEX □ KRONISOL

TOXICITY DATA WITH REFERENCE
orl-rat LD50:8380 mg/kg JIHTAB 30,63,48
orl-gpg LDLo:6000 mg/kg 29ZWAE -,336,68

CONSENSUS REPORTS: Reported in EPA TSCA Inventory.

SAFETY PROFILE: Mildly toxic by ingestion. When heated to decomposition it emits acrid smoke and irritating fumes. See also ESTERS.

BHK250 **CAS:15546-16-4** **HR: 3**
BIS(BUTOXYMALEOYLOXY)DIBUTYLSTANNANE
mf: $C_{24}H_{40}O_8Sn$ mw: 575.33

SYNS: DI-N-BUTYLTIN DI(MONOBUTYL)MALEATE □ DI-N-BUTYLZINN-DI(MONOBUTYL)MALEINAT (GERMAN)

TOXICITY DATA WITH REFERENCE
orl-rat LD50:120 mg/kg ARZNAD 19,934,69

CONSENSUS REPORTS: Reported in EPA TSCA Inventory.

OSHA PEL: TWA 0.1 mg(Sn)/m³ (skin)
ACGIH TLV: TWA 0.1 mg(Sn)/m³ (skin) (Proposed: TWA 0.1 mg(Sn)/m³; STEL 0.2 mg(Sn)/m³ (skin))
NIOSH REL: (Organotin Compounds) TWA 0.1 mg(Sn)/m³

SAFETY PROFILE: Poison by ingestion. See also TIN COMPOUNDS. When heated to decomposition it emits acrid smoke and irritating fumes.

For occupational chemical analysis use NIOSH: Organotin Compounds 5504.

BHK500 **CAS:29575-02-8** **HR: 2**
BIS(BUTOXYMALEOYLOXY)DIOCTYLSTANNANE
mf: $C_{32}H_{56}O_8Sn$ mw: 687.57

SYNS: DI-N-OCTYLTIN BIS(BUTYL MALEATE) □ DI-N-OCTYLTIN DIMONOBUTYLMALEATE □ DI-N-OCTYLZINN-DIMONOBUTYLMALEINAT (GERMAN)

TOXICITY DATA WITH REFERENCE
orl-rat LD50:2030 mg/kg ARZNAD 19,934,69
orl-mus LD50:3750 mg/kg FCTXAV 8,655,70

CONSENSUS REPORTS: Reported in EPA TSCA Inventory.

OSHA PEL: TWA 0.1 mg(Sn)/m³ (skin)
ACGIH TLV: TWA 0.1 mg(Sn)/m³ (skin) (Proposed: TWA 0.1 mg(Sn)/m³; STEL 0.2 mg(Sn)/m³ (skin))
NIOSH REL: (Organotin Compounds) TWA 0.1 mg(Sn)/m³

SAFETY PROFILE: Moderately toxic by ingestion. See also TIN COMPOUNDS. When heated to decomposition it emits acrid smoke and irritating fumes.

For occupational chemical analysis use NIOSH: Organotin Compounds 5504.

BHK750 **CAS:143-29-3** **HR: 2**
BIS(BUTYLCARBITOL)FORMAL
mf: $C_{17}H_{36}O_6$ mw: 336.53

SYNS: BUTYLCARBITOL FORMAL □ CRYOFLEX □ DIBUTYLCARBITOLFORMAL □ 5,8,11,13,16,19-HEXAOXATRICOSANE (9CI) □ TP 90B

TOXICITY DATA WITH REFERENCE
orl-rat LD50:1746 mg/kg NPIRI* 2,238,75
orl-mus LD50:2700 mg/kg GISAAA 46(5),87,81

CONSENSUS REPORTS: Reported in EPA TSCA Inventory.

SAFETY PROFILE: Moderately toxic by ingestion. When heated to decomposition it emits acrid smoke and irritating fumes.

BHL100 **CAS:25155-25-3** **HR: 1**
α-α′-BIS(tert-BUTYLPEROXY)DIISOPROPYLBENZENE
mf: $C_{20}H_{34}O_4$ mw: 338.54

SYNS: PEROXIDE, (PHENYLENEBIS(1-METHYLETHYLIDENE))BIS(1,1-DIMETHYLETHYL)- □ PEROXIDE, (PHENYLENEDIISOPROPYLIDENE)BIS(tert-BUTYL- □ (PHENYLENEDIISOPROPYLIDENE)BIS(tert-BUTYLPEROXIDE) □ VUL-CUP □ VUL-CUP 40KE □ VUL-CUP R

TOXICITY DATA WITH REFERENCE
skn-rbt 100%/24H MLD HERBU* PRC-304

CONSENSUS REPORTS: Reported in EPA TSCA Inventory.

SAFETY PROFILE: A skin irritant. When heated to decomposition it emits acrid smoke and irritating fumes.

BHL500 **CAS:1000-40-4** **HR: 3**
BIS(BUTYLTHIO)DIMETHYLTIN
mf: $C_{10}H_{24}S_2Sn$ mw: 327.15

SYN: BIS(BUTYLTHIO)DIMETHYL STANNANE

TOXICITY DATA WITH REFERENCE
ivn-mus LD50:320 mg/kg CSLNX* NX#01865

OSHA PEL: TWA 0.1 mg(Sn)/m³ (skin)
ACGIH TLV: TWA 0.1 mg(Sn)/m³ (skin) (Proposed: TWA 0.1 mg(Sn)/m³; STEL 0.2 mg(Sn)/m³ (skin))

NIOSH REL: (Organotin Compounds) TWA 0.1 mg(Sn)/m³

SAFETY PROFILE: Poison by intravenous route. See also TIN COMPOUNDS. When heated to decomposition it emits toxic fumes of SO_x.

For occupational chemical analysis use NIOSH: Organotin Compounds 5504.

BHL750 CAS:15263-52-2 HR: 3
1,3-BIS(CARBAMOYLTHIO)-2-(N,N-DIMETHYLAMINO) PROPANE HYDROCHLORIDE
mf: $C_7H_{15}N_3O_2S_2 \cdot ClH$ mw: 273.83

SYNS: CALDAN □ CARBAMOTHIOIC ACID-S,S′-(2-(DIMETHYLAMINO)-1,3-PROPANEDIYL) ESTER, MONOHYDROCHLORIDE (9CI) □ CARTAP HYDROCHLORIDE □ S,S′-(2-(DIMETHYLAMINO)TRIMETHYLENE)BIS(THIOCARBAMATE) HYDROCHLORIDE □ NTD 2 □PA-DAN □ PATAP □ SANVEX □ THIOBEL □ THIOCARBAMIC ACID-S,S-(2-(DIMETHYLAMINO)TRIMETHYLENE)ESTER HYDROCHLORIDE □ TI-1258 □ VEGETOX

TOXICITY DATA WITH REFERENCE
orl-rat TDLo:700 mg/kg (9-15D preg):TER TAKHAA 30,776,71
orl-rat LD50:250 mg/kg SPEADM 78-1,61,78
orl-mus LD50:165 mg/kg SPEADM 78-1,61,78
ivn-mus LD50:59 mg/kg JJPAAZ 17,491,67

CONSENSUS REPORTS: EPA Genetic Toxicology Program.

SAFETY PROFILE: Poison by ingestion and intravenous routes. An experimental teratogen. An insecticide. When heated to decomposition it emits very toxic fumes of NO_x, SO_x, and HCl. See also CARBAMATES.

BHM000 CAS:111-17-1 HR: 3
BIS(2-CARBOXYETHYL) SULFIDE
mf: $C_6H_{10}O_4S$ mw: 178.22

PROP: Very sol in alc, hot water, acetate; sltly sol in water. Mp: 134°.

SYNS: DIETHYL SULFIDE-2,2′-DICARBOXYLIC ACID □ KYSELINA-β,β′-THIODIPROPIONOVA (CZECH) □ TDPA □ 2-(2,3,5,6-TETRAMETHYLPHENOXY)PROPIONIC ACID □ 4-THIAHEPTANEDIOIC ACID □ THIODIPROPIONIC ACID □ β,β′-THIODIPROPIONIC ACID □ 3,3′-THIODIPROPIONIC ACID □ TYOX A

TOXICITY DATA WITH REFERENCE
skn-rbt 500 mg/24H MLD 28ZPAK -,171,72
eye-rbt 20 mg/24H MOD 28ZPAK -,171,72
orl-rat LD50:3980 mg/kg 28ZPAK -,171,72
ipr-rat LD50:500 mg/kg AFREAW 3,197,51
orl-mus LD50:2000 mg/kg AFREAW 3,197,51
ipr-mus LD50:250 mg/kg AFREAW 3,197,51
ivn-mus LD50:175 mg/kg AFREAW 3,197,51

CONSENSUS REPORTS: Reported in EPA TSCA Inventory.

SAFETY PROFILE: A poison by intraperitoneal and intravenous routes. Moderately toxic by ingestion. A skin and eye irritant. When heated to decomposition it emits toxic fumes of SO_x. See also SULFIDES.

BHM300 CAS:119-80-2 HR: 3
BIS(2-CARBOXYPHENYL) DISULFIDE
mf: $C_{14}H_{10}O_4S_2$ mw: 306.36

SYNS: BENZOIC ACID, 2,2′-DITHIOBIS-(9CI) □ BENZOIC ACID, 2,2′-DITHIODI- □ BIS(o-CARBOXYPHENYL) DISULFIDE □ 2,2′-DITHIOBIS(BENZOIC ACID) □ 2,2′-DITHIODIBENZOESAEURE □ 2,2′-DITHIODIBENZOIC ACID

TOXICITY DATA WITH REFERENCE
ipr-mus LD50:367 mg/kg ARZNAD 21,284,71

CONSENSUS REPORTS: Reported in EPA TSCA Inventory.

SAFETY PROFILE: Poison by intraperitoneal route. When heated to decomposition it emits toxic vapors of SO_x.

BHM500 HR: 3
BIS(3-CARBOXYPROPIONYL)PEROXIDE
mf: $C_8H_{10}O_8$ mw: 234.16

SAFETY PROFILE: Explodes on contact with flame. Commercial grade (dry 95%) is highly hazardous. When heated to decomposition it emits acrid smoke and fumes. See also PEROXIDES, ORGANIC.

BHM750 CAS:94-17-7 HR: 2
BIS(p-CHLOROBENZOYL) PEROXIDE
mf: $C_{14}H_8Cl_2O_4$ mw: 311.12

PROP: A white, granular material. Insol in water; sol in organic solvents.

SYNS: CADPX PS □ p-CHLOROBENZOYL PEROXIDE (DOT) □ p,p′-DICHLOROBENZOYL PEROXIDE □ DI-(4-CHLOROBENZOYL) PEROXIDE

TOXICITY DATA WITH REFERENCE
ipr-mus LDLo:500 mg/kg CBCCT• 4,110,52

CONSENSUS REPORTS: Reported in EPA TSCA Inventory.

SAFETY PROFILE: Moderately toxic by intraperitoneal route. Probably an irritant to skin and mucous membranes. Dangerous fire hazard; a powerful oxidizer. Store in a cool place away from fire hazards, sparks, open flames, and out of the direct rays of the sun. Dangerous explosion hazard; this material may explode by heat (over 38°) or contamination. Any contaminant that acts as an accelerator to the polymerization or decomposition of this material can cause an explosion. Heat or contact with certain fumes or mists can cause it to explode. To fight small fires, use CO_2 or foam extinguishers. Water spray or mist may also be used. Dry chemical is effective. When heated to decomposition it emits toxic fumes of Cl^-. See also PEROXIDES, ORGANIC.

BHN000 CAS:366-93-8 HR: 3
trans-N,N′-BIS(2-CHLOROBENZYL)-1,4-CYCLOHEXANEBIS(METHYLAMINE) DIHYDROCHLORIDE
mf: $C_{22}H_{28}Cl_2N_2 \cdot 2ClH$ mw: 464.34

SYNS: AY 9944 □ trans-1,4-BIS(2-DICHLOROBENZYLAMINOETHYL)

CYCLOHEXANE DICHLORHYDRATE (FRENCH) □ trans-N,N'-(1,4-CYCLOHEXYLENEDIMETHYLENE)BIS(2-CHLOROBENZYLAMINE) DIHYDROCHLORIDE

TOXICITY DATA WITH REFERENCE
unr-rat TDLo: 225 mg/kg (female 2-4D post): REP TJADAB 29(3),32A,84
orl-mus TDLo: 700 mg/kg (female 1-14D post): TER CRSBAW 163,327,69
orl-rbt TDLo: 650 mg/kg (female 1-13D post): REP CRSBAW 163,327,69
orl-rat TDLo: 50 mg/kg (female 4D post): TER PSEBAA 176,54,84
orl-rat TDLo: 150 mg/kg (female 2-4D post): TER CRSBAW 171,15,77
unr-rat TDLo: 225 mg/kg (female 2-4D post): TER TJADAB 29(3),32A,84
orl-rat TDLo: 150 mg/kg (2-4 Dpreg): TER CRSBAW 171(1),15,77
orl-mus LD50: 155 mg/kg PSEBAA 139,100,72

SAFETY PROFILE: Poison by ingestion. Experimental teratogenic and reproductive effects. Inhibits cholesterol synthesis. When heated to decomposition it emits very toxic fumes of NO_x and Cl^-.

BHN500 CAS:3374-04-7 HR: 3
N,N-BIS(β-CHLOROETHYL)-dl-ALANINE HYDROCHLORIDE
mf: $C_7H_{13}Cl_2NO_2{\cdot}ClH$ mw: 250.57

SYNS: ALANINE MUSTARD □ NSC 17663

TOXICITY DATA WITH REFERENCE
ice-rat LD50: 225 μg/kg JPPMAB 18,760,66
unk-man TDLo: 900 μg/kg: UNS CCROBU 50,219,66
ivn-dog LDLo: 1 mg/kg CCSUBJ 2,201,65
ivn-mky LDLo: 1 mg/kg CCSUBJ 2,201,65

SAFETY PROFILE: Deadly poison by intracerebral and intravenous routes. Human systemic effects by an unspecified route: bone marrow changes. When heated to decomposition it emits very toxic fumes of Cl^-, NO_x and HCl.

BHN750 CAS:334-22-5 HR: 3
BIS-β-CHLOROETHYLAMINE
mf: $C_4H_9Cl_2N$ mw: 142.04

SYNS: N,N-BIS-(β-CHLORAETHYL)-AMIN (GERMAN) □ NH-LOST □ NOR-NITROGEN MUSTARD □ NSC-10873

TOXICITY DATA WITH REFERENCE
mmo-sat 100 μmol/L CNREA8 41,2967,81
cyt-hmn: lym 1 mg/L CRNGDP 5,1637,84
ipr-mus TDLo: 50 mg/kg (female 12D post): REP TCMUD8 7,7,87
ipr-rat LD50: 97 mg/kg JMCMAR 8,167,65
ivn-rat LD50: 100 mg/kg ARZNAD 24,1149,74
scu-mus LD50: 20 mg/kg JPETAB 91,224,47
ivn-dog LDLo: 6 mg/kg CCSUBJ 2,201,65
ivn-mky LDLo: 11 mg/kg CCSUBJ 2,201,65

SAFETY PROFILE: Poison by intraperitoneal, subcutaneous, and intravenous routes. Experimental reproductive effects. Human mutation data reported. When heated to decomposition it emits very toxic NO_x and Cl^-.

BHO250 CAS:821-48-7 HR: 3
BIS(2-CHLOROETHYL)AMINE HYDROCHLORIDE
mf: $C_4H_9Cl_2N{\cdot}ClH$ mw: 178.50

SYNS: BIS(β-CHLOROETHYL)AMINE HYDROCHLORIDE □ N,N-BIS(2-CHLORO ETHYL)AMINE HYDROCHLORIDE □ BIS(2-CHLOROETHYL)AMMONIUM CHLORIDE □ 2-CHLORO-N-(2-CHLOROETHYL)ETHANAMINE HYDROCHLORIDE □ β,β'-DICHLORODIETHYLAMINE HYDROCHLORIDE □ 2,2'-DICHLORO DIETHYLAMINE HYDROCHLORIDE □ DI-2-CHLOROETHYLAMINE HYDROCHLORIDE □ LEO 72a □ NC 26 □ NOR-HN2 □ NOR-HN2 HYDROCHLORIDE □ NOR-LOST HYDROCHLORID (GERMAN) □ NORNITROGEN MUSTARD HYDROCHLORIDE □ NSC 10873 □ SK 555 □ TL 161

TOXICITY DATA WITH REFERENCE
hmn-lym 1 mg/L CRNGDP 5,163,84
hmn-lym 250 μg/L CRNGDP 5,163,84
mmo-sat 50 μg/plate PNASA6 72,979,75
ipr-mus TDLo: 2500 μg/kg (11D preg): TER TJADAB 4,141,71
ipr-mus TDLo: 40 mg/kg (11D preg): TER TJADAB 4,141,71
ipr-rat LD50: 100 mg/kg ARZNAD 11,143,61
ims-rat LD50: 160 mg/kg ZKKOBW 84,227,75
ihl-mus LCLo: 1000 mg/m³/10M NDRC** NDCrc-132,July,42
scu-mus LD50: 20 mg/kg JPETAB 91,224,47

CONSENSUS REPORTS: EPA Genetic Toxicology Program. Reported in EPA TSCA Inventory.

SAFETY PROFILE: A poison by inhalation, intraperitoneal, intramuscular, and subcutaneous routes. An experimental teratogen. Human mutation data reported. When heated to decomposition it emits toxic fumes of NH_3, NO_x, and Cl^-.

BHO500 CAS:1215-16-3 HR: 3
4'-(BIS(2-CHLOROETHYL)AMINO)ACETANILIDE
mf: $C_{12}H_{16}Cl_2N_2O$ mw: 275.20

SYNS: p-ACETYLAMINOPHENYL DERIVATIVE of NITROGEN MUSTARD □ LONIN 3

TOXICITY DATA WITH REFERENCE
mmo-sat 1 μmol/plate MUREAV 224,95,89
mmo-smc 808 μmol/L MUREAV 224,95,89
scu-mus TDLo: 36 mg/kg (female 7-9D post): TER JEEMAF 9,492,61
ipr-rat LD50: 28 mg/kg JMCMAR 8,167,65
ipr-mus LD50: 27 mg/kg JMCMAR 8,167,65

SAFETY PROFILE: Poison via intraperitoneal route. An experimental teratogen. Mutation data reported. When heated to decomposition it emits very toxic fumes of Cl^- and NO_x.

BHP125 CAS:1141-37-3 HR: 3
4-(BIS(2-CHLOROETHYL)AMINO)BENZOIC ACID
mf: $C_{11}H_{13}Cl_2NO_2$ mw: 262.15

SYN: p-(BIS(2-CHLOROETHYL)AMINO)BENZOIC ACID

TOXICITY DATA WITH REFERENCE
ipr-rat LD50: 96 mg/kg JMCMAR 8,167,65

unr-rat LD50:63 mg/kg NEOLA4 27,261,80
ipr-mus LD50:87 mg/kg JMCMAR 8,167,65

SAFETY PROFILE: Poison by intraperitoneal and possibly other routes. When heated to decomposition it emits toxic fumes of Cl^- and NO_x.

BHP150 CAS:24813-03-4 HR: 3
1-((BIS(2-CHLOROETHYL)AMINO)BENZOYL)PIPERIDINE
mf: $C_{16}H_{22}Cl_2N_2O$ mw: 329.30

SYNS: KETONE, m-(BIS(2-CHLOROETHYL)AMINO)PHENYL PIPERIDINO □ PIPERIDINE, 1-(m-(BIS(2-CHLOROETHYL)AMINO)BENZOYL)-

TOXICITY DATA WITH REFERENCE
unr-rat LD50:134 mg/kg NEOLA4 27,271,80

DOT CLASSIFICATION: 3; *Label:* Flammable Liquid

SAFETY PROFILE: A poison by an unspecified route. A flammable liquid. When heated to decomposition it emits toxic vapors of NO_x and Cl^-.

BHP500 CAS:63978-55-2 HR: 3
2-(BIS(2-CHLOROETHYL)AMINO)ETHYL VINYL SULFONE
mf: $C_8H_{15}Cl_2NO_2S$ mw: 260.20

SYN: VINYL(β-BIS(β-CHLOROETHYL)AMINO)ETHYL SULFONE

TOXICITY DATA WITH REFERENCE
scu-mus LD50:9 mg/kg JPETAB 93,1,48
ivn-rbt LD50:2550 µg/kg JPETAB 93,1,48

SAFETY PROFILE: Poison by subcutaneous and intravenous routes. See also SULFONATES. When heated to decomposition it emits very toxic fumes of Cl^-, NO_x, and SO_x.

BHP750 CAS:1492-93-9 HR: 3
4'-(BIS(2-CHLOROETHYL)AMINO)-2-FLUORO ACETANILIDE
mf: $C_{12}H_{15}Cl_2FN_2O$ mw: 293.19

SYN: p-FLUOROACETYLAMINOPHENYL DERIVATIVE of NITROGEN MUSTARD

TOXICITY DATA WITH REFERENCE
scu-mus TDLo:45 mg/kg (female 7-9D post):TER JEEMAF 9,492,61
ipr-rat LD50:7916 µg/kg JMCMAR 8,167,65
ipr-mus LD50:34 mg/kg JMCMAR 8,167,65

SAFETY PROFILE: Poison by intraperitoneal route. An experimental teratogen. When heated to decomposition it emits very toxic fumes of Cl^-, F^-, and NO_x.

BHQ750 CAS:7751-31-7 HR: 3
3-(BIS(2-CHLOROETHYL)AMINOMETHYL)-2-BENZOXAZOLINONE
mf: $C_{12}H_{14}Cl_2N_2O_2$ mw: 289.18

SYN: 3-(BIS-(2-CHLORAETHYL)AMINOMETHYL)BENZOXAZOLON-(2) (GERMAN)

TOXICITY DATA WITH REFERENCE
ims-rat LD50:8 mg/kg ZKKOBW 84,227,75
ipr-mus LD50:42 mg/kg ZKKOBW 84,227,75

SAFETY PROFILE: Poison by intramuscular and intraperitoneal routes. When heated to decomposition it emits very toxic fumes of Cl^- and NO_x.

BHQ760 CAS:21447-86-9 HR: 3
1-(3-(BIS(2-CHLOROETHYL)AMINO-4-METHYLBENZOYL)AZIRIDINE)
mf: $C_{14}H_{18}Cl_2N_2O$ mw: 301.24

SYNS: AZIRIDINE, 1-(3-(BIS(2-CHLOROETHYL)AMINO-p- TOLUOYL))- □ KETONE, 1-AZIRIDINYL 3-(BIS(2-CHLOROETHYL)AMINO)-p-TOLYL

TOXICITY DATA WITH REFERENCE
unr-rat LD50:17 mg/kg NEOLA4 27,271,80

DOT CLASSIFICATION: 3; *Label:* Flammable Liquid

SAFETY PROFILE: A poison by an unspecified route. A flammable liquid. When heated to decomposition it emits toxic vapors of NO_x and Cl^-.

BHR400 CAS:21447-39-2 HR: 3
1-(3-(BIS(2-CHLOROETHYL)AMINO)-4-METHYLBENZOYL)MORPHOLINE
mf: $C_{16}H_{22}Cl_2N_2O_2$ mw: 345.30

SYNS: KETONE, 3-(BIS(2-CHLOROETHYL)AMINO)-p-TOLYL MORPHOLINO- □ MORPHOLINE, 4-(3-(BIS(2-CHLOROETHYL)AMINO)-p-TOLUOYL)-

TOXICITY DATA WITH REFERENCE
unr-rat LD50:17 mg/kg NEOLA4 27,271,80

DOT CLASSIFICATION: 3; *Label:* Flammable Liquid

SAFETY PROFILE: A poison by an unspecified route. A flammable liquid. When heated to decomposition it emits toxic vapors of NO_x and Cl^-.

BHR500 CAS:10070-95-8 HR: 3
4-(BIS(2-CHLOROETHYL)AMINOMETHYL)-2,3-DIMETHYL-1-PHENYL-3-PYRAZOLIN-5-ONE HYDROCHLORIDE
mf: $C_{16}H_{21}Cl_2N_3O{\cdot}ClH$ mw: 378.76

SYN: 4-(BIS-(2-CHLORAETHYL)AMINOMETHYL)-1-PHENYL-2,3-DIMETHYLPYRAZOLON HYDROCHLORID (GERMAN)

TOXICITY DATA WITH REFERENCE
ims-rat LD50:30 mg/kg ZKKOBW 84,227,75
ipr-mus LD50:580 mg/kg ARZNAD 16,634,66

SAFETY PROFILE: Poison by intramuscular route. Moderately toxic by intraperitoneal route. When heated to decomposition it emits very toxic fumes of HCl and NO_x.

BHR750 CAS:2089-46-5 HR: 3
4-(BIS(2-CHLOROETHYL)AMINO)PHENOL
mf: $C_{27}H_{38}N_2O{\cdot}2BrH$ mw: 568.51

SYNS: 2,6-BIS(1-PIPERIDYLMETHYL)-4-(α,α-DIMETHYLBENZYL) PHENOL DIHYDROBROMIDE □ 4-α,α-DIMETHYLBENZYL-α,α'-DIPIPERIDINO-2,6-XYLENOL DIHYDROBROMIDE □ 4-(1-METHYL-1-PHE-

NYLETHYL)-2,6-BIS-(1-PIPERIDINYLMETHYL)PHENOL DIHYDROBROMIDE □ RO 2-5803 □ RYTHMOL

TOXICITY DATA WITH REFERENCE
ipr-rat LD50:17 mg/kg JMCMAR 8,167,65
orl-mus LD50:330 mg/kg AIPTAK 132,295,61
ipr-mus LD50:15 mg/kg JMCMAR 8,167,65
ivn-mus LD50:30 mg/kg AIPTAK 132,295,61
ims-gpg LD50:48,720 μg/kg ARPMAS 313,142,80

SAFETY PROFILE: Poison by ingestion, intravenous, intramuscular and intraperitoneal routes. When heated to decomposition it emits very toxic fumes of NO_x and HBr.

BHS250 CAS:66232-25-5 HR: 3
2-(N,N-BIS(2-CHLOROETHYL)AMINOPHENYL) ACETIC ACID BUTYL ESTER
mf: $C_{16}H_{23}Cl_2NO_2$ mw: 332.1

TOXICITY DATA WITH REFERENCE
orl-rat LD50:20 mg/kg PCJOAU 12,205,78
orl-mus LD50:15 mg/kg PCJOAU 12,205,78

SAFETY PROFILE: Poison by ingestion. See also ESTERS. When heated to decomposition it emits very toxic fumes of Cl^- and NO_x.

BHS500 CAS:66276-87-7 HR: 3
2-(N,N-BIS(2-CHLOROETHYL)AMINOPHENYL)ACETIC ACID DECYL ESTER
mf: $C_{22}H_{35}Cl_2NO_2$ mw: 416.2

TOXICITY DATA WITH REFERENCE
orl-rat LD50:150 mg/kg PCJOAU 12,205,78
orl-mus LD50:50 mg/kg PCJOAU 12,205,78

SAFETY PROFILE: Poison by ingestion. See also ESTERS. When heated to decomposition it emits very toxic fumes of Cl^- and NO_x.

BHS750 CAS:66232-30-2 HR: 3
2-(N,N-BIS(2-CHLOROETHYL)AMINOPHENYL)ACETIC ACID OCTADECYL ESTER
mf: $C_{30}H_{51}Cl_2NO_2$ mw: 529.2

TOXICITY DATA WITH REFERENCE
orl-rat LD50:200 mg/kg PCJOAU 12,205,78
orl-mus LD50:140 mg/kg PCJOAU 12,205,78

SAFETY PROFILE: Poison by ingestion. See also ESTERS. When heated to decomposition it emits very toxic fumes of Cl^- and NO_x.

BHT000 CAS:66232-28-8 HR: 3
2-(N,N-BIS(2-CHLOROETHYL)AMINOPHENYL)ACETIC ACID TETRADECYL ESTER
mf: $C_{26}H_{43}Cl_2NO_2$ mw: 472.2

TOXICITY DATA WITH REFERENCE
cyt-rat:oth 150 mg/L/24H-C TXAPA9 22,355,72
orl-rat LD50:46 mg/kg PCJOAU 12,205,78
orl-mus LD50:10 mg/kg PCJOAU 12,205,78

SAFETY PROFILE: Poison by ingestion. Mutation data reported. See also ESTERS. When heated to decomposition it emits very toxic fumes of Cl^- and NO_x.

BHT250 CAS:342-95-0 HR: 3
3-(o-(BIS-(β-CHLOROETHYL)AMINO)PHENYL)-dl-ALANINE
mf: $C_{13}H_{18}Cl_2N_2O_2$ mw: 305.23

SYNS: CB 1729 □ o-DI-2-CHLOROETHYLAMINO-dl-PHENYLALANINE □ FDA 0109 □ MEROPHAN □ o-MEROPHAN □ NSC-57199 □ ORTHOPHENYLALANINE MUSTARD □ (±)-o-PHENYLALANINE MUSTARD □ o-PHENYLALANINE MUSTARD □ o-dl-SARCOLYSIN

TOXICITY DATA WITH REFERENCE
ipr-rat LD50:3510 μg/kg BCPCA6 13,969,64
scu-mus LD50:18,480 μg/kg NCISP* JAN86
ivn-dog LDLo:190 μg/kg CCSUBJ 2,201,65
ivn-mky LDLo:380 μg/kg CCSUBJ 2,201,65

SAFETY PROFILE: Deadly poison by intraperitoneal, subcutaneous, and intravenous routes. When heated to decomposition it emits very toxic fumes of Cl^- and NO_x.

BHT750 CAS:531-76-0 HR: 3
dl-3-(p-(BIS(2-CHLOROETHYL)AMINO)PHENYL)ALANINE
mf: $C_{13}H_{18}Cl_2N_2O_2$ mw: 305.23

PROP: Needles from MeOH. Mp: 180–181°.

SYNS: 3-(p-(BIS(2-CHLOROETHYL)AMINO)PHENYL)ALANINE □ 4-(BIS(2-CHLOROETHYL)AMINO)-dl-PHENYLALANINE □ CB-3307 □ p-DI-(2-CHLORAETHYL)-AMINO-dl-PHENYL-ALANIN (GERMAN) □ p-DI (2-CHLOROETHYL)AMINO-dl-PHENYLALANINE □ MERFALAN □ MERPHALAN □ o-MERPHALAN □ NCI-C04944 □ NSC-14210 □ PHENYLALANIN-LOST (GERMAN) □ dl-PHENYLALANINE MUSTARD □ SAKOLYSIN (GERMAN) □ SARCOCLORIN □ dl-SARCOLYSIN □ dl-SARCOLYSINE

TOXICITY DATA WITH REFERENCE
mmo-omi 10 mmol/L MUREAV 23,5,74
pic-omi 5 mmol/L MUREAV 1,355,64
ipr-rat TDLo:5 mg/kg (female 6D post):TER DKBSAS 171,801,66
ipr-rat TDLo:8 mg/kg (female 4D post):REP DKBSAS 171,801,66
ipr-rat TDLo:8 mg/kg (6D preg):TER DKBSAS 171,801,66
ipr-mus TDLo:98 mg/kg/26W-I:ETA CANCAR 40S,1935,77
orl-rat LD50:105 mg/kg AICCA6 20,144,64
ipr-rat LD50:18 mg/kg DKBSAS 171,801,66
ivn-rat LD50:25 mg/kg ARZNAD 8,340,58
ice-rat LD50:250 μg/kg JPPMAB 18,760,66
orl-mus LD50:35 mg/kg XPHPAW 441,165,74
ipr-mus LD50:26 mg/kg ARZNAD 16,634,66

CONSENSUS REPORTS: IARC Cancer Review: Group 2B IMEMDT 7,56,87; Animal Limited Evidence IMEMDT 9,167,75; NCI Carcinogenesis Studies (ipr); Clear Evidence: mouse CANCAR 40,1935,77; No Evidence: rat CANCAR 40,1935,77.

SAFETY PROFILE: Suspected carcinogen with experimental tumorigenic data. A poison by ingestion, intraperitoneal, intravenous, and intracerebral routes. An experimental teratogen. Other experimental reproductive effects. Mutation data reported. An antineoplastic agent. When heated to decomposition it emits very toxic fumes of Cl^- and NO_x.

BHU500 **CAS:4213-34-7** **HR: 3**
3-(m-(BIS(β-CHLOROETHYL)AMINO)PHENYL)-dl-ALANINE HYDROCHLORIDE
mf: $C_{13}H_{18}Cl_2N_2O_2 \cdot ClH$ mw: 341.69

SYNS: METAPHENYLALANINE MUSTARD □ NCS-27381

TOXICITY DATA WITH REFERENCE
ivn-dog LDLo: 430 μg/kg CCSUBJ 2,201,65
ivn-mky LDLo: 860 μg/kg CCSUBJ 2,201,65

SAFETY PROFILE: Poison by intravenous route. When heated to decomposition it emits very toxic fumes of Cl^-, NO_x, and HCl.

BHU750 **CAS:4213-32-5** **HR: 3**
3-(p-(BIS(β-CHLOROETHYL)AMINO)PHENYL)-d-ALANINE HYDROCHLORIDE
mf: $C_{13}H_{18}Cl_2N_2O_2 \cdot ClH$ mw: 341.69

SYNS: 4-(BIS(2-CHLOROETHYL)AMINO)-d-PHENYLALANINE MONOHYDROCHLORIDE □ NSC-35051 □ PHENYLALANINE MUSTARD □ d-PHENYLALANINE MUSTARD

TOXICITY DATA WITH REFERENCE
ivn-dog LDLo: 2 mg/kg CCSUBJ 2,201,65
ivn-mky LDLo: 2 mg/kg CCSUBJ 2,201,65

SAFETY PROFILE: An intravenous poison. When heated to decomposition it emits very toxic fumes of Cl^-, NO_x, and HCl.

BHV000 **CAS:1465-26-5** **HR: 3**
3-(p-(BIS(β-CHLOROETHYL)AMINO)PHENYL)-dl-ALANINE HYDROCHLORIDE
mf: $C_{13}H_{18}Cl_2N_2O_2 \cdot ClH$ mw: 341.69

SYNS: ALKERAN (RUSSIAN) □ 4-(BIS(2-CHLOROETHYL)AMINO)-dl-PHENYLALANINE MONOHYDROCHLORIDE □ CB 3008 □ MELPHALAN (RUSSIAN) □ MERPHALAN HYDROCHLORIDE □ NCS-14210 □ dl-PHENYLALANINE MUSTARD HYDROCHLORIDE □ dl-SARCOLYSINE HYDROCHLORIDE □ SARCOLYSIN HYDROCHLORIDE □ SARKOKLORIN □ SKI 21739

TOXICITY DATA WITH REFERENCE
mmo-sat 100 μg/plate KHFZAN 16(10),11,82
mma-sat 100 μg/plate KHFZAN 16(10),11,82
dnd-esc 20 μmol/L MUREAV 89,95,81
pic-omi 5 mmol/L MUREAV 1,355,64
otr-mus:emb 1 μmol/L CBINA8 38,75,81
orl-rat LD50: 52 mg/kg GTPZAB 11(3),28,67
ipr-rat LD50: 16,200 μg/kg GTPZAB 11(3),28,67
orl-mus LD50: 44,600 μg/kg GTPZAB 11(3),28,67
ivn-dog LDLo: 430 μg/kg CCSUBJ 2,201,65
ivn-mky LDLo: 860 μg/kg CCSUBJ 2,201,65

SAFETY PROFILE: Poison by ingestion, intraperitoneal, and intravenous routes. Human mutation data reported. When heated to decomposition it emits very toxic fumes of Cl^- and NO_x.

BHV250 **CAS:3223-07-2** **HR: 3**
l-3-(p-(BIS(2-CHLOROETHYL)AMINO)PHENYL)ALANINE MONOHYDROCHLORIDE
mf: $C_{13}H_{18}Cl_2N_2O_2 \cdot ClH$ mw: 341.69

SYNS: ALANINE NITROGEN MUSTARD □ CB 3025 □ MELPHALAN HYDROCHLORIDE □ NSC-8806 □ l-PHENYLALANINE MUSTARD HYDROCHLORIDE □ l-SARCOLYSINE HYDROCHLORIDE

TOXICITY DATA WITH REFERENCE
ipr-mus TDLo: 1300 μg/kg/4W:CAR JNCIAM 36,915,66
orl-hmn TDLo: 1200 μg/kg/5D-I:GIT CCROBU 57,369,73
ivn-dog LDLo: 430 μg/kg CCSUBJ 2,201,65
ivn-mky LDLo: 430 μg/kg CCSUBJ 2,201,65

SAFETY PROFILE: Deadly poison by intravenous route. Human systemic effects by ingestion: nausea and vomiting. Questionable carcinogen with experimental carcinogenic data. When heated to decomposition it emits very toxic fumes of Cl^-, NO_x, and HCl.

BHV500 **CAS:1233-89-2** **HR: 3**
p-(BIS(2-CHLOROETHYL)AMINO)PHENYL BENZOATE
mf: $C_{17}H_{17}Cl_2NO_2$ mw: 338.25

SYN: p-(BIS(2-CHLOROETHYL)AMINO)PHENOL BENZOATE

TOXICITY DATA WITH REFERENCE
ipr-rat LD50: 20 mg/kg JMCMAR 12,491,69
ipr-mus LD50: 18,500 μg/kg JMCMAR 12,491,69

SAFETY PROFILE: Poison by intraperitoneal route. See also ESTERS. When heated to decomposition it emits very toxic fumes of Cl^- and NO_x.

BHV750 **CAS:22953-53-3** **HR: 3**
p-(BIS(2-CHLOROETHYL)AMINO)PHENYL-p-BROMOBENZOATE
mf: $C_{17}H_{16}BrCl_2NO_2$ mw: 417.15

SYN: p-(BIS(2-CHLOROETHYL)AMINO)PHENOL-p-BROMOBENZOATE

TOXICITY DATA WITH REFERENCE
ipr-rat LD50: 27 mg/kg JMCMAR 12,491,69
ipr-mus LD50: 19 mg/kg JMCMAR 12,491,69

SAFETY PROFILE: Poison by intraperitoneal route. When heated to decomposition it emits very toxic fumes of Br^-, Cl^-, and NO_x.

BHW000 **CAS:22953-54-4** **HR: 3**
p-(BIS(2-CHLOROETHYL)AMINO)PHENYL-m-CHLOROBENZOATE
mf: $C_{17}H_{16}Cl_3NO_2$ mw: 372.69

SYN: p-(BIS(2-CHLOROETHYL)AMINO)PHENOL-m-CHLOROBENZOATE

TOXICITY DATA WITH REFERENCE
ipr-rat LD50: 42 mg/kg JMCMAR 12,491,69
ipr-mus LD50: 21 mg/kg JMCMAR 12,491,69

SAFETY PROFILE: Poison by intraperitoneal route. When heated to decomposition it emits very toxic fumes of Cl^- and NO_x.

BHW250 **CAS:21667-01-6** **HR: 3**
p-(BIS(2-CHLOROETHYL)AMINO)PHENYL-2,6-DIMETHYLBENZOATE
mf: $C_{19}H_{21}Cl_2NO_2$ mw: 366.31

SYN: p-(BIS(2-CHLOROETHYL)AMINO)PHENOL-2,6-DIMETHYLBENZOATE

TOXICITY DATA WITH REFERENCE
ipr-rat LD50:290 mg/kg JMCMAR 12,491,69
ipr-mus LD50:500 mg/kg JMCMAR 12,491,69

SAFETY PROFILE: Poison by intraperitoneal route. When heated to decomposition it emits very toxic fumes of Cl^- and NO_x.

BHW300 **CAS:5185-77-3** **HR: 3**
2-(p-BIS(2-CHLOROETHYL)AMINOPHENYL)-1,3,2-DITHIARSENOLANE
mf: $C_{12}H_{16}AsCl_2NS_2$ mw: 384.23

SYN: 1,3,2-DITHIARSENOLANE, 2-(p-BIS(2-CHLOROETHYL)AMINOPHENYL)-

TOXICITY DATA WITH REFERENCE
ipr-mus LD50:15 mg/kg JMCMAR 9,221,66

OSHA PEL: TWA 0.5 mg(As)/m³

SAFETY PROFILE: Poison by intraperitoneal route. When heated to decomposition it emits toxic fumes of NO_x, SO_x, As, and Cl^-.

BHW500 **CAS:4465-92-3** **HR: 3**
4-(p-BIS(β-CHLOROETHYLAMINO)PHENYLETHYLAMINO)-7-CHLOROQUINOLINE MONOHYDROCHLORIDE

SYN: NSC-50982

TOXICITY DATA WITH REFERENCE
ivn-dog LDLo:11 mg/kg CCSUBJ 2,202,65
ivn-mky LDLo:11 mg/kg CCSUBJ 2,202,65

SAFETY PROFILE: Poison by intravenous route. When heated to decomposition it emits very toxic fumes of Cl^- and NO_x.

BHX250 **CAS:35849-41-3** **HR: 3**
l-3-(p-(BIS(2-CHLOROETHYL)AMINO)PHENYL)-N-FORMYLALANINE
mf: $C_{14}H_{18}Cl_2N_2O_3$ mw: 333.24

SYN: N-FORMYL-l-p-SARCOLYSIN

TOXICITY DATA WITH REFERENCE
orl-hmn TDLo:3 mg/kg:GIT XPHPAW 441,9,74
orl-rat LD50:700 mg/kg XPHPAW 441,9,74
ipr-rat LD50:80 mg/kg XPHPAW 441,9,74
orl-mus LD50:730 mg/kg XPHPAW 441,9,74
ipr-mus LD50:242 mg/kg NCISA* PH-43-63-1132

SAFETY PROFILE: Poison by intraperitoneal route. Moderately toxic by ingestion. Human gastrointestinal effects by ingestion. When heated to decomposition it emits very toxic fumes of Cl^- and NO_x.

BHX300 **CAS:4587-15-9** **HR: 3**
m-(BIS(2-CHLOROETHYL)AMINO)PHENYL MORPHOLINO KETONE
mf: $C_{15}H_{20}Cl_2N_2O_2$ mw: 331.27

TOXICITY DATA WITH REFERENCE
unr-rat LD50:40 mg/kg NEOLA4 27,261,80

DOT CLASSIFICATION: 3; *Label:* Flammable Liquid

SAFETY PROFILE: A poison by an unreported route. A flammable liquid. When heated to decomposition it emits toxic vapors of NO_x and Cl^-.

BHY500 **CAS:857-95-4** **HR: 3**
o-(4-(BIS(2-CHLOROETHYL)AMINO)PHENYL)-dl-TYROSINE
mf: $C_{19}H_{22}Cl_2N_2O_3$ mw: 397.26

SYN: PHENTYRIN

TOXICITY DATA WITH REFERENCE
orl-rat LD50:620 mg/kg FATOAO 43,100,80
ipr-rat LD50:115 mg/kg FATOAO 43,100,80
ivn-rat LD50:62 mg/kg FATOAO 43,100,80
orl-mus LD50:360 mg/kg FATOAO 43,100,80
ipr-mus LD50:110 mg/kg FATOAO 43,100,80
ivn-mus LD50:30 mg/kg FATOAO 43,100,80

SAFETY PROFILE: A poison by ingestion, intravenous, and intraperitoneal routes. When heated to decomposition it emits very toxic fumes of Cl^- and NO_x.

BHY625 **CAS:64508-90-3** **HR: 3**
5-(p-(BIS(2-CHLOROETHYL)AMINO)PHENYL)VALERIC ACID
mf: $C_{15}H_{21}Cl_2NO_2$ mw: 318.27

SYNS: 4-(BIS(2-CHLOROETHYL)AMINO)-BENZENEPENTANOIC ACID (9CI) □ 5-(4-BIS(2-CHLOROETHYL)AMINOPHENYL)PENTANOIC ACID □ CB 1356 □ p-N,N-DI-(2-CHLOROETHYL)AMINOPHENYLVALERIC ACID

TOXICITY DATA WITH REFERENCE
sln-dmg-par 5 mmol/L GENRA8 1,173,60
sln-dmg-unr 10 μmol/L ANYAA9 160,228,69
ipr-rat LDLo:50 mg/kg BCPCA6 5,192,60

CONSENSUS REPORTS: EPA Genetic Toxicology Program.

SAFETY PROFILE: Poison by intraperitoneal route. Mutation data reported. When heated to decomposition it emits toxic fumes of Cl^- and NO_x.

BHY750 **CAS:63815-37-2** **HR: 3**
β-(BIS(2-CHLOROETHYLAMINO))PROPIONITRILE
mf: $C_7H_{13}Cl_2N_3$ mw: 210.13

SYN: USAF UCTL-958

TOXICITY DATA WITH REFERENCE
ihl-mus LCLo:660 mg/m³/10M NDRC** No.9-4-1-19,44
ipr-mus LD50:10 mg/kg NTIS** AD277-689

SAFETY PROFILE: Poison by inhalation and intraperitoneal routes. When heated to decomposition it emits very toxic fumes of Cl^- and NO_x.

BHZ000 **CAS:38915-00-3** **HR: 3**
9-((3-(BIS(2-CHLOROETHYL)AMINO)PROPYL)AMINO) ACRIDINE DIHYDROCHLORIDE
mf: $C_{20}H_{23}Cl_2N_3 \cdot 2ClH$ mw: 449.28

SYN: ICR 220

TOXICITY DATA WITH REFERENCE
msc-ham:ovr 1 g/L CNREA8 39,4875,79
ipr-mus LD20:1 mg/kg JMCMAR 15,739,72

SAFETY PROFILE: Mutation data reported. Poison by intraperitoneal route. When heated to decomposition it emits very toxic fumes of HCl, NO_x, and Cl^-.

BIA000 **CAS:4213-40-5** **HR: 3**
o-(4-BIS(β-CHLOROETHYL)AMINO-o-TOLYLAZO)BENZOIC ACID
mf: $C_{18}H_{19}Cl_2N_3O_2$ mw: 380.30

SYN: NSC-16498

TOXICITY DATA WITH REFERENCE
sln-dmg-par 35 µg BCPCA6 5,206,60
ipr-rat LD50:45 mg/kg JNCIAM 50,243,73
ivn-dog LDLo:950 µg/kg CCSUBJ 2,202,65
ivn-mky LDLo:950 µg/kg CCSUBJ 2,202,65

SAFETY PROFILE: Poison by intraperitoneal and intravenous routes. Mutation data reported. When heated to decomposition it emits very toxic fumes of Cl^- and NO_x.

BIA100 **CAS:21447-87-0** **HR: 3**
3-(BIS(2-CHLOROETHYL)AMINO)-p-TOLYL PIPERIDYL KETONE
mf: $C_{17}H_{24}Cl_2N_2O$ mw: 343.33

SYNS: 1-(3-(BIS(2-CHLOROETHYL)AMINO)-p-TOLUOYL)PIPERIDINE □ KETONE, 3-(BIS(2-CHLOROETHYL)AMINO)-p-TOLYL PIPERIDYL- □ PIPERIDINE, 1-(3-(BIS(2-CHLOROETHYL)AMINO)-p-TOLUOYL)-

TOXICITY DATA WITH REFERENCE
unr-rat LD50:50 mg/kg CTRRDO 62,2045,78
unr-mus LD50:50 mg/kg JMCMAR 12,161,69

DOT CLASSIFICATION: 3; *Label:* Flammable Liquid

SAFETY PROFILE: A poison by an unreported route. A flammable liquid. When heated to decomposition it emits toxic vapors of NO_x and Cl^-.

BIA250 **CAS:66-75-1** **HR: 3**
5-(BIS(2-CHLOROETHYL)AMINO)URACIL
mf: $C_8H_{11}Cl_2N_3O_2$ mw: 252.12

PROP: Crystals from MeOH (aq). Mp: 206° (decomp).

SYNS: AMINOURACIL MUSTARD □ 5-(BIS(2-CHLOROETHYL)AMINO)-2,4(1H,3H)PYRIMIDINEDIONE □ 5-N,N-BIS(2-CHLOROETHYL) AMINOURACIL □ CB-4835 □ CHLORETHAMINACIL □ DEMETHYLDOPAN □ DESMETHYLDOPAN □ 5-(DI-(β-CHLOROETHYL)AMINO) URACIL □ 5-(DI-2-CHLOROETHYL)AMINOURACIL □ 2,6-DIHYDROXY-5-BIS(2-CHLOROETHYL)AMINOPYRAMIDINE □ ENT 50,439 □ NCI-C04820 □ NORDOPAN □ NSC-34462 □ RCRA WASTE NUMBER U237 □ SK-19849 □ U-8344 □ URACILLOST □ URACILMOSTAZA □ URACIL MUSTARD □ URAMUSTIN □ URAMUSTINE

TOXICITY DATA WITH REFERENCE
mmo-sat 125 µg/plate JNCIAM 62,893,79
msc-mus:lym 150 µg/L/2H MUREAV 59,61,79
ipr-rat TDLo:1000 µg/kg/26W-I:NEO RRCRBU 52,1,75
ipr-mus TDLo:240 µg/kg/4W:CAR JNCIAM 36,915,66
ipr-mus TD:9500 µg/kg/26W-I:NEO RRCRBU 52,1,75
ipr-mus TD:8 mg/kg/3W-I:NEO CNREA8 33,3069,73
ipr-mus TD:30 mg/kg/39W-I:CAR SCIEAS 147,1443,65
orl-rat LD50:3550 µg/kg NYKZAU 60,413,64
ipr-rat LD50:1250 µg/kg ADTEAS 3,181,68
ipr-mus LDLo:3 mg/kg TXAPA9 23,288,72

CONSENSUS REPORTS: IARC Cancer Review: Group 2B IMEMDT 7,370,87; Animal Sufficient Evidence IMEMDT 9,235,75; NCI Carcinogenesis Studies (ipr); Clear Evidence: mouse, rat RRCRBU 52,1,75. EPA Genetic Toxicology Program.

SAFETY PROFILE: Suspected carcinogen with experimental carcinogenic and neoplastigenic data. A deadly poison by ingestion and intraperitoneal routes. Mutation data reported. When heated to decomposition it emits very toxic fumes of Cl^- and NO_x.

BIA300 **CAS:5185-71-7** **HR: 3**
N,N-BIS(2-CHLOROETHYL)-p-ARSANILIC ACID
mf: $C_{10}H_{14}AsCl_2NO_3$ mw: 342.07

SYN: p-ARSANILIC ACID, N,N-BIS(2-CHLOROETHYL)-

TOXICITY DATA WITH REFERENCE
ipr-mus LD50:8789 µg/kg JMCMAR 9,221,66

OSHA PEL: TWA 0.5 mg(As)/m³

SAFETY PROFILE: Poison by intraperitoneal route. When heated to decomposition it emits toxic fumes of NO_x, As, and Cl^-.

BIA750 **CAS:55-51-6** **HR: 3**
N,N-BIS(2-CHLOROETHYL)BENZYLAMINE
mf: $C_{11}H_{15}Cl_2N$ mw: 232.17

SYNS: BENZYLBIS(β-CHLOROETHYL)AMINE □ BENZYL NORMECHLORETHAMINE □ N,N-BIS(2-CHLOROETHYL)BENZENEMETHANAMINE □ BIS(2-CHLOROETHYL)BENZYLAMINE □ DCBA □ DI-(2-CHLOROETHYL)BENZYLAMINE □ TL 965

TOXICITY DATA WITH REFERENCE
dni-mus-ivg 5000 ppm JIDEAE 62,378,74
mmo-asn 2500 µmol/L MUREAV 14,115,72
unr-rat LD50:10 mg/kg PHBUA9 1,297,53
scu-mus LDLo:80 mg/kg NDRC** No.9-4-1-9,43

CONSENSUS REPORTS: EPA Genetic Toxicology Program.

SAFETY PROFILE: Poison by ingestion and possibly other routes. Mutation data reported. When heated to decomposition it emits very toxic fumes of Cl^- and NO_x. See also AROMATIC AMINES.

BIB250 CAS:55112-89-5 HR: 3
N,N-BIS(2-CHLOROETHYL)BUTYLAMINE HYDROCHLORIDE
mf: $C_8H_{17}Cl_2N \cdot ClH$ mw: 234.62

SYNS: BUTYLBIS(β-CHLOROETHYL)AMINE HYDROCHLORIDE □ N-BUTYL-BIS(2-CHLOROETHYLAMINE) HYDROCHLORIDE □ TL 513 HYDROCHLORIDE

TOXICITY DATA WITH REFERENCE
ipr-mus LD50:4890 μg/kg CANCAR 2,1055,49
scu-mus LDLo:2 mg/kg NTIS** PB158-507

SAFETY PROFILE: Poison by intraperitoneal and subcutaneous routes. When heated to decomposition it emits very toxic fumes of NO_x and HCl.

BIC325 HR: 3
1,4-BIS(2-CHLOROETHYL)-1,4-DIAZONIABICYCLO (2.2.1)HEPTANE (Z)-2-BUTENEDIOATE (1:2)
mf: $C_9H_{18}Cl_2N_2 \cdot 2C_4H_4O_4$ mw: 457.35

SYN: NSC 262666

TOXICITY DATA WITH REFERENCE
orl-mus LD50:746 mg/kg NCISP* JAN86
ipr-mus LD50:210 mg/kg NCISP* JAN86
ivn-mus LD50:73,290 μg/kg NCISP* JAN86

SAFETY PROFILE: Poison by intraperitoneal and intravenous routes. Moderately toxic by ingestion. When heated to decomposition it emits toxic fumes of Cl^- and NO_x.

BIC500 CAS:63918-36-5 HR: 3
N,N′-BIS(2-CHLOROETHYL)-N,N′-DIETHYLETHYLENEDIAMINE DIHYDROCHLORIDE
mf: $C_{10}H_{22}Cl_2N_2 \cdot 2ClH$ mw: 314.16

SYN: N,N′-ETHYL-N,N′-(β-CHLOROETHYL)ETHYLENEDIAMINE DIHYDROCHLORIDE

TOXICITY DATA WITH REFERENCE
ipr-rat LD50:2300 μg/kg JPETAB 100,398,50
ipr-mus LD50:3141 μg/kg JPETAB 94,249,48

SAFETY PROFILE: Deadly poison by intraperitoneal route. When heated to decomposition it emits very toxic fumes of Cl^- and NO_x.

BIC600 CAS:4213-41-6 HR: 2
N,N-BIS(2-CHLOROETHYL)-2,3-DIMETHOXYANILINE
mf: $C_{12}H_{17}Cl_2NO_2$ mw: 278.20

SYNS: ANILINE, N,N-BIS(2-CHLOROETHYL)-2,3-DIMETHOXY- □ 2,3-DIMETHOXYANILINE MUSTARD □ NSC-18439

TOXICITY DATA WITH REFERENCE
ipr-mus TDLo:27 mg/kg/4W:CAR JNCIAM 36,915,66

SAFETY PROFILE: Questionable carcinogen with experimental carcinogenic data. When heated to decomposition it emits toxic fumes of NO_x.

BID000 CAS:6986-48-7 HR: 2
BIS(α-CHLOROETHYL) ETHER
mf: $C_4H_8Cl_2O$ mw: 143.02

PROP: Liquid. Bp: 112.5–114°.

SYN: 1,1′-OXYBIS(1-CHLOROETHANE)

TOXICITY DATA WITH REFERENCE
scu-mus TDLo:648 mg/kg/54W-I:ETA JNCIAM 48,1431,72

SAFETY PROFILE: Questionable carcinogen with experimental tumorigenic data. See also ETHERS. When heated to decomposition it emits toxic fumes of Cl^-.

BID250 CAS:538-07-8 HR: 3
BIS(2-CHLOROETHYL)ETHYLAMINE
mf: $C_6H_{13}Cl_2N$ mw: 170.10

SYNS: 2,2′-DICHLOROTRIETHYLAMINE □ ETHYLBIS(β-CHLOROETHYL)AMINE □ ETHYLBIS(2-CHLOROETHYL)AMINE □ ETHYL-S □ HN1 □ TL 329 □ TL 1149

TOXICITY DATA WITH REFERENCE
skn-rat LD50:17 mg/kg JPETAB 91,224,47
ivn-rat LD50:500 μg/kg NTIS** PB158-507
ihl-mus LC50:900 mg/m³/10M NTIS** PB158-508
skn-mus LD50:13 mg/kg JPETAB 91,224,47
ipr-mus LDLo:1030 μg/kg NTIS** PB158-507
scu-mus LDLo:1100 μg/kg NTIS** PB158-507
ihl-dog LC50:800 mg/m³/10M NTIS** PB158-508
skn-mus LD50:13 mg/kg JPETAB 91,224,47
ipr-mus LD50:1030 μg/kg NTIS** PB158-507
scu-mus LD50:1100 μg/kg NTIS** PB158-507
ihl-dog LC50:800 mg/m³/10M NTIS** PB158-508
skn-dog LDLo:40 mg/kg NTIS** PB158-507
ihl-mky LC50:1500 mg/m³/10M NTIS** PB158-508
ihl-cat LC50:400 mg/m³/10M NTIS** PB158-508
ihl-rbt LC50:900 mg/m³/20M NTIS** PB158-508
ihl-gpg LC50:1500 mg/m³/30M NTIS** PB158-508
ihl-dom LC50:1500 mg/m³/10M NTIS** PB158-508
skn-dog LDLo:40 mg/kg NTIS** PB158-507
skn-rbt LD50:15 mg/kg JPETAB 91,224,47
ivn-rbt LDLo:2 mg/kg NTIS** PB158-507

CONSENSUS REPORTS: Reported in EPA TSCA Inventory. EPA Extremely Hazardous Substances List.

SAFETY PROFILE: Deadly poison by inhalation, skin contact, ingestion, intravenous, subcutaneous, and intraperitoneal routes. When heated to decomposition it emits very toxic fumes of Cl^- and NO_x.

BID750 CAS:111-91-1 HR: 3
BIS(β-CHLOROETHYL)FORMAL
mf: $C_5H_{10}Cl_2O_2$ mw: 173.05

PROP: Liquid. Bp: 217.5°, flash p: 230°F (OC), d: 1.23, vap d: 5.9.

SYNS: BIS(2-CHLOROETHOXY)METHANE □ BIS(2-CHLOROETHYL)FORMAL □ DICHLOROETHYL FORMAL □ DI-2-CHLOROETHYL FORMAL □ FORMALDEHYDE BIS(β-CHLOROETHYL) ACETAL □ 1,1′-(METHYLENEBIS(OXY)BIS(2-CHLOROETHANE)) □ RCRA WASTE NUMBER U024

TOXICITY DATA WITH REFERENCE
skn-rbt 10 mg/24H open JIHTAB 30,63,48
eye-rbt 500 mg AJOPAA 29,1363,46
orl-rat LD50:65 mg/kg JIHTAB 30,63,48
ihl-rat LCLo:62 ppm/4H JIHTAB 31,343,49
skn-gpg LD50:170 mg/kg JIHTAB 30,63,48

CONSENSUS REPORTS: Reported in EPA TSCA Inventory.

SAFETY PROFILE: Poison by ingestion, inhalation, and skin contact. A skin and eye irritant. Combustible when exposed to heat or flame. Incompatible with oxidizers. To fight fire, use alcohol foam, foam, CO_2, dry chemical. When heated to decomposition it emits toxic fumes of Cl^-. See also CHLORIDES.

BIE250 CAS:51-75-2 HR: 3
BIS(β-CHLOROETHYL)METHYLAMINE
mf: $C_5H_{11}Cl_2N$ mw: 156.07

PROP: Dark liquid. Mp: 1° @ 10 mm, bp: 86–87° @ 11 mm, d: 1.09 @ 25°, vap press: 0.17 mm @ 25°, vap d: 5.9. Sltly sol in water.

SYNS: BIS(2-CHLOROETHYL)METHYLAMINE □ N,N-BIS(2-CHLOROETHYL)METHYLAMINE □ CARYOLYSIN □ CHLORMETHINE □ CLORAMIN □ DICHLORAMINE □ DICHLOREN (GERMAN) □ β,β′-DICHLORODIETHYL-N-METHYLAMINE □ DI(2-CHLOROETHYL)METHYLAMINE □ 2,2′-DICHLORO-N-METHYLDIETHYLAMINE □ EMBICHIN □ ENT 25,294 □ HN2 □ MBA □ MECHLORETHAMINE □ N-METHYL-BIS-CHLORAETHYLAMIN (GERMAN) □ METHYLBIS(β-CHLOROETHYL)AMINE □ N-METHYL-BIS(β-CHLOROETHYL)AMINE □ N-METHYL-BIS(2-CHLOROETHYL)AMINE (MAK) □ N-METHYL-2,2′-DICHLORODIETHYLAMINE □ METHYLDI(2-CHLOROETHYL)AMINE □ N-METHYL-LOST □ MUSTARGEN □ MUSTINE □ MUTAGEN □ NITROGEN MUSTARD □ N-LOST (GERMAN) □ NSC 762 □ TL 146

TOXICITY DATA WITH REFERENCE
eye-rbt 400 μg SEV AJOPAA 29,1553,46
eye-rbt 20 μg/30M INOPAO 15,308,76
mmo-sat 40 μg/plate CNREA8 37,2209,77
dnr-bcs 10 μg/plate TAKHAA 44,96,85
dns-hmn:fbr 160 μg/L TXCYAC 21,151,81
sce-hmn:lym 6250 ng/L CRNGDP 5,1637,84
ipr-mus TDLo:4 mg/kg (female 6-7D post):REP JPETAB 101,362,51
scu-rat TDLo:1 mg/kg (female 12D post):TER JEEMAF 18,215,67
par-mus TDLo:1 mg/kg (female 11D post):TER TCMUD8 3,281,83
ivn-rat TDLo:200 μg/kg (female 10D post):REP JLCMAK 36,833,50
ipr-mus TDLo:800 μg/kg (male 1D pre):REP CNREA8 42,122,82
ipr-rat TDLo:300 μg/kg (female 12D post):TER CANCAR 9,955,56
ipr-rat TDLo:800 μg/kg (female 4D post):TER ARZFAN 20,47,66
skn-man TDLo:153 mg/kg/3Y-C:CAR ADVEA4 58,421,78
skn-wmn TDLo:5840 mg/kg/8Y-I:CAR ADVEA4 58,421,78
ivn-rat TDLo:5720 μg/kg/1Y-I:CAR ARZNAD 20,1461,70
skn-mus TDLo:60 mg/kg/14W-I:ETA BJCAAI 9,177,55
ipr-mus TDLo:69 mg/kg/39W-I:ETA SCIEAS 147,1443,65
ivn-mus TDLo:10 mg/kg/42D-I:NEO CNREA8 25,20,65
skn-mus TD:276 mg/kg/23W-I:ETA EXPEAM 36,1211,80
skn-wmn TD:256 mg/kg/8.5Y-I:CAR ADVED7 111,127,84
orl-rat LD50:10 mg/kg NTIS** PB158-507
ihl-rat LC50:600 mg/m³/2M NTIS** PB158-508
skn-rat LD50:12 mg/kg FAATDF 5,S160,85
ivn-rat LD50:1100 μg/kg NTIS** PB158-507
orl-mus LD50:10 mg/kg NTIS** PB158-507
ihl-mus LC50:1500 mg/m³/30M NTIS** PB158-508
skn-mus LD50:29 mg/kg JPETAB 91,224,47
ihl-dog LC50:2 g/m³/10M NTIS** PB158-508
ihl-rbt LC50:1 g/m³/5M NTIS** PB158-508
skn-rbt LD50:12 mg/kg NTIS** PB158-507

CONSENSUS REPORTS: EPA Genetic Toxicology Program. Reported in EPA TSCA Inventory. EPA Extremely Hazardous Substances List. Community Right-To-Know List.

DFG MAK: Human Carcinogen

SAFETY PROFILE: Confirmed human carcinogen producing skin tumors by skin contact. Experimental carcinogenic, tumorigenic, and neoplastigenic data. A deadly poison by inhalation, ingestion, skin contact, and most other routes. Experimental teratogenic and reproductive effects. A powerful skin and eye irritant. Human mutation data reported. It has been used as a blistering agent in chemical warfare. When heated to decomposition it emits very toxic fumes of Cl^- and NO_x.

BIE500 CAS:55-86-7 HR: 3
BIS(2-CHLOROETHYL)METHYLAMINE HYDROCHLORIDE
mf: $C_5H_{11}Cl_2N{\cdot}ClH$ mw: 192.53

PROP: Leaflets from Me_2CO or $CHCl_3$. Mp: 119°.

SYNS: ANTIMIT □ AZOTOYPERITE □ C 6866 □ CAROLYSINE □ CARYOLYSINE □ CARYOLYSINE HYDROCHLORIDE □ CHLORAMIN □ CHLORAMINE □ CHLORAMIN HYDROCHLORIDE □ CHLORETHAMINE □ CHLORETHAZINE □ CHLORMETHINE HYDROCHLORIDE □ CHLORMETHINUM □ 2-CHLORO-N-(2-CHLOROETHYL)-N-METHYLETHANAMINE HYDROCHLORIDE □ DEMA □ DICHLOREN □ DICHLOREN HYDROCHLORIDE □ β,β′-DICHLORODIETHYL-N-METHYLAMINE HYDROCHLORIDE □ DI(2-CHLOROETHYL)METHYLAMINE HYDROCHLORIDE □ 1,5-DICHLORO-3-METHYL-3-AZAPENTANE HYDROCHLORIDE □ 2,2′-DICHLORO-N-METHYLDIETHYLAMINE HYDROCHLORIDE □ DIMITAN □ EMBECHINE □ EMBICHIN □ EMBICHIN HYDROCHLORIDE □ EMBIKHINE □ ERASOL □ ERASOL HYDROCHLORIDE □ ERASOL-IDO □ HN2.HCl □ HN2 HYDROCHLORIDE □ KLORAMIN □ N-LOST □ MBA HYDROCHLORIDE □ MEBICHLORAMINE □ MECHLORETHAMINE HYDROCHLORIDE □ MERCHLORETHANAMINE □ METHYLBIS(β-CHLOROETHYL)AMINE HYDROCHLORIDE □ N-METHYL-BIS-β-CHLORETHYLAMINE HYDROCHLORIDE □ METHYLBIS(2-CHLOROETHYL)AMINE HYDROCHLORIDE □ N-METHYLBIS(2-CHLOROETHYL)AMINE HYDROCHLORIDE □ N-METHYL-2,2′-DICHLORODIETHYLAMINE HYDROCHLORIDE □ N-METHYL-DI-2-CHLOROETHYLAMINE HYDROCHLORIDE □ METHYLDI(β-CHLOROETHYL)AMINE HYDROCHLORIDE □ METHYLDI(2-CHLOROETHYL)AMINE HYDROCHLORIDE □ MITOXINE □ N-MUSTARD (GERMAN) □ MUSTARGEN □ MUSTARGEN HYDROCHLORIDE □ MUSTINE HYDROCHLOR □ MUSTINE HYDROCHLORIDE □ NCI-C56382 □ NITOL □ NITOL "TAKEDA" □ NITROGEN MUSTARD HYDROCHLORIDE □ NITROGRANULOGEN □ NITROGRANULOGEN HYDROCHLORIDE □ NSC 762 □ NSC-762 HYDROCHLORIDE □ PLIVA □ STICKSTOFFLOST □ ZAGREB

TOXICITY DATA WITH REFERENCE
sln-dmg-orl 5 mmol/L MUREAV 95,237,82
dni-hmn:hla 1 µmol/L MUREAV 92,427,82
msc-mus:lym 20 µg/L FCTOD7 23,115,85
orl-rat TDLo:1130 mg/kg (male 8W pre):REP TJADAB 35,43A,87
scu-mus TDLo:2500 µg/kg (female 10D post):TER AMUK** 36,20,59
ipr-rat TDLo:700 µg/kg (female 12D post):TER PEDIAU 23,231,59
scu-rat TDLo:750 µg/kg (female 13D post):REP TJADAB 13,151,76
ivn-rat TDLo:500 µg/kg (male 1D pre):REP ZGEMAZ 117,467,51
ipr-mus TDLo:4 mg/kg (male 1D pre):REP CANCAR 2,1075,49
scu-rat TDLo:1 mg/kg (female 13D post):TER TJADAB 13,151,76
ipr-mus TDLo:6 mg/kg/4W:CAR JNCIAM 36,915,66
scu-mus TDLo:22 mg/kg/21W-I:CAR BJCAAI 3,118,49
ivn-mus TDLo:4 mg/kg/6D-I:NEO JNCIAM 11,415,50
scu-mus TD:6 mg/kg/6W-I:ETA JNCIAM 14,131,53
ivn-hmn TDLo:400 µg/kg:CNS,BLD CLPTAT 6,50,65
orl-rat LD50:10 mg/kg JPETAB 91,224,47
scu-rat LD50:1900 µg/kg JPETAB 91,224,47
ivn-rat LD50:1100 µg/kg MEIEDD 10,822,83
par-rat LD50:1700 µg/kg RRCRBU 52,76,75
orl-mus LD50:20 mg/kg JPETAB 91,224,47
ipr-mus LD50:2900 µg/kg NYKZAU 62,96,66
scu-mus LD50:2600 µg/kg JPETAB 91,224,47
ivn-mus LD50:2 mg/kg JPETAB 91,224,47
orl-rbt LD50:12,500 µg/kg NTIS** PB158-507
ivn-rbt LD50:1600 µg/kg JPETAB 91,224,47

CONSENSUS REPORTS: NTP 7th Annual Report On Carcinogens. IARC Cancer Review: Group 2A IMEMDT 7,269,87; Animal Sufficient Evidence IMEMDT 9,193,75. EPA Genetic Toxicology Program.

SAFETY PROFILE: Confirmed carcinogen with experimental carcinogenic, neoplastigenic, and tumorigenic data. Deadly poison by ingestion, intravenous, subcutaneous, intraperitoneal, and parenteral routes. Experimental teratogenic and reproductive effects. Human systemic effects by intravenous route: nausea or vomiting, reduction in the number of white blood cells and blood platelets. Other experimental reproductive effects. Human mutation data reported.

BIE750 CAS:63905-44-2 HR: 3
(N,N-BIS(2-CHLOROETHYL))-2-METHYLPROPYLAMINE HYDROCHLORIDE
mf: $C_8H_{17}Cl_2N{\cdot}ClH$ mw: 234.62

SYNS: N,N-BIS(2-CHLOROETHYL)ISOBUTYLAMINE HYDROCHLORIDE ☐ N,N-BIS(2-CHLOROETHYL)-2-METHYL-1-PROPANAMINE HYDROCHLORIDE ☐ 2,2′-DICHLORO-N-ISOBUTYL-DIETHYLAMINE HYDROCHLORIDE ☐ ISOBUTYLBIS(β-CHLOROETHYL)AMINE HYDROCHLORIDE ☐ ISOBUTYLBIS(2-CHLOROETHYL)AMINE HYDROCHLORIDE ☐ TL 525

TOXICITY DATA WITH REFERENCE
ipr-mus LD50:4420 µg/kg CANCAR 2,1075,49
scu-mus LDLo:5 mg/kg NDRC** No.9-4-1-9,43

SAFETY PROFILE: Poison by intraperitoneal and subcutaneous routes. When heated to decomposition it emits very toxic fumes of Cl^- and NO_x.

BIF250 CAS:494-03-1 HR: 3
N,N-BIS(2-CHLOROETHYL)-2-NAPHTHYLAMINE
mf: $C_{14}H_{15}Cl_2N$ mw: 268.20

PROP: Platelets from pet ether. Mp: 54–56°, bp: 210° @ 5 mm.

SYNS: 2-BIS(2-CHLOROETHYL)AMINONAPHTHALENE ☐ BIS(2-CHLOROETHYL)-β-NAPHTHYLAMINE ☐ CHLORNAFTINA ☐ CHLORNAPHAZIN ☐ CHLORNAPHTHIN ☐ CHLORONAFTINA ☐ CHLORONAPHTHINE ☐ CLORNAPHAZINE ☐ DICHLOROETHYL-β-NAPHTHYLAMINE ☐ DI(2-CHLOROETHYL)-β-NAPHTHYLAMINE ☐ N,N-DI(2-CHLOROETHYL)-β-NAPHTHYLAMINE ☐ 2-N,N-DI(2-CHLOROETHYL) NAPHTHYLAMINE ☐ ERYSAN ☐ NAPHTHYLAMINE MUSTARD ☐ β-NAPHTHYL-BIS-(β-CHLOROETHYL)AMINE ☐ 2-NAPHTHYLBIS(2-CHLOROETHYL)AMINE ☐ β-NAPHTHYL-DI-(2-CHLOROETHYL) AMINE ☐ NSC-62209 ☐ R48 ☐ RCRA WASTE NUMBER U026

TOXICITY DATA WITH REFERENCE
mmo-sat 40 µg/plate CNREA8 37,2209,77
mma-sat 10 µg/plate PNASA6 72,5135,75
dnd-dmg-orl 260 µmol/L CNREA8 30,195,70
orl-man TDLo:2468 mg/kg/6Y-I:CAR AMSVAZ 176,45,64
orl-wmn TDLo:3132 mg/kg/10Y-I:CAR AMSVAZ 175,721,64
unr-wmn TDLo:3200 mg/kg/11Y-I:CAR AMSVAZ 175,721,64
ipr-mus TDLo:330 mg/kg/4W:CAR JNCIAM 36,915,66
orl-wmn TD:4090 mg/kg/7Y-I:CAR AMSVAZ 176,45,64
orl-wmn TD:3200 mg/kg/11Y-I:CAR AMSVAZ 185,133,69
ipr-rat LD50:1086 mg/kg BCPCA6 13,969,64

CONSENSUS REPORTS: IARC Cancer Review: Group 1 IMEMDT 7,130,87; Animal Sufficient Evidence IMEMDT 4,119,74; Human Sufficient Evidence IMEMDT 4,119,74. EPA Genetic Toxicology Program.

SAFETY PROFILE: Confirmed human carcinogen producing bladder tumors. Human and experimental carcinogenic data. Moderately toxic by intraperitoneal route. When heated to decomposition it emits very toxic fumes of Cl^- and NO_x.

BIF500 CAS:67856-68-2 HR: 2
BIS(2-CHLOROETHYL)NITROSOAMINE
mf: $C_4H_8Cl_2N_2O$ mw: 171.04

SYNS: NITROSOBIS(2-CHLOROETHYL)AMINE ☐ N-NITROSO-2,2′-DICHLORODIETHYLAMINE

TOXICITY DATA WITH REFERENCE
mmo-sat-10 µg/plate MUREAV 66,1,79
orl-rat TDLo:345 mg/kg/30W-I:ETA CNREA8 38,2391,78

SAFETY PROFILE: Questionable carcinogen with experimental tumorigenic data. Mutation data reported. When heated to decomposition it emits toxic fumes of Cl^- and NO_x. See also NITROSAMINES and N-NITROSO COMPOUNDS.

BIF625 **CAS:77469-44-4** **HR: 3**
N,N′-BIS((2-CHLOROETHYL)-N-NITROSOCARBAMOYL)CYSTAMINE
mf: $C_{10}H_{18}Cl_2N_6O_4S_2$ mw: 421.36

SYNS: CNCC □ 13-CHLORO-N-(2-CHLOROETHYL)-N,11-DINITROSO-10-OXO-5,6-DITHIA-2,9,11-TRIAZATRIDECANAMIDE □ DI((CHLORO-2-ETHYL)-2-N-NITROSO-N-CARBAMOYL)-N,N-CYSTAMINE □ 1,1′-DITHIODIETHYLENEBIS(3-(2-(CHLOROETHYL)-3-NITROSOUREA)) □ I.C.I.G. 1325

TOXICITY DATA WITH REFERENCE
mmo-sat 200 μg/plate INSSDM 19,165,81
mma-sat 200 μg/plate INSSDM 19,165,81
oms-mus:oth 20 mg/L INSSDM 19,229,81
orl-mus LD50:280 mg/kg INSSDM 19,123,81
ipr-mus LD50:75 mg/kg INSSDM 19,123,81

SAFETY PROFILE: Poison by ingestion and intraperitoneal routes. Mutation data reported. When heated to decomposition it emits toxic fumes of Cl^-, SO_x and NO_x.

BIF750 **CAS:154-93-8** **HR: 3**
N,N′-BIS(2-CHLOROETHYL)-N-NITROSOUREA
mf: $C_5H_9Cl_2N_3O_2$ mw: 214.07

PROP: Light-yellow powder. Mp: 30–32°.

SYNS: BCNU □ BiCNU □ BIS(2-CHLOROETHYL)NITROSOUREA □ 1,3-BIS(β-CHLOROETHYL)-1-NITROSOUREA □ 1,3-BIS-(2-CHLOROETHYL)-1-NITROSOUREA □ BISCHLOROETHYLNITROSOUREA □ CARMUBRIS □ CARMUSTIN □ CARMUSTINE □ FDA 0345 □ NCI-C04773 □ NITRUMON □ NSC-409962 □ SK 27702 □ SRI 1720

TOXICITY DATA WITH REFERENCE
mmo-sat 33 μg/plate TCMUD8 5,319,85
sce-hmn:lym 25 μmol/L CNREA8 45,4708,85
ipr-rat TDLo:4 mg/kg (female 6-9D post):TER TXAPA9 30,422,74
ipr-rat TDLo:9 mg/kg (male 9W pre):REP TXAPA9 30,422,74
ipr-mus TDLo:11 mg/kg (male 1D pre):REP CANCAR 42,122,82
ipr-rat TDLo:8 mg/kg (female 6-9D post):TER TXAPA9 30,422,74
ipr-rat TDLo:15 mg/kg/7W-I:ETA CANCAR 40,1935,77
ivn-rat TDLo:16 mg/kg/60W-I:ETA DTESD7 8,273,80
skn-mus TDLo:276 mg/kg/23W-I:ETA EXPEAM 36,1211,80
ipr-mus TDLo:98 mg/kg/26W-I:ETA CANCAR 40,1935,77
ivn-rat TD:26 mg/kg/60W-I:ETA DTESD7 8,273,80
ivn-rat TD:45 mg/kg/60W-I:ETA DTESD7 8,273,80
ivn-rat TD:51 mg/kg/24W-I:ETA DTESD7 8,273,80
ivn-cld LDLo:78 mg/kg/52W I CANCAR 42,74,78
ivn-hmn TDLo:125 mg/kg:BLD,GIT ACRSAJ 16,273,72
ivn-hmn TDLo:6 mg/kg:BLD,GIT CTRRDO 60,709,76
par-wmn LDLo:1566 mg/kg:PUL JAMAAP 244,687,80
orl-rat LD50:20 mg/kg JPETAB 166,104,69
ipr-rat LD50:17,420 μg/kg NCISP* JAN86
scu-rat LD50:83,200 μg/kg IYKEDH 9,766,78
ivn-rat LD50:13,800 μg/kg ONCOBS 37,177,80
ims-rat LD50:79,600 μg/kg IYKEDH 9,766,78
orl-mus LD50:19 mg/kg TXAPA9 21,405,72
ipr-mus LD50:21,260 μg/kg NCISP* JAN86
scu-mus LD50:24 mg/kg TXAPA9 21,405,72
ivn-mus LD50:45 mg/kg PSEBAA 118,756,65

CONSENSUS REPORTS: NTP 7th Annual Report On Carcinogens. IARC Cancer Review: Group 2A IMEMDT 7,150,87; Human Limited Evidence IMEMDT 26,79,81; Animal Sufficient Evidence IMEMDT 26,79,81. NCI Carcinogenesis Studies (ipr); Some Evidence: rat CANCAR 40,1935,77; Clear Evidence: mouse CANCAR 40,1935,77. EPA Genetic Toxicology Program.

SAFETY PROFILE: Confirmed carcinogen with experimental carcinogenic and tumorigenic data. A human poison by parenteral route. An experimental poison by ingestion, intravenous, intraperitoneal, parenteral, and subcutaneous routes. Human systemic effects by parenteral, intravenous, and possibly other routes: nausea or vomiting, reduced white blood cell and blood platelet counts, bone marrow damage and potentially fatal respiratory system effects including lung fibrosis, dyspnea, and cyanosis. Experimental teratogenic and reproductive effects. Human mutation data reported. When heated to decomposition it emits very toxic fumes of Cl^- and NO_x. See also N-NITROSO COMPOUNDS.

BIG250 **CAS:2067-58-5** **HR: 3**
N,N-BIS(2-CHLOROETHYL)-p-PHENYLENEDIAMINE
mf: $C_{10}H_{14}Cl_2N_2$ mw: 233.16

SYN: p-AMINOPHENYL DERIVATIVE of NITROGEN MUSTARD

TOXICITY DATA WITH REFERENCE
scu-mus TDLo:12 mg/kg (female 7-9D post):TER JEEMAF 9,492,61
ipr-rat LD50:2200 μg/kg DBANAD 33,1005,80
ipr-mus LD50:7927 μg/kg JMCMAR 8,167,65

SAFETY PROFILE: Poison by intraperitoneal route. An experimental teratogen. When heated to decomposition it emits very toxic fumes of Cl^- and NO_x.

BIG500 **CAS:1070-42-4** **HR: 3**
BIS(2-CHLOROETHYL)PHOSPHITE
mf: $C_4H_9Cl_2O_3P$ mw: 207.00

PROP: Liquid. D: 1.40 @ 20°/4°, bp: 118–119° @ 4 mm.

TOXICITY DATA WITH REFERENCE
orl-rat LD50:260 mg/kg AIHAAP 30,470,69
ipr-mus LDLo:250 mg/kg CBCCT* 7,790,55
skn-rbt LD50:141 mg/kg AIHAAP 30,470,69

SAFETY PROFILE: Poison by ingestion and skin contact. When heated to decomposition it emits very toxic fumes of PO_x and Cl^-.

BIG750 **CAS:63980-44-9** **HR: 3**
N,N′-BIS(2-CHLOROETHYL)-1,4-PIPERAZINE HYDROCHLORIDE
mf: $C_8H_{16}Cl_2N_2 \cdot ClH$ mw: 247.62

TOXICITY DATA WITH REFERENCE
ipr-rat LD50:1100 μg/kg JPETAB 100,398,50
ipr-mus LD50:5700 μg/kg JPETAB 100,398,50

SAFETY PROFILE: Poison by intraperitoneal route. When heated to decomposition it emits very toxic fumes of HCl and NO_x.

BIH000 CAS:2045-41-2 HR: 3
N[4],N[4]-BIS(2-CHLOROETHYL)SULFANILAMIDE
mf: $C_{10}H_{14}Cl_2N_2O_2S$ mw: 297.22

TOXICITY DATA WITH REFERENCE
ipr-rat LD50:336 mg/kg JMCMAR 8,167,65
ipr-mus LD50:410 mg/kg JMCMAR 8,167,65

SAFETY PROFILE: Poison by intraperitoneal route. When heated to decomposition it emits very toxic fumes of Cl^-, NO_x, and SO_x.

BIH250 CAS:505-60-2 HR: 3
BIS(2-CHLOROETHYL)SULFIDE
mf: $C_4H_8Cl_2S$ mw: 159.08

PROP: Colorless (if pure) to light-yellow, oily liquid. Mp: 13–14°, bp: 215–217°, flash p: 221°F, d: 1.2741 @ 20°/4°, vap d: 5.4, vap press: 0.09 mm @ 30°.

SYNS: BIS(β-CHLOROETHYL)SULFIDE □ BIS(2-CHLOROETHYL) SULPHIDE □ 1-CHLORO-2-(β-CHLOROETHYLTHIO)ETHANE □ β,β-DICHLOR-ETHYL-SULPHIDE □ 2,2'-DICHLORODIETHYL SULFIDE □ DI-2-CHLOROETHYL SULFIDE □ β,β'-DICHLOROETHYL SULFIDE □ 2,2'-DICHLOROETHYL SULPHIDE (MAK) □ DISTILLED MUSTARD □ KAMPSTOFF "LOST" □ MUSTARD GAS □ MUSTARD HD □ MUSTARD VAPOR □ SCHWEFEL-LOST □ S-LOST □ S MUSTARD □ SULFUR MUSTARD □ SULFUR MUSTARD GAS □ SULPHUR MUSTARD GAS □ 1,1'-THIOBIS(2-CHLOROETHANE) □ YELLOW CROSS LIQUID □ YPERITE

TOXICITY DATA WITH REFERENCE
skn-man 2000 mg/m³/1H SEV NTIS** AD-A011-260
eye-man 100 mg/m³/6H MOD NTIS** AD-A011-260
eye-rbt 200 mg/m³ NTIS** AD-A011-260
eye-rbt 200 mg/m³/2M MLD NTIS** AD-A011-260
dnd-smc 500 μmol/L CBINA8 44,27,83
cyt-mam:lym 750 nmol/L CHRTBC 3,162,72
oms-hmn:hla 75 mg/L IUSMDJ 9,41,79
orl-rat TDLo:20 mg/kg (female 6-15D post):TER TJADAB 33,70C,86
ihl-rat TCLo:100 μg/m³/1Y-I:CAR NTIS** AD-A011-260
ihl-mus TCLo:1250 mg/m³/15M-C:NEO PSEBAA 82,457,53
scu-mus TDLo:6 mg/kg/6W-I:ETA JNCIAM 14,131,53
ivn-mus TDLo:600 mg/kg/6D-I:NEO JNCIAM 11,415,50
ihl-hmn LC50:1500 mg/m³/M NTIS** AD-A011-260
ihl-hmn LCLo:23 ppm/10M NTIS** PB214-270
skn-hmn LDLo:64 mg/kg WHOTAC -,24,70
ihl-rat LC50:100 mg/m³/10M NTIS** PB158-507
skn-rat LD50:5 mg/kg CNRMAW 25,141,47
scu-rat LD50:1500 μg/kg CNRMAW 25,141,47
ivn-rat LD50:700 μg/kg JPETAB 93,1,48
ihl-mus LC50:120 mg/m³/10M NTIS** PB158-507
skn-mus LD50:92 mg/kg JPETAB 93,1,48
scu-mus LD50:20 mg/kg NTIS** PB158-507
ivn-mus LD50:8600 μg/kg JPETAB 93,1,48
ihl-dog LC50:70 mg/m³/10M NTIS** PB158-507
skn-dog LD50:20 mg/kg NTIS** PB158-507
ivn-dog LD50:200 μg/kg NTIS** PB158-507
ihl-mky LC50:80 mg/m³/10M NTIS** PB158-507
skn-rbt LD50:40 mg/kg NTIS** PB158-507

CONSENSUS REPORTS: NTP 7th Annual Report On Carcinogens. IARC Cancer Review: Group 1 IMEMDT 7,259,87; Animal Sufficient Evidence IMEMDT 9,181,75; Human Limited Evidence IMEMDT 9,181,75. EPA Extremely Hazardous Substances List. Community Right-To-Know List. EPA Genetic Toxicology Program. Reported in EPA TSCA Inventory.

DFG MAK: Human Carcinogen

SAFETY PROFILE: Confirmed human carcinogen with experimental carcinogenic, neoplastigenic, and tumorigenic data. A human poison by inhalation and subcutaneous routes. An experimental poison by inhalation, skin contact, subcutaneous, and intravenous routes. An experimental teratogen. A severe human skin and eye irritant. Human mutation data reported. A military blistering gas. Strongly affects the skin, eyes, lungs, and gastric system. Pulmonary lesions are often fatal. It penetrates the skin deeply and injures blood vessels. Minute amounts can cause inflammation. Secondary infections are common. Combustible when exposed to heat or flame; can be ignited by a large explosive charge. It will react with water or steam to produce toxic and corrosive fumes. Vigorous reaction with oxidizing materials. Incompatible with bleaching powder. To fight fire, use water, foam, CO_2, dry chemical. Dangerous; when heated to decomposition or on contact with acid or acid fumes it emits highly toxic fumes of SO_x and Cl^-. See also SULFIDES and CHLORIDES.

BIH325 HR: 3
1,1-BIS(2-CHLOROETHYL)-2-SULFINYLHYDRAZINE
mf: $C_4H_8Cl_2N_2OS$ mw: 203.10

SYN: NSC 78409

TOXICITY DATA WITH REFERENCE
orl-mus LD50:37,240 μg/kg NCISP* JAN86
ipr-mus LD50:56,650 μg/kg NCISP* JAN86
scu-mus LD50:53,660 μg/kg NCISP* JAN86

SAFETY PROFILE: Poison by ingestion, subcutaneous, and intraperitoneal routes. When heated to decomposition it emits toxic fumes of Cl^-, SO_x, and NO_x.

BIH500 CAS:471-03-4 HR: 3
BIS(2-CHLOROETHYL)SULFONE
mf: $C_4H_8Cl_2O_2S$ mw: 191.08

PROP: Leaflets from H_2O or EtOH. Mp: 56°, bp: 183° @ 20 mm.

SYNS: BIS(β-CHLOROETHYL)SULFONE □ MUSTARD GAS SULFONE □ MUSTARD SULFONE □ YPERITE SULFONE

TOXICITY DATA WITH REFERENCE
scu-rat LD50:50 mg/kg JPETAB 93,1,48
scu-mus LD50:35 mg/kg JPETAB 93,1,48
ivn-mus LD50:50 mg/kg JPETAB 93,1,48
ihl-cat LCLo:1430 mg/m³/10M NDRC** NDCrc-132,JAN,42
ihl-rbt LCLo:1430 mg/m³/10M NDRC** NDCrc-132,JAN,42

SAFETY PROFILE: A poison via intravenous and subcutaneous routes. Moderately toxic via inhalation. See also SULFONATES. When heated to decomposition it emits very toxic fumes of Cl^- and SO_x.

BIH750 **HR: 3**
BIS(1-CHLOROETHYL THALLIUM CHLORIDE) OXIDE
mf: $C_4H_8Cl_4OTl_2$ mw: 622.66

CONSENSUS REPORTS: Thallium and its compounds are on the Community Right-To-Know List.

SAFETY PROFILE: An unstable explosive. When heated to decomposition it emits toxic fumes of Cl^-. See also THALLIUM COMPOUNDS.

BII250 **CAS:108-60-1** **HR: 3**
BIS(2-CHLOROISOPROPYL) ETHER
DOT: UN 2490
mf: $C_6H_{12}Cl_2O$ mw: 171.08

PROP: Colorless liquid. Bp: 187.8°, fp: > −20°, flash p: 185°F (OC), d: 1.11 @ 25°/25°, vap d: 6.0, vap press: 0.10 mm @ 20°.

SYNS: BIS(β-CHLOROISOPROPYL)ETHER □ BIS(2-CHLORO-1-METHYLETHYL) ETHER □ BIS(1-CHLORO-2-PROPYL) ETHER □ (2-CHLORO-1-METHYLETHYL) ETHER □ DCIP □ DCIP (nematocide) □ DICHLORODIISOPROPYL ETHER □ β,β′-DICHLORODIISOPROPYL ETHER □ DICHLOROISOPROPYL ETHER □ 2,2′-DICHLOROISOPROPYL ETHER □ DICHLOROISOPROPYL ETHER (DOT) □ NCI-C50044 □ NEMAMORT □ 2,2′-OXYBIS(1-CHLOROPROPANE) □ PROPANE, 2,2′-OXYBIS(1-CHLORO)- □ RCRA WASTE NUMBER U027

TOXICITY DATA WITH REFERENCE
eye-rbt 500 mg open AMIHBC 4,119,51
mmo-sat 1 mL/plate/3H DHEFDK FDA-78-1046,78
mma-sat 333 μg/plate ENMUDM 8(Suppl 7),1,86
orl-rat LD50:240 mg/kg AMIHBC 4,119,51
skn-rat LD50:>2 g/kg PEMNDP 9,83,91
orl-mus LD50:503 mg/kg FMCHA2-,C216,91
ihl-rat LCLo:700 ppm/5H BJIMAG 27,1,70
skn-rbt LD50:3000 mg/kg AMIHBC 4,119,51

CONSENSUS REPORTS: IARC Cancer Review: Group 3 IMEMDT 7,56,87, Animal Limited Evidence IMEMDT 41,149,86. NCI Carcinogenesis Bioassay (gavage); No Evidence: rat NCITR* NCI-CG-TR-191,79. Community Right-To-Know List. Reported in EPA TSCA Inventory.

DOT CLASSIFICATION: 6.1; *Label:* Poison

SAFETY PROFILE: Poison by ingestion. Moderately toxic by skin contact and inhalation. An eye irritant. Questionable carcinogen. Mutation data reported. A corrosive material. Moderate fire hazard when exposed to heat, flame, or powerful oxidizers. Incompatible with oxidizing materials. To fight fire, use water to blanket fire; foam, CO_2, dry chemical. When heated to decomposition it emits highly toxic fumes of Cl^-. See also ETHERS.

BII500 **CAS:67465-41-2** **HR: 3**
2,5-BIS(CHLOROMERCURI)FURAN
mf: $C_4H_2Cl_2Hg_2O$ mw: 538.14

SYN: USAF UCTL-974

TOXICITY DATA WITH REFERENCE
ipr-mus LD50:20 mg/kg NTIS** AD277-689

CONSENSUS REPORTS: Mercury and its compounds are on the Community Right-To-Know List.

OSHA PEL: CL 0.1 mg(Hg)/m³ (skin)
ACGIH TLV: TWA 0.1 mg(Hg)/m³ (skin)
NIOSH REL: (Mercury, Aryl And Inorganic): CL 0.1 mg/m³ (skin)

SAFETY PROFILE: Poison by intraperitoneal route. See also MERCURY COMPOUNDS and CHLORIDES. When heated to decomposition it emits very toxic fumes of Cl^- and Hg.

BII750 **HR: 3**
N,N-BIS(CHLOROMERCURI)HYDRAZINE
mf: $Cl_2H_2Hg_2N_2$ mw: 502.01

CONSENSUS REPORTS: Mercury and its compounds are on the Community Right-To-Know List.

SAFETY PROFILE: An explosive. When heated to decomposition it emits toxic fumes of Cl^-, NO_x, and Hg. See also MERCURY COMPOUNDS.

BIJ000 **CAS:64050-46-0** **HR: 3**
4,5-BIS(CHLOROMERCURI)-2-THIAZOLECARBAMIC ACID BENZYL ESTER
mf: $C_{11}H_8Cl_2Hg_2N_2O_2S$ mw: 704.35

TOXICITY DATA WITH REFERENCE
ipr-mus LDLo:125 mg/kg CBCCT* 8,752,56

CONSENSUS REPORTS: Mercury and its compounds are on the Community Right-To-Know List.

OSHA PEL: CL 0.1 mg(Hg)/m³ (skin)
ACGIH TLV: TWA 0.1 mg(Hg)/m³ (skin)
NIOSH REL: (Mercury, Aryl And Inorganic): CL 0.1 mg/m³ (skin)

SAFETY PROFILE: Poison by intraperitoneal route. See also CARBAMATES; MERCURY COMPOUNDS; and ESTERS. When heated to decomposition it emits very toxic fumes of Cl^-, Hg, NO_x, and SO_x.

BIJ250 **CAS:13483-18-6** **HR: 2**
BIS-1,2-(CHLOROMETHOXY)ETHANE
mf: $C_4H_8Cl_2O_2$ mw: 159.02

PROP: Viscous liquid. Bp: 99–100° @ 22 mm, d: 1.2879 @ 14°/15°.

SYN: ETHYLENE GLYCOL BIS(CHLOROMETHYL)ETHER

TOXICITY DATA WITH REFERENCE
skn-mus TDLo:8640 mg/kg/72W-I:NEO CNREA8 35,2553,75
ipr-mus TDLo:940 mg/kg/78W-I:NEO CNREA8 35,2553,75
scu-mus TDLo:970 mg/kg/81W-I:NEO CNREA8 35,2553,75

CONSENSUS REPORTS: IARC Cancer Review: Group 3 IMEMDT 7,56,87; Animal Sufficient Evidence IMEMDT 15,31,77. Reported in EPA TSCA Inventory. Glycol ethers are on the Community Right-To-Know List.

SAFETY PROFILE: Questionable carcinogen with experimental neoplastigenic data. See also GLYCOL

ETHERS. When heated to decomposition it emits toxic fumes of Cl^-.

For occupational chemical analysis use NIOSH: see 1,2-Dichlorodifluoroethane 1018.

BIJ500 CAS:56894-91-8 HR: 2
1,4-BIS(CHLOROMETHOXYMETHYL)BENZENE
mf: $C_{10}H_{12}Cl_2O_2$ mw: 235.12

PROP: Solid. Mp: 91.5–94.5°.

SYN: BIS-1,4-(CHLOROMETHOXY)-p-XYLENE

TOXICITY DATA WITH REFERENCE
skn-mus TDLo:2590 mg/kg/72W-I:NEO CNREA8 35,2553,75
ipr-mus TDLo:310 mg/kg/78W-I:ETA CNREA8 35,2553,75
scu-mus TDLo:970 mg/kg/81W-I:NEO CNREA8 35,2553,75

CONSENSUS REPORTS: IARC Cancer Review: Group 3 IMEMDT 7,56,87; Animal Sufficient Evidence IMEMDT 15,37,77.

SAFETY PROFILE: Questionable carcinogen with experimental neoplastigenic and tumorigenic data. When heated to decomposition it emits toxic fumes of Cl^-.

BIJ750 CAS:10387-13-0 HR: 3
9,10-BIS(CHLOROMETHYL)ANTHRACENE
mf: $C_{16}H_{12}Cl_2$ mw: 275.18

PROP: Crystals from xylene; yellow blades from toluene. Mp: 258–260° (decomp @ 2°).

SYNS: 9,10-DI(CHLOROMETHYL)ANTHRACENE □ ICR-450

TOXICITY DATA WITH REFERENCE
mma-sat 100 ng/plate PNASA6 72,5135,75
ivn-mus TDLo:1100 μg/kg:NEO CNREA8 36,2423,76
ivn-mus LD50:56 mg/kg CSLNX* NX#00245

CONSENSUS REPORTS: Reported in EPA TSCA Inventory. EPA Genetic Toxicology Program.

SAFETY PROFILE: Poison by intravenous route. Questionable carcinogen with experimental neoplastigenic data. Mutation data reported. When heated to decomposition it emits toxic fumes of Cl^-. See also CHLORINATED HYDROCARBONS, AROMATIC.

BIK000 CAS:542-88-1 HR: 3
BIS(CHLOROMETHYL) ETHER
DOT: UN 2249
mf: $C_2H_4Cl_2O$ mw: 114.96

PROP: Volatile liquid. Bp: 105°, d: 1.315 @ 20°, vap d: 4.0, flash p: <19°, fp: −41.5°.

SYNS: BCME □ BIS-CME □ CHLORO(CHLOROMETHOXY)METHANE □ DICHLORDIMETHYLAETHER (GERMAN) □ sym-DICHLORODIMETHYL ETHER (DOT) □ sym-DICHLOROMETHYL ETHER □ DIMETHYL-1,1'-DICHLOROETHER □ OXYBIS(CHLOROMETHANE) □ RCRA WASTE NUMBER P016

TOXICITY DATA WITH REFERENCE
otr-ham:kdy 80 μg/L BJCAAI 37,873,78
mma-sat 20 μg/plate BJCAAI 37,873,78
dns-hmn:fbr 160 μg/L TXCYAC 21,151,81
dns-mus-skn 360 μmol/kg/L CNREA8 33,769,73
ihl-rat TCLo:100 ppb/6H/4W-I:CAR AEHLAU 30,73,75
scu-rat TDLo:375 mg/kg/43W-I:CAR JNCIAM 43,481,69
ihl-mus TCLo:100 ppb/6H/26W-I:NEO TXAPA9 58,269,81
skn-mus TDLo:5520 mg/kg/23W-I:ETA AEHLAU 16,472,68
ipr-mus TDLo:50 mg/kg/61W-I:NEO CNREA8 35,2553,75
scu-mus TDLo:384 mg/kg/42W-I:NEO CNREA8 40,352,80
scu-mus TD:640 mg/kg/53W-I:NEO CNREA8 35,2553,75
ihl-rat TC:100 ppb/6H/6W-I:ETA PAACA3 21,106,80
scu-mus TD:12,500 μg/kg:NEO TXAPA9 15,92,69
ihl-rat TC:100 ppb/6H/26W-I:NEO TXAPA9 58,269,81
ihl-mus TC:1 ppm/82D-I:ETA AEHLAU 22,663,71
ihl-rat TC:75 ppb/6H/2Y-I:CAR JJIND8 68,597,82
skn-mus TD:11 g/kg/44W-I:CAR JNCIAM 43,481,69
ihl-man TCLo:3 ppm:EYE TJSGA8 51,596,73
ihl-man LCLo:100 ppm/3M:PUL TJSGA8 51,596,73
orl-rat LD50:210 mg/kg AIHAAP 30,470,69
ihl-rat LC50:7 ppm/7H AEHLAU 30,61,75
ihl-mus LC50:25 mg/m³/6H AEHLAU 22,663,71
skn-rbt LD50:280 mg/kg AIHAAP 30,470,69
ihl-ham LC50:7 ppm/7H AEHLAU 30,61,75

CONSENSUS REPORTS: NTP 7th Annual Report On Carcinogens. IARC Cancer Review: Group 1 IMEMDT 7,131,87; Animal Sufficient Evidence IMEMDT 4,231,74; Human Sufficient Evidence IMEMDT 4,231,74. Community Right-To-Know List. EPA Extremely Hazardous Substances List. Reported in EPA TSCA Inventory.

OSHA: Cancer Suspect Agent
ACGIH TLV: TWA 0.001 ppm; Confirmed Human Carcinogen
DFG MAK: Human Carcinogen
DOT CLASSIFICATION: 6.1; *Label:* Poison

SAFETY PROFILE: Confirmed human carcinogen with experimental carcinogenic, neoplastigenic, and tumorigenic data. Poison by inhalation, ingestion, and skin contact. Human systemic effects by inhalation: irritation of the conjunctiva, unspecified nasal and respiratory effects. Human mutation data reported. A dangerous fire hazard. When heated to decomposition it emits very toxic fumes of Cl^-. See also ETHERS.

For occupational chemical analysis use OSHA: #10.

BIK100 HR: 2
BIS(2-CHLORO-1-METHYLETHYL)ETHER mixed with 2-CHLORO-1-METHYLETHYL-(2-CHLOROPROPYL) ETHER
mf: $C_6H_{12}Cl_2O$ mw: 171.08

SYN: ETHER, BIS(2-CHLORO-1-METHYLETHYL), mixed with 2-CHLORO-1-METHYLETHYL-(2-CHLOROPROPYL)ETHER (7:3)

TOXICITY DATA WITH REFERENCE
orl-mus TDLo:51,500 mg/kg/2Y-I:CAR NTPTR* NTP-TR-239,82
orl-mus TD:103 g/kg/2Y-I:CAR NTPTR* NTP-TR-239,82

CONSENSUS REPORTS: NTP Carcinogenesis Bioassay

(gavage); Clear Evidence: mouse NTPTR* NTP-TR-239,83.

SAFETY PROFILE: Questionable carcinogen with experimental carcinogenic data. When heated to decomposition it emits toxic fumes of Cl^-.

BIK250 CAS:534-07-6 HR: 3
BIS(CHLOROMETHYL)KETONE
DOT: UN 2649
mf: $C_3H_4Cl_2O$ mw: 126.97

PROP: Crystals. Mp: 45°, bp: 173°, d: 1.3826 @ 46°/4°, vap d: 4.38. Sol in water.

SYNS: sym-DICHLOROACETONE □ α,α'-DICHLOROACETONE □ α,γ-DICHLOROACETONE □ 1,3-DICHLOROACETONE □ 1,3-DICHLOROACETONE (DOT) □ 1,3-DICHLORO-2-PROPANONE

TOXICITY DATA WITH REFERENCE
mmo-sat 1250 ng/plate MUREAV 157,111,85
mma-smc 5 μg/L MUREAV 155,53,85
ihl-rat LC50:29 mg/m³/2H 85GMAT -,44,82
ihl-mus LC50:27 mg/m³/2H 85GMAT -,44,82

CONSENSUS REPORTS: EPA Genetic Toxicology Program. Reported in EPA TSCA Inventory. EPA Extremely Hazardous Substances List.

DOT CLASSIFICATION: 6.1; *Label:* Poison

SAFETY PROFILE: Poison by inhalation. Mutation data reported. A sytemic irritant by ingestion and inhalation routes. See also KETONES. Dangerous; when heated to decomposition it emits highly toxic fumes of Cl^-.

BIK325 CAS:78-71-7 HR: 3
3,3-BIS(CHLOROMETHYL)OXETANE
mf: $C_5H_8Cl_2O$ mw: 155.03

PROP: Liquid. D: 1.295 @ 25°/25°. Mp: 18.7°, bp: 101° @ 27 mm.

SYN: 3,3-DICHLOROMETHYLOXYCYCLOBUTANE

TOXICITY DATA WITH REFERENCE
orl-rat LD50:600 mg/kg FATOBP 5,157,70
orl-mus LD50:420 mg/kg 85GMAT -,47,82
ihl-mus LC50:200 mg/m³/2H 85GMAT -,47,82

CONSENSUS REPORTS: EPA Extremely Hazardous Substances List.

SAFETY PROFILE: Poison by ingestion and inhalation. When heated to decomposition it emits toxic fumes of Cl^-.

BIK500 CAS:2209-86-1 HR: 2
2,2-BIS(CHLOROMETHYL)-1,3-PROPANEDIOL
mf: $C_5H_{10}Cl_2O_2$ mw: 173.05

PROP: Crystals. Mp: 79–80°, bp: 160° @ 12 mm. Sol in water.

SYN: DISPRANOL

TOXICITY DATA WITH REFERENCE
orl-rat LD50:1285 mg/kg BCFAAI 99,67,60
ipr-rat LD50:920 mg/kg BCFAAI 99,67,60
ipr-mus LD50:812 mg/kg BCFAAI 99,67,60
unr-mus LD50:1 g/kg RPTOAN 48,67,85

SAFETY PROFILE: Moderately toxic by ingestion, intraperitoneal, and possibly other routes. When heated to decomposition it emits toxic fumes of Cl^-.

BIK750 CAS:12712-28-6 HR: 3
2,2-BIS(CHLOROMETHYL)-1,3-PROPANEDIOL SULFATE
mf: $C_5H_{10}Cl_2O_2 \cdot H_2O_4S$ mw: 271.13

SYN: PHILIPS 2605

TOXICITY DATA WITH REFERENCE
orl-rat LD50:20 mg/kg TXAPA9 21,315,72
orl-bwd LD50:2400 μg/kg TXAPA9 21,315,72

SAFETY PROFILE: Poison by ingestion. See also SULFATES. When heated to decomposition it emits very toxic fumes of SO_x and Cl^-.

BIL250 CAS:14579-91-0 HR: 3
1,3-BIS(CHLOROMETHYL)-1,1,3,3-TETRAMETHYLDISILAZANE
mf: $C_6H_{17}Cl_2N\text{-}Si_2$ mw: 230.32

PROP: Liquid. D: 1.0543° @ 20 mm, bp: 103–105° @ 10 mm.

SYN: 1-(CHLOROMETHYL)-N-((CHLOROMETHYL)DIMETHYLSILYL)-1,1-DIMETHYL-SILANAMINE

TOXICITY DATA WITH REFERENCE
ipr-mus TDLo:40 mg/kg/I:NEO JNCIAM 54,495,75
ipr-mus LDLo:250 mg/kg StoGD# 27May75

CONSENSUS REPORTS: Reported in EPA TSCA Inventory.

SAFETY PROFILE: Poison by intraperitoneal route. Questionable carcinogen with experimental neoplastigenic data. When heated to decomposition it emits very toxic fumes of Cl^- and NO_x.

BIL500 CAS:83-05-6 HR: 2
BIS(p-CHLOROPHENYL)ACETIC ACID
mf: $C_{14}H_{10}Cl_2O_2$ mw: 281.14

SYNS: BIS(4-CHLOROPHENYL)ACETIC ACID □ BIS(p-CHLORPHENYL)ESSIGSAEURE (GERMAN) □ DICHLORODIPHENYLACETIC ACID □ p,p'-DICHLORODIPHENYLACETIC ACID □ DI(p-CHLOROPHENYL) ACETIC ACID

TOXICITY DATA WITH REFERENCE
sln-dmg-orl 3700 μmol/L MUREAV 16,157,72
cyt-rat:oth 150 mg/L/24H C TXAPA9 22,355,72
ipr-rat TDLo:350 mg/kg (7D pre):REP ENDOAO 91,1095,72
orl-rat TDLo:250 mg/kg (15-19D post):TER BNEOBV 26,283,75
orl-mus LD50:590 mg/kg AIPTAK 73,128,46

SAFETY PROFILE: Moderately toxic by ingestion. An experimental teratogen. Other experimental reproductive effects. Mutation data reported. When heated to decomposition it emits toxic fumes of Cl^-.

BIM000 CAS:4104-14-7 HR: 3
O,O-BIS(p-CHLOROPHENYL)ACETIMIDOYLPHOSPHORAMIDOTHIOATE
mf: $C_{14}H_{13}Cl_2N_2O_2PS$ mw: 375.22

SYNS: BAY 33819 ☐ BAYER 38819 ☐ DRC-714 ☐ GOPHACIDE ☐ PHOSAZETIM

TOXICITY DATA WITH REFERENCE
orl-rat LD50:3700 µg/kg FMCHA2 -,C117,83
skn-rat LD50:25 mg/kg FMCHA2 -,C117,83
ipr-rat LD50:3500 µg/kg AIPTAK 169,108,67
orl-mus LD50:12 mg/kg TXAPA9 25,42,73
ipr-mus LD50:5500 µg/kg AIPTAK 169,108,67
orl-dog LD50:23 mg/kg PCOC** -,107,66
orl-gpg LD50:20 mg/kg AIPTAK 169,108,67
ipr-gpg LD50:14 mg/kg AIPTAK 169,108,67

CONSENSUS REPORTS: EPA Extremely Hazardous Substances List.

SAFETY PROFILE: Poison by ingestion, skin contact, and intraperitoneal routes. A pesticide. When heated to decomposition it emits very toxic fumes of SO_x, PO_x, Cl^-, and NO_x. See also ESTERS.

BIM250 CAS:55-56-1 HR: 3
1,6-BIS(5-(p-CHLOROPHENYL)BIGUANIDINO)HEXANE
mf: $C_{22}H_{30}Cl_2N_{10}$ mw: 505.52

PROP: Solid. Mp: 134°.

SYNS: 1,6-BIS(p-CHLOROPHENYLDIGUANIDO)HEXANE ☐ CHLORHEXIDIN (CZECH) ☐ CHLORHEXIDINE ☐ 1,6-DI(4′-CHLOROPHENYLDIGUANIDO)HEXANE ☐ 1,1′-HEXAMETHYLENEBIS(5-(p-CHLOROPHENYL)BIGUANIDE) ☐ HIBITANE ☐ NOLVASAN ☐ ROTERSEPT ☐ STERIDO

TOXICITY DATA WITH REFERENCE
skn-hmn 1500 µg/3D-I MLD 85DKA8 -,127,77
mma-sat 400 nmol/L CBINA8 28,249,79
dnr-esc 7 µmol/disc CBINA8 28,249,79
orl-mus TDLo:1680 mg/kg (7D pre):REP MEXPAG 10,361,64
orl-rat TDLo:1500 mg/kg (30D male):REP YACHDS 6,2599,78
orl-rat LD50:9200 mg/kg YACHDS 6,2599,78
ipr-rat LD50:60 mg/kg VTYMAC 72,1330,77
ivn-rat LD50:21 mg/kg RMISDU 23,45,80
orl-mus LD50:2515 mg/kg RMISDU 23,45,80
ipr-mus LD50:44 mg/kg VTYMAC 72,1330,77

SAFETY PROFILE: Poison by intraperitoneal and intravenous routes. Mildly toxic by ingestion. Experimental reproductive effects. A human skin irritant. Mutation data reported. When heated to decomposition it emits very toxic fumes of Cl^- and NO_x.

BIM500 CAS:72-54-8 HR: 3
1,1-BIS(4-CHLOROPHENYL)-2,2-DICHLOROETHANE
mf: $C_{14}H_{10}Cl_4$ mw: 320.04

PROP: Crystalline solid from pet ether. Mp: 111°, vap d: 11.

SYNS: 1,1-BIS(p-CHLOROPHENYL)-2,2-DICHLOROETHANE ☐ 2,2-BIS(p-CHLOROPHENYL)-1,1-DICHLOROETHANE ☐ 2,2-BIS(4-CHLOROPHENYL)-1,1-DICHLOROETHANE ☐ DDD ☐ p,p′-DDD ☐ 1,1-DICHLOOR-2,2-BIS(4-CHLOOR FENYL)-ETHAAN (DUTCH) ☐ 1,1-DICHLOR-2,2-BIS(4-CHLOR-PHENYL)-AETHAN (GERMAN) ☐ 1,1-DICHLORO-2,2-BIS(p-CHLOROPHENYL)ETHANE ☐ 1,1-DICHLORO-2,2-BIS(p-CHLOROPHENYL)ETHANE (DOT) ☐ 1,1-DICHLORO-2,2-BIS(4-CHLOROPHENYL)-ETHANE (FRENCH) ☐ 1,1-DICHLORO-2,2-BIS(PARACHLOROPHENYL)ETHANE (DOT) ☐ 1,1-DICHLORO-2,2-DI(4-CHLOROPHENYL)ETHANE ☐ DICHLORODIPHENYL DICHLOROETHANE ☐ p,p′-DICHLORODIPHENYLDICHLOROETHANE ☐ 1,1-DICLORO-2,2-BIS(4-CLORO-FENIL)-ETANO (ITALIAN) ☐ DILENE ☐ ENT 4,225 ☐ ME-1700 ☐ NCI-C00475 ☐ RCRA WASTE NUMBER U060 ☐ RHOTHANE ☐ RHOTHANE D-3 ☐ ROTHANE ☐ p,p′-TDE ☐ TDE (DOT) ☐ TETRACHLORODIPHENYLETHANE

TOXICITY DATA WITH REFERENCE
cyt-rat:oth 10 µg/L 34LXAP -,555,76
otr-mus:emb 28,400 nmol/L JNCIAM 54,981,75
orl-rat TDLo:54 g/kg/78W-C:ETA NCITR* NCI-CG-TR-131,78
orl-mus TDLo:39 g/kg/2Y-C:NEO JNCIAM 52,883,74
orl-rat LD50:113 mg/kg GUCHAZ 6,154,73
orl-mus LDLo:600 mg/kg JPETAB 88,400,46
skn-rbt LD50:1200 mg/kg AFDOAQ 16,3,52

CONSENSUS REPORTS: IARC Cancer Review: Animal Sufficient Evidence IMEMDT 5,83,74. NCI Carcinogenesis Bioassay (feed); Clear Evidence: rat NCITR* NCI-CG-TR-131,78; No Evidence: mouse NCITR* NCI-CG-TR-131,78. EPA Genetic Toxicology Program.

SAFETY PROFILE: Confirmed carcinogen with experimental carcinogenic, neoplastigenic, and tumorigenic data. Poison by ingestion. Moderately toxic by skin contact. Mutation data reported. An insecticide. When heated to decomposition it emits toxic fumes of Cl^-. See also DDT.

BIM750 CAS:72-55-9 HR: 3
2,2-BIS(p-CHLOROPHENYL)-1,1-DICHLOROETHYLENE
mf: $C_{14}H_8Cl_4$ mw: 318.02

SYNS: DDE ☐ p,p′-DDE ☐ DDT DEHYDROCHLORIDE ☐ 1,1-DICHLORO-2,2-BIS(p-CHLOROPHENYL)ETHYLENE ☐ p,p′-DICHLORODIPHENYLDICHLOROETHYLENE ☐ 1,1′-DICHLOROETHENYLIDENE) BIS(4-CHLOROBENZENE) ☐ NCI-C00555

TOXICITY DATA WITH REFERENCE
sln-dmg-orl 1 pph ENMUDM 7,325,85
msc-mus:lym 40 mg/L/4H MUREAV 59,61,79
ipr-rat TDLo:3500 µg/kg (7D pre):REP ENDOAO 91,1095,72
orl-mus TDLo:9700 mg/kg/78W-C:CAR NCITR* NCI-TR-131,78
orl-ham TDLo:36 g/kg/86W-C:NEO CNREA8 43,776,83
orl-mus TD:28 g/kg/80W-C:NEO JNCIAM 52,883,74
orl-mus TD:17 g/kg/78W-C:CAR NCITR* NCI-TR-131,78
orl-ham TD:57 g/kg/68W-C:NEO CNREA8 43,776,83
orl-ham TD:41 g/kg/97W-C:NEO CNREA8 43,776,83
orl-ham TD:81 g/kg/97W-C:NEO CNREA8 43,776,83
orl-rat LD50:880 mg/kg TXAPA9 14,515,69
orl-mus LD50:700 mg/kg JPETAB 88,400,46

CONSENSUS REPORTS: IARC Cancer Review: Animal Limited Evidence IMEMDT 5,83,74. NCI Carcinogenesis Bioassay (feed); Clear Evidence: mouse NCITR* NCI-CG-TR-131,78; No Evidence: rat NCITR* NCI-CG-TR-131,78. EPA Genetic Toxicology Program.

SAFETY PROFILE: Suspected carcinogen with experimental carcinogenic and neoplastigenic data. Poison by ingestion. Experimental reproductive effects. Mutation data reported. An insecticide. When heated to decomposition it emits very toxic fumes of Cl^-. See also CHLORINATED HYDROCARBONS, ALIPHATIC.

BIM800 CAS:101-76-8 HR: 2
BIS(p-CHLOROPHENYL)METHANE
mf: $C_{13}H_{10}Cl_2$ mw: 237.13

SYNS: DI-(p-CHLOROPHENYL)METHANE □ DI-(4-CHLOROPHENYL)METHANE □ METHANE, BIS(4-CHLOROPHENYL)-

TOXICITY DATA WITH REFERENCE
orl-rat LD50:1 g/kg JPETAB 88,359,46
orl-mus LDLo:1500 mg/kg JPETAB 88,400,46

CONSENSUS REPORTS: Reported in EPA TSCA Inventory.

SAFETY PROFILE: Moderately toxic by ingestion. When heated to decomposition it emits toxic vapors of Cl^-.

BIN000 CAS:80-06-8 HR: 2
1,1-BIS(p-CHLOROPHENYL)METHYLCARBINOL
mf: $C_{14}H_{12}Cl_2O$ mw: 267.16

PROP: Crystals from pet ether. Mp: 68–69°.

SYNS: BCPE □ 1,1-BIS(p-CHLOROPHENYL)ETHANOL □ 1,1-BIS(4-CHLOROPHENYL)ETHANOL □ BIS(p-CHLOROPHENYL)METHYL CARBINOL □ 1,1-BIS(4-CHLORPHENYL) AETHANOL (GERMAN) □ CHLORFENETHOL □ DCPC □ DCPE □ DICHLORODIPHENYLETHANOL □ p,p'-DICHLORODIPHENYLMETHYLCARBINOL □ 4,4'-DICHLORO(METHYL BENZHYDROL) □ 4,4'-DICHLORO-α-METHYLBENZHYDROL □ 4,4'-DICHLORO-α-METHYLBENZOHYDROL □ DI-(p-CHLOROPHENYL)ETHANOL □ DI(p-CHLOROPHENYL) METHYLCARBINOL □ DIMITE □ DMC □ ENT 9,624 □ QIKRON

TOXICITY DATA WITH REFERENCE
orl-rat LD50:500 mg/kg ARSIM* 20,8,66
ipr-rat LD50:725 mg/kg OYYAA2 2,148,68

SAFETY PROFILE: Moderately toxic by ingestion and intraperitoneal routes. A pesticide. When heated to decomposition it emits toxic fumes of Cl^-.

BIN500 CAS:117-27-1 HR: 2
1,1-BIS(p-CHLOROPHENYL)-2-NITROPROPANE
mf: $C_{15}H_{13}Cl_2NO_2$ mw: 310.19

SYNS: C.I. AZOIC DIAZO COMPONENT 37 □ CS 645A □ DNP □ ENT 22,784 □ 2-NITRO-1,1-BIS(p-CHLOROPHENYL)PROPANE □ 1,1'-(2-NITROPORPYLIDENE)BIS(4-CHLOROBENZENE) □ PROLAN □ PROLAN (CSC)

TOXICITY DATA WITH REFERENCE
orl-rat LD50:750 mg/kg MEIEDD 10,946,83

SAFETY PROFILE: Moderately toxic by ingestion. An insecticide. When heated to decomposition it emits very toxic fumes of Cl^- and NO_x. See also AROMATIC AMINES.

BIN750 CAS:80-07-9 HR: 1
BIS(p-CHLOROPHENYL)SULFONE
mf: $C_{12}H_8Cl_2O_2S$ mw: 287.16

PROP: Crystals. Mp: 148°.

TOXICITY DATA WITH REFERENCE
orl-mus LD50:24 g/kg HCACAV 29,1317,46

CONSENSUS REPORTS: Reported in EPA TSCA Inventory.

SAFETY PROFILE: Mildly toxic by ingestion. See also SULFONATES. When heated to decomposition it emits very toxic fumes of Cl^- and SO_x.

BIO500 CAS:55216-04-1 HR: 3
BIS(p-CHLOROPHENYLTHIO)DIMETHYLTIN
mf: $C_{14}H_{14}Cl_2S_2Sn$ mw: 435.99

SYN: BIS(p-CHLOROPHENYLTHIO)DIMETHYL STANNANE

TOXICITY DATA WITH REFERENCE
ivn-mus LD50:56 mg/kg CSLNX* NX#01645

OSHA PEL: TWA 0.1 mg(Sn)/m³ (skin)
ACGIH TLV: TWA 0.1 mg(Sn)/m³ (skin) (Proposed: TWA 0.1 mg(Sn)/m³; STEL 0.2 mg(Sn)/m³ (skin))
NIOSH REL: (Organotin Compounds) TWA 0.1 mg(Sn)/m³

SAFETY PROFILE: Poison by intravenous route. See also TIN COMPOUNDS. When heated to decomposition it emits very toxic fumes of Cl^- and SO_x.

For occupational chemical analysis use NIOSH: Organotin Compounds 5504.

BIO625 CAS:789-02-6 HR: 2
2,2-BIS(o,p-CHLOROPHENYL)-1,1,1-TRICHLOROETHANE
mf: $C_{14}H_9Cl_5$ mw: 354.48

SYNS: o,p'-DDT □ 1,1,1-TRICHLORO-2-(o-CHLOROPHENYL)-2-(p-CHLOROPHENYL)ETHANE

TOXICITY DATA WITH REFERENCE
dns-rat-ipr 10 mg/kg JTEHD6 16,493,85
cyt-rat:oth 10 µg/L 34LXAP -,555,76
orl-rat TDLo:2150 mg/kg (female 1-22D post):REP BECTA6 12,373,74
scu-rat TDLo:15 mg/kg (female 3D pre):REP SCIEAS 173,642,71
ipr-rat TDLo:67,500 µg/kg (female 27D pre):REP ENDOAO 91,1095,72
orl-rat TDLo:250 mg/kg (female 15-19D post):TER BNEOBV 26,283,75
orl-mus LDLo:1000 mg/kg JPETAB 88,400,46
ipr-mus LD50:1577 mg/kg BECTA6 11,359,74

CONSENSUS REPORTS: EPA Genetic Toxicology Program.

SAFETY PROFILE: Moderately toxic by ingestion and intraperitoneal routes. An experimental teratogen. Other experimental reproductive effects. Mutation data reported. When heated to decomposition it emits toxic fumes of Cl^-. See also CHLORINATED HYDROCARBONS.

BIO750 **CAS:115-32-2** **HR: 3**
1,1-BIS(p-CHLOROPHENYL)-2,2,2-TRICHLOROETHANOL
mf: $C_{14}H_9Cl_5O$ mw: 370.48

PROP: Solid. Mp: 78.5°. Material used in cancer bioassay was 40–60% pure NCITR* NCI-CG-TR-90,78.

SYNS: ACARIN □ 1,1-BIS(CHLOROPHENYL)-2,2,2-TRICHLOROETHANOL □ 1,1-BIS(4-CHLOROPHENYL)-2,2,2-TRICHLOROETHANOL □ CARBAX □ CEKUDIFOL □ 4-CHLORO-α-(4-CHLOROPHENYL)-α-(TRICHLOROMETHYL)BENZENEMETHANOL □ CPCA □ DECOFOL □ DICHLOROKELTHANE □ DI-(p-CHLOROPHENYL)TRICHLOROMETHYLCARBINOL □ 4,4'-DICHLORO-α-(TRICHLOROMETHYL)BENZHYDROL □ DICOFOL □ DTMC □ ENT 23,648 □ FW 293 □ HIFOL □ KELTANE □ p,p'-KELTHANE □ KELTHANE (DOT) □ KELTHANE DUST BASE □ KELTHANETHANOL □ MILBOL □ MITIGAN □ NCI-C00486 □ 2,2,2-TRICHLOOR-1,1-BIS(4-CHLOOR FENYL)-ETHANOL (DUTCH) □ 1,1,1-TRICHLOR-2,2-BIS(4-CHLORPHENYL)-AETHANOL (GERMAN) □ 2,2,2-TRICHLOR-1,1-BIS(4-CHLOR-PHENYL)-AETHANOL (GERMAN) □ 2,2,2-TRICHLORO-1,1-BIS(4-CHLOROPHENYL)-ETHANOL (FRENCH) □ 2,2,2-TRICHLORO-1,1-BIS(4-CLORO-FENIL)-ETANOLO (ITALIAN) □ 2,2,2-TRICHLORO-1,1-DI-(4-CHLOROPHENYL)ETHANOL

TOXICITY DATA WITH REFERENCE
sce-hmn:lym 1 µmol/L ARTODN 52,221,83
orl-rat TDLo:430 mg/kg (female 6-15D post):TER CHYCDW 21,238,87
orl-rat TDLo:430 mg/kg (female 6-15D post):REP CHYCDW 21,238,87
orl-mus TDLo:17 g/kg/78W-C:CAR NCITR* NCI-CG-TR-90,78
orl-mus TD:35 g/kg/78W-C:CAR NCITR* NCI-CG-TR-90,78
orl-rat LD50:575 mg/kg WRPCA2 9,119,70
skn-rat LD50:100 mg/kg WRPCA2 9,119,70
ipr-rat LD50:1150 mg/kg TXAPA9 15,30,69
orl-mus LD50:420 mg/kg GTPZAB 19(9),55,75
orl-rbt LD50:1810 mg/kg TXAPA9 1,119,59
skn-rbt LD50:1870 mg/kg GUCHAZ 6,195,73
orl-gpg LD50:1810 mg/kg 85DPAN -,-,71/76
orl-ckn LD50:4365 mg/kg VETNAL 60(10),64,84

CONSENSUS REPORTS: IARC Cancer Review: Group 3 IMEMDT 7,56,87; Animal Limited Evidence IMEMDT 30,87,83. NCI Carcinogenesis Bioassay (feed); Clear Evidence: mouse NCITR* NCI-CG-TR-90,78; No Evidence: rat NCITR* NCI-CG-TR-90,78. Community Right-To-Know List.

SAFETY PROFILE: Poison by ingestion and skin contact. Moderately toxic by intraperitoneal route. Human mutation data reported. Questionable carcinogen with experimental carcinogenic data. An experimental teratogen. Other experimental reproductive effects. When heated to decomposition it emits toxic fumes of Cl^-.

BIP250 **HR: 3**
BIS-5-CHLORO TOLUENE DIAZONIUM ZINC TETRACHLORIDE
mf: $C_{14}H_{12}Cl_6N_4Zn$ mw: 514.37

CONSENSUS REPORTS: Zinc and its compounds are on the Community Right-To-Know List.

SAFETY PROFILE: A shock-sensitive explosive. When heated to decomposition it emits toxic fumes of Cl^-, NO_x, and ZnO. See also ZINC COMPOUNDS.

BIP500 **CAS:78371-84-3** **HR: 3**
1,3-BIS(6-CHLORO-o-TOLYL)-1-(2-(DIETHYLAMINO)ETHYL)UREA HYDROCHLORIDE
mf: $C_{21}H_{27}Cl_2N_3O{\cdot}ClH$ mw: 444.87

SYN: C 3183

TOXICITY DATA WITH REFERENCE
eye-rbt 2% MLD ARZNAD 8,664,58
ipr-rat LD50:65 mg/kg ARZNAD 8,664,58
scu-mus LD50:70 mg/kg ARZNAD 8,664,58

SAFETY PROFILE: Poison by intraperitoneal and subcutaneous routes. An eye irritant. When heated to decomposition it emits very toxic fumes of $Cl^-$and NO_x.

BIP750 **CAS:78371-85-4** **HR: 3**
1,3-BIS(6-CHLORO-o-TOLYL)-1-(2-PYRROLIDINYLETHYL)UREA HYDROCHLORIDE
mf: $C_{21}H_{25}Cl_2N_3O{\cdot}ClH$ mw: 442.85

SYN: C 3218

TOXICITY DATA WITH REFERENCE
eye-rbt 2% MLD ARZNAD 8,664,58
ipr-rat LD50:42 mg/kg ARZNAD 8,664,58
scu-mus LD50:71 mg/kg ARZNAD 8,664,58

SAFETY PROFILE: Poison by intraperitoneal and subcutaneous routes. An eye irritant. When heated to decomposition it emits very toxic fumes of Cl^-, NO_x, and HCl.

BIQ250 **CAS:40334-69-8** **HR: 3**
BIS(2-CHLOROVINYL)CHLOROARSINE
mf: $C_4H_4AsCl_3$ mw: 233.35

SYNS: DICHLOROVINYLARSINE CHLORIDE □ DICHLOROVINYLCHLOROARSINE (DOT) □ L-2 □ LEWISITE II

TOXICITY DATA WITH REFERENCE
skn-gpg LD50:8 mg/kg JPBAA7 58,411,46
scu-gpg LD50:200 µg/kg JPBAA7 58,411,46

CONSENSUS REPORTS: Arsenic and its compounds are on the Community Right-To-Know List.

OSHA PEL: TWA 0.5 mg(As)/m³
DOT CLASSIFICATION: Forbidden

SAFETY PROFILE: A poison by skin contact and subcutaneous routes. When heated to decomposition it emits very toxic fumes of As and Cl^-. See also ARSENIC COMPOUNDS.

BIQ500 **CAS:111-94-4** **HR: 3**
BIS(β-CYANOETHYL)AMINE
mf: $C_6H_9N_3$ mw: 123.18

$(NCC_2H_4)_2NH$

PROP: Liquid. Mp: −5.5°, bp: 135° @ 1 mm, d: 1.463 @ 25°, vap d: 3.3.

SYNS: BBCE □ BIS-(2-CYANOETHYL)AMINE □ N,N-BIS(2-CYANOETHYL)AMINE □ 2-CYANO-N-(2-CYANOETHYL)ETHANAMINE □ DI

(2-CIANOETIL)AMMINA (ITALIAN) □ 2,2'-DICYANODIETHYLAMINE □ DI-(2-CYANOETHYL)AMINE □ IDPN □ 3,3'-IMINOBISPROPANENITRILE □ IMINO-β,β'-DIPROPIONITRILE □ β,β-IMINODIPROPIONITRILE □ β,β'-IMINODIPROPIONITRILE □ 3,3'-IMINODIPROPIONITRILE □ 2341 I.S. □ USAF A-8564

TOXICITY DATA WITH REFERENCE
skn-rbt 500 mg/24H MLD 85JCAE-,924,86
eye-rbt 500 mg SEV AJOPAA 29,1363,46
eye-rbt 500 mg/24H MLD 85JCAE-,924,86
orl-rat TDLo:840 mg/kg (1-21D preg):REP SEIJBO 21,407,81
orl-rat TDLo:840 mg/kg (1-21D preg):TER SEIJBO 21,407,81
orl-rat LD50:2700 mg/kg JIHTAB 31,60,49
ipr-mus LD50:200 mg/kg NTIS** AD277-689
skn-rbt LD50:2520 mg/kg AMIHBC 10,61,54

CONSENSUS REPORTS: Reported in EPA TSCA Inventory. Cyanide and its compounds are on the Community Right-To-Know List.

SAFETY PROFILE: A poison by intraperitoneal route. Moderately toxic by ingestion and skin contact. Experimental teratogenic and reproductive effects. A skin and severe eye irritant. A storage hazard, may explode in a sealed container. When heated to decomposition it emits toxic fumes of NO_x and CN^-. See also NITRILES and AMINES.

BIR000 **HR: 3**
BIS(n-CYCLOOCTATETRANENE)URANIUM(O)
mf: $C_{16}H_{16}U$ mw: 446.33

SAFETY PROFILE: Ignites spontaneously in air. See also URANIUM.

BIR250 **HR: 3**
BIS(CYCLOPENTADIENYL)BIS(PENTAFLUOROPHENYL)ZIRCONIUM
mf: $C_{22}H_{16}F_{10}Zr$ mw: 555.52

SAFETY PROFILE: Explodes in air (but not nitrogen) above its melting point (219°). When heated to decomposition it emits toxic fumes of F^-. See also ZIRCONIUM COMPOUNDS.

BIR500 **CAS:12194-11-5** **HR: 3**
BIS(CYCLOPENTADIENYLCHROMIUM TRICARBONYL)MERCURY
mf: $C_{16}H_{10}Cr_2HgO_6$ mw: 602.85

PROP: Yellow crystals. Mp: 201–203°.

SYN: HEXACARBONYLDI-PI-CYCLOPENTADIENYL-MU-MERCURIODICHROMIUM

TOXICITY DATA WITH REFERENCE
ivn-mus LD50:56 mg/kg CSLNX* NX#04754

CONSENSUS REPORTS: Mercury and its compounds, as well as chromium and its compounds, are on the Community Right-To-Know List.

NIOSH REL: (Organomercury): TWA 0.01 mg/m³; STEL 0.03 mg/m³ (skin)

SAFETY PROFILE: Poison by intravenous route. See also CHROMIUM COMPOUNDS, MERCURY COMPOUNDS, and CARBONYLS. When heated to decomposition it emits toxic fumes of Hg.

BIR529 **CAS:1277-43-6** **HR: 3**
BIS(CYCLOPENTADIENYL)COBALT
mf: $C_{10}H_{10}Co$ mw: 189.13

PROP: Very air-sensitive, black-purple crystals. Mp: 173–174°.

SYNS: COBALTOCENE □ DICYCLOPENTADIENYLCOBALT

TOXICITY DATA WITH REFERENCE
mmo-sat 3333 μg/plate EMMUEG 19(Suppl 21),2,92
ims-rat TDLo:200 mg/kg/60W-I:ETA NCIUS* PH 42-64-886,SEPT,71
ipr-rat LD50:55 mg/kg NCIUS* PH 43-64-886,JAN,65
ipr-mus LD50:80 mg/kg NCIUS* PH 43-64-886,JAN,65

CONSENSUS REPORTS: Reported in EPA TSCA Inventory. Cobalt and its compounds are on the Community Right-To-Know List.

SAFETY PROFILE: Poison by intraperitoneal route. Questionable carcinogen with experimental tumorigenic data. Mutation data reported. When heated to decomposition it emits acrid smoke and fumes. See also COBALT.

BIR750 **HR: 3**
BIS(n-CYCLOPENTADIENYL)MAGNESIUM
mf: $C_{10}H_{10}Mg$ mw: 154.41

SAFETY PROFILE: Ignites spontaneously in air. When heated to decomposition it emits acrid smoke and fumes. See also MAGNESIUM COMPOUNDS.

BIS200 **CAS:15131-55-2** **HR: 2**
BIS(2-CYCLOPENTENYL)ETHER
mf: $C_{10}H_{14}O$ mw: 150.24

TOXICITY DATA WITH REFERENCE
orl-rat LD50:11,300 mg/kg AIHAAP 30,470,69
skn-rbt LD50:1590 mg/kg AIHAAP 30,470,69

SAFETY PROFILE: Moderately toxic by skin contact. Mildly toxic by ingestion. See also ETHERS. When heated to decomposition it emits acrid smoke and irritating fumes.

BIS250 **CAS:38780-36-8** **HR: 2**
cis-BIS(CYCLOPENTYLAMMINE)PLATINUM(II)
mf: $C_{10}H_{22}Cl_2N_2Pt$ mw: 436.33

SYNS: cis-DICHLOROBIS(CYCLOPENTYLAMMINE)PLATINUM(II) □ cis-DICYCLOPENTYLAMMINEDICHLOROPLATINUM(II)

TOXICITY DATA WITH REFERENCE
mma-sat 10 μg/plate MUREAV 95,79,82
dni-ham:ovr 26 mg/L CBINA8 14,217,76
scu-rat TDLo:109 mg/kg/6W-I:CAR CNREA8 39,913,79
ipr-mus TDLo:189 mg/kg/10W-I:CAR CNREA8 39,913,79
ipr-mus LD50:480 mg/kg CBINA8 11,145,75

SAFETY PROFILE: Moderately toxic by intraperitoneal route. Questionable carcinogen with experimental carcinogenic data. Mutation data reported. See also PLATINUM COMPOUNDS. When heated to decomposition it emits very toxic fumes of Cl^- and NO_x.

BIS500 CAS:3465-75-6 HR: 3
BIS(DECANOYLOXY)DI-n-BUTYLSTANNANE
mf: $C_{28}H_{56}O_4Sn$ mw: 575.53

SYN: BIS(DECANOYLOXY)DI-N-BUTYLTIN

TOXICITY DATA WITH REFERENCE
skn-rbt 500 mg/24H SEV 28ZPAK-,229,72
eye-rbt 100 mg/24H MOD 28ZPAK-,229,72
orl-rat LD50:153 mg/kg 28ZPAK -,229,72

OSHA PEL: TWA 0.1 mg(Sn)/m³ (skin)
ACGIH TLV: TWA 0.1 mg(Sn)/m³ (skin) (Proposed: TWA 0.1 mg(Sn)/m³; STEL 0.2 mg(Sn)/m³ (skin))
NIOSH REL: (Organotin Compounds) TWA 0.1 mg(Sn)/m³

SAFETY PROFILE: Poison by ingestion. A severe skin and eye irritant. See also TIN COMPOUNDS. When heated to decomposition it emits acrid and irritant fumes.

For occupational chemical analysis use NIOSH: Organotin Compounds 5504.

BIT000 CAS:5684-13-9 HR: 2
BISDEHYDROISYNOLIC ACID METHYL ESTER
mf: $C_{19}H_{22}O_3$ mw: 298.41

SYNS: DEHYDROFOLLICULINIC ACID □ DOISYNOESTROL □ 1-ETHYL-2-METHYL-7-METHOXY-1,2,3,4-TETRAHYDROPHENANTHRYL-2-CARBOXYLIC ACID □ FENOCYCLIN □ FENOCYCLINE □ 7-METHYL-BISDEHYDRODOISYNOLIC ACID □ METILESTER del ACIDO BISDEHIDROISYNOLICO (SPANISH) □ 16,17-SECO-13-α-ESTRA-1,3,5,6,7,9-PENTAEN-17-OIC ACID, METHYL ESTER □ SURESTRINE □ SURESTRYL □ TETRADEHYDRODOISYNOLIC ACID METHYL ETHER

TOXICITY DATA WITH REFERENCE
orl-rat TDLo:500 µg/kg (5D preg):REP CCPTAY 14,487,76
orl-rat TDLo:1 mg/kg (female 4-7D post):REP CCPTAY 14,487,76
imp-ham TDLo:160 mg/kg (male 54W pre):REP ENDOAO 57,255,55
imp-gpg TDLo:3952 µg/kg:ETA BSBSAS 8,142,51
imp-ham TDLo:437 mg/kg:ETA CNREA8 12,274,52

SAFETY PROFILE: Questionable carcinogen with experimental tumorigenic data. Experimental reproductive effects. See also ETHERS and ESTERS. When heated to decomposition it emits acrid smoke and irritating fumes.

BIT030 HR: 2
BISDEHYDRODOISYNOLIC ACID 7-METHYL ETHER
mf: $C_{19}H_{22}O_3$ mw: 298.41

SYNS: 7-METILETER del ACIDO BISDEHIDRODOISYNOLICO □ 16,17-SECOESTRA-1,3,5(10),6,8-PENTAEN-17-OIC ACID, 3-METHOXY-

TOXICITY DATA WITH REFERENCE
imp-gpg TDLo:21 mg/kg:ETA,REP BSBSAS 8,142,51

SAFETY PROFILE: Questionable carcinogen with experimental tumorigenic data. Experimental reproductive effects. When heated to decomposition it emits toxic fumes of NO_x.

BIT250 CAS:37333-40-7 HR: 3
BIS(DIALKYLPHOSPHINOTHIOYL)DISULFIDE
mf: $C_{11}H_{26}O_4P_2S_4 \cdot C_9H_{22}O_4P_2S_4$ mw: 797.04

SYNS: BIO 1,137 □ ENT 23,584 □ NIAGARA 1,137 □ PHOSTEX

TOXICITY DATA WITH REFERENCE
orl-rat LD50:265 mg/kg TXAPA9 14,515,69
skn-rat LD50:480 mg/kg TXAPA9 14,515,69

SAFETY PROFILE: Poison by ingestion. Moderately toxic by skin contact. When heated to decomposition it emits very toxic fumes of SO_x and PO_x.

BIT350 CAS:55870-36-5 HR: 3
BIS(1,2-DIAMINOETHANE)DIAQUACOBALT(III) PERCHLORATE
mf: $C_4H_{20}Cl_3CoN_4O_{14}$ mw: 514.51

$((C_2H_8N_2)_2Co(H_2O)_2)(ClO_4)_3$

PROP: Only observed in soln. Sol in H_2O.

CONSENSUS REPORTS: Cobalt and its compounds are on the Community Right-To-Know List.

SAFETY PROFILE: The dry perchlorate is violently explosive. Upon decomposition it emits toxic fumes of Cl^- and NO_x. See also COBALT COMPOUNDS and PERCHLORATES.

BIT500 CAS:26388-78-3 HR: 3
BIS-1,2-DIAMINO ETHANE DICHLORO COBALT(III) CHLORATE
mf: $C_4H_{16}Cl_3CoN_4O_3$ mw: 333.49

$((C_2H_8N_2)_2CoCl_2)ClO_3$

CONSENSUS REPORTS: Cobalt and its compounds are on the Community Right-To-Know List.

SAFETY PROFILE: Explodes when heated to 320°C. When heated to decomposition it emits toxic fumes of Cl^- and NO_x. See also COBALT COMPOUNDS and CHLORATES.

BIT750 CAS:14932-06-0 HR: 3
BIS-1,2-DIAMINO ETHANE DICHLORO COBALT(III) PERCHLORATE
mf: $C_4H_{16}Cl_3CoN_4O_4$ mw: 317.49

$((C_2H_8N_2)_2CoCl_2)ClO_4$

CONSENSUS REPORTS: Cobalt and its compounds are on the Community Right-To-Know List.

SAFETY PROFILE: The perchlorate has low impact sensitivity but explodes when heated to 300°C. When heated to decomposition it emits toxic fumes of Cl^- and

NO_x. See also COBALT COMPOUNDS and PERCHLORATES.

BIU000 **HR: 3**
cis-BIS-1,2-DIAMINO ETHANE DINITRO COBALT(III) IODATE
mf: $C_4H_{16}CoIN_6O_7$ mw: 446.04

CONSENSUS REPORTS: Cobalt and its compounds are on the Community Right-To-Know List.

SAFETY PROFILE: Explodes on heating. When heated to decomposition it emits toxic fumes of I^- and NO_x. See also COBALT COMPOUNDS.

BIU125 **CAS:14781-32-9** **HR: 3**
BIS(1,2-DIAMINOETHANE)DINITROCOBALT(III) PERCHLORATE
mf: $C_4H_{16}ClCoN_6O_8$ mw: 370.59

$((C_2H_8N_2)_2Co(NO_2)_2)ClO_4$

CONSENSUS REPORTS: Cobalt and its compounds are on the Community Right-To-Know List.

SAFETY PROFILE: A dangerous explosive. Upon decomposition it emits toxic fumes of Cl^- and NO_x. See also COBALT COMPOUNDS, PERCHLORATES, and EXPLOSIVES.

BIU250 **HR: 3**
BIS(1,2-DIAMINO ETHANE)HYDROXOOXO RHENIUM (V) DIPERCHLORATE
mf: $C_4H_{17}Cl_2N_4O_{10}Re$ mw: 538.30

SAFETY PROFILE: Explodes violently when dried at above room temperature. Shock sensitive. When heated to decomposition it emits toxic fumes of $Cl^-$and NO_x. See also RHENIUM and PERCHLORATES.

BIU260 **CAS:19267-68-6** **HR: 3**
BIS(1,2-DIAMINOETHANE)HYDROXOOXORHENIUM(V) PERCHLORATE
mf: $C_4H_{17}Cl_2N_4O_{10}Re$ mw: 538.31

$((C_2H_8N_2)_2Re(OH)O)(ClO_4)_2$

SAFETY PROFILE: A heat- and shock-sensitive explosive. Upon decomposition it emits toxic fumes of Cl^- and NO_x. See also RHENIUM and PERCHLORATES.

BIU500 **HR: 3**
BIS-1,2-DIAMINO PROPANE-cis-DICHLORO CHROMIUM(III) PERCHLORATE
mf: $C_6H_{20}Cl_3CrN_4O_4$ mw: 370.61

$((C_3H_{10}N_2)_2CrCl_2)ClO_4$

CONSENSUS REPORTS: Chromium and its compounds are on the Community Right-To-Know List.

SAFETY PROFILE: Mixture with concentrated perchloric acid explodes violently. When heated to decomposition it emits toxic fumes of Cl^- and NO_x. See CHROMIUM COMPOUNDS and PERCHLORATES.

BIU900 **HR: 3**
BIS(DI-n-BENZENE CHROMIUM(IV))DICHROMATE
mf: $C_{24}H_{24}Cr_4O_7$ mw: 632.44

CONSENSUS REPORTS: Chromium and its compounds are on the Community Right-To-Know List.

SAFETY PROFILE: An explosive catalyst. Upon decomposition it emits acrid smoke and fumes. See also CHROMIUM COMPOUNDS.

BIV000 **CAS:58451-87-9** **HR: 2**
2,6-BIS-(DIBENZYLHYDROXYMETHYL)PIPERIDINE
mf: $C_{35}H_{39}NO_2$ mw: 505.75

SYNS: 2,6-BIS-(DWUBENZYLOHYDROKSYMETYLO)-PIPERYDYNA (POLISH) □ α,α,α′,α′-TETRAKIS(PHENYLMETHYL)-2,6-PIPERIDINEDIMETHANOL

TOXICITY DATA WITH REFERENCE
orl-mus LD50:5000 mg/kg PJPPAA 27,549,75
ipr-mus LD50:3125 mg/kg PJPPAA 27,549,75

SAFETY PROFILE: Moderately toxic by intraperitoneal route. Mildly toxic by ingestion. When heated to decomposition it emits toxic fumes of NO_x.

BIV500 **CAS:3072-84-2** **HR: 1**
2,2-BIS(3,5-DIBROMO-4-(2,3-EPOXYPROPOXY)PHENYL)PROPANE
mf: $C_{21}H_{24}Br_4O_4$ mw: 660.09

SYN: 2,2-BIS-(3′,5′-DIBROM-4′-GLYCIDOXYFENYL)PROPAN (CZECH)

TOXICITY DATA WITH REFERENCE
skn-rbt 500 mg/24H MLD 28ZPAK -,137,72
eye rbt 500 mg/24H MOD 28ZPAK -,137,72
orl-rat LD50:7160 mg/kg 28ZPAK -,137,72

SAFETY PROFILE: Mildly toxic by ingestion. A skin and eye irritant. When heated to decomposition it emits toxic fumes of Br^-.

BIV900 **CAS:73926-85-9** **HR: 3**
BIS(DIBUTYLAMMONIUM)HEXACHLOROSTANNATE
mf: $C_{16}H_{40}N_2•Cl_6Sn$ mw: 591.97

SYNS: AMMONIUMYL, DIBUTYL-, HEXACHLOROSTANNATE(2-) (2:1) □ DIBUTYLAMINE, HEXACHLOROSTANNANE (2:1)

TOXICITY DATA WITH REFERENCE
ivn-mus LD50:56 mg/kg CSLNX* NX#06251

OSHA PEL: TWA 2 mg(Sn)/m³
ACGIH TLV: TWA 2 mg(Sn)/m³

SAFETY PROFILE: Poison by intravenous route. When heated to decomposition it emits toxic fumes of NO_x, Sn, and Cl^-.

BIW000 **HR: 3**
BIS(DIBUTYLBORINO)ACETYLENE
mf: $C_{18}H_{36}B_2$ mw: 468.71

SAFETY PROFILE: Ignites spontaneously in air. When heated to decomposition it emits acrid smoke and fumes. See also BORON COMPOUNDS and ACETYLENE COMPOUNDS.

BIW250 CAS:64653-03-8 HR: 3
BIS(DIBUTYLDITHIOCARBAMATO)DIBENZYLSTANNANE
mf: $C_{32}H_{50}N_2S_4Sn$ mw: 709.77

SYNS: BIS((DIBUTYLDITHIOCARBAMOYL)OXY)DIBENZYLSTANNANE ☐ DIBENZYLTIN BIS(DIBUTYLDITHIOCARBAMATE)

TOXICITY DATA WITH REFERENCE
ivn-mus LD50:180 mg/kg CSLNX* NX#02082

OSHA PEL: TWA 0.1 mg(Sn)/m^3 (skin)
ACGIH TLV: TWA 0.1 mg(Sn)/m^3 (skin) (Proposed: TWA 0.1 mg(Sn)/m^3; STEL 0.2 mg(Sn)/m^3 (skin))
NIOSH REL: (Organotin Compounds) TWA 0.1 mg(Sn)/m^3

SAFETY PROFILE: Poison by intravenous route. See also CARBAMATES and TIN COMPOUNDS. When heated to decomposition it emits very toxic fumes of NO_x and SO_x.

For occupational chemical analysis use NIOSH: Organotin Compounds 5504.

BIW500 CAS:66009-08-3 HR: 3
BIS(DIBUTYLDITHIOCARBAMATO)DIMETHYLSTANNANE
mf: $C_{20}H_{42}N_2S_4Sn$ mw: 557.57

SYNS: BIS((DIBUTYLDITHIOCARBAMOYL)OXY)DIMETHYLSTANNANE ☐ DIMETHYLTIN BIS(DIBUTYLDITHIOCARBAMATE)

TOXICITY DATA WITH REFERENCE
ivn-mus LD50:180 mg/kg CSLNX* NX#02075

OSHA PEL: TWA 0.1 mg(Sn)/m^3 (skin)
ACGIH TLV: TWA 0.1 mg(Sn)/m^3 (skin) (Proposed: TWA 0.1 mg(Sn)/m^3; STEL 0.2 mg(Sn)/m^3 (skin))
NIOSH REL: (Organotin Compounds) TWA 0.1 mg(Sn)/m^3

SAFETY PROFILE: Poison by intravenous route. See also CARBAMATES and TIN COMPOUNDS. When heated to decomposition it emits very toxic fumes of NO_x and SO_x.

For occupational chemical analysis use NIOSH: Organotin Compounds 5504.

BIW750 CAS:13927-77-0 HR: 1
BIS(DIBUTYLDITHIOCARBAMATO)NICKEL
mf: $C_{18}H_{36}N_2S_4{\cdot}Ni$ mw: 467.51

PROP: Green crystals from C_6H_6/EtOH. Mp: 91°. Sol in C_6H_6, Me_2CO.

SYNS: DIBUTYLDITHIOCARBAMIC ACID, NICKEL SALT ☐ NICKEL DIBUTYLDITHIOCARBAMATE ☐ UV CHEK AM 104 ☐ VANGUARD N

TOXICITY DATA WITH REFERENCE
orl-mus TDLo:22 mg/kg/78W-I:ETA NTIS** PB223-159
scu-mus TDLo:1000 mg/kg:ETA NTIS** PB223-159
orl-rat LD50:17 g/kg IPSTB3 3,93,76

CONSENSUS REPORTS: Reported in EPA TSCA Inventory. Nickel and its compounds are on the Community Right-To-Know List.

SAFETY PROFILE: Low toxicity by ingestion. Questionable carcinogen with experimental tumorigenic data. See also NICKEL COMPOUNDS and CARBAMATES. When heated to decomposition it emits very toxic fumes of SO_x and NO_x.

BIX000 CAS:136-23-2 HR: 3
BIS(DIBUTYLDITHIOCARBAMATO)ZINC
mf: $C_{18}H_{38}N_2S_4Zn$ mw: 476.19

PROP: White powder. Mp: 104–108°, d: 1.24 @ 20°/20°.

SYNS: ACETO ZDBD ☐ BUTAZATE ☐ BUTAZATE 50-D ☐ BUTYL ZIMATE ☐ BUTYL ZIRAM ☐ DIBUTYLDITHIO-CARBAMIC ACID ZINC COMPLEX ☐ DIBUTYLDITHIOCARBAMIC ACID ZINC SALT ☐ USAF GY-5 ☐ VULCACURE ☐ VULKACIT LDB/C ☐ ZINC-BIBUTYLDITHIOCARBAMATE ☐ ZINC-DIBUTYLDITHIOCARBAMATE ☐ ZINC-N,N-DIBUTYLDITHIOCARBAMATE

TOXICITY DATA WITH REFERENCE
orl-mus TDLo:290 g/kg/78W-I:ETA NTIS** PB223-159
scu-mus TDLo:1000 mg/kg:ETA NTIS** PB223-159
ipr-mus LD50:100 mg/kg NTIS** AD277-689

CONSENSUS REPORTS: Reported in EPA TSCA Inventory. Zinc and its compounds are on the Community Right-To-Know List.

SAFETY PROFILE: Poison by intraperitoneal route. Questionable carcinogen with experimental tumorigenic data. When heated to decomposition it emits very toxic fumes of NO_x, ZnO and SO_x. See also ZINC COMPOUNDS and CARBAMATES.

BIX250 CAS:1477-57-2 HR: 3
N,N′-BIS(DICHLOROACETYL)-1,8-DIAMINOOCTANE
mf: $C_{12}H_{20}Cl_4N_2O_2$ mw: 366.14

SYNS: N,N′-BIS(DICHLOROACETYL)-1,8-OCTAMETHYLENEDIAMINE ☐ FERTILYSIN ☐ N,N′-OCTAMETHYLENEBIS(2,2-DICHLOROACTAMIDE) ☐ R-010-TK ☐ WIN 18,441 ☐ WIN 18,446

TOXICITY DATA WITH REFERENCE
spm-hmn-orl 1150 mg/kg/23W TXAPA9 3,1,61
mnt-mus:oth 500 mg/kg NKEZA4 33,165,86
orl-man TDLo:150 mg/kg (42D male):REP TXAPA9 3,1,61
orl-rat TDLo:50 mg/kg (female 10-11D post):REP TJADAB 26,155,82
orl-mus TDLo:4 g/kg (female 10D post):TER NKEZA4 33,165,86
orl-rat TDLo:2250 mg/kg (3W male):REP EXMPA6 1,251,62
orl-rat TDLo:1313 mg/kg (male 21D pre):REP INJFA3 11,291,66
unr-rat TDLo:2 g/kg (female 9-10D post):TER TJADAB 28,10A,83
orl-rat TDLo:1 g/kg (female 11-12D post):TER TJADAB 18,5,78
orl-rat TDLo:1 g/kg (female 10D post):TER SEIJBO 25,29,85
orl-rat TDLo:1 g/kg (female 11-12D post):TER TJADAB 18,5,78

B

orl-mus TDLo:6400 mg/kg (female 8-9D post):TER TJADAB 30(1),13A,84
orl-mus TDLo:20 g/kg (female 14D post):TER NKEZA4 33,165,86
orl-rat TDLo:100 mg/kg (female 10-11D post):TER TJADAB 26,155,82
orl-rat TDLo:1 g/kg (female 11-12D post):TER TJADAB 18,5,78
unr-rat TDLo:2 g/kg (female 9-10D post):TER TJADAB 28,10A,83
orl-man TDLo:943 mg/kg/50D:GIT 15QWAW -93,65
ipr-mus LDLo:150 mg/kg TXAPA9 23,288,72

CONSENSUS REPORTS: EPA Genetic Toxicology Program.

SAFETY PROFILE: Poison by intraperitoneal route. Human systemic effects by ingestion: nausea and vomiting. Human reproductive effects by ingestion: changes in spermatogenesis. An experimental teratogen. Other experimental reproductive effects. Human mutation data reported. When heated to decomposition it emits very toxic fumes of Cl^- and NO_x.

BIX500 CAS:15442-77-0 HR: 3
BIS(3,4-DICHLOROBENZOATO)NICKEL
mf: $C_{14}H_6Cl_4NiO_4$ mw: 438.71

TOXICITY DATA WITH REFERENCE
ivn-mus LD50:100 mg/kg CSLNX* NX#03268

CONSENSUS REPORTS: Nickel and its compounds are on the Community Right-To-Know List.

OSHA PEL: TWA 0.1 mg (Ni)/m³
ACGIH TLV: TWA 0.1 mg(Ni)/m³; (Proposed: TWA 0.05 mg(Ni)/m³; Human Carcinogen)
NIOSH REL: (Inorganic Nickel) TWA 0.015 mg(Ni)/m³

SAFETY PROFILE: Suspected carcinogen. Poison by intravenous route. See also NICKEL COMPOUNDS and CHLORIDES. When heated to decomposition it emits toxic fumes of Cl^-.

BIX750 CAS:133-14-2 HR: 3
BIS(2,4-DICHLOROBENZOYL)PEROXIDE
mf: $C_{14}H_6Cl_4O_4$ mw: 380.00

SYNS: CADOX TS □ CADOX TS 40,50 □ DI-2,4-DICHLOROBENZOYL PEROXIDE, >75% with water (DOT) □ LUPERCO CST

TOXICITY DATA WITH REFERENCE
ipr-mus LD50:225 mg/kg IPSTB3 3,93,76

CONSENSUS REPORTS: Reported in EPA TSCA Inventory.

DOT CLASSIFICATION: Forbidden

SAFETY PROFILE: Poison by intraperitoneal route. Explosion Hazard: Pure compound is extremely shock sensitive and decomposes rapidly @ 80°. When heated to decomposition it emits toxic fumes of Cl^-. See also PEROXIDES, ORGANIC, and ESTERS.

BIY000 CAS:2589-02-8 HR: 2
2,2-BIS(3,5-DICHLORO-4-(2,3-EPOXYPROPOXY)PHENYL)PROPANE
mf: $C_{21}H_{24}Cl_4O_4$ mw: 482.25

SYN: 2,2-BIS-(3′,5′-DICHLOR-4′-GLYCIDOXYFENYL)PROPAN (CZECH)

TOXICITY DATA WITH REFERENCE
skn-rbt 500 mg/24H MOD 28ZPAK -,137,72
eye-rbt 500 mg/24H MOD 28ZPAK -,137,72

SAFETY PROFILE: A skin and eye irritant. When heated to decomposition it emits toxic fumes of Cl^-.

BIY250 CAS:19721-74-5 HR: 3
BIS(1,2-DICHLOROETHYL)SULFONE
mf: $C_4H_6Cl_4O_2S$ mw: 259.96

TOXICITY DATA WITH REFERENCE
orl-rat LD50:250 mg/kg AIHAAP 30,470,69
skn-rbt LD50:1000 mg/kg AIHAAP 30,470,69

SAFETY PROFILE: Poison by ingestion. Moderately toxic by skin contact. See also SULFONATES. When heated to decomposition it emits very toxic fumes of Cl^- and SO_x.

BIY500 HR: 3
BIS(1,3-DICHLORO-1,1,3,3-TETRAETHYLDISTANNOXANE)
mf: $C_{16}H_{40}Cl_4O_2 \cdot 2Sn$ mw: 643.72

SYN: DI-o-(CHLORODIETHYLSTANNYLOXO)BIS(CHLORO-DIETHYLTIN)

TOXICITY DATA WITH REFERENCE
ivn-mus LD50:22 mg/kg CSLNX* NX#03157

OSHA PEL: TWA 0.1 mg(Sn)/m³ (skin)
ACGIH TLV: TWA 0.1 mg(Sn)/m³ (skin) (Proposed: TWA 0.1 mg(Sn)/m³; STEL 0.2 mg(Sn)/m³ (skin))
NIOSH REL: (Organotin Compounds) TWA 0.1 mg(Sn)/m³

SAFETY PROFILE: Poison by intravenous route. See also TIN COMPOUNDS and CHLORIDES. When heated to decomposition it emits toxic fumes of Cl^-.

BIY600 CAS:1518-15-6 HR: 3
1,4-BIS(DICYANOMETHYLENE)CYCLOHEXANE
mf: $C_{12}H_8N_4$ mw: 208.24

SYN: $\Delta^{1,\alpha:4,\alpha'}$-CYCLOHEXANEDIMALONONITRILE

TOXICITY DATA WITH REFERENCE
ivn-mus LD50:56 mg/kg CSLNX* NX#05268

CONSENSUS REPORTS: Reported in EPA TSCA Inventory.

SAFETY PROFILE: Poison by intravenous route. When heated to decomposition it emits toxic vapors of NO_x.

BIZ000 **HR: 3**
1,3-BIS(DI-n-CYCLOPENTADIENYL IRON)-2-PROPEN-1-ONE
mf: $C_{23}H_{20}Fe_2O$ mw: 324.11

SAFETY PROFILE: The dry material is a powerful explosive and detonator. Incompatible with perchloric acid; acetic anhydride; ether; methanol.

BJA000 **CAS:73771-52-5** **HR: 1**
1,5-BIS(4-(2,3-DIDEHYDROTRIAZIRIDINYL)PHENYL)-1,4-PENTADIEN-3-ONE
mf: $C_{17}H_{12}N_6O$ mw: 316.35

SYN: DIAZIDODIBENZALACETON (CZECH)

TOXICITY DATA WITH REFERENCE
skn-rbt 500 mg/24H MLD 28ZPAK -,123,72
eye-rbt 500 mg/24H MLD 28ZPAK -,123,72

SAFETY PROFILE: A skin and eye irritant. See also KETONES. When heated to decomposition it emits toxic fumes of NO_x.

BJA200 **CAS:90466-79-8** **HR: 3**
BIS(2,2-DIETHOXYETHYL)DISELENIDE
mf: $C_{12}H_{26}O_4Se_2$ mw: 392.30

SYN: DISELENIDE, BIS(2,2-DIETHOXYETHYL)-

TOXICITY DATA WITH REFERENCE
ivn-mus LD50:1800 µg/kg CSLNX* NX#09262

OSHA PEL: TWA 0.2 mg(Se)/m^3
ACGIH TLV: TWA 0.2 mg(Se)/m^3

SAFETY PROFILE: Poison by intravenous route. When heated to decomposition it emits toxic fumes of Se.

BJA250 **CAS:105-18-0** **HR: 3**
1,4-BIS(DIETHYLAMINO)-2-BUTYNE
mf: $C_{12}H_{24}N_2$ mw: 196.38

SYNS: 2-BUTYNYLENEDIAMINE, N,N,N′N′-TETRAETHYL- □ N,N,N′,N′-TETRAETHYL-2-BUTYNYLENEDIAMINE

TOXICITY DATA WITH REFERENCE
ivn-mus LD50:56 mg/kg CSLNX* NX#04930

CONSENSUS REPORTS: Reported in EPA TSCA Inventory.

SAFETY PROFILE: Poison by intravenous route. When heated to decomposition it emits toxic vapors of NO_x.

BJA500 **CAS:35697-34-8** **HR: 3**
2,6-BIS(2-(DIETHYLAMINO)ETHOXY)-9,10-ANTHRACENEDIONE DIHYDROCHLORIDE
mf: $C_{26}H_{34}N_2O_4$•2ClH mw: 511.54

SYN: RMI 10024DA

TOXICITY DATA WITH REFERENCE
orl-mus LD50:1560 mg/kg ALACBI 12,77,79
scu-mus LD50:110 mg/kg ALACBI 12,77,79

SAFETY PROFILE: Poison by subcutaneous route. Moderately toxic by ingestion. When heated to decomposition it emits very toxic fumes of NO_x and HCl.

BJA750 **CAS:57665-49-3** **HR: 2**
1-(BIS(2-(DIETHYLAMINO)ETHYL)AMINO)-5-CHLORO-3-(p-CHLOROPHENYL)INDOLE DIHYDROCHLORIDE HEMIHYDRATE
mf: $C_{26}H_{36}Cl_2N_4$•2ClH•1/2H_2O mw: 557.49

TOXICITY DATA WITH REFERENCE
orl-rat LD50:780 mg/kg ARZNAD 30,919,80
orl-mus LD50:870 mg/kg ARZNAD 30,919,80

SAFETY PROFILE: Moderately toxic by ingestion. When heated to decomposition it emits very toxic fumes of Cl^- and NO_x.

BJA809 **CAS:57647-13-9** **HR: 2**
1-(BIS(2-(DIETHYLAMINO)ETHYL)AMINO)-3-PHENYLINDOLE DIHYDROCHLORIDE
mf: $C_{26}H_{38}N_4$•2ClH mw: 479.1

TOXICITY DATA WITH REFERENCE
orl-rat LD50:800 mg/kg ARZNAD 30,919,80
orl-mus LD50:540 mg/kg ARZNAD 30,919,80

SAFETY PROFILE: Moderately toxic by ingestion. When heated to decomposition it emits very toxic fumes of Cl^- and NO_x.

BJA825 **CAS:3572-35-8** **HR: 3**
3,6-BIS(3-DIETHYLAMINOPROPOXY)PYRIDAZINE BISMETHIODIDE
mf: $C_{20}H_{40}N_4O_2$•2I mw: 622.44

SYNS: 3,3′-(3,6-PYRIDAZINEDIYLBIS(OXY)BIS(N,N-DIETHYL-N-METHYL-1-PROPANAMINIUM)) DIIODIDE (9CI) □ (3,6-PYRIDAZINEDIYLBIS(OXYTRIMETHYLENE))BIS(DIETHYLMETHYLAMMONIUM IODIDE) □ WIN 4981

TOXICITY DATA WITH REFERENCE
orl-mus LD50:49 mg/kg JPETAB 118,395,56
ivn-mus LD50:610 µg/kg JPETAB 125,323,59
orl-cat LD50:5 mg/kg JPETAB 118,395,56
ivn-cat LD50:500 µg/kg JPETAB 118,395,56
ivn-rbt LD50:400 µg/kg JPETAB 118,395,56

SAFETY PROFILE: Poison by ingestion and intravenous routes. When heated to decomposition it emits toxic fumes of I^-, NO_x, and NH_3.

BJB500 **CAS:14239-68-0** **HR: 3**
BIS(DIETHYLDITHIOCARBAMATO)CADMIUM
mf: $C_{10}H_{20}CdN_2S_4$ mw: 408.96

SYNS: CADMIUM DIETHYL DITHIOCARBAMATE □ ETHYL CADMATE □ ETHYL TUADS

TOXICITY DATA WITH REFERENCE
mmo-sat 10 µg/plate MUREAV 68,313,79
dnd-esc 1 µmol/L ARTODN 46,277,80
orl-mus TDLo:7100 mg/kg/78W-I:ETA NTIS** PB223-159
scu-mus TDLo:1000 mg/kg:ETA NTIS** PB223-159

CONSENSUS REPORTS: Reported in EPA TSCA

Inventory. Cadmium and its compounds are on the Community Right-To-Know List.

OSHA PEL: TWA 5 µg(Cd)/m³
ACGIH TLV: TWA 0.05 mg(Cd)/m³ (Proposed: TWA 0.01 mg(Cd)/m³ (dust), Suspected Human Carcinogen; 0.002 mg(Cd)/m³ (respirable dust), Suspected Human Carcinogen); BEI: 10 µg/g creatinine in urine; 10 µg/L in blood
DFG BAT: Blood 1.5 µg/dL; Urine 15 µg/dL, Suspected Carcinogen
NIOSH REL: (Cadmium) Reduce to lowest feasible level

SAFETY PROFILE: Confirmed human carcinogen with experimental tumorigenic data. Mutation data reported. When heated to decomposition it emits very toxic fumes of NO_x and SO_x. See also CADMIUM COMPOUNDS and CARBAMATES.

BJB750 CAS:14239-51-1 HR: 3
BIS(DIETHYLDITHIOCARBAMATO)MERCURY
mf: $C_{10}H_{20}HgN_2S_4$ mw: 497.15

PROP: Yellow crystals from Me_2CO. Mp: 127–130°.

TOXICITY DATA WITH REFERENCE
ipr-rat LDLo:100 mg/kg NCNSA6 5,30,53
ivn-mus LD50:18 mg/kg CSLNX* NX#02505

CONSENSUS REPORTS: Mercury and its compounds are on the Community Right-To-Know List.

OSHA PEL: CL 0.1 mg(Hg)/m³ (skin)
ACGIH TLV: TWA 0.1 mg(Hg)/m³
NIOSH REL: (Organomercury): TWA 0.01 mg/m³; STEL 0.03 mg/m³ (skin)

SAFETY PROFILE: Poison by intravenous and intraperitoneal routes. See also MERCURY COMPOUNDS and CARBAMATES. When heated to decomposition it emits very toxic fumes of NO_x, SO_x, and Hg.

BJC000 CAS:14324-55-1 HR: 3
BIS(DIETHYLDITHIOCARBAMATO)ZINC
mf: $C_{10}H_{22}N_2S_4$•Zn mw: 363.95

PROP: White powder. D: 1.47 @ 20°/20°.

SYNS: DIETHYLDITHIOCARBAMIC ACID ZINC SALT ☐ ETHAZATE ☐ ETHYL CYMATE ☐ ETHYL ZIMATE ☐ ETHYL ZIRUM ☐ VULCACURE ☐ VULKACIT LDA ☐ ZINC DIETHYLDITHIOCARBAMATE ☐ ZINC-N,N-DIETHYLDITHIOCARBAMATE

TOXICITY DATA WITH REFERENCE
eye-rbt 100 mg/24H MOD 28ZPAK -,11,72
mmo-sat 25 µg/plate MUREAV 68,313,79
mma-sat 25 µg/plate MUREAV 68,313,79
scu-mus TDLo:464 mg/kg:CAR NTIS** PB223-159
orl-rat LD50:3340 mg/kg 28ZPAK -,11,72
ipr-mus LD50:142 mg/kg KOKABN 26,358,77
orl-rbt LD50:570 mg/kg INMEAF 16,473,47

CONSENSUS REPORTS: Reported in EPA TSCA Inventory. Zinc and its compounds are on the Community Right-To-Know List.

SAFETY PROFILE: Poison by intraperitoneal route. Moderately toxic by ingestion and subcutaneous routes. Severe irritant to eyes, nose, and throat. Questionable carcinogen with experimental carcinogenic and tumorigenic data. Mutation data reported. When heated to decomposition it emits very toxic fumes of NO_x and SO_x. See also ZINC COMPOUNDS and CARBAMATES.

BJC250 CAS:738-99-8 HR: 3
1,4-BIS(N,N′-DIETHYLENE PHOSPHAMIDE)PIPERAZINE
mf: $C_{12}H_{24}N_6O_2P_2$ mw: 346.36

PROP: Crystals from C_6H_6. Mp: 187–189°.

SYNS: 1,4-BIS(BIS(1-AZIRIDINYL)PHOSPHINYL)PIPERAZINE ☐ DIPIN ☐ DIPINE ☐ ENT 50,107 ☐ 1,4-PIPERAZINEDIYLBIS(BIS(1-AZIRIDINYL)PHOSPHINE) OXIDE ☐ TETRAETHYLENEIMIDEPIPERAZINE-N, N′-DIPHOSPHORIC ACID

TOXICITY DATA WITH REFERENCE
dlt-oin-unr 1 pph/3H-C AESAAI 62,790,69
cyt-hmn:lym 29 µmol/L SOGEBZ 10,1580,74
sce-hmn:lym 10 mg/L TGANAK 16(2),34,82
cyt-rat-ipr 60 mg/kg SOGEBZ 11,1347,75
ipr-mus LD50:90 mg/kg PCJOAU 14,363,80
orl-mus LD50:68 mg/kg RPTOAN 36,240,73
scu-mus LD50:58 mg/kg ANTBAL 21,262,76

SAFETY PROFILE: Poison by ingestion, intraperitoneal, and subcutaneous routes. Human mutation data reported. When heated to decomposition it emits very toxic fumes of PO_x and NO_x.

BJC500 HR: 3
BISDIETHYLENE TRIAMINE COBALT(III) PERCHLORATE
mf: $C_8H_{26}Cl_3CoN_6O_{12}$ mw: 562.54

CONSENSUS REPORTS: Cobalt and its compounds are on the Community Right-To-Know List.

SAFETY PROFILE: Very sensitive to impact. Explodes @ 325°. When heated to decomposition it emits toxic fumes of NO_x. See also COBALT COMPOUNDS and PERCHLORATES.

BJD000 CAS:34491-12-8 HR: 3
BIS(DIETHYLTHIO)CHLORO METHYL PHOSPHONATE
mf: $C_5H_{12}ClOPS_2$ mw: 218.71

SYNS: CHEMAGRO 5461 ☐ CHEMAGRO R-5461 ☐ S,S-DIETHYL (CHLOROMETHYL)PHOSPHONODITHIOATE ☐ ENT 27,267 ☐ R-5461

TOXICITY DATA WITH REFERENCE
orl-rat LD50:35 mg/kg ARSIM* 20,7,66
skn-rat LD50:79 mg/kg TXAPA9 12,286,68
ipr-rat LD50:23 mg/kg TXAPA9 12,286,68
orl-mus LDLo:210 mg/kg AECTCV 14,111,85
ipr-mus LD50:43 mg/kg TXAPA9 12,286,68
orl-gpg LD50:224 mg/kg TXAPA9 12,286,68
ipr-gpg LD50:109 mg/kg TXAPA9 12,286,68

SAFETY PROFILE: Poison by ingestion, skin contact, and intraperitoneal routes. When heated to decomposition it emits very toxic fumes of SO_x, PO_x, and Cl^-.

BJD250 CAS:4394-93-8 HR: 3
BIS(DIFLUOROAMINO)DIFLUOROMETHANE
mf: CF_6N_2 mw: 154.02

PROP: Gas. D: 1.50 @ 25°/4°, fp: −162° (to −1°), bp: −32°.

SAFETY PROFILE: An unstable explosive that may be initiated by phase changes. Upon decomposition it emits toxic fumes of F^- and NO_x. For preparation, handling, and storage, use protective equipment.

BJD375 CAS:30957-47-2 HR: 3
1,1-BIS(DIFLUOROAMINO)-2,2-DIFLUORO-2-NITRO-ETHYL METHYL ETHER
mf: $C_3H_3F_6N_3O_3$ mw: 243.07

SAFETY PROFILE: A shock-sensitive explosive. When heated to decomposition it emits toxic fumes of F^- and NO_x. See also ETHERS.

BJD500 CAS:13084-47-4 HR: 3
1,2-BIS(DIFLUOROAMINO)ETHANOL
mf: $C_2H_4F_4N_2O$ mw: 148.05

SAFETY PROFILE: An impact-sensitive explosive. When heated to decomposition it emits toxic fumes of F^- and NO_x. See also EXPLOSIVES.

BJD750 CAS:13084-45-2 HR: 3
1,2-BIS(DIFLUOROAMINO)ETHYL VINYL ETHER
mf: $C_4H_6F_4N_2O$ mw: 174.10

$F_2NCH_2CH(NF_2)OCH{=}CH_2$

SAFETY PROFILE: An impact-sensitive explosive. When heated to decomposition it emits toxic fumes of F^- and NO_x. See also EXPLOSIVES and ETHERS.

BJE000 CAS:33364-51-1 HR: 3
4,4-BIS(DIFLUOROAMINO)-3-FLUOROIMINO-1-PENTENE
mf: $C_5H_6F_5N_3$ mw: 203.06

$H_2C{=}CHC(N{:}F)C(NF_2)_2CH_3$

SAFETY PROFILE: May explode if heated. When heated to decomposition it emits toxic fumes of F^- and NO_x. See also EXPLOSIVES.

BJE250 CAS:18273-30-8 HR: 3
1,2-BIS(DIFLUOROAMINO)-N-NITROETHYLAMINE
mf: $C_2H_4F_4N_4O_2$ mw: 192.07

SAFETY PROFILE: May explode when heated above 75°C. When heated to decomposition it emits toxic fumes of F^- and NO_x. See also EXPLOSIVES.

BJE325 CAS:55124-14-6 HR: 3
BIS(DIFLUOROBORYL)METHANE
mf: $CH_2B_2F_4$ mw: 111.64

SAFETY PROFILE: Highly reactive. Explodes in air or on contact with water. When heated to decomposition it emits toxic fumes of F^-. See also BORON COMPOUNDS and BORANES.

BJE500 CAS:52578-56-0 HR: 2
BIS(DIHYDROXYPHENYL)SULFIDE
mf: $C_{12}H_{10}O_2S$ mw: 218.28

SYN: DIRESORCYL SULFIDE

TOXICITY DATA WITH REFERENCE
eye-rbt 500 mg SEV IHFCAY 6,1,67
orl-rat LD50:4290 mg/kg IHFCAY 6,1,67

SAFETY PROFILE: Mildly toxic by ingestion. A severe eye irritant. See also SULFIDES. When heated to decomposition it emits toxic fumes of SO_x.

BJE750 CAS:115-26-4 HR: 3
BIS(DIMETHYLAMIDO)FLUORO PHOSPHATE
mf: $C_4H_{12}FN_2OP$ mw: 154.15

PROP: Liquid. Misc in H_2O and most org solvs. D: 1.115 mm @ 20°, bp: 67° @ 4 mm.

SYNS: BFP ☐ BFPO ☐ BIS(DIMETHYLAMIDO)PHOSPHORYL FLUORIDE ☐ BIS(DIMETHYLAMINO)FLUOROPHOSPHATE ☐ BISDIMETHYLAMINOFLUOROPHOSPHINE OXIDE ☐ CR 409 ☐ DIFO ☐ DIMEFOX ☐ DMF ☐ ENT 19,109 ☐ FLUOPHOSPHORIC ACID DI(DIMETHYLAMIDE) ☐ FLUORURE de N,N,N',N'-TETRAMETHYLE PHOSPHORO-DIAMIDE (FRENCH) ☐ HANANE ☐ PESTOX IV ☐ PESTOX XIV ☐ PESTOX 14 ☐ T-2002 ☐ TERRA-SYSTAM ☐ TERRA-SYTAM ☐ TERRASYTUM ☐ N,N,N',N'-TETRAMETHYL-DIAMIDO-FOSFORZUUR-FLUORIDE (DUTCH) ☐ TETRAMETHYLDIAMIDOPHOSPHORIC FLUORIDE ☐ N,N,N',N'-TETRAMETHYL-DIAMIDO-PHOSPHORSAEURE-FLUORID (GERMAN) ☐ TETRAMETHYLPHOSPHORODIAMIDIC FLUORIDE ☐ N,N,N,N-TETRAMETHYLPHOSPHORODIAMIDIC FLUORIDE ☐ N,N,N',N'-TETRAMETIL-FOSFORODIAMMIDO-FLUORURO (ITALIAN) ☐ TETRA SYTAM ☐ TL 792 ☐ WACKER S 14/10

TOXICITY DATA WITH REFERENCE
orl-rat LD50:1 mg/kg NTIS** PB158-508
ihl-rat LC50:2 mg/m³/10M NTIS** PB158-508
skn-rat LD50:2 mg/kg WRPCA2 9,119,70
ipr-rat LD50:5 mg/kg AMIHBC 6,9,52
scu-rat LDLo:300 µg/kg NTIS** PB158-508
orl-mus LD50:2 mg/kg BESAAT 12,161,66
ihl-mus LC50:950 mg/m³/10M NTIS** PB158-508
ipr-mus LD50:1400 µg/kg JPETAB 112,231,54
scu-mus LD50:1 mg/kg NTIS** PB158-508
ivn-dog LD50:5 mg/kg JPETAB 112,231,54

CONSENSUS REPORTS: EPA Extremely Hazardous Substances List.

SAFETY PROFILE: Poison by ingestion, skin contact, intraperitoneal, subcutaneous, and intravenous routes. When heated to decomposition it emits very toxic fumes of F^-, NO_x, and PO_x.

BJF000 CAS:494-38-2 HR: 3
3,6-BIS(DIMETHYLAMINO)ACRIDINE
mf: $C_{17}H_{19}N_3$ mw: 265.39

PROP: Yellow needles from EtOH. Mp: 180–181°. Sol in EtOH and Me_2CO.

SYNS: ACRIDINE ORANGE □ ACRIDINE ORANGE FREE BASE □ BASIC ORANGE 3RN □ 2,8-BISDIMETHYLAMINOACRIDINE □ BRILLIANT ACRIDINE ORANGE E □ C.I. 46005 □ C.I. No. 46005:1 □ C.I. BASIC ORANGE 14 □ C.I. SOLVENT ORANGE 15 □ 3,6-DI(DIMETHYLAMINO)ACRIDINE □ EUCHRYSINE □ RHODULINE ORANGE □ SOLVENT ORANGE 15 □ N,N,N'-TETRAMETHYL-3,6-ACRIDINEDIAMINE □ WAXOLINE ORANGE A

TOXICITY DATA WITH REFERENCE
mmo-omi 10 µg/L MIBLAO 49,223,80
dns-rat:lvr 1 mmol/L ENMUDM 3,11,81
otr-ham:emb 1 µg/L NCIMAV 58,243,81
skn-mus TDLo:6630 mg/kg:ETA BJCAAI 23,587,69
scu-mus TDLo:657 mg/kg/63W-I:ETA BJCAAI 23,587,69
scu-mus LD50:250 mg/kg BJEPA5 28,1,47

CONSENSUS REPORTS: IARC Cancer Review: Group 3 IMEMDT 7,56,87; Animal Inadequate Evidence IMEMDT 16,145,78.

SAFETY PROFILE: Poison by subcutaneous route. Questionable carcinogen with experimental tumorigenic and carcinogenic data. Mutation data reported. When heated to decomposition it emits toxic fumes of NO_x.

BJF500 CAS:100-22-1 HR: 3
p-BIS(DIMETHYLAMINO)BENZENE
mf: $C_{10}H_{16}N_2$ mw: 164.28

PROP: Leaflets or crystals. Mp: 51°, bp: 260°. Sltly sol in cold water; more sol in hot water; freely sol in alc, chloroform, ether, and petroleum ether.

SYNS: 1,4-BIS(DIMETHYLAMINO)BENZENE □ TETRAMETHYL-p-PHENYLENEDIAMINE □ N,N,N',N'-TETRAMETHYL-p-PHENYLENEDIAMINE □ TL 85 □ TMPD □ WURSTER'S BLUE □ WURSTER'S REAGENT

TOXICITY DATA WITH REFERENCE
mmo-sat 333 µg/plate EMMUEG 11(Suppl 12),1,88
cyt-ham:lng 10 mg/L MUREAV 241,175,90
orl-rat LDLo:500 mg/kg JPETAB 90,260,47
ihl-mus LCLo:780 mg/m³/10M NDRC** NDCrc-132,Dec,42
orl-qal LD50:42 mg/kg EESADV 6,149,82
orl-bwd LD50:23,700 µg/kg AECTCV 12,355,83

CONSENSUS REPORTS: Reported in EPA TSCA Inventory.

SAFETY PROFILE: Poison by ingestion. Moderately toxic by inhalation. Mutation data reported. When heated to decomposition it emits toxic fumes of NO_x.

BJG000 CAS:39047-21-7 HR: 3
BIS(DIMETHYLAMINOBORANE)ALUMINUM TETRAHYDROBORATE
mf: $C_4H_{22}AlB_3N_2$ mw: 157.66

$((CH_3)_2NBH_3)_2AlBH_4$

SAFETY PROFILE: Ignites on contact with air. Violent reaction on contact with water. When heated to decomposition it emits toxic fumes of NO_x. See also ALUMINUM COMPOUNDS, BORANES, and BORON COMPOUNDS.

BJG100 CAS:17339-60-5 HR: 3
2,2'-BIS(DIMETHYLAMINO) DIETHYLSULPHIDE DIHYDROCHLORIDE
mf: $C_8H_{20}N_2S_2 \cdot 2ClH$ mw: 281.34

SYNS: ETHYLAMINE, 2,2'-DITHIOBIS(N,N-DIMETHYL), DIHYDROCHLORIDE □ 2,2'-DITHIOBIS(N,N-DIMETHYLETHYLAMINE) DIHYDROCHLORIDE

TOXICITY DATA WITH REFERENCE
ipr-mus LD50:310 mg/kg RPTOAN 33,127,70
ivn-mus LD50:310 mg/kg RPTOAN 33,127,70

CONSENSUS REPORTS: Reported in EPA TSCA Inventory.

SAFETY PROFILE: A poison by intraperitoneal and intravenous routes. When heated to decomposition it emits toxic vapors of SO_x, NO_x and HCl.

BJG125 CAS:993-74-8 HR: 3
BIS(DIMETHYLAMINO)DIMETHYLSTANNANE
mf: $C_6H_{18}N_2Sn$ mw: 236.93

$((CH_3)_2N)_2Sn(CH_3)_2$

PROP: Bp: 138°, d: 1.148 @ 20°/4°.

SAFETY PROFILE: Mixture with chloroform explodes when heated. When heated to decomposition it emits toxic fumes of NO_x. See also TIN COMPOUNDS.

BJG150 CAS:1007-22-3 HR: 3
3,5-BIS-DIMETHYLAMINO-1,2,4-DITHIAZOLIUM CHLORIDE
mf: $C_6H_{12}N_3S_2 \cdot Cl$ mw: 225.78

SYN: ORF 5513

TOXICITY DATA WITH REFERENCE
orl-rat TDLo:4 mg/kg (female 16-19D post):TER CCPTAY 21,529,80
orl-rat TDLo:150 µg/kg (3D pre):REP CCPTAY 21,529,80
orl-rat TDLo:4 mg/kg (female 10-13D post):REP CCPTAY 21,529,80
orl-rat TDLo:60 mg/kg (female 1-6D post):REP CCPTAY 21,529,80
ivn-mus LD50:56 mg/kg CSLNX* NX#07197
orl-qal LD50:24 mg/kg JRPFA4 48,371,76

SAFETY PROFILE: Poison by ingestion and intravenous routes. An experimental teratogen. Other experimental reproductive effects. When heated to decomposition it emits toxic fumes of Cl^-, SO_x and NO_x. See also CHLORIDES.

BJG250 CAS:3065-46-1 HR: 2
BIS(2-DIMETHYLAMINOETHOXY)ETHANE
mf: $C_{10}H_{24}N_2O_2$ mw: 204.36

TOXICITY DATA WITH REFERENCE
orl-rat LD50:2830 mg/kg AIHAAP 30,470,69
skn-rbt LD50:1200 mg/kg AIHAAP 30,470,69

SAFETY PROFILE: Moderately toxic by ingestion and skin contact. When heated to decomposition it emits toxic fumes such as NO_x.

BJG500 **CAS:54593-27-0** **HR: 3**
3,6-BIS(2-(DIMETHYLAMINO)ETHOXY)-9H-XANTHEN-9-ONE DIHYDROCHLORIDE
mf: $C_{21}H_{26}N_2O_4 \cdot 2ClH$ mw: 443.41

SYN: RMI 10874DA

TOXICITY DATA WITH REFERENCE
orl-mus LD50:1780 mg/kg ALACBI 12,77,79
scu-mus LD50:353 mg/kg ALACBI 12,77,79

SAFETY PROFILE: Poison by subcutaneous route. Moderately toxic by ingestion. When heated to decomposition it emits very toxic fumes of NO_x and HCl.

BJG750 **CAS:52673-65-1** **HR: 3**
1,3-BIS(2-DIMETHYLAMINOETHYL)ADAMANTANE DIHYDROCHLORIDE
mf: $C_{18}H_{34}N_2 \cdot 2ClH$ mw: 351.46

SYN: 2,2′-(1,3-ADAMANTYLENE)N,N,N′,N′-TETRAMETHYL-ETHYLAMINE DIHYDROCHLORIDE

TOXICITY DATA WITH REFERENCE
orl-mus LD50:1600 mg/kg JMCMAR 17,602,74
ipr-mus LD50:150 mg/kg JMCMAR 17,602,74

SAFETY PROFILE: Poison by intraperitoneal route. Moderately toxic by ingestion. When heated to decomposition it emits very toxic fumes of HCl and NO_x.

BJH000 **CAS:57647-53-7** **HR: 2**
1-(BIS(2-(DIMETHYLAMINO)ETHYL)AMINO)-5-METHYL-3-PHENYLINDOLE DIHYDROCHLORIDE
mf: $C_{23}H_{32}N_4 \cdot 2ClH$ mw: 437.3

TOXICITY DATA WITH REFERENCE
orl-rat LD50:980 mg/kg ARZNAD 30,919,80
orl-mus LD50:650 mg/kg ARZNAD 30,919,80

SAFETY PROFILE: Moderately toxic by ingestion. When heated to decomposition it emits very toxic fumes of Cl^- and NO_x.

BJH500 **CAS:74758-19-3** **HR: 2**
1-(BIS(2-(DIMETHYLAMINO)ETHYL)AMINO)-3-PHENYLINDOLE DIHYDROCHLORIDEHYDRATE
mf: $C_{22}H_{30}N_4 \cdot 2ClH \cdot H_2O$ mw: 441.2

TOXICITY DATA WITH REFERENCE
orl-rat LD50:1250 mg/kg ARZNAD 30,919,80
orl-mus LD50:810 mg/kg ARZNAD 30,919,80

SAFETY PROFILE: Moderately toxic by ingestion. When heated to decomposition it emits very toxic fumes of Cl^- and NO_x.

BJH750 **CAS:3033-62-3** **HR: 3**
BIS(2-DIMETHYLAMINOETHYL) ETHER
mf: $C_8H_{20}N_2O$ mw: 160.30

PROP: Bp: 180–182°.

SYN: NIAX CATALYST AL

TOXICITY DATA WITH REFERENCE
skn-rbt 500 mg open SEV UCDS** 12/27/71
skn-rbt 100 mg/24H SEV JTOTDO 5,3,86
eye-rbt 1 mg SEV UCDS** 12/27/71
eye-rbt 250 μg/24H SEV 85JCAE-,721,86
skn-rbt TDLo:650 mg/kg (female 6-18D post):REP JTOTDO 5,263,86
orl-rat LD50:1070 mg/kg JTOTDO 5,3,86
skn-rbt LD50:280 μL/kg AIHAAP 30,470,69

CONSENSUS REPORTS: Reported in EPA TSCA Inventory.

SAFETY PROFILE: Poison by skin contact. Moderately toxic by ingestion. Experimental reproductive effects. A severe skin and eye irritant. See also ETHERS. When heated to decomposition it emits toxic fumes of NO_x.

BJI000 **CAS:541-19-5** **HR: 3**
BIS(β-DIMETHYLAMINOETHYL)SUCCINATE BIS(METHYLIODIDE)
mf: $C_{14}H_{30}N_2O_4 \cdot 2I$ mw: 517.92

SYNS: ASCURON ☐ CELOCURINE ☐ CHOLINE IODIDE SUCCINATE (2:1) ☐ CURACIT ☐ DIACETYLCHOLINE DIIODIDE ☐ DITILIN IODIDE ☐ SUCCINIC ACID BIS(β-DIMETHYLAMINOETHYL) ESTER BISMETHIODIDE ☐ SUCCINIC ACID, DIESTER with CHOLINE IODIDE ☐ SUCCINYLDICHOLINE IODIDE ☐ o,o-SUCCINYLDICHOLINE IODIDE ☐ SUXAMETHONIUM IODIDE

TOXICITY DATA WITH REFERENCE
ipr-mus LD50:5 mg/kg AEPPAE 228,371,56
ivn-mus LD50:550 μg/kg JPETAB 99,458,50
ivn-rbt LD50:15 mg/kg AIPTAK 88,1,51

SAFETY PROFILE: Poison by intravenous and intraperitoneal routes. When heated to decomposition it emits very toxic fumes of NO_x and I^-. See also IODIDES and ESTERS.

BJI125 **CAS:21476-57-3** **HR: 3**
BIS(DIMETHYLAMINO)ISOPROPYLMETHACRYLATE
mf: $C_{11}H_{22}N_2O_2$ mw: 214.35

SYN: 2-METHYL-2-PROPENOIC ACID 2-(DIMETHYLAMINO)-1-((DIMETHYLAMINO)METHYL)ETHYL ESTER (9CI)

TOXICITY DATA WITH REFERENCE
orl-rat LD50:1605 mg/kg 85GMAT -,26,82
ihl-rat LC50:110 mg/m³/4H 85GMAT -,26,82
ihl-mus LC50:220 mg/m³/2H 85GMAT -,26,82

SAFETY PROFILE: Poison by inhalation. Moderately toxic by ingestion. When heated to decomposition it emits toxic fumes of Cl^-. See also ESTERS.

BJI250 **CAS:61-73-4** **HR: 3**
3,7-BIS(DIMETHYL AMINO)PHENAZA THIONIUM CHLORIDE
mf: $C_{16}H_{18}N_3S \cdot Cl$ mw: 319.88

PROP: Dark bronze-green crystals with bronze lustre. Sol in H_2O and EtOH.

SYNS: AIZEN METHYLENE BLUE BH ☐ BASIC BLUE 9 ☐ 3,7-BIS (DIMETHYLAMINO)PHENOTHIAZIN-5-IUM CHLORIDE ☐ CALCOZINE BLUE ZF ☐ CHROMOSMON ☐ C.I. 52015 (CZECH) ☐ C.I. BASIC BLUE 9 ☐ D&C BLUE NUMBER 1 ☐ EXTERNAL BLUE 1 ☐ HIDACO METHYLENE BLUE SALT FREE ☐ LEATHER PURE BLUE HB ☐ METH-

YLENE BLUE □ METHYLENE BLUE A □ METHYLENE BLUE BB □ METHYLENE BLUE BB ZINC FREE □ METHYLENE BLUE CHLORIDE □ METHYLENE BLUE CHLORIDE (biological stain) □ METHYLENE BLUE D □ METHYLENE BLUE (medicinal) □ METHYLENE BLUE I (medicinal) □ METHYLENE BLUE NF (medicinal) □ METHYLENE BLUE POLYCHROME □ METHYLENE BLUE USP (medicinal) □ METHYLENE BLUE USP XII (medicinal) □ METHYLENIUM CERULEUM □ METHYLTHIONINE CHLORIDE □ METHYLTHIONIUM CHLORIDE □ MITSUI METHYLENE BLUE □ MODR METHYLENOVA (CZECH) □ SANDOCRYL BLUE BRL □ SCHULTZ No. 1038 □ SWISS BLUE □ TETRAMETHYLTHIONINE CHLORIDE □ YAMAMOTO METHYLENE BLUE B

TOXICITY DATA WITH REFERENCE
mma-sat 20 µg/plate ABCHA6 45,327,81
mmo-sat 100 µmol/L AMACCQ 9,77,76
dnr-sat 10 pph AGACBH 4,286,74
mmo-esc 2 µmol/L MUREAV 137,1,84
mrc-smc 10 pph AGACBH 4,286,74
orl-rat TDLo: 2500 mg/kg (1-22D preg): REP AJANA2 110,29,62
unr-inf TDLo: 15 mg/kg: PUL,BLD 34ZIAG -,390,69
orl-rat LD50: 1180 mg/kg MarJV# 29MAR77
ipr-rat LD50: 180 mg/kg AEPPAE 204,288,47
ivn-rat LD50: 1250 mg/kg ARZNAD 18,678,68
orl-mus LD50: 3500 mg/kg CKFRAY 12,94,63
ipr-mus LD50: 150 mg/kg NTIS** AD691-490
ivn-mus LD50: 77 mg/kg CKFRAY 12,94,63
orl-dog LDLo: 500 mg/kg HBAMAK 4,1366,35
ivn-dog LDLo: 50 mg/kg HBAMAK 4,1366,35
ivn-mky LDLo: 10 mg/kg HBAMAK 4,1366,35

CONSENSUS REPORTS: EPA Genetic Toxicology Program. Reported in EPA TSCA Inventory.

SAFETY PROFILE: Poison by ingestion, intraperitoneal, intravenous, and subcutaneous routes. Human systemic effects: cyanosis, blood changes. Experimental reproductive effects. Mutation data reported. When heated to decomposition it emits very toxic fumes of NO_x, SO_x, and Cl^-.

BJI750 **CAS:52673-66-2** **HR: 3**
1,3-BIS(2-DIMETHYLAMINOPROPYL)ADAMANTANE DIHYDROCHLORIDE
mf: $C_{20}H_{38}N_2 \cdot 2ClH$ mw: 379.52

TOXICITY DATA WITH REFERENCE
orl-mus LD50: 1200 mg/kg JMCMAR 17,602,74
ipr-mus LD50: 75 mg/kg JMCMAR 17,602,74

SAFETY PROFILE: Poison by intraperitoneal route. Moderately toxic by ingestion. When heated to decomposition it emits very toxic fumes of NO_x and HCl.

BJJ000 **CAS:62778-13-6** **HR: 3**
N,N′-BIS(3-DIMETHYLAMINOPROPYL)DITHIOOXAMIDE
mf: $C_{12}H_{26}N_4S_2$ mw: 290.54

SYN: USAF MK-43

TOXICITY DATA WITH REFERENCE
ipr-mus LD50: 100 mg/kg NTIS** AD277-689

CONSENSUS REPORTS: Reported in EPA TSCA Inventory.

SAFETY PROFILE: Poison by intraperitoneal route. When heated to decomposition it emits very toxic fumes of NO_x and SO_x.

BJJ125 **CAS:3768-60-3** **HR: 3**
BIS(DIMETHYLAMINO)SULFOXIDE
mf: $C_4H_{12}N_2OS$ mw: 136.21

$(CH_3)_2NS(:O)N(CH_3)_2$

SAFETY PROFILE: Violent reaction with sulfinyl chloride; becomes explosive above 90°C. When heated to decomposition it emits toxic fumes of SO_x and NO_x.

BJJ200 **CAS:63382-64-9** **HR: 3**
BIS(DIMETHYLARSINYLDIAZOMETHYL)MERCURY
mf: $C_6H_{12}As_2HgN_4$ mw: 589.62

$((CH_3)_2AsCN_2)_2Hg$

CONSENSUS REPORTS: Arsenic and its compounds, as well as mercury and its compounds, are on the Community Right-To-Know List.

SAFETY PROFILE: An explosive. When heated to decomposition it emits toxic fumes of As, Hg, and NO_x. See also MERCURY COMPOUNDS and ARSENIC COMPOUNDS.

BJJ250 **CAS:503-80-0** **HR: 3**
BIS-DIMETHYL ARSINYL OXIDE
mf: $C_4H_{12}As_2O$ mw: 225.9

$((CH_3)_2As)_2O$

PROP: Liquid. D: 1.486 @ 15 mm, mp: −25°, bp: 149–151°. Sol in EtOH, Et_2O. Sltly sol in H_2O.

CONSENSUS REPORTS: Arsenic and its compounds are on the Community Right-To-Know List.

SAFETY PROFILE: Ignites spontaneously in air. When heated to decomposition it emits toxic fumes of As. See also ARSENIC COMPOUNDS.

BJJ500 **CAS:591-10-6** **HR: 3**
BIS-DIMETHYL ARSINYL SULFIDE
mf: $C_4H_{12}As_2S$ mw: 242

$((CH_3)_2As)_2S$

PROP: Bp: 211–220°. Sol in EtOH, Et_2O; sltly sol in H_2O.

CONSENSUS REPORTS: Arsenic and its compounds are on the Community Right-To-Know List.

SAFETY PROFILE: Ignites spontaneously in air. When heated to decomposition it emits toxic fumes of SO_x and As. See also ARSENIC COMPOUNDS and SULFIDES.

BJJ750 **CAS:69402-04-6** **HR: 1**
1,2-BIS(3,7-DIMETHYL-5-n-BUTOXY-1-AZA-5-BORA-4,6-DIOXOCYCLOOCTYL)ETHANE
mf: $C_{26}H_{46}B_2N_2O_6$ mw: 504.36

SYN: 1,1′-ETHYLENEBIS(5-BUTOXY-3,7-DIMETHYL-1,5-AZABOROCINE-4,6-DIONE)

TOXICITY DATA WITH REFERENCE
skn-rbt 500 mg IHFCAY 6,1,67
eye-rbt 100 mg IHFCAY 6,1,67
orl-rat LD50:5660 mg/kg IHFCAY 6,1,67

SAFETY PROFILE: Mildly toxic by ingestion. A skin and eye irritant. When heated to decomposition it emits toxic fumes of NO_x.

BJK250 CAS:15521-65-0 HR: 1
BIS(DIMETHYLDITHIOCARBAMATO)NICKEL
mf: $C_6H_{12}N_2S_4$•Ni mw: 299.15

TOXICITY DATA WITH REFERENCE
orl-rat LD50:17 g/kg RCTEA4 45(3),627,72
orl-mus LD50:5200 mg/kg JPIFAN (3),10,70

CONSENSUS REPORTS: Reported in EPA TSCA Inventory. Nickel and its compounds are on the Community Right-To-Know List.

SAFETY PROFILE: Mildly toxic by ingestion. See also NICKEL COMPOUNDS and CARBAMATES. When heated to decomposition it emits very toxic fumes of NO_x and SO_x.

BJK500 CAS:137-30-4 HR: 3
BIS(DIMETHYLDITHIOCARBAMATO)ZINC
mf: $C_6H_{12}N_2S_4$•Zn mw: 305.81

PROP: White powder. Mp: 248–250°, d: 1.65 @ 20°/20°.

SYNS: AAPROTECT ☐ AAVOLEX ☐ AAZIRA ☐ ACCELERATOR L ☐ ACETO ZDED ☐ ACETO ZDMD ☐ ALCOBAM ZM ☐ AMYL ZIMATE ☐ ANTENE ☐ BIS(DIMETHYLCARBAMODITHIOATO-S,S′) ZINC ☐ BIS(DIMETHYLDITHIOCARBAMATE de ZINC) (FRENCH) ☐ BIS(N,N-DIMETIL-DITIOCARBAMMATO) DI ZINCO (ITALIAN) ☐ CARBAMIC ACID, DIMETHYLDITHIO-, ZINC SALT (2:1) ☐ CARBAZINC ☐ CIRAM ☐ CORONA COROZATE ☐ COROZATE ☐ CUMAN ☐ CUMAN L ☐ CYMATE ☐ DIMETHYLCARBAMODITHIOIC ACID, ZINC COMPLEX ☐ DIMETHYLCARBAMODITHIOIC ACID, ZINC SALT ☐ DIMETHYLDITHIOCARBAMATE ZINC SALT ☐ DIMETHYLDITHIOCARBAMIC ACID, ZINC SALT ☐ DRUPINA 90 ☐ ENT 988 ☐ EPTAC 1 ☐ FUCLASIN ☐ FUCLASIN ULTRA ☐ FUKLASIN ☐ FUNGOSTOP ☐ HERMAT ZDM ☐ HEXAZIR ☐ KARBAM WHITE ☐ METHASAN ☐ METHAZATE ☐ METHYL ZIMATE ☐ METHYL ZINEB ☐ METHYL ZIRAM ☐ MEXENE ☐ MEZENE ☐ MILBAM ☐ MILBAN ☐ MOLURAME ☐ MYCRONIL ☐ NCI-C50442 ☐ ORCHARD BRAND ZIRAM ☐ POMARSOL Z FORTE ☐ PRODARAM ☐ RHODIACID ☐ SOXINAL PZ ☐ SOXINOL PZ ☐ TRICARBAMIX Z ☐ TSIMAT ☐ TSIRAM (RUSSIAN) ☐ USAF P-2 ☐ VANCIDE MZ-96 ☐ ZERLATE ☐ ZIMATE ☐ ZIMATE METHYL ☐ ZINC BIS(DIMETHYLDITHIOCARBAMATE) ☐ ZINC BIS(DIMETHYLDITHIOCARBAMOYL)DISULPHIDE ☐ ZINC DIMETHYLDITHIOCARBAMATE ☐ ZINC N,N-DIMETHYLDITHIOCARBAMATE ☐ ZINCMATE ☐ ZINK-BIS(N,N-DIMETHYL-DITHIOCARBAMAAT) (DUTCH) ☐ ZINK-BIS (N,N-DIMETHYL-DITHIOCARBAMAT) (GERMAN) ☐ZINKCARBA-MATE ☐ ZINK-(N,N-DIMETHYL-DITHIOCARBAMAT) (GERMAN) ☐ ZIRAM ☐ ZIRAMVIS ☐ ZIRASAN ☐ ZIRBERK ☐ ZIREX 90 ☐ ZIRIDE ☐ ZIRTHANE ☐ ZITOX

TOXICITY DATA WITH REFERENCE
mmo-sat 5 μg/plate MUREAV 68,313,79
cyt-hmn:lym 10 nmol/L TXCYAC 4,331,75
orl-rat TDLo:250 mg/kg (female 6-15D post):TER EESADV 7,531,83
orl-rat TDLo:500 mg/kg (female 6-15D post):REP EESADV 7,531,83
orl-rat TDLo:1 g/kg (female 6-15D post):TER EESADV 7,531,83
orl-rat TDLo:500 mg/kg (female 6-15D post):TER EESADV 7,531,83
orl-rat TDLo:12,978 mg/kg/2Y-I:CAR NTPTR* NTP-TR-238,82
imp-rat TDLo:60 mg/kg:ETA VPITAR 29,71,70
orl-mus TDLo:840 mg/kg/13W-I:ETA GISAAA 37(9),25,72
orl-rat TD:25,956 mg/kg/2Y-I:CAR NTPTR* NTP-TR-238,82
orl-rat TD:13,160 mg/kg/94W-I:CAR VPITAR 29,71,70
orl-rat LD50:267 mg/kg EPASR* 8EHQ-1090-1045
ihl-rat LC50:81 mg/m³/4H EPASR* 8EHQ-1090-1045
skn-rat LD50:>6 g/kg FMCHA2-,C329,91
ipr-rat LD50:23 mg/kg JAPMA8 41,662,52
ivn-mus LD50:18 mg/kg CSLNX* NX#04886

CONSENSUS REPORTS: IARC Cancer Review: Group 3 IMEMDT 7,56,87, Animal Inadequate Evidence IMEMDT 12,259,76; NTP Carcinogenesis Bioassay (feed); Clear Evidence: mouse, rat NTPTR* NTP-TR-238,83. EPA Genetic Toxicology Program. Reported in EPA TSCA Inventory. Zinc and its compounds are on the Community Right-To-Know List.

SAFETY PROFILE: Poison by ingestion, intraperitoneal, and intravenous routes. Moderately toxic by inhalation. Questionable carcinogen with experimental carcinogenic and tumorigenic data. An experimental teratogen. Other experimental reproductive effects. Human mutation data reported. See also ZINC COMPOUNDS and CARBAMATES. Severe irritant to eyes, nose, and throat. When heated to decomposition it emits very toxic fumes of NO_x and SO_x.

BJK550 CAS:36443-68-2 HR: 2
3,5-BIS(1,1-DIMETHYLETHYL)-4-HYDROXY-, 1,2-ETHANEDIYLBIS (OXY-2,1-ETHANEDIYL) ESTER BENZENEPROPANOIC ACID
mf: $C_{34}H_{50}O_8$ mw: 586.84

SYN: TK 12627

TOXICITY DATA WITH REFERENCE
orl-rat TDLo:36,500 mg/kg/2Y-C:ETA EPASR* 8EHQ-0388-0725

CONSENSUS REPORTS: Reported in EPA TSCA Inventory.

SAFETY PROFILE: Questionable carcinogen with experimental tumorigenic data. When heated to decomposition it emits acrid smoke and irritating vapors.

BJK650 CAS:63428-98-8 HR: 2
2,4-BIS(1,1-DIMETHYLETHYL)-6-(1-PHENYLETHYL) PHENOL
mf: $C_{22}H_{30}O$ mw: 310.52

SYNS: AI3-70736 ☐ PHENOL, 2,4-BIS(1,1-DIMETHYLETHYL)-6-(1-PHENYLETHYL)-

TOXICITY DATA WITH REFERENCE
orl-mus LD50:2510 mg/kg JAFCAU 27,1007,79

CONSENSUS REPORTS: Reported in EPA TSCA Inventory.

SAFETY PROFILE: Moderately toxic by ingestion. When heated to decomposition it emits acrid smoke and irritating vapors.

BJK750 CAS:4636-83-3 HR: 3
1,1′-BIS(3,5-DIMETHYLMORPHOLINOCARBONYLMETHYL)-4,4′-BIPYRIDYNIUM DICHLORIDE
mf: $C_{26}H_{36}N_4O_4{\cdot}2Cl$ mw: 539.56

SYNS: 1,1′-BIS(3,5-DIMETHYLMORPHOLINOCARBONYLMETHYL)-4,4′-BIPYRIDINIUM-DICHLORID (GERMAN) ☐ 1,1′-BIS(2-(3,5-DIMETHYL-4-MORPHOLINYL)-2-OXOETHYL)-4,4′-BIPYRIDINIUM DICHLORIDE ☐ CEROXONE ☐ MORFAMQUAT ☐ MORFOXONE ☐ MORPHANQUAT DICHLORIDE ☐ PP 745

TOXICITY DATA WITH REFERENCE
orl-rat LD50:345 mg/kg GUCHAZ 6,367,73
orl-mus LD50:325 mg/kg 28ZEAL 5,158,76
orl-cat LD50:160 mg/kg 28ZEAL 5,158,76
orl-ckn LD50:367 mg/kg 31ZOAD 1,311,68

SAFETY PROFILE: Poison by ingestion. When heated to decomposition it emits very toxic fumes of Cl^- and NO_x.

BJL000 CAS:3081-14-9 HR: 2
N,N′-BIS(1,4-DIMETHYLPENTYL)-p-PHENYLENEDIAMINE
mf: $C_{20}H_{36}N_2$ mw: 304.58

SYNS: N,N DI(1,4-DIMETHYLPENTYL)-p-PHENYLDIAMINE ☐ EASTOZONE ☐ EASTOZONE 33 ☐ NCI-C56337 ☐ SANTOFLEX 77 ☐ TENAMENE

TOXICITY DATA WITH REFERENCE
orl-rat LD50:750 mg/kg 85GMAT -,59,82
ipr-rat LDLo:800 mg/kg RCTEA4 45,627,72
orl-mus LD50:800 mg/kg IPSTB3 3,93,76
ipr-mus LDLo:400 mg/kg RCTEA4 45,627,72

CONSENSUS REPORTS: Reported in EPA TSCA Inventory.

SAFETY PROFILE: Moderately toxic by ingestion and intraperitoneal routes. When heated to decomposition it emits toxic fumes of NO_x.

BJL250 HR: 3
BISDIMETHYL STIBINYL OXIDE
mf: $C_4H_{12}OSb_2$ mw: 319.6

CONSENSUS REPORTS: Antimony and its compounds are on the Community Right-To-Know List.

SAFETY PROFILE: Antimony compounds are generally highly toxic. Ignites spontaneously in air. When heated to decomposition it emits acrid smoke and fumes. See also ANTIMONY COMPOUNDS.

BJL500 HR: 3
BIS(DIMETHYL THALLIUM)ACETYLIDE
mf: $C_6H_{12}Tl_2$ mw: 492.90

CONSENSUS REPORTS: Thallium and its compounds are on the Community Right-To-Know List.

SAFETY PROFILE: An extremely heat- and friction-sensitive explosive. When heated to decomposition it emits acrid smoke and fumes. See also THALLIUM COMPOUNDS and ACETYLIDES.

BJL600 CAS:97-74-5 HR: 3
BIS(DIMETHYLTHIOCARBAMOYL)SULFIDE
mf: $C_6H_{12}N_2S_3$ mw: 208.38

PROP: Yellow crystals from EtOH. Mp: 104°. Very sol in EtOH, $CHCl_3$; sltly sol in cold Et_2O.

SYNS: ACETO TMTM ☐ BIS(DIMETHYLTHIOCARBAMYL) MONOSULFIDE ☐ CARBAMIC ACID, DIMETHYLDITHIO-, ANHYDROSULFIDE ☐ MONEX ☐ MONO-THIURAD ☐ MONOTHIURAM ☐ PENNAC MS ☐ TETRAMETHYLTHIURAMMONIUM SULFIDE ☐ TETRAMETHYLTHIURAM MONOSULFIDE ☐ TETRAMETHYLTHIURAM SULFIDE ☐ TETRAMETHYLTRITHIO CARBAMIC ANHYDRIDE ☐ 1,1′-THIOBIS(N,N-DIMETHYLTHIO)FORMAMIDE ☐ THIONEX ☐ THIONEX RUBBER ACCELERATOR ☐ TMTM ☐ TMTMS ☐ UNADS ☐ USAF B-32 ☐ USAF EK-P-6255 ☐ VULKACIT THIURAM MS/C

TOXICITY DATA WITH REFERENCE
mmo-sat 100 μg/plate MUREAV 68,313,79
sce-ham:ovr 100 nmol/L SWEHDO 9(Suppl 2),27,83
scu-mus TDLo:900 mg/kg (6-14D preg):TER NTIS** PB223-160
scu-mus TDLo:100 mg/kg:ETA NTIS** PB223-159
ipr-rat LD50:383 mg/kg JNPHAG 9,35,78
orl-mus LD50:818 mg/kg ENVRAL 28(1),199,82
ipr-mus LD50:300 mg/kg NTIS** AD277-689
orl-dog LDLo:100 mg/kg RCTEA4 44,513,71

CONSENSUS REPORTS: Reported in EPA TSCA Inventory.

SAFETY PROFILE: Poison by ingestion and intraperitoneal routes. Questionable carcinogen with experimental tumorigenic data. Mutation data reported. An experimental teratogen. When heated to decomposition it emits very toxic fumes of NO_x and SO_x. See also SULFIDES.

BJM250 CAS:58451-85-7 HR: 2
2,6-BIS(DIPHENYLHYDROXYMETHYL)PIPERIDINE
mf: $C_{31}H_{31}NO_2$ mw: 449.63

TOXICITY DATA WITH REFERENCE
orl-mus LD50:5000 mg/kg PJPPAA 27,549,75
ipr-mus LD50:2000 mg/kg PJPPAA 27,549,75

SAFETY PROFILE: Moderately toxic by intraperitoneal route. Mildly toxic by ingestion. When heated to decomposition it emits toxic fumes of NO_x.

BJM500 CAS:58451-82-4 HR: 2
2,6-BIS(DIPHENYLHYDROXYMETHYL)PYRIDINE
mf: $C_{31}H_{25}NO_2$ mw: 443.57

TOXICITY DATA WITH REFERENCE
orl-mus LD50:5000 mg/kg PJPPAA 27,549,75
ipr-mus LD50:3000 mg/kg PJPPAA 27,549,75

SAFETY PROFILE: Moderately toxic by intraperitoneal route. Mildly toxic by ingestion. When heated to decomposition it emits toxic fumes of NO_x.

BJM625 **HR: 3**
3-(BIS(3,3-DIPHENYLPROPYL)AMINO)PROPANE-1-OL
mf: $C_{33}H_{37}NO$ mw: 463.71

SYN: 3-(BIS(3,3-DIPHENYLPROPYL)AMINO)-1-PROPANOL

TOXICITY DATA WITH REFERENCE
ipr-rat LD50:167 mg/kg ARZNAD 25,632,75
ivn-rat LD50:40,300 µg/kg ARZNAD 25,632,75
ipr-mus LD50:129 mg/kg ARZNAD 25,632,75
ivn-mus LD50:30,200 µg/kg ARZNAD 25,632,75

SAFETY PROFILE: Poison by intravenous and intraperitoneal routes. When heated to decomposition it emits toxic fumes of NO_x.

BJM700 **CAS:38998-91-3** **HR: 3**
BIS(1,3-DITHIOCYANATO-1,1,3,3-TETRABUTYLDISTANNOXANE)
mf: $C_{36}H_{72}N_4O_2S_4Sn_4$ mw: 1196.12

SYNS: DISTANNOXANE, BIS(1,3-DITHIOCYANATO-1,1,3,3-TETRABUTYL)- □ DI-µ-(THIOCYANATODI-n-BUTYLSTANNYLOXO)BIS(THIOCYANATODI-n-BUTYLTIN)

TOXICITY DATA WITH REFERENCE
ivn-mus LD50:180 mg/kg CSLNX* NX#03006

OSHA PEL: TWA 0.1 mg(Sn)/m^3
ACGIH TLV: TWA 0.1 mg(Sn)/m^3; STEL 0.2 mg/m^3 (skin)
NIOSH REL: 10H TWA 0.1 mg(Sn)/m^3

SAFETY PROFILE: Poison by intravenous route. When heated to decomposition it emits toxic fumes of NO_x, SO_x, and Sn.

For occupational chemical analysis use NIOSH: Organotin Compounds 5504.

BJM750 **CAS:10171-76-3** **HR: 3**
BIS(2,5-ENDOMETHYLENECYCLOHEXYLMETHYL) AMINE
mf: $C_{16}H_{27}N$ mw: 233.44

TOXICITY DATA WITH REFERENCE
skn-rbt 100 µg/24H open AIHAAP 23,95,62
orl-rat LD50:1410 mg/kg AIHAAP 23,95,62
skn-rbt LD50:110 mg/kg AIHAAP 23,95,62

SAFETY PROFILE: Poison by skin contact. Moderately toxic by ingestion. A skin irritant. When heated to decomposition it emits toxic fumes of NO_x.

BJN000 **CAS:10580-77-5** **HR: 3**
BIS(3,4-EPOXYBUTYL) ETHER
mf: $C_8H_{14}O_3$ mw: 158.22

TOXICITY DATA WITH REFERENCE
orl-rat LD50:1070 mg/kg AIHAAP 30,470,69
skn-rbt LD50:250 mg/kg AIHAAP 30,470,69

SAFETY PROFILE: Poison by skin contact. Moderately toxic by ingestion. See also ETHERS. When heated to decomposition it emits acrid smoke and fumes.

BJN250 **CAS:2386-90-5** **HR: 2**
BIS(2,3-EPOXYCYCLOPENTYL) ETHER
mf: $C_{10}H_{14}O_3$ mw: 182.24

SYNS: EP-205 □ ERR 4205 □ 2,2'-OXYBIS-6-OXABICYCLO-(3.1.0) HEXANE

TOXICITY DATA WITH REFERENCE
skn-rbt 500 mg open MLD UCDS** 12/13/63
mmo-sat 5700 µg/plate CIHPDR 6,210,84
mma-sat 5700 µg/plate CIHPDR 6,210,84
sce-hmn:lym 50 mg/L CIHPDR 6,210,84
mnt-mus-orl 1 g/kg CIHPDR 6,210,84
skn-mus TDLo:156 g/kg/2Y-I:CAR NTIS** ORNL-5375
skn-mus TD:312 g/kg/2Y-I:NEO,REP CNREA8 39,1718,79
skn-mus TD:395 g/kg/132W-I:ETA NTIS** ORNL-5762
orl-rat LDLo:2140 mg/kg AIHAAP 23,95,62
skn-mus LDLo:2000 mg/kg NTIS** ORNL-5375

CONSENSUS REPORTS: EPA Genetic Toxicology Program. Reported in EPA TSCA Inventory.

SAFETY PROFILE: Moderately toxic by ingestion. A systemic irritant by skin contact and ingestion. Experimental reproductive effects. Questionable carcinogen with experimental carcinogenic and neoplastigenic data. See also ETHERS. When heated to decomposition it emits acrid smoke and irritating fumes.

BJN500 **CAS:7487-28-7** **HR: 2**
BIS(2,3-EPOXY-2-METHYLPROPYL)ETHER
mf: $C_8H_{14}O_3$ mw: 158.22

SYN: BIS(2-METHYLGLYCIDYL) ETHER

TOXICITY DATA WITH REFERENCE
skn-rbt 500 mg open MLD UCDS** 4/21/67
orl-rat LD50:1680 mg/kg AIHAAP 24,305,63
skn-rbt LD50:1250 mg/kg UCDS** 4/21/67

SAFETY PROFILE: Moderately toxic by ingestion and skin contact. A skin irritant. See also ETHERS. When heated to decomposition it emits acrid and irritating fumes and smoke.

BJN750 **CAS:10043-09-1** **HR: 2**
2,3-BIS(2,3-EPOXYPROPOXY)-1,4-DIOXANE
mf: $C_{10}H_{16}O_6$ mw: 232.26

SYN: 2,3-BIS(GLYCIDYLOXY)-1,4-DIOXANE

TOXICITY DATA WITH REFERENCE
orl-rat LD50:1070 mg/kg AIHAAP 24,305,63
skn-rbt LD50:1590 mg/kg AIHAAP 23,95,62

SAFETY PROFILE: Moderately toxic by ingestion and skin contact. When heated to decomposition it emits acrid smoke and irritating fumes.

BJN850 **CAS:63951-08-6** **HR: 2**
N,N-BIS(2-(2,3-EPOXYPROPOXY)ETHOXY)ANILINE
mf: $C_{16}H_{23}NO_6$ mw: 325.40

SYNS: ANILINE, N,N-BIS(2-(2,3-EPOXYPROPOXY)ETHOXY)- □ DIGLYCIDYL ETHER of N,N-BIS(2-HYDROXYETHOXYETHYL)ANILINE

TOXICITY DATA WITH REFERENCE
scu-mus TDLo:5600 mg/kg/60W-I:ETA FCTXAV 4,365,66

SAFETY PROFILE: Questionable carcinogen with experimental tumorigenic data. When heated to decomposition it emits toxic fumes of NO_x.

BJN875 CAS:7329-29-5 HR: 2
N,N-BIS(2-(2,3-EPOXYPROPOXY)ETHYL)ANILINE
mf: $C_{16}H_{23}NO_4$ mw: 293.40

SYNS: ANILINE, N,N-BIS(2-(2,3-EPOXYPROPOXY)ETHYL)- □ DIGLYCIDYL ETHER of PHENYLDIETHANOLAMINE

TOXICITY DATA WITH REFERENCE
scu-rat TDLo:2820 mg/kg/43W-I:ETA BECCAN 42,37,64
scu-mus TD:22 g/kg/39W-I:ETA FCTXAV 4,365,66

SAFETY PROFILE: Questionable carcinogen with experimental tumorigenic data. When heated to decomposition it emits toxic fumes of NO_x.

BJO000 CAS:13561-08-5 HR: 2
BIS(2,6-(2,3-EPOXYPROPYL))PHENYL GLYCIDYL ETHER
mf: $C_{15}H_{18}O_4$ mw: 262.33

SYN: 2,6-BIS(2,3-EPOXYPROPYL)PHENYL 2,3-EPOXYPROPYLETHER

TOXICITY DATA WITH REFERENCE
skn-rbt 500 mg MOD SCCUR*-,2,61
orl-rat LD50:1620 µL/kg TXAPA9 28,313,74
skn-rbt LD50:2520 µL/kg TXAPA9 28,313,74

CONSENSUS REPORTS: Reported in EPA TSCA Inventory.

SAFETY PROFILE: Moderately toxic by ingestion and skin contact. See also ETHERS. When heated to decomposition it emits acrid smoke and irritating fumes.

BJO125 CAS:20539-85-9 HR: 3
BIS(ETHOXYCARBONYLDIAZOMETHYL)MERCURY
mf: $C_8H_{10}HgN_4O_4$ mw: 426.78

$(CH_3CH_2CO{\cdot}CN_2)_2Hg$

CONSENSUS REPORTS: Mercury and its compounds are on the Community Right-To-Know List.

SAFETY PROFILE: An impact-sensitive explosive that decomposes at its mp: 104°C. When heated to decomposition it emits toxic fumes of NO_x and Hg. See also MERCURY COMPOUNDS.

BJO225 CAS:109-44-4 HR: 2
BIS(2-ETHOXYETHYL) ADIPATE
mf: $C_{14}H_{26}O_6$ mw: 290.40

SYNS: ADIPIC ACID, BIS(2-ETHOXYETHYL) ESTER □ DIETHOXY ETHYL ADIPATE □ HEXANOIC ACID, BIS(2-ETHOXYETHYL) ESTER

TOXICITY DATA WITH REFERENCE
skn-man 50 mg/24H MLD CTOIDG 94(8),41,79
skn-rat 100 mg/24H MLD CTOIDG 94(8),41,79
skn-rbt 100 mg/24H SEV CTOIDG 94(8),41,79
skn-gpg 100 mg/24H MLD CTOIDG 94(8),41,79

CONSENSUS REPORTS: Reported in EPA TSCA Inventory.

SAFETY PROFILE: A severe skin irritant. When heated to decomposition it emits acrid smoke and irritating fumes.

BJO250 CAS:67856-66-0 HR: 2
BIS(2-ETHOXYETHYL)NITROSOAMINE
mf: $C_8H_{18}N_2O$ mw: 158.28

SYN: N-NITROSOBIS(2-ETHOXYETHYL)AMINE

TOXICITY DATA WITH REFERENCE
mma-sat 50 µg/plate MUREAV 66,1,79
orl-rat TDLo:3250 mg/kg/50W-I:ETA CNREA8 38,2391,78

SAFETY PROFILE: Questionable carcinogen with experimental tumorigenic data. Mutation data reported. When heated to decomposition it emits highly toxic fumes of NO_x. See also NITROSAMINES.

BJO500 CAS:101-93-9 HR: 3
N,N'-BIS(p-ETHOXYPHENYL)ACETAMIDINE
mf: $C_{18}H_{22}N_2O_2$ mw: 298.42

PROP: Solid. Mp: 117°.

SYNS: N',N(sup 2)-BIS(p-ETHOXYPHENYL)ACETAMIDINE □ N,N'-BIS(4-ETHOXYPHENYL)ETHANIMIDAMIDE □ FENACAINE □ HOLOCAINE □ PHENACAINE □ TANICAINE

TOXICITY DATA WITH REFERENCE
ivn-cat LDLo:10 mg/kg PHREA7 12,190,32
unr-rbt LDLo:5 mg/kg HBAMAK 4,1289,35
ipr-gpg LDLo:50 mg/kg PHREA7 12,190,32
scu-gpg LDLo:53 mg/kg PHREA7 12,190,32
ivn-gpg LDLo:15 mg/kg PHREA7 12,190,32

SAFETY PROFILE: Poison by subcutaneous and possibly other routes. When heated to decomposition it emits toxic fumes of NO_x.

BJP000 CAS:122-34-9 HR: 3
2,4-BIS(ETHYLAMINO)-6-CHLORO-s-TRIAZINE
mf: $C_7H_{12}ClN_5$ mw: 201.69

PROP: Crystals. Mp: 228–229°.

SYNS: AKTINIT S □ AQUAZINE □ BATAZINA □ 2,4-BIS(AETHYLAMINO)-6-CHLOR-1,3,5-TRIAZIN (GERMAN) □ BITEMOL □ BITEMOL S 50 □ CAT (herbicide) □ CDT □ CEKUSAN □ CEKUZINA-S □ CET □ 1-CHLORO-3,5-BISETHYLAMINO-2,4,6-TRIAZINE □ 2-CHLORO-4,6-BIS(ETHYLAMINO)-s-TRIAZINE □ 2-CHLORO-4,6-BIS(ETHYLAMINO)-1,3,5-TRIAZINE □ FRAMED □ GEIGY 27,692 □ GESARAN □ GESATOP □ HERBAZIN □ HERBEX □ HERBOXY □ HUNGAZIN DT □ PREMAZINE □ PRIMATOL S □ RADOCON □ RADOKOR □ SIMANEX □ SIMAZIN □ SIMAZINE (USDA) □ SIMAZINE 80W □ TAFAZINE □ TAPHAZINE □ ZEAPUR

TOXICITY DATA WITH REFERENCE
skn-rbt 500 mg open MLD CIGET* -,-,77
eye-rbt 80 mg MOD CIGET* -,-,77
sln-dmg-orl 2000 ppm JPFCD2 15,867,80
dlt-dmg-orl 6000 ppm JTEHD6 3,691,77
orl-rat TDLo:25 g/kg (female 6-15D post):TER CHYCDW 15,83,81

orl-rat TDLo: 3120 mg/kg (female 6-15D post): REP CHYCDW 15,83,81
orl-rat TDLo: 780 mg/kg (female 6-15D post): TER CHYCDW 15,83,81
scu-rat TDLo: 16 g/kg/61W-I: ETA VOONAW 16(1),82,70
scu-mus TDLo: 35 g/kg/87W-I: ETA VOONAW 16(1),82,70
orl-rat LD50: 971 mg/kg FAATDF 7,299,86
ihl-rat LC50: 9800 mg/m³/1H FMCHA2 -,C261,89
orl-mus LDLo: 5 g/kg GISAAA 27(8),22,62
ivn-mus LD50: 100 mg/kg CSLNX* NX#04003

CONSENSUS REPORTS: EPA Genetic Toxicology Program. Reported in EPA TSCA Inventory.

SAFETY PROFILE: Poison by intravenous route. Moderately toxic by ingestion. Questionable carcinogen with experimental tumorigenic data. An experimental teratogen. Other experimental reproductive effects. A skin and eye irritant. Mutation data reported. May cause weight loss and reduced red blood cell count. When heated to decomposition it emits very toxic fumes of Cl^- and NO_x.

BJP250 CAS:673-04-1 HR: 3
2,4-BIS(ETHYLAMINO)-6-METHOXY-s-TRIAZINE
mf: $C_8H_{15}N_5O$ mw: 197.28

SYNS: 4,6-BIS(ETHYLAMINO)-2-METHOXY-s-TRIAZINE □ GEIGY 30,044 □ GESADURAL □ 2-METHOXY-4,6-BIS(ETHYLAMINO)-s-TRIAZINE □ METHOXY SIMAZINE □ PIMETON □ SIMETON □ SIMETONE

TOXICITY DATA WITH REFERENCE
ivn-mus LD50: 180 mg/kg CSLNX* NX#03978
orl-rat LD50: 535 mg/kg RREVAH 10,97,65

SAFETY PROFILE: Poison by intravenous route. Moderately toxic by ingestion. When heated to decomposition it emits toxic fumes of NO_x.

BJP300 CAS:53213-78-8 HR: 3
1,2-BIS(ETHYLAMMONIO)ETHANE PERCHLORATE
mf: $C_6H_{18}Cl_2N_2O_8$ mw: 317.12

SAFETY PROFILE: An impact-sensitive mild explosive. When heated to decomposition it emits toxic fumes of NO_x, Cl^-, and NH_3. See also PERCHLORATES.

BJP425 CAS:29471-80-5 HR: 3
BIS(ETHYLENEDIAMINE)(MERCURICTETRATHIOCYANATO)COPPER
mf: $(C_4H_{16}CuN_4{\cdot}C_4HgN_4S_4)_x$

SYNS: COPPER, BIS(ETHYLENEDIAMINE)(MERCURICTETRATHIOCYANATO)- □ COPPER(2+), BIS(ETHYLENEDIAMINE)-, TETRAKIS (THIOCYANATO)MERCURATE(2-), POLYMERS

TOXICITY DATA WITH REFERENCE
ipr-mus LD50: 26,100 µg/kg IJEBA6 19,1187,81

ACGIH TLV: TWA 0.1 mg(Hg)/m³ (skin)
NIOSH REL: (Organomercury): TWA 0.01 mg/m³; STEL 0.03 mg/m³ (skin)

SAFETY PROFILE: Poison by intraperitoneal route. When heated to decomposition it emits toxic fumes of NO_x, SO_x, Hg, and Cl^-.

BJP450 CAS:1192-75-2 HR: 3
BISETHYLENEUREA
mf: $C_5H_8N_2O$ mw: 112.15

SYNS: AZIRIDINE, 1,1'-CARBONYLBIS- □ BIS(1-AZIRIDINYL)KETONE □ CARBONYLBIS(AZIRIDINE) □ CARBONYLBIS(1-AZIRIDINE) □ DIETHYLENEUREA □ N,N'-DIETHYLENEUREA

TOXICITY DATA WITH REFERENCE
ipr-rat LD50: 5800 µg/kg BJPCAL 25,223,65
ipr-mus LD50: 8500 µg/kg BJPCAL 25,223,65

DOT CLASSIFICATION: 3; *Label:* Flammable Liquid

SAFETY PROFILE: A poison by intraperitoneal route. A flammable liquid. When heated to decomposition it emits toxic vapors of NO_x.

BJP500 CAS:6708-69-6 HR: 3
2,6-BIS(ETHYLEN-IMINO)-4-AMINO-s-TRIAZINE
mf: $C_7H_{10}N_6$ mw: 178.23

TOXICITY DATA WITH REFERENCE
ipr-rat LD50: 700 µg/kg JPETAB 100,398,50
ipr-mus LD50: 1800 µg/kg JPETAB 100,398,50
ivn-dog LDLo: 400 µg/kg JPETAB 100,398,50

SAFETY PROFILE: A poison by intraperitoneal and intravenous routes. When heated to decomposition it emits toxic fumes of NO_x.

BJP899 CAS:19218-16-7 HR: 2
1,3-BIS(ETHYLENIMINOSULFONYL)PROPANE
mf: $C_7H_{14}N_2O_4S_2$ mw: 254.35

SYNS: BEP □ ω,ω'-BIS-(ETHYLENEIMINOSULPHONYL)PROPANE □ 1,3,-DI(ETHYLENESULPHAMOYL)PROPANE

TOXICITY DATA WITH REFERENCE
oms-rat-ipr 4 mg/kg BJPCAL 6,357,51
cyt-rat-ipr 4 mg/kg BJPCAL 6,357,51
scu-rat TDLo: 80 mg/kg/I: NEO ANYAA9 68,750,58

SAFETY PROFILE: Questionable carcinogen with experimental neoplastigenic data. When heated to decomposition it emits very toxic fumes of SO_x and NO_x.

BJQ250 CAS:2781-10-4 HR: 3
BIS(2-ETHYLHEXANOYLOXY)DIBUTYL STANNANE
mf: $C_{24}H_{48}O_4Sn$ mw: 519.41

SYNS: DIBUTYLBIS((2-ETHYLHEXANOYL)OXY)-STANNANE □ DIBUTYLBIS((2-ETHYL-1-OXOHEXYL)OXY)-STANNANE (9CI) □ DIBUTYLTIN BIS(α-ETHYLHEXANOATE) □ DIBUTYLTIN BIS(2-ETHYLHEXANOATE) □ DIBUTYLTIN DI(2-ETHYLHEXANOATE) □ DI-n-BUTYLTIN DI-2-ETHYLHEXANOATE □ DIBUTYLTIN DI(2-ETHYLHEXOATE)

TOXICITY DATA WITH REFERENCE
orl-rat LD50: 200 mg/kg JPMSAE 56,240,67
orl-mus LD50: 200 mg/kg JPMSAE 56,240,67
ivn-mus LD50: 178 mg/kg CSLNX* NX#00178

CONSENSUS REPORTS: Reported in EPA TSCA Inventory.

OSHA PEL: TWA 0.1 mg(Sn)/m³ (skin)
ACGIH TLV: TWA 0.1 mg(Sn)/m³ (skin) (Proposed: TWA 0.1 mg(Sn)/m³; STEL 0.2 mg(Sn)/m³ (skin))

NIOSH REL: (Organotin Compounds) TWA 0.1 mg(Sn)/m³

SAFETY PROFILE: Poison by ingestion and intravenous route. See also TIN COMPOUNDS. When heated to decomposition it emits acrid smoke and irritating fumes.

For occupational chemical analysis use NIOSH: Organotin Compounds 5504.

BJQ500 CAS:103-24-2 HR: 2
BIS(2-ETHYLHEXYL) AZELATE
mf: $C_{25}H_{48}O_4$ mw: 412.73

SYNS: AZELAIC ACID DI(2-ETHYLHEXYL)ESTER □ DIOCTYL AZELATE □ PLASTOLEIN 9058 □ PLASTOLEIN 9058 DOZ □ STAFLEX DOX □ TRUFLEX DOX

TOXICITY DATA WITH REFERENCE
skn-rbt 10 mg/24H open MLD AIHAAP 23,95,62
orl-rat LD50:8720 mg/kg AIHAAP 23,95,62
ivn-rat LD50:1060 mg/kg MRLR** No.256,54
skn-rbt LD50:20 g/kg AIHAAP 23,95,62
ivn-rbt LD50:640 mg/kg MRLR** No.256,54

CONSENSUS REPORTS: Reported in EPA TSCA Inventory.

SAFETY PROFILE: Moderately toxic by intravenous route. Mildly toxic by ingestion and skin contact. A skin irritant. See also ESTERS. When heated to decomposition it emits acrid smoke and irritating fumes.

BJQ709 CAS:3658-48-8 HR: 3
BIS(2-ETHYLHEXYL) HYDROGEN PHOSPHITE
mf: $C_{16}H_{35}O_3P$ mw: 306.42

PROP: Liquid. D: 0.93 @ 25°/25°, bp: 148–151° @ 1 mm.

TOXICITY DATA WITH REFERENCE
eye-rbt 25 mg MLD AMIHAB 18,464,58
orl-rat LD50:11,900 mg/kg ALBRW* #OPB-3,84
ipr-rat LD50:1500 mg/kg AMIHAB 18,464,58
ipr-mus LD50:620 mg/kg AMIHAB 18,464,58
skn-rbt LD50:4500 mg/kg ALBRW* #OPB-3,84
ivn-rbt LD50:100 mg/kg AMIHAB 18,464,58
ipr-gpg LD50:700 mg/kg AMIHAB 18,464,58

CONSENSUS REPORTS: Reported in EPA TSCA Inventory.

SAFETY PROFILE: Poison by intravenous route. Moderately toxic by intraperitoneal route. An eye irritant. When heated to decomposition it emits toxic fumes of PO_x. See also ESTERS.

BJQ750 CAS:137-89-3 HR: 1
BIS(2-ETHYLHEXYL) ISOPHTHALATE
mf: $C_{24}H_{38}O_4$ mw: 390.62

SYNS: DI-2-ETHYLHEXYL ISOPHTHALATE □ DIOCTYL ISOPHTHALATE

TOXICITY DATA WITH REFERENCE
skn-rbt 500 mg open MLD UCDS** 5/17/66
orl-rat LD50:17,300 mg/kg AIHAAP 23,95,62
skn-rbt LD50:7940 mg/kg AIHAAP 23,95,62

CONSENSUS REPORTS: Reported in EPA TSCA Inventory.

SAFETY PROFILE: Mildly toxic by ingestion and skin contact. When heated to decomposition it emits acrid smoke and irritating fumes.

BJR000 CAS:142-16-5 HR: 1
BIS(2-ETHYLHEXYL) MALEATE
mf: $C_{20}H_{36}O_4$ mw: 340.56

PROP: Liquid. Mp: −60°, bp: 164° @ 10 mm, flash p: 365°F, d: 0.9436 @ 20°/20°, vap d: 11.7.

SYNS: DI-(2-ETHYLHEXYL)MALEATE □ "DIOCTYL" MALEATE □ DOM □ RC COMONOMER DOM

TOXICITY DATA WITH REFERENCE
skn-rbt 10 mg/24H open MLD JIHTAB 31,60,49
eye-rbt 500 mg open JIHTAB 31,60,49
orl-rat LD50:14 g/kg JIHTAB 31,60,49
skn-rbt LD50:15 g/kg JIHTAB 31,60,49

CONSENSUS REPORTS: Reported in EPA TSCA Inventory.

SAFETY PROFILE: Mildly toxic by ingestion and skin contact. A skin and eye irritant. Combustible when exposed to heat or flame; can react with oxidizing materials. To fight fire, use alcohol foam, dry chemical, mist, or spray. When heated to decomposition it emits acrid smoke and irritating fumes. See also ESTERS.

BJR250 CAS:15546-12-0 HR: 3
BIS((2-(ETHYL)HEXYLOXY)MALEOYLOXY) DI(n-BUTYL)STANNANE
mf: $C_{32}H_{56}O_8Sn$ mw: 687.57

SYNS: BIS(HYDROGEN MALEATO)DIBUTYL-TIN BIS(2-ETHYLHEXYL) ESTER □ 2-ETHYLHEXYLMALEINAN DI-N-BUTYLCINICITY (CZECH)

TOXICITY DATA WITH REFERENCE
skn-rbt 500 mg/24H MOD 28ZPAK -,230,72
eye-rbt 100 mg/24H SEV 28ZPAK -,230,72
orl-rat LD50:284 mg/kg 28ZPAK -,230,72

OSHA PEL: TWA 0.1 mg(Sn)/m³ (skin)
ACGIH TLV: TWA 0.1 mg(Sn)/m³ (skin) (Proposed: TWA 0.1 mg(Sn)/m³; STEL 0.2 mg(Sn)/m³ (skin))
NIOSH REL: (Organotin Compounds) TWA 0.1 mg(Sn)/m³

SAFETY PROFILE: Poison by ingestion. A skin and eye irritant. See also TIN COMPOUNDS. When heated to decomposition it emits acrid smoke and irritating fumes.

For occupational chemical analysis use NIOSH: Organotin Compounds 5504.

BJR625 **CAS:16368-97-1** **HR: 3**
BIS(2-ETHYLHEXYL) PHENYL PHOSPHATE
mf: $C_{22}H_{39}O_4P$ mw: 398.58

SYNS: DAFF □ DEPP /S DI(2-ETHYLHEXYL)PHENYL PHOSPHATE

TOXICITY DATA WITH REFERENCE
ihl-rat LCLo:18 mg/m³/4H 85GMAT-,52,82
ipr-rat LD50:1178 mg/kg GTPZAB 15(8),30,71
orl-mus LD50:9333 mg/kg GTPZAB 15(8),30,71
ihl-mus LC50:5 g/m³ GTPZAB 15(8),30,71
ipr-mus LD50:473 mg/kg GTPZAB 15(8),30,71

SAFETY PROFILE: Poison by inhalation. Moderately toxic by intraperitoneal route. Mildly toxic by ingestion. When heated to decomposition it emits toxic fumes of PO_x. See also PHOSPHATES.

BJR750 **CAS:298-07-7** **HR: 3**
BIS(2-ETHYLHEXYL) PHOSPHATE
mf: $C_{16}H_{35}O_4P$ mw: 322.48

PROP: Viscous liquid. D: 0.975 @ 25 mm, bp: 155° @ 0.015 mm. Sol in C_6H_6, hexane, and 4-methyl-2-pentanone; sltly sol in H_2O.

SYNS: BIS(2-ETHYLHEXYL)HYDROGEN PHOSPHATE □ BIS(2-ETHYLHEXYL)ORTHOPHOSPHORIC ACID □ BIS(2-ETHYLHEXYL)PHOSPHORIC ACID □ DEHPA EXTRACTANT □ DI(2-ETHYLHEXYL)PHOSPHATE □ DI-2(ETHYLHEXYL)PHOSPHORIC ACID □ DI-(2-ETHYLHEXYL)PHOSPHORIC ACID (DOT) □ 2-ETHYL-1-HEXANOL HYDROGEN PHOSPHATE □ HDEHP □ KYSELINA DI-(2-ETHYLHEXYL)FOSFORECNA

TOXICITY DATA WITH REFERENCE
skn-rbt 500 mg open MOD UCDS** 5/18/72
skn-rbt 5 mg/24H SEV 85JCAE-,1130,86
eye-rbt 5 mg MOD UCDS** 5/18/72
eye-rbt 250 µg/24H SEV 85JCAE-,1130,86
orl-rat LD50:4940 mg/kg UCDS** 5/18/42
ipr-rat LD50:50 mg/kg HYDRDA 3,201,78
ipr-mus LDLo:63 mg/kg CBCCT* 9,132,57
skn-rbt LD50:1250 mg/kg UCDS** 5/18/72

CONSENSUS REPORTS: Reported in EPA TSCA Inventory.

SAFETY PROFILE: Poison by intraperitoneal route. A corrosive material. A severe eye and skin irritant. When heated to decomposition it emits toxic fumes of PO_x.

BJS250 **CAS:122-62-3** **HR: 2**
BIS(2-ETHYLHEXYL) SEBACATE
mf: $C_{26}H_{50}O_4$ mw: 426.76

PROP: Light, clear liquid; mild odor. Mp: −48°, fp: −55°, bp: 256° @ 5 mm, flash p: 410°F, d: 0.914 @ 20°/4°, vap d: 14.7.

SYNS: BISOFLEX DOS □ DECANEDIOIC ACID, BIS(2-ETHYLHEXYL) ESTER □ DI(2-ETHYLHEXYL)SEBACATE □ DIOCTYL SEBACATE □ DOS □ 2-ETHYLHEXYL SEBACATE □ MONOPLEX DOS □ OCTOIL S □ OCTYL SEBACATE □ PX 438 □ STALFLEX DOS □ UNIFLEX DOS

TOXICITY DATA WITH REFERENCE
ivn-rat LD50:900 mg/kg MRLR** No.256,54
orl-mus LD50:9500 mg/kg 85GMAT-,62,82
ivn-rbt LD50:540 mg/kg MRLR** No.256,54

CONSENSUS REPORTS: Reported in EPA TSCA Inventory.

SAFETY PROFILE: Moderately toxic by ingestion and intravenous routes. See also ESTERS. Combustible when exposed to heat or flame; can react with oxidizing materials. To fight fire, use foam, CO_2, dry chemical. When heated to decomposition it emits acrid and irritating fumes.

BJS500 **CAS:6422-86-2** **HR: 1**
BIS(2-ETHYLHEXYL) TEREPHTHALATE
mf: $C_{24}H_{38}O_4$ mw: 390.62

SYNS: 1,4-BENZENEDICARBOXYLIC ACID, BIS(2-ETHYLHEXYL)ESTER (9CI) □ KODAFLEX DOTP □ TEREPHTHALIC ACID, BIS(2-ETHYLHEXYL)ESTER

TOXICITY DATA WITH REFERENCE
orl-mus LDLo:20 g/kg GISAAA 47(8),91,82

CONSENSUS REPORTS: Reported in EPA TSCA Inventory.

SAFETY PROFILE: Low toxicity by ingestion. When heated to decomposition it emits acrid smoke and irritating vapors.

BJT250 **CAS:2440-45-1** **HR: 3**
BIS(ETHYLMERCURI) PHOSPHATE
mf: $C_4H_{11}Hg_2O_4P$ mw: 555.30

PROP: Solid.

SYNS: ETHYLMERCURIC PHOSPHATE □ ETHYLMERCURY PHOSPHATE □ LIGNASAN FUNGICIDE □ LIGNASAN-X □ NEW IMPROVED CERESAN □ NEW IMPROVED GRANOSAN

TOXICITY DATA WITH REFERENCE
scu-mus TDLo:40 mg/kg (10D preg):TER NISFAY 20,1479,68
orl-rat LD50:30 mg/kg PCOC** -,516,66
unk-rat LD50:30 mg/kg 30ZDA9 -,288,71
orl-mus LD50:56 mg/kg NYKZAU 58,235,62
scu-mus LD50:88 mg/kg KUMJAX 14,65,61

CONSENSUS REPORTS: Mercury and its compounds are on the Community Right-To-Know List.

OSHA PEL: TWA 0.01 mg(Hg)/m³; STEL 0.03 mg/m³ (skin)
ACGIH TLV: TWA 0.01 mg(Hg)/m³; STEL 0.03 mg(Hg)/m³

SAFETY PROFILE: Poison by ingestion and subcutaneous routes. See also MERCURY COMPOUNDS, ORGANIC. An experimental teratogen. When heated to decomposition it emits very toxic fumes of Hg and PO_x.

BJT500 **CAS:139-60-6** **HR: 2**
N,N′-BIS(1-ETHYL-3-METHYLPENTYL)-p-PHENYLENEDIAMINE
mf: $C_{22}H_{40}N_2$ mw: 332.64

SYNS: N,N′-BIS(5-METHYL-3-HEPTYL)-p-PHENYLENEDIAMINE □ N,N′-DI(1-ETHYL-3-METHYLPENTYL)-p-PHENYLENEDIAMINE □ EASTOZONE 31 □ ELASTOZONE 31 □ SANTOFLEX 17 □ TENAMENE 31 □ UOP 88

TOXICITY DATA WITH REFERENCE
orl-rat LD50:2400 mg/kg RCTEA4 45(3),627,72
skn-rbt LD50:1800 mg/kg RCTEA4 45(3),627,72

CONSENSUS REPORTS: Reported in EPA TSCA Inventory.

SAFETY PROFILE: Moderately toxic by ingestion and skin contact. When heated to decomposition it emits toxic fumes of NO_x.

BJT750 CAS:76-20-0 HR: 2
2,2-BIS(ETHYLSULFONYL)BUTANE
mf: $C_8H_{18}O_4S_2$ mw: 242.38

PROP: Lustrous bitter-tasting leaflet. Mp: 74–76°.

SYNS: DIETHYLSULFONMETHYLETHYLMETHANE □ ETHYLSULFONAL □ METHYLSULFONAL □ METHYLSULPHONAL □ SULFONETHYLMETHANE □ TIONAL □ TRIONAL

TOXICITY DATA WITH REFERENCE
skn-mus TDLo:1900 mg/kg/1W-I:ETA BJCAAI 9,177,55

SAFETY PROFILE: Questionable carcinogen with experimental tumorigenic data. When heated to decomposition it emits toxic fumes of SO_x.

BJU000 CAS:502-55-6 HR: 3
BIS(ETHYLXANTHOGEN) DISULFIDE
mf: $C_6H_{10}O_2S_4$ mw: 242.40

PROP: Yellow needles. Mp: 28–32°.

SYNS: AULIGEN □ BEK □ BEXIDE □ BEXT □ BIS(ETHYLXANTHIC)DISULFIDE □ BIETHYLXANTHOGENTRISULFIDE □ DEX □ DIETHYLDITHIO BIS(THIONOFORMATE) □ DIETHYL DIXANTHOGEN □ DIETHYL XANTHOGENATE □ DIETHYLXANTHOGEN DISULFIDE □ DITHIOBIS(THIOFORMIC ACID)-o,o-DIETHYL ESTER □ DIXANTHOGEN □ ETHYL XANTHOGEN DISULFIDE □ EXD □ K PREPARATION □ THIOPEROXYDICARBONIC ACID DIETHYL ESTER

TOXICITY DATA WITH REFERENCE
orl-rat LD50:480 mg/kg RREVAH 10,97,65
skn-rat LDLo:2100 mg/kg PCOC** -,578,66
orl-mus LD50:1200 mg/kg FATOAO 28,230,65
orl-rbt LD50:620 mg/kg PCOC** -,578,66
ipr-rbt LD50:320 mg/kg APTOA6 8,329,52
orl-gpg LD50:400 mg/kg PCOC** -,578,66
unk-mam LD50:600 mg/kg 30ZDA9 -,180,71

CONSENSUS REPORTS: Reported in EPA TSCA Inventory.

SAFETY PROFILE: Poison by ingestion and intraperitoneal routes. Moderately toxic by skin contact and possibly other routes. See also ESTERS and SULFIDES. When heated to decomposition it emits highly toxic fumes of SO_x.

BJU250 CAS:1851-71-4 HR: 3
BIS(ETHYLXANTHOGEN) TETRASULFIDE
mf: $C_6H_{10}O_2S_6$ mw: 306.52

SYN: TETRASULFIDE, BIS(ETHOXYTHIOCARBONYL)

TOXICITY DATA WITH REFERENCE
orl-rat LD50:275 mg/kg 28ZEAL 5,26,76
orl-mus LD50:275 mg/kg 28ZEAL 5,26,76

SAFETY PROFILE: Poison by ingestion. When heated to decomposition it emits toxic fumes such as SO_x. See also SULFIDES.

BJU350 CAS:73526-98-4 HR: 3
BIS(2-FLUORO-2,2-DINITROETHOXY)DIMETHYLSILANE
mf: $C_6H_{10}F_2N_4O_{10}Si$ mw: 364.25

$(F(O_2N)_2CCH_2O)_2Si(CH_3)_2$

SAFETY PROFILE: An explosive plasticizer sensitive to shock. When heated to decomposition it emits toxic fumes of F^- and NO_x.

BJU500 CAS:18139-03-2 HR: 3
BIS(2-FLUORO-2,2-DINITROETHYL)AMINE
mf: $C_4H_5F_2N_5O_8$ mw: 289.11

$(FC(NO_2)_2CH_2)_2NH$

SAFETY PROFILE: An explosive. When heated to decomposition it emits toxic fumes of F^- and NO_x. See also EXPLOSIVES and AMINES.

BJV625 CAS:72985-54-7 HR: 3
1,1-BIS(FLUOROOXY)HEXAFLUOROPROPANE
mf: $C_3F_8O_2$ mw: 220.02

SAFETY PROFILE: Decomposes explosively. When heated to decomposition it emits toxic fumes of F^-.

BJV630 CAS:16329-93-4 HR: 3
2,2-BIS(FLUOROOXY)HEXAFLUOROPROPANE
mf: $C_3F_8O_2$ mw: 220.02

SAFETY PROFILE: An unstable explosive. When heated to decomposition it emits toxic fumes of F^-.

BJV635 CAS:16329-92-3 HR: 3
1,1-BIS(FLUOROOXY)TETRAFLUOROETHANE
mf: $C_2F_6O_2$ mw: 170.01

SAFETY PROFILE: Potentially explosive at room temperature. Upon decomposition it emits toxic fumes of F^-.

BJV750 CAS:63698-38-4 HR: 3
trans-4-(4,4-BIS(p-FLUOROPHENYL)BUTYL)-1-(2-(4′-PHENYLCYCLOHEXYLAMINO)ETHYL)PIPERAZINE TRIHYDROCHLORIDE
mf: $C_{34}H_{43}F_2N_3$•3ClH mw: 641.18

SYN: M.G. 18001-3HCl

TOXICITY DATA WITH REFERENCE
orl-rat LD50:389 mg/kg FRPSAX 32,461,77
ipr-rat LD50:42 mg/kg FRPSAX 32,461,77

SAFETY PROFILE: Poison by ingestion and intraperitoneal routes. When heated to decomposition it emits very toxic fumes of F^-, HCl, and NO_x.

BJW000 CAS:63698-37-3 HR: 3
trans-2-(4-(4,4-BIS(p-FLUOROPHENYL)BUTYL)PIPERAZINYL)-N-(4′-PHENYLCYCLOHEXYL)ACETAMIDE DIHYDROCHLORIDE
mf: $C_{34}H_{41}F_2N_3O \cdot 2ClH$ mw: 618.70

SYN: M.G. 8948-2HCl

TOXICITY DATA WITH REFERENCE
orl-rat LD50:301 mg/kg FRPSAX 32,461,77
ipr-rat LD50:108 mg/kg FRPSAX 32,461,77

SAFETY PROFILE: Poison by ingestion and intraperitoneal routes. When heated to decomposition it emits very toxic fumes of F^-, NO_x, and HCl.

BJW250 CAS:20929-99-1 HR: 3
1,1-BIS(4-FLUOROPHENYL)-2-PROPYNYL-N-CYCLOHEPTYLCARBAMATE
mf: $C_{23}H_{23}F_2NO_2$ mw: 383.47

SYN: CYCLOHEPTANECARBAMIC ACID-1,1-BIS(p-FLUOROPHENYL)-2-PROPYNYL ESTER

TOXICITY DATA WITH REFERENCE
orl-rat TDLo:620 mg/kg/13W-C:CAR CNREA8 30,2881,70
ipr-rat LD50:215 mg/kg HarPN# 21OCT74
orl-mus LD50:405 mg/kg HarPN# 21OCT74
ipr-mus LD50:318 mg/kg HarPN# 21OCT74

SAFETY PROFILE: Poison by ingestion and intraperitoneal routes. Questionable carcinogen with experimental carcinogenic data. See also ESTERS. When heated to decomposition it emits very toxic fumes of F^- and NO_x.

BJW500 CAS:20930-00-1 HR: 2
1,1-BIS(4-FLUOROPHENYL)-2-PROPYNYL-N-CYCLOOCTYL CARBAMATE
mf: $C_{24}H_{25}F_2NO_2$ mw: 397.50

SYN: CYCLOOCTANECARBAMIC ACID-1,1-BIS(p-FLUOROPHENYL)-2-PROPYNYL ESTER

TOXICITY DATA WITH REFERENCE
orl-rat TDLo:1900 mg/kg/66D-C:CAR CNREA8 30,2881,70
ipr-rat LD50:617 mg/kg HarPN# 21OCT74
ipr-mus LD50:456 mg/kg HarPN# 21OCT74

SAFETY PROFILE: Moderately toxic by intraperitoneal route. Questionable carcinogen with experimental carcinogenic data. See also ESTERS and CARBAMATES. When heated to decomposition it emits very toxic fumes of F^- and NO_x.

BJW750 CAS:6784-25-4 HR: 3
BIS(N-FORMYL-p-AMINOPHENYL)SULFONE
mf: $C_{14}H_{12}N_2O_4S$ mw: 304.34

SYN: N,N′-DIFORMYL-p,p′-DIAMINODIPHENYLSULFONE

TOXICITY DATA WITH REFERENCE
ivn-rat LDLo:450 mg/kg TXAPA9 18,469,71
ipr-mus LD50:760 mg/kg EXPAAA 20,88,67
ivn-dog LDLo:98 mg/kg IJLEAG 36,432,68
ivn-cat LDLo:255 mg/kg IJLEAG 36,432,68

SAFETY PROFILE: Poison by intravenous route. Moderately toxic by intravenous and intraperitoneal routes. See also SULFONATES. When heated to decomposition it emits very toxic fumes of NO_x and SO_x.

BJW800 CAS:4387-13-7 HR: 3
BIS(FORMYLMETHYL) MERCURY
mf: $C_4H_6HgO_2$ mw: 286.69

PROP: Crystals from EtOH. Mp: 92–94°.

SYN: MERCURIDIACETALDEHYDE

TOXICITY DATA WITH REFERENCE
ivn-mus LD50:18 mg/kg CSLNX* NX#05651

OSHA PEL: CL 0.1 mg(Hg)/m³ (skin)
ACGIH TLV: TWA 0.01 mg(Hg)/m³; STEL 0.03 mg(Hg)/m³
NIOSH REL: (Organomercury): TWA 0.01 mg/m³; STEL 0.03 mg/m³ (skin)

SAFETY PROFILE: Poison by intravenous route. When heated to decomposition it emits toxic fumes of Hg.

BJW825 CAS:5188-42-1 HR: 3
BIS(GUANIDINIUM) CHROMATE
mf: $C_2H_{10}N_6 \cdot CrH_2O_4$ mw: 236.20

SYN: BIGUANIDINE, CHROMATE

TOXICITY DATA WITH REFERENCE
ivn-mus LD50:180 mg/kg CSLNX* NX#02828

OSHA PEL: CL 0.1 mg(CrO_3)/m³
ACGIH TLV: TWA 0.05 mg(Cr)/m³; Confirmed Human Carcinogen
NIOSH REL: (Chromium(VI)) TWA 0.025 mg/m³; CL 0.05 mg/15M

SAFETY PROFILE: A confirmed carcinogen. Poison by intravenous route. When heated to decomposition it emits toxic fumes of NO_x and Cr.

For occupational chemical analysis use NIOSH: Chromium Hexavalent 7024.

BJX750 CAS:19704-60-0 HR: 3
BIS(HEXANOYLOXY)DI-n-BUTYLSTANNANE
mf: $C_{20}H_{40}O_4Sn$ mw: 463.29

SYNS: BIS(HEXANOYLOXY)DI-n-BUTYL-TIN □ KAPRONAN DI-N-BUTYLCINICITY (CZECH)

TOXICITY DATA WITH REFERENCE
skn-rbt 500 mg/24H SEV 28ZPAK -,229,72
eye-rbt 20 mg/24H MOD 28ZPAK -,229,72
orl-rat LD50:94 mg/kg 28ZPAK -,229,72

OSHA PEL: TWA 0.1 mg(Sn)/m³ (skin)
ACGIH TLV: TWA 0.1 mg(Sn)/m³ (skin) (Proposed: TWA 0.1 mg(Sn)/m³; STEL 0.2 mg(Sn)/m³ (skin))
NIOSH REL: (Organotin Compounds) TWA 0.1 mg(Sn)/m³

SAFETY PROFILE: Poison by ingestion. A skin and eye irritant. See also TIN COMPOUNDS. When heated to decomposition it emits acrid smoke and fumes.

For occupational chemical analysis use NIOSH: Organotin Compounds 5504.

B

BJX800 **CAS:63270-67-7** **HR: 3**
BIS(l-HISTIDINATO)MANGANESE TETRAHYDRATE
mf: $C_{12}H_{16}MnN_6O_4{\cdot}4H_2O$ mw: 435.36

SYNS: MANGANESE, BIS(l-HISTIDINATO)-, TETRAHYDRATE □ MANGANESE, BIS(l-HISTIDINATO-N,O)-, TETRAHYDRATE

TOXICITY DATA WITH REFERENCE
uns-mus LD50:160 mg/kg FRMBAZ 29,215,81

OSHA PEL: CL 5 mg(Mn)/m³
ACGIH TLV: TWA 5 mg(Mn)/m³

SAFETY PROFILE: Poison by an unspecified route. When heated to decomposition it emits toxic fumes of NO_x and Mn.

BJY000 **CAS:14873-10-0** **HR: 3**
BIS(l-HISTIDINE)COBALT
mf: $C_{12}H_{14}N_6O_5{\cdot}Co$ mw: 365.25

SYNS: α-AMINOIMIDAZOLE-4-PROPIONIC ACID, COBALT(2+) SALT □ BIS(l-HISTIDINATO)COBALT □ COBALT-HISTIDINE □ KOBALT HISTIDIN (GERMAN)

TOXICITY DATA WITH REFERENCE
ipr-rat LD50:134 mg/kg AEPPAE 243,254,62
ivn-rat LD50:104 mg/kg AIPTAK 143,219,63
ivn-cat LD50:50 mg/kg AIPTAK 143,219,63

CONSENSUS REPORTS: Cobalt and its compounds are on the Community Right-To-Know List.

SAFETY PROFILE: Poison by intraperitoneal and intravenous routes. See also COBALT COMPOUNDS. When heated to decomposition it emits toxic fumes of NO_x.

BJY250 **HR: 3**
BISHYDRAZINE NICKEL(II)PERCHLORATE
mf: $Cl_2H_8N_4NiO_8$ mw: 323.7

CONSENSUS REPORTS: Nickel and its compounds are on the Community Right-To-Know List.

SAFETY PROFILE: Exploded by heat and dilute aqueous suspension. Upon decomposition it emits toxic fumes of Cl^- and NO_x. See also NICKEL COMPOUNDS and PERCHLORATES.

BJY500 **HR: 3**
BISHYDRAZINE TIN(II)CHLORIDE
mf: $Cl_2H_8N_4Sn$ mw: 253.69

SAFETY PROFILE: Explodes on heating. Upon decomposition it emits toxic fumes of Cl^-. See also TIN COMPOUNDS and CHLORIDES.

BJY750 **HR: 3**
BIS(1-HYDROPEROXY CYCLOHEXYL)PEROXIDE
mf: $C_{12}H_{22}O_6$ mw: 252.20

SAFETY PROFILE: Fire causes violent explosion. When heated to decomposition it emits acrid smoke and fumes. See also PEROXIDES, ORGANIC.

BJY825 **CAS:2614-76-8** **HR: 3**
2,2-BIS(HYDROPEROXY)PROPANE
mf: $C_3H_8O_4$ mw: 108.09

$(CH_3)_2C(OOH)_2$

PROP: Liquid.

SAFETY PROFILE: Ignites or explodes when heated. When heated to decomposition it emits acrid smoke and fumes. See also PEROXIDES.

BJZ000 **CAS:66-76-2** **HR: 3**
BISHYDROXYCOUMARIN
mf: $C_{19}H_{12}O_6$ mw: 336.31

PROP: Very small crystals from cyclohexanone with a slight pleasant odor and bitter taste. Mp: 288–289°. Sol in alkali.

SYNS: ACADYL □ ACAVYL □ ANTITROMBOSIN □ BARACOUMIN □ BHC □ BIS(4-HYDROXYCOUMARIN-3-YL)METHANE □ CUMA □ CUMID □ DICOUMARIN □ DICOUMAROL □ DICUMAN □ DICUMARINE □ DI-(4-HYDROXY-3-COUMARINYL)METHANE □ DI-4-HYDROXY-3,3′-METHYLENEDICOUMARIN □ DUFALONE □ KUMORAN □ MELITOXIN □ 3,3′-METHYLEEN-BIS(4-HYDROXY-CUMARINE) (DUTCH) □ 3,3′-METHYLEN-BIS(4-HYDROXY-CUMARIN) (GERMAN) □ 3,3′-METHYLENEBIS(4-HYDROXY-1,2-BENZOPYRONE) □ 3,3′-METHYLENEBIS(4-HYDROXYCOUMARIN) □ 3,3′-METHYLENE-BIS(4-HYDROXYCOUMARINE) (FRENCH) □ 3,3′-METILEN-BIS(4-IDROSSI-CUMARINA) (ITALIAN) □ TEMPARIN □ TROMBOSAN

TOXICITY DATA WITH REFERENCE
orl-wmn TDLo:110 mg/kg (31-40W preg):REP AJOGAH 57,965,49
orl-wmn TDLo:110 mg/kg (31-40W preg):TER AJOGAH 57,965,49
unr-wmn TDLo:28 mg/kg (28-30W preg):TER AJOGAH 77,1135,59
orl-rat LD50:250 mg/kg SMWOAS 83,471,53
ivn-rat LD50:52 mg/kg PSEBAA 50,228,42
orl-mus LD50:233 mg/kg PSEBAA 50,228,42
ipr-mus LD50:91 mg/kg DIPHAH 17,163,65
scu-mus LD50:50 mg/kg 85GDA2 8(1),360,82
ivn-mus LD50:42 mg/kg AEPPAE 222,107,54

CONSENSUS REPORTS: Reported in EPA TSCA Inventory.

SAFETY PROFILE: Poison by ingestion, subcutaneous, intravenous, and intraperitoneal routes. An experimental teratogen. Human reproductive effects by ingestion and possibly other routes: fetal death, unspecified developmental abnormalities, stillbirth, and unspecified neonatal effects. An anticoagulant. Excessive doses can cause hemorrhages. When heated to decomposition it emits acrid smoke and fumes. See also WARFARIN.

BKA000 **CAS:548-00-5** **HR: 3**
BIS(4-HYDROXY-3-COUMARIN) ACETIC ACID ETHYL ESTER
mf: $C_{22}H_{16}O_8$ mw: 408.38

PROP: Amorphous or crystalline from Me_2CO. Mp: 151° (amorphous), mp: 173° (crystalline).

SYNS: BIS-3,3′-(4-HYDROXYCOUMARINYL)ACETIC ACID ETHYL ESTER □ BIS-(4-HYDROXY-3-COUMARINYL)ETHYL ACETATE □ BIS(4-

HYDROXY-2-OXO-2H-1-BENZOPYRAN-3-YL)ACETIC ACID ETHYL ESTER □ BOEA □ B.O.E.A. □ 3,3'-(CARBOXYMETHYLENE)BIS(4-HYDROXYCOUMARIN) ETHYL ESTER □ DICUMACYL □ ETHYL BISCOUMACETATE □ ETHYL BIS(4-HYDROXYCOUMARINYL)ACETATE □ ETHYL BIS(4-HYDROXY-3-COUMARINYL)ACETATE □ ETHYLDICOUMAROL □ ETHYLDICOUMAROL ACETATE □ ETHYL-4,4'-DIHYDROXYDICOUMARINYL-3,3'-ACETATE □ NEODICOUMARIN □ NEODICOUMAROL □ NEODICUMARINUM □ PELENTAN □ STABILENE □ TROMBARIN □ TROMBIL □ TROMBOLYSAN □ TROMEXAN □ TROMEXAN ETHYL ACETATE

TOXICITY DATA WITH REFERENCE
orl-wmn TDLo:48 mg/kg (36-37W preg):REP BMJOAE 2,719,55
orl-wmn TDLo:48 mg/kg (36-37W preg):TER BMJOAE 2,719,55
orl-rat LD50:840 mg/kg FEPRA7 10,303,51
ipr-rat LD50:260 mg/kg AIPTAK 87,402,51
orl-mus LD50:750 mg/kg AEPPAE 222,107,54
scu-mus LD50:750 mg/kg LANCAO 2,611,51

SAFETY PROFILE: Poison by intraperitoneal route. Moderately toxic by ingestion and subcutaneous routes. An experimental teratogen. Human reproductive effects by ingestion: developmental abnormalities of the cardiovascular system, stillbirth, and unspecified neonatal effects. An anticoagulant. See also WARFARIN and ESTERS. When heated to decomposition it emits acrid and irritating fumes.

BKA250 HR: 3
BIS(1-HYDROXYCYCLOHEXYL)PEROXIDE
mf: $C_{22}H_{22}O_4$ mw: 230.3

SAFETY PROFILE: Explodes in vacuum. When heated to decomposition it emits acrid smoke and fumes. See also PEROXIDES, ORGANIC.

BKB000 CAS:21615-29-2 HR: 1
3'-(BIS(2-HYDROXYETHYL)AMINO)-p-ACETOPHENETIDIDE
mf: $C_{14}H_{22}N_2O_4$ mw: 282.38

SYNS: 2,2'-((5-ACETAMIDO-2-ETHOXYPHENYL)IMINO)DIETHANOL □ 2-BIS-HYDROXYETHYLAMINO-4-ACETAMINOFENETOL (CZECH)

TOXICITY DATA WITH REFERENCE
eye-rbt 100 mg/24H MOD 28ZPAK -,100,72

CONSENSUS REPORTS: Reported in EPA TSCA Inventory.

SAFETY PROFILE: An eye irritant. When heated to decomposition it emits toxic fumes of NO_x.

BKB250 CAS:63867-52-7 HR: 2
2-(BIS(β-HYDROXYETHYL)AMINO)-4,5-DIPHENYLOXAZOLE MONOHYDRATE
mf: $C_{19}H_{20}N_2O_3•H_2O$ mw: 342.43

SYNS: AGEROPLAS □ DIETHAMPHENAZOL MONOHYDRATE □ 2,2'-DIHYDROXY-N-(4,5-DIPHENYLOXAZOLE-2-YL)DIETHYLAMINE MONOHYDRATE □ N-(4,5-DIPHENYLOXAZOL-2-YL)DIETHANOLAMINE MONOHYDRATE □ 2,2'-((4,5-DIPHENYL-2-OXAZOLYL)IMINO)-DIETHANOLMONOHYDRATE □ DITAZOL MONOHYDRATE □ S 222

TOXICITY DATA WITH REFERENCE
orl-rat LD50:11,380 mg/kg ARZNAD 23,1283,73
ipr-rat LD50:7770 mg/kg ARZNAD 23,1283,73
orl-mus LD50:9621 mg/kg ARZNAD 23,1283,73
ipr-mus LD50:3390 mg/kg ARZNAD 23,1283,73

SAFETY PROFILE: Moderately toxic by intraperitoneal route. Mildly toxic by ingestion. An anti-inflammatory agent. When heated to decomposition it emits toxic fumes of NO_x.

BKB300 CAS:70711-40-9 HR: 3
1,4-BIS((2-((2-HYDROXYETHYL)AMINO)ETHYL)AMINO)-9,10-ANTHRACENEDIONE DIACETATE
mf: $C_{22}H_{28}N_4O_4•2C_2H_4O_2$ mw: 532.66

SYNS: AMETANTRONE ACETATE □ 1,4-BIS((2-((HYDROXYETHYL)AMINO)ETHYL)AMINO)-9,10-ANTHRACENEDIONE DIACETATE (SALT) (9CI) □ CI 881 □ HAQ □ NSC 287513

TOXICITY DATA WITH REFERENCE
oms-hmn:leu 400 μg/L CNREA8 39,2574,79
cyt-hmn:leu 50 μg/L CNREA8 39,2574,79
oms-ham:ovr 10 nmol/L CNREA8 39,2574,79
ipr-rbt TDLo:5200 μg/kg (6-18D preg):TER TJADAB 29(2),41A,84
orl-mus LD50:495 mg/kg NCISP* JAN86
ipr-mus LD50:62,830 μg/kg NCISP* JAN86
scu-mus LD50:297 mg/kg NCISP* JAN86

SAFETY PROFILE: Poison by subcutaneous and intraperitoneal routes. Moderately toxic by ingestion. An experimental teratogen. Human mutation data reported. When heated to decomposition it emits toxic fumes of NO_x.

BKB500 CAS:27464-23-9 HR: 3
3-(BIS(2-HYDROXYETHYL)AMINO)-6-HYDRAZINOPYRIDAZINEDIHYDROCHLORIDE
mf: $C_8H_{15}N_5O_2•2ClH$ mw: 286.20

PROP: Solid. Mp: 187.5–188.5°.

SYNS: 3-HYDRAZINO-6-(N,N-BIS-(2-HYDROXYETHYL)AMINO) PYRIDAZINE DIHYDROCHLORIDE □ 2-IDRAZINO-6-(N,N-BIS(2-IDROSSIETIL)-AMINO)-PIRIDAZINA CLORIDRATO (ITALIAN) □ L 6150

TOXICITY DATA WITH REFERENCE
ivn-hmn TDLo:29 μg/kg:CNS DRFUD4 2,172,77
orl-rat LD50:1800 mg/kg BCFAAI 111,480,72
ipr-rat LD50:335 mg/kg BCFAAI 111,480,72
orl-mus LD50:1520 mg/kg BCFAAI 111,480,72
ipr-mus LD50:263 mg/kg DRFUD4 2,172,77
orl-dog LD50:1 g/kg ARZNAD 23,1591,73
ivn-dog LD50:75 mg/kg ARZNAD 23,1591,73
orl-rbt LD50:5 g/kg ARZNAD 23,1591,73
orl-gpg LD50:188 mg/kg BCFAAI 111,480,72

SAFETY PROFILE: Poison by ingestion and intraperitoneal routes. Human systemic effects by intravenous route: somnolence and unspecified pulmonary system effects. When heated to decomposition it emits very toxic fumes of HCl and NO_x.

B

BKB750 **CAS:5055-20-9** **HR: 2**
4-BIS(2-HYDROXYETHYL)AMINO-2-(5-NITRO-2-FURYL)QUINAZOLINE
mf: $C_{16}H_{16}N_4O_5$ mw: 344.36

PROP: Solid. Mp: 167–168°.

TOXICITY DATA WITH REFERENCE
orl-rat TDLo:8437 mg/kg/22W-C:CAR JNCIAM 57,277,76

SAFETY PROFILE: Questionable carcinogen with experimental carcinogenic data. When heated to decomposition it emits toxic fumes of NO_x.

BKC250 **CAS:33372-39-3** **HR: 2**
4-BIS(2-HYDROXYETHYL)AMINO-2-(5-NITRO-2-THIENYL)QUINAZOLINE
mf: $C_{16}H_{16}N_4O_4S$ mw: 360.42

TOXICITY DATA WITH REFERENCE
mma-sat 1250 µg/plate CNREA8 35,3611,75
orl-rat TDLo:3800 mg/kg/15W-C:CAR JNCIAM 57,277,76

CONSENSUS REPORTS: EPA Genetic Toxicology Program.

SAFETY PROFILE: Questionable carcinogen with experimental carcinogenic data. Mutation data reported. When heated to decomposition it emits very toxic fumes of NO_x and SO_x.

BKC500 **CAS:78109-79-2** **HR: 3**
N-(3-(BIS(2-HYDROXYETHYL)AMINO)PROPYL)BENZAMIDE HYDROCHLORIDE
mf: $C_{18}H_{30}N_2O_4$•ClH mw: 374.96

SYN: D-695

TOXICITY DATA WITH REFERENCE
scu-mus LD50:800 mg/kg ARZNAD 10,743,60
ivn-mus LD50:80 mg/kg ARZNAD 10,743,60

SAFETY PROFILE: Poison by intravenous route. Moderately toxic by subcutaneous route. When heated to decomposition it emits very toxic fumes of HCl and NO_x.

BKC750 **CAS:101651-88-1** **HR: 3**
10-(3-(BIS(2-HYDROXYETHYL)AMINO)PROPYL)-7-CHLOROISOALLOXAZINE SULFATE
mf: $C_{17}H_{20}ClN_5O_4$•H_2O_4S mw: 491.95

TOXICITY DATA WITH REFERENCE
scu-mus LD50:80 mg/kg CMTRAG 2,96,61
ivn-mus LD50:67 mg/kg CMTRAG 2,96,61

SAFETY PROFILE: Poison by subcutaneous and intravenous routes. See also SULFATES. When heated to decomposition it emits very toxic fumes of SO_x, Cl^-, and NO_x.

BKD000 **CAS:78128-69-5** **HR: 3**
N-(3-(BIS(2-HYDROXYETHYL)AMINO)PROPYL)-o-PROPOXYBENZAMIDE HYDROCHLORIDE
mf: $C_{17}H_{28}N_2O_4$•ClH mw: 360.93

SYN: D-701

TOXICITY DATA WITH REFERENCE
ipr-mus LD50:330 mg/kg ARZNAD 10,743,60
scu-mus LD50:785 mg/kg ARZNAD 10,743,60
ivn-mus LD50:110 mg/kg ARZNAD 10,743,60

SAFETY PROFILE: Poison by intraperitoneal and intravenous routes. Moderately toxic by subcutaneous route. When heated to decomposition it emits very toxic fumes of NO_x and HCl.

BKD250 **CAS:20182-56-3** **HR: 1**
4,4′-BIS((4-(2-HYDROXYETHYL)AMINO-6-(p-SULFOANILINO)-s-TRIAZIN-2-YL)AMINO)-2,2′-STILBENEDISULFONIC ACID TETRASODIUM SALT
mf: $C_{36}H_{32}N_{12}O_{14}S_4$•4Na mw: 1077.00

TOXICITY DATA WITH REFERENCE
skn-rbt 500 mg/24H MLD MVCRB3 2,193,73
eye-rbt 100 mg MLD MVCRB3 2,193,73

CONSENSUS REPORTS: Reported in EPA TSCA Inventory.

SAFETY PROFILE: A skin and eye irritant. See also SULFONATES. When heated to decomposition it emits toxic fumes of NO_x, Na_2O and SO_x.

BKD500 **CAS:120-07-0** **HR: 2**
N,N-BIS(2-HYDROXYETHYL)ANILINE
mf: $C_{10}H_{15}NO_2$ mw: 181.26

SYNS: DIETHANOLAMINOBENZENE □ DIETHANOLANILINE □ N,N-DIETHANOLANILINE □ DIHYDROXYETHYLANILINE □ N,N-DI(β-HYDROXYETHYL)ANILINE □ N,N-DI(2-HYDROXYETHYL)ANILINE □ N,N-DIOXYETHYLANILINE □ EMERY 5703 □ 2,2′-(PHENYLAMINO)DIETHANOL □ PHENYL DIETHANOLAMINE □ N-PHENYLDIETHANOLAMINE □ 2,2′-(PHENYLIMINO)DIETHANOL

TOXICITY DATA WITH REFERENCE
skn-rbt 500 mg open MLD UCDS** 6/13/60
eye-rbt 100 mg SEV UCDS** 6/13/60
orl-rat LD50:980 mg/kg JIDHAN 23,259,41

CONSENSUS REPORTS: Reported in EPA TSCA Inventory.

SAFETY PROFILE: Moderately toxic by ingestion. A severe eye and mild skin irritant. When heated to decomposition it emits toxic fumes of NO_x. See also AROMATIC AMINES.

BKD600 **CAS:5185-70-6** **HR: 3**
N,N-BIS(2-HYDROXYETHYL)-p-ARSANILIC ACID
mf: $C_{10}H_{16}AsNO_5$ mw: 305.19

SYN: p-ARSANILIC ACID, N,N-BIS(2-HYDROXYETHYL)-

TOXICITY DATA WITH REFERENCE
ipr-mus LD50:1053 mg/kg JMCMAR 9,221,66

OSHA PEL: TWA 0.5 mg(As)/m³
ACGIH TLV: TWA 0.2 mg(As)/m³

SAFETY PROFILE: Moderately toxic by intraperitoneal route. When heated to decomposition it emits toxic fumes of NO_x and As.

BKD750 CAS:64036-91-5 HR: 3
BIS(2-HYDROXYETHYL)-2-((2-CHLORO ETHYL THIO) ETHYL SULFONIUM) CHLORIDE
mf: $C_8H_{18}ClO_2S_2 \cdot Cl$ mw: 281.28

SYNS: β-CHLOROETHYL-β-(BIS(β-HYDROXYETHYL)SULFONIUM) ETHYL SULFIDE CHLORIDE □ 2-(2-CHLOROETHYL)THIOETHYLBIS (2-HYDROXYETHYL)-CHLORIDE

TOXICITY DATA WITH REFERENCE
skn-rat LD50:10 mg/kg JPETAB 93,1,48
skn-mus LD50:15 mg/kg JPETAB 93,1,48
scu-mus LDLo:25 mg/kg NTIS** PB158-507
ivn-dog LD50:6 mg/kg JPETAB 93,1,48
ivn-rbt LD50:4500 μg/kg JPETAB 93,1,48

SAFETY PROFILE: A poison by skin contact, subcutaneous, and intravenous routes. See also SULFONATES. When heated to decomposition it emits very toxic fumes of Cl^- and SO_x.

BKD800 CAS:120-86-5 HR: 2
N,N′-BIS(2-HYDROXYETHYL)-DITHIOOXAMIDE
mf: $C_6H_{12}N_2O_2S_2$ mw: 208.32

SYNS: OXAMIDE, N,N′-BIS(2-HYDROXYETHYL)DITHIO- □ USAF MK-5

TOXICITY DATA WITH REFERENCE
ipr-mus LD50:750 mg/kg NTIS** AD277-689

CONSENSUS REPORTS: Reported in EPA TSCA Inventory.

SAFETY PROFILE: Moderately toxic by intraperitoneal route. When heated to decomposition it emits toxic vapors of NO_x and SO_x.

BKE500 CAS:120-40-1 HR: 2
N,N-BIS(2-HYDROXYETHYL)DODECAN AMIDE
mf: $C_{16}H_{33}NO_3$ mw: 287.50

PROP: Solid. Mp: 36°

SYNS: BIS(2-HYDROXYETHYL)LAURAMIDE □ N,N-BIS(HYDROXYETHYL)LAURAMIDE □ N,N-BIS(β-HYDROXYETHYL)LAURAMIDE □ N, N-BIS(2-HYDROXYETHYL)LAURAMIDE □ CLINDROL 101CG □ CLINDROL SUPERAMIDE 100L □ COCO DIETHANOLAMIDE □ COCONUT OIL AMIDE of DIETHANOLAMINE □ COMPERLAN LD □ CONDENSATE PL □ CRILLON L.D.E. □ DIETHANOLLAURAMIDE □ N,N-DIETHANOLLAURAMIDE □ N,N-DIETHANOLLAURIC ACID AMIDE □ EMID 6511 □ EMID 6541 □ ETHYLAN MLD □ HETAMIDE ML □ LAURAMIDE DEA □ LAURIC ACID DIETHANOLAMIDE □ LAURIC DIETHANOLAMIDE □ LAUROYL DIETHANOLAMIDE □ LAURYL DIETHANOLAMIDE □ LDA □ LDE □ MONAMID 150-LW □ NCI-C55323 □ NINOL AA-62 EXTRA □ NINOL 4821 □ NINOL AA62 □ ONYXOL 345 □ REWOMID DLMS □ RICHAMIDE 6310 □ ROLAMID CD □ STANDAMIDD LD □ STEINAMID DL 203 S □ SUPER AMIDE L-9A □ SYNOTOL L-60 □ UNAMIDE J-56 □ VARAMID ML 1

TOXICITY DATA WITH REFERENCE
orl-rat LD50:2700 mg/kg JSCCA5 13,469,62

CONSENSUS REPORTS: Reported in EPA TSCA Inventory.

SAFETY PROFILE: Moderately toxic by ingestion. When heated to decomposition it emits toxic fumes of NO_x. See also AMIDES.

BKE750 CAS:64058-26-0 HR: 3
1,1-BIS(β-HYDROXYETHYL)ETHYLENINONIUM CHLORIDE
mf: $C_6H_{14}NO_2 \cdot Cl$ mw: 167.66

SYN: 1,1-BIS(2-HYDROXYETHYL)AZIRIDINIUM CHLORIDE

TOXICITY DATA WITH REFERENCE
orl-hmn TDLo:342 μg/kg:CNS NTIS** PB158-507
ipr-mus LD50:5 mg/kg NTIS** PB158-507
ivn-rbt LD50:20 mg/kg NTIS** PB158-507

SAFETY PROFILE: Poison by intraperitoneal and intravenous routes. Human systemic effects by ingestion: nausea and vomiting. When heated to decomposition it emits very toxic fumes of NO_x and Cl^-.

BKF250 CAS:2784-94-3 HR: 3
N′,N′-BIS(2-HYDROXYETHYL)-N-METHYL-2-NITRO-p-PHENYLENEDIAMINE
mf: $C_{11}H_{17}N_3O_4$ mw: 255.31

SYNS: HC BLUE 1 □ NCI-C04159

TOXICITY DATA WITH REFERENCE
mmo-sat 333 μg/plate NTPTR* NTP-TR-271,85
mma-sat 100 μg/plate NTPTR* NTP-TR-271,85
dns-rat:lvr 50 mg/L NTPTR* NTP-TR-271,85
msc-mus:lym 30 mg/L NTPTR* NTP-TR-271,85
orl-rat TDLo:66 g/kg/2Y-C:CAR NTPTR* NTP-TR-271,85
orl-mus TDLo:98,280 mg/kg/39W-C:CAR FCTOD7 25,703,87
orl-mus TD:131 g/kg/2Y-C:CAR NTPTR* NTP-TR-271,85

CONSENSUS REPORTS: IARC Cancer Review: Group 2B IMEMDT 57,129,93; Animal Sufficient Evidence IMEMDT 57,129,93; Human Inadequate Evidence IMEMDT 57,129,93. NTP Carcinogenesis Studies (feed); Some Evidence: rat NTPTR* NTP-TR-271,85; (feed); Clear Evidence: mouse NTPTR* NTP-TR-271,85. Reported in EPA TSCA Inventory.

SAFETY PROFILE: Suspected carcinogen with experimental carcinogenic data. Mutation data reported. See also AMINES. When heated to decomposition it emits toxic fumes of NO_x.

BKF500 CAS:56863-02-6 HR: 2
N,N-BIS(2-HYDROXYETHYL)-9,12-OCTADECADIENAMIDE
mf: $C_{22}H_{41}NO_3$ mw: 367.64

SYNS: CLINDROL LT 15-73-1 □ CYCLOMIDE DIN 295/S □ LINOLEIC DIETHANOLAMIDE

TOXICITY DATA WITH REFERENCE
skn-rbt 500 mg/24H SEV TXAPA9 19,276,71
eye-rbt 100 mg TXAPA9 19,276,71

CONSENSUS REPORTS: Reported in EPA TSCA Inventory.

SAFETY PROFILE: A severe skin and eye irritant. When heated to decomposition it emits toxic fumes of NO_x.

BKF750 **CAS:63886-75-9** **HR: 3**
N,N-BIS(2-HYDROXYETHYL)-p-PHENYLENEDIAMINE SULFATE (1:1)
mf: $C_{10}H_{16}N_2O_2 \cdot H_2O_4S$ mw: 294.36

SYN: N,N-BIS-2-HYDROXYETHYL-p-FENYLENDIAMIN STRAN (CZECH)

TOXICITY DATA WITH REFERENCE
eye-rbt 500 mg/24H MOD 28ZPAK -,110,72
orl-rat LD50:131 mg/kg 28ZPAK -,110,72

SAFETY PROFILE: Poison by ingestion. An eye irritant. See also SULFATES. When heated to decomposition it emits very toxic fumes of SO_x and NO_x.

BKG250 **HR: 3**
BISHYDROXYL AMINE ZINC(II)CHLORIDE
mf: $Cl_2H_6N_2O_2Zn$ mw: 202.33

CONSENSUS REPORTS: Zinc and its compounds are on the Community Right-To-Know List.

SAFETY PROFILE: Explodes at 170°. When heated to decomposition it emits toxic fumes of Cl^-, NO_x, and ZnO. See also ZINC COMPOUNDS and CHLORIDES.

BKG500 **CAS:73118-23-7** **HR: 3**
3,5-BIS(3-HYDROXYMERCURI-2-METHOXYPROPYL) BARBITURIC ACID SODIUM SALT
mf: $C_{12}H_{19}Hg_2N_2O_7 \cdot 7Na$ mw: 865.44

SYN: (1,5-(2,4,6-TRIOXO-(1H,3H,5H)-PYRIMIDYLENE))BIS((2-METHOXYPROPYL)HYDROXYMERCURY SODIUM SALT

TOXICITY DATA WITH REFERENCE
ipr-rat LD50:13,500 µg/kg JAPMA8 39,297,50

CONSENSUS REPORTS: Mercury and its compounds are on the Community Right-To-Know List.

OSHA PEL: CL 0.1 mg(Hg)/m³ (skin)
ACGIH TLV: TWA 0.1 mg(Hg)/m³ (skin)
NIOSH REL: (Mercury, Aryl And Inorganic): CL 0.1 mg/m³ (skin)

SAFETY PROFILE: Poison by intraperitoneal route. See also MERCURY COMPOUNDS. When heated to decomposition it emits very toxic fumes of Na_2O, Hg, and NO_x.

BKG750 **CAS:73118-24-8** **HR: 3**
5,5-BIS(3-HYDROXYMERCURI-2-METHOXYPROPYL) BARBITURIC ACIDSODIUM SALT
mf: $C_{12}H_{19}Hg_2N_2O_7 \cdot xNa$ mw: 865.44

TOXICITY DATA WITH REFERENCE
ipr-rat LD50:30,500 µg/kg JAPMA8 39,297,50

CONSENSUS REPORTS: Merucury and its compounds are on the Community Right-To-Know List.

OSHA PEL: CL 0.1 mg(Hg)/m³ (skin)
ACGIH TLV: TWA 0.1 mg(Hg)/m³ (skin)
NIOSH REL: (Mercury, Aryl And Inorganic): CL 0.1 mg/m³ (skin)

SAFETY PROFILE: Poison by intraperitoneal route. See also MERCURY COMPOUNDS. When heated to decomposition it emits very toxic fumes of Na_2O, Hg and NO_x.

BKH000 **CAS:63951-09-7** **HR: 3**
2,6-BIS(HYDROXYMERCURI)-4-NITROANILINE
mf: $C_6H_6Hg_2N_2O_4$ mw: 571.32

TOXICITY DATA WITH REFERENCE
ipr-rat LDLo:250 mg/kg NCNSA6 5,12,53

CONSENSUS REPORTS: Mercury and its compounds are on the Community Right-To-Know List.

OSHA PEL: CL 0.1 mg(Hg)/m³ (skin)
ACGIH TLV: TWA 0.1 mg(Hg)/m³ (skin)
NIOSH REL: (Mercury, Aryl And Inorganic): CL 0.1 mg/m³ (skin)

SAFETY PROFILE: Poison by intraperitoneal route. See also MERCURY COMPOUNDS. When heated to decomposition it emits very toxic fumes of NO_x and Hg.

BKH125 **CAS:67536-44-1** **HR: 3**
1,2-BIS(HYDROXOMERCURIO)-1,1,2,2-BIS(OXYDIMERCURIO)ETHANE
mf: $C_2H_2Hg_6O_4$ mw: 1293.58

SYN: ETHANE HEXAMERCARBIDE

CONSENSUS REPORTS: Mercury and its compounds are on the Community Right-To-Know List.

SAFETY PROFILE: Explodes violently when heated to 230°C. When heated to decomposition it emits toxic fumes of Hg. See also MERCURY COMPOUNDS.

BKH325 **CAS:105-08-8** **HR: 2**
1,4-BIS(HYDROXYMETHYL)CYCLOHEXANE
mf: $C_8H_{16}O_2$ mw: 144.24

SYNS: 1,4-CHIDM □ HEXAHYDRO-2-OXO-1,4-CYCLOHEXANEDIMETHANOL

TOXICITY DATA WITH REFERENCE
orl-rat LDLo:3200 mg/kg KODAK* 21MAY71
ipr-rat LDLo:800 mg/kg 34ZIAG -,194,69
orl-mus LDLo:1600 mg/kg KODAK* 21MAY77
ipr-mus LDLo:1600 mg/kg 34ZIAG -,194,69

CONSENSUS REPORTS: Reported in EPA TSCA Inventory.

SAFETY PROFILE: Moderately toxic by intraperitoneal route. When heated to decomposition it emits acrid smoke and fumes.

BKH500 **CAS:794-93-4** **HR: 3**
BIS(HYDROXYMETHYL)FURATRIZINE
mf: $C_{11}H_{11}N_5O_5$ mw: 293.27

PROP: Yellow crystals. Mp: 161° (decomp).

SYNS: 3-BIS(HYDROXYMETHYL)AMINO-6-(5-NITRO-2-FURYLETHENYL)-1,2,4-TRIAZINE □ DHNT □ 3-DI(HYDROXYMETHYL)AMINO-6-(5-NITRO-2-FURYLETHENYL)-1,2,4-TRIAZINE □ 3-DI(HYDROXYMETHYL)AMINO-6-(2-(5-NITRO-2-FURYL)VINYL)-1,2,4-TRIAZINE □ DIHYDROXYMETHYL FURATRIZINE □ FURATONE □ FURATONE-S

□ N-(6-(5-NITROFURFURYLIDENEMETHYL)-1,2,4-TRIAZIN-3-YL)IMINODIMETHANOL □ 6-(5-NITRO-2-FURYLVINYL)-3-(DIHYDROXYDIMETHYLAMINO)-1,2,4-TRIAZENE □ N-(6-(2-(5-NITRO-2-FURYL)VINYL)-1,2,4-TRIAZIN-3-YL)IMINODIMETHANOL □ ((6-(2-(5-NITRO-2-FURYL)VINYL)-as-TRIAZIN-3-YL)IMINO)DIMETHANOL □ PANFURAN-S

TOXICITY DATA WITH REFERENCE
mmo-esc 125 μg/L MUREAV 146,243,85
pic-esc 800 μg/L MUREAV 146,243,85
orl-rat TDLo: 25,725 mg/kg/35W-C: CAR NAIZAM 31,31,80
orl-mus TDLo: 12,740 mg/kg/35W-C: CAR NAIZAM 31,31,80
orl-rat TD: 27 g/kg/36W-C: CAR JJIND8 60,1339,78
orl-rat TD: 27 g/kg/37W-C: CAR IGAYAY 108,96,79
orl-mus LD50: 2690 mg/kg PMDCAY 5,320,67
ipr-mus LD50: 1296 mg/kg PMDCAY 5,320,67
scu-mus LD50: 1602 mg/kg PMDCAY 5,320,67

CONSENSUS REPORTS: IARC Cancer Review: Group 3 IMEMDT 7,56,87; Animal Inadequate Evidence IMEMDT 24,77,80; Human No Adequate Data IMEMDT 24,77,80.

SAFETY PROFILE: Suspected carcinogen with experimental carcinogenic and tumorigenic data. Moderately toxic by ingestion, intraperitoneal, and subcutaneous routes. Mutation data reported. An antibacterial agent. When heated to decomposition it emits toxic fumes of NO_x.

BKH625 CAS:115-84-4 HR: 2
3,3-BIS(HYDROXYMETHYL)HEPTANE
mf: $C_9H_{20}O_2$ mw: 160.29

PROP: Solid. Mp: 40–42.5°, bp: 136–143° @ 8 mm.

SYNS: BEP □ 2-BUTYL-2-ETHYL-1,3-PROPANEDIOL □ 2-ETHYL-2-BUTYL-1,3-PROPANEDIOL

TOXICITY DATA WITH REFERENCE
orl-rat LD50: 5040 mg/kg 34ZIAG -,731,69
skn-rbt LD50: 3810 mg/kg 34ZIAG -,731,69

CONSENSUS REPORTS: Reported in EPA TSCA Inventory.

SAFETY PROFILE: Moderately toxic by skin contact. When heated to decomposition it emits acrid smoke and fumes.

BKH750 HR: 3
BIS HYDROXYMETHYL PEROXIDE
mf: $C_2H_6O_4$ mw: 94.06

SAFETY PROFILE: Highly explosive. Sensitive to friction. When heated to decomposition it emits acrid smoke and fumes. See also PEROXIDES, ORGANIC.

BKI250 CAS:620-92-8 HR: 1
BIS(p-HYDROXYPHENYL)METHANE
mf: $C_{13}H_{12}O_2$ mw: 200.25

PROP: Leaflets from H_2O. Mp: 160°.

SYNS: BIS(4-HYDROXYPHENYL)METHANE □ p,p′-BIS(HYDROXYPHENYL)METHANE □ 4,4′-METHYLENEBISPHENOL □ 4,4′-METHYLENE DIPHENOL

TOXICITY DATA WITH REFERENCE
orl-rat LD50: 4950 mg/kg AIHAAP 23,95,62

CONSENSUS REPORTS: Reported in EPA TSCA Inventory.

SAFETY PROFILE: Mildly toxic by ingestion. When heated to decomposition it emits acrid smoke and irritating fumes.

BKI500 CAS:2971-36-0 HR: 2
2,2-BIS(p-HYDROXYPHENYL)-1,1,1-TRICHLOROETHANE
mf: $C_{14}H_{11}Cl_3O_2$ mw: 317.60

SYN: 1,1,1-TRICHLORO-2,2-BIS(p-HYDROXYPHENYL)ETHANE

TOXICITY DATA WITH REFERENCE
orl-rat LDLo: 2 g/kg JAPMA8 36,349,47
orl-mus LD50: 3200 mg/kg THERAP 22,285,67

CONSENSUS REPORTS: Reported in EPA TSCA Inventory.

SAFETY PROFILE: Moderately toxic by ingestion. When heated to decomposition it emits very toxic fumes such as Cl^-.

BKI750 CAS:65-14-5 HR: 3
2,3-BIS(p-HYDROXYPHENYL)VALERONITRILE
mf: $C_{17}H_{17}NO_2$ mw: 267.35

SYN: SC-3402

TOXICITY DATA WITH REFERENCE
scu-rat TDLo: 7 mg/kg (35D pre): REP RPHRA6 20,395,64
par-rat TDLo: 2500 μg/kg (female 1D post): REP STEDAM 4,657,64
scu-rat TDLo: 250 μg/kg (female 1D post): REP PSEBAA 131,1326,69
scu-rat TDLo: 600 μg/kg (female 2-4D post): REP FESTAS 15,202,64
unr-rat TDLo: 3 mg/kg (female 30D pre): REP RPHRA6 20,395,64
ipr-rat LD50: 70 mg/kg JPETAB 112,176,54
orl-mus LD50: 2850 mg/kg JPETAB 112,176,54
ipr-mus LD50: 93 mg/kg JPETAB 112,176,54
ivn-dog LDLo: 100 mg/kg JPETAB 112,176,54

CONSENSUS REPORTS: Cyanide and its compounds are on the Community Right-To-Know List.

SAFETY PROFILE: Poison by intraperitoneal and intravenous routes. Moderately toxic by ingestion. Experimental reproductive effects. See also NITRILES. When heated to decomposition it emits toxic fumes of NO_x and CN^-.

BKJ250 CAS:62374-53-2 HR: 3
BIS(3-HYDROXY-1-PROPYNYL)MERCURY
mf: $C_6H_6HgO_2$ mw: 310.71

SYN: 3,3′-MERCURIDI-2-PROPYN-1-OL

TOXICITY DATA WITH REFERENCE
ivn-mus LD50: 4500 μg/kg CSLNX* NX#05895

CONSENSUS REPORTS: Mercury and its compounds are on the Community Right-To-Know List.

OSHA PEL: CL 0.1 mg(Hg)/m³ (skin)
ACGIH TLV: TWA 0.1 mg(Hg)/m³ (skin)
NIOSH REL: (Organomercury): TWA 0.01 mg/m³; STEL 0.03 mg/m³ (skin)

SAFETY PROFILE: Poison by intravenous route. See also MERCURY COMPOUNDS. When heated to decomposition it emits toxic vapors of Hg.

BKJ275 CAS:15702-65-5 HR: 3
BIS(8-HYDROXYQUINOLINE-5-SULFONIC ACID) MANGANESE(II)
mf: $C_{18}H_{12}N_2O_8S_2 \cdot Mn$ mw: 503.38

SYNS: BIS(5-SULFO-8-QUINOLINOLATO-N',O⁸) MANGANESE(II) □ MANGANESE, BIS(5-SULFO-8-QUINOLINOLATO)-

TOXICITY DATA WITH REFERENCE
ivn-mus LD50:56 mg/kg CSLNX* NX#01222

OSHA PEL: CL 5 mg(Mn)/m³
ACGIH TLV: TWA 5 mg(Mn)/m³

SAFETY PROFILE: Poison by intravenous route. When heated to decomposition it emits toxic fumes of NO_x, SO_x, and Mn.

BKJ325 CAS:3286-46-2 HR: 3
BISIBUTIAMINE
mf: $C_{32}H_{46}N_8O_6S_2$ mw: 702.98

SYN: O,O'-DIISOBUTYRYLTHIAMINE DISULFIDE

TOXICITY DATA WITH REFERENCE
ipr-rat LD50:660 mg/kg NIIRDN 6,606,82
scu-rat LD50:850 mg/kg NIIRDN 6,606,82
ivn-rat LD50:110 mg/kg NIIRDN 6,606,82

SAFETY PROFILE: Poison by intravenous route. Moderately toxic by some other routes. When heated to decomposition it emits toxic fumes of SO_x and NO_x. See also ESTERS.

BKJ500 CAS:73816-43-0 HR: 3
BIS(3-INDOLEMETHYLENEMORPHOLINIUM)HEXACHLOROSTANNATE
mf: $C_{26}H_{30}N_4O_2 \cdot Cl_6Sn$ mw: 761.99

SYN: MORPHOLINIUM, (3-INDOLYLMETHYLENE)-, HEXACHLOROSTANNATE(2-) (2:1)

TOXICITY DATA WITH REFERENCE
ivn-mus LD50:100 mg/kg CSLNX* NX#02753

OSHA PEL: TWA 2 mg(Sn)/m³
ACGIH TLV: TWA 2 mg(Sn)/m³

SAFETY PROFILE: Poison by intravenous route. When heated to decomposition it emits toxic fumes of NO_x, Sn, and Cl^-.

BKK250 CAS:25168-24-5 HR: 2
BIS(ISOOCTYLOXYCARBONYLMETHYLTHIO)DIBUTYL STANNANE
mf: $C_{28}H_{56}O_4S_2Sn$ mw: 639.65

SYNS: BIS(2-ETHYLHEXYLOXYCARBONYLMETHYLTHIO)DIBUTYLSTANNANE □ DIBUTYL-TIN BIS(ISOOCTYLTHIOGLYCOLLATE) □ DIBUTYLZINN-S,S'-BIS(ISOOCTYLTHIOGLYCOLAT) (GERMAN)

TOXICITY DATA WITH REFERENCE
orl-rat LD50:500 mg/kg TRIPA7 -,1,73

CONSENSUS REPORTS: Reported in EPA TSCA Inventory.

OSHA PEL: TWA 0.1 mg(Sn)/m³ (skin)
ACGIH TLV: TWA 0.1 mg(Sn)/m³ (skin) (Proposed: TWA 0.1 mg(Sn)/m³; STEL 0.2 mg(Sn)/m³ (skin))
NIOSH REL: (Organotin Compounds) TWA 0.1 mg(Sn)/m³

SAFETY PROFILE: Moderately toxic by ingestion. See also TIN COMPOUNDS. When heated to decomposition it emits toxic fumes of SO_x.

For occupational chemical analysis use NIOSH: Organotin Compounds, 5504.

BKK500 CAS:26636-01-1 HR: 2
BIS(ISOOCTYLOXYCARBONYLMETHYLTHIO)DIMETHYLSTANNANE
mf: $C_{22}H_{44}O_4S_2Sn$ mw: 555.47

SYNS: BIS(2-ETHYLHEXYLOXYCARBONYLMETHYLTHIO)DIMETHYLSTANNANE □ DIMETHYL-TIN BIS(ISOOCTYLTHIOGLYCOLLATE) □ DIMETHYLZINN-S,S'-BIS(ISOOCTYLTHIOGLYCOLAT) (GERMAN)

TOXICITY DATA WITH REFERENCE
orl-rat LD50:1380 mg/kg TRIPA7 -,1,73

CONSENSUS REPORTS: Reported in EPA TSCA Inventory.

OSHA PEL: TWA 0.1 mg(Sn)/m³ (skin)
ACGIH TLV: TWA 0.1 mg(Sn)/m³ (skin) (Proposed: TWA 0.1 mg(Sn)/m³; STEL 0.2 mg(Sn)/m³ (skin))
NIOSH REL: (Organotin Compounds) TWA 0.1 mg(Sn)/m³

SAFETY PROFILE: Moderately toxic by ingestion. See also TIN COMPOUNDS. When heated to decomposition it emits toxic fumes of SO_x.

For occupational chemical analysis use NIOSH: Organotin Compounds 5504.

BKK750 CAS:26401-97-8 HR: 2
BIS(ISOOCTYLOXYCARBONYLMETHYLTHIO)DIOCTYL STANNANE
mf: $C_{36}H_{72}O_4S_2Sn$ mw: 751.89

SYNS: ADVASTAB 17 MO □ BIS(MERCAPTOACETATE)DIOCTYLTIN BIS(ISOOCTYL) ESTER □ DIISOOCTYL ((DIOCTYLSTANNYLENE) DITHIO)DIACETATE □ DIOCTYLTIN BIS(ISOOCTYL MERCAPTOACETATE) □ DIOCTYLTIN-S,S'-BIS(ISOOCTYL MERCAPTOACETATE) □ DIOCTYLTIN BIS(ISOOCTYL THIOGLYCOLATE) □ DIOCTYL-TIN BIS (ISOOCTYLTHIOGLYCOLLATE) □ DI-n-OCTYLTIN DIISOOCTYL

THIOGLYCOLATE □ DI-n-OCTYL-ZINN-DI-ISOOCTYLTHIOGLYKOLAT (GERMAN) □ DOTG □ THERMOLITE 831

TOXICITY DATA WITH REFERENCE
orl-rat TDLo:420 mg/kg (1-21D preg):TER TXAPA9 26,253,73
orl-rat LD50:1277 mg/kg ARZNAD 19,934,69
skn-rat LD50:2250 mg/kg ARZNAD 19,934,69

CONSENSUS REPORTS: Reported in EPA TSCA Inventory.

OSHA PEL: TWA 0.1 mg(Sn)/m³ (skin)
ACGIH TLV: TWA 0.1 mg(Sn)/m³ (skin) (Proposed: TWA 0.1 mg(Sn)/m³; STEL 0.2 mg(Sn)/m³ (skin))
DFG MAK: 0.1 mg(Sn)/m³ calculated as total dust
NIOSH REL: (Organotin Compounds) TWA 0.1 mg(Sn)/m³

SAFETY PROFILE: Moderately toxic by ingestion and skin contact. An experimental teratogen. See also TIN COMPOUNDS and MERCAPTANS. When heated to decomposition it emits toxic fumes of SO_x.

For occupational chemical analysis use NIOSH: Organotin Compounds 5504.

BKL000 CAS:33568-99-9 HR: 2
BIS(ISOOCTYLOXYMALEOYLOXY)DIOCTYLSTANNANE
mf: $C_{40}H_{72}O_8Sn$ mw: 799.81

SYNS: (Z,Z)-BIS((3-CARBOXYACRYLOYL)OXY)DIOCTYL-STANNANE DIISOOCTYL ESTER (8CI) □ (Z,Z)-4,4′-((DIOCTYLSTANNYLENE)BIS (OXY))BIS(4-OXO-2-BUTANOIC ACID DIISOOCTYL ESTER □ DIOCTYLTINBIS(ISOOCTYL MALEATE)

TOXICITY DATA WITH REFERENCE
orl-rat LD50:2760 mg/kg TRIPA7 -,1,73

CONSENSUS REPORTS: Reported in EPA TSCA Inventory.

OSHA PEL: TWA 0.1 mg(Sn)/m³ (skin)
ACGIH TLV: TWA 0.1 mg(Sn)/m³ (skin) (Proposed: TWA 0.1 mg(Sn)/m³; STEL 0.2 mg(Sn)/m³ (skin))
DFG MAK: 0.1 mg(Sn)/m³ calculated as total dust
NIOSH REL: (Organotin Compounds) TWA 0.1 mg(Sn)/m³

SAFETY PROFILE: Moderately toxic by ingestion. See also TIN COMPOUNDS. When heated to decomposition it emits acrid smoke and irritating fumes.

For occupational chemical analysis use NIOSH: Organotin Compounds 5504.

BKL250 CAS:7287-19-6 HR: 2
2,4-BIS(ISOPROPYLAMINO)-6-METHYLMERCAPTO-s-TRIAZINE
mf: $C_{10}H_{19}N_5S$ mw: 241.40

PROP: Solid. Mp: 118–120°. Very sltly sol in H_2O.

SYNS: 4,6-BIS(ISOPROPYLAMINO)-2-METHYLMERCAPTO-s-TRIAZINE □ 2,4-BIS(ISOPROPYLAMINO)-6-METHYLTHIO-s-TRIAZINE □ 2,4-BIS(ISOPROPYLAMINO)-6-METHYLTHIO-1,3,5-TRIAZINE □ N,N′-BIS (1-METHYLETHYL)-6-METHYL-THIO-1,3,5-TRIAZINE-2,4-DIAMINE □ CAPAROL □ G 34161 □ GESAGARD □ MERKAZIN □ 2-METHYLMERCAPTO-4,6-BIS(ISOPROPYLAMINO)-s-TRIAZINE □ 2-METHYLTHIO-4,6-BIS(ISOPROPYLAMINO)-s-TRIAZINE □ POLISIN □ PRIMATOL Q □ PROMETREX □ PROMETRIN □ PROMETRYN □ PROMETRYNE (USDA) □ SELEKTIN □ SESAGARD

TOXICITY DATA WITH REFERENCE
eye-rbt 80 mg MLD CIGET* -,-,77
MUT mrc-smc 500 μg/L CYGEDX 21(2),59,87
orl-rat TDLo:2400 mg/kg (female 18-20D post):REP ZHYGAM 34,116,88
orl-rat LD50:2100 mg/kg GISAAA 34(3),94,69
orl-mus LD50:2138 mg/kg GISAAA 33(2),12,68

SAFETY PROFILE: Moderately toxic by ingestion. Experimental reproductive effects. An eye irritant. Mutation data reported. An herbicide. When heated to decomposition it emits very toxic fumes of NO_x and SO_x. See also MERCAPTANS.

BKL500 HR: 2
2,4-BIS(ISOPROPYLAMINO)-6-(METHYLTHIO)-s-TRIAZINE mixed with METHANEARSONIC ACID MONOSODIUM SALT (1:4)

TOXICITY DATA WITH REFERENCE
orl-rat LD50:2500 mg/kg CIGET* -,-,77
skn-rbt LD50:3700 mg/kg CIGET* -,-,77

SAFETY PROFILE: Moderately toxic by ingestion and skin contact. See also ARSENIC COMPOUNDS. When heated to decomposition it emits very toxic fumes of SO_x, As, and NO_x.

BKL750 CAS:3006-93-7 HR: 3
1,3-BISMALEIMIDO BENZENE
mf: $C_{14}H_8N_2O_4$ mw: 268.24

SYNS: 1,3-DIMALEIMIDOBENZENE □ HVA 2 □ HVA-2 CURING AGENT □ M-PHDM □ N,N′-(m-PHENYLENE)BISMALEIMIDE □ 1,1′-(m-PHENYLENE)BIS-1H-PYROLE-2,5-DIONE (9CI) □ N,N′-(m-PHENYLENEDIMALEIMIDE)

TOXICITY DATA WITH REFERENCE
orl-rat LD50:1370 mg/kg GISAAA 40(11),109,75
ihl-rat LC50:55 mg/m³/4H EPASR* 8EHQ-0790-1023S
ipr-rat LDLo:50 mg/kg NCNSA6 5,22,53
orl-rat LD50:1370 mg/kg GISAAA 40(11),109,75
orl-mus LD50:250 mg/kg GISAAA 40(11),109,75

CONSENSUS REPORTS: Reported in EPA TSCA Inventory.

SAFETY PROFILE: Poison by ingestion, inhalation, and intraperitoneal routes. When heated to decomposition it emits toxic fumes of NO_x.

BKL800 CAS:13676-54-5 HR: 3
BIS(4-MALEIMIDOPHENYL)METHANE
mf: $C_{21}H_{14}N_2O_4$ mw: 358.37

SYNS: 4,4′-BIPHENYLMETHANEBISMALEIMIDE □ BISMALEIMIDE S □ 4,4-BIS(MALEIMIDO)DIPHENYLMETHANE □ BIS(p-MALEIMIDOPHENYL)METHANE □ 4,4′-BIS(MALEIMIDOPHENYL)METHANE □ p,p′-DIMALEIMIDOPHENYLMETHANE □ 4,4′-DIMALEIMIDOPHENYLMETHANE □ DIPHENYLMETHANEBISMALEIMIDE □ 4,4′-DIPHENYLMETHANEBISMALEIMIDE □ 4,4′-DIPHENYLMETHANEDIMALEIMIDE □ MALEIMIDE, N,N′-(METHYLENEDI-p-PHENYLENE)DI- □ MB-3000

□ 4,4'-METHYLENEBIS(PHENYLMALEIMIDE) □ 1,1'-(METHYLENEDI-4,1-PHENYLENE)BIS-1H-PYRROLE-2,5-DIONE □ 1H-PYRROLE-2,5-DIONE, 1,1'-(METHYLENEDI-4,1-PHENYLENE)BIS-(9CI) □ XU 292A

TOXICITY DATA WITH REFERENCE
cyt-ham:ovr 2500 µg/L EPASR* 8EHQ-0491-1069
orl-rat LD50:>5 g/kg GISAAA 40(11),109,75
ihl-rat LC50:350 mg/m³/4H EPASR* 8EHQ-0790-1023S
orl-mus LD50:>5 g/kg GISAAA 40(11),109,75

CONSENSUS REPORTS: Reported in EPA TSCA Inventory.

SAFETY PROFILE: A poison by inhalation. Low toxicity by ingestion. Mutation data reported. When heated to decomposition it emits toxic vapors of NO_x.

BKM000 CAS:10193-95-0 HR: 3
BIS(MERCAPTOACETATE)-1,4-BUTANEDIOL
mf: $C_8H_{14}O_4S_2$ mw: 238.34

SYN: BUTYLENE GLYCOL BIS(MERCAPTOACETATE)

TOXICITY DATA WITH REFERENCE
orl-rat LD50:405 mg/kg TRIPA7 -,1,73

CONSENSUS REPORTS: Reported in EPA TSCA Inventory.

SAFETY PROFILE: Poison by ingestion. When heated to decomposition it emits toxic fumes of SO_x. See also MERCAPTANS.

BKM125 CAS:4672-49-5 HR: 3
1,2-BIS(MESYLOXY)ETHANE
mf: $C_4H_{10}O_6S_2$ mw: 218.26

SYNS: 1,2-ETHANEDIOL DIMETHANESULFONATE (9CI) □ 1,2-ETHANEDIYL DIMETHANESULFONATE □ ETHYLENE BIS(METHANESULFONATE) □ ETHYLENE DIMETHANESULFONATE □ ETHYLENE DIMETHANESULPHONATE

TOXICITY DATA WITH REFERENCE
orl-rat TDLo:250 mg/kg (10D male):REP 85GUAJ -,55,66
orl-rat TDLo:153 mg/kg (male 14D pre):REP JMCMAR 24,901,81
ipr-rat TDLo:75 mg/kg (male 1D pre):REP JRPFA4 71,393,84
ipr-rat LD50:150 mg/kg BJPCAL 24,24,65

SAFETY PROFILE: Poison by intraperitoneal route. Experimental reproductive effects. When heated to decomposition it emits toxic fumes of SO_x. See also SULFONATES.

BKM250 CAS:97-90-5 HR: 2
1,2-BIS(METHACRYLOYLOXY)ETHANE
mf: $C_{10}H_{14}O_4$ mw: 198.1

SYNS: AGEFLEX EGDM □ DIGLYCOL DIMETHACRYLATE □ ETHANEDIOL DIMETHACRYLATE □ 1,2-ETHANEDIOL DIMETHACRYLATE □ ETHYLDIOL METACRYLATE □ ETHYLENE GLYCOL BIS(METHACRYLATE) □ ETHYLENE GLYCOL DIMETHACRYLATE □ ETHYLENE METHACRYLATE □ GLYCOL DIMETHACRYLATE □ SARTOMER SR 206 □ SR 206

TOXICITY DATA WITH REFERENCE
msc-mus:lym 5820 µmol/L EMMUEG 17,264,91
orl-rat LD50:3300 mg/kg GTPZAB 24(4),58,80
ipr-rat LD50:2800 mg/kg AMPMAR 36,58,75
orl-mus LD50:2000 mg/kg GTPZAB 24(4),58,80

CONSENSUS REPORTS: Reported in EPA TSCA Inventory.

SAFETY PROFILE: Moderately toxic by ingestion and intraperitoneal routes. Mutation data reported. When heated to decomposition it emits acrid smoke and irritating fumes. See also ESTERS.

BKM500 CAS:1187-00-4 HR: 3
BIS(METHANE SULFONYL)-d-MANNITOL
mf: $C_8H_{18}O_{10}S_2$ mw: 338.38

SYNS: 1,6-BIS-o-METHYLSULFONYL-d-MANNITOL □ CB 2511 □ 1,6-DIMESYL-d-MANNITOL □ 1,6-DIMETHANESULFONATE-d-MANNITOL □ 1,6-DIMETHANE-SULFONOXY-d-MANNITOL □ 1,6-DIMETHANESULPHONOXY-1,6-DIDEOXY-d-MANNITOL □ DMM □ d-MANNITOL BUSULFAN □ MANNITOL MYLERAN □ MANNOGRANOL □ MM □ NSC-37538

TOXICITY DATA WITH REFERENCE
sln-dmg-unk 160 mmol/L ANYAA9 160,228,69
ipr-mus TDLo:1000 mg/kg/4W:NEO JNCIAM 36,915,66
ipr-rat LD50:2000 mg/kg EJCAAH 4,617,68
orl-mus LD50:6000 mg/kg ARZNAD 17,145,67
ivn-dog LDLo:135 mg/kg CCSUBJ 2,203,65
ivn-mky LDLo:135 mg/kg CCSUBJ 2,203,65

CONSENSUS REPORTS: EPA Genetic Toxicology Program.

SAFETY PROFILE: Poison by intravenous route. Moderately toxic by intraperitoneal route. Mildly toxic by ingestion. Questionable carcinogen with experimental neoplastigenic data. Mutation data reported. When heated to decomposition it emits toxic fumes of SO_x.

BKM750 CAS:7306-46-9 HR: 2
3,4-BIS(METHOXY)BENZYL CHLORIDE
mf: $C_9H_{11}ClO_2$ mw: 186.65

SYNS: 3,4-DIMETHOXYBENZYL CHLORIDE □ VERATRYL CHLORID (GERMAN) □ VERATRYL CHLORIDE

TOXICITY DATA WITH REFERENCE
scu-rat TDLo:2100 mg/kg/42W-I:ETA ZEKBAI 74,241,70
orl-rat LD50:4700 mg/kg GTPZAB 26(2),55,82
scu-rat LD50:3000 mg/kg ZEKBAI 74,241,70
orl-mus LD50:5g/kg GTPZAB 26(2),55,82

SAFETY PROFILE: Questionable carcinogen with experimental tumorigenic data. Moderately toxic by subcutaneous route. Mildly toxic by ingestion. When heated to decomposition it emits toxic fumes of Cl^-.

BKN000 CAS:3965-55-7 HR: 2
3,5-BIS(METHOXYCARBONYL)BENZENESULFONIC ACID, SODIUM SALT
mf: $C_{10}H_9O_7S \cdot Na$ mw: 296.24

SYN: 3,5-BIS-METHYLKARBOXY-BENZENSULFONAN SODNY (CZECH)

B

TOXICITY DATA WITH REFERENCE
eye-rbt 100 mg/24H SEV 28ZPAK -,185,72

CONSENSUS REPORTS: Reported in EPA TSCA Inventory.

SAFETY PROFILE: A severe eye irritant. When heated to decomposition it emits toxic fumes of SO_x and Na_2O. See also SULFONATES.

BKN250 CAS:58306-30-2 HR: 2
N-(2-(2,3-BIS-(METHOXYCARBONYL)-GUANIDINO)-5-(PHENYLTHIO)-PHENYL)-2-METHOXYACETAMIDE
mf: $C_{20}H_{22}N_4O_6S$ mw: 446.52

PROP: Solid. Mp: 129–130°.

SYN: FEBANTEL

TOXICITY DATA WITH REFERENCE
orl-rat LD50:10,605 mg/kg ARZNAD 28,2193,78
orl-rbt LD50:1250 mg/kg ARZNAD 28,2193,78

SAFETY PROFILE: Moderately toxic by ingestion. When heated to decomposition it emits very toxic fumes of NO_x and SO_x.

BKN500 CAS:60397-73-1 HR: 1
4,4′-BIS((4-(2-METHOXYETHOXY)-6-(N-METHYL-N-2-SULFOETHYL)AMINO-s-TRIAZIN-2-YL)AMINO)-2,2′-STILBENEDISULFONIC ACID
mf: $C_{32}H_{42}N_{10}O_{16}S_4$ mw: 951.08

TOXICITY DATA WITH REFERENCE
eye-rbt 100 mg MOD MVCRB3 2,193,73

SAFETY PROFILE: An eye irritant. See also SULFONATES. When heated to decomposition it emits very toxic fumes of NO_x and SO_x.

BKN750 CAS:111-96-6 HR: 3
BIS(2-METHOXY ETHYL)ETHER
mf: $C_6H_{14}O_3$ mw: 134.48

PROP: Liquid. Bp: 162°, d: 0.9451, mp: −68°, flash p: 158°F (70°C) (OC), n (20/D) 1.4097. Misc with water, alc, ether, and hydrocarbon solvents.

SYNS: DIETHYLENE GLYCOL DIMETHYL ETHER □ DIETHYL GLYCOL DIMETHYL ETHER □ DIGLYME

TOXICITY DATA WITH REFERENCE
sln-dmg-ihl 250 ppm/165M NTIS** PB83-138198
dlt-rat-ihl 1000 ppm/5D-C NTIS** PB83-138198
spm-mus-ihl 1000 ppm/5D-C NTIS** PB83-138198
orl-mus TDLo:1250 mg/kg (female 6-15D post):TER NTIS** PB86-135233
orl-mus TDLo:5 g/kg (female 6-15D post):REP NTIS** PB86-135233
orl-mus TDLo:2500 mg/kg (female 6-15D post):REP NTIS** PB86-135233
orl-mus TDLo:537 mg/kg (female 11D post):TER TJADAB 35,321,87

SAFETY PROFILE: An experimental teratogen. Other experimental reproductive effects. Mutation data reported. When heated to decomposition it emits toxic fumes of NO_x. Readily forms explosive peroxides upon exposure to air, light, or heat. Solution containing carbon dioxide may react with aluminum hydride to form an explosive product. Other metal hydrides may react similarly. See also ETHERS.

BKO000 CAS:67856-65-9 HR: 2
BIS(2-METHOXYETHYL)NITROSOAMINE
mf: $C_6H_{14}N_2O_3$ mw: 162.22

SYN: N-NITROSOBIS(2-METHOXYETHYL)AMINE

TOXICITY DATA WITH REFERENCE
orl-rat TDLo:2750 mg/kg/50W-I:ETA CNREA8 38,2391,78

SAFETY PROFILE: Questionable carcinogen with experimental tumorigenic data. When heated to decomposition it emits toxic fumes of NO_x. See also NITROSAMINES.

BKO250 CAS:15546-11-9 HR: 3
BIS(METHOXYMALEOYLOXY)DIBUTYLSTANNANE
mf: $C_{18}H_{28}O_8Sn$ mw: 491.15

SYNS: DIBUTYLBIS((3-CARBOXYACRYLOYL)OXY)-STANNANE DIMETHYL ESTER (Z,Z) (8CI) □ DIBUTYLTIN BIS(METHYL MALEATE) □ DIBUTYLTIN BIS(MONOMETHYL MALEATE) □ DIBUTYLTIN METHYL MALEATE □ 6,6-DIBUTYL-4,8,11-TRIOXO-5,7,12-TRIOXA-6-STANNATRIDECA-2,9-DIENOIC ACID METHYL ESTER □ DI-n-BUTYL-ZINN-DIMONOMETHYLMALEINAT (GERMAN) □ STAN-GUARD 156

TOXICITY DATA WITH REFERENCE
orl-rat LD50:62 mg/kg TRIPA7 -,1,73

CONSENSUS REPORTS: Reported in EPA TSCA Inventory.

OSHA PEL: TWA 0.1 mg(Sn)/m³ (skin)
ACGIH TLV: TWA 0.1 mg(Sn)/m³ (skin) (Proposed: TWA 0.1 mg(Sn)/m³; STEL 0.2 mg(Sn)/m³ (skin))
NIOSH REL: (Organotin Compounds) TWA 0.1 mg(Sn)/m³

SAFETY PROFILE: Poison by ingestion. See also TIN COMPOUNDS. When heated to decomposition it emits acrid smoke and irritating fumes.

For occupational chemical analysis use NIOSH: Organotin Compounds 5504.

BKO500 CAS:60494-19-1 HR: 2
BIS(METHOXYMALEOYLOXY)DIOCTYLSTANNANE
mf: $C_{26}H_{44}O_8Sn$ mw: 603.39

SYN: DI-n-OCTYLZINN-DIMONOMETHYLMALEINAT (GERMAN)

TOXICITY DATA WITH REFERENCE
orl-rat LD50:1673 mg/kg TRIPA7 -,1,73

OSHA PEL: TWA 0.1 mg(Sn)/m³ (skin)
ACGIH TLV: TWA 0.1 mg(Sn)/m³ (skin) (Proposed: TWA 0.1 mg(Sn)/m³; STEL 0.2 mg(Sn)/m³ (skin))
NIOSH REL: (Organotin Compounds) TWA 0.1 mg(Sn)/m³

SAFETY PROFILE: Moderately toxic by ingestion. See also TIN COMPOUNDS. When heated to decomposition it emits acrid smoke and irritating fumes.

For occupational chemical analysis use NIOSH: Organotin Compounds 5504.

BKO600 **CAS:101-70-2** **HR: 2**
BIS(4-METHOXYPHENYL)AMINE
mf: $C_{14}H_{15}NO_2$ mw: 229.30

SYNS: BENZENAMINE, 4-METHOXY-N-(4-METHOXYPHENYL)- □ 4-BIPHENYLAMINE, 4,4′-DIMETHOXY- □ BIS(p-ANISYLAMINE) □ BIS (p METHOXYPHENYL)AMINE □ DI-p-ANISYLAMINE □ p,p′-DIMETHOXYDIPHENYLAMINE □ 4,4′-DIMETHOXYDIPHENYLAMINE □ DI-p-METHOXYPHENYLAMINE □ TERMOFLEKS A

TOXICITY DATA WITH REFERENCE
cyt-ham:lng 30 mg/L MUREAV 241,175,90
orl-rat LD50:2470 mg/kg KCRZAE 26(9),28,67
orl-mus LD50:2500 mg/kg KCRZAE 26(9),28,67

CONSENSUS REPORTS: Reported in EPA TSCA Inventory.

SAFETY PROFILE: Moderately toxic by ingestion. Mutation data reported. When heated to decomposition it emits toxic vapors of NO_x.

BKO750 **CAS:7342-13-4** **HR: 2**
4,4′-BIS(4-METHOXY-6-PHENYLAMINO-2-s-TRIAZINYLAMINO)-2,2′-STILBENEDISULFONIC ACID
mf: $C_{34}H_{30}N_{10}O_8S_2$ mw: 770.86

SYN: RYLUX PRS (CZECH)

TOXICITY DATA WITH REFERENCE
eye-rbt 100 mg/24H SEV 28ZPAK ,251,72

CONSENSUS REPORTS: Reported in EPA TSCA Inventory.

SAFETY PROFILE: A severe eye irritant. When heated to decomposition it emits very toxic fumes of NO_x and SO_x.

BKP000 **CAS:69352-67-6** **HR: 3**
1,5-BIS(o-METHOXYPHENYL)-3,7-DIAZAADMANTAN-9-ONE
mf: $C_{22}H_{24}N_2O_3$ mw: 364.48

TOXICITY DATA WITH REFERENCE
ipr-mus LD50:20 mg/kg JMPCAS 5,1293,62
ivn-rbt LD50:66 mg/kg JMPCAS 5,1293,62

SAFETY PROFILE: Poison by intraperitoneal and intravenous routes. When heated to decomposition it emits toxic fumes of NO_x.

BKP200 **CAS:24407-55-4** **HR: 3**
BIS-(B-o-METHOXYPHENYL-ISOPROPYL)-AMINE LACTATE
mf: $C_{20}H_{27}NO_2 \cdot C_3H_6O_3$ mw: 403.57

SYNS: o,o′-DIMETHOXY-α-α′-DIMETHYL-DIPHENETHYLAMINE compounded with LACTIC ACID □ DIPHENETHYLAMINE, o,o′-DIMETHOXY-α-α′-DIMETHYL-, compounded with LACTIC ACID □ U-0045

TOXICITY DATA WITH REFERENCE
skn-rbt 2500 ppm MLD AIPTAK 137,410,62
eye-rbt 1 pph AIPTAK 137,410,62
ipr-mus LD50:51,200 μg/kg AIPTAK 137,410,62

SAFETY PROFILE: Poison by intraperitoneal route. A skin and eye irritant. When heated to decomposition it emits toxic fumes of NO_x.

BKP500 **CAS:2475-44-7** **HR: 3**
1,4-BIS(METHYLAMINO)-9,10-ANTHRACENEDIONE
mf: $C_{16}H_{14}N_2O_2$ mw: 266.32

SYNS: ANTHRAQUINONE, 1,4-BIS(METHYLAMINO)- □ C.I. DISPERSE BLUE 78 □ C.I. SOLVENT BLUE 78 □ C.I. SOLVENT BLUE 93 □ DIARESIN BLUE K □ DISPERSE BLUE 78 □ DISPERSE BLUE 110 □ MACROLEX BLUE FR □ SOLVENT BLUE 78 □ SOLVENT BLUE 93

TOXICITY DATA WITH REFERENCE
mma-sat 5 mg/plate EMMUEG 19(Suppl 20),8,92
ivn-mus LD50:180 mg/kg CSLNX* NX#01356

CONSENSUS REPORTS: Reported in EPA TSCA Inventory.

SAFETY PROFILE: Poison by intravenous route. Mutation data reported. When heated to decomposition it emits toxic vapors of NO_x.

BKQ250 **CAS:65210-37-9** **HR: 3**
N,N′-(BIS(2-(2-METHYL-1,3-BENZODIOXOL-2-YL)ETHYL))ETHYLENEDIAMINE DIHYDROCHLORIDE
mf: $C_{22}H_{28}N_2O_4 \cdot 2ClH$ mw: 457.44

TOXICITY DATA WITH REFERENCE
ivn-rat LD50:20 mg/kg EJMCA5 12,413,77
ipr-mus LD50:90 mg/kg EJMCA5 12,413,77

SAFETY PROFILE: Poison by intravenous and intraperitoneal routes. When heated to decomposition it emits very toxic fumes of HCl and NO_x.

BKQ500 **CAS:10024-74-5** **HR: 2**
BIS(α-METHYLBENZYL)AMINE
mf: $C_{16}H_{19}N$ mw: 225.36

PROP: Liquid. Mp: −65°, bp: 188.5°, flash p: 175°F (OC), d: 0.9535, vap press: 0.5 mm @ 20°, vap d: 4.18.

TOXICITY DATA WITH REFERENCE
skn-rbt 100 μg/24H open AIHAAP 23,95,62
orl-rat LD50:2930 mg/kg AIHAAP 23,95,62

SAFETY PROFILE: Moderately toxic by ingestion and skin contact. Combustible when exposed to heat or flame. To fight fire, use alcohol foam, CO_2, dry chemical. Incompatible with oxidizers. When heated to decomposition it emits toxic fumes of NO_x.

BKQ750 **CAS:74927-02-9** **HR: 3**
2,6-BIS(1-METHYLBUTYL)PHENOL
mf: $C_{16}H_{26}O$ mw: 234.2

TOXICITY DATA WITH REFERENCE
ivn-mus LD50:160 mg/kg JMCMAR 23,1350,80
ivn-rbt LDLo:30 mg/kg JMCMAR 23,1350,80

SAFETY PROFILE: Poison by intravenous route. When heated to decomposition it emits acrid smoke and irritating fumes.

BKR000 CAS:63982-52-5 HR: 3
1,4-BIS(METHYLCARBAMYLOXY)-2-ISOPROPYL-5-METHYLBENZENE
mf: $C_{14}H_{20}N_2O_4$ mw: 280.36

SYN: TL-1350

TOXICITY DATA WITH REFERENCE
orl-rat LD50:60 mg/kg 85ALAU -,107,76
orl-mus LD50:40 mg/kg 85ALAU -,107,76
scu-mus LDLo:20 mg/kg NTIS** PB158-508

SAFETY PROFILE: Poison by ingestion and subcutaneous routes. When heated to decomposition it emits toxic fumes of NO_x.

BKR100 CAS:68555-34-0 HR: 1
BIS((6-METHYL-3-CYCLOHEXEN-1-YL)METHYL) ESTER HEXANEDIOIC ACID
mf: $C_{22}H_{34}O_4$ mw: 362.56

SYNS: ADIPIC ACID, DIESTER with 6-METHYL-3-CYCLOHEXENE-1-METHANOL ☐ ADIPIC ACID, 6-METHYL-3-CYCLOHEXENYL-METHANOL DIESTER

TOXICITY DATA WITH REFERENCE
skn-rbt LD50:>16 g/kg TXAPA9 28,313,74

CONSENSUS REPORTS: Reported in EPA TSCA Inventory.

SAFETY PROFILE: Low toxicity by skin contact. When heated to decomposition it emits acrid smoke and irritating vapors.

BKR250 CAS:66903-23-9 HR: 2
BIS(3-METHYLCYCLOHEXYL PEROXIDE)
mf: $C_{14}H_{22}O_4$ mw: 254.36

SYN: 3-METHYLCYKLOHEXANONPEROXID (CZECH)

TOXICITY DATA WITH REFERENCE
skn-rbt 500 mg/24H SEV 28ZPAK -,142,72
eye-rbt 250 μg/24H SEV 28ZPAK -,142,72
orl-rat LD50:1500 mg/kg 28ZPAK -,142,72

SAFETY PROFILE: Moderately toxic by ingestion. A severe eye and skin irritant. When heated to decomposition it emits acrid smoke and irritating fumes. See also PEROXIDES, ORGANIC.

BKR500 CAS:64246-03-3 HR: 3
1,1-BIS((3,4-METHYLENEDIOXYPHENOXY)METHYL)-N,N-DIMETHYL-1-BUTANOL CITRATE
mf: $C_{22}H_{27}NO_7 \cdot C_6H_8O_7$ mw: 609.64

SYN: 1,3-BIS-(3,4-METILENDIOSSIFENOSSI)-2-(3-DIMETILAMINOPROPIL)PROPAN-2-OLO CITRATO (ITALIAN)

TOXICITY DATA WITH REFERENCE
orl-mus LD50:780 mg/kg FRPSAX 32,502,77
ivn-mus LD50:94 mg/kg FRPSAX 32,502,77

SAFETY PROFILE: Poison by intravenous route. Moderately toxic by ingestion. When heated to decomposition it emits toxic fumes of NO_x.

BKR750 CAS:64246-13-5 HR: 3
α,α-BIS((3,4-(METHYLENEDIOXY)PHENOXY)METHYL)-1-PIPERIDINEBUTANOLACETATE CITRATE
mf: $C_{27}H_{33}NO_8 \cdot C_6H_8O_7$ mw: 691.75

TOXICITY DATA WITH REFERENCE
orl-mus LD50:700 mg/kg FRPSAX 32,502,77
ivn-mus LD50:32 mg/kg FRPSAX 32,502,77

SAFETY PROFILE: Poison by intravenous route. Moderately toxic by ingestion. When heated to decomposition it emits toxic fumes of NO_x.

BKS500 HR: 3
1-(1,3-BIS(3,4-(METHYLENEDIOXY)PHENOXY)-2-PROPYL)PYRROLIDINE CITRATE
mf: $C_{21}H_{23}NO_6 \cdot C_6H_8O_7$ mw: 577.59

SYN: 1,3-BIS-(3,4-METILENDIOSSIFENOSSI)-2-PIRROLIDINOPROPANO CITRATO (ITALIAN)

TOXICITY DATA WITH REFERENCE
orl-mus LD50:195 mg/kg FRPSAX 32,502,77
ivn-mus LD50:58 mg/kg FRPSAX 32,502,77

SAFETY PROFILE: Poison by ingestion and intravenous routes. When heated to decomposition it emits toxic fumes of NO_x.

BKS750 CAS:26087-47-8 HR: 3
O,O-BIS(1-METHYLETHYL)-S-(PHENYLMETHYL) PHOSPHOROTHIOATE
mf: $C_{13}H_{21}O_3PS$ mw: 288.37

PROP: Yellow oil. Bp: 126° @ 0.04.

SYNS: O,O-DIISOPROPYL-S-BENZYL PHOSPHOROTHIOLATE ☐ O,O-DIISOPROPYL-S-BENZYL THIOPHOSPHATE ☐ IBP ☐IPROBEN-FOS ☐ KITAZIN L ☐ KITAZIN P ☐ RICID II ☐ RICID P

TOXICITY DATA WITH REFERENCE
orl-rat LD50:550 mg/kg GISAAA 50(11),75,85
ihl-rat LC50:2836 mg/m³ GISAAA 51(9),77,86
skn-rat LD50:3708 mg/kg GISAAA 51(9),77,86
ipr-rat LD50:220 mg/kg TOIZAG 29,51,82
scu-rat LD50:525 mg/kg TOIZAG 29,51,82
orl-mus LD50:435 mg/kg GISAAA 50(11),75,85
skn-mus LD50:4 g/kg PEMNDP 9,500,91
ipr-mus LD50:335 mg/kg TOIZAG 29,51,82
scu-mus LD50:1590 mg/kg TOIZAG 29,51,82

SAFETY PROFILE: Poison by intraperitoneal routes. Moderately toxic by ingestion, skin contact, and subcutaneous routes. When heated to decomposition it emits very toxic fumes of SO_x and PO_x. See also ESTERS.

BKS800 CAS:78313-59-4 HR: 3
BIS(2-METHYL-3-HYDROXY-4-METHOXYMETHYL-5-METHYLPYRIDYL)DISULFIDE DIHYDROCHLORIDE
mf: $C_{18}H_{24}N_2O_4S_2 \cdot 2ClH$ mw: 469.48

TOXICITY DATA WITH REFERENCE
orl-mus LD50:1450 mg/kg PCJOAU 15,79,81
ipr-mus LD50:434 mg/kg PCJOAU 15,79,81
ivn-mus LD50:111 mg/kg PCJOAU 15,79,81

SAFETY PROFILE: Poison by intravenous route. Moderately toxic by ingestion and other routes. When heated to decomposition it emits toxic fumes of NO_x, SO_x, and HCl.

BKS810 CAS:3810-81-9 HR: 3
BIS(METHYLMERCURIC)SULFATE
mf: $C_2H_6Hg_2O_4S$ mw: 527.32

PROP: Platelets from water. Sltly sol in EtOH. Mp: 255° (decomp).

SYNS: ARETAN-NIEUW □ B 4992 □ BIS-(METHYLMERCURY)-SULFATE □ BIS-(METHYLMERKURI)SULFAT □ CERESAN UNIVERSAL-FEUCHTBEIZE □ CEREWET □ COMPOUND-4992 □ MERCURY, SULFATOBIS(METHYL- □ METHYLMERCURIC SULFATE □ SULFURIC ACID, BIS(METHYLMERCURY) SALT

TOXICITY DATA WITH REFERENCE
orl-rat LD50:50 mg/kg FMCHA2-,C63,89

OSHA PEL: TWA 0.01 mg(Hg)/m^3; STEL 0.03 mg/m^3 (skin)
ACGIH TLV: TWA 0.01 mg(Hg)/m^3; STEL 0.03 mg(Hg)/m^3

SAFETY PROFILE: Poison by ingestion. When heated to decomposition it emits toxic fumes of SO_x and Hg.

BKS825 CAS:13018-50-3 HR: 3
N,N'-BIS(1-METHYL-4-PHENYL-4-PIPERIDYLMETHYL) SEBACAMIDE
mf: $C_{36}H_{54}N_4O_2$ mw: 574.94

SYNS: N,N'-BIS((1-METHYL-4-PHENYL-4-PIPERIDINYL)METHYL)-DECANEDIAMIDE (9CI) □ DIAMMIDE SEBACICA della 4-FENIL-4-AMMINOMETIL-N-METILPIPERIDINA (ITALIAN) □ 1665 I.S.

TOXICITY DATA WITH REFERENCE
ipr-rat LD50:17,500 mg/kg FRPSAX 17,24,62
ivn-rat LD50:12 mg/kg FRPSAX 17,24,62
ipr-mus LD50:5 mg/kg FRPSAX 17,24,62
ivn-mus LD50:2820 μg/kg CSLNX* NX#12224
ipr-gpg LD50:40 mg/kg FRPSAX 17,24,62
ivn-gpg LD50:4 mg/kg FRPSAX 17,24,62

SAFETY PROFILE: Poison by intravenous and intraperitoneal routes. When heated to decomposition it emits toxic fumes of NO_x.

BKT250 HR: 3
BIS(2-METHYL PYRIDINE)SODIUM
mf: $C_{12}H_{14}N_2Na$ mw: 209.25

SAFETY PROFILE: Ignites spontaneously in air. When heated to decomposition it emits toxic fumes of NO_x and Na_2O.

BKU000 CAS:73696-64-7 HR: 2
N,N'-BIS(3-METHYL-2-THIAZOLIDINYLIDENE)UREA
mf: $C_9H_{14}N_4S_2O$ mw: 258.39

TOXICITY DATA WITH REFERENCE
orl-mus LD50:2000 mg/kg JMCMAR 23,773,80
ipr-mus LD50:1140 mg/kg JMCMAR 23,773,80

SAFETY PROFILE: Moderately toxic by ingestion and intraperitoneal routes. When heated to decomposition it emits very toxic fumes of SO_x and NO_x.

BKU250 CAS:14024-75-0 HR: 3
BIS(4-MORPHOLINECARBODITHIOATO)MERCURY
mf: $C_{10}H_{16}HgN_2O_2S_4$ mw: 525.11

TOXICITY DATA WITH REFERENCE
ivn-mus LD50:56 mg/kg CSLNX* NX#02530

CONSENSUS REPORTS: Mercury and its compounds are on the Community Right-To-Know List.

OSHA PEL: CL 0.1 mg(Hg)/m^3 (skin)
ACGIH TLV: TWA 0.1 mg(Hg)/m^3 (skin)
NIOSH REL: (Mercury, Aryl And Inorganic): CL 0.1 mg/m^3 (skin)

SAFETY PROFILE: Poison by intravenous route. See also MERCURY COMPOUNDS. When heated to decomposition it emits very toxic fumes of SO_x and NO_x and Hg vapors.

BKU500 CAS:103-34-4 HR: 3
N,N'-BISMORPHOLINE DISULFIDE
mf: $C_8H_{16}N_2O_2S_2$ mw: 236.38

PROP: Tan to gray powder or crystals. Mp: 124–125°, d: 1.36 @ 25°.

SYNS: ACCEL R □ BISMORPHOLINO DISULFIDE □ DIMORPHOLINE DISULFIDE □ DIMORPHOLINO DISULFIDE □ DITHIOBISMORPHOLINE □ 4,4'-DITHIOBIS(MORPHOLINE) □ N,N-DITHIODIMORPHOLINE □ 4,4'-DITHIODIMORPHOLINE □ 4,4'-DITHIOMORPHOLINE □ MORPHOLINE DISULFIDE □ MORPHOLINODISULFIDE □ SULFASAN □ SULFASAN R POWDER □ USAF B-17 □ USAF EK-T-6645

TOXICITY DATA WITH REFERENCE
mma-sat 100 μg/plate PCBRD2 141,407,84
dnr-bcs 1 mg/disc SAIGBL 26,147,84
orl-rat LD50:4300 mg/kg GISAAA 51(12),67,86
orl-mus LD50:1660 mg/kg ARZNAD 11,797,61
ihl-mus LC50:1624 mg/m^3 GISAAA 53(3),90,88
ipr-mus LD50:50 mg/kg NTIS** AD277-689
ivn-mus LD50:100 mg/kg CSLNX* NX#02252

CONSENSUS REPORTS: Reported in EPA TSCA Inventory.

SAFETY PROFILE: Poison by intraperitoneal and intravenous routes. Moderately toxic by ingestion. Mutation data reported. See also MORPHOLINE. When heated to decomposition it emits very toxic fumes of NO_x and SO_x.

BKU750 CAS:7440-69-9 HR: 3
BISMUTH
af: Bi aw: 208.98

PROP: Hexagonal silver-white or reddish metallic crystals. Mp: 271.3°, bp: 1420–1560°, d: 9.80, vap press: 1 mm @ 1021°.

SYN: BISMUTH-209

TOXICITY DATA WITH REFERENCE
unr-man LDLo:221 mg/kg 85DCAI 2,73,70

CONSENSUS REPORTS: Reported in EPA TSCA Inventory.

SAFETY PROFILE: Poisonous to humans. See also BISMUTH COMPOUNDS. Flammable when exposed to flame. Reaction with $[Bi(OH)_3 + Al(OH)_3]$, coprecipitated and H_2 reduced produces a spontaneously flammable product. Moderately dangerous, can react with acid or acid fumes to emit toxic fumes. Incompatible with Al; BrF_3; acids; NOF; NH_4NO_3; $HClO_3$; Cl_2; IF_5; HNO_3; $HClO_4$.

BKV000 **HR: 3**
BISMUTH AMIDE OXIDE
mf: BiH_2NO mw: 241

SAFETY PROFILE: Stable in liquid NH_3. Very unstable when free of NH_3. Upon decomposition it emits toxic fumes of Bi and NO_x. See also BISMUTH COMPOUNDS.

BKV250 **CAS:12001-47-7** **HR: 3**
BISMUTH ARSPHENAMINE SULFONATE
mf: $C_{21}H_{24}As_3Bi_2N_3O_{12}S_3 \cdot 3Na$ mw: 1318.35

SYNS: BISMARSEN □ SULFARSPHENAMINE BISMUTH

TOXICITY DATA WITH REFERENCE
ims-rat LDLo:500 mg/kg ADSYAF 28,389,33
ipr-mus LDLo:128 mg/kg CBCCT* 2,241,50
ims-rbt LDLo:150 mg/kg ADSYAF 28,389,33

CONSENSUS REPORTS: Arsenic and its compounds are on the Community Right-To-Know List.

OSHA PEL: TWA 0.5 mg(As)/m^3

SAFETY PROFILE: A poison by intraperitoneal and intramuscular routes. See also ARSENIC COMPOUNDS and BISMUTH COMPOUNDS. When heated to decomposition it emits very toxic fumes of Na_2O, NO_x, SO_x, As, and Bi.

BKV750 **HR: 3**
BISMUTH COMPOUNDS

SAFETY PROFILE: Bismuth and its salts can cause kidney damage, although the degree of such damage is usually mild. Large doses can be fatal. Industrially it is considered one of the less toxic of the heavy metals, although intoxication has occurred from its use in medicine. The similarity between the pharmacologic and toxic behavior of lead and bismuth has been pointed out in the literature. Like lead, bismuth may be liberated from tissue deposits during periods of acidosis. Serious and sometimes fatal poisoning may occur from the injection of large doses into closed cavities and from extensive application to burns. Death of animals from bismuth nephritis following injections of soluble salts occurs within several hours to 24 days, the time being generally inversely proportional to the dose, and it appears to be in the order of 5–10 times higher than the dose by slow intravenous injection for rabbits. It is stated that the administration of bismuth should be stopped when gingivitis appears, for otherwise serious ulcerative stomatitis is likely to result. Other toxic results may develop, such as malaise, albuminuria, diarrhea, skin reactions, and sometimes serious exodermatitis. Industrial bismuth poisoning has not been reported, although bismuth absorbed in industrial cases may complicate a diagnosis of plumbism, since the dark line in the gums, which is often present in lead poisoning, is also produced by bismuth. All bismuth compounds do not have equal toxicity. See also individual entries.

Treatment and Antidotes: Personnel showing some of the symptoms noted above, which might indicate that they were absorbing too much bismuth into the body, should be removed from exposure as soon as possible. Get medical advice. Personnel should be cautioned against careless handling of these materials.

BKW000 **CAS:21260-46-8** **HR: 1**
BISMUTH DIMETHYL DITHIOCARBAMATE
mf: $C_9H_{18}N_3S_6 \cdot Bi$ mw: 569.64

SYNS: BISMATE □ TRIS(DIMETHYLDITHIOCARBAMATO)BISMUTH

TOXICITY DATA WITH REFERENCE
scu-mus TDLo:1000 mg/kg:ETA NTIS** PB223-159
orl-mus LD50:20 g/kg RCTEA4 44,512,71

CONSENSUS REPORTS: Reported in EPA TSCA Inventory.

SAFETY PROFILE: Low toxicity by ingestion. Questionable carcinogen with experimental tumorigenic data. See also BISMUTH COMPOUNDS and CARBAMATES. When heated to decomposition it emits very toxic fumes of SO_x and NO_x.

BKW100 **CAS:1304-85-4** **HR: 1**
BISMUTH HYDROXIDE NITRATE OXIDE
mf: $Bi_5H_9N_4O_{22}$ mw: 1462.03

SYNS: BASIC BISMUTH NITRATE □ BISMUTH MAGISTERY □ BISMUTH SUBNITRATE □ BISMUTH SUBNITRICUM □ BISMUTH WHITE □ BISMUTHYL NITRATE □ BLANC DE FARD □ C.I. 77169 □ C.I. PIGMENT WHITE 17 □ COSMETIC WHITE □ FLAKE WHITE □ MAGISTERY OF BISMUTH □ NOVISMUTH □ PAINT WHITE □ SNOWCAL 5SW □ SPANISH WHITE □ VICALIN

TOXICITY DATA WITH REFERENCE
orl-inf TDLo:259 mg/kg:BLD JAMAAP 133,1280,47
orl-inf LDLo:1 g/kg 34ZIAG -,134,69

CONSENSUS REPORTS: Reported in EPA TSCA Inventory.

SAFETY PROFILE: Human systemic effects by ingestion: methemoglobinemia and carboxyhemoglobin. When heated to decomposition it emits toxic vapors of NO_x and Bi.

BKW250 **CAS:10361-44-1** **HR: 3**
BISMUTH NITRATE
mf: BiN_3O_9 mw: 395.01

PROP: Triclinic, colorless, sltly hygroscopic crystals. Bp: $-5H_2O$ @ 80°, d: 2.83, mp: 30° (decomp).

SYN: NITRIC ACID, BISMUTH(3+) SALT

TOXICITY DATA WITH REFERENCE
itt-rat TDLo:31,601 μg/kg (male 1D pre):REP JRPFA4 7,21,64
ipr-mus LDLo:2500 mg/kg APFRAD 34,173,76
ivn-mus LDLo:21 mg/kg APFRAD 34,173,76

CONSENSUS REPORTS: Reported in EPA TSCA Inventory.

SAFETY PROFILE: Poison by intravenous route. Moderately toxic by intraperitoneal route. Experimental reproductive effects. When heated to decomposition it emits toxic fumes of Bi and NO_x. See also BISMUTH COMPOUNDS and NITRATES.

BKW500 CAS:12232-97-2 HR: 3
BISMUTH NITRIDE
mf: BiN mw: 222.99

SAFETY PROFILE: Very unstable; explodes when shaken, heated, or on contact with water and dilute acids. When heated to decomposition it emits toxic fumes of Bi and NO_x. See also BISMUTH COMPOUNDS and NITRIDES.

BKW600 CAS:1304-76-3 HR: 1
BISMUTH OXIDE
mf: Bi_2O_3 mw: 465.96

SYNS: BISMUTHOUS OXIDE ☐ BISMUTH(3+) OXIDE ☐ BISMUTH SESQUIOXIDE ☐ BISMUTH TRIOXIDE ☐ BISMUTH YELLOW ☐ C.I. 77160 ☐ DIBISMUTH TRIOXIDE

TOXICITY DATA WITH REFERENCE
orl-rat LD50:5 g/kg GTPZAB 30(6),16,86
orl-mus LD50:10 g/kg GTPZAB 30(6),16,86

CONSENSUS REPORTS: Reported in EPA TSCA Inventory.

SAFETY PROFILE: Slightly toxic by ingestion. When heated to decomposition it emits toxic vapors of Bi.

BKW750 CAS:7787-62-4 HR: 3
BISMUTH PENTAFLUORIDE
mf: BiF_5 mw: 303.98

PROP: Crystals or long white needles. Very sensitive to moisture. Mp: 154°, bp: 230°. Sol in FSO_3H. Sublimes @ 550°.

SAFETY PROFILE: An irritant poison via ingestion and inhalation routes. Decomposes vigorously and sometimes ignites on contact with moisture to yield O_3 and bismuth trifluoride. Very dangerous. Reacts violently with water and petrolatum above 50°, and acids at room temperature, liberating much heat and ozone. When heated to decomposition it emits highly toxic fumes of F^-. See also FLUORIDES and OZONE.

BKW850 CAS:32707-10-1 HR: 3
BISMUTH PERCHLORATE
mf: $BrClO_4$ mw: 308.43

PROP: Red liquid. Mp: −78°.

SAFETY PROFILE: A shock-sensitive explosive. When heated to decomposition it emits toxic fumes of Bi and Cl^-. See also BISMUTH COMPOUNDS and PERCHLORATES.

BKX000 HR: 3
BISMUTH PLUTONIDE
mf: BiPu mw: 451

SAFETY PROFILE: Ignites spontaneously in air. All plutonium compounds are extremely dangerous; when heated to decomposition it emits toxic fumes of Pu and Bi. See also BISMUTH COMPOUNDS and PLUTONIUM COMPOUNDS.

BKX250 CAS:63732-98-9 HR: 3
BISMUTH POTASSIUM SODIUM TARTRATE (SOLUBLE)

SYNS: SODIUM POTASSIUM BISMUTH TARTRATE (SOLUBLE) ☐ SOLUBLE TARTRO-BISMUTHATE ☐ TREPOL (FRENCH)

TOXICITY DATA WITH REFERENCE
ims-rbt LD50:55 mg/kg JPETAB 87,119,46
ims-mam LDLo:3000 mg/kg JPETAB 28,109,26

SAFETY PROFILE: Poison by intramuscular route. See also BISMUTH COMPOUNDS. When heated to decomposition it emits toxic fumes of oxides of Na_2O, K_2O, and Bi.

BKX500 CAS:5806-84-8 HR: 3
BISMUTH SODIUM-p-AMINOPHENYLARSONATE
mf: $C_6H_6AsNO_3$•BiO•Na mw: 463.02

SYNS: p-ARSANILIC ACID, BISMUTH, SODIUM SALT ☐ ARSENOBISMULAK

TOXICITY DATA WITH REFERENCE
ivn-rat LD50:631 mg/kg UCREAR 48,183,44
ims-rat LDLo:875 mg/kg UCREAR 48,183,44
ivn-rbt LD50:312 mg/kg UCREAR 48,183,44
ims-rbt LDLo:750 mg/kg UCREAR 48,183,44

CONSENSUS REPORTS: Arsenic and its compounds are on the Community Right-To-Know List.

OSHA PEL: TWA 0.5 mg(As)/m^3

SAFETY PROFILE: Poison by intravenous route. Moderately toxic by intramuscular route. See also BISMUTH COMPOUNDS and ARSENIC COMPOUNDS. When heated to decomposition it emits very toxic fumes of As, Bi, Na_2O and NO_x.

BKX750 CAS:150-49-2 HR: 3
BISMUTH SODIUM THIOGLYCOLLATE
mf: $C_6H_6BiNa_3O_6S_3$ mw: 548.25

SYNS: BISTRIMATE ☐ MERCAPTOACETIC ACID, SODIUM-BISMUTH SALT ☐ SODIUM BISMUTH THIOGLYCOLATE ☐ SODIUM BISMUTH THIOGLYCOLLATE ☐ THIOBISMOL

TOXICITY DATA WITH REFERENCE
orl-chd LD50:47,222 mg/kg/22W-I JAMAAP 198,187,66
ims-chd LDLo:2500 μg/kg JOPDAB 28,498,46

B

ims-chd TDLo:43 mg/kg:SYS JOPDAB 31,580,47
ims-chd TDLo:650 μg/kg AMDCA5 97,384,59

SAFETY PROFILE: Systemic toxic effects in children: somnolence, nausea or vomiting, kidney damage, and decreased urine volume. Poison by intramuscular route. See also BISMUTH COMPOUNDS. When heated to decomposition it emits very toxic fumes of SO_x and Na_2O.

BKY000 CAS:1304-82-1 HR: 3
BISMUTH TELLURIDE
mf: Bi_2Te_3 mw: 800.76

PROP: Gray crystals or solid. D: 7.7.

SYNS: BISMUTH SESQUITELLURIDE □ BISMUTH TELLURIDE, UNDOPED

CONSENSUS REPORTS: Reported in EPA TSCA Inventory.

OSHA PEL: Total Dust: TWA 0.1 mg(Te)/m³; Respirable Fraction: TWA 5 mg/m³; Se doped: 5 mg/m³
ACGIH TLV: TWA 10 mg/m³; (Se doped: 5 mg/m³)

SAFETY PROFILE: Moderate fire hazard by spontaneous chemical reaction with powerful oxidizers. Reacts with moisture to evolve a toxic gas. Slight explosion hazard by chemical reaction with powerful oxidizers; reacts with moisture. When heated to decomposition it emits toxic fumes of Te. See also BISMUTH COMPOUNDS and TELLURIUM COMPOUNDS.

BKY250 CAS:12010-67-2 HR: 3
BISMUTH TIN OXIDE
mf: $Bi_2O_9Sn_3 \cdot 5H_2O$ mw: 1008.13

SYN: BISMUTH STANNATE PENTAHYDRATE

TOXICITY DATA WITH REFERENCE
ivn-mus LD50:178 mg/kg CSLNX* NX#02286

OSHA PEL: TWA 2 mg(Sn)/m³
ACGIH TLV: TWA 2 mg(Sn)/m³
NIOSH REL: (Organotin Compounds) TWA 0.1 mg(Sn)/m³

SAFETY PROFILE: Poison by intravenous route. See also TIN COMPOUNDS and BISMUTH COMPOUNDS. When heated to decomposition it emits acrid smoke and irritating fumes.

BKY500 CAS:19025-95-7 HR: 3
BISMUTH TRISODIUM THIOGLYCOLLATE
mf: $C_6H_6BiO_6S_3 \cdot 3Na$ mw: 548.25

SYNS: THIOBISMOL □ TRIS(MERCAPTOACETATO(2-1))BISMUTHATE(3-) TRISODIUM

TOXICITY DATA WITH REFERENCE
unr-chd TDLo:8163 μg/kg AJSGA3 21,674,37
ipr-rat LDLo:26 mg/kg ADSYAF 15,550,27
ivn-rat LDLo:23 mg/kg ADSYAF 15,550,27
ims-rat LDLo:29 mg/kg ADSYAF 15,550,27
ipr-gpg LDLo:26 mg/kg ADSYAF 15,550,27

SAFETY PROFILE: Poison by intraperitoneal, intravenous, and intramuscular routes. Human systemic effects by an unspecified route: convulsions and kidney damage. See also BISMUTH COMPOUNDS. When heated to decomposition it emits toxic fumes of SO_x, Bi, and Na_2O.

BKZ000 HR: 3
2,5-BIS-(NITRATOMERCURIMETHYL)-1,4-DIOXANE
mf: $C_6H_{10}Hg_2N_2O_8$ mw: 639.36

SYN: 1,4-DIOXOLAN-2,5-DIYLDIMETHYLENEBIS(NITROMERCURY)

TOXICITY DATA WITH REFERENCE
ivn-mus LD50:18 mg/kg CSLNX* NX#06958

CONSENSUS REPORTS: Mercury and its compounds are on the Community Right-To-Know List.

OSHA PEL: CL 0.1 mg(Hg)/m³ (skin)
ACGIH TLV: TWA 0.1 mg(Hg)/m³ (skin)
NIOSH REL: (Mercury, Aryl And Inorganic): CL 0.1 mg/m³ (skin)

SAFETY PROFILE: Poison by intravenous route. See also MERCURY COMPOUNDS and NITRATES. When heated to decomposition it emits very toxic fumes of NO_x and Hg vapors.

BLA000 CAS:13826-66-9 HR: 2
BIS(NITRATO-O)OXOZIRCONIUM
mf: N_2O_7Zr mw: 231.24

SYN: ZIRCONYL NITRATE

TOXICITY DATA WITH REFERENCE
orl-rat LD50:2500 mg/kg AIHOAX 1,637,50
ipr-rat LD50:1250 mg/kg AIHOAX 1,637,50

CONSENSUS REPORTS: Reported in EPA TSCA Inventory.

OSHA PEL: TWA 5 mg(Zr)/m³; STEL 10 mg(Zr)/m³
ACGIH TLV: TWA 5 mg(Zr)/m³; STEL 10 mg(Zr)/m³
DFG MAK: 5 mg(Zr)/m³

SAFETY PROFILE: Moderately toxic by ingestion and intraperitoneal routes. See also ZIRCONIUM COMPOUNDS and NITRATES. When heated to decomposition it emits toxic fumes of NO_x.

BLA250 HR: 3
BIS-p-NITRO BENZENE DIAZO SULFIDE
mf: $C_{12}H_8N_6O_4S$ mw: 332.3

SAFETY PROFILE: Explosion Hazard: The dry material is extremely sensitive; avoid even light friction. When heated to decomposition it emits toxic fumes of SO_x and NO_x. See also SULFIDES and NITRO COMPOUNDS of AROMATIC HYDROCARBONS.

BLA600 CAS:645-15-8 HR: 2
BIS(4-NITROPHENYL) PHOSPHATE
mf: $C_{12}H_9N_2O_8P$ mw: 340.20

SYNS: BIS(p-NITROPHENYL) PHOSPHATE □ BNPP □ DI-p-NITROPHENYL PHOSPHATE □ PHENOL, p-NITRO-, HYDROGEN PHOSPHATE □ PHOSPHORIC ACID, BIS(p-NITROPHENYL) ESTER □ PHOSPHORSAEURE-BIS-(p-NITRO-PHENYLESTER)

TOXICITY DATA WITH REFERENCE
ipr-mus LD50:410 mg/kg HSZPAZ 348,609,67

CONSENSUS REPORTS: Reported in EPA TSCA Inventory.

SAFETY PROFILE: Moderately toxic by intraperitoneal route. When heated to decomposition it emits toxic vapors of NO_x and PO_x.

BLA750 CAS:1223-31-0 HR: 2
BIS(p-NITROPHENYL)SULFIDE
mf: $C_{12}H_8N_2O_4S$ mw: 276.28

PROP: Orange plates from AcOH. Mp: 156–157°.

TOXICITY DATA WITH REFERENCE
orl-rat LD50:1490 mg/kg MarJV# 29MAR77

CONSENSUS REPORTS: Reported in EPA TSCA Inventory.

SAFETY PROFILE: Moderately toxic by ingestion. When heated to decomposition it emits very toxic fumes of NO_x and SO_x.

BLA800 CAS:860-39-9 HR: 3
BIS(2-NITRO-4-TRIFLUOROMETHYLPHENYL) DISULFIDE
mf: $C_{14}H_6F_6N_2O_4S_2$ mw: 444.34

SYNS: DISULFIDE, BIS(2-NITRO-α-α-α-TRIFLUORO-p-TOLYL) □ USAF MA-9

TOXICITY DATA WITH REFERENCE
ipr-mus LD50:50 mg/kg NTIS** AD277-689

CONSENSUS REPORTS: Reported in EPA TSCA Inventory.

SAFETY PROFILE: Poison by intraperitoneal route. When heated to decomposition it emits toxic vapors of NO_x, SO_x, and F^-.

BLB250 CAS:4731-77-5 HR: 1
BIS(OCTANOYLOXY)DI-n-BUTYL STANNANE
mf: $C_{24}H_{48}O_4Sn$ mw: 519.41

SYNS: BIS(OCTANOYLOXY)DI-n-BUTYLTIN □ DIBUTYLBIS(OCTANOYLOXY)STANNANE □ DIBUTYLBIS((1-OXOOCTYL)OXY)STANNANE □ DIBUTYLTIN DICAPRYLATE □ DIBUTYLTIN DIOCTANOATE □ DIBUTYLTIN DIOCTATE □ DIBUTYLTIN OCTANOATE □ KAPRYLAN DI-N-BUTYLCINICITY (CZECH)

TOXICITY DATA WITH REFERENCE
skn-rbt 500 mg/24H SEV 28ZPAK -,229,72
eye-rbt 20 mg/24H MOD 28ZPAK -,229,72
ipr-mus LD50:14 g/kg JPMSAE 55,158,66

OSHA PEL: TWA 0.1 mg(Sn)/m³ (skin)
ACGIH TLV: TWA 0.1 mg(Sn)/m³ (skin) (Proposed: TWA 0.1 mg(Sn)/m³; STEL 0.2 mg(Sn)/m³ (skin))
NIOSH REL: (Organotin Compounds) TWA 0.1 mg(Sn)/m³

SAFETY PROFILE: Low toxicity. A severe skin and eye irritant. See also TIN COMPOUNDS. When heated to decomposition it emits acrid and irritating fumes. For occupational chemical analysis use NIOSH: Organotin Compounds 5504.

BLB500 CAS:19546-20-4 HR: 1
2,2-BIS(3′-tert-OCTYL)-4′-HYDROXYPHENYLPROPANE
mf: $C_{31}H_{48}O_2$ mw: 452.79

SYNS: ANTIOXIDANT TOD (CZECH) □ 2,2-BIS-3′-TERC. OKTYL-4′-HYDROXYFENYLPROPAN (CZECH)

TOXICITY DATA WITH REFERENCE
skn-rbt 500 mg/24H MOD 28ZPAK -,58,72
eye-rbt 100 mg/24H SEV 28ZPAK -,58,72
orl-rat LD50:4920 mg/kg 28ZPAK -,58,72

SAFETY PROFILE: Mildly toxic by ingestion. A skin and severe eye irritant. When heated to decomposition it emits acrid smoke and irritating fumes.

BLB750 CAS:131-15-7 HR: 1
BIS(2-OCTYL)PHTHALATE
mf: $C_{24}H_{38}O_4$ mw: 390.62

SYNS: BIS(2-OCTYL)PHTHALATE □ BIS-(2-OKTYL)ESTER KYSELINY FTALOVE □ CAPRYL o-PHTHALATE □ DICAPRYL 1,2-BENZENEDICARBOXYLATE □ DICAPRYL PHTHALATE □ DIOCTANOL-2-PHTHALATE □ MONOPLEX DCP □ PHTHALIC ACID, BIS(2-OCTYL) ESTER □ PHTHALIC ACID, DICAPRYL ESTER □ PHTHALIC ACID, DI-2-OCTYL ESTER

TOXICITY DATA WITH REFERENCE
ipr-mus LD50:14 g/kg JPMSAE 55,158,66

CONSENSUS REPORTS: Reported in EPA TSCA Inventory.

SAFETY PROFILE: Mildly toxic by intraperitoneal route. When heated to decomposition it emits acrid smoke and irritating fumes.

BLC000 CAS:868-18-8 HR: 2
BISODIUM TARTRATE
mf: $C_4H_4O_6$•2Na mw: 194.06

PROP: Transparent crystals; colorless and odorless. Sol in water.

SYNS: 2,3-DIHYDROXY-(R-(R*,R*))-BUTANEDIOIC ACID DISODIUM SALT (9CI) □ DISODIUM TARTRATE □ DISODIUM l-(+)-TARTRATE □ SODIUM TARTRATE (FCC) □ SODIUM l-(+)-TARTRATE

TOXICITY DATA WITH REFERENCE
orl-mus LDLo:3686 mg/kg JAPMA8 31,12,42
orl-rbt LDLo:5290 mg/kg FAONAU 53A,512,74

CONSENSUS REPORTS: Reported in EPA TSCA Inventory.

SAFETY PROFILE: Moderately toxic by ingestion. When heated to decomposition it emits acrid smoke and irritating fumes.

BLC250 **CAS:10380-28-6** **HR: 3**
BIS(8-OXYQUINOLINE)COPPER
mf: $C_{18}H_{12}CuN_2O_2$ mw: 351.86

PROP: Yellow-green powder or crystals. Insol in H_2O and common organic solvents.

SYNS: BIOQUIN □ BIOQUIN 1 □ BIS(8-QUINOLINATO)COPPER □ BIS(8-QUINOLINOLATO)COPPER □ BIS(8-QUINOLINOLATO-N', O[8])-COPPER □ CELLU-QUIN □ COPPER-8 □ COPPER HYDROXYQUINOLATE □ COPPER-8-HYDROXYQUINOLATE □ COPPER-8-HYDROXYQUINOLINATE □ COPPER-8-HYDROXYQUINOLINE □ COPPER OXINATE □ COPPER (2+) OXINATE □ COPPER OXINE □ COPPER OXYQUINOLATE □ COPPER OXYQUINOLINE □ COPPER QUINOLATE □ COPPER-8-QUINOLATE □ COPPER-8-QUINOLINOL □ COPPER QUINOLINOLATE □ COPPER-8-QUINOLINOLATE □ CUNILATE □ CUNILATE 2472 □ CUPRIC-8-HYDROXYQUINOLATE □ CUPRIC-8-QUINOLINOLATE □ DOKIRIN □ FRUITDO □ 8-HYDROXYQUINOLINE COPPER COMPLEX □ MILMER □ OXIME COPPER □ OXINE COPPER □ OXINE CUIVRE □ OXYQUINOLINOLEATE de CUIVRE (FRENCH) □ QUINONDO

TOXICITY DATA WITH REFERENCE
mma-sat 5 µg/plate MUREAV 116,185,83
scu-mus TDLo:156 mg/kg/39W-I:ETA JNCIAM 24,109,60
orl-rat LD50:9930 mg/kg GISAAA 51(1),85,86
ihl-rat LC50:820 mg/m³ NNGADV 16,563,91
ipr-rat LD50:22 mg/kg NNGADV 16,563,91
orl-mus LD50:3940 mg/kg GISAAA 51(1),85,86
ipr-mus LD50:67 mg/kg TXAPA9 5,599,63

CONSENSUS REPORTS: IARC Cancer Review: Group 3 IMEMDT 7,56,87, Animal Inadequate Evidence IMEMDT 15,103,77. Reported in EPA TSCA Inventory. Copper and its compounds are on the Community Right-To-Know List.

SAFETY PROFILE: Poison by intraperitoneal route. Moderately toxic by ingestion and inhalation. Questionable carcinogen with experimental tumorigenic data. Mutation data reported. See also COPPER COMPOUNDS. When heated to decomposition it emits toxic fumes of NO_x.

BLC500 **CAS:117-97-5** **HR: 2**
BIS(PENTACHLOROPHENOL), ZINC SALT
mf: $C_{12}Cl_{10}S_2Zn$ mw: 628.11

SYN: PENTACHLOROTHIOFENOLAT ZINECNATY (CZECH)

TOXICITY DATA WITH REFERENCE
skn-rbt 500 mg/24H SEV 28ZPAK -,11,72
eye-rbt 250 µg/24H SEV 28ZPAK -,11,72

CONSENSUS REPORTS: Reported in EPA TSCA Inventory. Zinc and its compounds are on the Community Right-To-Know List.

SAFETY PROFILE: A severe eye and skin irritant. See also ZINC COMPOUNDS and CHLORINATED HYDROCARBONS, AROMATIC. When heated to decomposition it emits very toxic fumes of ZnO, Cl^- and SO_x.

BLC750 **HR: 3**
BIS(PENTA FLUORO PHENYL)ALUMINUM BROMIDE
mf: $C_{12}AlBrF_{10}$ mw: 441

SAFETY PROFILE: Ignites spontaneously in air. Hydrolysis causes explosion. When heated to decomposition it emits toxic fumes of F^- and Br^-. See also ALUMINUM COMPOUNDS.

BLD000 **CAS:42310-84-9** **HR: 3**
BISPENTAFLUOROSULFUR OXIDE
mf: $F_{10}OS_2$ mw: 270.12

PROP: Colorless liquid. Mp: −118°, bp: 31°.

SYN: SULFUR FLUORIDE OXIDE

TOXICITY DATA WITH REFERENCE
ihl-rat LCLo:20 ppm/6H BJIMAG 27,1,70

OSHA PEL: TWA 2.5 mg(F)/m³
NIOSH REL: TWA 2.5 mg(F)/m³

SAFETY PROFILE: Poison by inhalation. See also FLUORIDES. When heated to decomposition it emits very toxic fumes of F^- and SO_x.

BLD250 **HR: 3**
BIS(2,4-PENTANEDIONATO)CHROMIUM
mf: $C_{10}H_{14}CrO_4$ mw: 250.21

CONSENSUS REPORTS: Chromium and its compounds are on the Community Right-To-Know List.

SAFETY PROFILE: Ignites spontaneously in air. When heated to decomposition it emits acrid smoke and fumes. See also CHROMIUM COMPOUNDS.

BLD325 **CAS:93431-23-3** **HR: 2**
4,5-BIS(4-PENTENYLOXY)-2-IMIDAZOLIDINONE
mf: $C_{13}H_{22}N_2O_3$ mw: 254.37

SYN: SRC-16

TOXICITY DATA WITH REFERENCE
orl-mus LD50:740 mg/kg CPBTAL 12,843,64
ipr-mus LD50:450 mg/kg CPBTAL 12,843,64
scu-mus LD50:621 mg/kg CPBTAL 12,843,64

SAFETY PROFILE: Moderately toxic by ingestion, subcutaneous, and intraperitoneal routes. When heated to decomposition it emits toxic fumes of NO_x.

BLD500 **CAS:80-05-7** **HR: 3**
BISPHENOL A
mf: $C_{15}H_{16}O_2$ mw: 228.31

PROP: White flakes; mild phenolic odor. Mp: 156–157°, bp: 250–252° @ 13 mm. Insol in water; sol in alcohol and dilute alkalies; sltly sol in CCl_4.

SYNS: BISFEROL A (GERMAN) □ 2,2-BIS-4'-HYDROXYFENYLPROPAN (CZECH) □ BIS(4-HYDROXYPHENYL) DIMETHYLMETHANE □ BIS(4-HYDROXYPHENYL)PROPANE □ 2,2-BIS(p-HYDROXYPHENYL) PROPANE □ 2,2-BIS(4-HYDROXYPHENYL)PROPANE □ DIAN □ p,p'-DIHYDROXYDIPHENYLDIMETHYLMETHANE □ 4,4'-DIHYDROXYDIPHENYLDIMETHYLMETHANE □ p,p'-DIHYDROXYDIPHENYLPROPANE □ 2,2-(4,4'-DIHYDROXYDIPHENYL)PROPANE □ 4,4'-DIHYDROXYDIPHENYLPROPANE □ 4,4'-DIHYDROXYDIPHENYL-2,2-PROPANE □ 4,4'-DIHYDROXY-2,2-DIPHENYLPROPANE □ β-DI-p-HYDROXYPHENYLPROPANE □ 2,2-DI(4-HYDROXYPHENYL)PROPANE □ DIMETHYL BIS(p-HYDROXYPHENYL)METHANE □ DIMETHYLME-

THYLENE-p,p'-DIPHENOL ☐ 2,2-DI(4-PHENYLOL)PROPANE ☐ p,p'-ISOPROPYLIDENEBISPHENOL ☐ 4,4'-ISOPROPYLIDENEBISPHENOL ☐ p,p'-ISOPROPYLIDENEDIPHENOL ☐ NCI-C50635

TOXICITY DATA WITH REFERENCE
skn-rbt 250 mg open MLD UCDS** 7/14/65
eye-rbt 20 mg/24H SEV 28ZPAK -,58,72
orl-mus TDLo:12,500 mg/kg (female 6-15D post):TER NTIS** PB85-205102
orl-rat TDLo:10 g/kg (6-15D preg):TER NTIS** PB85-20511-
ipr-rat TDLo:1275 mg/kg (female 1-15D post):REP SWEHDO 7(Suppl 4),66,81
orl-rat TDLo:15 g/kg (female 6-15D post):REP NTIS** PB85-205110
orl-mus TDLo:7500 mg/kg (female 6-15D post):REP NTIS** PB85-205102
ipr-rat TDLo:1275 mg/kg (female 1-15D post):TER SWEHDO 7(Suppl 4),66,81
ipr-rat TDLo:1875 mg/kg (female 1-15D post):TER SWEHDO 7(Suppl 4),66,81
orl rat LD50:3250 mg/kg AIHAAP 28,301,67
orl-mus LD50:2500 mg/kg AIHAAP 28,301,67
ipr-mus LD50:150 mg/kg NTIS** AD691-490
orl-rbt LD50:2230 mg/kg AIHAAP 28,301,67
skn-rbt LD50:3000 mg/kg AMIHBC 4,119,51

CONSENSUS REPORTS: NTP Carcinogenesis Bioassay (feed); Inadequate Studies: mouse, rat NTPTR* NTP-TR-215,82. Community Right-To-Know List. Reported in EPA TSCA Inventory.

SAFETY PROFILE: Poison by intraperitoneal route. Moderately toxic by ingestion, inhalation, and skin contact. Experimental teratogenic and reproductive effects. A skin and eye irritant. When heated to decomposition it emits acrid and irritating fumes.

BLD750 CAS:1675-54-3 HR: 3
BISPHENOL A DIGLYCIDYL ETHER
mf: $C_{21}H_{24}O_4$ mw: 340.45

SYNS: 2,2-BIS(4-(2,3-EPOXYPROPYLOXY)PHENYL)PROPANE ☐ BIS(4-GLYCIDYLOXYPHENYL)DIMETHYAMETHANE ☐ 2,2-BIS(p-GLYCIDYLOXYPHENYL)PROPANE ☐ BIS(4-HYDROXYPHENYL)DIMETHYLMETHANE DIGLYCIDYL ETHER ☐ 2,2-BIS(p-HYDROXYPHENYL)PROPANE, DIGLYCIDYL ETHER ☐ 2,2-BIS(4-HYDROXYPHENYL)PROPANE, DIGLYCIDYL ETHER ☐ D.E.R. 332 ☐ DIGLYCIDYL BISPHENOL A ETHER ☐ DIGLYCIDYL ETHER of 2,2-BIS(p-HYDROXYPHENYL)PROPANE ☐ DIGLYCIDYL ETHER of 2,2-BIS(4-HYDROXYPHENYL)PROPANE ☐ DIGLYCIDYL ETHER of BISPHENOL A ☐ DIGLYCIDYL ETHER of 4,4'-ISOPROPYLIDENEDIPHENOL ☐ 4,4'-DIHYDROXYDIPHENYLDIMETHYLMETHANE DIGLYCIDYL ETHER ☐ p,p'-DIHYDROXYDIPHENYLDIMETHYLMETHANE DIGLYCIDYL ETHER ☐ EPI-REZ 508 ☐ EPI-REZ 510 ☐ EPON 828 ☐ EPOXIDE A ☐ ERL-2774 ☐ 4,4'-ISOPROPYLIDENEDIPHENOL DIGLYCIDYL ETHER ☐ 2,2'-((1-METHYLETHYLIDENE)BIS(4,1-PHENYLENEOXYMETHYLENE))BISOXIRANE

TOXICITY DATA WITH REFERENCE
skn-rbt 500 mg open MLD UCDS** 4/21/67
eye-rbt 2 mg/24H SEV 28ZPAK -,137,72
mmo-sat 50 µg/plate MUREAV 66,367,79
mma-sat 50 µg/plate MUREAV 66,367,79
skn-mus TDLo:166 g/kg/2Y-I:CAR FCTOD7 26,611,88
skn-mus TD:312 g/kg/2Y-I:CAR,REP CNREA8 39,1718,79
orl-rat LD50:11,300 µL/kg UCDS** 4/21/67
ipr-rat LD50:2200 mg/kg 38MKAJ 2A,2219,81
orl-mus LD50:15,600 mg/kg 38MKAJ 2A,2219,81
ipr-mus LD50:4 g/kg 38MKAJ 2A,2219,81
orl-rbt LD50:1980 mg/kg 38MKAJ 2A,2219,81
skn-rbt LD50:20 mg/kg 38MKAJ 2A,2219,81

CONSENSUS REPORTS: EPA Genetic Toxicology Program. Reported in EPA TSCA Inventory.

SAFETY PROFILE: Poison by skin contact. Mildly toxic by ingestion. Mutation data reported. A skin and severe eye irritant. Experimental reproductive effects. Questionable carcinogen with experimental carcinogenic and tumorigenic data. See also ETHERS. When heated to decomposition it emits acrid and irritating fumes.

BLE000 HR: 2
BISPHENOL DIGLYCIDYL ETHER, MODIFIED

TOXICITY DATA WITH REFERENCE
skn-mus TDLo:470 g/kg/39W-I:ETA AIHAAP 24,305,63
skn-mus TD:1200 g/kg/2Y-I:ETA AIHAAP 24,305,63

SAFETY PROFILE: Questionable carcinogen with experimental tumorigenic data. See also BISPHENOL DIGLYCIDYL ETHER and ETHERS.

BLE250 CAS:17601-12-6 HR: 3
BIS(p-PHENOXYPHENYL)DIPHENYLTIN
mf: $C_{36}H_{28}O_2Sn$ mw: 611.33

SYN: BIS(p-PHENOXYPHENYL)DIPHENYLSTANNANE

TOXICITY DATA WITH REFERENCE
ivn-mus LD50:56 mg/kg CSLNX* NX#01351

OSHA PEL: TWA 0.1 mg(Sn)/m³ (skin)
ACGIH TLV: TWA 0.1 mg(Sn)/m³ (skin) (Proposed: TWA 0.1 mg(Sn)/m³; STEL 0.2 mg(Sn)/m³ (skin))
NIOSH REL: (Organotin Compounds) TWA 0.1 mg(Sn)/m³

SAFETY PROFILE: Poison by intravenous route. See also TIN COMPOUNDS. When heated to decomposition it emits acrid smoke and irritating fumes.

For occupational chemical analysis use NIOSH: Organotin Compounds 5504.

BLE500 CAS:74-31-7 HR: 3
1,4-BIS(PHENYL AMINO)BENZENE
mf: $C_{18}H_{16}N_2$ mw: 260.36

PROP: Gray crystals or solid. D: 1.20, mp: 147°, vap d: 9.0.

SYNS: AGERITE ☐ AGERITEDPPD ☐ N,N'-DIFENYL-p-FENYLENDIAMIN (CZECH) ☐ DIPHENYL-p-PHENYLENEDIAMINE ☐ N,N'-DIPHENYL-p-PHENYLENEDIAMINE ☐ DPPD ☐ FLEXAMINE G ☐ JZF ☐ NONOX DPPD ☐ p-PHENYLAMINODIPHENYLAMINE ☐ 4-PHENYLAMINODIPHENYLAMINE ☐ USAF GY-2

TOXICITY DATA WITH REFERENCE
eye-rbt 500 mg/24H SEV 28ZPAK -,73,72
mma-sat 10 µg/plate PCBRD2 141,407,84
msc-ham:lng 30 mg/L SWEHDO 9(Suppl 2),27,83

orl-rat TDLo:450 mg/kg (14D pre/1-22D preg):REP JAFCAU 4,796,56
orl-mus TDLo:4176 mg/kg (female 6-14D post):TER NTIS** PB223-160
orl-rat TDLo:2500 mg/kg (female 1-22D post):REP AJANA2 110,29,62
scu-mus TDLo:1000 mg/kg:ETA NTIS** PB223-159
orl-rat LD50:2370 mg/kg 28ZPAK -,73,72
orl-mus LD50:18 g/kg GTPZAB 10(3),49,66
ipr-mus LD50:300 mg/kg NTIS** AD277-689

CONSENSUS REPORTS: Reported in EPA TSCA Inventory.

SAFETY PROFILE: Poison by intraperitoneal route. Moderately toxic by ingestion. A weak allergen. Experimental teratogenic and reproductive effects. An eye irritant. Questionable carcinogen with experimental tumorigenic data. Mutation data reported. Combustible when exposed to heat or flame; can react with oxidizing materials. When heated to decomposition it emits toxic fumes of NO_x.

BLF250 **CAS:13754-23-9** **HR: 3**
BIS-N,N'-(3-PHENYLPROPYL-2)-PIPERAZINE DIHYDROCHLORIDE
mf: $C_{22}H_{30}N_2 \cdot 2ClH$ mw: 395.46

SYNS: N,N'-BIS(PHENYLISOPROPYL)PIPERAZINE DIHYDROCHLORIDE ☐ DIPHENAZINE DIHYDROCHLORIDE

TOXICITY DATA WITH REFERENCE
orl-mus LD50:230 mg/kg ARZNAD 7,225,57
scu-mus LD50:193 mg/kg 27ZQAG -,223,72
ivn-mus LD50:25 mg/kg ARZNAD 7,225,57

SAFETY PROFILE: Poison by ingestion, intravenous, and subcutaneous routes. When heated to decomposition it emits very toxic fumes of HCl and NO_x.

BLF500 **CAS:1666-13-3** **HR: 3**
BIS(PHENYLSELENIDE)
mf: $C_{12}H_{10}Se_2$ mw: 312.14

PROP: Yellow needles from hexane. Mp: 63–65°.

SYN: PHENYL DISELENIDE

TOXICITY DATA WITH REFERENCE
ivn-mus LD50:28 mg/kg CSLNX* NX#05657

CONSENSUS REPORTS: Reported in EPA TSCA Iventory. Selenium and its compounds are on the Community Right-To-Know List.

OSHA PEL: TWA 0.2 mg(Se)/m³
ACGIH TLV: TWA 0.2 mg(Se)/m³
DFG MAK: 0.1 mg(Se)/m³

SAFETY PROFILE: Poison by intravenous route. See also SELENIUM COMPOUNDS. When heated to decomposition it emits toxic fumes of Se.

BLF750 **CAS:4848-63-9** **HR: 3**
BIS(PHENYLTHIO)DIMETHYLTIN
mf: $C_{14}H_{16}S_2Sn$ mw: 367.11

SYN: DIMETHYLBIS(PHENYLTHIO)STANNANE

TOXICITY DATA WITH REFERENCE
ivn-mus LD50:56 mg/kg CSLNX* NX#01670

OSHA PEL: TWA 0.1 mg(Sn)/m³ (skin)
ACGIH TLV: TWA 0.1 mg(Sn)/m³ (skin) (Proposed: TWA 0.1 mg(Sn)/m³; STEL 0.2 mg(Sn)/m³ (skin))
NIOSH REL: (Organotin Compounds) TWA 0.1 mg(Sn)/m³

SAFETY PROFILE: Poison by intravenous route. See also TIN COMPOUNDS. When heated to decomposition it emits toxic fumes of SO_x.

For occupational chemical analysis use NIOSH: Organotin Compounds 5504.

BLG000 **HR: 3**
1,3-BIS((PHENYL)TRIAZENO)BENZENE
mf: $C_{18}H_{16}N_6$ mw: 316.37

SAFETY PROFILE: Explodes on rapid heating. When heated to decomposition it emits toxic fumes of NO_x.

BLG100 **CAS:52237-03-3** **HR: 1**
4,4'-BIS(4-PHENYL-2H-1,2,3-TRIAZOL-2-YL)-2,2'-STILBENEDISULFONIC ACID DIPOTASSIUM SALT
mf: $C_{30}H_{20}N_6O_6S_2 \cdot 2K$ mw: 702.88 2

SYNS: DIPOTASSIUM 4,4'-BIS(4-PHENYL-1,2,3-TRIAZOL-2-YL)STILBENE-2,2'-DISULFONATE ☐ DIPOTASSIUM 4,4'-BIS(4-PHENYL-1,2,3-TRIAZOL-2-YL)STILBENE-2,2'-SULFONATE ☐ 2,2'-STILBENEDISULFONIC ACID, 4,4'-BIS(4-PHENYL-2H-1,2,3-TRIAZOL-2-YL)-, DIPOTASSIUM SALT ☐ 2,2'-STILBENEDISULFONIC ACID, 4,4'-BIS(4-PHENYL-1,2,3-TRIAZOL-2-YL), DIPOTASSIUM SALT

TOXICITY DATA WITH REFERENCE
eye-rbt 100 mg MOD CTOXAO 13,171,78
orl-rat TDLo:3 g/kg (female 6-15D post):REP EQSFAP 4,223,75

SAFETY PROFILE: Experimental reproductive effects. An eye irritant. When heated to decomposition it emits toxic fumes of NO_x and SO_x.

BLG250 **CAS:2439-99-8** **HR: 2**
N,N-BIS(PHOSPHONOMETHYL)GLYCINE
mf: $C_4H_{11}NO_8P_2$ mw: 263.10

PROP: Crystals from EtOH (aq). Mp: 200° (decomp). Very sol in H_2O, sltly sol in EtOH, insol in C_6H_6.

SYNS: GLYPHOSINE ☐ POLARIS

TOXICITY DATA WITH REFERENCE
orl-rat LD50:3925 mg/kg 85AREA 3,60,76/77

CONSENSUS REPORTS: Reported in EPA TSCA Inventory.

SAFETY PROFILE: Moderately toxic by ingestion. When heated to decomposition it emits very toxic fumes of NO_x and PO_x.

B

BLG325 CAS:60012-89-7 HR: 2
3,8-BIS(1-PIPERIDINYLMETHYL)-2,7-DIOXASPIRO(4.4) NONANE-1,6-DIONE
mf: $C_{19}H_{30}N_2O_4$ mw: 350.51

TOXICITY DATA WITH REFERENCE
ipr-rat LD50:1160 mg/kg PJPPAA 28,157,76
orl-mus LD50:6 g/kg PJPPAA 28,157,76
ipr-mus LD50:820 mg/kg PJPPAA 28,157,76

SAFETY PROFILE: Moderately toxic by intraperitoneal route. When heated to decomposition it emits toxic fumes of NO_x.

BLG400 CAS:7652-64-4 HR: 2
N,N′-BISPROPYLENEISOPHTHALAMIDE
mf: $C_{14}H_{16}N_2O_2$ mw: 244.32

SYNS: AZIRIDINE, 1,1′-ISOPHTHALOYLBIS(2-METHYL)- □ AZIRIDINE, 1,1′-(1,3-PHENYLENEDICARBONYL)BIS(2-METHYL)-(9CI) □ HX 752 □ 1,1′-(1,3-PHENYLENEDICARBONYL)BIS(2-METHYLAZIRIDINE)

TOXICITY DATA WITH REFERENCE
eye-rbt 500 mg SEV WHYHAQ 14(5),1,85
mmo-sat 1500 ng/plate WHYHAQ 14(5),1,85
mma-sat 1500 ng/plate WHYHAQ 14(5),1,85
mnt-mus-orl 240 mg/kg WHYHAQ 14(5),1,85
cyt-rat-orl 240 mg/kg WHYHAQ 14(5),1,85
orl-rat LD50:1230 mg/kg WHYHAQ 14(5),1,85
orl-mus LD50:593 mg/kg WHYHAQ 14(5),1,85

CONSENSUS REPORTS: Reported in EPA TSCA Inventory.

SAFETY PROFILE: Moderately toxic by ingestion. A severe eye irritant. Mutation data reported. When heated to decomposition it emits toxic vapors of NO_x.

BLG500 CAS:1113-14-0 HR: 3
trans-1,2-BIS(n-PROPYLSULFONYL)ETHYLENE
mf: $C_8H_{16}O_4S_2$ mw: 240.36

SYNS: B-1843 □ C-272 □ CHEMAGRO B-1843 □ VANCIDE PA □ VANCIDE PA DISPERSION

TOXICITY DATA WITH REFERENCE
orl-rat LD50:200 mg/kg FMCHA2 -,C49,83
ipr-rat LD50:11,500 µg/kg 34ZIAG -,161,69
ipr-mus LDLo:11,500 mg/kg 34ZIAG -,162,69
ipr-gpg LDLo:11,500 mg/kg 34ZIAG -,162,69

SAFETY PROFILE: Poison by ingestion and intravenous routes. See also SULFONATES. When heated to decomposition it emits toxic fumes of SO_x.

BLH250 CAS:14167-18-1 HR: 3
BIS(SALICYLALDEHYDE)ETHYLENEDIIMINE COBALT(II)
mf: $C_{16}H_{14}CoN_2O_2$ mw: 325.25

PROP: Red crystals from DMF. Sol in C_6H_6, $CHCl_3$, and Py.

SYNS: N,N′-ETHYLENEBIS(SALICYLIDENEIMINATO)COBALT(II) □ SALCOMIN □ SALCOMINE POWDER □ SALICYLALDEHYDE ETHYLENEDIIMINE COBALT

TOXICITY DATA WITH REFERENCE
ihl-mus LCLo:390 mg/m³/5.5H AMRL** TR-74-78,74

CONSENSUS REPORTS: Reported in EPA TSCA Inventory. Cobalt and its compounds are on the Community Right-To-Know List. EPA Extremely Hazardous Substances List.

SAFETY PROFILE: Poison by inhalation. See also COBALT COMPOUNDS and ALDEHYDES. When heated to decomposition it emits toxic fumes of NO_x.

BLH309 CAS:28660-67-5 HR: 3
BIS(TETRADECANOYLOXY)DIBUTYLSTANNANE
mf: $C_{36}H_{72}O_4Sn$ mw: 687.77

SYNS: DI-n-BUTYL-TIN DI(TETRADECANOATE) □ MYRISTAN DI-n-BUTYLCINICITY (CZECH)

TOXICITY DATA WITH REFERENCE
skn-rbt 500 mg/24H SEV 28ZPAK -,230,72
eye-rbt 5 mg/24H SEV 28ZPAK -,230,72
orl-rat LD50:138 mg/kg 28ZPAK -,230,72

OSHA PEL: TWA 0.1 mg(Sn)/m³ (skin)
ACGIH TLV: TWA 0.1 mg(Sn)/m³ (skin) (Proposed: TWA 0.1 mg(Sn)/m³; STEL 0.2 mg(Sn)/m³ (skin))
NIOSH REL: (Organotin Compounds) TWA 0.1 mg(Sn)/m³

SAFETY PROFILE: Poison by ingestion. A severe skin and eye irritant. See also TIN COMPOUNDS. When heated to decomposition it emits smoke and acrid fumes.

For occupational chemical analysis use NIOSH: Organo tin Compounds 5504.

BLH325 CAS:62987-05-7 HR: 3
1,3-BIS(TETRAHYDRO-2-FURYL)-5-FLUOROURACIL
mf: $C_{12}H_{15}FN_2O_4$ mw: 270.29

SYNS: 1,3-BIS(TETRAHYDRO-2-FURANYL)-5-FLUORO-2,4-PYRIMIDINEDIONE □ FD-1 □ 5-FLUORO-1,3-BIS(TETRAHYDRO-2-FURANYL)-2,4(1H,3H)-PYRIMIDINEDIONE

TOXICITY DATA WITH REFERENCE
cyt-ham:fbr 100 nmol/L MUREAV 88,241,81
orl-rat LD50:1730 mg/kg GANNA2 71,30,80
orl-mus LD50:2664 mg/kg JMCMAR 21,738,78
orl-dog LD50:88,100 µg/kg OYYAA2 16,303,78

SAFETY PROFILE: Poison by ingestion. Mutation data reported. When heated to decomposition it emits toxic fumes of F^- and NO_x.

BLI000 CAS:68594-19-4 HR: 3
1,6-BIS(5-TETRAZOLYL)HEXAAZ-1,5-DIENE
mf: $C_2H_4N_{14}$ mw: 224.14

SAFETY PROFILE: An explosive extremely sensitive to pressure or heating to 90°C. Upon decomposition it emits toxic fumes of NO_x. See also EXPLOSIVES and AZIDES.

BLI250 CAS:1656-16-2 HR: 3
3,4-BIS(1,2,3,4-THIATRIAZOL-5-YL THIO) MALEIMIDE
mf: $C_4HN_7O_2S_4$ mw: 307.34

SAFETY PROFILE: Explodes on impact or when heated to its melting point. When heated to decomposition it emits toxic fumes of NO_x and SO_x. See also EXPLOSIVES.

BLI500 HR: 3
BIS(1,2,3,4-THIATRIAZOL-5-YL THIO)METHANE
mf: $C_3H_2N_6S_4$ mw: 250.33

SAFETY PROFILE: On impact or on heating to its melting point it explodes loudly with a flash. Upon decomposition it emits toxic fumes of SO_x and NO_x. See also EXPLOSIVES.

BLJ250 CAS:142-46-1 HR: 3
BIS(THIOUREA)
mf: $C_2H_6N_4S_2$ mw: 150.24

PROP: Needles from H_2O. Mp: 214–223° (decomp).

SYNS: BISTHIOCARBAMYL HYDRAZINE □ 2,5-DITHIOBIUREA □ 1,2-HYDRAZINEDICARBOTHIOAMIDE □ NCI-C03009 □ USAF B-44 □ USAF EK-P-6281

TOXICITY DATA WITH REFERENCE
sce-ham:ovr 145 mg/L EMMUEG 10(Suppl 10),1,87
orl-mus TDLo:655 g/kg/78W-C:ETA NCITR* NCI-CG-TR-132,79
orl-mus TD:1310 g/kg/78W-C:ETA NCITR* NCI-CG-TR-132,79
ipr-mus LD50:100 mg/kg NTIS** AD277-689

CONSENSUS REPORTS: NCI Carcinogenesis Bioassay (feed); No Evidence: mouse, rat NCITR* NCI-CG-TR-132,79. Reported in EPA TSCA Inventory.

SAFETY PROFILE: Poison by intraperitoneal route. Questionable carcinogen with experimental tumorigenic data. Mutation data reported. When heated to decomposition it emits very toxic fumes of NO_x and SO_x.

BLJ500 HR: 3
BISTOLUENE DIAZO OXIDE
mf: $C_{14}H_{14}N_4O$ mw: 254.29

SAFETY PROFILE: Ignites spontaneously. Explosion Hazard: Very unstable. Shock and friction sensitive. Incompatible with toluene. Upon decomposition it emits toxic fumes of NO_x. See also AZIDES.

BLK000 CAS:128-80-3 HR: 2
1,4-BIS(p-TOLYLAMINO)ANTHRAQUINONE
mf: $C_{28}H_{22}N_2O_2$ mw: 418.52

PROP: Dark green crystals or powder. Sol in C_6H_6 or acids; sltly sol in Me_2CO; insol in H_2O and EtOH.

SYNS: ALIZARINE CYANINE GREEN BASE □ AMAPLAST GREEN OZ □ ARLOSOL GREEN B □ BIS-1,4-p-TOLYLAMINOANTHRCHINON (CZECH) □ C-GREEN 10 □ C.I. 61565 □ C.I. SOLVENT GREEN 3 □ CYANINE GREEN G BASE □ D&C GREEN No. 6 □ 1,4-DI-p-TOLUIDINOANTHRAQUINONE □ FAT SOLUBLE GREEN ANTHRAQUINONE □ 11091 GREEN □ GREEN No. 2 □ MICRO-LEX GREEN 5B □ NITRO FAST GREEN GB □ ORGANOL FAST GREEN J □ QUINIZARINE GREEN BASE □ SUDAN GREEN 4B □ TOYO ORIENTAL OIL BLUE G □ WAXOLINE GREEN

TOXICITY DATA WITH REFERENCE
eye-rbt 20 mg/24H MOD 28ZPAK -,124,72
orl-rat LD50:3660 mg/kg 28ZPAK -,124,72

CONSENSUS REPORTS: Reported in EPA TSCA Inventory.

SAFETY PROFILE: Moderately toxic by ingestion. An eye irritant. When heated to decomposition it emits toxic fumes of NO_x.

BLK250 CAS:63869-05-6 HR: 3
N-BIS(p-TOLYLSULFONYL)AMIDOMETHYL MERCURY
mf: $C_{15}H_{17}HgNO_4S_2$ mw: 540.04

SYNS: N-METHYLMERCURI-BIS-p-TOLUENSULFONAMID (CZECH) □ METHYL(4-METHYL-N-((4-METHYLPHENYL)SULFONYL)BENZENESULFONAMIDATO-N)-MERCURY

TOXICITY DATA WITH REFERENCE
skn-rbt 500 mg/24H MLD 28ZPAK -,223,72
eye-rbt 50 μg/24H SEV 28ZPAK -,223,72
orl-rat LD50:98,900 μg/kg 85JCAE-,1211,86

CONSENSUS REPORTS: Mercury and its compounds are on the Community Right-To-Know List.

OSHA PEL: TWA 0.01 mg(Hg)/m³; STEL 0.03 mg/m³ (skin)
ACGIH TLV: TWA 0.01 mg(Hg)/m³; STEL 0.03 mg(Hg)/m³
NIOSH REL: (Mercury, Aryl And Inorganic): CL 0.1 mg/m³ (skin)

SAFETY PROFILE: Poison by ingestion. A skin and severe eye irritant. See also MERCURY COMPOUNDS and SULFONATES. When heated to decomposition it emits very toxic fumes of NO_x, SO_x, and Hg.

BLK750 CAS:10347-38-3 HR: 3
BIS(TRIBENZYLSTANNYL)SULFIDE
mf: $C_{42}H_{42}SSn_2$ mw: 816.28

SYNS: BIS(TRIBENZYLTIN) SULFIDE □ DISTANNATHIANE, HEXAKIS(PHENYLMETHYL)-(9CI) □ SIRNIK TRIBENZYLCINICTY (CZECH) □ THIOBIS(TRIBENZYL-TIN) (8CI)

TOXICITY DATA WITH REFERENCE
skn-rbt 500 mg/24H MOD 28ZPAK -,233,72
eye-rbt 500 mg/24H SEV 28ZPAK -,233,72
orl-rat LD50:314 mg/kg 28ZPAK -,233,72

OSHA PEL: TWA 0.1 mg(Sn)/m³ (skin)
ACGIH TLV: TWA 0.1 mg(Sn)/m³ (skin) (Proposed: TWA 0.1 mg(Sn)/m³; STEL 0.2 mg(Sn)/m³ (skin))
NIOSH REL: (Organotin Compounds) TWA 0.1 mg(Sn)/m³

SAFETY PROFILE: Poison by ingestion. A skin and eye irritant. See also TIN COMPOUNDS and SULFIDES. When heated to decomposition it emits toxic fumes of SO_x.

For occupational chemical analysis use NIOSH: Organotin Compounds 5504.

BLL000 **CAS:30099-72-0** **HR: 3**
BIS(TRIBUTYL(SEBACOYLDIOXY))TIN
mf: $C_{34}H_{70}O_4Sn_2$ mw: 780.42

SYN: SEBACOYLDIOXYBIS(TRIBUTYLSTANNANE)

TOXICITY DATA WITH REFERENCE
ivn-mus LD50:18 mg/kg CSLNX* NX#03600

OSHA PEL: TWA 0.1 mg(Sn)/m³ (skin)
ACGIH TLV: TWA 0.1 mg(Sn)/m³ (skin) (Proposed: TWA 0.1 mg(Sn)/m³; STEL 0.2 mg(Sn)/m³ (skin))
NIOSH REL: (Organotin Compounds) TWA 0.1 mg(Sn)/m³

SAFETY PROFILE: Poison by intravenous route. See also TIN COMPOUNDS. When heated to decomposition it emits acrid smoke and irritating fumes.

For occupational chemical analysis use NIOSH: Organotin Compounds 5504.

BLL250 **CAS:12291-11-1** **HR: 3**
BIS((TRI-n-BUTYLSTANNYL)CYCLOPENTADIENYL) IRON
mf: $C_{34}H_{62}FeSn_2$ mw: 764.19

SYN: 1,1′-BIS(TRIBUTYLSTANNYL)FERROCENE

TOXICITY DATA WITH REFERENCE
ivn-mus LD50:56 mg/kg CSLNX* NX#05870

OSHA PEL: TWA 0.1 mg(Sn)/m³ (skin)
ACGIH TLV: TWA 0.1 mg(Sn)/m³ (skin) (Proposed: TWA 0.1 mg(Sn)/m³; STEL 0.2 mg(Sn)/m³ (skin))
NIOSH REL: (Organotin Compounds) TWA 0.1 mg(Sn)/m³

SAFETY PROFILE: Poison by intravenous route. See also TIN COMPOUNDS. When heated to decomposition it emits acrid smoke and irritating fumes.

For occupational chemical analysis use NIOSH: Organotin Compounds 5504.

BLL500 **CAS:25711-26-6** **HR: 3**
BIS(TRIBUTYLTIN) ITACONATE
mf: $C_{29}H_{58}O_4Sn_2$ mw: 708.25

SYN: METHYLENESUCCINYLOXYBIS(TRIBUTYLSTANNANE)

TOXICITY DATA WITH REFERENCE
ivn-mus LD50:180 mg/kg CSLNX* NX#03635

OSHA PEL: TWA 0.1 mg(Sn)/m³ (skin)
ACGIH TLV: TWA 0.1 mg(Sn)/m³ (skin) (Proposed: TWA 0.1 mg(Sn)/m³; STEL 0.2 mg(Sn)/m³ (skin))
NIOSH REL: (Organotin Compounds) TWA 0.1 mg(Sn)/m³

SAFETY PROFILE: Poison by intravenous route. See also TIN COMPOUNDS. When heated to decomposition it emits acrid smoke and irritating fumes.

For occupational chemical analysis use NIOSH: Organotin Compounds 5504.

BLL750 **CAS:56-35-9** **HR: 3**
BIS(TRIBUTYL TIN)OXIDE
mf: $C_{24}H_{54}OSn_2$ mw: 596.16

PROP: Air-sensitive liquid. D: 1.17 @ 20°/4°, bp: 220–230° @ 10 mm.

SYNS: BIOMET TBTO □ BIS-(TRI-N-BUTYLCIN)OXID (CZECH) □ BIS(TRIBUTYLOXIDE) of TIN □ BIS(TRIBUTYLSTANNYL)OXIDE □ BIS(TRI-N-BUTYLZINN)-OXYD (GERMAN) □ BTO □ BUTINOX □ C-Sn-9 □ ENT 24,979 □ HEXABUTYLDISTANNOXANE □ HEXABUTYLDITIN □ KYSLICNIK TRI-N-BUTYLCINICITY (CZECH) □ L.S. 3394 □ OTBE (FRENCH) □ OXYBIS(TRIBUTYLTIN) □ OXYDE de TRIBUTYLETAIN □ TBOT □ TBTO □ TRI-n-BUTYL-STANNANE OXIDE □ TRIBUTYLTIN OXIDE

TOXICITY DATA WITH REFERENCE
eye-rbt 50 µg/24H SEV 28ZPAK -,232,72
eye-rbt 460 µg BJIMAG 26,165,69
dni-omi 56,200 ppb AEMIDF 45,48,83
orl-rat TDLo:150 mg/kg (female 6-20D post):REP TXAPA9 97,113,89
orl-rat TDLo:150 mg/kg (female 6-20D post):TER TXAPA9 97,113,89
orl-rat TDLo:37,100 mg/kg/2Y-C:CAR FCTOD7 28,179,90
orl-rat LD50:87 mg/kg MarJV# 29MAR77
ipr-rat LD50:7210 µg/kg FCTXAV 7,47,69
scu-rat LD50:11,700 mg/kg TRIPA7 -,1,73
orl-mus LD50:55 mg/kg GISAAA 41(5),10,76
ipr-mus LD50:12,500 µg/kg RPTOAN 42,73,79
ivn-mus LD50:6 mg/kg EJTXAZ 9,31,76
orl-rbt LDLo:50 mg/kg SAIGBL 15,3,73
skn-rbt LD50:900 mg/kg EJTXAZ 9,31,76

CONSENSUS REPORTS: Reported in EPA TSCA Inventory.

OSHA PEL: TWA 0.1 mg(Sn)/m³ (skin)
ACGIH TLV: TWA 0.1 mg(Sn)/m³ (skin) (Proposed: TWA 0.1 mg(Sn)/m³; STEL 0.2 mg(Sn)/m³ (skin))
DFG MAK: 0.002 ppm (0.05 mg/m³)
NIOSH REL: (Organotin Compounds) TWA 0.1 mg(Sn)/m³

SAFETY PROFILE: A poison by ingestion, intraperitoneal, and intravenous routes. Moderately toxic by skin contact. An experimental teratogen. Other experimental reproductive effects. Questionable carcinogen with experimental carcinogenic data. Mutation data reported. A severe eye irritant. See also TIN COMPOUNDS. When heated to decomposition it emits acrid and irritating fumes.

For occupational chemical analysis use NIOSH: Organotin Compounds 5504.

BLL825 **CAS:881-99-2** **HR: 2**
m-BIS(TRICHLORMETHYL)BENZENE
mf: $C_8H_4Cl_6$ mw: 312.82

SYNS: m-BIS(TRICHLOROMETHYL)BENZENE □ 1,3-BIS(TRICHLOROMETHYL)BENZENE □ 1,3-DI(TRICHLOROMETHYL)BENZENE □ α,α′-HEXACHLORO-m-XYLENE □ α,α,α,α′,α′,α′-HEXACHLOROXYLENE □ α,α,α,α′,α′,α′-HEXACHLORO-m-XYLENE

TOXICITY DATA WITH REFERENCE
skn-rbt 500 mg MOD 34ZIAG -,308,69

orl-rat TDLo:2129 mg/kg (26W male):REP GNAMAP 21,34,82
orl-rat LD50:2924 mg/kg GNAMAP 21,34,82

SAFETY PROFILE: Moderately toxic by ingestion. Experimental reproductive effects. A skin irritant. Violent reaction when heated with oxidants (e.g., potassium nitrate, selenium dioxide, and sodium chlorate). When heated to decomposition it emits toxic fumes of Cl^-. See CHLORINATED HYDROCARBONS, AROMATIC.

BLM000 CAS:2629-78-9 HR: 3
BIS(TRICHLOROACETYL)PEROXIDE
mf: $C_4Cl_6O_4$ mw: 324.76

$Cl_3CCO{\cdot}OOCO{\cdot}CCl_3$

PROP: Shock-sensitive crystals from CCl_3F.

SAFETY PROFILE: A very shock-sensitive explosive which may detonate at room temperature. Upon decomposition it emits toxic fumes of Cl^-. See also PEROXIDES.

BLM250 HR: 3
BIS-2,4,5-TRICHLORO BENZENE DIAZO OXIDE
mf: $C_{12}H_4Cl_6N_4O$ mw: 212.72

SAFETY PROFILE: Ignites spontaneously. Explodes on impact or on contact with benzene. Upon decomposition it emits toxic fumes of Cl^- and NO_x.

BLM500 CAS:3064-70-8 HR: 3
BIS(TRICHLOROMETHYL)SULFONE
mf: $C_2Cl_6O_2S$ mw: 300.78

SYN: N-1386 BIOCIDE

TOXICITY DATA WITH REFERENCE
orl-rat LD50:691 mg/kg TOXID9 4,16,84
ivn-mus LD50:18 mg/kg CSLNX* NX#04617

CONSENSUS REPORTS: Reported in EPA TSCA Inventory.

SAFETY PROFILE: Poison by intravenous route. Moderately toxic by ingestion. When heated to decomposition it emits very toxic fumes of Cl^- and SO_x.

BLM750 CAS:2532-50-5 HR: 3
BIS(TRICHLORO METHYL)TRISULFIDE
mf: $C_2Cl_6S_3$ mw: 332.90

SYNS: BISTRICHLOROMETHYLTRISULFID (CZECH) □ TRITHIOBIS (TRICHLOROMETHANE)

TOXICITY DATA WITH REFERENCE
skn-rbt 500 mg/24H MOD 28ZPAK -,170,72
eye-rbt 100 mg/24H SEV 28ZPAK -,170,72
orl-rat LD50:676 mg/kg 28ZPAK -,170,72
ivn-mus LD50:56 mg/kg CSLNX* NX#04597

SAFETY PROFILE: Poison by intravenous route. Moderately toxic by ingestion. A skin and eye irritant. See also SULFIDES. When heated to decomposition it emits very toxic fumes of Cl^- and SO_x.

BLN000 CAS:63885-02-9 HR: 2
BIS(2,3,5-TRICHLOROPHENYLTHIO)ZINC
mf: $C_{12}H_4Cl_6S_2Zn$ mw: 490.35

SYN: 2,3,5-TRICHLOROFENOLAT ZINECNATY (CZECH)

TOXICITY DATA WITH REFERENCE
skn-rbt 500 mg/24H MOD 28ZPAK -,11,72
eye-rbt 50 µg/24H SEV 28ZPAK -,11,72
orl-rat LD50:4260 mg/kg 28ZPAK -,11,72

CONSENSUS REPORTS: Zinc and its compounds are on the Community Right-To-Know List.

SAFETY PROFILE: Mildly toxic by ingestion. A severe eye and skin irritant. See also ZINC COMPOUNDS and CHLORINATED HYDROCARBONS, AROMATIC. When heated to decomposition it emits very toxic fumes of ZnO, Cl^- and SO_x.

BLN100 CAS:80660-68-0 HR: 3
BIS(TRIETHYLENETETRAMINE)TUNGSTATONICKEL
mf: $C_{12}H_{36}N_8O_4W{\cdot}Ni$ mw: 599.12

SYNS: NICKEL(2+), BIS(N,N'-BIS(2-AMINOETHYL)-1,2-ETHANEDIAMINE-N,N',N^N)-, (T-4)-TETRAOXOTUNGSTATE(2-) (1:1) □ NICKEL, BIS(TRIETHYLENETETRAMINE)TUNGSTATO-

TOXICITY DATA WITH REFERENCE
ipr-mus LD50:82,500 µg/kg IJEBA6 19,1187,81

OSHA PEL: TWA 1 mg(Ni)/m³
ACGIH TLV: TWA 5 mg(W)/m³; STEL 10 mg(W)/m³

SAFETY PROFILE: Poison by intraperitoneal route. When heated to decomposition it emits toxic fumes of NO_x, Ni, and W.

BLN250 HR: 3
BIS(TRIETHYL TIN)ACETYLENE
mf: $C_{14}H_{30}Sn_2$ mw: 435.77

SAFETY PROFILE: A sensitive, powerful explosive. Incompatible with stannic chloride. When heated to decomposition it emits acrid smoke and fumes. See also TIN COMPOUNDS and ACETYLENE COMPOUNDS.

BLN500 CAS:57-52-3 HR: 3
BIS(TRIETHYLTIN) SULFATE
mf: $C_{12}H_{30}O_4SSn_2$ mw: 507.86

SYNS: TRIAETHYLZINNSULFAT (GERMAN) □ TRIETHYLHYDROXYSTANNANE SULFATE (2:1) (8CI) □ TRIETHYLHYDROXYTIN SULFATE □ TRIETHYLTIN SULPHATE

TOXICITY DATA WITH REFERENCE
orl-rat LDLo:10 mg/kg BJPCAL 10,16,55
ipr-rat LD50:5700 µg/kg BJPCAL 10,16,55
scu-rat LDLo:25 mg/kg BJPCAL 10,16,55
ivn-rat LD50:9050 µg/kg AEPPAE 242,370,61
par-rat LD50:6 mg/kg BIJOAK 61,406,55
orl-rbt LDLo:10 mg/kg BJPCAL 10,16,55
ivn-rbt LDLo:3 mg/kg BJPCAL 10,16,55
ipr-gpg LD50:3 mg/kg BJIMAG 23,222,66
ivn-brd LDLo:3 mg/kg BJPCAL 10,16,55

OSHA PEL: TWA 0.1 mg(Sn)/m³ (skin)

ACGIH TLV: TWA 0.1 mg(Sn)/m³ (skin) (Proposed: TWA 0.1 mg(Sn)/m³; STEL 0.2 mg(Sn)/m³ (skin))
NIOSH REL: (Organotin Compounds) TWA 0.1 mg(Sn)/m³

SAFETY PROFILE: Poison by ingestion, intraperitoneal, subcutaneous, intravenous, and parenteral routes. See also TIN COMPOUNDS and SULFATES. When heated to decomposition it emits toxic fumes of SO_x.

For occupational chemical analysis use NIOSH: Organotin Compounds 5504.

BLN750 CAS:52112-09-1 HR: 3
BIS(TRIFLUOROACETOXY)DIBUTYLTIN
mf: $C_{12}H_{18}F_6O_4Sn$ mw: 458.99

SYNS: DIBUTYLTIN BIS(TRIFLUOROACETATE) □ DIBUTYLBIS (TRIFLUOROACETOXY)STANNANE □ DTBT □ STANNOUS DIBUTYLDITRIFLUOROACETATE □ TIN DIBUTYLDITRIFLUOROACETATE

TOXICITY DATA WITH REFERENCE
orl-rat LD50:55 mg/kg GISAAA 46(7),18,81
skn-rat LD50:1 g/kg GISAAA 46(7),18,81
orl-mus LD50:53,600 μg/kg GISAAA 41(5),10,76

OSHA PEL: TWA 0.1 mg(Sn)/m³ (skin)
ACGIH TLV: TWA 0.1 mg(Sn)/m³ (skin) (Proposed: TWA 0.1 mg(Sn)/m³; STEL 0.2 mg(Sn)/m³ (skin))
NIOSH REL: (Organotin Compounds) TWA 0.1 mg(Sn)/m³

SAFETY PROFILE: Poison by ingestion. Moderately toxic by skin contact. See also TIN COMPOUNDS and FLUORIDES. When heated to decomposition it emits toxic fumes of F^-.

For occupational chemical analysis use NIOSH: Organotin Compounds 5504.

BLO000 CAS:383-73-3 HR: 3
BIS(TRIFLUOROACETYL)PEROXIDE
mf: $C_4F_6O_4$ mw: 226.03

PROP: Crystals.

SAFETY PROFILE: A poison. May explode spontaneously at room temperature. Upon decomposition it emits toxic fumes of F^-. See also PEROXIDES.

BLO250 CAS:328-74-5 HR: 3
3,5-BIS(TRIFLUOROMETHYL)ANILINE
mf: $C_8H_5F_6N$ mw: 229.14

SYN: α,α,α,α,α,α-HEXAFLUORO-3,5-XYLIDINE

TOXICITY DATA WITH REFERENCE
ipr-mus LDLo:31 mg/kg CBCCT* 4,323,52
ivn-mus LD50:25 mg/kg CBCCT* 6,143,54

CONSENSUS REPORTS: Reported in EPA TSCA Inventory.

SAFETY PROFILE: Poison by intraperitoneal and intravenous routes. See also FLUORIDES. When heated to decomposition it emits very toxic fumes of $F^-$and NO_x.

BLO270 CAS:402-31-3 HR: 3
1,3-BIS(TRIFLUOROMETHYL)BENZENE
mf: $C_8H_4F_6$ mw: 214.11

PROP: Liquid. D: 1.378, bp: 116–116.3°.

SAFETY PROFILE: When heated to 90°C a mixture with nitric and sulfuric acids emits spark-sensitive explosive vapors. When heated to decomposition it emits toxic fumes of F^-.

BLO280 CAS:650-52-2 HR: 3
BIS(TRIFLUOROMETHYL)CHLOROPHOSPHINE
mf: C_2ClF_6P mw: 204.44

PROP: Liquid. Bp: 21–21.5°.

SAFETY PROFILE: Ignites spontaneously in air. When heated to decomposition it emits toxic fumes of F^-, Cl^-, and PO_x. See also PHOSPHINE.

BLO300 CAS:431-97-0 HR: 3
BIS(TRIFLUOROMETHYL)CYANOPHOSPHINE
mf: C_3F_6NP mw: 195.00

PROP: Liquid. Bp: 48°.

CONSENSUS REPORTS: Cyanide and its compounds are on the Community Right-To-Know List.

SAFETY PROFILE: Ignites spontaneously on contact with air. When heated to decomposition it emits toxic fumes of F^-, PO_x, CN^-, and NO_x. See also CYANIDE and PHOSPHINE.

BLO325 CAS:372-64-5 HR: 3
BIS(TRIFLUOROMETHYL)DISULFIDE
mf: $C_2F_6S_2$ mw: 202.13

PROP: Liquid. Bp: 34.6°.

SAFETY PROFILE: Mixtures of the solid with chlorine mono- or tri-fluorides are explosive. Dilute with halogenated solvents. When heated to decomposition it emits toxic fumes of F^- and SO_x. See also SULFIDES.

BLP250 CAS:30184-88-4 HR: 3
2,2-BIS(TRIFLUOROMETHYL)-4-METHYL-5-PHENYL-OXAZOLIDINE HYDRATE
mf: $C_{12}H_{11}F_6NO{\cdot}H_2O$ mw: 317.26

TOXICITY DATA WITH REFERENCE
orl-mus LD50:200 mg/kg JMCMAR 13,1215,70
ipr-mus LD50:300 mg/kg JMCMAR 13,1215,70

SAFETY PROFILE: Poison by ingestion and intraperitoneal routes. When heated to decomposition it emits very toxic fumes of F^- and NO_x.

BLP300 CAS:2154-71-4 HR: 3
BIS(TRIFLUOROMETHYL)NITROXIDE
mf: C_2F_6NO mw: 168.02

PROP: Purple gas, deep violet liquid or yellow crystals. Mp: −70°, bp: −25°.

SAFETY PROFILE: Explodes violently at room temperature. Upon decomposition it emits toxic fumes of F^- and NO_x.

BLP325 **CAS:24095-80-5** **HR: 3**
2-(3,5-BIS(TRIFLUOROMETHYL)PHENYL)-N-METHYL-HYDRAZINECARBOTHIOAMIDE (9CI)
mf: $C_{10}H_9F_6N_3S$ mw: 317.28

SYNS: 1-((3,5-BIS-TRIFLUOROMETHYL)PHENYL)-4-METHYL-THIOSEMICARBAZIDE □ CIBA 2696GO □ 1-(α,α,α,α',α',α'-HEXAFLUORO-3,5-XYLYL)-4-METHYL-3-THIO-SEMICARBAZIDE

TOXICITY DATA WITH REFERENCE
orl-rat TDLo:50 mg/kg (6-15D preg):TER ARZNAD 23,797,73
orl-rat TDLo:25 mg/kg (6-15D preg):REP ARZNAD 23,797,73
orl-rat TDLo:30 mg/kg (female 1-3D post):REP IJOCAP 23,1243,84
orl-rat LD50:414 mg/kg ARZNAD 23,797,73
ipr-rat LD50:212 mg/kg ARZNAD 23,797,73
orl-mus LD50:269 mg/kg ARZNAD 23,797,73
orl-dog LDLo:1500 mg/kg ARZNAD 23,797,73

SAFETY PROFILE: Poison by ingestion and intraperitoneal routes. An experimental teratogen. Other experimental reproductive effects. When heated to decomposition it emits toxic fumes of F^-, SO_x, and NO_x.

BLP500 **HR: 3**
BIS(TRIFLUOROMETHYL)PHOSPHORUS(III) AZIDE
mf: $C_2F_6N_3P$ mw: 211.01

SAFETY PROFILE: Explosive, very unstable even at −196°. Upon decomposition it emits toxic fumes of F^-, PO_x, and NO_x. See also AZIDES.

BLQ250 **CAS:30192-67-7** **HR: 3**
α,α-BIS(TRIFLUOROMETHYL)-1-PIPERIDINEMETHANOL HYDRATE
mf: $C_8H_{11}F_6NO•H_2O$ mw: 269.22

TOXICITY DATA WITH REFERENCE
orl-mus LD50:300 mg/kg JMCMAR 13,1215,70
ipr-mus LD50:300 mg/kg JMCMAR 13,1215,70

SAFETY PROFILE: Poison by ingestion and intraperitoneal routes. When heated to decomposition it emits very toxic fumes of F^- and NO_x.

BLQ325 **CAS:371-78-8** **HR: 3**
BIS(TRIFLUOROMETHYL)SULFIDE
mf: C_2F_6S mw: 170.07

PROP: Gas. Mp: −63.5°, bp: −22.2°.

SAFETY PROFILE: Mixtures of the solid with chlorine mono- or tri-fluorides are explosive. Dilute with halogenated solvents. When heated to decomposition it emits toxic fumes of F^- and SO_x. See also SULFIDES.

BLQ500 **CAS:28399-14-6** **HR: 3**
2,2-BIS(TRIFLUOROMETHYL)THIAZOLIDINE HYDRATE
mf: $C_5H_5F_6NS•H_2O$ mw: 243.19

TOXICITY DATA WITH REFERENCE
orl-mus LD50:300 mg/kg JMCMAR 13,1215,70
ipr-mus LD50:300 mg/kg JMCMAR 13,1215,70

SAFETY PROFILE: Poison by ingestion and intraperitoneal routes. When heated to decomposition it emits very toxic fumes of F^-, NO_x, and SO_x.

BLQ600 **HR: 2**
2,3-BISTRIMETHYLACETOXYMETHYL-1-METHYLPYRROLE
mf: $C_{17}H_{27}NO_4$ mw: 309.45

TOXICITY DATA WITH REFERENCE
skn-mus TDLo:291 mg/kg/47W-I:CAR CALEDQ 17,61,82

SAFETY PROFILE: Questionable carcinogen with experimental carcinogenic data. When heated to decomposition it emits toxic fumes of NO_x.

BLQ750 **CAS:64011-39-8** **HR: 3**
BIS(TRIMETHYLHEXYL)TIN DICHLORIDE
mf: $C_{18}H_{38}Cl_2Sn$ mw: 444.15

SYNS: DICHLORODIISONONYL STANNANE □ DIISONONYLTIN DICHLORIDE

TOXICITY DATA WITH REFERENCE
ivn-rat LDLo:10 mg/kg BJIMAG 15,15,58

OSHA PEL: TWA 0.1 mg(Sn)/m³ (skin)
ACGIH TLV: TWA 0.1 mg(Sn)/m³ (skin) (Proposed: TWA 0.1 mg(Sn)/m³; STEL 0.2 mg(Sn)/m³ (skin))
NIOSH REL: (Organotin Compounds) TWA 0.1 mg(Sn)/m³

SAFETY PROFILE: Poison by intravenous route. See also TIN COMPOUNDS. When heated to decomposition it emits toxic fumes of Cl^-.

For occupational chemical analysis use NIOSH: Organotin Compounds 5504.

BLQ850 **CAS:73452-31-0** **HR: 3**
N,N'-BIS(TRIMETHYLSILYL)AMINOBORANE
mf: $C_6H_{20}BNSi_2$ mw: 173.21

$((CH_3)_3SI)_2NBH_2$

SAFETY PROFILE: Ignites on contact with air. When heated to decomposition it emits toxic fumes of NO_x. See also BORANES and BORON COMPOUNDS.

BLQ900 **CAS:86045-52-5** **HR: 3**
cis-BIS(TRIMETHYLSILYLAMINO)TELLURIUM TETRAFLUORIDE
mf: $C_6H_{20}F_4N_2Si_2Te$ mw: 380.00

$((CH_3)_3SiNH)_2TeF_4$

SAFETY PROFILE: Explodes when heated to 100°C. During storage it converts to an explosive solid. When heated to decomposition it emits toxic fumes of F^-, SO_x, NO_x, and Te. See also TELLURIUM COMPOUNDS.

BLQ950 **CAS:1000-70-0** **HR: 2**
BIS(TRIMETHYLSILYL)CARBODIIMIDE
mf: $C_7H_{18}N_2Si_2$ mw: 186.45

SYNS: SILANAMINE, N,N′-METHANE TETRAYLBIS(1,1,1-TRIMETHYL)- □ CARBODIIMIDE, BIS(TRIMETHYLSILYL)-(7CI,8CI) □ N,N′-METHANE TETRAYLBIS(1,1,1-TRIMETHYLSILANAMINE)

TOXICITY DATA WITH REFERENCE
orl-rat LD50:1728 mg/kg GISAAA 55(6),86,90
orl-rbt LD50:1728 mg/kg GISAAA 55(6),86,90

CONSENSUS REPORTS: Reported in EPA TSCA Inventory.

SAFETY PROFILE: Moderately toxic by ingestion. When heated to decomposition it emits toxic vapors of NO_x.

BLR000 **CAS:1746-09-4** **HR: 3**
BIS(TRIMETHYLSILYL)CHROMATE
mf: $C_6H_{18}CrO_4Si_2$ mw: 262.57

$((CH_3)_3SiO)_2CrO_2$

CONSENSUS REPORTS: Chromium and its compounds are on the Community Right-To-Know List.

SAFETY PROFILE: May explode if heated above 75°C. When heated to decomposition it emits acrid smoke and fumes. See also CHROMIUM COMPOUNDS.

BLR125 **CAS:692-56-8** **HR: 3**
1,2-BIS(TRIMETHYLSILYL)HYDRAZINE
mf: $C_6H_{20}N_2Si_2$ mw: 176.41

$(CH_3)_3SiNHNHSi(CH_3)_3$

PROP: Bp: 149°.

SAFETY PROFILE: Hypergolic reaction with strong oxidants (e.g., fluorine or fuming nitric acid). When heated to decomposition it emits toxic fumes of NO_x. See also HYDRAZINE.

BLR140 **CAS:4656-04-6** **HR: 3**
BIS(TRIMETHYLSILYL)MERCURY
mf: $C_6H_{18}HgSi_2$ mw: 346.97

$((CH_3)_3Si)_2Hg$

PROP: Very light-sensitive yellow crystals. Mp: 102–104° (decomp). Sol in Et_2O, THF, C_6H_6, hexane, and CS_2.

CONSENSUS REPORTS: Mercury and its compounds are on the Community Right-To-Know List.

SAFETY PROFILE: May ignite spontaneously in air. When heated to decomposition it emits toxic fumes of Hg. See also MERCURY COMPOUNDS.

BLR250 **HR: 3**
BISTRIMETHYL SILYL OXIDE
mf: $C_6H_{18}OSi_2$ mw: 162.44

PROP: Flash p: −1°C.

SAFETY PROFILE: A very dangerous fire hazard. When heated to decomposition it emits acrid smoke and fumes.

BLR500 **CAS:23115-33-5** **HR: 3**
BIS(TRIMETHYLSILYL)PEROXOMONOSULFATE
mf: $C_6H_{18}O_5SSi_2$ mw: 258.61

$(CH_3)_3SiOSO_2OOSi(CH_3)_3$

PROP: Liquid.

SAFETY PROFILE: May decompose violently at room temperature and evolve toxic sulfur trioxide. See also PEROXIDES, ORGANIC; and SULFATES.

BLR625 **CAS:918-99-0** **HR: 3**
N,N′-BIS(2,2,2-TRINITROETHYL)UREA
mf: $C_5H_6N_8O_{13}$ mw: 386.15

$((O_2N)_3CCH_2NH)_2CO$

SAFETY PROFILE: Mixtures with sodium hydroxide are storage hazards due to the formation of unstable reaction products. When heated to decomposition it emits toxic fumes of NO_x.

BLR750 **CAS:28930-30-5** **HR: 2**
BIS(TRINITROPHENYL)SULFIDE
mf: $C_{12}H_4N_6O_{12}S$ mw: 456.20

SYNS: HEXANITRODIPHENYLSULFIDE □ PICRYL SULFIDE

TOXICITY DATA WITH REFERENCE
orl-rat LD50:1200 mg/kg TNICS* 13,132,73
orl-mus LD50:170 mg/kg TNICS* 13,132,73

SAFETY PROFILE: Moderately toxic by ingestion. See also SULFIDES and NITRO COMPOUNDS of AROMATIC HYDROCARBONS. See NITRATES for fire and explosion hazard. This material is a powerful explosive and has an added military advantage in that its explosive gases contain irritating and very toxic SO_x. See also EXPLOSIVES, HIGH.

BLS000 **HR: 3**
BISTRIPERCHLORATO SILICON OXIDE
mf: $Cl_6O_{25}Si_2$ mw: 508.88

SAFETY PROFILE: Heating to decomposition may form an explosive product. When heated to decomposition it emits toxic fumes of Cl^-. See also PERCHLORATES.

BLS250 **CAS:14264-16-5** **HR: 3**
BIS(TRIPHENYLPHOSPHINE)DICHLORONICKEL
mf: $C_{24}H_{54}P_2 \cdot Cl_2Ni$ mw: 534.33

SYNS: BIS(TRI-N-BUTYLPHOSPHINE)DICHLORONICKEL □ TRIBUTYL-PHOSPHINE compounded with NICKELCHLORIDE (2:1)

TOXICITY DATA WITH REFERENCE
ivn-mus LD50:56 mg/kg CSLNX* NX#03119

CONSENSUS REPORTS: Reported in EPA TSCA Inventory. Nickel and its compounds are on the Community Right-To-Know List.

OSHA PEL: TWA 0.1 mg (Ni)/m³
ACGIH TLV: TWA 0.1 mg(Ni)/m³; (Proposed: TWA 0.05 mg(Ni)/m³; Human Carcinogen)

SAFETY PROFILE: Suspected carcinogen. Poison by intravenous route. See also NICKEL COMPOUNDS. When heated to decomposition it emits very toxic fumes of Cl^- and PO_x.

BLS500 CAS:15709-62-3 HR: 3
BIS(TRIPHENYL PHOSPHINE)NICKEL DITHIOCYANATE
mf: $C_{38}H_{30}N_2NiP_2S_2$ mw: 699.47

SYN: NICKEL BISTRIPHENYLPHOSPHINE DITHIOCYANATE

TOXICITY DATA WITH REFERENCE
ivn-mus LD50:180 mg/kg CSLNX* NX#01983

CONSENSUS REPORTS: Nickel and its compounds are on the Community Right-To-Know List.

SAFETY PROFILE: Poison by intravenous route. See also NICKEL COMPOUNDS and THIOCYANATES. When heated to decomposition it emits very toxic fumes of SO_x, PO_x, NO_x, and CN^-.

BLS750 CAS:1624-02-8 HR: 2
BIS(TRIPHENYL SILYL)CHROMATE
mf: $C_{36}H_{30}CrO_4Si_2$ mw: 634.84

SYN: CHROMIC ACID, BIS(TRIPHENYLSILYL) ESTER

TOXICITY DATA WITH REFERENCE
orl-rat LD50:3360 mg/kg TXAPA9 28,313,74
skn-rbt LD50:710 mg/kg TXAPA9 28,313,74

CONSENSUS REPORTS: Reported in EPA TSCA Inventory. Chromium and its compounds are on the Community Right-To-Know List.

OSHA PEL: CL 0.1 mg(CrO_3)/m³
ACGIH TLV: TWA 0.05 mg(CrO_3)/m³
NIOSH REL: (Chromium(VI)): TWA 0.025 mg(Cr (VI))/m³; CL 0.05/15M

SAFETY PROFILE: Moderately toxic by ingestion and skin contact. See also CHROMIUM COMPOUNDS and ESTERS. When heated to decomposition it emits toxic fumes of CrO_3 particulates.

For occupational chemical analysis use NIOSH: Chromium Hexavalent 7024.

BLS900 CAS:73940-87-1 HR: 3
BIS(TRIPHENYLTIN)ACETYLENEDICARBOXYLATE
mf: $C_{40}H_{30}O_4Sn_2$ mw: 812.08

SYNS: ETHYNYLENEBIS(CARBONYLOXY)BIS(TRIPHENYLSTANNANE) □ STANNANE, ETHYNYLENEBIS(CARBONYLOXY)BIS(TRIPHENYL-

TOXICITY DATA WITH REFERENCE
ivn-mus LD50:18 mg/kg CSLNX* NX#05963

OSHA PEL: TWA 0.1 mg(Sn)/m³
ACGIH TLV: TWA 0.1 mg(Sn)/m³; STEL 0.2 mg/m³ (skin)
NIOSH REL: (Organotin compound): 10H TWA 0.1 mg(Sn)/m³

SAFETY PROFILE: Poison by intravenous route. When heated to decomposition it emits toxic fumes of Sn.

For occupational chemical analysis use NIOSH: Organotin Compounds 5504.

BLT000 CAS:3021-41-8 HR: 3
BIS(TRIPHENYLTIN)SULFATE
mf: $C_{36}H_{30}Sn_2 \cdot O_4S$ mw: 796.10

SYN: TRIPHENYLSTANNANE SULFATE (2:1)

TOXICITY DATA WITH REFERENCE
ipr-rat LD50:5700 μg/kg 85JCAE-,1253,86
ivn-mus LD50:18 mg/kg CSLNX* NX#04819

OSHA PEL: TWA 0.1 mg(Sn)/m³ (skin)
ACGIH TLV: TWA 0.1 mg(Sn)/m³ (skin) (Proposed: TWA 0.1 mg(Sn)/m³; STEL 0.2 mg(Sn)/m³ (skin))
NIOSH REL: (Organotin Compounds) TWA 0.1 mg(Sn)/m³

SAFETY PROFILE: Poison by intraperitoneal and intravenous routes. See also TIN COMPOUNDS and SULFATES. When heated to decomposition it emits toxic fumes of SO_x.

For occupational chemical analysis use NIOSH: Organotin Compounds 5504.

BLT250 CAS:77-80-5 HR: 3
BIS(TRIPHENYLTIN)SULFIDE
mf: $C_{36}H_{30}SSn_2$ mw: 732.10

PROP: Colorless crystals. Mp: 144°. Sol in organic solvents.

SYN: 1,1,1,3,3,3-HEXAPHENYLDISTANNTHIANE

TOXICITY DATA WITH REFERENCE
orl-mus LD50:710 mg/kg AECTCV 14,111,85
ivn-mus LD50:180 mg/kg CSLNX* NX#05814

OSHA PEL: TWA 0.1 mg(Sn)/m³ (skin)
ACGIH TLV: TWA 0.1 mg(Sn)/m³ (skin) (Proposed: TWA 0.1 mg(Sn)/m³; STEL 0.2 mg(Sn)/m³ (skin))
NIOSH REL: (Organotin Compounds) TWA 0.1 mg(Sn)/m³

SAFETY PROFILE: A poison via intravenous route. Moderately toxic by ingestion. See also TIN COMPOUNDS and SULFIDES. When heated to decomposition it emits toxic fumes of SO_x.

For occupational chemical analysis use NIOSH: Organotin Compounds 5504.

BLT300 **CAS:1067-29-4** **HR: 3**
BIS(TRIPROPYLTIN)OXIDE
mf: $C_{18}H_{42}OSn_2$ mw: 511.98

PROP: Air-sensitive liquid. Bp: 154.5° @ 3.5 mm.

SYNS: DISTANNOXANE, 1,1,1,3,3,3-HEXAPROPYL- □ 1,1,1,3,3,3-HEXAPROPYLDISTANNOXANE

TOXICITY DATA WITH REFERENCE
ivn-mus LD50:5600 μg/kg CSLNX* NX#03791

CONSENSUS REPORTS: Reported in EPA TSCA Inventory.

OSHA PEL: 8H TWA 0.1 mg(Sn)/m³
ACGIH TLV: TWA 0.1 mg(Sn)/m³; STEL 0.2 mg/m³ (skin)
NIOSH REL: (Organotin compound): 10H TWA 0.1 mg(Sn)/m³

SAFETY PROFILE: Poison by intravenous route. When heated to decomposition it emits toxic fumes of Sn.

For occupational chemical analysis use NIOSH: Organotin Compounds 5504.

BLT500 **CAS:74039-78-4** **HR: 3**
BIS(TRIS(p-CHLOROPHENYL)PHOSPHINE)MERCURIC CHLORIDE COMPLEX
mf: $C_{36}H_{24}Cl_6P_2{\cdot}Cl_2Hg$ mw: 1002.73

SYN: TRIS(p-CHLOROPHENYL)PHOSPHINE COMPLEX with MERCURIC CHLORIDE (2:1)

TOXICITY DATA WITH REFERENCE
ivn-mus LD50:56 mg/kg CSLNX* NX#02647

CONSENSUS REPORTS: Mercury and its compounds are on the Community Right-To-Know List.

OSHA PEL: CL 0.1 mg(Hg)/m³ (skin)
ACGIH TLV: TWA 0.1 mg(Hg)/m³ (skin)
NIOSH REL: (Mercury, Aryl And Inorganic): CL 0.1 mg/m³ (skin)

SAFETY PROFILE: Poison by intravenous route. See also MERCURY COMPOUNDS and CHLORIDES. When heated to decomposition it emits very toxic fumes of Cl^-, PO_x and Hg.

BLT750 **CAS:74039-79-5** **HR: 3**
BIS(TRIS(p-DIMETHYLAMINOPHENYL)PHOSPHINE) MERCURIC CHLORIDE COMPLEX
mf: $C_{48}H_{60}N_6P_2{\cdot}Cl_2Hg$ mw: 1054.57

SYN: TRIS(p-DIMETHYLAMINOPHENYL) PHOSPHINE COMPLEX with MERCURIC CHLORIDE (2:1)

TOXICITY DATA WITH REFERENCE
ivn-mus LD50:100 mg/kg CSLNX* NX#02644

CONSENSUS REPORTS: Mercury and its compounds are on the Community Right-To-Know List.

OSHA PEL: CL 0.1 mg(Hg)/m³ (skin)
ACGIH TLV: TWA 0.1 mg(Hg)/m³ (skin)
NIOSH REL: (Mercury, Aryl And Inorganic): CL 0.1 mg/m³ (skin)

SAFETY PROFILE: Poison by intravenous route. See also MERCURY COMPOUNDS and CHLORIDES. When heated to decomposition it emits very toxic fumes of NO_x, PO_x, Cl^-, and Hg.

BLT775 **CAS:38402-95-8** **HR: 3**
BIS(TRIS(p-DIMETHYLAMINOPHENYL)PHOSPHINE OXIDE)STANNIC CHLORIDE COMPLEX
mf: $C_{48}H_{60}N_6O_2P_2{\cdot}Cl_4Sn$ mw: 1075.57

SYN: PHOSPHINE OXIDE, TRIS(p-DIMETHYLAMINOPHENYL)-, compounded with STANNIC CHLORIDE (2:1)

TOXICITY DATA WITH REFERENCE
ivn-mus LD50:180 mg/kg CSLNX* NX#02651

OSHA PEL: TWA 2 mg(Sn)/m³
ACGIH TLV: TWA 2 mg(Sn)/m³

SAFETY PROFILE: Poison by intravenous route. When heated to decomposition it emits toxic fumes of NO_x, PO_x, Sn, and Cl^-.

BLU000 **CAS:13356-08-6** **HR: 2**
BIS(TRIS(β,β-DIMETHYLPHENETHYL)TIN)OXIDE
mf: $C_{60}H_{78}OSn_2$ mw: 1052.76

PROP: Crystals or powder. Sol in $CHCl_3$, C_6H_6. Sltly sol in Me_2CO; insol in H_2O.

SYNS: BENDEX □ BIS(TRIS(2-METHYL-2-PHENYLPROPYL)TIN)OXIDE □ DI(TRI-(2,2-DIMETHYL-2-PHENYLETHYL)TIN)OXIDE □ ENT 27,738 □ FENBUTATIN OXIDE □ HEXAKIS(β,β-DIMETHYLPHENETHYL)DISTANNOXANE □ HEXAKIS(2-METHYL-2-PHENYLPROPYL) DISTANNOXANE □ SD 14114 □ SHELL SD-14114 □ TORQUE □ VENDEX

TOXICITY DATA WITH REFERENCE
orl-rat LD50:2630 mg/kg 85ARAE 1,17,77
skn-rat LD50:1000 mg/kg TIUSAD 110,6,76

OSHA PEL: TWA 0.1 mg(Sn)/m³ (skin)
ACGIH TLV: TWA 0.1 mg(Sn)/m³ (skin) (Proposed: TWA 0.1 mg(Sn)/m³; STEL 0.2 mg(Sn)/m³ (skin))
NIOSH REL: (Organotin Compounds) TWA 0.1 mg(Sn)/m³

SAFETY PROFILE: Moderately toxic by ingestion and skin contact. See also TIN COMPOUNDS. When heated to decomposition it emits acrid smoke and irritating fumes.

For occupational chemical analysis use NIOSH: Organotin Compounds 5504.

BLU250 **CAS:74039-80-8** **HR: 3**
BIS(TRIS(p-METHOXYPHENYL)PHOSPHINE)MERCURIC CHLORIDE COMPLEX
mf: $C_{42}H_{42}O_6P_2{\cdot}Cl_2Hg$ mw: 976.27

SYN: TRIS(p-METHOXYPHENYL) PHOSPHINE COMPLEX with MERCURIC CHLORIDE (2:1)

TOXICITY DATA WITH REFERENCE
ivn-mus LD50:180 mg/kg CSLNX* NX#02645

CONSENSUS REPORTS: Mercury and its compounds are on the Community Right-To-Know List.

OSHA PEL: CL 0.1 mg(Hg)/m³ (skin)
ACGIH TLV: TWA 0.1 mg(Hg)/m³ (skin)
NIOSH REL: (Mercury, Aryl And Inorganic): CL 0.1 mg/m³ (skin)

SAFETY PROFILE: Poison by intravenous route. See also MERCURY COMPOUNDS and CHLORIDES. When heated to decomposition it emits very toxic vapors of PO_x, Cl^-, and Hg.

BLU500 CAS:74039-81-9 HR: 3
BIS(TRIS(p-METHYLTHIOPHENYL)PHOSPHINE)MERCURIC CHLORIDE COMPLEX
mf: $C_{42}H_{42}P_2S_6{\cdot}Cl_2Hg$ mw: 1072.63

SYN: TRIS(p-METHYLTHIOPHENYL) PHOSPHINE COMPLEX with MERCURIC CHLORIDE (2:1)

TOXICITY DATA WITH REFERENCE
ivn-mus LD50:180 mg/kg CSLNX* NX#02646

CONSENSUS REPORTS: Mercury and its compounds are on the Community Right-To-Know List.

OSHA PEL: CL 0.1 mg(Hg)/m³ (skin)
ACGIH TLV: TWA 0.1 mg(Hg)/m³ (skin)
NIOSH REL: (Mercury, Aryl And Inorganic): CL 0.1 mg/m³ (skin)

SAFETY PROFILE: Poison by intravenous route. See also MERCURY COMPOUNDS and CHLORIDES. When heated to decomposition it emits very toxic fumes of PO_x, SO_x, Cl^-, and Hg.

BLV000 CAS:5169-78-8 HR: 3
BITIODIN
mf: $C_{15}H_{17}NS_2$ mw: 275.45

PROP: Yellow crystals. Mp: 64–65°, bp: 178–184° @ 4–5 mm.

SYNS: AT 327 □ CR/662 □ 3-(DI-2-THIENYLMETHYLENE)-1-METHYLPIPERIDINE □ 1-METHYL-3-PIPERIDYLIDENEDI(2-THIENYL) METHANE □ TIPEDINE □ TIPEPIDINE

TOXICITY DATA WITH REFERENCE
orl-mus LD50:867 mg/kg MEIEDD 11,1490,89
ipr-mus LD50:294 mg/kg MEIEDD 11,1490,89
scu-mus LD50:222 mg/kg CPBTAL 7,372,59
ivn-mus LD50:55 mg/kg PCJOAU 10,1482,76
ims-mus LD50:308 mg/kg MEIEDD 11,1490,89
ivn-dog LD50:44 mg/kg CPBTAL 7,372,59

SAFETY PROFILE: A poison via subcutaneous, intraperitoneal, intravenous, and intramuscular routes. Moderately toxic by ingestion. An antitussive. When heated to decomposition it emits very toxic fumes of NO_x and SO_x.

BLV075 HR: 3
BITIS ARIETANS VENOM

SYNS: B. ARIETANS VENOM □ SNAKE VENOM BITIS ARIETANS

TOXICITY DATA WITH REFERENCE
ipr-mus LD50:560 µg/kg TOXIA6 18,384,80
scu-mus LD50:600 µg/kg TOXIA6 20,509,82
ivn-mus LD50:1055 µg/kg TOXIA6 2,5,64
ims-mus LD50:2 mg/kg TOXIA6 6,175,69
ivn-rbt LDLo:660 µg/kg TOXIA6 2,5,64
ipr-mam LD50:3680 µg/kg CLPTAT 8,849,67

SAFETY PROFILE: Deadly poison by subcutaneous, intramuscular, intravenous, and intraperitoneal routes.

BLV080 HR: 3
BITIS GABONICA VENOM

SYNS: B. GABONICA VENOM □ SNAKE VENOM BITIS GABONICA

TOXICITY DATA WITH REFERENCE
ipr-mus LD50:960 µg/kg TOXIA6 18,384,80
scu-mus LD50:5 mg/kg JOIMA3 67,299,51
ivn-mus LD50:550 µg/kg TOXIA6 14,146,76
ims-mus LD50:5200 µg/kg TOXIA6 6,175,69
ivn-rbt LDLo:1065 µg/kg SCIEAS 117,47,53

SAFETY PROFILE: Deadly poison by subcutaneous, intramuscular, intravenous, and intraperitoneal routes.

BLV125 CAS:30392-41-7 HR: 3
BITOLTEROL MESILATE
mf: $C_{29}H_{31}NO_5{\cdot}CH_4O_3S$ mw: 557.71

PROP: Solid. Mp: 170–172°.

SYNS: BITOLTEROL MESYLATE □ 4-(2-(tert-BUTYLAMINO)-1-HYDROXYETHYL)-o-PHENYLENE DI-p-TOLUATE MESILATE □ WIN 32784

TOXICITY DATA WITH REFERENCE
ivn-rat LD50:44 mg/kg NIIRDN 6,620,82
orl-mus LD50:4116 mg/kg IYKEDH 10,884,79
ivn-mus LD50:31,400 µg/kg IYKEDH 10,884,79

SAFETY PROFILE: Poison by intravenous route. When heated to decomposition it emits toxic fumes of SO_x and NO_x.

BLV250 CAS:13394-86-0 HR: 3
(m,o′-BITOLYL)-4-AMINE
mf: $C_{14}H_{15}N$ mw: 197.30

PROP: Oil. Bp: 201° @ 15 mm.

SYNS: 2′,3-DIMETHYL-4-AMINOBIPHENYL □ 3,2′-DIMETHYL-4-AMINOBIPHENYL □ 3,2′-DIMETHYL-4-AMINODIPHENYL □ 3,2′-DIMETHYL-4-BIPHENYLAMINE □ 3,2′-DMAB

TOXICITY DATA WITH REFERENCE
mma-sat 10 µg/plate PNASA6 72,5135,75
dns-rat:lvr 10 µmol/L CALEDQ 4,69,78
cyt-mus-orl 50 mg/kg JJIND8 71,133,83
otr-ham:emb 100 µg/L NCIMAV 58,243,81
scu-rat TDLo:680 mg/kg/I:CAR ARPAAQ 86,475,68
par-rat TDLo:280 mg/kg/14W-I:ETA 23HZAR-,280,70
scu-ham TDLo:2300 mg/kg/37W-I:CAR,REP JNCIAM 48,1733,72
scu-ham TD:1095 mg/kg/37W-I:CAR CALEDQ 20,349,83
scu-rat TD:960 mg/kg/21W-I:CAR ANZJA7 29,38,59
scu-rat TD:1000 mg/kg/20W-I:CAR JJIND8 71,419,83
scu-ham TD:4440 mg/kg/77W-I:CAR JJIND8 67,481,81
scu-rat TD:1000 mg/kg/20W-I:CAR CNREA8 41,1363,81
scu-rat TD:10,400 mg/kg/2Y-I:CAR CANCAR 28,29,71
scu-ham TD:2 g/kg/20W-I:NEO JJCREP 83,1286,92
scu-rat TD:2300 mg/kg/36W-I:CAR BJCAAI 9,170,55
scu-rat TD:1100 mg/kg/31W-I:CAR CNREA8 27,708,67

ipr-mus LD50:1130 mg/kg JJIND8 62,911,79

CONSENSUS REPORTS: EPA Genetic Toxicology Program.

SAFETY PROFILE: Suspected carcinogen with experimental carcinogenic and tumorigenic data. Moderately toxic by intraperitoneal route. Experimental reproductive effects. Mutation data reported. When heated to decomposition it emits toxic fumes of NO_x. See also AROMATIC AMINES.

BLV500 CAS:8013-76-1 HR: 3
BITTER ALMOND OIL

PROP: Volatile oil from dried ripe kernels of bitter almonds or from other kernels containing amygdalin, such as apricots, cherries, plums, and especially peaches. Colorless liquid; strong almond odor. Bp: 179°, d: 1.045–1.070 @ 15°. Sltly sol in water; sol in fixed oils and propylene glycol; insol in glycerin.

SYNS: ALMOND OIL BITTER, FFPA (FCC) □ OIL, BITTER ALMOND

TOXICITY DATA WITH REFERENCE
skn-rbt 500 mg/24H MOD FCTXAV 17,705,79
orl-hmn LDLo:107 mg/kg FCTXAV 17,705,79
orl-rat LD50:960 mg/kg FCTXAV 17,705,79
skn-rbt LD50:1220 mg/kg FCTXAV 17,705,79

CONSENSUS REPORTS: Reported in EPA TSCA Inventory.

SAFETY PROFILE: A human poison by ingestion. Moderately toxic by skin contact. A skin irritant. When heated to decomposition it emits toxic fumes of CN^-.

BLV750 CAS:68916-04-1 HR: 1
BITTER ORANGE OIL

PROP: Main constituent is d-limonene (FCTXAV 12,703,74). Pale yellow liquid, bitter taste. D: 0.842–0.848 @ 25°/25°. Very sltly sol in water; misc with abs alc; sol in 4 vols alc, in 1 vol glacial acetic acid. Keep well closed, cool, and protected from light.

TOXICITY DATA WITH REFERENCE
skn-mus 100% MLD FCTXAV 12,703,74
skn-rbt 500 mg/24H MOD FCTXAV 17,509,74

SAFETY PROFILE: A skin irritant. See also d-LIMONENE. When heated to decomposition it emits acrid smoke and irritating fumes.

BLW250 CAS:8006-82-4 HR: 1
BLACK PEPPER OIL

PROP: From steam distillation of dried fruit of *Piper nigrum L.* (Fam. Piperaceae). Main constituents include α- and β-pinene, β-caryophyllene, l-limonene, d-hydrocarveol, piperidine, and piperrine (FCTXAV 16,637,78). A colorless to greenish liquid; odor and taste of pepper. Sol in fixed oils, mineral oil, propylene glycol; sltly sol in glycerin.

TOXICITY DATA WITH REFERENCE
skn-rbt 500 mg/24H MOD FCTXAV 16,637,78
dnr-bcs 20 mg/disc TOFOD5 8,91,85

CONSENSUS REPORTS: Reported in EPA TSCA Inventory.

SAFETY PROFILE: A moderate skin irritant. Mutation data reported. When heated to decomposition it emits acrid smoke and irritating fumes.

BLW500 HR: 3
BLACK WIDOW SPIDER VENOM

SYN: LATRODECTUS M. MACTANS VENOM

TOXICITY DATA WITH REFERENCE
scu-mus LDLo:10 mg/kg SCNEBK 110,355,76
ivn-mus LDLo:5500 μg/kg SCNEBK 110,355,76

SAFETY PROFILE: Poison by subcutaneous and intravenous routes.

BLW750 CAS:21725-46-2 HR: 3
BLADEX
mf: $C_9H_{13}ClN_6$ mw: 240.73

PROP: A white, crystalline material. Mp: 167°.

SYNS: BLADEX 80WP □ 2-CHLORO-4-(1-CYANO-1-METHYLETHYLAMINO)-6-ETHYLAMINO-1,3,5-TRIAZINE □ 2-CHLORO-4-ETHYLAMINO-6-(1-CYANO-1-METHYL)ETHYLAMINO-s-TRIAZINE □ 2-(4-CHLORO-6-ETHYLAMINO-s-TRIAZINE-2-YLAMINO)-2-METHYLPROPIONITRILE □ 2-(4-CHLORO-6-ETHYLAMINO-1,3,5-TRIAZINE-2-YLAMINO)-2-METHYLPROPIONITRILE □ 2-((4-CHLORO-6-(ETHYLAMINO)-1,3,5-TRIAZIN-2-YL)AMINO)-2-METHYL-PROPANENITRILE □ 2-((4-CHLORO-6-(ETHYLAMINO)-s-TRIAZIN-2-YL)AMINO)-2-METHYLPROPIONITRILE □ CYANAZINE □ DW3418 □ FORTROL □ PAYZE □ SD 15418 □ WL 19805

TOXICITY DATA WITH REFERENCE
mma-sat 5170 μmol/L MUREAV 136,233,84
dlt-dmg-par 332 μmol/L JTEHD6 3,691,77
dlt-dmg-orl 100 ppm JTEHD6 3,691,77
sln-nsc 250 mg/L EVHPAZ 31,75,79
orl-rat TDLo:250 mg/kg (6-15D preg):TER TJADAB 25(2),59A,82
orl-rat LD50:149 mg/kg 85ARAE 2,132,77
skn-rat LD50:1200 mg/kg 28ZEAL 5,62,76
ipr-rat LD50:112 mg/kg NNGADV 11,127,86
scu-rat LD50:1738 mg/kg NNGADV 11,127,86
orl-mus LD50:380 mg/kg 28ZEAL 5,62,76
ihl-mus LC50:2470 mg/m³/4H NNGADV 11,127,86
orl-rbt LD50:141 mg/kg 85DPAN -,-,71/76
orl-qal LD50:400 mg/kg PEMNDP 8,198,87
orl-dck LD50:750 mg/kg PSSCBG 5,153,74

CONSENSUS REPORTS: EPA Genetic Toxicology Program. Cyanide and its compounds are on the Community Right-To-Know List.

SAFETY PROFILE: Poison by ingestion and intraperitoneal routes. Moderately toxic by skin contact. An experimental teratogen. Mutation data reported. See also NITRILES. An herbicide. When heated to decomposition it emits very toxic fumes of Cl^-, NO_x, and CN^-.

B

BLX000 **CAS:9084-06-4** **HR: 3**
BLANCOL
mf: $(C_{10}H_8O_3S{\cdot}CH_{2O})_x{\cdot}xNa$

SYNS: ATLOX 4862 ☐ BARRA SUPER ☐ BEVALOID 35 ☐ BLANCOL DISPERSANT ☐ DARVAN 1 ☐ DARVAN No. 1 ☐ DAXAD 11 ☐ DAXAD 15 ☐ DAXAD 18 ☐ DAXAD No. 11 ☐ DISPERGATOR NF ☐ DISPERSER NF ☐ DISPERSING AGENT NF ☐ DISPERSOL ACA ☐ FLUBE ☐ HUMIFEN NBL 85 ☐ LEUKANOL NF ☐ LISSATAN AC ☐ LOMAR D ☐ LOMAR LS ☐ LOMAR PW ☐ Na-CEMMIX ☐ NAPHTHALENESULFONIC ACID, POLYMER with FORMALDEHYDE, SODIUM SALT (9CI) ☐ NF ☐ NF (dispersant) ☐ NF-A ☐ POZZOLITH 400N ☐ QR 819 ☐ SODIUM SALT of SULFONATED NAPHTHALENEFORMALDEHYDE CONDENSATE ☐ SURFACTANT NF ☐ TAMOL L ☐ TAMOL SN

TOXICITY DATA WITH REFERENCE
orl-rat LD50:3800 mg/kg FMCHA2 -,D44,80
ipr-rat LD50: 460 mg/kg GISAAA 55(11),70,90
ivn-rat LD50: 435 mg/kg GISAAA 55(11),70,90
orl-mus LD50: 3400 mg/kg GISAAA 55(11),70,90
ipr-mus LD50: 315 mg/kg GISAAA 55(11),70,90
scu-mus LD50: 1275 mg/kg GISAAA 55(11),70,90

CONSENSUS REPORTS: Reported in EPA TSCA Inventory.

SAFETY PROFILE: Poison by intravenous and intraperitoneal routes. Moderately toxic by ingestion. See also ALDEHYDES and SULFONATES. When heated to decomposition it emits very toxic fumes of SO_x and Na_2O.

BLX250 **CAS:63732-07-0** **HR: 3**
BLASTICIDEN-S-LAURYLSULFONATE

TOXICITY DATA WITH REFERENCE
orl-rat LD50:39,500 µg/kg GUCHAZ 6,48,73
scu-rat LD50:220 mg/kg GUCHAZ 6,48,73
orl-rbt LD50:48,500 µg/kg GUCHAZ 6,48,73
orl-mam LD50:32 mg/kg GUCHAZ 6,48,73

SAFETY PROFILE: Poison by ingestion and subcutaneous routes. See also SULFONATES. When heated to decomposition it emits toxic fumes of SO_x.

BLX500 **CAS:2079-00-7** **HR: 3**
BLASTICIDIN S
mf: $C_{17}H_{26}N_8O_5$ mw: 422.51

PROP: Needles from H_2O. Mp: 235°. From *Streptomyces griseochromogenes* (JANTAJ 11,1,58).

SYNS: BABS ☐ BLA-S ☐ BLASTICIDIN ☐ CYTOVIRIN ☐ TOA BLA-S

TOXICITY DATA WITH REFERENCE
orl-rat LD50:16 mg/kg GUCHAZ 6,48,73
skn-rat LD50:3100 mg/kg 28ZEAL 5,27,76
orl-mus LD50:38 mg/kg JANTAJ 30,1022,77
skn-mus LD50:220 mg/kg 28ZEAL 5,27,76
ivn-mus LD50:2820 µg/kg JAJAAA 11,1,58

SAFETY PROFILE: Poison by ingestion, skin contact, and intravenous routes. When heated to decomposition it emits toxic fumes of NO_x. See other blastomycin entries.

BLX750 **CAS:522-70-3** **HR: 3**
BLASTOMYCIN
mf: $C_{26}H_{36}N_2O_9$ mw: 520.64

PROP: Needles from C_6H_6 pet ether. Mp: 174.5–175.0°.

SYNS: ANTIMYCIN A3 ☐ BLASTMYCIN

TOXICITY DATA WITH REFERENCE
ipr-mus LD50:1800 µg/kg JAJAAA 10,39,57
scu-mus LD50:1600 µg/kg JAJAAA 10,39,57

SAFETY PROFILE: Poison by intraperitoneal and subcutaneous routes. When heated to decomposition it emits toxic fumes of NO_x.

BLY000 **CAS:11056-06-7** **HR: 3**
BLEOMYCIN

PROP: A group of related glycopeptide antibiotics isolated from *Streptomyces verticillus.*

SYNS: BLENOXANE ☐ BLEO ☐ BLEOCIN ☐ BLM

TOXICITY DATA WITH REFERENCE
eye-rbt 1 mg MLD JJANAX 31,859,78
mnt-hmn:lym 1250 µg/L MUREAV 130,395,84
dnd-hmn:fbr 10 mg/L ENMUDM 7,267,84
dns-hmn:hla 110 µmol/L CRNGDP 7,77,86
cyt-mus:oth 4 nmol/L IPPABX 20,1,84
sce-ham-ipr 7500 µg/kg CNREA8 43,577,83
ipr-mus TDLo:200 mg/kg (1D male):REP CNREA8 39,3575,79
ivn-hmn LDLo:351 mg/kg:PUL AJCPAI 58,501,72
ims-hmn LDLo:418 mg/kg:PUL AJCPAI 58,501,72
ipr-rat LD50:168 mg/kg 40WDA5 -,311,78
ipr-mus LD50:35 mg/kg JANTAJ 37,239,84
ivn-mus LD50:53 mg/kg JANTAJ 31,667,78

CONSENSUS REPORTS: IARC Cancer Review: Group 2B IMEMDT 7,134,87, Human Inadequate Evidence IMEMDT 26,97,81. EPA Genetic Toxicology Program.

SAFETY PROFILE: A human poison by intravenous route; moderately toxic to humans by intramuscular route. Poison experimentally by intravenous and intraperitoneal routes. Human systemic effects by ingestion and intramuscular routes: dyspnea and fibrosing alveolitis (lung). Experimental reproductive effects. An eye irritant. Human mutation data reported. When heated to decomposition it emits toxic fumes of NO_x. See other bleomycin entries.

BLY250 **CAS:11116-31-7** **HR: 3**
BLEOMYCIN A2

SYN: ZHENGGUANGMYCIN A2 (CHINESE)

TOXICITY DATA WITH REFERENCE
mmo-sat 5 µg/plate MUREAV 117,9,83
dnd-esc 50 µg/L CNREA8 38,3900,78
dnd-rat:ast 13 mg/L PLCHB4 7,177,75
dnd-rat:lng 100 µmol/L CBINA8 45,65,83
dnd-mam:lym 100 µmol/L JPETAB 221,152,82
ivn-man TDLo:2143 µg/kg PUL CNREA8 36,1267,76
ipr-mus LD50:130 mg/kg YHHPAL 14,83,79
ivn-mus LD50:100 mg/kg YHHPAL 14,83,79

SAFETY PROFILE: Poison by intravenous and intraperitoneal routes. Noted for adverse pulmonary effects in humans. Mutation data reported. When heated to decomposition it emits toxic fumes of NO_x. See other bleomycin entries.

BLY500 CAS:11116-32-8 HR: 3
BLEOMYCIN A5

mf: $C_{57}H_{86}N_{18}O_{21}S_2$ mw: 1423.73

SYNS: N[1]-3-(((4-AMINOBUTYL)AMINO)PROPYL)BLEOMYCINAMIDE □ BLEOMYCETIN □ PINGYANGMYCIN (CHINESE) □ ZHENGGUANGMYCIN A5 (CHINESE)

TOXICITY DATA WITH REFERENCE
cyt-ham:ovr 400 μg/L HKXUDL 3,78,83
sce-ham:ovr 400 μg/L HKXUDL 3,78,83
unr-hmn TDLo:192 mg/kg:GIT,PUL,MET ANTBAL 28(8),632,83
ipr-rat LD50:117 mg/kg ANTBAL 24(5),363,79
ivn-rat LD50:100 mg/kg ANTBAL 24(5),363,79
ims-rat LD50:102 mg/kg ANTBAL 24(5),363,79
orl-mus LD50:840 mg/kg ANTBAL 24(5),363,79
ipr-mus LD50:88 mg/kg ANTBAL 24(5),363,79
scu-mus LD50:77 mg/kg ANTBAL 24(5),363,79
ivn-mus LD50:61,500 μg/kg ANTBAL 24(5),363,79

SAFETY PROFILE: Poison by intraperitoneal, intravenous, intramuscular, and subcutaneous routes. Moderately toxic by ingestion. Human systemic effects by an unspecified route: nausea or vomiting, dyspnea, and fever. Mutation data reported. When heated to decomposition it emits very toxic fumes of SO_x and NO_x. See other bleomycin entries.

BLY750 HR: 3
BLEOMYCIN A COMPLEX

PROP: Antibiotics produced by a strain of *Streptomyces verticillus* (JAJAAA 20,15,67).

TOXICITY DATA WITH REFERENCE
mmo-smc 10 mg/L/30M MUREAV 58,107,78
mrc-smc 30 mg/L/15M MUREAV 58,41,78
cyt-hmn-par 430 μg/kg MUREAV 56,341,78
cyt-hmn:lym 10 mg/L MUREAV 56,341,78
msc-ham:fbr 1 mg/L/24H MUREAV 40,325,76
ipr-mus LDLo:125 mg/kg JAJAAA 20,15,67
scu-mus LDLo:125 mg/kg JAJAAA 20,15,67
ivn-mus LDLo:125 mg/kg CANCAR 20,891,67
ivn-rbt LDLo:200 mg/kg JAJAAA 20,15,67

SAFETY PROFILE: Poison by intraperitoneal, subcutaneous, and intravenous routes. Human mutation data reported. When heated to decomposition it emits toxic fumes of NO_x. See other bleomycin entries.

BLY760 CAS:9060-10-0 HR: 3
BLEOMYCIN B2

mf: $C_{55}H_{84}N_{20}O_{21}S_2$ mw: 1425.71

SYNS: DEHYDROPHELOMYCIN D1 □ PHLEOMYCIN D2

TOXICITY DATA WITH REFERENCE
mmo-sat 1 μg/plate MUREAV 117,9,83
dnd-omi 100 mg/L JANTAJ 28,537,75
dnd-omi 10,700 pmol/L CNREA8 40,4173,80
itr-mus LDLo:14 μg/kg TXAPA9 56,326,80

SAFETY PROFILE: Poison by intratracheal route. Mutation data reported. When heated to decomposition it emits toxic fumes of SO_x and NO_x. See other bleomycin entries.

BLY770 CAS:68247-85-8 HR: 3
BLEOMYCIN PEP

mf: $C_{61}H_{88}N_{18}O_{21}S_2$ mw: 1473.79

SYNS: BLM-PEP □ NK 631 □ PEP □ PEPLEOMYCIN □ PEPLOMYCIN

TOXICITY DATA WITH REFERENCE
mmo-sat 1 μg/plate TAKHAA 44,96,85
pic-esc 56 ng/plate MUREAV 88,325,81
dnr-bcs 800 ng/plate TAKHAA 44,96,85
dnd-ham:ovr 25 mg/L JANTAJ 38,1257,85
scu-rat LD50: 199 mg/kg JJANAX 31,719,78
ivn-rat LD50: 215 mg/kg JJANAX 31,719,78
scu-mus LD50: 80 mg/kg JJANAX 31,719,78
ivn-mus LD50: 45 mg/kg JJANAX 31,719,78
ivn-dog LDLo:30 mg/kg JJANAX 31,719,78

SAFETY PROFILE: Poison by subcutaneous and intravenous routes. Mutation data reported. When heated to decomposition it emits toxic fumes of SO_x and NO_x. See other bleomycin entries.

BLY780 CAS:9041-93-4 HR: 3
BLEOMYCIN SULFATE

SYNS: BLENOXANE □ BLEOMYCIN, SULFATE (salt) (9CI) □ BLEXANE

TOXICITY DATA WITH REFERENCE
dnr-esc 20 μmol/L MUREAV 164,19,86
hma-mus/esc 10 mg/kg MUREAV 164,19,86
ipr-rat TDLo:8700 μg/kg (female 15-22D post):REP NTIS** PB261-972
ipr-rat TDLo:17,400 μg/kg (female 14D post): REP NTIS** PB261-972
ipr-rat TDLo:20,400 μg/kg (female 14D pre):REP NTIS** PB261-972
ipr-rat TDLo:8 mg/kg (female 6-9D post):TER NTIS** PB261-972
ivn-rbt TDLo:15,600 μg/kg (female 6-18D post):REP NTIS** PB261-972
ipr-rat TDLo:8 mg/kg (female 6-9D post):REP NTIS** PB261-972
scu-rat TDLo:14 mg/kg/68W-I:CAR ONCOBS 41,114,84
par-rat TDLo:18 mg/kg/52W-I:CAR PAACA3 24,96,83
par-rat TD:36 mg/kg/52W-I:CAR PAACA3 24,96,83
par-wmn TDLo:20 μg/kg:PUL,SKN ARHEAW 28,459,85
ipr-rat LD50:240 mg/kg IYKEDH 7,108,76
scu-rat LD50:86 mg/kg JJANAX 29,894,76
ipr-mus LD50:210 mg/kg YAKUD5 17,455,75
scu-mus LD50:103 mg/kg JJANAX 29,894,76

CONSENSUS REPORTS: IARC Cancer Review: Human Inadequate Evidence IMEMDT 26,97,81. EPA Genetic Toxicology Program.

SAFETY PROFILE: Poison by subcutaneous and intraperitoneal routes. Human systemic effects: cyanosis, allergic dermatitis. Questionable carcinogen with experimental carcinogenic data. An experimental teratogen. Other experimental reproductive effects. Mutation data reported. When heated to decomposition it emits toxic fumes of SO_x. See other bleomycin entries.

BMA000 CAS:2519-30-4 HR: 2
BLUE BLACK BN
mf: $C_{28}H_{21}N_5O_{14}S_4 \cdot 4Na$ mw: 871.74

SYNS: 1743 BLACK □ BLACK PN □ BRILLIANT ACID BLACK BNA EXPORT □ BRILLIANT ACID BLACK BN EXTRA PURE A □ BRILLIANT BLACK □ BRILLIANT BLACK A □ BRILLIANT BLACK BN □ BRILLIANT BLACK NAF □ BRILLIANT BLACK N.FQ □ BRILLIANTSCHWARZ BN (GERMAN) □ CERTICOL BLACK PNW □ C.I. 28440 □ C.I. FOOD BLACK 1, TETRASODIUM SALT □ CILEFA BLACK B □ E 151 □ EDICOL SUPRA BLACK BN □ HEXACOL BLACK PN □ MELAN BLACK □ NOIR BRILLANT BN (FRENCH) □ L-SCHWARZ 1 □ XYLENE BLACK F

TOXICITY DATA WITH REFERENCE
ipr-rat LD50:900 mg/kg FCTXAV 5,171,67
ivn-rat LD50:25,000 mg/kg APFRAD 15,402,57
orl-mus LD50:1100 mg/kg FCTXAV 5,171,67
ipr-mus LD50:500 mg/kg FCTXAV 5,171,67

SAFETY PROFILE: Moderately toxic by ingestion, intravenous, and intraperitoneal routes. When heated to decomposition it emits very toxic fumes of NO_x, Na_2O, and SO_x.

BMA125 CAS:7210-92-6 HR: 3
BLUECAIN
mf: $C_{15}H_{22}N_2O_3 \cdot ClH$ mw: 314.85

PROP: Solid. Mp: 139–140.5°.

SYNS: BAJKAIN □ BAYCAIN □ BAYCAINE □ BAYCALNE □ 2-(((DIETHYLAMINO)ACETYL)AMINO)-3-METHYL-BENZOIC ACID METHYL ESTER, MONOHYDROCHLORIDE □ TOLYCAINE HYDROCHLORIDE

TOXICITY DATA WITH REFERENCE
ivn-rat LD50:44 mg/kg NIIRDN 6,570,82
scu-mus LD50:450 mg/kg NIIRDN 6,570,82
ivn-mus LD50:60 mg/kg NIIRDN 6,570,82
ivn-rbt LD50:40 mg/kg NIIRDN 6,570,82

SAFETY PROFILE: Poison by intravenous route. Moderately toxic by other routes. When heated to decomposition it emits toxic fumes of NO_x and HCl.

BMA150 HR: 2
BLUE COHOSH

PROP: An erect herb, 1 to 3 feet tall, with clusters of small, yellow-green or purple-green flowers. It produces small, blue berries. It grows wild in damp woods in the region bounded by Alabama, Missouri, Manitoba, and New Brunswick.

SYNS: BLUEBERRY ROOT □ BLUE GINSENG □ CAULOPHYLLUM THALICTROIDES □ PAPOOSE ROOT □ SQUAW ROOT □ YELLOW GINSENG

SAFETY PROFILE: The berries and roots contain the poison N-methylcytisine (an alkaloid similar to nicotine) and saponins. The bitter taste usually limits ingestion, which could cause inflammation of the stomach and intestines. See also SAPONIN.

BMA600 CAS:8022-81-9 HR: 3
BOLDO LEAF OIL

TOXICITY DATA WITH REFERENCE
skn-rbt 500 mg/24H MOD FCTOD7 20(Suppl),643,82
orl-rat LD50:130 mg/kg FCTOD7 20(Suppl),643,82
ipr-mus LD50:420 mg/kg JPBEAJ 32,13,77
skn-rbt LD50:625 mg/kg FCTOD7 20(Suppl),643,82

CONSENSUS REPORTS: Reported in EPA TSCA Inventory.

SAFETY PROFILE: Poison by ingestion. Moderately toxic by skin contact and intraperitoneal routes. A skin irritant.

BMA625 CAS:21535-47-7 HR: 3
BOLVIDON
mf: $C_{18}H_{20}N_2 \cdot ClH$ mw: 300.86

PROP: Solid. Mp: 282–284°.

SYNS: ATHYMIL □ GB 94 □ 1,2,3,4,10,14b-HEXAHYDRO-2-METHYLDIBENZO(c,f)PYRAZINO(1,2-a)AZEPINE HYDROCHLORIDE □ MIANSERINE HYDROCHLORIDE □ MIANSERIN HYDROCHLORIDE □ NORVAL □ ORG GB 94 □ TOLUON □ TOLVIN □ TOLVON

TOXICITY DATA WITH REFERENCE
slt-dmg-orl 100 mmol/L MUREAV 286,155,93
orl-rat TDLo:5 mg/kg (1D pre):REP DCTODJ 7,41,84
orl-wmn TDLo:28 mg/kg/5W:CNS,CVS,PUL BMJOAE 284,1912,82
orl-wmn TDLo:18 mg/kg:BAH,BLD HUTODJ 6,401,87
orl-rat LD50:780 mg/kg PBPSDY 3,56,81
ipr-rat LD50:262 mg/kg SKKNAJ 31,112,79
ivn-rat LD50:31,850 µg/kg IYKEDH 14,484,83
orl-mus LD50:224 mg/kg SKKNAJ 31,112,79
ipr-mus LD50:117 mg/kg IYKEDH 14,484,83
scu-mus LD50:118 mg/kg SKKNAJ 31,112,79
ivn-mus LD50:31 mg/kg PBPSDY 3,56,81

SAFETY PROFILE: Poison by ingestion, subcutaneous, intravenous, and intraperitoneal routes. Experimental reproductive effects. Human systemic effects by ingestion: hallucinations and distorted perceptions, change in heart rate, and unspecified respiratory system effects. Experimental reproductive effects. A serotonin inhibitor and antihistamine. Mutation data reported. When heated to decomposition it emits toxic fumes of NO_x and HCl.

BMA750 CAS:8001-85-2 HR: 2
BONE OIL

PROP: Product of destructive distillation of bones in preparation of bone charcoal containing nitrogenous compounds such as pyridine, aniline, methylamine, and pyrrole (27ZTAP 3,25,69).

SYNS: ANIMAL OIL □ DIPPEL'S OIL □ OIL of HARTSHORN

TOXICITY DATA WITH REFERENCE
orl-rat LDLo:800 mg/kg 27ZTAP 3,25,69

CONSENSUS REPORTS: Reported in EPA TSCA Inventory.

SAFETY PROFILE: Moderately toxic by ingestion. When heated to decomposition it emits toxic fumes of NO_x.

BMB000 **CAS:1098-97-1** **HR: 3**
BONIFEN
mf: $C_{16}H_{20}N_2O_4S_2 \cdot 2ClH \cdot H_2O$ mw: 459.44

PROP: Solid. Mp: 218–220°.

SYNS: 3,3′-DITHIOBIS(METHYLENE)BIS(5-HYDROXY-6-METHYL-4-PYRIDINEMETHANOL) DIHYDROCHLORIDE □ 3,3′-DITHIODIMETHYLENEBIS(5-HYDROXY-6-METHYL-4-PYRIDINEMETHANOL) DIHYDROCHLORIDE HYDRATE □ 5,5′-DITHIODIMETHYLENEBIS(2-METHYL-3-HYDROXY-4-HYDROXYMETHYLPYRIDINE)DIHYDROCHLORIDE HYDRATE □ EPOCAN □ PYRIDOXIN-5′-DISULFID DIHYDROCHLORID HYDRAT (GERMAN) □ PYRITHIOXIN

TOXICITY DATA WITH REFERENCE
orl-rat LD50:6000 mg/kg ARZNAD 11,922,61
scu-rat LD50:3000 mg/kg ARZNAD 11,922,61
ivn-rat LD50:500 mg/kg ARZNAD 11,922,61
orl-mus LD50:5786 mg/kg TMPBAX 54,156,78
ipr-mus LD50:790 mg/kg ARZNAD 29,479,79
scu-mus LD50:3170 mg/kg ARZNAD 11,922,61
ivn-mus LD50:221 mg/kg TMPBAX 54,156,78
ivn-cat LD50:124 mg/kg ARZNAD 11,922,61

SAFETY PROFILE: Poison by intravenous route. Moderately toxic by subcutaneous and intraperitoneal routes. Mildly toxic by ingestion. When heated to decomposition it emits very toxic fumes of HCl, SO_x, and NO_x.

BMB125 **HR: 3**
BONNECOR
mf: $C_{21}H_{25}N_3O_3 \cdot ClH$ mw: 403.91

SYNS: AWD 19-166 □ 3-CARBETHOXYAMINO-5-DIMETHYLAMINOACETYL-10,11-DIHYDRODIBENZ(b,f)AZEPINE HYDROCHLORIDE □ GS 015

TOXICITY DATA WITH REFERENCE
orl-rat LD50:78 mg/kg PHARAT 40,871,85
ivn-rat LD50:10,900 µg/kg PHARAT 40,871,85
orl-mus LD50:48 mg/kg PHARAT 40,871,85
ivn-mus LD50:5400 µg/kg PHARAT 40,871,85

SAFETY PROFILE: Poison by ingestion and intravenous routes. When heated to decomposition it emits toxic fumes of NO_x and HCl. See also CARBAMATES and ESTERS.

BMB150 **CAS:17596-45-1** **HR: 3**
BORANE-AMMONIA
mf: BH_3-NH_3 mw: 30.86

SAFETY PROFILE: Complex may explode on rapid heating. When heated to decomposition it emits toxic fumes of NH_3. See also BORANES, BORON COMPOUNDS, and AMMONIA.

BMB250 **CAS:75-22-9** **HR: 3**
BORANE, COMPOUND with TRIMETHYLAMINE (1:1)
mf: $C_3H_9N \cdot BH_3$ mw: 72.97

PROP: White crystals or solid. Insol in hexane, very sol in most org solvs; sltly sol in H_2O and cyclohexane

SYNS: BORANE, COMPOUND with N,N-DIMETHYLMETHANAMINE (1:1) □ TMAB □ TRIMETHYLAMINE BORANE □ TRIMETHYLAMINE, COMPOUND with BORANE (1:1)

TOXICITY DATA WITH REFERENCE
dni-mus:ast 100 µmol/L JPMSAE 74,755,85
uns-mus:ast 100 µmol/L JPMSAE 74,755,85
ipr-rat LD50:175 mg/kg JOCMA7 1,46,59
ipr-mus LD50:740 mg/kg JPMSAE 19,1025,80

CONSENSUS REPORTS: Reported in EPA TSCA Inventory.

SAFETY PROFILE: Poison by intraperitoneal route. Mutation data reported. See also BORANES. When heated to decomposition it emits toxic fumes of NO_x.

BMB260 **HR: 3**
BORANE-HYDRAZINE
mf: BH_3-N_2H_4 mw: 45.88

SAFETY PROFILE: Complex is highly flammable and a shock-sensitive explosive. Upon decomposition it emits toxic fumes of NO_x. See also BORANES, BORON COMPOUNDS, and HYDRAZINE.

BMB270 **HR: 3**
BORANE-PHOSPHORUS TRIFLUORIDE
mf: BH_3-PF_3 mw: 101.80

SAFETY PROFILE: An unstable explosive complex which ignites spontaneously upon exposure to air. When heated to decomposition it emits toxic fumes of F^- and PO_x. See also BORANES, BORON COMPOUNDS, and PHOSPHORUS TRIFLUORIDE.

BMB280 **HR: 3**
BORANES

PROP: A series of boron hydrides (BH_3, B_2H_6,...,$B_{20}H_{26}$).

SAFETY PROFILE: Generally poisons. Most are unstable and react with water to produce explosive hydrogen gas. Many react violently with air. Many organoboranes are used as reducing agents. Haloboranes are highly reactive. Potentially explosive reaction with carbon tetrachloride.

BMB300 **CAS:14044-65-6** **HR: 3**
BORANE-TETRAHYDROFURAN
mf: BH_3-C_4H_8O mw: 85.93

PROP: Used only in soln, unstable in pure state, moisture-sensitive.

SAFETY PROFILE: The complex is an unstable explosive in tetrahydrofuran at room temperature. When heated to decomposition it emits acrid smoke and fumes. See also BORANES, BORON COMPOUNDS, and TETRAHYDROFURAN.

BMB325 **HR: 2**
BORASSUS FLABELLIFER Linn., extract

PROP: Indian plant belonging to the family Lalniae (IJEBA6 16,228,78).

TOXICITY DATA WITH REFERENCE
mmo-sat 530 μg/plate CALEDQ 26,113,85
mma-sat 1590 μg/plate CALEDQ 26,113,85
mmo-esc 530 μg/plate CALEDQ 26,113,85
ipr-rat LD50:850 mg/kg IJEBA6 16,228,78

SAFETY PROFILE: Moderately toxic by intraperitoneal route. Mutation data reported.

BMB500 **CAS:6569-51-3** **HR: 3**
BORAZINE
mf: $B_3H_6N_3$ mw: 80.5

SYN: BORAZOLE

PROP: Colorless liquid. Mp: −58°, bp: 55°, d: 0.824 @ 0°.

SAFETY PROFILE: A powerful irritant to skin, eyes, and mucous membranes. May explode spontaneously when stored in the light. Reacts with water to form toxic and flammable boron hydrides. A dangerous fire hazard. When heated to decomposition it emits toxic fumes of NO_x. See also BORON COMPOUNDS.

BMB750 **HR: 3**
BORDEAUX ARSENITE

CONSENSUS REPORTS: Arsenic and its compounds, as well as copper and its compounds, are on the Community Right-To-Know List.

OSHA PEL: TWA 0.01 mg(As)/m³
NIOSH REL: CL 0.002 mg(As)/m³/15M

SAFETY PROFILE: A poison. See also ARSENIC COMPOUNDS and COPPER COMPOUNDS. When heated to decomposition it emits toxic fumes of As.

BMC000 **CAS:10043-35-3** **HR: 3**
BORIC ACID
mf: BH_3O_3 mw: 61.84

PROP: White crystals, powder, or pearly scales. Mp: 171° (decomp), loses 1.5 H_2O @ 300°, d: 1.435 @ 15°.

SYNS: BORACIC ACID □ BOROFAX □ BORSAEURE (GERMAN) □ NCI-C56417 □ ORTHOBORIC ACID □ THREE ELEPHANT

TOXICITY DATA WITH REFERENCE
skn-hmn 15 mg/3D-I MLD 85DKA8 -,127,77
mmo-esc 17,000 ppm/24H AMNTA4 85,119,51
spm-rat-orl 6 mg/kg EVHPAZ 13,69,76
orl-rat TDLo:45 g/kg (90D male):REP TXAPA9 23,351,72
orl-rat TDLo:52 mg/kg (male 26W pre):REP EVHPAZ 13,69,76
orl-cld TDLo:500 mg/kg:GIT JTCTDW 24,269,86
orl-man LDLo:429 mg/kg:CVS,SYS JTCTDW 31,345,93
orl-cld TDLo: 500 mg/kg:GIT JTCTDW 24,269,86
orl-wmn LDLo:200 mg/kg LANCAO 2,162,17
orl-inf TDLo:800 mg/kg/4W-I ADCHAK 58,737,83
orl-inf LDLo:934 mg/kg JAMAAP 90,382,28
skn-inf LDLo:1200 mg/kg JAMAAP 129,332,45
skn-chd LDLo:4 g/kg/4D MMWOAU 52,763,05
skn-man LDLo:2430 mg/kg JAMAAP 128,266,45
skn-cld LDLo:1500 mg/kg QJPPAL 6,714,33
scu-inf LDLo:1100 mg/kg QJPPAL 6,714,33
unr-man TDLo:170 mg/kg:GIT RTPCAT 1,472,29
unr-man LDLo:147 mg/kg 85DCAI 2,73,70
orl-rat LD50:2660 mg/kg JAMAAP 128,266,45
ihl-rat LCLo:28 mg/m³/4H 85GMAT -,27,82
scu-rat LD50:1400 mg/kg 14KTAK -,694,64
ivn-rat LD50:1330 mg/kg MDSR** No.2,50
orl-mus LD50:3450 mg/kg JAMAAP 128,266,45
ipr-mus LDLo:800 mg/kg 14KTAK -,693,64
scu-mus LD50:1740 mg/kg JAMAAP 128,266,45
ivn-mus LD50:1240 mg/kg 14KTAK -,693,64
scu-dog LDLo:1000 mg/kg JAMAAP 128,266,45
par-dog LDLo:1 g/kg RTPCAT 1,472,29

CONSENSUS REPORTS: Reported in EPA TSCA Inventory.

SAFETY PROFILE: A human poison by ingestion and possibly other routes. Moderately toxic by skin contact and subcutaneous routes in humans. Poison experimentally by inhalation and subcutaneous routes. Moderately toxic experimentally by intraperitoneal and intravenous routes. Human systemic effects: anorexia, changes in kidney tubules, nausea or vomiting, wakefulness. Ingestion or absorption by other routes may also cause diarrhea, abdominal cramps, erythematous lesions on skin and mucous membranes, circulatory collapse, tachycardia, cyanosis, delirium, convulsions, and coma. Death has occurred from ingestion of less than 5 g in infants, and from 5 to 20 g in adults. Chronic exposure may result in borism (dry skin, eruptions, and gastrointestinal disturbances). Experimental reproductive effects. Mutation data reported. A human skin irritant. See also BORON COMPOUNDS. Incompatible with K; $(CH_3CO)_2O$.

BMC250 **CAS:34099-73-5** **HR: 3**
BORIC ACID, ETHYL ESTER
DOT: UN 1176
mf: $C_2H_7BO_3$ mw: 89.90

PROP: Colorless liquid, mild odor, decomp in water. Bp: 120°, flash p: 52°F (CC), d: 0.864 @ 26.5°, vap d: 5.04.

SYN: ETHYL BORATE (DOT)

TOXICITY DATA WITH REFERENCE
eye-rbt 5 mg SEV AJOPAA 29,1363,46

DOT CLASSIFICATION: 3; *Label:* Flammable Liquid

SAFETY PROFILE: A severe eye irritant. See also BORON COMPOUNDS and ESTERS. Dangerous fire hazard when exposed to heat or flame; will react with water or steam to produce flammable vapors. Incompatible with oxidizers, heat, and open flame. To fight fire, use CO_2, dry chemical.

B

BMC500 **CAS:5337-42-8** **HR: 1**
BORIC ACID, TRIOLEYL ESTER
mf: $C_{54}H_{108}BO_3$ mw: 816.43

SYN: TRIOLEYL BORATE

TOXICITY DATA WITH REFERENCE
eye-rbt 100 mg MLD 14KTAK -,693,64
orl-mus LD50:6200 mg/kg 14KTAK -,693,64

SAFETY PROFILE: Mildly toxic by ingestion. An eye irritant. When heated to decomposition it emits acrid smoke and irritating fumes. See also ESTERS and BORON COMPOUNDS.

BMC750 **CAS:5337-37-1** **HR: 2**
BORIC ACID, TRIS(4-METHYL-2-PENTYL) ESTER
mf: $C_{18}H_{39}BO_3$ mw: 314.38

SYN: TRI(METHYLISOBUTYLCARBINYL) BORATE

TOXICITY DATA WITH REFERENCE
eye-rbt 100 mg SEV 14KTAK -,706,64
orl-mus LD50:1320 mg/kg USBCC*

SAFETY PROFILE: Moderately toxic by ingestion. A severe eye irritant. See also ESTERS and BORON COMPOUNDS. When heated to decomposition it emits acrid smoke and irritating fumes.

BMD000 **CAS:507-70-0** **HR: 3**
BORNEOL
DOT: UN 1312
mf: $C_{10}H_{18}O$ mw: 154.28

PROP: Hexagonal crystals, peppery odor and burning taste. Mp: 208°, bp: 212°, flash p: 150°F, d: 1.01 @ 20°/4°, vap d: 5.31.

SYNS: 2-BORNANOL, endo- □ BAROS CAMPHOR □ BHIMSAIM CAMPHOR □ BICYCLO(2.2.1)HEPTAN-2-OL, 1,7,7-TRIMETHYL-, endo-(9CI) □ BORNEO CAMPHOR □ trans-BORNEOL □ BORNEOL (DOT) □ BORNYL ALCOHOL □ CAMPHANE, 2-HYDROXY- □ 2-CAMPHANOL □ CAMPHOL □ DRYOBALANOPS CAMPHOR □ 2-HYDROXYCAMPHANE □ MALAYAN CAMPHOR □ SUMATRA CAMPHOR □ endo-1,7,7-TRIMETHYL-BICYCLO(2.2.1)HEPTAN-2-OL

TOXICITY DATA WITH REFERENCE
dnr-bcs 10 mg/disc OIGZSE 34,267,85
cyt-smc 1 mmol/tube HEREAY 33,457,47
orl-rat LD50:500 mg/kg FRXXBL #2448856
orl-mus LD50:1059 mg/kg SHGKA3 75,934,75
orl-rbt LDLo:2000 mg/kg AEXPBL 17,363,1883

CONSENSUS REPORTS: Reported in EPA TSCA Inventory.

DOT CLASSIFICATION: 4.1; *Label:* Flammable Solid

SAFETY PROFILE: Moderately toxic by ingestion. Mutation data reported. A mild irritant. Flammable when exposed to heat or flame; can react with oxidizing materials. To fight fire, use water, CO_2, water spray, dry chemical. When heated to decomposition it emits acrid smoke and fumes.

BMD100 **CAS:76-49-3** **HR: 1**
BORNYL ACETATE
mf: $C_{12}H_{20}O_2$ mw: 196.29

PROP: Colorless liquid or white crystalline solid; sweet, piney odor. D: 0.981–0.985, refr index: 1.462, flash p: 192°F. Sol in alc, fixed oils; sltly sol in water; insol in glycerin, propylene glycol @ 226°.

SYNS: l-BORNYL ACETATE □ FEMA No. 2159

SAFETY PROFILE: Combustible liquid. When heated to decomposition it emits acrid smoke and irritating fumes.

BMD250 **CAS:40283-68-9** **HR: 3**
S-((N-BORNYLAMIDIN)METHYL) HYDROGEN THIOSULFATE
mf: $C_{12}H_{22}N_2O_3S_2$ mw: 306.48

TOXICITY DATA WITH REFERENCE
orl-mus LD50:225 mg/kg JMCMAR 15,1313,72
ipr-mus LD50:30 mg/kg JMCMAR 15,1313,72

SAFETY PROFILE: Poison by ingestion and intraperitoneal routes. See also THIOSULFATES. When heated to decomposition it emits very toxic fumes of NO_x and SO_x.

BMD500 **CAS:7440-42-8** **HR: 3**
BORON
af: B aw: 10.81

PROP: Monoclinic crystals, yellow or brown amorphous powder. Boron is very unreactive: reacts with NaOH at 5°, Na_2CO_3 at 8°. Mp: 2190°, bp: 3660°, d: 3.33 @ 20°.

TOXICITY DATA WITH REFERENCE
orl-rat LD50:650 mg/kg GISAAA 35(11),11,70
ipr-rat LD50:7 g/kg GTPZAB 35(2),42,91
orl-mus LD50:560 mg/kg GISAAA 35(11),11,70
ipr-mus LD50:11 g/kg GTPZAB 35(2),42,91
orl-dog LD50:310 mg/kg GISAAA 35(11),11,70
orl-cat LD50:250 mg/kg GISAAA 35(11),11,70
orl-rbt LD50:310 mg/kg GISAAA 35(11),11,70
orl-gpg LD50:310 mg/kg GISAAA 35(11),11,70

CONSENSUS REPORTS: Reported in EPA TSCA Inventory.

SAFETY PROFILE: A poison by ingestion. See also BORON COMPOUNDS. A relatively inert metal except in the form of powder or when exposed to highly oxidizing agents. Amorphous boron is very reactive, sometimes violently. Flammable in the form of dust when exposed to air, or by chemical reaction. An explosion hazard in the form of dust, which ignites on contact with air. Reacts explosively when ground with lead fluoride or silver fluoride. Ignites in contact with gaseous chlorine or fluorine at room temperature. Incompatible with NH_3, Br_2, BrF_3, Cs_2C_2, Cl_2, CuO, HIO_3, PbO_2, HNO_3, NO, NOF, N_2O, $KClO_3$, KNO_3, Rb_2C_2, S, BrF_5, IF_5, metal fluorides, interhalogens, nitryl fluoride (FNO_2), OF_2, KNO_2, NO_x, Na_2O_2, PbO, air. See also POWDERED METALS.

BMD750 **HR: 3**
BORON AZIDE DICHLORIDE
mf: BCl_2N_3 mw: 123.74

SAFETY PROFILE: Crust of sublimed compound explodes when crushed by spatula or upon removing solvent. When heated to decomposition it emits toxic fumes of Cl^- and NO_x. See also AZIDES, BORON COMPOUNDS, and CHLORIDES.

BMD825 **CAS:68533-38-0** **HR: 3**
BORON AZIDE DIIODIDE
mf: BI_2N_3 mw: 306.64

SAFETY PROFILE: Explodes on contact with water. When heated to decomposition it emits toxic fumes of I^- and NO_x. See also BORON COMPOUNDS, AZIDES, and IODIDES.

BME250 **CAS:14355-21-6** **HR: 3**
BORON BROMIDE DIIODIDE
mf: $BBrI_2$ mw: 344.53

PROP: Colorless liquid. Bp: 180°. Formed in mixtures of BBr_3 and BI_3. Sol in CH_2Cl_2; mod sol in methylcyclohexane.

SAFETY PROFILE: Dangerous. Violent reaction with water. When heated to decomposition it emits toxic fumes of Br^- and I^-. See BORON COMPOUNDS, BROMIDES, and IODIDES.

BME500 **HR: 3**
BORON COMPOUNDS

SAFETY PROFILE: Very toxic and therefore considered an industrial poison. Used in medicine as sodium borate, boric acid, or borax, which is a common cleanser. Fatal poisoning of children has been caused by the accidental substitution of boric acid for powdered milk. The medical literature reveals instances of accidental poisoning due to boric acid, ingestion of borates or boric acid, and, presumably, absorption of boric acid from wounds and burns. The fatal dose of orally ingested boric acid for an adult is somewhat greater than 15 to 20 g and, for an infant, 5 to 6 g. Boron is one of a group of elements, such as Pb, Mn, As, that affects the central nervous system. Boron poisoning causes depression of the circulation, persistent vomiting and diarrhea, followed by profound shock and coma. The temperature becomes subnormal and a scarlatina-form rash may cover the entire body. Containers of boric acid should be plainly labeled and should differ radically from those that contain powdered milk, particularly in institutions such as hospitals.

BME750 **HR: 3**
BORON DIBROMIDE IODIDE
mf: BBr_2I mw: 297.53

PROP: Colorless liquid. Bp: 125°, vap d: 10.3.

SAFETY PROFILE: Reaction with water or steam produces toxic and corrosive fumes. See BORON COMPOUNDS, BROMIDES, and IODIDES.

BMG000 **CAS:1303-86-2** **HR: 2**
BORON OXIDE
mf: B_2O_3 mw: 69.62

PROP: Vitreous or colorless. Two crystalline forms. Bp: 2250°. Mp: 450° (approx), d: 2.46.

SYNS: BORIC ANHYDRIDE □ BORON SESQUIOXIDE □ BORON TRIOXIDE □ FUSED BORIC ACID

TOXICITY DATA WITH REFERENCE
skn-rbt 1 g AIHAAP 20,284,59
eye-rbt 50 mg AIHAAP 20,284,59
orl-mus LD50:3163 mg/kg 85GMAT -,27,82
ipr-mus LD50:1868 mg/kg 85GMAT -,27,82

CONSENSUS REPORTS: Reported in EPA TSCA Inventory.

OSHA PEL: Total Dust: TWA 10 mg/m³; Respirable Fraction: TWA 5 mg/m³
ACGIH TLV: TWA 10 mg/m³
DFG MAK: 15 mg/m³

SAFETY PROFILE: Moderately toxic by ingestion and intraperitoneal routes. An eye and skin irritant. A pesticide. Mixed with CaO and put into fused $CaCl_2$, the mixture incandesces. See also BORON COMPOUNDS.

For occupational chemical analysis use NIOSH: Nuisance Dust, Total 0500; Nuisance Dust, Respirable, 0600.

BMG250 **HR: 3**
BORON PHOSPHIDE
mf: BP mw: 41.79

PROP: Maroon powder. Mp: 200°.

SAFETY PROFILE: A poison. Ignites @ 200°. Deflagrates with fused alkali nitrates. Incompatible with HNO_3; oxidants; i.e., nitrates. When heated to decomposition it emits toxic fumes of PO_x. See also BORON COMPOUNDS and PHOSPHIDES.

BMG325 **HR: 3**
BORON TRIAZIDE
mf: BN_9 mw: 136.87

SYN: TRIAZIDOBORANE

SAFETY PROFILE: An explosive that detonates by heat or contact with ether or water. See also BORON COMPOUNDS, AZIDES, and EXPLOSIVES.

BMG400 **CAS:10294-33-4** **HR: 3**
BORON TRIBROMIDE
DOT: UN 2692
mf: BBr_3 mw: 250.54

PROP: Colorless, fuming liquid. Very moisture-sensitive. Mp: −46°, bp: 91.3°, d: 2.650 @ 0°, vap press: 40 mm @ 14.0°, 100 mm @ 33.5°. Sol in CCl_4, SO_2 (l), SCl_2; mod sol in methylcyclohexane.

SYNS: BORON BROMIDE □ TRONA

CONSENSUS REPORTS: Reported in EPA TSCA Inventory.

OSHA PEL: CL 1 ppm
ACGIH TLV: CL 1 ppm
DOT CLASSIFICATION: 8; *Label:* Corrosive, Poison

SAFETY PROFILE: A poison. Corrosive. A skin, eye, and mucous membrane irritant. Dangerous; may explode when heated. This and other boron halides react with water or steam to produce toxic and corrosive fumes and may explode. Incompatible with K; Na. When heated to decomposition it emits toxic fumes of Br^-. See also BORON COMPOUNDS and HYDROBROMIC ACID.

BMG500 CAS:10294-34-5 HR: 3
BORON TRICHLORIDE
DOT: UN 1741
mf: BCl_3 mw: 117.16

PROP: Colorless gas, fuming liquid. Pungent, irritating odor. Very easily hydrolyzed. Mp: −107°, bp: 12.5°, d: 1.349 @ 11°/4°, vap press: 1 atm @ 12.7°, vap d: 4.03.

SYNS: BORON CHLORIDE □ CHLORURE de BORE (FRENCH)

TOXICITY DATA WITH REFERENCE
ihl-rat LCLo: 20 ppm/7H 14KTAK -,726,64
ihl-mus LCLo: 20 ppm/7H 14KTAK -,726,64

CONSENSUS REPORTS: Reported in EPA TSCA Inventory. EPA Extremely Hazardous Substances List.

DOT CLASSIFICATION: 2.3; *Label:* Poison Gas, Corrosive

SAFETY PROFILE: A poison by inhalation. A corrosive and severe irritant to skin, eyes, and mucous membranes. Reacts with water or steam to produce heat, toxic and corrosive fumes. Violent reaction with aniline or phosphine. Incompatible with hexafluorisopropylidene amino lithium, NO_2, grease, organic matter, O_2. When heated to decomposition it emits toxic fumes of Cl^-. See also BORON COMPOUNDS and HYDROCHLORIC ACID.

BMG700 CAS:7637-07-2 HR: 3
BORON TRIFLUORIDE
DOT: UN 1008
mf: BF_3 mw: 67.81

PROP: Colorless nonflammable gas; pungent, irritating odor. Mp: −128.4°, bp: −100.0°, d: 2.99 g/L. Sol in H_2O, and org solvs, e.g. alcohols, ethers (forming adducts).

SYNS: BORON FLUORIDE □ FLUORURE de BORE (FRENCH)

TOXICITY DATA WITH REFERENCE
ihl-rat LC50: 1180 mg/m³/4H 85GMAT -,27,82
ihl-mus LC50: 3460 mg/m³/2H FATOAO 35,369,72
ihl-gpg LC50: 109 mg/m³/4H FATOAO 35,369,72

CONSENSUS REPORTS: Reported in EPA TSCA Inventory. EPA Extremely Hazardous Substances List.

OSHA PEL: CL 1 ppm
ACGIH TLV: CL 1 ppm
DFG MAK: 1 ppm (3 mg/m³)
NIOSH REL: (Boron Trifluoride) No Exposure Limit

DOT CLASSIFICATION: 2.3; *Label:* Poison Gas

SAFETY PROFILE: A poison by inhalation. A strong irritant. See also BORON COMPOUNDS and FLUORIDES. A nonflammable gas. Dangerous; when heated to decomposition or upon contact with water or steam, will produce toxic and corrosive fumes of F^-. Incompatible with alkali metals, alkaline earth metals (except Mg), alkyl nitrates, and CaO.

BMG750 CAS:7578-36-1 HR: 3
BORON TRIFLUORIDE–ACETIC ACID COMPLEX
DOT: UN 1742

SYNS: ACETIC ACID, compd. with BORON FLUORIDE (BF3) (8CI) □ BORON FLUORIDE, compd. with ACETIC ACID

CONSENSUS REPORTS: Reported in EPA TSCA Inventory.

DOT CLASSIFICATION: 8; *Label:* Corrosive

SAFETY PROFILE: A very corrosive material. When heated to decomposition it emits very toxic fumes of F^-, B oxides. See BORON COMPOUNDS, ACETIC ACID, and FLUORIDES.

BMG800 CAS:13319-75-0 HR: 2
BORON TRIFLUORIDE DIHYDRATE
DOT: UN 2851
mf: $BF_3 \cdot 2H_2O$ mw: 103.85

SYNS: BORANE, TRIFLUORO-, DIHYDRATE □ BORON FLUORIDE DIHYDRATE □ BORON TRIFLUORIDE DIHYDRATE (DOT)

TOXICITY DATA WITH REFERENCE
ihl-rat LC50: 1210 mg/m³/4H TXAPA9 83,69,86

DOT CLASSIFICATION: 8; *Label:* Corrosive

SAFETY PROFILE: Moderately toxic by inhalation. A corrosive irritant. When heated to decomposition it emits toxic vapors of B and F^-.

BMH000 CAS:353-42-4 HR: 3
BORON TRIFLUORIDE-DIMETHYL ETHER
DOT: UN 2965
mf: $C_2H_6O \cdot BF_3$ mw: 113.89

PROP: Moisture-sensitive liquid. D: 1.239, mp: −14°, bp: 126–127°.

SYNS: BORON TRIFLUORIDE DIMETHYL ETHERATE (DOT) □ FLUORID BORITY-DIMETHYLETHER (1:1)

TOXICITY DATA WITH REFERENCE
ihl-gpg LCLo: 50 ppm/4H 14KTAK -,726,64

CONSENSUS REPORTS: Reported in EPA TSCA Inventory.

DOT CLASSIFICATION: 4.3; *Label:* Dangerous When Wet, Corrosive, Flammable Liquid

SAFETY PROFILE: Poison by inhalation. Corrosive. Flammable liquid. When heated to decomposition it emits toxic fumes of F^-. See also ETHERS and BORON COMPOUNDS.

BMH250 **CAS:109-63-7** **HR: 3**
BORON TRIFLUORIDE ETHERATE
mf: $C_4H_{10}BF_3O$ mw: 141.93

PROP: Moisture-sensitive liquid. D: 1.125 @ 25, mp: −58°, bp: 126°, flash p: <22°.

SYN: BORON TRIFLUORIDE DIETHYL ETHERATE

SAFETY PROFILE: Corrosive. A dangerous fire hazard. Peroxide containing etherate reacts explosively with solid lithium tetrahydroaluminate. Incompatible with water or steam to produce toxic, corrosive and flammable vapors; oxidizing materials. To fight fire, use dry chemical, CO_2, fog or mist. See BORON COMPOUNDS, FLUORIDES, and ETHER.

BMH500 **CAS:13517-10-7** **HR: 3**
BORON TRIIODIDE
mf: BI_3 mw: 391.52

PROP: Colorless crystals or hygroscopic plates. Mp: 2° (est.), d: 3.35 @ 50°. Sol in C_6H_6, CS_2, or CH_2Cl_2.

SAFETY PROFILE: A poison. Reacts violently with water. Incandescent reaction with red or white phosphorus. Exothermic reaction with ammonia. Incompatible with ethers, carbohydrates, POCl. When heated to decomposition it emits toxic fumes of I^-. See also BORON COMPOUNDS and IODIDES.

BMH659 **CAS:12007-33-9** **HR: 3**
BORON TRISULFIDE
mf: B_2S_3 mw: 117.80

PROP: Pale yellow solid, often glassy. Decomposes in moist air. Mp: 310°.

SAFETY PROFILE: Reacts violently with water. When heated to decomposition it emits toxic fumes of SO_x. See also BORON COMPOUNDS and SULFIDES.

BMH750 **CAS:7184-60-3** **HR: 3**
BORRELIDIN
mf: $C_{28}H_{43}NO_6$ mw: 489.72

PROP: Crystals. Mp: 145°.

TOXICITY DATA WITH REFERENCE
scu-rat LD50:1780 μg/kg JCINAO 28,1047,49
ivn-rat LD50:2 mg/kg 85ERAY 2,1198,78
scu-mus LD50:75 mg/kg JCINAO 28,1047,49
ivn-mus LD50:39 mg/kg JCINAO 28,1047,49
ims-ckn LD50:74 mg/kg 85ERAY 2,1198,78

SAFETY PROFILE: Poison by subcutaneous, intravenous, and intramuscular routes. When heated to decomposition it emits toxic fumes of NO_x.

BMI000 **HR: 3**
BOTHROPS ASPER VENOM

SYNS: B. ASPER VENOM □ VENOM, COSTA RICAN SNAKE, BOTHROPS ASPER

TOXICITY DATA WITH REFERENCE
ipr-mus LD50:469 μg/kg AJTHAB 21,360,72
ivn-mus LD50:1175 μg/kg AJTHAB 21,360,72

SAFETY PROFILE: Poison by intraperitoneal and intravenous routes.

BMI125 **HR: 3**
BOTHROPS ATROX VENOM

SYNS: B. ATROX VENOM □ VENOM, COSTA RICAN SNAKE, BOTHROPS ATROX

TOXICITY DATA WITH REFERENCE
scu-mus LD50:22,140 μg/kg AJTMAQ 31,489,51
ivn-mus LD50:1400 μg/kg TXAPA9 16,73,70
ivn-rbt LDLo:5 μg/kg SCIEAS 117,47,53
ipr-mam LD50:3800 μg/kg CLPTAT 8,849,67
ivn-mam LD50:4270 μg/kg CLPTAT 8,849,67

SAFETY PROFILE: Poison by subcutaneous, intravenous, and intraperitoneal routes.

BMI250 **HR: 3**
BOTHROPS COLOMIBIENSIS VENOM

SYN: VENOM, SNAKE, BOTHROPS COLOMBIENSIS

TOXICITY DATA WITH REFERENCE
ipr-mus LD50:4 mg/kg TOXIA6 17(Suppl 1),161,79
ivn-mus LD50:2 mg/kg TOXIA6 17(Suppl 1),161,79

SAFETY PROFILE: Poison by intraperitoneal and intravenous routes.

BMI500 **HR: 3**
BOTHROPS GODMANI VENOM

SYN: VENOM, COSTA RICAN SNAKE, BOTHROPS GODMANI

TOXICITY DATA WITH REFERENCE
ipr-mus LD50:375 μg/kg AJTHAB 21,360,72
ivn-mus LD50:4750 μg/kg AJTHAB 21,360,72

SAFETY PROFILE: A deadly poison by intraperitoneal and intravenous routes.

BMI750 **HR: 3**
BOTHROPS LATERALIS VENOM

SYN: VENOM, COSTA RICAN SNAKE, BOTHROPS LATERALIS

TOXICITY DATA WITH REFERENCE
ipr-mus LD50:644 μg/kg AJTHAB 21,360,72
ivn-mus LD50:5144 μg/kg AJTHAB 21,360,72

SAFETY PROFILE: A deadly poison by intraperitoneal and intravenous routes.

BMJ000 **HR: 3**
BOTHROPS NASUTUS VENOM

SYN: VENOM, COSTA RICAN SNAKE, BOTHROPS NASUTUS

TOXICITY DATA WITH REFERENCE
ipr-mus LD50:438 μg/kg AJTHAB 21,360,72
ivn-mus LD50:9063 μg/kg AJTHAB 21,360,72

SAFETY PROFILE: A deadly poison by intraperitoneal and intravenous routes.

BMJ250 **HR: 3**
BOTHROPS NIGROVIRIDIS NEGROVIRIDIS VENOM

SYN: VENOM, COSTA RICAN SNAKE, BOTHROPS NIGROVIRIDIS NIGROVIRIDIS

TOXICITY DATA WITH REFERENCE
ipr-mus LD50:875 μg/kg AJTHAB 21,360,72
ivn-mus LD50:4438 μg/kg AJTHAB 21,360,72

SAFETY PROFILE: A deadly poison by intraperitoneal and intravenous routes.

BMJ500 **HR: 3**
BOTHROPS NUMMIFER MEXICANUS VENOM

SYNS: VENOM, COSTA RICAN SNAKE, BOTHROPS NUMMIFER MEXICANUS □ B. N. MEXICANUS VENOM

TOXICITY DATA WITH REFERENCE
ipr-mus LD50:1063 μg/kg AJTHAB 21,360,72
ivn-mus LD50:5656 μg/kg AJTHAB 21,360,72

SAFETY PROFILE: A deadly poison by intraperitoneal and intravenous routes.

BMJ750 **HR: 3**
BOTHROPS OPHYOMEGA VENOM

SYN: VENOM, COSTA RICAN SNAKE, BOTHROPS OPHRYOMEGA

TOXICITY DATA WITH REFERENCE
ipr-mus LD50:719 μg/kg AJTHAB 21,360,72
ivn-mus LD50:6813 μg/kg AJTHAB 21,360,72

SAFETY PROFILE: A deadly poison by intraperitoneal and intravenous routes.

BMK000 **HR: 3**
BOTHROPS PICADOI VENOM

SYN: VENOM, COSTA RICAN SNAKE, BOTHROPS PICADOI

TOXICITY DATA WITH REFERENCE
ipr-mus LD50:375 μg/kg AJTHAB 21,360,72
ivn-mus LD50:1419 μg/kg AJTHAB 21,360,72

SAFETY PROFILE: A deadly poison by intraperitoneal and intravenous routes.

BMK250 **HR: 3**
BOTHROPS SCHLEGLII VENOM

SYN: VENOM, COSTA RICAN SNAKE, BOTHROPS SCHLEGLII

TOXICITY DATA WITH REFERENCE
ivn-mus LD50:1600 μg/kg TXAPA9 16,73,70
ipr-mus LD50:531 μg/kg AJTHAB 21,360,72
ivn-mus LD50:2125 μg/kg AJTHAB 21,360,72

SAFETY PROFILE: A deadly poison by intraperitoneal and intravenous routes.

BMK290 **CAS:27098-03-9** **HR: 3**
BOTRYODIPLODIN
mf: $C_7H_{12}O_3$ mw: 144.19

PROP: Crystals from Et_2O. Mp: 50–52°.

SYNS: (−)-BOTRYODIPLODIN □ 2-HYDROXY-3-METHYL-4-ACETYLTETRAHYDROFURANE □ METHYL TETRAHYDRO-5-HYDROXY-4-METHYL-3-FURYL KETONE □ 1-(TETRAHYDRO-5-HYDROXY-4-METHYL-3-FURANYL)-ETHANONE (9CI)

TOXICITY DATA WITH REFERENCE
dnd-hmn:fbr 30 μmol/L CRNGDP 5,1375,84
dns-rat:lvr 10 μmol/L CRNGDP 5,907,84
sce-ham:lng 300 μg/L CRNGDP 3,587,82
unr-mus LD50:40 mg/kg ENMUDM 3,287,81

DOT CLASSIFICATION: 3; *Label:* Flammable Liquid

SAFETY PROFILE: A poison. Human mutation data reported. A flammable liquid. When heated to decomposition it emits acrid smoke and fumes. See also KETONES.

BMM292 **HR: 3**
BOTULINUM NEUROTOXIN

SYNS: CLOSTRIDIUM BOTULINUM NEUROTOXIN □ NEUROTOXIN, CLOSTRIDIUM BOTULINUM

TOXICITY DATA WITH REFERENCE
ipr-mus LD50:200 pg/kg TOXIA6 24,123,86

SAFETY PROFILE: A poison by intraperitoneal route. When heated to decomposition it emits acrid smoke and irritating vapors.

BMK300 **CAS:18127-01-0** **HR: 2**
BOURGEONAL
mf: $C_{13}H_{18}O$ mw: 190.31

SYNS: BENZENEPROPANAL, 4-(1,1-DIMETHYLETHYL)- □ p-tert-BUTYLDIHYDROCINNAMALDEHYDE □ 4-(1,1-DIMETHYLETHYL)BENZENEPROPANAL

TOXICITY DATA WITH REFERENCE
orl-rat LD50:2700 mg/kg FCTOD7 26,287,88
skn-rbt LDLo:5 g/kg FCTOD7 26,287,88

CONSENSUS REPORTS: Reported in EPA TSCA Inventory.

SAFETY PROFILE: Moderately toxic by ingestion. When heated to decomposition it emits acrid smoke and irritating vapors.

BMK325 **CAS:64755-14-2** **HR: 3**
BOUVARDIN
mf: $C_{40}H_{48}N_6O_{10}$ mw: 772.94

PROP: Needles from $MeOH/CH_2Cl_2$.

SYN: NSC 259968

TOXICITY DATA WITH REFERENCE
dni-mus:lym 3160 nmol/L TUMOAB 71,261,85
oms-mus:lym 1 μmol/L TUMOAB 71,261,85
ipr-mus LD50:12,430 μg/kg NCISP* JAN86

SAFETY PROFILE: Poison by intraperitoneal route. Mutation data reported. When heated to decomposition it emits toxic fumes of NO_x.

BMK500 **CAS:774-64-1** **HR: 2**
BOVOLIDE
mf: $C_{11}H_{16}O_2$ mw: 180.27

TOXICITY DATA WITH REFERENCE
scu-rat TDLo: 2600 mg/kg/65W-I: NEO BJCAAI 19,392,65

SAFETY PROFILE: Questionable carcinogen with experimental neoplastigenic data. It is found in butter made from cow's milk and many other places. When heated to decomposition it emits acrid smoke and irritating fumes.

BMK620 **CAS:63323-31-9** **HR: 2**
(+)-BP-7-β,8-α-DIOL-9-α,10-α-EPOXIDE 2
mf: $C_{20}H_{14}O_3$ mw: 302.34

PROP: Solid. Mp: 226–228° (decomp).

SYNS: (+)-trans-7-β,8-α-DIHYDROXY-9-α,10-α-EPOXY-7,8,9,10-TETRAHDYROBENZO(a)PYRENE ☐ (+)-E-7,8,9,10-TETRAHYDRO-7-α,8-β-DIHYDROXY-9-β,19-β-EPOXY-BENZO(a)PYRENE

TOXICITY DATA WITH REFERENCE
mmo-sat 100 pmol/plate BBRCA9 77,1389,77
msc-ham: lng 300 nmol/L BBRCA9 77,1389,77
skn-mus TDLo: 1200 μg/kg: NEO CNREA8 39,67,79

SAFETY PROFILE: Questionable carcinogen with experimental neoplastigenic data by skin contact. Mutation data reported. When heated to decomposition it emits acrid smoke and fumes.

BMK630 **HR: 2**
B(a)P EPOXIDE II
mf: $C_{20}H_{14}O_3$ mw: 302.34

SYN: anti-(±)-7-β,8-α-DIHYDROXY-9-α,10-α-EPOXY-7,8,9,10-TETRAHYDROBENZO(a)PYRENE

TOXICITY DATA WITH REFERENCE
dnd-hmn: fbr 1500 nmol/L/15M-C CBINA8 38,261,82
dnd-hmn: lym 5 μmol/L PAACA3 24,70,83
msc-hmn: fbr 50 nmol/L MUREAV 94,435,82
dnd-mus: emb 1800 μg/L SCIEAS 209,297,80
skn-mus TDLo: 4830 μg/kg: NEO CCSUDL 3,371,78
scu-mus TDLo: 11 mg/kg: ETA JJIND8 64,617,80

CONSENSUS REPORTS: EPA Genetic Toxicology Program.

SAFETY PROFILE: Questionable carcinogen with experimental neoplastigenic and tumorigenic data by skin contact. Human mutation data reported. When heated to decomposition it emits acrid smoke and fumes.

BMK634 **CAS:75410-89-8** **HR: 2**
B(c)PH DIOL EPOXIDE-1
mf: $C_{18}H_{14}O_3$ mw: 278.32

SYNS: (±)-BENZO(c)PHENANTHRENE-3,4-DIOL-1,2-EPOXIDE-1 ☐ BENZO(c)PHENANTHRENE-3-α-4-β-DIOL, 1,2,3,4-TETRAHYDRO-1-β,2-β-EPOXY-, (±)- ☐ (±)-3-α-4-β-DIHYDROXY-1-β,2-β-EPOXY-1,2,3,4-TETRAHYDROBENZ O(c)PHENANTHRENE

TOXICITY DATA WITH REFERENCE
mmo-sat 100 pmol/plate CNREA8 40,2876,80
msc-ham: lng 200 nmol/L CNREA8 40,2876,80
skn-mus TDLo: 278 μg/kg: NEO CNREA8 46,2257,86

SAFETY PROFILE: Questionable carcinogen with experimental neoplastigenic data. Mutation data reported. When heated to decomposition it emits toxic and irritating fumes.

BMK635 **HR: 2**
B(c)PH DIOL EPOXIDE-2
mf: $C_{18}H_{14}O_3$ mw: 278.32

SYN: (±)-3-α,4-β-DIHYDROXY-1-α,2-α-EPOXY-1,2,3,4-TETRAHYDROBENZ(c)PHENANTHRACENE

TOXICITY DATA WITH REFERENCE
mmo-sat 100 pmol/plate CNREA8 40,2876,80
mma-sat 300 pmol/plate CRNGDP 4,1631,83
msc-ham: lng 200 nmol/L CNREA8 40,2876,80
skn-mus TDLo: 278 μg/kg: NEO CNREA8 46,2257,86
ipr-mus TDLo: 12 mg/kg/15D-I: NEO CNREA8 46,2257,86

SAFETY PROFILE: Questionable carcinogen with experimental neoplastigenic data. Mutation data reported. When heated to decomposition it emits toxic fumes of NO_x.

BMK750 **HR: 2**
BRACKEN FERN, CHLOROFORM FRACTION

PROP: Chloroform fraction of tannin isolated from Bracken Fern (*Pteridium aquilinum*).

TOXICITY DATA WITH REFERENCE
orl-rat TDLo: 1000 g/kg/56W-C: CAR JJIND8 65,131,80

SAFETY PROFILE: Questionable carcinogen with experimental carcinogenic data. Mutation data reported. See also SHIKIMIC ACID.

BML000 **HR: 3**
BRACKEN FERN, DRIED

SYNS: 1-CYCLOHEXENE-1-CARBOXYLIC ACID, 3,4,5 ☐ S. EGRELTRI ATUNUN (TURKISH) ☐ PTERIDIUM AQUILINUM ☐ PTERIS AQUALINA

TOXICITY DATA WITH REFERENCE
sln-dmg-orl 15 pph MUREAV 92,89,82
bfa-rat/sat 1000 g/kg JJIND8 65,131,80
bfa-ctl/sat 623 g/kg/2Y-C CNREA8 38,1556,78
orl-mus TDLo: 1782 g/kg (4W pre/1-7D preg): TER TXAPA9 28,264,74
orl-rat TDLo: 2209 g/kg/17W-C: CAR JNCIAM 45,179,70
orl-mus TDLo: 7140 g/kg/17W-C: ETA GANMAX 17,205,75
orl-ctl TDLo: 600 g/kg/2Y-C: CAR VTPHAK 13,110,76
orl-rat TD: 1350 g/kg/81D-C: NEO GANNA2 69,383,78
orl-rat TD: 8800 g/kg/58W-C: CAR NUCADQ 3,86,81
orl-ctl TD: 495 g/kg/68W-C: NEO CNREA8 27,917,67
orl-ctl TD: 3072 g/kg/3Y-C: CAR CNREA8 28,2247,68

CONSENSUS REPORTS: IARC Cancer Review: Human Inadequate Evidence IMEMDT 40,47,86; Animal Sufficient Evidence IMEMDT 40,47,86.

SAFETY PROFILE: Confirmed carcinogen with experimental carcinogenic, neoplastigenic, and tumorigenic data. Experimental teratogenic and reproductive effects. Mutation data reported.

BML250 **HR: 3**
BRACKEN FERN TANNIN

SYNS: PTERIDIUM AQUILINUM TANNIN □ TANNIN from BRACKEN FERN

TOXICITY DATA WITH REFERENCE
bfa-rat/sat 2000 g/kg/56W-C JJIND8 65,131,80
scu-rat TDLo: 1595 mg/kg/38W-I: NEO JJIND8 65,131,80
imp-mus TDLo: 50 mg/kg/1Y-C: NEO JNCIAM 56,33,76
ipr-mus LD50: 160 mg/kg JNCIAM 56,33,76

SAFETY PROFILE: Poison by intraperitoneal route. Questionable carcinogen with experimental neoplastigenic data. Mutation data reported.

BML750 **CAS:11011-73-7** **HR: 3**
BRAMYCIN
mf: $C_{32}H_{55}NO_{11}$ mw: 629.88

PROP: An antibiotic produced by *Streptomyces diastatochromogenes var. bracus.*

TOXICITY DATA WITH REFERENCE
orl-mus LD50: 4800 µg/kg 85ERAY 2,1134,78
ipr-mus LD50: 490 µg/kg 85ERAY 2,1134,78

SAFETY PROFILE: Poison by ingestion and intraperitoneal routes. When heated to decomposition it emits toxic fumes of NO_x.

BML825 **HR: 3**
BRAXORONE
mf: $C_{21}H_{27}BrO_3$ mw: 407.39

SYNS: 9-α-BROMO-11-KETOPROGESTERONE □ 9 BROMOPREGN-4-ENE-3,11,20-TRIONE □ 9-α-BROMOPREGN 4 ENE-3,11,20-TRIONE

TOXICITY DATA WITH REFERENCE
orl-wmn TDLo: 13 mg/kg (21D pre): REP AJOGAH 76,626,58
orl-wmn TDLo: 18 mg/kg (15D pre): REP OBGNAS 10,411,57
unr-mus TDLo: 200 µg/kg (female 1D pre): REP ANYAA9 71,494,58

SAFETY PROFILE: Human female reproductive effects by ingestion: disorders of the menstrual cycle and changes in the uterus, cervix, or vagina. When heated to decomposition it emits toxic fumes of Br^-.

BMM000 **CAS:50924-49-7** **HR: 2**
BREDININ
mf: $C_9H_{13}N_3O_6$ mw: 259.25

PROP: Crystals from MeOH.

SYNS: ANHYDRO-4-CARBAMOYL-5-HYDROXY-1-β-d-RIBOFURANOSYL-IMIDAZOLIUMHYDROXIDE □ BREDININE □ 4-CARBAMOYL-1-β-d-RIBOFURANOSYL-IMIDAZOLIUM-5-OLATE □ 5-HYDROXY-1-β-d-RIBOFURANOSYL-1H-IMIDAZOLE-4-CARBOXAMIDE □ MIZORIBINE

TOXICITY DATA WITH REFERENCE
mnt-mus-ipr 5 g/kg OYYAA2 24,703,82
cyt-mus: leu 20 µmol/L IDZAAW 51,61,76
cyt-mus: lym 20 µmol/L CNREA8 35,1643,75
sce-mus: lym 2500 nmol/L OYYAA2 24,703,82
dlt-mus-orl 160 mg/kg OYYAA2 24,711,82
cyt-ham: fbr 100 µmol/L CNREA8 35,1643,75
orl-rat TDLo: 48 mg/kg (17-22D preg): REP OYYAA2 26,397,83
orl-rat TDLo: 80 mg/kg (female 17-22D post): REP OYYAA2 26,397,83
orl-mus TDLo: 75 mg/kg (male 3D pre): TER OYYAA2 24,711,82
orl-rat TDLo: 11 mg/kg (7-17D preg): TER OYYAA2 26,377,83
orl-rat TDLo: 300 mg/kg (female 30D pre): REP OYYAA2 26,293,83
ipr-rat TDLo: 5 mg/kg (female 9D post): TER SEIJBO 18,227,78
ipr-rat TDLo: 5 mg/kg (female 11D post): TER SEIJBO 20,359,80
orl-uns TDLo: 44 mg/kg (female 8-18D post): TER JIDOAA 35,387,86
orl-rat LD50: 3100 mg/kg IYKEDH 15,668,84
scu-rat LD50: 4161 mg/kg IYKEDH 15,688,84
ivn-rat LD50: 1500 mg/kg DRFUD4 3,567,78
ipr-mus LD50: 5000 mg/kg 85GDA2 5,275,81
ivn-mus LD50: 500 mg/kg 85GDA2 5,275,81

CONSENSUS REPORTS: EPA Genetic Toxicology Program.

SAFETY PROFILE: Moderately toxic by ingestion and intravenous route. An experimental teratogen. Other experimental reproductive effects. Mutation data reported. An immunosuppressive agent. When heated to decomposition it emits toxic fumes of NO_x.

BMM500 **CAS:2580-78-1** **HR: 2**
BRILLIANT BLUE R
mf: $C_{22}H_{16}N_2O_{11}S_3 \cdot 2Na$ mw: 626.56

SYNS: CAVALITE BRILLIANT BLUE R □ C.I. 61200 □ C.I. REACTIVE BLUE 19 □ C.I. REACTIVE BLUE 19, DISODIUM SALT □ REACTIVE BLUE 19 □ REMALAN BRILLIANT BLUE R □ REMAZOL BRILLIANT BLUE R

TOXICITY DATA WITH REFERENCE
orl-rat TDLo: 87 g/kg/2Y-I: ETA TKORAS 3,53,67
scu-mus TDLo: 47 g/kg/39W-I: ETA TKORAS 3,53,67

CONSENSUS REPORTS: Reported in EPA TSCA Inventory.

SAFETY PROFILE: Questionable carcinogen with experimental tumorigenic data. When heated to decomposition it emits very toxic fumes of Na_2O, NO_x, and SO_x. See also SULFONATES.

BMM550 **CAS:10127-36-3** **HR: 3**
BRILLIANT CRESYL BLUE
mf: $C_{17}H_{21}N_4O \cdot Cl$ mw: 332.87

SYNS: BRILLIANT CRESYL BLUE BB □ C.I. 51010 □ 1,3 DIAMINO-7-(DIETHYLAMINO)-8-METHYLPHENOXAZIN-5-IUM CHLORIDE □ PHENOXAZIN-5-IUM, 1,3-DIAMINO-7-(DIETHYLAMINO)-8-METHYL-, CHLORIDE

TOXICITY DATA WITH REFERENCE
cyt-ham:ovr 20 µmol/L ENMUDM 1,27,79
ivn-mus LD50:32 mg/kg TXAPA9 44,225,78

CONSENSUS REPORTS: Reported in EPA TSCA Inventory.

SAFETY PROFILE: A poison by intravenous route. Mutation data reported. When heated to decomposition it emits toxic vapors of NO_x and Cl^-.

BMM625 CAS:63638-90-4 HR: 3
BROFAREMINE HYDROCHLORIDE
mf: $C_{14}H_{16}BrNO_2 \cdot ClH$ mw: 346.68

PROP: Crystals from MeOH/Et_2O. Mp: 242–243°.

SYNS: 4-(7-BROMO-5-METHOXY-2-BENZOFURANYL)PIPERIDINE HYDROCHLORIDE □ CGP-11305A

TOXICITY DATA WITH REFERENCE
orl-rat LD50:310 mg/kg DRFUD4 10,371,85
orl-mus LD50:190 mg/kg DRFUD4 10,371,85
orl-dog LD50:100 mg/kg DRFUD4 10,371,85

SAFETY PROFILE: Poison by ingestion. When heated to decomposition it emits toxic fumes of Br^-, NO_x, and HCl.

BMM650 CAS:314-40-9 HR: 2
BROMACIL
mf: $C_9H_{13}BrN_2O_2$ mw: 261.15

PROP: Crystals or solid from EtOH (aq). Sltly sol in H_2O; mod sol in strong aqueous bases from Me_2CO, MeCN, and EtOH.

SYNS: BOREA □ BROMAZIL □ 5-BROMO-3-sec-BUTYL-6-METHYLURACIL □ 5-BROMO-6-METHYL-3-(1-METHYLPROPYL)-2,4(1H,3H)-PYRIMIDINEDIONE □ 5-BROMO-6-METHYL-3-(1-METHYLPROPYL) URACIL □ 3-sek.BUTYL-5-BROM-6-METHYLURACIL (GERMAN) □ CYNOGAN □ DU PONT HERBICIDE 976 □ EEREX GRANULAR WEED KILLER □ EEREX WATER SOLUBLE CONCENTRATE WEED KILLER □ HERBICIDE 976 □ HYVAR □ HYVAREX □ HYVAR X □ HYVAR X BROMACIL □ HYVAR X WEED KILLER □ KROVAR II □ NALKIL □ URAGAN □ URAGON □ UROX B WATER SOLUBLE CONCENTRATE WEED KILLER □ UROX HX GRANULAR WEED KILLER

TOXICITY DATA WITH REFERENCE
sln-dmg-orl 2000 ppm JPFCD2 15,867,80
sln-nsc 10 mg/L MUREAV 167,35,86
msc-mus:lym 750 mg/L NTIS** PB84-138973
ihl-rat TCLo:38 mg/m³/2H (7-14D preg):TER NTIS** PB277-077
orl-rat LD50:641 mg/kg FAATDF 7,299,86
orl-mus LD50:3040 mg/kg JPFCD2 15,867,80

CONSENSUS REPORTS: EPA Genetic Toxicology Program.

OSHA PEL: TWA 1 ppm
ACGIH TLV: TWA 1 ppm

SAFETY PROFILE: Moderately toxic by ingestion. An experimental teratogen. Mutation data reported. An herbicide. When heated to decomposition it emits very toxic fumes of Br^- and NO_x.

BMN000 CAS:28772-56-7 HR: 3
BROMADIALONE
mf: $C_{30}H_{23}BrO_4$ mw: 527.11

SYNS: BROMADIOLONE □ 3-(3-(4'-BROMO(1,1'-BIPHENYL)-4-YL) 3-HYDROXY-1-PHENYLPROPYL)-4-HYDROXY-2H-1-BENZOPYRAN-2-ONE □ BROMONE □ 3-(α-(p-(p-BROMOPHENYL)-β-HYDROXYPHENETHYL)BENZYL)-4-HYDROXYCOUMARIN □ CANADIEN 2000 □ CONTRAC □ (HYDROXY-4 COUMARINYL 3)-3 PHENYL-3 (BROMO-4 BIPHENYLYL-4)-1 PROPANOL-1 (FRENCH) □ LM-637 □ MAKI □ RATIMUS □ SUPER-CAID □ SUPER-ROZOL □ SUP'OPERATS □ TEMUS

TOXICITY DATA WITH REFERENCE
orl-rat LD50:490 µg/kg MRBUDF 13,303,85
orl-mus LD50:1750 µg/kg PHPHA6 25,69,76
orl-rbt LD50:1 mg/kg PHPHA6 25,69,76

CONSENSUS REPORTS: EPA Extremely Hazardous Substances List.

SAFETY PROFILE: A deadly poison by ingestion. Used as a rodent poison. When heated to decomposition it emits toxic fumes of Br^-. See also WARFARIN.

BMN250 CAS:13977-28-1 HR: 3
BROMADRYL
mf: $C_{18}H_{22}BrNO \cdot ClH$ mw: 384.78

PROP: Crystals from Me_2CO/Et_2O. Mp: 152–155°.

SYNS: 2-(1-(4-BROMODIPHENYL)ETHOXY)-N,N-DIMETHYLETHYLAMINE HYDROCHLORIDE □ p-BROMO-α-METHYLBENZHYDRYL-2-DIMETHYLAMINOETHYL ETHER HYDROCHLORIDE □ 2-((p-BROMO-α-METHYL-α-PHENYLBENZYL)OXY)-N,N-DIMETHYLETHYLAMINE HYDROCHLORIDE □ 2-(1-(4-BROMOPHENYL)-1-PHENYLETHOXY)-N,N-DIMETHYLETHANAMINE HYDROCHLORIDE □ 1-(p-BROMOPHENYL)-1-PHENYL-1-(2-DIMETHYLAMINOETHOXY)ETHANE HYDROCHLORIDE □ (2-(1-p-BROMOPHENYL-1-PHENYLETHOXY)ETHYL)DIMETHYLETHYLAMINE HYDROCHLORIDE □ β-DIMETHYLAMINOETHYL-p-BROMO-α-METHYLBENZHYDRYL ETHER HYDROCHLORIDE □ EMBRAMINE HYDROCHLORIDE □ MEBROPHENHYDRAMINE □ MEBROPHENHYDRAMINE HYDROCHLORIDE □ MEBRYL

TOXICITY DATA WITH REFERENCE
orl-mus LD50:330 mg/kg MEIEDD 10,513,83
ivn-mus LD50:80 mg/kg MEIEDD 10,513,83

SAFETY PROFILE: Poison by ingestion and intravenous routes. An antihistaminic agent. When heated to decomposition it emits very toxic fumes of HCl, Br^-, and NO_x. See also ETHERS.

BMN350 CAS:332-69-4 HR: 2
BROMANYLPROMIDE
mf: $C_{11}H_{15}BrN_2O$ mw: 271.19

PROP: Solid. Mp: 114.5–116.5°.

SYNS: BROMAMID □ BROMAMIDE □ BROMAMIDE (pharmaceutical) □ N,N-DIMETHYL-β-(p-BROMOANILINO)PROPIONAMIDE □ 3-((4-BROMOPHENYL)AMINO)-N,N-DIMETHYL-PROPANAMIDE (9CI)

TOXICITY DATA WITH REFERENCE
orl-rat LD50:1810 mg/kg OYYAA2 2,70,68
orl-mus LD50:2431 mg/kg OYYAA2 2,70,68
scu-mus LD50:1375 mg/kg OYYAA2 2,70,68

SAFETY PROFILE: Moderately toxic by ingestion and

other routes. When heated to decomposition it emits toxic fumes of Br^- and NO_x.

BMN500 HR: 3
BROMATES

SAFETY PROFILE: Generally considered to be more toxic than chlorates; cause central nervous system paralysis. They may form methemoglobin, but less actively than chlorates. See also specific compounds. Flammable in the form of gas, vapor, or dust by chemical reaction with (powdered metals + acids); Al; As; CaH_2; C; Cu; powdered metals; metal sulfides; organic matter; PH_4I; P; SrH; S; (H_2SO_4 + metals). When heated to decomposition they emit toxic fumes of Br^-; can react with reducing materials.

BMN750 CAS:1812-30-2 HR: 3
BROMAZEPAM
mf: $C_{14}H_{10}BrN_3O$ mw: 316.18

PROP: Prisms from Me_2CO. Mp: 241–241° (decomp). Sol in dil acids, EtOH, Me_2CO.

SYNS: 7-BROMO-1,3-DIHYDRO-5-(2-PYRIDYL)-2H-1,4-BENZDIAZEPIN-2-ONE □ 7-BROMO-5-(2-PYRIDYL)-3H-1,4-BENZODIAZEPIN-2 (1H)-ONE □ COMPENDIUM □ 1,3-DIHYDRO-7-BROMO-5-(2-PYRIDYL)-2H-1,4-BENZODIAZEPIN-2-ONE □ KL-001 □ LA XVII □ LECTOPAM □ LEXOMIL □ LEXOTAN □ LEXOTANIL □ RO 4-9253 □ RO 5-3350

TOXICITY DATA WITH REFERENCE
skn-hmn 3 mg/3D-I MOD 85DKA8 -,127,77
rec-rat TDLo:11 mg/kg (female 7-17D post):REP OYYAA2 26,111,83
rec-rat TDLo:110 mg/kg (female 7-17D post):REP OYYAA2 26,111,83
rec-rat TDLo:180 mg/kg (female 17-22D post):REP OYYAA2 26,199,83
rec-rat TDLo:60 mg/kg (female 17-22D post):REP OYYAA2 26,199,83
rec-rbt TDLo:130 mg/kg (female 6-18D post):TER OYYAA2 26,99,83
orl-rbt TDLo:720 mg/kg (female 8-16D post):TER KSRNAM 7,2433,73
orl-mus TDLo:875 mg/kg (female 7-13D post):TER KSRNAM 7,2422,73
orl-mus TDLo:35 mg/kg (female 7-13D post):REP KSRNAM 7,2422,73
orl-rat TDLo:35 mg/kg (7-13D preg):TER KSRNAM 7,2422,73
orl-rat LD50:1950 mg/kg KSRNAM 7,2413,73
ipr-rat LD50:1660 mg/kg OYYAA2 17,115,79
scu-rat LD50:8800 mg/kg OYYAA2 17,115,79
orl-mus LD50:879 mg/kg OYYAA2 17,115,79
ipr-mus LD50:200 mg/kg JDGRAX 16,7,85
scu-mus LD50:6870 mg/kg OYYAA2 17,115,79
orl-rbt LD50:1690 mg/kg 27ZQAG -,158,72

SAFETY PROFILE: Poison by intraperitoneal route. Moderately toxic by ingestion. An experimental teratogen. Other experimental reproductive effects. Human skin irritant. A tranquilizer. When heated to decomposition it emits very toxic fumes of NO_x and HBr.

BMO000 CAS:9001-00-7 HR: 3
BROMELAIN

PROP: From pineapples *Ananas comosus* and *Ananas bracteatus* L. White to tan amorphous powder. Sol in water; insol in alc, chloroform, ether.

SYNS: ANANASE □ BROMELAINS □ BROMELIN □ E.C. 3.4.4.24 □ EXTRANASE □ INFLAMEN □ PLANT PROTEASE CONCENTRATE □ TRAUMANASE

TOXICITY DATA WITH REFERENCE
ipr-rat LD50:85,200 μg/kg AIPTAK 145,166,63
ipr-rat LD50:85 mg/kg AIPTAK 145,166,63
ipr-mus LD50:37 mg/kg AIPTAK 145,166,63
ivn-mus LD50:30 mg/kg AIPTAK 145,166,63

CONSENSUS REPORTS: Reported in EPA TSCA Inventory.

SAFETY PROFILE: A poison via intraperitoneal and intravenous routes. When heated to decomposition it emits acrid smoke and fumes.

BMO250 CAS:15086-94-9 HR: 1
BROMEOSIN
mf: $C_{20}H_8Br_4O_5$ mw: 647.92

PROP: Insol in H_2O; sltly sol in EtOH; sol in alkalis.

SYNS: BROMOEOSIN □ BROMOFLUORESCEIC ACID □ C.I. 45380:2 □ C.I. SOLVENT RED 43 □ D&C RED No. 21 □ EOSIN □ EOSINE □ 2,4,5,7-TETRABROMO-3,6-FLUORANDIOL □ TETRABROMOFLUORESCEIN □ 2′,4′,5′,7′-TETRABROMOFLUORESCEIN

TOXICITY DATA WITH REFERENCE
dnr-bcs 2 mg/disc TRENAF 27,153,76
scu-mus LDLo:450 mg/kg HBAMAK 4,1289,35
scu-frg LDLo:1 g/kg HBAMAK 4,1289,35

CONSENSUS REPORTS: IARC Cancer Review: Group 3 IMEMDT 7,56,87, Animal Inadequate Evidence IMEMDT 15,183,77. Reported in EPA TSCA Inventory.

SAFETY PROFILE: Mutation data reported. Incompatible with reducing agents. When heated to decomposition it emits very toxic fumes of Br^-. See also BROMIDES.

BMO325 CAS:611-75-6 HR: 3
BROMHEXINE CHLORIDE
mf: $C_{14}H_{20}Br_2N_2{\cdot}ClH$ mw: 412.64

PROP: Solid. Mp: 237–238° (decomp). Insol in Me_2CO and $CHCl_3$.

SYNS: 2-AMINO-3,5-DIBROMO-N-CYCLOHEXYL-N-METHYL-BENZENEMETHANAMINE MONOHYDROCHLORIDE (9CI) □ BISOLVON □ BISOLVON HYDROCHLORIDE □ BROMHEXINE HYDROCHLORIDE

TOXICITY DATA WITH REFERENCE
orl-rat LD50:6 g/kg GNRIDX 3,259,69
ipr-rat LD50:1680 mg/kg GNRIDX 3,259,69
orl-mus LD50:4800 mg/kg GNRIDX 3,259,69
ipr-mus LD50:2210 mg/kg GNRIDX 3,259,69
ivn-mus LD50:44 mg/kg GNRIDX 3,259,69

SAFETY PROFILE: Poison by intravenous route. Moder-

ately toxic by other routes. When heated to decomposition it emits toxic fumes of Br^-, NO_x, and HCl.

BMO750 **HR: 3**
BROMIDES

SAFETY PROFILE: The most common inorganic bromides are Na, K, NH_4, Ca, and Mg bromides. Methyl and ethyl bromides are among the most common organic bromides. The inorganic bromides produce depression, emaciation, and, in severe cases, psychosis and mental deterioration. Bromide rashes (bromoderma), especially of the face and resembling acne and furunculosis, often occur when bromide inhalation or administration is prolonged. Organic bromides, such as methyl bromide and ethyl bromide, are volatile liquids of relatively high toxicity. See also specific compounds. When strongly heated they emit highly toxic fumes of Br^-.

BMP000 **CAS:7726-95-6** **HR: 3**
BROMINE
DOT: UN 1744
mf: Br_2 mw: 159.82

PROP: Rhombic crystals or dark red-brown liquid with a strong disagreeable pungent odor. Strong oxidant. Fp: −7.3°, bp: 59.5°, d: 2.928 @ 59°, 3.12 @ 20°, vap press: 175 mm @ 21°, 1 atm @ 58.2°, vap d: 5.5. Sol in H_2O. Misc in most org solvs, although it may react.

SYNS: BROM (GERMAN) □ BROME (FRENCH) □ BROMINE, solution (DOT) □ BROMO (ITALIAN) □ BROOM (DUTCH)

TOXICITY DATA WITH REFERENCE
orl-hmn LDLo:14 mg/kg 34ZIAG -,645,69
ihl-hmn LCLo:1000 ppm 34ZIAG -,645,69
ihl-mus LC50:750 ppm/9M AIHAAP 39,129,78
ihl-cat LCLo:140 ppm/7H AHYGAJ 7,233,1887
ihl-rbt LCLo:180 ppm/6.5H HBTXAC 1,324,56
ihl-gpg LCLo:140 ppm/7H AHYGAJ 7,233,1887

CONSENSUS REPORTS: Reported in EPA TSCA Inventory. EPA Genetic Toxicology Program.

OSHA PEL: TWA 0.1 ppm; STEL 0.3 ppm
ACGIH TLV: TWA 0.1 ppm; STEL 0.2 ppm
DFG MAK: 0.1 ppm (0.7 mg/m³)
DOT CLASSIFICATION: 8; *Label:* Corrosive, Poison

SAFETY PROFILE: A human poison by ingestion and moderately toxic by inhalation. A poison by ingestion and inhalation experimentally. Corrosive. The action of bromine is essentially the same as that of chlorine, irritating the mucous membranes of the eyes and upper respiratory tract. Severe exposure may result in pulmonary edema. Usually, however, the irritant qualities of the chemical force the worker to leave the exposure area before serious poisoning can result. Chronic exposure is similar to the therapeutic ingestion of excessive bromides. See also BROMIDES. Regular physical examinations should be made of people who work with bromine or bromides. Flammable in the form of liquid or vapor by spontaneous chemical reaction with reducing materials. A very powerful oxidizer. Highly dangerous; when heated it emits highly toxic fumes; will react with water or steam to produce toxic and corrosive fumes. Reacts explosively with diethylzinc, germane, disilane, dimethylformamide, hydrogen, isobutyrophenone, metal azides (particularly silver or sodium azide), potassium, silane and homologs, praseodymium, antimony, trimethylamine, ammonia. Mixtures with lithium or sodium are shock-sensitive explosives. Ignition on contact with germanium, mono- or di-alkali metal acetylides, trialkyl boranes, copper acetylides. Violent reaction with carbonyl compounds (aldehydes, ketones, carboxylic acids), diethyl ether, phosphine, natural rubber, aluminum, mercury, titanium. Vigorous reaction with methanol and other alcohols, tetrahydrofuran, mixtures of ethanol and phosphorus. Incompatible with acetaldehyde, C_2H_2, acrylonitrile, NH_3, Sb, B, Ca_3N_2, Cs_2O, Cs_2C_2, CsC_2H, ClF_3C_2, CuH_2, dimethyl formamide, ethyl phosphine, F_2, Fe_2C, isobutyrophenone, Li_2C_2, Li_2Si_2, Mg_3P_2, $Ni(Co)_4$, NI_3, olefins, OF_2, O_3, P, PO_x, Rb_2C_2, RbC_2H, Na_2C_2, NaC_2H, Sr_3P, Sn, UC_2, ZrC2, reducing materials.

For occupational chemical analysis use OSHA: #ID-108 or NIOSH: Bromine and Chloride 6011.

BMP250 **CAS:13973-87-0** **HR: 3**
BROMINE AZIDE
mf: BrN_3 mw: 121.93

PROP: Crystals or red liquid. Mp: 45°, bp: explodes.

SYNS: BROMINE NITRIDE □ NITROGEN BROMIDE

DOT CLASSIFICATION: Forbidden

SAFETY PROFILE: A poison. Can explode spontaneously. The solid, liquid, and vapor are shock-sensitive explosives. Concentrated solutions in organic solvents may explode. Moderate fire hazard in the form of vapor by chemical reaction. A powerful oxidant. Moderately explosive when exposed to heat. The liquid explodes on contact with arsenic, sodium, silver foil, or phosphorus. Incompatible with Sb, ethyl ether, Ag, metals. When heated to decomposition it emits highly toxic fumes of $Br^-$and explodes. See also BROMINE and AZIDES.

BMP500 **CAS:21255-83-4** **HR: 3**
BROMINE DIOXIDE
mf: BrO_2 mw: 111.91

PROP: Light or dark yellow crystals.

CONSENSUS REPORTS: EPA Extremely Hazardous Substances List.

SAFETY PROFILE: Very unstable material. Flammable in the form of vapor by chemical reaction with reducing agents. Potentially explosive if heated rapidly. A strong oxidant. Reaction with water, steam, or reducing materials produces toxic and corrosive fumes. Must be stored at low temperatures. When heated to decomposition it emits toxic fumes of Br^-. See also BROMINE.

BMP750 **CAS:13863-59-7** **HR: 3**
BROMINE FLUORIDE
mf: BrF mw: 98.91

PROP: Red-brown gas. Mp: −33°, bp: 20°.

SAFETY PROFILE: A poison and powerful irritant. Very reactive. Ignites on contact with H_2. Incompatible with organic matter and water. When heated to decomposition it emits toxic fumes of F^- and Br^-. See also BROMINE and FLUORIDE.

BMQ000 CAS:7789-30-2 HR: 3
BROMINE PENTAFLUORIDE
DOT: UN 1745
mf: BrF_5 mw: 174.91

PROP: Colorless fuming liquid. An extremely vigorous fluorinating agent. Mp: −60.5, bp: 40.5, d: 2.466 @ 25°, vap d: 6.05.

OSHA PEL: TWA 0.1 ppm; TWA 2.5 mg(F)/m³
ACGIH TLV: TWA 0.1 ppm
NIOSH REL: (Inorganic Fluorides) TWA 2.5 mg(F)/m³
DOT CLASSIFICATION: 5.1; *Label:* Oxidizer, Poison, Corrosive

SAFETY PROFILE: A poisonous, corrosive, and extremely reactive gas. It is a powerful oxidizer. Will react with water or steam to produce toxic and corrosive fumes. The liquefied gas reacts violently with many organic compounds and some inorganic compounds. Explodes or ignites on contact with hydrogen-containing materials (e.g., acetic acid, ammonia, benzene, ethanol, hydrogen, hydrogen sulfide, methane, cork, grease, paper, wax, chloromethane). Reacts violently and may ignite on contact with acids, halogens, nonmetals, metal halides, metals, oxides, concentrated nitric or sulfuric acids, aluminum powder, ammonium chloride, antimony, arsenic, arsenic pentoxide, barium, bismuth, boron powder, boron trioxide, calcium oxide, carbon monoxide, charcoal, chlorine, chromium, chromium trioxide, cobalt powder, iodine, iodine pentoxide, iridium powder, iron powder, lithium powder, manganese, magnesium oxide, molybdenum, molybdenum trioxide, nickel powder, red phosphorus, phosphorus pentoxide, potassium iodide, rhodium powder, selenium, sulfur, sulfur dioxide, tellurium, tungsten, tungsten trioxide, water, zinc. When heated to decomposition it emits very toxic fumes of $F^-$and Br^-. See also BROMINE and FLUORIDES.

BMQ250 HR: 3
BROMINE PERCHLORATE
mf: $BrClO_4$ mw: 179.36

SAFETY PROFILE: A shock-sensitive explosive. Upon decomposition it emits toxic fumes of Cl^- and Br^-. See also PERCHLORATES and BROMIDES.

BMQ325 CAS:7787-71-5 HR: 3
BROMINE TRIFLUORIDE
DOT: UN 1746
mf: BrF_3 mw: 136.91

PROP: Colorless, fuming liquid. Mp: 8.8°, bp: 127°, d: 2.84.

CONSENSUS REPORTS: Reported in EPA TSCA Inventory.

OSHA PEL: TWA 2.5 mg(F)/m³
ACGIH TLV: TWA 2.5 mg(F)/m³
NIOSH REL: (Inorganic Fluorides) TWA 2.5 mg(F)/m³
DOT CLASSIFICATION: 5.1; *Label:* Oxidizer, Poison, Corrosive

SAFETY PROFILE: Poisonous and corrosive. Very reactive, a powerful oxidizer. Explosive or violent reaction with organic materials, water, acetone, ammonium halides, antimony, antimony trichloride oxide, arsenic, benzene, boron, bromine, carbon, carbon monoxide, carbon tetrachloride, carbon tetraiodide, chloromethane, cobalt, ether, halogens, iodine, powdered molybdenum, niobium, 2-pentanone, phosphorus, potassium hexachloroplatinate, pyridine, silicon, silicone grease, sulfur, tantalum, tin dichloride, titanium, toluene, vanadium, uranium, uranium hexafluoride. Incompatible with Sb_2O_3, $BaCl_2$, Bi_2O_5, $CdCl_2$, $CaCl_2$, CsCl, LiCl, $MnIO_3$, metals, Nb_2O_5, $PtBr_4$, $PtCl_4$, (Pt + KFO), KBr, KCl, KI, $RhBr_4$, RbCl, AgCl, NaBr, NaCl, NaI, Ta_2O_5, Sn, W, UO_x, rubber, plastics. The product of reaction with pyridine ignites when dry. When heated to decomposition it emits toxic fumes of F^- and Br^-. Very dangerous. See also BROMINE PENTAFLUORIDE, FLUORIDES, and BROMINE.

BMQ500 CAS:70142-16-4 HR: 3
BROMINE(1) TRIFLUOROMETHANESULFONATE
mf: $CBrF_3O_3S$ mw: 228.97

SAFETY PROFILE: A strong oxidizer that may react explosively with readily oxidizable materials. When heated to decomposition it emits toxic fumes of F^-, Br^-, and SO_x. See also SULFONATES.

BMQ750 HR: 3
BROMINE TRIOXIDE
mf: BrO_3 mw: 127.91

SAFETY PROFILE: The solid produced at −5° is only stable at −80° or in the presence of ozone. Decomposition can be violently explosive in the presence of trace impurities. Upon decomposition it emits toxic fumes of Br^-. See also BROMINE.

BMR100 CAS:103-88-8 HR: 3
4-BROMOACETANILIDE
mf: C_8H_8BrNO mw: 214.08

SYNS: ACETAMIDE, N-(4-BROMOPHENYL)- □ ACETANILIDE, p-BROMO- □ ACETANILIDE, 4'-BROMO- □ ANTISEPSIN □ ASEPSIN □ p-BROMOACETANILIDE □ 4'-BROMOACETANILIDE □ p-BROMO-N-ACETANILIDE □ BROMOANILIDE □ BROMOANTIFEBRIN □ USAF DO-40

TOXICITY DATA WITH REFERENCE
ipr-mus LD50:250 mg/kg NTIS** AD277-689

CONSENSUS REPORTS: Reported in EPA TSCA Inventory.

SAFETY PROFILE: Poison by intraperitoneal route. When heated to decomposition it emits toxic vapors of NO_x and Br^-.

BMR750 **CAS:79-08-3** **HR: 3**
α-BROMOACETIC ACID
DOT: UN 1938
mf: $C_2H_3O_2Br$ mw: 138.04

PROP: Hygroscopic crystals, sol in water and alc. D: 1.93, mp: 50°, bp: 208°. Sol in H_2O and EtOH.

SYNS: ACIDE BROMACETIQUE (FRENCH) □ BROMOACETIC ACID □ BROMOACETIC ACID, solid or solution (DOT) □ BROMOETHANOIC ACID □ α-BROMOETHANOIC ACID □ KYSELINA BROMOCTOVA □ MONOBROMESSIGSAEURE (GERMAN) □ MONOBROMOACETIC ACID □ TO NTU

TOXICITY DATA WITH REFERENCE
dnd-mus:leu 100 µmol/L BCPCA6 30,1497,81
orl-mus LD50:100 mg/kg JPETAB 86,336,46
ipr-mus LD50:66 mg/kg JNCIAM 31,297,63
ivn-rbt LDLo:45 mg/kg AEPPAE 160,551,31
ihl-rat LCLo:114 g/m³/30M RPTOAN 41,113,78
ipr-rat LD50:50 mg/kg RPTOAN 41,113,78

CONSENSUS REPORTS: Reported in EPA TSCA Inventory.

DOT CLASSIFICATION: 8; *Label:* Corrosive

SAFETY PROFILE: Poison by ingestion, intraperitoneal, and intravenous routes. Irritating and corrosive to skin and mucous membranes. Mutation data reported. When heated to decomposition it emits toxic fumes of Br^-. See also BROMIDES.

BMS000 **CAS:62116-25-0** **HR: 3**
BROMOACETONE OXIME
mf: C_3H_6BrNO mw: 151.98

$BrCH_2C(CH_3)=NOH$

PROP: Solid. Mp: 36.5°, bp: 83° @ 8 mm.

SYN: 1-BROMO-2-OXIMINOPROPANE

SAFETY PROFILE: Decomposes explosively during distillation. When heated to decomposition it emits toxic fumes of Br^- and NO_x. See also CHLOROACETONE and BROMIDES.

BMS250 **CAS:4189-47-3** **HR: 2**
1-BROMOACETOXY-2-PROPANOL
mf: $C_5H_9BrO_3$ mw: 197.05

SYN: NALCON 240

TOXICITY DATA WITH REFERENCE
orl-rat LD50:664 mg/kg PCOC** -,152,66
skn-rbt LD50:813 mg/kg PCOC** -,152,66

SAFETY PROFILE: Moderately toxic by ingestion and skin contact. When heated to decomposition it emits toxic fumes of Br^-.

BMS300 **CAS:143-84-0** **HR: 3**
1-BROMOACETYL-α-α-DIPHENYL-4-PIPERIDINEMETHANOL
mf: $C_{20}H_{22}BrNO_2$ mw: 388.34

SYNS: KETONE, BROMOMETHYL 4-(DIPHENYLHYDROXYMETHYL) PIPERIDINO □ 4-PIPERIDINEMETHANOL, 1-BROMOACETYL-α-α-DIPHENYL-

TOXICITY DATA WITH REFERENCE
ipr-mus LD50:20 mg/kg JPMSAE 55,529,66
unr-mus LD50:600 mg/kg JPMSAE 54,269,65

DOT CLASSIFICATION: 3; *Label:* Flammable Liquid

SAFETY PROFILE: A poison by intraperitoneal route. A flammable liquid. When heated to decomposition it emits toxic vapors of NO_x and Br^-.

BMS500 **HR: 3**
BROMOACETYLENE
mf: C_2HBr mw: 104.9

PROP: Gas. Bp: −2°, vap d: 4.684

SYNS: BROMACETYLENE □ BROMOETHYNE

SAFETY PROFILE: Toxicity is probably similar to dibromoacetylene. A dangerous fire hazard by spontaneous chemical reaction. A spontaneously flammable gas. Highly explosive. May explode or ignite on contact with air. Incompatible with oxidizing materials, even when solid at −196°. When heated to decomposition it burns and emits toxic fumes of Br^-. See also ACETYLENE COMPOUNDS and BROMIDES.

BMT000 **CAS:14925-39-4** **HR: 1**
2-BROMOACROLEIN
mf: C_3H_3BrO mw: 134.96

PROP: Bp: 46–48° @ 28 mm.

SYN: 2-BROMOPROPENALDEHYDE

TOXICITY DATA WITH REFERENCE
mmo-sat 1 nmol/plate MUREAV 78,113,80
mma-sat 1 nmol/plate MUREAV 78,113,80
dnd-rat:oth 1 µmol/L CRNGDP 6,705,85
otr-ham:emb 500 nmol/L CRNGDP 6,705,85
skn-mus TDLo:1200 mg/kg/2W-I:CAR CALEDQ 48,197,89

SAFETY PROFILE: Questionable carcinogen with experimental carcinogenic data. Mutation data reported. See also ALDEHYDES and BROMIDES. When heated to decomposition it emits toxic fumes of Br^-.

BMT100 **CAS:73599-95-8** **HR: 3**
1-BROMO-3-ADAMANTYL ETHOXYMETHYL KETONE
mf: $C_{14}H_{22}BrO_2$ mw: 302.27

SYN: ETHANONE, 1-(1-BROMO-3-ADAMANTYL)-2-ETHOXY-

TOXICITY DATA WITH REFERENCE
unr-mus LD50:740 mg/kg RPTOAN 43,73,80

DOT CLASSIFICATION: 3; *Label:* Flammable Liquid

SAFETY PROFILE: Moderately toxic by unspecified route. A flammable liquid. When heated to decomposition it emits toxic vapors of Br^-.

B

BMT130 **CAS:73599-91-4** **HR: 3**
1-BROMO-3-ADAMANTYL HYDROXYMETHYL KETONE
mf: $C_{12}H_{17}BrO_2$ mw: 273.20

SYN: ETHANONE, 1-(1-BROMO-3-ADAMANTYL)-2-HYDROXY-

TOXICITY DATA WITH REFERENCE
unr-mus LD50:700 mg/kg RPTOAN 43,73,80

DOT CLASSIFICATION: 3; *Label:* Flammable Liquid

SAFETY PROFILE: Moderately toxic by an unspecified route. A flammable liquid. When heated to decomposition it emits toxic vapors of Br^-.

BMT150 **CAS:101652-13-5** **HR: 3**
3-BROMOALLYL ISOCYANATE
DOT: UN 2206/UN 2207/UN 2478/UN 3080
mf: C_4H_4BrNO mw: 162.00

SYN: ISOCYANIC ACID, 3-BROMOALLYL ESTER

TOXICITY DATA WITH REFERENCE
ivn-mus LD50:56 mg/kg CSLNX* NX#09998

DOT CLASSIFICATION: 6.1; *Label:* KEEP AWAY FROM FOOD (UN2207); 6.1; *Label:* Poison (UN2206); 6.1; *Label:* Poison, Flammable Liquid (UN3080); 3; *Label:* Flammable Liquid, Poison (UN2478)

SAFETY PROFILE: A poison by intraperitoneal routes. A flammable liquid. When heated to decomposition it emits toxic vapors of NO_x and Br^-.

BMT250 **CAS:14519-10-9** **HR: 3**
BROMOAMINE
mf: BrH_2N mw: 95.93

PROP: Violet-black solid. Sol in Et_2O, insol in pentane.

SYN: BROMAMIDE

SAFETY PROFILE: Decomposes violently @ −70°. Upon decomposition it emits toxic fumes of Br^- and NO_x. See also BROMIDES and AMINES.

BMT300 **HR: 2**
3′-BROMO-trans-ANETHOLE
mf: $C_{10}H_{11}BrO$ mw: 227.12

SYNS: ANISOLE, p-(3-BROMOPROPENYL)-, (E)- □ (E)-p-(3-BROMOPROPENYL)ANISOLE

TOXICITY DATA WITH REFERENCE
mmo-sat 2 μmol/plate CRNGDP 7,2089,86
ipr-mus TDLo:69,600 μg/kg/4D-I:CAR CNREA8 47,2275,87

SAFETY PROFILE: Questionable carcinogen with experimental carcinogenic data. Mutation data reported. When heated to decomposition it emits toxic fumes of Br^-.

BMT325 **CAS:106-40-1** **HR: 3**
4-BROMOANILINE
mf: C_6H_6BrN mw: 172.04

PROP: Rhombic crystals from dil alc. Mp: 66–66.5°, d: 1.4970 (liq). Very sol in alc and ether; insol in cold water.

SYNS: 4-BROMANILINU (CZECH) □ p-BROMOANILINE □ 4-BROMO-BENZENAMINE (9CI) □ p-BROMOPHENYLAMINE

TOXICITY DATA WITH REFERENCE
dns-rat:lvr 50 μmol/L ENMUDM 3,11,81
orl-rat LD50:456 mg/kg CEHYAN 23,168,78
orl-mus LD50:289 mg/kg GISAAA 44(12),19,79
ipr-mus LD50:248 mg/kg GISAAA 44(12),19,79

CONSENSUS REPORTS: EPA Genetic Toxicology Program. Reported in EPA TSCA Inventory.

SAFETY PROFILE: Poison by ingestion and intraperitoneal routes. Mutation data reported. When heated to decomposition it emits toxic fumes of Br^- and NO_x. See also ANILINE DYES.

BMT400 **CAS:578-57-4** **HR: 2**
2-BROMOANISOLE
mf: C_7H_7BrO mw: 187.05

SYNS: ANISOLE, o-BROMO □ ANISYL BROMIDE □ BENZENE, 1-BROMO-2-METHOXY-(9CI) □ o-BROMOANISOLE □ 1-BROMO-2-METHOXYBENZENE □ o-BROMOPHENYL METHYL ETHER □ o-METHOXYBROMOBENZENE □ 2-METHOXYBROMOBENZENE □ o-METHOXYPHENYL BROMIDE □ 2-METHOXYPHENYL BROMIDE

TOXICITY DATA WITH REFERENCE
orl-mus LD50:2466 mg/kg GISAAA 44(12),19,79
ipr-mus LD50:1544 mg/kg GISAAA 44(12),19,79

CONSENSUS REPORTS: Reported in EPA TSCA Inventory.

SAFETY PROFILE: Moderately toxic by ingestion and intraperitoneal routes. When heated to decomposition it emits toxic vapors of Br^-.

BMT500 **CAS:19816-89-8** **HR: 3**
1-BROMOAZIRIDINE
mf: C_2H_4BrN mw: 121.96

SAFETY PROFILE: An unstable material that may spontaneously explode. When heated to decomposition it emits toxic fumes of Br^- and NO_x. See other aziridine compounds.

BMT700 **CAS:1122-91-4** **HR: 3**
4-BROMOBENZALDEHYDE
mf: C_7H_5BrO mw: 185.03

SYNS: BENZALDEHYDE, p-BROMO- □ BENZALDEHYDE, 4-BROMO-(9CI) □ p-BROMOBENZALDEHYDE

TOXICITY DATA WITH REFERENCE
orl-mus LD50:1230 mg/kg GISAAA 44(12),19,79
ipr-mus LD50:389 mg/kg GISAAA 44(12),19,79

CONSENSUS REPORTS: Reported in EPA TSCA Inventory.

SAFETY PROFILE: Poison by intraperitoneal route. Moderately toxic by ingestion. When heated to decomposition it emits toxic vapors of Br^-.

BMT750 CAS:32795-84-9 HR: 2
10-BROMO-1,2-BENZANTHRACENE
mf: $C_{18}H_{11}Br$ mw: 307.20

SYN: 10-BROM-1,2-BENZANTHRACEN (GERMAN)

TOXICITY DATA WITH REFERENCE
scu-rat TDLo:25 mg/kg:ETA SCPHA4 22,224,54

SAFETY PROFILE: Questionable carcinogen with experimental tumorigenic data. When heated to decomposition it emits toxic fumes of Br^-.

BMU000 CAS:81-96-9 HR: 3
3-BROMOBENZ(d,e)ANTHRONE
mf: $C_{17}H_9BrO$ mw: 309.17

SYNS: 3-BROMBENZANTHRONE □ 3-BROMO-7H-BENZ(DE)ANTHRACEN-7-ONE □ 7-BROMOMESOBENZANTHRONE

TOXICITY DATA WITH REFERENCE
eye-rbt 500 mg/24H MLD 28ZPAK -,89,72
ipr-rat LD50:2400 mg/kg RPTOAN 40,137,77
ipr-mus LD50:300 mg/kg RPTOAN 40,137,77

CONSENSUS REPORTS: Reported in EPA TSCA Inventory.

SAFETY PROFILE: Poison by intraperitoneal route. An eye irritant. When heated to decomposition it emits toxic fumes of Br^-.

BMU100 CAS:586-76-5 HR: 2
4-BROMOBENZOIC ACID
mf: $C_7H_5BrO_2$ mw: 201.03

SYNS: BENZOIC ACID, p-BROMO- □ BENZOIC ACID, 4-BROMO-(9CI) □ p-BROMOBENZOIC ACID □ p-CARBOXYBROMOBENZENE

TOXICITY DATA WITH REFERENCE
orl-mus LD50:1059 mg/kg GISAAA 44(12),19,79
ipr-mus LD50:536 mg/kg GISAAA 44(12),19,79

CONSENSUS REPORTS: Reported in EPA TSCA Inventory.

SAFETY PROFILE: Moderately toxic by ingestion and intraperitoneal routes. When heated to decomposition it emits toxic vapors of Br^-.

BMU500 CAS:21248-00-0 HR: 2
6-BROMOBENZO(a)PYRENE
mf: $C_{20}H_{11}Br$ mw: 331.22

PROP: Crystals from Me_2CO/MeOH. Mp: 223–224°.

TOXICITY DATA WITH REFERENCE
scu-mus TDLo:40 mg/kg:ETA BJCAAI 26,506,72

SAFETY PROFILE: Questionable carcinogen with experimental tumorigenic data. When heated to decomposition it emits toxic fumes of HBr. See also BROMIDES.

BMU750 CAS:14733-73-4 HR: 3
5-BROMO-2-BENZOXAZOLINONE
mf: $C_7H_4BrNO_2$ mw: 214.03

TOXICITY DATA WITH REFERENCE
orl-rat LD50:1050 mg/kg MDCHAG 4(1),308,64
orl-mus LD50:1440 mg/kg MDCHAG 4(1),308,64
ipr-mus LD50:262 mg/kg MDCHAG 4(1),308,64

SAFETY PROFILE: Poison by intraperitoneal route. Moderately toxic by ingestion. When heated to decomposition it emits very toxic fumes of Br^- and NO_x.

BMV000 CAS:19932-85-5 HR: 2
6-BROMO-2-BENZOXAZOLINONE
mf: $C_7H_4BrNO_2$ mw: 214.03

TOXICITY DATA WITH REFERENCE
orl-rat LD50:1000 mg/kg MDCHAG 4(1),308,64
orl-mus LD50:935 mg/kg MDCHAG 4(1),308,64
ipr-mus LD50:445 mg/kg MDCHAG 4(1),308,64

SAFETY PROFILE: Moderately toxic by ingestion and intraperitoneal routes. When heated to decomposition it emits very toxic fumes of Br^- and NO_x.

BMV250 CAS:14917-59-0 HR: 3
p-BROMOBENZOYL AZIDE
mf: $C_7H_4BrN_3O$ mw: 229.04

$BrC_6H_4CO{\bullet}N_3$

PROP: Mp: 46°.

SAFETY PROFILE: Explodes when heated above its melting point. When heated to decomposition it emits toxic fumes of Br^- and NO_x. See also AZIDES.

BMV750 CAS:61-75-6 HR: 3
(o-BROMOBENZYL)ETHYLDIMETHYLAMMONIUM-p-TOLUENESULFONATE
mf: $C_{11}H_{17}BrN{\bullet}C_7H_7O_3S$ mw: 414.40

PROP: Solid. Mp: 97–99°. Insol in Et_2O, sol in H_2O and EtOH.

SYNS: ASL-603 □ BRETYLAN □ BRETYLATE □ BRETYLIUM-p-TOLUENESULFONATE □ BRETYLIUM TOSYLATE □ BRETYLOL □ 2-BROMO-N-ETHYL-N,N-DIMETHYLBENZENEMETHANAMINIUM 4-METHYLBENZENESULFONATE □ DARENTHIN □ N-ETHYL-N-o-BROMOBENZYL-N,N-DIMETHYLAMMONIUM TOSYLATE □ ORNID

TOXICITY DATA WITH REFERENCE
ipr-rat LD50:57 mg/kg PHMGBN 21,256,80
ivn-rat LD50:17 mg/kg PBPSDY 2,148,79
ims-rat LD50:250 mg/kg TXAPA9 18,185,71
orl-mus LD50:400 mg/kg BJPCAL 14,536,59
ipr-mus LD50:39 mg/kg AIPTAK 155,69,65
scu-mus LD50:72 mg/kg BJPCAL 14,536,59
ivn-mus LD50:20 mg/kg BJPCAL 14,536,59

SAFETY PROFILE: A poison by ingestion, intraperitoneal, subcutaneous, intravenous, and intramuscular routes. An anti-adrenergic agent and antiarrhythmic cardiac depressant. When heated to decomposition it emits very toxic fumes of SO_x, NH_3, NO_x, and Br^-. See also SULFONATES.

B

BMW000 CAS:33855-47-9 HR: 3
N-p-BROMOBENZYL-N′-ETHYL-N′-METHYL-N-2-PYRIDYLETHYLENEDIAMINE MALEATE
mf: $C_{17}H_{22}BrN_3 \cdot C_4H_4O_4$ mw: 464.41

SYNS: N-p-BROMBENZYL-N-α-PYRIDYL-N′-METHYL-N′-AETHYL-AETHYLENDIAMIN-MALEINAT (GERMAN) □ WV 761

TOXICITY DATA WITH REFERENCE
orl-mus LD50:620 mg/kg ARZNAD 14,940,64
scu-mus LD50:119 mg/kg ARZNAD 14,940,64
ivn-mus LD50:16,700 μg/kg ARZNAD 14,940,64

SAFETY PROFILE: Poison by subcutaneous and intravenous routes. Moderately toxic by ingestion. When heated to decomposition it emits very toxic fumes of Br^- and NO_x.

BMW250 CAS:5798-79-8 HR: 3
BROMOBENZYLNITRILE
mf: C_8H_6BrN mw: 196.06

PROP: Pure: Yellowish-white crystals. Tech: brown, oily liquid with pungent odor of sour fruit; mp: 29°, bp: 242°, fp: 25.5°, flash p: none, d: 1.5160 @ 20°, vap d: 6.8, vap press: 0.011 mm @ 20°.

SYNS: BBC □ BBN □ BROMBENZYL CYANIDE □ α-BROMOBENZYL CYANIDE □ α-BROMOBENZYLNITRILE □ α-BROMOPHENYLACETONITRILE □ α-BROMO-α-TOLUNITRILE □ CA □ CAMITE

TOXICITY DATA WITH REFERENCE
ihl-hmn LC50:3500 mg/m³ SCJUAD 4,33,67
orl-rat LDLo:100 mg/kg NCNSA6 5,32,53

CONSENSUS REPORTS: Cyanide and its compounds are on the Community Right-To-Know List.

SAFETY PROFILE: Poison by ingestion. Moderately toxic to humans by inhalation. When heated to decomposition it emits very toxic fumes of NO_x, Br^-, and CN^-. See also NITRILES.

BMW290 CAS:2113-57-7 HR: 2
3-BROMOBIPHENYL
mf: $C_{12}H_9Br$ mw: 233.12

SYN: BIPHENYL, 3-BROMO-

TOXICITY DATA WITH REFERENCE
ipr-mus LDLo:500 mg/kg CBCCT* 6,217,54

CONSENSUS REPORTS: Reported in EPA TSCA Inventory.

SAFETY PROFILE: Moderately toxic by intraperitoneal route. When heated to decomposition it emits toxic vapors of Br^-.

BMX000 CAS:60883-74-1 HR: 2
α-BROMO-β,β-BIS(p-ETHOXYPHENYL)STYRENE
mf: $C_{24}H_{23}BrO_2$ mw: 423.38

SYN: α,α-DI(p-ETHOXYPHENYL)-β-BROMO-β-PHENYLETHYLENE

TOXICITY DATA WITH REFERENCE
scu-mus TDLo:94 mg/kg/26W-I:CAR MMJJAI 11,95,61

SAFETY PROFILE: Questionable carcinogen with experimental carcinogenic data. When heated to decomposition it emits toxic fumes of Br^-.

BMX250 CAS:34346-98-0 HR: 2
4-BROMO-7-BROMOMETHYLBENZ(a)ANTHRACENE
mf: $C_{19}H_{12}Br_2$ mw: 400.13

TOXICITY DATA WITH REFERENCE
skn-mus TDLo:16 mg/kg:ETA EJCAAH 7,473,71

SAFETY PROFILE: Questionable carcinogen with experimental tumorigenic data. When heated to decomposition it emits toxic fumes of Br^-. See also BROMIDES.

BMX500 CAS:109-65-9 HR: 3
1-BROMOBUTANE
DOT: UN 1126
mf: C_4H_9Br mw: 137.04

PROP: Colorless to pale straw-colored liquid. Mp: −112.3°, bp: 101.6°, flash p: 65°F (OC), d: 1.276 @ 20°/8°, autoign temp: 509°F, vap d: 4.72, lel: 2.8% @ 212°F, uel: 6.6% @ 212°F.

SYNS: BUTYL BROMIDE (DOT) □ n BUTYL BROMIDE (DOT)

TOXICITY DATA WITH REFERENCE
ihl-mam LC50:25,800 mg/m³ GTPZAB 18(4),55,74
ihl-rat LC50:237,000 mg/m³/30M FAVUAI 7,35,75
ipr-rat LD50:4450 mg/kg JPCEAO 320(1),133,78
ipr-mus LD50:6680 mg/kg JPCEAO 320(1),133,78
ipr-mam LD50:1424 mg/kg GTPZAB 18(4),55,74

CONSENSUS REPORTS: EPA Genetic Toxicology Program. Reported in EPA TSCA Inventory.

DOT CLASSIFICATION: 3; *Label:* Flammable Liquid

SAFETY PROFILE: Moderately toxic by intraperitoneal route. Mildly toxic by inhalation. Dangerous fire hazard when exposed to heat, flame, or oxidizers. Violent reaction with bromobenzene + sodium above 30°C. Can react with oxidizing materials. To fight fire, use CO_2, dry chemical, mist or spray. See also BROMIDES.

BMX750 CAS:78-76-2 HR: 2
2-BROMOBUTANE
DOT: UN 2339
mf: C_4H_9Br mw: 137.04

PROP: Colorless liquid. Fp: <−50°, bp: 91.4°, flash p: 70°F, d: 1.257 @ 25°/25°.

SYNS: sec-BUTYL BROMIDE □ METHYLETHYLBROMOMETHANE

TOXICITY DATA WITH REFERENCE
ipr-mus TDLo:3000 mg/kg/8W-I:NEO CNREA8 35,1411,75

CONSENSUS REPORTS: EPA Genetic Toxicology Program. Reported in EPA TSCA Inventory.

DOT CLASSIFICATION: 3; *Label:* Flammable Liquid

SAFETY PROFILE: Narcotic in high concentrations. Questionable carcinogen with experimental neoplastigenic data. See also BROMIDES and CHLORINATED HYDROCARBONS, ALIPHATIC. Flammable liquid.

Dangerous fire hazard when exposed to heat or flame. When heated to decomposition it emits toxic fumes of Br^-; can react with oxidizing materials. To fight fire, use water, spray or mist, foam, CO_2, dry chemical.

BMX825 **CAS:5162-44-7** **HR: 3**
4-BROMO-1-BUTENE
mf: C_4H_7Br mw: 135.00

$H_2C{=}CHCH_2CH_2Br$

PROP: Liquid. D: 1.32 @ 20°/4°, bp: 98.5°.

SAFETY PROFILE: A dangerous fire hazard (flash point <1°C). Violent reaction with chloromethylphenylsilane + chloroplatinic acid. When heated to decomposition it emits toxic fumes of Br^-. See also BROMIDES.

BMY250 **CAS:80-58-0** **HR: 3**
2-BROMOBUTYRIC ACID
mf: $C_4H_7BrO_2$ mw: 167.02

PROP: Colorless, oily liquid; sol in alc and ether; sparingly sol in water. D: 1.54, bp: 181° @ 250 mm, mp: −4°.

SYN: α-BROMOBUTYRIC ACID

TOXICITY DATA WITH REFERENCE
orl-mus LD50:310 mg/kg JPETAB 86,336,46

CONSENSUS REPORTS: Reported in EPA TSCA Inventory.

SAFETY PROFILE: Poison by ingestion. Dangerous; when heated to decomposition it emits toxic fumes of Br^-. See also BROMIDES.

BMY500 **CAS:5332-06-9** **HR: 3**
4-BROMOBUTYRONITRILE
mf: C_4H_6BrN mw: 148.02

SYN: USAF DO-6

TOXICITY DATA WITH REFERENCE
ipr-mus LD50:100 mg/kg NTIS** AD277-689

CONSENSUS REPORTS: Reported in EPA TSCA Inventory. Cyanide and its compounds are on the Community Right-To-Know List.

SAFETY PROFILE: Poison by intraperitoneal route. When heated to decomposition it emits very toxic fumes of NO_x, CN^-, and Br^-. See also BROMIDES and NITRILES.

BMY800 **CAS:83463-62-1** **HR: 2**
BROMOCHLOROACETONITRILE
mf: $C_2HBrClN$ mw: 154.40

PROP: Bp: 138–140°.

SYN: BROMOCHLOROMETHYL CYANIDE

TOXICITY DATA WITH REFERENCE
mmo-sat 1 nmol/plate ENMUDM 5,447,83
mma-sat 170 nmol/plate FAATDF 5,1065,85
dnd-hmn:lyms 2 μmol/L NTIS** PB84-246230
sce-ham:ovr 4200 nmol/L FAATDF 5,1065,85
orl-rat TDLo:825 mg/kg (7-21D post):REP TXCYAC 46,83,87
skn-mus TDLo:2400 mg/kg/2W-I:CAR FAATDF 5,1065,85

CONSENSUS REPORTS: IARC Cancer Review: Group 3 IMEMDT 52,269,91; Animal Inadequate Evidence IMEMDT 52,269,91; Human No Available Data IMEMDT 52,269,91.

SAFETY PROFILE: Experimental reproductive data. Questionable carcinogen with experimental carcinogenic data. Mutation data reported. When heated to decomposition it emits toxic fumes of Br^-, Cl^-, and NO_x.

BMY825 **CAS:25604-70-0** **HR: 3**
BROMOCHLOROACETYLENE
mf: C_2BrCl mw: 139.38

SAFETY PROFILE: An unstable high explosive. When heated to decomposition it emits toxic fumes of Cl^- and Br^-. See also ACETYLENE COMPOUNDS, BROMIDES, and CHLORIDES.

BMZ000 **CAS:5579-85-1** **HR: 2**
6-BROMO-5-CHLORO-2-BENZOXAZOLINONE
mf: $C_7H_3BrClNO_2$ mw: 248.47

PROP: Solid. Mp: 204–205°.

SYNS: BROMCHLORENONE □ 6-BROMO-5-CHLOROBENZOXAZOLONE □ NSC-24970

TOXICITY DATA WITH REFERENCE
orl-rat LD50:500 mg/kg MDCHAG 4(1),308,64
orl-mus LD50:871 mg/kg MDCHAG 4(1),308,64
skn-rbt LD50:3160 mg/kg HAZL** -,-,62

SAFETY PROFILE: Moderately toxic by ingestion and skin contact. When heated to decomposition it emits very toxic fumes of Cl^-, Br^-, and NO_x.

BNA000 **CAS:758-24-7** **HR: 2**
2-BROMO-2-CHLORO-1,1-DIFLUOROETHYLENE
mf: C_2BrClF_2 mw: 177.38

PROP: Liquid. Bp: 38° @ 625 mm.

SYNS: 1-BROMO-1-CHLORO-2,2-DIFLUOROETHENE □ 1-BROMO-1-CHLORO-2,2-DIFLUOROETHYLENE

TOXICITY DATA WITH REFERENCE
mma-sat 15 mmol/L ANESAV 51,424,79
oms-bcs 133 mol/L MUREAV 54,17,78
ihl-mus LC50:250 ppm/1H BJANAD 37,716,65

SAFETY PROFILE: Moderately toxic by inhalation. Mutation data reported. When heated to decomposition it emits very toxic fumes of Br^-, Cl^-, and F^-. See also CHLORINATED HYDROCARBONS, ALIPHATIC.

BNA250 **CAS:353-59-3** **HR: 1**
BROMOCHLORODIFLUOROMETHANE
DOT: UN 1974
mf: $CBrClF_2$ mw: 165.37

PROP: Colorless gas. Fp: −160.5°, bp: −4.

B

SYNS: CHLORODIFLUOROBROMOMETHANE (DOT) □ CHLORODIFLUOROMONOBROMOMETHANE □ FLUGEX 12B1 □ FLUOROCARBON 1211 □ FREON 12B1 □ HALON 1211 □ R12B1 (DOT)

TOXICITY DATA WITH REFERENCE
mmo-sat 10 pph MUREAV 142,187,85
mma-sat 5 pph MUREAV 142,187,85
ihl-rat LCLo:32 pph/15M FLCRAP 1,197,67

CONSENSUS REPORTS: Reported in EPA TSCA Inventory.

DOT CLASSIFICATION: 2.2; *Label:* Nonflammable Gas

SAFETY PROFILE: Mutation data reported. An asphyxiant. See also ARGON for description of inert gas asphyxiants. When heated to decomposition it emits very toxic fumes of Br^-, Cl^-, and F^-.

BNA325 CAS:126-06-7 HR: 2
3-BROMO-1-CHLORO-5,5-DIMETHYLHYDANTOIN
mf: $C_5H_6BrClN_2O_2$ mw: 241.49

SYN: 3-BROMO-1-CHLORO-5,5-DIMETHYL-2,4-IMIDAZOLIDINEDIONE

TOXICITY DATA WITH REFERENCE
skn-rbt 500 mg/24H SEV EPASR* 8EHQ-0181-0382
eye-rbt 100 mg/30S SEV EPASR* 8EHQ-0181-0382
orl-rat LD50:600 mg/kg EPASR* 8EHQ-0181-0382
orl-mus LD50:680 mg/kg EPASR* 8EHQ-0181-0382
skn-rbt LDLo:2 g/kg EPASR* 83HQ-0281-0382

CONSENSUS REPORTS: Reported in EPA TSCA Inventory.

SAFETY PROFILE: Moderately toxic by skin contact and ingestion. A severe eye and skin irritant. When heated to decomposition it emits toxic fumes of Cl^-, Br^-, and NO_x.

BNA750 CAS:41198-08-7 HR: 3
O-(4-BROMO-2-CHLOROPHENYL)-O-ETHYL-S-PROPYL PHOSPHOROTHIOATE
mf: $C_{11}H_{15}BrClO_3PS$ mw: 373.65

SYNS: CGA 15324 □ CURACRON □ POLYCRON □ PROFENOFOS □ SELECRON

TOXICITY DATA WITH REFERENCE
orl-rat LD50:400 mg/kg SPEADM 78-1,35,78
skn-rat LD50:300 mg/kg CIGET* -,-,77
orl-mus LD50:162 mg/kg TXAPA9 73,16,84
orl-mus LD50:298 mg/kg CIGET* -,-,77
orl-rbt LD50:700 mg/kg CIGET* -,-,77
skn-rbt LD50:192 mg/kg FMCHA2 -,C65,83
orl-ckn LD50:1900 μg/kg TXAPA9 73,16,84
skn-rbt LD50:472 mg/kg CIGET* -,-,77

SAFETY PROFILE: Poison by ingestion and skin contact. When heated to decomposition it emits very toxic SO_x, PO_x, Br^-, and Cl^-. See also ESTERS.

BNA825 CAS:109-70-6 HR: 2
1-BROMO-3-CHLOROPROPANE
DOT: UN 2688
mf: C_3H_6BrCl mw: 157.45

PROP: Bp: 142–143°.

SYNS: 3-BROMOPROPYL CHLORIDE □ 1,3-CHBP □ ω-CHLOROBROMOPROPANE □ 1-CHLORO-3-BROMOPROPANE (DOT) □ 3-CHLOROPROPYL BROMIDE □ TRIMETHYLENE BROMIDE CHLORIDE □ TRIMETHYLENE CHLOROBROMIDE

TOXICITY DATA WITH REFERENCE
orl-rat LD50:930 mg/kg TPKVAL 12,93,71
ihl-rat LC50:5668 mg/m³ GTPZAB 19(9),36,75
orl-mus LD50:1290 mg/kg 85GMAT -,35,82
ihl-mus LCLo:7270 mg/m³/2H 85GMAT -,35,82

CONSENSUS REPORTS: Reported in EPA TSCA Inventory.

DOT CLASSIFICATION: 6.1; *Label:* KEEP AWAY FROM FOOD

SAFETY PROFILE: Moderately toxic by ingestion. When heated to decomposition it emits toxic fumes of Cl^- and Br^-. See also CHLORINATED HYDROCARBONS, ALIPHATIC; and BROMIDES.

BNB250 CAS:25614-03-3 HR: 2
BROMOCRIPTINE
mf: $C_{32}H_{40}BrN_5O_5$ mw: 654.68

PROP: Crystals. Mp: 215–218° (decomp).

SYNS: BROMOCRIPTIN □ α BROMOERGOCRIPTINE □ BROMOERGOCRYPTINE □ 2-BROMOERGOCRYPTINE □ 2-BROMO-α-ERGOKRYPTIN □ 2-BROMO-12′-HYDROXY-2′-(1-METHYLETHYL)-5′-α-(2-METHYLPROPYL)ERGOTAMIN-3′,6′,18-TRIONE □ CB-154

TOXICITY DATA WITH REFERENCE
oms-hmn:lym 100 μmol/L MUREAV 117,163,83
dna-rat-ipr 4 mg/kg CNREA8 36,2223,76
unr-wmn TDLo:2800 μg/kg (24-52D preg):TER BCPHBM 5,227,78
unr-wmn TDLo:11,200 μg/kg (44D pre/1-12D preg):TER BCPHBM 5,227,78
unr-wmn TDLo:3500 μg/kg (14D pre/1-21D preg):TER BCPHBM 5,227,78
unr-wmn TDLo:21,700 μg/kg (28W pre/1-21D preg):TER BCPHBM 5,227,78
unr-wmn TDLo:11,900 μg/kg (77D pre/1-42D preg):TER BCPHBM 5,227,78
scu-rat TDLo:70 mg/kg (female 13-19D post):REP TJADAB 33,100C,86
scu-mus TDLo:84 mg/kg (lactating female 21D post):REP EXPEAM 30,1353,74
ims-rat TDLo:8125 μg/kg (female 6-18D post):TER ARCGDG 233,31,82
scu-rbt TDLo:8 mg/kg (female 27-30D post):TER PEREBL 15,183,81
ims-dog TDLo:600 μg/kg (female 42-47D post):REP JRPFA4 81,175,87
scu-mus TDLo:200 mg/kg (female 50D pre):REP EXPEAM 30,1353,74
scu-rat TDLo:1950 μg/kg (female 5-15D post):REP USXXAM #3752814

scu-rat TDLo: 88 mg/kg (female 1-22D post): REP ARCGDG 229,77,80
scu-mus TDLo: 212 mg/kg (female 32D pre): REP EXPEAM 30,1353,74
scu-rat TDLo: 5100 μg/kg (lactating female 3D post): REP USXXAM #3752814
scu-dom TDLo: 3 mg/kg (male 84D pre): REP BIREBV 17,192,77
scu-mus TDLo: 4 mg/kg (male 10D pre): REP BIREBV 11,319,74
scu-rat TDLo: 17,500 μg/kg (male 35D pre): REP IJEBA6 22,164,84
orl-rat TDLo: 7 g/kg/2Y-C: ETA,REP BMJOAE 2,1605,77
orl-wmn TDLo: 6 mg/kg/60D-I: NOSE NEJMAG 306,178,82
ivn-rat LD50: 72 mg/kg DRUGAY 17,313,78
ivn-rbt LD50: 12 mg/kg USXXAM #3752814
unr-mus LD50: 200 mg/kg BBIADT 43,1305,84

CONSENSUS REPORTS: EPA Genetic Toxicology Program.

SAFETY PROFILE: Poison by intravenous and possibly other routes. Human teratogenic effects by an unspecified route: developmental abnormalities of the respiratory system, musculoskeletal system, urogenital system, craniofacial area, and body wall. Human systemic effects by ingestion including: olfaction changes. An experimental teratogen. Other experimental reproductive effects. Human mutation data reported. Questionable carcinogen with experimental tumorigenic data. When heated to decomposition it emits very toxic fumes such as Br^- and NO_x.

BNB325 CAS:22260-51-1 HR: 3
BROMOCRIPTINE MESILATE
mf: $C_{32}H_4OBrN_5O_5{\cdot}CH_4O_3S$ mw: 714.43

PROP: Crystals from 2-butanone. Mp: 192–196° (decomp).

SYNS: 2-BROMO-α-ERGOCRYPTINE METHANESULFONATE □ 2-BROMO-α-ERGOKRYPTINE-MESILATE (GERMAN) □ CB-154 □ PARLODEL

TOXICITY DATA WITH REFERENCE
orl-wmn TDLo: 1400 mg/kg (lactating female 1-14D post): REP JRPMAP 33,630,88
par-rat TDLo: 35 mg/kg (female 14-20D post): TER GCENA5 39,118,79
orl-rat TDLo: 42 mg/kg (female 1-7D post): REP EXPEAM 30,1358,74
orl-rat TDLo: 60 mg/kg (female 6D pre): REP BIREBV 34,788,86
scu-rat TDLo: 10,500 μg/kg (female 6-8D post): REP EXPEAM 24,1130,68
ipr-mky TDLo: 20 mg/kg (male 10D pre): REP IJEBA6 23,679,85
orl-wmn TDLo: 650 μg/kg/9D-I: BAH,CVS AIMEAS 118,199,93
orl-cld TDLo: 375 μg/kg JOPDAB 105,838,84
orl-wmn TDLo: 1 mg/kg/20D-I: BAH AJPSAO 143,935,85
orl-man TDLo: 52 mg/kg/35W-I: CNS NEURAI 35,1193,85
ivn-rat LD50: 10,500 μg/kg YKYUA6 29,1231,78
orl-mus LD50: 2502 mg/kg YKYUA6 30,809,79
ivn-mus LD50: 189 mg/kg YKYUA6 30,809,79
ivn-rbt LD50: 8200 μg/kg IYKEDH 10,232,79

SAFETY PROFILE: Poison by intravenous route. Moderately toxic by ingestion. Human systemic effects by ingestion: cardiomyopathy. cerebral spinal fluid changes, distorted perceptions, hallucinations, headache, toxic psychosis. An experimental teratogen. Other experimental reproductive effects. When heated to decomposition it emits toxic fumes of Br^-, SO_x, and NO_x.

BNB750 HR: 3
1-BROMO-12-CYCLOTRIDECADIEN-4,8,10-TRIYNE
mf: $C_{13}H_9Br$ mw: 245.12

SAFETY PROFILE: Explodes @ 65° and decomposes @ 0° in the dark.

BNB800 CAS:112-29-8 HR: 2
1-BROMODECANE
mf: $C_{10}H_{21}Br$ mw: 221.22

SYNS: DECANE, 1-BROMO- □ DECYL BROMIDE □ n-DECYL BROMIDE □ 1-DECYL BROMIDE

TOXICITY DATA WITH REFERENCE
ipr-mus LD50: 4070 mg/kg GTPZAB 20(12),52,76
ihl-uns LC50: 4200 mg/m³ GTPZAB 18(4),55,74

CONSENSUS REPORTS: Reported in EPA TSCA Inventory.

SAFETY PROFILE: Moderately toxic by inhalation route. Mildly toxic by intraperitoneal routes. When heated to decomposition it emits toxic vapors of Br^-.

BNC750 CAS:59-14-3 HR: 2
5-BROMO-2′-DEOXYURIDINE
mf: $C_9H_{11}BrN_2O_5$ mw: 307.13

PROP: Solid. Mp: 187–189°.

SYNS: BDU □ 5-BDU □ BROMODEOXYURIDINE □ 5-BROMODEOXYURIDINE □ 5-BROMO-2-DEOXYURIDINE □ 5-BROMODESOXYURIDINE □ BROMOURACIL DEOXYRIBOSIDE □ 5-BROMOURACIL DEOXYRIBOSIDE □ 5-BROMOURACIL-2-DEOXYRIBOSIDE □ BROXURIDINE □ BRUDR □ BUDR □ 5-BUDR

TOXICITY DATA WITH REFERENCE
mnt-hmn: fbr 82 μmol/L MUREAV 4,353,67
cyt-hmn: leu 200 mg/L ECREAL 34,182,64
msc-hmn: fbr 15 mg/L CSHSAZ 29,151,65
ipr-rat TDLo: 700 mg/kg (female 9-15D post): REP TAKHAA 30,636,71
ipr-mus TDLo: 400 mg/kg (female 8D post): REP TCMUD8 4,403,84
ipr-rat TDLo: 700 mg/kg (female 9-15D post): REP TAKHAA 30,636,71
ipr-rat TDLo: 250 mg/kg (female 13D post): TER TJADAB 23,383,81
ivn-ham TDLo: 400 mg/kg (female 8D post): TER LIFSAK 4,633,65
ipr-rat TDLo: 200 mg/kg (female 13D post): TER TAKHAA 30,636,71
ivn-mus TDLo: 40 mg/kg (female 6D post): TER EXPEAM 33,448,77
ipr-ham TDLo: 640 mg/kg (female 8D post): REP JEEMAF 43,47,78

ipr-mus TDLo:200 mg/kg (female 8D post):TER TAKHAA 30,636,71
ipr-mus TDLo:300 mg/kg (female 10D post):TER TJADAB 4,87,71
scu-rat TDLo:16 mg/kg:CAR MUREAV 295,113,93
orl-rat LD50:8400 mg/kg IYKEDH 4,467,73
ipr-rat LD50:1500 mg/kg ADTEAS 3,181,68
scu-rat LD50:3900 mg/kg TAKHAA 30,530,71
ivn-rat LD50:2320 mg/kg TAKHAA 30,735,71
orl-mus LD50:9100 mg/kg TAKHAA 30,530,71
ipr-mus LD50:3050 mg/kg TAKHAA 30,530,71
scu-mus LD50:3500 mg/kg TAKHAA 30,530,71
ivn-mus LD50:2500 mg/kg TAKHAA 30,530,71

CONSENSUS REPORTS: Reported in EPA TSCA Inventory. EPA Genetic Toxicology Program.

SAFETY PROFILE: Moderately toxic by subcutaneous, intravenous, intraperitoneal, and possibly other routes. Mildly toxic by ingestion. Experimental teratogenic and reproductive effects. Human mutation data reported. When heated to decomposition it emits very toxic fumes of Br^- and NO_x.

BNC800 CAS:27312-17-0 HR: 1
2-BROMO-1,5-DIAMINO-4,8-DIHYDROXYANTHRAQUINONE
mf: $C_{14}H_9BrN_2O_4$ mw: 349.16

SYNS: ANTHRAQUINONE, 2-BROMO-1,5-DIAMINO-4,8-DIHYDROXY- □ MODR OSTACETOVA LR

TOXICITY DATA WITH REFERENCE
eye-rbt 500 mg/24H MLD 28ZPAK-,245,72

CONSENSUS REPORTS: Reported in EPA TSCA Inventory.

SAFETY PROFILE: An eye irritant. When heated to decomposition it emits toxic fumes of NO_x and Br^-.

BND250 CAS:65235-63-4 HR: 1
2-BROMO-1,8-DIAMINO-4,5-DIHYDROXYANTHRAQUINONE
mf: $C_{14}H_9BrN_2O_4$ mw: 349.16

SYN: MODR OSTACETOVA LG (CZECH)

TOXICITY DATA WITH REFERENCE
skn-rbt 500 mg/24H MLD 28ZPAK -,244,72
eye-rbt 500 mg/24H MLD 28ZPAK -,244,72
orl-rat LD50:12,500 mg/kg 28ZPAK -,244,72

CONSENSUS REPORTS: Reported in EPA TSCA Inventory.

SAFETY PROFILE: Mildly toxic by ingestion. A skin and eye irritant. When heated to decomposition it emits very toxic fumes of Br^- and NO_x.

BND325 CAS:23834-96-0 HR: 3
BROMODIBORANE
mf: B_2BrH_5 mw: 106.56

PROP: Gas.

SAFETY PROFILE: May ignite violently on exposure to air. When heated to decomposition it emits toxic fumes of Br^-. See also BORANES and BORON COMPOUNDS.

BND500 CAS:75-27-4 HR: 3
BROMODICHLOROMETHANE
mf: $CHBrCl_2$ mw: 163.83

PROP: Colorless liquid. Mp: −57.1°, bp: 88.4–88.6°, d: 1.971 @ 25°/25°.

SYNS: BDCM □ DICHLOROBROMOMETHANE □ NCI-C55243

TOXICITY DATA WITH REFERENCE
mmo-sat 50 μL/plate DHEFDK FDA-78-1046,78
sce-hmn:lym 400 μmol/L ENVRAL 32,72,83
sce-mus-orl 200 mg/kg/4D-I ENVRAL 32,72,83
orl-rat TDLo:25,500 mg/kg/2Y-C:CAR NTPTR* NTP-TR-321,87
orl-mus TDLo:25,500 mg/kg/2Y-C:CAR NTPTR* NTP-TR-321,87
orl-rat TD:51 g/kg/2Y-C:CAR NTPTR* NTP-TR-321,87
orl-mus TD:38,250 mg/kg/2Y-C:CAR NTPTR* NTP-TR-321,87
orl-rat LD50:916 mg/kg TXAPA9 52,351,80

CONSENSUS REPORTS: NTP 7th Annual Report On Carcinogens. NTP Carcinogenesis Studies (gavage): Clear Evidence: rat, mouse NTPTR* NTP-TR-321,87. EPA Genetic Toxicology Program. Community Right-To-Know List. Reported in EPA TSCA Inventory.

SAFETY PROFILE: Confirmed carcinogen with experimental carcinogenic data. Moderately toxic by ingestion. Human mutation data reported. When heated to decomposition it emits very toxic fumes of Br^- and Cl^-. See also CHLORINATED HYDROCARBONS, ALIPHATIC; and BROMIDES.

BND750 CAS:18936-66-8 HR: 3
o-(4-BROMO-2,5-DICHLOROPHENYL)-o-ETHYL PHENYLPHOSPHONOTHIOATE
mf: $C_{14}H_{12}BrCl_2O_2PS$ mw: 426.10

SYN: VELSICOL FCS-303

TOXICITY DATA WITH REFERENCE
orl-mus LD50:75 mg/kg JAFCAU 27,1197,79
orl-gpg LDLo:100 mg/kg JEENAI 61,1261,68
scu-gpg LDLo:100 mg/kg JEENAI 61,1261,68

SAFETY PROFILE: Poison by ingestion and subcutaneous routes. When heated to decomposition it emits very toxic fumes of SO_x, PO_x, Cl^-, and Br^-. See also ESTERS.

BNE250 CAS:53581-53-6 HR: 3
dl-4-BROMO-2,5-DIMETHOXYAMPHETAMINE HYDROBROMIDE
mf: $C_{11}H_{16}BrO_2{\cdot}BrH$ mw: 341.10

PROP: Crystals from EtOAc. Mp: 145–146°.

SYN: dl-4-BROMO-2,5-DIMETHOXY-α-METHYLPHENETHYLAMINE HYDROBROMIDE

TOXICITY DATA WITH REFERENCE
ipr-rat LD50:50 mg/kg TXAPA9 45(1),49,78

ivn-mus LD50: 80 mg/kg TXAPA9 45(1),49,78
ivn-dog LD50: 6400 µg/kg TXAPA9 45(1),49,78
orl-mky LD50: 2 mg/kg TXAPA9 45,49,78

SAFETY PROFILE: Poison by ingestion, intraperitoneal, and intravenous routes. See also BROMIDES and various amphetamine entries. When heated to decomposition it emits very toxic fumes of Br^-.

BNE325 CAS:70277-99-5 HR: 3
2-BROMO-3,5-DIMETHOXYANILINE
mf: $C_8H_{10}BrNO_2$ mw: 232.08

SAFETY PROFILE: May explode when heated. Upon decomposition it emits toxic fumes of Br^- and NO_x.

BNE500 CAS:66969-02-6 HR: 3
2-BROMO-N,N-DIMETHYL-1-ADAMANATANEMETHANAMINE HYDROCHLORIDEHEMIHYDRATE
mf: $C_{13}H_{22}BrN \cdot ClH \cdot 1/2H_2O$ mw: 317.74

SYN: 2-BROMO-1-(N,N-DIMETHYLAMINOMETHYL)ADAMANTANE HYDROCHLORIDEHEMIHYDRATE

TOXICITY DATA WITH REFERENCE
orl-mus LD50: 413 mg/kg JMCMAR 19,967,76
ipr-mus LD50: 159 mg/kg JMCMAR 19,967,76

SAFETY PROFILE: Poison by ingestion and intraperitoneal routes. When heated to decomposition it emits very toxic fumes of Br^-, NO_x, and HCl.

BNE600 CAS:17576-88-4 HR: 2
3′-BROMO-4-DIMETHYLAMINOAZOBENZENE
mf: $C_{14}H_{14}BrN_3$ mw: 304.22

SYNS: ANILINE, p-(m-BROMOPHENYLAZO)-N,N-DIMETHYL- □ BENZENAMINE, 4-((3-BROMOPHENYL)AZO)-N,N-DIMETHYL-(9CI) □ p-(m-BROMOPHENYLAZO)-N,N-DIMETHYLANILINE

TOXICITY DATA WITH REFERENCE
orl-rat TDLo: 7980 mg/kg/25W-C: CAR CBINA8 53,107,85
orl-rat LD50: 13 g/kg NEOLA4 27,237,80

SAFETY PROFILE: Questionable carcinogen with experimental carcinogenic data. Low oral toxicity. When heated to decomposition it emits toxic fumes of Br^- and NO_x.

BNE750 CAS:980-71-2 HR: 3
2-(p-BROMO-α-(2-(DIMETHYLAMINO)ETHYL)BENZYL) PYRIDINE MALEATE (1:1)
mf: $C_{16}H_{19}BrN_2 \cdot C_4H_4O_4$ mw: 435.36

SYNS: 2-((p-BROMO-α-(2-DIMETHYLAMINO)ETHYL)BENZYL)PYRIDINE BIMALEATE □ BROMOPHENIRAMINE MALEATE □ PARABROMODYLAMINE MALEATE

TOXICITY DATA WITH REFERENCE
orl-rat LD50: 318 mg/kg 29ZVAB -,19,69
ipr-rat LD50: 76 mg/kg 29ZVAB -,19,69

CONSENSUS REPORTS: Reported in EPA TSCA Inventory.

SAFETY PROFILE: Poison by ingestion and intraperitoneal routes. When heated to decomposition it emits very toxic fumes of Br^- and NO_x.

BNF000 CAS:52583-02-5 HR: 3
2-BROMO-1-(3-DIMETHYLAMINOPROPYL)ADAMANTANE HYDROCHLORIDE
mf: $C_{15}H_{26}BrN \cdot ClH$ mw: 336.79

SYN: 2-BROMO-N,N-DIMETHYL-1-ADAMANTANEPROPANAMINE HYDROCHLORIDE

TOXICITY DATA WITH REFERENCE
orl-mus LD50: 400 mg/kg JMCMAR 17,602,74
ipr-mus LD50: 150 mg/kg JMCMAR 17,602,74

SAFETY PROFILE: Poison by ingestion and intraperitoneal routes. When heated to decomposition it emits very toxic fumes of Br^-, NO_x, and HCl.

BNF250 CAS:586-77-6 HR: 3
4-BROMO-N,N-DIMETHYL ANILINE
mf: $C_8H_{10}BrN$ mw: 200.08

PROP: Crystals. Mp: 55°, bp: 264°.

SYNS: 4-BROMODIMETHYLANILINE □ N,N-DIMETHYL-4-BROMOANILINE

TOXICITY DATA WITH REFERENCE
orl-rat LDLo: 500 mg/kg JPETAB 90,260,47

CONSENSUS REPORTS: Reported in EPA TSCA Inventory.

SAFETY PROFILE: Moderately toxic by ingestion. May explode if heated. When heated to decomposition it emits toxic fumes of Br- and NO_x.

BNF300 HR: 2
3-BROMO-7,12-DIMETHYLBENZ(a)ANTHRACENE
mf: $C_{20}H_{15}Br$ mw: 335.26

SYN: 3-BROMO-DMBA

TOXICITY DATA WITH REFERENCE
mma-sat 5 µg/plate CRNGDP 4,1221,83
skn-mus TDLo: 520 mg/kg/50W-I: ETA CRNGDP 4,1221,83

SAFETY PROFILE: Questionable carcinogen with experimental tumorigenic data. Mutation data reported. When heated to decomposition it emits toxic fumes of Br^-.

BNF310 HR: 2
4-BROMO-7,12-DIMETHYLBENZ(a)ANTHRACENE
mf: $C_{20}H_{15}Br$ mw: 335.26

SYN: 4-BROMO-DMBA

TOXICITY DATA WITH REFERENCE
mma-sat 10 µg/plate CRNGDP 4,1221,83
skn-mus TDLo: 520 mg/kg/50W-I: CAR CRNGDP 4,1221,83

SAFETY PROFILE: Questionable carcinogen with experimental carcinogenic data. Mutation data reported. When heated to decomposition it emits toxic fumes of Br^-.

B

BNF315 CAS:63018-63-3 HR: 2
5-BROMO-9,10-DIMETHYL-1,2-BENZANTHRACENE
mf: $C_{20}H_{15}Br$ mw: 335.26

SYN: BENZ(a)ANTHRACENE, 8-BROMO-7,12-DIMETHYL-

TOXICITY DATA WITH REFERENCE
scu-mus TDLo:80 mg/kg:ETA CNREA8 6,454,46

SAFETY PROFILE: Questionable carcinogen with experimental tumorigenic data. When heated to decomposition it emits toxic fumes of Br^-.

BNF750 CAS:1463-08-7 HR: 3
p-BROMO-α,α-DIMETHYLPHENETHYLAMINE HYDROCHLORIDE
mf: $C_{10}H_{14}BrN{\cdot}ClH$ mw: 264.62

SYN: S 84

TOXICITY DATA WITH REFERENCE
ipr-rat LD50:172 mg/kg APTOA6 17,121,60
orl-mus LD50:325 mg/kg CHTPBA 6,453,71

SAFETY PROFILE: Poison by ingestion and intraperitoneal routes. When heated to decomposition it emits very toxic fumes of Br^-, HCl, and NO_x.

BNG125 CAS:65036-47-7 HR: 3
6-BROMO-2,4-DINITROBENZENEDIAZONIUM HYDROGEN SULFATE
mf: $C_6H_3BrN_4O_8S$ mw: 371.08

SAFETY PROFILE: Solution in sulfuric acid is explosive. When heated to decomposition it emits toxic fumes of Br^-, SO_x, and NO_x. See also SULFATES.

BNG250 HR: 3
3-BROMO-2,7-DINITRO-5-BENZO(b)-THIOPHENEDIAZONIUM-4-OLATE
mf: $C_8HBrN_4O_5S$ mw: 345.09

SAFETY PROFILE: An explosive. When heated to decomposition it emits toxic fumes of Br^-, SO_x, and NO_x.

BNG750 CAS:776-74-9 HR: 3
BROMODIPHENYLMETHANE
DOT: UN 1770
mf: $C_{13}H_{11}Br$ mw: 247.15

PROP: Solid. Mp: 45°, bp: 184° @ 20 mm. Decomp in hot water; sol in alc; very sol in benzene.

SYN: DIPHENYLMETHYL BROMIDE (DOT)

CONSENSUS REPORTS: Reported in EPA TSCA Inventory.

DOT CLASSIFICATION: 8; *Label:* Corrosive

SAFETY PROFILE: A corrosive poison. When heated to decomposition it emits toxic fumes of Br^-. See also BROMIDES.

BNH000 CAS:776-74-9 HR: 3
BROMODIPHENYLMETHANE (solution)
DOT: UN 1770
mf: $C_{13}H_{11}Br$ mw: 247.15

SYN: DIPHENYL METHYL BROMIDE, solution (DOT)

CONSENSUS REPORTS: Reported in EPA TSCA Inventory.

DOT CLASSIFICATION: 8; *Label:* Corrosive

SAFETY PROFILE: A corrosive, irritating liquid. When heated to decomposition it emits toxic fumes of Br^-. See also BROMODIPHENYLMETHANE and BROMIDES.

BNH100 CAS:728-84-7 HR: 3
2-BROMO-1,3-DIPHENYL-1,3-PROPANEDIONE
mf: $C_{15}H_{11}BrO_2$ mw: 303.17

SYN: 1,3-PROPANEDIONE, 2-BROMO-1,3-DIPHENYL-

TOXICITY DATA WITH REFERENCE
ipr-mus LDLo:31,200 μg/kg CBCCT* 4,232,52

CONSENSUS REPORTS: Reported in EPA TSCA Inventory.

SAFETY PROFILE: Poison by intraperitoneal route. When heated to decomposition it emits toxic vapors of Br^-.

BNH500 CAS:17372-87-1 HR: 3
BROMOEOSINE
mf: $C_{20}H_8Br_4O_5{\cdot}2Na$ mw: 693.90

PROP: Red crystals with bluish tinge, or brownish-red powder. Sol in H_2O; sltly sol in EtOH; insol in Et_2O.

SYNS: AIZEN EOSINE GH □ BROMO ACID □ BROMOFLUORESCEIC ACID □ BROMO FLUORESCEIN □ BRONZE BROMO □ CERTIQUAL EOSINE □ C.I. 45380 □ D&C RED No. 22 □ DISODIUM EOSIN □ EOSINE □ EOSINE SODIUM SALT □ EOSINE YELLOWISH □ EOSIN GELBLICH (GERMAN) □ FENAZO EOSINE XG □ HIDACID DIBROMO FLUORESCEIN □ IRGALITE BRONZE RED CL □ PHLOXINE TONER B □ PHLOX RED TONER X-1354 □ PURE EOSINE YY □ 11445 RED □ SODIUM EOSINATE □ SYMULER EOSIN TONER □ 2,4,5,7-TETRABROMO-9-o-CARBOXYPHENYL-6-HYDROXY-3-ISOXANTHONE, DISODIUM SALT □ 2,4,5,7-TETRABROMO-3,6-FLUORANDIOL □ TETRABROMOFLUORESCEIN □ 2′,4′,5′,7′-TETRABROMOFLUORESCEIN DISODIUM SALT □ TETRABROMOFLUORESCEIN S □ TETRABROMOFLUORESCEIN SOLUBLE □ 2-(2,4,5,7-TETRABROMO-6-HYDROXY-3-OXO-3H-XANTHENE-9-YL)BENZOIC ACID, DISODIUM SALT □ TOYO EOSINE G □ 1903 YELLOW PINK

TOXICITY DATA WITH REFERENCE
dnr-bcs 2 mg/disc TRENAF 27,153,76
scu-rat TDLo:13 g/kg/1Y-I:ETA GANNA2 47,51,56
ipr-rat LDLo:500 mg/kg IJLEAG 2,257,34
orl-mus LD50:2344 mg/kg EAPHA6 24,125,81
ivn-rbt LDLo:300 mg/kg IJLEAG 2,257,34

CONSENSUS REPORTS: IARC Cancer Review: Animal Inadequate Evidence IMEMDT 15,183,77. EPA Genetic Toxicology Program. Reported in EPA TSCA Inventory.

SAFETY PROFILE: Poison by intravenous and intraperitoneal routes. Moderately toxic by ingestion. Questionable carcinogen with experimental tumorigenic data.

When heated to decomposition it emits very toxic fumes of Br^- and Na_2O. See also BROMIDES.

BNI000 CAS:3132-64-7 HR: 3
3-BROMO-1,2-EPOXYPROPANE
DOT: UN 2558
mf: C_3H_5BrO mw: 136.99

PROP: Flash p: <22°.

SYNS: EPIBROMHYDRIN □ EPIBROMOHYDRIN (DOT) □ EPIBROMOHYDRINE

TOXICITY DATA WITH REFERENCE
mmo-sat 5 μmol/plate JTEHD6 5,1149,79
mmo-esc 20 μmol/L ARTODN 46,277,80

CONSENSUS REPORTS: EPA Genetic Toxicology Program. Reported in EPA TSCA Inventory.

DOT CLASSIFICATION: 6.1; *Label:* Poison

SAFETY PROFILE: Poison by intraperitoneal route. Human mutation data reported. A dangerous fire hazard when exposed to heat or flame. When heated to decomposition it emits toxic fumes of Br^-. See also BROMIDES.

BNI500 CAS:540-51-2 HR: 3
2-BROMO ETHANOL
mf: C_2H_5BrO mw: 124.98

PROP: D: 1.79 @ 0°/4°, bp: 149–150°.

SYNS: BE □ BROMOETHANOL □ ETHYLENEBROMOHYDRIN □ GLYCOL BROMOHYDRIN

TOXICITY DATA WITH REFERENCE
mmo-sat 10 μL/plate EVHPAZ 21,79,77
mma-sat 10 μL/plate EVHPAZ 21,79,77
dnr-esc 10 μmol/plate EVHPAZ 21,79,77
dnr-bcs 20 μL/disc AEMIDF 43,177,82
mmo-klp 15 mmol/L EXPEAM 25,85,69
orl-mus TDLo:43 g/kg/80W-C:ETA JACTDZ 2(2),246,83
ipr-mus TDLo:150 mg/kg/8W-I:NEO CNREA8 39,391,79
ipr-mus LDLo:80 mg/kg TXAPA9 23,288,72

CONSENSUS REPORTS: EPA Genetic Toxicology Program. Reported in EPA TSCA Inventory.

SAFETY PROFILE: Poison by intraperitoneal route. Questionable carcinogen with experimental neoplastigenic and tumorigenic data. Mutation data reported. When heated to decomposition it emits toxic fumes of Br^-. See also BROMIDES.

BNK000 CAS:77-65-6 HR: 3
2-BROMO-2-ETHYLBUTYRYLUREA
mf: $C_7H_{13}BrN_2O_2$ mw: 237.13

PROP: Solid. Mp: 116–119°.

SYNS: ADALIN □ ADDISOMNOL □ N-(AMINOCARBONYL)-2-BROMO-2-ETHYLBUTANAMIDE □ BROMACETOCARBAMIDE □ BROMADAL □ BROMADEL □ BROMODIETHYLACETYLCARBAMIDE □ BROMODIETHYLACETYLUREA □ (α-BROMO-α-ETHYLBUTYRYL) CARBAMIDE □ (α-BROMO-α-ETHYLBUTYRYL)UREA □ 1-BROMO-ETHYL-BUTYRYL-UREA □ 2-BROMO-2-ETHYLBUTYRLUREA □ CARBOMAL □ DIACID □ DORMITURIN □ FYDALIN □ HOGGAR □ KARBROMAL □ KARTRYL □ NCI-C03805 □ NENESIN □ NYCTAL □ PARKOSED □ PELIDORM □ PIANADALIN □ PLANADALIN □ TILDIN □ URADAL

TOXICITY DATA WITH REFERENCE
cyt-smc 10 mmol/tube HEREAY 33,457,47
ipr-rat LD50:427 mg/kg ITMZBJ 17,305,80
ivn-rat LD50:427 mg/kg ARTODN 40,211,78
unr-rat LDLo:350 mg/kg JPHAA3 23,788,34
orl-mus LD50:464 mg/kg NCILB* NIH-NCI-E-C-72-3252,73
orl-dog LD50:450 mg/kg MEIEDD 10,254,83
scu-dog LDLo:300 mg/kg HBAMAK 4,1292,35
orl-cat LDLo:350 mg/kg HBAMAK 4,1293,35
orl-rbt LDLo:600 mg/kg SAPHAO 28,193,13
scu-frg LDLo:1667 mg/kg HBAMAK 4,1293,35

CONSENSUS REPORTS: NCI Carcinogenesis Bioassay (feed); No Evidence: mouse, rat NCITR* NCI-CG-TR-173,79. Reported in EPA TSCA Inventory.

SAFETY PROFILE: Poison by ingestion, subcutaneous, and possibly other routes. Moderately toxic via intravenous and intraperitoneal routes. Mutation data reported. A sedative, hypnotic, and central nervous system depressant. When heated to decomposition it emits very toxic fumes of NO_x and Br^-.

BNK100 HR: 2
4′-BROMO-3′-ETHYL-4-DIMETHYLAMINOAZOBENZENE
mf: $C_{16}H_{18}BrN_3$ mw: 332.28

SYNS: ANILINE, p-((4-BROMO-3-ETHYLPHENYL)AZO)-N,N-DIMETHYL- □ BENZENAMINE, N,N-DIMETHYL-4′-BROMO-3′-ETHYL-4-(PHENYLAZO)- □ p-((4-BROMO-3-ETHYLPHENYL)AZO)-N,N-DIMETHYLANILINE

TOXICITY DATA WITH REFERENCE
orl-rat TDLo:14,414 mg/kg/52W-C:CAR CBINA8 53,107,85

SAFETY PROFILE: Questionable carcinogen with experimental carcinogenic data. When heated to decomposition it emits toxic fumes of Br^- and NO_x.

BNK250 HR: 3
2-BROMO ETHYL ETHYL ETHER
mf: C_4H_9BrO mw: 155

PROP: Liquid. Vap d: 5.25, flash p: 5°.

SAFETY PROFILE: An insecticide. A dangerous fire hazard when exposed to heat or flame. See also ETHERS and BROMIDES.

BNK275 HR: 2
p-((3-BROMO-4-ETHYLPHENYL)AZO)-N,N-DIMETHYLANILINE
mf: $C_{16}H_{18}BrN_3$ mw: 332.28

SYNS: ANILINE, p-((3-BROMO-4-ETHYLPHENYL)AZO)-N,N-DIMETHYL- □ BENZENAMINE, N,N-DIMETHYL-3′-BROMO-4′-ETHYL-4-(PHENYLAZO)- □ 3′-BROMO-4′-ETHYL-4-DIMETHYLAMINOAZOBENZENE

TOXICITY DATA WITH REFERENCE
orl-rat TDLo:6930 mg/kg/25W-C:CAR CBINA8 53,107,85

SAFETY PROFILE: Questionable carcinogen with experimental carcinogenic data. When heated to decomposition it emits toxic fumes of Br^- and NO_x.

BNK325 CAS:2758-06-7 HR: 3
(2-BROMOETHYL)TRIMETHYLAMMONIUM BROMIDE
mf: $C_5H_{13}BrN•Br$ mw: 247.01

PROP: Prisms from EtOAc. Mp: 251–253° (sinters at 2°).

TOXICITY DATA WITH REFERENCE
orl-rat LD50:190 mg/kg QJPPAL 20,81,47
scu-rat LD50:60 mg/kg QJPPAL 20,81,47
orl-mus LD50:450 mg/kg QJPPAL 20,81,47
ipr-mus LD50:55 mg/kg QJPPAL 20,81,47
scu-mus LD50:65 mg/kg QJPPAL 20,81,47
ims-mus LD50:60 mg/kg QJPPAL 20,81,47

SAFETY PROFILE: Poison by ingestion, subcutaneous, intramuscular, and intraperitoneal routes. When heated to decomposition it emits toxic fumes of Br^-, NH_3, and NO_x.

BNK350 CAS:2028-52-6 HR: 2
2-BROMOETHYNYL-2-BUTANOL
mf: C_6H_9BrO mw: 177.06

SYNS: BASON □ BROMOACETYLENYLETHYLMETHYLCARBINOL □ BROMOETHYNYLETHYLMETHYLCARBINOL □ 1-BROMO-3-METHYLPENTIN-3-OL □ 1-BROMO-3-METHYL-1-PENTYN-3-OL

TOXICITY DATA WITH REFERENCE
scu-rat LD50:940 mg/kg THERAP 10,56,55
orl-mus LD50:532 mg/kg THERAP 10,56,55
ipr-mus LD50:725 mg/kg AIPTAK 112,463,57
scu-mus LD50:910 mg/kg JPETAB 109,268,53

SAFETY PROFILE: Moderately toxic by ingestion and other routes. When heated to decomposition it emits toxic fumes of Br^-. See also ALCOHOLS.

BNK500 CAS:1940-57-4 HR: 3
9-BROMOFLUORENE
mf: $C_{13}H_9Br$ mw: 245.13

PROP: Crystals from ligroin. Mp: 104°.

TOXICITY DATA WITH REFERENCE
skn-man 2500 µg/48H SEV BJDEAZ 80,491,68
skn-hmn 2500 µg/24H SEV CHINAG (40),2080,67
ivn-mus LD50:180 mg/kg CSLNX* NX#01610

SAFETY PROFILE: Poison by intravenous route. A severe skin irritant in humans. When heated to decomposition it emits very toxic fumes of Br^-. See also BROMIDES.

BNK700 CAS:548-26-5 HR: 2
BROMOFLUORESCEIC ACID
mf: $C_{20}H_8Br_4O_5•2Na$ mw: 693.90

SYNS: AIZEN EOSINE GH □ BROMO ACID □ BROMO B □ BROMOEOSINE □ BROMO FLUORESCEIN □ BRONZE BROMO □ CERTIQUAL EOSINE □ C.I. 45380 □ C.I. ACID RED 87 □ EOSIN □ EOSINE B □ EOSINE FA □ EOSINE LAKE RED Y □ FENAZO EOSINE XG □ FLUORESCEIN, 2′,4′,5′,7′-TETRABROMO-, DISODIUM SALT □ HIDACID BROMO ACID REGULAR □ HIDACID DIBROMO FLUORESCEIN □ IRGALITE BRONZE RED CL □ PHLOXINE RED 20-7600

TOXICITY DATA WITH REFERENCE
scu-rat TDLo:13 g/kg/1Y-I:ETA GANNA2 47,51,56
ivn-mus LD50:550 mg/kg TXAPA9 44,225,78

SAFETY PROFILE: Moderately toxic by intravenous route. Questionable carcinogen with experimental tumorigenic data. When heated to decomposition it emits toxic fumes of Br^-.

BNL000 CAS:75-25-2 HR: 3
BROMOFORM
DOT: UN 2515
mf: $CHBr_3$ mw: 252.75

PROP: Colorless heavy liquid or hexagonal crystals. Mp: 6–7°, bp: 149°, flash p: none, d: 2.887 @ 20°/4°.

SYNS: BROMOFORME (FRENCH) □ BROMOFORMIO (ITALIAN) □ METHENYL TRIBROMIDE □ NCI-C55130 □ RCRA WASTE NUMBER U225 □ TRIBROMMETHAAN (DUTCH) □ TRIBROMMETHAN (GERMAN) □ TRIBROMOMETAN (ITALIAN) □ TRIBROMOMETHANE

TOXICITY DATA WITH REFERENCE
sln-dmg-orl 3000 ppm ENMUDM 7,677,85
sce-hmn:lym 80 µmol/L ENVRAL 32,72,83
sce-ham:ovr 290 µg/L ENMUDM 7,1,85
ipr-mus TDLo:1100 mg/kg/8W-I:NEO CNREA8 37,2717,77
orl-hmn LDLo:143 mg/kg 34ZIAG -,141,69
orl-rat LD50:1147 mg/kg TXAPA9 52,351,80
ihl-rat LCLo:45 g/m³/4H 85GMAT -,28,82
ipr-rat LD50:414 mg/kg TOLED5 15,251,83
orl-mus LD50:1400 mg/kg TXAPA9 44,213,78
scu-mus LD50:1820 mg/kg TXAPA9 4,354,62
scu-rbt LDLo:410 mg/kg AEXPBL 28,201,1891

CONSENSUS REPORTS: Reported in EPA TSCA Inventory. Community Right-To-Know List.

OSHA PEL: TWA 0.5 ppm (skin)
ACGIH TLV: TWA 0.5 ppm (skin)
DFG MAK: Suspected Carcinogen
DOT CLASSIFICATION: 6.1; *Label:* KEEP AWAY FROM FOOD

SAFETY PROFILE: Suspected carcinogen with experimental neoplastigenic data. A human poison by ingestion. Moderately toxic by intraperitoneal and subcutaneous routes. Human mutation data reported. A lachrymator. It can damage the liver to a serious degree and cause death. It has anesthetic properties similar to those of chloroform, but is not sufficiently volatile for inhalation purposes and is far too toxic for human use. As a sedative and antitussive its medicinal application has resulted in numerous poisonings. Inhalation of small amounts causes irritation, provoking the flow of tears and saliva, and reddening of the face. Abuse can lead to addiction and serious consequences. Explosive reaction with crown ethers or potassium hydroxide. Violent reaction with acetone or bases. Incompatible with Li or NaK alloys. When heated to decomposition it emits highly toxic fumes of Br^-. See also BROMIDES.

For occupational chemical analysis use NIOSH: Hydrocarbons, Halogenated, 1003.

BNL250 **CAS:2104-96-3** **HR: 2**
BROMOFOSMETHYL
mf: $C_8H_8BrCl_2O_3PS$ mw: 366.00

PROP: Yellowish crystals. Mp: 53–54°. Very sltly sol in H_2O, sol in CCl_4, Et_2O, and toluene.

SYNS: BROFENE □ O-(4-BROM-2,5-DICHLOR-PHENYL)-O,O-DI-METHYL-MONOTHIOPHOSPHAT (GERMAN) □ O-(4-BROMO-2,5-DI-CLORO-FENIL)-O,O-DIMETIL-MONOTIOFOSFATO (ITALIAN) □ 4-BROMO-2,5-DICHLOROPHENYL DIMETHYL PHOSPHOROTHIONATE □ O-(4-BROOM-2,5-DICHLOOR-FENYL)-O,O-DIMETHYL-MONOTHIO-FOSFAAT (DUTCH) □ BROMOFOS □ BROMOPHOS □ BRUOMO-PHOS (RUSSIAN) □ CELA S 1942 □ O,O-DIMETHYL-O-(4-BROMO-2,5-DICHLOROPHENYL) PHOSPHOROTHIOATE □ O,O-DIMETHYL-O-(2,5-DICHLOR-4-BROMPHENYL)-THIONOPHOSPHAT (GERMAN) □ O,O-DIMETHYL-O-(2,5-DICHLORO-4-BROMOPHENYL)PHOSPHOROTHIOATE □ O,O-DIMETHYL-O-(2,5-DICHLORO-4-BROMOPHENYL) THIOPHOSPHATE □ EL 400 □ ENT 27,162 □ MONSANTO CP 51969 □ NETAL □ NEXION □ NEXION 40 □ OMS-658 □ S 1942 □ THIOPHOSPHATE de O,O-DIMETHYLE et de O-4-BROMO-2,5-DICHLOROPHENYLE (FRENCH)

TOXICITY DATA WITH REFERENCE
cyt-mus-ipr 73,200 µg/kg RTOPDW 6,416,86
orl-mus TDLo: 293 mg/kg (female 6-12D post): REP RTOPDW 6,416,86
orl-wmn TDLo: 152 mg/kg: EYE,BAH,GIT JTCTDW 29,203,91
orl-rat LD50: 1600 mg/kg TXAPA9 14,515,69
ihl-rat LC50: 33 g/kg DOVEAA 32,40,78
ipr-rat LDLo: 1625 mg/kg ATXKA8 22,36,66
orl-mus LD50: 2829 mg/kg 28ZEAL 5,29,76
ipr-mus LD50: 1040 mg/kg ATXKA8 22,36,66
orl-cat LDLo: 750 mg/kg ATXKA8 22,36,66
skn-rbt LD50: 2181 mg/kg 28ZEAL 5,29,76
orl-gpg LD50: 1500 mg/kg ATXKA8 22,36,66
skn-mam LD50: 2820 mg/kg GTPZAB 21(7),34,77

CONSENSUS REPORTS: EPA Genetic Toxicology Program.

SAFETY PROFILE: Moderately toxic by ingestion, skin contact, intraperitoneal, and possibly other routes. Human systemic effects by ingestion: miosis, muscle contraction, hypermotility, diarrhea. Experimental reproductive effects. Mutation data reported. When heated to decomposition it emits very toxic fumes of SO_x, PO_x, Br^-, and Cl^-.

BNL260 **CAS:583-69-7** **HR: D**
2-BROMOHYDROQUINONE
mf: $C_6H_5BrO_2$ mw: 189.02

SYNS: 1,4-BENZENEDIOL, 2-BROMO- □ 2-BROMO-1,4-BENZENE-DIOL □ BROMOHYDROQUINONE □ 2-BROMOQUINOL □ HYDRO-QUINONE, BROMO-

TOXICITY DATA WITH REFERENCE
ipr-rat TDLo: 87,200 µg/kg (female 9D post): REP TXAPA9 120,1,93

CONSENSUS REPORTS: Reported in EPA TSCA Inventory.

SAFETY PROFILE: Experimental reproductive effects. When heated to decomposition it emits toxic vapors of Br^-.

BNL275 **CAS:23483-74-1** **HR: 3**
BROMO(2-HYDROXYETHYL)MERCURY AMMONIA SALT

SYNS: 2-(BROMOMERCURI) ETHANOL-AMMONIA (1:0.8 moles) compound □ MERCURY, BROMO(2-HYDROXYETHYL)-, compound with AMMONIA (1:0.8 moles)

TOXICITY DATA WITH REFERENCE
ivn-mus LD50: 56 mg/kg CSLNX* NX#05832

OSHA PEL: TWA 0.01 mg(Hg)/m³; CL 0.03 mg(Hg)/m³ (skin)
ACGIH TLV: TWA 0.01 mg(Hg)/m³; STEL 0.03 mg(Hg)/m³

SAFETY PROFILE: Poison by intravenous route. When heated to decomposition it emits toxic fumes of NH_3, Hg, and Br^-.

BNL750 **CAS:87-48-9** **HR: 2**
5-BROMOINDOLE-2,3-DIONE
mf: $C_8H_4BrNO_2$ mw: 226.04

PROP: Prisms from EtOH. Mp: 251–253°.

SYN: 5-BROMISATIN (CZECH)

TOXICITY DATA WITH REFERENCE
eye-rbt 5 mg/24H SEV 28ZPAK -,143,72
orl-rat LDLo: 4 g/kg 28ZPAK -,143,72
orl-mus LD50: 437 mg/kg RPTOAN 45,10,82
ipr-mus LD50: 437 mg/kg PCJOAU 15,858,81

SAFETY PROFILE: Moderately toxic by ingestion and intraperitoneal routes. A severe eye irritant. When heated to decomposition it emits very toxic fumes of Br^- and NO_x.

BNM000 **CAS:314-42-1** **HR: 2**
5-BROMO-3-ISOPROPYL-6-METHYLURACIL
mf: $C_8H_{11}BrN_2O_2$ mw: 247.12

PROP: Crystals from alc (aq). Mp: 158°. Sol in abs alc.

SYNS: 5-BROM-3-ISOPROPYL-6-METHYL-URACIL (GERMAN) □ 5-BROMO-3-ISOPROPYL-6-METHYL, 2,4-PYRIMIDINEDIONE (FRENCH) □ 5-BROMO-3-ISOPROPYL-6-METIL-URACIL (ITALIAN) □ 5-BROOM-3-ISOPROPYL-6-METHYL-URACIL DUTCH) □ HERBICIDE 82 □ HYVAR □ ISOCIL □ ISOPROCIL (FRENCH) □ 3-ISOPROPYL-5-BROMO-6-METHYLURACIL □ LOROX

TOXICITY DATA WITH REFERENCE
orl-rat LD50: 3400 mg/kg RREVAH 10,97,65
orl-mus LDLo: 3750 mg/kg TXAPA9 23,288,72

SAFETY PROFILE: Moderately toxic by ingestion. See also BROMIDES. When heated to decomposition it emits very toxic fumes of Br^- and NO_x.

BNM100 **CAS:565-74-2** **HR: 2**
2-BROMOISOVALERIC ACID
mf: $C_5H_9BrO_2$ mw: 181.05

SYNS: α-BROMOISOVALERIC ACID □ 2-BROMO-3-METHYLBUTANOIC ACID □ 2-BROMO-3-METHYLBUTYRIC ACID □ BUTANOIC ACID, 2-BROMO-3-METHYL-(9CI) □ BUTYRIC ACID, 2-BROMO-3-METHYL-

TOXICITY DATA WITH REFERENCE
orl-rat LD50:769 mg/kg EPASR* 8EHQ-0188-0714
skn-rat LD50:1410 mg/kg EPASR* 8EHQ-0188-0714

CONSENSUS REPORTS: Reported in EPA TSCA Inventory.

SAFETY PROFILE: Moderately toxic by ingestion and skin contact. When heated to decomposition it emits toxic vapors of Br^-.

BNM250 CAS:478-84-2 HR: 3
2-BROMO-d-LYSERGIC ACID DIETHYLAMIDE
mf: $C_{20}H_{26}BrN_3O$ mw: 404.40

SYNS: BOL □ BOL-148 □ d-2-BROM-DIETHYLAMIDE of LYSERGIC ACID □ BROM LSD □ BROMLYSERGAMIDE □ 2-BROM-d-LYSERGIC ACID DIETHYLAMINE □ 2-BROMO-9,10-DIDEHYDRO-N,N-DIETHYL-6-METHYLERGOLINE-8-β-CARBOXAMIDE □ BROMOLYSERGIDE □ 9, 10-DIDEHYDRO-N,N-DIETHYL-2-BROMO-6-METHYLERGOLINE-8-β-CARBOXAMIDE □ USAF SZ-1

TOXICITY DATA WITH REFERENCE
scu-ham TDLo:2 μg/kg (8D preg):TER SCIEAS 158,265,67
scu-ham TDLo:2 μg/kg (8D preg):REP SCIEAS 158,265,67
orl-hmn TDLo:75 μg/kg:CNS PSYPAG 1,20,59
ipr-mus LD50:25 mg/kg NTIS** AD277-689
ivn-mus LD50:20 mg/kg 28ZSAT -,-,64
ivn-rbt LD50:6 mg/kg ANYAA9 66,668,57

SAFETY PROFILE: Poison by intraperitoneal and intravenous routes. Experimental teratogenic and reproductive effects. Human systemic effects by ingestion: dilation of the arteries or veins. Many lysergic acid derivatives have central nervous system effects. When heated to decomposition it emits very toxic fumes such as Br^- and NO_x. See other lysergic acid derivatives.

BNM750 HR: 3
BROMOMETHANE mixed with DIBROMOETHANE
DOT: UN 1647

SYN: METHYL BROMIDE and ETHYLENE DIBROMIDE MIXTURE, liquid (DOT)

DOT CLASSIFICATION: 6.1; *Label:* Poison

SAFETY PROFILE: A poison. See also BROMIDES. When heated to decomposition it emits toxic fumes of Br^-.

BNN125 CAS:102433-83-0 HR: 3
2-(5-BROMO-2-METHOXYBENZYLOXY)TRIETHYLAMINE
mf: $C_{14}H_{22}BrNO_2$ mw: 316.28

TOXICITY DATA WITH REFERENCE
orl-rat LD50:220 mg/kg JPETAB 121,210,57
ipr-rat LD50:80 mg/kg JPETAB 121,210,57
orl-mus LD50:248 mg/kg JPETAB 121,210,57
ipr-mus LD50:123 mg/kg JPETAB 121,210,57

SAFETY PROFILE: Poison by ingestion and intraperitoneal routes. When heated to decomposition it emits toxic fumes of Br^- and NO_x.

BNN250 CAS:59177-64-9 HR: 3
2-BROMO-N-METHYL-1-ADAMANTANEETHYLAMINE MALEATE
mf: $C_{13}H_{22}BrN•C_4H_4O_4$ mw: 388.35

TOXICITY DATA WITH REFERENCE
orl-mus LD50:621 mg/kg JMCMAR 19,967,76
ipr-mus LD50:155 mg/kg JMCMAR 19,967,76

SAFETY PROFILE: Poison by intraperitoneal route. Moderately toxic by ingestion. When heated to decomposition it emits very toxic fumes of Br^- and NO_x.

BNN500 CAS:59177-85-4 HR: 3
2-BROMO-N-METHYL-1-ADAMANTANEMETHANAMINE HYDROCHLORIDE
mf: $C_{12}H_{20}BrN•ClH$ mw: 294.70

TOXICITY DATA WITH REFERENCE
orl-mus LD50:295 mg/kg JMCMAR 19,967,76
ipr-mus LD50:295 mg/kg JMCMAR 19,967,76

SAFETY PROFILE: Poison by ingestion and intraperitoneal routes. When heated to decomposition it emits very toxic fumes of Br^-, NO_x, and HCl.

BNN550 CAS:128-93-8 HR: 1
1-BROMO-4-(METHYLAMINO)ANTHRAQUINONE
mf: $C_{15}H_{10}BrNO_2$ mw: 316.17

PROP: Red brown needles from Py. Mp: 195–196°.

SYNS: 9,10-ANTHRACENEDIONE, 1-BROMO-4-(METHYLAMINO) □ ANTHRAQUINONE, 1-BROMO-4-(METHYLAMINO)- □ 1-METHYLAMINO-4-BROMANTHRACHINON □ 1-(METHYLAMINO)-4-BROMOANTHRAQUINONE

TOXICITY DATA WITH REFERENCE
eye-rbt 500 mg/24H MLD 05JCAE ,566,86

CONSENSUS REPORTS: Reported in EPA TSCA Inventory.

SAFETY PROFILE: An eye irritant. When heated to decomposition it emits toxic fumes of NO_x and Br^-.

BNO000 CAS:31897-92-4 HR: 3
2-BROMO-1-(2-METHYLAMINOPROPYL)ADAMANTANE HYDROCHLORIDE
mf: $C_{14}H_{24}BrN•ClH$ mw: 322.76

SYN: 1-(2-BROMO-1-ADAMANTYL)-N-METHYL-2-PROPYLAMINE HYDROCHLORIDE

TOXICITY DATA WITH REFERENCE
orl-mus LD50:300 mg/kg JMCMAR 17,602,74
ipr-mus LD50:150 mg/kg JMCMAR 17,602,74

SAFETY PROFILE: Poison by ingestion and intraperitoneal routes. When heated to decomposition it emits very toxic fumes of Br^-, NO_x, and HCl.

BNO250 CAS:31898-11-0 HR: 3
3-BROMO-1-(2-METHYLAMINOPROPYL)ADAMANTANE HYDROCHLORIDE
mf: $C_{14}H_{24}BrN \cdot ClH$ mw: 322.76

SYN: 1-(3-BROMO-1-ADAMANTYL)-N-METHYL-2-PROPYLAMINE HYDROCHLORIDE

TOXICITY DATA WITH REFERENCE
orl-mus LD50:400 mg/kg JMCMAR 17,602,74
ipr-mus LD50:150 mg/kg JMCMAR 17,602,74

SAFETY PROFILE: Poison by ingestion and intraperitoneal routes. When heated to decomposition it emits very toxic fumes of Br^-, NO_x, and HCl.

BNO500 CAS:2417-77-8 HR: 3
9-BROMOMETHYLANTHRACENE
mf: $C_{15}H_{11}Br$ mw: 271.17

PROP: Crystals from $CHCl_3$ or yellow needles from pet ether/C_6H_6. Mp: 145–147° (decomp).

SYN: ICR 506

TOXICITY DATA WITH REFERENCE
mma-sat 10 μg/plate PNASA6 72,5135,75
mmo-sat 10 μg/plate PNASA6 72,5135,75
ivn-mus TDLo:1350 μg/kg:NEO CNREA8 36,2423,76
ivn-mus LDLo:2700 μg/kg CNREA8 36,2423,76

CONSENSUS REPORTS: EPA Genetic Toxicology Program.

SAFETY PROFILE: Deadly poison by intravenous route. Mutation data reported. When heated to decomposition it emits toxic fumes of Br^-. See also BROMIDES.

BNO750 CAS:24961-39-5 HR: 3
7-BROMOMETHYLBENZ(a)ANTHRACENE
mf: $C_{19}H_{13}Br$ mw: 321.23

SYNS: 7-BMBA □ ICR 498

TOXICITY DATA WITH REFERENCE
dnr-esc 1 mg/L PNASA6 79,534,82
dns-hmn:fbr 1 μmol/L NARHAD 7,1343,79
dnr-ham:ovr 800 nmol/L PNASA6 79,534,82
dnd-ham:ovr 100 nmol/L SCMGDN 10,183,84
sce-ham:ovr 400 nmol/L PNASA6 79,534,82
msc-ham:ovr 50 nmol/L PNASA6 79,534,82
scu-rat TDLo:27 mg/kg:CAR CRNGDP 2,103,81
skn-mus TD:10 mg/kg/40W-I:ETA CNREA8 43,2034,83
ivn-mus TDLo:800 μg/kg:NEO CNREA8 36,2423,76
scu-rat TD:80 mg/kg:CAR CRNGDP 2,103,81
skn-mus TD:31 mg/kg/40W-I:ETA CNREA8 43,2034,83
ivn-mus LDLo:1600 μg/kg CNREA8 36,2423,76

CONSENSUS REPORTS: EPA Genetic Toxicology Program.

SAFETY PROFILE: A deadly poison by intravenous route. Questionable carcinogen with experimental carcinogenic, neoplastigenic, and tumorigenic data. Human mutation data reported. When heated to decomposition it emits toxic fumes of Br^-. See also BROMIDES.

BNP000 CAS:49852-85-9 HR: 2
6-BROMOMETHYLBENZO(a)PYRENE
mf: $C_{21}H_{13}Br$ mw: 345.25

TOXICITY DATA WITH REFERENCE
dnd-mam:lum 100 mg/L CBINA8 47,111,83
add-uns:lym 100 mg/L CBINA8 47,111,83
scu-rat TDLo:100 mg/kg/40D-I:ETA JMCMAR 16,714,73

SAFETY PROFILE: Questionable carcinogen with experimental tumorigenic data. Mutation data reported. See also BROMIDES. When heated to decomposition it emits toxic fumes of Br^-.

BNP250 CAS:107-82-4 HR: 3
1-BROMO-3-METHYL BUTANE
DOT: UN 2341
mf: $C_5H_{11}Br$ mw: 151.05

PROP: Colorless liquid. D: 1.210, mp: −112°, bp: 120–121°, flash p: 21°. Sltly sol in water; misc with alc and ether.

SYNS: ISOAMYL BROMIDE □ ISOPENTYL BROMIDE □ 3-METHYLBUTYL BROMIDE

TOXICITY DATA WITH REFERENCE
ipr-rat LD50:6150 mg/kg 85GMAT -,76,82
ipr-mus LD50:13,750 mg/kg 85GMAT -,76,82
ihl-mam LD50:21,300 mg/m³ GTPZAB 18(4),55,74
ipr-mam LD50:480 mg/kg GTPZAB 18(4),55,74

CONSENSUS REPORTS: EPA Genetic Toxicology Program. EPA TSCA Chemical Inventory.

DOT CLASSIFICATION: 3; *Label:* Flammable Liquid

SAFETY PROFILE: Moderately toxic by intraperitoneal route. Flammable liquid. Dangerous fire hazard when exposed to heat or flame. When heated to decomposition it emits toxic fumes of Br^-. See also BROMIDES.

BNP750 CAS:496-67-3 HR: 2
2-BROMO-3-METHYLBUTYRYLUREA
mf: $C_6H_{11}BrN_2O_2$ mw: 223.10

SYNS: ABROVAL □ ALLUVAL □ ALURAL □ N-(AMINOCARBONYL)-2-BROMO-3-METHYLBUTANAMIDE □ BROMARAL □ BROMCARBAMIDE □ BROMISOVAL □ BROMISOVALERYLUREA □ α-BROMISOVALERYLUREA □ BROMISOVALUM □ BROMIZOVAL □ BROMOCARBAMIDE □ α-BROMO-β-DIMETHYLPROPANOYLUREA □ α-BROMOISO VALERIC ACID UREIDE □ α-BROMOISOVALEROYLUREA □ (α-BROMOISOVALERYL)UREA □ BROMOVAL □BROMOVALEROCARBAMIDE □ BROMOVALERYLUREA □ BROMOXIL □ BROMURAL □ BROMUVAN □ BROMVALERYLUREA □ BROMVALETONE □ BROMVALETONUM □ BROMVALUREA □ BROMYL □ BROVALIN □ BROVALUREA □ BROVARIN □ BVU □ CALMOTIN □ DIAGRABROMYL □ DIBROLUUR □ DORMIGENE □ ISOBROMYL □ ISOVAL □ MONOBROMOISOVALERYLUREA □ 2-MONOBROMOISOVALERYLUREA □ PIVADORM □ PIVADORN □ SOMNUROL □ UPIOL □ UVALERAL

TOXICITY DATA WITH REFERENCE
orl-wmn TDLo:400 mg/kg:CNS BMJOAE 1,1238,55
orl-hmn LDLo:57 mg/kg TOIZAG 7,513,60
orl-rat LD50:1000 mg/kg FEPRA7 7,262,48
orl-mus LD50:2 g/kg OYYAA2 11,693,76
orl-cat LD50:450 mg/kg NIIRDN 6,738,82

orl-rbt LD50:1200 mg/kg MEIEDD 10,193,83

CONSENSUS REPORTS: Reported in EPA TSCA Inventory.

SAFETY PROFILE: Moderately toxic by ingestion. Human systemic effects by ingestion: nausea or vomiting, and coma. A sedative and hypnotic agent. When heated to decomposition it emits very toxic fumes of Br^- and NO_x.

BNP850 CAS:25855-92-9 HR: 2
9-(BROMOMETHYL)-10-CHLOROANTHRACENE
mf: $C_{15}H_{10}BrCl$ mw: 305.61

SYN: 10-BROMOMETHYL-9-CHLOROANTHRACENE

TOXICITY DATA WITH REFERENCE
ivn-mus TDLo:3066 μg/kg:NEO CNREA8 40,782,80

SAFETY PROFILE: Questionable carcinogen with experimental neoplastigenic data. When heated to decomposition it emits toxic fumes of Br^- and Cl^-.

BNQ000 CAS:34346-99-1 HR: 2
7-BROMOMETHYL-4-CHLOROBENZ(a)ANTHRACENE
mf: $C_{19}H_{12}BrCl$ mw: 355.67

SYN: 4-CHLORO-7-BROMOMETHYLBENZ(a)ANTHRACENE

TOXICITY DATA WITH REFERENCE
skn-mus TDLo:14 mg/kg:ETA EJCAAH 7,473,71

SAFETY PROFILE: Questionable carcinogen with experimental tumorigenic data by skin contact. When heated to decomposition it emits very toxic fumes of Br^- and Cl^-. See also CHLRORINATED HYDROCARBONS, AROMATIC.

BNQ100 HR: 2
3′-BROMO-4′-METHYL-4-DIMETHYLAMINOAZOBENZENE
mf: $C_{15}H_{16}BrN_3$ mw: 318.25

SYNS: ANILINE, p-((3-BROMO-p-TOLYL)AZO)-N,N-DIMETHYL- □ BENZENAMINE, N,N-DIMETHYL-3′-BROMO-4′-METHYL-4-(PHENYLAZO)- □ p-((3-BROMO-p-TOLYL)AZO)-N,N-DIMETHYLANILINE

TOXICITY DATA WITH REFERENCE
orl-rat TDLo:13,977 mg/kg/52W-C:CAR CBINA8 53,107,85

SAFETY PROFILE: Questionable carcinogen with experimental carcinogenic data. When heated to decomposition it emits toxic fumes of Br^- and NO_x.

BNQ110 HR: 2
4′-BROMO-3′-METHYL-4-DIMETHYLAMINOAZOBENZENE
mf: $C_{15}H_{16}BrN_3$ mw: 318.25

SYNS: ANILINE, p-((4-BROMO-m-TOLYL)AZO)-N,N-DIMETHYL- □ BENZENAMINE, N,N-DIMETHYL-4′-BROMO-3′-METHYL-4-(PHENYLAZO)- □ p-((4-BROMO-m-TOLYL)AZO)-N,N-DIMETHYLANILINE

TOXICITY DATA WITH REFERENCE
orl-rat TDLo:9677 mg/kg/36W-C:CAR CBINA8 53,107,85

SAFETY PROFILE: Questionable carcinogen with experimental carcinogenic data. When heated to decomposition it emits toxic fumes of Br^- and NO_x.

BNQ250 CAS:34346-97-9 HR: 2
7-BROMOMETHYL-6-FLUOROBENZ(a)ANTHRACENE
mf: $C_{19}H_{12}BrF$ mw: 339.22

SYN: 6-FLUORO-7-BROMOMETHYLBENZ(a)ANTHRACENE

TOXICITY DATA WITH REFERENCE
skn-mus TDLo:14 mg/kg:ETA EJCAAH 7,473,71

SAFETY PROFILE: Questionable carcinogen with experimental tumorigenic data by skin contact. When heated to decomposition it emits very toxic fumes of Br^- and F^-.

BNQ500 CAS:4437-18-7 HR: 3
2-BROMO METHYL FURAN
mf: C_5H_5BrO mw: 272.36

OCH=CHCH=CCH$_2$Br (ring closed O to C)

PROP: Unstable oil. D: 1.56 @ 20°/20°.

SAFETY PROFILE: A very unstable explosive. When heated to decomposition it emits toxic fumes of Br^-. See also BROMIDES.

BNQ750 CAS:34346-96-8 HR: 2
7-BROMOMETHYL-1-METHYLBENZ(a)ANTHRACENE
mf: $C_{20}H_{15}Br$ mw: 335.26

SYN: 1-METHYL-7-BROMOMETHYLBENZ(a)ANTHRACENE

TOXICITY DATA WITH REFERENCE
skn-mus TDLo:13 mg/kg:ETA EJCAAH 7,473,71

SAFETY PROFILE: Questionable carcinogen with experimental tumorigenic data by skin contact. When heated to decomposition it emits toxic fumes of Br^-.

BNR000 CAS:16238-56-5 HR: 2
7-BROMO METHYL-12-METHYLBENZ(a)ANTHRACENE
mf: $C_{20}H_{15}Br$ mw: 335.26

SYN: ICR 502

TOXICITY DATA WITH REFERENCE
mmo-sat 300 ng/plate ENMUDM 6(Suppl 2),1,84
mmo-esc 1 μg/plate ENMUDM 6(Suppl 2),1,84
otr-rat:emb 270 μg/L JJIND8 67,1303,81
otr-mus:fbr 16 μg/L JJIND8 67,1303,81
dnd-mus:emb 500 nmol/L CALEDQ 7,103,79
scu-rat TDLo:6 mg/kg:CAR CRNGDP 2,103,81
skn-mus TDLo:8040 mg/kg:NEO JJIND8 61,135,78
ivn-mus TDLo:1700 μg/kg:NEO CNREA8 36,2423,76
scu-rat TD:20 mg/kg/39D-I:NEO CNREA8 31,1951,71
scu-rat TD:13 mg/kg:NEO EJCAAH 6,417,70
scu-rat TD:188 mg/kg:CAR CRNGDP 2,103,81

CONSENSUS REPORTS: EPA Genetic Toxicology Program.

SAFETY PROFILE: Questionable carcinogen with experimental carcinogenic and neoplastigenic data. Mutation data reported. When heated to decomposition it emits toxic fumes of Br^-.

BNR250 CAS:59230-81-8 HR: 2
12-BROMOMETHYL-7-METHYLBENZ(a)ANTHRACENE
mf: $C_{20}F_{15}Br$ mw: 605.11

TOXICITY DATA WITH REFERENCE
mmo-sat 20 nmol/plate CBINA8 58,253,86
skn-mus TDLo:8040 mg/kg:NEO JJIND8 61,135,78

SAFETY PROFILE: Questionable carcinogen with experimental neoplastigenic data. Mutation data reported. When heated to decomposition it emits very toxic fumes of F^- and Br^-.

BNR325 CAS:57846-03-4 HR: 3
2-BROMOMETHYL-5-METHYLFURAN
mf: C_6H_7BrO mw: 175.02

SAFETY PROFILE: May decompose violently above 70°C. When heated to decomposition it emits toxic fumes of Br^-.

BNR750 CAS:78-77-3 HR: 2
1-BROMO-2-METHYLPROPANE
mf: C_4H_9Br mw: 137.04

PROP: Liquid. Flash p: 22°C, d: 1.253 @ 20°/4°, fp −117.4°, bp: 90.5–91°.

SYNS: 1-BUTYL BROMIDE □ iso-BUTYL BROMIDE □ ISOBUTYL BROMIDE

TOXICITY DATA WITH REFERENCE
ipr-mus TDLo:3000 mg/kg/8W-I:NEO CNREA8 35,1411,75
ipr-uns LD50:1660 mg/kg GTPZAB 18(4),55,74

CONSENSUS REPORTS: EPA Genetic Toxicology Program. Reported in EPA TSCA Inventory.

DOT CLASSIFICATION: 3; *Label:* Flammable Liquid

SAFETY PROFILE: Questionable carcinogen with experimental neoplastigenic data. Moderately toxic by intraperitoneal route. A dangerous fire hazard when exposed to heat or flame. When heated to decomposition it emits toxic fumes of Br^-. See also BROMIDES.

BNS000 HR: 3
2-BROMO-2-METHYL PROPANE
mf: C_4H_9Br mw: 137.04

PROP: Flash p: −18°C.

SAFETY PROFILE: A very dangerous fire hazard when exposed to heat or flame. When heated to decomposition it emits toxic fumes of Br^-. See also BROMIDES.

BNS200 CAS:90-11-9 HR: 2
1-BROMONAPHTHALENE
mf: $C_{10}H_7Br$ mw: 207.08

SYNS: α-BROMONAPHTHALENE □ NAPHTHALENE, 1-BROMO-

TOXICITY DATA WITH REFERENCE
ipr-mus LD50:810 mg/kg GTPZAB 20(12),52,76

CONSENSUS REPORTS: Reported in EPA TSCA Inventory.

SAFETY PROFILE: Moderately toxic by intraperitoneal route. When heated to decomposition it emits toxic vapors of Br^-.

BNS750 CAS:6954-48-9 HR: 3
6-BROMO-1,2-NAPHTHOQUINONE
mf: $C_{10}H_5BrO_2$ mw: 237.06

PROP: Golden needles from H_2O, orange-red crystals from C_6H_6. Mp: 168° (decomp), discolors at 1°, sinters at 1°. Mod sol in EtOH, Et_2O, AcOH, and ligroin.

SYN: BONAPHTHON

TOXICITY DATA WITH REFERENCE
orl-rat TDLo:700 mg/kg (female 12D post):TER FATOAO 41,109,78
orl-rat LD50:3900 mg/kg FATOAO 43,337,80
ipr-rat LD50:130 mg/kg FATOAO 39,628,76
orl-mus LD50:260 mg/kg FATOAO 39,628,76
ipr-mus LD50:9 mg/kg FATOAO 39,628,76
orl-gpg LD50:900 mg/kg FATOAO 39,628,76

SAFETY PROFILE: Poison by ingestion and intraperitoneal routes. An experimental teratogen. When heated to decomposition it emits toxic fumes of Br^-.

BNT000 CAS:30007-47-7 HR: 3
5-BROMO-5-NITRO-m-DIOXANE
mf: $C_4H_6BrNO_4$ mw: 212.02

SYNS: 5-BROM-5-NITRO-1,3-DIOXAN (GERMAN) □ 5-BROMO-5-NITRO-1,3-DIOXANE

TOXICITY DATA WITH REFERENCE
skn-rat 2500 μg/24H FSASAX 78,269,76
skn-mus 2500 μg/24H FSASAX 78,269,76
orl-rat LD50:455 mg/kg FSASAX 78,269,76
ipr-rat LD50:31 mg/kg FSASAX 78,269,76
orl-mus LD50:590 mg/kg FSASAX 78,269,76
scu-dog LDLo:500 mg/kg FSASAX 78,269,76

CONSENSUS REPORTS: Reported in EPA TSCA Inventory.

SAFETY PROFILE: Poison by intraperitoneal route. Moderately toxic by ingestion and subcutaneous routes. A skin irritant. When heated to decomposition it emits very toxic fumes of Br^- and NO_x.

BNT250 CAS:52-51-7 HR: 3
2-BROMO-2-NITRO-1,3-PROPANEDIOL
mf: $C_3H_6BrNO_4$ mw: 200.01

PROP: Crystals. Mp: 130–133°. Very sol in H_2O.

B

SYNS: 2-BROMO-2-NITROPANE-1,3-DIOL □ 2-BROMO-2-NITROPROPAN-1,3-DIOL □ β-BROMO-β-NITROTRIMETHYLENEGLYCOL □ BRONOCOT □ BRONOPOL □ BRONOSOL

TOXICITY DATA WITH REFERENCE
skn-hmn 10 mg MOD JSCCA5 29,3,78
skn-rbt 500 mg/24H MLD JEPTDQ 4(4),47,80
skn-rbt 80 mg MOD JEPTDQ 4(4),47,80
eye-rbt 5 mg JSCCA5 29,3,78
orl-rat LD50:180 mg/kg 28ZEAL 5,30,76
skn-rat LD50:1600 mg/kg 85JFAN A542,84
ipr-rat LD50:26 mg/kg JSCCA5 29,3,78
scu-rat LD50:170 mg/kg KSRNAM 8,1029,74
ivn-rat LD50:37,400 μg/kg IYKEDH 8,680,77
orl-mus LD50:270 mg/kg PEMNDP 9,103,91
skn-mus LD50:4750 mg/kg IYKEDH 8,680,77
ipr-mus LD50:15,500 μg/kg KHFZAN 11(1),73,77
scu-mus LD50:116 mg/kg IYKEDH 8,680,77
ivn-mus LD50:48 mg/kg IYKEDH 8,680,77
orl-dog LD50:250 mg/kg 28ZEAL 5,30,76

CONSENSUS REPORTS: Reported in EPA TSCA Inventory.

SAFETY PROFILE: Poison by ingestion, subcutaneous, intravenous, and intraperitoneal routes. Moderately toxic by skin contact. An eye and human skin irritant. An antiseptic. When heated to decomposition it emits very toxic fumes of NO_x and Br^-.

BNT500 CAS:14173-58-1 HR: 2
3-BROMO-4-NITROQUINOLINE-1-OXIDE
mf: $C_9H_5BrN_2O_3$ mw: 269.07

TOXICITY DATA WITH REFERENCE
cyt-omi 37 μmol GANNA2 60,155,69
scu-mus TDLo:60 mg/kg/I:ETA CPBTAL 17,544,69
scu-mus TD:120 mg/kg/50D-I:ETA BCPCA6 16,631,67

SAFETY PROFILE: Questionable carcinogen with experimental tumorigenic data. Mutation data reported. When heated to decomposition it emits very toxic fumes of Br^- and NO_x.

BNU000 CAS:111-83-1 HR: 1
1-BROMOOCTANE
mf: $C_8H_{17}Br$ mw: 193.16

PROP: Liquid. D: 1.11 @ 25°/4°, fp: −55°, bp: 201.5°.

SYN: n-OCTYL BROMIDE

TOXICITY DATA WITH REFERENCE
orl-rat LD50:5020 mg/kg AIHAAP 30,470,69
skn-rbt LD50:8944 mg/kg AIHAAP 30,470,69

CONSENSUS REPORTS: Reported in EPA TSCA Inventory.

SAFETY PROFILE: Mildly toxic by ingestion and skin contact. When heated to decomposition it emits toxic fumes of Br^-. See also BROMIDES.

BNU125 CAS:23753-67-5 HR: 3
1-BROMOPENTABORANE (9)
mf: B_5BrH_8 mw: 142.02

PROP: White solid. Mp: 34°, bp: 82° @ 34 mm.

SAFETY PROFILE: Ignites spontaneously in air. Explosive reaction with hexamine above 90°C. When heated to decomposition it emits toxic fumes of Br^-. See also BORANES and BORON COMPOUNDS.

BNU250 CAS:63867-64-1 HR: 3
4-BROMO-1,2,2,6,6-PENTAMETHYLPIPERIDINE
mf: $C_{10}H_{19}BrN$ mw: 233.21

TOXICITY DATA WITH REFERENCE
orl-mus LD50:172 mg/kg NATUAS 184,1707,59
ivn-mus LD50:51 mg/kg NATUAS 184,1707,59

SAFETY PROFILE: Poison by ingestion and intravenous routes. When heated to decomposition it emits very toxic fumes of NO_x and Br^-.

BNU500 CAS:107-81-3 HR: 3
2-BROMOPENTANE
DOT: UN 2343
mf: $C_5H_{11}Br$ mw: 151.07

PROP: Colorless to yellow liquid, strong odor. Bp: 120°, fp: <−30°, d: 1.211 @ 25°/25°, flash p: 90°F.

TOXICITY DATA WITH REFERENCE
ihl-rat TCLo:90 mg/m³/4H/17W-I GTPZAB 25(5),51,81
ihl-mus LC50:33 g/m³ GTPZAB 25(5),51,81
ihl-mus TCLo:90 mg/m³/4H/17W-I GTPZAB 25(5),51,81
ipr-mus LD50:150 mg/kg NTIS** AD691-490

CONSENSUS REPORTS: Reported in EPA TSCA Inventory.

DOT CLASSIFICATION: 3; *Label:* Flammable Liquid

SAFETY PROFILE: Poison by intraperitoneal route. Mildly toxic by inhalation. A local irritant and narcotic in high concentration. Ingestion can cause liver damage. A dangerous fire hazard when exposed to heat or flame. When heated to decomposition it emits toxic fumes of Br^-. See also BROMIDES, and CHLORINATED HYDROCARBONS, ALIPHATIC.

BNU660 HR: 3
4-(2-(5-BROMO-2-PENTYLOXYBENZYLOXY)ETHYL) MORPHOLINE
mf: $C_{18}H_{28}BrNO_3$ mw: 386.38

TOXICITY DATA WITH REFERENCE
orl-rat LD50:1200 mg/kg JPETAB 121,210,57
ipr-rat LD50:265 mg/kg JPETAB 121,210,57
orl-mus LD50:620 mg/kg JPETAB 121,210,57
ipr-mus LD50:400 mg/kg JPETAB 121,210,57

SAFETY PROFILE: Poison by intraperitoneal route. Moderately toxic by ingestion. When heated to decomposition it emits toxic fumes of Br^- and NO_x.

BNU700 **HR: 3**
2-(5-BROMO-2-PENTYLOXYBENZYLOXY)TRIETHYL-AMINE
mf: $C_{18}H_{30}BrNO_2$ mw: 372.40

TOXICITY DATA WITH REFERENCE
orl-rat LD50:320 mg/kg JPETAB 121,210,57
ipr-rat LD50:85 mg/kg JPETAB 121,210,57
orl-mus LD50:190 mg/kg JPETAB 121,210,57
ipr-mus LD50:100 mg/kg JPETAB 121,210,57

SAFETY PROFILE: Poison by ingestion and intraperitoneal routes. When heated to decomposition it emits toxic fumes of Br^- and NO_x.

BNU725 **CAS:10457-90-6** **HR: 3**
BROMOPERIDOL
mf: $C_{21}H_{23}BrFNO_2$ mw: 420.36

PROP: Off-white, amorphous or microcrystalline powder. Mp: 155–158°. Solubility in water: 0.09 mg/mL; in 0.1 M tartaric, lactic, citric, and acetic acids: about 10 mg/mL.

SYNS: AZURENE □ 4-(4-(p-BROMOPHENYL)-4-HYDROXYPIPERIDINO)-4′-FLUOROBUTYROPHENONE □ 4-(4-(4-BROMOPHENYL)-4-HYDROXYPIPERIDINO)-4′-FLUOROBUTYROPHENONE □ 4-(4-(p-BROMOPHENYL)-4-HYDROXYPIPERIDINOL)-4′-FLUOROBUTYROPHENONE □ 4-(4-(4-BROMOPHENYL)-4-HYDROXY-1-PIPERIDINYL)-1-(4-FLUOROPHENYL)-1-BUTANONE □ BROMPERIDOL □ IMPROMEN □ R 11333 □ TESOPREL

TOXICITY DATA WITH REFERENCE
orl-rat TDLo:110 mg/kg (female 7-17D post):REP JTSCDR 9(Suppl 1),109,84
orl-rat TDLo:16,500 µg/kg (female 7-17D post):REP JTSCDR 9(Suppl 1),109,84
orl-rat TDLo:110 mg/kg (female 7-17D post):TER JTSCDR 9(Suppl 1),109,84
orl-rat TDLo:5600 mg/kg (female 35D pre):REP IYKEDH 15,906,84
orl-rat TDLo:1400 mg/kg (35D male):REP IYKEDH 15,906,84
orl-rat LD50:359 mg/kg ARZNAD 24,45,74
ipr-rat LD50:323 mg/kg IYKEDH 16,1461,85
scu-rat LD50:84 mg/kg ARZNAD 24,45,74
ivn-rat LD50:10 mg/kg IYKEDH 16,1461,85
orl-mus LD50:174 mg/kg IYKEDH 16,1461,85
ipr-mus LD50:156 mg/kg IYKEDH 16,1461,85
scu-mus LD50:114 mg/kg ARZNAD 24,45,74
ivn-mus LD50:18,900 µg/kg ARZNAD 24,45,74

SAFETY PROFILE: Poison by ingestion, subcutaneous, intravenous, and intraperitoneal routes. An experimental teratogen. Other experimental reproductive effects. When heated to decomposition it emits toxic fumes of F^-, Br^- and NO_x.

BNU750 **CAS:106-41-2** **HR: 2**
p-BROMOPHENOL
mf: C_6H_5BrO mw: 173.02

PROP: Crystals. Mp: 66°, bp: 238°.

SYN: 4-BROMOPHENOL

TOXICITY DATA WITH REFERENCE
skn-mus TDLo:7200 mg/kg/18W-I:ETA CNREA8 19,413,59
orl-mus LD50:523 mg/kg GISAAA 44(12),19,79
ipr-mus LD50:411 mg/kg GISAAA 44(12),19,79

CONSENSUS REPORTS: Reported in EPA TSCA Inventory.

SAFETY PROFILE: Moderately toxic by ingestion and intraperitoneal routes. Questionable carcinogen with experimental tumorigenic data. When heated to decomposition it emits toxic fumes of Br^-.

BNU800 **CAS:32762-51-9** **HR: 2**
BROMOPHENOL
mf: C_6H_5BrO mw: 173.02

SYN: PHENOL, BROMO-

TOXICITY DATA WITH REFERENCE
orl-uns LD50:652 mg/kg GISAAA 45(10),16,80
skn-uns LD50:1620 mg/kg GISAAA 45(10),16,80

CONSENSUS REPORTS: Reported in EPA TSCA Inventory.

SAFETY PROFILE: Moderately toxic by ingestion and skin contact. When heated to decomposition it emits toxic vapors of Br^-.

BNV000 **CAS:95-56-7** **HR: 2**
o-BROMOPHENOL

PROP: Liquid. Bp: 194°.

TOXICITY DATA WITH REFERENCE
orl-mus LD50:652 mg/kg GISAAA 44(12),19,79
ipr-mus LD50:633 mg/kg GISAAA 44(12),19,79
scu-gpg LDLo:1500 mg/kg RMSRA6 16,449,1896

CONSENSUS REPORTS: Reported in EPA TSCA Inventory.

SAFETY PROFILE: Moderately toxic by ingestion, intraperitoneal, and subcutaneous routes. When heated to decomposition it emits toxic fumes of Br^-.

BNV250 **HR: 2**
BROMO PHENOLS
mf: $HO(C_6H_4)Br$ mw: 173

PROP: (m-) Crystals; insol in water; sol in alc, ether, and alkalis. (p-) Crystals; sltly sol in water; sol in alc, ether, chloroform, and glacial acetic acid. (o-) Yellow to oily, red liquid; unpleasant odor; insol in water; sol in alc, ether, and chloroform. D: (p-) 1.840 (15°), 1.5875 (80°); (o-) 1.5. Mp: (m-) 33°; (p-) 64°; (o-) 6°. Bp: (m-) 236°; (p-) 238°; (o-) 194°.

SAFETY PROFILE: Moderately toxic by several routes. Dangerous in a fire. When heated to decomposition it emits toxic fumes of Br^-. See also BROMIDES.

BNV750 CAS:16532-79-9 HR: 3
4-BROMOPHENYLACETONITRILE
mf: C_8H_6BrN mw: 196.06

SYNS: 4-BROMOBENZENEACETONITRILE □ p-BROMOBENZYL CYANIDE □ 4-BROMOBENZYLCYANIDE □ p-BROMOPHENYLACETONITRILE □ 2-(4-BROMOPHENYL)ACETONITRILE

TOXICITY DATA WITH REFERENCE
ivn-mus LD50:56 mg/kg CSLNX* NX#03252

CONSENSUS REPORTS: Reported in EPA TSCA Inventory. Cyanide and its compounds are on the Community Right-To-Know List.

SAFETY PROFILE: Poison by intravenous route. See also BROMIDES and NITRILES. When heated to decomposition it emits very toxic fumes of Br^-, NO_x, and CN^-.

BNV775 CAS:106-37-6 HR: 2
p-BROMOPHENYL BROMIDE
mf: $C_6H_4Br_2$ mw: 235.92

SYNS: BENZENE, p-DIBROMO- □ BENZENE, 1,4-DIBROMO-(9CI) □ p-DIBROMOBENZENE □ 1,4-DIBROMOBENZENE

TOXICITY DATA WITH REFERENCE
orl-mus LD50:3120 mg/kg GISAAA 44(12),19,79
ipr-mus LD50:1891 mg/kg GISAAA 44(12),19,79

CONSENSUS REPORTS: Reported in EPA TSCA Inventory.

SAFETY PROFILE: Moderately toxic by ingestion and intraperitoneal routes. When heated to decomposition it emits toxic vapors of Br^-.

BNW250 CAS:7239-21-6 HR: 2
1-(4-BROMOPHENYL)-3,3-DIMETHYLTRIAZENE
mf: $C_8H_{10}BrN_3$ mw: 228.12

SYNS: 1-p-BROMFENYL-3,3-DIMETHYLTRIAZEN (CZECH) □ 4-BROMO PDMT

TOXICITY DATA WITH REFERENCE
mma-sat 5 mmol/L MUREAV 36,1,76
sln-dmg-orl 100 μmol/L CBINA8 9,365,74
orl-rat LD50:423 mg/kg 28ZPAK -,98,72

SAFETY PROFILE: Moderately toxic by ingestion. Mutation data reported. When heated to decomposition it emits very toxic fumes of Br^- and NO_x.

BNW500 CAS:1808-12-4 HR: 3
BROMOPHENYL HYDRAMINE HYDROCHLORIDE
mf: $C_{17}H_{20}BrNO•ClH$ mw: 370.75

PROP: Crystals from 2-propanol. Mp: 144–145°.

SYNS: β-(p-BROMOBENZHYDRYLOXY)ETHYLDIMETHYLAMINE HYDROCHLORIDE □ 2-(4-BROMOBENZOHYDRYLOXY)ETHYLDIMETHYLAMINE HYDROCHLORIDE

TOXICITY DATA WITH REFERENCE
orl-rat LD50:602 mg/kg CLDND* 112,318,54
ivn-mus LD50:63 mg/kg CLDND*
ivn-dog LD50:21 mg/kg CLDND*

SAFETY PROFILE: Poison by intravenous route. Moderately toxic by ingestion. When heated to decomposition it emits very toxic fumes of Br^-, NO_x, and HCl.

BNW550 CAS:622-88-8 HR: 2
4-BROMOPHENYL HYDRAZINE HYDROCHLORIDE
mf: $C_6H_7BrN_2•ClH$ mw: 223.52

SYN: HYDRAZINE, 1-(p-BROMOPHENYL)-, HYDROCHLORIDE

TOXICITY DATA WITH REFERENCE
orl-rat LDLo:500 mg/kg JPETAB 90,260,47

CONSENSUS REPORTS: Reported in EPA TSCA Inventory.

SAFETY PROFILE: Moderately toxic by ingestion. When heated to decomposition it emits toxic vapors of NO_x, HCl, Cl^-, and Br^-.

BNW750 CAS:1470-37-7 HR: 3
4-BROMO-2-PHENYL-1,3-INDANDIONE
mf: $C_{15}H_9BrO_2$ mw: 301.15

SYN: 4-BROMO-2-FENILINDAN-1,3-DIONE (ITALIAN)

TOXICITY DATA WITH REFERENCE
orl-rat LD50:745 mg/kg FRPSAX 31,403,76
orl-mus LD50:114 mg/kg FRPSAX 31,403,76
ipr-mus LDLo:160 mg/kg FRPSAX 31,315,76

SAFETY PROFILE: Poison by ingestion and intraperitoneal routes. When heated to decomposition it emits toxic fumes of Br^-. See also BROMIDES.

BNW825 CAS:1985-12-2 HR: 3
p-BROMOPHENYL ISOTHIOCYANATE
mf: C_7H_4BrNS mw: 214.09

PROP: Yellow needles. Mp: 60–61°.

SYN: p-BROMOPHENYL ESTER ISOTHIOCYANIC ACID

TOXICITY DATA WITH REFERENCE
orl-rat LD50:400 mg/kg FCTXAV 5,741,67
ipr-rat LDLo:100 mg/kg ARZNAD 19,558,69
ipr-mus LDLo:100 mg/kg ARZNAD 21,121,71

SAFETY PROFILE: Poison by ingestion and intraperitoneal routes. When heated to decomposition it emits toxic fumes of SO_x, Br^-, and NO_x. See also THIOCYANATES and ESTERS.

BNX000 CAS:22480-64-4 HR: 3
p-BROMO PHENYL LITHIUM
mf: C_6H_4BrLi mw: 162.95

SAFETY PROFILE: Explodes on exposure to oxygen. When heated to decomposition it emits toxic fumes of Br^-. See also LITHIUM COMPOUNDS.

BNX125 CAS:23139-02-8 HR: 2
3-(p-BROMOPHENYL)-1-METHYL-1-NITROSOUREA
mf: $C_8H_8BrN_3O_2$ mw: 258.10

SYNS: 1-METHYL-3-(p-BROMOPHENYL)-1-NITROSOUREA □ 1-METHYL-3-(p-BROMPHENYL)-1-NITROSOHARNSTOFF (GERMAN) □ 1-METHYL-1-NITROSO-3-(p-BROMOPHENYL)UREA

TOXICITY DATA WITH REFERENCE
cyt-ham:lng 10 µmol/L IAPUDO 31,797,80
sce-ham:lng 10 µmol/L IAPUDO 31,797,80
orl-rat TDLo:774 mg/kg/88W-I:CAR ARGEAR 53,329,83
orl-rat TD:1080 mg/kg/30W-I:ETA IAPUDO 31,685,80

SAFETY PROFILE: Questionable carcinogen with experimental carcinogenic and tumorigenic data. Mutation data reported. When heated to decomposition it emits toxic fumes of Br^- and NO_x.

BNX250 CAS:60050-37-5 HR: 2
2-(m-BROMOPHENYL)-N-(4-MORPHOLINOMETHYL) SUCCINIMIDE
mf: $C_{15}H_{17}BrN_2O_3$ mw: 353.25

TOXICITY DATA WITH REFERENCE
orl-mus LD50:3012 mg/kg ARZNAD 29,290,79
ipr-mus LD50:443 mg/kg EJMCA5 13,465,78

SAFETY PROFILE: Moderately toxic by ingestion and intraperitoneal routes. When heated to decomposition it emits very toxic fumes of Br^- and NO_x.

BNX750 CAS:106-94-5 HR: 3
1-BROMOPROPANE
mf: C_3H_7Br mw: 123.01

PROP: Liquid. Mp: −110°, bp: 71°, d: 1.35 @ 20°/4°, autoign temp: 914°F, flash p: <22°, lel: 4.6%.

SYNS: 1-BROMOPROPANE (DOT) □ PROPYL BROMIDE

TOXICITY DATA WITH REFERENCE
orl-rat TDLo:2 g/kg (5D male):REP MUREAV 101,321,82
orl-rat LDLo:4000 mg/kg MUREAV 101,321,82
ihl-rat LC50:253,000 mg/m³/30M FAVUAI 7,35,75
ipr-rat LD50:2950 mg/kg 85GMAT -,102,82
ipr-mus LD50:2530 mg/kg 85GMAT -,102,82

CONSENSUS REPORTS: Reported in EPA TSCA Inventory.

SAFETY PROFILE: Moderately toxic by ingestion and intraperitoneal routes. Mildly toxic by inhalation. Experimental reproductive effects. Mutation data reported. Dangerous fire hazard when heated or exposed to flame or oxidizers. To fight fire, use water, foam, CO_2, dry chemical. When heated to decomposition it emits toxic fumes of Br^-. See also BROMIDES.

BNY000 CAS:75-26-3 HR: 3
2-BROMOPROPANE
DOT: UN 2344
mf: C_3H_7Br mw: 122.98

PROP: Liquid. Flash p: <14°, d: 1.31 @ 20°/4°, fp −89°, bp: 59.35°.

SYN: ISOPROPYL BROMIDE

TOXICITY DATA WITH REFERENCE
ipr-mus LD50:4837 mg/kg GTPZAB 20(12),52,76
ihl-uns LC50:36 g/m³ GTPZAB 18(4),55,74

CONSENSUS REPORTS: Reported in EPA TSCA Inventory.

DOT CLASSIFICATION: 3; *Label:* Flammable Liquid

SAFETY PROFILE: Moderately toxic by intraperitoneal route. A very flammable liquid and dangerous fire hazard. When heated to decomposition it emits toxic fumes of Br^-. See also BROMIDES.

BNZ000 CAS:598-31-2 HR: 3
BROMO-2-PROPANONE
DOT: UN 1569
mf: C_3H_5BrO mw: 136.99

PROP: Liquid that turns violet rapidly. D: 1.634°, fp −36.5°, bp: 136.5° @ 725 mm.

SYNS: ACETONYL BROMIDE □ ACETYL METHYL BROMIDE □ BROMOACETONE □ BROMOACETONE (DOT) □ BROMOACETONE, liquid (DOT) □ BROMOMETHYL METHYL KETONE □ 1-BROMO-2-PROPANONE □ MONOBROMOACETONE □ RCRA WASTE NUMBER P017

TOXICITY DATA WITH REFERENCE
ihl-hmn LCLo:572 ppm/10M NTIS** PB214-270

DOT CLASSIFICATION: 6.1; *Label:* Poison

SAFETY PROFILE: A poisonous gas. Moderately toxic to humans by inhalation. When heated to decomposition it emits toxic fumes of Br^-. See also BROMIDES.

BOA750 CAS:42461-89-2 HR: 2
5-(3-BROMO-1-PROPENYL)-1,3-BENZODIOXOLE
mf: $C_{10}H_9BrO_2$ mw: 241.10

SYNS: 3′-BROMOISOSAFROLE □ 1,2-(METHYLENEDIOXY)-4-(3-BROMO-1-PROPENYL)BENZENE

TOXICITY DATA WITH REFERENCE
scu-rat TDLo:359 mg/kg/10W-:ETA CNREA8 33,590,73

SAFETY PROFILE: Questionable carcinogen with experimental tumorigenic data. When heated to decomposition it emits toxic fumes such as Br^-. See also BROMIDES.

BOB000 CAS:598-72-1 HR: 2
α-BROMOPROPIONIC ACID
mf: $C_3H_5BrO_2$ mw: 152.99

TOXICITY DATA WITH REFERENCE
orl-mus LD50:250 mg/kg JPETAB 86,336,46

CONSENSUS REPORTS: Reported in EPA TSCA Inventory.

SAFETY PROFILE: Poison by ingestion. When heated to decomposition it emits toxic fumes of Br^-.

BOB250 CAS:590-92-1 HR: 2
3-BROMOPROPIONIC ACID
mf: $C_3H_5BrO_2$ mw: 152.99

PROP: Plates. D: 1.485, mp: 62.5°, bp: 140–142° @ 45 mm.

SYN: β-BROMOPROPIONIC ACID

TOXICITY DATA WITH REFERENCE
mmo-sat 25 μg/plate DHEFDK FDA-78-1046,78
ipr-mus TDLo:580 mg/kg/8W-I:ETA CNREA8 39,391,79
ipr-mus LDLo:500 mg/kg CBCCT* 6,228,54

CONSENSUS REPORTS: Reported in EPA TSCA Inventory.

SAFETY PROFILE: Moderately toxic by intraperitoneal route. Questionable carcinogen with experimental tumorigenic data. Mutation data reported. When heated to decomposition it emits toxic fumes of Br^-. See also BROMIDES.

BOB500 CAS:2417-90-5 HR: 3
3-BROMOPROPIONITRILE
mf: C_3H_4BrN mw: 133.99

PROP: Liquid. D: 1.62 @ 20°/4°, bp: 81–83° @ 15 mm.

SYN: USAF DO-51

TOXICITY DATA WITH REFERENCE
ipr-mus LD50:50 mg/kg NTIS** AD277-689
par-mus LDLo:80 mg/kg CBCCT* 7,692,55

CONSENSUS REPORTS: Reported in EPA TSCA Inventory. Cyanide and its compounds are on the Community Right-To-Know List.

SAFETY PROFILE: Poison by parenteral and intraperitoneal routes. See also NITRILES. When heated to decomposition it emits very toxic fumes of NO_x, CN^-, and Br^-.

BOB550 CAS:2114-00-3 HR: 3
2-BROMOPROPIOPHENONE
mf: C_9H_9BrO mw: 213.09

SYNS: α-BROMOPROPIOPHENONE □ PROPIOPHENONE, 2-BROMO- □ TL 336

TOXICITY DATA WITH REFERENCE
ihl-mus LCLo:1600 mg/m³/10M NDRC** NDCrc-132,AUG42
ivn-mus LD50:56 mg/kg CSLNX* NX#02729

CONSENSUS REPORTS: Reported in EPA TSCA Inventory.

SAFETY PROFILE: Poison by intravenous route. Moderately toxic by inhalation. When heated to decomposition it emits toxic vapors of Br^-.

BOB600 CAS:109-04-6 HR: 3
2-BROMOPYRIDINE
mf: C_5H_4BrN mw: 158.01

SYN: PYRIDINE, 2-BROMO-

TOXICITY DATA WITH REFERENCE
ipr-mus LDLo:31,300 μg/kg CBCCT* 4,322,52

CONSENSUS REPORTS: Reported in EPA TSCA Inventory.

SAFETY PROFILE: Poison by intraperitoneal route. When heated to decomposition it emits toxic vapors of NO_x and Br^-.

BOC510 CAS:626-55-1 HR: 3
3-BROMOPYRIDINE
mf: C_5H_4BrN mw: 158.00

HC=CHCH=CBrCH=N (ring)

PROP: Yellow liquid with strongly alkaline reaction. D: 1.645 @ 0°/4°, bp: 175°. Mod sol in H_2O.

SAFETY PROFILE: Mixture with acetic acid + hydrogen peroxide explodes when heated above 50°C. When heated to decomposition it emits toxic fumes of Br^- and NO_x.

BOD000 CAS:41287-72-3 HR: 3
3-(2-(5-BROMO-2-PYRIDYLOXY)ETHYL)THIAZOLIDINE HYDROCHLORIDE
mf: $C_{10}H_{13}BrN_2OS{\cdot}ClH$ mw: 325.68

SYN: 5-BROMO-2-(2-(3-THIAZOLIDINYL)ETHOXY)PYRIDINE HYDROCHLORIDE

TOXICITY DATA WITH REFERENCE
orl-mus LD50:400 mg/kg JMCMAR 16,319,73
ipr-mus LD50:150 mg/kg JMCMAR 16,319,73

SAFETY PROFILE: Poison by ingestion and intraperitoneal routes. When heated to decomposition it emits very toxic fumes of Br^-, HCl, NO_x, and SO_x.

BOD500 CAS:41287-56-3 HR: 3
2-(6-(5-BROMO-2-PYRIDYL OXY)HEXYL)AMINOETHANE THIOL HYDROCHLORIDE
mf: $C_{13}H_{21}BrN_2OS{\cdot}ClH$ mw: 369.79

TOXICITY DATA WITH REFERENCE
orl-mus LD50:350 mg/kg JMCMAR 16,319,73
ipr-mus LD50:140 mg/kg JMCMAR 16,319,73

SAFETY PROFILE: Poison by ingestion and intraperitoneal routes. When heated to decomposition it emits very toxic Br^-, Cl^-, SO_x, and NO_x.

BOD550 CAS:1113-59-3 HR: 3
3-BROMOPYRUVIC ACID
mf: $C_3H_3BrO_3$ mw: 166.97

SYNS: 3-BROMO-2-OXOPROPANOIC ACID □ 3-BROMOPYRUVATE □ BROMOPYRUVIC ACID □ β-BROMOPYRUVIC ACID □ PROPANOIC ACID, 3-BROMO-2-OXO-(9CI) □ PYRUVIC ACID, BROMO-

TOXICITY DATA WITH REFERENCE
ipr-mus LD50:72 mg/kg JPETAB 123,48,58

CONSENSUS REPORTS: Reported in EPA TSCA Inventory.

SAFETY PROFILE: Poison by intraperitoneal route. When heated to decomposition it emits toxic vapors of NO_x and Br^-.

BOD600 CAS:87-12-7 HR: 2
5-BROMOSALICYL-4-BROMOANILIDE
mf: $C_{13}H_9Br_2NO_2$ mw: 371.05

SYNS: BENZAMIDE, 5-BROMO-N-(4-BROMOPHENYL)-2-HYDROXY- □ p-BROMANILID KYSELINY 5-BROMSALICYLOVE □ 3-BROMO-6-HY-

DROXYBENZ-p-BROMANILIDE □ 4′,5-DIBROMOSALICYLANILIDE □ DIBROMSALAN □ NSC-20527 □ SALICYLANILIDE, 4′,5-DIBROMO- □ TEMASEPT

TOXICITY DATA WITH REFERENCE
orl-rat LD50:410 mg/kg IMSUAI 39,56,70

CONSENSUS REPORTS: Reported in EPA TSCA Inventory.

SAFETY PROFILE: Moderately toxic by ingestion. When heated to decomposition it emits toxic vapors of NO_x and Br^-.

BOE500 CAS:89-55-4 HR: 3
5-BROMOSALICYLIC ACID
mf: $C_7H_5BrO_3$ mw: 217.03

PROP: Needles. Mp: 168–169°.

TOXICITY DATA WITH REFERENCE
ivn-mus LD50:100 mg/kg CSLNX* NX#04478

CONSENSUS REPORTS: Reported in EPA TSCA Inventory.

SAFETY PROFILE: Poison by intravenous route. When heated to decomposition it emits toxic fumes of Br^-.

BOE750 CAS:13465-73-1 HR: 3
BROMOSILANE
mf: BrH_3Si mw: 111.02

SYN: SILYL BROMIDE

DOT CLASSIFICATION: Forbidden

SAFETY PROFILE: Ignites spontaneously upon exposure to air. When heated to decomposition it emits toxic fumes of Br^-.

BOF000 CAS:103-64-0 HR: 2
β-BROMOSTYRENE
mf: C_8H_7Br mw: 183.06

SYNS: α-BROMO-β-PHENYLETHYLENE □ ω-BROMOSTYRENE □ BROMOSTYROL □ BROMOSTYROLENE □ β-BROMSTYROL □ HYACINTH BASE

TOXICITY DATA WITH REFERENCE
orl-rat LD50:1250 mg/kg FCTXAV 11,1043,73

CONSENSUS REPORTS: Reported in EPA TSCA Inventory.

SAFETY PROFILE: Moderately toxic by ingestion. When heated to decomposition it emits toxic fumes of Br^-. See also BROMIDES.

BOF500 CAS:128-08-5 HR: 3
N-BROMOSUCCINIMIDE
mf: $C_4H_4BrNO_2$ mw: 178.00

$CH_2CO{\cdot}NBrCO{\cdot}CH_2$ (ring)

PROP: White to pale-buff, fine, orthorhombic, crystalline powder with faint odor of bromine. Mp: 173–175°, d: 2.098. Sol in Me_2CO, sltly sol in AcOH; sltly sol in H_2O and CCl_4; prac insol in hexane.

SYNS: 1-BROMO-2,5-PYRROLIDINEDIONE □ N-BROMOSUCCIMIDE □ SUCCINBROMIMIDE □ SUCCINIBROMIMIDE

TOXICITY DATA WITH REFERENCE
ipr-mus LDLo:256 mg/kg CBCCT* 2,244,50

CONSENSUS REPORTS: Reported in EPA TSCA Inventory.

SAFETY PROFILE: Poison by intraperitoneal route. An irritating poison to skin, eyes, and mucous membranes. Reacts explosively with aniline, diallyl sulfide, and hydrazine hydrate. Explosive reaction with propiononitrile after heating to 105°C for 24 hours. Violent reaction with dibenzoyl peroxide + 4-toluic acid. When heated to decomposition it emits toxic fumes of Br^- and NO_x. See also BROMIDES and NITROGEN MONOXIDE.

BOF750 CAS:679-84-5 HR: 2
3-BROMO-1,1,2,2-TETRAFLUOROPROPANE
mf: $C_3H_3BrF_4$ mw: 194.97

PROP: Liquid. Bp: 74°.

SYNS: FHD-3 □ HALOPROPANE

TOXICITY DATA WITH REFERENCE
ihl-hmn TCLo:40,000 ppm:CNS ANESAV 25,600,64
ihl-hmn TCLo:4000 ppm/30M:CVS ANESAV 25,600,64

SAFETY PROFILE: Human central nervous system and cardiovascular system effects by inhalation. When heated to decomposition it emits very toxic fumes of F^- and Br^-.

BOG000 CAS:14008-53-8 HR: 3
3-BROMOTETRAHYDROTHIOPHENE-1,1-DIOXIDE
mf: $C_4H_7BrO_2S$ mw: 199.08

SYN: TETRAHYDRO-3-BROMOTHIOPHENE-1,1-DIOXIDE

TOXICITY DATA WITH REFERENCE
orl-rat LD50:215 mg/kg AIPTAK 119,423,59
ipr-rat LD50:44 mg/kg AIPTAK 119,423,59
orl-mus LD50:121 mg/kg AIPTAK 119,423,59
ipr-mus LD50:59 mg/kg AIPTAK 119,423,59
ivn-mus LD50:25 mg/kg AIPTAK 119,423,59
ivn-dog LD50:29 mg/kg AIPTAK 119,423,59

SAFETY PROFILE: Poison by ingestion, intraperitoneal, and intravenous routes. When heated to decomposition it emits very toxic fumes of Br^- and SO_x.

BOG250 CAS:6926-40-5 HR: 3
N-BROMOTETRAMETHYL GUANIDINE
mf: $C_5H_{12}BrN_3$ mw: 194.07

$(CH_3)_2NC(:NBr)N(CH_3)_2$

SAFETY PROFILE: An unstable material that explodes when heated above 50°C. When heated to decomposition it emits toxic fumes of Br^- and NO_x.

BOG255 CAS:106-38-7 HR: 2
p-BROMOTOLUENE
mf: C_7H_7Br mw: 171.05

SYNS: PARABROMOTOLUENE □ TOLUENE, p-BROMO-

TOXICITY DATA WITH REFERENCE
ipr-mus LD50:1741 mg/kg GTPZAB 20(12),52,76
ihl-uns LC50:1300 mg/m³ GTPZAB 18(4),55,74

CONSENSUS REPORTS: Reported in EPA TSCA Inventory.

SAFETY PROFILE: Moderately toxic by inhalation and intraperitoneal routes. When heated to decomposition it emits toxic vapors of Br^-.

BOG260 CAS:95-46-5 HR: 2
2-BROMOTOLUENE
mf: C_7H_7Br mw: 171.05

SYNS: BENZENE, 1-BROMO-2-METHYL-(9CI) □ 1-BROMO-2-METHYLBENZENE □ o-BROMOTOLUENE □ 2-METHYLBROMOBENZENE □ o-METHYLPHENYL BROMIDE □ TOLUENE, o-BROMO- □ o-TOLYL BROMIDE □ 2-TOLYL BROMIDE

TOXICITY DATA WITH REFERENCE
orl-mus LD50:1864 mg/kg GISAAA 44(12),19,79
ipr-mus LD50:1358 mg/kg GISAAA 44(12),19,79

CONSENSUS REPORTS: Reported in EPA TSCA Inventory.

SAFETY PROFILE: Moderately toxic by ingestion and intraperitoneal route. When heated to decomposition it emits toxic vapors of Br^-.

BOG300 CAS:591-17-3 HR: 2
3-BROMOTOLUENE
mf: C_7H_7Br mw: 171.05

SYNS: BENZENE, 1-BROMO-3-METHYL □ m-BROMOTOLUENE □ 5-BROMOTOLUENE □ m-METHYLBROMOBENZENE □ 3-METHYLBROMOBENZENE □ TOLUENE, m-BROMO- □ m-TOLYL BROMIDE

TOXICITY DATA WITH REFERENCE
orl-mus LD50:1436 mg/kg GISAAA 44(12),19,79
ipr-mus LD50:1215 mg/kg GISAAA 44(12),19,79

CONSENSUS REPORTS: Reported in EPA TSCA Inventory.

SAFETY PROFILE: Moderately toxic by ingestion and intraperitoneal routes. When heated to decomposition it emits toxic vapors of Br^-.

BOH750 CAS:75-62-7 HR: 3
BROMOTRICHLOROMETHANE
mf: $CBrCl_3$ mw: 198.27

PROP: Colorless liquid. Fp −5.8°, bp: 104.2, d: 2.01 @ 20°/4°.

TOXICITY DATA WITH REFERENCE
dnd-mam:lym 1 mmol/L TOLED5 11,243,82
orl-rat LDLo:100 mg/kg IJMDAI 10,301,74
ipr-rat LD50:119 mg/kg FAATDF 2,161,82

CONSENSUS REPORTS: Reported in EPA TSCA Inventory.

SAFETY PROFILE: Poison by ingestion and intraperitoneal routes. Narcotic in high concentration. Mutation data reported. See also CHLOROFORM. Incompatible with ethylene. When heated to decomposition it emits very toxic fumes of Cl^- and Br^-.

BOI000 CAS:13749-37-6 HR: 2
3-BROMO-1,1,1-TRICHLORO PROPANE
mf: $C_3H_4BrCl_3$ mw: 224.31

SAFETY PROFILE: A preparative hazard. When heated to decomposition it emits toxic fumes of Cl^- and Br^-. See also BROMIDES; and CHLORINATED HYDROCARBONS, ALIPHATIC.

BOI250 CAS:63041-00-9 HR: 2
3-BROMOTRICYCLOQUINAZOLINE
mf: $C_{21}H_{11}BrN_4$ mw: 399.27

TOXICITY DATA WITH REFERENCE
skn-mus TDLo:1240 mg/kg/1Y-I:NEO BJCAAI 16,275,62

SAFETY PROFILE: Questionable carcinogen with experimental neoplastigenic data. When heated to decomposition it emits very toxic fumes of Br^- and NO_x.

BOI500 CAS:765-09-3 HR: 3
1-BROMOTRIDECANE
mf: $C_{13}H_{27}Br$ mw: 263.31

SYN: TRIDECANE, 1-BROMO-

TOXICITY DATA WITH REFERENCE
ivn-mus LD50:180 mg/kg CSLNX* NX#03504

CONSENSUS REPORTS: Reported in EPA TSCA Inventory.

SAFETY PROFILE: Poison by intravenous route. When heated to decomposition it emits toxic vapors of Br^-.

BOI750 CAS:2767-54-6 HR: 3
BROMOTRIETHYLSTANNANE
mf: $C_6H_{15}BrSn$ mw: 285.81

PROP: Colorless liquid. D: 1.630, mp: −13.5°, bp: 221.° Sol in organic solvents.

SYNS: TRIETHYLSTANNIUM BROMIDE □ TRIETHYL TIN BROMIDE

TOXICITY DATA WITH REFERENCE
ipr-rat TDLo:6 mg/kg (1D post):REP TXAPA9 72,557,84
ihl-mus LCLo:1640 mg/m³ NDRC** NDCrc-132,Feb,42

CONSENSUS REPORTS: Reported in EPA TSCA Inventory.

OSHA PEL: TWA 0.1 mg(Sn)/m³ (skin)
ACGIH TLV: TWA 0.1 mg(Sn)/m³ (skin) (Proposed: TWA 0.1 mg(Sn)/m³; STEL 0.2 mg(Sn)/m³ (skin))
NIOSH REL: (Organotin Compounds) TWA 0.1 mg(Sn)/m³

SAFETY PROFILE: Moderately toxic by inhalation. Experimental reproductive effects. See also TIN COM-

POUNDS and BROMIDES. When heated to decomposition it emits toxic fumes of Br^-.

For occupational chemical analysis use NIOSH: Organotin Compounds 5504.

BOJ000 CAS:598-73-2 HR: 3
BROMO TRIFLUOROETHYLENE
mf: BrF_3C_2 mw: 160.94

PROP: Gas. Bp: −2.5°.

SYNS: BROMOTRIFLUOROETHENE □ TRIFLUOROBROMOETHYLENE □ TRIFLUOROVINYLBROMIDE

CONSENSUS REPORTS: Reported in EPA TSCA Inventory.

SAFETY PROFILE: A poison. A flammable gas. Ignites spontaneously in air. Incompatible with powerful oxidizers, O_2. When heated to decomposition it emits highly toxic fumes of Br^-, F^-, and $COCF_2$.

BOJ500 CAS:401-78-5 HR: 1
m-BROMO-α,α,α-TRIFLUOROTOLUENE
mf: $C_7H_4BrF_3$ mw: 225.02

PROP: Oil. Bp: 44–48° @ 10 mm.

SYNS: 3-BROMBENZOTRIFLUORID (CZECH) □ m-BROMOBENZOTRIFLUORIDE □ 3-BROMOBENZOTRIFLUORIDE □ 3-BROMOBENZYLTRIFLUORIDE □ m-BROMO(TRIFLUOROMETHYL)BENZENE □ 3-BROMOTRIFLUOROMETHYLBENZENE □ m-(TRIFLUOROMETHYL) BROMOBENZENE □ 3-(TRIFLUOROMETHYL)BROMOBENZENE □ m-(TRIFLUOROMETHYL PHENYL BROMIDE □ 3-(TRIFLUOROMETHYL) PHENYL BROMIDE

TOXICITY DATA WITH REFERENCE
skn-rbt 500 mg/24H MLD 28ZPAK -,32,72
eye-rbt 500 mg/24H MOD 28ZPAK -,32,72

CONSENSUS REPORTS: Reported in EPA TSCA Inventory.

SAFETY PROFILE: A skin and eye irritant. See also FLUORIDES and BROMIDES. When heated to decomposition it emits very toxic fumes of Br^- and F^-.

BOJ750 CAS:392-83-6 HR: 2
o-BROMO-α,α,α-TRIFLUOROTOLUENE
mf: $C_7H_4BrF_3$ mw: 225.02

PROP: Oil. Bp: 167–168°.

SYNS: 2-BROMBENZOTRIFLUORID (CZECH) □ o-BROMOBENZOTRIFLUORIDE □ 2-BROMOBENZOTRIFLUORIDE □ o-BROMOBENZYLTRIFLUORIDE □ o-(TRIFLUOROMETHYL)BROMOBENZENE

TOXICITY DATA WITH REFERENCE
skn-rbt 500 mg/24H MOD 28ZPAK -,32,72
eye-rbt 500 mg/24H MOD 28ZPAK -,32,72
orl-rat LD50:2720 mg/kg 28ZPAK -,32,72

CONSENSUS REPORTS: Reported in EPA TSCA Inventory.

SAFETY PROFILE: Moderately toxic by ingestion. A skin and eye irritant. See also FLUORIDES and BROMIDES. When heated to decomposition it emits very toxic fumes of Br^- and F^-.

BOK250 CAS:3091-18-7 HR: 3
BROMOTRIPENTYLSTANNANE
mf: $C_{15}H_{33}BrSn$ mw: 412.08

SYN: TRI-N-PENTYLTIN BROMIDE

TOXICITY DATA WITH REFERENCE
ivn-mus LD50:56 mg/kg CSLNX* NX#05775

OSHA PEL: TWA 0.1 mg(Sn)/m³ (skin)
ACGIH TLV: TWA 0.1 mg(Sn)/m³ (skin) (Proposed: TWA 0.1 mg(Sn)/m³; STEL 0.2 mg(Sn)/m³ (skin))
NIOSH REL: (Organotin Compounds) TWA 0.1 mg(Sn)/m³

SAFETY PROFILE: Poison by intravenous route. See also BROMIDES and TIN COMPOUNDS. When heated to decomposition it emits toxic fumes of Br^-.

For occupational chemical analysis use NIOSH: Organotin Compounds 5504.

BOK750 CAS:2767-61-5 HR: 3
BROMOTRIPROPYLSTANNANE
mf: $C_9H_{21}BrSn$ mw: 327.90

PROP: Liquid. D: 1.426 @ 25°/4°, mp: −49°, bp: 133°.

SYN: TRI-N-PROPYLTIN BROMIDE

TOXICITY DATA WITH REFERENCE
ihl-mus LCLo:1650 mg/m³ NDRC** NDCrc-132,FEB,42
ivn-mus LD50:3600 μg/kg CSLNX* NX#02334

OSHA PEL: TWA 0.1 mg(Sn)/m³ (skin)
ACGIH TLV: TWA 0.1 mg(Sn)/m³ (skin) (Proposed: TWA 0.1 mg(Sn)/m³; STEL 0.2 mg(Sn)/m³ (skin))
NIOSH REL: (Organotin Compounds) TWA 0.1 mg(Sn)/m³

SAFETY PROFILE: Poison by intravenous route. Moderately toxic by inhalation. When heated to decomposition it emits toxic fumes of Br^-. See also BROMIDES and TIN COMPOUNDS.

For occupational chemical analysis use NIOSH: Organotin Compounds 5504.

BOL000 CAS:51-20-7 HR: 2
5-BROMOURACIL
mf: $C_4H_3BrN_2O_2$ mw: 191.00

PROP: Prisms from H_2O. Mp: 293°.

TOXICITY DATA WITH REFERENCE
mmo-esc 5000 ppm AGACBH 4,286,74
cyt-grh-ipr 10 mg IDZAAW 38,305,83
ipr-mus TDLo:400 mg/kg (female 12D post):REP TCMUD8 7,7,87
ipr-rat LD50:1700 mg/kg PSEBAA 93,124,56
ipr-mus LD50:1400 mg/kg PSEBAA 93,124,56

CONSENSUS REPORTS: Reported in EPA TSCA Inventory. EPA Genetic Toxicology Program.

SAFETY PROFILE: Moderately toxic by intraperitoneal

route. Experimental reproductive effects. Mutation data reported. When heated to decomposition it emits very toxic fumes of Br^- and NO_x.

BOL250 **CAS:584-93-0** **HR: 3**
α-BROMOVALERIC ACID
mf: $C_5H_9BrO_2$ mw: 181.05

TOXICITY DATA WITH REFERENCE
orl-mus LD50:380 mg/kg JPETAB 86,336,46

CONSENSUS REPORTS: Reported in EPA TSCA Inventory.

SAFETY PROFILE: Poison by ingestion. See also BROMIDES. When heated to decomposition it emits toxic fumes of Br^-.

BOL303 **CAS:2374-05-2** **HR: 2**
4-BROMO-2,6-XYLENOL
mf: C_8H_9BrO mw: 201.08

SYNS: 4-BROMO-2,6-DIMETHYLPHENOL □ 2,6-XYLENOL, 4-BROMO-

TOXICITY DATA WITH REFERENCE
ipr-mus LD50:650 mg/kg JMPCAS 2,201,60

CONSENSUS REPORTS: Reported in EPA TSCA Inventory.

SAFETY PROFILE: Moderately toxic by intraperitoneal route. When heated to decomposition it emits toxic vapors of Br^-.

BOL310 **CAS:22585-64-4** **HR: 3**
BROMYL FLUORIDE
mf: $BrFO_2$ mw: 130.90

PROP: White solid at low temp, liquid at room temperature. Mp: −10°. Sol in BrF_3.

SAFETY PROFILE: Reacts explosively with water. When heated to decomposition it emits toxic fumes of F^- and Br^-.

BOL325 **CAS:23233-88-7** **HR: 3**
BROTIANIDE
mf: $C_{15}H_{10}Br_2ClNO_2S$ mw: 463.59

PROP: Solid. Mp: 181°.

SYNS: BAY 4059 □ BAY-VA 4059 □ 2-(ACETYLOXY)-3-BROMO-N-(4-BROMOPHENYL)-5-CHLORO-BENZENECARBOTHIOAMIDE □ 2-BROMO-6-(N-(p-BROMOPHENYL)THIOCARBAMOYL)-4-CHLORO-BENZOIC ACID □ 3,4′-DIBROMO-5-CHLOROTHIOSALICYLANILIDE ACETATE (ESTER) □ DIRIAN

TOXICITY DATA WITH REFERENCE
orl-rat LD50:3000 mg/kg APFRAD 33,273,75
orl-mus LD50:184 mg/kg APFRAD 33,273,75
orl-rbt LD50:50 mg/kg APFRAD 33,273,75
orl-dom LD50:40 mg/kg FAZMAE 17,108,73

SAFETY PROFILE: Poison by ingestion. When heated to decomposition it emits toxic fumes of Cl^-, Br^-, SO_x, and NO_x.

BOL500 **CAS:41451-75-6** **HR: 3**
BRUCEANTIN
mf: $C_{28}H_{36}O_{11}$ mw: 548.64

PROP: Crystals from Et_2O. Mp: 225–226°. A quassinoid from the *Brucea antidysenterica* plant.

SYN: NSC-165563

TOXICITY DATA WITH REFERENCE
dni-hmn:hla 50 nmol/L FEPRA7 33,581,74
dni-mus:lym 15 μmol/L JPMSAE 68,883,79
oms-mus:lym 15 μmol/L JPMSAE 68,883,79
orl-mus LD50:7027 μg/kg NCISP* JAN86
ipr-mus LD50:2727 μg/kg NCISP* JAN86
scu-mus LD50:3359 μg/kg NCISP* JAN86
ivn-mus LD50:1950 μg/kg TXAPA9 41,192,77
ivn-dog LDLo:500 μg/kg TXAPA9 41,192,77

SAFETY PROFILE: A deadly poison by ingestion, subcutaneous, intravenous, and intraperitoneal routes. Human mutation data reported. When heated to decomposition it emits acrid smoke and irritating fumes.

BOL600 **HR: 3**
BRUCELLA MELITENSIS ENDOTOXIN

SYN: ENDOTOXIN, BRUCELLA MELITENSIS

TOXICITY DATA WITH REFERENCE
ipr-mus LD50:18 mg/kg CUMIDD 1,263,78
ivn-mus LDLo:125 mg/kg PSEBAA 112,463,63

SAFETY PROFILE: A poison by intraperitoneal and intravenous routes. When heated to decomposition it emits acrid smoke and irritating vapors.

BOL750 **CAS:357-57-3** **HR: 3**
BRUCINE
DOT: UN 1570
mf: $C_{23}H_{26}N_2O_4$ mw: 394.51

PROP: Crystals, powder, or monoclinic prisms. Mp: 105° (hydrate), mp: 78° (anhydrate). Sol in EtOH, and $CHCl_3$; sltly sol in C_6H_6, and Et_2O. An alkaloid extracted from *Strychnos* seeds (WQCHM* 4,-,74).

SYNS: BRUCINA (ITALIAN) □ (−)-BRUCINE □ BRUCINE (DOT) □ 2,3-DIMETHOXYSTRYCHNIDIN-10-ONE □ DIMETHOXY STRYCHNINE (DOT) □ 2,3-DIMETHOXYSTRYCHNINE □ 10,11-DIMETHYSTRYCHNINE □ STRYCHNIDIN-10-ONE, 2,3-DIMETHOXY-(9CI) □ STRYCHNINE, 2,3-DIMETHOXY- □ RCRA WASTE NUMBER P018

TOXICITY DATA WITH REFERENCE
ipr-rat LD50:91 mg/kg JPETAB 131,185,61
scu-mus LD50:60 mg/kg APSXAS 7,329,70
ivn-dog LDLo:8 mg/kg HBAMAK 4,1289,35
ivn-rbt LDLo:30 mg/kg NTIS** PB214-270
ivn-gpg LDLo:120 mg/kg NTIS** PB214-270
scu-pgn LDLo:58 mg/kg HBAMAK 4,1289,35

CONSENSUS REPORTS: Reported in EPA TSCA Inventory.

DOT CLASSIFICATION: 6.1; *Label:* Poison

SAFETY PROFILE: A poison by subcutaneous, intravenous, and intraperitoneal routes. An alkaloid-like strychnine, but one-sixth as toxic. When heated to decomposition it emits toxic fumes of NO_x. See also STRYCHNINE.

BOM000 CAS:60723-51-5 HR: 3
BRUCINE METHIODIDE
mf: $C_{23}H_{26}N_2O_4 \cdot CH_3I$ mw: 536.45

SYNS: BRUCINE IODOMETHYLATE □ BRUCINE IODOMETHYLE (FRENCH)

TOXICITY DATA WITH REFERENCE
ivn-mus LDLo:10 mg/kg CRSBAW 144,53,50
ivn-rbt LDLo:30 mg/kg CRSBAW 144,53,50
ivn-gpg LDLo:120 mg/kg CRSBAW 144,53,50

SAFETY PROFILE: A poison via intravenous route. See also BRUCINE. When heated to decomposition it emits very toxic fumes of NO_x and I^-.

BOM125 HR: 3
BUCKTHORN

PROP: A shrub that grows to 6 feet with small elliptical leaves 1 to 2 inches long. It produces a berry that turns black when mature and has a pit. It grows wild in western Texas and New Mexico.

SYNS: COYOTILLO □ KARWINSKIA HUMBOLDTIANA □ TULLIDORA

SAFETY PROFILE: The berry contains poisonous anthracenones. Ingestion may result (over a period of weeks or months) in loss of function in the peripheral nervous system including respiratory paralysis and death.

BOM250 CAS:129-74-8 HR: 3
BUCLIZINE DIHYDROCHLORIDE
mf: $C_{28}H_{33}ClN_2 \cdot 2ClH$ mw: 506.00

PROP: Crystals. Mp: 265–266°.

SYNS: BUCLODIN □ 1-(p-tert-BUTYLBENZYL)-4-(p-CHLORODIPHENYLMETHYL)PIPERAZINE DIHYDROCHLORIDE □ 1-(p-tert-BUTYLBENZYL-4-p-CHLORO-α-PHENYLBENZYL)PIPERAZINE DIHYDROCHLORIDE □ 1-(p-CHLOROBENZHYDRYL)-4-(p-tert-BUTYLBENZYL) DIETHYLENEDIAMINE DIHYDROCHLORIDE □ 1-p-CHLOROBENZHYDRYL-4-p-(tert)-BUTYLBENZYLPIPERAZINE DIHYDROCHLORIDE □ HISTABUTYZINE DIHYDROCHLORIDE □ LONGIFENE □ SOFTRAN □ UCB 4445 □ VIBAZINE

TOXICITY DATA WITH REFERENCE
orl-rat TDLo:360 mg/kg (10-15D preg):TER AJOGAH 95,109,66
ipr-mus LD50:430 mg/kg JAPMA8 43,653,54

CONSENSUS REPORTS: Reported in EPA TSCA Inventory.

SAFETY PROFILE: Poison by intraperitoneal route. An experimental teratogen. When heated to decomposition it emits very toxic fumes of NO_x and HCl.

BOM510 CAS:36556-75-9 HR: 3
BUCUMOLOL HYDROCHLORIDE
mf: $C_{17}H_{23}NO_4 \cdot ClH$ mw: 341.87

SYNS: dl-BUCUMOLOL HYDROCHLORIDE □ 8-(3-tert-BUTYLAMINO-2-HYDROXY)PROPOXY-5-METHYLCOUMARIN HYDROCHLORIDE □ CS 359

TOXICITY DATA WITH REFERENCE
orl-rat LD50:1259 mg/kg IYKEDH 13,349,82
ipr-rat LD50:74,200 μg/kg IYKEDH 13,349,82
scu-rat LD50:302 mg/kg IYKEDH 13,349,82
ivn-rat LD50:32,400 μg/kg IYKEDH 13,349,82
orl-mus LD50:676 mg/kg IYKEDH 13,349,82
ipr-mus LD50:59,200 μg/kg IYKEDH 13,349,82
scu-mus LD50:82,300 μg/kg IYKEDH 13,349,82
ivn-mus LD50:31,600 μg/kg JJPAAZ 23,497,73

SAFETY PROFILE: Poison by subcutaneous, intravenous, and intraperitoneal routes. Moderately toxic by ingestion. When heated to decomposition it emits toxic fumes of NO_x and HCl.

BOM520 CAS:51333-22-3 HR: 3
BUDESONIDE
mf: $C_{25}H_{34}O_6$ mw: 430.59

PROP: Crystals. Mp: 221–232° (decomp). It is a mixture of two isomers; the content of the S-isomer in the mixture varies between 40–51%.

SYNS: (11-β,16-α)-16,17-(BUTYLIDENEBIS(OXY))-11,21-DIHYDROXYPREGNA-1,4-DIENE-3,20-DIONE □ 16-α,17-α-BUTYLIDENEDIOXY-11-β,21-DIHYDROXY-1,4-PREGNADIENE-3,20-DIONE □ PREFERID □ PULMICORT □ RHINOCORT

TOXICITY DATA WITH REFERENCE
scu-rat TDLo:44 μg/kg (female 7-17D post):REP KSRNAM 19,5093,85
scu-rat TDLo:540 μg/kg (female 17-22D post):REP KSRNAM 19,5093,85
scu-rat TDLo:220 μg/kg (female 7-17D post):TER KSRNAM 19,5093,85
scu-rbt TDLo:1625 μg/kg (female 6-18D post):REP ARZNAD 37,43,87
scu-rat TDLo:21,600 ng/kg (female 17-22D post):REP KSRNAM 19,5093,85
ipr-rat LD50:138 mg/kg KSRNAM 19,4377,85
scu-rat LD50:58,400 μg/kg KSRNAM 19,4377,85
ivn-rat LD50:98,900 μg/kg KSRNAM 19,4377,85
orl-mus LD50:4750 mg/kg KSRNAM 19,4377,85
ipr-mus LD50:179 mg/kg KSRNAM 19,4377,85
scu-mus LD50:53,600 μg/kg KSRNAM 19,4377,85
ivn-mus LD50:124 mg/kg KSRNAM 19,4377,85
scu-dog LD50:173 mg/kg KSRNAM 19,4377,85

SAFETY PROFILE: Poison by subcutaneous, intravenous, and intraperitoneal routes. Moderately toxic by ingestion. An experimental teratogen. Other experimental reproductive effects. When heated to decomposition it emits acrid smoke and fumes.

B

BOM530 **CAS:57982-78-2** **HR: 3**
BUDIPINE
mf: $C_{21}H_{27}N$ mw: 293.49

PROP: Mp: 108–109°, bp: 160–165° @ 0.05 mm.

SYNS: BUDIPIN (GERMAN) □ 1-(1,1-DIMETHYLETHYL)-4,4-DIPHENYLPIPERIDINE

TOXICITY DATA WITH REFERENCE
orl-rat LD50:165 mg/kg ARZNAD 32,85,82
ivn-rat LD50:28 mg/kg ARZNAD 32,85,82
orl-mus LD50:120 mg/kg ARZNAD 32,85,82
ivn-mus LD50:33 mg/kg ARZNAD 32,85,82

SAFETY PROFILE: Poison by ingestion and intravenous routes. When heated to decomposition it emits toxic fumes of NO_x.

BOM600 **CAS:35543-24-9** **HR: 3**
BUFLOMEDIL HYDROCHLORIDE
mf: $C_{17}H_{25}NO_4$•ClH mw: 343.89

PROP: Solid. Mp: 192–193°.

SYNS: A-48257 □ BUFEDIL □ BUFLOMEDIL □ CHLORHYDRATE de (TRIMETHOXY-2-4-6) PHENYL-(PYRROLIDINE-3) PROPYLACETONE (FRENCH) □ FONZYLANE □ LL 1656 □ LOFTYL □ 4-(1-PYRROLIDINYL)-1-(2,4,6-TRIMETHOXYPHENYL)-1-BUTANONE HYDROCHLORIDE

TOXICITY DATA WITH REFERENCE
orl-rat TDLo:1500 mg/kg (female 17-22D post):REP KSRNAM 22,473,88
orl-wmn TDLo:50 mg/kg:BAH,SYS JTCTDW 30,305,92
orl-rat LD50:410 mg/kg:BAH,PUL KSRNAM 22,401,88
scu-rat LD50:796 mg/kg KSRNAM 22,401,88
ivn-rat LD50:58,500 μg/kg KSRNAM 22,401,88
orl-mus LD50:275 mg/kg THERAP 30,207,75
ivn-mus LD50:55 mg/kg THERAP 30,207,75
ims-mus LD50:250 mg/kg THERAP 30,207,75
orl-dog LDLo:500 mg/kg THERAP 30,207,75
ivn-dog LDLo:50 mg/kg THERAP 30,207,75

SAFETY PROFILE: Poison by ingestion, intramuscular, and intravenous routes. Experimental reproductive effects. Human systemic effects by ingestion: ataxia, coma, convulsions, metabolic acidosis, respiratory depression, somnolence. A vasodilator. When heated to decomposition it emits toxic fumes of NO_x and HCl.

BOM650 **CAS:465-39-4** **HR: 3**
BUFOGENIN
mf: $C_{24}H_{32}O_4$ mw: 384.56

PROP: Crystals from Me_2CO/hexane. Mp: 108–120°.

SYNS: 14,15-β-EPOXY-3-β-HYDROXY-5-β-BUFA-20,22-DIENOLIDE □ 3-β-HYDROXY-14,15-β-EPOXY-5-β-BUFA-20,22-DIENOLIDE □ RESIBUFOGENIN

TOXICITY DATA WITH REFERENCE
ivn-rat LD50:2200 μg/kg NIIRDN 6,899,82
ivn-mus LD50:4250 μg/kg NIIRDN 6,899,82
ivn-cat LD50:5 mg/kg JPETAB 111,365,54

SAFETY PROFILE: Deadly poison by intravenous route. When heated to decomposition it emits acrid smoke and fumes. See also BUFOGENIN B.

BOM655 **CAS:465-19-0** **HR: 3**
BUFOGENIN B
mf: $C_{24}H_{34}O_5$ mw: 402.58

PROP: Crystals from MeOH. Mp: 210–223° (decomp). Elongated prisms from methanol. Begins to sinter at 195°, decomp @ 210–223°. Very sparingly sol in chloroform, methanol, acetone.

SYNS: DESACETYLBUFOTALIN □ 3-β,14,16-β-TRIHYDROXY-5-β-BUFA-20,22-DIENOLIDE

TOXICITY DATA WITH REFERENCE
ivn-rat LDLo:2940 μg/kg OYYAA2 5,973,71
idu-rat LDLo:28,400 μg/kg OYYAA2 5,973,71
orl-mus LD50:24,500 μg/kg OYYAA2 5,973,71
scu-mus LD50:6950 μg/kg OYYAA2 5,973,71
ivn-mus LD50:10 μg/kg CPBTAL 24,1714,76
ivn-dog LDLo:580 μg/kg OYYAA2 5,973,71

SAFETY PROFILE: Deadly poison by ingestion, subcutaneous, intravenous, and intraduodenal routes. When heated to decomposition it emits acrid smoke and fumes. See also BUFOGENIN.

BOM750 **CAS:1190-53-0** **HR: 3**
BUFORMIN HYDROCHLORIDE
mf: $C_6H_{15}N_5$•ClH mw: 193.72

PROP: Solid. Mp: 174–177°. Sol in H_2O and EtOH.

SYNS: BUFONAMIN □ DIABRIN □ INSULAMIN

TOXICITY DATA WITH REFERENCE
cyt-hmn:emb 1900 μg/L SNSHBT (20),574,80
orl-rat LD50:320 mg/kg ARZNAD 12,314,62
orl-mus LD50:380 mg/kg ARZNAD 12,314,62
ipr-mus LD50:148 mg/kg PLRCAT 6,117,74

SAFETY PROFILE: A poison by intraperitoneal and ingestion routes. Mutation data reported. When heated to decomposition it emits very toxic fumes of HCl and NO_x.

BON000 **CAS:471-95-4** **HR: 3**
BUFOTALINE
mf: $C_{26}H_{36}O_6$ mw: 444.62

PROP: Crystals. Mp: 223° (decomp).

SYNS: BUFOTALIN □ 3-β,14,16-β-TRIHYDROXY-5-β-BUFA-20,22-DIENOLIDE-16-ACETATE

TOXICITY DATA WITH REFERENCE
scu-mus LD50:400 μg/kg CTOXAO 4,331,71
orl-dog LDLo:980 μg/kg CRSBAW 152,571,58
ivn-mus LD50:4130 μg/kg CPBTAL 24,1714,76
ivn-dog LDLo:360 μg/kg CRSBAW 152,571,58
ivn-cat LD50:130 μg/kg 85ELDJ -,189,63

SAFETY PROFILE: A deadly poison by ingestion, subcutaneous, and intravenous routes. When heated to decomposition it emits acrid and irritating fumes.

BON250 **HR: 2**
BULAN and PROLAN MIXTURE (2:1)

SYNS: 1,1-BIS(p-CHLOROPHENYL)-2-NITROPROPANE mixed with 1,

1-BIS(p-CHLOROPHENYL)-2-NITROBUTANE(1:2) ☐ CS 708 ☐ DILAN ☐ ENT 18,066

TOXICITY DATA WITH REFERENCE
orl-rat LD50:475 mg/kg FMCHA2 -,D103,80
skn-rat LD50:5900 mg/kg CMEP** -,1,56
orl-mus LD50:1100 mg/kg FEPRA7 12,368,53
ipr-mus LD50:950 mg/kg FEPRA7 12,368,53
orl-mam LD50:1100 mg/kg PCOC** -,929,66

SAFETY PROFILE: Moderately toxic by ingestion and intraperitoneal routes. Mildly toxic by skin contact. When heated to decomposition it emits very toxic fumes of Cl^- and NO_x. See also individual components.

BON300 **HR: 3**
BULKOSOL

TOXICITY DATA WITH REFERENCE
orl-rat LD50:3740 mg/kg NIIRDN 6,205,82
ivn-rat LD50:101 mg/kg NIIRDN 6,205,82
orl-mus LD50:2450 mg/kg NIIRDN 6,205,82
scu-mus LD50:4300 mg/kg NIIRDN 6,205,82
ivn-mus LD50:76,800 μg/kg NIIRDN 6,205,82

SAFETY PROFILE: Poison by intravenous route. Moderately toxic by ingestion.

BON325 **CAS:28395-03-1** **HR: 3**
BUMETANIDE
mf: $C_{17}H_{20}N_2O_5S$ mw: 364.45

PROP: Crystals from aq ethanol. Mp: 230–231°.

SYNS: 3-(AMINOSULFONYL)-5-(BUTYLAMINO)-4-PHENOXY-3-(AMINOSULFONYL)-5-(BUTYLAMINO)-4-PHENOXYBENZOIC ACID ☐ BUMEX ☐ BURINE ☐ BURINEX ☐ 3-(BUTYLAMINO)-4-PHENOXY-5-SULFAMOYLBENZOIC ACID ☐ FONTEGO ☐ FORDIURAN ☐ LIXIL ☐ LUNETORON ☐ PF 1593 ☐ RO 10-6338 ☐ SEGUREX

TOXICITY DATA WITH REFERENCE
ipr-rat LD50:1000 mg/kg ARZNAD 1,218,51
scu-rat LD50:22,500 μg/kg OYYAA2 9,413,75
ivn-rat LD50:4 mg/kg ARZNAD 1,218,51
orl-mus LD50:156 mg/kg OYYAA2 9,413,75
scu-mus LD50:140 mg/kg ARZNAD 1,218,51
ivn-mus LD50:4900 μg/kg OYYAA2 9,413,75
ivn-rbt LD50:2400 μg/kg OYYAA2 9,413,75

SAFETY PROFILE: Poison by ingestion, subcutaneous, and intravenous routes. Moderately toxic by intraperitoneal route. When heated to decomposition it emits toxic fumes of SO_x and NO_x.

BON350 **HR: 3**
BUNAZOCINE HYDROCHLORIDE
mf: $C_{19}H_{27}N_5O_3•ClH$ mw: 409.91

SYNS: 4-AMINO-2-(4-BUTYRYLHEXAHYDRO-1H-1,4-DIAZEPIN-1-YL)-6,7-DIMETHOXYQUINAZOLINE HYDROCHLORIDE ☐ BUNAZOSIN HYDROCHLORIDE

TOXICITY DATA WITH REFERENCE
orl-rat LD50:1280 mg/kg IYKEDH 16,866,85
scu-rat LD50:365 mg/kg IYKEDH 16,866,85
ivn-rat LD50:50 mg/kg IYKEDH 16,866,85
ims-rat LD50:152 mg/kg IYKEDH 16,866,85
orl-mus LD50:1201 mg/kg IYKEDH 16,866,85
scu-mus LD50:730 mg/kg IYKEDH 16,866,85
ivn-mus LD50:57 mg/kg IYKEDH 16,866,85
ims-mus LD50:660 mg/kg IYKEDH 16,866,85

SAFETY PROFILE: Poison by subcutaneous, intramuscular, intravenous, and intraperitoneal routes. Moderately toxic by ingestion. When heated to decomposition it emits toxic fumes of NO_x and HCl.

BON365 **HR: 3**
BUNGARUS CAERULEUS VENOM

SYN: VENOM, SNAKE, BUNGARUS CAERULEUS

TOXICITY DATA WITH REFERENCE
ipr-mus LD50:8 μg/kg TOXIA6 14,451,76
scu-mus LD50:450 μg/kg TOXIA6 5,47,67
ivn-mus LD50:96 μg/kg IJMRAQ 60,512,72
ivn-dog LDLo:120 μg/kg 19DDA6 1,269,67
ivn-rbt LDLo:40 μg/kg TOXIA6 2,5,64
ivn-mam LD50:90 μg/kg CLPTAT 8,849,67

SAFETY PROFILE: Deadly poison by subcutaneous, intravenous, and intraperitoneal routes.

BON367 **HR: 3**
BUNGARUS FASCIATUS VENOM

SYN: VENOM, SNAKE, BUNGARUS FASCIATUS

TOXICITY DATA WITH REFERENCE
ipr-mus LD50:150 μg/kg 85EGD4 5,161,78
scu-mus LD50:3580 μg/kg TOXIA6 5,47,67
ivn-mus LD50:170 μg/kg TOXIA6 21,681,83

SAFETY PROFILE: Deadly poison by subcutaneous, intravenous, and intraperitoneal routes.

BON370 **HR: 3**
BUNGARUS MULTICINCTUS VENOM

SYN: VENOM, FORMOSAN BANDED KRAIT, BUNGARUS MULTICINCTUS

TOXICITY DATA WITH REFERENCE
ipr-mus LD50:25 μg/kg JOBIAO 48,714,60
scu-mus LD50:160 μg/kg TIHHAH 61,239,62
ivn-mus LD50:71 μg/kg TOXIA6 9,131,71
ivn-rbt LDLo:1 mg/kg TOXIA6 3,281,66

SAFETY PROFILE: Deadly poison by subcutaneous, intravenous, and intraperitoneal routes.

BON400 **CAS:23093-74-5** **HR: 3**
BUNITROLOL HYDROCHLORIDE
mf: $C_{14}H_{20}N_2O_2•ClH$ mw: 284.82

PROP: Solid. Mp: 163–165°.

SYNS: BETRILOL ☐ o-(3-tert-BUTYLAMINO-2-HYDROXYPROPOXY)BENZONITRILE HYDROCHLORIDE ☐ 2-(3-((1,1-DIMETHYLETHYL)AMINO)-2-HYDROXYPROPOXY)-BENZONITRILE HYDROCHLORIDE ☐ o-(2-HYDROXY-3-(tert-BUTYLAMINO)PROPOXY)BENZONITRILE HYDROCHLORIDE ☐ KO 1366-CL ☐ KOE 1366 CHLORIDE ☐ STRESSON

TOXICITY DATA WITH REFERENCE
orl-rat TDLo:1100 mg/kg (female 7-17D post):REP IYKEDH 12,12,81
orl-rat TDLo:1350 mg/kg (17-22D preg/21D post):REP IYKEDH 12,976,81
orl-mus TDLo:840 mg/kg (female 7-13D post):REP OYYAA2 11,795,76
ipr-rat TDLo:70 mg/kg (female 7-13D post):TER OYYAA2 11,779,76
orl-mus TDLo:210 mg/kg (female 7-13D post):REP OYYAA2 11,795,76
orl-rat TDLo:140 mg/kg (female 28D pre):REP OYYAA2 9,465,75
orl-rat TDLo:18 g/kg (male 60D pre):REP IYKEDH 12,976,81
orl-rbt TDLo:325 mg/kg (female 6-18D post):TER IYKEDH 12,12,81
orl-rat TDLo:3300 mg/kg (7-17D preg):TER IYKEDH 12,12,81
orl-rat LD50:639 mg/kg OYYAA2 11,795,76
ipr-rat LD50:222 mg/kg IYKEDH 14,484,83
scu-rat LD50:902 mg/kg IYKEDH 14,484,83
ivn-rat LD50:69 mg/kg IYKEDH 14,484,83
orl-mus LD50:250 mg/kg IYKEDH 12,25,81
ipr-mus LD50:264 mg/kg IYKEDH 14,484,83
scu-mus LD50:542 mg/kg IYKEDH 14,484,83
ivn-mus LD50:264 mg/kg OYYAA2 9,457,75
orl-dog LD50:490 mg/kg IYKEDH 14,484,83
ivn-dog LD50:36 mg/kg IYKEDH 14,484,83

CONSENSUS REPORTS: Cyanide and its compounds are on the Community Right-To-Know List.

SAFETY PROFILE: Poison by ingestion, intravenous, and intraperitoneal routes. Moderately toxic by subcutaneous route. Experimental reproductive effects. An experimental teratogen. When heated to decomposition it emits toxic fumes of NO_x, CN^- and HCl. See also NITRILES.

BON750 CAS:27262-46-0 HR: 3
BUPICAINE HYDROCHLORIDE (+)
mf: $C_{18}H_{28}N_2O{\cdot}ClH$ mw: 324.94

SYN: 1-BUTYL-2′,6′-PIPECOLOXYLIDIDE HYDROCHLORIDE (+)

TOXICITY DATA WITH REFERENCE
scu-rat LD50:43 mg/kg AIPTAK 200,359,72
ivn-rat LD50:6 mg/kg AIPTAK 200,359,72
scu-mus LD50:58 mg/kg AIPTAK 200,359,72
ivn-mus LD50:7200 μg/kg AIPTAK 200,359,72
orl-rbt LD50:18 mg/kg AIPTAK 200,359,72
ivn-rbt LD50:3300 μg/kg AIPTAK 200,359,72
par-rbt LD50:185 mg/kg AIPTAK 200,359,72
itr-rbt LD50:12 mg/kg AIPTAK 200,359,72

SAFETY PROFILE: Poison by ingestion, subcutaneous, intravenous, parenteral, and intratracheal routes. When heated to decomposition it emits very toxic fumes of HCl and NO_x. See other bupicaine or bupivacaine entries.

BOO000 CAS:14252-80-3 HR: 3
BUPICAINE HYDROCHLORIDE (±)
mf: $C_{18}H_{28}N_2O{\cdot}ClH$ mw: 324.94

PROP: Crystals. Mp: 255–256°.

SYNS: BUPIVACAINE HYDROCHLORIDE □ 1-BUTYL-2′,6′-PIPECOLOXYLIDIDE (±) □ (±)-1-BUTYL-2′,6′-PIPECOLOXYLIDIDE MONOHYDROCHLORIDE, MONOHYDRATE □ CARBOSTESIN □ LAC-43 □ MARCAIN □ 2-PIPERIDINECARBOXAMIDE,1-BUTYL-N-(2,6-DIMETHYLPHENYL)MONOHYDROCHLORIDE MONOHYDRATE

TOXICITY DATA WITH REFERENCE
scu-rat LD50:43 mg/kg AIPTAK 200,359,72
ivn-rat LD50:6 mg/kg AIPTAK 200,359,72
scu-mus LD50:59 mg/kg AIPTAK 200,359,72
ivn-mus LD50:6400 μg/kg AIPTAK 200,359,72
orl-rbt LD50:18 mg/kg AIPTAK 200,359,72
ivn-rbt LD50:3400 μg/kg AIPTAK 200,359,72
par-rbt LD50:48 mg/kg AIPTAK 200,359,72
itr-rbt LD50:11 mg/kg AIPTAK 200,359,72
ipr-gpg LD50:50 mg/kg NIIRDN 6,680,82

SAFETY PROFILE: Poison by ingestion, subcutaneous, intravenous, intraperitoneal, parenteral, and intratracheal routes. A local anesthetic. When heated to decomposition it emits very toxic fumes of HCl and NO_x. See other bupicaine or bupivacaine entries.

BOO250 CAS:27262-45-9 HR: 3
d(+)-BUPIVACAINE
mf: $C_{18}H_{28}N_2O$ mw: 288.48

PROP: Crystals.

SYN: d-(+)-1-BUTYL-2′,6′-PIPECOLOXYLIDIDE

TOXICITY DATA WITH REFERENCE
scu-rat LD50:38 mg/kg APTOA6 31,273,72
ivn-rat LD50:3800 μg/kg APTOA6 31,273,72
scu-mus LD50:30 mg/kg APTOA6 31,273,72
ivn-mus LD50:7900 μg/kg APTOA6 31,273,72
ivn-rbt LDLo:5500 μg/kg APTOA6 31,273,72
itr-rbt LD50:10 mg/kg ARZNAD 26,78,76

SAFETY PROFILE: Poison by subcutaneous, intratracheal and intravenous routes. When heated to decomposition it emits toxic fumes of NO_x. See other bupicaine or bupivacaine entries.

BOO500 CAS:27262-47-1 HR: 3
l(−)-BUPIVACAINE
mf: $C_{18}H_{28}N_2O$ mw: 288.48

PROP: Crystals.

SYN: l-(−)-1-BUTYL-2′,6′-PIPECOLOXYLIDIDE

TOXICITY DATA WITH REFERENCE
scu-rat LD50:52 mg/kg APTOA6 31,273,72
ivn-rat LD50:7200 μg/kg APTOA6 31,273,72
scu-mus LD50:100 mg/kg APTOA6 31,273,72
ivn-mus LD50:9600 μg/kg APTOA6 31,273,72
ivn-rbt LDLo:9700 μg/kg APTOA6 31,273,72
itr-rbt LD50:14 mg/kg ARZNAD 26,78,76

SAFETY PROFILE: Poison by subcutaneous, intratracheal, and intravenous routes. When heated to decom-

B

position it emits toxic fumes of NO_x. See other bupicaine or bupivacaine entries.

BOO625 **HR: 2**
BUPLEURUM MARGINATUM WALL. EX. DC., EXTRACT

PROP: Indian plant belonging to the family Apiaceae (IJEBA6 22,312,84).

SYN: BURPLEURUM FALCATUM LINN. VAR. MARGINATUM (WALL. EX. DC.) CL., EXTRACT

TOXICITY DATA WITH REFERENCE
orl-rat TDLo:150 mg/kg (12-14D preg):REP IJEBA6 22,312,84
orl-ham TDLo:500 mg/kg (1-5D preg):REP IJEBA6 22,312,84
ipr-mus LD50:1 g/kg IJEBA6 22,312,84

SAFETY PROFILE: Moderately toxic by intraperitoneal route. Experimental reproductive effects.

BOO630 **CAS:53152-21-9** **HR: 3**
BUPRENORPHINE HYDROCHLORIDE
mf: $C_{29}H_{41}NO_4{\cdot}ClH$ mw: 504.17

SYNS: M-6029 ☐ MR 56

TOXICITY DATA WITH REFERENCE
ims-rat TDLo:1350 µg/kg (17-22D preg/21D post):REP IYKEDH 13,532,82
ims-rat TDLo:550 µg/kg (female 7-17D post):REP IYKEDH 13,509,82
ims-rat TDLo:300 µg/kg (female 17-22D post):REP IYKEDH 13,532,82
ims-rat TDLo:7 mg/kg (male 35D pre):REP IYKEDH 13,486,82
ims-rat TDLo:550 µg/kg (female 7-17D post):TER IYKEDH 13,509,82
orl-man TDLo:2857 µg/kg BMJOAE 296,214,88
ivn-rat LD50:62 mg/kg IYKEDH 13,486,82
orl-mus LD50:800 mg/kg IYKEDH 13,486,82
ivn-mus LD50:72 mg/kg IYKEDH 13,486,82
ivn-dog LD50:79 mg/kg YKYUA6 35,1351,84

SAFETY PROFILE: Poison by intravenous route. Moderately toxic by ingestion. An experimental teratogen. Other experimental reproductive effects. When heated to decomposition it emits toxic fumes of NO_x and HCl.

BOO635 **CAS:21564-17-0** **HR: 3**
BUSAN 72A
mf: $C_9H_6N_2S_3$ mw: 238.35

SYNS: TCMTB ☐ 2-(THIOCYANOMETHYLTHIO)BENZOTHIAZOLE, 60% ☐ THIOCYANIC ACID, 2-(BENZOTHIAZOLYLTHIO)METHYL ESTER

TOXICITY DATA WITH REFERENCE
orl-rat LD50:1590 mg/kg BUCKL* TCMTB,81
ipr-rat LD50:73 mg/kg NNGADV 12,343,87
scu-rat LD50:1300 mg/kg NNGADV 12,343,87
orl-mus LD50:445 mg/kg NNGADV 12,343,87
ipr-mus LD50:143 mg/kg NNGADV 12,343,87
scu-mus LD50:205 mg/kg NNGADV 12,343,87
skn-rbt LD50:10 g/kg BUCKL* TCMTB,81
orl-dck LD50:1310 mg/kg BUCKL* TCMTB,81

SAFETY PROFILE: Poison by intraperitoneal and subcutaneous routes. Moderately toxic by ingestion. When heated to decomposition it emits toxic fumes of SO_x and NO_x. See also THIOCYANATES.

BOO650 **CAS:8059-83-4** **HR: 3**
BUSCOPAN COMPOSITUM
mf: $C_{21}H_{30}NO_4{\cdot}C_{13}H_{17}N_3O_4S{\cdot}Br{\cdot}Na$ mw: 774.81

SYNS: N-BUTYLSCOPOLAMMONIUM BROMIDE combined with SODIUM SULPYRINE (1:25) ☐ SB 502

TOXICITY DATA WITH REFERENCE
ims-rat TDLo:22,500 mg/kg (female 30D pre):REP OYYAA2 9,615,75
ipr-rat TDLo:15 g/kg (30D male):REP OYYAA2 9,615,75
orl-rat LD50:4700 mg/kg KSRNAM 7,54,73
ipr-rat LD50:1400 mg/kg KSRNAM 7,54,73
scu-rat LD50:2850 mg/kg KSRNAM 7,54,73
orl-mus LD50:4350 mg/kg KSRNAM 7,54,73
ipr-mus LD50:2050 mg/kg KSRNAM 7,54,73
scu-mus LD50:2300 mg/kg KSRNAM 7,54,73
ivn-mus LD50:390 mg/kg KSRNAM 7,54,73
ivn-rbt LD50:480 mg/kg KSRNAM 7,54,73

SAFETY PROFILE: Poison by intravenous route. Moderately toxic by subcutaneous and intraperitoneal routes. Mildly toxic by ingestion. Experimental reproductive effects. When heated to decomposition it emits toxic fumes of Br^-, SO_x, NH_3, NO_x, and Na_2O.

BOO700 **HR: 3**
BUSHMAN'S POISON

PROP: An evergreen shrub or small tree native to Africa but also found in California, Florida, and Hawaii. They are used as ornamental shrubs in California and in greenhouses in the rest of the US. The plant has large leaves, fragrant flowers shaped like a flared tube, and a small, plum-like fruit that is red or purple-black when mature.

SYNS: ACOKANTHERA (VARIOUS SPECIES) ☐ A. LONGIFLORA ☐ A. OBLONGIFOLIA ☐ A. OPPOSITIFOLIA ☐ POISON BUSH ☐ POISON TREE ☐ WINTERSWEET

SAFETY PROFILE: The toxic agent is a cardiac glycoside similar to ouabain. It is found in all parts of the plant with the highest concentration in the seeds. The fruit of some species has low levels of toxin and is considered edible. Human systemic effects may include: nausea, vomiting, pain in the mouth and abdomen, cramps, diarrhea, slowed heartbeat, and high blood potassium levels. Symptoms develop after a delay period that is dependent upon the dose. See also OUABAIN.

BOO750 **CAS:149-16-6** **HR: 3**
BUTACAINE
mf: $C_{18}H_{30}N_2O_2$ mw: 306.50

PROP: Colorless, odorless powder. Mp: 98–100°, bp: 178–182° @ 0.11 mm.

B

SYNS: 3-(p-AMINOBENZOXY)-1-DI-n-BUTYLAMINOPROPANE □ p-AMINOBENZOYLDIBUTYLAMINOPROPANOL □ BUTYN □ 3-(DIBUTYLAMINO)-1-PROPANOL-p-AMINOBENZOATE □ 3-DIBUTYLAMINOPROPYL-p-AMINOBENZOATE

TOXICITY DATA WITH REFERENCE
scu-rat LDLo:150 mg/kg JPETAB 24,167,25
ivn-rat LDLo:7500 mg/kg PHREA7 12,190,32
scu-mus LDLo:100 mg/kg JPETAB 24,167,25
ivn-mus LDLo:12 mg/kg JAPMA8 39,4,50
scu-dog LDLo:55 mg/kg PHREA7 12,190,32
scu-cat LDLo:30 mg/kg JPETAB 24,167,25
ivn-cat LDLo:15 mg/kg AJPHAP 68,110,24
scu-rbt LDLo:50 mg/kg JPETAB 24,167,25
ivn-rbt LDLo:12 mg/kg PHREA7 12,190,32
scu-gpg LDLo:45 mg/kg JPETAB 62,69,38

SAFETY PROFILE: A poison via subcutaneous and intravenous routes. A weak allergen. Combustible. When heated to decomposition it emits toxic fumes of NO_x. See also BUTACAINE SULFATE.

BOP000 **CAS:149-15-5** **HR: 3**
BUTACAINE SULFATE
mf: $C_{36}H_{60}N_4O_4 \cdot H_2O_4S$ mw: 711.08

PROP: Solid. Mp: 138.5–139.5°.

SYNS: 3-(p-AMINOBENZOXY)-1-DI-n-BUTYLAMINOPROPANE SULFATE □ p-AMINOBENZOYLDIBUTYLAMINOPROPANOL SULFATE □ BUTELLINE □ BUTYN SULFATE □ 3-(DIBUTYLAMINO)-1-PROPANOL-p-AMINOBENZOATE (ESTER) SULFATE (2:1) □ 3-DIBUTYLAMINO-1-PROPANOL-4-AMINOBENZOATE (ESTER) SULFATE (SALT) (2:1) □ DIBUTYLAMINOPROPYL-p-AMINOBENZOATE SULFATE □ 3'-DIBUTYLAMINOPROPYL-4-AMINOBENZOATE SULFATE

TOXICITY DATA WITH REFERENCE
scu-rat LDLo:197 mg/kg PHREA7 12,262,32
ipr-mus LD50:80 mg/kg BJPCAL 1,90,46
scu-mus LDLo:100 mg/kg JPETAB 24,167,25
ivn-mus LD50:12 mg/kg JAPMA8 40,373,51
orl-bwd LD50:100 mg/kg TXAPA9 21,315,72

SAFETY PROFILE: A poison by ingestion, subcutaneous, intravenous, and intraperitoneal routes. A topical anesthetic. See also SULFATES. When heated to decomposition it emits very toxic fumes of SO_x and NO_2.

BOP100 **CAS:25339-57-5** **HR: 3**
BUTADIENE
DOT: UN 1010
mf: C_4H_6 mw: 54.10

SYNS: BUTADIENES, inhibited (DOT) □ PLIOLITE

DOT CLASSIFICATION: 2.1; *Label:* Flammable Gas

SAFETY PROFILE: A flammable gas. When heated to decomposition it emits acrid smoke and irritating vapors.

BOP250 **CAS:590-19-2** **HR: 2**
1,2-BUTADIENE
mf: C_4H_6 mw: 54.10

$H_2C{=}C{=}CHCH_3$

PROP: A gas. Flash p: <0°, fp: −136.19°, bp: 10.85°.

SAFETY PROFILE: A dangerous fire hazard. When heated to decomposition it emits acrid smoke and fumes. See also 1,3-BUTADIENE.

BOP500 **CAS:106-99-0** **HR: 3**
1,3-BUTADIENE
mf: C_4H_6 mw: 54.10

$H_2C{=}CHCH{=}CH_2$

PROP: Colorless gas; mild aromatic odor. Very reactive. Bp: −2.6°, mp: −113°, fp: −108.9°, flash p: −105°F, lel: 2.0%, uel: 11.5%, d: 0.621 @ 20°/4°, autoign temp: 788°F, vap d: 1.87, vap press: 1840 mm @ 21°.

SYNS: BIETHYLENE □ BIVINYL □ BUTADIEEN (DUTCH) □ BUTA-1,3-DIEEN (DUTCH) □ BUTADIEN (POLISH) □ BUTA-1,3-DIEN (GERMAN) □ BUTA-1,3-DIENE □ α-γ-BUTADIENE □ DIVINYL □ ERYTHRENE □ NCI-C50602 □ PYRROLYLENE □ VINYLETHYLENE

TOXICITY DATA WITH REFERENCE
mnt-mus:ihl 100 ppm/6H/2D-C ENMUDM 8(Suppl 6),18,86
msc-mus:lym 20 pph ENMUDM 8(Suppl 6),75,86
ihl-rat TCLo:8000 ppm/6H (6-15D preg):TER EPASR* 8EHQ-0382-0441
ihl-rat TCLo:625 ppm/6H/61W:CAR NTPTR* NTP-TR-288,84
ihl-mus TCLo:1250 ppm/6H/60W-I:CAR SCIEAS 227,548,85
ihl-rat TC:1000 ppm/6H/2Y-I:CAR AIHAAP 48,407,87
ihl-rat TC:8000 ppm/6H/2Y-I:NEO AIHAAP 48,407,87
ihl-rat TC:8000 ppm/6H/15W-I:CAR EPASR* 8EHQ-0482-0370
ihl-hmn TCLo:2000 ppm/7H:EYE JIHTAB 26,69,44
ihl-hmn TCLo:8000 ppm:EYE,PUL INMEAF 17,199,48
orl-rat LD50:5480 mg/kg 85JCAE -,14,86
ihl-rat LC50:285 g/m³/4H RPTOAN 31,162,68
ihl-mus LC50:270 g/m³/2H RPTOAN 31,162,68
ihl-rbt LCLo:25 pph/23M JIHTAB 26,69,44

CONSENSUS REPORTS: NTP 7th Annual Report On Carcinogens. IARC Cancer Review: Group 2A IMEMDT 54,237,92; Animal Sufficient Evidence IMEMDT 39,155,86; IARC Cancer Review: Animal Sufficent Evidence IMEMDT 54,237,92; Human Limited Evidence IMEMDT 54,237,92; Human Inadequate Evidence IMEMDT 39,155,86; NTP Carcinogenesis Studies (inhalation); Clear Evidence: mouse NTPTR* NTP-TR-288,84. Reported in EPA TSCA Inventory. Community Right-To-Know List.

OSHA PEL: TWA 1000 ppm
ACGIH TLV: TWA 2 ppm; Suspected Human Carcinogen
DFG MAK: Animal Carcinogen, Suspected Human Carcinogen
NIOSH REL: Reduce to lowest feasible level

SAFETY PROFILE: Confirmed carcinogen with experimental carcinogenic and neoplastigenic data. An experimental teratogen. Mutation data reported. Inhalation of high concentrations can cause unconsciousness and death. Human systemic effects by inhalation: cough, hallucinations, distorted perceptions, changes in the visual field and other unspecified eye effects. The vapors are irritating to eyes and mucous membranes. If spilled on skin or clothing, it can cause burns or frostbite (due to rapid vaporization). Chronic systemic poisoning in humans has not been reported.

Dangerous fire hazard when exposed to heat, flame, or powerful oxidizers. Upon exposure to air it forms explosive peroxides sensitive to heat, shock, or heating above 27°C. May decompose explosively when heated above 200°C/1.0 kbar. Explodes on contact with aluminum tetrahydroborate. Potentially explosive reaction with NO_x + O_2, ethanol + iodine + mercury oxide (at 35°C), ClO_2, crotonaldehyde (above 180°C), buten-3-yne (with heat and pressure). Reaction with sodium nitrite forms a spontaneously flammable product. Exothermic reaction with boron trifluoride etherate + phenol. To fight fire, stop flow of gas. When heated to decomposition it emits acrid smoke and fumes.

For occupational chemical analysis use OSHA: #ID-56 or NIOSH: 1,3-Butadiene, 1024.

BOP750 CAS:30031-64-2 HR: 3
l-BUTADIENE DIEPOXIDE
mf: $C_4H_6O_2$ mw: 86.10

PROP: Solid or liquid. Mp: 24–25.6°, bp: 144.5–145°.

SYNS: (S-(R*,R*))-2,2'-BIOXIRANE □ l-DIEPOXYBUTANE □ (2S, 3S)-DIEPOXYBUTANE □ l-1,2:3,4-DIEPOXYBUTANE □ (2S,3S)-1,2:3,4-DIEPOXYBUTANE □ NSC-32606

TOXICITY DATA WITH REFERENCE
dnd-omi 5 mmol/L BBACAQ 228,400,71
mmo-ssp 31 mmol/L ADWMAX -,193,62
ipr-mus TDLo:110 mg/kg/4W:NEO JNCIAM 36,915,66
par-mus TDLo:192 mg/kg/4W-I:NEO NCISA* PH-43-62-483

CONSENSUS REPORTS: IARC Cancer Review: Group 2B IMEMDT 7,56,87; Animal Sufficient Evidence IMEMDT 11,115,76. EPA Genetic Toxicology Program.

SAFETY PROFILE: Suspected carcinogen with experimental neoplastigenic data. Poison by intraperitoneal route. Mutation data reported. When heated to decomposition it emits acrid and irritating fumes.

BOQ250 HR: 3
BUTADIENE PEROXIDE
mf: $C_4H_6O_2$ mw: 86.09

SAFETY PROFILE: A shock-sensitive explosive formed by the peroxidation of butadiene upon prolonged exposure to air. Potentially explosive polymerization reaction with butadiene. Tank monitoring and a purging system are recommended to prevent explosion on contact with air over a long period of time. Concentration in butadiene as measured by standard methods of determining hydroperoxides may be only 5% of the true concentration. When heated to decomposition it emits acrid smoke and fumes. See also PEROXIDES.

BOQ500 CAS:16719-32-7 HR: 3
N-2,3-BUTADIENYL-N-METHYLBENZYLAMINE HYDROCHLORIDE
mf: $C_{12}H_{15}N \cdot ClH$ mw: 209.74

SYN: U-1247

TOXICITY DATA WITH REFERENCE
orl-mus LD50:339 mg/kg JPMSAE 57,430,68
ipr-mus LD50:156 mg/kg JPMSAE 57,430,68
ivn-mus LD50:32 mg/kg JPMSAE 57,430,68

SAFETY PROFILE: Poison by ingestion, intraperitoneal, and intravenous routes. When heated to decomposition it emits very toxic fumes as Cl^- and NO_x.

BOQ625 CAS:460-12-8 HR: 3
1,3-BUTADIYNE
mf: C_4H_2 mw: 50.06

PROP: A gas. Fp: −36°, bp: 10.3°.

SAFETY PROFILE: A dangerous explosive. Polymerizes violently above 0°C. Arsenic pentafluoride catalyzes explosive polymerization. Reaction with silver nitrate forms a very explosive friction-sensitive product. When heated to decomposition it emits acrid smoke and fumes.

BOQ750 CAS:125-88-2 HR: 3
BUTALBITAL SODIUM
mf: $C_{10}H_{14}N_2O_3 \cdot Na$ mw: 233.25

PROP: Hygroscopic powder.

SYNS: APROBARBITAL SODIUM □ APROBARBITONE SODIUM □ SODIUM-5-ALLYL-5-ISOPROPYLBARBITURATE

TOXICITY DATA WITH REFERENCE
ipr-mus LDLo:400 mg/kg NTIS** AD691-490
ipr-rat LD50:85 mg/kg APSCAX 18,204,49
scu-rat LDLo:125 mg/kg JACSAT 47,2236,25

SAFETY PROFILE: A poison via intraperitoneal and subcutaneous routes. When heated to decomposition it emits toxic fumes of NO_x and Na_2O. See also BARBITURATES.

BOR000 CAS:1142-70-7 HR: 3
BUTALLYLONAL
mf: $C_{11}H_{15}BrN_2O_3$ mw: 303.19

PROP: Crystals. Mp: 132–133°.

SYNS: 5-(2-BROMOALLYL)-5-sec-BUTYLBARBITURIC ACID □ 5-(2'-BROMOALLYL)-5-(1'-METHYL-N-PROPYL)BARBITURIC ACID □ BUTYLALYLONAL □ 5-sec-BUTYL-5-(β-BROMOALLYL)BARBITURIC ACID □ PERNOCTON □ PERNOSTON □ 2,4,6(1H,3H,5H)-PYRIMIDINETRIONE, 5-(2-BROMO-2-PROPENYL)-5-(1-METHYLPROPYL)-(9CI) □ SONBUTAL

TOXICITY DATA WITH REFERENCE
ipr-rat LDLo:65 mg/kg JPETAB 44,325,32
scu-rat LD50:90 mg/kg AEPPAE 152,341,30
scu-mus LDLo:150 mg/kg REDH** #3850
orl-rbt LDLo:350 mg/kg REDH** #3850
ipr-rbt LDLo:75 mg/kg JPETAB 41,465,31
scu-rbt LDLo:160 mg/kg REDH** #3850

ivn-rbt LDLo:70 mg/kg REDH•• #3850
scu-frg LDLo:150 mg/kg PHREA7 19,472,39
orl-mam LDLo:350 mg/kg JPETAB 42,253,31

SAFETY PROFILE: Poison by ingestion, intravenous, intraperitoneal, and subcutaneous routes. A central nervous system depressant (hypnotic) by ingestion. When heated to decomposition it emits very toxic fumes of Br^- and NO_x. See also BARBITURATES and ALLYL COMPOUNDS.

BOR250 CAS:3486-86-0 HR: 3
BUTALLYLONAL SODIUM
mf: $C_{11}H_{14}BrN_2O_3$•Na mw: 325.17

PROP: Powder. Sol in H_2O and EtOH.

SYNS: sec-BUTYL-BROM-ALLYL BARBITURIC ACID SODIUM SALT □ SODIUM-5-(2-BROMOALLYL)-5-sec-BUTYLBARBITURATE

TOXICITY DATA WITH REFERENCE
orl-cat LD50:135 mg/kg JPETAB 88,260,46
orl-rbt LD50:375 mg/kg JPETAB 42,253,31
ipr-rbt LD50:75 mg/kg JPETAB 42,253,31

SAFETY PROFILE: Poison by ingestion and intraperitoneal routes. When heated to decomposition it emits very toxic fumes of Br^- and NO_x. See also BARBITURATES.

BOR350 CAS:18109-81-4 HR: 2
BUTAMIRATE CITRATE
mf: $C_{18}H_{29}NO_3$•$C_6H_8O_7$ mw: 499.62

PROP: Crystals from Me_2CO. Mp: 75°.

SYNS: ABBOTT 36581 □ ACODEEN □ BUTAMYRATE CITRATE □ 2-((2-(DIETHYLAMINO)ETHOXY)ETHYL-2-PHENYLBUTYRATE CITRATE □ α-ETHYLBENZENEACETIC ACID-2-((2-DIETHYLAMINO)ETHOXY)ETHYL ESTER CITRATE □ HH-197 □ PHENYL ACETIC ACID DIETHYLAMINOETHOXYETHANOL ESTER CITRATE □ 2-PHENYLBUTYRIC ACID 2-(2-DIETHYLAMINO)ETHOXY)ETHYL ESTER CITRATE □ SINCODEEN □ SINCODEX □ SINCODIN □ SINCODIX □ SINECOD

TOXICITY DATA WITH REFERENCE
orl-rat TDLo:60 mg/kg (9-14D preg):REP TOIZAG 17,524,70
orl-mus TDLo:1800 mg/kg (7-12D preg):REP TOIZAG 17,524,70
orl-rat TDLo:60 mg/kg (9-14D preg):TER TOIZAG 17,524,70
orl-rat LD50:4164 mg/kg TOIZAG 18,115,71
scu-rat LD50:3638 mg/kg TOIZAG 18,115,71
orl-mus LD50:865 mg/kg TOIZAG 18,115,71

SAFETY PROFILE: Moderately toxic by ingestion and subcutaneous routes. Experimental reproductive effects. An experimental teratogen. When heated to decomposition it emits toxic fumes of NO_x. See also ESTERS.

BOR500 CAS:106-97-8 HR: 3
BUTANE
DOT: UN 1011
mf: C_4H_{10} mw: 58.14

PROP: Colorless gas; faint disagreeable odor. Bp: −0.5°, fp: −135°, lel: 1.9%, uel: 8.5%, flash p: −76°F (CC), d: 0.599, autoign temp: 761°F, vap press: 2 atm @ 18.8°, vap d: 2.046. Sltly sol in H_2O; mod sol in Et_2O and $CHCl_3$.

SYNS: n-BUTANE (DOT) □ BUTANE MIXTURES (DOT) □ BUTANEN (DUTCH) □ BUTANI (ITALIAN) □ DIETHYL □ METHYLETHYLMETHANE

TOXICITY DATA WITH REFERENCE
ihl-rat LC50:658 g/m³/4H FATOAO 30,102,67
ihl-mus LC50:680 g/m³/2H FATOAO 30,102,67

CONSENSUS REPORTS: Reported in EPA TSCA Inventory.

OSHA PEL: TWA 800 ppm
ACGIH TLV: TWA 800 ppm
DFG MAK: 1000 ppm (2350 mg/m³)
DOT CLASSIFICATION: 2.1; *Label:* Flammable Gas

SAFETY PROFILE: Mildly toxic by inhalation. Causes drowsiness. An asphyxiant. Very dangerous fire hazard when exposed to heat, flame, or oxidizers. Highly explosive when exposed to flame, or when mixed with $[Ni(CO)_4 + O_2]$. To fight fire, stop flow of gas. When heated to decomposition it emits acrid smoke and fumes.

BOR750 CAS:590-88-5 HR: 3
1,3-BUTANEDIAMINE
mf: $C_4H_{12}N_2$ mw: 88.18

PROP: Liquid. Bp: 142–150°, flash p: 125°F, d: 0.85, vap d: 3.04.

SYN: 1,3 DIAMINOBUTANE

TOXICITY DATA WITH REFERENCE
skn-rbt 10 mg/24H open SEV AMIHBC 4,119,51
eye-rbt 250 µg open SEV AMIHBC 4,119,51
orl-rat LD50:1350 mg/kg AMIHBC 4,119,51
skn-rbt LD50:430 mg/kg AMIHBC 4,119,51

SAFETY PROFILE: Moderately toxic by ingestion and skin contact. Severe skin and eye irritant. Flammable liquid when exposed to heat or flame. To fight fire, use alcohol foam, foam, CO_2, dry chemical. Incompatible with oxidizing materials. When heated to decomposition it emits toxic fumes of NO_x. See also 1,4-BUTANEDIAMINE and AMINES.

BOS000 CAS:110-60-1 HR: 3
1,4-BUTANEDIAMINE
mf: $C_4H_{12}N_2$ mw: 88.18

PROP: Crystals with strong odor. Mp: 27–28°, bp: 158–159°.

SYNS: BUTYLENEDIAMINE □ 1,4-BUTYLENEDIAMINE □ 1,4-DIAMINOBUTANE □ PUTRESCIN □ PUTRESCINE □ TETRAMETHYLENEDIAMINE □ 1,4-TETRAMETHYLENEDIAMINE

TOXICITY DATA WITH REFERENCE
cyt-hmn:hla 2 mmol/L JCLLAX 78,217,71
dns-mus:lvr 2 mmol/L AMOKAG 33,149,79
dni-mus:ast 10 mmol/L AMOKAG 33,149,79
dni-mus:lvr 20 mmol/L AMOKAG 33,149,79

ipr-mus TDLo:314 mg/kg (12D preg):TER TJADAB 28,237,83
orl-mus LDLo:1600 mg/kg AECTCV 14,111,85
orl-rbt LDLo:1600 mg/kg CRSBAW 83,481,20
orl-rbt LDLo:1600 mg/kg CRSBAW 83,481,20
scu-rbt LDLo:1 g/kg ZEPTAT 17,59,15
ivn-rbt LDLo:80 mg/kg CRSBAW 83,481,20
rec-rbt LDLo:400 mg/kg CRSBAW 83,481,20

CONSENSUS REPORTS: Reported in EPA TSCA Inventory.

SAFETY PROFILE: Poison by subcutaneous, intravenous, and rectal routes. Moderately toxic by ingestion. An experimental teratogen. Human mutation data reported. When heated to decomposition it emits toxic fumes of NO_x. See also 1,3-BUTANEDIAMINE and AMINES.

BOS100 CAS:2425-79-8 HR: 2
1,4-BUTANE DIGLYCIDYL ETHER
mf: $C_{10}H_{18}O_4$ mw: 202.28

SYNS: ARALDIT DY 026 □ 1,4-BIS(2,3-EPOXYPROPOXY)BUTANE □ 1,4-BIS(GLYCIDYLOXY)BUTANE □ BUTANE, 1,4-BIS(2,3-EPOXYPROPOXY)- □ BUTANEDIOL DIGLYCIDYL ETHER □ BUTANE-1:4-DIOL DIGLYCIDYL ETHER □ 1,4-BUTANEDIOL DIGLYCIDYL ETHER □ 2,2'-(1,4-BUTANEDIYLBIS(OXYMETHYLENE))BISOXIRANE □ CD 15006 A □ ChS-RR2 □ 1,4-DIGLYCIDLOXYBUTANE □ GRILONIT RV 1806 □ OXIRANE, 2,2'-(1,4-BUTANEDIYLBIS(OXYMETHYLENE))BIS-(9CI) □ TK 10352

TOXICITY DATA WITH REFERENCE
skn-rbt 10 mg/24H MOD AMIHAB 20,390,59
eye-rbt 100 mg MOD AMIHAB 20,390,59
mmo-sat 333 µg/plate MUREAV 172,105,86
uns-bac-esc 300 µmol/L MUREAV 231,205,90
slt-dmg-orl 28,400 ppm EMMUEG 23,51,94
trn-oin-dmg-orl 28 ppb EMMUEG 23,51,94
cyt-rat-ipr 100 mg/kg BJPCAL 6,235,51
sce-ham:lng 6250 nmol/L MUREAV 249,55,91
orl-rat LD50:1134 mg/kg TSCAT* OTS0206386
skn-rbt LD50:1130 mg/kg AMIHAB 20,390,59

CONSENSUS REPORTS: Reported in EPA TSCA Inventory.

SAFETY PROFILE: Moderately toxic by ingestion and skin contact. Mutation data reported. A skin and eye irritant. When heated to decomposition it emits acrid smoke and irritating vapors.

BOS250 CAS:584-03-2 HR: 2
1,2-BUTANEDIOL
mf: $C_4H_{10}O_2$ mw: 90.14

PROP: D: 1.0, vap d: 3.1, bp: 194°, flash p: 194°F.

SYN: 1,2-BUTYLENE GLYCOL

TOXICITY DATA WITH REFERENCE
orl-mus LD50:3720 mg/kg TXAPA9 49,385,79
orl-mus LD50:3720 mg/kg TXAPA9 49,385,79

CONSENSUS REPORTS: Reported in EPA TSCA Inventory.

SAFETY PROFILE: Moderately toxic by ingestion. Combustible when exposed to heat or flame. To fight fire, use alcohol foam. When heated to decomposition it emits acrid and irritating fumes.

BOS500 CAS:107-88-0 HR: 1
1,3-BUTANEDIOL
mf: $C_4H_{10}O_2$ mw: 90.14

PROP: Viscous liquid. Bp: 207.5°, fp: <−50°, flash p: 250°F, d: 1.006 @ 20°/20°, autoign temp: 741°F, vap press: 0.06 mm @ 20°, vap d: 3.2.

SYNS: 1,3-BUTANDIOL (GERMAN) □ BUTANE-1,3-DIOL □ β-BUTYLENE GLYCOL □ 1,3-BUTYLENE GLYCOL (FCC) □ 1,3-DIHYDROXYBUTANE □ METHYLTRIMETHYLENE GLYCOL

TOXICITY DATA WITH REFERENCE
skn-rbt 500 mg/24H MLD 85JCAE -,207,86
eye-rbt 505 mg AJOPAA 29,1363,46
eye-rbt 500 mg/24H MLD 85JCAE -,207,86
orl-rat TDLo:42,360 mg/kg (female 6-15D post):REP JACTDZ 5(4),189,86
orl-rat LD50:18,610 mg/kg JIDHAN 23,259,41
scu-rat LD50:20 g/kg NPIRI* 1,14,74
orl-mus LD50:12,980 mg/kg JAPMA8 45,669,56
orl-gpg LD50:11 g/kg JIHTAB 23,259,41

CONSENSUS REPORTS: Reported in EPA TSCA Inventory.

SAFETY PROFILE: Mildly toxic by ingestion and subcutaneous routes. A skin and eye irritant. See also ETHYLENE GLYCOL. Experimental reproductive effects. Combustible when exposed to heat or flame. Incompatible with oxidizing materials. To fight fire, use foam, alcohol foam, CO_2, dry chemical. When heated to decomposition it emits acrid smoke and irritating fumes.

BOS750 CAS:110-63-4 HR: 3
1,4-BUTANEDIOL
mf: $C_4H_{10}O_2$ mw: 90.14

PROP: Nearly odorless, colorless, viscous liquid or crystals; to needles on chilling. Mp: 16°, bp: 230°, flash p: 250°F (OC), d: 1.02 @ 20°, vap d: 3.1.

SYNS: BUTANE-1,4-DIOL □ 1,4-BUTYLENE GLYCOL □ 1,4-DIHYDROXYBUTANE □ 1,4-TETRAMETHYLENE GLYCOL

TOXICITY DATA WITH REFERENCE
unr-wmn LDLo:300 mg/kg:BAH PHARAT 3,110,48
rec-man LDLo:429 mg/kg PHARAT 3,110,48
orl-rat LD50:1525 mg/kg HYSAAV 33,41,68
ipr-rat LD50:1370 mg/kg TXAPA9 25,461,73
orl-mus LD50:2062 mg/kg HYSAAV 33,41,68
ipr-mus LDLo:500 mg/kg CBCCT* 3,363,51
orl-rbt LD50:2531 mg/kg HYSAAV 33,41,68
orl-gpg LD50:1200 mg/kg HYSAAV 33,41,68

CONSENSUS REPORTS: Reported in EPA TSCA Inventory.

SAFETY PROFILE: A human poison by an unspecified route. Moderately toxic by ingestion and intraperitoneal routes. Human systemic effects: altered sleep time. Combustible when exposed to heat or flame. To fight fire, use alcohol foam, mist, foam, CO_2, dry chemical.

Incompatible with oxidizing materials. When heated to decomposition it emits acrid smoke and fumes.

BOT000 CAS:513-85-9 HR: 1
2,3-BUTANEDIOL
mf: $C_4H_{10}O_2$ mw: 90.14

PROP: Colorless liquid or solid. Bp: 180°, fp: 19°, flash p: 185°F (TOC), d: 1.0095 @ 20°/20°, autoign temp: 756°F, vap press: 0.17 mm @ 20°, vap d: 3.1.

SYNS: 2,3-BUTYLENE GLYCOL □ 2,3-DIHYDROXYBUTANE □ DIMETHYLENE GLYCOL

TOXICITY DATA WITH REFERENCE
orl-mus LD50:5462 mg/kg TXAPA9 49,385,79

CONSENSUS REPORTS: Reported in EPA TSCA Inventory.

SAFETY PROFILE: Mildly toxic by ingestion. See also ETHYLENE GLYCOL. Flammable when exposed to heat or flame. Incompatible with oxidizing materials. To fight fire, use alcohol foam, CO_2, dry chemical. When heated to decomposition it emits acrid smoke and fumes.

BOT200 CAS:4437-85-8 HR: 1
1,2-BUTANEDIOL, CYCLIC CARBONATE
mf: $C_5H_8O_3$ mw: 116.13

SYNS: 1,2-BUTYLENE CARBONATE □ CARBONIC ACID, CYCLIC ETHYLETHYLENE ESTER □ 1,3-DIOXOLAN-2-ONE, 4-ETHYL-

TOXICITY DATA WITH REFERENCE
skn-rbt 500 mg MLD JACTDZ 1,12,90
eye-rbt 100 mg MLD JACTDZ 1,12,90
orl-rat LD50:>5 g/kg JACTDZ 1,11,90

SAFETY PROFILE: Low toxicity by ingestion. A skin and eye irritant. When heated to decomposition it emits acrid smoke and irritating fumes.

BOT250 CAS:55-98-1 HR: 3
1,4-BUTANEDIOL DIMETHYL SULFONATE
mf: $C_6H_{14}O_6S_2$ mw: 246.32

PROP: White crystals or needles. Mp: 116°.

SYNS: 1,4-BIS(METHANESULFONOXY)BUTANE □ (1,4-BIS(METHANESULFONYLOXY)BUTANE) □ BISULFAN □ BISULPHANE □ 1,4-BUTANEDIOL DIMETHANESULPHONATE □ BUZULFAN □ C.B. 2041 □ CITOSULFAN □ 1,4-DIMESYLOXYBUTANE □ 1,4-DIMETHANESULFONOXYBUTANE □ 1,4-DI(METHANESULFONYLOXY)BUTANE □ 1,4-DIMETHANESULPHONYLOXYBUTANE □ 1,4-DIMETHYLSULFONOXYBUTANE □ GT41 □ GT 2041 □ LEUCOSULFAN □ MABLIN □ METHANESULFONIC ACID TETRAMETHYLENE ESTER □ MIELUCIN □ MISULBAN □ MITOSTAN □ MYELOLEUKON □ MYLERAN □ NCI-C01592 □ NSC-750 □ SULPHABUTIN □ TETRAMETHYLENE BIS (METHANESULFONATE) □ TETRAMETHYLENE DIMETHANE SULFONATE □ X 149

TOXICITY DATA WITH REFERENCE
mmo-sat 333 µg/plate ENMUDM 8(Suppl 7),1,86
mma-esc 25 µg/plate TAKHAA 44,96,85
orl-wmn TDLo:17 mg/kg (4-36W preg):REP PEDIAU 25,85,60
unr-wmn TDLo:36,160 µg/kg (2.5Y pre):REP AJCPAI 44,385,65
orl-wmn TDLo:8460 µg/kg (22W pre):REP OBGNAS 21,466,63
orl-man TDLo:5400 µg/kg (90D male):REP ANYAA9 68,967,58
orl-wmn TDLo:17 mg/kg (4-36W preg):TER PEDIAU 25,85,60
unr-wmn TDLo:5280 µg/kg (14-26W preg):TER AJOGAH 92,580,65
ipr-rat TDLo:1 mg/kg (female 9D post):REP BEXBAN 68,1230,69
orl-rat TDLo:10 mg/kg (13D preg):REP TJADAB 30(1),14A,84
orl-mus TDLo:200 mg/kg (female 8-12D post):REP TCMUD8 6,361,86
orl-rat TDLo:48 mg/kg (female 7-14D post):TER PSDTAP 11,151,70
ipr-rat TDLo:10 mg/kg (female 2D post):TER BEXBAN 61,423,66
orl-rat TDLo:5600 µg/kg (female 7-14D post):TER PSDTAP 11,151,70
ipr-mus TDLo:10 mg/kg (female 1D pre):REP MUREAV 13,171,71
orl-rat TDLo:49 mg/kg (male 49D pre):REP 85GUAJ -,55,66
ipr-rat TDLo:10 mg/kg (female 2D post):REP BEXBAN 61,423,66
orl-mus TDLo:240 mg/kg (female 6-9D post):REP BTDCAV 13,103,72
orl-uns TDLo:8 mg/kg (female 1D pre):REP BIREBV 33,1237,85
ipr-mky TDLo:10 mg/kg (male 1D pre):REP JRPFA4 16,165,68
orl-rat TDLo:10 mg/kg (male 1D pre):REP ARTODN 7,147,84
ipr-mus TDLo:30 mg/kg (female 10D post):TER TJADAB 3,363,70
orl-rat TDLo:8 mg/kg (female 11-14D post):TER BTDCAV 13,103,72
orl-rbt TDLo:32 mg/kg (female 7-14D post):TER PSDTAP 10,227,69
orl-rat TDLo:5600 µg/kg (female 7-14D post):TER PSDTAP 11,151,70
orl-man TDLo:5684 µg/kg/21W-C:CAR BMJOAE 2,1513,77
orl-wmn TD:1140 mg/kg/9Y-I:CAR AIMDAP 124,66,69
orl-wmn TD:16,720 µg/kg/2Y-I:CAR JCROD7 108,362,84
ivn-mus TDLo:48 mg/kg/42D-I:NEO CNREA8 25,20,65
ipr-mus TD:1200 mg/kg/8W-I:ETA CNREA8 33,3069,73
ipr-mus TD:70 mg/kg/6W-I:ETA BLOOAW 44,49,74
orl-wmn TDLo:80 mg/kg/8Y:EYE,GIT JAMAAP 238,1951,77
orl-man TDLo:8 mg/kg/2D-I:CNS LANCAO 2,1463,84
ipr-rat LD50:18 mg/kg BCPCA6 1,39,58
scu-rat LD50:22 mg/kg KSRNAM 5,1894,71
ivn-rat LD50:1800 µg/kg ARZNAD 20,1467,70
orl-mus LD50:110 mg/kg KSRNAM 5,1894,71
ipr-mus LD50:86 mg/kg KSRNAM 5,1894,71
scu-mus LD50:63 mg/kg KSRNAM 5,1894,71
ivn dog LDLo:8 mg/kg CCSUBJ 2,203,65
ivn-mky LDLo:8 mg/kg CCSUBJ 2,203,65

CONSENSUS REPORTS: NTP 7th Annual Report On

B

Carcinogens. IARC Cancer Review: Group 1 IMEMDT 7,137,87; Animal Inadequate Evidence IMEMDT 4,247,74; Human Inadequate Evidence IMEMDT 4,247,74. EPA Genetic Toxicology Program.

SAFETY PROFILE: Confirmed carcinogen producing leukemia, kidney, and uterine tumors. Experimental neoplastigenic and tumorigenic data. Poison by ingestion, subcutaneous, intraperitoneal, intravenous, and possibly other routes. Ingestion by pregnant women can cause cancer of the reproductive system of the fetus including the uterus. Human teratogenic effects by ingestion and possibly other routes include developmental abnormalities of the eye, ear, craniofacial area including the nose and tongue, gastrointestinal system, endocrine system, urogenital system, and other unspecified areas. Other human reproductive effects by ingestion and possibly other routes include: impotence, changes in the uterus, cervix, and vagina, and menstrual-cycle disorders. Experimental reproductive effects. Human systemic effects by ingestion: eye general arteriolar or venous dilation, changes in structure or function of salivary glands. When heated to decomposition it emits toxic fumes of SO_x. See also SULFONATES.

BOT500 CAS:431-03-8 HR: 3
2,3-BUTANEDIONE
DOT: UN 2346
mf: $C_4H_6O_2$ mw: 86.10

$CH_3CO{\cdot}CO{\cdot}CH_3$

PROP: Greenish-yellow liquid; strong odor. Bp: 88°, flash p: 80°F, d: 0.9904 @ 15°/15°, refr index: 1.393–1.397, vap d: 3.00. Misc in alc, fixed oils, propylene glycol; sol in glycerin, alc, water.

SYNS: BIACETYL □ BUTADIONE □ BUTANEDIONE (DOT) □ DIACETYL (FCC) □ 2,3-DIKETOBUTANE □ DIMETHYL DIKETONE □ DIMETHYLGLYOXAL □ GLYOXAL, DIMETHYL- □ FEMA No. 2370

TOXICITY DATA WITH REFERENCE
skn-rbt 500 mg/24H MOD FCTXAV 17(Suppl.),695,79
mmo-sat 1 mg/plate MUREAV 67,367,79
oms-hmn:emb 20 mg/L BEXBAN 74,828,72
ipr-rat LD50:400 mg/kg FCTXAV 7,571,69
orl-mus LD50:250 mg/kg FRZKAP (1),44,83
orl-gpg LD50:990 mg/kg FCTXAV 2,327,64
orl-mam LD50:720 mg/kg RPTOAN 48,186,85

CONSENSUS REPORTS: Reported in EPA TSCA Inventory.

DOT CLASSIFICATION: 3; *Label:* Flammable Liquid

SAFETY PROFILE: A poison by ingestion and intraperitoneal route. Moderately toxic by ingestion. A skin irritant. Human mutation data reported. Flammable liquid. Dangerous fire hazard when exposed to heat or flame. To fight fire, use alcohol foam, CO_2, dry chemical. When heated to decomposition it emits acrid smoke and fumes. See also KETONES.

BOU250 CAS:1633-83-6 HR: 3
BUTANE SULTONE
mf: $C_4H_8O_3S$ mw: 136.18

PROP: Liquid. D: 1.33 @ 20°/4°, mp: 12.5–14.5°, bp: 134–136° @ 4 mm.

SYNS: BUTANESULFONE □ Δ-BUTANE SULTONE □ 1,4-BUTANESULTONE (MAK) □ 1,4-BUTYLENE SULFONE □ Δ-VALEROSULTONE

TOXICITY DATA WITH REFERENCE
mmo-sat 100 µg/plate JNCIAM 62,893,79
dnr-esc 10 µL/disc JNCIAM 62,873,79
hma-mus/sat 138 mg/kg JNCIAM 62,911,79
orl-rat TDLo:1300 mg/kg/1Y-I:ETA,REP ZEKBAI 75,69,70
scu-rat TDLo:2280 mg/kg/76W-I:ETA ZEKBAI 75,69,70
scu-mus TDLo:1680 mg/kg/42W-I:ETA JNCIAM 53,695,74
orl-rat LD50:500 mg/kg ZEKBAI 75,69,70
scu-rat LD50:350 mg/kg ZEKBAI 75,69,70
ivn-rat LD50:270 mg/kg ZEKBAI 75,69,70
ipr-mus LD50:138 mg/kg JJIND8 62,911,79

CONSENSUS REPORTS: EPA Genetic Toxicology Program. Reported in EPA TSCA Inventory.

DFG MAK: Suspected Carcinogen

SAFETY PROFILE: Suspected carcinogen with experimental tumorigenic data. Poison by subcutaneous, intravenous, and intraperitoneal routes. Moderately toxic by ingestion. Experimental reproductive effects. Human mutation data reported. See also SULFONATES. When heated to decomposition it emits toxic fumes of SO_x.

BOU500 CAS:1703-58-8 HR: 2
1,2,3,4-BUTANETETRACARBOXYLIC ACID
mf: $C_8H_{10}O_8$ mw: 234.18

SYN: BUTANETETRACARBOXYLIC ACID

TOXICITY DATA WITH REFERENCE
eye-rbt 100 µg SpiEW# 13FEB80
orl-rat LD50:1720 mg/kg SpiEW# 13FEB80
skn-rbt LDLo:8000 mg/kg SpiEW# 13FEB80

CONSENSUS REPORTS: Reported in EPA TSCA Inventory.

SAFETY PROFILE: Moderately toxic by ingestion. Mildly toxic by skin contact. An eye irritant. When heated to decomposition it emits acrid smoke and irritating fumes.

BOU550 CAS:36169-16-1 HR: 2
1-BUTANETHIOL, TIN(2+) SALT
mf: $C_4H_{10}S{\cdot}1/_2Sn$ mw: 149.54

SYN: ESTABEX S

TOXICITY DATA WITH REFERENCE
orl-mus LD50:690 mg/kg ERNFA7 11,424,66

OSHA PEL: TWA 2 mg(Sn)/m^3
ACGIH TLV: TWA 2 mg(Sn)/m^3

SAFETY PROFILE: Moderately toxic by ingestion. When heated to decomposition it emits toxic fumes of SO_x and Sn.

BOU700 **CAS:6659-60-5** **HR: 3**
1,2,4-BUTANETRIOL TRINITRATE
mf: $C_4H_7N_3O_9$ mw: 241.14

SYN: 1,2,4-BUTANETRIOL, TRINITRATE

CONSENSUS REPORTS: Reported in EPA TSCA Inventory.

DOT CLASSIFICATION: Forbidden

SAFETY PROFILE: An explosive. When heated to decomposition it emits toxic vapors of NO_x.

BOV000 **CAS:96-48-0** **HR: 2**
4-BUTANOLIDE
mf: $C_4H_6O_2$ mw: 86.10

PROP: Colorless liquid; mild caramel odor. Mp: −44°, bp: 203–204°, flash p: 209°F (OC), d: 1.441 @ 0°, refr index: 1.434–1.454 @ 25°, vap d: 3.0. Misc in H_2O.

SYNS: γ-6480 □ γ-BL □ BLO □ BLON □ BUTYRIC ACID LACTONE □ α-BUTYROLACTONE □ γ-BUTYROLACTONE (FCC) □ BUTYRYL LACTONE □ 4-DEOXYTETRONIC ACID □ DIHYDRO-2(3H)-FURANONE □ FEMA No. 3291 □ 4-HYDROXYBUTANOIC ACID LACTONE □ γ-HYDROXYBUTYRIC ACID CYCLIC ESTER □ 4-HYDROXYBUTYRIC ACID γ-LACTONE □ γ-HYDROXYBUTYROLACTONE □ NCI-C55878 □ TETRAHYDRO-2-FURANONE

TOXICITY DATA WITH REFERENCE
dnd-bcs 20 μL/disc PMRSDJ 1,175,81
otr-ham:kdy 25 mg/L PMRSDJ 1,638,81
orl-rat TDLo:500 mg/kg (female 6-15D post):TER PHTXA6 62,57,88
orl-rat TDLo:25 g/kg (20D male):REP ARANDR 10,239,83
skn-mus TDLo:50 g/kg/42W-I:ETA JNCIAM 31,41,63
orl-rat LD50:1540 mg/kg GTPZAB 31(1),49,87
ipr-rat LD50:1000 mg/kg AITEAT 13,70,65
orl-mus LD50:1720 mg/kg GTPZAB 31(1),49,87
ipr-mus LD50:1100 mg/kg AITEAT 13,70,65
ivn-rbt LDLo:500 mg/kg AITEAT 13,70,65

CONSENSUS REPORTS: IARC Cancer Review: Group 3 IMEMDT 7,56,87; Animal No Evidence IMEMDT 11,231,76. EPA Genetic Toxicology Program. Reported in EPA TSCA Inventory.

SAFETY PROFILE: Moderately toxic by ingestion, intravenous, and intraperitoneal routes. An experimental teratogen. Other experimental reproductive effects. Questionable carcinogen with experimental tumorigenic data by skin contact. Mutation data reported. Less acutely toxic than β-propiolactone. Combustible when exposed to heat or flame; can react with oxidizing materials. To fight fire, use foam, alcohol foam, CO_2, dry chemical. Potentially explosive reaction with butanol + 2,4-dichlorophenol + sodium hydroxide. When heated to decomposition it emits acrid and irritating fumes.

BOV625 **CAS:4154-69-2** **HR: 3**
2-BUTANONE OXIME HYDROCHLORIDE
mf: $C_4H_{10}ClNO$ mw: 123.58

SYN: (2-HYDROXYLIMINIOBUTANE CHLORIDE)

SAFETY PROFILE: Decomposes violently above 50°C. When heated to decomposition it emits toxic fumes of Cl^- and NO_x.

BOV750 **CAS:129-18-0** **HR: 3**
BUTAZOLIDINE SODIUM
mf: $C_{19}H_{20}N_2O_2{\cdot}Na$ mw: 331.40

SYNS: 4-BUTYL-1,2-DIPHENYL-3,5-PYRAZOLIDINEDIONE SODIUM SALT □ 3,5-DIOXO-1,2-DIPHENYL-4-N-BUTYLPYRAZOLIDIN SODIUM □ DIPHENYLDIOXOBUTYLPYRAZOLIDINE BUTAZOLIDINE-SODIUM □ PHENYLBUTAZONE SODIUM □ SODIUM BUTAZOLIDINE □ SODIUM PHENYLBUTAZONE □ SODIUM SALT of PHENYLBUTAZONE

TOXICITY DATA WITH REFERENCE
orl-wmn LDLo:16 mg/kg AIMEAS 39,1096,53
scu-rat LD50:360 mg/kg ARZNAD 8,229,58
orl-mus LD50:476 mg/kg RPOBAR 2,314,70
ipr-mus LD50:169 mg/kg RPOBAR 2,314,70
ivn-rat LD50:113 mg/kg FRPSAX 13,922,58
scu-mus LD50:271 mg/kg FRPSAX 12,521,57
ivn-mus LD50:94 g/kg FRPSAX 13,922,58

SAFETY PROFILE: A human poison by ingestion. Human systemic effects by ingestion: respiratory system damage, agranulocytosis, and dermatitis. An experimental poison via subcutaneous, intravenous, and intraperitoneal routes. An anti-inflammatory drug. When heated to decomposition it emits toxic fumes of NO_x and Na_2O.

BOV800 **HR: 3**
BUTEA FRONDOSA, seed extract

PROP: Indian plant belonging to the family Leguminosae (IJEBA6 11,43,73).

SYN: PALASH SEED EXTRACT

TOXICITY DATA WITH REFERENCE
orl-mus TDLo:50 mg/kg (female 1-5D post):REP IJPPAZ 13,239,69
orl-mus TDLo:4 g/kg (female 12-15D post):REP IJPPAZ 13,239,69
orl-rat TDLo:700 mg/kg (female 1-7D post):REP IJEBA6 16,1077,78
orl-mus LD50:7500 mg/kg IJPPAZ 13,239,69
ipr-mus LD50:20 mg/kg IJEBA6 11,43,73

SAFETY PROFILE: Poison by intraperitoneal route. Experimental reproductive effects.

BOV825 **CAS:5716-20-1** **HR: 3**
BUTEDRIN
mf: $C_{24}H_{38}N_2O_4{\cdot}H_2O_4S$ mw: 516.72

SYNS: BAMETAN SULFATE □ BAMETHAN SULFATE □ BASCURAT □ BUPATOL □ BUTIBATOL □ α-((BUTYLAMINO)METHYL)-p-HYDROXYBENZYL ALCOHOL SULFATE □ BUTYLNORSYMPATOL □ CYCLATE □ ECLERIN □ GARMIAN □ PERIPHETOL □ ROTESAR □ VASCULAT □ VASCULIT □ VASCUNICOL □ VASKULAT

TOXICITY DATA WITH REFERENCE
ipr-mus LD50:210 mg/kg NIIRDN 6,585,82
scu-mus LD50:422 mg/kg NIIRDN 6,585,82
ivn-mus LD50:72 mg/kg NIIRDN 6,585,82

SAFETY PROFILE: Poison by intravenous and intraperitoneal routes. Moderately toxic by subcutaneous route.

A vasodilator. When heated to decomposition it emits toxic fumes of SO_x and NO_x. See also SULFATES.

BOW250 **CAS:25167-67-3** **HR: 3**
1-BUTENE
mf: C_4H_8 mw: 56.11

PROP: A colorless, flammable gas; sltly aromatic odor. Bp: −6.3°, fp: −185.3°, lel: 1.6%, uel: 9.3%, flash p: −80° (−112°F), d: 0.668 @ 0°/1°, vap d: 1.93, vap press: 3480 mm @ 21°, autoign temp: 723°F.

SYNS: BUTYLENE □ α-BUTYLENE

CONSENSUS REPORTS: Reported in EPA TSCA Inventory.

SAFETY PROFILE: A simple asphyxiant. Very dangerous fire hazard when exposed to heat, flame, or oxidizers. To fight fire, stop flow of gas. Moderately explosive when exposed to flame. Mixtures with aluminum tetrahydroborate explode after an induction period. When heated to decomposition it emits acrid smoke and fumes.

BOW255 **CAS:107-01-7** **HR: 2**
2-BUTENE
mf: C_4H_8 mw: 56.12

SYNS: β-BUTYLENE □ PSEUDOBUTYLENE

TOXICITY DATA WITH REFERENCE
ihl-mus LC50:425 ppm 85JCAE -,12,86

CONSENSUS REPORTS: Reported in EPA TSCA Inventory.

SAFETY PROFILE: Moderately toxic by inhalation. When heated to decomposition it emits acrid smoke and irritating vapors.

BOW500 **HR: 3**
cis-2-BUTENE
mf: C_4H_8 mw: 56.11

PROP: Colorless, flammable gas; sltly aromatic odor. Bp: 1°, fp: −139°, flash p: −100°F, d: 0.627 @ 15.5°/15.5°, vap press: 1410 mm @ 21°, autoign temp: 615°F, lel: 1.7%, uel: 9.0%, vap d: 1.9.

SYNS: DIMETHYLETHYLENE □ PSEUDO-BUTYLENE

SAFETY PROFILE: A simple asphyxiant. Very dangerous fire hazard when exposed to heat or flame. Very likely to explode. Incompatible with oxidizing materials. To fight fire, stop flow of gas. When heated to decomposition it emits acrid smoke and fumes.

BOW750 **HR: 3**
trans-2-BUTENE
mf: C_4H_8 mw: 56.11

PROP: A colorless, flammable gas; sltly aromatic odor. Bp: 2.5°, fp: −105.6°, flash p: −100°F, d: 0.613 @ 15.5°/15.5°, vap d: 1.95, vap press: 1592 mm @ 21°, autoign temp: 615 F, lel: 1.8%, uel: 9.7%, vap d: 1.9.

CONSENSUS REPORTS: EPA Extremely Hazardous Substances List.

SAFETY PROFILE: A simple asphyxiant. Very dangerous fire hazard when exposed to heat or flame. Very likely to explode. To fight fire, stop flow of gas. Incompatible with oxidizing materials. When heated to decomposition it emits acrid smoke and fumes.

BOX250 **CAS:10099-70-4** **HR: 2**
2-BUTENEDIOIC ACID BIS(1-METHYLETHYL) ESTER
mf: $C_{10}H_{16}O_4$ mw: 200.26

TOXICITY DATA WITH REFERENCE
skn-rbt 10 mg/24H open MLD AMIHBC 10,61,54
eye-rbt 500 mg open AMIHBC 10,61,54
orl-rat LD50:2140 mg/kg AMIHBC 10,61,54

CONSENSUS REPORTS: Reported in EPA TSCA Inventory.

SAFETY PROFILE: Moderately toxic by ingestion. A skin and eye irritant. See also ESTERS. When heated to decomposition it emits acrid smoke and irritating fumes.

BOX300 **CAS:110-64-5** **HR: 3**
2-BUTENE-1,4-DIOL
mf: $C_4H_8O_2$ mw: 88.12

SYNS: AGRISYNTH B2D □ 2-BUTENE, 1,4-DIHYDROXY- □ 1,4-DIHYDROXY-2-BUTENE

TOXICITY DATA WITH REFERENCE
orl-rat LD50:1250 mg/kg GAFCC*
ipr-rat LD50:327 mg/kg JPPMAB 26,597,74

CONSENSUS REPORTS: Reported in EPA TSCA Inventory.

SAFETY PROFILE: Poison by intraperitoneal route. Moderately toxic by ingestion. When heated to decomposition it emits acrid smoke and irritating vapors.

BOX500 **CAS:109-75-1** **HR: 3**
3-BUTENE NITRILE
mf: C_4H_5N mw: 67.10

PROP: Colorless liquid, onion-like odor. Bp: 116–119°, d: 0.8341 @ 20°/4°, mp: −87°.

SYNS: ALLYL CYANIDE □ ALLYLNITRILE □ 1-BUTENE-4-NITRILE □ β-BUTENONITRILE □ TL 350 □ VINYLACETONITRILE

TOXICITY DATA WITH REFERENCE
skn-rbt 10 mg/24H open MLD AIHAAP 23,95,62
ihl-rat TCLo:50 ppm/6H (female 6-20D post):REP FAATDF 20,365,93
orl-rat LD50:115 mg/kg AIHAAP 30,470,69
scu-rat LD50:150 mg/kg 85GMAT-,18,82
skn-rbt LD50:1410 mg/kg AIHAAP 23,95,62
ihl-gpg LC50:2500 mg/m³/4H GISAAA 34(4),36,69

CONSENSUS REPORTS: Reported in EPA TSCA Inventory. Cyanide and its compounds are on the Community Right-To-Know List.

SAFETY PROFILE: A poison by ingestion and subcutaneous routes. Moderately toxic by inhalation and skin contact. Experimental reproductive effects. A skin irritant. See also NITRILES. Dangerous; emits highly toxic fumes of NO_x and CN^- when heated to decomposition or on contact with acids or acid fumes. To fight fire, use alcohol foam, mist.

BOX750 **CAS:106-88-7** **HR: 3**
1-BUTENE OXIDE
DOT: UN 3022
mf: C_4H_8O mw: 72.12

PROP: Colorless liquid. D: 0.8312 @ 20°/20°, bp: 63°, flash p: 5°F, lel: 1.5%, uel: 18.3%. Sol in water; misc with most organic solvents.

SYNS: BUTYLENE OXIDE □ 1,2-BUTYLENE OXIDE □ 1,2-BUTYLENE OXIDE, stabilized (DOT) □ EPOXYBUTANE □ 1,2-EPOXYBUTANE □ ETHYLENE OXIDE, ETHYL- □ ETHYLOXIRANE □ ETHYL ETHYLENE OXIDE □ NCI-C55527

TOXICITY DATA WITH REFERENCE
skn-rbt 500 mg/24H MLD 85JCAE-,770,86
eye-rbt 100 mg/24H MOD 85JCAE-,770,86
mmo-klp mmol/L MUREAV 89,269,81
trn-dmg-orl 5 pph ENMUDM 7,349,85
mma-ssp 1600 μmol/L TCMUD8 3,75,83
ihl-rbt TCLo:1000 ppm/7H (10-24D preg):REP NTIS** PB81-168510
ihl-rat TCLo:400 ppm/6H/5D/2Y-C:CAR NTPTR* NTP-TR-329,88
orl-rat LD50:500 mg/kg NTIS** PB81-168510
ihl-rat LCLo:4000 ppm/4H AIHAAP 23,95,62
skn-rbt LD50:2100 mg/kg AIHAAP 23,95,62

CONSENSUS REPORTS: NTP Carcinogenesis Studies (inhalation); Clear Evidence: rat; No Evidence: mouse NTPTR* NTP-TR-329,88. Community Right-To-Know List. EPA Genetic Toxicology Program. Reported in EPA TSCA Inventory.

DFG MAK: Animal Carcinogen, Suspected Human Carcinogen

SAFETY PROFILE: Confirmed carcinogen with experimental carcinogenic data. Moderately toxic by ingestion and skin contact. Mildly toxic by inhalation. Experimental reproductive effects. Mutation data reported. Dangerous fire hazard when exposed to heat, flame, or powerful oxidizers. To fight fire, use dry chemical, water spray, mist or fog, alcohol foam. When heated to decomposition it emits acrid smoke and fumes.

BOX825 **CAS:16187-15-8** **HR: 3**
trans-2-BUTENE OZONIDE
mf: $C_4H_8O_3$ mw: 104.11

SYN: (3,3-DIMETHYL-1,2,4-TRIOXOLANE)

SAFETY PROFILE: May explode when heated. When heated to decomposition it emits acrid smoke and fumes. See also OZONE.

BOY000 **CAS:6117-91-5** **HR: 2**
2-BUTEN-1-OL
mf: C_4H_8O mw: 72.12

PROP: Colorless liquid. Mp: <30°, bp: 118°, flash p: 92°F, d: 0.8726 @ 0°/4°, vap d: 2.49.

SYNS: 2-BUTENOL □ 2-BUTENYL ALCOHOL □ CROTONYL ALCOHOL □ CROTYL ALCOHOL

TOXICITY DATA WITH REFERENCE
mmo-sat 10 μg/plate TCMUD8 1,259,80
orl-rat LD50:930 mg/kg AIHAAP 23,95,62
skn-rbt LD50:1270 mg/kg AIHAAP 23,95,62

CONSENSUS REPORTS: Reported in EPA TSCA Inventory.

SAFETY PROFILE: Moderately toxic by ingestion and skin contact. Mutation data reported. Dangerous fire hazard when exposed to heat or flame; can react with oxidizing materials. To fight fire, use alcohol foam, CO_2, dry chemical. When heated to decomposition it emits acrid smoke and fumes. See also ALCOHOLS.

BOY250 **HR: 3**
1-BUTEN-3-ONE
mf: C_4H_6O mw: 70.10

PROP: Flash p: −7°C.

SAFETY PROFILE: A dangerous fire hazard. When heated to decomposition it emits acrid smoke and fumes. See also KETONES.

BOY500 **CAS:78-94-4** **HR: 3**
3-BUTEN-2-ONE
DOT: UN 1251
mf: C_4H_6O mw: 70.10

PROP: Colorless liquid, powerfully irritating odor. Bp: 81.4°, flash p: 20°F (CC), d: 0.8393 @ 25°/4°, vap d: 2.41.

SYNS: ACETYL ETHYLENE □ 3-BUTENE-2-ONE □ METHYLENE ACETONE □ METHYL-VINYL-CETONE (FRENCH) □ METHYLVINYLKETON (GERMAN) □ METHYL VINYL KETONE □ γ-OXO-α-BUTYLENE □ VINYL METHYL KETONE

TOXICITY DATA WITH REFERENCE
mmo-sat 250 μmol/L MUREAV 93,305,82
mma-sat 250 μmol/L MUREAV 93,305,82
orl-rat LD50:31 mg/kg 85GMAT -,88,82
ihl-rat LC50:7 mg/m³/4H 85GMAT -,88,82
orl-mus LD50:33 mg/kg 85GMAT -,88,82
ipr-mus LD50:76 mg/kg ZolH## 23OCT75

CONSENSUS REPORTS: Reported in EPA TSCA Inventory. EPA Extremely Hazardous Substances List.

DOT CLASSIFICATION: 3; *Label:* Flammable Liquid

SAFETY PROFILE: Poison by ingestion, inhalation, and intraperitoneal routes. A severe irritant to skin, eyes, and mucous membranes. A lachrymator. Mutation data reported. See also KETONES. Dangerous fire hazard when exposed to heat, flame, or oxidizers. To fight fire, use CO_2, dry chemical. When heated to decomposition it emits acrid smoke and fumes.

BPA250 CAS:14746-03-3 HR: 3
2-BUTEN-1-YL DIAZOACETATE
mf: $C_6H_8N_2O_2$ mw: 140.14

$N_2CHCO{\cdot}OCH_2CH{=}CHCH_3$

SAFETY PROFILE: Potentially explosive. When heated to decomposition it emits toxic fumes of NO_x. See also other diazo compounds.

BPA500 CAS:2237-92-5 HR: 3
5-(1-BUTENYL)-5-ETHYLBARBITURIC ACID
mf: $C_{10}H_{14}N_2O_3$ mw: 210.26

TOXICITY DATA WITH REFERENCE
orl-mus LD50:320 mg/kg JACSAT 62,1199,40
ipr-mus LD50:225 mg/kg JACSAT 62,1199,40

SAFETY PROFILE: Poison by ingestion and intraperitoneal routes. When heated to decomposition it emits toxic fumes of NO_x. See also BARBITURATES.

BPA750 CAS:67050-00-4 HR: 3
5-(2-BUTENYL)-5-ETHYL-2-THIOBARBITURIC ACID SODIUM SALT
mf: $C_{10}H_{13}N_2O_2S{\cdot}Na$ mw: 248.30

TOXICITY DATA WITH REFERENCE
ipr-rat LD50:123 mg/kg JAPMA8 34,183,45
ivn-rbt LD50:53 mg/kg JAPMA8 34,183,45

SAFETY PROFILE: Poison by intraperitoneal and intravenous routes. When heated to decomposition it emits very toxic fumes of NO_x, Na_2O, and SO_x. See also BARBITURATES.

BPB500 CAS:67050-04-8 HR: 3
5-(1-BUTENYL)-5-ISOPROPYLBARBITURIC ACID
mf: $C_{11}H_{16}N_2O_3$ mw: 224.29

TOXICITY DATA WITH REFERENCE
orl-mus LD50:300 mg/kg JACSAT 62,1199,40
ipr-mus LD50:250 mg/kg JACSAT 62,1199,40

SAFETY PROFILE: Poison by ingestion and intraperitoneal routes. When heated to decomposition it emits toxic fumes of NO_x. See also BARBITURATES.

BPC500 CAS:67050-11-7 HR: 3
5-(2-BUTENYL)-5-(1-METHYLBUTYL)-2-THIOBARBITURIC ACID SODIUM SALT
mf: $C_{13}H_{19}N_2O_2S{\cdot}Na$ mw: 290.39

TOXICITY DATA WITH REFERENCE
ipr-rat LD50:341 mg/kg JAPMA8 34,183,45
ivn-rbt LD50:49 mg/kg JAPMA8 34,183,45

SAFETY PROFILE: Poison by intraperitoneal and intravenous routes. When heated to decomposition it emits very toxic fumes of SO_x, Na_2O, and NO_x. See also BARBITURATES.

BPC600 HR: 2
3-(3-BUTENYLNITROSAMINO)-1-PROPANOL
mf: $C_7H_{14}N_2O_2$ mw: 158.23

SYN: BUTENYL(3-HYDROXYPROPYL)NITROSAMINE

TOXICITY DATA WITH REFERENCE
scu-ham TDLo:22,800 mg/kg/76W-I:CAR CDPRD4 4,79,81

SAFETY PROFILE: Questionable carcinogen with experimental carcinogenic data. When heated to decomposition it emits toxic fumes of NO_x.

BPC750 HR: 2
2-BUTENYLPHENOL (mixed isomers)
mf: $C_{10}H_{12}O$ mw: 148.22

TOXICITY DATA WITH REFERENCE
skn-rbt 10 mg/24H SEV AMIHBC 4,119,51
eye-rbt 50 µg SEV AMIHBC 4,119,51
orl-rat LD50:410 mg/kg AMIHBC 4,119,51

SAFETY PROFILE: Moderately toxic by ingestion. A severe skin and eye irritant. When heated to decomposition it emits acrid smoke and irritating fumes.

BPD000 CAS:54746-50-8 HR: 2
3-BUTENYL-(2-PROPENYL)-N-NITROSAMINE
mf: $C_7H_{12}N_2O$ mw: 140.21

SYN: N-ALLYL-N-NITROSO-3-BUTENYLAMINE

TOXICITY DATA WITH REFERENCE
mmo-sat 250 µg/plate MUREAV 68,195,79
scu-ham TDLo:15,300 mg/kg/51W-I:CAR CDPRD4 4,79,81

SAFETY PROFILE: Questionable carcinogen with experimental carcinogenic data. Mutation data reported. When heated to decomposition it emits acrid smoke and irritating fumes. See also NITROSAMINES.

BPE109 CAS:689-97-4 HR: 3
BUTEN-3-YNE
mf: C_4H_4 mw: 52.08

$HC{\equiv}CCH{=}CH_2$

PROP: Gas with acetylene-like odor. Flash p: $<-5°$, lel: 2%, uel: 100% d: 0.68 @ 1.7 atm, vap d: 1.8, bp: 2–3°.

SYN: VINYL ACETYLENE

TOXICITY DATA WITH REFERENCE
ihl-mus LC50:97,200 mg/m³/2H 85GMAT-,119,82

CONSENSUS REPORTS: Reported in EPA TSCA Inventory.

SAFETY PROFILE: Low toxicity by inhalation. Forms explosive peroxides with air or oxygen. Very exothermic decomposition when heated. Reacts explosively when heated with 1,3-butadiene or oxygen. Reacts with silver nitrate to form the explosive silver buten-3-ynide. When heated to decomposition it emits acrid smoke and irritating fumes. See also ACETYLENE COMPOUNDS.

BPE250 **HR: 3**
3-BUTEN-1-YNYL DIETHYL ALUMINUM
mf: $C_8H_{13}Al$ mw: 137.17

$H_2C{=}CH{=}CAl(CH_2CH_3)_2$

SAFETY PROFILE: Ignites spontaneously in air. When heated to decomposition it emits acrid smoke and fumes. See also ALUMINUM COMPOUNDS.

BPE500 **HR: 3**
3-BUTEN-1-YNYL DIISOBUTYL ALUMINUM
mf: $C_{12}H_{21}Al$ mw: 189.3

SAFETY PROFILE: Ignites spontaneously in air. When heated to decomposition it emits acrid smoke and fumes. See also ALUMINUM COMPOUNDS.

BPE750 **HR: 3**
2-BUTEN-1-YNYL TRIETHYL LEAD
mf: $C_{10}H_{18}Pb$ mw: 341.41

CONSENSUS REPORTS: Lead and its compounds are on the Community Right-To-Know List.

SAFETY PROFILE: Explodes when heated rapidly. See also LEAD COMPOUNDS.

BPF000 **CAS:125-40-6** **HR: 3**
BUTISOL
mf: $C_{10}H_{16}N_2O_3$ mw: 212.28

PROP: Bitter-tasting microcrystal powder. Mp: 165–168°. Sltly sol in H_2O.

SYNS: BUTABARB □ BUTABARBITAL □ BUTABARBITONE □ BUTATAB □ BUTATAL □ BUTICAPS □ BUTRATE □ 5-sec-BUTYL-5-ETHYLBARBITURIC ACID □ 5-sec-BUTYL-5-ETHYLMALONYL UREA □ 5-ETHYL-5-(1-METHYLPROPYL)BARBITURATE □ 5-ETHYL-5-(1-METHYLPROPYL)BARBITURIC ACID □ 5-ETHYL-5-(1-METHYLPROPYL)-2,4,6(1H,3H,5H)-PYRIMIDINETRIONE (9CI) □ MEDARSED □ NILOX □ SECBUBARBITAL □ SECBUTABARBITAL □ SECBUTOBARBITONE □ UNICELLES

TOXICITY DATA WITH REFERENCE
ipr-rat LD50:70 mg/kg JPETAB 44,325,32
scu-rat LDLo:140 mg/kg JACSAT 52,2440,30
ipr-mus LDLo:200 mg/kg JACSAT 58,731,36
ivn-mus LD50:175 mg/kg AIPTAK 132,164,61
orl-rbt LD50:140 mg/kg JPETAB 44,325,32
ipr-rbt LD50:75 mg/kg JPETAB 44,325,32
scu-rbt LDLo:200 mg/kg JACSAT 45,243,23
ivn-rbt LDLo:90 mg/kg JPPGAR 30,364,32

SAFETY PROFILE: Poison by ingestion, intravenous, intraperitoneal, and subcutaneous routes. A central nervous system depressant. When heated to decomposition it emits toxic fumes of NO_x. See also BARBITURATES.

BPF250 **CAS:143-81-7** **HR: 3**
BUTISOL SODIUM
mf: $C_{10}H_{16}N_2O_3{\cdot}Na$ mw: 235.27

PROP: Bitter powder. Very sol in H_2O; prac insol in Et_2O.

SYNS: BUTABARBITAL SODIUM □ 5-sec-BUTYL-5-ETHYLBARBITURIC ACID SODIUM SALT □ 5-ETHYL-5-(1-METHYLPROPYL)BARBITURIC ACID SODIUM SALT □ 5-ETHYL-5-(1-METHYLPROPYL)-2,4,6(1H,3H,5H)-PYRIMIDINETRIONE MONOSODIUM SALT □ SECBUBARBITAL SODIUM □ SODIUM BUTABARBITAL □ SODIUM-5-sec-BUTYL-5-ETHYLBARBITURATE □ SODIUM-5-ETHYL-5-sec-BUTYLBARBITURATE □ SODIUM-5-ETHYL-5-(1-METHYLPROPYL)BARBITURATE

TOXICITY DATA WITH REFERENCE
orl-hmn LDLo:125 mg/kg CTOXAO 10,327,77
orl-hmn TDLo:120 mg/kg:CNS,PSY BMJOAE 1,144,77
orl-rat LD50:78 mg/kg JPETAB 81,254,44
ipr-rat LD50:70 mg/kg JPETAB 81,254,44
ivn-rat LD50:70 mg/kg JPETAB 81,254,44
ipr-mus LD50:247 mg/kg JPETAB 81,254,44
ivn-dog LD50:90 mg/kg JPETAB 81,254,44
orl-rbt LD50:194 mg/kg JPETAB 81,254,44
ipr-rbt LD50:95 mg/kg JPETAB 81,254,44
ivn-rbt LD50:91 mg/kg JPETAB 81,254,44

SAFETY PROFILE: An experimental poison by ingestion, intraperitoneal, and intravenous routes. Human central nervous system and psychotropic effects by ingestion. When heated to decomposition it emits toxic fumes of NO_x and Na_2O. See also BARBITURATES.

BPF500 **CAS:77-28-1** **HR: 3**
BUTOBARBITAL
mf: $C_{10}H_{16}N_2O_3$ mw: 212.28

PROP: Crystals from EtOH (aq) with slightly bitter taste. Mp: 127–128°.

SYNS: BUDORM □ BUTETHAL □ BUTOBARBITONE □ BUTOBARBITURAL □ 5-BUTYL-5-ETHYLBARBITURIC ACID □ 5-BUTYL-5-ETHYL-2,4,6(1H,3H,5H)-PYRIMIDINETRIONE (9CI) □ 5-ETHYL-5-N-BUTYLBARBITURIC ACID □ ETOVAL □ HYPERBUTAL □ LONGANOCT □ MEONAL □ MONODORM □ NEONAL □ SONERILE □ SONERYL

TOXICITY DATA WITH REFERENCE
scu-rat TDLo:1300 mg/kg (7-19D preg):TER EXPEAM 33,499,77
ims-rat TDLo:28 mg/kg (7D male):REP IJEBA6 16,316,78
orl-wmn TDLo:120 mg/kg:BAH CTOXAO 2,133,69
orl-wmn TDLo:166 mg/kg:CNS BMJOAE 1,1238,55
ipr-rat LDLo:135 mg/kg JPETAB 44,325,32
scu-rat LDLo:190 mg/kg JPETAB 26,371,25
ipr-mus LD50:320 mg/kg JPETAB 89,356,47
orl-cat LDLo:80 mg/kg JPETAB 33,43,28
orl-rbt LDLo:100 mg/kg JPETAB 44,337,32
ipr-rbt LDLo:115 mg/kg JPETAB 44,325,32
scu-rbt LDLo:100 mg/kg JACSAT 45,243,23
ivn-rbt LDLo:90 mg/kg JACSAT 57,1961,35

SAFETY PROFILE: A poison by ingestion, intraperitoneal, subcutaneous, and intravenous routes. Experimental teratogenic and reproductive effects. Human systemic effects by ingestion: changes in motor activity, coma, and nausea or vomiting. A central nervous system depressant. When heated to decomposition it emits toxic fumes of NO_x. See also BARBITURATES.

BPF750 **CAS:35763-44-1** **HR: 3**
BUTOBARBITAL SODIUM
mf: $C_{10}H_{15}N_2O_3 \cdot Na$ mw: 234.26

SYNS: BUTETHAL SODIUM □ BUTOBARBITIONE SODIUM □ 5-BUTYL-5-ETHYL-2,4,6(1H,3H,5H)-PYRIMIDINETRIONE MONOSODIUM SALT (9CI) □ SODIUM ETHYL-N-BUTYL BARBITURATE

TOXICITY DATA WITH REFERENCE
ipr-rat LD50:197 mg/kg JPETAB 81,254,44
scu-rat LDLo:190 mg/kg JACSAT 47,2236,25
ipr-mus LDLo:275 mg/kg JPETAB 31,455,27

SAFETY PROFILE: Poison by intraperitoneal and subcutaneous routes. When heated to decomposition it emits toxic fumes of NO_x and Na_2O. See also BARBITURATES.

BPF825 **CAS:32838-28-1** **HR: 2**
BUTOCTAMIDE SEMISUCCINATE
mf: $C_{16}H_{29}NO_5$ mw: 315.46

SYNS: BUTANEDIOIC ACID MONO(3-((2-ETHYLHEXYL)AMINO)-1-METHYL-3-OXOPROPYL) ESTER (9CI) □ BUTOCTAMIDE HYDROGEN SUCCINATE □ N-(2-ETHYLHEXYL)-3-HYDROXYBUTYRAMIDE HYDROGEN SUCCINATE □ N-2-ETHYLHEXYL-β-OXYBUTYRAMIDE SEMISUCCINATE □ M-2H □ SUCCINIC ACID MONOESTER with N-(2-ETHYLHEXYL)-3-HYDROXYBUTYRAMIDE

TOXICITY DATA WITH REFERENCE
orl-mus TDLo:8 g/kg (7-14D preg):TER OYYAA2 8,1413,74
orl-mus TDLo:2400 mg/kg (7-14D preg):TER OYYAA2 8,1413,74
orl-rat LD50:12,100 mg/kg YAKUD5 24,2029,82
ipr-rat LD50:635 mg/kg TOIZAG 18,648,71
scu-rat LD50:3350 mg/kg TOIZAG 18,648,71
orl-mus LD50:5600 mg/kg TOIZAG 18,648,71
ipr-mus LD50:473 mg/kg TOIZAG 18,648,71
scu-mus LD50:3730 mg/kg TOIZAG 18,648,71

SAFETY PROFILE: Moderately toxic by intraperitoneal, ingestion, and subcutaneous routes. An experimental teratogen. When heated to decomposition it emits toxic fumes of NO_x.

BPG000 **CAS:126-22-7** **HR: 2**
BUTONATE
mf: $C_8H_{14}Cl_3O_5P$ mw: 327.54

PROP: Bp: 129°.

SYNS: BUTANOIC ACID 2,2,2-TRICHLORO-1-(DIMETHOXYPHOSPHINYL)ETHYL ESTER □ BUTILCHLOROFOS □ DIMETHOXY-2,2,2-TRICHLORO-1-N-BUTYRYLOXY-ETHYLPHOSPHINE OXIDE □ O,O-DIMETHYL-(1-BUTYRYLOXY-2,2,2-TRICHLOROETHYL) PHOSPHONATE □ O,O-DIMETHYL 2,2,2-TRICHLORO-1-(N-BUTYRYLOXY)ETHYLPHOSPHONATE □ ENT 20,852 □ F-139 □ T-113 □ TRIBUFON

TOXICITY DATA WITH REFERENCE
dnd-mus-ipr 200 mg/kg PCBPBS 6,101,76
orl-rat LD50:1100 mg/kg ARSIM* 20,6,66
skn-rat LD50:7000 mg/kg FMCHA2 -,C40,83
ipr-rat LD50:700 mg/kg ZHYGAM 25,512,79
scu-rat LD50:3000 mg/kg PAREAQ 11,636,59
orl-mus LD50:760 mg/kg ZHYGAM 25,512,79
skn-dog LD50:3080 mg/kg ZHYGAM 25,512,79

SAFETY PROFILE: Moderately toxic by ingestion, skin contact, intraperitoneal, subcutaneous, and possibly other routes. Mutation data reported. When heated to decomposition it emits highly toxic fumes of PO_x and Cl^-. See also ESTERS.

BPG250 **CAS:6365-83-9** **HR: 3**
BUTOPHEN
mf: $C_{10}H_{12}N_2O_5 \cdot H_3N$ mw: 257.28

SYNS: 2-sec-BUTYL-4,6-DINITROPHENOL AMMONIUM SALT □ CHEMOX SELECTIVE □ 4,6-DINITRO-2-sec.BUTYLFENOLATE AMMONY (CZECH) □ 4,6-DINITRO-o-sec-BUTYLPHENOL AMMONIUM SALT □ 4,6-DINITRO-2-sec-BUTYLPHENOL AMMONIUM SALT □ DINOSEB (AMINE) □ DNBP AMMONIUM SALT □ DOW SELECTIVE □ 2-(1-METHYL-N-PROPYL) 4,6-DINITROPHENOL AMMONIUM SALT □ SELECTIVE □ SINOX W

TOXICITY DATA WITH REFERENCE
eye-rbt 50 μg/24H SEV 28ZPAK -,108,72
orl-rat LD50:45 mg/kg 28ZPAK -,108,72
skn-rat LDLo:67 mg/kg BJIMAG 26,59,69

SAFETY PROFILE: A poison by ingestion and skin contact. A severe eye irritant. When heated to decomposition it emits very toxic fumes of NH_3 and NO_x.

BPG325 **CAS:58786-99-5** **HR: 3**
BUTORPHANOL TARTRATE
mf: $C_{21}H_{29}NO_2 \cdot C_4H_6O_6$ mw: 477.61

PROP: Solid. Mp: 217–219°.

SYNS: BT □ STADOL □ TORATE □ TORBUTROL

TOXICITY DATA WITH REFERENCE
ivn-rat TDLo:11 mg/kg (female 7-17D post):REP IYKEDH 13,446,82
scu-rat TDLo:26 mg/kg (female 17-22D post):REP IYKEDH 13,429,82
scu-rat TDLo:11 mg/kg (female 7-17D post):REP IYKEDH 13,401,82
scu-rat TDLo:55 mg/kg (female 7-17D post):REP IYKEDH 13,401,82
scu-rat TDLo:88 mg/kg (male 9W pre):REP PBPSDY 2,19,79
ivn-rat TDLo:55 mg/kg (female 7-17D post):TER IYKEDH 13,446,82
scu-rat TDLo:84 mg/kg (male 63D pre):TER IYKEDH 13,390,82
scu-rbt TDLo:13 mg/kg (female 6-18D post):REP IYKEDH 13,421,82
ipr-rat TDLo:9100 mg/kg (female 26W pre):REP IYKEDH 13,261,82
ipr-rat TDLo:1500 mg/kg (male 30D pre):REP IYKEDH 13,176,82
ipr-rat TDLo:3 g/kg (30D male):REP IYKEDH 13,176,82
scu-rat TDLo:55 mg/kg (female 7-17D post):TER IYKEDH 13,401,82
scu-rat TDLo:11 mg/kg (female 7-17D post):TER IYKEDH 13,401,82
orl-rat LD50:315 mg/kg IYKEDH 13,145,82
ipr-rat LD50:127 mg/kg IYKEDH 13,145,82
scu-rat LD50:425 mg/kg IYKEDH 13,145,82
ivn-rat LD50:17 mg/kg DRUGAY 16,474,78
ims-rat LD50:255 mg/kg IYKEDH 13,145,82
orl-mus LD50:395 mg/kg DRUGAY 16,474,78

ipr-mus LD50:192 mg/kg IYKEDH 13,145,82
scu-mus LD50:299 mg/kg PBPSDY 2,19,79
ivn-mus LD50:36 mg/kg IYKEDH 13,145,82
ims-mus LD50:208 mg/kg IYKEDH 13,145,82

SAFETY PROFILE: Poison by ingestion, subcutaneous, intramuscular, intravenous, and intraperitoneal routes. An experimental teratogen. Other experimental reproductive effects. An analgesic. When heated to decomposition it emits toxic fumes of NO_x.

BPG500 CAS:3329-56-4 HR: 3
BUTOXY ACETYLENE
mf: $C_6H_{10}O$ mw: 98.14

$(C_4H_9)OC{\equiv}CH$

PROP: Liquid. Bp: 106–108°.

SAFETY PROFILE: Explodes at 100°C when heated in a sealed container. When heated to decomposition it emits acrid smoke and fumes. See also ACETYLENE COMPOUNDS.

BPG750 CAS:60444-92-0 HR: 3
2-N-BUTOXYBENZAMIDE
mf: $C_{11}H_{15}NO_2$ mw: 193.27

SYNS: H.P. 165 □ o-BUTOXYBENZAMIDE

TOXICITY DATA WITH REFERENCE
orl-mus LD50:1300 mg/kg JPPMAB 4,872,52
ipr-mus LD50:360 mg/kg JPPMAB 4,872,52

SAFETY PROFILE: Poison by intraperitoneal route. Moderately toxic by ingestion. When heated to decomposition it emits toxic fumes of NO_x. See also AMIDES.

BPH750 CAS:67032-45-5 HR: 3
p-BUTOXYBENZOIC ACID-3-(2-METHYLPIPERIDINO) PROPYL ESTER HYDROCHLORIDE
mf: $C_{20}H_{31}NO_3{\cdot}ClH$ mw: 369.98

SYN: C-10

TOXICITY DATA WITH REFERENCE
skn-rbt 1% MLD AIPTAK 137,410,62
eye-rbt 2500 ppm MLD AIPTAK 137,410,62
ipr-mus LD50:73,600 μg/kg AIPTAK 137,410,62
scu-mus LD50:177 mg/kg JACSAT 68,2592,46
ivn-mus LD50:22 mg/kg JACSAT 68,2592,46

SAFETY PROFILE: Poison by intraperitoneal, subcutaneous and intravenous routes. A skin and eye irritant. When heated to decomposition it emits very toxic fumes of HCl and NO_x. See also ESTERS.

BPI125 CAS:29025-14-7 HR: 3
BUTOXYBENZYL HYOSCYAMINE BROMIDE
mf: $C_{28}H_{38}NO_4{\cdot}Br$ mw: 532.58

PROP: Crystals from ethanol-acetone. Mp: 166–168°. Also reported as white needles from isopropanol, mp: 158–160°. Freely sol in glacial acetic acid; sol in chloroform, DMF. Sparingly sol in ethanol; sltly sol in water, 0.1N HCl, 0.1N NaOH. Practically insol in acetone, ether, and benzene.

SYNS: BHB □ p-BUTOXYBENZYL HYOSCYAMINIUM BROMIDE □ (−)-8-(p-BUTOXYBENZYL)-3-α-HYDROXY-1-α-H,5-α-H-TROPANIUM BROMIDE TROPATE (ester) □ 1-(1-(p-n-BUTOXYBENZYL)HYOSCYAMINIUM) BROMIDE □ BUTROPIUM BROMIDE □ COLIOPAN

TOXICITY DATA WITH REFERENCE
ipr-rat TDLo:144 mg/kg (female 9-14D post):REP OYYAA2 8,319,74
orl-mus TDLo:360 mg/kg (female 7-12D post):REP OYYAA2 8,319,74
ipr-rat TDLo:36 mg/kg (female 9-14D post):REP OYYAA2 8,319,74
ipr-rat TDLo:36 mg/kg (female 9-14D post):TER OYYAA2 8,319,74
orl-rat TDLo:4375 mg/kg (25W male):REP OYYAA2 8,307,74
orl-rat TDLo:1440 mg/kg (9-14D preg):TER OYYAA2 8,319,74
ipr-rat LD50:113 mg/kg IYKEDH 5,106,74
ivn-rat LD50:21 mg/kg IYKEDH 5,106,74
orl-mus LD50:1500 mg/kg USXXAM #3696110
scu-mus LD50:370 mg/kg NIIRDN 6,355,82
ivn-mus LD50:6400 μg/kg IYKEDH 5,106,74
ims-mus LD50:285 mg/kg NIIRDN 6,355,82
ivn-rbt LD50:6800 μg/kg OYYAA2 8,285,74

SAFETY PROFILE: Poison by subcutaneous, intramuscular, intravenous, and intraperitoneal routes. Moderately toxic by ingestion. An experimental teratogen. Other experimental reproductive effects. When heated to decomposition it emits toxic fumes of Br^- and NO_x. See also BROMIDES.

BPI300 CAS:832-06-4 HR: 2
2-BUTOXYCARBONYLMETHYLENE-4-OXOTHIAZOLIDONE
mf: $C_9H_{13}NO_3S$ mw: 215.29

SYNS: ACETIC ACID, (4-OXO-2-THIAZOLIDINYLIDENE)-, BUTYL ESTER (9CI) □ 2-(n-BUTYLOXYCARBONYLMETHYLENE)THIAZOLID-4-ONE □ ICI 43823

TOXICITY DATA WITH REFERENCE
orl-rat TDLo:193 g/kg/2Y-C:ETA EXMDA4 (145),289,68

SAFETY PROFILE: Questionable carcinogen with experimental tumorigenic data. When heated to decomposition it emits toxic fumes of NO_x and SO_x.

BPI625 CAS:58763-31-8 HR: 3
4′-BUTOXY-3′-CHLORO-5′-METHYL-3-PIPERIDINO-PROPIOPHENONE HYDROCHLORIDE
mf: $C_{19}H_{28}ClNO_2{\cdot}ClH$ mw: 374.39

SYNS: 1-(4-BUTOXY-3-CHLORO-5-METHYLPHENYL)-3-(1-PIPERIDINYL)1-PROPANONE HYDROCHLORIDE (9CI) □ β-PIPERIDINOAETHYL-(3-CHLOR-4-n-BUTOXY-5-METHYLPHENYL)KETONHYDROCHLORID (GERMAN)

TOXICITY DATA WITH REFERENCE
orl-mus LD50:475 mg/kg PHARAT 31,21,76
scu-mus LD50:1000 mg/kg PHARAT 31,21,76
ivn-mus LD50:43 mg/kg PHARAT 31,21,76

SAFETY PROFILE: Poison by intravenous route. Moderately toxic by ingestion and subcutaneous routes. When heated to decomposition it emits toxic fumes of NO_x, Cl^-, and HCl.

BPI750 **HR: 3**
4'-BUTOXY-2'-CHLORO-2-PYRROLIDINYL ACETANILIDE HYDROCHLORIDE
mf: $C_{16}H_{23}ClN_2O_2 \cdot ClH$ mw: 347.32

SYN: C 3187

TOXICITY DATA WITH REFERENCE
eye-rbt 2% SEV ARZNAD 8,270,58
ipr-rat LD50:287 mg/kg ARZNAD 8,270,58
scu-mus LD50:550 mg/kg ARZNAD 8,270,58

SAFETY PROFILE: Poison by intraperitoneal route. Moderately toxic by subcutaneous route. A severe eye irritant. When heated to decomposition it emits very toxic fumes of Cl^-, NO_x, and HCl.

BPJ000 **CAS:41296-95-1** **HR: 3**
4'-BUTOXY-2-(DIETHYLAMINO)ACETANILIDE HYDROCHLORIDE
mf: $C_{16}H_{26}N_2O_2 \cdot ClH$ mw: 314.90

SYN: C 3121

TOXICITY DATA WITH REFERENCE
eye-rbt 2% MOD ARZNAD 8,270,58
ipr-rat LD50:220 mg/kg ARZNAD 8,270,58
scu-mus LD50:695 mg/kg ARZNAD 8,270,58

SAFETY PROFILE: Poison by intraperitoneal route. Moderately toxic by subcutaneous route. An eye irritant. When heated to decomposition it emits very toxic fumes of NO_x and HCl.

BPJ250 **CAS:77966-20-2** **HR: 3**
2-BUTOXY-N-(2-(DIETHYLAMINO)ETHYL)-N-(2,6-XYLYL)CINCHONINAMIDE HYDROCHLORIDE
mf: $C_{28}H_{37}N_3O_2 \cdot ClH$ mw: 484.14

SYN: 2-BUTOXY-N-((2-DIETHYLAMINO)ETHYL)-N-(2,6-XYLYL)-4-QUINOLINECARBOXAMIDE HYDROCHLORIDE

TOXICITY DATA WITH REFERENCE
eye-rbt 2% SEV ARZNAD 8,708,58
ipr-rat LD50:300 mg/kg ARZNAD 8,708,58
scu-mus-LD50:1175 mg/kg ARZNAD 8,708,58

SAFETY PROFILE: Poison by intraperitoneal route. Moderately toxic by subcutaneous route. A severe eye irritant. When heated to decomposition it emits very toxic fumes of HCl and NO_x.

BPJ500 **CAS:78109-80-5** **HR: 3**
o-BUTOXY-N-(5-(DIETHYLAMINO)-2-PENTYL)BENZAMIDE HYDROCHLORIDE
mf: $C_{20}H_{34}N_2O_2 \cdot ClH$ mw: 371.02

SYNS: 2-BUTOXYBENZOESAEURE-4'-DIAETHYLAMINO-L'-METHYLBUTYLAMID (1') HYDROCHLORID (GERMAN) □ D-649

TOXICITY DATA WITH REFERENCE
scu-mus LD50:130 mg/kg ARZNAD 10,743,60
ivn-mus LD50:15 mg/kg ARZNAD 10,743,60

SAFETY PROFILE: Poison by subcutaneous and intravenous routes. When heated to decomposition it emits very toxic fumes of HCl and NO_x.

BPJ750 **CAS:78109-81-6** **HR: 3**
o-BUTOXY-N-(3-(DIETHYLAMINO)PROPYL)BENZAMIDE HYDROCHLORIDE
mf: $C_{18}H_{30}N_2O_2 \cdot ClH$ mw: 342.96

SYNS: 2-BUTOXYBENZOESAEURE-3'-DIAETHYLAMINOPROPYLAMID-(1') HYDROCHLORID (GERMAN) □ D-638

TOXICITY DATA WITH REFERENCE
ipr-mus LD50:75 mg/kg ARZNAD 10,743,60
scu-mus LD50:160 mg/kg ARZNAD 10,743,60
ivn-mus LD50:30 mg/kg ARZNAD 10,743,60

SAFETY PROFILE: Poison by intraperitoneal, subcutaneous, and intravenous routes. When heated to decomposition it emits very toxic fumes of HCl and NO_x.

BPJ850 **CAS:111-76-2** **HR: 3**
2-BUTOXYETHANOL
DOT: UN 2369
mf: $C_6H_{14}O_2$ mw: 118.20

PROP: Clear, mobile liquid; pleasant odor. Fp: −74.8°, bp: 171–172°, flash p: 160°F (COC), d: 0.9012 @ 20°/20°, vap press: 300 mm @ 140°.

SYNS: BUCS □ BUTOKSYETYLOWY ALKOHOL (POLISH) □ 2-BUTOSSI-ETANOLO (ITALIAN) □ 2-BUTOXY-AETHANOL (GERMAN) □ BUTOXYETHANOL □ n-BUTOXYETHANOL □ 2-BUTOXY-1-ETHANOL □ BUTYL CELLOSOLVE □ o-BUTYL ETHYLENE GLYCOL □ BUTYL GLYCOL □ BUTYLGLYCOL (FRENCH, GERMAN) □ BUTYL OXITOL □ DOWANOL EB □ EKTASOLVE EB □ ETHYLENE GLYCOL-n-BUTYL ETHER □ ETHYLENE GLYCOL MONOBUTYL ETHER (MAK, DOT) □ GAFCOL EB □ GLYCOL BUTYL ETHER □ GLYCOL ETHER EB □ GLYCOL ETHER EB ACETATE □ GLYCOL MONOBUTYL ETHER □ JEFFERSOL EB □ MONOBUTYL GLYCOL ETHER □ 3-OXA-1-HEPTANOL □ POLY-SOLV EB

TOXICITY DATA WITH REFERENCE
skn-rbt 500 mg open MLD UCDS**
ihl-rat TCLo:200 ppm/6H (female 6-15D post):REP EVHPAZ 57,47,84
orl-mus TDLo:9440 mg/kg (female 7-14D post):REP EVHPAZ 57,141,84
ihl-rbt TCLo:200 ppm/6H (female 6-18D post):REP EVHPAZ 57,47,84
ihl-rbt TCLo:100 ppm/6H (female 6-18D post):TER EVHPAZ 57,47,84
ihl-rat TCLo:25 ppm/6H (female 6-15D post):TER EVHPAZ 57,47,84
orl-wmn TDLo:600 mg/kg HUTODJ 7,187,88
ihl-hmn TCLo:195 ppm/8H:GIT AMIHAB 14,114,56
ihl-hmn TCLo:100 ppm:NOSE,EYE,CNS NPIRI* 1,50,74
orl-rat LD50:470 mg/kg DOWCC* MSD-46
ihl-rat LC50:2900 mg/m³ GTPZAB 32(3),48,88
ipr-rat LD50:220 mg/kg 85GMAT -,67,82
ivn-rat LD50:340 mg/kg AMIHAB 14,114,56
ihl-mus LC50:700 ppm/7H JIHTAB 25,157,43

scu-mus LDLo:500 mg/kg JPETAB 42,355,31
orl-rbt LD50:300 mg/kg YKYUA6 32,1241,81
skn-gpg LD50:230 mg/kg TXAPA9 7,559,65

CONSENSUS REPORTS: Reported in EPA TSCA Inventory. Glycol ethers are on the Community Right-To-Know List.

OSHA PEL: TWA 25 ppm (skin)
ACGIH TLV: TWA 25 ppm (skin)
DFG MAK: 20 ppm (100 mg/m^3)
DOT CLASSIFICATION: 6.1; *Label:* KEEP AWAY FROM FOOD

SAFETY PROFILE: Poison by ingestion, skin contact, intraperitoneal, and intravenous routes. Moderately toxic via inhalation and subcutaneous routes. Human systemic effects by inhalation: nausea or vomiting, headache, nose tumors, unspecified eye effects. Experimental teratogenic and reproductive effects. A skin irritant. Combustible liquid when exposed to heat or flame. To fight fire, use foam, CO_2, and dry chemical. Incompatible with oxidizing materials, heat, and flame. When heated to decomposition it emits acrid smoke and irritating fumes.

For occupational chemical analysis use NIOSH: Alcohols IV, 1403.

BPK250 CAS:78-51-3 HR: 3
2-BUTOXYETHANOL PHOSPHATE
mf: $C_{18}H_{39}O_7P$ mw: 398.54

PROP: Light-colored liquid, butyl-like odor. Mp: −70°, bp: 200–230° @ 4 mm, flash p: 435°F, d: 1.02 @ 20°/20°, vap press: 0.03 mm @ 150°, vap d. 13.8.

SYNS: KP 140 □ KRONITEX KP-140 □ PHOSFLEX T-BEP □ TBEP □ TRI(2-BUTOXYETHANOL PHOSPHATE) □ TRIBUTOXYETHYL PHOSPHATE □ TRI(2-BUTOXYETHYL) PHOSPHATE □ TRIBUTYL CELLOSOLVE PHOSPHATE □ TRIS(2-BUTOXYETHYL) ESTER PHOSPHORIC ACID □ TRIS(2-BUTOXYETHYL) PHOSPHATE

TOXICITY DATA WITH REFERENCE
skn-rbt 500 mg/24H MLD 85JCAE-,1142,86
eye-rbt 500 mg/24H MLD 85JCAE-,1142,86
orl-rat LD50:3000 mg/kg NPIRI* 2,93,75
ivn-mus LD50:180 mg/kg CSLNX* NX#00391
orl-gpg LD50:3000 mg/kg 29ZWAE -,336,68

CONSENSUS REPORTS: Reported in EPA TSCA Inventory.

SAFETY PROFILE: A poison by intravenous route. Moderately toxic by ingestion. A skin and eye irritant. Combustible when exposed to heat or flame. Dangerous; see also PHOSPHATES; can react with oxidizing materials. To fight fire, use water, foam, CO_2, dry chemical. When heated to decomposition it emits toxic fumes of PO_x.

BPK500 CAS:7251-90-3 HR: 2
2-BUTOXYETHOXY ACRYLATE
mf: $C_9H_{16}O_4$ mw: 188.25

SYNS: BUTYL CELLOSOLVE ACRYLATE □ 2-PROPENOIC ACID 2-BUTOXYETHYL ESTER

TOXICITY DATA WITH REFERENCE
orl-rat LD50:6500 mg/kg AIHAAP 30,470,69
skn-rbt LD50:640 mg/kg AIHAAP 30,470,69

CONSENSUS REPORTS: Reported in EPA TSCA Inventory.

SAFETY PROFILE: Moderately toxic by skin contact. Mildly toxic by ingestion. See also ESTERS. When heated to decomposition it emits acrid smoke and irritating fumes.

BPK750 CAS:4413-13-2 HR: 2
1-BUTOXY-2-ETHOXYETHANE
mf: $C_8H_{18}O_2$ mw: 146.26

SYN: 1-(2-ETHOXYETHOXY)-BUTANE

TOXICITY DATA WITH REFERENCE
skn-rbt 10 mg/24H open MLD AMIHBC 10,61,54
skn-rbt 500 mg/24H MLD 85JCAE-,256,86
eye-rbt 20 mg open AMIHBC 10,61,54
eye-rbt 100 mg/24H MOD 85JCAE-,256,86
orl-rat LD50:2830 mg/kg AMIHBC 10,61,54
skn-rbt LD50:2120 mg/kg AMIHBC 10,61,54

SAFETY PROFILE: Moderately toxic by ingestion and skin contact. A skin and eye irritant. When heated to decomposition it emits acrid smoke and irritating fumes.

BPL250 CAS:112-56-1 HR: 3
2-(2-BUTOXY ETHOXY)ETHYL THIOCYANATE
mf: $C_9H_{17}NO_2S$ mw: 203.33

PROP: Liquid. Bp: 120–125° @ 0.25 mm.

SYNS: 2-(2-(BUTOXY)ETHOXY)ETHYL THIOCYANIC ACID ESTER □ BUTOXYRHODANODIETHYL ETHER □ β-BUTOXY-β′-THIOCYANODIETHYL ETHER □ 2-BUTOXY-2′-THIOCYANODIETHYL ETHER □ 1-BUTOXY-2-(2-THIOCYANOETHOXY)ETHANE □ 1-BUTOXY-2-(2-THIOCYANATOETHYXY)ETHANE □ BUTYL CARBITOL RHODANATE □ BUTYL CARBITOL THIOCYANATE □ ENT 6 □ ETHANOL-2-(2-BUTOXYETHOXY) THIOCYANATE □ LETHANE □ LETHANE 384 □ LETHANE 384 REGULAR

TOXICITY DATA WITH REFERENCE
orl-rat LD50:90 mg/kg FMCHA2 -,D180,80
skn-rat LD50:250 mg/kg WRPCA2 9,119,70
ipr-rat LD50:90 mg/kg INMEAF 11,-,42
scu-rat LD50:550 mg/kg INMEAF 11,-,42
ipr-mus LD50:41 mg/kg PCBPBS 2,95,72
scu-mus LDLo:200 mg/kg JIDHAN 18,310,36
ivn-mus LD50:56 mg/kg CSLNX* NX#02402
orl-dog LD50:30 mg/kg PCOC** -,657,66
scu-dog LD50:200 mg/kg INMEAF 11,-,42
orl-rbt LD50:35 mg/kg JPETAB 82,377,44
skn-rbt LD50:125 mg/kg SPEADM 78-1,20,78

SAFETY PROFILE: A poison by ingestion, skin contact, intraperitoneal, subcutaneous, and intravenous routes. Moderately toxic by an unspecified route. High concentrations can cause central nervous system depression. An insecticide. See also THIOCYANATES, ESTERS, and ETHERS. When heated to decomposition it emits very toxic fumes of SO_x, NO_x, and CN^-.

BPL500 **CAS:124-16-3** **HR: 2**
1-BUTOXY ETHOXY-2-PROPANOL
mf: $C_9H_{20}O_3$ mw: 176.29

PROP: D: 0.9310 @ 20°/20°, bp: 230.3°, fp: −90°, flash p: 250°F (OC). Sol in water.

SYN: 1-(2-BUTOXYETHOXY)-2-PROPANOL

TOXICITY DATA WITH REFERENCE
skn-rbt 485 mg open MLD UCDS** 12/29/71
orl-rat LD50:4 mL/kg AIHAAP 30,470,69
skn-rbt LD50:2830 μL/kg AIHAAP 30,470,69

CONSENSUS REPORTS: Reported in EPA TSCA Inventory.

SAFETY PROFILE: Moderately toxic by ingestion and skin contact. A skin and eye irritant. Combustible when exposed to heat or flame. To fight fire, use alcohol foam, dry chemical, spray, or mist. When heated to decomposition it emits acrid and irritating fumes.

BPL750 **CAS:10043-18-2** **HR: 2**
3-(2-BUTOXYETHOXY)PROPANOL
mf: $C_9H_{20}O_3$ mw: 176.29

TOXICITY DATA WITH REFERENCE
skn-rbt 10 mg/24H open MLD AMIHBC 10,61,54
eye-rbt 20 mg open SEV AMIHBC 10,61,54
orl-rat LD50:5160 mg/kg AMIHBC 10,61,54
skn-rbt LD50:3000 mg/kg AMIHBC 10,61,54

SAFETY PROFILE: Moderately toxic by skin contact. Mildly toxic by ingestion. A skin and severe eye irritant. When heated to decomposition it emits acrid smoke and irritating fumes.

BPM000 **CAS:112-07-2** **HR: 3**
2-BUTOXYETHYL ACETATE
mf: $C_8H_{16}O_3$ mw: 160.24

PROP: Colorless liquid; fruity odor. Bp: 192.3°, d: 0.9424 @ 20°/20°, fp: −63.5°, flash p: 190°F. Sol in hydrocarbons and organic solvents; insol in water.

SYNS: 2-BUTOXYETHANOL ACETATE □ 2-BUTOXYETHYL ESTER ACETIC ACID □ BUTYL CELLOSOLVE ACETATE □ EKTASOLVE EB ACETATE □ ETHYLENE GLYCOL MONOBUTYL ETHER ACETATE (MAK) □ GLYCOL MONOBUTYL ETHERACETATE

TOXICITY DATA WITH REFERENCE
skn-rbt 500 mg open MLD UCDS** 1/31/66
eye-rbt 500 mg/24H MLD 85JCAE-,713,86
orl-rat LD50:2400 mg/kg TXAPA9 51,117,79
orl-mus LD50:3200 mg/kg KODAK* 21MAY71
skn-rbt LD50:1500 mg/kg TXAPA9 51,117,79

CONSENSUS REPORTS: Reported in EPA ISCA Inventory. Glycol ethers are on the Community Right-To-Know List.

DFG MAK: 20 ppm (135 mg/m³)

SAFETY PROFILE: Moderately toxic by ingestion and skin contact. Mild skin irritant. Flammable when exposed to heat, flame, or oxidizers. To fight fire, use alcohol foam. When heated to decomposition it emits acrid smoke and irritating fumes. See also ESTERS.

BPM660 **CAS:1852-16-0** **HR: 2**
N-(BUTOXYMETHYL)-2-PROPENAMIDE
mf: $C_8H_{15}NO_2$ mw: 157.24

SYNS: ACRYLAMIDE, N-BUTOXYMETHYL- □ N-(BUTOXYMETHYL) ACRYLAMIDE □ N-BUTOXYMETHYLAKRYLAMID □ 2-PROPENAMIDE, N-(BUTOXYMETHYL)-(9CI)

TOXICITY DATA WITH REFERENCE
orl-rat LD50:1030 mg/kg 85JCAE -,706,86

CONSENSUS REPORTS: Reported in EPA TSCA Inventory.

SAFETY PROFILE: Moderately toxic by ingestion. When heated to decomposition it emits toxic vapors of NO_x.

BPM750 **CAS:27471-60-9** **HR: 3**
2-(p-BUTOXYPHENOXY)-N-(2-(DIETHYLAMINO)ETHYL)-2,5′-DIETHOXYACETANILIDE MONOHYDROCHLORIDE
mf: $C_{28}H_{42}N_2O_5$•ClH mw: 523.18

PROP: Solid. Mp: 140°.

SYNS: ANP 3548 □ CHLORHYDRATE de N-(DIETHOXY-2,5-PHENYL)-N-DIETHYLAMINO-2-ETHYL BUTOXY-4-PHENOXYACETAMIDE □ N,N-DIETHYL-N′-(2,5-DIETHOXYPHENYL)-N′-(4-BUTOXYPHENOXYACETYL) ETHYLENEDIAMINE HCl □ FENOXEDIL □ FENOXEDIL HYDROCHLORIDE □ SUPLEXEDIL

TOXICITY DATA WITH REFERENCE
orl-rat LD50:2400 mg/kg EJMCA5 10,291,75
ipr-rat LD50:175 mg/kg EJMCA5 10,291,75
scu-rat LD50:2065 mg/kg EJMCA5 10,291,75
ivn-rat LD50:10 mg/kg EJMCA5 10,291,75
orl-mus LD50:750 mg/kg EJMCA5 10,286,75
ipr-mus LD50:82 mg/kg EJMCA5 10,291,75
scu-mus LD50:341 mg/kg EJMCA5 10,291,75
ivn-mus LD50:17 mg/kg USXXAM #3818021
orl-rbt LD50:815 mg/kg EJMCA5 10,291,75

SAFETY PROFILE: Poison by intraperitoneal, intravenous, and subcutaneous routes. Moderately toxic by ingestion. A vasodilator. When heated to decomposition it emits very toxic fumes of NO_x and HCl.

BPN000 **CAS:27468-64-0** **HR: 3**
2-(p-BUTOXYPHENOXY)-N-(2-(DIETHYLAMINO)ETHYL)-N-(2,4-DIMETHOXYPHENYL)ACETAMIDE HYDROCHLORIDE
mf: $C_{26}H_{38}N_2O_5$•ClH mw: 495.12

TOXICITY DATA WITH REFERENCE
orl-mus LD50:400 mg/kg EJMCA5 10,286,75
ivn-mus LD50:40 mg/kg EJMCA5 10,286,75

SAFETY PROFILE: Poison by ingestion and intravenous routes. When heated to decomposition it emits very toxic fumes of NO_x and HCl.

BPN250 CAS:27468-66-2 **HR: 3**
2-(p-BUTOXYPHENOXY)-N-(2-(DIETHYLAMINO)ETHYL)-N-(2,5-DIMETHOXYPHENYL)ACETAMIDE HYDROCHLORIDE
mf: $C_{26}H_{38}N_2O_5 \cdot ClH$ mw: 495.12

TOXICITY DATA WITH REFERENCE
orl-mus LD50:400 mg/kg EJMCA5 10,286,75
ivn-mus LD50:25 mg/kg EJMCA5 10,286,75

SAFETY PROFILE: Poison by ingestion and intravenous routes. When heated to decomposition it emits very toxic fumes of NO_x and HCl.

BPO250 CAS:27468-71-9 **HR: 3**
2-(p-BUTOXYPHENOXY)-N-(2(DIMETHYLAMINO)ETHYL)-N-(2,6-DIMETHYLPHENYL)ACETAMIDE HYDROCHLORIDE
mf: $C_{24}H_{34}N_2O_3 \cdot ClH$ mw: 435.06

TOXICITY DATA WITH REFERENCE
orl-mus LD50:400 mg/kg EJMCA5 10,286,75
ivn-mus LD50:18 mg/kg EJMCA5 10,286,75

SAFETY PROFILE: Poison by ingestion and intravenous routes. When heated to decomposition it emits very toxic fumes of HCl and NO_x.

BPP250 CAS:3102-00-9 **HR: 3**
3-n-BUTOXY-1-PHENOXY-2-PROPANOL
mf: $C_{13}H_{20}O_3$ mw: 224.33

SYNS: (3-n-BUTOXY-2-HYDROXYPROPYL)PHENYL ETHER □ 1-BUTOXY-3-PHENOXY-2-PROPANOL □ FEBUPROL □ H-33 □ K-10033 □ VALBIL

TOXICITY DATA WITH REFERENCE
orl-rat LD50:2370 mg/kg DRFUD4 3,191,78
ipr-rat LD50:400 mg/kg DRFUD4 3,191,78
orl-mus LD50:3050 mg/kg DRFUD4 3,191,78
ipr-mus LD50:436 mg/kg DRFUD4 3,191,78
orl-dog LD50:500 mg/kg DRFUD4 3,191,78
ivn-dog LD50:150 mg/kg DRFUD4 3,191,78

SAFETY PROFILE: Poison by intraperitoneal and intravenous routes. Moderately toxic by ingestion. Stimulates the production of bile by the liver. See also ETHERS. When heated to decomposition it emits acrid smoke and irritating fumes.

BPP750 CAS:2438-72-4 **HR: 2**
p-BUTOXYPHENYLACETOHYDROXAMIC ACID
mf: $C_{12}H_{17}NO_3$ mw: 223.30

PROP: Needles from Me_2CO. Mp: 153–154°. Insol in H_2O.

SYNS: BUFEXAMIC ACID □ 4-BUTOXYPHENYLACETOHYDROXAMIC ACID □ CP 1044 J3 □ DROXAROL □ DROXARYL □ FLOGICID □ FLOGOCID N PLASTIGEL □ J3 □ PARFENAC □ PARFENAL

TOXICITY DATA WITH REFERENCE
sce-ham:ovr 20 μmol/L PAACA3 21,126,80
orl-rat TDLo:5 g/kg (6-15D preg):TER ARZNAD 20,565,70
orl-rat LD50:3370 mg/kg NIIRDN 6,681,82
ipr-rat LD50:805 mg/kg YKYUA6 28,253,77
orl-mus LD50:8000 mg/kg JMCMAR 13,211,70
ipr-mus LD50:1195 mg/kg YKYUA6 28,253,77

CONSENSUS REPORTS: EPA Genetic Toxicology Program.

SAFETY PROFILE: Moderately toxic by ingestion and intraperitoneal routes. Experimental teratogenic effects. Mutation data reported. When heated to decomposition it emits toxic fumes of NO_x.

BPR000 CAS:77791-53-8 **HR: 3**
4′-BUTOXY-2-PIPERIDINOACETANILIDE HYDROCHLORIDE
mf: $C_{17}H_{26}N_2O_2 \cdot ClH$ mw: 326.91

SYN: C 3125

TOXICITY DATA WITH REFERENCE
eye-rbt 2% SEV ARZNAD 8,407,58
ipr-rat LD50:200 mg/kg ARZNAD 8,407,58
scu-mus LD50:665 mg/kg ARZNAD 8,407,58

SAFETY PROFILE: Poison by intraperitoneal route. Moderately toxic by subcutaneous route. A severe eye irritant. When heated to decomposition it emits very toxic fumes of NO_x and HCl.

BPR250 **HR: 3**
4-BUTOXY-3-(PIPERIDINO)PROPIOPHENONE HYDROCHLORIDE
mf: $C_{18}H_{27}NO_2 \cdot ClH$ mw: 325.92

SYN: C 5422

TOXICITY DATA WITH REFERENCE
eye-rbt 2% SEV ARZNAD 8,708,58
ipr-rat LD50:33 mg/kg ARZNAD 8,708,58
scu-mus LD50:37 mg/kg ARZNAD 8,708,58

SAFETY PROFILE: Poison by intraperitoneal and subcutaneous routes. A severe eye irritant. When heated to decomposition it emits very toxic fumes of HCl and NO_x.

BPR500 CAS:536-43-6 **HR: 3**
4′-BUTOXY-3-PIPERIDINO PROPIOPHENONE HYDROCHLORIDE
mf: $C_{18}H_{27}NO_2 \cdot ClH$ mw: 325.92

PROP: Solid. Mp: 175–176°.

SYNS: 1-(2-(4-BUTOXYBENZOYL)ETHYL)PIPERIDINE HYDROCHLORIDE □ 4-n-BUTOXY-β-(1-PIPERIDYL)PROPIOPHENONE HYDROCHLORIDE □ DICLONIA □ DYCLOCAINUM □ DYCLONE HYDROCHLORIDE □ DYCLONINE HYDROCLORIDE □ DYCLOTHANE □ P-267 □ S 154

TOXICITY DATA WITH REFERENCE
skn-rbt 1% MLD AIPTAK 137,410,62
eye-rbt 1% MLD AIPTAK 137,410,62
eye-rbt 2% SEV ARZNAD 8,708,58
ipr-rat LD50:33 mg/kg ARZNAD 8,708,58
scu-rat LD50:201 mg/kg JPETAB 115,413,55
orl-mus LDLo:100 mg/kg TXAPA9 2,616,60
ipr-mus LD50:52 mg/kg AIPTAK 137,410,62
scu-mus LD50:42 mg/kg ARZNAD 5,559,55
ivn-mus LD50:20 mg/kg JPETAB 115,419,55
orl-dog LDLo:40 mg/kg TXAPA9 2,616,60

ivn-dog LD50:9500 µg/kg JPETAB 115,419,55
orl-rbt LDLo:200 mg/kg TXAPA9 2,616,60

SAFETY PROFILE: A poison by ingestion, intraperitoneal, subcutaneous, and intravenous routes. A skin and severe eye irritant. When heated to decomposition it emits very toxic fumes of HCl and NO_x.

BPS000 CAS:7420-06-6 HR: 2
3-BUTOXY PROPANOIC ACID
mf: $C_7H_{14}O_3$ mw: 146.21

SYN: 3-BUTOXYPROPIONIC ACID

TOXICITY DATA WITH REFERENCE
skn-rbt 10 mg/24H open MLD AMIHBC 10,61,54
eye-rbt 250 µg open SEV AMIHBC 10,61,54
orl-rat LD50:5190 mg/kg AMIHBC 10,61,54
skn-rbt LD50:630 mg/kg AMIHBC 10,61,54

SAFETY PROFILE: Moderately toxic by skin contact. Mildly toxic by ingestion. A skin and severe eye irritant. When heated to decomposition it emits acrid smoke and irritating fumes.

BPS250 CAS:5131-66-8 HR: 2
1-BUTOXY-2-PROPANOL
mf: $C_7H_{16}O_2$ mw: 132.23

SYNS: PROPASOL SOLVENT B ☐ PROPYLENE GLYCOL-n-BUTYL ETHER

TOXICITY DATA WITH REFERENCE
skn-rbt LD50:3100 mg/kg NPIRI* 1,102,74

CONSENSUS REPORTS: Reported in EPA TSCA Inventory. Glycol ethers are on the Community Right-To-Know List.

SAFETY PROFILE: Moderately toxic by skin contact. When heated to decomposition it emits acrid smoke and irritating fumes. See also ETHERS.

BPS500 CAS:10215-33-5 HR: 3
3-BUTOXY-1-PROPANOL
mf: $C_7H_{16}O_2$ mw: 132.23

SYN: PROPYLENE GLYCOL MONO-n-BUTYL ETHER

TOXICITY DATA WITH REFERENCE
skn-rbt 500 mg open MLD UCDS** 7/28/66
eye-rbt 15 mg SEV UCDS** 7/28/66
eye-rbt 2 mg/24H SEV 85JCAE-,633,86
orl-rat LD50:5950 µL/kg AIHAAP 30,470,69
ivn-mus LD50:320 mg/kg CSLNX* NX#02921
skn-rbt LD50:1590 mg/kg AIHAAP 30,470,69

CONSENSUS REPORTS: Glycol ethers are on the Community Right-To-Know List.

SAFETY PROFILE: Poison by intravenous route. Moderately toxic by skin contact. Mildly toxic by ingestion. A mild skin and severe eye irritant. When heated to decomposition it emits acrid smoke and irritating fumes.

BPS750 CAS:63716-40-5 HR: 2
n-BUTOXYPROPANOL (mixed isomers)
mf: $C_7H_{16}O_2$ mw: 132.23

SYN: BUTOXYPROPANOL (mixed isomers)

TOXICITY DATA WITH REFERENCE
skn-rbt 500 mg open MLD UCDS** 10/13/64
eye-rbt 15 mg SEV UCDS** 10/13/64
orl-rat LD50:2830 mg/kg UCDS** 10/13/64
skn-rbt LD50:3560 mg/kg TXAPA9 28,313,74

SAFETY PROFILE: Moderately toxic by ingestion and skin contact. A mild skin and severe eye irritant. When heated to decomposition it emits acrid smoke and irritating fumes.

BPT000 CAS:6959-71-3 HR: 2
3-BUTOXYPROPIONITRILE
mf: $C_7H_{13}NO$ mw: 127.21

SYN: 3-BUTOXYPROPANENITRILE

TOXICITY DATA WITH REFERENCE
eye-rbt 500 mg AMIHBC 10,61,54
eye-rbt 500 mg open AMIHBC 10,61,54
eye-rbt 500 mg/24H MLD 85JCAE-,918,86
orl-rat LD50:7460 mg/kg AMIHBC 10,61,54
ipr-mus LDLo:500 mg/kg CBCCT* 9,135,57
skn-rbt LD50:8980 mg/kg AMIHBC 10,61,54

CONSENSUS REPORTS: Reported in EPA TSCA Inventory. Cyanide and its compounds are on the Community Right-To-Know List.

SAFETY PROFILE: Moderately toxic by intraperitoneal route. Mildly toxic by ingestion and skin contact. An eye irritant. When heated to decomposition it emits toxic fumes of NO_x and CN^-. See also NITRILES.

BPT250 HR: 3
4′-BUTOXY-2-PYRROLIDINYLACETANILIDE HYDROCHLORIDE
mf: $C_{16}H_{24}N_2O_2 \cdot ClH$ mw: 312.88

SYN: C 3130

TOXICITY DATA WITH REFERENCE
eye-rbt 2% SEV ARZNAD 8,270,58
ipr-rat LD50:186 mg/kg ARZNAD 8,270,58
scu-mus LD50:545 mg/kg ARZNAD 8,270,58

SAFETY PROFILE: Poison by intraperitoneal route. Moderately toxic by subcutaneous route. A severe eye irritant. When heated to decomposition it emits very toxic fumes of NO_x and HCl.

BPT750 CAS:5585-73-9 HR: 3
BUTRIPTYLINE HYDROCHLORIDE
mf: $C_{21}H_{27}N \cdot ClH$ mw: 329.95

PROP: Solid. Mp: 188–190° (decomp).

SYNS: (±)-10,11-DIHYDRO-N,N,β-TRIMETHYL- 5H-DIBENZO(a,d,) CYCLOHEPTENE-5-PROPANAMINE HCl ☐ (±)-10,11-DIHYDRO-N,N,β-TRIMETHYL-5H-DIBENZO(a,d)-CYCLOHEPTENE-5-PROPYLAMINE HCl ☐ EVADYNE ☐ AY-62014

TOXICITY DATA WITH REFERENCE
orl-rat LD50:700 mg/kg 27ZQAG -,62,72
ipr-rat LD50:150 mg/kg 27ZQAG -,62,72
orl-mus LD50:345 mg/kg 27ZQAG -,62,72
ipr-mus LD50:120 mg/kg 27ZQAG -,62,72
ivn-mus LD50:48 mg/kg 27ZQAG -,62,72

SAFETY PROFILE: Poison by ingestion, intraperitoneal, and intravenous routes. An antidepressant. When heated to decomposition it emits toxic fumes of HCl and NO_x.

BPU000 CAS:16227-10-4 HR: 3
BUTRIZOL
mf: $C_6H_{11}N_3$ mw: 125.20

SYNS: BT □ 4-N-BUTYL-4H-1,2,4-TRIAZOLE □ 4-BUTYL-s-TRIAZOLE □ DITHANE R-24 □ INDAR □ RH-124

TOXICITY DATA WITH REFERENCE
orl-rat LD50:50 mg/kg 85ARAE 4,94,76/77
skn-rbt LD50:315 mg/kg FMCHA2 -,C131,83

SAFETY PROFILE: A poison by ingestion and skin contact. When heated to decomposition it emits toxic fumes of NO_x.

BPU500 CAS:91-49-6 HR: 3
N-BUTYLACETANILIDE
mf: $C_{12}H_{17}NO$ mw: 191.30

PROP: Yellowish liquid, mp: 20.8°, bp: 273–275° @ 718 mm, flash p: 286°F, vap d: 6.6, d: 0.992 @ 25°/25°.

TOXICITY DATA WITH REFERENCE
orl-mus LD50:800 mg/kg TXAPA9 19,20,71
orl-gpg LD50:300 mg/kg 28ZEAL 4,70,69

CONSENSUS REPORTS: Reported in EPA TSCA Inventory.

SAFETY PROFILE: Poison by ingestion. Combustible. To fight fire, use CO_2, dry chemical. When heated to decomposition it emits toxic fumes of NO_x.

BPU750 CAS:123-86-4 HR: 3
n-BUTYL ACETATE
DOT: UN 1123
mf: $C_6H_{12}O_2$ mw: 116.18

PROP: Colorless liquid; strong fruity odor. Fp: −77°, bp: 126°, ULC: 50–60, lel: 1.4%, uel: 7.5°, flash p: 72°F, d: 0.88 @ 20°/20°, refr index: 1.393–1.396, autoign temp: 797°F, vap press: 15 mm @ 25°. Misc with alc, ether, and propylene glycol. Sol in EtOH, Et_2CO and Me_2CO; insol in H_2O.

SYNS: ACETATE de BUTYLE (FRENCH) □ ACETIC ACID n-BUTYL ESTER □ BUTILE (ACETATI di) (ITALIAN) □ BUTYLACETAT (GERMAN) □ BUTYL ACETATE □ 1-BUTYL ACETATE □ BUTYLACETATEN (DUTCH) □ BUTYLE (ACETATE de) (FRENCH) □ BUTYL ETHANOATE □ FEMA No. 2174 □ OCTAN n-BUTYLU (POLISH)

TOXICITY DATA WITH REFERENCE
cyc-hmn 300 ppm JIHTAB 25,282,43
skn-rbt 500 mg/24H MOD FCTXAV 17,509,79
skn-rbt 500 mg/24H MLD 85JCAE -,355,86
eye-rbt 20 mg SEV AMIHBC 10,61,54
ihl-rat TCLo:1500 ppm/7H (female 7–16D post):TER NTIS** PB83-258038
ihl-hmn TCLo:200 ppm:NOSE,EYE,PUL JIHTAB 25,282,43
orl-rat LD50:13,100 mg/kg 85GMAT -,28,82
ihl-rat LC50:2000 ppm/4H NPIRI* 1,7,74
orl-mus LD50:7060 mg/kg YKYUA6 32,1241,81
ihl-mus LC50:6 g/m³/2H YKYUA6 32,1241,81
ipr-mus LD50:1230 mg/kg SCCUR* -,2,61
ihl-cat LCLo:68 g/m³/72M AGGHAR 5,1,33
orl-rbt LD50:3200 mg/kg 85GMAT -,28,82
orl-gpg LDLo:4700 mg/kg FCTXAV 17,509,79
ihl-gpg LCLo:67 g/m³/4H FCTXAV 17,515,79
ipr-gpg LDLo:1500 mg/kg AIHAAP 35,21,74

CONSENSUS REPORTS: Reported in EPA TSCA Inventory.

OSHA PEL: TWA 150 ppm; STEL 200 ppm
ACGIH TLV: TWA 150 ppm; STEL 200 ppm (Proposed: TWA 20 ppm; Not Classifiable as a Human Carcinogen)
DFG MAK: 200 ppm (950 mg/m³)
DOT CLASSIFICATION: 3; *Label:* Flammable Liquid

SAFETY PROFILE: Moderately toxic by intraperitoneal route. Mildly toxic by inhalation and ingestion. An experimental teratogen. A skin and severe eye irritant. Human systemic effects by inhalation: conjunctiva irritation, unspecified nasal and respiratory system effects. A mild allergen. High concentrations are irritating to eyes and respiratory tract and cause narcosis. Evidence of chronic systemic toxicity is inconclusive. Flammable liquid. Moderately explosive when exposed to flame. Ignites on contact with potassium tert-butoxide. To fight fire, use alcohol foam, CO_2, dry chemical. When heated to decomposition it emits acrid and irritating fumes. See also ESTERS.

For occupational chemical analysis use NIOSH: Esters I, 1450.

BPV000 CAS:105-46-4 HR: 3
sec-BUTYL ACETATE
DOT: UN 1123
mf: $C_6H_{12}O_2$ mw: 116.18

PROP: Colorless liquid, mild odor. Bp: 112°, flash p: 18°, d: 0.862–0.866 @ 20°/20°, vap d: 4.00. lel: 1.3%, uel: 7.5%.

SYNS: ACETATE de BUTYLE SECONDAIRE (FRENCH) □ ACETIC ACID-2-BUTOXY ESTER □ ACETIC ACID-1-METHYLPROPYL ESTER (9CI) □ 2-BUTANOL ACETATE □ sec-BUTYL ACETATE □ 2-BUTYL ACETATE □ sec-BUTYL ALCOHOL ACETATE

CONSENSUS REPORTS: Reported in EPA TSCA Inventory.

OSHA PEL: TWA 200 ppm
ACGIH TLV: TWA 200 ppm
DFG MAK: 200 ppm (950 mg/m³)
DOT CLASSIFICATION: 3; *Label:* Flammable Liquid

SAFETY PROFILE: An irritant and allergen. See also ESTERS. Flammable liquid. To fight fire, use alcohol foam, CO_2, dry chemical. When heated to decomposition it emits acrid and irritating fumes.

For occupational chemical analysis use NIOSH: Esters I, 1450.

BPV100 **CAS:540-88-5** **HR: 3**
tert-BUTYL ACETATE
DOT: UN 1123
mf: $C_6H_{12}O_2$ mw: 116.18

PROP: Liquid. Bp: 97–98°.

SYNS: ACETIC ACID-tert-BUTYL ESTER ☐ ACETIC ACID-1,1-DIMETHYLETHYL ESTER ☐ TEXACO LEAD APPRECIATOR ☐ TLA

CONSENSUS REPORTS: Reported in EPA TSCA Inventory.

OSHA PEL: TWA 200 ppm
ACGIH TLV: TWA 200 ppm
DFG MAK: 200 ppm (950 mg/m³)
DOT CLASSIFICATION: 3; *Label:* Flammable Liquid

SAFETY PROFILE: Poison by inhalation and ingestion. Flammable. To fight fire, use alcohol foam, CO_2, dry chemical. When heated to decomposition it emits acrid smoke and irritating fumes.

For occupational chemical analysis use NIOSH: Esters I, 1450.

BPV250 **CAS:591-60-6** **HR: 1**
BUTYL ACETOACETATE
mf: $C_8H_{14}O_3$ mw: 158.22

PROP: Bp: 214°, flash p: 185°F, d: 0.96, vap d: 5.55.

SYNS: ACETOACETIC ACID BUTYL ESTER ☐ 3-OXO-BUTANOIC ACID BUTYL ESTER

TOXICITY DATA WITH REFERENCE
skn-rbt 500 mg/24H MLD 85JCAE-,729,86
eye-rbt 500 mg open AMIHBC 10,61,54
orl-rat LD50:11,260 mg/kg AMIHBC 10,61,54

CONSENSUS REPORTS: Reported in EPA TSCA Inventory.

SAFETY PROFILE: Mildly toxic by ingestion. A skin and eye irritant. See also ESTERS. Flammable. To fight fire, use alcohol foam, CO_2, dry chemical. When heated to decomposition it emits acrid and irritating fumes.

BPV325 **CAS:56986-35-7** **HR: 2**
N-BUTYL-N-(1-ACETOXYBUTYL)NITROSAMINE
mf: $C_{10}H_{20}N_2O_3$ mw: 216.32

SYNS: ACETIC ACID-1-(BUTYLNITROSOAMINO)BUTYL ESTER ☐ N-(α-ACETOXY)BUTYL-N-BUTYLNITROSAMINE ☐ 1-ACETOXY-N-NITROSODIBUTYLAMINE ☐ BABN ☐ 1-(BUTYLNITROSOAMINO)BUTYL ACETATE

TOXICITY DATA WITH REFERENCE
mmo-sat 50 nmol/plate CNREA8 40,162,80
dnr-bcs 500 nmol/plate CNREA8 40,162,80
dns-rat:oth 10 µmol/L CBINA8 53,99,85
cyt-ham:lng 32 mg/L GMCRDC 27,95,81
scu-rat TDLo:82 mg/kg/10W-I:CAR IAPUDO 41,619,82
scu-rat TD:70,500 µg/kg/10W-I:ETA GANNA2 73,687,82

CONSENSUS REPORTS: EPA Genetic Toxicology Program.

SAFETY PROFILE: Questionable carcinogen with experimental carcinogenic and tumorigenic data. Mutation data reported. When heated to decomposition it emits toxic fumes of NO_x. See also NITROSAMINES and ESTERS.

BPW000 **CAS:66409-97-0** **HR: 3**
n-BUTYL-3,o-ACETYL-12-β-13-α-DIHYDROJERVINE
mf: $C_{33}H_{49}NO_4$ mw: 523.83

SYN: n-BUTYL-12-β-13-α-DIHYDROJERVINE-3-ACETATE

TOXICITY DATA WITH REFERENCE
orl-ham TDLo:170 mg/kg (7D preg):TER JAFCAU 26,564,78
orl-ham LDLo:170 mg/kg JAFCAU 26(3),564,78

SAFETY PROFILE: Poison by ingestion. An experimental teratogen. When heated to decomposition it emits toxic fumes of NO_x.

BPW050 **CAS:107-58-4** **HR: 2**
N-tert-BUTYLACRYLAMIDE
mf: $C_7H_{13}NO$ mw: 127.21

SYNS: ACRYLAMIDE, N-tert-BUTYL- ☐ N-(1,1-DIMETHYLETHYL)-2-PROPENAMIDE ☐ 2-PROPENAMIDE, N-(1,1-DIMETHYLETHYL)-(9CI)

TOXICITY DATA WITH REFERENCE
orl-mus LD50:941 mg/kg ARTODN 47,179,81

CONSENSUS REPORTS: Reported in EPA TSCA Inventory.

SAFETY PROFILE: Moderately toxic by ingestion. When heated to decomposition it emits toxic vapors of NO_x.

BPW100 **CAS:141-32-2** **HR: 3**
n-BUTYL ACRYLATE
DOT: UN 2348
mf: $C_7H_{12}O_2$ mw: 128.19

PROP: Water-white, extremely reactive monomer. Bp: 69° @ 50 mm, fp: −64.6°, flash p: 120°F (OC), d: 0.89 @ 25°/25°, vap press: 10 mm @ 35.5°, vap d: 4.42.

SYNS: ACRYLIC ACID BUTYL ESTER ☐ ACRYLIC ACID n-BUTYL ESTER (MAK) ☐ BUTYL ACRYLATE ☐ BUTYLACRYLATE, INHIBITED (DOT) ☐ BUTYL-2-PROPENOATE

TOXICITY DATA WITH REFERENCE
skn-rbt 10 mg/24H open MLD AMIHBC 4,119,51
skn-rbt 500 mg open MLD UCDS** 4/5/73
eye-rbt 50 mg MLD UCDS** 4/5/73
ihl-rat TCLo:135 ppm/6H (6-15D preg):REP FAATDF 3,443,83
orl-rat LD50:900 mg/kg 85GMAT -,28,82
ihl-rat LC50:2730 ppm/4H JTEHD6 16,811,85
skn-rat LDLo:1700 mg/kg PJPPAA 32,223,80
ipr-rat LD50:550 mg/kg AMPMAR 36,58,75
orl-mus LD50:7561 mg/kg TOLED5 11,125,82
ihl-mus LC50:7800 mg/m³/2H 85GMAT -,28,82
ipr-mus LD50:853 mg/kg JDREAF 51,526,72
skn-rbt LD50:2000 mg/kg TXAPA9 28,313,74

CONSENSUS REPORTS: IARC Cancer Review: Group 3 IMEMDT 7,56,87, Animal Inadequate Evidence IMEMDT 39,67,86. Reported in EPA TSCA Inventory. Community Right-To-Know List.

OSHA PEL: TWA 10 ppm
ACGIH TLV: TWA 10 ppm
DFG MAK: 10 ppm (55 mg/m³)
DOT CLASSIFICATION: 3; *Label:* Flammable Liquid

SAFETY PROFILE: Moderately toxic by ingestion, inhalation, skin contact, and intraperitoneal routes. Experimental reproductive effects. A skin and eye irritant. A flammable liquid when exposed to heat or flame. To fight fire, use foam, CO_2, dry chemical. Incompatible with oxidizing materials. When heated to decomposition it emits acrid and irritating fumes. See also ESTERS.

BPW250 **HR: 3**
tert-BUTYL-1-ADAMANTANE PEROXYCARBOXYLATE
mf: $C_{15}H_{24}O_3$ mw: 252.35

SAFETY PROFILE: Explodes on heating to 90-100°. When heated to decomposition it emits acrid smoke and fumes. See also PEROXIDES.

BPW500 **CAS:71-36-3** **HR: 3**
n-BUTYL ALCOHOL
mf: $C_4H_{10}O$ mw: 74.14

PROP: Colorless liquid; vinous odor. Bp: 117.4°, ULC: 40, lel: 1.4%, uel: 11.2%, fp: −90°, flash p: 95–100°F, d: 0.80978 @ 20°/4°, autoign temp: 689°F, vap press: 5.5 mm @ 20°, vap d: 2.55. Misc in alc, ether, and organic solvents. Mod sol in water.

SYNS: ALCOOL BUTYLIQUE (FRENCH) □ BUTANOL (FRENCH) □ n-BUTANOL □ BUTAN-1-OL □ 1-BUTANOL □ BUTANOL (DOT) □ BUTANOLEN (DUTCH) □ BUTANOLO (ITALIAN) □ BUTYL ALCOHOL (DOT) □ BUTYL HYDROXIDE □ BUTYLOWY ALKOHOL (POLISH) □ BUTYRIC or NORMAL PRIMARY BUTYL ALCOHOL □ CCS 203 □ FEMA No. 2178 □ 1-HYDROXYBUTANE □ METHYLOLPROPANE □ PROPYLCARBINOL □ PROPYLMETHANOL □ RCRA WASTE NUMBER U031

TOXICITY DATA WITH REFERENCE
eye-hmn 50 ppm JIHTAB 25,282,43
skn-rbt 405 mg/24H MOD BIOFX* 2-5/69
skn-rbt 20 mg/24H MOD 85JCAE -,193,86
eye-rbt 1620 μg SEV AJOPAA 29,1363,46
eye-rbt 2 mg/24H SEV 85JCAE -,193,86
cyt-smc 10 mmol/tube HEREAY 33,457,47
ihl-rat TCLo: 8000 ppm/7H (female 1-22D post): TER,REP TJADAB 35,56A,87
ihl-hmn TCLo: 25 ppm: IRR JIHTAB 25,282,43
orl-rat LD50: 790 mg/kg SAMJAF 43,795,69
ihl-rat LC50: 8000 ppm/4H NPIRI* 1,10,74
ivn-rat LD50: 310 mg/kg EVHPAZ 61,321,85
ipr-mus LD50: 603 mg/kg 85GMAT -,28,82
ivn-mus LD50: 377 mg/kg AIPTAK 135,330,62
orl-rbt LDLo: 4250 mg/kg JLCMAK 10,985,25
skn-rbt LD50: 3400 mg/kg NPIRI* 1,10,74

CONSENSUS REPORTS: Community Right-To-Know List. EPA Genetic Toxicology Program. Reported in EPA TSCA Inventory.

OSHA PEL: CL 50 ppm (skin)
ACGIH TLV: CL 50 ppm (skin) (Proposed: CL 25 ppm)
DFG MAK: 100 ppm (300 mg/m³)

SAFETY PROFILE: A poison by intravenous route. Moderately toxic by skin contact, ingestion, subcutaneous, and intraperitoneal routes. Human systemic effects by inhalation: conjunctiva irritation, unspecified respiratory system, and nasal effects. Experimental reproductive effects. A severe skin and eye irritant. Though animal experiments have shown the butyl alcohols to possess toxic properties, they have produced few cases of poisoning in industry probably because of their low volatility. The use of normal butyl alcohol is reported to have resulted in irritation of the eyes, with corneal inflammation, slight headache and dizziness, slight irritation of the nose and throat, and dermatitis about the fingernails and along the side of the fingers. Keratitis has also been reported. Mutation data reported. See also ALCOHOLS. Flammable liquid. Moderately explosive when exposed to flame. Incompatible with Al, chromium trioxide, oxidizing materials. To fight fire, use water spray, alcohol foam, CO_2, dry chemical. When heated to decomposition it emits acrid smoke and fumes.

For occupational chemical analysis use NIOSH: Alcohols II, 1401.

BPW750 **CAS:78-92-2** **HR: 3**
sec-BUTYL ALCOHOL
mf: $C_4H_{10}O$ mw: 74.14

PROP: Colorless liquid. Mp: −89°, bp: 99.5°, flash p: 14°, d: 0.808 @ 20°/4°, autoign temp: 763°F, vap press: 10 mm @ 20°, vap d: 2.55, lel: 1.7% @ 212°F, uel: 9.8% @ 212°F.

SYNS: ALCOOL BUTYLIQUE SECONDAIRE (FRENCH) □ sec-BUTANOL (DOT) □ BUTAN-2-OL □ 2-BUTANOL □ BUTANOL SECONDAIRE (FRENCH) □ 2-BUTYL ALCOHOL □ BUTYLENE HYDRATE □ CCS 301 □ ETHYLMETHYL CARBINOL □ 2-HYDROXYBUTANE □ METHYLETHYLCARBINOL □ S.B.A.

TOXICITY DATA WITH REFERENCE
skn-rbt 500 mg/24H MLD 85JCAE -,193,86
eye-rbt 16 mg open AMIHBC 10,61,54
eye-rbt 100 mg/24H MOD 85JCAE -,193,86
ihl-rat TCLo: 7000 ppm/7H (female 1-22D post): TER,REP TJADAB 35,56A,87
orl-rat LD50: 6480 mg/kg AMIHBC 10,61,54
ihl-rat LCLo: 16,000 ppm/4H AMIHBC 10,61,54
ipr-rat LD50: 1193 mg/kg EVHPAZ 61,321,85
ivn-rat LD50: 138 mg/kg EVHPAZ 61,321,85
ipr-mus LD50: 771 mg/kg SCCUR* -,2,61
ivn-mus LD50: 764 mg/kg AIPTAK 135,330,62
orl-rbt LD50: 4893 mg/kg IMSUAI 41,31,72
ipr-rbt LD50: 277 mg/kg EVHPAZ 61,321,85

CONSENSUS REPORTS: Community Right-To-Know List. Reported in EPA TSCA Inventory.

OSHA PEL: TWA 100 ppm
ACGIH TLV: TWA 100 ppm
DFG MAK: 100 ppm (300 mg/m³)

SAFETY PROFILE: Poison by intravenous and intraperi-

toneal routes. Mildly toxic by ingestion. Experimental reproductive effects. A skin and eye irritant. See also n-BUTYL ALCOHOL and ALCOHOLS. Dangerous fire hazard when exposed to heat or flame. Auto-oxidizes to an explosive peroxide. Ignites on contact with chromium trioxide. To fight fire, use water spray, alcohol foam, CO_2, dry chemical. Incompatible with oxidizing materials. When heated to decomposition it emits acrid smoke and fumes.

For occupational chemical analysis use NIOSH: Alcohols II, 1401.

BPX000 **CAS:75-65-0** **HR: 3**
tert-BUTYL ALCOHOL
mf: $C_4H_{10}O$ mw: 74.14

PROP: Colorless liquid or rhombic prisms or plates with camphoraceous odor. Mp: 25.5°, bp: 82.8°, flash p: 50°F (CC), d: 0.781 @ 25°/4°, autoign temp: 896°F, vap press: 40 mm @ 24.5°, vap d: 2.55, lel: 2.4%, uel: 8.0%. Misc in H_2O.

SYNS: ALCOOL BUTYLIQUE TERTIAIRE (FRENCH) □ tert-BUTANOL □ BUTANOL TERTIAIRE (FRENCH) □ tert-BUTYL HYDROXIDE □ 1,1-DIMETHYLETHANOL □ 2-METHYL-2-PROPANOL □ NCI-C55367 □ TRIMETHYLCARBINOL

TOXICITY DATA WITH REFERENCE
orl-mus TDLo: 103 g/kg (female 6–20D post): REP JPETAB 222,294,82
orl-mus TDLo: 135 g/kg (female 6–20D post): REP JPETAB 222,294,82
ihl-rat TCLo: 5000 ppm/7H (female 1–22D post): TER TJADAB 35,56A,87
orl-rat LD50: 3500 mg/kg SCIEAS 116,663,52
ipr-mus LD50: 933 mg/kg SCCUR* -,2,61
ivn-mus LD50: 1538 mg/kg AIPTAK 135,330,62
orl-rbt LD50: 3559 mg/kg IMSUAI 41,31,72
par-frg LDLo: 12 g/kg AIPTAK 50,296,35

CONSENSUS REPORTS: Community Right-To-Know List. Reported in EPA TSCA Inventory. EPA Genetic Toxicology Program.

OSHA PEL: TWA 100 ppm; STEL 150 ppm
ACGIH TLV: TWA 100 ppm; STEL 150 ppm (Proposed: TWA 100 ppm)
DFG MAK: 100 ppm (300 mg/m³)

SAFETY PROFILE: Moderately toxic by ingestion, intravenous, and intraperitoneal routes. An experimental teratogen. Other experimental reproductive effects. Dangerous fire hazard when exposed to heat or flame. Moderately explosive in the form of vapor when exposed to flame. Ignites on contact with potassium-sodium alloys. To fight fire, use alcohol foam, CO_2, dry chemical. Incompatible with oxidizing materials, H_2O_2. See also n-BUTYL ALCOHOL and ALCOHOLS.

For occupational chemical analysis use NIOSH: Alcohols I, 1400.

BPX500 **CAS:13449-22-4** **HR: 3**
n-BUTYL AMIDO SULFURYL AZIDE
mf: $C_4H_{10}N_4O_2S$ mw: 178.21

$(C_4H_9)NHSO_2N_3$

SAFETY PROFILE: May explode when heated. When heated to decomposition it emits toxic fumes of SO_x and NO_x. See also AZIDES.

BPX750 **CAS:109-73-9** **HR: 3**
n-BUTYLAMINE
DOT: UN 1125
mf: $C_4H_{11}N$ mw: 73.16

PROP: Liquid, ammonia-like odor. Mp: −50°, bp: 78°, flash p: 10°F (OC), 10°F (CC), d: 0.74–0.76 @ 20°/20°, autoign temp: 594°F, vap d: 2.52, lel: 1.7%, uel: 9.8%.

SYNS: 1-AMINO-BUTAAN (DUTCH) □ 1-AMINOBUTAN (GERMAN) □ 1-AMINOBUTANE □ 1-BUTANAMINE □ n-BUTILAMINA (ITALIAN) □ n-BUTYLAMIN (GERMAN) □ BUTYLAMINE (OSHA) □ MONOBUTILAMINA □ MONOBUTYLAMINE □ MONO-n-BUTYLAMINE □ NORVALAMINE

TOXICITY DATA WITH REFERENCE
cyt-rat-orl 110 mg/kg ZKKOBW 86,47,76
skn-rbt 10 mg/24H open JIHTAB 26,269,44
skn-rbt 500 mg open SEV UCDS** 7/19/65
ipr-mus TDLo: 800 mg/kg: ETA BCPCA6 2,168,59
orl-rat LD50: 366 mg/kg TXAPA9 63,150,82
par-rat LDLo: 600 mg/kg JPETAB 20,435,23
orl-mus LD50: 430 mg/kg GISAAA 40(11),21,75
ihl-mus LC50: 800 mg/m³/2H 85GMAT -,28,82
ipr-mus LD50: 629 mg/kg JPETAB 88,82,46
ivn-mus LD50: 198 mg/kg JPETAB 88,82,46
orl-gpg LD50: 430 mg/kg 85GMAT -,28,82
skn-rbt LD50: 850 mg/kg UCDS** 7/19/65
skn-gpg LD50: 370 mg/kg JIHTAB 26,269,44

CONSENSUS REPORTS: Reported in EPA TSCA Inventory.

OSHA PEL: CL 5 ppm (skin)
ACGIH TLV: CL 5 ppm
DFG MAK: 5 ppm (15 mg/m³)
DOT CLASSIFICATION: 3; *Label:* Flammable Liquid; DOT Class: 3; *Label:* Flammable Liquid, Corrosive

SAFETY PROFILE: Poison by ingestion, skin contact, and intravenous routes. Moderately toxic by inhalation, intraperitoneal, and parenteral routes. A corrosive and severe skin irritant. Questionable carcinogen with experimental tumorigenic data. Mutation data reported. A flammable liquid and dangerous fire hazard when exposed to heat, flame, or oxidizing materials. To fight fire, use alcohol foam, CO_2, dry chemical. Explodes on contact with perchloryl fluoride. When heated to decomposition it emits toxic fumes of NO_x. See also AMINES.

For occupational chemical analysis use NIOSH: n-Butylamine s138.

BPY000 CAS:13952-84-6 HR: 3
sec-BUTYLAMINE
DOT: UN 2733/UN 2734
mf: $C_4H_{11}N$ mw: 73.16

PROP: Liquid. Mp: −104°, bp: 63°, flash p: 15°F, d: 0.724 @ 20°.

SYNS: 2-AB ☐ 2-AMINOBUTANE ☐ BUTAFUME ☐ 2-BUTANAMINE ☐ DECCOTANE ☐ FRUCOTE ☐ 1-METHYLPROPYLAMINE ☐ TUTANE

TOXICITY DATA WITH REFERENCE
orl-rat LD50:152 mg/kg TXAPA9 63,150,82
orl-dog LD50:225 mg/kg PEMNDP 9,112,91
skn-rbt LD50:2500 mg/kg PEMNDP 9,112,91

CONSENSUS REPORTS: Reported in EPA TSCA Inventory.

DFG MAK: 5 ppm (15 mg/m³)
DOT CLASSIFICATION: 8; *Label:* Corrosive, Flammable Liquid (UN 2734); DOT Class: 3; *Label:* Flammable Liquid, Corrosive (UN 2733)

SAFETY PROFILE: A poison by ingestion. A powerful irritant. Moderately toxic by skin contact. Dangerous fire hazard when exposed to heat or flame. To fight fire, use alcohol foam, water spray or mist, dry chemical. Incompatible with oxidizing materials. When heated to decomposition it emits toxic fumes of NO_x. A fungicide.

BPY100 CAS:513-49-5 HR: 3
sec-BUTYLAMINE, (S)-
mf: $C_4H_{11}N$ mw: 73.16

SYNS: (+)-2-BUTYLAMINE ☐ (S)-2-BUTYLAMINE

TOXICITY DATA WITH REFERENCE
orl-rat LD50:380 mg/kg 28ZEAL 5,33,76
orl-dog LD50:225 mg/kg 28ZEAL 5,33,76
skn-rbt LD50:2500 mg/kg 28ZEAL 5,33,76

CONSENSUS REPORTS: Reported in EPA TSCA Inventory.

SAFETY PROFILE: Poison by ingestion. Moderately toxic by skin contact. When heated to decomposition it emits toxic vapors of NO_x.

BPY250 CAS:75-64-9 HR: 3
tert-BUTYLAMINE
DOT: UN 2733/UN 2734
mf: $C_4H_{11}N$ mw: 73.16

PROP: Colorless liquid. Mp: −67.5°, bp: 46.4°, fp: −72.65°, d: 0.700 @ 15°, lel: 1.7% @ 212°F, uel: 8.9% @ 212°F, vap d: 2.5, autoign temp: 716°F.

SYNS: 2-AMINOISOBUTANE ☐ 2-AMINO-2-METHYLPROPANE ☐ BUTYLAMINE, tertiary ☐ 1,1-DIMETHYLETHYLAMINE ☐ TRIMETHYLAMINOMETHANE

TOXICITY DATA WITH REFERENCE
ihl-man TCLo:40 mg/m³/8H-I BJIMAG 48,26,91
orl-rat LD50:78 mg/kg TXAPA9 63,150,82
orl-mus LD50:900 mg/kg WQCHM* 4,-,74

CONSENSUS REPORTS: Reported in EPA TSCA Inventory.

DFG MAK: 5 ppm (15 mg/m³)
DOT CLASSIFICATION: 8; *Label:* Corrosive, Flammable Liquid (UN 2734); DOT Class: 3; *Label:* Flammable Liquid, Corrosive (UN 2733)

SAFETY PROFILE: Poison by ingestion. Moderately toxic to humans by inhalation. A corrosive liquid. See also n-BUTYLAMINE and AMINES. Very dangerous fire hazard when exposed to heat or flame. Very exothermic reaction with 2,2-dibromo-1,3-dimethylcyclopropanoic acid. To fight fire, use alcohol foam. When heated to decomposition it emits toxic fumes of NO_x.

BPY500 CAS:77966-25-7 HR: 3
2-(BUTYLAMINO)-p-ACETOPHENETIDIDE HYDROCHLORIDE
mf: $C_{14}H_{22}N_2O_2$•ClH mw: 286.84

SYN: C 5414

TOXICITY DATA WITH REFERENCE
eye-rbt 2% MLD ARZNAD 8,407,58
ipr-rat LD50:220 mg/kg ARZNAD 8,407,58
scu-mus LD50:800 mg/kg ARZNAD 8,407,58

SAFETY PROFILE: Poison by intraperitoneal route. Moderately toxic by subcutaneous route. An eye irritant. When heated to decomposition it emits very toxic fumes of HCl and NO_x.

BPY625 CAS:78907-16-1 HR: 3
3-(tert-BUTYLAMINO)ACETYLINDOLE HYDROCHLORIDE HYDRATE
mf: $C_{14}H_{18}N_2O$•ClH•H_2O mw: 284.82

SYN: 3-((tert-BUTYLAMINO)ACETYL)INDOLE HYDROCHLORIDE HYDRATE

TOXICITY DATA WITH REFERENCE
orl-mus LD50:410 mg/kg PCJOAU 15,412,81
scu-mus LD50:275 µg/kg PCJOAU 15,412,81
ivn-mus LD50:90 mg/kg PCJOAU 15,412,81

SAFETY PROFILE: Poison by intravenous and subcutaneous routes. Moderately toxic by ingestion. When heated to decomposition it emits toxic fumes of NO_x and HCl.

BPZ000 CAS:94-25-7 HR: 3
BUTYL-p-AMINOBENZOATE
mf: $C_{11}H_{15}NO_2$ mw: 193.27

PROP: Yellow, amorphous powder; mp: 57–59°, bp: 174° @ 8 mm.

SYNS: p-AMINOBENZOIC ACID BUTYL ESTER ☐ BUTAMBEN

TOXICITY DATA WITH REFERENCE
ipr-mus LD50:67 mg/kg JMCMAR 17,900,74

CONSENSUS REPORTS: Reported in EPA TSCA Inventory.

SAFETY PROFILE: Poison by intraperitoneal route. An allergen. See also ESTERS and AMINES. Combustible when exposed to heat or flame. When heated to decomposition it emits toxic fumes such as NO_x.

BQA000 **CAS:16488-48-5** **HR: 3**
p-BUTYLAMINOBENZOIC ACID-2-(DIETHYLAMINO) ETHYL ESTER MONOHYDROCHLORIDE
mf: $C_{17}H_{28}N_2O_2$•ClH mw: 328.93

SYNS: BENZOE-DIAETHYL (GERMAN) □ HYDROCHLORID SALZ des p-N-n-BUTYLAMINO-BENZOESAURE-DIAETHYLAMINOAETHYLESTERS (GERMAN)

TOXICITY DATA WITH REFERENCE
ipr-rat LD50:27 mg/kg ARZNAD 1,218,51
scu-rat LD50:22,500 μg/kg OYYAA2 9,413,75
ivn-rat LD50:4 mg/kg ARZNAD 1,218,51
orl-mus LD50:156 mg/kg OYYAA2 9,413,75
scu-mus LD50:140 mg/kg ARZNAD 1,218,51
ivn-mus LD50:4900 μg/kg OYYAA2 9,413,75
ivn-rbt LD50:2400 μg/kg OYYAA2 9,413,75

SAFETY PROFILE: Poison by ingestion, subcutaneous, intravenous, and intraperitoneal routes. When heated to decomposition it emits toxic fumes of NO_x and HCl. See also AMINES and ESTERS.

BQA010 **CAS:94-24-6** **HR: 3**
p-(BUTYLAMINO)BENZOIC ACID-2-(DIMETHYLAMINO)ETHYL ESTER
mf: $C_{15}H_{24}N_2O_2$ mw: 264.41

PROP: Solid. Mp: 43°, bp: 210° @ 4 mm.

SYNS: AMETHOCAINE □ ANETAIN □ p-BUTYLAMINOBENZOYL-2-DIMETHYLAMINOETHANOL □ CONTRALGIN □ DICAIN □DI-CAINE □ DIKAIN □ DIMETHYLAMINOETHYL-p-BUTYL-AMINOBENZOATE □ 2-DIMETHYLAMINOETHYL-p-BUTYLAMINOBENZOATE □ FISSUCAIN □ INTERCAIN □ LANDOCAINE □ LAUDOCAINE □ MEDICAINE □ MEDIHALER-TETRACAINE □ MEETHOBALM □ METRASPRAY □ MUCAESTHIN □ NIPHANOID □ PANTOCAINE □ PONTOCAINE □ REXOCAINE □ TETRACAINE □ UROMUCAESTHIN

TOXICITY DATA WITH REFERENCE
dnd-esc 30 μmol/L MUREAV 89,95,81
dns-hmn:hla 1 μmol/L BCPCA6 14,205,65
par-man LDLo:1 mg/kg:CNS,PUL SAVEAB 10,50,39
ivn-rat LD50:6 mg/kg ARZNAD 8,539,58
ipr-mus LD50:20 mg/kg RPTOAN 35(3),114,72
scu-mus LD50:25 mg/kg PHTXA6 20,521,57
ivn-mus LD50:6 mg/kg EJMCA5 10,291,75
scu-rbt LDLo:20 mg/kg AEPPAE 160,53,31
ivn-rbt LDLo:6 mg/kg AEPPAE 160,53,31
par-rbt LD50:33,500 μg/kg ARZNAD 26,78,76
itr-rbt LD50:6500 μg/kg ARZNAD 26,78,76
par-frg LDLo:200 mg/kg AEPPAE 168,447,32

SAFETY PROFILE: A human poison by parenteral route with systemic effects including: muscle contractions, coma, and cyanosis. A poison experimentally by intravenous, parenteral, intratracheal, intraperitoneal, and subcutaneous routes. Human mutation data reported. A local anesthetic. See also ESTERS. When heated to decomposition it emits toxic fumes of NO_x.

BQA500 **CAS:77791-55-0** **HR: 2**
2-(BUTYLAMINO)-2′-CHLOROACETANILIDE HYDROCHLORIDE
mf: $C_{12}H_{17}ClN_2O$•ClH mw: 277.22

SYN: C 5413

TOXICITY DATA WITH REFERENCE
eye-rbt 2% MLD ARZNAD 8,407,58
ipr-rat LD50:670 mg/kg ARZNAD 8,407,58
scu-mus LD50:1075 mg/kg ARZNAD 8,407,58

SAFETY PROFILE: Moderately toxic by intraperitoneal and subcutaneous routes. An eye irritant. When heated to decomposition it emits very toxic fumes of Cl^-, NO_x, and HCl.

BQA750 **CAS:6027-28-7** **HR: 3**
2-(BUTYLAMINO)-6′-CHLORO-o-ACETOTOLUIDIDE MONOHYDROCHLORIDE
mf: $C_{13}H_{19}ClN_2O$•ClH mw: 291.25

PROP: Crystals from EtOH. Mp: 236–239°.

SYNS: BUTANILICAINE HYDROCHLORIDE □ 2-(BUTYLAMINO)-N-(2-CHLORO-6-METHYLPHENYL)ACETAMIDE HYDROCHLORIDE □ HOSTACAIN □ HOSTACAINE □ HOSTACAINE HYDROCHLORIDE

TOXICITY DATA WITH REFERENCE
eye-rbt 2% MLD ARZNAD 8,407,58
ipr-rat LD50:259 mg/kg ARZNAD 8,407,58
ipr-mus LD50:363 mg/kg ARZNAD 8,407,58
scu-mus LD50:570 mg/kg ARZNAD 8,181,58

SAFETY PROFILE: Poison by intraperitoneal route. Moderately toxic by subcutaneous route. An eye irritant. When heated to decomposition it emits very toxic fumes of Cl^-, NO_x, and HCl.

BQB000 **CAS:5915-41-3** **HR: 2**
2-tert-BUTYLAMINO-4-CHLORO-6-ETHYLAMINO-s-TRIAZINE
mf: $C_9H_{16}ClN_5$ mw: 229.75

PROP: Solid. Mp: 177–179°. Very sltly sol in H_2O; sltly sol in org solvs.

SYNS: 2-tert-BUTYLAMINO-4-AETHYLAMINO-6-CHLOR-1,3,5-TRIAZIN (GERMAN) □ GARDOPRIM □ GS 13529 □ PRIMATOL-M80 □ SORGOPRIM □ TERBUTHYLAZINE □ TURBULETHYLAZIN (GERMAN)

TOXICITY DATA WITH REFERENCE
orl-rat LD50:1845 mg/kg GUCHAZ 6,60,73
par-rat LD50:2160 mg/kg DOVEAA 26,5,72
unr-rat LD50:2500 mg/kg 30ZDA9 -,437,71

CONSENSUS REPORTS: Reported in EPA TSCA Inventory.

SAFETY PROFILE: Moderately toxic by ingestion and possibly other routes. When heated to decomposition it emits very toxic fumes of Cl^- and NO_x.

BQB250 CAS:15148-80-8 HR: 3
1-(tert-BUTYLAMINO)-3-(2-CHLORO-5-METHYLPHENOXY)-2-PROPANOL HYDROCHLORIDE
mf: $C_{14}H_{22}ClNO_2 \cdot ClH$ mw: 308.28

PROP: Crystals. Mp: 220–222°.

SYNS: BETADRENOL □ BETADRENOL HYDROCHLORIDE □ BUPRANOLOL HYDROCHLORIDE □ 1-(2-CHLORO-5-METHYLPHENOXY)-3-((1,1-DIMETHYLETHYL)AMINO)-2-PROPANOL HYDROCHLORIDE □ KL 255 □ (−)-KL 255 □ SKF 16805A

TOXICITY DATA WITH REFERENCE
orl-rat TDLo:240 mg/kg (9-14D preg):REP OYYAA2 7,76,73
orl-rat TDLo:900 mg/kg (female 9-14D post):TER OYYAA2 7,65,73
orl-rat LD50:518 mg/kg NIIRDN 6,682,82
ipr-rat LD50:96 mg/kg NIIRDN 6,682,82
scu-rat LD50:630 mg/kg OYYAA2 7,75,73
orl-mus LD50:329 mg/kg NIIRDN 6,682,82
scu-mus LD50:567 mg/kg OYYAA2 7,75,73
ivn-mus LD50:39 mg/kg NIIRDN 6,682,82
orl-dog LD50:438 mg/kg NIIRDN 6,682,82
orl-rbt LD50:895 mg/kg NIIRDN 6,682,82
ivn-rbt LD50:15,300 μg/kg NIIRDN 6,682,82

SAFETY PROFILE: Poison by ingestion, intraperitoneal, and intravenous routes. Moderately toxic by subcutaneous route. An experimental teratogen. Other experimental reproductive effects. When heated to decomposition it emits very toxic fumes of NO_x and Cl^-.

BQB825 CAS:81994-68-5 HR: 3
4-BUTYLAMINO-N-(2-(DIETHYLAMINO)ETHYL) PHTHALIMIDE HYDROCHLORIDE
mf: $C_{18}H_{27}N_3O_2 \cdot ClH$ mw: 353.94

SYN: 5-BUTYLAMINO-2-(2-DIETHYLAMINOETHYL)-1H-ISOINDOLE-1,3(2H)-DIONE HYDROCHLORIDE

TOXICITY DATA WITH REFERENCE
orl-rat LD50:580 mg/kg EJMCA5 16,59,81
ipr-rat LD50:66 mg/kg EJMCA5 16,59,81
scu-rat LD50:130 mg/kg EJMCA5 16,59,81
ivn-rat LD50:6200 μg/kg EJMCA5 16,59,81
orl-mus LD50:312 mg/kg EJMCA5 16,59,81
ipr-mus LD50:71 mg/kg EJMCA5 16,59,81
scu-mus LD50:67 mg/kg EJMCA5 16,59,81
ivn-mus LD50:3700 μg/kg EJMCA5 16,59,81

SAFETY PROFILE: Poison by ingestion, subcutaneous, intravenous, and intraperitoneal routes. When heated to decomposition it emits toxic fumes of HCl and NO_x.

BQC000 CAS:111-75-1 HR: 2
2-BUTYLAMINOETHANOL
mf: $C_6H_{15}NO$ mw: 117.22

PROP: Liquid. Bp: 200°, flash p: 170°F (OC), d: 0.89, vap d: 4.03. Sol in H_2O.

SYN: 2-n-BUTYLAMINOETHANOL

TOXICITY DATA WITH REFERENCE
skn-rbt 10 mg/24H open AMIHBC 10,61,54
eye-rbt 250 μg open SEV AMIHBC 10,61,54
orl-rat LD50:1150 mg/kg AMIHBC 10,61,54
ipr-rat LD50:840 mg/kg TXAPA9 12,486,68
orl-mam LD50:7100 mg/kg TXAPA9 8,344,66

CONSENSUS REPORTS: Reported in EPA TSCA Inventory.

SAFETY PROFILE: Moderately toxic by ingestion and intraperitoneal routes. A skin and severe eye irritant. See also AMINES. Combustible when exposed to heat or flame. To fight fire, use alcohol foam, foam, CO_2, dry chemical. Incompatible with oxidizing materials. When heated to decomposition it emits toxic fumes of NO_x.

BQC250 CAS:26259-45-0 HR: 2
2-sec-BUTYLAMINO-4-ETHYLAMINO-6-METHOXY-s-TRIAZINE
mf: $C_{10}H_{19}N_5O$ mw: 225.34

PROP: Powder. Mp: 86–88°. Sltly sol in H_2O.

SYNS: 2-sec-BUTYLAMINO-4-ETHYLAMINO-6-METHOXY-1,3,5-TRIAZINE □ ETAZIN □ ETAZINE □ GEIGY G.S. 14254 □ GS 15254 □ 2-METHOXY-4-sec-BUTYLAMINO-6-AETHYLAMINO-s-TRIAZIN (GERMAN) □ SUMITOL □ SUMITOL 80W

TOXICITY DATA WITH REFERENCE
eye-rbt 35 mg SEV CIGET* -,-,77
orl-rat LD50:1000 mg/kg FMCHA2 -,C224,83
skn-rbt LD50:1910 mg/kg CIGET* -,-,77

CONSENSUS REPORTS: Reported in EPA TSCA Inventory.

SAFETY PROFILE: Moderately toxic by ingestion and skin contact. A severe eye irritant. An herbicide. See also AMINES. When heated to decomposition it emits toxic fumes of NO_x.

BQC500 CAS:33693-04-8 HR: 2
2-tert-BUTYLAMINO-4-ETHYLAMINO-6-METHOXY-s-TRIAZINE
mf: $C_{10}H_{19}N_5O$ mw: 225.34

PROP: Solid. Mp: 123–124°. Very sltly sol in H_2O; sol in org solvs.

SYNS: 2-tert-BUTYLAMINO-4-ETHYLAMINO-6-METHOXY-1,3,5-TRIAZINE □ CARAGARD □ GS 14259 □ 2-METHOXY-4-tert-BUTYLAMINO-6-AETHYLAMINO-s-TRIAZIN (GERMAN) □ TERBUMETON

TOXICITY DATA WITH REFERENCE
orl-rat LD50:483 mg/kg GUCHAZ 6,62,73
skn-rat LD50:>3170 mg/kg PEMNDP 9,796,91
par-rat LD50:483 mg/kg DOVEAA 26,5,72

CONSENSUS REPORTS: Reported in EPA TSCA Inventory. EPA Genetic Toxicology Program.

SAFETY PROFILE: Moderately toxic by ingestion. An herbicide. See also AMINES. When heated to decomposition it emits toxic fumes of NO_x.

BQC750 **CAS:886-50-0** **HR: 2**
2-tert-BUTYLAMINO-4-ETHYLAMINO-6-METHYLMERCAPTO-s-TRIAZINE
mf: $C_{10}H_{19}N_5S$ mw: 241.40

PROP: Powder. Mp: 104–105°. Very sltly sol in H_2O; sol in most org solvs.

SYNS: 4-AETHYLAMINO-2-tert-BUTYLAMINO-6-METHYLTHIO-s-TRIAZIN (GERMAN) □ 2-tert-BUTYLAMINO-4-ETHYLAMINO-6-METHYLTHIO-s-TRIAZINE □ 2-METHYLTHIO-4-ETHYLAMINO-6-tert-BUTYLAMINO-s-TRIAZINE

TOXICITY DATA WITH REFERENCE
skn-rbt 380 mg open MLD CIGET* -,-,77
eye-rbt 76 mg MOD CIGET* -,-,77
orl-rat LD50:2045 mg/kg PESTD5 17,351,76
ipr-rat LD50:699 mg/kg PESTD5 17,351,76
orl-mus LD50:3884 mg/kg PESTD5 17,351,76
ipr-mus LD50:554 mg/kg PESTD5 17,351,76
orl-ckn LD50:4000 mg/kg 31ZOAD 1,56,68
unr-mam LD50:2900 mg/kg 30ZDA9 -,438,71

CONSENSUS REPORTS: EPA Extremely Hazardous Substances List.

SAFETY PROFILE: Moderately toxic by ingestion, intraperitoneal, and possibly other routes. A skin and eye irritant. An herbicide. When heated to decomposition it emits very toxic fumes of NO_x and SO_x. See also MERCAPTANS and AMINES.

BQD000 **CAS:54340-62-4** **HR: 3**
2-tert-BUTYLAMINO-1-(7-ETHYL-2-BENZOFURANYL) ETHANOL HYDROCHLORIDE
mf: $C_{16}H_{23}NO_2$•ClH mw: 297.86

SYNS: BUFURALOL □ 1-(7-ETHYLBENZOFURAN-2-YL)-2-tert-BUTYLAMINO-1-HYDROXYETHANE HYDROCHLORIDE

TOXICITY DATA WITH REFERENCE
orl-rat LD50:750 mg/kg ARZNAD 27,1410,77
scu-rat LD50:1400 mg/kg ARZNAD 27,1410,77
orl-mus LD50:177 mg/kg ARZNAD 27,1410,77
ipr-mus LD50:88 mg/kg ARZNAD 27,1410,77
ivn-mus LD50:30 mg/kg ARZNAD 27,1410,77

SAFETY PROFILE: Poison by ingestion, intraperitoneal, and intravenous routes. Moderately toxic by subcutaneous route. When heated to decomposition it emits very toxic fumes of Cl^- and NO_x. See also AMINES.

BQD125 **HR: 3**
3-(2-(tert-BUTYLAMINO)ETHYL)-6-HYDROXYBENZYL ALCOHOL SULFATE (2:1)
mf: $C_{26}H_{42}N_2O_4$•O_4S mw: 542.76

TOXICITY DATA WITH REFERENCE
ipr-rat LD50:295 mg/kg IYKEDH 9,222,78
ivn-rat LD50:59 mg/kg IYKEDH 9,222,78
orl-mus LD50:4750 mg/kg IYKEDH 9,222,78
ipr-mus LD50:239 mg/kg IYKEDH 9,222,78
scu-mus LD50:737 mg/kg IYKEDH 9,222,78
ivn-mus LD50:49 mg/kg IYKEDH 9,222,78

SAFETY PROFILE: Poison by intravenous and intraperitoneal routes. Moderately toxic by subcutaneous route. When heated to decomposition it emits toxic fumes of SO_x and NO_x. See also SULFATES.

BQD250 **CAS:3775-90-4** **HR: 3**
tert-BUTYL AMINO ETHYL METHACRYLATE
mf: $C_{10}H_{19}NO_2$ mw: 185.30

PROP: Liquid; bp: 100–105°; d: 0.914. flash p: 205°F (OC).

SYNS: AGEFLEX FM-4 □ 2-(tert-BUTYLAMINO)ETHYL METHACRYLATE

TOXICITY DATA WITH REFERENCE
ipr-mus LD50:174 mg/kg JDREAF 51,526,72

CONSENSUS REPORTS: Reported in EPA TSCA Inventory.

SAFETY PROFILE: Poison by intraperitoneal route. See also ESTERS and AMINES. Combustible when exposed to heat or flame. To fight fire, use alcohol foam, water spray or mist, dry chemical. When heated to decomposition it emits toxic fumes of NO_x.

BQD500 **CAS:34866-46-1** **HR: 3**
(5-(2-(tert-BUTYLAMINO)-1-HYDROXYETHYL)-2-HYDROXYPHENYL)UREA HYDROCHLORIDE
mf: $C_{13}H_{21}N_3O_3$•ClH mw: 303.83

PROP: Solid. Mp: 205–207° (decomp).

SYN: CARBUTEROL HYDROCHLORIDE

TOXICITY DATA WITH REFERENCE
ivn-rat LD50:87 mg/kg JPETAB 189,167,74
orl-mus LD50:3543 mg/kg JPETAB 189,167,74
ivn-mus LD50:37 mg/kg JPETAB 189,167,74
scu-gpg LD50:473 mg/kg JPETAB 189,167,74

SAFETY PROFILE: Poison by intravenous route. Moderately toxic by ingestion and subcutaneous routes. When heated to decomposition it emits very toxic fumes of HCl and NO_x. See also AMINES.

BQE000 **CAS:68377-91-3** **HR: 3**
(±)-2-(3′-tert-BUTYLAMINO-2′-HYDROXYPROPYLTHIO)-4-(5′-CARBAMOYL-2′-THIENYL)THIAZOLE HYDROCHLORIDE
mf: $C_{15}H_{21}N_3O_2S_3$•ClH mw: 408.03

PROP: Crystals from MeOH (aq). Mp: 234–235.5°.

SYN: S 596

TOXICITY DATA WITH REFERENCE
orl-rat LD50:86 mg/kg DRFUD4 4,442,79
orl-mus LD50:5000 mg/kg DRFUD4 4,442,79
ipr-mus LD50:360 mg/kg DRFUD4 4,442,79

SAFETY PROFILE: Poison by ingestion and intraperitoneal routes. When heated to decomposition it emits very toxic fumes of NO_x, SO_x, and HCl.

B

BQE250 CAS:56776-01-3 HR: 3
α-(tert-BUTYLAMINO)METHYL-2-CHLOROBENZYL ALCOHOL HYDROCHLORIDE
mf: $C_{12}H_{18}ClNO \cdot ClH$ mw: 264.22

SYNS: α-((tert-BUTYLAMINO)METHYL)-o-CHLOROBENZYL ALCOHOL HYDROCHLORIDE ☐ C 78 ☐ o-CHLORO-α-((tert-BUTYLAMINO)METHYL)BENZYLALCOHOL HYDROCHLORIDE ☐ 1-(o-CHLOROPHENYL)-2-tert-BUTYLAMINO ETHANOL HYDROCHLORIDE ☐ LOBUTEROL ☐ TOLUBUTEROL HYDROCHLORIDE

TOXICITY DATA WITH REFERENCE
orl-rat TDLo:825 mg/kg (female 7-17D post):REP KSRNAM 11,439,77
orl-rat TDLo:375 mg/kg (17-21D preg):REP KSRNAM 11,1917,77
orl-rat TDLo:4500 mg/kg (female 30D pre):REP OYYAA2 13,297,77
orl-rat LD50:780 mg/kg DRFUD4 1,217,76
ipr-rat LD50:104 mg/kg ARZNAD 25,1028,75
scu-rat LD50:349 mg/kg ARZNAD 25,1028,75
ivn-rat LD50:42 mg/kg YAKUD5 23,1107,81
orl-mus LD50:243 mg/kg DRFUD4 1,217,76
ipr-mus LD50:76 mg/kg ARZNAD 25,1028,75
scu-mus LD50:121 mg/kg IYKEDH 12,933,81
ivn-mus LD50:40 mg/kg ARZNAD 25,1028,75
orl-dog LD50:300 mg/kg ARZNAD 27,1439,77

SAFETY PROFILE: Poison by ingestion, intraperitoneal, subcutaneous, and intravenous routes. Experimental reproductive effects. A bronchodilator. When heated to decomposition it emits very toxic fumes of Cl^- and NO_x. See also AMINES.

BQF250 CAS:3703-79-5 HR: 3
α-((BUTYLAMINO)METHYL)-p-HYDROXYBENZYL ALCOHOL
mf: $C_{12}H_{19}NO_2$ mw: 209.32

SYNS: BAMETHANE ☐ BUTEDRINE ☐ 2-BUTYLAMINO-1-p-HYDROXYPHENYLETHANOL ☐ α-((BUTYLAMINO)METHYL)-4-HYDROXYBENZENEMETHANOL ☐ BUTYL-NOR-SYMPATOL ☐ n-BUTYLNORSYMPATHOL ☐ n-BUTYLNORSYNEPHRINE ☐ BUTYLSYMPATHOL ☐ 1-(p-HYDROXYPHENYL)-2-BUTYLAMINOETHANOL ☐ 1-(4-HYDROXYPHENYL)-1-HYDROXY-2-BUTYLAMINOETHANE

TOXICITY DATA WITH REFERENCE
ivn-rat LD50:80 mg/kg RPOBAR 2,272,70
orl-mus LD50:562 mg/kg RPOBAR 2,272,70
ipr-mus LD50:150 mg/kg JPETAB 89,297,47
ivn-mus LD50:72 mg/kg RPOBAR 2,271,70

SAFETY PROFILE: Poison by intravenous and intraperitoneal routes. Moderately toxic by ingestion. A vasodilator. When heated to decomposition it emits toxic fumes of NO_x. See also ALCOHOLS and AMINES.

BQF500 CAS:18559-94-9 HR: 3
α'-((tert-BUTYL AMINO)METHYL)-4-HYDROXY-m-XYLENE-α,α'-DIOL
mf: $C_{13}H_{21}NO_3$ mw: 239.35

SYNS: AEORLIN ☐ AH 3365 ☐ ALBUTEROL ☐ BRONCOVALEAS ☐ 2-(tert-BUTYLAMINO)-1-(4-HYDROXY-3-HYDROXYMETHYLPHENYL)ETHANOL ☐ α-1-((tert-BUTYLAMINO)METHYL)-4-HYDROXY-m-XYLENE-α,α-DIOL ☐ α-1-(((1,1-DIMETHYLETHYL)AMINO)METHYL)-4-HYDROXY-1,3-BENZENEDIMETHANOL ☐ 4-HYDROXY-3-HYDROXYMETHYL-α-((tert-BUTYLAMINO)METHYL)BENZYL ALCOHOL ☐ PROVENTIL ☐ SALBUTAMOL ☐ SOLBUTAMOL ☐ SULTANOL ☐ VENETLIN ☐ VENTOLIN

TOXICITY DATA WITH REFERENCE
orl-wmn TDLo:160 μg/kg (1D pre):REP RDCNBM 5,31,81
ivn-rat TDLo:3600 μg/kg (16-20D preg):TER DPTHDL 4(Suppl 1),150,82
orl-man TDLo:5714 μg/kg:BAH,CVS AEMED3 22,1474,93
orl-chd TDLo:1850 μg/kg:CNS,CVS BMJOAE 282,1932,81
orl-wmn TDLo:2240 μg/kg:BAH AEMED3 22,1474,93
ihl-man TCLo:36 μg/kg/6H BMJOAE 292,1430,86
ivn-hmn TDLo:6 μg/kg:CVS BMJOAE 1,365,76
orl-rat LD50:660 mg/kg USXXAM #4026897
ipr-rat LD50:295 mg/kg IYKEDH 4,193,73
ivn-rat LD50:57,100 μg/kg USXXAM #4026987
ipr-mus LD50:239 mg/kg IYKEDH 4,193,73
scu-mus LD50:737 mg/kg IYKEDH 4,193,73
ivn-mus LD50:48,700 μg/kg IYKEDH 4,193,73

SAFETY PROFILE: A poison by intraperitoneal and intravenous routes. Moderately toxic by ingestion and subcutaneous routes. Human systemic effects: change in heart rate and plasma or blood volume pulse rate increase, tremors. Human (child) behavioral and cardiac effects by ingestion including tremors, excitement, and change in heart rate. Human maternal effects of the uterus, cervix, and vagina by ingestion. An experimental teratogen. Other experimental reproductive effects. A bronchodilator. When heated to decomposition it emits toxic fumes of NO_x.

BQF750 CAS:86166-58-7 HR: 2
1-(tert-BUTYLAMINO)3-(3-METHYL-2-NITROPHENOXY)-2-PROPANOL
mf: $C_{13}H_{22}N_2O_4$ mw: 282.38

SYNS: dl-1-(2-NITRO-3-EMTHYLPHENOXY)-3-tert-BUTYLAMINO-PROPAN-2-OL ☐ ZAMI 1305 ☐ dl-ZAMI 1305

TOXICITY DATA WITH REFERENCE
dni-rat:lvr 14 mmol/L CBINA8 50,77,84
oms-rat:lvr 28 mmol/L CBINA8 50,77,84
oms-rat-ipr 300 mg/kg/6D CBINA8 52,203,84
dni-rat-ipr 100 mg/kg TOPADD 13,18,85
orl-rat TDLo:9 g/kg/26W-C:CAR JJIND8 68,669,82
orl-rat TD:18 g/kg/26W-C:CAR JJIND8 68,669,82

SAFETY PROFILE: Questionable carcinogen with experimental carcinogenic data. Mutation data reported. When heated to decomposition it emits toxic fumes of NO_x.

BQF825 CAS:102071-76-1 HR: 3
2-(BUTYLAMINO)-2-METHYL-1-PROPANOL BENZOATE HYDROCHLORIDE
mf: $C_{15}H_{23}NO_2 \cdot ClH$ mw: 285.85

SYN: 2-(BUTYLAMINO)-2-METHYL-1-PROPANOL BENZOATE (ester) HYDROCHLORIDE

TOXICITY DATA WITH REFERENCE
ipr-mus LD50:230 mg/kg AIPTAK 115,483,58
scu-mus LD50:305 mg/kg AIPTAK 115,483,58
ivn-mus LD50:21 mg/kg AIPTAK 115,483,58

SAFETY PROFILE: Poison by subcutaneous, intravenous, and intraperitoneal routes. When heated to decomposition it emits toxic fumes of NO_x and HCl.

BQG250 **HR: 3**
2-(BUTYLAMINO)-N-METHYL-N-(1-(2,6-XYLYLOXY)-2-PROPYL) ACETAMIDE HYDROCHLORIDE

SYN: C 6259

TOXICITY DATA WITH REFERENCE
eye-rbt 2% SEV ARZNAD 9,70,59
scu-mus LD50:170 mg/kg ARZNAD 9,70,59

SAFETY PROFILE: Poison by subcutaneous route. A severe eye irritant. When heated to decomposition it emits very toxic fumes of NO_x and HCl.

BQG500 **CAS:102585-37-5** **HR: 3**
2-(sec-BUTYLAMINO)-N-METHYL-N-(1-(2,4-XYLYLOXY)-2-PROPYL)ACETAMIDE HYDROCHLORIDE
mf: $C_{18}H_{30}N_2O_2 \cdot ClH$ mw: 342.96

SYN: C 6260

TOXICITY DATA WITH REFERENCE
eye-rbt 2% SEV ARZNAD 9,70,59
scu-mus LD50:180 mg/kg ARZNAD 9,70,59

SAFETY PROFILE: Poison by subcutaneous route. A severe eye irritant. When heated to decomposition it emits very toxic fumes of NO_x and HCl.

BQG750 **CAS:102585-38-6** **HR: 3**
2-(BUTYLAMINO)-N-(1-PHENOXY-2-PROPYL)ACETAMIDE HYDROCHLORIDE
mf: $C_{15}H_{24}N_2O_2 \cdot ClH$ mw: 300.87

SYN: C 6257

TOXICITY DATA WITH REFERENCE
eye-rbt 2% MLD ARZNAD 9,70,59
scu-mus LD50:245 mg/kg ARZNAD 9,70,59

SAFETY PROFILE: Poison by subcutaneous route. An eye irritant. When heated to decomposition it emits very toxic fumes of HCl and NO_x.

BQG850 **CAS:78907-15-0** **HR: 3**
3-(tert-BUTYLAMINO)PROPIONYLINDOLE HYDROCHLORIDE HYDRATE
mf: $C_{15}H_{20}N_2O \cdot ClH \cdot H_2O$ mw: 298.85

SYNS: 2-(tert-BUTYLAMINO)-1-(3-INDOLYL)-1-PROPANONE HYDROCHLORIDE HYDRATE □ 2-(tert-BUTYLAMINO)-1-(3-INDOLYL)-1-PROPANONE MONOHYDROCHLORIDE, MONOHYDRATE

TOXICITY DATA WITH REFERENCE
orl-mus LD50:515 µg/kg PCJOAU 15,412,81
scu-mus LD50:315 µg/kg PCJOAU 15,412,81
ivn-mus LD50:95 mg/kg PCJOAU 15,412,81

SAFETY PROFILE: Poison by subcutaneous and intravenous routes. Moderately toxic by ingestion. When heated to decomposition it emits toxic fumes of NO_x and HCl.

BQH250 **CAS:528-97-2** **HR: 3**
p-BUTYLAMINO SALICYLIC ACID-2-(DIETHYLAMINO) ETHYL ESTER HYDROCHLORIDE
mf: $C_{17}H_{28}N_2O_3 \cdot ClH$ mw: 344.93

SYNS: BRONCHIOCAIN □ BRONCHOCAIN □ BRONCHOCAINE □ 4-(BUTYLAMINO)SALICYLIC ACID 2-(DIETHYLAMINO)ETHYL ESTER HYDROCHLORIDE □ 4-(BUTYLAMINO)-SALICYLIC ACID 2-(DIETHYLAMINO)ETHYL ESTER MONOHYDROCHLORIDE □ C 4208 □ HCl SALZ DES p,N,N-BUTYLAMINOSALICYLSAEUREDIAETHYLAMINOAETHYLESTER (GERMAN) □ PARAESIN □ PHENOCAINE □ S 650 □ SALICYL-DIAETHYL (GERMAN) □ WOFACAIN A

TOXICITY DATA WITH REFERENCE
eye-rbt 2% SEV ARZNAD 8,708,58
ipr-rat LD50:62 mg/kg ARZNAD 1,218,51
ipr-mus LD50:12 mg/kg ARZNAD 1,218,51
scu-mus LD50:120 mg/kg ARZNAD 8,708,58
ivn-mus LD50:16 mg/kg ARZNAD 1,218,51

SAFETY PROFILE: A poison via intraperitoneal, subcutaneous, and intravenous routes. A severe eye irritant. See also AMINES and ESTERS. When heated to decomposition, it emits very toxic fumes of NO_x and HCl.

BQH500 **CAS:17284-75-2** **HR: 3**
p-BUTYLAMINOSALICYLIC ACID-2-(DIMETHYLAMINO)ETHYL ESTER HYDROCHLORIDE
mf: $C_{15}H_{24}N_2O_3 \cdot ClH$ mw: 316.87

PROP: Crystals from H_2O. Mp: 157°. Sltly sol in H_2O.

SYNS: C 4207 □ SALICYL-DIMETHYL (GERMAN)

TOXICITY DATA WITH REFERENCE
eye-rbt 2% SEV ARZNAD 8,708,58
ipr-rat LD50:90 mg/kg ARZNAD 1,218,51
ivn-rat LD50:12 mg/kg ARZNAD 1,218,51
scu-mus LD50:130 mg/kg ARZNAD 8,708,58
ivn-mus LD50:30 mg/kg ARZNAD 1,218,51

SAFETY PROFILE: Poison by subcutaneous, intraperitoneal, and intravenous routes. A severe eye irritant. See also AMINES and ESTERS. When heated to decomposition it emits very toxic fumes of NO_x and HCl.

BQH750 **CAS:78308-37-9** **HR: 3**
p-BUTYLAMINOSALICYLIC ACID-1-ETHYL-4-PIPERIDYL ESTER HYDROCHLORIDE
mf: $C_{18}H_{28}N_2O_3 \cdot ClH$ mw: 356.94

SYN: C 4211

TOXICITY DATA WITH REFERENCE
eye-rbt 2% SEV ARZNAD 8,708,58
scu-mus LD50:57 mg/kg ARZNAD 8,708,58

SAFETY PROFILE: Poison by subcutaneous route. A severe eye irritant. See also AMINES and ESTERS. When heated to decomposition it emits very toxic fumes of Cl^- and NO_x.

BQH800 **CAS:7532-60-7** **HR: 3**
1-(BUTYLAMINO)-3-p-TOLUIDINO-2-PROPANOL
mf: $C_{14}H_{24}N_2O$ mw: 236.40

SYN: 1-(BUTYLAMINO)-3-((4-METHYLPHENYL)AMINO)-2-PROPANOL (9CI)

TOXICITY DATA WITH REFERENCE
ipr-rat LDLo:7600 µg/kg JPETAB 107,250,53
orl-mus LDLo:20 mg/kg JPETAB 107,250,53
ipr-mus LD50:12,400 µg/kg JPETAB 109,407,53
ivn-mus LDLo:3 mg/kg JPETAB 107,250,53

SAFETY PROFILE: Poison by ingestion, intravenous, and intraperitoneal routes. When heated to decomposition it emits toxic fumes of NO_x.

BQH850 **CAS:1126-78-9** **HR: 3**
N-BUTYLANILINE
DOT: UN 2738
mf: $C_{10}H_{15}N$ mw: 149.26

PROP: Liquid. D: 0.936 @ 20°/4°, bp: 249°. Sol in acids, EtOH, C_6H_6, $CHCl_3$; insol in H_2O.

SYNS: BENZENAMINE, N-BUTYL-(9CI) ☐ N-(n-BUTYL)ANILINE ☐ N-n-BUTYLANILINE (DOT) ☐ N-BUTYLBENZENAMINE (9CI) ☐ 4-(PHENYLAMINO)BUTANE

TOXICITY DATA WITH REFERENCE
skn-rbt 20 mg/24H MOD 85JCAE-,465,86
eye-rbt 500 mg/24H MLD 85JCAE-,465,86
orl-rat LD50:1620 mg/kg AMIHBC 10,61,54
unr-mam LD50:282 mg/kg GISAAA 48(6),22,83

CONSENSUS REPORTS: Reported in EPA TSCA Inventory.

DOT CLASSIFICATION: 6.1; *Label:* Poison

SAFETY PROFILE: Poison by an unspecified route. Moderately toxic by skin contact and ingestion. A skin and eye irritant. When heated to decomposition it emits toxic fumes of NO_x. See also ANILINE DYES.

BQI000 **CAS:25013-16-5** **HR: 3**
BUTYLATED HYDROXYANISOLE
mf: $C_{11}H_{16}O_2$ mw: 180.27

PROP: White waxy solid; faint characteristic odor. Mp: 104–105°. Sol in alc and propylene glycol; insol in water.

SYNS: ANTRANCINE 12 ☐ BHA (FCC) ☐ BUTYLHYDROXYANISOLE ☐ tert-BUTYLHYDROXYANISOLE ☐ tert-BUTYL-4-HYDROXYANISOLE ☐ 2(3)-tert-BUTYL-4-HYDROXYANISOLE ☐ BUTYLOHYDROKSYANIZOL (POLISH) ☐ EMBANOX ☐ FEMA No. 2183 ☐ NIPANTIOX 1-F ☐ PREMERGE PLUS ☐ SUSTANE ☐ SUSTANE 1-F ☐ TENOX BHA ☐ VERTAC

TOXICITY DATA WITH REFERENCE
mmo-omi 12,500 µg/L FMLED7 14,183,82
sce-ham:fbr 100 µmol/L JNCIAM 58,1635,77
orl-mus TDLo:12,600 mg/kg (female 1-21D post):REP FEPRA7 31,596,72
orl-rat TDLo:30 g/kg (2W male/2W pre-2W post):REP NTOTDY 3,321,81
orl-rat TDLo:36 g/kg (male 2W pre):REP NETOD7 3,321,81
orl-rat TDLo:728 g/kg/2Y-C:CAR GANNA2 73,332,82
orl-mus TDLo:874 g/kg/1Y-C:ETA JJCREP 77,1083,86
orl-ham TDLo:437 g/kg/1Y-C:CAR JJCREP 77,1083,86
orl-rat TD:874 g/kg/2Y-C:CAR GANNA2 73,332,82
orl-rat TD:182 g/kg/2Y-C:ETA GANNA2 73,332,82
orl-rat TD:218 g/kg/2Y-C:ETA GANNA2 73,332,82
orl-rat TD:269 g/kg/32W-C:CAR CRNGDP 4,895,83
orl-rat TD:876 g/kg/2Y-C:CAR JJIND8 70,343,83
orl-ham TD:202 g/kg/24W-C:NEO GANNA2 74,459,83
orl-rat TD:4200 mg/kg/10W-C:NEO CNREA8 46,165,86
orl-rat TD:728 g/kg/2Y-C:CAR TOLED5 31(Suppl),207,86
orl-rat TD:874 g/kg/1Y-C:CAR JJCREP 77,1083,86
orl-rat TD:202 g/kg/24W-C:NEO JJCREP 77,854,86
orl-rat LD50:2 g/kg TRENAF 22,231,70
ipr-rat LD50:881 mg/kg TOLED5 27,15,85
orl-mus LD50:1100 mg/kg TRENAF 22,231,70
orl-rbt LD50:2100 mg/kg JAOCA7 54,239,77
orl-rat LDLo:2200 mg/kg AFREAW 3,197,51
orl-mus LD50:2000 mg/kg AFREAW 3,197,51

CONSENSUS REPORTS: NTP 7th Annual Report On Carcinogens. IARC Cancer Review: Group 2B IMEMDT 7,56,87; Animal Sufficient Evidence IMEMDT 40,123,86. Reported in EPA TSCA Inventory. EPA Genetic Toxicology Program.

SAFETY PROFILE: Confirmed carcinogen with experimental carcinogenic, neoplastigenic, and tumorigenic data. Moderately toxic by ingestion and intraperitoneal routes. Experimental reproductive effects. Mutation data reported. When heated to decomposition it emits acrid and irritating fumes.

BQI010 **CAS:88-32-4** **HR: 2**
3-tert-BUTYLATED HYDROXYANISOLE
mf: $C_{11}H_{16}O_2$ mw: 180.27

SYNS: 3-tert-BHA ☐ 3-tert-BUTYL-4-METHOXYPHENOL

TOXICITY DATA WITH REFERENCE
orl-ham TDLo:27 g/kg/3W-C:ETA JJIND8 76,143,86

SAFETY PROFILE: Questionable carcinogen with experimental tumorigenic data. When heated to decomposition it emits acrid and irritating fumes.

BQI125 **CAS:84928-98-3** **HR: 3**
N-BUTYL-N-2-AZIDOETHYLNITRAMINE
mf: $C_6H_{13}N_5O_2$ mw: 187.20

$(C_4H_9)N(NO_2)C_2H_4N_3$

SAFETY PROFILE: An impact-sensitive explosive. When heated to decomposition it emits toxic fumes of NO_x. See also AZIDES.

BQI250 **CAS:1070-19-5** **HR: 3**
tert-BUTYL AZIDOFORMATE
mf: $C_5H_9N_3O_2$ mw: 143.17

$(CH_3)_3COCO{\cdot}N_3$

PROP: Bp: 73–74° @ 70 mm.

SYNS: t-BUTOXYCARBONYL AZIDE ☐ tert-BUTOXYCARBONYL

AZIDE (DOT) □ tert-BUTYLOXYCARBONYL AZIDE □ CARBONAZIDIC ACID, 1,1-DIMETHYLETHYL ESTER □ FORMIC ACID, AZIDO-, tert-BUTYL ESTER

CONSENSUS REPORTS: Reported in EPA TSCA Inventory.

DOT CLASSIFICATION: Forbidden

SAFETY PROFILE: An unstable shock- and heat-sensitive explosive. It may explode above 100°C and ignites at 143°C. When heated to decomposition it emits toxic fumes of NO_x. See also AZIDES.

BQI300 CAS:64819-51-8 HR: 2
2-t-BUTYLAZO-2-HYDROXY-5-METHYLHEXANE
mf: $C_{11}H_{24}N_2O$ mw: 200.37

SYNS: 2-((1,1-DIMETHYLETHYL)AZO)-5-METHYL-2-HEXANOL □ 2-HEXANOL, 2-((1,1-DIMETHYLETHYL)AZO)-5-METHYL-

TOXICITY DATA WITH REFERENCE
ihl-rat LCLo:860 mg/m³/6H EPASR* 8EHQ-0491-1041
skn-rbt LD50:707 mg/kg EPASR* 8EHQ-0491-1041

CONSENSUS REPORTS: Reported in EPA TSCA Inventory.

SAFETY PROFILE: Moderately toxic by inhalation and skin contact routes. When heated to decomposition it emits toxic vapors of NO_x.

BQI500 CAS:63018-64-4 HR: 2
5-n-BUTYL-1,2-BENZANTHRACENE
mf: $C_{22}H_{20}$ mw: 284.42

SYN: 8-BUTYLBENZ(a)ANTHRACENE

TOXICITY DATA WITH REFERENCE
skn-mus TDLo:860 mg/kg/36W-I:ETA PRLBA4 129,439,40

SAFETY PROFILE: Questionable carcinogen with experimental tumorigenic data. When heated to decomposition it emits acrid smoke and irritating fumes.

BQI750 CAS:104-51-8 HR: 1
n-BUTYLBENZENE
mf: $C_{10}H_{14}$ mw: 134.24

PROP: Colorless liquid. Mp: −81.2°, bp: 182.1°, d: 0.875 @ 13°/4°, vap press: 1 mm @ 22.7°, autoign temp: 774°F, lel: 0.8%, uel: 5.8%, vap d: 4.6.

SYN: 1-PHENYLBUTANE

TOXICITY DATA WITH REFERENCE
orl-rat LDLo:5000 mg/kg AMIHAB 19,403,59

CONSENSUS REPORTS: Reported in EPA TSCA Inventory.

DOT CLASSIFICATION: 3; *Label:* Flammable Liquid

SAFETY PROFILE: Mildly toxic by ingestion. Flammable when exposed to heat or flame. To fight fire, use alcohol foam, CO_2, dry chemical. Incompatible with oxidizing materials. When heated to decomposition it emits acrid and irritating fumes.

BQJ000 CAS:135-98-8 HR: 3
sec-BUTYLBENZENE
mf: $C_{10}H_{14}$ mw: 134.24

PROP: Colorless liquid. Mp: −82.7°, bp: 173.5°, fp: −75.8°, flash p: 126°F (TOC), d: 0.8621 @ 20°, vap press: 1 mm @ 18.6°, vap d: 4.62, autoign temp: 788°F, lel: 0.8%, uel: 6.9%.

SYN: 2-PHENYLBUTANE

TOXICITY DATA WITH REFERENCE
skn-rbt 100 mg/24H MOD 85JCAE-,36,86
eye-rbt 500 mg/24H MLD 85JCAE-,36,86
orl-rat LD50:2240 mg/kg TXAPA9 28,313,74

CONSENSUS REPORTS: Reported in EPA TSCA Inventory.

DOT CLASSIFICATION: 3; *Label:* Flammable Liquid

SAFETY PROFILE: Moderately toxic by ingestion. A skin and eye irritant. Flammable liquid when exposed to heat or flame. To fight fire, use foam, CO_2, dry chemical, water spray or mist. Incompatible with oxidizing materials. When heated to decomposition it emits acrid smoke and fumes.

BQJ250 CAS:98-06-6 HR: 3
tert-BUTYLBENZENE
mf: $C_{10}H_{14}$ mw: 134.24

PROP: Colorless liquid. Bp: 170–171°, fp: −58°, flash p: 140°F (TOC), d: 0.8665 @ 20°, vap press: 1 mm @ 13.0°, vap d: 4.62, autoign temp: 842°F, lel: 0.7% @ 212°F, uel: 5.7% @ 212°F.

SYNS: 2-METHYL-2-PHENYLPROPANE □ PSEUDOBUTYLBENZENE □ TRIMETHYLPHENYLMETHANE

TOXICITY DATA WITH REFERENCE
orl-rat LDLo:5000 mg/kg AMIHAB 19,403,59

CONSENSUS REPORTS: Reported in EPA TSCA Inventory.

DOT CLASSIFICATION: 3; *Label:* Flammable Liquid

SAFETY PROFILE: Mildly toxic by ingestion. Flammable liquid when exposed to heat or flame. To fight fire, use foam, CO_2, dry chemical, water spray, fog, mist. Incompatible with oxidizing materials. When heated to decomposition it emits acrid smoke and fumes.

BQJ350 CAS:122-43-0 HR: 1
BUTYLBENZENEACETATE
mf: $C_{12}H_{16}O_2$ mw: 192.28

SYNS: ACETIC ACID, PHENYL-, BUTYL ESTER □ BENZENEACETIC ACID, BUTYL ESTER (9CI) □ BUTYL PHENYLACETATE □ n-BUTYL PHENYLACETATE □ PHENYLETHANOIC ACID BUTYL ESTER

TOXICITY DATA WITH REFERENCE
skn-rbt 500 mg/24H MLD FCTOD7 21,657,83
orl-rat LD50:>5 g/kg FCTOD7 21,657,83
skn-rbt LD50:>5 g/kg FCTOD7 21,657,83

CONSENSUS REPORTS: Reported in EPA TSCA Inventory.

SAFETY PROFILE: Low toxicity by ingestion. A skin irritant. When heated to decomposition it emits acrid smoke and irritating fumes.

BQJ500 **CAS:583-03-9** **HR: 3**
α-BUTYLBENZENEMETHANOL
mf: $C_{11}H_{16}O$ mw: 164.27

SYNS: α-BUTYLBENZYL ALCOHOL □ FENIPENTOL □ 1-HYDROXY-1-PHENYLPENTANE □ PANCORAL □ PC 1 □ PH BC □ PHENYLBUTYLCARBINOL □ 1-PHENYL-1-HYDROXYPENTANE □ PHENYLPENTANOL □ 1-PHENYLPENTANOL

TOXICITY DATA WITH REFERENCE
orl-rat TDLo:700 mg/kg (female 7-13D post):REP GNRIDX 5,357,71
orl-mus TDLo:3500 mg/kg (female 7-13D post):REP GNRIDX 5,357,71
orl-rat TDLo:3500 mg/kg (7-13D preg):TER GNRIDX 5,357,71
orl-rat TDLo:50 g/kg (14W pre):REP OYYAA2 6,15,72
orl-mus TDLo:7 g/kg (7-13D preg):TER GNRIDX 5,357,71
orl-rat LD50:5432 mg/kg IYKEDH 4,90,72
ipr-rat LD50:256 mg/kg NIIRDN 6,657,82
scu-rat LD50:6930 mg/kg IYKEDH 4,90,73
orl-mus LD50:2900 mg/kg OSDIAF 14,261,65
ipr-mus LD50:188 mg/kg NIIRDN 6,657,82
scu-mus LD50:3153 mg/kg IYKEDH 4,90,73

CONSENSUS REPORTS: Reported in EPA TSCA Inventory.

SAFETY PROFILE: Poison by intraperitoneal route. Moderately toxic by ingestion and subcutaneous routes. An experimental teratogen. Other experimental reproductive effects. Stimulates the production of bile by the liver. When heated to decomposition it emits acrid smoke and irritating fumes.

BQJ650 **CAS:3622-84-2** **HR: 2**
N-BUTYLBENZENESULFONAMIDE
mf: $C_{10}H_{15}NO_2S$ mw: 213.32

SYNS: BENZENESULFONAMIDE, N-BUTYL- □ BENZENESULFONIC ACID BUTYL AMIDE

TOXICITY DATA WITH REFERENCE
orl-rat LD50:2050 mg/kg TPKVAL 15,110,79
orl-mus LD50:2500 mg/kg TPKVAL 15,110,79
orl-uns LD50:2900 mg/kg GISAAA 39(4),86,74

CONSENSUS REPORTS: Reported in EPA TSCA Inventory.

SAFETY PROFILE: Moderately toxic by ingestion. When heated to decomposition it emits toxic vapors of NO_x and SO_x.

BQJ750 **CAS:24425-13-6** **HR: 3**
2-tert-BUTYLBENZIMIDAZOLE
mf: $C_{11}H_{14}N_2$ mw: 174.27

TOXICITY DATA WITH REFERENCE
mmo-sat 250 µg/plate CHIMAD 27,68,73
ivn-mus LD50:56 mg/kg CSLNX* NX#07472

SAFETY PROFILE: Poison by intravenous route. Mutation data reported. When heated to decomposition it emits toxic fumes of NO_x.

BQK000 **CAS:14255-87-9** **HR: 2**
5-BUTYL-2-BENZIMIDAZOLECARBAMIC ACID METHYL ESTER
mf: $C_{13}H_{17}N_3O_2$ mw: 247.33

PROP: Crystals from EtOH (aq). Mp: 225–227° (decomp).

SYNS: N-(BUTYL-5-BENZIMIDAZOLYL)-2-CARBAMATE de METHYLE (FRENCH) □ (4-BUTYL-1H-BENZIMIDAZOL-2-YL)-CARBAMIC ACID METHYL ESTER □ 5-BUTYL-2-(CARBOMETHOXYAMINO)BENZIMIDAZOLE □ HELMATAC □ METHYL-5-BUTYL-2-BENZIMIDAZOLECARBAMATE □ PARBENDAZOLE □ PBDZ □ SKF 29044 □ VERMINUM □ WORM GUARD

TOXICITY DATA WITH REFERENCE
oms-hmn:leu 1 mg/L THERAP 31,505,76
oms-dom:leu 1 mg/L THERAP 31,505,76
orl-dom TDLo:60 mg/kg (female 21D post):REP COVEAZ 64(Suppl 4),41,74
orl-rat TDLo:80 mg/kg (female 8-15D post):TER THERAP 31,505,76
orl-rat TDLo:150 mg/kg (female 8-15D post):TER RMVEAG 152,467,76
orl-rbt TDLo:130 mg/kg (female 6-18D post):REP COVEAZ 64(Suppl 4),104,74
orl-rbt TDLo:780 mg/kg (female 6-18D post):REP COVEAZ 64(Suppl 4),104,74
orl-dom TDLo:30 mg/kg (female 12D post):REP COVEAZ 64(Suppl 4),56,74
orl-rat TDLo:80 mg/kg (female 8-15D post):REP BSVMA8 76,147,74
orl-rat TDLo:80 mg/kg (female 8-15D post):TER THERAP 31,505,76
orl-rat TDLo:1200 mg/kg (7-18D preg):TER BSVMA8 75,117,73
orl-rat TDLo:150 mg/kg (female 8-15D post):TER RMVEAG 152,467,76
orl-mus TDLo:4 g/kg (female 6-15D post):TER BSVMA8 75,309,73
orl-mus TDLo:16 g/kg (female 6-15D post):TER BSVMA8 75,309,73
orl-mus LD50:1700 mg/kg BSVMA8 77,379,75
orl-dom LDLo:660 mg/kg AUVJA2 46,297,70

SAFETY PROFILE: Moderately toxic by ingestion. Experimental teratogenic and reproductive effects. Human mutation data reported. An anthelminthic agent. When heated to decomposition it emits toxic fumes of NO_x. See also CARBAMATES.

BQK250 **CAS:136-60-7** **HR: 2**
BUTYL BENZOATE
mf: $C_{11}H_{14}O_2$ mw: 178.25

PROP: Liquid. Mp: −21.5°, bp: 248–249°, flash p: 225°F (OC), d: 1.01 @ 15°/15°, vap press: <0.01 mm @ 20°, vap d: 6.15.

SYNS: ANTHRAPOLE AZ □ BENZOIC ACID-n-BUTYL ESTER □ n-BUTYL BENZOATE □ DAI CARI XBN

TOXICITY DATA WITH REFERENCE
skn-rbt 10 mg/24H open SEV AMIHBC 10,61,54
skn-rbt 500 mg open MOD UCDS** 10/15/58
eye-rbt 500 mg AMIHBC 10,61,54
orl-rat LD50:5140 mg/kg AMIHBC 10,61,54
skn-rbt LD50:4000 mg/kg NPIRI* 2,7,75

CONSENSUS REPORTS: Reported in EPA TSCA Inventory.

SAFETY PROFILE: Moderately toxic by skin contact. Mildly toxic by ingestion. Severe skin irritant and moderate eye irritant. Combustible when exposed to heat or flame; can react with oxidizing materials. To fight fire, use CO_2, dry chemical, water mist, fog, spray. When heated to decomposition it emits acrid and irritating fumes. See also ESTERS.

BQK500 CAS:98-73-7 HR: 2
p-tert-BUTYL BENZOIC ACID
mf: $C_{11}H_{14}O_2$ mw: 178.25

PROP: Colorless, fine, crystalline powder. Mp: 163–164.4°, d: 1.142 @ 20°/4°.

SYN: TBBA

TOXICITY DATA WITH REFERENCE
skn-rat TDLo:2450 mg/kg (male 7W pre):REP JACTDZ 6(2),233,87
ihl-rat TCLo:12,500 μg/m³/6H (male 7D pre):REP JACTDZ 6(2),233,87
ihl-rat TCLo:106 mg/m³/6H (male 7D pre):REP JACTDZ 6(2),233,87
orl-rat LD50:700 mg/kg TSCAT* OTS0510267

CONSENSUS REPORTS: Reported in EPA TSCA Inventory.

SAFETY PROFILE: Moderately toxic by ingestion. Experimental reproductive effects. An irritant. Combustible when exposed to heat or flame. Incompatible with oxidizing materials. To fight fire, use foam, CO_2, dry chemical. When heated to decomposition it emits acrid smoke and irritating fumes.

BQK750 CAS:95-31-8 HR: 3
N-tert-BUTYL-2-BENZOTHIAZOLESULFENAMIDE
mf: $C_{11}H_{14}N_2S$ mw: 206.33

PROP: Solid. Mp: 107.5–109°.

SYNS: PENNAC TBBS ☐ VANNAX NS

TOXICITY DATA WITH REFERENCE
mma-mus:lym 40 mg/L ENMUDM 5,193,83
otr-mus:emb 35 mg/L ENMUDM 5,193,83
orl-rat LDLo:7940 mg/kg JACTDZ 1,104,90
ipr-mus LD50:5 g/kg IPSTB3 3,93,76
ivn-mus LD50:180 mg/kg CSLNX* NX#02241

CONSENSUS REPORTS: Reported in EPA TSCA Inventory.

SAFETY PROFILE: Poison by intravenous route. Low toxicity by ingestion. Mutation data reported. When heated to decomposition it emits very toxic fumes of NO_x and SO_x.

BQK800 CAS:23902-88-7 HR: 3
4-(p-tert-BUTYLBENZYL)PIPERAZINYL β-(p-CHLOROPHENYL) KETONE
mf: $C_{30}H_{36}ClN_2O$ mw: 476.13

SYNS: 1-(p-tert-BUTYLBENZYL)-4-(3-(p-CHLOROPHENYL)-3-PHENYLPROPIONYL)PIPERAZINE ☐ KETONE, 4-(p-tert-BUTYLBENZYL)PIPERAZINYL β-(p-CHLOROPHENYL)PHENETHYL ☐ PIPERAZINE, 1-(p-tert-BUTYLBENZYL)-4-(3-(p-CHLOROPHENYL)-3-PHENYLPROPIONYL)-

TOXICITY DATA WITH REFERENCE
ipr-mus LD50:800 mg/kg JMCMAR 12,860,69

DOT CLASSIFICATION: 3; *Label:* Flammable Liquid

SAFETY PROFILE: Moderately toxic by intraperitoneal route. A flammable liquid. When heated to decomposition it emits toxic vapors of NO_x and Cl^-.

BQK830 CAS:17766-62-0 HR: 3
4-(p-tert-BUTYLBENZYL)PIPERAZINYL 3,4,5-TRIMETHOXYPHENYL KETONE
mf: $C_{25}H_{34}N_2O_4$ mw: 426.61

SYNS: 1-(p-tert-BUTYLBENZYL)-4-(3,4,5-TRIMETHOXYBENZOYL)PIPERAZINE ☐ KETONE, 4-(p-tert-BUTYLBENZYL)PIPERAZINYL 3,4,5-TRIMETHOXYPHENYL ☐ PIPERAZINE, 1-(p-tert-BUTYLBENZYL)-4-(3,4,5-TRIMETHOXYBENZOYL)-

TOXICITY DATA WITH REFERENCE
ipr-mus LD50:800 mg/kg JMCMAR 11,332,68

DOT CLASSIFICATION: 3; *Label:* Flammable Liquid

SAFETY PROFILE: Moderately toxic by intraperitoneal route. A flammable liquid. When heated to decomposition it emits toxic vapors of NO_x.

BQK850 CAS:61481-19-4 HR: 3
tert-BUTYL-BICYCLOPHOSPHATE
mf: $C_9H_{20}O_2$ mw: 206.20

PROP: Crystals from C_6H_6 or H_2O. Mp: 321–324°.

SYNS: 2-(tert-BUTYL)-2-(HYEROXYMETHYL)-1,3-PROPANEDIOL, CYCLIC PHOSPHATE (1:1) ☐ 4-tert-BUTYL-1-OXO-1-PHOSPHA-2,6,7-TRIOXABICYCLO(2.2.2)OCTANE ☐ 4-(tert-BUTYL)-2,6,7-TRIOXA-1-PHOSPHABICYCLO(2.2.2)OCTAN-1-ONE

TOXICITY DATA WITH REFERENCE
ipr-rat LD50:35 μg/kg TXAPA9 46,411,78
orl-mus LD50:45 μg/kg TXAPA9 46,411,78
ipr-mus LD50:35 μg/kg TXAPA9 46,411,78
ivn-mus LD50:120 μg/kg

SAFETY PROFILE: Poison by ingestion, intravenous, and intraperitoneal routes. When heated to decomposition it emits toxic fumes of PO_x. See also PHOSPHATES.

BQL000 CAS:1190-53-0 HR: 3
N-BUTYLBIGUANIDE HYDROCHLORIDE
mf: $C_6H_{15}N_5 \cdot ClH$ mw: 193.72

SYNS: ANDERE ☐ BIFORON ☐ BIGUNAL ☐ BUFONAMIN ☐ BUFORMIN HYDROCHLORIDE ☐ BULBONIN ☐ 1-BUTYLBIGUANIDE HYDROCHLORIDE ☐ 1-BUTYLDIGUANIDE HYDROCHLORIDE ☐ N-BUTYLIMIDODICARBONIMIDIC DIAMIDE MONOHYDROCHLORIDE (9CI) ☐ DIABRIN ☐ DIBETOS ☐ GLIBUTIDE ☐ GLIPORAL ☐ INSU-

LAMIN ☐ KREBON ☐ PANFORMIN ☐ SILUBIN ☐ SINDIATIL ☐ TIDEMOL ☐ ZIAVETINE

TOXICITY DATA WITH REFERENCE
orl-mus LD50:380 mg/kg ARZNAD 12,314,62
ipr-mus LD50:380 mg/kg JAJAAA 18,196,65
ivn-mus LD50:105 mg/kg ARZNAD 12,314,62

SAFETY PROFILE: A poison via ingestion, intravenous, and intraperitoneal routes. When heated to decomposition it emits very toxic fumes of HCl and NO_x.

BQL500 CAS:64037-56-5 HR: 3
sec-BUTYLBIS(2-CHLOROETHYL)AMINE HYDROCHLORIDE
mf: $C_8H_{17}Cl_2N{\cdot}ClH$ mw: 234.62

SYNS: sec-BUTYL-BIS(β-CHLOROETHYL)AMINE HYDROCHLORIDE ☐ N-sec-BUTYL-2,2′-DICHLORODIETHYLAMINE, HYDROCHLORIDE ☐ TL 524

TOXICITY DATA WITH REFERENCE
orl-rat LDLo:50 mg/kg NCNSA6 5,11,53
ipr-mus LD50:2800 µg/kg CANCAR 2,1055,49
scu-mus LDLo:2 mg/kg NDRC** No.9-4 1 9,43

SAFETY PROFILE: Poison by ingestion, intraperitoneal, and subcutaneous routes. When heated to decomposition it emits very toxic fumes of HCl and NO_x.

BQL750 CAS:64037-57-6 HR: 3
tert-BUTYLBIS(β-CHLOROETHYL)AMINE HYDROCHLORIDE
mf: $C_8H_{17}Cl_2N{\cdot}ClH$ mw: 234.62

SYNS: tert-BUTYLBIS(2-CHLOROETHYL)AMINE HYDROCHLORIDE ☐ N-tert-BUTYL-2,2′-DICHLORO-DIETHYLAMINE HYDROCHLORIDE ☐ TL 568

TOXICITY DATA WITH REFERENCE
orl-rat LDLo:75 mg/kg NCNSA6 5,11,53
ipr-rat LD50:3 mg/kg CPBTAL 8,99,60
ipr-mus LD50:1420 µg/kg CANCAR 2,1055,49
scu-mus LDLo:25 mg/kg NTIS** PB158-507

SAFETY PROFILE: Poison by ingestion, intraperitoneal, and subcutaneous routes. When heated to decomposition it emits very toxic fumes of HCl and NO_x. See also AMINES.

BQM000 CAS:102-79-4 HR: 2
N-BUTYL-N,N-BIS(HYDROXY ETHYL)AMINE
mf: $C_8H_{19}NO_2$ mw: 161.28

PROP: Liquid. Mp: −70, bp: 273–275°, flash p: 245°F (OC), d: 0.97, vap d: 5.55.

SYNS: N-BUTYLDIETHANOLAMINE ☐ N-BUTYL-2,2′-IMINODIETHANOL

TOXICITY DATA WITH REFERENCE
skn-rbt 10 mg/24H open MLD AMIHBC 10,61,54
eye-rbt 750 µg open SEV AMIHBC 10,61,54
orl-rat LD50:4250 mg/kg AMIHBC 10,61,54

CONSENSUS REPORTS: Reported in EPA TSCA Inventory.

SAFETY PROFILE: Mildly toxic via ingestion. A skin and severe eye irritant. Combustible when exposed to heat or flame. To fight fire, use alcohol foam, foam, CO_2, dry chemical. Incompatible with oxidizing materials. When heated to decomposition it emits toxic fumes of NO_x. See also AMINES.

BQM250 CAS:507-19-7 HR: 2
tert-BUTYL BROMIDE
mf: C_4H_9Br mw: 137.04

PROP: Colorless liquid. Mp: −20°, bp: 72.8°, fp: −16.3°, d: 1.20 @ 15°/4°.

SYNS: 2-BROMOISOBUTANE ☐ 2-BROMO-2-METHYLPROPANE (DOT) ☐ TRIMETHYLBROMOMETHANE

TOXICITY DATA WITH REFERENCE
ipr-mus TDLo:3000 mg/kg/8W-I:NEO CNREA8 35,1411,75
ipr-rat LD50:1250 mg/kg 85GMAT -,29,82
ipr-mus LD50:4400 mg/kg 85GMAT -,29,82

CONSENSUS REPORTS: EPA Genetic Toxicology Program. Reported in EPA TSCA Inventory.

DOT CLASSIFICATION: 3; *Label:* Flammable Liquid

SAFETY PROFILE: Moderately toxic by intraperitoneal route. Questionable carcinogen with experimental neoplastigenic data. When heated to decomposition it emits toxic fumes of Br^-. See also BROMIDES.

BQM309 CAS:1867-72-7 HR: 3
N-tert-BUTYL-1,4-BUTANEDIAMINE DIHYDROCHLORIDE
mf: $C_8H_{20}N_2{\cdot}2ClH$ mw: 217.22

SYNS: N-tert-BUTYL-1,4-DIAMINOBUTANE DIHYDROCHLORIDE ☐ CI-505 ☐ DIBUTADIAMIN DIHYDROCHLORIDE

TOXICITY DATA WITH REFERENCE
orl-rat LD50:810 mg/kg AIPTAK 154,263,65
ipr-rat LD50:349 mg/kg AIPTAK 154,263,65
scu-rat LD50:390 mg/kg AIPTAK 154,263,65
ivn-rat LD50:186 mg/kg AIPTAK 154,263,65
orl-mus LD50:1280 mg/kg AIPTAK 154,263,65
ipr-mus LD50:418 mg/kg AIPTAK 154,263,65
scu-mus LD50:1000 mg/kg AIPTAK 154,263,65
ivn-mus LD50:88 mg/kg AIPTAK 154,263,65

SAFETY PROFILE: Poison by subcutaneous, intravenous, and intraperitoneal routes. Moderately toxic by ingestion. When heated to decomposition it emits toxic fumes of NO_x and HCl. See also AMINES.

BQM500 CAS:109-21-7 HR: 3
n-BUTYL n-BUTANOATE
mf: $C_8H_{16}O_2$ mw: 144.24

PROP: Colorless liquid; pineapple odor. Bp: 166°, flash p: 128°F (OC), d: 0.67–0.871, refr index: 1.405, vap d: 5.0. Misc with alc, ether, vegetable oils; sltly sol in propylene glycol, water.

SYNS: BUTYL BUTYRATE (FCC) ☐ n-BUTYL BUTYRATE ☐ n-BUTYL n-BUTYRATE ☐ FEMA No. 2186

TOXICITY DATA WITH REFERENCE
skn-rbt 500 mg/24H MOD FCTXAV 17,521,79
ipr-rat LD50:2300 mg/kg FCTXAV 17,521,79
ipr-mus LD50:8900 mg/kg FCTXAV 17,521,79
orl-rbt LD50:9520 mg/kg IMSUAI 41,31,72

CONSENSUS REPORTS: Reported in EPA TSCA Inventory.

SAFETY PROFILE: Moderately toxic via intraperitoneal route. Mildly toxic by ingestion. Moderately irritating to eyes, skin, and mucous membranes by inhalation. Narcotic in high concentrations. Flammable liquid. To fight fire, use alcohol foam, foam, CO_2, dry chemical. Incompatible with oxidizing materials. When heated to decomposition it emits acrid and irritating fumes.

BQM750 CAS:63937-32-6 HR: 3
BUTYL-2-BUTOXYCYCLOPROPANE-1-CARBOXYLATE
mf: $C_{12}H_{22}O_3$ mw: 214.34

SYN: 2-BUTOXY-CYCLOPROPANECARBOXYLIC ACID BUTYL ESTER

TOXICITY DATA WITH REFERENCE
orl-rat LD50:24 mg/kg TXAPA9 28,313,74
skn-rbt LD50:110 mg/kg TXAPA9 28,313,74

SAFETY PROFILE: Poison by ingestion and skin contact. When heated to decomposition it emits acrid smoke and irritating fumes.

BQN250 CAS:78329-87-0 HR: 3
p-(N-BUTYL-2-(BUTYLAMINO)ACETAMIDO)BENZOIC ACID BUTYL ESTER HYDROCHLORIDE
mf: $C_{20}H_{32}N_2O_3$•ClH mw: 385.00

SYN: C 3192

TOXICITY DATA WITH REFERENCE
eye-rbt 2% SEV ARZNAD 8,609,58
ipr-rat LD50:260 mg/kg ARZNAD 8,609,58
scu-mus LD50:2825 mg/kg ARZNAD 8,609,58

SAFETY PROFILE: Poison by intraperitoneal route. Moderately toxic by subcutaneous route. See also ESTERS. A severe eye irritant. When heated to decomposition it emits very toxic fumes of HCl and NO_x.

BQN500 CAS:78218-43-6 HR: 3
N-BUTYL-2-(BUTYLAMINO)-2′,6′-PROPIONOXYLIDIDE HYDROCHLORIDE
mf: $C_{19}H_{32}N_2O$•ClH mw: 340.99

SYN: C 3160

TOXICITY DATA WITH REFERENCE
eye-rbt 2% SEV ARZNAD 8,609,58
ipr-rat LD50:68 mg/kg ARZNAD 8,609,58
scu-mus LD50:256 mg/kg ARZNAD 8,609,58

SAFETY PROFILE: Poison by intraperitoneal and subcutaneous routes. A severe eye irritant. When heated to decomposition it emits very toxic fumes of HCl and NO_x. See also AMINES.

BQN600 CAS:33629-47-9 HR: 3
N-sec-BUTYL-4-tert-BUTYL-2,6-DINITROANILINE
mf: $C_{14}H_{21}N_3O_4$ mw: 295.38

SYNS: A 820 ☐ 72-A34 ☐ AMCHEM 70-25 ☐ AMCHEM A-280 ☐ AMEX ☐ AMEX 820 ☐ ANILINE, N-sec-BUTYL-4-tert-BUTYL-2,6-DINITRO- ☐ 70-314B ☐ BENZENAMINE, 4-(1,1-DIMETHYLETHYL)-N-(1-METHYLPROPYL)-2,6-DINITRO-(9CI) ☐ BUTALIN ☐ BUTRALIN ☐ BUTRALINE ☐ DIBUTALIN ☐ 4-(1,1-DIMETHYLETHYL)-N-(1-METHYLPROPYL)-2,6-DINITROBENZENAMINE ☐ RUTRALIN ☐ TAMEX

TOXICITY DATA WITH REFERENCE
orl-rat LD50:2500 mg/kg SWSPBE 24,58,71
ihl-rat LC50:50 g/m³/4H SWSPBE 24,58,71
skn-rbt LD50:200 mg/kg FMCHA2-,C55,91

CONSENSUS REPORTS: Reported in EPA TSCA Inventory.

SAFETY PROFILE: A poison by skin contact. Moderately toxic by ingestion. When heated to decomposition it emits toxic vapors of NO_x.

BQP000 CAS:7492-70-8 HR: 1
BUTYL BUTYROLACTATE
mf: $C_{11}H_{20}O_4$ mw: 216.28

PROP: Colorless liquid; butter, creamlike odor. D: 0.970, refr index: 1.420, flash p: 212°F. Misc with alc, fixed oils; sol in propylene glycol; insol in water.

SYNS: BUTANOIC ACID-2-BUTOXY-1-METHYL-2-OXOETHYL ESTER (9CI) ☐ BUTYL BUTYRYL LACTATE ☐ BUTYRIC ACID ESTER with BUTYL LACTATE ☐ FEMA No. 2190 ☐ LACTIC ACID, BUTYL ESTER, BUTYRATE

TOXICITY DATA WITH REFERENCE
skn-rbt 500 mg/24H FCTXAV 17,241,79

CONSENSUS REPORTS: Reported in EPA TSCA Inventory.

SAFETY PROFILE: A skin irritant. See also ESTERS. Combustible liquid. When heated to decomposition it emits acrid smoke and irritating fumes.

BQP250 CAS:592-35-8 HR: 3
BUTYL CARBAMATE
mf: $C_5H_{11}NO_2$ mw: 117.17

SYNS: CARBAMIC ACID, BUTYL ESTER ☐ USAF EL-101 ☐ USAF FO-1

TOXICITY DATA WITH REFERENCE
mmo-esc 5000 ppm/3H AMNTA4 85,119,51
ipr-ham TDLo:492 mg/kg (8D preg):TER CNREA8 27,1696,67
ipr-mus TDLo:1980 mg/kg/6D-C:NEO PSEBAA 132,422,69
ipr-mus LD50:200 mg/kg NTIS** AD277-689
scu-mus LD50:540 mg/kg AJEBAK 45,507,67

CONSENSUS REPORTS: Reported in EPA TSCA Inventory.

SAFETY PROFILE: A poison via intraperitoneal route. Moderately toxic via subcutaneous route. Experimental teratogenic effects. Questionable carcinogen with experimental neoplastigenic data. Mutation data reported.

See also CARBAMATES. When heated to decomposition it emits toxic fumes of NO_x.

BQP500 **CAS:124-17-4** **HR: 2**
BUTYL CARBITOL ACETATE
mf: $C_{10}H_{20}O_4$ mw: 204.30

PROP: Colorless liquid. Fp: −32.2°, bp: 247°, flash p: 240°F (OC), d: 0.981 @ 20°/20°, autoign temp: 570°F, vap press: 0.01 mm @ 20°.

SYNS: 2-(2-BUTOXYETHOXY)ETHANOL ACETATE □ 2-(2-BUTOXYETHOXY)ETHYL ACETATE □ DIETHYLENE GLYCOL BUTYL ETHER ACETATE □ DIGLYCOL MONOBUTYL ETHER ACETATE □ EKTASOLVE DB ACETATE □ GLYCOL ETHER DB ACEATATE

TOXICITY DATA WITH REFERENCE
skn-rbt 500 mg open MLD UCDS** 12/29/71
eye-rbt 500 mg AJOPAA 29,1363,46
orl-rat LD50:6500 mg/kg 28ZEAL 5,32,76
orl-mus LD50:6600 mg/kg JPETAB 93,26,48
orl-rbt LD50:2600 mg/kg JPETAB 82,377,44
skn-rbt LD50:14,500 mg/kg NPIRI* 1,27,74
orl-gpg LD50:2340 mg/kg JIHTAB 23,259,41
orl-ckn LD50:5000 mg/kg JPETAB 93,26,48

CONSENSUS REPORTS: Reported in EPA TSCA Inventory. Glycol ethers are on the Community Right-To-Know List.

SAFETY PROFILE: Moderately toxic by ingestion. Mild skin and eye irritant. Combustible when exposed to heat or flame. To fight fire, use foam, CO_2, dry chemical. Incompatible with oxidizing materials; heat; flame. When heated to decomposition it emits acrid and irritating fumes.

BQP750 **CAS:85-70-1** **HR: 2**
BUTYL CARBOBUTOXYMETHYL PHTHALATE
mf: $C_{18}H_{24}O_6$ mw: 336.42

SYNS: BUTYL PHTHALATE BUTYL GLYCOLATE □ BUTYL PHTHALYL BUTYL GLYCOLATE □ DIBUTYL-o-(o-CARBOXYBENZOYL) GLYCOLATE □ DIBUTYL-o-CARBOXYBENZOYLOXYACETATE □ SANTICIZIER B-16

TOXICITY DATA WITH REFERENCE
eye-rbt 500 mg AJOPAA 29,1363,46
cyt-ham:fbr 125 mg/L/24H MUREAV 48,337,77
ipr-rat TDLo:689 mg/kg (5-15D preg):TER JPMSAE 61,51,72
ipr-rat TDLo:2296 mg/kg (5-15D preg):REP JPMSAE 61,51,72
orl-rat LD50:7 g/kg EVHPAZ 3,131,73
ipr-rat LD50:6889 mg/kg JPMSAE 61,51,72
orl-mus LD50:12,567 mg/kg IPSTB3 3,93,76
ipr-mus LD50:6880 mg/kg JSCCA5 28,667,77

CONSENSUS REPORTS: Reported in EPA TSCA Inventory.

SAFETY PROFILE: Mildly toxic via intraperitoneal route. Experimental teratogenic and reproductive effects. Mutation data reported. An eye irritant. When heated to decomposition it emits acrid and irritating fumes.

BQQ250 **CAS:38252-74-3** **HR: 3**
N-BUTYL-(3-CARBOXY PROPYL)NITROSAMINE
mf: $C_8H_{16}N_2O_3$ mw: 188.26

SYNS: BCPN □ 4-(BUTYLNITROSOAMINO)BUTANOIC ACID □ N-NITROSO-N-BUTYL-N-(3-CARBOXYPROPYL)AMINE

TOXICITY DATA WITH REFERENCE
mmo-sat 10 μmol/plate CNREA8 37,399,77
dnd-rat-par 50 mg/kg CBINA8 29,291,80
orl-rat TDLo:3760 mg/kg/12W-C:CAR CRNGDP 4,617,83
orl-mus TDLo:7 g/kg/20W-C:ETA GANNA2 67,175,76
orl-rat TD:4700 mg/kg/14W-C:ETA GANNA2 63,637,72
orl-rat TD:4375 mg/kg/20W-C:ETA GANNA2 67,175,76

CONSENSUS REPORTS: NTP 7th Annual Report On Carcinogens. IARC Cancer Review: Animal Limited Evidence IMEMDT 17,51,78. EPA Genetic Toxicology Program.

SAFETY PROFILE: Confirmed carcinogen with experimental carcinogenic and tumorigenic data. Mutation data reported. When heated to decomposition it emits toxic fumes of NO_x. See also NITROSAMINES.

BQQ750 **CAS:109-69-3** **HR: 3**
n-BUTYL CHLORIDE
mf: C_4H_9Cl mw: 92.58

PROP: Colorless liquid. Mp: −123.1°, bp: 78.5°, lel: 1.9%, uel: 10.1%, flash p: 15°F (OC), d: 0.892 @ 15°, autoign temp: 860°F, fp −123.1°, vap d: 3.20.

SYNS: BUTYL CHLORIDE (DOT) □ 1-CHLOROBUTANE (DOT) □ CHLORURE de BUTYLE (FRENCH) □ NCI-C06155 □ N-PROPYLCARBINYL CHLORIDE

TOXICITY DATA WITH REFERENCE
skn-rbt 10 mg/24H open MLD AMIHBC 10,61,54
eye-rbt 500 mg open AMIHBC 10,61,54
msc-mus:lym 500 mg/L NTPTR* NTP-TR-312,86
orl-rat LD50:2670 mg/kg AMIHBC 10,61,54
ihl-rat LCLo:8000 ppm/4H AMIHBC 10,61,54
skn-rbt LDLo:20 g/kg 34ZIAG -,745,69

CONSENSUS REPORTS: NTP Carcinogenesis Studies (gavage); No Evidence: mouse, rat NTPTR* NTP-TR-312,86. EPA Genetic Toxicology Program. Reported in EPA TSCA Inventory.

SAFETY PROFILE: Moderately toxic by ingestion. Mutation data reported. See CHLORINATED HYDROCARBONS, ALIPHATIC. Skin and eye irritant. Dangerous fire hazard when exposed to heat or flame. Moderately explosive when exposed to flame. When heated to decomposition it emits highly toxic fumes of phosgene and Cl^-. To fight fire, use foam, CO_2, dry chemical. Incompatible with oxidizing materials.

BQR000 **CAS:507-20-0** **HR: 3**
tert-BUTYL CHLORIDE
mf: C_4H_9Cl mw: 92.58

PROP: Liquid. Flash p: 32°F, d: 0.87, vap d: 3.2, bp: 51°, fp −27.1°.

SYNS: 2-CHLOROISOBUTANE □ 2-CHLORO-2-METHYLPROPANE □ TRIMETHYLCHLOROMETHANE

TOXICITY DATA WITH REFERENCE
ipr-mus TDLo:3000 mg/kg/8W-I:NEO CNREA8 35,1411,75

CONSENSUS REPORTS: Reported in EPA TSCA Inventory.

SAFETY PROFILE: Questionable carcinogen with experimental neoplastigenic data. Dangerous fire hazard when exposed to heat, flame (sparks), and oxidizers. To fight fire, use water, spray, fog, alcohol foam, dry chemical. When heated to decomposition it emits toxic fumes of Cl^-. See also CHLORINATED HYDROCARBONS, ALIPHATIC.

BQR250 CAS:27778-80-9 HR: 3
β-sec-BUTYL-3-CHLORO-N,N-DIMETHYL-4-ETHOXYPHENETHYLAMINE
mf: $C_{16}H_{26}ClNO$ mw: 283.88

TOXICITY DATA WITH REFERENCE
orl-rat LD50:400 mg/kg CHTPBA 6,453,71
ivn-mus LD50:30 mg/kg CHTPBA 6,453,71

SAFETY PROFILE: Poison by ingestion and intravenous routes. When heated to decomposition it emits very toxic fumes of Cl^- and NO_x. See also AMINES.

BQR750 CAS:27778-78-5 HR: 3
β-sec-BUTYL-3-CHLORO-N,N-DIMETHYL-4-METHOXYPHENETHYLAMINE
mf: $C_{15}H_{24}ClNO$ mw: 269.85

TOXICITY DATA WITH REFERENCE
orl-rat LD50:400 mg/kg CHTPBA 6,453,71
ivn-mus LD50:37 mg/kg CHTPBA 6,453,71

SAFETY PROFILE: Poison by ingestion and intravenous routes. When heated to decomposition it emits very toxic fumes of Cl^- and NO_x. See also AMINES.

BQS000 CAS:33132-85-3 HR: 3
β-sec-BUTYL-5-CHLORO-N,N-DIMETHYL-2-METHOXYPHENETHYLAMINE
mf: $C_{15}H_{24}ClNO$ mw: 269.85

TOXICITY DATA WITH REFERENCE
orl-mus LD50:115 mg/kg CHTPBA 6,453,71
ivn-mus LD50:25 mg/kg CHTPBA 6,453,71

SAFETY PROFILE: Poison by ingestion and intravenous routes. When heated to decomposition it emits very toxic fumes of Cl^- and NO_x. See also AMINES.

BQS250 CAS:33132-71-7 HR: 3
β-sec-BUTYL-p-CHLORO-N,N-DIMETHYLPHENETHYLAMINE
mf: $C_{14}H_{22}ClN$ mw: 239.82

TOXICITY DATA WITH REFERENCE
orl-mus LD50:145 mg/kg CHTPBA 6,453,71
ivn-mus LD50:40 mg/kg CHTPBA 6,453,71

SAFETY PROFILE: Poison by ingestion and intravenous routes. When heated to decomposition it emits very toxic fumes of Cl^- and NO_x. See also AMINES.

BQT000 CAS:29122-56-3 HR: 3
β-sec-BUTYL-5-CHLORO-2-ETHOXY-N,N-DIISOPROPYLPHENETHYLAMINE
mf: $C_{20}H_{34}ClNO$ mw: 340.00

TOXICITY DATA WITH REFERENCE
orl-mus LD50:220 mg/kg CHTPBA 6,453,71
ivn-mus LD50:31 mg/kg CHTPBA 6,453,71

SAFETY PROFILE: Poison by ingestion and intravenous routes. When heated to decomposition it emits very toxic fumes of Cl^- and NO_x. See also AMINES.

BQT250 CAS:29122-60-9 HR: 3
1-(β-sec-BUTYL-5-CHLORO-2-ETHOXYPHENETHYL) PIPERIDINE
mf: $C_{19}H_{30}ClNO$ mw: 323.95

TOXICITY DATA WITH REFERENCE
orl-rat LD50:400 mg/kg CHTPBA 6,453,71
ivn-mus LD50:27 mg/kg CHTPBA 6,453,71

SAFETY PROFILE: Poison by ingestion and intravenous routes. When heated to decomposition it emits very toxic fumes of NO_x and Cl^-.

BQT500 CAS:16224-33-2 HR: 2
BUTYL (3-CHLORO-2-HYDROXYPROPYL) ETHER
mf: $C_7H_{15}ClO_2$ mw: 166.67

SYN: BUTYL-CHLORHYDRINETHER (CZECH)

TOXICITY DATA WITH REFERENCE
skn-rbt 500 mg/24H MOD 28ZPAK -,81,72
eye-rbt 250 μg/24H SEV 28ZPAK -,81,72
orl-rat LD50:3520 mg/kg 28ZPAK -,81,72

CONSENSUS REPORTS: Reported in EPA TSCA Inventory.

SAFETY PROFILE: Moderately toxic by ingestion. A skin and severe eye irritant. See also ETHERS. When heated to decomposition it emits toxic fumes of Cl^-.

BQT600 CAS:12002-53-8 HR: 2
t-BUTYL-CHLORO-2-METHYL-CYCLOHEXANECARBOXYLATE
mf: $C_{12}H_{21}ClO_2$ mw: 232.78

SYNS: CYCLOHEXANECARBOXYLIC ACID, CHLORO-2-METHYL-, tert-BUTYL ESTER ☐ CYCLOHEXANECARBOXYLIC ACID, 4(or 5)-CHLORO-2-METHYL-, tert-BUTYL ESTER (8CI) ☐ CYCLOHEXANECARBOXYLIC ACID, 4(or 5)-CHLORO-2-METHYL-, 1,1-DIMETHYLETHYL ESTER (9CI) ☐ ENT 31560 ☐ PHEROCON MFF ☐ TRIMEDLURE

TOXICITY DATA WITH REFERENCE
orl-rat LD50:4556 mg/kg TXAPA9 31,421,75
ihl-rat LC50:>2900 mg/m³ TXAPA9 31,421,75
skn-rbt LDLo:2025 mg/kg TXAPA9 31,421,75

CONSENSUS REPORTS: Reported in EPA TSCA Inventory.

SAFETY PROFILE: Moderately toxic by skin contact. Low toxicity by ingestion and inhalation. When heated to decomposition it emits toxic vapors of Cl^-.

BQT750 **CAS:5902-51-2** **HR: 1**
3-tert-BUTYL-5-CHLORO-6-METHYLURACIL
mf: $C_9H_{13}ClN_2O_2$ mw: 216.69

PROP: Crystals or solid. Mp: 175–177°. Sltly sol in H_2O.

SYNS: 3-tert-BUTYL-5-CHLOR-6-METHYLURACIL (GERMAN) □ 5-CHLORO-3-tert-BUTYL-6-METHYLURACIL □ 5-CHLORO-3-(1,1-DIMETHYLETHYL)-6-METHYL-2,4(1H,3H)-PYRIMIDINEDIONE □ COMPOUNE 732 □ DU PONT 732 □ DU PONT HERBICIDE 732 □ EXPERIMENTAL HERBICIDE 732 □ SINBAR □ TERBACIL □ TURBSVIL

TOXICITY DATA WITH REFERENCE
orl-rat LD50:7500 mg/kg FMCHA2 -,D302,80
unk-mam LD50:5000 mg/kg 30ZDA9 -,421,71

SAFETY PROFILE: Mildly toxic by ingestion and possibly other routes. When heated to decomposition it emits very toxic fumes of Cl^- and NO_x.

BQU000 **CAS:56139-33-4** **HR: 3**
tert-BUTYL CHLOROPEROXYFORMATE
mf: $C_5H_9ClO_3$ mw: 152.58

$(CH_3)_3COOCO \cdot Cl$

SAFETY PROFILE: A storage hazard. May ignite or explode at room temperature. When heated to decomposition it emits toxic fumes of Cl^-. See also PEROXIDES, ORGANIC.

BQU500 **CAS:5902-52-3** **HR: 3**
o-(4-tert-BUTYL-2-CHLOROPHENYL)-o-METHYL PHOSPHORAMIDOTHIONATE
mf: $C_{11}H_{17}ClNO_2PS$ mw: 293.77

SYNS: DOWCO 109 □ METHYL-PHOSPHORAMIDOTHIOIC ACID o-(tert-BUTYL-2-CHLOROPHENYL)ESTER □ NARLENE

TOXICITY DATA WITH REFERENCE
orl-rat LD50:820 mg/kg TXAPA9 21,315,72
orl-bwd LD50:75 mg/kg TXAPA9 21,315,72

SAFETY PROFILE: Poison by ingestion. See also ESTERS. When heated to decomposition it emits very toxic fumes of SO_x, PO_x, NO_x, and Cl^-.

BQU750 **CAS:67195-50-0** **HR: 2**
tert-20-BUTYLCHOLANTHRENE
mf: $C_{24}H_{22}$ mw: 310.46

SYN: 3-tert-BUTYLCHOLANTHRENE

TOXICITY DATA WITH REFERENCE
scu-mus TDLo:600 mg/kg/39W-I:ETA JNCIAM 2,99,41

SAFETY PROFILE: Questionable carcinogen with experimental tumorigenic data. When heated to decomposition it emits acrid smoke and irritating fumes.

BQV000 **CAS:1189-85-1** **HR: 3**
tert-BUTYL CHROMATE
mf: $C_8H_{18}CrO_4$ mw: 230.26

$[(CH_3)_3CO]_2CrO_2$

PROP: Red crystals from pet ether.

SYN: CHROMIC ACID, DI-tert-BUTYL ESTER

CONSENSUS REPORTS: Chromium and its compounds are on the Community Right-To-Know List.

OSHA PEL: CL 0.1 $mg(CrO_3)/m^3$ (skin)
ACGIH TLV: CL 0.1 $mg(CrO_3)/m^3$ (skin)
NIOSH REL: (Chromium(VI)) CL 0.001 $Mg(Cr(VI))/m^3$

SAFETY PROFILE: A very flammable mixture. When heated to decomposition it emits acrid and irritating fumes. See CHROMIUM COMPOUNDS and ESTERS.

For occupational chemical analysis use NIOSH: Chromium Hexavalent 7024.

BQV250 **CAS:7492-44-6** **HR: 2**
α-BUTYLCINNAMALDEHYDE
mf: $C_{13}H_{16}O$ mw: 188.2

SYNS: BUTYL CINNAMIC ALDEHYDE □ α-BUTYLCINNAMIC ALDEHYDE □ α-n-BUTYL-β-PHENYLACROLEIN □ 2-(PHENYLMETHYLENE) HEXANAL

TOXICITY DATA WITH REFERENCE
skn-rbt 500 mg/24H SEV FCTXAV 18,649,80
orl-rat LD50:4400 mg/kg FCTXAV 18,649,80

SAFETY PROFILE: A severe skin irritant. Mildly toxic by ingestion. When heated to decomposition it emits acrid smoke and irritating fumes. See also ALDEHYDES.

BQV500 **CAS:538-65-8** **HR: 1**
n-BUTYL CINNAMATE
mf: $C_{13}H_{16}O_2$ mw: 204.27

SYNS: n-BUTYL PHENYLACRYLATE □ CINNAMIC ACID-n-BUTYL ESTER

TOXICITY DATA WITH REFERENCE
skn-rbt 500 mg/24H MOD FCTXAV 18,649,80
orl-rat LD50:>5 g/kg FCTXAV 18,655,80
orl-mus LD50:7 g/kg APFRAD 14,370,56
skn-rbt LD50:>5 g/kg FCTXAV 18,655,80

SAFETY PROFILE: Low toxicity by ingestion and skin contact. A skin irritant. See also ESTERS. When heated to decomposition it emits acrid smoke and irritating fumes.

BQV600 **CAS:88-60-8** **HR: 2**
6-tert-BUTYL-m-CRESOL
mf: $C_{11}H_{16}O$ mw: 164.27

SYNS: 2-(1,1-DIMETHYLETHYL)-5-METHYLPHENOL □ PHENOL, 2-tert-BUTYL-5-METHYL-

TOXICITY DATA WITH REFERENCE
orl-mus LD50:1080 mg/kg JAPMA8 38,366,49

SAFETY PROFILE: Moderately toxic by ingestion. When heated to decomposition it emits acrid smoke and irritating vapors.

BQV750 **CAS:2409-55-4** **HR: 3**
2-tert-BUTYL-p-CRESOL
mf: $C_{11}H_{16}O$ mw: 164.27

PROP: Clear liquid, sol in organic solvents and aqueous potassium hydroxide. Fp: 23.1°, bp: 118–119° @ 14 mm, d: 0.922, flash p: 116°F.

SYNS: 2-tert-BUTYL-p-KRESOL (CZECH) □ 2-tert-BUTYL-4-METHYLPHENOL

TOXICITY DATA WITH REFERENCE
skn-rbt 2 mg/24H SEV 85JCAE-,227,86
eye-rbt 50 µg/24H SEV 28ZPAK -,55,72
dni-hmn:lyms 25 µmol/L RCOCB8 54,133,86
orl-ham TDLo:84 g/kg/20W-C:NEO CRNGDP 7,1285,86
orl-rat LD50:2500 mg/kg TPKVAL 12,124,71
orl-mus LD50:700 mg/kg JAPMA8 38,366,49
ipr-mus LD50:144 mg/kg JMCMAR 18,868,75
ivn-mus LD50:10 mg/kg CSLNX* NX#03020
skn-rbt LD50:2200 mg/kg JAPMA8 38,366,49
orl-gpg LD50:1180 mg/kg TPKVAL 12,124,71

CONSENSUS REPORTS: Reported in EPA TSCA Inventory.

SAFETY PROFILE: A poison by intraperitoneal and intravenous routes. Moderately toxic by ingestion and skin contact. Questionable carcinogen with experimental neoplastigenic data. A severe skin and eye irritant. Mutation data reported. Flammable liquid when exposed to heat, flame, or oxidizers. To fight fire, use alcohol foam, foam, water spray, fog, dry chemical. When heated to decomposition it emits acrid and irritating fumes.

BQW000 **CAS:98-52-2** **HR: 3**
4-tert-BUTYLCYCLOHEXANOL
mf: $C_{10}H_{20}O$ mw: 156.30

SYNS: PADARYL □ USAF DO-20

TOXICITY DATA WITH REFERENCE
orl-rat LD50:4200 mg/kg FCTXAV 12,807,74
ipr-mus LD50:50 mg/kg NTIS** AD277-689

CONSENSUS REPORTS: Reported in EPA TSCA Inventory.

SAFETY PROFILE: Poison by intraperitoneal route. Moderately toxic by ingestion. When heated to decomposition it emits acrid smoke and irritating fumes. See also ALCOHOLS.

BQW250 **CAS:98-53-3** **HR: 1**
p-tert-BUTYLCYCLOHEXANONE
mf: $C_{10}H_{18}O$ mw: 154.28

PROP: Crystals. Mp: 49–50°, bp: 90–92° @ 9 mm.

TOXICITY DATA WITH REFERENCE
orl-rat LD50:5000 mg/kg FCTXAV 13,681,75
skn-rbt LD50:5000 mg/kg FCTXAV 13,681,75

CONSENSUS REPORTS: Reported in EPA TSCA Inventory.

SAFETY PROFILE: Mildly toxic by ingestion and skin contact. When heated to decomposition it emits acrid smoke and irritating fumes. See also KETONES.

BQW490 **CAS:88-41-5** **HR: 1**
2-tert-BUTYLCYCLOHEXYL ACETATE
mf: $C_{12}H_{22}O_2$ mw: 198.34

SYNS: 1-ACETOXY-2-tert-BUTYLCYCLOHEXANE □ 2-tert-BUTYLCYCLOHEXANOL ACETATE □ CYCLOHEXANOL, 2-(1,1-DIMETHYLETHYL)-, ACETATE □ 2-(1,1-DIMETHYLETHYL)CYCLOHEXANOL ACETATE □ GRUMEX □ VERDOX

TOXICITY DATA WITH REFERENCE
orl-rat LD50:4600 mg/kg FCTOD7 30,13S,92
skn-rbt LD50:>5 g/kg FCTOD7 30,13S,92

CONSENSUS REPORTS: Reported in EPA TSCA Inventory.

SAFETY PROFILE: Low toxicity by ingestion. When heated to decomposition it emits acrid smoke and irritating vapors.

BQW500 **CAS:32210-23-4** **HR: 1**
p-tert-BUTYLCYCLOHEXYL ACETATE
mf: $C_{12}H_{22}O_2$ mw: 198.34

SYNS: 4-tert-BUTYLCYCLOHEXYL ACETATE □ 4-tert-BUTYLHEXAHYDROPHENYL ACETATE □ VERTENEX

TOXICITY DATA WITH REFERENCE
skn-rbt 500 mg/24H MOD FCTXAV 16,637,78
orl-rat LD50:5000 mg/kg FCTXAV 16,637,78

CONSENSUS REPORTS: Reported in EPA TSCA Inventory.

SAFETY PROFILE: Mildly toxic by ingestion. A skin irritant. When heated to decomposition it emits acrid smoke and irritating fumes. See also ESTERS.

BQW750 **CAS:10108-56-2** **HR: 3**
N-BUTYL CYCLOHEXYL AMINE
mf: $C_{10}H_{21}N$ mw: 155.32

PROP: Liquid. Flash p: 200°F (OC), d: 0.8, bp: 207°.

TOXICITY DATA WITH REFERENCE
skn-rbt 100 µg/24H open AIHAAP 23,95,62
orl-rat LD50:330 mg/kg AIHAAP 23,95,62
skn-rbt LD50:530 mg/kg AIHAAP 23,95,62

SAFETY PROFILE: A poison by ingestion. Moderately toxic by skin contact. See also AMINES. A skin irritant. Combustible when exposed to heat or flame. To fight fire, use alcohol foam. When heated to decomposition it emits toxic fumes of NO_x.

BQW825 **CAS:841-73-6** **HR: 3**
5-BUTYL-1-CYCLOHEXYLBARBITURIC ACID
mf: $C_{14}H_{22}N_2O_3$ mw: 266.38

PROP: Needles from methanol. Mp: 84°, bp: 185–187°.

SYNS: BCP □ BUCOLOM □ BUCOLOME □ 5-BUTYL-1-CYCLOHEXYL-2,4,6(1H,3H,5H)-PYRIMIDINETRIONE □ 5-n-BUTYL-1-CYCLO-

HEXYL-2,4,6-TRIOXOPERHYDROPYRIMIDINE □ PARAMIDIN □ PARAMIDINE

TOXICITY DATA WITH REFERENCE
orl-rat TDLo:2800 mg/kg (7-13D preg):TER JJPAAZ 17,381,67
orl-rat TDLo:25,200 mg/kg (male 36W pre):REP NYKZAU 63,105,67
orl-rat LD50:1115 mg/kg NIIRDN 6,675,82
ipr-rat LD50:455 mg/kg NIIRDN 6,675,82
orl-mus LD50:1550 mg/kg NIIRDN 6,675,82
ipr-mus LD50:550 mg/kg ARZNAD 17,1519,67

SAFETY PROFILE: Poison by ingestion and intravenous routes. An experimental teratogen. Other experimental reproductive effects. When heated to decomposition it emits toxic fumes of NO_x. See also BARBITURATES.

BQX000 CAS:61925-70-0 HR: 3
N-(4-tert-BUTYL CYCLOHEXYL)-3,3-DIPHENYL PROPYLAMINE HYDROCHLORIDE
mf: $C_{25}H_{25}N{\cdot}ClH$ mw: 375.97

SYN: MG 18037

TOXICITY DATA WITH REFERENCE
orl-rat LD50:2550 mg/kg ARZNAD 26,2127,76
ipr-rat LD50:137 mg/kg ARZNAD 26,2127,76
orl-mus LD50:1850 mg/kg ARZNAD 26,2127,76
ipr-mus LD50:98 mg/kg ARZNAD 26,2127,76

SAFETY PROFILE: A poison by intraperitoneal route. Moderately toxic by ingestion. When heated to decomposition it emits very toxic fumes of HCl and NO_x.

BQX250 CAS:89-19-0 HR: 1
BUTYL DECYL PHTHALATE
mf: $C_{22}H_{34}O_4$ mw: 362.56

SYNS: DECYL BUTYL PHTHALATE □ PLASTICIZER BDP

TOXICITY DATA WITH REFERENCE
orl-rat LD50:21 g/kg AIHAAP 30,470,69
skn-rbt LD50:16 g/kg AIHAAP 30,470,69

SAFETY PROFILE: Mildly toxic by ingestion and skin contact. See also ESTERS. When heated to decomposition it emits acrid smoke and irritating fumes.

BQX750 HR: 3
tert-BUTYL DIAZOACETATE
mf: $C_6H_{10}N_2O_2$ mw: 142.16

SAFETY PROFILE: May explode during vacuum distillation. When heated to decomposition it emits toxic fumes of NO_x.

BQY000 CAS:10457-58-6 HR: 2
14-n-BUTYL DIBENZ(a,h)ACRIDINE
mf: $C_{25}H_{21}N$ mw: 335.47

SYN: 14-n-BUTYL-1,2,5,6-DIBENZACRIDINE (FRENCH)

TOXICITY DATA WITH REFERENCE
scu-mus TDLo:60 mg/kg/9W-I:ETA BAFEAG 42,186,55
scu-mus TD:60 mg/kg/9W-I:ETA ACRSAJ 4,315,56

SAFETY PROFILE: Questionable carcinogen with experimental tumorigenic data. When heated to decomposition it emits toxic fumes of NO_x. See also AROMATIC AMINES.

BQY250 CAS:2422-88-0 HR: 2
n-BUTYL-2-DIBUTYLTHIOUREA
mf: $C_{13}H_{28}N_2S$ mw: 244.49

TOXICITY DATA WITH REFERENCE
orl-rat LD50:3000 mg/kg TNICS* 13,78,73
orl-mus LD50:4300 mg/kg TNICS* 13,78,73

CONSENSUS REPORTS: Reported in EPA TSCA Inventory.

SAFETY PROFILE: Moderately toxic by ingestion. When heated to decomposition it emits very toxic fumes of NO_x and SO_x.

BQY300 CAS:684-82-2 HR: 3
sec-BUTYLDICHLOROARSINE
mf: $C_4H_9AsCl_2$ mw: 202.95

SYNS: ARSINE, sec-BUTYLDICHLORO- □ ARSONOUS DICHLORIDE, (1-METHYLPROPYL)-(9CI) □ sec-BUTYLDICHLORARSINE □ DICHLORO(1-METHYLPROPYL)ARSINE

TOXICITY DATA WITH REFERENCE
ihl-mus LC50:12 g/m³/10M NTIS** PB158 508

OSHA PEL: TWA 0.5 mg(As)/m³

SAFETY PROFILE: Poison by inhalation. When heated to decomposition it emits toxic fumes of As and Cl^-.

BQY500 CAS:14090-22-3 HR: 3
BUTYLDICHLOROBORANE
mf: $C_4H_9BCl_2$ mw: 138.7

PROP: Air and moisture-sensitive liquid. Bp: 106–108°.

SAFETY PROFILE: Explosive reaction on contact with water. Ignites in air after a delay period. When heated to decomposition it emits toxic fumes of Cl^-. See also BORANES and BORON COMPOUNDS.

BQZ000 CAS:94-80-4 HR: 2
BUTYL DICHLOROPHENOXYACETATE
mf: $C_{12}H_{14}Cl_2O_3$ mw: 277.16

PROP: Bp: 146–147° @ 1 mm.

SYNS: BUTYL 2,4-D □ BUTYL (2,4-DICHLOROPHENOXY)ACETATE □ 2,4-D BUTYL ESTER □ BUTYL ESTER 2,4-D □ (2,4-DICHLOROPHENOXY)ACETIC ACID, BUTYL ESTER □ ESSO HERBICIDE 10 □ FERNESTA □ LIRONOX □ SHELL 40

TOXICITY DATA WITH REFERENCE
orl-rat TDLo:100 µg/kg (female 10D post):TER GISAAA 44(4),70,79
scu-mus TDLo:900 mg/kg (female 6-14D post):TER NTIS** PB223-160
scu-mus TDLo:414 mg/kg (female 6-14D post):REP NTIS** PB223-160
orl-mus TDLo:1109 mg/kg (female 12-15D post):TER AECTCV 6,33,77

B

orl-mus TDLo:251 mg/kg (female 11D post):TER TJADAB 33,15,86
orl-rat TDLo:1500 mg/kg (6-15D preg):TER TXAPA9 22,14,72
orl-rat LD50:600 mg/kg FAATDF 9,423,87
orl-mus LD50:425 mg/kg 85GMAT -,29,82
orl-cat LD50:780 mg/kg 85GMAT -,29,82

CONSENSUS REPORTS: IARC Cancer Review: Animal Inadequate Evidence IMEMDT 15,111,77.

SAFETY PROFILE: Moderately toxic by ingestion. Experimental teratogenic and reproductive effects. Questionable carcinogen. An herbicide. See also ESTERS. When heated to decomposition it emits toxic fumes of Cl^-.

BRA250 CAS:555-37-3 HR: 3
1-BUTYL-3-(3,4-DICHLOROPHENYL)-1-METHYLUREA
mf: $C_{12}H_{16}Cl_2N_2O$ mw: 275.20

PROP: White or colorless crystals from dioxan (aq). Mp: 101.5–103°. Sltly sol in hydrocarbon solvents: practically insol in water.

SYNS: N-BUTYL-N′-(3,4-DICHLOROPHENYL)-N-METHYLUREA □ 3-(3,4-DICHLORPHENYL)-1-N-BUTYL-HARNSTOFF (GERMAN) □ 3-(3,4-DICHLOROPHENYL)-1-METHYL-1-BUTYLUREA □ GRANUREX □ KLOBEN □ KLOBEN NEBURON □ NEBUREA □ NEBUREX □ NEBURON

TOXICITY DATA WITH REFERENCE
orl-rat LD50:11,000 mg/kg 85ARAE 2,144,77
ivn-mus LD50:180 mg/kg CSLNX* NX#03862

SAFETY PROFILE: Poison by intravenous route. Mildly toxic by ingestion. See also CHLORIDES and NITROGEN MONOXIDE. When heated to decomposition it emits toxic fumes of Cl^- and NO_x.

BRA500 CAS:102489-47-4 HR: 3
2-(BUTYL(2-(DIETHYLAMINO)ETHYL)AMINO)-6′-CHLORO-o-ACETOTOLUIDIDE HYDROCHLORIDE
mf: $C_{19}H_{32}ClN_3O•ClH$ mw: 390.45

SYN: C 5388

TOXICITY DATA WITH REFERENCE
eye-rbt 2% MLD ARZNAD 9,167,59
ipr-rat LD50:31 mg/kg ARZNAD 9,167,59
scu-mus LD50:47 mg/kg ARZNAD 9,167,59

SAFETY PROFILE: Poison by intraperitoneal and subcutaneous routes. An eye irritant. When heated to decomposition it emits very toxic fumes of $Cl^-$and NO_x.

BRA550 CAS:17563-48-3 HR: 3
n-BUTYLDIETHYLTIN IODIDE
mf: $C_8H_{19}ISn$ mw: 360.86

SYN: STANNANE, BUTYLDIETHYLIODO-

TOXICITY DATA WITH REFERENCE
ivn-mus LD50:7100 μg/kg CSLNX* NX#05977

OSHA PEL: TWA 0.1 mg(Sn)/m³
ACGIH TLV: TWA 0.1 mg(Sn)/m³; STEL 0.2 mg/m³ (skin)
NIOSH REL: (Organotin compound): 10H TWA 0.1 mg(Sn)/m³

SAFETY PROFILE: Poison by intravenous route. When heated to decomposition it emits toxic fumes of Sn and I^-.

For occupational chemical analysis use NIOSH: Organotin Compounds 5504.

BRA600 CAS:29149-32-4 HR: 3
tert-BUTYLDIFLUOROPHOSPHINE
mf: $C_4H_9F_2P$ mw: 126.09

$(CH_3)_3CPF_2$

PROP: Liquid. Bp: 54°.

SAFETY PROFILE: Ignites spontaneously in air. When heated to decomposition it emits toxic fumes of F^- and PO_x. See also PHOSPHINE.

BRA625 CAS:692-13-7 HR: 3
1-BUTYLDIGUANIDE
mf: $C_6H_{15}N_5$ mw: 157.26

PROP: Strong base. Very sol in water.

SYNS: BUFORMIN □ BUFORMINE □ BUTFORMIN □ BUTYLBIGUANIDE □ BUTYLDIGUANIDE □ DBV □ GLYBIGID □ H 224 □ W 37

TOXICITY DATA WITH REFERENCE
orl-mus LD50:300 mg/kg JMCMAR 24,1521,81
ipr-mus LD50:140 mg/kg JMCMAR 24,1521,81
scu-gpg LD50:18 mg/kg MEXPAG 8,237,63

SAFETY PROFILE: Poison by ingestion, subcutaneous, and intraperitoneal routes. When heated to decomposition it emits toxic fumes of NO_x.

BRB450 CAS:24596-39-2 HR: 2
4′-n-BUTYL-4-DIMETHYLAMINOAZOBENZENE
mf: $C_{18}H_{23}N_3$ mw: 281.44

SYNS: ANILINE, p-((p-BUTYLPHENYL)AZO)-N,N-DIMETHYL- □ p-((p-BUTYLPHENYL)AZO)-N,N-DIMETHYLANILINE

TOXICITY DATA WITH REFERENCE
orl-rat TDLo:13 mg/kg/Y-C:ETA JNCIAM 27,663,61

SAFETY PROFILE: Questionable carcinogen with experimental tumorigenic data. When heated to decomposition it emits toxic fumes of NO_x.

BRB460 CAS:24596-41-6 HR: 2
4′-tert-BUTYL-4-DIMETHYLAMINOAZOBENZENE
mf: $C_{18}H_{23}N_3$ mw: 281.44

SYNS: ANILINE, p-((p-(tert-BUTYL)PHENYL)AZO)-N,N-DIMETHYL- □ p-((p-tert-BUTYLPHENYL)AZO)-N,N-DIMETHYLANILINE

TOXICITY DATA WITH REFERENCE
orl-rat TDLo:12,852 mg/kg/Y-C:ETA JNCIAM 27,663,61

SAFETY PROFILE: Questionable carcinogen with experimental tumorigenic data. When heated to decomposition it emits toxic fumes of NO_x.

BRB500 **CAS:69745-66-0** **HR: 3**
4-(1-sec-BUTYL-2-(DIMETHYLAMINO)ETHYL)PHENOL
mf: $C_{14}H_{23}NO$ mw: 221.38

TOXICITY DATA WITH REFERENCE
orl-mus LD50:375 mg/kg CHTPBA 6,453,71
ivn-mus LD50:73 mg/kg CHTPBA 6,453,71

SAFETY PROFILE: Poison by ingestion and intravenous routes. When heated to decomposition it emits toxic fumes of NO_x.

BRB750 **CAS:33098-26-9** **HR: 3**
2-(1-sec-BUTYL-2-(DIMETHYLAMINO)ETHYL)QUINOLINE
mf: $C_{17}H_{24}N_2$ mw: 256.43

TOXICITY DATA WITH REFERENCE
orl-mus LD50:60 mg/kg CHTPBA 6,453,71
ivn-mus LD50:20 mg/kg CHTPBA 6,453,71

SAFETY PROFILE: Poison by ingestion and intravenous routes. When heated to decomposition it emits toxic fumes of NO_x.

BRC000 **CAS:33098-27-0** **HR: 3**
2-(1-sec-BUTYL-2-(DIMETHYLAMINO)ETHYL)QUINOXALINE
mf: $C_{16}H_{23}N_3$ mw: 257.42

TOXICITY DATA WITH REFERENCE
orl-mus LD50:102 mg/kg CHTPBA 6,453,71
ivn-mus LD50:35 mg/kg CHTPBA 6,453,71

SAFETY PROFILE: Poison by ingestion and intravenous routes. When heated to decomposition it emits toxic fumes of NO_x.

BRC250 **CAS:34548-72-6** **HR: 3**
2-(1-sec-BUTYL-2-(DIMETHYLAMINO)ETHYL)THIOPHENE
mf: $C_{12}H_{21}NS$ mw: 211.40

TOXICITY DATA WITH REFERENCE
orl-mus LD50:260 mg/kg CHTPBA 6,453,71
ivn-mus LD50:50 mg/kg CHTPBA 6,453,71

SAFETY PROFILE: Poison by ingestion and intravenous routes. When heated to decomposition it emits very toxic fumes of NO_x and SO_x.

BRC500 **CAS:51003-83-9** **HR: 3**
2-n-BUTYL-3-DIMETHYLAMINO-5,6-METHYLENEDIOXYINDENE HYDROCHLORIDE
mf: $C_{16}H_{21}NO_2 \cdot ClH$ mw: 295.84

SYNS: 6-BUTYL-5-DIMETHYLAMINO-5H-INDENO(5,6-d)-1,3-DIOXOLE HYDROCHLORIDE □ bu-MDI

TOXICITY DATA WITH REFERENCE
ipr-rat LD50:240 mg/kg RCOCB8 26,85,79
ipr-mus LD50:185 mg/kg RCOCB8 26,85,79
ivn-mus LD50:32 mg/kg RCOCB8 26,85,79

SAFETY PROFILE: A poison by intraperitoneal and intravenous routes. When heated to decomposition it emits very toxic fumes of NO_x and HCl.

BRC750 **CAS:6279-54-5** **HR: 2**
BUTYL-3-((DIMETHYLAMINO)METHYL)-4-HYDROXYBENZOATE
mf: $C_{14}H_{21}NO_3$ mw: 251.36

TOXICITY DATA WITH REFERENCE
orl-mus LDLo:2000 mg/kg ARZNAD 11,85,61
scu-mus LD50:475 mg/kg ARZNAD 11,85,61

SAFETY PROFILE: Moderately toxic by ingestion and subcutaneous routes. When heated to decomposition it emits toxic fumes of NO_x. See also ESTERS.

BRD000 **CAS:5221-53-4** **HR: 3**
5-BUTYL-2-(DIMETHYLAMINO)-6-METHYL-4(1H)-PYRIMIDINONE
mf: $C_{11}H_{19}N_3O$ mw: 209.33

SYNS: 5-n-BUTYL-2-DIMETHYLAMINO-4-HYDROXY-6-METHYLPYRIMIDINE □ 5-BUTYL-2-(DIMETHYLAMINO)-6-METHYL-4-PYRIMIDINOL □ DIMETHIRIMOL □ 2-DIMETHYLAMINO-4-HYDROXY-5-n-BUTYL-6-METHYLPYRIMIDINE □ 2-DIMETHYLAMINO-4-METHYL-5-n-BUTYL-6-HYDROXYPYRIMIDINE □ METHYRIMOL □ MILCURB □ PP 675

TOXICITY DATA WITH REFERENCE
orl-rat LD50:2350 mg/kg WRPCA2 9,119,70
ipr-rat LDLo:200 mg/kg NATUAS 219,1160,68
orl-mus LD50:800 mg/kg 28ZEAL 5,79,76

SAFETY PROFILE: Poison by intraperitoneal route. Moderately toxic by ingestion and possibly other routes. When heated to decomposition it emits toxic fumes of NO_x.

BRD500 **CAS:27778-82-1** **HR: 3**
β-sec-BUTYL-N,N-DIMETHYL-2-ETHOXY-5-FLUOROPHENETHYLAMINE
mf: $C_{16}H_{26}FNO$ mw: 267.43

TOXICITY DATA WITH REFERENCE
orl-rat LD50:285 mg/kg CHTPBA 6,453,71
ivn-mus LD50:11 mg/kg CHTPBA 6,453,71

SAFETY PROFILE: Poison by ingestion and intravenous routes. When heated to decomposition it emits very toxic fumes of F^- and NO_x. See also AMINES.

BRD750 **CAS:27684-90-8** **HR: 3**
β-sec-BUTYL-N,N-DIMETHYL-5-FLUORO-2-METHOXYPHENETHYLAMINE
mf: $C_{15}H_{24}FNO$ mw: 253.40

TOXICITY DATA WITH REFERENCE
orl-rat LD50:300 mg/kg CHTPBA 6,453,71
ivn-mus LD50:18 mg/kg CHTPBA 6,453,71

SAFETY PROFILE: Poison by ingestion and intravenous routes. When heated to decomposition it emits very toxic fumes of F^- and NO_x. See also AMINES.

BRE000 CAS:56654-53-6 HR: 2
1-BUTYL-3,3-DIMETHYL-1-NITROSOUREA
mf: $C_7H_{15}N_3O_2$ mw: 173.25

TOXICITY DATA WITH REFERENCE
mmo-esc 4 mmol/L CPBTAL 34,5056,86
orl-rat TDLo:3140 mg/kg/45W-C:ETA,REP JNCIAM 56,1177,76

SAFETY PROFILE: Questionable carcinogen with experimental tumorigenic data. Experimental reproductive effects. Mutation data reported. When heated to decomposition it emits toxic fumes of NO_x.

BRE250 CAS:33132-61-5 HR: 3
β-sec-BUTYL-N,N-DIMETHYLPHENETHYLAMINE
mf: $C_{14}H_{23}N$ mw: 205.38

TOXICITY DATA WITH REFERENCE
orl-mus LD50:170 mg/kg CHTPBA 6,453,71
ivn-mus LD50:31 mg/kg CHTPBA 6,453,71

SAFETY PROFILE: Poison by ingestion and intravenous routes. When heated to decomposition it emits toxic fumes of NO_x. See also AMINES.

BRE255 HR: 3
β-sec-BUTYL-N,N-DIMETHYLPHENETHYLAMINE HYDROCHLORIDE
mf: $C_{14}H_{23}N{\cdot}ClH$ mw: 241.84

SYNS: 1-DIMETHYLAMINO-2-PHENYL-3-METHYLPENTANE HYDROCHLORIDE □ Z-134

TOXICITY DATA WITH REFERENCE
orl-rat LD50:357 mg/kg JPETAB 117,451,56
ipr-rat LD50:93 mg/kg JPETAB 117,451,56
ims-rat LD50:119 mg/kg JPETAB 117,451,56
orl-mus LD50:237 mg/kg JPETAB 117,451,56
ipr-mus LD50:110 mg/kg JPETAB 117,451,56

SAFETY PROFILE: Poison by ingestion, intraperitoneal, and intramuscular routes. When heated to decomposition it emits toxic fumes of NO_x and HCl.

BRE500 CAS:88-85-7 HR: 3
2-sec-BUTYL-4,6-DINITROPHENOL
mf: $C_{10}H_{12}N_2O_5$ mw: 240.24

PROP: Crystals from pentane (tech grade usually liquid). Vap d: 7.73, mp: 40–41°.

SYNS: ARETIT □ BASANITE □ BNP 30 □ BUTAPHENE □ CALDON □ CHEMOX GENERAL □ CHEMOX P.E. □ DINITRO □ DINITRO-3 □ 4,6-DINITRO-2-sec.BUTYLFENOL (CZECH) □ DINITROBUTYLPHENOL □ 2,4-DINITRO-6-sec-BUTYLPHENOL □ 4,6-DINITRO-o-sec-BUTYLPHENOL □ 4,6-DINITRO-2-sec-BUTYLPHENOL □ 4,6-DINITRO-2-(1-METHYL-N-PROPYL)PHENOL □ 2,4-DINITRO-6-(1-METHYLPROPYL)PHENOL (FRENCH) □ DINOSEB □ DINOSEBE (FRENCH) □ DN 289 □ DNBP □ DNOSBP □ DNSBP □ DOW GENERAL □ DOW GENERAL WEED KILLER □ DOW SELECTIVE WEED KILLER □ ELGETOL □ ELGETOL 318 □ ENT 1,122 □ GEBUTOX □ HEL-FIRE □ KILOSEB □ 6-(1-METHYL-PROPYL)-2,4-DINITROFENOL (DUTCH) □ 2-(1-METHYLPROPYL)-4,6-DINITROPHENOL □ 6-(1-METIL-PROPIL)-2,4-DINITRO-FENOLO (ITALIAN) □ NITROPONE C □ PHENOTAN □ PREMERGE □ PREMERGE 3 □ RCRA WASTE NUMBER P020 □ SINOX GENERAL □ SPARIC □ SPURGE □ SUBITEX □ UNICROP DNBP □ VERTAC DINITRO WEED KILLER □ VERTAC GENERAL WEED KILLER □ VERTAC SELECTIVE WEED KILLER

TOXICITY DATA WITH REFERENCE
eye-rbt 50 µg/24H SEV 28ZPAK -,108,72
mrc-smc 185 ppm MUREAV 21,83,73
orl-rat TDLo:820 mg/kg (male/60D pre/1-22D preg):REP TXAPA9 45,235,78
orl-rat TDLo:150 mg/kg (female 6-15D post):TER TJADAB 33,19A,86
scu-mus TDLo:53,100 µg/kg (female 14-16D post):REP FCTXAV 11,31,73
orl-rat TDLo:1201 mg/kg (77D male):REP AECTCV 11,475,82
orl-rat TDLo:78 mg/kg (male 20D pre):REP AECTCV 11,475,82
ipr-mus TDLo:30 mg/kg (female 10-12D post):TER FCTXAV 11,31,73
ipr-mus TDLo:47,400 µg/kg (female 10-12D post):TER TJADAB 12,147,75
orl-rat TDLo:100 mg/kg (female 6-15D post):TER TJADAB 33,19A,86
orl-mus TDLo:26 mg/kg (female 8D post):TER TCMUD8 5,3,85
orl-mus TDLo:764 mg/kg/78W-I:ETA NTIS** PB223-159
orl-rat LD50:25 mg/kg TXAPA9 7,353,65
skn-rat LD50:80 mg/kg WRPCA2 9,119,70
scu-rat LD50:20,368 µg/kg JPPMAB 4,1062,52
orl-mus LD50:16 mg/kg 85GMAT -,61,82
ihl-cat LCLo:45 mg/m³/3H 85GMAT -,61,82
skn-rbt LD50:80 mg/kg 31ZOAD 1,178,68

CONSENSUS REPORTS: EPA Genetic Toxicology Program. EPA Extremely Hazardous Substances List.

SAFETY PROFILE: A poison by ingestion, inhalation, skin contact, subcutaneous, and intraperitoneal routes. Experimental teratogenic and reproductive effects. A severe eye irritant. Questionable carcinogen with experimental tumorigenic data. Mutation data reported. An herbicide. When heated to decomposition it emits toxic fumes of NO_x.

BRE750 CAS:6420-47-9 HR: 3
o-sec-BUTYL-4,6-DINITROPHENOLTRIETHANOLAMINE SALT
mf: $C_{16}H_{27}N_3O_8$ mw: 389.46

SYNS: 2-sec-BUTYL-4,6-DINITROPHENOL- 2,2′,2″-NITRILOTRIETHANOL SALT □ DINITROBUTYLPHENOL-2,2′,2″-NITRILOTRIETHANOL SALT □ 2-(1-METHYL-N-PROPYL)-4,6-DINITROPHENOL TRIETHANOLAMINE SALT

TOXICITY DATA WITH REFERENCE
orl-rat LD50:37 mg/kg SPEADM 74-1,-,74
skn-rat LD50:80 mg/kg SPEADM 74-1,-,74

SAFETY PROFILE: Poison by ingestion and skin contact. When heated to decomposition it emits toxic fumes of NO_x.

BRF500 CAS:50-33-9 HR: 3
4-BUTYL-1,2-DIPHENYL-3,5-DIOXO PYRAZOLIDINE
mf: $C_{19}H_{20}N_2O_2$ mw: 308.41

PROP: Crystals from EtOH. Mp: 105.5–106.5°.

SYNS: ALINDOR □ ALQOVERIN □ ANERVAL □ ANTADOL □ ANUSPIRAMIN □ ARTIZIN □ ARTRIZONE □ ARTROPAN □ AZDID □ AZOLID □ BENZONE □ BETAZED □ BUSONE □ BUTACOMPREN □ BUTACOTE □ BUTALAN □ BUTALGINA □ BUTALIDON □ BUTAPIRAZOL □ BUTAPYRAZOLE □ BUTARECBON □ BUTARTRINA □ BUTAZINA □ BUTAZONA □ BUTAZONE □ BUTIDIONA □ BUTONE □ BUTOZ □ 4-BUTYL-1,2-DIPHENYLPYRAZOLIDINE-3,5-DIONE □ BUTYLPYRIN □ BUVETZONE □ BUZON □ DIGIBUTINA □ DIOSSIDONE □ 3,5-DIOXO-1,2-DIPHENYL-4-N-BUTYLPYRAZOLIDENE □ DIOZOL □ DIPHEBUZOL □ DIPHENYLBUTAZONE □ 1,2-DIPHENYL-4-BUTYL-3,5-DIOXOPYRAZOLIDINE □ ELMEDAL □ EQUI BUTE □ ERIBUTAZONE □ ESTEVE □ FENARTIL □FENIBUTAZO-NA □ FENIBUTOL □ FENILBUTINE □ FENILIDINA □ FENOTONE □ FENYLBUTAZON □ FLEXAZONE □ INTALBUT □ IPSOFLAME □ LINGEL □ MALGESIC □ MEPHABUTAZONE □ MERIZONE □ NADOZONE □ NCI-C56531 □ NOVOPHENYL □ PHEBUZIN □ PHENBUTAZOL □ PHENOPYRINE □ PHENYLBUTAZON (GERMAN) □ PHENYLBUTAZONE □ PIRARREUMOL 'B" □ PRAECIRHEUMIN □ PYRAZOLIDIN □ REUDO □ REUMASYL □ REUMAZOL □ REUPOLAR □ RUBATONE □ SCANBUTAZONE □ SHIGRODIN □ TAZONE □ TEVCODYNE □ THERAZONE □ TODALGIL □ UZONE □ WESCOZONE □ ZOLAPHEN □ ZOLIDINUM □ ZORANE

TOXICITY DATA WITH REFERENCE
eye-rbt 100 mg MOD CMROCX 4,17,76
oms-hmn:emb 20 mg/L BEXBAN 74,828,72
mnt-mus-ipr 50 mg/kg IJEBA6 18,869,80
orl-rat TDLo:1100 mg/kg (female 7-17D post):REP OYYAA2 20,377,80
orl-mus TDLo:8 g/kg (female 6-15D post):REP OYYAA2 20,289,80
orl-rat TDLo:1140 mg/kg (female 14D pre):REP TXAPA9 15,46,69
orl-rbt TDLo:780 mg/kg (female 6-18D post):TER IYKEDH 10,149,79
orl-rat TDLo:600 mg/kg (female 5-16D post):TER RCOCB8 7,701,74
orl-rbt TDLo:1560 mg/kg (female 6-18D post):REP IYKEDH 10,149,79
orl-rbt TDLo:1050 mg/kg (female 1-7D post):REP JRPFA4 10,129,65
orl-rat TDLo:25,200 mg/kg (female 84D pre):REP OYYAA2 19,1,80
orl-rat TDLo:360 mg/kg (female 17-22D post):REP IYKEDH 10,164,79
orl-rat TDLo:21 g/kg (42D male):REP YACHDS 7,1257,79
orl-rat TDLo:30 mg/kg (female 21D post):TER OYYAA2 27,117,84
orl-rat TDLo:1100 mg/kg (female 7-17D post):TER OYYAA2 20,377,80
orl-man TDLo:4368 mg/kg/4Y-C:CAR,BLD BMJOAE 1,744,64
mul-wmn TDLo:4200 mg/kg/77W-I:CAR BMJOAE 2,1569,61
orl-man TD:140 mg/kg/3W-C:CAR,BLD BMJOAE 2,1552,60
par-hmn LDLo:168 mg/kg/2W-I:SYS 27ZXA3 -,448,63
orl-man TDLo:17,500 µg/kg/3W-I:KID AIMEAS 41,1075,54
unr-man TDLo:200 mg/kg/5W-I:CVS,KID BMJOAE 282,950,81
unr-wmn TDLo:40 mg/kg/4D-I:BLD,MET PGPKA8 4(5),48,59
orl-rat LD50:245 mg/kg AIPTAK 123,48,59
ipr-rat LD50:142 mg/kg FRPSAX 14,347,59
scu-rat LD50:230 mg/kg OYYAA2 6,1285,72
ivn-rat LD50:100 mg/kg ARZNAD 10,665,60
ims-rat LD50:220 mg/kg ARZNAD 10,665,60
orl-mus LD50:270 mg/kg BCFAAI 111,293,72
ipr-mus LD50:128 mg/kg PCJOAU 19,33,85
scu-mus LD50:230 mg/kg JPPMAB 7,1022,55
ivn-mus LD50:90 mg/kg ARZNAD 19,36,69
ims-mus LD50:430 mg/kg OYYAA2 13,97,77
orl-dog LD50:332 mg/kg OYYAA2 20,265,80
ivn-dog LD50:121 mg/kg AIPTAK 149,571,64
ivn-cat LD50:100 mg/kg ARZNAD 19,36,69
orl-rbt LD50:781 mg/kg OYYAA2 20,265,80
ivn-rbt LD50:146 mg/kg ARZNAD 10,129,60
orl-gpg LD50:250 mg/kg ARZNAD 19,1207,69
orl-ham LD50:1260 mg/kg ATSUDG 7,365,84

CONSENSUS REPORTS: IARC Cancer Review: Group 3 IMEMDT 7,316,87; Human Inadequate Evidence IMEMDT 13,183,77. EPA Genetic Toxicology Program. Reported in EPA TSCA Inventory.

SAFETY PROFILE: Suspected human carcinogen producing leukemia. A human poison by parenteral route. An experimental poison by ingestion, intraperitoneal, subcutaneous, intravenous, and intramuscular routes. Human systemic effects by ingestion and possibly other routes: fever, blood pressure increase, other unspecified vascular effects, damage to kidney tubules and glomeruli, decreased urine volume, blood in the urine, reduction in the number of white blood cells, and agranulocytosis. Experimental teratogenic and reproductive effects. Human mutation data reported. An eye irritant. An anti-inflammatory agent. When heated to decomposition it emits toxic fumes of NO_x.

BRF550 **CAS:20333-40-8** **HR: 3**
BUTYL DISELENIDE
mf: $C_8H_{18}Se_2$ mw: 272.18

SYNS: DIBUTYL DISELENIDE □ DI-n-BUTYL-DISELENIDE □ DIBUTYLDISELENIUM □ DISELENIDE, DIBUTYL-(9CI)

TOXICITY DATA WITH REFERENCE
ivn-mus LD50:75 mg/kg CSLNX* NX#09252

OSHA PEL: TWA 0.2 mg(Se)/m^3
ACGIH TLV: TWA 0.2 mg(Se)/m^3

SAFETY PROFILE: Poison by intravenous route. When heated to decomposition it emits toxic fumes of Se.

BRG000 **CAS:110-57-6** **HR: 3**
2-BUTYLENE DICHLORIDE
mf: $C_4H_6Cl_2$ mw: 125.00

PROP: Colorless liquid. Mp: 1–3°, bp: 156°, d: 1.183 @ 25°/4°.

SYNS: 1,4-DICHLOROBUTENE-2 (trans) □ 1,4-DICHLORO-2-BUTENE

TOXICITY DATA WITH REFERENCE
ipr-mus TDLo:150 mg/kg/77W-I:ETA CNREA8 35,2553,75
scu-mus TDLo:150 mg/kg/77W-I:NEO CNREA8 35,2553,75
ihl-rat LC50:86 ppm/4H AIHAM* -,-,68

CONSENSUS REPORTS: IARC Cancer Review: Group 3

IMEMDT 7,56,87; Animal Inadequate Evidence IMEMDT 15,149,77. Reported in EPA TSCA Inventory. EPA Extremely Hazardous Substances List.

SAFETY PROFILE: A poison by inhalation. Questionable carcinogen with experimental neoplastigenic and tumorigenic data. When heated to decomposition it emits toxic fumes of Cl^-. See also CHLORINATED HYDROCARBONS, ALIPHATIC.

BRG500 CAS:19485-03-1 HR: 2
1,3-BUTYLENE GLYCOL DIACRYLATE
mf: $C_{10}H_{14}O_4$ mw: 198.24

SYNS: ACRYLIC ACID-1-METHYLTRIMETHYLENE ESTER □ 1,3-BUTANEDIOL DIACRYLATE □ 1,3-BUTYLENE DIACRYLATE □ 2-PROPENOIC ACID-1-METHYL-13-PROPANEDIYL ESTER

TOXICITY DATA WITH REFERENCE
orl-rat LD50:3540 mg/kg TXAPA9 28,313,74
skn-rbt LD50:450 mg/kg TXAPA9 28,313,74

CONSENSUS REPORTS: Reported in EPA TSCA Inventory.

SAFETY PROFILE: Moderately toxic by ingestion and skin contact. See also ESTERS. When heated to decomposition it emits acrid smoke and irritating fumes.

BRH250 CAS:106-83-2 HR: 2
BUTYL-9,10-EPOXYSTEARATE
mf: $C_{22}H_{42}O_3$ mw: 354.64

SYN: 9,10-EPOXYOCTADECANOIC ACID BUTYL ESTER

TOXICITY DATA WITH REFERENCE
unr-mus TDLo:24 g/kg:ETA RARSAM 3,193,63

CONSENSUS REPORTS: Reported in EPA TSCA Inventory.

SAFETY PROFILE: Questionable carcinogen with experimental tumorigenic data. See also ESTERS. When heated to decomposition it emits acrid smoke and irritating fumes.

BRH750 CAS:142-96-1 HR: 3
n-BUTYL ETHER
DOT: UN 1149
mf: $C_8H_{18}O$ mw: 130.26

PROP: Colorless liquid. Mp: −98°, bp: 142°, flash p: 77°F, d: 0.784 @ 0°/4°, autoign temp: 382°F, vap d: 4.48, lel: 1.5%, uel: 7.6%.

SYNS: 1-BUTOXYBUTANE □ BUTYL ETHER (DOT) □ DI-n-BUTYL ETHER (DOT) □ DIBUTYL OXIDE □ ETHER BUTYLIQUE (FRENCH) □ 1,1′-OXYBIS(BUTANE)

TOXICITY DATA WITH REFERENCE
eye-hmn 200 ppm/15M JIHTAB 28,262,46
skn-rbt 100 mg/24H MOD 85JCAE-,250,86
eye-rbt 500 mg open AMIHBC 10,61,54
eye-rbt 500 mg/24H MLD 85JCAE-,250,86
ihl-hmn TCLo:200 ppm:NOSE,EYE JIHTAB 28,262,46
orl-rat LD50:7400 mg/kg AMIHBC 10,61,54
ihl-rat LCLo:4000 ppm/4H AMIHBC 10,61,54
skn-rbt LD50:10 g/kg AMIHBC 10,61,54

CONSENSUS REPORTS: Reported in EPA TSCA Inventory.

DOT CLASSIFICATION: 3; *Label:* Flammable Liquid

SAFETY PROFILE: Mildly toxic by inhalation, ingestion, and skin contact. Human systemic effects by inhalation: conjunctiva irritation and unspecified nasal effects. An experimental skin and human eye irritant. See also ETHERS. Dangerous fire hazard when exposed to heat, flame, or oxidizers. Incompatible with NCl_3 and oxidizing materials. To fight fire, use alcohol foam, dry chemical. When heated to decomposition it emits acrid smoke and fumes.

BRH760 CAS:6863-58-7 HR: 3
sec-BUTYL ETHER
mf: $C_8H_{18}O$ mw: 130.26

SYNS: BIS(2-BUTYL)ETHER □ BUTANE, 2,2′-OXYBIS-(9CI) □ DI-sec-BUTYL ETHER □ 2,2′-OXYBISBUTANE

TOXICITY DATA WITH REFERENCE
ihl-mus LC50:130 mg/m³/15M ANESAV 11,455,50

CONSENSUS REPORTS: Reported in EPA TSCA Inventory.

DOT CLASSIFICATION: 3; *Label:* Flammable Liquid

SAFETY PROFILE: Poison by inhalation. A flammable liquid. When heated to decomposition it emits acrid smoke and irritating vapors.

BRI000 CAS:123-05-7 HR: 3
BUTYL ETHYL ACETALDEHYDE
mf: $C_8H_{16}O$ mw: 128.24

PROP: Bp: 163.4°, flash p: 125°F (OC), autoign temp: 387°F, d: 0.8205, vap press: 1.8 mm @ 20°, vap d: 4.42.

SYNS: ETHYLBUTYLACETALDEHYDE □ α-ETHYLCAPROALDEHYDE □ 2-ETHYLHEXALDEHYDE □ ETHYLHEXALDEHYDE (DOT) □ 2-ETHYLHEXANAL □ β-PROPYL-α-ETHYLACROLEIN

TOXICITY DATA WITH REFERENCE
skn-rbt 20 mg/24H MOD 85JCAE-,274,86
skn-rbt 425 mg open MLD UCDS** 7/21/65
eye-rbt 500 mg open AMIHBC 4,119,51
orl-rat LD50:3730 mg/kg AMIHBC 4,119,51
ihl-rat LCLo:4000 ppm/4H AMIHBC 4,119,51
ipr-rat LD50:500 mg/kg HYDRDA 3,201,78
orl-mus LD50:3550 mg/kg 85GMAT -,103,82
skn-rbt LD50:5040 mg/kg AMIHBC 4,119,51

CONSENSUS REPORTS: Reported in EPA TSCA Inventory.

SAFETY PROFILE: Moderately toxic by ingestion and intraperitoneal routes. Mildly toxic by inhalation and skin contact. An eye and severe skin irritant. See also ALDEHYDES. Dangerous fire hazard; spontaneously flammable in air. To fight fire, use foam, CO_2, dry chemical, water spray, mist, fog. Incompatible with oxidizing materials. When heated to decomposition it emits acrid and irritating fumes.

BRI250 CAS:149-57-5 HR: 2
BUTYL ETHYL ACETIC ACID
mf: $C_8H_{16}O_2$ mw: 144.24

PROP: Flash p: 260°F (OC), bp: 225–228°.

SYNS: α-ETHYLCAPROIC ACID □ 2-ETHYLHEXANOIC ACID □ 2-ETHYLHEXOIC ACID

TOXICITY DATA WITH REFERENCE
skn-rbt 10 mg/24H open JIHTAB 26,269,44
skn-rbt 450 mg open MLD UCDS** 11/4/71
eye-rbt 20 mg SEV AJOPAA 29,1363,46
orl-rat TDLo: 1803 mg/kg (female 12D post): TER TJADAB 35,41,87
orl-rat LD50: 3000 mg/kg JIHTAB 26,269,44
skn-rbt LD50: 1260 mg/kg UCDS** 11/4/71

CONSENSUS REPORTS: Reported in EPA TSCA Inventory.

SAFETY PROFILE: Moderately toxic by ingestion and skin contact. An experimental teratogen. A skin and severe eye irritant. Combustible when exposed to heat or flame. When heated to decomposition, it emits acrid and irritating fumes.

BRI500 CAS:77966-77-9 HR: 3
n-BUTYL-2-(ETHYLAMINO)-2′,6′-ACETOXYLIDIDE HYDROCHLORIDE
mf: $C_{16}H_{26}N_2O{\cdot}ClH$ mw: 298.90

SYN: C 3164

TOXICITY DATA WITH REFERENCE
eye-rbt 2% MLD ARZNAD 8,609,58
ipr-rat LD50: 72 mg/kg ARZNAD 8,609,58
scu-mus LD50: 125 mg/kg ARZNAD 8,609,58

SAFETY PROFILE: Poison by intraperitoneal and subcutaneous routes. An eye irritant. When heated to decomposition it emits very toxic fumes of HCl and NO_x.

BRI750 CAS:23947-60-6 HR: 2
5-n-BUTYL-2-ETHYLAMINO-4-HYDROXY-6-METHYLPYRIMIDINE
mf: $C_{11}H_{19}N_3O$ mw: 209.33

PROP: Solid. Mp: 159°. Very sltly sol in H_2O; sltly sol in EtOH; sol in $CHCl_3$.

SYNS: 5-BUTYL-2-(ETHYLAMINO)-6-METHYL-4(1H)-PYRIMIDINONE □ ETHIRIMOL □ 2-ETHYLAMINO-4-METHYL-5-n-BUTYL-6-HYDROXYPYRIMIDINE □ MILCURB □ MILCURB SUPER □ MILGO □ MILGO E □ MILSTEM □ MILSTEM SEED DRESSING □ NEW MILSTEM □ PP149

TOXICITY DATA WITH REFERENCE
mmo-smc 50 ppm RSTUDV 6,161,76
orl-rat LD50: 4000 mg/kg 28ZEAL 5,106,76
skn-rat LD50: >1 g/kg PEMNDP 9,345,91
par-rat LD50: 4 g/kg DOVEAA 26,5,72
orl-mus LD50: 4 g/kg 85JFAN A183,83
ivn-mus LD50: 800 mg/kg CHINAG (42),1512,69
unr-mus LD50: 4 g/kg TGANAK 16(1),45,82
orl-cat LD50: 1000 mg/kg CHINAG (42),1512,69
orl-rbt LD50: 1000 mg/kg CHINAG (42),1512,69
orl-gpg LD50: 500 mg/kg CHINAG (42),1512,69

SAFETY PROFILE: Moderately toxic by ingestion, intravenous, and possibly other routes. Mutation data reported. When heated to decomposition it emits toxic fumes of NO_x.

B

BRJ000 CAS:41483-43-6 HR: 2
5-BUTYL-2-ETHYLAMINO-6-METHYLPYRIMIDIN-4-YL DIMETHYLSULPHAMATE
mf: $C_{13}H_{24}N_4O_3S$ mw: 316.47

PROP: Pale-tan waxy solid. Mp: 50–51°. Very sltly sol in H_2O; sol in most org solvs.

SYNS: 2-AETHYLAMINO-5-BUTYL-4-YL-DIMETHYLSULFAMAT (GERMAN) □ DIMETHYLSULFAMIC ACID 5-BUTYL-2-(ETHYLAMINO)-6-METHYL-4-PYRIMIDINYL ESTER □ NIMROD □ NIMROD T □ PP588 □ SULFAMIC ACID, DIMETHYL-, 5-BUTYL-2-(ETHYLAMINO)-6-METHYL-4-PYRIMIDINYL ESTER

TOXICITY DATA WITH REFERENCE
orl-rat LD50: 4000 mg/kg 85ARAE 4,115,76
skn-rat LD50: 500 mg/kg DOVEAA 30,200,76
orl-mus LD50: 4000 mg/kg 85DPAN -,-,71/76
orl gpg LD50: 4000 mg/kg 85DPAN -,-,71/76

SAFETY PROFILE: Moderately toxic by ingestion and skin contact. When heated to decomposition it emits very toxic fumes of NO_x and SO_x. See also ESTERS.

BRJ125 CAS:13080-06-3 HR: 3
BUTYLETHYLMALONIC ACID-2-(DIETHYLAMINO) ETHYL ETHYL ESTER
mf: $C_{17}H_{33}NO_4$ mw: 315.51

SYNS: BUTYLAETHYLMALONSAEURE-AETHYL-DIAETHYLAMINOAETHYL-DI-ESTER (GERMAN) □ BUTYLETHYL-PROPANEDIOIC ACID-2-(DIETHYLAMINO)ETHYL ETHYL ESTER (9CI) □ Sch 5712

TOXICITY DATA WITH REFERENCE
orl-rat LD50: 638 mg/kg AEPPAE 237,264,59
ipr-rat LD50: 225 mg/kg AEPPAE 237,264,59
orl-mus LD50: 412 mg/kg AEPPAE 237,264,59
ipr-mus LD50: 258 mg/kg AEPPAE 237,264,59

SAFETY PROFILE: Poison by intraperitoneal route. Moderately toxic by ingestion. When heated to decomposition it emits toxic fumes of NO_x. See also ESTERS.

BRJ250 CAS:67050-26-4 HR: 3
5-sec-BUTYL-5-ETHYL-1-METHYLBARBITURIC ACID
mf: $C_{11}H_{18}N_2O_3$ mw: 226.31

SYN: N-METHYLBUTABARBITAL

TOXICITY DATA WITH REFERENCE
ipr-rat LDLo: 120 mg/kg JACSAT 58,1358,36
ivn-mus LD50: 75 mg/kg AIPTAK 132,164,61
ivn-rbt LDLo: 85 mg/kg JACSAT 58,1354,36

SAFETY PROFILE: Poison by intraperitoneal and intravenous routes. When heated to decomposition it emits toxic fumes of NO_x. See also BARBITURATES.

BRJ325 **CAS:67330-25-0** **HR: 2**
BUTYL FLUFENAMATE
mf: $C_{18}H_{18}F_3NO_2$ mw: 337.37

PROP: Bp: 169–170° @ 1 mm.

SYNS: BUTYL-o-((m-(TRIFLUOROMETHYL)PHENYL)AMINO)BENZOATE □ BUTYL-2-((3-(TRIFLUOROMETHYL)PHENYL)AMINO)BENZOATE □ HF 264 □ N-(α,α,α-TRIFLUORO-m-TOLYL)ANTHRANILIC ACID BUTYL ESTER

TOXICITY DATA WITH REFERENCE
orl-rat TDLo:990 mg/kg (7-17D preg):REP KSRNAM 13,3288,79
orl-rat TDLo:150 mg/kg (female 17-21D post):REP KSRNAM 13,3302,79
orl-rbt TDLo:1560 mg/kg (female 6-18D post):TER AMBNAS 27,33,79
orl-rat TDLo:3720 mg/kg (male 31D pre):REP OYYAA2 18,597,79
orl-rat TDLo:660 mg/kg (7-17D preg):TER KSRNAM 13,3288,79
orl-rat LD50:510 mg/kg OYYAA2 18,845,79
ipr-rat LD50:4550 mg/kg OYYAA2 18,845,79
scu-rat LD50:7800 mg/kg OYYAA2 18,845,79
ivn-rat LD50:650 mg/kg OYYAA2 18,845,79
orl-mus LD50:3100 mg/kg OYYAA2 18,845,79
ipr-mus LD50:4100 mg/kg OYYAA2 18,845,79
ivn-mus LD50:610 mg/kg IYKEDH 14,297,83
ipr-dog LD50:1500 mg/kg OYYAA2 18,845,79
scu-dog LD50:9300 mg/kg OYYAA2 18,845,79
ipr-rbt LD50:11,500 mg/kg OYYAA2 18,845,79

SAFETY PROFILE: Moderately toxic by ingestion and other routes. An experimental teratogen. Experimental reproductive effects. When heated to decomposition it emits toxic fumes of F^- and NO_x. See also ESTERS.

BRJ750 **CAS:2425-74-3** **HR: 3**
tert-BUTYL FORMAMIDE
mf: $C_5H_{11}NO$ mw: 101.17

TOXICITY DATA WITH REFERENCE
ivn-mus LD50:180 mg/kg CSLNX* NX#04680

CONSENSUS REPORTS: Reported in EPA TSCA Inventory.

SAFETY PROFILE: A poison by intravenous route. When heated to decomposition it emits toxic fumes of NO_x.

BRK000 **CAS:592-84-7** **HR: 3**
n-BUTYL FORMATE
DOT: UN 1128
mf: $C_5H_{10}O_2$ mw: 102.15

PROP: Colorless liquid. Mp: −90°, bp: 106.0°, flash p: 64°F (CC), d: 0.911, autoign temp: 612°F, vap press: 40 mm @ 31.6°, vap d: 3.52, lel: 1.7%, uel: 8%.

SYNS: BUTYLESTER KYSELINY MRAVENCI □ BUTYL FORMATE (DOT)

TOXICITY DATA WITH REFERENCE
ihl-hmn TCLo:10,418 ppm:EYE,CNS,PUL AMIHAB 20,517,59
ihl-cat LCLo:10,418 ppm/70M AMIHAB 20,517,59
orl-rbt LD50:2656 mg/kg IMSUAI 41,31,72

CONSENSUS REPORTS: Reported in EPA TSCA Inventory.

DOT CLASSIFICATION: 3; *Label:* Flammable Liquid

SAFETY PROFILE: Moderately toxic by ingestion. Mildly toxic by inhalation. Human systemic effects by inhalation: muscle contractions and spasticity, conjunctiva irritation, and unspecified respiratory changes. An irritant and narcotic in high concentrations. See also ESTERS, n-BUTYL ALCOHOL, and FORMIC ACID. Dangerous fire hazard when exposed to heat or flame. To fight fire, use alcohol foam, foam, CO_2, dry chemical. Incompatible with oxidizing materials. When heated to decomposition it emits acrid and irritating fumes.

BRK100 **CAS:16120-70-0** **HR: 2**
N-n-BUTYL-N-FORMYLHYDRAZINE
mf: $C_5H_{12}N_2O$ mw: 116.19

SYNS: BFH □ FORMIC ACID, 1-BUTYLHYDRAZIDE

TOXICITY DATA WITH REFERENCE
orl-mus TDLo:70 g/kg/84W-C:CAR CRNGDP 1,589,80

SAFETY PROFILE: Questionable carcinogen with experimental carcinogenic data. When heated to decomposition it emits toxic fumes of NO_x.

BRK250 **CAS:64441-42-5** **HR: 3**
1-BUTYL-3-(2-FUROYL)UREA
mf: $C_{10}H_{14}NO_3$ mw: 196.25

SYNS: n-BUTYL-N'-(2-FUROYL) □ N-FUROYL-N'-n-BUTYLHARNSTOFF (GERMAN)

TOXICITY DATA WITH REFERENCE
orl-mus LD50:730 mg/kg ARZNAD 10,686,60
ipr-mus LD50:230 mg/kg ARZNAD 10,686,60

SAFETY PROFILE: Poison by intraperitoneal route. Moderately toxic by ingestion. When heated to decomposition it emits toxic fumes of NO_x.

BRK750 **CAS:2426-08-6** **HR: 2**
n-BUTYL GLYCIDYL ETHER
mf: $C_7H_{14}O_2$ mw: 130.21

SYNS: AGEFLEX BGE □ BGE □ BGE (OSHA) □ BUTYL GLYCIDYL ETHER □ 2,3-EPOXYPROPYL BUTYL ETHER □ ETHER, BUTYL 2,3-EPOXYPROPYL □ ETHER, BUTYL GLYCIDYL □ GLYCIDYL BUTYL ETHER □ TK 10408

TOXICITY DATA WITH REFERENCE
skn-rbt 454 mg/3D MLD AMIHAB 14,250,56
skn-rbt 20 mg/24H MOD 85JCAE-,774,86
eye-rbt 91 mg MLD AMIHAB 14,250,56
eye-rbt 750 μg/24H SEV 28ZPAK -,135,72
mmo-esc 20 μmol/L ARTODN 46,277,80
dnd-esc 1 μmol/L ARTODN 46,277,80
skn-mus TDLo:36 g/kg (24D male):TER MUREAV 124,225,83
orl-rat LD50:2050 mg/kg AIHAAP 23,95,62
ihl-rat LCLo:670 ppm AMIHAB 14,250,56
ipr-rat LD50:1140 mg/kg AMIHAB 14,250,56

orl-mus LD50:1520 mg/kg AMIHAB 14,250,56
ipr-mus LD50:700 mg/kg AMIHAB 14,250,56
skn-rbt LD50:2520 mg/kg AIHAAP 23,95,62

CONSENSUS REPORTS: Reported in EPA TSCA Inventory.

OSHA PEL: TWA 25 ppm
ACGIH TLV: TWA 25 ppm
DFG MAK: Suspected Carcinogen
NIOSH REL: (Glycidyl Ethers) CL 30 mg/m³/15M

SAFETY PROFILE: Moderately toxic by ingestion, skin contact, and intraperitoneal routes. Mildly toxic by inhalation. An experimental teratogen. Mutation data reported. A skin and severe eye irritant. See also ETHERS. When heated to decomposition it emits acrid and irritating fumes.

For occupational chemical analysis use NIOSH: n-Butyl Glycidyl Ether S81.

BRK900 CAS:626-82-4 HR: 1
BUTYL HEXANOATE
mf: $C_{10}H_{20}O_2$ mw: 172.30

SYNS: BUTYL CAPROATE □ n-BUTYL HEXANOATE □ HEXANOIC ACID, BUTYL ESTER

TOXICITY DATA WITH REFERENCE
skn-rbt 500 mg/24H MLD FCTOD7 21,653,83
orl-rat LD50:>5 g/kg FCTOD7 21,653,83
skn-rbt LD50:>5 g/kg FCTOD7 21,653,83

CONSENSUS REPORTS: Reported in EPA TSCA Inventory.

SAFETY PROFILE: Low toxicity by ingestion and skin contact. A skin irritant. When heated to decomposition it emits acrid smoke and irritating fumes.

BRL500 CAS:56795-65-4 HR: 2
n-BUTYLHYDRAZINE HYDROCHLORIDE
mf: $C_4H_{12}N_2 \cdot ClH$ mw: 124.64

TOXICITY DATA WITH REFERENCE
orl-mus TDLo:14/g/kg/8W-C:NEO EJCAAH 11,473,75

SAFETY PROFILE: Questionable carcinogen with experimental neoplastigenic data. When heated to decomposition it emits very toxic fumes of NO_x and HCl. See also HYDRAZINE.

BRM000 HR: 3
O,O-tert-BUTYL HYDROGEN MONOPEROXY MALEATE
mf: $C_8H_{12}O_5$ mw: 188.18

SAFETY PROFILE: Slightly shock-sensitive. Commercial grade 95% dry is very hazardous. When heated to decomposition it emits acrid smoke and fumes. See also PEROXIDES, ORGANIC.

BRM250 CAS:75-91-2 HR: 2
tert-BUTYLHYDROPEROXIDE
mf: $C_4H_{10}O_2$ mw: 90.14

CH_3COOH

PROP: Water-white liquid. Flash p: 80°F or above, fp: −35°, d: 0.860, mp: −8°, bp: 40° @ 23 mm, vap d: 2.07. Sltly sol in water; very sol in esters and alc.

SYNS: terc. BUTYLHYDROPEROXID (CZECH) □ CADOX TBH □ 1,1-DIMETHYLETHYL HYDROPEROXIDE □ HYDROPEROXYDE de BUTYLE TERTIAIRE (FRENCH) □ 2-HYDROPEROXY-2-METHYLPROPANE □ PERBUTYL H □ TBHP-70 □ TRIGONOX A-75 (CZECH)

TOXICITY DATA WITH REFERENCE
dnd-ham:lng 500 µmol/L MUREAV 213,243,89
cyt-ham:lng 150 µmol/L MUREAV 213,243,89
skn-rbt 500 mg AIHAAP 19,205,58
skn-rbt 500 mg/24H SEV 28ZPAK -,39,72
eye-rbt 7 mg AIHAAP 19,205,58
eye-rbt 100 mg/24H MOD 28ZPAK -,39,72
eye-rbt 150 mg/1M rns SEV ZAARAM 8,25,58
mmo-sat 17 µg/plate ENMUDM 5(Suppl 1),3,83
mma-sat 17 µg/plate ENMUDM 5(Suppl 1),3,83
pic-esc 25 mg/L VIRLAX 99,257,79
orl rat LD50:406 mg/kg AIHAAP 19,205,58
ihl-rat LC50:500 ppm/4H AIHAAP 19,205,58
skn-rat LD50:790 mg/kg BSPII* 1/75-19B
ipr-rat LD50:87 mg/kg AIHAAP 19,205,58
orl-mus LD50:710 mg/kg BSPII* 1/75-19B
ihl-mus LC50:350 ppm/4H AIHAAP 19,205,58

CONSENSUS REPORTS: EPA Genetic Toxicology Program. Reported in EPA TSCA Inventory.

DFG MAK: Moderate skin effects
DOT CLASSIFICATION: Forbidden

SAFETY PROFILE: Moderately toxic by ingestion and inhalation. A severe skin and eye irritant. Mutation data reported. At highest dosage levels, symptoms noted were severe depression, incoordination, and cyanosis. Death was due to respiratory arrest. Very dangerous fire hazard when exposed to heat or flame, or by spontaneous chemical reaction such as with reducing materials. Moderately explosive; may explode during distillation. Violent reaction with traces of acid. Concentrated solutions may ignite spontaneously on contact with molecular sieve. Mixtures with transition metal salts may react vigorously and release oxygen. Forms an unstable solution with 1,2-dichloroethane. To fight fire, use alcohol foam, CO_2, dry chemical. When heated to decomposition it emits acrid smoke and fumes. See also PEROXIDES, ORGANIC.

BRM500 CAS:1948-33-0 HR: 3
tert-BUTYLHYDROQUINONE
mf: $C_{10}H_{14}O_2$ mw: 166.24

PROP: White crystalline solid; characteristic odor. Mp: 126.5–128.5°. Sol in alc, ether; insol in water.

SYNS: MONO-tert-BUTYLHYDROQUINONE □ MTBHQ □ SUSTANE □ TBHQ (FCC) □ TENOX TBHQ

TOXICITY DATA WITH REFERENCE
cyt-mus-ipr 200 mg/kg FCTOD7 22,459,84

orl-rat LD50:700 mg/kg JAOCA7 52,53,75
ihl-rat LCLo:2900 mg/m³/4H JACTDZ 1,753,92
ipr-rat LD50:300 mg/kg JAOCA7 52,53,75
orl-mus LD50:1000 mg/kg KODAK* 21MAY71
ipr-mus LD50:144 mg/kg DCTODJ 7,335,84

CONSENSUS REPORTS: Reported in EPA TSCA Inventory.

SAFETY PROFILE: Poison by intraperitoneal route. Moderately toxic by ingestion and inhalation. Mutation data reported. When heated to decomposition it emits acrid smoke and irritating fumes.

BRM750 CAS:21070-33-7 HR: 2
6-BUTYL-4-HYDROXYAMINOQUINOLINE-1-OXIDE
mf: $C_{13}H_{16}N_2O_2$ mw: 232.31

TOXICITY DATA WITH REFERENCE
scu-mus TDLo:60 mg/kg/I:ETA CPBTAL 17,544,69

SAFETY PROFILE: Questionable carcinogen with experimental tumorigenic data. When heated to decomposition it emits toxic fumes of NO_x.

BRN000 CAS:121-00-6 HR: 3
3-tert-BUTYL-4-HYDROXYANISOLE
mf: $C_{11}H_{16}O_2$ mw: 180.27

PROP: Solid. Mp: 62–63°.

SYNS: 2-tert-BUTYL-4-METHOXYPHENOL ☐ 4-METHOXY-2-tert-BUTYLPHENOL

TOXICITY DATA WITH REFERENCE
cyt-ham:lng 125 mg/L MUREAV 241,125,90
orl-ham TDLo:168 g/kg/20W-C:NEO CRNGDP 7,1285,86
orl-rat LD50:2910 mg/kg PLRCAT 16,1041,84
ipr-rat LD50:32 mg/kg PLRCAT 16,1041,84
orl-mus LD50:1583 mg/kg PLRCAT 16,1041,84
ipr-mus LD50:29 mg/kg PLRCAT 16,1041,84

CONSENSUS REPORTS: Reported in EPA TSCA Inventory.

SAFETY PROFILE: Poison by intraperitoneal route. Moderately toxic by ingestion. Questionable carcinogen with experimental neoplastigenic data. Mutation data reported. When heated to decomposition it emits acrid smoke and irritating fumes.

BRO000 CAS:51938-14-8 HR: 2
BUTYL(2-HYDROXYETHYL)NITROSOAMINE
mf: $C_6H_{14}N_2O_2$ mw: 146.22

SYNS: BHEN ☐ 2-(BUTYLNITROSAMINO)ETHANOL

TOXICITY DATA WITH REFERENCE
mmo-sat 100 µg/plate MUREAV 56,219,78
mma-sat 5 µmol/plate CNREA8 37,399,77
orl-rat TDLo:4800 mg/kg/20W-C:ETA GANNA2 65,13,74

CONSENSUS REPORTS: EPA Genetic Toxicology Program.

SAFETY PROFILE: Questionable carcinogen with experimental tumorigenic data. Mutation data reported. When heated to decomposition it emits toxic fumes of NO_x. See also NITROSAMINES.

BRO250 CAS:78128-80-0 HR: 2
3-BUTYL-4-HYDROXY-2(5H)FURANONE
mf: $C_8H_{12}O_3$ mw: 156.20

SYN: α-n-BUTYL-β-HYDROXY-$\Delta^{\alpha,\beta}$-BUTENOLID (GERMAN)

TOXICITY DATA WITH REFERENCE
scu-mus LD50:1750 mg/kg ARZNAD 11,277,61
ivn-mus LD50:1187 mg/kg ARZNAD 11,277,61

SAFETY PROFILE: Moderately toxic by subcutaneous and intravenous routes. When heated to decomposition it emits acrid smoke and irritating fumes.

BRO750 CAS:67590-46-9 HR: 3
2-(tert-BUTYL)-2-(HYDROXYMETHYL)-1,3-PROPANEDIOL, CYCLIC PHOSPHITE (1:1)
mf: $C_8H_{15}O_3P$ mw: 190.20

SYN: 4-(tert-BUTYL)-2,6,7-TRIOXA-1-PHOSPHABICYCLO(2.2.2)OCTANE

TOXICITY DATA WITH REFERENCE
ipr-mus LD50:40 µg/kg TXAPA9 47,287,79
ivn-mus LD50:210 µg/kg EJMCA5 13,207,78

SAFETY PROFILE: A deadly poison by intraperitoneal and intravenous routes. When heated to decomposition it emits toxic fumes of PO_x.

BRP250 CAS:9003-13-8 HR: 1
α-BUTYL-ω-HYDROXYPOLY(OXY(METHYL-1,2-ETHANEDIYL))
mf: $(C_3H_6O)_n \cdot C_4H_{10}O$

SYNS: BUTOXYPOLYPROPYLENE GLYCOL ☐ BUTOXYPROPANEDIOL POLYMER ☐ CRAG FLY REPELLENT ☐ ENT 8286 ☐ EXP. MITICIDE No. 7 ☐ NEWPOL LB3000 ☐ OPSB ☐ POLY(OXYPROPYLENE) BUTYL ETHER ☐ POLYOXYPROPYLENE MONOBUTYL ETHER ☐ POLYPROPYLENE GLYCOL MONOBUTYL ETHER ☐ PPG-14 BUTYL ETHER ☐ PPG-16 BUTYL ETHER ☐ PPG-33 BUTYL ETHER ☐ STABILENE ☐ STABILENE FLY REPELLENT ☐ UCON LB-250 ☐ UCON LB 1145 ☐ UCON LB 1800X

TOXICITY DATA WITH REFERENCE
skn-rbt 500 mg open MLD UCDS** 5/23/68
orl-rat LD50:9100 mg/kg ARSIM* 20,6,66
orl-rbt LD50:23,900 mg/kg SPEADM 78-1,53,78
skn-rbt LD50:21 g/kg UCDS** 1/16/58

CONSENSUS REPORTS: Reported in EPA TSCA Inventory. Glycol ethers are on the Community Right-To-Know List.

SAFETY PROFILE: Mildly toxic by ingestion and by skin contact. A skin irritant. An insect repellent. When heated to decomposition it emits acrid smoke and irritating fumes.

BRP500 CAS:507-40-4 HR: 3
tert-BUTYL HYPOCHLORITE
mf: C_4H_9OCl mw: 106.6

PROP: Pale-yellow liquid. Bp: 77–78°.

SAFETY PROFILE: A storage hazard. Ultraviolet light causes exothermic decomposition. Reacts violently with rubber. Reaction with sodium hydrogen cyanamide forms the explosive cyanonitrene. When heated to decomposition it emits toxic fumes of Cl^-. See also HYPOCHLORITES.

BRP750 CAS:85-60-9 HR: 1
4,4'-BUTYLIDENEBIS(3-METHYL-6-tert-BUTYLPHENOL)
mf: $C_{26}H_{38}O_2$ mw: 382.64

SYNS: 1,1-BIS(2-METHYL-4-HYDROXY-5-tert-BUTYLPHENYL)BUTANE □ 4,4'-BUTYLIDENEBIS(6-tert-BUTYL-m-CRESOL) □ 4,4'-BUTYLIDENEBIS(6-tert-BUTYL-3-METHYLPHENYL) □ SANTOWHITE POWDER □ SUMILIT BBM □ SWP (ANTIOXIDANT)

TOXICITY DATA WITH REFERENCE
orl-rat LDLo:17 g/kg RCTEA4 45(3),627,72

CONSENSUS REPORTS: Reported in EPA TSCA Inventory.

SAFETY PROFILE: Mildly toxic by ingestion. An antioxidant. When heated to decomposition it emits acrid smoke and irritating fumes.

BRQ000 CAS:3772-23-4 HR: 1
6,6'-BUTYLIDENEBIS(2,4-XYLENOL)
mf: $C_{20}H_{22}O$ mw: 278.42

TOXICITY DATA WITH REFERENCE
eye-rbt 100 mg IHFCAY 6,1,67
orl-rat LD50:5500 mg/kg IHFCAY 6,1,67

SAFETY PROFILE: Mildly toxic by ingestion. An eye irritant. When heated to decomposition it emits smoke and acrid, irritating fumes.

BRQ050 CAS:541-33-3 HR: 1
BUTYLIDENE CHLORIDE
mf: $C_4H_8Cl_2$ mw: 127.02

SYNS: BUTANE, 1,1-DICHLORO- □ 1,1 DICHLOROBUTANE

TOXICITY DATA WITH REFERENCE
orl-mus LD50:4859 mg/kg JPPMAB 3,169,51

CONSENSUS REPORTS: Reported in EPA TSCA Inventory.

SAFETY PROFILE: Slightly toxic by ingestion. When heated to decomposition it emits toxic vapors of Cl^-.

BRQ100 CAS:551-08-6 HR: 2
3-BUTYLIDENE PHTHALIDE
mf: $C_{12}H_{12}O_2$ mw: 188.24

PROP: Needles from $CHCl_3$. Mp: 82–83°.

SYNS: BUTYLIDENE PHTHALIDE □ n-BUTYLIDENE PHTHALIDE □ 1(3H)-ISOBENZOFURANONE, 3-BUTYLIDENE-(9CI) □ PHTHALIDE, 3-BUTYLIDENE-

TOXICITY DATA WITH REFERENCE
skn-rbt 500 mg/24H MLD FCTOD7 21,659,83
orl-rat LD50:1850 mg/kg FCTOD7 21,659,83

CONSENSUS REPORTS: Reported in EPA TSCA Inventory.

SAFETY PROFILE: Moderately toxic by ingestion. A skin irritant. When heated to decomposition it emits acrid smoke and irritating fumes.

BRQ250 CAS:542-69-8 HR: 3
n-BUTYL IODIDE
mf: C_4H_9I mw: 184.03

PROP: Liquid. D: 1.6166 @ 20°/4°, fp: −103°, bp: 130.4–131°.

SYN: 1-IODOBUTANE

TOXICITY DATA WITH REFERENCE
ipr-mus TDLo:480 mg/kg/8W-I:NEO CNREA8 35,1411,75
ihl-rat LC50:6100 mg/m³/4H 34ZIAG -,756,69
ipr-rat LD50:692 mg/kg 85GMAT -,30,82
ipr-mus LD50:101 mg/kg 85GMAT -,30,82

CONSENSUS REPORTS: EPA Genetic Toxicology Program. Reported in EPA TSCA Inventory.

SAFETY PROFILE: A poison by intraperitoneal route. Moderately toxic by inhalation. Questionable carcinogen with experimental neoplastigenic data. See also IODIDES. When heated to decomposition it emits toxic fumes of I^-.

BRQ350 HR: 3
BUTYL ISOBUTYRATE
mf: $C_8H_{16}O_2$ mw: 44.44

PROP: Colorless liquid; apple-pineapple odor. D: 0.859–0.864, refr index: 1.401, flash p: 113°F. Misc with alc, ether, fixed oils; insol in glycerin, propylene glycol, water @ 166°.

SYN: FEMA No. 2188

SAFETY PROFILE: Flammable liquid. When heated to decomposition it emits acrid smoke and irritating fumes.

BRQ500 CAS:111-36-4 HR: 3
n-BUTYL ISOCYANATE
DOT: UN 2485
mf: C_5H_9NO mw: 99.15

PROP: Colorless liquid. Bp: 115°, d: 0.880 @ 20°/4°.

SYNS: BIC □ ISOCYANIC ACID, BUTYL ESTER

TOXICITY DATA WITH REFERENCE
orl-rat LD50:600 mg/kg GTPZAB 20(3),53,76
ihl-rat LC50:3000 mg/m³ GTPZAB 20(3),53,76
orl-mus LD50:150 mg/kg GTPZAB 20(3),53,76
ihl-mus LC50:680 mg/m³ GTPZAB 20(3),53,76
ivn-mus LD50:1 mg/kg CSLNX* NX#05701
orl-gpg LD50:250 mg/kg GTPZAB 20(3),53,76

CONSENSUS REPORTS: Reported in EPA TSCA Inventory.

DOT CLASSIFICATION: 3; *Label:* Flammable Liquid, Poison; DOT Class: 6.1; *Label:* Poison; DOT Class: 6.1;

Label: Poison, Flammable Liquid; DOT Class: 3; *Label:* Flammable Liquid, Poison

SAFETY PROFILE: A poison by ingestion and intravenous routes. Mildly toxic by inhalation. A powerful irritant to eyes, skin, and mucous membranes. A flammable liquid. See also CYANATES and NITROGEN MONOXIDE.

BRQ750 CAS:7188-38-7 HR: 2
tert-BUTYL ISOCYANIDE
mf: C_5H_9N mw: 83.15

PROP: Liquid with very unpleasant odor. Bp: 92–93° @ 725 mm.

SYN: tert-BUTYLISONITRILE

TOXICITY DATA WITH REFERENCE
ihl-rat LC50:710 mg/m³/4H ARTODN 33,241,75
ihl-mus LC50:377 mg/m³/4H ARTODN 33,241,75

CONSENSUS REPORTS: Cyanide and its compounds are on the Community Right-To-Know List.

SAFETY PROFILE: Moderately toxic by inhalation. When heated to decomposition it emits toxic fumes of NO_x and CN^-.

BRQ800 CAS:73791-40-9 HR: 3
BUTYL(ISOPROPYL)ARSINIC ACID
mf: $C_7H_{17}AsO_2$ mw: 208.16

SYNS: ARSINE OXIDE, BUTYLHYDROXYISOPROPYL- □ BUTYLHYDROXYISOPROPYLARSINE OXIDE

TOXICITY DATA WITH REFERENCE
ivn-mus LD50:56 mg/kg CSLNX* NX#05105

OSHA PEL: TWA 0.5 mg(As)/m³

SAFETY PROFILE: Poison by intravenous route. When heated to decomposition it emits toxic fumes of As.

BRR250 CAS:30026-92-7 HR: 3
tert-BUTYL ISOPROPYL BENZENE HYDROPEROXIDE
mf: $C_{13}H_{20}O_2$ mw: 208.33

PROP: Crystals.

SYN: tert-BUTYL ISOPROPYL BENZENE HYDROPEROXIDE (DOT)

SAFETY PROFILE: Powerful irritant. See also PEROXIDES, ORGANIC. Dangerous fire hazard when exposed to heat or flame or by chemical reaction. Incompatible with oxidizing or reducing materials. When heated to decomposition it emits acrid smoke and fumes.

BRR500 CAS:74926-97-9 HR: 3
2-sec-BUTYL-6-ISOPROPYLPHENOL
mf: $C_{13}H_{20}O$ mw: 192.2

TOXICITY DATA WITH REFERENCE
ivn-mus LD50:50 mg/kg JMCMAR 23,1350,80
ivn-rbt LDLo:15 mg/kg JMCMAR 23,1350,80

SAFETY PROFILE: Poison by intravenous route. When heated to decomposition it emits acrid smoke and irritating fumes.

BRR600 CAS:138-22-7 HR: 3
n-BUTYL LACTATE
mf: $C_7H_{14}O_3$ mw: 146.21

PROP: Liquid. Sltly sol in water; misc in alc and ether. Mp: −43°, bp: 188°, flash p: 160°F (OC), d: 0.968, autoign temp: 720°F, vap d: 5.04, vap press: 0.4 mm @ 20°.

SYNS: BUTYL α-HYDROXYPROPIONATE □ BUTYL LACTATE □ 2-HYDROXYPROPANOIC ACID, BUTYL ESTER □ LACTIC ACID, BUTYL ESTER

TOXICITY DATA WITH REFERENCE
skn-rbt 500 mg/24H MOD FCTXAV 17,727,79
scu-rat LD50:12 g/kg NPIRI* 1,15,74
ipr-mus LDLo:200 mg/kg CBCCT* 7,690,55
scu-mus LD50:11,000 mg/kg FCTXAV 17,727,79

CONSENSUS REPORTS: Reported in EPA TSCA Inventory.

OSHA PEL: TWA 5 ppm
ACGIH TLV: TWA 5 ppm

SAFETY PROFILE: Poison by intraperitoneal route. A skin irritant. Toxic concentration in air for humans is about 4 ppm. Flammable when exposed to heat or flame; can react with oxidizing materials. To fight fire, use alcohol foam, foam, CO_2, dry chemical. When heated to decomposition it emits acrid smoke and irritating fumes. See also ESTERS, n-BUTYL ALCOHOL, and LACTIC ACID.

BRR700 CAS:2052-15-5 HR: 1
n-BUTYL LEVULINATE
mf: $C_9H_{16}O_3$ mw: 172.25

SYNS: BUTYL LAEVULINATE □ n-BUTYL LAEVULINATE □ BUTYL LEVULINATE □ BUTYL 4-OXOPENTANOATE □ 4-KETOPENTANOIC ACID BUTYL ESTER □ LEVULINIC ACID, BUTYL ESTER □ PENTANOIC ACID, 4-OXO-, BUTYL ESTER (9CI)

TOXICITY DATA WITH REFERENCE
skn-rbt 500 mg/24H MLD FCTOD7 21,655,83
orl-rat LD50:>5 g/kg FCTOD7 21,655,83
skn-rbt LD50:>5 g/kg FCTOD7 21,655,83

CONSENSUS REPORTS: Reported in EPA TSCA Inventory.

SAFETY PROFILE: Low toxicity by ingestion and skin contact. A skin irritant. When heated to decomposition it emits acrid smoke and irritating vapors.

BRR739 CAS:109-72-8 HR: 3
BUTYL LITHIUM
mf: C_4H_9Li mw: 64.06

PROP: Liquid. Eliminates LiH on heating. D: 0.765 @ 25°, mp: −76°, bp: 80–90° @ 0.0001 mm. Sol in ethers or hydrocarbons.

SAFETY PROFILE: Probably very toxic. Solutions of greater than 20% will ignite spontaneously in air. Ignites on contact with water or CO_2. May cause potentially explosive polymerization of styrene. Extremely flammable. To fight fire, use dry chemical; see special instructions of manufacturer. See also LITHIUM COMPOUNDS and BUTYL LITHIUM.

BRR750 **CAS:594-19-4** **HR: 3**
tert-BUTYL LITHIUM
mf: C_4H_9Li mw: 64.06

PROP: Colorless crystals. Decomposes to LiH and $(H_3C)_2CHCH_2$. Sublimes at 0.1°.

SAFETY PROFILE: Probably very toxic. Solutions in heptane may ignite spontaneously in air. Potentially violent reaction with 2,2,2,4,4,4-hexafluoro-1,3-dimethyl-1,3,2,4-diazadiphosphetidine. Extremely flammable. To fight fire, use dry chemical; see special instructions of manufacturer. See also LITHIUM COMPOUNDS and BUTYL LITHIUM.

BRR900 **CAS:109-79-5** **HR: 3**
n-BUTYL MERCAPTAN
DOT: UN 2347
mf: $C_4H_{10}S$ mw: 90.20

PROP: Colorless liquid, skunk-like odor. Mp: −116°, bp: 98°, d: 0.8365 @ 25°/4°, flash p: 35°F, vap d: 3.1.

SYNS: BUTANETHIOL (OSHA) □ BUTYL MERCAPTAN □ n-BUTYL MERCAPTAN (ACGIH,DOT) □ NCI-C60866

TOXICITY DATA WITH REFERENCE
eye-rbt 83 mg AIHAAP 19,171,58
ihl-mus TCLo:68 ppm/6H (female 6-16D post):REP FAATDF 8,170,87
ihl-mus TCLo:68 ppm/6H (female 6-16D post):TER FAATDF 8,170,87
orl-rat LD50:1500 mg/kg AIHAAP 19,171,58
ihl-rat LC50:4020 ppm/4H AIHAAP 19,171,58
ipr-rat LD50:399 mg/kg AIHAAP 19,171,58
orl-mus LD50:3 g/kg 85JCAE -,982,86
ihl-mus LC50:2500 ppm/4H AIHAAP 19,171,58
orl-rat LD50:3800 mg/kg

CONSENSUS REPORTS: Reported in EPA TSCA Inventory.

OSHA PEL: TWA 0.5 ppm
ACGIH TLV: TWA 0.5 ppm
DFG MAK: 0.5 ppm (1.5 mg/m³)
NIOSH REL: (n-Alkane Mono Thiols) CL 0.5 ppm/15M
DOT CLASSIFICATION: 3; *Label:* Flammable Liquid

SAFETY PROFILE: Poison by intraperitoneal route. Moderately toxic by ingestion. An eye irritant. Dangerous fire hazard by exposure to heat, flame, sparks, or powerful oxidizers. Reacts violently with HNO_3. Incompatible with acids, acid fumes, oxidizing materials, heat, flame, and sparks. To fight fire, use alcohol foam. When heated to decomposition it emits toxic SO_x. See also MERCAPTANS.

For occupational chemical analysis use NIOSH: 1-Butanethiol 2525.

BRS000 **CAS:486-17-9** **HR: 3**
p-BUTYLMERCAPTOBENZHYDRYL-β-DIMETHYLAMINOETHYLSULPHIDE
mf: $C_{21}H_{29}NS_2$ mw: 359.63

SYNS: 2-((p-(BUTYLTHIO)-α-PHENYLBENZYL)THIO)-N,N-DIMETHYLETHYLAMINE □ CAPTODIAME □ CAPTODIAMIN □ CAPTODIAMINE □ COVATIN □ COVATIX □ N 68 □ SUVREN

TOXICITY DATA WITH REFERENCE
orl-rat LD50:3800 mg/kg ARZNAD 8,154,58
ipr-rat LD50:343 mg/kg ARZNAD 8,154,58
orl-mus LD50:1630 mg/kg ARZNAD 8,154,58
ipr-mus LD50:116 mg/kg JPETAB 108,201,53
scu-mus LD50:1750 mg/kg AIPTAK 136,440,62

SAFETY PROFILE: Poison by intraperitoneal route. Moderately toxic by ingestion and subcutaneous routes. See also MERCAPTANS and SULFIDES. When heated to decomposition it emits very toxic fumes of NO_x and SO_x.

BRS250 **CAS:6192-29-6** **HR: 3**
BUTYLMERCAPTOMETHYLPENICILLIN
mf: $C_{14}H_{22}N_2O_4S_2$ mw: 346.50

SYNS: n-BUTYLTHIOMETHYLPENICILLIN □ PENICILLIN BT

TOXICITY DATA WITH REFERENCE
ice-mus LD50:101 mg/kg JLCMAK 24,126,49
ice-dog LD50:11,500 mg/kg JLCMAK 24,126,49
isp-dog LD50:56 mg/kg JLCMAK 24,126,49
ice-rbt LD50:15,600 mg/kg JLCMAK 24,126,49

SAFETY PROFILE: Poison by intracerebral and intraspinal routes. When heated to decomposition it emits very toxic fumes of NO_x and SO_x. See also MERCAPTANS and other penicillin entries.

BRS500 **CAS:6165-01-1** **HR: 3**
9-BUTYL-6-MERCAPTOPURINE
mf: $C_9H_{12}N_4S$ mw: 208.31

SYNS: 9-BUTYL-1,9-DIHYDRO-6H-PURINE-6-THIONE □ 9-BUTYL-6-MP □ 9-BUTYL-9H-PURINE-6-THIOL □ NSC 19488 □ SRI 753

TOXICITY DATA WITH REFERENCE
ipr-rat LDLo:300 mg/kg CPCHAO 18,307,62
ipr-mus LD50:270 mg/kg NCISP* JAN86

SAFETY PROFILE: Poison by intraperitoneal route. When heated to decomposition it emits very toxic fumes of NO_x and SO_x. See also MERCAPTANS.

BRS750 **CAS:543-63-5** **HR: 3**
n-BUTYLMERCURIC CHLORIDE
mf: C_4H_9ClHg mw: 293.17

PROP: Plates or needles from EtOH. Mp: 128.3–128.8°. Sol in $CHCl_3$; sltly sol in EtOH; insol in H_2O.

SYN: BMC

TOXICITY DATA WITH REFERENCE
dnr-esc 2 mmol/L MJDHDW 28,F39,80
cyt-hmn:hla 1 mg/L JJEMAG 39,47,69
scu-rat LDLo:73 mg/kg JJEMAG 39,47,69

CONSENSUS REPORTS: Mercury and its compounds are on the Community Right-To-Know List.

OSHA PEL: TWA 0.01 mg(Hg)/m³; STEL 0.03 mg/m³ (skin)
ACGIH TLV: TWA 0.01 mg(Hg)/m³; STEL 0.03 mg(Hg)/m³
NIOSH REL: (Organomercury): TWA 0.01 mg/m³; STEL 0.03 mg/m³ (skin)

SAFETY PROFILE: A poison by subcutaneous route. Mutation data reported. See also MERCURY COMPOUNDS, ORGANIC, and CHLORIDES. When heated to decomposition it emits very toxic fumes of Cl^- and Hg.

BRT000 **CAS:532-34-3** **HR: 2**
n-BUTYL MESITYL OXIDE OXALATE
mf: $C_{12}H_{18}O_4$ mw: 226.30

PROP: Yellow to pale-red liquid. Bp: 256–270°, d: 1.052–1.060 @ 20°/4°, flash p: 315°F.

SYNS: BMOO □ BUTOPYRONOXYL □ BUTYL-3,4-DIHYDRO-2,2-DIMETHYL-4-OXO-2H-PYRAN-6-CARBOXYLATE □ n-BUTYL ESTER of 3,4-DIHYDRO-2,2-DIMETHYL-4-OXO-2H-PYRAN-6-CARBOXYLIC ACID □ n-BUTYLMESITYLOXID OXALATE □ 2-CARBO-n-BUTOXY-6,6-DIMETHYL-5,6-DIHYDRO-1,4-PYRONE □ 3,4-DIHYDRO-2,2-DIMETHYL-4-OXO-2H-PYRAN-6-CARBOXYLIC ACID-n-BUTYL ESTER □ DIHDYRO-PYRONE □ α,α-DIMETHYL-α′-CARBOBUTOXY-DIHYDRO-γ-PYRONE □ 2,2-DIMETHYL-6-CARBOBUTOXY-2,3-DIHYDRO-4-PYRONE □ ENT 9 □ INDALONE

TOXICITY DATA WITH REFERENCE
orl-rat LD50:7400 mg/kg JPETAB 93,26,48
orl-mus LD50:11,600 mg/kg JPETAB 93,26,48
orl-rbt LD50:5400 mg/kg JPETAB 93,26,48
orl-gpg LD50:3200 mg/kg JPETAB 93,26,48

CONSENSUS REPORTS: Reported in EPA TSCA Inventory.

SAFETY PROFILE: Moderately toxic by ingestion. Produces liver necrosis in experimental animals. A mild skin irritant. See also OXALATES and ESTERS. Combustible when exposed to heat or flame. When heated to decomposition it emits acrid fumes.

BRU250 **CAS:5412-64-6** **HR: 3**
n-BUTYL-α-METHYLBENZYLAMINE
mf: $C_{12}H_{19}N$ mw: 177.32

TOXICITY DATA WITH REFERENCE
skn-rbt 10 mg/24H open MLD AIHAAP 23,95,62
orl-rat LD50:360 mg/kg AIHAAP 23,95,62
skn-rbt LD50:570 mg/kg AIHAAP 23,95,62

SAFETY PROFILE: Poison by ingestion. Moderately toxic by skin contact. A skin irritant. When heated to decomposition it emits toxic fumes of NO_x. See also AMINES.

BRU300 **CAS:464-07-3** **HR: 1**
tert-BUTYL METHYL CARBINOL
mf: $C_6H_{14}O$ mw: 102.20

SYNS: 2-BUTANOL, 3,3-DIMETHYL- □ 3,3-DIMETHYL-2-BUTANOL □ PINACOLYL ALCOHOL (6CI)

TOXICITY DATA WITH REFERENCE
ihl-rat LCLo:3600 ppm/2.3H JJATDK 7,307,87

CONSENSUS REPORTS: Reported in EPA TSCA Inventory.

SAFETY PROFILE: Slightly toxic by inhalation. When heated to decomposition it emits acrid smoke and irritating vapors.

BRU500 **CAS:83-66-9** **HR: 3**
6-tert-BUTYL-3-METHYL-2,4-DINITRO ANISOLE
mf: $C_{12}H_{16}N_2O_5$ mw: 268.30

SYNS: 2,6-DINITRO-3-METHOXY-4-tert-BUTYLTOLUENE □ MUSK AMBRETTE

TOXICITY DATA WITH REFERENCE
skn-rbt 500 mg/24H MOD FCTXAV 13,681,75
mmo-sat 2 μmol/plate FCTOD7 21,707,83
mma-sat 100 μg/plate FCTOD7 24,27,86
sln-dmg-orl 10 mmol/L FCTOD7 21,707,83
orl-rat LD50:339 mg/kg FCTXAV 2,327,64

CONSENSUS REPORTS: Reported in EPA TSCA Inventory.

SAFETY PROFILE: A poison by ingestion. Mutation data reported. A skin irritant. When heated to decomposition it emits toxic fumes of NO_x. See also AROMATIC AMINES.

BRU750 **CAS:2487-01-6** **HR: 3**
2-tert-BUTYL-5-METHYL-4,6-DINITROPHENYL ACETATE
mf: $C_{13}H_{16}N_2O_6$ mw: 296.31

SYNS: ACETIC ACID-2-(tert-BUTYL)-4,6-DINITRO-m-TOLYL ESTER □ 6-(1,1-DIMETHYLETHYL)-3-METHYL-2,4-DINITROPHENYL ACETATE □ 2,4-DINITRO-3-METHYL-6-tert-BUTYLPHENYLACETAT (GERMAN) □ 2,4-DINITRO-3-METHYL-6-tert-BUTYLPHENYL ACETATE □ MC 1488 □ MEDINOTERB ACETATE □ P 1488

TOXICITY DATA WITH REFERENCE
orl-rat LD50:42 mg/kg FMCHA2 -,D191,80
skn-rat LD50:1300 mg/kg GUCHAZ 6,326,73
orl-mus LD50:90 mg/kg 85GYAZ -,75-71
orl-rbt LD50:80 mg/kg 28ZEAL 4,82,69
orl-gpg LD50:55 mg/kg 28ZEAL 4,82,69
skn-gpg LD50:7200 mg/kg 85GYAZ -,75,71
orl-ckn LD50:560 mg/kg 28ZEAL 5,144,76

SAFETY PROFILE: Poison by ingestion. Moderately toxic by skin contact. See also ESTERS. When heated to decomposition it emits toxic fumes of NO_x.

B

BRU780 CAS:628-28-4 HR: 3
BUTYL METHYL ETHER (DOT)
DOT: UN 2350
mf: $C_5H_{12}O$ mw: 88.17

SYNS: BUTANE, 1-METHOXY-(9CI) □ ETHER, BUTYL METHYL □ α-METHOXYBUTANE □ 1-METHOXYBUTANE □ METHYL BUTYL ETHER □ METHYL n-BUTYL ETHER

TOXICITY DATA WITH REFERENCE
ihl-mus LC50:176 mg/m³/15M ANESAV 11,455,50

DOT CLASSIFICATION: 3; *Label:* Flammable Liquid

SAFETY PROFILE: Poison by inhalation. A flammable liquid. When heated to decomposition it emits acrid smoke and irritating vapors.

BRU790 CAS:2219-82-1 HR: 3
2-tert-BUTYL-6-METHYLPHENOL
mf: $C_{11}H_{16}O$ mw: 164.27

SYN: PHENOL, 2-tert-BUTYL-6-METHYL-

TOXICITY DATA WITH REFERENCE
ivn-mus LD50:120 mg/kg JMCMAR 23,1350,80

CONSENSUS REPORTS: Reported in EPA TSCA Inventory.

SAFETY PROFILE: Poison by intravenous route. When heated to decomposition it emits acrid smoke and irritating vapors.

BRU800 CAS:98-27-1 HR: 3
4-tert-BUTYL-2-METHYLPHENOL
mf: $C_{11}H_{16}O$ mw: 164.27

SYN: PHENOL, 4-tert-BUTYL-2-METHYL-

TOXICITY DATA WITH REFERENCE
ipr-mus LD50:81 mg/kg JMCMAR 18,868,75
ivn-mus LD50:180 mg/kg JMCMAR 23,1350,80

CONSENSUS REPORTS: Reported in EPA TSCA Inventory.

SAFETY PROFILE: Poison by intravenous and intraperitoneal routes. When heated to decomposition it emits acrid smoke and irritating vapors.

BRV000 CAS:100836-63-3 HR: 3
tert-BUTYL-N-(3-METHYL-2-THIAZOLIDINYLIDENE) CARBAMATE
mf: $C_9H_{16}N_2O_2S$ mw: 216.33

TOXICITY DATA WITH REFERENCE
orl-mus LD50:306 mg/kg JMCMAR 23,773,80
ivn-mus LD50:68 mg/kg JMCMAR 23,773,80

SAFETY PROFILE: Poison by ingestion and intravenous routes. See also CARBAMATES. When heated to decomposition it emits very toxic fumes of NO_x and SO_x.

BRV100 CAS:1005-67-0 HR: 3
4-BUTYLMORPHOLINE
mf: $C_8H_{17}NO$ mw: 143.26

PROP: Bp: 110–115°

SYNS: N-BUTYLMORPHOLINE □ N-(n-BUTYL)MORPHOLINE □ MORPHOLINE, 4-BUTYL-

TOXICITY DATA WITH REFERENCE
skn-rbt 500/24H MOD JACTDZ 1,13,90
eye-rbt 100 mg MOD JACTDZ 1,13,90
orl rat LD50:338 mg/kg JACTDZ 1,13,90
skn-rbt LD50:1800 mg/kg JACTDZ 1,13,90

SAFETY PROFILE: Poison by ingestion. Moderately toxic and corrosive to skin. A skin and eye irritant. When heated to decomposition it emits toxic fumes of NO_x.

BRV325 CAS:928-45-0 HR: 3
BUTYL NITRATE
mf: $C_4H_9NO_3$ mw: 119.12

PROP: Liquid. Bp: 136°.

SAFETY PROFILE: An explosive. Reacts explosively with Lewis acids (e.g., boron trifluoride; aluminum chloride; etc.). When heated to decomposition it emits toxic fumes of NO_x. See also NITRATES.

BRV500 CAS:544-16-1 HR: 3
n-BUTYL NITRITE
mf: $C_4H_9NO_2$ mw: 103.14

PROP: Oily liquid, characteristic odor, misc in alc and ether. Bp: 78°, d: 0.9114 @ 0°/4°, vap d: 3.5, flash p: 10°.

SYNS: BUTYL NITRITE (DOT) □ NBN □ NCI-C56553 □ NITROUS ACID-n-BUTYL ESTER

TOXICITY DATA WITH REFERENCE
mmo-sat 1 mg/plate PSEBAA 157,688,78
orl-man TDLo:153 mg/kg:BLD AIMEAS 92,570,80
orl-rat LD50:83 mg/kg JJATDK 1,30,81
ihl-rat LC50:420 ppm/4H FAATDF 8,101,87
orl-mus LD50:171 mg/kg RCSADO 3,233,82
ihl-mus LC50:567 ppm/1H FAATDF 1,448,81
ipr-mus LD50:169 mg/kg TXAPA9 48,A43,79

CONSENSUS REPORTS: Reported in EPA TSCA Inventory.

SAFETY PROFILE: A poison by ingestion and intraperitoneal routes. Mildly toxic by inhalation. An irritant. Human systemic effects by ingestion: methemoglobinemia-carboxhemoglobinemia. Resembles amyl nitrite in causing fall in blood pressure, headache, pulse throbbing, and weakness. Mutation data reported. Flammable when exposed to heat or flame or by spontaneous chemical reaction. When heated to decomposition it emits toxic fumes of NO_x. See also NITRITES, n-BUTYL ALCOHOL, and ESTERS.

BRV750 **CAS:924-43-6** **HR: 3**
sec-BUTYL NITRITE
mf: $C_4H_9NO_2$ mw: 103.14

PROP: Liquid. Bp: 68°, d: 0.8981 @ 0°/4°, vap d: 3.5.

SYNS: NITROUS ACID-sec-BUTYL ESTER □ NITROUS ACID-1-METHYL PROPYL ESTER

TOXICITY DATA WITH REFERENCE
mma-sat 1 mg/plate BSIBAC 56,816,80
orl-mus LD50:423 mg/kg RCSADO 3,233,82
ihl-mus LD50:1753 ppm/1H FAATDF 1,448,81
ipr-mus LD50:592 mg/kg TXAPA9 48,A43,79

CONSENSUS REPORTS: Reported in EPA TSCA Inventory.

SAFETY PROFILE: Moderately toxic by ingestion, inhalation, and intraperitoneal routes. Mutation data reported. Flammable when exposed to heat or flame or by spontaneous chemical reaction. An oxidizer. Potentially explosive. To fight fire, use water, spray, foam, dry chemical. When heated to decomposition it emits toxic fumes of NO_x. See also n-BUTYL NITRITE, NITRITES, and ESTERS.

BRV760 **CAS:540-80-7** **HR: 3**
tert-BUTYL NITRITE
mf: $C_4H_9NO_2$ mw: 103.14

PROP: Yellow liquid, agreeable odor. D: 0.8941, bp: 63°, n (20/D) 1.3687. Very sol in alc, ether, chloroform, carbon disulfide; sltly sol in water; practically insol in glycerol.

SYNS: α,α-DIMETHYLETHYL NITRITE □ NITROUS ACID-1,1-DIMETHYLETHYL ESTER

TOXICITY DATA WITH REFERENCE
mmo-sat 10 µmol/plate BCPCA6 35,3847,86
orl-mus LD50:308 mg/kg RCSADO 3,233,82
ihl-mus LC50:10,852 ppm/1H FAATDF 1,448,81
ipr-mus LD50:625 mg/kg TXAPA9 48,A43,79

SAFETY PROFILE: Poison by ingestion. Moderately toxic by intraperitoneal route. Mutation data reported. A jet propellant. When heated to decomposition it emits toxic fumes of NO_x. See also n-BUTYL NITRITE, NITRITES, and ESTERS.

BRW000 **HR: 3**
tert-BUTYL NITROACETYLENE
mf: $C_6H_9NO_2$ mw: 127.14

SAFETY PROFILE: When ignited in absence of a solvent, the primary, secondary and tertiary amines explode. Incompatible with amines. See also ACETYLENE COMPOUNDS.

BRW250 **HR: 3**
tert-BUTYL-p-NITRO PEROXY BENZOATE
mf: $C_{11}H_{13}NO_5$ mw: 239.2

SAFETY PROFILE: Explodes in contact with flame. When heated to decomposition it emits toxic fumes of NO_x. See also PEROXIDES, ORGANIC.

BRW500 **CAS:71002-67-0** **HR: 3**
BUTYL-p-NITROPHENYL ESTER of ETHYLPHOSPHONIC ACID
mf: $C_{12}H_{18}NO_5P$ mw: 287.28

SYN: ETHYLPHOSPHONIC ACID BUTYL-p-NITROPHENYL ESTER

TOXICITY DATA WITH REFERENCE
scu-mus LD50:1500 µg/kg RPTOAN 42,106,79
ivn-mus LD50:1300 µg/kg RPTOAN 42,106,79
scu-rat LD50:1500 µg/kg FATOAO 42(3),299,79
ivn-rat LD50:1300 µg/kg FATOAO 42(3),299,79

SAFETY PROFILE: Deadly poison by subcutaneous and intravenous routes. See also ESTERS. When heated to decomposition it emits very toxic fumes of PO_x and NO_x.

BRW750 **CAS:21070-32-6** **HR: 2**
6-BUTYL-4-NITROQUINOLINE-1-OXIDE
mf: $C_{13}H_{14}N_2O_3$ mw: 246.29

TOXICITY DATA WITH REFERENCE
dns-ham:oth 4 µmol/L NATUAS 229,416,71
dnd-mus:fbr 100 µmol/L CNREA8 35,521,75
scu-mus TDLo:60 mg/kg/I:ETA CPBTAL 17,544,69

CONSENSUS REPORTS: EPA Genetic Toxicology Program.

SAFETY PROFILE: Questionable carcinogen with experimental tumorigenic data. Mutation data reported. When heated to decomposition it emits toxic fumes of NO_x.

BRX500 **CAS:56986-36-8** **HR: 2**
BUTYLNITROSOAMINOMETHYL ACETATE
mf: $C_7H_{14}N_2O_3$ mw: 174.23

SYNS: ACETOXYMETHYLBUTYLNITROSAMINE □ N-(ACETOXY)METHYL-N,N-BUTYLNITROSAMINE □ BAMN □ BUTYL ACETOXYMETHYLNITROSAMINE □ N-BUTYL-N-(ACETOXYMETHYL)NITROSAMINE □ N-NITROSO-N-(1-ACETOXYMETHYL)BUTYLAMINE

TOXICITY DATA WITH REFERENCE
mmo-sat 1 µmol/plate MUREAV 49,187,78
mmo-esc 1 µmol/plate GANNA2 71,124,80
dnr-bcs 500 nmol/plate GANNA2 66,457,75
dns-rat:oth 10 µmol/L CBINA8 53,99,85
dnd-mus:fbr 260 nmol/L GANNA2 73,565,82
cyt-ham:fbr 16 mg/L/24H MUREAV 48,337,77
msc-ham:lng 100 µmol/L GANNA2 75,531,81
orl-rat TDLo:555 mg/kg/90D-I:ETA ZKKOBW 91,317,78
scu-rat TDLo:50 mg/kg/10W-I:CAR JCROD7 104,13,82
scu-rat TD:66 mg/kg/10W-I:CAR IAPUDO 41,619,82
orl-rat LD50:1500 mg/kg ZKKOBW 91,317,78

CONSENSUS REPORTS: EPA Genetic Toxicology Program.

SAFETY PROFILE: Moderately toxic by ingestion. Questionable carcinogen with experimental carcinogenic and tumorigenic data. Mutation data reported. When heated to decomposition it emits toxic fumes of NO_x. See also NITROSAMINES.

BRY000 CAS:51938-15-9 HR: 2
1-(BUTYLNITROSOAMINO)-2-PROPANONE
mf: $C_7H_{14}N_2O_2$ mw: 158.23

SYNS: BUTYL(2-OXOPROPYL)NITROSOAMINE □ N-NITROSO-1-BUTYLAMINO-2-PROPANONE □ N-NITROSO-(2-OXOPROPYL)-N-BUTYLAMINE

TOXICITY DATA WITH REFERENCE
mmo-sat 31 μmol/plate CNREA8 37,399,77
mma-sat 4 μmol/plate CNREA8 37,399,77
orl-rat TDLo:2000 mg/kg/13W-C:ETA GANNA2 65,13,74

CONSENSUS REPORTS: EPA Genetic Toxicology Program.

SAFETY PROFILE: Questionable carcinogen with experimental tumorigenic data. Mutation data reported. When heated to decomposition it emits toxic fumes of NO_x. See also NITROSAMINES.

BRY250 CAS:16339-05-2 HR: 2
N-BUTYL-N-NITROSO AMYL AMINE
mf: $C_9H_{20}N_2O$ mw: 172.31

SYNS: BUTYLAMYLNITROSAMIN (GERMAN) □ N-BUTYL-N-NITROSOPENTYLAMINE □ N-BUTYL-N-PENTYLNITROSAMINE □ N-NITROSO-N-BUTYLPENTYLAMINE □ N-NITROSO-N-BUTYL-N-PENTYLAMINE

TOXICITY DATA WITH REFERENCE
scu-mus TDLo:17 g/kg/21W-I:ETA ZEKBAI 69,103,67
scu-rat LD50:2500 mg/kg ZEKBAI 69,103,67

SAFETY PROFILE: Moderately toxic by subcutaneous route. Questionable carcinogen with experimental tumorigenic data. See also N-NITROSO COMPOUNDS and NITROSAMINES. When heated to decomposition it emits toxic fumes of NO_x.

BRY500 CAS:924-16-3 HR: 3
n-BUTYL-N-NITROSO-1-BUTAMINE
mf: $C_8H_{18}N_2O$ mw: 158.28

PROP: Pale-yellow liquid. Bp: 235°.

SYNS: DBN □ DBNA □ DI-n-BUTYLNITROSAMIN (GERMAN) □ DIBUTYLNITROSOAMINE □ DI-n-BUTYLNITROSAMINE □ N,N-DI-n-BUTYLNITROSAMINE □ N,N-DIBUTYLNITROSOAMINE □ NDBA □ N-NITROSODIBUTYLAMINE □ N-NITROSODI-n-BUTYLAMINE (MAK) □ RCRA WASTE NUMBER U172

TOXICITY DATA WITH REFERENCE
mma-esc 1 μmol/plate GANNA2 75,8,84
dnd-esc 100 nmol/tube CRNGDP 3,781,82
dns-hmn:hla 10 μmol/L CNREA8 38,2621,78
dnd-rat:lvr 100 μmol/L CNREA8 42,2592,82
bfa-rat-sat 158 mg/kg CRNGDP 6,967,85
hma-rat-smc 2912 mg/kg TCMUD8 3,41,83
scu-ham TDLo:30 mg/kg (female 15D post):REP ZEKBAI 86,69,76
orl-rat TDLo:1200 mg/kg (12D preg):TER BEXBAN 78,1308,74
orl-rat TDLo:140 mg/kg/4W-C:CAR CNREA8 46,6160,86
scu-rat TDLo:8 g/kg/20W-I:ETA XENOBH 3,271,73
orl-mus TDLo:1200 mg/kg/8W-I:NEO TXAPA9 82,19,86
ipr-mus TDLo:240 mg/kg/8W-I:NEO TXAPA9 82,19,86
scu-mus TDLo:800 mg/kg/40W-I:CAR EJCAAH 6,433,70
ivn-mus TDLo:12 mg/kg/25W-I:CAR IDZAAW 45,71,70
scu-rbt TDLo:34 g/kg/78W-I:NEO INURAQ 12,262,75
orl-gpg TDLo:24 g/kg/2Y-I:ETA ZEKBAI 71,183,68
orl-ham TDLo:9 g/kg:ETA,TER PSEBAA 136,1007,71
orl-ham TDLo:3900 mg/kg/13W-I:ETA PSEBAA 136,168,71
ipr-ham TDLo:200 mg/kg:ETA ZEKBAI 79,85,73
scu-ham TDLo:240 mg/kg:CAR,TER ZEKBAI 86,69,76
scu-ham TDLo:240 mg/kg/8D-I:CAR ZEKBAI 86,69,76
scu-rat TD:7400 mg/kg/37W-I:ETA ARZNAD 19,1077,69
scu-ham TD:2162 mg/kg/46W-I:ETA JNCIAM 57,401,76
orl-rat TD:64 g/kg/20W-C:ETA GANNA2 67,825,76
scu-mus TD:3200 mg/kg/30W-I:ETA BECCAN 46,271,68
orl-rat TD:3900 mg/kg/56W-C:ETA ARZNAD 19,1077,69
orl-rat TD:12,600 mg/kg/30W-C:ETA IAPUDO 41,649,82
orl-rat TD:3640 mg/kg/26W-C:ETA EVHPAZ 50,169,83
orl-mus TD:2190 mg/kg/1Y-C:NEO GANNA2 60,353,69
orl-rat TD:648 mg/kg/2Y-I:CAR CALEDQ 19,207,83
orl-rat TD:882 mg/kg/2W-C:ETA JJCREP 78,227,87
orl-rat LD50:1200 mg/kg NATWAY 50,735,63
scu-rat LD50:1200 mg/kg XENOBH 3,271,73
orl-ham LD50:2150 mg/kg ZKKOBW 79,85,73
ipr-ham LD50:1200 mg/kg ZKKOBW 79,85,73
scu-ham LD50:561 mg/kg PSEBAA 136,168,71

CONSENSUS REPORTS: NTP 7th Annual Report On Carcinogens. IARC Cancer Review: Group 2B IMEMDT 7,56,87; Animal Sufficient Evidence IMEMDT 28,151,82; IMEMDT 17,51,78; IMEMDT 4,197,74; Human Limited Evidence IMEMDT 17,51,78. Community Right-To-Know List. EPA Genetic Toxicology Program. Reported in EPA TSCA Inventory.

DFG MAK: Animal Carcinogen, Suspected Human Carcinogen

SAFETY PROFILE: Confirmed carcinogen with experimental carcinogenic, tumorigenic, and neoplastigenic data. Moderately toxic by ingestion, subcutaneous, and intraperitoneal routes. Experimental teratogenic effects. Human mutation data reported. When heated to decomposition it emits toxic fumes of NO_x. See also NITROSAMINES.

For occupational chemical analysis use OSHA: #27 or NIOSH: Nitrosamines 2522.

BRZ000 CAS:6558-78-7 HR: 3
N-BUTYL-N-NITROSO ETHYL CARBAMATE
mf: $C_7H_{14}N_2O_3$ mw: 174.23

SYNS: N-BUTYL-N-NITROSOURETHAN □ 1-BUTYL-1-NITROSOURETHAN □ TL 478

TOXICITY DATA WITH REFERENCE
mmo-bcs 5 g/L MUREAV 42,19,77
dnr-bcs 5 g/L MUREAV 42,19,77
cyt-ham:fbr 120 mg/L/48H MUREAV 48,337,77
cyt-ham:lng 35 mg/L GMCRDC 27,95,81
sce-ham:fbr 100 μmol/L JNCIAM 58,1635,77
orl-rat TDLo:500 mg/kg (20D preg):ETA,TER GANNA2 71,811,80
orl-rat TDLo:2240 mg/kg/8W-C:ETA GANNA2 65,227,74
scu-rat TDLo:150 mg/kg (15-21D preg):ETA,TER GANNA2 71,811,80
orl-mus TDLo:5300 mg/kg/20W-C:NEO GANNA2 67,231,76

orl-rat TD:5040 mg/kg/18W-C:ETA NIPAA4 78,157,81
orl-mam TD:5 g/kg/16W-C:ETA AMBNAS 28,85,81
orl-rat TD:5 g/kg/16W-C:ETA AMBNAS 28,85,81
orl-rat LD50:900 mg/kg GANNA2 65,227,74
ihl-mus LCLo:300 mg/m^3/10M NDRC** NDCrc-132,Nov,42

CONSENSUS REPORTS: EPA Genetic Toxicology Program.

SAFETY PROFILE: A poison by inhalation. Moderately toxic by ingestion. Experimental teratogenic data. Questionable carcinogen with experimental neoplastigenic and tumorigenic data. Mutation data reported. See also N-nitroso compounds and CARBAMATES. When heated to decomposition it emits toxic fumes of NO_x.

BRZ200 CAS:17721-94-7 HR: 2
4-tert-BUTYL-1-NITROSOPIPERIDINE
mf: $C_9H_{18}N_2O$ mw: 170.29

SYN: N-NITROSO-4-tert-BUTYLPIPERIDINE

TOXICITY DATA WITH REFERENCE
mma-sat 250 µg/plate MUREAV 111,135,83
orl-rat TDLo:4500 mg/kg/2Y-I:CAR CRNGDP 2,1045,81

SAFETY PROFILE: Questionable carcinogen with experimental carcinogenic data. Mutation data reported. When heated to decomposition it emits toxic fumes of NO_x.

BSA250 CAS:869-01-2 HR: 3
n-BUTYLNITROSOUREA
mf: $C_5H_{11}N_3O_2$ mw: 145.19

PROP: Solid. Mp: 82.5–84°.

SYNS: BNU □ BUTYLNITROSOHARNSTOFF (GERMAN) □ N-n-BUTYL-N-NITROSOUREA □ 1-BUTYL-1-NITROSOUREA □ N-NITROSOBUTYLUREA

TOXICITY DATA WITH REFERENCE
pic-esc 2 mg/L TCMUE9 1,91,84
sce-ham:fbr 500 µmol/L CNREA8 44,3270,84
orl-rat TDLo:4867 mg/kg/24W-I:CAR GANNA2 67,33,76
orl-rat TDLo:120 mg/kg (22D preg):ETA,TER ARGEAR 48,9,78
orl-rat TDLo:16,512 mg/kg/50W-I:CAR,REP JCROD7 107,32,84
ipr-rat TDLo:300 mg/kg:ETA GANMAX 12,283,72
scu-rat TDLo:300 mg/kg:ETA GMCRDC 12,283,72
scu-rat TDLo:120 mg/kg (22D preg):ETA ARGEAR 48,9,78
ivn-rat TDLo:120 mg/kg (22D preg):ETA ARGEAR 48,9,78
orl-mus TDLo:2800 mg/kg/10W-C:CAR GANNA2 68,281,77
skn-mus TDLo:581 mg/kg/50W-I:ETA JCROD7 102,13,81
orl-dog TDLo:960 mg/kg/81W-I:ETA PARPDS 164,216,79
orl-rat LD:1400 mg/kg/5W-C:CAR ESKHA5 (102),66,84
orl-mus TD:2800 mg/kg/10W-C:ETA PPTCBY 6,57,76
orl-rat TD:200 mg/kg:ETA GANNA2 66,615,75
orl-rat TD:300 mg/kg:ETA GMCRDC 12,283,72
orl-rat TD:1050 mg/kg/5W-C:ETA BIHAA2 (40),107,73
orl-mus TDLo:3360 mg/kg/12W-C:ETA GANNA2 61,287,70
ivn-rat LD:250 mg/kg (female 15D post):ETA KFIZAO 84,23,75
orl-rat LD:2400 mg/kg/17W-I:ETA GANNA2 62,557,71
orl-rat LD:2505 mg/kg/24W-C:ETA 25NJAN -,24,70
scu-rat TD:450 mg/kg:ETA ANYAA9 381,250,82
orl-rat LD50:400 mg/kg PPTCBY 2,73,72
scu-rat LD50:1200 mg/kg ZEKBAI 69,103,67

CONSENSUS REPORTS: EPA Genetic Toxicology Program.

SAFETY PROFILE: Suspected carcinogen with experimental carcinogenic and tumorigenic data. A poison by ingestion. Moderately toxic by subcutaneous route. Experimental teratogenic and reproductive effects. Mutation data reported. When heated to decomposition it emits toxic fumes of NO_x. See also NITROSAMINES.

BSA500 CAS:3913-02-8 HR: 1
2-BUTYL-1-OCTANOL
mf: $C_{12}H_{26}O$ mw: 186.38

PROP: Liquid. Mp: −80°, flash p: 230°F(OC), bp: 253.3°, d: 0.8355 @ 20°/20°, vap d: 6.42.

SYN: 2-BUTYLOCTYL ALCOHOL

TOXICITY DATA WITH REFERENCE
skn-rbt 10 mg/24H open MLD AMIHBC 4,119,51
eye-rbt 500 mg open AMIHBC 4,119,51
orl-rat LD50:13 g/kg AMIHBC 4,119,51

CONSENSUS REPORTS: Reported in EPA TSCA Inventory.

SAFETY PROFILE: Mildly toxic by ingestion. A skin and eye irritant. See also ALCOHOLS. Combustible when exposed to heat or flame. Incompatible with oxidizing materials. To fight fire, use CO_2, dry chemical. When heated to decomposition it emits acrid and irritating fumes.

BSA750 CAS:10097-26-4 HR: 1
2-BUTYLOCTYL ESTER METHACRYLIC ACID
mf: $C_{16}H_{30}O_2$ mw: 254.46

TOXICITY DATA WITH REFERENCE
skn-rbt 10 mg/24H open MLD AMIHBC 10,61,54
eye-rbt 500 mg open AMIHBC 10,61,54
orl-rat LD50:26 g/kg AMIHBC 10,61,54

SAFETY PROFILE: Mildly toxic by ingestion. A skin and eye irritant. See also ESTERS. When heated to decomposition it emits smoke and irritating fumes.

BSB000 CAS:142-77-8 HR: 1
BUTYL OLEATE
mf: $C_{22}H_{42}O_2$ mw: 338.64

PROP: Liquid. Bp: 173°, flash p: 356°F(OC), d: 0.873, vap d: 11.3.

SYNS: (Z)-9-OCTADECENOIC ACID BUTYL ESTER □ OLEIC ACID, BUTYL ESTER □ PLASTHALL 503 □ UNIFLEX BYO

TOXICITY DATA WITH REFERENCE
skn-rbt 500 mg/24H MOD FCTXAV 17,241,79

B

CONSENSUS REPORTS: Reported in EPA TSCA Inventory.

SAFETY PROFILE: A skin irritant. Combustible when exposed to heat or flame. To fight fire, use CO_2, dry chemical. Incompatible with oxidizing materials. When heated to decomposition it emits acrid smoke and irritating fumes. See also ESTERS; n-BUTYL ALCOHOL; and OLEIC ACID.

BSB500 CAS:61734-89-2 HR: 2
N-BUTYL-N-(2-OXOBUTYL)NITROSAMINE
mf: $C_8H_{16}N_2O_2$ mw: 172.26

SYN: N-NITROSO-N-(2-OXOBUTYL)BUTYLAMINE

TOXICITY DATA WITH REFERENCE
mma-sat 4 μmol/plate CNREA8 37,399,77
orl-rat TDLo:69 g/kg/20W-C:ETA GANNA2 67,825,76

CONSENSUS REPORTS: EPA Genetic Toxicology Program.

SAFETY PROFILE: Questionable carcinogen with experimental tumorigenic data. Mutation data reported. When heated to decomposition it emits toxic fumes of NO_x. See also NITROSAMINES.

BSB750 CAS:61734-90-5 HR: 2
N-BUTYL-N-(3-OXOBUTYL)NITROSAMINE
mf: $C_8H_{16}N_2O_2$ mw: 172.26

SYN: N-NITROSO-N-(3-OXOBUTYL)BUTYLAMINE

TOXICITY DATA WITH REFERENCE
mma-sat 4 μmol/plate CNREA8 37,399,77
orl-rat TDLo:69 g/kg/20W-C:ETA GANNA2 67,825,76

CONSENSUS REPORTS: EPA Genetic Toxicology Program.

SAFETY PROFILE: Questionable carcinogen with experimental tumorigenic data. Mutation data reported. When heated to decomposition it emits toxic fumes of NO_x. See also NITROSAMINES.

BSC000 CAS:94-26-8 HR: 3
BUTYL PARABEN
mf: $C_{11}H_{14}O_3$ mw: 194.25

PROP: Solid. Mp: 68–69°.

SYNS: BUTOBEN ☐ BUTYL CHEMOSEPT ☐ BUTYL-p-HYDROXYBENZOATE ☐ n-BUTYL PARAHYDROXYBENZOATE ☐ BUTYL PARASEPT ☐ BUTYL TEGOSEPT ☐ p-HYDROXYBENZOIC ACID BUTYL ESTER ☐ NIPABUYL ☐ PARASEPT ☐ SOLBROL B ☐ TEGOSEPT B

TOXICITY DATA WITH REFERENCE
skn-gpg 5%/48H MLD JSCCA5 28,357,77
orl-mus LD50:13,200 mg/kg NEZAAQ 28,463,73
ipr-mus LD50:230 mg/kg JSCCA5 28,357,77

CONSENSUS REPORTS: Reported in EPA TSCA Inventory.

SAFETY PROFILE: Poison by intraperitoneal route. A skin irritant. When heated to decomposition it emits acrid smoke and irritating fumes. See also ESTERS.

BSC250 CAS:107-71-1 HR: 3
tert-BUTYL PERACETATE
mf: $C_6H_{12}O_3$ mw: 132.18

PROP: Clear, colorless, benzene solution; insol in water; sol in organic solvents. D: 0.923, vap press: 50 mm @ 26°, flash p: <80°F (COC).

SYNS: t-BUTYL PERACETATE ☐ t-BUTYL PEROXYACETATE ☐ tert-BUTYL PEROXYACETATE, >76% in solution (DOT) ☐ ETHANEPEROXOIC ACID, 1,1-DIMETHYLETHYL ESTER ☐ LUPERSOL 70 ☐ TRIGONOX F-C50

TOXICITY DATA WITH REFERENCE
orl-rat LD50:675 mg/kg 85GMAT -,30,82
ihl-rat LC33:8200 mg/m³/4H 85GMAT -,30,82
orl-mus LD50:632 mg/kg TPKVAL 9,78,67
ihl-mus LCLo:6000 mg/m³ TPKVAL 9,78,67

CONSENSUS REPORTS: Reported in EPA TSCA Inventory.

DFG MAK: Moderate skin irritant
DOT CLASSIFICATION: Forbidden

SAFETY PROFILE: Moderately toxic by ingestion. Mildly toxic by inhalation. Moderate skin and eye irritant. A shock- and heat sensitive explosive. Dangerous fire hazard when exposed to heat, flame, reducing agents. To fight fire, use dry chemical, alcohol foam, spray, and mist. When heated to decomposition it emits acrid smoke and fumes. See also PEROXIDES, ORGANIC; and ESTERS.

BSC500 CAS:614-45-9 HR: 3
tert-BUTYL PERBENZOATE
mf: $C_{11}H_{14}O_3$ mw: 194.25

PROP: Colorless to slightly yellow liquid; mild aromatic odor. Bp: 112° (decomp), flash p: 19°, fp: 8°, vap press: 0.33 mm @ 50°, d: 1.0. Insol in water; sol in organic solvents.

SYNS: terc.BUTYLESTER KYSELINY PEROXYBENZOOVE (CZECH) ☐ terc.BUTYLPERBENZOAN (CZECH) ☐ t-BUTYL PERBENZOATE ☐ t-BUTYL PEROXY BENZOATE ☐ ESPEROX 10 ☐ NOVOX ☐ PERBENZOATE de BUTYLE TERTIAIRE (FRENCH) ☐ TRIGONOX C

TOXICITY DATA WITH REFERENCE
skn-rbt 500 mg/24H MLD 28ZPAK -,52,72
eye-rbt 100 mg/1M rns MLD ZAARAM 8,25,58
eye-rbt 500 mg/24H MLD 28ZPAK -,52,72
mma-sat 67 μg/plate ENMUDM 8(Suppl 7),52,72
unr-mus TDLo:311 mg/kg:ETA RARSAM 3,193,63
orl-rat LD50:1012 mg/kg 85GMAT -,30,82
orl-mus LD50:914 mg/kg 85GMAT -,30,82

CONSENSUS REPORTS: Reported in EPA TSCA Inventory.

SAFETY PROFILE: Moderately toxic by ingestion. A skin and eye irritant. Questionable carcinogen with experimental tumorigenic data. Mutation data reported. See also PEROXIDES, ORGANIC. Potentially explosive when heated above 115°C. Explosive reaction on contact with organic matter or copper(I) bromide + limonene. When heated to decomposition it emits acrid smoke and fumes.

BSC600 CAS:109-13-7 HR: 3
tert-BUTYL PERISOBUTYRATE
mf: $C_8H_{16}O_3$ mw: 160.24

SYNS: tert-BUTYL PEROXYISOBUTYRATE ☐ tert-BUTYL PEROXYISOBUTYRATE, >77% in solution (DOT) ☐ ESPEROX 24M ☐ LUPERSOL 8 ☐ PEROXYISOBUTYRIC ACID, tert-BUTYL ESTER ☐ PROPANEPEROXOIC ACID, 2-METHYL-, 1,1-DIMETHYLETHYL ESTER

CONSENSUS REPORTS: Reported in EPA TSCA Inventory.

DOT CLASSIFICATION: Forbidden

SAFETY PROFILE: An explosive and flammable peroxide. Handle very carefully in concentrated solutions. When heated to decomposition it emits acrid smoke and irritating vapors.

BSC750 CAS:110-05-4 HR: 3
tert-BUTYL PEROXIDE
mf: $C_8H_{18}O_2$ mw: 146.26

PROP: Clear, water-white liquid. Mp: −40°, bp: 80° @ 284 mm, flash p: 65°F (OC), d: 0.79, vap press: 19.51 mm @ 20°, vap d: 5.03. Very sltly sol in H_2O.

SYNS: CADOX ☐ DI-tert-BUTYLPEROXID (GERMAN) ☐ DI-tert-BUTYL PEROXIDE (MAK) ☐ DI-tert-BUTYL PEROXYDE (DUTCH) ☐ DTBP ☐ PEROSSIDO di BUTILE TERZIARIO (ITALIAN) ☐ PEROXYDE de BUTYLE TERTIAIRE (FRENCH) ☐ (TRIBUTYL)PEROXIDE

TOXICITY DATA WITH REFERENCE
skn-rbt 500 mg AIHAAP 19,205,58
eye-rbt 500 mg/24H MLD 28ZPAK -,40,72
eye-rbt 200 mg/1M rns MLD ZAARAM 8,25,58
unr-mus TDLo: 585 mg/kg: ETA RARSAM 3,193,63
orl-rat LC50: 10,200 mg/kg 28ZPAK -,40,72
ipr-rat LD50: 3210 mg/kg AIHAAP 19,205,58
orl-mus LD50: 20 g/kg FEPRA7 7,252,48

CONSENSUS REPORTS: Reported in EPA ISCA Inventory.

DFG MAK: Mild skin irritant

SAFETY PROFILE: Moderately toxic by intraperitoneal route. A powerful irritant by ingestion and inhalation. A mild skin and eye irritant. Questionable carcinogen with experimental tumorigenic data. Flammable liquid; see PEROXIDES, ORGANIC, for fire and explosion hazards. Warning: Water may not work to fight fire. When heated to decomposition it emits acrid smoke and fumes.

BSC800 CAS:16215-49-9 HR: 3
BUTYL PEROXYDICARBONATE
mf: $C_{10}H_{18}O_6$ mw: 234.28

SYNS: n-BUTYL PEROXYDICARBONATE, >52% in solution (DOT) ☐ DIBUTYL PEROXYDICARBONATE ☐ DI-n-BUTYL PEROXYDICARBONATE, >52% in solution (DOT) ☐ PEROXYDICARBONIC ACID, DIBUTYL ESTER

DOT CLASSIFICATION: Forbidden

SAFETY PROFILE: A highly unstable peroxide. When heated to decomposition it emits acrid smoke and irritating vapors.

BSD000 CAS:19910-65-7 HR: 2
sec-BUTYL PEROXYDICARBONATE
mf: $C_{10}H_{18}O_6$ mw: 234.28

SYNS: DI-sec-BUTYL PEROXYDICARBONATE ☐ DI-sec-BUTYL PEROXYDICARBONATE, not more than 52% in solution (DOT) ☐ DI-sec-BUTYL PEROXYDICARBONATE, technically pure (DOT)

TOXICITY DATA WITH REFERENCE
skn-rbt LD50: 1200 mg/kg BSPII* 1/75-19B

CONSENSUS REPORTS: Reported in EPA TSCA Inventory.

SAFETY PROFILE: Moderately toxic by skin contact. See also PEROXIDES, ORGANIC. When heated to decomposition it emits acrid smoke and irritating fumes.

BSD250 CAS:927-07-1 HR: 3
tert-BUTYL PEROXYPIVALATE
mf: $C_9H_{18}O_3$ mw: 174.27

PROP: Colorless liquid. D: 0.854 @ 25°/25°, fp: <19°, flash p: >155°F (OC), rapid decomp @ 21°. Insol in water and ethylene glycol; sol in most organic solvents.

SYNS: t-BUTYL PEROXYPIVALATE ☐ tert-BUTYL PERPIVALATE ☐ tert-BUTYL TRIMETHYLPEROXYACETATE ☐ ESPEROX 31M ☐ LUPERSOL 11 ☐ TRIGONOX 25/75 ☐ TRIGONOX 25-C75

TOXICITY DATA WITH REFERENCE
orl-rat LD50: 4300 mg/kg BSPII* 1/75-19B

CONSENSUS REPORTS: Reported in EPA TSCA Inventory.

SAFETY PROFILE: Mildly toxic by ingestion. Moderately flammable by heat, flame (sparks), oxidizers. Can explode on heating. To fight fire, use water, fog, mist, alcohol foam, dry chemical. When heated to decomposition it emits acrid smoke and fumes. See also PEROXIDES, ORGANIC.

BSE000 CAS:89-72-5 HR: 3
o-sec-BUTYLPHENOL
mf: $C_{10}H_{14}O$ mw: 150.24

PROP: Colorless liquid. Bp: 226–228° @ 25 mm, fp: 12°, flash p: 225°F, d: 0.981 @ 25°/25°.

SYN: 2-sec.-BUTYLFENOL (CZECH)

TOXICITY DATA WITH REFERENCE
skn-rbt 500 mg/24H SEV 28ZPAK -,55,72
eye-rbt 50 µg/24H SEV 28ZPAK -,55,72
ipr-mus LD50: 63 mg/kg JMCMAR 18,868,75
ivn-mus LD50: 60 mg/kg JMCMAR 23,1350,80
orl-gpg LD50: 600 mg/kg DTLVS* 4,58,80
skn-gpg LD50: 600 mg/kg DTLVS* 4,58,80

CONSENSUS REPORTS: Reported in EPA TSCA Inventory.

OSHA PEL: TWA 5 ppm (skin)
ACGIH TLV: TWA 5 ppm (skin)

SAFETY PROFILE: A poison by intraperitoneal and intravenous routes. Moderately toxic by ingestion and skin contact. A severe skin and eye irritant. Combustible when exposed to heat or flame. To fight fire, use foam, spray, CO_2, dry chemical. When heated to decomposition it emits acrid and irritating fumes. See also PHENOL and other butyl phenols.

BSE250 **CAS:99-71-8** **HR: 3**
p-sec-BUTYLPHENOL
mf: $(CH_3CHC_2H_5)C_6H_4OH$ mw: 150.2

PROP: Nearly white flakes. Bp: 135.4–136.5° @ 25 mm, fp: 51°, flash p: 240°F, d: 0.963 @ 60°/60°, mp: 60°.

SYN: 4-sec BUTYL PHENOL

TOXICITY DATA WITH REFERENCE
orl-rat LD50:2450 mg/kg SCIEAS 36(1-4),10,89
ipr-mus LD50:66 mg/kg JMCMAR 18,868,75
ivn-mus LD50:40 mg/kg JMCMAR 23,1350,80

CONSENSUS REPORTS: Reported in EPA TSCA Inventory.

SAFETY PROFILE: Poison by intravenous and intraperitoneal routes. Moderately toxic by ingestion. Combustible when exposed to heat or flame. When heated to decomposition it emits toxic fumes. To fight fire, use foam, CO_2, dry chemical. Incompatible with oxidizing materials. See also PHENOL and other butyl phenols.

BSE440 **CAS:3180-09-4** **HR: 2**
2-n-BUTYLPHENOL
mf: $C_{10}H_{14}O$ mw: 150.24

PROP: D: 0.975 @ 20°/4°, bp: 234–237°.

TOXICITY DATA WITH REFERENCE
skn-mus TDLo:3800 mg/kg/12W-I:NEO CNREA8 19,413,59
orl-rat LD50:634 mg/kg PSEBAA 32,592,35

SAFETY PROFILE: Questionable carcinogen with experimental neoplastigenic data. When heated to decomposition it emits acrid smoke and irritating fumes. See also PHENOL and other butyl phenols.

BSE450 **CAS:1638-22-8** **HR: 2**
4-n-BUTYLPHENOL
mf: $C_{10}H_{14}O$ mw: 150.24

PROP: Solid or liquid. D: 0.976 @ 22°/4°, mp: 22°, bp: 248°.

TOXICITY DATA WITH REFERENCE
skn-mus TDLo:3840 mg/kg/12W-I:ETA CNREA8 19,413,59

CONSENSUS REPORTS: Reported in EPA TSCA Inventory.

SAFETY PROFILE: A poison. Questionable carcinogen with experimental tumorigenic data. When heated to decomposition it emits acrid smoke and irritating fumes. See also PHENOL and other butyl phenols.

BSE460 **CAS:88-18-6** **HR: 3**
2-t-BUTYLPHENOL
mf: $C_{10}H_{14}O$ mw: 150.24

SYN: PHENOL, o-(tert-BUTYL)-

TOXICITY DATA WITH REFERENCE
ipr-mus LD50:82 mg/kg JMCMAR 18,868,75

CONSENSUS REPORTS: Reported in EPA TSCA Inventory.

SAFETY PROFILE: Poison by intraperitoneal route. When heated to decomposition it emits acrid smoke and irritating vapors.

BSE500 **CAS:98-54-4** **HR: 3**
4-t-BUTYLPHENOL
mf: $C_{10}H_{14}O$ mw: 150.24

PROP: Crystals, needles, or practically white flakes. Mp: 99°, bp: 236–238°, d: 0.9081 @ 114°/4°, vap press: 1 mm @ 70.0°, vap d: 5.1.

SYNS: p-tert-BUTYLFENOL (CZECH) □ BUTYLPHEN □ p-tert-BUTYLPHENOL (MAK) □ 4-(1,1-DIMETHYLETHYL)PHENOL □ 1-HYDROXY-4-tert-BUTYLBENZENE □ UCAR BUTYLPHENOL 4-T

TOXICITY DATA WITH REFERENCE
skn-rbt 500 mg/4H MLD DCTODJ 11,43,88
skn-rbt 500 mg/24H MLD 85JCAE-,224,86
eye-rbt 10 mg SEV DCTODJ 11,43,88
eye-rbt 50 µg/24H SEV 85JCAE-,224,86
orl-ham TDLo:252 g/kg/20W-C:NEO CRNGDP 7,1285,86
orl-rat LD50:3250 µL/kg AIHAAP 30,470,69
ihl-rat LCLo:5600 mg/m³/4H DCTODJ 11,43,88
ipr-mus LD50:78 mg/kg JMCMAR 18,868,75
skn-rbt LD50:2520 µL/kg AIHAAP 30,470,69
orl-mam LD50:1500 mg/kg GISAAA 45(10),16,80
skn-uns LD50:1580 mg/kg GISAAA 45(10),16,80

CONSENSUS REPORTS: Reported in EPA TSCA Inventory.

DFG MAK: 0.08 ppm (0.5 mg/m³)

SAFETY PROFILE: Poison by intraperitoneal route. Moderately toxic by skin contact and ingestion. A skin and severe eye irritant. Questionable carcinogen with experimental neoplastigenic data. Combustible when exposed to heat or flame; can react with oxidizing materials. To fight fire, use foam, CO_2, dry chemical. When heated to decomposition it emits acrid and irritating fumes. See also PHENOL and other butyl phenols.

BSE750 **CAS:56488-59-6** **HR: 3**
4′-(3-(4′-tert-BUTYLPHENOXY)-2-HYDROXYPROPOXY)BENZOIC ACID
mf: $C_{20}H_{24}O_5$ mw: 344.44

SYN: 4-(3-(4-(1,1-DIMETHYLETHYL)PHENOXY)-2-HYDROXYPROPOXY)BENZOIC ACID

TOXICITY DATA WITH REFERENCE
orl-rat LD50:2400 mg/kg DRFUD4 4,140,79
ipr-rat LD50:500 mg/kg DRFUD4 4,140,79
orl-mus LD50:2100 mg/kg DRFUD4 4,140,79

ipr-mus LD50:335 mg/kg DRFUD4 4,140,79
orl-rbt LD50:1800 mg/kg DRFUD4 4,140,79
orl-gpg LD50:320 mg/kg DRFUD4 4,140,79
orl-mam LD50:5000 mg/kg DRFUD4 4,140,79

SAFETY PROFILE: Poison by ingestion and intraperitoneal routes. When heated to decomposition it emits acrid smoke and irritating fumes.

BSF250 CAS:61005-12-7 HR: 3
o-sec-BUTYLPHENYL CARBAMATE
mf: $C_{12}H_{17}NO_2$ mw: 207.30

TOXICITY DATA WITH REFERENCE
orl-rat LD50:410 mg/kg OYYAA2 3,74,69
orl-mus LD50:340 mg/kg OYYAA2 3,74,69

SAFETY PROFILE: A poison by ingestion. See also CARBAMATES. When heated to decomposition it emits toxic fumes of NO_x.

BSF750 CAS:1126-79-0 HR: 2
BUTYL PHENYL ETHER
mf: $C_{10}H_{14}O$ mw: 150.24

PROP: Liquid. Flash p: 180°F (OC), d: 0.9, vap d: 5.2, mp: −19°, bp: 210°.

SYN: BUTOXYPHENYL

TOXICITY DATA WITH REFERENCE
orl-mus LD50:3200 mg/kg JPETAB 88,400,46

CONSENSUS REPORTS: Reported in EPA TSCA Inventory.

SAFETY PROFILE: Moderately toxic by ingestion. See also ETHERS. When heated to decomposition it emits acrid and irritating fumes.

BSG000 CAS:329-21-5 HR: 3
S-p-tert-BUTYLPHENYL-o-ETHYL ETHYLPHOSPHONODITHIOATE
mf: $C_{14}H_{23}OPS_2$ mw: 302.46

SYNS: S-(4-(1,1-DIMETHYLETHYL)PHENYL)-o-ETHYL ETHYLPHOSPHONODITHIOATE □ ENT 25,765 □ N 3051 □ STAUFFER N-3051

TOXICITY DATA WITH REFERENCE
orl-rat LD50:141 mg/kg ARSIM* 20,22,66
orl-ckn LD50:64 mg/kg TXAPA9 7,606,65

SAFETY PROFILE: Poison by ingestion. When heated to decomposition it emits very toxic fumes of SO_x and PO_x. See also ESTERS.

BSG250 CAS:673-19-8 HR: 3
m-sec-BUTYLPHENYL-N-METHYLCARBAMATE
mf: $C_{12}H_{17}NO_2$ mw: 207.30

SYNS: 3-sec-BUTYLPHENYL-N-METHYLCARBAMATE □ CALIFORNIA CHEMICAL COMPANY RE5305 □ CHEVRON RE5305 □ ENT 27,039 □ H-28 □ m-(1-METHYLPROPYL)PHENYLMETHYLCARBAMATE □ RE 5305 (CALIFORNIA CHEMICAL)

TOXICITY DATA WITH REFERENCE
orl-rat LD50:10 mg/kg ARSIM* 20,7,66
orl-ckn LD50:14 mg/kg TXAPA9 11,49,67
orl-bwd LD50:4600 μg/kg TXAPA9 21,315,72

SAFETY PROFILE: Poison by ingestion. See also CARBAMATES. When heated to decomposition it emits toxic fumes of NO_x.

BSG300 CAS:780-11-0 HR: 2
3-tert-BUTYLPHENYL N-METHYLCARBAMATE
mf: $C_{12}H_{17}NO_2$ mw: 207.30

SYNS: CARBAMIC ACID, METHYL-, 3-tert-BUTYLPHENYL ESTER □ H-22 □ KNOCKBAL □ PHENOL, 3-(1,1-DIMETHYLETHYL)-, METHYLCARBAMATE (9CI) □ RE 5030 □ TBPMC □ TERBAM

TOXICITY DATA WITH REFERENCE
orl-mus LD50:470 mg/kg 85ARAE 1,44,77
skn-mus LD50:2660 mg/kg JPIFAN (4),28,70
orl-rbt LD50:505 mg/kg JPIFAN (4),28,70

CONSENSUS REPORTS: Reported in EPA TSCA Inventory.

SAFETY PROFILE: Moderately toxic by ingestion and skin contact. When heated to decomposition it emits toxic vapors of NO_x.

BSH000 HR: 3
2-tert-BUTYL-3-PHENYL OXAZIRANE
mf: $C_{11}H_{15}NO$ mw: 177.05

SAFETY PROFILE: May explode in vacuum. When heated to decomposition it emits toxic fumes of NO_x.

BSH100 CAS:87-18-3 HR: 2
p-tert-BUTYLPHENYL SALICYLATE
mf: $C_{17}H_{18}O_3$ mw: 270.13

SYNS: BENZOIC ACID, 2-HYDROXY-, 4-(1,1-DIMETHYLETHYL)PHENYL ESTER □ p-terc.BUTYLFENYLESTER KYSELINY SALICYLOVE

TOXICITY DATA WITH REFERENCE
orl-mus LD50:2900 mg/kg 85JCAE-,672,86

CONSENSUS REPORTS: Reported in EPA TSCA Inventory.

SAFETY PROFILE: Moderately toxic by ingestion. When heated to decomposition it emits acrid smoke and irritating fumes.

BSH250 CAS:78-48-8 HR: 3
BUTYL PHOSPHOROTRITHIOATE
mf: $C_{12}H_{27}OPS_3$ mw: 314.54

PROP: Liquid. Bp: 167–170° @ 1 mm, d: 1.06 @ 20 mm. Insol in water; sol in aliphatic, aromatic, and chlorinated hydrocarbons.

SYNS: B-1,776 □ BUTIFOS □ BUTIPHOS □ CHEMAGRO 1,776 □ CHEMAGRO B-1776 □ DEF □ DEF DEFOLIANT □ DE-GREEN □ E-Z-OFF D □ FOS-FALL "A" □ ORTHO PHOSPHATE DEFOLIANT □ S,S,S-TRIBUTYL PHOSPHOROTRITHIOATE □ S,S,S-TRIBUTYL TRITHIOPHOSPHATE

TOXICITY DATA WITH REFERENCE
orl-rat TDLo:216 mg/kg (8W pre):REP MZUZA8 (2),48,80

orl-rat LD50: 150 mg/kg TXAPA9 14,515,69
skn-rat LD50: 168 mg/kg WRPCA2 9,119,70
ipr-rat LD50: 210 mg/kg 34ZIAG -,199,69
orl-mus LD50: 77 mg/kg 85JCAE -,1188,86
ihl-mus LCLo: 3804 mg/m³/1H 34ZIAG -,199,69
ipr-mus LD50: 290 mg/kg 34ZIAG -,199,69
skn-rbt LD50: 97 mg/kg 85GMAT -,44,82

CONSENSUS REPORTS: Reported in EPA TSCA Inventory.

SAFETY PROFILE: A poison by ingestion, skin contact, and intraperitoneal routes. Experimental reproductive effects. Animal experiments show an anti-cholinesterase effect. When heated to decomposition it emits toxic fumes of PO_x and SO_x. See also PARATHION, PHOSPHATES, ESTERS, and SULFATES.

BSH500 CAS:6066-49-5 HR: 2
3-n-BUTYLPHTHALIDE
mf: $C_{12}H_{14}O_2$ mw: 190.26

SYNS: BUTYLPHTHALIDE □ 3-BUTYLPHTHALIDE

TOXICITY DATA WITH REFERENCE
skn-rbt 500 mg/24H MOD FCTXAV 17,241,79
orl-rat LD50: 2450 mg/kg FCTXAV 17,241,79

CONSENSUS REPORTS: Reported in EPA TSCA Inventory.

SAFETY PROFILE: Moderately toxic by ingestion. A skin irritant. When heated to decomposition it emits acrid smoke and irritating fumes.

BSI000 CAS:536-69-6 HR: 3
5-BUTYL PICOLINIC ACID
mf: $C_{10}H_{13}NO_2$ mw: 179.24

PROP: Plates from pet ether. Mp: 108–109°.

SYNS: 5-BUTYL-2-PYRIDINECARBOXYLIC ACID □ FUSARIC ACID □ FUSARINIC ACID

TOXICITY DATA WITH REFERENCE
orl-rat LD50: 480 mg/kg JJANAX 29,439,76
ipr-rat LD50: 250 mg/kg JJANAX 29,439,76
scu-rat LD50: 300 mg/kg JJANAX 29,439,76
ivn-rat LD50: 210 mg/kg JJANAX 29,439,76
orl-mus LD50: 180 mg/kg JOPHDQ 6,922,83
ipr-mus LD50: 75 mg/kg JOPHDQ 6,922,83
ivn-mus LD50: 100 mg/kg 85ERAY 3,1873,78

CONSENSUS REPORTS: EPA Genetic Toxicology Program. Reported in EPA TSCA Inventory.

SAFETY PROFILE: A poison by ingestion, intraperitoneal, subcutaneous, and intravenous routes. When heated to decomposition it emits toxic fumes of NO_x.

BSI250 CAS:2180-92-9 HR: 3
1-BUTYL-2′,6′-PIPECOLOXYLIDIDE
mf: $C_{18}H_{28}N_2O$ mw: 288.48

SYNS: BUPIVACAINE □ dl-BUPIVACAINE

TOXICITY DATA WITH REFERENCE
isp-wmn TDLo: 2396 mg/kg (female 39W post): REP BJOGAS 88,407,81
icv-wmn TDLo: 500 μg/kg (female 39W post): TER JPEMAO 12,75,84
ivn-hmn TDLo: 4300 μg/kg: BPR,PSY AANEAB 21,521,77
scu-rat LD50: 48 mg/kg APTOA6 31,273,72
ivn-rat LD50: 5600 μg/kg APTOA6 31,273,72
ipr-mus LD50: 58,700 μg/kg TXAPA9 54,501,80
scu-mus LD50: 53 mg/kg APTOA6 31,273,72
ivn-mus LD50: 7300 μg/kg APTOA6 31,273,72
ivn-rbt LD50: 1620 μg/kg AACRAT 64,209,85
par-rbt LD50: 64 mg/kg ARZNAD 26,78,76
itr-rbt LD50: 12,500 μg/kg ARZNAD 26,78,76

SAFETY PROFILE: A poison by subcutaneous, intraperitoneal, intratracheal, parenteral, and intravenous routes. An experimental teratogen. Other experimental reproductive effects. Human systemic effects by intravenous route: changes in regional blood flow rates and euphoria. When heated to decomposition it emits toxic fumes of NO_x.

BSI750 CAS:78329-88-1 HR: 2
p-(N-BUTYL-2-(PIPERIDINO)ACETAMIDO)BENZOIC ACID BUTYL ESTER HYDROCHLORIDE
mf: $C_{22}H_{34}N_2O_3 \cdot ClH$ mw: 411.04

SYN: C 3181

TOXICITY DATA WITH REFERENCE
eye-rbt 2% SEV ARZNAD 8,609,58
ipr-rat LD50: 480 mg/kg ARZNAD 8,609,58
scu-mus LD50: 3750 mg/kg ARZNAD 8,609,58

SAFETY PROFILE: Moderately toxic by intraperitoneal and subcutaneous routes. A severe eye irritant. See also ESTERS. When heated to decomposition it emits very toxic fumes of NO_x and HCl.

BSJ500 CAS:590-01-2 HR: 3
BUTYL PROPANOATE
DOT: UN 1914
mf: $C_7H_{14}O_2$ mw: 130.2

PROP: Water-white liquid, apple-like odor. Mp: −89.6°, bp: 145.4°, flash p: 90°F, d: 0.893 @ 0°/0°, autoign temp: 800°F, vap d: 4.49.

SYNS: BUTYL PROPIONATE □ n-BUTYL PROPIONATE □ PROPANOIC ACID BUTYLESTER (9CI)

TOXICITY DATA WITH REFERENCE
skn-rbt 500 mg/24H MOD FCTXAV 18,661,80
eye-rbt 100 mg SEV JACTDZ 1,192,92
orl-rat LD50: 5 g/kg FCTXAV 18,661,80

CONSENSUS REPORTS: Reported in EPA TSCA Inventory.

DOT CLASSIFICATION: 3; *Label:* Flammable Liquid

SAFETY PROFILE: Mildly toxic by ingestion. A skin irritant. Dangerously flammable when exposed to heat or flame. To fight fire, use foam, CO_2, dry chemical. Incompatible with oxidizing materials. See also ESTERS, n-BUTYL ALCOHOL, and PROPIONIC ACID.

BSJ550 **CAS:539-32-2** **HR: 3**
3-BUTYLPYRIDINE
mf: $C_9H_{13}N$ mw: 135.23

SYNS: 3-n-BUTYLPYRIDINE □ PYRIDINE, 3-BUTYL- □ 1-(3-PYRIDYL)BUTANE

TOXICITY DATA WITH REFERENCE
ipr-mus LD50:270 mg/kg JPETAB 88,82,46
ivn-mus LD50:59 mg/kg JPETAB 88,82,46

CONSENSUS REPORTS: Reported in EPA TSCA Inventory.

SAFETY PROFILE: Poison by intravenous and intraperitoneal routes. When heated to decomposition it emits toxic vapors of NO_x.

BSK000 **CAS:98-29-3** **HR: 3**
4-tert-BUTYLPYROCATECHOL
mf: $C_{10}H_{14}O_2$ mw: 166.24

PROP: Crystals. Fp: 52°, bp: 285°, flash p: 265°F, d: 1.049 @ 60°/25°.

SYNS: 4-tert-BUTYLCATECHOL □ p-tert-BUTYLPYROCATECHOL □ 4-tert-BUTYLPYROKATECHIN (CZECH) □ 4-(1,1-DIMETHYLETHYL)-1,2-BENZENEDIOL □ SYNOX TBC

TOXICITY DATA WITH REFERENCE
skn-rbt 10 mg/24H open AMIHBC 10,61,54
skn-rbt 750 µg/24H SEV 85JCAE-,236,86
eye-rbt 50 µg open SEV AMIHBC 10,61,54
skn-gpg 0.1%/3W MLD JIDEAE 55,190,70
skn-gpg 1%/3W MOD JIDEAE 55,190,70
msc-mus:lym 80 µg/L EMMUEG 11,523,88
orl-rat LD50:2820 mg/kg AMIHBC 10,61,54
ivn-mus LD50:32 mg/kg CSLNX* NX#07874
skn-rbt LD50:630 mg/kg AMIHBC 10,61,54

CONSENSUS REPORTS: Reported in EPA TSCA Inventory.

SAFETY PROFILE: A poison by intravenous route. Moderately toxic by ingestion and skin absorption. A severe skin and eye irritant. Mutation data reported. Combustible when exposed to heat or flame. To fight fire, use CO_2, dry chemical, fog, mist. When heated to decomposition it emits acrid and irritating fumes.

BSK250 **CAS:767-10-2** **HR: 3**
n-BUTYLPYRROLIDINE
mf: $C_8H_{17}N$ mw: 127.26

TOXICITY DATA WITH REFERENCE
orl-mus LD50:51 mg/kg INHEAO 4,63,66
skn-mus LD50:1000 mg/kg INHEAO 4,63,66
ipr-mus LD50:37 mg/kg INHEAO 4,63,66
scu-mus LD50:57 mg/kg INHEAO 4,63,66

SAFETY PROFILE: Poison by ingestion, intraperitoneal, and subcutaneous routes. Moderately toxic by skin contact. When heated to decomposition it emits toxic fumes of NO_x.

BSL250 **CAS:2052-14-4** **HR: 2**
n-BUTYL SALICYLATE
mf: $C_{11}H_{14}O_3$ mw: 194.25

SYNS: BUTYL-o-HYDROXYBENZOATE □ n-BUTYL-o-HYDROXYBENZOATE □ BUTYL SALICYLATE □ 2-HYDROXYBENZOIC ACID BUTYL ESTER

TOXICITY DATA WITH REFERENCE
orl-rat LD50:1700 mg/kg FCTXAV 16,637,78

CONSENSUS REPORTS: Reported in EPA TSCA Inventory.

SAFETY PROFILE: Moderately toxic by ingestion. When heated to decomposition it emits acrid smoke and irritating fumes. See also ESTERS.

BSL325 **HR: 3**
n-BUTYLSCOPOLAMINE TANNATE

TOXICITY DATA WITH REFERENCE
orl-rat TDLo:164 g/kg (91D male):REP KSRNAM 7,442,73
ipr-rat LD50:343 mg/kg OYYAA2 5,599,71
ipr-mus LD50:146 mg/kg OYYAA2 5,599,71
scu-mus LD50:228 mg/kg OYYAA2 5,599,71

SAFETY PROFILE: Poison by subcutaneous and intraperitoneal routes. Experimental reproductive effects. When heated to decomposition it emits toxic fumes of NO_x. See also AMINES.

BSL450 **CAS:52670-52-7** **HR: 3**
17-BUTYLSPARTEIN
mf: $C_{19}H_{34}N_2$ mw: 290.55

SYN: 6-BUTYLDODECAHYDRO-7,14-METHANO-2H,6H-DIPYRIDO(1,2-a:1′,2′-e)(1,5)DIAZOCINE

TOXICITY DATA WITH REFERENCE
orl-mus LD50:1820 mg/kg ARZNAD 30,1497,80
ipr-mus LD50:160 mg/kg ARZNAD 30,1497,80
ivn-mus LD50:27,300 µg/kg ARZNAD 30,1497,80

SAFETY PROFILE: Poison by intravenous and intraperitoneal routes. Moderately toxic by ingestion. When heated to decomposition it emits toxic fumes of NO_x.

BSL500 **CAS:2273-43-0** **HR: 3**
BUTYL STANNOIC ACID
mf: $C_4H_{10}O_2Sn$ mw: 208.83

PROP: White infusible solid. Sol in Me_2CO.

SYN: BUTYLHYDROXYOXOSTANNANE

TOXICITY DATA WITH REFERENCE
ivn-mus LD50:180 mg/kg CSLNX* NX#03474

CONSENSUS REPORTS: Reported in EPA TSCA Inventory.

OSHA PEL: TWA 0.1 mg(Sn)/m^3 (skin)
ACGIH TLV: TWA 0.1 mg(Sn)/m^3 (skin) (Proposed: TWA 0.1 mg(Sn)/m^3; STEL 0.2 mg(Sn)/m^3 (skin))
NIOSH REL: (Organotin Compounds) TWA 0.1 mg(Sn)/m^3

SAFETY PROFILE: A poison by intravenous route. See

also TIN COMPOUNDS. When heated to decomposition it emits acrid smoke and irritating fumes.

For occupational chemical analysis use NIOSH: Organotin Compounds 5504.

BSL600 CAS:123-95-5 HR: 1
BUTYL STEARATE
mf: $C_{22}H_{44}O_2$ mw: 340.57

PROP: Crystals from alcohol, propanol, or ether.

SYNS: APEX 4 ☐ BS ☐ BUTYL OCTADECANOATE ☐ n-BUTYL OCTADECANOATE ☐ n-BUTYL STEARATE ☐ EMEREST 2325 ☐ GROCO 5810 ☐ KESSCO BSC ☐ KESSCOFLEX BS ☐ POLYCIZER 332 ☐ OCTADECANOIC ACID, BUTYL ESTER (9CI) ☐ RC PLASTICIZER B-17 ☐ STARFOL BS-100 ☐ TEGESTER BUTYL STEARATE ☐ UNIFLEX BYS ☐ WICKENOL 122 ☐ WITCIZER 200 ☐ WITCIZER 201

TOXICITY DATA WITH REFERENCE
skn-rbt 500 mg MOD JACTDZ 4(5),107,85
orl-rat TDLo:418 g/kg (male 10W pre):REP AMIHBC 7,310,53
orl-rat LD50:32 g/kg IPSTB3 3,93,76

CONSENSUS REPORTS: Reported in EPA TSCA Inventory.

SAFETY PROFILE: Low toxicity by ingestion. Experimental reproductive data. A skin irritant. When heated to decomposition it emits acrid smoke and irritating fumes.

BSL750 CAS:63979-65-7 HR: 3
n-BUTYL-k-STROPHANTHIDIN
mf: $C_{27}H_{38}O_7$ mw: 474.65

TOXICITY DATA WITH REFERENCE
ivn-cat LDLo:350 μg/kg AEPPAE 185,329,37
ivn-rat LDLo:500 μg/kg AEPPAE 185,329,37

SAFETY PROFILE: Deadly poison by intravenous route. When heated to decomposition it emits acrid smoke and irritating fumes.

BSM000 CAS:339-43-5 HR: 3
1-BUTYL-3-SULFANILYL UREA
mf: $C_{11}H_{17}N_3O_3S$ mw: 271.37

PROP: Solid. Mp: 144–145°.

SYNS: ALENTIN ☐ N-(4-AMINOBENZENESULFONYL)-N'-BUTYLUREA ☐ 4-AMINO-N-((BUTYLAMINO)CARBONYL)BENZENESULFONAMIDE ☐ AMINOPHENUROBUTANE ☐ BUCARBAN ☐ BUCROL ☐ BUKARBAN ☐ BURCOL ☐ BUTISULFINA ☐ N'-(BUTYLCARBAMOYL) SULFANILAMIDE ☐ N[1]-(BUTYLCARBAMOYL)SULFANILAMIDE ☐ N-BUTYLSULFANILYLUREA ☐ CARBUTAMID ☐ CARBUTAMIDE ☐ CICLORAL ☐ DIABORAL ☐ EMEDAN ☐ GLUCIDORAL ☐ GLUCOFREN ☐ GLYBUTAMIDE ☐ INBUTON ☐ INVENOL ☐ NADISAN ☐ NADIZAN ☐ NORBORAL ☐ ORANIL ☐ ORANYL ☐ ORASULIN ☐ N[1]-SULFANILYL-N[2]-BUTYLCARBAMIDE ☐ N[1]-SULFANILYL-N[2]-BUTYLUREA ☐ N-SULFANILYL-N'BUTYLUREE (FRENCH) ☐ U 6987

TOXICITY DATA WITH REFERENCE
unr-rat TDLo:1 g/kg (female 1D post):TER AKGIAO 42(12),35,66
orl-rat TDLo:3200 mg/kg (female 7-14D post):TER PSDTAP 11,151,70
unr-mus TDLo:8800 mg/kg (female 5-15D post):REP BSVMA8 71,289,69
unr-mus TDLo:4400 mg/kg (female 5-15D post):TER BSVMA8 71,289,69
unr-rat TDLo:3 g/kg (female 9D post):TER AKGIAO 42(12),35,66
orl-rat TDLo:5 g/kg (female 9-10D post):TER ANENAG 19,167,58
orl-rat TDLo:1056 mg/kg (7-14D preg):TER PSDTAP 11,151,70
orl-rat LD50:7800 mg/kg FATOAO 25,93,62
ivn-rat LD50:980 mg/kg DIAEAZ 6,2,57
orl-mus LD50:2800 mg/kg FATOAO 25,93,62
ipr-mus LD50:250 mg/kg NTIS** AD691-490
scu-mus LD50:2640 mg/kg DIAEAZ 6,2,57

CONSENSUS REPORTS: Reported in EPA TSCA Inventory.

SAFETY PROFILE: A poison by intraperitoneal route. Moderately toxic by ingestion and subcutaneous routes. An experimental teratogen. Other experimental reproductive effects. When heated to decomposition it emits very toxic fumes of NO_x and SO_x.

BSM125 CAS:544-40-1 HR: 2
BUTYL SULFIDE
mf: $C_8H_{18}S$ mw: 146.32

PROP: Liquid. Mp: −79.7°, bp: 185–185.5°, d: 0.839. Insol in water; very sol in alc and ether.

SYNS: BUTYL MONOSULFIDE ☐ n-BUTYL-SULFIDE ☐ BUTYLTHIOBUTANE ☐ n-DIBUTYL SULFIDE ☐ DI-n-BUTYLSULFIDE ☐ DIBUTYL SULPHIDE ☐ DIBUTYL THIOETHER ☐ 5 THIANONANE ☐ THIANONANE-5

TOXICITY DATA WITH REFERENCE
skn-rbt 500 mg/24H MOD FCTXAV 17,769,79
orl-rat LD50:2220 mg/kg FCTXAV 17,769,79
ihl-mus LCLo:1800 mg/m^3 FCTXAV 17,769,79

CONSENSUS REPORTS: Reported in EPA TSCA Inventory.

SAFETY PROFILE: Moderately toxic by ingestion. A skin irritant. When heated to decomposition it emits toxic fumes of SO_x. See also SULFIDES.

BSM250 CAS:64910-63-0 HR: 3
1-BUTYLSULFONIMIDOCYCLOHEXAMETHYLENE
mf: $C_{10}H_{21}NO_2S$ mw: 219.38

SYN: N-CYCLOHEXYL-1-BUTANESULFONAMIDE

TOXICITY DATA WITH REFERENCE
skn-rbt 175 mg/14D MLD NTIS** AD-A022-909
orl-rat LD50:2816 mg/kg NTIS** AD-A022-909
ipr-rat LD50:1074 mg/kg NTIS** AD-A022-909
ivn-rat LDLo:225 mg/kg NTIS** AD-A022-909
orl-mus LD50:5400 mg/kg NTIS** AD-A022-909
skn-mus LD50:7560 mg/kg NTIS** AD-A022-909
scu-mus LD50:519 mg/kg NTIS** AD-A022-909

SAFETY PROFILE: Poison by intravenous route. Moderately toxic by ingestion, intraperitoneal, and subcutaneous routes. A skin irritant. When heated to decomposition it emits very toxic fumes of SO_x and NO_x.

BSM825 CAS:63906-57-0 HR: 3
1-BUTYL THEOBROMINE
mf: $C_{11}H_{14}N_4O_2$ mw: 234.29

SYNS: 1-(2'-BUTENYL)THEOBROMINE □ 1-CROTYL THEOBROMINE

TOXICITY DATA WITH REFERENCE
orl-mus LD50:667 mg/kg JPETAB 116,343,56
ipr-mus LD50:230 mg/kg JPETAB 116,343,56
ivn-mus LD50:95 mg/kg JPETAB 86,113,46

SAFETY PROFILE: Poison by intravenous and intraperitoneal routes. Moderately toxic by ingestion. When heated to decomposition it emits toxic fumes of NO_x.

BSN000 CAS:34014-18-1 HR: 3
1-(5-(tert-BUTYL)-1,3,4-THIADIAZOL-2-YL)-1,3-DIMETHYLUREA
mf: $C_9H_{16}N_4OS$ mw: 228.35

PROP: Solid. Mp: 161.5–164°. Sltly sol in H_2O; sol in Me_2CO, MeOH; sltly sol in hexane.

SYNS: BRULAN □ 1-(5-tert-BUTYL-1,3,4-THIADIAZOL-2-YL)-3-DIMETHYLHARNSTOFF (GERMAN) □ N-(5-(1,1-DIMETHYLAETHYL)-1,3,4-THIADIAZOL-2-YL)-N,N'-DIMETHYLHARNSTOFF (GERMAN) □ E-103 □ EI-103 □ EL-103 □ GRASLAN □ PERFMID □ PREFLAN □ PREFMID □ SPIKE □ TEBULAN □ TEBUTHIURON □ TIUROLAN

TOXICITY DATA WITH REFERENCE
orl-rat LD50:644 mg/kg FMCHA2 -,D286,80
ipr-rat LD50:480 mg/kg NNGADV 17,S35,92
scu-rat LD50:500 mg/kg NNGADV 17,S35,92
orl-mus LD50:579 mg/kg 85DPAN -,-,71/76
ipr-mus LD50:505 mg/kg NNGADV 17,S35,92
scu-mus LD50:545 mg/kg NNGADV 17,S35,92
orl-rbt LD50:286 mg/kg 85DPAN -,-,71/76

SAFETY PROFILE: Poison by ingestion. Moderately toxic by intraperitoneal and subcutaneous routes. When heated to decomposition it emits very toxic fumes of SO_x and NO_x.

BSN325 CAS:2314-17-2 HR: 2
2-BUTYLTHIOBENZOTHIAZOLE
mf: $C_{11}H_{13}NS_2$ mw: 223.37

SYN: BUTYLCAPTAX

TOXICITY DATA WITH REFERENCE
orl-rat LD50:1270 mg/kg 85GMAT -,29,82
unr-rat LD50:1300 mg/kg GISAAA 47(2),63,82
orl-mus LD50:1610 mg/kg 85GMAT -,29,82
orl-rbt LD50:2344 mg/kg 85GMAT -,29,82

SAFETY PROFILE: Moderately toxic by ingestion and possibly other routes. When heated to decomposition it emits toxic fumes of SO_x and NO_x.

BSN500 CAS:628-83-1 HR: 3
n-BUTYL THIOCYANATE
mf: C_5H_9NS mw: 115.21

SYNS: n-BUTYL RHODANATE □ BUTYRHODANID (GERMAN) □ 1-THIOCYANOBUTANE

TOXICITY DATA WITH REFERENCE
orl-rat LDLo:250 mg/kg JIHTAB 18,310,36
scu-rat LDLo:70 mg/kg JIHTAB 18,310,36
ipr-mus LD50:13 mg/kg PCBPBS 2,95,72
scu-mus LDLo:130 mg/kg JIHTAB 18,310,36

CONSENSUS REPORTS: Reported in EPA TSCA Inventory.

SAFETY PROFILE: A poison by ingestion and subcutaneous routes. When heated to decomposition it emits very toxic fumes of NO_x and SO_x. See also THIOCYANATES.

BSO000 CAS:13071-79-9 HR: 3
S-((tert-BUTYLTHIO)METHYL)-O,O-DIETHYLPHOSPHORODITHIOATE
mf: $C_9H_{21}O_2PS_3$ mw: 288.45

PROP: Pale-yellow liquid. D: 1.105 @ 24 mm, mp: −29°, bp: 69° @ 0.01 mm. Very sltly sol in H_2O; sol in most org solvs.

SYNS: AC 921000 □ COUNTER □ COUNTER 15G SOIL INSECTICIDE □ COUNTER 15G SOIL INSECTICIDE-NEMATICIDE □ S-(((1,1-DIMETHYLETHYL)THIO)METHYL)-O,O-DIETHYL PHOSPHORODITHIOATE □ PHOSPHORODITHIOIC ACID S-((tert-BUTYLTHIO)METHYL)-O,O-DIETHYL ESTER □ PHOSPHORODITHIOIC ACID S-(((1,1-DIMETHYLETHYL)THIO)METHYL)-O,O-DIETHYL ESTER □ TERBUFOS

TOXICITY DATA WITH REFERENCE
orl-rat LD50:1600 μg/kg MEIEDD 10,1310,83
orl-mus LD50:3500 μg/kg FMCHA2 -,C63,83
orl-dog LD50:4500 μg/kg FMCHA2 -,C63,83
skn-rbt LD50:1100 μg/kg/24H FMCHA2 -,C63,83
orl-qal LD50:15 mg/kg EESADV 8,551,84

CONSENSUS REPORTS: EPA Genetic Toxicology Program. EPA Extremely Hazardous Substances List.

SAFETY PROFILE: Deadly poison by ingestion and skin contact. An insecticide. When heated to decomposition it emits very toxic fumes of SO_x and PO_x. See also ESTERS.

BSO200 CAS:70303-47-8 HR: 3
(BUTYLTHIO)TRIOCTYLSTANNANE
mf: $C_{28}H_{60}SSn$ mw: 547.63

SYNS: STANNANE, (BUTYLTHIO)TRIOCTYL- □ TRIOCTYL(BUTYLTHIO)STANNANE

TOXICITY DATA WITH REFERENCE
ipr-mus LD50:389 mg/kg RPTOAN 42,73,79

OSHA PEL: TWA 0.1 mg(Sn)/m^3
ACGIH TLV: TWA 0.1 mg(Sn)/m^3; STEL 0.2 mg/m^3 (skin)
NIOSH REL: (Organotin compound): 10H TWA 0.1 mg(Sn)/m^3

SAFETY PROFILE: Poison by intraperitoneal route.

When heated to decomposition it emits toxic fumes of SO_x and Sn.

For occupational chemical analysis use NIOSH: Organotin Compounds 5504.

BSO500 CAS:1516-32-1 HR: 3
n-BUTYL THIOUREA
mf: $C_5H_{12}N_2S$ mw: 132.25

SYN: USAF D-5

TOXICITY DATA WITH REFERENCE
ipr-mus LD50:300 mg/kg NTIS** AD277-689

CONSENSUS REPORTS: Reported in EPA TSCA Inventory.

SAFETY PROFILE: A poison by intraperitoneal route. When heated to decomposition it emits very toxic fumes of NO_x and SO_x.

BSO750 CAS:25151-00-2 HR: 3
BUTYLTIN TRILAURATE
mf: $C_{40}H_{72}O_6Sn$ mw: 767.81

SYNS: BTT □ n-BUTYLTIN TRICHLORIDE □ BUTYLTIN TRI (DODECANOATE) □ BUTYLTRI(LAUROYLOXY)STANNANE □ MONOBUTYLTIN TRICHLORIDE □ MONOBUTYLTIN TRILAURATE

TOXICITY DATA WITH REFERENCE
orl-mus LD50:325 mg/kg GISAAA 41(5),10,76

OSHA PEL: TWA 0.1 mg(Sn)/m³ (skin)
ACGIH TLV: TWA 0.1 mg(Sn)/m³ (skin) (Proposed: TWA 0.1 mg(Sn)/m³; STEL 0.2 mg(Sn)/m³ (skin))
NIOSH REL: (Organotin Compounds) TWA 0.1 mg(Sn)/m³

SAFETY PROFILE: Poison by ingestion. See also TIN COMPOUNDS. When heated to decomposition it emits acrid smoke and irritating fumes.

For occupational chemical analysis use NIOSH: Organotin Compounds 5504.

BSP000 CAS:73927-88-5 HR: 3
n-BUTYLTIN TRIS(DIBUTYLDITHIOCARBAMATE)
mf: $C_{31}H_{63}N_3S_6Sn$ mw: 789.02

SYN: BUTYLTRIS(DIBUTYLDITHIOCARBAMATO)STANNANE

TOXICITY DATA WITH REFERENCE
ivn-mus LD50:180 mg/kg CSLNX* NX#02083

OSHA PEL: TWA 0.1 mg(Sn)/m³ (skin)
ACGIH TLV: TWA 0.1 mg(Sn)/m³ (skin) (Proposed: TWA 0.1 mg(Sn)/m³; STEL 0.2 mg(Sn)/m³ (skin))
NIOSH REL: (Organotin Compounds) TWA 0.1 mg(Sn)/m³

SAFETY PROFILE: Poison by intravenous route. See also CARBAMATES and TIN COMPOUNDS. When heated to decomposition it emits very toxic fumes of NO_x and SO_x.

For occupational chemical analysis use NIOSH: Organotin Compounds 5504.

BSP250 CAS:5593-70-4 HR: 3
BUTYL TITANATE
mf: $C_{16}H_{36}O_4 \cdot Ti$ mw: 340.42

PROP: Colorless to light yellow liquid or oil with the odor of butanol. Mp: −55°, bp: 155° @ 1 mm, d: 0.993 @ 25°/4°, flash p: 170°F, vap d: 11.5.

SYN: TETRABUTYLTITANATE (CZECH)

TOXICITY DATA WITH REFERENCE
orl-rat LD50:3122 mg/kg MarJV# 29MAR77
ivn-mus LD50:180 mg/kg CSLNX* NX#01650

CONSENSUS REPORTS: Reported in EPA TSCA Inventory.

SAFETY PROFILE: A poison by intravenous route. Moderately toxic by ingestion. See n-BUTYL ALCOHOL and TITANIUM COMPOUNDS. Flammable when exposed to heat or flame. To fight fire, use water, spray, foam, dry chemical. Incompatible with oxidizing materials. When heated to decomposition it emits acrid and irritating fumes.

BSP500 CAS:98-51-1 HR: 2
p-tert-BUTYLTOLUENE
mf: $C_{11}H_{16}$ mw: 148.27

PROP: Colorless liquid. D: 0.861 @ 20°/4°, mp: −54°, bp: 189–192°.

SYNS: p-METHYL-tert-BUTYLBENZENE □ 1-METHYL-4-tert-BUTYLBENZENE □ TBT

TOXICITY DATA WITH REFERENCE
eye-hmn 5 ppm/2H AMIHBC 9,227,54
skn-rbt 500 mg/24H MLD AMIHBC 9,227,54
eye-rbt 100 mg AMIHBC 9,227,54
ihl-hmn TCLo:10 ppm/3M:GIT AMIHBC 9,227,54
ihl-hmn TCLo:20 ppm/5M:EYE,IRR,GIT 28ZRAQ -,156,60
orl-rat LD50:1500 mg/kg AMIHBC 9,227,54
ihl-rat LC50:165 ppm/8H AMIHBC 9,227,54
orl-mus LD50:778 mg/kg AMIHBC 9,227,54
ihl-mus LC50:248 ppm/2H AMIHBC 9,227,54
orl-rbt LD50:2000 mg/kg AMIHBC 9,227,54

CONSENSUS REPORTS: Reported in EPA TSCA Inventory.

OSHA PEL: TWA 10 ppm; STEL 20 ppm
ACGIH TLV: TWA 10 ppm; STEL 20 ppm. (Proposed: TWA 1 ppm)
DFG MAK: 10 ppm (60 mg/m³)

SAFETY PROFILE: Moderately toxic by inhalation and ingestion. A skin and human eye irritant. Human systemic effects by inhalation: nausea or vomiting, conjunctiva irritation, unspecified effects on the sense of taste. Inhalation of vapors causes irritation of lungs and depression of central nervous system. Prolonged exposure may result in damage to liver and kidneys. Flammable when exposed to heat or flame. Incompatible with oxidizing materials. When heated to decomposition it emits acrid smoke and fumes.

For occupational chemical analysis use NIOSH: Hydrocarbons, Aromatic, 1501.

BSP750 CAS:778-28-9 HR: 3
n-BUTYL-p-TOLUENESULFONATE
mf: $C_{11}H_{16}O_3S$ mw: 228.33

SYNS: BUTYL-p-METHYLBENZENESULFONATE □ BUTYL-p-TOLUENESULFONATE □ BUTYL TOSYLATE □ 4-METHYL-BENZENESULFONIC ACID BUTYL ESTER (9CI)

TOXICITY DATA WITH REFERENCE
scu-rat LD50:5000 mg/kg ZEKBAI 74,241,70
ivn-mus LD50:320 mg/kg CSLNX* NX#01764

CONSENSUS REPORTS: Reported in EPA TSCA Inventory.

SAFETY PROFILE: Poison by intravenous route. Mildly toxic by subcutaneous route. See also SULFONATES. When heated to decomposition it emits toxic fumes of SO_x.

BSQ000 CAS:64-77-7 HR: 2
1-BUTYL-3-(p-TOLYL SULFONYL)UREA
mf: $C_{12}H_{18}N_2O_3S$ mw: 270.38

PROP: Crystals. Mp: 128.5–129.5°. Insol in H_2O; sol in $CHCl_3$, dil acids and alkalis.

SYNS: AGLICID □ ARKOZAL □ ARTOSIN □ ARTOZIN □ BUTAMID □ N-((BUTYLAMINO)CARBONYL)-4-METHYLBENZENESULFONAMIDE □ 1-BUTYL-3-(p-METHYLPHENYLSULFONYL)UREA □ n-BUTYL-N'-p-TOLUENESULFONYLUREA □ N-n-BUTYL-N'-TOSYLUREA □ 1-BUTYL-3-TOSYLUREA □ BZ 55 □ D 860 □ DIABEN □ DIABETAMID □ DIABETOL □ DIABUTON □ DOLIPOL □ DRABET □ HLS 831 □ IPOGLICONE □ MOBENOL □ NCI-CO1763 □ ORABET □ ORALIN □ OREZAN □ ORINASE □ ORINAZ □ OTERBEN □ RASTINON □ SK-TOLBUTAMIDE □ N-(SULFONYL-p-METHYLBENZENE)-N'-N-BUTYLUREA □ TOLBUSAL □ TOLBUTAMID □ TOLBUTAMIDE □ 1-p-TOLUENESULFONYL-3-BUTYLUREA □ TOLUINA □ TOLUMID □ TOLUVAN □ N-(p-TOLYLSULFONYL)-N'-BUTYLCARBAMIDE □ 3-(p-TOLYL-4-SULFONYL)-1-BUTYLUREA □ TOLYLSULFONYLBUTYLUREA □ WILLBUTAMIDE

TOXICITY DATA WITH REFERENCE
sce-mus-orl 28,600 μg/kg MUREAV 77,349,80
sce-ham-ipr 28,600 μg/kg MUREAV 77,349,80
orl-wmn TDLo:2 mg/kg (1-13W preg):REP CMAJAX 87,193,62
unr-wmn TDLo:1800 mg/kg (26-39W preg):REP BMJOAE 2,187,64
orl-wmn TDLo:2 mg/kg (1-13W preg):TER CMAJAX 87,193,62
orl-wmn TDLo:2600 mg/kg(1-37W preg):TER JOPDAB 77,457,70
ipr-rat TDLo:1650 mg/kg (female 1-22D post):REP METAAJ 22,1389,73
orl-rat TDLo:70 mg/kg (female 7-13D post):REP KSRNAM 6,1969,72
orl-rat TDLo:12 g/kg (female 1-12D post):TER COREAF 247,1134,58
orl-mus TDLo:1250 mg/kg (female 2D post):TER ARZNAD 36,219,86
orl-rat TDLo:12 g/kg (female 1-12D post):REP COREAF 247,1134,58
orl-rbt TDLo:1800 mg/kg (female 8-16D post):REP TXAPA9 10,244,67
orl-rbt TDLo:1800 mg/kg (female 8-16D post):TER TXAPA9 10,244,67
scu-rat TDLo:5400 mg/kg (female 9-14D post):TER SEIJBO 17,31,77
orl-rat TDLo:8400 mg/kg (male 3D pre):TER TXAPA9 7,409,65
orl-wmn LDLo:1 g/kg:GIT:SYS ATXKA8 23,153,68
orl-rat LD50:2490 mg/kg PMDCAY 1,187,61
ipr-rat LD50:860 mg/kg FRPSAX 12,268,57
ivn-rat LD50:700 mg/kg PMDCAY 1,187,61
orl-mus LD50:490 mg/kg IJCREE 26,81,88
ipr-mus LD50:650 mg/kg TXAPA9 4,631,62
scu-mus LD50:980 mg/kg NATUAS 193,891,62
ivn-mus LD50:770 mg/kg PMDCAY 1,187,61
ipr-mus LD50:700 mg/kg PCJOAU 14,107,80

CONSENSUS REPORTS: NCI Carcinogenesis Bioassay (feed); No Evidence: mouse, rat NCITR* NCI-CG-TR-31,77. Reported in EPA TSCA Inventory. EPA Genetic Toxicology Program.

SAFETY PROFILE: Moderately toxic by ingestion and several other routes. A human teratogen. Human reproductive effects by ingestion and possibly other routes: stillbirth, developmental abnormalities of the cardiovascular (circulatory) system and urogenital system, and unspecified neonatal effects. Human systemic effects by ingestion: nausea or vomiting, hypoglycemia. Other experimental teratogenic and reproductive effects. Mutation data reported. Implicated in aplastic anemia. When heated to decomposition it emits very toxic fumes of NO_x and SO_x.

BSQ500 CAS:4872-26-8 HR: 3
BUTYLTRICHLOROGERMANE
mf: $C_4H_9Cl_3Ge$ mw: 236.07

TOXICITY DATA WITH REFERENCE
ipr-rat LDLo:48 mg/kg CHDDAT 262,1302,66
ipr-mus LD50:190 mg/kg CHDDAT 262,1302,66

SAFETY PROFILE: Poison by intraperitoneal route. When heated to decomposition it emits very toxic fumes of Cl^-. See also GERMANIUM COMPOUNDS and CHLORIDES.

BSQ750 CAS:93-79-8 HR: 1
BUTYL-2,4,5-TRICHLOROPHENOXYACETATE
mf: $C_{12}H_{13}Cl_3O_3$ mw: 311.60

SYNS: ARBORICID □ BUTYL-2,4,5-T □ BUTYLATE-2,4,5-T □ N-BUTYLESTER KYSELINI-2,4,5-TRICHLORFENOXYOCTOVE (CZECH) □ N-BUTYL (2,4,5-TRICHLOROPHENOXY)ACETATE □ FLOMORE □ KILEX 3 □ KRZEWOTOKS □ 2,4,5-T-N-BUTYL ESTER □ TORMONA □ 2,4,5-TRICHLOROPHENOXYACETIC ACID, BUTYL ESTER □ TRIOXONE □ U46KW

TOXICITY DATA WITH REFERENCE
skn-rbt 500 mg/24H MOD 28ZPAK -,85,72
eye-rbt 100 mg/24H MOD 28ZPAK -,85,72
cyt-dmg-orl 250 ppm/24H HEREAY 68,115,71
cyt-rat-orl 10 μg/kg GTPZAB 18(4),24,74
orl-rat TDLo:2 mg/kg (1-20D preg):REP GTPZAB 20(8),5,76
orl-mus TDLo:1246 mg/kg (female 12-15D post):TER AECTCV 6,33,77

orl-rat TDLo: 2 mg/kg (1-20D preg): TER GTPZAB 20(8),5,76
orl-mus TDLo: 748 mg/kg (female 11-13D post): TER AECTCV 6,33,77
orl-rat TDLo: 200 μg/kg (7-8D preg): TER GTPZAB 20(8),5,76

CONSENSUS REPORTS: EPA Genetic Toxicology Program.

SAFETY PROFILE: Moderately toxic by ingestion. Experimental teratogenic and reproductive effects. A skin and eye irritant. Mutation data reported. See also ESTERS. When heated to decomposition it emits toxic fumes of Cl^-.

BSR000 CAS:7521-80-4 HR: 3
BUTYLTRICHLOROSILANE
DOT: UN 1747
mf: $C_4H_9Cl_3Si$ mw: 191.57

PROP: Colorless liquid. Vap d: 6.4, flash p: 130°F (OC), d: 1.16 @ 20°/4°, bp: 148–149°.

SYN: TRICHLOROBUTYLSILANE

CONSENSUS REPORTS: Reported in EPA TSCA Inventory.

DOT CLASSIFICATION: 8; *Label:* Corrosive

SAFETY PROFILE: A corrosive poison. See also CHLOROSILANE. Flammable liquid when exposed to heat, flame (sparks), or oxidizers. To fight fire, use water to blanket fire, fog, mist, dry chemical, alcohol foam. Reacts with water or steam to produce heat and toxic and corrosive fumes. When heated to decomposition it emits highly toxic fumes of Cl^-.

BSR250 CAS:1118-46-3 HR: 2
BUTYL TRICHLORO STANNANE
mf: $C_4H_9Cl_3Sn$ mw: 282.17

PROP: Liquid. D: 0.85 @ 20°/4°, bp: 93° @ 10 mm.

SYN: CHLORID-N-BUTYLCINICITY (CZECH)

TOXICITY DATA WITH REFERENCE
skn-rbt 750 μg/24H SEV 85JCAE-,1245,86
eye-rbt 50 μg/24H SEV 85JCAE-,1245,86
mmo-sat 100 μg/tube MUREAV 300,265,93
uns-bac-esc 5 mg/tube MUREAV 280,195,92
orl-rat LD50: 2140 mg/kg 28ZPAK -,225,72

CONSENSUS REPORTS: Reported in EPA TSCA Inventory.

OSHA PEL: TWA 0.1 mg(Sn)/m³ (skin)
ACGIH TLV: TWA 0.1 mg(Sn)/m³ (skin) (Proposed: TWA 0.1 mg(Sn)/m³; STEL 0.2 mg(Sn)/m³ (skin))
NIOSH REL: (Organotin Compounds) TWA 0.1 mg(Sn)/m³

SAFETY PROFILE: Moderately toxic by ingestion. A severe skin and eye irritant. Mutation data reported. See also TIN COMPOUNDS. When heated to decomposition it emits toxic fumes of Cl^-.
For occupational chemical analysis use NIOSH: Organotin Compounds 5504.

BSR500 CAS:313-94-0 HR: 2
3-tert-BUTYLTRICYCLOQUINAZOLINE
mf: $C_{25}H_{21}N_4$ mw: 377.50

TOXICITY DATA WITH REFERENCE
skn-mus TDLo: 1200 mg/kg/50W-I: CAR BCPCA6 14,323,65

SAFETY PROFILE: Questionable carcinogen with experimental carcinogenic data. When heated to decomposition it emits toxic fumes of NO_x.

BSR600 CAS:54546-26-8 HR: 1
2-BUTYL-4,4,6-TRIMETHYL-1,3-DIOXANE
mf: $C_{11}H_{22}O_2$ mw: 186.33

SYN: 1,3-DIOXANE, 2-BUTYL-4,4,6-TRIMETHYL-

TOXICITY DATA WITH REFERENCE
orl-rat LD50: >5 g/kg FCTOD7 30,15S,92
skn-rbt LD50: >5 g/kg FCTOD7 30,15S,92

CONSENSUS REPORTS: Reported in EPA TSCA Inventory.

SAFETY PROFILE: Low toxicity by ingestion and skin contact. When heated to decomposition it emits acrid smoke and irritating vapors.

BSR825 CAS:73452-32-1 HR: 3
N-tert-BUTYL-N-TRIMETHYLSILYLAMINOBORANE
mf: C_7H_2OBNSi mw: 154.99

$(C_4H_9)N(Si(CH_3)_3)BH_2$

SAFETY PROFILE: Ignites spontaneously on contact with air. When heated to decomposition it emits toxic fumes of NO_x. See also BORANES and BORON COMPOUNDS.

BSR900 CAS:1779-51-7 HR: 3
n-BUTYLTRIPHENYLPHOSPHONIUM BROMIDE
mf: $C_{22}H_{24}P \cdot Br$ mw: 399.34

SYN: PHOSPHONIUM, BUTYLTRIPHENYL-, BROMIDE

TOXICITY DATA WITH REFERENCE
ivn-mus LD50: 56 mg/kg CSLNX* NX#06771

CONSENSUS REPORTS: Reported in EPA TSCA Inventory.

SAFETY PROFILE: Poison by intravenous route. When heated to decomposition it emits toxic vapors of PO_x and Br^-.

BSS000 CAS:25852-70-4 HR: 2
BUTYLTRIS(ISOOCTYLOXYCARBONYLMETHYLTHIO) STANNANE
mf: $C_{34}H_{66}O_6S_3Sn$ mw: 785.87

SYN: BUTYLTRIS(2-ETHYLHEXYLOXYCARBONYLMETHYLTHIO) STANNANE

TOXICITY DATA WITH REFERENCE
orl-rat LD50:1063 mg/kg TRIPA7-,1,73

CONSENSUS REPORTS: Reported in EPA TSCA Inventory.

OSHA PEL: TWA 0.1 mg(Sn)/m³ (skin)
ACGIH TLV: TWA 0.1 mg(Sn)/m³ (skin) (Proposed: TWA 0.1 mg(Sn)/m³; STEL 0.2 mg(Sn)/m³ (skin))
NIOSH REL: (Organotin Compounds) TWA 0.1 mg(Sn)/m³

SAFETY PROFILE: Moderately toxic by ingestion. See also TIN COMPOUNDS. When heated to decomposition it emits toxic fumes of SO_x.

For occupational chemical analysis use NIOSH: Organotin Compounds 5504.

BSS100 CAS:109-42-2 HR: 1
BUTYL 10-UNDECENOATE
mf: $C_{15}H_{28}O_2$ mw: 240.43

SYNS: BUTYL UNDECYLENATE □ 10-UNDECENOIC ACID, BUTYL ESTER

TOXICITY DATA WITH REFERENCE
skn-rbt 500 mg/24H MLD FCTXAV 17,729,79
orl-rat LD50:5000 mg/kg FCTXAV 17,729,79

CONSENSUS REPORTS: Reported in EPA TSCA Inventory.

SAFETY PROFILE: Mildly toxic by ingestion. A skin irritant. When heated to decomposition it emits acrid smoke and irritating fumes.

BSS250 CAS:592-31-4 HR: 2
N-BUTYLUREA
mf: $C_5H_{12}N_2O$ mw: 116.19

PROP: Needles from C_6H_6. Mp: 96°.

SYN: NCI-CO2131

TOXICITY DATA WITH REFERENCE
cyt-rat-orl 100 mg/kg ZKKOBW 86,47,76
cyt-ham:fbr 4 g/L/48H MUREAV 48,337,77
orl-rat LD:>500 mg/kg NCNSA6 5,47,53
par-mus LDLo:1627 mg/kg JPETAB 51,217,34

CONSENSUS REPORTS: Reported in EPA TSCA Inventory. EPA Genetic Toxicology Program.

SAFETY PROFILE: Moderately toxic by parenteral route. Mutation data reported. When heated to decomposition it emits toxic fumes of NO_x.

BSS300 CAS:689-11-2 HR: 2
sec-BUTYLUREA
mf: $C_5H_{12}N_2O$ mw: 116.19

SYN: UREA, sec-BUTYL-

TOXICITY DATA WITH REFERENCE
par-mus LDLo:2789 mg/kg JPETAB 52,216,34

CONSENSUS REPORTS: Reported in EPA TSCA Inventory.

SAFETY PROFILE: Moderately toxic by parenteral route. When heated to decomposition it emits toxic vapors of NO_x.

BSS310 CAS:1118-12-3 HR: 2
tert-BUTYLUREA
mf: $C_5H_{12}N_2O$ mw: 116.19

SYNS: (1,1-DIMETHYLETHYL)UREA □ UREA, tert-BUTYL- □ UREA, (1,1-DIMETHYLETHYL)-(9CI)

TOXICITY DATA WITH REFERENCE
orl-mus LD50:3050 mg/kg AIPTAK 219,103,76

CONSENSUS REPORTS: Reported in EPA TSCA Inventory.

SAFETY PROFILE: Moderately toxic by ingestion. When heated to decomposition it emits toxic vapors of NO_x.

BSS500 HR: 3
1-BUTYLUREA and SODIUM NITRITE (2:1)

TOXICITY DATA WITH REFERENCE
orl-rat TDLo:126 g/kg/33W-C:CAR IJCNAW 23,253,79
orl-rat TDLo:1350 mg/kg (13-21D preg):ETA,TER GANNA2 68,81,77
orl-mus TDLo:380 g/kg/42W-C:CAR IJCNAW 23,253,79
orl-rat TD:170 g/kg/37W-C:CAR IJCNAW 23,253,79
orl-mus TD:421 g/kg/46W-C:CAR IJCNAW 23,253,79
orl-rat TD:27,500 mg/kg/50D-C:NEO ZAPPAN 121,61,77

SAFETY PROFILE: Suspected carcinogen with experimental carcinogenic, neoplastigenic, and tumorigenic data. An experimental teratogen. When heated to decomposition it emits toxic fumes of NO_x. See also NITRITES.

BSS550 CAS:105-77-1 HR: 2
BUTYLXANTHIC DISULFIDE
mf: $C_{10}H_{18}O_2S_4$ mw: 298.52

SYNS: BIS-BUTYLXANTHOGEN □ CPB □ DI(BUTOXYTHIOCARBONYL) DISULFIDE □ DIBUTYL DIXANTHOGEN □ DIBUTYLDIXANTOGENATE □ DIBUTYL XANTHOGEN DISULFIDE □ DITHIOBIS (THIOFORMIC ACID) O,O-DIBUTYL ESTER) □ DXG □ FORMIC ACID, DITHIOBIS(THIO-, O,O-DIBUTYL ESTER) □ THIOPEROXYDICARBONIC ACID, DIBUTYL ESTER

TOXICITY DATA WITH REFERENCE
orl-mus LD50:2700 mg/kg GISAAA 47(3),88,82

CONSENSUS REPORTS: Reported in EPA TSCA Inventory.

SAFETY PROFILE: Moderately toxic by ingestion. When heated to decomposition it emits toxic vapors of SO_x.

BST000 CAS:1879-09-0 HR: 3
6-tert-BUTYL-2,4-XYLENOL
mf: $C_{12}H_{18}O$ mw: 178.30

SYNS: 6-tert-BUTYL-2,4-DIMETHYLPHENOL □ PRODOX 340

TOXICITY DATA WITH REFERENCE
orl-rat LDLo:1400 mg/kg JAPMA8 38,366,49
orl-mus LD50:530 mg/kg JAPMA8 38,366,49

orl-rbt LDLo:55 mg/kg JAPMA8 38,366,49
skn-rbt LDLo:55 mg/kg JAPMA8 38,366,49
orl-gpg LDLo:420 mg/kg JAPMA8 38,366,49
skn-gpg LDLo:7100 mg/kg JAPMA8 38,366,49

CONSENSUS REPORTS: Reported in EPA TSCA Inventory.

SAFETY PROFILE: Poison by skin contact. Moderately toxic by ingestion. When heated to decomposition it emits smoke and acrid, irritating fumes.

BST500 CAS:110-65-6 HR: 3
2-BUTYNE-1,4-DIOL
DOT: UN 2716
mf: $C_4H_6O_2$ mw: 86.10

$HOCH_2C{=}CCH2OH$

PROP: Plates from EtOAc or C_6H_6. Mp: 57–57°, bp: 145° @ 15 mm. Very sol in H_2O, EtOH; sltly sol in $CHCl_3$.

SYN: 1,4-BUTYNEDIOL (DOT)

TOXICITY DATA WITH REFERENCE
orl-rat LD50:104 mg/kg HYSAAV 33,41,68
ihl-rat LCLo:150 mg/m³/2H 85GMAT -,30,82
orl-mus LD50:105 mg/kg HYSAAV 33,41,68
ihl-mus LCLo:150 mg/m³/2H 85GMAT -,30,82
orl-rbt LD50:150 mg/kg HYSAAV 33,41,68
orl-gpg LD50:130 mg/kg HYSAAV 33,41,68
orl-bwd LD50:75 mg/kg AECTCV 12,355,83

CONSENSUS REPORTS: Reported in EPA TSCA Inventory.

DOT CLASSIFICATION: 6.1; *Label:* KEEP AWAY FROM FOOD

SAFETY PROFILE: A poison by ingestion. A skin sensitizer upon long or repeated contact. Moderately explosive. When heated to decomposition it emits acrid smoke and fumes and may explode. Explosive reaction with traces of alkalies, alkali earth hydroxides, halide salts, strong acids, mercury salts + strong acids. See also ACETYLENE COMPOUNDS.

BST750 HR: 3
2-BUTYNE-1-THIOL
mf: C_4H_6S mw: 86.16

SAFETY PROFILE: Forms an explosive polymer on exposure to air. Store at −20° in the presence of a stabilizer under nitrogen. When heated to decomposition it emits toxic fumes of SO_x. See also ACETYLENE COMPOUNDS.

BSU000 HR: 3
3-BUTYN-1-YL-p-TOLUENE SULFONATE
mf: $C_{11}H_{12}O_3S$ mw: 224.18

SAFETY PROFILE: Explodes in vacuum at 0.65 mbar. May be safe in small amounts below 0.01 mbar. When heated to decomposition it emits toxic fumes of SO_x. See also SULFONATES and ACETYLENE COMPOUNDS.

BSU250 CAS:123-72-8 HR: 3
n-BUTYRALDEHYDE
DOT: UN 1129
mf: C_4H_8O mw: 72.12

PROP: Colorless, mobile liquid; pungent, nutty odor. Mp: −100°, bp: 74.7°, flash p: 20°F (CC), (−6°), d: 0.7988 @ 25°, autoign temp: 446°F, lel: 2.5%, uel: 12.5%, vap d: 2.5. Sol in water; misc with ether @ 74.8°.

SYNS: ALDEHYDE BUTYRIQUE (FRENCH) □ ALDEIDE BUTIRRICA (ITALIAN) □ BUTAL □ BUTALDEHYDE □ BUTALYDE □ BUTANAL □ n-BUTANAL (CZECH) □ n-BUTYL ALDEHYDE □ BUTYRAL □ BUTYRALDEHYD (GERMAN) □ BUTYRALDEHYDE (CZECH) □ BUTYRIC ALDEHYDE □ FEMA No. 2219 □ NCI-C56291

TOXICITY DATA WITH REFERENCE
skn-rbt 2 mg/24H SEV 85JCAE-,270,86
skn-rbt 500 mg/24H SEV 28ZPAK -,40,72
eye-rbt 20 mg/24H MOD 85JCAE-,270,86
eye-rbt 20 mg/24H MOD 28ZPAK -,40,72
skn-gpg 100% MOD FCTXAV 17,731,79
spm-mus-ipr 30 mg/kg MUREAV 39,317,77
spm-mus-orl 15 g/kg/50D MUREAV 39,317,77
ihl-hmn TCLo:580 mg/m³:IMM BMJOAE 2,913,56
orl-rat LD50:2490 mg/kg 28ZPAK -,40,72
ihl-rat LCLo:8000 ppm/4H AMIHBC 4,119,51
ipr-rat LD50:800 mg/kg FCTXAV 17,731,79
scu-rat LDLo:10 g/kg ARZNAD 11,73,61
ihl-mus LC50:44,610 mg/m³/2H 85GMAT -,30,82
ipr-mus LD50:1140 mg/kg FCTXAV 17,731,79
scu-mus LD50:2700 mg/kg APTOA6 6,299,50
skn-rbt LD50:3560 mg/kg UCDS** 7/20/67
ihl-mam LC50:64 g/m³ GTPZAB 12(7),16,68

CONSENSUS REPORTS: Community Right-To-Know List. Reported in EPA TSCA Inventory.

DOT CLASSIFICATION: 3; *Label:* Flammable Liquid

SAFETY PROFILE: Moderately toxic by ingestion, inhalation, skin contact, intraperitoneal, and subcutaneous routes. Severe skin and eye irritant. Human immunological effects by inhalation: delayed hypersensitivity. See also ALDEHYDES. Highly flammable liquid. To fight fire, use foam, CO_2, dry chemical. Incompatible with oxidizing materials. Reacts vigorously with chlorosulfonic acid, HNO_3, oleum, H_2SO_4. When heated to decomposition it emits acrid smoke and fumes.

BSU500 CAS:110-69-0 HR: 3
n-BUTYRALDEHYDE OXIME
DOT: UN 2840
mf: C_4H_9NO mw: 87.14

PROP: Liquid. Mp: −29.5°, bp: 152°, flash p: 136°F (CC), d: 0.923, vap d: 3.01.

SYNS: BUTANAL OXIME □ BUTYRALDOXIME (DOT) □ N-BUTYRALDOXIME □ SKINO #1 □ TROYKYD ANTI-SKIN BTO □ USAF AM-6

TOXICITY DATA WITH REFERENCE
msc-mus:lyms 1700 mg/L MUREAV 204,149,88
ipr-mus LD50:200 mg/kg NTIS** AD277-689

CONSENSUS REPORTS: Reported in EPA TSCA Inventory.

DOT CLASSIFICATION: 3; *Label:* Flammable Liquid

SAFETY PROFILE: A poison by intraperitoneal route. Mutation data reported. Flammable liquid when exposed to heat or flame. To fight fire, use alcohol foam, dry chemical. Highly explosive. Can explode during vacuum distillation. Incompatible with oxidizing materials, metallic impurities. When heated to decomposition it emits toxic fumes of NO_x. See also ALDEHYDES.

BSV250 CAS:29067-70-7 HR: 3
2-(3-BUTYRAMIDO-2,4,6-TRIIODOPHENYL)PROPIONIC ACID
mf: $C_{13}H_{14}I_3NO_3$ mw: 612.98

TOXICITY DATA WITH REFERENCE
orl-mus LD50:100 mg/kg JMCMAR 13,559,70
ivn-mus LD50:300 mg/kg JMCMAR 13,559,70

SAFETY PROFILE: Poison by ingestion and intravenous routes. When heated to decomposition it emits very toxic fumes of NO_x and I^-.

BSV500 CAS:1129-50-6 HR: 2
n-BUTYRANILIDE
mf: $C_{10}H_{13}NO$ mw: 163.24

PROP: Solid. Mp: 96°.

TOXICITY DATA WITH REFERENCE
orl-mus LD50:1630 mg/kg TXAPA9 19,20,71

CONSENSUS REPORTS: Reported in EPA TSCA Inventory.

SAFETY PROFILE: Moderately toxic by ingestion. When heated to decomposition it emits toxic fumes of NO_x.

BSV750 CAS:2440-29-1 HR: 3
(BUTYRATO)PHENYLMERCURY
mf: $C_{10}H_{12}HgO_2$ mw: 364.81

SYN: PHENYL(BUTYRATE)MERCURY

CONSENSUS REPORTS: Mercury and its compounds are on the Community Right-To-Know List.

OSHA PEL: CL 0.1 mg(Hg)/m³ (skin)
ACGIH TLV: TWA 0.1 mg(Hg)/m³ (skin)
NIOSH REL: (Mercury, Aryl And Inorganic): CL 0.1 mg/m³ (skin)

SAFETY PROFILE: Probably a poison. See also MERCURY COMPOUNDS. When heated to decomposition it emits toxic Hg vapors.

BSW000 CAS:107-92-6 HR: 2
n-BUTYRIC ACID
DOT: UN 2820
mf: $C_4H_8O_2$ mw: 88.12

PROP: Colorless liquid; strong, rancid-butter odor. Mp: −7.9°, bp: 163.5°, flash p: 161°F, d: 0.9590 @ 20°/20°, refr index: 1.397, autoign temp: 846°F, vap press: 0.43 mm @ 20°, vap d: 3.04, lel: 2.0%, uel: 10.0%. Misc in H_2O, EtOH, Et_2O.

SYNS: BUTANOIC ACID □ BUTTERSAEURE (GERMAN) □ ETHYLACETIC ACID □ FEMA No. 2221 □ 1-PROPANECARBOXYLIC ACID □ PROPYLFORMIC ACID

TOXICITY DATA WITH REFERENCE
skn-rbt 10 mg/24H open SEV AMIHBC 10,61,54
skn-rbt 20 mg/24H MOD 85JCAE-,306,86
eye-rbt 250 µg open SEV AMIHBC 10,61,54
dnd-hmn:hla 3 mmol/L CELLB5 12,855,77
dni-hmn:lym 4 mmol/L HAONDL 2,381,84
orl-rat LD50:2 g/kg 85GMAT-,30,82
orl-mus LDLo:500 mg/kg TPKVAL 4,19,62
ipr-mus LD50:3180 mg/kg JPPMAB 21,85,69
scu-mus LD50:3180 mg/kg JPPMAB 21,85,69
ivn-mus LD50:800 mg/kg APTOA6 18,141,61
skn-rbt LD50:530 mg/kg UCDS** 4/10/68

CONSENSUS REPORTS: Reported in EPA TSCA Inventory.

DOT CLASSIFICATION: 8; *Label:* Corrosive

SAFETY PROFILE: Moderately toxic by ingestion, skin contact, subcutaneous, intraperitoneal, and intravenous routes. Human mutation data reported. Severe skin and eye irritant. A corrosive material. Combustible liquid. Could react with oxidizing materials. Incandescent reaction with chromium trioxide above 100°. To fight fire, use alcohol foam, CO_2, dry chemical. When heated to decomposition it emits acrid smoke and irritating fumes.

BSW500 CAS:539-90-2 HR: 1
BUTYRIC ACID ISOBUTYL ESTER
mf: $C_8H_{16}O_2$ mw: 144.24

PROP: Colorless liquid; apple-pineapple odor. D: 0.858–0863, refr index: 1.402. Sol in alc, fixed oils; sltly sol in water; insol in glycerin.

SYNS: FEMA No. 2187 □ ISOBUTYL BUTANOATE □ ISOBUTYL BUTYRATE (FCC) □ 2-METHYLPROPYL BUTYRATE

TOXICITY DATA WITH REFERENCE
skn-rbt 500 mg/24H MLD FCTXAV 17,833,79
orl-rbt LD50:9520 mg/kg IMSUAI 41,31,72
idu-rbt LD50:9500 mg/kg FCTXAV 17(Suppl),695,79

CONSENSUS REPORTS: Reported in EPA TSCA Inventory.

SAFETY PROFILE: Mildly toxic by ingestion and intraduodenal routes. A skin irritant. See also ESTERS. When heated to decomposition it emits acrid smoke and irritating fumes.

BSW550 CAS:106-31-0 HR: 1
BUTYRIC ANHYDRIDE
DOT: UN 2739
mf: $C_8H_{14}O_3$ mw: 158.22

SYNS: ANHYDRID KYSELINY MASELNE □ BUTANOIC ACID, ANHYDRIDE (9CI) □ BUTANOIC ANHYDRIDE □ BUTYRANHYDRID □ BUTYRIC ACID ANHYDRIDE □ n-BUTYRIC ACID ANHYDRIDE □ n-BUTYRIC ANHYDRIDE □ BUTYRYL OXIDE

TOXICITY DATA WITH REFERENCE
orl-rat LD50:8790 mg/kg 85JCAE -,321,86
orl-mus LD30: 2 g/kg 85GMAT -,31,82

CONSENSUS REPORTS: Reported in EPA TSCA Inventory.

DOT CLASSIFICATION: 8; *Label:* Corrosive

SAFETY PROFILE: Mildly toxic by ingestion. A corrosive liquid. When heated to decomposition it emits acrid smoke and irritating vapors.

BSX000 CAS:3068-88-0 HR: 3
β-BUTYROLACTONE
mf: $C_4H_6O_2$ mw: 86.10

SYNS: 3-HYDROXYBUTANOIC ACID-β-LACTONE □ HYDROXYBUTYRIC ACID LACTONE □ 3-HYDROXYBUTYRIC ACID LACTONE □ 4-METHYL-2-OXETANONE

TOXICITY DATA WITH REFERENCE
skn-rbt 500 mg open MOD UCDS** 1/20/66
dnd-mam:lym 10 mmol/L BBACAQ 138,611,67
oms-mam:lym 286 nmol/L CBINA8 34,323,81
orl-rat TDLo:31 g/kg/61W-I:ETA JNCIAM 37,825,66
scu-rat TDLo:38 g/kg/78W-I:CAR JNCIAM 39,1213,67
skn-mus TDLo:59 g/kg/49W-I:CAR JNCIAM 35,707,65
scu-mus TDLo:12 g/kg/30W-I:NEO JNCIAM 37,825,66
skn-mus TD:80 g/kg/67W-I:CAR 14JTAF -,275,64
skn-mus TD:43 g/kg/36W-I:ETA JNCIAM 39,1217,67
orl-rat LD50:17,200 µL/kg AIHAAP 30,470,69

CONSENSUS REPORTS: IARC Cancer Review: Group 2B IMEMDT 7,56,87; Animal Sufficient Evidence IMEMDT 11,225,76. Reported in EPA TSCA Inventory.

SAFETY PROFILE: Suspected carcinogen with experimental carcinogenic, neoplastigenic, and tumorigenic data. Mildly toxic by ingestion. A moderate skin irritant. Mutation data reported. When heated to decomposition it emits acrid and irritating fumes. See also 4-BUTYROLACTONE.

BSX250 CAS:109-74-0 HR: 3
BUTYRONITRILE
DOT: UN 2411
mf: C_4H_7N mw: 69.12

PROP: Colorless liquid. D: 0.796 @ 15°, mp: −112.6°, bp: 117°, flash p: 79°F (OC). Sltly sol in water; sol in alc and ether.

SYNS: BUTANENITRILE □ n-BUTANENITRILE □ BUTYRIC ACID NITRILE □ BUTYRONITRILE (DOT) □ 1-CYANOPROPANE □ PROPYL CYANIDE

TOXICITY DATA WITH REFERENCE
eye-rbt 500 mg/24H MLD 85JCAE-,900,86
ihl-rat TCLo:200 ppm/6H (female 6-20D post):REP FAATDF 20,365,93
orl-rat LD50:50 mg/kg 38MKAJ 2C,4873,82
ipr-rat LD50:50 mg/kg 38MKAJ 2C,4873,82
skn-rbt 395 mg open MLD UCDS** 5/17/60
orl-rat LD50:140 mg/kg AIHAAP 23,95,62
ihl-rat LCLo:1000 ppm/4H AIHAAP 23,95,62
orl-mus LD50:27,689 µg/kg NEZAAQ 39,123,84
ihl-mus LC50:249 ppm/1H CTOXAO 18,991,81
ipr-mus LD50:38 mg/kg TXAPA9 59,589,81
skn-rbt LD50:500 mg/kg AIHAAP 23,95,62
scu-rbt LDLo:10 mg/kg AIPTAK 5,161,1899
ivn-rbt LDLo:980 mg/kg COREAF 153,895,11
skn-gpg LDLo:100 mg/kg KODAK* 21MAY71
scu-gpg LDLo:100 mg/kg COREAF 153,895,11
scu-frg LDLo:3100 mg/kg AIPTAK 5,161,1899

CONSENSUS REPORTS: Reported in EPA TSCA Inventory. Cyanide and its compounds are on the Community Right-To-Know List.

NIOSH REL: (Nitriles) TWA 22 mg/m³
DOT CLASSIFICATION: 3; *Label:* Flammable Liquid, Poison

SAFETY PROFILE: A poison by ingestion, skin contact, intraperitoneal, and subcutaneous routes. Moderately toxic by inhalation. Experimental reproductive data. A skin irritant. Dangerous fire hazard when exposed to heat, flame, or oxidizers. To fight fire, use alcohol foam. When heated to decomposition it emits toxic fumes of NO_x and CN^-.

BSX325 CAS:34291-02-6 HR: 3
BUTYROSIN A
mf: $C_{21}H_{41}N_5O_{12}$ mw: 555.67

PROP: Amorphous solid with a broad melting point.

SYNS: AMBUTYROSIN A □ AMBUYROSIN A □ BUTIROSIN A □ o-2,6-DIAMINO-2,6-DIDEOXY-α-d-GLUCOPYRANOSYL-(1-4)-o-(β-d-XYLOFURANOSYL-(1-5))-N¹-(4-AMINO-2-HYDROXY-1-OXOBUTYL)-2-DEOXY-d-STREPTAMINE

TOXICITY DATA WITH REFERENCE
ipr-mus LD50:2198 mg/kg 85GDA2 1,145,80
scu-mus LD50:3050 mg/kg 85GDA2 1,145,80
ivn-mus LD50:50 mg/kg 38KLAC -,239,77

SAFETY PROFILE: Poison by intravenous route. Moderately toxic by subcutaneous and intraperitoneal routes. When heated to decomposition it emits toxic fumes of NO_x. See also AMINES.

BSX500 CAS:67557-56-6 HR: 2
N-(1-BUTYROXYMETHYL)METHYLNITROSAMINE
mf: $C_6H_{12}N_2O_3$ mw: 160.20

SYNS: N-(1-BUTYROXYMETHYL)-N-NITROSOMETHYLAMINE □ N-NITROSO-N-(1-BUTYROXYMETHYL)METHYL AMINE

TOXICITY DATA WITH REFERENCE
mmo-sat 1 µmol/plate ARTODN 39,51,77
orl-rat TDLo:60 mg/kg/90D-I:ETA ZKKOBW 91,317,78
orl-rat LD50:800 mg/kg ZKKOBW 91,317,78

SAFETY PROFILE: Moderately toxic by ingestion. Questionable carcinogen with experimental tumorigenic data. Mutation data reported. When heated to decomposition it emits toxic fumes of NO_x. See also NITROSAMINES.

BSX750 CAS:37415-56-8 HR: 2
12-o-BUTYROYL-PHORBOLDODECANOATE
mf: $C_{36}H_{57}O_8$ mw: 617.93

SYN: PHORBOL-12-o-BUTYROYL-13-DODECANOATE

TOXICITY DATA WITH REFERENCE
skn-mus 3 ng MLD 85CVA2 5,213,70

C

CAB125 CAS:20064-38-4 HR: 3
C-666
mf: $C_{18}H_{20}N_4O_3 \cdot 2ClH$ mw: 413.34

SYNS: N,N-DIMETHYL-N'-(1-NITRO-9-ACRIDINYL)-1,3-PROPANEDIAMINE-N-OXIDE, DIHYDROCHLORIDE (9CI) □ 1-NITRO-9-(3-DIMETHYLAMINOPROPYLAMINE)ACRIDINE-N^{10}-OXIDE DIHYDROCHLORIDE □ N^{10}-OXIDE-1-NITRO-9-(3-DIMETHYLAMINOPROPYLAMINO)-DIHYDROCHLORIDE ACRIDINE

TOXICITY DATA WITH REFERENCE
orl-rat LD50:110 mg/kg MMDPA6 8,252,76
ivn-rat LD50:5400 µg/kg MMDPA6 8,252,76
orl-mus LD50:108 mg/kg MMDPA6 8,252,76
ivn-mus LD50:9 mg/kg MMDPA6 8,252,76
ivn-pgn LD50:9000 µg/kg AITEAT 28,777,80

SAFETY PROFILE: Poison by ingestion and intravenous routes. When heated to decomposition it emits toxic fumes of NO_x and HCl. See also AMINES.

CAB250 CAS:78265-91-5 HR: 3
C 3206
mf: $C_{17}H_{25}ClN_2O_2 \cdot ClH$ mw: 361.35

SYN: 6'-CHLORO-2-PYRROLIDINYL-o-HEXANOTOLUIDIDE HYDROCHLORIDE

TOXICITY DATA WITH REFERENCE
eye-rbt 2% MOD ARZNAD 8,544,58
ipr-rat LD50:42 mg/kg ARZNAD 8,544,58
ipr-mus LD50:37 mg/kg ARZNAD 8,544,58
scu-mus LD50:70 mg/kg ARZNAD 8,544,58

SAFETY PROFILE: Poison by intraperitoneal and subcutaneous routes. An eye irritant. When heated to decomposition it emits very toxic fumes of $Cl^-$and NO_x.

CAB500 CAS:78265-89-1 HR: 3
C 3207
mf: $C_{17}H_{27}ClN_2O \cdot ClH$ mw: 347.37

SYN: 2-(BUTYLAMINO)-6'-CHLORO-o-HEXANOTOLUIDIDE HYDROCHLORIDE

TOXICITY DATA WITH REFERENCE
eye-rbt 2% SEV ARZNAD 8,544,58
ipr-rat LD50:33 mg/kg ARZNAD 8,544,58
scu-mus LD50:210 mg/kg ARZNAD 8,544,58

SAFETY PROFILE: Poison by intraperitoneal and subcutaneous routes. A severe eye irritant. When heated to decomposition it emits very toxic fumes of Cl^- and NO_x.

CAB750 CAS:78265-90-4 HR: 3
C 3208
mf: $C_{15}H_{23}ClN_2O \cdot ClH$ mw: 319.31

SYN: 6'-CHLORO-2-(ETHYLAMINO)-o-HEXANOTOLUIDIDE HYDROCHLORIDE

TOXICITY DATA WITH REFERENCE
eye-rbt 2% MOD ARZNAD 8,544,58
ipr-rat LD50:25 mg/kg ARZNAD 8,544,58
scu-mus LD50:62 mg/kg ARZNAD 8,544,58

SAFETY PROFILE: Poison by intraperitoneal and subcutaneous routes. An eye irritant. When heated to decomposition it emits very toxic fumes of $Cl^-$and NO_x.

CAB800 CAS:68188-03-4 HR: 1
CABREUVA OIL

TOXICITY DATA WITH REFERENCE
skn-rbt 500 mg/24H MOD FCTOD7 20,645,82
orl-rat LD50:>5 g/kg FCTOD7 20,645,82
skn-rbt LD50:>5 g/kg FCTOD7 20,645,82

CONSENSUS REPORTS: Reported in EPA TSCA Inventory.

SAFETY PROFILE: Low toxicity by ingestion and skin contact. A skin irritant. When heated to decomposition it emits acrid smoke and irritating vapors.

CAC250 HR: 3
CACODYL SULFIDE
mf: $((CH_3)_2As)_2S$ mw: 242

PROP: Oily liquid. Bp: 211°. Sltly sol in water.

SYN: DICACODYL SULFIDE

CONSENSUS REPORTS: Arsenic and its compounds are on the Community Right-To-Know List.

SAFETY PROFILE: Poison by most routes. See also ARSENIC COMPOUNDS and SULFIDES. Dangerous fire hazard when exposed to heat or by spontaneous chemical reaction, i.e., in air. Vigorous reaction with oxidizing materials. When heated to decomposition it emits toxic fumes of As.

CAC500 HR: 2
CADIA DEL PERRO

PROP: Aqueous extract from the dried leaves of the plant (JNCIAM 46,1131,71).

SYNS: K. IXINA □ KRAMERIA IXINA

TOXICITY DATA WITH REFERENCE
scu-rat TDLo:300 mg/kg/1Y-I:NEO JNCIAM 46,1131,71

ims-rat TDLo:45 g/kg/1Y-I:ETA JNCIAM 46,1131,71
skn-ham TDLo:53,950 mg/kg/65W-I:CAR JNCIAM 53,1259,74
scu-rat TD:990 mg/kg/55W-I:NEO JNCIAM 52,1579,74

SAFETY PROFILE: Questionable carcinogen with experimental carcinogenic, tumorigenic, and neoplastigenic data. When heated to decomposition it emits acrid smoke and fumes.

CAD000 CAS:7440-43-9 HR: 3
CADMIUM
mf: Cd mw: 112.40

PROP: Hexagonal, ductile, crystals or soft, silver-white, lustrous, malleable metal. Tarnishes in air, particularly moist air. Mp: 321°, bp: 767°, d: 8.642, vap press: 1 mm @ 394°. Sol dil acids (H_2 evolved).

SYNS: C.I. 77180 □ COLLOIDAL CADMIUM □ KADMIUM (GERMAN)

TOXICITY DATA WITH REFERENCE
mnt-mus:emb 6 μmol/L TXCYAC 4,57,90
cyt-ham:ovr 1 μmol/L CGCGBR 26,251,80
orl-rat TDLo:155 mg/kg (male 13W pre):REP BECTA6 20,96,78
scu-rat TDLo:250 μg,/kg (female 19D post):REP APTOD9 19,A122,80
orl-rat TDLo:21,500 μg/kg (multi):TER ENVRAL 22,466,80
orl-mus TDLo:1700 mg/kg (female 8–12D post):REP TCMUD8 6,361,86
orl-mus TDLo:448 mg/kg (multi):TER AEHLAU 23,102,71
ivn-rat TDLo:8 mg/kg (female 8–15D post):TER JJATDK 1,264,81
orl-rat TDLo:220 mg/kg (female 1–22D post):TER TOLED5 11,233,82
orl-rat TDLo:21,500 μg/kg (multi):REP ENVRAL 22,466,80
ipr-rat TDLo:1124 μg/kg (male 1D pre):REP TXAPA9 41,194,77
orl-rat TDLo:23 mg/kg (female 1–22D post):TER PSEBAA 158,614,78
ivn-rat TDLo:1250 μg/kg (female 9D post):TER JJATDK 1,264,81
par-ham TDLo:2 mg/kg (female 8D post):TER TJADAB 31,52A,85
ihl-wmn TDLo:129 μg/m³/20Y-C:CAR AJIM08 10,153,86
scu-rat TDLo:3372 μg/kg:CAR ENVRAL 55,40,91
ims-rat TDLo:40 mg/kg/4W-I:CAR JEPTDQ 1(1),51,77
ims-rat TD:70 mg/kg:ETA BJCAAI 18,124,64
ims-rat TD:63 mg/kg:ETA NATUAS 193,592,62
ims-rat TD:45 mg/kg/4W-I:NEO NCIUS* PH-43-64-886,SEPT,71
ihl-man TCLo:88 μg/m³/8.6Y:KID AEHLAU 28,147,74
ihl-hmn LCLo:39 mg/m³/20M AIHAAP 31,180,70
unk-man LDLo:15 mg/kg 85DCAI 2,73,70
orl-rat LD50:225 mg/kg TXAPA9 41,667,77
ihl-rat LC50:25 mg/m³/30M SAIGBL 16,212,74
orl-mus LD50:890 mg/kg 41HTAH-,14,78
ihl-mus LCLo:170 mg/m³ NTIS** PB158-508
ipr-mus LD50:5700 μg/kg TXAPA9 37,403,76
unr-mus LD50:890 mg/kg GTPZAB 22(5),6,78
orl-rbt LDLo:70 mg/kg AMPMAR 34,127,73
scu-rbt LDLo:6 mg/kg PROTA*-,-,55
ivn-rbt LDLo:5 mg/kg JOGBAS 35,693,28

CONSENSUS REPORTS: NTP 7th Annual Report on Carcinogens. IARC Cancer Review: Group 1 IMEMDT 58,119,93; Animal Sufficient Evidence IMEMDT 2,74,73; Animal Sufficient Evidence IMEMDT 11,39,76; Human Sufficient Evidence IMEMDT 58,119,93; Human Limited Evidence IMEMDT 7,139,87; Animal Limited Evidence IMEMDT 58,119,93. Cadmium and its compounds are on the Community Right-To-Know List. Reported in EPA TSCA Inventory. EPA Genetic Toxicology Program.

OSHA PEL: TWA 5 μg(Cd)/m³
ACGIH TLV: Dust and Salts: TWA 0.05 mg(Cd)/m³ (Proposed: TWA 0.01 mg(Cd)/m³ (dust), Suspected Human Carcinogen; 0.002 mg(Cd)/m³ (respirable dust), Suspected Human Carcinogen); BEI: 10 μg/g creatinine in urine; 10 μg/L in blood. (Proposed: 5 μg/g creatinine in urine; 5 μg/L in blood)
DFG BAT: Blood 1.5 μg/dL; Urine 15 μg/dL. MAK: Suspected Carcinogen
NIOSH REL: (Cadmium) Reduce to lowest feasible level

SAFETY PROFILE: Confirmed human carcinogen with experimental carcinogenic, tumorigenic, and neoplastigenic data. A human poison by inhalation and possibly other routes. Poison experimentally by ingestion, inhalation, intraperitoneal, subcutaneous, and intravenous routes. In humans inhalation causes an excess of protein in the urine. Experimental teratogenic and reproductive effects. Mutation data reported. The dust ignites spontaneously in air and is flammable and explosive when exposed to heat, flame, or by chemical reaction with oxidizing agents, metals, HN_3, Zn, Se, and Te. Explodes on contact with hydrazoic acid. Violent or explosive reaction when heated with ammonium nitrate. Vigorous reaction when heated with nitryl fluoride. When heated to a high temperature it emits toxic fumes of Cd. See also CADMIUM COMPOUNDS.

For occupational chemical analysis use OSHA: #ID-125G or NIOSH: Cadmium, 7048; welding and Brazing Fume, 7200; elements, 7300.

CAD250 CAS:543-90-8 HR: 3
CADMIUM(II) ACETATE
mf: $C_2H_4O_2 \cdot 1/2Cd$ mw: 116.25

PROP: Monoclinic, colorless crystals; odor of acetic acid. Mp: 256°, bp: decomp, d: 2.341.

SYNS: ACETIC ACID, CADMIUM SALT □ BIS(ACETOXY)CADMIUM □ CADMIUM ACETATE (DOT) □ CADMIUM DIACETATE □ C.I. 77185

TOXICITY DATA WITH REFERENCE
cyt-hmn:lym 10 nmol/L MUREAV 85,236,81
otr-ham:emb 1 μmol/L CNREA8 39,193,79
dnd-ham:emb 1 μmol/L CNREA8 39,193,79
ipr-rat TDLo:2 mg/kg (female 20D post):REP BECTA6 23,25,79
ipr-rat TDLo:1 mg/kg (14D preg):TER BECTA6 23,25,79
ipr-rat TDLo:2371 μg/kg (12D preg):REP BECTA6 20,206,78
scu-rat TDLo:2325 ng/kg (female 1D pre):REP JRPFA4 17,559,68

ipr-rat TDLo: 2371 μg/kg (12D preg): TER BECTA6 20,206,78
ipr-mus LD50: 14 mg/kg TXAPA9 49,41,79

CONSENSUS REPORTS: IARC Cancer Review: Group 1 IMEMDT 58,119,93; Human Sufficient Evidence IMEMDT 58,119,93; Animal Sufficient Evidence IMEMDT 58,119,93. Reported in EPA TSCA Inventory. EPA Genetic Toxicology Program. Cadmium and its compounds are on the Community Right-To-Know List.

OSHA PEL: TWA 5 μg(Cd)/m³
ACGIH TLV: TWA 0.05 mg(Cd)/m³ (Proposed: TWA 0.01 mg(Cd)/m³ (dust), Suspected Human Carcinogen; 0.002 mg(Cd)/m³ (respirable dust), Suspected Human Carcinogen); BEI: 10 μg/g creatinine in urine; 10 μg/L in blood
NIOSH REL: (Cadmium) Reduce to lowest feasible level

SAFETY PROFILE: Confirmed human carcinogen. Poison by intraperitoneal route. An experimental teratogen. Other experimental reproductive effects. Human mutation data reported. When heated to decomposition it emits toxic fumes of Cd. See also CADMIUM COMPOUNDS.

CAD275 CAS:5743-04-4 HR: 3
CADMIUM ACETATE DIHYDRATE
mf: $C_4H_6O_4 \cdot Cd \cdot 2H_2O$ mw: 266.54

PROP: Crystals, becoming anhydrous at 130°; slt acetic acid odor. D: 2.01, 2.341 (anhydrous), mp: 255° (anhydrous). Sol in water and alc.

SYNS: ACETIC ACID, CADMIUM SALT, DIHYDRATE □ CADMIUM DIACETATE DIHYDRATE

TOXICITY DATA WITH REFERENCE
cyt-hmn:lyms 1 mg/L CYGEDX 12(3),46,78

OSHA PEL: TWA 5 μg(Cd)/m³
ACGIH TLV: TWA 0.01 mg(Cd)/m³; Suspected Carcinogen

SAFETY PROFILE: Confirmed human carcinogen. Mutation data reported. When heated to decomposition it emits toxic fumes of Cd.

CAD325 CAS:22750-53-4 HR: 3
CADMIUM AMIDE
mf: CdH_4N_2 mw: 144.46

PROP: White solid, which turns brown in air.

SYN: CADMIUM DIAMIDE

CONSENSUS REPORTS: Cadmium compounds are on the Community Right-To-Know List.

OSHA PEL: TWA 5 μg(Cd)/m³
ACGIH TLV: TWA 0.05 mg(Cd)/m³ (Proposed: TWA 0.01 mg(Cd)/m³ (dust), Suspected Human Carcinogen; 0.002 mg(Cd)/m³ (respirable dust), Suspected Human Carcinogen); BEI: 10 μg/g creatinine in urine; 10 μg/L in blood
NIOSH REL: (Cadmium) Reduce to lowest feasible level

SAFETY PROFILE: Confirmed human carcinogen. May explode if heated. Reacts violently with water. When heated to decomposition it emits toxic fumes of Cd and NO_x. See also CADMIUM COMPOUNDS and AMIDES.

CAD350 CAS:14215-29-3 HR: 3
CADMIUM AZIDE
mf: CdN_6 mw: 196.45

$Cd(N_3)_2$

PROP: White crystals.

SYN: CADMIUM DIAZIDE

CONSENSUS REPORTS: Cadmium compounds are on the Community Right-To-Know List.

OSHA PEL: TWA 5 μg(Cd)/m³
ACGIH TLV: TWA 0.05 mg(Cd)/m³ (Proposed: TWA 0.01 mg(Cd)/m³ (dust), Suspected Human Carcinogen; 0.002 mg(Cd)/m³ (respirable dust), Suspected Human Carcinogen); BEI: 10 μg/g creatinine in urine; 10 μg/L in blood
NIOSH REL: (Cadmium) Reduce to lowest feasible level

SAFETY PROFILE: Confirmed human carcinogen. The dry solid is an unstable heat- and friction-sensitive explosive. When heated to decomposition it emits toxic fumes of NO_x and Cd. See also CADMIUM COMPOUNDS and AZIDES.

CAD500 CAS:7495-93-4 HR: 3
CADMIUM BIS(2-ETHYLHEXYL) PHOSPHITE
mf: $C_{32}H_{68}O_6P_2 \cdot Cd$ mw: 723.34

SYN: BIS(2-ETHYLHEXYL) ESTER PHOSPHOROUS ACID CADMIUM SALT

TOXICITY DATA WITH REFERENCE
ipr-mus LDLo: 250 mg/kg CBCCT* 7,790,55

CONSENSUS REPORTS: Cadmium and its compounds are on the Community Right-To-Know List.

OSHA PEL: TWA 5 μg(Cd)/m³
ACGIH TLV: TWA 0.05 mg(Cd)/m³ (Proposed: TWA 0.01 mg(Cd)/m³ (dust), Suspected Human Carcinogen; 0.002 mg(Cd)/m³ (respirable dust), Suspected Human Carcinogen); BEI: 10 μg/g creatinine in urine; 10 μg/L in blood
NIOSH REL: (Cadmium) Reduce to lowest feasible level

SAFETY PROFILE: Confirmed human carcinogen. Poison by intraperitoneal route. When heated to decomposition it emits toxic fumes of PO_x and Cd. See also CADMIUM COMPOUNDS.

CAD600 CAS:7789-42-6 HR: 3
CADMIUM BROMIDE
mf: Br_2Cd mw: 272.22

PROP: Pearly or colorless hexagonal crystals; hygroscopic. Mp: 570°, bp: 863°, d: 5.192. Sol in water, alc, and Me_2CO; moderately sol in acetone.

SYN: CADMIUM DIBROMIDE

CONSENSUS REPORTS: Reported in EPA TSCA Inventory.

OSHA PEL: TWA 5 µg(Cd)/m³
ACGIH TLV: TWA 0.01 mg(Cd)/m³; Suspected Carcinogen.

SAFETY PROFILE: Confirmed human carcinogen. When heated to decomposition it emits toxic fumes of Cd and Br⁻.

CAD750 CAS:2191-10-8 HR: 3
CADMIUM CAPRYLATE
mf: $C_{16}H_{30}O_4$•Cd mw: 398.86

SYN: OCTANOIC ACID, CADMIUM SALT (2:1)

TOXICITY DATA WITH REFERENCE
orl-rat LD50:950 mg/kg JHEMA2 18,144,74
itr-rat LDLo:9 mg/kg JOHYAY 18,144,74
orl-mus LD50:300 mg/kg JHEMA2 18,144,74

CONSENSUS REPORTS: Reported in EPA TSCA Inventory. Cadmium and its compounds are on the Community Right-To-Know List.

OSHA PEL: TWA 5 µg(Cd)/m³
ACGIH TLV: TWA 0.05 mg(Cd)/m³ (Proposed: TWA 0.01 mg(Cd)/m³ (dust), Suspected Human Carcinogen; 0.002 mg(Cd)/m³ (respirable dust), Suspected Human Carcinogen); BEI: 10 µg/g creatinine in urine; 10 µg/L in blood
NIOSH REL: (Cadmium) Reduce to lowest feasible level

SAFETY PROFILE: Confirmed human carcinogen. Poison by ingestion and intratracheal routes. When heated to decomposition it emits toxic fumes of Cd. See also CADMIUM COMPOUNDS.

CAD800 CAS:513-78-0 HR: 3
CADMIUM CARBONATE
mf: CO_3•Cd mw: 172.41

SYNS: CADMIUM MONOCARBONATE □ CARBONIC ACID, CADMIUM SALT □ CHEMCARB □ KALCIT □ MIKROKALCIT □ SUPERMIKROKALCIT

TOXICITY DATA WITH REFERENCE
sce-ham:ovr 870 nmol/L ENMUDM 7,381,85
orl-mus LD50:310 mg/kg GTPZAB 25(2),42,81

OSHA PEL: TWA 5 µg(Cd)/m³
ACGIH TLV: TWA 0.01 mg(Cd)/m³; Suspected Carcinogen
NIOSH REL: (Cadmium, dust and fume): lowest feasible concentration

CONSENSUS REPORTS: NTP 7th Annual Report On Carcinogens. IARC Cancer Review: Group 1 IMEMDT 58,119,93; Human Sufficient Evidence IMEMDT 58,119,93; Animal Sufficient Evidence IMEMDT 58,119,93. Reported in EPA TSCA Inventory.

SAFETY PROFILE: Confirmed human carcinogen. Poison by ingestion. Mutation data reported. When heated to decomposition it emits toxic fumes of cadmium.

CAE000 HR: 3
CADMIUM CHLORATE
mf: $CdCl_2O_6$ mw: 279.31

$Cd(ClP_3)_2$

PROP: Colorless, deliquescent prisms. Mp: 80°, d: 2.28 @ 18°.

CONSENSUS REPORTS: Cadmium and its compounds are on the Community Right-To-Know List.

OSHA PEL: TWA 5 µg(Cd)/m³
ACGIH TLV: TWA 0.05 mg(Cd)/m³ (Proposed: TWA 0.01 mg(Cd)/m³ (dust), Suspected Human Carcinogen; 0.002 mg(Cd)/m³ (respirable dust), Suspected Human Carcinogen); BEI: 10 µg/g creatinine in urine; 10 µg/L in blood
NIOSH REL: (Cadmium) Reduce to lowest feasible level

SAFETY PROFILE: Confirmed human carcinogen. A powerful oxidizing agent. Flammable by chemical reaction with reducing agents. Moderate explosion hazard when shocked or exposed to heat. Violent or explosive reaction with sulfides (e.g., copper(II) sulfide (explodes); antimony(II) sulfide; arsenic(III) sulfide; tin(II) sulfide; tin(IV) sulfide). When heated to decomposition it emits toxic fumes of Cd and Cl⁻. See also CHLORATES.

CAE250 CAS:10108-64-2 HR: 3
CADMIUM CHLORIDE
mf: $CdCl_2$ mw: 183.30

PROP: Hexagonal, colorless crystals. Mp: 568°, bp: 960.6°, d: 4.047 @ 25°, vap press: 10 mm @ 656°. Sol in H_2O; sltly sol in EtOH.

SYNS: CADDY □ CADMIUM DICHLORIDE □ KADMIUMCHLORID (GERMAN) □ VI-CAD

TOXICITY DATA WITH REFERENCE
dni-hmn:hla 250 µmol/L MUREAV 92,427,82
cyt-ofs-mul 630 µg/L/4W-C BECTA6 36,199,86
orl-mus TDLo:6 mg/kg (female 15-19D post):REP JEPTDQ 1(3),187,78
orl-rat TDLo:48 mg/kg (female 90D pre):REP NTIS** CONF-771017
ihl-rat TCLo:130 µg/m³ (female 1–19D post):TER GTPZAB 31(8),25,87
orl-rat TDLo:14,677 µg/kg (female 6–14D post):TER TXAPA9 23,222,72
ipr-rat TDLo:2933 µg/kg (female 10D post):TER TJADAB 7,237,73
scu-rat TDLo:7 mg/kg (female 12D post):TER TJADAB 35,74A,87
ihl-rat TCLo:200 µg/m³/24H (female 1–21D post):TER TXCYAC 10,297,78
scu-ham TDLo:5 mg/kg (female 1D pre):REP BIREBV 29,249,83
orl-rat TDLo:652 mg/kg (female 7–16D post):REP JJATDK 2,255,82
ipr-mus TDLo:1 mg/kg (male 1D pre):REP JPETAB 187,641,73
orl-rbt TDLo:990 mg/kg (female 1–6D post):REP JRPFA4 70,323,84

orl-rat TDLo:652 mg/kg (female 7–16D post):REP JJATDK 2,255,82
scu-rbt TDLo:5 mg/kg (female 1D pre):REP CCPTAY 26,181,82
scu-mus TDLo:3 mg/kg (female 1D pre):REP TJADAB 16,127,77
orl-rat TDLo:4890 ng/kg (male 30D pre):REP EVHPAZ 13,59,76
scu-rat TDLo:1250 μg/kg (male 1D pre):REP ABMGAJ 5,153,60
orl-mus TDLo:248 g/kg (female 1–19D post):TER JONUAI 109,1640,79
ipr-rat TDLo:2933 μg/kg (female 10D post):TER TJADAB 7,237,73
orl-rat TDLo:280 mg/kg (female 6–19D post):TER NATUAS 239,231,72
ipr-rat TDLo:2 mg/kg (female 10D post):TER ARTODN 59,443,87
ipr-mus TDLo:1833 μg/kg (female 7D post):TER NTIS** CONF-771017
orl-mus TDLo:6 mg/kg (female 15–19D post):TER JEPTDQ 1(3),187,78
ipr-rat TDLo:2933 μg/kg (female 10D post):TER TJADAB 7,237,73
scu-rat TDLo:4892 μg/kg (female 19D post):TER TXCYAC 18,103,80
orl-rat TDLo:17 mg/kg (male 6W pre):TER EESADV 4,51,80
scu-rat TDLo:32 mg/kg (female 12–15D post):TER TJADAB 23,75,81
orl-rat TDLo:14,677 μg/kg (female 6–14D post):TER TXAPA9 23,222,72
ihl-rat TCLo:20 μg/m³/23H/78W-C:CAR JJIND8 70,367,83
scu-rat TDLo:3666 μg/kg:CAR PAACA3 24,84,83
ims-rat TDLo:4500 μg/kg:ETA ARPAAQ 83,493,67
par-rat TDLo:1700 μg/kg:ETA ARPAAQ 83,493,67
scu-mus TDLo:5499 μg/kg:ETA JNCIAM 31,745,63
scu-rat TD:4500 μg/kg:ETA ARPAAQ 83,493,67
ihl-rat TC:41 μg/m³/23H/78-W-C:CAR JJIND8 70,367,83
ihl-rat TC:82 μg/m³/23H/78W-C:CAR JJIND8 70,367,83
scu-rat TD:7332 μg/kg:CAR PAACA3 24,84,83
scu-rat TD:5499 μg/kg:NEO PSEBAA 115,653,64
scu-rat TD:3666 mg/kg:NEO CNREA8 43,4575,83
scu-rat TD:7332 mg/kg:NEO CNREA8 43,4575,83
scu-rat TD:40,770 ng/kg/2D-I:CAR . CNREA8 48,4656,88
orl-wmn LDLo:3 g/kg:BPR,GIT BMJOAE 292,1559,86
orl-rat LD50:88 mg/kg AFDOAQ 15,122,51
ipr-rat LD50:1800 μg/kg EVHPAZ 28,89,79
orl-mus LD50:60 mg/kg APTOA6 48,108,81
ihl-mus LC50:2300 mg/m³ NTIS** PB158-508
ipr-mus LD50:9300 μg/kg NEZAAQ 32,472,77
scu-mus LD50:3200 μg/kg APTOA6 48,108,81
ivn-mus LD50:3500 μg/kg TXAPA9 53,510,80
ihl-dog LC90:420 mg/m³/30M JIHTAB 29,302,47
ivn-dog LDLo:5 mg/kg EQSSDX 1,1,75
scu-cat LDLo:25 mg/kg EQSSDX 1,1,75
ivn-cat LDLo:5 mg/kg HBAMAK 4,1289,35
orl-rbt LDLo:70 mg/kg EQSSDX 1,1,75

CONSENSUS REPORTS: NTP 7th Annual Report on Carcinogens. IARC Cancer Review: Group 1 IMEMDT 58,119,93; Animal Sufficient Evidence IMEMDT 2,74,73; Animal Sufficient Evidence IMEMDT 11,39,76; Human Sufficient Evidence IMEMDT 58,119,93; IARC Cancer Review: Animal Sufficient Evidence IMEMDT 58,119,93; EPA Genetic Toxicology Program. Cadmium and its compounds are on the Community Right-To-Know List. Reported in EPA TSCA Inventory.

OSHA PEL: TWA 5 μg(Cd)/m³
ACGIH TLV: TWA 0.05 mg(Cd)/m³ (Proposed: TWA 0.01 mg(Cd)/m³ (dust), Suspected Human Carcinogen; 0.002 mg(Cd)/m³ (respirable dust), Suspected Human Carcinogen); BEI: 10 μg/g creatinine in urine; 10 μg/L in blood
DFG MAK: Animal Carcinogen, Suspected Human Carcinogen
NIOSH REL: (Cadmium) Reduce to lowest feasible level

SAFETY PROFILE: Confirmed human carcinogen with experimental carcinogenic and tumorigenic data. Poison by ingestion, inhalation, skin contact, intraperitoneal, subcutaneous, intravenous, and possibly other routes. Human systemic effects by ingestion: blood pressure, acute pulmonary edema, hypermotility, diarrhea. Experimental teratogenic and reproductive effects. Human mutation data reported. Reacts violently with BrF_3 and K. When heated to decomposition it emits very toxic fumes of Cd and Cl^-. See also CADMIUM COMPOUNDS and CHLORIDES.

CAE375 CAS:72589-96-9 HR: 3
CADMIUM CHLORIDE, DIHYDRATE
mf: $CdCl_2 \cdot 2H_2O$ mw: 219.34

TOXICITY DATA WITH REFERENCE
orl-mus TDLo:15 mg/kg (5D male):TER AXVMAW 34,399,80
itt-mam TDLo:29 μg/kg (1D male):REP BECTA6 26,233,81
scu-rat TDLo:6580 μg/kg:ETA ARGEAR 36,119,70

CONSENSUS REPORTS: Cadmium and its compounds are on the Community Right-To-Know List.

OSHA PEL: TWA 5 μg(Cd)/m³
ACGIH TLV: TWA 0.05 mg(Cd)/m³ (Proposed: TWA 0.01 mg(Cd)/m³ (dust), Suspected Human Carcinogen; 0.002 mg(Cd)/m³ (respirable dust), Suspected Human Carcinogen); BEI: 10 μg/g creatinine in urine; 10 μg/L in blood
DFG MAK: Animal Carcinogen, Suspected Human Carcinogen
NIOSH REL: (Cadmium) Reduce to lowest feasible level

SAFETY PROFILE: Confirmed human carcinogen with experimental tumorigenic data. An experimental teratogen. Other experimental reproductive effects. When heated to decomposition it emits toxic fumes of Cl^- and Cd. See also CADMIUM CHLORIDE, CADMIUM COMPOUNDS, and CHLORIDES.

CAE425 CAS:7790-78-5 HR: 3
CADMIUM CHLORIDE, HYDRATE (2:5)
mf: $CdCl_2 \cdot 5/2H_2O$ mw: 228.35

PROP: Crystals.

TOXICITY DATA WITH REFERENCE
dni-hmn:lym 28 μmol/L IAAAAM 79,83,86
scu-rat TDLo:2 mg/kg (male 1D pre):REP CCPTAY 27,521,83
scu-rat TDLo:7 mg/kg (male 1D pre):REP JOENAK 34,329,66
ipr-rat TDLo:6237 μg/kg:REP ESKGA2 29,P-46,83
orl-rat LD50:665 mg/kg TXAPA9 103,28,90
orl-mus LD50:194 mg/kg JTEHD6 22,35,87
ipr-mus LD50:4567 μg/kg TXAPA9 63,461,82

CONSENSUS REPORTS: Cadmium and its compounds are on the Community Right-To-Know List.

OSHA PEL: TWA 5 μg(Cd)/m³
ACGIH TLV: TWA 0.05 mg(Cd)/m³ (Proposed: TWA 0.01 mg(Cd)/m³ (dust), Suspected Human Carcinogen; 0.002 mg(Cd)/m³ (respirable dust), Suspected Human Carcinogen); BEI: 10 μg/g creatinine in urine; 10 μg/L in blood
DFG MAK: Animal Carcinogen, Suspected Human Carcinogen
NIOSH REL: (Cadmium) Reduce to lowest feasible level

SAFETY PROFILE: Confirmed human carcinogen. Poison by ingestion and intraperitoneal routes. Experimental reproductive effects. Human mutation data reported. When heated to decomposition it emits toxic fumes of Cl^- and Cd. See also CADMIUM CHLORIDE, CADMIUM COMPOUNDS, and CHLORIDES.

CAE500 **CAS:35658-65-2** **HR: 3**
CADMIUM CHLORIDE, MONOHYDRATE
mf: $CdCl_2 \cdot H_2O$ mw: 201.32

TOXICITY DATA WITH REFERENCE
orl-rat TDLo:179 mg/kg (male 1D pre):REP CALEDQ 36,307,87
orl-rat TDLo:65 mg/kg/2Y-C:ETA CALEDQ 9,191,80
scu-rat TDLo:4478 μg/kg:CAR CALEDQ 36,307,87

CONSENSUS REPORTS: Cadmium and its compounds are on the Community Right-To-Know List.

OSHA PEL: TWA 5 μg(Cd)/m³
ACGIH TLV: TWA 0.05 mg(Cd)/m³ (Proposed: TWA 0.01 mg(Cd)/m³ (dust), Suspected Human Carcinogen; 0.002 mg(Cd)/m³ (respirable dust), Suspected Human Carcinogen); BEI: 10 μg/g creatinine in urine; 10 μg/L in blood
DFG MAK: Animal Carcinogen, Suspected Human Carcinogen
NIOSH REL: (Cadmium) Reduce to lowest feasible level

SAFETY PROFILE: Confirmed human carcinogen with experimental carcinogenic and tumorigenic data. Experimental reproductive effects. When heated to decomposition it emits very toxic fumes of Cd and Cl^-. See also CADMIUM CHLORIDE, CADMIUM COMPOUNDS, and CHLORIDES.

CAE750 **HR: 3**
CADMIUM COMPOUNDS

TOXICITY DATA WITH REFERENCE
ihl-hmn TCLo:1500 μg/m³/14Y-I:CAR,PUL ANYAA9 271,273,76

CONSENSUS REPORTS: Cadmium and its compounds are on the Community Right-To-Know List.

OSHA PEL: TWA 5 μg(Cd)/m³
ACGIH TLV: Dust and Salts: TWA 0.05 mg(Cd)/m³ (Proposed: TWA 0.01 mg(Cd)/m³ (dust), Suspected Human Carcinogen; 0.002 mg(Cd)/m³ (respirable dust), Suspected Human Carcinogen); BEI: 10 μg/g creatinine in urine; 10 μg/L in blood
DFG BAT: Blood 1.5 μg/dL; Urine 15 μg/dL. MAK: Suspected Carcinogen.
NIOSH REL: (Cadmium) Reduce to lowest feasible level

SAFETY PROFILE: Confirmed human carcinogens producing lung tumors. Poison by ingestion. The irritating and emetic action is so violent, however, that little of the cadmium has time to be absorbed and fatal poisoning rarely ensues. Experimental carcinogens and teratogens. Cases of human poisoning have been reported from ingestion of food or beverages prepared or stored in cadmium-plated containers. Inhalation of fumes or dusts affects the respiratory tract and the kidneys. Brief exposure to high concentrations may result in pulmonary edema and death. Fatal concentrations may be breathed without sufficient discomfort to warn a worker to leave the exposure site. Cadmium oxide fumes can cause metal fume fever resembling that caused by zinc oxide fumes. When heated to decomposition cadmium compounds emit toxic fumes of Cd.

CAF500 **HR: 3**
CADMIUM DICYANIDE
mf: C_2CdN_2 mw: 164.44

CONSENSUS REPORTS: Cadmium and its compounds and Cyanide and its compounds are on the Community Right-To-Know List.

OSHA PEL: TWA 5 μg(Cd)/m³
ACGIH TLV: TWA 0.05 mg(Cd)/m³ (Proposed: TWA 0.01 mg(Cd)/m³ (dust), Suspected Human Carcinogen; 0.002 mg(Cd)/m³ (respirable dust), Suspected Human Carcinogen); BEI: 10 μg/g creatinine in urine; 10 μg/L in blood
NIOSH REL: (Cadmium) Reduce to lowest feasible level

SAFETY PROFILE: Confirmed human carcinogen. A poison. Incompatible with magnesium. When heated to decomposition it emits toxic fumes of Cd and CN^-. See also CADMIUM COMPOUNDS and CYANIDE.

CAF750 **CAS:15954-91-3** **HR: 3**
CADMIUM(II) EDTA COMPLEX

SYN: (ETHYLENEDINITRILO)TETRAACETIC ACID CADMIUM(II) COMPLEX

TOXICITY DATA WITH REFERENCE
ipr-mus LD50:7800 μg,(Cd)/kg PABIAQ 11,853,63
ivn-mus LD50:21,400 μg(Cd)/kg ABMGAJ 16,149,66

CONSENSUS REPORTS: Cadmium and its compounds are on the Community Right-To-Know List.

OSHA PEL: TWA 5 µg(Cd)/m³
ACGIH TLV: TWA 0.05 mg(Cd)/m³ (Proposed: TWA 0.01 mg(Cd)/m³ (dust), Suspected Human Carcinogen; 0.002 mg(Cd)/m³ (respirable dust), Suspected Human Carcinogen); BEI: 10 µg/g creatinine in urine; 10 µg/L in blood
NIOSH REL: (Cadmium) Reduce to lowest feasible level

SAFETY PROFILE: Confirmed human carcinogen. Poison by intraperitoneal and intravenous routes. When heated to decomposition it emits toxic fumes of NO_x and Cd.

CAG000 CAS:14486-19-2 HR: 3
CADMIUM FLUOBORATE
mf: B_2CdF_8 mw: 286.02

SYNS: BORATE(1-), TETRAFLUORO-, CADMIUM (2:1) (9CI) □ CADMIUM FLUOROBORATE □ CADMIUM TETRAFLUOROBORATE (7CI) □ TL 1026

TOXICITY DATA WITH REFERENCE
orl-rat LDLo:250 mg/kg NCNSA6 5,27,53
ihl-mus LCLo:650 mg/m³/10M NDRC** No.9-4-1-19,44

CONSENSUS REPORTS: NTP 7th Annual Report on Carcinogens. Reported in EPA TSCA Inventory. Cadmium and its compounds are on the Community Right-To-Know List.

OSHA PEL: TWA 5 µg(Cd)/m³
ACGIH TLV: TWA 0.05 mg(Cd)/m³ (Proposed: TWA 0.01 mg(Cd)/m³ (dust), Suspected Human Carcinogen; 0.002 mg(Cd)/m³ (respirable dust), Suspected Human Carcinogen); BEI: 10 µg/g creatinine in urine; 10 µg/L in blood
NIOSH REL: (Cadmium) Reduce to lowest feasible level

SAFETY PROFILE: Confirmed human carcinogen. Poison by ingestion and inhalation. When heated to decomposition it emits very toxic fumes of Cd and F^-. See TETRAFLUOROBORATE.

CAG250 CAS:7790-79-6 HR: 3
CADMIUM FLUORIDE
mf: CdF_2 mw: 150.40

PROP: Cubic, white, non-hygroscopic, non-volatile crystals. Mp: 1078°, bp: 1748°, d: 6.64, vap press: 1 mm @ 1112°. Sltly sol in H_2O.

SYN: CADMIUM FLUORURE (FRENCH)

TOXICITY DATA WITH REFERENCE
scu-frg LDLo:280 mg/kg CRSBAW 124,133,37

CONSENSUS REPORTS: Reported in EPA TSCA Inventory. Cadmium and its compounds are on the Community Right-To-Know List.

OSHA PEL: TWA 5 µg(Cd)/m³
ACGIH TLV: TWA 0.05 mg(Cd)/m³ (Proposed: TWA 0.01 mg(Cd)/m³ (dust), Suspected Human Carcinogen; 0.002 mg(Cd)/m³ (respirable dust), Suspected Human Carcinogen); BEI: 10 µg/g creatinine in urine; 10 µg/L in blood.
NIOSH REL: (Cadmium) Reduce to lowest feasible level

SAFETY PROFILE: Confirmed human carcinogen. Poison by subcutaneous route. Violent reaction with K. When heated to decomposition it emits very toxic fumes of Cd and F^-. See also FLUORIDES and CADMIUM COMPOUNDS.

CAG500 CAS:17010-21-8 HR: 3
CADMIUM FLUOSILICATE
mf: CdF_6Si mw: 254.49

PROP: Hexagonal, colorless crystals.

SYNS: CADMIUM FLUOROSILICATE □ CADMIUM HEXAFLUOROSILICATE (7CI) □ CADMIUM SILICON FLUORIDE □ SILICATE(2-), HEXAFLUORO-, CADMIUM (8CI,9CI) □ TL 1070

TOXICITY DATA WITH REFERENCE
orl-rat LDLo:100 mg/kg NCNSA6 5,27,53
ihl-mus LCLo:670 mg/m³/10M NDRC** No.9-4-1-19,44

CONSENSUS REPORTS: Cadmium and its compounds are on the Community Right-To-Know List.

OSHA PEL: TWA 5 µg(Cd)/m³
ACGIH TLV: TWA 0.05 mg(Cd)/m³ (Proposed: TWA 0.01 mg(Cd)/m³ (dust), Suspected Human Carcinogen; 0.002 mg(Cd)/m³ (respirable dust), Suspected Human Carcinogen); BEI: 10 µg/g creatinine in urine; 10 µg/L in blood
NIOSH REL: (Cadmium) Reduce to lowest feasible level
DOT CLASSIFICATION: 6.1; *Label:* KEEP AWAY FROM FOOD

SAFETY PROFILE: Confirmed human carcinogen. Poison by ingestion and inhalation. When heated to decomposition it emits very toxic fumes of Cd and F^-.

CAG750 CAS:16039-55-7 HR: 3
CADMIUM LACTATE
mf: $C_6H_{10}O_6{\cdot}Cd$ mw: 290.56

PROP: Needles.

SYN: LACTIC ACID, CADMIUM SALT

CONSENSUS REPORTS: Cadmium and its compounds are on the Community Right-To-Know List.

OSHA PEL: TWA 5 µg(Cd)/m³
ACGIH TLV: TWA 0.05 mg(Cd)/m³ (Proposed: TWA 0.01 mg(Cd)/m³ (dust), Suspected Human Carcinogen; 0.002 mg(Cd)/m³ (respirable dust), Suspected Human Carcinogen); BEI: 10 µg/g creatinine in urine; 10 µg/L in blood
NIOSH REL: (Cadmium) Reduce to lowest feasible level

SAFETY PROFILE: Confirmed human carcinogen. A poison. When heated to decomposition it emits toxic fumes of Cd. See also CADMIUM COMPOUNDS.

CAG775 **CAS:2605-44-9** **HR: 3**
CADMIUM LAURATE
mf: $C_{24}H_{46}O_4$•Cd mw: 511.10

SYNS: CADMIUM DILAURATE □ CADMIUM DODECANOATE □ DODECANOIC ACID, CADMIUM SALT (9CI) □ LAURIC ACID, CADMIUM SALT (2:1)

TOXICITY DATA WITH REFERENCE
orl-rat LD50:2370 mg/kg 41HTAH -,14,78
orl-mus LD50:1060 mg/kg 41HTAH -,14,78

CONSENSUS REPORTS: Reported in EPA TSCA Inventory.

OSHA PEL: TWA 5 µg(Cd)/m³
ACGIH TLV: TWA 0.01 mg(Cd)/m³; Suspected Carcinogen

SAFETY PROFILE: Confirmed human carcinogen. Moderately toxic by ingestion. When heated to decomposition it emits toxic fumes of Cd.

CAH000 **CAS:10325-94-7** **HR: 3**
CADMIUM NITRATE
mf: CdN_2O_6 mw: 236.42

PROP: Strongly hygroscopic, white, prismatic needles. Mp: 350–360°. Very sol in H_2O: sol in EtOAc.

SYNS: CADMIUM DINITRATE □ CADMIUM(II) NITRATE □ NITRIC ACID, CADMIUM SALT

TOXICITY DATA WITH REFERENCE
mrc-bcs 5 mmol/L MUREAV 77,109,80
orl-rat LD50:300 mg/kg YAKUD5 22,455,80
unr-rat LD50:200 mg/kg GISAAA 50(3),57,85
orl-mus LD50:100 mg/kg 41HTAH -,14,78
ihl-mus LC50:3850 mg/m³ NTIS** PB1580508

CONSENSUS REPORTS: NTP 7th Annual Report on Carcinogens. IARC Cancer Review: Group 1 IMEMDT 58,119,93; Human Sufficient Evidence IMEMDT 58,119,93; Animal Sufficient Evidence IMEMDT 58,119,93. Reported in EPA TSCA Inventory. EPA Genetic Toxicology Program. Cadmium and its compounds are on the Community Right-To-Know List.

OSHA PEL: TWA 5 µg(Cd)/m³
ACGIH TLV: TWA 0.05 mg(Cd)/m³ (Proposed: TWA 0.01 mg(Cd)/m³ (dust), Suspected Human Carcinogen; 0.002 mg(Cd)/m³ (respirable dust), Suspected Human Carcinogen); BEI: 10 µg/g creatinine in urine; 10 µg/L in blood
NIOSH REL: (Cadmium) Reduce to lowest feasible level

SAFETY PROFILE: Confirmed human carcinogen. Poison by ingestion and possibly other routes. Moderately toxic by inhalation. Mutation data reported. When heated to decomposition it emits very toxic fumes of Cd and NO_x. See also CADMIUM COMPOUNDS and NITRATES.

CAH250 **CAS:10022-68-1** **HR: 3**
CADMIUM(II) NITRATE TETRAHYDRATE (1:2:4)
mf: N_2O_6•Cd•$4H_2O$ mw: 308.50

PROP: Crystals in H_2O. Mp: 59.4°.

SYNS: DUSICNAN KADEMNATY (CZECH) □ NITRIC ACID, CADMIUM SALT, TETRAHYDRATE

TOXICITY DATA WITH REFERENCE
skn-rbt 500 mg/24H SEV 28ZPAK -,12,72
eye-rbt 20 mg/24H MOD 28ZPAK -,12,72
mmo-esc 6 µmol/L ENVRAL 26,279,85
orl-rat LD50:300 mg/kg 28ZPAK -,12,72

CONSENSUS REPORTS: Cadmium and its compounds are on the Community Right-To-Know List.

OSHA PEL: TWA 5 µg(Cd)/m³
ACGIH TLV: TWA 0.05 mg(Cd)/m³ (Proposed: TWA 0.01 mg(Cd)/m³ (dust), Suspected Human Carcinogen; 0.002 mg(Cd)/m³ (respirable dust), Suspected Human Carcinogen); BEI: 10 µg/g creatinine in urine; 10 µg/L in blood
NIOSH REL: (Cadmium) Reduce to lowest feasible level

SAFETY PROFILE: Confirmed human carcinogen. Poison by ingestion. A severe skin and moderate eye irritant. Mutation data reported. See also CADMIUM COMPOUNDS, CADMIUM NITRATE, and NITRATES. When heated to decomposition it emits very toxic fumes of Cd and NO_x.

CAH500 **CAS:1306-19-0** **HR: 3**
CADMIUM OXIDE
mf: CdO mw: 128.40

PROP: (1) Amorphous, brown powder; (2) cubic, brown crystals. Changes color on heating. Mp (1): <1426°, mp (2): decomp @ 950°, bp: 1559°, d (1): 6.95, d (2): 8.15, vap press: 1 mm @ 1000°. Subl at 7°. Insol in H_2O; sol in acids and NH_3.

SYNS: CADMIUM MONOXIDE □ KADMU TLENEK (POLISH) □ NCI-C02551

TOXICITY DATA WITH REFERENCE
ihl-rat TCLo:23 µg/m³/5H (15W pre/1-20D preg):REP TOLED5 22,53,84
ihl-rat TCLo:183 µg/m³/5H (15W pre/1-20D preg):REP TOLED5 22,53,84
ihl-rat TCLo:91 µg/m³ (female 1–19D post):TER GTPZAB 31(8),25,87
scu-rat TDLo:90 mg/kg:NEO BJCAAI 20,190,66
ihl-hmn TCLo:8630 µg/m³/5H YAKUD5 22,455,80
ihl-man TCLo:500 µg/m³/5Y-I:NOSE,KID QJMEA7 38,425,69
ihl-man TCLo:40 µg/m³:CVS,KID GISAAA 45(10)22,80
orl-rat LD50:72 mg/kg YAKUD5 22,455,80
ihl-rat LC50:780 mg/m³/10M NTIS** PB158-508
ipr-rat LD50:12 mg/kg ZDKAA8 38(9),18,78
orl-mus LD50:72 mg/kg 41HTAH -,14,78
ihl-mus LC50:340 mg/m³/10M NTIS** PB158-508
ihl-dog LC50:400 mg/m³/10M YAKUD5 22,455,80
ihl-mky LC50:15 g/m³/10M NTIS** PB158-508
ihl-rbt LC50:3 g/m³/15M NTIS** PB158-508
ihl-gpg LC50:3 g/m³/15M NTIS** PB158-508

CONSENSUS REPORTS: NTP 7th Annual Report on Carcinogens. IARC Cancer Review: Group 1 IMEMDT 58,119,93; Animal Sufficient Evidence IMEMDT 2,74,73; Animal Sufficient Evidence IMEMDT 11,39,76; Animal Sufficient Evidence IMEMDT 58,119,93; Human Limited

Evidence IMEMDT 11,39,76; Human Sufficient Evidence IMEMDT 58,119,93. Reported in EPA TSCA Inventory. EPA Extremely Hazardous Substances List. Cadmium and its compounds are on the Community Right-To-Know List.

OSHA PEL: TWA 5 μg(Cd)/m³
ACGIH TLV: TWA 0.05 mg(Cd)/m³ (Proposed: TWA 0.01 mg(Cd)/m³ (dust), Suspected Human Carcinogen; 0.002 mg(Cd)/m³ (respirable dust), Suspected Human Carcinogen); BEI: 10 μg/g creatinine in urine; 10 μg/L in blood
DFG MAK: Suspected Carcinogen
NIOSH REL: (Cadmium) Reduce to lowest feasible level

SAFETY PROFILE: Confirmed human carcinogen with experimental neoplastigenic data. Poison by ingestion, inhalation, and intraperitoneal routes. An experimental teratogen. Other experimental reproductive effects. Human systemic effects by inhalation include: change in the sense of smell, change in heart rate, blood pressure increase, an excess of protein in the urine, and other kidney or bladder changes. Mixtures with magnesium explode when heated. When heated to decomposition it emits toxic fumes of Cd. See also CADMIUM COMPOUNDS.

For occupational chemical analysis use NIOSH: Cadmium, 7048; welding and Brazing Fume, 7200; Elements, 7300.

CAH750 **HR: 3**
CADMIUM OXIDE FUME
mf: CdO mw: 128.40

SYN: CADMIUM FUME

TOXICITY DATA WITH REFERENCE
ihl-hmn LCLo:2500 mg/m³ JIHTAB 29,279,47
ihl-man TCLo:8630 μg/m³/5H:PUL BJIMAG 23,292,66
ihl-rat LC50:500 mg/m³/10M JIHTAB 29,279,47
ihl-mus LCLo:700 mg/m³/10M JIHTAB 29,279,47
ihl-dog LC50:4000 mg/m³/10M JIHTAB 29,279,47
ihl-mky LC50:15,000 mg/m³/10M JIHTAB 29,279,47
ihl-rbt LC50:2500 mg/m³/10M JIHTAB 29,279,47
ihl-gpg LC50:3500 mg/m³/10M JIHTAB 29,279,47

CONSENSUS REPORTS: Reported in EPA TSCA Inventory. Cadmium and its compounds are on the Community Right-To-Know List.

OSHA PEL: TWA 5 μg(Cd)/m³
ACGIH TLV: TWA 0.05 mg(Cd)/m³ (Proposed: TWA 0.01 mg(Cd)/m³ (dust), Suspected Human Carcinogen; 0.002 mg(Cd)/m³ (respirable dust), Suspected Human Carcinogen); BEI: 10 μg/g creatinine in urine; 10 μg/L in blood
NIOSH REL: (Cadmium) Reduce to lowest feasible level

SAFETY PROFILE: Confirmed human carcinogen. Poison by inhalation. Moderately toxic to humans by inhalation. Human pulmonary system effects by inhalation, including: coughing, difficult breathing, and cyanosis. A strong irritant via inhalation. When heated to decomposition it emits toxic fumes of Cd. See also CADMIUM OXIDE and CADMIUM COMPOUNDS.

CAI000 **CAS:13477-17-3** **HR: 3**
CADMIUM PHOSPHATE
mf: $Cd_3O_8P_2•4H_2O$ mw: 599.22

PROP: Amorphous or colorless crystals. Mp: 1180°.

SYN: TL 1182

TOXICITY DATA WITH REFERENCE
ihl-mus LCLo:650 mg/m³/10M NDRC** No.9-4-1-19,44

CONSENSUS REPORTS: Reported in EPA TSCA Inventory. Cadmium and its compounds are on the Community Right-To-Know List.

OSHA PEL: TWA 5 μg(Cd)/m³
ACGIH TLV: TWA 0.05 mg(Cd)/m³ (Proposed: TWA 0.01 mg(Cd)/m³ (dust), Suspected Human Carcinogen; 0.002 mg(Cd)/m³ (respirable dust), Suspected Human Carcinogen); BEI: 10 μg/g creatinine in urine; 10 μg/L in blood
NIOSH REL: (Cadmium) Reduce to lowest feasible level

SAFETY PROFILE: Confirmed human carcinogen. Poison by inhalation. When heated to decomposition it emits toxic fumes of Cd and PO_x. See CADMIUM COMPOUNDS and PHOSPHATES.

CAI125 **CAS:12014-28-7** **HR: 3**
CADMIUM PHOSPHIDE
mf: Cd_3P_2 mw: 399.18

PROP: Grey needles or platelets.

CONSENSUS REPORTS: Cadmium compounds are on the Community Right-To-Know List.

OSHA PEL: TWA 5 μg(Cd)/m³
ACGIH TLV: Dust and Salts: TWA 0.05 mg(Cd)/m³ (Proposed: TWA 0.01 mg(Cd)/m³ (dust), Suspected Human Carcinogen; 0.002 mg(Cd)/m³ (respirable dust), Suspected Human Carcinogen); BEI: 10 μg/g creatinine in urine; 10 μg/L in blood
DFG BAT: Blood 1.5 μg/dL; Urine 15 μg/dL. MAK: Suspected Carcinogen.
NIOSH REL: (Cadmium) Reduce to lowest feasible level

SAFETY PROFILE: Confirmed human carcinogen. Explosive reaction with concentrated nitric acid. When heated to decomposition it emits toxic fumes of PO_x and Cd. See also CADMIUM COMPOUNDS and PHOSPHIDES.

CAI250 **HR: 3**
CADMIUM PROPIONATE
mf: $C_6H_{10}CdO_5$ mw: 258.55

CONSENSUS REPORTS: Cadmium and its compounds are on the Community Right-To-Know List.

OSHA PEL: TWA 5 μg(Cd)/m³
ACGIH TLV: Dust and Salts: TWA 0.05 mg(Cd)/m³ (Proposed: TWA 0.01 mg(Cd)/m³ (dust), Suspected Human Carcinogen; 0.002 mg(Cd)/m³ (respirable dust), Suspected Human Carcinogen); BEI: 10 μg/g creatinine in urine; 10 μg/L in blood
DFG BAT: Blood 1.5 μg/dL; Urine 15 μg/dL. MAK: Suspected Carcinogen.

NIOSH REL: (Cadmium) Reduce to lowest feasible level

SAFETY PROFILE: Confirmed human carcinogen. The salt has exploded. Incompatible with 3-pentanone vapor. When heated to decomposition it emits toxic fumes of Cd. See also CADMIUM COMPOUNDS.

CAI350 CAS:18897-36-4 HR: 3
CADMIUM 2-PYRIDINETHIONE
mf: $C_{10}H_8CdN_2O_2S_2$ mw: 364.72

SYNS: CADMIUM, BIS(1-HYDROXY-2(1H)-PYRIDINETHIONATO)- □ CADMIUM PT □ CdPT

TOXICITY DATA WITH REFERENCE
orl-rat LD50:240 mg/kg TOANDB 3,1,79
ivn-rbt LD50:1340 μg/kg TOANDB 3,1,79

OSHA PEL: TWA 5 μg(Cd)/m³
ACGIH TLV: TWA 0.01 mg(Cd)/m³; Suspected Carcinogen
NIOSH REL: (Cadmium): TWA reduce to lowest feasible level

SAFETY PROFILE: Confirmed human carcinogen. Poison by ingestion and intravenous routes. When heated to decomposition it emits toxic fumes of NO_x, SO_x, and Cd.

CAI400 CAS:19010-79-8 HR: 3
CADMIUM SALICYLATE
mf: $C_{14}H_{10}CdO_6$ mw: 386.64

PROP: Monohydrate small needles or plates. Mp: 242°. Sltly sol in cold water, methanol, eth; very sol in boiling water.

SYNS: BIS(2-HYDROXYBENZOATO-O¹O²-), (T-4)-CADMIUM (9CI) □ CADMIUM, BIS(SALICYLATO)-

TOXICITY DATA WITH REFERENCE
orl-rat LD50:1200 mg/kg 41HTAH -,14,78
orl-mus LD50:164 mg/kg 41HTAH -,14,78

OSHA PEL: TWA 5 μg(Cd)/m³
ACGIH TLV: TWA 0.01 mg(Cd)/m³; Suspected Carcinogen

SAFETY PROFILE: Confirmed human carcinogen. Poison by ingestion. When heated to decomposition it emits toxic fumes of Cd.

CAI500 HR: 3
CADMIUM SELENIDE
mf: CdSe mw: 191.36

PROP: Preparative hazard.

CONSENSUS REPORTS: Cadmium and its compounds as well as selenium and its compounds are on the Community Right-To-Know List.

OSHA PEL: TWA 5 μg(Cd)/m³
ACGIH TLV: Dust and Salts: TWA 0.05 mg(Cd)/m³ (Proposed: TWA 0.01 mg(Cd)/m³ (dust), Suspected Human Carcinogen; 0.002 mg(Cd)/m³ (respirable dust), Suspected Human Carcinogen); BEI: 10 μg/g creatinine in urine; 10 μg/L in blood
DFG BAT: Blood 1.5 μg/dL; Urine 15 μg/dL. MAK: Suspected Carcinogen; 0.1 mg(Se)/m³.
NIOSH REL: (Cadmium) Reduce to lowest feasible level

SAFETY PROFILE: Confirmed human carcinogen. Selenium compounds are considered to be poisons. When heated to decomposition it emits toxic fumes of Cd and Se. See also CADMIUM COMPOUNDS and SELENIUM COMPOUNDS.

CAI600 CAS:12626-36-7 HR: 2
CADMIUM SELENIDE SULFIDE

SYNS: CADMIUM SULFIDE SELENIDE □ CADMIUM SULFOSELENIDE □ CADMIUM SULPHOSELENIDE

TOXICITY DATA WITH REFERENCE
orl-mus LD50:2425 mg/kg GTPZAB 25(2),42,81

CONSENSUS REPORTS: Reported in EPA TSCA Inventory.

OSHA PEL: TWA 0.2 mg(Se)/m³
ACGIH TLV: TWA 0.2 mg(Se)/m³; TWA 0.05 mg(Cd)/m³
NIOSH REL: (Cadmium, dust and fume): lowest feasible concentration

SAFETY PROFILE: Moderately toxic by ingestion. When heated to decomposition it emits toxic vapors of Cd, Se, and SO_x.

CAI750 CAS:141-00-4 HR: 3
CADMIUM SUCCINATE
mf: $C_4H_4O_4$•Cd mw: 228.48

SYNS: CADMINATE □ SUCCINIC ACID, CADMIUM SALT (1:1)

TOXICITY DATA WITH REFERENCE
orl-rat LD50:660 mg/kg FMCHA2 -,D53,80
orl-mus LD50:312 mg/kg 28ZEAL 5,35,76
ipr-mus LD50:270 mg/kg AIPTAK 128,391,60

CONSENSUS REPORTS: Reported in EPA TSCA Inventory. Cadmium and its compounds are on the Community Right-To-Know List.

OSHA PEL: TWA 5 μg(Cd)/m³
ACGIH TLV: Dust and Salts: TWA 0.05 mg(Cd)/m³ (Proposed: TWA 0.01 mg(Cd)/m³ (dust), Suspected Human Carcinogen; 0.002 mg(Cd)/m³ (respirable dust), Suspected Human Carcinogen); BEI: 10 μg/g creatinine in urine; 10 μg/L in blood
DFG BAT: Blood 1.5 μg/dL; Urine 15 μg/dL. MAK: Suspected Carcinogen.
NIOSH REL: (Cadmium) Reduce to lowest feasible level

SAFETY PROFILE: Confirmed human carcinogen. Poison by ingestion and intraperitoneal routes. Moderately toxic by ingestion. When heated to decomposition it emits toxic fumes of Cd. See also CADMIUM COMPOUNDS.

CAJ000 CAS:10124-36-4 HR: 3
CADMIUM SULFATE (1:1)
mf: O_4S•Cd mw: 208.46

PROP: Rhombic, white crystals or prisms. Mp: 1000°, d:

4.691. Sol in H_2O; very sltly sol in MeOH, EtOH, and EtOAc.

SYNS: CADMIUM SULFATE □ CADMIUM SULPHATE □ SULFURIC ACID, CADMIUM(2+) SALT □ SULPHURIC ACID, CADMIUM SALT (1:1)

TOXICITY DATA WITH REFERENCE
mrc-bcs 5 mmol/L MUREAV 77,109,80
dnd-rat:lvr 30 µmol/L MUREAV 113,357,83
msc-mus:lym 150 µg/L JTEHD6 9,367,82
par-mus TDLo:28 µg/kg (female 12D post):TER IGSBAL 92,65,76
ipr-mus TDLo:1030 µg/kg (female 9D post):TER TJADAB 29,427,84
ipr-mus TDLo:5150 µg/kg (female 9D post):TER TJADAB 32,407,85
ipr-mus TDLo:5150 µg/kg (female 9D post):REP TJADAB 29,427,84
ipr-mus TDLo:3 mg/kg (female 9D post):TER TJADAB 27,54A,83
par-mus TDLo:28 µg/kg (female 12D post):TER IGSBAL 92,65,76
ivn-ham TDLo:2 mg/kg (female 8D post):TER BECTA6 22,175,79
ipr-mus TDLo:2570 µg/kg (female 9D post):TER TJADAB 29,427,84
ipr-ham TDLo:2800 µg/kg (female 8D post):TER TJADAB 29(2),30A,84
orl-rat LD50:280 mg/kg 41HTAH -,14,78
orl-mus LD50:88 mg/kg 41HTAH -,14,78
ipr-mus LD50:12,760 µg/kg COREAF 256,1043,63
orl-dog LDLo:105 mg/kg EQSSDX 1,1,75
scu-dog LDLo:27 mg/kg EQSSDX 1,1,75
scu-frg LDLo:105 mg/kg HBAMAK 4,1317,35

CONSENSUS REPORTS: NTP 7th Annual Report on Carcinogens. IARC Cancer Review: Group 1 IMEMDT 58,119,93; Animal Sufficient Evidence IMEMDT 2,74,73; Animal Sufficient Evidence IMEMDT 11,39,76; Animal Sufficient Evidence IMEMDT 58,119,93; Human Sufficient Evidence IMEMDT 58,119,93. Reported in EPA TSCA Inventory. EPA Genetic Toxicology Program. Cadmium and its compounds are on the Community Right-To-Know List.

OSHA PEL: TWA 5 µg(Cd)/m³
ACGIH TLV: TWA 0.05 mg(Cd)/m³ (Proposed: TWA 0.01 mg(Cd)/m³ (dust), Suspected Human Carcinogen; 0.002 mg(Cd)/m³ (respirable dust), Suspected Human Carcinogen); BEI: 10 µg/g creatinine in urine; 10 µg/L in blood
DFG MAK: Suspected Carcinogen
NIOSH REL: (Cadmium) Reduce to lowest feasible level

SAFETY PROFILE: Confirmed human carcinogen with experimental carcinogenic data. Poison by ingestion, subcutaneous, and intraperitoneal routes. Experimental teratogenic and reproductive effects. Mutation data reported. See also CADMIUM COMPOUNDS and SULFATES. When heated to decomposition it emits very toxic fumes of Cd and SO_x.

CAJ250 CAS:7790-84-3 HR: 3
CADMIUM SULFATE (1:1) HYDRATE (3:8)
mf: $O_4S{\cdot}Cd{\cdot}8/3H_2O$ mw: 256.51

PROP: Crystals from aq soln @ 174°.

SYNS: CADMIUM SULFATE OCTAHYDRATE □ SULFURIC ACID, CADMIUM SALT, HYDRATE

TOXICITY DATA WITH REFERENCE
dnd-esc 3 µmol/L JOBAAY 133,75,78
cyt-ham:fbr 10 µmol/L/1H MUREAV 40,125,76
ivn-ham TDLo:2 mg/kg (8D preg):REP TJADAB 21,181,80
ivn-ham TDLo:2 mg/kg (8D preg):TER TJADAB 21,181,80
ivn-ham TDLo:2 mg/kg (female 8D post):TER TJADAB 21,181,80
scu-rat TDLo:60 mg/kg/2Y-I:NEO AOHYA3 16,111,73
scu-rat TD:15 mg/kg/2Y-I:ETA AOHYA3 16,111,73

CONSENSUS REPORTS: IARC Cancer Review: Animal Sufficient Evidence IMEMDT 2,74,73. Cadmium and its compounds are on the Community Right-To-Know List.

OSHA PEL: TWA 5 µg(Cd)/m³
ACGIH TLV: TWA 0.05 mg(Cd)/m³ (Proposed: TWA 0.01 mg(Cd)/m³ (dust), Suspected Human Carcinogen; 0.002 mg(Cd)/m³ (respirable dust), Suspected Human Carcinogen); BEI: 10 µg/g creatinine in urine; 10 µg/L in blood
NIOSH REL: (Cadmium) Reduce to lowest feasible level

SAFETY PROFILE: Confirmed human carcinogen with experimental tumorigenic and neoplastigenic data. Experimental teratogenic and reproductive effects. Mutation data reported. When heated to decomposition it emits very toxic fumes of Cd and SO_x. See also CADMIUM SULFATE, CADMIUM COMPOUNDS, and SULFATES.

CAJ500 CAS:13477-21-9 HR: 3
CADMIUM SULFATE TETRAHYDRATE
mf: $O_4S{\cdot}Cd{\cdot}4H_2O$ mw: 280.54

SYN: SULFURIC ACID, CADMIUM SALT, TETRAHYDRATE

TOXICITY DATA WITH REFERENCE
scu-rat TDLo:20 mg/kg/10W-I:NEO BJCAAI 18,667,64

CONSENSUS REPORTS: Cadmium and its compounds are on the Community Right-To-Know List.

OSHA PEL: TWA 5 µg(Cd)/m³
ACGIH TLV: TWA 0.05 mg(Cd)/m³ (Proposed: TWA 0.01 mg(Cd)/m³ (dust), Suspected Human Carcinogen; 0.002 mg(Cd)/m³ (respirable dust), Suspected Human Carcinogen); BEI: 10 µg/g creatinine in urine; 10 µg/L in blood
NIOSH REL: (Cadmium) Reduce to lowest feasible level

SAFETY PROFILE: Confirmed human carcinogen with experimental neoplastigenic data. When heated to decomposition it emits very toxic fumes of Cd and SO_x. See also CADMIUM COMPOUNDS.

CAJ750 **CAS:1306-23-6** **HR: 3**
CADMIUM SULFIDE
mf: CdS mw: 144.46

PROP: Hexagonal, lemon yellow to orange crystals. Mp: 1750° @ 100 atm, bp: subl in N_2, subl @ 9°, d: 4.82. Sltly sol in H_2O.

SYNS: AURORA YELLOW □ CADMIUM GOLDEN 366 □ CADMIUM LEMON YELLOW 527 □ CADMIUM MONOSULFIDE □ CADMIUM ORANGE □ CADMIUM PRIMROSE 819 □ CADMIUM SULPHIDE □ CADMIUM YELLOW □ CADMIUM YELLOW 000 □ CADMIUM YELLOW 892 □ CADMIUM YELLOW CONC. DEEP □ CADMIUM YELLOW CONC. GOLDEN □ CADMIUM YELLOW CONC. LEMON □ CADMIUM YELLOW CONC. PRIMROSE □ CADMIUM YELLOW 10G CONC. □ CADMIUM YELLOW OZ DARK □ CADMIUM YELLOW PRIMROSE 47-4100 □ CADMOPUR GOLDEN YELLOW N □ CADMOPUR YELLOW □ CAPSEBON □ C.I. 77199 □ C.I. PIGMENT ORANGE 20 □ C.I. PIGMENT YELLOW 37 □ FERRO LEMON YELLOW □ FERRO ORANGE YELLOW □ FERRO YELLOW □ GREENOCKITE □ NCI-C02711

TOXICITY DATA WITH REFERENCE
cyt-hmn:leu 62 µg/L PJACAW 48,133,72
otr-ham:emb 1 mg/L CNREA8 42,2757,82
dnd-ham:ovr 10 mg/L CRNGDP 3,657,82
scu-rat TDLo:90 mg/kg:CAR BJCAAI 20,190,66
ims-rat TDLo:120 mg/kg:ETA BJCAAI 20,190,66
scu-rat TD:135 mg/kg:ETA PBPHAW 14,47,78
scu-rat TD:250 mg/kg:ETA NATUAS 198,1213,63
orl-rat LD50:7080 mg/kg 41HTAH -,14,78
orl-mus LD50:1166 mg/kg 41HTAH -,14,78
ihl-mus LCLo:1350 mg/m³ NTIS** PB158-508

CONSENSUS REPORTS: NTP 7th Annual Report on Carcinogens. IARC Cancer Review: Group 1 IMEMDT 58,119,93; Animal Sufficient Evidence IMEMDT 2,74,73; Animal Sufficient Evidence IMEMDT 11,39,76; Animal Sufficient Evidence IMEMDT 58,119,93, Human Sufficient Evidence IMEMDT 58,119,93. EPA Genetic Toxicology Program. Cadmium and its compounds are on the Community Right-To-Know List. Reported in EPA TSCA Inventory.

OSHA PEL: TWA 5 µg(Cd)/m³
ACGIH TLV: TWA 0.05 mg(Cd)/m³ (Proposed: TWA 0.01 mg(Cd)/m³ (dust), Suspected Human Carcinogen; 0.002 mg(Cd)/m³ (respirable dust), Suspected Human Carcinogen); BEI: 10 µg/g creatinine in urine; 10 µg/L in blood
DFG MAK: Suspected Carcinogen
NIOSH REL: (Cadmium) Reduce to lowest feasible level

SAFETY PROFILE: Confirmed human carcinogen with experimental carcinogenic and tumorigenic data. Moderately toxic by ingestion and inhalation. Human mutation data reported. When heated to decomposition it emits very toxic fumes of Cd and SO_x. See also CADMIUM COMPOUNDS and SULFIDES.

CAJ800 **CAS:1306-25-8** **HR: 2**
CADMIUM TELLURIDE
mf: CdTe mw: 240.00

SYNS: CADMIUM MONOTELLURIDE □ IRTRAN 6

TOXICITY DATA WITH REFERENCE
ipr-rat LD50:2820 mg/kg GTPZAB 25(2),42,81
ipr-mus LD50:2100 mg/kg GTPZAB 25(2),42,81

CONSENSUS REPORTS: Reported in EPA TSCA Inventory.

ACGIH TLV: TWA 0.05 mg(Cd)/m³, TWA 0.1 mg(Te)/m³
NIOSH REL: (Cadmium, dust and fume): Lowest feasible concentration

SAFETY PROFILE: Moderately toxic by intraperitoneal route. When heated to decomposition it emits toxic vapors of Cd and Te.

CAK000 **HR: 3**
CADMIUM THERMOVACUUM AEROSOL
mf: Cd mw: 112.40

SYN: AEROSOL of THERMOVACUUM CADMIUM

TOXICITY DATA WITH REFERENCE
unr-rat LD50:1365 mg/kg GTPZAB 22(5),6,78
unr-mus LD50:815 mg/kg GTPZAB 22(5),6,78

CONSENSUS REPORTS: Cadmium and its compounds are on the Community Right-To-Know List.

OSHA PEL: TWA 5 µg(Cd)/m³
ACGIH TLV: TWA 0.05 mg(Cd)/m³ (Proposed: TWA 0.01 mg(Cd)/m³ (dust), Suspected Human Carcinogen; 0.002 mg(Cd)/m³ (respirable dust), Suspected Human Carcinogen); BEI: 10 µg/g creatinine in urine; 10 µg/L in blood
NIOSH REL: (Cadmium) Reduce to lowest feasible level

SAFETY PROFILE: Confirmed human carcinogen. Moderately toxic by an unspecified route. When heated to decomposition it emits very toxic fumes of Cd. See also CADMIUM and CADMIUM COMPOUNDS.

CAK250 **CAS:73419-42-8** **HR: 3**
CADMIUM-THIONEINE
mf: $C_{18}H_{30}N_6O_4S_2$•Cd mw: 571.06

PROP: Cadmium(II) is bound to the protein thioneine from rat or rabbit liver (BCPCA6 26,25,77).

TOXICITY DATA WITH REFERENCE
ivn-rat LD50:280 µg/kg BCPCA6 26,25,77

CONSENSUS REPORTS: Cadmium and its compounds are on the Community Right-To-Know List.

OSHA PEL: TWA 5 µg(Cd)/m³
ACGIH TLV: TWA 0.05 mg(Cd)/m³ (Proposed: TWA 0.01 mg(Cd)/m³ (dust), Suspected Human Carcinogen; 0.002 mg(Cd)/m³ (respirable dust), Suspected Human Carcinogen); BEI: 10 µg/g creatinine in urine; 10 µg/L in blood
NIOSH REL: (Cadmium) Reduce to lowest feasible level

SAFETY PROFILE: Confirmed human carcinogen. Deadly poison by intravenous route. When heated to decomposition it emits very toxic fumes of NO_x, SO_x, and Cd. See also CADMIUM COMPOUNDS.

CAK275 **CAS:64241-34-5** **HR: 3**
CADRALAZINE
mf: $C_{12}H_{21}N_5O_3$ mw: 283.38

SYNS: ETHYL-6-(ETHYL(2-HYDROXYPROPYL)AMINO)-3-PYRIDA-

C

ZINECARBAZATE □ ETHYL-2-(6(ETHYL(2-HYDROXYPROPYL)AMINO)-3-PYRIDAZINYL)HYDRAZINECARBOXYLATE □ 3-(6-(ETHYL-(2-HYDROXYPROPYL)AMINO)PYRIDAZIN-3-YL)CARBAZIC ACID ETHYL ESTER □ 2-(6-ETHYL(2-HYDROXYPROPYL)AMINO)-3-PYRIDAZINYL)-HYDRAZINECARBOXYLIC ACID ETHYL ESTER □ ISF 2469

TOXICITY DATA WITH REFERENCE
orl-rat TDLo:132 mg/kg (female 7-17D post):TER YACHDS 15,3913,87
orl-rat TDLo:3300 mg/kg (female 7-17D post):TER YACHDS 15,3913,87
orl-rat LD50:2060 mg/kg JCPCDT 3,455,81
ipr-rat LD50:440 mg/kg DRFUD4 7,382,82
ivn-rat LD50:269 mg/kg JCPCDT 3,455,81
orl-mus LD50:825 mg/kg DRFUD4 7,382,82
ipr-mus LD50:362 mg/kg DRFUD4 7,382,82
ivn-mus LD50:162 mg/kg DRFUD4 7,382,82
ivn-dog LD50:400 mg/kg JCPCDT 3,455,81

SAFETY PROFILE: Poison by intravenous and intraperitoneal routes. Moderately toxic by ingestion. Experimental reproductive effects. When heated to decomposition it emits toxic fumes of NO_x.

CAK325 HR: 2
CAESALPINIA (various species)

PROP: Several species of shrubs of various sizes producing yellow flowers with red filaments. *C. gilliesti* (Bird of Paradise) is a tall, thornless shrub (15 ft) with compound leaves and a fruit pod about 4 inches long containing 6 to 8 seeds. It is common as a cultivated plant in the southern United States from Florida to Arizona. *C. pulcherrima* (Dwarf poinciana) is a vertical shrub with some thorns. Seed pods are similar to *C. gilliesti*. It is common as a cultivated plant in the West Indies and frost-free regions of the United States. *C. bonduc* is a ground-hugging shrub with many thorns. The seed pods are about 3 inches long, covered with thorns and contain 2 large seeds. It is common in the West Indies. *C. vesicaria* is a small, thorny tree native to the West Indies except Puerto Rico.

SYNS: BARBADOS PRIDE □ BIRD of PARADISE □ BRASIL (CUBA) □ BRASILETTO (BAHAMAS) □ BRIER (BAHAMAS) □ CARZAZO (DOMINICAN REPUBLIC) □ C. BONDUC □ C. DRUMMONDII □ C. GILLIESII □ CLAVELLINA (PUERTO RICO) □ C. MEXICANA □ C. PULCHERRIMA □ C. VESICARIA □ DODDLE-DO (PUERTO RICO) □ DUL-DUL (PUERTO RICO) □ DWARF POINCIANA □ ESPIGA de AMOR (PUERTO RICO) □ FLOR de CAMARON (MEXICO) □ FLOWER FENCE □ FRANCILLADE (HAITI) □ GREY NICKER □ GUACALOTE AMARILLO (CUBA) □ GUACAMAYA (CUBA) □ HABA de SAN ANTONIO (PUERTO RICO) □ HORSE NICKER □ INDIAN SAVIN TREE (JAMAICA) □ MARAVILLA (MEXICO) □ MATO AZUL (PUERTO RICO) □ MATO de PLAYA (PUERTO RICO) □ 'OHAI-ALI'I (HAWAII) □ SPANISH CARNATION □ TABACHIN (MEXICO)

SAFETY PROFILE: The seeds usually contain toxic tannins except the immature seeds of *C. pulcherrima* and cooked seeds of *C. bonduc*. Ingestion of the seeds may cause persistent vomiting and diarrhea after a delay of 30 minutes to 6 hours.

CAK375 CAS:331-39-5 HR: 2
CAFFEIC ACID
mf: $C_9H_8O_4$ mw: 180.17

PROP: Constituent of plants, probably occurs in plants only in conjugated forms, e.g., chlorogenic acid. Yellow crystals from concentrated aq solns. Monohydrate from dil solns. Decomp 223–225° (softens at 194°). Sparingly sol in cold water; freely sol in hot water and cold alc. Alkaline solns turn from yellow to orange.

SYNS: 3,4-DIHYDROXYBENZENEACRYLIC ACID □ 3,4-DIHYDROXYCINNAMIC ACID □ 3-(3,4-DIHYDROXYPHENYL)-2-PROPENOIC ACID (9CI)

TOXICITY DATA WITH REFERENCE
mrc-smc 300 mg/L MUREAV 135,109,84
cyt-ham:ovr 200 mg/L CALEDQ 14,251,81
ipr-rat TDLo:480 mg/kg (5-12D preg):TER TXAPA9 36,227,76
ipr-rat LDLo:1500 mg/kg TXAPA9 36,227,76

SAFETY PROFILE: Moderately toxic by intraperitoneal route. An experimental teratogen. Mutation data reported. When heated to decomposition it emits acrid smoke and fumes.

CAK500 CAS:58-08-2 HR: 3
CAFFEINE
mf: $C_8H_{10}N_4O_2$ mw: 194.22

PROP: White, fleecy masses; odorless with bitter taste. Mp: 235° (anhyd). Sol in water, alc, chloroform, ether.

SYNS: CAFFEIN □ COFFEIN (GERMAN) □ COFFEINE □ 3,7-DIHYDRO-1,3,7-TRIMETHYL-1H-PURINE-2,6-DIONE □ ELDIATRIC C □ FEMA No. 2224 □ GUARANINE □ KOFFEIN (GERMAN) □ METHYLTHEOBROMIDE □ 1-METHYLTHEOBROMINE □ 7-METHYLTHEOPHYLLINE □ NCI-C02733 □ NO-DOZ □ ORGANEX □ THEIN □ THEINE □ 1,3,7-TRIMETHYL-2,6-DIOXOPURINE □ 1,3,7-TRIMETHYLXANTHINE

TOXICITY DATA WITH REFERENCE
dns-hmn:oth 1 mmol/L BIOJAU 35,665,81
dni-hmn:oth 4 mmol/L BIOJAU 35,665,81
cyt-hmn:lym 100 µg/L/24H MUREAV 46,205,77
orl-wmn TDLo:3276 mg/kg (1-39W preg):REP POMDAS 62(3),64,77
orl-wmn TDLo:1092 mg/kg (1-91D preg):REP POMDAS 62,(3),64,77
orl-wmn TDLo:6750 mg/kg (1-39W preg):TER LANCAO 1,1415,81
orl-rat TDLo:85 mg/kg (female 3-19D post):REP NETOD7 8,29,86
orl-rat TDLo: 627 mg/kg (1-22D preg):REP EXPEAM 36,1105,80
orl-mus TDLo:1 g/kg (female 8-12D post):REP JTEHD6 10,541,82
orl-ham TDLo:8160 mg/kg (male 60D pre):REP JRPFA4 43,141,75
scu-rat TDLo:200 mg/kg (male 4D pre):REP PCBRD2 36,49,80
orl-rat TDLo:660 mg/kg (female 1-22D post):REP TXAPA9 44,1,78
ipr-mus TDLo:250 mg/kg (female 10D post):TER KAIZAN 36,521,61

orl-rat TDLo:200 mg/kg (female 13-14D post):TER TJADAB 26(3),8A,82
orl-rat TDLo:120 mg/kg (female 12D post):TER FAATDF 4,240,84
ipr-rat TDLo:420 mg/kg (female 1-21D post):REP JRPFA4 34,495,73
orl-mus TDLo:1650 mg/kg (female 6-16D post):REP TXCYAC 23,57,82
orl-rat TDLo:1750 mg/kg (female 15-21D post):TER TXAPA9 22,449,72
ipr-rat TDLo:75 mg/kg (female 12D post):TER TJADAB 25,95,82
orl-mus TDLo:1650 mg/kg (female 6-16D post):TER TXCYAC 23,57,82
scu-mus TDLo:400 mg/kg (female 13D post):TER FOMOAJ 29,316,70
orl-rat TDLo:114 mg/kg (female 1-19D post):TER FCHNA9 22(26),35,80
orl-mus TDLo:2691 mg/kg (female 5-18D post):TER TXCYAC 23,57,82
mul-mus TDLo:408 mg/kg (female 9W pre):TER SEIJBO 6,171,66
orl-rat TDLo:660 mg/kg (female 1-22D post):TER TXAPA9 44,1,78
orl-mus TDLo:30,800 mg/kg/44W-C:CAR CNREA8 48,2078,88
orl-man TDLo:51 mg/kg:CVS,SYS,NEO AEMED3 18,94,89
orl-wmn TDLo:96 mg/kg/1D-I:PSY,GIT JOPDAB 105,493,84
orl-man TDLo:13 mg/kg:PSY AJPSAO 143,1320,86
orl-hmn LDLo:192 mg/kg JNDRAK 5,252,65
orl-cld LDLo:320 mg/kg FNSCA6 3,275,74
orl-wmn LDLo:1 g/kg:GIT BIATDR-,6,73
ivn-hmn TDLo:7 mg/kg:PSY APTOA6 15,331,59
orl-inf TDLo:14,700 µg/kg:CNS CLBIAS 10,148,77
ivn-inf TDLo:68 mg/kg:PSY AJDCAI 134,495,80
ivn-wmn LDLo:57 mg/kg:CNS,BLD APTOA6 15,331,59
ims-inf TDLo:36 mg/kg:PSY AJDCAI 134,495,80
orl-rat LD50:192 mg/kg JNDRAK 5,252,65
ipr-rat LD50:260 mg/kg ZERNAL 15,64,76
scu-rat LD50:170 mg/kg JCPHB8 7,131,67
ivn-rat LD50:105 mg/kg JPETAB 82,89,44
rec-rat LD50:300 mg/kg JCPHB8 7,131,67
orl-mus LD50:127 mg/kg TXAPA9 44,1,78
ipr-mus LD50:168 mg/kg CPBTAL 22,1459,74
scu-mus LD50:270 mg/kg AEPPAE 241,182,61
ivn-mus LD50:62 mg/kg TOLED5 29,25,85
orl-dog LD50:140 mg/kg NIIRDN 6,174,82

CONSENSUS REPORTS: Reported in EPA TSCA Inventory. EPA Genetic Toxicology Program.

SAFETY PROFILE: A human poison by ingestion. An experimental poison by ingestion, subcutaneous, intraperitoneal, intramuscular, rectal, and intravenous routes. Human systemic effects by: ataxia, blood pressure elevation, change in heart rate, changes in tubules, convulsions or effect on seizure threshold, diarrhea, distorted perceptions, hallucinations, hypermotility, muscle contraction, musculoskeletal tumors, nausea or vomiting, toxic psychosis, tremors. A human teratogen causing developmental abnormalities of the craniofacial and musculoskeletal systems, pregnancy termination (abortion) and stillbirth. Human maternal effects include an unspecified effect on labor or childbirth. Human mutation data reported. An experimental teratogen. Other experimental reproductive effects. Questionable carcinogen with experimental carcinogenic data. Large doses (above 1.0 g) cause palpitation, excitement, insomnia, dizziness, headache, and vomiting. Continued excessive use of caffeine in tea or coffee may lead to digestive disturbances, constipation, palpitations, shortness of breath, and depressed mental states. It is also implicated in cardiac disorders under those conditions. When heated to decomposition it emits toxic fumes of NO_x.

CAK750 CAS:5743-18-0 HR: 3
CAFFEINE HYDROBROMIDE
mf: $C_8H_{10}N_4O_2 \cdot BrH$ mw: 275.14

SYNS: CAFFEINE BROMIDE □ 3,7-DIHYDRO-1,3,7-TRIMETHYL-1H-PURINE-2,6-DIONE MONOHYDROBROMIDE

TOXICITY DATA WITH REFERENCE
orl-rbt LDLo:400 mg/kg HBAMAK 4,1289,35
scu-rbt LDLo:150 mg/kg HBAMAK 4,1289,35
ivn-rbt LDLo:100 mg/kg HBAMAK 4,1289,35

SAFETY PROFILE: Poison by ingestion, subcutaneous, and intravenous routes. See also CAFFEINE and BROMIDES. When heated to decomposition it emits very toxic fumes of NO_x and HBr.

CAK800 CAS:8000-95-1 HR: 2
CAFFEINE and SODIUM BENZOATE
mf: $C_8H_{10}N_4O_2 \cdot C_7H_5NaO_2$ mw: 338.33

SYN: SODIUM BENZOATE and CAFFEINE

TOXICITY DATA WITH REFERENCE
ipr-rat TDLo:20 mg/kg (1D male):REP SCIEAS 127,84,58
orl-rat LD50:860 mg/kg 85GMAT -,31,82
orl-mus LD50:800 mg/kg 85GMAT -,31,82
ipr-mus LD50:525 mg/kg JPETAB 116,343,56

SAFETY PROFILE: Moderately toxic by ingestion and intraperitoneal routes. Experimental reproductive effects. When heated to decomposition it emits toxic fumes of NO_x and Na_2O. See also CAFFEINE and SODIUM BENZOATE.

CAL000 CAS:470-82-6 HR: 3
CAJEPUTOL
mf: $C_{10}H_{18}O$ mw: 154.28

PROP: Colorless liquid or oil with characteristic camphoraceous odor; pungent, cooling taste. D: 0.921–0.924, refr index: 1.455–1.460, flash p: 122°F, mp: 1.5°, bp: 176–177°. Sol in alc, fixed oils, glycerin, and propylene glycol.

SYNS: 1,8-CINEOL □ CINEOLE □ 1,8-CINEOLE □ 1,8-EPOXY-p-MENTHANE □ EUCALYPTOL (FCC) □ EUCALYPTOLE □ FEMA No. 2465 □ LIMONENE OXIDE □ NCI-C56575 □ 1,8-OXIDO-p-MENTHANE □ 1,3,3-TRIMETHYL-2-OXABICYCLO(2.2.2)OCTANE

TOXICITY DATA WITH REFERENCE
sce-ham:ovr 200 mg/L EMMUEG 10(Suppl 10),1,87
scu-rat TDLo:2 g/kg (19-22D preg):REP BCPCA6 22,543,73

orl-rat LD50:2480 mg/kg FCTXAV 2,327,64
scu-mus LD50:1070 mg/kg SIZSAR 3,73,52
ims-mus LD50:100 mg/kg JSICAZ 21,342,62
scu-dog LDLo:1500 mg/kg TFAKA4 1,134,55
ims-gpg LDLo:2250 mg/kg TFAKA4 1,134,55

CONSENSUS REPORTS: Reported in EPA TSCA Inventory.

SAFETY PROFILE: Poison by intramuscular route. Moderately toxic by ingestion and subcutaneous routes. Experimental reproductive effects. Mutation data reported. Flammable liquid. When heated to decomposition it emits acrid smoke and fumes. See also LIMONENE.

CAL075 CAS:60996-85-2 HR: 3
CALACIDOL
mf: $C_{20}H_{39}N_2 \cdot Cl$ mw: 343.06

SYNS: 1-(2-(DICYCLOHEXYLAMINO)ETHYL)-1-METHYL-PIPERIDINIUM CHLORIDE ☐ I.U. 7

TOXICITY DATA WITH REFERENCE
orl-mus LD50:892 mg/kg FRPSAX 16,773,61
ipr-mus LD50:125 mg/kg FRPSAX 16,773,61
scu-mus LD50:151 mg/kg FRPSAX 16,773,61

SAFETY PROFILE: Poison by subcutaneous and intraperitoneal routes. Moderately toxic by ingestion. When heated to decomposition it emits toxic fumes of NO_x and Cl^-.

CAL125 HR: 2
CALADIUM

PROP: The various species of this genus have variegated, heart-shaped leaves. The leaf coloration may be green with white, orange, or red. They are popular house plants and may be cultivated all year in subtropical gardens and in the summer in temperate zones.

SYNS: ANGEL WINGS ☐ CALADIO (PUERTO RICO) ☐ CANANGA ☐ CAPOTILLO (MEXICO) ☐ C. BICOLOR ☐ CORAZON de CABRITO (CUBA) ☐ COUER SAIGNANT (HAITI) ☐ ELEPHANT'S EAR ☐ HEART-OF-JESUS ☐ LAGRIMAS de MARIA ☐ MOTHER-IN-LAW PLANT ☐ PALETA de PINTOR (PUERTO RICO)

SAFETY PROFILE: The whole plant contains toxic calcium oxalate raphides. Chewing any part of the plant results in burning pain in the lips, mouth and throat, possibly followed by inflammation and blistering. Systemic effects are usually not seen because of the insolubility of calcium oxalate; however, ingestion may cause inflammation of the stomach and intestines. See also OXALATES.

CAL250 CAS:7440-70-2 HR: 3
CALCIUM
DOT: UN 1401
af: Ca aw: 40.08

PROP: Silvery white, relatively soft metal. The bulk metal tarnishes in air, forming a white coating of Ca_3N_2. Mp: 849° @ 8°, bp: 1494°, d: 1.54 @ 20°, vap press: 10 mm @ 983°.

SYN: CALCICAT

CONSENSUS REPORTS: Reported in EPA TSCA Inventory.

DOT CLASSIFICATION: 4.3; *Label:* Dangerous When Wet

SAFETY PROFILE: See CALCIUM COMPOUNDS. Flammable when heated or in intimate contact with moisture or acids. Moderate explosion hazard in intimate contact with very powerful oxidizing agents. Reacts with moisture or acids to liberate large quantities of hydrogen; can develop explosive pressure in containers. To fight fire, use special mixtures of dry chemical. Violent reaction with water may evolve explosive hydrogen gas. Potentially explosive reaction with alkali metal hydroxides or carbonates; dinitrogen tetraoxide; lead chloride + heat; phosphorus(V) oxide + heat; sulfur + heat. Molten calcium reacts explosively with asbestos cement. Hypergolic reaction with chlorine fluorides (e.g., chlorine trifluoride; chlorine pentafluoride). Ignition on contact with halogens (e.g., fluorine, chlorine); sulfur + vanadium(V) oxide. Violent reaction with mercury (at 390°C); silicon (above 1050°C); sodium + mixed oxides + heat. Incompatible with air.

For occupational chemical analysis use NIOSH: Calcium 7020.

CAL500 CAS:64046-96-4 HR: 3
CALCIUM ACETARSONE
mf: $C_8H_{10}AsNO_5 \cdot 7Ca$ mw: 555.67

SYN: N-ACETYL-4-HYDROXY-m-ARSANILIC ACID, CALCIUM SALT

TOXICITY DATA WITH REFERENCE
orl-cat LDLo:135 mg/kg PSEBAA 27,267,30

CONSENSUS REPORTS: Arsenic and its compounds are on the Community Right-To-Know List.

OSHA PEL: TWA 0.5 mg(As)/m^3
ACGIH TLV: TWA 0.2 mg(As)/m^3

SAFETY PROFILE: Poison by ingestion. See also ARSENIC COMPOUNDS and CALCIUM COMPOUNDS. When heated to decomposition it emits very toxic fumes of As and NO_x.

CAL750 CAS:62-54-4 HR: 3
CALCIUM ACETATE
mf: $C_4H_6O_4 \cdot Ca$ mw: 158.18

PROP: Fine, white, hygroscopic, bulky powder. Very sol in water; sltly sol in alc.

SYNS: ACETATE of LIME ☐ BROWN ACETATE ☐ CALCIUM DIACETATE ☐ GRAY ACETATE ☐ LIME ACETATE ☐ LIME PYROLIGNITE ☐ SORBO-CALCIAN ☐ SORBO-CALCION ☐ TELTOZAN ☐ VINEGAR SALTS

TOXICITY DATA WITH REFERENCE
dns-rat-rat 1290 μmol/kg/5D-I CRNGDP 6,1819,85
ivn-rat LDLo:147 mg/kg JPETAB 71,1,41
ipr-mus LD50:75 mg/kg ABMGAJ 6,447,61
ivn-mus LD50:52 mg/kg JLCMAK 29,809,44

CONSENSUS REPORTS: Reported in EPA TSCA Inventory.

SAFETY PROFILE: Poison by intravenous and intraperitoneal routes. Mutation data reported. See also CALCIUM COMPOUNDS. When heated to decomposition it emits acrid smoke and fumes.

CAM000 **CAS:5902-95-4** **HR: 3**
CALCIUM ACID METHYL ARSONATE
mf: $C_2H_8As_2O_6{\cdot}Ca$ mw: 318.02

SYNS: CALAR □ CALCIUM ACID METHANEARSONATE □ CALCIUM HYDROGEN METHANEARSONATE □ CALCIUM METHANEARSONATE □ CAMA □ SUPER CRAB-E-RAD-CALAR □ SUPER DAL-E-RAD □ SUPER DAL-E-RAD-CALAR □ USAF AN-11

TOXICITY DATA WITH REFERENCE
ipr-mus LD50:500 mg/kg NTIS** AD414-344
unr-mam LD50:4000 mg/kg FMCHA2 -,C241,83

CONSENSUS REPORTS: Arsenic and its compounds are on the Community Right-To-Know List.

OSHA PEL: TWA 0.5 mg(As)/m^3
ACGIH TLV: TWA 0.2 mg(As)/m^3 (Proposed: 0.01 mg(As)/m^3; Human Carcinogen)

SAFETY PROFILE: Moderately toxic by intraperitoneal and possibly other routes. Arsenic compounds are considered to be poisons. An herbicide. When heated to decomposition it emits toxic fumes of As. See also ARSENIC COMPOUNDS and CALCIUM COMPOUNDS.

CAM200 **CAS:9005-35-0** **HR: 3**
CALCIUM ALGINATE
mf: $[(C_6H_70_6)_2Ca]_n$ mw: 195.16

PROP: White to yellow, granular powder. Insol in water, organic solvents.

SYNS: ALGIN □ CA 33 □ CALGINATE □ COMBINACE □ KALTOSTAT

TOXICITY DATA WITH REFERENCE
ipr-rat LD50:1407 mg/kg FAONAU 53A,381,74
ivn-rat LD50:64 mg/kg FAONAU 53A,381,74

CONSENSUS REPORTS: Reported in EPA TSCA Inventory.

SAFETY PROFILE: Poison by intravenous route. Moderately toxic by intraperitoneal route. When heated to decomposition it emits acrid smoke and irritating fumes.

CAM300 **CAS:15194-98-6** **HR: 3**
CALCIUM ARSENITE
DOT: NA 1574
mf: $AsO_4{\cdot}Ca$ mw: 179.00

SYNS: ARSENENOUS ACID, CALCIUM SALT (2:1) □ CALCIUM ARSENITE, solid (DOT) □ PROTARS

CONSENSUS REPORTS: Reported in NTP 7th Annual Report on Carcinogens, 1992.

DOT CLASSIFICATION: 6.1; *Label:* Poison

SAFETY PROFILE: A carcinogen and poison. When heated to decomposition it emits toxic vapors of As.

CAM500 **CAS:27152-57-4** **HR: 3**
CALCIUM ARSENITE
DOT: NA 1574
mf: $As_2O_6{\cdot}3Ca$ mw: 366.08

PROP: White, granular powder.

SYNS: ARSENIOUS ACID, CALCIUM SALT □ CALCIUM ARSENITE, solid (DOT) □ MONOCALCIUM ARSENITE

TOXICITY DATA WITH REFERENCE
orl-hmn LDLo:1666 μg/kg YKYUA6 28,329,77
orl-mus LD50:1 mg/kg YKYUA6 28,329,77
orl-dog LDLo:85 mg/kg YKYUA6 28,329,77
orl-pig LDLo:5 mg/kg YKYUA6 28,329,77

CONSENSUS REPORTS: NTP 7th Annual Report on Carcinogens. Arsenic and its compounds are on the Community Right-To-Know List.

OSHA PEL: Cancer Hazard
ACGIH TLV: TWA 0.2 mg(As)/m^3 (Proposed: 0.01 mg(As)/m^3; Human Carcinogen)
NIOSH REL: (Inorganic Arsenic) CL 0.002 mg(As)/m^3/15M
DOT CLASSIFICATION: 6.1; *Label:* Poison

SAFETY PROFILE: Confirmed carcinogen. A poison by inhalation and ingestion. When heated to decomposition it emits toxic fumes of As. See also ARSENIC COMPOUNDS and CALCIUM COMPOUNDS.

CAM675 **CAS:21059-46-1** **HR: 2**
CALCIUM ASPARTATE
mf: $C_4H_7NO_4{\cdot}7Ca$ mw: 413.68

SYNS: ASPARAGINATE CALCIUM □ CALCIRETARD □ CALCIUM-l-ASPARTATE

TOXICITY DATA WITH REFERENCE
orl-mus LD50:10 g/kg NIIRDN 6,12,82
ipr-mus LD50:1059 mg/kg NIIRDN 6,12,82
ivn-mus LD50:646 mg/kg NIIRDN 6,12,82

SAFETY PROFILE: Moderately toxic by intraperitoneal and intravenous routes. When heated to decomposition it emits toxic fumes of NO_x. See also CALCIUM COMPOUNDS.

CAM680 **HR: 2**
CALCIUM BENZOATE
mf: $C_{14}H_{10}O_4{\cdot}3H_2O$ mw: 374.26

PROP: Orthorhombic crystals or powder. D: 1.44. Sol in water.

SAFETY PROFILE: Combustible when exposed to heat or flame. When heated to decomposition it emits acrid smoke and irritating fumes.

CAM750 **CAS:6485-34-3** **HR: 3**
CALCIUM-o-BENZOSULFIMIDE
mf: $C_{14}H_{10}N_2O_6S_2{\cdot}Ca$ mw: 406.46

PROP: White, crystalline powder; odorless or faint aromatic odor; sol in water.

SYNS: 1,2-BENZISOTHIAZOL-3(2H)-ONE-1,1-DIOXIDE, CALCIUM

SALT □ CALCIUM-o-BENZOSULPHIMIDE □ CALCIUM-2-BENZOSULPHIMIDE □ CALCIUM SACCHARIN □ CALCIUM SACCHARINA □ CALCIUM SACCHARINATE □ DARAMIN □ SACCHARIN CALCIUM □ SULPHOBENZOIC IMIDE CALCIUM SALT

TOXICITY DATA WITH REFERENCE
dns-rat:lvr 100 mg/L CNREA8 40,4541,80
cyt-ham:lng 8 g/L MUREAV 163,63,86

CONSENSUS REPORTS: NTP 7th Annual Report On Carcinogens. Reported in EPA TSCA Inventory.

SAFETY PROFILE: Confirmed carcinogen. Mutagenic data reported. When heated to decomposition it emits toxic fumes of SO_x and NO_x.

CAN000 CAS:13780-03-5 HR: 3
CALCIUM BISULFITE
DOT: UN 1923

PROP: Colorless or sltly yellowish liquid, strong sulfur dioxide odor. D: 1.06.

SYNS: CALCIUM DITHIONITE (DOT) □ CALCIUM HYDROSULFITE (DOT) □ SULFUROUS ACID, CALCIUM SALT (2:1) (8CI,9CI)

TOXICITY DATA WITH REFERENCE
eye-rbt 250 mg/5D MLD AMIHAB 14,265,56

CONSENSUS REPORTS: Reported in EPA TSCA Inventory.

DOT CLASSIFICATION: 4.2; *Label:* Spontaneously Combustible

SAFETY PROFILE: A poison via ingestion. Strong irritant via skin and eye contact, ingestion, and inhalation. Spontaneously combustible. When heated to decomposition it emits toxic fumes of SO_x. See also SULFITES and SULFUROUS ACID.

CAN250 CAS:12007-56-6 HR: 2
CALCIUM BORATE
mf: B_4CaO_7 mw: 195.32

PROP: Colorless, rhombic or long, flat plates. Mp: 1154°.

SYNS: BORIC ACID (H2-B4-O7), CALCIUM SALT (1:1) (8CI) □ BORON CALCIUM OXIDE □ CALCIUM TETRABORATE □ COLEMANITE

TOXICITY DATA WITH REFERENCE
orl-rat LD50:5600 mg/kg GTPZAB 25(6),53,81
orl-mus LD50:5900 mg/kg GTPZAB 25(6),53,81
ipr-mus LD50:3900 mg/kg GTPZAB 25(6),53,81

CONSENSUS REPORTS: Reported in EPA TSCA Inventory.

SAFETY PROFILE: Moderately toxic by intraperitoneal route. Mildly toxic by ingestion. See also CALCIUM COMPOUNDS and BORON COMPOUNDS.

CAN400 HR: 1
CALCIUM BROMATE
mf: $Ca(BrO_3)_2 \cdot H_2O$ mw: 313.90

PROP: White crystalline powder. Very sol in water.

SAFETY PROFILE: A nuisance dust.

CAN750 CAS:75-20-7 HR: 3
CALCIUM CARBIDE
DOT: UN 1402
mf: C_2Ca mw: 64.10

PROP: Rhombic, moisture-sensitive, gray crystals. Mp: approx 2300°, d: 2.222.

SYNS: ACETYLENOGEN □ CALCIUM ACETYLIDE □ CALCIUM DICARBIDE

CONSENSUS REPORTS: Reported in EPA TSCA Inventory.

DOT CLASSIFICATION: 4.3; *Label:* Dangerous When Wet

SAFETY PROFILE: Reaction on contact with moisture forms explosive acetylene gas. Flammable on contact with moisture, acid, or acid fumes; evolves heat or flammable vapors. Moderate explosion hazard. Incandescent reaction with Cl_2 (245°C), Br_2 (350°C), I_2 (305°C), HCl gas + heat, PbF_2, Mg + heat. Incompatible with Se; (KOH + Cl_2), $AgNO_3$, Na_2O_2, $SnCl_2$, S, water. Mixtures with iron(III) chloride, iron(III) oxide, tin(II) chloride are easily ignited and burn fiercely. Vigorous reaction with methanol after an induction period. Addition to silver nitrate solutions precipitates the dangerously explosive silver acetylide. Copper salt solutions behave similarly. See also CALCIUM HYDROXIDE and ACETYLENE.

CAO000 CAS:1317-65-3 HR: 1
CALCIUM CARBONATE
mf: $CO_3 \cdot Ca$ mw: 100.09

PROP: White microcrystalline powder. Mp: 825° (α), 1339° (β) @ 102.5 atm; d: 2.7–2.95. Found in nature as the minerals limestone, marble, aragonite, calcite, and vaterite. Odorless, tasteless powder or crystals. Two crystalline forms are of commercial importance: aragonite, orthorhombic, mp: 825° (decomp), d: 2.83, formed at temperatures above 30°; calcite, hexagonal-rhombohedral, mp: 1339° (102.5 atm), d: 2.711, formed at temperatures below 30°. At about 825° it decomposes into CaO and CO_2. Practically insol in water, alc; sol in dilute acids.

SYNS: AGRICULTURAL LIMESTONE □ AGSTONE □ ARAGONITE □ ATOMIT □ BELL MINE PULVERIZED LIMESTONE □ CALCITE □ CARBONIC ACID, CALCIUM SALT (1:1) □ CHALK □ DOLOMITE □ FRANKLIN □ LIMESTONE (FCC) □ LITHOGRAPHIC STONE □ MARBLE □ NATURAL CALCIUM CARBONATE □ PORTLAND STONE □ SOHNHOFEN STONE □ VATERITE

CONSENSUS REPORTS: Reported in EPA TSCA Inventory.

OSHA PEL: Total Dust: 15 mg/m^3; Respirable Fraction: 5 mg/m^3
ACGIH TLV: TWA (nuisance particulate) 10 mg/m^3 of total dust (when toxic impurities are not present, e.g., quartz <1%)

SAFETY PROFILE: A nuisance dust. An eye and skin irritant. Ignites on contact with F_2. Incompatible with acids, alum, ammonium salts, ($Mg + H_2$). Calcium carbonate is a common air contaminant. See also CALCIUM COMPOUNDS.

For occupational chemical analysis use NIOSH: Nuisance Dust, Total, 0500; Nuisance Dust, Respirable, 0600.

CAO250 CAS:9049-05-2 HR: 2
CALCIUM CARRAGHEENATE

PROP: A mixture of highly sulfated polygalactosides. It is extracted from seaweed (FAONAU 53A,398,74).

SYNS: ALGIN GUM □ CALCIUM CARAGEENIN □ CALCIUM CARRAGEENAN □ CARRAGEENAN, CALCIUM(II) SALT □ VISCARIN 402

TOXICITY DATA WITH REFERENCE
orl-rat TDLo: 13 g/kg (14D pre-21D post):REP TXAPA9 50,267,79
orl-rat TDLo: 105 g/kg (MGN):REP FCTXAV 15,533,77
orl-rat LD50:5140 mg/kg FAONAU 53A,386,74
orl-mus LD50:8710 mg/kg FAONAU 53A,398,74
orl-rbt LD50:2280 mg/kg FAONAU 53A,398,74
orl-ham LD50:6180 mg/kg FAONAU 53A,398,74

CONSENSUS REPORTS: Reported in EPA TSCA Inventory.

SAFETY PROFILE: Moderately toxic by ingestion. Experimental reproductive effects. When heated to decomposition it emits toxic fumes of SO_x. See also CALCIUM COMPOUNDS.

CAO500 CAS:10137-74-3 HR: 2
CALCIUM CHLORATE
DOT: UN 1452/UN 2429
mf: $Cl_2O_6{\cdot}Ca$ mw: 206.98

PROP: Monoclinic, yellowish-white, deliquescent crystals. Mp: 340° loses H_2O @ >100°, d: 2.711. Very sol in H_2O.

SYNS: CALCIUM CHLORATE, aqueous solution (DOT) □ CHLORATE de CALCIUM (FRENCH)

TOXICITY DATA WITH REFERENCE
orl-rat LDLo:4500 mg/kg JPETAB 35,1,29
ipr-rat LDLo:625 mg/kg JPETAB 35,1,29

DOT CLASSIFICATION: 5.1; *Label:* Oxidizer

SAFETY PROFILE: Moderately toxic by ingestion and intraperitoneal routes. A powerful oxidant. Incompatible with Al, As, C, Cu, charcoal, MnO_2, metal sulfides, S, dibasic organic acids, organic matter, P. When heated to decomposition it emits toxic fumes of Cl^-. See also CHLORATES for fire, disaster, and explosion hazards.

CAO750 CAS:10043-52-4 HR: 2
CALCIUM CHLORIDE
mf: $CaCl_2$ mw: 110.98

PROP: Cubic, colorless, deliq crystals. Mp: 782°, bp: >1600°, d: 2.512 @ 25°. Very sol H_2O; sol in EtOH, Me_2CO, and AcOH.

SYNS: CALCIUM CHLORIDE, anhydrous □ CALPLUS □ CALTAC □ DOWFLAKE □ LIQUIDOW □ PELADOW □ SNOMELT □ SUPERFLAKE ANHYDROUS

TOXICITY DATA WITH REFERENCE
dns-rat-ipr 2500 μmol/kg JOENAK 65,45,75
cyt-rat:ast 3500 mg/kg GANNA2 7,165,87
orl-rat TDLo: 112 g/kg/20W-C:ETA AJCAA7 23,550,35
ivn-wmn TDLo: 20 mg/kg/1H-C:SKN,GLN ARDEAC 124,922,88
orl-rat LD50:1000 mg/kg CJCMAV 12,216,48
ipr-rat LD50:264 mg/kg OYYAA2 14,963,77
scu-rat LD50:2630 mg/kg OYYAA2 14,963,77
ivn-rat LDLo:161 mg/kg JLCMAK 15,35,29
ims-rat LD50:25 mg/kg EMSUA8 4,223,46
orl-mus LD50:1940 mg/kg OYYAA2 14,963,77
ipr-mus LD50:210 mg/kg GTPZAB 34(5),51,90
scu-mus LD50:823 mg/kg OYYAA2 14,963,77
ivn-mus LD50:42 mg/kg TXAPA9 22,150,72
ipr-dog LDLo:110 mg/kg AVERAG 44,555,37
scu-dog LDLo:274 mg/kg HBAMAK 4,1316,35

CONSENSUS REPORTS: Reported in EPA TSCA Inventory. EPA Genetic Toxicology Program.

SAFETY PROFILE: Moderately toxic by ingestion. Poison by intravenous, intramuscular, intraperitoneal, and subcutaneous routes. Human systemic effects: dermatitis, changes in calcium. Questionable carcinogen with experimental tumorigenic data. Mutation data reported. Reacts violently with (B_2O_3 + CaO), BrF_3. Reaction with zinc releases explosive hydrogen gas. Catalyzes exothermic polymerization of methyl vinyl ether. Exothermic reaction with water. When heated to decomposition it emits toxic fumes of Cl^-. See also CALCIUM COMPOUNDS and CHLORIDES.

CAP000 CAS:14674-72-7 HR: 3
CALCIUM CHLORITE
DOT: UN 1453
mf: $CaCl_2O_4$ mw: 174.98

$Ca(ClO_2)_2$

PROP: White solid.
DOT CLASSIFICATION: 5.1; *Label:* Oxidizer

SAFETY PROFILE: A strong oxidizer. Ignites on contact with potassium thiocyanate. Reaction with Cl_2 yields explosive ClO_2. When heated to decomposition it emits toxic fumes of Cl^-. See also CHLORITES and CALCIUM COMPOUNDS.

CAP250 CAS:85721-24-0 HR: 2
CALCIUM-4-(p-CHLOROPHENYL)-2-PHENYL-5-THIAZOLEACETATE
mf: $C_{17}H_{11}ClNO_2S{\cdot}Ca$ mw: 368.88

SYNS: CALCIUM-2-PHENYL-4-(p-CHLOROPHENYL)-5-THIAZOLEACETATE □ 4-(p-CHLOROPHENYL)-2-PHENYL-5-THIAZOLEACETIC ACID CALCIUM SALT □ FENTIAZAC CALCIUM SALT

TOXICITY DATA WITH REFERENCE
orl-rat LD50:860 mg/kg CMROCX 6,53,79
orl-mus LD50:1353 mg/kg CMROCX 6,53,79

SAFETY PROFILE: Moderately toxic by ingestion. When

heated to decomposition it emits very toxic fumes of SO_x, NO_x, and Cl^-. See also CALCIUM COMPOUNDS.

CAP500 **CAS:13765-19-0** **HR: 3**
CALCIUM CHROMATE
mf: $CrO_4 \cdot Ca$ mw: 156.08

PROP: Monoclinic prisms; yellow colored crystals. Sltly sol in H_2O; insol in EtOH and Me_2CO.

SYNS: CALCIUM CHROMATE (VI) □ CALCIUM CHROME YELLOW □ CALCIUM CHROMIUM OXIDE ($CaCrO_4$) □ CALCIUM MONOCHROMATE □ CHROMIC ACID, CALCIUM SALT (1:1) □ C.I. 77223 □ C.I. PIGMENT YELLOW 33 □ GELBIN □ RCRA WASTE NUMBER U032 □ YELLOW ULTRAMARINE

TOXICITY DATA WITH REFERENCE
mmo-sat 50 nmol/plate CRNGDP 2,283,81
mma-esc 100 μg/plate ENMUDM 6(Suppl 2),1,84
otr-rat:emb 58 μg/L JJIND8 67,1303,81
dlt-mus-unr 40 mg/kg MUREAV 97,180,82
dnd-ham:ovr 25 μmol/L/1H-C PAACA3 24,74,83
sce-ham:ovr 100 μg/L MUREAV 156,219,85
unr-mus TDLo:1400 mg/kg (35W male):REP MUREAV 97,180,82
ims-rat TDLo:76 mg/kg/19W-I:NEO BJCAAI 23,172,69
itr-rat TDLo:163 mg/kg/130W-I:CAR EXPADD 30,129,86
imp-rat TDLo:8 mg/kg:CAR CRNGDP 7,831,86
imp-rat TD:50 mg/kg:ETA AEHLAU 5,445,62
ims-rat TD:216 mg/kg/1Y-I:NEO CNREA8 36,1779,76
imp-rat TD:125 mg/kg:CAR AIHAAP 20,274,59
imp-rat TD:10,866 μg/kg:CAR BJIMAG 43,243,86

CONSENSUS REPORTS: NTP 7th Annual Report on Carcinogens. IARC Cancer Review: Group 1 IMEMDT 49,49,90; Human Sufficient Evidence IMEMDT 23,205,80; Animal Sufficient Evidence IMEMDT 23,205,80. Reported in EPA TSCA Inventory. EPA Genetic Toxicology Program. Chromium and its compounds are on the Community Right-To-Know List.

OSHA PEL: CL 0.1 $mg(CrO_3)/m^3$
ACGIH TLV: TWA 0.001 $mg(Cr)/m^3$; Suspected Human Carcinogen
DFG TRK: 0.1 mg/m^3 calculated as CrO_3 in that portion of dust that can possibly be inhaled; 0.2 mg/m^3 arc-welding by hand; others 0.1 mg/m^3. Animal Carcinogen, Suspected Human Carcinogen.
NIOSH REL: (Chromium(VI)) TWA 0.001 $mg(Cr(VI))/m^3$

SAFETY PROFILE: Confirmed human carcinogen with experimental carcinogenic, neoplastigenic, and tumorigenic data. Experimental reproductive effects. Mutation data reported. A powerful oxidizer. Mixture with boron burns violently if ignited. See also CHROMIUM COMPOUNDS and CALCIUM COMPOUNDS.

For occupational chemical analysis use NIOSH: Chromium Hexavalent 7024.

CAP750 **CAS:8012-75-7** **HR: 3**
CALCIUM CHROMATE(VI) DIHYDRATE
mf: $CrO_4 \cdot Ca \cdot 2H_2O$ mw: 192.12

SYNS: CALCIUM CHROME YELLOW □ CHROMIC ACID, CALCIUM SALT (1:1), DIHYDRATE □ C.I. 77223 □ C.I. PIGMENT YELLOW 33 □ GELBIN YELLOW ULTRAMARINE □ PIGMENT YELLOW 33 □ STEINBUHL YELLOW

TOXICITY DATA WITH REFERENCE
otr-ham:kdy 250 mg/L CNREA8 35,1058,75
scu-mus TDLo:400 mg/kg:ETA AMIHAB 21,530,60
imp-mus TDLo:400 mg/kg:CAR AMIHAB 21,530,60
orl-rat LD50:327 mg/kg TXAPA9 42,417,77
imp-rat LDLo:112 mg/kg AMIHAB 21,530,60

CONSENSUS REPORTS: IARC Cancer Review: Animal Sufficient Evidence IMEMDT 2,100,72. Chromium and its compounds are on the Community Right-To-Know List.

OSHA PEL: CL 0.1 $mg(CrO^3)/m^3$
ACGIH TLV: TWA 0.05 $mg(Cr)/m^3$; Confirmed Human Carcinogen
NIOSH REL: (Chromium(VI)) TWA 0.001 $mg(Cr(VI))/m^3$

SAFETY PROFILE: Confirmed human carcinogen with experimental tumorigenic and carcinogenic data. Poison by ingestion and implant routes. Mutation data reported. A powerful oxidizer. See also CHROMIUM COMPOUNDS and CALCIUM COMPOUNDS.

For occupational chemical analysis use NIOSH: Chromium Hexavalent 7024.

CAQ000 **HR: 1**
CALCIUM COMPOUNDS

SAFETY PROFILE: The fumes evolved by burning calcium in air are composed of calcium oxide (quicklime), which is an irritant to the skin, eyes, and mucous membranes. Generally speaking, calcium compounds should be considered toxic only when they contain toxic components (such as arsenic, etc.) or as calcium oxide or hydroxide. Calcium compounds are common air contaminants.

CAQ250 **CAS:156-62-7** **HR: 3**
CALCIUM CYANAMIDE
DOT: UN 1403
mf: $CN_2 \cdot Ca$ mw: 80.11

PROP: Hexagonal, rhombohedral, colorless, moisture-sensitive, crystals. Mp: 1300°, subl >1500°. Decomposes in water. Compound not hydrated; compound contains more than 0.1% calcium (FEREAC 41,15972,76).

SYNS: AERO-CYANAMID □ AERO CYANAMID GRANULAR □ AERO CYANAMID SPECIAL GRADE □ ALZODEF □ CALCIUM CARBIMIDE □ CALCIUM CYANAMID □ CCC □ CYANAMIDE □ CYANAMIDE CALCIQUE (FRENCH) □ CYANAMIDE, CALCIUM SALT (1:1) □ CYANAMID GRANULAR □ CYANAMID SPECIAL GRADE □ CY-L 500 □ LIME-NITROGEN (DOT) □ NCI-C02937 □ NITROGEN LIME □NITROLIME □ USAF CY-2

TOXICITY DATA WITH REFERENCE
mmo-sat 1 mg/plate ENMUDM 5(Suppl 1),3,83
mma-sat 100 μg/plate ENMUDM 5(Suppl 1),3,83
orl-mus TDLo:170 g/kg/2Y-C:ETA NCITR* NCI-CG-TR-163,79
orl-hmn LDLo:571 mg/kg 34ZIAG -,149,69
orl-rat LD50:158 mg/kg NIIRDN 6,304,82

ihl-rat LCLo:86 mg/m³/4H 85GMAT -,40,82
skn-rat LD50:84 mg/kg 85GMAT -,40,82
ivn-rat LD50:125 mg/kg NIIRDN 6,304,82
unr-rat LD50:1000 mg/kg GUCHAZ 6,73,73
orl-mus LD50:334 mg/kg NIIRDN 6,304,82
ipr-mus LD50:100 mg/kg NTIS** AD277-689
ivn-mus LD50:282 mg/kg NIIRDN 6,304,82
orl-cat LD50:100 mg/kg 85GMAT -,40,82
orl-rbt LD50:1400 mg/kg PCOC** -,174,66
skn-rbt LD50:590 mg/kg 37ASAA 7,291,79

CONSENSUS REPORTS: NCI Carcinogenesis Bioassay (feed); No Evidence: mouse, rat NCITR* NCI-CG-TR-163,79. Community Right-To-Know List. Reported in EPA TSCA Inventory.

OSHA PEL: TWA 0.5 mg/m³
ACGIH TLV: TWA 0.5 mg/m³
DFG MAK: 1 mg/m³
DOT CLASSIFICATION: 4.3; *Label:* Dangerous When Wet

SAFETY PROFILE: Poison by ingestion, inhalation, skin contact, intravenous, and intraperitoneal routes. Moderately toxic to humans by ingestion. Questionable carcinogen with experimental tumorigenic data. Mutation data reported. The fatal dose, by ingestion, is probably around 20 to 30 g for an adult. It does not have a cyanide effect. Calcium cyanamide is not believed to have a cumulative action. Flammable. Reaction with water forms the explosive acetylene gas. When heated to decomposition it emits toxic fumes of NO_x and CN^-. See also CALCIUM COMPOUNDS; AMIDES; and CYANIDE.

CAQ500 CAS:592-01-8 HR: 3
CALCIUM CYANIDE
DOT: UN 1575
mf: C_2CaN_2 mw: 92.12

PROP: Rhombohedral crystals or white powder. Mp: decomp >350°.

SYNS: CALCID ☐ CALCIUM CYANIDE MIXTURE, solid (DOT) ☐ CALCYAN ☐ CALCYANIDE ☐ CYANOGAS ☐ CYANURE de CALCIUM (FRENCH) ☐ RCRA WASTE NUMBER P021

TOXICITY DATA WITH REFERENCE
orl-rat LD50:39 mg/kg AIHAAP 30,470,69

CONSENSUS REPORTS: Cyanide and its compounds are on the Community Right-To-Know List. Reported in EPA TSCA Inventory.

OSHA PEL: TWA 5 mg(CN)/m³
ACGIH TLV: CL 5 mg(CN)/m³ (skin)
DFG MAK: 5 mg/m³
NIOSH REL: (Cyanide) CL 5 mg(CN)/m³/10M
DOT CLASSIFICATION: 6.1; *Label:* Poison

SAFETY PROFILE: A deadly poison by ingestion and probably other routes. When heated to decomposition it emits toxic fumes of NO_x and CN^-. See also CALCIUM COMPOUNDS and CYANIDE.

CAR000 CAS:139-06-0 HR: 3
CALCIUM CYCLOHEXYLSULPHAMATE
mf: $C_{12}H_{24}N_2O_6S_2{\cdot}Ca$ mw: 396.58

PROP: White, crystalline powder; almost odorless; freely sol in water; practically insol in alc, benzene, chloroform, and ether.

SYNS: CALCIUM CYCLAMATE ☐ CALCIUM CYCLOHEXANESULFAMATE ☐ CALCIUM CYCLOHEXANE SULPHAMATE ☐ CALCIUM CYCLOHEXYLSULFAMATE ☐ CYCLAMATE CALCIUM ☐ CYCLAMATE, CALCIUM SALT ☐ CYCLAN ☐ CYCLOHEXANESULFAMIC ACID, CALCIUM SALT ☐ CYCLOHEXYLSULPHAMIC ACID, CALCIUM SALT ☐ CYLAN ☐ DIETIL ☐ KALZIUMZYKLAMATE (GERMAN) ☐ SUCARYL CALCIUM

TOXICITY DATA WITH REFERENCE
sln-dmg-orl 5 mmol/L DRISAA 46,114,71
dni-hmn:lng 100 mg/L JCLBA3 47,30a,70
cyt-hmn:leu 250 mg/L SCIEAS 164,568,69
cyt-ham:fbr 10 mg/L MUREAV 39,1,76
cyt-ham:lng 100 mg/L HEREAY 70,271,72
cyt-grb-ipr 150 mg/kg CNJGA8 13,189,71
orl-rat TDLo:55 mg/kg (1-22D preg):REP AJCNAC 23,782,70
orl-rat TDLo:3465 g/kg/88W-C:NEO JNCIAM 49,751,72
scu-rat TDLo:45 g/kg/66W-I:ETA FCTXAV 9,463,71
orl-rat LDLo:10 mg/kg CLDND* 7,178,87

CONSENSUS REPORTS: IARC Cancer Review: Group 3 IMEMDT 7,178,87; Animal Limited Evidence IMEMDT 22,55,80; Human Inadequate Evidence IMEMDT 22,55,80. Reported in EPA TSCA Inventory. EPA Genetic Toxicology Program.

SAFETY PROFILE: Poison by ingestion and intravenous routes. Experimental reproductive effects. Questionable carcinogen with experimental tumorigenic and neoplastigenic data. Human mutation data reported. When heated to decomposition it emits very toxic fumes of SO_x and NO_x. See also CALCIUM COMPOUNDS.

CAR375 CAS:7789-41-5 HR: 2
CALCIUM DIBROMIDE
mf: Br_2Ca mw: 199.90

PROP: Colorless, orthorhombic, deliquescent crystals. The N.F. grade is a hydrated salt, containing not less than 84% and not more than 94% $CaBr_2$. Odorless, deliquescen granules or rhombic crystals; sharp, saline taste. Becomes yellow on long exposure to air. Mp: 742° (anhydrous), d: (25/4) 3.353. When strongly heated in air, becomes alkaline due to loss of bromine and formation of lime. Very sol in water, methanol, ethanol; sol in acetone; practically insol in dioxane, chloroform, ether.

SYN: CALCIUM BROMIDE

TOXICITY DATA WITH REFERENCE
cyt-rat/ast 2300 mg/kg GANNA2 54,155,63
ipr-rat LD50:437 mg/kg OYYAA2 16,229,78
ipr-mus LD50:740 mg/kg OYYAA2 16,229,78
scu-mus LD50:1580 mg/kg OYYAA2 20,693,80

CONSENSUS REPORTS: Reported in EPA TSCA Inventory.

C

SAFETY PROFILE: Moderately toxic by intraperitoneal and subcutaneous routes. Mutation data reported. Incompatible with potassium. When heated to decomposition it emits toxic fumes of Br^-. See also BROMIDES and CALCIUM COMPOUNDS.

CAR750 CAS:12013-56-8 HR: 3
CALCIUM DISILICIDE
mf: $CaSi_2$ mw: 96.25

PROP: Gray hexagonal crystals with metallic luster. Mp: 1033°. Insol in water.

SAFETY PROFILE: Mixture with CCl_4 is a friction-sensitive explosive. Ignites on close contact with alkali metal fluorides. Mixture with iron(III) oxide (silicon thermite) reacts violently when heated producing molten iron as with the normal thermite mixture. Mixtures with potassium nitrate are easily ignited and burn at a very high temperature. See also CALCIUM COMPOUNDS.

CAR780 CAS:62-33-9 HR: 2
CALCIUM DISODIUM ETHYLENEDIAMINETETRAACETATE
mf: $C_{10}H_{12}CaN_2O_8 \cdot 2Na$ mw: 374.30

SYNS: ACETIC ACID, (ETHYLENEDINITRILO)TETRA-, CALCIUM DISODIUM SALT □ ADSORBONAC □ ANTALLIN □ CALCIATE(2^-), ((ETHYLENEDINITRILO)TETRAACETATO)-, DISODIUM □ CALCITETRACEMATE DISODIUM □ CALCIUM DISODIUM EDATHAMIL □ CALCIUM DISODIUM EDETATE □ CALCIUM DISODIUM EDTA □ CALCIUM DISODIUM (ETHYLENEDINITRILO)TETRAACETATE □ CALCIUM DISODIUM VERSENATE □ CALCIUM EDTA □ CALCIUM TITRIPLEX □ DISODIUM CALCIUM EDTA □ DISODIUM CALCIUM ETHYLENEDIAMINETETRAACETATE □ EDATHAMIL CALCIUM DISODIUM □ EDETAMIN □ EDETAMINE □ EDETATE CALCIUM □ EDETIC ACID CALCIUM DISODIUM SALT □ EDTACAL □ EDTA CALCIUM DISODIUM SALT □ ETHYLENEDIAMINETETRAACETIC ACID, CALCIUM DISODIUM CHELATE □ LEDCLAIR □ MONOCALCIUM DISODIUM EDTA □ MOSATIL □ RIKELATE CALCIUM □ SODIUM CALCIUM EDETATE □ SORMETAL □ TETACIN □ TETACIN-CALCIUM □ TETAZINE □ VERSENE CA

TOXICITY DATA WITH REFERENCE
orl-rat LD50:10 g/kg TXAPA9 5,142,63
ipr-rat LD50:3850 mg/kg JPETAB 117,20,56
ivn-rat LD50:3 g/kg CLDND* -,188,90
ipr-mus LD50:4500 mg/kg CLDND* 5,142,63
orl-rbt LD50:7 g/kg TXAPA9 5,142,63
ipr-rbt LD50:6 g/kg DRUGAY -,188,90
ivn-rbt LDLo:4 g/kg FEPRA7 11,321,52

CONSENSUS REPORTS: Reported in EPA TSCA Inventory.

SAFETY PROFILE: Moderately toxic by intraperitoneal route. Mildly toxic by ingestion and intravenous routes. When heated to decomposition it emits toxic vapors of NO_x.

CAR800 CAS:12264-18-5 HR: 2
CALCIUM EDTA COMPLEX
mf: $C_{10}H_{12}CaN_2O_8 \cdot 2H$ mw: 330.34

SYNS: ACETIC ACID, (ETHYLENEDINITRILO)TETRA-, CALCIUM (II) COMPLEX □ CALCIATE(2-), ((ETHYLENEDINITRILO)TETRAACETATO)-, DIHYDROGEN (8CI) □ VERSENE CA

TOXICITY DATA WITH REFERENCE
scu-rat TDLo:1344 mg/kg (female 11-15D post):TER TXAPA9 82,426,86
scu-rat TDLo:2016 mg/kg (female 11-15D post):REP TXAPA9 82,426,86
scu-rat TDLo:2687 mg/kg (female 11-15D post):REP TXAPA9 82,426,86
scu-rat TDLo:2016 mg/kg (female 11-15D post):TER TXAPA9 82,426,86
ipr-mus LD50:573 mg(Ca)/kg PABIAQ 11,853,63

SAFETY PROFILE: Moderately toxic by intraperitoneal route. An experimental teratogen. Other experimental reproductive effects. When heated to decomposition it emits toxic fumes of NO_x.

CAR875 CAS:32266-82-3 HR: 2
CALCIUM-N-2-ETHYLHEXYL-β-OXYBUTYRAMIDE SEMISUCCINATE
mf: $C_{32}H_{58}N_2O_{10} \cdot Ca$ mw: 671.00

SYN: M-2

TOXICITY DATA WITH REFERENCE
orl-mus TDLo:300 mg/kg (female 7-14D post):REP TOIZAG 18,88,71
orl-rat TDLo:3 g/kg (9-14D preg):TER TOIZAG 18,88,71
orl-rat TDLo:3 g/kg (9-14D preg):REP TOIZAG 18,88,71
orl-rat TDLo:600 mg/kg (9-14D preg):TER TOIZAG 18,88,71
orl-rat LD50:5746 mg/kg TOIZAG 17,579,70
ipr-rat LD50:741 mg/kg TOIZAG 17,579,70
scu-rat LD50:3037 mg/kg TOIZAG 17,579,70
orl-mus LD50:2129 mg/kg TOIZAG 17,579,70
ipr-mus LD50:549 mg/kg TOIZAG 17,579,70
scu-mus LD50:1187 mg/kg TOIZAG 17,579,70

SAFETY PROFILE: Moderately toxic by intraperitoneal, subcutaneous, and ingsetion routes. An experimental teratogen. Experimental reproductive effects. When heated to decomposition it emits toxic fumes of NO_x. See also CALCIUM COMPOUNDS and ESTERS.

CAS000 CAS:7789-75-5 HR: 2
CALCIUM FLUORIDE
mf: CaF_2 mw: 78.08

PROP: Hygroscopic, cubic, colorless crystals; luminous with heat. Mp: 1418°, d: 3.180. Practically insol in H_2O; insol in Me_2CO; sol in acids.

SYNS: ACID-SPAR □ CALCIUM DIFLUORIDE □ FLUORITE □ FLUORSPAR □ IRTRAN 3 □ LIPARITE □ MET-SPAR

TOXICITY DATA WITH REFERENCE
cyt-rat/ast 1 g/kg GANNA2 54,155,63
ipr-mus TDLo:3200 mg/kg (9D preg):REP DZZEA7 34,124,79
ipr-mus TDLo:67,200 mg/kg (1-21D preg):TER DZZEA7 34,484,79
orl-rat LD50:4250 mg/kg VAMNAQ 32,28,77
ipr-mus LD50:2638 mg/kg DZZEA7 34,484,79

CONSENSUS REPORTS: Reported in EPA TSCA Inventory.

OSHA PEL: TWA 2.5 mg(F)/m^3
ACGIH TLV: TWA 2.5 mg(F)/m^3
NIOSH REL: (Inorganic Fluorides) TWA 2.5 mg(F)/m^3

SAFETY PROFILE: Moderately toxic by intraperitoneal route. Mildly toxic by ingestion. An experimental teratogen. Other experimental reproductive effects. Mutation data reported. See also FLUORIDES and CALCIUM COMPOUNDS. When heated to decomposition it emits toxic fumes of F^-.

CAS250 CAS:544-17-2 HR: 3
CALCIUM FORMATE
mf: $C_2H_2O_4{\cdot}Ca$ mw: 130.12

PROP: Colorless, orthorhombic crystals. Also exists in several other polymorphic forms. Very sol in H_2O; insol in EtOH.

SYNS: FORMIC ACID, CALCIUM SALT ☐ MRAVENCAN VAPENATY (CZECH)

TOXICITY DATA WITH REFERENCE
eye-rbt 100 mg/24H MOD 28ZPAK -,9,72
orl-rat LD50:2650 mg/kg 28ZPAK -,9,72
orl-mus LD50:1920 mg/kg ZERNAL 9,332,69
ivn-mus LD50:154 mg/kg ZERNAL 9,332,69

CONSENSUS REPORTS: Reported in EPA TSCA Inventory.

SAFETY PROFILE: Poison by intravenous route. Moderately toxic by ingestion. An eye irritant. When heated to decomposition it emits acrid smoke and fumes. See also CALCIUM COMPOUNDS.

CAS750 CAS:299-28-5 HR: 2
CALCIUM GLUCONATE
mf: $C_{12}H_{22}O_{14}{\cdot}Ca$ mw: 430.42

PROP: White, fluffy powder or granules; odorless and tasteless. Sol in hot water; less sol in cold water; insol in alc, acetic acid, and other organic solvents. Mp: loses H_2O @ 120°.

SYNS: CALCICOL ☐ CALCIOFON ☐ CALCIPUR ☐ CALCIUM d-GLUCONATE ☐ CALCIUM HEXAGLUCONATE ☐ CALGLUCOL ☐ CALGLUCON ☐ DRAGOCAL ☐ EBUCIN ☐ GLUCAL ☐ GLUCOBIOGEN ☐ GLUCONATE de CALCIUM (FRENCH) ☐ GLUCONATO di CALCIO ☐ d-GLUCONIC ACID, CALCIUM SALT (2:1) (9CI) ☐ KALPREN ☐ NOVOCAL

TOXICITY DATA WITH REFERENCE
ims-inf TDLo:143 mg/kg:SKN,MET JAMAAP 129,347,45
ims-inf LDLo:10 g/kg JAMAAP 129,347,45
orl-rat LDLo:10 g/kg FRPPAO 26,144,71
ivn-rat LD50:950 mg/kg NIIRDN 6,226,82
orl-mus LDLo:10 g/kg FRPPAO 26,144,71
ipr-mus LD50:2200 mg/kg JDGRAX 15(1-2),121,84
scu-mus LD50:2890 mg/kg JAPMA8 45,47,56
ivn-mus LD50:950 mg/kg TXAPA9 4,492,62
ivn-gpg LDLo:1810 mg/kg AIPTAK 191,44,71

CONSENSUS REPORTS: Reported in EPA TSCA Inventory.

SAFETY PROFILE: Moderately toxic by subcutaneous, intraperitoneal, and intravenous routes. Human systemic effects in infants by intramuscular route: dermatitis and fever. When heated to decomposition it emits acrid smoke and fumes. See also CALCIUM COMPOUNDS.

C

CAS825 HR: 1
CALCIUM HEXAMETAPHOSPHATE

SAFETY PROFILE: A nuisance dust.

CAT125 CAS:17097-76-6 HR: 2
CALCIUM HOMOPANTOTHENATE
mf: $C_{20}H_{38}N_2O_6{\cdot}Ca$ mw: 442.68

PROP: Solid. Mp: 155–165°.

SYNS: CALCIUM-d-HOMOPANTOTHENATE ☐ CALCIUM HOPANTENATE ☐ (R)-4-((2,4-DIHYDROXY-3,3-DIMETHYL-1-OXOBUTYL) AMINO)-BUTANOIC ACID CALCIUM SALT (2:1) ☐ HOPANTENATE CALCIUM ☐ PANTOGAM

TOXICITY DATA WITH REFERENCE
orl-rat LD50:13,800 mg/kg NIIRDN 6,788,82
scu-rat LD50:5600 mg/kg NIIRDN 6,788,82
orl-mus LD50:6000 mg/kg NIIRDN 6,788,82
scu-mus LD50:2600 mg/kg NIIRDN 6,788,82

SAFETY PROFILE: Moderately toxic by subcutaneous route. Mildly toxic by ingestion. When heated to decomposition it emits toxic fumes of NO_x. See also CALCIUM COMPOUNDS.

CAT175 CAS:1990-07-4 HR: 2
CALCIUM HOPANTENATE HEMIHYDRATE
mf: $C_{20}H_{36}N_2O_{10}{\cdot}Ca{\cdot}1/2H_2O$ mw: 513.61

SYNS: CALCIUM-d-(+)-4-(2,4-DIHYDROXY-3,3-DIMETHYLBUTYRAMIDE)BUTYRATE HEMIHYDRATE ☐ HOPA ☐ HOPANTENATE CALCIUM HEMIHDYRATE

TOXICITY DATA WITH REFERENCE
orl-rat TDLo:405 mg/kg (17-22D preg/21D post):REP OYYAA2 19,1011,80
orl-rat TDLo:810 mg/kg (female 17-22D post):REP OYYAA2 19,1011,80
orl-rat TDLo:1620 mg/kg (female 17-22D post):REP OYYAA2 19,1011,80
orl-rat LD50:13,900 mg/kg IYKEDH 9,829,78
ipr-rat LD50:13,500 mg/kg IYKEDH 9,829,78
scu-rat LD50:5600 mg/kg IYKEDH 9,829,78
orl-mus LD50:6000 mg/kg IYKEDH 9,829,78
ipr-mus LD50:850 mg/kg YAKUD5 20,259,78
scu-mus LD50:2600 mg/kg IYKEDH 9,829,78

SAFETY PROFILE: Moderately toxic by subcutaneous and intraperitoneal routes. Experimental reproductive effects. When heated to decomposition it emits toxic fumes of NO_x. See also CALCIUM COMPOUNDS.

CAT200 CAS:7789-78-8 HR: 3
CALCIUM HYDRIDE
mf: CaH_2 mw: 60.24

PROP: Moisture sensitive, white, orthorhombic crystals. Mp: 816° (in water).

SYN: CALCIUM DIHYDRIDE

SAFETY PROFILE: Explosive reaction on heating with tetrahydrofuran. Mixtures with potassium chlorate and other metal oxohalogenates (e.g., chlorates; bromates; and perchlorates) are heat- and friction-sensitive explosives. Vigorous or incandescent reaction on heating with halogens (chlorine; bromine; or iodine); manganese dioxide; and silver halides (e.g., silver fluoride; silver iodide). See also CALCIUM COMPOUNDS and HYDRIDES.

CAT225 CAS:1305-62-0 HR: 2
CALCIUM HYDROXIDE
mf: CaH_2O_2 mw: 74.10

PROP: Rhombic, trigonal, colorless crystals or white power; sltly bitter taste. Mp: loses H_2O @ 580°, bp: decomp, d: 2.343. Sltly sol in water and glycerin; insol in alc.

SYNS: BELL MINE ☐ BIOCALC ☐ CALCIUM DIHYDROXIDE ☐ CALCIUM HYDRATE ☐ CALCIUM HYDROXIDE (ACGIH, OSHA) ☐ CALVIT ☐ CARBOXIDE ☐ HYDRATED LIME ☐ KALKHYDRATE ☐ KEMIKAL ☐ LIMBUX ☐ LIME MILK ☐ LIME WATER ☐ MILK OF LIME ☐ SLAKED LIME

TOXICITY DATA WITH REFERENCE
eye-rbt 10 mg SEV TXAPA9 55,501,80
cyt-rat/ast 1200 mg/kg GANNA2 54,155,62
orl-rat LD50:7340 mg/kg AIHAAP 30,470,69
orl-mus LD50:7300 mg/kg YKYUA6 32,1477,81

CONSENSUS REPORTS: Reported in EPA TSCA Inventory.

OSHA PEL: TWA 5 mg/m^3
ACGIH TLV: TWA 5 mg/m^3

SAFETY PROFILE: Mildly toxic by ingestion. A severe eye irritant. A skin, mucous membrane, and respiratory system irritant. Mutation data reported. Causes dermatitis. Dust is considered to be a significant industrial hazard. A common air contaminant. Violent reaction with maleic anhydride; nitroethane; nitromethane; nitroparaffins; nitropropane; phosphorus. Reaction with polychlorinated phenols + potassium nitrate forms extremely toxic products. See also CALCIUM COMPOUNDS.

For occupational chemical analysis use NIOSH: Calcium, 7020; Elements, 7300.

CAT250 CAS:7789-79-9 HR: 3
CALCIUM HYPOPHOSPHITE
mf: $CaH_4O_4P_2$ mw: 170.06

$Ca(OP(O)H_2)_2$

PROP: Monoclinic crystals from aq Me_2CO.

SYN: CALCIUM PHOSPHINATE

SAFETY PROFILE: Mixture with potassium chlorate is a friction-sensitive explosive. When heated to decomposition it emits toxic fumes of PO_x. See also CALCIUM COMPOUNDS.

CAT500 CAS:7789-80-2 HR: 1
CALCIUM IODATE
mf: $Ca(IO_3)_2 \cdot H_2O$ mw: 407.90

PROP: White powder or colorless monoclinic crystals. Decomposes on heating. Sltly sol in water; insol in alc.

SAFETY PROFILE: A nuisance dust.

CAT600 CAS:814-80-2 HR: 3
CALCIUM LACTATE
mf: $C_6H_{10}CaO_6 \cdot xH_2O$ mw: 218.22

PROP: White crystalline powder with up to 5 H_2O. Sol in water; insol in alc.

SYNS: CALPHOSAN ☐ CONCLYTE CALCIUM ☐ 2-HYDROXYPROPANOIC ACID CALCIUM SALT ☐ PROPANOIC ACID, 2-HYDROXY-, CALCIUM SALT

TOXICITY DATA WITH REFERENCE
ivn-mus LDLo:140 mg/kg JAPMA8 27,484,38

CONSENSUS REPORTS: Reported in EPA TSCA Inventory.

SAFETY PROFILE: Poison by intravenous route. When heated to decomposition it emits acrid smoke and irritating fumes.

CAT700 CAS:819-17-0 HR: 3
CALCIUM METHIONATE
mf: $CH_2O_6S_2 \cdot Ca$ mw: 214.23

SYN: METHANEDISULFONIC ACID, CALCIUM SALT (1:1)

TOXICITY DATA WITH REFERENCE
ivn-rat LD50:329 mg/kg JAPMA8 45,47,56
scu-mus LD50:1085 mg/kg JAPMA8 45,47,56
ivn-mus LD50:422 mg/kg JAPMA8 45,47,56

SAFETY PROFILE: Poison by intravenous route. Moderately toxic by subcutaneous route. When heated to decomposition it emits toxic fumes of SO_x. See also CALCIUM COMPOUNDS and SULFONATES.

CAT750 CAS:7789-82-4 HR: 3
CALCIUM MOLYBDATE
mf: $MoO_4 \cdot Ca$ mw: 200.02

PROP: White crystals. An electrical conductor. Mp: 965° (decomp). Insol in H_2O.

SYNS: CALCIUM MOLYBDENUM OXIDE ($CaMoO_4$) ☐ MOLYBDATE, CALCIUM ☐ MOLYBDIC ACID (H_2MoO_4), CALCIUM SALT (1:1)

TOXICITY DATA WITH REFERENCE
ipr-rat LD50:208 mg/kg EQSSDX 1,1,75

CONSENSUS REPORTS: Reported in EPA TSCA Inventory.

OSHA PEL: TWA Total Dust: 10 mg/m^3; Respirable Fraction: 5 mg/m^3
ACGIH TLV: TWA 10 $mg(Mo)/m^3$

SAFETY PROFILE: Poison by intraperitoneal route. See also MOLYBDENUM and CALCIUM COMPOUNDS.

CAT775 CAS:471-34-1 HR: 1
CALCIUM MONOCARBONATE
mf: $CO_3 \cdot Ca$ mw: 100.09

SYNS: AEROMATT ☐ AKADAMA ☐ ALBACAR ☐ ALBACAR 5970 ☐ ALBAFIL ☐ ALBAGLOS ☐ ALBAGLOS SF ☐ ALLIED WHITING ☐ ATOMIT ☐ ATOMITE ☐ AX 363 ☐ BF 200 ☐ BRILLIANT 15 ☐ BRITOMYA M ☐ CALCENE CO ☐ CALCICOLL ☐ CALCIDAR 40 ☐ CALCILIT 8 ☐ CALCIUM CARBONATE (1:1) ☐ CALIBRITE ☐ CAL-LIGHT SA ☐ CALMOS ☐ CALMOTE ☐ CALOFIL A 4 ☐ CALOFORT S ☐ CALOFORT U ☐ CALOFOR U 50 ☐ CALOPAKE F ☐ CALOPAKE HIGH OPACITY ☐ CALSEEDS ☐ CALTEC ☐ CAMEL-CARB ☐ CAMEL-TEX ☐ CAMEL-WITE ☐ CARBITAL 90 ☐ CARBIUM ☐ CARBONIC ACID, CALCIUM SALT (1:1) ☐ CARBIUM MM ☐ CARBOREX 2 ☐ CARUSIS P ☐ CCC G-WHITE ☐ CCC No. AA OOLITIC ☐ CCR ☐ CCW ☐ CHEMCARB ☐ C.I. PIGMENT WHITE 18 ☐ CLEFNON ☐ CRYSTIC PREFIL S ☐ DACOTE ☐ DOMAR ☐ DURAMITE ☐ DURCAL 10 ☐ EGRI M 5 ☐ ESKALON 100 ☐ FILTEX WHITE BASE ☐ FINNCARB 6002 ☐ GAROLITE SA ☐ GILDER'S WHITING ☐ HAKUENKA CC ☐ HAKUENKA R 06 ☐ HOMOCAL D ☐ HYDROCARB 60 ☐ K 250 ☐ KOTAMITE ☐ KREDAFIL 150 EXTRA ☐ KREDAFIL RM 5 ☐ KS 1300 ☐ KULU 40 ☐ LEVIGATED CHALK ☐ MARBLEWHITE 325 ☐ MARFIL ☐ MC-T ☐ MICROCARB ☐ MICROMIC CR 16 ☐ MICROMYA ☐ MICROWHITE 25 ☐ MONOCALCIUM CARBONATE ☐ MSK-C ☐ MULTIFLEX MM ☐ N 34 ☐ NCC 45 ☐ NEOANTICID ☐ NEOLITE F ☐ NON-FER-AL ☐ NS (carbonate) ☐ NS 100 (carbonate) ☐ NS 200 (filler) ☐ NZ ☐ OA-A 1102 ☐ OMYA ☐ OMYA BLH ☐ OMYACARB F ☐ OMYALENE G 200 ☐ OMYALITE 90 ☐ OS-CAL ☐ PIGMENT WHITE 18 ☐ P-LITE 500 ☐ POLCARB ☐ PREPARED CHALK ☐ PS 100 (carbonate) ☐ PURECAL ☐ PURECALO ☐ PZ ☐ QUEENSGATE WHITING ☐ RED BALL ☐ R JUTAN ☐ ROYAL WHITE LIGHT ☐ RX 2557 ☐ SHIPRON A ☐ SILVER W ☐ SL 700 ☐ SMITHKO KALKARB WHITING ☐ SNOWCAL ☐ SNOWFLAKE WHITE ☐ SNOW TOP ☐ SOCAL ☐ SOCAL E 2 ☐ SOFTON 1000 ☐ SS 30 (carbonate) ☐ SS 50 (carbonate) ☐ SSB 100 ☐ STANWHITE 500 ☐ STURCAL D ☐ SUNLIGHT 700 ☐ SUPER 1500 ☐ SUPERCOAT ☐ SUPERMITE ☐ SUPER MULTIFEX ☐ SUPER-PFLEX ☐ SUPER 3S ☐ SUPER SSS ☐ SURFEX MM ☐ SURFIL S ☐ SUSPENSO ☐ SYLACAUGA 88B ☐ T 130-2500 ☐ TAMA PEARL TP 121 ☐ TANCAL 100 ☐ TM 1 (filler) ☐ TONASO ☐ TOYOFINE TF-X ☐ TP 121 (filler) ☐ TP 222 ☐ ULTRA-PFLEX ☐ UNIBUR 70 ☐ VEVETONE ☐ VICRON ☐ VICRON 31-6 ☐ VIENNA WHITE ☐ VIGOT 15 ☐ WHICA BA ☐ WHITCARB W ☐ WHITE-POWDER ☐ WHITING ☐ WHITON 450 ☐ WINNOFIL 3 ☐ WITCARD ☐ WITCARB P ☐ WITCARB REGULAR ☐ YORK WHITE ☐ ZG 301

TOXICITY DATA WITH REFERENCE
skn-rbt 500 mg/24H MOD 28ZPAK -,267,72
eye-rbt 750 µg/24H SEV 28ZPAK -,267,72
orl-rat LD50:6450 mg/kg 28ZPAK -,267,72

CONSENSUS REPORTS: Reported in EPA TSCA Inventory.

SAFETY PROFILE: Mildly toxic by ingestion. A skin and severe eye irritant. When heated to decomposition it emits acrid smoke and irritating vapors.

CAU000 CAS:10124-37-5 HR: 3
CALCIUM(II) NITRATE (1:2)
DOT: UN 1454
mf: $N_2O_6 \cdot Ca$ mw: 164.10

PROP: Hygroscopic, colorless, cubic crystals. Mp: 561°. Decomposes on heating. Very sol in H_2O, EtOH; sol in MeOH, and Me_2CO; insol in Et_2O.

SYNS: CALCIUM NITRATE (DOT) ☐ CALCIUM DINITRATE ☐ CALCIUM SALTPETER ☐ NITRIC ACID, CALCIUM SALT (8CI,9CI) ☐ NORGE SALTPETER ☐ NORWAY SALTPETER ☐ NORWEGIAN SALTPETER ☐ SYNFAT 1006 ☐ UN1454 (DOT)

TOXICITY DATA WITH REFERENCE
orl-rat LD50:302 mg/kg GISAAA 46(12),66,81

CONSENSUS REPORTS: Reported in EPA TSCA Inventory.

DOT CLASSIFICATION: 5.1; *Label:* Oxidizer

SAFETY PROFILE: A poison by ingestion. An irritant. A strong oxidant. Forms powerfully explosive mixtures with aluminum + ammonium nitrate + formamide + water, ammonium nitrate + hydrocarbon oils, ammonium nitrate + water-soluble fuels, and organic materials. When heated to decomposition it emits toxic fumes of NO_x. See also NITRATES and CALCIUM COMPOUNDS.

CAU250 CAS:13477-34-4 HR: 2
CALCIUM(II) NITRATE TETRAHYDRATE (1:2:4)
mf: $N_2O_6 \cdot Ca \cdot 4H_2O$ mw: 236.18

PROP: Cubic, colorless, hygroscopic, monoclinic, deliquescent, crystals. Mp: 43°, d: 2.36. Decomposes on heating with water loss. Decomp on heating with H_2O loss. Very sol in H_2O; sol in Me_2CO and EtOH.

SYNS: DUSICNAN VAPENATY (CZECH) ☐ NITRIC ACID, CALCIUM SALT, TETRAHYDRATE

TOXICITY DATA WITH REFERENCE
skn-rbt 500 mg/24H MLD 28ZPAK -,9,72
eye-rbt 500 mg/24H MLD 28ZPAK -,9,72
orl-rat LD50:3900 mg/kg 28ZPAK -,9,72

SAFETY PROFILE: Moderately toxic by ingestion. A skin and eye irritant. See also CALCIUM COMPOUNDS, and NITRATES. When heated to decomposition it emits toxic fumes of NO_x.

CAU500 CAS:1305-78-8 HR: 3
CALCIUM OXIDE
DOT: UN 1910
mf: CaO mw: 56.08

PROP: Cubic, colorless, white crystals. Mp: 2580°, d: 3.37, bp: 2850°. Sol in water and glycerin; insol in alc.

SYNS: AIRLOCK ☐ BELL CML(E) ☐ BURNT LIME ☐ CALCIA ☐ CALOXOL CP2 ☐ CALOXOL W3 ☐ CALX ☐ CALXYL ☐ CML 21 ☐ CML 31 ☐ DESICAL P ☐ LIME ☐ LIME, BURNED ☐ LIME, UNSLAKED (DOT) ☐ OXYDE de CALCIUM (FRENCH) ☐ QUICKLIME (DOT) ☐ RHENOSORB C ☐ RHENOSORB F ☐ WAPNIOWY TLENEK (POLISH)

CONSENSUS REPORTS: Reported in EPA TSCA Inventory.

OSHA PEL: TWA 5 mg/m³
ACGIH TLV: TWA 2 mg/m³
DFG MAK: 5 mg/m³
DOT CLASSIFICATION: 8; *Label:* Corrosive

SAFETY PROFILE: A caustic and irritating material. See also CALCIUM COMPOUNDS. A common air contaminant. A powerful caustic to living tissue. The powdered oxide may react explosively with water. Mixtures with ethanol may ignite if heated and thus can cause an air-vapor explosion. Violent reaction with (B_2O_3 + $CaCl_2$) interhalogens (e.g., BF_3, ClF_3), F_2, HF, P_2O_5 + heat, water. Incandescent reaction with liquid HF. Incompatible with phosphorus(V) oxide.

For occupational chemical analysis use OSHA: #ID-125G or NIOSH: Calcium, 7020; elements, 7300.

CAU750 **CAS:137-08-6** **HR: 2**
CALCIUM-d-PANTOTHENATE
mf: $C_{19}H_{34}N_2O_{10}$•Ca mw: 490.63

PROP: White, sltly hygroscopic powder; odorless; bitter taste; crystals from MeOH. Mp: 195–196° decomp at 195–196°. Sol in water and glycerin; insol in alc, chloroform, and ether.

SYNS: CALCIUM d(+)-N-(α,γ-DIHYDROXY-β,β-DIMETHYLBUTYRYL)-β-ALANINATE □ CALCIUM PANTHOTHENATE (FCC) □ CALCIUM PANTOTHENATE □ d-CALCIUM PANTOTHENATE □ CALPANATE □ DEXTRO CALCIUM PANTOTHENATE □ N-(2,4-DIHYDROXY-3,3-DIMETHYLBUTYRYL)-β-ALANINE CALCIUM □ PANCAL □ PANTHOJECT □ PANTHOLIN □ PANTOTHENATE CALCIUM □ PANTOTHENIC ACID, CALCIUM SALT □ (+)-PANTOTHENIC ACID, CALCIUM SALT □ VITAMIN B-5

TOXICITY DATA WITH REFERENCE
ipr-rat LD50:820 mg/kg PSEBAA 45,311,40
scu-rat LD50:3400 mg/kg PSEBAA 45,311,40
ivn-rat LD50:830 mg/kg NIIRDN 6,599,82
orl-mus LD50:10 g/kg NIIRDN 6,599,82
ipr-mus LD50:920 mg/kg PSEBAA 45,311,40
scu-mus LD50:2700 mg/kg PSEBAA 45,311,40
ivn-mus LD50:910 mg/kg PSEBAA 45,311,40

CONSENSUS REPORTS: Reported in EPA TSCA Inventory.

SAFETY PROFILE: Moderately toxic by intraperitoneal, subcutaneous, and intravenous routes. Mildly toxic by ingestion. A vitamin. See also CALCIUM COMPOUNDS. When heated to decomposition it emits toxic fumes of NO_x.

CAU780 **HR: 1**
CALCIUM PANTOTHENATE, CALCIUM CHLORIDE DOUBLE SALT
mf: $C_{19}H_{34}N_2O_{10}$•Ca_2Cl_2 mw: 601.61

PROP: White, sltly hygroscopic powder; odorless with bitter taste. Sol in water and glycerin; insol in alc, chloroform, and ether.

SAFETY PROFILE: Moderately toxic by intraperitoneal, subcutaneous, and intravenous routes. Mildly toxic by ingestion. A vitamin. See also CALCIUM COMPOUNDS. When heated to decomposition it emits toxic fumes of NO_x.

CAV000 **CAS:7563-42-0** **HR: 3**
CALCIUM PENTOBARBITAL
mf: $C_{11}H_{18}N_2O_3$•7Ca mw: 506.87

SYNS: CALCIUM NEMBUTAL □ INSOM-RAPIDO □ NEMBUTAL CALCIUM □ PENTOBARBITAL CALCIUM □ 2,4,6(1H,3H,5H)-PYRIMIDINETRIONE, 5-ETHYL-5-(1-METHYLBUTYL)-, CALCIUM SALT (9CI) □ RAVONA □ REPOCAL □ SCHLAFEN

TOXICITY DATA WITH REFERENCE
orl-dog LDLo:60 mg/kg CRAAA7 20,350,41
ivn-dog LDLo:70 mg/kg CRAAA7 20,350,41

SAFETY PROFILE: Poison by ingestion and intravenous routes. When heated to decomposition it emits toxic fumes of NO_x. See also BARBITURATES.

CAV250 **CAS:10118-76-0** **HR: 3**
CALCIUM PERMANGANATE
DOT: UN 1456
mf: Mn_2O_8•Ca mw: 277.96

$Ca(MnO_4)_2$

PROP: Violet, deliquescent crystals. Mp: decomp, d: 2.4.

SYNS: ACERDOL □ KALIUMPERMANGANAT (GERMAN) □ PERMANGANIC ACID($HMnO_4$), CALCIUM SALT (8CI,9CI)

TOXICITY DATA WITH REFERENCE
ivn-rbt LDLo:50 mg/kg TDBU** -,-,33

CONSENSUS REPORTS: Manganese and its compounds are on the Community Right-To-Know List.

OSHA PEL: CL 5 mg(Mn)/m³
ACGIH TLV: TWA 5 mg(Mn)/m³
DOT CLASSIFICATION: 5.1; *Label:* Oxidizer

SAFETY PROFILE: Poison by intravenous route. See also CALCIUM COMPOUNDS, MANGANESE COMPOUNDS, and PERMANGANATES. A strong oxidant. May explode on contact with acetic acid or acetic anhydride. Ignites on contact with cellulose. Incompatible with hydrogen peroxide.

CAV500 **CAS:1305-79-9** **HR: 3**
CALCIUM PEROXIDE
DOT: UN 1457
mf: CaO_2 mw: 72.08

PROP: Yellow crystals or powder or white crystals, decomposes in air. Mp: decomp @ 275°. Insol in water; sol in acids, forming hydrogen peroxide.

SYNS: CALCIUM DIOXIDE □ CALCIUM SUPEROXIDE

CONSENSUS REPORTS: Reported in EPA TSCA Inventory.

DOT CLASSIFICATION: 5.1; *Label:* Oxidizer

SAFETY PROFILE: Irritating in concentrated form. Will react with moisture to form slaked lime. Flammable if hot and mixed with finely divided combustible material. Mixtures with oxidizable materials can also be ignited by grinding and are explosion hazards. A strong alkali. An oxidizer. Mixtures with polysulfide polymers may ignite. See also CALCIUM COMPOUNDS, CALCIUM HYDROXIDE, and PEROXIDES, INORGANIC.

CAW000 **CAS:13235-16-0** **HR: 3**
CALCIUM PEROXODISULPHATE
mf: CaO_8S_2 mw: 232.21

SAFETY PROFILE: A powerful shock-sensitive explosive. Upon decomposition it emits toxic fumes of SO_x. See also CALCIUM COMPOUNDS and PEROXIDES.

CAW100 **CAS:7757-93-9** **HR: 1**
CALCIUM PHOSPHATE, DIBASIC
mf: $CaHPO_4{\cdot}2H_2O$ mw: 172.09

PROP: White powder or crystals. Sol in dilute acid; insol in water, alc.

SYN: DICALCIUM PHOSPHATE

SAFETY PROFILE: Skin and eye irritant. A nuisance dust.

CAW110 **CAS:7758-23-8** **HR: 1**
CALCIUM PHOSPHATE, MONOBASIC
mf: $Ca(H_2PO_4)_2$ mw: 234.05

PROP: White crystals or granular powder. Sltly sol in water; insol in alc.

SYNS: ACID CALCIUM PHOSPHATE □ CALCIUM BIPHOSPHATE □ MONOCALCIUM PHOSPHATE

TOXICITY DATA WITH REFERENCE
orl-rat LD50:17,500 mg/kg GISAAA 52(12),87,87
orl-mus LD50:15,250 mg/kg GISAAA 52(12),87,87

SAFETY PROFILE: Low toxicity by ingestion. A nuisance dust.

CAW120 **CAS:12167-74-7** **HR: 1**
CALCIUM PHOSPHATE, TRIBASIC
mf: $10CaO{\cdot}3P_2O_5{\cdot}H_2O$ mw: 1004.64

PROP: White powder or clear colorless hexagonal crystals. Sol in dilute HCl; practically insol in water, alc.

SYNS: PRECIPITATED CALCIUM PHOSPHATE □ TRICALCIUM PHOSPHATE

SAFETY PROFILE: Skin and eye irritant. A nuisance dust.

CAW250 **CAS:1305-99-3** **HR: 3**
CALCIUM PHOSPHIDE
DOT: UN 1360
mf: Ca_3P_2 mw: 182.18

PROP: Red-brown crystals. Mp: >1600°, d: 2.238 @ 25°. Insol in EtOH, Et_2O, and C_6H_6.

SYNS: CALCIUM PHOTOPHOR □ PHOTOPHOR

CONSENSUS REPORTS: Reported in EPA TSCA Inventory.

DOT CLASSIFICATION: 4.3; *Label:* Dangerous When Wet, Poison

SAFETY PROFILE: Highly toxic due to phosphide, which in presence of moisture emits phosphine. The phosphine may ignite spontaneously in air. Incandescent reaction with oxygen at 300°C. Incompatible with dichlorine oxide. When heated to decomposition it emits toxic fumes of PO_x. See also CALCIUM COMPOUNDS and PHOSPHIDES.

CAW376 **CAS:26016-98-8** **HR: 2**
CALCIUM PHOSPHONOMYCIN HYDRATE
mf: $C_3H_5O_4P{\cdot}Ca{\cdot}H_2O$ mw: 194.15

SYNS: CALCIUM (−)-(1R,2S)-(1,2-EPOXYPROPYL)PHOSPHONATE HYDRATE □ CALCIUM FOSFOMYCIN HYDRATE □ FOM-Ca HYDRATE □ FOSFOMYCIN-Ca HYDRATE □ FOSFOMYCIN CALCIUM HYDRATE

TOXICITY DATA WITH REFERENCE
orl-rat TDLo:1540 mg/kg (female 7-17D post):REP JJANAX 32,546,79
orl-rat TDLo:84 g/kg (female 14-22D post):REP JJANAX 33,733,80
orl-rat TDLo:11 g/kg (male 9W pre):TER JJANAX 33,613,80
orl-rat TDLo:113 g/kg (male 9W pre):REP JJANAX 33,613,80
orl-rat TDLo:15,400 mg/kg (female 7-17D post):REP JJANAX 32,546,79
orl-rat TDLo:15,400 mg/kg (female 7-17D post): TER JJANAX 32,546,79
ipr-rat LD50:1036 mg/kg DRUGAY 6,785,82
ipr-mus LD50:994 mg/kg IYKEDH 11,811,80

SAFETY PROFILE: Moderately toxic by intraperitoneal route. An experimental teratogen. Other experimental reproductive effects. When heated to decomposition it emits toxic fumes of PO_x. See also CALCIUM COMPOUNDS.

CAW400 **CAS:4075-81-4** **HR: 2**
CALCIUM PROPIONATE
mf: $C_6H_{10}CaO_4$ mw: 186.22

PROP: White crystals; faint odor of propionic acid. Sol in water.

SYNS: BIOBAN-C □ CALCIUM DIPROPIONATE □ CALCIUM PROPIONATE □ PROPANOIC ACID, CALCIUM SALT (9CI)

TOXICITY DATA WITH REFERENCE
orl-rat LD50:3920 mg/kg TRENAF 27,159,76
orl-mus LD50:2350 mg/kg TRENAF 27,159,76

SAFETY PROFILE: Moderately toxic by ingestion. When heated to decomposition it emits acrid smoke and irritating fumes.

CAW450 **CAS:7790-76-3** **HR: 1**
CALCIUM PYROPHOSPHATE
mf: $Ca_2P_2O_7$ mw: 254.10

PROP: Fine white powder. Sol in dilute HCl; insol in water.

SAFETY PROFILE: A nuisance dust.

CAW500 **CAS:9007-13-0** **HR: 1**
CALCIUM RESINATE
DOT: UN 1313/UN 1314
mf: $Ca(C_{44}H_{62}O_4)_2$ mw: 1349.50

PROP: Yellowish-white, amorphous powder or lumps.

SYNS: CALCIUM RESINATE (UN 1313) (DOT) □ CALCIUM RESIN-

ATE, fused (UN 1314) (DOT) □ LIMED ROSIN □ RESIN ACIDS and ROSIN ACIDS, CALCIUM SALTS □ URAPRINT 62-126

CONSENSUS REPORTS: Reported in EPA TSCA Inventory.

DOT CLASSIFICATION: 4.1; *Label:* Flammable Solid

SAFETY PROFILE: Flammable solid when heated; can react with oxidizing materials. When heated to decomposition it emits acrid smoke and fumes. See also CALCIUM COMPOUNDS.

CAW850 CAS:1344-95-2 HR: 1
CALCIUM SILICATE

PROP: Varying proportions of CaO and SiO_2. White powder. Insol in water.

SYNS: CALCIUM HYDROSILICATE □ CALCIUM MONOSILICATE □ CALCIUM POLYSILICATE □ CALCIUM SILICATE, synthetic nonfibrous (ACGIH) □ CALFLO E □ CALSIL □ CS LAFARGE □ FLORITE R □ MARIMET 45 □ MICROCAL 160 □ MICROCAL ET □ MICRO-CEL □ MICRO-CEL A □ MICRO-CEL B □ MICRO-CEL C □ MICRO-CEL E □ MICRO-CEL T □ MICRO-CEL T26 □ MICRO-CEL T38 □ MICRO-CEL T41 □ PROMAXON P60 □ SILENE EF □ SILMOS T □ SOLEX □ STABINEX NW 7PS □ STARLEX L □ SW 400 □ TOYOFINE A

OSHA PEL: Total Dust: 15 mg/m^3; Respirable Fraction: 5 mg/m^3
ACGIH TLV: TWA (nuisance particulate) 10 mg/m^3 of total dust (when toxic impurities are not present, e.g., quartz <1%)

SAFETY PROFILE: A nuisance dust.

CAX250 CAS:16925-39-6 HR: 3
CALCIUM SILICOFLUORIDE
mf: CaF_6Si mw: 182.17

PROP: White, crystalline powder. Hydrolyzes in H_2O forming complex mixtures of products. Decomp on heating with formation of CaF_2 and SiF_4 at 225°. Sltly sol in EtOHD.

SYNS: CALCIUM FLUOROSILICATE □ CALCIUM FLUOSILICATE □ CALCIUM HEXAFLUOROSILICATE □ SILICATE(2-), HEXAFLUORO-, CALCIUM (1:1) (9CI)

CONSENSUS REPORTS: Reported in EPA TSCA Inventory.

OSHA PEL: TWA 2.5 $mg(F)/m^3$
NIOSH REL: (Inorganic Fluorides) TWA 2.5 $mg(F)/m^3$
DOT CLASSIFICATION: 6.1; *Label:* KEEP AWAY FROM FOOD

SAFETY PROFILE: Poison by ingestion and subcutaneous routes. See also CALCIUM COMPOUNDS. When heated to decomposition it emits toxic fumes of F^-.

CAX260 CAS:23209-59-8 HR: 2
CALCIUM SODIUM METAPHOSPHATE
mf: $HO_3P{\cdot}Ca{\cdot}Na$ mw: 143.05

SYN: METAPHOSPHORIC ACID, CALCIUM SODIUM SALT

TOXICITY DATA WITH REFERENCE
ipl-rat TDLo: 200 mg/kg/2Y-C:ETA EPASR* 8EHQ-0386-0619

CONSENSUS REPORTS: Reported in EPA TSCA Inventory.

SAFETY PROFILE: Questionable carcinogen with experimental tumorigenic data. When heated to decomposition it emits toxic fumes of PO_x.

CAX350 CAS:1592-23-0 HR: 1
CALCIUM STEARATE

PROP: Variable proportions of calcium stearate and calcium palmitate. Fine white powder; sltly characteristic odor. Insol in water, alc, ether.

SYNS: AQUACAL □ CALCIUM DISTEARATE □ CALSTAR □ FLEXICHEM □ FLEXICHEM CS □ G 339 S □ NOPCOTE C 104 □ OCTADECANOIC ACID, CALCIUM SALT □ STAVINOR 30 □ SYNPRO STEARATE □ WITCO G 339S

CONSENSUS REPORTS: Reported in EPA TSCA Inventory.

ACGIH TLV: TWA 10 mg/m^3, total dust

SAFETY PROFILE: A nuisance dust. When heated to decomposition it emits acrid smoke and irritating fumes.

CAX500 CAS:7778-18-9 HR: 1
CALCIUM SULFATE
mf: $CaSO_4$ mw: 136.14

PROP: Pure anhydrous, colorless or white powder or odorless crystals. D: 2.964; mp: 1570°. Dissolves in acids. Sltly sol in H_2O.

SYNS: ANHYDROUS CALCIUM SULFATE □ CRYSALBA □DRIER-ITE □ GIBS □ PLASTER of PARIS □ THIOLITE

OSHA PEL: Total Dust: 15 mg/m^3; Respirable Fraction: 5 mg/m^3
ACGIH TLV: TWA (nuisance particulate) 10 mg/m^3 of total dust (when toxic impurities are not present, e.g., quartz <1%).

SAFETY PROFILE: A nuisance dust. Reacts violently with aluminum when heated. Mixtures with diazomethane react exothermically and eventually explode. Mixtures with phosphorus ignite at high temperatures. When heated to decomposition it emits toxic fumes of SO_x. See also CALCIUM COMPOUNDS and SULFATES.

CAX750 CAS:10101-41-4 HR: 1
CALCIUM(II) SULFATE DIHYDRATE (1:1:2)
mf: $O_4S{\cdot}Ca{\cdot}2H_2O$ mw: 172.18

PROP: Colorless, monoclinic, hygroscopic crystals. D: 2.32, mp: 128°, bp: 163°. Sltly sol in water.

SYNS: ALABASTER □ ANNALINE □ C.I. 77231 □ C.I. PIGMENT WHITE 25 □ GYPSUM □ GYPSUM STONE □ LAND PLASTER □ LIGHT SPAR □ MAGNESIA WHITE □ MINERAL WHITE □ NATIVE CALCIUM SULFATE □ PRECIPITATED CALCIUM SULFATE □SATIN-

ITE □ SATIN SPAR □ SULFURIC ACID, CALCIUM(2+) SALT, DIHYDRATE □ TERRA ALBA

TOXICITY DATA WITH REFERENCE
ipr-rat TDLo:450 mg/kg/3W-I:CAR ZHPMAT 162,467,76
ihl-hmn TCLo:194 g/m³/10Y-I:NOSE,PUL GTPZAB 11(10),23,67

OSHA PEL: Total Dust: 15 mg/m³; Respirable Fraction: 5 mg/m³
ACGIH TLV: TWA (nuisance particulate) 10 mg/m³ of total dust (when toxic impurities are not present, e.g., quartz <1%)

SAFETY PROFILE: Human systemic effects by inhalation: fibrosing alveolitis (growth of fibrous tissue in the lung); unspecified respiratory system effects and unspecified effects on the nose. Questionable carcinogen with experimental carcinogenic data. Long considered a nuisance dust (depending on silica content). When heated to decomposition it emits toxic fumes of SO_x. See also CALCIUM SULFATE, CALCIUM COMPOUNDS, and SULFATES.

CAY000 **CAS:20548-54-3** **HR: 3**
CALCIUM SULFIDE
mf: CaS mw: 72.14

PROP: Cubic, colorless crystals. Mp: 2525, bp: decomp, d: 218 @ 15°. Insol in MeOH.

SYNS: CALCIC LIVER of SULFUR □ HEPAR CALCIS □ OLDHAMITE

SAFETY PROFILE: A poison via inhalation. Reacts violently with chromyl chloride, lead dioxide, potassium chlorate (mild explosion), potassium nitrate (violent explosion). Incompatible with oxidants. When heated to decomposition it emits toxic fumes of SO_x. See also CALCIUM COMPOUNDS and SULFIDES.

CAY250 **CAS:2092-16-2** **HR: 3**
CALCIUM THIOCYANATE
mf: $C_2N_2S_2$•Ca mw: 156.24

PROP: White, deliquescent crystals. Very sol in H_2O; sol in MeOH and EtOH.

SYNS: CALCIUM DITHIOCYANATE □ CALCIUM RHODANID (GERMAN) □ CALCIUMRHODANID □ CALCIUM SULFOCYANATE □ THIOCYAN

TOXICITY DATA WITH REFERENCE
orl-mus LDLo:120 mg/kg AEPPAE 169,429,33
ivn-rbt LDLo:250 mg/kg AEPPAE 169,429,33

CONSENSUS REPORTS: Reported in EPA TSCA Inventory.

SAFETY PROFILE: Poison by ingestion and intravenous routes. See also THIOCYANATES and CALCIUM COMPOUNDS. When heated to decomposition it emits toxic fumes of NO_x and SO_x.

CAY500 **CAS:12111-24-9** **HR: 2**
CALCIUM TRISODIUM DIETHYLENE TRIAMINE PENTAACETATE
mf: $C_{14}H_{18}N_3O_{10}$•$CaNa_3$ mw: 497.40

SYNS: Ba 2797 □ CALCIUM CHEL-330 □ CALCIUM-DTPA □ CALCIUM TRISODIUM CHEL 330 □ CALCIUM TRISODIUM DTPA □ CALCIUM TRISODIUM PENTETATE □ CALCIUM TRISODIUM SALT of DIETHYLENETRIAMINEPENTAACETIC ACID □ DIETHYLENETRIAMINE PENTAACETIC ACID, CALCIUM TRISODIUM SALT □ DITRIPENTAT □ DTPA CALCIUM TRISODIUM SALT □ PENTACIN □ PENTACINE □ PENTETATE TRISODIUM CALCIUM □ PENTHAMIL

TOXICITY DATA WITH REFERENCE
dni-rat-scu 4 mmol/kg BCPCA6 23,901,74
dni-rat:lvr 20 mmol/L BCPCA6 23,901,74
ipr-rat TDLo:659 mg/kg (9D preg):REP JANSAG 23,908,64
scu-mus TDLo:3580 mg/kg (female 7-11D post):REP HLTPAO 33,624,77
scu-mus TDLo:30 g/kg (female 1-21D post):REP HLTPAO 29,780,75
scu-rat TDLo:1790 mg/kg (9-13D preg):TER HOKBAQ 18,37,83
ipr-rat TDLo:659 mg/kg (9D preg):TER JANSAG 23,908,64
ivn-rat LD50:2512 mg/kg AAJRDX 142,619,84
ipr-mus LD50:7269 mg/kg PHTXA6 64,247,89

CONSENSUS REPORTS: Reported in EPA TSCA Inventory.

SAFETY PROFILE: Moderately toxic by intravenous and intraperitoneal routes. Experimental teratogenic and reproductive effects. Mutation data reported. When heated to decomposition it emits toxic fumes of Na_2O and NO_x. See also CALCIUM COMPOUNDS.

CAY675 **HR: 3**
CALCIUM VALPROATE
mf: $C_{16}H_{30}O_4$•Ca mw: 326.54

SYNS: DIPROPYLACETIC ACID CALCIUM SALT □ 2-PROPYLVALERIC ACID CALCIUM SALT (2:1) □ VALONTIN □ VALPROIC ACID CALCIUM SALT □ VALPROIC ACID HEMI-CALCIUM SALT

TOXICITY DATA WITH REFERENCE
orl-rat TDLo:6 g/kg (6-15D preg):TER FAATDF 3,121,83
orl-rat TDLo:6 g/kg (6-15D preg):REP FAATDF 3,121,83
orl-rat TDLo:1500 mg/kg (6-15D preg):TER FAATDF 3,121,83
ipr-rat LD50:375 mg/kg JNPHAG 2,313,71
idu-rat LD50:1065 mg/kg JNPHAG 2,313,71
ipr-mus LD50:320 mg/kg JNPHAG 2,313,71
idu-mus LD50:673 mg/kg JNPHAG 2,313,71

SAFETY PROFILE: Poison by intraperitoneal route. Moderately toxic by intraduodenal route. Experimental teratogenic and reproductive effects. When heated to decomposition it emits acrid smoke and fumes. See also CALCIUM COMPOUNDS.

CAY800 **HR: 2**
CALLA

PROP: A commonly cultivated ornamental. The leaves are shaped like an arrowhead and are sometimes mottled with white. The lily-type flower may be white, green, pink, or yellow. It is grown outdoors in mild climates and indoors elsewhere.

SYNS: CALLA LILY □ LIRIO CALA (SPANISH) □ ZANTEDESCHIA AETHIOPICA

SAFETY PROFILE: The leaves contain poisonous crystals of calcium oxalate. Chewing the leaves results in burning pain in the lips, mouth, and throat, possibly followed by inflammation and blistering. Systemic effects are usually not seen because of the insolubility of calcium oxalate. The sap can cause contact dermatitis. See also OXALATES.

CAY875 CAS:42839-36-1 HR: 3
CALNEGYT
mf: $C_9H_{20}N_4 \cdot H_2O_4S \cdot H_2O$ mw: 300.43

PROP: Crystals. Mp: 239-241°.

SYNS: EGYT 739 □ GUANAZODINE SULFATE MONOHYDRATE □ ((OCTAHYDRO-2-AZOCINYL)METHYL)GUANIDINE SULFATE HYDRATE □ SANEGYT

TOXICITY DATA WITH REFERENCE
orl-rat LD50:3550 mg/kg OYYAA2 14,235,77
ipr-rat LD50:970 mg/kg OYYAA2 14,235,77
ivn-rat LD50:136 mg/kg NYKZAU 72,837,76
ims-rat LD50:1080 mg/kg OYYAA2 14,235,77
orl-mus LD50:2450 mg/kg USXXAM #3856778
scu-mus LD50:700 mg/kg USXXAM #3856778
ivn-mus LD50:100 mg/kg OYYAA2 14,235,77
ims-mus LD50:1240 mg/kg OYYAA2 14,235,77

SAFETY PROFILE: Poison by intravenous route. Moderately toxic by ingestion, intraperitoneal, intramuscular, and subcutaneous routes. When heated to decomposition it emits toxic fumes of SO_x and NO_x. An antihypertensive agent. See also SULFATES.

CAY950 CAS:8065-83-6 HR: 3
CALO-CLOR
mf: $Cl_2Hg_2 \cdot Cl_2Hg$ mw: 743.57

TOXICITY DATA WITH REFERENCE
orl-rat LD50:55,200 μg/kg FMCHA2-,C56,89

ACGIH TLV: TWA 0.1 mg(Hg)/m³ (skin)
NIOSH REL: (Mercury, Aryl and Inorganic): CL 0.1 mg/m³ (skin)

SAFETY PROFILE: Poison by ingestion. When heated to decomposition it emits toxic fumes of Hg and Cl^-.

CAZ000 HR: 2
CALOMEL and MAGNESIUM SULFATE (5:8)

SYN: MAGNESIUM SULFATE and CALOMEL (8:5)

TOXICITY DATA WITH REFERENCE
orl-mus TDLo:44 g/kg/69W-I:ETA CNREA8 28,2272,68

CONSENSUS REPORTS: Mercury and its compounds are on the Community Right-To-Know List.

SAFETY PROFILE: Questionable carcinogen with experimental tumorigenic data. See also MERCUROUS CHLORIDE, MERCURY COMPOUNDS, MAGNESIUM COMPOUNDS, and SULFATES. When heated to decomposition it emits very toxic fumes of Hg, Cl^-, and SO_x.

CAZ075 HR: 1
CALOTROPIS PROCERA (Ait.) R.Br., flower extract

PROP: Indian plant belonging to the family *Asclepiadaceae* (JOETD7 22,211,88).

SYNS: AK, flower extract □ AKRA, flower extract

TOXICITY DATA WITH REFERENCE
orl-grb TDLo:3 g/kg (male 15D pre):REP IJEBA6 17,859,79
orl-grb TDLo:3 g/kg (male 15D pre):REP IJEBA6 17,859,79
orl-mus LDLo:3 g/kg JOETD7 22,211,88

SAFETY PROFILE: Slightly toxic by ingestion. Experimental reproductive effects. When heated to decomposition it emits acrid smoke and irritating fumes.

CAZ125 CAS:6874-80-2 HR: 3
CALPURNINE
mf: $C_{20}H_{27}N_3O_3$ mw: 357.50

PROP: Prisms from EtOAc. Mp: 152–154°.

SYNS: HOE 933 □ 13-HYDROXYLUPANINE-2-PYRROLE CARBOXYLIC ACID ESTER □ (2S-(2-α,7-β,7A-β,14-β,14a-α))-1H-PYRROLE-2-CARBOXYLIC ACID-DODECAHYDRO-11-OXO-7,14-METHANO-2H,6H-DIPYRIDO(1,2-α:1′,2′-e)(1,5)DIAZOCIN-2-YL) ESTER

TOXICITY DATA WITH REFERENCE
orl-rat LD50:132 mg/kg DRFUD4 2,365,77
scu-rat LD50:41 mg/kg DRFUD4 2,365,77
ivn-rat LD50:3 mg/kg DRFUD4 2,365,77
orl-mus LD50:32 mg/kg DRFUD4 2,365,77
ivn-mus LD50:3100 μg/kg DRFUD4 2,365,77

SAFETY PROFILE: Poison by ingestion, subcutaneous, and intravenous routes. When heated to decomposition it emits toxic fumes of NO_x.

CBA000 CAS:9012-59-3 HR: 3
CALVACIN

PROP: High molecular weight glycopeptide from the giant puffball mushroom *Calvatia Gigantea* (CNREA8 23,1036,63).

TOXICITY DATA WITH REFERENCE
ipr-rat LD50:65 mg/kg CNREA8 23,1036,63
ipr-mus LD50:138 mg/kg CNREA8 23,1036,63
ivn-rbt LDLo:13 mg/kg CNREA8 23,1036,63

SAFETY PROFILE: Poison by intraperitoneal and intravenous routes. When heated to decomposition it emits acrid smoke and fumes. See also MUSHROOMS.

CBA075 HR: 3
CALYCANTHINE, HYDROCHLORIDE
mf: $C_{22}H_{26}N_4 \cdot ClH$ mw: 382.98

TOXICITY DATA WITH REFERENCE
ivn-rat LD50:17,160 μg/kg JAPMA8 31,513,42

ivn-mus LD50:43,790 µg/kg JAPMA8 31,513,42
ivn-rbt LDLo:10 mg/kg JAPMA8 31,513,42

SAFETY PROFILE: Poison by intravenous route. When heated to decomposition it emits toxic fumes of NO_x and HCl.

CBA125 CAS:2752-65-0 HR: 3
CAMBOGIC ACID
mf: $C_{38}H_{44}O_8$ mw: 628.82

PROP: Crystals or amorphous mass. Mp: 86–91°.

SYNS: β-GUTTIFERIN □ B''-GUTTIFERIN

TOXICITY DATA WITH REFERENCE
ipr-rat LD50:88 mg/kg IJEBA6 5,96,67
ivn-rat LD50:107 mg/kg IJEBA6 5,96,67
scu-mus LD50:354 mg/kg 85DGAU 8(1),331,82

SAFETY PROFILE: Poison by subcutaneous, intravenous, and intraperitoneal routes. When heated to decomposition it emits acrid smoke and fumes.

CBA200 CAS:68916-73-4 HR: 1
CAMELIA OIL

SYN: KAMILLEN OEL

TOXICITY DATA WITH REFERENCE
orl-rat LD50:8560 mg/kg PHARAT 14,435,59

CONSENSUS REPORTS: Reported in EPA TSCA Inventory.

SAFETY PROFILE: Low toxicity by ingestion. When heated to decomposition it emits acrid smoke and irritating vapors.

CBA375 CAS:54063-28-4 HR: 3
CAMIVERINE
mf: $C_{19}H_{30}N_2O_2$ mw: 318.51

PROP: Pale yellow oil. Bp: 184-188 @ 2 mm.

SYNS: ESTERE ISOAMILICO dell'ACIDO α-(N-(PIRROLIDINOETIL))-AMINOFENILACETICO (ITALIAN) □ FC 4/58 □ 2-PHENYL-N-(2-(1-PYRROLIDINYL)ETHYL)GLYCINE ISOPENTYL ESTER □ SANASPASMINA

TOXICITY DATA WITH REFERENCE
ipr-rat LD50:140 mg/kg FRPSAX 17,914,62
ivn-rat LD50:21 mg/kg FRPSAX 17,914,62
orl-mus LD50:920 mg/kg FRPSAX 17,914,62
ipr-mus LD50:175 mg/kg FRPSAX 17,914,62
ivn-mus LD50:28 mg/kg FRPSAX 17,914,62
ivn-rbt LD50:13 mg/kg FRPSAX 17,914,62

SAFETY PROFILE: Poison by intravenous and intraperitoneal routes. Moderately toxic by ingestion. When heated to decomposition it emits toxic fumes of NO_x. See also ESTERS.

CBA500 CAS:79-92-5 HR: 1
CAMPHENE
mf: $C_{10}H_{16}$ mw: 136.26

PROP: Colorless cubic crystals; oily odor. Mp: 50–51°, bp: 159°, d: 0.842 @ 54°/4°, refr index: 1.452 @ 55°. Sol in alc; misc in fixed oils; insol in water.

SYNS: BICYCLO(2.2.1)HEPTANE, 2,2-DIMETHYL-3-METHYLENE-(9CI) □ FEMA No. 2229

TOXICITY DATA WITH REFERENCE
bfa-rat/sat 2500 mg/kg NUCADQ 1,10,79

CONSENSUS REPORTS: Reported in EPA TSCA Inventory.

SAFETY PROFILE: Mutation data reported. Combustible; yields flammable vapors when heated and can react with oxidizing materials. To fight fire, use water spray, foam, fog, CO_2. When heated to decomposition it emits acrid smoke and irritating fumes.

CBA750 CAS:76-22-2 HR: 3
CAMPHOR
DOT: UN 2717
mf: $C_{10}H_{16}O$ mw: 152.26

PROP: White, transparent, crystalline masses; penetrating odor; pungent, aromatic taste. Mp: 180°, bp: 204°, lel: 0.6%, uel: 3.5%, flash p: 150°F (CC), d: 0.992 @ 25°/4°, autoign temp: 871°F, vap d: 5.24.

SYNS: 2-BORNANONE □ 2-CAMPHANONE □ CAMPHOR, synthetic (ACGIH, DOT) □ CAMPHOR-natural □ FORMOSA CAMPHOR □ GUM CAMPHOR □ HUILE de CAMPHRE (FRENCH) □ JAPAN CAMPHOR □ KAMPFER (GERMAN) □ 2-KETO-1,7,7-TRIMETHYLNORCAMPHANE □ LAUREL CAMPHOR □ MATRICARIA CAMPHOR □ 2-OXOBORNANE □ 1,7,7-TRIMETHYLBICYCLO(2.2.1)-2-HEPTANONE □ 1,7,7-TRIMETHYLNORCAMPHOR

TOXICITY DATA WITH REFERENCE
cyt-smc 2 mmol/tube HEREAY 33,457,47
orl-inf LDLo:70 mg/kg AJPAA4 30,857,54
unk-man LDLo:29 mg/kg 85DCAI 2,73,70
ipr-rat LDLo:900 mg/kg JPETAB 65,275,39
scu-rat LD50:70 mg/kg CDGU** -,-,34
orl-mus LD50:1310 mg/kg SHCKA3 75,934,75
ihl-mus LCLo:400 mg/m³/3H 85GMAT -,31,82
ipr-mus LD50:3000 mg/kg AJPAA4 30,857,54
scu-mus LDLo:200 mg/kg HDTU** -,-,33
orl-dog LDLo:800 mg/kg HBAMAK 4,1289,35
ipr-cat LDLo:400 mg/kg HBAMAK 4,1289,35
orl-rbt LDLo:2000 mg/kg AJPAA4 30,857,54
scu-frg LDLo:240 mg/kg AEXPBL 50,199,1903

CONSENSUS REPORTS: Reported in EPA TSCA Inventory.

OSHA PEL: TWA 2 mg/m³
ACGIH TLV: TWA 2 ppm; STEL 3 ppm
DFG MAK: 2 ppm (13 mg/m³)
DOT CLASSIFICATION: 4.2; *Label:* Flammable Solid

SAFETY PROFILE: A human poison by ingestion, and possibly other routes. An experimental poison by inhalation, subcutaneous, and intraperitoneal routes. A local irritant. Ingestion causes nausea, vomiting, dizziness, excitation, and convulsions. Mutation data reported. Used as a topical anti-infective and anti-itching agent. Flammable liquid when exposed to heat or flame; can react with oxidizing materials. Vapor is explosive when exposed to heat or flame or CrO_3. To fight fire, use foam, carbon dioxide, dry chemical. See also KETONES and other camphor entries.

For occupational chemical analysis use NIOSH: Ketones II (desorption in 99:1 CS_2:methanol) 1301.

CBA800 **CAS:21368-68-3** **HR: 2**
dl-CAMPHOR
mf: $C_{10}H_{16}O$ mw: 152.26

SYN: (±)-CAMPHOR

TOXICITY DATA WITH REFERENCE
ipr-rat LD50:956 mg/kg KHFZAN 16(7),108,82
scu-rat LD50:3040 mg/kg KHFZAN 16(7),108,82
ipr-mus LD50:884 mg/kg KHFZAN 16(7),108,82
scu-mus LD50:3020 mg/kg KHFZAN 16(7),108,82

CONSENSUS REPORTS: Reported in EPA TSCA Inventory.

SAFETY PROFILE: Moderately toxic by subcutaneous and intraperitoneal routes. When heated to decomposition it emits acrid smoke and fumes. See other camphor entries.

CBB000 **CAS:464-48-2** **HR: 3**
l-(−)-CAMPHOR
mf: $C_{10}H_{16}O$ mw: 152.26

PROP: Crystals. Mp: 179°.

SYN: l-CAMPHOR

TOXICITY DATA WITH REFERENCE
orl-rat LDLo:800 µg/kg JPETAB 1,445,09
ivn-mus LD50:320 mg/kg CSLNX* NX#02534

CONSENSUS REPORTS: Reported in EPA TSCA Inventory.

SAFETY PROFILE: Deadly poison by ingestion. Poison by intravenous route. When heated to decomposition it emits acrid smoke and irritating fumes. See also (IR,4R)-(+)-CAMPHOR and CAMPHOR.

CBB250 **CAS:464-49-3** **HR: 3**
(1R,4R)-(+)-CAMPHOR
mf: $C_{10}H_{16}O$ mw: 152.26

PROP: Rhombohedra or cubic crystals. Mp: 204°, bp: 204°. Sltly sol in water.

SYNS: ALCANFOR □ (+)-2-BORNANONE □ d-2-BORNANONE □ d-2-CAMPHANONE □ (+)-CAMPHOR □ d-CAMPHOR □ d-(+)-CAMPHOR □ CAMPHOR USP □ JAPANESE CAMPHOR □ (1R)-1,7,7-TRIMETHYL-BICYCLO(2.2.1)HEPTAN-2-ONE

TOXICITY DATA WITH REFERENCE
skn-rbt 500 mg/24H MLD FCTXAV 16,665,78
scu-rat LDLo:1700 mg/kg FCTXAV 16,665,78
ipr-rat LDLo:3500 mg/kg FCTXAV 16,665,78
orl-mus LD50:1310 mg/kg FCTXAV 16,665,78
scu-mus LDLo:2200 mg/kg FCTXAV 16,665,78
ivn-mus LD90:525 mg/kg FCTXAV 16,665,78
ipr-cat LDLo:400 mg/kg FCTXAV 16,665,78

CONSENSUS REPORTS: Reported in EPA TSCA Inventory.

SAFETY PROFILE: Poison by intraperitoneal route. Moderately toxic by ingestion, subcutaneous, and intravenous routes. A skin irritant. When heated to decomposition it emits acrid and irritating fumes. See other camphor entries.

CBB375 **CAS:8011-47-0** **HR: 2**
CAMPHORATED OIL

SYN: CAMPHOR LINIMENT

TOXICITY DATA WITH REFERENCE
orl-wmn TDLo:240 mg/kg (40W preg):REP OBGNAS 25,255,65
orl-wmn TDLo:240 mg/kg (female 40W post):TER OBGNAS 25,255,65
orl-wmn TDLo:148 mg/kg:CNS,GIT PEDIAU 52,713,73
orl-wmn TDLo:1180 µL/kg:BAH,SYS CTOXAO 11,151,77
orl-man TDLo:843 µL/kg:EYE,BAH,GIT CTOXAO 11,151,77
unr-wmn TDLo:900 mg/kg:CNS,GIT JFMAAQ 43,999,57
ivn-mus LD50:1600 mg/kg THERAP 20,321,65

SAFETY PROFILE: Moderately toxic by intravenous route. Human systemic effects: coma, convulsions, excitement, liver function tests impaired, muscle weakness, nausea or vomiting, visual field changes. Human teratogenic effects by ingestion include these developmental abnormalities: extra embryonic structures, homeostasis, reduced viability and other neonatal effects. Other experimental reproductive effects. When heated to decomposition it emits acrid smoke and fumes. See other camphor entries.

CBB500 **CAS:8008-51-3** **HR: 3**
CAMPHOR OIL
DOT: UN 1130

PROP: Colorless or yellowish, oily, fragrant liquid. Bp: 175–200°, flash p: 117°F (CC), d: 0.875–0.900 @ 20°/20°. Insol in water; sol in chloroform, ether, oils, and in approx 3 vols alc. Found in the trees and bark of *Cinnamomum carphora sieb* (Fam. Lauraceae) and prepared by fractional distillation of crude camphor oil after the camphor has been crystallized out; a white, viscous liquid with cineole as the principal ingredient along with monoterpenes (FCTXAV 11,1011,73).

SYNS: CAMPHOR OIL, RECTIFIED □ CAMPHOR OIL WHITE □ CAMPHOR OIL YELLOW □ FORMOSA CAMPHOR OIL □ FORMOSE OIL of CAMPHOR □ JAPANESE CAMPHOR OIL □ JAPANESE OIL of CAMPHOR □ LIGHT CAMPHOR OIL □ LIGHT OIL of CAMPHOR □ LIQUID CAMPHOR □ OIL of CAMPHOR RECTIFIED □ OIL CAMPHOR SASSAFRASSY □ OIL of CAMPHOR WHITE □ WHITE CAMPHOR OIL □ WHITE OIL of CAMPHOR

TOXICITY DATA WITH REFERENCE
skn-rbt 500 mg MLD FCTXAV 11,1047,73
orl-hmn TDLo:29 mg/kg:CNS,PUL 34ZIAG -,150,69
orl-chd LDLo:50 mg/kg 34ZIAG -,150,69
orl-rat LD50:3730 mg/kg FCTXAV 13,739,75

CONSENSUS REPORTS: Reported in EPA TSCA Inventory.

DOT CLASSIFICATION: 3; *Label:* Flammable Liquid

SAFETY PROFILE: A human poison by ingestion. Human systemic effects by ingestion: convulsions, tremors, and unspecified respiratory system effects. A skin irritant. Flammable liquid when exposed to heat or flame; can react with oxidizing materials. To fight fire, use foam, CO_2, dry chemical, mist, fog. See also SAFROL and CAMPHOR.

CBB870 CAS:7689-03-4 HR: 3
CAMPTOTHECINE
mf: $C_{20}H_{16}N_2O_4$ mw: 348.38

PROP: Pale-yellow needles from methanol + acetonitrile. Decomp 264–267°. Does not form stable salts with acids.

SYNS: CAMPTOTHECIN □ 20(S)-CAMPTOTHECINE □ (S)-4-ETHYL-4-HYDROXY 1H-PYRANO(3′,4′:6,7)INDOLIZINO(1,2-b)QUINOLINE-3,14(4H,12H)-DIONE □ NSC 94600 □ NSC 100880 □ 21,22-SECO-CAMPTOTHECIN-21-OIC ACID LACTONE

TOXICITY DATA WITH REFERENCE
dnd-omi 100 mg/L/30M NATUAS 248,226,74
dnd-hmn:hla 20 μmol/L CNREA8 33,2834,73
dni-hmn:hla 5 μmol/L HXPHAU 38(Pt 2),649,75
oms-hmn:hla 5 μmol/L HXPHAU 38(Pt 2),649,75
oms-mus:lym 1 mg/L BCPCA6 21,1977,72
dni-ckn:emb 500 μg/L CJBIAE 55,1180,77
ipr-mus LD50:64 mg/kg CNREA8 39,2204,79
ivn-mus LD50:38 mg/kg NCISP* JAN86

SAFETY PROFILE: Poison by intravenous and intraperitoneal routes. Human mutation data reported. When heated to decomposition it emits toxic fumes of NO_x. See also CAMPTOTHECIN, SODIUM SALT.

CBB875 CAS:25387-67-1 HR: 3
CAMPTOTHECIN, SODIUM SALT
mf: $C_{20}H_{15}N_2O_4 \cdot Na$ mw: 370.36

SYN: NSC-100880

TOXICITY DATA WITH REFERENCE
ivn-hmn TDLo:2500 μg/kg/7D-I:BLD CCROBU 56,515,72
orl-mus LD50:27 mg/kg PMDCAY 9,1,73
ivn-mus LD50:57 mg/kg PMDCAY 9,1,73

SAFETY PROFILE: Poison by ingestion and intravenous routes. Human systemic effects by intravenous route: reduction in the number of white blood cells (leukopenia), reduction in the number of blood platelets (thrombocytopenia), and changes in blood cell count. When heated to decomposition it emits toxic fumes of NO_x and Na_2O.

CBB900 CAS:8021-28-1 HR: 1
CANADIAN FIR NEEDLE OIL

SYNS: ABIES OIL □ BALSAM FIR OIL

TOXICITY DATA WITH REFERENCE
orl-rat LD50:>5 g/kg FCTXAV 13,449,75
skn-rbt LD50:>5 g/kg FCTXAV 13,449,75

CONSENSUS REPORTS: Reported in EPA TSCA Inventory.

SAFETY PROFILE: Low toxicity by ingestion and skin contact. When heated to decomposition it emits acrid smoke and irritating vapors.

CBC375 HR: 3
CANDIDA ALBICANS GLYCOPROTEINS

PROP: Glycoprotein complex isolated from the cell walls of the 29–3–109 strain of *Candida albicans.* 40YJAX -,35,76

TOXICITY DATA WITH REFERENCE
ivn-rat TDLo:15 mg/kg (8D preg):TER 40YJAX -,35,79
orl-rat TDLo:130 mg/kg (4-14D preg):TER 40YJAX -,91,76
ivn-rat TDLo:30 mg/kg (female 8D post):TER 40YJAX -,83,76
ivn-mus LD50:290 mg/kg TOXIA6 12,103,74

SAFETY PROFILE: Poison by intravenous route. An experimental teratogen. Other experimental reproductive effects. When heated to decomposition it emits toxic fumes of NO_x.

CBC500 CAS:1405-90-9 HR: 3
CANDIDIN
mf: $C_{46}H_{75}NO_{17}$ mw: 914.22

SYN: CANDIDINE

TOXICITY DATA WITH REFERENCE
orl-mus LD50:100 mg/kg 85GDA2 2,288,80
ipr-mus LD50:7 mg/kg MEIEDD 10,240,83
scu-mus LD50:30 mg/kg MEIEDD 10,240,83
ivn-mus LD50:1500 μg/kg MEIEDD 10,240,83

SAFETY PROFILE: Poison by ingestion, subcutaneous, intraperitoneal, and intravenous routes. When heated to decomposition it emits toxic fumes of NO_x.

CBD250 CAS:64854-99-5 HR: 1
CANDLETOXIN A
mf: $C_{35}H_{44}O_9$ mw: 608.8

PROP: Glassy resin.

TOXICITY DATA WITH REFERENCE
skn-mus 290 ng OPEN ARTODN 44,279,80

SAFETY PROFILE: A skin irritant. When heated to decomposition it emits acrid smoke and irritating fumes.

CBD500 CAS:64854-98-4 HR: 1
CANDLETOXIN B
mf: $C_{33}H_{42}O_8$ mw: 566.8

PROP: Resin.

TOXICITY DATA WITH REFERENCE
skn-mus 110 ng OPEN ARTODN 44,279,80

SAFETY PROFILE: A skin irritant. When heated to

decomposition it emits acrid smoke and irritating fumes.

CBD599 **CAS:13956-29-1** **HR: 3**
CANNABIDIOL
mf: $C_{21}H_{30}O_2$ mw: 314.51

PROP: Pale yellow resin or crystals from pet ether. Mp: 66–67°, bp: 160–180° @ 0.001 mm, d: 1.040, n (20/D) 1.5404. Practically insol in water or 10% NaOH; sol in ethanol, methanol, ether, benzene, chloroform, and petr ether.

SYNS: (−)-CANNABIDIOL ☐ (−)-trans-CANNABIDIOL ☐ CBD ☐ (−)-trans-2-p-MENTHA-1,8-DIEN-3-YL-5-PENTYLRESORCINOL ☐ (1R-trans)-2-(3-METHYL-6-(1-METHYLETHENYL)-2-CYCLOHEXEN-1-YL)-5-PENTYL-1,3-BENZENEDIOL

TOXICITY DATA WITH REFERENCE
mnt-mus-ipr 50 mg/kg/5D-I PHMGBN 21,277,80
dni-mus-ipr 200 mg/kg RCOCB8 17,703,77
dni-mus:lng 33,700 nmol/L CNREA8 36,95,76
dni-mus:bmr 489 μmol/L CNREA8 36,95,76
cyt-mus-ipr 50 mg/kg/5D-I PHMGBN 21,277,80
orl-mus TDLo:50 mg/kg (female 12D post):REP TJADAB 33,195,86
orl-mus TDLo:50 mg/kg (female 12D post):REP NETOD7 8,391,86
orl-mus TDLo:750 mg/kg (15D male):REP SCIEAS 216,315,82
orl-mus TDLo:750 mg/kg (15D male):TER SCIEAS 216,315,82
orl-mus TDLo:50 mg/kg (female 12D post):TER TJADAB 33,195,86
ivn-mus LD50:50 mg/kg JMCMAR 18,213,75
ivn-mky LD50:212 mg/kg TXAPA9 58,118,81

CONSENSUS REPORTS: EPA Genetic Toxicology Program.

SAFETY PROFILE: Poison by intravenous route. An experimental teratogen. Other experimental reproductive effects. Mutation data reported. When heated to decomposition it emits acrid smoke and fumes.

CBD625 **CAS:521-35-7** **HR: 1**
CANNABINOL
mf: $C_{21}H_{26}O_2$ mw: 310.42

PROP: Leaflets or crystals from petr ether. Mp: 76–77°, bp: 185° at 0.05 mm. Insol in water; sol in methanol, ethanol, and aq alkaline solns.

SYNS: 3-AMYL-1-HYDROXY-6,6,9-TRIMETHYL-6H-DIBENZO(b,d)PYRAN ☐ CBN ☐ 6,6,9-TRIMETHYL-3-PENTYL-6H-DIBENZO(b,d)PYRAN-1-OL

TOXICITY DATA WITH REFERENCE
dni-hmn:hla 10 μmol/L ANTRD4 3,211,83
mnt-mus-ipr 50 mg/kg/5D-I PHMGBN 21,277,80
dni-mus-ipr 200 mg/kg RCOCB8 17,703,77
dni-mus:lng 2300 nmol/L CNREA8 36,95,76
cyt-mus-ipr 50 mg/kg/5D-I PHMGBN 21,277,80
spm-mus-ipr 50 mg/kg/5D-C PHMGBN 18,143,79
orl-mus TDLo:50 mg/kg (female 12D post):REP TJADAB 33,195,86
orl-mus TDLo:50 mg/kg (female 20D post):REP SCIEAS 205,1420,79
orl-mus TDLo:50 mg/kg (female 12D post):REP NETOD7 8,391,86
orl-mus TDLo:400 mg/kg (male 8D pre):REP PBBHAU 12,143,80
scu-rat TDLo:100 mg/kg (male 10D pre):REP BIREBV 20,1039,79
orl-mus TDLo:450 mg/kg (male 9D pre):REP PBBHAU 12,143,80
orl-mus LD50:13,500 mg/kg JPETAB 88,154,46

CONSENSUS REPORTS: EPA Genetic Toxicology Program.

SAFETY PROFILE: Low toxicity by ingestion. An experimental teratogen. Other experimental reproductive effects. Human mutation data reported. When heated to decomposition it emits acrid smoke and fumes. See also CANNABIS.

CBD750 **CAS:8063-14-7** **HR: 3**
CANNABIS

PROP: A greenish-black, resinous, bitter substance from *Cannabis sativa.*

SYNS: BHANG ☐ CANNABIS RESIN ☐ CHARAS ☐ CME ☐ GANJA ☐ HASACH ☐ HASHISH ☐ INDIAN CANNABIS ☐ INDIAN HEMP ☐ MARIHUANA ☐ MARIJUANA

TOXICITY DATA WITH REFERENCE
sln-dmg-orl 1 pph 48NTAS 7,101,81
dlt-dmg-orl 5000 ppm 48NTAS 7,101,81
oms-hmn:lym 500 mg/L JAINAA 24,71,75
cyt-hmn:lym 500 mg/L JAINAA 24,71,75
orl-rat TDLo:3 g/kg (female 2-21D post):REP NETOD7 1,285,79
orl-rat TDLo:1600 mg/kg (15-22D preg):REP NTOTDY 3,351,81
orl-rbt TDLo:19,500 μg/kg (female 6-18D post):REP TXAPA9 38,223,76
orl-rat TDLo:3 g/kg (female 3-22D post):REP PSCHDL 71,71,80
ihl-rbt TCLo:1440 μg/kg (female 6-18D post):TER FAATDF 7,236,86
orl-rbt TDLo:79,300 μg/kg (female 6-18D post):TER BANMAC 164,276,80
orl-rat TDLo:3 g/kg (female 3-22D post):TER PSCHDL 71,71,80
orl-rat TDLo:44 mg/kg (male 9W pre):REP TXAPA9 38,223,76
ipr-rat TDLo:30 mg/kg (male 1D pre):REP RCOCB8 7,779,74
orl-rbt TDLo:195 mg/kg (female 6-18D post):REP TXAPA9 38,223,76
orl-mus TDLo:25 mg/kg (male 1D pre):REP SCIEAS 216,315,82
ipr-mus TDLo:2 g/kg (male 25D pre):REP EJPHAZ 26,111,74
ipr-rat TDLo:25,200 μg/kg (female 1-6D post):TER LANCAO 2,406,68
scu-rbt TDLo:250 mg/kg (female 7-10D post):TER TXAPA9 14,276,69
orl-hmn TDLo:60 mg/kg/20D:CVS BMJOAE 1,460,78
orl-rat LD50:1380 mg/kg TXAPA9 25,363,73

ipr-mus LDLo:5 g/kg NATUAS 228,134,70

CONSENSUS REPORTS: EPA Genetic Toxicology Program.

SAFETY PROFILE: Moderately toxic by ingestion. An experimental teratogen. Experimental reproductive effects. Human systemic effects by ingestion include: change in heart rate, change in cardiac resting or action potential, and blood pressure decrease. Human mutation data reported. An allergen. When ingested or inhaled as smoke, it can cause euphoria, delirium, hallucinations, drowsiness, weakness, and hyporeflexia. An overdose can cause coma and death. Dried material can burn; can react with oxidizing materials. When heated to decomposition it emits toxic fumes of NO_x. See also THC.

CBD760 HR: 2
CANNABIS SMOKE RESIDUE

SYN: MARIJUANA, SMOKE RESIDUE

TOXICITY DATA WITH REFERENCE
dnd-esc 10 ppm MUREAV 89,95,81
scu-rat TDLo:11,640 mg/kg/18D-I:ETA VHTODE 21(Suppl),148,79

SAFETY PROFILE: Questionable carcinogen with experimental tumorigenic data. Mutation data reported. When heated to decomposition it emits acrid smoke and irritating fumes.

CBE250 HR: 2
CANTHARIDES
mf: $C_{10}H_{12}O_4$ mw: 196.15

PROP: Brown to black powder or scales. Mp: 218°, bp: subl @ 90°.

SYNS: BLISTERING BEETLES ☐ BLISTERING FLIES ☐ SPANISH FLY

SAFETY PROFILE: Strong irritant via skin contact, ingestion, inhalation, and contact with eyes. An allergen. Can cause conjunctivitis, keratitis, blepharitis, slight swelling of cornea and inflammation of iris. It is often mistakenly used as an aphrodisiac, but it is much too dangerous and irritating a material for this purpose. When heated to decomposition it emits acrid smoke and fumes.

CBE750 CAS:56-25-7 HR: 3
CANTHARIDINE
mf: $C_{10}H_{12}O_4$ mw: 196.22

PROP: Plates. Mp: 218°.

SYNS: CANTHARIDES CAMPHOR ☐ CANTHARIDIN ☐ CANTHARONE ☐ exo-1,2-cis-DIMETHYL-3,6-EPOXYHEXAHYDROPHTHALIC ANHYDRIDE ☐ 2,3-DIMETHYL-7-OXABICYCLO(2.2.1)HEPTANE-2,3-DICARBOXYLIC ANHYDRIDE ☐ HEXAHYDRO-3A,7A-DIMETHYL-4,7-EPOXYISOBENZOFURAN-1,3-DIONE

TOXICITY DATA WITH REFERENCE
skn-mus TDLo:25 mg/kg/14W-I:NEO BJCAAI 9,177,55
skn-mus TD:70 mg/kg/52W-I:ETA CNREA8 32,1463,72
orl-hmn LDLo:428 μg/kg 34ZIAG -,646,69
ipr-mus LD50:1 mg/kg JAFCAU 35,823,87
orl-dog LDLo:50 mg/kg FDWU** -,-,31

CONSENSUS REPORTS: IARC Cancer Review: Group 3 IMEMDT 7,56,87; Animal Limited Evidence IMEMDT 10,79,76. EPA Extremely Hazardous Substances List. Reported in EPA TSCA Inventory.

SAFETY PROFILE: A deadly human poison by ingestion. Questionable carcinogen with experimental tumorigenic and neoplastigenic data. See also CANTHARIDES. When heated to decomposition it emits acrid and irritating fumes.

CBF000 CAS:76-90-4 HR: 3
CANTRIL
mf: $C_{21}H_{26}NO_3 \cdot Br$ mw: 420.39

PROP: Mp: 228–229° (decomp).

SYNS: BENZILIC ACID ester with 3-HYDROXY-1,1-DIMETHYLPIPERIDINIUM BROMIDE ☐ CANTIL ☐ GASTROPIDIL ☐ 3-HYDROXY-1,1-DIMETHYLPIPERIDINIUM BROMIDE BENZILATE ☐ 3-((HYDROXYDIPHENYLACETYL)OXY)-1,1-DIMETHYLPIPERIDINIUM BROMIDE ☐ JB 340 ☐ MEPENZOLATE ☐ MEPENZOLATE BROMIDE ☐ N-METHYL-3-PIPERIDYL BENZILATE METHOBROMIDE ☐ N-METHYL-3-PIPERIDYL-DIPHENYLGLYCOLATE METHOBROMIDE ☐ 1-METHYL-3-PIPERIDYL ESTER METHOBROMIDE BENZILIC ACID ☐ TRANCOLON

TOXICITY DATA WITH REFERENCE
orl-rat LD50:742 mg/kg JOPDAB 69,663,66
ipr-rat LD50:158 mg/kg DRUGAY 6,358,82
scu-rat LD50:740 mg/kg TXAPA9 18,185,71
ivn-rat LD50:22 mg/kg 27ZIAQ -,148,73
orl-mus LD50:900 mg/kg 27ZIAQ -,-,65
ivn-mus LD50:9800 μg/kg 27ZIAQ -,-,65
scu-mus LD50:455 mg/kg NIIRDN 6,358,82

SAFETY PROFILE: Poison by intravenous and intraperitoneal routes. Moderately toxic by ingestion and subcutaneous routes. When heated to decomposition it emits very toxic fumes of Br^- and NO_x.

CBF250 CAS:302-22-7 HR: 3
CAP
mf: $C_{23}H_{29}ClO_4$ mw: 404.97

PROP: Crystals from Me_2CO/Et_2O. Mp: 211–212°.

SYNS: 17-ACETOXY-6-CHLORO-6-DEHYDROPROGESTERONE ☐ 17-α-ACETOXY-6-CHLORO-6-DEHYDROPROGESTERONE ☐ 17-α-ACETOXY-6-CHLORO-6,7-DEHYDROPROGESTERONE ☐ 17-α-ACETOXY-6-CHLOROPREGNA-4,6-DIENE-3,20-DIONE ☐ 17-α-ACETOXY-6-CHLORO-4,6-PREGNADIENE-3,20-DIONE ☐ 17-(ACETYLOXY)-6-CHLOROPREGNA-4,6-DIENE-3,20-DIONE ☐ CHLORMADINON ACETATE ☐ CHLORMADINONE ACETATE ☐ CHLORMADINONU (POLISH) ☐ 6-CHLORO-17-α-ACETOXY-4,6-PREGNADIENE-3,20-DIONE ☐ Δ^6-6-CHLORO-17-α-ACETOXYPROGESTERONE ☐ 6-CHLORO-Δ^6-17-ACETOXYPROGESTERONE ☐ 6-CHLORO-Δ^6-(17-α)ACETOXYPROGESTERONE ☐ 6-CHLORO-Δ^6-DEHYDRO-17-ACETOXYPROGESTERONE ☐ 6-CHLORO-6-DEHYDRO-17-α-ACETOXYPROGESTERONE ☐ 6-CHLORO-6-DEHYDRO-17 α HYDROXYPROGESTERONE ACETATE ☐ 6-CHLORO-17-α-HYDROXYPREGNA-4,6-DIENE 3,20-DIONE ACETATE ☐ 6-CHLORO-17-α-HYDROXY-Δ^6-PROGESTERONE ACETATE ☐ CHLOROMADINONE ACETATE ☐ 6-CHLORO-$\Delta^{4,6}$-PREGNADIENE-17-α-OL-3,20-DIONE-17-ACETATE ☐ 6-CHLORO-PREGNA-4,6-DIEN-17-α-OL-3,20-DIONE ACETATE ☐ CLORDION ☐ CMA ☐ C-QUENS ☐ 6-DEHYDRO-6-CHLORO-

17-α-ACETOXYPROGESTERONE □ LORMIN □ LUTINYL □ NSC-92338 □ RS 1280 □ SKEDULE □ ST 155

TOXICITY DATA WITH REFERENCE

orl-wmn TDLo:1200 μg/kg (female 17W pre):REP FESTAS 17,49,66
orl-wmn TDLo:400 μg/kg (female 20D pre):REP INJFA3 9,57,64
orl-wmn TDLo:600 μg/kg (female 60D pre):REP BMJOAE 2,263,68
orl-wmn TDLo:80 μg/kg (80D pre):REP FESTAS 18,57,67
orl-wmn TDLo:3650 μg/kg (female 52W pre):REP CCPTAY 7,503,73
unr-wmn TDLo:3640 μg/kg (female 52W pre):TER CCPTAY 3,45,71
orl-rat TDLo:120 mg/kg (female 7-18D post):REP OYYAA2 16,153,78
orl-mus TDLo:80 mg/kg (female 8-15D post):TER PSEBAA 121,455,66
orl-rat TDLo:3600 mg/kg (female 7-18D post):TER OYYAA2 16,153,78
scu-rat TDLo:350 μg/kg (female 7D pre):REP JRPFA4 10,105,65
orl-rbt TDLo:416 mg/kg (female 6-18D post):REP OYYAA2 16,153,78
scu-rbt TDLo:5 μg/kg (female 1D pre):REP ACEDAB 73,3,63
orl-rbt TDLo:300 μg/kg (female 3D pre):REP BIREBV 1,372,69
orl-mus TDLo:200 mg/kg (female 6-7D post):REP OYYAA2 15,955,78
scu-mus TDLo:2 mg/kg (female 1-4D post):REP CCPTAY 21,537,80
orl-mus TDLo:910 mg/kg (female 26W pre):REP IJEBA6 11,292,73
scu-rat TDLo:28 mg/kg (female 14D pre):REP CCPTAY 5,57,72
scu-rat TDLo:25 mg/kg (female 2-6D post):REP JRPFA4 39,119,74
ims-rat TDLo:80 mg/kg (male 40D pre):REP JRPFA4 48,177,76
orl-rat TDLo:378 mg/kg (male 63D pre):REP OYYAA2 15,1211,78
scu-rbt TDLo:2 mg/kg (male 1D pre):REP JRPFA4 20,105,69
orl-rbt TDLo:130 mg/kg (female 8-20D post):TER PSEBAA 121,455,66
orl-mus TDLo:10 mg/kg (female 8-17D post):TER PSEBAA 121,455,66
orl-rat TDLo:3600 mg/kg (female 7-18D post):TER OYYAA2 16,153,78
orl-rat TDLo:1260 mg/kg (female 8-21D post):TER ANENAG 28,433,67
orl-dog TDLo:182 mg/kg/2Y-C:ETA JAMAAP 219,1601,72
orl-dog TDLo:639 mg/kg/7Y-C:ETA JTEHD6 3,167,77
ipr-mus LD50:3 g/kg KSRNAM 11,571,77

CONSENSUS REPORTS: IARC Cancer Review: Animal Limited Evidence IMEMDT 21,365,79; Animal Sufficient Evidence IMEMDT 6,149,74.

SAFETY PROFILE: Suspected carcinogen with experimental carcinogenic and tumorigenic data. Moderately toxic by intraperitoneal route. Human maternal and reproductive effects by ingestion, intramuscular, and possibly other routes: ovary, uterus, cervix, vagina, and fallopian tube changes; menstrual cycle changes or disorders; changes in fertility; and other unspecified female effects. A human teratogen that causes developmental abnormalities of the endocrine system in the fetus. Experimental teratogenic and reproductive effects. An oral contraceptive. When heated to decomposition it emits toxic fumes of Cl^-.

CBF625 CAS:27276-25-1 HR: 2
CAPOBENATE
mf: $C_{16}H_{22}NO_6 \cdot Na$ mw: 347.38

SYNS: CAPOBENATE SODIUM □ C-3 SODIUM SALT □ epsilon-(3,4,5-TRIMETHOXYBENZAMIDO)CAPROIC ACID SODIUM SALT □ epsilon-(3,4,5-TRIMETHOXYBENZAMIDO)CAPRONSAEURE NATRIUM (GERMAN) □ 6-((3,4,5-TRIMETHOXYBENZOYL)AMINO)HEXANOIC ACID SODIUM SALT

TOXICITY DATA WITH REFERENCE

ipr-rat LD50:2500 mg/kg USXXAM #3697563
orl-mus LD50:5 g/kg USXXAM #3697563
ipr-mus LD50:3 g/kg USXXAM #3697563
ivn-mus LD50:2500 mg/kg USXXAM #3697563

SAFETY PROFILE: Moderately toxic by intravenous and intraperitoneal routes. Mildly toxic by ingestion. Used as a cardiac anti-arrhythmic. When heated to decomposition it emits toxic fumes of NO_x and Na_2O.

CBF675 CAS:1405-36-3 HR: 3
CAPREOMYCIN DISULFATE

SYNS: CAPROCIN □ OGOSTAL

TOXICITY DATA WITH REFERENCE

ipr-rat LD50:157 mg/kg ANYAA9 135,960,66
scu-rat LD50:1191 mg/kg ANYAA9 135,960,66
ivn-rat LD50:325 mg/kg ANYAA9 135,960,66
scu-mus LD50:514 mg/kg ANYAA9 135,960,66
ivn-mus LD50:250 mg/kg ANYAA9 135,960,66

SAFETY PROFILE: Poison by intravenous and intraperitoneal routes. Moderately toxic by subcutaneous route. When heated to decomposition it emits toxic fumes of SO_x. See also SULFATES.

CBF680 CAS:37280-35-6 HR: 3
CAPREOMYCIN IA
mf: $C_{25}H_{44}N_{14}O_8$ mw: 668.83

PROP: Crystals. Mp: 246–248°.

SYNS: A-250-II □ ANTIBIOTIC 29275 □ ANTIBIOTIC A-250-II □ CAPROMYCIN □ CAPSTAT

TOXICITY DATA WITH REFERENCE

orl-mus LD50:10 g/kg 85GDA2 4(1),288,80
scu-mus LD50:514 mg/kg 85GDA2 4(1),288,80
ivn-mus LD50:250 mg/kg 85GDA2 4(1),288,80

SAFETY PROFILE: Poison by intravenous route. Moderately toxic by subcutaneous route. When heated to decomposition it emits toxic fumes of NO_x.

C

CBF700 **CAS:105-60-2** **HR: 3**
CAPROLACTAM
mf: $C_6H_{11}NO$ mw: 113.18

$HN(CH_2)_5CO$

PROP: White crystals or leaflets from ligroin. Mp: 69°, bp: 139° @ 12 mm, vap press: 6 mm @ 120°.

SYNS: AMINOCAPROIC LACTAM ☐ 6-AMINOHEXANOIC ACID CYCLIC LACTAM ☐ 2-AZACYCLOHEPTANONE ☐ 6-CAPROLACTAM ☐ ω-CAPROLACTAM (MAK) ☐ CAPROLATTAME (FRENCH) ☐ CYCLOHEXANONE ISO-OXIME ☐ EPSYLON KAPROLAKTAM (POLISH) ☐ HEXAHYDRO-2-AZEPINONE ☐ HEXAHYDRO-2H-AZEPIN-2-ONE ☐ 6-HEXANELACTAM ☐ HEXANONE ISOXIME ☐ HEXANONISOXIM (GERMAN) ☐ 1,6-HEXOLACTAM ☐ e-KAPROLAKTAM (CZECH) ☐ 2-KETOHEXAMETHYLENIMINE ☐ NCI-C50646 ☐ 2-OXOHEXAMETHYLENIMINE ☐ 2-PERHYDROAZEPINONE

TOXICITY DATA WITH REFERENCE
skn-rbt 500 mg/24H MLD 28ZPAK -,149,72
eye-rbt 20 mg/24H MOD 28ZPAK -,149,72
slt-dmg-orl 5 mmol/L PMRSDJ 5,313,85
mmo-smc 100 mg/L PMRSDJ 5,271,85
cyt hmn:lym 270 mg/L PMRSDJ 5,457,85
orl-rbt TDLo:3450 mg/kg (female 6–28D post):TER JJATDK 7,317,87
orl-rat TDLo:10 g/kg (female 6–15D post):REP JJATDK 7,317,87
ihl-rat TCLo:125 mg/m³/24H (76D male):REP GTPZAB 19(10),40,75
ihl-hmn TCLo:100 ppm:PUL AIHAAP 34,384,73
orl-rat LD50:1210 mg/kg NTPTR* NTP-TR-214,82
ihl-rat LC50:300 mg/m³/2H 85GMAT -,32,82
ipr-rat LDLo:800 mg/kg BJIMAG 11,1,54
orl-mus LD50:930 mg/kg GTPZAB 10(10),54,66
ihl-mus LC50:450 mg/m³ GTPZAB 10(10),54,66
ipr-mus LD50:650 mg/kg JPMSAE 60,1058,71
scu-mus LDLo:750 mg/kg AEXPBL 50,199,1903
skn-rbt LDLo:1438 mg/kg AIHAAP 30,470,69
scu-frg LDLo:2800 mg/kg AEXPBL 50,199,1903

CONSENSUS REPORTS: IARC Cancer Review: Group 4 IMEMDT 7,56,87; Animal No Evidence IMEMDT 39,247,86. Reported in EPA TSCA Inventory.

OSHA PEL: Dust: 1 mg/m³; STEL 3 mg/m³; Vapor: 5 ppm; STEL 10 ppm
ACGIH TLV: TWA Dust: 1 mg/m³; Vapor: 5 ppm; STEL 10 ppm
DFG MAK: 5 mg/m³

SAFETY PROFILE: Moderately toxic by ingestion, skin contact, intraperitoneal, and subcutaneous routes. Human systemic effects by inhalation: cough. An experimental teratogen. Other experimental reproductive effects. Human mutation data reported. A skin and eye irritant. Potentially explosive reaction with acetic acid + dinitrogen trioxide. When heated to decomposition it emits toxic fumes of NO_x.

CBF705 **CAS:762-16-3** **HR: 2**
CAPROLYL PEROXIDE
mf: $C_{16}H_{30}O_4$ mw: 286.46

SYNS: CAPRYL PEROXIDE ☐ CAPRYLYL PEROXIDE ☐ CAPRYLYL PEROXIDE (DOT) ☐ CAPRYLYL PEROXIDE SOLUTION (DOT) ☐ DICAPRYLYL PEROXIDE ☐ DIOCTANOYL PEROXIDE ☐ n-OCTANOYL PEROXIDE (DOT) ☐ PERKADOX SE 8 ☐ PEROXIDE, BIS(1-OXOOCTYL) (9CI) ☐ PEROXIDE, OCTANOYL

CONSENSUS REPORTS: Reported in EPA TSCA Inventory.

SAFETY PROFILE: A peroxide. Handle carefully. When heated to decomposition it emits acrid smoke and irritating vapors.

CBF710 **CAS:52622-27-2** **HR: 1**
CAPRYLIC/CAPRIC TRIGLYCERIDE

SYNS: CAPTEX 300 ☐ MIGLYOL 810 NEUTRAL OIL ☐ MIGLYOL 812 NEUTRAL OIL ☐ MYRITOL 318 ☐ NEOBEE M-5 ☐ NEOBEE O ☐ OCTANOIC/DECANOIC ACID TRIGLYCERIDE ☐ VEGETABLE OIL 1400

TOXICITY DATA WITH REFERENCE
eye-rbt 100 mg/24H MLD JEPTDQ 4(4),105,80

SAFETY PROFILE: An eye irritant. When heated to decomposition it emits acrid smoke and irritating fumes.

CBF725 **CAS:5299-65-0** **HR: 3**
4-CAPRYLMORPHOLINE
mf: $C_{14}H_{27}NO_2$ mw: 241.42

SYNS: AI3-18285 ☐ 4-DECANOYLMORPHOLINE ☐ MORPHOLINE, 4-DECANOYL- ☐ MORPHOLINE, 4-(1-OXODECYL) (9CI)

TOXICITY DATA WITH REFERENCE
eye-rbt 100 mg MLD NTIS** AD-A002-053
ivn-mus LD50:18 mg/kg CSLNX* NX#08964

SAFETY PROFILE: Poison by intravenous route. An eye irritant. When heated to decomposition it emits toxic fumes of NO_x.

CBF750 **CAS:404-86-4** **HR: 3**
CAPSAICIN
mf: $C_{18}H_{27}NO_3$ mw 305.46

PROP: Monoclinic, rectangular plates, crystals, and scales. Mp: 65°C, bp: 210–220°C. Freely soluble in ethanol, ether, benzene, chloroform; sltly soluble in carbon disulfide; insoluble in water. Highly volatile with a pungent odor.

SYNS: CAPSAICINE ☐ N-((4-HYDROXY-3-METHOXYPHENYL) METHYL)-8-METHYL-6-NONENAMIDE ☐ trans-N-((4-HYDROXY-3-METHOXYPHENYL)METHYL)-8-METHYL-6-NONEAMIDE ☐ trans-8-METHYL-N-VANILLYL-6-NONEAMIDE ☐ NCI-C56564

TOXICITY DATA WITH REFERENCE
mma-sat 10 µg/plate ENMUDM 7,881,85
mnt-mus-ipr 7500 µg/kg ENMUDM 7,881,85
dni-mus-ipr 1800 µg/kg ENMUDM 7,881,85
ipr-rat LD50:9500 µg/kg TOXIA6 18,215,80
orl-mus LD50:47,200 µg/kg YAHOA3 25,191,81
ipr-mus LD50:6500 µg/kg TOXIA6 18,215,80
scu-mus LD50:9000 µg/kg TOXIA6 18,215,80
ivn-mus LD50:400 µg/kg YAHOA3 25,101,81
ims-mus LD50:7800 µg/kg TOXIA6 18,215,80
itr-mus LD50:1600 µg/kg TOXIA6 18,215,80
ipr-gpg LD50:1100 µg/kg TOXIA6 18,215,80

CONSENSUS REPORTS: Reported in EPA TSCA Inventory.

SAFETY PROFILE: Deadly poison by intravenous and intraperitoneal routes. Poison by ingestion, subcutaneous, intramuscular, and intratracheal routes. Mutation data reported. Capsaicin produced erythema and burning without blistering the human skin. Capsicum is considered a moderate irritant to human skin and a strong irritant to gastric mucosa. Irritating to mucous membranes; produces severe gastritis and diarrhea. Intragastric infusion of capsaicin in humans increased the DNA content of the gastric aspirate. Capsaicin inhibits transplanted tumors in mice. Capsicum chiles fed to rats produced tumors in 15 of 26 animals. It is the component in peppers which makes them hot. When heated to decomposition it emits toxic fumes of NO_x.

CBF800 CAS:2425-06-1 HR: 3
CAPTAFOL
mf: $C_{10}H_9Cl_4NO_2S$ mw: 349.06

PROP: Crystals. Mp: 160–161°.

SYNS: CAPTOFOL ☐ DIFOLATAN ☐ DIFOSAN ☐ FOLCID ☐ ORTHO 5865 ☐ SANSPOR ☐ SULFONIMIDE ☐ SULPHEIMIDE ☐ N-(1,1,2,2-TETRACHLORAETHYLTHIO)CYCLOHEX-4-EN-1,4-DIACARBOXIMID (GERMAN) ☐ N-(1,1,2,2-TETRACHLORAETHYLTHIO)TETRAHYDROPHTHALAMID (GERMAN) ☐ N-1,1,2,2-TETRACHLOROETHYLMERCAPTO-4-CYCLOHEXENE-1,2-CARBOXIMIDE ☐ N-((1,1,2,2-TETRACHLOROETHYL)SULFENYL)-cis-4-CYCLOHEXENE-1,2-DICARBOXIMIDE ☐ N-(1,1,2,2-TETRACHLOROETHYLTHIO)-4-CYCLOHEXENE-1,2-DICARBOXIMIDE

TOXICITY DATA WITH REFERENCE
mmo-esc 50 µg/plate MUREAV 40,19,76
mma-esc 50 µg/plate MUREAV 116,185,83
mrc-bcs 100 ng/disc/24H MUREAV 40,19,76
dlt-rat-ipr 25 mg/kg/5D FCTXAV 10,353,72
cyt-ham:lng 10 µmol/L MUREAV 78,177,80
sce-ham:lng 2 µmol/L MUREAV 78,177,80
orl-rat TDLo:11 g/kg (multi) :REP TXAPA9 13,420,68
orl-ham TDLo:200 mg/kg (female 8D post):TER TXAPA9 16,24,70
orl-rbt TDLo:975 mg/kg (female 6-18D post):REP TXAPA9 13,420,68
orl-ham TDLo:400 mg/kg (female 7D post):TER TXAPA9 16,24,70
orl-mus TDLo:60,480 mg/kg/96W-C:CAR GANNA2 75,853,84
orl-rat LD50:2500 mg/kg WRPCA2 9,119,70
orl-mus TD:15,120 mg/kg/96W-C:ETA JTSCDR 7,278,82
orl-rat LD50:2500 mg/kg WRPCA2 9,119,70
ipr-mus LDLo:3 mg/kg FCTXAV 13,55,75

CONSENSUS REPORTS: IARC Cancer Review: Group 2A IMEMDT 53,353,91; Animal Sufficient Evidence IMEMDT 53,353,91; Human No Available Data IMEMDT 53,353,91. EPA Genetic Toxicology Program.

OSHA PEL: TWA 0.1 mg/m³
ACGIH TLV: TWA 0.1 mg/m³

SAFETY PROFILE: Confirmed carcinogen with experimental carcinogenic data. Poison by intraperitoneal route. Moderately toxic by ingestion. An experimental teratogen. Other experimental reproductive effects. Mutation data reported. A fungicide. When heated to decomposition it emits very toxic fumes of Cl^-, NO_x, and SO_x.

CBF825 CAS:1892-80-4 HR: 3
CAPTAGON HYDROCHLORIDE
mf: $C_{18}H_{23}N_5O_2{\cdot}ClH$ mw: 377.92

PROP: Mp: 227–229° and 237–239°.

SYNS: AMFETYLINE HYDROCHLORIDE ☐ BZT ☐ FENETHYLLINE HYDROCHLORIDE ☐ 7-(2-((α-METHYLPHENETHYL)AMINO)ETHYL) THEOPHYLLINE HYDROCHLORIDE ☐ 7-(2-(1-METHYL-2-PHENETHYLAMINO)ETHYL)THEOPHYLLINE HYDROCHLORIDE ☐ 7-(PHENYLISOPROPYL-AMINO-AETHYL)-THEOPHYLLIN-HYDROCHLORID (GERMAN)

TOXICITY DATA WITH REFERENCE
orl-rat LD50:100 mg/kg 27ZQAG -,230,72
ipr-rat LD50:57 mg/kg 27ZQAG -,230,72
scu-rat LD50:196 mg/kg 27ZQAG -,230,72
orl-mus LD50:347 mg/kg 27ZQAG -,230,72
ipr-mus LD50:347 mg/kg ARZNAD 8,190,58
scu-mus LD50:80 mg/kg 27ZQAG -,230,72
ivn-mus LD50:55 mg/kg 27ZQAG -,230,72

SAFETY PROFILE: Poison by ingestion, subcutaneous, intravenous, and intraperitoneal routes. When heated to decomposition it emits toxic fumes of NO_x and HCl. See also THEOPHYLLINE and other theophylline entries.

CBG000 CAS:133-06-2 HR: 3
CAPTAN
mf: $C_9H_8Cl_3NO_2S$ mw: 300.59

PROP: Odorless crystals from CCl_4 or C_6H_6. Mp: 172–173°, d: 1.745. Practically insol in water; sol in benzene, alcohol and chloroform.

SYNS: AACAPTAN ☐ AGROSOL S ☐ AGROX 2-WAY and 3-WAY ☐ AMERCIDE ☐ BANGTON ☐ BEAN SEED PROTECTANT ☐ CAPTAF ☐ CAPTANCAPTENEET 26,538 ☐ CAPTANE ☐ CAPTAN-STREPTOMYCIN 7.5-0.1 POTATO SEED PIECE PROTECTANT ☐ CAPTEX ☐ ENT 26,538 ☐ ESSO FUNGICIDE 406 ☐ FLIT 406 ☐ FUNGUS BAN TYPE II ☐ GLYODEX 3722 ☐ GRANOX PPM ☐ GUSTAFSON CAPTAN 30-DD ☐ HEXACAP ☐ KAPTAN ☐ LE CAPTANE (FRENCH) ☐ MALIPUR ☐ MERPAN ☐ MICRO-CHECK 12 ☐ NCI-C00077 ☐ NERACID ☐ ORTHOCIDE ☐ OSOCIDE ☐ SR406 ☐ STAUFFER CAPTAN ☐ 3a,4,7,7a-TETRAHYDRO-N-(TRICHLOROMETHANESULPHENYL)PHTHALIMIDE ☐ 3a,4,7,7a-TETRAHYDRO-2-((TRICHLOROMETHYL)THIO)-1H-ISOINDOLE-1,3(2H)-DIONE ☐ 1,2,3,6-TETRAHYDRO-N-(TRICHLOROMETHYLTHIO)PHTHALIMIDE ☐ N-(TRICHLOR-METHYLTHIO)-PHTHALIMID (GERMAN) ☐ N-TRICHLOROMETHYLMERCAPTO-4-CYCLOHEXENE-1,2-DICARBOXIMIDE ☐ N-(TRICHLOROMETHYLMERCAPTO)-Δ^4-TETRAHYDROPHTHALIMIDE ☐ N-TRICHLOROMETHYLTHIOCYCLOHEX-4-ENE-1,2-DICARBOXIMIDE ☐ N-TRICHLOROMETHYLTHIO-cis-Δ^4-CYCLOHEXENE-1,2-DICARBOXIMIDE ☐ N-((TRICHLOROMETHYL)THIO)-4-CYCLOHEXENE-1,2-DICARBOXIMIDE ☐ TRICHLOROMETHYLTHIO-1,2,5,6-TETRAHYDROPHTHALAMIDE ☐ N-((TRICHLOROMETHYL)THIO)TETRAHYDROPHTHALIMIDE ☐ N-TRICHLOROMETHYLTHIO-3A,4,7,7A-TETRAHYDROPHTHALIMIDE ☐ VANCIDE 89 ☐ VANGARD K ☐ VANICIDE ☐ VONDCAPTAN

TOXICITY DATA WITH REFERENCE
mmo-sat 310 ng/plate MUREAV 130,79,84
cyt-hmn:lng 10 mg/L ANYAA9 160,344,69
sce-hmn:lym 30 µmol/L MUREAV 79,53,80

oms-ctl:lvr 1 mmol/L CBINA8 56,289,85
orl-mus TDLo:250 mg/kg (male 5D pre):REP TXAPA9 23,277,72
orl-rat TDLo:25 g/kg (female 6-10D post):REP BECTA6 33,84,84
unr-rat TDLo:105 mg/kg (female 10-16D post):REP SAKNAH 64,737,71
orl-dog TDLo:1890 mg/kg (female 1-63D post):REP 32OAAP -,253,73
orl-ham TDLo:200 mg/kg (female 8D post):TER TXAPA9 16,24,70
ipr-rat TDLo:25 mg/kg (female 14D post):TER CRSBAW 168,1173,74
scu-mus TDLo:900 mg/kg (female 9-14D post):TER NTIS** PB223-160
scu-mus TDLo:900 mg/kg (female 9-14D post):REP NTIS** PB223-160
orl-rbt TDLo:244 mg/kg (female 6-18D post):REP TXAPA9 13,420,68
orl-rat TDLo:2500 mg/kg (female 6-10D post):REP BECTA6 33,84,84
orl-ham TDLo:750 mg/kg (female 7D post):TER TXAPA9 16,24,70
scu-mus TDLo:900 mg/kg (female 9-14D post):TER NTIS** PB223-160
orl-ham TDLo:300 mg/kg (female 8D post):TER TXAPA9 16,24,70
orl-mus TD:540 g/kg/80W-C:ETA NCITR* NCI-TR-15,77
orl-hmn LDLo:1071 mg/kg 34ZIAG -,151,69
orl-rat LD50:9 g/kg ARSIM* 20,6,66
ihl-mus LC50:5000 mg/m³/2H TXAPA9 45,320,78
ipr-rat LDLo:25 mg/kg CRSBAW 168,1173,74

CONSENSUS REPORTS: IARC Cancer Review: Group 3 IMEMDT 7,56,87; Animal Limited Evidence IMEMDT 30,295,83. NCI Carcinogenesis Bioassay (feed); Clear Evidence: mouse NCITR* NCI-CG-TR-15,77; No Evidence: rat NCITR* NCI-CG-TR-15,77. EPA Genetic Toxicology Program. Community Right-To-Know List. Reported in EPA TSCA Inventory.

OSHA PEL: TWA 5 mg/m³
ACGIH TLV: TWA 5 mg/m³

SAFETY PROFILE: Poison by intraperitoneal route. Moderately toxic to humans by ingestion. Moderately toxic experimentally by ingestion and inhalation routes. Experimental teratogenic and reproductive effects. Questionable carcinogen with experimental tumorigenic and neoplastigenic data. Human mutation data reported. When heated to decomposition it emits toxic fumes of Cl^-, SO_x, and NO_x.

CBG075 CAS:81424-67-1 HR: 3
CARACEMIDE
mf: $C_6H_{11}N_3O_4$ mw: 189.20

PROP: Crystals. Mp: 121–123.5°.

SYNS: N-ACETYL N-(METHYLCARBAMOYLOXY)-N'-METHYLUREA □ N-((METHYLAMINO)CARBONYL) N (((METHYLAMINO)CARBONYL) OXY)ACETAMIDE □ NSC-253272

TOXICITY DATA WITH REFERENCE
mmo-sat 5 μmol/plate MUREAV 172,199,86
dni-hmn:leu 50 μmol/L NEOLA4 35,27,88
orl-mus LD50:388 mg/kg NCISP* JAN86
ipr-mus LD50:167 mg/kg NCISP* JAN86
ivn-mus LD50:238 mg/kg NTIS** PB84-152032

SAFETY PROFILE: Poison by ingestion, intravenous, and intraperitoneal routes. Human mutation data reported. When heated to decomposition it emits toxic fumes of NO_x.

C

CBG250 CAS:125-86-0 HR: 3
CARAMIPHEN ETHANE DISULFONATE
mf: $C_{18}H_{27}NO_2 \cdot 1/2C_2H_6O_6S_2$ mw: 479.66

PROP: Crystals from Me_2CO. Mp: 115–116°. Sol in H_2O and EtOH.

SYNS: BIS(1-(CARBO-β-DIETHYLAMINOETHOXY)-1-PHENYLCYCLOPENTANE)ETHANE DISULFONATE □ BIS(1-(2-DIETHYLAMINOETHOXYCARBONYL)-1-PHENYLCYCLOPENTANE)ETHANE DISULFONATE □ DIETHYLAMINOETHYL-1-PHENYLCYCLOPENTANE-1-CARBOXYLATE ETHANE DISULFONATE □ PARANIT ETHANE DISULFONATE □ 1-PHENYLCYCLOPENTANECARBOXYLIC ACID 1-DIETHYLAMINOETHYL ESTER, 1,2-ETHANE DISULFONATE □ SKF No. 769-J² □ TAORYL □ TORYN

TOXICITY DATA WITH REFERENCE
orl-rat LDLo:1400 mg/kg CLDND* -,363,72
orl-mus LD50:485 mg/kg 27ZQAG -,363,72
ipr-mus LD50:240 mg/kg 27ZQAG -,363,72
ivn-mus LD50:67 mg/kg 27ZQAG -,363,72
ivn-rbt LD50:12 mg/kg 27ZQAG -,363,72

SAFETY PROFILE: Poison by intraperitoneal and intravenous routes. Moderately toxic by ingestion. When heated to decomposition it emits very toxic fumes of SO_x and NO_x. See also SULFONATES.

CBG375 CAS:57554-34-4 HR: 3
CARAMIPHEN HYDROCHLORIDE
mf: $C_{18}H_{27}NOS \cdot ClH$ mw: 341.98

SYN: CARAMIFENE (ITALIAN)

TOXICITY DATA WITH REFERENCE
ims-rat LD50:1148 μg/kg BJPCBM 39,822,70
orl-mus LD50:180 mg/kg BCFAAI 111,293,72
ipr-mus LD50:339 mg/kg EJMCA5 10,262,75
ims-mus LD50:651 μg/kg BJPCBM 39,822,70
ims-gpg LD50:115 μg/kg BJPCBM 39,822,70

SAFETY PROFILE: Poison by ingestion, intramuscular, and intraperitoneal routes. When heated to decomposition it emits toxic fumes of SO_x, NO_x, and HCl.

CBG500 CAS:8000-42-8 HR: 2
CARAWAY OIL

PROP: The main constituent of caraway oil is 1-carvone; found in the fruits of *Carum carvi L.* (Fam. Umbelliferae) (FCTXAV 11,1011,73.) Colorless liquid; odor and taste of caraway.

SYNS: KUEMMEL OIL (GERMAN) □ OIL of CARAWAY

TOXICITY DATA WITH REFERENCE
skn-rbt 500 mg/24H FCTXAV 11,1051,73
mmo-sat 5 μg/plate KEKHB8 (9),11,79

orl-rat LD50:3500 mg/kg FCTXAV 11,1051,73
skn-rbt LD50:1780 mg/kg FCTXAV 11,1051,73

CONSENSUS REPORTS: Reported in EPA TSCA Inventory.

SAFETY PROFILE: Moderately toxic by ingestion and skin contact. A skin irritant. Mutation data reported. When heated to decomposition it emits acrid smoke and irritating fumes. See also l(−)-CARVONE.

CBH250 **CAS:51-83-2** **HR: 3**
CARBACHOL CHLORIDE
mf: $C_6H_{15}N_2O_2 \cdot Cl$ mw: 182.68

PROP: Hard prisms. Mp: 204–205°. Insol in $CHCl_3$ and Et_2O.

SYNS: 2-((AMINOCARBONYL)OXY)-N,N,N-TRIMETHYLETHANAMINIUM CHLORIDE □ CARBACHOL □ CARBACHOLIN □ CARBACHOLINE CHLORIDE □ CARBACOLINA □ CARBAMIC ACID, ESTER with CHOLINE CHLORIDE □ CARBAMINOCHOLINE CHLORIDE □ CARBAMINOYLCHOLINE CHLORIDE □ CARBAMIOTIN □ CARBAMOYLCHOLINE CHLORIDE □ γ-CARBAMOYL CHOLINE CHLORIDE □ CARBAMYLCHOLINE CHLORIDE □ CARBOCHOL □ CARBOCHOLIN □ CARBYL □ CARCHOLIN □ CHOLINE CARBAMATE CHLORIDE □ CHOLINE CHLORINE CARBAMATE □ CHOLINE, CHLORIDE CARBAMATE(ESTER) □ COLEYTL □ DORYL (PHARMACEUTICAL) □ (2-HYDROXYETHYL)TRIMETHYL AMMONIUM CHLORIDE CARBAMATE □ ISOPTO CARBACHOL □ JESTRYL □ LENTIN □ LENTINE (FRENCH) □ MIOSTAT □ MISTURA C □ MORYL □ P.V. CARBACHOL □ TL 457 □ VASOPERIF

TOXICITY DATA WITH REFERENCE
ivn-man TDLo:1428 ng/kg:CVS,GIT CRSBAW 113,79,33
ims-hmn TDLo:6 μg/kg:EYE,CVS,SKN SCALA9 36,1,33
ims-man TDLo:2857 ng/kg:CVS CRSBAW 113,79,33
orl-rat LD50:40 mg/kg JPETAB 58,337,36
ipr-rat LD50:2 mg/kg AIPTAK 149,560,64
scu-rat LD50:4 mg/kg JPETAB 58,337,36
ivn-rat LD50:100 μg/kg JPETAB 58,337,36
orl-mus LD50:15 mg/kg NIIRDN 6,182,82
ipr-mus LD50:370 μg/kg ATXKA8 29,39,72
scu-mus LD50:3 mg/kg JPETAB 58,337,36
ivn-mus LD50:300 μg/kg JPETAB 58,337,36
orl-dog LDLo:3 mg/kg AEPPAE 164,346,32

CONSENSUS REPORTS: EPA Extremely Hazardous Substances List. Reported in EPA TSCA Inventory.

SAFETY PROFILE: Deadly poison by subcutaneous, intravenous, and intraperitoneal routes. Poison by ingestion and possibly other routes. Human systemic effects by intravenous and intramuscular routes including: lowered blood pressure, venous dilation, nausea or vomiting, sweating and lacrimation (increased flow of tears). A cholinergic agent (parasympathetic nerve stimulant). When heated to decomposition it emits very toxic fumes of Cl^-, NH_3, and NO_x. See also CARBAMATES.

CBH750 **HR: 3**
CARBAMATES

PROP: Compounds based upon carbamic acid, NH_2COOH. Used only in the form of its numerous salts and derivatives.

SAFETY PROFILE: Many carbamates are poisons or moderately toxic, and some are carcinogenic, teratogenic, or mutagenic. They are used as insecticides, fungicides, herbicides, and as accelerators in the vulcanization of rubber. There is little data on persistence or breakdown in the environment.

The N-alkylcarbamates and thiocarbamates can react with nitrite under mildly acid conditions to form N-nitroso compounds. Nitrite is found in soils, in human saliva, and in cured meats. N-nitrosodimethylamine is formed by soil microorganisms from thiram. Other N-nitroso compounds could similarly be formed from other carbamate pesticides. However, the extent of the reaction of carbamates and nitrite in humans is not known. The N-nitrosodialkylamines formed from dialkylthiocarbamate pesticides and nitrite are potent animal carcinogens and mutagens. The N-nitroso derivatives of several N-alkylcarbamates produce cancers in experimental animals at small doses.

Carbaryl, semicarbazide hydrochloride, n-propyl carbamate, Maneb, Zineb, Ferbam, and Thiram are experimental teratogens.

Many of the carbamates have central nervous system effects. Carbaryl and Zectran are acetylcholinesterase inhibitors.

Ethylenethiourea, which produces thyroid carcinomas in rats and liver cell tumors in mice by ingestion, is formed from ethylenebisdithiocarbamates such as Maneb and Zineb by metabolic processes and cooking.

See also individual compounds, NITROSAMINES, and N-NITROSO COMPOUNDS.

CBI250 **CAS:120-02-5** **HR: 3**
4-CARBAMIDOPHENYL BIS(CARBOXYMETHYLTHIO) ARSENITE
mf: $C_{11}H_{13}AsN_2O_5S_2$ mw: 392.30

SYNS: 2,2′-((4-((AMINOCARBONYL)AMINO)PHENYL)ARSINIDENE) BIS(THIO)BISACETIC ACID □ BIS(CARBOXYMETHYLMERCAPTO)(p-UREIDOPHENYL)ARSINE □ BIS(CARBOXYMETHYLTHIO)(p-UREIDOPHENYL)ARSINE □ (p-CARBAMOYLAMINO)PHENYLARSINOBIS(2-THIO-ACETIC ACID) □ CC 914 □ C.C. No. 914 □ MERCAPTOACETIC ACID, DIESTER with DITHIO-p-UREIDOBENZENEARSONOUS ACID □ PHENYL UREA-p-DI(CARBOXYMETHYL) THIOARSENITE □ THIOCARBARSONE □ (p-UREIDOPHENYLARSYLENEDITHIO)DIACETIC ACID

TOXICITY DATA WITH REFERENCE
orl-rat LD50:1000 mg/kg JPETAB 91,112,47
ipr-rat LD50:75 mg/kg JPETAB 91,112,47
ivn-rat LD50:29 mg/kg JPETAB 91,112,47
ipr-mus LD50:100 mg/kg JPETAB 91,112,47
ivn-mus LD50:43 mg/kg JPETAB 91,112,47
ivn-rbt LDLo:100 mg/kg JPETAB 91,112,47

CONSENSUS REPORTS: Arsenic and its compounds are on the Community Right-To-Know List.

OSHA PEL: TWA 0.5 mg(As)/m^3
ACGIH TLV: TWA 0.2 mg(As)/m^3

SAFETY PROFILE: Poison by intraperitoneal and intravenous routes. Moderately toxic by ingestion. See also ARSENIC COMPOUNDS, MERCAPTANS, and ESTERS. When heated to decomposition it emits very toxic fumes of As and SO_x.

CBI500 CAS:2490-89-3 HR: 3
4-CARBAMIDOPHENYLOXOARSINE
mf: $C_7H_7AsN_2O_2$ mw: 226.08

SYNS: 1-(p-ARSENOPHENYL)UREA □ p-CARBAMIDOPHENYL ARSENOUS ACID □ p-CARBAMIDOPHENYL ARSENOUS OXIDE □ CARBARSONE OXIDE □ CHEMOTHERAPY CENTER No. 606

TOXICITY DATA WITH REFERENCE
eye-rbt 2 mg JPETAB 82,377,44
orl-rat LD50:510 mg/kg FEPRA7 5,162,46
ipr-rat LD50:55 mg/kg JPETAB 91,112,47
ivn-rat LD50:17 mg/kg JPETAB 91,112,47
ipr-mus LD50:59 mg/kg JPETAB 91,112,47
ivn-mus LD50:41 mg/kg JPETAB 91,112,47
ivn-rbt LDLo:20 mg/kg JPETAB 91,112,47

CONSENSUS REPORTS: Arsenic and its compounds are on the Community Right-To-Know List.

OSHA PEL: TWA 0.5 mg(As)/m^3

SAFETY PROFILE: Poison by intraperitoneal and intravenous routes. Moderately toxic by ingestion. An eye irritant. When heated to decomposition it emits very toxic fumes of As and NO_x. See also ARSENIC COMPOUNDS.

CBI675 CAS:21704-46-1 HR: 3
CARBAMIMIDOTHIOIC ACID, ETHYL ESTER, MONO (DIETHYL PHOSPHATE)
mf: $C_4H_{11}O_4P{\cdot}C_3H_8N_2S$ mw: 258.31

SYN: S-ETHYLISOTHIURONIUM DIETHYL PHOSPHATE

TOXICITY DATA WITH REFERENCE
ipr-rat LD50:113 mg/kg FATOAO 43,212,80
orl-mus LD50:2380 mg/kg FATOAO 43,212,80
ipr-mus LD50:680 mg/kg FATOAO 43,212,80
scu-mus LD50:705 mg/kg FATOAO 43,212,80
ims-mus LD50:772 mg/kg FATOAO 43,212,80

SAFETY PROFILE: Poison by intraperitoneal route. Moderately toxic by ingestion, subcutaneous, and intramuscular routes. When heated to decomposition it emits toxic fumes of NO_x, PO_x, and SO_x. See also ESTERS and PHOSPHATES.

CBJ000 CAS:121-59-5 HR: 3
N-CARBAMOYLARSANILIC ACID
mf: $C_7H_9AsN_2O_4$ mw: 260.10

PROP: White, nearly odorless powder or needles from water. Sltly acid taste. Mp: 174°. Sltly sol in cold H_2O and org solvs; sol in hot H_2O.

SYNS: AMABEVAN □ AMEBAN □ AMEBARSONE □ AMIBIARSON □ AMINARSON □ AMINARSONE □ AMINOARSON □ (4-((AMINOCARBONYL)AMINO)PHENYL)ARSONIC ACID □ ARSAMBIDE □ p-ARSONOPHENYLUREA □ p-CARBAMIDOBENZENEARSONIC ACID □ CARBAMINOPHENYL-p-ARSONIC ACID □ p-CARBAMINO PHENYL ARSONIC ACID □ 4-CARBAMYLAMINOPHENYLARSONIC ACID □ N-CARBAMYL ARSANILIC ACID □ CARBARSONE (USDA) □ CARBASONE □ FENARSONE □ HISTOCARB □ LEUCARSONE □ p-UREIDOBENZENEARSONIC ACID □ 4-UREIDO-1-PHENYLARSONIC ACID

TOXICITY DATA WITH REFERENCE
orl-rat TDLo:5000 mg/kg:ETA CNREA8 26,619,66
orl-rat LD50:510 mg/kg MEIEDD 10,246,83
ipr-rat LDLo:1000 mg/kg JPETAB 80,393,44
orl-cat LDLo:250 mg/kg PSEBAA 29,125,31
orl-rbt LDLo:200 mg/kg PSEBAA 29,125,31
orl-gpg LDLo:200 mg/kg PSEBAA 29,125,31

CONSENSUS REPORTS: Arsenic and its compounds are on the Community Right-To-Know List.

OSHA PEL: TWA 0.5 mg(As)/m^3
ACGIH TLV: TWA 0.2 mg(As)/m^3

SAFETY PROFILE: Poison by ingestion. Moderately toxic by intraperitoneal route. Questionable carcinogen with experimental tumorigenic data. See also ARSENIC COMPOUNDS. When heated to decomposition it emits very toxic fumes of As and NO_x.

CBJ750 CAS:618-25-7 HR: 3
N-(CARBAMOYLMETHYL)ARSANILIC ACID
mf: $C_8H_{11}AsN_2O_4$ mw: 274.13

PROP: White, crystalline powder.

SYNS: (4-((2-AMINO-2-OXOETHYL)AMINO)PHENYL)ARSONIC ACID □ 4-ARSONOPHENYLGLYCINAMIDE □ p-((CARBAMOYLMETHYL)AMINO)-BENZENEARSONIC ACID □ SODIUM-N-PHENYLGLYCINAMIDE-p-ARSONATE □ TRYPARSAMIDE

TOXICITY DATA WITH REFERENCE
ivn-rat LDLo:2000 mg/kg JPETAB 63,122,38
ims-rat LDLo:2500 mg/kg JPETAB 63,122,38
ivn-mus LD50:4 g/kg THERAP 2,28,47
ivn-rbt LD50:700 mg/kg JPETAB 80,93,44
orl-gpg LDLo:150 mg/kg PSEBAA 29,125,31

CONSENSUS REPORTS: Arsenic and its compounds are on the Community Right-To-Know List.

OSHA PEL: TWA 0.5 mg(As)/m^3
ACGIH TLV: TWA 0.2 mg(As)/m^3

SAFETY PROFILE: Poison by ingestion and intramuscular route. Moderately toxic by intravenous route. See also ARSENIC COMPOUNDS. When heated to decomposition it emits very toxic fumes of As and NO_x.

CBK000 CAS:817-99-2 HR: 2
N-(CARBAMOYLMETHYL)-2-DIAZOACETAMIDE
mf: $C_4H_6N_4O_2$ mw: 142.14

SYNS: N-(2-AMINO-2-OXOETHYL)-2-DIAZOACETAMIDE □ N-DIAZOACETILGLICINA-AMIDE (ITALIAN) □DIAZOACETYLGLYCINAMIDE □ N-(DIAZOACETYL)GLYCINAMIDE □ DIAZOACETYLGLYCINE AMIDE □ N-DIAZOACETYLGLYCINE AMIDE

TOXICITY DATA WITH REFERENCE
mmo-sat 10 μg/plate AMACCQ 6,655,74
mma-sat 10 μg/plate PNASA6 72,5135,75
dnd-rat-ipr 3700 μg/kg BSIBAC 57,414,81
dnd-mus:fbr 620 μmol/L TOLED5 1,115,77
dni-mus/ast 1500 mg/kg BCPCA6 23,289,74
ipr-mus TDLo:720 mg/kg/4D-I:CAR BSIBAC 45,227,69
ipr-mus LD50:2630 mg/kg ARZNAD 23,690,73

CONSENSUS REPORTS: EPA Genetic Toxicology Program.

SAFETY PROFILE: Moderately toxic by intraperitoneal

C

route. Questionable carcinogen with experimental carcinogenic data. Mutation data reported. When heated to decomposition it emits toxic fumes of NO_x. See also AMIDES.

CBK125 CAS:475-08-1 HR: 3
2-CARBAMOYL-2-NITROACETONITRILE
mf: $C_3H_3N_3O_3$ mw: 129.08

$H_2NCO{\cdot}CH(NO_2)C{\equiv}N$

PROP: Prisms from EtOH or crystals from Et-OAc/ligroin. Mp: 136–149° (decomp). Sol in H_2O and EtOH; sltly sol in Et_2O; insol in $CHCl_3$, C_6H_6, and ligroin.

CONSENSUS REPORTS: Cyanide and its compounds are on the Community Right-To-Know List.

SAFETY PROFILE: A heat-sensitive explosive. When heated to decomposition it emits toxic fumes of NO_x and CN^-. See also NITRILES.

CBK500 CAS:533-06-2 HR: 3
1-CARBAMOYLOXY-2-HYDROXY-3(o-METHYLPHENOXY)PROPANE
mf: $C_{11}H_{15}NO_4$ mw: 225.27

SYNS: 2-HYDROXY-3-o-TOLYLOXYPROPYL-1-CARBAMATE □ KIMAVOXYL □ MC 2303 □ MEPHENESIN CARBAMATE □ 3-(2-METHYLPHENOXY)-1,2-PROPANEDIOL 1-CARBAMATE □ SQ 2303 □ 3-o-TOLOXY-2-HYDROXYPROPYL-1-CARBAMATE □ 3-o-TOLOXY-1,2-PROPANEDIOL-1-CARBAMIC ACID ESTER □ TOLSERAM □ 3-o-TOLYLOXY-2-HYDROXYPROPYL-1-CARBAMATE

TOXICITY DATA WITH REFERENCE
orl-rat LD50:1050 mg/kg JPETAB 129,75,60
ipr-rat LD50:413 mg/kg JPETAB 129,75,60
orl-mus LD50:1050 mg/kg JPETAB 129,75,60
ipr-mus LD50:490 mg/kg JPETAB 129,75,60
orl-ham LD50:982 mg/kg JPETAB 129,75,60
ipr-ham LD50:385 mg/kg JPETAB 129,75,60

SAFETY PROFILE: Poison by intraperitoneal route. Moderately toxic by ingestion. A skeletal muscle relaxant. When heated to decomposition it emits toxic fumes of NO_x. See also CARBAMATES and ESTERS.

CBK750 CAS:64046-99-7 HR: 3
N-(1-CARBAMOYLPROPYL)ARSANILIC ACID
mf: $C_{10}H_{15}AsN_2O_4$ mw: 302.19

SYNS: n-BUTARSAMIDE □ PHENYL-α-AMINO-n-BUTYRAMIDE-p-ARSONIC ACID

TOXICITY DATA WITH REFERENCE
orl-rbt LDLo:50 mg/kg PSEBAA 29,125,31
orl-gpg LDLo:100 mg/kg PSEBAA 29,125,31

CONSENSUS REPORTS: Arsenic and its compounds are on the Community Right-To-Know List.

OSHA PEL: TWA 0.5 mg(As)/m^3

SAFETY PROFILE: Poison by ingestion. See also ARSENIC COMPOUNDS. When heated to decomposition it emits very toxic fumes of As and NO_x.

CBL000 CAS:103-03-7 HR: 3
1-CARBAMYL-2-PHENYLHYDRAZINE
mf: $C_7H_9N_3O$ mw: 151.19

PROP: Crystals or leaflets from water or alc. Mp: 172°.

SYNS: CPH □ CRYOGENINE □ KRYOGENIN □ 2-PHENYLDIAZENECARBOXAMIDE □ 2-PHENYLHYDRAZIDE, CARBAMIC ACID □ 1-PHENYLHYDRAZINE CARBOXAMIDE □ 2-PHENYLHYDRAZINECARBOXAMIDE □ PHENYLSEMICARBAZIDE □ 1-PHENYLSEMICARBAZIDE

TOXICITY DATA WITH REFERENCE
dnd-esc 250 μg/well MUREAV 133,161,84
orl-mus TDLo:394 mg/kg/62W-C:NEO JNCIAM 52,241,74
ipr-mus LD50:198 mg/kg CNREA8 41,1469,81

CONSENSUS REPORTS: IARC Cancer Review: Group 3 IMEMDT 7,56,87; Animal Limited Evidence IMEMDT 12,177,76. Reported in EPA TSCA Inventory.

SAFETY PROFILE: Poison by intraperitoneal route. Questionable carcinogen with experimental neoplastigenic data. Mutation data reported. When heated to decomposition it emits toxic fumes of NO_x. See also CARBAMATES.

CBL500 CAS:16118-49-3 HR: 2
d-(−)-CARBANILIC ACID (1-ETHYLCARBAMOYL) ETHYL ESTER
mf: $C_{12}H_{16}N_2O_3$ mw: 236.30

SYNS: CARBETAMEX □ CARBETAMID (GERMAN) □ CARBETAMIDE □ d-N-ETHYLACETAMIDE CARBANILATE □ d-(−)-1-(ETHYLCARBAMOYL)ETHYL PHENYLCARBAMATE □ d-N-ETHYLLACTAMIDE CARBANILATE (ESTER) □ (R)-N-ETHYL-2-(((PHENYLAMINO)CARBONYL)OXY)PROPANAMIDE □ LEGURAME □ 2-PHENYL-CARBAMOYLOXY-N-AETHYL-PROPIONAMID (GERMAN) □ (PHENYLCARBAMOYLOXY)-2-N-ETHYLPROPIONAMIDE □ N-PHENYL-1-(ETHYLCARBAMOYL-1)-ETHYLCARBAMATE, D ISOMER □ 11,561 RP

TOXICITY DATA WITH REFERENCE
orl-rat LD50:11,000 mg/kg 85ARAE 2,83,77
orl-mus LD50:1200 mg/kg GUCHAZ 6,80,73
orl-dog LD50:900 mg/kg GUCHAZ 6,80,73

SAFETY PROFILE: Moderately toxic by ingestion. An herbicide. When heated to decomposition it emits toxic fumes of NO_x. See also CARBAMATES.

CBL750 CAS:101-99-5 HR: 3
CARBANILIC ACID ETHYL ESTER
mf: $C_9H_{11}NO_2$ mw: 165.21

PROP: Crystals from water. Mp: 53°, bp: 238° (sltly decomp), d: 1.106.

SYNS: EPC (the plant regulator) □ ETHYL CARBANILATE □ ETHYL-N-PHENYLCARBAMATE □ EUPHORIN □ KEIMSTOP □ PHENYLETHYL CARBAMATE □ PHENYLURETHAN □ PHENYLURETHAN(E) □ N-PHENYLURETHANE

TOXICITY DATA WITH REFERENCE
skn-mus TDLo:20 g/kg/2W-I:NEO BJCAAI 9,177,55
skn-mus TD:72 g/kg/15W-I:NEO BJCAAI 10,363,56
unr-rat LDLo:500 mg/kg BJPCAL 7,142,52
ipr-mus LD50:350 mg/kg HBTXAC 5,45,59
scu-mus LDLo:1 g/kg HDTU** -,-,33

ivn-mus LD50:400 mg/kg HBTXAC 5,45,59

CONSENSUS REPORTS: Reported in EPA TSCA Inventory.

SAFETY PROFILE: Poison by intraperitoneal and intravenous routes. Moderately toxic by subcutaneous and possibly other routes. Questionable carcinogen with experimental neoplastigenic data. When heated to decomposition it emits toxic fumes of NO_x. See also CARBAMATES.

CBM000 CAS:122-42-9 HR: 3
CARBANILIC ACID ISOPROPYL ESTER
mf: $C_{10}H_{13}NO_2$ mw: 179.24

PROP: A white, crystalline solid; sol in acetone and benzene. Mp: 90°.

SYNS: BAN-HOE □ BEET-KLEEN □ CHEM-HOE □ IFC □ IPPC □ ISOPROPIL-N-FENIL-CARBAMMATO (ITALIAN) □ ISOPROPYL CARBANILATE □ ISOPROPYL CARBANILIC ACID ESTER □ ISOPROPYL-N-FENYL-CARBAMAAT (DUTCH) □ ISOPROPYL-N-PHENYL-CARBAMAT (GERMAN) □ ISOPROPYL PHENYLCARBAMATE □ ISOPROPYL-N-PHENYLCARBAMATE □ o-ISOPROPYL-N-PHENYL CARBAMATE □ ISOPROPYL-N-PHENYLURETHAN (GERMAN) □ ORTHO GRASS KILLER □ N-PHENYLCARBAMATE D'ISOPROPYLE (FRENCH) □ PHENYLCARBAMIC ACID-1-METHYLETHYL ESTER □ N-PHENYL ISOPROPYL CARBAMATE □ PREMALOX □ PROFAM □ PROPHAM □ TRIHERBIDE □ TRIHERBIDE-IPC □ TUBERIT □ TUBERITE □ USAF D-9 □ Y 2

TOXICITY DATA WITH REFERENCE
cyt-omi 550 μmol/L JCLBA3 63,84,74
sce-hmn:lym 2 mg/L MUREAV 147,296,85
scu-mus TDLo:7650 mg/kg (6-14D preg):TER NTIS** Pb 223-160
orl-mus TDLo:6 g/kg/10W-I:NEO BJCAAI 12,355,58
orl-hmn LDLo:714 mg/kg CRSBAW 175,496,81
orl-rat LD50:1000 mg/kg RREVAH 10,97,65
ipr-rat LD50:600 mg/kg CRSBAW 175,496,81
orl-mus LD50:2160 mg/kg 85GMAT -,79,82
ipr-mus LD50:200 mg/kg NTIS** AD277-689
unr-mam LD50:1000 mg/kg 30ZDA9 -,199,71

CONSENSUS REPORTS: IARC Cancer Review: Group 3 IMEMDT 7,56,87; Animal Inadequate Evidence IMEMDT 12,189,76. Reported in EPA TSCA Inventory. EPA Genetic Toxicology Program.

SAFETY PROFILE: Poison by intraperitoneal route. Moderately toxic to humans by ingestion. Moderately toxic experimentally by ingestion and possibly other routes. An experimental teratogen. Human mutation data reported. Questionable carcinogen with experimental neoplastigenic and teratogenic data. An herbicide. When heated to decomposition it emits toxic fumes of NO_x. See also CARBAMATES.

CBM250 CAS:102-07-8 HR: 3
CARBANILIDE
mf: $C_{13}H_{12}N_2O$ mw: 212.27

PROP: Prisms. Mp: 239–240°, bp: 260°.

SYNS: N,N'-DIPHENYLUREA □ sym-DIPHENYLUREA □ 1,3-DIPHENYLUREA □ USAF EK-534

TOXICITY DATA WITH REFERENCE
orl-rat LDLo:500 mg/kg JPETAB 90,260,47
ipr-mus LD50:200 mg/kg NTIS** AD277-689

CONSENSUS REPORTS: Reported in EPA TSCA Inventory.

SAFETY PROFILE: Poison by intraperitoneal route. Moderately toxic by ingestion. When heated to decomposition it emits toxic fumes of NO_x.

CBM500 CAS:116-06-3 HR: 3
CARBANOLATE
mf: $C_7H_{14}N_2O_2S$ mw: 190.29

PROP: A solid material or crystals. Mp: 98–100°. Sltly sol in water.

SYNS: ALDECARB □ ALDICARB (USDA) □ ALDICARBE (FRENCH) □ AMBUSH □ ENT 27,093 □ 2-METHYL-2-(METHYLTHIO)PROPANAL-O-((METHYLAMINO)CARBONYL)OXIME □ 2-METHYL-2-(METHYLTHIO)PROPIONALDEHYDE-O-(METHYLCARBAMOYL)OXIME □ 2-METHYL-2-(METHYLTHIO)PROPIONALDEHYDE OXIME □ 2-METHYL-2-METHYLTHIO-PROPIONALDEHYD-O-(N-METHYL-CARBAMOYL)-OXIM (GERMAN) □ 2-METIL-2-TIOMETIL-PROPIONALDEID-O-(N-METIL-CARBAMOIL)-OSSIMA (ITALIAN) □ NCI C08640 □ OMS-771 □ RCRA WASTE NUMBER P070 □ TEMIC □ TEMIK □ TEMIK G10 □ UC-21149

TOXICITY DATA WITH REFERENCE
sce-hmn:lym 10 mg/L MUREAV 138,175,84
otr-rat:emb 117 μg/plate IJATDK 1,190,81
orl-rat LD50:650 μg/kg TXAPA9 14,515,69
ihl-rat LC50:200 mg/m³/5H 85JCAE-,999,86
skn-rat LD50:2500 μg/kg TXAPA9 14,515,69
scu-rat LDLo:666 μg/kg TXAPA9 25,569,73
orl-mus LD50:300 μg/kg JAFCAU 18,793,70
skn-rbt LD50:1400 mg/kg GUCHAZ 6,4,73
skn-gpg LD50:2400 mg/kg 85DPAN -,-,71/76
orl-pgn LD50:3160 μg/kg ASTTA8 (680),157,79
orl-ckn LD50:8 mg/kg 85GYAZ -,62,71
orl-qal LD50:2 mg/kg EESADV 8,551,84
orl-dck LD50:3400 μg/kg TXAPA9 47,451,79
orl-bwd LD50:750 μg/kg ASTTA8 (680),157,79

CONSENSUS REPORTS: IARC Cancer Review: Group 3 IMEMDT 53,93,91; Animal Inadequate Evidence IMEMDT 53,93,91; Human No Available Data IMEMDT 53,93,91. NCI Carcinogenesis Bioassay (feed); No Evidence: mouse, rat NCITR* NCI-CG-TR-136,79. Reported in EPA TSCA Inventory. EPA Extremely Hazardous Substances List.

SAFETY PROFILE: Deadly poison by ingestion, skin contact, subcutaneous, and possibly other routes. Human mutation data reported. A powerful systemic poison. In 1985 over 150 people in California exhibited toxic effects from eating watermelons contaminated with aldicarb. When heated to decomposition it emits very toxic fumes of NO_x and SO_x.

CBM750 CAS:63-25-2 HR: 3
CARBARYL
mf: $C_{12}H_{11}NO_2$ mw: 201.24

PROP: White crystals. Mp: 142°, d: 1.232 @ 20°/20°.

SYNS: ARILAT □ ARILATE □ ARYLAM □ ATOXAN □ BERCEMA NMC50 □ BUG MASTER □ CAPROLIN □ CARBAMINE □ CARBARYL (ACGIH,DOT,OSHA) □ CARBATOX □ CARBATOX-60 □ CARBATOX-75 □ CARBAVUR □ CARBOMATE □ CARPOLIN □ CARYLDERM □ CEKUBARYL □ COMPOUND 7744 □ CRUNCH □ DENAPON □ DEVICARB □ DICARBAM □ DYNA-CARBYL □ CRAG SEVIN □ ENT 23,969 □ EXPERIMENTAL INSECTICIDE 7744 □ GAMONIL □ GERMAIN'S □ HEXAVIN □ KARBARYL □ KARBASPRAY □ KARBATOX □ KARBATOX 75 □ KARBATOX ZAWIESINOWY □ KARBOSEP □ LATKA 7744 □ MENAPHTAM □ N-METHYLCARBAMATE DE 1-NAPHTYLE □ METHYLCARBAMATE-1-NAPHTHALENOL □ METHYLCARBAMATE-1-NAPHTHOL □ KARBARYL (POLISH) □ N-METHYLCARBAMATE de 1-NAPHTYLE (FRENCH) □ METHYLCARBAMIC ACID-1-NAPHTHYL ESTER □ N-METHYL-1-NAFTYL-CARBAMAAT (DUTCH) □ N-METHYL-1-NAPHTHYL-CARBAMAT (GERMAN) □ N-METHYL-α-NAPHTHYLCARBAMATE □ N-METHYL-1-NAPHTHYL CARBAMATE □ N-METHYL-α-NAPHTHYLURETHAN □ N-METIL-1-NAFTIL-CARBAMMATO (ITALIAN) □ MONSUR □ MUGAN □ MURVIN □ NAC □ 1-NAFTYL-ESTER KYSELINY METHYLKARBAMINOVE □ α-NAFTYL-N-METHYLKARBAMAT □ 1-NAPHTHALENOL, METHYLCARBAMATE (9CI) □ α-NAPHTHALENYL METHYLCARBAMATE □ 1-NAPHTHALENYL METHYLCARBAMATE □ 1-NAPHTHOL N-METHYLCARBAMATE □ α-NAPHTHYL METHYLCARBAMATE □ α-NAPHTHYL N-METHYLCARBAMATE □ 1-NAPHTHYL METHYLCARBAMATE □ 1-NAPHTHYL N-METHYLCARBAMATE □ 1-NAPHTHYL-N-METHYL-KARBAMAT □ NMC 50 □ OLTITOX □ OMS-29 □ PANAM □ POMEX □ PROSEVOR 85 □ RAVYON □ RYLAM □ SAVIT □ SEFFEIN □ SEPTENE □ SEVIMOL □ SEVIN □ SEVIN 4 □ SEVIN (OSHA) □ SEWIN □ SOK □ TERCYL □ TOXAN □ TRICARNAM □ UC 7744 □ UNION CARBIDE 7,744 □ VETOX □ VIOXAN

TOXICITY DATA WITH REFERENCE

skn-rbt 12 mg/24H SEV JAFCAU 9,30,61
eye-rbt 500 mg/24H MOD 28ZPAK -,164,72
mmo-sat 250 µg/plate RPZHAW 30,81,79
mma-hmn:fbr 1 µmol/L MUREAV 42,161,77
dns-hmn:fbr 1 µmol/L MUREAV 42,161,77
cyt-hmn:emb 40 µg/kg ZDVKAP 20(4),14,77
orl-rat TDLo:27,500 µg/kg (multi) :REP TXAPA9 21,390,72
orl-dog TDLo:197 mg/kg (female 1-63D post):REP 32OAAP -,253,73
orl-grb TDLo:20,300 mg/kg (multi) :REP TXAPA9 19,202,71
orl-rat TDLo:220 mg/kg (female 5-15D post):REP TXAPA9 21,390,72
orl-ham TDLo:250 mg/kg (female 8D post):TER TXAPA9 15,152,69
scu-mus TDLo:418 mg/kg (female 6-14D post):TER NTIS** PB223-160
orl-mus TDLo:11,660 mg/kg (female 6-15D post):TER TXAPA9 51,81,79
orl-grb TDLo:20,300 mg/kg (multi) :REP TXAPA9 19,202,71
orl-rat TDLo:5500 mg/kg (multi) :REP TXAPA9 17,273,70
orl-dog TDLo:197 mg/kg (female 1-63D post):REP 32OAAP -,253,73
orl-rat TDLo:1370 mg/kg (female 39W pre):REP VPITAR 27(6),49,68
orl-rbt TDLo:1950 mg/kg (female 6-18D post):REP TXAPA9 51,81,79
orl-rat TDLo:5475 mg/kg (female 52W pre):REP VPITAR 27(6),49,68
orl-dog TDLo:375 mg/kg (female 3-62D post):TER TXAPA9 13,392,68
scu-mus TDLo:900 mg/kg (female 6-15D post):TER NTIS** PB223-160
orl-dog TDLo:394 mg/kg (female 1-63D post):TER 32OAAP -,253,73
orl-rat TDLo:5640 mg/kg/94W-I:ETA VPITAR 29,71,70
imp-rat TDLo:80 mg/kg:CAR VPITAR 29,71,70
orl-man TDLo:500 mg/kg:PNS NEURAI 37,1229,87
orl-rat LD50:230 mg/kg TXAPA9 11,546,67
skn-rat LD50:4000 mg/kg 85DPAN -,-,71/76
ipr-rat LD50:64 mg/kg PSEBAA 114,509,63
scu-rat LD50:1400 mg/kg 34ZIAG-,528,69
ivn-rat LD50:41,900 µg/kg BWHOA6 44,241,71
orl-mus LD50:128 mg/kg JPETAB 181,576,72
ipr-mus LD50:25 mg/kg TXAPA9 6,402,64
scu-mus LD50:6717 mg/kg TOIZAG 17,60,70
skn-rbt LD50:2000 mg/kg 85DPAN-,-,71/76

CONSENSUS REPORTS: IARC Cancer Review: Group 3 IMEMDT 7,56,87; Animal Inadequate Evidence IMEMDT 12,37,76. Community Right-To-Know List.

OSHA PEL: TWA 5 mg/m^3
ACGIH TLV: TWA 5 mg/m^3
DFG MAK: 5 mg/m^3
NIOSH REL: (Carbaryl) TWA 5 mg/m^3

SAFETY PROFILE: Poison by ingestion, intravenous, intraperitoneal, and possibly other routes. Human systemic effects by ingestion: sensory change involving peripheral nerves and muscle weakness. Experimental teratogenic and reproductive effects. Questionable carcinogen with experimental carcinogenic and tumorigenic data. Human mutation data reported. An eye and severe skin irritant. Absorbed by all routes, although skin absorption is slow. No accumulation in tissue. Symptoms include blurred vision, headache, stomach ache, vomiting. Symptoms similar to but less severe than those due to parathion. A reversible cholinesterase inhibitor. See also CARBAMATES and ESTERS. When heated to decomposition it emits toxic fumes of NO_x.

For occupational chemical analysis use OSHA: #ID-63 or NIOSH: Carbaryl, 5006.

CBM875 CAS:33060-69-4 HR: 2
CARBAVINE
mf: $C_6H_9NO_2$ mw: 127.16

SYN: METHYL-3-BUTYN-2-OL CARBAMATE

TOXICITY DATA WITH REFERENCE

ipr-rat LDLo:700 mg/kg RPTOAN 33,191,70
unr-mus LD50:1500 mg/kg RPTOAN 33,191,70
ipr-rbt LDLo:700 mg/kg RPTOAN 33,191,70

SAFETY PROFILE: Moderately toxic by intraperitoneal and possibly other routes. When heated to decomposition it emits toxic fumes of NO_x. See also CARBAMATES.

CBN000 CAS:86-74-8 HR: 3
CARBAZOLE
mf: $C_{12}H_9N$ mw: 167.22

PROP: White crystals or plates from xylene. Mp: 244.8°, bp: 354.7°, d: 1.10 @ 18°/4°, vap press: 400 mm @ 323.0°. Sltly sol in most org solvs; sol in hot EtOH.

SYNS: 9-AZAFLUORENE □ 9H-CARBAZOLE □ DIBENZOPYRROLE □ DIBENZO(b,d)PYRROLE □ DIPHENYLENEIMINE □ DIPHENYLENIMIDE □ DIPHENYLENIMINE □ USAF EK-600

TOXICITY DATA WITH REFERENCE
mor-rat-orl 504 mg/kg/6W CRNGDP 9,387,88
orl-rat LDLo:500 mg/kg JPETAB 90,260,47
ipr-mus LD50:200 mg/kg NTIS** AD277-689

CONSENSUS REPORTS: IARC Cancer Review: Group 3 IMEMDT 7,56,87; Animal Limited Evidence IMEMDT 32,239,83. Reported in EPA TSCA Inventory.

SAFETY PROFILE: Poison by intraperitoneal route. Questionable carcinogen. Moderately toxic by ingestion. Mutation data reported. A pesticide. When heated to decomposition it emits toxic fumes of NO_x.

CBN100 CAS:86-72-6 HR: 1
4-(3-CARBAZOLYLAMINO)PHENOL
mf: $C_{18}H_{14}N_2O$ mw: 274.34

SYNS: CARBAZOLE, 3-(p-HYDROXYANILINO)- □ 3-(4'-HYDROXYFENYL)AMINOKARBAZOL □ PHENOL, 4-(3-CARBAZOLYLAMINO)- □ R-BASE

TOXICITY DATA WITH REFERENCE
skn-rbt 500 mg/24H MLD 85JCAE-,825,86
eye-rbt 100 mg/24H MOD 85JCAE-,825,86

CONSENSUS REPORTS: Reported in EPA TSCA Inventory.

SAFETY PROFILE: A skin and eye irritant. When heated to decomposition it emits toxic fumes of NO_x.

CBN375 HR: 3
CARBENDAZIM and SODIUM NITRITE (5:1)

SYNS: METHYL-2-BENZIMIDAZOLE CARBAMATE and SODIUM NITRITE □ SODIUM NITRITE and CARBENDAZIM (1:5) □ SODIUM NITRITE and METHYL-2-BENZIMIDAZOLE CARBAMATE

TOXICITY DATA WITH REFERENCE
orl-mus TDLo:31 g/kg/26W-I:CAR IJCNAW 15,830,75
orl-mus TDLo:3000 mg/kg (7-14D preg):TER IJCNAW 17,742,76
orl-mus TD:88 g/kg/12W-I:CAR MGONAD 19,175,75

SAFETY PROFILE: Suspected carcinogen with experimental carcinogenic data. An experimental teratogen. When heated to decomposition it emits toxic fumes of Na_2O and NO_x. See also NITRITES and CARBAMATES.

CBN750 CAS:27025-49-6 HR: 2
CARBENICILLIN PHENYL
mf: $C_{23}H_{22}N_2O_6S$ mw: 454.53

SYNS: CARBENICILLIN PHENYL ESTER □ CARFECILLIN

TOXICITY DATA WITH REFERENCE
ipr-rat LD50:980 mg/kg ANTBAL 25,513,80
orl-mus LD50:3924 mg/kg ANTBAL 25,513,80
ivn-mus LD50:728 mg/kg ANTBAL 25,513,80

SAFETY PROFILE: Moderately toxic by ingestion, intraperitoneal, and intravenous routes. When heated to decomposition it emits very toxic fumes of NO_x and SO_x. See also CARBENICILLIN PHENYL SODIUM and ESTERS.

CBO000 CAS:21649-57-0 HR: 2
CARBENICILLIN PHENYL SODIUM
mf: $C_{23}H_{21}N_2NaO_6S$ mw: 476.51

PROP: Crystals from EtOH. Sol in H_2O.

SYNS: BRL 3475 □ CARBOXYBENZYLPENICILLIN PHENYL ESTER SODIUM SALT □ CARFECILLIN SODIUM □ SODIUM-α-PHENOXYCARBONYLBENZYLPENICILLIN □ UTICILLIN

TOXICITY DATA WITH REFERENCE
ipr-rat LD50:572 mg/kg NIIRDN 6,186,82
scu-rat LD50:4530 mg/kg NIIRDN 6,186,82
ivn-rat LD50:710 mg/kg NIIRDN 6,186,82
orl-mus LD50:3040 mg/kg ANTBAL 23(5),450,78
ipr-mus LD50:942 mg/kg NIIRDN 6,186,82
scu-mus LD50:2010 mg/kg NIIRDN 6,186,82
ivn-mus LD50:717 mg/kg NIIRDN 6,186,82
ivn-dog LD50:625 mg/kg NIIRDN 6,186,82
orl-rbt LD50:10 g/kg NIIRDN 6,186,82
ivn-rbt LD50:625 mg/kg NIIRDN 6,186,82

SAFETY PROFILE: Moderately toxic by ingestion, subcutaneous, intravenous, and intraperitoneal routes. When heated to decomposition it emits very toxic fumes of NO_x, Na_2O, and SO_x. See also ESTERS and other penicillin entries.

CBO250 CAS:4800-94-6 HR: 1
CARBENICILLIN SODIUM
mf: $C_{17}H_{18}N_2O_6S•2Na$ mw: 424.41

SYNS: ANABACTYL □ BRL-2064 □ CARBECIN □ CARBENICILLIN DISODIUM SALT □ CARBOXYBENZYLPENICILLIN SODIUM □ N-(2-CARBOXY-3,3-DIMETHYL-7-OXO-4-THIA-1-AZABICYCLO(3.2.0)HEPT-6-YL) 2-PHENYL-MALONAMIC ACID DISODIUM SALT □ CBPC □ CP-15-639-2 □ FUGACILLIN □ GEOPEN □ GRIPENIN □ MICROCILLIN □ NSC-111071 □ PIOPEN □ PYOPEN □ PYOPENE □ SODIUM CARBENICILLIN

TOXICITY DATA WITH REFERENCE
ipr-rat LD50:10 g/kg NIIRDN 6,187,82
ivn-rat LD50:6800 mg/kg NIIRDN 6,187,82
ipr-mus LD50:7600 mg/kg NIIRDN 6,187,82
scu-mus LD50:9 g/kg NIIRDN 6,187,82
ivn-mus LD50:4500 mg/kg NKRZAZ 23,572,75
ivn-mky LD50:9800 mg/kg TAKHAA 34,405,75

SAFETY PROFILE: Mildly toxic by subcutaneous, intravenous, and intraperitoneal routes. When heated to decomposition it emits very toxic fumes of NO_x, Na_2O, and SO_x. See also CARBENICILLIN PHENYL SODIUM.

CBO500 CAS:7421-40-1 HR: 3
CARBENOXALONE, DISODIUM SALT
mf: $C_{34}H_{48}O_7•2Na$ mw: 614.80

SYNS: BIOGASTRONE □ BIORAL □ CARBENOXOLONE, DISODIUM SALT □ CARBENOXOLONE SODIUM □ 3-(3-CARBOXY-1-OXOPROPOXY)-11-OXOOLEAN-12-EN-29-OIC ACID, DISODIUM SALT (3-β, 20-β) □ 3-o-(β-CARBOXYPROPIONYL)-11-OXO-18-β-OLEAN-12-EN-30-OIC ACID, DISODIUM SALT □ DUOGASTRONE □ GLYCYRRHETINIC ACID HYDROGEN SUCCINATE DISODIUM SALT □ 18-β-GLYCYRRHE-

TINIC ACID HYDROGEN SUCCINATE DISODIUM SALT □ 3-β-HYDROXY-11-OXOOLEAN-12-EN-30-OIC ACID HYDROGEN SUCCINATE DISODIUM SALT □ NEOGEL □ PYROGASTRONE □ SANODIN □ SODIUM-3-β-HYDROXY-11-OXO-12-OLEANEN-30-OATE SODIUM SUCCINATE □ ULCUS-TABLINEN

TOXICITY DATA WITH REFERENCE
orl-dog TDLo:3 g/kg (30D male):REP OYYAA2 11,831,76
orl-dog TDLo:900 mg/kg (30D male):REP OYYAA2 11,831,76
orl-hmn TDLo:120 mg/kg/6W:CVS,MET CMAJAX 117,1155,77
orl-rat LD50:2450 mg/kg OYYAA2 11,263,76
ipr-rat LD50:112 mg/kg IYKEDH 10,710,79
scu-rat LD50:1515 mg/kg IYKEDH 10,710,79
ipr-mus LD50:120 mg/kg 21NDAB -,6,68
ivn-mus LD50:198 mg/kg 21NDAB -,6,68
orl-dog LD50:3900 mg/kg IYKEDH 10,710,79
ipr-dog LD50:371 mg/kg IYKEDH 10,710,79
scu-dog LD50:1060 mg/kg OYYAA2 11,263,76
ivn-dog LD50:371 mg/kg OYYAA2 11,263,76
orl-rbt LD50:2 g/kg 21NDAB -,6,68

SAFETY PROFILE: Poison by intravenous and intraperitoneal routes. Moderately toxic by ingestion and subcutaneous routes. Human systemic effects by ingestion: blood pressure increase, change in blood potassium levels. Experimental reproductive effects. An anti-inflammatory agent used to treat gastric ulcers. When heated to decomposition it emits toxic fumes of Na_2O.

CBO625 CAS:1755-52-8 HR: 2
CARBESTROL
mf: $C_{17}H_{22}O_3$ mw: 274.39

PROP: A solid. Mp: 157–158°.

SYNS: 3-ETHYL-4-(p-METHOXYPHENYL)-2-METHYL-3-CYCLOHEXENE-1-CARBOXYLIC ACID □ 2-METHYL-3-ETHYL-4-p-METHOXYPHENYL-Δ^3-CYCLOHEXENE CARBOXYLIC ACID □ NSC-19962 □ ORF 2166

TOXICITY DATA WITH REFERENCE
orl-rat TDLo:30 μg/kg (1D preg):REP JPETAB 167,105,69
orl-rat TDLo:16 mg/kg (8D male):REP JRPFA4 12,381,66
orl-hmn TDLo:126 mg/kg/6W-I:CNS,GIT CCROBU 56,641,72

SAFETY PROFILE: Human systemic effects by ingestion: anorexia, diarrhea, and nausea or vomiting. Experimental reproductive effects. When heated to decomposition it emits acrid smoke and fumes.

CBP250 CAS:21600-51-1 HR: 3
1(4-CARBETHOXYPHENYL)-3,3-DIMETHYLTRIAZENE
mf: $C_{11}H_{15}N_3O_2$ mw: 221.29

SYNS: 1-(p-CARBOXYAETHYLPHENYL)-3,3-DIMETHYLTRIAZEN (GERMAN) □ 1-(p-ETHYLCARBOXYPHENYL)-3,3-DIMETHYLTRIAZENE

TOXICITY DATA WITH REFERENCE
sln-dmg-orl 1 mmol/L CBINA8 9,365,74
mrc-smc 10 mmol/L CBINA8 9,365,74
hma-mus/smc 1 mmol/L CBINA8 9,365,74
scu-rat TDLo:760 mg/kg/19W-I:CAR ARZNAD 23,800,73
ivn-rat TDLo:805 mg/kg/23W-I:CAR ARZNAD 23,800,73
scu-mus TDLo:760 mg/kg/19W-I:ETA ARZNAD 23,800,73
scu-rat LD50:450 mg/kg ARZNAD 23,800,73
ivn-rat LD50:150 mg/kg ARZNAD 23,800,73

CONSENSUS REPORTS: EPA Genetic Toxicology Program.

SAFETY PROFILE: Poison by intravenous route. Moderately toxic by subcutaneous route. Questionable carcinogen with experimental carcinogenic and tumorigenic data. Mutation data reported. When heated to decomposition it emits toxic fumes of NO_x. See also ESTERS.

CBP325 HR: 3
2-(N-(4-CARBETHOXY-4-PHENYL)PIPERIDINO)PROPIOPHENONE HYDROCHLORIDE
mf: $C_{23}H_{27}NO_3{\cdot}ClH$ mw: 401.97

SYNS: 1-(2-BENZOYLETHYL)-4-PHENYLISONIPECOTIC ACID ETHYL ESTER HYDROCHLORIDE □ R 951

TOXICITY DATA WITH REFERENCE
orl-rat LD50:145 mg/kg APPNAH 7,373,58
scu-rat LD50:360 mg/kg APPNAH 7,373,58
ivn-rat LD50:3300 μg/kg APPNAH 7,373,58
orl-mus LD50:610 mg/kg APPNAH 7,373,58
scu-mus LD50:215 mg/kg APPNAH 7,373,58
ivn-mus LD50:13,800 μg/kg APPNAH 7,373,58

SAFETY PROFILE: Poison by ingestion, subcutaneous, and intravenous routes. When heated to decomposition it emits toxic fumes of NO_x and HCl. See also ESTERS.

CBQ125 CAS:3811-06-1 HR: 3
CARBIDIUM ETHANESULFONATE
mf: $C_{23}H_{22}N_3O_2{\cdot}C_2H_5O_3S$ mw: 481.61

PROP: Deep yellow prisms from H_2O. Mp: 288–290° (decomp).

SYNS: 3-AMINO-9-p-CARBETHOXYAMINOPHENYL-10-METHYLPHENANTHRIDINIUM ETHANESULPHONATE □ 2-AMINO-6-(p-CARBOXYAMINOPHENYL)-5-METHYLPHENANTHRIDINIUM ETHANESULFONATE ETHYL ESTER □ 74C48 □ CARBIDIUM ETHANESULPHONATE

TOXICITY DATA WITH REFERENCE
ipr-mus LD50:40 mg/kg BJPCAL 5,287,50
scu-mus LD50:130 mg/kg BJPCAL 5,287,50
ivn-mus LD50:10 mg/kg BJPCAL 5,287,50
ivn-rbt LD50:10 mg/kg BJPCAL 5,287,50

SAFETY PROFILE: Poison by subcutaneous, intravenous, and intraperitoneal routes. When heated to decomposition it emits toxic fumes of SO_x and NO_x. See also SULFONATES and ESTERS.

CBQ500 CAS:28860-95-9 HR: 2
CARBIDOPA
mf: $C_{10}H_{14}N_2O_4$ mw: 226.26

PROP: Crystals from hot water. Mp: 203–205° (decomp). dl-Form: Tan, fluffy crystals; mp: 206–208° (decomp).

SYNS: N-AMINOMETHYLDOPA □ (S)-α-HYDRAZINO-3,4-DIHYDROXY-α-METHYL-BENZENEPROPANOIC ACID (9CI) □ HYDRAZINO-α-METHYLDOPA □ LODOSIN □ LODOSYN □ l-α-METHYLDOPAHYDRAZINE □ MK 486

TOXICITY DATA WITH REFERENCE
mma-sat 1 mg/plate RCOCB8 49,415,85
orl-rat TDLo:2880 mg/kg (90D pre):REP TXAPA9 66,201,82
orl-rat TDLo:5850 mg/kg (90D male):REP TXAPA9 66,201,82
orl-rat TDLo:2100 mg/kg (1-21D preg):TER TXAPA9 38,251,76
ipr-rat LD50:2804 mg/kg YKYUA6 31,1127,80
ipr-mus LD50:468 mg/kg YKYUA6 31,1127,80

SAFETY PROFILE: Moderately toxic by intraperitoneal route. An experimental teratogen. Other experimental reproductive effects. Mutation data reported. An antihypertensive agent. When heated to decomposition it emits toxic fumes of NO_x.

CBQ529 **HR: 3**
CARBIDOPA MONOHYDRATE
mf: $C_{10}H_{14}N_2O_4 \cdot H_2O$ mw: 244.28

SYNS: (−)-l-α-HYDRAZINO-3,4-DIHYDROXY-α-METHYLHYDROCINNAMIC ACID MONOHYDRATE □ S(−)-α-HYDRAZINO-3,4-DIHYDROXY-α-METHYLHYDROCINNAMIC ACID MONOHYDRATE

TOXICITY DATA WITH REFERENCE
orl-rat LD50:4810 mg/kg TXAPA9 29,181,74
ipr-rat LD50:352 mg/kg TXAPA9 29,181,74
scu-rat LD50:3428 mg/kg YKYUA6 31,237,80
orl-mus LD50:1750 mg/kg TXAPA9 29,181,74
ipr-mus LD50:148 mg/kg TXAPA9 29,181,74
scu-mus LD50:4955 mg/kg YKYUA6 31,237,80
ivn-mus LD50:519 mg/kg TXAPA9 29,181,74

SAFETY PROFILE: Poison by intraperitoneal route. Moderately toxic by ingestion, subcutaneous, and intravenous routes. When heated to decomposition it emits toxic fumes of NO_x.

CBQ575 **HR: 3**
CARBINOXAMINE DIPHENYLDISULFONATE
mf: $C_{16}H_{19}ClN_2O \cdot C_{12}H_{10}O_6S_2$ mw: 605.16

SYNS: 2-(p-CHLORO-α-(2-(DIMETHYLAMINO)ETHOXY)BENZYL)-PYRIDINE DIPHENYLDISULFONATE □ CXA-DPS

TOXICITY DATA WITH REFERENCE
orl-mus LD50:630 mg/kg TOIZAG 15,367,68
ipr-mus LD50:220 mg/kg TOIZAG 15,367,68
scu-mus LD50:560 mg/kg TOIZAG 15,367,68

SAFETY PROFILE: Poison by intraperitoneal route. Moderately toxic by ingestion and subcutaneous routes. When heated to decomposition it emits toxic fumes of Cl^-, NO_x, and SO_x. See also SULFONATES and AMINES.

CBQ625 **CAS:467-22-1** **HR: 3**
CARBIPHENE HYDROCHLORIDE
mf: $C_{28}H_{34}N_2O_2 \cdot ClH$ mw: 467.10

PROP: A solid. Mp: 163–165°.

SYNS: BANDOL □ 2-ETHOXY-N-METHYL-N-(2-(METHYLPHENETHYLAMINO)ETHYL)-2,2-DIPHENYLACETAMIDE HYDROCHLORIDE □ ETOMIDE HYDROCHLORIDE □ NSC-106959 □ SQ 10269

TOXICITY DATA WITH REFERENCE
orl-man LDLo:2 mg/kg JMCMAR 6,547,63
orl-mus LD50:370 mg/kg AIPTAK 154,484,65
ipr-mus LD50:190 mg/kg AIPTAK 154,484,65
ivn-mus LD50:40 mg/kg AIPTAK 154,484,65

SAFETY PROFILE: A human poison by ingestion. An experimental poison by ingestion, intravenous, and intraperitoneal routes. When heated to decomposition it emits toxic fumes of NO_x and HCl.

CBQ750 **CAS:112-15-2** **HR: 2**
CARBITOL ACETATE
mf: $C_8H_{16}O_4$ mw: 176.24

PROP: Liquid. Bp: 217.4°, fp: −25°, flash p: 230°F (OC), d: 1.0114 @ 20°/20°, vap press: 0.05 mm @ 20°, vap d: 6.07.

SYNS: DIETHYLENE GLYCOL MONOETHYL ETHER ACETATE □ DIGLYCOL MONOETHYL ETHER ACETATE □ EKTASOLVE de ACETATE □ 2-(2-ETHOXYETHOXY)ETHANOL ACETATE □ GLYCOL ETHER de ACETATE

TOXICITY DATA WITH REFERENCE
skn-rbt 500 mg open MLD UCDS** 7/20/65
eye-rbt 505 mg AJOPAA 29,1363,46
orl-rat LD50:11 g/kg UCDS** 7/20/65
skn-rbt LD50:15,100 µL/kg UCDS** 7/20/65
orl-gpg LD50:3930 mg/kg JIHTAB 23,259,41

CONSENSUS REPORTS: Reported in EPA TSCA Inventory. Glycol ether compounds are on the Community Right-To-Know List.

SAFETY PROFILE: Moderately toxic by ingestion. A skin and eye irritant. See also GLYCOL ETHERS. Combustible when exposed to heat; can react with oxidizing materials. To fight fire, use alcohol foam, water, CO_2, dry chemical. When heated to decomposition it emits acrid smoke and fumes.

CBR000 **CAS:111-90-0** **HR: 2**
CARBITOL CELLOSOLVE
mf: $C_6H_{14}O_3$ mw: 134.20

PROP: Very hygroscopic, colorless liquid; mild pleasant odor. Bp: 201.9°, flash p: 201°F (OC), d: 0.986 @ 25°/4°, vap d: 4.62. Misc in water.

SYNS: APV □ CARBITOL □ CARBITOL SOLVENT □ DIETHYLENE GLYCOL ETHYL ETHER □ DIETHYLENE GLYCOL MONOETHYL ETHER □ DIGLYCOL MONOETHYL ETHER □ DIOXITOL □ DOWANOL □ DOWANOL DE □ ETHOXY DIGLYCOL □ 2-(2-ETHOXYETHOXY)ETHANOL □ ETHYL CARBITOL □ ETHYL DIETHYLENE GLYCOL □ ETHYLENE DIGLYCOL MONOETHYL ETHER □ LOSUNGSMITTEL APV □ MONOETHYL ETHER of DIETHYLENE GLYCOL □ POLY-SOLV □ SOLVOSOL

TOXICITY DATA WITH REFERENCE
skn-rbt 500 mg/24H MLD JPETAB 82,377,44
eye-rbt 500 mg MOD UCDS** 11/22/68
eye-rbt 125 mg MLD ADSYAF 45,553,42
mmo-sat 986 mg/plate BCFAAI 125,401,86
orl-mus TDLo:44 g/kg (7-14D preg):REP EVHPAZ 57,141,84

orl-mus TDLo: 32 g/kg (female 6-13D post): REP TCMUD8 7,29,87
orl-rat LD50: 5500 mg/kg JIDHAN 21,173,39
skn-rat LD50: 6000 mg/kg JIHTAB 29,190,47
ipr-rat LD50: 6310 mg/kg TXAPA9 21,454,72
ivn-rat LD50: 2200 mg/kg ARZNAD 28,1571,78
skn-mus LD50: 6000 mg/kg JIHTAB 29,190,47
ipr-mus LD50: 2300 mg/kg PHTHDT 5,467,79
scu-mus LD50: 5500 mg/kg JPETAB 65,89,39
ivn-dog LD50: 3000 mg/kg JIHTAB 29,190,47
ivn-cat LDLo: 1 g/kg ARZNAD 28,1571,78
orl-rbt LD50: 3620 mg/kg JIHTAB 23,259,41
skn-rbt LD50: 8500 mg/kg JIHTAB 29,325,47

CONSENSUS REPORTS: Reported in EPA TSCA Inventory. Glycol ether compounds are on the Community Right-To-Know List.

SAFETY PROFILE: Moderately toxic by ingestion, intravenous, intraperitoneal, and possibly other routes. Mildly toxic by skin contact. A skin and eye irritant. Experimental reproductive effects. Mutation data reported. Combustible when exposed to heat; can react with oxidizing materials. To fight fire, use alcohol foam, CO_2, dry chemical. When heated to decomposition it emits acrid smoke and irritating fumes.

CBR125 **CAS:1138-80-3** **HR: 3**
CARBOBENZOXYLGLYCINE
mf: $C_{10}H_{11}NO_4$ mw: 209.22

PROP: A solid. Mp: 119–120°.

SYNS: BENZYLOXYCARBONYLGLYCINE □ N-BENZYLOXYCARBONYLGLYCINE □ N-CARBOBENZOYLGLYCINE □ CARBOBENZOYL GLYCINE □ CARBOBENZYLOXYGLYCINE □ N-CARBOBENZYLOXYGLYCINE □ (CBZ)GLY □ Z-GLY

TOXICITY DATA WITH REFERENCE
ipr-mus TDLo: 280 mg/kg (28D preg): REP JPMSAE 70,60,81
ivg-mus LD50: 380 mg/kg JPMSAE 68,696,79

CONSENSUS REPORTS: Reported in EPA TSCA Inventory.

SAFETY PROFILE: Poison by intravaginal route. Experimental reproductive effects. When heated to decomposition it emits toxic fumes of NO_x. See also ESTERS.

CBR175 **CAS:64187-25-3** **HR: 3**
N-CARBOBENZOXYGLYCINE-1,2-DIBROMOETHYL ESTER
mf: $C_{12}H_{13}Br_2NO_4$ mw: 395.08

PROP: Colorless, odorless gas. D: 1.529, mp: 57° (under 5 atm), bp: 78.2°. Sltly sol in H_2O; forming H_2CO_3.

TOXICITY DATA WITH REFERENCE
ipr-mus TDLo: 280 mg/kg (28D pre): REP JPMSAE 70,60,81
ivg-mus LD50: 148 mg/kg JPMSAE 68,696,79

SAFETY PROFILE: Poison by intraperitoneal and intravaginal routes. Experimental reproductive effects. When heated to decomposition it emits toxic fumes of Br^- and NO_x. See also ESTERS.

CBR200 **CAS:64187-24-2** **HR: 2**
N-CARBOBENZOXYGLYCINE VINYL ESTER
mf: $C_{12}H_{13}NO_4$ mw: 235.26

TOXICITY DATA WITH REFERENCE
ipr-mus TDLo: 280 mg/kg (28D pre): REP JPMSAE 70,60,81
ipr-mus LD50: 501 mg/kg JMCMAR 20,1584,77
ivg-mus LD50: 500 mg/kg JPMSAE 68,696,79

SAFETY PROFILE: Moderately toxic by intraperitoneal and intravaginal routes. Experimental reproductive effects. When heated to decomposition it emits toxic fumes of NO_x. See also ESTERS.

CBR210 **HR: 3**
N-CARBOBENZOXY-l-LEUCINE-1,2-DIBROMOETHYL ESTER
mf: $C_{16}H_{21}Br_2NO_4$ mw: 451.20

SYN: l-N-CARBOXYLEUCINE-N-BENZYL-1-(1,2-DIBROMOETHYL) ESTER

TOXICITY DATA WITH REFERENCE
ipr-mus TDLo: 280 mg/kg (28D pre): REP JPMSAE 68,696,79
ivg-mus LD50: 81 mg/kg JPMSAE 68,696,79

SAFETY PROFILE: Poison by intravaginal route. Experimental reproductive effects. When heated to decomposition it emits toxic fumes of Br^- and NO_x. See also ESTERS.

CBR215 **CAS:64187-27-5** **HR: 2**
N-CARBOBENZOXY-l-LEUCINE VINYL ESTER
mf: $C_{16}H_{21}NO_4$ mw: 291.38

SYNS: l-N-CARBOXYLEUCINE N-BENZYL 1-VINYL ESTER □ LEUCINE, N-CARBOXY-, N-BENZYL 1-VINYL ESTER □ l-LEUCINE, N-((PHENYLMETHOXY)CARBONYL)-, ETHENYL ESTER

TOXICITY DATA WITH REFERENCE
ivg-mus TDLo: 280 mg/kg (female 28D pre): REP JPMSAE 68,696,79
ivg-mus LD50: 500 mg/kg JPMSAE 68,696,79

SAFETY PROFILE: Moderately toxic by intravaginal route. Experimental reproductive effects. When heated to decomposition it emits toxic fumes of NO_x.

CBR220 **CAS:1161-13-3** **HR: 3**
CARBOBENZOXYPHENYLALANINE
mf: $C_{17}H_{17}NO_4$ mw: 299.35

PROP: A solid. Mp: 88–89°.

SYNS: (BENZYLOXYCARBONYL)PHENYLALANINE □ CARBOBENZOXY-l-PHENYLALANINE □ N-CARBOBENZOXY-l-PHENYLALANINE □ l-N-CARBOXY-3-PHENYLALANINE-N-BENZYL ESTER

TOXICITY DATA WITH REFERENCE
ipr-mus TDLo: 280 mg/kg (28D pre): REP JPMSAE 68,696,79
unr-mus LD50: 251 mg/kg JPMSAE 67,1726,78
ivg-mus LD50: 250 mg/kg JPMSAE 68,696,79

SAFETY PROFILE: Poison by intravaginal route. Experimental reproductive effects. When heated to decomposition it emits toxic fumes of NO_x.

CBR225 **CAS:64187-43-5** **HR: 3**
N-CARBOBENZOXY-l-PHENYLALANINE-1,2-DIBROMOETHYL ESTER
mf: $C_{19}H_{19}Br_2NO_4$ mw: 485.21

SYN: 1-N-BENZYLOXYCARBONYL-3-PHENYLALANINE-1,2-DIBROMOETHYL ESTER

TOXICITY DATA WITH REFERENCE
ivg-mus TDLo:280 mg/kg (28D pre):REP JPMSAE 68,696,79
ipr-mus LD50:73 mg/kg JMCMAR 20,1578,77
ivg-mus LD50:74 mg/kg JPMSAE 68,696,79

SAFETY PROFILE: Poison by intraperitoneal and intravaginal routes. Experimental reproductive effects. When heated to decomposition it emits toxic fumes of Br^- and NO_x. See also ESTERS.

CBR235 **CAS:64187-42-4** **HR: 2**
N-CARBOBENZOXY-l-PHENYLALANINE VINYL ESTER
mf: $C_{19}H_{19}NO_4$ mw: 325.39

SYNS: ALANINE, N-BENZYLOXYCARBONYL-3-PHENYL-, VINYL ESTER, l- □ ALANINE, N-CARBOXY-3-PHENYL-, N-BENZYL 1-VINYL ESTER, l- □ N-BENZYLOXYCARBONYL-l-PHENYLALANINE VINYL ESTER □ l-N-CARBOXY-3-PHENYLALANINE N-BENZYL 1-VINYL ESTER □ l-PHENYLALANINE, N-((PHENYLMETHOXY)CARBONYL)-, ETHENYL ESTER □ N-((PHENYLMETHOXY)CARBONYL)-l-PHENYLALANINE ETHENYL ESTER

TOXICITY DATA WITH REFERENCE
ivg-mus TDLo:280 mg/kg (female 28D pre):REP JPMSAE 68,696,79
unr-mus LD50:2001 mg/kg JPMSAE 67,1726,78

SAFETY PROFILE: Moderately toxic. Experimental reproductive effects. When heated to decomposition it emits toxic fumes of NO_x.

CBR245 **HR: 3**
N-CARBOBENZOXY-l-PROLINE-1,2-DIBROMOETHYL ESTER
mf: $C_{15}H_{17}Br_2NO_4$ mw: 435.15

SYN: 1,2-PYRROLIDINEDICARBOXYLIC ACID-1-BENZYL 2-(1,2-DIBROMOETHYL) ESTER

TOXICITY DATA WITH REFERENCE
ipr-mus TDLo:280 mg/kg (28D pre):REP JPMSAE 68,696,79
ivg-mus LD50:225 mg/kg JPMSAE 68,696,79

SAFETY PROFILE: Poison by intravaginal route. Experimental reproductive effects. When heated to decomposition it emits toxic fumes of Br^- and NO_x. See also ESTERS.

CBR247 **HR: 2**
N-CARBOBENZOXY-l-PROLINE VINYL ESTER
mf: $C_{15}H_{17}NO_4$ mw: 275.33

SYN: 1,2-PYRROLIDINECARBOXYLIC ACID, 1-BENZYL-2-VINYL ESTER

TOXICITY DATA WITH REFERENCE
ivg-mus TDLo:280 mg/kg (female 28D pre):REP JPMSAE 68,696,79
ivg-mus LD50:500 mg/kg JPMSAE 68,696,79

SAFETY PROFILE: Moderately toxic by intravaginal route. Experimental reproductive effects. When heated to decomposition it emits toxic fumes of NO_x.

CBR250 **CAS:1722-62-9** **HR: 3**
CARBOCAINE HYDROCHLORIDE
mf: $C_{15}H_{22}N_2O\cdot ClH$ mw: 282.85

PROP: A solid. Mp: 262–264°. Sol in H_2O.

SYNS: CHLOROCAIN □ N-(2,6-DIMETHYLPHENYL)-1-METHYL-2-PIPERIDINECARBOXAMIDE-MONOHYDROCHLORIDE □ MEAVERIN □ MEPIVACAINE HYDROCHLORIDE □ dl-MEPIVACAINE HYDROCHLORIDE □ MEPIVASTESIN □ 1-METHYL-2′,6′-PIPECOLOXYLIDIDE HYDROCHLORIDE □ dl-1-METHYL-2′,6′ PIPECOLOXYLIDIDE HYDROCHLORIDE □ (1-METHYL-dl-PIPERIDINE-2-CARBOXYLIC ACID)-2,6-DIMETHYLANILIDE HYDROCHLORIDE □ SCANDICAIN

TOXICITY DATA WITH REFERENCE
ipr-rat TDLo:160 mg/kg (1D male):REP JRPFA4 51,477,77
ipr-mus LD50:117 mg/kg JPPMAB 40,592,88
scu-mus LD50:260 mg/kg NIIRDN 6,846,82
ivn-mus LD50:35 mg/kg APTOA6 42,88,78
imp-mus LD50:260 mg/kg 29ZVAB -,69,69
scu-rbt LD50:110 mg/kg TXAPA9 2,295,60
scu-rbt LD50:110 mg/kg NIIRDN 6,846,82
ivn-rbt LD50:22 mg/kg 29ZVAB -,69,69
imp-rbt LD50:110 mg/kg 29ZVAB -,69,69
scu-gpg LD50:94 mg/kg NIIRDN 6,846,82
ivn-gpg LD50:20 mg/kg NIIRDN 6,846,82

CONSENSUS REPORTS: Reported in EPA TSCA Inventory.

SAFETY PROFILE: Poison by subcutaneous, intravenous, subcutaneous, and implant routes. Experimental reproductive effects. An anesthetic. When heated to decomposition it emits very toxic fumes of HCl and NO_x.

CBR500 **CAS:655-35-6** **HR: 3**
CARBOCHROMENE HYDROCHLORIDE
mf: $C_{20}H_{27}NO_5\cdot ClH$ mw: 397.94

PROP: A solid. Mp: 159–160°.

SYNS: A-27053 □ AG 3 □ ANTIANGOR □ CARBOCROMENE □ CASSELLA 4489 □ CHROMONAR HYDROCHLORIDE □ 3-(β-DIETHYLAMINOETHYL)-4-METHYL-7-(CARBETHOXYMETHOXY)-COUMARIN HYDROCHLORIDE □ INTENKORDIN □ INTENSAIN □ INTENSAIN HYDROCHLORIDE □ KARBOKROMEN (RUSSIAN) □ NSC-110430

TOXICITY DATA WITH REFERENCE
orl-mus LD50:6300 mg/kg ARZNAD 13,243,63
ipr-mus LD50:528 mg/kg ARZNAD 13,243,63
ivn-mus LD50:34 mg/kg KHFZAN 9,57,75

SAFETY PROFILE: Poison by intravenous route. Moderately toxic by intraperitoneal route. Mildly toxic by ingestion. A coronary vasodilator. When heated to decomposition it emits very toxic fumes of NO_x and HCl.

CBR675 CAS:638-23-3 HR: 2
CARBOCISTEINE
mf: $C_5H_9NO_4S$ mw: 179.21

PROP: l-Form: Mp: 204–207°. dl-Form: Spherical aggregates of needles.

SYNS: CARBOCIT □ CARBOCYSTEINE □ S-(CARBOXYMETHYL) CYSTEINE □ l-CARBOXYMETHYLCYSTEINE □ 3-(CARBOXYMETHYLTHIO)ALANINE □ l-3-((CARBOXYMETHYL)THIO)ALANINE □ FLUIFORT □ L.J. 206 □ LOVISCOL □ MUCICLAR □ MUCOCIS □ MUCODYNE □ MUCOLASE □ MUCOLEX □ MUCOPRONT □ PECTOX □ PULMOCLASE □ REOMUCIL □ RHINATHIOL □ RINATIOL □ THIODRIL □ TRANSBRONCHIN

TOXICITY DATA WITH REFERENCE
orl-rat TDLo:625 mg/kg (17-21D preg):REP KSRNAM 13,1311,79
ipr-rat LD50:7800 mg/kg NIIRDN 6,190,82
scu-rat LD50:10,300 mg/kg NIIRDN 6,190,82
orl-mus LD50:8400 mg/kg NIIRDN 6,190,82
ipr-mus LD50:1433 mg/kg YKKZAJ 94,1419,74
scu-mus LD50:9 g/kg NIIRDN 6,190,82

CONSENSUS REPORTS: Reported in EPA TSCA Inventory.

SAFETY PROFILE: Moderately toxic by intraperitoneal route. Mildly toxic by ingestion. Experimental reproductive effects. When heated to decomposition it emits toxic fumes of SO_x and NO_x.

CBS000 CAS:14679-73-3 HR: 3
N[1]-CARBOETHOXY-N[2]-PHTHALAZINO HYDRAZINE
mf: $C_{11}H_{12}N_4O_2$ mw: 232.27

SYNS: BT 621 □ CARBOETHOXYPHTHALAZINO HYDRAZINE □ 3-(1-PHTHALAZINYL)CARBAZIC ACID ETHYL ESTER □ TODRALAZINA (ITALIAN) □ TODRALAZINE

TOXICITY DATA WITH REFERENCE
orl-rat LD50:318 mg/kg PJPPAA 31,127,79
ipr-rat LD50:337 mg/kg BJPCAL 32,104,68
ivn-rat LD50:110 mg/kg BJPCAL 32,104,68
ims-rat LD50:333 mg/kg BJPCAL 32,104,68
ipr-mus LD50:382 mg/kg BJPCAL 32,104,68
ivn-mus LD50:360 mg/kg BJPCAL 32,104,68
ims-mus LD50:417 mg/kg BJPCAL 32,104,68
orl-frg LD50:650 mg/kg BJPCAL 32,104,68
par-frg LD50:636 mg/kg BJPCAL 32,104,68

SAFETY PROFILE: Poison by ingestion, intraperitoneal, intravenous, and intramuscular routes. Moderately toxic by ingestion and parenteral routes. When heated to decomposition it emits toxic fumes of NO_x.

CBS250 CAS:4425-78-9 HR: 3
CARBOFLUORENE AMINO ESTER
mf: $C_{20}H_{23}NO_2$ mw: 309.44

SYNS: FLUORENE-9-CARBOXYLIC ACID-2-(DIETHYLAMINO)ETHYL ESTER □ PAVATRIN □ PAVATRINEAT

TOXICITY DATA WITH REFERENCE
orl-mus LD50:900 mg/kg CLDND* 91,103,47
ivn-rbt LD50:16 mg/kg CLDND* 91,103,47

SAFETY PROFILE: Poison by intravenous routes. Moderately toxic by ingestion. See also AMINES and ESTERS. When heated to decomposition it emits toxic fumes of NO_x.

CBS275 CAS:1563-66-2 HR: 3
CARBOFURAN
mf: $C_{12}H_{15}NO_3$ mw: 221.28

PROP: White, crystalline solid; odorless. Mp: 150–152°, d: 1.180 @ 20°/20°, vap press: 2×10^{-5} mm @ 33°. Sltly sol in water.

SYNS: BAY 70143 □ CURATERR □ D 1221 □ 2,3-DIHYDRO-2,2-DIMETHYLBENZOFURANYL-7-N-METHYLCARBAMATE □ 2,3-DIHYDRO-2,2-DIMETHYL-7-BENZOFURANYL METHYLCARBAMATE □ 2,2-DIMETHYL-7-COUMARANYL-N-METHYLCARBAMATE □ 2,2-DIMETHYL-2,3-DIHYDROBENZOFURAN-7-YL ESTER, METHYLCARBAMIC ACID □ 2,2-DIMETHYL-2,3-DIHYDRO-7-BENZOFURANYL-N-METHYLCARBAMATE □ ENT 27,164 □ FMC 10242 □ FURADAN □ FURODAN □ METHYL CARBAMIC ACID 2,3-DIHYDRO-2,2-DIMETHYL-7-BENZOFURANYL ESTER □ NIA 10242 □ NIAGRA 10242 □ YALTOX

TOXICITY DATA WITH REFERENCE
mmo-sat 10 mg/plate MUREAV 116,185,83
cyt-hmn:lym 100 mg/L TGANAK 18(1),17,84
sce-hmn:lym 5 mg/L MUREAV 147,296,85
orl-mus TDLo:210 μg/kg (female 1-21D post):REP JEPTDQ 2(2),357,78
orl-mus TDLo:10,500 μg/kg (female 1-21D post):REP JEPTDQ 4(5-6),53,80
orl-mus TDLo:110 mg/kg (female 6-16D post):TER JESEDU 20,373,85
orl-rat TDLo:70 mg/kg (female 6-19D post):REP JESEDU 20,373,85
orl-mus TDLo:10,500 μg/kg (female 1-21D post):TER JEPTDQ 4(5-6),53,80
orl-mus TDLo:220 mg/kg (female 6-16D post):TER JESEDU 20,373,85
orl-rat LD50:5 mg/kg PSSCBG 1,117,70
ihl-rat LC50:85 mg/m³ JOCMA7 12,16,70
skn-rat LD50:120 mg/kg WRPCA2 9,119,70
orl-mus LD50:2 mg/kg JAFCAU 18,793,70
ivn-mus LD50:450 μg/kg CSLNX* NX#11280
orl-dog LD50:19 mg/kg JOCMA7 12,16,70
ihl-dog LC50:52 mg/m³ JOCMA7 12,16,70
skn-rbt LD50:885 mg/kg GUCHAZ 6,81,73
ihl-gpg LC50:43 mg/m³/4H TobJS# 9NOV73

CONSENSUS REPORTS: EPA Extremely Hazardous Substances List. Reported in EPA TSCA Inventory. EPA Genetic Toxicology Program.

OSHA PEL: TWA 0.1 mg/m³
ACGIH TLV: TWA 0.1 mg/m³

SAFETY PROFILE: Poison by inhalation, ingestion, skin contact, and intravenous routes. Experimental teratogenic and reproductive effects. Human mutation data reported. When heated to decomposition it emits toxic fumes of NO_x. See also CARBAMATES.

CBS500 **CAS:497-18-7** **HR: 3**
CARBOHYDRAZIDE
mf: CH_6N_4O mw: 90.11

PROP: Crystals from EtOH (aq). Mp: 153–154°. Sol in H_2O; practically insol in org solvs.

SYNS: 4-AMINOSEMICARBAZIDE □ CARBAZIC ACID HYDRAZIDE □ CARBAZIDE □ CARBODIHYDRAZIDE □ CARBONIC ACID DIHYDRAZIDE □ CARBONIC DIHYDRAZIDE □ CARBONOHYDRAZIDE □ CARBONYLDIHYDRAZINE □ 1,3-DIAMINOUREA

TOXICITY DATA WITH REFERENCE
ivn-mus LD50:120 mg/kg JPETAB 122,110,58
ipr-mus LD50:167 mg/kg JMPCAS 4,259,61

DOT CLASSIFICATION: Forbidden

CONSENSUS REPORTS: Reported in EPA TSCA Inventory.

SAFETY PROFILE: Poison by intravenous and intraperitoneal routes. Reacts with nitrous acid to form the explosive carbonic diazide. When heated to decomposition it emits toxic fumes of NO_x.

CBS750 **CAS:63042-08-0** **HR: 2**
4′-CARBOMETHOXY-2,3′-DIMETHYLAZOBENZENE
mf: $C_{16}H_{16}N_2O_3$ mw: 284.34

SYNS: 4′-CARBOMETHOXY-2,3′-DIMETHYLAZOBENZOL □ CARBONIC ACID METHYL-4-(o-TOLYLAZO)-o-TOLYL ESTER □ 2,3′-DIMETHYLAZOBENZENE-4′-METHYLCARBONATE

TOXICITY DATA WITH REFERENCE
orl-rat TDLo:27 g/kg/43W-C:ETA GANNA2 33,196,39

SAFETY PROFILE: Questionable with experimental tumorigenic data. When heated to decomposition it emits toxic fumes of NO_x.

CBT125 **CAS:25147-05-1** **HR: 3**
N-CARBOMETHOXYMETHYLIMINOPHOSPHORYL CHLORIDE
mf: $C_2H_3Cl_3NO_2P$ mw: 310.38

SAFETY PROFILE: Violent or explosive spontaneous decomposition. Upon decompositon it emits toxic fumes of Cl^-, PO_x and NO_x. See also CHLORIDES.

CBT175 **CAS:89022-11-7** **HR: 2**
2′-CARBOMETHOXYPHENYL 4-GUANIDINOBENZOATE
mf: $C_{16}H_{15}N_3O_4$ mw: 313.34

SYNS: 4-((AMINOIMINOMETHYL)AMINO)BENZOIC ACID 2-(METHOXYCARBONYL)PHENYL ESTER □ BENZOIC ACID, 4-((AMINOIMINOMETHYL)AMINO)-, 2-(METHOXYCARBONYL)PHENYL ESTER □ SALICYLIC ACID, METHYL ESTER, ESTER with p-GUANIDINOBENZOIC ACID

TOXICITY DATA WITH REFERENCE
ivg-rbt TDLo:100 μg/kg (female 1D pre):REP CCPTAY 32,183,85
ipr-mus LD50:750 mg/kg JMCMAR 29,514,86

SAFETY PROFILE: Moderately toxic by intraperitoneal route. Experimental reproductive effects. When heated to decomposition it emits toxic fumes of NO_x.

CBT250 **CAS:4564-87-8** **HR: 3**
CARBOMYCIN
mf: $C_{42}H_{67}NO_{16}$ mw: 842.10

PROP: Laths from MeOH (aq). Mp: 210–214° (decomp).

SYNS: CARBOMYCIN A □ DELTAMYCIN A □ 9-DEOXY-12,13-EPOXY-9-OXOLEUCOMYCIN V 3-ACETATE 4[B]-(3-METHYLBUTANOATE) □ M-4209 □ MAGNAMYCIN □ MAGNAMYCIN A

TOXICITY DATA WITH REFERENCE
scu-mus LD50:295 mg/kg ANTCAO 3,55,53
ivn-mus LD50:550 mg/kg MEIEDD 10,250,83
ims-mus LD50:1000 mg/kg ANTCAO 3,55,53
ivn-rbt LD50:700 mg/kg ANTCAO 3,55,53

SAFETY PROFILE: Poison by subcutaneous route. Moderately toxic by intravenous and intramuscular routes. When heated to decomposition it emits toxic fumes of NO_x.

CBT500 **CAS:7440-44-0** **HR: 1**
CARBON
DOT: UN 1361/UN 1362
af: C aw: 12.01

PROP: Black crystals, powder or diamond form. Mp: 3652–3697° (subl), bp: approx 4200°, d (amorph): 1.8–2.1, d (graphite): 2.25, d (diamond): 3.51, vap press: 1 mm @ 3586°.

SYNS: ACTICARBONE □ ACTIVATED CARBON □ AG 3 □ AG 5 □ AG 3 (ADSORBENT) □ AG 5 (ADSORBENT) □ AR (ADSORBENT) □ ANTHRASORB □ AR 3 □ ART 2 □ AU 3 □ BAU □ BG 6080 □ BLACK LEAD □ CARBON, activated (DOT) □ CARBON-12 □ CARBON, animal or vegetable origin (DOT) □ CARBOPOL EXTRA □ CARBOPOL M □ CARBOPOL Z 4 □ CARBOPOL Z EXTRA □CARBO-SIEVE □ CARBOSORBIT R □ CECARBON □ CF 8 □ CF 8 (CARBON) □ C.I. 77265 □ C.I. PIGMENT BLACK 10 □ CLF II □ CMB 50 □ CMB 200 □ COKE POWDER □ COLUMBIA LCK □ CONDUCTEX □ CUZ 3 □ CWN 2 □ DARCO □ FILTRASORB □ FILTRASORB 200 □ FILTRASORB 400 □ GRAPHITE □ GRAPHITE SYNTHETIC (ACGIH,OSHA) □ GROSAFE □ HYDRODARCO □ IRGALITE 1104 □ JADO □ K 257 □ MA 100 (CARBON) □ NORIT □ NUCHAR □ OU-B □ PELIKAN C 11/1431a □ PLUMBAGO □ SKG □ SKT □ SKT (ADSORBENT) □ SU 2000 □ SUCHAR 681 □ SUPERSORBON IV □ SUPERSORBON S 1 □ U 02 □ WATERCARB □ WITCARB 940 □ XE 340 □ XF 4175L

TOXICITY DATA WITH REFERENCE
scu-rat TDLo:167 mg/kg (8D preg):REP TJADAB 4,327,71
ivn-mus LD50:440 mg/kg TXAPA9 24,497,73

CONSENSUS REPORTS: Reported in EPA TSCA Inventory.

OSHA PEL: (Natural graphite) TWA 2.5 mg/m³; (Synthetic graphite) TWA Total Dust: 10 mg/m³; Respirable Fraction: 5 mg/m³
ACGIH TLV: TWA 2 mg/m³ (respirable dust)
DFG MAK: 6 mg/m³
DOT CLASSIFICATION: 4.2; *Label:* Spontaneously Combustible

SAFETY PROFILE: Moderately toxic by intravenous route. Experimental reproductive effects. It can cause a dust irritation, particularly to the eyes and mucous membranes. See also CARBON BLACK, SOOT. Combustible when exposed to heat. Dust is explosive when exposed to heat or flame or oxides, peroxides, oxosalts, halogens, interhalogens, O_2, (NH_4NO_3 + heat), (NH_4ClO_4@ 240°), bromates, $Ca(OCl)_2$, chlorates, (Cl_2+ $Cr(OCl)_2$), ClO, iodates, IO_5, ($Pb(NO_3)_2$, $HgNO_3$, HNO_3, (oils + air), (K + air), Na_2S, $Zn(NO_3)_2$. Incompatible with air, metals, oxidants, unsaturated oils.

CBT750 CAS:1333-86-4 HR: 1
CARBON BLACK

PROP: A generic term applied to a family of high-purity colloidal carbons commercially produced by carefully controlled pyrolysis of gaseous or liquid hydrocarbons. Carbon blacks, including commercial colloidal carbons such as furnace blacks, lamp blacks, and acetylene blacks, usually contain less than several tenths percent of extractable organic matter and less than one percent ash.

SYNS: ACETYLENE BLACK □ ARO □ AROFLOW □ AROGEN □ AROMEX □ AROTONE □ AROVEL □ ARROW □ ATLANTIC □ BLACK PEARLS □ CANCARB □ CARBODIS □ CARBOLAC □ CARBOLAC 1 □ CARBOMET □ CARBON BLACK, ACETYLENE □ CARBON BLACK BV and V □ CARBON BLACK, CHANNEL □ CARBON BLACK, FURNACE □ CARBON BLACK, LAMP □ CARBON BLACK, THERMAL □ CHANNEL BLACK □ C.I. 77266 □ C.I. PIGMENT BLACK 6 □ C.I. PIGMENT BLACK 7 □ CK3 □ COLLOCARB □ COLUMBIA CARBON □ CONDUCTEX □ CONTINENTAL □ CONTINEX □ CORAX □ CORAX P □ CROFLEX □ CROLAC □ DEGUSSA □ DELUSSA BLACK FW □ DIXIE □ DIXIECELL □ DIXIEDENSED □ DIXITHERM □DU-REX □ EAGLE GERMANTOWN □ ELF □ ELFTEX □ ESSEX □EXCELSIOR □ EXPLOSION BLACK □ EXPLOSION ACETYLENE BLACK □ FARBRUSS □ FECTO □ FLAMRUSS □ FURNAL □ FURNEX □ FURNEX N 765 □ GAS-FURNACE BLACK □ GASTEX □ HUBER □HUMENEG-RO □ IMPINGEMENT BLACK □ KETJENBLACK EC □ KOSMINK □ KOSMOBIL □ KOSMOLAK □ KOSMOS □ KOSMOTHERM □KOSMO-VAR □ MAGECOL □ METANEX □ MICRONEX □ MIIKE 20 □ MODULEX □ MOGUL □ MOGUL L □ MOLACCO □ MONARCH □ NEO-SPECTRA □ NEO SPECTRA II □ NEOTEX □ OIL-FURNACE BLACK □ P-33 □ P68 □ P1250 □ PEERLESS □ PELLETEX □ PHILBLACK □ PHILBLACK N 550 □ PHILBLACK N 765 □ PHILBLACK O □ PIGMENT BLACK 7 □ PRINTEX □ PRINTEX 60 □ RAVEN □ RAVEN 30 □ RAVEN 420 □ RAVEN 500 □ RAVEN 8000 □ REBONEX □ REGAL □ REGAL 99 □ REGAL 300 □ REGAL 330 □ REGAL 600 □ REGAL 400R □ REGAL SRF □ REGENT □ ROYAL SPECTRA □ SEVACARB □ SEVAL □ SHAWINIGAN ACETYLENE BLACK □ SHELL CARBON □ SPECIAL BLACK 1V & V □ SPECIAL SCHWARZ □ SPHERON □ SPHERON 6 □ STATEX □ STATEX N 550 □ STERLING □ STERLING N 765 □ STERLING NS □ STERLING SO 1 □ SUPERBA □ SUPER-CARBOVAR □ SUPER-SPECTRA □ TEXAS □ THERMA-ATOMIC BLACK □ THERMAL ACETYLENE BLACK □ THERMATOMIC □ THERMAX □ THERMBLACK □ TINOLITE □ TM 30 □ TORCH BRAND □ TRIANGLE □ UCET □ UKARB □ UNITED □ VELVETEX □ VULCAN □ WITCO □ WITCOBLAK NO. 100 □ WYEX

TOXICITY DATA WITH REFERENCE
mmo-sat 1 mg/plate EVSRBT 27,297,83
add-mus-ihl 6200 μg/m³/16H/12W-I EMMUEG 16,64,90

CONSENSUS REPORTS: IARC Cancer Review: Group 3 IMEMDT 7,142,87; Human Inadequate Evidence IMEMDT 33,35,84; Animal Inadequate Evidence IMEMDT 33,35,84

OSHA PEL: TWA 3.5 mg/m³
ACGIH TLV: TWA 3.5 mg/m³
NIOSH REL: (Carbon Black) TWA 3.5 mg/m³

SAFETY PROFILE: Mildly toxic by ingestion, inhalation, and skin contact. Questionable carcinogen. Mutation data reported. See also CARBON. A nuisance dust in high concentrations. While it is true that the tiny particulates of carbon black contain some molecules of carcinogenic materials, the carcinogens are apparently held tightly and are not eluted by hot or cold water, gastric juices, or blood plasma.

For occupational chemical analysis use NIOSH: Carbon Black, 5000.

CBU250 CAS:124-38-9 HR: 1
CARBON DIOXIDE
DOT: UN 1013/UN 1845/UN 2187
mf: CO_2 mw: 44.01

PROP: Colorless, odorless gas. Mp: 57° (sublimes @ −78.5°), vap d: 1.53, 78.2°. Sltly sol in water, forming H_2CO_3.

SYNS: ANHYDRIDE CARBONIQUE (FRENCH) □ CARBON DIOXIDE, refrigerated liquid (UN 2187) (DOT) □ CARBON DIOXIDE, solid (UN 1845) (DOT) □ CARBONIC ACID ANHYDRIDE □ CARBONIC ACID Gas □ CARBONIC ANHYDRIDE □ CARBON OXIDE □ DRY ICE □ DRY ICE (UN 1845) (DOT) □ KHLADON 744 □ KOHLENDIOXYD (GERMAN) □ KOHLENSAEURE (GERMAN) □ R 744

TOXICITY DATA WITH REFERENCE
ihl-rat TCLo:6 pph/24H (10D preg):REP CIRUAL 8,1218,60
ihl-mus TCLo:55 pph/4H (male 6D pre):REP JRPFA4 13,165,67
ihl-mus TCLo:2 pph/8H (female 10D post):REP TJADAB 30,187,84
ihl-rat TCLo:6 pph/24H (10D preg):TER CIRUAL 8,1218,60
ihl-hmn LCLo:9 pph/5M TABIA2 3,231,33
ihl-mam LCLo:90,000 ppm/5M AEPPAE 138,65,28

CONSENSUS REPORTS: Reported in EPA TSCA Inventory.

OSHA PEL: TWA 10,000 ppm; STEL 30,000 ppm
ACGIH TLV: TWA 5000 ppm; STEL 30,000 ppm
DFG MAK: 5000 ppm (9000 mg/m³)
NIOSH REL: (Carbon Dioxide) TWA 10,000 ppm; CL 30,000 ppm/10M
DOT CLASSIFICATION: 2.2; *Label:* Nonflammable Gas; DOT Class: 9; *Label:* None (UN 1845)

SAFETY PROFILE: An asphyxiant. See discussion of simple asphyxiants under ARGON. Experimental teratogenic and reproductive effects. Contact of solid carbon dioxide snow with the skin can cause burns. Dusts of magnesium, zirconium, titanium, and some magnesium-aluminum alloys ignite and then explode in CO_2 atmospheres. Dusts of aluminum, chromium, and manganese ignite and then explode when heated in CO_2. Several bulk metals will burn in CO_2. Reacts vigorously with (Al + Na_2O_2), Cs_2O, $Mg(C_2H_5)_2$, Li, (Mg + Na_2O_2), K, KHC,

Na, Na_2C_2, NaK, Ti. CO_2 fire extinguishers can produce highly incendiary sparks of 5-15 mJ at 10–20 kV by electrostatic discharge. Incompatible with acrylaldehyde, aziridine, metal acetylides, sodium peroxide.

For occupational chemical analysis use OSHA: #ID-172 or NIOSH: Carbon Dioxide S249.

CBV000 CAS:53569-62-3 HR: 2
CARBON DIOXIDE mixed with NITROUS OXIDE
DOT: UN 1015
mf: $CO_2 \cdot N_2O$ mw:88.03

SYNS: CARBON DIOXIDE, mixture with NITROGEN OXIDE (N_2O) □ CARBON DIOXIDE–NITROUS OXIDE mixture (DOT)

NIOSH REL: (Carbon Dioxide) TWA 10000 ppm; CL 30000 ppm/10M; (N_2O as Anesthetic Agent) TWA 25 ppm/1H
DOT CLASSIFICATION: 2.2; *Label:* Nonflammable Gas

SAFETY PROFILE: See components as listed. An anesthetic mixture. Combustible. An oxidizing mixture. Can react with reducing materials.

CBV250 CAS:8063-77-2 HR: 1
CARBON DIOXIDE mixed with OXYGEN
DOT: UN 1014

SYNS: CARBOGEN (8CI) □ CARBON DIOXIDE-OXYGEN mixture (DOT)

NIOSH REL: (Carbon Dioxide) TWA 10,000 ppm; CL 30,000 ppm/10M
DOT CLASSIFICATION: 2.2; *Label:* Nonflammable Gas

SAFETY PROFILE: Possible asphyxiant.

CBV500 CAS:75-15-0 HR: 3
CARBON DISULFIDE
DOT: UN 1131
mf: CS_2 mw: 76.13

PROP: Highly refracting, clear, colorless liquid; nearly odorless when pure. Mp: −111.6°, d: 1.293 @ 0°/4°, bp: 46.5°, lel: 1.3%, uel: 50%, flash p: −22°F (CC), autoign temp: 257°F, vap press: 400 mm @ 28°, vap d: 2.64. Misc in EtOH, Et_2O, and C_6H_6; sltly sol in H_2O.

SYNS: CARBON BISULFIDE (DOT) □ CARBON BISULPHIDE □ CARBON DISULPHIDE □ CARBONE (SUFURE de) (FRENCH) □ CARBONIO (SOLFURO di) (ITALIAN) □ CARBON SULFIDE □ CARBON SULPHIDE (DOT) □ DITHIOCARBONIC ANHYDRIDE □ KOHLENDISULFID (SCHWEFELKOHLENSTOFF) (GERMAN) □ KOOLSTOFDISULFIDE (ZWAVELKOOLSTOF) (DUTCH) □ NCI-C04591 □ RCRA WASTE NUMBER P022 □ SCHWEFELKOHLENSTOFF (GERMAN) □ SOLFURO di CARBONIO (ITALIAN) □ SULPHOCARBONIC ANHYDRIDE □ WEEVILTOX □ WEGLA DWUSIARCZEK (POLISH)

TOXICITY DATA WITH REFERENCE
mmo-sat 100 μL/plate NIOSH* 5AUG77
sce-hmn:lym 10,200 μg/L BCTKAG 14,115,81
ihl-man TCLo:40 mg/m³ (91W male):REP MELAAD 60,566,69
ihl-rat TCLo:30 μg/m³/8H (1-22D preg):REP ATSUDG 4,252,80
ihl-rat TCLo:100 mg/m³/8H (1-22D preg):REP TOLED5 2,129,78
ihl-rat TCLo:10 mg/m³/8H (1-22D preg):REP ATSUDG 4,252,80
ihl-rat TCLo:100 mg/m³/8H (1-21D preg):TER TJADAB 14,374,76
orl-rat TDLo:2 g/kg (6-15D preg):TER TOXID9 4,86,84
ihl-mus TCLo:2000 mg/m³/2H (1-21D preg):REP BEXBAN 68,1158,69
orl-rbt TDLo:350 mg/kg (6-19D preg):REP TOXID9 4,86,84
ihl-rat TCLo:200 mg/m³/24H (1-21D preg):REP KHZDAN 21,257,78
ihl-rat TCLO:600 ppm/6H (50D male):REP TXAPA9 73,275,84
ihl-rat TCLo:10 mg/m³/8H (1-22D preg):TER ATSUDG 4,252,80
orl-rbt TDLo:2100 mg/kg (6-19D preg):TER TOXID9 4,86,84
ihl-hmn LCLo:4000 ppm/30M 29ZWAE -,118,68
ihl-hmn LCLo:2000 ppm/5M TABIA2 3,231,33
unr-man LDLo:186 mg/kg 85DCAI 2,73,70
orl-rat LD50:3188 mg/kg GISAAA 31(1),13,66
ihl-rat LC50:25 g/m³/2H 85GMAT -,32,82
orl-mus LD50:2780 mg/kg GISAAA 31(1),13,66
ihl-mus LC50:10 g/m³/2H 85GMAT -,32,82 GISAAA 31(1),13,66
orl-gpg LD50:2125 mg/kg GISAAA 31(1),13,66
ipr-gpg LDLo:400 mg/kg AIHAAP 35,21,74

CONSENSUS REPORTS: Reported in EPA TSCA Inventory. EPA Genetic Toxicology Program. Community Right-To-Know List. EPA Extremely Hazardous Substances List.

OSHA PEL: TWA 4 ppm; STEL 12 (skin)
ACGIH TLV: TWA 10 ppm (skin); BEI: 5 mg(2-thiothiazolidine-4-carboxylic acid (TTCA))/g creatinine in urine
DFG MAK: 10 ppm (30 mg/m³); BAT: 8 mg/L of 4-thio-4-thiazolidine carboxylic acid (TTCA) at end of shift.
NIOSH REL: (Carbon Disulfide) TWA 1 ppm; CL 10 ppm/15M
DOT CLASSIFICATION: 3; *Label:* Flammable Liquid, Poison

SAFETY PROFILE: A human poison by unspecified route. Mildly toxic to humans by inhalation. An experimental poison by intraperitoneal route. Human reproductive effects on spermatogenesis by inhalation. Experimental teratogenic and reproductive effects. Human mutation data reported. The main toxic effect is on the central nervous system, acting as a narcotic and anesthetic in acute poisoning with death following from respiratory failure. In chronic poisoning, the effect on the nervous system is one of central and peripheral damage, which may be permanent if the damage has been severe.

Flammable liquid. A dangerous fire hazard when exposed to heat, flame, sparks, friction, or oxidizing materials. Severe explosion hazard when exposed to heat or flame. Ignition and potentially explosive reaction when heated in contact with rust or iron. Mixtures with sodium or potassium-sodium alloys are powerful, shock-sensitive explosives. Explodes on contact with permanganic acid. Potentially explosive reaction with nitrogen oxide; chlorine (catalyzed by iron). Mixtures with dinitrogen tetraoxide are heat-, spark- and shock-sensitive explosives. Reacts with metal azides to produce shock- and heat-sensitive, explosive metal azidodithioformates. Aluminum powder ignites in CS_2 vapor. The vapor ignites on contact with fluorine. Reacts violently with azides, CsN_3, ClO, ethylamine diamine, ethylene imine, $Pb(N_3)_2$, LiN_3, (H_2SO_4 + permanganates), KN_3, RbN_3, NaN_3, phenylcopper-triphenylphosphine complexes. Incompatible with air, metals, oxidants. To fight fire, use water, CO_2, dry chemical, fog, mist. When heated to decomposition it emits highly toxic fumes of SO_x.

For occupational chemical analysis use NIOSH: Carbon Disulfide, 1600.

CBW000 **CAS:973-21-7** **HR: 3**
CARBONIC ACID-2-sec-BUTYL-4,6-DINITROPHENYL ISOPROPYL ESTER
mf: $C_{14}H_{18}N_2O_7$ mw: 326.34

SYNS: ACREX □ 2-sec-BUTYL-4,5-DINITROPHENOL ISOPROPYL CARBONATE □ 2-sec-BUTYL-4,6-DINITROPHENYL ISOPROPYL CARBONATE □ DESSIN □ 2,4-DINITRO-6-sek.BUTYL-ISOPROPYLPHENYL-CARBONAT (GERMAN) □ 2,4-DINITRO-6-sec-BUTYLPHENYL ISOPROPYL CARBONATE □ DINOBUTON □ DINOFEN □ DRAWINOL □ DS 18302 □ ENT 27,244 □ ISOPHEN □ ISOPHEN (pesticide) □ ISOPROPYL-2,4-DINITRO-6-SEC-BUTYLPHENYL CARBONATE □ ISOPROPYL-2-(1-METHYL-N-PROPYL)-4,6-DINITROPHENYL CARBONATE □ KASEBON □ MC 1053 □ 1-METHYLETHYL-2-(1-ETHYLPROPYL)-4,6-DINITROPHENYL CARBONATE □ 1-METHYLETHYL-2-(1-METHYLPROPYL)-4,5-DINITROPHENYLESTER CARBONIC ACID □ 2-(1-METHYL-2-PROPYL)-4,6-DINITROPHENYL ISOPROPYLCARBONATE □ SYTASOL □ TALAN □ UC 19786 □ UNION CARBIDE 19786

TOXICITY DATA WITH REFERENCE
cyt-mus-orl 25 mg/kg CYGEDX 14(6),38,80
orl-rat LD50:59 mg/kg TXAPA9 14,515,69
ihl-rat LC50:80 mg/m^3/4H 85JFAN A157,83
skn-rat LDLo:1500 mg/kg TXAPA9 14,515,69
unk-rat LD50:140 mg/kg 30ZDA9 -,100,71
orl-mus LD50:170 mg/kg GTPZAB 19(9),55,75
ipr-mus LD50:125 mg/kg BCPCA6 18,1389,69
unk-mus LD50:2540 mg/kg 30ZDA9 -,100,71
skn-rbt LD50:3200 mg/kg FMCHA2 -,C82,83
orl-ckn LD50:235 mg/kg VETNAL 63(1),59,87

SAFETY PROFILE: Poison by ingestion, inhalation, intraperitoneal, and possibly other routes. Moderately toxic by skin contact. Mutation data reported. A miticide. See also ESTERS. When heated to decomposition it emits toxic fumes of NO_x.

CBW200 **CAS:1184-64-1** **HR: 3**
CARBONIC ACID, COPPER(2+) SALT (1:1)
mf: CO_3•Cu mw: 123.55

SYNS: COPPER CARBONATE □ COPPER CARBONATE (1:1) □ COPPER(II) CARBONATE □ COPPER MONOCARBONATE □ CUPRIC CARBONATE □ CUPRIC CARBONATE (1:1) □ XANTHIC ACID, COPPER(II) SALT

TOXICITY DATA WITH REFERENCE
orl-hmn LDLo:200 mg/kg FAONAU 53A,43,74
orl-mus LDLo:320 mg/kg AECTCV 14,111,85

CONSENSUS REPORTS: Reported in EPA TSCA Inventory.

SAFETY PROFILE: A human poison by ingestion. When heated to decomposition it emits toxic vapors of Cu.

CBW400 **CAS:2463-45-8** **HR: 3**
CARBONIC ACID, CYCLIC 3-CHLOROPROPYLENE ESTER
mf: $C_4H_5ClO_3$ mw: 136.54

SYN: 1,3-DIOXOLAN-2-ONE, 4-(CHLOROMETHYL)-

TOXICITY DATA WITH REFERENCE
orl-rat TDLo:70 mg/kg (male 14D pre):REP CCPTAY 9,451,74
orl-rat LD50:80 mg/kg CCPTAY 9,451,74

CONSENSUS REPORTS: Reported in EPA TSCA Inventory.

SAFETY PROFILE: Poison by ingestion. Experimental reproductive effects. When heated to decomposition it emits toxic fumes of Cl^-.

CBW500 **CAS:108-32-7** **HR: 1**
CARBONIC ACID CYCLIC PROPYLENE ESTER
mf: $C_4H_6O_3$ mw: 102.10

PROP: A clear liquid. Bp: 242.1°, fp: −48.8°, flash p: 275°F (OC), d: 1.2069 @ 20°/20°, vap press: 0.03 mm @ 20°.

SYNS: CYCLIC METHYLETHYLENE CARBONATE □ CYCLIC PROPYLENE CARBONATE □ CYCLIC-1,2-PROPYLENE CARBONATE □ 1-METHYLETHYLENE CARBONATE □ 1,2-PROPANEDIOL CARBONATE □ 1,2-PROPANEDIOL CYCLIC CARBONATE □ 1,2-PROPANEDIYL CARBONATE □ 1,2-PROPYLENE CARBONATE □ PROPYLENE GLYCOL CYCLIC CARBONATE

TOXICITY DATA WITH REFERENCE
skn-hmn 100 mg/3D-I MOD 85DKA8 -,127,77
eye-rbt 60 mg MOD UCDS** 4/25/58
orl-rat LD50:29,100 µL/kg UCDS** 4/25/58
scu-rat LD50:11,100 mg/kg SKIZAB 28,276,72
orl-mus LD50:20,700 mg/kg JACTDZ 6(1),23,87
scu-mus LD50:15,800 mg/kg SKIZAB 28,276,72

CONSENSUS REPORTS: Reported in EPA TSCA Inventory.

SAFETY PROFILE: Mildly toxic by ingestion and other routes. A human skin irritant. An eye irritant. See also ESTERS. Combustible when exposed to heat or flame. To fight fire, use alcohol foam. Can react with oxidizing materials. When heated to decomposition it emits acrid smoke and irritating fumes. See also ESTERS.

CBW750 **CAS:630-08-0** **HR: 3**
CARBON MONOXIDE
DOT: UN 1016/NA 9202
mf: CO mw: 28.01

PROP: Colorless, odorless, tasteless gas. Mp: −213°, bp: −190°, lel: 12.5%, uel: 74.2%, d: (gas) 1.250 g/L @ 0°, (liquid) 0.793, autoign temp: 1128°F. Very sltly sol in H_2O; sol in AcOH, MeOH, and EtOH.

SYNS: CARBONE (OXYDE de) (FRENCH) ☐ CARBONIC OXIDE ☐ CARBONIO (OSSIDO di) (ITALIAN) ☐ CARBON MONOXIDE (ACGIH, OSHA) ☐ CARBON MONOXIDE (UN 1016) (DOT) ☐ CARBON MONOXIDE, refrigerated liquid (cryogenic liquid) (NA 9202) (DOT) ☐ CARBON OXIDE (CO) ☐ EXHAUST GAS ☐ FLUE GAS ☐ KOHLENMONOXID (GERMAN) ☐ KOHLENOXYD (GERMAN) ☐ KOOLMONOXYDE (DUTCH) ☐ OXYDE de CARBONE (FRENCH) ☐ WEGLA TLENEK (POLISH)

TOXICITY DATA WITH REFERENCE

ihl-mus TCLo:65 ppm/24H (female 7–18D post):REP TJADAB 29(2),8B,84
ihl-rat TCLo:1 mg/m³/24H (female 72D pre):REP HYSAAV 35(4-6),277,70
ihl-rat TCLo:150 ppm/24H (1–22D preg):REP NETOD7 2,7,80
ihl-rbt TCLo:180 ppm/24H (female 1–30D post):REP LANCAO 2,1220,72
ihl-mus TCLo:8 pph/1H (female 8D post):TER FPNJAG 11,301,58
ihl-mus TCLo:125 ppm/24H (female 7–18D post):TER TJADAB 30,253,84
ihl-rat TCLo:1 mg/m³/24H (72D pre):REP HYSAAV 35(4-6),277,70
ihl-mus TCLo:8 pph/1H (female 8D post):REP FPNJAG 11,301,58
ihl-mus TCLo:250 ppm/7H (female 6–15D post):REP TJADAB 19,385,79
ihl-rat TCLo:1 mg/m³/24H (72D pre):REP HYSAAV 35(4-6),277,70
ihl-rat TDLo:150 ppm/24H (1–22D preg):TER TXAPA9 56,370,80
ihl-mus TCLo:8 pph/1H (female 8D post):TER FPNJAG 11,301,58
ihl-mus TCLo:250 ppm/7H (female 6–15D post):TER TJADAB 19,385,79
ihl-hmn TCLo:600 mg/m³/10M GTPZAB 31(4),34,87
ihl-man LCLo:4000 ppm/30M 29ZWAE -,207,68
ihl-man TCLo:650 ppm/45M:CNS,BLD AIHAAP 34,212,73
ihl-hmn LCLo:5000 ppm/5M TABIA2 3,231,33
ihl-rat LC50:1807 ppm/4H TXAPA9 17,752,70
ihl-mus LC50:2444 ppm/4H TXAPA9 17,752,70
ihl-dog LCLo:4000 ppm/46M HBAMAK 4,1360,35
ihl-rbt LCLo:4000 ppm HBAMAK 4,1360,35
ihl-gpg LC50:5718 ppm/4H TXAPA9 17,752,70
ihl-mam LCLo:5000 ppm/5M AEPPAE 138,65,28
ihl-bwd LD50:1334 ppm AECTCV 12,355,83

CONSENSUS REPORTS: Reported in EPA TSCA Inventory.

OSHA PEL: TWA 35; CL 200 ppm
ACGIH TLV: 25 ppm; BEI: less than 8% carboxyhemoglobin in blood at end of shift; less than 40 ppm CO in end-exhaled air at end of shift. (Proposed: less than 3.5% carboxyhemoglobin in blood at end of shift; less than 20 ppm CO in end-exhaled air at end of shift.)
DFG MAK: 30 ppm (33 mg/m³); BAT: 5% carboxyhemoglobin in blood at end of shift
NIOSH REL: (Carbon Monoxide) TWA 35 ppm; CL 200 ppm
DOT CLASSIFICATION: 2.3; *Label:* Poison Gas, Flammable Gas

SAFETY PROFILE: Mildly toxic by inhalation in humans but has caused many fatalities. Experimental teratogenic and reproductive effects. Human systemic effects by inhalation: changes in psychophysiological tests and methemoglobinemia-carboxhemoglobinemia. Can cause asphyxiation by preventing hemoglobin from binding oxygen. After removal from exposure, the half-life of elimination from the blood is one hour. Chronic exposure effects can occur at lower concentrations. A common air contaminant. Acute cases of poisoning resulting from brief exposures to high concentrations seldom result in any permanent disability if recovery takes place. Chronic effects as the result of repeated exposure to lower concentrations have been described, particularly in the Scandinavian literature. Auditory disturbances and contraction of the visual fields have been demonstrated. Glycosuria does occur, and heart irregularities have been reported. Other workers have found that where the poisoning has been relatively long and severe, cerebral congestion and edema may occur, resulting in long-lasting mental or nervous damage. Repeated exposure to low concentration of the gas, up to 100 ppm in air, is generally believed to cause no signs of poisoning or permanent damage. Industrially, sequelae are rare, as exposure, though often severe, is usually brief. It is a common air contaminant.

A dangerous fire hazard when exposed to flame. Severe explosion hazard when exposed to heat or flame. Violent or explosive reaction on contact with bromine trifluoride, bromine pentafluoride, chlorine dioxide, or peroxodisulfuryl difluoride. Mixture of liquid CO with liquid O_2 is explosive. Reacts with sodium or potassium to form explosive products sensitive to shock, heat, or contact with water. Mixture with copper powder + copper(II) perchlorate + water forms an explosive complex. Mixture of liquid CO with liquid difluorgen oxide is a rocket propellant combination. Ignites on warming with iodine heptafluoride. Ignites on contact with cesium oxide + water. Potentially explosive reaction with iron(III) oxide between 0° and 150°C. Exothermic reaction with ClF_3, (Li + H_2O), NF_3, OF_2, (K + O_2), Ag_2O, (Na + NH_3). To fight fire, stop flow of gas.

For occupational chemical analysis use NIOSH: Carbon Monoxide S340.

CBX109 **CAS:1885-14-9** **HR: 3**
CARBONOCHLORIDIC ACID PHENYL ESTER
DOT: UN 2746
mf: $C_7H_5ClO_2$ mw: 156.57

PROP: Bp: 68–71° @ 9 mm.

SYNS: CHLOROFORMIC ACID PHENYL ESTER ☐ FENYLESTER KYSELINY CHLORMRAVENCI (CZECH) ☐ PHENYL CHLOROCARBONATE ☐ PHENYL CHLOROFORMATE ☐ PHENYLCHLOROFORMATE (DOT)

TOXICITY DATA WITH REFERENCE

skn-rbt 500 mg/24H MLD 85JCAE-,940,86
eye-rbt 50 µg/24H SEV 85JCAE-,940,86
orl-rat LD50:1410 mg/kg AIHAAP 30,470,69
ihl-rat LCLo:44 ppm/4H AIHAAP 30,470,69
skn-rbt LD50:3970 mg/kg AIHAAP 30,470,69

CONSENSUS REPORTS: Reported in EPA TSCA Inventory.

DOT CLASSIFICATION: 6.1; *Label:* Poison, Corrosive

SAFETY PROFILE: Poison by inhalation. Moderately toxic by ingestion and skin contact. A corrosive skin and eye irritant. See also ESTERS. When heated to decomposition it emits toxic fumes of Cl^-.

CBX750 CAS:558-13-4 HR: 3
CARBON TETRABROMIDE
DOT: UN 2516
mf: CBr_4 mw: 331.65

PROP: Colorless, monoclinic tablets. Mp: (α) 48.4°, (β) 90.1°, bp: 102° @ 50 mm, d: 2.961 @ 99.5°/4°, vap press: 40 mm @ 96.3°. Sol in EtOH, Et_2O, and $CHCl_3$; insol in H_2O.

SYNS: BROMID UHLICITY □ CARBON BROMIDE □ METHANE, TETRABROMIDE □ METHANE, TETRABROMO- □ TETRABROMIDE METHANE □ TETRABROMOMETHANE

TOXICITY DATA WITH REFERENCE
scu-mus LD50:298 mg/kg TXAPA9 4,354,62
ivn-mus LD50:56 mg/kg CSLNX* NX#01612

CONSENSUS REPORTS: Reported in EPA TSCA Inventory.

OSHA PEL: TWA 0.1 ppm; STEL 0.3 ppm
ACGIH TLV: TWA 0.1 ppm; STEL 0.3 ppm
DOT CLASSIFICATION: 6.1; *Label:* KEEP AWAY FROM FOOD

SAFETY PROFILE: Poison by subcutaneous and intravenous routes. Narcotic in high concentration. Mixture with Li particles is an impact-sensitive explosive. Explodes on contact with hexacyclohexyldilead. When heated to decomposition it emits toxic fumes of Br^-. See also CHLORINATED HYDROCARBONS, ALIPHATIC.

CBY000 CAS:56-23-5 HR: 3
CARBON TETRACHLORIDE
DOT: UN 1846
mf: CCl_4 mw: 153.81

PROP: Colorless liquid; heavy, ethereal odor. Mp: −22.6°, bp: 76.8°, flash p: none, d: 1.632 @ 0°/4°, vap press: 100 mm @ 23.0°. Sol in EtOH, and Et_2O; prac insol in H_2O.

SYNS: BENZINOFORM □ CARBONA □ CARBON CHLORIDE □ CARBON TET □ CZTEROCHLOREK WEGLA (POLISH) □ ENT 4,705 □ FASCIOLIN □ FLUKOIDS □ METHANE TETRACHLORIDE □ NECATORINA □ NECATORINE □ PERCHLOROMETHANE □ R 10 □ RCRA WASTE NUMBER U211 □ TETRACHLOORKOOLSTOF (DUTCH) □ TETRACHLOORMETAAN □ TETRACHLORKOHLENSTOFF, (GERMAN) □ TETRACHLORMETHAN (GERMAN) □ TETRACHLOROCARBON □ TETRACHLOROMETHANE □ TETRACHLORURE de CARBONE (FRENCH) □ TETRACLOROMETANO (ITALIAN) □ TETRACLORURO di CARBONIO (ITALIAN) □ TETRAFINOL □ TETRAFORM □ TETRASOL □ UNIVERM □ VERMOESTRICID

TOXICITY DATA WITH REFERENCE
skn-rbt 4 mg MLD XEURAQ MDDC-1715
skn-rbt 500 mg/24H MLD 85JCAE-,91,86
eye-rbt 2200 μg/30S MLD XEURAQ MDDC-1715
eye-rbt 500 mg/24H MLD 85JCAE -,91,86
mmo-sat 20 μL/L EJMBA2 18,213,83
mmo-asn 5000 ppm MUREAV 147,288,85
ihl-rat TCLo:250 ppm/8H (female 10–15D post):REP DABBBA 32,2021,71
orl-rat TDLo:3 g/kg (14D preg):TER BEXBAN 82,1262,76
ihl-rat TCLo:300 ppm/7H (female 6–15D post):TER TXAPA9 28,452,74
orl-rat TDLo:2 g/kg (7–8D preg):REP 85DJA5 -,95,71
ipr-rat TDLo:5 g/kg (male 1D pre):REP TXCYAC 10,39,78
ipr-rat TDLo:71,500 mg/kg (male 15D pre):REP EXPEAM 22,395,66
orl-rat TDLo:7691 mg/kg (male 10D pre):REP ESKHA5 (99),156,81
par-rat TDLo:2384 mg/kg (female 18D post):TER BEXBAN 76,1467,73
ihl-rat TCLo:300 ppm/7H (female 6–15D post):TER TXAPA9 28,452,74
scu-rat TDLo:15,600 mg/kg/12W-I:ETA JJIND8 38,891,67
orl-mus TDLo:4400 mg/kg/19W-I:NEO JJIND8 20,431,58
par-mus TDLo:305 g/kg/30W-I:ETA BEXBAN 89,845,80
orl-ham TDLo:9250 mg/kg/30W-I:ETA JJIND8 26,855,61
orl-mus TD:12 g/kg/88D-I:NEO JJIND8 4,385,44
scu-rat TD:100 g/kg/25W-I:ETA KRMJAC 12,37,65
scu-rat TD:31 g/kg/12W-I:ETA JJIND8 45,1237,70
scu-rat TD:182 g/kg/70W-I:CAR JJIND8 44,419,70
orl-mus TD:8580 mg/kg/9W-I:NEO JJIND8 4,385,44
orl-mus TD:57,600 mg/kg/12W-I:NEO JJIND8 2,197,41
ihl-hmn TCLo:20 ppm:GIT 85CYAB 2,136,59
orl-wmn TDLo:1800 mg/kg:EYE,CNS TXMDAX 69,86,73
orl-man TDLo:1700 mg/kg:CNS,PUL,GIT SAMJAF 49,635,75
orl-man LDLo:429 mg/kg:CNS,PUL,GIT ZHYGAM 19,781,73
ihl-hmn LCLo:1000 ppm PCOC** -,198,66
ihl-hmn TCLo:45 ppm/3D:CNS,GIT LANCAO 1,360,60
ihl-hmn TCLo:317 ppm/30M:GIT JAMAAP 103,962,34
ihl-hmn LCLo:5 pph/5M TABIA2 3,231,33
unk-man LDLo:93 mg/kg 85DCAI 2,73,70
orl-rat LD50:2350 mg/kg ARTODN 54,275,83
ihl-rat LC50:8000 ppm/4H NPIRI* 1,16,74
skn-rat LD50:5070 mg/kg SPEADM 78-1,16,78
ipr-rat LD50:1500 mg/kg XEURAQ MDDC-1715
orl-mus LD50:8263 mg/kg JPPMAB 3,169,51
ihl-mus LC50:9526 ppm/8H JIDHAN 29,382,47
ipr-mus LD50:572 mg/kg PHMCAA 10,172,68
orl-dog LDLo:1000 mg/kg QJPPAL 7,205,34
ihl-dog LCLo:14,620 ppm/8H NIHBAZ 191,1,49
ipr-dog LD50:1500 mg/kg TXAPA9 10,119,67
ivn-dog LDLo:125 mg/kg QJPPAL 7,205,34
ihl-cat LCLo:38,110 ppm/2H HBAMAK 4,1405,35
scu-cat LDLo:300 mg/kg JPETAB 63,153,38

CONSENSUS REPORTS: NTP 7th Annual Report on Carcinogens. IARC Cancer Review: Group 2B IMEMDT 7,143,87; Animal Sufficient Evidence IMEMDT 20,371,79; IMEMDT 1,53,72; Human Inadequate Evidence IMEMDT 1,53,72; Human Limited Evidence IMEMDT 20,371,79. Community Right-To-Know List. EPA Genetic Toxicology Program. Reported in EPA TSCA Inventory.

OSHA PEL: TWA 2 ppm

ACGIH TLV: TWA 5 ppm; STEL 30 (skin); Suspected Human Carcinogen (Proposed: TWA 5; STEL 10 (skin); Animal Carcinogen)
DFG MAK: 10 ppm (65 mg/m³); BEI: 1.6 mL/m³ in alveolar air 1 hour after exposure; Suspected Carcinogen
NIOSH REL: (Carbon Tetrachloride) CL 2 ppm/60M
DOT CLASSIFICATION: 6.1; *Label:* Poison

SAFETY PROFILE: Confirmed carcinogen with experimental carcinogenic, neoplastigenic, and tumorigenic data. A human poison by ingestion and possibly other routes. Poison by subcutaneous and intravenous routes. Mildly toxic by inhalation. Human systemic effects by inhalation and ingestion: nausea or vomiting, pupillary constriction, coma, antipsychotic effects, tremors, somnolence, anorexia, unspecified respiratory system, and gastrointestinal system effects. Experimental teratogenic and reproductive effects. An eye and skin irritant. Damages liver, kidneys, and lungs. Mutation data reported. A narcotic. Individual susceptibility varies widely. Contact dermatitis can result from skin contact.

Carbon tetrachloride has a narcotic action resembling that of chloroform, though not as strong. Following exposure to high concentrations, the victim may become unconscious, and, if exposure is not terminated, death can follow from respiratory failure. The aftereffects following recovery from narcosis are more serious than those of delayed chloroform poisoning, usually taking the form of damage to the kidneys, liver, and lungs. Exposure to lower concentrations, insufficient to produce unconsciousness, usually results in severe gastrointestinal upset and may progress to serious kidney and hepatic damage. The kidney lesion is an acute nephrosis; the liver involvement consists of an acute degeneration of the central portions of the lobules. When recovery takes place, there may be no permanent disability. Marked variation in individual susceptibility to carbon tetrachloride exists; some persons appear to be unaffected by exposures that seriously poison their fellow workers. Alcoholism and previous liver and kidney damage seem to render the individual more susceptible. Concentrations on the order of 1000 to 1500 ppm are sufficient to cause symptoms if exposure continues for several hours. Repeated daily exposure to such concentration may result in poisoning.

Though the common form of poisoning following industrial exposure is usually one of gastrointestinal upset, which may be followed by renal damage, other cases have been reported in which the central nervous system has been affected, resulting in the production of polyneuritis, narrowing of the visual fields, and other neurological changes. Prolonged exposure to small amounts of carbon tetrachloride has also been reported as causing cirrhosis of the liver.

Locally, a dermatitis may be produced following long or repeated contact with the liquid. The skin oils are removed and the skin becomes red, cracked, and dry. The effect of carbon tetrachloride on the eyes, either as a vapor or as a liquid, is one of irritation with lacrimation and burning.

Industrial poisoning is usually acute with malaise, headache, nausea, dizziness, and confusion, which may be followed by stupor and sometimes loss of consciousness. Symptoms of liver and kidney damage may follow later with development of dark urine, sometimes jaundice and liver enlargement, followed by scanty urine, albuminuria and renal casts; uremia may develop and cause death. Where exposure has been less acute, the symptoms are usually headache, dizziness, nausea, vomiting, epigastric distress, loss of appetite, and fatigue. Visual disturbances (blind spots, spots before the eyes, a visual "haze" and restriction of the visual fields), secondary anemia, and occasionally a slight jaundice may occur. Dermatitis may be noticed on the exposed parts.

Forms impact-sensitive explosive mixtures with particulates of many metals, e.g., aluminum (when ball milled or heated to 152° in a closed container); barium (bulk metal also reacts violently); beryllium; potassium (200 times more shock-sensitive than mercury fulminate); potassium-sodium alloy (more sensitive than potassium); lithium; sodium; zinc (burns readily). Also forms explosive mixtures with chlorine trifluoride; calcium hypochlorite (heat sensitive); calcium disilicide (friction and pressure sensitive); triethyldialuminum trichloride (heat sensitive); decaborane(14) (impact sensitive); dinitrogen tetraoxide. Violent or explosive reaction on contact with fluorine. Forms explosive mixtures with ethylene between 25° and 105° and between 30 and 80 bar. Potentially explosive reaction on contact with boranes. 9.1 mixtures of methanol and CCl_4 react exothermically with aluminum, magnesium, or zinc. Potentially dangerous reaction with dimethyl formamide, 1,2,3,4,5,6-hexachlorocyclohexane, or dimethylacetamide when iron is present as a catalyst. CCl_4 has caused explosions when used as a fire extinguisher on wax and uranium fires. Incompatible with aluminum trichloride, dibenzoyl peroxide, potassium-tert-butoxide. Vigorous exothermic reaction with allyl alcohol, $Al(C_2H_5)_3$, (benzoyl peroxide + C_2H_4), BrF_3, diborane, disilane, liquid O_2, Pu, ($AgClO_4$ + HCl), potassium-tert-butoxide, tetraethylenepentamine, tetrasilane, trisilane, Zr. When heated to decomposition it emits toxic fumes of Cl^- and phosgene. It has been banned from household use by the FDA. See also CHLORINATED HYDROCARBONS, ALIPHATIC.

For occupational chemical analysis use NIOSH: Hydrocarbons, halogenated, 1003.

CBY250 CAS:75-73-0 HR: 2
CARBON TETRAFLUORIDE
DOT: UN 1982
mf: CF_4 mw: 88.01

PROP: Colorless gas. Mp: −184°, bp: −127.7°, d: 1.96 @ −184°. Sltly sol in H_2O.

SYNS: ARCTON 0 □ CARBON FLUORIDE □ F 14 □ FC 14 □ FREON 14 □ HALOCARBON 14 □ HALON 14 □ METHANE, TETRAFLUORO- □ PERFLUOROMETHANE □ R 14 □ R14 (DOT) □ REFRIGERANT 14 □ REFRIGERANT R 14 □ R 14 (REFRIGERANT) □ TETRAFLUOROCARBON □ TETRAFLUOROMETHANE □ TETRAFLUOROMETHANE (DOT)

TOXICITY DATA WITH REFERENCE
ihl-rat LCLo:89,5000 ppm/15M MRLR** No.23,50

CONSENSUS REPORTS: Reported in EPA TSCA Inventory.

DOT CLASSIFICATION: 2.2; *Label:* Nonflammable Gas

SAFETY PROFILE: Mildly toxic by inhalation. Less chronically toxic than carbon tetrachloride. Violent reaction with Al. When heated to decomposition it emits toxic fumes of F^-. See also FLUORIDES.

CBY500 CAS:507-25-5 HR: 3
CARBON TETRAIODIDE
mf: CI_4 mw: 519.61

PROP: Octahedral, dark-red crystals. Mp: 171°, d: 4.32. Sol in C_6H_6, and $CHCl_3$.

SYNS: CARBON IODIDE □ TETRAIODOMETHANE

TOXICITY DATA WITH REFERENCE
ivn-mus LD50:178 mg/kg CSLNX* NX#02298

CONSENSUS REPORTS: Reported in EPA TSCA Inventory.

SAFETY PROFILE: Poison by intravenous route. See also IODOFORM. Explodes on contact with bromine trifluoride. Mixtures with lithium particles are impact-sensitive explosives. Vigorous reaction with BrF_2. When heated to decomposition it emits toxic fumes of I^-.

CBY750 CAS:75-46-7 HR: 2
CARBON TRIFLUORIDE
DOT: UN 1984/UN 3136
mf: CHF_3 mw: 70.02

PROP: Colorless, odorless gas. Mp: −163°, bp: −82.2°, d: 1.52 (liquid) @ −100°. Sol in water.

SYNS: ARCTON □ CARBON TRIFLUORIDE □ FLUOROFORM □ FLUORYL □ FREON 23 □ FREON F-23 □ GENETRON-23 □ HALOCARBON 23 □ METHYL TRIFLUORIDE □ R 23 □ TRIFLUOROMETHANE □ TRIFLUOROMETHANE, refrigerated, liquid (UN 3136) (DOT) □ TRIFLUOROMETHANE (UN 1984) (DOT)

TOXICITY DATA WITH REFERENCE
sln-dmg-ihl 98 pph/10M ENVRAL 7,275,74

CONSENSUS REPORTS: EPA Genetic Toxicology Program. Reported in EPA TSCA Inventory.

DOT CLASSIFICATION: 2.2; *Label:* Nonflammable Gas

SAFETY PROFILE: Narcotic in high concentration. A mild respiratory irritant. Mutation data reported. See also FLUORIDES. When heated to decomposition it emits toxic fumes of F^-.

CCA000 CAS:14435-92-8 HR: 3
CARBONYL DIAZIDE
mf: CN_6O mw: 112.05

$O:C(N_3)_2$

PROP: Needles.

SYNS: CARBONIC DIAZIDE □ CARBONYL AZIDE

SAFETY PROFILE: A very dangerous high explosive. May explode violently in ice water or on exposure to light. When heated to decomposition it emits toxic fumes of CO and NO_x. See also AZIDES and CARBONYLS.

CCA125 CAS:6470-09-3 HR: 2
CARBONYL DIISOTHIOCYANATE
mf: $C_3N_2OS_2$ mw: 144.17

$O:C(N{=}C{=}S)_2$

PROP: Oil. Bp: 27–32° @ 0.5 mm.

SAFETY PROFILE: A strong Lewis acid. It reacts explosively with dimethyl sulfoxide. When heated to decomposition it emits toxic fumes of CO, SO_x, NO_x, and CN^-. See also CARBONYLS and THIOCYANATES.

CCA500 CAS:353-50-4 HR: 3
CARBONYL FLUORIDE
DOT: UN 2417
mf: CF_2O mw: 66.01

PROP: Colorless gas; pungent; hygroscopic. Readily hydrolyzes to CO_2 and HF. Mp: −114°, bp: −83°, d: 1.139 @ −114°.

SYNS: CARBON DIFLUORIDE OXIDE □ CARBON FLUORIDE OXIDE □ CARBONIC DIFLUORIDE □ CARBON OXYFLUORIDE □ CARBONYL DIFLUORIDE □ DIFLUOROFORMALDEHYDE □ FLUOPHOSGENE □ FLUOROFORMYL FLUORIDE □ FLUOROPHOSGENE □ RCRA WASTE NUMBER U033

TOXICITY DATA WITH REFERENCE
ihl-rat LC50:360 ppm/1H AIHAAP 29,41,68

CONSENSUS REPORTS: Reported in EPA TSCA Inventory.

OSHA PEL: TWA 2 ppm; STEL 5 ppm
ACGIH TLV: TWA 2 ppm; STEL 5 ppm
DOT CLASSIFICATION: 2.3; *Label:* Poison Gas

SAFETY PROFILE: A poison. Moderately toxic by inhalation. A powerful irritant. Hydrolyzes instantly to form HF on contact with moisture. See also CARBONYLS, HYDROFLUORIC ACID, and FLUORINE. Incompatible with hexafluoroisopropylideneamino-lithium. When heated to decomposition it emits toxic fumes of CO and F^-. See CARBON MONOXIDE for fire and explosion hazard.

CCB500 CAS:12397-35-2 HR: 3
CARBONYL POTASSIUM
mf: CKO mw: 67.11

SYN: POTASSIUM CARBONYL (DOT)

DOT CLASSIFICATION: Forbidden

SAFETY PROFILE: Explodes on heating in air or contact with water. Incompatible with oxygen. When heated to decomposition it emits toxic fumes of CO and K_2O. See also CARBONYLS.

CCB609 HR: 3
CARBONYLS

PROP: The (CO) group with a metal (M). They may

exist as dimeric acetylene derivatives (MOC≡COM) or as salts of hexahydroxybenzene.

SAFETY PROFILE: Most carbonyls are highly toxic. The toxicity of carbonyls depends in part, but not always entirely, on their ready decomposition, which releases carbon monoxide. Symptoms are due in part to carbon monoxide and in part to the direct irritating action of the carbonyl. See specific carbonyl in question. Many carbonyl metals ignite spontaneously in air, some with a delay period. Others are moderate fire and explosion hazards when exposed to heat or flame. Carbonyls of alkali metals are potentially explosive. Hypergolic reaction with dinitrogen tetraoxide. They react with water or steam to produce toxic and flammable vapors; can react vigorously with oxidizing materials. When heated to decomposition they emit highly toxic fumes of carbon monoxide. See also CARBON MONOXIDE and POWDERED METALS.

CCC000 CAS:463-58-1 HR: 3
CARBONYL SULFIDE
DOT: UN 2204
mf: COS mw: 60.07

PROP: Gas or liquid. Hydrolyzed by water. Mp: −138.2°, bp: 50.2°, lel: 12%, ucl: 28.5%, d: liq 1.24 @ −87°, vap d: 2.1, d: 1.19 @ 50 mm. Very sltly sol in water, alc, and toluene.

SYNS: CARBON OXIDE SULFIDE □ CARBON OXYSULFIDE □ CARBONYL SULFIDE-^{35}S □ OXYCARBON SULFIDE

TOXICITY DATA WITH REFERENCE
ipr-rat LD50:23 mg/kg TXAPA9 55,198,80
ihl-mus LCLo:1200 ppm/35M BDCGAS 76,299,43

CONSENSUS REPORTS: Community Right-To-Know List. Reported in EPA TSCA Inventory.

DOT CLASSIFICATION: 2.3, *Label:* Poison Gas, Flammable Gas

SAFETY PROFILE: Poison by intraperitoneal route. Mildly toxic by inhalation. Narcotic in high concentration. An irritant. May liberate highly toxic hydrogen sulfide upon decomposition. A very dangerous fire hazard and moderate explosion hazard when exposed to heat or flame. Can react vigorously with oxidizing materials. To fight fire, stop flow of gas or use CO_2, dry chemical, or water spray. When heated to decomposition it emits toxic fumes of CO. See also CARBONYLS and SULFIDES.

CCC075 CAS:41575-94-4 HR: 3
CARBOPLATIN
mf: $C_6H_{12}N_2O_4Pt$ mw: 371.29

PROP: White solid. Sol in H_2O.

SYNS: CBDCA □ cis-(1,1-CYCLOBUTANEDICARBOSYLATO)DIAMMINEPLATINUM(II) □ 1,1-CYCLOBUTANEDICARBOXYLATE DIAMMINE PLATINUM(II) □ cis-(1,1-CYCLOBUTANEDICARBOXYLATO)DIAMMINEPLATINUM(II) □ DIAMMINE(1,1-CYCLOBUTANEDICARBOXYLATO)PLATINUM (II) □ cis-DIAMMINE(1,1-CYCLOBUTANEDICARBOXYLATO)PLATINUM(II) □ JM 8 □ NSC-241240

TOXICITY DATA WITH REFERENCE
mmo-esc 300 μmol/L MUREAV 173,13,86
dnd-mus:leu 200 μmol/L CNREA8 45,4043,85
mnt-ham:lng 8250 nmol/L NEOLA4 31,655,84
sce-ham:lng 8250 nmol/L NEOLA4 31,655,84
ivn-rat TDLo:24 mg/kg (female 6-9D post):REP JTSCDR 14,115,89
orl-rat LD50:343 mg/kg DRUGAY-,288,90
scu-rat LD50:72 mg/kg DRUGAY-,288,90
ivn-rat LD50:61 mg/kg JJIND8 67,201,81
ipr-mus LD50:118 mg/kg EJMCA5 27,611,92
ivn-mus LD50:89,360 μg/kg NCISP* JAN86
unr-mus LD50:180 mg/kg RRCRBU 48,12,74

SAFETY PROFILE: Poison by ingestion, intravenous, intraperitoneal, and possibly other routes. Experimental reproductive effects. Mutation data reported. When heated to decomposition it emits toxic fumes of NO_x. See also AMINES and PLATINUM COMPOUNDS.

CCC100 CAS:35700-23-3 HR: 3
CARBOPROST
mf: $C_{21}H_{36}O_5$ mw: 368.57

SYNS: 15-M3-PGF2-α □ METHYL-PGF2-α □ 15(S)-15-METHYL PGF2-α □ 15-METHYLPROSTAGLANDIN F2-α □ 15(S)-METHYLPROSTAGLANDIN F2-α □ 15(S)-15-METHYL-PROSTAGLANDIN F2-α □ (15S) 15-METHYLPROSTAGLANDIN F2-α □ PROSTIN □ (5Z,9-α,11-α,13E,15S) 9,11,15-TRIHYDROXY-15-METHYLPROSTA-5,13-DIEN-1-OIC ACID □ U 32921E

TOXICITY DATA WITH REFERENCE
ims-wmn TDLo:12 μg/kg (30D preg):REP PRGLBA 14,785,77
ivg-wmn TDLo:60 μg/kg (42D preg):REP CCPTAY 21,273,80
ims-rat TDLo:100 μg/kg (female 10D post):REP CYLPDN 8,540,87
ivg-rbt TDLo:500 μg/kg (1D preg):REP PSEBAA 151,575,76

SAFETY PROFILE: In humans very small amounts cause abortion by intramuscular, intravaginal and intraplacental routes. Experimental reproductive effects. When heated to decomposition it emits acrid smoke and fumes.

CCC110 CAS:58551-69-2 HR: 3
CARBOPROST TROMETHAMINE
mf: $C_{21}H_{36}O_5 \cdot C_4H_{11}NO_3$ mw: 489.73

SYNS: 15(2)15-METHYL PGF2-α TROMETHAMINE SALT □ 15(S)15-METHYL PROSTAGLANDIN F2-α TROMETHAMINE □ 9,11,15-TRIHYDROXY-15-METHYL-PROSTA-5,13-DIEN-1-OIC ACID, (5Z,9-α,11-α,13E,15S)-compounded with 2-AMINO-2-(HYDROXYMETHYL)-1,3-PROPANEDIOL (1:1)

TOXICITY DATA WITH REFERENCE
ims-wmn TDLo:30 μg/kg (14W preg):REP CCPTAY 11,533,75
scu-rat TDLo:6 mg/kg (male 3D pre):REP APTRDI 4,157,78
scu-rat TDLo:87 μg/kg (15-22D preg/21D post):REP APTRDI 4,157,78
scu-rat TDLo:300 μg/kg (female 9-11D post):TER APTRDI 4,157,78

ivn-rat LD50:25,100 μg/kg APTRDI 4,157,78
ivn-mus LD50:131 mg/kg APTRDI 4,157,78

SAFETY PROFILE: Poison by intravenous route. In humans, very small amounts cause abortion by intramuscular route. An experimental teratogen. Other experimental reproductive effects. When heated to decomposition it emits toxic fumes of NO_x.

CCC250 CAS:59-31-4 HR: 3
CARBOSTYRIL
mf: C_9H_7NO mw: 145.17

PROP: White crystals or powder. Mp: 199–200° (anhyd), bp: 267°. Very sltly sol in water.

SYN: 2-QUINOLINOL

TOXICITY DATA WITH REFERENCE
ipr-mus LD50:150 mg/kg NTIS** AD607-952

CONSENSUS REPORTS: Reported in EPA TSCA Inventory.

SAFETY PROFILE: Poison by intraperitoneal route. A fungicide. A central nervous system stimulant. When heated to decomposition it emits toxic fumes of NO_x.

CCC325 CAS:33330-91-5 HR: 3
1-p-(CARBOXAMIDOPHENYL)-3,3-DIMETHYLTRIAZINE
mf: $C_9H_{12}N_4O$ mw: 192.25

SYNS: 1-(4′-CARBOXYLAMIDOPHENYL)-3,3-DIMETHYLTRIAZINE □ CB 10286 □ p-(3,3-DIMETHYLTRIAZENO)BENZAMIDE □ p-(3,3-DIMETHYL-1-TRIAZENYL)BENZAMIDE □ 4-(3,3-DIMETHYL-1-TRIAZENYL)BENZAMIDE □ 1-(4′-KARBOXYLAMIDOFENYL)-3,3-DIMETHYLTRIAZENU (CZECH) □ 1-(4′-KARBOXYLAMIDOPHENYL)-3,3-DIMETHYLTRIAZEN (GERMAN)

TOXICITY DATA WITH REFERENCE
mma-sat 91 μmol/L JMCMAR 22,473,79
orl-rat LD50:54 mg/kg CKFRAY 27,384,78
ipr-mus LD50:356 mg/kg CTRRDO 62,721,78

SAFETY PROFILE: Poison by ingestion and intraperitoneal routes. Mutation data reported. When heated to decomposition it emits toxic fumes of NO_x. See also AMIDES.

CCC500 CAS:5234-68-4 HR: 3
CARBOXINE
mf: $C_{12}H_{13}NO_2S$ mw: 235.32

PROP: Solid. Sltly sol in H_2O; sol in C_6H_6, EtOH, and MeOH. Very sol in Me_2CO.

SYNS: 5-CARBOXANILIDO-2,3-DIHYDRO-6-METHYL-1,4-OXATHIIN □ CARBOXIN (USDA) □ D 735 □ DCMO □ 2,3-DIHYDRO-5-CARBOXANILIDO-6-METHYL-1,4-OXATHIIN □ 5,6-DIHYDRO-2-METHYL-3-CARBOXANILIDO-1,4-OXATHIIN (GERMAN) □ 2,3-DIHYDRO-6-METHYL-1,4-OXATHIIN-5-CARBOXANILIDE □ 5,6-DIHYDRO-2-METHYL-1,4-OXATHIIN-3-CARBOXANILIDE □ 5,6-DIHYDRO-2-METHYL-N-PHENYL-1,4-OXATHIIN-3-CARBOXAMIDE □ F 735 □ FLO PRO V SEED PROTECTANT □ VITAVAX

TOXICITY DATA WITH REFERENCE
cyt-rat-ipr 382 mg/kg/48H-C EMMUEG 12,235,88
orl-rat LD50:430 mg/kg GTPZAB 23(2),55,79
skn-rat LD50:1050 mg/kg GTPZAB 23(2),55,79
orl-mus LD50:3200 mg/kg GTPZAB 23(2),55,79
orl-ckn LD50:24 g/kg VETNAL 54(6),85,78
orl-bwd LD50:42,200 μg/kg AECTCV 12,355,83

SAFETY PROFILE: Poison by ingestion. Moderately toxic by skin contact and possibly other routes. Mutation data reported. When heated to decomposition it emits very toxic fumes of NO_x and SO_x.

CCC750 CAS:141-82-2 HR: 3
CARBOXYACETIC ACID
mf: $C_3H_4O_4$ mw: 104.07

PROP: Crystals. Mp: 135.6°. Sol in H_2O, EtOH, and Et_2O; mod sol in Py.

SYNS: DICARBOXYMETHANE □ METHANEDICARBOXYLIC ACID □ PROPANEDIOIC ACID □ USAF EK-695

TOXICITY DATA WITH REFERENCE
skn-rbt 500 mg/24H MLD BIOFX* 22-3/71
eye-rbt 100 mg SEV BIOFX* 22-3/71
mor-rat-orl 10,080 mg/kg/6W CRNGDP 9,387,88
orl-rat LD50:1310 mg/kg BIOFX* 22-3/71
ipr-rat LD50:1500 mg/kg 38MKAJ 2C,4937,82
orl-mus LD50:4000 mg/kg BIJOAK 34,1196,40
ipr-mus LD50:300 mg/kg NTIS** AD277-689

CONSENSUS REPORTS: Reported in EPA TSCA Inventory.

SAFETY PROFILE: Poison by intraperitoneal route. Moderately toxic by ingestion and intraperitoneal routes. A skin and severe eye irritant. Mutation data reported. When heated to decomposition it emits acrid smoke and irritating fumes.

CCD625 CAS:56743-33-0 HR: 3
CARBOXYBENZENESULFONYL AZIDE
mf: $C_7H_5N_3O_4S$ mw: 227.19

$HOCO{\cdot}C_6H_4SO_2N_3$

SAFETY PROFILE: Decomposes explosively at 120°C. When heated to decomposition it emits toxic fumes of SO_x and NO_x. See also AZIDES.

CCD750 CAS:69365-73-7 HR: 3
N-(2-CARBOXYCAPROYL)HYDRAZOBENZENE CALCIUM SALT HEMIHYDRATE
mf: $C_{38}H_{42}N_4O_6{\cdot}Ca{\cdot}1/2H_2O$ mw: 699.93

SYNS: BUMADIZON CALCIUM SALT HEMIHYDRATE □ BUTYLMALONIC ACID MONO(1,2-DIPHENYLHYDRAZIDE) CALCIUM SALT HEMIHYDRATE □ BUTYL-MALONSAEURE-MONO-(1,2-DIPHENYL-HYDRAZID)-CALCIUM-SEMIHYDRAT (German) □ BUTYLPROPANEDIOIC ACID MONO(1,2-DIPHENYLHYDRAZIDE) CALCIUM SALT HEMIHYDRATE □ α-CARBOXYCAPROYL-N,N′-DIPHENYLHYDRAZINE CALCIUM SALT HEMIHYDRATE □ EUMOTOL □ RHEUMATOL

TOXICITY DATA WITH REFERENCE
orl-rat LD50:1250 mg/kg ARZNAD 23,1215,73
ivn-mus LD50:263 mg/kg ARZNAD 23,1215,73
orl-mus LD50:2500 mg/kg ARZNAD 23,1215,73
ivn-mus LD50:258 mg/kg ARZNAD 23,1215,73

SAFETY PROFILE: Poison by intravenous route. Moderately toxic by ingestion. An analgesic, antipyretic, and antirheumatic. When heated to decomposition it emits toxic fumes of NO_x.

CCE500 CAS:493-52-7 HR: 2
2-CARBOXY-4′-(DIMETHYLAMINO)AZOBENZENE
mf: $C_{15}H_{15}N_3O_2$ mw: 269.33

PROP: Shiny violet crystals.

SYNS: C.I. 13020 □ C.I. ACID RED 2 □ p-(DIMETHYLAMINO)AZOBENZENE-o-CARBOXYLIC ACID □ 4′-DIMETHYLAMINOAZOBENZENE-2-CARBOXYLIC ACID □ o-((p-(DIMETHYLAMINO)PHENYL)AZO)BENZOIC ACID □ 2-((4-DIMETHYLAMINO)PHENYLAZO)BENZOIC ACID □ METHYL RED

TOXICITY DATA WITH REFERENCE
mma-sat 50 µg/plate MUREAV 56,249,78
dnr-bcs 2 mg/disc TRENAF 27,153,76
dns-rat:lvr 10 µmol/L CNREA8 46,1654,86
orl-rat TDLo:12 g/kg/57W-C:ETA BJCAAI 9,310,55

CONSENSUS REPORTS: IARC Cancer Review: Group 3 IMEMDT 7,56,87; Animal Inadequate Evidence IMEMDT 8,161,75. Reported in EPA TSCA Inventory. EPA Genetic Toxicology Program.

SAFETY PROFILE: Questionable carcinogen with experimental tumorigenic data. Mutation data reported. When heated to decomposition it emits toxic fumes of NO_x.

CCE750 CAS:20691-84-3 HR: 2
3′-CARBOXY-4-DIMETHYLAMINOAZOBENZENE
mf: $C_{15}H_{15}N_3O_2$ mw: 269.33

SYN: 3-((p-(DIMETHYLAMINO)PHENYL)AZO)BENZOIC ACID

TOXICITY DATA WITH REFERENCE
mma-sat 1 µmol/plate CRNGDP 1,121,80
dns-rat:lvr 10 µmol/L CNREA8 46,1654,86
orl-rat TDLo:12 g/kg/57W-C:ETA BJCAAI 9,310,55
orl-rat LD50:3757 mg/kg NEOLA4 27,237,80

SAFETY PROFILE: Moderately toxic by ingestion. Questionable carcinogen with experimental tumorigenic data. Mutation data reported. When heated to decomposition it emits toxic fumes of NO_x.

CCF125 CAS:12758-40-6 HR: 2
CARBOXYETHYLGERMANIUM SESQUIOXIDE
mf: $C_6H_{10}Ge_2O_7$ mw: 339.34

SYNS: BIS-β-CARBOXYETHYLGERMANIUM SESQUIOXIDE □ 2-CARBOXYETHYLGERMASESQUIOXANE □ 3,3′-(DIOXODIGERMOXANYLENE) DIPROPANOIC ACID □ DIPROPANOIC ACID GERMANIUM SESQUIOXIDE □ Ge 132 □ GERMANATE(2-), BIS(2-CARBOXYLATOETHYL)TRIOXODI-, DIHYDROGEN (9CI) □ 3,3′-(GERMANOIC ANHYDRIDE) DIPROPANOIC ACID

TOXICITY DATA WITH REFERENCE
ipr-rat TDLo:94 g/kg (male 90D pre):REP OYYAA2 20,271,80
orl-rat LD50:9500 mg/kg SIGZAL 46,227,86
ipr-rat LD50:3200 mg/kg SIGZAL 46,227,86
scu-rat LD50:16,300 mg/kg SIGZAL 46,227,86
ivn-rat LD50:3200 mg/kg DRFUD4 5,545,80
orl-mus LD50:11,400 mg/kg SIGZAL 46,227,86
ivn-mus LD50:2110 mg/kg DRFUD4 5,548,80

CONSENSUS REPORTS: Reported in EPA TSCA Inventory.

SAFETY PROFILE: Moderately toxic by intravenous route. Experimental reproductive effects. See also GERMANIUM COMPOUNDS.

CCF250 CAS:4033-46-9 HR: 2
3-((2-CARBOXYETHYL)THIO)ALANINE
mf: $C_6H_{11}NO_4S$ mw: 193.24

PROP: A solid. Mp: 218°.

SYN: S-2-CARBOXYETHYL-l-CYSTEINE

TOXICITY DATA WITH REFERENCE
scu-rat TDLo:520 mg/kg/52W-I:ETA BJCAAI 15,85,61

SAFETY PROFILE: Questionable carcinogen with experimental tumorigenic data. When heated to decomposition it emits very toxic fumes of SO_x and NO_x. See also AMINES.

CCF500 CAS:63907-33-5 HR: 3
(3-(4-(CARBOXYLATOMETHOXY)PHENYL)-2-HYDROXYPROPYL)HYDROXY-MERCURATE(1-), SODIUM
mf: $C_{11}H_{13}HgO_5$•Na mw: 448.82

SYN: (p-(2-HYDROXY-3-HYDROXYMERCURI)PROPYL)PHENOXY)) ACETIC ACID, SODIUM SALT

TOXICITY DATA WITH REFERENCE
ivn-rbt LDLo:7 mg/kg JPETAB 41,21,31

CONSENSUS REPORTS: Mercury and its compounds are on the Community Right-To-Know List.

OSHA PEL: CL 0.1 mg(Hg)/m³ (skin)
ACGIH TLV: TWA 0.1 mg(Hg)/m³ (skin)
NIOSH REL: (Mercury, Aryl and Inorganic): CL 0.1 mg/m³ (skin)

SAFETY PROFILE: Poison by intravenous route. See also MERCURY COMPOUNDS. When heated to decomposition it emits toxic fumes of Hg and Na_2O.

CCF750 CAS:13442-14-3 HR: 2
6-CARBOXYL-4-HYDROXYLAMINOQUINOLINE-1-OXIDE
mf: $C_{10}H_8N_2O_4$ mw: 220.20

SYN: 4-(HYDROXYAMINO)-6-QUINOLINECARBOXYLIC ACID-1-OXIDE

TOXICITY DATA WITH REFERENCE
scu-mus TDLo:120 mg/kg/50D-I:ETA BCPCA6 16,631,67

SAFETY PROFILE: Questionable carcinogen with experimental tumorigenic data. When heated to decomposition it emits toxic fumes of NO_x.

CCG000 CAS:1425-67-8 HR: 2
6-CARBOXYL-4-NITROQUINOLINE-1-OXIDE
mf: $C_{10}H_6N_2O_5$ mw: 234.18

SYNS: 6-CARBOXY-4-NITROQUINOLINE-1-OXIDE □ 4-NITROQUINOLINE-6-CARBOXYLIC ACID-1-OXIDE □ 4-NITRO-6-QUINOLINE-CARBOXYLIC ACID-1-OXIDE

TOXICITY DATA WITH REFERENCE
mmo-esc 500 µg/plate CNREA8 32,2369,72
mrc-esc 500 µg/well CNREA8 32,2369,72
mmo-smc 100 mg/L IGSBAL 85,127,72
dnd-mus:fbr 100 µmol/L CNREA8 35,521,75
dns-ham:oth 4 µmol/L NATUAS 229,416,71
dnd-mam:lym 5 mg BIPMAA 4,409,66
scu-rat TDLo:90 mg/kg/20W-I:ETA GANNA2 58,397,67
scu-mus TDLo:120 mg/kg/50D-I:ETA BCPCA6 16,631,67

CONSENSUS REPORTS: EPA Genetic Toxicology Program.

SAFETY PROFILE: Questionable carcinogen with experimental tumorigenic data. Mutation data reported. When heated to decomposition it emits toxic fumes of NO_x.

CCG500 CAS:36568-91-9 HR: 3
(4-(CARBOXY METHOXY)-3-CHLOROPHENYL)(5,5-DIETHYL-2,4,6(1H,3H,5H)-PYRIMIDINETRIONATO)-O²-MERCURY, MONOSODIUM SALT
mf: $C_{16}H_{18}ClHgN_2O_6 \cdot Na$ mw: 593.39

SYNS: MERBAPHEN □ NOVASUROL

TOXICITY DATA WITH REFERENCE
ivn-rbt LDLo:20 mg/kg JPETAB 41,21,31

CONSENSUS REPORTS: Mercury and its compounds are on the Community Right-To-Know List.

OSHA PEL: CL 0.1 mg(Hg)/m³ (skin)
ACGIH TLV: TWA 0.1 mg(Hg)/m³ (skin)
NIOSH REL: (Mercury, Aryl and Inorganic): CL 0.1 mg/m³ (skin)

SAFETY PROFILE: Poison by intravenous route. See also MERCURY COMPOUNDS. When heated to decomposition it emits very toxic fumes of Cl^-, NO_x, and Hg.

CCH000 CAS:9086-60-6 HR: 2
CARBOXYMETHYLCELLULOSE NORDIC

SYNS: AMMONIUM CARBOXYMETHYL CELLULOSE □ CARBOXYMETHYL CELLULOSE, AMMONIUM SALT

TOXICITY DATA WITH REFERENCE
scu-rat TDLo:6600 mg/kg/73W-I:NEO RCBIAS 20,701,61

CONSENSUS REPORTS: Reported in EPA TSCA Inventory.

SAFETY PROFILE: Questionable carcinogen with experimental neoplastigenic data. When heated to decomposition it emits toxic fumes of NO_x and NH_3. See also CARBOXYMETHYLCELLULOSE.

CCH125 CAS:2387-59-9 HR: 2
S-CARBOXYMETHYLCYSTEINE
mf: $C_5H_9NO_4S$ mw: 179.21

PROP: A solid. Mp: 204–207°.

SYNS: AHR-3053 □ 3-((CARBOXYMETHYL)THIO)ALANINE □ LJ 206 □ S-CMC

TOXICITY DATA WITH REFERENCE
ipr-rat LD50:7800 mg/kg OYYAA2 14,567,77
scu-rat LD50:10,300 mg/kg OYYAA2 14,567,77
ipr-mus LD50:2980 mg/kg IYKEDH 12,668,81
scu-mus LD50:9000 mg/kg IYKEDH 12,668,81

SAFETY PROFILE: Moderately toxic by intraperitoneal route. Mildly toxic by subcutaneous route. When heated to decomposition it emits toxic fumes of SO_x and NO_x. See also AMINES.

CCH199 HR: 2
2-CARBOXYMETHYLISOTHIOURONIUM CHLORIDE
mf: $C_3H_7ClN_2O_2S$ mw: 170.61

$HOCO \cdot CH_2SC(:N^+H_2)NH_2Cl^-$

SYN: CARBOXYMETHYL CARBAMIMONIOTHIOATE CHLORIDE

SAFETY PROFILE: Reaction with chlorine may form the dangerously explosive nitrogen trichloride. When heated to decomposition it emits toxic fumes of Cl^-, SO_x, and NO_x. See also CHLORIDES.

CCH250 CAS:63938-93-2 HR: 3
2-(CARBOXYMETHYLMERCAPTO)PHENYLSTIBONIC ACID
mf: $C_8H_9O_5SSb$ mw: 338.98

SYNS: 2-(CARBOXYMETHYLMERCAPTO)PHENYL-STIBONSAEURE (GERMAN) □ RO 2-1160 □ ((2-STIBONOPHENYL)THIO)ACETIC ACID

TOXICITY DATA WITH REFERENCE
orl-rat LD50:5000 mg/kg ARZNAD 4,116,54
orl-mus LD50:5000 mg/kg AIPTAK 85,100,51
scu-mus LD50:2520 mg/kg AIPTAK 85,100,51
ivn-mus LD50:965 mg/kg AIPTAK 85,100,51
ivn-rbt LD50:186 mg/kg AIPTAK 85,100,51
ipr-gpg LD50:350 mg/kg AIPTAK 85,100,51
ipr-ham LD50:550 mg/kg AIPTAK 85,100,51

CONSENSUS REPORTS: Antimony and its compounds are on the Community Right-To-Know List.

OSHA PEL: TWA 0.5 mg(Sb)/m³
ACGIH TLV: TWA 0.5 mg(Sb)/m³
NIOSH REL: (Antimony) TWA 0.5 mg(Sb)/m³

SAFETY PROFILE: Poison by intravenous and intraperitoneal routes. Moderately toxic by subcutaneous route. Mildly toxic by ingestion. When heated to decomposition it emits very toxic fumes of antimony and SO_x. See also ANTIMONY COMPOUNDS and MERCAPTANS.

CCH500 **HR: 3**
CARBOXYMETHYLNITROSOUREA
mf: $C_3H_5N_3O_4$ mw: 147.11

TOXICITY DATA WITH REFERENCE
orl-rat TDLo:4 g/kg/74W-I:NEO JJIND8 62,1523,79
ipr-rat LD50:210 mg/kg JJIND8 62,1523,79

SAFETY PROFILE: Poison by intraperitoneal route. Questionable carcinogen with experimental neoplastigenic data. When heated to decomposition it emits toxic fumes of NO_x. See also N-NITROSO COMPOUNDS.

CCH800 **HR: 2**
5-CARBOXYMETHYL-3-p-TOLYL-THIAZOLIDINE-2,4-DIONE-2-ACETOPHENONE HYDRAZONE
mf: $C_{20}H_{16}N_3O_4S$ mw: 394.45

SYN: 5-KARBOKSIMETIL-3-p-TOLIL-TIAZOLIDIN-2,4-DION-2-ACETOFENONHIDRAZON (CZECH)

TOXICITY DATA WITH REFERENCE
orl-rat LD50:600 mg/kg ZDVEA7 39(Suppl 1),20,70
ipr-rat LD50:2160 mg/kg ZDVEA7 39(Suppl 1),20,70
orl-mus LD50:820 mg/kg ZDVEA7 39(Suppl 1),20,70
ipr-mus LD50:1870 mg/kg ZDVEA7 39(Suppl 1),20,70

SAFETY PROFILE: Moderately toxic by ingestion and intraperitoneal routes. When heated to decomposition it emits toxic fumes of SO_x and NO_x.

CCH850 **CAS:590-46-5** **HR: 1**
(CARBOXYMETHYL)TRIMETHYLAMMONIUM CHLORIDE
mf: $C_5H_{12}NO_2{\cdot}Cl$ mw: 153.63

SYNS: AMMONIUM, (CARBOXYMETHYL)TRIMETHYL-, CHLORIDE □ GLYKOKOLLBETAIN-CHLORID □ METHANAMINIUM, 1-CARBOXY-N,N,N-TRIMETHYL-, CHLORIDE

TOXICITY DATA WITH REFERENCE
scu-mus LD50:8 g/kg ABMGAJ 3,28,59

CONSENSUS REPORTS: Reported in EPA TSCA Inventory.

SAFETY PROFILE: Slightly toxic by subcutaneous route. When heated to decomposition it emits toxic vapors of NO_x and Cl^-.

CCI250 **CAS:62-23-7** **HR: 2**
1-CARBOXY-4-NITROBENZENE
mf: $C_7H_5NO_4$ mw: 167.13

PROP: Crystals or leaflets from water. Mp: 241.5°, bp: sublimes, d: 1.550 @ 32°/4°.

SYNS: KYSELINA-p-NITROBENZOOVA (CZECH) □ p-NITROBENZOIC ACID □ 4-NITROBENZOIC ACID □ 4-NITRODRACYLIC ACID

TOXICITY DATA WITH REFERENCE
eye-rbt 20 mg/24H MOD 28ZPAK -,129,72
mmo-sat 100 µg/plate MUREAV 137,71,84
mma-sat 10 µmol/plate MUREAV 58,11,78
bfa-rat/sat 400 mg/kg/4D PNASA6 72,4607,75
orl-rat LD50:1960 mg/kg CRSBAW 160,1097,66
ipr-rat LD50:1210 mg/kg CRSBAW 160,1097,66
par-rat LD50:1960 mg/kg CRSBAW 160,1097,66
ipr-mus LD50:880 mg/kg CRSBAW 160,1097,66
ivn-mus LD50:770 mg/kg CRSBAW 160,1097,66
par-mus LD50:1470 mg/kg CRSBAW 160,1097,66

CONSENSUS REPORTS: Reported in EPA TSCA Inventory. EPA Genetic Toxicology Program.

SAFETY PROFILE: Moderately toxic by ingestion, intravenous, parenteral, and intraperitoneal routes. An eye irritant. Mutation data reported. When heated to decomposition it emits toxic fumes of NO_x. See also NITRO COMPOUNDS of AROMATIC HYDROCARBONS.

CCI500 **CAS:41956-77-8** **HR: 2**
2-(5-CARBOXYPENTYL)-4-THIAZOLIDONE
mf: $C_9H_{15}NO_3S$ mw: 217.31

SYNS: ACIDOMYCIN □ ACTITHIAZIC ACID □ CINNAMONIN □ 1-MYCOBACIDIN □ 1-4-OXO-2-THIAZOLIDINEHEXANOIC ACID □ 4-THIAZOLIDONE-2-CAPROIC ACID □ epsilon-(2-(4-THIAZOLIDONE)) HEXANOIC ACID

TOXICITY DATA WITH REFERENCE
scu-mus LD50:20 g/kg PHBUA9 1,84,53
ivn-mus LD50:3500 mg/kg PHBUA9 1,84,53

SAFETY PROFILE: Moderately toxic by intravenous route. Mildly toxic by subcutaneous route. When heated to decomposition it emits very toxic fumes of NO_x and SO_x.

CCI550 **CAS:1197-16-6** **HR: 3**
p-CARBOXY PHENYLARSENOXIDE
mf: $C_7H_5AsO_3$ mw: 212.04

SYNS: ARSINE, OXO(4-CARBOXY)PHENYL- □ BENZOIC ACID, 4-ARSENOSO-

TOXICITY DATA WITH REFERENCE
ivn-rbt LD50:2800 µg/kg JPETAB 80,93,44

OSHA PEL: TWA 0.5 mg(As)/m³

SAFETY PROFILE: Poison by intravenous route. When heated to decomposition it emits toxic fumes of As.

CCJ000 **CAS:64050-44-8** **HR: 2**
4′-CARBOXYPHENYLMETHANESULFONANILIDE, SODIUM SALT
mf: $C_{14}H_{12}NO_4S{\cdot}Na$ mw: 313.32

TOXICITY DATA WITH REFERENCE
orl-mus LD50:2450 mg/kg JPETAB 91,263,47
scu-mus LD50:1650 mg/kg JPETAB 91,263,47
ivn-mus LD50:1300 mg/kg JPETAB 91,263,47
ivn-dog LD50:1693 mg/kg JPETAB 91,263,47
ivn-rbt LD50:1419 mg/kg JPETAB 91,263,47

SAFETY PROFILE: Moderately toxic by ingestion, subcutaneous, and intravenous routes. When heated to decomposition it emits very toxic fumes of SO_x, Na_2O, and NO_x.

CCJ350 CAS:65296-81-3 HR: 3
4-CARBOXYPHTHALATO(1,2-DIAMINOCYCLOHEXANE)PLATINUM(II)
mf: $C_{15}H_{18}N_2O_6Pt$ mw: 517.44

PROP: Readily sol in 1% $NaHCO_3$ soln.

SYNS: (CYCLOHEXANE-1,2-DIAMMINE)(4-CARBOXYPHTHLATO) PLATINUM(II) □ NSC 271674

TOXICITY DATA WITH REFERENCE
ivn-rat LD50:84 mg/kg JJIND8 67,201,81
ipr-mus LD50:46,100 μg/kg NCISP* JAN86
ivn-mus LD50:40,130 μg/kg NCISP* JAN86

SAFETY PROFILE: Poison by intravenous and intraperitoneal routes. When heated to decomposition it emits toxic fumes of NO_x. See also PLATINUM COMPOUNDS.

CCJ375 HR: 2
3-CARBOXYPROPYL(2-PROPENYL)NITROSAMINE
mf: $C_7H_{12}N_2O_3$ mw: 172.21

SYN: 4-(ALLYLNITROSOAMINO)BUTRIC ACID

TOXICITY DATA WITH REFERENCE
scu-ham TDLo:23,100 mg/kg/77W-I:CAR CDPRD4 4,79,81

SAFETY PROFILE: Questionable carcinogen with experimental carcinogenic data. When heated to decomposition it emits toxic fumes of NO_x.

CCJ400 HR: 2
CARBOXY VINYL POLYMER

PROP: A finely divided white powder disperses in water to yield a low viscosity acid solution. When neutralized, the solution is changed into a clear, stable gel (AIPTAK 114,258,58).

SYN: CP

TOXICITY DATA WITH REFERENCE
orl-rat LD50:4000 mg/kg AIPTAK 114,258,58
orl-mus LD50:4300 mg/kg AIPTAK 114,258,58
orl-gpg LD50:2000 mg/kg AIPTAK 114,258,58

SAFETY PROFILE: Moderately toxic by ingestion. When heated to decomposition it emits acrid smoke and fumes. See also POLYMERS, SOLUBLE.

CCJ500 CAS:19477-24-8 HR: 2
CARCINOLIPIN
mf: $C_{44}H_{78}O_2$ mw: 639.22

PROP: Crystals. Mp: 75°.

SYNS: CHOLESTERYL-14-METHYLHEXADECANOATE □ 3-β-14-METHYLHEXADECANOATE-CHOLEST-5-EN-3-OL

TOXICITY DATA WITH REFERENCE
scu-mus TDLo:720 mg/kg/(14-21D preg):ETA,TER NEOLA4 20,347,73

SAFETY PROFILE: Questionable carcinogen with experimental tumorigenic data. An experimental teratogen. When heated to decomposition it emits acrid smoke and irritating fumes.

CCJ625 CAS:8000-66-6 HR: 1
CARDAMON OIL

PROP: From the seed of *Elettaria cardamomun* (L.) Maton (Fam. *Zingiberazeae*). Colorless liquid; aromatic penetrating odor of cardamom, pungent taste. Misc with alc.

SYNS: CARDAMON □ OIL of CARDAMON

TOXICITY DATA WITH REFERENCE
mmo-sat 2500 ng/plate KEKHB8 (9),11,79
mmo-esc 2500 ng/plate KEKHB8 (9),11,79
dnr-bcs 19 mg/disc SKEZAP 25,378,84
orl-rat LD50:5 g/kg FCTXAV 12,837,74

CONSENSUS REPORTS: Reported in EPA TSCA Inventory.

SAFETY PROFILE: Mildly toxic by ingestion. Mutation data reported. When heated to decomposition it emits acrid smoke and fumes.

CCJ825 HR: 3
CARDINAL FLOWER

PROP: Annual weeds with distinctive flowers. They have 2 small petals opposed by 3 large petals and may be blue, pink, white, red, or yellow. The various species grow wild across the United States. Indian tobacco is cultivated as a drug plant. Cardinalis is grown as an ornamental.

SYNS: ASTHMA WEED □ BLADDERPOD LOBELIA □ BLUE CARDINAL FLOWER □ CARDENAL de MACETA (MEXICO) □ EMETIC WEED □ EYE BRIGHT □ GAG ROOT □ GREAT BLUE LOBELIA □ HIGH BELIA □ HOG PHYSIC □ INDIAN PINK □ INDIAN TOBACCO □ KINNIKINNIK □ LOBELIA INFLATA □ LOBELIA SIPHILITICA □ LOBELLOA CARDINALIS □ LOUISIANA LOBELIA □ LOW BELIA □ PUKE WEED □ RED LOBELIA □ SCARLET LOBELIA □ WILD TOBACCO

SAFETY PROFILE: The whole plant contains the poisonous lobeline and related alkaloids. Poisonings are most common when the plant is used in home medicine. The leaves are sold for use in tea and tobacco as a psychoactive ingredient. Ingestion of the leaves may cause nausea, vomiting, sensory disturbances, dizziness, and convulsions. See also LOBELINE.

CCK000 CAS:3599-32-4 HR: 3
CARDIO-GREEN
mf: $C_{43}H_{48}N_2O_6S_2$•Na mw: 776.04

PROP: Green powder. Mp: 243–245° (decomp).

SYNS: ICG □ INDOCYANINE GREEN □ IR 125 □ UJOVIRIDIN □ WOFAVERDIN

TOXICITY DATA WITH REFERENCE
ivn-mus LD50:60 mg/kg TXAPA9 44,225,78

CONSENSUS REPORTS: Reported in EPA TSCA Inventory.

SAFETY PROFILE: Poison by intravenous route. When heated to decomposition it emits very toxic fumes of SO_x, Na_2O, and NO_x.

CCK125 **CAS:87-33-2** **HR: 2**
CARDIS
DOT: UN 2907
mf: $C_6H_8N_2O_8$ mw: 236.16

PROP: Hard, colorless crystals. Mp: 71°. Sparingly sol in water. Freely sol in organic solvents, such as acetone, alc, and ether.

SYNS: ASTRIDINE □ CARDIO □ CARVANIL □ CARVASIN □ CEDOCARD □ CLAODICAL □ COROSORBIDE □ COROVLISS □ 1,4:3,6-DIANHYDROSORBITOL-2,5-DINITRATE □ DINITROSORBIDE □ DISORLON □ DURANITRAT □ EURECOR □ FLINDIX □ GLENTONIN-RETARD □ HARRICAL □ IBD □ ISDIN □ ISO-BID □ ISOKET □ ISOMACK □ ISO-PUREN □ ISORBID □ ISORDIL □ ISORDIL TEMBIDS □ ISOSORBIDE DINITRATE □ ISOSTENASE □ ISOTRATE □ KORODIL □ LANGORAN □ LASERDIL □ MAYCOR □ MONOCLAIR □ MYOREXON □ NITROSORBID □ NITROSORBIDE □ NITROSORBON □ NOSIM □ RESOIDAN □ RIFLOC RETARD □ RIGEDAL □ SORBANGIL □ SORBID □ SORBIDE NITRATE □ SORBIDILAT □ SORBIDINITRATE □ SORBISLO □ SORBITRATE □ SORBONIT □ SORQUAD □ SORQUAT □ VASCARDIN □ VASORBATE □ VASOTRATE

TOXICITY DATA WITH REFERENCE
cyt-mus:mmr 1 mmol/L/48H-C JTSCDR 5,141,80
ivn-rat TDLo:260 mg/kg (6-22D preg/20D post):REP KSRNAM 19,5021,85
orl-rat LD50:747 mg/kg YAKUD5 26,309,84
ipr-rat LD50:620 mg/kg NIIRDN 6,72,82
scu-rat LD50:1237 mg/kg YACHDS 10,2109,82
orl-mus LD50:1050 mg/kg NIIRDN 6,72,82
ipr-mus LD50:960 mg/kg NIIRDN 6,72,82
scu-mus LD50:1050 mg/kg NIIRDN 6,72,82
ims-mus LD50:1080 mg/kg NIIRDN 6,72,82

CONSENSUS REPORTS: Reported in EPA TSCA Inventory.

DOT CLASSIFICATION: 4.1; *Label:* Flammable Solid

SAFETY PROFILE: Moderately toxic by ingestion, intraperitoneal, intramuscular, and subcutaneous routes. Experimental reproductive effects. Mutation data reported. A flammable solid. When heated to decomposition it emits toxic fumes of NO_x. A coronary vasodilator. See also NITRATES.

CCK250 **CAS:959-24-0** **HR: 3**
β-CARDONE
mf: $C_{12}H_{20}N_2O_3S{\cdot}ClH$ mw: 308.86

PROP: A solid. Mp: 206.5–207° (decomp). Sol in H_2O; sltly sol in $CHCl_3$.

SYNS: 4'-(1-HYDROXY-2-(ISOPROPYLAMINO)ETHYL)METHANESULFOANILIDE HYDROCHLORIDE □ 4'-(1-HYDROXY-2-ISOPROPYLAMINO)ETHYL)METHANESULFONANILIDE MONOHYDROCHLORIDE □ 4-(2-ISOPROPYLAMINE-1-HYDROXYETHYL)METHANESULFOANILIDE HYDROCHLORIDE □ 4-(2-ISOPROPYLAMINO-1-HYDROXYAETHYL) METHANESULFONALID HYDROCHLORID (GERMAN) □ ISOPROPYLAMINOHYDROXYETHYLMETHANESULFONALIDE HYDROCHLORIDE □ N-ISOPROPYL-β-(4-METHANESULFONAMIDOPHENYL) ETHANOLAMINE HYDROCHLORIDE □ MEAD JOHNSON 1999 □ MJ 1999 □ MJ 1999 HYDROCHLORIDE □ SOTACOR □ SOTALEX □ SOTALOL □ SOTALOL HYDROCHLORIDE

TOXICITY DATA WITH REFERENCE
orl-man LDLo:45,714 μg/kg ARTODN 43,221,80
orl-rat LD50:3450 mg/kg JPETAB 149,161,65
ipr-rat LD50:680 mg/kg JPETAB 149,161,65
orl-mus LD50:2600 mg/kg JPETAB 149,161,65
ipr-mus LD50:670 mg/kg JPETAB 149,161,65
ivn-mus LD50:166 mg/kg ARZNAD 27,1022,77
ipr-dog LD50:330 mg/kg JPETAB 149,161,65
orl-rbt LD50:1000 mg/kg JPETAB 149,161,65

SAFETY PROFILE: A human poison by ingestion. Poison experimentally by intravenous and intraperitoneal routes. Moderately toxic by ingestion. Human systemic effects by ingestion: excitement, dyspnea, and convulsions. When heated to decomposition it emits very toxic fumes of HCl, SO_x, and NO_x.

CCK500 **CAS:13466-78-9** **HR: 1**
3-CARENE
mf: $C_{10}H_{16}$ mw: 136.26

PROP: Colorless, mobile liquid; found in many volatile oils (such as Swedish and Finnish turpentine oils, galanga root oil and in German pine needle oils such as those from *Pinus pumilio* and *Pinus sylvestris*) and isolated from turpentine fractions (FCTXAV 11,1011,73).

SYNS: Δ^3-CARENE □ S 3 CARENE □ ISODIPRENE □ 3,7,7-TRIMETHYLBICYCLO(4.1.0)-3-HEPTENE □ 3,7,7-TRIMETHYL-3-NORCARENE □ 4,7,7-TRIMETHYL-3-NORCARENE

TOXICITY DATA WITH REFERENCE
skn-rbt 500 mg/24H FCTXAV 11,1053,73
orl-rat LD50:4800 mg/kg FCTXAV 11,1053,73

CONSENSUS REPORTS: Reported in EPA TSCA Inventory.

SAFETY PROFILE: Mildly toxic by ingestion. A skin irritant. When heated to decomposition it emits acrid smoke and fumes.

CCK625 **CAS:50935-04-1** **HR: 3**
CARMINOMYCIN I
mf: $C_{26}H_{27}NO_{10}$ mw: 513.54

SYNS: CARMINOMICIN I □ CARUBICIN □ NSC-180024

TOXICITY DATA WITH REFERENCE
dni-mus:leu 390 nmol/L JANTAJ 34,1596,81
oms-mus:leu 490 nmol/L JANTAJ 34,1596,81
orl-mus LD50:7300 μg/kg ANTBAL 19,57,74
ipr-mus LD50:1100 μg/kg ANTBAL 29,666,84
scu-mus LD50:3800 μg/kg ANTBAL 19,57,74
ivn-mus LD50:3700 μg/kg ANTBAL 19,57,74

SAFETY PROFILE: Deadly poison by ingestion, subcutaneous, intravenous, and intraperitoneal routes. Mutation data reported. When heated to decomposition it emits toxic fumes of NO_x.

CCK630 **CAS:61422-45-5** **HR: 3**
CARMOFUR
mf: $C_{11}H_{16}FN_3O_3$ mw: 257.30

PROP: White crystals from ethanol. Mp: 110–111°.

SYNS: 2,4-DIOXO-5-FLUORO-N-HEXYL-3,4-DIHYDRO-1(2H)-PYRI-

MIDINECARBOXAMIDME □ 2,4-DIOXO-5-FLUORO-N-HEXYL-1,2,3,4-TETRAHYDRO-1-PYRIMIDINECARBOXAMIDE □ 5-FLUORO-1-HEXYL-CARBAMOYL-URACIL □ HCFU □ 1-HEXYLCARBAMOYL-5-FLUOROURACIL □ MIFUROL □ 1,2,3,4-TETRAHYDRO-2,4-DIOXO-5-FLUORO-N-HEXYL-1-PYRIMIDINECARBOXAMIDE □ YAMAFUL

TOXICITY DATA WITH REFERENCE
dnr-bcs 40 μg/plate TAKHAA 44,96,85
cyt-mus-orl 400 mg/kg OYYAA2 19,363,80
orl-rat TDLo:550 mg/kg (female 7-17D post):REP KSRNAM 14,1373,80
orl-rat TDLo:275 mg/kg (7-17D preg):TER KSRNAM 14,1373,80
orl-rat TDLo:110 mg/kg (7-17D preg):REP KSRNAM 14,1373,80
orl-rat TDLo:3 g/kg (male 4W pre):REP OYYAA2 17,575,79
orl-wmn TDLo:1152 mg/kg/14W-I:CNS,PSY JNRYA9 234,365,87
orl-man TDLo:1749 mg/kg/29W-I JNRYA9 234,365,87
orl-rat LD50:268 mg/kg NIIRDN 6,191,82
ipr-rat LD50:93 mg/kg NIIRDN 6,191,82
scu-rat LD50:260 mg/kg NIIRDN 6,191,82
orl-mus LD50:1260 mg/kg NIIRDN 6,191,82
ipr-mus LD50:96 mg/kg NIIRDN 6,191,82
scu-mus LD50:532 mg/kg NIIRDN 6,191,82
orl-dog LD50:65 mg/kg NIIRDN 6,191,82
orl-rbt LD50:55 mg/kg NIIRDN 6,191,82

SAFETY PROFILE: Poison by ingestion, subcutaneous, and intraperitoneal routes. Human systemic effects by ingestion: encephalitis, hallucinations, distorted perceptions, ataxia. Experimental reproductive effects. An experimental teratogen. Mutation data reported. When heated to decomposition it emits toxic fumes of $F^-$and NO_x.

CCK650 CAS:56-99-5 HR: 2
CARNITINE CHLORIDE
mf: $C_7H_{16}NO_3•Cl$ mw: 197.69

SYNS: (3-CARBOXY-2-HYDROXYPROPYL)TRIMETHYLAMMONIUM CHLORIDE □ 3-CARBOXY-2-HYDROXY-N,N,N-TRIMETHYL-1-PROPANAMINIUM CHLORIDE (9CI)

TOXICITY DATA WITH REFERENCE
orl-mus LD50:6690 mg/kg NIIRDN 6,135,82
scu-mus LD50:4030 mg/kg NIIRDN 6,135,82
ivn-mus LD50:1150 mg/kg NIIRDN 6,135,82

SAFETY PROFILE: Moderately toxic by intravenous route. Mildly toxic by ingestion and subcutaneous routes. When heated to decomposition it emits toxic fumes of Cl^-, NH_3, and NO_x. See also CHLORIDES.

CCK655 CAS:461-05-2 HR: 1
dl-CARNITINE CHLORIDE
mf: $C_7H_{15}NO_3•Cl$ mw: 196.68

SYNS: AMMONIUM, (3-CARBOXY-2-HYDROXYPROPYL)TRIMETHYL-, CHLORIDE, (±)- □ BICARNESINE □ (±)-(3-CARBOXY-2-HYDROXYPROPYL)TRIMETHYLAMMONIUM CHLORIDE □ (±)-CARNITINE CHLORIDE □ (±)-CARNITINE HYDROCHLORIDE □ d,l-CARNITINE HYDROCHLORIDE □ dl-CARNITINE HYDROCHLORIDE □ 1-PROPANAMINIUM, 3-CARBOXY-2-HYDROXY-N,N,N-TRIMETHYL-, CHLORIDE, (±)-(9CI)

TOXICITY DATA WITH REFERENCE
scu-rat LD50:10 g/kg ABMGAJ 3,28,59
scu-mus LD50:6 g/kg ABMGAJ 3,28,59

CONSENSUS REPORTS: Reported in EPA TSCA Inventory.

SAFETY PROFILE: Mildly toxic by subcutaneous route. When heated to decomposition it emits toxic vapors of NO_x and Cl^-.

CCK660 CAS:6645-46-1 HR: 2
l-CARNITINE HYDROCHLORIDE
mf: $C_7H_{15}NO_3•Cl$ mw: 196.68

SYNS: AMMONIUM, (3-CARBOXY-2-HYDROXYPROPYL)TRIMETHYL-, CHLORIDE, (−)- □ (−)-(3-CARBOXY-2-HYDROXYPROPYL)TRIMETHYLAMMONIUM CHLORIDE □ l-(3-CARBOXY-2-HYDROXYPROPYL)TRIMETHYLAMMONIUM CHLORIDE □ (R)-3-CARBOXY-2-HYDROXY-N,N,N-TRIMETHYL-1-PROPANAMINIUM CHLORIDE □ l-CARNITINE CHLORIDE □ (R)-CARNITINE HYDROCHLORIDE □ LC-80 □ 1-PROPANAMINIUM, 3-CARBOXY-2-HYDROXY-N,N,N-TRIMETHYL-, CHLORIDE, (R)- (9CI)

TOXICITY DATA WITH REFERENCE
orl-rat TDLo:33 g/kg (female 7-17D post):REP IYKEDH 19,465,88
orl-rat TDLo:33 g/kg (female 7-17D post):TER IYKEDH 19,465,88
orl-rat LD50:6890 mg/kg IYKEDH 19,191,88
ipr-rat LD50:1920 mg/kg IYKEDH 19,191,88
ivn-rat LD50:1440 mg/kg IYKEDH 19,191,88
orl-mus LD50:8 g/kg IYKEDH 19,446,88
ipr-mus LD50:1690 mg/kg IYKEDH 19,446,88
scu-mus LD50:4320 mg/kg IYKEDH 19,446,88
ivn-dog LD50:2272 mg/kg IYKEDH 19,238,88

SAFETY PROFILE: Moderately toxic by intraperitoneal and intravenous routes. An experimental teratogen. Other experimental reproductive effects. When heated to decomposition it emits toxic fumes of NO_x and Cl^-.

CCK665 CAS:305-84-0 HR: 1
CARNOSINE
mf: $C_9H_{14}N_4O_3$ mw: 226.27

PROP: Needles. Mp: 246–250° (decomp).

SYNS: β-ALANYL-l-HISTIDINE □ l-CARNOSINE □ l-HISTIDINE, N-β-ALANYL- □ IGNOTINE □ KARNOZZN □ N-2-M

TOXICITY DATA WITH REFERENCE
ipr-rat TDLo:21 mg/kg (female 8-14D post):TER OYYAA2 8,1219,74
ipr-rat TDLo:21 mg/kg (female 8-14D post):REP OYYAA2 8,1219,74
ipr-mus LD50:9087 mg/kg USXXAM #4446149

CONSENSUS REPORTS: Reported in EPA TSCA Inventory.

SAFETY PROFILE: Mildly toxic by intraperitoneal route. An experimental teratogen. Other experimental reproductive effects. When heated to decomposition it emits toxic fumes of NO_x.

C

CCK675 **HR: 2**
CAROLINA ALLSPICE

PROP: A large shrub (to 12 feet tall) with large (2- to 3-inch), fruity smelling, brownish-red or purple flowers. The fruit is fig-shaped and contains large glossy brown seeds. Various species are native to the eastern states from Pennsylvania, through northern Florida to Alabama and in California.

SYNS: AMERICAN ALLSPICE □ BUBBIE BLOSSOMS □ BUBBY BUSH □ CALYCANTH □ CALYCANTHUS (VARIOUS SPECIES) □ C. FERTILIS □ C. FLORIDUS □ C. OCCIDENTALIS □ PINEAPPLE SHRUB □ SPICEBUSH □ STRAWBERRY BUSH □ SWEET BETTIE □ SWEET SHRUB

SAFETY PROFILE: The seeds contain the toxin calycanthin and some related alkaloids. No human poisonings have been reported, but ingestion of the seeds could cause symptoms similar to strychnine poisoning: convulsions, weak contractions of the heart, and low blood pressure.

CCK775 **CAS:7075-03-8** **HR: 3**
CARPIPRAMINE DIHYDROCHLORIDE
mf: $C_{28}H_{38}N_4O•2ClH$ mw: 519.62

PROP: Crystals. Mp: 260°.

SYNS: CARPIPRAMINE HYDROCHLORIDE □ DEFEKTON

TOXICITY DATA WITH REFERENCE
orl-rat LD50:1025 mg/kg NIIRDN 6,185,82
ipr-rat LD50:76 mg/kg NIIRDN 6,185,82
ivn-rat LD50:37 mg/kg NIIRDN 6,185,82
orl-mus LD50:2180 mg/kg NIIRDN 6,185,82
ipr-mus LD50:136 mg/kg NIIRDN 6,185,82
ivn-mus LD50:28,200 µg/kg NIIRDN 6,185,82

SAFETY PROFILE: Poison by intravenous and intraperitoneal routes. Moderately toxic by ingestion. When heated to decomposition it emits toxic fumes of NO_x and HCl. See also CARPIPRAMINE DIHYDROCHLORIDE MONOHYDRATE and AMINES.

CCK780 **HR: 3**
CARPIPRAMINE DIHYDROCHLORIDE MONOHYDRATE
mf: $C_{28}H_{38}N_4O•2ClH•H_2O$ mw: 537.64

SYNS: DEFEKTON □ PRAZINIL □ PZ 1511

TOXICITY DATA WITH REFERENCE
orl-rat LD50:1025 mg/kg MEIEDD 10,260,83
ipr-rat LD50:76 mg/kg MEIEDD 10,260,83
ivn-rat LD50:37 mg/kg MEIEDD 10,260,83
orl-mus LD50:2180 mg/kg MEIEDD 10,260,83
ipr-mus LD50:136 mg/kg MEIEDD 10,260,83
ivn-mus LD50:28,200 µg/kg MEIEDD 10,260,83

SAFETY PROFILE: Poison by intravenous and intraperitoneal routes. Moderately toxic by ingestion. When heated to decomposition it emits toxic fumes of NO_x and HCl. A psychotropic agent. See other carpipramine entries.

CCK790 **CAS:100482-23-3** **HR: 3**
CARPIPRAMINE MALEATE
mf: $C_{28}H_{38}N_4O•C_4H_4O_4$ mw: 562.78

SYN: CARBADIPIMIDINE MALEATE

TOXICITY DATA WITH REFERENCE
ipr-rat LD50:169 mg/kg NIIRDN 6,185,82
orl-mus LD50:2055 mg/kg NIIRDN 6,185,82
ipr-mus LD50:147 mg/kg NIIRDN 6,185,82

SAFETY PROFILE: Poison by intraperitoneal route. Moderately toxic by ingestion. When heated to decomposition it emits toxic fumes of NO_x. See other carpipramine entries.

CCK800 **CAS:53716-49-7** **HR: 3**
CARPROFEN
mf: $C_{15}H_{12}ClNO_2$ mw: 273.72

PROP: Crystals from chloroform. Mp: 197–198°.

SYNS: dl-6-CHLORO-α-METHYLCARBAZOLE-2-ACETIC ACID □ IMADYL □ RIMADYL

TOXICITY DATA WITH REFERENCE
orl rat TDLo:2360 mg/kg (9W male/2W pre-3W post):REP TXAPA9 56,376,80
orl-rat TDLo:708 mg/kg (male 9W pre):REP TXAPA9 56,376,80
orl-rat LD50:74 mg/kg OYYAA2 14,251,77
orl-mus LD50:186 mg/kg MDACAP 18,170,82

SAFETY PROFILE: Poison by ingestion. Experimental reproductive effects. When heated to decomposition it emits toxic fumes of Cl^- and NO_x.

CCL109 **CAS:23734-06-7** **HR: 3**
CARQUEJOL
mf: $C_{10}H_{14}O$ mw: 150.24

SYN: (1S-cis)-5-METHYLENE-6-(1-METHYLETHENYL) 2 CYCLOHEXEN-1-OL

TOXICITY DATA WITH REFERENCE
ipr-rat LD50:410 mg/kg APFRAD 18,715,60
orl-mus LD50:1800 mg/kg APFRAD 18,715,60
ipr-mus LD50:456 mg/kg APFRAD 18,715,60
ipr-dog LDLo:250 mg/kg APFRAD 18,715,60

SAFETY PROFILE: Poison by intraperitoneal route. Moderately toxic by ingestion. When heated to decomposition it emits acrid smoke and fumes.

CCL250 **CAS:9000-07-1** **HR: 2**
CARRAGEEN

PROP: A sulfated polysaccharide. Dried plant of seaweed *Chondrus crispus, Chondrus ocellatus, Eucheuma cottonil, Eucheuma spinosum, Gigartina acicularis, Gigartina pistillata, Gigartina radula, Gigartina stellata.* Yellow-white when powdered. Sol in water @ 80°; insol in organic solvents. Dried, bleached *Chondrus crispus* containing salts of sulfated polygalactose esters.

SYNS: 3,6-ANHYDRO-d-GALACTAN □ AUBYGEL GS □ AUBYGUM DM □ BURTONITE-V-40-E □ CARASTAY □ CARASTAY G □ CARRAGEENAN (FCC) □ CARRAGEENAN GUM □ CARRAGHEANIN □ CAR-

RAGHEEN □ CARRAGHEENAN □ CHONDRUS □ CHONDRUS EXTRACT □ COLLOID 775 □ COREINE □ EUCHEUMA SPINOSUM GUM □ FLANOGEN ELA □ GALOZONE □ GELCARIN □ GELCARIN HMR □ GELOZONE □ GENU □ GENUGEL □ GENUGEL CJ □ GENUGOL RLV □ GENUVISCO J □ GUM CARRAGEENAN □ GUM CHON 2 □ GUM CHROND □ IRISH GUM □ IRISH MOSS EXTRACT □ IRISH MOSS GELOSE □ KILLEEN □ LYGOMME CDS □ PEARLPUSS □ PELLUGEL □ PENCOGEL □ PIG-WRACK □ SATIAGEL GS 350 □ SATIAGUM 3 □ SATIAGUM STANDARD □ SEAKEM CARRAGEENIN □ SEATREM □ SELF ROCK MOSS □ VISCARIN

TOXICITY DATA WITH REFERENCE
orl-rat TDLo:2100 g/kg/40W-C:ETA CNREA8 38,4427,78
scu-rat TDLo:525 mg/kg/21W-I:NEO 13BYAH -,83,62
par-rat TDLo:430 mg/kg:ETA BJCAAI 15,607,61
par-rat TD:320 mg/kg:ETA OYYAA2 32,711,86
ivn-rbt LDLo:5 mg/kg JPPMAB 17,647,65
ivn-gpg LDLo:20 mg/kg NATUAS 202,401,64

CONSENSUS REPORTS: IARC Cancer Review: Group 3 IMEMDT 7,56,87; Animal Limited Evidence IMEMDT 10,181,76. Reported in EPA TSCA Inventory.

SAFETY PROFILE: Poison by intravenous route. Questionable carcinogen with experimental neoplastigenic and tumorigenic data. When heated to decomposition it emits acrid smoke and fumes.

CCL350 CAS:11114-20-8 HR: 2
kappa-CARRAGEENAN

SYNS: kappa-CARRAGEEN □ kappa-CARRAGEENIN □ SATIAGEL GS 350

TOXICITY DATA WITH REFERENCE
par-rat TDLo:320 mg/kg:ETA OYYAA2 32,711,86
orl-rbt LDLo:3 mg/kg JPPMAB 17,647,65

SAFETY PROFILE: Poison by ingestion. Questionable carcinogen with experimental tumorigenic data. When heated to decomposition it emits acrid smoke and irritating fumes.

CCL500 HR: 3
CARRAGEENAN, DEGRADED

PROP: Carrageenan derived from *Eucheuma spinosum*, degraded by acid hydrolysis; average molecular weight 20,000–40,000 (CALEDQ 4,171,78).

TOXICITY DATA WITH REFERENCE
orl-rat TDLo:360 g/kg/9W-C:CAR CALEDQ 14,267,81
orl-rat TD:1700 g/kg/52W-C:CAR CALEDQ 4,171,78
orl-rat TD:6834 g/kg/73W-C:CAR PPTCBY 9,127,79
orl-rat TD:3116 g/kg/77W-C:NEO PPTCBY 9,127,79
orl-rat TD:4860 g/kg/64W-C:CAR PPTCBY 9,127,79
orl-rat TD:2250 g/kg/64W-C:ETA PPTCBY 9,127,79
orl-rat TD:1500 g/kg/43W-C:CAR CALEDQ 4,171,78
orl-rat TD:1080 g/kg/26W-C:CAR CALEDQ 14,267,81
orl-rat TD:1620 g/kg/39W-C:CAR CALEDQ 14,267,81

CONSENSUS REPORTS: IARC Cancer Review: Animal Sufficient Evidence IMEMDT 31,79,83.

SAFETY PROFILE: Confirmed carcinogen with experimental carcinogenic, neoplastigenic, and tumorigenic data. See also CARRAGEEN. When heated to decomposition it emits toxic fumes of SO_x.

CCL750 CAS:8015-88-1 HR: 1
CARROT SEED OIL

PROP: Distilled from the seeds of *Daucus carota L.* (Fam. Umbelliferae). (FCTXAV 14,659,76). Light-yellow to amber liquid; aromatic odor. Sol in fixed oils, mineral oil; insol in glycerin, propylene glycol.

SYNS: DAUCUS OIL □ OILS, CARROT

TOXICITY DATA WITH REFERENCE
skn-rbt 500 mg/24H MLD FCTXAV 14,659,76
skn-gpg 100% MLD FCTXAV 14,705,76

CONSENSUS REPORTS: Reported in EPA TSCA Inventory.

SAFETY PROFILE: A skin irritant. When heated to decomposition it emits acrid smoke and irritating fumes.

CCL800 CAS:51781-06-7 HR: 3
CARTEOLOL
mf: $C_{16}H_{24}N_2O_3$ mw: 292.42

SYN: 5-(3-((1,1-DIMETHYLETHYL)AMINO)-2-HYDROXYPROPOXY)-3,4-DIHYDRO-2(1H)-QUINOLINONE

TOXICITY DATA WITH REFERENCE
orl-dog LD50:830 mg/kg OYYAA2 19,323,80
orl-rbt LD50:740 mg/kg OYYAA2 19,323,80
ivn-rbt LD50:112 mg/kg OYYAA2 19,323,80

SAFETY PROFILE: Poison by intravenous route. Moderately toxic by ingestion. When heated to decomposition it emits toxic fumes of NO_x.

CCM000 CAS:499-75-2 HR: 3
CARVACROL
mf: $C_{10}H_{14}O$ mw: 150.24

PROP: Colorless to pale-yellow liquid; spicy thymol odor. D: 0.974–0.980, mp: 3.5°, bp: 237–238°, refr index: 1.521–1.526, flash p: 212°F. Sol in alc, ether; insol in water.

SYNS: 2-p-CYMENOL □ FEMA No. 2245 □ 2-HYDROXY-p-CYMENE □ ISOPROPYL-o-CRESOL □ 5-ISOPROPYL-2-METHYLPHENOL □ ISOTHYMOL □ 2-METHYL-5-ISOPROPYLPHENOL □ o-THYMOL

TOXICITY DATA WITH REFERENCE
skn-rbt 500 mg/24H SEV FCTXAV 17(suppl)695,79
orl-rat LD50:810 mg/kg FCTXAV 2,327,64
scu-mus LD50:680 mg/kg SIZSAR 3,73,52
ivn-mus LD50:80 mg/kg JMCMAR 23,1350,80
ivn-dog LDLo:310 mg/kg THERAP 3,109,48
orl-cat LDLo:100 mg/kg HBTXAC 5,46,59
orl-rbt LDLo:100 mg/kg AEPPAE 161,196,31
skn-rbt LDLo:2700 mg/kg JAPMA8 38,366,49
scu-rbt LDLo:1000 mg/kg HBTXAC 5,46,59
scu-frg LDLo:75 mg/kg HBTXAC 5,46,59

CONSENSUS REPORTS: Reported in EPA TSCA Inventory.

SAFETY PROFILE: Poison by ingestion, intravenous, and subcutaneous routes. Moderately toxic by skin contact. A severe skin irritant. Combustible liquid.

When heated to decomposition it emits acrid smoke and irritating fumes.

CCM100 CAS:2244-16-8 HR: 3
d-CARVONE
mf: $C_{10}H_{14}O$ mw: 150.24

PROP: Colorless liquid or oil; caraway odor. D: 0.956–0.960, bp: 230°, refr index: 1.96–1.499. Sol in propylene glycol, fixed oils; misc in alc; insol in glycerin.

SYNS: (+)-CARVONE □ d(+)-CARVONE □ (S)-CARVONE □ (S)-(+)-CARVONE □ FEMA No. 2249 □ d-1-METHYL-4-ISOPROPENYL-6-CYCLOHEXEN-2-ONE □ d-p-MENTHA-6,8,(9)-DIEN-2-ONE □ (S)-2-METHYL-5-(1-METHYLETHENYL)-2-CYCLOHEXEN-1-ONE

TOXICITY DATA WITH REFERENCE
skn-rbt 500 mg/24H MLD FCTXAV 16,673,78
orl-rat LD50:3710 µg/kg FCTXAV 16,673,78
skn-rbt LD50:4 mg/kg FCTXAV 16,673,78

CONSENSUS REPORTS: Reported in EPA TSCA Inventory.

SAFETY PROFILE: Poison by ingestion and skin contact. A skin irritant. When heated to decomposition it emits acrid smoke and irritating fumes.

CCM120 CAS:6485-40-1 HR: 3
l(−)-CARVONE
mf: $C_{10}H_{14}O$ mw: 150.22

PROP: Colorless liquid or oil; spearmint odor. D: 0.956–0.960, bp: 230–231°, refr index: 1.495–1.499. Sol in propylene glycol, fixed oils; misc in alc, insol in glycerin.

SYNS: (−)-CARVONE □ 1-CARVONE □ (R)-CARVONE □ FEMA No. 2249 □ 1-6,8(9)-p-MENTHADIEN-2-ONE □ (R)-(−)-p-MENTHA-6,8-DIEN-2-ONE □ 1-1-METHYL-4-ISOPROPENYL-6-CYCLOHEXEN-2-ONE □ (R)-2-METHYL-5-(1-METHYLETHENYL)-2-CYCLOHEXEN-1-ONE (9CI)

TOXICITY DATA WITH REFERENCE
orl-rat LD50:1640 mg/kg FCTXAV 11,1057,73
ivn-mus LD50:56 mg/kg CSLNX* NX#02834
orl-gpg LD50:766 mg/kg FCTXAV 11,1057,73

SAFETY PROFILE: Poison by intravenous route. Moderately toxic by ingestion. When heated to decomposition it emits acrid smoke and irritating fumes.

CCM750 CAS:97-42-7 HR: 1
1-CARVYL ACETATE
mf: $C_{12}H_{18}O_2$ mw: 194.30

SYNS: 1-p-MENTHA-6(8,9)-DIEN-2-YL ACETATE □ 2-METHYL-5-(1-METHYLETHENYL)-2-CYCLOHEXEN-1-OL ACETATE

TOXICITY DATA WITH REFERENCE
skn-rbt 500 mg/24H MLD FCTXAV 16,637,78

CONSENSUS REPORTS: Reported in EPA TSCA Inventory.

SAFETY PROFILE: A skin irritant. When heated to decomposition it emits acrid smoke and fumes. See also ESTERS.

CCN000 CAS:87-44-5 HR: 1
CARYOPHYLLENE
mf: $C_{15}H_{26}$ mw: 206.41

PROP: Colorless to sltly yellow oily liquid; clove odor. Found in oil of clove, cinnamon leaves, and copaiba balsam, and in minor quantities in various other essential oils, especially lavender; prepared by isolation from clove leaf oil, clove stem oil, cinnamon leaf oil, or pine oil fractions (FCTXAV 11,1011,73). D: 0.897–0.910, refr index: 1.498–1.504, bp: 118–119° @ 9.7 mm, flash p: 206°F. Sol in alc, ether; insol in water.

SYNS: β-CARYOPHYLLENE (FCC) □ FEMA No. 2252 □ 8-METHYLENE-4,11,11-(TRIMETHYL)BICYCLO(7.2.0)UNDEC-4-ENE

TOXICITY DATA WITH REFERENCE
skn-rbt 500 mg/24H FCTXAV 11,1059,73

CONSENSUS REPORTS: Reported in EPA TSCA Inventory.

SAFETY PROFILE: A skin irritant. Combustible liquid. When heated to decomposition it emits acrid smoke and irritating fumes.

CCN050 CAS:57082-24-3 HR: 1
CARYOPHYLLENE ACETATE
mf: $C_{17}H_{28}O_2$ mw: 264.45

SYN: TRICYCLO(6.3.1.0^{2,5}))DODECAN-1-OL, 4,4,8-TRIMETHYL-, ACETATE, (1R-(1-α-2-α-5-β,8-β))-

TOXICITY DATA WITH REFERENCE
orl-rat LD50:>5 g/kg FCTXAV 12,839,74
skn-rbt LD50:>5 g/kg FCTXAV 12,839,74

CONSENSUS REPORTS: Reported in EPA TSCA Inventory.

SAFETY PROFILE: Low toxicity by ingestion and skin contact. When heated to decomposition it emits acrid smoke and irritating vapors.

CCN100 CAS:1139-30-6 HR: 1
β-CARYOPHYLLENE EPOXIDE
mf: $C_{15}H_{24}O$ mw: 220.39

PROP: Crystals. Mp: 63.5–64°.

SYNS: CARYOPHYLLENE OXIDE □ CARYOPHYLLENE EPOXIDE □ (−)-CARYOPHYLLENE OXIDE □ β-CARYOPHYLLENE OXIDE □ EPOXYCARYOPHYLLENE □ (−)-EPOXYDIHYDROCARYOPHYLLENE □ 5-OXATRICYCLO(8.2.0.0^{4,6})DODECANE, 4,12,12-TRIMETHYL-9-METHYLENE-, (1R,4R,6R,10S)- □ 4,11,11-TRIMETHYL-8-METHYLENE-5-OXATRICYCLO(8.2.0.0(4,6))DODECANE

TOXICITY DATA WITH REFERENCE
skn-rbt 500 mg/24H MOD FCTOD7 21,661,83
orl-rat LD50:>5 g/kg FCTOD7 21,661,83
skn-rbt LD50:>2 g/kg FCTOD7 21,661,83

CONSENSUS REPORTS: Reported in EPA TSCA Inventory.

SAFETY PROFILE: Low toxicity by ingestion and skin contact. A skin irritant. When heated to decomposition it emits acrid smoke and irritating fumes.

CCN250 CAS:1403-27-6 HR: 3
CARZINOCIDIN

PROP: An antitumor substance from *Streptomyces sahachiroi.*

SYN: CARCINOCIDIN

TOXICITY DATA WITH REFERENCE
ipr-mus LD50:43,500 µg/kg JAJAAA 9,9,56
scu-mus LD50:20 mg/kg JAJAAA 9,9,56
ivn-mus LD50:4700 µg/kg JAJAAA 9,6,56

SAFETY PROFILE: Poison by subcutaneous, intravenous, and intraperitoneal routes. When heated to decomposition it emits acrid smoke and irritating fumes.

CCN500 CAS:1403-28-7 HR: 3
CARZINOPHILIN

SYNS: CARDINOPHILLIN ☐ CARDINOPHYLLIN

TOXICITY DATA WITH REFERENCE
mmo-esc 10 mg/disc ANYAA9 76,475,58
scu-rat TDLo:50 µg/kg (female 6-10D post):TER OSDIAF 14,107,65
scu-rat TDLo:50 µg/kg (female 6-10D post):REP OSDIAF 14,107,65
ipr-rat TDLo:500 iu/kg (male 1D pre):REP HIKYAJ 12,1339,66
ipr-mus LD50:8000 unit/kg JAJAAA 13,27,60
scu-mus LD50:3 mg/kg 85GDA2 6,300,81
ivn-mus LD50:500 µg/kg 85GDA2 6,300,81

SAFETY PROFILE: Poison by subcutaneous and intravenous routes. An experimental teratogen. Other experimental reproductive effects. Mutation data reported. When heated to decomposition it emits acrid smoke and irritating fumes.

CCN750 CAS:1403-29-8 HR: 3
CARZINOPHILIN A
mf: $C_{31}H_{33}N_3O_{12}$ mw: 639.67

PROP: Needles. Mp: 217–222° (decomp). Active fraction of antitumor substance *Carzinophilin* obtained from *Streptomyces sahachiroi* .

TOXICITY DATA WITH REFERENCE
mmo-esc 500 µg/disc APMBAY 6,23,58
ivn-mus LD50:15 µg/kg 85ERAY 2,1356,78

SAFETY PROFILE: Deadly poison by intravenous route. Mutation data reported.

CCO675 CAS:33445-03-3 HR: 3
CASSAINE HYDROCHLORIDE
mf: $C_{24}H_{39}NO_4$•ClH mw: 442.10

SYN: (E)-7-OXO-3-β-HYDROXY-14-α-METHYL-8-β-PODOCARPANE-Δ^{13}-α-ACETIC ACID-2-(DIMETHYLAMINO)ETHYL ESTER HYDROCHLORIDE

TOXICITY DATA WITH REFERENCE
ivn-dog LDLo:400 µg/kg JMCMAR 10,582,67
ivn-cat LDLo:806 µg/kg JPHAA3 27,9,38
ivn-gpg LDLo:2640 µg/kg APSXAS 13,35,76

SAFETY PROFILE: Deadly poison by intravenous route. When heated to decomposition it emits toxic fumes of NO_x and HCl.

CCO680 HR: 3
CASSAVA

PROP: A bushy shrub up to 9 feet tall that grows long, tuberous roots. The alternate leaves have 3 to 7 lobes. It is cultivated for food in the United States Gulf Coast states, Hawaii, Guam, and the West Indies.

SYNS: JUCA ☐ MANIHOT ESCULENTA ☐ MANIOC ☐ MANIOKA ☐ SWEET POTATO PLANT ☐ TAPIOCA ☐ YUCA ☐ YUCA BRAVA

SAFETY PROFILE: The leaves, and especially the tubers, contain the cyanogenetic glycosides linamarin and lotaustralin. Cyanogenetic glycosides release cyanide when exposed to stomach acid. Ingestion may cause after a delay period of several hours: abdominal pain, vomiting, lack of muscle control, coma, and convulsions. See also CYANIDE.

CCO750 CAS:8007-80-5 HR: 3
CASSIA OIL

PROP: Chief constituent is cinnamic aldehyde, found in the leaves and twigs of *Cinnamomum cassia blume* (FCTXAV 13,91,75). Yellow liquid; cinnamon odor, spicy burning taste. Sol in fixed oils, propylene glycol; insol in glycerin, mineral oil.

SYNS: ARTIFICIAL CINNAMON OIL ☐ CINNAMON BARK OIL ☐ CINNAMON BARK OIL, CEYLON TYPE (FCC) ☐ CINNAMON OIL ☐ KASSIA OEL (GERMAN) ☐ OIL of CASSIA ☐ OIL of CHINESE CINNAMON ☐ OIL of CINNAMON ☐ OIL of CINNAMON, CEYLON ☐ OILS, CINNAMON

TOXICITY DATA WITH REFERENCE
skn-hmn 100% FCTXAV 13,109,75
skn-mus 100% MLD FCTXAV 13,109,75
skn-rbt 500 mg/24H SEV FCTXAV 13,91,75
dnr-bcs 600 µg/disc TOFOD5 8,91,85
orl-rat LD50:2800 mg/kg FCTXAV 13,91,75
orl-mus LD50:2670 mg/kg TOFOD5 8,91,85
ipr-mus LD50:500 mg/kg PHMCAA 3,62,61
skn-rbt LD50:320 mg/kg FCTXAV 13,91,75

CONSENSUS REPORTS: Reported in EPA TSCA Inventory.

SAFETY PROFILE: Poison by skin contact. Moderately toxic by ingestion and intraperitoneal routes. A human skin irritant. Mutation data reported. See also CINNAMALDEHYDE and ALDEHYDES. When heated to decomposition it emits acrid smoke and irritating fumes.

CCO800 HR: 3
CASSIA TORA Linn., leaf extract

PROP: Indian plant belonging to the family *Leguminosae* INDRBA 15,49,78).

TOXICITY DATA WITH REFERENCE
orl-mus LDLo:200 mg/kg INDRBA 15,49,78
ipr-mus LDLo:100 mg/kg INDRBA 15,49,78
ivn-mus LDLo:20 mg/kg INDRBA 15,49,78

SAFETY PROFILE: Poison by ingestion, intravenous, and intraperitoneal routes.

CCP000 **HR: 3**
CASTOR BEAN
DOT: UN 2969

PROP: An annual may grow higher than 15 feet. The large, lobed leaves may be 3 feet across. The spiny seed pods grow in clusters and contain plump seeds that are white with brown or black mottling. The seeds have a pleasant taste.

SYNS: AFRICAN COFFEE TREE □ CASTOR BEANS (DOT) □ CASTOR FLAKE (DOT) □ CASTOR MEAL (DOT) □ CASTOR OIL PLANT □ CASTOR POMACE (DOT) □ HIGUERETA (CUBA, PUERTO RICO) □ HIGUERILLA (MEXICO) □ KOLI (HAWAII) □ LA'AU-'AILA (HAWAII) □ MAN'S MOTHERWORT □ MEXICO WEED □ PA'AILA (HAWAII) □ PALMA CHRISTI (HAITI) □ RICIN (HAITI) □ RICINO (PUERTO RICO) □ RICINUS COMMUNIS □ STEADFAST □ WONDER TREE

TOXICITY DATA WITH REFERENCE
orl-chd LDLo:500 μg/kg 34ZIAG -,158,69

DOT CLASSIFICATION: 9; *Label:* None

SAFETY PROFILE: Deadly poison by ingestion in humans. The seeds contain the deadly poison ricin, a plant lectin (toxalbumin) that inhibits protein synthesis in the intestinal wall. Ingestion of the seeds can cause after a delay period of several hours: nausea, vomiting, diarrhea, and intestinal dysfunction. There may be massive fluid and electrolyte loss. Ingestion of as few as 2 seeds could be fatal. A potent allergen. When heated to decomposition it emits toxic fumes of NO_x. See also RICIN.

CCP250 **CAS:8001-79-4** **HR: 1**
CASTOR OIL

PROP: From seeds of *Ricinus communis L.* (Fam. Euphorbiaceae). A colorless to pale yellow, viscous liquid; bland taste, characteristic odor. Mp: −12°, bp: 313°, flash p: 445°F (CC), d: 0.96, autoign temp: 840°F. Sol in alc; misc in abs alc, glacial acetic acid, chloroform, and ether.

SYNS: AROMATIC CASTOR OIL □ CASTOR OIL AROMATIC □ COSMETOL □ CRYSTAL O □ GOLD BOND □ NCI-C55163 □ NEOLOID □ OIL of PALMA CHRISTI □ PHORBYOL □ RICINUS OIL □ RICIRUS OIL □ TANGANTANGAN OIL

TOXICITY DATA WITH REFERENCE
skn-man 50 mg/48H MLD CTOIDG 94(8),41,79
skn-rat 100 mg/24H MLD CTOIDG 94(8),41,79
skn-rbt 100 mg/24H SEV CTOIDG 94(8),41,79
eye-rbt 500 mg MLD AJOPAA 29,1363,46
skn-gpg 100 mg/24H MLD CTOIDG 94(8),41,79

CONSENSUS REPORTS: Reported in EPA TSCA Inventory.

SAFETY PROFILE: An allergen. A human skin and eye irritant. Combustible when exposed to heat. Spontaneous heating may occur. To fight fire, use CO_2, dry chemical, fog, mist. See also CASTOR BEAN.

CCP500 **CAS:535-89-7** **HR: 3**
CASTRIX
mf: $C_7H_{10}ClN_3$ mw: 171.65

PROP: Solid, sltly water-sol crystals. Mp: 87°, bp: 140–147° @ 4 mm.

SYNS: 2-CHLOOR-4-DIMETHYLAMINO-6-METHYL-PYRIMIDINE (DUTCH) □ 2-CHLOR-4-DIMETHYLAMINO-6-METHYLPYRIMIDIN (GERMAN) □ 2-CHLORO-4-DIMETHYLAMINO-6-METHYL-PYRIMIDINE □ 2-CHLORO-4-METHYL-6-DIMETHYLAMINOPYRIMIDINE □ 2-CLORO-4-DIMETILAMINO-6-METIL-PIRIMIDINA (ITALIAN) □ CRIMIDIN (GERMAN) □ CRIMIDINA (ITALIAN) □ CRIMIDINE □ W 491

TOXICITY DATA WITH REFERENCE
orl-rat LD50:1250 μg/kg GUCHAZ 6,139,73
ipr-rat LD50:1 mg/kg JAPMA8 27,307,48
orl-mus LD50:1200 μg/kg MEIEDD 11,405,89
ipr-mus LD50:420 μg/kg JAPMA8 37,307,48
ipr-dog LD50:500 μg/kg JAPMA8 37,307,48
orl-rbt LD50:5 mg/kg 28ZEAL 5,59,76
ipr-rbt LD50:5 mg/kg JAPMA8 37,307,48
orl-gpg LD50:2660 μg/kg PCOC** -,202,66
ipr-gpg LD50:2660 μg/kg JAPMA8 37,307,48

CONSENSUS REPORTS: EPA Extremely Hazardous Substances List.

SAFETY PROFILE: Deadly poison by ingestion and intraperitoneal routes. Can cause central nervous system damage and convulsions. Intensely poisonous to mammals. A pesticide. When heated to decomposition it emits very toxic fumes of Cl^- and NO_x.

CCP525 **CAS:9001-05-2** **HR: D**
CATALASE from MICROCOCCUS LYSODEIKTICUS

PROP: Derived from *Micrococcus lysodeikticus.*

SYNS: CAPERASE □ EQUILASE □ OPTIDASE

TOXICITY DATA WITH REFERENCE
mic-mic-uns 5 pph POASAD 34,114,53

CONSENSUS REPORTS: Reported in EPA TSCA Inventory.

SAFETY PROFILE: Mutation data reported. When heated to decomposition it emits acrid smoke and irritating fumes.

CCP675 **CAS:3758-54-1** **HR: 1**
CATANAC SP ANTISTATIC AGENT
mf: $C_{25}H_{53}N_2O_2 \cdot H_2O_4P$ mw:510.79

SYNS: CATANAC SP □ CATIONIC SP □ (2-HYDROXYETHYL)DIMETHYL(3-STEARAMIDOPROPYL)-AMMONIUM PHOSPHATE (1:1) (SALT)

TOXICITY DATA WITH REFERENCE
orl-rat LD50:8100 mg/kg 34ZIAG-,158,69

CONSENSUS REPORTS: Reported in EPA TSCA Inventory.

SAFETY PROFILE: Mildly toxic by ingestion. When heated to decomposition it emits toxic fumes of NH_3, NO_x, and PO_x.

CCP850 **CAS:120-80-9** **HR: 3**
CATECHOL
mf: $C_6H_6O_2$ mw: 110.12

PROP: Colorless crystals or needles from water. Mp: 105°, bp: 240°, flash p: 261°F (CC), d: 1.341 @ 15°, vap press: 10 mm @ 118.3°, vap d: 3.79. Sol in water, chloroform, and benzene; very sol in alc and ether.

SYNS: o-BENZENEDIOL □ 1,2-BENZENEDIOL □ CATECHIN □ C.I. 76500 □ C.I. OXIDATION BASE 26 □ o-DIHYDROXYBENZENE □ 1,2-DIHYDROXYBENZENE □ o-DIOXYBENZENE □ o-DIPHENOL □ DURAFUR DEVELOPER C □ FOURAMINE PCH □ FOURRINE 68 □ o-HYDROQUINONE □ o-HYDROXYPHENOL □ 2-HYDROXYPHENOL □ NCI-C55856 □ OXYPHENIC ACID □ PELAGOL GREY C □ o-PHENYLENEDIOL □ PYROCATECHIN □ PYROCATECHINIC ACID □ PYROCATECHOL □ PYROCATECHUIC ACID

TOXICITY DATA WITH REFERENCE
mrc-smc 300 mg/L MUREAV 135,109,84
dni-hmn:hla 200 μmol/L MUREAV 92,427,82
dns-rat-orl 1 g/kg JJIND8 74,1283,85
scu-rat TDLo:5 mg/kg (1D pre):REP ENDOAO 57,466,55
orl-rat TDLo:437 g/kg/2Y-C:CAR JJCREP 81,207,90
orl-mus TDLo:645 g/kg/96W-C:CAR JJCREP 81,207,90
orl-rat LD50:260 mg/kg AFREAW 3,197,51
scu-rat LDLo:110 mg/kg AIPTAK 176,193,68
orl-mus LD50:260 mg/kg AFREAW 3,197,51
ipr-mus LD50:68 mg/kg PHTXA6 64,247,89
scu-mus LD50:247 mg/kg INHEAO 5,143,67
ivn-dog LDLo:40 mg/kg HBTXAC 1,62,56
skn-rbt LD50:800 mg/kg AIHAAP 37,596,76
ipr-gpg LDLo:150 mg/kg HBTXAC 1,62,55
par-frg LDLo:160 mg/kg AEPPAE 166,437,32

CONSENSUS REPORTS: IARC Cancer Review: Group 3 IMEMDT 7,56,87; Animal Inadequate Evidence IMEMDT 15,155,77. EPA Extremely Hazardous Substances List. Reported in EPA TSCA Inventory. EPA Genetic Toxicology Program.

OSHA PEL: TWA 5 ppm (skin)
ACGIH TLV: TWA 5 ppm (skin)

SAFETY PROFILE: Poison by ingestion, subcutaneous, intraperitoneal, intravenous, and parenteral routes. Moderately toxic by skin contact. Experimental reproductive effects. Can cause dermatitis on skin contact. An allergen. Human mutation data reported. Questionable carcinogen. Systemic effects similar to those of phenol. Combustible when exposed to heat or flame; can react vigorously with oxidizing materials. Hypergolic reaction with concentrated nitric acid. To fight fire, use water, CO_2, dry chemical. When heated to decomposition it emits acrid smoke and irritating fumes. See also PHENOL.

CCP875 **CAS:154-23-4** **HR: 2**
d-CATECHOL
mf: $C_{15}H_{14}O_6$ mw: 290.29

PROP: dl-Form: Needles from water + acetic acid. Mp: 212–216°. Sltly sol in cold water, ether; sol in hot water, alc, glacial acetic acid, acetone. Practically insol in benzene, chloroform, petr ether. Hydrated d-form: Needles from water + acetic acid. Mp: 93–96° (175–177° when anhydrous). Hydrated l-form: Needles from water + acetic acid. Mp: 93–96° (175–177° when anhydrous).

SYNS: CATECHIN □ (+)-CATECHIN □ d-CATECHIN □ d-(+)-CATECHIN □ CATECHIN (FLAVAN) □ CATECHINIC ACID □ CATECHOL □ (+)-CATECHOL □ CATECHOL (FLAVAN) □ CATECHUIC ACID □ CATERGEN □ CIANIDANOL □ KB-53

TOXICITY DATA WITH REFERENCE
oms-hmn:lym 5 μmol/L CNREA8 45,2471,85
sce-hmn:lym 5 μmol/L CNREA8 45,2471,85
orl-rat TDLo:12,150 mg/kg (17-22D preg/21D post):REP OYYAA2 24,509,82
orl-rat TDLo:55 g/kg (7-17D preg):TER OYYAA2 24,495,82
ipr-rat LD50:1084 mg/kg OYYAA2 24,361,82
ipr-mus LD50:1 g/kg PLMEAA 42,75,81

SAFETY PROFILE: Moderately toxic by intraperitoneal route. An experimental teratogen. Other experimental reproductive effects. Human mutation data reported. When heated to decomposition it emits acrid smoke and fumes.

CCP900 **CAS:2050-46-6** **HR: 1**
CATECHOL DIETHYL ETHER
mf: $C_{10}H_{14}O_2$ mw: 166.24

PROP: Crystals. Mp: 43–45°, bp: 219°, d: 1.0.

SYNS: BENZENE, o-DIETHOXY- □ BENZENE, 1,2-DIETHOXY-(9CI) □ o-DIETHOXYBENZENE □ 1,2-DIETHOXYBENZENE

TOXICITY DATA WITH REFERENCE
eye-rbt 100 mg MLD FCTOD7 20,573,82
eye-rbt 100 mg/30S RNS MLD FCTOD7 20,573,82

CONSENSUS REPORTS: Reported in EPA TSCA Inventory.

SAFETY PROFILE: An eye irritant. When heated to decomposition it emits acrid smoke and irritating fumes.

CCQ125 **HR: 3**
CAULOPHYLLUM THALICTROIDES, glycoside extract

PROP: Crystalline glycoside isolated from *Caulophyllum thalictroides* blue cohosh) (JAPMA8 43,16,54).

TOXICITY DATA WITH REFERENCE
eye-rbt 5000 ppm/1M JAPMA8 43,16,54
ivn-rat LDLo:20,300 μg/kg JAPMA8 43,16,54
ivn-mus LD50:11,800 μg/kg JAPMA8 43,16,54

SAFETY PROFILE: Poison by intravenous route. An eye irritant. When heated to decomposition it emits acrid smoke and fumes.

CCQ500 **CAS:8007-20-3** **HR: 2**
CEDAR LEAF OIL

PROP: Constituent is d-α-thujone, found in leaves of *Thuja occidentalis L.* (Fam. Cupressaaceae) (FCTXAV 12,807,74). Yellowish, volatile oil; strong sage odor. D: 0.910–0.920. Sol in fixed oils, mineral oil, propylene glycol; insol in glycerin.

SYNS: OIL of ARBOR VITAE □ OIL of CEDAR LEAF □ OILS, CEDAR LEAF □ OIL THUJA □ OIL of THUJA □ OIL of WHITE CEDAR □ THUJA OIL □ WHITE CEDAR OIL

TOXICITY DATA WITH REFERENCE
skn-rbt 500 mg/24H MOD FCTXAV 12,807,74
orl-rat LD50:830 mg/kg FCTXAV 12,807,74
skn-rbt LD50:4100 mg/kg FCTXAV 12,843,74

CONSENSUS REPORTS: Reported in EPA TSCA Inventory.

SAFETY PROFILE: Moderately toxic by ingestion and skin contact. A skin irritant. Ingestion of large quantities causes hypertension, bradycardia, tachypnea, convulsions, death. When heated to decomposition it emits acrid smoke and fumes. See also ARTEMISIA OIL.

CCQ750 CAS:8023-85-6 HR: 1
CEDARWOOD OIL ATLAS

PROP: From *Cedrus atlantica*, contains α- and β-atalantone (FCTXAV 14,659,76).

SYNS: CEDARWOOD OIL MOROCCAN □ CEDRUS ATLANTICA OIL

TOXICITY DATA WITH REFERENCE
skn-rbt 500 mg/24H MLD FCTXAV 14,659,76

CONSENSUS REPORTS: Reported in EPA TSCA Inventory.

SAFETY PROFILE: A skin irritant. When heated to decomposition it emits acrid smoke and irritating fumes.

CCR000 CAS:8000-27-9 HR: 1
CEDARWOOD OIL (VIRGINIA)

PROP: Colorless or sltly yellow, viscid liquid. Composition. Cedrene and cedrol. D: 0.940–0.950 @ 20°/20°. From steam distillation of the wood of *Juniperus virginiana L.* The main constituents are cedrene, thujopsene, and cedrol (FCTXAV 12,807,74).

SYNS: OIL CEDAR □ RED CEDARWOOD OIL

TOXICITY DATA WITH REFERENCE
skn-rbt 500 mg/24H MOD FCTXAV 12,807,74
orl-rat LD50:>5 g/kg FCTXAV 12,845,74
skn-rbt LD50:>5 g/kg FCTXAV 12,845,74

CONSENSUS REPORTS: Reported in EPA TSCA Inventory.

SAFETY PROFILE: Low toxicity by ingestion and skin contact. A skin irritant and allergen. Combustible when exposed to heat or flame. When heated to decomposition it emits acrid smoke and irritating fumes.

CCR250 CAS:77-54-3 HR: 1
8-β-H-CEDRAN-8-OL ACETATE
mf: $C_{17}H_{28}O_2$ mw: 264.45

SYNS: ACETIC ACID, CEDROL ESTER □ CEDRANYL ACETATE □ CEDRYL ACETATE □ OCTAHYDRO-3,6,8,8-TETRAMETHYL 1H-3a,7-METHANOAZULEN-6-OL ACETATE

TOXICITY DATA WITH REFERENCE
skn-rbt 500 mg/24H MOD FCTXAV 12,847,74
orl-rat LD50:44,750 mg/kg FCTXAV 2,327,64
skn-rbt LD50:>5 g/kg FCTXAV 12,847,74

CONSENSUS REPORTS: Reported in EPA TSCA Inventory.

SAFETY PROFILE: Very low toxicity by ingestion and skin contact. A skin irritant. When heated to decomposition it emits acrid smoke and irritating fumes.

CCR500 CAS:469-61-4 HR: 1
α-CEDRENE
mf: $C_{15}H_{24}$ mw: 204.39

PROP: Oil. Bp: 262–263° @ 760 mm.

SYN: CEDR-8-ENE

TOXICITY DATA WITH REFERENCE
skn-rbt 500 mg/24H MLD FCTXAV 16,637,78

CONSENSUS REPORTS: Reported in EPA TSCA Inventory.

SAFETY PROFILE: A skin irritant. When heated to decomposition it emits acrid smoke and irritating fumes.

CCR510 CAS:29597-36-2 HR: 1
CEDR-8-ENE EPOXIDE

SYNS: ANDRANE □ CEDRANE, 8,9 EPOXIDE

TOXICITY DATA WITH REFERENCE
skn-rbt 500 mg/24H MOD FCTXAV 18,663,80
orl-rat LD50:>5 g/kg FCTXAV 18,663,80
skn-rbt LD50:>5 g/kg FCTXAV 18,663,80

CONSENSUS REPORTS: Reported in EPA TSCA Inventory.

SAFETY PROFILE: Very low toxicity by ingestion and skin contact. A skin irritant. When heated to decomposition it emits acrid smoke and irritating fumes.

CCR524 CAS:39900-38-4 HR: 1
CEDROL FORMATE
mf: $C_{16}H_{26}O_2$ mw: 250.42

SYNS: CEDRYL FORMATE □ 1H-3-α-7-METHANOAZULEN-6-OL, OCTAHYDRO-3,6,8,8-TETRAMETHYL-, FORMATE, (3R-(3-α-3a-β,6-α-7-β, 8aα-))-

TOXICITY DATA WITH REFERENCE
skn-rbt 500 mg/24H MOD FCTOD7 20,647,82
orl-rat LD50:>5 g/kg FCTOD7 20,647,82
skn-rbt LD50:>5 g/kg FCTOD7 20,647,82

CONSENSUS REPORTS: Reported in EPA TSCA Inventory.

SAFETY PROFILE: Very low toxicity by ingestion and skin contact. A skin irritant. When heated to decomposition it emits acrid smoke and irritating fumes.

CCR525 CAS:67874-81-1 HR: 1
CEDROL METHYL ETHER
mf: $C_{16}H_{28}O$ mw: 236.44

SYNS: CEDRAMBER □ 1H-3a,7-METHANOAZULENE, OCTAHYDRO-6-METHOXY-3,6,8,8-TETRAMETHYL-,(3R-(3-α-3a-β, 6-α-7-β,8aα-)- □ METHYL CEDRYL ETHER

TOXICITY DATA WITH REFERENCE
skn-rbt 500 mg/24H MOD FCTXAV 17,747,79

CONSENSUS REPORTS: Reported in EPA TSCA Inventory.

SAFETY PROFILE: A skin irritant. When heated to decomposition it emits acrid smoke and irritating fumes.

CCR850 CAS:70356-03-5 HR: 2
CEFACLOR HYDRATE
mf: $C_{15}H_{14}ClN_3O_4S•H_2O$ mw: 385.85

PROP: Crystalline solid. Sol in water; practically insol in methanol, chloroform, benzene.

SYNS: ALFATIL □ CECLOR □ CEFACLOR □ DISTACLOR □ LILLY 99638 HYDRATE □ PANACEF □ PANORAL □ PANORAL HYDRATE

TOXICITY DATA WITH REFERENCE
ipr-rat LD50:1259 mg/kg IYKEDH 13,637,82
scu-rat LD50:4838 mg/kg IYKEDH 13,637,82
ipr-mus LD50:1227 mg/kg IYKEDH 13,637,82
scu-mus LD50:4180 mg/kg IYKEDH 13,637,82

SAFETY PROFILE: Moderately toxic by intraperitoneal route. Mildly toxic by subcutaneous route. When heated to decomposition it emits toxic fumes of Cl^-, SO_x, and NO_x.

CCR875 CAS:3254-89-5 HR: 3
CEFADOL
mf: $C_{21}H_{27}NO•ClH$ mw: 345.95

PROP: Crystals from $CHCl_3$/EtOAc. Mp: 212–214°.

SYNS: CELMIDOL □ DEPHENIDOL HYDROCHLORIDE □ DIFENIDOL HYDROCHLORIDE □ DIFENIDOLIN □ α,α-DIPHENYL-1-PIPERIDINEBUTANOL HYDROCHLORIDE □ MANIOL □ MECALMIN □ PINERORO □ SATANOLON □ TENESDOL □ WANSAR □ YESDOL

TOXICITY DATA WITH REFERENCE
orl-rat TDLo:4950 mg/kg (11D pre):REP GNRIDX 5,430,71
orl-rat LD50:515 mg/kg IYKEDH 4,193,73
ipr-rat LD50:82 mg/kg IYKEDH 4,193,73
scu-rat LD50:670 mg/kg IYKEDH 4,193,73
ivn-rat LD50:29 mg/kg IYKEDH 4,193,73
ims-rat LD50:635 mg/kg IYKEDH 4,193,73
orl-mus LD50:400 mg/kg IYKEDH 4,193,73
ipr-mus LD50:105 mg/kg IYKEDH 4,193,73
scu-mus LD50:163 mg/kg IYKEDH 4,193,73
ivn-mus LD50:37 mg/kg IYKEDH 4,193,73

SAFETY PROFILE: Poison by ingestion, subcutaneous, intravenous, and intraperitoneal routes. Moderately toxic by intramuscular route. Experimental reproductive effects. When heated to decomposition it emits toxic fumes of NO_x and HCl.

CCR890 CAS:3577-01-3 HR: 2
CEFALOGLYCIN
mf: $C_{18}H_{19}N_3O_6S$ mw: 405.46

PROP: Dihydrate, Kafocin. Crystalline powder. Mp: 223–250° (decomp).

SYNS: 7-(d-α-AMINOPHENYL-ACETAMIDO)CEPHALOSPORANIC ACID □ CEPHALOGLYCIN □ CEPHALOGLYCINE □ d-CEPHALOGLYCINE □ CEPHAOGLYCIN ACID □ KAFOCIN □ KEFGLYCIN □ LILLY 39435

TOXICITY DATA WITH REFERENCE
orl-rat TDLo:1200 mg/kg (7-12D preg):REP NKRZAZ 18,39,70
ipr-rat LD50:1300 mg/kg NKRZAZ 18,22,70
scu-rat LD50:2800 mg/kg NKRZAZ 18,22,70
ipr-mus LD50:1030 mg/kg NKRZAZ 18,22,70
scu-mus LD50:3700 mg/kg NKRZAZ 18,22,70

SAFETY PROFILE: Moderately toxic by subcutaneous and intraperitoneal routes. Experimental reproductive effects. When heated to decomposition it emits toxic fumes of SO_x and NO_x.

CCR925 CAS:30034-03-8 HR: 2
CEFAMANDOLE SODIUM
mf: $C_{18}H_{17}N_6O_5S_2•Na$ mw: 484.52

SYN: SODIUM CEFAMANDOLE

TOXICITY DATA WITH REFERENCE
ivn-rat TDLo:5500 mg/kg (female 7-17D post):REP NKRZAZ 27(Suppl 5),658,79
ivn-rat TDLo:5500 mg/kg (7-17D preg):REP NKRZAZ 27(Suppl 5),658,79
ipr-rat TDLo:11 g/kg (7-17D preg):REP NKRZAZ 27(Suppl 5),658,79
ipr-rat TDLo:13,500 mg/kg (female 17-22D post):REP NKRZAZ 27(Suppl 5),682,79
ipr-rbt TDLo:195 mg/kg (female 6-18D post):TER NKRZAZ 27(Suppl 5),658,79
ipr-rbt TDLo:780 mg/kg (female 6-18D post):REP NKRZAZ 27(Suppl 5),658,79
ivn-rat TDLo:27 g/kg (female 17-22D post):REP NKRZAZ 27(Suppl 5),682,79
ipr-rat TDLo:11 g/kg (7-17D preg):TER NKRZAZ 27(Suppl 5),658,79
scu-rat LD50:12,100 mg/kg YAKUD5 26,115,84
ivn-rat LD50:4410 mg/kg YAKUD5 26,115,84
scu-mus LD50:10,300 mg/kg YAKUD5 26,115,84
ivn-mus LD50:4460 mg/kg YAKUD5 26,115,84

SAFETY PROFILE: Moderately toxic by intravenous route. An experimental teratogen. Experimental reproductive effects. When heated to decomposition it emits toxic fumes of SO_x, NO_x, and Na_2O.

CCR950 CAS:64485-93-4 HR: 2
CEFATOXIME SODIUM
mf: $C_{16}H_{16}N_5O_7S_2•Na$ mw: 477.48

PROP: A solid. Mp: 162–163° (decomp).

SYNS: (6R-(6-α,7-β(Z)))-3-((ACETYLOXY)METHYL)-7-(((2-AMINO-4-THIAZOLYL)(METHOXYIMIO)ACETYL)AMINO)-8-OXO-5-THIA-1-AZABICYCLO(4,2,0)OCT-2-ENE-2-CARBOXYLIC ACID, SODIUM SALT □

CEFOTAXIME SODIUM □ CTX □ HR 756 □ RU 24756 □ SODIUM-7-(2-(2-AMINO-4-THIAZOLYL)-2-METHOXYIMINOACETAMIDO) CEPHALOSPORANATE

TOXICITY DATA WITH REFERENCE
ivn-rbt TDLo:325 mg/kg (6-18D preg):TER OYYAA2 21,375,81
ivn-cld TDLo:1800 mg/kg/18D-I:BLD DICPBB 17,739,83
orl-rat LD50:20 g/kg NKRZAZ 28(Suppl 1),98,80
ipr-rat LD50:10 g/kg NKRZAZ 28(Suppl 1),98,80
scu-rat LD50:18,400 mg/kg NKRZAZ 28(Suppl 1),98,80
ivn-rat LD50:7000 mg/kg NIIRDN 6,APP-8,82
ims-rat LD50:2000 mg/kg NKRZAZ 28(Suppl 1),98,80
orl-mus LD50:20 g/kg NKRZAZ 28(Suppl 1),98,80
ipr-mus LD50:10 g/kg NIIRDN 6,APP-8,82
scu-mus LD50:12,950 mg/kg NIIRDN 6,APP-8,82
ivn-mus LD50:8350 mg/kg NKRZAZ 28(Suppl 1),98,80
ivn-rbt LD50:1880 mg/kg NKRZAZ 28(Suppl 1),98,80

SAFETY PROFILE: Moderately toxic by intravenous and intramuscular routes. Mildly toxic by ingestion and intraperitoneal routes. Human systemic effects by intravenous route: agranulocytosis. An experimental teratogen. When heated to decomposition it emits toxic fumes of SO_x, NO_x and Na_2O.

CCS250 CAS:27164-46-1 HR: 2
CEFAZOLIN SODIUM SALT
mf: $C_{14}H_{13}N_8O_4S_3{\cdot}Na$ mw: 476.52

PROP: Mp: 185–186°. Sol in H_2O.

SYNS: ACEF □ ANCEF □ ATIRIN □ BIAZOLINA □ CEFACIDAL □ CEFAMEDIN □ CEFAMEZIN □ CEFAZIL □ CEFAZINA □ CEFAZOLIN □ CEFAZOLINE SODIUM □ CEZ SODIUM □ ELZOGRAM □ FIRMACEF □ GRAMAXIN □ KEFZOL □ LIVICLINA □ MONOSODIUM CEFAZOLIN □ SKF 41588 □ SODIUM CEFAZOLIN □ SODIUM CEPHAZOLIN □ SODIUM CEZ □ TOTACEF □ ZOLICEF

TOXICITY DATA WITH REFERENCE
ivn-rat TDLo:5500 mg/kg (female 7-17D post):REP NKRZAZ 28(Suppl 7),1119,80
ivn-rat TDLo:5500 mg/kg (7-17D preg):TER NKRZAZ 28(Suppl 7),1119,80
ivn-rat TDLo:5500 mg/kg (female 7-17D post):TER NKRZAZ 28(Suppl 7),1119,80
ivn-rat TDLo:21 g/kg (female 17-22D post):REP NKRZAZ 35(Suppl 1),404,87
scu-rat TDLo:28 g/kg (28D male):REP JZKEDZ 2,1,76
ivn-rat TDLo:21 g/kg (female 17-22D post):REP NKRZAZ 35(Suppl 1),404,87
ivn-wmn TDLo:660 mg/kg/11D-I NPRNAY 45,72,87
ims-hmn TDLo:14 mg/kg/D:GIT,SKN JMGZAI 8(8),10,71
scu-rat LD50:7400 mg/kg NIIRDN 6,404,82
scu-rat LD50:10 g/kg MEIEDD 10,269,83
ivn-rat LD50:2760 mg/kg NKRZAZ 35(Suppl 1),207,87
ipr-mus LD50:6200 mg/kg NIIRDN 6,404,82
scu-mus LD50:7600 mg/kg JIDIAQ 128,S379,73
ivn-mus LD50:3900 mg/kg JIDIAQ 128,S379,73
scu-dog LD50:4 g/kg NKRZAZ 18,528,70
ivn-dog LD50:2200 mg/kg ARZNAD 29,424,79
ivn-rbt LD50:2500 mg/kg ARZNAD 29,424,79

SAFETY PROFILE: Moderately toxic by subcutaneous and intravenous routes. Mildly toxic by intraperitoneal route. Human systemic effects by intramuscular route: changes in structure or function of the salivary glands, nausea or vomiting, and allergic dermatitis. An experimental teratogen. Other experimental reproductive effects. When heated to decomposition it emits very toxic fumes of NO_x, Na_2O, and SO_x.

CCS300 CAS:75738-58-8 HR: 2
CEFMENOXIME HEMIHYDROCHLORIDE
mf: $C_{16}H_{17}N_9O_5S_3{\cdot}1/2ClH$ mw: 529.77

SYNS: AB 50912 HEMIHYDROCHLORIDE □ SCE 1365 HYDROCHLORIDE

TOXICITY DATA WITH REFERENCE
ipr-rat TDLo:10,500 mg/kg (35D male):REP TAKHAA 42,104,83
scu-rat LD50:13,150 mg/kg JJANAX 25,2615,82
ivn-rat LD50:2680 mg/kg JJANAX 35,2615,82
orl-mus LD50:17,540 mg/kg IYKEDH 14,297,83
scu-mus LD50:11,830 mg/kg JJANAX 35,2615,82
ivn-mus LD50:7830 mg/kg YAKUD5 22,1605,80

SAFETY PROFILE: Moderately toxic by intravenous route. Mildly toxic by ingestion. Experimental reproductive effects. When heated to decomposition it emits toxic fumes of SO_x, NO_x, and HCl.

CCS350 CAS:56796-20-4 HR: 1
CEFMETAZOLE
mf: $C_{15}H_{17}N_7O_5S_2$ mw: 471.57

SYNS: CS 1170 □ SKF 83088

TOXICITY DATA WITH REFERENCE
ivn-dog TDLo:9 g/kg (18-35D preg):TER JZKEDZ 6,289,80
ipr-mus LD50:10,233 mg/kg SKKNAJ 31,49,79
scu-mus LD50:12,190 mg/kg SKKNAJ 31,49,79
ivn-mus LD50:8690 mg/kg SKKNAJ 31,49,79

SAFETY PROFILE: Mildly toxic by subcutaneous, intravenous, and intraperitoneal routes. An experimental teratogen. When heated to decomposition it emits toxic fumes of SO_x and NO_x.

CCS360 HR: 2
CEFMETAZOLE SODIUM
mf: $C_{15}H_{16}N_7O_5S_3{\cdot}Na$ mw: 493.55

SYNS: CMZ SODIUM □ CS 1170 SODIUM □ SKF 83088 SODIUM

TOXICITY DATA WITH REFERENCE
ivn-mus TDLo:5 g/kg (female 6-15D post):REP SKKNAJ 30,148,78
ivn-rat TDLo:5500 mg/kg (female 7-17D post):TER SKKNAJ 30,148,78
ivn-mus TDLo:5 g/kg (female 6-15D post):TER SKKNAJ 30,148,78
orl-rat LD50:3204 mg/kg JOPHDQ 8,633,85
orl-mus LD50:3228 mg/kg JOPHDQ 8,633,85
ipr-mus LD50:10,233 mg/kg SKKNAJ 30,112,78
scu-mus LD50:12,190 mg/kg SKKNAJ 30,112,78
ivn-mus LD50:8690 mg/kg SKKNAJ 30,112,78

SAFETY PROFILE: Moderately toxic by ingestion. Mildly toxic by subcutaneous, intravenous, and intraperito-

C

neal routes. An experimental teratogen. Other experimental reproductive effects. When heated to decomposition it emits toxic fumes of SO_x, NO_x, and Na_2O.

CCS365 **HR: 1**
CEFMINOX
mf: $C_{16}H_{20}N_7O_7S3{\cdot}Na$ mw: 541.60

SYNS: (6R-(6-α,7-α))-7-((((2-AMINO-2-CARBOXYETHYL)THIO)ACETYL)AMINO)-7-METHOXY-3-(((1-METHYL-1H-TETRAZOL-5-YL)THIO)METHYL)-8-OXO-5-THIA-1-AZABICYCLO(4.2.0)OCT-2-ENE-2-CARB OXYLIC ACID MONOSODIUM SALT □ CEPHAMYCIN □ MEICELIN □ MT-141

TOXICITY DATA WITH REFERENCE
unr-rat TDLo:36 g/kg (male 36D pre):REP TOLED5 31(Suppl),66,86
ipr-rat LD50:8550 mg/kg JJANAX 37,847,84
ivn-rat LD50:5700 mg/kg JJANAX 37,847,84
ims-rat LD50:9600 mg/kg JJANAX 37,847,84
ivn-mus LD50:5200 mg/kg JJANAX 37,847,84
ims-mus LD50:8200 mg/kg JJANAX 37,847,84

SAFETY PROFILE: Mildly toxic by intravenous, intramuscular, and intraperitoneal routes. Experimental reproductive effects. When heated to decomposition it emits toxic fumes of SO_x, NO_x, and Na_2O.

CCS369 **CAS:62893-20-3** **HR: 2**
CEFOPERAZONE SODIUM
mf: $C_{25}H_{27}N_9O_8S_2{\cdot}Na$ mw: 668.72

SYNS: CPZ □ T-1551

TOXICITY DATA WITH REFERENCE
ivn-man TDLo:57 mg/kg/4D-I:SYS,BLD DICPBB 18,314,84
ivn-wmn TDLo:220 mg/kg/5D-I:GIT,BLD DICPBB 20,281,86
par-hmn TDLo:622 mg/kg/10D-I:GIT,BLD SMJOAV 80,1360,87
unr-man TDLo:229 mg/kg/4D-I:BLD AIMEAS 102,721,85
ivn-rat LD50:4260 mg/kg NKRZAZ 28(Suppl 6),179,80
ipr-mus LD50:8200 mg/kg NKRZAZ 28(Suppl 6),179,80
scu-mus LDLo:15 g/kg NKRZAZ 28(Suppl 6),179,80
ivn-mus LD50:3840 mg/kg NKRZAZ 28(Suppl 6),179,80
ivn-dog LDLo:6 g/kg NKRZAZ 28(Suppl 6),179,80

SAFETY PROFILE: Moderately toxic by intravenous routes. Mildly toxic by subcutaneous and intraperitoneal routes. Human systemic effects by an unspecified route: change in clotting factors, hematuria, hemorrhage, ulceration, or bleeding from large intestine. When heated to decomposition it emits toxic fumes of SO_x, NO_x, and Na_2O.

CCS371 **CAS:74356-00-6** **HR: 1**
CEFOTAN
mf: $C_{17}H_{17}N_7O_8S_4{\cdot}2Na$ mw: 621.63

SYNS: (6R-cis)-7-(((4-(2-AMINO-1-CARBOXY-2-OXOETHYL)-1,3-DITHIETAN-2-YL)CARBONYL)AMINO)-7-METHOXY-3-(((1-METHYL-1H-TETRAZOL-5-YL)THIO)METHYL)-8-OXO-5-THIA-1-AZABICYCLO(4.2.0)OCT-2-ENE-2-CARBO XYLIC ACID MONOSODIUM SALT □ CEFOTETAN DISODIUM SALT □ ICI 156834 DISODIUM □ YM 09330

TOXICITY DATA WITH REFERENCE
ivn-rat TDLo:2700 mg/kg (17-22D preg/21D post):REP OYYAA2 23,767,82
ipr-rat TDLo:5460 mg/kg (26W male):REP KSRNAM 16,98,82
ipr-rat LD50:8250 mg/kg NKRZAZ 30(Suppl 1),212,82
ivn-rat LD50:6790 mg/kg NKRZAZ 30(Suppl 1),212,82
ipr-mus LD50:8120 mg/kg NKRZAZ 30(Suppl 1),212,82
ivn-mus LD50:4990 mg/kg NKRZAZ 30(Suppl 1),212,82

SAFETY PROFILE: Mildly toxic by intravenous and intraperitoneal routes. Experimental reproductive effects. When heated to decomposition it emits toxic fumes of SO_x, NO_x, and Na_2O.

CCS373 **CAS:69712-56-7** **HR: 1**
CEFOTETAN
mf: $C_{17}H_{17}N_7O_8S_4$ mw: 575.65

PROP: Crystals from $MeOH/CH_2Cl_2$.

SYNS: CTT □ 5-THIA-1-AZABICYCLO(4.2.0)OCT-2-ENE-2-CARBOXYLIC ACID, 7-(((4-(2-AMINO-1-CARBOXY- 2-OXOETHYLIDENE)-1,3-DITHIETAN-2-YL)CARBONYL)AMINO)-7-METHOXY-3-(((1-METHY L-1H-TETRAZOL- 5-YL)THIO)METHYL)-8-OXO-, (6R-(6-α-7-α))- □ YM 09330

TOXICITY DATA WITH REFERENCE
ivn-rat TDLo:1100 mg/kg (female 7-17D post):REP NKRZAZ 30(Suppl 1),278,82
ivn-rat LD50:5 g/kg 43MKAT 1,273,80

SAFETY PROFILE: Slightly toxic by intravenous route. Experimental reproductive effects. When heated to decomposition it emits toxic fumes of NO_x and SO_x.

CCS375 **CAS:66309-69-1** **HR: 2**
CEFOTIAM DIHYDROCHLORIDE
mf: $C_{18}H_{23}N_9O_4S_2{\cdot}2ClH$ mw: 598.60

PROP: Light-yellow crystals. Sol in MeOH; sltly sol in EtOH.

SYN: CEFOTIAM HYDROCHLORIDE

TOXICITY DATA WITH REFERENCE
ivn-rat LD50:3680 mg/kg IYKEDH 12,668,81
scu-mus LD50:7800 mg/kg NIIRDN 6,411,82
ivn-mus LD50:3840 mg/kg IYKEDH 12,668,81

SAFETY PROFILE: Moderately toxic by intravenous route. Mildly toxic by subcutaneous route. When heated to decomposition it emits toxic fumes of SO_x, NO_x, and HCl.

CCS500 **CAS:35607-66-0** **HR: 2**
CEFOXITIN
mf: $C_{16}H_{17}N_3O_7S_2$ mw: 427.48

PROP: A solid. Mp: 149–150° (decomp).

SYNS: CEPHOXITIN □ CFX □ REPHOXITIN

TOXICITY DATA WITH REFERENCE
ivn-wmn TDLo:75 mg/kg/18H-I:BLD AIMEAS 92,874,80
ivn-rat LD50:8580 mg/kg NKRZAZ 26(Suppl 1),150,78
scu-mus LD50:9250 mg/kg NKRZAZ 26(Suppl 1),150,78
ivn-mus LD50:4970 mg/kg NKRZAZ 26(Suppl 1),150,78

SAFETY PROFILE: Mildly toxic by subcutaneous and intravenous routes. Human systemic effects by intravenous route: reduction in the white blood cell count. When heated to decomposition it emits very toxic fumes of NO_x and SO_x. See also AMINES.

CCS510 CAS:33564-30-6 HR: 1
CEFOXOTIN SODIUM
mf: $C_{16}H_{16}N_3O_7S_2$•Na mw: 449.46

PROP: Crystals.

SYNS: (6R-cis)-3-(((AMINOCARBONYL)OXY)METHYL)-7-METHYOXY-8-OXO-7-((2-THIENYLACETYL)AMINO)-5-THIA-1-AZABICYCLO (4.2.0)OCT-2-ENE-2-CARBOXYLIC ACID MONOSODIUM SALT □ CEFOXITIN SODIUM SALT □ CENOMYCIN □ MEFOXIN □MEFOXITIN □ MERXIN □ MONOSODIUM CEROXITIN

TOXICITY DATA WITH REFERENCE
ivn-man TDLo:100 mg/kg/2D-I:SYS SMJOAV 80,274,87
ivn-man TDLo:229 mg/kg/4D-I:BLD DICPBB 17,816,83
orl-rat LD50:>10 g/kg YAKUD5 22,123,80
scu-rat LD50:>10 g/kg YKYUA6 31,629,80
ivn-rat LD50:8580 mg/kg IYKEDH 11,181,80
scu-mus LD50:9250 mg/kg YKYUA6 31,629,80
ivn-mus LD50:4970 mg/kg IYKEDH 11,181,80
ivn-dog LD50:10,000 mg/kg NIIRDN 6,410,82

SAFETY PROFILE: Mildly toxic by ingestion subcutaneous, and intravenous routes. Human systemic effects: aplastic anemia, interstitial nephritis, normocytic anemia. When heated to decomposition it emits toxic fumes of SO_x, NO_x, and Na_2O. See also CEFOXITIN.

CCS525 CAS:85287-61-2 HR: 2
CEFPIMIZOLE SODIUM
mf: $C_{28}H_{26}N_6O_{10}S_2$•Na mw: 693.66

PROP: Crystals.

SYNS: AC 1370 □ AC 1370 SODIUM □ (6R-(6-α,7-β(R*)))-1-((2-CARBOXY-7-(((((5-CARBOXY-1H-IMIDAZOL-4-YL)CARBONYL)AMINO) PHENYLACETYLE)AMINO)-8-OXO-5-THIA-1-AZABICYCLO(4.2.0)OCT-2-EN-3-YL)METHYL)-4-(2-SULFOETHYL)-PYRI DINIUM HYDROXODIE, inner salt, MONOSODIUM SALT □ U 631963

TOXICITY DATA WITH REFERENCE
orl-rat LD50:>15 g/kg TOLED5 23,135,84
scu-rat LD50:11,500 mg/kg TOLED5 23,135,84
ivn-rat LD50:3500 mg/kg TOLED5 23,135,84
scu-mus LD50:6800 mg/kg TOLED5 23,135,84
ivn-mus LD50:2700 mg/kg TOLED5 23,135,84

SAFETY PROFILE: Moderately toxic by intravenous route. Mildly toxic by subcutaneous route. When heated to decomposition it emits toxic fumes of NO_x, SO_x, and Na_2O.

CCS530 CAS:51762-05-1 HR: 1
CEFROXADIN
mf: $C_{16}H_{19}N_3O_5S$ mw: 365.44

PROP: Internal salt. Mp: 170° (decomp).

SYNS: (6R-(6-α,7-β(R*)))-7-((AMINO-1,4-CYCLOHEXADIEN-1-YLACETAL)AMINO)-3-METHYL-8-OXO-5-THIA-1-AZABICYCLO(4.2.0) OCT-2-ENE-2-CARBOXYLIC ACID □ 7-(D-2-AMINO-2-(1,4-CYCLOHEXADIENYL)ACETAMIDE)-3-METHOXY-3-CEPHEM-4-CARBOXYLIC ACID □ ANTIBIOTIC CGP 9000 □ CEFROXADINE □ CGP 9000 □ CXD □ ORASPOR

TOXICITY DATA WITH REFERENCE
orl-rat TDLo:2700 mg/kg (female 17-22D post):REP OYYAA2 19,615,80
orl-rat TDLo:27 g/kg (17-22D preg/21D post):REP OYYAA2 19,615,80
orl-rat TDLo:29,120 mg/kg (91D pre):REP NKRZAZ 28(Suppl 3),103,80
orl-rat TDLo:182 g/kg (male 91D pre):REP NKRZAZ 28(Suppl 3),103,80
orl-rat TDLo:27 g/kg (17-22D preg/21D post):TER OYYAA2 19,615,80
ipr-rat LD50:6 g/kg NKRZAZ 28(Suppl 3),98,80
ipr-mus LD50:7090 mg/kg JANTAJ 29,653,76
orl-rbt LD50:10 g/kg NIIRDN 6,APP-11,82

SAFETY PROFILE: Mildly toxic by ingestion and intraperitoneal routes. An experimental teratogen. Other experimental reproductive effects. When heated to decomposition it emits toxic fumes of SO_x and NO_x.

CCS535 HR: 2
CEFROXADIN DIHYDRATE
mf: $C_{16}H_{19}N_3O_5S$•$2H_2O$ mw: 401.48

SYNS: (6R-(6-α,7-β(R*)))-7-((AMINO-1,4-CYCLOHEXADIEN-1-YLACETAL)AMINO)-3-METHYL-8-OXO-5-THIA-1-AZABICYCLO(4.2.0) OCT-2-ENE-2-CARBOXYLIC ACID DIHYDRATE □ CGP 9000 DIHYDRATE

TOXICITY DATA WITH REFERENCE
orl-rat TDLo:11 g/kg (7-17D preg):REP IYKEDH 10,802,79
orl-rat TDLo:440 mg/kg (7-17D preg):REP IYKEDH 10,802,79
orl-rat TDLo:81 g/kg (male 8W pre):REP IYKEDH 10,825,79
ipr-rat LD50:3320 mg/kg IYKEDH 13,349,82

SAFETY PROFILE: Moderately toxic by intraperitoneal route. Experimental reproductive effects. When heated to decomposition it emits toxic fumes of SO_x and NO_x. See also CEFROXADIN.

CCS550 CAS:52152-93-9 HR: 2
CEFSULODIN SODIUM
mf: $C_{22}H_{19}N_4O_8S_2$•Na mw: 554.56

PROP: Needles from EtOH (aq). Mp: 175° (decomp).

SYNS: ABBOTT-468 11 □ (6R-(6-α,7-β(R*)))-4-(AMINOCARBONYL)-1-((2-CARBOXY-8-OXO-7-((PHENYLSULFOACETYL)AMINO)-5-THIO-1-AZABICYCLO(4.2.0)OCT-2-EN-3-YL)METHYL)-PYRIDINIUM HYDROXIDE, inner salt, MONOSODIUM SALT □ CEFSULODIN SODIUM □ CGP 7174E □ MONASPOR □ PSEUDOCEF □ PSEUDOMONIL □ PYOCEFAL □ SCE 129 □ SULCEPHALOSPORIN □ TAKESULIN □TILMAPOR □ ULFARET

TOXICITY DATA WITH REFERENCE
orl-rat LD50:>15 g/kg LYPHAD 34,343,83
ipr-rat LD50:3030 mg/kg IYKEDH 12,668,81
scu-rat LD50:5550 mg/kg IYKEDH 12,668,81
ivn-rat LD50:3030 μg/kg YAKUD5 23,439,81
ims-rat LD50:5530 mg/kg IYKEDH 12,668,81

ipr-mus LD50:6350 mg/kg IYKEDH 12,668,81
scu-mus LD50:6940 mg/kg IYKEDH 12,668,81
ivn-mus LD50:3780 mg/kg NIIRDN 6,412,82
ims-mus LD50:3800 mg/kg IYKEDH 12,668,81

SAFETY PROFILE: Moderately toxic by intramuscular, intravenous, and intraperitoneal routes. Mildly toxic by subcutaneous route. When heated to decomposition it emits toxic fumes of NO_x, SO_x, and Na_2O.

CCS560 CAS:41136-22-5 HR: 2
CEFTEZOLE SODIUM
mf: $C_{13}H_{11}N_8O_4S_3$•Na mw: 462.49

TOXICITY DATA WITH REFERENCE
scu-rat TDLo:56,400 mg/kg (26W male):REP NKRZAZ 24,671,76
ivn-rat LD50:3800 mg/kg NIIRDN 6,413,82
ipr-mus LD50:8900 mg/kg NIIRDN 6,413,82
ivn-mus LD50:4700 mg/kg NIIRDN 6,413,82

SAFETY PROFILE: Moderately toxic by intravenous route. Mildly toxic by intraperitoneal route. Experimental reproductive effects. When heated to decomposition it emits toxic fumes of SO_x, NO_x and Na_2O.

CCS600 CAS:55268-75-2 HR: 1
CEFUROXIM
mf: $C_{16}H_{16}N_4O_8S$ mw: 424.42

PROP: White, crystalline solid.

SYNS: (6R-(6-α,7-β(Z)))-3-(((AMINOCARBONYL)OXY)METHYL)-7-((2-FURANYL(METHYOXYIMINO)ACETYL)AMINO)-8-OXO-5-THIA-1-AZABICYCLO(4.2.0)OCT-2-ENE-2-CARBOXYLIC ACID □ CEFUROXIME □ CEPHUROXIME □ CXM □ ZINACEF

TOXICITY DATA WITH REFERENCE
scu-rat TDLo:8800 mg/kg (7-17D preg):REP NKRZAZ 27(Suppl 6),245,79
scu-rat TDLo:4400 mg/kg (female 7-17D post):TER NKRZAZ 27(Suppl 6),245,79
scu-rat TDLo:8800 mg/kg (7-17D preg):TER NKRZAZ 27(Suppl 6),245,79
scu-rat TDLo:32 g/kg (male 9W pre):REP NKRZAZ 27(Suppl 6),245,79
scu-rat TDLo:24 g/kg (60D male):REP NKRZAZ 27(Suppl 6),245,79
scu-rat TDLo:8800 mg/kg (7-17D preg):TER NKRZAZ 27(Suppl 6),245,79
ivn-man TDLo:64 mg/kg/16H-I LANCAO 1,965,84
ivn-mus LD50:10,400 mg/kg DRUGAY 17,233,79

SAFETY PROFILE: Mildly toxic by intravenous route. An experimental teratogen. Other experimental reproductive effects. When heated to decomposition it emits toxic fumes of SO_x and NO_x.

CCS625 CAS:64544-07-6 HR: 3
CEFUROXIME AXETIL
mf: $C_{20}H_{22}N_4O_{10}S$ mw: 510.52

SYNS: CXM-AX □ SN 407 □ 5-THIA-1-AZABICYCLO(4.2.0)OCT-2-ENE-2-CARBOXYLIC ACID, 3-(((AMINOCARBONYL)OXY)METHYL)-7-((2- FURANYL(METHOXYIMINO)ACETYL)AMINO)-8-OXO-, 1-(ACETYLOXY)ETHYL ESTER, (6R-(6-α-7-β (Z)))-

TOXICITY DATA WITH REFERENCE
orl-rat TDLo:26 g/kg (female 17-22D post):REP NKRZAZ 34(Suppl 5),251,86
orl-rbt TDLo:195 mg/kg (female 6-18D post):REP NKRZAZ 34(Suppl 5),271,86
orl-rat TDLo:7800 mg/kg (female 17-22D post):REP NKRZAZ 34(Suppl 5),251,86
ipr-rat LD50:950 mg/kg NKRZAZ 34(Suppl 5),64,86
scu-rat LD50:2500 mg/kg NKRZAZ 34(Suppl 5),64,86
ipr-mus LD50:510 mg/kg NKRZAZ 34(Suppl 5),64,86
scu-mus LD50:1840 mg/kg NKRZAZ 34(Suppl 5),68,86
orl-rbt LD50:200 mg/kg NKRZAZ 34(Suppl 5),64,86

SAFETY PROFILE: Poison by ingestion. Moderately toxic by intraperitoneal and subcutaneous routes. Experimental reproductive effects. When heated to decomposition it emits toxic fumes of NO_x and SO_x.

CCS635 CAS:82219-81-6 HR: 2
CEFZONAME SODIUM
mf: $C_{16}H_{14}N_7O_5S_4$•Na mw: 535.60

SYNS: CL 251931 SODIUM SALT □ CZON □ L-105 □ 5-THIA-1-AZABICYCLO(4.2.0)OCT-2-ENE-2-CARBOXYLIC ACID, 7-(((2-AMINO-4-THIAZOLYL) (METHOXYIMINO)ACETYL)AMINO)-8-OXO-3-((1,2,3-THIADIAZOL-5-YLTHIO)METHYL)-, SODIUM SALT, (6R-(6-α-7-β(Z)))-

TOXICITY DATA WITH REFERENCE
ipr-rat TDLo:233 g/kg (female 26W pre):REP YACHDS 14,4013,86
ipr-rat TDLo:233 g/kg (male 26W pre):REP YACHDS 14,4013,86
ivn-rat LD50:4222 mg/kg NKRZAZ 34(Suppl 3),96,86
ipr-mus LD50:6424 mg/kg NKRZAZ 34(Suppl 3),96,86
scu-mus LD50:8 g/kg NKRZAZ 34(Suppl 3),96,86
ivn-mus LD50:4117 mg/kg NKRZAZ 34(Suppl 3),96,86
ivn-dog LD50:2500 mg/kg NKRZAZ 34(Suppl 3),96,86

SAFETY PROFILE: Moderately toxic by intravenous route. Experimental reproductive effects. When heated to decomposition it emits toxic fumes of NO_x and SO_x.

CCS650 HR: 3
CELANDINE

PROP: A low (1 to 3 feet) herb which produces small yellow flowers from March to August. Most of the plant is covered with fine white hairs and its sap is a red-orange color. It grows in wet soil in the region bounded by Georgia, Missouri, British Columbia, and Nova Scotia.

SYNS: CHELIDONIUM MAJUS L. □ ELON WORT □ FELONWORT □ SWALLOW WORT □ TETTERWORT □ WORT-WEED

SAFETY PROFILE: The whole plant contains poisonous isoquinoline alkaloids some of which are adrenergic blockers. Ingestion (rare because of the unpleasant taste) can cause headache and sleepiness within 14 hours, followed by fever, vomiting, diarrhea, coma, and circulatory collapse within 6 hours.

CCS675 CAS:8064-08-2 HR: D
CELESTAN-DEPOT
mf: $C_{24}H_{31}FO_6 \cdot C_{22}H_{30}FO_8P \cdot 2Na$ mw: 953.02

SYNS: BETAMETHASONE ACETATE mixed with BETAMETHASONE SODIUM PHOSPHATE □ BETAMETHASONE SODIUM PHOSPHATE mixed with BETAMETHASONE ACETATE □ CELESTONE CHRONODOSE □ CELESTONE SOLOSPAN □ CELESTONE SOLUSPAN □ 21-(PHOSPHONOOXY)PREGNA-1,4-DIENE-3,20-DIONE DISODIUM SALT

TOXICITY DATA WITH REFERENCE
ims-rbt TDLo: 800 µg/kg (female 24-26D post): REP AJOGAH 136,234,80
ims-rbt TDLo: 200 µg/kg (female 26D post): REP BBACAQ 574,197,79
scu-rat TDLo: 4200 µg/kg (female 12-13D post): TER TJADAB 23,15,81
scu-rat TDLo: 1800 µg/kg (12-13D preg): TER TJADAB 23,15,81
ims-rbt TDLo: 200 µg/kg (female 26D post): TER BBACAQ 574,197,79

SAFETY PROFILE: An experimental teratogen. Experimental reproductive effects. When heated to decomposition it emits toxic fumes of F^-, PO_x, and Na_2O.

CCT250 CAS:9005-81-6 HR: 2
CELLOPHANE
mf: $(C_6H_{10}O_5)_n$

SYN: VISKING CELLOPHANE

TOXICITY DATA WITH REFERENCE
imp-rat TDLo: 18 mg/kg: ETA CNREA8 15,333,55
imp-mus TDLo: 720 mg/kg: ETA CNREA8 15,333,55
imp-rat TD: 4200 mg/kg: ETA PSEBAA 67,33,48

CONSENSUS REPORTS: Reported in EPA TSCA Inventory.

SAFETY PROFILE: Questionable carcinogen with experimental tumorigenic data by implant. See also POLYMERS. When heated to decomposition it emits acrid smoke and irritating fumes.

CCT825 HR: 1
CELLRYL

PROP: Protein-free extract of calf blood which comprises various kinds of amino acids, peptides, nucleosides, electrolytes, and unidentified organic substances; exerts healing effect on experimentally induced ulcer and wound (UsuT## 29JUN79).

TOXICITY DATA WITH REFERENCE
scu-rat TDLo: 100 g/kg (8-17D preg): REP YACHDS 4,1114,76
scu-rat TDLo: 25 g/kg (8-17D preg): TER YACHDS 4,1114,76
scu-rat TDLo: 114 g/kg (male 91D pre): REP YACHDS 4,1123,76
ivn-mus LD50: 43 g/kg YACHDS 4,74,76

SAFETY PROFILE: Mildly toxic by intravenous route. An experimental teratogen. Other experimental reproductive effects. When heated to decomposition it emits toxic fumes of NO_x.

CCT900 HR: 2
CELLULASE AP3

TOXICITY DATA WITH REFERENCE
ipr-rat LD50: 2650 mg/kg KSRNAM 8,3751,74
scu-rat LD50: 11,920 mg/kg KSRNAM 8,3751,74
orl-mus LD50: 30,900 mg/kg KSRNAM 8,3751,74
ipr-mus LD50: 3660 mg/kg KSRNAM 8,3751,74
scu-mus LD50: 6710 mg/kg KSRNAM 8,3751,74

SAFETY PROFILE: Moderately toxic by intraperitoneal route. Mildly toxic by ingestion and subcutaneous routes. When heated to decomposition it emits toxic fumes of NO_x.

CCU100 HR: 1
CELLULOSE, MICROCRYSTALLINE

PROP: Fine white crystalline powder from treatment of α-cellulose with mineral acids. Insol in water, most organic solvents.

SYN: CELLULOSE GEL

SAFETY PROFILE: A nuisance dust. When heated to decomposition it emits acrid smoke and irritating fumes.

CCU150 CAS:9004-34-6 HR: 1
CELLULOSE, POWDERED

PROP: Fine white fibrous particles from treatment of bleached cellulose from wood or cotton. Insol in water and most organic solvents.

SYNS: ABICEL □ β-AMYLOSE □ ARBOCEL □ ARBOCEL BC 200 □ ARBOCELL B 600/30 □ AVICEL □ AVICEL 101 □ AVICEL 102 □ AVICEL PH 101 □ AVICEL PH 105 □ CELLEX MX □ α-CELLULOSE □ CELLULOSE 248 □ CELLULOSE (ACGIH,OSHA) □ CELLULOSE CRYSTALLINE □ CELUFI □ CEPO □ CEPO CFM □ CEPO S 20 □ CEPO S 40 □ CHROMEDIA CC 31 □ CHROMEDIA CF 11 □CUPRICELLULOSE □ ELCEMA F 150 □ ELCEMA G 250 □ ELCEMA P 050 □ ELCEMA P 100 □ FRESENIUS D 6 □ HEWETEN 10 □ HYDROXYCELLULOSE □ KINGCOT □ LA 01 □ MN-CELLULOSE □ ONOZUKA P 500 □ PYROCELLULOSE □ RAYOPHANE □ RAYWEB Q □ REXCEL □ SIGMACELL □ SOLKA-FIL □ SOLKA-FLOC □ SOLKA-FLOC BW □ SOLKA-FLOC BW 20 □ SOLKA-FLOC BW 100 □ SOLKA-FLOC BW 200 □ SOLKA-FLOC BW 2030 □ SPARTOSE OM-22 □ SULFITE CELLULOSE □ TOMOFAN □ TUNICIN □ WHATMAN CC-31

OSHA PEL: Total Dust: 15 mg/m^3; Respirable Fraction: 5 mg/m^3
ACGIH TLV: TWA (nuisance particulate) 10 mg/m^3 of total dust (when toxic impurities are not present, e.g., quartz $<1\%$).

SAFETY PROFILE: A nuisance dust. When heated to decomposition it emits acrid smoke and irritating fumes.

For occupational chemical analysis use NIOSH: Nuisance Dust, Total, 0500; Nuisance Dust, Respirable, 0600.

CCU250 CAS:9004-70-0 HR: 3
CELLULOSE TETRANITRATE
DOT: UN 0340/UN 0341/UN 0342/UN 0343/UN 2059/UN 2555/UN 2556/UN 2557
mf: $C_{12}H_{16}(ONO_2)_4O_6$ mw: 504.3

PROP: White, amorphous solid. D: 1.66, flash p: 55°F.

SYNS: AS ☐ C 2018 ☐ CA 80-15 ☐ CELEX ☐ CELLOIDIN ☐ CELLULOSE NITRATE ☐ CELLULOSE, NITRATE (9CI) ☐ COLLODION ☐ COLLODION COTTON ☐ COLLODION WOOL ☐ COLLOXYLIN ☐ CORIAL EM FINISH F ☐ E 1440 ☐ FLEXIBLE COLLODION ☐ FM-NTS ☐ GUNCOTTON ☐ HX 3/5 ☐ KODAK LR 115 ☐ LR 115 ☐ NITROCELLULOSE, dry or wetted with <25% water (or alcohol), by weight (UN 0340) (DOT) ☐ NITROCELLULOSE, plasticized with not <18% plasticizing substance, by weight (UN 0343) (DOT) ☐ NITROCELLULOSE, solution, flammable with not >12.6% nitrogen, by weight (UN 2059) (DOT) ☐ NITROCELLULOSE, unmodified or plasticized with <18% plasticizing substance (UN 0341) (DOT) ☐ NITROCELLULOSE, wetted with not <25% alcohol, by weight (UN 0342) (DOT) ☐ NITROCELLULOSE with alcohol not <25% alcohol by weight, and not >12.6% nitrogen (UN 2556) (DOT) ☐ NITROCELLULOSE with plasticizing not <18% plasticizing substance, by weight (UN 2557) (DOT) ☐ NITROCELLULOSE with water not <25% water, by weight (UN 2555) (DOT) ☐ NITROCELLULOSE E950 ☐ NITROCOTTON ☐ NITRON ☐ NITRON (NITROCELLULOSE) ☐ NIXON N/C ☐ NTs 62 ☐ NTs 218 ☐ NTs 222 ☐ NTs 539 ☐ NTs 542 ☐ PARLODION ☐ PYRALIN ☐ PYROXYLIN ☐ RF 10 ☐ RS ☐ R.S.NITROCELLULOSE ☐ SOLUBLE GUN COTTON ☐ SS ☐ SYNPOR ☐ TSAPOLAK 964 ☐ XYLOIDIN

TOXICITY DATA WITH REFERENCE
orl-rat LD50:>5 g/kg TXAPA9 33,159,75
orl-mus LD50:>5 g/kg TXAPA9 33,159,75

CONSENSUS REPORTS: Reported in EPA TSCA Inventory.

DOT CLASSIFICATION: EXPLOSIVE 1.1D; *Label:* EXPLOSIVE 1.1D (UN 0340, UN 0341); DOT Class: EXPLOSIVE 1.3C; *Label:* EXPLOSIVE 1.3C (UN 0343, UN 0342); DOT Class: 3; *Label:* Flammable Liquid (UN 2059); DOT Class: 4.1; *Label:* Flammable Solid (UN 2556, UN 2557, UN 2555)

SAFETY PROFILE: Very low oral toxicity. Flammable solid. Highly dangerous fire hazard in the dry state when exposed to heat, flame, or powerful oxidizers. When wet with 35% of denatured ethanol it is about as hazardous as ethanol alone or gasoline. Dry cellulose tetranitrate burns rapidly with intense heat and ignites easily. Moderately dangerous explosion hazard. To fight fire, use copious volumes of water; alcohol foam. CO_2 is effective in extinguishing fires of nitrocellulose solvents. See also EXPLOSIVES, HIGH.

CCW250 HR: 3
CEMENT (rubber)

PROP: Flash p: 50°F or less.

SYNS: CEMENT, RUBBER ☐ RUBBER CEMENT

SAFETY PROFILE: May contain benzene or other toxic solvents. See specific constituent. Dangerous fire hazard when exposed to heat or flame; can react with oxidizing materials.

CCW375 CAS:82636-28-0 HR: 3
CENTBUCRIDINE HYDROCHLORIDE
mf: $C_{17}H_{22}N{\cdot}7ClH$ mw: 495.62

SYN: 1,2,3,4-TETRAHYDRO-4-(N-BUTYLAMINO)ACRIDINE HYDROCHLORIDE

TOXICITY DATA WITH REFERENCE
scu-mus TDLo:400 mg/kg (6-15D preg):TER IJEBA6 20,337,82
scu-rat LD50:45 mg/kg IJEBA6 20,330,82
ipr-mus LD50:25 mg/kg INJPD2 19,44,87
scu-mus LD50:26 mg/kg IJEBA6 20,330,82
scu-mky LD50:10,500 µg/kg IJEBA6 20,330,82

SAFETY PROFILE: Poison by subcutaneous and intraperitoneal routes. Experimental reproductive effects. When heated to decomposition it emits toxic fumes of NO_x and HCl.

CCW500 CAS:41510-23-0 HR: 3
CENTBUTINDOLE
mf: $C_{24}H_{26}FN_3O$ mw: 391.53

SYN: 1,2,3,4,6,7,12A-OCTAHYDRO-2-(1-(p-FLUOROPHENYL)-1-OXO-4-BUTYL)-PYRAZINO(2,1:6,1)PYRIDO(3,4-B)INDOLE

TOXICITY DATA WITH REFERENCE
orl-rat LD50:700 mg/kg DRFUD4 3,803,78
ipr-mus LD50:180 mg/kg DRFUD4 3,803,78

SAFETY PROFILE: Poison by intraperitoneal route. Moderately toxic by ingestion. When heated to decomposition it emits very toxic fumes of F^- and NO_x.

CCW750 CAS:51023-56-4 HR: 3
CENTCHROMAN HYDROCHLORIDE
mf: $C_{30}H_{35}NO_2{\cdot}ClH$ mw: 478.12

SYNS: 67/20CDRI ☐ 3,4-trans-2,2-DIMETHYL-3-PHENYL-4-p-(β-PYRROLIDINOETHOXY)PHENYL-7-METHOXYCHROMAN HCl

TOXICITY DATA WITH REFERENCE
scu-rat TDLo:1250 µg/kg (9D preg):REP CCPTAY 9,279,74
scu-rat TDLo:1 mg/kg (1D pre):REP CCPTAY 13,597,76
orl-rat LDLo:1600 mg/kg IJEBA6 15,1159,77
ipr-mus LD50:400 mg/kg IJEBA6 15,1159,77

SAFETY PROFILE: Poison by intraperitoneal route. Moderately toxic by ingestion. Experimental reproductive effects. When heated to decomposition it emits very toxic fumes of NO_x and HCl.

CCW800 CAS:98459-16-6 HR: 3
CENTPHENAQUIN
mf: $C_{24}H_{27}N_3{\cdot}2ClH$ mw: 429.41

SYN: 7,8,9,10-TETRAHYDRO-11-(4-PHENYL-1-PIPERAZINYL)-6H-CYCLOHEPTA(b)QUINOLINE DIHYDROCHLORIDE

TOXICITY DATA WITH REFERENCE
ipr-rat LD50:493 mg/kg IJEBA6 23,214,85
ipr-mus LD50:494 mg/kg IJEBA6 23,214,85
ivn-mus LD50:56 mg/kg IJEBA6 23,214,85

SAFETY PROFILE: Poison by intravenous route. Moder-

ately toxic by intraperitoneal route. When heated to decomposition it emits toxic fumes of NO_x and HCl.

CCW925 **HR: 3**
CENTRUROIDES SUFFUSUS SUFFUSUS VENOM

SYNS: C. SUFFUSUS SUFFUSUS VENOM □ VENOM, SCORPION, CENTRUROIDES SUFFUSUS SUFFUSUS

TOXICITY DATA WITH REFERENCE
ipr-mus LD50:78 µg/kg TOXIA6 22,308,84
ice-mus LD50:1600 ng/kg TOXIA6 22,308,84
unr-mus LD50:25 µg/kg TOXIA6 20,9,82

SAFETY PROFILE: Deadly poison by intraperitoneal, intracerebral, and possibly other routes.

CCX000 **CAS:123-03-5** **HR: 3**
CEPACOL CHLORIDE
mf: $C_{21}H_{38}N{\bullet}Cl$ mw: 340.05

PROP: A solid. Mp: 87–88°. Sol in water.

SYNS: ACETOQUAT CPC □ AKTIVEX □ AMMONYX CPC □ BIO-SEPT □ CEEPRYN □ CEEPRYN CHLORIDE □ CEPRIM □CETAMI-UM □ CETYLPYRIDINIUM CHLORIDE □ N-CETYLPYRIDINIUM CHLO-RIDE □ 1-CETYLPYRIDINIUM CHLORIDE □ DOBENDAN □ HEXADECYLPYRIDINIUM CHLORIDE □ n-HEXADECYLPYRIDINIUM CHLORIDE □ 1-HEXADECYLPYRIDINIUM CHLORIDE □ INTEXSAN CPC □ PRISTACIN □ PYRISEPT □ QUATERNARIO CPC

TOXICITY DATA WITH REFERENCE
skn-rbt 50 mg/24H MOD 33NFA8 -,2,75
eye-rbt 1% ARZNAD 18,137,68
eye-rbt 100 mg JPMSAE 59,188,70
orl-rat LD50:200 mg/kg SDSTBT 5R,24,72
ipr-rat LD50:6 mg/kg JAPMA8 35,89,46
scu-rat LD50:250 mg/kg JAPMA8 35,89,46
ivn-rat LD50:30 mg/kg AFDOAQ 18,43,54
orl-mus LD50:108 mg/kg PSEBAA 120,511,65
ipr-mus LD50:10 mg/kg JMCMAR 23,469,80
orl-rbt LD50:400 mg/kg PCOC** -,208,66
skn-rbt LDLo:2 g/kg JPMSAE 59,188,70
ivn-rbt LD50:36 mg/kg PCOC** -,208,66

CONSENSUS REPORTS: Reported in EPA TSCA Inventory.

SAFETY PROFILE: Poison by ingestion, intraperitoneal, subcutaneous, and intravenous routes. Moderately toxic by skin contact. A skin and eye irritant. When heated to decomposition it emits very toxic fumes of NO_x and Cl^-.

CCX125 **CAS:5853-29-2** **HR: 3**
CEPHAELINE HYDROCHLORIDE
mf: $C_{28}H_{38}N_2O_4{\bullet}2ClH$ mw: 539.60

SYN: (−)-CEPHAELINE DIHYDROCHLORIDE

TOXICITY DATA WITH REFERENCE
ipr-rat LD50:10 mg/kg JPETAB 104,421,52
orl-mus LD50:74,970 µg/kg NCISP* JAN86
ipr-mus LD50:20,530 µg/kg NCISP* JAN86

SAFETY PROFILE: Poison by ingestion and intraperitoneal routes. When heated to decomposition it emits toxic fumes of NO_x and HCl.

CCX175 **CAS:11005-92-8** **HR: 3**
CEPHALOMYCIN

TOXICITY DATA WITH REFERENCE
orl-mus LD50:1000 mg/kg 85GDA2 4(2),235,80
ipr-mus LD50:55 mg/kg 85FZAT -,204,67
scu-mus LD50:161 mg/kg 85FZAT -,204,67
ivn-mus LD50:31 mg/kg 85ERAY 2,1237,78

SAFETY PROFILE: Poison by subcutaneous, intravenous, and intraperitoneal routes. Moderately toxic by ingestion.

CCX250 **CAS:153-61-7** **HR: 3**
CEPHALOTHIN
mf: $C_{16}H_{16}N_2O_6S_2$ mw: 396.46

PROP: A solid. Mp: 160–160.5°.

SYNS: CEFALOTIN □ CEPHALOTIN □ CET □ CT □ 7-(2-THIENY-LACETAMIDO)CEPHALOSPORANIC ACID □ 7-(THIOPHENE-2-ACET-AMIDO)CEPHALOSPORANIC ACID

TOXICITY DATA WITH REFERENCE
pic-omi 25 µg/plate ZMMPAO 231,369,75
scu-mus TDLo:18 g/kg (9-11D preg):TER SEIJBO 16,250,76
scu-rat TDLo:8800 mg/kg (7-17D preg):TER NKRZAZ 27(Suppl 6),245,79
unr-mus TDLo:9 g/kg (female 9-11D post):TER TJA-DAB 14,250,76
scu-mus TDLo:18 g/kg (9-11D preg):TER SEIJBO 16,250,76
ipr-rat LD50:4296 mg/kg ANTBAL 26(1),44,82
scu-rat LDLo:10 g/kg NKRZAZ 27(Suppl 6),124,79
par-rat LD50:23 mg/kg AACHAX -,863,65
ivn-mus LD50:4990 mg/kg NKRZAZ 26(Suppl 1),150,78
ims-mus LD50:7 g/kg BYYADW 3,220,78
ice-mus LD50:81 mg/kg AACHAX -,863,65

SAFETY PROFILE: Poison by parenteral and intracerebral routes. Moderately toxic by intravenous route. Mildly toxic by subcutaneous and intraperitoneal routes. An experimental teratogen. Mutation data reported. See also ESTERS. When heated to decomposition it emits very toxic fumes of NO_x and SO_x.

CCX500 **CAS:21593-23-7** **HR: 2**
CEPHAPIRIN
mf: $C_{17}H_{17}N_3O_6S_2$ mw: 423.49

PROP: Crystals from Me_2CO (aq). Mp: 155°.

SYNS: CEFAPIRIN (GERMAN) □ 3-(HYDROXYMETHYL)-8-OXO-7-(2-(4-PYRIDYLTHIO)ACETAMIDO)-5-THIA-1-AZABICYCLO(4.2.0)OCT-2-ENE-2-CARBOXYLIC ACID, ACETATE (ESTER)

TOXICITY DATA WITH REFERENCE
ivn-man TDLo:514 µg/kg/9D-I:SYS DICPBB 19,553,85
orl-rat LD50:16,356 mg/kg TOIZAG 21,279,74
ipr-rat LD50:7850 mg/kg TOIZAG 21,279,74
orl-mus LD50:26,088 mg/kg TOIZAG 21,279,74
ipr-mus LD50:8899 mg/kg TOIZAG 21,279,74
scu-mus LD50:13,556 mg/kg TOIZAG 21,279,74

SAFETY PROFILE: Moderately toxic by intraperitoneal route. Human systemic effects by intravenous route:

jaundice. Experimental reproductive effects. When heated to decomposition it emits very toxic fumes of NO_x and SO_x.

CCX550 **CAS:481-49-2** **HR: 3**
CEPHARANTHINE
mf: $C_{37}H_{38}N_2O_6$ mw: 606.77

PROP: Yellow amorphous powder. Mp: 145–155°. From tubers of *Stephania cephalantha Hayata*, and *Stephania sasahii Hayata, Menispermaceae*. Yellow powder. Mp: 145–155°. Obtained by drying solvated needles from acetone + benzene. Soluble in the usual organic solvents except petr ether.

SYNS: CEPHARANTHIN □ 6′,12′-DIMETHOXY-2,2′-DIMETHYL-6,7-(METHYLENEBIS(OXY)OXYACANTHAN)

TOXICITY DATA WITH REFERENCE
ipr-rat TDLo:2400 mg/kg (female 30D pre):REP KSRNAM 16,3855,82
ipr-rat TDLo:1200 mg/kg (female 30D pre):REP KSRNAM 16,3855,82
orl-rat TDLo:182 mg/kg (female 26W pre):REP KSRNAM 16,3855,82
orl-rat TDLo:182 mg/kg (male 26W pre):REP KSRNAM 16,3855,82
orl-rat TDLo:546 mg/kg (91D male):REP KSRNAM 16,3855,82
orl-rat LD50:2 g/kg KSRNAM 16,3855,82
scu-rat LD50:100 mg/kg KSRNAM 16,3855,82
ivn-rat LD50:57 mg/kg KSRNAM 16,3855,82
orl-mus LD50:1900 mg/kg KSRNAM 16,3855,82
ipr-mus LD50:125 mg/kg CPBTAL 24,2413,76
scu-mus LD50:100 mg/kg KSRNAM 16,3855,82
ivn-mus LD50:43,500 µg/kg KSRNAM 16,3855,82

SAFETY PROFILE: Poison by subcutaneous, intravenous, and intraperitoneal routes. Moderately toxic by ingestion. Experimental reproductive effects. When heated to decomposition it emits toxic fumes of NO_x.

CCX600 **CAS:67055-59-8** **HR: 2**
CEPHEDRINE
mf: $C_{13}H_{18}N_2O$ mw: 218.33

SYNS: CEFEDRIN □ 3-((1-HYDROXY-1-PHENYL-2-PROPYL)METHYLAMINO)PROPIONITRILE □ 1-PHENYL-2-(METHYL-(β-CYANOETHYL)AMINO)PROPAN-1-OL

TOXICITY DATA WITH REFERENCE
orl-rat LD50:940 mg/kg PCJOAU 14,773,80
scu-rat LD50:600 mg/kg PCJOAU 14,773,80
orl-mus LD50:465 mg/kg PCJOAU 14,773,80
scu-mus LD50:410 mg/kg PCJOAU 14,773,80
orl-gpg LD50:1200 mg/kg FATOAO 41,345,78

CONSENSUS REPORTS: Cyanide and its compounds are on the Community Right-To-Know List.

SAFETY PROFILE: Moderately toxic by ingestion and subcutaneous routes. When heated to decomposition it emits toxic fumes of NO_x and CN^-. See also NITRILES.

CCX620 **HR: 3**
CERASTES CERASTES VENOM

SYNS: C. CERASTES VENOM □ VENOM, SNAKE, CERASTES CERASTES

TOXICITY DATA WITH REFERENCE
ipr-mus LD50:1080 µg/kg TOXIA6 18,384,80
ivn-mus LD50:375 µg/kg TOXIA6 14,146,76
ivn-dog LDLo:100 µg/kg TOXIA6 6,221,69

SAFETY PROFILE: Deadly poison by intravenous and intraperitoneal routes.

CCX625 **CAS:11005-70-2** **HR: 3**
CERBEROSIDE
mf: $C_{42}H_{66}O_{18}$ mw: 859.08

SYNS: CERBEROSID (GERMAN) □ CERBROSIDE □ THEVETIN B

TOXICITY DATA WITH REFERENCE
ivn-cat LD50:810 µg/kg 85ELDJ -,189,63
unr-cat LDLo:636 µg/kg 85ELDJ 134,63
ivn-gpg LDLo:3539 µg/kg AEPPAE 252,314,66

SAFETY PROFILE: Deadly poison by intravenous and possibly other routes. When heated to decomposition it emits acrid smoke and fumes.

CCX725 **CAS:55467-31-7** **HR: 3**
CEREXIN A
mf: $C_{63}H_{103}N_{15}O_{19}$ mw: 1374.81

PROP: Amorphous powder.

SYN: ANTIBIOTIC 60-6

TOXICITY DATA WITH REFERENCE
ipr-mus LD50:50 mg/kg 85GDA2 4(1),261,80
scu-mus LD50:500 mg/kg 85GDA2 4(1),261,80
ivn-mus LD50:25 mg/kg 85GDA2 4(1),261,80

SAFETY PROFILE: Poison by intravenous and intraperitoneal routes. Moderately toxic by subcutaneous route. When heated to decomposition it emits toxic fumes of NO_x.

CCY000 **CAS:1306-38-3** **HR: 1**
CERIC OXIDE
mf: CeO_2 mw: 172.12

PROP: Pale yellow solid (white when pure). Mp: 2600°. Insol in H_2O; sol in H_2SO_4, and HNO_3 with difficulty.

SYNS: CERIA □ CERIC DIOXIDE □ CERIUM DIOXIDE □ CERIUM (4+) OXIDE □ NIDORAL

TOXICITY DATA WITH REFERENCE
orl-rat LD50:>5 g/kg JACTDZ 12,617,93

SAFETY PROFILE: Low toxicity by ingestion. See also CERIUM COMPOUNDS.

CCY250 **CAS:7440-45-1** **HR: 3**
CERIUM
af: Ce aw: 140.13

PROP: Malleable gray metal, forms lustrous crystals that tarnish in air. Cubic or hexagonal, steel-gray crystals. Mp: 804°, bp: 3433°, d: (cubic form): 6.90, hexagonal form 6.75. Reacts with moist air readily and with H_2O (slow in cold), acids and alkalis.

CONSENSUS REPORTS: Reported in EPA TSCA inventory.

SAFETY PROFILE: Cerium resembles aluminum in its pharmacological action as well as in its chemical properties. The insoluble salts such as the oxalates are stated to be nontoxic even in large doses. It is used to prevent vomiting in pregnancy. The average dose is from 0.05 to 0.5 g.

The effect on the central nervous system of the rare-earth metals following inhalation may preclude welding operations with these materials to any large extent. Cerium is stated to produce polycythemia but is useless in the treatment of anemia owing to its toxic effects. The salts of cerium increase the blood coagulation rate. See also RARE EARTHS. A strong reducing agent. Moderate fire hazard; ignites spontaneously in air at 150–180°. Moderate explosion hazard in the form of dust when exposed to flame. The metal or its alloys spark with friction. Many alloys are pyrophoric in air. See also IRON DUST. Explosive reaction with zinc. Very exothermic reaction with antimony or bismuth. Ignites when heated in atmospheres of $CO_2 + N_2$, Cl_2, or Br_2. Violent reaction when heated with phosphorus (400°C), silicon (1400°C).

CCY500 **CAS:537-00-8** **HR: 3**
CERIUM ACETATE
mf: $C_6H_9O_6 \cdot Ce$ mw: 317.27

SYNS: CERIUM TRIACETATE □ CEROUS ACETATE

TOXICITY DATA WITH REFERENCE
ivn-hmn TDLo:2 mg/kg:CNS JCINAO 21,447,42

CONSENSUS REPORTS: Reported in EPA TSCA Inventory.

SAFETY PROFILE: Human central nervous system effects. See also CERIUM COMPOUNDS. When heated to decomposition it emits acrid and irritating fumes.

CCY699 **HR: 3**
CERIUM AZIDE
mf: CeN_9 mw:266.18

$Ce(N_3)_3$

SAFETY PROFILE: An explosive. Upon decomposition it emits toxic fumes of NO_x. See also CERIUM COMPOUNDS and AZIDES.

CCY750 **CAS:7790-86-5** **HR: 3**
CERIUM CHLORIDE
mf: $CeCl_3$ mw: 246.47

PROP: Colorless or white solid or deliquescent crystals. Mp: 722°, bp: 1705°, d: 3.92. Sol in water and THF.

SYNS: CERIUM(III) CHLORIDE □ CERIUM TRICHLORIDE □ CEROUS CHLORIDE

TOXICITY DATA WITH REFERENCE
orl-rat LD50:2111 mg/kg EQSSDX 1,1,75
scu-rat LDLo:4000 mg/kg AEXPBL 100,230,23
ivn-rat LD50:5096 µg/kg APYPAY 32,205,81
orl-mus LD50:5277 mg/kg EQSSDX 1,1,75
ipr-mus LD50:172 mg/kg COREAF 256,1043,63
scu-mus LDLo:4000 mg/kg AEPPAE 188,465,38
ivn-dog LDLo:60 mg/kg HBAMAK 4,1289,35
ipr-gpg LD50:56 mg/kg AMIHAB 15,9,57
scu-gpg LDLo:2 g/kg AEXPBL 72,228,13
scu-frg LDLo:211 mg/kg EQSSDX 1,1,75

CONSENSUS REPORTS: Reported in EPA TSCA Inventory. EPA Genetic Toxicology Program.

SAFETY PROFILE: Poison by intravenous, intraperitoneal, and subcutaneous routes. Moderately toxic by ingestion. See also CERIUM COMPOUNDS. When heated to decomposition it emits toxic fumes of Cl^-.

CCZ000 **CAS:512-24-3** **HR: 3**
CERIUM CITRATE
mf: $C_6H_8O_7 \cdot Ce$ mw: 332.26

SYNS: CERIUM(III) CITRATE □ CEROUS CITRATE □ 2-HYDROXY-1,2,3-PROPANETRISCARBOXYLIC ACID CERIUM(3+) SALT (1:1) (9CI)

TOXICITY DATA WITH REFERENCE
scu-mus TDLo:190 mg/kg (12D preg):REP JTEHD6 10,449,82
ipr-mus LD50:149 mg/kg AEHLAU 5,437,62
ipr-gpg LD50:83 mg/kg AEHLAU 5,437,62

SAFETY PROFILE: Poison by intraperitoneal route. Experimental reproductive effects. See also CERIUM COMPOUNDS. When heated to decomposition it emits acrid and irritating fumes.

CDA250 **HR: 2**
CERIUM COMPOUNDS

PROP: Compounds of cerium and the other rare-earth elements are generally of low toxicity. The greatest exposures are likely to be during manufacture of cerium. Exposed workers have experienced sensitivity to heat, itching, and skin lesions. Large doses to experimental animals have caused writhing, ataxia (loss of muscle coordination), labored respiration, sedation, hypotension, and death by cardiovascular collapse. The chloride, bromide, nitrate, bromate, and perchlorate salts are water soluble and thus are more likely to cause systemic effects when ingested. The sulfates, iodides, and iodates are less water soluble. Oxides, oxalates, sulfides, carbonates, fluorides, and phosphates are insoluble. The salts of cerium increase the blood coagulation rate. Cerium tartrate has been found to produce a direct injurious action on the hearts of small animals. Cerium oxalate has been used to suppress motion sickness and to suppress vomiting during pregnancy (by ingestion of 1 g/24 hr). The toxicity of cerium compounds may be taken to be that of cerium, except when the anion has a toxicity of its own. See also CERIUM and RARE EARTHS.

CDA500 CAS:15158-67-5 HR: 3
CERIUM EDETATE

TOXICITY DATA WITH REFERENCE
ipr-mus LD50:37.6 mg/kg AEHLAU 5,437,62
ipr-gpg LD50:129 mg/kg AEHLAU 5,437,62

SAFETY PROFILE: Poison by intraperitoneal route. See also CERIUM COMPOUNDS. When heated to decomposition it emits acrid smoke and irritating fumes.

CDA750 CAS:7758-88-5 HR: 1
CERIUM FLUORIDE
mf: CeF_3 mw: 197.12

PROP: White, hexagonal crystals or solid. D: 6.16, mp: 1460°, bp: 2300°. Insol in water; sol in H_2SO_4.

SYNS: CERIUM FLUORURE (FRENCH) □ CERIUM TRIFLUORIDE □ CEROUS FLUORIDE

TOXICITY DATA WITH REFERENCE
orl-rat LD50:>5 g/kg JACTDZ 12,632,93

CONSENSUS REPORTS: Reported in EPA TSCA Inventory.

OSHA PEL: TWA 2.5 mg(F)/m³
ACGIH TLV: TWA 2.5 mg(F)/m³
NIOSH REL: (Inorganic Fluorides) TWA 2.5 mg(F)/m³

SAFETY PROFILE: Low toxicity by ingestion. See FLUORIDES and CERIUM COMPOUNDS. When heated to decomposition it emits toxic fumes of F^-.

CDB000 CAS:10108-73-3 HR: 3
CERIUM(III) NITRATE
mf: $N_3O_9 \cdot Ce$ mw: 326.15

SYNS: CERIUM NITRATE □ CERIUM(3+) NITRATE □ CERIUM TRINITRATE □ CEROUS NITRATE □ DUSICNAN CERITY (CZECH) □ NITRIC ACID, CERIUM(3+) SALT (8CI, 9CI)

TOXICITY DATA WITH REFERENCE
itt-rat TDLo:26,092 μg/kg (1D male):REP JRPFA4 7,21,64
orl-rat LD50:3154 mg/kg EQSSDX 1,1,75
ipr-rat LD50:216 mg/kg EQSSDX 1,1,75
ivn-rat LD50:37 mg/kg EQSSDX 1,1,75

CONSENSUS REPORTS: Reported in EPA TSCA Inventory.

SAFETY PROFILE: Poison by intravenous and intraperitoneal routes. Moderately toxic by ingestion. Experimental reproductive effects. See also CERIUM COMPOUNDS and NITRATES. When heated to decomposition it emits toxic fumes of NO_x.

CDB250 CAS:10294-41-4 HR: 3
CERIUM(III) NITRATE, HEXAHYDRATE (1:3:6)
mf: $N_3O_9 \cdot Ce \cdot 6H_2O$ mw: 434.27

SYNS: CERIUM NITRATE, HEXAHYDRATE □ CERIUM TRINITRATE HEXAHYDRATE □ CEROUS NITRATE HEXAHYDRATE □ NITRIC ACID, CERIUM(3+) SALT, HEXAHYDRATE

TOXICITY DATA WITH REFERENCE
skn-rbt 500 mg/24H MLD JACTDZ 12,615,93
eye-rbt 100 mg SEV JACTDZ 12,615,93
orl-rat LD50:4200 mg/kg TXAPA9 5,750,63
ipr-rat LD50:290 mg/kg TXAPA9 5,750,63
ivn-rat LD50:4 mg/kg TXAPA9 5,750,63
ipr-mus LD50:470 mg/kg TXAPA9 5,750,63

SAFETY PROFILE: Poison by intraperitoneal and intravenous routes. Moderately toxic by ingestion. A skin and eye irritant. See also CERIUM COMPOUNDS and NITRATES. When heated to decomposition it emits toxic fumes of NO_x.

CDB325 CAS:25764-08-3 HR: 3
CERIUM NITRIDE
mf: CeN mw: 154.13

SAFETY PROFILE: Reaction with water or dilute acids may cause ignition and the release of toxic ammonia gas and explosive hydrogen gas. When heated to decomposition it emits toxic fumes of NO_x. See also CERIUM COMPOUNDS and NITRIDES.

CDB500 HR: 3
CERIUM(III) TETRAHYDROALUMINATE
mf: Al_3CeH_{12} mw: 236.46

PROP: Decomp @ −80°C.

SAFETY PROFILE: A dangerous fire hazard. Ignites spontaneously in air. Unstable. See also CERIUM COMPOUNDS and ALUMINUM COMPOUNDS.

CDB760 CAS:8054-43-1 HR: 1
CERNILTON

SYN: CN 009

TOXICITY DATA WITH REFERENCE
orl-rat TDLo:576 g/kg (male 26W pre):REP TOIZAG 15,201,68
orl-rat TDLo:441 g/kg (male 35D pre):REP TOIZAG 15,201,68
ipr-rat LD50:6660 mg/kg TOIZAG 15,201,68
orl-mus LD50:27,610 mg/kg TOIZAG 15,201,68
ipr-mus LD50:6940 mg/kg TOIZAG 15,201,68
scu-mus LD50:13,060 mg/kg TOIZAG 15,201,68

SAFETY PROFILE: Mildly toxic. Experimental reproductive effects. When heated to decomposition it emits acrid smoke and irritating fumes.

CDB770 CAS:106440-54-4 HR: 2
CERNITIN GBX

SYN: GBX

TOXICITY DATA WITH REFERENCE
orl-rat TDLo:700 g/kg (male 35D pre):REP TOIZAG 15,201,68
ipr-rat LD50:3310 mg/kg TOIZAG 15,201,68
orl-mus LD50:52,250 mg/kg TOIZAG 15,201,68
ipr-mus LD50:1720 mg/kg TOIZAG 15,201,68
scu-mus LD50:26,130 mg/kg TOIZAG 15,201,68

SAFETY PROFILE: Moderately toxic by intraperitoneal route. Experimental reproductive effects. When heated to decomposition it emits acrid smoke and irritating fumes.

CDB772 CAS:106440-55-5 HR: 1
CERNITIN T-60

TOXICITY DATA WITH REFERENCE
orl-rat TDLo:420 g/kg (male 35D pre):REP TOIZAG 15,201,68
ipr-rat LD50:7580 mg/kg TOIZAG 15,201,68
orl-mus LD50:27,750 mg/kg TOIZAG 15,201,68
ipr-mus LD50:8310 mg/kg TOIZAG 15,201,68
scu-mus LD50:9470 mg/kg TOIZAG 15,201,68

SAFETY PROFILE: Mildly toxic. Experimental reproductive effects. When heated to decomposition it emits acrid smoke and irritating fumes.

CDB775 CAS:64318-79-2 HR: 3
CERVAGEM
mf: $C_{23}H_{38}O_5$ mw: 394.61

SYNS: 16,16-DIMETHYL-trans-Δ^2-PGE1 METHYL ESTER □ 16,16-DIMETHYL-trans-Δ^2-PROSTAGLANDIN E1 METHYL ESTER □GEMEPROST □ ONO 802 □ PREGLANDIN

TOXICITY DATA WITH REFERENCE
ivg-wmn TDLo:20 µg/kg (10W preg):REP CCPTAY 27,51,83
ivg-rat TDLo:2 mg/kg (female 17-20D post):REP GEIRDK 14,593,82
ivg-rat TDLo:80 µg/kg (female 17-20D post):REP GEIRDK 14,593,82
ivg-rat TDLo:438 µg/kg (female 1-7D post):TER GEIRDK 14,809,82
ivg-rat TDLo:5 mg/kg (female 6-15D post):TER TCMUD8 4,233,84
ipr-rbt TDLo:200 µg/kg (female 24-25D post):REP NYKZAU 79,15,82
ivg rat TDLo:438 µg/kg (female 1-7D post):REP GEIRDK 14,809,82
ivg-rbt TDLo:162 µg/kg (female 6-18D post):REP TCMUD8 4,225,84
ipr-rat TDLo:4 mg/kg (female 2D post):REP NYKZAU 79,15,82
scu-rat TDLo:960 µg/kg (female 30D pre):REP GEIRDK 14,188,82
scu-rat TDLo:15 µg/kg (female 30D pre):REP GEIRDK 14,188,82
scu-rat TDLo:60 mg/kg (male 30D pre):REP GEIRDK 14,188,82
ivg-rat TDLo:2 mg/kg (female 17-20D post):TER GEIRDK 14,593,82
orl-rat LD50:56,500 µg/kg GEIRDK 14,188,82
scu-rat LD50:22,600 µg/kg GEIRDK 14,188,82
ivn-rat LD50:28,600 µg/kg GEIRDK 14,188,82
ivg-rat LD50:32,500 µg/kg GEIRDK 14,188,82
orl-mus LD50:59 mg/kg GEIRDK 14,188,82
scu-mus LD50:32,500 µg/kg GEIRDK 14,188,82
ivn-mus LD50:29,500 µg/kg GEIRDK 14,188,82
ivg-mus LD50:36 mg/kg GEIRDK 14,188,82

SAFETY PROFILE: Poison by ingestion, subcutaneous, intravenous, and intravaginal routes. Human reproductive effects by intravaginal route: abortion, changes in the uterus, cervix, and vagina. An experimental teratogen. Other experimental reproductive effects. When heated to decomposition it emits acrid smoke and fumes. See other prostaglandin entries.

CDB800 HR: 3
CESALIN

SYN: NCS 110435

TOXICITY DATA WITH REFERENCE
ipr-rat LD50:1838 µg/kg NCISP* JAN86
scu-rat LD50:7777 µg/kg NCISP* JAN86
ipr-mus LD50:2286 µg/kg NCISP* JAN86

SAFETY PROFILE: Deadly poison by subcutaneous and intraperitoneal routes.

CDC000 CAS:7440-46-2 HR: 3
CESIUM
DOT: UN 1407
af: Cs aw: 132.91

PROP: Bright, shiny, hexagonal crystals; silver-white, ductile metal; or possibly a silvery liquid. Golden when ultra pure. Spontaneously ignites in the atmosphere forming caesium oxides, carbonates and hydroxide. Mp: 28.5°, bp: 668°, d: 1.873, vap press: 1 mm @ 279°. Reacts violently with H_2O forming CsOH and dihydrogen.

SYN: CESIUM-133

TOXICITY DATA WITH REFERENCE
ipr-mus LD50:1700 mg/kg 85IXA4-,704,48

CONSENSUS REPORTS: Reported in EPA TSCA Inventory.

DOT CLASSIFICATION: 4.3; *Label:* Dangerous When Wet

SAFETY PROFILE: Moderately toxic by intraperitoneal route. Cesium is quite similar to potassium in its elemental state. It has been shown, however, to have pronounced physiological action in experimentation with animals. Hyper-irritability, including marked spasms, has been shown to follow the administration of cesium in amounts equal to the potassium content of the diet. It has been found that replacing the potassium in the diet of rats with cesium caused death after 10–17 days. Ignites spontaneously in air. Violent reaction with water, moisture, or steam releases hydrogen gas which explodes. Violent reaction with acids, halogens, and other oxidizing materials. Incandescent reaction with nonmetals (e.g., sulfur, phosphorus). See also SODIUM.

CDC125 CAS:22750-56-7 HR: 3
CESIUM ACETYLIDE
mf: C_2Cs_2 mw: 289.83

SAFETY PROFILE: Explosive reaction on contact with nitric acid. Ignition on contact with fluorine, chlorine, bromine, iodine, and hydrogen chloride. Vigorous or incandescent reaction on heating with iron(III) choride, boron, or silicon. See also CESIUM and ACETYLIDES.

CDC375 CAS:61136-62-7 HR: 3
CESIUM ARSENATE
mf: $AsO_4 \cdot 3Cs$ mw: 537.65

SYN: ARSENIC ACID, TRICESIUM SALT

TOXICITY DATA WITH REFERENCE
ihl-rat TCLo:430 µg/m³/24H (1-22D preg):TER VAMNAQ (8),10,78
ihl-rat TCLo:4600 µg/m³/24H (1-22D preg):TER VAMNAQ (8),10,78
orl-mus LD50:116 mg/kg VAMNAQ (8),10,78

CONSENSUS REPORTS: Arsenic and its compounds are on the Community Right-To-Know List.

OSHA: Cancer Hazard

SAFETY PROFILE: Poison by ingestion. Experimental teratogenic effects by inhalation. When heated to decomposition it emits toxic fumes of As. See also ARSENIC COMPOUNDS and CESIUM.

CDC500 CAS:7787-69-1 HR: 2
CESIUM BROMIDE
mf: BrCs mw: 212.82

PROP: Deliquescent colorless cubic crystals. Mp: 636°, bp: 1300°. Very sol in water.

TOXICITY DATA WITH REFERENCE
ipr-rat LD50:1400 mg/kg AIHOAX 1,637,50

CONSENSUS REPORTS: Reported in EPA TSCA Inventory.

SAFETY PROFILE: Moderately toxic by intraperitoneal route. See also CESIUM and BROMIDES. When heated to decomposition it emits toxic fumes of Br^-.

CDC699 HR: 1
CESIUM BROMOXENATE
mf: $BrCsO_3Xe$ mw: 392.10

SAFETY PROFILE: Solution in water is extremely unstable. When heated to decomposition it emits toxic fumes of Br^-. See also CESIUM and BROMIDES.

CDC750 CAS:534-17-8 HR: 2
CESIUM CARBONATE
mf: $CO_3 \cdot 2Cs$ mw: 325.83

PROP: Deliquescent colorless monoclinic crystals. Very sol in H_2O; sol in EtOH and Et_2O.

SYNS: CARBONIC ACID, DICESIUM SALT □ DICESIUM CARBONATE

TOXICITY DATA WITH REFERENCE
mrc-bcs 5 mol/L MUREAV 77,109,80
orl-rat LD50:2333 mg/kg VAMNAQ (8),10,78
orl-mus LD50:2170 mg/kg VAMNAQ (8),10,78

CONSENSUS REPORTS: EPA Genetic Toxicology Program. Reported in EPA TSCA Inventory.

SAFETY PROFILE: Moderately toxic by ingestion. Mutation data reported. When heated to decomposition it emits acrid smoke and fumes. See also CESIUM.

CDD000 CAS:7647-17-8 HR: 2
CESIUM CHLORIDE
mf: ClCs mw: 168.36

PROP: Deliquescent cubic crystals. Undergoes transition to high temp polymorph at 4°. D: 3.99, mp: 646°, bp: 1209°. Very sol in H_2O, MeOH, and EtOH; insol in Me_2CO.

SYNS: CESIUM MONOCHLORIDE □ DICESIUM DICHLORIDE □ TRICESIUM TRICHLORIDE

TOXICITY DATA WITH REFERENCE
mrc-bcs 5 mol/L MUREAV 77,109,80
sln-smc 20 mmol/L MUTAEX 1,21,86
orl-mus TDLo:262 g/kg (female 1-21D post):REP PHBHA4 46,89,89
orl-rat LD50:2600 mg/kg VAMNAQ (8),10,78
ipr-rat LD50:1500 mg/kg AIHOAX 1,637,50
orl-mus LD50:2306 mg/kg VAMNAQ (8),10,78
ipr-mus LD50:1849 mg/kg COREAF 256,1043,63

CONSENSUS REPORTS: Reported in EPA TSCA Inventory. EPA Genetic Toxicology Program.

SAFETY PROFILE: Moderately toxic by ingestion and intraperitoneal routes. Experimental reproductive effects. Mutation data reported. Reacts violently with BF_3. See also CESIUM. When heated to decomposition it emits toxic fumes of Cl^-.

CDD325 CAS:71250-00-5 HR: 3
CESIUM CYANOTRIDECAHYDRODECABORATE (2-)
mf: $CH_{13}B_{10}Cs_2N$ mw: 413.03

PROP: A solid.

CONSENSUS REPORTS: Cyanide compounds are on the Community Right-To-Know List.

SAFETY PROFILE: A poison. Violent reaction with concentrated hydrochloric acid. When heated to decomposition it emits toxic fumes of NO_x. See also CESIUM, CYANIDE, and BORON COMPOUNDS.

CDD500 CAS:13400-13-0 HR: 3
CESIUM FLUORIDE
mf: CsF mw: 151.91

PROP: Deliquescent colorless cubic crystals. Mp: 703°, bp: 1251°. Very sol in H_2O and MeOH; insol in Py.

SYNS: CESIUM MONOFLUORIDE □ DICESIUM DIFLUORIDE □ TRICESIUM TRIFLUORIDE

CONSENSUS REPORTS: Reported in EPA TSCA Inventory.

OSHA PEL: TWA 2.5 mg(F)m³
ACGIH TLV: TWA 2.5 mg(F)/m³
NIOSH REL: (Inorganic Fluorides) TWA 2.5 mg(F)/m³

SAFETY PROFILE: A poison. Incompatible with benzenediazonium tetrafluoroborate and difluoroamine. When heated to decomposition it emits toxic fumes of F^-.

CDD625 **CAS:12079-66-2** **HR: 3**
CESIUM GRAPHITE
mf: C_8Cs mw: 228.99

PROP: Moisture-sensitive shiny black powder.

SAFETY PROFILE: Explodes on contact with water. Ignites spontaneously in air. See also CESIUM.

CDD750 **CAS:21351-79-1** **HR: 3**
CESIUM HYDROXIDE
DOT: UN 2681/UN 2682
mf: CsHO mw: 149.92

PROP: Colorless to yellowish, very deliquescent crystals. Undergoes transition from orthorhombic to cubic at 2°. Mp: 315°, d: 3.675. Very sol in H_2O and EtOH.

SYNS: CAESIUM HYDROXIDE, solid (UN 2682) (DOT) □ CAESIUM HYDROXIDE, solution (UN 2681) (DOT) □ CESIUM HYDRATE □ CESIUM HYDROXIDE (ACGIH, OSHA) □ CESIUM HYDROXIDE DIMER

TOXICITY DATA WITH REFERENCE
skn-rbt 5 mg/24H MLD TXAPA9 32,239,75
eye-rbt 5 mg/5M rns SEV TXAPA9 32,239,75
orl-rat LD50:570 mg/kg GTPZAB 21(1),29,77
ipr-rat LD50:100 mg/kg AIHOAX 1,637,50
orl-mus LD50:800 mg/kg 20PKA3 -,-,67

CONSENSUS REPORTS: Reported in EPA TSCA Inventory.

OSHA PEL: TWA 2 mg/m³
ACGIH TLV: TWA 2 mg/m³
DOT CLASSIFICATION: 8; *Label:* Corrosive

SAFETY PROFILE: Poison by intraperitoneal route. Moderately toxic by ingestion. A powerful caustic. A corrosive skin and eye irritant. See also CESIUM.

CDE000 **CAS:7789-17-5** **HR: 2**
CESIUM IODIDE
mf: CsI mw: 259.81

PROP: Deliquescent colorless orthorhombic crystals. Mp: 626°, bp: 1280°. Very sol in H_2O; sol in EtOH.

SYNS: CESIUM MONOIODIDE □ DICESIUM DIIODIDE □ TRICESIUM TRIIODIDE

TOXICITY DATA WITH REFERENCE
orl-rat LD50:2386 mg/kg NIOSH* TR-74,1,72
ipr-rat LD50:1400 mg/kg AIHOAX 1,637,50

CONSENSUS REPORTS: Reported in EPA TSCA Inventory.

SAFETY PROFILE: Moderately toxic by ingestion and intraperitoneal routes. See also CESIUM and IODIDES. When heated to decomposition, it emits toxic fumes of I^-.

CDE125 **CAS:12430-27-2** **HR: 3**
CESIUM LITHIUM TRIDECAHYDRONONABORATE
mf: $H_{13}B_9CsLi$ mw: 250.24

SAFETY PROFILE: Ignites spontaneously in air. See also CESIUM, LITHIUM COMPOUNDS, and BORON COMPOUNDS.

CDE250 **CAS:7789-18-6** **HR: 2**
CESIUM(I) NITRATE (1:1)
DOT: UN 1451
mf: NO_3•Cs mw: 194.92

PROP: Colorless, hexagonal or cubic, glittering crystalline powder. Undergoes hexagonal to cubic transition at 1°. Piezoelectric. Mp: 414°, bp: decomp, d: 3.685, 2.71 @ 500° (liq). Very sol in H_2O; sol in Me_2CO; sltly sol in EtOH.

SYNS: CESIUM NITRATE (DOT) □ NITRIC ACID, CESIUM SALT

TOXICITY DATA WITH REFERENCE
mrc-bcs 5 mol/L MUREAV 77,109,80
orl-rat LD50:2390 mg/kg VAMNAQ (8),10,78
ipr-rat LD50:1200 mg/kg AIHOAX 1,637,50
orl-mus LD50:2300 mg/kg VAMNAQ (8),10,78

CONSENSUS REPORTS: Reported in EPA TSCA Inventory. EPA Genetic Toxicology Program.

DOT CLASSIFICATION: 5.1; *Label:* Oxidizer

SAFETY PROFILE: Moderately toxic by ingestion and intraperitoneal routes. Mutation data reported. When heated to decomposition it emits toxic fumes of NO_x. See also CESIUM and NITRATES.

CDE325 **CAS:20281-00-9** **HR: 3**
CESIUM OXIDE
mf: Cs_2O mw: 297.81

PROP: Orange rhombohedral crystals; moisture sensitive.

SAFETY PROFILE: Ignition or incandescent reaction on contact with water; ethanol; moisture + carbon monoxide or carbon dioxide; sulfur dioxide + heat; or halogens (fluorine; chlorine; or iodine) above 150°C. Reacts with H_2O. with formation of CsOH. See also CESIUM.

CDE400 **CAS:78937-12-9** **HR: 3**
CESIUM PENTACARBONYLVANADATE (3-)
mf: $C_5Cs_3O_5V$ mw: 589.71

SAFETY PROFILE: Ignites spontaneously in air or when scratched under a non-reactive gas. Explodes on contact with water or alcohols. When heated to decomposition it emits toxic fumes of VO_x. See also CESIUM and VANADIUM COMPOUNDS.

CDE500 **CAS:10294-54-9** **HR: 2**
CESIUM SULFATE
mf: Cs_2O_4S mw: 361.88

PROP: Colorless orthorhombic crystals. Hygroscopic. Undergoes orthorhombic to hexagonal transition at 6°. Mp: 1005°. Very sol in H_2O; prac insol in EtOH, and Me_2CO.

SYNS: DICESIUM SULFATE □ SULFURIC ACID, DICESIUM SALT

TOXICITY DATA WITH REFERENCE
mrc-bcs 5 mol/L MUREAV 77,109,80
orl-rat LD50:2830 mg/kg VAMNAQ (8),10,78
orl-mus LD50:3180 mg/kg VAMNAQ (8),10,78

CONSENSUS REPORTS: Reported in EPA TSCA Inventory. EPA Genetic Toxicology Program.

SAFETY PROFILE: Moderately toxic by ingestion. Mutation data reported. When heated to decomposition it emits toxic fumes of SO_x.

CDF250 CAS:29144-42-1 HR: 3
CETOCYLINE
mf: $C_{22}H_{21}NO$ mw: 315.44

PROP: Bright-yellow needles.

SYNS: β-CHELOCARDIN ☐ 2-DECARBOXAMIDO-2-ACETYL-4-DES-DIMETHYLAMINO-4-AMINO-9-METHYL-5A,6-ANHYDROTETRACYCLINE

TOXICITY DATA WITH REFERENCE
orl-mus LD50:2500 μg/kg 85ERAY 1,534,78
ipr-mus LD50:140 mg/kg 85ERAY 1,534,78
ivn-mus LD50:88 mg/kg 85ERAY 1,534,78

SAFETY PROFILE: Poison by ingestion, intraperitoneal, and intravenous routes. When heated to decomposition it emits toxic fumes of NO_x.

CDF375 CAS:34675-84-8 HR: 3
CETRAXATE
mf: $C_{17}H_{23}NO_4$ mw: 305.41

PROP: Crystals from methanol, melts over a range of 200–280°.

SYNS: trans-4-(((4-(AMINOMETHYL)CYCLOHEXYL)CARBONYL) OXY)BENZENEPROPANOIC ACID ☐ trans-p-HYDROXY HYDROCINNAMIC ACID-4-(AMINOMETHYL)CYCLOHEXANE CARBOXYLATE

TOXICITY DATA WITH REFERENCE
ipr-rat LD50:716 mg/kg OYYAA2 19,323,80
scu-rat LD50:1503 mg/kg OYYAA2 19,323,80
ivn-rat LD50:345 mg/kg OYYAA2 19,323,80
ipr-mus LD50:1520 mg/kg OYYAA2 19,323,80
scu-mus LD50:4310 mg/kg OYYAA2 19,323,80
ivn-mus LD50:681 mg/kg OYYAA2 19,323,80

SAFETY PROFILE: Poison by intravenous route. Moderately toxic by intraperitoneal and subcutaneous routes. When heated to decomposition it emits toxic fumes of NO_x.

CDF380 CAS:27724-96-5 HR: 3
CETRAXATE HYDROCHLORIDE
mf: $C_{17}H_{23}NO_4$•ClH mw: 341.87

PROP: Crystals from MeOH/Et_2O. Mp: 238–240°.

SYNS: trans-4-(((4-(AMINOMETHYL)CYCLOHEXYL)CARBONYL) OXY)-BENZENEPROPANOIC ACID HYDROCHLORIDE ☐ 4′-(2-CARBOXYETHYL)PHENYL-trans-4-AMINOMETHYLCYCLOHEXANE CARBOXYLATE HYDROCHLORIDE ☐ CV 1006

TOXICITY DATA WITH REFERENCE
ipr-rat LD50:716 mg/kg OYYAA2 12,265,76
scu-rat LD50:1415 mg/kg IYKEDH 10,710,79
ivn-rat LD50:298 mg/kg IYKEDH 10,710,79
ipr-mus LD50:1520 mg/kg OYYAA2 12,265,76
scu-mus LD50:4210 mg/kg IYKEDH 10,710,79
ivn-mus LD50:666 mg/kg IYKEDH 10,710,79

SAFETY PROFILE: Poison by intravenous route. Moderately toxic by subcutaneous and intraperitoneal routes. When heated to decomposition it emits toxic fumes of NO_x and HCl.

CDF400 CAS:3151-59-5 HR: 3
CETYLAMINE HYDROFLUORIDE
mf: $C_{16}H_{35}N$•FH mw: 261.53

SYNS: CETYLAMINE-HF ☐ CETYLAMINHYDROFLUORID (GERMAN) ☐ GA 242 ☐ HEPTAFLUR ☐ HEXADECYLAMINE HYDROFLUORIDE ☐ 1-HEXADECANAMINE HYDROFLUORIDE (9CI) ☐ SKF 2208K

TOXICITY DATA WITH REFERENCE
orl-mus TDLo:40 mg/kg (9D preg):REP DZZEA7 32,861,77
orl-mus TDLo:40 mg/kg (9D preg):TER DZZEA7 32,861,77
ipr-mus LD50:45,246 μg/kg DZZEA7 35,1070,80

OSHA PEL: TWA 2.5 mg(F)/m^3
ACGIH TLV: TWA 2.5 mg(F)/m^3

SAFETY PROFILE: Poison by intraperitoneal route. An experimental teratogen. Other experimental reproductive effects. When heated to decomposition it emits toxic fumes of NO_x and HF.

CDF450 CAS:693-33-4 HR: 3
CETYL BETAINE
mf: $C_{20}H_{41}NO_2$ mw: 327.62

SYNS: AMMONIUM, (CARBOXYMETHYL)HEXADECYLDIMETHYL-, HYDROXIDE, inner salt (8CI) ☐ N-(CARBOXYMETHYL)-N,N-DIMETHYL-1-HEXADECANAMINIUM HYDROXIDE inner salt ☐ (CARBOXYMETHYL)HEXADECYLDIMETHYLAMMONIUM HYDROXIDE, inner salt (7CI) ☐ C16BET ☐ N,N-DIMETHYL-N-HEXADECYLGLYCINE ☐ 1-HEXADECANAMINIUM, N-(CARBOXYMETHYL)-N,N-DIMETHYL-, HYDROXIDE, inner salt ☐ HEXADECYLBETAINE ☐ LONZAINE 16S ☐ PRODUCT HDN

TOXICITY DATA WITH REFERENCE
orl-rat LD50:1620 mg/kg FAATDF 16,41,91
ipr-rat LD50:150 mg/kg FAATDF 16,41,91

CONSENSUS REPORTS: Reported in EPA TSCA Inventory.

SAFETY PROFILE: A poison by intraperitoneal route. Moderately toxic by ingestion. When heated to decomposition it emits toxic vapors of NO_x.

CDF500 CAS:13316-70-6 HR: 3
CETYLDIETHYLETHYLAMMONIUM BROMIDE
mf: $C_{22}H_{48}N$•Br mw: 406.62

SYNS: CETYLTRIETHYLAMMONIUM BROMIDE ☐ HEXADECYLTRIETHYLAMMONIUM BROMIDE ☐ TRIETHYLHEXADECYLAMMONIUM BROMIDE

TOXICITY DATA WITH REFERENCE
eye-rbt 1% SEV JAPMA8 38,428,49

orl-mus TDLo:900 mg/kg (female 8-12D post):REP TCMUD8 7,7,87
orl-mus LD50:60 mg/kg JAPMA8 38,428,49
ivn-mus LD50:50 mg/kg JAPMA8 38,428,49

SAFETY PROFILE: Poison by ingestion and intravenous routes. Experimental reproductive effects. A severe eye irritant. When heated to decomposition it emits very toxic fumes of Br^-, NH_3 and NO_x. See also BROMIDES.

CDF750 CAS:6004-24-6 HR: 3
CETYLPYRIDINIUM CHLORIDE MONOHYDRATE
mf: $C_{21}H_{38}N{\cdot}Cl{\cdot}H_2O$ mw: 358.07

SYNS: CEEPRYN □ CEPACOL □ 1-HEXADECYLPYRIDINIUM CHLORIDE MONOHYDRATE

TOXICITY DATA WITH REFERENCE
ipr-rat LDLo:15 mg/kg JPETAB 74,401,42
ipr-mus LDLo:3 mg/kg JPETAB 74,401,42
ivn-dog LDLo:100 mg/kg JPETAB 74,401,42
orl-rbt LDLo:400 mg/kg JPETAB 74,401,42
ipr-rbt LDLo:5 mg/kg JPETAB 74,401,42
scu-rbt LDLo:200 mg/kg JPETAB 74,401,42
ivn-rbt LDLo:20 mg/kg JPETAB 74,401,42
ipr-gpg LDLo:5 mg/kg JPETAB 74,401,42

SAFETY PROFILE: Poison by ingestion, intraperitoneal, intravenous, and subcutaneous routes. When heated to decomposition it emits very toxic fumes of Cl^- and NO_x.

CDG000 CAS:62-59-9 HR: 3
CEVADINE
mf: $C_{32}H_{49}NO_9$ mw: 591.82

PROP: A solid. Mp: 213–214.5°.

SYNS: CEVADENE □ CEVADIN □ (Z)-3-((Z) 2-METHYLCROTONATE)4,9-EPOXYCEVANE-3-β,4-β,12,14,16-β,17,20-HEPTOL □ VERATRINE □ VERATRINE (crystallized)

TOXICITY DATA WITH REFERENCE
ipr-mus LD50:3500 μg/kg PSEBAA 76,847,51
scu-mus LD50:4900 μg/kg JPETAB 113,89,55
ivn-mus LD50:1 mg/kg PHREA7 26,383,46
scu-gpg LDLo:1 mg/kg PHREA7 26,383,46

SAFETY PROFILE: Poison by intravenous, subcutaneous, and intraperitoneal routes. When heated to decomposition it emits toxic fumes of NO_x.

CDG250 CAS:5205-82-3 HR: 3
CG 201
mf: $C_{22}H_{28}NO_3{\cdot}CH_3O_4S$ mw: 465.61

PROP: Crystals from petr ether. Mp: 134–135°.

SYNS: ACABEL □ BENZILIC ACID ester with 2-(HYDROXYMETHYL)-1,1-DIMETHYLPIPERIDINIUM METHYL SULFATE □ BENZILSAEURE-(N,N-DIMETHYL-2-HYDROXYMETHYL-PIPERIDINIUM)-ESTER-METHYLSULFAT (GERMAN) □ BEVONIUM METHYL SULFATE □ BEVONIUM METILSULFATE □ ESTER d'ACIDE BENZILIQUE et DU-1-METHYLSULFATE de 1,1-DIMETHYL-(2-HYDROXY METHYL)PIPERIDINIUM □ ESTER del ACIDO BENCILICO del-1,1-DIMETIL-2-OXIMETILPIPERIDINIO-METILSULFATO (SPANISH) □ 2-(((HYDROXYDIPHENYLACETYL)OXY)METHYL)-1,1-DIMETHYLPIPERIDINIUM METHYL SULFATE (SALT) □ 2-(HYDROXYMETHYL)-1,1-DIMETHYLPIPERIDINIUM METHYL SULFATE BENZILATE □ L-99 □ α-PHENYLMANDELIC ACID-N,N-DIMETHYLPIPERIDINIUM-2-METHYL ESTER METHYLSULFATE □ PIRIBENZIL METHYL SULFATE

TOXICITY DATA WITH REFERENCE
orl-rat LD50:5080 mg/kg ARZNAD 16,901,66
scu-rat LD50:2400 mg/kg ARZNAD 16,901,66
ivn-rat LD50:26 mg/kg ARZNAD 16,901,66
scu-mus LD50:436 mg/kg ARZNAD 16,901,66
orl-mus LD50:1360 mg/kg ARZNAD 16,901,66
ivn-mus LD50:17,400 μg/kg ARZNAD 16,901,66
orl-dog LD50:1000 mg/kg ARZNAD 16,901,66
orl-rbt LD50:1000 mg/kg ARZNAD 16,901,66
ims-rbt LD50:500 mg/kg ARZNAD 16,901,66
orl-gpg LD50:3860 mg/kg ARZNAD 16,901,66
scu-gpg LD50:182 mg/kg ARZNAD 16,901,66

SAFETY PROFILE: Poison by subcutaneous and intravenous routes. Moderately toxic by ingestion and intramuscular routes. When heated to decomposition it emits toxic fumes of NO_x and SO_x. See also ESTERS.

CDG300 HR: 3
CGS 10787B
mf: $C_{15}H_{13}N_3O_2{\cdot}C_6H_{15}NO_3$ mw: 416.53

SYN: α-CYANO-1-METHYL-β-OXO-PYRROLE-2-PROPIONANILIDE compounded with 2,2′,2″-NITRILOTRIETHANOL

TOXICITY DATA WITH REFERENCE
orl-rat LD50:1071 mg/kg JACTDZ 2(2),249,83
ivn-rat LD50:243 mg/kg JACTDZ 2(2),249,83
orl-mus LD50:674 mg/kg JACTDZ 2(2),249,83
ivn-mus LD50:144 mg/kg JACTDZ 2(2),249,83

CONSENSUS REPORTS: Cyanide and its compounds are on the Community Right-To-Know List.

SAFETY PROFILE: Poison by intravenous route. Moderately toxic by ingestion. When heated to decomposition it emits toxic fumes of NO_x and CN^-.

CDG500 CAS:20562-03-2 HR: 3
α-CHACONINE

PROP: A solid. Mp: 243°.

TOXICITY DATA WITH REFERENCE
ipr-rat TDLo:20 mg/kg (female 5-12D post):TER TXAPA9 36,227,76
ipr-rat TDLo:40 mg/kg (female 5-12D post):REP TXAPA9 36,227,76
orl-ham TDLo:165 mg/kg (female 8D post):TER TJADAB 30,371,84
ipr-rat LD50:84 mg/kg TXAPA9 36,227,76
ipr-mus LD50:19 mg/kg TOLED5 3,349,79
ipr-rbt LDLo:50 mg/kg RCOCB8 12,657,75

SAFETY PROFILE: Poison by intraperitoneal route. An experimental teratogen. Other experimental reproductive effects.

CDG750 CAS:50335-03-0 HR: 3
CHAETOGLOBOSIN A
mf: $C_{32}H_{36}N_2O_5$ mw: 528.70

PROP: Pale-yellow prisms from CH_2Cl_2. Mp: 188°. Isolated from cultures of *Chaetonium globosum* (JJEMAG 48,105,78).

TOXICITY DATA WITH REFERENCE
ipr-mus TDLo: 700 μg/kg (8D preg): REP MAIKD3 (10),17,80
orl-rat LD50: 400 mg/kg JJEMAG 48,105,78
ipr-rat LDLo: 2000 μg/kg JJEMAG 48,105,78
orl-mus LD50: 400 mg/kg JJEMAG 48,105,78
scu-mus LD50: 6500 μg/kg JJEMAG 48,105,78

SAFETY PROFILE: Poison by ingestion, intraperitoneal, and subcutaneous routes. Experimental reproductive effects. When heated to decomposition it emits toxic fumes of NO_x.

CDH000 CAS:94-41-7 HR: 3
CHALCONE
mf: $C_{15}H_{12}O$ mw: 208.27

SYNS: 2-BENZALACETOPHENONE ☐ 1-BENZOYL-1-PHENYLETHENE ☐ β-BENZOYLSTYRENE ☐ 2-BENZYLIDENEACETOPHENONE ☐ CINNAMOPHENONE ☐ 1,3-DIPHENYL-1-PROPEN-3-ONE ☐ 3-PHENYLACRYLOPHENONE ☐ β-PHENYLACRYLOPHENONE ☐ 1-PHENYL-2-BENZOYLETHYLENE ☐ PHENYL STYRYL KETONE

TOXICITY DATA WITH REFERENCE
ivn-mus LD50: 56 mg/kg CSLNX* NX#04476

CONSENSUS REPORTS: Reported in EPA TSCA Inventory.

DOT CLASSIFICATION: 3; *Label:* Flammable Liquid

SAFETY PROFILE: Poison by intravenous route. See also KETONES. When heated to decomposition it emits acrid smoke and irritating fumes.

CDH125 HR: 2
CHALICE VINE

PROP: Climbing or erect vines with large yellow or creamy yellow trumpet-shaped flowers and elongated berries. They are native to Mexico and the subtropical areas of the United States, and are cultivated in Florida, Hawaii, and the West Indies.

SYNS: BEJUCO DO PEO (PUERTO RICO) ☐ CHAMICO BEJUCO (CUBA) ☐ CUP-OF-GOLD ☐ PALO GUACO (CUBA) ☐ SILVER CUP ☐ SOLANDRA (VARIOUS SPECIES) ☐ TRUMPET PLANT

SAFETY PROFILE: All parts of the plant including the nectar contain poisonous atropine alkaloids. Ingestion of any part of the plant can cause rapid heartbeat, fever, blurred vision, dilated pupils, excitement, headache, delirium, and hallucinations. See also ATROPINE.

CDH250 CAS:520-36-5 HR: 1
CHAMOMILE
mf: $C_{15}H_{10}O_5$ mw: 270.25

PROP: Blue liquid, turning brownish-yellow. Yellow needles from Py (aq). Composed of amyl and butyl esters of angelic and tiglic acids, butyric acid, etc. Mp: 352°, d: 0.905–0.915 @ 15°/15°.

SYNS: APIGENIN ☐ APIGENINE ☐ APIGENOL ☐ C.I. NATURAL YELLOW 1 ☐ 5,7-DIHYDROXY-2-(4-HYDROXYPHENYL)-4H-1-BENZOPYRAN-4-ONE ☐ 2-(p-HYDROXYPHENYL)-5,7-DIHYDROXYCHROMONE ☐ PELARGIDENON 1449 ☐ 4′,5,7-TRIHYDROXYFLAVONE ☐ VERSULIN

TOXICITY DATA WITH REFERENCE
mmo-sat 100 μg/plate BCSTB5 5,1489,77
mma-sat 100 μg/plate BCSTB5 5,1489,77

SAFETY PROFILE: Mutation data reported. A mild allergen. When heated to decomposition it emits acrid smoke and irritating fumes. See also ESTERS.

CDH500 CAS:8002-66-2 HR: 1
CHAMOMILE OIL

PROP: By steam distillation of the flowers and stalks of *Matrilaria chamomilla* L. (FCTXAV 12,807,74). Blue–yellowish–brown liquid; strong odor and bitter aromatic taste. Composed of amyl and butyl esters of angelic, tiglic acids, and butyric acid. D: 0.905–0.915 @ 15°/15°. Sol in fixed oils, propylene glycol; insol in mineral oil, glycerin.

SYNS: BLUE CHAMOMILE OIL ☐ CAMOMILE OIL GERMAN ☐ CHAMOMILE-GERMAN OIL ☐ GERMAN CHAMOMILE OIL ☐ HUNGARIAN CHAMOMILE OIL ☐ KAMILLENOEL ☐ OILS, CHAMOMILE, GERMAN

TOXICITY DATA WITH REFERENCE
skn-rbt 500 mg/24H MOD FCTXAV 12,851,74
orl-rat LD50: 10 g/kg ARZNAD 19,615,69
skn-rbt LD50: >5 g/kg FCTXAV 12,851,74

CONSENSUS REPORTS: Reported in EPA TSCA Inventory.

SAFETY PROFILE: Low toxicity by ingestion and skin contact. A mild allergen. A skin irritant. See also ESTERS. When heated to decomposition it emits acrid and irritating fumes.

CDH750 CAS:8015-92-7 HR: 1
CHAMOMILE OIL (ROMAN)

PROP: Obtained by the steam distillation of the dried flowers of *Anthemis nobilis* L. (FCTXAV 12,807,74). Blue liquid, turning brownish-yellow; strong aromatic odor. Composition: Amyl and butyl esters of angelic and tiglic acids, butyric acid, etc. D: 0.905–0.915 @ 15°/15°. Sol in fixed oils, mineral oil, propylene glycol; insol in glycerin.

SYN: CAMOMILE OIL, ENGLISH TYPE (FCC)

TOXICITY DATA WITH REFERENCE
skn-rbt 500 mg/24H MOD FCTXAV 12,853,74
orl-rat LD50: >5 g/kg FCTXAV 12,853,74
skn-rbt LD50: >5 g/kg FCTXAV 12,853,74

CONSENSUS REPORTS: Reported in EPA TSCA Inventory.

SAFETY PROFILE: Low toxicity by ingestion and skin contact. A mild allergen. A skin irritant. See also ESTERS. Combustible when heated. When heated to decomposition it emits acrid smoke and irritating fumes.

CDI000 **CAS:64365-11-3** **HR: 1**
CHARCOAL, ACTIVATED (DOT)
DOT: NA 1361
af: C aw: 12.01

PROP: Black porous solid, coarse granules or powder. Insol in water, organic solvents.

SYNS: ACTIVATED CARBON □ CARBON, ACTIVATED □ CARBORAFFIN □ CARBORAFINE □ KARBORAFIN □ NUCHAR 722

CONSENSUS REPORTS: Reported in EPA TSCA Inventory.

DOT CLASSIFICATION: 4.2; *Label:* Spontaneously Combustible

SAFETY PROFILE: It can cause a dust irritation, particularly to the eyes and mucous membranes. Combustible when exposed to heat. Dust is flammable and explosive when exposed to heat or flame or oxides.

CDI250 **CAS:16291-96-6** **HR: 2**
CHARCOAL (BRIQUETTES)
DOT: NA 1361

PROP: Black amorphous solid. Composition: carbon + impurities. Mw: 12.0, mp: >3500°, bp: 4200°, d: 3.51.

SYNS: CHARCOAL □ CHARCOAL SCREENINGS (DOT) □ CHARCOAL WOOD (DOT)

CONSENSUS REPORTS: Reported in EPA TSCA Inventory.

DOT CLASSIFICATION: 4.2; *Label:* Spontaneously Combustible

SAFETY PROFILE: Carbon itself has no toxic action, but it contains impurities that may be toxic. Fire hazard: reacts with liquid air, $Ba(ClO_3)_2$, BrF_5, ClO, $Ca(ClO_3)_2$, ClF_3, F_2, H_2O_2, $Mg(ClO_3)_2$, (O_2 + wood), perchlorates, peroxides, (P + air), K + $KClO_4$, KNO_3, RuO_4, $AgNO_3$, $NaClO_3$, (AgCl + NaO_2), S, (S + $NaNO_3$), $Zn(ClO_3)_2$. Heats spontaneously, particularly when wet, freshly calcined, or tightly packed, and it can ignite and burn. Slight explosion hazard when exposed to heat or flame. To fight fire, use water, mist, foam or dry chemical. When heated to decomposition it emits acrid smoke and fumes.

CDJ000 **HR: 3**
CHARCOAL (SHELL)
DOT: NA 1361

SYN: CHARCOAL, SHELL (DOT)

DOT CLASSIFICATION: 4.2; *Label:* Spontaneously Combustible

SAFETY PROFILE: A flammable solid. See also CHARCOAL (BRIQUETTES).

CDK250 **CAS:6377-18-0** **HR: 3**
CHARTREUSIN
mf: $C_{32}H_{32}O_{14}$ mw: 640.64

PROP: Yellow plates from Me_2CO (aq) or CH_2Cl_2/EtOH. Mp: 184–186°. Antibiotic substances produced by *Streptomyces chartreusis* from soil (JACSAT 75,4011,53).

SYNS: ANTIBIOTIC X-465A □ LAMBDAMYCIN □ NSC 5159 □ U-7257

TOXICITY DATA WITH REFERENCE
mmo-sat 200 μg/plate ABCHA6 44(4),919,80
pic-esc 150 ng/plate CNREA8 43,2819,83
dni-hmn:oth 5 mg/L RCOCB8 34,173,81
oms-hmn:oth 5 mg/L RCOCB8 34,173,81
ipr-mus LD50:300 mg/kg CNCRA6 30,9,63
scu-mus LD50:500 mg/kg 85GDA2 6,335,81

SAFETY PROFILE: Poison by intraperitoneal route. Moderately toxic by subcutaneous route. Human mutation data reported. When heated to decomposition it emits acrid smoke and irritating fumes.

CDK500 **CAS:1393-72-2** **HR: 3**
CHARTREUSIN, SODIUM SALT
mf: $C_{18}H_{17}O_8$•Na mw: 384.34

SYNS: ANTIBIOTIC X465A SODIUM SALT □ X465A SODIUM SALT

TOXICITY DATA WITH REFERENCE
ipr-mus LD50:600 mg/kg 85ERAY 1,768,78
ivn-mus LD50:250 mg/kg JACSAT 75,4011,53

SAFETY PROFILE: Poison by intravenous route. Moderately toxic by intraperitoneal route. See also CHARTREUSIN. When heated to decomposition it emits toxic fumes of Na_2O.

CDL000 **CAS:476-32-4** **HR: 3**
CHELIDONINE
mf: $C_{20}H_{19}NO_5$ mw: 353.40

PROP: White, crystalline powder. Mp: 135–136°.

TOXICITY DATA WITH REFERENCE
scu-rat LDLo:300 mg/kg JAMAAP 75,1324,20
ivn-mus LD50:35 mg/kg FEPRA7 5,163,46
scu-mus LDLo:300 mg/kg JAMAAP 75,1324,20
scu-rbt LDLo:300 mg/kg JAMAAP 75,1324,20
scu-gpg LDLo:300 mg/kg JAMAAP 75,1324,20
scu-frg LDLo:300 mg/kg JAMAAP 75,1324,20

SAFETY PROFILE: Poison by intravenous and subcutaneous routes. A central nervous system depressant causing sleepiness, depression, slowing of the pulse, and, in large doses, coma and circulatory failures. Combustible when exposed to heat or flame. When heated to decomposition it emits toxic fumes of NO_x.

CDL325 **CAS:474-25-9** **HR: 3**
CHENODESOXYCHOLIC ACID
mf: $C_{24}H_{40}O_4$ mw: 392.64

PROP: Needles from ethyl acetate + heptane. Mp: 119°. Freely sol in methanol, alc, acetone, acetic acid; more sol in ether and ethyl acetate than deoxycholic acid. Practically insol in water, petr ether, benzene. Forms beautiful crystalline salts of Na, K, and Ba. While the acid is tasteless, the Na salt tastes slightly sweet at first, then bitter.

C

SYNS: ANTHROPODEOXYCHOLIC ACID □ ANTHROPODESOXYCHOLIC ACID □ ANTHROPODODESOXYCHOLIC ACID □ CDC □ CDCA □ CHENDAL □ CHENDOL □ CHENIC ACID □ CHENIX □ CHENOCEDON □ CHENODEOXYCHOLIC ACID □ CHENODESOXYCHOLSAEURE (GERMAN) □ CHENODEX □ CHENODIOL □ CHENOFALK □ CHENOSAURE □ CHENOSSIL □ CHOLANORM □ 3-α,7-α-DIHYDROXYCHOLANIC ACID □ 3-α,7-α-DIHYDROXY-5-β-CHOLAN-24-OIC ACID □ FLUIBIL □ GALLODESOXYCHOLIC ACID □ HEKBILIN □ KEBILIS □ ULMENIDE

TOXICITY DATA WITH REFERENCE

mmo-sat 20 mg/L MUREAV 158,45,85
sln-smc 100 mg/L CRNGDP 5,447,84
orl-rat TDLo:3480 mg/kg (female 15-22D post):REP NTIS** PB81-127581
orl-mky TDLo:1500 mg/kg (female 21-45D post):REP TXCYAC 2,239,74
orl-rat TDLo:100 mg/kg (male 1D pre):TER NTIS** PB81-127581
orl-rbt TDLo:260 mg/kg (female 6-18D post):REP OYYAA2 16,33,78
orl-rat TDLo:550 mg/kg (female 7-17D post):REP OYYAA2 16,51,78
orl-rat TDLo:1638 mg/kg (26W pre):REP YACHDS 10,4529,82
orl-rat TDLo:1820 mg/kg (male 91D pre):REP OYYAA2 15,975,78
orl-mky TDLo:1500 mg/kg (female 21-45D post):TER LANCAO 2,1021,73
orl-rat TDLo:1785 mg/kg (female 4-20D post):TER AIPTAK 246,149,80
orl-rat TDLo:1375 mg/kg (female 7-17D post):TER OYYAA2 16,51,78
orl-mky TDLo:1500 mg/kg (female 21-45D post):TER LANCAO 2,1021,73
orl-wmn TDLo:24 g/kg/5Y-C:CAR CLONEA 7,245,81
orl-rat LD50:4000 mg/kg IYKEDH 13,1128,82
ipr-rat LD50:105 mg/kg OYYAA2 15,915,78
ivn-rat LD50:106 mg/kg OYYAA2 15,915,78
orl-mus LD50:3000 mg/kg IYKEDH 13,1128,82
ipr-mus LD50:86 mg/kg OYYAA2 15,915,78
ivn-mus LD50:100 mg/kg ARZNAD 20,323,70

SAFETY PROFILE: Poison by intravenous and intraperitoneal routes. Moderately toxic by ingestion. An experimental teratogen. Experimental reproductive effects. Questionable human carcinogen producing liver tumors. Mutation data reported. When heated to decomposition it emits acrid smoke and fumes.

CDL375 CAS:2646-38-0 HR: 3
CHENODESOXYCHOLIC ACID SODIUM SALT

mf: $C_{24}H_{39}O_4{\cdot}Na$ mw: 414.62

SYNS: CHENODEOXYCHOLIC ACID SODIUM SALT □ SODIUM CHENODEOXYCHOLATE □ SODIUM CHENODESOXYCHOLATE

TOXICITY DATA WITH REFERENCE

ivn-rat LD50:100 mg/kg KSRNAM 11,2499,77
scu-mus LD50:1450 mg/kg KSRNAM 11,2499,77
ivn-mus LD50:114 mg/kg KSRNAM 11,2499,77

SAFETY PROFILE: Poison by intravenous route. Moderately toxic by subcutaneous route. When heated to decomposition it emits toxic fumes of Na_2O. See also CHENODESOXYCHOLIC ACID.

CDL500 CAS:8006-99-3 HR: 3
CHENOPODIUM OIL

PROP: American wormseed. Ingredients are ascaridol, cymene, camphor, and saponins (27ZTAP 3,33,69). Colorless or pale yellow liquid, characteristic disagreeable odor and taste. Composition: 60–70% ascaridol. D: 0.950–0.980 @ 25°/25°. Insol in water; sol in 8 vols 70% alc; sltly sol in glacial acetic acid. Keep well closed, cool, and protected from light.

SYNS: OIL of AMERICAN WORMSEED □ OIL of CHENOPODIUM

TOXICITY DATA WITH REFERENCE

skn-mus 100% FCTXAV 14,713,76
skn-rbt 500 mg/24H MLD FCTXAV 14,713,76
skn-pig 100% FCTXAV 14,713,76
orl-rat LD50:255 mg/kg FCTXAV 14,713,76
skn-rbt LD50:415 mg/kg FCTXAV 14,713,76

SAFETY PROFILE: Poison by ingestion. Moderately toxic by skin contact. A skin irritant. See also ASCARIDOL, CAMPHOR, SAPONINE. When heated to decomposition it emits acrid smoke and irritating fumes.

CDL750 HR: 2
CHERRY BARK OAK

PROP: Tannin containing fraction of bark used (JNCIAM 57,207,76).

SYNS: QUERCUS FALCATA PAGODAEFOLIA □ TANNIN from CHERRY BARK OAK

TOXICITY DATA WITH REFERENCE

scu-rat TDLo:720 mg/kg/45W-I:NEO JNCIAM 57,207,76

SAFETY PROFILE: Questionable carcinogen with experimental neoplastigenic data. See also TANNIN. When heated to decomposition it emits acrid and irritating fumes.

CDM000 HR: 3
CHERRY LAUREL OIL

PROP: Volatile oil from leaves of *Prunus laurocerasus L., Rosacene.* Pale yellow liquid, odor and taste similar to oil of bitter almond. D: 1.054–1.066 @ 20°/20°. Sltly sol in water; sol in 2 vols 70% alc, benzene, chloroform, and ether.

SAFETY PROFILE: Very poisonous. Hydrogen cyanide component is responsible for highly toxic properties. Keep well closed, cool, and protected from light. See also CYANIDE. When heated to decomposition it emits toxic fumes of CN^-.

CDM250 CAS:1401-55-4 HR: 3
CHESTNUT TANNIN

SYNS: CASTANEA SATIVA MILL TANNIN □ TANNIN from CHESTNUT

TOXICITY DATA WITH REFERENCE

scu-mus TDLo:750 mg/kg/12W-I:ETA BJCAAI 14,147,60
ipr-mus LD50:150 mg/kg JPPMAB 9,98,57
scu-mus LD50:140 mg/kg JPPMAB 9,98,57
ivn-mus LD50:50 mg/kg JPPMAB 9,98,57

ims-mus LD50:120 mg/kg JPPMAB 9,98,57

SAFETY PROFILE: Poison by subcutaneous, intramuscular, intravenous, and intraperitoneal routes. Questionable carcinogen with experimental tumorigenic data. See also TANNIN. When heated to decomposition it emits acrid and irritating fumes.

CDM325 **HR: 3**
CHINABERRY

PROP: A tree that may grow to 50 feet. The leaves are compound with 2-inch long serrated leaflets. The flowers are purple, fragrant, and form clusters. The yellow berries remain on the tree after the leaves fall. They grow wild and are cultivated in the US coastal states from Virginia to Texas, Hawaii, Guam, and the West Indies.

SYNS: AFRICAN LILAC TREE □ ALELAILA (PUERTO RICO) □ ARBOL DEL QUITASOL (CUBA) □ BEAD TREE □ CHINA TREE □ FALSE SYCAMORE □ HOG BUSH □ 'INIA (HAWAII) □ INDIAN LILAC □ JAPANESE BEAD TREE □ LILAILA □ LILAS (HAITI, DOMINICAN REPUBLIC) □ MELIA AZEDARACH □ PARADISE TREE □ PARAISO (MEXICO) □ PASILLA (PUERTO RICO) □ PERSIAN LILAC □ PRIDE of CHINA □ PRIDE of INDIA □ SYRIAN BEAD TREE □ TEXAS UMBRELLA TREE □ WEST INDIAN LILAC □ WHITE CEDAR

SAFETY PROFILE: The fruit and bark contain tetranortriterpene neurotoxins and gastroenteric toxins. In some areas the fruit is edible. Ingestion may cause, after an indefinite delay period, poor muscle coordination, confusion, stupor, intense gastritis, vomiting, diarrhea, difficult breathing, convulsions, and partial to complete paralysis.

CDM500 **CAS:71392-29-5** **HR: 3**
CHINOIN-127
mf: $C_{11}H_{17}N_3O_2$ mw: 223.31

SYN: 1,6-DIMETHYL-4-OXO-1,6,7,8,9,9a-HEXAHYDRO-4H-PYRIDO(1,2-a)PYRIMIDINE-3-CARBOXAMIDE

TOXICITY DATA WITH REFERENCE
orl-rat LD50:370 mg/kg ARZNAD 29,766,79
scu-rat LD50:280 mg/kg ARZNAD 29,266,79
ivn-rat LD50:210 mg/kg ARZNAD 29,766,79
orl-mus LD50:360 mg/kg ARZNAD 29,766,79

SAFETY PROFILE: Poison by ingestion, subcutaneous, and intravenous routes. When heated to decomposition it emits toxic fumes of NO_x.

CDM575 **CAS:88338-63-0** **HR: 3**
CHINOIN-170
mf: $C_{11}H_{12}N_6O_3$ mw: 276.29

SYNS: 3,7-DIHYDRO-1,3-DIMETHYL-7-((5-METHYL-1,2,4-OXADIAZOL-3-YL)METHYL)-1H-PURINE-2,6-DIONE □ 3-((1,3-DIMETHYLXANTHIN-7-YL)METHYL)-5-METHYL-1,2,4-OXADIAZOLE

TOXICITY DATA WITH REFERENCE
orl-rat LD50:1549 mg/kg DRFUD4 10,624,85
ivn-rat LD50:397 mg/kg DRFUD4 10,624,85
orl-mus LD50:1243 mg/kg DRFUD4 10,624,85
ivn-mus LD50:417 mg/kg DRFUD4 10,624,85
ims-mus LD50:721 mg/kg DRFUD4 10,624,85

SAFETY PROFILE: Poison by intravenous route. Moderately toxic by ingestion and intramuscular routes. When heated to decomposition it emits toxic fumes of NO_x.

CDM625 **CAS:41024-90-2** **HR: 3**
CHIRAL BINAPHTHOL
mf: $C_{20}H_{14}O_2$ mw: 286.33

SYN: (±)-1,1-BI-2-NAPHTHOL

TOXICITY DATA WITH REFERENCE
eye-rbt 50 mg MOD DCTODJ 8,451,85
orl-rat LD50:113 mg/kg DCTODJ 8,451,85
ipr-rat LD50:20 mg/kg DCTODJ 8,451,85
ipr-mus LD50:6 mg/kg DCTODJ 8,451,85

SAFETY PROFILE: Poison by ingestion and intraperitoneal routes. An eye irritant. When heated to decomposition it emits acrid smoke and fumes. See also various napthol entries.

CDM700 **HR: 3**
CHIRONEX FLECKERI TOXIN

SYN: TOXIN, JELLYFISH, CHIRONEX FLECKERI

TOXICITY DATA WITH REFERENCE
ivn-mus LDLo:1 mg/kg BJPCBM 35,510,69
par-mus LDLo:167 µg/kg TOXIA6 9,145,71
unr-mam LD50:5 mg/kg CLPTAT 8,849,67

SAFETY PROFILE: Poison by intravenous, parenteral, and possibly other routes.

CDN000 **CAS:53-19-0** **HR: 2**
CHLODITHANE
mf: $C_{14}H_{10}Cl_4$ mw: 320.04

PROP: A solid. Mp: 76–78°.

SYNS: CHLODITAN □ 1-CHLORO-2-(2,2-DICHLORO-1-(4-CHLOROPHENYL)ETHYL)BENZENE □ 2-(o-CHLOROPHENYL)-2-(p-CHLOROPHENYL)-1,1-DICHLOROETHANE □ o,p'-DDD □ 2,4'-DDD □ 1,1-DICHLORO-2,2-BIS(2,4'-DICHLOROPHENYL)ETHANE □ 1,1-DICHLORO-2-(o-CHLOROPHENYL)-2-(p-CHLOROPHENYL)ETHANE □ o,p'-DICHLORODIPHENYLDICHLOROETHANE □ 2,4'-DICHLOROPHENYLDICHLOROETHANE □ MITOTANE □ NCI-C04933 □ NSC 38721 □ o,p-TDE □ o,p'-TDE

TOXICITY DATA WITH REFERENCE
cyt-rat:oth 10 µg/L 34LXAP -,555,76
orl-man TDLo:16 g/kg (male 15W pre):REP JTCTDW 25,463,87
scu-mus TDLo:900 mg/kg (female 6-14D post):TER NTIS** PB223-160
scu-mus TDLo:900 mg/kg (female 6-14D post):REP NTIS** PB223-160
orl-rat TDLo:250 mg/kg (female 15-19D post):TER BNEOBV 26,283,75
orl-rat TDLo:250 mg/kg (female 15-19D post):TER BNEOBV 26,283,75
orl-rat TDLo:10 g/kg/52W-C:ETA BAFEAG 52,89,65
ipr-rat TDLo:2500 mg/kg/7W-I:ETA,TER CANCAR 40(Suppl 4),1935,77

ipr-mus TDLo: 9750 mg/kg/26W-I: ETA CANCAR 40(Suppl 4),1935,77
orl-man TDLo: 17 g/kg/35W: CNS CANCAR 42,2177,78
orl-wmn TDLo: 800 mg/kg/4D: SKN CANCAR 42,2177,78
orl-wmn TDLo: 11 g/kg/15W: BLD,CVS CANCAR 42,2177,78
orl-wmn TDLo: 14 g/kg/22W: GIT,CNS CANCAR 42,2177,78

CONSENSUS REPORTS: NCI Carcinogenesis Studies (ipr); Equivocal Evidence: mouse, rat CANCAR 40,1935,77. EPA Genetic Toxicology Program.

SAFETY PROFILE: Human systemic effects by ingestion: somnolence, blood pressure depression, diarrhea, nausea or vomiting, normocytic anemia (decrease in the number of red blood cells), and pigmented or nucleated red blood cells. Experimental teratogenic and reproductive effects. Questionable carcinogen with experimental carcinogenic and tumorigenic data. Mutation data reported. When heated to decomposition it emits toxic fumes of Cl^-.

CDN200 CAS:78-95-5 HR: 3
CHLORACETONE
DOT: UN 1695
mf: C_3H_5ClO mw: 92.53

PROP: Colorless, lachrymatory liquid with pungent odor. Mp: −44.5°, bp: 119°, d: 1.162.

SYNS: ACETONYL CHLORIDE □ A-STOFF □ CHLORACETONE □ CHLOROACETONE □ CHLOROACETONE, stabilized (DOT) □ CHLOROPROPANONE □ 1-CHLORO-2-PROPANONE □ MONOCHLORACETONE □ MONOCHLOROACETONE □ MONOCHLOROACETONE, inhibited (DOT) □ MONOCHLOROACETONE, stabilized (DOT) □ MONOCHLOROACETONE, unstabilized (DOT) □ TONITE

TOXICITY DATA WITH REFERENCE
sln-oin-dmg-ihl 100 pph/6M PREBA3 62,284,46/47
ihl-hmn LCLo: 605 ppm/10M NTIS** PB214-270
orl-rat LD50: 100 mg/kg AIHAAP 47,375,86
ihl-rat LC50: 262 ppm/1H AIHAAP 47,375,86
skn-rat LDLo: 100 mg/kg KODAK* 21MAY71
ipr-rat LD50: 80 mg/kg OYYAA2 33,695,87
orl-mus LD50: 127 mg/kg AIHAAP 47,375,86
ipr-mus LD50: 92 mg/kg OYYAA2 33,695,87
skn-rbt LD50: 141 mg/kg AIHAAP 47,375,86

CONSENSUS REPORTS: Reported in EPA TSCA Inventory.

ACGIH TLV: CL 1 ppm (skin)
DOT CLASSIFICATION: 6.1; *Label:* Poison (UN 1695); DOT Class: Forbidden

SAFETY PROFILE: Poison by inhalation, ingestion, and skin contact. Mutation data reported. A lachrymator poison gas. See also CHLORINATED HYDROCARBONS, ALIPHATIC; ACETONE. Flammable when exposed to heat or flame, or oxidizers. Old material can explode. When heated to decomposition it emits highly toxic fumes.

CDN500 CAS:107-14-2 HR: 3
CHLORACETONITRILE
DOT: UN 2668
mf: C_2H_2ClN mw: 75.50

PROP: Fuming liquid. D: 1.193 @ 20°, bp: 126–127°.

SYNS: CHLOROACETONITRILE (DOT) □ α-CHLOROACETONITRILE □ 2-CHLOROACETONITRILE □ CHLOROMETHYL CYANIDE □ MONOCHLOROACETONITRILE □ MONOCHLOROMETHYL CYANIDE □ USAF KF-5

TOXICITY DATA WITH REFERENCE
skn-rbt 14 mg/24H open MLD AIHAAP 23,95,62
dnd-hmn: lym 50 μmol/L AIHAAP 23,95,62
sce-ham: ovr 79,100 nmol/L FAATDF 5,1065,85
skn-mus TDLo: 4800 mg/kg/2W-I: CAR FAATDF 5,1065,85
orl-rat LD50: 220 mg/kg AIHAAP 23,95,62
ihl-rat LCLo: 250 ppm/4H AIHAAP 23,95,62
orl-mus LD50: 139 mg/kg ARTODN 55,47,84
ipr-mus LD50: 100 mg/kg NTIS** AD277-689
skn-rbt LDLo: 71 mg/kg AIHAAP 23,95,62

CONSENSUS REPORTS: Reported in EPA TSCA Inventory. Cyanide and its compounds are on the Community Right-To-Know List.

DOT CLASSIFICATION: 6.1; *Label:* Poison

SAFETY PROFILE: Poison by ingestion, skin contact, and intraperitoneal route. Moderately toxic by inhalation. A skin irritant. Human mutation data reported. Questionable carcinogen with experimental tumorigenic data. Flammable liquid. See also NITRILES. When heated to decomposition it emits very toxic fumes of Cl^-, NO_x, and CN^-.

CDN525 CAS:6310-09-4 HR: 3
5-CHLOR-2-ACETYL THIOPHEN
mf: C_6H_5ClOS mw: 160.62

SYNS: BA 11044 □ 5-CHLORO-2-ACETYL THIOPHEN □ KETONE, 5-CHLORO-2-THIENYL METHYL

TOXICITY DATA WITH REFERENCE
orl-rat TDLo: 500 mg/kg (male 20D pre): REP ANREAK 125,312,56

DOT CLASSIFICATION: 3; *Label:* Flammable Liquid

SAFETY PROFILE: Experimental reproductive effects. A flammable liquid. When heated to decomposition it emits toxic vapors of SO_x and Cl^-.

CDN550 CAS:75-87-6 HR: 3
CHLORAL
DOT: UN 2075
mf: C_2HCl_3O mw: 147.38

SYNS: ACETALDEHYDE, TRICHLORO-(9CI) □ ANHYDROUS CHLORAL □ CHLORAL, anhydrous, inhibited (DOT) □ CLORALIO □ GRASEX □ RCRA WASTE NUMBER U034 □ TRICHLOROACETALDEHYDE □ 2,2,2-TRICHLOROACETALDEHYDE □ TRICHLOROETHANAL

TOXICITY DATA WITH REFERENCE
mmo-sat 1 mg/plate STEVA8 46,229,85
mma-sat 10 mg/plate STEVA8 46,229,85
mma-smc 1 g/L MUREAV 155,53,85

ipr-mus LD50:600 mg/kg BCFAAI 111,293,72

CONSENSUS REPORTS: Reported in EPA TSCA Inventory.

DOT CLASSIFICATION: 6.1; *Label:* Poison

SAFETY PROFILE: A poison. Mutation data reported.

CDO000 CAS:302-17-0 HR: 3
CHLORAL HYDRATE
mf: $C_2HCl_3O \cdot H_2O$ mw: 165.40

PROP: Transparent, colorless crystals; aromatic, penetrating, sltly acrid odor and sltly bitter, caustic taste. Mp: 52°, bp: 97.5°, d: 1.9.

SYNS: AQUACHLORAL □ Bi 3411 □ CHLORALDURAT □DOR-MAL □ FELSULES □ HYDRAL □ HYDRAL de CHLORAL □ KESSODRATE □ LORINAL □ NOCTEC □ NORTEC □ NYCOTON □ PHALDRONE □ RECTULES □ SK-CHLORAL HYDRATE □ SOMNI SED □ SOMNOS □ SONTEC □ TOSYL □ TRAWOTOX □ TRICHLORACETALDEHYD-HYDRAT (GERMAN) □ TRICHLOROACETALDEHYDE HYDRATE □ TRICHLOROACETALDEHYDE MONOHYDRATE □ 2,2,2-TRICHLORO-1, 1-ETHANEDIOL

TOXICITY DATA WITH REFERENCE
mrc-smc 15 mmol/L MUREAV 141,19,84
sce-hmn:lym 54 mg/L AGTQAH 24,105,81
orl-mus TDLo:13 g/kg (female 3W pre-3W post):REP NETOD7 6,137,84
orl-mus TDLo:10 mg/kg:CAR CDPRD4 9,279,86
skn-mus TDLo:960 mg/kg/1W-I:ETA BJCAAI 9,177,55
orl-wmn TDLo:465 mg/kg:EYE,BPR AMSVAZ 223,269,88
orl-hmn LDLo:4 mg/kg PHBTH* 3,132,67
orl-hmn TDLo:300 mg/kg:CNS,CVS BMJOAE 2,994,77
unr-cld TDLo:48 mg/kg/6H-I:BPR AACRAT 65,691,86
unk-man LDLo:103 mg/kg 85DCAI 2,73,70
unr-inf TDLo:160 mg/kg/3D-I:CVS AACRAT 65,691,86
rec-cld TDLo:97 mg/kg:CVS AACRAT 65,691,86
orl-rat LD50:479 mg/kg TXAPA9 18,185,71
skn-rat LD50:3030 mg/kg 85JFAN A065,84
ipr-rat LD50:472 mg/kg JAPMA8 41,100,52
par-rat LD50:710 mg/kg NIIRDN 6,784,82
orl-mus LD50:1100 mg/kg JPETAB 106,444,52
ipr-mus LD50:580 mg/kg ARZNAD 10,665,60
scu-mus LDLo:800 mg/kg AEPPAE 166,437,32
orl-dog LDLo:1000 mg/kg JPETAB 78,340,43
orl-cat LDLo:400 mg/kg JPETAB 78,340,43

CONSENSUS REPORTS: Reported in EPA TSCA Inventory. EPA Genetic Toxicology Program.

SAFETY PROFILE: A human poison by ingestion and possibly other routes. Poison experimentally by ingestion, intravenous, and rectal routes. Moderately toxic by subcutaneous, parenteral, and intraperitoneal routes. Experimental reproductive effects. Human systemic effects by ingestion: general anesthetic; cardiac arrhythmias, blood pressure depression, eye effects, coma, pulse rate increase, arrhythmias. Human mutation data reported. Questionable carcinogen with experimental carcinogenic and tumorigenic data by skin contact. A sedative, anesthetic, and narcotic. Combustible when exposed to heat or flame. When heated to decomposition it emits toxic fumes of Cl^-.

CDO250 CAS:95-06-7 HR: 3
2-CHLORALLYL DIETHYLDITHIOCARBAMATE
mf: $C_8H_{14}ClNS_2$ mw: 223.80

PROP: Amber liquid or oil. Bp: 129° @ 1 mm, d: 1,088 @ 25°. Very sltly sol in H_2O; sol in most org solvs.

SYNS: CDEC □ CHLORALLYL DIETHYLDITHIOCARBAMATE □ 2-CHLOROALLYL DIETHYLDITHIOCARBAMATE □ 2-CHLOROALLYL-N,N-DIETHYLDITHIOCARBAMATE □ 2-CHLORO-2-PROPENE-1-THIOL DIETHYLDITHIOCARBAMATE □ 2-CHLORO-2-PROPENYL DIETHYLCARBAMODITHIOATE □ CP 4572 □ DIETHYLCARBAMODITHIOIC ACID 2-CHLORO-2-PROPENYL ESTER □ DIETHYLDITHIOCARBAMIC ACID-2-CHLOROALLYL ESTER □ NCI-C00453 □ SULFALLATE □ THIOALLATE □ VEGADEX □ VEGADEX SUPER

TOXICITY DATA WITH REFERENCE
mmo-sat 10 µL/plate PMRSDJ 2,87,81
mma-sat 10 µL/plate PMRSDJ 2,87,81
mmo-omi 1100 µg/plate PMRSDJ 2,87,81
mmo-asn 20 µL/plate JSFAAE 32,826,81
orl-rat TDLo:6825 mg/kg/78W-C:CAR NCITR* NCI-CG-TR-115,78
orl-mus TDLo:59 g/kg/78W-C:CAR NCITR* NCI-CG-TR-115,78
orl-rat TD:11 g/kg/78W-I:CAR NCITR* NCI-CG-TR-115,78
orl-mus TD:120 g/kg/78W-I:CAR NCITR* NCI-CG-TR-115,78
orl-rat LD50:850 mg/kg RREVAH 10,97,65
skn-rbt LD50:2200 mg/kg 85DPAN -,-,71/76

CONSENSUS REPORTS: NTP 7th Annual Report on Carcinogens. IARC Cancer Review: Group 2B IMEMDT 7,56,87; Animal Sufficient Evidence IMEMDT 30,283,83. NCI Carcinogenesis Bioassay (feed); Clear Evidence: mouse, rat NCITR* NCI-CG-TR-115,78. EPA Genetic Toxicology Program.

SAFETY PROFILE: Confirmed carcinogen with experimental carcinogenic data. Moderately toxic by ingestion and skin contact. Mutation data reported. An herbicide. When heated to decomposition it emits very toxic fumes of Cl^-, NO_x, and SO_x. See also ALLYL COMPOUNDS; CARBAMATES; and ESTERS.

CDO500 CAS:305-03-3 HR: 3
CHLORAMBUCIL
mf: $C_{14}H_{19}Cl_2NO_2$ mw: 304.24

PROP: Flattened needles from pet ether. Mp: 64–66°. Sol in Et_2O.

SYNS: AMBOCHLORIN □ AMBOCLORIN □ 4-(BIS(2-CHLOROETHYL)AMINO)BENZENEBUTANOIC ACID □ γ-(p-BIS(2-CHLOROETHYL)AMINOPHENYL)BUTYRIC ACID □ 4-(p-BIS(β-CHLOROETHYL) AMINOPHENYL)BUTYRIC ACID □ 4-(p-(BIS(2-CHLOROETHYL)AMINO)PHENYL)BUTYRIC ACID □ CB 1348 □ CHLORAMINOPHEN □ CHLORAMINOPHENE □ CHLOROAMBUCIL □ CHLOROBUTIN □ CHLOROBUTINE □ N,N-DI-2-CHLOROETHYL-γ-p-AMINOPHENYLBUTYRIC ACID □ p-N,N-DI-(β-CHLOROETHYL)AMINOPHENYL BUTYRIC ACID □ p-(N,N-DI-2-CHLOROETHYL)AMINOPHENYL BUTYRIC ACID □ γ-(p-DI(2-CHLOROETHYL)AMINOPHENYL)BUTYRIC ACID □ ECLORIL □ ELCORIL □ LEUKERAN □ LEUKERSAN □ LEUKORAN □ LINFOLIZIN □ LINFOLYSIN □ NCI-C03485 □ NSC-3088 □ PHENYLBUTYRIC ACID NITROGEN MUSTARD □ RCRA WASTE NUMBER U035

TOXICITY DATA WITH REFERENCE
mma-sat 100 µg/plate ENMUDM 8(Suppl 7),1,86

dns-hmn:lym 1 µmol/L JTEHD6 6,1059,80
sce-hmn:lym 150 µg/L MUREAV 143,225,85
sln-dmg-orl 2000 ppm ENMUDM 7,677,85
trn-dmg-orl 3000 ppm ENMUDM 7,677,85
orl-wmn TDLo: 13 mg/kg (56D pre): REP JAMAAP 191,444,65
unr-man TDLo: 143 µg/kg (6D male): REP CANCAR 25,1026,70
orl-wmn TDLo: 5160 µg/kg (33-75D preg): TER JAMAAP 186,74,63
orl-mus TDLo: 80 mg/kg (female 8-12D post): REP TCMUD8 6,361,86
ipr-rat TDLo: 3 mg/kg (female 3D post): TER TCMUD8 4,341,84
orl-rat TDLo: 10 mg/kg (female 15D post): TER 40YJAX -,107,79
ipr-rat TDLo: 6 mg/kg (female 12D post): TER EXPEAM 27,1329,71
orl-rat TDLo: 8 mg/kg (female 11D post): REP 40YJAX -,107,79
ipr-mus TDLo: 4 mg/kg (male 1D pre): REP CNREA8 42,122,82
scu-mus TDLo: 5 mg/kg (female 10D post): TER RFGOAO 70,419,75
orl-rat TDLo: 8 mg/kg (15D preg): TER 40YJAX -,107,76
ipr-rat TDLo: 6 mg/kg (female 13D post): TER PSEBAA 139,62,72
ipr-ham TDLo: 13,500 µg/kg (female 8D post): TER TJADAB 29(2),30A,84
ipr-rat TDLo: 6 mg/kg (female 10D post): TER ANREAK 139,145,61
orl-man TDLo: 84 mg/kg/2.5Y-C: CAR,BLD AMSVAZ 199,373,76
orl-wmn TDLo: 101 mg/kg/82W-C: CAR,BLD GYNOA3 6,115,78
unr-wmn TDLo: 161 mg/kg/3Y-I: CAR,BLD SJHAAQ 13,179,74
ipr-rat TDLo: 120 mg/kg/26W-I: CAR RRCRBU 52,1,75
skn-mus TDLo: 108 mg/kg/9W-I: NEO BJCAAI 10,363,56
ipr-mus TDLo: 18 mg/kg/4W: CAR JNCIAM 36,915,66
ipr-rat TD: 170 mg/kg/26W-I: CAR RRCRBU 52,1,75
orl-man TD: 59 mg/kg/96W-C: CAR,BLD NPMDAD 10,1717,81
orl-wmn TD: 200 mg/kg/6Y-C: CAR,BLD ACHAAH 62,283,80
orl-wmn TD: 307 mg/kg/7Y-C: CAR,BLD AIMDAP 134,728,74
orl-wmn TD: 141 mg/kg/5Y-I: CAR,BLD NEJMAG 304,441,81
orl-hmn TD: 180 mg/kg/3Y-I: CAR,BLD AIMDAP 137,355,77
orl-wmn TD: 135 mg/kg/4Y-C: CAR,BLD AIMDAP 137,355,77
orl-hmn TD: 84 mg/kg/3Y-C: CAR,BLD ACCBAT 38,228,83
orl-wmn TD: 70 mg/kg/94W-C: CAR,BLD ACCBAT 38,228,83
orl-chd TD: 108 mg/kg/77W-C: CAR,BLD AFPEAM 36,592,79
orl-wmn TDLo: 82,600 µg/kg: PUL CANCAR 41,455,78
orl-man TDLo: 3571 µg/kg: CNS JTCTDW 20,361,83
orl-rat LD50: 76 mg/kg FCTOD7 22,665,84
ipr-rat LD50: 14 mg/kg BCPCA6 13,969,64
scu-rat LDLo: 32 mg/kg EJCAAH 13,873,77
orl-mus LD50: 101 mg/kg NCISP* JAN86
ipr-mus LD50: 30 mg/kg PHMGBN 11,231,74
scu-mus LD50: 115 mg/kg NCISP* JAN86
ivn-dog LDLo: 3 mg/kg CCSUBJ 2,201,65
ivn-mky LDLo: 3 mg/kg CCSUBJ 2,201,65

CONSENSUS REPORTS: NTP 7th Annual Report on Carcinogens. IARC Cancer Review: Group 1 IMEMDT 7,144,87; Human Inadequate Evidence IMEMDT 9,125,75; Human Limited Evidence IMEMDT 26,115,81; Animal Limited Evidence IMEMDT 26,115,81; Animal Sufficient Evidence IMEMDT 9,125,75. EPA Genetic Toxicology Program.

SAFETY PROFILE: Confirmed carcinogen producing leukemia. Experimental carcinogenic and neoplastigenic data. Poison by ingestion, intravenous, intraperitoneal, and subcutaneous routes. Human system effects by ingestion: convulsions, cough, dyspnea, and interstitial fibrosis. Human reproductive effects by ingestion and possibly other routes: changes in spermatogenesis; menstrual cycle changes or disorders; and teratogenic effects of the fetal urogenital system. Experimental teratogenic and reproductive effects. Human mutation data reported. An anti-neoplastic agent. When heated to decomposition it emits very toxic fumes of Cl^- and NO_x.

CDO625 CAS:1030-06-4 HR: 3
CHLORAMBUCIL SODIUM SALT
mf: $C_{14}H_{19}Cl_2NO_2{\bullet}Na$ mw: 327.23

SYNS: CB 1348 SODIUM SALT □ p-(DI-2-CHLOROETHYLAMINE) PHENYL BUTYRIC ACID SODIUM SALT □ SODIUM CHLORAMBUCIL

TOXICITY DATA WITH REFERENCE
sln-oin-dmg-par 2 mmol/L GENRA8 1,173,60
ipr-mus TDLo: 10 mg/kg (12D preg): REP BJPCAL 11,437,56
ipr-mus TDLo: 20 mg/kg (female 1D post): REP BJPCAL 11,437,56
ipr-mus TDLo: 20 mg/kg (female 7D post): REP BJPCAL 11,437,56
unr-mus LD50: 45 mg/kg BJPCAL 11,437,56

CONSENSUS REPORTS: EPA Genetic Toxicology Program.

SAFETY PROFILE: Poison by an unspecified route. Experimental reproductive effects. Mutation data reported. When heated to decomposition it emits toxic fumes of Cl^-, NO_x, and Na_2O. See also CHLORAMBUCIL.

CDO750 CAS:10599-90-3 HR: 2
CHLORAMIDE
mf: ClH_2N mw: 51.48

PROP: A liquid formed as a solution in Et_2O. Mp: −66°. Sol in H_2O, Et_2O, and NH_3.

SYNS: CHLORAMINE □ CHLORAMINE (inorganic compound) □ CHLOROAMINE □ MONOCHLORAMIDE □ MONOCHLORAMINE □ MONOCHLOROAMINE □ MONOCHLOROAMMONIA

TOXICITY DATA WITH REFERENCE
mmo-bcs 18 µmol/L SCIEAS 192,1141,76
dnr-bcs 18 µmol/L SCIEAS 192,1141,76
orl-rat TDLo: 7280 mg/kg/2Y-C: ETA NTPTR* NTP-TR-392,92

CONSENSUS REPORTS: Reported in EPA TSCA Inventory.

SAFETY PROFILE: Questionable carcinogen with experimental tumorigenic data. Mutation data reported. The dry material decomposes violently at −50°C. When heated to decomposition it emits very toxic fumes of NO_x, NH_3, and Cl^-.

CDP000 CAS:127-65-1 HR: 3
CHLORAMINE T
mf: $C_7H_8ClNO_2S{\cdot}Na$ mw: 228.66

PROP: Faintly yellow crystals or powder (as Na salt). Mp: 167–170° (as Na salt). Mod sol in H_2O; insol in C_6H_6, $CHCl_3$; decomp in EtOH (Na salt).

SYNS: ACTI-CHLORE □ AKTIVIN □ ANEXOL □ BENZENESULFONAMIDE, N-CHLORO-4-METHYL-, SODIUM SALT (9CI) □ BERKENDYL □ CHLORALONE □ CHLORASAN □ CHLORASEPTINE □ CHLORAZAN □ CHLORAZENE □ CHLORAZONE □ CHLOROZONE □ CHLORSEPTOL □ CLORINA □ CLOROSAN □ DESINFECT □ EUCLORINA □ GANSIL □ GYNECLORINA □ HALAMID □ HELIOGEN □ KLORAMIN □ KLORAMINE-T □ MULTICHLOR □ SODIUM CHLORAMINE T □ SODIUM p-TOLUENESULFONYLCHLORAMIDE □ SODIUM TOSYLCHLORAMIDE □ TAMPULES □ TOCHLORINE □ TOLAMINE □ TOSYLCHLORAMIDE SODIUM

TOXICITY DATA WITH REFERENCE
cyt-hmn:lyms 100 ppm/24H ARMCAH 21,409,70
par-mus LDLo:300 mg/kg JPETAB 14,259,20
ivn-rbt LDLo:25 mg/kg JPETAB 14,259,20
scu-gpg LDLo:900 mg/kg JPETAB 14,259,20
par-frg LDLo:200 mg/kg JPETAB 14,259,20

CONSENSUS REPORTS: Reported in EPA TSCA Inventory.

SAFETY PROFILE: Poison by parenteral and intravenous routes. Human mutagenic data reported. When heated to decomposition it emits toxic fumes of Cl^-, SO_x, Na_2O, and NO_x. See also SULFONATES and CHLORIDES.

CDP250 CAS:56-75-7 HR: 3
CHLORAMPHENICOL
mf: $C_{11}H_{12}Cl_2N_2O_5$ mw: 323.15

PROP: Pale-yellow needles (H_2O or 1,2-dichloroethane) or crystals. Mp: 151°. Sltly sol in water. Sol in EtOH, EtOAc, and Me_2CO; insol in C_6H_6 and pet ether.

SYNS: ALFICETYN □ AMBOFEN □ AMPHENICOL □ AMPHICOL □ AMSECLOR □ ANACETIN □ AQUAMYCETIN □ AUSTRACIL □ AUSTRACOL □ BIOCETIN □ BIOPHENICOL □ CAF □ CAM □ CAP □ CATILAN □ CHEMICETIN □ CHEMICETINA □ CHLOMIN □ CHLOMYCOL □ CHLORAMEX □ CHLORAMFICIN □CHLORAMFILIN □ d-CHLORAMPHENICOL □ d-threo-CHLORAMPHENICOL □ CHLORAMSAAR □ CHLORASOL □ CHLORA-TABS □ CHLORICOL □ CHLORNITROMYCIN □ CHLOROCAPS □ CHLOROCID □ CHLOROCIDIN C TETRAN □ CHLOROCOL □ CHLOROJECT L □ CHLOROMAX □ CHLOROMYCETIN □ CHLORONITRIN □ CHLOROPTIC □ CHLOROVULES □ CIDOCETINE □ CIPLAMYCETIN □CLORAMIDINA □ CLOROAMFENICOLO (ITALIAN) □ CLOROMISAN □ CLOROSINTEX □ COMYCETIN □ CPH □ CYLPHENICOL □ DESPHEN □ DETREOMYCINE □ DEXTROMYCETIN □ d-(−)-threo-2-DICHLOROACETAMIDO-1-p-NITROPHENYL-1,3-PROPANEDIOL □ d-threo-N-DICHLOROACETYL-1-p-NITROPHENYL-2-AMINO-1,3-PROPANEDIOL □ d-(−)-2,2-DICHLORO-N-(β-HYDROXY-α-(HYDROXYMETHYL)-p-NITROPHENYLETHYL)ACETAMIDE □ d-(−)-threo-2,2-DICHLORO-N-(β-HYDROXY-α-(HYDROXYMETHYL))-p-NITROPHENETHYLACETAMIDE □ d-threo-N-(1,1′-DIHYDROXY-1-p-NITROPHENYLISOPROPYL)DICHLOROACETAMIDE □ DOCTAMICINA □ ECONOCHLOR □ EMBACETIN □ EMETREN □ ENICOL □ ENTEROMYCETIN □ ERBAPLAST □ ERTILEN □ FARMICETINA □ FENICOL □ GLOBENICOL □ GLOROUS □ HALOMYCETIN □ HORTFENICOL □ I 337A □ INTRAMYCETIN □ ISMICETINA □ ISOPHENICOL □ ISOPTO FENICOL □ KAMAVER □ KEMICETINE □ LEUKOMYAN □ LEVOMYCETIN □ LOROMISIN □ MASTIPHEN □ MEDIAMYCETINE □ MICOCHLORINE □ MICROCETINA □ MYCHEL □ MYCINOL □ NCI-C55709 □ d-(−)-threo-1-p-NITROPHENYL-2-DICHLORACETAMIDO-1,3-PROPANEDIOL □ d-threo-1-(p-NITROPHENYL)-2-(DICHLOROACETYLAMINO)-1,3-PROPANEDIOL □ NORIMYCIN V □ NOVOCHLOROCAP □ NOVOMYCETIN □ NOVOPHENICOL □ NSC 3069 □ OFTALENT □ OLEOMYCETIN □ OPTHOCHLOR □ OTOPHEN □ PANTOVERNIL □ PARAXIN □ PETNAMYCETIN □ QUEMICETINA □ RIVOMYCIN □ ROMPHENIL □ SEPTICOL □ SINTOMICETINA □ STANOMYCETIN □ SYNTHOMYCINE □ TEVCOCIN □ TIFOMYCINE □ TREOMICETINA □ U-6062 □ UNIMYCETIN □ VETICOL

TOXICITY DATA WITH REFERENCE
dni-hmn:bmr 1500 μmo/L 46GFA5 -,17,81
cyt-hmn:lym 500 mg/L HUMAA7 7,305,69
orl-mus TDLo:175 mg/kg (female 15-21D post):REP NEPHBW 13,233,74
ivn-rat TDLo:2 g/kg (female 10-14D post):TER AEMBAP 27,291,72
orl-mus TDLo:6 g/kg (female 8-10D post):TER TXAPA9 19,667,71
orl-rat TDLo:23 g/kg (1-21D preg):TER TJADAB 12,291,75
par-rbt TDLo:2700 mg/kg (female 11-19D post):REP JJANAX 23,353,70
par-rbt TDLo:2700 mg/kg (female 2-10D post):REP JJANAX 23,353,70
orl-mus TDLo:7 g/kg (female 6-12D post):REP TXAPA9 19,667,71
ipr-rat TDLo:250 mg/kg (female 3D post):REP EXPEAM 35,1649,79
scu-rat TDLo:2400 mg/kg (female 12-14D post):REP VHAGA3 71,623,77
orl-rat TDLo:2 g/kg (female 8D post):TER TXAPA9 19,667,71
orl-rat TDLo:2500 mg/kg (female 9D post):TER AAGEAA 60(5),25,71
scu-rat TDLo:3500 mg/kg (female 6-10D post):TER OSDIAF 14,107,65
orl-rat TDLo:23 g/kg (1-21D preg):TER TJADAB 12,291,75
orl-mus TDLo:7 g/kg (female 6-12D post):TER TXAPA9 19,667,71
orl-wmn TDLo:300 mg/kg/60W-I:CAR,BLD NEJMAG 277,1003,67
ipr-mus TDLo:2500 mg/kg/5W-I:ETA,BLD CNREA8 41,3478,81
orl-wmn TD:1680 mg/kg/6W-I:CAR,BLD NEJMAG 277,1003,67
orl-man TD:434 mg/kg/W-C:CAR,BLD ACHAAH 66,267,81
orl-inf TDLo:440 mg/kg:CNS,GIT,MET JAMAAP 234,149,75
orl-wmn LDLo:400 mg/kg JAMAAP 234,149,75
unr-chd TDLo:250 mg/kg/10D:BLD,LIV CPEDAM 14,499,75
ims-inf TDLo:250 mg/kg/2D:CVS NEJMAG 262,787,60

orl-rat TDLo:2500 mg/kg FRPSAX 10,3,55
ipr-rat LD50:1811 mg/kg TXAPA9 18,185,71
scu-rat LD50:5 g/kg TXAPA9 9,445,66
ivn-rat LD50:171 mg/kg JCINAO 28,943,49
orl-mus LD50:1500 mg/kg ARZNAD 5,1,55
ipr-mus LD50:1100 mg/kg DIPHAH 14,21,62
scu-mus LD50:400 mg/kg 85ERAY 1,493,78
ivn-mus LD50:110 mg/kg JCINAO 28,943,49
ivn-dog LDLo:150 mg/kg JOBAAY 55,425,48
ivn-rbt LD50:117 mg/kg JCINAO 28,943,49
orl-gpg LD50:500 mg/kg FRPSAX 10,3,55
ivn-gpg LD50:560 mg/kg FRPSAX 9,21,54

CONSENSUS REPORTS: IARC Cancer Review: Group 2B IMEMDT 7,145,87; Human Limited Evidence IMEMDT 10,85,76. Reported in EPA TSCA Inventory. EPA Genetic Toxicology Program.

SAFETY PROFILE: Suspected human carcinogen producing leukemia, aplastic anemia, and other bone marrow changes. Experimental tumorigenic data. Poison by intravenous and subcutaneous routes. Moderately toxic by ingestion and intraperitoneal routes. Human systemic effects by an unknown route: changes in plasma or blood volume, unspecified liver effects, and hemorrhaging. Experimental teratogenic and reproductive effects. Human mutation data reported. An antibiotic. When heated to decomposition it emits very toxic fumes of NO_x and Cl^-. See also other chloramphenicol entries.

CDP500 CAS:982-57-0 HR: 2
CHLORAMPHENICOL MONOSUCCINATE SODIUM SALT
mf: $C_{15}H_{15}Cl_2N_2O_8 \cdot Na$ mw: 445.21

SYNS: CHLORAMPHENICOL SODIUM MONOSUCCINATE ☐ CHLORAMPHENICOL SODIUM SUCCINATE ☐ CHLORAMPHENICOL SUCCINATE SODIUM ☐ CHLORAMPHENICOL-SUKZINAT-NATRIUM (GERMAN) ☐ PROTOPHENICOL ☐ SODIUM CHLORAMPHENICOL SUCCINATE

TOXICITY DATA WITH REFERENCE
mic-mus:lym 1048 mg/L EMMUEG 12(Suppl 13),37,88
msc-mus:lym 2953 mg/L EMMUEG 12(Suppl 13),37,88
scu-rat TDLo:6 g/kg (female 11-14D post):TER DEGEA3 31,1181,76
ivn-rbt TDLo:1200 mg/kg (female 7-12D post):REP VHAGAS 71,623,77
scu-rat TDLo:35 g/kg (female 35D pre):REP KSRNAM 4,301,70
ivn-inf LDLo:135 mg/kg/3D-I JOPDAB 103,485,83
ipr-rat LD50:1400 mg/kg NIIRDN 6,248,82
ivn-rat LD50:1500 mg/kg NIIRDN 6,248,82

SAFETY PROFILE: Moderately toxic by intravenous and intraperitoneal routes. An experimental teratogen. Other experimental reproductive effects. Mutation data reported. When heated to decomposition it emits very toxic fumes of Cl^-, Na_2O, and NO_x. See also other chloramphenicol entries.

CDP700 CAS:530-43-8 HR: 2
CHLORAMPHENICOL PALMITATE
mf: $C_{27}H_{42}Cl_2N_2O_6$ mw: 561.61

PROP: Crystals from C_6H_6. Mp: 90°.

SYNS: CAP-P ☐ CAP-PALMITATE ☐ CHLORAMPHENICOL MONOPALMITATE ☐ DETREOPAL ☐ α-ESTER PALMITIC ACID with D-threo-(−)-2,2-DICHLORO-N-(β-HYDROXY-α-(HYDROXYMETHYL)-p-NITROPHENETHYL)ACETAMIDE

TOXICITY DATA WITH REFERENCE
orl-rbt TDLo:1200 mg/kg (7-12D preg):TER VHAGAS 71,623,77
orl-mus LD50:2640 mg/kg NIIRDN 6,248,82

SAFETY PROFILE: Moderately toxic by oral route. An experimental teratogen. Other experimental reproductive effects. An antibiotic. When heated to decomposition it emits very toxic fumes of NO_x and Cl^-. See also other chloramphenicol entries.

CDP725 CAS:3544-94-3 HR: 2
CHLORAMPHENICOL SUCCINATE
mf: $C_{15}H_{16}Cl_2N_2O_8$ mw: 423.23

SYNS: CHLORAMPHENICOL ACID SUCCINATE ☐ CHLORAMPHENICOL HEMISUCCINATE ☐ CHLORAMPHENICOL HYDROGEN SUCCINATE ☐ CHLORAMPHENICOL MONOSUCCINATE ☐ CHLOROMYCETIN SUCCINATE ☐ CHRONICIN FOAM ☐ CPSA ☐ KEMICETINE SUCCINATE ☐ LEVOMYCETIN HEMISUCCINATE ☐ LEVOMYCETIN SUCCINATE ☐ PARAXIN SUCCINATE ☐ SUCCINATO de CLORANFENICOL (SPANISH) ☐ SUCCINIC ACID-α-MONOESTER with d-threo-(−)-2,2-DICHLORO-N-(β-HYDROXY-α-(HYDROXYMETHYL)-p-NITROPHENETHYL)ACETAMIDE

TOXICITY DATA WITH REFERENCE
scu-rat TDLo:750 mg/kg (7-21D preg):REP SCIEAS 213,238,81
orl-rat LD50:8300 mg/kg KSRNAM 4,135,70
scu-rat LD50:2400 mg/kg THERAP 22,1405,67
ivn-rat LD50:1720 mg/kg THERAP 22,1405,67
orl-mus LD50:11 g/kg KSRNAM 4,135,70
ipr-mus LD50:1400 mg/kg FRPSAX 9,21,54
scu-mus LD50:4200 mg/kg KSRNAM 4,135,70
ivn-mus LD50:1500 mg/kg FRPSAX 9,21,54
ivn-gpg LD50:1000 mg/kg FRPSAX 9,21,54

CONSENSUS REPORTS: EPA Genetic Toxicology Program.

SAFETY PROFILE: Moderately toxic by subcutaneous, intravenous, and intraperitoneal routes. Experimental reproductive effects. When heated to decomposition it emits toxic fumes of Cl^- and NO_x. See also other chloramphenicol entries.

CDP750 CAS:20856-57-9 HR: 3
CHLORANIFORMETHANE
mf: $C_9H_7Cl_5N_2O$ mw: 336.43

SYNS: BAY 79770 ☐ CHLORANIFORMETHAN ☐ 1-(3,4-DICHLORANILINO)-1-FORMYLAMINO-2,2,2-TRICHLORAETHAN (GERMAN) ☐ N-FORMYL-N'-(3',4'-DICHLORPHENYL)-2,2,2-TRICHLORACETALDEHYDAM (GERMAN) ☐ IMUGAN ☐ MILFARON ☐ N-(2,2,2-TRICHLORO-1-(3,4-DICHLOROANILINO))ETHYLFORMAMIDE

TOXICITY DATA WITH REFERENCE
orl-rat LD50:2500 mg/kg 85ARAE 4,129,76/77
orl-gpg LD50:250 mg/kg 85DPAN -,-,71/76

SAFETY PROFILE: Poison by ingestion. A pesticide. When heated to decomposition it emits very toxic fumes of Cl^- and NO_x. See also ALDEHYDES.

CDQ000 **HR: 3**
CHLORATES

PROP: Chlorates are a combination of a metal or hydrogen and $^-ClO_3$ monovalent radical. They are crystalline and somewhat deliquescent.

SAFETY PROFILE: The principal toxic effects of chlorates are the production of methemoglobin in the blood and destruction of red blood corpuscles. The latter may lead to irritation of the kidneys. Damage to heart muscle has been reported.

Dangerous fire hazard in contact with flammable matter. When contaminated with oxidizable materials, they are particularly sensitive to friction, heat, and shock. They are powerful oxidizing agents and can undergo violent reactions with reducing materials. Dangerous explosion hazard when shocked, exposed to heat, or rubbed, particularly when contaminated with sugar, charcoal, shellac, sulfur, starch, sawdust, sulfuric acid, ammonium compounds, cyanides, phosphorous or antimony sulfide, Al, (metals + acids), As_2S_3, CaH_2, MnO_2, metal sulfides, organic acids, powdered metals, Hg_3P_4, PHI_4, SCN, (S + Cu), Se, NaH_2PO_2, SrH, SO_2. Chlorates when mixed with combustible materials may form explosive mixtures. For instance, potassium chlorate, when mixed with sulfur or with other combustible substances explodes on friction. Pure chlorates which have been spilled on the floor, or mixed with small amounts of impurities, become very sensitive to shock and friction. Water is considered the best agent for fighting fires involving chlorates. When heated to decomposition they can emit toxic fumes of Cl^- and explode.

CDQ250 **CAS:57109-90-7** **HR: 3**
CHLORAZEPATE DIPOTASSIUM
mf: $C_{16}H_{10}ClN_2O_3•K•HKO$ mw: 408.94

PROP: Powder. Sol in H_2O; insol in Et_2O and $CHCl_3$.

SYNS: AB 35616 □ ABBOTT-35616 □ AH 3232 □ BELSEREN □ BIPOTASSIUM CHLORAZEPATE □ CB 4306 □ CHLORAZEPAM □ CLORAZEPATE DIPOTASSIUM □ DIPOTASSIUM CHLORAZEPATE □ DIPOTASSIUM CLORAZEPATE □ MENDON □ NEVRACTEN □ POTASSIUM 7-CHLORO-2,3-DIHYDRO-2-OXO-5-PHENYL-1H-1,4-BENZODIAZEPINE-3-CARBOXYLATE KOH □ TENCILAN □ TRANSENE □ TRANSILIUM □ TRANXENE □ TRANXILEN □ TRANXILENE □ TRANXILIUM

TOXICITY DATA WITH REFERENCE
sln-asn 1 mg/L MUREAV 26,159,74
unr-wmn TDLo:18 mg/kg (25-39W preg):REP THERAP 36,305,81
ims-rat TDLo:96 mg/kg (female 8 10D post):REP ARNEAS 37,350,80
orl-rat TDLo:36,400 mg/kg (26W pre):REP OYYAA2 8,89,74
orl-rat LD50:880 mg/kg IYKEDH 10,710,79
ipr-rat LD50:31,200 μg/kg IYKEDH 10,710,79
scu-rat LD50:1478 mg/kg NIIRDN 6,245,82
ivn-rat LD50:279 mg/kg NIIRDN 6,245,82
orl-mus LD50:700 mg/kg ARZNAD 20,123,70
ipr-mus LD50:290 mg/kg ARZNAD 20,123,70
scu-mus LD50:443 mg/kg IYKEDH 10,710,79
ivn-mus LD50:157 mg/kg IYKEDH 10,710,79

SAFETY PROFILE: Poison by intravenous and intraperitoneal routes. Moderately toxic by ingestion and subcutaneous routes. Experimental reproductive effects. Mutation data reported. A tranquilizer. When heated to decomposition it emits very toxic fumes of Cl^-, NO_x and K_2O.

CDQ325 **CAS:580-48-3** **HR: 2**
CHLORAZINE
mf: $C_{11}H_{20}ClN_5$ mw: 257.81

PROP: A solid or liquid. Mp: 27°, bp: 104–106° @ 0.05 mm.

SYNS: 2-CHLORO-4,6-BIS(DIETHYLAMINO)-s-TRIAZINE □ 6-CHLORO-N,N,N',N'-TETRAETHYL-1,3,5-TRIAZINE-2,4-DIAMINE

TOXICITY DATA WITH REFERENCE
orl-rat LD50:850 mg/kg FMCHA2 -,C50,83
unr-rat LD50:3500 mg/kg 30ZDA9 -,420,71
orl-mus LD50:743 mg/kg 85GMAT -,35,82

SAFETY PROFILE: Moderately toxic by ingestion and possibly other routes. When heated to decomposition it emits toxic fumes of Cl^- and NO_x.

CDQ500 **CAS:5576-62-5** **HR: 3**
CHLORBENZOXYETHAMINE DIHYDROCHLORIDE
mf: $C_{27}H_{31}ClN_2O•2ClH$ mw: 507.97

PROP: Bitter crystals. Mp: 197–200°.

SYNS: ANTIULCERA MASTER □ CHLORBENZOSAMINE DIHYDROCHLORIDE □ CHLORBENZOXAMINE DIHYDROCHLORIDE □ GASTOMAX □ LIBRATAR □ U.C.B. 1474

TOXICITY DATA WITH REFERENCE
orl-rat LD50:3350 mg/kg AIPTAK 118,167,59
ivn-rat LD50:66 mg/kg AIPTAK 118,167,59
orl-mus LD50:1400 mg/kg AIPTAK 118,167,59

SAFETY PROFILE: Poison by intravenous route. Moderately toxic by ingestion. When heated to decomposition it emits very toxic fumes of NO_x and Cl^-.

CDQ750 **CAS:95-25-0** **HR: 3**
5-CHLORBENZOZAZOLIN-2-ON
mf: $C_7H_4ClNO_2$ mw: 169.57

PROP: Crystals from Me_2CO. Mp: 191–191.5°. Sltly sol in H_2O.

SYNS: BIOMIORAN □ 5-CHLORBENZOXAZOLIN-2-ON □ 5-CHLOROBENZOXAZOLIDONE □ 5-CHLORO-2-BENZOXAZOLINONE □ 6-CHLORO-2-BENZOXAZOLINONE □ 5-CHLOROBENZOXAZOL-2-ONE □ 5-CHLORO-3(H)-2-BENZOXAZOLONE □ CHLOROXAZONE □ CHLORZOXAZONE □ MYOFLEXINE □ PARAFLEX □ SOLAXIN □ USAF MA-10

TOXICITY DATA WITH REFERENCE
orl-rat LD50:763 mg/kg JPETAB 129,75,60
ipr-rat LD50:150 mg/kg JPETAB 129,75,60
orl-mus LD50:440 mg/kg ARZNAD 17,242,67
ipr-mus LD50:50 mg/kg NTIS** AD277-689
scu-mus LD50:170 mg/kg APTOA6 19,247,62
orl-ham LD50:662 mg/kg JPETAB 129,75,60
ipr-ham LD50:166 mg/kg JPETAB 129,75,60
ipr-mam LD50:550 mg/kg CHTPBA 6,65,71

CONSENSUS REPORTS: Reported in EPA TSCA Inventory.

SAFETY PROFILE: Poison by intraperitoneal and subcutaneous routes. Moderately toxic by ingestion. A skeletal muscle relaxant. When heated to decomposition it emits very toxic fumes of Cl^- and NO_x.

CDR000 CAS:129-71-5 HR: 3
CHLORCYCLIZINE DIHYDROCHLORIDE
mf: $C_{18}H_{21}ClN_2•2ClH$ mw: 373.78

PROP: Prisms from $EtOH/Et_2O$. Mp: 216–216.5°.

SYNS: AH 289 □ 1-(4-CHLOROBENZHYDRYL)-4-METHYLPIPERAZINE DIHYDROCHLORIDE □ 1-(p-CHLORO-α-PHENYLBENZYL)-4-METHYL-PIPERAZINE DIHYDROCHLORIDE □ DI-PARALENE-2-HYDROCHLORIDE □ HISTANTINE DIHYDROCHLORIDE □ N-METHYL-N'-(4-CHLOROBENZHYDRYL)PIPERAZINE DIHYDROCHLORIDE □ PERAZIL □ PERAZIL DIHYDROCHLORIDE □ TRIHISTAN

TOXICITY DATA WITH REFERENCE
ipr-rat LD50:100 mg/kg AIPTAK 80,378,49
orl-mus LDLo:150 mg/kg AIPTAK 80,378,49
ipr-mus LD50:137 mg/kg JPETAB 96,388,49
scu-mus LDLo:150 mg/kg AIPTAK 80,378,49
ivn-mus LD50:35 mg/kg AIPTAK 80,378,49
ipr-dog LDLo:100 mg/kg AIPTAK 80,378,49
ipr-cat LD50:75 mg/kg AIPTAK 80,378,49
ipr-gpg LD50:100 mg/kg AIPTAK 80,378,49

SAFETY PROFILE: Poison by ingestion, subcutaneous, intravenous, and intraperitoneal routes. An antihistamine. When heated to decomposition it emits very toxic fumes of Cl^- and NO_x. See also CHLORCYCLIZINE DIHYDROCHLORIDE.

CDR250 CAS:14362-31-3 HR: 3
CHLORCYCLIZINE HYDROCHLORIDE
mf: $C_{18}H_{21}ClN_2•xClH$ mw: 556.08

SYNS: AH-289 HYDROCHLORIDE □ CHLORCYCLIZINIUM CHLORIDE □ 1-(p-CHLOROBENZHYDRYL)-4-METHYLPIPERAZINE HYDROCHLORIDE □ DIPARALENE HYDROCHLORIDE □ ERAMIDE

TOXICITY DATA WITH REFERENCE
ims-mus TDLo:200 mg/kg (female 12-13D post):TER AIPTAK 194,168,71
ims-mus TDLo:200 mg/kg (female 12-13D post):REP AIPTAK 194,168,71
orl-rat TDLo:3024 mg/kg (male 28D pre):REP BIREBV 7,398,72
orl-rat TDLo:100 mg/kg (female 12-15D post):TER JPETAB 147,391,65
ipr-rat LD50:100 mg/kg CLDND*
orl-mus LD50:300 mg/kg CLDND*
ipr-mus LD50:100 mg/kg CLDND*
scu-mus LD50:200 mg/kg CLDND*
ivn-mus LD50:35 mg/kg CLDND*
ipr-dog LD50:125 mg/kg CLDND*
ipr-cat LD50:100 mg/kg CLDND*
ipr-gpg LD50:100 mg/kg CLDND*

SAFETY PROFILE: Poison by ingestion, subcutaneous, intravenous, and intraperitoneal routes. An experimental teratogen. Other experimental reproductive effects. When heated to decomposition it emits very toxic fumes of HCl and NO_x.

CDR500 CAS:894-56-4 HR: 3
CHLORCYCLIZINE HYDROCHLORIDE A
mf: $C_{18}H_{21}ClN_2•ClH$ mw: 337.32

SYNS: CHLOROCYCLIZINE HYDROCHLORIDE □ 1-(p-CHLORO-α-PHENYLBENZYL)-4-METHYLPIPERAZINE HYDROCHLORIDE □ PERAZIL

TOXICITY DATA WITH REFERENCE
orl-rat TDLo:650 mg/kg (10-22D preg):REP VAAZA2 39,59,82
orl-rat TDLo:500 mg/kg (6-15D preg):TER TXCYAC 8,87,77
orl-mus TDLo:800 mg/kg (11-14D preg):TER AJOGAH 95,109,66
ipr-rat LD50:100 mg/kg 27ZQAG -,213,72
orl-mus LD50:300 mg/kg 27ZQAG -,213,72
ipr-mus LD50:137 mg/kg JPETAB 96,388,49
scu-mus LD50:200 mg/kg 27ZQAG -,213,72
ivn-mus LD50:50,800 μg/kg TXAPA9 1,454,56
ipr-dog LD50:125 mg/kg 27ZQAG -,213,72
ipr-cat LD50:75 mg/kg 27ZQAG -,213,72
ipr-gpg LD50:100 mg/kg 27ZQAG -,213,72

SAFETY PROFILE: Poison by ingestion, subcutaneous, intravenous, and intraperitoneal routes. An experimental teratogen. Other experimental reproductive effects. When heated to decomposition it emits very toxic fumes of HCl and NO_x.

CDR550 CAS:963-03-1 HR: 2
CHLORCYCLOHEXAMIDE
mf: $C_{13}H_{17}ClN_2O_3S$ mw: 316.83

SYNS: BENZENESULFONAMIDE, 4-CHLORO-N-((CYCLOHEXYLAMINO)CARBONYL)- □ CHLORHEXAMIDE □ CHLOROCYCLAMIDE-R □ 1-((p-CHLOROPHENYL)SULFONYL)-3-CYCLOHEXYLUREA □ K 694 □ ORADIAN □ UREA, 1-((p-CHLOROPHENYL)SULFONYL)-3-CYCLOHEXYL-

TOXICITY DATA WITH REFERENCE
orl-rat TDLo:1 g/kg (female 10D post):TER PROEAS 14,89,68
orl-rat LD50:1525 mg/kg FATOAO 25,93,62
orl-mus LD50:1525 mg/kg FATOAO 25,93,62

SAFETY PROFILE: Moderately toxic by ingestion. Experimental teratogenic effects. When heated to decomposition it emits toxic fumes of SO_x, NO_x, and Cl^-.

CDR575 **CAS:5566-34-7** **HR: 3**
trans-CHLORDAN
mf: $C_{10}H_6Cl_8$ mw: 409.76

SYNS: γ-CHLORDAN ☐ γ(trans)-CHLORDANE ☐ 4,7-METHANOINDAN, 2,2,4,5,6,7,8,8-OCTACHLORO-3a,4,7,7a-TETRAHYDRO- ☐ 4,7-METHANO-1H-INDENE, 2,2,4,5,6,7,8,8-OCTACHLORO-2,3,3a,4,7,7a-HEXAHYDRO- (9CI) ☐ 2,2,4,5,6,7,8,8-OCTACHLORO-3a,4,7,7a-TETRAHYDRO-4,7-METHANOINDAN

TOXICITY DATA WITH REFERENCE
ipr-mus TDLo:100 mg/kg (female 1D post):REP BECTA6 17,559,77
scu-mus TDLo:18 mg/kg (female 3D pre):REP JTEHD6 3,713,77
orl-rat LD50:500 mg/kg NTIS** PB85-143766

CONSENSUS REPORTS: IARC Cancer Review: Group 2B IMEMDT 53,115,91; Animal Sufficient Evidence IMEMDT 53,115,91; Human Inadequate Evidence IMEMDT 53,115,91.

SAFETY PROFILE: Confirmed carcinogen. Moderately toxic by ingestion. Experimental reproductive effects. When heated to decomposition it emits toxic fumes of Cl^-.

CDR675 **CAS:5103-71-9** **HR: 3**
α-CHLORDAN
mf: $C_{10}H_6Cl_8$ mw: 409.76

SYNS: cis-CHLORDAN ☐ α-CHLORDANE ☐ α(cis)-CHLORDANE ☐ cis-CHLORDANE

TOXICITY DATA WITH REFERENCE
ipr-mus TDLo:50 mg/kg (1D preg):REP BECTA6 17,559,77
scu-mus TDLo:18 mg/kg (female 3D pre):REP JTEHD6 3,713,77
scu-mus TDLo:9 mg/kg (3D pre):REP JTEHD6 3,713,77
orl-rat LD50:500 mg/kg NTIS** PB85-143766
orl-mus LD50:125 mg/kg JAFCAU 21,1113,73
ipr-mus LDLo:290 mg/kg TXAPA9 23,288,72

SAFETY PROFILE: Poison by ingestion and intraperitoneal routes. Experimental reproductive effects. When heated to decomposition it emits toxic fumes of Cl^-. See also CHLORDANE.

CDR750 **CAS:57-74-9** **HR: 3**
CHLORDANE
mf: $C_{10}H_6Cl_8$ mw: 409.76

PROP: Colorless to amber; odorless, viscous liquid. Bp: 175°, d: 1.57–1.63 @ 15.5°/15.5°.

SYNS: ASPON-CHLORDANE ☐ BELT ☐ CD 68 ☐ CHLOORDAAN (DUTCH) ☐ CHLORDAN ☐ γ-CHLORDAN ☐ CHLORDANE, liquid (DOT) ☐ CHLORINDAN ☐ CHLOR KIL ☐ CHLORODANE ☐ CHLORTOX ☐ CLORDAN (ITALIAN) ☐ CLORODANE ☐ CORTILAN-NEU ☐ DICHLOROCHLORDENE ☐ DOWCHLOR ☐ ENT 9,932 ☐ ENT 25,552-X ☐ HCS 3260 ☐ KYPCHLOR ☐ M 140 ☐ M 410 ☐ NCI-C00099 ☐ NIRAN ☐ 1,2,4,5,6,7,8,8-OCTACHLOOR-3a,4,7,7a-TETRAHYDRO-4,7-endo-METHANO-INDAAN (DUTCH) ☐ OCTACHLOR ☐ OCTACHLORODIHYDRODICYCLOPENTADIENE ☐ 1,2,4,5,6,7,8,8-OCTACHLORO-2,3,3a,4,7,7a-HEXAHYDRO-4,7-METHANOINDENE ☐ 1,2,4,5,6,7,8,8-OCTACHLORO-2,3,3a,4,7,7a-HEXAHYDRO-4,7-METHANO-1H-INDENE ☐ 1,2,4,5,6,7,8,8-OCTACHLORO-3a,4,7,7a-HEXAHYDRO-4,7-METHYLENE INDANE ☐ OCTACHLORO-4,7-METHANOHYDROINDANE ☐ OCTACHLORO-4,7-METHANOTETRAHYDROINDANE ☐ 1,2,4,5,6,7,8,8-OCTACHLORO-4,7-METHANO-3a,4,7,7a-TETRAHYDROINDANE ☐ 1,2,4,5,6,7,8,8-OCTACHLORO-3a,4,7,7a-TETRAHYDRO-4,7-METHANOINDAN ☐ 1,2,4,5,6,7,10,10-OCTACHLORO-4,7,8,9-TETRAHYDRO-4,7-METHYLENEINDANE ☐ 1,2,4,5,6,7,8,8-OCTACHLOR-3a,4,7,7a-TETRAHYDRO-4,7-endo-METHANO-INDAN (GERMAN) ☐ OCTA-KLOR ☐ OKTATERR ☐ ORTHO-KLOR ☐ 1,2,4,5,6,7,8,8-OTTOCHLORO-3A,4,7,7A-TETRAIDRO-4,7-endo-METANO-INDANO (ITALIAN) ☐ RCRA WASTE NUMBER U036 ☐ SD 5532 ☐ SHELL SD-5532 ☐ SYNKLOR ☐ TAT CHLOR 4 ☐ TOPICHLOR 20 ☐ TOPICLOR ☐ TOPICLOR 20 ☐ TOXICHLOR ☐ VELSICOL 1068

TOXICITY DATA WITH REFERENCE
sce-ofs-mul 54 pmol/L MUREAV 118,61,83
sce-hmn:lym 10 µmol/L ARTODN 52,221,83
orl-mus TDLo:7 mg/kg (15-21D preg):REP BJPCBM 49,311,73
unr-mus TDLo:168 mg/kg (female 1-21D post):REP APTOD9 19,A18,80
orl-mus TDLo:3360 µg/kg (1-21D preg):TER JEPTDQ 2(2),357,78
orl-mus TDLo:152 mg/kg (1-19D preg):TER TXAPA9 62,402,82
orl-mus TDLo:2020 mg/kg/80W-C:CAR NCITR* NCI-CG-TR-8,77
orl-mus TD:3780 mg/kg/80W-C:CAR NCITR* NIC-CG-TR-8,77
orl-man TDLo:3071 µg/kg JTCTDW 20,291,83
orl-hmn LDLo:29 mg/kg:LIV CMEP** -,1,56
orl-wmn LDLo:120 µg/kg:CNS,GIT CMEP** -,1,56
skn-hmn LDLo:428 mg/kg:CNS 34ZIAG -,648,69
unr-man LDLo:118 mg/kg 85DCAI 2,73,70
orl-rat LD50:200 mg/kg ARZNAD 17,614,67
skn-rat LD50:690 mg/kg JAVMA4 157,1835,70
ipr-rat LD50:343 mg/kg TXAPA9 32,443,75
orl-mus LD50:145 mg/kg ARSIM* 20,19,66
ipr-mus LDLo:240 mg/kg TXAPA9 23,288,72
ivn-mus LD50:100 mg/kg CSLNX* NX#04876
ihl-cat LC50:100 mg/m³/4H GTPZAB 8(4),30,64
orl-rbt LD50:100 mg/kg PCOC** -,226,66
skn-rbt LD50:780 mg/kg 85DPAN -,-,71/76
ivn-rbt LDLo:10 mg/kg AIHOAX 1,13,50

CONSENSUS REPORTS: IARC Cancer Review: Group 2B IMEMDT 53,115,91; Animal Sufficient Evidence IMEMDT 20,45,79; Animal Limited Evidence IARC Monographs, Supplement. IMEMDT 7,146,87; Animal Sufficient Evidence IMEMDT 53,115,91; Human Inadequate Evidence IMEMDT 20,45,79; Human Inadequate Evidence IMEMDT 53,115,91. NCI Carcinogenesis Bioassay (feed); Clear Evidence: mouse NCITR* NCI-CG-TR-8,77; No Evidence: rat NCITR* NCI-CG-TR-8,77. EPA Genetic Toxicology Program. Community Right-To-Know List. EPA Extremely Hazardous Substances List.

OSHA PEL: TWA 0.5 mg/m³ (skin)
ACGIH TLV: TWA 0.5 mg/m³ (skin)
DFG MAK: 0.5 mg/m³; Suspected Carcinogen

SAFETY PROFILE: Confirmed carcinogen with experimental carcinogenic data. Poison to humans by ingestion and possibly other routes. An experimental poison by ingestion, inhalation, intravenous, and intraperitoneal routes. Moderately toxic by skin contact. Human systemic effects by ingestion or skin contact: tremors, convulsions, excitement, ataxia (loss of muscle coordination), and gastritis. Experimental teratogenic and reproductive effects. Human mutation data reported. Combustible liquid. It is no longer permitted for use as a termiticide in homes.

A central nervous system stimulant whose exact mode of action is unknown, but it may involve microsomal enzyme stimulation. Animals poisoned by this and related compounds show an extremely marked loss of appetite and neurological symptoms. The fatal dose to humans is unknown. It has been estimated to be between 6 and 60 g (0.2 and 2 ounces). One person receiving an accidental skin application of 25% solution (amounting to something over 30 g of technical chlordane) developed symptoms within about 40 minutes and died, apparently of respiratory failure, before medical attention was obtained. In two patients, death followed exposure to low ingestion doses of chlordane (2–4 g). On microscopic examination, both patients showed severe chronic fatty degeneration of the liver, characteristic of chronic alcoholism. Although these two fatalities cannot be attributed exclusively to chlordane, they are entirely consistent with previous observations that the toxicity of other chlorinated hydrocarbons is much enhanced in the presence of chronic liver damage. The dangerous chronic dose in humans is unknown.

Experimental animals exposed to repeated small doses exhibit hyperexcitability, tremors, and convulsions, and those that survive long enough show marked anorexia and loss of weight. Symptoms in animals frequently occur within an hour of the administration of a large dose, but death often is delayed for several days depending on the dosage and route of administration. In any event, symptoms are of longer duration with chlordane than with DDT under similar conditions.

Laboratory analyses on poisoned animals are essentially normal, except that the insecticide is found in tissues by means of bioassay. A method for specific, quantitative chemical analysis for chlordane is now available using small amounts of subcutaneous fat. Chronically poisoned animals show degenerative changes in the liver and kidney tubules.

When heated to decomposition chlordane emits toxic fumes of Cl^-.

CDS000 CAS:115-28-6 HR: 3
CHLORENDIC ACID
mf: $C_9H_4Cl_6O_4$ mw:388.83

SYNS: HET ACID □ 1,4,5,6,7,7-HEXACHLORO-5-NORBORNENE-2,3-DICARBOXYLIC ACID □ KYSELINA 3,6-ENDOMETHYLEN-3,4,5,6,7,7-HEXACHLOR-Δ^4-TETRAHYDROFTALOVA (CZECH) □ KYSELINA HET (CZECH) □ NCI-C55072

TOXICITY DATA WITH REFERENCE
skn-rbt 500 mg/24H MLD 28ZPAK -,92,72
eye-rbt 250 µg/24H SEV 28ZPAK -,92,72
msc-mus:lyms 1700 mg/L NTPTR* NTP-TR-304,87
orl-mus TDLo:84 g/kg/14D-C:REP NTPTR* NTP-TR-304,87
orl-rat TDLo:45,063 mg/kg/2Y-C:CAR NTPTR* NTP-TR-304,87
orl-mus TDLo:108 g/kg/2Y-C:CAR NTPTR* NTP-TR-304,87

CONSENSUS REPORTS: NTP 7th Annual Report on Carcinogens. NTP Carcinogenesis Studies (feed): Clear Evidence: mouse, rat NTPTR* NTP-TR-304,87. Reported in EPA TSCA Inventory.

SAFETY PROFILE: Confirmed carcinogen with experimental carcinogenic data. A severe eye and mild skin irritant. When heated to decomposition it emits toxic fumes of Cl^-.

CDS025 CAS:1770-80-5 HR: 1
CHLORENDIC ACID DIBUTYL ESTER
mf: $C_{17}H_{20}Cl_6O_4$ mw: 501.07

SYNS: BICYCLO(2.2.1)HEPT-5-ENE-2,3-DICARBOXYLIC ACID, 1,4,5,6,7,7-HEXACHLORO-, DIBUTYL ESTER □ DIBUTYL CHLORENDATE □ DIBUTYL 1,4,5,6,7,7-HEXACHLOROBICYCLO(2.2.1)HEPT-5-ENE-2,3-DICARBOXYLATE □ 5-NORBORNENE-2,3-DICARBOXYLIC ACID, 1,4,5,6,7,7-HEXACHLORO-, DIBUTYL ESTER (8CI)

TOXICITY DATA WITH REFERENCE
orl-rat LD50:>10 g/kg NTIS** PB85-143766

CONSENSUS REPORTS: Reported in EPA TSCA Inventory.

SAFETY PROFILE: Low toxicity by ingestion. When heated to decomposition it emits toxic vapors of Cl^-.

CDS050 CAS:115-27-5 HR: 3
CHLORENDIC ANHYDRIDE
mf: $C_9H_2Cl_6O_3$ mw: 370.81

SYN: 5-NORBORNENE-2,3-DICARBOXYLIC ANHYDRIDE, 1,4,5,6,7,7-HEXACHLORO-

TOXICITY DATA WITH REFERENCE
orl-rat LD50:2300 mg/kg GTPZAB 26(4),49,82
ihl-rat LC50:>1 g/m³ GTPZAB 26(4),49,82
orl-mus LD50:2400 mg/kg GTPZAB 26(4),49,82
ihl-cat LC50:>1 g/m³ GTPZAB 26(4),49,82
ihl-rbt LC50:>1 g/m³ GTPZAB 26(4),49,82

CONSENSUS REPORTS: Reported in EPA TSCA Inventory.

SAFETY PROFILE: Poison by inhalation. Moderately toxic by ingestion. When heated to decomposition it emits toxic vapors of Cl^-.

CDS100 CAS:6889-41-4 HR: 3
CHLORENDIC IMIDE
mf: $C_9H_3Cl_6NO_2$ mw: 369.83

SYNS: 1,4,5,6,7,7-HEXACHLOROBICYCLO(2.2.1)HEPTENE-2,3-DICARBOXIMIDE □ 1,4,5,6,7,7-HEXACHLORO-5-NORBORNENE-2,3-DICARBOXIMIDE

TOXICITY DATA WITH REFERENCE
orl-rat LD50:2300 mg/kg GTPZAB 26(4),49,82
orl-mus LD50:2400 mg/kg GTPZAB 26(4),49,82
ivn-mus LD50:320 mg/kg CSLNX* NX#00789

SAFETY PROFILE: Poison by intravenous route. Moderately toxic by ingestion. When heated to decomposition it emits toxic fumes of Cl^- and NO_x.

CDS125 **CAS:16672-87-0** **HR: 2**
CHLORETHEPHON
mf: $C_2H_6ClO_3P$ mw: 144.50

PROP: Very hygroscopic needles from benzene. Mp: 74–75°. Freely sol in water, methanol, acetone, ethylene glycol, propylene glycol; sltly sol in benzene, toluene; practically insol in petr ether.

SYNS: AMCHEM 68-250 □ BROMOFLOR □ CAMPOSAN □ CEP □ 2-CEPA □ CEPHA □ CEPHA 10LS □ 2-CHLORAETHYL-PHOSPHONSAEURE (GERMAN) □ 2-CHLORETHYLPHOSPHONIC ACID □ 2-CHLOROETHANEPHOSPHONIC ACID □ ETHEFON □ ETHEL □ ETHEPHON □ ETHEVERSE □ ETHREL □ FLORDIMEX □ FLOREL □ G 996 □ KAMPOSAN □ ROLL-FRUCT □ TOMATHREL

TOXICITY DATA WITH REFERENCE
orl-rat LD50:3400 mg/kg ZKMAAX 20,274,80
ihl-rat LC50:90 mg/m³/4H 85JCAE-,1120,86
orl-mus LD50:2850 mg/kg PHARAT 32,181,77
orl-rbt LD50:5000 mg/kg GISAAA 48(8),79,83
skn-rbt LD50:5730 mg/kg 85DPAN -,-,71/76
orl-gpg LD50:4200 mg/kg GISAAA 48(8),79,83

CONSENSUS REPORTS: EPA Genetic Toxicology Program.

SAFETY PROFILE: Moderately toxic by ingestion. Mildly toxic by skin contact. A plant growth regulator. Caution: Spray formulations are quite acidic, about pH 1.0. May be irritating to exposed skin and eyes, or if inhaled. When heated to decomposition it emits toxic fumes of Cl^- and PO_x.

CDS250 **CAS:132-89-8** **HR: 2**
CHLORETHYLBENZMETHOXAZONE
mf: $C_{10}H_{10}ClNO_2$ mw: 211.66

PROP: Cryst from EtOH. Mp: 146–147° (decomp).

SYNS: 2-(2-CHLOROETHYL)-3-AZA-4-CHROMANONE □ 2-(2-CHLOROETHYL)-2,3-DIHYDRO-4H-1,3-BENZOXAZIN-4-ONE □ 2-(2-CHLOROETHYL)-2,3-DIHYDRO-4-OXO-1,3-BENZOXAZINE □ 2-(β-CHLOROETHYL)-2,3-DIHYDRO-4-OXO(BENZO-1,3-OXAZINE) □ 4-OXO-2-(β-CHLOROETHYL)-2,3-DIHYDROBENZO-1,3-OXAZINE

TOXICITY DATA WITH REFERENCE
ipr-rat LD50:730 mg/kg BMJOAE 1,36,60
orl-rat LD50:10 g/kg ARZNAD 7,651,57

SAFETY PROFILE: Moderately toxic by intraperitoneal route. Mildly toxic by ingestion. When heated to decomposition it emits very toxic fumes of Cl^- and NO_x.

CDS275 **CAS:21267-72-1** **HR: 3**
CHLORETIN
mf: $C_{12}H_{12}ClNO$ mw: 221.70

SYNS: 2903 H □ 2-CHLORO-N-(1-METHYL-2-PROPYNYL)-ACETANILIDE (8CI) □ 2-CHLORO-N-(1-METHYL-2-PROPYNYL)-N-PHENYLACETAMIDE □ BASAMAIZE □ BAS-290-H □ BAS 2900H □ BAS 2903H □ BUTISAN □ BUTISANE □ CHLORESSIGSAEURE-N-ISOBUTINYLANILID (GERMAN) □ 2-CHLORO-N-(1-METHYL-2-PROPYNYL)ACETANILIDE □ PRYNACHLOR

TOXICITY DATA WITH REFERENCE
orl-rat LD50:1170 mg/kg GUCHAZ 6,440,73
orl-mus LD50:150 mg/kg GUCHAZ 6,440,73
skn-rbt LD50:1926 mg/kg GUCHAZ 6,440,73

CONSENSUS REPORTS: EPA Genetic Toxicology Program.

SAFETY PROFILE: Poison by ingestion. Moderately toxic by skin contact. When heated to decomposition it emits toxic fumes of Cl^- and NO_x.

CDS500 **CAS:2274-74-0** **HR: 2**
CHLORFENSULFIDE
mf: $C_{12}H_6Cl_4N_2S$ mw: 352.06

SYNS: CHLORFENSULFID (GERMAN) □ 4-CHLOROPHENYL-2,4,5-TRICHLOROPHENYLAZOSULFIDE □ 4-CHLORPHENYL-2′,4′,5′-TRICHLORPHENYLAZOSULFID (GERMAN) □ CPAS □ MICASIN □ 2,4,5-TRICHLOROBENZENEDIAZO p-CHLOROPHENYL SULFIDE □ 2,4,5-TRICHLOROPHENYLAZO-4′-CHLOROPHENYL-SULFIDE

TOXICITY DATA WITH REFERENCE
orl-rat LD50:4000 mg/kg 85DPAN -,-,71/76
unr-mus LD50:3000 mg/kg 30ZDA9 -,281,71

SAFETY PROFILE: Moderately toxic by ingestion and possibly other routes. See also SULFIDES. When heated to decomposition it emits very toxic fumes of Cl^-, NO_x, and SO_x.

CDS750 **CAS:470-90-6** **HR: 3**
CHLORFENVINFOS
mf: $C_{12}H_{14}Cl_3O_4P$ mw: 359.58

PROP: Amber liquid. Mp: −23°, bp: 124–126° @ 0.008 mm.

SYNS: APACHLOR □ BIRLANE □ C-10015 □ CFV □ CGA 26351 □ CHLOFENVINPHOS □ O-2-CHLOOR-1-(2,4-DICHLOOR FENYL)-VINYL-O,O-DIETHYLFOSFAAT (DUTCH) □ O-2-CHLOR-1-(2,4-DICHLORPHENYL)-VINYL-O,O-DIAETHYLPHOSPHAT (GERMAN) □ CHLORFENVINFOS □ CHLORFENVINPHOS □ 2-CHLORO-1-(2,4-DICHLOROPHENYL)VINYL DIETHYL PHOSPHATE □ β-2-CHLORO-1-(2′,4′-DICHLOROPHENYL) VINYL DIETHYLPHOSPHATE □ CHLOROFENVINPHOS □ CHLORPHENVINFOS □ CHLORPHENVINPHOS □ O-2-CLORO-1-(2,4-DICLORO-FENIL)-VINYL-O,O-DIETILFOSFATO (ITALIAN) □ COMPOUND 4072 □ CVP □ DERMATON □ O,O-DIAETHYL-O-1-(4,5-DICHLORPHENYL)-2-CHLOR-VINYL-PHOSPHAT (GERMAN) □ 2,4 DICHLORO-α-(CHLOROMETHYLENE)BENZYL ALCOHOL DIETHYL PHOSPHATE □ O,O-DIETHYL-O-(2-CHLORO-1-(2′,4′-DICHLOROPHENYL)VINYL) PHOSPHATE □ ENT 24,969 □ GC 4072 □ OMS 1328 □ PHOSPHATE de O,O-DIETHYLE et de O-2-CHLORO-1-(2,4-DICHLOROPHENYL) VINYLE (FRENCH) □ SAPECRON □ SHELL 4072 □ STELADONE □ SUPONA □ SUPONE □ UNITOX □ VINYLPHATE

TOXICITY DATA WITH REFERENCE
mmo-sat 500 μg/plate MUREAV 116,185,83
skn-hmn TDLo:10 mg/kg:BLD,BIO IMSUAI 38,25,69
orl-rat LD50:10 mg/kg FMCHA2 -,C52,83
skn-rat LD50:26,400 μg/kg APYPAY 32,507,81
ipr-rat LD50:8500 μg/kg VETRAX 77,1140,65
scu-rat LD50:7 mg/kg APYPAY 32,507,81
ivn-rat LD50:6600 μg/kg TXAPA9 17,323,70
orl-mus LD50:117 mg/kg GUCHAZ 6,100,73
ipr-mus LD50:87 mg/kg JPPMAB 19,612,67
scu-mus LD50:339 mg/kg JPPMAB 19,612,67
ivn-mus LD50:87 mg/kg JPPMAB 19,612,67
orl-dog LD50:1200 mg/kg 85GYAZ -,18,71
ivn-dog LD50:51 mg/kg TXAPA9 17,323,70

CONSENSUS REPORTS: EPA Extremely Hazardous Substances List.

SAFETY PROFILE: Poison by ingestion, skin contact, intraperitoneal, subcutaneous, and intravenous routes. Human systemic effects by skin contact: unspecified blood system effects. Mutation data reported. A cholinesterase inhibitor. An insecticide. See also PARATHION. When heated to decomposition it emits very toxic fumes of Cl^- and PO_x.

CDT000 CAS:2536-31-4 HR: 2
CHLORFLURENOL METHYL ESTER
mf: $C_{15}H_{11}ClO_3$ mw: 274.71

PROP: A solid. Mp: 152°.

SYNS: CF 125 □ CHLORFLURECOL □ CHLORFLURECOL-METHYL □ CHLORFLURECOL-METHYL ESTER □ CHLORFLURENOL □ 2-CHLOR-9-HYDROXYFLUOREN-CARBONSAEURE-(9)-METHYLESTER (GERMAN) □ CHLOROFLURENOL-METHYL ESTER □ 2-CHLORO-9-HYDROXY-9-METHYLCARBOXYLATEFLUORENE □ CURBISET □ IT 3456 □ MAINTAIN A □ MAINTAIN CF125 □ METHYL-2-CHLORO-9-HYDROXYFLUORENE-9-CARBOXYLATE □ MORPHACTIN □ MULTIPROP

TOXICITY DATA WITH REFERENCE
orl-rat LD50:3100 mg/kg GUCHAZ 6,102,73

CONSENSUS REPORTS: Reported in EPA TSCA Inventory.

SAFETY PROFILE: Moderately toxic by ingestion. An herbicide. When heated to decomposition it emits very toxic fumes of Cl^-. See also ESTERS.

CDT125 CAS:56-95-1 HR: 3
CHLORHEXIDINE DIACETATE
mf: $C_{22}H_{30}Cl_2N_{10}{\cdot}2C_2H_4O_2$ mw: 625.64

PROP: A solid. Mp: 154–155°.

SYNS: 1,6-BIS(5-(p-CHLOROPHENYL)BIGUANIDINO)HEXANE DIACETATE □ CHLORHEXIDINE ACETATE □ 10,040 DIACETATE □ 1,6-DI(4'-CHLOROPHENYLDIGUANIDINO)HEXANE DIACETATE □ 1,1'-HEXAMETHYLENEBIS(5-(p-CHLOROPHENYL)BIGUANIDE) DIACETATE □ HIBITANE DIACETATE

TOXICITY DATA WITH REFERENCE
skn-rbt 500 mg/24H MLD JACTDZ 4(6),309,85
orl-mus LD50:2 g/kg BJPCAL 9,192,54
ipr-mus LD50:38 mg/kg BJPCAL 9,192,54
scu-mus LD50:325 mg/kg BJPCAL 9,192,54
ivn-mus LD50:25 mg/kg BJPCAL 9,192,54

SAFETY PROFILE: Poison by subcutaneous, intravenous, and intraperitoneal routes. Moderately toxic by ingestion. A skin irritant. When heated to decomposition it emits toxic fumes of Cl^- and NO_x.

CDT250 CAS:18472-51-0 HR: 3
CHLORHEXIDINE DIGLUCONATE
mf: $C_{22}H_{30}Cl_2N_{10}{\cdot}2C_6H_{12}O_7$ mw: 897.88

SYNS: ABACIL □ ARLACIDE G □ BACTICLENS □ BIGUANIDE, 1,1'-HEXAMETHYLENEBIS(5-(p-CHLOROPHENYL))-, DIGLUCONATE □ 1,6-BIS(5-(p-CHLOROPHENYL)BIGUANDINO)HEXANE DIGLUCONATE □ CASWELL NO. 481G □ CHLORHEXIDINE GLUCONATE □ CHLORHEXIDIN GLUKONATU □ CORSODYL □ DISTERYL □ GLUCONIC ACID, compd. with 1,1'-HEXAMETHYLENE BIS(5-(p-CHLOROPHENYL)BIGUANIDE) (2:1), D-(8CI) □ 1,1'-HEXAMETHYLENEBIS(5-(p-CHLOROPHENYL)BIGUANIDE)DIGLUCONATE □ HIBICLENS □ HIBIDIL □ HIBISCRUB □ HIBITANE □ ORAHEXAL □ PERIDEX □ PLAC OUT □ PLUREXID □ ROTERSEPT □ SEPTEAL □ UNISEPT

TOXICITY DATA WITH REFERENCE
dnr-bcs 1 mg/L ESKGA2 32,171,86
orl-rat LD50:2 g/kg DRUGAY 6,255,82
scu-rat LD50:3320 mg/kg DRUGAY 6,255,82
ivn-rat LD50:24,200 μg/kg DRUGAY 6,255,82
orl-mus LD50:1260 mg/kg DRUGAY 6,255,82
scu-mus LD50:1140 mg/kg DRUGAY 6,255,82
ivn-mus LD50:12,900 μg/kg DRUGAY 6,255,82

CONSENSUS REPORTS: Reported in EPA TSCA Inventory.

SAFETY PROFILE: Poison by intravenous route. Moderately toxic by ingestion and subcutaneous routes. Mutation data reported. When heated to decomposition it emits very toxic fumes of Cl^- and NO_x.

CDT750 CAS:96-24-2 HR: 3
CHLORHYDRIN
DOT: UN 2689
mf: $C_3H_7ClO_2$ mw: 110.55

$ClCH_2CHOHCH_2$

PROP: Colorless liquid. Bp: 213° decomp, d: 1.326.

SYNS: α-CHLORHYDRIN □ CHLORODEOXYGLYCEROL □ 1-CHLORO-2,3-DIHYDROXYPROPANE □ 3-CHLORO-1,2-DIHYDROXYPROPANE □ α-CHLOROHYDRIN □ 1-CHLOROPROPANE-2,3-DIOL □ 1-CHLORO-2,3-PROPANEDIOL □ 3-CHLOROPROPANE-1,2-DIOL □ 3-CHLORO-1,2-PROPANEDIOL □ 3-CHLOROPROPYLENE GYLCOL □ β,β'-DIHYDROXYISOPROPYL CHLORIDE □ 2,3-DIHYDROXYPROPYL CHLORIDE □ EPIBLOC □ GLYCERIN-α-MONOCHLORHYDRIN □ GLYCEROL CHLOROHYDRIN □ GLYCEROL-α-CHLOROHYDRIN □ GLYCEROL-α-MONOCHLOROHYDRIN (DOT) □ GLYCERYL-α-CHLOROHYDRIN □ MONOCHLORHYDRIN □ MONOCHLOROHYDRIN □ α-MONOCHLOROHYDRIN □ U-5897

TOXICITY DATA WITH REFERENCE
eye-rbt 100 mg SEV AJOPAA 29,1363,46
mmo-ssp 100 mmol/L MUREAV 118,213,83
mma-ssp 300 mmol/L MUREAV 118,213,83
spm-rat-orl 600 mg/kg/24D-C CUSCAM 44,193,75
msc-mus:lym 10 mmol/L PAACA3 21,74,80
orl-mky TDLo:1260 mg/kg (male 42D pre):REP JRPFA4 21,275,70
scu-rat TDLo:750 mg/kg (female 15D pre):REP AEFTAA 10,23,79
orl-mky TDLo:5600 mg/kg (male 40D pre):REP ENDKAC 69,157,77
ipr-rat TDLo:150 mg/kg (male 10D pre):REP IJEBA6 16,1278,78
orl-rat TDLo:2500 μg/kg (1D male):REP JRPFA4 38,1,74
orl-rat TDLo:85,500 μg/kg (male 9D pre):REP PSEBAA 132,656,69
orl-rat TDLo:34,580 mg/kg/72W-C:ETA JJIND8 67,75,81
orl-rat TDLo:26 mg/kg IPCLBZ 24,20,82
ihl-rat LCLo:125 ppm/4H JIHTAB 31,343,49
ipr-rat LDLo:10 mg/kg NCNSA6 5,9,53
orl-mus LD50:160 mg/kg AMIHAB 14,250,56

ipr-mus LD50:73 mg/kg JMCMAR 18,116,75
orl-bwd LD50:23,700 μg/kg AECTCV 12,355,83

CONSENSUS REPORTS: Reported in EPA TSCA Inventory. EPA Genetic Toxicology Program.

DOT CLASSIFICATION: 6.1; *Label:* KEEP AWAY FROM FOOD

SAFETY PROFILE: Poison by ingestion and intraperitoneal routes. Moderately toxic by inhalation. Experimental reproductive effects. A severe eye irritant. Questionable carcinogen with experimental tumorigenic data. Mutation data reported. A chemosterilant for rodents. Combustible when exposed to heat or flame. Reaction with perchloric acid forms a sensitive explosive product more powerful than glyceryl nitrate. When heated to decomposition it emits toxic fumes of Cl^-.

CDU000 **CAS:7790-93-4** **HR: 3**
CHLORIC ACID
DOT: UN 2626
mf: $ClHO_3$ mw: 84.46

PROP: Colorless solution. Fairly stable in cold H_2O up to 30%. Strong oxidant, stable as alkali metal salts. Mp: $<-20°$, bp: decomp @ 40°, d: 1.282 @ 14.2°.

SYN: CHLORIC ACID, solution, containing not more than 10% acid (DOT)

CONSENSUS REPORTS: Reported in EPA TSCA Inventory.

DOT CLASSIFICATION: 5.1; *Label:* Oxidizer

SAFETY PROFILE: A poison. A strong irritant by ingestion and inhalation. Dangerous fire hazard; ignites organic matter upon contact. A very powerful oxidizing agent. Violent or explosive reaction with oxidizable materials. Aqueous solutions decompose explosively during evaporation. Solutions greater than 40% are unstable. Reacts violently with NH_3, Sb, Sb_2S_3, As_2S_3, Bi, CuS, PHI_4, SnS_2, SnS. Reaction with cellulose causes ignition after a delay period. Dangerous reaction with metal sulfides and metal chlorides (e.g., incandescent reaction with antimony trisulfide, arsenic trisulfide, tin(II)sulfide, tin(IV) sulfide, explosion on contact with copper sulfide). Reaction with metals (e.g., antimony, bismuth, iron) forms explosive products. When heated to decomposition it emits toxic fumes of Cl^-. See also CHLORATES and CHLORINE.

CDU250 **HR: D**
CHLORIDES

SAFETY PROFILE: Varies widely. Sodium chloride (table salt) has very low toxicity, while carbonyl chloride (phosgene) is lethal in small doses. Therefore, see specific entries. When heated to decomposition or on contact with acids or acid fumes, they evolve highly toxic chloride fumes. Some organic chlorides decompose to yield phosgene.

CDU325 **HR: 3**
CHLORIERTES CAMPHEN

PROP: Consists of a mixture of chlorinated camphene with 67–69% chlorine (85GYAZ-,50,71).

TOXICITY DATA WITH REFERENCE
orl-rat LD50:60 mg/kg 85GYAZ -,50,71
skn-rat LD50:1 g/kg 85GYAZ -,50,71
skn-rbt LD50:250 mg/kg 85GYAZ -,50,71

SAFETY PROFILE: Poison by ingestion and skin contact. When heated to decomposition it emits toxic fumes of Cl^-. See also CAMPHENE.

CDU750 **CAS:303-49-1** **HR: 3**
CHLORIMIPRAMINE
mf: $C_{19}H_{23}ClN_2$ mw: 314.89

PROP: Bp: 160–170° @ 0.3 mm.

SYNS: ANAFRANIL □ 3-CHLORO-5-(3-(DIMETHYLAMINO)PROPYL)-10,11-DIHYDRO-5H-DIBENZ(b,f)AZEPINE □ 3-CHLOROIMIPRAMINE □ CIM □ CLOMIPRAMINE □ MONOCHLORIMIPRAMINE

TOXICITY DATA WITH REFERENCE
ivn-wmn TDLo:50 mg/kg (38-47D preg):TER TJADAB 24(1),42A,81
orl-rat TDLo:420 mg/kg (lactating female 21D post):REP PSCHDL 56,93,78
orl-rat TDLo:400 mg/kg (female 20D pre):REP PSCHDL 56,93,78
orl-man TDLo:357 μg/kg:GIT JCPYDR 2,215,82
orl-wmn TDLo:10 mg/kg/5D-I:CVS BMJOAE 1,406,71
ivn-wmn TDLo:3400 μg/kg/47M-I:CNS,CVS BMJOAE 3,698,72
orl-rat LD50:610 mg/kg TXCYAC 24,335,82
ipr-rat LD50:149 mg/kg TXCYAC 24,335,82
orl-mus LD50:380 mg/kg GWXXBX #2618152
ipr-mus LD50:150 mg/kg JMCMAR 21,448,78
ivn-mus LD50:27 mg/kg APSXAS 13,485,76

SAFETY PROFILE: Poison by ingestion, intraperitoneal, and intravenous routes. Human systemic effects by ingestion and intravenous routes: convulsions, heart damage, and blood pressure increase, nausea or vomiting. Human teratogenic effects on the cardiovascular system of the fetus. Experimental reproductive effects. When heated to decomposition it emits very toxic fumes of Cl^- and NO_x.

CDV000 **CAS:17321-77-6** **HR: 3**
CHLORIMIPRAMINE HYDROCHLORIDE
mf: $C_{19}H_{23}ClN_2 \cdot ClH$ mw: 351.35

PROP: A solid. Mp: 189–192°.

SYNS: ANAFRANIL □ ANAPHRANIL □ 3-CHLORO-10,11-DIHYDRO-N,N-DIMETHYL-5H-DIBENZ(b,f)AZEPINE-5-PROPANAMINE MONOHYDROCHLORIDE □ 3-CHLORO-5-(3-(DIMETHYLAMINO)PROPYL)-10,11-DIHYDRO-5H-DIBENZ(b,f)AZEPINE MONOHYDROCHLORIDE □ 3-CHLOROIMIPRAMINE HYDROCHLORIDE □ CHLOROIMIPRAMINE MONOHYDROCHLORIDE □ CLOMIPRAMINE HYDROCHLORIDE □ G 34586

TOXICITY DATA WITH REFERENCE
sln-dmg-orl 200 mg SOGEBZ 7,1042,71

scu-mus TDLo: 110 mg/kg (9D preg): REP DGDFA5 22,61,80
orl-wmn TDLo: 360 mg/kg/13W-I: BPR JCLPDE 46,290,85
orl-rat LD50: 1150 mg/kg NIIRDN 6,243,82
ipr-rat LD50: 135 mg/kg NIIRDN 6,243,82
scu-rat LD50: 1750 mg/kg NIIRDN 6,243,82
ivn-rat LD50: 26 mg/kg NIIRDN 6,243,82
orl-mus LD50: 470 mg/kg NIIRDN 6,243,82
ipr-mus LD50: 90 mg/kg NIIRDN 6,243,82
scu-mus LD50: 400 mg/kg KSRNAM 4,2105,70
ivn-mus LD50: 26 mg/kg NIIRDN 6,243,82
orl-dog LD50: 383 mg/kg KSRNAM 4,2105,70
ivn-dog LD50: 32 mg/kg KSRNAM 4,2105,70
orl-rbt LD50: 792 mg/kg KSRNAM 4,2105,70
scu-rbt LD50: 217 mg/kg KSRNAM 4,2105,70
ivn-rbt LD50: 12 mg/kg KSRNAM 4,2105,70

SAFETY PROFILE: Poison by ingestion, subcutaneous, intravenous, and intraperitoneal routes. Human systemic effects by ingestion: pulse rate increase, no fall in blood pressure. Experimental reproductive effects. Mutation data reported. When heated to decomposition it emits very toxic fumes of Cl^- and NO_x. See also CHLORIMIPRAMINE.

CDV100 CAS:8001-35-2 HR: 3
CHLORINATED CAMPHENE
mf: $C_{10}H_{10}Cl_8$ mw: 413.80

PROP: Yellow, waxy solid; pleasant piney odor. Mp: 65–90°. Almost insol in water; very sol in aromatic hydrocarbons.

SYNS: AGRICIDE MAGGOT KILLER (F) □ ALLTEX □ ALLTOX □ ATTAC 6 □ ATTAC 6-3 □ CAMPHECHLOR □ CAMPHOCHLOR □ CAMPHOCLOR □ CAMPHOFENE HUILEUX □ CHEM-PHENE □ CHLOROCAMPHENE □ CLOR CHEM T-590 □ COMPOUND 3956 □ CRESTOXO □ CRISTOXO 90 □ ENT 9,735 □ ESTONOX □ FASCO-TERPENE □ GENIPHENE □ GY-PHENE □ HERCULES 3956 □ HERCULES TOXAPHENE □ KAMFOCHLOR □ M 5055 □ MELIPAX □ MOTOX □ NCI-C00259 □ OCTACHLOROCAMPHENE □ PCC □ PENPHENE □ PHENACIDE □ PHENATOX □ POLYCHLORCAMPHENE □ POLYCHLORINATED CAMPHENES □ POLYCHLOROCAMPHENE □ RCRA WASTE NUMBER P123 □ STROBANE-T-90 □ SYNTHETIC 3956 □ TOXADUST □ TOXAFEEN (DUTCH) □ TOXAKIL □ TOXAPHEN (GERMAN) □ TOXAPHENE □ TOXON 63 □ TOXYPHEN □ VERTAC 90% □ VERTAC TOXAPHENE 90

TOXICITY DATA WITH REFERENCE
skn-mam 500 mg MOD JAMAAP 149,1135,52
mmo-sat 100 μg/plate ENMUDM 8(Suppl 7),1,86
mma-sat 500 μg/plate SCIEAS 205,591,79
sce-hmn:lym 10 μmol/L ARTODN 52,221,83
otr-mus:emb 12,400 μg/L PMRSDJ 5,659,85
orl-rat TDLo: 900 μg/kg (female 5-22D post): REP AECTCV 9,247,80
orl-mus TDLo: 375 mg/kg (female 8-12D post): REP TCMUD8 4,403,84
orl-rat TDLo: 250 mg/kg (female 7-16D post): TER BECTA6 15,660,76
orl-rat TDLo: 280 mg/kg (10W male): REP GISAAA 45(5),14,80
orl-rat TDLo: 150 mg/kg (female 7-16D post): TER BECTA6 15,660,76
orl-rat TDLo: 30 g/kg/80W-C: ETA NCITR* NCI-CG-TR-37,79
orl-mus TDLo: 6600 mg/kg/80W-C: CAR NCITR* NCI-CG-TR-37,79
orl-mus TD: 13 g/kg/80W-C: CAR NCITR* NCI-CG-TR-37,79
orl-hmn LDLo: 28 mg/kg: CNS 34ZIAG -,598,69
orl-man LDLo: 29 mg/kg CMEP** -,-,56
skn-hmn TDLo: 657 mg/kg: SKN CMEP** -,1,56
unr-man LDLo: 44 mg/kg 85DCAI 2,73,70
orl-rat LD50: 50 mg/kg ARZNAD 17,614,67
skn-rat LD50: 600 mg/kg SPEADM 74-1,-,74
ipr-rat LDLo: 70 mg/kg BECTA6 19,47,78
orl-mus LD50: 112 mg/kg SPEADM 74-1,-,74
ihl-mus LCLo: 2000 mg/m³/2H JAMAAP 149,1135,52
ipr-mus LD50: 47 mg/kg JAFCAU 25,1394,77
unr-mus LD50: 45 mg/kg CYGEDX 8(1),23,74
orl-dog LD50: 15 mg/kg SPEADM 74-1,-,74
orl-rbt LD50: 75 mg/kg 85JFAN A054,84
skn-rbt LD50: 1025 mg/kg JEENAI 46,702,53
orl-gpg LD50: 250 mg/kg 85DPAN -,-,71/76
orl-ham LD50: 200 mg/kg TXAPA9 48,A192,79
orl-dck LD50: 31 mg/kg DOEAAH 35,25,79

CONSENSUS REPORTS: NTP 7th Annual Report on Carcinogens. IARC Cancer Review: Group 2B IMEMDT 7,56,87; Human Limited Evidence IMEMDT 20,327,79; Animal Sufficient Evidence IMEMDT 20,327,79. NCI Carcinogenesis Bioassay (feed); Clear Evidence: mouse, rat NCITR* NCI-CG-TR-37,79.

OSHA PEL: TWA 0.5 mg/m³; STEL 1 mg/m³ (skin)
ACGIH TLV: TWA 0.5 mg/m³; STEL 1 mg/m³ (skin)
DFG MAK: 0.5 mg/m³

SAFETY PROFILE: Confirmed carcinogen with experimental carcinogenic and tumorigenic data. Human poison by ingestion and possibly other routes. Experimental poison by ingestion, intraperitoneal, and possibly other routes. Moderately toxic experimentally by inhalation and skin contact. Human systemic effects by ingestion and skin contact: somnolence, convulsions or effect on seizure threshold, coma, and allergic skin dermatitis. A skin irritant; absorbed through the skin. Experimental teratogenic and reproductive effects. Human mutation data reported. Liver injury has been reported. Lethal amounts of toxaphene can enter the body through the mouth, lungs, and skin. Systemic absorption of the insecticide is increased by the presence of digestible oils, and liquid preparations of the insecticide, which penetrate the skin more readily than do dusts and wettable powders.

A toxic mixture of organochlorine pesticides stored to some extent in body fat. It resembles chlordane and, to some extent, camphor in its physiological action. It causes diffuse stimulation of the brain and spinal cord, resulting in generalized convulsions of a tonic or clonic character. Death usually results from respiratory failure. Detoxification appears to occur in the liver. The lethal ingestion dose for humans is estimated to be 2–7 g, a toxicity of about four times that of DDT. At least seven human deaths have been reported due to toxaphene, all in children. Two families have been made ill by eating vegetables containing a large residue of toxaphene. When heated to decomposition it emits toxic fumes of Cl^-.

For occupational chemical analysis use NIOSH: Chlorinated Camphene S67.

CDV125 **HR: 3**
CHLORINATED DIBENZO DIOXINS

PROP: See individual entries for physical properties.

SYNS: 1-CHLORODIBENZO-p-DIOXIN □ DIBENZO-p-DIOXIN □ 1,2,3,8-TETRACHLORODIBENZO-p-DIOXIN □ 1,2,4-TRICHLORO DIBENZO-p-DIOXIN

SAFETY PROFILE: The chlorinated dibenzo dioxins are not manufactured on a commercial basis, but some are present as impurities in herbicide and fungicide formulations, such as 2,4,5-T, the pentachlorophenols, and hexachlorphene (from trichlorophenol). The chlorinated dibenzo dioxins include some with antibacterial, flame-proofing, insecticidal, and fungicidal actions. Their acute toxicity ranges from moderate to high and some are carcinogens, mutagens, and teratogens. They tend to accumulate in living organisms. When heated to decomposition they emit toxic fumes of Cl^-. See also SPECIFIC COMPOUNDS.

CDV175 **CAS:31242-93-0** **HR: 2**
CHLORINATED DIPHENYL OXIDE
mf: $C_{12}H_4Cl_6O$ mw: 376.86

PROP: Light-yellow, very viscous liquid. Bp: 230–260° @ 8 mm, d: 1.60 @ 20°/60°, autoign temp: 1148°F, vap d: 13.0.

SYNS: BENZENE, 1,1'-OXYBIS-, HEXACHLORO derivatives (9CI) □ ETHER, HEXACHLOROPHENYL □ HEXACHLORODIPHENYL ETHER □ HEXACHLORO DIPHENYL OXIDE □ PHENYL ETHER, HEXACHLORO derivative (8CI) □ TRICHLORO DIPHENYL ETHER □ TRICHLORO DIPHENYL OXIDE

TOXICITY DATA WITH REFERENCE
orl-rat LD:>500 mg/kg NCNSA6 5,16,53

OSHA PEL: TWA 0.5 mg/m^3
ACGIH TLV: TWA 0.5 mg/m^3
DFG MAK: 0.5 mg/m^3

SAFETY PROFILE: Moderately toxic by ingestion and probably by inhalation. Combustible when exposed to heat, flame, or oxidizing materials. To fight fire, use water spray, fog, foam, dry chemical, CO_2. When heated to decomposition it emits toxic fumes of Cl^-. See also ETHERS and ALDRIN (a closely related compound).

For occupational chemical analysis use NIOSH: Chlorinated Diphenyl Ether, 5025.

CDV250 **HR: 2**
CHLORINATED HYDROCARBONS, ALIPHATIC

SYNS: ALIPHATIC CHLORINATED HYDROCARBONS □ CHLORINATED HC, ALIPHATIC

SAFETY PROFILE: Suspected carcinogen with experimental tumors of the liver, lung, skin, and blood-forming tissues. The substitution of a chlorine (or other halogen) atom for a hydrogen greatly increases the anesthetic action of the aliphatic hydrocarbons and increases the range of their systemic effects. In many cases, the chlorine derivative is quite toxic. In general, the unsaturated chlorine derivatives are more narcotic but less toxic than the saturated derivatives. In the saturated group, the narcotic effect is proportional to the number of chlorine atoms. This relationship is not true for toxicity.

In dealing with these chlorinated hydrocarbons, it must be remembered that a toxic action may result from repeated exposure to concentrations that are too low to produce a narcotic effect, and that, consequently, are too low to give warning of danger. Individual susceptibility varies widely. Certain workmen may be seriously affected by concentrations that seem to have no effect on fellow employees at the same exposure.

In general, reactivity decreases with greater substitution of halogen for hydrogen atoms. Halogenated (i.e., fluorine-, chlorine-, or bromine-containing) acetylene compounds are unstable and should be treated as explosives. Lightly substituted haloalkanes are highly flammable and can react with divalent light metals to form dangerously reactive products. Lightly substituted haloalkenes are highly flammable, peroxidizable, and may polymerize violently. When heated to decomposition they emit highly toxic fumes of phosgene. They may react violently with Al, liquid O_2, K, and Na.

CDV500 **HR: 3**
CHLORINATED HYDROCARBONS, AROMATIC

SYN: CHLORINATED HC AROMATIC

SAFETY PROFILE: In most instances, it is difficult to predict the toxicity of these compounds. However, in the case of most aromatic chlorine compounds, their toxicity is usually no greater, and frequently is less, than that of the corresponding aromatic hydrocarbons, with the notable exception of naphthalene and the various biphenyls. They can react with oxidizing materials. React violently with Al, liquid O_2, K, or Na. When heated to decomposition they emit toxic fumes of Cl^-.

CDV575 **HR: 3**
CHLORINATED NAPHTHALENES

SAFETY PROFILE: Questionable carcinogens that can cause tumors of the liver. Severe irritants by ingestion, inhalation, and skin contact. The action of the chlorinated naphthalenes on the body is quite similar to that of the chlorinated biphenyls, the chief effects being the production of chloracne of the skin, and systemically an acute yellow atrophy of the liver. When heated to decomposition they emit toxic fumes of Cl^-.

CDV625 **CAS:56641-03-3** **HR: 3**
CHLORINATED POLYETHER POLYURETHAN

PROP: Polymer formed from toluene diisocyanate and 1,4-butanediol and cured with 4,4'-methylenebis(o-chloroaniline) (CNREA8 36,3973,76).
mf: $(C_{13}H_{12}Cl_2N_2{\cdot}C_9H_6N_2O_2{\cdot}(C_4H_8O)_nH_2O)_x$

SYNS: OSTAMER □ POLYURETHANE Y-238 □ Y-238

TOXICITY DATA WITH REFERENCE
imp-rat TDLo:20 mg/kg:ETA CNREA8 36,3973,76
imp-rat TD:293 mg/kg:ETA JNCIAM 33,1005,64
imp-rat TD:6750 mg/kg:ETA CNREA8 35,1591,75

CONSENSUS REPORTS: IARC Cancer Review: Animal Sufficient Evidence IMEMDT 19,303,79.

SAFETY PROFILE: Confirmed carcinogen with experimental tumorigenic data. When heated to decomposition it emits toxic fumes of Cl^- and NO_x.

CDV700 **CAS:145-94-8** **HR: 3**
CHLORINDANOL
mf: C_9H_9ClO mw: 168.63

PROP: Needles from petr ether. Mp: 91–93°.

SYNS: CLORINDANOL □ 7-CHLORO-4-INDANOL □ 2,3-DIHYDRO-7-CHLORO-1H-INDEN-4-OL (9CI) □ LANESTA

TOXICITY DATA WITH REFERENCE
orl-mus LD50:920 mg/kg JAPMA8 48,212,59
ipr-mus LD50:96 mg/kg JAPMA8 48,212,59
ivn-mus LD50:48 mg/kg JAPMA8 48,212,59

SAFETY PROFILE: Poison by intravenous and intraperitoneal routes. Moderately toxic by ingestion. When heated to decomposition it emits toxic fumes of Cl^-.

CDV750 **CAS:7782-50-5** **HR: 3**
CHLORINE
DOT: UN 1017
mf: Cl_2 mw: 70.90

PROP: Greenish-yellow gas, liquid, or rhombic crystals. Mp: −101°, bp: −34.9°, d: (liquid) 1.47 @ 0° (3.65 atm), vap press: 4800 mm @ 20°, vap d: 2.49. Sol in water.

SYNS: BERTHOLITE □ CHLOOR (DUTCH) □ CHLOR (GERMAN) □ CHLORE (FRENCH) □ CHLORINE MOL. □ CLORO (ITALIAN) □ MOLECULAR CHLORINE

TOXICITY DATA WITH REFERENCE
mma-sat 1800 µg/L OZSEDS 8,217,86
cyt-hmn:lym 20 ppm CBINA8 6,375,73
spm-mus-orl 20 mg/kg/5D-C ENMUDM 7,201,85
ihl-hmn LCLo:2530 mg/m³/30M:PUL 28ZOAH -,150,37
ihl-hmn LCLo:500 ppm/5M TABIA2 3,231,33
ihl-rat LC50:293 ppm/1H NTIS** PB214-270
ihl-mus LC50:137 ppm/1H NTIS** PB214-270
ihl-dog LCLo:800 ppm/30M JPETAB 14,65,19
ihl-cat LCLo:660 ppm/4H AHYGAJ 7,233,1887
ihl-rbt LDLo:660 ppm/4H AHYGAJ 7,233,1887

CONSENSUS REPORTS: Reported in EPA TSCA Inventory. Community Right-To-Know List. EPA Extremely Hazardous Substances List.

OSHA PEL: TWA 0.5 ppm; STEL 1 ppm
ACGIH TLV: TWA 0.5 ppm; STEL 1 ppm
DFG MAK: 0.5 ppm (1.5 mg/m³)
NIOSH REL: (Chlorine) CL 0.5 ppm/15M
DOT CLASSIFICATION: 2.3; *Label:* Poison Gas

SAFETY PROFILE: Moderately toxic to humans by inhalation. Very irritating by inhalation. Human mutation data reported. Human respiratory system effects by inhalation: changes in the trachea or bronchi, emphysema, chronic pulmonary edema or congestion. A strong irritant to eyes and mucous membranes.

Chlorine is extremely irritating to the mucous membranes of the eyes and the respiratory tract at 3 ppm. Combines with moisture to form HCl. Both these substances, if present in quantity, cause inflammation of the tissues with that they come in contact. A concentration of 3.5 ppm produces a detectable odor; 15 ppm causes immediate irritation of the throat. Concentrations of 50 ppm are dangerous for even short exposures; 1000 ppm may be fatal, even when exposure is brief. Because of its intensely irritating properties, severe industrial exposure seldom occurs, as the worker is forced to leave the exposure area before he can be seriously affected. In cases where this is impossible, the initial irritation of the eyes and mucous membranes of the nose and throat is followed by coughing, a feeling of suffocation, and, later, pain and a feeling of constriction in the chest. If exposure has been severe, pulmonary edema may follow, with rales being heard over the chest. It is a common air contaminant.

Explodes on contact with acetylene + heat or UV light, air + ethylene, molten aluminum, ammonia, amidosulfuric acid, antimony trichloride + tetramethyl silane (at 100°), benzene + light, biuret, bromine pentafluoride + heat, tert-butanol, butyl rubber + naphtha, carbon disulfide + iron catalyst, chlorinated pyridine + iron powder, 3-chloropropyne, cobalt(II) chloride + methanol, diborane, dibutyl phthalate (at 118°), dichloro(methyl)arsine (in a sealed container), diethyl ether, dimethyl phosphoramidiate, dioxygen difluoride, disilyl oxide, 4,4′-dithiodimorpholine, ethane over activated carbon (at 350°), fluorine + sparks, gasoline, glycerol (above 70° in a sealed container), hexachlorodisilane (above 300°), hydrocarbon oils or waxes, iron(III) chloride + monomers (e.g., styrene), methane over mercury oxide, methanol, methanol + tetrapyridine cobalt(II) chloride, naphtha + sodium hydroxide, nitrogen triiodide, oxygen difluoride, white phosphorus (in liquid Cl_2), phosphorus compounds, polypropylene + zinc oxide, propane (at 300°), silicones when heated in a sealed container [e.g., polydimethyl siloxane (above 88°), polymethyl trifluoropropylsiloxane (above 68°)], stibine, synthetic rubber (in liquid Cl_2), tetraselenium tetranitride, trimethyl thionophosphate. Explosive products are formed on reaction with alkylthiouronium salts, amidosulfuric acid, acidic ammonium chloride solutions, aziridine, bis(2,4-dinitrophenyl)disulfide, cyanuric acid, phenyl magnesium bromide. Mixtures with ethylene are explosives initiated by light, heat, or by the presence or mercury, mercury oxide, silver oxide, lead oxide (at 100°). Mixtures with hydrogen are explosives initiated by sparks, light, heating to over 280°, or the presence of yellow mercuric oxide or nitrogen trichloride. Mixtures with hydrogen and other gases (e.g., air, hydrogen chloride, oxygen) are also explosive.

Ignition or explosive reaction with metals (e.g., aluminum, antimony powder, bismuth powder, brass, calcium powder, copper, germanium, iron, manganese, potassium, tin, vanadium powder). Reaction with some metals requires moist Cl_2 or heat. Ignites with diethyl zinc (on contact), polyisobutylene (at 130°), metal acetylides, metal carbides, metal hydrides (e.g., potassium hydride, sodium hydride, copper hydride), metal phosphides (e.g., copper(II) phosphide), methane + oxygen, hydrazine, hydroxylamine, calcium nitride, nonmetals (e.g., boron, active carbon, silicon, phosphorus), nonmetal hydrides (e.g., arsine, phosphine, silane), steel (above 200° or as low as 50° when impurities are present), sulfides (e.g., arsenic disulfide, boron trisulfide, mercuric sulfide), trialkyl boranes.

Violent reaction with alcohols, N-aryl sulfinamides, dimethyl formamide, polychlorobiphenyl, sodium hydroxide, hydrochloric acid + dinitroanilines. Incandescent reaction when warmed with cesium oxide (above 150°), tellurium, arsenic, tungsten dioxide. Potentially dangerous reaction with hydrocarbons + Lewis acids releases toxic and reactive HCl gas.

Can react to cause fires or explosions upon contact with turpentine, illuminating gas, polypropylene, rubber, sulfamic acid, $As_2(CH_3)_4$, UC_2, acetaldehyde, alcohols, alkylisothiourea salts, alkyl phosphines, Al, Sb, As, AsS_2, AsH_3, Ba_3P_2, C_6H_6, Bi, B, BPI_2, B_2S_3, brass BrF_5, Ca, (CaC_2 + KOH), $Ca(ClO_2)_2$, Ca_3N_2 Ca_3P_2, C, CS_2, Cs, $CsHC_2$, Co_2O, Cs_3N, (C + $Cr(OCl)_2$), CuH_2, CuC_2, dialklyl phosphines, diborane, dibutyl phthalate, $Zn(C_2H_5)_2$, C_2H_6, C_2H_4, ethylene imine, $C_2H_5PH_2$, F_2, Ge, glycerol, $(NH_2)_2$, (H_2O + KOH), I_2, hydroxylamine, Fe, FeC_2, Li, Li_2C_2, Li_6C_2, Mg, Mg_2P_3, Mn, Mn_4P_2, HgO, HgS, Hg, Hg_3P_2, CH_4, Nb, NI_3, OF_2, H_2SiO, (OF_2+ Cu), PH_3, P, $P(SNC)_3$, P_2O_3, PCB's, K, KHC_2, KH, Ru, $RuHC_2$, Si, SiH_2, Ag_2O, Na, $NaHC_2$, Na_2C_2, SnF_2, SbH_3, Sr_3P, Te, Th, Sn, WO_2, U, V, Zn, ZrC_2.

For occupational chemical analysis use OSHA: #ID-101 or NIOSH: Bromine and Chloride 6011.

CDW000 CAS:13973-88-1 HR: 3
CHLORINE AZIDE
mf: ClN_3 mw: 77.48

PROP: An explosive gas. Bp: 15°. Sol in org solvs.

SYN: NITROGEN CHLORIDE

DOT CLASSIFICATION: Forbidden

SAFETY PROFILE: Strong irritant by inhalation. An extremely unstable explosive. Reacts with liquid ammonia to form an explosive liquid. Explosive reaction with 1,3-butadiene, C_2H_6, C_2H_4, CH_4, C_3H_8, phophorus, silver azide, sodium. Reacts with water or steam to produce toxic and corrosive fumes of HCl. Has been used as an initiator in chemical gas lasers. When heated to decomposition it emits toxic fumes of Cl^- and NO_x. See also CHLORINE and AZIDES.

CDW450 CAS:10049-04-4 HR: 3
CHLORINE DIOXIDE
mf: ClO_2 mw: 67.45

PROP: Red-yellow or orange-green gas or orange-red crystals. Unstable in light, stable in the dark. Mp: −59°, bp: 11°, d: 3.09 g/L @ 11°. Insol in water.

SYNS: ALCIDE □ ANTHIUM DIOXIDE □ CHLORINE DIOXIDE, not hydrated (DOT) □ CHLORINE OXIDE □ CHLORINE(IV) OXIDE □ CHLORINE PEROXIDE □ CHLOROPEROXYL □ CHLORYL RADICAL □ DOXCIDE 50

TOXICITY DATA WITH REFERENCE
eye-rbt 100 mg MLD JJATDK 2,160,82
mma-sat 400 µg/plate FCTOD7 22,623,84
orl-rat TDLo:570 mg/kg (female 14D pre-21D post):REP NTIS** PB85-245983
orl-rat TDLo:1130 mg/kg (male 8W pre):REP TJADAB 35,43A,87
orl-mus TDLo:840 mg/kg (female 1-21D post):REP EVHPAZ 46,31,82
orl-rat TDLo:960 mg/kg (76D pre/1-20D preg):REP JJATDK 3,75,83
orl-rat LD50:292 mg/kg JJATDK 2,160,82
ihl-rat LCLo:500 ppm/15M TXAPA9 27,527,74
unr-rat LD50:140 mg/kg GISAAA 36(11),18,71

CONSENSUS REPORTS: Reported in EPA TSCA Inventory. Community Right-To-Know List.

OSHA PEL: TWA 0.1 ppm; STEL 0.3 ppm
ACGIH TLV: TWA 0.1 ppm; STEL 0.3 ppm
DFG MAK: 0.1 ppm (0.3 mg/m³)
DOT CLASSIFICATION: Forbidden

SAFETY PROFILE: Moderately toxic by inhalation. Experimental reproductive effects. Mutation data reported. An eye irritant. A powerful explosive sensitive to spark, impact, sunlight, or heating rapidly to 100°C. A powerful oxidizer. Concentrations of greater than 10% in air are explosive. Explodes on mixing with carbon monoxide, hydrocarbons (e.g., butadiene, ethane, ethylene, methane, propane), fluoramines (e.g., difluoramine, trifluoramine). Mixtures with hydrogen explode with sparking or contact with platinum. Explodes on contact with mercury, potassium hydroxide, phosphorus pentachloride + chlorine. Ignites or explodes on contact with non-metals (e.g., phosphorus, sulfur, sugar). Reacts violently with F_2, NHF_2. Reacts with water or steam to produce toxic and corrosive fumes of HCl. When heated to decomposition it emits toxic fumes of Cl^-. See also CHLORINE.

CDX250 CAS:13637-63-3 HR: 3
CHLORINE PENTAFLUORIDE
DOT: UN 2548
mf: ClF_5 mw: 130.45

PROP: Colorless gas, extremely vigorous fluorinating agent. D: 2.105 @ −80°, mp: −103°, bp: −13.1°.

SYNS: CHLORINE FLUORIDE (ClF_5) □ CHLORINE PENTAFLUORIDE (DOT)

TOXICITY DATA WITH REFERENCE
ihl-rat LC50:122 ppm/1H AIHAAP 33,661,72
ihl-mus LC50:57 ppm/1H AIHAAP 33,661,72
ihl-dog LC50:122 ppm/1H AIHAAP 33,661,72
ihl-mky LC50:173 ppm/1H AIHAAP 33,661,72

OSHA PEL: TWA 2.5 mg(F)/m³
ACGIH TLV: TWA 2.5 mg(F)/m³
NIOSH REL: (Inorganic Fluorides) TWA 2.5 mg(F)/m³
DOT CLASSIFICATION: 2.3; *Label:* Poison Gas, Oxidizer, Corrosive

SAFETY PROFILE: Poison by inhalation. A corrosive material. Vigorous reaction in contact with water or anhydrous nitric acid. Violent reaction on contact with metals. When heated to decomposition it emits very toxic fumes of Cl^- and F^-. See also CHLORINE, FLUORINE, FLUORIDES, and CHLORINE TRIFLUORIDE.

CDX500 **CAS:27218-16-2** **HR: 3**
CHLORINE PERCHLORATE
mf: Cl_2O_4 mw: 134.91

PROP: Yellow liquid. Mp: −117°, bp: 44.5°.

SAFETY PROFILE: A shock-sensitive explosive. Potentially explosive reaction with chlorotrifluoroethylene, perfluoroalkyl iodides (e.g., perfluoromethyl iodide, 1,2-diiodoperfluoroethane, 1,3-diiodoperfluoropropane). When heated to decomposition it emits toxic fumes of Cl^-. See also CHLORINE and PERCHLORATES.

CDX750 **CAS:7790-91-2** **HR: 3**
CHLORINE TRIFLUORIDE
DOT: UN 1749
mf: ClF_3 mw: 92.45

PROP: Colorless gas to yellow liquid; sweet odor. One of the most reactive chemical compds known. Mp: −83°, bp: 11.8°, d: 1.77 @ 13°.

SYNS: CHLORINE FLUORIDE □ CHLOROTRIFLUORIDE □ TRIFLUORURE de CHLORE (FRENCH)

TOXICITY DATA WITH REFERENCE
eye-rat 21 ppm/12H-I AMIHAB 12,515,55
eye-dog 21 ppm/12H-I AMIHAB 12,515,55
ihl-hmn LCLo:50 ppm 34ZIAG -,651,66
ihl-rat LCLo:400 ppm/30M TXAPA9 27,527,74
ihl-mus LC50:178 ppm/1H AMRL** TR-70-55/70
ibl-mky LC50:230 ppm/1H AMRL** TR-70-77/70

CONSENSUS REPORTS: Reported in EPA TSCA Inventory.

OSHA PEL: CL 0.1 ppm
ACGIH TLV: CL 0.1 ppm
DFG MAK: 0.1 ppm (0.4 mg/m³)
DOT CLASSIFICATION: 2.3; *Label:* Poison Gas, Oxidizer, Corrosive

SAFETY PROFILE: Human poison by inhalation. An eye irritant. See also FLUORIDES; CHLORINE; and FLUORINE. Spontaneously flammable. A powerful oxidant which may react violently with oxidizable materials. A rocket propellant.

Explosive reaction with water, bis(trifluoromethyl)sulfide or -disulfide, polychlorotrifluoroethylene, trifluoromethanesulfenyl chloride, and other hydrogen containing materials (e.g., ammonia, coal gas, hydrogen, hydrogen sulfide, methane, acetic acid, benzene, ether, cotton, paper, wood). Forms shock-sensitive explosive mixtures with highly chlorinated compounds (e.g., carbon tetrachloride), nitroaryl compounds (e.g., trinitrotoluene, hexanitrobiphenyl, hexanitrodiphenyl amine, hexanitrodiphenyl sulfide, hexanitrodiphenyl ether). Reaction with ammonium fluoride or ammonium hydrogen fluoride forms explosive gaseous products.

Ignition on contact with boron-containing materials, iodine, finely divided refractory materials (e.g., asbestos, glass wool, sand, tungsten carbide), fluorinated polymers (with flowing trifluoride).

Violent reaction with acids (e.g., nitric or sulfuric), chromium trioxide, ruthenium, selenium tetrafluoride (above 106°C), metals, metal oxides, metal salts, nonmetals, nonmetal salts, organic matter, glass wool, acetic acid, Al, Sb, As, Cu, Ir, Fe, Pb, Mg, Mo, Os, P, K, Rh, Se, Si, Ag, Na, S, Te, Sn, W, Zn, oxides, CO, graphite, HgI_2, HNO_3, K_2CO_3, KI, rubber, AgN_3, $AgNO_3$, NaOH, V_2P_5, WO_3. Incompatible with fuels, nitro compounds. When heated to decomposition or in reaction with water or steam it emits toxic fumes of F^- and Cl^-.

CDX800 **CAS:65597-24-2** **HR: 3**
CHLORINE(1)TRIFLUOROMETHANESULFONATE
mf: $CClF_3O_3S$ mw: 184.52

SAFETY PROFILE: Explodes on contact with easily oxidizable materials (e.g., organic materials). When heated to decomposition it emits toxic fumes of F^-, Cl^- and SO_x. See also SULFONATES.

CDY000 **CAS:69-27-2** **HR: 3**
CHLORISONDAMINE CHLORIDE
mf: $C_{14}H_{20}Cl_4N_2{\cdot}2Cl$ mw: 429.06

PROP: Crystals. Mp: 258–265°.

SYNS: CHLORISONDAMINE □ CHLORISONDAMINE DIMETHOCHLORIDE □ N-((2-DIMETHYLAMMONIUM)ETHYL)-4,5,6,7-TETRACHLOROISOINDOLINIUM DIMETHOCHLORIDE □ ECOLID □ ECOLID CHLORIDE □ HISINDAMONE A □ ISOINDOLINE, 2-(2-DIMETHYLAMINOETHYL)-4,5,6,7-TETRACHLORODIMETHO CHLORIDE □ SU 3088 □ 4,5,6,7-TETRACHLORO-2-(2-DIMETHYLAMINOETHYL)-ISOINDOLINE DIMETHOCHLORIDE

TOXICITY DATA WITH REFERENCE
orl-rat LD50:300 mg/kg SKNEA7 10,15,60
ivn-rat LD50:28 mg/kg JPETAB 115,172,55
orl-mus LD50:380 mg/kg BCFAAI 103,490,64
ipr-mus LD50:62 mg/kg AIPTAK 155,69,65
scu-mus LD50:240 mg/kg FRPSAX 20,482,65
ivn-mus LD50:28 mg/kg CSLNX* NX#00058

SAFETY PROFILE: Poison by ingestion, subcutaneous, intravenous and intraperitoneal routes. When heated to decomposition it emits very toxic fumes of Cl^-, NH_3 and NO_x.

CDY100 **CAS:2281-78-9** **HR: 3**
CHLORISOPROPAMIDE
mf: $C_{10}H_{13}ClN_2O_3S$ mw: 276.76

SYNS: BENZENESULFONAMIDE, 4-CHLORO-N-(((1-METHYLETHYL)AMINO)CARBONYL)- (9CI) □ UREA, 1-((p-CHLOROPHENYL)SULFONYL)-3-ISOPROPYL-

TOXICITY DATA WITH REFERENCE
orl-rat TDLo:1 g/kg (female 10D post):TER PROEAS 14,89,68
ipr-mus LD50:153 mg/kg FRZKAP (6),26,87

SAFETY PROFILE: Poison by intraperitoneal route. An experimental teratogen. When heated to decomposition it emits toxic fumes of SO_x, NO_x, and Cl^-.

CDY250 **HR: 3**
CHLORITES

SAFETY PROFILE: Many chlorite salts are heat- and

impact-sensitive explosives. The metal salts are powerful oxidants. They are much less stable than the analogous chlorates. React violently with NH_3, organic matter, or metals. See individual chlorites.

CDY275 **CAS:1961-77-9** **HR: 2**
CHLORMADINON
mf: $C_{21}H_{27}ClO_3$ mw: 362.93

PROP: Crystals from MeOH. Mp: 212–214°.

SYNS: CHLORMADINONE ◻ 6-CHLORO-17-HYDROXYPREGNA-4,6-DIENE-3,20-DIONE

TOXICITY DATA WITH REFERENCE
orl-wmn TDLo:200 µg/kg (20D pre):REP FESTAS 16,158,65
scu-rbt TDLo:300 mg/kg (female 1-3D post):TER FESTAS 20,211,69
orl-rbt TDLo:150 µg/kg (female 3D pre):REP ENDOAO 81,1251,67
scu-rbt TDLo:300 µg/kg (female 2D pre):REP AJOGAH 117,167,73
scu-rat TDLo:16,800 µg/kg (21D male):REP ACENA7 49,145,65
scu-rat TDLo:50,400 µg/kg (male 21D pre):REP ACENA7 49,145,65

SAFETY PROFILE: Human reproductive effects by ingestion: changes in the uterus, cervix, vagina, and female fertility. An experimental teratogen. Other experimental reproductive effects. When heated to decomposition it emits toxic fumes of Cl^-.

CDY299 **CAS:24934-91-6** **HR: 3**
CHLORMEPHOS
mf: $C_5H_{12}ClO_2PS_2$ mw: 234.70

PROP: Oil. Bp: 81–85°C @ 0.1 mm Hg, d: 1.260, vap. press: 0.0056 (mm Hg) @ 30°C. Sltly water sol (60 mg/L @ 20C).

SYNS: DOTAN ◻ MC 2188 ◻ S-(CHLOROMETHYL)-O,O-DIETHYL PHOSPHORODITHIOATE ◻ S-CHLOROMETHYL-O,O-DIETHYL PHOSPHOROTHIOLOTHIOATE ◻ S-CHLOROMETHYL-O,O-DIETHYL PHOSPHOROTHIOLOTHIONATE ◻ S-(CHLOROMETHYL)-O,O-DIMETHYL PHOSPHORODITHIOIC ACID, ESTER

TOXICITY DATA WITH REFERENCE
orl-rat LD50:7 mg/kg FMCHA2 -,C53,83
skn-rat LD50:27 mg/kg GUCHAZ 6,103,73

CONSENSUS REPORTS: EPA Extremely Hazardous Substances List.

SAFETY PROFILE: Poison by inhalation, ingestion, and skin contact. Symptoms may include nausea, vomiting, abdominal cramps, diarrhea, excessive salivation, headache, giddiness, weakness, muscle twitching, difficult breathing, blurring or dimness of vision, and loss of muscle coordination. Death may occur from failure of the respiratory center, paralysis of the respiratory muscles, or intense bronchoconstriction. An organophosphorus pesticide. Combustible. For small fires, use dry chemical, carbon dioxide, water spray, or foam. For large fires, use water spray, fog, or foam. When heated to decomposition it emits toxic fumes of Cl^-, PO_x, and SO_x.

CDY325 **CAS:3689-76-7** **HR: 3**
CHLORMIDAZOLE
mf: $C_{15}H_{13}ClN_2$ mw: 256.75

PROP: Crystals. Mp: 67–68°, bp: 240–242°.

SYNS: 1-p-CHLORBENZYL-2-METHYL-BENZIMIDAZOL (GERMAN) ◻ 1-((4-CHLOROPHENYL)METHYL)-2-METHYL-1H-BENZIMIDAZOLE (9CI) ◻ CLOMIDAZOLE ◻ DIAMYCELINE ◻ FUNGO-POLYCID ◻ FUTRICAN ◻ H115 ◻ MYCO-POLYCID

TOXICITY DATA WITH REFERENCE
orl-rat LD50:2200 mg/kg ARZNAD 9,489,59
ipr-rat LD50:58 mg/kg ARZNAD 9,489,59
orl-mus LD50:200 mg/kg ARZNAD 9,489,59
ipr-mus LD50:90 mg/kg ARZNAD 9,489,59
scu-mus LD50:250 mg/kg ARZNAD 9,489,59

SAFETY PROFILE: Poison by ingestion, subcutaneous, and intraperitoneal routes. When heated to decomposition it emits toxic fumes of Cl^- and NO_x.

CDY500 **CAS:107-20-0** **HR: 3**
CHLOROACETALDEHYDE
DOT: UN 2232
mf: C_2H_3ClO mw: 78.50

PROP: Clear, colorless liquid; pungent odor. Bp: 90.0–100.1° (40% soln), fp: −16.3° (40% soln), flash p: 190°F, d: 1.19 @ 25°/25° (40% soln), vap press: 100 mm @ 45° (40% soln).

SYNS: CHLOROACETALDEHYDE MONOMER ◻ 2-CHLOROACETALDEHYDE ◻ 2-CHLOROETHANAL ◻ 2-CHLORO-1-ETHANAL ◻ MONOCHLOROACETALDEHYDE ◻ RCRA WASTE NUMBER P023

TOXICITY DATA WITH REFERENCE
mmo-omi 250 µL/plate CBINA8 30,9,80
mmo-omi 1 mmol/L MUREAV 73,1,80
mmo-asn 30 µL/plate CBINA8 30,9,80
dnd-sat:tes 3 g CRNGDP 3,663,82
orl-rat LD50:75 mg/kg JPMSAE 61,19,72
ipr-rat LD50:6 mg/kg JPMSAE 61,19,72
orl-mus LD50:69 mg/kg JPMSAE 61,19,72
ipr-mus LD50:6 mg/kg JPMSAE 61,19,72
skn-rbt LD50:224 mg/kg JPMSAE 61,19,72
ipr-rbt LD50:4640 mg/kg JPMSAE 61,19,72
ipr-gpg LD50:2 mg/kg JPMSAE 61,19,72

CONSENSUS REPORTS: Reported in EPA TSCA Inventory. EPA Genetic Toxicology Program.

OSHA PEL: CL 1 ppm
ACGIH TLV: CL 1 ppm
DFG MAK: 1 ppm (3 mg/m^3)
DOT CLASSIFICATION: 6.1; *Label:* Poison

SAFETY PROFILE: Poison by ingestion, skin contact, and intraperitoneal routes. Mutation data reported. Combustible when exposed to heat or flame. Reacts with oxidizing materials. To fight fire, use water, foam, CO_2, dry chemical. When heated to decomposition it emits toxic fumes of Cl^-. See also ALDEHYDES and CHLORIDES.

For occupational chemical analysis use NIOSH: Chloroacetaldehyde S11.

CDY825 CAS:598-49-2 HR: 3
N-CHLOROACETAMIDE
mf: C_2H_4ClNO mw: 93.51

PROP: Plates from C_6H_6. Mp: 111–112°.

SAFETY PROFILE: Solutions of the acetamide may explode during drying or concentration operations. When heated to decomposition it emits toxic fumes of Cl^- and NO_x.

CDY850 CAS:79-07-2 HR: 3
2-CHLOROACETAMIDE
mf: C_2H_4ClNO mw: 93.52

PROP: Crystals. Mp: 120°, bp: 225° (decomp). Sol in H_2O, EtOH; sltly sol in Et_2O.

SYNS: CHLORACETAMID (GERMAN) □ CHLOROACETAMIDE □ α-CHLOROACETAMIDE □ 2-CHLOROETHANAMIDE □ USAF DO-29

TOXICITY DATA WITH REFERENCE
mmo-esc 2 mmol/L MUREAV 210,255,89
orl-mus LD50:155 mg/kg CHHTAT 58,462,78
ipr-mus LD50:100 mg/kg NTIS** AD277-689
ivn-mus LD50:180 mg/kg CSLNX* NX#03764

CONSENSUS REPORTS: Reported in EPA TSCA Inventory.

SAFETY PROFILE: Poison by ingestion, intravenous, and intraperitoneal routes. Mutation data reported. When heated to decomposition it emits very toxic Cl^- and NO_x. See also N-CHLOROACETAMIDE.

CDZ000 CAS:3272-96-6 HR: 3
CHLOROACETAMIDE OXIME
mf: $C_2H_5ClN_2O$ mw: 108.53

SAFETY PROFILE: An unstable heat-sensitive explosive. When heated to decomposition it emits toxic fumes of Cl^- and NO_x. See also EXPLOSIVES.

CDZ050 CAS:588-07-8 HR: 3
3′-CHLOROACETANILIDE
mf: C_8H_8ClNO mw: 169.62

SYN: m-CHLOROACETANILIDE

TOXICITY DATA WITH REFERENCE
ipr-rat LD50:350 mg/kg JAPMA8 48,204,59
ipr-mus LD50:610 mg/kg JAPMA8 48,204,59

CONSENSUS REPORTS: Reported in EPA TSCA Inventory.

SAFETY PROFILE: Poison by intraperitoneal route. When heated to decomposition it emits toxic vapors of NO_x and Cl^-.

CDZ100 CAS:539-03-7 HR: 3
4′-CHLOROACETANILIDE
mf: C_8H_8ClNO mw: 169.62

PROP: Needles or plates. Mp: 178.4°. Sol in C_6H_6.

SYNS: ACETIC-4-CHLOROANILIDE □ N-(4-CHLOROPHENYL)ACETAMIDE

TOXICITY DATA WITH REFERENCE
mmo-sat 2500 μg/plate PCBPBS 10,174,79
ipr-rat LD50:245 mg/kg JAPMA8 48,204,59
ipr-mus LD50:730 mg/kg 49RQAC 1,140,82

CONSENSUS REPORTS: Reported in EPA TSCA Inventory.

SAFETY PROFILE: Poison by intraperitoneal route. Mutation data reported. When heated to decomposition it emits toxic fumes of Cl^- and NO_x.

CEA000 CAS:79-11-8 HR: 3
CHLOROACETIC ACID
DOT: UN 1750/UN 1751
mf: $C_2H_3ClO_2$ mw: 94.50

PROP: Colorless crystals in three forms. Mp: (α) 61.3°, (β) 56.2°, (τ) 52.5°, bp: 189°, flash p: 259°F, d: 1.58 @ 20°/20°, vap d: 3.26. Very sol in H_2O; sol org solvs.

SYNS: ACIDE CHLORACETIQUE (FRENCH) □ ACIDE MONOCHLORACETIQUE (FRENCH) □ ACIDOMONOCLOROACETICO (ITALIAN) □ CHLORACETIC ACID □ CHLOROACETIC ACID, liquid (UN 1750) (DOT) □ CHLOROACETIC ACID, solid (UN 1751) (DOT) □ α-CHLOROACETIC ACID □ CHLOROETHANOIC ACID □ KYSELINA CHLOROCTOVA □ MCA □ MONOCHLOORAZIJNZUUR (DUTCH) □ MONOCHLORACETIC ACID □ MONOCHLORESSIGSAEURE (GERMAN) □ MONOCHLOROACETIC ACID □ MONOCHLOROETHANOIC ACID □ NCI-C60231

TOXICITY DATA WITH REFERENCE
mma-mus:lym 548 mg/L MUREAV 97,49,82
scu-mus TDLo:100 mg/kg:ETA NTIS** PB223-159
scu-mus TD:1300 mg/kg/65W-I:ETA JNCIAM 53,695,74
ihl-rat LC50:180 mg/m³ GTPZAB 18(9),32,74
scu-rat LD50:5 mg/kg TXAPA9 22,303,72
ivn-rat LD50:55 mg/kg GTPZAB 18(9),32,74
orl-mus LD50:165 mg/kg JPETAB 86,336,46
scu-mus LD50:250 mg/kg AIPTAK 116,154,58

CONSENSUS REPORTS: Reported in EPA TSCA Inventory. EPA Genetic Toxicology Program. EPA Extremely Hazardous Substances List. Community Right-To-Know List.

DOT CLASSIFICATION: 8; *Label:* Corrosive, Poison (UN 1750)

SAFETY PROFILE: Poison by ingestion, inhalation, subcutaneous, and intravenous route. A corrosive skin, eye, and mucous membrane irritant. Questionable carcinogen with experimental tumorigenic data. Mutation data reported. Combustible liquid when exposed to heat or flame. To fight fire, use water spray, fog, mist, dry chemical, foam. When heated to decomposition it emits toxic fumes of Cl^-. See also CHLORIDES.

For occupational chemical analysis use NIOSH: Chloroacetic Acid, 2008.

CEA050 CAS:39082-00-3 HR: 3
4-CHLOROACETOACETANILIDE
mf: $C_{10}H_{10}ClNO_2$ mw: 211.66

SYNS: ACETOACETANILIDE, 4-CHLORO- □ BUTANAMIDE, 4-CHLORO-3-OXO-N-PHENYL- □ γ-CHLOROACETOACETANILIDE □ γ-CHLO-

ROACETO ACETIC ACID ANILIDE □ 4-CHLORO-3-OXO-N-PHENYLBUTANAMIDE

TOXICITY DATA WITH REFERENCE
orl-rat LD50:1825 mg/kg GTPZAB 31(1),49,87
ipr-rat LD50:59 mg/kg OYYAA2 33,695,87
scu-rat LD50:4500 mg/kg GTPZAB 31(1),49,87
orl-mus LD50:1780 mg/kg GTPZAB 31(1),49,87
ipr-mus LD50:71 mg/kg OYYAA2 33,695,87
orl-rbt LD50:1780 mg/kg GTPZAB 31(1),49,87

CONSENSUS REPORTS: Reported in EPA TSCA Inventory.

SAFETY PROFILE: A poison by intraperitoneal route. Moderately toxic by ingestion. When heated to decomposition it emits toxic vapors of NO_x and Cl^-.

CEA100 CAS:4116-10-3 HR: 2
α-CHLOROACETOACETIC ACID MONOETHYLAMIDE
mf: $C_6H_{10}ClNO_2$ mw: 163.62

SYNS: ACETOACETAMIDE, 2-CHLORO-N-ETHYL- □ α-CHLORACETESSIGSAEUREAETHYLAMID □ BUTANAMIDE, 2-CHLORO-N-METHYL-3-OXO-(9CI) □ 2-CHLORO-N-METHYL-3-OXOBUTANAMIDE

TOXICITY DATA WITH REFERENCE
orl-rat LD50:700 mg/kg ATXKA8 18,316,60
orl-mus LD50:750 mg/kg ATXKA8 18,316,60

CONSENSUS REPORTS: Reported in EPA TSCA Inventory.

SAFETY PROFILE: Moderately toxic by ingestion. When heated to decomposition it emits toxic vapors of NO_x and Cl^-

CEA750 CAS:532-27-4 HR: 3
α-CHLOROACETOPHENONE
DOT: UN 1697
mf: C_8H_7ClO mw: 154.60

PROP: Leaflets from pet ether. Mp: 54°, bp: 139–141° @ 14 mm.

SYNS: CAF □ CAP □ CHEMICAL MACE □ CHLORACETOPHENONE □ ω-CHLOROACETOPHENONE □ 1-CHLOROACETOPHENONE □ CHLOROACETOPHENONE, liquid or solid (DOT) □ CHLOROMETHYL PHENYL KETONE □ 2-CHLORO-1-PHENYLETHANONE □ CN □ ETHANONE, 2-CHLORO-1-PHENYL- □ MACE (lachrymator) □ NCI-C55107 □ PHENACYL CHLORIDE □ PHENYLCHLOROMETHYLKETONE

TOXICITY DATA WITH REFERENCE
skn-rat 12%/6H open MOD ARTODN 40,75,78
skn-rbt 5 mg/24H MLD TXAPA9 17,295,70
skn-rbt 12%/6H open MOD ARTODN 40,75,78
eye-rbt 1 mg MLD TXAPA9 17,295,70
eye-rbt 3 mg SEV TXAPA9 17,295,70
skn-gpg 12%/6H open MOD ARTODN 40,75,78
skn-mus TDLo:2400 mg/kg/27W-I:NEO BJCAAI 7,482,53
ihl-hmn LCLo:159 mg/m³/20M 34ZIAG -,163,69
ihl-hmn TCLo:93 mg/m³/3M:EYE AIHAAP 23,199,62
ihl-hmn TCLo:20 mg/m³:EYE BJIMAG 29,298,72
orl-rat LD50:127 mg/kg ARTODN 40,75,78
ihl-rat LCLo:417 mg/m³/15M ARTODN 40,75,78
ipr-rat LD50:36 mg/kg ARTODN 40,75,78
ivn-rat LD50:41 mg/kg ARTODN 40,75,78
orl-mus LD50:139 mg/kg NTIS** AD837-111
ihl-mus LCLo:600 mg/m³/15M ARTODN 40,75,78
ipr-mus LD50:60 mg/kg NTIS** AD837-111
ivn-mus LD50:81 mg/kg ARTODN 40,75,78
orl-rbt LD50:118 mg/kg ARTODN 40,75,78
ihl-rbt LCLo:465 mg/m³/20M ARTODN 40,75,78
ivn-rbt LD50:30 mg/kg ARTODN 40,75,78
orl-gpg LD50:158 mg/kg ARTODN 40,75,78
ihl-gpg LCLo:490 mg/m³/30M ARTODN 40,75,78
ipr-gpg LD50:17 mg/kg ARTODN 40,75,78

CONSENSUS REPORTS: Reported in EPA TSCA Inventory. Community Right-To-Know List.

OSHA PEL: TWA 0.05 ppm
ACGIH TLV: TWA 0.05 ppm
DOT CLASSIFICATION: 6.1; *Label:* Poison

SAFETY PROFILE: A human poison by inhalation. An experimental poison by ingestion, inhalation, intraperitoneal, and intravenous routes. Human systemic effects by inhalation: lacrimation, conjunctiva irritation, and unspecified eye effects, cough, and dyspnea. A severe eye and moderate skin irritant. Questionable carcinogen with experimental neoplastigenic data by skin contact. A riot control agent. When heated to decomposition it emits toxic fumes of Cl^-. See also KETONES.

CEB000 CAS:2142-68-9 HR: 2
2′-CHLOROACETOPHENONE
mf: C_8H_7ClO mw: 154.60

PROP: D: 1.19 @ 25°/25°, bp: 227–228° @ 738 mm.

TOXICITY DATA WITH REFERENCE
orl-rat LD50:1020 mg/kg CroHP# 02SEPT80
orl-mus LD50:880 mg/kg CroHP# 02SEPT80

CONSENSUS REPORTS: Reported in EPA TSCA Inventory.

SAFETY PROFILE: Moderately toxic by ingestion. When heated to decomposition it emits toxic fumes of Cl^-.

CEB250 CAS:99-91-2 HR: 3
p-CHLOROACETOPHENONE
mf: C_8H_7ClO mw: 154.60

PROP: Pale straw-colored liquid or white crystals; fragrant, non-persistent odor. Mp: 20°, bp: 236°, fp: 59°, d: 1.19 @ 20°, vap press: 0.012 mm @ 0°, vap d: 5.2, flash p: 244°F.

SYNS: 4-CHLOROACETOPHENONE □ 4′-CHLOROACETOPHENONE □ 1-(4-CHLOROPHENYL)ETHANONE □ USAF DO-1

TOXICITY DATA WITH REFERENCE
ihl-hmn TCLo:1 mg/m³/1M:NOSE,EYE 27ZZA9 8,192,57
orl-mus LD50:1207 mg/kg PHARAT 31,317,76
ipr-mus LD50:100 mg/kg NTIS** AD277-689

CONSENSUS REPORTS: Reported in EPA TSCA Inventory.

SAFETY PROFILE: Poison by intraperitoneal route. Moderately toxic by ingestion. A powerful irritant and lachrymator. Human systemic effects by inhalation:

C

lachrimation and unspecified effects on the eye and sense of smell. Combustible when exposed to heat or flame. To fight fire, use water, foam, alcohol foam, dry chemical. When heated to decomposition or on contact with water or steam it emits toxic fumes of Cl^-.

CEB750 **CAS:7149-79-3** **HR: 3**
3′-CHLORO-p-ACETOTOLUIDIDE
mf: $C_9H_{10}ClNO$ mw: 183.65

PROP: Crystals from EtOH or hexane. Mp: 85° from EtOH, mp: 104–105° from Hexane. Sol in C_6H_6.

TOXICITY DATA WITH REFERENCE
orl-bwd LD50:1300 µg/kg TXAPA9 21,315,72

CONSENSUS REPORTS: Reported in EPA TSCA Inventory.

SAFETY PROFILE: A deadly poison by ingestion. When heated to decomposition it emits very toxic fumes of Cl^- and NO_x.

CEC000 **CAS:140-49-8** **HR: 2**
4′-CHLOROACETYL ACETANILIDE
mf: $C_{10}H_{10}ClNO_2$ mw: 211.66

SYNS: p-ACETAMIDOPHENACYL CHLORIDE □ p-(ACETYLAMINO) PHENACYL CHLORIDE □ 4′-(CHLOROACETYL)ACETANILIDE □ NCI-C03770

TOXICITY DATA WITH REFERENCE
mmo-sat 3300 ng/plate ENMUDM 7(Suppl 5),1,85
orl-rat LD50:2150 mg/kg NCILB* NIH-NCI-E-C-72-3252
orl-mus LD50:1470 mg/kg NCILB* NIH-NCI-E-C-72-3252

CONSENSUS REPORTS: NCI Carcinogenesis Bioassay (feed); No Evidence: mouse, rat NCITR* NCI- CG-TR-177,79. Reported in EPA TSCA Inventory.

SAFETY PROFILE: Moderately toxic by ingestion. Mutation data reported. When heated to decomposition it emits very toxic fumes of Cl^- and NO_x. See also CHLORIDES.

CEC100 **CAS:1132-20-3** **HR: 3**
N-(CHLOROACETYL)-3-AZABICYCLO(3.2.1)NONANE
mf: $C_{10}H_{16}ClNO$ mw: 201.72

SYNS: 3-AZABICYCLO(3.2.2)NONANE, 3-(CHLOROACETYL)- □ KETONE, 3-AZABICYCLO(3.2.2)NONYL CHLOROMETHYL

TOXICITY DATA WITH REFERENCE
ivn-mus LD50:56 mg/kg CSLNX* NX#03484

DOT CLASSIFICATION: 3; *Label:* Flammable Liquid

SAFETY PROFILE: A poison by intravenous route. A flammable liquid. When heated to decomposition it emits toxic vapors of NO_x and Cl^-.

CEC250 **CAS:79-04-9** **HR: 3**
CHLOROACETYL CHLORIDE
DOT: UN 1752
mf: $C_2H_2Cl_2O$ mw: 112.94

PROP: Water-white or sltly yellow liquid. Bp: 108–110°, fp: −22.5°, flash p: none, d: 1.495 @ 0°.

SYNS: CHLORACETYL CHLORIDE □ CHLORID KYSELINY CHLOROCTOVE □ CHLOROACETIC ACID CHLORIDE □ CHLOROACETIC CHLORIDE □ CHLORURE de CHLORACETYLE (FRENCH) □ MONOCHLOROACETYL CHLORIDE

TOXICITY DATA WITH REFERENCE
orl-rat LD50:208 mg/kg 85INA8 5,123(89),86
orl-mus LD50:220 mg/kg GISAAA 54(2),90,89
ihl-mus LC50:1300 ppm/2H 85INA8 5,123(89),86
ihl-rat LCLo:1000 ppm/4H 34ZIAG -,607,69
ivn-mus LD50:32 mg/kg CSLNX* NX#04538

CONSENSUS REPORTS: Reported in EPA TSCA Inventory.

OSHA PEL: TWA 0.05 ppm
ACGIH TLV: TWA 0.05 ppm; STEL: 0.15 ppm
DOT CLASSIFICATION: 8; *Label:* Corrosive, Poison

SAFETY PROFILE: Poison by ingestion and intravenous routes. Mildly toxic by inhalation. Corrosive. A lachrymator. When heated to decomposition it emits toxic fumes of Cl^-.

CEC300 **CAS:143-85-1** **HR: 3**
1-CHLOROACETYL-α-α-DIPHENYL-4-PIPERIDINEMETHANOL
mf: $C_{20}H_{22}ClNO_2$ mw: 343.88

SYNS: KETONE, CHLOROMETHYL 4-(DIPHENYLHYDROXYMETHYL) PIPERIDINO □ 4-PIPERIDINEMETHANOL, 1-CHLOROACETYL-α-α-DIPHENYL-

TOXICITY DATA WITH REFERENCE
ipr-mus LD50:20 mg/kg JPMSAE 55,529,66

DOT CLASSIFICATION: 3; *Label:* Flammable Liquid

SAFETY PROFILE: A poison by intraperitoneal route. A flammable liquid. When heated to decomposition it emits toxic vapors of NO_x and Cl^-.

CEC500 **CAS:593-63-5** **HR: 3**
CHLOROACETYLENE
mf: C_2HCl mw: 60.47

PROP: Very unstable gas with a nauseating odor. Bp: −32°.

SAFETY PROFILE: Explodes or ignites on contact with air. Its reactivity and volatility make it extremely dangerous. Probably a strong irritant by inhalation. When heated to decomposition it emits toxic fumes of Cl^-. See also ACETYLENE COMPOUNDS.

CEC700 **CAS:24040-34-4** **HR: 3**
2-CHLOROACETYLFLUORENE
mf: $C_{15}H_{11}ClO$ mw: 242.71

SYNS: ETHANONE, 2-CHLORO-1-(9H-FLUOREN-2-YL)- □ FLUO-

RENE, 2-(CHLOROACETYL)- □ KETONE, CHLOROMETHYL 2-FLUORENYL

TOXICITY DATA WITH REFERENCE
ipr-mus LD50:340 mg/kg RPTOAN 48,143,85

DOT CLASSIFICATION: 3; *Label:* Flammable Liquid

SAFETY PROFILE: A poison by intraperitoneal route. A flammable liquid. When heated to decomposition it emits toxic vapors of Cl^-.

CEE500 CAS:598-79-8 HR: 3
2-CHLOROACRYLIC ACID
mf: $C_3H_3ClO_2$ mw: 106.51

PROP: A solid. Mp: 65°.

SYNS: CHLOROACRYLIC ACID □ α-CHLOROACRYLIC ACID

TOXICITY DATA WITH REFERENCE
ivn-mus LD50:42 mg/kg CSLNX* NX#07795

CONSENSUS REPORTS: Reported in EPA TSCA Inventory.

SAFETY PROFILE: Poison by intravenous route. When heated to decomposition it emits toxic fumes of Cl^-.

CEE750 CAS:920-37-6 HR: 3
2-CHLOROACRYLONITRILE
mf: C_4H_2ClN mw: 87.51

PROP: Liquid. Bp: 88°, flash p: 46.4°F.

SYNS: CHLOROACRYLONITRILE □ α-CHLOROACRYLONITRILE

TOXICITY DATA WITH REFERENCE
ivn-mus LD50:100 mg/kg CSLNX* NX#02164

CONSENSUS REPORTS: Reported in EPA TSCA Inventory. Cyanide and its compounds are on the Community Right-To-Know List.

SAFETY PROFILE: Poison by intravenous route. A powerful irritant. A dangerous fire hazard when exposed to heat or flame. To fight fire, use water, dry chemical, CO_2, foam. When heated to decomposition it emits very toxic fumes of Cl^-, NO_x, and CN^-. See also NITRILES.

CEE800 CAS:52917-86-9 HR: 3
3-CHLOROADAMANTYL DIAZOMETHYL KETONE
mf: $C_{12}H_{15}ClN_2O$ mw: 238.74

SYNS: (3-CHLORO-1-ADAMANTANOYL)DIAZOMETHANE □ 1-ETHANONE, 1-(3-CHLOROADAMANTYL)-2-DIAZO- □ KETONE, 3-CHLORO-1-ADAMANTYL DIAZOMETHYL

TOXICITY DATA WITH REFERENCE
ivn-rat LD50:440 mg/kg PCJOAU 10,454,76

DOT CLASSIFICATION: 3; *Label:* Flammable Liquid

SAFETY PROFILE: Moderately toxic by intravenous route. A flammable liquid. When heated to decomposition it emits toxic vapors of NO_x and Cl^-.

CEE825 CAS:73599-94-7 HR: 3
1-CHLORO-3-ADAMANTYL ETHOXYMETHYL KETONE
mf: $C_{14}H_{22}ClNO_2$ mw: 271.82

SYN: ETHANONE, 1-(1-CHLORO-3-ADAMANTYL)-2-ETHOXY-

TOXICITY DATA WITH REFERENCE
unr-mus LD50:600 mg/kg RPTOAN 43,73,80

DOT CLASSIFICATION: 3; *Label:* Flammable Liquid

SAFETY PROFILE: Moderately toxic by an unreported route. A flammable liquid. When heated to decomposition it emits toxic vapors of NO_x and Cl^-.

CEE850 CAS:73599-90-3 HR: 3
1-CHLORO-3-ADAMANTYL HYDROXYMETHYL KETONE
mf: $C_{12}H_{17}ClO_2$ mw: 228.74

SYN: ETHANONE, 1-(1-CHLORO-3-ADAMANTYL)-2-HYDROXY-

TOXICITY DATA WITH REFERENCE
unr-mus LD50:795 mg/kg RPTOAN 43,73,80

DOT CLASSIFICATION: 3; *Label:* Flammable Liquid

SAFETY PROFILE: Moderately toxic by an unreported route. A flammable liquid. When heated to decomposition it emits toxic vapors of Cl^-.

CEF100 CAS:146-77-0 HR: 3
2-CHLOROADENOSINE
mf: $C_{10}H_{12}ClN_5O_4$ mw: 301.72

PROP: Crystals. Mp: 133–136° (decomp).

SYNS: ADENOSINE, 2-CHLORO- □ Cl-ADO

TOXICITY DATA WITH REFERENCE
ipr-mus TDLo:603 μg/kg (female 11-12D post):TER TJADAB 25(2),35A,82
ivn-mus LDLo:50 mg/kg AIPTAK 118,95,59

SAFETY PROFILE: Poison by intravenous route. An experimental teratogen. When heated to decomposition it emits toxic fumes of NO_x and Cl^-.

CEF125 HR: 3
2-CHLOROADENOSINE-5′-SULFAMATE
mf: $C_{10}H_{13}ClN_6O_6S$ mw: 380.80

PROP: Isolated from a subspecies of *Streptomyces rishiriensis* (JANTAJ 35,939,82).

SYN: 5′-SULFAMOYL-2-CHLOROADENOSINE

TOXICITY DATA WITH REFERENCE
dni-esc 10 mg/L JANTAJ 35,939,82
oms-esc 10 mg/L JANTAJ 35,939,82
orl-mus LD50:3400 μg/kg JANTAJ 35,939,82
ipr-mus LD50:200 μg/kg JANTAJ 35,939,82

SAFETY PROFILE: Deadly poison by ingestion and intraperitoneal routes. Mutation data reported. When heated to decomposition it emits toxic fumes of Cl^-, SO_x, and NO_x.

CEF250 CAS:5976-47-6 HR: 3
β-CHLORO ALLYL ALCOHOL
mf: C_3H_5ClO mw: 92.53

PROP: Liquid. D: 1.166, bp: 136–140°.

SYN: 2-CHLORO-2-PROPEN-1-OL

TOXICITY DATA WITH REFERENCE
skn-rbt 500 mg SEV SCCUR* -,2,61
mmo-sat 1 nmol/plate JAFCAU 28,880,80
mma-sat 1 nmol/plate JAFCAU 28,880,80
orl-mus LDLo:63 mg/kg SCCUR* -,2,61
ihl-mus LCLo:198 ppm/6H SCCUR* -,2,61
ivn-mus LD50:56 mg/kg CSLNX* NX#02539
skn-rbt LDLo:25 mg/kg SCCUR* -,2,61

SAFETY PROFILE: Poison by ingestion, skin contact, and intravenous routes. Moderately toxic by inhalation. A severe skin irritant. Mutation data reported. When heated to decomposition it emits toxic fumes of Cl^-. See also pi-CHLORO ALLYL ALCOHOL, ALLYL COMPOUNDS, CHLORIDES, and ALCOHOLS.

CEF500 CAS:29560-84-7 HR: 3
pi-CHLORO ALLYL ALCOHOL
mf: C_3H_5ClO mw: 92.53

SYN: 3-CHLORO-2-PROPEN-1-OL

TOXICITY DATA WITH REFERENCE
skn-rbt 500 mg SEV SCCUR* -,2,61
orl-rat LD50:102 mg/kg SCCUR* -,2,61
ihl-rat LC50:370 ppm/1H SCCUR* -,2,61
orl-mus LD50:175 mg/kg SCCUR* -,2,61
ihl-mus LC50:540 ppm/1H SCCUR* -,2,61
skn-mus LDLo:170 mg/kg SCCUR* -,2,61

SAFETY PROFILE: Poison by ingestion and skin contact. Moderately toxic by inhalation. A severe skin irritant. When heated to decomposition it emits toxic fumes of Cl^-. See also β-CHLORO ALLYL ALCOHOL, ALLYL COMPOUNDS, CHLORIDES, and ALCOHOLS.

CEG550 CAS:4080-31-3 HR: 3
1-(3-CHLOROALLYL)-3,5,7-TRIAZA-1-AZONIAADAMANTANE CHLORIDE
mf: $C_9H_{16}ClN_4{\cdot}Cl$ mw: 251.19

SYNS: DOWCO 184 □ DOWICIDE Q □ DOWICIL 75 □ DOWICIL 100 □ QUATERNIUM 15 □ 3,5,7-TRIAZA-1-AZONIAADAMANTANE, 1-(3-CHLOROALLYL)-, CHLORIDE

TOXICITY DATA WITH REFERENCE
skn-rbt 500 mg/24H MLD JACTDZ 5(3),61,86
mma-sat 333 μg/plate EMMUEG 11(Suppl 12),1,88
orl-rat TDLo:250 mg/kg (female 6-15D post):TER JACTDZ 5(3),61,86
orl-rat LD50:500 mg/kg PCOC** -,455,66
orl-rbt LD50:78,500 μg/kg JACTDZ 5(3),61,86
skn-rbt LD50:565 mg/kg JACTDZ 5(3),61,86

CONSENSUS REPORTS: Reported in EPA TSCA Inventory.

SAFETY PROFILE: Poison by ingestion. Moderately toxic by skin contact. Experimental teratogenic effects. A skin irritant. Mutation data reported. When heated to decomposition it emits toxic fumes of NO_x and Cl^-.

CEG600 CAS:615-66-7 HR: 1
3-CHLORO-4-AMINOANILINE
mf: $C_6H_7ClN_2$ mw: 142.60

PROP: Needles. Mp: 64°.
$Cl(H_2N)C_6H_3NH_2$

SYNS: 3-CHLOR-p-FENYLENDIAMIN (CZECH) □ 2-CHLORO-1,4-BENZENEDIAMINE □ o-CHLORO-p-PHENYLENEDIAMINE □ 2-CHLORO-p-PHENYLENEDIAMINE □ C.I. 76065 □ URSOL BROWN O

TOXICITY DATA WITH REFERENCE
eye-rbt 20 mg/24H MOD 28ZPAK -,96,72
orl-rat TDLo:4 g/kg (6-15D preg):TER FCTOD7 22,147,84
orl-rat TDLo:4 g/kg (6-15D preg):REP FCTOD7 22,147,84

CONSENSUS REPORTS: Reported in EPA TSCA Inventory.

SAFETY PROFILE: An experimental teratogen. Other experimental reproductive effects. An eye irritant. Decomposes explosively at 165°C/33 mbar. When heated to decomposition it emits toxic fumes of Cl^- and NO_x. See also AROMATIC AMINES.

CEG625 CAS:6219-71-2 HR: 2
3-CHLORO-4-AMINOANILINE SULFATE
mf: $C_6H_5ClN_2{\cdot}H_2O_4S$ mw: 238.66

SYNS: 2-CHLORO-1,4-BENZENEDIAMINE SULFATE □ 2-CHLORO-p-PHENYLENEDIAMINE SULFATE □ C.I. 76066 □ C.I. OXIDATION BASE 13A □ 2-Cl-P-PD □ FOURRINE 81 □ FOURRINE SO □ NCI-C03316 □ RENAL SO

TOXICITY DATA WITH REFERENCE
orl-rat TDLo:41 g/kg/77W-C:ETA CRNGDP 1,495,80
orl-mus TDLo:194 g/kg/77W-C:ETA CRNGDP 1,495,80
orl-mus TD:12,150 g/kg/96W-C:ETA CRNGDP 1,495,80

CONSENSUS REPORTS: NCI Carcinogenesis Bioassay (Feed); Results Negative: mouse, rat NCITR* NCI-CG-TR-113,78

SAFETY PROFILE: Questionable carcinogen with experimental tumorigenic data. When heated to decomposition it emits toxic fumes of Cl^-, SO_x, and NO_x. See also AROMATIC AMINES and SULFATES.

CEG750 CAS:2457-76-3 HR: 2
2-CHLORO-4-AMINOBENZOIC ACID
mf: $C_7H_6ClNO_2$ mw: 171.59

PROP: A solid. Mp: 213°.

SYNS: 4-AMINO-2-CHLOROBENZOIC ACID □ USAF NB-1

TOXICITY DATA WITH REFERENCE
mmo-esc 500 mg/L JGMIAN 18,543,58
ipr-mus LDLo:500 mg/kg NTIS** AD277-689

CONSENSUS REPORTS: Reported in EPA TSCA Inventory.

SAFETY PROFILE: Moderately toxic by intraperitoneal route. Mutation data reported. When heated to decomposition it emits very toxic fumes of Cl^- and NO_x. See also AROMATIC AMINES.

CEG800 **CAS:121-50-6** **HR: 3**
4-CHLORO-3-AMINOBENZOTRIFLUORIDE
mf: $C_7H_5ClF_3N$ mw: 195.58

SYNS: 3-AMINO-4-CHLOROBENZOTRIFLUORIDE □ 3-AMINO-4-CHLORO-α-α-α-TRIFLUOROTOLUENE □ AZOENE FAST ORANGE RD SALT □ 2-CHLORO-5-(TRIFLUOROMETHYL)ANILINE □ 6-CHLORO-α-α-α-TRIFLUORO-m-TOLUIDINE □ C.I. 37050 □ DAITO ORANGE SALT RD □ DIAZO FAST ORANGE RD □ FAST ORANGE RD OIL □ FAST ORANGE RD SALT □ FAST ORANGE SALT RD □ FAST ORANGE SALT RDA □ FAST ORANGE SALT RDN □ HILTOSAL FAST ORANGE RD SALT □ ORANGE SALT NRD □ SANYO FAST ORANGE SALT RD □ m-TOLUIDINE, 6-CHLORO-α-α-α-TRIFLUORO- □ USAF MA-13

TOXICITY DATA WITH REFERENCE
ipr-mus LD50:100 mg/kg NTIS** AD277-689

CONSENSUS REPORTS: Reported in EPA TSCA Inventory.

SAFETY PROFILE: Poison by intraperitoneal route. When heated to decomposition it emits toxic vapors of NO_x, F^-, and Cl^-.

CEH000 **CAS:5730-85-8** **HR: 2**
3-CHLORO-4-AMINODIPHENYL
mf: $C_{12}H_{10}ClN$ mw: 203.68

SYN: 3-CHLOROBIPHENYLAMINE

TOXICITY DATA WITH REFERENCE
scu-rat TDLo:7000 mg/kg/W-I:ETA BMBUAQ 14,141,58

SAFETY PROFILE: Questionable carcinogen with experimental tumorigenic data. When heated to decomposition it emits very toxic fumes of Cl^- and NO_x.

CEH125 **CAS:101-79-1** **HR: 2**
4-CHLORO-4′-AMINODIPHENYL ETHER
mf: $C_{12}H_{10}ClNO$ mw: 219.68

SYNS: 4′-CHLORO-4-AMINOBIPHENYL ETHER □ p-(p-CHLOROPHENOXY)ANILINE □ 4-(4-CHLOROPHENOXY)ANILINE □ 4-(4-CHLOROPHENOXY)-BENZENAMINE (9CI)

TOXICITY DATA WITH REFERENCE
mma-sat 100 μg/plate CBINA8 44,133,83
orl-rat TDLo:38 g/kg/78W-C:CAR JEPTDQ 2(2),325,78
orl-mus TDLo:150 g/kg/78W-C:NEO JEPTDQ 2(2),325,78
orl-mus TD:302 g/kg/78W-C:ETA JEPTDQ 2(2),325,78

CONSENSUS REPORTS: Reported in EPA TSCA Inventory.

SAFETY PROFILE: Questionable carcinogen with experimental carcinogenic, neoplastigenic, and tumorigenic data. Mutation data reported. When heated to decomposition it emits toxic fumes of Cl^- and NO_x. See also ETHERS.

CEH250 **CAS:95-85-2** **HR: 3**
p-CHLORO-o-AMINOPHENOL
DOT: UN 2673
mf: C_6H_6ClNO mw: 143.58

PROP: Plates from H_2O. Mp: 140–141°.

SYN: 2-AMINO-4-CHLOROPHENOL (DOT)

TOXICITY DATA WITH REFERENCE
mma-sat 333 μg/plate EMMUEG 11(Suppl 12),1,88
orl-rat LD50:690 mg/kg GTPZAB 25(8),50,81
orl-mus LD50:1030 mg/kg GTPZAB 25(8),50,81

CONSENSUS REPORTS: Reported in EPA TSCA Inventory.

DOT CLASSIFICATION: 6.1; *Label:* Poison

SAFETY PROFILE: A poison. Moderately toxic by ingestion. Mutation data reported. When heated to decomposition it emits very toxic fumes of Cl^- and NO_x. See also AROMATIC AMINES and CHLORIDES.

CEH500 **CAS:2047-30-5** **HR: 3**
CHLOROAMITRIPTYLINE HYDROCHLORIDE
mf: $C_{20}H_{22}ClN{\cdot}ClH$ mw: 348.34

SYNS: CHLORPROHEPTADIEN □ CHLORPROHEPTADIENE HYDROCHLORIDE

TOXICITY DATA WITH REFERENCE
orl-mus LD50:210 mg/kg AIPTAK 144,481,63
ivn-mus LD50:43 mg/kg AIPTAK 144,481,63

SAFETY PROFILE: Poison by ingestion and intravenous routes. When heated to decomposition it emits very toxic fumes of Cl^- and NO_x.

CEH670 **CAS:95-51-2** **HR: 3**
2-CHLOROANILINE
mf: C_6H_6ClN mw: 127.58

PROP: Liquid. Mp: −1.94°, bp: 208.84°, d: 1.21 @ 20°/4°. Practically insol in water; sol in most organic solvents, also in acids.

SYNS: 1-AMINO-2-CHLOROBENZENE □ o-CHLORANILINE □ o-CHLOROANILINE □ o-CHLOROANILINE, liquid □ o-CHLOROANILINE, solid □ 2-CHLORO-BENZENAMINE (9CI) □ FAST YELLOW GC BASE

TOXICITY DATA WITH REFERENCE
dnr-esc 500 μg/L JJIND8 62,873,79
mmo-asn 200 mg/L CJMIAZ 16,369,70
orl-mus LD50:256 mg/kg GTPZAB 13(5),29,69
skn-cat LD50:222 mg/kg GTPZAB 13(5),29,69
scu-cat LDLo:310 mg/kg AHBAAM 110,12,33

CONSENSUS REPORTS: EPA Genetic Toxicology Program. Reported in EPA TSCA Inventory.

SAFETY PROFILE: Poison by skin contact, ingestion, and subcutaneous routes. Mutation data reported. When heated to decomposition it emits toxic fumes of Cl^- and NO_x. See also ANILINE DYES.

C

CEH675 **CAS:108-42-9** **HR: 3**
3-CHLOROANILINE
mf: C_6H_6ClN mw: 127.58

PROP: Liquid. Bp: 230.5°, mp: −10.4°, d: 1.2225 @ 15°/15°. Practically insol in water; sol in most common organic solvents.

SYNS: m-AMINOCHLOROBENZENE □ 1-AMINO-3-CHLOROBENZENE □ 3-CHLOORANILINEN (DUTCH) □ m-CHLORANILINE □ m-CHLOROANILINE □ 3-CHLOROANILINE (ITALIAN) □ m-CHLOROANILINE, liquid □ m-CHLOROANILINE, solid □ 3-CHLOROBENZENAMINE □ m-CHLOROPHENYLAMINE □ 3-CHLOROPHENYLAMINE □ FAST ORANGE GC BASE □ ORANGE GC BASE

TOXICITY DATA WITH REFERENCE
mmo-asn 200 mg/L CJMIAZ 16,369,70
orl-rat LD50:256 mg/kg GISAAA 31(12),6,66
skn-rat LD50:250 mg/kg 85GMAT-,34,82
orl-mus LD50:334 mg/kg GTPZAB 13(5),29,69
ihl-mus LC50:550 mg/m³/4H 85GMAT-,34,82
ivn-dog LDLo:50 mg/kg AEPPAE 244,387,63
skn-cat LD50:223 mg/kg GTPZAB 13(5),29,69
scu-cat LDLo:125 mg/kg AHBAAM 110,12,33
orl-gpg LD50:250 mg/kg GISAAA 31(12),6,66

CONSENSUS REPORTS: EPA Genetic Toxicology Program. Reported in EPA TSCA Inventory.

SAFETY PROFILE: Poison by ingestion, skin contact, subcutaneous, and intravenous routes. Mutation data reported. When heated to decomposition it emits toxic fumes of Cl^- and NO_x. See also ANILINE DYES.

CEH680 **CAS:106-47-8** **HR: 3**
4-CHLOROANILINE
mf: C_6H_6ClN mw: 127.58

PROP: Orthorhombic crystals from alc or petr ether; needles from toluene. Mp: 70–71°, bp: 232°, d: 1.169. Sol in hot water; freely sol in alc, ether, acetone, carbon disulfide.

SYNS: 1-AMINO-4-CHLOROBENZENE □ 4-CHLORANILIN (CZECH) □ p-CHLORANILINE □ p-CHLOROANILINE □ p-CHLOROANILINE, liquid □ p-CHLOROANILINE, solid □ 4-CHLOROBENZENAMINE □ 4-CHLOROBENZENEAMINE □ 4-CHLOROPHENYLAMINE □ NCI-C02039 □ RCRA WASTE NUMBER P024

TOXICITY DATA WITH REFERENCE
skn-rbt 500 mg/24H MLD 28ZPAK -,96,72
eye-rbt 250 µg/24H SEV 28ZPAK -,96,72
mma-sat 100 µg/plate ENMUDM 7(Suppl 5),1,85
otr-rat:emb 14,500 ng/plate JJATDK 1,190,81
dns-rat:lvr 5 mg/L MUREAV 97,359,82
orl-rat TDLo:9270 mg/kg/2Y-C:CAR FCTOD7 29,119,91
orl-mus TDLo:5150 mg/kg/2Y-C:CAR FCTOD7 29,119,91
orl-rat TD:14 g/kg/78W-C:ETA NCITR* NCI-TR-189,79
orl-rat TD:18,200 mg/kg/2Y-C:NEO FCTOD7 25,619,87
orl-mus TD:328 g/kg/78W-C:ETA NCITR* NCI-TR-189,79
orl-rat LD50:310 mg/kg AIHAAP 23,95,62
skn-rat LD50:3200 mg/kg AGGHAR 15,447,57
ipr-rat LD50:420 mg/kg AGGHAR 15,447,57
orl-mus LD50:100 mg/kg NCILB* NCI-E-C-72-3252,73
ihl-mus LC12:250 mg/m³/6H 85GMAT -,34,82
ivn-dog LDLo:100 mg/kg AEPPAE 244,387,63
skn-cat LD50:239 mg/kg GTPZAB 13(5),29,69
scu-cat LDLo:125 mg/kg AHBAAM 110,12,33
skn-rbt LD50:360 mg/kg AIHAAP 23,95,62
orl-gpg LD50:350 mg/kg 85GMAT -,34,82

CONSENSUS REPORTS: EPA Genetic Toxicology Program. Reported in EPA TSCA Inventory.

DFG MAK: Confirmed Carcinogen

SAFETY PROFILE: Confirmed carcinogen with experimental neoplastigenic and tumorigenic data. Poison by ingestion, inhalation, skin contact, subcutaneous, and intravenous routes. Moderately toxic by inhalation and intraperitoneal routes. A skin and severe eye irritant. Mutation data reported. When heated to decomposition it emits toxic fumes of Cl^- and NO_x. See also ANILINE DYES.

CEH700 **CAS:10238-21-8** **HR: 1**
1-((p-(2-(CHLORO-o-ANISAMIDO)ETHYL)PHENYL) SULFONYL)-3-CYCLOHEXYL UREA
mf: $C_{23}H_{28}ClN_3O_5S$ mw: 494.05

PROP: Crystals from EtOH/DMF or MeOH. Mp: 172–174°.

SYNS: 5-CHLORO-N-(2-(4-((((CYCLOHEXYLAMINO) CARBONYL) AMINO)SULFONYL)PHENYL)ETHYL)-2-METHOXYBENZAMIDE □ N-(4-(2-(5-CHLORO-2-METHOXYBENZAMIDO)ETHYL)PHENYLSULFONYL)-N′-CYCLOHEXYLUREA □ DAONIL □ DIABETA □ EUGLUCAN □ EUGLUCON □ EUGLUCON 5 □ EUGLYKON □ GILEMAL □ GLIBENCLAMIDE □ GLYBENZCYCLAMIDE □ GLYBURIDE □ HB 419 □ HD 419 □ MANINIL □ MICRONASE □ U 26452 □ UR 606

TOXICITY DATA WITH REFERENCE
ipr-rat TDLo:4 mg/kg (5-9D preg):REP ARZNAD 21,846,71
scu-rat TDLo:1800 mg/kg (9-14D preg):TER SEIJBO 17,31,77
scu-rat TDLo:180 mg/kg (female 9-14D post):TER SEIJBO 17,31,77
orl-rat TDLo:20 mg/kg (6-15D preg):TER OYYAA2 4,271,70
orl-wmn TDLo:147 µg/kg DICPBB 18,142,84
ipr-rat LD50:3750 mg/kg ARZNAD 19,1413,69
orl-mus LD50:3250 mg/kg ARZNAD 16,1640,66
ipr-mus LD50:5900 mg/kg OYYAA2 4,247,70
scu-mus LD50:20 g/kg OYYAA2 4,247,70

SAFETY PROFILE: Mildly toxic by ingestion, intraperitoneal, and subcutaneous routes. An experimental teratogen. Other experimental reproductive effects. When heated to decomposition it emits very toxic fumes of SO_x, NO_x, and Cl^-.

CEH750 **CAS:5345-54-0** **HR: 2**
3-CHLOROANISIDINE
mf: C_7H_8ClNO mw: 157.61

PROP: Needles. Mp: 62°.

SYNS: 3-CHLORO-4-METHOXY-BENZENAMINE (9CI) □ OCPA □ ORTHOCHLOROPARANISIDINE

TOXICITY DATA WITH REFERENCE
skn-rbt 500 mg/24H MLD TOERD9 2,77,79
eye-rbt 100 mg MLD TOERD9 2,77,79
orl-rat LD50:550 mg/kg TOERD9 2,77,79

ipr-rat LD50:510 mg/kg TOERD9 2,77,79
orl-mus LD50:650 mg/kg TOERD9 2,77,79
ipr-mus LD50:670 mg/kg TOERD9 2,77,79

SAFETY PROFILE: Moderately toxic by ingestion and intraperitoneal routes. A skin and eye irritant. When heated to decomposition it emits toxic fumes of Cl^- and NO_x. See also AROMATIC AMINES.

CEI000 CAS:82-44-0 HR: 3
1-CHLOROANTHRAQUINONE
mf: $C_{14}H_8ClO_2$ mw: 243.67

PROP: Yellow needles from EtOH. Mp: 162°.

SYNS: 1-CHLORANTHRACHINON (CZECH) □ 1-CHLORO-9,10-ANTHRACENEDIONE □ α-CHLOROANTHRAQUINONE □ 1-CHLORO-9,10-ANTHRAQUINONE □ α-MONOCHLOROANTHRAQUINONE

TOXICITY DATA WITH REFERENCE
eye-rbt 500 mg/24H MLD 85JCAE-,565,86
ivn-mus LD50:56 mg/kg CSLNX* NX#03287

CONSENSUS REPORTS: Reported in EPA TSCA Inventory.

SAFETY PROFILE: Poison by intravenous route. An eye irritant. When heated to decomposition it emits very toxic fumes of Cl^- and NO_x.

CEI100 CAS:131-09-9 HR: 1
2-CHLOROANTHRAQUINONE
mf: $C_{14}H_7ClO_2$ mw: 242.66

SYNS: 9,10-ANTHRACENEDIONE, 2-CHLORO- □ ANTHRAQUINONE, 2-CHLORO- □ 2-CHLORO-9,10-ANTHRACENEDIONE

TOXICITY DATA WITH REFERENCE
ipr-rat LD50:4310 mg/kg GISAAA 49(4),90,84

CONSENSUS REPORTS: Reported in EPA TSCA Inventory.

SAFETY PROFILE: Mildly toxic by intraperitoneal route. When heated to decomposition it emits toxic vapors of Cl^-.

CEI250 CAS:300-88-9 HR: 3
CHLOROARSENOL
mf: $C_7H_{14}AsClO_3{\cdot}H_3N$ mw: 273.62

SYNS: AMMONIUM CHLOROHEPTENE ARSONATE □ ARSION □ CHLORARSENOL □ (2-CHLORO-1-HEPTENYL)ARSONIC ACID MONOAMMONIUM SALT □ SOLARSON □ SOLASON

TOXICITY DATA WITH REFERENCE
cyt-hmn:hla 1000 ppm/6H IDZAAW 40,135,65
cyt-hmn:lng 1000 ppm/12H IDZAAW 40,135,65
scu-rbt LDLo:1 g/kg HBAMAK 4,1289,35
ivn-rbt LDLo:1 g/kg HBAMAK 4,1289,35

CONSENSUS REPORTS: Arsenic and its compounds are on the Community Right-To-Know List.

OSHA PEL: TWA 0.5 mg(As)/m^3
ACGIH TLV: TWA 0.2 mg(As)/m^3

SAFETY PROFILE: Arsenic compounds are generally poisons. Moderately toxic by subcutaneous and intravenous routes. Human mutation data reported. When heated to decomposition it emits very toxic fumes of As, Cl^-, NO_x, and NH_3. See also ARSENIC COMPOUNDS.

CEI325 CAS:25167-31-1 HR: 3
1-CHLOROAZIRIDINE
mf: C_2H_4ClN mw: 77.51

$CH_2NCl\,CH_2$ (ring)

SYN: CHLOROETHYLENIMINE

TOXICITY DATA WITH REFERENCE
mic-uns 4000 ppm/24H-C DKBSAS 183,668,68

SAFETY PROFILE: Mutation data reported. A dangerous storage hazard, it may explode at room temperature. When heated to decomposition it emits toxic fumes of Cl^- and NO_x. See also various aziridine compounds.

CEI500 CAS:89-98-5 HR: 3
o-CHLOROBENZALDEHYDE
mf: C_7H_5ClO mw: 140.57

PROP: Liquid or needles with strong odor. Fp: 11°, bp: 213–214°.

SYNS: o-CHLOORBENZALDEHYDE (DUTCH) □ 2-CHLOORBENZALDEHYDE (DUTCH) □ 2-CHLORBENZALDEHYD (GERMAN) □ 2-CHLOROBENZALDEHYDE □ 2-CLOROBENZALDEIDE (ITALIAN) □ o-CHLOROBENZENECARBOXALDEHYDE □ USAF M-7

TOXICITY DATA WITH REFERENCE
ipr-mus LD50:10 mg/kg NTIS** AD277-689
ivn-rbt LD50:8500 μg/kg PJPPAA 31,563,79

CONSENSUS REPORTS: Reported in EPA TSCA Inventory.

SAFETY PROFILE: Poison by intraperitoneal and intravenous routes. When heated to decomposition it emits toxic fumes of Cl^-. See also ALDEHYDES and CHLORIDES.

CEI600 CAS:104-88-1 HR: 2
p-CHLOROBENZALDEHYDE
mf: C_7H_5ClO mw: 140.57

SYNS: BENZALDEHYDE, p-CHLORO- □ 4-CHLOROBENZALDEHYDE □ p-CHLOROBENZENECARBOXALDEHYDE

TOXICITY DATA WITH REFERENCE
orl-rat LD50:1575 mg/kg GTPZAB 31(2),50,87
orl-mus LD50:1400 mg/kg GTPZAB 31(2),50,87

CONSENSUS REPORTS: Reported in EPA TSCA Inventory.

SAFETY PROFILE: Moderately toxic by ingestion. When heated to decomposition it emits toxic vapors of Cl^-.

CEJ000 CAS:20268-52-4 HR: 2
10-CHLORO-1,2-BENZANTHRACENE
mf: $C_{18}H_{11}Cl$ mw: 262.74

PROP: Yellow plates from C_6H_6/EtOH; crystals from Me_2CO. Mp: 146–148°.

SYN: 7-CHLOROBENZ(a)ANTHRACENE

TOXICITY DATA WITH REFERENCE
mma-sat 20 µg/plate MUREAV 155,91,85
skn-mus TDLo: 530 mg/kg/22W-I: ETA COREAF 226,1852,48

SAFETY PROFILE: Questionable carcinogen with experimental tumorigenic data. Mutation data reported. When heated to decomposition it emits toxic fumes of Cl^-.

CEJ125 CAS:108-90-7 HR: 3
CHLOROBENZENE
DOT: UN 1134
mf: C_6H_5Cl mw: 112.56

PROP: Clear, colorless liquid with faint odor. Bp: 131.7°, lel: 1.3%, uel: 7.1%, @ 150°, mp: −45°, flash p: 85°F (CC), d: 1.11 @ 20°/4°, autoign temp: 1180°F, vap press: 10 mm @ 22.2°, vap d: 3.88.

SYNS: BENZENE CHLORIDE □ CHLOORBENZEEN (DUTCH) □ CHLORBENZENE □ CHLORBENZOL □ CHLOROBENZEN (POLISH) □ CHLOROBENZOL (DOT) □ CLOROBENZENE (ITALIAN) □ MCB □ MONOCHLOORBENZEEN (DUTCH) □ MONOCHLORBENZENE □ MONOCHLORBENZOL (GERMAN) □ MONOCHLOROBENZENE □ MONOCLOROBENZENE (ITALIAN) □ NCI-C54886 □ PHENYL CHLORIDE □ RCRA WASTE NUMBER U037

TOXICITY DATA WITH REFERENCE
mrc-smc 1000 ppm NTIS** PB84-138973
ihl-rbt TCLo: 590 ppm/6H (6-18D preg): REP TXAPA9 76,365,84
ihl-rat TCLo: 210 ppm/6H (6-15D preg): TER TXAPA9 76,365,84
ihl-rat TCLo: 75 ppm/6H (6-15D preg): TER TXAPA9 76,365,84
orl-rat LD50: 2290 mg/kg 38MKAJ 2B,3603,81
ipr-rat LD50: 1655 mg/kg FAATDF 5,105,85
scu-rat LDLo: 7000 mg/kg RMSRA6 16,449,1896
orl-mus LD50: 2300 mg/kg 85GMAT -,34,82
ihl-mus LCLo: 15 g/m³ GISAAA 20(8),19,55
ipr-mus LD50: 515 mg/kg PHMCAA 10,172,68
orl-rbt LD50: 2830 mg/kg 14CYAT 2,1394,63
ipr-gpg LDLo: 4100 mg/kg RMSRA6 16,449,1896

CONSENSUS REPORTS: NTP Carcinogenesis Studies (gavage); Some Evidence: rat NTPTR* NTP-TR-261,85; No Evidence: mouse NTPTR* NTP-TR-261,85. Reported in EPA TSCA Inventory. Community Right-To-Know List.

OSHA PEL: TWA 75 ppm
ACGIH TLV: TWA 10 ppm
DFG MAK: 50 ppm (230 mg/m³)
DOT CLASSIFICATION: 3; *Label:* Flammable Liquid

SAFETY PROFILE: Moderately toxic by ingestion and intraperitoneal routes. Experimental teratogenic and reproductive effects. Mutation data reported. Strong narcotic with slight irritant qualities. Dichlorobenzols are strongly narcotic. Little is known of the effects of repeated exposures at lower concentrations, but it may cause kidney and liver damage. The industrial illnesses reported may possibly be due to nitrobenzol. Dangerous fire hazard when exposed to heat or flame. Moderate explosion hazard when exposed to heat or flame. Potentially explosive reaction with powdered sodium or phosphorus trichloride + sodium. Violent reaction with $AgClO_4$ or dimethyl sulfoxide. Reacts vigorously with oxidizers. See also CHLORINATED HYDROCARBONS, AROMATIC. To fight fire, use foam, CO_2, dry chemical, water to blanket fire. Associated with EPA Superfund sites.

For occupational chemical analysis use NIOSH: Hydrocarbons, halogenated, 1003.

CEJ250 CAS:17333-84-5 HR: 3
m-CHLOROBENZENEDIAZONIUM SALTS
mf: $C_6H_4ClN_2X$

SAFETY PROFILE: Violent explosion on contact with sodium disulfide. Reaction with potassium O,O-diphenylphosphorodithioates; o-alkyldithiocarbonate (xanthate) solutions; thiophenoxide solutions (e.g., sodium-2-chlorothiophenoxide; potassium thiophenoxide); various sulfides or derivatives (e.g., hydrogen sulfide; sodium hydrogen sulfide; sodium mono-, di- or polysulfides) forms highly explosive products. When heated to decomposition it emits toxic fumes of Cl^- and NO_x.

CEJ500 CAS:17333-83-4 HR: 3
o-CHLOROBENZENEDIAZONIUM SALTS
mf: $C_6H_4ClN_2X$

SYN: 2-CHLOROBENZENEDIAZONIUM SALTS

SAFETY PROFILE: Reaction with potassium O,O-diphenylphosphorodithioates; o-alkyldithiocarbonate (xanthate) solutions; thiophenoxide solutions (e.g., sodium-2-chlorothiophenoxide; potassium thiophenoxide); various sulfides or derivatives (e.g., hydrogen sulfide; sodium hydrogen sulfide; sodium mono-, di- or polysulfides) forms highly explosive products. Incompatible with potassium-2-chlorothiophenolate. When heated to decomposition it emits toxic fumes of Cl^- and NO_x.

CEJ600 CAS:100-03-8 HR: 3
p-CHLOROBENZENESULFINIC ACID
mf: $C_6H_5ClO_2S$ mw: 176.62

SYN: BENZENESULFINIC ACID, p-CHLORO-

TOXICITY DATA WITH REFERENCE
orl-rat LD50: >500 mg/kg JAFCAU 25,501,77
ivn-mus LD50: 56 mg/kg CSLNX* NX#04978

CONSENSUS REPORTS: Reported in EPA TSCA Inventory.

SAFETY PROFILE: Poison by intravenous route. Moderately toxic by ingestion. When heated to decomposition it emits toxic vapors of SO_x and Cl^-.

CEK000 CAS:98-64-6 HR: 3
p-CHLOROBENZENESULFONAMIDE
mf: $C_6H_6ClNO_2S$ mw: 191.64

PROP: Leaflets. Mp: 143–144°.

SYN: USAF MA-3

TOXICITY DATA WITH REFERENCE
orl-rat LDLo:500 mg/kg NCNSA6 5,20,53
ipr-mus LD50:200 mg/kg NTIS** AD277-689

CONSENSUS REPORTS: Reported in EPA TSCA Inventory.

SAFETY PROFILE: Poison by intraperitoneal route. Moderately toxic by ingestion. When heated to decomposition it emits very toxic fumes of SO_x, NO_x, and Cl^-.

CEK250 CAS:5138-90-9 HR: 1
p-CHLOROBENZENESULFONIC ACID, SODIUM SALT
mf: $C_6H_4ClO_3S{\cdot}Na$ mw: 214.60

SYN: p-CHLORBENZENSULFONAN SODNY (CZECH)

TOXICITY DATA WITH REFERENCE
skn-rbt 500 mg/24H MLD 28ZPAK -,178,72
eye-rbt 100 mg/24H MOD 28ZPAK -,178,72
orl-rat LD50:10 g/kg 28ZPAK -,178,72

CONSENSUS REPORTS: Reported in EPA TSCA Inventory.

SAFETY PROFILE: Mildly toxic by ingestion. A skin and eye irritant. When heated to decomposition it emits very toxic fumes of SO_x, Na_2O, and Cl^-. See also SULFONATES.

CEK375 CAS:98-60-2 HR: 3
p-CHLOROBENZENESULFONYL CHLORIDE
mf: $C_6H_4Cl_2O_2S$ mw: 211.06

PROP: Prisms. Mp: 53°, bp: 141° @ 15 mm.

SYNS: p-CHLORBENZENESULFOCHLORID (CZECH) □ CHLORID KYSELINY-p-CHLORBENSULFONOVE (CZECH)

TOXICITY DATA WITH REFERENCE
skn-rbt 20 mg/24H MOD 85JCAE-,1071,86
eye-rbt 50 μg/24H SEV 28ZPAK -,198,72
orl-rat LD50:4250 mg/kg 28ZPAK -,198,72
ipr-mus LDLo:250 mg/kg CBCCT* 7,774,55

CONSENSUS REPORTS: Reported in EPA TSCA Inventory.

SAFETY PROFILE: Poison by intraperitoneal route. Mildly toxic by ingestion. A severe eye and moderate skin irritant. When heated to decomposition it emits toxic fumes of Cl^- and SO_x. See also CHLORIDES.

CEK425 CAS:106-54-7 HR: 3
4-CHLOROBENZENETHIOL
mf: C_6H_5ClS mw: 144.62

PROP: Prisms or plates from EtOH. Mp: 54°, bp: 205–207°.

SYNS: p-CHLORO-PHENYL MERCAPTAN □ p-CHLOROTHIOPHENOL □ 4-CHLOROTHIOPHENOL □ p-CHLORTHIOFENOL (CZECH)

TOXICITY DATA WITH REFERENCE
skn-rbt 20 mg/24H MOD 85JCAE-,989,86
eye-rbt 50 μg/24H SEV 28ZPAK -,167,72
skn-mus TDLo:8000 mg/kg/20W-I:ETA CNREA8 19,413,59
ipr-mus LD50:75 mg/kg NTIS** AD691-490

CONSENSUS REPORTS: Chlorophenol compounds are on the Community Right-To-Know List. Reported in EPA TSCA Inventory.

SAFETY PROFILE: Poison by intraperitoneal route. A severe eye and moderate skin irritant. Questionable carcinogen with experimental tumorigenic data by skin contact. When heated to decomposition it emits toxic fumes of Cl^- and SO_x. See also CHLOROPHENOLS and MERCAPTANS

CEK500 CAS:21248-01-1 HR: 2
6-CHLOROBENZENO(a)PYRENE
mf: $C_{20}H_{11}Cl$ mw: 286.69

PROP: Crystals from MeOH/C_6H_6. Mp: 209–210°.

TOXICITY DATA WITH REFERENCE
unk-mus TDLo:80 mg/kg/8D-I:ETA BEBMAE 88(11),592,79

SAFETY PROFILE: Questionable carcinogen with experimental tumorigenic data. When heated to decomposition it emits toxic fumes of Cl^-. See also CHLORINATED HYDROCARBONS, AROMATIC; and PYRENE.

CEK875 CAS:55981-23-2 HR: 3
N[1]-(4′-CHLOROBENZHYDRYL)-N[4]-SPIROMORPHOLINO-PIPERAZINIUM CHLORIDE HYDROCHLORIDE
mf: $C_{21}H_{26}ClN_2O{\cdot}Cl{\cdot}ClH$ mw: 429.85

SYNS: 9-((4-CHLOROPHENYL)PHENYLMETHYL)3-OXA-9-AZA-6-AZONIASPIRO(5.5)UNDECANE CHLORIDE HCl □ CRC 7001

TOXICITY DATA WITH REFERENCE
ipr-rat LD50:40,500 μg/kg APJUA8 28,21,78
ivn-rat LD50:2400 μg/kg APJUA8 28,21,78
ims-rat LD50:320 mg/kg APJUA8 28,21,78
ipr-mus LD50:51,300 μg/kg APJUA8 28,21,78
ims-mus LD50:281 mg/kg APJUA8 28,21,78
orl-dog LD50:315 mg/kg APJUA8 28,21,78

SAFETY PROFILE: Poison by ingestion, intramuscular, intravenous, and intraperitoneal routes. When heated to decomposition it emits toxic fumes of NO_x and Cl^-.

CEL000 CAS:32226-65-6 HR: 2
2-CHLOROBENZO(e)(1)BENZOTHIOPYRANO(4,3-b)INDOLE
mf: $C_{19}H_{10}ClNS$ mw: 319.81

TOXICITY DATA WITH REFERENCE
scu-mus TDLo:378 mg/kg/27W-I:NEO JNCIAM 46,1257,71

SAFETY PROFILE: Questionable carcinogen with experimental neoplastigenic data. When heated to decomposition it emits very toxic fumes of Cl^-, NO_x, and SO_x.

CEL250 CAS:118-91-2 HR: 2
2-CHLOROBENZOIC ACID
mf: $C_7H_5ClO_2$ mw: 156.57

PROP: Crystals from toluene. Mp: 142°. Sol in H_2O, EtOH, and Et_2O.

SYNS: 2-CBA □ o-CHLOROBENZOIC ACID □ KYSELINA o-CHLORBENZOOVA (CZECH)

TOXICITY DATA WITH REFERENCE
skn-rbt 500 mg/24H MLD 28ZPAK -,91,72
eye-rbt 20 mg/24H MOD 28ZPAK -,91,72
orl-rat LD50:6460 mg/kg 28ZPAK -,91,72
ipr-rat LD50:2300 mg/kg BCFAAI 112,53,73

CONSENSUS REPORTS: Reported in EPA TSCA Inventory.

SAFETY PROFILE: Moderately toxic by intraperitoneal route. Mildly toxic by ingestion. An eye and skin irritant. When heated to decomposition it emits toxic fumes of Cl^-.

CEL290 CAS:535-80-8 HR: 2
3-CHLOROBENZOIC ACID
mf: $C_7H_5ClO_2$ mw: 156.57

SYNS: ACIDO m-CLOROBENZOICO □ BENZOIC ACID, 3-CHLORO- □ m-CHLOROBENZOIC ACID □ BENZOIC ACID, m-CHLORO-

TOXICITY DATA WITH REFERENCE
ipr-rat LD50:750 mg/kg BCFAAI 112,53,73

CONSENSUS REPORTS: Reported in EPA TSCA Inventory.

SAFETY PROFILE: Moderately toxic by intraperitoneal route. When heated to decomposition it emits toxic vapors of Cl^-.

CEL300 CAS:74-11-3 HR: 2
4-CHLOROBENZOIC ACID
mf: $C_7H_5ClO_2$ mw: 156.57

SYNS: ACIDO p-CLOROBENZOICO □ BENZOIC ACID, p-CHLORO- □ BENZOIC ACID, 4-CHLORO-(9CI) □ p-CARBOXYCHLOROBENZENE □ p-CHLORBENZOIC ACID □ p-CHLOROBENZOIC ACID □ CHLORODRACYLIC ACID

TOXICITY DATA WITH REFERENCE
orl-rat LD50:1170 mg/kg JAFCAU 21,794,73
ipr-rat LD50:1000 mg/kg BCFAAI 112,53,73
orl-mus LD50:1170 mg/kg JAFCAU 23,724,75

CONSENSUS REPORTS: Reported in EPA TSCA Inventory.

SAFETY PROFILE: Moderately toxic by ingestion and intraperitoneal routes. When heated to decomposition it emits toxic vapors of Cl^-.

CEL500 CAS:7250-60-4 HR: 3
o-CHLOROBENZOIC ACID NICKEL(II) SALT
mf: $C_{14}H_8Cl_2O_4 \cdot Ni$ mw: 369.83

TOXICITY DATA WITH REFERENCE
ipr-mus LDLo:125 mg/kg CBCCT* 4,317,52

CONSENSUS REPORTS: Nickel and its compounds are on the Community Right-To-Know List.

NIOSH REL: (Inorganic Nickel) TWA 0.015 mg(Ni)/m³

SAFETY PROFILE: Poison by intraperitoneal route. See also NICKEL COMPOUNDS. When heated to decomposition it emits toxic fumes of Cl^-.

CEM000 CAS:873-32-5 HR: 3
o-CHLOROBENZONITRILE
mf: C_7H_4ClN mw: 137.57

PROP: Crystals. Mp: 42–43°, bp: 232°.

SYNS: o-CHLORBENZONITRIL (CZECH) □ NITRIL KYSELINY-o-CHLORBENZOOVE (CZECH)

TOXICITY DATA WITH REFERENCE
eye-rbt 100 mg/24H MOD 28ZPAK -,160,72
orl-mus LD50:>300 mg/kg JMCMAR 21,906,78
ipr-mus LD50:150 mg/kg NTIS** AD691-490

CONSENSUS REPORTS: Reported in EPA TSCA Inventory. Cyanide and its compounds are on the Community Right-To-Know List.

SAFETY PROFILE: Poison by intraperitoneal route. Moderately toxic by ingestion. An eye irritant. When heated to decomposition or on contact with water, steam, acid, or acid fumes it emits toxic fumes of Cl^-, and CN^-. See also NITRILES.

CEM250 CAS:623-03-0 HR: 3
p-CHLOROBENZONITRILE
mf: C_7H_4ClN mw: 137.57

PROP: Crystals from EtOH (aq). Mp: 94–96°, bp: 223°.

SYN: NITRIL KYSELINY p-CHLORBENZOOVE (CZECH)

TOXICITY DATA WITH REFERENCE
eye-rbt 100 mg/24H MOD 28ZPAK -,161,72
orl-mus LD50:>300 mg/kg JMCMAR 21,906,78
ipr-mus LD50:150 mg/kg NTIS** AD691-490

CONSENSUS REPORTS: Reported in EPA TSCA Inventory. Cyanide and its compounds are on the Community Right-To-Know List.

SAFETY PROFILE: Poison by intraperitoneal route. Moderately toxic by ingestion. An eye irritant. When heated to decomposition it emits very toxic fumes of NO_x, Cl^-, and CN^-. See also NITRILES.

CEM500 CAS:615-20-3 HR: 3
2-CHLOROBENZOTHIAZOLE
mf: C_7H_4C1NS mw: 169.63

SYN: USAF EK-2784

TOXICITY DATA WITH REFERENCE
orl-rat LDLo:250 mg/kg NCNSA6 5,24,53
ipr-mus LD50:200 mg/kg NTIS** AD277-689

CONSENSUS REPORTS: Reported in EPA TSCA Inventory.

SAFETY PROFILE: Poison by ingestion and intraperito-

neal routes. When heated to decomposition it emits very toxic fumes of SO_x, NO_x, and Cl^-.

CEM625 **HR: 3**
1-CHLOROBENZOTRIAZOL
mf: $C_6H_4ClN_3$ mw: 153.57

$C_6H_4N(Cl)N{=}N$

SAFETY PROFILE: May ignite spontaneously. When heated to decomposition it emits toxic fumes of Cl^- and NO_x.

CEM825 **CAS:98-56-6** **HR: 1**
p-CHLOROBENZOTRIFLUORIDE
mf: $C_7H_4ClF_3$ mw: 180.56

PROP: Bp: 135–136°.

SYNS: (p-CHLOROPHENYL)TRIFLUOROMETHANE □ p-CHLOROTRIFLUOROMETHYLBENZENE □ 4-CHLOROTRIFLUOROMETHYLBENZENE □ 1-CHLORO-4-(TRIMETHYL)-BENZENE (9CI) □ α,α,α-TRIFLUORO-4-CHLOROTOLUENE □ p-(TRIFLUOROMETHYL)CHLOROBENZENE □ p-TRIFLUOROMETHYLPHENYL CHLORIDE

TOXICITY DATA WITH REFERENCE
dns-hmn:emb 1 g/L AISSAW 18,123,82
orl-rat LD50:13 g/kg GTPZAB 28(5),49,84
ihl-rat LC50:22 g/m³ GTPZAB 28(5),49,84
orl-mus LD50:11,500 mg/kg GTPZAB 28(5),49,84
ihl-mus LC50:20 g/m³ GTPZAB 28(5),49,84

CONSENSUS REPORTS: Reported in EPA TSCA Inventory.

SAFETY PROFILE: Mildly toxic by ingestion and inhalation. Human mutation data reported. Flammable. Strongly exothermic reaction with sodium dimethylsulfinate. When heated to decomposition it emits toxic fumes of F^- and Cl^-. See also CHLORINATED HYDROCARBONS, AROMATIC; and FLUORIDES.

CEM850 **CAS:615-18-9** **HR: 1**
2-CHLOROBENZOXAZOLE
mf: C_7H_4ClNO mw: 153.57

SYN: BENZOXAZOLE, 2-CHLORO-

TOXICITY DATA WITH REFERENCE
orl-mus LD50:2 g/kg MDCHAG 4(1),337,64

CONSENSUS REPORTS: Reported in EPA TSCA Inventory.

SAFETY PROFILE: Slightly toxic by ingestion. When heated to decomposition it emits toxic vapors of NO_x and Cl^-.

CEO100 **CAS:14415-66-8** **HR: 3**
2-(p-CHLOROBENZOYL)-1-(2-MORPHOLINOETHYL)PYRROLE MONOHYDROCHLORIDE
mf: $C_{17}H_{19}ClN_2O_2{\cdot}ClH$ mw: 355.29

SYN: KETONE, p-CHLOROPHENYL 1-(2-MORPHOLINOETHYL)PYRROL-2-YL, HYDROCHLORIDE

TOXICITY DATA WITH REFERENCE
ivn-mus LD50:140 mg/kg CHTPBA 1,127,66

DOT CLASSIFICATION: 3; *Label:* Flammable Liquid

SAFETY PROFILE: A poison by intravenous route. A flammable liquid. When heated to decomposition it emits toxic vapors of NO_x, HCl, and Cl^-.

CEP000 **CAS:103-17-3** **HR: 2**
p-CHLOROBENZYL-p-CHLOROPHENYL SULFIDE
mf: $C_{13}H_{10}Cl_2S$ mw: 269.19

PROP: Crystals, almond-like odor. Mp: 75–76°, d: 1.4210 @ 25°/4°, vap press: 1.21 × 10^{-5} mm @ 30°. Insol in water, sol in most organic solvents.

SYNS: CHLOORBENZIDE (DUTCH) □ (4-CHLOOR-BENZYL)-(4-CHLOOR-FENYL)-SULFIDE (DUTCH) □ CHLORBENSID (GERMAN) □ CHLORBENSIDE □ CHLORBENXIDE □ CHLORBENZIDE □ (4-CHLOR-BENZYL)-(4-CHLOR-PHENYL)-SULFID (GERMAN) □ p-CHLOROBENZYL-p-CHLOROPHENYL SULPHIDE □ 4-CHLOROBENZYL-4-CHLOROPHENYL SULPHIDE □ 1-CHLORO-4-(((4-CHLOROPHENYL)METHYL)THIO)BENZENE □ CHLOROCIDE □ CHLOROPARACIDE □ 4-CHLOROPHENYL-4′-CHLOROBENZYL SULFIDE □ CHLOROSULFACIDE □ CHLORPARACIDE □ CHLORSULPHACIDE □ (4-CLORO-BENZIL)-(4-CLORO-FENIL) SOLFURO (ITALIAN) □ p,p′-DICHLORODIPHENYL SULFIDE □ ENT 20,696 □ HRS 860 □ METOX □ MITOX □ RD 2195 □ SULFURE de 4-CHLOROBENZYLE et de 4-CHLOROPHENYLE (FRENCH)

TOXICITY DATA WITH REFERENCE
orl-rat LD50:2000 mg/kg WRPCA2 9,119,70
unr-mus LD50:3000 mg/kg 30ZDA9 -,242,71

SAFETY PROFILE: Moderately toxic by ingestion and possibly other routes. Has caused liver and kidney injury and skin irritation in experimental animals. When heated to decomposition it emits toxic fumes of Cl^- and SO_x. See also CHLORIDES and SULFIDES.

CEP300 **CAS:140-53-4** **HR: 3**
4-CHLOROBENZYL CYANIDE
mf: C_8H_6ClN mw: 151.60

SYNS: ACETONITRILE, (p-CHLOROPHENYL)- □ BENZENEACETONITRILE, 4-CHLORO- □ 4-CHLOR-BENZYL-CYANID □ 4-CHLOROBENZENEACETONITRILE □ p-CHLOROBENZYL CYANIDE □ p-CHLOROPHENYLACETONITRILE □ (4-CHLOROPHENYL)ACETONITRILE □ 2-(4-CHLOROPHENYL)ACETONITRILE

TOXICITY DATA WITH REFERENCE
ipr-mus LD50:27 mg/kg PCBPBS 2,95,72
ivn-mus LD50:56 mg/kg CSLNX* NX#07883

CONSENSUS REPORTS: Reported in EPA TSCA Inventory.

SAFETY PROFILE: Poison by intravenous and intraperitoneal routes. When heated to decomposition it emits toxic vapors of NO_x and Cl^-.

CEP675 **CAS:13835-15-9** **HR: 3**
4-(p-CHLOROBENZYL)-2-((2-DIMETHYLAMINO)ETHYL)-1(2H)-PHTHALAZINONE HYDROCHLORIDE
mf: $C_{19}H_{20}ClN_3O{\cdot}ClH$ mw: 378.33

SYN: HL 2197

C

TOXICITY DATA WITH REFERENCE
orl-mus LD50:300 mg/kg ARZNAD 8,219,58
ipr-mus LD50:157 mg/kg ARZNAD 7,678,57
scu-mus LD50:100 mg/kg ARZNAD 8,219,58
ivn-mus LD50:75 mg/kg ARZNAD 8,219,58

SAFETY PROFILE: Poison by ingestion, subcutaneous, intravenous, and intraperitoneal routes. When heated to decomposition it emits toxic fumes of NO_x and Cl^-.

CEQ500 CAS:3309-77-1 HR: 3
p-CHLOROBENZYL-3-HYDROXYCROTONATE DIMETHYL PHOSPHATE
mf: $C_{13}H_{16}ClO_6P$ mw: 334.71

SYNS: BAS 4239 □ 1-(p-CHLOROBENZYLOXYCARBONYL)-1-PROPEN-2-YL-DIMETHYLPHOSPHATE □ DIMETHYL PHOSPHATE-3-HYDROXY-CROTONIC ACID, p-CHLOROBENZYL ESTER □ ENT 24716 □ PHOSPHORIC ACID, DIMETHYL ESTER with p-CHLOROBENZYL-3-HYDROXYCROTONATE □ SD 4,239 □ SHELL SD 4,239

TOXICITY DATA WITH REFERENCE
orl-rat LD50:139 mg/kg 28ZEAL 4,98,69
orl-mus LD50:212 mg/kg 28ZEAL 4,98,69

SAFETY PROFILE: Poison by ingestion. When heated to decomposition it emits very toxic fumes of PO_x and Cl^-. See also ESTERS and PHOSPHATES.

CEQ600 CAS:2698-41-1 HR: 3
o-CHLOROBENZYLIDENE MALONONITRILE
mf: $C_{10}H_5ClN_2$ mw: 188.62

PROP: White crystals. Mp: 95°, bp: 313°.

SYNS: o-CHLOROBENZAL MALONONITRILE □ 2-CHLOROBENZAL MALONONITRILE □ o-CHLOROBENZYLIDENE MALONITRILE □ 2-CHLOROBENZYLIDENE MALONONITRILE □ 2-CHLOROBMN □ CS □ β,β-DICYANO-o-CHLOROSTYRENE □ NCI-C55118 □ PROPANEDINITRILE((2-CHLOROPHENYL)METHYLENE) □ USAF KF-11

TOXICITY DATA WITH REFERENCE
skn-hmn 10 mg/1H MLD BJDEAZ 90,657,74
eye-man 5 mg/m³/20S SEV MMEDA9 134,219,69
eye-man 624 ng APTOA6 35,412,74
skn-rat 12%/6H open MLD ARTODN 40,75,78
skn-rbt 12%/6H open MLD ARTODN 40,75,78
eye-rbt 5 mg TXAPA9 4,656,62
eye-rbt 1 mg MLD TXAPA9 17,295,70
eye-rbt 1150 ng APTOA6 35,412,74
skn-gpg 12%/6H open MLD ARTODN 40,75,78
eye-gpg 429 ng APTOA6 35,412,74
mmo-sat 100 μg/plate ARTODN 49,15,81
mma-sat 100 μg/plate ARTODN 49,15,81
ihl-rat TCLo:6 mg/m³/5M (female 6-15D post):TER 37QLAZ -,79,77
ihl-hmn TCLo:1500 μg/m³/90M:EYE,PUL AEHLAU 6,366,63
orl-rat LD50:178 mg/kg AEHLAU 24,449,72
ihl-rat LCLo:1806 mg/m³/45M ARTODN 40,75,78
ipr-rat LD50:48 mg/kg TXAPA9 24,45,73
ivn-rat LD50:28 mg/kg ARTODN 40,75,78
orl-mus LD50:282 mg/kg NTIS** AD837-111
ihl-mus LCLo:2753 mg/m³/20M ARTODN 40,75,78
ipr-mus LD50:32,320 μg/kg TOLED5 8,73,81
ivn-mus LD50:47,700 μg/kg TXAPA9 25,111,73
orl-rbt LD50:143 mg/kg ARTODN 40,75,78
ihl-rbt LCLo:1802 mg/m³/10M ARTODN 40,75,78
ivn-rbt LDLo:8 mg/kg TXAPA9 4,656,62
orl-gpg LD50:212 mg/kg ARTODN 40,75,78
ihl-gpg LCLo:2326 mg/m³/10M ARTODN 40,75,78
ipr-gpg LD50:73 mg/kg ARTODN 40,75,78

CONSENSUS REPORTS: Reported in EPA TSCA Inventory. Cyanide and its compounds are on the Community Right-To-Know List.

OSHA PEL: CL 0.05 ppm (skin)
ACGIH TLV: CL 0.05 ppm (skin)

OSHA PEL: TWA 0.05 ppm

SAFETY PROFILE: Poison by ingestion, intraperitoneal, and intravenous routes. Moderately toxic by inhalation. Human systemic effects by inhalation: conjuntiva irritation, cough, and unspecified respiratory system effects. A human skin and eye irritant. Human exposure data suggest relatively low systematic toxicity, but intense irritation of eyes, skin, and mucous membranes. Mutation data reported. A teargas used for riot control. When heated to decomposition it emits very toxic fumes of Cl^-, NO_x, and CN^-. See also NITRILES.

CEQ625 CAS:50264-86-3 HR: 3
1-p-CHLOROBENZYL-1H-INDAZOLE-3-CARBOXYLIC ACID
mf: $C_{15}H_{11}ClN_2O_2$ mw: 286.73

SYNS: AF 1312/TS □ 1-(4-CHLOROBENZYL)-1H-INDAZOLE-3-CARBOXYLIC ACID □ 1-((4-CHLOROPHENYL)METHYL)-1H-INDOLE-3-CARBOXYLIC ACID

TOXICITY DATA WITH REFERENCE
spm-rat-orl 200 mg/kg EXMPA6 23,288,75
spm-mky-orl 14,400 mg/kg/24W-I EXMPA6 23,357,75
orl-rat TDLo:1 g/kg (female 6-15D post):TER ARTODN 5,197,82
orl-rat TDLo:19 g/kg (63D male):REP JRPFA4 50,159,77
orl-rat TDLo:3 g/kg (male 20D pre):REP EMPSAL 23,308,75
orl-rat TDLo:500 mg/kg (male 1D pre):REP EMPSAL 23,321,75
orl-rat TDLo:400 mg/kg (male 1D pre):REP JMCMAR 19,778,76
orl-rat LD50:1140 mg/kg DRFUD4 3,87,78
ipr-rat LD50:58 mg/kg EXMPA6 23,288,75

SAFETY PROFILE: Poison by intraperitoneal route. Moderately toxic by ingestion. An experimental teratogen. Other experimental reproductive effects. Mutation data reported. When heated to decomposition it emits toxic fumes of Cl^- and NO_x.

CEQ750 CAS:3694-45-9 HR: 3
4-CHLOROBENZYL ISOTHIOCYANATE
mf: C_8H_6ClNS mw: 183.66

SYN: ISOTHIOCYANIC ACID,-p-CHLOROBENZYL ESTER

TOXICITY DATA WITH REFERENCE
ipr-rat LDLo:100 mg/kg ARZNAD 16,870,66
ipr-mus LDLo:100 mg/kg ARZNAD 21,121,71
ivn-mus LD50:56 mg/kg CSLNX* NX#02757

SAFETY PROFILE: Poison by intraperitoneal and intravenous routes. See also ESTERS and THIOCYANATES. When heated to decomposition it emits very toxic fumes of SO_x, Cl^- and CN^-.

CEQ800 CAS:544-47-8 HR: 2
p-CHLOROBENZYLPSEUDOTHIURONIUM CHLORIDE
mf: $C_8H_9ClN_2S{\cdot}ClH$ mw: 237.16

SYNS: 2-(4-CHLOROBENZYL)-2-THIOPSEUDOUREA HYDROCHLORIDE □ PSEUDOUREA, 2-(p-CHLOROBENZYL)-2-THIO-, MONOHYDROCHLORIDE

TOXICITY DATA WITH REFERENCE
orl-mus LDLo:1600 mg/kg AECTCV 14,111,85

CONSENSUS REPORTS: Reported in EPA TSCA Inventory.

SAFETY PROFILE: Moderately toxic by ingestion. When heated to decomposition it emits toxic vapors of NO_x, SO_x, HCl, and Cl^-.

CER825 CAS:72985-56-9 HR: 3
2-CHLORO-1,1-BIS(FLUOROOXY)TRIFLUOROETHANE
mf: $C_2ClF_5O_2$ mw: 186.47

SAFETY PROFILE: An extremely unstable explosive. Upon decomposition it emits toxic fumes of F^- and Cl^-. See also EXPLOSIVES.

CES250 CAS:64037-53-2 HR: 3
1-CHLORO-3-BROMO-BUTENE-1
mf: C_4H_6BrCl mw: 169.46

TOXICITY DATA WITH REFERENCE
orl-rat LD50:74 mg/kg SCCUR* -,3,61
orl-mus LD50:56,500 μg/kg SCCUR* -,3,61

SAFETY PROFILE: Poison by ingestion. When heated to decomposition it emits very toxic fumes of Br^- and Cl^-. See also CHLORINATED HYDROCARBONS, ALIPHATIC; and BROMIDES.

CES500 CAS:107-04-0 HR: 3
1-CHLORO-2-BROMOETHANE
mf: C_2H_4BrCl mw: 143.42

PROP: Colorless, volatile liquid; sweet chloroform-like odor. Bp: 107°, fp: −16.7°, flash p: none, d: 1.74 @ 20°/4°, vap press: 40 mm @ 29.7°, vap d: 4.94.

SYNS: sym-CHLOROBROMOETHANE □ ETHYLENE CHLOROBROMIDE

TOXICITY DATA WITH REFERENCE
mmo-sat 1 mmol/L CRNGDP 2,499,81
mma-sat 10 μmol/plate EVHPAZ 21,79,77
dnr-esc 10 μL/plate EVHPAZ 21,79,77
dnd-mus-ipr 500 μmol/kg CRNGDP 4,1491,83
msc-ham:ovr 200 μmol/L MUREAV 90,183,81
orl-rat LD50:64 mg/kg 28ZEAL 4,73,69

CONSENSUS REPORTS: Reported in EPA TSCA Inventory. EPA Genetic Toxicology Program.

SAFETY PROFILE: Poison by ingestion and probably inhalation routes. An irritant to the skin, eyes, and mucous membranes. May cause injury to liver and kidneys. Mutation data reported. When heated to decomposition it emits toxic fumes of Cl^- and Br^-. See also CHLORINATED HYDROCARBONS, ALIPHATIC; and BROMIDES.

C

CES650 CAS:74-97-5 HR: 3
CHLOROBROMOMETHANE
DOT: UN 1887
mf: CH_2BrCl mw: 129.39

PROP: Clear, colorless liquid; sweet odor. Fp: −88°, bp: 67.8°, flash p: none, d: 1.930 @ 25°/25°, vap d: 4.46. Sol in organic solvs; insol in water.

SYNS: BROMOCHLOROMETHANE □ HALON 1011 □ METHYLENE CHLOROBROMIDE □ MIL-B-4394-B □ MONO-CHLORO-MONO-BROMO-METHANE

TOXICITY DATA WITH REFERENCE
mma-sat 10 μg/plate TECSDY 13,205,87
orl-rat LD50:5000 mg/kg 34ZIAG -,390,69
ihl-rat LCLo:28,800 ppm/15M MRLR** #113,52
orl-mus LD50:4300 mg/kg JIHTAB 29,382,47
ihl-mus LC50:15,850 mg/m³/8H JIHTAB 29,382,47

CONSENSUS REPORTS: Reported in EPA TSCA Inventory.

OSHA PEL: TWA 200 ppm
ACGIH TLV: TWA 200 ppm
DFG MAK: 200 ppm (1050 mg/m³)
DOT CLASSIFICATION: 6.1; *Label:* KEEP AWAY FROM FOOD

SAFETY PROFILE: A poison. Mildly toxic by ingestion and inhalation. Mutation data reported. This material has a narcotic action of moderate intensity, although of prolonged duration. Animals exposed for several weeks to 1000 ppm had blood bromide levels as high as 350 mg/100 g. Therefore, until further data are available, it should be considered at least as toxic as carbon tetrachloride and more than minimal exposure to its vapors should be avoided. Dangerous; when heated to decomposition it emits highly toxic fumes of Br^- and Cl^-. See also BROMIDES and CHLORINATED HYDROCARBONS, ALIPHATIC.

For occupational chemical analysis use NIOSH: Hydrocarbons, halogenated, 1003.

CES750 CAS:13360-45-7 HR: 2
1-(3-CHLORO-4-BROMOPHENYL)-3-METHYL-3-METHOXYUREA
mf: $C_9H_{10}BrClN_2O_2$ mw: 293.57

PROP: Tan crystals. Mp: 98°. Very sltly sol in H_2O; sol in org solvs.

SYNS: BROMEX □ N-(4-BROMO-3-CHLOROPHENYL)-N′-METHOXY-N′-METHYLUREA □ N′-(4-BROMO-3-CHLOROPHENYL)-N-METHOXY-N-METHYLUREA □ 3-(4-BROMO-3-CHLOROPHENYL)-1-METHOXY-1-METHYLUREA □ C-6313 □ CHLORBROMURON □ CHLOROBROMURON □ CIBA 6313 □ MALORAN

TOXICITY DATA WITH REFERENCE
cye-rbt 50 mg MOD CIGET* -,-,77

orl-rat LD50: 2150 mg/kg PHARAT 37,370,82

SAFETY PROFILE: Moderately toxic by ingestion. An eye irritant. An herbicide. When heated to decomposition it emits very toxic fumes of NO_x, Cl^-, and Br^-.

CET000 CAS:73926-87-1 HR: 3
trans-CHLORO(2-(3-BROMOPROPIONAMIDO)CYCLOHEXYL)MERCURY
mf: $C_9H_{15}BrClHgNO$ mw: 469.20

SYNS: 3-BROMO-N-(2-CHLOROMERCURICYCLOHEXYL)PROPIONAMIDE □ MERCURY (E)-CHLORO(2-(3-BROMOPROPIONAMIDO)CYCLOHEXYL)

TOXICITY DATA WITH REFERENCE
ivn-mus LD50: 32 mg/kg CSLNX* NX#04830

CONSENSUS REPORTS: Mercury and its compounds are on the Community Right-To-Know List.

OSHA PEL: TWA 0.01 mg(Hg)/m³; STEL 0.03 mg/m³ (skin)
ACGIH TLV: TWA 0.1 mg(Hg)/m³ (skin)
NIOSH REL: (Organomercury): TWA 0.01 mg/m³; STEL 0.03 mg/m³ (skin)

SAFETY PROFILE: Poison by intravenous route. See also MERCURY COMPOUNDS. When heated to decomposition it emits very toxic fumes of Br^-, Cl^-, NO_x, and Hg.

CET250 CAS:627-22-5 HR: 3
1-CHLOROBUTADIENE
mf: C_4H_5Cl mw: 88.54

PROP: Colorless liquid. Bp: 68°, d: 0.961 @ 20°/4°, flash p: −4°F, lel: 4.0%, uel: 20.0%, vap d: 3.0.

SYN: 1-CHLORO-1,3-BUTADIENE

TOXICITY DATA WITH REFERENCE
mmo-sat 8 pph/4H ARTODN 41,249,79
sln-dmg-orl 5700 µmol/L/3D-I 35WYAM -,63,76

CONSENSUS REPORTS: EPA Genetic Toxicology Program.

SAFETY PROFILE: Probably a poison. Mutation data reported. Dangerous fire hazard when exposed to heat or flame. To fight fire, use alcohol foam. When heated to decomposition it emits toxic fumes of Cl^-. See also CHLORINATED HYDROCARBONS, ALIPHATIC.

CEU250 CAS:78-86-4 HR: 3
2-CHLOROBUTANE
mf: C_4H_9Cl mw: 92.58

PROP: Flash p: 14°F, d: 0.87, vap d: 3.2, bp: 68.50.

SYN: sec-BUTYL CHLORIDE

TOXICITY DATA WITH REFERENCE
ipr-mus TDLo: 3240 mg/kg/8W-I: NEO CNREA8 35,1411,75
orl-rat LD50: 17,460 mg/kg AIHAAP 30,470,69
ihl-rat LCLo: 8000 ppm/4H AIHAAP 30,470,69
skn-rbt LD50: 20 g/kg AIHAAP 30,470,69

CONSENSUS REPORTS: Reported in EPA TSCA Inventory.

SAFETY PROFILE: Mildly toxic by ingestion, inhalation, and skin contact. Questionable carcinogen with experimental neoplastigenic data. Dangerous fire hazard when exposed to heat, open flame (sparks), or oxidizers. To fight fire, use water, water spray, fog, mist, dry chemical, alcohol foam. When heated to decomposition it emits toxic fumes of Cl^-. See also CHLORINATED HYDROCARBONS, ALIPHATIC.

CEU300 CAS:628-20-6 HR: 3
4-CHLOROBUTANENITRILE
mf: C_4H_6ClN mw: 103.56

SYNS: BUTANENITRILE, 4-CHLORO-(9CI) □ BUTYRONITRILE, 4-CHLORO- □ γ-CHLOROBUTYRONITRILE □ 4-CHLOROBUTYRONITRILE

TOXICITY DATA WITH REFERENCE
orl-mus LD50: 53,380 µg/kg ARTODN 55,47,84

CONSENSUS REPORTS: Reported in EPA TSCA Inventory.

SAFETY PROFILE: Poison by ingestion. When heated to decomposition it emits toxic vapors of NO_x and Cl^-.

CEU500 CAS:928-51-8 HR: 2
4-CHLORO-1-BUTANOL
mf: C_4H_9ClO mw: 108.58

PROP: D: 1.088 @ 20°/4°, bp: 84–85° @ 16 mm.

SYNS: 4-CHLORBUTAN-1-OL (GERMAN) □ 4-CHLORO-1-BUTANEOL □ 4-CHLOROBUTANOL □ TETRAMETHYLENE CHLOROHYDRIN

TOXICITY DATA WITH REFERENCE
ipr-mus TDLo: 3650 mg/kg/8W-I: NEO CNREA8 39,391,79
orl-mus LD50: 990 mg/kg ZHYGAM 26,17,80

CONSENSUS REPORTS: Reported in EPA TSCA Inventory.

SAFETY PROFILE: Moderately toxic by ingestion. Questionable carcinogen with experimental neoplastigenic data. Mutation data reported. When heated to decomposition it emits toxic fumes of Cl^-. See also CHLORIDES and ALCOHOLS.

CEU750 CAS:616-27-3 HR: 3
1-CHLORO-2-BUTANONE
mf: C_4H_7ClO mw: 106.55

PROP: Bp: 137.5°.

SAFETY PROFILE: A dangerously unstable explosive; may explode spontaneously. Upon decomposition it emits toxic fumes of Cl^-.

CEU825 CAS:591-97-9 HR: 1
1-CHLORO-2-BUTENE
mf: C_4H_7Cl mw: 90.56

PROP: trans-Isomer: Liquid. Bp: 84.8°, n (20/D) 1.4350,

n (25/D) 1.4327, d: 0.9295. cis-Isomer: Liquid. Bp: 84.1°, n (20/D) 1.4390, d: 0.9426.

SYNS: 2-BUTENYL CHLORIDE □ CROTYL CHLORIDE □ KROTYLCHLORID □ γ-METHALLYL CHLORIDE □ γ-METHYLALLYL CHLORIDE

TOXICITY DATA WITH REFERENCE
mmo-sat 5 µmol/plate BCPCA6 29,2611,80
dns-hmn:hla 100 µmol/L CALEDQ 20,263,83

SAFETY PROFILE: Human mutation data reported. An eye and mucous membrane irritant. When heated to decomposition it emits toxic fumes of Cl^-. See also CHLORINATED HYDROCARBONS, ALIPHATIC.

CEV000 CAS:4461-41-0 HR: 3
2-CHLORO-2-BUTENE
mf: C_4H_7Cl mw: 90.55

$CH_3CH{=}CClCH_3$

PROP: Flash p: −13°F; lel: 2.3%; uel: 9.3%.

TOXICITY DATA WITH REFERENCE
mmo-sat 10 µmol/plate BCPCA6 29,2611,80
ihl-mus LCLo:48,800 ppm/15M UCPHAQ 2,89,42

SAFETY PROFILE: Low toxicity by inhalation. Mutation data reported. A dangerous fire hazard when exposed to heat or flame. When heated to decomposition it emits toxic fumes of Cl^-. See also CHLORINATED HYDROCARBONS, ALIPHATIC.

CEV250 CAS:563-52-0 HR: 3
3-CHLORO-1-BUTENE
mf: C_4H_7Cl mw: 90.55

$H_2C{=}CHCHClCH_3$

PROP: dl-Form: Liquid. Bp: 63.9–64.2°, d: 0.9001, n (20/D) 1.4150. d(−)-Form: Liquid. Bp: −5°. l(+)-Form: Liquid. Flash p.: −16.6°F.

SYNS: 3-CHLORO-1-BUTENE □ α-METHALLYL CHLORIDE □ α-METHYLALLYL CHLORIDE □ 1-METHYLALLYL CHLORIDE

TOXICITY DATA WITH REFERENCE
mmo-sat 10 µmol/plate BCPCA6 29,2611,80
dns-hmn:hla 500 µmol/L CALEDQ 20,263,83

SAFETY PROFILE: Human mutation data reported. Irritates eyes, respiratory passages. A dangerous fire hazard when exposed to heat or flame. When heated to decomposition it emits toxic fumes of Cl^-. See also CHLORINATED HYDROCARBONS, ALIPHATIC.

CEV500 CAS:7119-27-9 HR: 3
1-CHLORO-1-BUTEN-3-ONE
mf: C_4H_5ClO mw: 104.54

$ClCH{=}CHCO{\cdot}CH_3$

SAFETY PROFILE: An unstable explosive. Upon decomposition it emits toxic fumes of Cl^-.

CEV750 CAS:55477-20-8 HR: 3
4-CHLORO-2-(tert-BUTYLAMINO)-6-(4-METHYLPIPERAZINO)-5-METHYLTHIOPYRIMIDINE
mf: $C_{14}H_{24}ClN_5S$ mw: 329.94

TOXICITY DATA WITH REFERENCE
orl-mus LD50:600 mg/kg JMCMAR 18,553,75
ivn-mus LD50:67 mg/kg JMCMAR 18,553,75

SAFETY PROFILE: Poison by intravenous route. Moderately toxic by ingestion. When heated to decomposition it emits very toxic fumes of SO_x, NO_x, and Cl^-.

CEV800 CAS:13280-07-4 HR: 3
4-CHLORO-2-BUTYNOL
mf: C_4H_5ClO mw: 104.54

$ClCH_2C{\equiv}CCH_2OH$

SAFETY PROFILE: A heat-sensitive explosive. When heated to decomposition it emits toxic fumes of Cl^-.

CEW000 CAS:1951-12-8 HR: 2
β-CHLOROBUTYRIC ACID
mf: $C_4H_7ClO_2$ mw: 122.56

SYNS: 3-CHLOROBUTANOIC ACID □ 3-CHLOROBUTYRIC ACID

TOXICITY DATA WITH REFERENCE
ipr-mus TDLo:1180 mg/kg/8W-I:ETA CNREA8 39,391,79

CONSENSUS REPORTS: Reported in EPA TSCA Inventory.

SAFETY PROFILE: Questionable carcinogen with experimental tumorigenic data. When heated to decomposition it emits toxic fumes of Cl^-. See also CHLORIDES.

CEW500 CAS:101-27-9 HR: 3
m-CHLORO CARBANILIC ACID-4-CHLORO-2-BUTYNYL ESTER
mf: $C_{11}H_9Cl_2NO_2$ mw: 258.11

PROP: Crystals. Mp: 75–76°. Very sltly sol in H_2O.

SYNS: A-980 □ BARBAMATE □ BARBAN □ BARBANE □ 2-BUTYNYL-4-CHLORO-m-CHLOROCARBANILATE □ C-847 □ CARBIN □ CARBYNE □ CARYNE □ CBN □ (4-CHLOOR-BUT-2-YN-YL)-N-(3-CHLOOR-FENYL)-CARBAMAAT (DUTCH) □ (4-CHLOR-BUT-2-IN-YL)-N-(3-CHLOR-PHENYL)-CARBAMAT (GERMAN) □ CHLORINAT □ CHLORO-2-BUTYNYL-m-CHLOROCARBAMATE □ 4-CHLOROBUT-2-YNYL-m-CHLOROCARBANILATE □ 4-CHLORO-2-BUTYNYL-m-CHLOROCARBANILATE □ 4-CHLOROBUT-2-YNYL-3-CHLOROPHENYLCARBAMATE □ 4-CHLORO-2-BUTYNYL-N-(3-CHLOROPHENYL)CARBAMATE □ N-(3-CHLORO PHENYL) CARBAMATE de 4-CHLORO 2-BUTYNYLE (FRENCH) □ (3-CHLOROPHENYL)CARBAMIC ACID 4-CHLORO-2-BUTYNYL ESTER □ (4-CLORO-BUT-2-IN-IL)-N-(3-CLORO-FENIL)-CARBAMMATO (ITALIAN) □ CS-847 □ FISONS B25 □ NEOBAN □ S-847

TOXICITY DATA WITH REFERENCE
mrc-bcs 20 µg/disc/24H MUREAV 40,19,76
orl-rat LD50:600 mg/kg RREVAH 10,97,65
ihl-rat LD50:527 mg/kg EQSFAP 3,618,75
orl-mus LD50:322 mg/kg FATOAO 35,356,72
orl-rbt LD50:600 mg/kg PAREAQ 14,225,62
skn-rbt LD50:23,000 mg/kg 85DPAN -,-,71/76
orl-gpg LD50:240 mg/kg PAREAQ 14,225,62

unr-mam LD50:240 mg/kg 30ZDA9 -,202,71

CONSENSUS REPORTS: EPA Genetic Toxicology Program.

SAFETY PROFILE: Poison by ingestion, inhalation, and possibly other routes. Mildly toxic by skin contact. Mutation data reported. An herbicide. See also CARBAMATES and ESTERS. When heated to decomposition it emits very toxic fumes of Cl^- and NO_x.

CEX250 CAS:1967-16-4 HR: 3
m-CHLOROCARBANILIC ACID-1-METHYL-2-PROPYNYL ESTER
mf: $C_{11}H_{10}ClNO_2$ mw: 223.67

PROP: Crystals. Mp: 45–46°. Sltly sol in H_2O; sol in MeOH, EtOH, and Me_2CO.

SYNS: BICP □ BIPC (the herbicide) □ BUTYN-1-OL-3-ESTER of m-CHLOROPHENYLCARBAMIC ACID □ 3-BUTYNYL-m-CHLOROCARBANILATE □ 1-BUTYN-3-YL-m-CHLOROPHENYLCARBAMATE □ CHLORBUFAM □ CHLORBUPHAM □ CHLOROBUFAM □ 3-CHLOROPHENYLCARBAMIC ACID-1-METHYLPROPYNYL ESTER □ 3-CHLORPHENYLCARBAMIDSAURE-BUTIN-(1)-YL(3)-ESTER (GERMAN) □ GRISEMIN □ GRISIN □ IEM-1-15 □ ISOBUTINYL-N-(3-CHLORPHENYL)-CARBAMAT (GERMAN) □ 1-METHYL-2-PROPYNYL-m-CHLOROCARBANILATE □ 1-METHYL-2-PROPYNYL-m-CHLOROPHENYLCARBAMATE □ 1-METHYLPROPYNYL 3-CHLOROPHENYLCARMATE □ 1-METHYLPROPYNYL ESTER of 3-CHLOROPHENYLCARBAMIC ACID

TOXICITY DATA WITH REFERENCE
orl-rat LD50:2380 mg/kg 85JFAN A071,83
ipr-mus LD50:250 mg/kg 85DPAN -,-,71/76
unr-mus LD50:15,500 μg/kg 85GDA2 1,262,80

SAFETY PROFILE: Poison by intraperitoneal and possibly other routes. Moderately toxic by ingestion. A pesticide. See also CARBAMATES and ESTERS. When heated to decomposition it emits very toxic fumes of Cl^- and NO_x.

CEX255 CAS:52716-12-8 HR: 3
N-(CHLOROCARBONYLOXY)TRIMETHYLUREA
mf: $C_5H_9ClN_2O_3$ mw: 180.59

$ClCO{\cdot}ONCH_3CO{\cdot}N(CH_3)_2$

SAFETY PROFILE: An unstable explosive. When heated to decomposition it emits toxic fumes of Cl^- and NO_x. See also EXPLOSIVES.

CEX275 CAS:5665-94-1 HR: 3
5-CHLOROCARVACROL
mf: $C_{10}H_{13}ClO$ mw: 184.68

SYNS: CARVASEPT □ CHLORCARVACROL □ MCIT □ MONOCHLOROISOTHYMOL

TOXICITY DATA WITH REFERENCE
ipr-mus LDLo:250 mg/kg CBCCT* 5,338,53
scu-mus LDLo:1500 mg/kg AEPPAE 161,196,31
orl-cat LDLo:1000 mg/kg AEPPAE 161,196,31
orl-rbt LDLo:750 mg/kg AEPPAE 161,196,31
scu-frg LDLo:400 mg/kg AEPPAE 161,196,31

SAFETY PROFILE: Poison by subcutaneous and intraperitoneal routes. Moderately toxic by ingestion. When heated to decomposition it emits toxic fumes of Cl^-.

CEX500 CAS:77966-38-2 HR: 3
6′-CHLORO-2-(p-CHLOROBENZYL(2-(DIETHYLAMINO)ETHYL)AMINO)-o-ACETOLUIDIDE DIHYDROCHLORIDE
mf: $C_{22}H_{29}Cl_2N_3O{\cdot}2ClH$ mw: 495.36

SYN: C 5364

TOXICITY DATA WITH REFERENCE
eye-rbt 2% SEV ARZNAD 9,167,59
ipr-rat LD50:103 mg/kg ARZNAD 9,167,59
scu-mus LD50:240 mg/kg ARZNAD 9,167,59

SAFETY PROFILE: Poison by intraperitoneal and subcutaneous routes. A severe eye irritant. When heated to decomposition it emits very toxic fumes of Cl^- and NO_x.

CEY250 CAS:78110-37-9 HR: 3
6-CHLORO-9-((2-((2-CHLOROETHYL)AMINO)ETHYL)AMINO)-2-METHOXYACRIDINE 2-HYDROCHLORIDE SESQUIHYDRATE
mf: $C_{18}H_{19}Cl_2N_3O{\cdot}2ClH{\cdot}3/2H_2O$ mw: 464.21

TOXICITY DATA WITH REFERENCE
mmo-sat 5 μg/plate JMCMAR 15,739,72
ipr-mus LD20:186 mg/kg JMCMAR 15,739,72

SAFETY PROFILE: Poison by intraperitoneal route. Mutation data reported. When heated to decomposition it emits very toxic fumes of NO_x and Cl^-.

CFA250 HR: 3
6-CHLORO-9-(3-(2-CHLOROETHYL)MERCAPTOPROPYLAMINO)-2-METHOXYACRIDINE HYDROCHLORIDE
mf: $C_{19}H_{20}Cl_2N_2OS{\cdot}ClH$ mw: 431.83

SYNS: ICR 342 □ 2-METHOXY-6-CHLORO-9-(3-(2-CHLOROETHYL) MERCAPTO PROPYLAMINO) ACRIDINE HYDROCHLORIDE

TOXICITY DATA WITH REFERENCE
ivn-mus TDLo:6500 μg/kg:NEO CNREA8 36,2423,76
ivn-mus LDLo:12 mg/kg CNREA8 36,2423,76

SAFETY PROFILE: Poison by intravenous route. Questionable carcinogen with experimental neoplastigenic data. When heated to decomposition it emits very toxic fumes of Cl^-, SO_x, and NO_x. See also MERCAPTANS.

CFA500 CAS:126-85-2 HR: 3
2-CHLORO-N-(2-CHLOROETHYL)-N-METHYLETHANAMINE-N-OXIDE
mf: $C_5H_{11}Cl_2NO$ mw: 172.07

SYNS: 2,2′-DICHLORO-N-METHYLDIETHYLAMINE-N-OXIDE □ DIETHYLAMINE, 2,2′-DICHLORO-N-METHYL-, OXIDE □ HN_2 AMINE OXIDE □ HN_2 OXIDE MUSTARD □ MBAO □ MECHLORETHAMINE OXIDE □ METHYL-BIS(β-CHLOROETHYL)AMINE OXIDE □ METHYLBIS (β-CHLOROETHYL)AMINE N-OXIDE □ N-METHYL-DI-2-CHLOROETHYLAMINE-N-OXIDE □ MITOMEN □ MITOMIN □ NITROGEN MUSTARD OXIDE □ NITROGEN MUSTARD-N-OXIDE □ NITROMIN □ NMO □ N-OXYD-LOST □ N-OXYD-MUSTARD □ NSC-10107 □ OXY-NH2

TOXICITY DATA WITH REFERENCE
mmo-esc 50 μg/plate TAKHAA 44,96,85
dlt-mus-ipr-20 mg/kg MUREAV 26,285,74
ipr-rat TDLo: 20 mg/kg (male 1D pre): REP 85GUAJ -,37,66
ipr-rat TDLo: 80 mg/kg (1D male): REP HIKYAJ 12,1339,66
scu-mus TDLo: 60 mg/kg (male 1D pre): REP JJPAAZ 10,64,60
ivn-rat TDLo: 218 mg/kg/1Y-I: CAR ARZNAD 20,1461,70
skn-mus TDLo: 204 g/kg/17W-I: ETA GANNA2 57,295,66
ipr-rat LD50: 50 mg/kg ARZNAD 11,143,61
ipr-mus LD50: 100 mg/kg NTIS** PB158-507
ivn-mus LD50: 80 mg/kg JAJAAA 13,19,60

SAFETY PROFILE: Poison by intravenous and intraperitoneal routes. Experimental reproductive effects. Questionable carcinogen with experimental carcinogenic and tumorigenic data. Mutation data reported. When heated to decomposition it emits toxic fumes of Cl^- and NO_x.

CFA750 CAS:302-70-5 HR: 3
2-CHLORO-N-(2-CHLOROETHYL)-N-METHYLETHANAMINE-N-OXIDE HYDROCHLORIDE
mf: $C_5H_{11}Cl_2NO{\cdot}ClH$ mw: 208.53

PROP: A solid. Mp: 109–110°.

SYNS: CHLORMETHINE-N-OXIDE HYDROCHLORIDE □ 2,2′-DICHLORO-N-METHYLDIETHYLAMINE N-OXIDE HYDROCHLORIDE □ HN_2 OXIDE HYDROCHLORIDE □ MBAO HYDROCHLORIDE □ MECHLORETHAMINE OXIDE HYDROCHLORIDE □ METHYL-BIS-(β-CHLORAETHYL)-AMIN-N-OXYD-HYDROCHLORID (GERMAN) □ METHYLBIS(β-CHLOROETHYL)AMINE-N-OXIDE HYDROCHLORIDE □ N-METHYLBIS(2-CHLOROETHYL)AMINE-N-OXIDE HYDROCHLORIDE □ N-METHYL-2,2′-DICHLORODIETHYLAMINE-N-OXIDE HYDROCHLORIDE □ METHYLDI(2-CHLOROETHYL)AMINE-N-OXIDE HYDROCHLORIDE □ MITOMEN □ MUSTRON □ NITROGEN MUSTARD OXIDE □ NITROGEN MUSTARD-N-OXIDE □ NITROGEN MUSTARD-N-OXIDE HYDROCHLORIDE □ NITROMIM □ NITROMIN HYDROCHLORIDE □ N-OXYD-LOST □ NSC-10107 □ OSSIAMINA □ OSSICHLORIN □ OXYAMINE □ SK-598 □ XA 2

TOXICITY DATA WITH REFERENCE
mmo-sat 500 μg/plate URLRA5 7,119,79
pic-esc 200 mg/L ARMKA7 51,9,65
cyt-slw-par 1670 μL ZEVBA5 89,216,58
bfa-rat/sat 250 mg/kg URLRA5 7,119,79
mul-rat TDLo: 1300 mg/kg/93W-I: ETA ZEKBAI 62,112,57
orl-rat LD50: 60 mg/kg YAKUD5 21,359,79
ipr-rat LD50: 79,500 μg/kg NIIRDN 6,256,82
scu-rat LD50: 76,100 μg/kg NIIRDN 6,256,82
ipr-mus LD50: 125 mg/kg NCISP* JAN86
ivn-dog LDLo: 4 mg/kg CCSUBJ 2,201,65
ivn-mky LDLo: 8 mg/kg CCSUBJ 2,201,65

CONSENSUS REPORTS: IARC Cancer Review: Group 2B IMEMDT 7,56,87; Animal Sufficient Evidence IMEMDT 9,209,75; EPA Genetic Toxicology Program.

SAFETY PROFILE: Suspected carcinogen with experimental tumorigenic data. Poison by ingestion, subcutaneous, intravenous, and intraperitoneal routes. Mutation data reported. An antineoplastic agent. When heated to decomposition it emits toxic fumes of Cl^- and NO_x.

CFB500 CAS:19996-03-3 HR: 3
9-CHLORO-10-CHLOROMETHYL ANTHRACENE
mf: $C_{15}H_{10}Cl_2$ mw: 261.15

SYNS: 10-CHLOROMETHYL-9-CHLOROANTHRACENE □ ICR 486

TOXICITY DATA WITH REFERENCE
mma-sat 1 μg/plate PNASA6 72,5135,75
pic-esc 50 ng/plate CNREA8 43,2819,83
ivn-mus TDLo: 2600 μg/kg: NEO CNREA8 36,2423,76
ivn-mus TD: 2612 μg/kg: NEO CNREA8 40M782,80

CONSENSUS REPORTS: EPA Genetic Toxicology Program.

SAFETY PROFILE: Questionable carcinogen with experimental neoplastigenic data. Mutation data reported. When heated to decomposition it emits toxic fumes of Cl^-. See also CHLORINATED HYDROCARBONS, AROMATIC; and ANTHRACENE.

CFB600 CAS:104-83-6 HR: 2
1-CHLORO-4-CHLOROMETHYLBENZENE
mf: $C_7H_6Cl_2$ mw: 161.03

SYN: TOLUENE, p,α-DICHLORO-

TOXICITY DATA WITH REFERENCE
unr-rat LD50: 1075 mg/kg GISAAA 55(5),13,90
unr-mus LD50: 1156 mg/kg GISAAA 55(5),13,90
unr-gpg LD50: 5625 mg/kg GISAAA 55(5),13,90

CONSENSUS REPORTS: Reported in EPA TSCA Inventory.

SAFETY PROFILE: Moderately toxic by unspecified routes. When heated to decomposition it emits toxic vapors of Cl^-.

CFB750 CAS:7205-90-5 HR: 2
1-CHLORO-4-(CHLOROMETHYLTHIO)BENZENE
mf: $C_7H_6Cl_2S$ mw: 193.09

SYN: p-CHLORFENYLMERKAPTOMETHYLCHLORID (CZECH)

TOXICITY DATA WITH REFERENCE
skn-rbt 500 mg/24H MOD 28ZPAK -,172,72
eye-rbt 50 μg/24H SEV 28ZPAK -,172,72

CONSENSUS REPORTS: Reported in EPA TSCA Inventory.

SAFETY PROFILE: A skin and severe eye irritant. When heated to decomposition it emits very toxic fumes of SO_x and Cl^-. See also MERCAPTANS.

CFB825 CAS:23784-96-5 HR: 3
2-CHLORO-5-CHLOROMETHYLTHIOPHENE
mf: $C_5H_4Cl_2S$ mw: 167.05

$CH{=}C(Cl)SC(CH_2Cl){=}CH$ (ring)

SAFETY PROFILE: A storage hazard, it may decompose explosively at room temperature. When heated to decomposition it emits toxic fumes of Cl^- and SO_x.

CFC000 **CAS:1018-71-9** **HR: 3**
3-CHLORO-4-(3-CHLORO-2-NITROPHENYL)PYRROLE
mf: $C_{10}H_6Cl_2N_2O_2$ mw: 257.08

PROP: Pale-yellow crystals from C_6H_6. Mp: 125°.

SYNS: 3-CHLORO-4-(2′-NITRO-3′-CHLOROPHENYL)PYRROLE □ NSC-107654 □ PN □ PYROACE □ PYROLLNITRIN □ PYRROLNITRIN

TOXICITY DATA WITH REFERENCE
ipr-rat LD50:68 mg/kg 85ERAY 3,1479,78
orl-mus LD50:1000 mg/kg 85ERAY 3,1479,78
ipr-mus LD50:680 mg/kg MEIEDD 10,1157,83
scu-mus LD50:2000 mg/kg 85GDA2 5,44,81
ipr-rbt LD50:105 mg/kg 85ERAY 3,1479,78

SAFETY PROFILE: Poison by intraperitoneal route. Moderately toxic by ingestion and subcutaneous routes. When heated to decomposition it emits very toxic fumes of Cl^- and NO_x.

CFC100 **CAS:57808-65-8** **HR: 3**
N-(5-CHLORO-4-((4-CHLOROPHENYL)CYANOMETHYL)-2-METHYLPHENYL)-2-HYDROXY-3, 5-DIIODOBENZAMIDE
mf: $C_{22}H_{14}Cl_2I_2N_2O_2$ mw: 663.08

PROP: A solid. Mp: 217°.

SYNS: BENZAMIDE, N-(5-CHLORO-4-((4-CHLOROPHENYL)CYANOMETHYL)-2-METHYLPHENYL)-2-HYDROXY-3,5- DIIODO- □ CLOSANTEL □ R 31520

TOXICITY DATA WITH REFERENCE
orl-rat TDLo:3600 mg/kg (male 90D pre):REP DCTODJ 8,101,85
orl-rat LD50:262 mg/kg DCTODJ 8,101,85
ims-rat LD50:28,400 μg/kg DCTODJ 8,101,85
orl-mus LD50:331 mg/kg DCTODJ 8,101,85
ims-mus LD50:56,800 μg/kg DCTODJ 8,101,85

SAFETY PROFILE: Poison by ingestion and intramuscular routes. Experimental reproductive effects. When heated to decomposition it emits toxic fumes of NO_x, I^-, and Cl^-.

CFC250 **CAS:846-49-1** **HR: 2**
7-CHLORO-5-(o-CHLOROPHENYL)-1,3-DIHYDRO-3-HYDROXY-2H-1,4-BENZODIAZEPIN-2-ONE
mf: $C_{15}H_{10}Cl_2N_2O_2$ mw: 321.17

SYNS: ALMAZINE □ ATIVAN □ 7-CHLORO-5-(2-CHLOROPHENYL)-1,3-DIHYDRO-3-HYDROXY-2H-1,4-BENZODIAZEPIN-2-ONE □ 7-CHLORO-5-(2-CHLOROPHENYL)-3-HYDROXY-1H-1,4-BENZODIAZEPIN-2(3H)-ONE □ EMOTIVAL □ LORAX □ LORAZEPAM □ LORSILAN □ PSICOPAX □ TAVOR □ TEMESTA □ WY 4036 □ WYPAX

TOXICITY DATA WITH REFERENCE
orl-wmn TDLo:2420 μg/kg (36-39W preg/2D post):REP BMJOAE 282,1106,81
orl-rat TDLo:2800 μg/kg (9-15D preg):TER JZKEDZ 1,25,75
orl-chd TDLo:71 μg/kg:CNS TOLED5 2,109,78
orl-wmn TDLo:380 μg/kg/19D-I:BLD BMJOAE 296,1332,88
orl-hmn TDLo:21 μg/kg CLPTAT 39,526,86
orl-rat LD50:4500 mg/kg IYKEDH 8,680,77
ipr-rat LD50:870 mg/kg PHMGBN 10,345,73
orl-mus LD50:1850 mg/kg IYKEDH 8,680,77
ipr-mus LD50:1810 mg/kg JZKEDZ 1,5,74

SAFETY PROFILE: Moderately toxic by ingestion and intraperitoneal routes. Human systemic effects by ingestion: general anesthetic, hallucinations or distorted perceptions, loss of muscle control (ataxia), aplastic anemia, bone marrow changes. Human reproductive effects by ingestion: unspecified neonatal effects. An experimental teratogen. Other experimental reproductive effects. An anticonvulsant and tranquilizer. When heated to decomposition it emits very toxic fumes of $Cl^-$and NO_x. See also DIAZEPAM.

CFC500 **CAS:24283-57-6** **HR: 3**
5-CHLORO-3-(4-CHLOROPHENYL)-4′-FLUORO-2′-METHYLSALICYLANILIDE
mf: $C_{20}H_{14}Cl_2FNO_2$ mw: 390.25

SYNS: CP 48985 □ 4′,5-DICHLORO-N-(4-FLUORO-2-METHYLPHENYL)-2-HYDROXY-(1,1′-BIPHENYL)-3-CARBOXAMIDE □ ENT 27,349 □ MONSANTO CP-48985 □ NSC 190947

TOXICITY DATA WITH REFERENCE
orl-rat LD50:2510 mg/kg ARSIM* 20,16,66
ipr-mus LD50:151 mg/kg BCPCA6 18,1389,69

SAFETY PROFILE: Poison by intraperitoneal route. Moderately toxic by ingestion. When heated to decomposition it emits very toxic fumes of Cl^-, F^-, and NO_x.

CFC750 **CAS:14437-17-3** **HR: 2**
2-CHLORO-3-(4-CHLOROPHENYL)METHYLPROPIONATE
mf: $C_{10}H_{10}Cl_2O_2$ mw: 233.10

PROP: Liquid. Bp: 110–113° @ 0.1 mm.

SYNS: BAY 70533 □ BAYER 70533 □ BIDISIN □ CHLORFENPROP-METHYL □ 2-CHLORO-3-(4-CHLOROPHENYL)PROPIONIC ACID METHYL ESTER □ CHLORPHENPROP-METHYL □ 3-(4-CHLORPHENYL)-2-CHLORPROPIONSAEUREMETHYLESTER (GERMAN) □ METHACHLORPHENPROP □ METHYL-2-CHLORO-3-(4-CHLOROPHENYL)PROPIONATE □ METHYL α,4-DICHLOROPHENYLPROPANOATE □ W5769

TOXICITY DATA WITH REFERENCE
orl-rat LD50:1072 mg/kg WRPCA2 9,119,70
orl-mus LD50:1000 mg/kg 85DPAN -,-,71,76
orl-rbt LD50:500 mg/kg 28ZEAL 5,44,76
skn-rbt LD50:756 mg/kg WRPCA2 9,119,70
orl-gpg LD50:500 mg/kg 28ZEAL 5,44,76

SAFETY PROFILE: Moderately toxic by ingestion and skin contact. An herbicide. See also ESTERS. When heated to decomposition it emits toxic fumes of Cl^-.

CFD250 **CAS:5980-86-9** **HR: 3**
CHLORO(2-CHLOROVINYL)MERCURY
mf: $C_2H_2Cl_2Hg$ mw: 297.53

SYN: (2-CHLOROVINYL)MERCURIC CHLORIDE

TOXICITY DATA WITH REFERENCE
ivn-mus LD50:56,200 μg/kg CSLNX* NX#05831

CONSENSUS REPORTS: Mercury and its compounds are on the Community Right-To-Know List.

OSHA PEL: CL 0.1 mg(Hg)/m³ (skin)
ACGIH TLV: TWA 0.1 mg(Hg)/m³
NIOSH REL: (Organomercury): TWA 0.01 mg/m³; STEL 0.03 mg/m³ (skin)

SAFETY PROFILE: Poison by intravenous route. See also MERCURY COMPOUNDS. When heated to decomposition it emits very toxic fumes of Cl^- and Hg.

CFE000 CAS:1570-64-5 HR: 3
4-CHLORO-o-CRESOL
mf: C_7H_7ClO mw: 142.59

PROP: Crystals from pet ether. Mp: 49°, bp: 222–225°.

TOXICITY DATA WITH REFERENCE
mmo-sat 100 μg/plate TECSDY 14,143,87
orl-mus LD50:1320 mg/kg PHARAT 30,147,75
ivn-mus LD50:56 mg/kg CSLNX* NX#03270

CONSENSUS REPORTS: Reported in EPA TSCA Inventory. Chlorophenol compounds are on the Community Right-To-Know List.

SAFETY PROFILE: Poison by intravenous route. Moderately toxic by ingestion. Mutation data reported. When heated to decomposition it emits toxic fumes of Cl^- and phosgene. See also CRESOL and CHLOROPHENOLS.

CFE250 CAS:59-50-7 HR: 3
4-CHLORO-m-CRESOL
mf: C_7H_7ClO mw: 142.59

PROP: Odorless crystals (when pure) from pet ether. Mp: 55.5°, bp: 235°. Somewhat sol in water; very sol in organic solvents.

SYNS: APTAL □ BAKTOL □ BAKTOLAN □ CANDASEPTIC □ p-CHLOR-m-CRESOL □ CHLOROCRESOL □ p-CHLOROCRESOL □ p-CHLORO-m-CRESOL □ 6-CHLORO-m-CRESOL □ 2-CHLORO HYDROXYTOLUENE □ 6-CHLORO-3-HYDROXYTOLUENE □ 4-CHLORO-3-METHYLPHENOL □ 3-METHYL-4-CHLOROPHENOL □ OTTAFACT □ PARMETOL □ PAROL □ PCMC □ PREVENTOL CMK □ RASCHIT □ RASEN-ANICON □ RCRA WASTE NUMBER U039

TOXICITY DATA WITH REFERENCE
mmo-sat 25 μg/plate TECSDY 14,143,87
uns-bac-esc 100 μmol/L MUREAV 307,141,94
orl-rat LD50:1830 mg/kg SchP## 04APR86
scu-rat LD50:400 mg/kg QJPPAL 12,212,39
orl-mus LD50:600 mg/kg SCIEAS 36(1-4),10,89
scu-mus LD50:360 mg/kg QJPPAL 12,212,39
ivn-mus LD50:70 mg/kg QJPPAL 12,212,39

CONSENSUS REPORTS: Reported in EPA TSCA Inventory. Chlorophenol compounds are on the Community Right-To-Know List.

SAFETY PROFILE: Poison by intravenous, subcutaneous, and intraperitoneal routes. Moderately toxic by ingestion. An allergen. Mutation data reported. Incompatible with sodium hydroxide. When heated to decomposition it emits toxic fumes of Cl^- and phosgene. See also CRESOL and CHLOROPHENOLS.

CFE500 CAS:615-74-7 HR: 2
6-CHLORO-m-CRESOL
mf: C_7H_7ClO mw: 142.59

SYNS: PHENOL, 2-CHLORO-5-METHYL- □ 2-CHLORO-5-METHYLPHENOL

TOXICITY DATA WITH REFERENCE
orl-qal LD50:562 mg/kg AECTCV 12,355,83
orl-brd LD50:562 mg/kg AECTCV 12,355,83

CONSENSUS REPORTS: Reported in EPA TSCA Inventory.

SAFETY PROFILE: Moderately toxic by ingestion. When heated to decomposition it emits toxic vapors of Cl^-.

CFE750 CAS:2003-31-8 HR: 3
CHLOROCYANOACETYLENE
mf: C_3ClN mw: 85.49

ClC≡CC≡N

PROP: Crystals. Mp: 42–42.5°.

SYN: CHLOROPROPYNENITRILE

CONSENSUS REPORTS: Cyanide and its compounds are on the Community Right-To-Know List.

SAFETY PROFILE: Ignites spontaneously in air. A dangerous storage hazard, it may explode in a sealed container. When heated to decomposition it emits toxic fumes of Cl^-, NO_x, and CN^-. See also ACETYLENE COMPOUNDS, CYANIDE, NITRILES, and CHLORINATED HYDROCARBONS, ALIPHATIC.

CFF100 HR: 3
2-CHLORO-α-CYANO-6-METHYLERGOLINE-8-PROPIONAMIDE
mf: $C_{19}H_{21}ClN_4O$ mw: 356.89

SYN: ERGOLINE-8-PROPIONAMIDE, 2-CHLORO-α-CYANO-6-METHYL-

TOXICITY DATA WITH REFERENCE
orl-rat TDLo:8 mg/kg (female 5D post):REP ARZNAD 33,1094,83
orl-mus LD50:12,500 μg/kg ARZNAD 33,1094,83

SAFETY PROFILE: Poison by ingestion. Experimental reproductive effects. When heated to decomposition it emits toxic fumes of NO_x and Cl^-.

CFF250 CAS:15271-41-7 HR: 3
3-CHLORO-6-CYANO-2-NORBORNANONE-o-(METHYLCARBAMOYL)OXIME
mf: $C_{10}H_{12}ClN_3O_2$ mw: 241.70

SYNS: endo-3-CHLORO-exo-6-CYANO-2-NORBORNANONE-o-(METHYLCARBAMOYL)OXIME □ 2-exo-CHLORO-6-endo-CYANO-2-NORBORNANONE-o-(METHYLCARBAMOYL)OXIME2-CARBONITRILE □ 3-CHLORO-6-CYANONORBORNANONE-2-OXIME-o,N-METHYLCARBAMATE □ 5-CHLORO-6-((((METHYLAMINO)CARBONYL)OXY)IMINO) BICYCLO(2.2.1)HEPTANE □ exo 5 CHLORO-6-OXO-endo-2-NORBORNANECARBONITRILE-o-(METHYLCARBAMOYL)OXIME □ COMPOUND UC-20047 A □ ENT 25,962 □ TRANID □ UC 20047 □ UC 26089 □ UC 20,047A □ UNION CARBIDE UC 20047

TOXICITY DATA WITH REFERENCE
orl-rat LD50: 19 mg/kg TXAPA9 14,515,69
skn-rat LD50: 303 mg/kg WRPCA2 9,119,70
unk-rat LD50: 26 mg/kg 30ZDA9 -,198,71
skn-rbt LDLo: 303 mg/kg BESAAT 12,161,66

CONSENSUS REPORTS: EPA Extremely Hazardous Substances List. Cyanide and its compounds are on the Community Right-To-Know List.

SAFETY PROFILE: Poison by ingestion, skin contact, and possibly other routes. A pesticide. When heated to decomposition it emits very toxic fumes of Cl^- and NO_x. See also CARBAMATES.

CFF500 **CAS:82-93-9** **HR: 3**
CHLOROCYCLINE
mf: $C_{18}H_{21}ClN_2$ mw: 300.86

SYNS: CHLORCYCLINE □ CHLORCYCLIZINE □ 1-(4-CHLOROBENZHYDRYL)-4-METHYLPIPERAZINE □ CHLOROCYCLIZINE □ 1-(p-CHLORO-α-PHENYLBENZYL)-4-METHYLPIPERAZINE □ DI-PARALEN □ DIPARALENE □ HISTANTIN □ HISTANTINE

TOXICITY DATA WITH REFERENCE
dni-rat-ipr 50 mg/kg JPETAB 171,109,70
orl-rat TDLo: 330 mg/kg (female 7-17D post): TER ARZNAD 31,1225,81
orl-uns TDLo: 240 mg/kg (female 18-25D post): REP AOBIAR 13,1281,68
orl-rat TDLo: 1890 mg/kg (female 21D pre): REP JRPFA4 31,463,72
orl-rat TDLo: 755 mg/kg (female 7-17D post): TER ARZNAD 31,1225,81
orl-rat TDLo: 160 mg/kg (13-16D preg): TER TXAPA9 17,76,70
orl-rat TDLo: 200 mg/kg (female 13-16D post): TER TXAPA9 17,67,70
orl-rat TDLo: 330 mg/kg (female 7-17D post): TER ARZNAD 31,1225,81
orl-rat TDLo: 160 mg/kg (13-16D preg): TER TXAPA9 17,76,70

SAFETY PROFILE: Unspecified human reproductive effects. Experimental teratogenic and reproductive effects. Mutation data reported. When heated to decomposition it emits very toxic fumes of Cl^- and NO_x.

CFF600 **CAS:766-66-5** **HR: 2**
2-CHLOROCYCLOHEPTANONE
mf: $C_7H_{11}ClO$ mw: 146.63

SYNS: α-CHLOROCYCLOHEPTANONE □ CYCLOHEPTANONE, 2-CHLORO-

TOXICITY DATA WITH REFERENCE
ipr-mus LD50: 980 mg/kg COREAF 254,2683,62

CONSENSUS REPORTS: Reported in EPA TSCA Inventory.

SAFETY PROFILE: Moderately toxic by intraperitoneal route. When heated to decomposition it emits toxic vapors of Cl^-.

CFG250 **CAS:822-87-7** **HR: 2**
α-CHLOROCYCLOHEXANONE
mf: C_6H_9ClO mw: 132.60

SYN: 2-CHLOROCYCLOHEXANONE

TOXICITY DATA WITH REFERENCE
mmo-sat 370 nmol/plate CBINA8 45,305,83
ipr-mus LD50: 830 mg/kg COREAF 254,2683,62

CONSENSUS REPORTS: Reported in EPA TSCA Inventory.

SAFETY PROFILE: Moderately toxic by intraperitoneal route. Mutation data reported. When heated to decomposition it emits toxic fumes of Cl^-. See also KETONES and CHLORIDES.

CFG500 **CAS:14737-08-7** **HR: 2**
4-CHLORO-4-CYCLOHEXENE-1,2-DICARBOXYLIC ANHYDRIDE
mf: $C_8H_7ClO_3$ mw: 186.60

SYNS: ANHYDRIDE KYSELINY 4-CHLOR-1,2,3,6-TETRAHYDROFTALOVE (CZECH) □ 4-CHLORTETRAHYDROFTALANHYDRID (CZECH)

TOXICITY DATA WITH REFERENCE
skn-rbt 500 mg/24H MLD 28ZPAK -,140,72
eye-rbt 20 mg/24H SEV 28ZPAK -,140,72
orl-rat LD50: 3390 mg/kg 28ZPAK -,140,72

SAFETY PROFILE: Moderately toxic by ingestion. A severe eye and mild skin irritant. When heated to decomposition it emits toxic fumes of Cl^-. See also ANHYDRIDES.

CFG750 **CAS:10379-14-3** **HR: 2**
7-CHLORO-5-(CYCLOHEXEN-1-YL)-1,3-DIHYDRO-1-METHYL-2H-1,4-BENZODIAZEPIN-2-ONE
mf: $C_{16}H_{17}ClN_2O$ mw: 288.80

PROP: Yellow-brown crystals from EtOAc. Mp: 144°.

SYNS: CB 4261 □ 4361 CB □ 7-CHLORO-5-(1-CYCLOHEXENYL)-1-METHYL-2-OXO-2,3-DIHYDRO-1H-(1,4)-BENZO(f)DIAZEPINE □ CLINOXAN □ MUSARIL □ MYOLASTAN □ TETRAZEPAM

TOXICITY DATA WITH REFERENCE
orl-mus LD50: 2000 mg/kg CHTPBA 2,254,67
ipr-mus LD50: 415 mg/kg 27ZQAG -,171,72

SAFETY PROFILE: Moderately toxic by ingestion and intraperitoneal routes. A tranquilizer and muscle relaxant. When heated to decomposition it emits very toxic fumes of Cl^- and NO_x. See also DIAZEPAM and KETONES.

CFH000 **CAS:77966-40-6** **HR: 3**
6′-CHLORO-2-(CYCLOHEXYLAMINO)-o-ACETOTOLUIDIDE HYDROCHLORIDE
mf: $C_{15}H_{21}ClN_2O{\cdot}ClH$ mw: 317.29

SYN: C 3115

TOXICITY DATA WITH REFERENCE
eye-rbt 2% MLD ARZNAD 8,407,58
ipr-rat LD50: 83 mg/kg ARZNAD 8,407,58
scu-mus LD50: 122 mg/kg ARZNAD 8,407,58

SAFETY PROFILE: Poison by intraperitoneal and subcutaneous routes. An eye irritant. When heated to decomposition it emits very toxic fumes of Cl^- and NO_x.

CFH500 CAS:13909-11-0 HR: 3
cis-3-(2-CHLOROCYCLOHEXYL)-1-(2-CHLOROETHYL)-1-NITROSOUREA
mf: $C_9H_{15}Cl_2N_3O_2$ mw: 268.17

SYNS: cis-N'-(2-CHLOROCYCLOHEXYL)-N-(2-CHLOROETHYL)-N-NITROSOUREA □ (Z)-3-(2-CHLOROCYCLOHEXYL)-1-(2-CHLOROETHYL)-1-NITROSOUREA □ NSC 84954

TOXICITY DATA WITH REFERENCE
ipr-mus LD50:52,100 µg/kg NCISP* JAN86

SAFETY PROFILE: Poison by intraperitoneal route. Many N-nitroso compounds are carcinogens. When heated to decomposition it emits very toxic fumes of Cl^- and NO_x. See also N-NITROSO COMPOUNDS.

CFH750 CAS:13909-12-1 HR: 3
trans-3-(2-CHLOROCYCLOHEXYL)-1-(2-CHLOROETHYL)-1-NITROSOUREA
mf: $C_9H_{15}Cl_2N_3O_2$ mw: 268.17

SYNS: trans-N'-(2-CHLOROCYCLOHEXYL)-N-(2 CHLOROETHYL)-N-NITROSOUREA □ NSC 88104 □ SRI 2656

TOXICITY DATA WITH REFERENCE
orl-mus LD50:51,100 µg/kg NCISP* JAN86
ipr-mus LD50:62,230 µg/kg NCISP* JAN86
scu-mus LD50:74,600 µg/kg NCISP* JAN86

SAFETY PROFILE: Poison by ingestion, subcutaneous, and intraperitoneal routes. Many N-nitroso compounds are carcinogens. When heated to decomposition it emits very toxic fumes of NO_x and Cl^-. See also N-NITROSO COMPOUNDS.

CFH825 CAS:28968-07-2 HR: 3
6-CHLORO-5-CYCLOHEXYL-1-INDANCARBOXYLIC ACID
mf: $C_{16}H_{19}ClO_2$ mw: 278.80

PROP: Colorless crystals from petr ether. Mp: 150.5–152.5°. (S)-(+)-Form: Colorless needles from petr ether. Mp: 135–136°. (R)-(−)-Form: Colorless crystals from petr ether. Mp: 134–135°. Sltly sol in water.

SYNS: BRITAI □ (±)-6-CHLORO-5-CYCLOHEXYL-2,3-DIHYDRO-1H-INDENE-1-CARBOXYLIC ACID (9CI) □ (±)-6-CHLORO-5-CYCLOHEXYLINDAN-1-CARBOXYLIC ACID □ (±)-6-CHLORO-5-CYCLOHEXYL-1-INDANCARBOXYLIC ACID □ CLIDANAC □ (±)-2,3-DIHYDRO-6-CHLORO-5-CYCLOHEXYL-1H-INDENE-1-CARBOXYLIC ACID □ INDANAL □ TAI-284 □ (±)-TAI 284 □ dl-TAI 284

TOXICITY DATA WITH REFERENCE
orl-mus TDLo:810 mg/kg (7-15D preg):TER SEIJBO 17,177,77
orl-rat TDLo:90 mg/kg (30D male):REP OYYAA2 7,333,73
orl-mus TDLo:90 mg/kg (7D preg):TER SEIJBO 17,177,77
orl-rat LD50:41 mg/kg JMCMAR 15,1297,72
ipr-rat LD50:50 mg/kg OYYAA2 7,333,73
scu-rat LD50:60 mg/kg OYYAA2 7,333,73
ivn-rat LD50:45 mg/kg OYYAA2 7,333,73
orl-mus LD50:750 mg/kg OYYAA2 7,333,73
ipr-mus LD50:500 mg/kg OYYAA2 7,333,73
scu-mus LD50:800 mg/kg OYYAA2 7,333,73
ivn-mus LD50:150 mg/kg OYYAA2 7,333,73
orl-rbt LD50:250 mg/kg OYYAA2 7,333,73
orl-gpg LD50:400 mg/kg OYYAA2 7,333,73

SAFETY PROFILE: Poison by ingestion, subcutaneous, intravenous, and intraperitoneal routes. Experimental teratogenic and reproductive effects. A non-steroidal anti-inflammatory agent. When heated to decomposition it emits toxic fumes of Cl^-.

CFI000 CAS:77966-41-7 HR: 3
6'-CHLORO-2-(N-CYCLOHEXYL-N-METHYLAMINO)-o-ACETOTOLUIDIDE HYDROCHLORIDE
mf: $C_{16}H_{23}ClN_2O{\cdot}ClH$ mw: 331.32

SYN: C 3120

TOXICITY DATA WITH REFERENCE
eye-rbt 2% SEV ARZNAD 8,407,58
ipr-rat LD50:105 mg/kg ARZNAD 8,407,58
scu-mus LD50:337 mg/kg ARZNAD 8,407,58

SAFETY PROFILE: Poison by intraperitoneal and subcutaneous routes. A severe eye irritant. When heated to decomposition it emits very toxic fumes of Cl^- and NO_x.

CFI625 CAS:96-40-2 HR: 3
3-CHLOROCYCLOPENTENE
mf: C_5H_7Cl mw: 102.56

$\overline{CH{=}CHC_2H_4\,CHCl}$

PROP: Unstable oil. Bp: 18–25° @ 5 mm.

SAFETY PROFILE: A dangerous storage hazard; it may decompose explosively at room temperature. When heated to decomposition it emits toxic fumes of Cl^-. See also CHLORINATED HYDROCARBONS, ALIPHATIC.

CFI750 CAS:13347-42-7 HR: 2
4-CHLORO-2-CYCLOPENTYL PHENOL
mf: $C_{11}H_{13}ClO$ mw: 196.69

SYN: DOWICIDE 9

TOXICITY DATA WITH REFERENCE
skn-hmn 25 mg MLD DOWCC* Apr.55
eye-rbt 100 mg SEV DOWCC* Apr.55
orl-rat LDLo:420 mg/kg DOWCC*

CONSENSUS REPORTS: Chlorophenol compounds are on the Community Right-To-Know List.

SAFETY PROFILE: Moderately toxic by ingestion and skin contact. A human skin irritant. A severe eye irritant. When heated to decomposition it emits toxic fumes of Cl^-. See also CHLOROPHENOLS.

CFJ000 CAS:63007-70-5 HR: 2
3-CHLORO-4-CYCLO-PROPYLMETHOXYPHENYLACETIC ACID LYSINE SALT (d,l)
mf: $C_{12}H_{13}ClO_3 \cdot C_6H_{14}N_2O_2$ mw: 386.92

SYNS: 2-(3-CHLORO-4-CYCLOPROPYLMETHOXYPHENYL)ACETIC ACID LYSINE SALT (d,l) □ ISF 2508

TOXICITY DATA WITH REFERENCE
orl-rat LD50:895 mg/kg FRPSAX 32,286,77
ipr-rat LD50:429 mg/kg FRPSAX 32,286,77

SAFETY PROFILE: Moderately toxic by ingestion and intraperitoneal routes. When heated to decomposition it emits very toxic fumes of NO_x and Cl^-.

CFJ100 CAS:447-31-4 HR: 3
α-CHLORODEOXYBENZOIN
mf: $C_{14}H_{11}ClO$ mw: 230.70

SYNS: ACETOPHENONE, α-CHLORO-α-PHENYL- □ ACETOPHENONE, 2-CHLORO-2-PHENYL-(8CI) □ α-CHLOROBENZYL PHENYL KETONE □ 2-CHLORO-1,2-DIPHENYLETHANONE □ 2-CHLORO-2-PHENYLACETOPHENONE □ DESYL CHLORIDE □ ETHONE, 2-CHLORO-1, 2-DIPHENYL-(9CI)

TOXICITY DATA WITH REFERENCE
ipr-mus LDLo:500 mg/kg CBCCT* 5,284,53

CONSENSUS REPORTS: Reported in EPA TSCA Inventory.

DOT CLASSIFICATION: 3; *Label:* Flammable Liquid

SAFETY PROFILE: Moderately toxic by intraperitoneal route. Flammable liquid. When heated to decomposition it emits toxic vapors of Cl^-.

CFJ750 CAS:50-90-8 HR: 2
5-CHLORO-2′-DEOXYURIDINE
mf: $C_9H_{11}ClN_2O_5$ mw: 262.67

PROP: Fine needles from MeOH/Et_2O. Mp: 178–179.5°.

SYNS: 5-CHLORODEOXYURIDINE □ CLUDR

TOXICITY DATA WITH REFERENCE
msc-hmn:lym 100 µmol/L LIFSAK 19,563,76
cyt-ham:ovr 10 µmol/L BLFSBY 29A,69,84
sce-ham:ovr 10 µmol/L MUREAV 91,395,81
msc-ham:ovr 500 nmol/L ENMUDM 4,301,82
ipr-rat LD50:2000 mg/kg ADTEAS 3,181,68

SAFETY PROFILE: Moderately toxic by intraperitoneal route. Human mutation data reported. When heated to decomposition it emits very toxic fumes of Cl^- and NO_x.

CFK000 CAS:93-71-0 HR: 3
2-CHLORO-N,N-DIALLYLACETAMIDE
mf: $C_8H_{12}ClNO$ mw: 173.66

PROP: Amber liquid. Bp: 74° @ 0.3 mm. Sltly sol in water; sol in alc, hexane, and xylene.

SYNS: ALIDOCHLOR □ ALLIDOCHLOR □ CDAA □ CDAAT □ α-CHLORO-N,N-DIALLYLACETAMIDE □ 2-CHLORO-N,N-DI-2-PROPENYLACETAMIDE □ CP 6,343 □ DIALLYLCHLOROACETAMIDE □ N,N-DIALLYLCHLOROACETAMIDE □ N,N-DIALLYL-α-CHLOROACETAMIDE □ N,N-DIALLYL-2-CHLOROACETAMIDE □ NCI-CO4035 □ RADOX □ RANDOX □ RANTOX T

TOXICITY DATA WITH REFERENCE
orl-rat LD50:700 mg/kg RREVAH 10,97,65
skn-rat LD50:360 mg/kg WRPCA2 9,119,70

CONSENSUS REPORTS: Reported in EPA TSCA Inventory.

SAFETY PROFILE: Poison by skin contact. Moderately toxic by ingestion. An herbicide. When heated to decomposition it emits very toxic fumes of $Cl^-$and NO_x. See also ALLYL COMPOUNDS.

CFK125 CAS:95-83-0 HR: 3
4-CHLORO-1,2-DIAMINOBENZENE
mf: $C_6H_7ClN_2$ mw: 142.60

PROP: Leaflets from H_2O. Mp: 76°.

SYNS: p-CHLORO-o-PHENYLENEDIAMINE □ 4-CHLORO-o-PHENYLENEDIAMINE □ 4-CHLORO-1,2-PHENYLENEDIAMINE □ 4-Cl-o-PD □ 1,2-DIAMINO-4-CHLOROBENZENE □ 3,4-DIAMINOCHLOROBENZENE □ 3,4-DIAMINO-1-CHLOROBENZENE □ NCI-C03292 □ URSOL OLIVE 6G

TOXICITY DATA WITH REFERENCE
mma-sat 10 µg/plate ENMUDM 7(Suppl 5),1,85
mma-esc 1 mg/plate ENMUDM 7(Suppl 5),1,85
dnd-hmn:fbr 50 µmol/L MUREAV 127,107,84
cyt-mus-ipr 100 mg/kg ENMUDM 8(Suppl 6),53,86
orl-rat TDLo:135 g/kg/77W-C:CAR CRNGDP 1,495,80
orl-mus TDLo:324 g/kg/77W-C:CAR CRNGDP 1,495,80
orl-rat TD:273 g/kg/78W-C:CAR NCITR* NCI-CG-TR-63,78
orl-mus TD:917 g/kg/78W-C:CAR NCITR* NCI-CG-TR-63,78
orl-rat TD:136 g/kg/78W-C:CAR NCITR* NCI-CG-TR-63,78
orl-mus TD:460 g/kg/78W-C:CAR NCITR* NCI-CG-TR-63,78
orl-mus TD:40,500 g/kg/96W-C:NEO CRNGDP 1,495,80
orl-rat TD:18,750 g/kg/2Y-C:CAR CRNGDP 1,495,80

CONSENSUS REPORTS: NTP 7th Annual Report on Carcinogens. IARC Cancer Review: Group 2B IMEMDT 7,56,87; Human Limited Evidence IMEMDT 27,81,82; Animal Sufficient Evidence IMEMDT 27,81,82. NCI Carcinogenesis Bioassay (feed); Clear Evidence: mouse, rat NCITR* NCI-CG-TR-63,78. Reported in EPA TSCA Inventory.

SAFETY PROFILE: Confirmed with experimental carcinogenic and neoplastigenic data. Human mutation data reported. When heated to decomposition it emits toxic fumes of Cl^- and NO_x. See also AROMATIC AMINES.

CFK325 CAS:17927-57-0 HR: 3
CHLORODIBORANE
mf: B_2ClH_5 mw: 62.11

PROP: Colorless gas. Mp: −143.4°.

SAFETY PROFILE: A dangerous fire hazard. It ignites spontaneously in air and is a gas above −11°C. When heated to decomposition it emits toxic fumes of Cl^-. See also BORANES.

CFK500 CAS:124-48-1 HR: 2
CHLORODIBROMOMETHANE
mf: $CHBr_2Cl$ mw: 208.29

PROP: Colorless to pale-yellow, heavy liquid. Bp: 121–122°, fp: −22°, d: 2.440 @ 25°/25°.

SYNS: CDBM □ DIBROMOCHLOROMETHANE □ NCI-C55254

TOXICITY DATA WITH REFERENCE
mmo-sat 10 µL/plate DHEFDK FDA-78-1046,78
mrc-smc 100 µL/L MUREAV 155,53,85
sce-hmn:lym 400 µmol/L ENVRAL 32,72,83
sce-mus-orl 100 mg/kg/4D-I ENVRAL 32,72,83
orl-mus TDLo:52,500 mg/kg/2Y-C:CAR NTPTR* NTP-TR-282,85
orl-rat LD50:370 mg/kg SCIEAS 36(1-4),10,89
orl-mus LD50:800 mg/kg TXAPA9 44,213,78

CONSENSUS REPORTS: IARC Cancer Review: Group 3 IMEMDT 52,243,91; Animal Limited Evidence IMEMDT 52,243,91; Human Inadequate Evidence IMEMDT 52,243,91. NTP Carcinogenesis Studies (gavage); Some Evidence: mouse NTPTR* NTP-TR-282,86; No Evidence: rat NTPTR* NTP-TR-282,85; Reported in EPA TSCA Inventory.

SAFETY PROFILE: Moderately toxic by ingestion. Questionable carcinogen with experimental carcinogenic data. Human mutation data reported. Compounds of this type are generally irritating and narcotic. See also BROMOFORM and CHLOROFORM. When heated to decomposition it emits toxic fumes of Cl^- and Br^-.

CFK750 CAS:63869-01-2 HR: 3
CHLORO(DIBUTOXYPHOSPHINYL)MERCURY
mf: $C_8H_{18}ClHgO_3P$ mw: 429.27

SYN: (DIBUTOXYPHOSPHINYL)MERCURY CHLORIDE

TOXICITY DATA WITH REFERENCE
ipr-mus LDLo:31,300 µg/kg CBCCT* 8,104,56

CONSENSUS REPORTS: Mercury and its compounds are on the Community Right-To-Know List.

OSHA PEL: CL 0.1 mg(Hg)/m³ (skin)
ACGIH TLV: TWA 0.1 mg(Hg)/m³ (skin)
NIOSH REL: (Organomercury): TWA 0.01 mg/m³; STEL 0.03 mg/m³ (skin)

SAFETY PROFILE: Poison by intraperitoneal route. When heated to decomposition it emits very toxic fumes of PO_x, Hg, and Cl^-. See also MERCURY COMPOUNDS and CHLORIDES.

CFL000 CAS:77966-42-8 HR: 3
6′-CHLORO-2-(DIBUTYLAMINO)-o-ACETOTOLUIDIDE, HYDROCHLORIDE
mf: $C_{17}H_{27}ClN_2O{\cdot}ClH$ mw: 347.37

SYN: C 3072

TOXICITY DATA WITH REFERENCE
eye-rbt 2% SEV ARZNAD 8,407,58
ipr-rat LD50:250 mg/kg ARZNAD 8,407,58
scu-mus LD50:2295 mg/kg ARZNAD 8,407,58

SAFETY PROFILE: Poison by intraperitoneal route. Moderately toxic by subcutaneous route. A severe eye irritant. When heated to decomposition it emits very toxic fumes of Cl^- and NO_x.

CFL200 CAS:68786-66-3 HR: 3
6-CHLORO-5-(2,3-DICHLOROPHENOXY)-2-METHYLTHIO-BENZIMIDAZOLE
mf: $C_{14}H_9Cl_3N_2OS$ mw: 359.66

PROP: Crystals from EtOH (aq). Mp: 175–176°.

SYNS: 1H-BENZIMIDAZOLE, 5-CHLORO-6-(2,3-DICHLOROPHENOXY)-2-(METHYLTHIO)- □ CGA 89317 □ 5-CHLORO-6-(2,3-DICHLOROPHENOXY)-2-(METHYLTHIO)-1H-BENZIMIDAZOLE □ FASINEX □ TRICLABENDAZOLE

TOXICITY DATA WITH REFERENCE
orl-rat TDLo:800 mg/kg (female 8-15D post):TER TXCYAC 43,283,87
orl-rbt LD50:206 mg/kg MDACAP 21,227,85

SAFETY PROFILE: Poison by ingestion. Experimental teratogenic effects. When heated to decomposition it emits toxic fumes of SO_x, NO_x, and Cl^-.

CFL500 CAS:77791-58-3 HR: 2
2′-CHLORO-2-(DIETHYLAMINO)ACETANILIDE HYDROCHLORIDE
mf: $C_{12}H_{17}ClN_2O{\cdot}ClH$ mw: 277.22

SYN: C 3070

TOXICITY DATA WITH REFERENCE
eye-rbt 2% MOD ARZNAD 8,270,58
ipr-rat LD50:550 mg/kg ARZNAD 8,270,58
ipr-mus LD50:600 mg/kg ARZNAD 8,270,58
scu mus LD50:2150 mg/kg ARZNAD 8,270,58

SAFETY PROFILE: Moderately toxic by intraperitoneal and subcutaneous routes. An eye irritant. When heated to decomposition it emits very toxic fumes of Cl^- and NO_x.

CFL750 CAS:55489-49-1 HR: 3
3′-CHLORO-2-(DIETHYLAMINO)ACETANILIDE HYDROCHLORIDE
mf: $C_{12}H_{17}ClN_2O{\cdot}ClH$ mw: 277.22

SYN: C 3191

TOXICITY DATA WITH REFERENCE
eye-rbt 2% MLD ARZNAD 8,170,58
ipr-rat LD50:370 mg/kg ARZNAD 8,170,58
ipr-mus LD50:175 mg/kg JAPMA8 49,80,60
scu-mus LD50:375 mg/kg JAPMA8 49,80,60

SAFETY PROFILE: Poison by subcutaneous and intraperitoneal routes. An eye irritant. When heated to decomposition it emits very toxic fumes of Cl^- and NO_x.

CFM000 CAS:74816-28-7 HR: 3
4′-CHLORO-2-(DIETHYLAMINO)ACETANILIDE HYDROCHLORIDE
mf: $C_{12}H_{17}ClN_2O{\cdot}ClH$ mw: 277.22

SYN: C 3061

TOXICITY DATA WITH REFERENCE
eye-rbt 2% MLD ARZNAD 8,170,58
ipr-rat LD50:318 mg/kg ARZNAD 8,170,58
ipr-mus LD50:375 mg/kg ARZNAD 8,170,58
scu-mus LD50:800 mg/kg ARZNAD 8,170,58

SAFETY PROFILE: Poison by intraperitoneal route. Moderately toxic by subcutaneous route. An eye irritant. When heated to decomposition it emits very toxic fumes of Cl^- and NO_x.

CFM250 CAS:77966-46-2 HR: 3
3′-CHLORO-2-(DIETHYLAMINO)-o-ACETOTOLUIDIDE HYDROCHLORIDE
mf: $C_{13}H_{19}ClN_2O{\cdot}ClH$ mw: 291.25

SYNS: C 3037 □ 2′-CHLORO-2-(DIETHYLAMINO)-2′-METHYLACETANILIDE HYDROCHLORIDE

TOXICITY DATA WITH REFERENCE
eye-rbt 2% MLD ARZNAD 8,270,58
ipr-rat LD50:245 mg/kg ARZNAD 8,270,58
ipr-mus LD50:200 mg/kg JAPMA8 49,80,60
scu-mus LD50:375 mg/kg JAPMA8 49,80,60

SAFETY PROFILE: Poison by subcutaneous and intraperitoneal routes. An eye irritant. When heated to decomposition it emits very toxic fumes of $Cl^-$and NO_x.

CFM500 CAS:77966-47-3 HR: 3
4′-CHLORO-2-(DIETHYLAMINO)-o-ACETOTOLUIDIDE HYDROCHLORIDE
mf: $C_{13}H_{19}ClN_2O{\cdot}ClH$ mw: 291.25

SYNS: C 3101 □ 4′-CHLORO-2-(DIETHYLAMINO)-2′-METHYLACETANILIDE HYDROCHLORIDE

TOXICITY DATA WITH REFERENCE
eye-rbt 2% MLD ARZNAD 8,270,58
ipr-rat LD50:350 mg/kg ARZNAD 8,270,58
scu-mus LD50:860 mg/kg ARZNAD 8,270,58

SAFETY PROFILE: Poison by intraperitoneal route. Moderately toxic by subcutaneous route. An eye irritant. When heated to decomposition it emits very toxic fumes of Cl^- and NO_x.

CFM750 CAS:77966-48-4 HR: 3
5′-CHLORO-2-(DIETHYLAMINO)-o-ACETOTOLUIDIDE HYDROCHLORIDE
mf: $C_{13}H_{19}ClN_2O{\cdot}ClH$ mw: 291.25

SYNS: C 3152 □ 3′-CHLORO-2-(DIETHYLAMINO)-6′-METHYLACETANILIDE HYDROCHLORIDE

TOXICITY DATA WITH REFERENCE
eye-rbt 2% MLD ARZNAD 8,270,58
ipr-rat LD50:261 mg/kg ARZNAD 8,270,58
scu-mus LD50:1125 mg/kg ARZNAD 8,270,58

SAFETY PROFILE: Poison by intraperitoneal route. Moderately toxic by subcutaneous route. An eye irritant. When heated to decomposition it emits very toxic fumes of Cl^- and NO_x.

CFN000 CAS:77966-49-5 HR: 3
6′-CHLORO-2-(DIETHYLAMINO)-m-ACETOTOLUIDIDE HYDROCHLORIDE
mf: $C_{13}H_{19}ClN_2O{\cdot}ClH$ mw: 291.25

SYNS: C 3201 □ 2′-CHLORO-2-(DIETHYLAMINO)-5′-METHYLACETANILIDE HYDROCHLORIDE

TOXICITY DATA WITH REFERENCE
eye-rbt 2% MOD ARZNAD 8,270,58
ipr-rat LD50:370 mg/kg ARZNAD 8,270,58
scu-mus LD50:2450 mg/kg ARZNAD 8,270,58

SAFETY PROFILE: Poison by intraperitoneal route. Moderately toxic by subcutaneous route. An eye irritant. When heated to decomposition it emits very toxic fumes of Cl^- and NO_x.

CFN500 CAS:77985-16-1 HR: 3
6′-CHLORO-3-(DIETHYLAMINO)-o-BUTYROTOLUIDIDE HYDROCHLORIDE
mf: $C_{15}H_{23}ClN_2O{\cdot}ClH$ mw: 319.31

SYN: C 5126

TOXICITY DATA WITH REFERENCE
ipr-rat LD50:35 mg/kg ARZNAD 8,544,58
ipr-mus LD50:30 mg/kg ARZNAD 8,544,58
scu-mus LD50:40 mg/kg ARZNAD 8,544,58

SAFETY PROFILE: Poison by intraperitoneal and subcutaneous routes. When heated to decomposition it emits very toxic fumes of NO_x and Cl^-.

CFO000 CAS:102489-48-5 HR: 3
6′-CHLORO-2-(2-(DIETHYLAMINO)ETHOXY)-o-ACETOTOLUIDIDE HYDROCHLORIDE
mf: $C_{15}H_{23}ClN_2O_2{\cdot}ClH$ mw: 335.31

SYN: C 3068

TOXICITY DATA WITH REFERENCE
eye-rbt 2% MLD ARZNAD 9,113,59
ipr-rat LD50:175 mg/kg ARZNAD 9,113,59
scu-mus LD50:465 mg/kg ARZNAD 9,113,59

SAFETY PROFILE: Poison by intraperitoneal route. Moderately toxic by subcutaneous route. An eye irritant. When heated to decomposition it emits very toxic fumes of Cl^- and NO_x.

CFO250 CAS:7432-27-1 HR: 3
5-CHLORO-2-(2-(DIETHYLAMINO)ETHOXY)BENZANILIDE
mf: $C_{19}H_{23}ClN_2O_2$ mw: 346.89

SYN: o-DIAETHYLAMINOAETHOXY-5-CHLOR-BENZANILID (GERMAN)

TOXICITY DATA WITH REFERENCE
scu-mus TDLo:20 mg/kg (1-19D preg):REP ARZNAD 18,658,68

scu-mus TDLo: 20 mg/kg (1-19D preg): TER ARZNAD 18,658,68
orl-mus LD50: 480 mg/kg ARZNAD 16,1127,66
scu-mus LD50: 720 mg/kg ARZNAD 16,1127,66
ivn-mus LD50: 49 mg/kg ARZNAD 16,1127,66

SAFETY PROFILE: Poison by intravenous route. Moderately toxic by ingestion and subcutaneous routes. An experimental teratogen. Other experimental reproductive effects. When heated to decomposition it emits very toxic fumes of Cl^- and NO_x.

CFO750 CAS:56287-41-3 HR: 3
5-CHLORO-2-(2-(2-(DIETHYLAMINO)ETHOXY)ETHYL)-2-METHYL-1,3-BENZODIOXOLE
mf: $C_{16}H_{24}ClNO_3$ mw: 313.86

SYNS: 2-(2-(5-CHLORO-2-METHYL-1,3-BENZODIOXOL-2-YL)ETHOXY)-N,N-DIETHYLETHANAMINE □ LR-529 □ 2 METHYL-5-CHLORO-2-(N,N-DIETHYLAMINOETHOXYETHYL)-1,3-BENZODIOXOLE

TOXICITY DATA WITH REFERENCE
ivn-rat LD50: 18 mg/kg DRFUD4 3,379,78
ipr-mus LD50: 111 mg/kg DRFUD4 3,379,78

SAFETY PROFILE: Poison by intravenous and intraperitoneal routes. When heated to decomposition it emits very toxic fumes of Cl^- and NO_x.

CFP000 CAS:102489-49-6 HR: 3
6'-CHLORO-2-(2-(DIETHYLAMINO)ETHYL)AMINO-o-ACETOTOLUIDIDE HYDROCHLORIDE
mf: $C_{15}H_{24}ClN_3O \cdot ClH$ mw: 334.33

SYN: C 3173

TOXICITY DATA WITH REFERENCE
ipr-rat LD50: 305 mg/kg ARZNAD 9,167,59
scu-mus LD50: 1010 mg/kg ARZNAD 9,167,59

SAFETY PROFILE: Poison by intraperitoneal route. Moderately toxic by subcutaneous route. When heated to decomposition it emits very toxic fumes of Cl^- and NO_x.

CFP250 CAS:52400-77-8 HR: 3
5-CHLORO-2-(2-(2-(DIETHYLAMINO)ETHYLAMINO) ETHYL)-2-METHYL-1,3-BENZODIOXOLE DIHYDROCHLORIDE
mf: $C_{16}H_{25}ClN_2O_2 \cdot 2ClH$ mw: 385.80

TOXICITY DATA WITH REFERENCE
ivn-rat LD50: 35 mg/kg EJMCA5 12,413,77
ipr-mus LD50: 132 mg/kg EJMCA5 12,413,77

SAFETY PROFILE: Poison by intravenous and intraperitoneal routes. When heated to decomposition it emits very toxic fumes of NO_x and Cl^-.

CFP750 CAS:43047-59-2 HR: 3
8-CHLORO-2-(2-(DIETHYLAMINO)ETHYL)-2H-(1)-BENZOTHIOPYRANO(4,3,2-cd)INDAZOLE-5-METHANOL MONOMETHANE SULFONATE
mf: $C_{20}H_{22}ClN_3OS \cdot CH_4O_3S$ mw: 484.07

SYN: IA-4

TOXICITY DATA WITH REFERENCE
mmo-sat 465 nmol/plate JPETAB 200,1,77
mma-sat 465 nmol/plate JPETAB 200,1,77
mmo-nsc 20 µmol/L JTEHD6 1,271,75
bfa-mus/sat 100 mg/kg JPETAB 200,1,77
hma-mus/sat 100 mg/kg JPETAB 200,1,77
ivn-mus LD50: 131 mg/kg JPETAB 186,402,73
ims-mus LD50: 1560 mg/kg JPETAB 186,402,73

CONSENSUS REPORTS: EPA Genetic Toxicology Program.

SAFETY PROFILE: Poison by intravenous route. Moderately toxic by intramuscular route. Mutation data reported. When heated to decomposition it emits very toxic fumes of SO_x, Cl^-, and NO_x. See also SULFONATES.

CFQ000 CAS:54484-91-2 HR: 2
8-CHLORO-2-(2-(DIETHYLAMINO)ETHYL)-2H-(1)BENZOTHIOPYRANO(4,3,2-cd)INDAZOLE-5-METHANOL-N-OXIDE
mf: $C_{20}H_{22}ClN_3O_2S$ mw: 403.96

SYN: IA-4 N-OXIDE

TOXICITY DATA WITH REFERENCE
mmo-sat 492 nmol/plate JPETAB 200,1,77
mma-sat 492 nmol/plate JPETAB 200,1,77
sln-dmg-par 2800 µmol/L MUREAV 82,111,81
bfa-mus/sat 200 mg/kg CNREA8 38,4478,78
hma-mus/sat 100 mg/kg JPETAB 200,1,77
ims-mus LD50: 3500 mg/kg JPETAB 200,1,77

CONSENSUS REPORTS: EPA Genetic Toxicology Program.

SAFETY PROFILE: Moderately toxic by intramuscular route. Mutation data reported. When heated to decomposition it emits very toxic fumes of SO_x, NO_x, and Cl^-.

CFQ250 CAS:101651-60-9 HR: 3
2'-CHLORO-2-((2-(DIETHYLAMINO)ETHYL)ETHYLAMINO)ACETANILIDE DIHYDROCHLORIDE
mf: $C_{16}H_{26}ClN_3O \cdot 2ClH$ mw: 384.82

SYN: C 5412

TOXICITY DATA WITH REFERENCE
eye-rbt 2% MLD ARZNAD 9,262,59
ipr-rat LD50: 100 mg/kg ARZNAD 9,262,59
scu-mus LD50: 460 mg/kg ARZNAD 9,262,59

SAFETY PROFILE: Poison by intraperitoneal route. Moderately toxic by subcutaneous route. An eye irritant. When heated to decomposition it emits very toxic fumes of Cl^- and NO_x.

CFQ500 CAS:102489-50-9 HR: 3
6'-CHLORO-2-((2-(DIETHYLAMINO)ETHYL)ETHYLAMINO)-o-ACETOTOLUIDIDE HYDROCHLORIDE
mf: $C_{17}H_{28}ClN_3O \cdot ClH$ mw: 362.39

SYN: C 3253

TOXICITY DATA WITH REFERENCE
eye-rbt 2% MLD ARZNAD 9,167,59
ipr-rat LD50: 55 mg/kg ARZNAD 9,167,59

scu-mus LD50:210 mg/kg ARZNAD 9,167,59

SAFETY PROFILE: Poison by subcutaneous and intraperitoneal routes. An eye irritant. When heated to decomposition it emits very toxic fumes of $Cl^-$and NO_x.

CFQ750 CAS:102489-51-0 HR: 3
6′-CHLORO-2-((2-(DIETHYLAMINO)ETHYL)ISOPROPYLAMINO)-o-ACETOTOLUIDIDE HYDROCHLORIDE
mf: $C_{18}H_{30}ClN_3O{\bullet}ClH$ mw: 376.42

SYN: C 5384

TOXICITY DATA WITH REFERENCE
eye-rbt 2% MLD ARZNAD 9,167,59
ipr-rat LD50:22 mg/kg ARZNAD 9,167,59
scu-mus LD50:60 mg/kg ARZNAD 9,167,59

SAFETY PROFILE: Poison by subcutaneous and intraperitoneal routes. An eye irritant. When heated to decomposition it emits very toxic fumes of $Cl^-$and NO_x.

CFR000 CAS:55489-49-1 HR: 3
2′-CHLORO-2-(2-(DIETHYLAMINO)ETHYL)METHYLAMINOACETANILIDE DIHYDROCHLORIDE
mf: $C_{15}H_{24}ClN_3O{\bullet}2ClH$ mw: 370.79

SYN: C 5366

TOXICITY DATA WITH REFERENCE
ipr-rat LD50:160 mg/kg ARZNAD 9,262,59
scu-mus LD50:1040 mg/kg ARZNAD 9,262,59

SAFETY PROFILE: Poison by intraperitoneal route. Moderately toxic by subcutaneous route. When heated to decomposition it emits very toxic fumes of Cl^- and NO_x.

CFR250 CAS:77791-57-2 HR: 3
4′-CHLORO-2-(2-(DIETHYLAMINO)ETHYL)METHYLAMINOACETANILIDE DIHYDROCHLORIDE
mf: $C_{15}H_{24}ClN_3O{\bullet}2ClH$ mw: 370.79

SYN: C 5400

TOXICITY DATA WITH REFERENCE
eye-rbt 2% MLD ARZNAD 8,262,59
ipr-rat LD50:166 mg/kg ARZNAD 9,262,59
scu-mus LD50:435 mg/kg ARZNAD 9,262,59

SAFETY PROFILE: Poison by intraperitoneal route. Moderately toxic by subcutaneous route. An eye irritant. When heated to decomposition it emits very toxic fumes of Cl^- and NO_x.

CFR500 CAS:77984-94-2 HR: 3
3′-CHLORO-2-(2-(DIETHYLAMINO)ETHYL)METHYLAMINO-o-ACETOTOLUIDIDE DIHYDROCHLORIDE
mf: $C_{16}H_{26}ClN_3O{\bullet}2ClH$ mw: 384.82

SYN: C 5397

TOXICITY DATA WITH REFERENCE
ipr-rat LD50:72 mg/kg ARZNAD 9,262,59
scu-mus LD50:500 mg/kg ARZNAD 9,262,59

SAFETY PROFILE: Poison by intraperitoneal route. Moderately toxic by subcutaneous route. When heated to decomposition it emits very toxic fumes of Cl^- and NO_x.

CFR750 CAS:77966-43-9 HR: 3
4′-CHLORO-2-(2-(DIETHYLAMINO)ETHYL)METHYLAMINO-o-ACETOTOLUIDIDE DIHYDROCHLORIDE
mf: $C_{16}H_{26}ClN_3O{\bullet}2ClH$ mw: 384.82

SYN: C 5401

TOXICITY DATA WITH REFERENCE
eye-rbt 2% MLD ARZNAD 9,262,59
ipr-rat LD50:135 mg/kg ARZNAD 9,262,59
scu-mus LD50:350 mg/kg ARZNAD 9,262,59

SAFETY PROFILE: Poison by subcutaneous and intraperitoneal routes. An eye irritant. When heated to decomposition it emits very toxic fumes of $Cl^-$and NO_x.

CFS000 CAS:77966-44-0 HR: 3
5′-CHLORO-2-(2-(DIETHYLAMINO)ETHYL)METHYLAMINO-o-ACETOTOLUIDIDE DIHYDROCHLORIDE
mf: $C_{16}H_{26}ClN_3O{\bullet}2ClH$ mw: 384.82

SYN: C 5402

TOXICITY DATA WITH REFERENCE
ipr-rat LD50:114 mg/kg ARZNAD 9,262,59
scu-mus LD50:340 mg/kg ARZNAD 9,262,59

SAFETY PROFILE: Poison by subcutaneous and intraperitoneal routes. When heated to decomposition it emits very toxic fumes of Cl^- and NO_x.

CFS250 CAS:77966-45-1 HR: 3
6′-CHLORO-2-(2-(DIETHYLAMINO)ETHYL)METHYLAMINO-m-ACETOTOLUIDIDE DIHYDROCHLORIDE
mf: $C_{16}H_{26}ClN_3O{\bullet}2ClH$ mw: 384.82

SYN: C 5398

TOXICITY DATA WITH REFERENCE
eye-rbt 2% MLD ARZNAD 9,262,59
ipr-rat LD50:104 mg/kg ARZNAD 9,262,59
scu-mus LD50:750 mg/kg ARZNAD 9,262,59

SAFETY PROFILE: Poison by intraperitoneal route. Moderately toxic by subcutaneous route. An eye irritant. When heated to decomposition it emits very toxic fumes of Cl^- and NO_x.

CFS500 CAS:102489-52-1 HR: 3
6′-CHLORO-2-((2-(DIETHYLAMINO)ETHYL)METHYLAMINO)-o-ACETOTOLUIDIDE HYDROCHLORIDE
mf: $C_{16}H_{26}ClN_3O{\bullet}ClH$ mw: 348.36

SYN: C 3249

TOXICITY DATA WITH REFERENCE
eye-rbt 2% MLD ARZNAD 9,167,59
ipr-rat LD50:110 mg/kg ARZNAD 9,167,59
scu-mus LD50:460 mg/kg ARZNAD 9,167,59

SAFETY PROFILE: Poison by intraperitoneal route. Moderately toxic by subcutaneous route. An eye irritant.

When heated to decomposition it emits very toxic fumes of Cl^- and NO_x.

CFS750 CAS:102489-53-2 HR: 3
6'-CHLORO-2-((2-(DIETHYLAMINO)ETHYL)OCTYLAMINO)-o-ACETOTOLUIDIDE HYDROCHLORIDE
mf: $C_{23}H_{40}ClN_3O{\cdot}ClH$ mw: 446.57

SYN: C 5347

TOXICITY DATA WITH REFERENCE
eye-rbt 2% SEV ARZNAD 9,167,59
ipr-rat LD50:98 mg/kg ARZNAD 9,167,59
scu-mus LD50:260 mg/kg ARZNAD 9,167,59

SAFETY PROFILE: Poison by intraperitoneal and subcutaneous routes. A severe eye irritant. When heated to decomposition it emits very toxic fumes of Cl^- and NO_x.

CFT000 CAS:102489-54-3 HR: 3
6'-CHLORO-2-((2-(DIETHYLAMINO)ETHYL)(2-PHENOXYETHYL)AMINO)-o-ACETOTOLUIDIDE HYDROCHLORIDE
mf: $C_{23}H_{32}ClN_3O_2{\cdot}ClH$ mw: 454.49

SYN: C 5290

TOXICITY DATA WITH REFERENCE
eye-rbt 2% SEV ARZNAD 9,113,59
scu-mus LD50:92 mg/kg ARZNAD 9,113,59

SAFETY PROFILE: Poison by subcutaneous route. A severe eye irritant. When heated to decomposition it emits very toxic fumes of Cl^- and NO_x.

CFT250 CAS:102489-55-4 HR: 3
6'-CHLORO-2-((2-(DIETHYLAMINO)ETHYL)PROPYLAMINO)-o-ACETOTOLUIDIDE HYDROCHLORIDE
mf: $C_{18}H_{30}ClN_3O{\cdot}ClH$ mw: 376.42

SYN: C 5385

TOXICITY DATA WITH REFERENCE
eye-rbt 2% MLD ARZNAD 9,167,59
ipr-rat LD50:28 mg/kg ARZNAD 9,167,59
scu-mus LD50:74 mg/kg ARZNAD 9,167,59

SAFETY PROFILE: Poison by intraperitoneal and subcutaneous routes. An eye irritant. When heated to decomposition it emits very toxic fumes of $Cl^-$and NO_x.

CFT500 CAS:101651-61-0 HR: 2
2'-CHLORO-2-(2-(DIETHYLAMINO)ETHYLTHIO)ACETANILIDE HYDROCHLORIDE
mf: $C_{14}H_{21}ClN_2OS{\cdot}ClH$ mw: 337.34

SYN: C 4920

TOXICITY DATA WITH REFERENCE
eye-rbt 2% MLD ARZNAD 9,683,59
ipr-rat LD50:490 mg/kg ARZNAD 9,683,59
scu-mus LD50:1750 mg/kg ARZNAD 9,683,59

SAFETY PROFILE: Moderately toxic by intraperitoneal and subcutaneous routes. An eye irritant. When heated to decomposition it emits very toxic fumes of Cl^-, SO_x, and NO_x.

CFT750 CAS:102489-56-5 HR: 3
6'-CHLORO-2-(2-(DIETHYLAMINO)ETHYLTHIO)-o-ACETOTOLUIDIDE HYDROCHLORIDE
mf: $C_{15}H_{23}ClN_2OS{\cdot}ClH$ mw: 351.37

SYN: C 4926

TOXICITY DATA WITH REFERENCE
eye-rbt 2% MLD ARZNAD 9,683,59
ipr-rat LD50:118 mg/kg ARZNAD 9,683,59
scu-mus LD50:350 mg/kg ARZNAD 9,683,59

SAFETY PROFILE: Poison by intraperitoneal and subcutaneous routes. An eye irritant. When heated to decomposition it emits very toxic fumes of Cl^-, SO_x, and NO_x.

CFU000 CAS:101651-90-5 HR: 3
7-CHLORO-10-(3-(DIETHYLAMINO)-2-HYDROXYPROPYL)ISOALLOXAZINE SULFATE
mf: $C_{17}H_{20}ClN_5O_3{\cdot}H_2O_4S$ mw: 475.95

TOXICITY DATA WITH REFERENCE
ipr-rat LD50:50 mg/kg CMTRAG 2,96,61
scu-mus LD50:48 mg/kg CMTRAG 2,96,61
ivn-mus LD50:132 mg/kg CMTRAG 2,96,61
ims-mus LD50:54 mg/kg CMTRAG 2,96,61

SAFETY PROFILE: Poison by intraperitoneal, subcutaneous, intravenous, and intramuscular routes. When heated to decomposition it emits very toxic fumes of SO_x, NO_x, and Cl^-.

CFU250 CAS:74816-32-3 HR: 3
4'-CHLORO-2-(DIETHYLAMINO)-N-METHYLACETANILIDE HYDROCHLORIDE
mf: $C_{13}H_{19}ClN_2O{\cdot}ClH$ mw: 291.25

SYN: C 3049

TOXICITY DATA WITH REFERENCE
eye-rbt 2% MLD ARZNAD 8,609,58
ipr-rat LD50:220 mg/kg ARZNAD 8,609,58
scu-mus LD50:350 mg/kg ARZNAD 8,609,58

SAFETY PROFILE: Poison by intraperitoneal and subcutaneous routes. An eye irritant. When heated to decomposition it emits very toxic fumes of NO_x and Cl^-.

CFU500 CAS:77966-51-9 HR: 3
6'-CHLORO-2-(DIETHYLAMINO)-N-METHYL-o-ACETOTOLUIDIDE HYDROCHLORIDE
mf: $C_{14}H_{21}ClN_2O{\cdot}ClH$ mw: 305.28

SYN: V 316

TOXICITY DATA WITH REFERENCE
ipr-rat LD50:136 mg/kg ARZNAD 8,609,58
scu-mus LD50:305 mg/kg ARZNAD 8,609,58

SAFETY PROFILE: Poison by intraperitoneal and subcutaneous routes. When heated to decomposition it emits very toxic fumes of Cl^- and NO_x.

CFU750 CAS:69-05-6 HR: 3
6-CHLORO-9-((4-(DIETHYLAMINO)-1-METHYLBUTYL) AMINO)-2-METHOXYACRIDINE DIHYDROCHLORIDE
mf: $C_{23}H_{30}ClN_3O{\cdot}2ClH$ mw: 472.93

PROP: Crystals from H_2O. Mp: 248–250°.

SYNS: ACRICHINE □ ARICHIN □ ATABRINE DIHYDROCHLORIDE □ ATABRINE HYDROCHLORIDE □ CHEMIOCHIN □ CHINACRIN HYDROCHLORIDE □ 2-CHLORO-5-(ω-DIETHYLAMINO-α-METHYLBUTYLAMINO)-7-METHOXYACRIDINE DIHYDROCHLORIDE □ 3-CHLORO-9-(4′-DIETHYLAMINO-1′-METHYLBUTYLAMINO)-7-METHOXYACRIDINE DIHYDROCHLORIDE □ 3-CHLORO-7-METHOXY-9-(1-METHYL-4-DIETHYLAMINOBUTYLAMINO)ACRIDINE DIHYDROCHLORIDE □ CRINODORA □ DIAL □ ERION □ ITALCHIN □ MALARICIDA □ MECRYL □ MEPACRINE DIHYDROCHLORIDE □ MEPACRINE HYDROCHLORIDE □ METHOQUINE □ 2-METHOXY-6-CHLORO-9-(4-DIETHYLAMINO-1-METHYLBUTYLAMINO)ACRIDINE-DIHYDROCHLORIDE □ METOQUINE □ PALACRIN □ PENTILEN □ QUINACRINE DIHYDROCHLORIDE □ QUINACRINE HYDROCHLORIDE □ 866 R.P. □ SN 390

TOXICITY DATA WITH REFERENCE
mma-sat 1 mg/plate MUREAV 22,295,74
sln-dmg-orl 5 mmol/L MUREAV 158,177,85
icv-wmn TDLo:20 mg/kg (1D pre):REP CCPTAY 14,75,76
iut-wmn TDLo:15 mg/kg (3D pre):REP FESTAS 21,581,70
iut-rat TDLo:2 mg/kg (female 8D post):REP 49VWAG -,71,83
iut-mus TDLo:80 mg/kg (female 1D post):REP CCPTAY 17,231,78
unr-grb TDLo:375 mg/kg (female 25D pre):REP IJEBA6 16,984,78
iut-rat TDLo:50 mg/kg (female 1D pre):REP IJEBA6 16,1074,78
iut-rat TDLo:200 mg/kg (female 1D pre):REP CCPTAY 29,553,84
itt-mky TDLo:20 mg/kg (male 1D pre):REP FESTAS 39(Suppl),441,83
orl-man TDLo:34 mg/kg/8D-I SMJOAV 75,359,82
orl-wmn TDLo:18 mg/kg/3D-I SMJOAV 75,359,82
orl-rat LD50:660 mg/kg JPETAB 91,157,47
ivn-rat LD50:29 mg/kg JPETAB 91,157,47
iut-rat LD50:100 mg/kg IJEBA6 1074,78
orl-mus LD50:557 mg/kg JPETAB 91,157,47
ipr-mus LD50:189 mg/kg JPETAB 91,133,47
scu-mus LD50:212 mg/kg ABEMAV 1,317,41
ivn-mus LD50:38 mg/kg JPETAB 91,157,47
ivn-dog LDLo:20 mg/kg JAPMA8 34,20,45
orl-cat LDLo:200 mg/kg AEPPAE 170,328,33

CONSENSUS REPORTS: EPA Genetic Toxicology Program.

SAFETY PROFILE: Poison by ingestion, subcutaneous, intrauterine, intravenous, and intraperitoneal routes. Human reproductive effects by intrauterine and intracervical routes: changes in fertility and unspecified effects on the uterus, cervix and vagina. Experimental reproductive effects. Mutation data reported. Used as a treatment for parasitic worms. When heated to decomposition it emits very toxic fumes of Cl^- and NO_x.

CFW000 CAS:77966-53-1 HR: 2
2′-CHLORO-2-(DIETHYLAMINO)-5′-TRIFLUOROMETHYLACETANILIDE HYDROCHLORIDE
mf: $C_{13}H_{16}ClF_3N_2O{\cdot}ClH$ mw: 345.22

SYN: C 3078

TOXICITY DATA WITH REFERENCE
eye-rbt 2% SEV ARZNAD 8,270,58
ipr-rat LD50:590 mg/kg ARZNAD 8,270,58
scu-mus LD50:4250 mg/kg ARZNAD 8,270,58

SAFETY PROFILE: Moderately toxic by intraperitoneal route. Mildly toxic by subcutaneous route. A severe eye irritant. When heated to decomposition it emits very toxic fumes of Cl^-, F^-, and NO_x.

CFW250 CAS:77966-52-0 HR: 3
4′-CHLORO-2-(DIETHYLAMINO)-3′-TRIFLUOROMETHYLACETANILIDE HYDROCHLORIDE
mf: $C_{13}H_{16}ClF_3N_2O{\cdot}ClH$ mw: 345.22

SYN: C 3074

TOXICITY DATA WITH REFERENCE
eye-rbt 2% SEV ARZNAD 8,270,58
ipr-rat LD50:300 mg/kg ARZNAD 8,270,58
scu-mus LD50:1175 mg/kg ARZNAD 8,270,58

SAFETY PROFILE: Poison by intraperitoneal route. Moderately toxic by subcutaneous route. A severe eye irritant. See also FLUORIDES. When heated to decomposition it emits very toxic fumes of Cl^-, F^-, and NO_x.

CFW625 CAS:5314-83-0 HR: 3
CHLORODIETHYLBORANE
mf: $C_4H_{10}BCl$ mw: 104.39

PROP: Air and moisture-sensitive liquid. Bp: 25° @ 100 mm.

SAFETY PROFILE: Ignites spontaneously in air. When heated to decomposition it emits toxic fumes of Cl^-. See also BORANES.

CFW750 CAS:23184-66-9 HR: 2
2-CHLORO-2′,6′-DIETHYL-N-(BUTOXYMETHYL)ACETANILIDE
mf: $C_{17}H_{26}ClNO_2$ mw: 311.89

PROP: Light yellow or amber oily liquid. D: 1.070 @ 30°/4°, bp: 196° @ 0.5 mm. Sltly sol in H_2O; sol in most org solvs.

SYNS: BUTACHLOR □ BUTANEX □ N-BUTOXYMETHYL-2-CHLORO-2′,6′-DIETHYLACETANILIDE □ N-(BUTOXYMETHYL)-2-CHLORO-N-(2,6-DIETHYLPHENYL)ACETAMIDE □ CP 53619 □ LAMBAST □ MACHETE □ MACHETE (herbicide) □ MACHETTE

TOXICITY DATA WITH REFERENCE
mma-sat 500 µg/plate MUREAV 116,185,83
mmo-omi 4 mg/L JASIAB 104,571,85
orl-rat LD50:1740 mg/kg EKMMA8 13,123,74
skn-rbt LD50:4080 mg/kg FMCHA2 -,C39,83

SAFETY PROFILE: Moderately toxic by ingestion. Mildly toxic by skin contact. Mutation data reported. An herbicide. When heated to decomposition it emits very toxic fumes of Cl^- and NO_x.

CFX000 **CAS:15972-60-8** **HR: 2**
2-CHLORO-2′,6′-DIETHYL-N-(METHOXYMETHYL)ACETANILIDE
mf: $C_{14}H_{20}ClNO_2$ mw: 269.80

PROP: Crystals. Sltly sol in H_2O; sol in Me_2CO, C_6H_6, EtOH, and EtOAc.

SYNS: ALACHLOR (USDA) □ ALANEX □ ALOCHLOR □ CHLORESSIGSAEURE-N-(METHOXYMETHYL)-2,6-DIAETHYLANILID (GERMAN) □ 2-CHLORO-N-(2,6-DIETHYLPHENYL)-N-(METHOXYMETHYL)ACETAMIDE □ CP 50144 □ LASSO □ LAZO □ METACHLOR □ METHACHLOR □ PILLARZO

TOXICITY DATA WITH REFERENCE
mmo-omi 90 mg/L JASIAB 104,571,85
mrc-smc 33 μg/plate MUREAV 136,233,84
cyt-hmn:lym 4 mg/L MUREAV 116,341,83
cyt-rat-ipr 1250 μg/kg MUREAV 116,341,83
orl-mus TDLo:142 g/kg/78W-C:CAR TXCYAC 3,383,87
orl-rat LD50:930 mg/kg TXCYAC 3,383,87
orl-rat LD50:1200 mg/kg WRPCA2 9,119,70
orl-mus LD50:462 mg/kg GTPZAB 21(12),30,77
skn-rbt LD50:3500 mg/kg GUCHAZ 6,3,73
orl-mam LD50:3000 mg/kg GUCHAZ 6,3,73
unk-mam LD50:1200 mg/kg 30ZDA9 -,126,71

CONSENSUS REPORTS: EPA Genetic Toxicology Program.

SAFETY PROFILE: Moderately toxic by ingestion, skin contact, and possibly other routes. Questionable carcinogen with experimental carcinogenic data. Human mutation data reported. When heated to decomposition it emits very toxic fumes of Cl^- and NO_x.

CFX125 **CAS:68674-44-2** **HR: 3**
CHLORODIFLUOROACETYL HYPOCHLORITE
mf: $C_2Cl_2F_2O_2$ mw: 164.92

SAFETY PROFILE: An unstable explosive above 22°C. When heated to decomposition it emits toxic fumes of F^- and Cl^-. See also HYPOCHLORITES.

CFX250 **CAS:75-68-3** **HR: 1**
1-CHLORO-1,1-DIFLUOROETHANE
DOT: UN 2517
mf: $C_2H_3ClF_2$ mw: 100.50

PROP: Gas. Mp: −131°, bp: −9.5°, d: 1.19, lel: 9.0%, uel: 14.8%. Insol in water.

SYNS: CFC 142b □ CHLORODIFLUOROETHANES (DOT) □ CHLOROETHYLIDENE FLUORIDE □ α-CHLOROETHYLIDENE FLUORIDE □ DIFLUOROCHLOROETHANES (DOT) □ 1,1-DIFLUORO-1-CHLOROETHANE □ FC142b □ FLUOROCARBON FC142b □ FREON 142 □ FREON 142b □ GENETRON 101 □ GENETRON 142b □ GENTRON 142B □ HYDROCHLOROFLUOROCARBON 142b □ R142B (DOT)

TOXICITY DATA WITH REFERENCE
mma-sat 50 pph/24H TXAPA9 72,15,84
ihl-rat LC50:2050 g/m³/4H 85GMAT -,53,82
ihl-mus LC50:1758 g/m³/2H 85GMAT -,53,82

CONSENSUS REPORTS: Reported in EPA TSCA Inventory.

DFG MAK: 100 ppm (4170 mg/m³)
DOT CLASSIFICATION: 2.1; *Label:* Flammable Gas

SAFETY PROFILE: Very mildly toxic by inhalation. Mutation data reported. A very dangerous fire hazard when exposed to heat, flame, or oxidizing materials. To fight fire, stop flow of gas. Can react vigorously with oxidizing materials. When heated to decomposition it emits toxic fumes of F^- and Cl^-.

CFX500 **CAS:75-45-6** **HR: 1**
CHLORODIFLUOROMETHANE
DOT: UN 1018
mf: $CHClF_2$ mw: 86.47

PROP: Gas. D: 1.49 @ 69°/4°, mp: −146°, bp: −40.8°, fp −160, autoign temp: 1170°F. Sltly sol in water.

SYNS: ALGEON 22 □ ALGOFRENE 22 □ ALGOFRENE TYPE 6 □ ARCTON 4 □ ARCTON 22 □ CFC 22 □ CHLORODIFLUOROMETHANE □ CHLORODIFLUOROMETHANE (ACGIH,DOT,OSHA) □ DAIFLON 22 □ DIFLUOROCHLOROMETHANE □ DIFLUOROMONOCHLOROMETHANE □ DYMEL 22 □ ELECTRO-CF 22 □ ESKIMON 22 □ F 22 □ FC 22 □ FLUGENE 22 □ FLUOROCARBON-22 □ FORANE 22 □ FREON □ FREON 22 □ FRIGEN □ FRIGEN 22 □ GENETRON 22 □ HALTRON 22 □ ISCEON 22 □ ISOTRON 22 □ KHALADON 22 □ MONOCHLORODIFLUOROMETHANE □ PROPELLANT 22 □ R-22 □ R22 (DOT) □ REFRIGERANT 22 □ UCON 22 □ UCON 22/HALOCARBON 22

TOXICITY DATA WITH REFERENCE
mmo-sat 33 pph/24H-C TOLED5 2,1,78
mma-sat 33 pph/24H-C TOLED5 2,1,78
ihl-rat TCLo:50,000 ppm/5H (56D male):REP FAATDF 1,266,81
ihl-rat LC50:35 pph/15M HUTODJ 1,239,82
ihl-mus LC50:28 pph/20M TXAPA9 59,64,81
ihl-dog LCLo:70 pph TXAPA9 2,363,60

CONSENSUS REPORTS: IARC Cancer Review: Group 3 IMEMDT 7,149,87; Human Inadequate Evidence IMEMDT 41,237,86; Animal Limited Evidence IMEMDT 41,237,86. Reported in EPA TSCA Inventory. EPA Genetic Toxicology Program.

OSHA PEL: TWA 1000 ppm
ACGIH TLV: TWA 1000 ppm
DFG MAK: 500 ppm (1800 mg/m³)
DOT CLASSIFICATION: 2.2; *Label:* Nonflammable Gas

SAFETY PROFILE: Mildly toxic by inhalation. Experimental reproductive effects. Mutation data reported. An asphyxiant in high concentrations. At elevated pressures, 50% mixtures with air are combustible although ignition is difficult. When heated to decomposition it emits toxic fumes of F^- and Cl^-. See also CHLORINATED HYDROCARBONS, ALIPHATIC; and FLUORIDES.

CFX625 CAS:59034-34-3 HR: 3
1-CHLORO-3,3-DIFLUORO-2-METHOXYCYCLOPROPENE
mf: $C_4H_3ClF_2O$ mw: 140.52

$ClC{=}C(OCH_3)\,CF_2$

SAFETY PROFILE: Explosive reaction on contact with water or methanol. When heated to decomposition it emits toxic fumes of F^- and Cl^-. See also CHLORINATED HYDROCARBONS, ALIPHATIC.

CFY000 CAS:58-93-5 HR: 3
6-CHLORO-3,4-DIHYDRO-2H-1,2,4-BENZOTHIADIAZINE-7-SULFONAMIDE- 1,1-DIOXIDE
mf: $C_7H_8ClN_3O_4S_2$ mw: 297.75

PROP: A solid. Mp: 273–275°.

SYNS: AQUARILLS □ AQUARIUS □ BREMIL □ 6-CHLORO-3,4-DIHYDRO-7-SULFAMOYL-2H-1,2,4-BENZOTHIADIAZINE-1,1-DIOXIDE □ 6-CHLORO-7-SULFAMOYL-3,4-DIHYDRO-2H-1,2,4-BENZOTHIADIAZINE-1,1-DIOXIDE □ CHLOROSULTHIADIL □ CHLORSULFONAMIDO DIHYDROBENZOTHIADIAZINE DIOXIDE □ CHLORZIDE □ CIDREX □ DICHLOROSAL □ DICHLOTIAZID □ DICHLOTRIDE □DICLOTRIDE □ 3,4-DIHYDRO-6-CHLORO-7-SULFAMYL-1,2,4-BENZOTHIADIAZINE-1,1-DIOXIDE □ DIHYDROCHLOROTHIAZID □ DIHYDROCHLOROTHIAZIDE □ 3,4-DIHYDROCHLOROTHIAZIDE □ DIHYDROXYCHLOROTHIAZIDUM □ DIREMA □ DISALUNIL □ DRENOL □ DYAZIDE □ ESIDREX □ ESIDRIX □ FLUVIN □ HCTZ □ HCZ □ HIDRIL □ HIDROCHLORTIAZID □ HIDRORONOL □ HIDROTIAZIDA □ HYDRO-AQUIL □ HYDROCHLORTHIAZID □ HYDRODIURETIC □ HYDRO-DIURIL □ HYDROSALURIC □ HYDROTHIDE □ HYPOTHIAZIDE □ IDROTIAZIDE □ IVAUGAN □ JEN-DIRIL □ MASCHITT □ MEGADIURIL □ NCI-C55925 □ NEFRIX □ NEOCODEMA □ NEOFLUMEN □ ORETIC □ PANURIN □ RO-HYDRAZIDE □ SU 5879 □ THIARETIC □ THIURETIC □ THLARETIC □ URODIAZIN □ VETIDREX □ ZIDE

TOXICITY DATA WITH REFERENCE
dnd-esc 5 mg/L MUREAV 89,95,81
cyt-ham:lng 500 mg/L/48H GMCRDC 27,95,81
orl-rat TDLo:582 g/kg (26W pre):REP IYKEDH 11,294,80
orl-mus TDLo:309 g/kg/2Y-C:ETA NTPTR* NTP-TR-357,89
orl-wmn TDLo:2 mg/kg/12H-I:SYS SMJOAV 76,1363,83
orl-wmn TDLo:500 µg/kg:PUL,GIT DICPBB 18,238,84
orl-rat LD50:2750 mg/kg TXAPA9 1,333,59
ipr-rat LD50:234 mg/kg 27ZIAQ -,124,73
scu-rat LD50:1270 mg/kg 27ZIAQ -,124,73
ivn-rat LD50:990 mg/kg JPETAB 140,249,63
orl-mus LD50:1175 mg/kg FRZKAP (1),44,83
ipr-mus LD50:578 mg/kg 27ZIAQ -,77,65
scu-mus LD50:1470 mg/kg 27ZIAQ -,124,73
ivn-mus LD50:590 mg/kg JPETAB 134,273,61
ivn-dog LD50:250 mg/kg 27ZIAQ -,124,73
ivn-rbt LD50:461 mg/kg 27ZIAQ -,124,73

CONSENSUS REPORTS: Reported in EPA TSCA Inventory. EPA Genetic Toxicology Program.

SAFETY PROFILE: Poison by intraperitoneal and intravenous routes. Moderately toxic by ingestion and subcutaneous routes. Human systemic effects by ingestion: sodium level changes, chlorine level changes, acute pulmonary edema, nausea or vomiting. Experimental reproductive effects. Questionable carcinogen with experimental tumorigenic data. Mutation data reported. A diuretic. When heated to decomposition it emits very toxic fumes of SO_x, Cl^-, and NO_x.

CFY250 CAS:36104-80-0 HR: 2
7-CHLORO-1,3-DIHYDRO-3-(N,N-DIMETHYLCARBAMOYL)-1-METHYL-5-PHENYL-2H-1,4-BENZODIAZEPIN-2-ONE
mf: $C_{19}H_{18}ClN_3O_3$ mw: 371.85

SYNS: ALBEGO □ B 5333 □ CAMAZEPAM □ 7-CHLORO-1,3-DIHYDRO-3-HYDROXY-1-METHYL-5-PHENYL-1,4-BENZODIAZEPIN-2-ONE DIMETHYLCARBAMATE □ SB 5833

TOXICITY DATA WITH REFERENCE
orl-mus LD50:970 mg/kg DRFUD4 1,458,76
ipr-mus LD50:800 mg/kg DRFUD4 1,458,76

SAFETY PROFILE: Moderately toxic by ingestion and intraperitoneal routes. A tranquilizer. See also CARBAMATES and DIAZEPAM. When heated to decomposition it emits very toxic fumes of Cl^- and NO_x.

CFY500 CAS:4700-56-5 HR: 3
7-CHLORO-1,3-DIHYDRO-3-HEMISUCCINYLOXY-2H-1,4-BENZODIAZEPIN-2-ONE
mf: $C_{19}H_{15}ClN_2O_5$ mw: 386.81

SYNS: BUTANEDIOIC ACID MONO(7-CHLORO-2,3-DIHYDRO-2-OXO-5-PHENYL-1H-1,4-BENZODIAZEPIN-3-YL) ESTER □ NULANS □ SAS 538 □ SUCCINIC ACID MONOESTER with 7-CHLORO-1,3-DIHYDRO-3-HYDROXY-5-PHENYL-2H-1,4-BENZODIAZEPIN-2-ONE

TOXICITY DATA WITH REFERENCE
orl-mus LD50:1148 mg/kg FRPPAO 22,506,67
ipr-mus LD50:375 mg/kg FRPPAO 22,506,67
ivn-mus LD50:285 mg/kg FRPPAO 22,506,67

SAFETY PROFILE: Poison by intraperitoneal and intravenous routes. Moderately toxic by ingestion. See also DIAZEPAM and ESTERS. When heated to decomposition it emits very toxic fumes of Cl^- and NO_x.

CFY750 CAS:846-50-4 HR: 3
7-CHLORO-1,3-DIHYDRO-3-HYDROXY-1-METHYL-5-PHENYL-2H-1,4-BENZODIAZEPIN-2-ONE
mf: $C_{16}H_{13}ClN_2O_2$ mw: 300.76

PROP: Crystals from cyclohexane. Mp: 119–121°.

SYNS: CEREPAX □ CRISONAR □ 1,3-DIHYDRO-7-CHLORO-3-HYDROXY-1-METHYL-5-PHENYL-2H-1,4-BENZODIAZEPIN-2-ONE □ ER 115 □ EUHYPNOS □ HYDROXYDIAZEPAM □ 3-HYDROXYDIAZEPAM □ K3917 □ LEVANXENE □ LEVANXOL □ MABERTIN □ METHYLOXAZEPAM □ N-METHYLOXAZEPAM □ NORMISON □ OXYDIAZEPAM □ PLANUM □ REMESTAN □ RESTORIL □ RO 5-5345 □ SIGNOPAM □ TEMAZEPAM □ WY 2917 □ WY 3917

TOXICITY DATA WITH REFERENCE
orl-hmn TDLo:15,430 µg/kg/12D-I:CNS,GIT ARZNAD 22,93,72
orl-rat LD50:2000 mg/kg DRUGAY 21,321,81
ipr-rat LD50:600 mg/kg DRUGAY 21,321,81
orl-mus LD50:370 mg/kg EJPHAZ 4,467,68
ipr-mus LD50:85 mg/kg AIPTAK 185,135,70
orl-dog LD50:3620 mg/kg DRUGAY 21,321,81

SAFETY PROFILE: Poison by ingestion and intraperito-

neal routes. Human systemic effects by ingestion: muscle weakness and changes in the structure or function of the salivary glands. A tranquilizer. When heated to decomposition it emits toxic fumes of Cl^- and NO_x. See also DIAZEPAM.

CFZ000 CAS:604-75-1 HR: 2
7-CHLORO-1,3-DIHYDRO-3-HYDROXY-5-PHENYL-2H-1,4-BENZODIAZEPINE-2-ONE
mf: $C_{15}H_{11}ClN_2O_2$ mw: 286.73

SYNS: ADUMBRAN □ ANSIOLISINA □ ANSIOXACEPAM □ ANXIOLIT □ APLAKIL □ ASTRESS □ BONARE □ 7-CHLORO-3-HYDROXY-5-PHENYL-1,3-DIHYDRO-2H-1,4-BENZODIAZEPIN-2-ONE □ ENIDREL □ HILONG □ ISODIN □ LIMBIAL □ NESONTIL □ NOCTAZEPAM □ NOTARAL □ OX □ OXAZEPAM □ PACIENX □ PRAXITEN □ PROPAX □ PSICOPAX □ QUEN □ QUILIBREX □ RO 5-6789 □ RONDAR □ SERAX □ SERENAL □ SERENID □ SERENID-D □ SEREPAX □ SERESTA □ SERPAX □ SIGACALM □ SOBRIL □ TAZEPAM □ TRANQUO-BUSCOPAN-WIRKSTOFF □ VABEN □ WY-3498 □ Z10-TR

TOXICITY DATA WITH REFERENCE
mma-sat 5200 pmol/plate CNREA8 38,4478,78
orl-mus TDLo:2520 mg/kg (1-21D preg/21D post):REP PLRCAT 9,325,77
orl-mus TDLo:4140 mg/kg (male 5D pre):REP PLRCAT 9,187,77
orl-mus TDLo:2 g/kg (male 5D pre):REP RCOCB8 13,601,76
orl-mus TDLo:400 mg/kg (female 14D post):REP TXAPA9 25,453,73
orl-mus TDLo:1560 mg/kg (female 26D pre):REP RCOCB8 11,155,75
orl-mus TDLo:120 g/kg/57W-C:ETA TOXID9 13,373,93
orl-chd TDLo:8 mg/kg:CNS JAMAAP 196,662,66
ipr-rat LD50:1535 mg/kg PHMGBN 10,345,73
ipr-mus LD50:767 mg/kg JMCMAR 11,777,68

CONSENSUS REPORTS: IARC Cancer Review: Group 3 IMEMDT 7,56,87; Animal Limited Evidence IMEMDT 13,57,77

SAFETY PROFILE: Moderately toxic by intraperitoneal route. Human (child) systemic effects by ingestion: somnolence, changes in REM sleep, and loss of muscle control (ataxia). Experimental reproductive effects. Questionable carcinogen with experimental tumorigenic data. Mutation data reported. Used to treat anxiety and tension. When heated to decomposition it emits very toxic fumes of NO_x and HCl. See also DIAZEPAM.

CGA000 CAS:2898-12-6 HR: 3
7-CHLORO-2,3-DIHYDRO-1-METHYL-5-PHENYL-1H-1,4-BENZODIAZEPINE
mf: $C_{16}H_{15}ClN_2$ mw: 270.78

PROP: Prisms from Et_2O/pet ether. Mp: 95–97°.

SYNS: ANSILAN □ DIEPIN □ 2,3-DIHYDRO-7-CHLORO-1-METHYL-5-PHENYL-1H-1,4-BENZODIAZEPINE □ ELBRUS □ ESMAIL □ MEDAZEPAM □ MEDAZEPOL □ MEGASEDAN □ MEZEPAN □ NARSIS □ NOBRIUM □ PAZITAL □ PSIQUIM □ RESMIT □ RUDOTEL □ SERENIUM □ SIMAN □ TRANQUILAX

TOXICITY DATA WITH REFERENCE
orl-rat TDLo:860 mg/kg (1-22D preg/21D post):REP PSYPAG 41,113,75
orl-rat LD50:900 mg/kg 26RAAN -,35,73
orl-mus LD50:475 mg/kg ARZNAD 24,2029,74
ipr-mus LD50:360 mg/kg AIPTAK 185,135,70

SAFETY PROFILE: Poison by intraperitoneal route. Moderately toxic by ingestion. Experimental reproductive effects. A tranquilizer. When heated to decomposition it emits very toxic fumes of Cl^- and NO_x. See also DIAZEPAM.

C

CGA500 CAS:1088-11-5 HR: 3
7-CHLORO-1,3-DIHYDRO-5-PHENYL-2H-1,4-BENZODIAZEPIN-2-ONE
mf: $C_{15}H_{11}ClN_2O$ mw: 270.73

PROP: Plates from Me_2CO. Mp: 216–217°.

SYNS: A-101 □ CALMDAY □ DEALKYLPRAZEPAM □ DEMETHYLDIAZEPAM □ N-DEMETHYLDIAZEPAM □ 1-DEMETHYLDIAZEPAM □ N-DEOXYDEMOXAPAM □ DESALKYLPRAZEPAM □ N-DESCYCLOPROPYLMETHYLPRAZEPAM □ DESMETHYLDIAZEPAM □ N-DESMETHYLDIAZEPAM □ DMDZ □ MADAR □ NDD □ NORDIAZEPAM □ NORPRAZEPAM □ RO 5-2180 □ STILNY

TOXICITY DATA WITH REFERENCE
mma-sat 57 nmol/plate CNREA8 38,4478,78
orl-mus TDLo:400 mg/kg/(14D preg):TER TXAPA9 25,453,73
orl-mus LD50:670 mg/kg EJPHAZ 4,467,68
ipr-mus LD50:290 mg/kg EJPHAZ 16,311,71

SAFETY PROFILE: Poison by intraperitoneal route. Moderately toxic by ingestion. An experimental teratogen. Mutation data reported. A tranquilizer. When heated to decomposition it emits very toxic fumes of NO_x and Cl^-. See also DIAZEPAM.

CGB000 CAS:55299-24-6 HR: 2
7-CHLORO-1,3-DIHYDRO-5-PHENYL-1-TRIMETHYLSILYL-2H-1,4-BENZODIAZEPIN-2-ONE
mf: $C_{18}H_{18}ClOSi$ mw: 313.90

SYNS: ST 720 (FRENCH) □ TRIMETHYL SILYL-1-CHLORO-7-DIHYDRO-1,3-PHENYL-5,2H-BENZODIAZEPINE-1,4-ONE-2 (FRENCH)

TOXICITY DATA WITH REFERENCE
orl-mus LD50:1000 mg/kg APFRAD 36,621,78
ipr-mus LD50:600 mg/kg APFRAD 36,621,78

SAFETY PROFILE: Moderately toxic by ingestion and intraperitoneal routes. When heated to decomposition it emits toxic fumes of Cl^-. See also DIAZEPAM.

CGB250 CAS:14437-41-3 HR: 2
4′-CHLORO-3,5-DIIODOSALICYLANILIDE ACETATE
mf: $C_{15}H_{10}ClI_2NO_3$ mw: 541.51

PROP: Needles from Me_2CO. Mp: 215–216°.

SYNS: ACETOXY-4′-CHLORO-3,5-DIIODOBENZANILIDE □ 2-(ACETYLOXY)-N-(4-CHLOROPHENYL)-3,5-DIIODOBENZAMIDE □ C.I. 633 □ CLIOXANIDE □ CN 59,567 □ SYD 230 □ TREMERAD

TOXICITY DATA WITH REFERENCE
ipr-mus LD50:720 mg/kg AUVJA2 46,297,70
orl-dom LD50:414 mg/kg AUVJA2 46,297,70

CONSENSUS REPORTS: Reported in EPA TSCA Inventory.

SAFETY PROFILE: Moderately toxic by ingestion and intraperitoneal routes. Used in treatment against parasitic worms. When heated to decomposition it emits very toxic fumes of Cl^-, I^-, and NO_x.

CGB500 CAS:1779-25-5 HR: 1
CHLORO DIISOBUTYL ALUMINUM
mf: $C_8H_{18}AlCl$ mw: 176.69

SYNS: ALLUMINIO DIISOBUTIL-MONOCLORURO (ITALIAN) □ BIS (ISOBUTYL)ALUMINUM CHLORIDE □ CHLOROBIS(2-METHYLPROPYL)ALUMINUM □ DIISOBUTYLALUMINUM CHLORIDE □ DIISOBUTYLALUMINUM MONOCHLORIDE □ DIISOBUTYLCHLOROALUMINUM

TOXICITY DATA WITH REFERENCE
ihl-rat LC50:67 ppm/1H 85JCAE-,1216,86
ihl-mus LDLo:680 g/kg/15M MELAAD 58,290,67

CONSENSUS REPORTS: Reported in EPA TSCA Inventory.

ACGIH TLV: TWA 2 mg(Al)/m³

SAFETY PROFILE: Mildly toxic by inhalation. See also ALUMINUM COMPOUNDS and CHLORIDES. Ignites spontaneously in air. When heated to decomposition it emits toxic fumes of Cl^-.

CGB750 CAS:63869-02-3 HR: 3
CHLORO(DIISOPROPOXYPHOSPHINYL)MERCURY
mf: $C_6H_{14}ClHgO_3P$ mw: 401.21

TOXICITY DATA WITH REFERENCE
ipr-mus LDLo:15,600 µg/kg CBCCT* 8,104,56

CONSENSUS REPORTS: Mercury and its compounds are on the Community Right-To-Know List.

OSHA PEL: CL 0.1 mg(Hg)/m³ (skin)
ACGIH TLV: TWA 0.1 mg(Hg)/m³ (skin)
NIOSH REL: (Organomercury): TWA 0.01 mg/m³; STEL 0.03 mg/m³ (skin)

SAFETY PROFILE: Poison by intraperitoneal route. See also MERCURY COMPOUNDS. When heated to decomposition it emits very toxic fumes of PO_x, Cl^-, and Hg.

CGC000 CAS:77966-54-2 HR: 3
6′-CHLORO-2-(DIISOPROPYLAMINO)-o-ACETOTOLUIDIDE HYDROCHLORIDE
mf: $C_{15}H_{23}ClN_2O{\cdot}ClH$ mw: 319.31

TOXICITY DATA WITH REFERENCE
eye-rbt 2% MLD ARZNAD 8,407,58
ipr-rat LD50:100 mg/kg ARZNAD 8,407,58
scu-mus LD50:515 mg/kg ARZNAD 8,407,58

SAFETY PROFILE: Poison by intraperitoneal route. Moderately toxic by subcutaneous route. An eye irritant. When heated to decomposition it emits very toxic fumes of Cl^- and NO_x.

CGC100 CAS:119-21-1 HR: 3
1-CHLORO-2,4-DIMETHOXY-5-NITROBENZENE
mf: $C_8H_8ClNO_4$ mw: 217.62

SYN: BENZENE, 5-CHLORO-1-NITRO-2,4-DIMETHOXY-

TOXICITY DATA WITH REFERENCE
orl-brd LD50:100 mg/kg TXAPA9 21,315,72

CONSENSUS REPORTS: Reported in EPA TSCA Inventory.

SAFETY PROFILE: Poison by ingestion. When heated to decomposition it emits toxic vapors of NO_x and Cl^-.

CGC200 CAS:1585-74-6 HR: 3
N-CHLORODIMETHYLAMINE
mf: C_2H_6ClN mw: 79.53

SAFETY PROFILE: Products of reaction with antimony chlorides (e.g. antimony trichloride and antimony pentachloride) are dangerous heat- and shock-sensitive explosives which may explode at room temperature. When heated to decomposition it emits toxic fumes of Cl^- and NO_x. See also AMINES.

CGD000 CAS:77966-55-3 HR: 3
6′-CHLORO-2-(DIMETHYLAMINO)-o-ACETOTOLUIDIDE HYDROCHLORIDE
mf: $C_{11}H_{15}ClN_2O{\cdot}ClH$ mw: 263.19

SYN: V 252

TOXICITY DATA WITH REFERENCE
ipr-rat LD50:218 mg/kg ARZNAD 8,407,58
ipr-mus LD50:243 mg/kg ARZNAD 8,407,58
scu-mus LD50:600 mg/kg ARZNAD 8,407,58

SAFETY PROFILE: Poison by intraperitoneal route. Moderately toxic by subcutaneous route. When heated to decomposition it emits very toxic fumes of Cl^- and NO_x.

CGD250 CAS:2491-76-1 HR: 2
p-CHLORO DIMETHYLAMINOAZOBENZENE
mf: $C_{14}H_{14}ClN_3$ mw: 259.76

SYNS: 4′-CHLORO-4-DIMETHYLAMINOAZOBENZENE □ N,N-DIMETHYL-p-((p-CHLOROPHENYL)AZO)ANILINE

TOXICITY DATA WITH REFERENCE
scu-mus TDLo:500 mg/kg (9D preg):TER OFAJAE 36,195,60
orl-rat TDLo:6100 mg/kg/21W-C:NEO JEMEAV 87,139,48
scu-mus LDLo:500 mg/kg OFAJAE 36,195,60

SAFETY PROFILE: Moderately toxic by subcutaneous route. Questionable carcinogen with experimental neoplastigenic data. Experimental teratogenic effects. When heated to decomposition it emits very toxic fumes of Cl^- and NO_x.

CGD399 **HR: 3**
β-CHLORODIMETHYLAMINO DIBORANE
mf: $C_2H_{10}B_2ClN$ mw: 105.18

$ClHB{:}H_2{:}BHN(CH_3)_2$

SYN: B-CHLORO-N,N-DIMETHYLAMINODIBORANE

SAFETY PROFILE: Ignites spontaneously in air. When heated to decomposition it emits toxic fumes of Cl^- and NO_x. See also BORANES, CHLORIDES, and AMINES.

CGD500 **CAS:3505-38-2** **HR: 3**
2-(p-CHLORO-α-(2-(DIMETHYLAMINO)ETHOXY)BENZYL)PYRIDINE BIMALEATE
mf: $C_{16}H_{19}ClN_2O{\bullet}C_4H_4O_4$ mw: 406.90

SYNS: ALLERGEFON MALEATE □ CARBINOXAMINE MALEATE □ p-CARBINOXAMINE MALEATE □ 2-(p-CHLORO-α-(2-(DIMETHYLAMINO) ETHOXY)BENZYL)PYRIDINE MALEATE □ 2-((4-CHLOROPHENYL)-2-PYRIDINYLMETHOXY)-N,N-DIMETHYLETHANAMINE-(Z)-2-BUTENEDIOATE (1:1) □ CLISTIN □ CLISTIN MALEATE □ CLISTINE MALEATE

TOXICITY DATA WITH REFERENCE
orl-cld TDLo:1880 μg/kg/4D-I:BAH JTCTDW 25,161,87
orl-mus LD50:162 mg/kg CLDND* 15,367,68
scu-mus LD50:350 mg/kg TOIZAG 15,367,68
ivn-mus LD50:32 mg/kg CLDND*
ivn-dog LDLo:36 mg/kg CLDND*
orl-gpg LD50:411 mg/kg CLDND*
scu-gpg LD50:120 mg/kg APFRAD 20,463,62

CONSENSUS REPORTS: Reported in EPA TSCA Inventory.

SAFETY PROFILE: Poison by ingestion, subcutaneous, and intravenous routes. Human systemic effects by ingestion: distorted perceptions, excitement hallucinations. When heated to decomposition it emits very toxic fumes of Cl^- and NO_x.

CGD750 **CAS:101651-62-1** **HR: 3**
2′-CHLORO-2-((2-(DIMETHYLAMINO)ETHYL)ETHYLAMINO)ACETANILIDE DIHYDROCHLORIDE
mf: $C_{14}H_{22}ClN_3O{\bullet}2ClH$ mw: 356.76

SYN: C 5417

TOXICITY DATA WITH REFERENCE
eye-rbt 2% MLD ARZNAD 9,262,59
ipr-rat LD50:148 mg/kg ARZNAD 9,262,59
scu-mus LD50:575 mg/kg ARZNAD 9,262,59

SAFETY PROFILE: Poison by intraperitoneal route. Moderately toxic by subcutaneous route. An eye irritant. When heated to decomposition it emits very toxic fumes of Cl^- and NO_x.

CGE000 **CAS:101651-94-9** **HR: 3**
7-CHLORO-10-(2-(DIMETHYLAMINO)ETHYL)ISOALLOXAZINE SULFATE
mf: $C_{14}H_{14}ClN_5O_2{\bullet}H_2O_4S$ mw: 417.86

TOXICITY DATA WITH REFERENCE
orl-mus LD50:1900 mg/kg CMTRAG 2,96,61
scu-mus LD50:38 mg/kg CMTRAG 2,96,61
ivn-mus LD50:60 mg/kg CMTRAG 2,96,61

SAFETY PROFILE: Poison by subcutaneous and intravenous routes. Moderately toxic by ingestion. See also SULFATES. When heated to decomposition it emits very toxic fumes of SO_x, Cl^-, and NO_x.

CGE250 **CAS:95770-03-9** **HR: 2**
2′-CHLORO-2-(2-(DIMETHYLAMINO)ETHYLTHIO)ACETANILIDE HYDROCHLORIDE
mf: $C_{12}H_{17}ClN_2OS{\bullet}ClH$ mw: 309.28

SYN: C 5501

TOXICITY DATA WITH REFERENCE
eye-rbt 2% MOD ARZNAD 9,683,59
ipr-rat LD50:750 mg/kg ARZNAD 9,683,59
scu-mus LD50:1400 mg/kg ARZNAD 9,683,59

SAFETY PROFILE: Moderately toxic by intraperitoneal and subcutaneous routes. An eye irritant. When heated to decomposition it emits very toxic fumes of Cl^-, NO_x, and SO_x.

CGE500 **CAS:100620-36-8** **HR: 3**
6′-CHLORO-2-(2-(DIMETHYLAMINO)ETHYLTHIO)-o-ACETOTOLUIDIDE
mf: $C_{13}H_{19}ClN_2OS$ mw: 286.85

SYN: C 5458

TOXICITY DATA WITH REFERENCE
ipr-rat LD50:140 mg/kg ARZNAD 9,683,59
scu-mus LD50:460 mg/kg ARZNAD 9,683,59

SAFETY PROFILE: Poison by intraperitoneal route. Moderately toxic by subcutaneous route. When heated to decomposition it emits very toxic fumes of SO_x, NO_x, and Cl^-.

CGE750 **CAS:77966-56-4** **HR: 3**
6′-CHLORO-2-(DIMETHYLAMINO)-N-METHYL-o-ACETOTOLUIDIDE HYDROCHLORIDE
mf: $C_{12}H_{17}ClN_2O{\bullet}ClH$ mw: 277.22

SYN: C 3133

TOXICITY DATA WITH REFERENCE
ipr-rat LD50:390 mg/kg ARZNAD 8,609,58
scu-mus LD50:445 mg/kg ARZNAD 8,609,58

SAFETY PROFILE: Poison by intraperitoneal route. Moderately toxic by subcutaneous route. When heated to decomposition it emits very toxic fumes of Cl^- and NO_x.

CGF000 **CAS:19986-35-7** **HR: 3**
5-CHLORO-3-(DIMETHYLAMINOMETHYL)-2-BENZOXAZOLINONE
mf: $C_{10}H_{11}ClN_2O_2$ mw: 226.68

TOXICITY DATA WITH REFERENCE
orl-mus LD50:1500 mg/kg MDCHAG 4(1),308,64
ipr-mus LD50:400 mg/kg MDCHAG 4(1),308,64

SAFETY PROFILE: Poison by intraperitoneal route. Moderately toxic by ingestion. When heated to decomposition it emits very toxic fumes of Cl^- and NO_x.

CGF250 CAS:101651-96-1 HR: 3
7-CHLORO-10-(4-(DIMETHYLAMINO)-1-METHYLBUTYL)ISOALLOXAZINE SULFATE
mf: $C_{19}H_{24}ClN_5O_2 \cdot H_2O_4S$ mw: 488.01

TOXICITY DATA WITH REFERENCE
ipr-rat LD50:55 mg/kg CMTRAG 2,96,61
scu-mus LD50:120 mg/kg CMTRAG 2,96,61
ivn-mus LD50:28 mg/kg CMTRAG 2,96,61

SAFETY PROFILE: Poison by intraperitoneal, subcutaneous, and intravenous routes. See also SULFATES. When heated to decomposition it emits very toxic fumes of SO_x, NO_x and Cl^-.

CGF500 CAS:78218-37-8 HR: 3
6′-CHLORO-3-(DIMETHYLAMINO)-o-PROPIONOTOLUIDIDE HYDROCHLORIDE
mf: $C_{12}H_{17}ClN_2O \cdot ClH$ mw: 277.22

TOXICITY DATA WITH REFERENCE
ipr-rat LD50:114 mg/kg ARZNAD 8,544,58
ipr-mus LD50:175 mg/kg ARZNAD 8,544,58
scu-mus LD50:445 mg/kg ARZNAD 8,544,58

SAFETY PROFILE: Poison by intraperitoneal route. Moderately toxic by subcutaneous route. When heated to decomposition it emits very toxic fumes of NO_x and Cl^-.

CGG500 CAS:28907-45-1 HR: 3
7-CHLORO-10-(3-DIMETHYLAMINOPROPYL)-BENZO-(b)(1,8)-5(10H)-NAPHTHAPYRIDONE HYDROCHLORIDE
mf: $C_{17}H_{18}ClN_3O \cdot ClH$ mw: 352.29

SYNS: C 45 □ C 45 (pharmaceutical) □ CHLOROWODORKU 10-γ-DWUMETYLOAMINOPROPYLO-7-CHLOROBENZO(b)-(1,8)-NAFTYRYDONU-5 (POLISH) □ IFC-45

TOXICITY DATA WITH REFERENCE
orl-rat LD50:500 mg/kg PJPPAA 27,503,75
ipr-rat LD50:105 mg/kg PJPPAA 27,503,75
ivn-rat LD50:33 mg/kg DRFUD4 3,303,78
orl-mus LD50:200 mg/kg DRFUD4 3,303,78
ipr-mus LD50:106 mg/kg PJPPAA 27,503,75
ivn-mus LD50:33 mg/kg DRFUD4 3,303,78
ivn-rbt LD50:12 mg/kg PJPPAA 27,503,75

SAFETY PROFILE: Poison by ingestion, intravenous, and intraperitoneal routes. When heated to decomposition it emits very toxic fumes of Cl^- and NO_x.

CGG600 CAS:14051-55-9 HR: 3
3-CHLORO-5-(3-(DIMETHYLAMINO)PROPYL)-10,11-DIHYDRO-5H-DIBENZ(b,f)AZEPINE HYDROCHLORIDE
mf: $C_{19}H_{23}ClN_2 \cdot ClH$ mw: 351.35

TOXICITY DATA WITH REFERENCE
orl-rat LD50:1150 mg/kg IYKEDH 4,193,73
ipr-rat LD50:135 mg/kg IYKEDH 4,193,73
scu-rat LD50:1750 mg/kg IYKEDH 4,193,73
ivn-rat LD50:26 mg/kg IYKEDH 4,193,73
orl-mus LD50:470 mg/kg IYKEDH 4,193,73
ipr-mus LD50:90 mg/kg IYKEDH 4,193,73
scu-mus LD50:400 mg/kg IYKEDH 4,193,73
ivn-mus LD50:26 mg/kg IYKEDH 4,193,73

SAFETY PROFILE: Poison by subcutaneous, intravenous, and intraperitoneal routes. Moderately toxic by ingestion. When heated to decomposition it emits toxic fumes of NO_x and HCl.

CGG750 CAS:101651-97-2 HR: 3
7-CHLORO-10-(3-(DIMETHYLAMINO)PROPYL)ISOALLOXAZINE HYDROCHLORIDE
mf: $C_{15}H_{16}ClN_5O_2 \cdot ClH$ mw: 370.27

TOXICITY DATA WITH REFERENCE
orl-mus LD50:1250 mg/kg CMTRAG 2,96,61
ipr-mus LD50:70 mg/kg CMTRAG 2,96,61
scu-mus LD50:24 mg/kg CMTRAG 2,96,61
ivn-mus LD50:60 mg/kg CMTRAG 2,96,61

SAFETY PROFILE: Poison by intraperitoneal, subcutaneous, and intravenous routes. Moderately toxic by ingestion. When heated to decomposition it emits very toxic fumes of Cl^- and NO_x.

CGH250 CAS:63019-52-3 HR: 2
9-CHLORO-8,12-DIMETHYLBENZ(a)ACRIDINE
mf: $C_{19}H_{14}ClN$ mw: 291.79

SYNS: 2-CHLORO-1,10-DIMETHYL-5,6-BENZACRIDINE (FRENCH) □ 1,10-DIMETHYL-2-CHLORO-5,6-BENZACRIDINE □ 8,12-DIMETHYL-9-CHLOROBENZ(a)ACRIDINE

TOXICITY DATA WITH REFERENCE
skn-mus TDLo:500 mg/kg/41W-I:ETA AICCA6 11,736,55
skn-mus TD:540 mg/kg/45W-I:ETA ACRSAJ 4,315,56

SAFETY PROFILE: Questionable carcinogen with experimental tumorigenic data by skin contact. When heated to decomposition it emits very toxic fumes of NO_x and Cl^-.

CGH500 CAS:64050-23-3 HR: 2
10-CHLORO-6,9-DIMETHYL-5,10-DIHYDRO-3,4-BENZOPHENARSAZINE
mf: $C_{18}H_{15}AsClN$ mw: 355.71

SYN: 12-CHLORO-7,12-DIHYDRO-8,11-DIMETHYLBENZO(a)PHENARSAZINE

TOXICITY DATA WITH REFERENCE
skn-mus TDLo:380 mg/kg/16W-I:ETA CRSBAW 145,1451,51

CONSENSUS REPORTS: Arsenic and its compounds are on the Community Right-To-Know List.

OSHA PEL: TWA 0.5 mg(As)/m^3

SAFETY PROFILE: Questionable carcinogen with experimental tumorigenic data by skin contact. When heated to decomposition it emits very toxic fumes of As, Cl^-, and NO_x. See also ARSENIC COMPOUNDS.

CGH675 CAS:10140-91-7 HR: 3
p-CHLORO-5,10-DIMETHYL-2,4-DIOXA-p-THIONO-3-PHOSPHABICYCLO(4.4.0)DECANE

TOXICITY DATA WITH REFERENCE
skn-rbt 10 mg/24H open MLD AIHAAP 23,95,62
orl-rat LD50:110 mg/kg AIHAAP 23,95,62
skn-rbt LD50:200 mg/kg AIHAAP 23,95,62

SAFETY PROFILE: Poison by ingestion and skin contact. When heated to decomposition it emits toxic fumes of Cl^-, PO_x, and SO_x.

CGI125 CAS:26096-99-1 HR: 3
N-(p-CHLORO-α,α-DIMETHYLPHENETHYL)-2-(DIETHLAMINO)PROPIONAMIDE HYDROCHLORIDE
mf: $C_{17}H_{27}ClN_2O•ClH$ mw: 347.37

SYN: N-(2-(4-CHLOROPHENYL)-1,1-DIMETHYLETHYL)-2-(DIETHYLAMINO)-PROPANAMIDE HYDROCHLORIDE

TOXICITY DATA WITH REFERENCE
orl-mus LD50:690 mg/kg APSXAS 15,87,78
ipr-mus LD50:220 mg/kg APSXAS 15,87,78
ivn-mus LD50:35 mg/kg APSXAS 15,87,78

SAFETY PROFILE: Poison by intravenous and intraperitoneal routes. Moderately toxic by ingestion. When heated to decomposition it emits toxic fumes of Cl^- and NO_x.

CGI500 CAS:671-04-5 HR: 3
2-CHLORO-4,5-DIMETHYLPHENYL METHYLCARBAMATE
mf: $C_{10}H_{12}ClNO_2$ mw: 213.68

SYNS: BANOL □ BANOL TUCO SOK □ CARBANOLATE □ 2-CHLORO-4,5-DIMETHYLPHENOL, METHYL CARBAMATE □ (2-CHLORO-4,5-DIMETHYL)PHENYL ESTER, CARBAMIC ACID □ CHLOROXYLAM □ 6-CHLORO-3,4-XYLENYL N-METHYLCARBAMATE □ 2-CHLORO-4,5-XYLYL ESTER, CARBAMIC ACID □ 6-CHLORO-3,4-XYLYL N-METHYLCARBAMATE □ OMS-174 □ U 12927 □ U-17004 □ UPJOHN U-12,927

TOXICITY DATA WITH REFERENCE
orl-rat LD50:30 mg/kg WRPCA2 9,119,70
ipr-rat LD50:11,200 μg/kg BWHOA6 44(1-3),241,71
ivn-rat LD50:3 mg/kg BJIMAG 22,317,65
ims-rat LD50:24 mg/kg BJIMAG 22,317,65
unk-rat LD50:293 mg/kg 30ZDA9 -,190,71
orl-mus LD50:300 mg/kg ARSIM* 20,26,66
orl-pgn LD50:4200 μg/kg TXAPA9 21,315,72
orl-qal LD50:7500 μg/kg ASTTA8 (680),157,79
orl-dck LD50:2400 μg/kg TXAPA9 21,315,72
orl-bwd LD50:1780 μg/kg ASTTA8 (680),157,79

CONSENSUS REPORTS: Chlorophenol compounds are on the Community Right-To-Know List.

SAFETY PROFILE: Poison by ingestion, intraperitoneal, intravenous, intramuscular, and possibly other routes. See also CARBAMATES; CHLOROPHENOLS; and ESTERS. A pesticide. When heated to decomposition it emits very toxic fumes of Cl^- and NO_x.

CGI625 CAS:811-62-1 HR: 3
CHLORODIMETHYLPHOSPHINE
mf: C_2H_6ClP mw: 96.50

PROP: Pale-yellow liquid. D: 1.23 @ 25°/4°, mp: −2°, bp: 73–74° @ 749 mm.

SAFETY PROFILE: Ignites spontaneously in air. When heated to decomposition it emits toxic fumes of Cl^- and PO_x. See also PHOSPHINE.

CGI750 CAS:77966-93-9 HR: 3
6′-CHLORO-2-(2,6-DIMETHYLPIPERIDINO)-o-ACETOTOLUIDIDE HYDROCHLORIDE
mf: $C_{16}H_{23}ClN_2O•ClH$ mw: 331.32

SYN: V 375

TOXICITY DATA WITH REFERENCE
eye-rbt 2% MLD ARZNAD 8,407,58
ipr-rat LD50:72 mg/kg ARZNAD 8,407,58
ipr-mus LD50:77 mg/kg ARZNAD 8,407,58
scu-mus LD50:180 mg/kg ARZNAD 8,407,58

SAFETY PROFILE: Poison by intraperitoneal and subcutaneous routes. An eye irritant. When heated to decomposition it emits very toxic fumes of Cl^- and NO_x.

CGJ000 CAS:102504-64-3 HR: 3
6′-CHLORO-3-(2,6-DIMETHYLPIPERIDINO)-o-PROPIONOTOLUIDIDE HYDROCHLORIDE
mf: $C_{16}H_{25}ClN_2O•ClH$ mw: 333.34

SYN: C 3140

TOXICITY DATA WITH REFERENCE
eye-rbt 2% MLD ARZNAD 8,544,58
ipr-rat LD50:33 mg/kg ARZNAD 8,544,58
scu-mus LD50:25 mg/kg ARZNAD 8,544,58

SAFETY PROFILE: Poison by intraperitoneal and subcutaneous routes. An eye irritant. When heated to decomposition it emits very toxic fumes of Cl^- and NO_x.

CGJ250 CAS:24358-29-0 HR: 2
2-CHLORO-5-(3,5-DIMETHYLPIPERIDINO SULPHONYL)BENZOIC ACID
mf: $C_{14}H_{18}ClNO_4S$ mw: 331.84

SYN: TIBRIC ACID

TOXICITY DATA WITH REFERENCE
dni-mus:oth 500 μmol/L CNREA8 40,36,80
orl-rat TDLo:39 g/kg/71W-C:CAR NATUAS 283,397,80

SAFETY PROFILE: Questionable carcinogen with experimental carcinogenic data. Mutation data reported. When heated to decomposition it emits very toxic fumes of SO_x, NO_x, and Cl^-.

CGK500 CAS:63020-91-7 HR: 2
2′-CHLORO-N,N-DIMETHYL-4-STILBENAMINE
mf: $C_{16}H_{16}ClN$ mw: 257.78

SYNS: 2′-CHLORO-4-DIMETHYLAMINOSTILBENE □ 2′-CHLORO-4-STILBENYL-N,N-DIMETHYLAMINE

TOXICITY DATA WITH REFERENCE
orl-rat TDLo:625 mg/kg/60W-C:ETA ABMGAJ 9,87,62
ipr-rat TDLo:300 mg/kg/8W-I:CAR BJCAAI 10,123,56
ipr-rat TD:420 mg/kg/9W-I:ETA BJCAAI 6,392,52

SAFETY PROFILE: Questionable carcinogen with experimental carcinogenic and tumorigenic data. When heated to decomposition it emits very toxic fumes of Cl^- and NO_x.

CGK750 CAS:63040-27-7 HR: 2
3′-CHLORO-N,N-DIMETHYL-4-STILBENAMINE
mf: $C_{16}H_{16}ClN$ mw: 257.78

SYNS: 3′-CHLORO-N,N-DIMETHYLAMINOSTIBEN (GERMAN) □ 3′-CHLORO-4-DIMETHYLAMINOSTILBENE □ 3′-CHLORO-4-STILBENYL-N,N-DIMETHYLAMINE

TOXICITY DATA WITH REFERENCE
orl-rat TDLo:440 mg/kg/42W-C:ETA ABMGAJ 9,87,62

SAFETY PROFILE: Questionable carcinogen with experimental tumorigenic data. When heated to decomposition it emits very toxic fumes of Cl^- and NO_x.

CGL000 CAS:7378-50-9 HR: 2
4′-CHLORO-N,N-DIMETHYL-4-STILBENAMINE
mf: $C_{16}H_{16}ClN$ mw: 257.78

SYNS: 4′-CHLORO-N,N-DIMETHYLAMINOSTIBEN (GERMAN) □ 4′-CHLORO-4-DIMETHYLAMINOSTILBENE □ 4′-CHLORO-4-STILBENYL-N,N-DIMETHYLAMINE

TOXICITY DATA WITH REFERENCE
orl-rat TDLo:625 mg/kg/59W-C:ETA ABMGAJ 9,87,62

SAFETY PROFILE: Questionable carcinogen with experimental tumorigenic data. When heated to decomposition it emits very toxic fumes of Cl^- and NO_x.

CGL125 CAS:72040-09-6 HR: 3
N-CHLORO-4,5-DIMETHYLTRIAZOLE
mf: $C_4H_6ClN_3$ mw: 131.56

$ClNN{=}NC(CH_3){=}CCH_3$

SAFETY PROFILE: The solid triazole and its concentrated solutions are unstable and may decompose violently at room temperature. When heated to decomposition it emits toxic fumes of Cl^- and NO_x.

CGL250 CAS:7287-36-7 HR: 2
4′-CHLORO-2,2-DIMETHYLVALERANILIDE
mf: $C_{13}H_{18}ClNO$ mw: 239.77

PROP: Crystals. Mp: 87–88°.

SYNS: N-(4-CHLOROPHENYL)-2,2-DIMETHYLPENTANAMIDE □ N-(4-CHLOROPHENYL)-2,2-DIMETHYLVALEROAMIDE □ N-(4-CHLORPHENYL)-2,2-DIMETHYLPENTAMID (GERMAN) □ N-(4-CHLOR-PHENYL)-2,2-DIMETHYL-VALERIANSAEUREAMID (GERMAN) □ D-90-A □ MONALIDE □ POTABLAN □ SCHERING-35830 □ SN 35830

TOXICITY DATA WITH REFERENCE
orl-rat LD50:2600 mg/kg 85ARAE 2,220,77
skn-rat LD50:>800 mg/kg PEMNDP 9,596,91
skn-rbt LD50:800 mg/kg 85JFAN A283,83

CONSENSUS REPORTS: EPA Genetic Toxicology Program.

SAFETY PROFILE: Moderately toxic by ingestion and skin contact. A pesticide. When heated to decomposition it emits very toxic fumes of Cl^- and NO_x.

CGL325 CAS:3531-19-9 HR: 3
2-CHLORO-4,6-DINITROANILINE
mf: $C_6H_4ClN_3O_4$ mw: 217.57

$Cl(NO_2)_2C_6H_2NH_2$

PROP: Yellow crystals from DMF (aq). Mp: 157°.

SYNS: BENZENAMINE, 2-CHLORO-4,6-DINITRO- □ 6-CHLORO-2,4-DINITROANILINE

TOXICITY DATA WITH REFERENCE
mmo-sat 500 μg/plate SAIGBL 29,34,87

CONSENSUS REPORTS: Reported in EPA TSCA Inventory.

SAFETY PROFILE: Mutation data reported. An explosive sensitive to heat or an initiating charge. Solution in nitrosylsulfuric acid explodes between 50-160°C depending on the concentration. When heated to decomposition it emits toxic fumes of Cl^- and NO_x. See also EXPLOSIVES and ANILINE DYES.

CGL500 CAS:5388-62-5 HR: 3
4-CHLORO-2,6-DINITROANILINE
mf: $C_6H_4ClN_3O_4$ mw: 217.57

$Cl(O_2N)_2C_6H_2NH_2$

PROP: Orange-yellow cryst from EtOH. Mp: 146°.

SYN: 2,6-DINITRO-4-CHLOROANILINE

TOXICITY DATA WITH REFERENCE
orl-rat LD50:400 mg/kg TSCAT* OTS 206512

CONSENSUS REPORTS: Reported in EPA TSCA Inventory.

SAFETY PROFILE: A poison by ingestion. Solution in nitrosylsulfuric acid may explode when heated. When heated to decomposition it emits toxic fumes of Cl^- and NO_x. See also 2-CHLORO-4,6-DINITROANILINE; ANILINE; and ANILINE DYES.

CGL750 CAS:25567-67-3 HR: 3
CHLORODINITROBENZENE
DOT: UN 1577
mf: $C_6H_3ClN_2O_4$ mw: 202.56

SYNS: CHLORODINITROBENZENE (mixed isomers) (DOT) □ DINITROCHLOROBENZENE □ DINITROCHLOROBENZENE (DOT)

TOXICITY DATA WITH REFERENCE
dnd-mus-ipr 60 mg/kg BSIBAC 56,1680,80
orl-rat LD50:300 mg/kg 85GMAT-,61,82

DOT CLASSIFICATION: 6.1; *Label:* Poison

SAFETY PROFILE: A poison by ingestion. Mutation data

reported. Potentially explosive. When heated to decomposition it emits very toxic fumes of Cl^- and NO_x. See also other chloro-dinitrobenzenes.

CGM000 CAS:97-00-7 HR: 3
1-CHLORO-2,4-DINITROBENZENE
mf: $C_6H_3ClN_2O_4$ mw: 202.56

PROP: Yellow rhombic crystals from Et_2O; insol in water. Mp(α): 51°, mp(β): 43°, mp(γ): 27°, bp: 315° (sltly decomp), lel: 2.0%, uel: 22%, flash p: 382°F (CC), d(α): 1.687 @ 22°, d(β): 1.680 @ 20°/4°, vap d: 6.98.

SYNS: 1-CHLOOR-2,4-DINITROBENZEEN (DUTCH) ☐ 1-CHLOR-2,4-DINITROBENZENE ☐ 4-CHLORO-1,3-DINITROBENZENE ☐ 6-CHLORO-1,3-DINITROBENZENE ☐ 1-CHLORO-2,4-DINITROBENZOL (GERMAN) ☐ 1-CLORO-2,4-DINITROBENZENE (ITALIAN) ☐ 2,4-DINITROCHLOROBENZENE ☐ 1,3-DINITRO-4-CHLOROBENZENE ☐ 2,4-DINITRO-1-CHLOROBENZENE ☐ DINITROCHLOROBENZOL ☐ DNCB

TOXICITY DATA WITH REFERENCE
skn-hmn 30 μg CODEDG 2,247,76
skn-rbt 100 μg/24H open AIHAAP 23,95,62
skn-rbt 2 mg/24H SEV 85JCAE-,600,86
eye-rbt 50 μg/24H SEV 85JCAE-,600,86
mmo-sat 3 μg/plate ARDEAC 121,348,85
mma-sat 50 μg/plate ADREDL 266,315,79
dnd-mus-ipr 30 mg/kg MUREAV 116,239,83
otr-ham:kdy 10 mg/L ARTODN 45,307,80
orl-rat LD50:780 mg/kg GTPZAB 32(2),48,88
ipr-rat LD50:280 mg/kg AGGHAR 17,217,59
skn-rbt LD50:130 mg/kg AIHAAP 23,95,62

CONSENSUS REPORTS: Reported in EPA TSCA Inventory.

SAFETY PROFILE: Poison by skin contact and intraperitoneal routes. Moderately toxic by ingestion. A severe human skin and eye irritant. Acts as a primary irritant as well as a sensitizer of skin. An allergen. Mutation data reported. Combustible when exposed to heat or flame. A moderate explosion hazard when exposed to flame, sparks, heated to 150°, or when shocked in a sealed container. Explosive reaction with ammonia at 170°C/40 bar. To fight fire use CO_2, dry chemical. Reacts violently with hydrazine sulfate or hydrazine hydrate. See also NITRO COMPOUNDS of AROMATIC HYDROCARBONS.

CGM199 HR: 3
4-CHLORO-2,5-DINITROBENZENE DIAZONIUM-6-OXIDE
mf: $C_6HClN_4O_5$ mw: 244.55

SAFETY PROFILE: A very shock-sensitive explosive solid. When heated to decomposition it emits toxic fumes of Cl^- and NO_x. See also NITRO COMPOUNDS of AROMATIC HYDROCARBONS.

CGM225 CAS:393-75-9 HR: 2
4-CHLORO-3,5-DINITRO-α-α-α-TRIFLUOROTOLUENE
mf: $C_7H_2ClF_3N_2O_4$ mw: 270.56

SYNS: BENZENE, 2-CHLORO-1,3-DINITRO-5-(TRIFLUOROMETHYL)- ☐ BENZOTRIFLUORIDE, 4-CHLORO-3,5-DINITRO- ☐ 4-CHLORO-3,5-DINITROBENZOTRIFLUORIDE ☐ 3,5-DINITRO-4-CHLORO-α-α-α-TRIFLUOROTOLUENE ☐ TOLUENE, 4-CHLORO-3,5-DINITRO-α-α-α-TRIFLUORO-

TOXICITY DATA WITH REFERENCE
orl-rat LD50:930 mg/kg AISSAW 19,351,83

CONSENSUS REPORTS: Reported in EPA TSCA Inventory.

SAFETY PROFILE: Moderately toxic by ingestion. When heated to decomposition it emits toxic vapors of NO_x, Cl^-, and F^-.

CGM375 CAS:63886-82-8 HR: 3
p-CHLORO-2,4-DIOXA-5-ETHYL-p-THIONO-3-PHOSPHABICYCLO(4.4.0)DECANE

TOXICITY DATA WITH REFERENCE
skn-rbt 10 mg/24H open MLD AIHAAP 23,95,62
orl-rat LD50:140 mg/kg AIHAAP 23,95,62
skn-rbt LD50:400 mg/kg AIHAAP 23,95,62

SAFETY PROFILE: Poison by ingestion and skin contact. A skin irritant. When heated to decomposition it emits toxic fumes of Cl^-, PO_x, and SO_x.

CGM400 CAS:2921-31-5 HR: 3
p-CHLORO-2,4-DIOXA-5-METHYL-p-THIONO-3-PHOSPHABICYCLO(4.4.0)DECANE
mf: $C_8H_{14}ClO_2PS$ mw: 240.70

SYNS: ENT 23,970 ☐ UC 8305 ☐ UNION CARBIDE UC-8305

TOXICITY DATA WITH REFERENCE
skn-rbt 10 mg/24H open MLD AIHAAP 23,95,62
orl-rat LD50:120 mg/kg ARSIM* 20,25,66
skn-rbt LD50:360 mg/kg AIHAAP 23,95,62
orl-ckn LD50:26 mg/kg TXAPA9 11,49,67

SAFETY PROFILE: Poison by ingestion and skin contact. A skin irritant. When heated to decomposition it emits toxic fumes of Cl^-, SO_x, and PO_x.

CGM450 CAS:3367-31-5 HR: 2
CHLORO((3-(2,4-DIOXO-5-IMIDAZOLIDINYL)-2-METHOXY)PROPYL) MERCURY
mf: $C_7H_{11}ClHgN_2O_3$ mw: 407.24

TOXICITY DATA WITH REFERENCE
orl-mus LD50:3320 mg/kg JMPCAS 5,168,62

ACGIH TLV: TWA 0.1 mg(Hg)/m^3 (skin)
NIOSH REL: (Mercury, Aryl and Inorganic): CL 0.1 mg/m^3 (skin)

SAFETY PROFILE: Moderately toxic by ingestion. When heated to decomposition it emits toxic fumes of NO_x, Hg, and Cl^-.

CGM500 CAS:3861-99-2 HR: 2
6-CHLORO-1,3-DIOXO-5-ISOINDOLINESULFONAMIDE
mf: $C_8H_5ClN_2O_4S$ mw: 260.66

SYNS: 4-CHLORO-5-SULPHAMOYLPHTHALIMIDE ☐ 1H-ISOINDOLE-5-SULFONAMIDE, 6-CHLORO-2,3-DIHYDRO-1,3-DIOXO- ☐ 5-ISOINDOLINESULFONAMIDE, 6-CHLORO-1,3-DIOXO-

C

TOXICITY DATA WITH REFERENCE
skn-rbt 500 mg MLD FCTOD7 20,573,82
eye-rbt 100 mg SEV FCTOD7 20,573,82
eye-rbt 100 mg/4S RNS SEV FCTOD7 20,573,82

SAFETY PROFILE: A severe eye and mild skin irritant. When heated to decomposition it emits toxic fumes of NO_x, SO_x, and Cl^-.

CGM750 CAS:2051-60-7 HR: 2
2-CHLORODIPHENYL
mf: $C_{12}H_9Cl$ mw: 188.66

PROP: Crystals. Mp: 33.5°, bp: 273–274°.

SYNS: 2-CHLORO-1,1'-BIPHENYL □ o-CHLORODIPHENYL

TOXICITY DATA WITH REFERENCE
orl-mam LDLo: 2500 mg/kg JIDHAN 13,87,31

SAFETY PROFILE: Moderately toxic by ingestion. When heated to decomposition it emits toxic fumes of Cl^-.

CGN000 CAS:712-48-1 HR: 3
CHLORODIPHENYLARSINE
DOT: UN 1699
mf: $C_{12}H_{10}AsCl$ mw: 264.59

PROP: Colorless crystals when pure, technical product is dark-brown liquid. Mp: 44°, bp: 333° (decomp), d: 1.333 @ 40° (solid): 1.358 @ 45° (liquid), vap press: 0.00049 mm @ 20°, vap d: 9.15. Insol in H_2O; sol org solvs.

SYNS: BLUE CROSS □ CLARK I □ DA □ DIPHENYLARSINOUS CHLORIDE □ DIPHENYLCHLOORARSINE (DUTCH) □ DIPHENYL-CHLOROARSINE (DOT) □ SNEEZING GAS

TOXICITY DATA WITH REFERENCE
ihl-hmn LCLo: 55 ppm/30M NTIS** PB214-270
ihl-dog LCLo: 31 ppm/50M ZGEMAZ 13,523,21
ihl-cat LCLo: 6 ppm/12M ZGEMAZ 13,523,21
scu-cat LDLo: 500 μg/kg ZGEMAZ 13,523,21

CONSENSUS REPORTS: Arsenic and its compounds are on the Community Right-To-Know List.

OSHA PEL: TWA 0.5 mg(As)/m³
DOT CLASSIFICATION: 6.1; *Label:* Poison

SAFETY PROFILE: A human poison by inhalation. Poison experimentally by inhalation and skin contact. A powerfully irritating military poison. Exposure yields cold-like symptoms, plus headache, vomiting and nausea. A nonpersistent gas. Decontamination is by use of chlorine or caustic soda in confined spaces. When heated to decomposition it emits toxic fumes of As and Cl^-. See also ARSENIC COMPOUNDS.

CGN250 CAS:77966-58-6 HR: 3
6'-CHLORO-2-(DIPROPYLAMINO)-o-ACETOTOLUIDIDE HYDROCHLORIDE
mf: $C_{15}H_{23}ClN_2O{\cdot}ClH$ mw: 319.31

SYN: C 3071

TOXICITY DATA WITH REFERENCE
eye-rbt 2% SEV ARZNAD 8,407,58
ipr-rat LD50: 260 mg/kg ARZNAD 8,407,58
ipr-mus LD50: 245 mg/kg ARZNAD 8,407,58
scu-mus LD50: 1780 mg/kg ARZNAD 8,407,58

SAFETY PROFILE: Poison by intraperitoneal route. Moderately toxic by subcutaneous route. A severe eye irritant. When heated to decomposition it emits very toxic fumes of Cl^- and NO_x.

CGN325 CAS:22086-53-9 HR: 3
CHLORODIPROPYLBORANE
mf: $C_6H_{14}BCl$ mw: 132.44

PROP: Air and moisture-sensitive colorless liquid. D: 0.848 @ 20°/4°, bp: 127°, mp: −125°.

SAFETY PROFILE: Ignites spontaneously in air. When heated to decomposition it emits toxic fumes of Cl^-. See also BORANES and CHLORIDES.

CGO125 CAS:1622-32-8 HR: 3
2-CHLOROETHANE SULFOCHLORIDE
mf: $C_2H_4Cl_2O_2S$ mw: 163.02

SYNS: β-CHLOROETHANESULFONYL CHLORIDE □ 2-CHLOROETHANESULFONYL CHLORIDE □ 2-CHLOROETHYLSULFONYL CHLORIDE

TOXICITY DATA WITH REFERENCE
orl-rat LD50: 240 mg/kg GTPZAB 15(6),59,71
ihl-rat LC50: 420 mg/m³/4H GTPZAB 15(6),59,71
ihl-mus LC50: 250 mg/m³/4H GTPZAB 15(6),59,71

CONSENSUS REPORTS: EPA Extremely Hazardous Substances List.

SAFETY PROFILE: Poison by inhalation and ingestion. When heated to decomposition it emits toxic fumes of Cl^- and SO_x.

CGO500 CAS:115-96-8 HR: 3
2-CHLOROETHANOL PHOSPHATE
mf: $C_6H_{12}Cl_3O_4P$ mw: 285.50

PROP: Flash p: 421°F (COC), boiling range: 210–220° @ 20 mm, d: 1.425 @ 20°/20°, autoign temp: 1115°F, vap press: 0.5 mm @ 145°.

SYNS: CELLUFLEX □ FYROL CEF □ NCI-C60128 □ NIAX FLAME RETARDANT 3 CF □ TRICHLORETHYL PHOSPHATE □ TRI-β-CHLOROETHYL PHOSPHATE □ TRI(2-CHLOROETHYL)PHOSPHATE □ TRIS (2-CHLOROETHYL)ESTER PHOSPHORIC ACID □ TRIS(β-CHLOROETHYL) PHOSPHATE □ TRIS(2-CHLOROETHYL) PHOSPHATE

TOXICITY DATA WITH REFERENCE
skn-rbt 10 mg/24H open MLD AMIHBC 4,119,51
eye-rbt 500 mg open AMIHBC 4,119,51
eye-rbt 500 mg/24H MLD 85JCAE-,1135,86
mmo-sat 333 μg/plate ENMUDM 5(Suppl 1),3,83
mma-sat 1 μmol/plate MUREAV 66,373,79
mnt-ham-ipr 125 mg/kg EJCODS 18,1337,82
otr-ham: emb 400 mg/L EJCODS 18,1337,82
sce-ham: lng 700 mg/L EJCODS 18,1337,82
orl-rat TDLo: 1800 mg/kg (7-15D preg): REP ESKHA5 101,55,83
orl-rat TDLo: 45,760 mg/kg/2Y-C: CAR NTPTR* NTP-TR-391,91

orl-mus TDLo: 182 g/kg/2Y-C: ETA NTPTR* NTP-TR-391,91
orl-rat LD50: 1230 mg/kg BECTA6 17,720,77
orl-mus LD50: 1866 mg/kg NTIS** AD-A067-313
ipr-mus LDLo: 250 mg/kg CBCCT* 7,396,55

CONSENSUS REPORTS: IARC Cancer Review: Group 3 IMEMDT 48,109,90; Animal Inadequate Evidence IMEMDT 48,109,90. Reported in EPA TSCA Inventory. EPA Genetic Toxicology Program.

SAFETY PROFILE: Poison by intraperitoneal route. Moderately toxic by ingestion. Experimental reproductive effects. Questionable carcinogen with experimental tumorigenic data. A skin and eye irritant. Combustible when exposed to heat or flame. When heated to decomposition it emits very toxic fumes of PO_x and Cl^-. See also PHOSPHATES, CHLORIDES, and ESTERS.

CGO600 CAS:542-58-5 HR: 3
2-CHLOROETHYL ACETATE
mf: $C_4H_7ClO_2$ mw: 122.56

SYNS: ACETOXYETHYL CHLORIDE □ 2-CHLORETHYLACETAT □ 2-CHLOROETHANOL ACETATE □ β-CHLOROETHYL ACETATE □ ETHANOL, 2 CHLORO-, ACETATE

TOXICITY DATA WITH REFERENCE
orl-mus LD50: 250 mg/kg ZHYGAM 26,17,80

CONSENSUS REPORTS: Reported in EPA TSCA Inventory.

SAFETY PROFILE: Poison by ingestion. When heated to decomposition it emits toxic vapors of Cl^-.

CGP125 CAS:689-98-5 HR: 3
2-CHLOROETHYLAMINE
mf: C_2H_6ClN mw: 79.53

PROP: Oil.

SAFETY PROFILE: Unstable and may polymerize explosively. When heated to decomposition it emits toxic fumes of Cl^- and NO_x. See also AMINES and CHLORIDES.

CGP250 CAS:870-24-6 HR: 3
2-CHLOROETHYLAMINE HYDROCHLORIDE
mf: $C_2H_6ClN \cdot ClH$ mw: 116.00

SYNS: β-CHLOROETHYLAMINE HYDROCHLORIDE □ 2-CHLOROETHYLAMMONIUM CHLORIDE

TOXICITY DATA WITH REFERENCE
mmo-sat 8600 μmol/L ENMUDM 3,11,81
mmo-esc 8600 μmol/L ENMUDM 3,11,81
dns-rat: lvr 100 μmol/L ENMUDM 3,11,81
ipr-mus LD50: 2204 mg/kg JPETAB 94,249,48

CONSENSUS REPORTS: Reported in EPA TSCA Inventory.

SAFETY PROFILE: Moderately toxic by intraperitoneal route. Mutation data reported. Explosive reaction with concentrated alkali above 50°C. When heated to decomposition it emits toxic fumes of Cl^-, NH_3, and NO_x.

CGP375 CAS:110335-28-9 HR: 3
3′-CHLORO-2-ETHYLAMINO-o-ACETOTOLUIDIDE HYDROCHLORIDE
mf: $C_{11}H_{15}ClN_2O \cdot ClH$ mw: 263.19

TOXICITY DATA WITH REFERENCE
eye-rbt 100 mg MOD JAPMA8 49,80,60
ipr-mus LD50: 125 mg/kg JAPMA8 49,80,60
scu-mus LD50: 300 mg/kg JAPMA8 49,80,60

SAFETY PROFILE: Poison by subcutaneous and intraperitoneal routes. An eye irritant. When heated to decomposition it emits toxic fumes of NO_x and Cl^-.

CGP380 CAS:77966-59-7 HR: 3
6′-CHLORO-2-(ETHYLAMINO)-o-ACETOTOLUIDIDE HYDROCHLORIDE
mf: $C_{11}H_{15}ClN_2O \cdot ClH$ mw: 263.19

SYNS: C 3063 □ 2′-CHLORO-2-(ETHYLAMINO)-6′-METHYLACETANILIDE, HYDROCHLORIDE

TOXICITY DATA WITH REFERENCE
ipr-rat LD50: 235 mg/kg ARZNAD 8,407,58
ipr-mus LD50: 270 mg/kg ARZNAD 8,407,58
scu-mus LD50: 655 mg/kg ARZNAD 8,407,58

SAFETY PROFILE: Poison by intraperitoneal route. Moderately toxic by subcutaneous route. When heated to decomposition it emits toxic fumes of NO_x and Cl^-.

CGP500 CAS:38915-49-0 HR: 3
7-((2-((2-CHLOROETHYL)AMINO)ETHYL)AMINO)BENZ(c)ACRIDINE DIHYDROCHLORIDE HYDRATE
mf: $C_{21}H_{20}ClN_3 \cdot 2ClH \cdot H_2O$ mw: 440.83

TOXICITY DATA WITH REFERENCE
mmo-sat 5 μg/plate JMCMAR 15,739,72
ipr-mus LD20: 170 mg/kg JMCMAR 15,739,72

SAFETY PROFILE: Poison by intraperitoneal route. Mutation data reported. When heated to decomposition it emits very toxic fumes of Cl^- and NO_x.

CGQ250 CAS:40713-31-3 HR: 3
2-CHLOROETHYLAMINOETHYL DEHYDROABIETATE HYDROCHLORIDE
mf: $C_{24}H_{36}ClNO_2 \cdot ClH$ mw: 442.52

SYN: DEHYDRO-ABIETIC ACID-2-(2-(CHLOROETHYL)AMINO)ETHYL ESTER

TOXICITY DATA WITH REFERENCE
ipr-rat LD50: 400 mg/kg PCJOAU 6,647,72
ipr-mus LD50: 400 mg/kg PCJOAU 6,647,72

SAFETY PROFILE: Poison by intraperitoneal route. See also ESTERS. When heated to decomposition it emits very toxic fumes of Cl^- and NO_x.

C

CGQ280 **CAS:56538-00-2** **HR: 3**
N-(2-CHLOROETHYL)AMINOMETHYL-4-HYDROXYNITROBENZENE
mf: $C_9H_{11}ClN_2O_3$ mw: 230.67

SYNS: 2-(((2-CHLOROETHYL)AMINO)METHYL)-4-NITROPHENOL □ PHENOL, 2-(((2-CHLOROETHYL)AMINO)METHYL)-4-NITRO-

TOXICITY DATA WITH REFERENCE
orl-rat TDLo:35 mg/kg (female 1-7D post):REP IJMRAQ 66,987,77
orl-rat LD50:25 mg/kg IJMRAQ 66,987,77

SAFETY PROFILE: Poison by ingestion. Experimental reproductive effects. When heated to decomposition it emits toxic fumes of NO_x and Cl^-.

CGQ400 **CAS:56538-01-3** **HR: 3**
N-(2-CHLOROETHYL)AMINOMETHYL-4-METHOXYNITROBENZENE
mf: $C_{10}H_{13}ClN_2O_3$ mw: 244.70

SYNS: BENZYLAMINE, N-(2-CHLOROETHYL)-2-METHOXY-5-NITRO- □ N-(2-CHLOROETHYL)-2-METHOXY-5-NITROBENZYLAMINE

TOXICITY DATA WITH REFERENCE
orl-rat TDLo:35 mg/kg (female 1-7D post):REP IJMRAQ 66,987,77
orl-rat LD50:28 mg/kg IJMRAQ 66,987,77

SAFETY PROFILE: Poison by ingestion. Experimental reproductive effects. When heated to decomposition it emits toxic fumes of NO_x and Cl^-.

CGQ500 **CAS:21715-46-8** **HR: 3**
6-CHLORO-2-ETHYLAMINO-4-METHYL-4-PHENYL-4H-3,1-BENZOXAZINE
mf: $C_{17}H_{17}ClN_2O$ mw: 300.81

SYNS: 2-AETHYLAMINO-6-CHLOR-4-METHYL-4-PHENYL-4H-3,1-BENZOXAZIN (GERMAN) □ 6-CHLORO-N-ETHYL-4-METHYL-4-PHENYL-4H-3,1-BENZOXAZIN-2-AMINE □ ETIFOXIN □ ETIFOXINE □ HOE 36801

TOXICITY DATA WITH REFERENCE
orl-rat LD50:1500 mg/kg THERAP 27,325,72
ipr-rat LD50:292 mg/kg THERAP 27,325,72
ivn-rat LD50:55 mg/kg THERAP 27,325,72
orl-mus LD50:1388 mg/kg THERAP 27,325,72
ipr-mus LD50:450 mg/kg THERAP 27,325,72
ivn-mus LD50:120 mg/kg THERAP 27,325,72
ivn-gpg LDLo:133 mg/kg THERAP 27,325,72

SAFETY PROFILE: Poison by intravenous and intraperitoneal routes. Moderately toxic by ingestion. A tranquilizer. When heated to decomposition it emits very toxic fumes of Cl^- and NO_x.

CGR000 **CAS:38914-96-4** **HR: 2**
9-((3-((2-CHLOROETHYL)AMINO)PROPYL)AMINO)ACRIDINE DIHYDROCHLORIDE HYDRATE
mf: $C_{18}H_{20}ClN_3•2ClH•H_2O$ mw: 404.80

TOXICITY DATA WITH REFERENCE
mmo-sat 5 µg/plate JMCMAR 15,739,72
ipr-mus LD20:243 mg/kg JMCMAR 15,739,72

SAFETY PROFILE: Moderately toxic by intraperitoneal route. Mutation data reported. When heated to decomposition it emits very toxic fumes of Cl^- and NO_x.

CGR250 **CAS:38915-50-3** **HR: 3**
7-((3-((2-CHLOROETHYL)AMINO)PROPYL)AMINO)BENZ(c)ACRIDINE DIHYDROCHLORIDE SESQUIHYDRATE
mf: $C_{22}H_{22}ClN_3•2ClH•3/2H_2O$ mw: 463.87

TOXICITY DATA WITH REFERENCE
mmo-sat 5 µg/plate JMCMAR 15,739,72
ipr-mus LD20:70 mg/kg JMCMAR 15,739,72

SAFETY PROFILE: Poison by intraperitoneal route. Mutation data reported. When heated to decomposition it emits very toxic fumes of NO_x and Cl^-.

CGR500 **CAS:38915-61-6** **HR: 3**
7-((3-((2-CHLOROETHYL)AMINO)PROPYL)AMINO)BENZO(b)(1,10)PHENANTHROLINE DIHYDROCHLORIDE
mf: $C_{21}H_{21}ClN_4•2ClH$ mw: 437.83

SYN: ICR 395

TOXICITY DATA WITH REFERENCE
mmo-sat 500 ng/plate MUREAV 136,185,84
ipr-mus LD20:68 mg/kg JMCMAR 15,739,72

SAFETY PROFILE: Poison by intraperitoneal route. Mutation data reported. When heated to decomposition it emits very toxic fumes of Cl^- and NO_x.

CGR750 **CAS:38915-61-6** **HR: 3**
7-((3-((2-CHLOROETHYL)AMINO)PROPYL)AMINO)BENZO(b)(1,8)PHENANTHROLINE DIHYDROCHLORIDE HYDRATE
mf: $C_{21}H_{21}ClN_4•2ClH•H_2O$ mw: 455.85

TOXICITY DATA WITH REFERENCE
mmo-sat 5 µg/plate JMCMAR 15,739,72
ipr-mus LD20:68 mg/kg JMCMAR 15,739,72

SAFETY PROFILE: Poison by intraperitoneal route. Mutation data reported. When heated to decomposition it emits very toxic fumes of Cl^- and NO_x.

CGS500 **CAS:36167-69-8** **HR: 3**
10-((2-CHLOROETHYLAMINO)PROPYLAMINO)-2-METHOXY-7-CHLOROBENZO(b)-(1,5)-NAPHTHYRIDINE
mf: $C_{18}H_{20}Cl_2N_4O$ mw: 379.32

SYNS: N-(2-CHLOROETHYL)-N'-(7-CHLORO-2-METHOXYBENZO(b)-1,5-NAPHTHYRIDIN-10-YL)-1,3-PROPANEDIAMINE □ ICR 372

TOXICITY DATA WITH REFERENCE
mmo-sat 500 ng/plate MUREAV 136,185,84
mmo-esc 5 µg/plate GENTAE 78,823,74
msc-ham:ovr 1 µmol/L CNREA8 38,506,78
ipr-mus LD20:94 mg/kg JMCMAR 15,739,72

CONSENSUS REPORTS: EPA Genetic Toxicology Program.

SAFETY PROFILE: Poison by intraperitoneal route. Mutation data reported. When heated to decomposition it emits very toxic fumes of NO_x and Cl^-.

CGS750 CAS:38925-90-5 HR: 3
4-((3-((2-CHLOROETHYL)AMINO)PROPYL)AMINO)-6-METHOXYQUINOLINE HYDROCHLORIDE
mf: $C_{15}H_{20}ClN_3O{\cdot}2ClH$ mw: 366.75

SYN: ICR 180

TOXICITY DATA WITH REFERENCE
mmo-sat 5 µg/plate JMCMAR 15,739,72
ipr-mus LD20:92 mg/kg JMCMAR 15,739,72

SAFETY PROFILE: Poison by intraperitoneal route. Mutation data reported. When heated to decomposition it emits very toxic fumes of Cl^- and NO_x.

CGT000 CAS:102584-97-4 HR: 3
6′-CHLORO-2-(ETHYLAMINO)-o-VALEROTOLUIDIDE HYDROCHLORIDE
mf: $C_{14}H_{21}ClN_2O{\cdot}ClH$ mw: 305.28

SYN: C 3205

TOXICITY DATA WITH REFERENCE
eye-rbt 2% MLD ARZNAD 8,544,58
ipr-rat LD50:35 mg/kg ARZNAD 8,544,58
scu-mus LD50:65 mg/kg ARZNAD 8,544,58

SAFETY PROFILE: Poison by intraperitoneal and subcutaneous routes. An eye irritant. When heated to decomposition it emits very toxic fumes of $Cl^-$and NO_x.

CGU000 CAS:2114-18-3 HR: 2
β-CHLOROETHYL CARBAMATE
mf: $C_3H_6ClNO_2$ mw: 123.55

TOXICITY DATA WITH REFERENCE
orl-mus LD50:950 mg/kg THERAP 16,678,61
ipr-mus LD50:1000 mg/kg JNCIAM 8,99,47

CONSENSUS REPORTS: Reported in EPA TSCA Inventory.

SAFETY PROFILE: Moderately toxic by ingestion and intraperitoneal routes. See also CARBAMATES. When heated to decomposition it emits very toxic fumes of Cl^- and NO_x.

CGU199 CAS:627-11-2 HR: 3
2-CHLOROETHYL CHLOROFORMATE
mf: $C_3H_4Cl_2O_2$ mw: 142.97

PROP: Bp: 155.7°, d: 1.3847, insoluble in water.

SYNS: 2-CHLORETHYLESTER KYSELINY CHLORMRAVENCI □ CHLOROFORMIC ACID 2-CHLOROETHYL ESTER □ FORMIC ACID, CHLORO-, 2-CHLOROETHYL ESTER □ TL 207

TOXICITY DATA WITH REFERENCE
ihl-mus LCLo:200 mg/m³/10M NDRC** NCDrc-132,June,42

CONSENSUS REPORTS: EPA Extremely Hazardous Substances List. EPA Genetic Toxicology Program. Reported in EPA TSCA Inventory.

DOT CLASSIFICATION: 6.1; *Label:* Poison, Corrosive

SAFETY PROFILE: Poison by inhalation. May also cause fatalities by ingestion or skin contact. May cause burns to the eyes and skin. When heated to decomposition it emits toxic fumes of Cl^-.

CGV000 CAS:13909-14-3 HR: 3
1-(2-CHLOROETHYL)-3-CYCLODODECYL-1-NITROSOUREA
mf: $C_{15}H_{28}ClN_3O_2$ mw: 317.91

SYNS: N-(2-CHLOROETHYL)-N′-CYCLODODECYL-N-NITROSOUREA □ NSC 91726

TOXICITY DATA WITH REFERENCE
ipr-mus LD50:1297 mg/kg NCISP* JAN86

SAFETY PROFILE: Poison by intraperitoneal route. Many N-nitroso compounds are carcinogens. When heated to decomposition it emits very toxic fumes of Cl^- and NO_x. See also N-NITROSO COMPOUNDS.

CGV250 CAS:13010-47-4 HR: 3
1-(2-CHLOROETHYL)-3-CYCLOHEXYL-1-NITROSOUREA
mf: $C_9H_{16}ClN_3O_2$ mw: 233.73

PROP: Yellow powder. Mp: 90°.

SYNS: BELUSTINE □ CCNU □ CECENU □ CEENU □ CHLOROETHYLCYCLOHEXYLNITROSOUREA □ N-(2-CHLOROETHYL)-N′-CYCLOHEXYL-N-NITROSOUREA □ (CHLORO-2-ETHYL)-1-CYCLOHEXYL-3-NITROSOUREA □ CINU □ (CLORO-2-ETIL)-1-CICLOESIL-3-NITROSOUREA (ITALIAN) □ ICIG 1109 □ LOMUSTINE □ NCI-C04740 □ NSC-79037 □ RB 1509 □ SRI 2200

TOXICITY DATA WITH REFERENCE
mma-sat 100 nmol/plate JJIND8 65,149,80
mmo-esc 500 µmol/L GANNA2 71,674,80
dnd-hmn:emb 100 µmol/L CNREA8 44,1352,84
dns-hmn:lym 10 mg/L FRPSAX 36,947,81
dni-hmn:leu 1 µmol/L BBACAQ 425,463,76
sce-hmn:lym 50 µmol/L CGCYDF 9,261,83
msc-ham:lng 10 µmol/L CNREA8 40,2719,80
ivn-rbt TDLo:39 mg/kg (female 6-18D post):REP TXAPA9 34,456,75
ipr-rat TDLo:8 mg/kg (9-12D preg):TER TXAPA9 34,456,75
ipr-rat TDLo:22,500 µg/kg (9W male):REP TXAPA9 34,456,75
ipr-rat TDLo:16 mg/kg (6-9D preg):REP TXAPA9 34,456,75
ipr-mus TDLo:9 mg/kg (male 1D pre):REP CNREA8 42,122,82
ipr-rat TDLo:16 mg/kg (6-9D preg):TER TXAPA9 34,456,75
ipr-rat TDLo:8 mg/kg (9-12D preg):TER TXAPA9 34,456,75
orl-rat TDLo:50 mg/kg:ETA PTBIAN 35,139,87
ipr-rat TDLo:60 mg/kg/7W-I:ETA CANCAR 40(Suppl 4),1935,77
ivn-rat TDLo:64 mg/kg/60W-I:ETA DTESD7 8,273,80
skn-mus TDLo:276 mg/kg/23W-I:ETA EXPEAM 36,1211,80
ipr-mus TDLo:98 mg/kg/26W-I:ETA CANCAR 40S,1935,77

ivn-rat TD: 127 mg/kg/60W-I: ETA DTESD7 8,273,80
ivn-rat TD: 178 mg/kg/42W-I: ETA DTESD7 8,273,80
orl-hmn TDLo: 30 mg/kg: CNS,GIT CCYPBY 3,33,72
orl-hmn TDLo: 3 mg/kg: GIT,BLD CTRRDO 60,709,76
orl-rat LD50: 70 mg/kg TXAPA9 21,405,72
ipr-rat LD50: 50,350 μg/kg NCISP* JAN86
orl-mus LD50: 38 mg/kg TXAPA9 21,405,72
ipr-mus LD50: 53 mg/kg TXAPA9 21,405,72
scu-mus LD50: 54 mg/kg TXAPA9 21,405,72
ivn-mus LD10: 40 mg/kg ANBCB3 23,64,78
orl-dog LDLo: 10 mg/kg ACRSAJ 16,273,72
ivn-dog LDLo: 5 mg/kg CCYPBY 4(3),13,73

CONSENSUS REPORTS: NTP 7th Annual Report on Carcinogens. IARC Cancer Review: Group 2A IMEMDT 7,150,87; Human Limited Evidence IMEMDT 26,137,81; Animal Sufficient Evidence IMEMDT 26,137,81. NCI Carcinogenesis Studies (ipr); Clear Evidence: mouse CANCAR 40,1935,77; No Evidence: rat CANCAR 40,1935,77. EPA Genetic Toxicology Program.

SAFETY PROFILE: Confirmed carcinogen with experimental carcinogenic and tumorigenic data. Poison by ingestion, intraperitoneal, subcutaneous, intravenous, and possibly other routes. Human systemic effects by ingestion: anorexia, nausea or vomiting, leukopenia (decrease in the white blood cell count), and thrombocytopenia (decrease in the number of blood platelets). Experimental teratogenic and reproductive effects. Human mutation data reported. When heated to decomposition it emits very toxic fumes of Cl^- and NO_x. See also N-NITROSO COMPOUNDS.

CGV275 CAS:30077-45-3 HR: 1
β-CHLOROETHYLDICHLOROARSINE
mf: $C_2H_4AsCl_3$ mw: 209.33

PROP: Oil. D: 1.840, bp: 92–93° @ 32 mm.

SYNS: ARSINE, (2-CHLOROETHYL)DICHLORO- □ β-CHLOROETHYLDICHLORARSINE

TOXICITY DATA WITH REFERENCE
ihl-mus LC50: 13 g/m³/10M NTIS** PB158-508

OSHA PEL: TWA 0.5 mg(As)/m³

SAFETY PROFILE: Toxic by inhalation. When heated to decomposition it emits toxic fumes of Asand Cl^-.

CGV500 CAS:100-35-6 HR: 3
N-(2-CHLORO ETHYL)DIETHYLAMINE
mf: $C_6H_{14}ClN$ mw: 135.66

SYNS: (2-CHLOROETHYL)DIETHYLAMINE □ β-CHLOROTRIETHYLAMINE □ 2-CHLOROTRIETHYLAMINE □ 2-(DIETHYLAMINO)CHLOROETHANE □ DIETHYLAMINOETHYL CHLORIDE □ β-(DIETHYLAMINO)ETHYL CHLORIDE □ N-DIETHYLAMINOETHYL CHLORIDE □ 2-(DIETHYLAMINO)ETHYL CHLORIDE □ DIETHYL(2-CHLOROETHYL) AMINE

TOXICITY DATA WITH REFERENCE
skn-rbt 10 mg/24H MOD AMIHBC 4,119,51
eye-rbt 2 mg SEV AMIHBC 4,119,51
dni-mus-ivg 5000 ppm JIDEAE 62,378,74
orl-rat LD50: 17 mg/kg AMIHBC 4,119,51
unr-rat LD50: 30 mg/kg PHBUA9 1,297,53
skn-rbt LD50: 300 mg/kg AMIHBC 4,119,51

CONSENSUS REPORTS: EPA Genetic Toxicology Program.

SAFETY PROFILE: Poison by ingestion, skin contact and possibly other routes. A severe eye and moderate skin irritant. When heated to decomposition it emits very toxic fumes of Cl^- and NO_x. See also AMINES and CHLORIDES.

CGW000 CAS:107-99-3 HR: 3
N-(2-CHLOROETHYL)DIMETHYLAMINE
mf: $C_4H_{10}ClN$ mw: 107.60

PROP: Liquid. Vap d: 3.72.

SYNS: CHLORO(DIMETHYLAMINO)ETHANE □ β-CHLOROETHYLDIMETHYLAMINE □ (2-CHLOROETHYL)DIMETHYLAMINE □ DIMETHYLAMINOETHYL CHLORIDE □ β-(DIMETHYLAMINO)ETHYL CHLORIDE □ 2-DIMETHYLAMINOETHYLCHLORIDE □ DIMETHYL(2-CHLOROETHYL)AMINE □ HN 1 □ NITROGEN HALF MUSTARD

TOXICITY DATA WITH REFERENCE
mmo-sat 100 nmol/plate ARTODN 56,267,85
mma-sat 100 nmol/plate ARTODN 56,267,85
mmo-esc 100 nmol/plate ARTODN 56,267,85
mma-esc 100 nmol/plate ARTODN 56,267,85
oms-omi 2 mol EXPEAM 29,1344,73
mmo-smc 20 mmol/L GENTAE 92,83,79
dni-mus-ivg 2 pph JIDEAE 62,378,74
unr-rat LD50: 30 mg/kg PHBUA9 1,297,53

CONSENSUS REPORTS: EPA Genetic Toxicology Program. Reported in EPA TSCA Inventory.

SAFETY PROFILE: Poison by an unspecified route. A systemic irritant. Mutation data reported. When heated to decomposition it emits highly toxic fumes of Cl^- and NO_x.

CGW100 HR: 2
3′-CHLORO-4′-ETHYL-4-DIMETHYLAMINOAZOBENZENE
mf: $C_{16}H_{18}ClN_3$ mw: 287.82

SYNS: BENZENAMINE, N,N-DIMETHYL-3′-CHLORO-4′-ETHYL-4-(PHENYLAZO)- □ p-((3-CHLORO-4-ETHYLPHENYL)AZO)-N,N-DIMETHYLANILINE

TOXICITY DATA WITH REFERENCE
orl-rat TDLo: 4515 mg/kg/25W-C: CAR CBINA8 53,107,85

SAFETY PROFILE: Questionable carcinogen with experimental carcinogenic data. When heated to decomposition it emits toxic fumes of Cl^- and NO_x.

CGW105 HR: 2
4′-CHLORO-3′-ETHYL-4-DIMETHYLAMINOAZOBENZENE
mf: $C_{16}H_{18}ClN_3$ mw: 287.82

SYNS: BENZENAMINE, N,N-DIMETHYL-4′-CHLORO-3′-ETHYL-4-(PHENYLAZO)- □ p-((4-CHLORO-3-ETHYLPHENYL)AZO)-N,N-DIMETHYLANILINE

TOXICITY DATA WITH REFERENCE
orl-rat TDLo: 15,725 mg/kg/52W-C: CAR CBINA8 53,107,85

SAFETY PROFILE: Questionable carcinogen with experimental carcinogenic data. When heated to decomposition it emits toxic fumes of Cl^- and NO_x.

CGW250 CAS:13909-02-9 HR: 3
1-(2-CHLOROETHYL)-3-(2,6-DIOXO-3-PIPERIDYL)-1-NITROSOUREA
mf: $C_8H_{10}ClN_4O_4$ mw: 261.67

SYNS: N-(2-CHLOROETHYL)-N′-(2,6-DIOXO-3-PIPERIDINYL)-N-NITROSOUREA □ NSC-95466 □ PCNU

TOXICITY DATA WITH REFERENCE
mmo-sat 200 μg/plate TCMUD8 5,319,85
dnd-mus: leu 200 μmol/L PAACA3 24,249,83
dni-mus/oth 16 mg/kg INSSDM 19,85,81
oms mus/leu 16 mg/kg INSSDM 19,85,81
orl-mus LD50: 35,700 μg/kg NTIS** PB282-250
ipr-mus LD50: 15,210 μg/kg NCISP* JAN86
ivn-mus LD50: 22 mg/kg NTIS** PB282-250
ivn-dog LDLo: 3 mg/kg NTIS** PB282-250
ivn-mky LDLo: 10 mg/kg NTIS** PB282-250

SAFETY PROFILE: Poison by ingestion, intravenous, and intraperitoneal routes. Many N-nitroso compounds are carcinogens. Mutation data reported. When heated to decomposition it emits very toxic fumes of Cl^- and NO_x. See also N-NITROSO COMPOUNDS.

CGW300 CAS:9011-06-7 HR: D
CHLOROETHYLENE-1,1-DICHLOROETHYLENE POLYMER
mf: $(C_2H_3Cl{\cdot}C_2H_2Cl_2)_x$

SYNS: BREON 202 □ BREON CS 100/30 □ DARAN □ DARAN CR 6795H □ 1,1-DICHLOROETHENE POLYMER with CHLOROETHENE □ 1,1-DICHLOROETHYLENE-MONOCHLOROETHYLENE POLYMER □ 1,1-DICHLOROETHYLENE POLYMER with CHLOROETHYLENE □ DOW 874 □ DOW LATEX 874 □ ET 67 □ ETHENE, 1,1-DICHLORO-, POLYMER with CHLOROETHENE (9CI) □ ETHYLENE, 1,1-DICHLORO-, POLYMER with CHLOROETHYLENE □ GEON 222 □ GEON 652 □ IKhS 1 □ KhS 596 □ KUREHALON A0 □ LAPLEN □ LATEX SVKh □ POLYCO 2611 □ QX 2168 □ SARAN 683 □ SARAN 746 □ SARAN RESIN 683 □ SP 489 □ SVKh 1 □ SVKh 40 □ UP 925 □ VELON □ VIKh 65 □ VINIDEN 60 □ VINYL CHLORIDE COPOLYMER with VINYLIDENE CHLORIDE □ VINYL CHLORIDE-1,1-DICHLOROETHYLENE COPOLYMER □ VINYL CHLORIDE-VINYLIDENE CHLORIDE COPOLYMER □ VINYL CHLORIDE-VINYLIDENE CHLORIDE POLYMER □ VINYLIDENE CHLORIDE-VINYL CHLORIDE POLYMER □ VKhVD 40 □ WINIDEN 60

CONSENSUS REPORTS: IARC Cancer Review: Group 3 IMEMDT 7,56,87; Human No Adequate Data IMEMDT 19,439,79; Animal No Adequate Data IMEMDT 19,439,79

CONSENSUS REPORTS: Reported in EPA TSCA Inventory.

SAFETY PROFILE: Questionable carcinogen. When heated to decomposition it emits toxic vapors of Cl^-.

CGW750 CAS:63019-51-2 HR: 2
4-CHLORO-6-ETHYLENEIMINO-2-PHENYLPYRIMIDINE
mf: $C_{12}H_{10}ClN_3$ mw: 231.70

SYN: 6-(1-AZIRIDINYL)-4-CHLORO-2-PHENYLPYRIMIDINE

TOXICITY DATA WITH REFERENCE
scu-rat TDLo: 960 mg/kg/16W-I: NEO BJPCAL 9,306,54

SAFETY PROFILE: Questionable carcinogen with experimental neoplastigenic data. When heated to decomposition it emits very toxic fumes of Cl^- and NO_x.

CGX000 CAS:7763-77-1 HR: 2
CHLOROETHYLENE OXIDE
mf: C_2H_3ClO mw: 78.50

SYNS: CHLOROEPOXYETHANE □ CHLOROOXIRANE □ MONOCHLOROETHYLENE OXIDE

TOXICITY DATA WITH REFERENCE
mma-sat 400 μmol/L MUREAV 58,217,78
mmo-esc 500 μmol/L MUREAV 152,147,85
mrc-smc 1 mmol/L TOERD9 3,131,81
dns-rat-ivn 5 g/kg CBINA8 17,239,77
msc-ham: lng 6 μmol/L IJCNAW 16,639,75
scu mus TDLo: 128 mg/kg/42W-I: NEO CNREA8 40,352,80

CONSENSUS REPORTS: EPA Genetic Toxicology Program.

SAFETY PROFILE: Questionable carcinogen with experimental neoplastigenic data. Mutation data reported. When heated to decomposition it emits very toxic fumes of Cl^-. See also CHLORIDES.

CGX250 CAS:4535-87-9 HR: 3
N-(2-CHLOROETHYL)ETHANAMINE HYDROCHLORIDE
mf: $C_4H_{10}ClN{\cdot}ClH$ mw: 144.06

SYNS: N-(2-CHLOROETHYL)ETHYLAMINE HYDROCHLORIDE □ ETHYL-β-CHLOROETHYLAMINE HYDROCHLORIDE

TOXICITY DATA WITH REFERENCE
ipr-rat LD50: 400 mg/kg ARZNAD 11,143,61
ipr-mus LD50: 1120 mg/kg CANCAR 2,1055,49
scu-mus LD50: 1 g/kg NTIS** PB158-507
ivn-mus LD50: 100 mg/kg JPETAB 91,224,47

SAFETY PROFILE: Poison by intravenous and intraperitoneal routes. Moderately toxic by subcutaneous route. When heated to decomposition it emits toxic fumes of Cl^- and NO_x.

CGX325 CAS:56538-02-4 HR: 3
N-(2-CHLOROETHYL)-2-ETHOXY-5-NITROBENZYLAMINE
mf: $C_{11}H_{15}ClN_2O_3$ mw: 258.73

SYNS: BENZENEMETHANAMINE-N-(2-CHLOROETHYL)-2-ETHOXY-5-NITRO □ N-(2-CHLOROETHYL)AMINOETHYL-4-ETHOXYNITROBENZENE

TOXICITY DATA WITH REFERENCE
orl-rat TDLo: 20 mg/kg (female 6-9D post): REP IJMRAQ 66,987,77

orl-rat TDLo: 10 mg/kg (4-5D preg): REP IJMRAQ 66,987,77
orl-rat LD50: 25 mg/kg IJMRAQ 66,987,77

SAFETY PROFILE: Poison by ingestion. Experimental reproductive effects. When heated to decomposition it emits toxic fumes of Cl^- and NO_x.

CGX500 CAS:38915-22-9 HR: 3
2-((2-CHLOROETHYL)ETHYLAMINO))-N-(3-((6-CHLORO-2-METHOXY-9-ACRIDINYL)AMINO)PROPYL) PROPYL)ACETAMIDE DIHYDROCHLORIDE, HEMIHYDRATE
mf: $C_{23}H_{28}Cl_2N_4O_2{\cdot}2ClH{\cdot}1/2H_2O$ mw: 545.38

SYN: ICR 290

TOXICITY DATA WITH REFERENCE
mmo-sat 5 µg/plate JMCMAR 15,739,72
ipr-mus LD20: 5454 µg/kg JMCMAR 15,739,72

SAFETY PROFILE: Poison by intraperitoneal route. Mutation data reported. When heated to decomposition it emits very toxic fumes of Cl^- and NO_x.

CGX625 CAS:51775-17-8 HR: 3
2-(N-(2-CHLOROETHYL)-N-ETHYLAMINOMETHYL)-1, 4-BENZODIOXAN HYDROCHLORIDE
mf: $C_{13}H_{18}ClNO_2{\cdot}ClH$ mw: 292.23

SYNS: CHLORHYDRATE de (N-ETHYL,N,β-CHLORETHYL)AMINOMETHYLBENZODIOXANE (FRENCH) □ N-(2-CHLOROETHYL)-N-ETHYL-1,4-BENZODIOXAN-2-METHYLAMINE HYDROCHLORIDE □ 3718 RP

TOXICITY DATA WITH REFERENCE
scu-rat LDLo: 40 mg/kg AIPTAK 95,285,53
scu-mus LDLo: 40 mg/kg AIPTAK 95,285,53
ivn-mus LD50: 56 mg/kg CSLNX* NX#07473
ivn-dog LDLo: 15 mg/kg AIPTAK 95,285,53
scu-rbt LDLo: 30 mg/kg AIPTAK 95,285,53
ivn-rbt LDLo: 15 mg/kg AIPTAK 95,285,53
scu-frg LDLo: 50 mg/kg AIPTAK 95,285,53

SAFETY PROFILE: Poison by subcutaneous and intravenous routes. When heated to decomposition it emits toxic fumes of NO_x and Cl^-.

CGX750 CAS:38914-97-5 HR: 3
9-((3-((2-CHLOROETHYL)ETHYLAMINO)PROPYL)AMINO)ACRIDINE DIHYDROCHLORIDE
mf: $C_{20}H_{24}ClN_3{\cdot}2ClH$ mw: 414.84

SYNS: N'-9-ACRIDINYL-N-(2-CHLOROETHYL)-N-ETHYL-1,3-PROPANEDIAMINE DIHYDROCHLORIDE □ ICR 217

TOXICITY DATA WITH REFERENCE
mmo-sat 500 ng/plate MUREAV 136,185,84
msc-ham: ovr 1 µmol/L CNREA8 39,4875,79
ipr-mus LD20: 1659 µg/kg JMCMAR 15,739,72

SAFETY PROFILE: A poison by intraperitoneal route. Mutation data reported. When heated to decomposition it emits very toxic fumes of Cl^- and NO_x.

CGY000 CAS:39013-93-9 HR: 3
7-((3-((2-CHLOROETHYL)ETHYLAMINO)PROPYL)AMINO)BENZO(b) (1,10)PHENANTHROLINE DIHYDROCHLORIDE
mf: $C_{23}H_{25}ClN_4{\cdot}2ClH$ mw: 465.89

SYNS: 7-((3-((2-CHLOROETHYL)ETHYLAMINO)PROPYL)AMINO) BENZO(b)(1,10)PHENANTHROLINE 2HCl □ ICR 368

TOXICITY DATA WITH REFERENCE
mmo-sat 500 ng/plate MUREAV 136,185,84
msc-ham: ovr 250 nmol/L CNREA8 39,4875,79
ipr-mus LD20: 932 µg/kg JMCMAR 15,739,72

SAFETY PROFILE: A poison by intraperitoneal route. Mutation data reported. When heated to decomposition it emits very toxic fumes of Cl^- and NO_x.

CGY500 CAS:78218-16-3 HR: 3
9-((3-((2-CHLOROETHYL)ETHYLAMINO)PROPYL)AMINO)-4-METHOXYACRIDINE DIHYDROCHLORIDE HEMIHYDRATE
mf: $C_{21}H_{26}ClN_3O{\cdot}2ClH{\cdot}1/2H_2O$ mw: 453.84

SYNS: 9-((3-((2-CHLOROETHYL)ETHYLAMINO)PROPYL)AMINO)-4-METHOXYACRIDINE 2HCl HEMIHYDRATE □ N-(2-CHLOROETHYL)-N-ETHYL-N'-(4-METHOXY-9-ACRIDINYL)-1,3-PROPANEDIAMINE DIHYDROCHLORIDE HEMIHYDRATE

TOXICITY DATA WITH REFERENCE
mmo-sat 5 µg/plate JMCMAR 15,739,72
ipr-mus LD20: 1362 µg/kg JMCMAR 15,739,72

SAFETY PROFILE: Poison by intraperitoneal route. Mutation data reported. When heated to decomposition it emits very toxic fumes of Cl^- and NO_x.

CGY750 CAS:693-07-2 HR: 3
CHLOROETHYL ETHYL SULFIDE
mf: C_4H_9ClS mw: 124.64

PROP: Liquid with penetrating odor. D: 1.07 @ 25°/4°, bp: 156.5°.

SYNS: 2-CHLOROETHYL ETHYL SULFIDE □ 2-CHLOROETHYL ETHYL THIOETHER □ 1-CHLORO-2-(ETHYLTHIO)ETHANE □ ETHYL-β-CHLOROETHYL SULFIDE □ ETHYL-2-CHLOROETHYL SULFIDE □ β-ETHYLMERKAPTOETHYLCHLORID (CZECH) □ 2-(ETHYLTHIO)CHLOROETHANE □ 2-ETHYLTHIOETHYL CHLORIDE □ HALF-MUSTARD GAS □ h-MG

TOXICITY DATA WITH REFERENCE
skn-rbt 500 mg/24H SEV 28ZPAK -,170,72
eye-rbt 250 µg/24H SEV 28ZPAK -,170,72
mmo-esc 500 µmol/L MUREAV 28,257,75
dnd-esc 500 µmol/L MUREAV 28,257,75
slt-dmg-par 5 mmol/L MUREAV 13,19,71
sln-dmg-par 5 mmol/L CNREA8 32,550,72
orl-rat LD50: 252 mg/kg 28ZPAK -,170,72
scu-mus LDLo: 25 mg/kg NTIS** PB158-507

CONSENSUS REPORTS: Reported in EPA TSCA Inventory. EPA Genetic Toxicology Program.

SAFETY PROFILE: Poison by ingestion and subcutaneous routes. Mutation data reported. A severe skin and eye irritant. See also ETHERS and SULFIDES. When heated to decomposition it emits very toxic fumes of Cl^- and SO_x.

CGY825 CAS:1537-62-8 HR: 3
2-CHLOROETHYL FLUOROACETATE
mf: $C_4H_5ClFO_2$ mw: 140.55

SYNS: β-CHLOROETHYL FLUOROACETATE □ TL 671

TOXICITY DATA WITH REFERENCE
ihl-rat LC50:200 mg/m^3/10M NTIS** PB158-508
ihl-mus LC50:700 mg/m^3/10M NTIS** PB158-508
ihl-rbt LC50:100 mg/m^3/10M NTIS** PB158-508
ihl-gpg LC50:150 mg/m^3/10M NTIS** PB158-508

SAFETY PROFILE: Poison by inhalation. When heated to decomposition it emits toxic fumes of F^- and Cl^-.

CGZ000 CAS:371-28-8 HR: 3
2-CHLOROETHYL-γ-FLUOROBUTYRATE
mf: $C_6H_{10}ClFO_2$ mw: 168.61

SYN: 4-FLUORO-BUTYRIC ACID-2-CHLOROETHYL ESTER

TOXICITY DATA WITH REFERENCE
ihl-mus LC50:54 mg/m^3/10M NTIS** PB158-508
ihl-gpg LC50:100 mg/m^3/10M NTIS** PB158-508

SAFETY PROFILE: Poison by inhalation. See also ESTERS. When heated to decomposition it emits very toxic fumes of Cl^- and F^-.

CHA000 CAS:58484-07-4 HR: 3
1-(2-CHLOROETHYL)-3-(β-d-GLUCOPYRANOSYL)-1-NITROSOUREA
mf: $C_9H_{16}ClN_3O_7$ mw: 313.73

SYNS: GANU □ NSC D 254157

TOXICITY DATA WITH REFERENCE
dnd-mus:leu 100 μmol/L INSSDM 19,49,81
dni-mus/leu 10 mg/kg CNREA8 37,783,77
ivn-rat LD50:40 mg/kg GANNA2 68,247,77
ipr-mus LD50:15 mg/kg CNREA8 37,783,77
ipr-mus LDLo:10 mg/kg CNREA8 37,2615,77
ivn-mus LD10:10 mg/kg ANBCB3 23,64,78

SAFETY PROFILE: Poison by intravenous and intraperitoneal routes. Many N-nitroso compounds are carcinogens. Mutation data reported. When heated to decomposition it emits very toxic fumes of NO_x and Cl^-. See also N-NITROSO COMPOUNDS.

CHA250 CAS:60784-48-7 HR: 3
1-(2-CHLOROETHYL)-3-(4-HYDROXYBUTYL)-1-NITROSOUREA
mf: $C_7H_{14}ClN_3$ mw: 175.69

TOXICITY DATA WITH REFERENCE
mrc-smc 1 mmol/L/16H MUREAV 42,45,77
ipr-rat LD50:32 mg/kg EJCAAH 13,937,77

SAFETY PROFILE: Poison by intraperitoneal route. Mutation data reported. Many N-nitroso compounds are carcinogens. When heated to decomposition it emits very toxic fumes of NO_x and Cl^-. See also N-NITROSO COMPOUNDS.

CHA500 CAS:52049-26-0 HR: 3
1-(2-CHLOROETHYL)-3-(cis-4-HYDROXYCYCLOHEXYL)-1-NITROSOUREA
mf: $C_9H_{16}ClN_3O_3$ mw: 249.73

SYNS: cis-4-HYDROXY-CCNU □ cis-N-(2-CHLOROETHYL)-N'-(4-HYDROXYCYCLOHEXYL)-N-NITROSOUREA □ cis-4-OH-CCNU □ NSC 239724

TOXICITY DATA WITH REFERENCE
ipr-mus LD50:60,180 μg/kg NCISP* JAN86
ivn-mus LD10:25 mg/kg ANBCB3 23,64,78
ipl-mus LDLo:33 mg/kg JMCMAR 18,634,75

SAFETY PROFILE: Poison by intravenous, implant, and intraperitoneal routes. Many N-nitroso compounds are carcinogens. When heated to decomposition it emits very toxic fumes of Cl^- and NO_x. See also N-NITROSO COMPOUNDS.

CHA750 CAS:58494-43-2 HR: 3
1-(2-CHLOROETHYL)-3-(trans-2-HYDROXYCYCLOHEXYL)-1-NITROSOUREA
mf: $C_9H_{16}ClN_3O_4$ mw: 265.73

SYNS: N-(2-CHLOROETHYL)-N'-(trans-2-HYDROXYCYCLOHEXYL)-N-NITROSOUREA □ NSC 253947 □ trans-2-OH-CCNU □ trans-N-(2-CHLOROETHYL)-N'-(2-HYDROXYCYCLOHEXYL)-N-NITROSOUREA

TOXICITY DATA WITH REFERENCE
mmo-sat 100 nmol/plate JJIND8 65,149,80
ipr-mus LD50:90,140 μg/kg NCISP* JAN86

SAFETY PROFILE: Poison by intraperitoneal route. Mutation data reported. Many N-nitroso compounds are carcinogens. When heated to decomposition it emits very toxic fumes of Cl^- and NO_x. See also N-NITROSO COMPOUNDS.

CHB000 CAS:56239-24-8 HR: 3
1-(2-CHLOROETHYL)-3-(trans-4-HYDROXYCYCLOHEXYL)-1-NITROSOUREA
mf: $C_9H_{16}ClN_3O_3$ mw: 249.73

SYNS: N-(2-CHLOROETHYL)-N'-(trans-4-HYDROXYCYCLOHEXYL)-N-NITROSOUREA □ trans-N-(2-CHLOROETHYL)-N'-(4-HYDROXYCYCLOHEXYL)-N-NITROSOUREA □ trans-4-HYDROXY-CCNU □ NSC-239717 □ trans-4-OH CCNU

TOXICITY DATA WITH REFERENCE
mmo-sat 100 nmol/plate JJIND8 65,149,80
mma-sat 100 nmol/plate JJIND8 65,149,80
dnd-hmn:emb 100 μmol/L CNREA8 44,1352,84
ipr-mus LD50:52,420 μg/kg NCISP* JAN86
ivn-mus LD10:12 mg/kg ANBCB3 23,64,78
ipl-mus LDLo:35 mg/kg JMCMAR 18,634,75

SAFETY PROFILE: Poison by implant, intravenous and intraperitoneal routes. Human mutation data reported. Many N-nitroso compounds are carcinogens. When heated to decomposition it emits very toxic fumes of NO_x and Cl^-. See also N-NITROSO COMPOUNDS.

CHB250 CAS:101651-98-3 HR: 3
7-CHLORO-10-(3-(N-ETHYL-N-(2-HYDROXYETHYL) AMINO)PROPYL)ISOALLOXAZINE HYDROCHLORIDE
mf: $C_{17}H_{20}ClN_5O_3{\cdot}ClH$ mw: 414.33

TOXICITY DATA WITH REFERENCE
ipr-rat LD50:75 mg/kg CMTRAG 2,96,61
orl-mus LD50:5000 mg/kg CMTRAG 2,96,61
ipr-mus LD50:90 mg/kg CMTRAG 2,96,61
scu-mus LD50:145 mg/kg CMTRAG 2,96,61
ivn-mus LD50:60 mg/kg CMTRAG 2,96,61
ims-mus LD50:90 mg/kg CMTRAG 2,96,61

SAFETY PROFILE: Poison by intraperitoneal, subcutaneous, intravenous, and intramuscular routes. Mildly toxic by ingestion. When heated to decomposition it emits very toxic fumes of Cl^- and NO_x.

CHB750 CAS:60784-46-5 HR: 3
1-(2-CHLOROETHYL)-3-(2-HYDROXYETHYL)-1-NITROSOUREA
mf: $C_5H_{10}ClN_3O_3$ mw: 195.63

PROP: A solid. Mp: 56°. Sol in H_2O.

SYNS: CNU-ETHANOL □ HECNU □ 1-(2-HYDROXYETHYL)-3-(2-CHLOROETHYL)-3-NITROSOUREA □ HYDROXYETHYL CNU □ NSC 294895

TOXICITY DATA WITH REFERENCE
sln-dmg-orl 5 mmol/L DRISAA 52,20,77
sln-dmg-par 5 mmol/L DRISAA 52,20,77
mrc-smc 1 mmol/L/16H MUREAV 42,45,77
dnd-rat-ipr 100 μmol/kg CNREA8 44,514,84
cyt-mus:lng 1 mg/L/1H MUREAV 44,87,77
ivn-rat TDLo:16 mg/kg/60W-I:ETA DTESD7 8,273,80
ivn-rat LD50:32 mg/kg:ETA DTESD7 8,273,80
ivn-rat TD:64 mg/kg/60W-I:ETA DTESD7 8,273,80
ivn-rat TD:96 mg/kg/60W-I:ETA DTESD7 8,273,80
ipr-rat LD50:25,300 μg/kg ONCOBS 37,177,80
ipr-mus LD50:25 mg/kg INSSDM 19,123,81

CONSENSUS REPORTS: EPA Genetic Toxicology Program.

SAFETY PROFILE: Poison by intraperitoneal route. Questionable carcinogen with experimental tumorigenic data. Mutation data reported. Many N-nitroso compounds are carcinogens. When heated to decomposition it emits very toxic fumes of Cl^- and NO_x. See also N-NITROSO COMPOUNDS.

CHC000 CAS:693-30-1 HR: 3
2-CHLOROETHYL-2-HYDROXYETHYL SULFIDE
mf: C_4H_9ClOS mw: 140.64

SYNS: β-CHLOROETHYL-β-HYDROXYETHYL SULFIDE □ 2-((2-CHLOROETHYL)THIO)ETHANOL □ HALF MUSTARD GAS □ HALF SULFUR MUSTARD □ 2-HYDROXYETHYL-2-CHLOROETHYL SULFIDE □ MUSTARD CHLOROHYDRIN □ SULFUR HALF-MUSTARD

TOXICITY DATA WITH REFERENCE
dni-hmn:hla 1500 mg/L IUSMDJ 9,41,79
oms-hmn:hla 1500 mg/L IUSMDJ 9,41,79
dnd-rat:lvr 100 μmol/L BIJOAK 80,496,61
dnd-mus/ast 4 mg/kg BIJOAK 80,496,61
ims-rat LD50:500 μg/kg NTIS** PB158-507
skn-mus LD50:600 mg/kg JPETAB 93,1,48
ivn-mus LD50:35 mg/kg JPETAB 93,1,48

SAFETY PROFILE: Poison by intravenous and intramuscular routes. Moderately toxic by skin contact. Human mutation data reported. When heated to decomposition it emits very toxic fumes of Cl^- and SO_x. See also SULFIDES.

CHC250 CAS:60784-47-6 HR: 3
3-(2-CHLOROETHYL)-1-(3-HYDROXYPROPYL)-3-NITROSOUREA
mf: $C_6H_{12}ClN_3O_3$ mw: 209.66

SYN: 1-(3-HYDROXYPROPYL)-CNU

TOXICITY DATA WITH REFERENCE
mrc-smc 1 mmol/L/16H MUREAV 42,45,77
ipr-rat LDLo:16 mg/kg JNCIAM 60,345,78

SAFETY PROFILE: Poison by intraperitoneal route. Mutation data reported. Many N-nitroso compounds are carcinogens. When heated to decomposition it emits very toxic fumes of Cl^- and NO_x. See also N-NITROSO COMPOUNDS.

CHC500 CAS:107-27-7 HR: 3
CHLOROETHYL MERCURY
mf: C_2H_5ClHg mw: 265.11

PROP: Silvery, iridescent leaflets from EtOH. Mp: 196–198°. Sltly sol in Et_2O and EtOH; sol in $CHCl_3$; insol in H_2O.

SYNS: CERESAN □ EMC □ ETHYLMERCURIC CHLORIDE □ ETHYLMERCURY CHLORIDE □ GANOZAN □ GRANOSAN

TOXICITY DATA WITH REFERENCE
mmo-esc 20 nmol/L MJDHDW 28,F39,80
dnr-esc 3 mmol/L MJDHDW 28,F39,80
orl-rat TDLo:5600 μg/kg (28D male):REP TIVSAI 46,157,73
orl-rat TDLo:9 mg/kg (14D pre/1-22D preg):TER TIVSAI 46,157,73
orl-rat LD50:40 mg/kg TAGTBR 9,25,74
ihl-rat LC50:689 mg/m^3/4H GISAAA 56(2),80,91
skn-rat LD50:200 mg/kg PHJOAV 185,361,60
orl-mus LD50:56 mg/kg 85GMAT -,68,82
ihl-mus LC50:5 mg/m^3 85JCAE -,1199,86
ipr-mus LD50:16 mg/kg OCRAAH 3,137,68

CONSENSUS REPORTS: Mercury and its compounds are on the Community Right-To-Know List.

OSHA PEL: TWA 0.01 mg(Hg)/m^3; STEL 0.03 mg/m^3 (skin)
ACGIH TLV: TWA 0.01 mg(Hg)/m^3; STEL 0.03 mg(Hg)/m^3
NIOSH REL: (Organomercury): TWA 0.01 mg/m^3; STEL 0.03 mg/m^3 (skin)

SAFETY PROFILE: Poison by ingestion, inhalation, skin contact, and intraperitoneal routes. An experimental teratogen. Other experimental reproductive effects. Human mutation data reported. See also MERCURY

COMPOUNDS, ORGANIC. When heated to decomposition it emits very toxic fumes of Cl^- and Hg.

CHC675 **CAS:1888-94-4** **HR: 3**
CHLOROETHYL METHACRYLATE
mf: $C_6H_9ClO_2$ mw: 148.60

SYNS: β-CHLOROETHYL METHACRYLATE □ 2-CHLOROETHYL METHACRYLATE

TOXICITY DATA WITH REFERENCE
orl-rat LD50:200 mg/kg 85GMAT -,36,82
ihl-rat LC50:550 mg/m³/4H 85GMAT -,36,82
ihl-mus LC50:700 mg/m³/2H 85GMAT -,36,82

SAFETY PROFILE: Poison by inhalation and ingestion. When heated to decomposition it emits toxic fumes of Cl^-. See also ESTERS.

CHC750 **CAS:3570-58-9** **HR: 3**
2-CHLOROETHYL METHANESULFONATE
mf: $C_3H_7ClO_3S$ mw: 158.61

PROP: Bp: 125–126° @ 9 mm.

SYNS: CB 1506 □ β-CHLOROETHYLMETHANESULFONATE □ CHLOROETHYL METHANESULPHONATE □ CHLOROMETHANE SULFONATE d'ETHYLE (FRENCH) □ METHANESULFONIC ACID CHLOROETHYL ESTER □ NSC 18016

TOXICITY DATA WITH REFERENCE
dnd-dmg-par 10,500 µmol/L CNREA8 30,195,70
sln-dmg-par 15 mmol/L ANYAA9 68,731,58
mmo-nsc 100 mmol/L MGBUA3 17,5,60
mmo-esp 8 mmol/L ADWMAX -,193,62
ipr-rat TDLo:250 mg/kg (5D male):REP 85GUAJ -,55,66
ipr-rat LD50:135 mg/kg CNCRA6 9,56,60
ivn-rat LD50:143 mg/kg CNCRA6 9,56,60
ipr-mus LD50:182 mg/kg CNCRA6 9,56,60
ivn-mus LD50:182 mg/kg CNCRA6 9,56,60
ivn-dog LDLo:32 mg/kg CCSUBJ 2,203,65
ivn-mky LD50:127 mg/kg CNCRA6 9,56,60
orl-qal LD50:208 mg/kg JRPFA4 48,271,76

CONSENSUS REPORTS: EPA Genetic Toxicology Program.

SAFETY PROFILE: Poison by ingestion, intravenous, and intraperitoneal routes. Experimental reproductive effects. Mutation data reported. See also SULFONATES. When heated to decomposition it emits very toxic fumes of Cl^- and SO_x.

CHD250 **CAS:13909-09-6** **HR: 3**
1-(2-CHLOROETHYL)-3-(4-METHYL-CYCLOHEXYL)-1-NITROSOUREA
mf: $C_{10}H_{18}ClN_3O_2$ mw: 247.76

PROP: Mp: 64° (decomp).

SYNS: N-(2-CHLOROETHYL)-N'-(trans-4-METHYLCYCLOHEXYL)-N-NITROSOUREA □ 1-(2-CHLOROETHYL)-3-(trans-4-METHYL-CYCLOHEXYL)-1-NITROSOUREA □ ICIG 1110 □ ME-CCNU □ METHYL-CCNU □ trans-METHYL-CCNU □ METHYL-LOMUSTINE □ NCI-C04955 □ NSC-95441 □ SEMUSTINE

TOXICITY DATA WITH REFERENCE
skn-rbt 5 mg/24H rns TXCYAC 14,117,79
mmo-sat 100 nmol/plate JJIND8 65,149,80
mma-sat 100 nmol/plate JJIND8 65,149,80
dnd-esc 50 µmol/L MUREAV 89,95,81
dni-mus:oth 10 µmol/L CNREA8 43,5837,83
orl-hmn TDLo:22 mg/kg/60W-C:CAR NEJMAG 309,1079,83
ipr-rat TDLo:30 mg/kg/7W-I:CAR CANCAR 40(Suppl 4),1935,77
ivn-rat TDLo:32 mg/kg/60W-I:ETA DTESD7 8,273,80
ipr-mus TDLo:117 mg/kg/26W-I:CAR CANCAR 40(Suppl 4),1935,77
ivn-rat TD:64 mg/kg/60W-I:ETA DTESD7 8,273,80
ivn-rat TD:127 mg/kg/60W-I:ETA DTESD7 8,273,80
ivn-rat TD:203 mg/kg/48W-I:ETA DTESD7 8,273,80
orl-wmn TD:73 mg/kg/3Y-I:CAR NEJMAG 302,120,80
orl-hmn TDLo:90 mg/kg:BLD,GIT CCYPBY 4,257,73
orl-hmn TDLo:6 mg/kg:GIT,BLD CTRRDO 60,709,76
orl-chd LDLo:5550 mg/kg/30W-I NEJMAG 300,1200,79
orl-chd TDLo:37,950 mg/kg/2Y-I NEJMAG 300,1200,79
orl-mus LD50:49,900 µg/kg NTIS** PB269-473
ipr-mus LD10:37 mg/kg CNREA8 34,194,74
orl-dog LDLo:25 mg/kg ACRSAJ 16,273,72
ivn-dog LDLo:14 mg/kg CTRRDO 60,1559,76
orl-mky LDLo:100 mg/kg ACRSAJ 16,273,72
ivn-mky LDLo:45 mg/kg CTRRDO 60,1559,76

CONSENSUS REPORTS: NTP 7th Annual Report on Carcinogens. IARC Cancer Review: Group 1 IMEMDT 7,150,87; Animal Limited Evidence 7,150,87; Human Sufficient Evidence IMEMDT 7,150,87. NCI Carcinogenesis Studies (ipr); Clear Evidence: rat CANCAR 40,1935,77; No Evidence: mouse CANCAR 40,1935,77.

SAFETY PROFILE: Confirmed human carcinogen producing leukemia. Experimental carcinogenic and tumorigenic data. Poison by ingestion, intraperitoneal, intravenous, and possibly other routes. Mutation data reported. Human systemic effects by ingestion: nausea or vomiting, damage to kidney tubules and glomeruli, and hematuria (blood in the urine). When heated to decomposition it emits very toxic fumes of Cl^- and NO_x. See also N-NITROSO COMPOUNDS.

CHD500 **CAS:61137-63-1** **HR: 3**
trans-1-(2-CHLOROETHYL)-3-(3-METHYLCYCLOHEXYL)-1-NITROSOUREA
mf: $C_{10}H_{18}ClN_3O_2$ mw: 247.76

SYNS: Me-CCNU □ trans-N-(2-CHLOROETHYL)-N'-(3-METHYLCYCLOHEXYL)-N-NITROSOUREA

TOXICITY DATA WITH REFERENCE
msc-ham:ovr 4 mg/L CNREA8 35,460,75
ipr-mus LD50:71,460 µg/kg NCISP* JAN86

SAFETY PROFILE: Poison by intraperitoneal route. Mutation data reported. Many N nitroso compounds are carcinogens. When heated to decomposition it emits very toxic fumes of Cl^- and NO_x. See also N-NITROSO COMPOUNDS.

CHD675 CAS:1755-67-5 HR: 3
5-(2-CHLOROETHYL)-4-METHYLOXAZOLE-1,2-ETHANEDISULFONATE (2:1)
mf: $C_{12}H_{16}Cl_2O_2 \cdot C_2H_6O_6S_2$ mw: 453.38

TOXICITY DATA WITH REFERENCE
orl-mus LD50:380 mg/kg APSXAS 4,269,67
ipr-mus LD50:315 mg/kg APSXAS 4,269,67
scu-mus LD50:380 mg/kg APSXAS 4,269,67
ivn-mus LD50:250 mg/kg APSXAS 4,269,67

SAFETY PROFILE: Poison by ingestion, subcutaneous, intravenous, and intraperitoneal routes. When heated to decomposition it emits toxic fumes of $Cl^-$and SO_x. See also SULFONATES.

CHD700 CAS:5959-98-8 HR: 3
3-(2-CHLOROETHYL)-2-METHYLPYRIDINE HYDROCHLORIDE
mf: $C_8H_{10}ClN \cdot ClH$ mw: 192.10

SYN: 3-(2-CHLOROETHYL)-2-PICOLINE HYDROCHLORIDE

TOXICITY DATA WITH REFERENCE
orl-mus LD50:1300 mg/kg APSXAS 4,269,67
ipr-mus LD50:350 mg/kg APSXAS 4,269,67
ivn-mus LD50:150 mg/kg APSXAS 4,269,67

SAFETY PROFILE: Poison by intravenous and intraperitoneal routes. Moderately toxic by ingestion. When heated to decomposition it emits toxic fumes of Cl^- and NO_x.

CHD750 CAS:533-45-9 HR: 3
5-(2-CHLOROETHYL)-4-METHYLTHIAZOLE
mf: C_6H_8ClNS mw: 161.66

PROP: Oily, viscous liquid. D: 1.233 @ 25°/4°, bp: 92° @ 7 mm.

SYNS: CHLORETHIAZOL □ CHLORMETHIAZOLE □ CHLORO-S.C.T.Z. □ CLOMETHIAZOLE □ CLOMETHIAZOLUM □ DISTRANEURIN □ EMINEURINA □ HEMINEVRIN □ 4-METHYL-5-(β-CHLOROETHYL)THIAZOLE □ SCTZ □ SOMNEVRIN □ WY 1485

TOXICITY DATA WITH REFERENCE
orl-mus LD50:2110 mg/kg JMCMAR 7,167,64
ipr-mus LD50:190 mg/kg APSXAS 8,39,71
ivn-mus LD50:94 mg/kg APSXAS 7,423,70

SAFETY PROFILE: Poison by intravenous and intraperitoneal routes. Moderately toxic by ingestion. When heated to decomposition it emits very toxic fumes of Cl^-, NO_x, and SO_x.

CHD800 HR: 2
5-(2-CHLOROETHYL)-4-METHYLTHIAZOLE ETHANE DISULFONATE
mf: $C_6H_8ClNS \cdot C_2H_6O_6S_2$ mw: 351.86

TOXICITY DATA WITH REFERENCE
orl-mus LD50:835 mg/kg APSXAS 4,269,67
ipr-mus LD50:590 mg/kg APSXAS 4,269,67
scu-mus LD50:620 mg/kg APSXAS 4,269,67
ivn-mus LD50:620 mg/kg APSXAS 4,269,67

SAFETY PROFILE: Moderately toxic by ingestion, subcutaneous, intravenous and intraperitoneal routes. When heated to decomposition it emits toxic fumes of Cl^-, SO_x, and NO_x. See also SULFONATES.

CHD875 CAS:3240-94-6 HR: 3
4-(2-CHLOROETHYL)MORPHOLINE
mf: $C_6H_{12}ClNO$ mw: 149.64

SYN: TL 401

TOXICITY DATA WITH REFERENCE
dni-mus ivg 1 pph JIDEAE 62,378,74
unr-rat LD50:70 mg/kg PHBUA9 1,297,53
ihl-mus LCLo:370 mg/m³/10M NDRC** NDCrc-132,Nov,42

SAFETY PROFILE: Poison by inhalation and possibly other routes. Mutation data reported. When heated to decomposition it emits toxic fumes of Cl^- and NO_x.

CHE000 CAS:3647-69-6 HR: D
4-(2-CHLOROETHYL)MORPHOLINE HYDROCHLORIDE
mf: $C_6H_{12}ClNO \cdot ClH$ mw: 186.10

TOXICITY DATA WITH REFERENCE
mmo-sat 21,500 nmol/L ENMUDM 3,11,81
mmo-esc 21,500 nmol/L ENMUDM 3,11,81
dns-rat:lvr 100 μmol/L ENMUDM 3,11,81
msc-mus:lym 44 μmol/L ENMUDM 3,33,81
ipr-mus LD50:150 mg/kg (1D male):REP CANCAR 2,1075,49

CONSENSUS REPORTS: Reported in EPA TSCA Inventory.

SAFETY PROFILE: Experimental reproductive effects. Mutation data reported. When heated to decomposition it emits very toxic fumes of Cl^- and NO_x.

CHE250 CAS:64057-51-8 HR: D
N-(β-CHLOROETHYL)-N-NITROSOACETAMIDE
mf: $C_4H_7ClN_2O_2$ mw: 150.58

SYN: N-(2-CHLOROETHYL)-N-NITROSOACETAMIDE

TOXICITY DATA WITH REFERENCE
mmo-sat 1 nmol/plate CNREA8 43,175,83
mma-sat 2 nmol/plate CNREA8 39,1328,79
pic-esc 2 mg/L CNREA8 39,1328,79
dnd-mus:leu 500 nmol/L CNREA8 43,175,83

SAFETY PROFILE: Mutation data reported. Many N-nitroso compounds are carcinogens. When heated to decomposition it emits very toxic fumes of Cl^- and NO_x. See also N-NITROSO COMPOUNDS.

CHE325 CAS:60784-40-9 HR: 3
N-(2-CHLOROETHYL)-N-NITROSOCARBOMOYL AZIDE
mf: $C_3H_4ClN_5O_2$ mw: 177.55

$ClC_2H_4N(N:O)CO \cdot N_3$

SAFETY PROFILE: A potentially explosive material. Many N-nitroso compounds are experimental carcinogens. When heated to decomposition it emits toxic fumes of Cl^- and NO_x. See also N-NITROSO COMPOUNDS and AZIDES.

CHE500 CAS:13909-13-2 HR: 3
1-(2-CHLOROETHYL)-1-NITROSO-3-(2-NORBORNYL) UREA
mf: $C_{10}H_{16}ClN_3O_2$ mw: 245.74

SYNS: N'-BICYCLO(2.2.1)HEPT-2-YL-N-(2-CHLOROETHYL)-N-NITROSOUREA □ 1-(2-CHLOROETHYL)-3-(2-NORBORNYL)-1-NITROSOUREA □ NSC-88106 □ SRI 2638

TOXICITY DATA WITH REFERENCE
orl-rat LD50:83 mg/kg NCIMR* -,372,68
orl-mus LD50:56 mg/kg NCIMR* -,372,68
ipr-mus LD50:54,760 µg/kg NCISP* JAN86

SAFETY PROFILE: Poison by ingestion and intraperitoneal routes. Many N-nitroso compounds are carcinogens. When heated to decomposition it emits very toxic fumes of Cl^- and NO_x. See also N-NITROSO COMPOUNDS.

CHE750 CAS:2365-30-2 HR: 3
1-(2-CHLOROETHYL)-1-NITROSOUREA
mf: $C_3H_6ClN_3O_2$ mw: 151.57

SYNS: N-(2-CHLOROETHYL)-N-NITROSOUREA □ CNU □ MP 655 □ N-NITROSO-2-CHLOROETHYLUREA □ 1-NITROSO-1-(2-CHLOROETHYL)UREA □ NSC 47547 □ SKI 28404

TOXICITY DATA WITH REFERENCE
mmo-sat 1 µg/plate MUREAV 68,1,79
mma-sat 20 nmol/plate CNREA8 39,1328,79
dnr-esc 1 mmol/L CNREA8 45,6471,85
dnd-hmn:ovr 50 µmol/L INSSDM 19,33,81
dnd-mus:leu 10 µmol/L CNREA8 43,175,83
msc-ham:lng 10 µmol/L CNREA8 40,2719,80
ipr-mus LD50:4368 µg/kg NCISP* JAN86
ivn-mus LD10:4700 µg/kg ANBCB3 23,64,78

CONSENSUS REPORTS: EPA Genetic Toxicology Program.

SAFETY PROFILE: Poison by intraperitoneal and intravenous routes. Human mutation data reported. Many N nitroso compounds are carcinogens. When heated to decomposition it emits very toxic fumes of Cl^- and NO_x. See also N-NITROSO COMPOUNDS.

CHF000 CAS:33073-60-8 HR: 3
trans-4-(3-(2-CHLOROETHYL))-3-NITROSOUREIDO-CYCLOHEXANE CARBOXYLIC ACID ETHYL ESTER
mf: $C_{12}H_{20}ClN_3O_4$ mw: 305.80

SYNS: 4-((((2-CHLOROETHYL)NITROSOAMINO)CARBONYL)AMINO)CYCLOHEXANE CARBOXYLIC ACID, ETHYL ESTER □ NSC 103548

TOXICITY DATA WITH REFERENCE
ipr mus LD50:24,150 µg/kg NCISP* JAN86

SAFETY PROFILE: Poison by intraperitoneal route. Many N-nitroso compounds are carcinogens. When heated to decomposition it emits very toxic fumes of Cl^- and NO_x. See also N-NITROSO COMPOUNDS.

CHF250 CAS:61866-12-4 HR: 3
2-(3-(2-CHLOROETHYL)3-NITROSOUREIDO)ETHYL METHANE SULFONATE
mf: $C_5H_{10}ClN_3O_5S$ mw: 259.69

SYNS: 2-(3-(2-CHLORATHYL)-3-NITROSOUREIDO)ATHYLMETHANSULFONAT (GERMAN) □ CNUEMS □ ETHYLMETHANESULFONATO-CNU □ NSC-294896

TOXICITY DATA WITH REFERENCE
mrc-smc 500 µmol/L/16H MUREAV 42,45,77
ipr-rat LD50:23 mg/kg VDGIA2 85,1293,79
ivn-rat LD50:13,800 µg/kg ONCOBS 37,177,80

SAFETY PROFILE: Poison by intraperitoneal and intravenous routes. Mutation data reported. Many N-nitroso compounds are carcinogens. When heated to decomposition it emits very toxic fumes of Cl^-, NO_x, and SO_x. See also N-NITROSO COMPOUNDS, SULFONATES, and ESTERS.

CHF500 CAS:6296-45-3 HR: 2
2-CHLOROETHYL-N-NITROSOURETHANE
mf: $C_5H_9ClN_2O_3$ mw: 180.61

SYNS: N-(2-CHLOROETHYL)-N-NITROSOETHYLCARBAMATE □ N (β-CHLOROETHYL)-N-NITROSOURETHAN □ ETHYL-N-(β-CHLOROETHYL)-N-NITROSOCARBAMATE □ TL 154

TOXICITY DATA WITH REFERENCE
mmo-sat 1 nmol/plate CNREA8 43,175,83
dnd-mus:leu 500 nmol/L CNREA8 43,175,83
orl-rat TDLo:6 mg/kg:ETA CNREA8 31,573,71
orl-rat LDLo:10 mg/kg CNREA8 31,573,71
ihl-rat LCLo:330 mg/m³/10M NDRC** NDCrc-132,Apr,42
ipr rat LDLo:6500 µg/kg CNREA8 31,573,71
ihl-dog LCLo:330 mg/m³/10M NDRC** NDCrc-132,Apr,42
ihl-cat LCLo:330 mg/m³/10M NDRC** NDCrc-132,Apr,42

SAFETY PROFILE: Poison by inhalation, ingestion, and intraperitoneal routes. Questionable carcinogen with experimental tumorigenic data. Mutation data reported. Many N-nitroso compounds are carcinogens. See also CARBAMATES and N-NITROSO COMPOUNDS. When heated to decomposition it emits very toxic fumes of Cl^- and NO_x.

CHG000 CAS:113-18-8 HR: 3
1-CHLORO-3-ETHYL-1-PENTEN-4-YN-3-OL
mf: C_7H_9ClO mw: 144.61

PROP: Liquid with pungent, aromatic odor, slowly darkening on exposure to light and air. D: 1.065–1.070 @ 25°/4°, bp: 173–174°. Misc most org solvents; immisc in H_2O.

SYNS: A 71 □ AETHYL-CHLORVYNOL □ ALVINOL □ ARVYNOL □ β-CHLOROVINYL ETHYLETHYNYL CARBINOL □ 3-(β-CHLOROVINYL)-1-PENTYN-3-OL □ ETCHLORVINOLO □ ETHCHLOROVYNOL □ ETHCHLORVINYL □ ETHCLORVYNOL □ ETHOCHLORVYNOL □ ETHYL-β-CHLOROVINYLETHYNYL CARBINOL □ETHYLCHLORVYN-OL □ NORMONSON □ NORMOSAN □ NORMOSON □ NOSTEL □ PLACIDIL □ PLACIDYL □ ROERIDORM □ SERENIL □ SERENSIL

TOXICITY DATA WITH REFERENCE
unr-wmn TDLo:643 mg/kg (26-39W preg):REP PEDIAU 52,714,73

orl-rat TDLo:400 mg/kg (1-20D preg):REP EESADV 5,494,81
orl-rat TDLo:1680 mg/kg (1-21D preg):REP EVHPAZ 21,85,77
orl-rat TDLo:400 mg/kg (1-20D preg):TER EESADV 5,494,81
orl-wmn TDLo:10 mg/kg:CNS BMJOAE 2,1610,62
orl-wmn TDLo:15 mg/kg/2D:BLD AIMEAS 77,73,72
scu-rat LD50:200 mg/kg 27ZQAG -,420,72
orl-mus LD50:290 mg/kg JPETAB 114,326,55
ipr-mus LD50:275 mg/kg 27ZQAG -,420,72
scu-mus LD50:240 mg/kg JPETAB 114,326,55
ivn-dog LD50:55 mg/kg 27ZQAG -,420,72
orl-bwd LD50:42 mg/kg TXAPA9 21,315,72

SAFETY PROFILE: Poison by ingestion, subcutaneous, intraperitoneal, and intravenous routes. Human systemic effects by ingestion: general anesthesia and thrombocytopenia (reduction in the number of blood platelets). Human effects on newborn by an unspecified route: drug dependency and apgar score (condition of newborn). Experimental teratogenic and reproductive effects. When heated to decomposition it emits toxic fumes of Cl^-.

CHG250 CAS:73816-74-7 HR: 1
4-(4-CHLORO-6-ETHYLPHENYLAMINO)-2-s-TRIAZINYLAMINO-5-HYDROXY-6-(4-METHYL-2-SULFOPHENYLAZO)-2,7-NAPHTHALENE DISULFONIC ACID TRISODIUM SALT
mf: $C_{28}H_{21}ClN_7O_{10}S_3{\cdot}3Na$ mw: 816.16

SYN: CERVEN BRILANTNI OSTAZINOVA H-3B (CZECH)

TOXICITY DATA WITH REFERENCE
eye-rbt 20 mg/24H MOD 28ZPAK -,234,72
orl-rat LD50:6730 mg/kg 28ZPAK -,234,72

SAFETY PROFILE: Mildly toxic by ingestion. An eye irritant. When heated to decomposition it emits very toxic fumes of Cl^-, NO_x, Na_2O, and SO_x.

CHG375 CAS:10419-79-1 HR: 2
(2-CHLOROETHYL)PHOSPHONIC ACID DIETHYL ESTER
mf: $C_6H_{14}ClO_3P$ mw: 200.62

PROP: Liquid. D: 1.16 @ 20°/4°, bp: 92–93° @ 2.5 mm.

TOXICITY DATA WITH REFERENCE
orl-rat LD50:1 g/kg GISAAA 48(8),79,83
orl-rbt LD50:2 g/kg GISAAA 48(8),79,83
orl-gpg LD50:1450 mg/kg GISAAA 48(8),79,83

SAFETY PROFILE: Moderately toxic by ingestion. When heated to decomposition it emits toxic fumes of Cl^- and PO_x. See also ESTERS.

CHG400 CAS:23510-39-6 HR: 2
(2-CHLOROETHYL)PHOSPHONIC ACID MONOETHYL ESTER
mf: $C_4H_{10}ClO_3P$ mw: 172.56

TOXICITY DATA WITH REFERENCE
orl-rat LD50:1250 mg/kg GISAAA 48(8),79,83
orl-rbt LD50:2 g/kg GISAAA 48(8),79,83
orl-gpg LD50:1800 mg/kg GISAAA 48(8),79,83

SAFETY PROFILE: Moderately toxic by ingestion. When heated to decomposition it emits toxic fumes of Cl^- and PO_x. See also ESTERS.

CHG500 CAS:2008-75-5 HR: 3
1-(2-CHLOROETHYL)PIPERIDINE HYDROCHLORIDE
mf: $C_7H_{14}ClN{\cdot}ClH$ mw: 184.13

PROP: A solid. Mp: 233–236°.

SYNS: β-CHLOROETHYLPIPERIDINE HYDROCHLORIDE □ PIPERIDINOETHYL CHLORIDE, HYDROCHLORIDE

TOXICITY DATA WITH REFERENCE
mmo-sat 5400 nmol/L ENMUDM 3,11,81
mmo-esc 5400 nmol/L ENMUDM 3,11,81
dns-rat:lvr 500 µmol/L ENMUDM 3,11,81
msc-mus:lym 44 µmol/L ENMUDM 3,33,81
ipr-mus LD50:93 mg/kg JPETAB 94,249,48
scu-mus LD50:125 mg/kg JPETAB 97,25,49
ivn-mus LD50:56 mg/kg CSLNX* NX#07326

SAFETY PROFILE: Poison by subcutaneous, intravenous, and intraperitoneal routes. Mutation data reported. When heated to decomposition it emits very toxic fumes of Cl^- and NO_x.

CHG750 CAS:101651-64-3 HR: 3
2′-CHLORO-2-(ETHYL(2-PIPERIDINOETHYL)AMINO)ACETANILIDE DIHYDROCHLORIDE
mf: $C_{17}H_{26}ClN_3O{\cdot}2ClH$ mw: 396.83

SYN: C 5410

TOXICITY DATA WITH REFERENCE
eye-rbt 2% MLD ARZNAD 9,262,59
ipr-rat LD50:89 mg/kg ARZNAD 9,262,59
scu-mus LD50:400 mg/kg ARZNAD 9,262,59

SAFETY PROFILE: Poison by intraperitoneal and subcutaneous routes. An eye irritant. When heated to decomposition it emits very toxic fumes of $Cl^-$and NO_x.

CHH000 CAS:38915-59-2 HR: 3
7-(3-(2-CHLOROETHYL-n-PROPYLAMINO)PROPYLAMINO)BENZO(b)(1,10)-PHENATHROLINE HYDROCHLORIDE
mf: $C_{24}H_{27}ClN_4{\cdot}3ClH$ mw: 516.38

SYN: ICR 394

TOXICITY DATA WITH REFERENCE
mmo-sat 5 µg/plate JMCMAR 15,739,72
ivn-mus TDLo:1550 µg/kg:NEO CNREA8 36,2423,76
ipr-mus LD20:1 mg/kg JMCMAR 15,739,72

SAFETY PROFILE: Poison by intraperitoneal route. Questionable carcinogen with experimental neoplastigenic data. Mutation data reported. When heated to decomposition it emits very toxic fumes of Cl^- and NO_x.

CHH125 CAS:10140-94-0 HR: 3
2-CHLORO-5-ETHYL-4-PROPYL-2-THIONO-1,3,2-DIOXAPHOSPHORINANE
mf: $C_8H_{16}ClO_2PS$ mw: 242.72

SYN: PHOSPHOROCHLORIDOTHIOIC ACID, cyclic O,O-(2-ETHYL-1-PROPYLTRIMETHYLENE) ESTER

TOXICITY DATA WITH REFERENCE
skn-rbt 10 mg/24H open SEV AIHAAP 23,95,62
orl-rat LD50:300 mg/kg AIHAAP 23,95,62
skn-rbt LD50:1000 mg/kg AIHAAP 23,95,62

SAFETY PROFILE: Poison by ingestion. Moderately toxic by skin contact. A severe skin irritant. When heated to decomposition it emits toxic fumes of Cl^-, PO_x, and SO_x. See also ESTERS.

CHI125 CAS:80-41-1 HR: 2
2-CHLOROETHYL TOSYLATE
mf: $C_9H_{11}ClO_3S$ mw: 234.71

PROP: Bp: 153° @ 0.3 mm.

SYNS: 2-CHLORO-ETHANOL-4-METHYLBENZENESULFONATE (9CI) □ 2-CHLORO-ETHANOL-p-TOLUENESULFONATE (8CI) □ β-CHLOROETHYLESTER KYSELINY-p-TOLUENSULFONOVE (CZECH)

TOXICITY DATA WITH REFERENCE
eye-rbt 500 mg/24H MLD 28ZPAK -,197,72
orl-rat LD50:498 mg/kg 28ZPAK -,197,72

CONSENSUS REPORTS: Reported in EPA TSCA Inventory.

SAFETY PROFILE: Moderately toxic by ingestion. An eye irritant. When heated to decomposition it emits toxic fumes of Cl^- and SO_x. See also SULFONATES.

CHI250 CAS:110-75-8 HR: 3
2-CHLOROETHYL VINYL ETHER
mf: C_4H_7ClO mw: 106.56

PROP: Liquid. Mp: −70.3°, bp: 108°, d: 1.05 @ 20°/4°, flash p: 80°F (OC).

SYNS: 2-CHLORETHYL VINYL ETHER □ (2-CHLOROETHOXY)ETHENE □ RCRA WASTE NUMBER U042 □ VINYL-β-CHLOROETHYL ETHER □ VINYL-2-CHLOROETHYL ETHER

TOXICITY DATA WITH REFERENCE
skn-rbt 525 mg open SEV UCDS** 11/15/71
eye-rbt 500 mg open JIHTAB 31,60,49
eye-rbt 500 mg/24H MLD 85JCAE-,541,86
orl-rat LD50:210 mg/kg UCDS** 11/15/71
orl-rat LD50:250 mg/kg JIHTAB 31,60,49
ihl-rat LCLo:250 ppm/4H JIHTAB 31,343,49
skn-rbt LD50:3354 mg/kg JIHTAB 31,60,49

CONSENSUS REPORTS: Reported in EPA TSCA Inventory.

SAFETY PROFILE: Poison by ingestion. Moderately toxic by inhalation and skin contact. A severe eye and skin irritant. See also ETHERS. Dangerous fire hazard when exposed to heat, flame, or oxidizers. Potentially explosive. May form dangerous peroxides on exposure to air. To fight fire, use alcohol foam, dry chemical. When heated to decomposition it emits toxic fumes of Cl^-. See also CHLORIDES and ETHERS.

CHI750 HR: 2
CHLOROETHYNYL NORGESTREL mixed with MESTRANOL (20:1)

SYNS: MESTRANOL mixed with CHLOROETHYNYL NORGESTREL (1:20) □ WY-4355 mixed with MESTRANOL (20:1)

TOXICITY DATA WITH REFERENCE
orl-dog TDLo:463 mg/kg/90W-I:CAR JJIND8 65,137,80
orl-dog TD:423 mg/kg/4Y-I:ETA JTEHD6 3,179,77
orl-dog TD:383 mg/kg/4Y-I:ETA JNCIAM 59,933,77

SAFETY PROFILE: Questionable carcinogen with experimental carcinogenic and tumorigenic data. When heated to decomposition it emits very toxic fumes of Cl^- and NO_x.

CHI825 CAS:5096-17-3 HR: 2
N-(7-CHLORO-2-FLUORENYL)ACETAMIDE
mf: $C_{15}H_{11}ClNO$ mw: 256.72

SYN: N-2-(7-CHLORO)FLUORENYLACETAMIDE

TOXICITY DATA WITH REFERENCE
orl-rat TDLo:1265 mg/kg/11W-C:ETA JNCIAM 24,149,60
orl-rat TD:5000 mg/kg:ETA CNREA8 26,619,66
orl-rat LDLo:5000 mg/kg CNREA8 26,619,66

SAFETY PROFILE: Questionable carcinogen with experimental tumorigenic data. Mildly toxic by ingestion. When heated to decomposition it emits toxic fumes of Cl^- and NO_x.

CHI900 CAS:593-70-4 HR: 3
CHLOROFLUOROMETHANE
mf: CH_2ClF mw: 68.48

PROP: Gas. Bp: −9°.

SYNS: CFC 31 □ FC 31 □ FREON 31 □ MONOCHLOROMONOFLUOROMETHANE □ R 31 □ R 31 (refrigerant)

TOXICITY DATA WITH REFERENCE
mmo-sat 5 pph MUREAV 118,277,83
mma-sat 5 pph MUREAV 118,277,83
otr-ham:kdy 100 μmol/L TXAPA9 72,15,84
msc-ham:ovr 10 pph EVSRBT 25,91,82
orl-rat TDLo:78 g/kg/1Y-I:CAR TXAPA9 72,15,84
ihl-mky LCLo:1000 ppm/4H TXAPA9 48,A109,79

CONSENSUS REPORTS: IARC Cancer Review: Group 3 IMEMDT 7,56,87; Animal Limited Evidence IMEMDT 41,229,86.

DFG MAK: Animal Carcinogen, Suspected Human Carcinogen.

SAFETY PROFILE: Confirmed carcinogen with experimental carcinogenic data. Moderately toxic by inhalation. Mutation data reported. When heated to decomposition it emits very toxic fumes of Cl^- and F^-. See also CHLORINATED HYDROCARBONS, ALIPHATIC; and FLUORIDES.

C

CHI950 CAS:350-30-1 HR: D
2-CHLORO-1-FLUORO-4-NITROBENZENE
mf: $C_6H_3ClFNO_2$ mw: 175.55

SYNS: BENZENE, 2-CHLORO-1-FLUORO-4-NITRO- □ 3-CHLORO-4-FLUORONITROBENZENE

TOXICITY DATA WITH REFERENCE
mmo-sat 102 μg/plate MUREAV 116,217,83

SAFETY PROFILE: Mutation data reported. When heated to decomposition it emits toxic vapors of NO_x, F^-, and Cl^-.

CHJ000 CAS:34214-51-2 HR: 2
3-(2-CHLORO-6-FLUOROPHENYL)-5-METHYL-4-ISOXAZOLYLPENICILLIN SODIUM MONOHYDRATE
mf: $C_{19}H_{17}ClFN_3O_5S•Na•H_2O$ mw: 494.91

SYNS: 6-(3-(2-CHLORO-6-FLUOROPHENYL)-5-METHYL-4-ISOXAZOLECARBOXAMIDO)PENICILLANIC ACID SODIUM SALT □ CULPEN □ FLOXACILLIN SODIUM MONOHYDRATE □ FLOXAPEN □ FLUCLOXACILLIN SODIUM MONOHYDRATE □ STAPHYLEX

TOXICITY DATA WITH REFERENCE
orl-mus LD50:3800 mg/kg MEIEDD 10,589,83
scu-mus LD50:2200 mg/kg MEIEDD 10,589,83

SAFETY PROFILE: Moderately toxic by ingestion and subcutaneous routes. When heated to decomposition it emits very toxic fumes of Cl^-, F^-, NO_x, Na_2O, and SO_x. See other penicillin entries.

CHJ250 CAS:6186-91-0 HR: 3
3-CHLORO-2-FLUOROPROPENE
mf: C_3H_4ClF mw: 94.52

PROP: Bp: 53°.

SYN: 3-CHLORO-2-FLUORO-1-PROPENE

TOXICITY DATA WITH REFERENCE
orl-rat LD50:280 mg/kg AIHAAP 23,95,62
ihl-rat LCLo:1000 ppm/4H AIHAAP 23,95,62
skn-rbt LD50:200 mg/kg AIHAAP 23,95,62

SAFETY PROFILE: Poison by ingestion and skin contact. Mildly toxic by inhalation. When heated to decomposition it emits very toxic fumes of F^- and Cl^-. See also CHLORINATED HYDROCARBONS, ALIPHATIC, and FLUORIDES.

CHJ300 CAS:63284-71-9 HR: 3
2-CHLORO-4′-FLUORO-α-(PYRIMIDIN-5-YL)DIPHENYLMETHANOL
mf: $C_{17}H_{12}ClFN_2O$ mw: 314.76

SYNS: (+-)-2-CHLORO-4′-FLUORO-α-(PYRIMIDIN-5-YL)BENZHYDRYL ALCOHOL □ 2-CHLORO-4′-FLUORO-α-(PYRIMIDIN-5-YL)BENZHYDRYL ALCOHOL □ α-(2-CHLOROPHENYL)-α-(4-FLUOROPHENYL)-5-PYRIMIDINEMETHANOL □ EL 228 □ EL 2289 □ GUANTLET □ MUROX □ NUARIMOL □ 5-PYRIMIDINEMETHANOL, α-(2-CHLOROPHENYL)-α-(4-FLUOROPHENYL)- □ TRIMIDAL □ TRIMIFRUIT SC □ TRIMINOL

TOXICITY DATA WITH REFERENCE
orl-rat LD50:1250 mg/kg PEMNDP 9,625,91
orl-mus LD50:2500 mg/kg PEMNDP 9,625,91
orl-dog LD50:500 mg/kg PEMNDP 9,625,91
skn-rbt LD50:>2 g/kg PEMNDP 9,625,91
orl-qal LD50:200 mg/kg PEMNDP 9,625,91
orl-brd LD50:200 mg/kg 85JFAN A303,86

CONSENSUS REPORTS: Reported in EPA TSCA Inventory.

SAFETY PROFILE: A poison by ingestion. Low toxicity by skin contact. When heated to decomposition it emits toxic vapors of NO_x, F^-, and Cl^-.

CHJ500 CAS:67-66-3 HR: 3
CHLOROFORM
DOT: UN 1888
mf: $CHCl_3$ mw: 119.37

PROP: Colorless liquid; heavy, ethereal odor. Mp: −63.2°, bp: 61.3°, flash p: none, d: 1.481 @ 25°/4°, vap press: 100 mm @ 10.4°, vap d: 4.12. Sltly sol in H_2O.

SYNS: CHLOROFORME (FRENCH) □ CLOROFORMIO (ITALIAN) □ FORMYL TRICHLORIDE □ METHANE TRICHLORIDE □ METHENYL TRICHLORIDE □ METHYL TRICHLORIDE □ NCI-C02686 □ R 20 (refrigerant) □ RCRA WASTE NUMBER U044 □ TCM □ TRICHLOORMETHAAN (DUTCH) □ TRICHLORMETHAN (CZECH) □ TRICHLOROFORM □ TRICHLOROMETHANE □ TRICLOROMETANO (ITALIAN)

TOXICITY DATA WITH REFERENCE
skn-rbt 10 mg/24H open MLD AIHAAP 23,95,62
skn-rbt 500 mg/24H MLD 85JCAE-,89,86
eye-rbt 148 mg AIHAAP 37,697,76
eye-rbt 20 mg/24H MOD 85JCAE-,89,86
sce-hmn:lym 10 mmol/L ENVRAL 32,72,83
dns-mus-ipr 50 mg/kg TOLED5 21,357,84
orl-mus TDLo:2177 mg/kg (male 3W pre):REP NETOD7 1,199,79
orl-mus TDLo:2115 mg/kg (male 3W pre):REP EVHPAZ 46,127,82
ihl-rat TCLo:20,100 μg/m³/1H (female 7-14D post):TER NTIS** PB277-077
orl-rat TDLo:1260 mg/kg (6-15D preg):TER TXAPA9 29,348,74
ihl-mus TCLo:100 ppm/7H (female 1-7D post):REP TXAPA9 50,515,79
ihl-rat TCLo:30 ppm/7H (6-15D preg):REP TXAPA9 28,442,74
ihl-rat TCLo:300 ppm/7H (female 6-15D post):REP TXAPA9 28,442,74
ihl-mus TCLo:100 ppm/7H (female 8-15D post):TER TXAPA9 50,515,79
ihl-rat TCLo:100 ppm/7H (female 6-15D post):TER TXAPA9 28,442,74
orl-rat TDLo:1260 mg/kg (6-15D preg):TER TXAPA9 29,348,74
orl-rat TDLo:13,832 mg/kg/2Y-C:CAR FAATDF 5,760,85
orl-mus TDLo:127 g/kg/92W-I:CAR NCITR* NCI-CG-TR-0,76
orl-rat TD:98 g/kg/78W-I:NEO NCITR* NCI-CG-TR-0,76
orl-mus TD:18 g/kg/17W-I:NEO JNCIAM 5,251,45
orl-rat TD:7020 mg/kg/78W-I:CAR EVHPAZ 31,171,79
orl-rat TD:70 g/kg/78W-I:NEO NCITR* NCI-CG-TR-0,76
orl-mus TD:24,752 mg/kg/2Y-C:ETA FAATDF 5,760,85
orl-rat TD:58,968 mg/kg/2Y-C:NEO FAATDF 5,760,85
orl-mus LD:130 g/kg/2Y-I:NEO VOONAW 33(8),81,87
ihl-hmn TCLo:10 mg/m³/1Y:CNS,GIT IRGGAJ 24,127,67

ihl-hmn LCLo:25,000 ppm/5M TABIA2 3,231,33
ihl-hmn TCLo:5000 mg/m³/7M:CNS AHBAAM 116,131,36
unr-man LDLo:546 mg/kg 85DCAI 2,73,70
orl-rat LD50:908 mg/kg JPFCD2 17,205,82
ihl-rat LC50:47,702 mg/m³/4H ENVRAL 40,411,86
orl-mus LD50:36 mg/kg ATSUDG 2,371,79
ihl-mus LC50:28 g/m³ PCOC** -,230,66
ipr-mus LD50:623 mg/kg AGGHAR 18,109,60
scu-mus LD50:704 mg/kg JPETAB 123,224,58
orl-dog LDLo:1000 mg/kg QJPPAL 7,205,34
ihl-dog LC50:100 g/m³ PCOC** -,230,66
ipr-dog LD50:1000 mg/kg TXAPA9 10,119,67
ivn-dog LDLo:75 mg/kg QJPPAL 7,205,34
ihl-cat LCLo:35,000 mg/m³/4H AHBAAM 116,131,36
orl-rbt LDLo:500 mg/kg AEXPBL 97,86,23

CONSENSUS REPORTS: NTP 7th Annual Report on Carcinogens. IARC Cancer Review: Group 2B IMEMDT 7,152,87; Animal Limited Evidence IMEMDT 1,61,72; Human Limited Evidence IMEMDT 20,401,79; Animal Sufficient Evidence IMEMDT 20,401,79. NCI Carcinogenesis Bioassay (gavage); Clear Evidence: mouse, rat NCITR* NCI-CG-TR,1976. EPA Genetic Toxicology Program. EPA Extremely Hazardous Substances List. Community Right-To-Know List. Reported in EPA TSCA Inventory.

OSHA PEL: TWA 2 ppm
ACGIH TLV: TWA 10 ppm; Suspected Human Carcinogen
DFG MAK: 10 ppm (50 mg/m³; Suspected Carcinogen.
NIOSH REL: (Waste Anesthetic Gases and Vapors) CL 2 ppm/1H; (Chloroform) CL 2 ppm/60M
DOT CLASSIFICATION: 6.1; *Label:* Poison

SAFETY PROFILE: Confirmed carcinogen with experimental carcinogenic, neoplastigenic, and tumorigenic data. A human poison by ingestion and inhalation. An experimental poison by ingestion and intravenous routes. Moderately toxic experimentally by intraperitoneal and subcutaneous routes. Human systemic effects by inhalation: hallucinations and distorted perceptions, nausea, vomiting, and other unspecified gastrointestinal effects. Human mutation data reported. Experimental teratogenic and reproductive effects.

Inhalation of the concentrated vapor causes dilation of the pupils with reduced reaction to light, as well as reduced intraocular pressure (experimental). In the initial stages there is a feeling of warmth of the face and body, then an irritation of the mucous membranes, conjunctiva, and skin; followed by excitation, loss of reflexes, sensation, and consciousness. Prolonged inhalation will bring on paralysis accompanied by cardiac-respiratory failure and finally death.

Chloroform has been widely used as an anesthetic. However, due to its toxic effects, this use is being abandoned. Concentrations of 68,000–82,000 ppm in air can kill most animals in a few minutes. 14,000 ppm may cause death after an exposure of from 30 to 60 minutes. 5,000–6,000 ppm can be tolerated by animals for 1 hour without serious disturbances. The maximum concentration tolerated for several hours or for prolonged exposure with slight symptoms is 2,000–2,500 ppm. Prolonged administration as an anesthetic may lead to such serious effects as profound toxemia and damage to the liver, heart, and kidneys. Experimental prolonged but light anesthesia in dogs produces a typical hepatitis.

Explosive reaction with sodium + methanol or sodium methoxide + methanol. Mixtures with sodium or potassium are impact sensitive explosives. Reacts violently with acetone + alkali (e.g., sodium hydroxide, potassium hydroxide, or calcium hydroxide), Al, disilane, Li, Mg, methanol + alkali, nitrogen tetroxide, perchloric acid + phosphorus pentoxide, potassium-tert-butoxide, sodium methylate, NaK. Incompatible with dinitrogen tetraoxide, fluorine, metals, or triisopropylphosphine. Nonflammable. When heated to decomposition it emits toxic fumes of Cl^-.

See also CHLORINATED HYDROCARBONS, ALIPHATIC.

For occupational chemical analysis use OSHA: #05 or NIOSH: Hydrocarbons, halogenated, 1003.

CHJ599 CAS:29671-92-9 HR: 3
CHLOROFORMAMIDINIUM CHLORIDE
mf: CH_4Cl2N_2 mw: 102.95

$ClC(:NH)N^+H_3Cl^-$

PROP: A solid. Mp: 175°

SAFETY PROFILE: Reaction with perchloric acid forms highly explosive products. When heated to decomposition it emits toxic fumes of Cl^- and NO_x. See also CHLORIDES.

CHJ625 CAS:75524-40-2 HR: 3
CHLOROFORMAMIDINIUM NITRATE
mf: $CH_3ClN_3O_3$ mw: 140.51

SAFETY PROFILE: A powerful explosive and a strong oxidant. Mixtures with wet magnesium powder, powdered aluminum, or powdered iron ignite and then explode. Reacts violently with ammonia or amines. When heated to decomposition it emits toxic fumes of Cl^- and NO_x. See also NITRATES and EXPLOSIVES.

CHJ750 CAS:54-31-9 HR: 3
4-CHLORO-N-FURFURYL-5-SULFAMOYLANTHRANILIC ACID
mf: $C_{12}H_{11}ClN_2O_5S$ mw: 330.76

PROP: Crystals from EtOH (aq). Mp: 206°.

SYNS: AISEMIDE □ ALUZINE □ 5-(AMINOSULFONYL)-4-CHLORO-2-((2-FURNAYLMETHYL)AMINO)BENZOIC ACID □ BERONALD □ CHLOR-N-(2-FURYLMETHYL)-5-SULFAMYLANTHRANILSAEURE (GERMAN) □ 4-CHLORO-N-(2-FURYLMETHYL)-5-SULFAMOYLANTHRANILIC ACID □ DESDEMIN □ DIURAL □ DRYPTAL □ ERROLON □ EUTENSIN □ FRUSEMIDE □ FRUSEMIN □ FRUSID □ FULSIX □ FULUVAMIDE □ FURANTHRIL □ FURANTHRYL □ FURANTRIL □ FURESIS □ FUROSEDON □ FUROSEMID □ FUROSEMIDE □ FUROSEMIDE "MITA" □ FURSEMID □ FURSEMIDE □ FUSID □ HYDRO-RAPID □ KATLEX □ LASEX □ LASIX □ LB 502 □ LOWPSTRON □ MACASIROOL □ NICOROL □ NCI-C55936 □ PREFEMIN □ PROFEMIN □ RADONNA □ ROSEMIDE □ SALIX □ SEGURIL □ TRANSIT □ TROFURIT □ UREX □ UROSEMIDE

TOXICITY DATA WITH REFERENCE
cyt-hmn:leu 200 mg/L/24H MUREAV 66,69,79

cyt-ham:lng 1 g/L ATSUDG (4),41,80
orl-rat TDLo:300 mg/kg (female 16D post):TER TJADAB 34,452,86
orl-mus TDLo:12,500 mg/kg (female 6-15D post):TER SEIJBO 24,111,84
orl-rat TDLo:122 mg/kg (male 6W pre):REP JRPFA4 81,259,87
orl-rat TDLo:150 mg/kg (12-14D preg):TER TJADAB 31,401,85
orl-wmn TDLo:120 mg/kg/21W-I:SYS
orl-rat TDLo:21,112 mg/kg/2Y-C:ETA NTPTR* NTP-TR-356,89
orl-mus TDLo:15,652 mg/kg/2Y-C:CAR NTPTR* NTP-TR-356,89
orl-mus TD:122 g/kg/2Y-C:CAR JJATDK 10,369,90 JRSMD9 79,239,86
ivn-man TDLo:29 mg/kg:EAR,KID NEJMAG 282,1413,70
ivn-hmn TDLo:1300 µg/kg:CVS AIMEAS 103,1,85
ivn-inf TDLo:1 mg/kg/4H-I:SYS ADCHAK 59,907,84
ivn-wmn TDLo:2500 µg/kg/2M-C:BPR ICMED9 12,54,86
orl-rat LD50:2600 mg/kg TXAPA9 18,185,71
ipr-rat LD50:800 mg/kg APPHAX 42,199,85
ivn-rat LD50:800 mg/kg NIIRDN 6,725,82
orl-mus LD50:2200 mg/kg PCJOAU 19,706,85
ivn-mus LD50:308 mg/kg ARZNAD 14,44,64
orl-dog LD50:2000 mg/kg NIIRDN 6,725,82
orl-rbt LD50:800 mg/kg NIIRDN 6,725,82
ivn-rbt LD50:400 mg/kg NIIRDN 6,725,82

CONSENSUS REPORTS: IARC Cancer Review: Group 3 IMEMDT 50,277,90; Human Inadequate Evidence IMEMDT 50,277,90; Animal Inadequate Evidence IMEMDT 50,277,90. EPA Genetic Toxicology Program.

SAFETY PROFILE: Poison by intravenous route. Moderately toxic by ingestion and intraperitoneal routes. Human systemic effects by intravenous route: change in the sensitivity of the ear to sound, tinnitus, unspecified effects on the heart, constriction of the arteries, and a decrease in urine volume, interstitial nephritis, metabolic alkalosis, pulse rate decrease, fall in BP. Ingestion can damage the liver. Experimental teratogenic and reproductive effects. Questionable carcinogen with experimental carcinogenic effects. Human mutation data reported. When heated to decomposition it emits very toxic fumes of Cl^-, NO_x, and SO_x.

CHK000 CAS:5857-37-4 HR: 3
CHLORO(2-FURYL)MERCURY
mf: C_4H_3ClHgO mw: 303.11

PROP: Solid from EtOH (aq). Mp: 152–153°.

SYNS: CHLORO-2-FURANYL MERCURY □ 2-CHLOROMERCURIFURAN □ 2-FURYLMERCURIC CHLORIDE □ 2-FURYLMERCURY CHLORIDE

TOXICITY DATA WITH REFERENCE
ipr-mus LDLo:20 mg/kg HBTXAC 5,81,59
ivn-mus LD50:56 mg/kg CSLNX* NX#03269

CONSENSUS REPORTS: Mercury and its compounds are on the Community Right-To-Know List.

OSHA PEL: CL 0.1 mg(Hg)/m³ (skin)
ACGIH TLV: TWA 0.1 mg(Hg)/m³ (skin)
NIOSH REL: (Mercury, Aryl and Inorganic): CL 0.1 mg/m³ (skin)

SAFETY PROFILE: Poison by intravenous and intraperitoneal routes. See also MERCURY COMPOUNDS, ORGANIC. When heated to decomposition it emits very toxic fumes of Hg and Cl^-.

CHK125 CAS:102489-58-7 HR: 3
6′-CHLORO-2-(2-FURYLMETHYL)AMINO-o-ACETOTOLUIDIDE HYDROCHLORIDE
mf: $C_{14}H_{15}ClN_2O_2 \cdot ClH$ mw: 315.22

SYN: C 3211

TOXICITY DATA WITH REFERENCE
eye-rbt 2% MLD ARZNAD 8,407,58
ipr-rat LD50:330 mg/kg ARZNAD 8,407,58
scu-mus LD50:2350 mg/kg ARZNAD 8,407,58

SAFETY PROFILE: Poison by intraperitoneal route. Moderately toxic by subcutaneous route. When heated to decomposition it emits toxic fumes of NO_x and Cl^-.

CHK175 CAS:327-97-9 HR: 2
CHLOROGENIC ACID
mf: $C_{16}H_{18}O_9$ mw: 354.34

PROP: Needles. Mp: 208°.

SYNS: 3-CAFFEOYLQUINIC ACID □ 3-o-CAFFEOYLQUINIC ACID

TOXICITY DATA WITH REFERENCE
mrc-smc 1 g/L MUREAV 135,109,84
cyt-ham:ovr 250 mg/L MUREAV 111,209,83
ipr-rat TDLo:40 mg/kg (5-12D preg):TER TXAPA9 36,227,76
ipr-rat LDLo:4000 mg/kg TXAPA9 36,337,76

SAFETY PROFILE: Moderately toxic by intraperitoneal route. An experimental teratogen. Mutation data reported. When heated to decomposition it emits acrid smoke and fumes.

CHK250 CAS:13637-65-5 HR: 3
CHLOROGERMANE
mf: $ClGeH_3$ mw: 111.07

PROP: Liquid. D: 2.147 @ −52°, mp: −52°, bp: 28°.

SAFETY PROFILE: Reaction with ammonia forms heat-sensitive explosive products. When heated to decomposition it emits toxic fumes of Cl^-. See also GERMANIUM COMPOUNDS.

CHK750 CAS:400-44-2 HR: 3
2-CHLORO-1,1,1,4,4,4-HEXAFLUOROBUTENE-2
mf: C_4HClF_6 mw: 198.50

PROP: Liquid. D: 1.5482 @ 20°/4°, bp: 32.2°.

SYNS: CHFB □ 1,1,1,4,5,5-HEXAFLUORO-2-CHLORO-2-BUTENE

TOXICITY DATA WITH REFERENCE
ihl-hmn TCLo:10 ppm/1H:PUL CENEAR 44,6,66
ihl-rat LC50:3 ppm/6H 34ZIAG -,310,69

SAFETY PROFILE: Poison by inhalation. Human respiratory system effects by inhalation. When heated to decomposition it emits very toxic fumes of $F^-$and Cl^-. See also CHLORINATED HYDROCARBONS, ALIPHATIC, and FLUORIDES.

CHK825 **CAS:73803-48-2** **HR: D**
endo-4-CHLORO-N-(HEXAHYDRO-4,7-METHANOISOINDOL-2-YL)-3-SULFAMOYLBENZAMIDE

SYNS: TDS □ N-(4-AZA-endo-TRICYCLO(5.2.1.$5^{2,6}$)-DECAN-4-YL)-4-CHLORO-3-SULFAMOYLBENZAMIDE

TOXICITY DATA WITH REFERENCE
orl-rat TDLo:5 g/kg (8-17D preg):REP YKRYAH 12,651,79
orl-rat TDLo:5 g/kg (female 8-17D post):REP YKRYAH 12,651,79
orl-rat TDLo:40 g/kg (8-17D preg):TER YKRYAH 12,651,79
orl-rat TDLo:5 g/kg (8-17D preg):TER YKRYAH 12,651,79
orl-rat TDLo:40 g/kg (8-17D preg):REP YKRYAH 12,651,79

SAFETY PROFILE: An experimental teratogen. Other experimental reproductive effects. When heated to decomposition it emits toxic fumes of Cl^-, SO_x, and NO_x.

CHL000 **CAS:73926-88-2** **HR: 3**
trans-CHLORO(2-HEXANAMIDOCYCLOHEXYL)MERCURY
mf: $C_{12}H_{22}ClHgNO$ mw: 432.39

SYNS: CHLORO(2-HEXANAMIDOCYCLOHEXYL)MERCURY, (E)- □ N-(2-CHLOROMERCURICYCLOHEXYL) HEXANAMIDE, (E)-

TOXICITY DATA WITH REFERENCE
ivn-mus LD50:56 mg/kg CSLNX* NX#04829

CONSENSUS REPORTS: Mercury and its compounds are on the Community Right-To-Know List.

OSHA PEL: CL 0.1 mg(Hg)/m^3 (skin)
ACGIH TLV: TWA 0.1 mg(Hg)/m^3 (skin)
NIOSH REL: (Organomercury): TWA 0.01 mg/m^3; STEL 0.03 mg/m^3 (skin)

SAFETY PROFILE: Poison by intravenous route. See also MERCURY COMPOUNDS, ORGANIC. When heated to decomposition it emits very toxic fumes of Cl^-, NO_x, and Hg.

CHL250 **CAS:13654-91-6** **HR: 3**
CHLOROHEXYL ISOCYANATE
DOT: UN 2206/UN 2207/UN 2478/UN 3080
mf: $C_7H_{10}ClNO$ mw: 159.63

SYNS: 6-CHLORHEXYLISOKYANAT □ ISOCYANIC ACID, 6-CHLOROHEXYL ESTER

TOXICITY DATA WITH REFERENCE
ihl-rat LCLo:60 mg/m^3/4H GTPZAB 12(10),40,68
ihl-mus LC50:69 mg/m^3/2H GTPZAB 12(10),40,68

DOT CLASSIFICATION: 6.1; *Label:* KEEP AWAY FROM FOOD (UN 2207); DOT Class: 6.1; *Label:* Poison (UN 2206); DOT Class: 6.1; *Label:* Poison, Flammable Liquid (UN 3080); DOT Class: 3; *Label:* Flammable Liquid, Poison (UN 2478)

SAFETY PROFILE: Poison by inhalation. See also THIOCYANATES and ESTERS. A flammable liquid. When heated to decomposition it emits very toxic fumes of Cl^- and NO_x.

CHL500 **CAS:18979-94-7** **HR: 2**
4-CHLORO-2-HEXYLPHENOL
mf: $C_{12}H_{17}ClO$ mw: 212.74

SYN: 2-HEXYL-4-CHLOROPHENOL

TOXICITY DATA WITH REFERENCE
skn-mus TDLo:8400 mg/kg/21W-I:NEO CNREA8 19,413,59

CONSENSUS REPORTS: Chlorophenol compounds are on the Community Right-To-Know List.

SAFETY PROFILE: Questionable carcinogen with experimental neoplastigenic data by skin contact. When heated to decomposition it emits toxic fumes of Cl^- and NO_x. See also CHLOROPHENOLS.

CHL875 **CAS:52340-46-2** **HR: D**
dl-α-CHLOROHYDRIN
mf: $C_3H_7ClO_2$ mw: 110.55

PROP: Liquid with sweet taste. Bp: 139° @ 18 mm.

SYNS: (±)-3-CHLORO-1,2-PROPANEDIOL □ dl-3-CHLORO-1,2-PROPANEDIOL □ (±)-2,3-DIHYDROXYCHLOROPROPANE

TOXICITY DATA WITH REFERENCE
orl-rat TDLo:50 mg/kg (male 1D pre):REP CCPTAY 13,639,76
orl-rat TDLo:50 mg/kg (5D male):REP CCPTAY 13,639,76

SAFETY PROFILE: Experimental reproductive effects. When heated to decomposition it emits toxic fumes of Cl^-.

CHM000 **CAS:615-67-8** **HR: 3**
CHLOROHYDROQUINONE
mf: $C_6H_5ClO_2$ mw: 144.56

PROP: Leaflets from $CHCl_3$; needles from toluene. Mp: 106°, bp: 263°.

TOXICITY DATA WITH REFERENCE
scu-rat TDLo:5 mg/kg (female 1D pre):REP ENDOAO 57,466,55
orl-rat LDLo:200 mg/kg KODAK* 21MAY71
skn-rat LDLo:500 mg/kg KODAK* 21MAY71
ipr-rat LDLo:100 mg/kg KODAK* 21MAY71

CONSENSUS REPORTS: Reported in EPA TSCA Inventory.

SAFETY PROFILE: Poison by ingestion and intraperitoneal routes. Moderately toxic by skin contact. Experimental reproductive effects. When heated to decomposition it emits toxic fumes of Cl^-. See also CHLORIDES.

CHM500 **CAS:13442-11-0** **HR: 2**
5-CHLORO-4-(HYDROXYAMINO)QUINOLINE-1-OXIDE
mf: $C_9H_7ClN_2O_2$ mw: 210.63

TOXICITY DATA WITH REFERENCE
scu-mus TDLo: 120 mg/kg/50D-I: ETA BCPCA6 16,631,67

SAFETY PROFILE: Questionable carcinogen with experimental tumorigenic data. When heated to decomposition it emits very toxic fumes of Cl^- and NO_x.

CHM750 **CAS:14076-05-2** **HR: 2**
6-CHLORO-4-(HYDROXYAMINO)QUINOLINE-1-OXIDE
mf: $C_9H_7ClN_2O_2$ mw: 210.63

TOXICITY DATA WITH REFERENCE
scu-mus TDLo: 120 mg/kg/50D-I: ETA BCPCA6 16,631,67

SAFETY PROFILE: Questionable carcinogen with experimental tumorigenic data. When heated to decomposition it emits very toxic fumes of Cl^- and NO_x.

CHN000 **CAS:13442-12-1** **HR: 2**
7-CHLORO-4-(HYDROXYAMINO)QUINOLINE-1-OXIDE
mf: $C_9H_7ClN_2O_2$ mw: 210.63

TOXICITY DATA WITH REFERENCE
scu-mus TDLo: 120 mg/kg/50D-I: ETA BCPCA6 16,631,67

SAFETY PROFILE: Questionable carcinogen with experimental tumorigenic data. When heated to decomposition it emits very toxic fumes of Cl^- and NO_x.

CHN500 **CAS:92-04-6** **HR: 2**
3-CHLORO-4-HYDROXYBIPHENYL
mf: $C_{12}H_9ClO$ mw: 204.66

PROP: White flakes or prisms. Bp: 322° (decomp), fp: 74.2°, flash p: 345°F, d: <1, mp: 76–76.5°.

SYNS: 3-CHLOR-4-HYDROXYBIFENYL (CZECH) □ 3-CHLORO-4-HYDROXYDIPHENYL □ 2-CHLORO-4-PHENYLPHENOL □ DOWICIDE 4 □ 4-PHENYL-2-CHLOROPHENOL

TOXICITY DATA WITH REFERENCE
skn-rbt 500 mg/24H MLD 28ZPAK -,82,72
eye-rbt 50 μg/24H SEV 28ZPAK -,82,72

CONSENSUS REPORTS: Reported in EPA TSCA Inventory. Chlorophenol compounds are on the Community Right-To-Know List.

SAFETY PROFILE: A severe eye and mild skin irritant. A pesticide. Combustible when exposed to heat or flame. To fight fire, use alcohol foam, CO_2, dry chemical. When heated to decomposition it emits toxic fumes of Cl^-. See also CHLOROPHENOLS.

CHN750 **CAS:24579-91-7** **HR: 2**
CHLORO(2-HYDROXY-3,5-DINITROPHENYL)MERCURY
mf: $C_6H_3ClHgN_2O_5$ mw: 419.15

SYN: 2-(CHLOROMERCURI)-4,6-DINITROPHENOL

TOXICITY DATA WITH REFERENCE
orl-rat LDLo: 500 mg/kg NCNSA6 5,37,53

CONSENSUS REPORTS: Mercury and its compounds are on the Community Right-To-Know List.

OSHA PEL: CL 0.1 mg(Hg)/m³ (skin)
ACGIH TLV: TWA 0.1 mg(Hg)/m³ (skin)
NIOSH REL: (Mercury, Aryl and Inorganic): CL 0.1 mg/m³ (skin)

SAFETY PROFILE: Moderately toxic by ingestion. See also MERCURY COMPOUNDS, ORGANIC. When heated to decomposition it emits very toxic fumes of Cl^-, NO_x, and Hg.

CHO125 **CAS:94-87-1** **HR: 3**
2-CHLORO-N-(2-HYDROXYETHYL)ANILINE
mf: $C_8H_{10}ClNO$ mw: 171.63

$ClC_6H_4NHC_2H_4OH$

SAFETY PROFILE: Potentially explosive decomposition above 210°C, catalyzed by the presence of mild steel. This reaction has caused a violent explosion during an industrial scale distillation. Upon decomposition it emits toxic fumes of Cl^-, HCl, and NO_x. Decomposition also produces primary amines, ethylene, methane, carbon monoxide, and carbon dioxide. See also ANILINE DYES.

CHO250 **CAS:55477-27-5** **HR: 3**
4-CHLORO-6-(2-HYDROXYETHYLPIPERAZINO-2-METHYLAMINO-5)-METHYLTHIOPYRIMIDINE
mf: $C_{12}H_{20}ClN_5OS$ mw: 317.88

TOXICITY DATA WITH REFERENCE
orl-mus LD50: 525 mg/kg JMCMAR 18,553,75
ivn-mus LD50: 124 mg/kg JMCMAR 18,553,75

SAFETY PROFILE: Poison by intravenous route. Moderately toxic by ingestion. When heated to decomposition it emits very toxic fumes of Cl^-, NO_x, and SO_x.

CHO750 **CAS:538-04-5** **HR: 3**
2-CHLORO-4-(HYDROXY MERCURI)PHENOL
mf: $C_6H_5ClHgO_2$ mw: 345.15

PROP: Insol solid. Contains 20% mercury (27ZTAP 3,36,69).

SYNS: (3-CHLORO-4-HYDROXYPHENYL)HYDROXYMERCURY □ SEMESAN

TOXICITY DATA WITH REFERENCE
ipr-mus LDLo: 25 mg/kg JPETAB 31,87,27

CONSENSUS REPORTS: Mercury and its compounds and chlorophenol compounds are on the Community Right-To-Know List.

ACGIH TLV: TWA 0.1 mg(Hg)/m³ (skin)
NIOSH REL: (Mercury, Aryl and Inorganic): CL 0.1 mg/m³ (skin)

SAFETY PROFILE: Poison by ingestion, inhalation, intraperitoneal, and intravenous routes. See also MERCURY COMPOUNDS and CHLOROPHENOLS. When heated to decomposition it emits very toxic fumes of Cl^- and Hg.

CHP250 **CAS:303-47-9** **HR: 3**
(−)-N-((5-CHLORO-8-HYDROXY-3-METHYL-1-OXO-7-ISOCHROMANYL)CARBONYL)-3-PHENYLALANINE
mf: $C_{29}H_{18}ClNO_6$ mw: 403.84

PROP: Crystals from xylene. Mp: 169°.

SYNS: (R)N-((5-CHLORO-3,4-DIHYDRO-8-HYDROXY-3-METHYL-1-OXO-1H-2-BENZOPYRAN-7-YL))PHENYLALANINE ☐ NCI-C56586 ☐ OCHRATOXIN A

TOXICITY DATA WITH REFERENCE
cyt-mky:kdy 20 mg/L TXAPA9 32,198,75
orl-mus TDLo:3750 µg/kg (female 15-17D post):REP TJADAB 27,293,83
orl-rat TDLo:5 mg/kg (female 8-9D post):REP ARCVBP 6,379,75
orl-mus TDLo:3 mg/kg (female 15D post):TER TXAPA9 57,127,81
orl-mus TDLo:4 mg/kg (female 9D post):TER ACVTA8 22,535,81
orl-rat TDLo:8 mg/kg (female 11-14D post):TER TJADAB 28,37A,83
orl-rat TDLo:7500 µg/kg (female 6-15D post):TER TXAPA9 37,331,76
scu-rat TDLo:1750 µg/kg (female 7D post):REP TXCYAC 32,277,84
orl-rat TDLo:8 mg/kg (female 7-10D post):REP TJADAB 28,37A,83
ipr-rat TDLo:35 mg/kg (female 6-12D post):REP ENDKAC 77,152,81
ipr-rat TDLo:40 mg/kg (male 8D pre):REP IJEBA6 17,121,79
scu-rat TDLo:1750 µg/kg (female 7D post):TER TXCYAC 25,175,82
orl-rat TDLo:10 mg/kg (female 6-15D post):TER TXAPA9 37,331,76
orl-rat TDLo:5 mg/kg (female 6-15D post):TER TXAPA9 37,331,76
orl-rat TDLo:8 mg/kg (female 7-10D post):TER TJADAB 28,37A,83
orl-rat TDLo:36,050 µg/kg/2Y-C:CAR NTPTR* NTP-TR-358,89
orl-mus TDLo:2216 mg/kg/44W-C:CAR GANNA2 69,599,78
orl-mus TD:1478 mg/kg/44W-C:NEO MAIKD3 (18),15,83
orl-mus TD:1478 mg/kg/44W-C:ETA GANRAE 30,1445,84
orl-mus TD:3504 mg/kg/2Y-C:CAR JJIND8 75,733,85
orl-mus TD:2 g/kg/30W-C:ETA TOLED5 31(Suppl),206,86
orl-rat LD50:20 mg/kg FCTXAV 6,479,68
ipr-rat LD50:12,600 µg/kg ARCVBP 5(2),233,74
ivn-rat LD50:12,750 µg/kg ARCVBP 5(2),233,74
orl-mus LD50:46 mg/kg TOLED5 25,1,85
ipr-mus LD50:22 mg/kg APTOA6 2,109,46
ivn-mus LD50:25,710 µg/kg ARCVBP 5(2),233,74
orl-dog LD50:200 µg/kg CRTXB2 2,499,74
orl-pig LD50:1 mg/kg CRTXB2 2,499,74
orl-ckn LD50:3300 µg/kg APMBAY 21,492,71

CONSENSUS REPORTS: NTP 7th Annual Report On Carcinogens. IARC Cancer Review: Group 2B IMEMDT 56,489,93; Animal Limited Evidence IMEMDT 31,191,83; Animal Sufficient Evidence IMEMDT 56,489,93; Human Inadequate Evidence; IMEMDT 31,191,83; Human Inadequate Evidence IMEMDT 56,489,93.

SAFETY PROFILE: Confirmed carcinogen with carcinogenic and neoplastigenic data. Poison by ingestion, intraperitoneal, intravenous, and subcutaneous routes. Experimental teratogenic and reproductive effects. Questionable carcinogen with experimental carcinogenic, neoplastigenic, tumorigenic data. Mutation data reported. When heated to decomposition it emits very toxic fumes of Cl^- and NO_x.

CHP375 **HR: D**
6-CHLORO-17-α-HYDROXY-16-α-METHYLPREGNA-4,6-DIENE-3,20-DIONE
mf: $C_{24}H_{31}ClO_4$ mw: 419.00

SYN: 6-CHLORO-16-α-METHYL-Δ^6-DEHYDRO-17-α-ACETOXYPROGESTERONE

TOXICITY DATA WITH REFERENCE
scu-rbt TDLo:10 µg/kg (female 1D pre):REP ACEDAB 73,3,63
scu-rbt TDLo:20 µg/kg (1D pre):REP ACEDAB 73,3,63
scu-mus TDLo:2 mg/kg (1D pre):REP 85GRAA -,150,65

SAFETY PROFILE: Experimental reproductive effects. When heated to decomposition it emits toxic fumes of Cl^-.

CHP500 **CAS:5160-02-1** **HR: 2**
5-CHLORO-2-((2-HYDROXY-1-NAPHTHYL)AZO)-p-TOLUENE SULFONIC ACID, BARIUM SALT
mf: $C_{17}H_{12}ClN_2O_4S•1/2Ba$ mw: 444.49

SYNS: BRIGHT RED ☐ BRILLIANT RED ☐ BRILLIANT SCARLET ☐ BRILLIANT TONER Z ☐ BRONZE RED RO ☐ BRONZE SCARLET ☐ 5-CHLORO-2-((2-HYDROXY-1-NAPHTHALENYL)AZO)-4-METHYLBENZENE SULFONIC ACID, BARIUM SALT (2:1) ☐ 5-CHLORO-2-((2-HYDROXY-1-NAPHTHALENYL)AZO)-4-METHYLBENZENE SULPHONIC ACID, BARIUM SALT ☐ 1-(4-CHLORO-o-SULFO-5-TOLYLAZO)-2-NAPHTHOL,BARIUM SALT ☐ C.I. PIGMENT RED ☐ COSMETIC CORAL RED KO BLUISH ☐ DAINICHI LAKE RED C ☐ D&C RED No. 9 ☐ DESERT RED ☐ ELJON LAKE RED C ☐ HAMILTON RED ☐ HELIO RED TONER LCLL ☐ IRGALITE RED CBN ☐ ISOL LAKE RED LCS 12527 ☐ LAKE RED C ☐ LATEXOL SCARLET R ☐ LD RUBBER RED 16913 ☐ LUTETIA RED CLN ☐ MICROTEX LAKE RED CR ☐ MOHICAN RED A-8008 ☐ NCI-C53792 ☐ No. 3 CONC. SCARLET ☐ PARIDINE RED LCL ☐ PIGMENT RED CD ☐ POTOMAC RED ☐ RECOLITE RED LAKE C ☐ 1860 RED ☐ RED SCARLET ☐ SANYO LAKE RED C ☐ SEGNALE RED LC ☐ SICO LAKE RED 2L ☐ SUPEROL RED C RT-265 ☐ SYMULER LAKE RED C ☐ TERMOSOLIDO RED LCG ☐ TEXAN RED TONER D ☐ TONER LAKE RED C ☐ TRANSPARENT BRONZE SCARLET ☐ VULCAFIX SCARLET R ☐ VULCAN RED LC ☐ VULCOL FAST RED L ☐ WAYNE RED X-2486

TOXICITY DATA WITH REFERENCE
mmo-sat 1 mg/plate SCIEAS 236,933,87
orl-rat TDLo:130 g/kg/2Y-C:CAR NTPTR* NTP-TR-225,82
orl-rat TD:109 g/kg/2Y-C:NEO FCTOD7 25,619,87

CONSENSUS REPORTS: IARC Cancer Review: Group 3 IMEMDT 57,203,93; Animal Inadequate Evidence IMEMDT 8,107,75; Human No Adequate Data IMEMDT 8,107,75; Human Inadequate Evidence IMEMDT

57,203,93. NTP Carcinogenesis Bioassay (feed); Clear Evidence: rat NTPTR* NTP-TR-225,82; No Evidence: mouse NTPTR* NTP-TR-225,82. Reported in EPA TSCA Inventory.

SAFETY PROFILE: Questionable carcinogen with experimental carcinogenic and tumorigenic data. Mutation data reported. When heated to decomposition it emits very toxic fumes of SO_x, NO_x, and Cl^-. See also SULFONATES.

CHP750 CAS:3124-93-4 HR: 2
21-CHLORO-17-HYDROXY-19-NOR-17-α-PREGNA-4,9-DIEN-20-YN-3-ONE
mf: $C_{20}H_{23}ClO_2$ mw: 330.88

PROP: A solid. Mp: 151°.

SYNS: 17-α-CHLOROETHINYL-17-β-HYDROXYESTRA-4,9-DIEN-3-ONE □ 17-α-CHLOROETHYNYL-17-β-HYDROXY-19-NOR-4,9-ANDROSTADIEN-3-ONE □ 17-α-CHLOROETHYNLY-19-NOR-4,9-ANDROSTADIEN-17-β-OL-3-ONE □ ETHYNERONE □ MK 665

TOXICITY DATA WITH REFERENCE
orl-dog TDLo:1008 mg/kg/4Y-I:ETA JTEHD6 3,179,77

SAFETY PROFILE: Questionable carcinogen with experimental tumorigenic data. When heated to decomposition it emits toxic fumes of Cl^-.

CHQ000 CAS:101652-00-0 HR: 3
7-CHLORO-10-(2-HYDROXY-3-PIPERIDINOPROPYL)ISOALLOXAZINE SULFATE
mf: $C_{18}H_{20}ClN_5O_3 \cdot H_2O_4S$ mw: 487.96

TOXICITY DATA WITH REFERENCE
ipr-rat LD50:38 mg/kg CMTRAG 2,96,61
scu-mus LD50:60 mg/kg CMTRAG 2,96,61
ivn-mus LD50:115 mg/kg CMTRAG 2,96,61

SAFETY PROFILE: Poison by intraperitoneal, subcutaneous and intravenous routes. See also SULFATES. When heated to decomposition it emits very toxic fumes of SO_x, Cl^-, and NO_x.

CHQ500 CAS:637-61-6 HR: 3
4-CHLOROIMINO-2,5-CYCLOHEXADIENE-1-ONE
mf: C_6H_4ClNO mw: 141.56

$O:C_6H_4:NCl$

PROP: Yellow crystals from pet ether. Mp: 86°. Sltly sol in cold H_2O; sol in hot H_2O, EtOH, Et_2O, C_6H_6, and $CHCl_3$.

SYN: QUINONE CHLORIMIDE

TOXICITY DATA WITH REFERENCE
ipr-mus LD50:12 mg/kg JMCMAR 21,11,78

CONSENSUS REPORTS: Reported in EPA TSCA Inventory.

SAFETY PROFILE: Poison by intraperitoneal route. Explodes on heating. Upon decomposition it emits toxic fumes of Cl^-.

CHQ750 CAS:537-45-1 HR: 3
4-CHLOROIMINO-2,6-DIBROMO-2,5-CYCLOHEXADIENE-1-ONE
mf: $C_6H_2Br_2ClNO$ mw: 299.36

HC=CBrCO•CBr=CH C:NCl

PROP: Yellow crystals from pet ether. Mp: 85–86°. Sol in EtOH, $CHCl_3$, Me_2CO, and Et_2O.

TOXICITY DATA WITH REFERENCE
ipr-mus LD50:63 mg/kg JMCMAR 21,11,78

CONSENSUS REPORTS: Reported in EPA TSCA Inventory.

SAFETY PROFILE: Poison by intraperitoneal route. A storage hazard. May explode at room temperature. Explodes when heated above 50°C. When heated to decomposition it emits very toxic fumes of Br^-, Cl^-, and NO_x.

CHR000 CAS:101-38-2 HR: 3
4-CHLOROIMINO-2,6-DICHLORO-2,5-CYCLOHEXADIENE-1-ONE
mf: $C_6H_2Cl_3NO$ mw: 210.44

HC=CClCO•CCl=CH C:NCl

PROP: Yellow needles from EtOH. Mp: 66.5–66.7°.

SYNS: 2,6-DICHLOROQUINONE CHLOROIMIDE □ N,2,6-TRICHLORO-p-BENZOQUINONE IMINE

TOXICITY DATA WITH REFERENCE
ipr-mus LD50:20 mg/kg JMCMAR 21,11,78
ivn-mus LD50:56 mg/kg CSLNX* NX#00254

CONSENSUS REPORTS: Reported in EPA TSCA Inventory.

SAFETY PROFILE: Poison by intraperitoneal and intravenous routes. A storage hazard. It may explode at room temperature. When heated to decomposition it emits very toxic fumes of Cl^- and NO_x.

CHR325 CAS:25604-71-1 HR: 3
CHLOROIODOACETYLENE
mf: C_2ClI mw: 186.38

SYN: CHLOROIODOETHYNE

SAFETY PROFILE: A very unstable, explosive material. When heated to decomposition it emits toxic fumes of Cl^- and I^-. See also ACETYLENE COMPOUNDS and EXPLOSIVES.

CHR400 CAS:109-71-7 HR: 3
3-CHLORO-1-IODOPROPYNE
mf: C_3H_2ClI mw: 200.41

SAFETY PROFILE: Reacts explosively with air when heated to 47°C. When heated to decomposition it emits toxic fumes of Cl^- and I^-. See also CHLORINATED HYDROCARBONS, ALIPHATIC.

CHR500 **CAS:130-26-7** **HR: 3**
5-CHLORO-7-IODO-8-QUINOLINOL
mf: C_9H_5ClINO mw: 305.50

PROP: Brownish-yellow powder, darkens when exposed to light. Sltly sol in Et_2O.

SYNS: ALCHLOQUIN □ AMEBIL □ AMOENOL □ BACTOL □ BARQUINOL □ BUDOFORM □ CHINOFORM □ 5-CHLOR-7-JOD-8-8HYDROXY-CHINOLIN (GERMAN) □ 5-CHLORO-8-HYDROXY-7-IODOQUINOLINE □ 5-CHLORO-7-IODO-8-HYDROXYQUINOLINE □ CHLOROIODOQUINE □ CLIOQUINOL □ CLIQUINOL □ ECZECIDIN □ EMAFORM □ ENTERO-BIO FORM □ ENTEROQUINOL □ ENTEROSEPTOL □ ENTERO-VIOFORM □ ENTEROZOL □ ENTERUM LOCORTEN □ ENTROKIN □ HI-ENTEROL □ HYDRIODIDE-ENTROL □ IODENTEROL □ IODOCHLORHYDROXYQUINOL □ IODOCHLORHYDROXYQUINOLINE □ 7-IODO-5-CHLORO-8-HYDROXYQUINOLINE □ 7-IODO-5-CHLOROXINE □ IODOENTEROL □ NIOFORM □ QUINAMBICIDE □ ROMETIN □ VIOFORM □ VIOFORM N.N.R.

TOXICITY DATA WITH REFERENCE
sln-asn 1 g/L MUREAV 26,159,74
dnd-hmn:hla 40 µmol/L ANYAA9 284 525,77
orl-rat TDLo:528 mg/kg (7-17D preg):REP OYYAA2 14,211,77
orl-rat TDLo:1320 mg/kg (7-17D preg):REP OYYAA2 14,211,77
orl-rat TDLo:1320 mg/kg (7-17D preg):TER OYYAA2 14,211,77
orl-rat TDLo:9000 mg/kg (30D pre):REP OYYAA2 14,211,77
orl-rat TDLo:1560 mg/kg (male 22W pre):REP OYYAA2 14,191,77
orl-rat TDLo:900 mg/kg (male 30D pre):REP OYYAA2 14,75,77
orl-rat TDLo:528 mg/kg (7-17D preg):TER OYYAA2 14,211,77
orl-wmn TDLo:11 g/kg/36W:EYE LANCAO 1,1015,72
orl-hmn TDLo:1400 mg/kg/20D-I:BRN JJMCAQ 24,195,71
ipr-rat LD50:3400 mg/kg OYYAA2 14,75,77
orl-mus LD50:69 mg/kg ATSUDG 2,371,79
orl-cat LD50:400 mg/kg AJTMAQ 24,29,44
orl-rbt LDLo:250 mg/kg JAMAAP 100,1658,33
orl-gpg LDLo:175 mg/kg AJTMAQ 24,29,44

CONSENSUS REPORTS: Reported in EPA TSCA Inventory. EPA Genetic Toxicology Program.

SAFETY PROFILE: Poison by ingestion. Moderately toxic by intraperitoneal route. Human systemic effects by ingestion: change in central nervous system electrical function, optic nerve damage, and changes in vision. Experimental teratogenic and reproductive effects. Human mutation data reported. When heated to decomposition it emits very toxic fumes of Cl^-, I^-, and NO_x.

CHR700 **CAS:109651-74-3** **HR: 3**
3′-CHLORO-2-ISOBUTYLAMINO-p-ACETOTOLUIDIDE HYDROCHLORIDE
mf: $C_{13}H_{19}ClN_2O{\cdot}ClH$ mw: 291.25

TOXICITY DATA WITH REFERENCE
eye-rbt 100 mg MOD JAPMA8 49,80,60
ipr-mus LD50:75 mg/kg JAPMA8 49,80,60
scu-mus LD50:225 mg/kg JAPMA8 49,80,60

SAFETY PROFILE: Poison by subcutaneous and intraperitoneal routes. An eye irritant. When heated to decomposition it emits toxic fumes of NO_x and Cl^-.

CHR750 **CAS:102489-59-8** **HR: 3**
6′-CHLORO-2-(ISOBUTYLAMINO)-o-ACETOTOLUIDIDE HYDROCHLORIDE
mf: $C_{13}H_{19}ClN_2O{\cdot}ClH$ mw: 291.25

SYN: C 3156

TOXICITY DATA WITH REFERENCE
ipr-rat LD50:345 mg/kg ARZNAD 8,407,58
ipr-mus LD50:390 mg/kg ARZNAD 8,407,58
scu-mus LD50:700 mg/kg ARZNAD 8,407,58

SAFETY PROFILE: Poison by intraperitoneal route. Moderately toxic by subcutaneous route. When heated to decomposition it emits very toxic fumes of Cl^- and NO_x.

CHR850 **CAS:109509-25-3** **HR: 3**
3′-CHLORO-3-ISOBUTYLAMINO-o-PROPIONOTOLUIDIDE HYDROCHLORIDE
mf: $C_{14}H_{21}ClN_2O{\cdot}ClH$ mw: 305.28

TOXICITY DATA WITH REFERENCE
eye-rbt 100 mg MOD JAPMA8 49,80,60
ipr-mus LD50:325 mg/kg JAPMA8 49,80,60
scu-mus LD50:1500 mg/kg JAPMA8 49,80,60

SAFETY PROFILE: Poison by intraperitoneal route. Moderately toxic by subcutaneous route. An eye irritant. When heated to decomposition it emits toxic fumes of NO_x and HCl. See also CHLORINATED HYDROCARBONS, AROMATIC.

CHS500 **CAS:1918-16-7** **HR: 3**
2-CHLORO-N-ISOPROPYLACETANILIDE
mf: $C_{11}H_{14}ClNO$ mw: 211.71

PROP: Light-tan solid. Mp: 67–76°. Sltly sol in H_2O.

SYNS: BEXTON □ CHLORESSIGSAEURE-N-ISOPROPYLANILID (GERMAN) □ α-CHLORO-N-ISOPROPYLACETANILIDE □ 2-CHLORO-N-ISOPROPYL-N-PHENYLACETAMIDE □ 2-CHLORO-N-(1-METHYLETHYL)-N-PHENYLACETAMIDE □ CP 31393 □ N-ISOPROPYL-α-CHLOROACETANILIDE □ N-ISOPROPYL-2-CHLOROACETANILIDE □ NITICID □ PROPACHLOR □ PROPACHLORE □ RAMROD □ SATECID

TOXICITY DATA WITH REFERENCE
cyt-mus-unr 10 mg/kg TGANAK 14(6),41,80
cyt-mus-orl 10 mg/kg CYGEDX 14(6),38,80
orl-rat LD50:710 mg/kg 85ARAE 2,63,77
orl-mus LD50:290 mg/kg EKMMA8 13,123,74
orl-rbt LD50:710 mg/kg KHZDAN 17,90,74
skn-rbt LD50:380 mg/kg WRPCA2 9,119,70
orl-dck LD50:512 mg/kg DOEAAH 35,25,79

CONSENSUS REPORTS: EPA Genetic Toxicology Program.

SAFETY PROFILE: Poison by ingestion and skin contact. Mutation data reported. A selective herbicide. When heated to decomposition it emits very toxic fumes of Cl^- and NO_x.

C

CHS750 **CAS:77966-61-1** **HR: 3**
6′-CHLORO-2-(ISOPROPYLAMINO)-o-ACETOTOLUIDIDE HYDROCHLORIDE
mf: $C_{12}H_{17}ClN_2O{\bullet}ClH$ mw: 277.22

SYNS: C 3059 □ 2′-CHLORO-2-(ISOPROPYLAMINO)-6′-METHYLACETANILIDE HYDROCHLORIDE

TOXICITY DATA WITH REFERENCE
ipr-rat LD50:330 mg/kg ARZNAD 8,407,58
ipr-mus LD50:275 mg/kg ARZNAD 8,407,58
scu-mus LD50:730 mg/kg ARZNAD 8,407,58

SAFETY PROFILE: Poison by intraperitoneal route. Moderately toxic by subcutaneous route. When heated to decomposition it emits very toxic fumes of NO_x and Cl^-.

CHU000 **CAS:33965-80-9** **HR: 3**
3-CHLORO-LACTONITRILE
mf: C_3H_4ClNO mw: 105.53

$ClCH_2CH(CN)OH$

SYN: 2-CHLORO-1-CYANOETHANOL

TOXICITY DATA WITH REFERENCE
ipr-mus LDLo:4 mg/kg CBCCT* 2,241,50
ivn-dog LD50:7 mg/kg CBCCT* 2,299,50
skn-rbt LDLo:25 mg/kg CBCCT* 2,299,50
ivn-rbt LDLo:10 mg/kg CBCCT* 2,299,50

CONSENSUS REPORTS: Cyanide and its compounds are on the Community Right-To-Know List.

SAFETY PROFILE: Poison by skin contact, intravenous, and intraperitoneal routes. Heating above 110°C may cause explosive decomposition to 2-chloroacetaldehyde and HCN. When heated to decomposition it emits very toxic fumes of Cl^-, CN^-, and NO_x. See also NITRILES.

CHU100 **CAS:615-48-5** **HR: 3**
5-CHLORO-2-MERCAPTOANILINE HYDROCHLORIDE
mf: $C_6H_6ClNS{\bullet}ClH$ mw: 196.10

SYN: ANILINE, 5-CHLORO-2-MERCAPTO-, HYDROCHLORIDE

TOXICITY DATA WITH REFERENCE
ivn-mus LD50:56 mg/kg CSLNX* NX#00461

CONSENSUS REPORTS: Reported in EPA TSCA Inventory.

SAFETY PROFILE: Poison by intravenous route. When heated to decomposition it emits toxic vapors of NO_x, SO_x, HCl, and Cl^-.

CHU500 **CAS:59-85-8** **HR: 3**
p-CHLOROMERCURIC BENZOIC ACID
mf: $C_7H_5ClHgO_2$ mw: 357.16

PROP: A solid. Mp: 273°.

SYNS: (p-CARBOXYPHENYL)CHLOROMERCURY □ p-(CHLOROMERCURI)BENZOIC ACID □ USAF D-3

TOXICITY DATA WITH REFERENCE
ipr-mus LD50:25 mg/kg NTIS** AD277-689

CONSENSUS REPORTS: Reported in EPA TSCA Inventory. Mercury and its compounds are on the Community Right-To-Know List.

OSHA PEL: CL 0.1 mg(Hg)/m³ (skin)
ACGIH TLV: TWA 0.1 mg(Hg)/m³ (skin)
NIOSH REL: (Mercury, Aryl and Inorganic): CL 0.1 mg/m³ (skin)

SAFETY PROFILE: Poison by intraperitoneal route. See also MERCURY COMPOUNDS. When heated to decomposition it emits very toxic fumes of Cl^- and Hg.

CHU750 **CAS:73940-90-6** **HR: 3**
N-(CHLOROMERCURI)FORMANILIDE
mf: $C_7H_6ClHgNO$ mw: 356.18

SYN: CHLORO(N-PHENYLFORMAMIDO)MERCURY

TOXICITY DATA WITH REFERENCE
ivn-mus LD50:32 mg/kg CSLNX* NX#05982

CONSENSUS REPORTS: Mercury and its compounds are on the Community Right-To-Know List.

OSHA PEL: CL 0.1 mg(Hg)/m³ (skin)
ACGIH TLV: TWA 0.1 mg(Hg)/m³ (skin)
NIOSH REL: (Mercury, Aryl and Inorganic): CL 0.1 mg/m³ (skin)

SAFETY PROFILE: Poison by intravenous route. See also MERCURY COMPOUNDS. When heated to decomposition it emits very toxic fumes of Cl^-, NO_x, and Hg.

CHV250 **CAS:3477-28-9** **HR: 3**
3-(3-CHLOROMERCURI-2-METHOXY-1-PROPYL)-5,5-DIMETHYLHYDANTOIN
mf: $C_9H_{15}ClN_2O_3Hg$ mw: 435.21

SYN: CHLORO((3-(5,5-DIMETHYL-2,4-DIOXO-3-IMIDAZOLIDINYL)-2-METHOXY)PROPYL)MERCURY

TOXICITY DATA WITH REFERENCE
orl-mus LD50:346 mg/kg JMPCAS 5,168,62

CONSENSUS REPORTS: Mercury and its compounds are on the Community Right-To-Know List.

OSHA PEL: CL 0.1 mg(Hg)/m³ (skin)
ACGIH TLV: TWA 0.1 mg(Hg)/m³ (skin)
NIOSH REL: (Mercury, Aryl and Inorganic): CL 0.1 mg/m³ (skin)

SAFETY PROFILE: Poison by ingestion. See also MERCURY COMPOUNDS. When heated to decomposition it emits very toxic fumes of Cl^-, NO_x and Hg.

CHV500 **CAS:3367-32-6** **HR: 2**
1-(3-CHLOROMERCURI-2-METHOXY)PROPYLHYDANTOIN
mf: $C_7H_{10}ClHgN_2O_3$ mw: 406.23

SYNS: 1-(3-CHLOROMERCURI-2-METHOXY-1-PROPYL)-HYDANTOIN □ 1-(3-(CHLOROMERCURY)-2-METHOXYPROPYL)HYDANTOIN

TOXICITY DATA WITH REFERENCE
orl-mus LD50:1580 mg/kg AIPTAK 149,415,64

CONSENSUS REPORTS: Mercury and its compounds are on the Community Right-To-Know List.

OSHA PEL: CL 0.1 mg(Hg)/m³ (skin)
ACGIH TLV: TWA 0.1 mg(Hg)/m³ (skin)
NIOSH REL: (Mercury, Aryl and Inorganic): CL 0.1 mg/m³ (skin)

SAFETY PROFILE: Moderately toxic by ingestion. See also MERCURY COMPOUNDS. When heated to decomposition it emits very toxic fumes of Cl^-, NO_x and Hg.

CHV750 CAS:3367-29-1 HR: 3
3-(3-CHLOROMERCURI-2-METHOXY-1-PROPYL)HYDANTOIN
mf: $C_7H_{11}ClHgN_2O_3$ mw: 407.24

SYN: CHLORO((3-(2,4-DIOXO-3-IMIDAZOLIDINYL)-2-METHOXY)PROPYL)MERCURY

TOXICITY DATA WITH REFERENCE
orl-mus LD50:358 mg/kg JMPCAS 5,168,62

CONSENSUS REPORTS: Mercury and its compounds are on the Community Right-To-Know List.

OSHA PEL: CL 0.1 mg(Hg)/m³ (skin)
ACGIH TLV: TWA 0.1 mg(Hg)/m³ (skin)
NIOSH REL: (Mercury, Aryl and Inorganic): CL 0.1 mg/m³ (skin)

SAFETY PROFILE: Poison by ingestion. See also MERCURY COMPOUNDS. When heated to decomposition it emits very toxic fumes of Cl^-, NO_x, and Hg.

CHW000 CAS:67465-39-8 HR: 3
1-(3-CHLOROMERCURI-2-METHOXY-1-PROPYL)-3-METHYLHYDANTOIN
mf: $C_8H_{13}ClHgN_2O_3$ mw: 421.27

SYN: CHLORO((3-(2,4-DIOXO-3-METHYL-1-IMIDAZOLIDINYL)-2-METHOXY)PROPYL)MERCURY

TOXICITY DATA WITH REFERENCE
orl-mus LD50:298 mg/kg JMPCAS 5,168,62

CONSENSUS REPORTS: Mercury and its compounds are on the Community Right-To-Know List.

OSHA PEL: CL 0.1 mg(Hg)/m³ (skin)
ACGIH TLV: TWA 0.1 mg(Hg)/m³ (skin)
NIOSH REL: (Mercury, Aryl and Inorganic): CL 0.1 mg/m³ (skin)

SAFETY PROFILE: Poison by ingestion. See also MERCURY COMPOUNDS. When heated to decomposition it emits very toxic fumes of Cl^-, Hg, and NO_x.

CHW250 CAS:3367-28-0 HR: 3
3-(3-CHLOROMERCURI-2-METHOXY-1-PROPYL)-1-METHYLHYDANTOIN
mf: $C_8H_{13}ClHgN_2O_3$ mw: 421.27

SYNS: CHLORO((3-(2,4-DIOXO-1-METHYL-3-IMIDAZOLIDINYL)-2-METHOXY)PROPYL)MERCURY ☐ 3-(3-(CHLOROMERCURI)-2-METHOXYPROPYL)-1-METHYLHYDANTOIN

TOXICITY DATA WITH REFERENCE
orl-mus LD50:264 mg/kg JMPCAS 5,168,62

CONSENSUS REPORTS: Mercury and its compounds are on the Community Right-To-Know List.

OSHA PEL: CL 0.1 mg(Hg)/m³ (skin)
ACGIH TLV: TWA 0.1 mg(Hg)/m³
NIOSH REL: (Organomercury): TWA 0.01 mg/m³; STEL 0.03 mg/m³ (skin)

SAFETY PROFILE: Poison by ingestion. See also MERCURY COMPOUNDS. When heated to decomposition it emits very toxic fumes of Cl^-, NO_x, and Hg.

CHW500 CAS:3367-30-4 HR: 3
5-(3-CHLOROMERCURI-2-METHOXY-1-PROPYL)-3-METHYLHYDANTOIN
mf: $C_7H_{11}ClHgN_2O_3$ mw: 407.24

SYN: CHLORO((3-(2,4-DIOXO-3-METHYL-5-IMIDAZOLIDINYL)-2-METHOXY)PROPYL)MERCURY

TOXICITY DATA WITH REFERENCE
orl-mus LD50:715 mg/kg JMPCAS 5,168,62

CONSENSUS REPORTS: Mercury and its compounds are on the Community Right-To-Know List.

OSHA PEL: CL 0.1 mg(Hg)/m³ (skin)
ACGIH TLV: TWA 0.1 mg(Hg)/m³ (skin)
NIOSH REL: (Mercury, Aryl and Inorganic): CL 0.1 mg/m³ (skin)

SAFETY PROFILE: A poison. Moderately toxic by ingestion. See also MERCURY COMPOUNDS. When heated to decomposition it emits very toxic fumes of Cl^-, NO_x, and Hg.

CHW675 CAS:90-03-9 HR: 3
o-CHLOROMERCURIPHENOL
mf: C_6H_5ClHgO mw: 329.15

PROP: Crystals from H_2O. Mp: 152.5°.

SYNS: CHLORO(o-HYDROXYPHENYL)MERCURY ☐ o-HYDROXYFENYLMERKURICHLORID ☐ o-HYDROXYPHENYLMERCURIC CHLORIDE ☐ MERCUFENOL CHLORIDE ☐ MERCURY, CHLORO(2-HYDROXYPHENYL)- ☐ MYRINGACAINE DROPS ☐ PHENOL, o-(CHLOROMERCURI)- ☐ SALICRESIN FLUID ☐ U-7743 ☐ USPULUM

TOXICITY DATA WITH REFERENCE
orl-rat LDLo:100 mg/kg NCNSA6 5,36,53
ipr-rat LDLo:25 mg/kg NCNSA6 5,36,53
ipr-mus LDLo:48 mg/kg JPETAB 31,87,27
scu-mus LD50:36 mg/kg HBTXAC 5,114,59
ivn-mus LD50:23 mg/kg HBTXAC 5,114,59

CONSENSUS REPORTS: Reported in EPA TSCA Inventory. Mercury and its compounds as well as chlorophenol compounds are on the Community Right-To-Know List.

OSHA PEL: CL 0.1 mg(Hg)/m³ (skin)
ACGIH TLV: TWA 0.1 mg(Hg)/m³ (skin)
NIOSH REL: (Mercury, Aryl and Inorganic): CL 0.1 mg/m³ (skin)

SAFETY PROFILE: Poison by ingestion, subcutaneous, intravenous, and intraperitoneal routes. An antiseptic. See also MERCURY COMPOUNDS, ORGANIC; and

CHLOROPHENOLS. When heated to decomposition it emits toxic fumes of Cl^- and Hg.

CHW750 **CAS:623-07-4** **HR: 3**
p-CHLOROMERCURIPHENOL
mf: C_6H_5ClHgO mw: 329.15

PROP: Plates from Me_2CO. Mp: 226–227°.

SYNS: CHLORO(p-HYDROXYPHENYL)MERCURY ☐ p-(CHLOROMERCURI)PHENOL

TOXICITY DATA WITH REFERENCE
ipr-rat LDLo:50 mg/kg NCNSA6 5,36,53
ipr-mus LDLo:55 mg/kg JPETAB 31,87,27

CONSENSUS REPORTS: Reported in EPA TSCA Inventory. Mercury and its compounds as well as chlorophenol compounds are on the Community Right-To-Know List.

OSHA PEL: CL 0.1 mg(Hg)/m^3 (skin)
ACGIH TLV: TWA 0.1 mg(Hg)/m^3 (skin)
NIOSH REL: (Mercury, Aryl and Inorganic): CL 0.1 mg/m^3 (skin)

SAFETY PROFILE: Poison by intraperitoneal route. When heated to decomposition it emits very toxic fumes of Cl^- and Hg. See also MERCURY COMPOUNDS, ORGANIC; and CHLOROPHENOLS.

CHX250 **CAS:62-37-3** **HR: 3**
CHLOROMERODRIN
mf: $C_5H_{11}ClHgN_2O_2$ mw: 367.22

PROP: Air and light-stable crystals from EtOH. Mp: 152–153°. Sltly sol in H_2O and MeOH.

SYNS: (3-((AMINOCARBONYL)AMINO)-2-METHOXYPROPYL)CHLOROMERCURY ☐ CHLORMEROPRIN ☐ (3-(CHLOROMERCURI)-2-METHOXYPROPYL)UREA ☐ 1-(3-(CHLOROMERCURI)-2-METHOXYPROPYL)UREA ☐ CHLOROMERIDIN ☐ CHLOROMERODRIN ☐ DIURONE ☐ HG-203 ☐ KATONIL ☐ MERCLORAN ☐ MERCORAL ☐ MERILID ☐ (2-METHOXYPROPYL)UREA, MERCURY COMPLEX ☐ NEOHYDRIN ☐ ORICUR ☐ PERCAPYL ☐ PROMERAN

TOXICITY DATA WITH REFERENCE
orl-rat LD50:150 mg/kg TXAPA9 18,185,71
orl-mus LD50:215 mg/kg AIPTAK 143,181,63
ipr-mus LDLo:62,500 µg/kg CBCCT* 5,144,53

CONSENSUS REPORTS: Mercury and its compounds are on the Community Right-To-Know List.

OSHA PEL: CL 0.1 mg(Hg)/m^3 (skin)
ACGIH TLV: TWA 0.1 mg(Hg)/m^3 (skin)
NIOSH REL: (Organomercury): TWA 0.01 mg/m^3; STEL 0.03 mg/m^3 (skin)

SAFETY PROFILE: Poison by ingestion and intraperitoneal routes. A diuretic. See also MERCURY COMPOUNDS. When heated to decomposition it emits very toxic fumes of Cl^-, NO_x, and Hg.

CHX750 **HR: 3**
CHLOROMETHANE mixed with DICHLOROMETHANE

SYN: METHYL CHLORIDE–METHYLENE CHLORIDE MIXTURE

SAFETY PROFILE: Flammable when exposed to heat or flame. When heated to decomposition it emits toxic fumes of Cl^-.

CHY000 **CAS:3518-65-8** **HR: 3**
CHLOROMETHANE SULFONYL CHLORIDE
mf: $CH_2Cl_2O_2S$ mw: 148.99

PROP: Liquid. Bp: 80–81° @ 25 mm.

SYNS: CHLORID KYSELINY CHLORMETHANSULFONOVE (CZECH) ☐ CHLORMETHANSULFOCHLORID (CZECH)

TOXICITY DATA WITH REFERENCE
skn-rbt 500 mg/24H SEV 28ZPAK -,198,72
eye-rbt 50 µg/24H SEV 28ZPAK -,198,72
orl-rat LD50:372 mg/kg 28ZPAK -,198,72

SAFETY PROFILE: Poison by ingestion. A severe skin and eye irritant. When heated to decomposition it emits very toxic fumes of Cl^- and SO_x.

CHY250 **CAS:148-65-2** **HR: 3**
CHLOROMETHAPYRILENE
mf: $C_{14}H_{18}ClN_3S$ mw: 295.86

PROP: Bp: 155–156° @ 10 mm.

SYNS: CHLOROPYRILENE ☐ CHLOROTHEN ☐ 2-((5-CHLORO-2-THENYL)(2-DIMETHYLAMINOETHYL)AMINO)PYRIDINE ☐ CHLOROTHENYLPYRAMINE ☐ N,N-DIMETHYL-N′-(2-PYRIDYL)-N′-(5-CHLORO-2-THENYL)ETHYLENEDIAMINE ☐ ETHYLENEDIAMINE, N-(5-CHLORO-2-THENYL)-N′,N′-DIMETHYL-N-2-PYRIDYL- ☐ NCI-C60559 ☐ PYRITHEN ☐ TAGATHEN ☐ 2-THENYLAMINE, 5-CHLORO-N-(2-(DIMETHYLAMINO)ETHYL)-N-2-PYRIDYL-

TOXICITY DATA WITH REFERENCE
ipr-mus LD50:105 mg/kg JPETAB 96,388,49

SAFETY PROFILE: Poison by intraperitoneal route. When heated to decomposition it emits very toxic fumes of Cl^-, NO_x, and SO_x.

CIB500 **CAS:5185-84-2** **HR: 3**
CHLORO(trans-2-METHOXYCYCLOOCTYL)MERCURY
mf: $C_9H_{17}ClHgO$ mw: 377.30

TOXICITY DATA WITH REFERENCE
ivn-mus LD50:14 mg/kg CSLNX* NX#02812

CONSENSUS REPORTS: Mercury and its compounds are on the Community Right-To-Know List.

OSHA PEL: CL 0.1 mg(Hg)/m^3 (skin)
ACGIH TLV: TWA 0.1 mg(Hg)/m^3 (skin)
NIOSH REL: (Organomercury): TWA 0.01 mg/m^3; STEL 0.03 mg/m^3 (skin)

SAFETY PROFILE: Poison by intravenous route. See also MERCURY COMPOUNDS. When heated to decomposition it emits very toxic fumes of Cl^- and Hg.

CIB625 **CAS:4222-27-9** **HR: 3**
3-CHLORO-3-METHOXYDIAZIRINE
mf: $C_2H_3ClN_2O$ mw: 106.51

SAFETY PROFILE: The liquid is a dangerously unstable

explosive. Upon decomposition it emits toxic fumes of Cl^- and NO_x. See also EXPLOSIVES.

CIB700 CAS:91-38-3 HR: 3
4-CHLORO-N-(p-METHOXYPHENYL)ANTHRANILIC ACID
mf: $C_{14}H_{12}ClNO_3$ mw: 277.72

SYN: 5-CHLORO-4-METHOXYDIPHENYLAMINE-2-CARBOXYLIC ACID

TOXICITY DATA WITH REFERENCE
orl-mus LD50:450 mg/kg QJPPAL 21,10,48
ipr-mus LD50:150 mg/kg QJPPAL 21,10,48
scu-mus LD50:250 mg/kg QJPPAL 21,10,48

SAFETY PROFILE: Poison by subcutaneous and intraperitoneal routes. Moderately toxic by ingestion. When heated to decomposition it emits toxic fumes of Cl^- and NO_x.

CIB725 CAS:4222-26-8 HR: 3
CHLORO-(4-METHOXYPHENYL)DIAZIRINE
mf: $C_8H_7ClN_2O$ mw: 182.61

$CH_3OC_6H_4(Cl)\overline{CN{=}N}$

SAFETY PROFILE: Explodes at room temperature. When heated to decomposition it emits toxic fumes of Cl^- and NO_x.

CIC000 CAS:73926-89-3 HR: 3
CHLORO(2-(3-METHOXYPROPIONAMIDO)CYCLOHEXYL)MERCURY
mf: $C_{10}H_{18}ClHgNO_2$ mw: 420.33

SYN: N-(2-CHLOROMERCURICYCLOHEXYL)PROPIONAMIDE

TOXICITY DATA WITH REFERENCE
ivn-mus LD50:18 mg/kg CSLNX* NX#04831

CONSENSUS REPORTS: Mercury and its compounds are on the Community Right-To-Know List.

OSHA PEL: CL 0.1 mg(Hg)/m³ (skin)
ACGIH TLV: TWA 0.1 mg(Hg)/m³ (skin)
NIOSH REL: (Organomercury): TWA 0.01 mg/m³; STEL 0.03 mg/m³ (skin)

SAFETY PROFILE: Poison by intravenous route. See also MERCURY COMPOUNDS. When heated to decomposition it emits very toxic fumes of Cl^-, Hg, and NO_x.

CIC500 CAS:99999-42-5 HR: 3
7-CHLORO-8-METHOXY-10-(2-PYRROLIDINYLETHYL) ISOALLOXAZINE ACETATE
mf: $C_{17}H_{18}ClN_5O_3 \cdot C_2H_4O_2$ mw: 435.91

TOXICITY DATA WITH REFERENCE
ipr-rat LD50:155 mg/kg CMTRAG 2,96,61
scu-mus LD50:1000 mg/kg CMTRAG 2,96,61
ivn-mus LD50:105 mg/kg CMTRAG 2,96,61

SAFETY PROFILE: Poison by intraperitoneal and intravenous routes. Moderately toxic by subcutaneous route. When heated to decomposition it emits very toxic fumes of Cl^- and NO_x.

CID250 CAS:77966-62-2 HR: 3
6′-CHLORO-2-(METHYLAMINO)-o-ACETOTOLUIDIDE HYDROCHLORIDE
mf: $C_{10}H_{13}ClN_2O \cdot ClH$ mw: 249.16

SYNS: C 3167 □ 2′-CHLORO-6′-METHYL-2-(METHYLAMINO)ACETANILIDE HYDROCHLORIDE

TOXICITY DATA WITH REFERENCE
eye-rbt 2% MLD ARZNAD 8,407,58
ipr-rat LD50:330 mg/kg ARZNAD 8,407,58
ipr-mus LD50:305 mg/kg ARZNAD 8,407,58
scu-mus LD50:775 mg/kg ARZNAD 8,407,58

SAFETY PROFILE: Poison by intraperitoneal route. Moderately toxic by subcutaneous route. An eye irritant. When heated to decomposition it emits very toxic fumes of Cl^- and NO_x.

CID825 CAS:27683-73-4 HR: 3
4′-CHLORO-2-((METHYLAMINO)METHYL)BENZHYDROL HYDROCHLORIDE
mf: $C_{15}H_{16}ClNO \cdot ClH$ mw: 298.23

SYNS: α-(4-CHLOROPHENYL)-2-((METHYLAMINO)METHYL)-BENZENEMETHANOL HYDROCHLORIDE (9CI) □ PR-F 36 Cl

TOXICITY DATA WITH REFERENCE
orl-rat LD50:560 mg/kg AIPTAK 211,253,74
ipr-rat LD50:115 mg/kg AIPTAK 211,253,74
scu-rat LD50:328 mg/kg AIPTAK 211,253,74
orl-mus LD50:320 mg/kg AIPTAK 211,253,74
ipr-mus LD50:54 mg/kg AIPTAK 211,253,74
scu-mus LD50:83 mg/kg AIPTAK 211,253,74
ivn-mus LD50:53 mg/kg AIPTAK 211,253,74

SAFETY PROFILE: Poison by ingestion, subcutaneous, intravenous, and intraperitoneal routes. When heated to decomposition it emits toxic fumes of NO_x and Cl^-.

CIE250 CAS:102504-65-4 HR: 3
6′-CHLORO-2-(METHYLAMINO)-o-PROPIONOTOLUIDIDE HYDROCHLORIDE
mf: $C_{11}H_{15}ClN_2O \cdot ClH$ mw: 263.19

SYN: C 3158

TOXICITY DATA WITH REFERENCE
eye-rbt 2% MLD ARZNAD 8,544,58
ipr-rat LD50:183 mg/kg ARZNAD 8,544,58
scu-mus LD50:340 mg/kg ARZNAD 8,544,58

SAFETY PROFILE: Poison by intraperitoneal and subcutaneous routes. An eye irritant. When heated to decomposition it emits very toxic fumes of NO_x and Cl^-.

CIE500 CAS:78218-38-9 HR: 3
6′-CHLORO-3-(METHYLAMINO)-o-PROPIONOTOLUIDIDE HYDROCHLORIDE
mf: $C_{11}H_{15}ClN_2O \cdot ClH$ mw: 263.19

SYN: C 3162

TOXICITY DATA WITH REFERENCE
eye-rbt 2% MLD ARZNAD 8,544,58
ipr-rat LD50:365 mg/kg ARZNAD 8,544,58
scu-mus LD50:700 mg/kg ARZNAD 8,544,58

SAFETY PROFILE: Poison by intraperitoneal route. Moderately toxic by subcutaneous route. An eye irritant. When heated to decomposition it emits very toxic fumes of Cl^- and NO_x.

CIF250 **CAS:1199-85-5** **HR: 3**
p-CHLORO-N-METHYLAMPHETAMINE
mf: $C_{10}H_{14}ClN$ mw: 183.70

SYNS: p-CHLORO-N-α-DIMETHYLPHENETHYLAMINE □ d-1-p-CHLORO-METHYLAMPHETAMINE (FRENCH) □ CMA □ pCMA □ RO 4-6861 □ S-33

TOXICITY DATA WITH REFERENCE
orl-rat LD50:110 mg/kg THERAP 26,219,71
scu-rat LD50:55 mg/kg AIPTAK 159,442,66
ivn-rat LD50:52 mg/kg THERAP 26,219,71
orl-mus LD50:100 mg/kg THERAP 26,219,71
ipr-mus LD50:20 mg/kg ISYAM* -,729,70
ivn-mus LD50:50 mg/kg THERAP 26,219,71

SAFETY PROFILE: Poison by ingestion, intravenous, intraperitoneal, and subcutaneous routes. When heated to decomposition it emits very toxic fumes of Cl^- and NO_x. See also BENZEDRINE and other amphetamine entries.

CIG250 **CAS:6325-54-8** **HR: 3**
7-CHLOROMETHYL BENZ(a)ANTHRACENE
mf: $C_{19}H_{13}Cl$ mw: 276.77

SYN: ICR 451

TOXICITY DATA WITH REFERENCE
mma-sat 1 µg/plate PNASA6 72,5135,75
ivn-mus TDLo:700 µg/kg:NEO CNREA8 36,2423,76
ivn-mus LDLo:1384 µg/kg CNREA8 36,2423,76

SAFETY PROFILE: Poison by intravenous route. Questionable carcinogen with experimental neoplastigenic data. Mutation data reported. When heated to decomposition it emits toxic Cl^-. See also CHLORINATED HYDROCARBONS, AROMATIC.

CIG750 **CAS:27165-08-8** **HR: 3**
4-CHLORO-2-METHYLBENZENEDIAZONIUM SALTS

SAFETY PROFILE: Reaction with sulfides (e.g., hydrogen sulfide, ammonium sulfide, sodium hydrogen sulfide, disodium sulfide, or disodium polysulfide) forms explosive products. When heated to decomposition it emits toxic fumes of Cl^- and NO_x.

CIH100 **CAS:1006-99-1** **HR: 2**
5-CHLORO-2-METHYLBENZOTHIAZOLE
mf: C_8H_6ClNS mw: 183.66

SYNS: BENZOTHIAZOLE, 5-CHLORO-2-METHYL- □ USAF EK-P-4382

TOXICITY DATA WITH REFERENCE
ipr-mus LD50:500 mg/kg NTIS** AD277-689

CONSENSUS REPORTS: Reported in EPA TSCA Inventory.

SAFETY PROFILE: Moderately toxic by intraperitoneal route. When heated to decomposition it emits toxic vapors of NO_x, SO_x, and Cl^-.

CIH900 **CAS:65313-33-9** **HR: 3**
CHLOROMETHYL BISMUTHINE
mf: C_2H_6BiCl mw: 274.50

SAFETY PROFILE: Ignites spontaneously in air. When heated to decomposition it emits toxic fumes of Cl^- and Bi. See also BISMUTH COMPOUNDS.

CII000 **CAS:107-84-6** **HR: 3**
1-CHLORO-3-METHYLBUTANE
mf: $C_5H_{11}Cl$ mw: 106.60

PROP: Liquid. Flash p: 16°, mp: −104°; bp: 99°, d: 0.8704 @ 20°/4°, lel: 1.5%, uel: 7.4%.

SAFETY PROFILE: Very dangerous fire hazard when exposed to heat, flame, or powerful oxidizers. Reaction with divalent metals may form reactive products. When heated to decomposition it emits toxic fumes of Cl^-. See also CHLORINATED HYDROCARBONS, ALIPHATIC.

CII250 **CAS:594-36-5** **HR: 3**
2-CHLORO-2-METHYLBUTANE
mf: $C_5H_{11}Cl$ mw: 106.60

PROP: Flash p: 16°, d: 0.869 @ 15°/15°, mp: −73.°, bp: 86°, lel: 1.5%, uel: 7.4%.

SAFETY PROFILE: Very dangerous fire hazard when exposed to heat, flame, or powerful oxidizers. Reaction with divalent metals may form very reactive products. When heated to decomposition it emits toxic fumes of Cl^-. See also CHLORINATED HYDROCARBONS, ALIPHATIC.

CIJ250 **CAS:20228-97-1** **HR: 3**
2-CHLORO-6-METHYLCARBANILIC ACID-2-(DIETHYLAMINO)ETHYL ESTER, HYDROCHLORIDE
mf: $C_{14}H_{21}ClN_2O_2{\cdot}ClH$ mw: 321.28

SYN: C 3069

TOXICITY DATA WITH REFERENCE
eye-rbt 2% MLD ARZNAD 8,664,58
ipr-rat LD50:55 mg/kg ARZNAD 8,664,58
scu-mus LD50:112 mg/kg ARZNAD 8,664,58

SAFETY PROFILE: Poison by intraperitoneal and subcutaneous routes. An eye irritant. See also ESTERS. When heated to decomposition it emits very toxic fumes of Cl^- and NO_x.

CIK250 **CAS:33531-34-9** **HR: 3**
2-CHLORO-6-METHYLCARBANILIC ACID-N-METHYL-4-PIPERIDINYL ESTER
mf: $C_{14}H_{19}ClN_2O_2$ mw: 282.80

TOXICITY DATA WITH REFERENCE
scu-mus LD50:63 mg/kg JMCMAR 14,710,71
ivn-mus LD50:14 mg/kg JMCMAR 14,710,71

SAFETY PROFILE: Poison by subcutaneous and intravenous routes. See also ESTERS. When heated to decomposition it emits very toxic fumes of Cl^- and NO_x.

CIK500 CAS:77944-89-9 HR: 3
2-CHLORO-6-METHYLCARBANILIC ACID-2-(PYRROLIDINYL)ETHYL ESTER HYDROCHLORIDE
mf: $C_{14}H_{19}ClN_2O_2 \cdot ClH$ mw: 319.26

SYNS: C 3067 ☐ 2-(PYRROLIDINYL)ETHYL-2-CHLORO-6-METHYLCARBANILATE HYDROCHLORIDE

TOXICITY DATA WITH REFERENCE
eye-rbt 2% MLD ARZNAD 8,664,58
ipr-rat LD50:72 mg/kg ARZNAD 8,664,58
scu-mus LD50:160 mg/kg ARZNAD 8,664,58
ivn-mus LD50:36 mg/kg ARZNAD 10,475,60

SAFETY PROFILE: Poison by intraperitoneal, subcutaneous, and intravenous routes. An eye irritant. See also ESTERS. When heated to decomposition it emits very toxic fumes of NO_x and Cl^-.

CIK750 CAS:321-54-0 HR: 3
3-CHLORO-4-METHYL-7-COUMARINYL DIETHYLPHOSPHATE
mf: $C_{14}H_{16}ClO_6P$ mw: 346.72

SYNS: COROXON ☐ COUMAPHOS-O-ANALOG ☐ COUMAPHOS OXYGEN ANALOG (USDA) ☐ O,O-DI(2-CHLOROETHYL)-7-(3-CHLORO-4-METHYLCOUMARINYL)PHOSPHATE ☐ O,O-DIETHYL-O-(3-CHLORO-4-METHYLCOUMARIN-7-YL)PHOSPHATE ☐ DIETHYL-3-CHLORO-4-METHYL-7-COUMARINYL PHOSPHATE ☐ PHOSPHORIC ACID, DIETHYL ESTER, with 3-CHLORO-7-HYDROXY-4-METHYLCOUMARIN

TOXICITY DATA WITH REFERENCE
orl-ckn LD50:2200 μg/kg BCPCA6 16,1183,67

SAFETY PROFILE: Deadly poison by ingestion. When heated to decomposition it emits very toxic fumes of PO_x and Cl^-. See also ESTERS and PHOSPHATES.

CIK825 CAS:4222-21-3 HR: 3
3-CHLORO-3-METHYLDIAZIRINE
mf: $C_2H_3ClN_2$ mw: 90.51

SAFETY PROFILE: A powerful, extremely shock-sensitive explosive. Upon decomposition it emits toxic fumes of Cl^- and NO_x. See also EXPLOSIVES.

CIL000 CAS:102129-02-2 HR: 3
o-CHLORO-2-(METHYL(2-(DIETHYLAMINO)ETHYL)AMINO)PROPIONANILIDE DIHYDROCHLORIDE
mf: $C_{16}H_{26}ClN_3O \cdot 2ClH$ mw: 389.54

SYN: C 5405

TOXICITY DATA WITH REFERENCE
ipr-rat LD50:100 mg/kg ARZNAD 9,262,59
scu-mus LD50:700 mg/kg ARZNAD 9,262,59

SAFETY PROFILE: Poison by intraperitoneal route. Moderately toxic by subcutaneous route. When heated to decomposition it emits very toxic fumes of Cl^- and NO_x.

CIL500 CAS:29053-27-8 HR: 2
7-CHLORO-2-METHYL-3,3a-DIHYDRO-2H,9H-ISOXAZOLO(3,2-b)(1,3)BENZOXAZIN-9-ONE
mf: $C_{11}H_9ClNO_3$ mw: 238.66

PROP: Crystals from EtOAc. Mp: 147–149°.

SYNS: 3,3a-DIHYDRO-7-CHLORO-2-METHYL-2H,9H-ISOXAZOLO(3,2-b)(1,3)BENZOXAZIN-9-ONE ☐ MESECLAZONE ☐ W-2395

TOXICITY DATA WITH REFERENCE
orl-rat LD50:1160 mg/kg TXAPA9 33,147,75
orl-mus LD50:2250 mg/kg TXAPA9 33,147,75

SAFETY PROFILE: Moderately toxic by ingestion. When heated to decomposition it emits very toxic fumes of NO_x and Cl^-.

CIL750 CAS:101651-65-4 HR: 3
o-CHLORO-2-(METHYL(2-(DIMETHYLAMINO)ETHYL)AMINO)ACETANILIDE DIHYDROCHLORIDE
mf: $C_{13}H_{20}ClN_3O \cdot 2ClH$ mw: 342.73

SYN: C 5415

TOXICITY DATA WITH REFERENCE
ipr-rat LD50:240 mg/kg ARZNAD 9,262,59
scu-mus LD50:1280 mg/kg ARZNAD 9,262,59

SAFETY PROFILE: Poison by intraperitoneal route. Moderately toxic by subcutaneous route. When heated to decomposition it emits very toxic fumes of Cl^- and NO_x.

CIL775 CAS:73639-62-0 HR: 2
4-(CHLOROMETHYL)-2,2-DIMETHYL-1,3-DIOXA-2-SILACYCLOPENTANE
mf: $C_5H_{11}ClO_2Si$ mw: 166.70

SYNS: 2,2-DIMETHYL-4-(CHLOROMETHYL)-1,3-DIOXA-2-SILACYCLOPENTANE ☐ 1,3-DIOXA-2-SILACYCLOPENTANE, 4-(CHLOROMETHYL)-2,2-DIMETHYL- ☐ SOC

TOXICITY DATA WITH REFERENCE
ipr-rat TDLo:525 mg/kg (male 21D pre):REP CMBID4 29,299,83
ipr-rat LD50:500 mg/kg CMBID4 29,299,83

SAFETY PROFILE: Moderately toxic by intraperitoneal route. Experimental reproductive effects. When heated to decomposition it emits toxic fumes of Cl^-.

CIL800 CAS:4362-40-7 HR: 3
4-(CHLOROMETHYL)-2,2-DIMETHYL-1,3-DIOXOLANE
mf: $C_6H_{11}ClO_2$ mw: 150.62

TOXICITY DATA WITH REFERENCE
orl-rat TDLo:140 mg/kg (14D male):REP CCPTAY 9,451,74
orl-rat TDLo:32,500 μg/kg (13D male):REP JRPFA4 30,117,72
orl-rat LD50:115 mg/kg CCPTAY 9,451,74

CONSENSUS REPORTS: Reported in EPA TSCA Inventory.

SAFETY PROFILE: Poison by ingestion. Experimental

C

reproductive effects. When heated to decomposition it emits toxic fumes of Cl^-.

CIL850 CAS:53460-80-3 HR: 3
1-(4-CHLOROMETHYL-1,3-DIOXOLAN-2-YL)-2-PROPANONE
mf: $C_7H_{11}ClO_3$ mw: 178.63

SYN: 2-PROPANONE, 1-(4-(CHLOROMETHYL)-1,3-DIOXOLAN-2-YL)-

TOXICITY DATA WITH REFERENCE
orl-rat TDLo:140 mg/kg (male 14D pre):REP CCPTAY 9,451,74
orl-rat LD50:120 mg/kg CCPTAY 9,451,74

SAFETY PROFILE: Poison by ingestion. Experimental reproductive effects. When heated to decomposition it emits toxic fumes of Cl^-.

CIL900 CAS:869-50-1 HR: 3
1-CHLOROMETHYL-1,2-ETHANEDIOL DIACETATE
mf: $C_7H_{11}ClO_4$ mw: 194.63

PROP: Bp: 145–150° @ 40 mm.

SYNS: ACETIC ACID, 3-CHLOROPROPYLENE ESTER □ CHLORODEOXYGLYCEROL DIACETATE □ α-CHLOROHYDRIN DIACETATE □ 1-CHLORO-2,3-PROPANEDIOL DIACETATE □ 1,2-DIACETOXY-3-CHLOROPROPANE □ 1,2-PROPANEDIOL, 3-CHLORO-, DIACETATE

TOXICITY DATA WITH REFERENCE
orl-rat TDLo:77,500 μg/kg (male 31D pre):REP JRPFA4 30,117,72
ipr-mus LD50:340 mg/kg JMCMAR 20,644,77

SAFETY PROFILE: Poison by intraperitoneal route. Experimental reproductive effects. When heated to decomposition it emits toxic fumes of Cl^-.

CIM000 CAS:3188-13-4 HR: 3
CHLOROMETHYL ETHYL ETHER
mf: C_3H_7ClO mw: 94.54

PROP: Flash p: <−2.2°F, bp: 83°.

SYNS: CHLOROMETHOXY ETHANE □ ETHOXY CHLOROMETHANE □ ETHOXY METHYL CHLORIDE

CONSENSUS REPORTS: Reported in EPA TSCA Inventory.

SAFETY PROFILE: A poison by inhalation and ingestion. A very dangerous fire and explosion hazard when exposed to heat or flame. See also ETHERS.

CIM300 HR: 3
3-CHLOROMETHYLFURAN
mf: C_5H_5ClO mw: 116.55

SAFETY PROFILE: A storage hazard. It may explode spontaneously at room temperature. Violent reaction with lithium aluminum hydride, ethylacetate. When heated to decomposition it emits toxic fumes of Cl^-.

CIM399 HR: 3
N-CHLORO-4-METHYL-2-IMIDAZOLINONE
mf: $C_4H_5ClN_2O$ mw: 132.55

ClNCO•NHCCH$_3$= CH

SAFETY PROFILE: Explodes spontaneously at room temperature. Upon decomposition it emits toxic fumes of Cl^- and NO_x.

CIN000 CAS:16781-80-9 HR: 3
α-(CHLOROMETHYL)-5-IODO-2-METHYL-4-NITROIMIDAZOLE-2-ETHANOL
mf: $C_7H_9ClIN_3O_3$ mw: 345.54

SYN: 1-(3-CHLORO-2-HYDROXYPROPYL)-5-IODO-2-METHYL-4-NITROIMIDAZOLE

TOXICITY DATA WITH REFERENCE
orl-mus LD50:901 mg/kg JMCMAR 17,1019,74
ipr-mus LD50:354 mg/kg JMCMAR 17,1019,74

SAFETY PROFILE: Poison by intraperitoneal route. Moderately toxic by ingestion. When heated to decomposition it emits very toxic fumes of I^-, Cl^-, and NO_x.

CIN750 CAS:13345-62-5 HR: 3
7-CHLOROMETHYL-12-METHYL BENZ(a)ANTHRACENE
mf: $C_{20}H_{15}Cl$ mw: 290.80

SYN: IRC 453

TOXICITY DATA WITH REFERENCE
mma-sat 1 μg/plate PNASA6 72,5135,75
scu-rat TDLo:150 mg/kg/39D-I:NEO CNREA8 31,1951,71
ivn-mus TDLo:1150 μg/kg:NEO CNREA8 36,2423,76
ivn-mus LDLo:2 mg/kg CNREA8 36,2423,76

CONSENSUS REPORTS: EPA Genetic Toxicology Program.

SAFETY PROFILE: Poison by intravenous route. Questionable carcinogen with experimental neoplastigenic data. Mutation data reported. When heated to decomposition it emits toxic Cl^-. See also CHLORINATED HYDROCARBONS, AROMATIC.

CIO000 CAS:2212-10-4 HR: 2
CHLOROMETHYLMETHYLDIETHOXY SILANE
mf: $C_6H_{15}ClO_2Si$ mw: 182.75

PROP: Liquid. D: 1.00 @ 20°/4°, bp: 69–70° @ 20 mm.

SYN: CHLORMETHYL-METHYL-DIETHOXYSILAN (CZECH)

TOXICITY DATA WITH REFERENCE
skn-rbt 20 mg/24H MOD 85JCAE-,1226,86
eye-rbt 100 mg/24H SEV 28ZPAK -,218,72
orl-rat LD50:1300 mg/kg 28ZPAK -,218,72
ihl-rat LCLo:2270 ppm/4H 28ZPAK -,218,72

CONSENSUS REPORTS: Reported in EPA TSCA Inventory.

SAFETY PROFILE: Moderately toxic by ingestion and inhalation. A skin and severe eye irritant. When heated to decomposition it emits toxic fumes of Cl^-.

CIO250 CAS:107-30-2 HR: 3
CHLOROMETHYL METHYL ETHER
DOT: UN 1239
mf: C_2H_5ClO mw: 80.52

$ClCH_2OCH_3$

PROP: Liquid. Flash p: <73.4°F. D: 1.070 @ 25 mm, bp: 59.5°.

SYNS: CHLORDIMETHYLETHER (CZECH) □ CMME □ DIMETHYLCHLOROETHER □ ETHER METHYLIQUE MONOCHLORE (FRENCH) □ METHYLCHLOROMETHYL ETHER (DOT) □ METHYL CHLOROMETHYL ETHER, anhydrous (DOT) □ MONOCHLORODIMETHYL ETHER (MAK) □ RCRA WASTE NUMBER U046

TOXICITY DATA WITH REFERENCE
dni-hmn:lym 5 mL/L CALEDQ 13,213,81
ihl-rat TCLo:1 ppm/6H/72W:ETA AEHLAU 30,70,75
ihl-mus TCLo:2 ppm/82D-I:ETA AEHLAU 22,663,71
scu-mus TDLo:312 mg/kg/26W-I:NEO JNCIAM 46,143,71
ihl-ham TCLo:1 ppm/6H/90D:ETA AEHLAU 30,70,75
scu-mus TD:852 mg/kg/71W-I:NEO JNCIAM 48,1431,72
scu-mus TD:125 mg/kg:ETA TXAPA9 15,92,69
ihl-rat LC50:55 ppm/7H AEHLAU 30,61,75
ihl-mus LC50:1030 mg/m³/2H 85GMAT -,89,82
ihl-ham LC50:65 ppm/7H AEHLAU 30,61,75

CONSENSUS REPORTS: NTP 7th Annual Report On Carcinogens. IARC Cancer Review: Group 1 IMEMDT 7,131,87; Animal Sufficient Evidence IMEMDT 4,239,74; Human Limited Evidence IMEMDT 4,239,74. EPA Genetic Toxicology Program. Reported in EPA TSCA Inventory. Community Right-To-Know List. EPA Extremely Hazardous Substances List.

OSHA: Cancer Suspect Agent
ACGIH TLV: Suspected Human Carcinogen.
DFG MAK: Human Carcinogen.
NIOSH REL: (Methyl Chloromethyl Ether) TWA use 29 CFR 1910.1006
DOT CLASSIFICATION: 6.1; *Label:* Poison, Flammable Liquid

SAFETY PROFILE: Confirmed human carcinogen with experimental carcinogenic, tumorigenic, and neoplastigenic data. Poison by inhalation. Moderately toxic by ingestion. Human mutation data reported. A very dangerous fire hazard when exposed to heat or flame. To fight fire, use alcohol foam, water, CO_2, or dry chemical. Reaction with divalent metals forms a very reactive product. When heated to decomposition it emits toxic fumes of Cl^-. See also ETHERS and CHLORINATED HYDROCARBONS, ALIPHATIC.

For occupational chemical analysis use OSHA: #10.

CIO275 CAS:52157-57-0 HR: 3
2-CHLOROMETHYL-5-METHYLFURAN
mf: C_6H_7ClO mw: 130.57

$OC(CH_3)=CHCH=CH_2Cl$ (ring)

SAFETY PROFILE: A storage hazard. It may explode spontaneously at room temperature. Less stable than 2-chloromethylfuran. When heated to decomposition it emits toxic fumes of Cl^-. See also CHLORINATED HYDROCARBONS, AROMATIC.

CIO500 CAS:67293-64-5 HR: 3
2-CHLORO-10-((2-METHYL-3-(4-METHYL-1-PIPERAZINYL)PROPYL)-PHENOTHIAZINE
mf: $C_{21}H_{26}ClN_3S$ mw: 388.01

SYN: 6710 RP

TOXICITY DATA WITH REFERENCE
orl-mus LD50:430 mg/kg CRSBAW 152,1371,58
ipr-mus LD50:120 mg/kg CRSBAW 152,1371,58
scu-mus LD50:420 mg/kg CRSBAW 152,1371,58
ivn-mus LD50:95 mg/kg CRSBAW 152,1371,58

SAFETY PROFILE: Poison by intraperitoneal and intravenous routes. Moderately toxic by ingestion and subcutaneous routes. When heated to decomposition it emits very toxic fumes of Cl^-, NO_x, and SO_x.

CIO750 CAS:50308-83-3 HR: 3
7-CHLORO-1-METHYL-4-((p-((1-METHYLPYRIDINIUM-4-YL)AMINO)PHENYL)CARBAMOYL)ANILINO)QUINOLINIUM DIBROMIDE
mf: $C_{29}H_{26}ClN_5O\cdot 2Br$ mw: 655.87

TOXICITY DATA WITH REFERENCE
dnd-mus:lym 1 µmol/L JMCMAR 22,134,79
ipr-mus LD10:39 mg/kg JMCMAR 22,134,79

SAFETY PROFILE: Poison by intraperitoneal route. Mutation data reported. When heated to decomposition it emits very toxic fumes of Br^-, NO_x, and Cl^-.

CIP000 CAS:50308-82-2 HR: 3
6-CHLORO-1-METHYL-4-(p-((p-((1-METHYLPYRIDINIUM-4-YL)AMINO)PHENYL)CARBAMOYL)ANILINO) QUINOLINIUM, DI-p-TOLUENESULFONATE
mf: $C_{29}H_{26}ClN_5O\cdot 2C_7H_7O_3S$ mw: 838.45

TOXICITY DATA WITH REFERENCE
dnd-mus:lym 790 nmol/L JMCMAR 22,134,79
ipr-mus LD10:40 mg/kg JMCMAR 22,134,79

SAFETY PROFILE: Poison by intraperitoneal route. Mutation data reported. When heated to decomposition it emits very toxic fumes of SO_x, NO_x and Cl^-.

CIP250 CAS:50308-84-4 HR: 3
8-CHLORO-1-METHYL-4-((p-((p-((1-METHYLPYRIDINIUM-4-YL)AMINO)PHENYL)CARBAMOYL)ANILINO) QUINOLINIUM) DI-p-TOLUENESULFONATE
mf: $C_{29}H_{26}ClN_5O\cdot 2C_7H_7O_3S$ mw: 838.45

TOXICITY DATA WITH REFERENCE
dnd-mus:lym 1100 nmol/L JMCMAR 22,134,79
ipr-mus LD10:55 mg/kg JMCMAR 22,134,79

SAFETY PROFILE: Poison by intraperitoneal route. Mutation data reported. See also SULFONATES. When heated to decomposition it emits very toxic fumes of SO_x, NO_x, and Cl^-.

CIP500 **CAS:3688-85-5** **HR: 2**
4-CHLORO-N-METHYL-3-(METHYLSULFAMOYL) BENZAMIDE
mf: $C_9H_{11}ClN_2O_3S$ mw: 262.73

PROP: Crystals from AcOH (aq). Mp: 165–166°.

SYNS: 4-CHLORO-N-METHYL-3-((METHYLAMINO)SULFONYL)BENZAMIDE ☐ C.I. 456 ☐ CN-36337 ☐ D 1593 ☐ DIAPAMIDE ☐ THIAMIZIDE ☐ TIAMIZID ☐ TIAMIZIDE ☐ VECTREN

TOXICITY DATA WITH REFERENCE
orl-rat LD50:1400 mg/kg JNDRAK 3,302,63
orl-mus LD50:2580 mg/kg JNDRAK 3,302,63
ipr-mus LD50:520 mg/kg JNDRAK 3,302,63

SAFETY PROFILE: Moderately toxic by ingestion and intraperitoneal routes. When heated to decomposition it emits very toxic SO_x, NO_x, and Cl^-.

CIP750 **CAS:86-52-2** **HR: 2**
1-CHLOROMETHYL NAPHTHALENE
mf: $C_{11}H_9Cl$ mw: 176.65

PROP: Prisms. Flash p: 270°F (OC), autoign temp: 1036°F, d: 1.19382 @ 20°/4°, mp: 31–32°, bp: 291–292°. Sol in benzene, petr ether, alc; insol in water.

SYN: α-CHLOROMETHYLNAPHTHALENE

TOXICITY DATA WITH REFERENCE
orl-rat LD50:890 mg/kg IHFCAY 6,1,67
skn-rbt LD50:2000 mg/kg IHFCAY 6,1,67

CONSENSUS REPORTS: Reported in EPA TSCA Inventory.

SAFETY PROFILE: Moderately toxic by ingestion and skin contact. See also CHLORINATED HYDROCARBONS, AROMATIC. Combustible when exposed to heat or flame. To fight fire, use dry chemical, spray or mist, CO_2. When heated to decomposition it emits toxic fumes of Cl^-.

CIQ000 **CAS:64059-42-3** **HR: 2**
2-(8-CHLOROMETHYL-1-NAPHTHYLTHIO)ACETIC ACID
mf: $C_{13}H_{11}ClO_2S$ mw: 266.75

SYN: KYSELINA-S-(8-CHLORMETHYL-1-NAFTYL)THIOGLYKOLOVA (CZECH)

TOXICITY DATA WITH REFERENCE
eye-rbt 5 mg/24H SEV 28ZPAK -,173,72
orl-rat LDLo:2500 mg/kg 28ZPAK -,173,72

SAFETY PROFILE: Moderately toxic by ingestion. A severe eye irritant. When heated to decomposition it emits very toxic fumes of Cl^- and SO_x.

CIQ400 **CAS:53460-81-4** **HR: 3**
4-(CHLOROMETHYL)-2-(o-NITROPHENYL)-1,3-DIOXOLANE
mf: $C_{10}H_{10}ClNO_4$ mw: 243.66

SYN: 1,3-DIOXOLANE, 4-(CHLOROMETHYL)-2-(o-NITROPHENYL)-

TOXICITY DATA WITH REFERENCE
orl-rat TDLo:280 mg/kg (male 14D pre):REP CCPTAY 9,451,74
orl-rat LD50:65 mg/kg CCPTAY 9,451,74

SAFETY PROFILE: Poison by ingestion. Experimental reproductive effects. When heated to decomposition it emits toxic fumes of NO_x and Cl^-.

CIQ500 **CAS:16339-16-5** **HR: 3**
2-CHLORO-N-METHYL-N-NITROSOETHYLAMINE
mf: $C_3H_7ClN_2O$ mw: 122.57

SYNS: 2-CHLORO-2-METHYL-N-NITROSOETHANAMINE ☐ METHYL-2-CHLORAETHYLNITROSAMIN (GERMAN) ☐ METHYL(2-CHLOROETHYL)NITROSAMINE ☐ N-NITROSOMETHYL-2-CHLOROETHYLAMINE

TOXICITY DATA WITH REFERENCE
orl-rat TDLo:111 mg/kg/53W-C:ETA ZEKBAI 69,103,67
orl-rat LD50:22 mg/kg ZEKBAI 69,103,67
ivn-rat LD50:22 mg/kg ZEKBAI 69,103,67
unr-mam LD50:22 mg/kg GMCRDC 17,107,75

SAFETY PROFILE: Poison by ingestion, intravenous, and possibly other routes. Questionable carcinogen with experimental tumorigenic data. Many nitrosamine compounds are carcinogens. When heated to decomposition it emits very toxic fumes of Cl^- and NO_x. See also NITROSAMINES.

CIQ625 **CAS:25480-76-6** **HR: 3**
N-CHLORO-5-METHYL-2-OXAZOLIDINONE
mf: $C_4H_6ClNO_2$ mw: 135.55

$OCO•NClCH_2 CHCH_3$ (ring)

SAFETY PROFILE: Potentially explosive above 160°C. When heated to decomposition it emits toxic fumes of Cl^- and NO_x.

CIR000 **HR: 2**
4-CHLORO-2-METHYLPHENOL
mf: C_7H_7ClO mw: 142.59

$Cl(CH_3)C_6H_3OH$

CONSENSUS REPORTS: Chlorophenol compounds are on the Community Right-To-Know List.

SAFETY PROFILE: Very exothermic reaction with concentrated sodium hydroxide; releases explosive fumes. When heated to decomposition it emits toxic fumes of Cl^-. See also CHLOROPHENOLS.

CIR250 **CAS:94-74-6** **HR: 3**
(4-CHLORO-2-METHYLPHENOXY)ACETIC ACID
mf: $C_9H_9ClO_3$ mw: 200.63

PROP: Crystals from C_6H_6 or toluene. Mp: 120°.

SYNS: AGRITOX ☐ AGROXONE ☐ ANICON KOMBI ☐ ANICON M ☐ BH MCPA ☐ BORDERMASTER ☐ BROMINAL M & PLUS ☐ B-SELEKTONON M ☐ CHIPTOX ☐ 4-CHLORO-o-CRESOXYACETIC ACID ☐ 4-CHLORO-o-TOLOXYACETIC ACID ☐ ((4-CHLORO-o-TOLYL)OXY) ACETIC ACID ☐ CHWASTOX ☐ CORNOX-M ☐ DED-WEED ☐ DICOPUR-M ☐ DICOTEX ☐ DOW MCP AMINE WEED KILLER ☐EMCEP-AN

□ EMPAL □ HEDAPUR M 52 □ HERBICIDE M □ HORMOTUHO □ 4K-2M □ KILSEM □ KREZONE □ LEGUMEX DB □ LEUNA M □ LEYSPRAY □ LINORMONE □ M 40 □ 2M-4C □ MCP □ MCPA □ MEPHANAC □ METAXON □ METHOXONE □ 2-METHYL-4-CHLOROPHENOXYACETIC ACID □ 2-METHYL-4-CHLORPHENOXYESSIGSAEURE (GERMAN) □ 2M-4KH □ NETAZOL □ OKULTIN M □ PHENOXYLENE SUPER □ RAPHONE □ RAZOL DOCK KILLER □ RHOMENE □ RHONOX □ SEPPIC MMD □ SHAMROX □ SOVIET TECHNICAL HERBICIDE 2M-4C □ TRASAN □ U 46 M-FLUID □ USTINEX □ VACATE □ VERDONE □ VESAKONTUHO MCPA □ WEEDAR MCPA CONCENTRATE □ WEEDONE MCPA ESTER □ WEED-RHAP □ ZELAN

TOXICITY DATA WITH REFERENCE
sln-dmg-orl 5 mmol/L EXPEAM 30,621,74
mmo-smc 30 μmol/L/3H MUREAV 60,291,79
dns-mus-orl 200 mg/kg MUREAV 55,197,78
hma-mus/sat 200 mg/kg ECBUDQ 27,182,78
sce-ham:ovr 10 μmol/L/1H CRNGDP 5,703,84
orl-mus TDLo:1 g/kg (6-15D preg):TER ARZNAD 33,1479,83
orl-mus TDLo:3 g/kg (6-15D preg):REP ARZNAD 33,1479,83
orl-mus TDLo:2 g/kg (6-15D preg):TER ARZNAD 33,1479,83
orl-man LDLo:814 mg/kg:CNS,CVS BMJOAE 2,629,65
orl-rat LD50:700 mg/kg AJVRAH 15,622,54
orl-mus LD50:439 mg/kg RPZHAW 31,373,80
scu-mus LDLo:28 mg/kg PCOC** -,711,66
ivn-mus LD50:28 mg/kg PCOC** -,711,66

CONSENSUS REPORTS: IARC Cancer Review: Group 2B IMEMDT 7,156,87; Animal Inadequate Evidence IMEMDT 30,255,83; Human Inadequate Evidence IMEMDT 30,255,83; Human Limited Evidence IMEMDT 41,357,86. Reported in EPA TSCA Inventory. EPA Genetic Toxicology Program.

SAFETY PROFILE: Suspected carcinogen. Poison by subcutaneous and intravenous routes. Moderately toxic by ingestion. Human systemic effects by ingestion: blood pressure decrease and coma. Experimental teratogenic and reproductive effects. Mutation data reported. An herbicide. When heated to decomposition it emits toxic fumes of Cl^-.

CIR325 CAS:16484-77-8 HR: 2
2-(4-CHLORO-2-METHYLPHENOXY)PROPANOIC ACID (R) (9CI)
mf: $C_{10}H_{11}ClO_3$ mw: 214.66

SYN: 2M-4XP

TOXICITY DATA WITH REFERENCE
orl-mus TDLo:3 g/kg (6-15D preg):TER ARZNAD 33,1479,83
orl-mus TDLo:5 g/kg (6-15D preg):REP ARZNAD 33,1479,83
orl-mus TDLo:5 g/kg (6-15D preg):TER ARZNAD 33,1479,83
orl-mus TDLo:4 g/kg (6-15D preg):TER ARZNAD 33,1479,83
orl-rat LD50:1050 mg/kg FMCHA2 -,C185,89
ipr-rat LD50:680 mg/kg ZDBEA9 23(11),83,77

SAFETY PROFILE: Moderately toxic by ingestion and intraperitoneal routes. An experimental teratogen. Experimental reproductive effects. When heated to decomposition it emits toxic fumes of Cl^-.

CIR500 CAS:93-65-2 HR: 3
4-CHLORO-2-METHYLPHENOXY-α-PROPIONIC ACID
mf: $C_{10}H_{11}ClO_3$ mw: 214.66

SYNS: ACIDE 2-(4-CHLORO-2-METHYL-PHENOXY)PROPIONIQUE (FRENCH) □ ACIDO 2-(4-CLORO-2-METIL-FENOSSI)-PROPIONICO (ITALIAN) □ BH MECOPROP □ CHIPCO TURF HERBICIDE MCPP □ 2-(4-CHLOOR-2-METHYL-FENOXY)-PROPIONZUUR (DUTCH) □ 2-(4-CHLOR-2-METHYL-PHENOXY)-PROPIONSAEURE (GERMAN) □ (+)-α-(4-CHLORO 2 METHYLPHENOXY) PROPIONIC ACID □ 2-(4-CHLORO-2-METHYLPHENOXY)PROPIONIC ACID □ 2-(4-CHLOROPHENOXY-2-METHYL)PROPIONIC ACID □ 2-(p-CHLORO-o-TOLYLOXY)PROPIONIC ACID □ CMPP □ COMPITOX □ FBC CMPP □ HEDONAL MCPP □ ISO-CORNOX □ KILPROP □ LIRANOX □ 2M-4CP □ MCPP □ 2-MCPP □ MCPP 2,4-D □ MCPP-D-4 □ MCPP-K-4 □ MECOMEC □ MECOPEOP □ MECOPER □ MECOPEX □ MECOPROP □ MECOTURF □ MECPROP □ MEPRO □ METHOXONE □ α (2 METHYL-4-CHLOROPHENOXY)PROPIONIC ACID □ 2-(2-METHYL-4-CHLOROPHENOXY)PROPIONIC ACID □ 2-METHYL 4 CHLOROPHENOXY-α-PROPIONIC ACID □ 2-(2-METHYL-4-CHLORPHENOXY)-PROPIONSAEURE (GERMAN) □ 2M 4KHP □ N.B. MECOPROP □ PROPAL □ PROPONEX-PLUS □ RANKOTEX □ RD 4593 □ RUNCATEX □ U 46 □ U 46 KV-ESTER □ U 46 KV-FLUID □ VI-PAR □ VI-PEX

TOXICITY DATA WITH REFERENCE
mrc-smc 742 ppm MUREAV 21,83,73
dns-mus-orl 100 mg/kg MUREAV 55,197,78
orl-mus TDLo:3 g/kg (female 6-15D post):TER ARZNAD 33,1479,83
orl-mus TDLo:7 g/kg (6-15D preg):REP ARZNAD 33,1479,83
orl-mus TDLo:5 g/kg (female 6-15D post):TER ARZNAD 33,1479,83
orl-mus TDLo:7 g/kg (6-15D preg):TER ARZNAD 33,1479,83
orl-rat LD50:650 mg/kg WRPCA2 9,119,70
ipr-rat LD50:402 mg/kg TXCYAC 3,349,75
orl-mus LD50:369 mg/kg RPZHAW 31,373,80
skn-rbt LD50:900 mg/kg PCOC** -,683,66

CONSENSUS REPORTS: IARC Cancer Review: Group 2B IMEMDT 7,156,87; Human Limited Evidence IMEMDT 41,357,86. EPA Genetic Toxicology Program. Reported in EPA TSCA Inventory.

SAFETY PROFILE: Suspected carcinogen. Poison by ingestion. Moderately toxic by skin contact and intraperitoneal routes. Experimental teratogenic and reproductive effects. Mutation data reported. An herbicide. When heated to decomposition it emits toxic fumes of Cl^-.

CIR750 CAS:22316-47-8 HR: 3
7-CHLORO-1-METHYL-5-PHENYL-1H-1,5-BENZODIAZEPINE-2,4(3H,5H)-DIONE
mf: $C_{16}H_{13}ClN_2O_2$ mw: 300.76

PROP: A solid. Mp: 180–182°.

SYNS: CHLOREPIN □ CLOBAZAM □ CLOREPIN □ FRISIUM □ H-4723 □ HR 376 □ LM-2717 □ 1-PHENYL 5 METHYL-8-CHLORO-1,2,4,5-TETRAHYDRO-2,4-DIOXO-3H-1,5-BENZODIAZEPINE □ RU-4723 □ URBANYL

TOXICITY DATA WITH REFERENCE
orl-rat TDLo:6500 mg/kg (female 17-21D post):REP OYYAA2 25,917,83
orl-rbt TDLo:650 mg/kg (female 7-17D post):REP OYYAA2 25,1055,83
orl-rat TDLo:3750 mg/kg (female 17-21D post):REP OYYAA2 25,917,83
orl-rbt TDLo:650 mg/kg (female 7-17D post):TER OYYAA2 25,1055,83
orl-rat TDLo:1575 mg/kg (63D male):REP OYYAA2 25,907,83
orl-rat TDLo:2750 mg/kg (7-17D preg):TER OYYAA2 25,1039,83
orl-man TDLo:104 mg/kg/1Y-C:CNS,GIT BMJOAE 282,1931,81
orl-rat LD50:6 g/kg MDACAP 16,9,80
ipr-rat LD50:740 mg/kg BCPHBM 7,33S,79
scu-rat LD50:8700 mg/kg OYYAA2 25,663,83
orl-mus LD50:580 mg/kg ARZNAD 35,133,85
ipr-mus LD50:289 mg/kg BCPHBM 7,33S,79
scu-mus LD50:2250 mg/kg BCPHBM 7,33S,79
orl-rbt LD50:320 mg/kg BCPHBM 7,33S,79
orl-gpg LD50:109 mg/kg BCPHBM 7,33S,79

SAFETY PROFILE: Poison by ingestion and intraperitoneal routes. Moderately toxic by subcutaneous route. Human systemic effects by ingestion: wakefulness, withdrawal, nausea and vomiting. An experimental teratogen. Other experimental reproductive effects. A tranquilizer. When heated to decomposition it emits very toxic fumes of NO_x and Cl^-. See also DIAZEPAM.

CIS000 CAS:562-09-4 HR: 3
2-((p-CHLORO-α-METHYL-α-PHENYLBENZYL)OXY)-N,N-DIMETHYLAMINE HYDROCHLORIDE
mf: $C_{18}H_{22}ClNO{\cdot}ClH$ mw: 340.32

PROP: Needles. Mp: 128°. Sol in H_2O.

SYNS: 2-(α-(p-CHLOROPHENYL)-α-METHYLBENZYLOXY)-N,N-DIMETHYLAMINE □ (1-(p-CHLOROPHENYL)-1-PHENYL)ETHYL (β-DIMETHYLAMINOETHYL) ETHER HYDROCHLORIDE □ 2-(1-(4-CHLOROPHENYL)-1-PHENYLETHOXY)-N,N-DIMETHYLETHANAMINE HYDROCHLORIDE □ CHLORPHENOXAMINE HYDROCHLORIDE □ CONTRISTAMINE HYDROCHLORIDE □ β-DIMETHYLAMINOETHYL (p-CHLORO-α-METHYLBENZHYDRYL) ETHER HYDROCHLORIDE □ PHENOXENE HYDROCHLORIDE □ SUBSTANZ NR. 1766 (GERMAN) □ SYSTRAL

TOXICITY DATA WITH REFERENCE
orl-rat LD50:1000 mg/kg 29ZVAB -,31,69
orl-mus LD50:345 mg/kg CLDND* 2,83,60
scu-mus LD50:159 mg/kg CLDND* 4,638,62
scu-gpg LD50:140 mg/kg ARZNAD 4,189,54

SAFETY PROFILE: Poison by ingestion and subcutaneous routes. Moderately toxic by ingestion. When heated to decomposition it emits very toxic fumes of Cl^- and NO_x.

CIS250 CAS:15545-48-9 HR: 2
N-(3-CHLORO-4-METHYLPHENYL)-N′,N′-DIMETHYLUREA
mf: $C_{10}H_{13}ClN_2O$ mw: 212.70

PROP: Crystals. Mp: 147–148°. Very sltly sol in H_2O: sol most org solvs.

SYNS: C 2242 □ 3-(3-CHLOR-4-METHYLPHENYL)-1,1-DIMETHYLHARNSTOFF (GERMAN) □ 3-(3-CHLORO-4-METHYLPHENYL)-1,1-DIMETHYL-UREA □ CHLOROTOLURON □ CHLORTOLURON □ CLORTOKEM □ DICURAN

TOXICITY DATA WITH REFERENCE
mma-sat 1 μg/plate MUREAV 58,353,78
dni-mus-orl 200 mg/kg MUREAV 58,353,78
orl-rat TDLo:2 g/kg (4D preg):REP KHZDAN 22,362,79
orl-rat LD50:5800 mg/kg KHZDAN 22,362,79
ihl-rat LC50:1300 mg/m³ 85DPAN -,-,71/76

SAFETY PROFILE: Moderately toxic by inhalation. Mildly toxic by ingestion. Experimental reproductive effects. Mutation data reported. A pesticide. When heated to decomposition it emits very toxic fumes of Cl^- and NO_x.

CIS325 CAS:65039-20-5 HR: 3
2-CHLORO-5-METHYLPHENYLHYDROXYLAMINE
mf: C_7H_8ClNO mw: 157.60

$Cl(CH_3)C_6H_3NHOH$

SAFETY PROFILE: Explodes when heated above 120°C. When heated to decomposition it emits toxic fumes of Cl^- and NO_x. See also AROMATIC AMINES.

CIS625 CAS:1631-82-9 HR: 2
CHLOROMETHYLPHENYLSILANE
mf: C_7H_9ClSi mw: 156.69

PROP: Liquid. D: 1.04 @ 20°/4°, bp: 66° @ 20 mm.

SAFETY PROFILE: Vigorous reaction above 100°C with 4-bromobutene + chloroplatinic acid. When heated to decomposition it emits toxic fumes of Cl^-.

CIS750 CAS:2058-52-8 HR: 3
2-CHLORO-11-(4-METHYLPIPERAZINO)DIBENZO(b,f)(1,4)THIAZEPINE
mf: $C_{18}H_{18}ClN_3S$ mw: 343.90

PROP: Crystals from Et_2O/pet ether. Mp: 118–120°.

SYNS: 2-CHLORO-11-(4-METHYL-1-PIPERAZINYL)DIBENZO(b,f)(1,4)THIAZEPINE □ DIBENZOTHIAZEPINE

TOXICITY DATA WITH REFERENCE
orl-rat TDLo:72 mg/kg (female 9-14D post):REP OYYAA2 4,305,70
orl-rat TDLo:6 mg/kg (9-14D preg):TER OYYAA2 4,305,70
orl-rat TDLo:6 mg/kg (9-14D preg):REP OYYAA2 4,305,70
orl-rat LD50:280 mg/kg DRUGAY 6,234,82
orl-mus LD50:270 mg/kg ARZNAD 15,841,65
orl-gpg LD50:150 mg/kg TXAPA9 14,657,69

SAFETY PROFILE: Poison by ingestion. An experimental teratogen. Other experimental reproductive effects.

When heated to decomposition it emits very toxic fumes of Cl^-, NO_x, and SO_x.

CIT000 CAS:4956-31-4 HR: 3
2-CHLORO-11-(4′-METHYL)PIPERAZINO-DIBENZO(b, f)(1,4)THIAZEPINE HYDROCHLORIDE
mf: $C_{18}H_{18}ClN_3S \cdot ClH$ mw: 380.36

SYNS: 2-CHLORO-11-(4-METHYL-1-PIPERAZINYL)DIBENZO(b,f)(1,4)THIAZEPINE HYDROCHLORIDE □ HF-2159 HYDROCHLORIDE

TOXICITY DATA WITH REFERENCE
orl-rat LD50:430 mg/kg IJNEAQ 4,375,65
ivn-rat LD50:42 mg/kg IJNEAQ 4,375,65
orl-mus LD50:270 mg/kg IJNEAQ 4,375,65
ivn-mus LD50:46 mg/kg IJNEAQ 4,375,65
orl-gpg LD50:154 mg/kg IJNEAQ 4,375,65

SAFETY PROFILE: Poison by ingestion and intravenous routes. When heated to decomposition it emits very toxic fumes of Cl^-, NO_x, and SO_x.

CIT625 CAS:59943-31-6 HR: 3
7-CHLORO-3-(4-METHYL-1-PIPERAZINYL)-4H-1,2,4-BENZOTHIADIAZINE-1,1-DIOXIDE
mf: $C_{12}H_{15}ClN_4O_2S$ mw: 314.82

SYN: DU-717

TOXICITY DATA WITH REFERENCE
ivn-rat LD50:551 mg/kg IYKEDH 11,294,80
ivn-mus LD50:242 mg/kg IYKEDH 11,294,80
ivn-dog LD50:400 mg/kg IYKEDH 11,294,80

SAFETY PROFILE: Poison by intravenous route. When heated to decomposition it emits toxic fumes of Cl^-, SO_x, and NO_x.

CIT750 CAS:77966-63-3 HR: 3
6′-CHLORO-2-(2-METHYLPIPERIDINO)-o-ACETOTOLUIDIDE HYDROCHLORIDE
mf: $C_{15}H_{21}ClN_2O \cdot ClH$ mw: 317.29

SYN: V 315

TOXICITY DATA WITH REFERENCE
eye-rbt 2% MLD ARZNAD 8,407,58
ipr-rat LD50:77 mg/kg ARZNAD 8,407,58
ipr-mus LD50:110 mg/kg ARZNAD 8,407,58
scu-mus LD50:139 mg/kg ARZNAD 8,407,58

SAFETY PROFILE: Poison by intraperitoneal and subcutaneous routes. An eye irritant. When heated to decomposition it emits very toxic fumes of $Cl^-$and NO_x.

CIU000 CAS:101651-66-5 HR: 3
o-CHLORO-2-(METHYL(2-(PIPERIDINO)ETHYL)AMINO)ACETANILIDE DIHYDROCHLORIDE
mf: $C_{16}H_{24}ClN_3O \cdot 2ClH$ mw: 382.80

SYN: C 5106

TOXICITY DATA WITH REFERENCE
ipr-rat LD50:124 mg/kg ARZNAD 9,262,59
scu-mus LD50:590 mg/kg ARZNAD 9,262,59

SAFETY PROFILE: Poison by intraperitoneal route. Moderately toxic by subcutaneous route. When heated to decomposition it emits very toxic fumes of Cl^- and NO_x.

CIU250 CAS:78218-42-5 HR: 3
6′-CHLORO-3-(2-METHYLPIPERIDINO)-o-PROPIONOTOLUIDIDE HYDROCHLORIDE
mf: $C_{16}H_{23}ClN_2O \cdot ClH$ mw: 331.32

SYN: C 3139

TOXICITY DATA WITH REFERENCE
ipr-rat LD50:39 mg/kg ARZNAD 8,544,58
scu-mus LD50:36 mg/kg ARZNAD 8,544,58

SAFETY PROFILE: Poison by intraperitoneal and subcutaneous routes. When heated to decomposition it emits very toxic fumes of Cl^- and NO_x.

CIU325 CAS:58763-27-2 HR: 3
3′-CHLORO-5′-METHYL-3-PIPERIDINO-4′-PROPOXYPROPIOPHENONE HYDROCHLORIDE
mf: $C_{18}H_{26}ClNO_2 \cdot ClH$ mw: 360.36

SYNS: 1-(3-CHLORO-5-METHYL-4-PROPOXYPHENYL)-3-(1-PIPERIDINYL)1-PROPANONE HYDROCHLORIDE (9CI) □ β-PIPERIDINOAETHYL-(3-CHLOR-4-PROPOXY-5-METHYLPHENYL)-KETONHYDROCHLORID (GERMAN)

TOXICITY DATA WITH REFERENCE
orl-mus LD50:460 mg/kg PHARAT 31,21,76
scu-mus LD50:330 mg/kg PHARAT 31,21,76
ivn-mus LD50:50 mg/kg PHARAT 31,21,76

SAFETY PROFILE: Poison by subcutaneous and intravenous routes. Moderately toxic by ingestion. When heated to decomposition it emits toxic fumes of NO_x and Cl^-.

CIU500 CAS:513-36-0 HR: 3
1-CHLORO-2-METHYLPROPANE
mf: C_4H_9Cl mw: 92.57

PROP: Liquid. Flash p: 21.2°F, lel: 2.0%, uel: 8.7%, d: 0.881 @ 20°/4°, mp: −130.3°, bp: 68.8°.

SYN: ISOBUTYL CHLORIDE

SAFETY PROFILE: A poison by ingestion and inhalation. A very dangerous fire and explosion hazard when exposed to heat or flame. When heated to decomposition it emits toxic fumes of Cl^-. See also CHLORINATED HYDROCARBONS, ALIPHATIC.

CIU750 CAS:563-47-3 HR: 3
3-CHLORO-2-METHYLPROPENE
DOT: UN 2554
mf: C_4H_7Cl mw: 90.56

$H_2C{=}C(CH_3)CH_2Cl$

PROP: Colorless, volatile liquid, disagreeable odor. Bp: 72.17°, lel: 2.3%, uel: 9.3%, fp: <−80°, d: 0.9257 @ 20°/4°, vap press: 101.7 mm @ 20°, vap d: 3.12, flash p: −10°. Misc in alc and ether.

SYNS: 3-CHLOR-2-METHYL-PROP-1-EN (GERMAN) □ γ-CHLOROI-

SOBUTYLENE ☐ 3-CHLORO-2-METHYL-1-PROPENE ☐ CHLORURE de METHALLYLE (FRENCH) ☐ 3-CLORO-2-METIL-PROP-1-ENE (ITALIAN) ☐ CLORURO di METALLILE (ITALIAN) ☐ ISOBUTENYL CHLORIDE ☐ METHALLYL CHLORIDE ☐ α-METHALLYL CHLORIDE ☐ 2-METHYL-ALLYLCHLORID (GERMAN) ☐ β-METHYLALLYL CHLORIDE ☐ 2-METHYLALLYL CHLORIDE ☐ METHYL ALLYL CHLORIDE (DOT) ☐ NCI-C54820

TOXICITY DATA WITH REFERENCE
mmo-sat 6 μmol/plate BCPCA6 29,2611,80
dns-hmn:hla 1 mmol/L CALEDQ 20,263,83
orl-rat TDLo:77,250 mg/kg/2Y-I:CAR NTPTR* NTP-TR-300,86
orl-rat TDLo:38,625 mg/kg/2Y-I:ETA PAACA3 26,95,85
orl-mus TDLo:51,500 mg/kg/2Y-I:CAR NTPTR* NTP-TR-300,86
orl-mus TDLo:51,500 mg/kg/2Y-I:ETA PAACA3 26,95,85
orl-rat TD:77,250 mg/kg/2Y-I:NEO,REP NTPTR* NTP-TR-300,86

CONSENSUS REPORTS: NTP 7th Annual Report On Carcinogens. NTP Carcinogenesis Studies (gavage); Clear Evidence: mouse, rat NTPTR* NTP-TR-300,86. Reported in EPA TSCA Inventory.

DFG MAK: Suspected Carcinogen.
DOT CLASSIFICATION: 3; *Label:* Flammable Liquid

SAFETY PROFILE: Confirmed carcinogen with experimental carcinogenic, neoplastigenic, and tumorigenic data. Experimental reproductive effects. An irritant. Human mutation data reported. Very dangerous fire hazard when exposed to heat, flame, or oxidizers. Moderately explosive when exposed to heat or flame. Can react vigorously with oxidizing materials. To fight fire, use alcohol foam, CO_2, dry chemical. When heated to decomposition it emits toxic fumes of Cl^-. See also CHLORINATED HYDROCARBONS, ALIPHATIC; and ALLYL COMPOUNDS.

CIV000 CAS:6959-48-4 HR: 3
3-(CHLOROMETHYL) PYRIDINE HYDROCHLORIDE
mf: $C_6H_6ClN•ClH$ mw: 164.04

SYN: NCI-C03838

TOXICITY DATA WITH REFERENCE
mmo-sat 333 μg/plate IARCCD 27,283,80
mma-sat 33,300 ng/plate ENMUDM 7(Suppl 5),1,85
mma-esc 1 mg/plate ENMUDM 7(Suppl 5),1,85
otr-rat:emb 640 ng/plate JJATDK 1,190,81
orl-rat TDLo:37 g/kg/83W-I:CAR NCITR* NCI-CG-TR-95,78
orl-mus TDLo:49 g/kg/81W-I:CAR NCITR* NCI-CG-TR-95,78
orl-rat LD50:316 mg/kg NCILB* NIH-NCI-E-C-72-3252
orl-mus LD50:316 mg/kg NCILB* NIH-NCI-E-C-72-3252

CONSENSUS REPORTS: NCI Carcinogenesis Bioassay (gavage); Clear Evidence: mouse, rat NCITR* NCI-CG-TR-95,78. EPA Genetic Toxicology Program.

SAFETY PROFILE: Suspected carcinogen with experimental carcinogenic data. Poison by ingestion. Mutation data reported. When heated to decomposition it emits very toxic fumes of NO_x and Cl^-.

CIW250 HR: 3
5′-CHLORO-2-(METHYL(2-(PYRROLIDINYL)ETHYL) AMINO)-O-ACETOTOLUIDIDE DIHYDROCHLORIDE
mf: $C_{16}H_{24}ClN_3O•2ClH$ mw: 382.80

SYN: C 5420

TOXICITY DATA WITH REFERENCE
eye-rbt 2% MLD ARZNAD 9,262,59
ipr-rat LD50:118 mg/kg ARZNAD 9,262,59
scu-mus LD50:385 mg/kg ARZNAD 9,262,59

SAFETY PROFILE: Poison by intraperitoneal and subcutaneous routes. An eye irritant. When heated to decomposition it emits very toxic fumes of $Cl^-$and NO_x.

CIX000 CAS:102129-03-3 HR: 3
o-CHLORO-2-(METHYL(2-(PYRROLIDINYL)ETHYL) AMINO)PROPIONANILIDE DIHYDROCHLORIDE
mf: $C_{16}H_{24}ClN_3O•2ClH$ mw: 382.80

SYN: C 5407

TOXICITY DATA WITH REFERENCE
eye-rbt 2% MLD ARZNAD 9,262,59
ipr-rat LD50:112 mg/kg ARZNAD 9,262,59
scu-mus LD50:855 mg/kg ARZNAD 9,262,59

SAFETY PROFILE: Poison by intraperitoneal route. Moderately toxic by subcutaneous route. An eye irritant. When heated to decomposition it emits very toxic fumes of NO_x and Cl^-.

CIX200 CAS:13244-35-4 HR: 2
2-CHLORO-4-(METHYLSULFONYL)ANILINE
mf: $C_7H_8ClNO_2S$ mw: 205.67

SYNS: 4-AMINO-3-CHLOROFENYLOMETHYLOSULFON ☐ 4-AMINO-3-CHLOROPHENYLMETHYLSULFONE ☐ 4-AMINO-3-CHLOROPHENYL-METHYLSULPHONE ☐ ANILINE, 2-CHLORO-4-(METHYLSULFONYL)- ☐ BENZENAMINE, 2-CHLORO-4-(METHYLSULFONYL)-

TOXICITY DATA WITH REFERENCE
orl-rat LD50:9500 mg/kg BCTKAG 13,107,80
ipr-rat LDLo:1800 mg/kg BCTKAG 13,107,80

CONSENSUS REPORTS: Reported in EPA TSCA Inventory.

SAFETY PROFILE: Moderately toxic by intraperitoneal route. When heated to decomposition it emits toxic vapors of NO_x, SO_x, and Cl^-.

CIY250 CAS:617-88-9 HR: 3
2-CHLOROMETHYLTHIOPHENE
mf: C_5H_5ClS mw: 132.61

SCH=CHCH=CCH$_2$Cl (ring closure from S to C)

PROP: D: 1.178 @ 20°/4°.

SAFETY PROFILE: Flammable when exposed to heat or flame. A storage hazard. It decomposes at room temperature to release hydrogen chloride, and may explode in a sealed container. Highly explosive when shocked, exposed to heat, or by spontaneous chemical reaction.

Can react vigorously with oxidizing materials. See also THIOPHENE.

CIY325 CAS:1558-25-4 HR: 3
(CHLOROMETHYL)TRICHLOROSILANE
mf: CH_2Cl_4Si mw: 183.92

PROP: Liquid. D 1.465 @ 20°/4°, mp: 111–112° mm.

SYNS: CHLOROMETHYL(TRICHLORO)SILANE □ TRICHLORO (CHLOROMETHYL)SILANE (9CI)

TOXICITY DATA WITH REFERENCE
orl-mus LDLo: 100 mg/kg 85GMAT -,37,82
ihl-mus LD50: 30 mg/m³/2H 85GMAT -,37,82
skn-mus LDLo: 100 mg/kg 85GMAT -,37,82
ipr-mus LDLo: 100 mg/kg 85GMAT -,37,82

SAFETY PROFILE: Poison by inhalation, skin contact, ingestion, and intraperitoneal routes. When heated to decomposition it emits toxic fumes of Cl^-.

CIY500 CAS:15267-95-5 HR: 2
(CHLOROMETHYL)TRIETHOXYSILANE
mf: $C_7H_{17}ClO_3Si$ mw: 212.78

PROP: Liquid. Bp: 91–93° @ 27 mm.

SYN: CHLORMETHYL TRIETHOXYSILAN (CZECH)

TOXICITY DATA WITH REFERENCE
skn-rbt 500 mg/24H MOD 28ZPAK -,219,72
eye-rbt 500 mg/24H MOD 28ZPAK -,219,72
orl-rat LD50: 2400 mg/kg 28ZPAK -,219,72

CONSENSUS REPORTS: Reported in EPA TSCA Inventory.

SAFETY PROFILE: Moderately toxic by ingestion. A skin and eye irritant. When heated to decomposition it emits toxic fumes of Cl^-.

CIY899 HR: 3
N-CHLORO-3-MORPHOLINONE
mf: $C_4H_6ClNO_2$ mw: 135.55

$ClNCH_2CO{\cdot}OCH_2CH_2$ (ring)

SAFETY PROFILE: May explode when heated to 115°C. When heated to decomposition it emits toxic fumes of Cl^- and NO_x.

CIZ000 CAS:90-13-1 HR: 2
1-CHLORONAPHTHALENE
mf: $C_{10}H_7Cl$ mw: 162.62

PROP: Oil. D: 1.194 @ 20°/4°, mp: −20°, bp: 263°, flash p: 121°. Sol in pet ether and EtOH.

SYNS: α-CHLORNAPHTHALENE □ α-CHLORONAPHTHALENE

TOXICITY DATA WITH REFERENCE
mmo-sat 200 μg/plate EMMUEG 19(Suppl 21),2,92
orl-rat LD50: 1540 mg/kg NTIS** PB214-270
orl-mus LD50: 1091 mg/kg NTIS** PB214-270
orl-gpg LD50: 2000 mg/kg GISAAA 47(11),78,82

CONSENSUS REPORTS: Reported in EPA TSCA Inventory.

SAFETY PROFILE: Moderately toxic by ingestion. Mutation data reported. When heated to decomposition it emits toxic fumes of Cl^-. See also CHLORINATED HYDROCARBONS, AROMATIC.

CJA000 CAS:91-58-7 HR: 2
2-CHLORONAPHTHALENE
mf: $C_{10}H_7Cl$ mw: 162.62

PROP: Leaflets from EtOH (aq). D: 1.377 @ 71°, mp: 61°, bp: 256°. Insol in water; sol in alc, benzene, chloroform, ether, CS_2.

SYNS: β-CHLORONAPHTHALENE □ RCRA WASTE NUMBER U047

TOXICITY DATA WITH REFERENCE
orl-rat LD50: 2078 mg/kg NTIS** PB214-270
orl-mus LD50: 886 mg/kg NTIS** PB214-270

CONSENSUS REPORTS: Reported in EPA TSCA Inventory.

SAFETY PROFILE: Moderately toxic by ingestion. When heated to decomposition it emits toxic fumes of Cl^-. See also CHLORINATED HYDROCARBONS, AROMATIC.

CJA100 CAS:2675-77-6 HR: 1
CHLORONEB
mf: $C_8H_8Cl_2O_2$ mw: 207.06

PROP: A solid. Mp: 134°, bp: 262° @ 744 mm.

SYNS: CHLORONEBE (FRENCH) □ DEMOSAN □ 1,4-DICHLORO-2, 5-DIMETHOXYBENZENE □ SOIL FUNGICIDE 1823 □ TERSAN SP

TOXICITY DATA WITH REFERENCE
mmo-asn 24 μmol/L PHYTAJ 66,217,76
sln-asn 48 μmol/L EVHPAZ 31,81,79
orl-rat LD50: 11 g/kg 85ARAE 4,82,76/77

CONSENSUS REPORTS: EPA Genetic Toxicology Program.

SAFETY PROFILE: Very mildly toxic by ingestion. Mutation data reported. When heated to decomposition it emits toxic fumes of Cl^-.

CJA150 CAS:59483-61-3 HR: 3
N-CHLORO-4-NITROANILINE
mf: $C_6H_5ClN_2O_2$ mw: 172.57

SAFETY PROFILE: Decomposes explosively at room temperature. Upon decomposition it emits toxic fumes of Cl^- and NO_x. See also ANILINE DYES and NITRO COMPOUNDS of AROMATIC HYDROCARBONS.

CJA175 CAS:121-87-9 HR: 3
2-CHLORO-4-NITROANILINE
mf: $C_6H_5ClN_2O_2$ mw: 172.58

PROP: Yellow needles from water. Mp: 108°. Sltly sol in water; very sol in alc, ether.

SYNS: 1-AMINO-2-CHLORO-4-NITROBENZENE □ o-CHLORO-p-NITROANILINE □ 4-NITRO-2-CHLOROANILINE □ OCPNA

TOXICITY DATA WITH REFERENCE
mmo-sat 1 mg/plate SAIGBL 29,34,87
orl-rat LD50:6430 mg/kg GTPZAB 25(8),50,81
orl-mus LD50:1250 mg/kg GTPZAB 25(8),50,81
ipr-mus LDLo:500 mg/kg CBCCT* 5,337,53
ivn-mus LDLo:50 mg/kg CBCCT* 6,138,54

CONSENSUS REPORTS: Reported in EPA TSCA Inventory.

SAFETY PROFILE: Poison by intravenous route. Moderately toxic by ingestion and intraperitoneal routes. Mutation data reported. When heated to decomposition it emits toxic fumes of Cl^- and NO_x. See also ANILINE DYES.

CJA180 CAS:6283-25-6 HR: 3
2-CHLORO-5-NITROANILINE
mf: $C_6H_5ClN_2O_2$ mw: 172.58

SYN: ANILINE, 2-CHLORO-5-NITRO-

TOXICITY DATA WITH REFERENCE
orl-rat LD50:2015 mg/kg TSCAT* OTS 206512
ipr-rat LD50:400 mg/kg TSCAT* OTS 206512
orl-mus LD50:1600 mg/kg TSCAT* OTS 206512
ipr-mus LD50:800 mg/kg TSCAT* OTS 206512
ivn-mus LD50:180 mg/kg CSLNX* NX#03227

CONSENSUS REPORTS: Reported in EPA TSCA Inventory.

SAFETY PROFILE: A poison by intravenous and intraperitoneal routes. Moderately toxic by ingestion. When heated to decomposition it emits toxic vapors of NO_x and Cl^-.

CJA185 CAS:635-22-3 HR: 3
4-CHLORO-3-NITROANILINE
mf: $C_6H_5ClN_2O_2$ mw: 172.58

PROP: Yellow needles from H_2O. Mp: 103°.

SYN: ANILINE, 4-CHLORO-3-NITRO-

TOXICITY DATA WITH REFERENCE
eye-rbt 100 mg MLD EPASR* 8EHQ-0882-0452
skn-gpg 500 mg/24H MLD EPASR* 8EHQ-0882-0452
orl-rat TDLo:5850 mg/kg (male 90D pre):REP FAATDF 6,551,86
orl-rat LD50:400 mg/kg EPASR* 8EHQ-0882-0452
ipr-rat LD50:200 mg/kg EPASR* 8EHQ-0882-0452
orl-mus LD50:800 mg/kg EPASR* 8EHQ-0882-0452
ipr-mus LD50:200 mg/kg EPASR* 8EHQ-0882-0452
orl-brd LD50:100 mg/kg TXAPA9 21,315,72

CONSENSUS REPORTS: Reported in EPA TSCA Inventory.

SAFETY PROFILE: Poison by ingestion and intraperitoneal routes. Experimental reproductive effects. Skin and eye irritant. When heated to decomposition it emits toxic fumes of NO_x and Cl^-.

CJA200 CAS:4920-79-0 HR: 2
2-CHLORO-4-NITRO-ANISOLE
mf: $C_7H_6ClNO_3$ mw: 187.59

PROP: Needles or prisms from MeOH. Mp: 98°.

SYNS: 2-CHLORO-1-METHOXY-4-NITROBENZENE (9CI) □ CHLORONITROANISOLE □ OCNA □ o-CHLORO-p-NITROANISOLE

TOXICITY DATA WITH REFERENCE
skn-rbt 500 mg/24H MLD TOERD9 2,77,79
eye-rbt 100 mg MLD TOERD9 2,77,79
orl-rat LD50:1180 mg/kg TOERD9 2,77,79
ipr-rat TDLo:445 mg/kg TOERD9 2,77,79
orl-mus LD50:1550 mg/kg TOERD9 2,77,79
ipr-mus LD50:815 mg/kg TOERD9 2,77,79

SAFETY PROFILE: Moderately toxic by ingestion and intraperitoneal routes. An eye and skin irritant. When heated to decomposition it emits toxic fumes of Cl^- and NO_x. See also NITRO COMPOUNDS of AROMATIC HYDROCARBONS.

CJA250 CAS:129-40-8 HR: 1
1-CHLORO-5-NITROANTHRAQUINONE
mf: $C_{14}H_6ClNO_4$ mw: 287.66

PROP: Yellow felted needles from nitrobenzene. Mp: 314°.

SYNS: 1-CHLOR-5-NITROANTHRACHINON (CZECH) □ 1-CHLORO-5-NITRO-9,10-ANTHRACENEDIONE □ 5-CHLORO-1-NITROANTHRAQUINONE □ 1-NITRO-5-CHLOROANTHRAQUINONE

TOXICITY DATA WITH REFERENCE
eye-rbt 500 mg/24H MLD 28ZPAK -,86,72

CONSENSUS REPORTS: Reported in EPA TSCA Inventory.

SAFETY PROFILE: An eye irritant. When heated to decomposition it emits very toxic fumes of Cl^- and NO_x. See also NITRO COMPOUNDS of AROMATIC HYDROCARBONS.

CJA950 CAS:25167-93-5 HR: 3
CHLORONITROBENZENE
DOT: UN 1578
mf: $C_6H_4ClNO_2$ mw: 157.56

SYNS: CHLORONITROBENZENE, ortho, liquid (DOT) □ MONONITROCHLOROBENZENE □ NITROCHLOROBENZENE

TOXICITY DATA WITH REFERENCE
dnd-mus-ipr 60 mg/kg ARTODN (5),355,82

DOT CLASSIFICATION: 6.1; *Label:* Poison

SAFETY PROFILE: A poison. Mutation data reported. When heated to decomposition it emits toxic fumes of Cl^- and NO_x. See also other chloronitrobenzene entries and NITRO COMPOUNDS of AROMATIC HYDROCARBONS.

CJB250 **CAS:121-73-3** **HR: 3**
1-CHLORO-3-NITROBENZENE
DOT: UN 1578
mf: $C_6H_4ClNO_2$ mw: 157.56

PROP: Pale-yellow crystals or prisms. Mp: 46°, flash p: 103°, bp: 236°, d: 1.534 @ 20°/4°.

SYNS: CHLORO-m-NITROBENZENE □ m-CHLORONITROBENZENE □ m-CHLORONITROBENZENE (DOT) □ m-NITROCHLOROBENZENE □ m-NITROCHLOROBENZENE, solid (DOT)

TOXICITY DATA WITH REFERENCE
orl-rat LD50:420 mg/kg 85GMAT-,92,82
orl-mus LD50:380 mg/kg GTPZAB 25(8),50,81

CONSENSUS REPORTS: Reported in EPA TSCA Inventory.

DOT CLASSIFICATION: 6.1; *Label:* Poison

SAFETY PROFILE: Poison by ingestion and inhalation. It forms methemoglobin in the body and gives rise to cyanosis and blood changes. Its effects are cumulative and analogous to those of nitrobenzene. The para compound is thought to be somewhat less toxic than the ortho compound. Chemically, it is probably converted in the body to chloroaniline, which is also poisonous. In industry, it is the dust of this material that is most often the source of intoxication. Flammable liquid and dangerous fire hazard when exposed to heat or flame. It can react with oxidizing materials. When heated to decomposition it emits toxic fumes of Cl^-, NO_x, and phosgene. See also other chloronitrobenzene entries and NITRO COMPOUNDS of AROMATIC HYDROCARBONS.

CJB750 **CAS:88-73-3** **HR: 3**
CHLORO-o-NITROBENZENE
DOT: UN 1578
mf: $C_6H_4ClNO_2$ mw: 157.56

PROP: Yellow crystals or needles. Mp: 32–33°, bp: 245–246°, d: 1.348, flash p: 123°.

SYNS: o-CHLORONITROBENZENE □ o-CHLORONITROBENZENE, liquid (DOT) □ 1-CHLORO-2-NITROBENZENE □ 2-CHLORONITROBENZENE □ 2-CHLORO-1-NITROBENZENE □ o-NITROCHLOROBENZENE □ ONCB

TOXICITY DATA WITH REFERENCE
mmo-sat 205 μg/plate MUREAV 116,217,83
mma-sat 100 μg/plate ENMUDM 5(Suppl 1),3,83
orl-rat TDLo:22 g/kg/78W-C:NEO JEPTDQ 2(2),325,78
orl-mus TDLo:140 g/kg/78W-C:CAR JEPTDQ 2(2),325,78
orl-mus TD:280 g/kg/78W-C:CAR JEPTDQ 2(2),325,78
orl-rat LD50:288 mg/kg NTIS** PB214-270
orl-mus LD50:135 mg/kg NTIS** PB214-270
orl-rbt LD50:280 mg/kg 85GMAT -,92,82
skn-rbt LD50:400 mg/kg FAATDF 7,609,86

CONSENSUS REPORTS: Reported in EPA TSCA Inventory.

DFG MAK: Suspected Carcinogen
DOT CLASSIFICATION: 6.1; *Label:* Poison

SAFETY PROFILE: Suspected carcinogen with experimental carcinogenic and neoplastigenic data. Poison by ingestion, skin contact, and probably inhalation. Combustible when exposed to heat or flame. To fight fire, use water, foam. Potentially explosive reaction with ammonia at 160°C/30 bar. When heated to decomposition it emits toxic fumes of Cl^-, NO_x, and phosgene. See also other chloronitrobenzene entries and NITRO COMPOUNDS of AROMATIC HYDROCARBONS.

CJB825 **CAS:96-73-1** **HR: 3**
2-CHLORO-5-NITROBENZENESULFONIC ACID
mf: $C_6H_4ClNO_5S$ mw: 237.61

$Cl(NO_2)C_6H_3SO_2OH$

PROP: Crystals. Mp: 168–169° (decomp).

SAFETY PROFILE: Decomposes violently at 150°C. When heated to decomposition it emits toxic fumes of Cl^-, SO_x and NO_x. See also SULFONATES and NITRO COMPOUNDS of AROMATIC HYDROCARBONS.

CJC000 **CAS:4515-30-4** **HR: 2**
2-CHLORO-3,5-NITROBENZENESULFONIC ACID, SODIUM SALT
mf: $C_6H_2ClN_2O_7S \cdot Na$ mw: 304.60

SYN: 2,4-DINITROCHLORBENZEN-6-SULFONAN SODNY (CZECH)

TOXICITY DATA WITH REFERENCE
skn-rbt 500 mg/24H MLD 28ZPAK -,181,72
eye-rbt 20 mg/24H MOD 28ZPAK -,181,72
orl-rat LD50:1640 mg/kg 28ZPAK -,181,72

SAFETY PROFILE: Moderately toxic by ingestion. A skin and eye irritant. When heated to decomposition it emits very toxic fumes of Cl^-, Na_2O, NO_x, and SO_x. See also SULFONATES.

CJC250 **CAS:99-60-5** **HR: 2**
2-CHLORO-4-NITROBENZOIC ACID
mf: $C_7H_4ClNO_4$ mw: 201.57

PROP: Crystals from water. Mp: 142–143°. Sol in hot water and hot benzene.

SYN: KYSELINA-2-CHLORO-4-NITROBENZOOVA (CZECH)

TOXICITY DATA WITH REFERENCE
eye-rbt 250 μg/24H SEV 28ZPAK -,91,72

CONSENSUS REPORTS: Reported in EPA TSCA Inventory.

SAFETY PROFILE: A severe eye irritant. When heated to decomposition it emits very toxic fumes of Cl^- and NO_x.

CJC500 **CAS:96-99-1** **HR: 3**
4-CHLORO-3-NITROBENZOIC ACID
mf: $C_7H_4ClNO_4$ mw: 201.57

PROP: Crystals from alc. D: 1.645 @ 18°, mp: 183°. Sltly sol in alc; sol in hot water.

SYN: KYSELINA 4-CHLORO-3-NITROBENZOOVA (CZECH)

TOXICITY DATA WITH REFERENCE
mmo-sat 500 μg/plate SAIGBL 29,34,87
orl-rat LD50:3150 mg/kg MarJV# 29MAR77
orl-bwd LD50:75 mg/kg TXAPA9 21,315,72

CONSENSUS REPORTS: Reported in EPA TSCA Inventory.

SAFETY PROFILE: Poison by ingestion. Mutation data reported. When heated to decomposition it emits very toxic fumes of NO_x and Cl^-.

CJC515 **CAS:6307-82-0** **HR: 1**
2-CHLORO-5-NITROBENZOIC ACID METHYL ESTER
mf: $C_8H_6ClNO_4$ mw: 215.60

PROP: Crystals from MeOH. Mp: 73°.

SYNS: BENZOIC ACID, 2-CHLORO-5-NITRO-, METHYL ESTER □ METHYLESTER KYSELINY 2-CHLOR-5-NITROBENZOOVE

TOXICITY DATA WITH REFERENCE
skn-rbt 500 mg/24H MLD 85JCAE-,591,86
eye-rbt 500 mg/24H MLD 85JCAE-,591,86
orl-rat LD50:5360 mg/kg 85JCAE-,591,86

SAFETY PROFILE: Mildly toxic by ingestion. A skin and eye irritant. When heated to decomposition it emits toxic fumes of NO_x and Cl^-.

CJC549 **HR: 3**
2-CHLORO-5-NITROBENZYL ALCOHOL
mf: $C_7H_6ClNO_3$ mw: 187.58

$O_2N(Cl)C_6H_3CH_2OH$

PROP: Mp: 79°.

SAFETY PROFILE: Decomposes exothermically at 211°C. When heated to decomposition it emits toxic fumes of Cl^- and NO_x. See also NITRO COMPOUNDS of AROMATIC HYDROCARBONS.

CJC600 **CAS:56433-01-3** **HR: 3**
6-CHLORO-2-NITROBENZYL BROMIDE
mf: $C_7H_5BrClNO_2$ mw: 250.48

SAFETY PROFILE: Potentially dangerous exothermic reaction when heated above 190°C in a sealed container. When heated to decomposition it emits toxic fumes of Cl^-, Br^-, and NO_x. See also NITRO COMPOUNDS of AROMATIC HYDROCARBONS and BROMIDES.

CJC610 **CAS:50274-95-8** **HR: 3**
2-CHLORO-4-NITROBENZYL CHLORIDE
mf: $C_7H_5Cl_2NO_2$ mw: 206.03

SAFETY PROFILE: Potentially dangerous exothermic reaction when heated above 190°C in a sealed container. When heated to decomposition it emits toxic fumes of Cl^- and NO_x. See also NITRO COMPOUNDS of AROMATIC HYDROCARBONS.

CJC625 **CAS:938-71-6** **HR: 3**
4-CHLORO-2-NITROBENZYL CHLORIDE
mf: $C_7H_5Cl_2NO_2$ mw: 206.03

SAFETY PROFILE: Potentially dangerous exothermic reaction when heated above 190°C in a sealed container. When heated to decomposition it emits toxic fumes of Cl^- and NO_x. See also NITRO COMPOUNDS of AROMATIC HYDROCARBONS.

CJC750 **CAS:22236-53-9** **HR: 3**
2-CHLORO-2-NITROBUTANE
mf: $C_4H_8ClNO_2$ mw: 137.58

TOXICITY DATA WITH REFERENCE
orl-mus LD50:970 mg/kg HYSAAV 34(10-12),429,69
ihl-mus LC50:135 g/m³/3H HYSAAV 34(10-12),429,69
scu-mus LD50:250 mg/kg HYSAAV 30,169,65

SAFETY PROFILE: Poison by subcutaneous route. Moderately toxic by ingestion. Very mildly toxic by inhalation. When heated to decomposition it emits very toxic fumes of Cl^- and NO_x.

CJC800 **CAS:598-92-5** **HR: 3**
1-CHLORO-1-NITROETHANE
mf: $C_2H_4ClNO_2$ mw: 109.52

PROP: Oil. D: 1.258 @ 20°/20°, bp: 125°, flash p: 56°. Sol in alkalis; insol in H_2O.

TOXICITY DATA WITH REFERENCE
orl-mus LD50:620 mg/kg HYSAAV 34(10-12),429,69
ihl-mus LC50:21 g/m³/3H HYSAAV 34(10-12),429,69
ipr-mus LD50:560 mg/kg KHFZAN 10(6),53,76
scu-mus LD50:185 mg/kg HYSAAV 34(10-12),429,69
orl-rbt LDLo:100 mg/kg JIHTAB 22,315,40

CONSENSUS REPORTS: Reported in EPA TSCA Inventory.

SAFETY PROFILE: Poison by ingestion and subcutaneous route. Moderately toxic by intraperitoneal route. Very mildly toxic by inhalation. When heated to decomposition it emits toxic fumes of Cl^- and NO_x.

CJD250 **CAS:619-08-9** **HR: 3**
2-CHLORO-4-NITROPHENOL
mf: $C_6H_4ClNO_3$ mw: 173.56

PROP: Needles from water. Mp: 111°.

TOXICITY DATA WITH REFERENCE
orl-rat LD50:900 mg/kg 34ZIAG -,169,69
ipr-mus LD50:80 mg/kg JMCMAR 18,868,75

CONSENSUS REPORTS: Reported in EPA TSCA Inventory. Chlorophenol compounds are on the Community Right-To-Know List.

SAFETY PROFILE: Poison by intraperitoneal route. Moderately toxic by ingestion. When heated to decomposition it emits very toxic fumes of Cl^- and NO_x. See also CHLOROPHENOLS and NITRO COMPOUNDS of AROMATIC HYDROCARBONS.

CJD600 **CAS:135-12-6** **HR: 1**
4-CHLORO-2-NITROPHENYL p-CHLOROPHENYL ETHER
mf: $C_{12}H_7Cl_2NO_3$ mw: 284.10

SYNS: BENZENE, 4-CHLORO-1-(4-CHLOROPHENOXY)-2-NITRO- □

C

4,4'-DICHLOR-2-NITRODIFENYLETHER □ ETHER, 4-CHLOROPHENYL (4'-CHLORO-2'-NITRO)PHENYL

TOXICITY DATA WITH REFERENCE
skn-rbt 500 mg/24H MLD 28ZPAK-,84,72
eye-rbt 100 mg/24H MOD 28ZPAK-,84,72

CONSENSUS REPORTS: Reported in EPA TSCA Inventory.

SAFETY PROFILE: A skin and eye irritant. When heated to decomposition it emits toxic fumes of NO_x and Cl^-.

CJD625 CAS:64046-47-5 HR: 1
2-CHLORO-5-NITROPHENYL ESTER ACETIC ACID
mf: $C_8H_6ClNO_4$ mw: 215.60

SYN: METHYLESTER KYSELINY-2-CHLOR-5-NITROBENZOOVE (CZECH)

TOXICITY DATA WITH REFERENCE
skn-rbt 500 mg/24H MLD 28ZPAK -,92,72
eye-rbt 500 mg/24H MLD 28ZPAK -,92,72
orl-rat LD50:5360 mg/kg 28ZPAK -,92,72

SAFETY PROFILE: Mildly toxic by ingestion. An eye and skin irritant. When heated to decomposition it emits toxic fumes of Cl^- and NO_x. See also ESTERS.

CJD630 CAS:42069-72-7 HR: 3
1-(4-CHLORO-3-NITROPHENYL)-2-ETHOXY-2-((4-(METHYLTHIO)PHENYL)AMINO)ETHAN ONE
mf: $C_{17}H_{17}ClN_2O_4S$ mw: 380.87

SYNS: ETHANONE, 1-(4-CHLORO-3-NITROPHENYL)-2-ETHOXY-2-((4-(METHYLTHIO)PHENYL)AMINO)- □ KETONE, 1-(4-CHLORO-3-NITROPHENYL)-2-ETHOXY-2-((4-(METHYLTHIO)PHENYL)AMINO)-

TOXICITY DATA WITH REFERENCE
ipr-mus LD50:1200 mg/kg ARZNAD 23,573,73

DOT CLASSIFICATION: 3; *Label:* Flammable Liquid

SAFETY PROFILE: Moderately toxic by intraperitoneal route. A flammable liquid. When heated to decomposition it emits toxic vapors of NO_x, SO_x, and Cl^-.

CJD650 CAS:328-04-1 HR: 3
o-(2-CHLORO-4-NITROPHENYL)-o-ISOPROPYL ETHYLPHOSPHONOTHIOATE
mf: $C_{11}H_{15}ClNO_4PS$ mw: 323.75

SYNS: ENT 25,755 □ N 2404 □ STAUFFER N-2404

TOXICITY DATA WITH REFERENCE
orl-rat LD50:32 mg/kg ARSIM* 20,22,66
orl-mus LDLo:62 mg/kg AECTCV 14,111,85
orl-ckn LD50:4 mg/kg TXAPA9 7,606,65

SAFETY PROFILE: Poison by ingestion. When heated to decomposition it emits toxic fumes of Cl^-, NO_x, PO_x, and SO_x.

CJD750 CAS:2425-66-3 HR: 3
CHLORONITROPROPANE
mf: $C_3H_6ClNO_2$ mw: 123.55

SYNS: CHLORONITROPROPAN (POLISH) □ 1-CHLORO-2-NITROPROPANE

TOXICITY DATA WITH REFERENCE
orl-rat LD50:197 mg/kg PCOC** -,647,66
ihl-rat LC50:1070 mg/m³ GUCHAZ 6,111,73
skn-rat LD50:362 mg/kg 31ZOAD 1,89,68
orl-mus LD50:105 mg/kg GUCHAZ 6,111,73
skn-rbt LD50:362 mg/kg WRPCA2 9,119,70

SAFETY PROFILE: Poison by ingestion and skin contact. Moderately toxic by inhalation. When heated to decomposition it emits very toxic fumes of Cl^- and NO_x. See also other chloronitropropane entries.

CJE000 CAS:600-25-9 HR: 3
1-CHLORO-1-NITROPROPANE
mf: $C_3H_6ClNO_2$ mw: 123.55

PROP: Liquid. Bp: 139.5°, flash p: 144°F (OC), d: 1.209 @ 20°/20°, vap d: 4.26.

SYN: CHLORONITROPROPANE

TOXICITY DATA WITH REFERENCE
mmo-sat 333 µg/plate EMMUEG 19(Suppl 21),2,92
orl-rbt LDLo:50 mg/kg JIHTAB 27,95,45
orl-mus LD50:510 mg/kg HYSAAV 34(10-12),429,69
ihl-mus LC50:66 g/m³/3H HYSAAV 34(10-12),429,69
scu-mus LD50:165 mg/kg HYSAAV 34(10-12),429,69
ihl-rbt LCLo:2000 mg/m³/6H JIHTAB 27,95,45
ihl-gpg LCLo:580 mg/m³/1H JIHTAB 27,95,45

OSHA PEL: TWA 2 ppm
ACGIH TLV: TWA 2 ppm
DFG MAK: 20 ppm (100 mg/m³)

SAFETY PROFILE: Poison by ingestion, subcutaneous, and possibly other routes. Moderately toxic by inhalation. Causes injury to kidneys, liver, and cardiovascular system. Mutation data reported. Flammable liquid when exposed to heat, flame (sparks), and oxidizers. Moderately explosive when exposed to heat. To fight fire, use alcohol foam, water, CO_2, or dry chemical. When heated to decomposition it emits toxic fumes of Cl^- and NO_x. See also other chloropropane entries and CHLORIDES.

CJE250 CAS:594-71-8 HR: 3
2-CHLORO-2-NITROPROPANE
mf: $C_3H_6ClNO_2$ mw: 123.55

PROP: Liquid. Bp: 134° (slt decomp), flash p: 135°F (OC), d: 1.179 @ 16°, vap d: 4.26.

TOXICITY DATA WITH REFERENCE
orl-rbt LDLo:500 mg/kg JIHTAB 27,95,45
orl-mus LD50:580 mg/kg HYSAAV 34(10-12),429,69
ihl-mus LC50:55 g/m³/3H HYSAAV 34(10-12),429,69
scu-mus LD50:195 mg/kg HYSAAV 34(10-12),429,69

SAFETY PROFILE: Poison by subcutaneous route. Moderately toxic by ingestion. Mildly toxic by inhalation. Flammable liquid when exposed to heat, flame, or oxidizers. Explodes on rapid heating. When heated to decomposition it emits toxic fumes of Cl^- and NO_x. See also other chloronitropropane entries and CHLORIDES.

CJF500 CAS:6505-75-5 HR: 3
3′-CHLORO-5-NITROSALICYLANILIDE
mf: $C_{13}H_9ClN_2O_4$ mw: 292.69

SYN: USAF BV-8

TOXICITY DATA WITH REFERENCE
orl-mus LDLo:710 mg/kg AECTCV 14,111,85
ipr-mus LD50:15 mg/kg NTIS** AD610-947
ivn-mus LD50:100 mg/kg CSLNX* NX#00478

SAFETY PROFILE: Poison by intraperitoneal and intravenous routes. Moderately toxic by ingestion. When heated to decomposition it emits very toxic fumes of Cl^- and NO_x.

CJF825 CAS:695-64-7 HR: 3
1-CHLORO-1-NITROSOCYCLOHEXANE
mf: $C_6H_{10}ClNO$ mw: 147.60

PROP: Crystals from Me_2CO or cyclohexane. D: 1.114 @ 20°/4°, mp: 116–117°.

SAFETY PROFILE: May explode during vacuum distillation. When heated to decomposition it emits toxic fumes of Cl^- and NO_x.

CJG750 CAS:10140-97-3 HR: 3
α-CHLORO-p-NITROSTYRENE
mf: $C_8H_6ClNO_2$ mw: 183.60

PROP: Pale-yellow needles from ligroin. Mp: 63–64°.

TOXICITY DATA WITH REFERENCE
orl-rat LDLo:710 mg/kg AIHAAP 23,95,62
skn-rbt LD50:390 mg/kg AIHAAP 23,95,62

SAFETY PROFILE: Poison by skin contact. Moderately toxic by ingestion. When heated to decomposition it emits very toxic fumes of Cl^- and NO_x.

CJG800 CAS:121-86-8 HR: 1
2-CHLORO-4-NITROTOLUENE
mf: $C_7H_6ClNO_2$ mw: 171.59

PROP: Needles. Fp: 62°, mp: 65°.

SYN: TOLUENE, 2-CHLORO-4-NITRO-

TOXICITY DATA WITH REFERENCE
skn-rbt 500 mg/24H MLD 85JCAE-,603,86
eye-rbt 500 mg/24H MLD 85JCAE-,603,86

CONSENSUS REPORTS: Reported in EPA TSCA Inventory.

SAFETY PROFILE: A skin and eye irritant. When heated to decomposition it emits toxic fumes of NO_x and Cl^-.

CJG825 CAS:83-42-1 HR: 3
2-CHLORO-6-NITROTOLUENE
mf: $C_7H_6ClNO_2$ mw: 171.59

SYNS: BENZENE, 1-CHLORO-2-METHYL-3-NITRO- □ 6-CHLORO-2-NITROTOLUENE □ TOLUENE, 2-CHLORO-6-NITRO-

TOXICITY DATA WITH REFERENCE
ivn-mus LD50:56 mg/kg CSLNX* NX#03505

CONSENSUS REPORTS: Reported in EPA TSCA Inventory.

SAFETY PROFILE: Poison by intravenous route. When heated to decomposition it emits toxic vapors of NO_x and Cl^-.

CJH500 CAS:1120-10-1 HR: 2
9-CHLORONONANOIC ACID
mf: $C_9H_{17}ClO_2$ mw: 192.71

SYNS: CHLOROPERALGONIC ACID □ CHLORPERALGONIC

TOXICITY DATA WITH REFERENCE
orl-rat LD50:3000 mg/kg GISAAA 27(10),12,62
orl-mus LD50:2 g/kg TPKVAL 5,89,63
ihl-mus LCLo:500 mg/m³/2H TPKVAL 5,89,63

SAFETY PROFILE: Moderately toxic by ingestion and inhalation. When heated to decomposition it emits toxic fumes of Cl^-.

CJH750 CAS:35941-71-0 HR: 3
4-((5-CHLORO-2-OXO-3(2H)-BENZOTHIAZOLYL)ACETYL)-1-PIPERAZINEETHANOL
mf: $C_{15}H_{18}ClN_3O_3S•ClH$ mw: 392.33

PROP: A solid. Mp: 159–161°. Very sol in water: sltly sol in org solvs.

SYNS: 5-CHLORO-3-(4-(2-HYDROXYETHYL)-1-PIPERAZINYL)CARBONYLMETHYL-2-BENZOTHIAZOLINONE □ FK 1160 □ NTA-194 □ SOLANTAL □ TIARAMIDE HYDROCHLORIDE

TOXICITY DATA WITH REFERENCE
eye-rbt 200 mg rns MLD ARZNAD 22,724,72
orl-rat TDLo:6 g/kg (9-14D preg):TER ARZNAD 23,504,73
orl-mus TDLo:300 mg/kg (7-12D preg):TER ARZNAD 32,225,82
orl-rat LD50:2300 mg/kg ARZNAD 32,225,82
ipr-rat LD50:530 mg/kg NIIRDN 6,451,82
scu-rat LD50:930 mg/kg ARZNAD 23,504,73
ivn-rat LD50:203 mg/kg ARZNAD 23,504,73
orl-mus LD50:564 mg/kg ARZNAD 23,504,73
ipr-mus LD50:298 mg/kg ARZNAD 23,504,73
scu-mus LD50:375 mg/kg ARZNAD 23,504,73
ivn-mus LD50:178 mg/kg ARZNAD 23,504,73

SAFETY PROFILE: Poison by subcutaneous, intravenous, and intraperitoneal routes. Moderately toxic by ingestion. An experimental teratogen. An eye irritant. When heated to decomposition it emits very toxic fumes of Cl^-, NO_x, and SO_x.

CJI100 CAS:7203-90-9 HR: 3
CHLORO-PDMT
mf: $C_8H_{10}ClN_3$ mw: 183.66

SYNS: 1-p-CHLORFENYL-3,3-DIMETHYLTRIAZEN (CZECH) □ 1-(p-CHLOROPHENYL)-3,3-DIMETHYL-TRIAZENE □ 1-(4-CHLOROPHENYL)-3,3-DIMETHYLTRIAZENE □ 1-(p-CHLOR-PHENYL)-3,3-DIMETHYL-TRIAZEN (GERMAN)

TOXICITY DATA WITH REFERENCE
mma-sat 5 mmol/L MUREAV 36,1,76
sln-dmg-orl 100 μmol/L CBINA8 9,365,74
mnt-mus-ipr 50 mg/kg/24H MUREAV 56,319,78
scu-rat TDLo:3200 mg/kg/70W-I:CAR ZKKOBW 81,285,74
scu-rat TD:300 mg/kg:NEO ZKKOBW 81,285,74
orl-rat LD50:362 mg/kg 28ZPAK -,98,72
scu-rat LD50:330 mg/kg ZKKOBW 81,285,74

CONSENSUS REPORTS: EPA Genetic Toxicology Program.

SAFETY PROFILE: Poison by ingestion and subcutaneous routes. Questionable carcinogen with experimental carcinogenic and neoplastigenic data. Mutation data reported. When heated to decomposition it emits toxic fumes of Cl^- and NO_x.

CJI500 CAS:76-15-3 HR: 1
CHLOROPENTAFLUOROETHANE
DOT: UN 1020
mf: C_2ClF_5 mw: 154.47

PROP: Colorless gas. Bp: −37.7°, mp: −106°, d: 1.5678 @ -42°. Insol in water; sol in alc and ether.

SYNS: F-115 □ FLUOROCARBON-115 □ FREON 115 □ GENETRON 115 □ HALOCARBON 115 □ MONOCHLOROPENTAFLUOROETHANE (DOT)

CONSENSUS REPORTS: Reported in EPA TSCA Inventory.

OSHA PEL: TWA 1000 ppm
ACGIH TLV: TWA 1000 ppm
DOT CLASSIFICATION: 2.2; *Label:* Nonflammable Gas

SAFETY PROFILE: Mildly toxic by inhalation. A nonflammable gas. When heated to decomposition it emits toxic fumes of F^- and Cl^-.

CJI609 HR: 3
CHLOROPEPTIDE

PROP: Chlorine-containing cyclic pentapeptide from metabolites of *Penicillium islandicum Sopp.* (JTSCDR 2,349,77).

TOXICITY DATA WITH REFERENCE
orl-rat LD50:5 mg/kg JTSCDR 2,349,77
scu-rat LD50:400 μg/kg JTSCDR 2,349,77
orl-mus LD50:7 mg/kg JTSCDR 2,349,77
ipr-mus LD50:450 μg/kg JTSCDR 2,349,77
scu-mus LD50:500 μg/kg JTSCDR 2,349,77
ivn-mus LD50:400 μg/kg JTSCDR 2,349,77
scu-gpg LD50:500 mg/kg JTSCDR 2,349,77

SAFETY PROFILE: Poison by subcutaneous, intravenous, and intraperitoneal routes. Mildly toxic by ingestion. When heated to decomposition it emits toxic fumes of Cl^- and NO_x.

CJI809 CAS:32755-26-3 HR: 3
CHLOROPEROXYTRIFLUOROMETHANE
mf: $CClF_3O_2$ mw: 98.46

PROP: Pale-yellow liquid. Mp: −132°, bp: −22°.

SAFETY PROFILE: Initiates explosive polymerization of tetrafluoroethylene. When heated to decomposition it emits toxic fumes of F^- and Cl^-. See also PEROXIDES.

CJJ000 CAS:3691-35-8 HR: 3
CHLOROPHACINONE
mf: $C_{23}H_{15}ClO_3$ mw: 374.83

SYNS: AFNOR □ CAID □ CHLOORFACINON (DUTCH) □ 2(2-(4-CHLOOR-FENYL-2-FENYL)-ACETYL)-INDAAN-1,3-DION (DUTCH) □ CHLORFACINON (GERMAN) □ 2-(α-p-CHLOROPHENYLACETYL)INDANE-1,3-DIONE □ 2 ((p-CHLOROPHENYL)PHENYLACETYL)-1,3-INDANDIONE □ 2(2-(4-CHLOROPHENYL)-2-PHENYLACETYL)INDAN-1,3-DIONE □ 2-((4-CHLOROPHENYL)PHENYLACETYL)-1H-INDENE-1,3(2H)-DIONE □ CHLORPHACINON (ITALIAN) □ 2(2-(4-CHLOR-PHENYL-2 PHENYL)ACETYL)INDAN-1,3-DION (GERMAN) □ ((4-CHLOR PHENYL)-1-PHENYL)-ACETYL-1,3-INDANDION (GERMAN) □ 1-(4-CHLORPHENYL)-1-PHENYL-ACETYL-INDAN-1,3-DION (GERMAN) □ 2 (2-(4-CLORO-FENIL-2-FENIL)-ACETIL)INDAN 1,3-DIONE (ITALIAN) □ DELTA □ DRAT □ LIPHADIONE □ LM 91 □ MICROZUL □MU-RIOL □ 2-(2-PHENYL-2-(4-CHLOROPHENYL)ACETYL)-1,3-INDANDIONE □ QUICK □ RAMUCIDE □ RANAC □ RATOMET □ RAVIAC □ ROZOL □ TOPITOX

TOXICITY DATA WITH REFERENCE
orl-wmn TDLo:5 mg/kg JTCTDW 27,79,89
orl-rat LD50:2100 μg/kg GUCHAZ 6,112,73
orl-mus LD50:1060 μg/kg TXAPA9 25,42,73
orl-rbt LD50:50 mg/kg GUCHAZ 6,112,73
skn-rbt LD50:200 mg/kg GUCHAZ 6,112,73
orl-dck LD50:100 mg/kg GUCHAZ 6,112,73
orl mam LD50:7500 μg/kg GUCHAZ 6,112,73
orl-brd LD50:430 mg/kg GUCHAZ 6,112,73

CONSENSUS REPORTS: EPA Extremely Hazardous Substances List.

SAFETY PROFILE: Poison by ingestion and skin contact. Human systemic effects by ingestion: vascular changes. A pesticide. When heated to decomposition it emits toxic fumes of Cl^-.

CJJ100 CAS:536-38-9 HR: 3
p-CHLOROPHENACYL BROMIDE
mf: C_8H_6BrClO mw: 233.50

SYNS: ACETOPHENONE, 2-BROMO-4′-CHLORO- □ BROMOMETHYL p-CHLOROPHENYL KETONE □ 4′-CHLOROPHENACYL BROMIDE □ ETHANONE, 2-BROMO-1-(4-CHLOROPHENYL)-

TOXICITY DATA WITH REFERENCE
orl-mus LD50:>2 g/kg MEXPAG 11,137,64

DOT CLASSIFICATION: 3; *Label:* Flammable Liquid

SAFETY PROFILE: Low toxicity by ingestion. A flammable liquid. When heated to decomposition it emits toxic vapors of Br^- and Cl^-.

CJJ250 **CAS:6164-98-3** **HR: 3**
CHLOROPHENAMIDINE
mf: $C_{10}H_{13}ClN_2$ mw: 196.70

PROP: Crystals. Mp: 32°, bp: 163–165° @ 14 mm.

SYNS: ACARON □ BERMAT □ C 8514 □ CARZOL □ CDM □ CHLORDIMEFORM □ CHLORFENAMIDINE □ N'-(4-CHLORO-2-METHYLPHENYL)-N,N-DIMETHYLMETHANIMIDAMIDE □ CHLOROPHENAMADIN □ N'-(4-CHLORO-o-TOLYL)-N,N-DIMETHYLFORMAMIDINE □ CHLORPHENAMIDINE □ N'-(4-CHLOR-o-TOLYL)-N,N-DIMETHYLFORMAMIDIN (GERMAN) □ CIBA 8514 □ N,N-DIMETHYL-N'-(2-METHYL-4-CHLOROPHENYL)-FORMAMIDINE □ N,N-DIMETHYL-N'-(2-METHYL-4-CHLORPHENYL)-FORMADIN (GERMAN) □ ENT 27,335 □ ENT 27,567 □ EP-333 □ FUNDAL □ FUNDAL 500 □ FUNDEX □ GALECRON □ N'-(2-METHYL-4-CHLOROPHENYL)-N,N-DIMETHYLFORMAMIDINE □ N'-(2-METHYL-4-CHLORPHENYL)-FORMAMIDIN-HYDROCHLORID (GERMAN) □ NSC 190935 □ RS 141 □ SCHERING 36268 □ SN 36268 □ SPANON □ SPANONE

TOXICITY DATA WITH REFERENCE
eye-rbt 100 mg MLD CIGET* 6/2/75
mmo-smc 5 ppm RSTUDV 6,161,76
dni-hmn:hla 1 mmol/L BECTA6 11,184,74
oms-hmn:hla 1 mmol/L BECTA6 11,184,74
orl-rat TDLo:1800 μg/kg (5-22D preg):REP BECTA6 20,760,78
orl-mus TDLo:6552 mg/kg/78W-C:CAR CHYCDW 19,154,85
orl-rat LD50:160 mg/kg KSKZAN 16(2),59,78
skn-rat LD50:263 mg/kg FAATDF 7,299,86
ipr-rat LD50:90 mg/kg PSDTAP 15,239,74
orl-mus LD50:160 mg/kg SPEADM 78-1,-,78
skn-mus LD50:225 mg/kg SPEADM 78-1,21,78
ipr-mus LD50:71 mg/kg OYYAA2 1,78,67
orl-rbt LD50:625 mg/kg 28ZEAL 5,43,76
skn-rbt LD50:640 mg/kg 85DPAN -,-,71/76

CONSENSUS REPORTS: IARC Cancer Review: Group 3 IMEMDT 7,56,87. EPA Genetic Toxicology Program.

SAFETY PROFILE: Poison by ingestion, skin contact, and intraperitoneal routes. Experimental reproductive effects. Human mutation data reported. An eye irritant. Questionable carcinogen with experimental carcinogenic data. When heated to decomposition it emits very toxic fumes of NO_x and Cl^-.

CJJ500 **CAS:19750-95-9** **HR: 3**
CHLOROPHENAMIDINE HYDROCHLORIDE
mf: $C_{10}H_{13}ClN_2 \cdot ClH$ mw: 233.16

PROP: A solid. Mp: 225–227° (decomp).

SYNS: CHLORDIMEFORM HYDROCHLORIDE □ N'-(4-CHLORO-2-METHYLPHENYL)-METHANIMIDAMIDE MONOHYDROCHLORIDE □ N'-(4-CHLORO-o-TOLYL)-N,N-DIMETHYLFORMAMIDINE HYDROCHLORIDE □ N,N-DIMETHYL-N'-(2-METHYL-4-CHLOROPHENYL)-FORMAMIDINE HYDROCHLORIDE □ ENT 27,567 □ EP 333 □ FUNDAL SP □ GALECRON SP □ MORTON EP 333 □ NOR-AM EP 333 □ NSC 195102 □ SCHERING 36268

TOXICITY DATA WITH REFERENCE
eye-rbt 100 mg MLD CIGET* 78-1,22,78
skn-rat LD50:4 g/kg SPEADM 78-1,22,78
orl-mus LD50:290 mg/kg BESAAT 15,103,69
ipr-mus LD50:85,500 μg/kg PCBPBS 9,148,78
orl-rbt LD50:625 mg/kg BESAAT 15,103,69
skn-rbt LD50:>4 g/kg SPEADM 78-1,22,78

SAFETY PROFILE: Poison by ingestion and intraperitoneal routes. Low toxicity by skin contact. An eye irritant. A pesticide. When heated to decomposition it emits very toxic fumes of NO_x and Cl^-.

CJK000 **CAS:2598-25-6** **HR: 3**
1-(p-CHLOROPHENETHYL)HYDRAZINE HYDROGEN SULFATE
mf: $C_8H_{11}ClN_2 \cdot H_2O_4S$ mw: 268.74

SYNS: p-CHLORO-β-PHENYLETHYLHYDRAZINE DIHYDROGEN SULFATE □ WL 28

TOXICITY DATA WITH REFERENCE
scu-mus TDLo:360 mg/kg (1-6D preg):REP JOENAK 30,205,64
orl-mus LD50:175 mg/kg JMPCAS 5,221,62
scu-mus LD50:250 mg/kg JOENAK 30,205,64

SAFETY PROFILE: Poison by ingestion and subcutaneous routes. Experimental reproductive effects. When heated to decomposition it emits very toxic fumes of Cl^-, NO_x, and SO_x. See also HYDRAZINE and SULFATES.

CJK100 **CAS:2598-75-6** **HR: 2**
2-(o-CHLOROPHENETHYL)-3-THIOSEMICARBAZIDE
mf: $C_9H_{12}ClN_3S$ mw: 229.75

SYNS: SEMICARBAZIDE, 2-(o-CHLOROPHENETHYL)-3-THIO- □ WL 34

TOXICITY DATA WITH REFERENCE
scu-mus TDLo:1500 mg/kg (female 1-6D post):REP JOENAK 30,205,64
scu-mus LD50:500 mg/kg JOENAK 30,205,64

SAFETY PROFILE: Moderately toxic by subcutaneous route. Experimental reproductive effects. When heated to decomposition it emits toxic fumes of SO_x, NO_x, and Cl^-.

CJK250 **CAS:95-57-8** **HR: 3**
2-CHLOROPHENOL
mf: C_6H_5ClO mw: 128.56

PROP: Light-amber liquid. Mp: 9°, fp: 7°, bp: 174.9°, d: 1.263 @ 20°/4°, flash p: 147°F, vap press: 1 mm @ 12.1°. Sltly water sol; very sol in alc, ether, and alkali.

SYNS: o-CHLOROPHENOL □ o-CHLOROPHENOL, liquid □ o-CHLOROPHENOL, solid □ o-CHLORPHENOL (GERMAN) □ RCRA WASTE NUMBER U048

TOXICITY DATA WITH REFERENCE
sln-ham:lng 800 μmol/L MUREAV 182,135,87
orl-rat TDLo:4550 mg/kg (70 pre/1-21D preg):REP EVHPAZ 46,137,82
skn-mus TDLo:4800 mg/kg/12W-I:ETA CNREA8 19,413,59
orl-rat LD50:670 mg/kg FEPRA7 2,76,43
ipr-rat LD50:230 mg/kg BJPCAL 13,20,58
scu-rat LD50:950 mg/kg FEPRA7 2,76,43
orl-mus LD50:345 mg/kg TOLED5 29 39,85
ipr-mus LD50:235 mg/kg JMCMAR 18,868,75
scu-rbt LDLo:950 mg/kg HBAMAK 4,1361,35

ivn-rbt LDLo: 120 mg/kg HBTXAC 5,112,59

CONSENSUS REPORTS: Reported in EPA TSCA Inventory. Chlorophenol compounds are on the Community Right-To-Know List.

SAFETY PROFILE: Poison by ingestion, intraperitoneal, and intravenous routes. Experimental reproductive effects. Questionable carcinogen with experimental tumorigenic data. Mutation data reported. Flammable liquid when exposed to heat, flame, or oxidizers. To fight fire, use alcohol foam. When heated to decomposition it emits toxic fumes of Cl^-. See also CHLOROPHENOLS and CHLORIDES.

CJK500 CAS:108-43-0 HR: 3
3-CHLOROPHENOL
mf: C_6H_5ClO mw: 128.56

PROP: Crystals or needles from pet ether. Mp: 33°, bp: 210–214°, d: 1.245 @ 45°/4°, vap press: 1 mm @ 44.2°, flash p: >112°.

SYN: m-CHLOROPHENOL

TOXICITY DATA WITH REFERENCE
mmo-sat 10 μg/plate TECSDY 14,143,87
skn-mus TDLo: 6000 mg/kg/15W-I: ETA CNREA8 19,413,59
orl-rat LD50: 570 mg/kg FEPRA7 2,76,43
ipr-rat LD50: 355 mg/kg BJPCAL 13,20,58
scu-rat LD50: 1390 mg/kg FEPRA7 2,76,43
orl-mus LD50: 521 mg/kg TOLED5 29,39,85

CONSENSUS REPORTS: Reported in EPA TSCA Inventory. Chlorophenol compounds are on the Community Right-To-Know List.

SAFETY PROFILE: Poison by intraperitoneal route. Moderately toxic by ingestion and subcutaneous routes. Questionable carcinogen with experimental tumorigenic data by skin contact. Mutation data reported. Flammable or combustible liquid. When heated to decomposition it emits toxic fumes of Cl^-. See also CHLOROPHENOLS.

CJK750 CAS:106-48-9 HR: 3
4-CHLOROPHENOL
DOT: UN 2020/UN 2021
mf: C_6H_5ClO mw: 128.56

PROP: Needle-like, white to straw-colored crystals; unpleasant odor. Flash p: 250°F, d: 1.246 @ 60°/25°, vap press: 1 mm @ 49.8°, mp: 43.5 (α-form) mp: 43.5 (γ-form), bp: 220°. Sltly water sol; very sol in alc, chloroform, and ether.

SYNS: p-CHLORFENOL (CZECH) □ p-CHLOROPHENOL □ PARA-CHLOROPHENOL □ PHENOL, 4-CHLORO-

TOXICITY DATA WITH REFERENCE
skn-rbt 500 mg/24H SEV 28ZPAK -,78,72
eye-rbt 250 μg/24H SEV 28ZPAK -,78,72
mmo-sat 200 μg/plate PCBPBS 10,174,79
ihl-mus TCLo: 760 μg/m³/24H (male 17W pre): REP GISAAA 57(7-8),9,92
ihl-hmn TCLo: 10 g/m³/8H: BAH GISAAA 29(10),37,64
ihl-rat LC50: 11 mg/m³ GISAAA 29(10),37,64
skn-rat LD50: 1500 mg/kg GISAAA 29(10),37,64
ipr-rat LD50: 281 mg/kg BJPCAL 13,20,58
scu-rat LD50: 1030 mg/kg FEPRA7 2,76,43
orl-mus LD50: 1373 mg/kg TOLED5 29,39,85
ipr-mus LD50: 332 mg/kg JMCMAR 18,868,75
skn-mam LD50: 1000 mg/kg GISAAA 45(10),16,80

CONSENSUS REPORTS: Reported in EPA TSCA Inventory. Chlorophenol compounds are on the Community Right-To-Know List.

SAFETY PROFILE: Poison by inhalation and intraperitoneal routes. Moderately toxic by ingestion, skin contact, and subcutaneous routes. A severe skin and eye irritant. Human systemic effects by inhalation: excitement, irritability. Mutation data reported. Combustible when exposed to heat or flame. To fight fire, use water, spray, mist, fog, foam, dry chemical. When heated to decomposition it emits toxic fumes of Cl^-. See also CHLOROPHENOLS and CHLORIDES.

For occupational chemical analysis use NIOSH: o-Chlorophenol P&CAM 337.

CJL000 HR: 3
CHLOROPHENOLS

CONSENSUS REPORTS: Chlorophenol compounds are on the Community Right-To-Know List.

SAFETY PROFILE: Many are suspected experimental carcinogens. Most are strong eye and skin irritants. They are systemic irritants by inhalation, ingestion, and skin contact. Generally mutagenic.

Trichlorophenols are generally poisons and may be carcinogens. They may contain 2,3,7,8-tetrachlorodibenzo-p-dioxin (TCDD) as a contaminant. Some trichlorophenols are used as herbicides (e.g., 2,4,5-T and silvex). Human exposure may cause chloracne, liver dysfunction, muscle weakness, and prophyria.

Pentachlorophenol is a poison by several routes. Human exposure causes increased respiration, fever, tachycardia, muscle weakness, and cardiac failure. Many toxic effects are due to impurities in commercial-grade material. A teratogen and mutagen. Pentachlorophenol and 2,4,6-trichlorophenol may interfere with mitochondrial oxidative phosphorylation. When heated to decomposition they emit toxic fumes of Cl^-. See also specific compounds, PHENOL, and CHLORIDES.

CJL100 CAS:92-39-7 HR: 3
2-CHLOROPHENOTHIAZINE
mf: $C_{12}H_8ClNS$ mw: 233.72

SYNS: PHENOTHIAZINE, 2-CHLORO- □ 10H-PHENOTHIAZINE, 2-CHLORO-

TOXICITY DATA WITH REFERENCE
dni-bac-esc 160 mg/L BCPCA6 26,1205,77
oth-esc 160 mg/L BCPCA6 26,1205,77
ivn-mus LD50: 18 mg/kg CSLNX* NX#00590

CONSENSUS REPORTS: Reported in EPA TSCA Inventory.

SAFETY PROFILE: Poison by intravenous route. Mutation data reported. When heated to decomposition it emits toxic vapors of NO_x, SO_x, and Cl^-.

CJL409 **CAS:19142-68-8** **HR: 3**
2-(2-(4-(2-((2-CHLORO-10-PHENOTHIAZINYL)METHYL)PROPYL)-1-PIPERAZINYL)ETHOXY)ETHANOL
mf: $C_{24}H_{32}ClN_3O_2S$ mw: 462.10

SYN: UCB 2493

TOXICITY DATA WITH REFERENCE
orl-rat LD50:400 mg/kg ANPBAZ 61,669,61
ivn-rat LD50:50 mg/kg ANPBAZ 61,669,61

SAFETY PROFILE: Poison by ingestion and intravenous routes. When heated to decomposition it emits very toxic fumes of SO_x, NO_x, and Cl^-.

CJL500 **CAS:84-04-8** **HR: 3**
1-(3-(3-CHLOROPHENOTHIAZIN-10-YL)PROPYL)-ISONIPECOTAMIDE
mf: $C_{21}H_{24}ClN_3OS$ mw: 401.99

PROP: A solid. Mp: 139°.

SYNS: MOMETINE □ MORNIDINE □ NAUSIDOL □ NOMETINE □ PIPAMAZINE □ SC 8016 □ SC 9387

TOXICITY DATA WITH REFERENCE
orl-chd TDLo:500 µg/kg:CNS 34ZIAG -,478,69
orl-rat LD50:620 mg/kg 27ZQAG -,39,72
orl-mus LD50:370 mg/kg 27ZQAG -,39,72
ipr-mus LD50:80 mg/kg 27ZQAG -,39,72

SAFETY PROFILE: Poison by ingestion and intraperitoneal routes. Human (child) systemic effects by ingestion: somnolence, sleep, and coma. When heated to decomposition it emits very toxic fumes of SO_x, NO_x, and Cl^-.

CJM250 **CAS:58-39-9** **HR: 3**
4-(3-(2-CHLOROPHENOTHIAZIN-10-YL)PROPYL)-1-PIPERAZINEETHANOL
mf: $C_{21}H_{26}ClN_3OS$ mw: 404.01

PROP: Crystals from EtOAc. Mp: 100–101°, bp: 278–281° @ 1 mm.

SYNS: 2-CHLORO-10-3-(1-(2-HYDROXYETHYL)-4-PIPERAZINYL) PROPYL PHENOTHIAZINE □ DECENTAN □ ETAPERAZIN □ ETAPERAZINE □ ETHAPERAZINE □ FENTAZIN □ 1-(2-HYDROXYETHYL)-4-(3-(2-CHLORO-10-PHENOTHIAZINYL)PROPYL)PIPERAZINE □ γ-(4-(β-HYDROXYETHYL)PIPERAZIN-1-YL)PROPYL-2-CHLOROPHENOTHIAZINE □ 1′,1-(2-IDROSSIETIL)4-(3-(2-CLORO-10-FENOTIAZIL)PROPILPIPERAZINA (ITALIAN) □ PERFENAZINA (ITALIAN) □ PERPHENAZIN □ PERPHENAZINE □ TRIFARON □ TRILAFON

TOXICITY DATA WITH REFERENCE
cyt-hmn:leu 270 mg/kg BMJOAE 3,634,69
orl-mus TDLo:150 mg/kg (female 10-12D post):TER TOIZAG 28,621,81
unr-rat TDLo:70 mg/kg (female 6-15D post):TER TJADAB 8,214,73
unr-mus TDLo:4 mg/kg (female 1D pre):REP 85GRAA -,57,65
orl-rat TDLo:90 mg/kg (14D preg):REP ABAHAU 27,15,76
par-rat TDLo:80 mg/kg (female 1-8D post):REP 15QWAW -,290,65
orl-rat TDLo:90 mg/kg (14D preg):TER ABAHAU 27,15,76
orl-rat TDLo:150 mg/kg (female 10D post):TER ABAHAU 27,15,76
ims-hmn TDLo:71,428 ng/kg:CNS BMJOAE 3,867,67
orl-rat LD50:318 mg/kg TXAPA9 21,315,72
ipr-rat LD50:146 mg/kg ARZNAD 24,917,74
ivn-rat LD50:34 mg/kg FRPPAO 26,585,71
orl-mus LD50:120 mg/kg 27ZIAQ -,-,65
ipr-mus LD50:64 mg/kg JPMSAE 59,976,70
ivn-mus LD50:19 mg/kg ARZNAD 11,932,61
ivn-dog LD50:51 mg/kg ARZNAD 10,638,60

SAFETY PROFILE: Poison by ingestion, intravenous, subcutaneous, intraperitoneal, and intramuscular routes. Human systemic effects by intramuscular route: muscle spasms. Experimental teratogenic and reproductive effects. Human mutation data reported. When heated to decomposition it emits very toxic fumes of SO_x, NO_x, and Cl^-.

CJM750 **CAS:2865-70-5** **HR: 3**
10-CHLOROPHENOXARSINE
mf: $C_{12}H_8AsClO$ mw: 278.57

PROP: Colorless prisms from $CHCl_3$/octane. Mp: 122°. Very sol in C_6H_6, Me_2CO, $CHCl_3$; mod sol EtOH, Et_2O, and AcOH.

SYNS: 10-CHLORO-10H-PHENOXARSINE □ 10-CHLOROPHENOXYARSINE □ DID 95

TOXICITY DATA WITH REFERENCE
orl-mus LDLo:42 mg/kg AECTCV 14,111,85
orl-bwd LD50:100 mg/kg TXAPA9 21,315,72

CONSENSUS REPORTS: Reported in EPA TSCA Inventory. Arsenic and its compounds are on the Community Right-To-Know List.

OSHA PEL: TWA 0.5 mg(As)/m^3

SAFETY PROFILE: Poison by ingestion. See also ARSINE and ARSENIC COMPOUNDS. When heated to decomposition it emits very toxic fumes of As and Cl^-.

CJN000 **CAS:122-88-3** **HR: 2**
p-CHLOROPHENOXYACETIC ACID
mf: $C_8H_7ClO_3$ mw: 186.60

PROP: Needles or prisms from H_2O. Mp: 159°.

SYNS: (4-CHLOROPHENOXY)ACETIC ACID □ 4-CP □ CPA □ MARKS 4-CPA □ PCPA □ SURE-SET □ TOMATO FIX CONCENTRATE □ TOMATO HOLD □ TOMATOTONE

TOXICITY DATA WITH REFERENCE
dns-hmn:oth 10 mmol/L ZYDXDM 13(2),74,84
orl-rat LD50:850 mg/kg RREVAH 10,97,65
ipr-mus LD50:680 mg/kg CHTPBA 5,211,70

CONSENSUS REPORTS: Reported in EPA TSCA Inventory.

SAFETY PROFILE: Moderately toxic by ingestion and intraperitoneal routes. Human mutation data reported.

When heated to decomposition it emits toxic fumes of Cl^-.

CJN250 CAS:3544-35-2 HR: 3
p-CHLOROPHENOXYACETIC ACID-2-ISOPROPYLHYDRAZIDE
mf: $C_{11}H_{15}ClN_2O_2$ mw: 242.73

PROP: A solid. Mp: 93–94°.

SYNS: 1-(p-CHLOROPHENOXYACETYL)-2-ISOPROPYL HYDRAZINE ☐ IPROCLOZIDE ☐ ISOPROPILIDRAZIDE dell'ac. p-CLORO-FENOSSIACETICO (ITALIAN) ☐ PC 603 ☐ PU 603 ☐ SINDERESIN ☐ SOG-4 ☐ SURSUM

TOXICITY DATA WITH REFERENCE
orl-rat TDLo:600 mg/kg (15D preg):REP AOGNAX 71,153,66
orl-rat LD50:550 mg/kg 27ZQAG -,394,72
orl-mus LD50:278 mg/kg 27ZQAG -,394,72
ipr-mus LD50:219 mg/kg PCJOAU 11,887,77

SAFETY PROFILE: Poison by ingestion and intraperitoneal routes. Experimental reproductive effects. When heated to decomposition it emits very toxic fumes of Cl^- and NO_x. See also HYDRAZINE.

CJN750 CAS:1223-36-5 HR: 3
2-(p-CHLOROPHENOXY)-N-(2-(DIETHYLAMINO)ETHYL)ACETAMIDE
mf: $C_{14}H_{21}ClN_2O_2$ mw: 284.82

SYNS: AMICHLOPHENE ☐ ANP 246 ☐ CHLOFEXAMIDE ☐ 2-(4-CHLOROPHENOXY)-N-(2-(DIETHYLAMINO)ETHYL)ACETAMIDE ☐ CLOFEXAMIDE ☐ IEM 455 ☐ NP 246

TOXICITY DATA WITH REFERENCE
orl-rat LD50:1040 mg/kg OYYAA2 17,935,79
orl-mus LD50:1200 μg/kg CHTPBA 1,444,66
ivn-mus LD50:215 mg/kg ARZNAD 14,556,64
ivn-rbt LDLo:90 mg/kg ARZNAD 14,556,64

SAFETY PROFILE: Poison by ingestion and intravenous routes. When heated to decomposition it emits very toxic fumes of Cl^- and NO_x.

CJO250 CAS:43121-43-3 HR: 3
1-(4-CHLOROPHENOXY)-3,3-DIMETHYL-1-(1,2,4-TRIAZOL-1-YL)-2-BUTAN-2-ONE
mf: $C_{14}H_{16}ClN_3O_2$ mw: 293.78

PROP: A solid. Mp: 82–83°.

SYNS: AMIRAL ☐ BAY 6681 F ☐ BAYLETON ☐ BAY-MEB-6447 ☐ 1-((tert-BUTYLCARBONYL-4-CHLOROPHENOXY)METHYL)-1H-1,2,4-TRIAZOLE ☐ 1-(4-CHLOROPHENOXY)-3,3-DIMETHYL-1-(1H-1,2,4-TRIAZOL-1-YL)-2-BUTANONE ☐ MEB 6447 ☐ TRIADIMEFON

TOXICITY DATA WITH REFERENCE
mnt-rat-ipr 3 mg/kg MUREAV 321,103,94
cyt-rat-ipr 3 mg/kg MUREAV 321,103,94
orl-rat LD50:363 mg/kg DOVEAA 31,97,77
ihl-rat LC50:2450 mg/m³ GTPZAB 34(5),64,90
skn-rat LD50:>5 g/kg FMCHA2-,C38,91
orl-mus LD50:1 g/kg PEMNDP 9,831,91
ihl-mus LC50:2337 mg/m³ GTPZAB 34(5),64,90
orl-rbt LD50:500 mg/kg 85DPAN -,-,71/76

SAFETY PROFILE: Poison by ingestion. Mutation data reported. When heated to decomposition it emits very toxic fumes of Cl^- and NO_x. See also KETONES.

CJO500 CAS:1892-43-9 HR: 2
2-(p-CHLOROPHENOXY)ETHANOL
mf: $C_8H_9ClO_2$ mw: 172.62

SYNS: 2-(4'-CHLORFENOXY)ETHANOL (CZECH) ☐ p-CHLORFENYLMONOGLYKOLETHER (CZECH) ☐ 2-(4-CHLOROPHENOXY)ETHANOL

TOXICITY DATA WITH REFERENCE
skn-rbt 10 mg/24H open MLD AIHAAP 23,95,62
skn-rbt 500 mg/24H MOD 28ZPAK -,81,72
eye-rbt 250 μg/24H SEV 28ZPAK -,81,72
skn-rbt LD50:500 mg/kg AIHAAP 23,95,62

CONSENSUS REPORTS: Reported in EPA TSCA Inventory.

SAFETY PROFILE: Moderately toxic by skin contact. A severe eye and moderate skin irritant. When heated to decomposition it emits very toxic fumes of Cl^-.

CJP250 CAS:69782-24-7 HR: 3
(2-(p-CHLOROPHENOXY)ETHYL)HYDRAZINE HYDROCHLORIDE
mf: $C_8H_{11}ClN_2O{\cdot}ClH$ mw: 223.12

TOXICITY DATA WITH REFERENCE
orl-mus LD50:300 mg/kg JMCMAR 6,63,63
ipr-mus LD50:300 mg/kg JMCMAR 6,63,63

SAFETY PROFILE: Poison by ingestion and intraperitoneal routes. When heated to decomposition it emits very toxic fumes of Cl^- and NO_x. See also HYDRAZINE.

CJP500 CAS:2598-73-4 HR: 3
1-(2-(o-CHLOROPHENOXY)ETHYL)HYDRAZINE HYDROGEN SULFATE
mf: $C_8H_{11}ClN_2O{\cdot}H_2O_4S$ mw: 284.74

SYNS: HYDRAZINE, 1-(2-(o-CHLOROPHENOXY)ETHYL)-, HYDROGEN SULFATE (1:1) ☐ HYDRAZINE, 1-(2-(o-CHLOROPHENOXY)ETHYL)-, SULFATE (1:1) ☐ WEG 147

TOXICITY DATA WITH REFERENCE
scu-mus TDLo:180 mg/kg (female 1-6D post):REP JOENAK 30,205,64
scu-mus LD50:150 mg/kg JOENAK 30,205,64

SAFETY PROFILE: Poison by subcutaneous route. Experimental reproductive effects. When heated to decomposition it emits toxic fumes of SO_x, NO_x, and Cl^-.

CJP750 CAS:26129-32-8 HR: 3
2-(4-(4-CHLOROPHENOXY)PHENOXY)PROPIONIC ACID
mf: $C_{15}H_{13}ClO_4$ mw: 292.73

SYNS: ACIDO FENOFIBRICO ☐ 2-(4-(4'-CHLORPHENOXY)-PHENOXY)-PROPIONSAEURE (GERMAN) ☐ FENOFIBRIC ACID ☐ HCG-004

TOXICITY DATA WITH REFERENCE
orl-rat LD50:1200 mg/kg DRFUD4 4,326,79

ivn-rat LD50:190 mg/kg DRFUD4 4,326,79
ipr-mus LD50:500 mg/kg NPMDAD 9,3737,80

SAFETY PROFILE: Poison by intravenous route. Moderately toxic by ingestion and intraperitoneal routes. When heated to decomposition it emits toxic fumes of Cl^-.

CJQ000 CAS:1982-47-4 HR: 2
3-(p-(p-CHLOROPHENOXY)PHENYL)-1,1-DIMETHYLUREA
mf: $C_{15}H_{15}ClN_2O_2$ mw: 290.77

PROP: Crystals. Mp: 151–152°.

SYNS: C 1983 □ 3-((4-(4-CHLOOR-FENOXY)-FENOXY)-FENYL)-1,1-DIMETHYLUREUM (DUTCH) □ 3-(4-(4-CHLORO-FENOSSIL))-1,1-DIMETIL-UREA (ITALIAN) □ N'-4-(4-CHLOROPHENOXY)PHENYL-N,N-DIMETHYLUREA □ 1-(4-(4-CHLORO-PHENOXY)PHENYL)-3,3-D'METHYLUREE (FRENCH) □ CHLOROXIFENIDIM □ CHLOROXURON □ 3-(4-(4-CHLOR-PHENOXY)-PHENYL)-1,1-DIMETHYLHARNSTOFF (GERMAN) □ CIBA 1983 □ NOREX □ TENORAN

TOXICITY DATA WITH REFERENCE
orl-rat LD50:3700 mg/kg WRPCA2 9,119,70
unk-rat LD50:1000 mg/kg 30ZDA9 -,232,71

CONSENSUS REPORTS: EPA Genetic Toxicology Program. EPA Extremely Hazardous Substances List.

SAFETY PROFILE: Moderately toxic by ingestion and possibly other routes. An herbicide. When heated to decomposition it emits very toxic fumes of Cl^- and NO_x.

CJQ250 CAS:886-74-8 HR: 3
3-(p-CHLOROPHENOXY)-1,2-PROPANEDIOL-1-CARBAMATE
mf: $C_{10}H_{12}ClNO_4$ mw: 245.68

PROP: Crystals from C_6H_6/toluene. Mp: 89–91°. Insol in H_2O, C_6H_6, and cyclohexane; sol in dioxan.

SYNS: 3-(p-CHLOROPHENOXY)-2-HYDROXYPROPYL CARBAMATE □ 3-(4-CHLOROPHENOXY)-1,2-PROPANEDIOL-1-CARBAMATE □ CHLORPHENESIN CARBAMATE □ MAOLATE □ U-19,646

TOXICITY DATA WITH REFERENCE
orl-dog TDLo:40 g/kg (14W pre):REP OYYAA2 14,27,77
orl-dog TDLo:40 g/kg (male 14W pre):REP OYYAA2 14,27,77
orl-rat TDLo:60 g/kg (24W male):REP OYYAA2 15,725,78
orl-hmn TDLo:23 mg/kg/D:CNS 34ZIAG -,171,69
orl-rat LD50:632 mg/kg IYKEDH 10,710,70
ipr-rat LD50:354 mg/kg 27ZQAG -,386,72
scu-rat LD50:952 mg/kg OYYAA2 13,659,77
ivn-rat LD50:237 mg/kg YKYUA6 31,363,80
orl-mus LD50:741 mg/kg IYKEDH 10,710,79
ipr-mus LD50:452 mg/kg IYKEDH 10,710,79
scu-mus LD50:798 mg/kg IYKEDH 10,710,79
ivn-mus LD50:239 mg/kg 27ZQAG -,386,72

SAFETY PROFILE: Poison by intraperitoneal and intravenous routes. Moderately toxic by ingestion and subcutaneous routes. Human systemic effects by ingestion: somnolence, hallucinations or distorted perceptions, and muscle weakness. Experimental reproductive effects. When heated to decomposition it emits very toxic fumes of Cl^- and NO_x. See also CARBAMATES.

CJQ300 CAS:101-10-0 HR: 2
2-(3-CHLOROPHENOXY)PROPIONIC ACID
mf: $C_9H_9ClO_3$ mw: 200.63

SYNS: CLOPROP □ 2-(3-CHLOROPHENOXY)PROPANOIC ACID □ 2-(m-CHLOROPHENOXY)PROPIONIC ACID □ 3CP □ 3-CPA □ FRUITONE CPA □ METACHLORPHENPROP □ PROPIONIC ACID, 2-(m-CHLOROPHENOXY)-

TOXICITY DATA WITH REFERENCE
orl-rat LD50:>750 mg/kg FMCHA2-,C147,91
skn-rbt LD50:>2 g/kg PEMNDP 9,180,91

CONSENSUS REPORTS: Reported in EPA TSCA Inventory.

SAFETY PROFILE: Moderately toxic by ingestion. Low toxicity by skin contact. When heated to decomposition it emits toxic vapors of Cl^-.

CJQ500 CAS:102585-42-2 HR: 3
N-(1-(o-CHLOROPHENOXY)-2-PROPYL)-2-(DIETHYLAMINO)-N-ETHYLACETAMIDE HYDROCHLORIDE
mf: $C_{17}H_{27}ClN_2O_2 \cdot ClH$ mw: 363.37

SYN: C 2096

TOXICITY DATA WITH REFERENCE
eye-rbt 2% MLD ARZNAD 9,70,59
scu-mus LD50:200 mg/kg ARZNAD 9,70,59

SAFETY PROFILE: Poison by subcutaneous route. An eye irritant. When heated to decomposition it emits very toxic fumes of Cl^- and NO_x.

CJQ750 CAS:102585-43-3 HR: 3
N-(1-(o-CHLOROPHENOXY)-2-PROPYL)-2-(DIETHYLAMINO)-N-METHYLACETAMIDE HYDROCHLORIDE
mf: $C_{16}H_{25}ClN_2O_2 \cdot ClH$ mw: 349.34

SYN: C 2095

TOXICITY DATA WITH REFERENCE
eye-rbt 2% MLD ARZNAD 9,70,59
scu-mus LD50:330 mg/kg ARZNAD 9,70,59

SAFETY PROFILE: Poison by subcutaneous route. An eye irritant. When heated to decomposition it emits very toxic fumes of Cl^- and NO_x.

CJR200 CAS:20265-96-7 HR: 3
4-CHLOROPHENYLAMINE HYDROCHLORIDE
mf: $C_6H_6ClN \cdot ClH$ mw: 164.04

SYNS: 1-AMINO-4-CHLOROBENZENE HYDROCHLORIDE □ ANILINE, p-CHLORO-, HYDROCHLORIDE □ BENZENAMINE, 4-CHLORO-, HYDROCHLORIDE □ p-CHLOROANILINE HYDROCHLORIDE □ 4-CHLOROANILINE HYDROCHLORIDE □ p-CHLOROANILINIUM CHLORIDE □ 4-CHLOROBENZENAMINE HYDROCHLORIDE □ p-CHLOROPHENYLAMINE HYDROCHLORIDE

TOXICITY DATA WITH REFERENCE
mma-sat 1 mg/plate NTPTR* NTP-TR-351,89
msc-mus:lyms 375 mg/L NTPTR* NTP-TR-351,89
cyt-ham:ovr 900 mg/L NTPTR* NTP-TR-351,89
sce-ham:ovr 200 mg/L NTPTR* NTP-TR-351,89
orl-rat TDLo:9270 mg/kg/2Y-C:CAR NTPTR* NTP-TR-351,89

orl-mus TDLo:15,450 mg/kg/2Y-C:CAR NTPTR* NTP-TR-351,89

CONSENSUS REPORTS: NTP Carcinogenesis Studies (gavage): Clear Evidence: Rat NTPTR* NTP-TR-351,89; (gavage): Some Evidence: Mouse NTPTR* NTP-TR-351,89. Reported in EPA TSCA Inventory.

SAFETY PROFILE: Suspected carcinogen with carcinogenic data. Mutation data reported. When heated to decomposition it emits toxic vapors of NO_x, HCl, and Cl^-.

CJR250 CAS:3647-19-6 HR: 3
N-(3-CHLOROPHENYL)-1-AZIRIDINECARBOXAMIDE
mf: $C_9H_9ClN_2O$ mw: 196.65

SYNS: 1-(1-AZIRIDINYL)-N-(m-CHLOROPHENYL)FORMAMIDE ☐ 3-CHLOROPHENYL-N-CARBAMOYLAZIRIDINE

TOXICITY DATA WITH REFERENCE
ipr-mus TDLo:120 mg/kg/4W-I:NEO CNREA8 29,2184,69
ivn-mus LD50:180 mg/kg CSLNX* NX#03943

SAFETY PROFILE: Poison by intravenous route. Questionable carcinogen with experimental neoplastigenic data. When heated to decomposition it emits very toxic fumes of Cl^- and NO_x.

CJR500 CAS:80-38-6 HR: 2
4-CHLOROPHENYL BENZENESULFONATE
mf: $C_{12}H_9ClO_3S$ mw: 268.72

PROP: Colorless crystals. Mp: 62°. Insol in water; sol in organic solvents.

SYNS: ARACID ☐ BENZENESULFONATE de 4-CHLOROPHENYLE (FRENCH) ☐ BENZENESULFONIC ACID, 4-CHLOROPHENYL ESTER ☐ (4-CHLOOR-FENYL)-BENZEEN-SULFONAAT (DUTCH) ☐ p-CHLOROFENYLESTER KYSELINY BENZENSULFONOVE (CZECH) ☐ p-CHLOROPHENYL BENZENESULFONATE ☐ p-CHLOROPHENYL BENZENESULPHONATE ☐ 4-CHLOROPHENYL BENZENESULPHONATE ☐ (4-CHLOR-PHENYL)-BENZOLSULFONAT (GERMAN) ☐ (4-CLORO-FENIL)-BENZOL-SOLFONATO (ITALIAN) ☐ CPB ☐ CPBS ☐ ENT 4,585 ☐ FENIZON (FRENCH) ☐ FENSON ☐ GC 928 ☐ MURVESCO ☐ PCBS ☐ PCI ☐ PCPBS ☐ TRIFENSON

TOXICITY DATA WITH REFERENCE
skn-rbt 500 mg/24H MLD 28ZPAK -,197,72
eye-rbt 100 mg/24H MOD 28ZPAK -,197,72
orl-rat LD50:1350 mg/kg ARSIM* 20,10,66
unk-mam LD50:1300 mg/kg 30ZDA9 -,274,71

SAFETY PROFILE: Moderately toxic by ingestion and possibly other routes. An eye and skin irritant. See also ESTERS and SULFONATES. An acaricide. When heated to decomposition it emits toxic fumes of Cl^- and SO_x.

CJR809 CAS:1982-36-1 HR: 3
1-(p-CHLORO-α-PHENYLBENZYL)HEXAHYDRO-4-METHYL-1H-1,4-DIAZEPINE DIHYDROCHLORIDE
mf: $C_{19}H_{23}ClN_2 \cdot 2ClH$ mw: 387.81

PROP: Very bitter crystals from EtOH. Mp: 227–228°.

SYNS: HOMOCHLORCYCLIZINE DIHYDROCHLORIDE ☐ HOMOCHLOROCYCLIZINE DIHYDROCHLORIDE ☐ SA 97 DIHYDROCHLORIDE

TOXICITY DATA WITH REFERENCE
orl-rat LD50:490 mg/kg JOALAS 31,237,60
ipr-rat LD50:80 mg/kg JOALAS 31,237,60
ivn-rat LD50:36 mg/kg JOALAS 31,237,60
orl-mus LD50:390 mg/kg JOALAS 31,237,60
ipr-mus LD50:125 mg/kg JOALAS 31,237,60
scu-mus LD50:135 mg/kg JOALAS 31,237,60
ivn-mus LD50:47 mg/kg JOALAS 31,237,60
ipr-dog LD50:50 mg/kg JOALAS 31,237,60

SAFETY PROFILE: Poison by ingestion, subcutaneous, intravenous, and intraperitoneal routes. When heated to decomposition it emits very toxic fumes of Cl^- and NO_x.

CJR909 CAS:68-88-2 HR: 3
1-(p-CHLORO-α-PHENYLBENZYL)-4-(2-((2-HYDROXYETHOXY)ETHYL)PIPERAZINE)
mf: $C_{21}H_{27}ClN_2O_2$ mw: 374.95

SYNS: ATARA ☐ ATARAX ☐ ATARAXOID ☐ ATARAZOID ☐ ATAZINA ☐ ATERAX ☐ 1-(p-CHLOROBENZHYDRYL)-4-(2-(2-HYDROXYETHOXY)ETHYL)DIETHYLENEDIAMINE ☐ 1-(p-CHLOROBENZHYDRYL)-4-(2-(2-HYDROXYETHOXY)ETHYL)PIPERAZINE ☐ N-(4-CHLOROBENZHYDRYL)-N'-(HYDROXYETHOXYETHYL)PIPERAZINE ☐ 1-(p-CHLORODIPHENYLMETHYL)-4-(2-(2-HYDROXYETHOXY)ETHYL)PIPERAZINE ☐ 2-(2-(4-(p-CHLORO-α-PHENYLBENZYL)-1-PIPERAZINYL)ETHOXY)ETHANOL ☐ DEINAIT ☐ EQUIPOISE ☐ FENAROL ☐ HYCHOTINE ☐ HYDROXINE ☐ HYDROXYCINE ☐ HYDROXYZINE ☐ IDROSSIZINA ☐ NEO-CALMA ☐ NEUROZINA ☐ NP 212 ☐ PAMAZONE ☐ PARENTERAL ☐ PAXISTIL ☐ PLACIDOL ☐ PLAXIDOL ☐ TRAN-Q ☐ TRAQUIZINE ☐ UCB 492 ☐ U.CB 4492 ☐ VESPARAZ-WIRKSTOFF

TOXICITY DATA WITH REFERENCE
orl-dog TDLo:1500 mg/kg (female 1-60D post):REP 28QFAD -,233,73
unr-mus TDLo:40 mg/kg (female 1D pre):REP FESTAS 12,346,61
orl-dog TDLo:350 mg/kg (female 15-21D post):REP 28QFAD ,233,73
orl-rat TDLo:400 mg/kg (13-16D preg):TER TXAPA9 17,67,70
orl-dog TDLo:350 mg/kg (female 15-21D post):TER 28QFAD -,233,73
orl-rat LD50:840 mg/kg CHTPBA 3,210,68
ipr-rat LD50:160 mg/kg CHTPBA 3,210,68
ivn-rat LD50:45 mg/kg ANPBAZ 61,669,61
orl-mus LD50:480 mg/kg AANEAB 7,87,63
ipr-mus LD50:81,300 μg/kg DPHFAK 23,281,71
ivn-mus LD50:137 mg/kg 27ZQAG -,237,72

SAFETY PROFILE: Poison by intravenous and intraperitoneal routes. Moderately toxic by ingestion. Experimental teratogenic and reproductive effects. When heated to decomposition it emits very toxic fumes of Cl^- and NO_x.

CJR959 HR: 3
2-(α-(p-CHLOROPHENYL)BENZYLOXY)-N,N-DIMETHYLETHYLAMINE HYDROCHLORIDE
mf: $C_{17}H_{20}ClNO \cdot ClH$ mw: 326.29

SYN: SUBSTANZ NR. 1602 (GERMAN)

TOXICITY DATA WITH REFERENCE
ivn-mus LD50:46 mg/kg ARZNAD 4,189,54

scu-gpg LD50:94 mg/kg ARZNAD 4,189,54

SAFETY PROFILE: Poison by intravenous and subcutaneous routes. When heated to decomposition it emits very toxic fumes of NO_x and Cl^-.

CJT125 CAS:15842-89-4 HR: 3
1-(4-CHLOROPHENYL)BIGUANIDINIUM HYDROGEN DICHROMATE
mf: $C_8H_{12}ClCr_2N_5O_7$ mw: 429.66

$ClC_6H_4NHC(:NH)NHC(:NH)N^+H_3HCr_2O_7^-$

CONSENSUS REPORTS: Chromium compounds are on the Community Right-To-Know List.

SAFETY PROFILE: Decomposes violently above 130°C. When heated to decomposition it emits toxic fumes of Cl^- and NO_x. See also CHROMIUM COMPOUNDS.

CJT750 CAS:80-33-1 HR: 2
4-CHLOROPHENYL-4-CHLOROBENZENESULFONATE
mf: $C_{12}H_8Cl_2O_3S$ mw: 303.16

PROP: Crystals. Mp: 86.5°.

SYNS: ACARICYDOL E 20 □ C-854 □ C 1,006 □ CCS □ CHLOORFENSON (DUTCH) □ (4-CHLOOR-FENYL)-4-CHLOOR-BENZEEN-SULFONAAT (DUTCH) □ CHLOREFENIZON (FRENCH) □ CHLORFENSON □ CHLORFENSONE □ 4-CHLOROBENZENESULFONATE de 4-CHLOROPHENYLE (FRENCH) □ p-CHLOROBENZENESULFONIC ACID-p-CHLOROPHENYL ESTER □ CHLOROFENIZON □ p-CHLOROPHENYL-p-CHLOROBENZENE SULFONATE □ 4-CHLOROPHENYL-4-CHLOROBENZENESULPHONATE □ 4-CHLORPHENYL-4'-CHLORBENZOLSULFONAT (GERMAN) □ (4-CHLOR-PHENYL)-4-CHLOR-BENZOL-SULFONATE (GERMAN) □ (4-CLORO-FENIL)-4-CLORO-VENZOL-SOLFONATO (ITALIAN) □ COROTRAN □ CPCBS □ D 854 □DIFEN-SON □ ENT 16,358 □ EPHIRSULPHONATE □ ESTER SULFONATE □ ESTONMITE □ ETHERSULFONATE □ GENITE 883 □ K 6451 □ LETHALAIRE G-58 □ MITICIDE K-101 □ NIAGARATRAN □ ONEX □ ORTHOTRAN □ OTRACID □ OVATRAN □ OVEX □ OVOCHLOR □ OVOTOX □ OVOTRAN □ PCPCBS □ SAPPILAN □ SAPPIRAN □ TRICHLORFENSON (OBS.)

TOXICITY DATA WITH REFERENCE
orl-mus TDLo:115 g/kg/78W-I:ETA NTIS** PB223-159
orl-rat LD50:2000 mg/kg GUCHAZ 6,381,73
unr-rat LD50:2050 mg/kg DABBBA 32,4116,72
orl-mus LD50:2000 mg/kg FMCHA2 -,C167,83
orl-rbt LD50:5660 mg/kg PCOC** -,841,66
orl-gpg LD50:640 mg/kg PCOC** -,841,66
orl-ckn LD50:3780 mg/kg PCOC** -,841,66

SAFETY PROFILE: Moderately toxic by ingestion and possibly other routes. Questionable carcinogen with experimental tumorigenic data. A pesticide. When heated to decomposition it emits very toxic fumes of Cl^- and SO_x. See also SULFONATES and CHLORINATED HYDROCARBONS, AROMATIC.

CJU125 CAS:17710-62-2 HR: 2
p-CHLOROPHENYL-N-(4'-CHLOROPHENYL)THIOCARBAMATE
mf: $C_{13}H_9Cl_2NOS$ mw: 298.19

SYN: p-CHLOROTHIOCARBANILIC ACID-o-(p-CHLOROPHENYL) ESTER

TOXICITY DATA WITH REFERENCE
orl-mus LD50:2161 mg/kg YKKZAJ 88,465,68
ipr-mus LD50:426 mg/kg YKKZAJ 88,465,68
scu-mus LD50:462 mg/kg YKKZAJ 88,465,68

SAFETY PROFILE: Moderately toxic by ingestion, subcutaneous, and intraperitoneal routes. When heated to decomposition it emits toxic fumes of Cl^-, SO_x, and NO_x. See also CARBAMATES and ESTERS.

CJU200 CAS:92-00-2 HR: 1
N-(3-CHLOROPHENYL)DIETHANOLAMINE
mf: $C_{10}H_{14}ClNO_2$ mw: 215.70

SYNS: N,N-BIS(2-HYDROXYETHYL)CHLOROANILIDE □ N-(m-CHLOROPHENYL)DIETHANOLAMINE □ DIETHANOLAMINOCHLOROBENZENE □ N,N-DIETHANOLANILIDE, 3-CHLORO- □ DIETHANOLCHLOROANILIDE □ N,N-DIHYDROXYETHYL-m-CHLOROANILINE □ N,N-DIHYDROXYETHYL-3-CHLOROANILINE □ EMERY 5715 □ EMERY 5717 □ ETHANOL, 2,2'-((3-CHLOROPHENYL)IMINO)BIS-

TOXICITY DATA WITH REFERENCE
orl-rat LD50:5150 mg/kg LONZA# 13JUL81

CONSENSUS REPORTS: Reported in EPA TSCA Inventory.

SAFETY PROFILE: Mildly toxic by ingestion. When heated to decomposition it emits toxic vapors of NO_x and Cl^-.

CJU250 CAS:120-32-1 HR: 3
4-CHLORO-α-PHENYL-o-CRESOL
mf: $C_{13}H_{11}ClO$ mw: 218.69

PROP: Nearly colorless flakes. Mp: 48.5°, bp: 160–162° @ 3.5 mm, d: 1.2 @ 55°/25°.

SYNS: o-BENZYL-p-CHLOROPHENOL □ 2-BENZYL-4-CHLOROPHENOL □ BIO-CLAVE □ 5-CHLORO-2-HYDROXYDIPHENYLMETHANE □ CHLOROPHENE □ 4-CHLORO-2-(PHENYLMETHYL)PHENOL □ CLOROPHENE □ KETOLIN-H □ NCI-C61201 □ NEOSABENYL □ SANTOPHEN □ SANTOPHEN I GERMICIDE □ SENTIPHENE

TOXICITY DATA WITH REFERENCE
msc-hmn:lym 40 mg/L MUREAV 196,61,88
msc-mus:lym 25 mg/L MUREAV 196,61,88
skn-mus TDLo:14 g/kg/34W-I:ETA CNREA8 19,413,59
orl-rat LD50:1700 mg/kg JPMSAE 63,1068,74
orl-mus LD50:65 mg/kg PHTXA6 22,270,59
scu-mus LD50:350 mg/kg PHTXA6 22,270,59

CONSENSUS REPORTS: Reported in EPA TSCA Inventory. Chlorophenol compounds are on the Community Right-To-Know List.

SAFETY PROFILE: A poison by ingestion and subcutaneous routes. Questionable carcinogen with experimental tumorigenic data. Mutation data reported. When heated to decomposition it emits toxic fumes of Cl^-. See also CHLOROPHENOLS.

CJU300 CAS:27695-55-2 HR: 3
1-(3-CHLOROPHENYL)-2-((4-(DICHLOROACETYL) PHENYL)AMINO)-2-HYDROXYETHANONE
mf: $C_{16}H_{12}Cl_3NO_3$ mw: 372.64

SYNS: ETHANONE, 1-(3-CHLOROPHENYL)-2-((4-(DICHLOROACETYL)PHENYL)AMINO)-2-HYDROXY- ☐ KETONE, 1-(3-CHLOROPHENYL)-2-((4-(DICHLOROACETYL)PHENYL)AMINO)-2-HYDROXY-

TOXICITY DATA WITH REFERENCE
ipr-mus LD50:2400 mg/kg ARZNAD 23,573,73

DOT CLASSIFICATION: 3; *Label:* Flammable Liquid

SAFETY PROFILE: Moderately toxic by intraperitoneal route. A flammable liquid. When heated to decomposition it emits toxic vapors of NO_x and Cl^-.

CJV250 CAS:35367-38-5 HR: 2
1-(4-CHLOROPHENYL)-3-(2,6-DIFLUOROBENZOYL) UREA
mf: $C_{14}H_9ClF_2N_2O_2$ mw: 310.70

PROP: Crystals. Mp: 230–232°.

SYNS: N-(((4-CHLOROPHENYL)AMINO)CARBONYL)-2,6-DIFLUOROBENZAMIDE ☐ DIFLUBENZURON ☐ DIFLURON ☐ DIMILIN ☐ DU 112307 ☐ ENT 29,054 ☐ OMS 1804 ☐ PDD 6040I ☐ PH 60-40 ☐ PHILIPS-DUPHAR PH 60-40 ☐ TH 6040 ☐ THOMPSON-HAYWARD TH6040

TOXICITY DATA WITH REFERENCE
mmo-sat 1 mg/plate PCBPBS 10,174,79
cyt-mus-unr 500 mg/kg TGANAK 16(1),45,82
orl-rat LD50:4640 mg/kg 85JCAE-,968,86
orl-mus LD50:4640 mg/kg SPEADM 78-1,20,78
skn-rbt LD50:2000 mg/kg SPEADM 74-1,-,74

CONSENSUS REPORTS: Reported in EPA TSCA Inventory. EPA Genetic Toxicology Program.

SAFETY PROFILE: Moderately toxic by skin contact. Mildly toxic by ingestion. Mutation data reported. When heated to decomposition it emits very toxic fumes of Cl^-, F^-, and NO_x.

CJW500 CAS:54708-51-9 HR: 3
1-(m-CHLOROPHENYL)-3-N,N-DIMETHYLCARBAMOYL-5-METHOXYPYRAZOLE
mf: $C_{13}H_{14}ClN_3O_2$ mw: 279.75

SYNS: 1-(3-CHLOROPHENYL)-3,N,N-DIMETHYLCARBAMOYL-5-METHOXYPYRAZOLE ☐ PZ 177

TOXICITY DATA WITH REFERENCE
orl-rat TDLo:2750 mg/kg (7-17D preg):TER IYKEDH 9,538,78
orl-rbt TDLo:5200 mg/kg (female 6-18D post):REP IYKEDH 9,558,78
orl-rat TDLo:3640 mg/kg (26W pre):REP IYKEDH 8,524,77
orl-rat TDLo:15 g/kg (male 30D pre):REP IYKEDH 8,494,77
orl-rat LD50:790 mg/kg NYKZAU 72,31,76
ipr-rat LD50:235 mg/kg NYKZAU 72,31,76
scu-rat LD50:1100 mg/kg NYKZAU 72,31,76
orl-mus LD50:955 mg/kg NYKZAU 72,31,76
ipr-mus LD50:490 mg/kg NYKZAU 72,31,76
scu-mus LD50:1190 mg/kg NYKZAU 72,31,76
orl-rbt LD50:1650 mg/kg IYKEDH 8,494,77
ipr-rbt LD50:1040 mg/kg IYKEDH 8,494,77

SAFETY PROFILE: Poison by intraperitoneal route. Moderately toxic by ingestion and subcutaneous routes. An experimental teratogen. Other experimental reproductive effects. An analgesic and anti-inflammatory agent. When heated to decomposition it emits very toxic fumes of NO_x and Cl^-.

CJX000 CAS:60719-83-7 HR: 3
2-(4-CHLOROPHENYL)-1,1-DIMETHYLETHYL 2-AMINOPROPANOATE HYDROCHLORIDE
mf: $C_{13}H_{18}ClNO_2 \cdot ClH$ mw: 292.23

SYN: p-CHLORO-α,α-DIMETHYL-2-AMINOPROPIONATE-PHENETHYL ALCOHOL HYDROCHLORIDE

TOXICITY DATA WITH REFERENCE
orl-mus LD50:901 mg/kg JMCMAR 21,448,78
ipr-mus LD50:200 mg/kg JMCMAR 21,448,78
ivn-mus LD50:54 mg/kg JMCMAR 21,448,78

SAFETY PROFILE: Poison by intraperitoneal and intravenous routes. Moderately toxic by ingestion. When heated to decomposition it emits very toxic fumes of Cl^- and NO_x.

CJX750 CAS:150-68-5 HR: 2
3-(p-CHLOROPHENYL)-1,1-DIMETHYLUREA
mf: $C_9H_{11}ClN_2O$ mw: 198.67

PROP: Crystals or thin rectangular prisms from MeOH; slight odor. Mp: 170.5–171.51°, vap press: 0.002 mm @ 100°. Sltly sol in water and hydrocarbons.

SYNS: 3-(4-CHLOOR-FENYL)-1,1-DIMETHYLUREUM (DUTCH) ☐ CHLORFENIDIM ☐ N (p CHLOROPHENYL)-N′,N′-DIMETHYLUREA ☐ N′-(4-CHLOROPHENYL)-N,N-DIMETHYLUREA ☐ 1-(p-CHLOROPHENYL)-3,3-DIMETHYLUREA ☐ 3-(4-CHLOROPHENYL)-1,1 DIMETHYLUREA ☐ 1-(4-CHLORO PHENYL)-3,3-DIMETHYLUREE (FRENCH) ☐ 3-(4-CHLOR-PHENYL)-1,1-DIMETHYL-HARNSTOFF (GERMAN) ☐ 3-(4-CLORO-FENIL)-1,1-DIMETIL-UREA (ITALIAN) ☐ CMU ☐ N,N-DIMETHYL-N′-(4-CHLOROPHENYL)UREA ☐ 1,1-DIMETHYL-3-(p-CHLOROPHENYL)UREA ☐ HERBICIDES, MONURON ☐ KARMEX MONURON HERBICIDE ☐ KARMEX W. MONURON HERBICIDE ☐ LIROBETAREX ☐ MONUREX ☐ MONURON ☐ MONUROX ☐ MONURUON ☐ MONUURON ☐ NCI-C02846 ☐ TELVAR ☐ TELVAR MONURON WEEDKILLER ☐ USAF P-8 ☐ USAF XR-41

TOXICITY DATA WITH REFERENCE
mma-sat 1 μg/plate MUREAV 58,353,78
dni-mus-orl 200 mg/kg MUREAV 58,353,78
otr-ham:emb 5 mg/L CRNGDP 4,291,83
orl-mus TDLo:1935 mg/kg (6-15D preg):TER NTIS** PB223-160
scu-mus TDLo:1935 mg/kg (female 6-14D post):TER NTIS** PB223-160
orl-mus TDLo:1935 mg/kg (6-15D preg):REP NTIS** PB223-160
orl-rat TDLo:46,350 mg/kg/2Y-C:CAR NTPTR* NTP-TR 266,88
orl-mus TDLo:4320 mg/kg/16W-I:CAR VOONAW 16(10),51,70
orl-rat LD:58 g/kg/18W-C:CAR VOONAW 16(10),51,70
orl-rat LD50:1053 mg/kg FAATDF 7,299,86

unk-rat LD50:3600 mg/kg JPFCD2 B15,929,80
orl-mus LD50:1920 mg/kg NTPTR* NTP-TR-266,88
ipr-mus LD50:1000 mg/kg NTIS** AD277-689
orl-gpg LDLo:670 mg/kg PAREAQ 14,225,62
unr-mam LD50:3500 mg/kg 30ZDA9 -,231,71

CONSENSUS REPORTS: IARC Cancer Review: Group 3 IMEMDT 7,56,87; Animal Sufficient Evidence IMEMDT 12,167,76. Reported in EPA TSCA Inventory. EPA Genetic Toxicology Program.

SAFETY PROFILE: Moderately toxic by ingestion, intraperitoneal, and possibly other routes. Experimental teratogenic and reproductive effects. Questionable carcinogen with experimental carcinogenic data. Mutation data reported. An herbicide. When heated to decomposition it emits very toxic fumes of NO_x, and Cl^-.

CJY000 CAS:140-41-0 HR: 2
3-(p-CHLOROPHENYL)-1,1-DIMETHYLUREA TRICHLOROACETATE
mf: $C_2HCl_3O_2 \cdot C_9H_{11}ClN_2O$ mw: 362.05

PROP: A solid. Mp: 78–81°. Sol in MeOH, xylene; sltly sol in H_2O.

SYNS: 3-(p-CHLOROPHENYL)-1,1-DIMETHYLUREA compounded with TRICHLOROACETIC ACID (1:1) □ GC-2996 □ MONURON-TCA □ TRICHLOROACETIC ACID compounded with N'-(4-CHLOROPHENYL)-N,N-DIMETHYLUREA (1:1) □ UROX 379 □ XORU-OX

TOXICITY DATA WITH REFERENCE
orl-rat LD50:2300 mg/kg 28ZEAL 4,292,69
scu-rbt LD50:1000 mg/kg FMCHA2 -,C249,83

SAFETY PROFILE: Moderately toxic by ingestion and subcutaneous routes. When heated to decomposition it emits very toxic fumes of Cl^- and NO_x. See also 3-(p-CHLOROPHENYL)-1,1-DIMETHYLUREA and TRICHLOROACETIC ACID.

CJY120 CAS:5131-60-2 HR: 3
4-CHLORO-m-PHENYLENEDIAMINE
mf: $C_6H_7ClN_2$ mw: 142.60

PROP: Needles. Mp: 91°.

SYNS: C.I. 76027 □ 4-CHLORO-1,3-BENZENEDIAMINE □ 1-CHLORO-2,4-DIAMINOBENZENE □ 4-CHLOROPHENE-1,3-DIAMINE □ 4-CHLOROPHENYLENE-1,3-DIAMINE □ 4-CHLORO-1,3-PHENYLENEDIAMINE □ 4-Cl-m-PD □ NCI-C03305

TOXICITY DATA WITH REFERENCE
mmo-sat 1 mg/plate ENMUDM 7(Suppl 5),1,85
mma-sat 10 µg/plate ENMUDM 7(Suppl 5),1,85
orl-rat TDLo:164 g/kg/78W-C:CAR NCITR* NCI-CG-TR-85,78
orl-mus TDLo:648 g/kg/77W-C:CAR CRNGDP 1,495,80
orl-mus TD:917 g/kg/78W-C:ETA NCITR* NCI-CG-TR-85,78
orl-mus TD:764 g/kg/78W-C:CAR NCITR* NCI-CG-TR-85,78
orl-rat TD:108 g/kg/77W-C:NEO CRNGDP 1,495,80
orl-mus TD:16,200 g/kg/96W-C:CAR CRNGDP 1,495,80
orl-rat TD:7500 g/kg/2Y-C:CAR CRNGDP 1,495,80
orl-rat TD:1092 g/kg/78W-C:CAR,REP IARC** 27,81,82
orl-mus TD:3745 g/kg/78W-C:CAR IARC** 27,81,82
orl-rat TD:2184 g/kg/78W-C:CAR,REP IARC** 27,81,82
orl-rat TD:5460 g/kg/78W-C:CAR IARC** 27,81,82
orl-mus TD:7140 g/kg/78W-C:CAR IARC** 27,81,82

CONSENSUS REPORTS: IARC Cancer Review: Group 3 IMEMDT 7,56,87; Animal Inadequate Evidence IMEMDT 27,81,82. NCI Carcinogenesis Bioassay (feed); Clear Evidence: mouse, rat NCITR* NCI-CG-TR-85,78. Reported in EPA TSCA Inventory.

SAFETY PROFILE: Suspected carcinogen with experimental carcinogenic, neoplastigenic, and tumorigenic data. Experimental reproductive effects. Mutation data reported. When heated to decomposition it emits toxic fumes of Cl^- and NO_x. See also AROMATIC AMINES.

CJZ000 CAS:155-00-0 HR: 3
o-CHLORO-β-PHENYLETHYLHYDRAZINE DIHYDROGEN SULPHATE
mf: $C_8H_{11}ClN_2 \cdot H_2O_4S$ mw: 268.74

TOXICITY DATA WITH REFERENCE
scu-mus TDLo:240 mg/kg (female 1-6D post):REP JOENAK 49,635,71
scu-mus TDLo:120 mg/kg (female 1-6D post):REP JOENAK 30,205,64
scu-mus TDLo:640 mg/kg (female 16D pre):REP JOENAK 30,205,64
scu-mus TDLo:1120 mg/kg (female 28D pre):REP JOENAK 30,205,64
scu-mus LD50:182 mg/kg JOENAK 30,205,64

SAFETY PROFILE: Poison by acute subcutaneous route. Experimental reproductive effects. See also SULFATES. When heated to decomposition it emits very toxic fumes of SO_x, Cl^-, and NO_x.

CKA000 CAS:33671-46-4 HR: 2
5-(2-CHLOROPHENYL)-7-ETHYL-1-METHYL-1,3-DIHYDRO-2H-THIENO(2,3-e)(1,4)DIAZEPIN-2-ONE
mf: $C_{16}H_{15}ClN_2OS$ mw: 318.84

PROP: A solid. Mp: 243–246°.

SYNS: 5-(o-CHLOROPHENYL)-7-ETHYL-1,3-DIHYDRO-1-METHYL-2H-THIENO(2,3-e)-1,4-DIAZEPIN-2-ONE □ CLOTIAZEPAM □ RISE □ TRECALMO □ Y 6047

TOXICITY DATA WITH REFERENCE
orl-rat TDLo:7 g/kg (35D pre):REP KSRNAM 6,2228,72
orl-rat TDLo:175 mg/kg (female 35D pre):REP KSRNAM 6,2228,72
orl-mus TDLo:600 mg/kg (7-12D preg):TER KSRNAM 6,2264,72
orl-rat LD50:1461 mg/kg NIIRDN 6,233,82
ipr-rat LD50:682 mg/kg NIIRDN 6,233,82
orl-mus LD50:636 mg/kg JMCMAR 16,214,73
ipr-mus LD50:440 mg/kg JMCMAR 16,214,73
scu-mus LD50:2837 mg/kg NIIRDN 6,233,82

SAFETY PROFILE: Moderately toxic by ingestion, subcutaneous, and intraperitoneal routes. An experimental teratogen. Other experimental reproductive effects. A tranquilizer. When heated to decomposition it emits very toxic fumes of SO_x, NO_x, and Cl^-. See also DIAZEPAM.

CKA500 **CAS:13822-05-4** **HR: 3**
1-(p-CHLOROPHENYL)-1,2,3,4,5,6-HEXAHYDRO-2,5-BENZODIAZOCINO DIHYDROCHLORIDE
mf: $C_{16}H_{17}ClN_2$•2ClH mw: 345.72

SYNS: 1-(p-CHLOROPHENYL)-1,2,3,4,5,6-HEXAHYDRO-2,5-BENZODIAZOCINE, DIHYDROCHLORIDE ☐ WY 5244

TOXICITY DATA WITH REFERENCE
orl-rat LD50:100 mg/kg TXAPA9 18,185,71
orl-bwd LD50:100 mg/kg TXAPA9 21,315,72

SAFETY PROFILE: Poison by ingestion. When heated to decomposition it emits very toxic fumes of NO_x and Cl^-.

CKA550 **CAS:555-60-2** **HR: 3**
((3-CHLOROPHENYL)HYDRAZONO)PROPANEDINITRILE
mf: $C_9H_5ClN_4$ mw: 204.63

SYNS: CARBONYL CYANIDE, 3-CHLOROPHENYLHYDRAZONE ☐ CCCP ☐ CCP ☐ MESOXALONITRILE, (m-CHLOROPHENYL)HYDRAZONE (8CI) ☐ PROPANEDINITRILE, (3-CHLOROPHENYL)HYDRAZONO)-(9CI)

TOXICITY DATA WITH REFERENCE
scu-rat LDLo:50 mg/kg EJMCA5 12,361,77
ipr-mus LD50:8 mg/kg BCPCA6 18,1389,69

CONSENSUS REPORTS: Reported in EPA TSCA Inventory.

SAFETY PROFILE: Poison by subcutaneous and intraperitoneal routes. When heated to decomposition it emits toxic vapors of NO_x and Cl^-.

CKA575 **CAS:21905-40-8** **HR: 3**
(m-CHLOROPHENYL)HYDROXY(β-HYDROXYPHENETHYL)ARSINE OXIDE
mf: $C_{14}H_{14}AsClO_3$ mw: 340.65

SYNS: ARSINE OXIDE, (m-CHLOROPHENYL)HYDROXY(β-HYDROXYPHENETHYL)- ☐ 2-PHENYL-2-HYDROXYETHYL, m-CHLOROPHENYL ARSINIC ACID

TOXICITY DATA WITH REFERENCE
ivn-mus LD50:100 mg/kg CSLNX* NX#06919

OSHA PEL: TWA 0.5 mg(As)/m^3

SAFETY PROFILE: Poison by intravenous route. When heated to decomposition it emits toxic fumes of As and Cl^-.

CKA600 **CAS:24671-21-4** **HR: 3**
4-(4-(p-CHLOROPHENYL)-4-HYDROXYPIPERIDINO)-4′-(DIMETHYLAMINO)BUTYROPHENO NE
mf: $C_{23}H_{29}ClN_2O_2$ mw: 400.99

SYNS: BUTYROPHENONE, 4-(4-(p-CHLOROPHENYL)-4-HYDROXYPIPERIDINO)-4′-(DIMETHYLAMINO)- ☐ 4-(DIMETHYLAMINO)PHENYL-3-(4-(4 CHLOROPHENYL)-4-HYDROXYPIPERIDINO)-PROPYL KETONE

TOXICITY DATA WITH REFERENCE
ivn-mus LD50:75 mg/kg CSLNX* NX#11083

DOT CLASSIFICATION: 3; *Label:* Flammable Liquid

SAFETY PROFILE: A poison by intravenous route. A flammable liquid. When heated to decomposition it emits toxic vapors of NO_x and Cl^-.

CKA750 **CAS:2909-38-8** **HR: 3**
m-CHLOROPHENYL ISOCYANATE
DOT: UN 2206/UN 2207/UN 2478/UN 3080
mf: C_7H_4ClNO mw: 153.57

PROP: Water-white liquid with irritating odors. Mp: −4°, bp: 101° @ 30 mm. Sol in organic solvents.

SYNS: m-CHLORFENYLISOKYANAT ☐ ISOCYANIC ACID-m-CHLOROPHENYL ESTER

TOXICITY DATA WITH REFERENCE
ihl-mam LD50:63 mg/kg GTPZAB 11(4),23,67

CONSENSUS REPORTS: Reported in EPA TSCA Inventory.

DOT CLASSIFICATION: 6.1; *Label:* KEEP AWAY FROM FOOD (UN 2207); DOT Class: 6.1; *Label:* Poison (UN 2206); DOT Class: 6.1; *Label:* Poison, Flammable Liquid (UN 3080); DOT Class: 3; *Label:* Flammable Liquid, Poison (UN 2478)

SAFETY PROFILE: Poison by inhalation and ingestion. A flammable liquid when exposed to heat or flame. When heated to decomposition it emits toxic fumes of Cl^-, CN^-, and NO_x. See also ESTERS and THIOCYANATES.

CKB000 **CAS:104-12-1** **HR: 3**
p-CHLOROPHENYL ISOCYANATE
DOT: UN 2206/UN 2207/UN 2478/UN 3080
mf: C_7H_4ClNO mw: 153.57

PROP: White solid or crystals. Mp: 31°, bp: 204°. Sol in org solvs.

SYNS: p-CHLORFENYLISOKYANAT (CZECH) ☐ ISOCYANIC ACID-p-CHLOROPHENYL ESTER ☐ PCPI

TOXICITY DATA WITH REFERENCE
skn-rbt 20 mg/24H MOD 85JCAE-,931,86
eye-rbt 250 μg/24H SEV 85JCAE-,931,86
ihl-man TCLo:800 μg/m^3/1M HYSAAV 31(7-9),481,66
orl-rat LD50:4710 mg/kg 28ZPAK -,165,72
orl-mus LD50:450 mg/kg 85GMAT-,38,82
ihl-mus LCLo:40 mg/m^3 HYSAAV 31(7-9),481,66

CONSENSUS REPORTS: Reported in EPA TSCA Inventory.

DOT CLASSIFICATION: 6.1; *Label:* KEEP AWAY FROM FOOD (UN 2207); DOT Class: 6.1; *Label:* Poison (UN 2206); DOT Class: 6.1; *Label:* Poison, Flammable Liquid (UN 3080); DOT Class: 3; *Label:* Flammable Liquid, Poison (UN 2478)

SAFETY PROFILE: Poison by ingestion and inhalation. Moderately toxic by ingestion. Unspecified human systemic effects. A severe eye and moderate skin irritant. A flammable liquid when exposed to heat or flame. Dangerous, can explode on distillation. When heated to decomposition it emits toxic fumes of Cl^-, CN^-, and NO_x.

C

CKB250 **CAS:500-92-5** **HR: 3**
1-(p-CHLOROPHENYL)-5-ISOPROPYLBIGUANIDE
mf: $C_{11}H_{16}ClN_5$ mw: 253.77

PROP: White powder or rectangular plates from toluene. Mp: 130–131°.

SYNS: BIGUMAL □ CHLORGUANIDE □ CHLOROGUANIDE □ PALUDRINE □ PROGUANIL

TOXICITY DATA WITH REFERENCE
orl-rat TDLo: 700 mg/kg (7D pre): REP MEXPAG 10,361,64
orl-mus TDLo: 504 mg/kg (female 42D pre): REP MEXPAG 10,361,64
orl-mus TDLo: 252 mg/kg (female 21D pre): REP MEXPAG 10,361,64
ipr-rat LD50: 15 mg/kg 14XBAV -,367,64
ivn-rat LDLo: 40 mg/kg CLDND* 4,14,49
orl-rbt LD50: 150 mg/kg CLDND* 4,14,49
ipr-rbt LD50: 50 mg/kg CLDND* 4,14,49
orl-ckn LDLo: 400 mg/kg CLDND* 4,14,49
ivn-ckn LDLo: 60 mg/kg CLDND* 4,14,49

SAFETY PROFILE: Poison by ingestion, intravenous, and intraperitoneal routes. Experimental reproductive effects. When heated to decomposition it emits toxic fumes of Cl^- and NO_x.

CKB500 **CAS:637-32-1** **HR: 3**
1-(p-CHLOROPHENYL)-5-ISOPROPYLBIGUANIDE HYDROCHLORIDE
mf: $C_{11}H_{16}ClN_5 \cdot ClH$ mw: 290.23

SYNS: CHLORGUANIDE HYDROCHLORIDE □ CHLOROGUANIDE HYDROCHLORIDE □ CHLOROGUANIDINE HYDROCHLORIDE □ N-4-CHLOROPHENYL-N[5]-ISOPROPYLDIGUANIDE HYDROCHLORIDE □ DIGUANYL □ DRINUPAL HYDROCHLORIDE □ GUANATOL HYDROCHLORIDE □ M 4888 □ PALUDRINE HYDROCHLORIDE □ PALUSIL HYDROCHLORIDE □ PROGUANIL HYDROCHLORIDE □ SN 12,837 □ TIRIAN HYDROCHLORIDE □ 3359 RP

TOXICITY DATA WITH REFERENCE
orl-mus TDLo: 168 mg/kg (7D pre/1-7D preg): REP ANTCAO 12,671,62
orl-rat LD50: 58 mg/kg JPETAB 91,157,47
ivn-rat LD50: 33 mg/kg JPETAB 91,157,47
orl-mus LD50: 27 mg/kg JPETAB 91,157,47
ivn-mus LD50: 23 mg/kg JPETAB 91,157,47
orl-rbt LD50: 172 mg/kg BJPCAL 2,181,47
ivn-rbt LD50: 44,850 μg/kg JPETAB 91,157,47
ivn-gpg LD50: 39,510 μg/kg JPETAB 91,157,47
orl-ckn LD50: 400 mg/kg BJPCAL 2,181,47

SAFETY PROFILE: Poison by ingestion and intravenous routes. Experimental reproductive effects. An antimalarial drug. When heated to decomposition it emits very toxic fumes of Cl^- and NO_x.

CKC000 **CAS:101-21-3** **HR: 2**
N-3-CHLOROPHENYLISOPROPYLCARBAMATE
mf: $C_{10}H_{12}ClNO_2$ mw: 213.68

PROP: Light-brown, crystalline solid; faint characteristic odor. Mp: 41°, bp: 149° @ 2 mm, d: 1,180 @ 30°. Very sparingly sol in water.

SYNS: BEET-KLEEN □ BUD-NIP □ N-(3-CHLOOR-FENYL)-ISOPROPYL CARBAMAAT (DUTCH) □ CHLOR-IFC □ CHLOR-IPC □ m-CHLOROCARBANILIC ACID, ISOPROPYL ESTER □ 3-CHLOROCARBANILIC ACID, ISOPROPYL ESTER □ N-(3-CHLORO PHENYL) CARBAMATE D'ISOPROPYLE (FRENCH) □ N-(3-CHLOROPHENYL)CARBAMIC ACID, ISOPROPYL ESTER □ (3-CHLOROPHENYL)CARBAMIC ACID, 1-METHYLETHYL ESTER □ CHLOROPROPHAM □ N-(3-CHLOR-PHENYL)-ISOPROPYL-CARBAMAT (GERMAN) □ CHLORPROPHAM □ CHLORPROPHAME (FRENCH) □ CICP □ CI-IPC □ CIPC □ N-(3-CLOROFENIL)-ISOPROPIL-CARBAMMATO (ITALIAN) □ ELBANIL □ ENT 18,060 □ FASCO WY-HOE □ FURLOE □ FURLOE 4EC □ ISOPROPYL-m-CHLOROCARBANILATE □ ISOPROPYL-3-CHLOROCARBANILATE □ ISOPROPYL-3-CHLOROPHENYLCARBAMATE □ ISOPROPYL-N-(3-CHLOROPHENYL)CARBAMATE □ o-ISOPROPYL-N-(3-CHLOROPHENYL)CARBAMATE □ ISOPROPYL-N-(3-CHLORPHENYL)-CARBAMAT (GERMAN) □ JACK WILSON CHLORO 51 (oil) □ LIRO CIPC □ METOXON □ NEXOVAL □ PREVENOL □ PREVENOL 56 □ PREVEN-TOL □ PREVENTOL 56 □ PREWEED □ SPROUT NIP □ SPROUT-NIP EC □ SPUD-NIC □ SPUD-NIE □ STOPGERME-S □ TATERPEX □ TRI-HERBICIDE CIPC □ UNICROP CIPC □ Y 3

TOXICITY DATA WITH REFERENCE
mmo-smc 100 mg/L PMRSDJ 1,414,81
dns-hmn:fbr 4 mg/L PMRSDJ 1,528,81
scu-mus TDLo: 9 g/kg (6-14D preg): TER NTIS** PB223-160
orl-mus TDLo: 600 mg/kg: NEO BJCAAI 12,355,58
orl-rat LD50: 1200 mg/kg AEHLAU 19,621,69
ipr-rat LD50: 700 mg/kg CRSBAW 175,496,81
ipr-mus LD50: 2600 mg/kg CRSBAW 175,496,81
orl-rbt LD50: 5000 mg/kg PAREAQ 14,225,62

CONSENSUS REPORTS: IARC Cancer Review: Group 3 IMEMDT 7,56,87; Animal Inadequate Evidence IMEMDT 12,55,76. EPA Genetic Toxicology Program.

SAFETY PROFILE: Moderately toxic by ingestion and intraperitoneal routes. Questionable carcinogen with experimental neoplastigenic and teratogenic data. Human mutation data reported. An herbicide. When heated to decomposition it emits highly toxic fumes of Cl^-, NO_x, and phosgene. See also CARBAMATES.

CKC325 **CAS:14774-78-8** **HR: 3**
4-CHLOROPHENYLLITHIUM
mf: C_6H_4ClLi mw: 118.49

PROP: A solid. Sol in ethers; insol in hexane.

SAFETY PROFILE: Explodes on contact with oxygen. When heated to decomposition it emits toxic fumes of Cl^-. See also LITHIUM COMPOUNDS.

CKC500 **CAS:1631-29-4** **HR: 3**
N-(p-CHLOROPHENYL)MALEIMIDE
mf: $C_{10}H_6ClNO_2$ mw: 207.62

TOXICITY DATA WITH REFERENCE
skn-rbt 500 mg MLD SCCUR* -,7,61
orl-rat LD50: 580 mg/kg SCCUR* -,7,61
orl-mus LD50: 290 mg/kg SCCUR* -,7,61
orl-rbt LDLo: 350 mg/kg SCCUR* -,7,61

SAFETY PROFILE: Poison by ingestion. A skin irritant. When heated to decomposition it emits very toxic fumes of Cl^- and NO_x.

CKD250 CAS:54708-68-8 HR: 2
1-(3-CHLOROPHENYL-5-METHOXY-N-METHYL)-1H-PYRAZOLE-3-CARBOXAMIDE
mf: $C_{12}H_{12}ClN_3O_2$ mw: 265.72

SYNS: PC 222 □ PZ 222

TOXICITY DATA WITH REFERENCE
orl-rat LD50:1540 mg/kg NYKZAU 72,31,76
ipr-rat LD50:410 mg/kg NYKZAU 72,31,76
orl-mus LD50:1540 mg/kg NYKZAU 72,31,76
ipr-mus LD50:910 mg/kg NYKZAU 72,31,76
orl-rbt LD50:4800 mg/kg NYKZAU 72,31,76

SAFETY PROFILE: Moderately toxic by ingestion and intraperitoneal routes. When heated to decomposition it emits very toxic fumes of Cl^- and NO_x.

CKD500 CAS:1746-81-2 HR: 3
3-(4-CHLOROPHENYL)-1-METHOXY-1-METHYLUREA
mf: $C_9H_{11}ClN_2O_2$ mw: 214.67

PROP: A solid. Mp: 76–78°.

SYNS: AFESIN □ ARESIN □ AREZIN □ AREZINE □ ARRESIN □ N-(4-CHLOROPHENYL)-N'-METHOXY-N-METHYLUREA □ N'-(4-CHLOROPHENYL)-N-METHOXY-N-METHYLUREA □ 3-(4-CHLORPHENYL)-1-METHOXY-1-METHYLHARNSTOFF (GERMAN) □ HOE 2747 □ MONOLINURON □ PREMALIN

TOXICITY DATA WITH REFERENCE
orl-mus TDLo:8 g/kg (6-15D preg):TER ARTODN 38,261,77
orl-mus TDLo:8 g/kg (6-15D preg):REP ARTODN 38,261,77
orl-mus TDLo:4 g/kg (female 10-13D post):REP ARTODN 38,261,77
orl-mus TDLo:4 g/kg (female 10-13D post):TER ARTODN 38,261,77
orl-mus TDLo:8 g/kg (female 6-15D post):TER ARTODN 38,261,77
orl-rat LD50:1800 mg/kg WRPCA2 9,119,70
orl-dog LD50:500 mg/kg 28ZEAL 5,158,76

CONSENSUS REPORTS: Reported in EPA TSCA Inventory.

SAFETY PROFILE: Moderately toxic by ingestion. Experimental teratogenic and reproductive effects. When heated to decomposition it emits very toxic fumes of Cl^- and NO_x.

CKD750 CAS:1867-66-9 HR: 3
2-(o-CHLOROPHENYL)-2-(METHYLAMINO)CYCLOHEXANONE HYDROCHLORIDE
mf: $C_{13}H_{16}ClNO \cdot ClH$ mw: 274.21

SYNS: CI 581 □ CL 369 □ CN-52,372-2 □ KETAJECT □ KETALAR □ KETAMINE □ KETAMINE HYDROCHLORIDE □ KETANEST □ KETASET □ KETAVET □ KETOLAR □ VETALAR

TOXICITY DATA WITH REFERENCE
ipr-rat TDLo:1700 mg/kg (1-17D preg):TER REANDJ 26,137,79
ipr-rat TDLo:375 mg/kg (female 1-15D post):TER RCOCB8 54,413,86
unr-man TDLo:1 mg/kg:CNS CSLNX* NX#03551
ivn-hmn TDLo:2 mg/kg:CNS BMJOAE 2,943,76
mul-inf TDLo:11,587 μg/kg:PUL BJANAD 58,573,86
orl-rat LD50:447 mg/kg NIIRDN 6,264,82
ipr-rat LD50:224 mg/kg TXAPA9 18,185,71
ivn-rat LD50:58,880 μg/kg NIIRDN 6,264,82
orl-mus LD50:617 mg/kg NIIRDN 6,164,82
ivn-mus LD50:224 mg/kg NIIRDN 6,262,82
ivn-mus LD50:55,900 μg/kg NIIRDN 6,264,82
ims-gpg LD50:361 mg/kg VHTODE 24,410,82
ivn-ckn LD50:67,900 μg/kg AJVRAH 45,531,84

CONSENSUS REPORTS: EPA Genetic Toxicology Program.

SAFETY PROFILE: Poison by intramuscular, intraperitoneal, and intravenous routes. Moderately toxic by ingestion. Human systemic effects by intravenous and possibly other routes: analgesia, coma, hallucinations and distorted perceptions, dyspnea. An experimental teratogen. An anesthetic. When heated to decomposition it emits very toxic fumes of Cl^- and NO_x.

CKD800 CAS:73791-42-1 HR: 3
(4-CHLOROPHENYL)METHYLARSINIC ACID
mf: $C_7H_8AsClO_2$ mw: 234.52

SYNS: ARSINE OXIDE, (p-CHLOROPHENYL)HYDROXYMETHYL- □ (p-CHLOROPHENYL)HYDROXYMETHYLARSINE OXIDE

TOXICITY DATA WITH REFERENCE
ivn-mus LD50:100 mg/kg CSLNX* NX#01207

OSHA PEL: TWA 0.5 mg(As)/m³

SAFETY PROFILE: Poison by intravenous route. When heated to decomposition it emits toxic fumes of As and Cl^-.

CKE000 CAS:511-46-6 HR: 3
2-(α-(p-CHLOROPHENYL)-α-METHYLBENZYLOXY)-N,N DIETHYLETHYLAMINE
mf: $C_{20}H_{26}ClNO$ mw: 331.92

SYNS: ETHYLAMINE, 2-(α-(p-CHLOROPHENYL)-α-METHYLBENZYLOXY)-N,N-DIETHYL □ SUBSTANZ NR. 1925 (GERMAN)

TOXICITY DATA WITH REFERENCE
ivn-mus LD50:32 mg/kg ARZNAD 4,189,54
scu-gpg LD50:120 mg/kg ARZNAD 4,189,54

SAFETY PROFILE: Poison by intravenous and subcutaneous routes. See also AMINES. When heated to decomposition it emits very toxic fumes of Cl^- and NO_x.

CKE250 CAS:102584-42-9 HR: 3
1-(2-(α-(p-CHLOROPHENYL)-α-METHYLBENZYLOXY)) ETHYL PYRROLIDINE
mf: $C_{20}H_{24}ClNO$ mw: 329.90

SYN: SUBSTANZ NR. 2135

TOXICITY DATA WITH REFERENCE
scu-mus LD50:90 mg/kg ARZNAD 4,189,54
ivn-mus LD50:50 mg/kg ARZNAD 4,189,54

SAFETY PROFILE: Poison by subcutaneous and intrave-

C

nous routes. When heated to decomposition it emits very toxic fumes of Cl^- and NO_x.

CKE750 CAS:79-93-6 HR: 3
2-p-CHLOROPHENYL-3-METHYL-2,3-BUTANEDIOL
mf: $C_{11}H_{15}ClO_2$ mw: 214.71

SYNS: ACALMID □ ACALO □ ALTERTON □ ATADIOL □ B586 □ 2-(4-CHLOROPHENYL)-3-METHYL-2,3-BUTANEDIOL □ ENT 15,208 □ FELIXYN □ FENAGLICODOLO □ PAUSITAL □ PHENAGLYCODOL □ PHENGLYKODOL □ REMIN □ SEDAPSIN □ SINFORIL □ STESIL □ ULTRAN □ USAF EL-44

TOXICITY DATA WITH REFERENCE
orl-rat LD50:832 mg/kg TXAPA9 21,315,72
orl-mus LD50:514 mg/kg 27ZQAG -,401,72
ipr-mus LD50:200 mg/kg NTIS** AD277-689
scu-mus LD50:358 mg/kg 27ZQAG -,401,72
ivn-mus LD50:254 mg/kg 27ZQAG -,401,72
orl-bwd LD50:32 mg/kg TXAPA9 21,315,72

SAFETY PROFILE: Poison by ingestion, intraperitoneal, subcutaneous, and intravenous routes. A tranquilizer. When heated to decomposition it emits toxic fumes of Cl^-.

CKF000 CAS:3942-54-9 HR: 3
o-CHLOROPHENYL METHYLCARBAMATE
mf: $C_8H_8ClNO_2$ mw: 185.62

SYNS: 2-CHLOROPHENYL-N-METHYLCARBAMATE □ CPMC □ ETROFOL □ HOPCIDE

TOXICITY DATA WITH REFERENCE
orl-rat LD50:648 mg/kg FMCHA2 -,D132,80
orl-mus LD50:150 mg/kg GUCHAZ 6,115,73

CONSENSUS REPORTS: EPA Genetic Toxicology Program. Reported in EPA TSCA Inventory.

SAFETY PROFILE: A poison by ingestion and possibly other routes. An insecticide. See also CARBAMATES. When heated to decomposition it emits very toxic fumes of Cl^- and NO_x.

CKF500 CAS:80-77-3 HR: 3
2-(4-CHLOROPHENYL)-3-METHYL-4-METATHIAZA-NONE-1,1-DIOXIDE
mf: $C_{11}H_{12}ClNO_3S$ mw: 273.75

SYNS: BANABIN □ BANABIN-SINTYAL □ BISINA □ CHLORMETHAZANONE □ CHLORMETHAZONE □ CHLORMEZANONE □ 2-(p-CHLOROPHENYL)TETRAHYDRO-3-METHYL-4H-1,3-THIAZIN-4-ONE 1,1-DIOXIDE □ 2-(p-CHLORPHENYL)-3-METHYL-1,3-PERHYDROTHIAZIN-4-ON-1,1-DIOXIDE □ CLORILAX □ CLORMETAZANONE □ CLORMETHAZON □ DICHLOROMETHAZANONE □ FENAROL □ LOBAK □ MIORILAX □ MIO-SED □ MUSKEL □ MUSKEL-TRANCOPAL □ PHENAROL □ REXAN □ RILANSYL □ RILAQUIL □ RILASSOL □ RILAX □ RILLASOL □ SUPOTRAN □ SUPROTAN □ TANAFOL □ TETRAHYDRO-2-(p-CHLOROPHENYL)-3-METHYL-4H-1,3-THIAZIN-4-ONE-1,1-DIOXIDE □ TRANCOPAL

TOXICITY DATA WITH REFERENCE
orl-man TDLo:157 mg/kg:EYE,CNS BMJOAE 292,732,86
orl-rat LD50:605 mg/kg AIPTAK 130,280,61
ipr-rat LD50:370 mg/kg ARZNAD 17,242,67
orl-mus LD50:600 mg/kg OYYAA2 9,601,75
ipr-mus LD50:570 mg/kg ARZNAD 17,242,67
scu-mus LD50:322 mg/kg APTOA6 19,247,62
orl-dog LD50:500 mg/kg TXAPA9 1,168,59
ipr-dog LD50:500 mg/kg TXAPA9 1,168,59
ipr-gpg LD50:600 mg/kg TXAPA9 1,168,59

SAFETY PROFILE: Poison by intraperitoneal and subcutaneous routes. Moderately toxic by ingestion. Human systemic effects by ingestion: dilation of the pupils, ataxia (loss of muscle coordination), and coma. When heated to decomposition it emits very toxic fumes of SO_x, NO_x, and Cl^-.

CKF750 CAS:3766-60-7 HR: 2
3-(p-CHLOROPHENYL)-1-METHYL-1-(1-METHYL-2-PROPYNYL)UREA
mf: $C_{12}H_{13}ClN_2O$ mw: 236.72

PROP: A solid. Very sltly sol in H_2O; sltly sol in C_6H_6; sol in Me_2CO and MeOH.

SYNS: ARISAN □ BUTURON □ BUTYRON □ N'-(4-CHLOROPHENYL)-N-ISOBUTINYL-N-METHYLUREA □ N'-(4-CHLOROPHENYL)-N-METHYL-N-(1-METHYL-2-PROPYNYL)-UREA □ N-(4-CHLORPHENYL)-N'-METHYL-N'-ISOBUTINYLHARNSTOFF (GERMAN) □ 3-(4-CHLORPHENYL)-1-METHYL-1-ISOBUTINYLHARNSTOFF (GERMAN) □ EPTAPUR □ H 95

TOXICITY DATA WITH REFERENCE
orl-mus TDLo:700 mg/kg (12-13D preg):REP ARTODN 38,261,77
orl-mus TDLo:3 g/kg (6-15D preg):TER ARTODN 38,261,77
orl-mus TDLo:3500 mg/kg (6-15D preg):TER ARTODN 38,261,77
orl-rat LD50:1791 mg/kg ARTODN 38,261,77
ipr-mus LD50:500 mg/kg 85DPAN -,-,71/76

CONSENSUS REPORTS: EPA Genetic Toxicology Program.

SAFETY PROFILE: Moderately toxic by ingestion and intraperitoneal routes. Experimental teratogenic and reproductive effects. An herbicide. When heated to decomposition it emits very toxic fumes of Cl^- and NO_x.

CKG000 CAS:15687-18-0 HR: 2
2-(p-CHLOROPHENYL)-4-METHYLPENTANE-2,4-DIOL
mf: $C_{12}H_{17}ClO_2$ mw: 228.74

SYNS: 2-(p-CHLOROPHENYL)-4-METHYL-2,4-PENTANEDIOL □ FENPENTADIOL □ 2-METHYL-4-(p-CHLOROPHENYL)-2,4-PENTANEDIOL □ RD 292 □ TREDUM

TOXICITY DATA WITH REFERENCE
orl-rat LD50:1200 mg/kg ARZNAD 21,9,71
orl-mus LD50:940 mg/kg ARZNAD 21,9,71

SAFETY PROFILE: Moderately toxic by ingestion. A tranquilizer and an analeptic agent (stimulant). When heated to decomposition it emits toxic fumes of Cl^-.

CKG500 CAS:123-09-1 HR: 3
p-CHLOROPHENYL METHYL SULFIDE
mf: C_7H_7ClS mw: 159.56

PROP: D: 1.222, bp: 169°.

SYNS: p-CHLOROTHIOANISOLE □ 4-CHLOROTHIOANISOLE □ METHYL-p-CHLOROPHENYL SULFIDE □ METHYL-4-CHLOROPHENYL SULFIDE

TOXICITY DATA WITH REFERENCE
eye-rbt 100 mg MLD NTIS** AD-A082-824
orl-rat LD50:400 mg/kg TOLED5 1000(Sp Iss 1),32,80
skn-rat LDLo:5630 mg/kg NTIS** AD-A082-824
orl-mus LD50:672 mg/kg NTIS** AD-A082-824

SAFETY PROFILE: Poison by ingestion. Mildly toxic by skin contact. An eye irritant. When heated to decomposition it emits very toxic fumes of Cl^- and SO_x. See also CHLORINATED HYDROCARBONS, AROMATIC; and SULFIDES.

CKG750 CAS:98-57-7 HR: 3
p-CHLOROPHENYL METHYL SULFONE
mf: $C_7H_7ClO_2S$ mw: 191.56

SYNS: 4-CHLOROPHENYL METHYL SULFONE □ METHYL 4 CHLOROPHENYL SULFONE

TOXICITY DATA WITH REFERENCE
skn-rbt 500 mg/24H MLD NTIS** AD-A082-824
orl-rat LD50:400 mg/kg NTIS** AD-A082-824
orl-mus LD50:606 mg/kg NTIS** AD-A082-824

SAFETY PROFILE: Poison by ingestion. A skin irritant. When heated to decomposition it emits very toxic fumes of Cl^- and SO_x.

CKH000 CAS:934-73-6 HR: 3
p-CHLOROPHENYL METHYL SULFOXIDE
mf: C_7H_7ClOS mw: 175.56

SYNS: 4-CHLOROPHENYL METHYL SULFOXIDE □ METHYL-4-CHLOROPHENYL SULFOXIDE

TOXICITY DATA WITH REFERENCE
skn-rbt 500 mg/24H SEV NTIS** AD-A082-824
eye-rbt 100 mg SEV NTIS** AD-A082-824
orl-rat LD50:463 mg/kg NTIS** AD-A082-824
orl-mus LD50:328 mg/kg JACTDZ 12,369,93

SAFETY PROFILE: A poison by ingestion. A severe eye and skin irritant. When heated to decomposition it emits very toxic fumes of Cl^- and SO_x.

CKI000 CAS:3818-90-4 HR: 3
1-p-CHLOROPHENYL PENTYL SUCCINATE
mf: $C_{15}H_{19}ClO_4$ mw: 298.79

SYNS: AF 425 □ α-BUTYL-p-CHLOROBENZYL ESTER of SUCCINIC ACID □ SUCCINATO ACIDO DI 1-p-CLOROFENILPENTILE (ITALIAN) □ SUCCINIC ACID-α-BUTYL-p-CHLOROBENZYL ESTER

TOXICITY DATA WITH REFERENCE
orl-rat LD50:1130 mg/kg BCFAAI 100,504,61
ipr-mus LD50:259 mg/kg BCFAAI 100,855,61
ivn-mus LD50:76 mg/kg BCFAAI 100,504,61

SAFETY PROFILE: Poison by intravenous and intraperitoneal routes. Moderately toxic by ingestion. When heated to decomposition it emits toxic fumes of Cl^-.

CKI020 CAS:25174-66-7 HR: 3
β-(p-CHLOROPHENYL)PHENETHYL 4-(o-CHLOROPHENYL)PIPERAZINYL KETONE
mf: $C_{25}H_{24}Cl_2N_2O$ mw: 439.41

SYNS: 1-(o-CHLOROPHENYL)-4-(3-(p-CHLOROPHENYL)-3-PHENYLPROPIONYL)PIPERAZINE □ KETONE, β-(p-CHLOROPHENYL)PHENETHYL 4-(o-CHLOROPHENYL)PIPERAZINYL □ PIPERAZINE, 1-(o-CHLOROPHENYL)-4-(3-(p-CHLOROPHENYL)-3-PHENYLPROPIONYL)-

TOXICITY DATA WITH REFERENCE
ipr-mus LD50:800 mg/kg JMCMAR 12,860,69

DOT CLASSIFICATION: 3; *Label:* Flammable Liquid

SAFETY PROFILE: Moderately toxic by intraperitoneal route. A flammable liquid. When heated to decomposition it emits toxic vapors of NO_x and Cl^-.

CKI030 CAS:23917-55-7 HR: 3
β-(p-CHLOROPHENYL)PHENETHYL 4-(m-METHYLBENZYL)PIPERAZINYL KETONE
mf: $C_{27}H_{29}ClN_2O$ mw: 433.03

SYNS: 1-(3-(p-CHLOROPHENYL)-3-PHENYLPROPIONYL)-4-(m-METHYLBENZYL)PIPERAZINE □ KETONE, β-(p-CHLOROPHENYL)PHENETHYL 4-(m-METHYLBENZYL)PIPERAZINYL □ PIPERAZINE, 1-(3-(p-CHLOROPHENYL)-3-PHENYLPROPIONYL)-4-(m-METHYLBENZYL)-

TOXICITY DATA WITH REFERENCE
ipr-mus LD50:800 mg/kg JMCMAR 12,860,69

DOT CLASSIFICATION: 3; *Label:* Flammable Liquid

SAFETY PROFILE: Moderately toxic by intraperitoneal route. A flammable liquid. When heated to decomposition it emits toxic vapors of NO_x and Cl^-.

CKI040 CAS:23902-89-8 HR: 3
β-(p-CHLOROPHENYL)PHENETHYL 4-PHENETHYLPIPERAZINYL KETONE
mf: $C_{27}H_{29}ClN_2O$ mw: 433.03

SYNS: 1-(3-(p-CHLOROPHENYL)-3-PHENYLPROPIONYL)-4-PHENETHYLPIPERAZINE □ KETONE, β-(p-CHLOROPHENYL)PHENETHYL 4-PHENETHYLPIPERAZINYL □ PIPERAZINE, 1-(3-(p-CHLOROPHENYL)-3-PHENYLPROPIONYL)-4-PHENETHYL-

TOXICITY DATA WITH REFERENCE
ipr-mus LD50:800 mg/kg JMCMAR 12,860,69

DOT CLASSIFICATION: 3; *Label:* Flammable Liquid

SAFETY PROFILE: Moderately toxic by intraperitoneal route. A flammable liquid. When heated to decomposition it emits toxic vapors of NO_x and Cl^-.

CKI050 CAS:23904-74-7 HR: 3
β-(p-CHLOROPHENYL)PHENETHYL 4-(2-PYRIDYL)PIPERAZINYL KETONE
mf: $C_{24}H_{24}ClN_3O$ mw: 405.96

SYNS: 1-(3-(p-CHLOROPHENYL)-3-PHENYLPROPIONYL)-4-(2-PYRIDYL)PIPERAZINE □ KETONE, β-(p-CHLOROPHENYL)PHENETHYL 4-(2-PYRIDYL)PIPERAZINYL □ PIPERAZINE, 1-(3-(p-CHLOROPHENYL)-3-PHENYLPROPIONYL)-4-(2-PYRIDYL)-

TOXICITY DATA WITH REFERENCE
ipr-mus LD50:800 mg/kg JMCMAR 12,860,69

DOT CLASSIFICATION: 3; *Label:* Flammable Liquid

SAFETY PROFILE: Moderately toxic by intraperitoneal route. A flammable liquid. When heated to decomposition it emits toxic vapors of NO_x and Cl^-.

CKI060 CAS:33656-20-1 HR: 3
β-(p-CHLOROPHENYL)PHENETHYL 4-(2-PYRIMIDYL) PIPERAZINYL KETONE
mf: $C_{23}H_{23}ClN_4O$ mw: 406.95

SYNS: 1-(3-(p-CHLOROPHENYL)-3-PHENYLPROPIONYL)-4-(2-PYRIMIDYL)PIPERAZINE □ KETONE, β-(p-CHLOROPHENYL)PHENETHYL 4-(2-PYRIMIDYL)PIPERAZINYL □ PIPERAZINE, 1-(3-(p-CHLOROPHENYL)-3-PHENYLPROPIONYL)-4-(2-PYRIMIDYL)-

TOXICITY DATA WITH REFERENCE
ipr-mus LD50:800 mg/kg JMCMAR 12,860,69

DOT CLASSIFICATION: 3; *Label:* Flammable Liquid

SAFETY PROFILE: Moderately toxic by intraperitoneal route. A flammable liquid. When heated to decomposition it emits toxic vapors of NO_x and Cl^-.

CKI070 CAS:23920-57-2 HR: 3
β-(p-CHLOROPHENYL)PHENETHYL 4-(2-THIAZOLYL) PIPERAZINYL KETONE
mf: $C_{22}H_{22}ClN_3OS$ mw: 411.98

SYNS: 1-(3-(p-CHLOROPHENYL)-3-PHENYLPROPIONYL)-4-(2-THIAZOLYL)PIPERAZINE □ KETONE, β-(p-CHLOROPHENYL)PHENETHYL 4-(2-THIAZOLYL)PIPERAZINYL □ PIPERAZINE, 1-(3-(p-CHLOROPHENYL)-3-PHENYLPROPIONYL)-4-(2-THIAZOLYL)-

TOXICITY DATA WITH REFERENCE
ipr-mus LD50:800 mg/kg JMCMAR 12,860,69

DOT CLASSIFICATION: 3; *Label:* Flammable Liquid

SAFETY PROFILE: Moderately toxic by intraperitoneal route. A flammable liquid. When heated to decomposition it emits toxic vapors of NO_x, SO_x, and Cl^-.

CKI080 CAS:23904-88-3 HR: 3
β-(p-CHLOROPHENYL)PHENETHYL 4-(m-TOLYL)PIPERAZINYL KETONE
mf: $C_{26}H_{27}ClN_2O$ mw: 419.00

SYNS: 1-(3-(p-CHLOROPHENYL)-3-PHENYLPROPIONYL)-4-(m-TOLYL)PIPERAZINE □ KETONE, β-(p-CHLOROPHENYL)PHENETHYL 4-(m-TOLYL)PIPERAZINYL □ PIPERAZINE, 1-(3-(p-CHLOROPHENYL)-3-PHENYLPROPIONYL)-4-(m-TOLYL)-

TOXICITY DATA WITH REFERENCE
ipr-mus LD50:800 mg/kg JMCMAR 12,860,69

DOT CLASSIFICATION: 3; *Label:* Flammable Liquid

SAFETY PROFILE: Moderately toxic by intraperitoneal route. A flammable liquid. When heated to decomposition it emits toxic vapors of NO_x and Cl^-.

CKI090 CAS:23904-87-2 HR: 3
β-(p-CHLOROPHENYL)PHENETHYL 4-(o-TOLYL)PIPERAZINYL KETONE
mf: $C_{26}H_{27}ClN_2O$ mw: 419.00

SYNS: 1-(3-(p-CHLOROPHENYL)-3-PHENYLPROPIONYL)-4-(o-TOLYL)PIPERAZINE □ KETONE, β-(p-CHLOROPHENYL)PHENETHYL 4-(o-TOLYL)PIPERAZINYL □ PIPERAZINE, 1-(3-(p-CHLOROPHENYL)-3-PHENYLPROPIONYL)-4-(o-TOLYL)-

TOXICITY DATA WITH REFERENCE
ipr-mus LD50:800 mg/kg JMCMAR 12,860,69

DOT CLASSIFICATION: 3; *Label:* Flammable Liquid

SAFETY PROFILE: Moderately toxic by intraperitoneal route. A flammable liquid. When heated to decomposition it emits toxic vapors of NO_x and Cl^-.

CKI175 CAS:102071-30-7 HR: 3
α-(p-CHLOROPHENYL)-α-PHENYL-2-PIPERIDINEMETHANOL HYDROCHLORIDE
mf: $C_{18}H_{20}ClNO•ClH$ mw: 338.30

SYN: α-(p-CLOROFENIL)-α-FENIL-2-PIPERIDILMETANOLO CLORIDRATO (ITALIAN)

TOXICITY DATA WITH REFERENCE
orl-rat LD50:400 mg/kg FRPSAX 12,853,57
orl-mus LD50:145 mg/kg FRPSAX 12,853,57
ipr-mus LD50:78 mg/kg FRPSAX 12,853,57

SAFETY PROFILE: Poison by ingestion and intraperitoneal routes. When heated to decomposition it emits toxic fumes of NO_x and Cl^-.

CKI180 CAS:23902-87-6 HR: 3
1-(3-(p-CHLOROPHENYL)-3-PHENYLPROPIONYL)-4-(2-HYDROXYPROPYL)PIPERAZINE
mf: $C_{22}H_{27}ClN_2O_2$ mw: 386.96

SYNS: β-(p-CHLOROPHENYL)PHENETHYL 4-(2-HYDROXYPROPYL) PIPERAZINYL KETONE □ KETONE, β-(p-CHLOROPHENYL)PHENETHYL 4-(2-HYDROXYPROPYL)PIPERAZINYL □ PIPERAZINE, 1-(3-(p-CHLOROPHENYL)-3-PHENYLPROPIONYL)-4-(2-HYDROXYPROPYL)-

TOXICITY DATA WITH REFERENCE
ipr-mus LD50:800 mg/kg JMCMAR 12,860,69

DOT CLASSIFICATION: 3; *Label:* Flammable Liquid

SAFETY PROFILE: Moderately toxic by intraperitoneal route. A flammable liquid. When heated to decomposition it emits toxic vapors of NO_x and Cl^-.

CKI185 CAS:23902-91-2 HR: 3
1-(3-(p-CHLOROPHENYL)-3-PHENYLPROPIONYL)-4-(o-METHOXYPHENYL)PIPERAZINE
mf: $C_{26}H_{27}ClN_2O_2$ mw: 435.00

SYNS: β-(p-CHLOROPHENYL)PHENETHYL 4-(o-METHOXYPHENYL) PIPERAZINYL KETONE □ KETONE, β-(p-CHLOROPHENYL)PHENETHYL 4-(o-METHOXYPHENYL)PIPERAZINYL □ PIPERAZINE, 1-(3-(p-CHLOROPHENYL)-3-PHENYLPROPIONYL)-4-(o-METHOXYPHENYL)-

TOXICITY DATA WITH REFERENCE
ipr-mus LD50:800 mg/kg JMCMAR 12,860,69

DOT CLASSIFICATION: 3; *Label:* Flammable Liquid

SAFETY PROFILE: Moderately toxic by intraperitoneal route. A flammable liquid. When heated to decomposition it emits toxic vapors of NO_x and Cl^-.

CKI190 CAS:23904-72-5 HR: 3
1-(3-(p-CHLOROPHENYL)-3-PHENYLPROPIONYL)-4-(p-TOLYL)PIPERAZINE
mf: $C_{26}H_{27}ClN_2O$ mw: 419.00

SYNS: β-(p-CHLOROPHENYL)PHENETHYL 4-(p-TOLYL)PIPERAZINYL KETONE □ KETONE, β-(p-CHLOROPHENYL)PHENETHYL 4-(p-TOLYL)PIPERAZINYL □ PIPERAZINE, 1-(3-(p-CHLOROPHENYL)-3-PHENYLPROPIONYL)-4-(p-TOLYL)-

TOXICITY DATA WITH REFERENCE
ipr-mus LD50:800 mg/kg JMCMAR 12,860,69

DOT CLASSIFICATION: 3; *Label:* Flammable Liquid

SAFETY PROFILE: Moderately toxic by intraperitoneal route. A flammable liquid. When heated to decomposition it emits toxic vapors of NO_x and Cl^-.

CKI250 HR: 3
1-(p-CHLOROPHENYL)-1-PHENYL-2-PROPYN-1-OL CARBAMATE
mf: $C_{16}H_{12}ClNO_2$ mw: 285.74

SYNS: 4-CHLORO-α-ETHYNYL-α-PHENYLBENZENEMETHANOL CARBAMATE □ 1-(4-CHLOROPHENYL)-1-PHENYL-2-PROPYNYL ESTER CARBAMIC ACID

TOXICITY DATA WITH REFERENCE
orl-rat TDLo:4040 mg/kg/42W-C:CAR TXAPA9 21,414,72
orl-rat TD:14 g/kg/77W-C:ETA JJIND8 71,211,83
ipr-mus LD50:347 mg/kg JMCMAR 11,1155,68

SAFETY PROFILE: Poison by intraperitoneal route. Questionable carcinogen with experimental carcinogenic and tumorigenic data. When heated to decomposition it emits toxic fumes of Cl^- and NO_x. See also CARBAMATES; and CHLORINATED HYDROCARBONS, AROMATIC.

CKI625 CAS:80-00-2 HR: 2
p-CHLOROPHENYL PHENYL SULFONE
mf: $C_{12}H_9ClO_2S$ mw: 252.72

PROP: Crystals, sltly aromatic odor, no taste, insol in water. Mp: 90–94°.

SYNS: 4-CHLORODIPHENYL SULFONE □ R-CHLORODIPHENYL SULPHONE □ p-CHLOROPHENYL PHENYL SULPHONE □ 1-CHLORO-4-(PHENYLSULFONYL)BENZENE □ COMPOUND R-242 □ ENT 17,941 □ p-MONOCHLOROPHENYL PHENYL SULFONE □ R-242 □ R-242-B □ SULFENONE □ SULPHENONE

TOXICITY DATA WITH REFERENCE
orl-rat LD50:1400 mg/kg GUCHAZ 5,95,68
ipr-rat LDLo:500 mg/kg JAFCAU 3,836,55
orl-mus LD50:2700 mg/kg JAFCAU 3,836,55
ipr-mus LD50:1000 mg/kg JAFCAU 3,836,55

CONSENSUS REPORTS: Reported in EPA TSCA Inventory.

SAFETY PROFILE: Moderately toxic by ingestion and intraperitoneal routes. An acaricide. When heated to decomposition it emits very toxic fumes of Cl^- and SO_x.

CKI750 CAS:18046-21-4 HR: 3
4-(p-CHLOROPHENYL)-2-PHENYL-5-THIAZOLEACETIC ACID
mf: $C_{17}H_{12}ClNO_2S$ mw: 329.81

PROP: Needles from C_6H_6. Mp: 161–162°.

SYNS: BR 700 □ CH 800 □ 4-(p-CHLOROPHENYL)-2-PHENYLTHIAZOLE-5-ACETIC ACID □ 4-((4-CHLOROPHENYL)-2-PHENYL)-5-THIAZOLEACETIC ACID (9CI) □ DONOREST □ FENTIAZAC □ FLOGENE □ NORVEDAN

TOXICITY DATA WITH REFERENCE
orl-rat TDLo:1350 mg/kg (female 17-22D post):REP KSRNAM 13,1929,79
orl-rat TDLo:3500 mg/kg (male 35D pre):REP KSRNAM 13,1901,79
orl-rat TDLo:1750 mg/kg (male 35D pre):REP KSRNAM 13,1901,79
orl-rat TDLo:525 mg/kg (14D pre/1-7D preg):TER KSRNAM 13,1929,79
orl-rat LD50:409 mg/kg NIIRDN 6,APP-17,82
ipr-rat LD50:325 mg/kg NIIRDN 6,APP-17,82
scu-rat LD50:543 mg/kg IYKEDH 13,637,82
ivn-rat LD50:160 mg/kg KSRNAM 13,1895,79
orl-mus LD50:520 mg/kg NIIRDN 6,APP-17,82
ipr-mus LD50:385 mg/kg IYKEDH 13,637,82
scu-mus LD50:655 mg/kg IYKEDH 13,637,82
ivn-mus LD50:161 mg/kg KSRNAM 13,1895,79
ivn-dog LD50:103 mg/kg KSRNAM 13,1895,79
orl-rbt LD50:625 mg/kg KSRNAM 13,1895,79
ivn-rbt LD50:71 mg/kg KSRNAM 13,1895,79

SAFETY PROFILE: Poison by intravenous and intraperitoneal routes. Moderately toxic by ingestion and subcutaneous routes. An experimental teratogen. Other experimental reproductive effects. An anti-inflammatory agent. When heated to decomposition it emits very toxic fumes of SO_x, NO_x, and Cl^-.

CKI825 CAS:4415-51-4 HR: 3
4-(p-CHLOROPHENYL)-1-PIPERAZINEETHANOL-2-PYRIDINEACRYLATE(ESTER)
mf: $C_{20}H_{22}ClN_3O_2$ mw: 371.90

SYN: 1-(4-CHLOROPHENYL)-4-(2-(3-(2-PYRIDYL)ACRYLOXY)ETHYL)-PIPERAZINE

TOXICITY DATA WITH REFERENCE
orl-rat LD50:478 mg/kg TXAPA9 10,444,67
ipr-rat LD50:311 mg/kg JPMSAE 55,1105,66
orl-mus LD50:529 mg/kg JPMSAE 55,1105,66
ipr-mus LD50:280 mg/kg JPMSAE 55,290,66

SAFETY PROFILE: Poison by intraperitoneal route. Moderately toxic by ingestion. When heated to decomposition it emits toxic fumes of Cl^- and NO_x.

CKJ000 CAS:25332-39-2 HR: 3
2-(3-(4-(3-CHLOROPHENYL)-1-PIPERAZINYL)PROPYL)-1,2,4-TRIZOLO(4,3-a)PYRIDIN-3(2H)-ONE HYDROCHLORIDE
mf: $C_{19}H_{22}ClN_5O \cdot ClH$ mw: 408.37

PROP: Plates from EtOH. Mp: 223°. Sol in $CHCl_3$; sltly sol in H_2O and EtOH.

SYNS: AF 1161 □ 2-(3-(4-(m-CHLOROPHENYL)-1-PIPERAZINYL) PROPYL)-s-TRIAZOLE-(4,3-α)-PIRIDIN-3(WH)-ONE HCl □ DESYREL □ MOLIPAXIN □ PRAGMAZONE □ THOMBRAN □ TOMBRAN □ TRAZODONE HYDROCHLORIDE □ TRITTICO

TOXICITY DATA WITH REFERENCE
orl-man TDLo:159 mg/kg (male 7W pre):REP AJPSAO 140,1256,83
orl-man TDLo:112 mg/kg (male 6W pre):REP JCLPDE 45,232,84
orl-wmn TDLo:720 mg/kg/34W-I:SYS AIMEAS 118,791,93
orl-wmn TDLo:750 μg/kg AJPSAO 141,434,84
orl-wmn TDLo:7500 μg/kg/5D-I AJPSAO 140,642,83
orl-man TDLo:46 mg/kg/8D-I JCPYDR 6,117,86
orl-man TDLo:667 μg/kg:CVS AJPSAO 141,1472,84
orl-rat LD50:690 mg/kg MPPPBK 9,76,74
ipr-rat LD50:178 mg/kg MPPPBK 9,76,74
ivn-rat LD50:91 mg/kg MPPPBK 9,76,74
orl-mus LD50:610 mg/kg MPPPBK 9,76,74
ivn-mus LD50:91 mg/kg MPPPBK 9,76,74
orl-dog LD50:500 mg/kg MPPPBK 9,76,74
ivn-mky LD50:25 mg/kg PBPSDY 3,94,81
orl-rbt LD50:560 mg/kg MPPPBK 9,76,74
ivn-rbt LD50:52 mg/kg MPPPBK 9,76,74

SAFETY PROFILE: Poison by intravenous route. Moderately toxic by ingestion. Human systemic effects by ingestion: cardiomyopathy including infarction, hepatitis, jaundice. Experimental reproductive effects. When heated to decomposition it emits very toxic fumes of Cl^- and NO_x. A tranquilizer and hypotensive agent. See also KETONES.

CKJ100 CAS:17766-66-4 HR: 3
4-(p-CHLOROPHENYL)PIPERAZINYL 3,4,5-TRIMETHOXYPHENYL KETONE
mf: $C_{20}H_{23}ClN_2O_4$ mw: 390.90

SYNS: 1-(p-CHLOROPHENYL)-4-(3,4,5-TRIMETHOXYBENZOYL)PIPERAZINE □ KETONE, 4-(p-CHLOROPHENYL)PIPERAZINYL 3,4,5-TRIMETHOXYPHENYL □ PIPERAZINE, 1-(p-CHLOROPHENYL)-4-(3,4,5-TRIMETHOXYBENZOYL)-

TOXICITY DATA WITH REFERENCE
ipr-mus LD50:300 mg/kg JMCMAR 11,332,68

DOT CLASSIFICATION: 3; *Label:* Flammable Liquid

SAFETY PROFILE: A poison by intraperitoneal route. A flammable liquid. When heated to decomposition it emits toxic vapors of NO_x and Cl^-.

CKJ825 CAS:20064-40-8 HR: 3
5-((p-CHLOROPHENYL)SULFONYL)-3-METHYL-1,2,4-THIADIAZOLE
mf: $C_9H_7ClN_2O_2S_2$ mw: 274.75

TOXICITY DATA WITH REFERENCE
eye-rbt 200 mg/1M YKKZAJ 88,1437,68
ivn-rat LD50:132 mg/kg YKKZAJ 88,1437,68
orl-mus LD50:3 g/kg YKKZAJ 88,1437,68
ipr-mus LD50:347 mg/kg YKKZAJ 88,1437,68
scu-mus LD50:646 mg/kg YKKZAJ 88,1437,68
ivn-mus LD50:151 mg/kg YKKZAJ 88,1437,68

SAFETY PROFILE: Poison by intravenous and intraperitoneal routes. Moderately toxic by ingestion and subcutaneous routes. An eye irritant. When heated to decomposition it emits toxic fumes of Cl^-, SO_x, and NO_x.

CKK000 CAS:94-20-2 HR: 3
1-(p-CHLOROPHENYLSULFONYL)-3-PROPYLUREA
mf: $C_{10}H_{13}ClN_2O_3S$ mw: 276.76

PROP: Crystals from EtOH (aq). Mp: 127–129°. Sltly sol in Et_2O and C_6H_6.

SYNS: ADIABEN □ ASUCROL □ CATANIL □ 1-(p-CHLOROBENZENESULFONYL)-3-PROPYLUREA □ N-(p-CHLOROBENZENESULFONYL)-N′-PROPYLUREA □ CHLORODIABINA □ CHLORONASE □ 1-p-CHLOROPHENYL-3-(PROPYLSULFONYL)UREA □ CHLOROPROPAMIDE □ 4-CHLORO-4-((PROPYLAMINO)CARBONYL)BENZENESULFONAMIDE □ CHLORPROPAMID □ CHLORPROPAMIDE □ CLORPROPAMIDE (ITALIAN) □ DIABARIL □ DIABECHLOR □ DIABENAL □ DIABENESE □ DIABENEZA □ DIABETORAL □ DIABET-PAGES □ DIABINESE □ GLISEMA □ MELDIAN □ MELITASE □ MELLINESE □ NCI-CO1752 □ ORADIAN □ N-PROPYL-N′-(p-CHLOROBENZENESULFONYL)UREA □ 1-PROPYL-3-(p-CHLOROBENZENESULFONYL)UREA □ N-PROPYL-N′-p-CHLORPHENYLSULFONYLCARBAMIDE □ STABINOL □ U-3818

TOXICITY DATA WITH REFERENCE
sce-mus-orl 177 mg/kg MUREAV 77,349,80
sce-ham-ipr 7100 μg/kg MUREAV 77,349,80
orl-wmn TDLo:840 mg/kg (9-39W preg):REP ADCHAK 49,283,74
unr-wmn TDLo:1358 mg/kg (1-26W preg):REP BMJOAE 2,187,64
unr-wmn TDLo:180 mg/kg (26-39W preg):REP BMJOAE 2,187,64
orl-wmn TDLo:884 mg/kg (1-35W preg):TER ADCHAK 49,283,74
orl-rat TDLo:1 g/kg (9D preg):TER PROEAS 14,89,68
orl-rat TDLo:12 g/kg (female 1-12D post):REP COREAF 249,1160,59
orl-wmn TDLo:25 mg/kg/5D-I:BLD DICPBB 18,981,84
orl-wmn TDLo:300 mg/kg:CNS CTOXAO 11,13,77
orl-man TDLo:83 mg/kg/25D-I:GIT AJGAAR 80,381,85
orl-rat LD50:2150 mg/kg FATOAO 25,93,62
ipr-rat LD50:580 mg/kg TXAPA9 18,185,71
ivn-rat LD50:590 mg/kg PMDCAY 1,187,61
orl-mus LD50:1546 mg/kg AIPTAK 153,379,65
ivn-mus LD50:500 mg/kg PMDCAY 1,187,61

CONSENSUS REPORTS: NCI Carcinogenesis Bioassay (feed); No Evidence: mouse, rat NCITR* NCI-CG-TR-45,78. EPA Genetic Toxicology Program.

SAFETY PROFILE: Moderately toxic by ingestion, intraperitoneal, and intravenous routes. Human systemic effects by ingestion: hemolysis with or without anemia, ulceration or bleeding from large intestine, cholestatic jaundice. Human teratogenic effects by ingestion: fetal death. Human reproductive effects by ingestion and possibly other routes: stillbirth, reduced viability and abnormal characteristics of newborn at birth. An experimental teratogen. Experimental reproductive effects. Mutation data reported. An antidiabetic agent. When heated to decomposition it emits very toxic fumes of Cl^-, NO_x, and SO_x.

CKK099 **HR: 3**
N-CHLORO-5-PHENYLTETRAZOLE
mf: $C_7H_6ClN_3$ mw: 167.60

SAFETY PROFILE: May explode when heated. When heated to decomposition it emits toxic fumes of Cl^- and NO_x.

CKK250 **CAS:17969-20-9** **HR: 3**
2-(p-CHLOROPHENYL)-4-THIAZOLE ACETIC ACID
mf: $C_{11}H_8ClNO_2S$ mw: 253.71

PROP: Crystals from EtOAc. Mp: 155–156°. Sltly sol in H_2O.

SYNS: ACIDE p-CHLOROPHENYL-2-THIAZOLE-ACETIQUE-4 (FRENCH) □ ACIDO FENCLOZICO (ITALIAN) □ ACIDUM FENCLOZICUM □ 2-(4-CHLOROPHENYL)-4-THIAZOLEACETIC ACID (9CI) □ 2-(p-CHLOROPHENYL)THIAZOL-4-YLACETIC ACID □ FENCLOZIC ACID □ ICI 54,450 □ MIALEX □ MYALEX

TOXICITY DATA WITH REFERENCE
orl-rat TDLo:50 mg/kg (20-21D preg):REP NATUAS 240,37,72
orl-rat TDLo:200 mg/kg (5D male):REP CCPTAY 19,129,79
orl-rat LD50:850 mg/kg NATUAS 221,582,69
ivn-rat LD50:300 mg/kg NATUAS 221,582,69
orl-mus LD50:400 mg/kg FRPSAX 40,875,85
ipr-mus LD50:250 mg/kg FRPSAX 40,875,85
ivn-mus LD50:250 mg/kg NATUAS 221,582,69

SAFETY PROFILE: Poison by ingestion, intraperitoneal, and intravenous routes. Experimental reproductive effects. An anti-inflammatory agent. When heated to decomposition it emits very toxic fumes of Cl^-, NO_x, and SO_x.

CKK500 **CAS:15446-08-9** **HR: 2**
4-(p-CHLOROPHENYL)THIO-1-BUTANOL
mf: $C_9H_{13}ClOS$ mw: 204.73

SYN: 4-(p-CHLORPHENYLTHIO)-BUTANOL (GERMAN)

TOXICITY DATA WITH REFERENCE
orl-rat TDLo:18,200 mg/kg (91D male):REP ARZNAD 29,1382,79
orl-rat TDLo:6 g/kg (30D male):REP ARZNAD 29,1141,79
ipr-rat LD50:885 mg/kg ARZNAD 29,1141,79
orl-mus LD50:1145 mg/kg ARZNAD 19,1141,79
ipr-mus LD50:880 mg/kg ARZNAD 29,1141,79

SAFETY PROFILE: Moderately toxic by ingestion and intraperitoneal routes. Experimental reproductive effects. When heated to decomposition it emits very toxic fumes of Cl^- and SO_x.

CKL000 **CAS:5344-82-1** **HR: 3**
2-CHLOROPHENYL THIOUREA
mf: $C_7H_7ClN_2S$ mw: 186.67

SYN: RCRA WASTE NUMBER P026

TOXICITY DATA WITH REFERENCE
orl-rat LD50:4600 μg/kg JPETAB 90,260,47

CONSENSUS REPORTS: EPA Extremely Hazardous Substances List. Reported in EPA TSCA Inventory.

SAFETY PROFILE: Poison by ingestion. When heated to decomposition it emits very toxic fumes of Cl^-, NO_x, and SO_x.

CKL250 **CAS:29975-16-4** **HR: 3**
8-CHLORO-6-PHENYL-4H-s-TRIAZOLO(4,3-a)(1,4) BENZODIAZEPINE
mf: $C_{16}H_{11}ClN_4$ mw: 294.76

PROP: A solid. Mp: 228–229°.

SYNS: 8-CHLORO-6-PHENYL-4H-(1,2,4)TRIAZOLO(4,3-a)(1,4-BENZODIAZEPINE) (9CI) □ D-40TA □ ESILGAN □ ESTAZOLAM □ EURODIN □ JULODIN □ NUCTALON

TOXICITY DATA WITH REFERENCE
orl-rat TDLo:300 mg/kg (30D pre):REP TAKHAA 32,158,73
orl-rat TDLo:15 g/kg (30D male):REP TAKHAA 32,158,73
orl-rat LD50:2500 mg/kg NIIRDN 6,103,82
ipr-rat LD50:339 mg/kg IYKEDH 6,530,75
scu-rat LD50:3580 mg/kg IYKEDH 6,530,75
ims-rat LD50:868 mg/kg IYKEDH 6,530,75
orl-mus LD50:600 mg/kg CCCCAK 48,2395,83
ipr-mus LD50:438 mg/kg JJPAAZ 21,497,71
orl-rbt LD50:300 mg/kg TAKHAA 32,152,73

SAFETY PROFILE: Poison by ingestion and intraperitoneal routes. Moderately toxic by intramuscular and subcutaneous routes. Experimental reproductive effects. An hypnotic and sedative agent. When heated to decomposition it emits very toxic fumes of Cl^- and NO_x. See also DIAZEPAM.

CKL325 **CAS:66535-86-2** **HR: 3**
2-(p-CHLOROPHENYL)-s-TRIAZOLO(5,1-a)ISOQUINOLINE
mf: $C_{16}H_{10}ClN_3$ mw: 279.74

PROP: Crystals. Mp: 238–240°.

SYNS: CANOCENTA □ 2-(4-CHLOROPHENYL)-(1,2,4)TRIAZOLO(5,1-a)ISOQUINOLINE (9CI) □ DL 717-IT □ L 12717 □ LOTRIFEN □ PRIVAPROL

TOXICITY DATA WITH REFERENCE
orl-ham TDLo:40 mg/kg (female 4-8D post):REP RDMIDP 4,237,82
scu-rat TDLo:4500 μg/kg (female 7D post):REP RDMIDP 4,237,82
ipr-mus LD50:1 g/kg MDACAP 23,75,87
ims-dog LDLo:75 mg/kg MDACAP 23,75,87

SAFETY PROFILE: Poison by intramuscular route. Experimental reproductive effects. When heated to decomposition it emits toxic fumes of Cl^- and NO_x.

CKL500 **CAS:8072-20-6** **HR: 2**
4-CHLOROPHENYL-2,4,5-TRICHLOROPHENYLAZO-SULFIDE mixed with 1,1-BIS(4-CHLOROPHENYL) ETHANOL
mf: $C_{14}H_{12}Cl_2O \cdot C_{12}H_6Cl_4N_2S$ mw: 619.22

SYNS: ANILIX □ BCPE mixed with SPAS □CHLOROFENSUL-PHIDE □ CPAS mixed with BCPE □ DANINON □ MICASIN □ MILBEX □ MILSAR □ MISASIN

TOXICITY DATA WITH REFERENCE
orl-mus TDLo:1200 mg/kg (1-12D preg):TER OYYAA2 2,355,68
orl-rat LD50:1856 mg/kg OYYAA2 2,161,68
orl-mus LD50:2000 mg/kg OYYAA2 2,355,68

SAFETY PROFILE: Moderately toxic by ingestion. An experimental teratogen. A miticide. When heated to decomposition it emits very toxic fumes of Cl^-, SO_x, and NO_x. See also SULFIDES.

CKL750 **CAS:2227-13-6** **HR: 2**
p-CHLOROPHENYL-2,4,5-TRICHLOROPHENYL SULFIDE
mf: $C_{12}H_6Cl_4S$ mw: 324.04

PROP: Crystals. Sltly sol in H_2O; mod sol in Me_2CO and Et_2O; sol in C_6H_6 and $CHCl_3$.

SYNS: ANIMERT □ ANIMERT V-10 □ ANIMERT V-101 □ ANIMERT V-10K □ 4-CHLOROPHENYL 2,4,5-TRICHLOROPHENYL SULFIDE □ ENT 27,115 □ PHILIPS-DUPHAR V-101 □ 3,4,6,R'-TETRACHLOR-DI-PHENYLSULFID (GERMAN) □ 2,4,4',5-TETRACHLORODIPHENYL SULFIDE □ 2,4,5,4'-TETRACHLORODIPHENYL SULFIDE □ TETRASUL □ 1,2,4-TRICHLORO-5-((4-CHLOROPHENYL)THIO)-BENZENE

TOXICITY DATA WITH REFERENCE
orl-rat TDLo:18 g/kg (17W male):REP TXCYAC 1,63,73
orl-rat LD50:3960 mg/kg BESAAT 15,129,69
orl-mus LD50:5010 mg/kg GUCHAZ 6,495,73
orl-rbt LDLo:1350 mg/kg TXCYAC 1,63,73
skn-rbt LD50:2000 mg/kg SPEADM 78-1,55,78
orl-gpg LD50:500 mg/kg TXCYAC 1,63,73
skn-gpg LD50:8200 mg/kg 85DPAN -,-,71/76
ipr-gpg LD50:550 mg/kg TXCYAC 1,63,73

SAFETY PROFILE: Moderately toxic by ingestion, skin contact, and intraperitoneal routes. Experimental reproductive effects. A pesticide. When heated to decomposition it emits very toxic fumes of Cl^- and SO_x. See also SULFIDES.

CKM000 **CAS:116-29-0** **HR: 2**
p-CHLOROPHENYL-2,4,5-TRICHLOROPHENYL SULFONE
mf: $C_{12}H_6Cl_4O_2S$ mw: 356.04

PROP: Crystals. Mp: 148–149°. Nearly water-insol.

SYNS: AKARITOX □ AREDION □ 4-CHLOROPHENYL-2,4,5-TRICHLOROPHENYL SULFONE □ p-CHLOROPHENYL-2,4,5-TRICHLOROPHENYL SULPHONE □ DUPHAR □ ENT 23,737 □ FMC 5488 □ MITION □ NIA 5488 □ POLACARITOX □ ROZTOZOL □ SULFONE-2,4,4',5-TETRACHLORODIPHENYL □ TEDION □ TEDION V-18 □ 2,4,4',5-TETRACHLOOR-DIFENYL-SULFON (DUTCH) □ 2,4,4',5-TETRACHLOR-DIPHENYL-SULFON (GERMAN) □ 2,4,4',5-TETRACHLORODIPHENYL SULFONE □ 2,4,5,4'-TETRACHLORODIPHENYLSULPHONE □ 2,4,4',5-TETRACLORO-DIFENIL-SOLFONE (ITALIAN) □TETRADI-CHLONE □ TETRADIFON □ TETRADIPHON □ TETRAFIDON □ 1,2,4-TRICHLORO-5-((4-CHLOROPHENYL)SULFONYL)-BENZENE □ V-18

TOXICITY DATA WITH REFERENCE
scu-mus TDLo:1953 mg/kg (6-14D preg):TER NTIS** PB223,-160
orl-rat LD50:566 mg/kg WRPCA2 9,119,70
orl-dog LD50:2000 mg/kg SPEADM 78-1,55,78
skn-rbt LD50:10 g/kg 28ZEAL 5,220,76

CONSENSUS REPORTS: EPA Genetic Toxicology Program.

SAFETY PROFILE: Moderately toxic by ingestion. Mildly toxic by skin contact. An experimental teratogen. Used to control worms in crops. When heated to decomposition it emits highly toxic fumes of Cl^- and SO_x.

CKM250 **CAS:26571-79-9** **HR: 3**
CHLOROPHENYLTRICHLOROSILANE
DOT: UN 1753
mf: $C_6H_4Cl_4Si$ mw: 245.99

PROP: Colorless to pale-yellow liquid, readily hydrolyzed by moisture with the liberation of HCl (a mixture of 3 isomers). Bp: 230°, d: 1.439 @ 25°/25°, flash p: 255°F (COC).
DOT CLASSIFICATION: 8; *Label:* Corrosive

SAFETY PROFILE: A poison irritant by ingestion and inhalation. A corrosive irritant to the skin, eyes, and mucous membranes. Combustible when exposed to heat or flame. In contact with water it readily hydrolyzes to HCl and evolves heat. When heated to decomposition it emits toxic fumes of Cl^-. See also CHLOROSILANES.

CKM750 **CAS:29025-67-0** **HR: 3**
1-(p-CHLOROPHENYL)-2,8,9-TRIOXA-5-AZA-1-SILABICYCLO(3.3.3) UNDECANE
mf: $C_{12}H_{16}ClNO_3Si$ mw: 285.83

SYNS: p-CHLORFENYLSILATRAN (CZECH) □ 1-(p-CHLOROPHENYL)SILATRANE

TOXICITY DATA WITH REFERENCE
orl-rat LD50:520 mg/kg MarJV# 29MAR77
orl-mus LDLo:8 mg/kg AECTCV 14,111,85
ipr-mus LD50:1050 mg/kg RCRVAB 38(12),975,69

SAFETY PROFILE: Poison by ingestion. Moderately toxic by intraperitoneal route. When heated to decomposition it emits very toxic fumes of Cl^- and NO_x.

CKN000 **CAS:1406-65-1** **HR: 3**
CHLOROPHYLL

PROP: Dark-green liquid.

SYNS: CHLOROFOLIN □ CHLOROFYL □ CHLOROPHYLLS □ C.I. 1956 □ DEODOPHYLL □ E 140 □ L-GRUEN No. 1 □ L-GRUEN No. 1 (GERMAN)

TOXICITY DATA WITH REFERENCE
ivn-mus LD50:285 mg/kg ARZNAD 4,19,54
ivn-gpg LD50:85 mg/kg ARZNAD 7,357,57

SAFETY PROFILE: Poison by intravenous and intraperitoneal routes. When heated to decomposition it emits toxic fumes of NO_x.

CKN375 CAS:4684-94-0 HR: 2
6-CHLOROPICOLINIC ACID
mf: $C_6H_4ClNO_2$ mw: 157.56

PROP: Plates. Mp: 190°. Sol in EtOH and $CHCl_3$; sltly sol in Et_2O; insol in pet ether.

SYNS: 6-CHLORO-2-PYRIDINECARBOXYLIC ACID □ 6-CPA

TOXICITY DATA WITH REFERENCE
orl-rat LD50:2180 mg/kg DOEAAH 32,22,76
unr-rat LD50:1487 mg/kg GISAAA 48(4),52,83
unr-mus LD50:777 mg/kg GISAAA 48(4),52,83

SAFETY PROFILE: Moderately toxic by ingestion and possibly other routes. When heated to decomposition it emits toxic fumes of Cl^- and NO_x.

CKN500 CAS:76-06-2 HR: 3
CHLOROPICRIN
DOT: UN 1580/UN 1583
mf: CCl_3NO_2 mw: 164.37

PROP: Sltly oily, colorless liquid. D: 1.692 @ 0.4°, Fp −69, mp: −64°, bp: 112.3 @ 766 mm, vap press: 40 mm @ 33.80, vap d: 6.69. Sol in water, alc, and ether.

SYNS: ACQUINITE □ CHLOORPIKRINE (DUTCH) □ CHLOR-O-PIC □ CHLOROPICRIN, liquid (DOT) □ CHLOROPICRIN, ABSORBED (DOT) □ CHLOROPICRINE (FRENCH) □ CHLORPIKRIN (GERMAN) □ CLOROPICRINA (ITALIAN) □ DOJYOPICRIN □ DOLOCHLOR □ LARVACIDE □ MICROLYSIN □ NCI-C00533 □ NITROCHLOROFORM □ NITROTRICHLOROMETHANE □ PIC-CLOR □ PICFUME □ PICRIDE □ PROFUME A □ PS □ TRICHLOORNITROMETHAAN (DUTCH) □ TRICHLORNITROMETHAN (GERMAN) □ TRICHLORONITROMETHANE □ TRI-CLOR □ TRICLORO-NITRO-METANO (ITALIAN)

TOXICITY DATA WITH REFERENCE
mma-sat 50 µg/plate MUREAV 116,185,83
orl-mus TDLo:26 g/kg/78W-I:ETA NCITR* NCI-CG-TR-65,78
ihl-hmn TCLo:2 mg/m³:EYE,PUL ZGEMAZ 106,296,39
ihl-hmn TCLo:2000 mg/m³/10M 34ZIAG -,169,69
orl-rat LD50:250 mg/kg DOWCC* -,-,72
ihl-mus LC50:1600 mg/m³/10M NDRC** NDCrc-132, an,42
ipr-mus LD50:25 mg/kg KHFZAN 10(6),53,75
ihl-cat LCLo:800 mg/m³/20M ZGEMAZ 106,296,39
ihl-rbt LC50:800 mg/m³/20M FMCHA2-,C72,91
ihl-gpg LCLo:800 mg/m³/20M ZGEMAZ 106,296,39
ivn-gpg LD50:4200 µg/kg AIPTAK 114,146,58

CONSENSUS REPORTS: NCI Carcinogenesis Bioassay (gavage); No Evidence: mouse NCITR* NCI-GC-TR-65,78. Reported in EPA TSCA Inventory

OSHA PEL: TWA 0.1 ppm
ACGIH TLV: TWA 0.1 ppm
DFG MAK: 0.1 ppm (0.7 mg/m³)
DOT CLASSIFICATION: 6.1; *Label:* Poison (UN 1580); DOT Class: 6.1; *Label:* Poison, KEEP AWAY FROM FOOD (UN 1583)

SAFETY PROFILE: Poison by ingestion, intravenous, and intraperitoneal routes. Moderately toxic by inhalation. Human systemic effects by inhalation: lacrimation, conjunctiva irritation, and pulmonary changes. Mutation data reported. A powerful irritant that affects all body surfaces. It causes lacrymation, vomiting, bronchitis, pulmonary edema, irritation to gastrointestinal and respiratory tracts. Questionable carcinogen with experimental tumorigenic data. An additional toxic effect is its reaction with SH-groups in hemoglobin thus interfering with oxygen transport. Photochemical transformation of chloropicrin into phosgene (carboxy chloride, $COCl_2$) has been reported. A concentration of 1 ppm causes a smarting pain in the eyes and therefore in itself constitutes a good warning of exposure. Inhalation causes vomiting, probably due to swallowing saliva in which small amounts of chloropicrin have dissolved. Its primary lethal effect is to produce lung injury and it is a difficult gas to protect oneself against because it is chemically inert and does not react with the usual chemicals used in gas masks. Four ppm is sufficient to render a worker unfit for action and 20 ppm, when breathed from 1 to 2 minutes, causes definite bronchial or pulmonary lesions. Industrially it is used as a warning agent in commercial fumigants. It is more toxic than chlorine but less so than phosgene.

Above a critical volume it can be shocked into detonation. Mixtures with 3-bromopropyne are shock- and heat-sensitive explosives. Violent reaction with aniline + heat, alcoholic sodium hydroxide, sodium methoxide, and propargyl bromide. When heated to decomposition it emits very toxic fumes of Cl^- and NO_x.

Used for insect and rodent control in grain elevators and bins and as a soil fumigant and fungicide. See also NITRO COMPOUNDS.

CKN675 CAS:2156-71-0 HR: 3
1-CHLOROPIPERIDINE
mf: $C_5H_{10}ClN$ mw: 119.61

PROP: Oil. Bp: 53–54° @ 29 mm.

$ClN(CH_2)_4CH_2$ (ring)

SYN: N-CHLOROPIPERIDINE

TOXICITY DATA WITH REFERENCE
mmo-sat 32 µg/plate JEPTDQ 4(2-3),345,80
bfa-mus/sat 50 mg/kg JEPTDQ 4(2-3),345,80
oms-ham:ovr 1200 µg/L JEPOEC 6,241,85
cyt-ham:ovr 400 µg/L JEPOEC 6,241,85
orl-mus TDLo:2800 mg/kg (female 7D pre):REP JACTDZ 2(2),209,83
orl-mus TDLo:2800 mg/kg (male 7D pre):REP JACTDZ 2(2),209,83
orl-mus TDLo:16 g/kg (40D male):REP JACTDZ 2(2),209,83
ipr-mus LD50:100 mg/kg JEPTDQ 4(2-3),345,80

SAFETY PROFILE: Poison by intraperitoneal route. Mutation data reported. Experimental reproductive effects. Can undergo rapid spontaneous decomposition.

When heated to decomposition it emits toxic fumes of Cl^- and NO_x.

CKN750 **CAS:77966-67-7** **HR: 3**
6′-CHLORO-2-PIPERIDINO-o-ACETOTOLUIDIDE HYDROCHLORIDE
mf: $C_{14}H_{19}ClN_2O \cdot ClH$ mw: 303.26

SYN: V 255

TOXICITY DATA WITH REFERENCE
eye-rbt 2% MLD ARZNAD 8,407,58
ipr-rat LD50:161 mg/kg ARZNAD 8,407,58
ipr-mus LD50:170 mg/kg ARZNAD 8,407,58
scu-mus LD50:265 mg/kg ARZNAD 8,407,58

SAFETY PROFILE: Poison by intraperitoneal and subcutaneous routes. An eye irritant. When heated to decomposition it emits very toxic fumes of $Cl^-$and NO_x.

CKO000 **CAS:78218-40-3** **HR: 3**
6′-CHLORO-3-(PIPERIDINO)-o-PROPIONOTOLUIDIDE HYDROCHLORIDE
mf: $C_{15}H_{21}ClN_2O \cdot ClH$ mw: 317.29

SYN: C 3138

TOXICITY DATA WITH REFERENCE
ipr-rat LD50:58 mg/kg ARZNAD 8,544,58
scu-mus LD50:67 mg/kg ARZNAD 8,544,58

SAFETY PROFILE: Poison by intraperitoneal and subcutaneous routes. When heated to decomposition it emits very toxic fumes of NO_x and Cl^-.

CKO750 **CAS:16941-12-1** **HR: 3**
CHLOROPLATINIC ACID
DOT: UN 2507
mf: $Cl_6Pt \cdot 2H$ mw: 409.81

PROP: Reddish-brownish-yellow deliquescent solid or crystalline mass. D: 2.431, mp: 60°. Easily sol in water and alc.

SYNS: CHLOROPLATINIC(IV) ACID □ DIHYDROGEN HEXACHLOROPLATINATE □ DIHYDROGEN HEXACHLOROPLATINATE(2-) □ HEXACHLOROPLATINIC ACID □ HEXACHLOROPLATINIC(IV) ACID □ HEXACHLOROPLATINIC(4+) ACID, HYDROGEN- □ HYDROGEN HEXACHLOROPLATINATE(4+) □ PLATINIC CHLORIDE

TOXICITY DATA WITH REFERENCE
mrc-bcs 10 mmol/L MUREAV 77,109,80
ivn-rat LD50:49 mg/kg NTIS** PB291-731
ipr-mus LD50:61 mg/kg COREAF 256,1043,63

CONSENSUS REPORTS: Reported in EPA TSCA Inventory. EPA Genetic Toxicology Program.

OSHA PEL: TWA 0.002 mg(Pt)/m³
ACGIH TLV: TWA 0.002 mg(Pt)/m³
DOT CLASSIFICATION: 8; *Label:* Corrosive

SAFETY PROFILE: Poison by intravenous and intraperitoneal routes. Mutation data reported. See PLATINUM COMPOUNDS and CHLORIDES. Incompatible with BrF_3. When heated to decomposition it emits toxic fumes of Cl^-.

CKP250 **CAS:50-53-3** **HR: 3**
CHLOROPROMAZINE
mf: $C_{17}H_{19}ClN_2S$ mw: 318.89

PROP: Oily liquid. Bp: 200–205° @ 0.8 mm. Insol in H_2O.

SYNS: AMINASINE □ AMINAZIN □ AMINAZINE □ AMPLIACTIL □ AMPLICITIL □ 2-CHLORO-10-(3-(DIMETHYLAMINO)PROPYL)PHENOTHIAZINE □ CHLORO-3-(DIMETHYLAMINO-3-PROPYL)-10 PHENOTHIAZINE (FRENCH) □ CHLORPROMAZIN □ CHLORPROMAZINE □ CHLOR-PZ □ CLOPROMAZINA (ITALIAN) □ 2-CLORO-10-(3-DIMETILAMINOPROPIL)FENOTIAZINA (ITALIAN) □ CPZ □ FENACTIL □ FENAKTYL □ FRACTION AB □ HIBANIL □ HIBERNAL □ LARGACTIL □ LARGACTILOTHIAZINE □ LARGACTYL □ MEGAPHEN □ NOVOMAZINA □ PHENACTYL □ PHENATHYL □ PLEGOMAZIN □ PRAZIL □ PROMACTIL □ PROMAZIL □ PROPAPHENIN □ PROZIL □ PROZIN □ SKF-2601 □ THORAZINE □ TORAZINA □ WINTERMIN

TOXICITY DATA WITH REFERENCE
add-bac-esc 3 µmol/L MUREAV 89,95,81
mmo-smc 1 mg/L ATPNAB 10,463,74
unr-wmn TDLo:540 mg/kg (1-39W preg):REP AGPSA3 2,606,60
unr-wmn TDLo:480 mg/kg (12-17W preg):REP AGPSA3 2,606,60
unr-wmn TDLo:242 mg/kg (1-17W preg):REP AGPSA3 2,606,60
orl-rat TDLo:30 mg/kg (6-15D preg):REP TXAPA9 53,541,80
orl-mus TDLo:336 mg/kg (female 1-21D post):REP ANREAK 157,311,67
orl-rat TDLo:30 mg/kg (6-15D preg):REP TXAPA9 53,541,80
unr-rat TDLo:245 mg/kg (female 8-22D post):REP PHMCAA 18,231,76
scu-rat TDLo:24 mg/kg (female 4-7D post):REP EXNEAC 14,33,66
ims-rat TDLo:270 mg/kg (female 18-22D post):TER TJADAB 26,21,82
orl-mus TDLo:50 mg/kg (female 10D post):TER TOIZAG 28,621,81
orl-rat TDLo:119 mg/kg (female 13D post):TER PJPPAA 32,199,80
orl-mus TDLo:448 mg/kg (female 1-28D post):REP PSEBAA 113,833,63
ipr-rat TDLo:1 mg/kg (male 1D pre):REP SCIEAS 127,84,58
orl-rat TDLo:4 mg/kg (female 1D pre):REP DCTODJ 7,41,84
par-mus TDLo:19,800 µg/kg (female 6-16D post):REP CMJODS 101,339,88
ims-rat TDLo:20 mg/kg (female 5D post):REP PSEBAA 100,555,59
scu-rat TDLo:225 mg/kg (female 15D pre):REP AJPHAP 199,1073,60
scu-rat TDLo:50 mg/kg (female 1D pre):REP ENDOAO 65,563,59
ims-mus TDLo:160 mg/kg (female 10D pre):REP FATOAO 38,473,75
orl-mus TDLo:448 mg/kg (female 1-28D post):REP PSEBAA 113,833,63
ipr-uns TDLo:96 mg/kg (male 6D pre):REP IJEBA6 18,149,80
ims-dom TDLo:12 mg/kg (male 8D pre):REP AJVRAH 25,523,64

ims-rat TDLo: 28 mg/kg (male 7D pre): REP IJEBA6 16,316,78
ims-mus TDLo: 176 mg/kg (female 10D pre): TER FATOAO 38,473,75
orl-mus TDLo: 50 mg/kg (female 8D post): TER TOIZAG 28,621,81
orl-inf TDLo: 20 mg/kg: CVS,CNS AJDCAI 130,507,76
orl-wmn TDLo: 200 μg/kg: CNS,LIV NYSJAM 57,1922,57
orl-hmn TDLo: 8570 μg/kg/12D-I: CNS ARZNAD 22,93,72
orl-rat LD50: 142 mg/kg ARZNAD 18,261,68
ihl-rat LC50: 209 mg/m³/2H TPKVAL 10,73,68
ihl-rat LC50: 209 mg/m³/2H TPKVAL 10,73,68
ipr-rat LD50: 58 mg/kg 27ZTAP 3,69,69
scu-rat LD50: 75 mg/kg NYKZAU 56,377,60
ivn-rat LD50: 23 mg/kg FRPPAO 26,585,71
orl-mus LD50: 135 mg/kg AIPTAK 118,358,59
ihl-mus LC50: 209 mg/m³/2H TPKVAL 10,73,68
ipr-mus LD50: 14 mg/kg FRPSAX 14,269,59
scu-mus LD50: 33 mg/kg ARZNAD 11,932,61
ivn-mus LD50: 16 mg/kg ARZNAD 11,932,61

CONSENSUS REPORTS: EPA Genetic Toxicology Program.

SAFETY PROFILE: A human poison by an unspecified route. Poison experimentally by ingestion, inhalation, intravenous, intraperitoneal, and subcutaneous routes. Human systemic effects by ingestion: decrease in blood pressure, somnolence, sleep, wakefulness, tremors, muscle weakness, and jaundice. Human teratogenic effects by an unspecified route: stillbirth, physical abnormalities, and other neonatal effects. Human reproductive effects by an unspecified route: terminates pregnancy. Experimental reproductive effects. An experimental teratogen. Mutation data reported. Has been implicated in aplastic anemia. Mutation data reported. When heated to decomposition it emits very toxic fumes of Cl^-, NO_x, and SO_x.

CKP500 CAS:69-09-0 HR: 3
CHLOROPROMAZINE HYDROCHLORIDE
mf: $C_{17}H_{19}ClN_2S{\cdot}ClH$ mw: 355.35

SYNS: AMINAZIN MONOHYDROCHLORIDE □ AMPLIACTIL MONOHYDROCHLORIDE □ CHLORACTIL □ CHLORAZIN □ 2-CHLORO-10-(3-DIMETHYLAMINOPROPYL) PHENOTHIAZINE MONOHYDROCHLORIDE □ CHLOROPROMAZINE MONOHYDROCHLORIDE □ CPZ □ 10-(3-DIMETHYLAMINOPROPYL)-2-CHLOROPHENOTHIAZINE MONOHYDROCHLORIDE □ HEBANIL □ HIBANIL □ HIBERNAL □ HYBERNAL □ KLORPROMAN □ KLORPROMEX □ LARGACTIL MONOHYDROCHLORIDE □ LARGAKTYL □ MEGAPHEN □ NCI-CO5210 □ NEURAZINE □ NORCOZINE □ PHENOTHIAZINE HYDROCHLORIDE □ PLEGOMAZIN □ PROMACID □ PROMAPAR □ PROPAPHEN □ PROPAPHENIN HYDROCHLORIDE □ PSYCHOZINE □ 4560 RP HYDROCHLORIDE □ SONAZINE □ TAROCTYL □ THORAZINE □ THORAZINE HYDROCHLORIDE □ TORAZINA □ TRANZINE □ UNITENSEN

TOXICITY DATA WITH REFERENCE
mmo-sat 5 mg/L MUREAV 208,33,88
mmo esc 5 mg/L GENRA8 30,13,77
orl-rat TDLo: 300 mg/kg (female 6-20D post): REP TJADAB 37,185,88
orl-rat TDLo: 1980 mg/kg (male 4W pre): REP JONRA9 20,625,73
ipr-rat TDLo: 100 mg/kg (female 14D post): REP IJEBA6 16,542,78
scu-rat TDLo: 24 mg/kg (female 17-20D post): REP JNEUAY 4,87,62
scu-mus TDLo: 10 mg/kg (female 3D post): TER JRPFA4 76,527,86
orl-mus TDLo: 300 mg/kg (female 6-15D post): TER NTIS** PB83-179846
ipr-rat TDLo: 25 mg/kg (female 8D post): TER IJEBA6 18,344,80
orl-rat TDLo: 50 mg/kg (female 7-16D post): TER TXAPA9 21,230,72
orl-rat TDLo: 350 mg/kg (female 7-16D post): REP TXAPA9 21,230,72
ipr-rat TDLo: 2500 μg/kg (male 1D pre): REP PSEBAA 103,392,60
orl-rat TDLo: 250 mg/kg (female 7-16D post): REP TXAPA9 21,230,72
scu-rat TDLo: 24 mg/kg (female 17-20D post): REP JNEUAY 4,87,62
scu-rat TDLo: 125 mg/kg (male 5D pre): REP JRPFA4 28,177,72
ipr-rat TDLo: 50 mg/kg (female 12D post): TER TJADAB 18,167,78
ipr-rat TDLo: 100 mg/kg (female 14D post): TER IJMRAQ 67,300,78
orl-wmn TDLo: 6 mg/kg: BAH,GIT PGMJAO 60,564,84
orl-wmn TDLo: 35 g/kg/16Y-I: BAH,CVS BIPCBF 18,1441,83
ivn-man TDLo: 1786 μg/kg/2D-I: BAH,PUL, SKN BIPCBF 18,1441,83
ivn-wmn TDLo: 822 μg/kg: BAH,CVS AEMED3 17,380,88
orl-rat LD50: 145 mg/kg JPETAB 148,151,65
ihl-rat LC50: 40 mg/m³/2H 85JCAE -,1106,86
ipr-rat LD50: 62 mg/kg PSYPAG 12,142,68
ivn-rat LD50: 26 mg/kg FAZMAE 5,269,63
orl-mus LD50: 135 mg/kg CPBTAL 24,1179,76
ihl-mus LC50: 40 mg/m³/2H 85JCAE -,1106,86
ipr-mus LD50: 115 mg/kg AIPTAK 134,328,61
ivn-mus LD50: 20 mg/kg ARZNAD 21,808,71

CONSENSUS REPORTS: Reported in EPA TSCA Inventory.

SAFETY PROFILE: Poison by ingestion, intraperitoneal, intravenous, and subcutaneous routes. An experimental teratogen. Experimental reproductive effects. An antiemetic and antipsychotic drug. Human systemic effects: anorexia (human), excitement, gastrointestinal changes, irritability, pulse rate increase, respiratory stimulation, rigidity, somnolence, sweating. Mutation data reported. When heated to decomposition it emits very toxic fumes of Cl^-, NO_x, and SO_x.

CKP725 CAS:627-30-5 HR: 2
3-CHLOROPROPANOL
DOT: UN 2849
mf: C_3H_7ClO mw: 94.55

SYNS: 3-CHLORPROPAN-1-OL □ 1-PROPANOL, 3-CHLORO- □ TRIMETHYLENE CHLOROHYDRIN

TOXICITY DATA WITH REFERENCE
mma-sat 33 µg/plate ENMUDM 9(Suppl 9),1,87
orl-mus LD50:2300 mg/kg ZHYGAM 26,17,80

CONSENSUS REPORTS: Reported in EPA TSCA Inventory.

DOT CLASSIFICATION: 6.1; *Label:* KEEP AWAY FROM FOOD

SAFETY PROFILE: Moderately toxic by ingestion. Mutation data reported.

CKP750 CAS:540-54-5 HR: 3
1-CHLOROPROPANE
DOT: UN 1278
mf: C_3H_7Cl mw: 78.55

PROP: Colorless liquid, chloroform-like odor. Mp: −122.8°, d: 0.897 @ 15°/4°, bp: 46.60°, lel: 2.6%, uel: 11.1%, flash p: <0°F, d: 0.890, vap d: 2.71, autoign temp: 968°F.

SYN: N-PROPYL CHLORIDE

CONSENSUS REPORTS: Reported in EPA TSCA Inventory.

DOT CLASSIFICATION: 3; *Label:* Flammable Liquid

SAFETY PROFILE: A moderately poisonous irritant to skin, eyes, and mucous membranes. Narcotic in high concentrations. Flammable liquid and dangerous fire hazard when exposed to heat, flame, or oxidizers. Moderately explosive when exposed to flame. Keep away from heat and open flame; can react vigorously with oxidizing materials. To fight fire, use CO_2, dry chemical. When heated to decomposition it emits toxic fumes of Cl^-. See also CHLORINATED HYDROCARBONS, ALIPHATIC.

CKQ000 CAS:75-29-6 HR: 3
2-CHLOROPROPANE
DOT: UN 2356
mf: C_3H_7Cl mw: 78.54

$CH_3CHClCH_3$

PROP: Flash p.: −25.6°F; lel: 2.8%; uel: 10.7%, d: 0.868 @ 15°, mp: −117°, bp: 34.8°.

SYN: ISOPROPYL CHLORIDE

TOXICITY DATA WITH REFERENCE
mmo-sat 1 g/plate DTESD7 2,249,77
mma-sat 1 g/plate DTESD7 2,249,77

DOT CLASSIFICATION: Flammable Liquid; *Label:* Flammable Liquid

SAFETY PROFILE: Mutation data reported. A flammable liquid. A very dangerous fire hazard when exposed to heat, flame, or oxidizers. When heated to decomposition it emits toxic fumes of Cl^-. See also 1-CHLOROPROPANE.

CKQ250 CAS:1331-07-3 HR: 3
CHLOROPROPANE DIOL-1,3
mf: $C_3H_7ClO_2$ mw: 110.55

SYN: CHLORO-1,3-PROPANEDIOL

TOXICITY DATA WITH REFERENCE
skn-rbt 10 mg/24H JIHTAB 30,63,48
eye-rbt 20 mg JIHTAB 30,63,48
orl-rat LD50:110 mg/kg JIHTAB 30,63,48
ihl-rat LCLo:1000 ppm/1H JIHTAB 30,63,48
skn-rbt LD50:450 mg/kg JIHTAB 30,63,48

SAFETY PROFILE: Poison by ingestion. Moderately toxic by skin contact. Mildly toxic by inhalation. A skin and eye irritant. Reaction with 70% perchloric acid forms a dangerously unstable explosive product. When heated to decomposition it emits toxic fumes of Cl^-.

CKQ500 CAS:3477-94-9 HR: 2
3-CHLORO-1,2-PROPANEDIOL 1-BENZOATE
mf: $C_{10}H_{11}ClO_3$ mw: 214.66

SYNS: BENZOIC ACID, 3-CHLORO-2-HYDROXYPROPYL ESTER □ 1-BENZOYLOXY-3-CHLOROPROPAN-2-OL □ 1,2-PROPANEDIOL, 3-CHLORO-, 1-BENZOATE □ U 27,574

TOXICITY DATA WITH REFERENCE
orl-rat TDLo:960 mg/kg (male 8D pre):REP JRPFA4 21,263,70
ipr-mus LD50:450 mg/kg JMCMAR 20,644,77

SAFETY PROFILE: Moderately toxic by intraperitoneal route. Experimental reproductive effects. When heated to decomposition it emits toxic fumes of Cl^-.

CKQ750 CAS:15121-11-6 HR: 3
CHLOROPROPANEDIOL CYCLIC SULFITE
mf: $C_3H_5ClO_3S$ mw: 156.69

SYN: 1-CHLOROMETHYLETHYLENE GLYCOL CYCLIC SULFITE

TOXICITY DATA WITH REFERENCE
orl-rat TDLo:112 mg/kg (male 14D pre):REP CCPTAY 9,451,74
ivn-mus LD50:178 mg/kg CSLNX* NX#02154

CONSENSUS REPORTS: Reported in EPA TSCA Inventory.

SAFETY PROFILE: Poison by intravenous route. See also SULFITES. Experimental reproductive effects. When heated to decomposition it emits very toxic fumes of Cl^- and SO_x.

CKR500 CAS:78-89-7 HR: 3
2-CHLORO-1-PROPANOL
DOT: UN 2611
mf: C_3H_7ClO mw: 94.55

PROP: Colorless liquid, mild non-residual odor. Bp: 133.5°, flash p: 125°F (CC), d: 1.103 @ 20°, vap d: 3.26.

SYNS: 2-CHLOROPROPANOL □ 2-CHLOROPROPYL ALCOHOL □ PROPYLENECHLOROHYDRIN

TOXICITY DATA WITH REFERENCE
skn-rbt 500 mg open MLD UCDS** 7/23/71

eye-rbt 2230 μg SEV AJOPAA 29,1363,46
orl-rat LD50:218 mg/kg FAONAU 53A,359,74
ihl-rat LCLo:500 ppm/4H AIHAAP 30,470,69
orl-dog LDLo:200 mg/kg FAONAU 53A,359,74
skn-rbt LD50:529 mg/kg AIHAAP 30,470,69
orl-gpg LD50:720 mg/kg JIHTAB 23,259,41

CONSENSUS REPORTS: Reported in EPA TSCA Inventory.

DOT CLASSIFICATION: 6.1; *Label:* Poison

SAFETY PROFILE: Poison by ingestion. Moderately toxic by inhalation and skin contact. A skin and severe eye irritant. Flammable liquid when exposed to heat, flame, or powerful oxidizers. To fight fire, use alcohol foam, CO_2, dry chemical. When heated to decomposition it emits toxic fumes of Cl^-.

CKR750 CAS:127-00-4 HR: 1
1-CHLORO-2-PROPANOL with 2-CHLORO-1-PROPANOL
mf: C_3H_7ClO mw: 94.55

TOXICITY DATA WITH REFERENCE
mmo-sat 40 μmol/plate FCTXAV 18,115,80
mma-sat 1100 μg/plate MUREAV 30,303,75
ihl-rat LC50:1000 ppm/4H 85JCAE ,517,86

SAFETY PROFILE: Low toxicity by inhalation. Mutation data reported. When heated to decomposition it emits toxic fumes of Cl^-. See also individual components.

CKS000 CAS:557-98-2 HR: 3
2-CHLORO-1-PROPENE
DOT: UN 2456
mf: C_3H_5Cl mw: 76.53

PROP: Colorless liquid or gas. Bp: 22.65°, fp: −137.4°, d: 0.918 @ 9°, flash p: −4°, lel: 4.5%, uel: 16%, mp: −138.6°.

SYN: 2-CHLOROPROPENE (DOT)

TOXICITY DATA WITH REFERENCE
mma-sat 100 μmol/plate BCPCA6 29,2611,80
ihl-mus LC50:267 g/m³ UCPHAQ 2,39,41

CONSENSUS REPORTS: Reported in EPA TSCA Inventory.

DOT CLASSIFICATION: 3; *Label:* Flammable Liquid

SAFETY PROFILE: Mildy toxic by inhalation. Mutation data reported. Very dangerous fire hazard when exposed to heat, flame, sparks, or powerful oxidizers. To fight fire, use water, spray, mist, fog, dry chemical, alcohol foam. When heated to decomposition it emits toxic fumes of Cl^-. See also CHLORIDES.

CKS325 CAS:62861-56-7 HR: 3
2-CHLORO-2-PROPENYL TRIFLUOROMETHANE SULFONATE
mf: $C_4H_4ClF_3O_3S$ mw: 224.58

$H_2C{=}CClCH_2OSO_2CF_3$

SAFETY PROFILE: A dangerous storage hazard. Store in a vented container at −78°C. Reacts violently with aprotic solvents (e.g., DMF and DMSO). When heated to decomposition it emits toxic fumes of F^-, Cl^-, and SO_x. See also SULFONATES.

CKS500 CAS:107-94-8 HR: 2
3-CHLOROPROPIONIC ACID
mf: $C_3H_5ClO_2$ mw: 108.53

PROP: Leaflets from H_2O; crystals from ligroin. Mp: 41°, bp: 204° (part decomp).

SYNS: β-CHLOROPROPIONIC ACID □ β-MONOCHLOROPROPIONIC ACID

TOXICITY DATA WITH REFERENCE
mmo-sat 100 μg/plate DHEFDK FDA-78-1046,78
ipr-mus TDLo:730 mg/kg/4W-I:NEO CNREA8 39,391,79
skn-mus LDLo:1040 mg/kg CNREA8 28,653,68

CONSENSUS REPORTS: Reported in EPA TSCA Inventory.

SAFETY PROFILE: Moderately toxic by skin contact. Questionable carcinogen with experimental neoplastigenic data. Mutation data reported. When heated to decomposition it emits toxic fumes of Cl^-. See also α-CHLOROPROPIONIC ACID.

CKS750 CAS:598-78-7 HR: 2
α-CHLOROPROPIONIC ACID
DOT: UN 2511
mf: $C_3H_5ClO_2$ mw: 108.53

PROP: Sol in water. D: 1.260–1.268 @ 20°, bp: 183–187°, flash p: 225°F.

TOXICITY DATA WITH REFERENCE
skn-gpg LD50:126 mg/kg 85INA8 5,136(89),86

CONSENSUS REPORTS: Reported in EPA TSCA Inventory.

ACGIH TLV: TWA 0.1 ppm (skin) 85INA8 5,136(89),86
DOT CLASSIFICATION: 8; *Label:* Corrosive

SAFETY PROFILE: Poison by skin contact. A corrosive. Combustible when exposed to heat or flame. To fight fire, use water, foam, alcohol foam. When heated to decomposition it emits toxic fumes of Cl^-. See also 3-CHLOROPROPIONIC ACID.

CKT000 CAS:17639-93-9 HR: 3
2-CHLOROPROPIONIC ACID METHYL ESTER
DOT: UN 2933
mf: $C_4H_7ClO_2$ mw: 122.56

SYN: METHYL-2-CHLOROPROPIONATE (DOT)

TOXICITY DATA WITH REFERENCE
ipr-mus LDLo:250 mg/kg CBCCT* 6,228,54

CONSENSUS REPORTS: Reported in EPA TSCA Inventory.

DOT CLASSIFICATION: 3; *Label:* Flammable Liquid

SAFETY PROFILE: Poison by intraperitoneal route. See also ESTERS. A flammable liquid when exposed to heat of flame. When heated to decomposition it emits toxic fumes of Cl^-.

CKT100 CAS:16987-02-3 HR: 2
2-CHLOROPROPIONIC ACID SODIUM SALT
mf: $C_3H_4ClO_2$•Na mw: 130.51

SYNS: 2-CHLOROPROPANOIC ACID SODIUM SALT □ 2-CHLOROPROPIONATE SODIUM SALT □ α-CHLOROPROPIONIC ACID SODIUM SALT □ PROPANOIC ACID, 2-CHLORO-, SODIUM SALT □ SODIUM-2-CHLOROPROPIONATE

TOXICITY DATA WITH REFERENCE
orl-mus TDLo:44 g/kg (male 84D pre):REP JPETAB 222,501,82
orl-mus LD50:1671 mg/kg JPETAB 222,501,82

SAFETY PROFILE: Moderately toxic by ingestion. Experimental reproductive effects. When heated to decomposition it emits toxic fumes of Cl^-.

CKT250 CAS:542-76-7 HR: 3
3-CHLOROPROPIONITRILE
mf: C_3H_4ClN mw: 89.53

PROP: Colorless liquid. Mp: −51°, bp: 176° decomp, flash p: 168°F (CC), d: 1.1363 @ 25°, vap press: 6 mm @ 50°, vap d: 3.09.

SYNS: 3-CHLOROPROPANENITRILE □ 3-CHLOROPROPANONITRILE □ β-CHLOROPROPIONITRILE □ RCRA WASTE NUMBER P027 □ USAF A-8798

TOXICITY DATA WITH REFERENCE
orl-rat LD50:100 mg/kg 14CYAT 2,2025,62
orl-mus LD50:9 mg/kg 14CYAT 2,2025,62
ipr-mus LD50:100 mg/kg NTIS** AD277-689
ivn-mus LD50:56 mg/kg CSLNX* NX#01996

CONSENSUS REPORTS: EPA Extremely Hazardous Substances List. Cyanide and its compounds are on the Community Right-To-Know List. Reported in EPA TSCA Inventory.

SAFETY PROFILE: Poison by ingestion, intravenous, and intraperitoneal routes. Flammable in its liquid form when exposed to heat or flame. To fight fire, use alcohol foam, water, foam, CO_2, or dry chemical. When heated to decomposition it emits very toxic fumes of Cl^-, CN^-, and NO_x. See also NITRILES.

CKT500 CAS:6285-05-8 HR: 3
p-CHLOROPROPIOPHENONE
mf: C_9H_9ClO mw: 168.63

PROP: A solid. Mp: 37–38°, bp: 134–137° @ 31 mm.

SYN: USAF EK-5296

TOXICITY DATA WITH REFERENCE
ipr-mus LD50:200 mg/kg NTIS** AD277-689
ivn-mus LD50:100 mg/kg CSLNX* NX#04477

CONSENSUS REPORTS: Reported in EPA TSCA Inventory.

SAFETY PROFILE: Poison by intraperitoneal and intravenous routes. When heated to decomposition it emits toxic fumes of Cl^-.

CKT750 CAS:77966-68-8 HR: 3
6′-CHLORO-2-(PROPYLAMINO)-o-ACETOTOLUIDIDE HYDROCHLORIDE
mf: $C_{12}H_{17}ClN_2O$•ClH mw: 277.22

SYNS: C 3058 □ 2′-CHLORO-6′-METHYL-2-(PROPYLAMINO)ACETANILIDE HYDROCHLORIDE

TOXICITY DATA WITH REFERENCE
ipr-rat LD50:400 mg/kg ARZNAD 8,407,58
ipr-mus LD50:400 mg/kg ARZNAD 8,407,58
scu-mus LD50:1125 mg/kg ARZNAD 8,407,58

SAFETY PROFILE: Poison by intraperitoneal route. Moderately toxic by subcutaneous route. When heated to decomposition it emits very toxic fumes of Cl^- and NO_x.

CKU000 CAS:77985-17-2 HR: 3
6′-CHLORO-2-(PROPYLAMINO)-o-BUTYROTOLUIDIDE HYDROCHLORIDE
mf: $C_{14}H_{21}ClN_2O$•ClH mw: 305.28

SYN: C 3189

TOXICITY DATA WITH REFERENCE
eye-rbt 2% MLD ARZNAD 8,544,58
ipr-rat LD50:34 mg/kg ARZNAD 8,544,58
scu-mus LD50:57 mg/kg ARZNAD 8,544,58

SAFETY PROFILE: Poison by intraperitoneal and subcutaneous routes. An eye irritant. When heated to decomposition it emits very toxic fumes of $Cl^-$and NO_x.

CKU250 CAS:77846-96-9 HR: 3
9-((2-((2-CHLOROPROPYL)AMINO)ETHYL)AMINO)-2-METHOXYACRIDINE DIHYDROCHLORIDE HEMIHYDRATE
mf: $C_{19}H_{22}ClN_3O$•2ClH•1/2H_2O mw: 425.79

SYNS: ACRIDINE, 9-((2-((2-CHLOROPROPYL)AMINO)ETHYL)AMINO)-2-METHOXY-, DIHYDROCHLORIDE, HEMIHYDRATE □ ICR 442

TOXICITY DATA WITH REFERENCE
mmo-sat 5 μg/plate JMCMAR 15,739,72
ipr-mus LD20:64 mg/kg JMCMAR 15,739,72

SAFETY PROFILE: Poison by intraperitoneal route. Mutation data reported. When heated to decomposition it emits very toxic fumes of Cl^- and NO_x.

CKU625 CAS:2612-33-1 HR: 3
1-CHLORO-2,3-PROPYLENE DINITRATE
mf: $C_3H_5ClN_2O_6$ mw: 200.54

$ClCH_2CH(ONO_2)CH_2ONO_2$

PROP: Pale-yellow liquid. D: 1.51 @ 9°/0°, bp: 190–195° (decomp).

SAFETY PROFILE: A viscous liquid explosive. Upon decomposition it emits toxic fumes of Cl^- and NO_x. See also NITRATES and EXPLOSIVES.

CKV275 CAS:624-65-7 HR: 3
3-CHLOROPROPYNE
mf: C_3H_3Cl mw: 74.51

PROP: Liquid. D: 1.045 @ 5°, bp: 65°.

SYN: PROPARGYL CHLORIDE

SAFETY PROFILE: A pressure-sensitive explosive. Reacts explosively with ammonia in a closed container. When heated to decomposition it emits toxic fumes of Cl^-. See also ACETYLENE COMPOUNDS; CHLORINATED HYDROCARBONS, ALIPHATIC; and EXPLOSIVES.

CKV500 CAS:87-42-3 HR: 3
6-CHLOROPURINE
mf: $C_5H_3ClN_4$ mw: 154.57

PROP: Crystals from H_2O.

SYNS: 6-CHLORO-9H-PURINE □ 6-CHLORO-1H-PURINE (9CI) □ ClP □ NSC 744 □ SK 6048

TOXICITY DATA WITH REFERENCE
mmo-sat 10 µL/plate ANYAA9 76,475,58
mmo-esc 10 µL/disc ANYAA9 76,475,58
mmo-omi 500 mg/L SOGEBZ 6,1509,70
hma-mus/sat 25 mg/kg MUREAV 26,455,74
ipr-rat TDLo:100 mg/kg (7D preg):REP JRPFA4 4,291,62
ipr-rat LD50:400 mg/kg ADTEAS 3,181,68
orl-mus LD50:720 mg/kg NCISP* JAN86
ipr-mus LD50:132 mg/kg NCISP* JAN86
scu-mus LD50:514 mg/kg NCISP* JAN86
unr-mus LD50:230 mg/kg PMDCAY 7,69,70

SAFETY PROFILE: Poison by intraperitoneal and possibly other routes. Moderately toxic by ingestion and subcutaneous routes. Experimental reproductive effects. Mutation data reported. When heated to decomposition it emits very toxic fumes of Cl^- and NO_x.

CKV625 CAS:59-32-5 HR: 3
CHLOROPYRAMINE
mf: $C_{16}H_{20}ClN_3$ mw: 289.84

PROP: Light-yellow, viscous, oily liquid; pungent odor. Bp: 154–155°.

SYNS: ALLERGAN □ AVAPENA □ 2-((p-CHLOROBENZYL)(2-(DIMETHYLAMINO)ETHYL)AMINO)PYRIDINE □ N-(p-CHLOROBENZYL)-N′,N′-DIMETHYL-N-(2-PYRIDYL)ETHYLENEDIAMINE □ p-CHLORBENZYL-α-PYRIDYL-DIMETHYL-AETHYLENDIAMIN (GERMAN) □ CHLORONEOANTERGAN □ N-((4-CHLOROPHENYL)METHYL)-N′,N′-DIMETHYL-N-2-PYRIDINYL-1,2-ETHANEDIAMINE (9CI) □ CHLOROPYRIBENZAMINE □ HALOPYRAMINE □ SUPRASTIN □ SYNOPEN □ SYNOPEN R □ SYNPEN

TOXICITY DATA WITH REFERENCE
orl-rat LD50:920 mg/kg ARZNAD 25,1723,75
ipr-rat LD50:104 mg/kg ARZNAD 25,1723,75
ivn-rat LD50:32,500 µg/kg ARZNAD 25,1723,75
orl-mus LD50:354 mg/kg ARZNAD 25,1723,75
ipr-mus LD50:79,200 µg/kg ARZNAD 25,1723,75
ivn-mus LD50:24,100 µg/kg ARZNAD 25,1723,75
ipr-gpg LD50:108 mg/kg ARZNAD 25,1723,75
scu-gpg LD50:142 mg/kg ARZNAD 7,131,57

SAFETY PROFILE: Poison by ingestion, subcutaneous, intravenous, and intraperitoneal routes. When heated to decomposition it emits toxic fumes of $Cl^-$and NO_x.

CKW000 CAS:109-09-1 HR: 3
2-CHLOROPYRIDINE
DOT: UN 2822
mf: C_5H_4ClN mw: 113.55

PROP: Colorless, oily liquid or crystals. Mp: 65°, bp: 170°, d: 1.205 @ 15°, vap press: 1 mm @ 13.3°, vap d: 3.93.

SYNS: o-CHLOROPYRIDINE □ α-CHLOROPYRIDINE

TOXICITY DATA WITH REFERENCE
mma-sat 5 mg/plate MUREAV 176,185,87
sln-smc 4000 ppm MUREAV 163,23,86
ihl-rat LCLo:100 ppm/4H TXAPA9 11,361,67
orl-mus LD50:110 mg/kg TXAPA9 11,361,67
ipr-mus LD50:130 mg/kg TXAPA9 11,361,67
skn-rbt LD50:64 mg/kg TXAPA9 11,361,67
ipr-rbt LD50:48 mg/kg TXAPA9 11,361,67
orl-bwd LD50:1 g/kg AECTCV 12,355,83

CONSENSUS REPORTS: Reported in EPA TSCA Inventory.

DOT CLASSIFICATION: 6.1; *Label:* Poison

SAFETY PROFILE: Poison by ingestion, inhalation, skin contact, and intraperitoneal routes. Combustible when exposed to heat or flame. Can react with oxidizing materials. When heated to decomposition it emits very toxic fumes of Cl^-, NO_x, and phosgene.

CKW250 CAS:626-60-8 HR: 3
3-CHLOROPYRIDINE
mf: C_5H_4ClN mw: 113.55

PROP: D: 1.194, bp: 148° @ 744 mm.

SYN: m-CHLOROPYRIDINE

TOXICITY DATA WITH REFERENCE
mnt-mus:lyms 1433 mg/L MUREAV 301,57,93
cyt-mus:lyms 1433 mg/L MUREAV 301,57,93
ipr-mus LD50:235 mg/kg TXAPA9 11,361,67
orl-bwd LD50:750 mg/kg AECTCV 12,355,83

CONSENSUS REPORTS: Reported in EPA TSCA Inventory.

SAFETY PROFILE: Poison by intraperitoneal route. Moderately toxic by ingestion. Mutation data reported. When heated to decomposition it emits very toxic fumes of Cl^- and NO_x.

CKW325 CAS:2402-95-1 HR: 3
2-CHLOROPYRIDINE-N-OXIDE
mf: C_5H_4ClNO mw: 129.55

CH=CHCH=CHCCl=N:O

SAFETY PROFILE: Decomposes violently when heated above 90°C. Upon decomposition it emits toxic fumes of Cl^- and NO_x.

C

CKW500 **CAS:5428-90-0** **HR: 3**
CHLORO-3-PYRIDYLMERCURY
mf: C_5H_4ClHgN mw: 314.14

PROP: Needles from H_2O. Mp: 279.5–280°.

SYN: 3-(CHLOROMERCURI)PYRIDINE

CONSENSUS REPORTS: Mercury and its compounds are on the Community Right-To-Know List.

OSHA PEL: CL 0.1 mg(Hg)/m³ (skin)
ACGIH TLV: TWA 0.1 mg(Hg)/m³ (skin)
NIOSH REL: (Mercury, Aryl and Inorganic): CL 0.1 mg/m³ (skin)

SAFETY PROFILE: Probably a poison. See also MERCURY COMPOUNDS. When heated to decomposition it emits very toxic fumes of Cl^-, Hg, and NO_x.

CLB250 **CAS:78110-10-8** **HR: 3**
6′-CHLORO-2-(PYRROLIDINYL)-o-DIACETOTOLUIDIDE HYDROCHLORIDE
mf: $C_{15}H_{19}ClN_2O_2\cdot ClH$ mw: 294.81

SYNS: N-ACETYL-N-(2-CHLORO-6-METHYLPHENYL)-1-PYRROLIDINEACETAMIDE MONOHYDROCHLORIDE □ C 3199

TOXICITY DATA WITH REFERENCE
eye-rbt 2% MLD ARZNAD 8,609,58
ipr-rat LD50:180 mg/kg ARZNAD 8,609,58
ipr-mus LD50:120 mg/kg ARZNAD 8,609,58
scu-mus LD50:325 mg/kg ARZNAD 8,609,58

SAFETY PROFILE: Poison by intraperitoneal and subcutaneous routes. An eye irritant. When heated to decomposition it emits toxic fumes of Cl^- and NO_x.

CLC100 **HR: 3**
2′-CHLORO-2-PYRROLIDINYL-5′-TRIFLUOROMETHYLACETANILIDE HYDROCHLORIDE
mf: $C_{13}H_{14}ClF_3N_2O\cdot ClH$ mw: 343.20

SYNS: C 3078 □ 6′-CHLORO-2-PYRROLIDINYL-α,α,α-TRIFLUORO-m-ACETOTOLUIDINE, HYDROCHLORIDE

TOXICITY DATA WITH REFERENCE
eye-rbt 2% MOD ARZNAD 8,270,58
ipr-rat LD50:157 mg/kg ARZNAD 8,270,58
scu-mus LD50:232 mg/kg ARZNAD 8,270,58

SAFETY PROFILE: Poison by subcutaneous and intraperitoneal routes. An eye irritant. When heated to decomposition it emits toxic fumes of F^-, NO_x, and Cl^-.

CLC125 **HR: 3**
4′-CHLORO-2-PYRROLIDINYL-3′-TRIFLUOROMETHYLACETANILIDE HYDROCHLORIDE
mf: $C_{13}H_{14}ClF_3N_2O\cdot ClH$ mw: 343.20

SYNS: C 3073 □ 4′-CHLORO-2-PYRROLIDINYL-α,α,α-TRIFLUORO-m-ACETOTOLUIDIDE, HYDROCHLORIDE

TOXICITY DATA WITH REFERENCE
eye-rbt 2% SEV ARZNAD 8,270,58
ipr-rat LD50:174 mg/kg ARZNAD 8,270,58
scu-mus LD50:460 mg/kg ARZNAD 8,270,58

SAFETY PROFILE: Poison by intraperitoneal route. Moderately toxic by subcutaneous route. A severe eye irritant. When heated to decomposition it emits toxic fumes of F^-, NO_x, and Cl^-.

CLC500 **CAS:72-80-0** **HR: 3**
CHLOROQUINALDOL
mf: $C_{10}H_7Cl_2NO$ mw: 228.08

PROP: Yellow needles from EtOH. Mp: 114–115° (decomp). Insol in H_2O; sol in $CHCl_3$, EtOH, C_6H_6, and hexane.

SYNS: CHLORQUINALDOL □ 5,7-DICHLORO-8-HYDROXYQUINALDINE □ 5,7-DICHLORO-2-METHYL-8-HYDROXYQUINOLINE □ 5,7-DICHLORO-2-METHYL-8-QUINOLINOL □ 5,7-DICHLORO-8-QUINALDINOL □ HYDROXYDICHLOROQUINALDINE

TOXICITY DATA WITH REFERENCE
orl-rat LD50:660 mg/kg 29ZVAB -,33,69
orl-dog LD50:2250 mg/kg 29ZVAB -,33,69
orl-rbt LD50:160 mg/kg 29ZVAB -,33,69

SAFETY PROFILE: Poison by ingestion. When heated to decomposition it emits very toxic fumes of Cl^- and NO_x.

CLC750 **CAS:16064-14-5** **HR: 3**
6-CHLORO-4-QUINAZOLINONE
mf: $C_8H_5ClN_2O$ mw: 180.60

SYN: 6-CHLORO-4(3H)-QUINAZOLINONE

TOXICITY DATA WITH REFERENCE
orl-mus LD50:404 mg/kg ARZNAD 12,1204,62
ipr-mus LD50:340 mg/kg ARZNAD 12,1204,62

SAFETY PROFILE: Poison by intraperitoneal route. Moderately toxic by ingestion. When heated to decomposition it emits very toxic fumes of Cl^- and NO_x.

CLD000 **CAS:54-05-7** **HR: 3**
CHLOROQUINE
mf: $C_{18}H_{26}ClN_3$ mw: 319.92

SYNS: AMOKIN □ ARALEN □ ARTHROCHIN □ AVLOCLOR □ BEMACO □ BEMAPHATE □ BEMASULPH □ CHEMOCHIN □ CHINGAMIN □ CHLORAQUINE □ CHLOROCHIN □ 7-CHLORO-4-(4-DIETHYLAMINO-1-METHYLBUTYLAMINO)QUINOLINE □ CHLOROQUINIUM □ N⁴-(7-CHLORO-4-QUINOLINYL)-N¹,N¹-DIETHYL-1,4-PENTANEDIAMINE □ CIDANCHIN □ CLOROCHINA □ COCARTRIT □ DELAGIL □ DICHINALEX □ ELESTOL □ GONTOCHIN □ HELIOPAR □ IMAGON □ IROQUINE □ KLOROKIN □ LAPAQUIN □ MALAQUIN □ MALAREN □ MALAREX □ MESYLITH □ NEOCHIN □ NIVACHINE □ NIVAQUINE B □ QUINACHLOR □QUINAGA-MINE □ QUINERCYL □ QUINILON □ QUINOSCAN □ RESOCHIN □ RESOQUINA □ RESOQUINE □ REUMACHLOR □ REUMAQUIN □ ROQUINE □ RP 3377 □ SANOQUIN □ SENAQUIN □ SILBESAN □ SIRAGAN □ SN 6718 □ SN 7618 □ SOLPRINA □ SOPAQUIN □ TANAKAN □ TRESOCHIN □ TROCHIN □ W 7618 □ WIN 244

TOXICITY DATA WITH REFERENCE
mmo-sat 100 μmol/L AMACCQ 9,77,76
cyt-hmn:lym 100 mg/L BEXBAN 82,1095,76
unr-wmn TDLo:1155 mg/kg (6-39W preg):REP JOPDAB 69,1150,66
unr-wmn TDLo:840 mg/kg (1-84D preg):REP JOPDAB 69,1150,66

unr-wmn TDLo:1155 mg/kg (6-39W preg):TER JOPDAB 69,1150,66
unr-wmn TDLo:420 mg/kg (1-42D preg):TER JOPDAB 69,1150,66
orl-wmn LDLo:110 mg/kg:CVS,GIT NEJMAG 318,1,88
orl-wmn TDLo:3600 mg/kg/3Y TGMEAJ 32,216,80
orl-man LDLo:86 mg/kg:CVS,GIT NEJMAG 318,1,88
orl-hmn LDLo:20 mg/kg JETOAS 6,86,73
orl-rat LD50:330 mg/kg JTCTDW 20,271,83
ipr-rat LD50:102 mg/kg PHMGBN 13,401,75
orl-mus LD50:311 mg/kg OYYAA2 7,753,73
ipr-mus LD50:66 mg/kg ARZNAD 32,1219,82
scu-mus LD50:150 mg/kg JETOAS 6,86,73
ivn-mus LD50:21,600 μg/kg CYLPDN 4,69,83
ims-mus LD50:71 mg/kg CYLPDN 4,69,83
scu-rbt LD50:75 mg/kg JETOAS 6,86,73
ivn-rbt LD50:8 mg/kg JETOAS 6,86,73
ims-rbt LDLo:80 mg/kg YHHPAL 15,630,80

CONSENSUS REPORTS: IARC Cancer Review: Group 3 IMEMDT 7,56,87; Animal Inadequate Evidence IMEMDT 13,47,77. EPA Genetic Toxicology Program.

SAFETY PROFILE: Poison by ingestion, intraperitoneal, intravenous, intramuscular, and subcutaneous routes. Human systemic effects by ingestion: heart rate changes, nausea or vomiting. Human teratogenic effects by an unspecified route include developmental abnormalities of the urogenital system, eyes and ears, other unspecified areas, and postnatal effects. Human reproductive effects by an unspecified route: terminates pregnancy. Human mutation data reported. Questionable carcinogen. An antimalarial agent. When heated to decomposition it emits very toxic fumes of Cl^- and NO_x.

CLD250 CAS:50-63-5 HR: 3
CHLOROQUINE DIPHOSPHATE
mf: $C_{18}H_{26}ClN_3{\cdot}2H_3O_4P$ mw: 515.92

PROP: Dimorphic crystals. Mp: 193–195°.

SYNS: ALERMINE □ ARALEN DIPHOSPHATE □ ARALEN PHOSPHATE □ ARECHIN □ AROCLOR 54 □ AVLOCLOR □ BEMAPHATE □ CHINGAMIN □ 7-CHLOR-4-(4-(DIAETHYLAMINO)-1-METHYLBUTYLAMINO)-CHINOLINDIPHOSPHAT (GERMAN) □ 7-CHLORO-4-((4′-DIETHYLAMINO-1-METHYLBUTYL)AMINO)QUINOLINE DIPHOSPHATE □ CHLOROIN □ 2-((p-CHLORO-α-(2-DIMETHYLAMINO)ETHYL)BENZYL)PYRIDINE MALEATE (1:1) □ CHLOROQUINE PHOSPHATE □ CHLOR-TRIMETON □ CQ □ DELAGIL □ GONTOCHIN PHOSPHATE □ HISTASPAN □ H-STADUR □ KHINGAMIN □ NOSCOSED □ RESOCHIN □ RESOCHIN DIPHOSPHATE □ RESOQUINE □ SANOQUIN □ TANAKAN □ TELDRIN □ TELODRON

TOXICITY DATA WITH REFERENCE
mmo-sat 100 mg/L MUREAV 68,41,79
scu-rat TDLo:475 mg/kg (female 10-14D post):REP VAAZA2 39,59,82
orl-rat TDLo:750 mg/kg (female 9D post):TER RPTOAN 30,114,67
orl-rat TDLo:1200 mg/kg (5-16D preg):TER RCOCB8 7,701,74
orl-rat TDLo:550 mg/kg (female 9D post):TER RPTOAN 30,114,67
orl-man TDLo:8571 μg/kg:GIT HUTODJ 8,387,89
orl-cld LDLo:250 mg/kg:BAH ATXKA8 23,204,68
orl-wmn TDLo:167 mg/kg:CVS,BPR,PUL JTCTDW 19,1067,82/83
orl-man LDLo:179 mg/kg ATXKA8 23,204,68
orl-rat LDLo:600 mg/kg 85GLAQ 1,390,46
orl-mus LD50:500 mg/kg TMPRAD 30,308,79
ipr-mus LD50:68 mg/kg 85GLAQ 1,390,46
scu-mus LD60:200 mg/kg ATMPA2 74,393,80
ivn-brd LD50:64,500 μg/kg ARZNAD 20,1775,70

SAFETY PROFILE: A human poison by ingestion. Poison by intravenous, subcutaneous, and intraperitoneal routes. Human systemic effects by ingestion: EKG changes, blood pressure lowering, respiratory depression. An experimental teratogen. Experimental reproductive effects. Human systemic effects: blood pressure lowering, coma, EKG changes, nausea or vomiting, nausea or vomiting, respiratory depression, ulceration or bleeding from duodenum. Mutation data reported. When heated to decomposition it emits very toxic fumes of Cl^-, NO_x, and PO_x. See also CHLOROQUINE.

CLD500 CAS:4213-44-9 HR: 3
CHLOROQUINE MUSTARD
mf: $C_{18}H_{24}Cl_3N_3{\cdot}2ClH$ mw: 461.72

SYNS: 4-((4-(BIS(2-CHLOROETHYL)AMINO)-1-METHYLBUTYL)AMINO)-7-CHLOROQUINOLINE, DIHYDROCHLORIDE □ ICR-25A □ NSC-17118

TOXICITY DATA WITH REFERENCE
dnd-mus:lvr 70 μmol/L CNREA8 21,1124,61
dnd-mus:oth 70 μmol/L CNREA8 21,1124,61
ipr-mus TDLo:8 mg/kg/4W:CAR JNCIAM 36,915,66
ipr-rat LD10:1100 μg/kg CCROBU 17,63,62
ivn-dog LDLo:200 μg/kg CCSUBJ 2,202,65
ivn-mky LDLo:410 μg/kg CCSUBJ 2,202,65

SAFETY PROFILE: A deadly poison by intraperitoneal and intravenous routes. Questionable carcinogen with experimental carcinogenic data. Mutation data reported. When heated to decomposition it emits very toxic fumes of Cl^- and NO_x. See also CHLOROQUINE.

CLD600 CAS:130-16-5 HR: 2
5-CHLORO-8-QUINOLINOL
mf: C_9H_6ClNO mw: 179.61

SYNS: 5-CHLORO-8-HYDROXYQUINOLINE □ CHLOROXYQUINOLINE □ 8-QUINOLINOL, 5-CHLORO-

TOXICITY DATA WITH REFERENCE
mma-sat 50 nmol/plate MUREAV 42,335,77
orl-gpg LDLo:1200 mg/kg PSEBAA 28,484,31

CONSENSUS REPORTS: Reported in EPA TSCA Inventory.

SAFETY PROFILE: Moderately toxic by ingestion. Mutation data reported. When heated to decomposition it emits toxic vapors of NO_x and Cl^-.

CLD750 CAS:95-88-5 HR: 3
4-CHLORORESORCINOL
mf: $C_6H_5ClO_2$ mw: 144.56

PROP: Crystals from C_6H_6. Mp: 89°, bp: 147°.

TOXICITY DATA WITH REFERENCE
eye-rbt 5% MLD JAPMA8 46,185,57
orl-rat TDLo: 2 g/kg (6-15D preg): REP JACTDZ 2(4),325,83
orl-rat LD50: 369 mg/kg FCTXAV 15,607,77
ipr-mus LD50: 195 mg/kg JAPMA8 46,185,57

CONSENSUS REPORTS: Reported in EPA TSCA Inventory. EPA Genetic Toxicology Program. Chlorophenols are on the Community Right-To-Know List.

SAFETY PROFILE: Poison by ingestion and intraperitoneal routes. Experimental reproductive effects. An eye irritant. A hair dye component. When heated to decomposition it emits toxic fumes of Cl^-. See also CHLOROPHENOLS and RESORCINOL.

CLD800 CAS:635-93-8 HR: 3
5-CHLOROSALICYLALDEHYDE
mf: $C_7H_5ClO_2$ mw: 156.57

SYN: SALICYLALDEHYDE, 5-CHLORO-

TOXICITY DATA WITH REFERENCE
ivn-mus LD50: 56 mg/kg CSLNX* NX#05073

CONSENSUS REPORTS: Reported in EPA TSCA Inventory.

SAFETY PROFILE: Poison by intravenous route. When heated to decomposition it emits toxic vapors of NO_x and Cl^-.

CLD825 CAS:321-14-2 HR: 3
5-CHLOROSALICYLIC ACID
mf: $C_7H_5ClO_3$ mw: 172.57

SYNS: BENZOIC ACID, 5-CHLORO-2-HYDROXY- □ SALICYLIC ACID, 5-CHLORO-

TOXICITY DATA WITH REFERENCE
orl-rat LD50: 250 mg/kg PHMGBN 9,164,73
ipr-mus LDLo: 250 mg/kg CBCCT* 7,792,55

CONSENSUS REPORTS: Reported in EPA TSCA Inventory.

SAFETY PROFILE: Poison by ingestion and intraperitoneal routes. When heated to decomposition it emits toxic vapors of Cl^-.

CLE250 HR: 3
CHLOROSILANES

PROP: Compounds of Si, Cl, and H where the total number of atoms of Cl and H add up to 4. SiH_xCl_{4x}.

SAFETY PROFILE: Poison by ingestion and inhalation, and a poisonous irritant to skin, eyes, and mucous membranes. Toxicity is based on HCl which is formed upon hydrolysis of a chlorosilane. Self-ignites in air. With a little ammonia, it forms a self-igniting product. They react with water or steam to produce heat and toxic and corrosive fumes of HCl. When heated to decomposition they emit highly toxic fumes of Cl^-.

CLE600 CAS:1331-28-8 HR: 3
CHLOROSTYRENE
mf: C_8H_7Cl mw: 138.60

SYN: STYRENE, CHLORO-

TOXICITY DATA WITH REFERENCE
skn-rbt 10 mg/24H open JIDHAN 30,63,48
eye-rbt 500 mg open JIDHAN 30,63,48
orl-rat LD50: 5200 mg/kg JIDHAN 30,63,48
orl-mus LD50: 1230 mg/kg SAIGBL 15,544,73
ipr-mus LD50: 1090 mg/kg SAIGBL 15,544,73
skn-rbt LD50: 20 g/kg JIDHAN 30,63,48

CONSENSUS REPORTS: Reported in EPA TSCA Inventory.

SAFETY PROFILE: Moderately toxic by ingestion and intraperitoneal routes. A skin and eye irritant. A flammable liquid. When heated to decomposition it emits toxic vapors of Cl^-.

CLE750 CAS:2039-87-4 HR: 1
o-CHLOROSTYRENE
mf: C_8H_7Cl mw: 138.60

PROP: A solid. Mp: −63.15, bp: 188.6°, d: 1.100 @ 20°/4°.

CONSENSUS REPORTS: Reported in EPA TSCA Inventory.

OSHA PEL: TWA 50 ppm; STEL: 75 ppm
ACGIH TLV: TWA 50 ppm; STEL: 75 ppm

SAFETY PROFILE: A skin and eye irritant. When heated to decomposition it emits toxic fumes of Cl^-. See also CHLORINATED HYDROCARBONS, AROMATIC.

CLF100 CAS:956-04-7 HR: 3
4-CHLOROSTYRYL PHENYL KETONE
mf: $C_{15}H_{11}ClO$ mw: 242.71

SYNS: CHALCONE, 4-CHLORO-(6CI,7CI,8CI) □ (4-CHLOROBENZYLIDENE)ACETOPHENONE □ p-CHLOROCHALCONE □ 4-CHLOROCHALCONE □ 3-(4-CHLOROPHENYL)-1-PHENYL-2-PROPEN-1-ONE □ p-CHLOROSTYRYL PHENYL KETONE □ 2-PROPEN-1-ONE, 3-(4-CHLOROPHENYL)-1-PHENYL-

TOXICITY DATA WITH REFERENCE
orl-mus LD:>1 g/kg PHARAT 46,542,91
ipr-mus LD50:>1 g/kg PHARAT 46,542,91

DOT CLASSIFICATION: 3; *Label:* Flammable Liquid

SAFETY PROFILE: Low toxicity by ingestion and intraperitoneal routes. A flammable liquid. When heated to decomposition it emits toxic vapors of Cl^-.

CLF150 CAS:3300-67-2 HR: 3
o-CHLOROSTYRYL PHENYL KETONE
mf: $C_{15}H_{11}ClO$ mw: 242.71

SYNS: CHALCONE, 2-CHLORO-(CI,7CI,8CI) □ 2-CHLOROBENZYLIDENEACETOPHENONE □ 2-CHLOROCHALCONE □ 3-(2-CHLOROPHENYL)-1-PHENYL-2-PROPEN-1-ONE □ 2-PROPEN-1-ONE, 3-(2-CHLOROPHENYL)-1-PHENYL-

TOXICITY DATA WITH REFERENCE
orl-mus LD:>1 g/kg PHARAT 46,542,91
ipr-mus LD50:>750 mg/kg PHARAT 46,542,91

DOT CLASSIFICATION: 3; *Label:* Flammable Liquid

SAFETY PROFILE: Moderately toxic by intraperitoneal route. A flammable liquid. When heated to decomposition it emits toxic vapors of Cl^-.

CLF325 CAS:14293-44-8 HR: 3
4-CHLORO-5-SULFAMOYL-2′,6′-SALICYLOXYLIDIDE
mf: $C_{15}H_{15}ClN_2O_4S$ mw: 354.83

PROP: Crystals from methanol-water. Mp: 256°.

SYNS: 5-(AMINOSULFONYL)-4-CHLORO-N-(2,6-DIMETHYLPHENYL)-2-HYDROXY BENZAMIDE (9CI) ☐ AQUAPHOR ☐ BE 1293 ☐ BEI-1293 ☐ 4-CHLOR-5-SULFAMOYL-2′,6′-SALICYLOXYLIDID (GERMAN) ☐ DIUREXAN ☐ XIPAMID ☐ XIPAMIDE

TOXICITY DATA WITH REFERENCE
orl-rat LD50:1640 mg/kg ARZNAD 25,245,75
ipr-rat LD50:320 mg/kg ARZNAD 25,245,75
orl-mus LD50:1810 mg/kg ARZNAD 25,245,75
ipr-mus LD50:520 mg/kg ARZNAD 25,245,75
scu-mus LD50:1480 mg/kg ARZNAD 25,245,75

SAFETY PROFILE: Poison by intraperitoneal route. Moderately toxic by ingestion and subcutaneous routes. When heated to decomposition it emits toxic fumes of Cl^-, SO_x, and NO_x.

CLF500 CAS:25081-01-0 HR: 3
N-CHLOROSULFINYLIMIDE
mf: ClNOS mw: 97.52

O:S:NCl

PROP: Colorless liquid with suffocating smell. Mp: −80°, bp: 65.5°. Reacts vigorously with H_2O and Hg.

SAFETY PROFILE: May explode if melted in a sealed container. Reacts with chlorine fluoride to form an explosive, powerfully oxidizing product. When heated to decomposition it emits toxic fumes of Cl^-, SO_x, and NO_x.

CLG000 CAS:73926-94-0 HR: 2
4-CHLORO-4′-(6-SULFO-2H-NAPHTHO(1,2-d)TRIAZOL-2-YL)-2,2′-STILBENEDISULFONIC ACID TRISODIUM SALT
mf: $C_{24}H_{13}ClN_3O_9S_3 \cdot 3Na$ mw: 688.00

SYN: 2-(4″-CHLOR-4′-STILBYL)NAFTOTRIAZOL-6,2′,2″-TRISULFONAN SODNY (CZECH)

TOXICITY DATA WITH REFERENCE
eye-rbt 5 mg/24H SEV 28ZPAK -,250,72
orl-rat LD50:19,900 mg/kg 28ZPAK -,250,72

SAFETY PROFILE: Mildly toxic by ingestion. A severe eye irritant. When heated to decomposition it emits very toxic fumes of Cl^-, NO_x, Na_2O, and SO_x.

CLG200 HR: 2
5-(CHLOROSULFONYL)-2,4-DICHLOROBENZOIC ACID
mf: $C_7H_4Cl_3O_4S$ mw: 290.52

SYN: 2,4-DICHLORO-5-CHLOROSULPHONYLBENZOIC ACID

TOXICITY DATA WITH REFERENCE
skn-rbt 500 mg MLD FCTOD7 20,563,82
eye-rbt 100 mg SEV FCTOD7 20,573,82
eye-rbt 100 mg/4S rns SEV FCTOD7 20,573,82

SAFETY PROFILE: A skin and severe eye irritant. When heated to decomposition it emits toxic fumes of Cl^- and SO_x.

CLG250 CAS:1189-71-5 HR: 3
CHLOROSULFONYLISOCYANATE
mf: $CClNO_3S$ mw: 141.54

$ClSO_2N:C:O$

PROP: Liquid that fumes in moist air. Fp: −43°, bp: 107°.

SAFETY PROFILE: A very strong irritant. Reacts violently with water or moist air forming CO_2, H_2NS_3OH, and HCl. When heated to decomposition it emits toxic fumes of Cl^-, SO_x, and NO_x

CLG500 CAS:7790-94-5 HR: 3
CHLOROSULFURIC ACID
DOT: UN 1754/UN 2240
mf: $ClHO_3S$ mw: 116.52

PROP: Strong acid; clear to cloudy or colorless to pale-yellow liquid; sharp odor. Fumes in moist air. Mp: −80°, bp: 155°, d. 1.77 @ 28°, vap press: 1 mm @ 32°, vap d: 4.02. Sol in $CHCl_3$, CH_2Cl_2, and Py, insol in CS_2 and CCl_4.

SYNS: CHLOROSULFONIC ACID ☐ CHROMOSULFURIC ACID (UN 2240) (DOT) ☐ CHLOROSULFONIC ACID (with or without sulfur trioxide) (UN 1754) (DOT) ☐ MONOCHLOROSULFURIC ACID ☐ SULFONIC ACID, MONOCHLORIDE ☐ SULFURIC CHLOROHYDRIN

CONSENSUS REPORTS: Reported in EPA TSCA Inventory.

DOT CLASSIFICATION: 8; *Label:* Corrosive, Poison (UN 1754); DOT Class: 8; *Label:* Corrosive (UN 2240)

SAFETY PROFILE: A poison irritant. See also SULFURIC ACID. Chlorosulfonic acid is corrosive, can cause severe acid burns and is very irritating to the eyes, lungs, and mucous membranes. It can cause acute toxic effects either in the liquid or vapor state. Inhalation of concentrated vapor may cause loss of consciousness with serious damage to lung tissue. Contact of liquid with the eyes can cause severe burns if the liquid is not immediately and completely removed. It also causes severe skin burns due to its highly corrosive action. Upon ingestion it will irritate the mouth, esophagus, and stomach to a serious degree and on contact with skin cause dermatitis. It may cause conjunctivitis even in the vapor form. If spilled on a person, remove all contaminated clothing, wash contaminated skin with copious amounts of water, followed by a baking soda solution. Irrigate eyes with warm water for 15 minutes. Consult a physician.

Stored drums should be vented two times per month to control the hydrogen pressure, which is produced by the action of acid on the drum metal. Decomposes explosively on contact with water, alcohol, or acids. Explosive reaction with phosphorus. Violent reaction with silver nitrate. Potentially violent reaction with sulfuric acid or diphenyl ether. Incompatible with acetic acid, acetic anhydride, acetonitrile, acrolein, acrylic acid, acrylonitrile, allyl alcohol, allyl chloride, 2-amino ethanol, ammonium hydroxide, aniline, n-butyraldehyde, creosote oil, cresol, cumene, dichloroethyl ether, diethylene glycol monomethyl ether, diisobutylene, diisopropylether, epichloro hydrin, ethyl acetate, ethyl acrylate, ethylene chlorohydrin, ethylene cyanohydrin, ethylene diamine, ethylene glycol, ethylene glycol monoethyl ether acetate, ethylene imine, glyoxal, HCl, HF, H_2O_2, isoprene, mesityl oxide, metal powders, methyl ethyl ketone, HNO_3, 2-nitropropane, β-propiolactone, propylene oxide, pyridene, NaOH, sulfolane, styrene monomer, vinyl acetate, vinylidene chloride, water, organic matter, combustibles. Dangerous. To fight fire, avoid water, use dry chemicals. When heated to decomposition it emits toxic fumes of Cl^- and SO_x. See SULFURIC ACID, HYDROCHLORIC ACID, and SULFONATES.

CLG825 **HR: 2**
4-CHLORO-5-SULPHAMOYLPHTHALIMIDE
mf: $C_8H_5ClN_2O_4$ mw: 228.60

SYN: 6-CHLORO-1,3-DIOXO-5-ISOINDOLINESULFONAMIDE

TOXICITY DATA WITH REFERENCE
skn-rbt 500 mg MLD FCTOD7 20,573,82
eye-rbt 100 mg SEV FCTOD7 20,573,82
eye-rbt 100 mg/4S rns SEV FCTOD7 20,573,82

SAFETY PROFILE: A severe eye and mild skin irritant. When heated to decomposition it emits toxic fumes of Cl^- and NO_x.

CLH000 **CAS:63938-10-3** **HR: 1**
CHLOROTETRAFLUOROETHANE
mf: C_2HClF_4 mw: 136.48

PROP: Colorless gas.

SYN: MONOCHLOROTETRAFLUOROETHANE (DOT)

SAFETY PROFILE: Probably acts as a simple asphyxiant. See also CHLORINATED HYDROCARBONS, ALIPHATIC; and FLUORIDES. When heated to decomposition it emits highly toxic fumes of F^- and Cl^-.

CLH625 **CAS:4113-57-9** **HR: 3**
5-CHLORO-1,2,3-THIADIAZOLE
mf: C_2HClN_2S mw: 120.56

PROP: Bp: 72–74° @ 27 mm.

SAFETY PROFILE: A heat- and impact-sensitive explosive. When heated to decomposition it emits toxic fumes of Cl^-, SO_x, and NO_x. See also EXPLOSIVES.

CLH750 **CAS:58-94-6** **HR: 2**
CHLOROTHIAZIDE
mf: $C_7H_6ClN_3O_4S_2$ mw: 295.73

PROP: Mp: 342.5–343 (decomp). Sol in alkali.

SYNS: ALURENE □ CHLORIAZID □ 6-CHLORO-2H-1,2,4-BENZOTHIADIAZINE-7-SULFONAMIDE-1,1-DIOXIDE □ 6-CHLORO-7-SULFAMOYL-2H-1,2,4-BENZOTHIADIAZINE-1,1-DIOXIDE □CHLOROTHIA-ZID □ CHLORSAL □ CHLORTHIAZIDE □ CHLORURIT □ CHLOTRIDE □ CLOTRIDE □ CT □ DIURESAL □ DIURIL □ DIURILIX □ DIURITE □ DIUTRID □ FLUMEN □ MINZIL □ NEO-DEMA □ SALISAN □ SALUNIL □ SALURETIL □ SALURIC □ SK-CHLOROTHIAZIDE □ THIAZIDE □ URINEX □ WARDUZIDE □ YADALAN

TOXICITY DATA WITH REFERENCE
orl-rat TDLo: 12,450 mg/kg (8-22D preg): REP JCINAO 41,710,62
unr-rat TDLo: 3520 mg/kg (1-22D preg): REP ANREAK 193,174,79
orl-rat LD50: 10 g/kg YAKUD5 21,775,79
ipr-rat LD50: 1386 mg/kg 27ZIAQ -,77,73
ivn-rat LD50: 200 mg/kg TXAPA9 1,333,59
orl-mus LD50: 8 g/kg AIPTAK 118,467,59
ipr-mus LD50: 1400 mg/kg JPETAB 134,273,61
ivn-mus LD50: 940 mg/kg JPETAB 134,273,61
ivn-dog LD50: 1000 mg/kg 27ZIAQ -,77,73

CONSENSUS REPORTS: Reported in EPA TSCA Inventory. EPA Genetic Toxicology Program.

SAFETY PROFILE: Moderately toxic by intraperitoneal and intravenous routes. Mildly toxic by ingestion. Experimental reproductive effects. Has been implicated in aplastic anemia. When heated to decomposition it emits very toxic fumes of SO_x, NO_x, and Cl^-.

CLJ750 **CAS:2812-73-9** **HR: 3**
CHLOROTHIOFORMIC ACID ETHYL ESTER
DOT: UN 2826
mf: C_3H_5ClOS mw: 124.59

PROP: Bp: 52–55° @ 40 mm.

SYN: ETHYL CHLOROTHIOFORMATE (DOT)

DOT CLASSIFICATION: 8; *Label:* Corrosive, Poison

SAFETY PROFILE: Probably a poison by inhalation and ingestion. A corrosive irritant to skin, eyes, and mucous membranes. See also ESTERS and CHLORIDES. Flammable when exposed to heat or flame. When heated to decomposition it emits very toxic fumes of Cl^- and SO_x.

CLJ800 **CAS:89-68-9** **HR: 2**
6-CHLOROTHYMOL
mf: $C_{10}H_{13}ClO$ mw: 184.68

SYNS: 4-CHLORO-5-METHYL-2-(1-METHYLETHYL)PHENOL □ CHLOROTHYMOL □ CHLORTHYMOL □ PHENOL, 4-CHLORO-5-METHYL-2-(1-METHYLETHYL)-(9CI) □ THYMOL, 6-CHLORO-

TOXICITY DATA WITH REFERENCE
scu-mus LD50: 2460 mg/kg SIZSAR 3,73,52

CONSENSUS REPORTS: Reported in EPA TSCA Inventory.

SAFETY PROFILE: Moderately toxic by subcutaneous

route. When heated to decomposition it emits toxic vapors of Cl^-.

CLJ875 CAS:21923-23-9 HR: 3
CHLORTHIOPHOS
mf: $C_{11}H_{15}Cl_2O_3PS_2$ mw: 361.25

SYNS: CELAMERCK S-2957 □ CELA S-2957 □ CELATHION □ CM S 2957 □ O-(DICHLORO(METHYLTHIO)PHENYL) O,O-DIETHYL PHOSPHOROTHIOATE (3 isomers) □ O,O-DIETHYL-O-2,4,5-DICHLORO-(METHYLTHIO)PHENYL THIONOPHOSPHATE □ ENT 27,635 □ NSC 195164 □ OMS 1342 □ S 2957

TOXICITY DATA WITH REFERENCE
orl-rat LD50:7800 μg/kg FMCHA2 -,C56,8
skn-rat LD50:58 mg/kg AHRTAN 27,3,76
orl-mus LD50:141 mg/kg AHRTAN 27,3,76
orl-rbt LD50:20 mg/kg AHRTAN 27,3,76
skn-rbt LD50:48 mg/kg AHRTAN 27,3,76
orl-gpg LD50:58 mg/kg AHRTAN 27,3,76
orl-ckn LD50:45 mg/kg AHRTAN 27,3,76
orl-qal LD50:45 mg/kg AHRTAN 27,3,76

CONSENSUS REPORTS: EPA Extremely Hazardous Substances List.

SAFETY PROFILE: Poison by ingestion and skin contact. When heated to decomposition it emits toxic fumes of Cl^-, PO_x, and SO_x.

CLK100 CAS:95-49-8 HR: 3
o-CHLOROTOLUENE
DOT: UN 2238
mf: C_7H_7Cl mw: 126.59

PROP: Liquid. Mp: −34°, bp: 159°, d: 1.08 @ 20°/4°. Volatile with steam. Sltly sol in water; freely sol in alc, benzene, chloroform, ether.

SYNS: 2-CHLORO-1-METHYLBENZENE (9CI) □ 2-CHLOROTOLUENE □ HALSO 99 □ 1-METHYL-2-CHLOROBENZENE □ 2-METHYLCHLOROBENZENE □ o-TOLYL CHLORIDE

TOXICITY DATA WITH REFERENCE
ihl-rat LCLo:17,500 ppm DTLVS* 4,95,80
unr-rat LD50:5700 mg/kg GISAAA 45(12),64,80
unr-mus LD50:4400 mg/kg GISAAA 45(12),64,80
unr-gpg LD50:3000 mg/kg GISAAA 46(2),14,81

CONSENSUS REPORTS: Reported in EPA TSCA Inventory.

OSHA PEL: TWA 50 ppm
ACGIH TLV: TWA 50 ppm
DOT CLASSIFICATION: 3; *Label:* Flammable Liquid

SAFETY PROFILE: Moderately toxic by unspecified routes. Flammable when exposed to heat or flame. When heated to decomposition it emits toxic fumes of Cl^-. See also TOLYL CHLORIDE, and CHLORINATED HYDROCARBONS, AROMATIC.

CLK130 CAS:25168-05-2 HR: 3
CHLOROTOLUENES
DOT: UN 2238
mf: C_7H_7Cl mw: 126.59

SYNS: BENZENE, CHLOROMETHYL-(9CI) □ CHLOROMETHYLBENZENE □ CHLOROTOLUENE □ ar-CHLOROTOLUENE □ TOLUENE, ar-CHLORO-

TOXICITY DATA WITH REFERENCE
ihl-uns LC50:48 g/m³ GTPZAB 30(3),6,86

CONSENSUS REPORTS: Reported in EPA TSCA Inventory.

DOT CLASSIFICATION: 3; *Label:* Flammable Liquid

SAFETY PROFILE: Low toxicity by inhalation. A flammable liquid. When heated to decomposition it emits toxic vapors of Cl^-.

CLK200 CAS:87-60-5 HR: 3
3-CHLORO-o-TOLUIDINE
mf: C_7H_8ClN mw: 141.61

PROP: Bp: 245°.

SYNS: 1-AMINO-2-CHLORO-6-METHYLBENZENE □ 1-AMINO-3-CHLORO-2-METHYLBENZENE □ 2-AMINO-6-CHLOROTOLUENE □ AZOIC DIAZO COMPONENT 46 □ 3-CHLORO-2-METHYLANILINE □ 3-CHLOR 2 TOLUIDIN (CZECH) □ FAST SCARLET TR BASE □ SCARLET TR BASE

TOXICITY DATA WITH REFERENCE
dni-mus-orl 200 mg/kg MUREAV 46,305,77
orl-rat LD50:574 mg/kg 85JCAE-,613,86
orl-rat LD50:574 mg/kg MarJV# 29MAR77
orl-bwd LD50:237 mg/kg AECTCV 12,355,83

CONSENSUS REPORTS: Reported in EPA TSCA Inventory.

DOT CLASSIFICATION: 6.1; *Label:* KEEP AWAY FROM FOOD

SAFETY PROFILE: Poison by ingestion. Mutation data reported. When heated to decomposition it emits toxic fumes of Cl^- and NO_x. See also other chloro toluidine entries.

CLK210 CAS:615-65-6 HR: 3
2-CHLORO-p-TOLUIDINE
mf: C_7H_8ClN mw: 141.61

PROP: Liquid. D: 1.151 @ 20°, mp: 7°, bp: 219° @ 732 mm.

SYNS: BENZENAMINE, 2-CHLORO-4-METHYL- □ 2-CHLORO-4-METHYLANILINE □ 2-CHLOR-4-TOLUIDIN (CZECH) □ 4-METHYL-2-CHLOROANILINE

TOXICITY DATA WITH REFERENCE
skn-rbt 2 mg/24H SEV 85JCAE-,612,86
eye-rbt 250 μg/24H SEV 85JCAE-,612,86
mma-sat 1 μmol/plate MUREAV 77,317,80
orl-rat LD50:367 mg/kg 28ZPAK -,97,72

CONSENSUS REPORTS: Reported in EPA TSCA Inventory.

DOT CLASSIFICATION: 6.1; *Label:* KEEP AWAY FROM FOOD

SAFETY PROFILE: Poison by ingestion. A severe eye and skin irritant. Mutation data reported. When heated to decomposition it emits toxic fumes of Cl^- and NO_x. See also other chloro toluidine entries.

CLK215 CAS:95-74-9 HR: 3
3-CHLORO-p-TOLUIDINE
mf: C_7H_8ClN mw: 141.61

PROP: A solid or liquid. Mp: 26°, bp: 237–238.5°.

SYNS: 1-AMINO-3-CHLORO-4-METHYLBENZENE □ 4-AMINO-2-CHLOROTOLUENE □ 2-CHLORO-4-AMINOTOLUENE □ 3-CHLORO-4-METHYLANILINE □ CPT □ DKC 1347 □ DRC 1339 □ NCI-C02040

TOXICITY DATA WITH REFERENCE
dni-mus-orl 200 mg/kg MUREAV 46,305,77
orl-rat LD50:1500 mg/kg TXAPA9 21,315,72
ipr-rat LD50:325 mg/kg TXAPA9 18,517,71
ivn-rat LD50:48 mg/kg TXAPA9 18,517,71
orl-mus LD50:316 mg/kg NCILB* NCI-E-C-72-3252,73
orl-pgn LD50:13 mg/kg TXAPA9 21,315,72
orl-qal LD50:1 mg/kg AECTCV 12,355,83
orl-bwd LD50:2400 µg/kg TXAPA9 21,315,72

CONSENSUS REPORTS: Reported in EPA TSCA Inventory. NCI Carcinogenogenesis Bioassay (Feed); Results Negative: Mouse, Rat NCITR* NCI-CG-TR-145,78

DOT CLASSIFICATION: 6.1; *Label:* KEEP AWAY FROM FOOD

SAFETY PROFILE: Poison by ingestion, intravenous, and intraperitoneal routes. Mutation data reported. When heated to decomposition it emits toxic fumes of Cl^- and NO_x. See also other chloro toluidine entries.

CLK220 CAS:95-69-2 HR: 3
4-CHLORO-o-TOLUIDINE
mf: C_7H_8ClN mw: 141.61

PROP: Leaflets from EtOH. Mp: 29–30°, bp: 236–238° @ 730 mm.

SYNS: AMARTHOL FAST RED TR BASE □ 2-AMINO-5-CHLOROTOLUENE □ AZOENE FAST RED TR BASE □ AZOGENE FAST RED TR □ AZOIC DIAZO COMPONENT 11 BASE □ BRENTAMINE FAST RED TR BASE □ 5-CHLORO-2-AMINOTOLUENE □ 4-CHLORO-2-METHYLANILINE □ 4-CHLORO-6-METHYLANILINE □ 4-CHLORO-2-METHYLBENZENEAMINE □ 4-CHLORO-2-TOLUIDINE □ DAITO RED BASE TR □ DEVAL RED K □ DEVAL RED TR □ DIAZO FAST RED TRA □ FAST RED BASE TR □ FAST RED 5CT BASE □ FAST RED TR □ FAST RED TR11 □ FAST RED TR BASE □ FAST RED TRO BASE □ KAKO RED TR BASE □ KAMBAMINE RED TR □ 2-METHYL-4-CHLOROANILINE □ MITSUI RED TR BASE □ RED BASE CIBA IX □ RED BASE IRGA IX □ RED BASE NTR □ RED TR BASE □ SANYO FAST RED TR BASE □ TULABASE FAST RED TR

TOXICITY DATA WITH REFERENCE
mmo-sat 400 µg/plate JPFCD2 19,95,84
dnr-sat 250 mg/disc JPFCD2 19,95,84
dnr-esc 2 g/disc JPFCD2 19,95,84
oms-hmn:hla 1 mmol/L BECTA6 11,184,74
slt-mus-orl 12 g/kg/3D-I MUREAV 135,219,84
dnd-ham:lng 3 mmol/L/2H MUREAV 77,317,80
scu-cat LDLo:310 mg/kg AHBAAM 110,12,33
orl-brd LD50:75 mg/kg AECTCV 12,355,83

CONSENSUS REPORTS: IARC Cancer Review: Group 2A IMEMDT 7,56,87; Human Inadequate Evidence IMEMDT 16,277,78; Animal Sufficient Evidence IMEMDT 30,61,83. Reported in EPA TSCA Inventory.

DFG MAK: Human Carcinogen.
DOT CLASSIFICATION: 6.1; *Label:* KEEP AWAY FROM FOOD

SAFETY PROFILE: Confirmed carcinogen. Poison by ingestion and subcutaneous routes. Human mutation data reported. In the presence of copper(II) chloride catalyst decomposition occurs above 239°C. When heated to decomposition it emits toxic fumes of Cl^- and NO_x. See also other chloro toluidine entries.

CLK225 CAS:95-79-4 HR: 3
5-CHLORO-o-TOLUIDINE
mf: C_7H_8ClN mw: 141.61

PROP: Solid. Bp: 237° @ 722 mm, mp: 21–22°.

SYNS: ACCO FAST RED KB BASE □ 1-AMINO-3-CHLORO-6-METHYLBENZENE □ 2-AMINO-4-CHLOROTOLUENE □ ANSIBASE RED KB □ AZOENE FAST RED KB BASE □ AZOIC DIAZO COMPONENT 32 □ 4-CHLORO-2-AMINOTOLUENE □ 3-CHLORO-6-METHYLANILINE □ 5-CHLORO-2-METHYLANILINE □ FAST RED KB AMINE □ FAST RED KB BASE □ FAST RED KB SALT □ FAST RED KB SALT SUPRA □ FAST RED KBS SALT □ GENAZO RED KB SOLN □ HILTONIL FAST RED KB BASE □ LAKE RED KB BASE □ METROGEN RED FORMER KB SOLN □ NAPHTHOSOL FAST RED KB BASE □ NCI-C02051 □ PHARMAZOID RED KB □ RED KB BASE □ SPECTROLENE RED KB □ STABLE RED KB BASE

TOXICITY DATA WITH REFERENCE
dni-mus-orl 200 mg/kg MUREAV 46,305,77
orl-rat TDLo:164 g/kg/78W-C:ETA NCITR* NCI-CG-TR-187,79
orl-mus TDLo:131 g/kg/78W-C:CAR NCITR* NCI-CG-TR-187,79
orl-mus TD:262 g/kg/78W-C:CAR NCITR* NCI-CG-TR-187,79
orl-rat LD50:464 mg/kg NCILB* NIH-NCI-E-C-72-3252

CONSENSUS REPORTS: NTP Carcinogenesis Bioassay (feed): Clear Evidence: mouse NCITR* NCI-TR-187,79; (feed): Inadequate Studies: rat NCITR* NCI-TR-187,79. Reported in EPA TSCA Inventory. EPA Genetic Toxicology Program.

DFG MAK: Suspected Carcinogen.
DOT CLASSIFICATION: 6.1; *Label:* KEEP AWAY FROM FOOD

SAFETY PROFILE: Suspected carcinogen with experimental carcinogenic and tumorigenic data. Moderately toxic by ingestion. When heated to decomposition it emits very toxic fumes of Cl^- and NO_x. See also AROMATIC AMINES.

CLK227 CAS:87-63-8 HR: 3
6-CHLORO-o-TOLUIDINE
mf: C_7H_8ClN mw: 141.61

SYNS: 2-AMINO-3-CHLOROTOLUENE ☐ 3-CHLORO-2-AMINOTOLUENE ☐ 6-CHLORO-2-METHYLANILINE ☐ 6-CHLORO-2-TOLUIDINE ☐ o-TOLUIDINE, 6-CHLORO-

TOXICITY DATA WITH REFERENCE
dni-mus-orl 200 mg/kg MUREAV 46,305,77
scu-cat LDLo:200 mg/kg AHBAAM 110,12,33

CONSENSUS REPORTS: Reported in EPA TSCA Inventory.

DOT CLASSIFICATION: 6.1; *Label:* KEEP AWAY FROM FOOD

SAFETY PROFILE: Poison by subcutaneous route. Mutation data reported. When heated to decomposition it emits toxic vapors of NO_x and Cl^-.

CLK230 CAS:7745-89-3 HR: 3
3-CHLORO-p-TOLUIDINE HYDROCHLORIDE
mf: $C_7H_8ClN{\cdot}ClH$ mw: 178.07

SYNS: CTH ☐ DRC-1,339 ☐ 4-METHYL-3-CHLOROANILINE HYDROCHLORIDE ☐ STARLICIDE

TOXICITY DATA WITH REFERENCE
orl-rat LD50:655 mg/kg TXAPA9 18,517,71
ipr-mus LD50:338 mg/kg TXAPA9 29,135,74
orl-pgn LD50:18 mg/kg TXAPA9 21,315,72
orl-ckn LD50:4 mg/kg PCOC** -,457,66
ipr-ckn LDLo:100 mg/kg TXAPA9 22,458,72
orl-dck LD50:18 mg/kg TXAPA9 21,315,72
orl-bwd LD50:2400 μg/kg TXAPA9 21,315,72

SAFETY PROFILE: Poison by ingestion and intraperitoneal routes. When heated to decomposition it emits toxic fumes of NO_x and Cl^-. See also other chloro toluidine entries.

CLK235 CAS:3165-93-3 HR: 3
4-CHLORO-2-TOLUIDINE HYDROCHLORIDE
DOT: UN 1579
mf: $C_7H_8ClN{\cdot}ClH$ mw: 178.07

SYNS: AMARTHOL FAST RED TR BASE ☐ AMARTHOL FAST RED TR SALT ☐ 2-AMINO-5-CHLOROTOLUENE HYDROCHLORIDE ☐ AZANIL RED SALT TRD ☐ AZOENE FAST RED TR SALT ☐ AZOGENE FAST RED TR ☐ AZOIC DIAZO COMPONENT 11 BASE ☐ BRENTAMINE FAST RED TR SALT ☐ CHLORHYDRATE de 4-CHLOROORTHOTOLUIDINE (FRENCH) ☐ 5-CHLORO-2-AMINOTOLUENE HYDROCHLORIDE ☐ 4-CHLORO-2-METHYLANILINE HYDROCHLORIDE ☐ 4-CHLORO-6-METHYLANILINE HYDROCHLORIDE ☐ 4-CHLORO-2-METHYLBENZENEAMINE HYDROCHLORIDE ☐ 4-CHLORO-o-TOLUIDINE HYDROCHLORIDE ☐ 4-CHLORO-o-TOLUIDINE HYDROCHLORIDE (DOT) ☐ C.I. 37085 ☐ C.I. AZOIC DIAZO COMPONENT 11 ☐ DAITO RED SALT TR ☐ DEVOL RED K ☐ DEVOL RED TA SALT ☐ DEVOL RED TR ☐ DIAZO FAST RED TR ☐ DIAZO FAST RED TRA ☐ FAST RED 5CT SALT ☐ FAST RED SALT TR ☐ FAST RED SALT TRA ☐ FAST RED SALT TRN ☐ FAST RED TR SALT ☐ HINDASOL RED TR SALT ☐ KROMON GREEN B ☐ 2-METHYL-4-CHLOROANILINE HYDROCHLORIDE ☐ NATASOL FAST RED TR SALT ☐ NCI-C02368 ☐ NEUTROSEL RED TRVA ☐ OFNA-PERL SALT RRA ☐ RCRA WASTE NUMBER U049 ☐ RED BASE CIBA IX ☐ RED BASE IRGA IX ☐ RED SALT CIBA IX ☐ RED SALT IRGA IX ☐ RED TRS SALT ☐ SANYO FAST RED SALT TR

TOXICITY DATA WITH REFERENCE
orl-mus TDLo:49 g/kg/78W-C:CAR JEPTDQ 2(2),325,78
orl-mus TD:104 g/kg/99W-C:CAR NCITR* NCI-CG-TR-165,78
orl-mus TD:108 g/kg/78W-C:CAR JEPTDQ 2(2),325,78
orl-mus TD:216 g/kg/78W-C:CAR JEPTDQ 2(2),325,78
orl-mus TD:97 g/kg/78W-C:CAR JEPTDQ 2(2),325,78
ipr-rat LD50:560 mg/kg NCIBR* NCI-E-68-1311,73
ipr-mus LD50:680 mg/kg NCIBR* NCI-E-68-1311,73

CONSENSUS REPORTS: IARC Cancer Review: Group 2A IMEMDT 48,123,90; Animal Inadequate Evidence, Human Inadequate Evidence IMEMDT 16,277,78. NCI Carcinogenesis Bioassay (Feed); Clear Evidence: Mouse; No Evidence: Rat NCITR* NCI-CG-TR-165,79. Reported in EPA TSCA Inventory.

DOT CLASSIFICATION: 6.1; *Label:* KEEP AWAY FROM FOOD

SAFETY PROFILE: Suspected carcinogen with experimental carcinogenic data. Moderately toxic by intraperitoneal route. When heated to decomposition it emits toxic fumes of NO_x and Cl^-. See also other chloro toluidine entries.

CLK325 CAS:13710-19-5 HR: 3
N-(3-CHLORO-o-TOLYL)ANTHRANILIC ACID
mf: $C_{14}H_{12}ClNO_2$ mw: 261.72

PROP: Crystals from abs ethanol. Mp: 207–207.5°.

SYNS: N-(3-CHLORO-2-METHYLPHENYL)ANTHRANILIC ACID ☐ CLOTAM ☐ GEA 6414 ☐ N-(2-METHYL-3-CHLOROPHENYL)ANTHRANILIC ACID ☐ TOLFENAMIC ACID

TOXICITY DATA WITH REFERENCE
orl-rat TDLo:648 mg/kg (female 17-22D post):REP TOIZAG 29,889,83
orl-rat TDLo:1 mg/kg (21D preg):TER OYYAA2 27,117,84
orl-rat TDLo:264 mg/kg (female 7-17D post):TER TOIZAG 29,889,83
orl-rat LD50:225 mg/kg TOIZAG 28,99,81
ipr-rat LD50:238 mg/kg TOIZAG 29,851,83
scu-rat LD50:246 mg/kg IYKEDH 14,838,83
orl-mus LD50:280 mg/kg IYKEDH 14,838,83
ipr-mus LD50:185 mg/kg IYKEDH 14,838,83
scu-mus LD50:267 mg/kg IYKEDH 14,838,83

SAFETY PROFILE: Poison by ingestion, subcutaneous, and intraperitoneal routes. An experimental teratogen. Experimental reproductive effects. When heated to decomposition it emits toxic fumes of Cl^- and NO_x.

CLK500 CAS:78371-90-1 HR: 3
1-(6-CHLORO-o-TOLYL)-3-CYCLOHEXYL-3-(2-(DIETHYLAMINO)ETHYL)UREA HYDROCHLORIDE
mf: $C_{20}H_{32}ClN_3O{\cdot}ClH$ mw: 402.46

TOXICITY DATA WITH REFERENCE
ipr-rat LD50:41 mg/kg ARZNAD 8,664,58
scu-mus LD50:87 mg/kg ARZNAD 8,664,58

SAFETY PROFILE: Poison by intraperitoneal and subcutaneous routes. When heated to decomposition it emits very toxic fumes of Cl^- and NO_x.

CLK750 **CAS:78371-91-2** **HR: 3**
1-(6-CHLORO-o-TOLYL)-3-(3-(DIBUTYLAMINO)PROPYL)UREA HYDROCHLORIDE
mf: $C_{19}H_{32}ClN_3O \cdot ClH$ mw: 390.45

TOXICITY DATA WITH REFERENCE
eye-rbt 2% MOD ARZNAD 8,664,58
ipr-rat LD50:275 mg/kg ARZNAD 8,664,58
scu-mus LD50:450 mg/kg ARZNAD 8,664,58

SAFETY PROFILE: Poison by intraperitoneal route. Moderately toxic by subcutaneous route. An eye irritant. When heated to decomposition it emits very toxic fumes of Cl^- and NO_x.

CLL000 **CAS:78371-93-4** **HR: 3**
1-(6-CHLORO-o-TOLYL)-3-(2-(DIETHYLAMINO)ETHYL)-3-METHYLUREA
mf: $C_{15}H_{24}ClN_3O$ mw: 297.87

SYN: C 3247

TOXICITY DATA WITH REFERENCE
eye-rbt 2% MLD ARZNAD 8,664,58
ipr-rat LD50:108 mg/kg ARZNAD 8,664,58
scu-mus LD50:262 mg/kg ARZNAD 8,664,58

SAFETY PROFILE: Poison by intraperitoneal and subcutaneous routes. An eye irritant. When heated to decomposition it emits very toxic fumes of Cl^- and NO_x.

CLL250 **CAS:78371-92-3** **HR: 3**
1-(6-CHLORO-o-TOLYL)-3-(2-(DIETHYLAMINO)ETHYL)UREA HYDROCHLORIDE
mf: $C_{14}H_{22}ClN_3O \cdot ClH$ mw: 320.30

SYN: C 3182

TOXICITY DATA WITH REFERENCE
eye-rbt 2% MLD ARZNAD 8,664,58
ipr-rat LD50:212 mg/kg ARZNAD 8,664,58
scu-mus LD50:500 mg/kg ARZNAD 8,664,58

SAFETY PROFILE: Poison by intraperitoneal route. Moderately toxic by subcutaneous route. An eye irritant. When heated to decomposition it emits very toxic fumes of Cl^- and NO_x.

CLL500 **CAS:78371-95-6** **HR: 3**
1-(6-CHLORO-o-TOLYL)-3-(2-(DIETHYLAMINO)ETHYL)-3-(2,6-XYLYL)UREA HYDROCHLORIDE
mf: $C_{22}H_{30}ClN_3O \cdot ClH$ mw: 424.46

SYN: C 3184

TOXICITY DATA WITH REFERENCE
eye-rbt 2% MLD ARZNAD 8,664,58
ipr-rat LD50:84 mg/kg ARZNAD 8,664,58
scu-mus LD50:75 mg/kg ARZNAD 8,664,58

SAFETY PROFILE: Poison by intraperitoneal and subcutaneous routes. An eye irritant. When heated to decomposition it emits very toxic fumes of $Cl^-$and NO_x.

CLL750 **CAS:78371-94-5** **HR: 3**
1-(6-CHLORO-o-TOLYL)-1-(2-(DIETHYLAMINO)ETHYL)-3-(2,6-XYLYL)UREA HYDROCHLORIDE
mf: $C_{22}H_{30}ClN_3O \cdot ClH$ mw: 424.46

SYN: C 3186

TOXICITY DATA WITH REFERENCE
ipr-rat LD50:62 mg/kg ARZNAD 8,664,58
scu-mus LD50:90 mg/kg ARZNAD 8,664,58

SAFETY PROFILE: Poison by intraperitoneal and subcutaneous routes. When heated to decomposition it emits very toxic fumes of Cl^- and NO_x.

CLM000 **CAS:78371-96-7** **HR: 3**
1-(6-CHLORO-o-TOLYL)-3-(3-(DIETHYLAMINO)PROPYL)UREA
mf: $C_{15}H_{24}ClN_3O$ mw: 297.87

SYN: C 3214

TOXICITY DATA WITH REFERENCE
eye-rbt 2% MLD ARZNAD 8,664,58
ipr-rat LD50:275 mg/kg ARZNAD 8,664,58
scu-mus LD50:450 mg/kg ARZNAD 8,664,58

SAFETY PROFILE: Poison by intraperitoneal route. Moderately toxic by subcutaneous route. An eye irritant. When heated to decomposition it emits very toxic fumes of Cl^- and NO_x.

CLM250 **CAS:78371-98-9** **HR: 3**
1-(6-CHLORO-o-TOLYL)-3-(2-(DIMETHYLAMINO)ETHYL)-3-ISOPROPYLUREA HYDROCHLORIDE
mf: $C_{15}H_{24}ClN_3O \cdot ClH$ mw: 334.33

SYN: C 3246

TOXICITY DATA WITH REFERENCE
eye-rbt 2% MLD ARZNAD 8,664,58
ipr-rat LD50:22 mg/kg ARZNAD 8,664,58
scu-mus LD50:30 mg/kg ARZNAD 8,664,58

SAFETY PROFILE: Poison by intraperitoneal and subcutaneous routes. An eye irritant. When heated to decomposition it emits very toxic fumes of Cl^- and NO_x.

CLM500 **CAS:78371-97-8** **HR: 3**
1-(6-CHLORO-o-TOLYL)-3-(2-(DIMETHYLAMINO)ETHYL)UREA HYDROCHLORIDE
mf: $C_{12}H_{18}ClN_3O \cdot ClH$ mw: 292.24

SYN: C 3213

TOXICITY DATA WITH REFERENCE
eye-rbt 2% MLD ARZNAD 8,664,58
ipr-rat LD50:362 mg/kg ARZNAD 8,664,58
scu-mus LD50:1025 mg/kg ARZNAD 8,664,58

SAFETY PROFILE: Poison by intraperitoneal route. Moderately toxic by subcutaneous route. An eye irritant. When heated to decomposition it emits very toxic fumes of NO_x and Cl^-.

CLM750 CAS:78371-99-0 HR: 3
1-(6-CHLORO-o-TOLYL)-3-(3-(DIMETHYLAMINO)PROPYL)UREA HYDROCHLORIDE
mf: $C_{13}H_{20}ClN_3O•ClH$ mw: 306.27

SYN: C 3229

TOXICITY DATA WITH REFERENCE
ipr-rat LD50:300 mg/kg ARZNAD 8,664,58
scu-mus LD50:1375 mg/kg ARZNAD 8,664,58

SAFETY PROFILE: Poison by intraperitoneal route. Moderately toxic by subcutaneous route. When heated to decomposition it emits very toxic fumes of Cl^- and NO_x.

CLN000 CAS:78372-00-6 HR: 3
1-(6-CHLORO-o-TOLYL)-3-(4-METHOXYBENZYL)-3-(2-PIPERIDINOETHYL)UREA
mf: $C_{23}H_{30}ClN_3O_2$ mw: 416.01

SYN: C 5320

TOXICITY DATA WITH REFERENCE
eye-rbt 2% MLD ARZNAD 8,664,58
ipr-rat LD50:70 mg/kg ARZNAD 8,664,58
scu-mus LD50:130 mg/kg ARZNAD 8,664,58

SAFETY PROFILE: Poison by intraperitoneal and subcutaneous routes. An eye irritant. When heated to decomposition it emits very toxic fumes of $Cl^-$and NO_x.

CLN250 CAS:78393-39-2 HR: 3
1-(6-CHLORO-o-TOLYL)-3-(4-METHOXYBENZYL)-3-(2-(PYRROLIDINYL)ETHYL)UREA HYDROCHLORIDE
mf: $C_{22}H_{28}ClN_3O_2•ClH$ mw: 438.44

SYN: C 5324

TOXICITY DATA WITH REFERENCE
ipr-rat LD50:72 mg/kg ARZNAD 8,664,58
scu-mus LD50:145 mg/kg ARZNAD 8,664,58

SAFETY PROFILE: Poison by intraperitoneal and subcutaneous routes. When heated to decomposition it emits very toxic fumes of Cl^- and NO_x.

CLN500 CAS:78372-01-7 HR: 3
1-(4-CHLORO-o-TOLYL)-3-(p-METHYLBENZYL)-3-(2-PYRROLIDINYLETHYL)UREA HYDROCHLORIDE
mf: $C_{22}H_{28}ClN_3O•ClH$ mw: 422.44

SYN: C 5326

TOXICITY DATA WITH REFERENCE
eye-rbt 2% MLD ARZNAD 8,664,58
ipr-rat LD50:85 mg/kg ARZNAD 8,664,58
scu-mus LD50:175 mg/kg ARZNAD 8,664,58

SAFETY PROFILE: Poison by intraperitoneal and subcutaneous routes. An eye irritant. When heated to decomposition it emits very toxic fumes of $Cl^-$and NO_x.

CLN750 CAS:94-81-5 HR: 2
4-((4-CHLORO-o-TOLYL)OXY)BUTYRIC ACID
mf: $C_{11}H_{13}ClO_3$ mw: 228.69

PROP: Crystals. Mp: 100. Very sltly sol in H_2O.

SYNS: BEXANE □ BEXONE □ CAN-TROL □ 4-(4-CHLOR-2-METHYLPHENOXY)-BUETTERSAEURE (GERMAN) □ 4-(4-CHLOR-2-METHYLPHENOXY)-BUTTERSAEURE (GERMAN) □ 4-(4-CHLORO-2-METHYLPHENOXY)BUTANOIC ACID □ γ-(4-CHLORO-2-METHYLPHENOXY)BUTYRIC ACID □ 4-(4-CHLORO-2-METHYLPHENOXY)BUTYRIC ACID □ (4-CHLORO-o-TOLYLOXY)BUTYRIC ACID □ LEGUMEX □ 4-(MCB) □ MCPB □ MCP-BUTYRIC □ 2-METHYL-4-CHLOROPHENOXYBUTYRIC ACID □ γ-2-METHYL-4-CHLOROPHENOXYBUTYRIC ACID □ 4-(2-METHYL-4-CHLOROPHENOXY)BUTYRIC ACID □ 4-(2-METHYL-4-CHLORPHENOXY)-BUTTERSAEURE (GERMAN) □ PDQ □ THISTROL □ TRIFOLEX □ TRITROL □ TROPOTOX □ TROTOX □ U46 MCPB

TOXICITY DATA WITH REFERENCE
sln-dmg-orl 4400 μmol/L EXPEAM 30,621,74
mrc-smc 13,500 μmol/L IARCCD 10,161,74
orl-rat LD50:680 mg/kg WRPCA2 4,36,65
orl-mus LD50:800 mg/kg FMCHA2 -,C43,83

CONSENSUS REPORTS: EPA Genetic Toxicology Program.

SAFETY PROFILE: Moderately toxic by ingestion. Mutation data reported. An herbicide. When heated to decomposition it emits toxic fumes of Cl^-.

CLO000 CAS:6062-26-6 HR: 2
(4-CHLORO-o-TOLYLOXY)BUTYRIC ACID SODIUM SALT
mf: $C_{11}H_{12}ClO_3•Na$ mw: 250.67

SYNS: CANTROL □ 4-(4-CHLOR-2-METHYL-PHENOXY)-BUTTERSAEURE NATRIUMSALZ (GERMAN) □ CHLOROMETHYLPHENOXYBUTYRIC ACID SODIUM SALT □ 4-(4-CHLORO-2-METHYLPHENOXY)BUTYRIC ACID SODIUM SALT □ 4-(4-CHLORO-2-METHYLPHENOXY)BUTANOIC ACID, SODIUM SALT □ M&B 3046 □ MCPB □ 4-(MCPD) □ 4-(2-METHYL-4-CHLOROPHENOXY)BUTYRIC ACID, SODIUM SALT □ THISTROL □ TROPOTOX

TOXICITY DATA WITH REFERENCE
orl-rat LD50:700 mg/kg PCOC** -,715,66
skn-rat LD50:1000 mg/kg WRPCA2 9,119,70
orl-mus LD50:600 mg/kg GTPZAB 10(3),50,66

SAFETY PROFILE: Moderately toxic by ingestion and skin contact. A pesticide. When heated to decomposition it emits toxic fumes of Cl^- and Na_2O.

CLO200 CAS:1929-86-8 HR: 2
2-((4-CHLORO-o-TOLYL)OXY)PROPIONIC ACID POTASSIUM SALT
mf: $Cl_0H_{10}ClO_3•K$ mw: 168.10

SYNS: GORDON'S MECOMEC □ HEDONAL MCPP □ MCPP POTASSIUM SALT □ MECOPEX □ MECOPROP POTASSIUM SALT □ METHOXONE M □ PROPANOIC ACID, 2-(4-CHLORO-2-METHYLPHENOXY)-, POTASSIUM SALT (9CI) □ PROPIONIC ACID, 2-((4-CHLORO-o-TOLYL)OXY)-, POTASSIUM SALT □ SYS 67MPROP □ VI PEX

TOXICITY DATA WITH REFERENCE
orl-rat TDLo:520 mg/kg (female 4-18D post):REP TJADAB 33,11A,86

orl-rat TDLo: 1320 mg/kg (female 4-18D post): REP TJADAB 33,11A,86
orl-rat TDLo: 1320 mg/kg (female 4-18D post): TER TJADAB 33,11A,86
orl-rat LD50: 930 mg/kg FMCHA2 -,C184,89

SAFETY PROFILE: Moderately toxic by ingestion. An experimental teratogen. Other experimental reproductive effects. When heated to decomposition it emits toxic fumes of Cl^-.

CLO500 CAS:78372-02-8 HR: 3
1-(6-CHLORO-o-TOLYL)-3-(2-PYRROLIDINYLETHYL) UREA HYDROCHLORIDE
mf: $C_{14}H_{20}ClN_3O{\cdot}ClH$ mw: 318.28

SYN: C 3193

TOXICITY DATA WITH REFERENCE
eye-rbt 2% MLD ARZNAD 8,664,58
ipr-rat LD50: 210 mg/kg ARZNAD 8,664,58
scu-mus LD50: 550 mg/kg ARZNAD 8,664,58

SAFETY PROFILE: Poison by intraperitoneal route. Moderately toxic by subcutaneous route. An eye irritant. When heated to decomposition it emits very toxic fumes of Cl^- and NO_x.

CLO600 CAS:94-76-8 HR: 3
CHLOROTOLYLTHIOGLYCOLIC ACID
mf: $C_9H_9ClO_2S$ mw: 216.69

SYNS: ACETIC ACID, ((4-CHLORO-2-METHYL)PHENYL)THIO- □ 4-CHLORO-2-METHYLPHENYLTHIOGLYCOLIC ACID □ RED 3B ACID

TOXICITY DATA WITH REFERENCE
ipr-mus LD50: 150 mg/kg NTIS** AD691-490

CONSENSUS REPORTS: Reported in EPA TSCA Inventory.

SAFETY PROFILE: Poison by intraperitoneal route. When heated to decomposition it emits toxic vapors of SO_x and Cl^-.

CLO750 CAS:569-57-3 HR: 3
CHLOROTRIANISENE
mf: $C_{23}H_{21}ClO_3$ mw: 380.89

PROP: Crystals from MeOH. Mp: 114–116°.

SYNS: ANISENE □ CHLORESTROLO □ 1,1',1''-(1-CHLORO-1-ETHENYL-2-YLIDENE)-TRIS(4-METHOXYBENZENE) □CHLOROTRIANIZ-EN □ CHLOROTRISIN □ CHLOROTRIS(p-METHOXYPHENYL)ETHYLENE □ CHLORTRIANISEN □ CLORESTROLO □ CLOROTRISIN □ CTA □ HORMONISENE □ KHLORTRIANIZEN □ MERBENTUL □ METACE □ NSC-10108 □ RIANIL □ TACE □ TACE-FN □ TRI-p-ANISYLCHLOROETHYLENE □ TRIS(p-METHOXYPHENYL)CHLOROETHYLENE

TOXICITY DATA WITH REFERENCE
orl-wmn TDLo: 48 mg/kg (14W pre): REP OBGNAS 8,399,56
orl-rat TDLo: 37 mg/kg/2Y-C: ETA TXAPA9 11,489,67
scu-mus TDLo: 180 mg/kg/89W-I: ETA AMPLAO 50,750,50

CONSENSUS REPORTS: IARC Cancer Review: Animal Inadequate Evidence IMEMDT 21,139,79; Human Limited Evidence IMEMDT 21,139,79.

SAFETY PROFILE: Suspected human carcinogen with experimental tumorigenic data. Human reproductive effects by ingestion: changes in fertility. Used in cancer treatment. When heated to decomposition it emits very toxic fumes of Cl^-.

CLP000 CAS:3151-41-5 HR: 3
CHLOROTRIBENZYLSTANNANE
mf: $C_{21}H_{21}ClSn$ mw: 427.56

PROP: Colorless needles from EtOAc. Mp: 142–144°.

SYNS: CHLORID TRIBENZYLCINICITY (CZECH) □ TRIBENZYLCHLOROSTANNANE □ TRIBENZYLTIN CHLORIDE

TOXICITY DATA WITH REFERENCE
skn-rbt 500 mg/24H MLD 28ZPAK -,232,72
eye-rbt 20 mg/24H MOD 85JCAE-,1254,86
orl-rat LD50: 175 mg/kg 28ZPAK -,232,72

OSHA PEL: TWA 0.1 mg(Sn)/m³ (skin)
ACGIH TLV: TWA 0.1 mg(Sn)/m³ (skin) (Proposed: TWA 0.1 mg(Sn)/m³; STEL 0.2 mg(Sn)/m³ (skin))
NIOSH REL: (Organotin Compounds) TWA 0.1 mg(Sn)/m³

SAFETY PROFILE: Poison by ingestion. A skin and severe eye irritant. See also TIN COMPOUNDS. When heated to decomposition it emits toxic fumes of Cl^-.

For occupational chemical analysis use NIOSH: Organotin Compounds 5504.

CLP250 CAS:2117-36-4 HR: 2
CHLOROTRIBUTYLGERMANIUM
mf: $C_{12}H_{27}ClGe$ mw: 279.43

SYN: TRIBUTYLCHLOROGERMANE

TOXICITY DATA WITH REFERENCE
ipr-rat LDLo: 1970 mg/kg CHDDAT 262,1302,66
orl-mus LD50: 1280 mg/kg 85JCAE-,1243,86
ipr-mus LDLo: 1280 mg/kg CHDDAT 262,1302,66

CONSENSUS REPORTS: Reported in EPA TSCA Inventory.

SAFETY PROFILE: Moderately toxic by ingestion and intraperitoneal routes. When heated to decomposition it emits very toxic fumes of Cl^-. See also GERMANIUM COMPOUNDS.

CLP500 CAS:1461-22-9 HR: 3
CHLOROTRIBUTYLSTANNANE
mf: $C_{12}H_{27}ClSn$ mw: 325.53

PROP: Liquid. D: 1.2105 @ 20°, bp: 171–173° @ 25 mm.

SYNS: CHLORID TRI-n-BUTYLCINICITY (CZECH) □ TRIBUTYLCHLOROSTANNANE □ TRI-n-BUTYLTIN CHLORIDE □ TRI-n-BUTYLZINN-CHLORID (GERMAN)

TOXICITY DATA WITH REFERENCE
eye-rbt 50 µg/24H SEV 28ZPAK -,231,72
mmo-sat 100 ng/tube MUREAV 300,265,93
dnd-bcs 10 µg/disk MUREAV 280,195,92
orl-rat TDLo: 225 mg/kg (female 7-15D post): REP DCTODJ 13,283,90

orl-rat TDLo:45 mg/kg (female 7-15D post):REP DCTODJ 13,283,90
orl-rat TDLo:15 mg/kg (female 6-20D post):REP NRTXDN 13,99,91
orl-rat LD50:129 mg/kg 28ZPAK -,231,72
orl-mus LD50:60 mg/kg YKYUA6 30,505,79
orl-rbt LDLo:30 mg/kg SAIGBL 15,3,73
skn-rbt LDLo:70 mg/kg SAIGBL 15,3,73

CONSENSUS REPORTS: Reported in EPA TSCA Inventory.

OSHA PEL: TWA 0.1 mg(Sn)/m^3 (skin)
ACGIH TLV: TWA 0.1 mg(Sn)/m^3 (skin) (Proposed: TWA 0.1 mg(Sn)/m^3; STEL 0.2 mg(Sn)/m^3 (skin))
DFG MAK: 0.002 ppm (0.05 mg/m^3)
NIOSH REL: (Organotin Compounds) TWA 0.1 mg(Sn)/m^3

SAFETY PROFILE: Poison by ingestion and skin contact. A severe eye irritant. Mutation data reported. Tributyl tin compounds are extremely toxic to marine life. See also TIN COMPOUNDS. When heated to decomposition it emits toxic fumes of Cl^-.

For occupational chemical analysis use NIOSH: Organotin Compounds, 5504.

CLP625 **HR: 3**
3-CHLORO-3-TRICHLOROMETHYLDIAZIRINE
mf: $C_2Cl_4N_2$ mw: 193.85

SAFETY PROFILE: An extremely shock-sensitive explosive. Upon decomposition it emits toxic fumes of Cl^- and NO_x. See also EXPLOSIVES.

CLP750 **CAS:1929-82-4** **HR: 3**
2-CHLORO-6-(TRICHLOROMETHYL)PYRIDINE
mf: $C_6H_3Cl_4N$ mw: 230.90

PROP: Crystals. Mp: 62–63°. Very sltly sol in H_2O.

SYNS: DOWCO-163 □ NITRAPYRIN (ACGIH) □ N-SERVE NITROGEN STABILIZER

TOXICITY DATA WITH REFERENCE
mmo-sat 100 µg/plate EMMUEG 11(Suppl 12),1,88
orl-rbt TDLo:390 mg/kg (female 6-18D post):REP FAATDF 11,464,88
orl-rat LD50:940 mg/kg PCOC** -,819,66
orl-mus LD50:710 mg/kg GUCHAZ 6,122,73
orl-rbt LD50:713 mg/kg FAATDF 11,464,88
skn-rbt LD50:850 mg/kg PCOC** -,819,66
orl-ckn LD50:235 mg/kg 28ZEAL 5,166,76

CONSENSUS REPORTS: NCI Carcinogenesis Studies (ipr); Clear Evidence: mouse, rat RRCRBU 52,1,75. Reported in EPA TSCA Inventory.

OSHA PEL: Total Dust: 15 mg/m^3; Respirable Fraction: 5 mg/m^3
ACGIH TLV: TWA 10 mg/m^3; STEL 20 mg/m^3

SAFETY PROFILE: Poison by ingestion. Moderately toxic by skin contact. Experimental reproductive effects. Mutation data reported. When heated to decomposition it emits very toxic fumes of Cl^- and NO_x.

CLQ250 **CAS:869-24-9** **HR: 3**
2-CHLOROTRIETHYLAMINE HYDROCHLORIDE
mf: $C_6H_{14}ClN \cdot ClH$ mw: 172.12

SYNS: β-CHLOROETHYLDIETHYLAMINE HYDROCHLORIDE □ (2-CHLOROETHYL)DIETHYLAMINE HYDROCHLORIDE □ DIETHYLAMINOETHYL CHLORIDE HYDROCHLORIDE □ β-DIETHYLAMINOETHYL CHLORIDE HYDROCHLORIDE □ DIETHYL-β-CHLOROETHYLAMINEHYDROCHLORIDE

TOXICITY DATA WITH REFERENCE
mmo-sat 500 µmol/L ENMUDM 3,11,81
mmo-esc 1 µmol/L JPPMAB 31,67P,79
dns-rat:lvr 100 µmol/L ENMUDM 3,11,81
msc-mus:lym 22 µmol/L ENMUDM 3,33,81
ipr-rat LD50:30 mg/kg CPBTAL 8,807,60
orl-mus LDLo:320 mg/kg AECTCV 14,111,85
ipr-mus LD50:71 mg/kg JPETAB 94,249,48
scu-mus LD50:100 mg/kg JPETAB 91,224,47
ivn-mus LD50:100 mg/kg JPETAB 91,224,47
ivn-rbt LDLo:40 mg/kg JPETAB 91,224,47
orl-bwd LD50:42 mg/kg TXAPA9 21,315,72

CONSENSUS REPORTS: Reported in EPA TSCA Inventory. EPA Genetic Toxicology Program.

SAFETY PROFILE: Poison by ingestion, intraperitoneal, subcutaneous, and intravenous routes. Mutation data reported. When heated to decomposition it emits very toxic fumes of Cl and NO_x. See also AMINES and CHLORIDES.

CLQ500 **CAS:15529-90-5** **HR: 3**
CHLORO(TRIETHYLPHOSPHINE)GOLD
mf: $C_6H_{15}AuClP$ mw: 350.60

PROP: Crystals from EtOH. Mp: 78°, bp: 210° @ 0.03 mm. Sol in $CHCl_3$ and EtOH.

SYNS: SK&F 36914 □ TRIETHYLPHOSPHINEAUROUS CHLORIDE

TOXICITY DATA WITH REFERENCE
dni-hmn:oth 37,500 nmol/L BCPCA6 34,3243,85
orl-rbt TDLo:46,280 µg/kg (6-18D preg):TER VTPHAK 15(Suppl 5),97,78
orl-rbt TDLo:92,560 µg/kg (6-18D preg):REP VTPHAK 15(Suppl 5),97,78
orl-rat LD50:79 mg/kg VTPHAK 15(Suppl 5),1,78
orl-mus LD50:68 mg/kg VTPHAK 15(Suppl 5),1,78

SAFETY PROFILE: Poison by ingestion. Experimental teratogenic and reproductive effects. Human mutation data reported. When heated to decomposition it emits very toxic fumes of Cl^- and PO_x. See also PHOSPHINE and GOLD.

CLQ750 **CAS:79-38-9** **HR: 3**
CHLOROTRIFLUOROETHYLENE
DOT: UN 1082
mf: C_2ClF_3 mw: 116.47

PROP: A gas. Fp: −157.5°, bp: −26.2°, flash p: −18°F, lel. 24%, uel: 40.3%.

SYNS: 1-CHLORO-1,2,2-TRIFLUOROETHYLENE □ 2-CHLORO-1,1,2-TRIFLUOROETHYLENE □ CHLORTRIFLUORAETHYLEN (GERMAN) □ CTFE □ DAIFLON □ FLUOROPLAST 3 □ GENETRON 1113 □ MONOCHLOROTRIFLUOROETHYLENE □ TRIFLUOROCHLOROETHY-

LENE (DOT) □ 1,1,2-TRIFLUORO-2-CHLOROETHYLENE □ TRIFLUOROMONOCHLOROETHYLENE □ TRIFLUOROVINYL CHLORIDE □ TRITHENE

TOXICITY DATA WITH REFERENCE
ihl-rat LC50:1000 ppm/4H FLCRAP 1,197,67
orl-mus LD50:268 mg/kg ABMGAJ 21,377,68
ihl-mus LC50:3000 ppm/7H ABMGAJ 21,377,68
ipr-mus LD50:175 mg/kg ABMGAJ 21,377,68
ihl-gpg LC50:4300 mg/m³/4H GTPZAB 21(5),36,77

CONSENSUS REPORTS: Reported in EPA TSCA Inventory.

DOT CLASSIFICATION: 2.1; *Label:* Flammable Gas

SAFETY PROFILE: Poison by ingestion and intraperitoneal routes. Moderately toxic by inhalation. Very dangerous fire hazard when exposed to heat, flames (sparks), or oxidizers. To fight fire, stop flow of gas. Violent reaction when mixed with ($Br_2 + O_2$) or (ClF_3 + water). Potentially explosive polymerization reaction with ethylene. Incompatible with 1,1-dichloroethylene; oxygen. When heated to decomposition it emits toxic fumes of F^- and Cl^-. See also CHLORINATED HYDROCARBONS, ALIPHATIC; and FLUORIDES.

CLR000 CAS:425-87-6 HR: 3
2-CHLORO-1,1,2-TRIFLUOROETHYL METHYL ETHER
mf: $C_3H_4ClF_3O$ mw: 148.52

PROP: A liquid. D: 1.364 @ 20°/4°. Mp: −92.6°, mp: −109.4° (dimorph), bp: 70.6°.

TOXICITY DATA WITH REFERENCE
eye-rbt 2 mg open SEV AMIHBC 4,119,51
orl-rat LD50:5130 mg/kg AMIHBC 4,119,51
skn-rbt LD50:200 mg/kg AMIHBC 4,119,51

SAFETY PROFILE: Poison by skin contact. Mildly toxic by ingestion. Severe eye irritant. See also ETHERS. When heated to decomposition it emits very toxic fumes of Cl^- and F^-. See also ETHERS, CHLORIDES, and FLUORIDES.

CLR250 CAS:75-72-9 HR: 1
CHLOROTRIFLUOROMETHANE
DOT: UN 1022
mf: $CClF_3$ mw: 104.46

PROP: Colorless gas, ethereal odor. Mp: −181°, bp: −81.4°, fp: −181°.

SYNS: ARCTON 3 □ F 13 □ FREON 13 □ GENETRON 13 □ HALOCARBON 13/UCON 13 □ MONOCHLOROTRIFLUOROMETHANE (DOT) □ R 13 □ R13 (DOT) □ TRIFLUOROCHLOROMETHANE (DOT) □ TRIFLUOROMETHYL CHLORIDE □ TRIFLUOROMONOCHLOROCARBON

CONSENSUS REPORTS: Reported in EPA TSCA Inventory.

DFG MAK: 1000 ppm (4330 mg/m³)
DOT CLASSIFICATION: 2.2; *Label:* Nonflammable Gas

SAFETY PROFILE: A mild irritant. Narcotic in high concentrations. Reacts violently with Al. When heated to decomposition it emits highly toxic fumes of F^- and Cl^-.

CLR825 CAS:58911-30-1 HR: 3
3-CHLORO-3-TRIFLUOROMETHYLDIAZIRINE
mf: $C_2ClF_3N_2$ mw: 144.48

PROP: Gas. Bp: −19°.

SAFETY PROFILE: Potentially explosive. When heated to decomposition it emits toxic fumes of F^-, Cl^-, and NO_x.

CLS075 CAS:50594-66-6 HR: 3
5-(2-CHLORO-4-(TRIFLUOROMETHYL)PHENOXY)-2-NITROBENZOIC ACID
mf: $C_{14}H_7ClF_3NO_5$ mw: 361.67

SYNS: ACIFLUORFEN □ ACIFLUORFENE □ BENZOIC ACID, 5-(2-CHLORO-4-(TRIFLUOROMETHYL)PHENOXY)-2-NITRO- □ 5-(2-CHLORO-α-α-α-TRIFLUORO-p-TOLYLOXY)-2-NITROBENZOIC ACID (IUPAC) □ TACKLE

TOXICITY DATA WITH REFERENCE
orl-rat LD50:1370 mg/kg PEMNDP 9,6,91
ihl-rat LC50:>6900 mg/m³/4H PEMNDP 9,6,91
orl-mus LD50:1370 mg/kg PEMNDP 9,6,91
skn-rbt LD50:3680 mg/kg PEMNDP 9,6,91
orl-qal LD50:325 mg/kg PEMNDP 9,6,91
orl-dck LD50:2821 mg/kg PEMNDP 9,6,91

CONSENSUS REPORTS: Reported in EPA TSCA Inventory.

SAFETY PROFILE: A poison by ingestion. Moderately toxic by skin contact. When heated to decomposition it emits toxic vapors of NO_x, F^-, and Cl^-.

CLS125 CAS:25238-02-2 HR: 3
2-CHLORO-N,N,N′-TRIFLUOROPROPIONAMIDINE
mf: $C_3H_4ClF_3N_2$ mw: 160.53

SAFETY PROFILE: A shock-sensitive explosive. When heated to decomposition it emits toxic fumes of F^-, Cl^-, and NO_x. See also EXPLOSIVES.

CLS250 CAS:17230-87-4 HR: 3
4-(4-(4-CHLORO-α,α,α-TRIFLUORO-m-TOLYL)-4-HYDROXYPIPERIDINO)BUTYROPHENONE-4′-FLUOROHYDROCHLORIDE
mf: $C_{22}H_{22}ClF_4NO_2 \cdot ClH$ mw: 480.36

PROP: Crystals from EtOH/Et_2O. Mp: 203.5–206°.

SYNS: CLOFLUPEROL HYDROCHLORIDE □ R 9298 □ SEPERIDOL □ SEPEROL

TOXICITY DATA WITH REFERENCE
orl-rat LD50:195 mg/kg 27ZQAG -,186,72
scu-rat LD50:69 mg/kg 27ZQAG -,186,72
ivn-rat LD50:17 mg/kg 27ZQAG -,186,72
scu-mus LD50:47 mg/kg 27ZQAG -,186,72
ivn-mus LD50:19 mg/kg 27ZQAG -,186,72

SAFETY PROFILE: Poison by ingestion, subcutaneous, and intravenous routes. When heated to decomposition it emits very toxic fumes of Cl^-, F^-, and NO_x.

CLS500 **CAS:7342-38-3** **HR: 3**
CHLORO(TRIISOBUTYL)STANNANE
mf: $C_{12}H_{27}ClSn$ mw: 325.53

PROP: Solid. D: 1.1290 @ 34°, mp: 30.2°, bp: 174° @ 13 mm.

SYN: TRIISOBUTYLTIN CHLORIDE

TOXICITY DATA WITH REFERENCE
ivn-mus LD50:5 mg/kg CSLNX* NX#05523

OSHA PEL: TWA 0.1 mg(Sn)/m³ (skin)
ACGIH TLV: TWA 0.1 mg(Sn)/m³ (skin) (Proposed: TWA 0.1 mg(Sn)/m³; STEL 0.2 mg(Sn)/m³ (skin))
NIOSH REL: (Organotin Compounds) TWA 0.1 mg(Sn)/m³

SAFETY PROFILE: Poison by intravenous route. Tributyl tin compounds are very toxic to marine life. See also TIN COMPOUNDS. When heated to decomposition it emits toxic fumes of Cl^-.

For occupational chemical analysis use NIOSH: Organotin Compounds 5504.

CLT000 **CAS:1066-45-1** **HR: 3**
CHLOROTRIMETHYLSTANNANE
mf: C_3H_9ClSn mw: 199.26

PROP: Colorless needles. Mp: 42°, bp: 154–156°.

SYNS: CHLOROTRIMETHYLTIN ◻ TRIMETHYLCHLOROSTANNANE ◻ TRIMETHYLCHLOROTIN ◻ TRIMETHYLSTANNYL CHLORIDE ◻ TRIMETHYLTIN CHLORIDE

TOXICITY DATA WITH REFERENCE
dni-rbt:oth 10 μg/L JTEHD6 16,229,85
orl-rat TDLo:9066 μg/kg (14D pre-21D post):REP NTOTDY 4,539,82
ipr-rat TDL0:9 mg/kg (17D preg):REP TJADAB 29(2),50A,84
orl-rat LD50:12,600 μg/kg AJPAA4 97,59,79
ipr-rat LD50:7450 μg/kg NETOD7 4,127,82
ivn-mus LD50:1800 μg/kg CSLNX* NX#02983

CONSENSUS REPORTS: EPA Extremely Hazardous Substances List. Reported in EPA TSCA Inventory.

OSHA PEL: TWA 0.1 mg(Sn)/m³ (skin)
ACGIH TLV: TWA 0.1 mg(Sn)/m³ (skin) (Proposed: TWA 0.1 mg(Sn)/m³; STEL 0.2 mg(Sn)/m³ (skin))
NIOSH REL: (Organotin Compounds) TWA 0.1 mg(Sn)/m³

SAFETY PROFILE: A deadly poison by intravenous route. Experimental reproductive effects. See also TIN COMPOUNDS. When heated to decomposition it emits toxic fumes of Cl^-.

For occupational chemical analysis use NIOSH: Organotin Compounds 5504.

CLT250 **CAS:1943-16-4** **HR: 3**
CHLOROTRINITROMETHANE
mf: $CClN_3O_6$ mw: 185.49

PROP: D: 1.68 @ 20°/4°, bp: 32–32.5° @ 11 mm.

TOXICITY DATA WITH REFERENCE
ipr-mus LD50:29,300 μg/kg KHFZAN 10(6),53,76
ipr-mam LDLo:500 mg/kg COREAF 171,1396,20
ihl-mam LCLo:5 g/m³ COREAF 171,1396,20

SAFETY PROFILE: Poison by intraperitoneal route. Mildly toxic by inhalation. Potentially explosive. When heated to decomposition it emits very toxic fumes of Cl^- and NO_x.

CLT500 **CAS:76-83-5** **HR: 3**
CHLOROTRIPHENYLMETHANE
mf: $C_{19}H_{15}Cl$ mw: 278.79

PROP: Crystals from C_6H_6 or pet ether. Mp: 111–112°.

SYN: TRITYL CHLORIDE

TOXICITY DATA WITH REFERENCE
ivn-mus LD50:180 mg/kg CSLNX* NX#04021

CONSENSUS REPORTS: Reported in EPA TSCA Inventory.

SAFETY PROFILE: Poison by intravenous route. When heated to decomposition it emits toxic fumes of Cl^-. See also CHLORINATED HYDROCARBONS, AROMATIC.

CLU000 **CAS:639-58-7** **HR: 3**
CHLOROTRIPHENYLSTANNANE
mf: $C_{18}H_{15}ClSn$ mw: 385.47

PROP: Colorless crystals from alc. Mp: 106°, bp: 240° @ 13.5 mm. Insol in water; sol in organic solvents.

SYNS: AQUATIN ◻ BRESTANOL ◻ CHLOROTRIPHENYLTIN ◻ FENTIN CHLORIDE ◻ GC 8993 ◻ GENERAL CHEMICALS 8993 ◻ HOE 2872 ◻ LS 4442 ◻ TINMATE ◻ TPTC ◻ TRIPHENYLCHLOROSTANNANE ◻ TRIPHENYLCHLOROTIN ◻ TRIPHENYLTIN CHLORIDE

TOXICITY DATA WITH REFERENCE
sln-hmn:lyms 30 nmol/L MUREAV 246,109,91
oth-ham:ovr 60 μg/L MUREAV 300,5,93
orl-rat TDLo:380 mg/kg (19D male):REP JEENAI 61,32,68
orl-rat LD50:135 mg/kg FMCHA2 -,C245,83
orl-mus LD50:18 mg/kg FMCHA2 -,C245,83
ipr-mus LD50:21,500 μg/kg JICSAH 67,740,90
ivn-mus LD50:18 mg/kg CSLNX* NX#01649

CONSENSUS REPORTS: Reported in EPA TSCA Inventory. EPA Extremely Hazardous Substances List.

OSHA PEL: TWA 0.1 mg(Sn)/m³ (skin)
ACGIH TLV: TWA 0.1 mg(Sn)/m³ (skin) (Proposed: TWA 0.1 mg(Sn)/m³; STEL 0.2 mg(Sn)/m³ (skin))
NIOSH REL: (Organotin Compounds) TWA 0.1 mg(Sn)/m³

SAFETY PROFILE: Poison by ingestion, intraperitoneal, and intravenous routes. Experimental reproductive effects. Mutation data reported. An insect chemosterilant. See also TIN COMPOUNDS. When heated to decomposition it emits toxic fumes of Cl^-.

For occupational chemical analysis use NIOSH: Organotin Compounds 5504.

CLU250 **CAS:2279-76-7** **HR: 3**
CHLOROTRIPROPYLSTANNANE
mf: $C_9H_{21}ClSn$ mw: 283.44

PROP: Colorless liquid. D: 1.2678 @ 28°, mp: −23.5°. Sol in organic solvents.

SYNS: TRIPROPYLTIN CHLORIDE □ TRI-n-PROPYLTIN CHLORIDE

TOXICITY DATA WITH REFERENCE
ivn-mus LD50:4 mg/kg CSLNX* NX#02220

OSHA PEL: TWA 0.1 mg(Sn)/m³ (skin)
ACGIH TLV: TWA 0.1 mg(Sn)/m³ (skin) (Proposed: TWA 0.1 mg(Sn)/m³; STEL 0.2 mg(Sn)/m³ (skin))
NIOSH REL: (Organotin Compounds) TWA 0.1 mg(Sn)/m³

SAFETY PROFILE: Poison by intravenous route. See also TIN COMPOUNDS. When heated to decomposition it emits toxic fumes of Cl^-.

For occupational chemical analysis use NIOSH: Organotin Compounds 5504.

CLU500 **CAS:10008-90-9** **HR: 3**
CHLORO(TRIVINYL)STANNANE
mf: C_6H_9ClSn mw: 235.29

PROP: Colorless liquid. Bp: 59–60° @ 6 mm.

SYN: TRIVINYLTIN CHLORIDE

TOXICITY DATA WITH REFERENCE
ivn-mus LD50:40 mg/kg CSLNX* NX#05524

OSHA PEL: TWA 0.1 mg(Sn)/m³ (skin)
ACGIH TLV: TWA 0.1 mg(Sn)/m³ (skin) (Proposed: TWA 0.1 mg(Sn)/m³; STEL 0.2 mg(Sn)/m³ (skin))
NIOSH REL: TWA (Organotin Compounds) 0.1 mg(Sn)/m³

SAFETY PROFILE: Poison by intravenous route. See also TIN COMPOUNDS and CHLORIDES. When heated to decomposition it emits toxic fumes of Cl^-.

For occupational chemical analysis use NIOSH: Organotin Compounds 5504.

CLV000 **CAS:541-25-3** **HR: 3**
CHLOROVINYLARSINE DICHLORIDE
mf: $C_2H_2AsCl_3$ mw: 207.31

PROP: Liquid, faint odor of geranium. Bp: 190° decomp, fp: −13°, d: 1.888 @ 20°/4°, vap press: 0.4 mm @ 20°, vap d: 7.15.

SYNS: (2-CHLOROETHENYL) ARSONOUS DICHLORIDE □ β-CHLOROVINYLBICHLOROARSINE □ 2-CHLOROVINYLDICHLOROARSINE □ (2-CHLOROVINYL)DICHLOROARSINE □ DICHLORO(2-CHLOROVINYL)ARSINE □ LEWISITE □ LEWISITE (ARSENIC COMPOUND)

TOXICITY DATA WITH REFERENCE
skn-hmn 95 µg NTIS** PB158-508
orl-rat TDLo:20 mg/kg (female 6-15D post):TER NTIS** DE88-008303
ihl-hmn LCLo:6 ppm/30M NTIS** PB214-270
ihl-hmn LC50:1500 mg/kg/M:PUL YKYUA6 30,355,79
ihl-rat LC50:580 mg/m³/1H NTIS** PB158-508
skn-rat LD50:15 mg/kg NTIS** PB158-508
scu-rat LD50:1 mg/kg JPBAA7 58,411,46
ihl-mus LC50:500 mg/m³/9M NTIS** PB158-508
skn-mus LD50:15 mg/kg NTIS** PB158-508
ihl-dog LC50:1400 mg/m³/15M NTIS** PB158-508
skn-dog LD50:15 mg/kg JPBAA7 58,411,46
scu-dog LD50:2 mg/kg JPBAA7 58,411,46
ivn-dog LD50:2 mg/kg NTIS** PB158-508
ihl-cat LC50:30 g/m³/30M NTIS** PB158-508
ihl-rbt LC50:1200 mg/m³/8M NTIS** PB158-508

CONSENSUS REPORTS: Reported in EPA TSCA Inventory. EPA Genetic Toxicology Program. EPA Extremely Hazardous Substances List. Arsenic and its compounds are on the Community Right-To-Know List.

SAFETY PROFILE: A human poison by inhalation. Poison experimentally by inhalation, skin contact, subcutaneous, intraperitoneal, and intravenous routes. An experimental teratogen. A blistering-type military poison. Lewisite is absorbed through skin; as little as 2 mL on the skin can cause death. Has a delayed action similar to distilled mustard gas. This gas exhibits a systemic poisoning effect on humans. When heated to decomposition it emits toxic fumes of Cl^- and As. See also ARSENIC COMPOUNDS.

CLV250 **CAS:64049-11-2** **HR: 3**
(2-CHLOROVINYL)DIETHOXYARSINE
mf: $C_6H_{12}AsClO_2$ mw: 226.55

TOXICITY DATA WITH REFERENCE
ihl-mus LCLo:500 mg/m³ NDRC** -,7,43
skn-mus LDLo:80 mg/kg NDRC** -,24,42

CONSENSUS REPORTS: Arsenic and its compounds are on the Community Right-To-Know List.

SAFETY PROFILE: Poison by skin contact. Moderately toxic by inhalation. See also ARSENIC COMPOUNDS. When heated to decomposition it emits very toxic fumes of As and Cl^-.

CLV375 **CAS:311-47-7** **HR: 3**
2-CHLOROVINYL DIETHYL PHOSPHATE
mf: $C_6H_{12}ClO_4P$ mw: 214.60

SYNS: COMPOUND 1836 □ DIETHYL-2-CHLOROVINYL PHOSPHATE □ O,O-DIETHYL-O-(2-CHLOROVINYL) PHOSPHATE □ OS 1836 □ SD 1836 □ SHELL OS 1836

TOXICITY DATA WITH REFERENCE
orl-rat LD50:10 mg/kg AMIHBC 9,45,54
ihl-rat LC50:22 ppm AMIHBC 9,45,54
ipr-rat LD50:9 mg/kg AMIHBC 9,45,54
orl-mus LD50:32 mg/kg PAREAQ 11,636,59
orl-rbt LD50: 3 mg/kg AMIHBC 9,45,54
skn-rbt LD50:18 mg/kg AMIHBC 9,45,54

SAFETY PROFILE: Poison by inhalation, skin contact, ingestion, and intraperitoneal routes. When heated to decomposition it emits toxic fumes of $Cl^-$and PO_x.

CLW000 CAS:88-04-0 HR: 3
4-CHLORO-3,5-XYLENOL
mf: C_8H_9ClO mw: 156.62

PROP: Crystals or prisms from (C_6H_6); phenolic odor. Mp: 115.5°, bp: 246°. Sltly water-sol.

SYNS: BENZYTOL □ 4-CHLORO-3,5-DIMETHYLPHENOL □ CHLORO-XYLENOL □ p-CHLORO-m-XYLENOL □ DESSON □ DETTOL □ ESPADOL □ HUSEPT EXTRA □ OTTASEPT □ OTTASEPT EXTRA □ PCMX □ RBA 777

TOXICITY DATA WITH REFERENCE
eye-rbt 100 mg MOD JACTDZ 4(5),147,85
orl-rat TDLo:17,100 mg/kg (1-19D preg):TER AOISDR (45),100,83
orl-rat LD50:3830 mg/kg JACTDZ 4(5),147,85
orl-mus LDLo:1600 mg/kg AECTCV 14,111,85
ipr-mus LD50:115 mg/kg JAPMA8 41,595,52

CONSENSUS REPORTS: Reported in EPA TSCA Inventory. Chlorophenols are on the Community Right-To-Know List.

SAFETY PROFILE: Poison by intraperitoneal route. Moderately toxic by ingestion. An experimental teratogen. An eye irritant. An antimicrobial agent. See also CHLOROPHENOLS; and CHLORINATED HYDROCARBONS, AROMATIC. When heated to decomposition it emits toxic fumes of Cl^-.

CLW250 CAS:50892-23-4 HR: 3
(4-CHLORO-6-(2,3-XYLIDINO)-2-PYRIMIDINYLTHIO) ACETIC ACID
mf: $C_{14}H_{14}ClN_3O_2S$ mw: 323.82

PROP: Crystals from EtOAc. Mp: 150–153°.

SYNS: ((4-CHLORO-6-((2,3-DIMETHYLPHENYL)AMINO) 2-PYRIMIDINYL)THIO)ACETIC ACID □ WY-14,643

TOXICITY DATA WITH REFERENCE
dns-rat:lvr 1 mmol/L CALEDQ 24,147,84
dni-mus:oth 100 μmol/L CNREA8 40,36,80
orl-rat TDLo:29 g/kg/69W-C:CAR CNREA8 39,152,79
orl-mus TDLo:37 g/kg/62W-C:CAR CNREA8 39,152,79
orl-rat TD:46 g/kg/65W-C:CAR CRNGDP 2,645,81
orl-rat TD:27 g/kg/64W-C:ETA CALEDQ 32,33,86
orl-rat LD50:4150 mg/kg ATHSBL 30,45,78
orl-mus LD50:1600 mg/kg ATHSBL 30,45,78

SAFETY PROFILE: Moderately toxic by ingestion. Suspected carcinogen with experimental carcinogenic and tumorigenic data. Mutation data reported. When heated to decomposition it emits very toxic fumes of Cl^-, NO_x, and SO_x.

CLW500 CAS:65089-17-0 HR: 3
2-((4-CHLORO-6-(2,3-XYLIDINO)-2-PYRIMIDINYL) THIO)-N-(2-HYDROXYETHYL)ACETAMIDE
mf: $C_{16}H_{19}ClN_4O_2S$ mw: 366.90

PROP: Crystals from Me_2CO. Mp: 144–146°.

SYNS: BR-931 □ PIRINIXIL

TOXICITY DATA WITH REFERENCE
dns-rat:lvr 100 μmol/L CALEDQ 24,147,84
dni-mus:oth 25 μmol/L CNREA8 40,36,80
orl-rat TDLo:17 g/kg/81W-C:CAR NATUAS 283,397,80
orl-mus TDLo:137 g/kg/81W-C:CAR NATUAS 283,397,80

SAFETY PROFILE: Suspected carcinogen with experimental carcinogenic data. Mutation data reported. When heated to decomposition it emits very toxic fumes of Cl^-, NO_x, and SO_x.

C

CLW625 CAS:30544-72-0 HR: 3
4-(p-CHLORO-N-2,6-XYLYLBENZAMIDO)BUTYRIC ACID
mf: $C_{19}H_{29}ClNO_3$ mw: 345.85

SYNS: B 66347 □ B 67347 □ N-(p-CHLORBENZOYL)-γ-(2,6-DIMETHYLANILINO)-BUTTERSAEURE (GERMAN)

TOXICITY DATA WITH REFERENCE
orl-rat LD50:1900 mg/kg GWXXBX #1917036
ivn-rat LD50:300 mg/kg GWXXBX #1917036
orl-mus LD50:710 mg/kg GWXXBX #1917036
ipr-mus LD50:305 mg/kg GWXXBX #1917036

SAFETY PROFILE: Poison by intravenous and intraperitoneal routes. Moderately toxic by ingestion. When heated to decomposition it emits toxic fumes of Cl^- and NO_x.

CLX000 CAS:54749-90-5 HR: 3
CHLOROZOTOCIN
mf: $C_9H_{16}ClN_3O_7$ mw: 313.73

SYNS: 1-(2-CHLOROETHYL)-3-(d-GLUCOPYRANOS-2-YL)-1-NITROSOUREA □ 2-((((2-CHLOROETHYL)NITROSOAMINO)CARBONYL) AMINO)-2-DEOXY-d-GLUCOPYRANOSE □ 2-((((2-CHLOROETHYL)NITROSOAMINO)CARBONYL)AMINO)-2-DEOXY-d-GLUCOSE □ 2-(3-(2-CHLOROETHYL)-3-NITROSOUREIDO)-2-DEOXY-d-GLUCOSOPYRANOSE □ 2-(3-(2-CHLOROETHYL)-3-NITROSOUREIDO)-d-GLUCO-PYRANOSE □ CHLZ □ CZT □ DCNU □ NSC 178248 □ NSC D 254157

TOXICITY DATA WITH REFERENCE
mmo-sat 100 nmol/plate JJIND8 65,149,80
mma-sat 100 mg/L/1H MUREAV 40,281,76
dnd-rat-ipr 100 μmol/kg CNREA8 44,514,84
sce-rat:oth 1 μmol/L CNREA8 43,473,83
dnd-mam:lym 10 mmol/L CNREA8 44,1887,84
ipr-rat TDLo:34 mg/kg/85W-I:CAR CALEDQ 8,133,79
ivn-rat TDLo:16 mg/kg/60W-I:ETA DTESD7 8,273,80
ivn-rat TD:64 mg/kg/60W-I:ETA DTESD7 8,273,80
ivn-hmn TDLo:2027 μg/kg/5D:BLD CTRRDO 63,17,79
ivn-man TDLo:500 mg/kg:CNS,GIT,BLD CANCAR 46,2365,80
ipr-rat LD50:28 mg/kg CALEDQ 8,133,79
scu-rat LDLo:40 mg/kg TXAPA9 82,540,86
ivn-rat LD50:22,500 μg/kg ONCOBS 37,177,80
ipr-mus LD50:35 mg/kg INSSDM 19,123,81
scu-mus LD50:66,230 μg/kg NCISP* JAN86
ivn-mus LD10:15 mg/kg GANNA2 71,686,80

CONSENSUS REPORTS: IARC Cancer Review: Group 2A IMEMDT 50,65,90; Animal Sufficient Evidence IMEMDT 50,65,90; Human No Adequate Data IMEMDT 50,65,90. EPA Genetic Toxicology Program.

SAFETY PROFILE: Suspected carcinogen with experimental carcinogenic and tumorigenic data. Poison by subcutaneous, intravenous, and intraperitoneal routes.

Human systemic effects by intravenous route: anorexia, leukopenia, nausea or vomiting, thrombocytopenia. Mutation data reported. When heated to decomposition it emits very toxic fumes of Cl^- and NO_x. See also NITROSAMINES.

CLX250 CAS:633-59-0 HR: 3
CHLORPERPHENTHIXENE DIHYDROCHLORIDE
mf: $C_{22}H_{25}ClN_2OS \cdot 2ClH$ mw: 473.92

PROP: Crystals from MeOH/Et_2O. Mp: 257–258°. Very sol in H_2O; insol in org solvents.

SYNS: AY 62021 □ 4-(3-(2-CHLOROTHIOXANTHEN-9-YLIDENE) PROPYL)-1-PIPERAZINEETHANOL DIHYDROCHLORIDE □ CHLORPENTHIXOL DIHYDROCHLORIDE □ CIATYL □ CLOPENTHIXOL DIHYDROCHLORIDE □ CLOPIXOL □ N-746 □ SORDENAC □ SORDINOL

TOXICITY DATA WITH REFERENCE
orl-rat LD50:660 mg/kg 27ZQAG -,67,72
ipr-rat LD50:105 mg/kg 27ZQAG -,67,72
ivn-rat LD50:125 mg/kg 27ZQAG -,67,72
orl-mus LD50:560 mg/kg 27ZQAG -,67,72
ipr-mus LD50:222 mg/kg 27ZQAG-,67,72
ivn-mus LD50:111 mg/kg USXXAM #3996211

SAFETY PROFILE: Poison by intraperitoneal and intravenous routes. Moderately toxic by ingestion. When heated to decomposition it emits very toxic fumes of SO_x, NO_x, and Cl^-.

CLX300 CAS:132-22-9 HR: 3
CHLORPHENIRAMINE
mf: $C_{16}H_{19}ClN_2$ mw: 274.82

SYNS: ALLERGICAN □ ALLERGISAN □ 2-(p-CHLORO-α-(2-(DIMETHYLAMINO)ETHYL)BENZYL)PYRIDINE □ CHLOROPHENYLPYRIDAMINE □ 1-(p-CHLOROPHENYL)-1-(2-PYRIDYL)-3-DIMETHYLAMINOPROPANE □ 1-(p-CHLOROPHENYL)-1-(2-PYRIDYL)-3-N,N-DIMETHYLPROPYLAMINE □ 4-CHLOROPHENIRAMINE □ CHLOROPIRIL □ CHLOROPROPHENPYRIDAMINE □ CHLORPHENAMINE □ CHLORPROPHENPYRIDAMINE □ CHLOR-TRIMETON □ CHLOR-TRIPOLON □ CLORFENIRAMINA □ CLOROPIRIL □ HAYNON □ HISTADUR □ PIRITON □ POLARONIL □ PYRIDINE, 2-(p-CHLORO-α-(2-(DIMETHYLAMINO)ETHYL)BENZYL)- □ 2-PYRIDINEPROPANAMINE, γ-(4-CHLOROPHENYL)-N,N-DIMETHYL-(9CI)

TOXICITY DATA WITH REFERENCE
orl-rat LD50:118 mg/kg MEWEAC 17,2791,66
orl-mus LD50:121 mg/kg MEWEAC 17,2791,66
ipr-mus LD50:125 mg/kg YKKZAJ 92,1339,72
scu-mus LD50:160 mg/kg BCFAAI 111,293,72
ivn-mus LD50:20 mg/kg MEWEAC 17,2791,66
ivn-rbt LD50:22 mg/kg MEWEAC 17,2791,66

CONSENSUS REPORTS: Reported in EPA TSCA Inventory.

SAFETY PROFILE: Poison by ingestion, intraperitoneal, subcutaneous, and intravenous routes. When heated to decomposition it emits toxic vapors of NO_x and Cl^-.

CLY250 CAS:461-78-9 HR: 3
CHLORPHENTERMINE
mf: $C_{10}H_{14}ClN$ mw: 183.70

PROP: A liquid. Bp: 231°.

SYNS: p-CHLORO-α,α-DIMETHYLPHENETHYLAMINE □ CHLOROPHENTERMINE □ β-(p-CHLOROPHENYL)-α,α-DIMETHYLETHYLAMINE □ 1-(p-CHLOROPHENYL)-2-METHYL-2-AMINOPROPANE □ CHLORPHENTERAMINE

TOXICITY DATA WITH REFERENCE
dns-rat-unr 300 mg/kg/5D-C 40QBA3 -,459,78
oms-rat-unr 300 mg/kg/5D-C 40QBA3 -,459,78
scu-rat TDLo:480 mg/kg (10-12D preg):TER VAAZA2 12,295,73
orl-man TDLo:5357 μg/kg:ANS,CVS THERAP 34,205,79
orl-rat LD50:250 mg/kg NYKZAU 65(6),218S,69
orl-mus LD50:270 mg/kg APPHAX 26,598,69
ipr-mus LD50:150 mg/kg APSXAS 15,87,78
scu-mus LD50:260 mg/kg APPHAX 26,598,69
ivn-mus LD50:56 mg/kg CSLNX* NX#00697

SAFETY PROFILE: Poison by ingestion, intraperitoneal, subcutaneous, and intravenous routes. Human systemic effects by ingestion: blood pressure elevation and sympathetic nervous system stimulation. An experimental teratogen. Mutation data reported. An anorectic drug which diminishes the appetite. When heated to decomposition it emits very toxic fumes of Cl^- and NO_x.

CLY500 CAS:52-86-8 HR: 3
γ-(4-(p-CHLORPHENYL)-4-HYDROXPIPERIDINO)-p-FLUORBUTYROPHENONE
mf: $C_{21}H_{23}ClFNO_2$ mw: 375.90

PROP: A solid. Mp: 148–149.4°.

SYNS: ALDO □ ALOPERIDIN □ ALOPERIDOLO □ BROTOPON □ 4-(4-(4-CHLOROPHENYL)-4-HYDROXY-1-PIPERIDINYL)-1-(4-FLUOROPHENYL)-1-BUTANONE □ EINALON S □ EUKYSTOL □ 1-(3-p-FLUOROBENZOYLPROPYL)-4-p-CHLOROPHENYL-4-HYDROXYPIPERIDINE □ 4'-FLUORO-4-(4-HYDROXY-4-(4'-CHLOROPHENYL)PIPERIDINO)BUTYROPHENONE □ GALOPERIDOL □ HALDOL □ HALOPERIDOL □ HALOSTEN □ 4-(4-HYDROXY-4'-CHLORO-4-PHENYLPIPERIDINO)-4'-FLUOROBUTYROPHENONE □ KESELAN □ LEALGIN COMPOSITUM □ LINTON □ PELUCES □ PERNOX □ R 1625 □ SERENACE □ SERNAS □ SERNEL □ ULCOLIND □ ULIOLIND □ VESALIUM

TOXICITY DATA WITH REFERENCE
mma-sat 100 nmol/plate CRNGDP 3,223,82
cyt-hmn:fbr 10 g/L AMBUCH 6,42,79
orl-wmn TDLo:14,700 μg/kg (1-7W preg):REP JAMAAP 231,62,75
ims-man TDLo:71 μg/kg (male 13D pre):REP TXMDAX 81(9),47,85
orl-wmn TDLo:14,700 μg/kg (1-7W preg):TER JAMAAP 231,62,75
scu-rat TDLo:7500 μg/kg (female 4-18D post):REP NETOD7 7,489,85
scu-rat TDLo:10,750 μg/kg (female 1-22D post):REP PSCHDL 70,47,80
orl-mus TDLo:25 mg/kg (female 8D post):TER TOIZAG 28,621,81
orl-rat TDLo:24 mg/kg (female 1-8D post):TER JRPFA4 22,591,70

orl-mus TDLo: 21 mg/kg (female 6-12D post): REP ARZNAD 18,1420,68
orl-rat TDLo: 17,500 µg/kg (female 8-14D post): REP ARZNAD 18,1420,68
ims-rat TDLo: 672 mg/kg (female 12W pre): REP KSRNAM 19,6731,85
orl-rat TDLo: 60 mg/kg (female 1-12D post): REP CHDDAT 264,114,67
scu-rat TDLo: 6750 µg/kg (female 1-22D post): REP PSCHDL 70,47,80
ims-rat TDLo: 84 mg/kg (male 12W pre): REP KSRNAM 19,6731,85
orl-rat TDLo: 15 mg/kg (female 8-12D post): TER ARZNAD 18,1420,68
orl-rat TDLo: 5040 µg/kg (female 14-22D post): TER DPTHDL 7,188,84
ipr-ham TDLo: 140 mg/kg (female 8D post): TER DPTHDL 4,1,82
scu-rat TDLo: 70 mg/kg (female 6-12D post): TER ARZNAD 18,1420,68
ipr-mus TDLo: 25 mg/kg/5D-C: CAR CRNGDP 3,223,82
ipr-mus TD: 50 mg/kg/10D-C: CAR CRNGDP 3,223,82
orl-man TDLo: 9 mg/kg/30W-I: PNS BIPCBF 22,111,87
orl-man TDLo: 480 µg/kg/6D-I: BAH AJPSAO 142,389,85
orl-cld TDLo: 72 µg/kg: BAH AJPSAO 143,1176,85
orl-wmn TDLo: 100 µg/kg/10D-I: BLD JAGSAF 35,248,87
orl-hmn TDLo: 71 µg/kg: BAH JCPYDR 5,120,85
unr-cld TDLo: 375 µg/kg/3D: CNS LANCAO 2,479,80
unr-hmn TDLo: 9800 µg/kg/28D: BAH ARZNAD 32,911,82
unr-man TDLo: 500 µg/kg/5D-I: BAH JAMAAP 250,485,83
mul-man TDLo: 343 µg/kg SMJOAV 76,546,83
mul-man TDLo: 1 mg/kg/1D-I JCPYDR 3,338,83
orl-rat LD50: 128 mg/kg ARZNAD 24,45,74
ipr-rat LD50: 27 mg/kg 27ZQAG -,190,72
scu-rat LD50: 60 mg/kg NIIRDN 6,594,82
ivn-rat LD50: 15 mg/kg NIIRDN 6,594,82
orl-mus LD50: 71 mg/kg FRPSAX 31,442,76
ipr-mus LD50: 30 mg/kg BCFAAI 111,293,72
scu-mus LD50: 41 mg/kg OYYAA2 1,74,67
ivn-mus LD50: 13 mg/kg ARZNAD 11,932,61

CONSENSUS REPORTS: EPA Genetic Toxicology Program.

SAFETY PROFILE: Poison by ingestion, intraperitoneal, intravenous, and subcutaneous routes. Human systemic effects: change in motor activity, distorted perceptions, excitement, fasciculations, hallucinations, muscle contraction or spasticity, muscle weakness, rigidity, somnolence, tremors. A human teratogen by ingestion which causes developmental abnormalites of the musculoskeletal and cardiovascular (circulatory) systems, and abnormal conditions of newborn at birth. Human mutation data reported. An experimental teratogen. Experimental reproductive effects. Questionable carcinogen with experimental carcinogenic data. A tranquilizer used in the treatment of schizophrenia and agitated psychoses. When heated to decomposition it emits very toxic fumes of F^-, Cl^-, and NO_x.

CLY600 CAS:77-36-1 HR: 1
CHLORPHTHALIDOLONE
mf: $C_{14}H_{11}ClN_2O_4S$ mw: 338.78

SYNS: BENZENESULFONAMIDE, 2-CHLORO-5-(2,3-DIHYDRO-1-HYDROXY-3-OXO-1H-ISOINDOL-1-YL)- (9CI) □ BENZENESULFONAMIDE, 2-CHLORO-5-(1-HYDROXY-3-OXO-1-ISOINDOLINYL)- □ CHLOROTHALIDONE □ CHLORPHTHALIDONE □ CHLORTALIDONE □ CHLORTHALIDON □ CHLORTHALIDONE □ G 33182 □HYGRO-TON □ IGROTON □ ISOREN □ NATRIURAN □ ORADIL □ OXODOLIN □ PHTHALAMODINE □ PHTHALAMUDINE □ RENON □SALURE-TIN □ ZAMBESIL

TOXICITY DATA WITH REFERENCE
orl-man TDLo: 30 mg/kg (male 3W pre): REP BMJOAE 281,714,80
orl-mus TDLo: 2200 mg/kg (female 2-12D post): REP MPHEAE 12,245,65
orl-hmn TDLo: 2587 µg/kg/8D-I: EYE,GIT JAMAAP 258,484,87
orl-man TDLo: 5714 µg/kg/4D-I: BPR,GLN JAMAAP 220,1592,72
orl-wmn TDLo: 12 mg/kg/6D-I: SYS SMJOAV 79,629,86

SAFETY PROFILE: Human systemic effects by ingestion: BP elevation, hyperglycemia, sodium and chlorine level changes, headache, nausea or vomiting. Experimental reproductive effects. When heated to decomposition it emits toxic fumes of SO_x, NO_x, and Cl^-.

CLY750 CAS:84-01-5 HR: 3
CHLORPROETHAZINE
mf: $C_{19}H_{23}ClN_2S$ mw: 346.95

PROP: Bp: 225–240° @ 1 mm.

SYNS: 2-CHLORO-10-(3-DIETHYLAMINOPROPYL)PHENOTHIAZINE □ NEURIPLEGE □ RP 4909

TOXICITY DATA WITH REFERENCE
orl-mus LD50: 300 mg/kg PSCBAY 2,17,63
ipr-mus LD50: 90 mg/kg PSCBAY 2,17,63
scu-mus LD50: 325 mg/kg PSCBAY 2,17,63
ivn-mus LD50: 80 mg/kg PSCBAY 2,17,63

SAFETY PROFILE: Poison by ingestion, intraperitoneal, subcutaneous, and intravenous routes. When heated to decomposition it emits very toxic fumes of Cl^-, NO_x, and SO_x.

CLZ000 CAS:4611-02-3 HR: 3
CHLORPROETHAZINE HYDROCHLORIDE
mf: $C_{19}H_{23}ClN_2S \cdot ClH$ mw: 383.41

PROP: A solid. Mp: 178°.

SYN: 2-CHLORO-10-(3′-DIETHYLAMINOPROPYL)PHENOTHIAZINE HYDROCHLORIDE

TOXICITY DATA WITH REFERENCE
orl-mus LD50: 300 mg/kg 27ZQAG -,14,72
ipr-mus LD50: 90 mg/kg 27ZQAG -,14,72
scu-mus LD50: 325 mg/kg 27ZQAG -,14,72
ivn-mus LD50: 80 mg/kg 27ZQAG -,14,72

SAFETY PROFILE: Poison by ingestion, intravenous, intraperitoneal, and subcutaneous routes. When heated to decomposition it emits very toxic fumes of Cl^-, NO_x, and SO_x.

CMA100 **CAS:2921-88-2** **HR: 3**
CHLORPYRIFOS
mf: $C_9H_{11}Cl_3NO_3PS$ mw: 350.59

PROP: Crystals with mild mercaptan odor. Mp: 42–43.5°. Very sltly sol in H_2O; sol in most org solvs.

SYNS: BRODAN □ CHLORPYRIFOS-ETHYL □ CHLORPYRIPHOS □ CHLORPYRIPHOS-ETHYL □ DETMOL U.A. □ O,O-DIAETHYL-O-3,5,6-TRICHLOR-2-PYRIDYLMONOTHIOPHOSPHAT □ O,O-DIETHYL O-3,5,6-TRICHLORO-2-PYRIDYL PHOSPHOROTHIOATE □ DOWCO 179 □ DURSBAN □ DURSBAN F □ ENT 27311 □ ERADEX □ ETHION, dry □ LORSBAN □ OMS-0971 □ PIRIDANE □ 2-PYRIDINOL, 3,5,6-TRICHLORO-, O-ESTER with O,O-DIETHYL PHOSPHOROTHIOATE □ PYRINEX □ STIPEND

TOXICITY DATA WITH REFERENCE
cyt-dmg-orl 50 ppb/3S ENMUDM 5,835,83
orl-mus TDLo:250 mg/kg (female 6-15D post):REP TXAPA9 54,31,80
orl-man TDLo:300 mg/kg:PNS ARTODN 59,176,86
orl-rat LD50:82 mg/kg TXAPA9 14,515,69
ihl-rat LC50:>200 mg/m³/4H PEMNDP 9,166,91
skn-rat LD50:202 mg/kg TXAPA9 14,515,69
orl-mus LD50:60 mg/kg JESEDU 13,11,78
ipr-mus LD50:192 mg/kg TXAPA9 65,144,82
orl-rbt LD50:1000 mg/kg SPEADM 78-1,45,78
skn-rbt LD50:2000 mg/kg GUCHAZ 6,203,73

CONSENSUS REPORTS: EPA Genetic Toxicology Program.

OSHA PEL: TWA 0.2 mg/m³ (skin)
ACGIH TLV: TWA 0.2 mg/m³ (skin)

SAFETY PROFILE: Poison by ingestion, intraperitoneal, skin contact, and inhalation routes. Human systemic effects by ingestion: paresthesia, muscle weakness, coma. Experimental reproductive effects: developmental toxicity. Mutation data reported. When heated to decomposition it emits very toxic fumes of Cl^-, NO_x, PO_x, and SO_x.

For occupational chemical analysis use OSHA: #ID-62.

CMA250 **CAS:5598-13-0** **HR: 2**
CHLORPYRIFOS-METHYL
mf: $C_7H_7Cl_3NO_3PS$ mw: 322.53

PROP: A solid. Mp: 44.5–45.5°. Sltly sol in H_2O; very sol in org solvs.

SYNS: O,O-DIMETHYL-O-(3,5,6-TRICHLORO-2-PYRIDYL)PHOSPHOROTHIOATE □ DOWCO 217 □ DURSBAN METHYL □ ENT 27,520 □ METHYL CHLORPYRIFOS □ METHYL DURSBAN □ NOLTRAN □ NSC 60380 □ OMS-1155 □ RELDAN □ ZERTELL

TOXICITY DATA WITH REFERENCE
skn-rbt 500 mg/24H MLD TXAPA9 21,369,72
orl-mus TDLo:2184 mg/kg (male 26W pre):REP HOEKAN 23,71,73
orl-rat LD50:1828 mg/kg HOEKAN 23,57,73
skn-rat LD50:3713 mg/kg YKYUA6 35,1315,84
orl-mus LD50:2032 mg/kg HOEKAN 23,57,73
ipr-mus LD50:2325 mg/kg TXAPA9 65,144,82
scu-mus LD50:23,800 mg/kg YKYUA6 30,409,79
orl-rbt LD50:2000 mg/kg BESAAT 15,123,69

SAFETY PROFILE: Moderately toxic by ingestion, intraperitoneal, and skin contact. A skin irritant. Experimental reproductive effects. When heated to decomposition it emits very toxic fumes of Cl^-, NO_x, PO_x, and SO_x. A pesticide.

CMA500 **CAS:3495-42-9** **HR: 3**
CHLORQUINOX
mf: $C_8H_2Cl_4N_2$ mw: 267.92

SYNS: LUCEL □ 5,6,7,8-TETRACHLOROQUINOXALINE

TOXICITY DATA WITH REFERENCE
orl-rat LD50:6400 mg/kg 28ZEAL 5,50,76
orl-rbt LD50:3000 mg/kg 28ZEAL 5,50,76
orl-brd LD50:400 mg/kg 28ZEAL 5,50,76

SAFETY PROFILE: Poison by ingestion. When heated to decomposition it emits very toxic fumes of Cl^- and NO_x.

CMA600 **CAS:97919-22-7** **HR: 3**
CHLORSULFAQUINOXALINE
mf: $C_{14}H_{11}ClN_4O_2S$ mw: 334.80

SYNS: 4-AMINO-N-(5-CHLORO-2-QUINOXALINYL)BENZENESULFONAMIDE □ BENZENESULFONAMIDE, 4-AMINO-N-(5-CHLORO-2-QUINOXALINYL)- □ NSC-339004

TOXICITY DATA WITH REFERENCE
ivn-rat TDLo:600 mg/kg (male 1D pre):REP NTIS** PB87-128658
ivn-rat LDLo:600 mg/kg NTIS** PB87-128658
ivn-mus LD50:607 mg/kg NTIS** PB87-128658
ivn-dog LDLo:12 mg/kg NTIS** PB87-128658

SAFETY PROFILE: Poison by intravenous route. Experimental reproductive effects. When heated to decomposition it emits toxic fumes of SO_x, NO_x, and Cl^-.

CMA750 **CAS:57-62-5** **HR: 3**
CHLORTETRACYCLINE
mf: $C_{22}H_{23}ClN_2O_8$ mw: 478.92

PROP: Golden-yellow crystals. Mp: 168–169°. Sltly sol in water; very sol in aq soln pH 7.65; freely sol in the "cellosolves," dioxane, "Carbitol"; sol in methanol, ethanol, butanol, acetone, ethyl acetate, and benzene; insol in ether and petr ether.

SYNS: ACRONIZE □ AUREOCINA □ AUREOMYCIN □ AUREOMYCIN A-377 □ AUREOMYKOIN □ BIOMITSIN □ BIOMYCIN □ 7-CHLORO-4-(DIMETHYLAMINO)-1,4,4a,5,5a,6,11,12a-OCTAHYDRO-2-NAPHTHACENECARBOXAMIDE □ 7-CHLOROTETRACYCLINE □ CHRYSOMYKINE □ CTC □ DUOMYCIN □ FLAMYCIN

TOXICITY DATA WITH REFERENCE
scu-mus TDLo:372 mg/kg (1-6D preg):REP ASPHAK 23,481,69
orl-rat LDLo:3 g/kg JAFCAU 17,497,69
ipr-rat LDLo:335 mg/kg CLDND*
orl-dog LDLo:750 mg/kg AAGAAW -,595,60
ivn-dog LD50:150 mg/kg HBTXAC 5,52,59
ipr-gpg LDLo:1800 mg/kg ANTBAL 20,793,75
ivn-gpg LDLo:100 mg/kg ANYAA9 51,182,48

CONSENSUS REPORTS: Reported in EPA TSCA Inventory.

SAFETY PROFILE: Poison by intravenous and intraperitoneal routes. Moderately toxic by ingestion. Experimental reproductive effects. When heated to decomposition it emits toxic fumes of Cl^- and NO_x. See also TETRACYCLINE.

CMB000 CAS:64-72-2 HR: 3
CHLORTETRACYCLINE HYDROCHLORIDE
mf: $C_{22}H_{23}ClN_2O_8 \cdot ClH$ mw: 515.38

PROP: Yellow crystals.

SYNS: AUREOCICLINA □ AUREOCYCLINE □ AUREOMYCIN HYDROCHLORIDE □ AUXEOMYCIN □ CHLOROTETRACYCLINE HYDROCHLORIDE □ CLOROTETRACICLINA CLORIDRATO (ITALIAN) □ ISPHAMYCIN □ NSC-13252 □ U-6780

TOXICITY DATA WITH REFERENCE
scu-mus TDLo:1125 mg/kg (10-18D preg):REP CRSBAW 161,300,67
orl-rat LD50:10,300 mg/kg TXAPA9 18,185,71
ivn-rat LD50:100 mg/kg BCFAAI 102,660,63
orl-mus LD50:2740 mg/kg FRPSAX 10,197,55
ipr-mus LD50:197 mg/kg RPOBAR 2,278,70
scu-mus LDLo:1000 mg/kg ANYAA9 51,254,48
ivn-mus LD50:101 mg/kg RPOBAR 2,278,70

SAFETY PROFILE: Poison by intraperitoneal and intravenous routes. Moderately toxic by ingestion and subcutaneous routes. Experimental reproductive effects. When heated to decomposition it emits very toxic fumes of Cl^- and NO_x. See also TETRACYCLINE.

CMB125 CAS:14008-79-8 HR: 3
CHLORTROPBENZYL
mf: $C_{21}H_{21}ClNO \cdot ClH$ mw: 378.37

PROP: Crystals from 2-propanol. Mp: 215–217°.

SYNS: 3-α-((p-CHLORO-α-PHENYLBENZYL)OXY)-1-α-H,5-α-H-TROPANE HYDROCHLORIDE □ FC-1 □ CL 6057 □ TROPINE-4-CHLOROBENZHYDRYL ETHER HYDROCHLORIDE □ WY 2149

TOXICITY DATA WITH REFERENCE
orl-rat LD50:364 mg/kg JPETAB 114,192,55
ipr-rat LD50:58 mg/kg JPETAB 114,192,55
orl-mus LD50:174 mg/kg JPETAB 114,192,55
ipr-mus LD50:32 mg/kg JPETAB 114,192,55
ivn-dog LDLo:28 mg/kg JPETAB 114,192,55

SAFETY PROFILE: Poison by ingestion, intravenous, and intraperitoneal routes. When heated to decomposition it emits toxic fumes of NO_x and HCl.

CMB500 CAS:12442-63-6 HR: 3
CHLORYL PERCHLORATE
mf: Cl_2O_6 mw: 166.91

PROP: A red liquid. Mp: 3.5°, bp: 203°.

SAFETY PROFILE: Probably a poison and irritant due to its reactivity. A very powerful oxidant. Explodes when heated or on contact with water or thionyl chloride. Violent or explosive reaction with organic matter (e.g., ethanol; stopcock grease; wood). The least explosive of the chlorine oxide compounds. When heated to decomposition it emits toxic fumes of Cl^-. See also CHLORIDES and PERCHLORATES.

CMB675 CAS:35317-79-4 HR: 2
CHLOTAZOLE
mf: $C_5H_5Cl_3N_2OS$ mw: 247.53

SYNS: KHLOTAZOL □ 2,2,2-TRICHLORO-1-(2-THIAZOLYLAMINO) ETHANOL

TOXICITY DATA WITH REFERENCE
orl-rat LD50:2500 mg/kg PCJOAU 17,519,83
orl-mus LD50:1 g/kg RPTOAN 46,213,83
ipr-mus LD50:708 mg/kg FRXXBL #2400361
orl-cat LD50:700 mg/kg PCJOAU 17,519,83
orl-rbt LD50:700 mg/kg PCJOAU 17,519,83

SAFETY PROFILE: Moderately toxic by ingestion and intraperitoneal routes. When heated to decomposition it emits toxic fumes of Cl^-, SO_x, and NO_x.

CMC750 CAS:67-97-0 HR: 3
CHOLECALCIFEROL
mf: $C_{27}H_{44}O$ mw: 384.71

PROP: White crystals; odorless. Mp: 87–88°. Insol in water; sol in alc, chloroform, and fatty oils.

SYNS: COLECALCIFEROL □ 7-DEHYDROCHOLESTROL, ACTIVATED □ DELSTEROL □ DEPARAL □ D3-VIGANTOL □ OLEOVITAMIN D3 □ RICKETON □ 9,10-SECOCHOLESTA-5,7,10(19)-TRIEN-3-β-OL □ TRIVITAN □ VIGORSAN □ VITAMIN D3 □ VITINC DAN-DEE-3

TOXICITY DATA WITH REFERENCE
scu-rat TDLo:90 mg/kg (12-20D preg):TER FOMOAJ 29,333,70
orl-inf TDLo:39 mg/kg/34W-I BMJOAE 295,1173,87
orl-rat LD50:42 mg/kg TXAPA9 43,125,78
orl-mus LD50:42,500 μg/kg DOVEAA 43(255-256),14,89
orl-dog LD50:80 mg/kg JAVMA4 193,211,88

CONSENSUS REPORTS: Reported in EPA TSCA Inventory.

SAFETY PROFILE: Poison by ingestion. An experimental teratogen. When heated to decomposition it emits acrid smoke and irritating fumes.

CMC800 HR: 3
CHOLERA ENTEROTOXIN

SYNS: CHOLERA ENTERO-EXOTOXIN □ CHOLERAGEN □ ENTERO-EXOTOXIN, CHOLERA □ ENTEROTOXIN, CHOLERA

TOXICITY DATA WITH REFERENCE
ivn-mus TDLo:8 μg/kg (female 4-6D post):REP IMLCAV 1,223,72
ivn-mus LD50:260 μg/kg IMLCAV 1,223,72
ivn-mky LDLo:10 μg/kg TOXIA6 18,309,80
ivn-rbt LDLo:100 μg/kg TOXIA6 19,701,81

SAFETY PROFILE: Poison by intravenous route. Experimental reproductive effects. When heated to decomposition it emits acrid smoke and irritating fumes.

CMD750 **CAS:57-88-5** **HR: 2**
CHOLESTEROL
mf: $C_{27}H_{46}O$ mw: 386.73

PROP: White or faint yellow, pearly leaflets from aq alc. Mp: 148.5° (anhyd), bp: 360° (decomp).

SYNS: CHOLEST-5-EN-3-β-OL □ Δ^5-CHOLESTEN-3-β-OL □ 5-CHOLESTEN-3-β-OL □ 5:6-CHOLESTEN-3-β-OL □ CHOLESTERIN □ CHOLESTEROL BASE H □ CHOLESTERYL ALCOHOL □ CHOLESTRIN □ CHOLESTROL □ CORDULAN □ DUSOLINE □ DUSORAN □DYTHOL □ HYDROCERIN □ 3-β-HYDROXYCHOLEST-5-ENE □ KATHRO □ LANOL □ NIMCO CHOLESTEROL BASE H □ PROVITAMIN D □ SUPER HARTOLAN □ TEGOLAN

TOXICITY DATA WITH REFERENCE
mmo-sat 500 µg/plate FCTOD7 20,35,82
dnd-mus:oth 1 µmol/L CJBBDU 62,94,84
orl-rbt TDLo:6 g/kg (1-31D preg):REP JONUAI 102,1681,72
orl-rbt TDLo:6 g/kg (1-31D preg):TER JONUAI 102,1681,72
scu-rat TDLo:1900 mg/kg (1-19D preg):REP JDREAF 51,1421,72
scu-rat TDLo:175 mg/kg (8-14D preg):TER AOBIAR 12,1221,67
ipr-rat TDLo:800 mg/kg/43W-I:ETA EMSUA8 3,95,45
scu-mus TDLo:15 g/kg/47W-I:CAR NATUAS 160,270,47
imp-mus TDLo:800 mg/kg:ETA SCIEAS 167,996,70
scu-mus TDLo:60 g/kg/2W-I:CAR NATWAY 60,525,73

CONSENSUS REPORTS: IARC Cancer Review: Group 3 IMEMDT 7,161,87; Human Inadequate Evidence IMEMDT 31,95,83; Animal Inadequate Evidence IMEMDT 10,99,76. Reported in EPA TSCA Inventory. EPA Genetic Toxicology Program.

SAFETY PROFILE: Experimental teratogenic and reproductive effects. Questionable carcinogen with experimental carcinogenic and tumorigenic data. Mutation data reported. Used in pharmaceutical and dermal preparations as an emulsifying agent. When heated to decomposition it emits acrid smoke and irritating fumes.

CME250 **CAS:3546-10-9** **HR: 3**
CHOLESTERYL-p-BIS(2-CHLOROETHYL)AMINO PHENYLACETATE
mf: $C_{39}H_{59}Cl_2NO_2$ mw: 644.89

SYNS: (p-(BIS(2-CHLOROETHYL)AMINO)PHENYL)ACETATE CHOLESTEROL □ (p-(BIS(2-CHLOROETHYL)AMINO)PHENYL)ACETIC ACID CHOLESTEROL ESTER □ (4-(BIS(2-CHLOROETHYL)AMINO) PHENYL)ACETIC ACID CHOLESTERYL ESTER □ 5-CHOLESTEN-3-β-OL 3-(p-(BIS(2-CHLOROETHYL)AMINO)PHENYL)ACETATE □ FENESTERIN □ FENESTRIN □ NCI-C01558 □ NSC 104469 □ PHENESTERINE □ PHENESTRIN

TOXICITY DATA WITH REFERENCE
orl-rat TDLo:780 mg/kg/52W-I:CAR NCITR* NCI-CG-TR-60,78
orl-mus TDLo:1092 mg/kg/1Y-I:CAR NCITR* NCI-CG-TR-60,78
ipr-mus TDLo:2400 mg/kg/8W-I:NEO CNREA8 33,3069,73
orl-rat TD:1560 mg/kg/1Y-I:CAR NCITR* NCI-CG-TR-60,78
orl-mus TD:2340 mg/kg/1Y-I:CAR NCITR* NCI-CG-TR-60,78
orl-mus TD:4680 mg/kg/1Y-I:CAR NCITR* NCI-CG-TR-60,78

CONSENSUS REPORTS: NCI Carcinogenesis Bioassay (gavage); Clear Evidence: mouse, rat NCITR* NCI-CG-TR-60,78.

SAFETY PROFILE: Suspected carcinogen with experimental carcinogenic and neoplastigenic data. When heated to decomposition it emits very toxic fumes of Cl^- and NO_x.

CME400 **CAS:11041-12-6** **HR: 2**
CHOLESTYRAMINE

SYNS: CHOLESTYRAMINE CHLORIDE □ CHOLESTYRAMINE RESIN □ COLESTYRAMIN □ CUEMID □ QUANTALAN □ QUESTRAN

TOXICITY DATA WITH REFERENCE
orl-rat TDLo:6 g/kg (17-22D post):REP KSRNAM 16,2078,82
orl-rat TDLo:42 g/(9 W male pre/female 2 W pre/1-7D post):TER KSRNAM 16,2040,82
orl-rat TDLo:11 g/kg (7-17D post):REP KSRNAM 16,2050,82
orl-rbt TDLo:13 g/kg (6-18D post):REP KSRNAM 16,2070,82
orl-rat TDLo:5500 mg/kg (7-17D post):TER KSRNAM 16,2050,82
orl-man TDLo:112 g/kg/W-C:CAR NEJMAG 301,1007,79
orl-inf TDLo:4 g/kg/2D-I:SYS AJDCAI 141,479,87
orl-cld TDLo:46 g/kg/39W-I:NOSE CMAJAX 134,609,86
orl-rat LD50:>4 g/kg DRUGAY-,412,90

CONSENSUS REPORTS: Reported in EPA TSCA Inventory.

SAFETY PROFILE: Low toxicity by ingestion. Questionable human carcinogen producing colon tumors. An experimental teratogen. Other experimental reproductive effects. Toxic effects by ingestion: acidosis and nose bleeds. When heated to decomposition it emits acrid smoke and irritating fumes.

CME675 **CAS:27959-26-8** **HR: 1**
CHOLEXAMIN
mf: $C_{34}H_{32}N_4O_9$ mw: 640.70

PROP: Crystals from dilute acetic acid and aq alc; practically odorless and tasteless. Mp: 177–180°. Sltly sol in water; ethanol, and ether.

SYNS: CHOLEXAMINE □ 2-HYDROXYCYCLOHEXANE-1,1,3,3-TETRAMETHANOL TETRAESTER with NICOTINIC ACID □ K 31 (pharmaceutical) □ NICOMOL □ 3-PYRIDINECARBOXYLIC ACID, (2-HYDROXY-1,3-CYCLOHEXANEDIYLIDENE)TETRAKIS(METHYLENE) ESTER □ 2,2,6,6-TETRAKIS(NICOTINOYLOXYMETHYL)CYCLOHEXANOL □ TETRANICOTINIC ACID-2-HYDROXYCYCLOHEXA-1,1,3,3-TETRAMETHYL ESTER

TOXICITY DATA WITH REFERENCE
orl-rat TDLo:10,500 mg/kg (female 35D pre):REP OYYAA2 14,755,77
orl-rat TDLo:42 g/kg (female 35D pre):REP OYYAA2 14,755,77

orl-rat TDLo: 168 g/kg (male 35D pre): REP OYYAA2 14,755,77
orl-rat TDLo: 10,500 mg/kg (35D male): REP OYYAA2 14,755,77
orl-rat LD50: 10 g/kg OYYAA2 14,741,77

SAFETY PROFILE: Mildly toxic by ingestion. Experimental reproductive effects. When heated to decomposition it emits toxic fumes of NO_x. An anticholesteremic agent that reduces the blood cholesterol level. See also ESTERS.

CME750 CAS:81-25-4 HR: 3
CHOLIC ACID
mf: $C_{24}H_{40}O_5$ mw: 408.64

PROP: Crystals. Mp: 197° (anhyd). The most abundant bile acid; the monohydrate crystallizes in plates from dilute acetic acid; sol in glacial acetic acid, acetone, and alc. Sltly sol in chloroform, practically insol in water and benzene.

SYNS: CHOLALIN □ CHOLSAEURE (GERMAN) □ COLALIN □ 3-α,7-α,12-α-TRIHYDROXY-5-β-CHOLAN-24-OIC ACID □ 3,7,12-TRIHYDROXY-CHOLAN-24-OIC ACID (3-α,5-β,7-α,12-α) □ 3-α,7-α,12-α-TRIHYDROXYCHOLANSAEURE (GERMAN)

TOXICITY DATA WITH REFERENCE
mmo sat 50 mg/L MUREAV 158,45,85
sln-smc 400 mg/L CRNGDP 5,447,84
orl-mus LD50: 4950 mg/kg ESKHA5 (103),29,85
ipr-mus LD50: 330 mg/kg ARZNAD 20,323,70

CONSENSUS REPORTS: Reported in EPA TSCA Inventory.

SAFETY PROFILE: Poison by intraperitoneal route. Moderately toxic by ingestion. Mutation data reported. When heated to decomposition it emits acrid smoke and irritating fumes.

CMF000 CAS:62-49-7 HR: 3
CHOLINE
mf: $C_5H_{14}NO$ mw: 104.20

SYNS: BILINEURINE □ CHOLINE ION □ (2-HDYROXYETHYL)TRIMETHYLAMMONIUM □ 2-HYDROXY-N,N,N-TRIMETHYLETHANAMINIUM

TOXICITY DATA WITH REFERENCE
scu-cat LDLo: 150 mg/kg HBAMAK 4,1289,35
ivn-cat LDLo: 35 mg/kg 85IXA4-,358,48
scu-rbt LDLo: 800 mg/kg CRSBAW 83,481,20
ivn-rbt LDLo: 70 mg/kg CRSBAW 83,481,20
rec-rbt LDLo: 460 mg/kg CRSBAW 83,481,20
scu-frg LDLo: 1500 mg/kg HBAMAK 4,1289,35

CONSENSUS REPORTS: Reported in EPA TSCA Inventory.

SAFETY PROFILE: Poison by subcutaneous and intravenous routes. Moderately toxic by rectal route. Mildly toxic by ingestion. When heated to decomposition it emits toxic fumes of NO_x and NH_3.

CMF250 CAS:51-84-3 HR: 3
CHOLINE ACETATE (ESTER)
mf: $C_7H_{16}NO_2$ mw: 146.24

SYNS: ACECOLINE □ ACETYLCHOLINE □ ACETYL CHOLINE ION □ 2-(ACETYLOXY)-N,N,N-TRIMETHYLETHANAMINIUM □ ACH □ ARTEROCOLINE □ CHOLINE ACETATE □ OVISOT

TOXICITY DATA WITH REFERENCE
scu-rat LD50: 250 mg/kg 27ZIAQ -,29,73
ivn-rat LD50: 22 mg/kg JPETAB 58,337,36
orl-mus LD50: 3000 mg/kg JPETAB 58,337,36
ipr-mus LD50: 170 mg/kg AIPTAK 192,88,71
scu-mus LD50: 170 mg/kg JPETAB 58,337,36
ivn-mus LD50: 11 mg/kg ATXKA8 29,39,72
ivn-rbt LD50: 300 µg/kg 27ZIAQ -,39,73

SAFETY PROFILE: Poison by subcutaneous, intravenous, and intraperitoneal routes. Moderately toxic by ingestion. When heated to decomposition it emits toxic fumes of NO_x. See also CHOLINE and ESTERS.

CMF260 CAS:66-23-9 HR: 3
CHOLINE ACETATE (ESTER), BROMIDE
mf: $C_7H_{16}NO_2 \cdot Br$ mw: 226.15

SYNS: ACETOXYETHYL-TRIMETHYLAMMONIUM BROMIDE □ ACETYLCHOLINE BROMHYDRATE □ ACETYLCHOLINE BROMIDE □ ACETYLCHOLINE HYDROBROMIDE □ 2-(ACETYLOXY)-N,N,N-TRIMETHYLETHANAMINIUM BROMIDE □ CHOLINE, ACETYL-, BROMIDE □ ETHANAMINIUM, 2-(ACETYLOXY)-N,N,N-TRIMETHYL-, BROMIDE (9CI) □ PRAGMOLINE □ TONOCHOLIN B

TOXICITY DATA WITH REFERENCE
scu-mus LD50: 170 mg/kg JPETAB 103,62,51

CONSENSUS REPORTS: Reported in EPA TSCA Inventory.

SAFETY PROFILE: Poison by subcutaneous route. When heated to decomposition it emits toxic vapors of NO_x and Br^-.

CMF350 CAS:987-78-0 HR: 1
CHOLINE CYTIDINE DIPHOSPHATE
mf: $C_{14}H_{26}N_4O_{11}P_2$ mw: 488.38

PROP: Hygroscopic powder.

SYNS: CHOLINE, HYDROXIDE, 5'-ESTER with CYTIDINE 5'-(TRIHYDROGEN PYROPHOSPHATE), inner salt □ CDP-CHOLIN □ CDP-CHOLINE □ CDP-COLINA □ CEREB □ CHOLINE 5'-CYTIDINE DIPHOSPHATE □ CITICHOLINE □ CITICOLINE □ CITIDIN DIFOSFATO de COLINA □ CITIDOLINE □ COLITE □CYTIDINDIPHOSPHOCHO-LIN □ CYTIDINE CHOLINE DIPHOSPHATE □ CYTIDINE 5'-(CHOLINE DIPHOSPHATE) □ CYTIDINE DIPHOSPHATE CHOLINE □ CYTIDINE 5'-DIPHOSPHATE CHOLINE □ CYTIDINE DIPHOSPHATE CHOLINE ESTER □ CYTIDINE DIPHOSPHATE CHOLIN ESTER □ CYTIDINE DIPHOSPHOCHOLINE □ CYTIDINE 5'-DIPHOSPHOCHOLINE □ CYTIDINE DIPHOSPHORYLCHOLINE □ CYTIDOLINE □ ENSIGN □ NICHOLIN □ NICOLIN □ NITICOLIN □ RECOFNAN □RECOG-NAN □ SOMAZINA □ SUNCHOLIN

TOXICITY DATA WITH REFERENCE
ipr-rat TDLo: 7500 mg/kg (male 30D pre): REP OYYAA2 20,109,80
ipr-rat TDLo: 60 g/kg (male 30D pre): REP OYYAA2 20,109,80

orl-rat LD50:18 g/kg DRUGAY 6,322,82
ipr-rat LD50:5344 mg/kg OYYAA2 20,109,80
scu-rat LD50:8218 mg/kg OYYAA2 20,109,80
ivn-rat LD50:2973 mg/kg OYYAA2 20,109,80
orl-mus LD50:12,500 mg/kg DRUGAY 6,322,82
ipr-mus LD50:5393 mg/kg OYYAA2 20,109,80
scu-mus LD50:5800 mg/kg DRUGAY 6,322,82
ivn-mus LD50:4600 mg/kg ARZNAD 33,1033,83

SAFETY PROFILE: Moderately toxic by intravenous route. Experimental reproductive effects. When heated to decomposition it emits toxic fumes of NO_x and PO_x.

CMF400 CAS:999-81-5 HR: 3
CHOLINE DICHLORIDE
mf: $C_5H_{13}ClN•Cl$ mw: 158.09

PROP: Crystals. Mp: 245° (decomp). Very sol in H_2O.

SYNS: ANTYWYLEGACZ □ CCC PLANT GROWTH REGULANT □ CE CE CE □ 2-CHLORAETHYL-TRIMETHYLAMMONIUMCHLORID □ CHLORCHOLINCHLORID □ CHLORCHOLINE CHLORIDE □ CHLORMEQUAT □ CHLORMEQUAT CHLORIDE □ CHLOROCHOLINE CHLORIDE □ (β-CHLOROETHYL)TRIMETHYLAMMONIUM CHLORIDE □ (2-CHLOROETHYL)TRIMETHYLAMMONIUM CHLORIDE □ 2-CHLORO-N,N,N-TRIMETHYLETHANAMINIUM CHLORIDE □ 60-CS-16 □ CYCLOCEL □ CYCOCEL □ CYCOCEL-EXTRA □ CYCOGAN □ CYCOGAN EXTRA □ CYOCEL □ EI 38,555 □ ETHANAMINIUM, 2-CHLORO-N,N,N-TRIMETHYL-, CHLORIDE (9CI) □ HICO CCC □ HORMOCEL-2CCC □ INCRECEL □ LIHOCIN □ NCI-C02960 □ RETACEL □ STABILAN □ TRIMETHYL-β-CHLORETHYLAMMONIUMCHLORID □ TUR

TOXICITY DATA WITH REFERENCE
skn-rbt 500 mg/24H MLD 85JCAE-,616,86
dni-mus-ivg 5 pph JIDEAE 62,378,74
orl-mus TDLo:7100 mg/kg/78W-I:NEO NTIS** PB223-159
orl-hmn LDLo:10 mg/kg:PUL AXVMAW 31,527,77
ivn-hmn LDLo:1 mg/kg:PUL
orl-rat LD50:600 mg/kg GISAAA 36(11),33,71 AXVMAW 31,527,77
skn-rat LD50:4000 mg/kg FMCHA2-,C53,83
ipr-rat LD50:64 mg/kg ABMGAJ 33,89,74
ivn-rat LD50:12,500 µg/kg AXVMAW 31,527,77
unr-rat LD50:780 mg/kg VINIT* #4758-80
orl-mus LD50:54 mg/kg ABMGAJ 27,663,71
ipr-mus LD50:62 mg/kg ABMGAJ 27,663,71
ivn-mus LD50:7 mg/kg AXVMAW 31,527,77
unr-mus LD50:560 mg/kg VINIT* #4758-80
orl-dog LD50:50 mg/kg AXVMAW 24,1049,70
orl-cat LD50:7 mg/kg AXVMAW 24,1049,70
ivn-rbt LDLo:4 mg/kg AXVMAW 31,527,77

CONSENSUS REPORTS: NTP Carcinogenesis Bioassay (feed): No Evidence: mouse, rat NCITR* NCI-TR-158,79. Reported in EPA TSCA Inventory.

SAFETY PROFILE: Human poison by ingestion and intravenous routes. Moderately toxic by skin contact. Human systemic effects: respiratory depression. Questionable carcinogen with experimental neoplastigenic data. Mutation data reported. When heated to decomposition it emits toxic fumes of Cl^-.

CMF500 CAS:4499-40-5 HR: 3
CHOLINE, with THEOPHYLLINE (1:1)
mf: $C_7H_7N_4O_2•C_5H_{14}NO$ mw: 283.38

PROP: Granules.

SYNS: CHOLEDYL □ CHOLEGYL □ CHOLINE THEOPHYLLINATE □ CHOLINE THEOPHYLLINE SALT □ CHOLINOPHYLLINE □ FILORAL □ (2-HYDROXYETHYL)TRIMETHYLAMMONIUM with THEOPHYLLINE □ 2-HYDROXY-N,N,N-TRIMETHYLETHANAMINIUM SALT with 3,7-DIHYDRO-1,3-DIMETHYLPURINE-2,6-DIONE □OXTRIMETHYL-LINE □ OXTRIPHYLLINE □ SOLIPHYLLINE □ TEOFILCOLINA □ TEOKOLIN □ THEOPHYLLINE CHOLINATE □ THEOPHYLLINE SALT of CHOLINE □ THEOXYLLINE □ THIOPHYLLINE CHOLINATE □ THIOPHYLLINE with CHOLLINE

TOXICITY DATA WITH REFERENCE
orl-man TDLo:429 mg/kg:SYS AJEMEN 3,408,85
orl-inf TDLo:113 mg/kg/31D:END,KID ADCHAK 53,757,78
orl-wmn TDLo:420 mg/kg:GIT,CNS SMJOAV 71,965,78
orl-rat LD50:600 mg/kg NIIRDN 6,278,82
ipr-rat LD50:185 mg/kg NIIRDN 6,278,82
ims-rat LD50:240 mg/kg NIIRDN 6,278,82
orl-mus LD50:770 mg/kg CLDND*
ivn-mus LD50:112 mg/kg CLDND*
ims-mus LD50:360 mg/kg CLDND*
orl-gpg LD50:210 mg/kg CLDND*
ivn-gpg LD50:118 mg/kg CLDND*
ims-gpg LD50:185 mg/kg CLDND*

SAFETY PROFILE: Poison by ingestion, intravenous, intraperitoneal, and intramuscular routes. Human systemic effects by ingestion: changes in potassium, changes in urine composition, hyperglycemia, metabolic acidosis, nausea or vomiting, tremors, and excitement. When heated to decomposition it emits toxic fumes of NO_x and NH_3. See also THEOPHYLLINE and CHOLINE.

CMF750 CAS:67-48-1 HR: 3
CHOLINE HYDROCHLORIDE
mf: $C_5H_{14}NO•Cl$ mw: 139.65

PROP: Colorless to white, deliquescent, hygroscopic crystals; slt odor of trimethylamine. Sol in water and alc.

SYNS: BIOCOLINA □ CHLORIDE de CHOLINE (FRENCH) □ CHOLINE CHLORHYDRATE □ CHOLINE CHLORIDE (FCC) □ CHOLINIUM CHLORIDE □ HEPACHOLINE □ (2-HYDROXYETHYL)TRIMETHYLAMMONIUM CHLORIDE □ LIPOTRIL

TOXICITY DATA WITH REFERENCE
cyt-ham:ovr 500 µg/L ENMUDM 7,1,85
sce-ham:ovr 500 µg/L ENMUDM 7,1,85
orl-rat LD50:3400 mg/kg PSEBAA 58,87,45
ipr-rat LD50:400 mg/kg TXAPA9 12,486,68
orl-mus LD50:3900 mg/kg ARZNAD 33,1016,83
ipr-mus LD50:320 mg/kg PSEBAA 51,281,42
scu-mus LDLo:735 mg/kg JPETAB 6,477,14/15
ivn-mus LD50:53 mg/kg ARZNAD 33,1016,83
ivn-dog LDLo:5 mg/kg HBAMAK 4,1289,35
ivn-cat LDLo:25 mg/kg HBAMAK 4,1289,35
ipr-rbt LDLo:500 mg/kg JIDIAQ 42,473,28
scu-rbt LDLo:1 g/kg PSEBAA 51,281,42
ivn-rbt LDLo:1100 µg/kg PSEBAA 51,281,42

CONSENSUS REPORTS: Reported in EPA TSCA Inventory.

SAFETY PROFILE: Poison by intraperitoneal and intravenous routes. Moderately toxic experimentally by ingestion and subcutaneous routes. Mutation data reported. A lipotropic agent which induces the reduction in fats contained in the liver. When heated to decomposition it emits toxic fumes of Cl^-, SO_x, and NO_x. See also CHOLINE.

CMF800 **CAS:123-41-1** **HR: 3**
CHOLINE HYDROXIDE
mf: $C_5H_{14}NO{\cdot}HO$ mw: 121.21

SYNS: BURSINE □ FAGINE □ GOSSYPINE □ LURIDINE □ SINCALINE □ SINKALIN □ SINKALINE □ VIDINE

TOXICITY DATA WITH REFERENCE
ivn-mus LD50:21,400 μg/kg THERAP 23,1357,68

CONSENSUS REPORTS: Reported in EPA TSCA Inventory.

SAFETY PROFILE: Poison by intravenous route. When heated to decomposition it emits toxic vapors of NO_x.

CMG000 **CAS:2016-36-6** **HR: 2**
CHOLINE SALICYLATE
mf: $C_{12}H_{19}NO_4$ mw: 241.32

PROP: Very hygroscopic. Mp: 49.5–50°.

SYNS: ACTASAL □ ARRET □ ARTHROPAN □ ARTROBIONE □ CHOLINE SALICYLATE B □ CHOLINE, SALICYLATE (SALT) □ CHOLINE SALICYLIC ACID SALT □ (2-HYDROXYETHYL)TRIMETHYLAMMONIUM SALICYLATE □ 2-HYDROXY-N,N,N-TRIMETHYLETHANAMINIUM SALT with 2-HYDROXYBENZOIC ACID (1:1) □ MUNDISAL □ SALICOL □ SALICYLIC ACID CHOLINE SALT □ SYRAP

TOXICITY DATA WITH REFERENCE
orl-mus LD50:2690 NIIRDN 6,291,82
ipr-mus LD50:410 mg/kg NIIRDN 6,291,82
scu-mus LD50:1 g/kg NIIRDN 6,291,82

CONSENSUS REPORTS: Reported in EPA TSCA Inventory.

SAFETY PROFILE: Moderately toxic by ingestion, intraperitoneal, and subcutaneous routes. An analgesic and antipyretic. When heated to decomposition it emits toxic fumes of NO_x and NH_3. See also CHOLINE.

CMG250 **CAS:306-40-1** **HR: 3**
CHOLINE SUCCINATE (2:1) (ESTER)
mf: $C_{14}H_{30}N_2O_4$ mw: 290.46

SYNS: ANECTINE □ CHOLINE SUCCINATE (ester) □ DIACETYLCHOLINE □ DICHOLINE SUCCINATE □ 2,2'-((1,4-DIOXO-1,4-BUTANEDIYL)BIS(OXY))BIS(N,N,N-TRIMETHYLETHANAMINIUM) □ DITILIN □ DITILINE □ QUELICIN □ SUCCINIC ACID DIESTER with CHOLINE □ SUCCINOCHOLINE □ SUCCINOYLCHOLINE □ SUCCINYLBISCHOLINE □ SUCCINYLDICHOLINE □ SUXAMETHONIUM □ SUXEMETHONIUM

TOXICITY DATA WITH REFERENCE
ivn-hmn TDLo:1430 μg/kg:PUL ANATAE 21,27,72
orl-mus LD50:125 mg/kg 27ZIAQ -,-,65
ipr-mus LD50:2140 μg/kg AIPTAK 152,277,64
scu-mus LD50:7500 μg/kg ARZNAD 15,126,65
ivn-mus LD50:280 μg/kg RCOCB8 1,141,70
ivn-dog LDLo:300 μg/kg AIPTAK 88,1,51
ivn-rbt LD50:800 μg/kg AIPTAK 88,1,51

SAFETY PROFILE: Poison by ingestion, intraperitoneal, intravenous,and subcutaneous routes. Human systemic effects by intravenous route: changes in the trachea or bronchi. When heated to decomposition it emits toxic fumes of NO_x.

CMG700 **HR: 3**
CHRISTMAS ROSE

PROP: An evergreen perennial herb that grows to 2 feet. The white or pink-white flower has 5 petals and is 2 to 3 inches across. It is native to Europe and grows wild in the northern United States and Canada.

SYNS: HELLEBORE □ HELLEBORUS NIGER

SAFETY PROFILE: The whole plant contains the poisons hellebrin, helleborin, and helleborein (cardiac glycosides), and the direct irritants saponin and protoanemonin. Ingestion may cause mouth and abdominal pain, nausea, vomiting, and diarrhea. Cardiac glycosides may cause death due to their effect on heart function. See also DIGITALIS and SAPONIN.

CMG800 **CAS:15005-90-0** **HR: 3**
CHROMALUM HEXAHYDRATE
mf: $Cr_2O_{12}S_3{\cdot}6H_2O$ mw: 500.30

SYN: CHROMIUM(III) SULFATE, HEXAHYDRATE (2:3:6)

TOXICITY DATA WITH REFERENCE
ivn-rat LDLo:144 mg/kg AJPHAP 209,489,65

OSHA PEL: TWA 0.5 mg(Cr)/m^3
ACGIH TLV: TWA 0.5 mg(Cr)/m^3; Not Classifiable as a Carcinogen

SAFETY PROFILE: Poison by intravenous route. When heated to decomposition it emits toxic fumes of SO_x and Cr^-.

CMG850 **CAS:7788-99-0** **HR: 2**
CHROME ALUM (DODECAHYDRATE)
mf: $CrKO_8S_2{\cdot}12H_2O$ mw: 499.41

PROP: Deep purple crystals. Mp: 89°. Sol in H_2O; virtually insol in EtOH.

SYNS: CHROME ALUM □ POTASSIUM CHROMIUM ALUM □ SULFURIC ACID, CHROMIUM(3+)POTASSIUM SALT(2:1:1), DODECAHYDRATE

TOXICITY DATA WITH REFERENCE
dnr-esc 125 μg/well MUREAV 133,161,84
scu-rat TDLo:135 mg/kg:ETA PBPHAW 14,47,78
ivn-rat LD50:112 mg/kg EQSFAP 1,1,75

OSHA PEL: TWA 0.5 mg(Cr)/m^3
ACGIH TLV: TWA 0.5 mg(Cr)/m^3; Not Classifiable as a Carcinogen

SAFETY PROFILE: Poison by intravenous route. Questionable carcinogen with experimental tumorigenic data. Mutation data reported. When heated to decomposition it emits toxic fumes of Cr^-.

CMH000 CAS:1066-30-4 HR: 3
CHROMIC ACETATE
mf: $C_6H_9O_6$•Cr mw: 229.15

PROP: Gray-green powder or bluish-green pasty mass.

SYNS: CHROMIC ACETATE(III) □ CHROMIUM ACETATE □ CHROMIUM(III) ACETATE □ CHROMIUM TRIACETATE

TOXICITY DATA WITH REFERENCE
mmo-esc 16 mmol/L MUREAV 58,175,78
cyt-hmn:leu 16 µmol/L MUREAV 58,175,78
imp-rat TDLo:1000 mg/kg/56W-I:ETA AEHLAU 5,445,62
ivn-mus LDLo:2290 mg/kg EQSSDX 1,1,75
ivn-rbt LDLo:1604 mg/kg EQSSDX 1,1,75
ivn-frg LDLo:6185 mg/kg AIPTAK 62,330,39

CONSENSUS REPORTS: NTP 7th Annual Report on Carcinogens. IARC Cancer Review: Group 3 IMEMDT 7,165,87; Animal Inadequate Evidence IMEMDT 2,100,73; IMEMDT 23,205,80. Chromium and its compounds are on the Community Right-To-Know List. Reported in EPA TSCA Inventory.

OSHA PEL: TWA 0.5 mg(Cr)/m^3
ACGIH TLV: TWA 0.5 mg(Cr)/m^3; Not Classifiable as a Carcinogen

SAFETY PROFILE: Confirmed carcinogen with experimental tumorigenic data. Moderately toxic by intravenous route. Human mutation data reported. See also CHROMIUM COMPOUNDS. When heated to decomposition it emits acrid smoke and irritating fumes.

CMH250 CAS:7738-94-5 HR: 3
CHROMIC(VI) ACID
mf: CrH_2O_4 mw: 118.02

PROP: Found in solution.

SYNS: ACIDE CHROMIQUE (FRENCH) □ CHROMIC ACID

TOXICITY DATA WITH REFERENCE
mmo-sat 80 µg/plate MUREAV 54,139,78
dnr-smc 1200 nmol/L CNJGA8 24,771,82
dnr-ssp 1200 nmol/L CNJGA8 24,771,82
scu-dog LDLo:320 mg/kg EQSSDX 1,1,75

CONSENSUS REPORTS: Reported in EPA TSCA Inventory. Chromium and its compounds are on the Community Right-To-Know List.

OSHA PEL: CL 0.1 mg(CrO_3)/m^3
ACGIH TLV: TWA 0.05 mg(Cr)/m^3, Confirmed Human Carcinogen.
DFG MAK: Animal Carcinogen, Suspected Human Carcinogen.
NIOSH REL: (Chromium(VI)) TWA 0.025 mg(Cr(VI))/m^3; CL 0.05/15M

SAFETY PROFILE: Confirmed human carcinogen. Poison by subcutaneous route. Mutation data reported. A powerful oxidizer. A powerful irritant of skin, eyes, and mucous membranes. Can cause a dermatitis, bronchoasthma, "chrome holes," damage to the eyes. Dangerously reactive. Incompatible with acetic acid, acetic anhydride, tetrahydronaphthalene, acetone, alcohols, alkali metals, ammonia, arsenic, bromine penta fluoride, butyric acid, n,n-dimethylformamide, hydrogen sulfide, peroxyformic acid, phosphorus, potassium hexacyanoferrate, pyridine, selenium, sodium, sulfur and many other materials. See also CHROMIUM COMPOUNDS.

For occupational chemical analysis use NIOSH: Chromium Hexavalent 7024.

CMH260 CAS:1308-14-1 HR: 3
CHROMIC(III) ACID
mf: CrH_3O_3 mw: 103.03

PROP: Green flocculent solid or crystals. Readily hydrolyzed. Practically insol in H_2O; sol in mineral acids.

SYNS: CHROMIC (III) HYDROXIDE □ CHROMIUM(III) HYDROXIDE □ CHROMIUM TRIHYDROXIDE

CONSENSUS REPORTS: Reported in EPA TSCA Inventory.

ACGIH TLV: TWA 0.05 mg(Cr)/m^3; Confirmed Human Carcinogen.

CONSENSUS REPORTS: IARC Cancer Review: Group 3 IMEMDT 49,49,90; Human Inadequate Evidence IMEMDT 49,49,90; Animal Inadequate Evidence IMEMDT 49,49,90

OSHA PEL: CL 0.1 mg(CrO_3)/m^3
ACGIH TLV: TWA 0.05 mg(Cr)/m^3; Confirmed Human Carcinogen.
NIOSH REL: (Chromium(VI)) TWA 0.001 mg(Cr)/m^3

SAFETY PROFILE: A confirmed carcinogen. A poison. A powerful oxidizer. A powerful irritant of skin, eyes, and mucous membranes.

For occupational chemical analysis use NIOSH: Chromium Hexavalent 7024.

CMH300 CAS:15242-96-3 HR: 3
CHROMIC CHLORIDE STEARATE
mf: $C_{18}H_{36}Cl_4Cr_2O_3$ mw: 546.34

SYNS: CHROMIUM, TETRACHLORO-µ-HYDROXY(µ-(OCTADECANOATO-O: O'))DI- □ CHROMIUM, TETRACHLORO-µ-HYDROXY(µ-STEARATO)DI- □ KHROMOLAN □ NCI-C60800 □ QUILON S □ STEARATE CHROMIC CHLORIDE □ STEARATO CHROMIC CHLORIDE □ STEARATO-CHROMIC CHLORIDE COMPLEX □ STEARATO-CHROMIUM CHLORIDE

TOXICITY DATA WITH REFERENCE
mmo-sat 333 µg/plate EMMUEG 11(Suppl 12),1,88
orl-mus LD50:1280 mg/kg ESKHA5 (103),37,85
skn-mus LD50:>2500 mg/kg ESKHA5 (103),37,85
ivn-mus LD50:180 mg/kg CSLNX* NX#03305

CONSENSUS REPORTS: Reported in EPA TSCA Inventory.

SAFETY PROFILE: A poison by intravenous route. Moderately toxic by ingestion. Mutation data reported. When heated to decomposition it emits toxic vapors of Cr and Cl^-.

CMI250 **CAS:24613-89-6** **HR: 3**
CHROMIC CHROMATE
mf: Cr_3O_{12}•2Cr mw: 452.00

SYNS: CHROMIC ACID, CHROMIUM(3+) SALT (3:2) □ CHROMIUM CHROMATE (MAK)

TOXICITY DATA WITH REFERENCE
imp-rat TDLo:112 mg/kg:NEO AIHAAP 20,274,59

CONSENSUS REPORTS: IARC Cancer Review: Animal Sufficient Evidence IMEMDT 2,100,73. Reported in EPA TSCA Inventory. Chromium and its compounds are on the Community Right-To-Know List.

OSHA PEL: CL 0.1 mg(CrO_3)/m³
ACGIH TLV: TWA 0.05 mg(Cr)/m³; Confirmed Human Carcinogen.
DFG MAK: Animal Carcinogen, Suspected Human Carcinogen.
NIOSH REL: (Chromium(VI)) TWA 0.001 mg(Cr(VI))/m³

SAFETY PROFILE: Confirmed carcinogen with experimental carcinogenic and neoplastigenic data. Very powerful oxidizer. See also CHROMIUM COMPOUNDS.

For occupational chemical analysis use NIOSH: Chromium Hexavalent 7024.

CMI300 **CAS:13537-21-8** **HR: 3**
CHROMIC PERCHLORATE
mf: Cl_3O_{12}•Cr mw: 350.35

SYNS: CHROMIUM PERCHLORATE □ CHROMIUM TRIPERCHLORATE □ PERCHLORIC ACID, CHROMIUM(3+) SALT

CONSENSUS REPORTS: IARC Cancer Review: Group 3 IMEMDT 49,49,90; Animal Inadequate Evidence IMEMDT 49,49,90; Human Inadequate Evidence IMEMDT 49,49,90.

OSHA PEL: TWA 0.5 mg(Cr)/m³
ACGIH TLV: TWA 0.5 mg(Cr)/m³; Not Classifiable as a Carcinogen.
DOT CLASSIFICATION: 5.1; *Label:* Oxidizer

SAFETY PROFILE: A toxic and reactive oxidizing solid. When heated to decomposition it emits toxic vapors of Cr.

CMI500 **CAS:1308-31-2** **HR: 3**
CHROMITE (mineral)
mf: Cr_2FeO_4 mw: 223.85

PROP: Black or brown-black cubic crystals. Relatively insol in acids.

SYNS: CHROME ORE □ CHROMITE □ CHROMITE ORE □ IRON CHROMITE

TOXICITY DATA WITH REFERENCE
mma-sat 2 mg/plate CRNGDP 3,1331,82
cyt-hmn:oth 500 mg/L BJCAAI 44,219,81
sce-ham:ovr 10 mg/L CRNGDP 3,1331,82

CONSENSUS REPORTS: NTP 7th Annual Report on Carcinogens. IARC Cancer Review: Group 3 IMEMDT 7,165,87; Animal Inadequate Evidence IMEMDT 23,205,80. Chromium and its compounds are on the Community Right-To-Know List.

OSHA PEL: TWA 0.5 mg(Cr)/m³
ACGIH TLV: TWA 0.05 mg/m³ (ore processing); Confirmed Human Carcinogen (ore processing)

SAFETY PROFILE: Confirmed human carcinogen during ore processing. Human mutation data reported. See also CHROMIUM COMPOUNDS and IRON.

CMI750 **CAS:7440-47-3** **HR: 3**
CHROMIUM
mf: Cr mw: 52.00

PROP: Hard ductile blue-white metal. Resists oxidation in air. Bp: 26° @ 2690 mm. More reactive to acids than Mo or W and can be rendered passive. Rapidly attacked by fused NaOH + KNO_3 or $KClO_4$.

SYNS: CHROME □ CHROMIUM METAL (OSHA)

TOXICITY DATA WITH REFERENCE
ivn-rat TDLo:2160 μg/kg/6W-I:ETA JNCIAM 16,447,55
imp-rat TDLo:1200 μg/kg/6W-I:ETA JNCIAM 16,447,55
imp-rbt TDLo:75 mg/kg:ETA ZEKBAI 52,425,42

CONSENSUS REPORTS: NTP 7th Annual Report on Carcinogens. IARC Cancer Review: Group 3 IMEMDT 7,165,87; Animal Inadequate Evidence IMEMDT 23,205,80. Chromium and its compounds are on the Community Right-To-Know List. Reported in EPA TSCA Inventory.

OSHA PEL: TWA 1 mg/m³
ACGIH TLV: TWA 0.5 (Cr)mg/m³; Not Classifiable As Carcinogen

SAFETY PROFILE: Confirmed human carcinogen with experimental tumorigenic data. Powder will explode spontaneously in air. Ignites and is potentially explosive in atmospheres of carbon dioxide. Violent or explosive reaction when heated with ammonium nitrate. May ignite or react violently with bromine pentafluoride. Incandescent reaction with nitrogen oxide or sulfur dioxide. Incompatible with oxidants. See also CHROMIUM COMPOUNDS.

For occupational chemical analysis use OSHA: #ID-125G or NIOSH: Chromium, 7024; Welding and Brazing Fume, 7200; Elements, 7300.

CMJ000 **CAS:628-52-4** **HR: 3**
CHROMIUM ACETATE HYDRATE
mf: $C_4H_6O_4$•Cr•H_2O mw: 188.12

PROP: Air-sensitive dark red crystals. Stable in air for short period. Sltly sol in H_2O and EtOH; sol in hot H_2O.

SYNS: ACETIC ACID, CHROMIUM (2+) SALT (8CI, 9CI) □ CHROMIUM(2+) ACETATE □ CHROMIUM(II) ACETATE □ CHROMIUM DIACETATE □ CHROMOUS ACETATE □ CHROMOUS ACETATE MONOHYDRATE

TOXICITY DATA WITH REFERENCE
orl-rat LD50:11,260 mg/kg AIHAAP 30,470,69

CONSENSUS REPORTS: Reported in EPA TSCA

Inventory. Chromium and its compounds are on the Community Right-To-Know List.

OSHA PEL: TWA 0.5 mg(Cr)/m^3
ACGIH TLV: TWA 0.5 mg(Cr)/m^3; Not Classifiable as a Carcinogen.

SAFETY PROFILE: Mildly toxic by ingestion. The anhydrous acetate ignites spontaneously in air. See also CHROMIUM COMPOUNDS. When heated to decomposition it emits acrid smoke and irritating fumes.

CMJ100 **CAS:29689-14-3** **HR: 3**
CHROMIUM CARBONATE
mf: $CH_2O_3 \cdot xCr$ mw: 426.03

SYNS: BASIC CHROMIUM CARBONATE □ CARBONIC ACID, CHROMIUM SALT

CONSENSUS REPORTS: NTP 7th Annual Report on Carcinogens. Reported in EPA TSCA Inventory.

SAFETY PROFILE: Confirmed carcinogen. When heated to decomposition it emits toxic fumes of Cr.

CMJ250 **CAS:10025-73-7** **HR: 3**
CHROMIUM CHLORIDE
mf: Cl_3Cr mw: 158.35

PROP: Red-violet flaky crystals. Mp: 1152°, bp: 1300° (subl). Insol in cold H_2O; sltly sol in hot H_2O.

SYNS: CHROMIC CHLORIDE □ CHROMIUM(III) CHLORIDE (1:3) □ CHROMIUM CHLORIDE, anhydrous □ CHROMIUM TRICHLORIDE □ C.I. 77295 □ PURATRONIC CHROMIUM CHLORIDE □ TRICHLOROCHROMIUM

TOXICITY DATA WITH REFERENCE
cyt-hmn:oth 500 mg/L BJCAAI 44,219,81
sce-ham:lng 39 mg/L CRNGDP 4,605,83
ipr-mus TDLo:59,500 µg/kg (female 8D post):TER JTSCDR 1(2),1,76
ipr-mus TDLo:30 mg/kg (female 8D post):TER JTSCDR 1(2),1,76
ipr-mus TDLo:59,500 µg/kg (female 9D post):REP JTSCDR 1(2),1,76
itt-rat TDLo:12,668 µg/kg (male 1D pre):REP JRPFA4 7,21,64
ipr-mus TDLo:44,600 µg/kg (female 8D post):TER JTSCDR 1(2),1,76
orl-rat LD50:1870 mg/kg YAKUD5 22,291,80
ihl-mus LC50:31,500 µg/m^3/2H 85GMAT -,39,82
ipr-mus LD50:434 mg/kg COREAF 256,1043,63
skn-gpg LDLo:202 mg/kg AEHLAU 11,201,65*
ipr-gpg LDLo:200 mg/kg AEHLAU 11,201,65

CONSENSUS REPORTS: IARC Cancer Review: Group 3; IMEMDT 49,49,90; Human Inadequate Evidence IMEMDT 49,49,90; Animal Inadequate Evidence IMEMDT 49,49,90. Reported in EPA TSCA Inventory. EPA Genetic Toxicology Program. Chromium and its compounds are on the Community Right-To-Know List. EPA Extremely Hazardous Substances List.

OSHA PEL: TWA 0.5 mg(Cr)/m^3
ACGIH TLV: TWA 0.5 mg(Cr)/m^3; Not Classifiable as a Carcinogen

SAFETY PROFILE: Poison by skin contact, inhalation, and intraperitoneal routes. Experimental teratogenic and reproductive effects. Human mutation data reported. Reacts violently with lithium under nitrogen atmosphere. When heated to decomposition it emits toxic fumes of Cl^-.

CMJ300 **CAS:10049-05-5** **HR: 2**
CHROMIUM(II) CHLORIDE (1:2)
mf: Cl_2Cr mw: 122.90

SYNS: CHROMIUM CHLORIDE □ CHROMIUM(II) CHLORIDE □ CHROMIUM DICHLORIDE □ CHROMOUS CHLORIDE

TOXICITY DATA WITH REFERENCE
mmo-sat 1 mg/L BECTA6 40,597,88
pic-esc 33,300 ng/well MUREAV 260,349,91
orl-rat LD50:1870 mg/kg AIHAAP 30,470,69

CONSENSUS REPORTS: Reported in EPA TSCA Inventory.

OSHA PEL: 8H TWA 0.5 mg(Cr)/m^3
ACGIH TLV: TWA 0.5 mg(Cr)/m^3

SAFETY PROFILE: Moderately toxic by ingestion. Mutation data reported. When heated to decomposition it emits toxic vapors of Cr.

CMJ500 **HR: 3**
CHROMIUM COMPOUNDS

CONSENSUS REPORTS: Chromium and its compounds are on the Community Right-To-Know List.

SAFETY PROFILE: Chromate salts are suspected human carcinogens producing tumors of the lungs, nasal cavity, and paranasal sinus. Chromic acid and its salts have a corrosive action on the skin and mucous membranes. The lesions are confined to the exposed parts, affecting chiefly the skin of the hands and forearms and the mucous membranes of the nasal septum. The characteristic lesion is a deep, penetrating ulcer, which, for the most part, does not tend to suppurate, and which is slow in healing. Small ulcers, about the size of a matchhead, may be found, chiefly around the base of the nails, on the knuckles, dorsum of the hands and forearms. These ulcers tend to be clean and progress slowly. They are frequently painless, even though quite deep. They heal slowly and leave scars. On the mucous membranes of the nasal septum, the ulcers are usually accompanied by purulent discharge and crusting. If exposure continues, perforation of the nasal septum may result but produces no deformity of the nose. Hexavalent compounds are more toxic than the trivalent. Eczematous dermatitis due to trivalent chromium compounds has been reported.

CMJ530 **CAS:7788-97-8** **HR: 3**
CHROMIUM(III) FLUORIDE
DOT: UN 1756/UN 1757
mf: CrF_3 mw: 109.00

SYNS: CHROME FLUORURE □ CHROMIC FLUORIDE □ CHROMIC FLUORIDE, solid (UN1756) (DOT) □ CHROMIC FLUORIDE, solution (UN1757) (DOT) □ CHROMIC TRIFLUORIDE □ CHROMIUM TRIFLUORIDE

TOXICITY DATA WITH REFERENCE
orl-gpg LDLo:150 mg/kg YAKUD5 22,291,80
scu-frg LDLo:420 mg/kg CRSBAW 124,133,37 85INA8 5,139,86

CONSENSUS REPORTS: Reported in EPA TSCA Inventory.

OSHA PEL: 8H TWA 0.5 mg(Cr)/m³; TWA 2.5 mg(F)/m³
ACGIH TLV: TWA 0.5 mg(Cr)/m³; TWA 2.5 mg(F)/m³
NIOSH REL: (Fluorides, Inorganic): TWA 2.5 mg(F)/m³
DOT CLASSIFICATION: 8; *Label:* Corrosive

SAFETY PROFILE: A poison by ingestion and subcutaneous routes. A corrosive. When heated to decomposition it emits toxic vapors of CR and F^-.

CMJ560 CAS:7788-97-8 HR: 3
CHROMIUM(III) FLUORIDE
DOT: UN 1756/UN 1757
mf: CrF_3 mw: 109.00

SYNS: CHROME FLUORURE ☐ CHROMIC FLUORIDE ☐ CHROMIC FLUORIDE, solid (UN1756) (DOT) ☐ CHROMIC FLUORIDE, solution (UN1757) (DOT) ☐ CHROMIC TRIFLUORIDE ☐ CHROMIUM TRIFLUORIDE

TOXICITY DATA WITH REFERENCE
orl gpg LDLo:150 mg/kg YAKUD5 22,291,80
scu-frg LDLo:420 mg/kg CRSBAW 124,133,37

CONSENSUS REPORTS: Reported in EPA TSCA Inventory.

OSHA PEL: 8H TWA 0.5 mg(Cr)/m³; TWA 2.5 mg(F)/m³
ACGIH TLV: TWA 0.5 mg(Cr)/m³; TWA 2.5 mg(F)/m³
NIOSH REL: (Fluorides, Inorganic): TWA 2.5 mg(F)/m³
DOT CLASSIFICATION: 8; *Label:* Corrosive

SAFETY PROFILE: A poison by ingestion. A corrosive. When heated to decomposition it emits toxic vapors of Cr and F^-.

CMJ600 CAS:13548-38-4 HR: 3
CHROMIUM(III) NITRATE
DOT: UN 2720
mf: CrN_3O_9 mw: 238.03

PROP: Very deliquescent, pale green powder. Nonvolatile. Sol in H_2O, EtOAc, MeCN, and DMSO; insol in C_6H_6, CCl_4, and $CHCl_3$.

SYNS: CHROMIC NITRATE ☐ CHROMIUM NITRATE ☐ CHROMIUM (3+) NITRATE ☐ CHROMIUM NITRATE (DOT) ☐ CHROMIUM TRINITRATE ☐ NITRIC ACID, CHROMIUM (3+) SALT

TOXICITY DATA WITH REFERENCE
dnr-bcs 160 mmol/L MUREAV 58,175,78
orl-rat LD50:3250 mg/kg YAKUD5 22,291,80
orl-mus LD50:2976 mg/kg SAIGBL 20,590,78
ipr-mus LD50:110 mg/kg SAIGBL 20,590,78
scu-mus LD50:3232 mg/kg SAIGBL 20,590,78

CONSENSUS REPORTS: IARC Cancer Review: Group 3 IMEMDT 49,49,90; Human Inadequate Evidence IMEMDT 49,49,90; Animal Inadequate Evidence IMEMDT 49,49,90. Reported in EPA TSCA Inventory.

OSHA PEL: TWA 0.5 mg(Cr)/m³
ACGIH TLV: TWA 0.5 mg(Cr)/m³; Not Classifiable as a Carcinogen.
DOT CLASSIFICATION: 5.1; *Label:* Oxidizer

SAFETY PROFILE: Poison by intraperitoneal route. Moderately toxic by subcutaneous and ingestion routes. Mutation data reported. Questionable carcinogen. When heated to decomposition it emits toxic fumes of NO_x and Cr.

CMJ850 CAS:24094-93-7 HR: 3
CHROMIUM NITRIDE
mf: CrN mw: 66.00

PROP: Dark gray powder; bronze yellow crystals with metallic lustre. Sol in conc HCl.

CONSENSUS REPORTS: Chromium compounds are on the Community Right-To-Know List.

SAFETY PROFILE: Mixture with potassium nitrate ignites when heated. When heated to decomposition it emits toxic fumes of NO_x. See also CHROMIUM COMPOUNDS and NITRIDES.

CMJ900 CAS:1308-38-9 HR: 3
CHROMIUM(III) OXIDE (2:3)
mf: Cr_2O_3 mw: 152.00

PROP: Green crystals. Mp: 2275°.

SYNS: ANADOMIS GREEN ☐ ANIDRIDE CROMIQUE (FRENCH) ☐ CASALIS GREEN ☐ CHROME GREEN ☐ CHROME OCHER ☐ CHROME OXIDE ☐ CHROME OXIDE GREEN ☐ CHROMIA ☐ CHROMIC ACID ☐ CHROMIC ACID GREEN ☐ CHROMIC OXIDE ☐ CHROMIUM OXIDE ☐ CHROMIUM(III) OXIDE ☐ CHROMIUM(3+) OXIDE ☐ CHROMIUM SESQUIOXIDE ☐ CHROMIUM(3+) TRIOXIDE ☐ C.I. 77288 ☐ C.I. No. 77278 ☐ C.I. PIGMENT GREEN 17 ☐ DICHROMIUM TRIOXIDE ☐ 11661 GREEN ☐ GREEN CHROME OXIDE ☐ GREEN CHROMIC OXIDE ☐ GREEN CINNABAR ☐ GREEN ROUGE ☐ GUIGNER'S GREEN ☐ LEAF GREEN ☐ LEVANOX GREEN GA ☐ OIL GREEN ☐ OXIDE of CHROMIUM ☐ ULTRAMARINE GREEN

TOXICITY DATA WITH REFERENCE
mmo-sat 1 mmol/L TOLED5 8,195,81
dnr-sat 50 mmol/L TOLED5 7,439,81
dnd-esc 5 mmol/L CNREA8 40,2455,80
sce-ham:lng 34 mg/L CRNGDP 4,605,83
ipr-rat TDLo:90 mg/kg:ETA VOONAW 13(11),57,67
ipl-rat TDLo:45 mg/kg:ETA VOONAW 13(11),57,67
itr-rat TDLo:90 mg/kg:ETA VOONAW 13(11),57,67

CONSENSUS REPORTS: NTP 7th Annual Report on Carcinogens. IARC Cancer Review: Group 3 IMEMDT 7,165,87; Animal Inadequate Evidence IMEMDT 23,205,80. Reported in EPA TSCA Inventory. Chromium and its compounds are on the Community Right-To-Know List.

OSHA PEL: TWA 0.5 mg(Cr)/m³
ACGIH TLV: TWA 0.5 mg(Cr)/m³; Not Classifiable as a Carcinogen.
DFG MAK: Suspected Carcinogen.

SAFETY PROFILE: Confirmed carcinogen with experimental tumorigenic data. Mutation data reported. Probably a severe eye, skin, and mucous membrane irritant. A powerful oxidizer. Reacts violently with CLF_3. See also CHROMIUM COMPOUNDS.

For occupational chemical analysis use NIOSH: Chromium, 7024; Welding and Brazing Fume, 7200; Elements, 7300.

CMK000 **CAS:1333-82-0** **HR: 3**
CHROMIUM(VI) OXIDE (1:3)
DOT: UN 1463/NA 1463/UN 1755
mf: CrO_3 mw: 100.00

PROP: Dark orange-red, rhombic, deliquescent crystals. D: 2.70, mp: 190°, bp: decomp, sol: 61.7 g/100 cc @ 0°, 67.45 g/100 cc @ 100°. Very sol in H_2O; sol in H_2SO_4, and org solvs.

SYNS: ANHYDRIDE CHROMIQUE (FRENCH) □ ANIDRIDE CROMICA (ITALIAN) □ CHROME (TRIOXYDE de) (FRENCH) □ CHROMIC ACID □ CHROMIC ACID, solution (UN 1755) (DOT) □ CHROMIC ACID, solid (NA 1463) (DOT) □ CHROMIC ANHYDRIDE □ CHROMIC TRIOXIDE □ CHROMIUM TRIOXIDE, anhydrous (UN 1463) (DOT) □ CHROMIC(VI) ACID □ CHROMIUM OXIDE □ CHROMIUM(VI) OXIDE □ CHROMIUM TRIOXIDE □ CHROMIUM(6+) TRIOXIDE □ CHROMIUM TRIOXIDE, anhydrous (DOT) □ CHROMO (TRIOSSIDO di) (ITALIAN) □ CHROMSAEUREANHYDRID (GERMAN) □ CHROMTRIOXID (GERMAN) □ CHROOMTRIOXYDE (DUTCH) □ CHROOMZUURANHYDRIDE (DUTCH) □ MONOCHROMIUM OXIDE) □ MONOCHROMIUM TRIOXIDE □ PURATRONIC CHROMIUM TRIOXIDE

TOXICITY DATA WITH REFERENCE
mmo-sat 1 mmol/L TOLED5 8,195,81
cyt-hmn:leu 2 mg/L MUREAV 58,175,78
scu-mus TDLo:20 mg/kg (8D preg):TER SEIJBO 19,171,79
ivn-ham TDLo:7500 µg/kg (female 8D post):REP ENVRAL 16,101,78
ivn-ham TDLo:8 mg/kg (female 8D post):TER ANREAK 199,89A,81
ivn-ham TDLo:5 mg/kg (female 8D post):TER ENVRAL 16,101,78
ivn-ham TDLo:7500 µg/kg (female 8D post):TER ENVRAL 16,101,78
ihl-hmn TCLo:110 µg/m³/3Y-C:CAR AGGHAR 13,528,55
imp-rat TDLo:125 mg/kg:CAR AIHAAP 20,274,59
ihl-mus TCLo:3480 µg/m³/2H/1Y-I:ETA SAIGBL 29,17,87
ihl-hmn TCLo:110 µg/m³ YAKUD5 22,291,80
orl-rat LD50:80 mg/kg TRENAF 27(2),119,76
orl-mus LD50:127 mg/kg CHYCDW 14,86,80
ipr-mus LD50:14 mg/kg NEZAAQ 34,193,79
scu-mus LDLo:20 mg/kg SEIJBO 19,171,79

CONSENSUS REPORTS: NTP 7th Annual Report on Carcinogens. IARC Cancer Review: Group 1 IMEMDT 7,165,87; Animal Sufficient Evidence IMEMDT 23,205,80. EPA Genetic Toxicology Program. Chromium and its compounds are on the Community Right-To-Know List. Reported in EPA TSCA Inventory.

OSHA PEL: CL 0.1 mg(CrO_3)/m³
ACGIH TLV: TWA 0.05 mg(Cr)/m³; Confirmed Human Carcinogen
DFG MAK: 0.1 mg/m³, Suspected Carcinogen.
NIOSH REL: (Chromium(VI)) TWA 0.025 mg(Cr(VI))/m³; CL 0.05/15M
DOT CLASSIFICATION: 5.1; *Label:* Oxidizer, Corrosive (NA 1463, UN 1463); DOT Class: 8; *Label:* Corrosive (UN 1755)

SAFETY PROFILE: Confirmed human carcinogen producing nasal and lung tumors. Experimental carcinogenic and tumorigenic data. Poison by ingestion, intraperitoneal, and subcutaneous routes. Experimental teratogenic and reproductive effects. Human mutation data reported. Corrosive. Probably a severe eye, skin, and mucous membrane irritant. See also CHROMIUM COMPOUNDS.

A powerful oxidizer. Explosive reaction with acetaldehyde, acetic acid + heat, acetic anhydride + heat, benzaldehyde, benzene, benzylthylaniline, butyraldehyde, 1,3-dimethylhexahydropyrimidone, diethyl ether, ethylacetate, isopropylacetate, methyl dioxane, pelargonic acid, pentyl acetate, phosphorus + heat, propionaldehyde and other organic materials or solvents. Forms a friction- and heat-sensitive explosive mixture with potassium hexacyanoferrate. Ignites on contact with alcohols, acetic anhydride + tetrahydronaphthalene, acetone, butanol, chromium(II) sulfide, cyclohexanol, dimethyl formamide, ethanol, ethylene glycol, methanol, 2-propanol, pyridine. Violent reaction with acetic anhydride + 3-methylphenol (above 75°C), acetylene, bromine pentafluoride, glycerol, hexamethylphosphoramide, peroxyformic acid, selenium, sodium amide. Incandescent reaction with alkali metals (e.g., sodium, potassium), ammonia, arsenic, butyric acid (above 100°C), chlorine trifluoride, hydrogen sulfide + heat, sodium + heat, and sulfur. Incompatible with N,N-dimethylformamide.

For occupational chemical analysis use NIOSH: Chromium, Hexavalent, 7600.

CMK275 **CAS:14884-42-5** **HR: 3**
CHROMIUM PENTAFLUORIDE
mf: CrF_5 mw: 146.99

CONSENSUS REPORTS: Chromium compounds are on the Community Right-To-Know List.

PROP: Volatile crimson solid. Readily hydrol. Mp: 30°, bp: 117° (decomp).

SAFETY PROFILE: Undergoes violent redox and halogen exchange reactions. Mixtures with phosphorus trichloride react violently on slight heating. When heated to decomposition it emits toxic fumes of F^-. See also CHROMIUM COMPOUNDS.

CMK300 **CAS:7789-04-0** **HR: 3**
CHROMIUM PHOSPHATE
mf: $Cr{\cdot}H_3O_4P$ mw: 150.00

SYNS: ARNAUDON'S GREEN □ ARNAUDON'S GREEN (HEMIHEPTAHYDRATE) □ CHROMIC PHOSPHATE □ CHROMIUM MONOPHOSPHATE □ CHROMIUM ORTHOPHOSPHATE □ PHOSPHORIC ACID CHROMIUM (III) SALT □ PHOSPHORIC ACID, CHROMIUM(3+) SALT (1:1) □ PLESSY'S GREEN (HEMIHEPTAHYDRATE)

CONSENSUS REPORTS: NTP 7th Annual Report on Carcinogens. IARC Cancer Review: Group 3 IMEMDT 49,49,90; Human Inadequate Evidence IMEMDT 49,49,

90; IARC Cancer Review: Animal Inadequate Evidence IMEMDT 49,49,90. Reported in EPA TSCA Inventory.

OSHA PEL: TWA 0.5 mg(Cr)/m^3
ACGIH TLV: TWA 0.5 mg(Cr)/m^3; Not Classifiable as a Carcinogen.

SAFETY PROFILE: Confirmed carcinogen. When heated to decomposition it emits toxic fumes of PO_x and Cr.

CMK400 CAS:37224-57-0 HR: 2
CHROMIUM POTASSIUM ZINC OXIDE

SYNS: POTASSIUM ZINC CHROMATE □ ZINC POTASSIUM CHROMATE

TOXICITY DATA WITH REFERENCE
imp-rat TDLo: 10,746 µg/kg: CAR BJIMAG 43,243,86

CONSENSUS REPORTS: IARC Cancer Review: Human Sufficient Evidence IMEMDT 23,205,80; Animal Sufficient Evidence IMEMDT 2,100,73. Chromium and its compounds, as well as zinc and its compounds, are on the Community Right-To-Know List.

OSHA PEL: CL 0.1 mg(CrO_3)/m^3
ACGIH TLV: TWA 0.01 mg(Cr)/m^3; Confirmed Human Carcinogen
DFG MAK: Human Carcinogen.
NIOSH REL: (Chromium (VI)) TWA 0.001 mg(Cr(VI))/m^3

SAFETY PROFILE: Confirmed carcinogen with experimental carcinogenic data. When heated to decomposition it emits toxic fumes of Cr^- and Zn .

CMK415 CAS:10101-53-8 HR: 3
CHROMIUM (III) SULFATE (2:3)
mf: $O_{12}S_3$•2Cr mw: 392.18

SYNS: BAYCHROM A □ BAYCHROM F □ CHROMIC SULFATE □ CHROMIC SULPHATE □ CHROMITAN B □ CHROMITAN MS □ CHROMITAN NA □ CHROMIUM III SULFATE □ CHROMIUM SULFATE (2:3) □ CHROMIUM SULPHATE □ CHROMIUM SULPHATE (2:3) □ C.I. 77305 □ DICHROMIUM SULFATE □ DICHROMIUM SULPHATE □ DICHROMIUM TRISULFATE □ DICHROMIUM TRISULPHATE □ KOREON □ SULFURIC ACID, CHROMIUM(3+) SALT (3:2)

TOXICITY DATA WITH REFERENCE
mmo-sat 10 mg/plate BECTA6 32,400,84
oth-hmn: oth 500 mg/L BJCAAI 44,219,81
dni-ham: kdy 500 mg/L BJCAAI 44,219,81
ivn-mus LDLo:85 mg/kg AQMOAC #70-15,70
ivn-rbt LDLo:215 mg/kg EQSFAP 1,1,75

CONSENSUS REPORTS: IARC Cancer Review: Group 3 IMEMDT 49,49,90; Animal Inadequate Evidence IMEMDT 23,205,80; Animal Inadequate Evidence IMEMDT 49,49,90. Reported in EPA TSCA Inventory.

OSHA PEL: 8H TWA 0.5 mg(Cr)/m^3
ACGIH TLV: TWA 0.5 mg(Cr)/m^3

SAFETY PROFILE: A poison by intravenous route. Questionable carcinogen. Human mutation data reported. When heated to decomposition it emits toxic vapors of SO_x and Cr.

CMK425 CAS:10031-37-5 HR: 3
CHROMIUM SULFATE, PENTADECAHYDRATE
mf: $O_{12}S_3$•2Cr•15H_2O mw: 662.38

SYNS: SULFURIC ACID, CHROMIUM(3+) SALT (3:2), PENTADECAHYDRATE □ WOOL MORDANT

TOXICITY DATA WITH REFERENCE
ipr-mus LD50:258 mg/kg CRNGDP 4,1535,83

ACGIH TLV: TWA 0.5 mg(Cr)/m^3; Not Classifiable as a Carcinogen

SAFETY PROFILE: Poison by intraperitoneal route. When heated to decomposition it emits toxic fumes of SO_x and Cr.

CMK450 CAS:10060-12-5 HR: 3
CHROMIUM TRICHLORIDE HEXAHYDRATE
mf: Cl_3Cr•6HO_2 mw: 356.41

SYNS: CHLORID CHROMITY HEXAHYDRAT □ CHROMIC CHLORIDE HEXAHYDRATE □ CHROMIUM CHLORIDE, HEXAHYDRATE (8CI,9CI) □ CHROMIUM(III) CHLORIDE, HEXAHYDRATE (1:3:6) □ CHROMIUM SESQUICHLORIDE □ HEXAAQUACHROMIUM CHLORIDE □ HEXAAQUACHROMIUM (III) CHLORIDE

TOXICITY DATA WITH REFERENCE
oth-hmn: oth 500 mg/L BJCAAI 44,219,81
cyt-hmn: leu 400 mg/L SAIGBL 18,136,76
dnd-ham: kdy 500 mg/L CBINA8 37,309,81
sce-ham: fbr 32 mg/L MUREAV 104,141,82
ipr-mus TDLo: 134 mg/kg (female 8D post): TER TXCYAC 26,257,83
orl-rat LD50: 1790 mg/kg SinJF# 29MAR77
ipr-mus LD50: 285 mg/kg TXAPA9 63,461,82
ivn-mus LDLo: 1602 mg/kg AIPTAK 62,330,39
ivn-rbt LDLo: 576 mg/kg AIPTAK 62,330,39
ivn-frg LDLo: 374 mg/kg AIPTAK 62,330,39

OSHA PEL: TWA 0.5 mg(Cr)/m^3
ACGIH TLV: TWA 0.5 mg(Cr)/m^3; Not Classifiable as a Carcinogen

SAFETY PROFILE: Poison by intraperitoneal and intravenous routes. Moderately toxic by ingestion. An experimental teratogen. Human mutation data reported. When heated to decomposition it emits acrid smoke and toxic fumes.

CMK500 CAS:15930-94-6 HR: 3
CHROMIUM(6+)ZINC OXIDE HYDRATE (1:2:6:1)
mf: CrO_4•H_2O_2•Zn_2•H_2O mw: 298.78

SYNS: BUTTERCUP YELLOW □ CHROMIC ACID, ZINC SALT (1:2) □ ZINC CHROMATE HYDROXIDE □ ZINC CHROMATE(VI) HYDROXIDE □ ZINC HYDROXYCHROMATE □ ZINC YELLOW

TOXICITY DATA WITH REFERENCE
sce-ham: ovr 100 µg/L MUREAV 156,219,85

CONSENSUS REPORTS: IARC Cancer Review: Human Sufficient Evidence IMEMDT 23,205,80; Animal Sufficient Evidence IMEMDT 2,100,73. Chromium and its compounds, as well as zinc and its compounds, are on the Community Right-To-Know List.

OSHA PEL: CL 0.1 mg(CrO_3)/m^3

ACGIH TLV: TWA 0.01 mg(Cr)/m^3; Confirmed Human Carcinogen
DFG MAK: Human Carcinogen.
NIOSH REL: (Chromium (VI)) TWA 0.001 mg(Cr(VI))/m^3

SAFETY PROFILE: Confirmed human carcinogen. Mutation data reported. When heated to decomposition it emits toxic fumes of ZnO. See also CHROMIUM and ZINC COMPOUNDS.

For occupational chemical analysis use NIOSH: Chromium Hexavalent 7024.

CMK650 CAS:7059-24-7 HR: 3
CHROMOMYCIN A3
mf: $C_{57}H_{82}O_{26}$ mw: 1183.39

PROP: Pale-yellow crystals. Mp: 185° (decomp).

SYNS: ABURAMYCIN B □ 3B-o-(4-o-ACETYL-2,6-DIDEOXY-3-C-METHYL-α-l-ARABINOHEXOPYRANOSYL)-7-METHYL-OLIVOMYCIN D □ ANTIBIOTIC B 599 □ CHROMOMYSIN A_3 □ NSC-58514 □ TOYOMYCIN

TOXICITY DATA WITH REFERENCE
dnr-bcs 800 ng/plate TAKHAA 44,96,85
dnd-hmn:hla 400 μg/L CNREA8 45,2813,85
msc-hmn:hla 8 μg/L CNREA8 45,2813,85
msc-mus:emb 25 μg/L CNREA8 45,2813,85
scu-rat TDLo:125 μg/kg (6-10D preg):TER OSDIAF 14,107,65
scu-rat TDLo:125 μg/kg (6-10D preg):REP OSDIAF 14,107,65
ipr-rat TDLo:250 μg/kg (1D male):REP HIKYAJ 12,1339,66
scu-mus TDLo:500 μg/kg (1D male):REP JJPAAZ 10,64,60
ipr-rat LDLo:250 μg/kg 85ERAY 2,1401,78
orl-mus LD50:1431 μg/kg NCISP* JAN86
ipr-mus LD50:800 μg/kg JAJAAA 16,22,63
scu-mus LD50:2800 μg/kg NIIRDN 6,245,82
ivn-mus LD50:1 mg/kg JAJAAA 16,22,63
ipr-dog LDLo:250 μg/kg 85ERAY 2,1401,78
ivn-dog LDLo:200 μg/kg TXAPA9 27,259,74
ivn-mky LDLo:330 μg/kg TXAPA9 27,259,74
ipr-cat LDLo:250 μg/kg 85ERAY 2,1041,78

SAFETY PROFILE: A deadly poison by ingestion, subcutaneous, intravenous, and intraperitoneal routes. An experimental teratogen. Experimental reproductive effects. Human mutation data reported. When heated to decomposition it emits acrid smoke and fumes.

CMK750 CAS:12622-79-6 HR: 3
CHROMOMYCIN SODIUM

PROP: Produced by a strain of *Actinomyces olivoreticuli* (85ERAY 2,1322,78).

SYN: OLIVOMYCIN, SODIUM SALT

TOXICITY DATA WITH REFERENCE
ipr-rat LDLo:1 mg/kg ANTBAL 7,53,62
ivn-rat LDLo:1 mg/kg ANTBAL 7,53,62
orl-mus LDLo:250 mg/kg ANTBAL 7,53,62
ipr-mus LD50:12,700 μg/kg ANTBAL 7,53,62
scu-mus LD50:15,600 μg/kg ANTBAL 7,53,62
ivn-mus LD50:138 mg/kg 85ERAY 2,1322,78
ivn-dog LDLo:300 μg/kg ANTBAL 7,53,62
ivn-rbt LDLo:2500 μg/kg ANTBAL 7,53,62
ipr-gpg LDLo:2 mg/kg ANTBAL 7,53,62

SAFETY PROFILE: Poison by ingestion, intraperitoneal, intravenous, and subcutaneous routes. When heated to decomposition it emits toxic fumes including Na_2O.

CML000 CAS:14259-67-7 HR: 3
CHROMYL AZIDE CHLORIDE
mf: $ClCrN_3O_2$ mw: 161.47

$CrO_2(N_3)Cl$

PROP: Dark-green, amorphous solid.

CONSENSUS REPORTS: Chromium and its compounds are on the Community Right-To-Know List.

SAFETY PROFILE: An explosive. When heated to decomposition it emits toxic fumes of Cl^- and NO_x. See also CHROMIUM COMPOUNDS, AZIDES, and CHLORIDES.

CML125 CAS:14977-61-8 HR: 3
CHROMYL CHLORIDE
DOT: UN 1758
mf: Cl_2CrO_2 mw: 154.90

PROP: Dark-red liquid; yellow-red vapor; musty burning odor. Readily hydrol to HCl and CrO_3. Fumes in air. Sol in org solvs and inorganic acid halides. Mp: −96.5°, bp: 115.7°, d: 1.9145 @ 25°/4°, vap press: 20 mm @ 20°.

SYNS: CHLORURE de CHROMYLE (FRENCH) □ CHROMIC OXYCHLORIDE □ CHROMIUM CHLORIDE OXIDE □ CHROMIUM DICHLORIDE DIOXIDE □ CHROMIUM DIOXIDE DICHLORIDE □ CHROMIUM(VI) DIOXYCHLORIDE □ CHROMIUM OXYCHLORIDE □ CHROMOXYCHLORID (GERMAN) □ CHROMYLCHLORID (GERMAN) □ CHROOMOXYLCHLORIDE (DUTCH) □ CROMILE, CLORURO di (ITALIAN) □ CROMO, OSSICLORURO di (ITALIAN) □ DICHLORODIOXOCHROMIUM □ DIOXODICHLOROCHROMIUM □ OXYCHLORURE CHROMIQUE (FRENCH)

TOXICITY DATA WITH REFERENCE
mmo-sat 50 μg/plate CRNGDP 1,583,80
mma-sat 100 μg/plate CRNGDP 1,583,80

CONSENSUS REPORTS: Reported in EPA TSCA Inventory. Chromium and its compounds are on the Community Right-To-Know List.

OSHA PEL: TWA 0.5 mg(Cr)/m^3
ACGIH TLV: TWA 0.025 ppm
DFG MAK: Suspected Carcinogen.
NIOSH REL: (Chromium(VI)) TWA 0.001 mg(Cr(VI))/m^3
DOT CLASSIFICATION: 8; *Label:* Corrosive

SAFETY PROFILE: Probably a poison by various routes. Mutation data reported. Corrosive. A strong irritant. Hydrolyzes to form chromic and hydrochloric acids. A strong oxidizer and chlorinating agent. Violent reaction with water. Reacts violently with alcohol, ether, acetone, turpentine. Ignites or explodes on contact with nonmetal halides (e.g., disulfur dichloride, phosphorus trichloride, and phosphorus tribromide), nonmetal hydrides (e.g., hydrogen sulfide, and hydrogen phosphide), flowers of sulfur, moist phosphorus, sodium azide, and urea. During preparation can violently explode. Incompatible with ammonia, disulfur dichloride, organic solvents, phosphorus, phosphorus trichloride, sodium azide, and sulfur. When heated to decomposition it emits toxic fumes of Cl^-. See also CHROMIUM COMPOUNDS.

For occupational chemical analysis use NIOSH: Chromium Hexavalent 7024.

CML325 CAS:16017-38-2 HR: 3
CHROMYL NITRATE
mf: CrN_2O_8 mw: 208.00

CONSENSUS REPORTS: Chromium compounds are on the Community Right-To-Know List.

SAFETY PROFILE: A powerful oxidant and nitrating agent. Ignites on contact with many organic materials (e.g., hydrocarbons, orgnic solvents, paper, rubber and wood). When heated to decomposition it emits toxic fumes of NO_x. See also CHROMIUM COMPOUNDS.

CML500 HR: 3
CHROMYL PERCHLORATE
mf: Cl_2CrO_{10} mw: 282.90

$CrO_2(ClO_4)_2$

CONSENSUS REPORTS: Chromium and its compounds are on the Community Right-To-Know List.

SAFETY PROFILE: A powerful oxidant. Explodes when heated above 80°C. Ignites on contact with organic solvents. When heated to decomposition it emits toxic fumes of Cl^-. See also CHROMIUM COMPOUNDS and PERCHLORATES.

CML600 CAS:517-92-0 HR: 3
CHRYSAMMINIC ACID
mf: $C_{14}H_4N_4O_{12}$ mw: 420.22

SYNS: 9,10-ANTHRACENEDIONE, 1,8-DIHYDROXY-2,4,5,7-TETRANITRO- □ ANTHRAQUINONE, 1,8-DIHYDROXY-2,4,5,7-TETRANITRO- □ CHRYSAMMIC ACID □ 1,8-DIHYDROXY-2,4,5,7-TETRANITROANTHRAQUINONE (chrysamminic acid) (DOT) □ 2,4,5,7-TETRANITROCHRYSAZIN

DOT CLASSIFICATION: Forbidden

SAFETY PROFILE: An unstable acid forbidden for transport. When heated to decomposition it emits toxic vapors of NO_x.

CML620 CAS:39067-39-5 HR: 1
CHRYSANTHAL
mf: $C_{11}H_{16}O$ mw: 164.27

SYNS: BICYCLO(2.2.1)HEPT-5-ENE-2-CARBOXALDEHYDE, 3 PROPYL- □ 2,5-METHYLENE-6-PROPYL-3-CYCLOHEXENECARBOXALDEHYDE □ 3-PROPYLBICYCLO(2.2.1)HEPT-5-ENE-2-CARBOXALDEHYDE

TOXICITY DATA WITH REFERENCE
orl-rat LD50:4600 mg/kg FCTOD7 26,401,88
skn-rbt LD50:>5 g/kg FCTOD7 26,401,88

CONSENSUS REPORTS: Reported in EPA TSCA Inventory.

SAFETY PROFILE: Low toxicity by ingestion and skin contact. When heated to decomposition it emits acrid smoke and irritating vapors.

CML650 CAS:10453-89-1 HR: 3
CHRYSANTHEMIC ACID
mf: $C_{10}H_{16}O_2$ mw: 168.26

SYNS: CHRYSANTHEMUMIC ACID □ CHRYSANTHEMUMMONOCARBOXYLIC ACID □ CYCLOPROPANECARBOXYLIC ACID, 2,2-DIMETHYL-3-(2-METHYLPROPENYL)-

TOXICITY DATA WITH REFERENCE
cyt-mus-ihl 1248 mg/m³ GISAAA 51(1),16,86
orl-rat LD50:2500 mg/kg GISAAA 51(1),16,86
ihl-rat LC:>400 mg/m³ GTPZAB 31(7),53,87
orl-mus LD50:1250 mg/kg GISAAA 51(1),16,86
ihl-mus LC:>400 mg/m³ GTPZAB 31(7),53,87
skn-mus LD50:>5 g/kg GISAAA 51(1),16,86
ipr-mus LD50:150 mg/kg NTIS** AD691-490
ivn-mus LD50:56 mg/kg CSLNX* NX#00083
orl-gpg LD50:2500 mg/kg GISAAA 51(1),16,86

CONSENSUS REPORTS: Reported in EPA TSCA Inventory.

SAFETY PROFILE: A poison by intravenous route. Moderately toxic by ingestion. Mutation data reported. When heated to decomposition it emits acrid smoke and irritating vapors.

CML800 CAS:2642-98-0 HR: 2
6-CHRYSENAMINE
mf: $C_{18}H_{13}N$ mw: 243.32

PROP: Leaflets from alc. Mp: 210–211°. Sltly sol in alc, benzene, ethyl acetate.

SYNS: 6-AMC □ 6-AMINOCHRYSENE □ CHRYSENEX □ CHRYSONEX

TOXICITY DATA WITH REFERENCE
mmo-sat 2500 ng/plate CNREA8 44,3408,84
mma-sat 500 ng/plate MUREAV 155,7,85
dnr-bcs 20 μL/disc MUREAV 97,1,82
dns-rat:lvr 500 nmol/L ENMUDM 3,11,81
msc-ham:ovr 50 mg/L JTEHD6 13,531,84
skn-mus TDLo:1100 g/kg/39W-I:CAR EJCAAH 11,327,75

CONSENSUS REPORTS: EPA Genetic Toxicology Program.

SAFETY PROFILE: Questionable carcinogen with experimental carcinogenic data by skin contact. Mutation data reported. When heated to decomposition it emits toxic fumes of NO_x. See also AROMATIC AMINES and CHRYSENE.

CML810 CAS:218-01-9 HR: 3
CHRYSENE
mf: $C_{18}H_{12}$ mw: 228.30

PROP: Plates from C_6H_6 or AcOH with reddish-violet fluorescence. Occurs in coal tar. Is formed during distillation of coal, in very small amount during distillation or pyrolysis of many fats and oils. Orthorhombic bipyramidal plates from benzene. D: 1.274, mp: 255–256°. Sublimes easily in vacuum, bp: 448°. Sltly sol in alc, ether, carbon disulfide, and glacial acetic acid; moderately sol in boiling benzene; insol in water. Chrysene is generally only sltly sol in cold organic solvents, but fairly sol in these solvents when hot, including glacial acetic acid.

SYNS: BENZO(a)PHENANTHRENE ☐ 1,2-BENZOPHENANTHRENE ☐ BENZ(a)PHENANTHRENE ☐ 1,2-BENZPHENANTHRENE ☐ 1,2,5,6-DIBENZONAPHTHALENE ☐ RCRA WASTE NUMBER U050

TOXICITY DATA WITH REFERENCE
mma-sat 5 µg/plate MUREAV 156,61,85
msc-hmn:lym 6 µmol/L DTESD7 10,227,82
msc-hmn:oth 12 µmol/L MUREAV 130,127,84
skn-mus TDLo:3600 µg/kg:NEO CNREA8 38,1831,78
scu-mus TDLo:200 mg/kg:ETA CNREA8 15,632,55
skn-mus TD:99 mg/kg/31W-I:ETA CNREA8 11,301,51
skn-mus TD:40 mg/kg/3W-I:ETA CCSUDL 1,325,76
skn-mus TD:3600 mg/kg/30W-I:ETA CANCAR 12,1079,59
skn-mus TD:23 mg/kg:NEO CNREA8 40,642,80

CONSENSUS REPORTS: IARC Cancer Review: Group 3 IMEMDT 7,56,87; Animal Limited Evidence IMEMDT 32,247,83; Human No Adequate Data IMEMDT 32,247,83. EPA Genetic Toxicology Program. Reported in EPA TSCA Inventory.

OSHA PEL: 0.2 mg/m^3
ACGIH TLV: Suspected Human Carcinogen
DFG MAK: Animal Carcinogen, Suspected Human Carcinogen
NIOSH REL: (Chrysene) To be controlled as a carcinogen

SAFETY PROFILE: Confirmed carcinogen with experimental carcinogenic, neoplastigenic, and tumorigenic data by skin contact. Human mutation data reported. When heated to decomposition it emits acrid smoke and fumes.

For occupational chemical analysis use OSHA: #ID-58 or NIOSH: Polynuclear Aromatic Hydrocarbons (HPLC), 5506; (GC), 5515.

CML820 CAS:63339-68-4 HR: 2
CHUANGHSINMYCIN
mf: $C_{12}H_{11}NO_2S$ mw: 233.30

SYNS: CHUANGXINMYCIN ☐ cis-(−)-3,5-DIHYDRO-3-METHYL-2H-THIOPYRANO(4,3,2-cd)INDOLE-2-CARBOXYLIC ACID

TOXICITY DATA WITH REFERENCE
orl-mus LD50:1770 mg/kg 85GDA2 5,133,81
ipr-mus LD50:875 mg/kg 85GDA2 5,133,81
ivn-mus LD50:600 mg/kg 85GDA2 5,133,81

SAFETY PROFILE: Moderately toxic by ingestion, intravenous, and intraperitoneal routes. When heated to decomposition it emits toxic fumes of SO_x and NO_x.

CML822 CAS:102419-73-8 HR: 2
CHUANGHSINMYCIN SODIUM
mf: $C_{12}H_{10}NO_2$•Na mw: 223.22

SYN: CHUANGXIMYCIN SODIUM

TOXICITY DATA WITH REFERENCE
orl-mus LD50:1770 mg/kg JANTAJ 32,79-23,79
ipr-mus LD50:875 mg/kg JANTAJ 32,79-23,79
ivn-mus LD50:600 mg/kg JANTAJ 32,79-23,79

SAFETY PROFILE: Moderately toxic by ingestion, intravenous, and intraperitoneal routes. When heated to decomposition it emits toxic fumes of NO_x and Na_2O.

CML825 CAS:58812-37-6 HR: 3
CHUANLIANSU
mf: $C_{30}H_{38}O_{11}$ mw: 574.68

SYN: TOOSENDANIN

TOXICITY DATA WITH REFERENCE
ipr-rat LD50:98 mg/kg CTYAD8 13,317,82
orl-mus LD50:244 mg/kg CTYAD8 13,317,82
ipr-mus LD50:13,800 µg/kg CTYAD8 13,317,82
scu-mus LD50:14,300 µg/kg CTYAD8 13,317,82
ivn-mus LD50:14,600 µg/kg CTYAD8 13,317,82
ivn-rbt LD50:4200 µg/kg CTYAD8 13,317,82

SAFETY PROFILE: Poison by ingestion, subcutaneous, intravenous and intraperitoneal routes. When heated to decomposition it emits acrid smoke and fumes.

CML835 CAS:37106-97-1 HR: 2
CHYMEX
mf: $C_{23}H_{20}N_2O_5$ mw: 404.45

PROP: Crystals from methanol/water. Mp: 240–242°.

SYNS: BENTIROMIDE ☐ (S)-p-(α-BENZAMIDO-p-HYDROXYHYDROCINNAMAMIDO)BENZOIC ACID ☐ 4-((2-(BENZOYLAMINO)-3-(4-HYDROXYPHENYL)-1-OXOPROPYL)AMINO)BENZOIC ACID ☐ (S)-4-(((2-BENZOYLAMINO)-3-(4-HYDROXYPHENYL)-1-OXOPROPYL)AMINO)BENZOIC ACID ☐ p-((N-BENZOYL-l-TYROSIN)AMIDO)BENZOIC ACID ☐ BENZOYLTYROSYL-p-AMINOBENZOIC ACID ☐ N-BENZOYL-l-TYROSYL-p-AMINOBENZOIC ACID ☐ BTPABA ☐ E-2663 ☐ PFT

TOXICITY DATA WITH REFERENCE
ipr-rat LD50:2000 mg/kg IYKEDH 11,181,80
ivn-rat LD50:485 mg/kg IYKEDH 11,181,80
ipr-mus LD50:1650 mg/kg IYKEDH 11,181,80
ivn-mus LD50:1020 mg/kg IYKEDH 11,181,80

SAFETY PROFILE: Moderately toxic by intravenous and intraperitoneal routes. When heated to decomposition it emits toxic fumes of NO_x.

CML850 CAS:9004-07-3 HR: 3
α-CHYMOTRYPSIN

SYNS: ALPHA CHYMAR ☐ ALPHA-CHYMAR OPHTH ☐ AVAZYME ☐ CHYMAR ☐ CHYMOTEST ☐ CHYMOTRYPSIN A ☐ CHYMOTRYPSIN B ☐ E.C. 3.4.4.5 ☐ E.C. 3.4.4.6 ☐ E.C. 3.4.21.1 ☐ ENZEON ☐ QUIMAR ☐ QUIMOTRASE

TOXICITY DATA WITH REFERENCE
ipr-rat LD50:65,100 μg/kg JAPMA8 39,42,50
scu-rat LD50:250 mg/kg NIIRDN 6,202,82
ivn-rat LD50:84 mg/kg NIIRDN 6,202,82
ims-rat LD50:116 mg/kg NIIRDN 6,202,82
scu-mus LD50:185 mg/kg NIIRDN 6,202,82
ivn-mus LD50:89 mg/kg NIIRDN 6,202,82
ims-mus LD50:104 mg/kg NIIRDN 6,202,82
ivn-rbt LD50:24,000 units/kg AIPTAK 106,164,56
ivn-gpg LDLo:50,000 units/kg AIPTAK 106,164,56

CONSENSUS REPORTS: Reported in EPA TSCA Inventory.

SAFETY PROFILE: Poison by subcutaneous, intramuscular, intravenous and intraperitoneal routes. When heated to decomposition it emits toxic fumes of NO_x.

CML890 **HR: 3**
CI-914
mf: $C_{13}H_{12}N_4O{\bullet}ClH$ mw: 276.75

SYN: 4,5-DIHYDRO-6-((4-1H-IMIDAZOL-1-YL)PHENYL)-3(2H)-PYRIDAZINONE HYDROCHLORIDE

TOXICITY DATA WITH REFERENCE
orl-rat LD50:125 mg/kg TOXID9 4,7,84
ivn-rat LD50:98 mg/kg TOXID9 4,6,84
orl-mus LD50:250 mg/kg TOXID9 4,7,84
ivn-mus LD50:226 mg;kg TOXID9 4,6,84

SAFETY PROFILE: Poison by ingestion and intravenous routes. When heated to decomposition it emits toxic fumes of NO_x and HCl.

CMM000 **CAS:6441-77-6** **HR: 2**
C.I. 45405
mf: $C_{20}H_6Br_4Cl_2O_5{\bullet}2K$ mw: 795.00

SYNS: C.I. ACID RED 98 □ 4,6-DICHLORO-2′,4′,5′,7′-TETRABROMOFLUORESCEIN DIPOTASSIUM SALT □ FOOD DYE RED No. 104 □ PHLOXIN □ PHLOXINE □ PHLOXINE K □ 2′,4′,5′,7′-TETRABROMO-4,7-DICHLORO-FLUORESCEIN DIPOTASSIUM SALT □ 2,4,5,7-TETRABROMO-12,15-DICHLOROFLUORESCEIN, DIPOTASSIUM SALT □ TOYO ACID PHLOXINE

TOXICITY DATA WITH REFERENCE
dnr-esc 1 mg/disc MUREAV 88,1,81
dnd-esc 1 mg/disc FCTXAV 18,215,80
dni-hmn:leu 500 mg/L NEZAAQ 30,574,75
cyt-hmn:leu 1 mg/L NEZAAQ 30,574,75
cyt-hmn:fbr 1 mg/L NEZAAQ 30,574,75
orl-rat TDLo:75 g/kg (1-20D preg):REP SKEZAP 16,34,75
orl-rat TDLo:75 g/kg (1-20D preg):TER SKEZAP 16,34,75
orl-rat LD50:2870 mg/kg SKEZAP 16,34,75

CONSENSUS REPORTS: EPA Genetic Toxicology Program.

SAFETY PROFILE: Moderately toxic by ingestion. An experimental teratogen. Other experimental reproductive effects. Human mutation data reported. When heated to decomposition it emits toxic fumes of Cl^-, Br^-, and K_2O.

CMM062 **CAS:3536-49-0** **HR: 2**
C.I. ACID BLUE 3, CALCIUM SALT (2:1) (8CI)
mf: $C_{54}H_{62}N_4O_{14}S_4{\bullet}Ca$ mw: 1159.52

SYNS: ACIDAL CARMINE V □ ACID BLUE 3 □ BLEU PATENTE V □ BLUE ZN 3 □ CARMINE BLUE V □ C.I. 712 □ C.I. 42051 □ C.I. ACID BLUE 3 □ C.I. FOOD BLUE 5 □ DAI-EI ACID PURE BLUE VX □ E 131 □ ETHANAMINIUM, N-(4-((4-(DIETHYLAMINO)PHENYL)(5-HYDROXY-2,4-DISULFOPHENYL)METHYLENE)-2,5-CYCLOHEXADIEN-1-YLIDENE)-N-ETHYL-, HYDROXIDE, inner salt, CALCIUM SALT (2:1) □ m-HYDROXYTETRAETHYLDIAMINOTRIPHENYLCARBINOL ANHYDRIDE DISULFONIC ACID CALCIUM SALT □ L-BLAU 3 □ MERANTINE BLUE V □ MITSUI ACID PURE BLUE VX □ NEW PATENT BLUE A-CE EXTRA □ NEW PATENT BLUE EXTRA PURE A □ PATENT BLUE V □ PATENT BLUE V CARMINE BLUE V □ SCHULTZ No. 826 □ SOLAR PURE BLUE VX

TOXICITY DATA WITH REFERENCE
ivn-rat LD50:5 g/kg FAONAU 38B,71,66
ivn-mus LD50:1200 mg/kg FAONAU 38B,71,66

CONSENSUS REPORTS: Reported in EPA TSCA Inventory.

SAFETY PROFILE: Moderately toxic by intravenous route. When heated to decomposition it emits toxic vapors of NO_x and SO_x.

CMM070 **CAS:2666-17-3** **HR: 1**
C.I. ACID BLUE 41
mf: $C_{23}H_{18}N_3O_6S{\bullet}Na$ mw: 487.49

SYNS: ACID BLUE 41 □ ALIZARINE BLUE A □ ALIZARINE BLUE AR □ ALIZARINE DIRECT BLUE AR □ ALIZARINE DIRECT BLUE ARA □ ALIZARINE SAPPHIRE AR □ 2-ANTHRACENESULFONIC ACID, 1-AMINO-9,10-DIHYDRO-4-(p-(N-METHYLACETAMIDO)ANILINO)-9,10-DIOXO-, MONOSODIUM SALT □ ANTHRALAN BLUE B □ C.I. 62130 □ C.I. ACID BLUE 71 □ DIACID LIGHT BLUE BR □ ERIO FAST BLUE BRL □ FENALAN BLUE B □ KAYACYL BLUE BR □ LANAPERL BLUE B □ LISSAMINE BLUE AR □ LISSAMINE ULTRA BLUE AR □ MODR KYSELA 41 □ NYLOMINE ACID BLUE B-B □ SUPERAN BLUE AR □ UNITERTRACID LIGHT BLUE AB

TOXICITY DATA WITH REFERENCE
eye-rbt 500 mg/24H MLD 85JCAE-,1327,86

CONSENSUS REPORTS: Reported in EPA TSCA Inventory.

SAFETY PROFILE: An eye irritant. When heated to decomposition it emits toxic vapors of NO_x, NaO_2 and SO_x.

CMM080 **CAS:4368-56-3** **HR: 2**
C.I. ACID BLUE 62 (8CI)
mf: $C_{20}H_{19}N_2O_5S{\bullet}Na$ mw: 422.46

SYNS: ACID ALIZARINE PURE BLUE R □ ACID BLUE 62 □ ALIZARINE BRILLIANT SAPPHIRE R □ ALIZARINE BRILLIANT SKY BLUE R □ ALIZARINE DIRECT PURE BLUE R □ ALIZARINE FAST BLUE RFE □ ALIZARINE SKY BLUE R □ ALIZARINE SUPRA BLUE R □ ALIZARINE SUPRA SKY RA □ 2-ANTHRACENESULFONIC ACID, 9,10-DIHYDRO-1-AMINO-4-(CYCLOHEXYLAMINO)-9,10-DIOXO-, MONOSODIUM SALT □ BRILLIANT ALIZARINE CYANINE R □ BRILLIANT ALIZARINE LIGHT BLUE 3FR □ C.I. 62045 □ 9,10-DIHYDRO-1-AMINO-4-(CYCLOHEXYLAMINO)-9,10-DIOXO-2-ANTHRACENESULFONIC ACID SODIUM SALT □ ERIO ANTHRACENE BRILLIANT BLUE RFF □ ERIONYL BLUE E-RFF □ ERIOSIN FAST BLUE RFF □ FENAZO LIGHT BLUE RA □ KAYACYL

C

SKY BLUE R □ KAYAKU ALIZARINE SKY BLUE R □ POLAN NAVY BLUE E 2R □ SOL. SULFUR BLUE 10 □ SUMINOL LEVELLING SKY BLUE R □ TELON BLUE RRL □ TERTRACID BRILLIANT LIGHT BLUE R

TOXICITY DATA WITH REFERENCE
mnt-mus-ipr 87,500 μg/kg/24H-I TOLED5 40,183,88
ipr-mus LD50:983 mg/kg TOLED5 40,183,88

CONSENSUS REPORTS: Reported in EPA TSCA Inventory.

SAFETY PROFILE: Moderately toxic by intraperitoneal route. Mutation data reported. When heated to decomposition it emits toxic vapors of NO_x and SO_x.

CMM090 CAS:6424-75-5 HR: 1
C.I. ACID BLUE 78
mf: $C_{21}H_{14}BrN_2O_5S{\cdot}Na$ mw: 509.33

SYNS: ACID ALIZARINE PURE BLUE B □ ACID ALIZARINE SKY BLUE B □ ACID ANTHRAQUINONE PURE BLUE □ ACID BLUE 78 □ ACID PURE SKY BLUE ANTHRAQUINONE □ ACID SKY BLUE ANTHRAQUINONE □ AHCOQUINONE SKY BLUE B □ ALIZARINE ACID BLUE B □ ALIZARINE BLUE GRL □ ALIZARINE FAST BLUE 2B □ ALIZARINE FAST LIGHT BLUE C □ ALIZARINE LIGHT BLUE AR □ ALIZARINE PURE BLUE B □ ALIZARINE SKY BLUE B □ ALIZARINE SKY BLUE BS-CF □ ANTHRAQUINONE BLUE SKY □ ATLANTIC ACID FAST BLUE B □ CALCOCID ALIZARINE BLUE SKY □ C.I. 62105 □ CUROL PURE BLUE B □ DIACID ALIZARINE SKY BLUE B □ DIACID ALIZARIN SKY BLUE B □ ERIO ANTHRACENE BRILLIAN BLUE B □ ERIO ANTHRACENE FAST BLUE BB □ ERIO FAST BLUE BS □ ERIONYL BLUE E-B □ ERIOSIN FAST BLUE B □ FENAZO LIGHT BLUE AC □ MERPACYL BLUE SK □ MITSUI NYLON FAST SKY BLUE B □ MODR KYSELA 78 □ NYLOMINE BLUE A 2B □ SANDOLAN BLUE P-ARL □ SOLWAY SKY BLUE B □ SOLWAY SKY BLUE BA □ SUMINOL FAST SKY BLUE B □ SUPERNYLITE BLUE BR □ TELON BLUE BL □ m-TOLUENESULFONIC ACID, 6-((4-AMINO-3-BROMO-1-ANTHRAQUINONYL)AMINO)-, MONOSODIUM SALT □ VONDACID FAST BLUE BR

TOXICITY DATA WITH REFERENCE
eye-rbt 500 mg/24H MLD 85JCAE-,1327,86

CONSENSUS REPORTS: Reported in EPA TSCA Inventory.

SAFETY PROFILE: An eye irritant. When heated to decomposition it emits toxic vapors of NO_x, SO_x, Br^-, and Cl^-.

CMM100 CAS:6397-02-0 HR: 1
C.I. ACID BLUE 129
mf: $C_{23}H_{19}N_2O_5S{\cdot}Na$ mw: 458.49

SYNS: ACID BLUE 129 □ ACID FAST BLUE BS □ ALIZARINE FAST BLUE BE □ 2-ANTHRACENESULFONIC ACID, 1-AMINO-9,10-DIHYDRO-9,10-DIOXO-4-(2,4,6-TRIMETHYLANILINO)-, MONOSODIUM SALT □ BRILLIANT ALIZARINE SKY BLUE BS □ C.I. 62059 □ ERIONYL BLUE BFF □ ERIONYL BLUE E-BFF □ ERIOSIN FAST BLUE BFF □ FAST ACID LIGHT BLUE BRL □ MODR KYSELA 129 □ NEONYL BLUE 4R □ NYLOMINE BLUE A 3R □ SOLWAY BLUE RB □ SUMINOL FAST BLUE PR □ TERTRACID FAST BLUE TR

TOXICITY DATA WITH REFERENCE
skn-rbt 500 mg/24H MLD 85JCAE-,1327,86
eye-rbt 100 mg/24H MOD 85JCAE-,1327,86

CONSENSUS REPORTS: Reported in EPA TSCA Inventory.

SAFETY PROFILE: A skin and eye irritant. When heated to decomposition it emits toxic vapors of NO_x and SO_x.

CMM200 CAS:12219-87-3 HR: 1
C.I. ACID GREEN 40

SYNS: ACID GREEN 40 □ LANASYN GREEN BL □ NYLOSAN GREEN F-BL □ SILK FAST GREEN B □ XYLENE ACID MILLING GREEN BL □ ZELEN KYSELA 40

TOXICITY DATA WITH REFERENCE
eye-rbt 500 mg/24H MLD 85JCAE-,1330,86
orl-rat LD50:8680 mg/kg 85JCAE-,1330,86

SAFETY PROFILE: Mildly toxic by ingestion. An eye irritant. When heated to decomposition it emits acrid smoke and irritating fumes.

CMM220 CAS:633-96-5 HR: D
C.I. ACID ORANGE 7, MONOSODIUM SALT □ 75025-97-7
mf: $C_{16}H_{12}N_2O_4S{\cdot}Na$ mw: 351.35

SYNS: ACID LEATHER ORANGE EXTRA □ ACID LEATHER ORANGE EXTRA G □ ACID LEATHER ORANGE EXTRA PRW □ ACID ORANGE □ ACID ORANGE 7 □ ACID ORANGE A □ ACID ORANGE II □ ACILAN ORANGE II □ AIREDALE ORANGE II □ AMACID ORANGE Y □ ATUL ACID ORANGE II □ BENZENESULFONIC ACID, 4-((2-HYDROXY-1-NAPHTHALENYL)AZO)-, MONOSODIUM SALT □ BETANAPHTHOL ORANGE □ BRASILAN ORANGE A □ BUCACID ORANGE A □ CALCOCID ORANGE Y □ CERTIQUAL ORANGE II □ C.I. 15510 □ C.I. ACID ORANGE 7 □ COLACID ORANGE □ CUROL ORANGE □ D and C ORANGE No. 4 □ DIACID ORANGE II □ ERIO ORANGE II □ FENAZO ORANGE □ HIDACID ORANGE II □ HISPACID ORANGE AF □ p-((2-HYDROXY-1-NAPHTHYL)AZO)BENZENESULFONIC ACID SODIUM SALT □ JAVA ORANGE II □ KITON ORANGE II □ KROMON LAKE ORANGE TONER □ LAKE ORANGE A □ LAKE ORANGE II YS □ LEATHER ORANGE EXTRA □ LURAZOL ORANGE E □ LUTETIA ORANGE 3JR □ MANDARIN G □ NAPHTHALENE LAKE ORANGE G □ NAPHTHALENE ORANGE G □ NAPHTHOL ORANGE □ β-NAPHTHOL ORANGE □ 2-NAPHTHOL ORANGE II □ β-NAPHTHYL ORANGE □ NAPHTOCARD ORANGE II □ NEKLACID ORANGE II □ No. 177 ORANGE LAKE □ NUBILON ORANGE R □ 11550 ORANGE □ ORANGE EXTRA N □ ORANGE EXTRA P □ ORANGE II □ ORANGE IIC □ ORANGE II for LAKES □ ORANGE IIP □ ORANGE IIS □ ORANGE IISM □ ORANGE II SPECIAL FOR LACQUER □ ORANGE No. 205 □ ORANGE 2 SODIUM SALT □ ORANGE TONER GRT □ ORANGE Y □ ORANGE YA □ ORANGE YZ □ ORANZ KYSELA 7 □ PEERACID ORANGE II □ PERCA ORANGE GR □ PERSIAN ORANGE □ PERSIAN ORANGE LAKE □ PERSIAN ORANGE X □ PURE ORANGE II S □ RIFA ACID ORANGE II □ SANYO GUM ORANGE A □ SOLAR ORANGE □ SPECIAL ORANGE GR □ SPECIAL ORANGE H □ SYMULER ORANGE LAKE 43 □ SYMULON ACID ORANGE II □ TANGARINE LAKE X-917 □ TERTRACID ORANGE II □ TROPAEOLIN OOO □ TROPAEOLIN OOO 2 □ VONDACID ORANGE II □ WOOL ORANGE A

TOXICITY DATA WITH REFERENCE
msc-hmn:lyms 12 nmol/L MUREAV 249,265,91
itt-rat TDLo:10 mg/kg (male 1D pre):REP IJEBA6 15,1215,77
orl-rat TDLo:150 g/kg (male 43W pre):REP IJEBA6 17,1100,79

CONSENSUS REPORTS: Reported in EPA TSCA Inventory.

SAFETY PROFILE: Experimental reproductive effects.

Mutation data reported. When heated to decomposition it emits toxic vapors of NO_x and SO_x.

CMM300 CAS:3734-67-6 HR: D
C.I. ACID RED 1, DISODIUM SALT
mf: $C_{18}H_{15}N_3O_8S_2{\cdot}2Na$ mw: 511.46

SYNS: ACETYL RED G □ ACETYL RED J □ ACETYL ROSE 2GL □ ACIDAL BRILLIANT RED 2G □ ACID BRIGHT RED □ ACID BRILLIANT RED □ ACID FAST RED EGG □ ACID FAST RED 3G □ ACIDINE RED G □ ACID LEATHER RED KG □ ACID NAFTOL RED G □ ACID PHLOXINE GA □ ACID RED 1 □ ACID RED 2G □ ACID RED GA □ ACID ROSE 2GL □ ACILAN NAPHTHOL RED G □ ACILAN NAPHTOL RED G □ AHCOCID CARMINE 2G □ AMACID PHLOXINE □ AMIDO NAPHTHOL RED G □ AMIDO NAPHTHOL RED 2G □ AMIDO NAPHTHOL RED GA □ AMIDO RED 2G □ ARIAVIT RED 2G □ ATUL ACID GERANINE G □ AZO GERANINE 2G □ AZO GERANINE 2GA □ AZONAPHTHOL RED J □ AZOPHLOXIN □ AZOPHLOXINE □ AZO PHLOXINE GA □ AZO PHLOXINE GA-CF □ AZO RHODINE 2G □ BELACID PHLOXINE G □ BRILLIANT ACID RED G<01> □ BRILLIANT ACID ROSAMINE 2G □ BRILLIANT COLACID RED G □ BUCACID FAST CRIMSON □ CALCOCID PHLOXINE 2G □ CETIL LIGHT RED GG □ CERVEN 2G □ CERVEN KYSELA 1 □ CERVEN POTRAVINARSKA 10 □ C.I. 18050 □ C.I. ACID RED 1 □ C.I. FOOD RED 10 □ EDICOL SUPRA GERANINE 2G □ EDICOL SUPRA GERANINE 2GS □ EGACID RED G □ ENIACID LIGHT RED 3G □ ERIO FLOXINE 2G □ ERIO FLOXINE 2GN □ EXT D and C RED NO. 11 □ FAST CRIMSON GR □ FAST DRIMSON GR □ FENAZO RED B □ GERANINE 2GS □ HASTINGS CARMINE 2G □ HEXACOL RED 2G □ HEXALAN RED 2G □ HIDACID FAST CRIMSON □ HISPACID FAST CARMOISINE G □ INK RED JSN □ JAVA NAPHTOL RED G □ KITON RED G □ KITON RED 2G □ LEATHER RED G □ LISSAMINE RED 2G □ 2,7-NAPHTHALENEDISULFONIC ACID, 5-(ACETYLAMINO)-4-HYDROXY-3-(PHENYLAZO)-, DISODIUM SALT □ NAPHTHAZINE ROSE 2G □ NAPHTOCARD RED 2G □ PHLOXINE G □ PHLOXINE 2G □ PONTACYL CARMINE 2G □ 1379 RED □ RED 2G □ SOLAR FAST RED 3G □ UNITERTRACID RED 2G □ VONDACID LIGHT RED NG

TOXICITY DATA WITH REFERENCE
mmo-esc 10 g/L MUREAV 117,127,83
bfa-rat:sat 800 mg/kg FCTOD7 22,593,84

CONSENSUS REPORTS: Reported in EPA TSCA Inventory.

SAFETY PROFILE: Mutation data reported. When heated to decomposition it emits toxic vapors of NO_x and SO_x.

CMM320 CAS:3567-65-5 HR: D
C.I. ACID RED 85, DISODIUM SALT
mf: $C_{35}H_{26}N_4O_{10}S_3{\cdot}2Na$ mw: 804.81

SYNS: ACIDINE SCARLET GD □ ACID LEATHER RED GR □ ACID LEATHER SCARLET G □ ACID RED 85 □ ACID RED PG □ AIREDALE RED PGM □ ALTOCHROME MILLING SCARLET G □ AMACID MILLING RED PGS □ APOCID MILLING RED G □ BENZYL FAST RED GRG □ BENZYL RED GR □ C.I. 22245 □ C.I. ACID RED 85 □ COOMASSIE RED PG □ COOMASSIE RED PGP □ CUTAMIN BRILLIANT RED CG □ ELITE FAST RED G □ ELITE FAST RED GRS □ ERIONYL RED G □ FAST SCARLET S □ FENAFOR RED PG □ FOLAN YELLOW G □ KAYANOL MILLING RED PG □ KAYANOL RED PG □ KCA SILK RED G □ LANAPERL FAST RED 3G □ LANAPERL RED G □ MIDLON RED PG □ MILLING BRILLIANT SCARLET GN □ MILLING FAST RED G □ MILLING FAST RED GL □ MILLING FAST RED PG □ MILLING RED J □ MILLING RED SWG □ MILLING SCARLET G □ MITSUI MILLING SCARLET G □ 1,3-NAPHTHALENEDISULFONIC ACID, 7-HYDROXY-8-((4′-((4-(((4-METHYLPHENYL)SULFONYL)OXY)PHENYL)AZO)(1,1′-BIPHENYL)-4-YL)AZO)-, DISODIUM SALT □ NEONYL SCARLET R □ NEUTRAL RED PG □ NITTO ACID RED PG □ NYLOMINE ACID SCARLET C-R □ NYLOMINE ACID SCARLET P-R □ OPTANOL FAST SCARLET GN □ PHARMANIL SCARLET Y □ POLAR RED G □ POLAR RED G SUPRA □ SHIKISO ACID RED PG □ SULFONINE RED G □ SULFONINE RED GN □ SULFONINE RED GS □ SULFONINE RED SG □ SULPHONOL RED PG □ SUMINOL RED PG □ SUPRANOL FAST SCARLET GN □ SUPRANOL RED PG-CF □ SUPRANOL SCARLET BN □ SYMULON ACID RED PG □ TELON FAST SCARLET N □ TERTRACID MILLING RED G □ VONDAMOL FAST RED G

TOXICITY DATA WITH REFERENCE
mma-sat 100 nmol/plate MUREAV 136,33,84

CONSENSUS REPORTS: Reported in EPA TSCA Inventory.

SAFETY PROFILE: Mutation data reported. When heated to decomposition it emits toxic vapors of NO_x and SO_x.

CMM325 CAS:10169-02-5 HR: D
C.I. ACID RED 97, DISODIUM SALT (8CI)
mf: $C_{32}H_{20}N_4O_8S_2{\cdot}2Na$ mw: 698.66

SYNS: ACID ANTHRACENE RED G □ ACID ANTHRACENE RED GA-CF □ ACID RED 97 □ AIREDALE SCARLET GM □ ALIZARINE CHROME RED G □ ALTOCHROME SCARLET G □ AMACID MILLING SCARLET G □ ANTHRA RED G □ AZO MILLING RED G □ BELACID MILLING RED G □ BENZYL RED GS □ BENZYL RED MG □ (1,1′-BIPHENYL)-2,2′-DISULFONIC ACID, 4,4′-BIS((2-HYDROXY-1-NAPHTHALENYL)AZO)-, DISODIUM SALT □ CALCOCID MILLING RED G □ C.I. 22890 □ C.I. ACID RED 97 □ COOMASSIE MILLING SCARLET G □ COOMASSIE MILLING SCARLET GP □ CORIACID SCARLET R □ CRISPIN RED GM □ CYANINE FAST SCARLET G □ FENAZO RED FG □ HEXADERM RED MRG □ KOROSTAN RED G □ MILLING RED A □ MILLING SCARLET DH □ MILLING SCARLET 2G □ MILLING SCARLET R □ NAPHTHALENE LEATHER SCARLET G □ OPTANOL SCARLET GS □ PHARMAGLO RED G □ POLYCOR RED GS □ SHIKISO ACID ANTHRACENE RED G □ SUMINOL BRILLIANT SCARLET DH □ SUMITOMO FAST SCARLET G □ SUPRANOL SCARLET GS □ TERTRACID MILLING RED AGE □ VONDAMOL BRILLIANT RED G □ XYLENE MILLING RED G

TOXICITY DATA WITH REFERENCE
mma-sat 125 μg/plate VHTODE 22,413,80

CONSENSUS REPORTS: Reported in EPA TSCA Inventory.

SAFETY PROFILE: Mutation data reported. When heated to decomposition it emits toxic vapors of NO_x and SO_x.

CMM330 CAS:6459-94-5 HR: 3
C.I. ACID RED 114, DISODIUM SALT
mf: $C_{37}H_{28}N_4O_{10}S_3{\cdot}2Na$ mw: 830.85

SYNS: ACID LEATHER RED BG □ ACID RED 114 □ AMACID MILLING RED PRS □ BENZYL FAST RED BG □ BENZYL RED BR □ CERVEN KYSELA 114 □ C.I. 23635 □ C.I. ACID RED 114 □ ELCACID MILLING FAST RED RS □ ERIONYL RED RS □ FENAFOR RED PB □ FOLAN RED B □ INTRAZONE RED BR □ KAYANOL MILLING RED RS □ LEATHER FAST RED B □ LEVANOL RED GG □ MIDLON RED PRS □ MILLING FAST RED B □ MILLING RED B □ MILLING RED BB □ MILLING RED SWB □ 1,3-NAPHTHALENEDISULFONIC ACID, 8-((3,3′-DIMETHYL-4′-((4-(((4-METHYLPHENYL)SULFONYL)OXY) PHENYL)

AZO)(1,1'-BIPHENYL)-4-YL)AZO)-7-HYDROXY-, DISODIUM SALT □ NCI C61096 □ POLAR RED RS □ SANDOLAN RED N-RS □ SELLA FAST RED RS □ SULPHONOL FAST RED R □ SULPHONOL RED R □ SUMINOL MILLING RED RS □ SUPRANOL FAST RED 3G □ SUPRANOL FAST RED GG □ SUPRANOL RED PBX-CF □ SUPRANOL RED R □ TELON FAST RED GG □ TETRACID MILLING RED B □ TETRACID MILLING RED G □ VONDAMOL FAST RED RS

TOXICITY DATA WITH REFERENCE
mmo-sat 300 nmol/plate MUREAV 136,33,84
mma-sat 125 μg/plate VHTODE 22,413,80
orl-rat TDLo: 15 mg/kg/2Y-C:CAR NTPTR* NTP-TR-405,91

CONSENSUS REPORTS: IARC Cancer Review: Group 2B IMEMDT 57,247,93; Animal Sufficient Evidence IMEMDT 57,247,93; Human Inadequate Evidence IMEMDT 57,247,93. NTP Carcinogenesis Studies (drinking); Clear Evidence: rat NTPTR* NTP-TR-405,91. Reported in EPA TSCA Inventory.

SAFETY PROFILE: Confirmed carcinogen with experimental carcinogenic data. Mutation data reported. When heated to decomposition it emits toxic vapors of NO_x and SO_x.

CMM510 CAS:8004-92-0 HR: 1
C.I. ACID YELLOW 3
mf: $C_{18}H_9NO_8S_2 \cdot 2Na$ mw: 477.38

SYNS: ACID YELLOW 3 □ CHINOGELB □ CHINOGELB EXTRA □ CHINOGELB WASSERLOESLICH □ C.I. 801 □ C.I. 918 □ C.I. 47005 □ C.I. FOOD YELLOW 3 □ C.I. FOOD YELLOW 13 □ D and C YELLOW No. 10 □ DYE QUINOLINE YELLOW □ E 104 □ FD and C YELLOW No. 10 □ FOOD YELLOW 13 □ JAPAN YELLOW 203 □ JAUNE de QUINOLEINE □ LEMON YELLOW ZN 3 □ L-GELB 3 □ QUINIDINE YELLOW KT □ QUINOLINE YELLOW □ QUINOLINE YELLOW EXTRA □ 2-(2-QUINOLYL)-1,3-INDANDIONE DISULFONIC ACID DISODIUM SALT □ SCHULTZ No. 918 □ ZLUT CHINOLONOVA □ ZLUT KYSELA 3 □ ZLUT POTRAVINARSKA 13

TOXICITY DATA WITH REFERENCE
orl-rat LD50: 2 g/kg SCPHA4 47,39,79

CONSENSUS REPORTS: Reported in EPA TSCA Inventory.

SAFETY PROFILE: Low toxicity by ingestion. When heated to decomposition it emits toxic vapors of NO_x and SO_x.

CMM758 CAS:2706-28-7 HR: 2
C.I. ACID YELLOW 9, DISODIUM SALT (8CI)
mf: $C_{12}H_9N_3O_6S_2 \cdot 2Na$ mw: 401.34

SYNS: ACETYL YELLOW G □ ACID YELLOW □ ACID YELLOW AT □ ACID YELLOW G □ ACID YELLOW GEIGY □ ACID YELLOW G KOND □ ACILAN YELLOW EXTRA □ AMACID YELLOW RG □ 4-AMINOAZOBENZENE-3,4'-DISULFONIC ACID DISODIUM SALT □ BENZENESULFONIC ACID, 2-AMINO-5-((4-SULFOPHENYL)AZO)-, DISODIUM SALT □ C.I. 13015 □ C.I. ACID YELLOW 9 □ C.I. FOOD YELLOW 2 □ CILEFA YELLOW R □ FAST YELLOW AB □ FAST YELLOW EXTRA SPECIALLY PURE □ FAST YELLOW S □ FAST YELLOW S EXTRA SPECIALLY PURE □ FAST YELLOW Y □ FOOD YELLOW 2 □ GOLDEN YELLOW RUAF □ HEXACOL ACID YELLOW G □ KITON YELLOW EXTRA □ LACQUER YELLOW T □ WOOL YELLOW G □ YELLOW ACID □ ZLUT KYSELA 9 □ ZLUT POTRAVINARSKA 2

TOXICITY DATA WITH REFERENCE
ipr-rat LD50: 2 g/kg 85JCAE-,1312,86
ivn-rat LD50: 2500 mg/kg 85JCAE-,1312,86

CONSENSUS REPORTS: Reported in EPA TSCA Inventory.

SAFETY PROFILE: Moderately toxic by intravenous route. When heated to decomposition it emits toxic vapors of NO_x and SO_x.

CMM759 CAS:6375-55-9 HR: D
C.I. ACID YELLOW 42 DISODIUM SALT
mf: $C_{32}H_{24}N_8O_8S_2 \cdot 2Na$ mw: 758.74

SYNS: ACID ANTHRACENE YELLOW GR □ ACID FAST YELLOW MR □ ACID LEATHER YELLOW CRS □ ACID YELLOW 42 □ ACID YELLOW K □ AIREDALE YELLOW 3GM □ AMACID FAST YELLOW RS □ BELACID MILLING YELLOW R □ BENZYL FAST YELLOW RS □ (1, 1'-BIPHENYL)-2,2'-DISULFONIC ACID, 4,4'-BIS((4,5-DIHYDRO-3-METHYL-5-OXO-1-PHENYL-1H-PYRAZOL-4-YL)AZO)-, DISODIUM SALT □ CALCOCID MILLING YELLOW R □ C.I. 22910 □ C.I. ACID YELLOW 42 □ COOMASSIE YELLOW R □ COOMASSIE YELLOW RP □ CYANINE FAST YELLOW M □ ERIO FAST YELLOW RL □ FAST SILK YELLOW SH □ KCA ACID MILLING YELLOW M □ MIDLON YELLOW PROPYL □ MILLING YELLOW 3G □ MILLING YELLOW 3J □ MILLING YELLOW RX □ NAPHTHALENE LEATHER YELLOW 2G □ OPTANOL YELLOW R □ PHARMACINE YELLOW R □ PHARMATEX YELLOW G □ SHIKISO ACID FAST YELLOW MR □ SULFONINE YELLOW CSR □ SULPHON YELLOW RS-CF □ SUMINOL MILLING YELLOW MR □ SUPRANOL YELLOW R □ SYMULON ACID FAST YELLOW MR □ TERTRACID MILLING YELLOW R □ XYLENE MILLING YELLOW SH

TOXICITY DATA WITH REFERENCE
mmo-sat 500 μg/plate VHTODE 22,413,80

CONSENSUS REPORTS: Reported in EPA TSCA Inventory.

SAFETY PROFILE: Mutation data reported. When heated to decomposition it emits toxic vapors of NO_x and SO_x.

CMM760 CAS:5042-54-6 HR: D
C.I. BASIC ORANGE 1
mf: $C_{13}H_{14}N_4$ mw: 226.31

SYNS: 1,3-BENZENEDIAMINE, 4-METHYL-6-(PHENYLAZO)- □ BRILLIANT OIL ORANGE R BASE □ CHRYSOIDINE R BASE □ CHRYSOIDINE 3RN BASE □ C.I. 11320B □ 2,4-DIAMINO-5-METHYLAZOBENZENE □ 4-METHYL-6-(PHENYLAZO)-1,3-BENZENEDIAMINE □ m-PHENYLENEDIAMINE, 4-METHYL-6-(PHENYLAZO)- □ TOLUENE-2,4-DIAMINE, 5-(PHENYLAZO)-(6CI)

TOXICITY DATA WITH REFERENCE
mma-sat 20 μg/plate MUREAV 240,227,90
dns-rat:lvr 500 ng/well MUREAV 240,227,90

CONSENSUS REPORTS: Reported in EPA TSCA Inventory.

SAFETY PROFILE: Mutation data reported. When heated to decomposition it emits toxic vapors of NO_x.

CMM764 **CAS:3056-93-7** **HR: 2**
C.I. BASIC ORANGE 21
mf: $C_{22}H_{22}N_2 \cdot ClH$ mw: 350.92

SYNS: AIZEN CATHILON ORANGE GL □ AIZEN CATHILON ORANGE GLH □ ASTRAZON ORANGE □ ASTRAZON ORANGE G □ BASIC ORANGE 21 □ CATIONIC ORANGE ZH □ C.I. 48035 □ GENACRYL ORANGE G □ 3H-INDOLIUM, 2-(2-(2-METHYLINDOL-3-YL) VINYL)-1,3,3-TRIMETHYL-, CHLORIDE □ NABOR ORANGE G □ ORANGE ZH □ SANDOCRYL ORANGE B-G □ SEVRON ORANGE G □ SUMIACRYL ORANGE G

TOXICITY DATA WITH REFERENCE
orl-rat LD50:2450 mg/kg GISAAA 51(1),61,86
orl-mus LD50:490 mg/kg GISAAA 51(1),61,86
orl-rbt LD50:4500 mg/kg GISAAA 51(1),61,86

CONSENSUS REPORTS: Reported in EPA TSCA Inventory.

SAFETY PROFILE: Moderately toxic by ingestion. When heated to decomposition it emits toxic vapors of NO_x, HCl, and Cl^-.

CMM765 **CAS:6320-14-6** **HR: 3**
C.I. BASIC RED 12
mf: $C_{25}H_{29}N_2 \cdot Cl$ mw: 393.01

SYNS: ACRONOL PHLOXINE FF □ AIZEN ASTRA PHLOXINE FF □ ASTRAPHLOXIN □ ASTRA PHLOXINE □ ASTRA PHLOXINE G □ ASTRA PHLOXINE G EXTRA □ ASTRAPHLOXIN G □ BRILLIANT PINK AS □ CALCOZINE RED BG Liquid □ C.I. 48070 □ COSMOPHLOXINE F □ 3H-INDOLIUM, 1,3,3-TRIMETHYL-2-(3-(1,3,3-TRIMETHYL-2-INDOLINYLIDENE)PROPENYL)-, CHLORIDE □ TOKYO ANILINE ASTRAPHLOXINE FF □ VERONA PAPER RED

TOXICITY DATA WITH REFERENCE
orl-rat LD50:18 mg/kg EPASR* 8EHQ-1285-0581

CONSENSUS REPORTS: Reported in EPA TSCA Inventory.

SAFETY PROFILE: A poison by ingestion. When heated to decomposition it emits toxic vapors of NO_x and Cl^-.

CMM768 **CAS:3648-36-0** **HR: 2**
C.I. BASIC RED 13
mf: $C_{22}H_{26}ClN_2 \cdot Cl$ mw: 389.40

SYNS: AIZEN CATHILON PINK FG □ AIZEN CATHILON PINK FGH □ ASTRAZON PINK FG □ BASIC RED 13 □ BASIC ROSE 2S □ CATHILON PINK FGH □ CATIONIC PINK 2S □ CATIONIC ROSE 2S □ C.I. 48015 □ GENACRYL PINK G □ 3H-INDOLIUM, 2-(p-((2-CHLOROETHYL)METHYLAMINO)STYRYL)-1,3,3-TRIMETHYL-, CHLORIDE □ NABOR BRILLIANT PINK 28 □ PINK 2S

TOXICITY DATA WITH REFERENCE
orl-rat LD50:2260 mg/kg GISAAA 51(1),61,86
orl-mus LD50:430 mg/kg GISAAA 51(1),61,86
orl-rbt LD50:3750 mg/kg GISAAA 51(1),61,86

CONSENSUS REPORTS: Reported in EPA TSCA Inventory.

SAFETY PROFILE: Moderately toxic by ingestion. When heated to decomposition it emits toxic vapors of NO_x and Cl^-.

CMM770 **CAS:42373-04-6** **HR: D**
C.I. BASIC RED 29
mf: $C_{19}H_{17}N_4S \cdot Cl$ mw: 368.91

SYNS: BASACRYL RED GL □ BASIC RED 29 □ 3-METHYL-2-((1-METHYL-2-PHENYL-1H-INDOL-3-YL)AZO)THIAZOLIUM CHLORIDE □ THIAZOLIUM, 3-METHYL-2-((1-METHYL-2-PHENYL-1H-INDOL-3-YL) AZO)-, CHLORIDE

TOXICITY DATA WITH REFERENCE
mmo-sat 1 mg/plate MUREAV 155,17,85
msc-mus:lym 6 mg/L MUREAV 189,223,87

CONSENSUS REPORTS: Reported in EPA TSCA Inventory.

SAFETY PROFILE: Mutation data reported. When heated to decomposition it emits toxic vapors of NO_x, SO_x, and Cl^-.

CMN000 **CAS:505-75-9** **HR: 3**
CICUTOXIN
mf: $C_{17}H_{22}O_2$ mw: 258.39

PROP: Crystals from Et_2O/pet ether. Mp: 54°.

SYN: 8,10,12-HEPTADECATRIENE-4,6-DIYNE-1,14-DIOL, (E,E,E)-(−)-

TOXICITY DATA WITH REFERENCE
ipr-mus LD50:48,300 µg/kg NYKZAU 52,29S,56
orl-cat LDLo:7 mg/kg HBAMAK 4,1289,35
ivn-cat LDLo:5360 µg/kg NYKZAU 52,29S,56
unr-frg LDLo:61 mg/kg HBAMAK 4,1289,35

SAFETY PROFILE: Poison by ingestion, intravenous, and possibly other routes. When heated to decomposition it emits acrid smoke and irritating fumes.

CMN125 **CAS:64440-87-5** **HR: 3**
CIDEFERRON

TOXICITY DATA WITH REFERENCE
ivn-rat LD50:385 mg/kg NIIRDN 6,923,82
ivn-mus LD50:403 mg/kg NIIRDN 6,923,82
ivn-rbt LD50:280 mg/kg NIIRDN 6,923,82

SAFETY PROFILE: Poison by intravenous route.

CMN230 **CAS:2945-96-2** **HR: D**
C.I. DIRECT BLACK 17, MONOSODIUM SALT
mf: $C_{24}H_{22}N_6O_5S \cdot Na$ mw: 529.57

SYNS: AZINE FAST BLACK D □ AZOGEN BLACK D □ BENZAMIN BLACK DS □ BENZANIL FAST BLACK D □ BENZO GREY LBGV □ CARBIDE BLACK D □ CARBIDE BLACK DU □ CERN PRIMA 17 □ CHLORAMINE BLACK SD □ CHLORAZOL DIAZO BLACK SD □ CHROME LEATHER BLACK D □ C.I. 27700 □ C.I. DIRECT BLACK 17 □ DIACOTTON FAST BLACK D □ DIAPHTOGENE BLACK D □ DIAZAMINE BLACK D □ DIAZINE BLACK DR □ DIAZO BLACK D □ DIAZO FAST BLACK D □ DIAZO FAST BLACK MD □ DIAZOL BLACK SD □ DIAZOPHENYL BLACK D □ DIAZOPHENYL BLACK DC □ DIRECT BLACK 17 □ DURAZOL GRAY B □ FENAMIN DIAZO BLACK D □ GIUBA BLACK D □ HISPADIAZO BLACK D □ INDIGENE BLACK D □ JAPANOL FAST BLACK D □ KAYAKU DIRECT FAST BLACK D □ MITSUI DIRECT BLACK D □ 2-NAPHTHALENESULFONIC ACID, 6-AMINO-3-((4-((4-AMINOPHENYL)AZO)-2-METHOXY-5-METHYLPHENYL)AZO)-4-HYDROXY-, MONOSODIUM SALT □ PHENAZO BLACK D □ TELON FAST BLACK PE □ TERTRODIRECT BLACK ZD □ TRIAMINE BLACK

C

D □ VONDACEL BLACK D □ ZAMBESI BLACK D □ ZAMBESI BLACK DA-CF

TOXICITY DATA WITH REFERENCE
mmo-sat 100 μg/plate MUREAV 68,307,79

CONSENSUS REPORTS: Reported in EPA TSCA Inventory.

SAFETY PROFILE: Mutation data reported. When heated to decomposition it emits toxic vapors of NO_x and SO_x.

CMN240 CAS:6428-31-5 HR: D
C.I. DIRECT BLACK 19, DISODIUM SALT
mf: $C_{34}H_{29}N_{13}O_7S_2{\bullet}2Na$ mw: 841.86

SYNS: ARTIFICIAL SILK BLACK G □ ARTIFICIAL SILK BLACK GN □ ARTIFICIAL SILK BLACK GR □ ATLANTIC ARTIFICIAL SILK BLACK G □ BALI VISCOSE BLACK G □ BALI VISCOSE BLACK N □ BENCIDAL FAST BLACK G □ BENZANIL FAST BLACK G □ BENZO FAST BLACK G □ BENZOFORM BLACK RRA-CF □ CERN PRIMA 19 □ CHLORAZOL VISCOSE BLACK B □ CHROME LEATHER BLACK GNA □ C.I. 35255 □ C.I. DIRECT BLACK 19 □ COLUMBIA FAST BLACK G □ COLUMBIA FAST BLACK GB □ CORANIL DIRECT BLACK B □ CUTAMIN BLACK CG □ DIAMINE FAST BLACK B □ DIAPHTAMINE FAST BLACK KG □ DIRECT ARTIFICIAL SILK BLACK G □ DIRECT BLACK 19 □ DIRECT FAST BLACK G □ DIRECT FAST BLACK GU □ DIRECT FAST BLACK SA □ DIRECT RAYON BLACK KSG □ FENAMIN BLACK GR □ FORMAL FAST BLACK 2B □ HISPAMIN FAST BLACK CG □ INDOBLACK GR □ KAYARUS BLACK G □ KAYARUS BLACK G CONC. □ ORIENT WATER BLACK 100L □ ORIENT WATER BLACK 200L □ PHENAMINE VISCOSE BLACK RR □ PYRAZOL FAST BLACK GS □ RAYON BLACK G □ RAYON BLACK GSN □ RAYON BLACK M □ RAYON FAST BLACK B □ SOLAR BLACK G □ SUMILIGHT BLACK G □ TETRODIRECT BLACK V □ TETRAZO DEEP BLACK GC EXTRA □ VISCO BLACK N □ VISCOSE BLACK G □ VISCOSE BLACK GNA □ VISCOSE BLACK J □ VISCOSE BLACK N □ VISCOSE BLACK NG □ VONDACEL BLACK VG □ VONDACEL BLACK VN □ WATER BLACK 100 □ WATER BLACK 200L □ WATER BLACK P 200

TOXICITY DATA WITH REFERENCE
mma-sat 100 nmol/plate MUREAV 156,131,85
bfa-rat:sat 500 mg/kg MUREAV 156,131,85

CONSENSUS REPORTS: Reported in EPA TSCA Inventory.

SAFETY PROFILE: Mutation data reported. When heated to decomposition it emits toxic vapors of NO_x and SO_x.

CMN800 CAS:2429-73-4 HR: 1
C.I. DIRECT BLUE 2
mf: $C_{32}H_{24}N_6O_{11}S_3{\bullet}3Na$ mw: 833.77

SYNS: AIREDALE BLACK BHD □ AIZEN DIRECT BLACK BH □ ALTAZINE BLACK BH □ AMANIL DEVELOPED BLACK BHSW □ ATUL DEVELOPED BLACK BT □ AZINE DIAZO BLACK BHK □ AZOCARD BLUE BH □ AZOMINE BLACK BH □ BELAMINE DIAZO BLACK BH □ BENCIDAL NAVY BLUE BH □ BENZANIL BLACK BH □ BENZO BLACK BLUE BH □ BENZO BLACK BLUE FBH □ BLUE BH □ BRASILAZOL BLACK BH □ CALCOLOID DIAZO BLACK BHL □ CALCOMINE DIAZO BLACK BHD □ CALCOMINE DIAZO BLACK BTCW □ CHLORAMINE BLACK BH □ CHLORAZOL BLACK BH □ CHLORAZOL LEATHER BLACK BH □ CHROME LEATHER BLACK BH □ CHROME LEATHER BLACK CR □ CHROME LEATHER BLACK DS □ CHROME LEATHER DARK BLUE BHM □ C.I. 22590 □ C.I. DIRECT BLUE 2, TRISODIUM SALT □ CUTAMIN DARK BLUE CB □ DIACOTTON BLACK BH □ DIAMINE BLACK BH □ DIAMINE BLACK BHM □ DIAMINOGENE VELOUR BLACK B □ DIANIL DARK BLUE H □ DIAPHTAMINE BLACK BH □ DIAZINE BLACK BHC □ DIAZINE BLACK H □ DIAZINE BLACK HDW □ DIAZINE BLACK HNJ □ DIAZO BLACK BH □ DIAZO BLACK BHN-CF □ DIAZO BLACK BHSW □ DIAZO BLACK BHSWK □ DIAZO BLACK CR □ DIAZO DIRECT BLACK N □ DIAZO FAST BLACK BH □ DIAZO FAST BLACK MBH □ DIAZOL BLACK BH □ DIAZO NAVY BLUE BH □ DIAZOPHENYL BLACK BH □ DIPHENYL BLUE BLACK GHS □ DIPHENYL BLUE BLACK MBH □ DIRECT BLACK BH □ DIRECT BLUE 2 □ DIRECT BLUE BLACK BH □ DIRECT DARK BLUE BH □ DIRECT DIAZO BLACK □ DIRECT DIAZO BLACK C □ DIRECT DIAZO BLACK N □ DIRECT DIAZO BLACK S □ DIRECT NAVY BLUE BH □ ENIAZOL BLUE BLACK BHN □ FENAMIN NAVY BLUE H □ FIXANOL BLUE BH □ INDOXINE KL □ JAPANOL BLACK BHK □ KAYAKU DIRECT BLACK BH □ MELANTHERINE BH □ MELANTHERINE BHX □ MITSUI DIRECT BLACK BH □ 2,7-NAPHTHALENEDISULFONIC ACID, 5-AMINO-3-((4′-((7-AMINO-1-HYDROXY-3-SULFO-2-NAPHTHALENYL) AZO)(1,1′-BIPHENYL)-4-YL)AZO)-4-HYDROXY-, TRISODIUM SALT □ NAVY BLUE EMBL □ NCI-C61110 □ NEKLAMIN BLACK BH □ PARAMINE BLACK BH □ PHENAZO BLACK BH □ PHENO NAVY BLUE □ PONTAMINE DEEP BLUE BH □ PONTAMINE DIAZO BLACK BHSW □ SYMULON DIRECT BLACK BH □ TERTRODIRECT BLACK BH □ TERTRODIRECT BLACK BHS □ UNION FAST NAVY BLUE DS □ VONDACEL DARK BLUE BH □ ZAMBESI DARK BLUE BH

TOXICITY DATA WITH REFERENCE
eye-rbt 100 mg MLD TSCAT* OTS 215154
mmo-sat 1 mg/plate ENMUDM 8(Suppl 7),1,86
mma-sat 1 mg/plate ENMUDM 8(Suppl 7),1,86
orl-rat LD50:6450 mg/kg TSCAT* OTS 215154
orl-rbt LDLo:10 g/kg TSCAT* OTS 215154

CONSENSUS REPORTS: Reported in EPA TSCA Inventory.

SAFETY PROFILE: Low toxicity by ingestion. An eye irritant. Mutation data reported. When heated to decomposition it emits toxic vapors of NO_x and SO_x.

CMO000 CAS:2602-46-2 HR: 3
C.I. DIRECT BLUE 6, TETRASODIUM SALT
mf: $C_{32}H_{20}N_6O_{14}S_4{\bullet}4Na$ mw: 932.78

PROP: A dye.

SYNS: AIREDALE BLUE 2BD □ AIZEN DIRECT BLUE 2BH □ AMANIL BLUE 2BX □ ATLANTIC BLUE 2B □ ATUL DIRECT BLUE 2B □ AZOCARD BLUE 2B □ AZOMINE BLUE 2B □ BELAMINE BLUE 2B □ BENCIDAL BLUE 2B □ BENZANIL BLUE 2B □ BENZO BLUE GS □ BLUE 2B □ BRASILAMINA BLUE 2B □ CALCOMINE BLUE 2B □ CHLORAMINE BLUE 2B □ CHLORAZOL BLUE B □ CHROME LEATHER BLUE 2B □ C.I. 22610 □ CRESOTINE BLUE 2B □ DIACOTTON BLUE BB □ DIAMINE BLUE 2B □ DIAPHTAMINE BLUE BB □ DIAZINE BLUE 2B □ DIAZOL BLUE 2B □ DIPHENYL BLUE 2B □ DIRECT BLUE 6 □ ENIANIL BLUE 2BN □ FENAMIN BLUE 2B □ FIXANOL BLUE 2B □ HISPAMIN BLUE 2B □ INDIGO BLUE 2B □ KAYAKU DIRECT □ MITSUI DIRECT BLUE 2BN □ NAPHTAMINE BLUE 2B □ NB2B □ NCI-C54579 □ NIAGARA BLUE 2B □ NIPPON BLUE BB □ PARAMINE BLUE 2B □ PHENAMINE BLUE BB □ PHENO BLUE 2B □ PONTAMINE BLUE BB □ SODIUM DIPHENYL-4,4′-BIS-AZO-2″-8″-AMINO-1″-NAPHTHOL-3″,6″ DISULPHONATE □ TERTRODIRECT BLUE 2B □ VONDACEL BLUE 2B

TOXICITY DATA WITH REFERENCE
mma-sat 100 nmol/plate MUREAV 136,33,94
dnd-rat-ipr 61,200 μg/kg TXCYAC 32,315,84

ipr-rat TDLo: 200 mg/kg (8D preg): TER PSEBAA 127,215,68
ipr-rat TDLo: 200 mg/kg (female 8D post): TER TJADAB 2,85,69
ipr-rat TDLo: 140 mg/kg (8D preg): REP PSEBAA 127,215,68
scu-rat TDLo: 150 mg/kg (female 8D post): TER BIJOAK 91,14P,64
orl-rat TDLo: 5250 mg/kg/5W-C: CAR NCITR* NCI-CG-TR-108,78
scu-rat TDLo: 750 mg/kg/27W-I: ETA BJEPA5 38,291,57
orl-mus TDLo: 34 g/kg/60W-C: ETA TJIDAH 88,467,73
imp-mus TDLo: 80 mg/kg: ETA TJIDAH 88,467,73
orl-rat TD: 2310 mg/kg/4W-C: NEO NCITR* NCI-CG-TR-108,78

CONSENSUS REPORTS: NTP 7th Annual Report on Carcinogens. IARC Cancer Review: Animal Sufficient Evidence IMEMDT 29,311,82; Human Inadequate Evidence IMEMDT 29,311,82. NCI Carcinogenesis Bioassay (feed); Clear Evidence: rat NCITR* NCI-CG-TR-108,78; No Evidence: mouse NCITR* NCI-CG-TR-108,78. Reported in EPA TSCA Inventory. Community Right-To-Know List.

SAFETY PROFILE: Confirmed carcinogen with experimental carcinogenic, neoplastigenic, and tumorigenic data. Experimental teratogenic and reproductive effects. Mutation data reported. When heated to decomposition it emits very toxic fumes of NO_x, Na_2O, and SO_x.

For occupational chemical analysis use NIOSH: Dyes, 5013.

CMO250 CAS:72-57-1 HR: 3
C.I. DIRECT BLUE 14, TETRASODIUM SALT
mf: $C_{34}H_{28}N_6O_{14}S_4$•4Na mw: 964.88

PROP: Dark blue crystals or powder. Sol in H_2O and acids.

SYNS: AMANIL SKY BLUE □ AMIDINE BLUE 4B □ AZIDINE BLUE 3B □ AZZURRO DIRETTO 3B □ BENCIDAL BLUE 3B □ BENZAMINE BLUE □ BENZO BLUE □ BLEU DIAMINE □ BLUE EMB □ BRASILAMINA BLUE 3B □ CENTRALINE BLUE 3B □ CHLORAMINE BLUE □ CHLORAZOL BLUE 3B □ CHROME LEATHER BLUE 3B □ C.I. 23850 □ C.I. DIRECT BLUE 14 □ CONGOBLAU 3B □ CONGO BLUE □ CRESOTINE BLUE 3B □ DIAMINE BLUE 3B □ DIANILBLAU □ DIANIL BLUE □ DIAZINE BLUE 3B □ DIPHENYL BLUE 3B □ DIRECT BLUE 14 □ HISPAMIN BLUE 3BX □ NAPHTAMINE BLUE 2B □ NAPHTHYLAMINE BLUE □ NCI-C61289 □ NIAGARA BLUE □ ORION BLUE 3B □ PARAMINE BLUE 3B □ PARKIBLEU □ PARKIPAN □ PONTAMINE BLUE 3BX □ PYRAZOL BLUE 3B □ PYROTROPBLAU □ RCRA WASTE NUMBER U236 □ RENOLBLAU 3B □ SODIUM DITOLYLDIAZOBIS-8-AMINO-1-NAPHTHOL-3,6-DISULFONATE □ SODIUM DITOLYLDIAZOBIS-8-AMINO-1-NAPHTHOL-3,6-DISULPHONATE □ TB □ TRIANOL DIRECT BLUE 3B □ TRIPAN BLUE □ TRYPANBLAU (GERMAN) □ TRYPAN BLUE □ TRYPAN BLUE SODIUM SALT

TOXICITY DATA WITH REFERENCE
mma-sat 250 μg/plate MUREAV 56,249,78
dnd-esc 20 μmol/L MUREAV 89,95,81
dns-rat: lvr 10 μmol/L MUREAV 136,255,84
bfa-rat/sat 500 mg/kg MUREAV 156,131,85
dns-ham: lvr 100 μmol/L MUREAV 136,255,84
scu-rat TDLo: 10 mg/kg (female 7D post): REP ESKHA5 (103),75,85
ipr-mus TDLo: 1 g/kg (female 8-12D post): REP TCMUD8 7,7,87
scu-ham TDLo: 20 μg/kg (female 8D post): TER LIFSAK 8,525,69
ipr-rat TDLo: 30 mg/kg (female 8D post): TER TJADAB 37,43,88
scu-gpg TDLo: 40 mg/kg (female 11D post): REP ANREAK 141,173,61
orl-rat TDLo: 150 mg/kg (7-9D preg): REP MEXPAG 10,201,64
ipr-rat TDLo: 12,500 μg/kg (female 8D post): TER JPETAB 144,429,64
ipr-rat TDLo: 40 mg/kg (female 8D post): TER TJADAB 37,43,88
orl-rat TDLo: 150 mg/kg (7-9D preg): TER MEXPAG 10,201,64
scu-mus TDLo: 45 mg/kg (female 6-14D post): TER NTIS** PB223-160
ipr-rat TDLo: 30 mg/kg (female 8D post): TER TJADAB 37,43,88
scu-rbt TDLo: 150 mg/kg (female 1D pre): TER ANREAK 133,513,59
ipr-mus TDLo: 100 mg/kg (female 7D post): TER JEZOAO 171,343,69
scu-uns TDLo: 50 mg/kg (female 13D post): TER TJADAB 18,187,78
scu-rat TDLo: 630 mg/kg/43W-I: CAR BJEPA5 33,524,52
par-rat TDLo: 250 mg/kg/10W-I: ETA CANCAR 5,792,52
scu-rat TD: 7500 mg/kg/86W-I: NEO LAINAW 12,1221,63
scu-mus TD: 1275 mg/kg/34W-I: ETA APHGBP 33,1,53
scu-rat TD: 1300 mg/kg/1Y-I: ETA AJPAA4 106,326,82
scu-rat TD: 300 mg/kg/6W-I: ETA AJPAA4 106,326,82
par-rat TD: 620 mg/kg/31W-I: ETA GANNA2 48,573,57
unr-rat TD: 700 mg/kg/4W-I: ETA FZPAAZ 75,74,66
orl-rat LD50: 6200 mg/kg TSCAT* OTS 215154
ipr-rat LDLo: 300 mg/kg PSEBAA 31,825,34
scu-rat LDLo: 300 mg/kg PSEBAA 31,825,34
ivn-rat LDLo: 300 mg/kg PSEBAA 31,825,34
ipr-mus LDLo: 350 mg/kg PSEBAA 31,825,34
scu-mus LD50: 267 mg/kg NNAPBA 267,31,70
ivn-mus LD50: 328 mg/kg TXAPA9 23,537,72
ipr-rbt LDLo: 400 mg/kg PSEBAA 31,825,34
ivn-rbt LDLo: 100 mg/kg PSEBAA 31,825,34
ipr-gpg LDLo: 250 mg/kg PSEBAA 31,825,34
scu-gpg LDLo: 300 mg/kg PSEBAA 31,825,34

CONSENSUS REPORTS: IARC Cancer Review: Group 2B IMEMDT 7,56,87; Animal Sufficient Evidence IMEMDT 8,267,75. EPA Genetic Toxicology Program. Reported in EPA TSCA Inventory.

SAFETY PROFILE: Suspected carcinogen with experimental carcinogenic, neoplastigenic, and tumorigenic data. Poison by intraperitoneal, intravenous, and subcutaneous routes. Experimental teratogenic and reproductive effects. Mutation data reported. When heated to decomposition it emits very toxic fumes of NO_x, Na_2O, and SO_x.

CMO500 CAS:2429-74-5 HR: 3
C.I. DIRECT BLUE 15, TETRASODIUM SALT
mf: $C_{34}H_{28}N_6O_{16}S_4$•4Na mw: 996.88

SYNS: AIREDALE BLUE D □ AIZEN DIRECT SKY BLUE 5BH □ AMANIL SKY BLUE □ ATLANTIC SKY BLUE A □ ATUL DIRECT SKY

BLUE □ AZINE SKY BLUE 5B □ BELAMINE SKY BLUE A □ BENZANIL SKY BLUE □ BENZO SKY BLUE S □ BENZO SKY BLUE A-CF □ CHLORAMINE SKY BLUE A □ CHLORAMINE SKY BLUE 4B □ CHROME LEATHER PURE BLUE □ C.I. 24400 □ C.I. DIRECT BLUE 15 □ CRESOTINE PURE BLUE □ DIACOTTON SKY BLUE 5B □ DIAMINE SKY BLUE CI □ DIAPHTAMINE PURE BLUE □ DIAZOL PURE BLUE 4B □ DIPHENYL BRILLIANT BLUE □ DIPHENYL SKY BLUE 6B □ DIRECT BLUE 10G □ DIRECT BLUE 15 □ DIRECT BLUE HH □ DIRECT PURE BLUE □ DIRECT PURE BLUE M □ DIRECT SKY BLUE A □ ENIANIL PURE BLUE AN □ FENAMIN SKY BLUE □ HISPAMIN SKY BLUE 3B □ KAYAKU DIRECT SKY BLUE 5B □ MITSUI DIRECT SKY BLUE 5B □ MODR PRIMA 15 □ NAPHTAMINE BLUE 10G □ NCI C61290 □ NIAGARA BLUE 4B □ NIAGARA SKY BLUE □ NIPPON DIRECT SKY BLUE □ NITTO DIRECT SKY BLUE 5B □ PHENAMINE SKY BLUE A □ PONTAMINE SKY BLUE 5BX □ PONTACYL SKY BLUE 4BX □ SHIKISO DIRECT SKY BLUE 5B □ SKY BLUE 4B □ SKY BLUE 5B □ TERTRODIRECT BLUE F □ VONDACEL BLUE HH

TOXICITY DATA WITH REFERENCE
mmo-sat 500 µg/plate VHTODE 22,413,80
mma-sat 300 nmol/plate MUREAV 116,305,83
ipr-rat TDLo: 140 mg/kg (female 8D post): TER TJADAB 2,85,69
ipr-rat TDLo: 70 mg/kg (female 8D post): REP PSEBAA 127,215,68
ipr-rat TDLo: 200 mg/kg (female 8D post): REP PSEBAA 127,215,68
orl-rat TDLo: 168 g/kg/96W-C: CAR NTPTR* NTP-TR-397,92

CONSENSUS REPORTS: IARC Cancer Review: Group 2B IMEMDT 57,235,93; Animal Sufficient Evidence IMEMDT 57,235,93; Human Inadequate Evidence IMEMDT 57,235,93. Reported in EPA TSCA Inventory.

SAFETY PROFILE: A confirmed carcinogen with experimental carcinogenic data reported. An experimental teratogen. Other experimental reproductive effects. Mutation data reported. When heated to decomposition it emits very toxic fumes of NO_x, Na_2O, and SO_x.

CMO600 CAS:2586-57-40 HR: 1
C.I. DIRECT BLUE 22, DISODIUM SALT (8CI)
mf: $C_{34}H_{27}N_5O_{10}S_2{\cdot}2Na$ mw: 775.76

SYNS: AIREDALE BLUE RWD □ AMANIL BLUE RW □ ATLANTIC BLUE RW □ AZINE BRILLIANT BLUE RW □ BENZANIL BLUE RW □ BENZANOL BLUE RW □ BENZO BLUE RWA □ BENZO BLUE RWS □ BRASILAMINA BLUE RW □ CHICAGO BLUE RW □ CHLORAZOL BLUE RW □ C.I. 24280 □ C.I. DIRECT BLUE 22 (7CI) □ CUTAMIN BLUE CR □ DIACOTTON BRILLIANT BLUE RW □ DIAPHTAMINE BLUE RW □ DIAZOL BLUE RW □ DIPHENYL BLUE G □ DIRECT BLUE 22 □ DIRECT BLUE BR □ DIRECT BLUE MRW □ DIRECT BLUE RW □ DIRECT BLUE RWN □ DURAZOL BLUE 5G □ ENIANIL BLUE RW □ FENAMIN BLUE RW □ HISPAMIN BLUE RW □ JAPANOL BRILLIANT BLUE RWL (6CI) □ KAYAKU DIRECT BRILLIANT BLUE RW □ MITSUI DIRECT BLUE RW □ NAPHTAMINE BLUE RW □ 1,3-NAPHTHALENEDISULFONIC ACID, 4-AMINO-5-HYDROXY-6-((4'-((2-HYDROXY-1-NAPHTHALENYL)AZO)-3,3'-DIMETHOXY(1, 1'-BIPHENYL)-4-YL)AZO)-, DISODIUM SALT □ NIAGARA BLUE RW □ NYANZA BLUE RW □ PHENAMINE BLUE RW □ PONTAMINE BLUE RW □ SHIKISO DIRECT BRILLIANT BLUE RW □ TERTRODIRECT BLUE RW □ TRISULFON BLUE RW

TOXICITY DATA WITH REFERENCE
orl-rat LD50: 6700 mg/kg TSCAT* OTS 215154

CONSENSUS REPORTS: Reported in EPA TSCA Inventory.

SAFETY PROFILE: Low toxicity by ingestion. When heated to decomposition it emits toxic vapors of NO_x and SO_x.

CMO650 CAS:28407-37-6 HR: 2
C.I. DIRECT BLUE 218
mf: $C_{32}H_{16}Cu_2N_6O_{16}S_4{\cdot}4Na$ mw: 1087.82

SYNS: AMANIL SUPRA BLUE 9GL □ C.I. 24401 □ CUPRATE(4-), (mu-((3,3'-((3,3'-DIHYDROXY(1,1'-BIPHENYL)-4,4'-DIYL)BIS(AZO)) BIS(5-AMINO-4-HYDROXY-2,7-NAPHTHALENEDISULFONATO))(8-))) DI-, TETRASODIUM □ FASTUSOL BLUE 9GLP □ NCI C60877 □ PONTAMINE BOND BLUE B □ PONTAMINE FAST BLUE 7GLN □ SOLANTINE BLUE 10GL

TOXICITY DATA WITH REFERENCE
mmo-sat 1 mg/plate VHTODE 22,413,80
orl-rat LD50: 3290 mg/kg TSCAT* OTS 215154
orl-rbt LDLo: 3920 mg/kg TSCAT* OTS 215154
skn-rbt LDLo: 8 g/kg TSCAT* OTS 215154

CONSENSUS REPORTS: Reported in EPA TSCA Inventory.

SAFETY PROFILE: Moderately toxic by ingestion. Low toxicity by skin contact. Mutation data reported. When heated to decomposition it emits toxic vapors of NO_x, SO_x, and Cu.

CMO750 CAS:16071-86-6 HR: 3
C.I. DIRECT BROWN
mf: $C_{31}H_{20}N_6O_9S{\cdot}Cu{\cdot}2Na$ mw: 762.15

SYNS: AIZEN PRIMULA BROWN BRLH □ AMANIL SUPRA BROWN LBL □ ATLANTIC RESIN FAST BROWN BRL □ BENZAMIL SUPRA BROWN BRLL □ CALCODUR BROWN BRL □ CHLORAMINE FAST BROWN BRL □ CHROME LEATHER BROWN BRLL □ C.I. 30145 □ DERMA FAST BROWN W-GL □ DIPHENYL FAST BROWN BRL □ DIRECT BROWN 95 □ NCI-C54568 □ SATURN BROWN LBR □ SOLAR BROWN PL □ TETRAMINE FAST BROWN BRS

TOXICITY DATA WITH REFERENCE
mmo-sat 100 nmol/plate MUREAV 136,147,84
mma-sat 30 nmol/plate MUREAV 136,147,84
dns-rat-orl 100 mg/kg MUREAV 136,147,84
orl-rat TDLo: 2625 mg/kg/5W-C: NEO NCITR* NCI-CG-TR-108,78
orl-rat TD: 6825 mg/kg/13W-C: ETA TXAPA9 54,431,80
orl-rat LD50: 12,060 mg/kg TSCAT* OTS 215154

CONSENSUS REPORTS: IARC Cancer Review: Animal Limited Evidence IMEMDT 29,321,82; Human Limited Evidence IMEMDT 29,321,82; NCI Carcinogenesis Bioassay (feed); Clear Evidence: rat NCITR* NCI-CG-TR-108,78; No Evidence: mouse NCITR* NCI-CG-TR-108,78. Reported in EPA TSCA Inventory. Community Right-To-Know List.

SAFETY PROFILE: Low toxicity by ingestion. Suspected carcinogen with experimental carcinogenic and neoplastigenic data. Mutation data reported. When heated to decomposition it emits very toxic fumes of Na_2O, SO_x, and NO_x.

For occupational chemical analysis use NIOSH: Dyes, 5013.

CMO810 CAS:2586-58-5 HR: D
C.I. DIRECT BROWN 1A, DISODIUM SALT
mf: $C_{32}H_{24}N_8O_6S \cdot 2Na$ mw: 694.68

SYNS: ATLANTIC BROWN D 3Y □ ATUL DIRECT BROWN CN □ BENZOIC ACID, 5-((4'-((2,6-DIAMINO-3-METHYL-5-((4-SULFOPHENYL) AZO)PHENYL)AZO)(1,1'-BIPHE NYL)-4-YL)AZO)-2-HYDROXY-, DISODIUM SALT □ BENZO BROWN D 3GA-CF □ CHLORAZOL ORANGE BROWN X □ C.I. 30110 □ C.I. DIRECT BROWN 1:2 □ DIPHENYL BROWN PT □ DIRECT BROWN 1:2 □ DIRECT BROWN 1A □ DIRECT BROWN 5C □ DIRECT BROWN CGN □ DIRECT BROWN 5G □ DIRECT BROWN 2GS □ ENIANIL BROWN 2GS □ FIXANOL ORANGE BROWN X □ HONEY YELLOW 3GNT □ OXYDIAMINE BROWN 3GN □ PHENAMINE BROWN D 3G □ PONTAMINE BROWN D 3GN □ PONTAMINE BROWN NCR

TOXICITY DATA WITH REFERENCE
mmo-sat 30 nmol/plate MUREAV 136,33,84

CONSENSUS REPORTS: Reported in EPA TSCA Inventory.

SAFETY PROFILE: Mutation data reported. When heated to decomposition it emits toxic vapors of NO_x and SO_x.

CMO820 CAS:2429-81-4 HR: D
C.I. DIRECT BROWN 31
mf: $C_{46}H_{30}N_{10}O_{13}S_3 \cdot 4Na$ mw: 1119.00

SYNS: AIREDALE BROWN BSD □ AMANIL FAST BROWN HP □ AMANIL RAYON BROWN D □ ATLANTIC BROWN BCW □ ATLANTIC BROWN BP □ BELAMINE FAST BROWN BP □ BENZANIL BROWN BS □ BENZO DEEP BROWN NZ □ BENZOIC ACID, 5-((4'-((2,6-DIAMINO-3-((8-HYDROXY-3,6-DISULFO-7-((4-SULFO-1-NAPHTHALENYL) AZO)-2-NAPHTHALENYL)AZO)-5 METHYLPHENYL)AZO)(1,1'-BIPHENYL)-4-YL) AZO)-2-HYDROXY-, TETRASODIUM SALT □ CALCOMINE BROWN B □ CALCOMINE CATECHU 2B □ CHLORAZOL BROWN LF □ CHOCOLATE EMBL □ CHROME LEATHER BROWN DS □ C.I. 35660 □ C.I. DIRECT BROWN 31, TETRASODIUM SALT □ CUPRANIL BROWN BCW □ CUPRANIL BROWN BCWR □ DIAPHTAMINE FAST BROWN TB □ DIAZOL CUTCH F □ DIAZOL CUTCH FB □ DIPHENYL BROWN BS □ DIPHENYL BROWN TB □ DIPHENYL FAST BROWN F □ DIRECT BROWN 31 □ DIRECT BROWN B □ DIRECT BROWN 3B □ DIRECT BROWN BS □ DIRECT BROWN BSB □ DIRECT BROWN FS □ DIRECT BROWN TRB □ DIRECT FAST BROWN BP □ DIRECT FAST BROWN TSN □ DIRECT FAST BROWN TWC □ ERIE FAST BROWN B □ FENAMIN BROWN PBL □ FIXANOL BROWN LF □ HISPAMIN FAST BROWN NZ □ PHENAMINE FAST BROWN T □ PHENAMINE FAST BROWN TWC □ PONTAMINE BROWN BCW □ PONTAMINE BROWN BT □ TERTRODIRECT BROWN TB □ TRIAZOL BROWN B □ TRISULPHONE BROWN B □ VEGENTINE FAST BROWN B □ VONDACEL BROWN S □ VONDACEL BROWN SP

TOXICITY DATA WITH REFERENCE
mmo-sat 300 nmol/plate MUREAV 136,33,84
mma-sat 10 nmol/plate MUREAV 136,33,84

CONSENSUS REPORTS: Reported in EPA TSCA Inventory.

SAFETY PROFILE: Mutation data reported. When heated to decomposition it emits toxic vapors of NO_x and SO_x.

CMO825 CAS:6360-54-9 HR: 1
C.I. DIRECT BROWN 154, DISODIUM SALT (8CI)
mf: $C_{33}H_{28}N_8O_6S \cdot 2Na$ mw: 710.73

SYNS: BENZOIC ACID, 5-((4'-((2,6-DIAMINO-3-METHYL-5-((4-SULFOPHENYL)AZO)PHENYL)AZO)(1,1'-BIPHE NYL)-4-YL)AZO)-2-HYDROXY-3-METHYL-, DISODIUM SALT □ C.I. 30120 □ C.I. DIRECT BROWN 154 □ DIAMINE BROWN 3GN-CF □ DIAPHTAMINE BROWN 3GC □ DIPHENYL BROWN 3GT □ DIRECT BROWN 154 □ DIRECT BROWN CMD □ DIRECT BROWN D3Y □ DIRECT BROWN 5GR □ ERIE BROWN 3GN □ METADIAZOL BROWN 450 □ PHENAMINE BROWN 3G □ PONTAMINE BROWN N3G

TOXICITY DATA WITH REFERENCE
eye-rbt 100 mg MLD TSCAT* OTS 215154
orl-rat LD50:9960 mg/kg TSCAT* OTS 215154
orl-rbt LDLo:6330 mg/kg TSCAT* OTS 215154

CONSENSUS REPORTS: Reported in EPA TSCA Inventory.

SAFETY PROFILE: Low toxicity by ingestion. An eye irritant. When heated to decomposition it emits toxic vapors of NO_x and SO_x.

CMO830 CAS:3626-28-6 HR: D
C.I. DIRECT GREEN 1
mf: $C_{34}H_{25}N_7O_8S_2 \cdot 2Na$ mw: 769.76

SYNS: AIREDALE GREEN BWD □ AIZEN DIRECT DARK GREEN BH □ AMANIL GREEN LT □ ATLANTIC DARK GREEN □ ATLANTIC GREEN WT □ ATUL DIRECT DARK GREEN P □ AZINE DARK GREEN BH/C □ AZOCARD DARK GREEN B □ BENCIDAL DARK GREEN B □ BENZANIL DARK GREEN BW □ BENZO DARK GREEN B □ BENZO DARK GREEN BA-CF □ BRASILAMINA GREEN B □ CALCOMINE DARK GREEN BG □ CHLORAZOL DARK GREEN PL □ CHROME LEATHER DARK GREEN N □ CHROME LEATHER DARK GREEN S □ CHROME LEATHER GREEN B □ C.I. 30280 □ CRESOTINE DARK GREEN B □ DARK GREEN EMBL □ DIACOTTON DARK GREEN □ DIAMINE DARK GREEN B □ DIAMINE DARK GREEN N □ DIAPHTHAMINE FAST BLACK FE □ DIAZINE DARK GREEN BO □ DIAZINE DARK GREEN P □ DIAZOL GREEN BLACK N □ DIPHENYL DARK GREEN B □ DIPHENYL DARK GREEN BN □ DIRECT BLACK GREEN □ DIRECT DARK GREEN A □ DIRECT DARK GREEN B □ DIRECT DARK GREEN BF □ DIRECT DARK GREEN BG □ DIRECT DARK GREEN MB □ DIRECT DARK GREEN S □ DIRECT DARK GREEN SUPRA □ DIRECT DARK GREEN WS □ DIRECT DEEP GREEN A □ DIRECT GREEN WAC □ ENIANIL DARK GREEN BG □ ERIE GREEN WT □ FENAMIN GREEN M □ HISPAMIN GREEN WT □ KAYAKU DIRECT DARK GREEN B □ MITSUI DIRECT DARK GREEN BX □ 2,6-NAPHTHALENEDISULFONIC ACID, 4-AMINO-5-HYDROXY-3-((4'-((4-HYDROXYPHENYL)AZO)(1,1'-BIPHENYL)-4-YL)AZO)-6-(PHENYLAZO)-, DISODIUM SALT □ NAPHTHAMINE DARK GREEN B □ NIPPON DARK GREEN B □ PHENAMINE DARK GREEN B □ POLYCOR DARK GREEN S □ PONTAMINE GREEN S □ SANDOPEL DARK GREEN B □ SHIKISO DIRECT DARK GREEN B □ TERTRODIRECT GREEN BG □ UNION DARK GREEN B □ VONDACEL GREEN DB

TOXICITY DATA WITH REFERENCE
mmo-sat 3333 μg/plate ENMUDM 9(Suppl 9),1,87
bfa-rat:sat 1500 mg/kg SAIGBL 22,194,80

CONSENSUS REPORTS: Reported in EPA TSCA Inventory.

SAFETY PROFILE: Mutation data reported. When heated to decomposition it emits toxic vapors of NO_x and SO_x.

CMO840 CAS:4335-09-5 HR: 2
C.I. DIRECT GREEN 6, DISODIUM SALT
mf: $C_{34}H_{24}N_8O_{10}S_2 \cdot 2Na$ mw: 814.76

SYNS: AIREDALE GREEN BD □ AIZEN DIRECT GREEN BH □ AMANIL GREEN B □ APOMINE GREEN B □ ATLANTIC GREEN 2B □ ATUL DIRECT GREEN B □ AZINE GREEN BX □ AZOCARD GREEN B □ AZOMINE GREEN B □ BELAMINE GREEN BX □ BENCIDAL GREEN B □ BENZANIL GREEN B □ BENZANIL GREEN BN □ BENZO GREEN B □ BENZO GREEN BG-CF □ BENZO GREEN CA-CF □ BENZO GREEN GA-CF □ BRASILAMINA GREEN G □ CALCOMINE GREEN BY □ CHLORAMINE GREEN B □ CHLORAMINE GREEN 2B □ CHLORAMINE GREEN BC □ CHLORAMINE GREEN 3G □ CHLORAZOL GREEN BN □ CHLORAZOL GREEN BNP □ CHLORAZOL PAPER GREEN BN □ C.I. 30295 □ C.I. DIRECT GREEN 6 (7CI) □ COTTON GREEN B □ CRESOTINE GREEN B □ DIACOTTON GREEN B □ DIAMINE GREEN B □ DIAPHTAMINE GREEN B □ DIAZINE GREEN B □ DIAZINE GREEN DB □ DIAZOL GREEN B □ DIAZOL GREEN BJ □ DIPHENYL GREEN BB □ DIPHENYL GREEN BY □ DIPHENYL GREEN C □ DIPHENYL GREEN GPD □ DIPHENYL GREEN KG □ DIPHENYL GREEN MB □ DIRECT BRILLIANT GREEN BB □ DIRECT BRILLIANT GREEN C □ DIRECT BRILLIANT GREEN CBM □ DIRECT GREEN □ DIRECT GREEN 6 □ DIRECT GREEN A □ DIRECT GREEN 2B □ DIRECT GREEN BN □ DIRECT GREEN BP □ DIRECT GREEN BX □ DIRECT GREEN MB □ ENIANIL GREEN B □ ENIANIL GREEN BBN □ ERIE GREEN GPD □ ERIE GREEN MT □ ERIE GREEN TCM □ FENAMIN GREEN A □ FENAMIN GREEN B □ FENAMIN GREEN G □ FIXANOL GREEN BN □ HISPAMIN GREEN B □ KAYAKU DIRECT GREEN B □ MITSUI DIRECT GREEN BC □ NAPHTAMINE GREEN B □ 2,7-NAPHTHALENEDISULFONIC ACID, 4-AMINO-5-HYDROXY-6-((4'-((4-HYDROXYPHENYL)AZO) (1,1'-BIPHENYL)-4-YL)AZO)-3-((4-NITROPHENYL)AZO)-, DISODIUM SALT □ NAPHTHALENE LEATHER GREEN BL □ NIPPON GREEN B □ ORBAMIN GREEN B □ PARAMINE GREEN B □ PARAMINE GREEN BN □ PHENAMINE GREEN BG □ PHENAMINE GREEN C □ PHENAMINE GREEN G □ PHENO BRIGHT GREEN □ PONTAMINE GREEN BXN □ PONTAMINE GREEN GXN □ TERTRODIRECT GREEN B □ VONDACEL GREEN B

TOXICITY DATA WITH REFERENCE
orl-rat LD50:14,730 mg/kg TSCAT* OTS 215154
orl-rbt LDLo:3990 mg/kg TSCAT* OTS 215154

CONSENSUS REPORTS: Reported in EPA TSCA Inventory.

SAFETY PROFILE: Moderately toxic by ingestion. When heated to decomposition it emits toxic vapors of NO_x and SO_x.

CMO870 CAS:2429-84-7 HR: 1
C.I. DIRECT RED 1, DISODIUM SALT (8CI)
mf: $C_{29}H_{21}N_5O_7S \cdot 2Na$ mw: 629.59

SYNS: AIREDALE RED FD □ AIZEN DIRECT FAST RED FH □ AMANIL FAST RED FS □ ATLANTIC FAST RED F □ AZINE FAST RED FC □ AZOCARD FAST RED F □ BELAMINE FAST RED FC □ BENCIDAL FAST RED F □ BENZANIL FAST RED F □ BENZANOL FAST RED F □ BENZO FAST RED F □ BENZOIC ACID, 5-((4'-((2-AMINO-8-HYDROXY-6-SULFO-1-NAPHTHALENYL)AZO)(1,1'-BIPHENYL)-4-YL)AZO)-2-HYDROXY-, DISODIUM SALT □ BRASILAMINA FAST RED F □ CALCOMINE RED FC □ CHLORAMINE FAST RED F □ CHLORAMINE FAST RED FS □ CHLORAZOL FAST RED FP □ CHLORAZOL FAST RED FS □ CHROME FAST RED F □ CHROME FAST RED FB □ CHROME FAST RED FW □ CHROME LEATHER FAST RED N □ CHROME LEATHER RED F □ CHROME LEATHER RED F EXTRA □ C.I. 22310 □ C.I. DIRECT RED 1 (6CI,7CI) □ C.I. MORDANT RED 57 □ COLUMBIA FAST RED F □ CRESOTINE FAST RED F □ CUTAMIN RED CF □ DIACOTTON FAST RED F □ DIAMINE FAST RED F □ DIAMINE FAST RED FACF □ DIAMINE FAST RED N □ DIAMINE FAST RED OJCD □ DIAPHTAMINE FAST RED FC □ DIAZINE FAST RED F □ DIAZOL FAST RED F □ DIAZOL FAST RED FS □ DIPHENYL FAST RED B □ DIPHENYL FAST RED F □ DIPHENYL RED B □ DIRECT FAST RED B □ DIRECT FAST RED F □ DIRECT FAST RED FN □ DIRECT FAST RED FR □ DIRECT FAST RED G □ DIRECT FAST RED MF □ DIRECT RED 1 □ DIRECT RED F □ DIRECT RED FR □ DIRECT RED Kh □ DIRECT RED M □ DIRECT RED MN □ DIRECT ROSE MN □ ENIANIL FAST RED F □ ERIE FAST RED FD □ FAST RED F □ FENAMIN FAST RED F □ FIXANOL RED FS □ HISPAMIN FAST RED FN □ JAPANOL FAST RED 1 □ JAPANOL FAST RED F □ KAYAKU DIRECT FAST RED F □ METACHROME RED F □ MITSUI DIRECT FAST RED F □ NAPHTAMINE FAST RED F □ NEONYL RED 2B □ NYANZA FAST RED FA □ NYLOMINE ACID RED C-R □ PARAMINE FAST RED F □ PHENAMINE FAST RED F □ PHENO FAST RED F □ POLAN RED FS □ PONTAMINE FAST RED F □ PONTAMINE FAST RED FCB □ TERTRODIRECT RED F □ UNION FAST RED 3B □ VONDACEL RED FN

TOXICITY DATA WITH REFERENCE
eye-rbt 100 mg MOD TSCAT* OTS 215154
orl-rat LD50:22,400 mg/kg TSCAT* OTS 215154
orl-rbt LDLo:6 g/kg TSCAT* OTS 215154

CONSENSUS REPORTS: Reported in EPA TSCA Inventory.

SAFETY PROFILE: Low toxicity by ingestion. An eye irritant. When heated to decomposition it emits toxic vapors of NO_x and SO_x.

CMO875 CAS:6358-29-8 HR: D
C.I. DIRECT RED 39
mf: $C_{32}H_{28}N_4O_8S_2 \cdot 2Na$ mw: 706.74

SYNS: AIREDALE SCARLET 3BD □ AMANIL FAST SCARLET 3B □ ATLANTIC SCARLET 3B □ BENZANIL SCARLET 3B □ BENZANOL BRILLIANT SCARLET 3B □ BENZO RED 3B □ BENZYL SCARLET 3BS □ CALCOMINE SCARLET 3B □ CHLORAMINE RED 3B □ CHROME LEATHER SCARLET 3BS □ C.I. 23630 □ DIAMINE SCARLET 3BA-CF □ DIAPHTAMINE FAST SCARLET □ DIAZOL SCARLET 3B □ DIPHENYL RED 3BS □ DIPHENYL SCARLET 3BS □ DIRECT FAST SCARLET 3B □ DIRECT RED 39 □ DIRECT SCARLET 3BS □ ERIE SCARLET 3B □ FENAMIN SCARLET 3B □ KAYAKU DIRECT SCARLET 3B □ MITSUI DIRECT SCARLET 3BX □ 1,3-NAPHTHALENEDISULFONIC ACID, 8-((4'-((4-ETHOXYPHENYL)AZO)-3,3'-DIMETHYL(1,1'-BIPHENYL)-4-YL) AZO)-7-HYDROXY-, DISODIUM SALT □ PAPER SCARLET 3BX □ PARAMINE FAST SCARLET 3B □ PHENAMINE SCARLET 3B □ PHENO FAST SCARLET 4B □ PONTAMINE SCARLET 3B □ SHIKISO DIRECT SCARLET 3B □ TRIAZOL FAST SCARLET 3B □ UNION FAST SCARLET 3B

TOXICITY DATA WITH REFERENCE
mmo-sat 125 μg/plate VHTODE 22,413,80

CONSENSUS REPORTS: Reported in EPA TSCA Inventory.

SAFETY PROFILE: Mutation data reported. When heated to decomposition it emits toxic vapors of NO_x and SO_x.

CMO880 CAS:6548-29-4 HR: D
C.I. DIRECT RED 46, TETRASODIUM SALT
mf: $C_{32}H_{22}Cl_2N_6O_{12}S_4 \cdot 4Na$ mw: 973.70

SYNS: ACETOPURPURINE 8B □ AMANIL CHLORAMINE RED 8BS □ BENZO BRILLIANT RED 8BS □ CHLORAMINE BRILLIANT RED 8B □ CHLORAMINE RED 8B □ CHLORAZOL BRILLIANT PURPURINE 8B □ C.I. 23050 □ C.I. DIRECT RED 46 □ DIAPHTAMINE FAST PURPU-

RINE 8B ☐ DIAZOL FAST PURPURINE 8B ☐ DIPHENYL RED 8B ☐ DIRECT FAST PURPURINE 8B ☐ DIRECT RED 8BS ☐ FAST PURPURINE ☐ MITSUI DIRECT BRILLIANT SCARLET 8B ☐ NIPPON PURPURINE 8B ☐ TRIAZOL FAST RED 8B

TOXICITY DATA WITH REFERENCE
dns-rat-orl 100 mg/kg ENMUDM 5,488,83
dns-rat:lvr 500 mg/L ENMUDM 5,488,83

CONSENSUS REPORTS: Reported in EPA TSCA Inventory.

SAFETY PROFILE: Mutation data reported. When heated to decomposition it emits toxic vapors of NO_x, SO_x and Cl^-.

CMO885 CAS:2610-11-9 HR: 2
C.I. DIRECT RED 81
mf: $C_{29}H_{19}N_5O_8S_2 \cdot 2Na$ mw: 675.63

SYNS: AIREDALE RED KD ☐ AIZEN PRIMULA RED 4BH ☐ AMANIL FAST RED 8BL ☐ AMANIL FAST RED 8BLW ☐ BELAMINE FAST RED 8 BL ☐ BENZANIL FAST RED K ☐ BENZO FAST RED 8BL ☐ BORDEAUX EMBL ☐ CALCODUR RED 8BL ☐ CHLORAMINE FAST RED 5BL ☐ CHLORAMINE FAST RED K ☐ CHLORANTINE FAST RED ☐ CHLORANTINE FAST RED 5B (6CI) ☐ CHROME LEATHER RED 5B ☐ C.I. 28160 ☐ DIALUMINOUS RED 4B ☐ DIAPHTAMINE LIGHT RED 4B ☐ DIAZINE FAST RED 8BK ☐ DIAZO LIGHT RED 8BD ☐ DIAZO LIGHT RED 8B ☐ DIPHENYL FAST 5BL SUPRA I RED ☐ DIPHENYL FAST RED 5BL ☐ DIPHENYL FAST RED 5BLN ☐ DIRECT FAST RED 5B ☐ DIRECT FAST RED 8BL ☐ DIRECT FAST RED 2S ☐ DIRECT LIGHTFAST RED 2S ☐ DIRECT LIGHT RED 4B ☐ DIRECT LIGHT RED 8B ☐ DIRECT LIGHT RED M 8BL ☐ DIRECT RED 81 ☐ DURAZOL RED 2B ☐ DURAZOL RED 2BP ☐ ELIAMINA RED 8BL ☐ FASTOLITE RED 8BL ☐ FASTUSOL RED 4BA-CF ☐ FENALUZ RED 4B ☐ HELION RED 8B ☐ HISPALUZ RED 8BL ☐ JAPAN RED NO. 505 ☐ 2-NAPHTHALENESULFONIC ACID, 7-(BENZOYLAMINO)-4-HYDROXY-3-((4-((4-SULFOPHENYL)AZO)PHENYL)AZO)-,DISODIUM SALT ☐ PARANOL FAST RED 8BL ☐ PHENO FAST SCARLET 9B ☐ PONTAMINE FAST RED 8BLX ☐ PYRAZOL FAST RED 5BL ☐ PYRAZOL FAST RED 8BL ☐ PYRAZOLINE RED 8BL ☐ SATURN RED B ☐ SIRIUS RED 4B ☐ SIRIUS RED 4BA ☐ SOLANTINE RED 8BL ☐ SOLAR RED B ☐ SOLIUS RED 4B ☐ SUMILIGHT RED 4B ☐ SUPRAZO RED 4B ☐ SUPREXCEL RED 8BL ☐ TERTRODIRECT FAST RED 5B ☐ TETRAMINE FAST RED 8B ☐ TRIANTINE FAST RED 4BN ☐ TRIANTINE LIGHT RED 4BN

TOXICITY DATA WITH REFERENCE
mnt-mus-ipr 91,250 μg/kg/24H-I TOLED5 40,183,88
ipr-mus LD50:1048 mg/kg TOLED5 40,183,88

CONSENSUS REPORTS: Reported in EPA TSCA Inventory.

SAFETY PROFILE: Moderately toxic by intraperitoneal route. Mutation data reported. When heated to decomposition it emits toxic vapors of NO_x and SO_x.

CMP000 CAS:2586-60-9 HR: 2
C.I. DIRECT VIOLET 1, DISODIUM SALT
mf: $C_{32}H_{22}N_6O_8S_2 \cdot 2Na$ mw: 728.70

SYNS: AIREDALE VIOLET ND ☐ AMANIL FAST VIOLET N ☐ ATLANTIC VIOLET N ☐ ATUL DIRECT VIOLET N ☐ AZOCARD VIOLET N ☐ BENCIDAL FAST VIOLET N ☐ BENZANIL VIOLET N ☐ BENZO VIOLET N ☐ BRASILAMINA VIOLET 3R ☐ CALCOMINE VIOLET N ☐ CHLORAZOL VIOLET N ☐ C.I. 22570 ☐ COTTON VIOLET R ☐ DIAMINE VIOLET N ☐ DIAPHTAMINE VIOLET N ☐ DIAZINE VIOLET N ☐ DIAZOL VIOLET N ☐ DIRECT FAST VIOLET N ☐ DIRECT VIOLET C ☐ ERIE VIOLET 3R ☐ FIXANOL VIOLET N ☐ HISPAMIN VIOLET 3R ☐ JAPANOL VIOLET J ☐ NAPHTAMINE VIOLET N ☐ PARAMINE FAST VIOLET N ☐ PHENO VIOLET N ☐ PONTAMINE VIOLET N ☐ TERTRODIRECT VIOLET N ☐ TRISULFON VIOLET N

TOXICITY DATA WITH REFERENCE
orl-rat TDLo:1125 g/kg/71W-C:ETA VOONAW 23(7),72,77
unk-rat TDLo:125 g/kg/71W-C:ETA GTPZAB 22(10),22,78

SAFETY PROFILE: Questionable carcinogen with experimental tumorigenic data. When heated to decomposition it emits very toxic fumes of NO_x, Na_2O, and SO_x.

CMP050 CAS:1829-00-1 HR: 2
C.I. DIRECT YELLOW 9, DISODIUM SALT
mf: $C_{28}H_{19}N_5O_6S_4 \cdot 2Na$ mw: 695.74

SYNS: ATLANTIC BRILLIANT YELLOW MN ☐ 7-BENZOTHIAZOLESULFONIC ACID, 2,2'-(1-TRIAZENE-1,3-DIYLDI-4,1-PHENYLENE)BIS(6-METHYL-), DISODIUM SALT ☐ BENZO YELLOW TZ ☐ CHLORAZOL YELLOW DP ☐ CHLORAZOL YELLOW 2G ☐ C.I. 19540 ☐ C.I. DIRECT YELLOW 9 ☐ CLAYTON YELLOW ☐ DIAPHTAMINE BRILLIANT YELLOW 6GS ☐ DIAZAMINE GOLDEN YELLOW T ☐ DIAZOL YELLOW J ☐ DIRECT YELLOW MTZ ☐ DIRECT YELLOW TZ ☐ HISPAMIN PURE YELLOW T2G ☐ MIMOSA Z ☐ PEERAMINE BRIGHT YELLOW MN ☐ PONTAMINE PURE YELLOW ☐ PONTAMINE PURE YELLOW MN ☐ THIAZOLE YELLOW ☐ THIAZOLE YELLOW G ☐ THIAZOL YELLOW ☐ THIAZOL YELLOW G ☐ THIAZOL YELLOW GGM ☐ THIAZOL YELLOW R ☐ THIAZOL YELLOW Z ☐ TITAN YELLOW ☐ TITAN YELLOW DYE ☐ TITAN YELLOW G

TOXICITY DATA WITH REFERENCE
orl-rat LD:>500 mg/kg NCNSA6 5,24,53

CONSENSUS REPORTS: Reported in EPA TSCA Inventory.

SAFETY PROFILE: Moderately toxic by ingestion. When heated to decomposition it emits toxic vapors of NO_x and SO_x.

CMP060 CAS:15791-78-3 HR: D
C.I. DISPERSE BLUE 27
mf: $C_{22}H_{16}N_2O_7$ mw: 420.40

SYNS: 9,10-ANTHRACENEDIONE, 1,8-DIHYDROXY-4-((4-(2-HYDROXYETHYL)PHENYL)AMINO)-5-NITRO- ☐ ANTHRAQUINONE, 1,8-DIHYDROXY-4-(p-(2-HYDROXYETHYL)ANILINO)-5-NITRO- ☐ C.I. 60767 ☐ 1,8-DIHYDROXY-4-(p-(2-HYDROXYETHYL)ANILINO)-5-NITROANTHRAQUINONE ☐ 1,8-DIHYDROXY-4-(4'-β-HYDROXYETHYL) ANILINO-6-NITROANTHROQUINONE ☐ EASTMAN FAST BLUE B-GLF ☐ SERISOL FAST BLUE BGLW

TOXICITY DATA WITH REFERENCE
mmo-sat 500 μg/plate CRNGDP 2,879,81

CONSENSUS REPORTS: Reported in EPA TSCA Inventory.

SAFETY PROFILE: Mutation data reported. When heated to decomposition it emits toxic vapors of NO_x.

CMP070 CAS:12217-79-7 HR: 1
C.I. DISPERSE BLUE 56
mf: $C_{14}H_9ClN_2O_4$ mw: 304.70

SYNS: 9,10-ANTHRACENEDIONE, 1,5-DIAMINOCHLORO-4,8-DIHYDROXY- ☐ C.I. 63285 ☐ C.I. DISPERSE BLUE 59 ☐ C.I. DISPERSE

BLUE 71 □ DISPERSE BLUE 56 □ DISPERSOL BLUE B-R □ DURANOL BLUE TR □ KAYALON POLYESTER BLUE EBL-E □ LATYL BLUE BCN □ MIKETON POLYESTER BLUE FBL □ MODR DISPERZNI 56 □ PALANIL BLUE R □ RESOLIN BLUE FBL □ SAMARON BLUE FBL □ SUMIKARON BLUE E-BL □ SUMIKARON BLUE E-FBL □ SUMIKARON BLUE R □ TERASIL BRILLIANT BLUE 3RL □ TERSETILE BLUE RBL

TOXICITY DATA WITH REFERENCE
eye-rbt 500 mg/24H MLD 85JCAE-,1326,86

CONSENSUS REPORTS: Reported in EPA TSCA Inventory.

SAFETY PROFILE: An eye irritant. When heated to decomposition it emits toxic vapors of NO_x and Cl^-.

CMP080 CAS:3180-81-2 HR: D
C.I. DISPERSE RED 13
mf: $C_{16}H_{17}ClN_4O_3$ mw: 348.82

SYNS: ACETAMINE RUBINE B □ ACETATE FAST RUBINE B □ ACETOQUINONE LIGHT RUBINE BLZ □ AMACEL RUBINE B □ CELLITON DISCHARGING RUBINE BL □ CELLITON FAST RUBINE B □ CELLITON FAST RUBINE BA-CF □ CELLITON RUBINE B □ CELLITON RUBY B □ CIBACET RUBINE BS □ CIBACET RUBINE R □ C.I. 11115 □ CILLA FAST RUBINE B □ DIACELLITON FAST BLUE BORDEAUZ B □ DIACELLITON FAST BORDEAUX B □ DISPERSE BORDEAUX S □ DISPERSE RED 13 □ DISPERSOL FAST CRIMSON B □ DISPERSOL RUBINE B □ DURGACET RUBINE B □ ETHANOL, 2-((4-((2-CHLORO-4-NITROPHENYL)AZO)PHENYL)ETHYLAMINO)- □ INTERCHEM ACETATE BORDEAUX B □ FENACET FAST RUBINE B □ KAYALON FAST RUBINE B □ KCA ACETATE CRIMSON B □ MICROSETILE RUBINE 2B □ NYLOQUINONE BORDEAUX B □ PALACET SCARLET B □ PERLITON RUBINE 4B □ SERISOL FAST CRIMSON BD □ SETACYL RED 2B □ SILOTRAS RUBINE TSB □ SUPRACET FAST CRIMSON B

TOXICITY DATA WITH REFERENCE
mma-sat 5 mg/plate EMMUEG 19(Suppl 20),8,92

CONSENSUS REPORTS: Reported in EPA TSCA Inventory.

SAFETY PROFILE: Mutation data reported. When heated to decomposition it emits toxic vapors of NO_x and Cl^-.

CMP090 CAS:6300-37-4 HR: 2
C.I. DISPERSE YELLOW 7
mf: $C_{19}H_{16}N_4O$ mw: 316.39

SYNS: ACETATE FAST YELLOW 5RL □ AZOFENOL 4K □ BENZENE-1-AZOBENZENE-4-AZO-o-CRESOL □ CELLITON DISCHARGE YELLOW 5RL □ CELLITON FAST YELLOW 5R □ CELLITON YELLOW 5R □ C.I. 26090 □ CILLA FAST YELLOW 5R □ DIANIX FAST YELLOW 5R □ DIANIX YELLOW 5R-E □ DISPERSE DYE FAST YELLOW 4K □ DISPERSE FAST YELLOW 4K □ DISPERSE YELLOW 7 □ DZhp-4K □ KAYALON FAST YELLOW 4R □ KAYALON POLYESTER YELLOW 4R-E □ KAYALON POLYESTER YELLOW RF □ MIKETON POLYESTER YELLOW 5R □ PALANIL YELLOW 5R □ PALANIL YELLOW 5RX □ PHENOL, 2-METHYL-4-((4-(PHENYLAZO)PHENYL)AZO)- □ RESOLIN YELLOW 5R □ SAMARON YELLOW 5RL □ SERISOL FAST YELLOW N 5RD □ SUPRACET FAST YELLOW 4R □ TERSETILE YELLOW 5R □ TERSETILE YELLOW 5RL □ TULASTERON FAST YELLOW 5R-B □ VONTERYL YELLOW 3R □ YELLOW FAST DYE 4K □ YELLOW STABLE 4K

TOXICITY DATA WITH REFERENCE
unr-rat LD50:2870 mg/kg GISAAA 48(7),6,83
unr-mus LD50:1864 mg/kg GISAAA 48(7),6,83
unr-rbt LD50:>10 g/kg GISAAA 48(7),6,83

CONSENSUS REPORTS: Reported in EPA TSCA Inventory.

SAFETY PROFILE: Moderately toxic by an unspecified route. When heated to decomposition it emits toxic vapors of NO_x.

CMP200 CAS:16090-02-1 HR: 1
C.I. FLUORESCENT BRIGHTENER 260
mf: $C_{40}H_{38}N_{12}O_8S_2 \cdot 2Na$ mw: 925.00

SYNS: 4,4′-BIS((4-ANILINO-6-MORPHOLINO-1,3,5-TRIAZINE-2-YL) AMINO)STILBENE 2,2-DISULFONATE 2Na □ BLANKOPHOR BBH □ BLANKOPHOR MBBH □ DISODIUM 4,4′-BIS((4-ANILINO-6-MORPHOLINO-1,3,5-TRIAZIN-2-YL)AMINO)STILBENE-2,2′-DISULFONATE □ MBBH 766 □ MIKEPHOR TB □ 2,2′-STILBENEDISULFONIC ACID, 4,4′-BIS((4-ANILINO-6-MORPHOLINO-s-TRIAZIN-2-YL)AMINO)-, DISODIUM SALT □ TINOPAL AMS □ TINOPAL EMS

TOXICITY DATA WITH REFERENCE
eye-rbt 100 mg MLD TXAPA9 27,494,74
eye-rbt 100 mg/2S RNS MLD TXAPA9 27,494,74
orl-rat LD50:>8 g/kg CCECAU 7,91,77
orl-mus LD50:>20 g/kg SCIEAS 36(1-4),10,89
orl-cat LDLo:>500 mg/kg CCECAU 7,91,77
orl-rbt LDLo:>1 g/kg CCECAU 7,91,77
skn-rbt LD:>10 g/kg TXAPA9 27,494,74
orl-gpg LDLo:>1 g/kg CCECAU 7,91,77

CONSENSUS REPORTS: Reported in EPA TSCA Inventory.

SAFETY PROFILE: Low toxicity by ingestion. An eye irritant. When heated to decomposition it emits toxic vapors of NO_x and Cl^-.

CMP250 CAS:8062-14-4 HR: 2
C.I. FOOD BROWN 1
mf: $C_{18}H_{14}N_6O_6S_2 \cdot C_{13}H_{13}N_4O_3S \cdot 3Na$ mw: 848.83

SYNS: 1545 BROWN □ BROWN FK □ 4,4′-((4,6-DIAMINO-m-PHENYLENE)BIS(AZO))DIBENZENESULFONIC ACID, DISODIUM SALT mixed with p-((4,6-DIAMINO-m-TOLYL)AZO)-BENZENESULFONIC ACID, SODIUM SALT □ EEC SERIAL No. 124 □ GOLDEN BROWN RK-FQ

TOXICITY DATA WITH REFERENCE
mmo-esc 1 g/L FCTXAV 18,215,80
mma-esc 10 g/L FCTXAV 18,215,80
orl-rat LDLo:4000 mg/kg FCTXAV 6,1,68
ipr-rat LD50:750 mg/kg FCTXAV 6,737,68
ipr-mus LDLo:1500 mg/kg FCTXAV 6,1,68

CONSENSUS REPORTS: EPA Genetic Toxicology Program.

SAFETY PROFILE: Moderately toxic by ingestion and intraperitoneal routes. Mutation data reported. When heated to decomposition it emits very toxic fumes of NO_x, Na_2O, and SO_x.

CMP500 **CAS:4553-89-3** **HR: 3**
C.I. FOOD BROWN 3, DISODIUM SALT
mf: $C_{27}H_{20}N_4O_9S_2 \cdot 2Na$ mw: 654.61

SYNS: C.I. 20285 ☐ 2,4-DIHYDROXY-3,5-DI(4-SULPHO-1-NAPHTHYLAZO)BENZYL ALCOHOL, DISODIUM SALT

TOXICITY DATA WITH REFERENCE
ipr-rat LD50:375 mg/kg FCTXAV 4,143,66
ipr-mus LD50:220 mg/kg FCTXAV 4,143,66

SAFETY PROFILE: Poison by intraperitoneal route. When heated to decomposition it emits very toxic fumes of NO_x, Na_2O, and SO_x.

CMP600 **CAS:2051-85-6** **HR: 2**
C.I. FOOD ORANGE 3
mf: $C_{12}H_{10}N_2O_2$ mw: 214.24

SYNS: BENZENEAZORESORCINOL ☐ 1,3-BENZENEDIOL, 4-(PHENYLAZO)- ☐ CERES ORANGE G ☐ CERES ORANGE GN ☐ CERISOL YELLOW GR ☐ C.I. 11920 ☐ C.I. SOLVENT ORANGE 1 (8CI) ☐ 2,4-DIBENZENEAZORESORCINOL ☐ 2,4-DIHYDROXYAZOBENZENE ☐ FAST OIL ORANGE T ☐ FAST OIL YELLOW G ☐ FAST OIL YELLOW 2G ☐ FAT ORANGE A ☐ FAT ORANGE G ☐ FAT ORANGE GS ☐ FAT ORANGE RG ☐ FAT VICTORIA YELLOW D ☐ GRASOL YELLOW RSF ☐ HEXACOL OIL YELLOW GG ☐ LACQUER ORANGE V 3G ☐ OIL ORANGE G ☐ OIL ORANGE 4G ☐ OIL ORANGE MO ☐ OIL ORANGE MON ☐ OIL ORANGE MON EXTRA ☐ OIL-SOL. YELLOW ZH ☐ OIL YELLOW G EXTRA ☐ OIL YELLOW GG ☐ ORANZ POTRAVINARSKA 3 ☐ ORANZ ROZPOUSTEDLOVA 1 ☐ ORGANOL ORANGE 2J ☐ 4-(PHENYLAZO)RESORCINOL ☐ PLASTORESIN ORANGE F 3A ☐ RESINOL ORANGE G ☐ SOLVENT ORANGE 1 ☐ SUDAN G ☐ SUDAN ORANGE G ☐ SUDAN YELLOW AR ☐ TERTROGRAS ORANGE SG ☐ 1504 YELLOW ☐ YELLOW M SOLUBLE IN GREASE

TOXICITY DATA WITH REFERENCE
ipr-rat LD50:600 mg/kg 85JCAE-,1307,86

CONSENSUS REPORTS: Reported in EPA TSCA Inventory

SAFETY PROFILE: Moderately toxic by intraperitoneal route. When heated to decomposition it emits toxic vapors of NO_x.

CMP620 **CAS:3257-28-1** **HR: 1**
C.I. FOOD RED 2, DISODIUM SALT
mf: $C_{18}H_{14}N_2O_7S_2 \cdot 2Na$ mw: 480.44

SYNS: ACID SCARLET GNA ☐ ACID SCARLET JN EXTRA PURE A ☐ CERVEN POTRAVINARSKA 2 ☐ C.I. 14815 ☐ C.I. FOOD RED 2 ☐ CILEFA RED G ☐ E 125 ☐ ECARLATE GN ☐ EUROCERT SCARLET GN ☐ FOOD RED 2 ☐ L RED Z 3000 ☐ 1-NAPHTHALENESULFONIC ACID, 6-((2,4-DIMETHYL-6-SULFOPHENYL)AZO)-5-HYDROXY-, DISODIUM SALT ☐ SARLACH GN ☐ SCARLET GN ☐ SCARLET GN SPECIALLY PURE

TOXICITY DATA WITH REFERENCE
ivn-rat LD50:1 g/kg APFRAD 15,402,57

CONSENSUS REPORTS: Reported in EPA TSCA Inventory.

SAFETY PROFILE: Low toxicity by intravenous route. When heated to decomposition it emits toxic vapors of NO_x and SO_x.

CMP800 **HR: 3**
CIGARETTE REFINED TAR

SYNS: CIGARETTE TAR ☐ COLOMBIAN BLACK TOBACCO CIGARETTE REFINED TAR ☐ TAR, from tobacco ☐ TOBACCO REFINED TAR ☐ TOBACCO TAR ☐ U.S. BLENDED LIGHT TOBACCO CIGARETTE REFINED TAR

TOXICITY DATA WITH REFERENCE
mmo-sat 1 mg/L GANNA2 69,85,78
otr-ham:emb 1 mg/L GANNA2 69,85,78
otr-ham:lng 10 mg/L BJCAAI 25,574,71
cyt-ham:lng 10 mg/L BJCAAI 25,574,71
skn-mus TDLo:118 g/kg/52W-I:CAR CANCAR 21,376,68
ivg-mus TDLo:1320 mg/kg/44W-I:CAR,REP JNCIAM 23,1,59
skn-mus TD:235 g/kg/52W-I:CAR CANCAR 21,376,68
skn-mus TD:390 g/kg/65W-I:CAR AJPAA4 102,381,81

DOT CLASSIFICATION: 3; *Label:* Flammable Liquid

SAFETY PROFILE: Suspected carcinogen with experimental carcinogenic data. Experimental reproductive effects. Mutation data reported. See also TOBACCO and NICOTINE.

CMP825 **CAS:73963-72-1** **HR: 1**
CILOSTAZOL
mf: $C_{20}H_{27}N_5O_2$ mw: 369.52

PROP: Needles from MeOH aq. Mp: 158–159°.

SYNS: 6-(4-(1-CYCLOHEXYL-1H-TETRAZOL-5-YL)BUTOXY)-3,4-DIHYDRO-2(1H)-QUINOLINONE ☐ 3,4-DIHYDRO-6-(4-(1-CYCLOHEXYL-1H-TETRAZOL-5-YL)BUTOXY)-2(1H)-QUINOLINONE ☐ OPC-13013

TOXICITY DATA WITH REFERENCE
orl-rat TDLo:1650 mg/kg (female 17-22D post):REP IYKEDH 16,1073,85
orl-rat TDLo:11 g/kg (female 7-17D post):TER IYKEDH 16,1053,85
orl-rat TDLo:5250 mg/kg (5W pre):REP IYKEDH 16,1204,85
orl-rat TDLo:137 g/kg (male 13W pre):REP IYKEDH 16,1245,85
orl-rat TDLo:11 g/kg (7-17D preg):TER IYKEDH 16,1053,85
orl-man TDLo:1248 μg/kg:CNS ARZNAD 35,1173,85

SAFETY PROFILE: Human systemic effects by ingestion: headache. An experimental teratogen. Experimental reproductive effects. A vasodilator. When heated to decomposition it emits toxic fumes of NO_x.

CMP880 **CAS:3564-14-5** **HR: D**
C.I. MORDANT BLACK 3, MONOSODIUM SALT
mf: $C_{20}H_{13}N_2O_5S \cdot Na$ mw: 416.40

SYNS: ALIZARINE BLUE BLACK OCBN ☐ ALIZARINE BLUE BLACK OCGN ☐ ALIZARINE BLUE BLACK OCGP ☐ ALIZARINE BLUE OCB ☐ AZOCHROMOL BLUE BLACK EB ☐ CALCOCHROME BLUE BLACK BC ☐ CHROMAZINE BLUE BLACK B ☐ CHROME BLUE BLACK BF ☐ CHROME FAST CYANINE BP ☐ CHROME FAST CYANIDE G ☐ CHROME FAST CYANINE GN ☐ CHROME FAST CYANINE GNN ☐ CHROME FAST CYANINE GP ☐ CHROME FAST CYANINE GSS ☐ CHROMOCARD BLUE BLACK B ☐ C.I. 14640 ☐ C.I. MORDANT BLACK 3 ☐ DIACROMO BLUE G ☐ DIAMOND BLUE BLACK AE ☐ DIAMOND

BLUE BLACK EB ☐ DIAMOND BLUE BLACK EBS-CF ☐ DUROCHROME BLUE OCG ☐ DUROCHROME CYANINE G ☐ DUROCHROME FAST CYANINE 6BN ☐ ERIOCHROME BLUE BLACK ☐ ERIOCHROME BLUE BLACK B ☐ ERIOCHROME BLUE BLACK 2B ☐ ERIOCHROME BLUE BLACK BC ☐ ERIOCHROME BLUE BLACK 2BP ☐ ERIOCHROME BLUE BLACK BSS ☐ ERIOCHROME BLUE BLACK 2G ☐ FAST CHROME CYANINE 6B ☐ FAST CHROME CYANINE G ☐ FENAKRON BLUE BLACK EB ☐ HISPACROM BLUE BG ☐ JAVA CHROME BLUE BLACK BN ☐ KENACHROME BLACK 6B ☐ LIGHTHOUSE CHROME BLACK 6B ☐ MITSUI CHROME BLUE BLACK BC ☐ 1-NAPHTHALENESULFONIC ACID, 3-HYDROXY-4-((1-HYDROXY-2-NAPHTHALENYL)AZO)-, MONOSODIUM SALT ☐ OMEGA CHROME BLUE BLACK B ☐ PONTACHROME BLUE BLACK BB ☐ SALICINE BLUE BLACK AEF ☐ SOLOCHROMATE BLACK 6BN ☐ SOLOCHROME BLACK 6BN ☐ SUNCHROMINE BLUE BLACK B ☐ SUPERCHROME BLUE BC ☐ SUPERCHROME BLUE BG ☐ TERTROCHROME BLUE B

TOXICITY DATA WITH REFERENCE
mmo-sat 300 nmol/plate MUREAV 116,305,83

CONSENSUS REPORTS: Reported in EPA TSCA Inventory.

SAFETY PROFILE: Mutation data reported. When heated to decomposition it emits toxic vapors of NO_x and SO_x.

CMP885 CAS:539-35-5 HR: 2
CINAMONIN
mf: $C_9H_{15}NO_3S$ mw: 217.31

SYNS: ACIDOMYCIN ☐ ACTITHIAZIC ACID ☐ MYCOBACIDIN

TOXICITY DATA WITH REFERENCE
ipr-mus LD50:2500 mg/kg TDKNAF 14,60,55
scu-mus LD50:20 g/kg 85GDA2 4(1),143,80
ivn-mus LD50:1500 mg/kg 85GDA2 4(1),143,80

SAFETY PROFILE: Moderately toxic intravenous and intraperitoneal routes. Mildly toxic by subcutaneous routes. When heated to decomposition it emits toxic fumes of SO_x and NO_x.

CMP900 CAS:54-84-2 HR: 3
CINANSERIN HYDROCHLORIDE
mf: $C_{20}H_{24}N_2OS{\cdot}ClH$ mw: 376.98

PROP: Crystals from MeCN. Mp: 146–148°.

SYNS: n-(2-((3-(DIMETHYLAMINO)PROPYL)THIO)PHENYL)-3-PHENYL-2-PROPENAMIDE MONOHYDROCHLORIDE ☐ 2'-((3-(DIMETHYLAMINO)PROPYL)THIO)CINNAMANILIDE HYDROCHLORIDE ☐ MAPTC ☐ SQ 10,643

TOXICITY DATA WITH REFERENCE
orl-rat LD50:1500 mg/kg AIPTAK 152,132,64
orl-mus LD50:480 mg/kg AIPTAK 152,132,64
ivn-mus LD50:35 mg/kg AIPTAK 152,132,64
ivn-dog LDLo:8 mg/kg AIPTAK 152,132,64

SAFETY PROFILE: Poison by intravenous route. Moderately toxic by ingestion. When heated to decomposition it emits toxic fumes of SO_x, NO_x, and HCl.

CMP910 CAS:485-71-2 HR: 3
CINCHONIDINE
mf: $C_{19}H_{22}N_2O$ mw: 294.43

SYNS: CINCHONAN-9-OL, (8-α-9R)-(9CI) ☐ (−)-CINCHONIDINE ☐ (8S,9R)-CINCHONIDINE ☐ CINCHOVATINE ☐ α-QUINIDINE ☐ 2-QUINUCLIDINEMETHANOL, α-4-QUINOLYL-5-VINYL-

TOXICITY DATA WITH REFERENCE
ipr-rat LD50:206 mg/kg APTOA6 4,265,48

CONSENSUS REPORTS: Reported in EPA TSCA Inventory.

SAFETY PROFILE: Poison by intraperitoneal route. When heated to decomposition it emits toxic vapors of NO_x.

CMP925 CAS:118-10-5 HR: 3
d-CINCHONINE
mf: $C_{19}H_{22}N_2O$ mw: 294.43

PROP: Crystals, prisms, or needles from alcohol or ether. Mp: 264°. Occurs in most varieties of cinchona bark, especially in bark of *Cinchona micrantha R. & P., Rubiaceae.* Prisms, needles from alcohol or ether. Mp: about 265°; begins to sublime at 220°. One gram dissolves in 60 mL alc, 25 mL boiling alc, 110 mL chloroform, 500 mL ether. Practically insol in water.

SYNS: α-4-QUINOLYL-5-VINYL-2-QUINUCLIDINEMETHANOL ☐ α-(5-VINYL-2-QUINOLYL)-2-QUINUCLIDINEMETHANOL

TOXICITY DATA WITH REFERENCE
ipr-rat LD50:152 mg/kg APTOA6 4,265,48
scu-mus LDLo:400 mg/kg AEPPAE 205,129,48
scu-frg LDLo:200 mg/kg AEPPAE 205,129,48

CONSENSUS REPORTS: Reported in EPA TSCA Inventory.

SAFETY PROFILE: Poison by intraperitoneal and subcutaneous routes. When heated to decomposition it emits toxic fumes of NO_x.

CMP950 CAS:20168-99-4 HR: 3
CINDOMET
mf: $C_{21}H_{19}NO_4$ mw: 349.39

PROP: Yellow crystals from acetone or Me_2CO (aq). Loses water. Mp: 170–172°.

SYNS: CINMETACIN ☐ CINMETHACIN ☐ 1-CINNAMOYL-5-METHOXY-2-METHYLINDOLE-3-ACETIC ACID ☐ 1-CINNAMOYL-2-METHOXY-5-METHOXY-3-INDOLYLACETIC ACID ☐ INDOLACIN ☐ 5-METHOXY-2-METHYL-1-(1-OXO-3-PHENYL-2-PROPENYL)-1H-INDOLE-3-ACETIC ACID (9CI)

TOXICITY DATA WITH REFERENCE
orl-rat LD50:1020 mg/kg ARZNAD 23,1690,73
ipr-rat LD50:590 mg/kg ARZNAD 23,1690,73
orl-mus LD50:750 mg/kg ARZNAD 23,1690,73
ipr-msu LD50:360 mg/kg ARZNAD 23,1690,73

SAFETY PROFILE: Poison by intraperitoneal route. Moderately toxic by ingestion. An anti-inflammatory agent. When heated to decomposition it emits toxic fumes of NO_x.

CMP969 CAS:104-55-2 HR: 3
CINNAMALDEHYDE
mf: C_9H_8O mw: 132.17

PROP: Found in Ceylon and Chinese cinnamon oils. Yellowish, oily liquid; strong odor of cinnamon. D: 1.048–1.052, mp: −7.5°, bp: 246.0° (some decomp), d: 1.048–1.052 @ 25°/25°, refr index: 1.619–1.623, flash p: 248°F. Very sltly sol in water; misc with alc, ether, chloroform, fixed oils.

SYNS: BENZYLIDENEACETALDEHYDE □ CASSIA ALDEHYDE □ CINNAMAL □ CINNAMYL ALDEHYDE □ CINNIMIC ALDEHYDE □ FEMA No. 2286 □ NCI-C56111 □ PHENYLACROLEIN □ 3-PHENYLACROLEIN □ 3-PHENYLPROPENAL □ 3-PHENYL-2-PROPENAL □ ZIMTALDEHYDE

TOXICITY DATA WITH REFERENCE
skn-hmn 40 mg/48H SEV FCTXAV 17,253,79
mma-sat 500 µg/plate FCTOD7 22,623,84
sln-dmg-par 2 pph ENMUDM 7,677,85
dni-mus:leu 31,500 µg/L DCTODJ 6,521,83
oms-mus:leu 31,500 µg/L DCTODJ 6,521,83
cyt-ham:fbr 15 mg/L FCTOD7 22,623,84
orl-rat TDLo:55 mg/kg (female 7-17D post):REP FCTOD7 27,781,89
orl-rat TDLo:275 mg/kg (female 7-17D post):REP FCTOD7 27,781,89
orl-rat LD50:2220 mg/kg FCTXAV 2,327,64
orl-mus LD50:2225 mg/kg YKKZAJ 92,135,72
ipr-mus LD50:200 mg/kg FAONAU 44A,16,67
ivn-mus LD50:75 mg/kg CSLNX* NX#07571
par-mus LDLo:200 mg/kg CBCCT* 7,687,55
orl-gpg LD50:1160 mg/kg FCTXAV 2,327,64

CONSENSUS REPORTS: Reported in EPA TSCA Inventory.

SAFETY PROFILE: Poison by intravenous and parenteral routes. Moderately toxic by ingestion and intraperitoneal routes. A severe human skin irritant. Mutation data reported. Combustible liquid. May ignite after a delay period in contact with NaOH. When heated to decomposition it emits acrid smoke and fumes. See also ALDEHYDES.

CMP970 CAS:621-79-4 HR: 3
CINNAMAMIDE
mf: C_9H_9NO mw: 147.19

SYNS: 2-BENZYLIDENEACETAMIDE □ 3-PHENYLACRYLAMIDE □ 3-PHENYLPROPENAMIDE □ 3-PHENYL-2-PROPENAMIDE □ 2-PROPENAMIDE, 3-PHENYL-(9CI)

TOXICITY DATA WITH REFERENCE
orl-mus LD50:1600 mg/kg CHTPBA 8,202,73
ivn-rbt LDLo:62 mg/kg COREAF 153,895,11

CONSENSUS REPORTS: Reported in EPA TSCA Inventory.

SAFETY PROFILE: Poison by intravenous route. Moderately toxic by ingestion. When heated to decomposition it emits toxic vapors of NO_x.

CMP975 CAS:621-82-9 HR: 3
CINNAMIC ACID
mf: $C_9H_8O_2$ mw: 148.17

PROP: Occurs free and partly esterified in storax, balsam Peru or Tolu, oil of cinnamon, coca leaves. White monoclinic crystals; honey floral odor. D: (4/4) 1.2475, mp: 133°, bp: 300°, flash p: 212°F. One gram dissolves in about 2000 mL water at 25° (more sol in hot water), in 6 mL alc, 5 mL methanol, 12 mL chloroform. Freely sol in benzene, ether, acetone, glacial acetic acid, carbon disulfide, fixed oils.

SYNS: FEMA No. 2288 □ PHENYLACRYLIC ACID □ tert-β-PHENYLACRYLIC ACID □ 3-PHENYLACRYLIC ACID □ 3-PHENYLPROPENOIC ACID □ 3-PHENYL-2-PROPENOIC ACID □ ZIMTSAEURE (GERMAN)

TOXICITY DATA WITH REFERENCE
skn-rbt 500 mg/24H MLD FCTXAV 16,687,78
orl-rat LD50:2500 mg/kg FCTXAV 16,687,78
ipr-rat LD50:1600 mg/kg BCFAAI 112,53,73
orl-mus LD50:5 g/kg GISAAA 46(1),94,81
ipr-mus LD50:160 mg/kg FCTXAV 16,687,78
ivn-mus LD50:380 mg/kg ARZNAD 19,617,69

CONSENSUS REPORTS: Reported in EPA TSCA Inventory.

SAFETY PROFILE: Poison by intravenous and intraperitoneal routes. Moderately toxic by ingestion. A skin irritant. Combustible liquid. When heated to decomposition it emits acrid smoke and fumes.

CMP980 CAS:140-10-3 HR: 3
trans-CINNAMIC ACID
mf: $C_9H_8O_2$ mw: 148.17

SYNS: trans-β-CARBOXYSTYRENE □ CINNAMIC ACID, (E)- □ (E)-CINNAMIC ACID □ trans-3-PHENYLACRYLIC ACID □ (E)-3-PHENYL-2-PROPENOIC ACID □ 2-PROPENOIC ACID, 3-PHENYL-, (E)-(9CI)

TOXICITY DATA WITH REFERENCE
orl-brd LD50:100 mg/kg AECTCV 12,355,83

CONSENSUS REPORTS: Reported in EPA TSCA Inventory.

SAFETY PROFILE: Poison by ingestion. When heated to decomposition it emits acrid smoke and irritating vapors.

CMQ000 CAS:63938-16-9 HR: 3
CINNAMIC ACID, NICKEL(II) SALT
mf: $C_{18}H_{14}O_4 \cdot Ni$ mw: 353.03

TOXICITY DATA WITH REFERENCE
par-mus LDLo:40 mg/kg CBCCT* 7,687,55

CONSENSUS REPORTS: Nickel and its compounds are on the Community Right-To-Know List.

NIOSH REL: (Nickel, Inorganic) TWA 0.015 mg(Ni)/m³

SAFETY PROFILE: Poison by parenteral route. See also NICKEL COMPOUNDS. When heated to decomposition it emits acrid smoke and irritating fumes.

CMQ100 **CAS:90-50-6** **HR: 2**
CINNAMIC ACID, 3,4,5-TRIMETHOXY-
mf: $C_{12}H_{14}O_5$ mw: 238.26

SYNS: O-METHYLSINAPIC ACID □ 2-PROPENOIC ACID, 3-(3,4,5-TRIMETHOXYPHENYL)-(9CI) □ 3,4,5-TRIMETHOXYCINNAMIC ACID □ 3,4,5-TRIMETHOXYPHENYLACRYLIC ACID □ 3-(3,4,5-TRIMETHOXYPHENYL)-2-PROPENOIC ACID

TOXICITY DATA WITH REFERENCE
orl-brd LD50:422 mg/kg AECTCV 12,355,83

CONSENSUS REPORTS: Reported in EPA TSCA Inventory.

SAFETY PROFILE: Moderately toxic by ingestion. When heated to decomposition it emits acrid smoke and irritating vapors.

CMQ475 **CAS:3669-32-7** **HR: 3**
CINNAMOHYDROXAMIC ACID
mf: $C_9H_9NO_2$ mw: 163.19

PROP: Crystals. Sol in EtOH and Me_2CO.

SYNS: CINNAMOYLHYDROXAMIC ACID □ N-HYDROXY-3-PHENYL-2-PROPENAMIDE (9CI)

TOXICITY DATA WITH REFERENCE
orl-rat LD50:2140 mg/kg TXAPA9 28,313,74
orl-mus LD50:1350 mg/kg BJPCAL 26,41,66
scu-mus LD50:86 mg/kg BJPCAL 26,41,66

SAFETY PROFILE: Poison by subcutaneous route. Moderately toxic by ingestion. When heated to decomposition it emits toxic fumes of NO_x.

CMQ500 **CAS:4360-47-8** **HR: 3**
CINNAMONITRILE
mf: C_9H_7N mw: 129.17

SYN: CINNAMYL NITRILE

TOXICITY DATA WITH REFERENCE
skn-rbt 500 mg/24H MOD FCTXAV 14,659,76
orl-rat LD50:4150 mg/kg FCTXAV 14,659,76
ivn-rbt LDLo:34 mg/kg COREAF 153,895,11
scu-gpg LDLo:130 mg/kg COREAF 153,895,11

CONSENSUS REPORTS: Cyanide and its compounds are on the Community Right-To-Know List. Reported in EPA TSCA Inventory.

SAFETY PROFILE: Poison by subcutaneous and intravenous routes. Mildly toxic by ingestion. A skin irritant. See also NITRILES. When heated to decomposition it emits toxic fumes of NO_x and CN^-.

CMQ625 **CAS:751-01-9** **HR: 3**
14-CINNAMOYLOXYCODEINONE
mf: $C_{27}H_{25}NO_5$ mw: 443.53

TOXICITY DATA WITH REFERENCE
orl-mus LD50:1100 mg/kg JPPMAB 17,759,65
scu-mus LD50:530 mg/kg JPPMAB 17,759,65
ivn-mus LD50:31 mg/kg JPPMAB 17,759,65

SAFETY PROFILE: Poison by intravenous route. Moderately toxic by ingestion and subcutaneous route. When heated to decomposition it emits toxic fumes of NO_x.

CMQ725 **CAS:1405-39-6** **HR: 3**
CINNAMYCIN

SYNS: ANTIBIOTIC NSC-71936 □ NSC-71936

TOXICITY DATA WITH REFERENCE
ipr-mus LD50:2 mg/kg 85GDA2 4(1),249,80
scu-mus LD50:400 mg/kg 85GDA2 4(1),249,80
ivn-mus LD50:2500 μg/kg 85GDA2 4(1),249,80

SAFETY PROFILE: Poison by subcutaneous, intraperitoneal, and intravenous routes.

CMQ730 **CAS:103-54-8** **HR: 2**
CINNAMYL ACETATE
mf: $C_{11}H_{12}O_2$ mw: 176.23

PROP: Colorless liquid; sweet floral odor. D: 1.047–1.051, refr index: 1.539–1.543, flash p: 244°F. Misc with chloroform, ether, fixed oils; insol in glycerin, water @ 264°.

SYNS: ACETIC ACID, CINNAMYL ESTER □ FEMA No. 2293 □ γ-PHENYLALLYL ACETATE □ 3-PHENYL-2-PROPEN-1-YL ACETATE

TOXICITY DATA WITH REFERENCE
skn-man 16 mg/48H MLD CTOIDG 94(8),41,79
skn-rbt 100 mg/24H MOD CTOIDG 94(8),41,79
skn-gpg 100 mg/24H MLD CTOIDG 94(8),41,79
orl-rat LD50:3300 mg/kg FCTXAV 11,1065,73
orl-mus LD50:4750 mg/kg VPITAR 33(5),48,74
ipr-mus LD50:1200 mg/kg PHMCAA 3,62,61
orl-gpg LD50:4750 mg/kg VPITAR 33(5),48,74

CONSENSUS REPORTS: Reported in EPA TSCA Inventory.

SAFETY PROFILE: Moderately toxic by ingestion and intraperitoneal routes. A skin irritant. Combustible liquid. When heated to decomposition it emits acrid smoke and fumes. See also ALLYL COMPOUNDS.

CMQ740 **CAS:104-54-1** **HR: 2**
CINNAMYL ALCOHOL
mf: $C_9H_{10}O$ mw: 134.19

PROP: Occurs (in the esterified form) in storax and in balsam Peru, cinnamon leaves, hyacinth oil. Needles or crystalline mass; odor of hyacinth. Mp: 33°, d: 1.0397, bp: 250.0°, n (20/D) 1.58190. Sol in water, glycerol, and propylene glycol; freely sol in alc, ether, other common organic solvents.

SYNS: CINNAMIC ALCOHOL □ CINNAMYL ALCOHOL, SYNTHETIC □ FEMA No. 2294 □ γ-PHENYLALLYL ALCOHOL □ 3-PHENYLALLYL ALCOHOL □ 3-PHENYL-2-PROPEN-1-OL □ STYRONE □ STYRYL CARBINOL

TOXICITY DATA WITH REFERENCE
skn-rbt 500 mg/24H MOD FCTXAV 12,855,74
dnr-bcs 10 mg/disc OIGZSE 34,267,85
orl-rat LD50:2000 mg/kg FCTXAV 12,855,74
orl-mus LD50:2675 mg/kg VPITAR 33(5),48,74
orl-gpg LD50:2675 mg/kg VPITAR 33(5),48,74

CONSENSUS REPORTS: Reported in EPA TSCA Inventory.

SAFETY PROFILE: Moderately toxic by ingestion. A skin irritant. Mutation data reported. When heated to decomposition it emits acrid smoke and fumes. See also ALCOHOLS and ALLYL COMPOUNDS.

CMQ750 CAS:5320-75-2 HR: 2
CINNAMYL BENZOATE
mf: $C_{16}H_{14}O_2$ mw: 238.30

SYNS: BENZOIC ACID, CINNAMYL ESTER □ CINNAMYL ALCOHOL, BENZOATE

TOXICITY DATA WITH REFERENCE
skn-rbt 500 mg/24H MOD FCTXAV 14,659,76
orl-rat LD50:4 g/kg FCTXAV 14,659,76

CONSENSUS REPORTS: Reported in EPA TSCA Inventory.

SAFETY PROFILE: Moderately toxic by ingestion. A skin irritant. See also ESTERS. When heated to decomposition it emits acrid smoke and irritating fumes.

CMQ800 CAS:103-61-7 HR: 1
CINNAMYL BUTYRATE
mf: $C_{13}H_{16}O_2$ mw: 204.29

SYNS: BUTYNOIC ACID, 3-PHENYL-2-PROPENYL ESTER □ BUTYRIC ACID, CINNAMYL ESTER □ PHENYLPROPENYL n-BUTYRATE

TOXICITY DATA WITH REFERENCE
skn-rbt 500 mg/24H FCTXAV 16,691,78

CONSENSUS REPORTS: Reported in EPA TSCA Inventory.

SAFETY PROFILE: A skin irritant. When heated to decomposition it emits acrid smoke and irritating fumes.

CMQ850 CAS:122-69-0 HR: 1
CINNAMYL CINNAMATE
mf: $C_{18}H_{16}O_2$ mw: 264.34

SYNS: CINNAMIC ACID, CINNAMYL ESTER □ CINNAMYL ALCOHOL, CINNAMATE □ CINNAMYLESTER KYSELINY SKORICOVE □ PHENYLALLYL CINNAMATE □ 3-PHENYL-2-PROPEN-1-YL CINNAMATE □ 3-PHENYL-2-PROPENYL 3-PHENYL-2-PROPENOATE □ 2-PROPENOIC ACID, 3-PHENYL-, 3-PHENYL-2-PROPENYL ESTER (9CI) □ STYRACIN

TOXICITY DATA WITH REFERENCE
orl-rat LD50:5000 mg/kg FCTXAV 13,753,75

CONSENSUS REPORTS: Reported in EPA TSCA Inventory.

SAFETY PROFILE: Mildly toxic by ingestion. When heated to decomposition it emits acrid smoke and irritating vapors.

CMR100 CAS:298-57-7 HR: 3
1-CINNAMYL-4-(DIPHENYLMETHYL)PIPERAZINE
mf: $C_{26}H_{28}N_2$ mw: 368.56

PROP: Very sltly sol in H_2O.

SYNS: trans-1-CINNAMYL-(4-DIPHENYLMETHYL)PIPERAZINE □ CINNARIZINE □ DIMITRON □ DIMITRONAL □ FOLCODAL □ GLANIL □ LABYRIN □ LAZETA □ MARISAN □ 516 MD □ MIDRONAL □ MITRONAL □ PIPERAZINE, 1-CINNAMYL-4-(DIPHENYLMETHYL)- □ PIPERAZINE, 1-(DIPHENYLMETHYL)-4-(3-PHENYL-2-PROPENYL)-(9CI) □ R 516 □ R 1575 □ SEPAN □ STUGERON □ STUTGERON □ STUTGIN □ TOLIMAN

TOXICITY DATA WITH REFERENCE
orl-rat TDLo:3750 mg/kg (female 30D pre):REP KSRNAM 16,1748,82
orl-wmn TDLo:252 mg/kg/12W-I:SKN BJDEAZ 112,607,85
ipr-rat LD50:1050 mg/kg MDACAP 21,443,85
ivn-rat LD50:24 mg/kg KSRNAM 16,1735,82
ipr-mus LD50:730 mg/kg MDACAP 21,443,85
ivn-mus LD50:22 mg/kg KSRNAM 16,1735,82

SAFETY PROFILE: Poison by intravenous route. Moderately toxic by intraperitoneal route. Experimental reproductive effects. Human systemic effects by ingestion: allergic dermatitis. When heated to decomposition it emits toxic fumes of NO_x.

CMR250 CAS:64043-53-4 HR: 3
d-CINNAMYLEPHEDRINE HYDROCHLORIDE
mf: $C_{19}H_{23}NO{\cdot}ClH$ mw: 317.89

SYN: CINNAMYLEPHEDRINE HYDROCHLORIDE, DEXTRO

TOXICITY DATA WITH REFERENCE
idr-hmn TDLo:143 ng/kg:SKN JPETAB 76,295,42
scu-mus LD50:75 mg/kg JPETAB 76,295,42

SAFETY PROFILE: Poison by subcutaneous route. Human systemic effects by very small amounts administered intradermally: unspecified skin effects. When heated to decomposition it emits very toxic fumes of HCl and NO_x.

CMR500 CAS:104-65-4 HR: 2
CINNAMYL FORMATE
mf: $C_{10}H_{10}O_2$ mw: 162.20

PROP: Colorless liquid; balsamic odor. D: 1.077–1.082, refr index: 1.550–1.556, flash p: 212°F. Misc with alc, chloroform, ether, fixed oils; insol in water @ 250°.

SYNS: CINNAMYL ALCOHOL, FORMATE □ CINNAMYL METHANOATE □ FEMA No. 2299 □ FORMIC ACID, CINNAMYL ESTER □ 3-PHENYL-2-PROPEN-1-YL FORMATE

TOXICITY DATA WITH REFERENCE
orl-rat LD50:2900 mg/kg FCTXAV 14,659,76

CONSENSUS REPORTS: Reported in EPA TSCA Inventory.

SAFETY PROFILE: Moderately toxic by ingestion. See also ESTERS. Combustible liquid. When heated to decomposition it emits acrid smoke and irritating fumes.

C

CMR750 CAS:103-59-3 HR: 1
CINNAMYL ISOBUTYRATE
mf: $C_{13}H_{16}O_2$ mw: 204.29

SYNS: ISOBUTYRIC ACID, CINNAMYL ESTER □ 2-METHYL-PROPANOIC ACID-3-PHENYL-2-PROPENYL ESTER

TOXICITY DATA WITH REFERENCE
skn-rbt 500 mg/24H MLD FCTXAV 17,509,79
orl-rat LD50:>5 g/kg FCTXAV 17,523,79
skn-rbt LD50:>5 g/kg FCTXAV 17,523,79

CONSENSUS REPORTS: Reported in EPA TSCA Inventory.

SAFETY PROFILE: Low toxicity by ingestion and skin contact. A skin irritant. When heated to decomposition it emits acrid smoke and irritating fumes. See also ESTERS.

CMR800 HR: 1
CINNAMYL ISOVALERATE
mf: $C_{14}H_{18}O_2$ mw: 218.30

PROP: Colorless to sltly yellow liquid; spicy, floral, fruity odor. D: 0.991–0.996, refr index: 1.518–1.524, flash p: 212°F. Misc in alc, chloroform, ether, most oils; insol in glycerin, propylene glycol, and water @ 313°.

SYN: FEMA No. 2302

SAFETY PROFILE: Combustible liquid. When heated to decomposition it emits acrid smoke and irritating fumes.

CMR850 CAS:103-56-0 HR: 2
CINNAMYL PROPIONATE
mf: $C_{12}H_{14}O_2$ mw: 190.24

PROP: Colorless to pale-yellow liquid; spicy, fruity, balsamic odor. D: 1.029–1.033, refr index: 1.523–1.537, flash p: 212°F. Misc in alc, chloroform, ether, most oils; insol in glycerin, propylene glycol, and water @ 289°.

SYNS: FEMA No. 2301 □ 3-PHENYL-2-PROPENYL PROPIONATE □ 3-PHENYL-2-PROPEN-1-YL PROPIONATE □ PROPIONIC ACID, CINNAMYL ESTER

TOXICITY DATA WITH REFERENCE
skn-rbt 500 mg/24H MLD FCTXAV 12,859,74
orl-rat LD50:3400 mg/kg FCTXAV 12,859,74

CONSENSUS REPORTS: Reported in EPA TSCA Inventory.

SAFETY PROFILE: Moderately toxic by ingestion. A skin irritant. Combustible liquid. When heated to decomposition it emits acrid smoke and irritating fumes.

CMS125 CAS:60763-49-7 HR: 2
CINNARIZINE CLOFIBRATE
mf: $C_{26}H_{28}N_2 \cdot C_{10}H_{11}ClO_3$ mw: 583.22

SYNS: CLOFIBRATO de CINARIZINA (SPANISH) □ LM-16

TOXICITY DATA WITH REFERENCE
orl-rat LD50:6800 mg/kg DRFUD4 3,572,78
ipr-rat LD50:2000 mg/kg DRFUD4 3,572,78
orl-mus LD50:4300 mg/kg DRFUD4 3,572,78

SAFETY PROFILE: Moderately toxic by ingestion and intraperitoneal routes. When heated to decomposition it emits toxic fumes of Cl^- and NO_x.

CMS130 CAS:28657-80-9 HR: 2
CINOXACIN
mf: $C_{12}H_{10}N_2O_5$ mw: 262.24

PROP: Light tan crystals. Mp: 261–262° (decomp). Sol in most polar organic solvents.

SYNS: CINOBAC □ CINX □ COMPOUND 64716 □ 1-ETHYL-1,4-DIHYDRO-4-OXO(1,3)DIOXOLO(4,5-g)CINNOLINE-3-CARBOXYLIC ACID □ 1-ETHYL-6,7-METHYLENEDIOXY-4(1H)-OXOCINNOLINE-3-CARBOXYLIC ACID

TOXICITY DATA WITH REFERENCE
orl-rat TDLo:1100 mg/kg (female 7-17D post):TER NKRZAZ 28,484,80
orl-rat TDLo:1050 mg/kg (female 14D pre):REP NKRZAZ 28,484,80
orl-rat TDLo:3150 mg/kg (63D male):REP NKRZAZ 28,484,80
orl-rat TDLo:550 mg/kg (female 7-17D post):TER NKRZAZ 28,484,80
orl-rat LD50:4160 mg/kg NKRZAZ 28(Suppl 4),406,80
scu-rat LD50:1380 mg/kg IYKEDH 14,297,83
ivn-rat LD50:860 mg/kg IYKEDH 14,297,83
orl-mus LD50:2330 mg/kg IYKEDH 14,297,83
scu-mus LD50:900 mg/kg IYKEDH 14,297,83
ivn-mus LD50:850 mg/kg IYKEDH 14,297,83

SAFETY PROFILE: Moderately toxic by ingestion, subcutaneous and intravenous routes. An experimental teratogen. Other experimental reproductive effects. When heated to decomposition it emits toxic fumes of NO_x.

CMS135 CAS:71631-15-7 HR: 1
C.I. PIGMENT BLACK 30

SYNS: CHROME IRON NICKEL BLACK SPINEL □ C.I. 77504 □ DCMA-13-50-9 □ NICKEL IRON CHROMITE BLACK SPINEL

CONSENSUS REPORTS: IARC Cancer Review: Animal Limited Evidence IMEMDT 49,257,90. Reported in EPA TSCA Inventory.

SAFETY PROFILE: A questionable carcinogen. When heated to decomposition it emits toxic vapors of Cr, Fe, and Ni.

CMS140 CAS:14302-13-7 HR: D
C.I. PIGMENT GREEN 36
mf: $C_{32}Br_6Cl_{10}CuN_8$ mw: 1393.90

SYNS: C.I. 74265 □ C.I. PIGMENT GREEN 38 □ C.I. PIGMENT GREEN 41 □ COPPER, (1,3,8,16,18,24-HEXABROMO-2,4,9,10,11,15,17,22,23,25-DECACHLOROPHTHALOCYAN INATO(2-))- □ FASTOGEN GREEN Y □ FASTOGEN GREEN 2YK □ HELIO FAST GREEN GN □ HELIO FAST GREEN GT □ HELIOGEN GREEN 9360 □ HELIOGEN GREEN 6G □ HELIOGEN GREEN 6GA □ HELIOGEN GREEN 8GA □ HOSTAPERM GREEN 8G □ MONASTRAL FAST GREEN 3Y □ MONASTRAL FAST GREEN 6Y □ MONASTRAL FAST GREEN 3YA □ MONASTRAL FAST GREEN 6YA □ MONASTRAL GREEN Y-GT 805D □ PHTHAL-

OCYANINE GREEN 6G □ PIGMENT GREEN 38 □ SANDORIN GREEN 8GLS □ VYNAMON GREEN 6Y

TOXICITY DATA WITH REFERENCE
mmo-sat 1 mg/plate ENMUDM 8(Suppl 7),1,86

CONSENSUS REPORTS: Reported in EPA TSCA Inventory.

SAFETY PROFILE: Mutation data reported. When heated to decomposition it emits toxic vapors of NO_x, Cu, Br^- and Cl^-.

CMS145 CAS:3520-72-7 HR: 1
C.I. PIGMENT ORANGE 13
mf: $C_{32}H_{24}Cl_2N_8O_2$ mw: 623.54

SYNS: ATUL VULCAN FAST PIGMENT ORANGE G □ BENZIDINE ORANGE □ BENZIDINE ORANGE 45-2850 □ BENZIDINE ORANGE 45-2880 □ BENZIDINE ORANGE TONER □ BENZIDINE ORANGE WD 265 □ CALCOTONE ORANGE R □ CARNELIO ORANGE G □ C.I. 21110 □ DAINICHI FAST ORANGE RR □ DALTOLITE FAST ORANGE G □ DIARYLIDE ORANGE □ ELJON FAST ORANGE G □ FAST BENZIDENE ORANGE YB 3 □ FASTONA ORANGE G □ FAST ORANGE G □ GRAPHTOL ORANGE GP □ IRGALITE ORANGE P □ IRGALITE ORANGE PG □ IRGALITE ORANGE PX □ IRGAPLAST ORANGE G □ KROMON ORANGE G □ LATEXOL FAST ORANGE J □ LUTETIA ORANGE J □ MONOLITE FAST ORANGE G □ MONOLITE FAST ORANGE GA □ NO. 56 CONC. PERMANENT ORANGE G □ NO. 59 FORTHFAST BENZIDINE YELLOW □ ORALITH ORANGE PG □ ORANGE G □ ORANGE Y □ OSWEGO ORANGE X 2065 □ PERMANENT ORANGE G □ PERMANENT ORANGE G EXTRA □ PIGMENT FAST ORANGE G □ PIGMENT ORANGE 13 □ PIGMENT ORANGE ERH □ PIGMENT ORANGE G □ PIGMENT ORANGE ZH □ PLASTOL ORANGE G □ POLYMO ORANGE GR □ PONOLITH ORANGE Y □ PV-ORANGE G □ PYRAZALONE ORANGE NP 215 □ PYRAZOLONE ORANGE □ PYRAZOLONE ORANGE YB 3 □ RECOLITE ORANGE G □ RESAMINE FAST ORANGE G □ SANYO BENZIDINE ORANGE □ SEGNALE LIGHT ORANGE G □ SEGNALE LIGHT ORANGE PG □ SIEGLE ORANGE S □ SILOGOMMA ORANGE G □ SILOTERMO ORANGE G □ SILOTON ORANGE GT □ SYMULER FAST PYRAZOLONE ORANGE G □ SYTON FAST ORANGE G □ TERTROPIGMENT ORANGE PG □ VULCAFIX ORANGE J □ VULCAFIX ORANGE JV □ VULCAFOR FAST ORANGE G □ VULCAFOR FAST ORANGE GA □ VULCAN FAST ORANGE G □ VULCAN FAST ORANGE GA □ VULCAN FAST ORANGE GN □ VULCOL FAST ORANGE G □ VYNAMON ORANGE G

TOXICITY DATA WITH REFERENCE
mmo-sat 1 mg/plate GTPZAB 27(10),52,83
mma-sat 500 µg/plate GTPZAB 27(10),52,83
orl-rat LD50:>5 g/kg EPASR* 8EHQ-0690-0962

CONSENSUS REPORTS: Reported in EPA TSCA Inventory.

SAFETY PROFILE: Low toxicity by ingestion. Mutation data reported. When heated to decomposition it emits toxic vapors of NO_x and Cl^-.

CMS148 CAS:7585-41-3 HR: 1
C.I. PIGMENT RED 48:1
mf: $C_{18}H_{12}ClN_2O_6S \cdot Ba$ mw: 557.17

SYNS: BON RED YELLOW SHADE □ BRIGHT RED G TONER □ C.I. PIGMENT RED 48, BARIUM SALT (1:1) (8CI) □ ELJON RUBINE BS □ ISOL BONA RED NR BARIUM SALT □ ISOL BONA RED N 5R BARIUM SALT □ LITHOL SCARLET K 3700 □ 2-NAPHTHALENECARBOXYLIC ACID, 4-((5-CHLORO-4-METHYL-2-SULFOPHENYL)AZO)-3-HYDROXY-, BARIUM SALT (1:1) □ PERMANENT RED BB □ PERMANENT RED BBa □ PIGMENT RED 48:1 □ RESINO RED K □ RUBINE TONER B □ RUBINE TONER BA □ RUBINE TONER BT □ SANYO FAST RED 2B □ SANYO FAST RED 2BE □ SEGNALE RED GS □ SEIKAFAST RED 8040 □ SYMULER RED 3023 □ SYMULER RED NRY □ WATCHUNG RED Y

TOXICITY DATA WITH REFERENCE
ihl-rat LCLo:5420 mg/m³/4H JACTDZ 1,745,92

CONSENSUS REPORTS: Reported in EPA TSCA Inventory.

SAFETY PROFILE: Low toxicity by inhalation. When heated to decomposition it emits toxic vapors of NO_x, SO_x, and Cl^-.

CMS150 CAS:2092-56-0 HR: 1
C.I. PIGMENT RED 53
mf: $C_{17}H_{12}ClN_2O_4S \cdot Na$ mw: 398.81

SYNS: BENZENESULFONIC ACID, 5-CHLORO-2-((2-HYDROXY-1-NAPHTHALENYL)AZO)-4-METHYL-, MONOSODIUM SALT □ BRIGHT RED □ BRONZE ORANGE □ BRONZE ORANGE TONER □ CERVEN PIGMENT 53 □ 5-CHLORO-2-((2-HYDROXY-1-NAPHTHALENYL)AZO)-4-METHYLBENZENESULFONIC ACID SODIUM SALT □ C.I. 15585 □ D and C RED No. 8 □ DUPLEX RED LAKE CD 20-5925 □ IRGALITE RED C □ ISOL LAKE RED LCL □ ISOL LAKE RED LCR □ ISOL LAKE RED LCT □ JAPAN RED 203 □ JAPAN RED NO. 203 □ KROMON LAKE RED C □ LAKE RED C □ LAKE RED C 18958 □ LAKE RED CY □ LITHOSOL RED C □ LITHOSOL RED CLM □ LUTETIA RED C □ MONOLITE RED CN □ PIGMENT RED 53 □ PIGMENT RED GG □ RED 203 □ 11938 RED □ RED FOR LAKE C TONER □ RESAMINE RED GB □ SEGNALE BRONZE RED P □ SEGNALE RED C □ SICO LAKE RED CU □ TERTROPIGMENT RED LC □ TONING ORANGE □ VULCAN RED LCN □ VULCOL FAST RED C

TOXICITY DATA WITH REFERENCE
orl-mus LD50:>12 g/kg SCIEAS 36(1-4),10,89

CONSENSUS REPORTS: Reported in EPA TSCA Inventory.

SAFETY PROFILE: Low toxicity by ingestion. When heated to decomposition it emits toxic vapors of NO_x, SO_x, and Cl^-.

CMS155 CAS:5858-81-1 HR: 1
C.I. PIGMENT RED 57, DISODIUM SALT (8CI)
mf: $C_{18}H_{14}N_2O_6S \cdot 2Na$ mw: 432.38

SYNS: CERVEN PIGMENT 57 □ C.I. 15850 □ C.I. PIGMENT RED 57 (7CI) □ D and C RED NO. 6 □ IRGALITE RED 4B □ IRGALITE RUBINE PB □ ISOL BONA RUBINE BK □ ISOL BONA RUBINE BKS □ ISOL BONA RUBINE KBK □ ISOL RUBY BK □ ISOL RUBY BKS □ JAPAN RED 201 □ KROMON PERMANENT RED 4B □ LITHOL RUBINE □ LITHOL RUBIN B □ LITHOL RUBINE BNA □ PERMANENT RED 4B □ PERMANENT RED F 6R □ PIGMENT RED 57 □ PIGMENT RUBINE B □ PIGMENT RUBINE BCL □ PLASTOL RUBINE BC □ 2-NAPHTHALENECARBOXYLIC ACID, 3-HYDROXY-4-((4-METHYL-2-SULFOPHENYL)AZO)-, DISODIUM SALT □ RED 102 □ 11070 RED □ RESAMINE RUBINE BC □ RUBINE RED RR 1253 □ SEGNALE RED 3R □ SILOTERMO CARMINE G □ SILOTON RUBINE B □ SILOTON RUBINE 2B □ VYNAMON CLARET Y

TOXICITY DATA WITH REFERENCE
orl-rat LD50:>10,800 mg/kg SCIEAS 36(1-4),10,89

CONSENSUS REPORTS: Reported in EPA TSCA Inventory.

SAFETY PROFILE: Low toxicity by ingestion. When heated to decomposition it emits toxic vapors of NO_x and SO_x.

CMS160 CAS:6371-76-2 HR: 2
C.I. PIGMENT RED 64, CALCIUM SALT (2:1) (8CI)
mf: $C_{34}H_{22}N_4O_6•Ca$ mw: 622.68

SYNS: BRILLIANT LAKE M □ BRILLIANT LAKERED R □ BRILLIANT RED TONER RA □ BRILLIANT SCARLET G □ C.I. 15800 CA SALT □ C.I. PIGMENT RED 64:1 □ DAINICHI BRILLIANT SCARLET G □ DAINICHI BRILLIANT SCARLET RG □ D and C RED NO.31 □ JAPAN RED 219 □ 2-NAPHTHALENECARBOXYLIC ACID, 3-HYDROXY-4-(PHENYLAZO)-, CALCIUM SALT (2:1) □ PIGMENT RED 64:1 □ PIGMENT SCARLET TONER RB □ R 219 □ RED 219 □ 11067 RED □ RED NO. 219 □ SYMULER BRILLIANT SCARLET G □ TOPAZ TONER R 5

TOXICITY DATA WITH REFERENCE
orl-rat LD50:920 mg/kg SCIEAS 36(1-4),10,89

CONSENSUS REPORTS: Reported in EPA TSCA Inventory.

SAFETY PROFILE: Moderately toxic by ingestion. When heated to decomposition it emits toxic vapors of NO_x.

CMS208 CAS:5102-83-0 HR: 1
C.I. PIGMENT YELLOW 13
mf: $C_{36}H_{34}Cl_2N_6O_4$ mw: 685.66

SYNS: BENZIDINE YELLOW 20544 □ BENZIDINE YELLOW GE □ BENZIDINE YELLOW GR □ BENZIDINE YELLOW LEMON 12221 □ BUTANAMIDE, 2,2′-((3,3′-DICHLORO(1,1′-BIPHENYL)-4,4′-DIYL)BIS(AZO))BIS(N-(2,4-DIMETHYLPHENYL)-3-OXO)- □ C.I. 21100 □ DAINICHI BENZIDINE YELLOW 2GR □ DIARYLIDE YELLOW □ ELKON FAST YELLOW GR □ GRAPHTOL YELLOW RGS □ HELIO FAST YELLOW GRF □ HELIO FAST YELLOW GRN □ HOSTAPERM YELLOW GR □ IRGALITE YELLOW BAW □ IRGALITE YELLOW BAWX □ IRGAPLAST YELLOW IRS □ ISOL BENZIDINE FAST YELLOW GRX □ ISOL BENZIDINE YELLOW GRX 2548 □ ISOL DIARYL YELLOW GRF □ KROMON YELLOW GXR □ LATEXOL FAST YELLOW JR □ LIGHT YELLOW JBR □ LIONOL YELLOW FG 1310 □ MONOLITE FAST YELLOW GLV □ MONOLITE YELLOW GL □ MONOLITE YELLOW GLA □ PERMANENT YELLOW GR □ PERMANENT YELLOW GR01 □ PIGMENT YELLOW 13 □ PIGMENT YELLOW MH □ POLYMO YELLOW GR □ RECOLITE FAST YELLOW BLF □ RECOLITE FAST YELLOW BLT □ RUBBER FAST YELLOW GRA □ SARANAC YELLOW X 2838 □ SEGNALE LIGHT YELLOW GRX □ SICO FAST YELLOW D 1355 □ SYMULER FAST YELLOW GRF □ SYMULER FAST YELLOW GRTF □ TERTROPIGMENT FAST YELLOW VGR □ TERTROPIGMENT PGR □ VULCAN FAST YELLOW GR □ VULCAN FAST YELLOW GRA □ VULCAN FAST YELLOW GRN □ VYNAMON YELLOW GRE □ VYNAMON YELLOW GRES □ YELLOW AAMX □ YELLOW TONER YB5

TOXICITY DATA WITH REFERENCE
orl-rat LD50:>5 g/kg EPASR* 8EHQ-0690-0962

CONSENSUS REPORTS: Reported in EPA TSCA Inventory.

SAFETY PROFILE: Low toxicity by ingestion. When heated to decomposition it emits toxic vapors of NO_x and Cl^-.

CMS212 CAS:5468-75-7 HR: 1
C.I. PIGMENT YELLOW 14
mf: $C_{34}H_{30}Cl_2N_6O_4$ mw: 657.60

SYNS: AAOT YELLOW □ BENZIDINE YELLOW AAOT □ BENZIDINE YELLOW ABZ 249 □ BENZIDINE YELLOW G □ BENZIDINE YELLOW GGT □ BENZIDINE YELLOW L □ BENZIDINE YELLOW OT (6CI) □ BENZIDINE YELLOW OTYA 8055 □ BUTANAMIDE, 2,2′-((3,3′-DICHLORO(1,1′-BIPHENYL)-4,4′-DIYL)BIS(AZO))BIS(N-(2-METHYLPHEN YL)-3-OXO)- □ CALCOTONE YELLOW GP □ C.I. 21095 □ DIARYLIDE YELLOW AAOT □ GRAPHTOL YELLOW GXS □ HOSTAPERM YELLOW GT □ HOSTAPERM YELLOW GTT □ IRGALITE YELLOW BR □ IRGALITE YELLOW BRE □ ISOL BENZIDINE YELLOW GO □ LAKE YELLOW GA □ LIGHT YELLOW JBV □ LIGHT YELLOW JBVT □ LIONOL YELLOW GGR □ No. 55 Conc. PALE YELLOW SF □ PERMAGEN YELLOW □ PERMAGEN YELLOW GA □ PERMANENT YELLOW G □ PERMANENT YELLOW LIGHT □ PIGMENT FAST YELLOW GP □ PIGMENT FAST YELLOW 2GP □ PIGMENT YELLOW 14 □ PIGMENT YELLOW 2G □ PIGMENT YELLOW GGP □ PIGMENT YELLOW GPP □ PLASTOL YELLOW GG □ PLASTOL YELLOW GP □ RADIANT YELLOW □ RECOLITE FAST YELLOW B 2T □ RESAMINE FAST YELLOW GGP □ RESAMINE YELLOW GP □ RUBBER FAST YELLOW GA □ SACANDAGA YELLOW X 2476 □ SANYO BENZIDINE YELLOW □ SEGNALE YELLOW 2GR □ SEIKAFAST YELLOW 2200 □ SILOGOMMA FAST YELLOW 2G □ SILOTERMO YELLOW G □ SUMATRA YELLOW X 1940 □ SUMIKAPRINT YELLOW GFN □ SYMULER FAST YELLOW 4090G □ SYMULER FAST YELLOW 5GF □ TETROPIGMENT FAST YELLOW VG □ TETROPIGMENT FAST YELLOW BG □ VERSAL FAST YELLOW PG □ VULCAFOR FAST YELLOW 2G □ VULCAN FAST YELLOW G □ VYNAMON YELLOW 2G □ VYNAMON YELLOW 2GE

TOXICITY DATA WITH REFERENCE
orl-rat LD50:>5 g/kg EPASR* 8EHQ-0690-0962

CONSENSUS REPORTS: Reported in EPA TSCA Inventory.

SAFETY PROFILE: Low toxicity by ingestion. When heated to decomposition it emits toxic vapors of NO_x and Cl^-.

CMS215 CAS:62865-26-3 HR: 3
C.I. PIGMENT YELLOW 35

SYNS: B-3-Zh □ CADMIUM GOLDEN □ CADMIUM LEMON □ CADMIUM PRIMROSE □ CADMIUM SULFIDE mixed with ZINC SULFIDE (1:1) □ C.I. 77205

OSHA PEL: TWA 5 μg(Cd)/m³
ACGIH TLV: TWA 0.01 mg(Cd)/m³; Suspected Carcinogen
NIOSH REL: (Cadmium): TWA reduce to lowest feasible level

SAFETY PROFILE: Confirmed human carcinogen. When heated to decomposition it emits toxic fumes of Cd.

CMS225 CAS:9008-54-2 HR: 3
CIRCULIN
mf: $C_{56}H_{96}N_{12}O_{13}$ mw: 1145.64

SYN: POLYPEPTIN

TOXICITY DATA WITH REFERENCE
ivn-rat LD50:20 mg/kg JCINAO 28,1032,49
ims-rat LD50:23 mg/kg JCINAO 28,1032,49

ipr-mus LD50:15 mg/kg JOBAAY 56,749,48
scu-mus LD50:77 mg/kg JCINAO 29,1032,49
ivn-mus LD50:10 mg/kg JCINAO 28,1032,49

SAFETY PROFILE: Poison by subcutaneous, intramuscular, intravenous, and intraperitoneal routes. When heated to decomposition it emits toxic fumes of NO_x.

CMS227 CAS:12225-26-2 HR: 1
C.I. REACTIVE BLACK 8

SYNS: CERN REAKTIVNI 8 □ CIBACRON BLACK B-D □ HELAKTYN BLACK DN □ OSTAZIN BLACK H-N □ PROCION BLACK H-N □ REACTIVE BLACK 8

TOXICITY DATA WITH REFERENCE
skn-rbt 500 mg/24H MLD 85JCAE-,1293,86
eye-rbt 500 mg/24H MLD 85JCAE-,1293,86
orl-rat LD50:9120 mg/kg 85JCAE-,1293,86

SAFETY PROFILE: Mildly toxic by ingestion. A skin and eye irritant. When heated to decomposition it emits acrid smoke and irritating fumes.

CMS228 CAS:13324-20-4 HR: 1
C.I. REACTIVE BLUE 4
mf: $C_{23}H_{14}Cl_2N_6O_8S_2$ mw: 637.45

SYNS: 2-ANTHRACENESULFONIC ACID, 1-AMINO-4-(3-((4,6-DICHLORO-s-TRIAZIN-2-YL)AMINO)-4-SULFOANILINO)-9,10-DIHYDRO-9,10-DIOXO- □ C.I. 61205 □ HELAKTYN PURE BLUE FR □ MIKACION BRILLIANT BLUE RS □ MODR REAKTIVNI 4 □ OSTAZIN BRILLIANT BLUE S-R □ PROCION BLUE MX-R □ PROCION BRILLIANT BLUE MR □ PROCION BRILLIANT BLUE MX-R □ PROCION BRILLIANT BLUE RS □ REACTIVE BLUE 4

TOXICITY DATA WITH REFERENCE
eye-rbt 100 mg/24H MOD 85JCAE -,1328,86
orl-rat LD50:8980 mg/kg 85JCAE-,1328,86

CONSENSUS REPORTS: Reported in EPA TSCA Inventory.

SAFETY PROFILE: Low toxicity by ingestion. An eye irritant. When heated to decomposition it emits toxic vapors of NO_x, SO_x, and Cl^-.

CMS230 CAS:58128-20-4 HR: 2
C.I. REACTIVE YELLOW 73

SYNS: POLACTINE G YELLOW □ POLAKTYN YELLOW G □ YELLOW POLAKTIN G □ ZOLCIENI POLACTYNOWEJ G (POLISH)

TOXICITY DATA WITH REFERENCE
eye-rbt 20 mg MLD MEPAAX 30,157,79
orl-rat LD50:8200 mg/kg MEPAAX 30,157,79
ipr-rat LD50:1330 mg/kg MEPAAX 30,157,79

SAFETY PROFILE: Moderately toxic by intraperitoneal route. Mildly toxic by ingestion. An eye irritant.

CMS231 CAS:3567-66-6 HR: D
C.I. RED 33, DISODIUM SALT □ 64553-75-9
mf: $C_{16}H_{13}N_3O_7S_2{\cdot}2Na$ mw: 469.42

SYNS: ACETYL RED B □ ACID FUCHSINE D □ ACID FUCHSIN FAST B □ ACID RED 33 □ ACID RED 2A □ ACID RED B □ AMACID FUCHSINE 4B □ AZO FUCHSINE □ AZO GRENADINE □ AZO MAGENTA G □ BRASILAN FUCHSINE D □ CERTICOL RED B □ C.I. ACID RED 33 □ C.I. FOOD RED 12 □ C.I. REDUCING AGENT 6 □ COLACID RED 2A □ D and C RED No. 33 □ EDICOL SUPRA 10B □ EDICOL SUPRA 10BS □ ENIACID FUCHSINE BN □ FAST ACID MAGENTA □ FAST ACID MAGENTA B □ HEXACOL RED 10B □ HEXALAN RED B □ 2,7-NAPHTHALENEDISULFONIC ACID, 4-AMINO-5-HYDROXY-6-PHENYLAZO-, DISODIUM SALT □ NAPHTHALENE RED B □ 1424 RED □ 11427 RED □ RED 10B □ RED No. 227

TOXICITY DATA WITH REFERENCE
mmo-esc 10 g/L FCTXAV 18,223,80
dnr-esc 10 g/L FCTXAV 18,223,80

CONSENSUS REPORTS: Reported in EPA TSCA Inventory.

SAFETY PROFILE: Mutation data reported. When heated to decomposition it emits toxic vapors of NO_x and SO_x.

CMS232 CAS:11056-12-5 HR: 3
CIROLEMYCIN

SYNS: ANTIBIOTIC U-12,241 □ U 12241

TOXICITY DATA WITH REFERENCE
dnd-sat 217 μg ABBIA4 120,292,67
dnd-esc 5 mg/L MUREAV 89,95,81
dnd-omi 217 μg ABBIA4 120,292,67
dnd-sal:spr 100 mmol ABBIA4 120,292,67
dnd-mam:lym 217 μg ABBIA4 120,292,67
ipr-mus LD50:5600 μg/kg 85GDA2 3,161,80
scu-mus LD50:1250 μg/kg 85GDA2 3,161,80

SAFETY PROFILE: Poison by subcutaneous and intraperitoneal routes. Mutation data reported.

CMS237 CAS:81098-60-4 HR: 2
CISAPRIDE
mf: $C_{23}H_{29}ClFN_3O_4{\cdot}H_2O$ mw: 484.02

SYNS: BENZAMIDE, 4-AMINO-5-CHLORO-N-(1-(3-(4-FLUOROPHENOXY)PROPYL)-3-METHOXY-4-PIPERIDINYL)-2-METHOXY-, MONOHYDRATE, cis- □ R 51619

TOXICITY DATA WITH REFERENCE
orl-rat LD50:4166 mg/kg IYKEDH 19,599,88
ipr-rat LD50:3435 mg/kg IYKEDH 19,599,88
ivn-rat LD50:27,400 μg/kg IYKEDH 19,599,88
orl-mus LD50:8715 mg/kg IYKEDH 19,599,88
ivn-mus LD50:32,200 μg/kg IYKEDH 19,599,88

SAFETY PROFILE: Moderately toxic by intraperitoneal and intravenous routes. Mildly toxic by ingestion. Experimental reproductive effects. When heated to decomposition it emits toxic fumes of NO_x, F^-, and Cl^-.

CMS242 CAS:1229-55-6 HR: 1
C.I. SOLVENT RED
mf: $C_{17}H_{14}N_2O_2$ mw: 278.33

SYNS: BRILLIANT FAT SCARLET R □ CERES RED G □ CERES RED G 102 □ C.I. 12150 □ C.I. FOOD RED 16 □ C RED 2 □ FAT RED BG □ FAT RED G □ FAT RED RS □ FAT SOLUBLE RED S □ FOOD RED 16 □ LACQUER RED V 2G □ OIL PINK □ 2-NAPHTHALENOL, 1-((2-METHOXYPHENYL)AZO)-(9CI) □ OIL RED □ OIL RED 113 □

OIL RED OG □ OIL SCARLET 389 □ OIL SOLUBLE RED S □ OIL VERMILION □ OIL VERMILION LP □ OLEAL RED G □ ORGANOL VERMILION □ ORIENT OIL RED OG □ PLASTORESIN RED FR □ RESINOL RED G □ SICO FAT RED BG NEW □ SILOTRAS RED TG □ SOLVENT RED 1 □ SOMALIA RED PG □ SUDAN R □ SUDAN RED 290 □ SUDAN RED G (6CI)

TOXICITY DATA WITH REFERENCE
eye-rbt 100 mg NTIS** AD-A172-758

CONSENSUS REPORTS: Reported in EPA TSCA Inventory.

SAFETY PROFILE: An eye irritant. When heated to decomposition it emits toxic fumes of NO_x.

CMS245 CAS:8003-22-3 HR: 1
C.I. SOLVENT YELLOW 33

SYNS: ARLOSOL YELLOW S □ CHINOLINE YELLOW D SOL. IN SPIRITS □ CHINOLINE YELLOW ZSS □ CI 47000 □ D and C YELLOW NO. 11 □ NITRO FAST YELLOW SL □ OIL YELLOW SIS □ PETROL YELLOW C □ QUINOLINE YELLOW A SPIRIT SOLUBLE □ QUINOLINE YELLOW BASE □ QUINOLINE YELLOW SPIRIT SOLUBLE □ QUINOLINE YELLOW SS □ SOLVANT YELLOW 33 □ WAXOLINE YELLOW T □ YELLOW 204

TOXICITY DATA WITH REFERENCE
mmo-sat 333 μg/plate EMMUEG 11(Suppl 12),1,88
mic-mus:lym 5 mg/L EMMUEG 12,219,88
cyt-mus:lym 1 mg/L ENMUDM 8(Suppl 6),24,86
sce-mus:lym 20 mg/L EMMUEG 12,219,88
msc-mus:lym 15 mg/L EMMUEG 12,219,88
orl-rat TDLo:67 g/kg (female 4W pre):REP NTPTR* NIH-91-3127
orl-mus TDLo:5384 mg/kg (male 13W pre):REP NTPTR* NIH-91-3127
orl-rat LD50:>13,650 mg/kg SCIEAS 36(1-4),10,89

CONSENSUS REPORTS: Reported in EPA TSCA Inventory.

SAFETY PROFILE: Low toxicity by ingestion. Experimental reproductive effects. Mutation data reported. When heated to decomposition it emits acrid smoke and irritating vapors.

CMS248 HR: 3
CISTANCHE TUBULOSA Wight (extract)

PROP: Indian plant belonging to the family *Orobanchaceae* (IJEBA6 18,594,80).

SYN: PHELIPAEA CALOTROPIDIS Walp., extract

TOXICITY DATA WITH REFERENCE
orl-rat TDLo:150 mg/kg (12-14D preg):REP IJEBA6 18,594,80
orl-ham TDLo:500 mg/kg (1-5D preg):REP IJEBA6 18,594,80
ipr-mus LD50:250 mg/kg IJEBA6 18,594,80

SAFETY PROFILE: Poison by intraperitoneal route. Experimental reproductive effects.

CMS250 CAS:1786-81-8 HR: 3
CITANEST HYDROCHLORIDE
mf: $C_{13}H_{20}N_2O{\cdot}ClH$ mw: 256.81

PROP: Crystals from EtOH/isopropyl ether. Mp: 167–168°.

SYNS: CITANEST □ L-67 HYDROCHLORIDE □ N-(2-METHYLPHENYL)-2-(PROPYLAMINO)-PROPANAMIDE MONOHYDROCHLORIDE □ PRILOCAINE HYDROCHLORIDE □ PROPITOCAINE HYDROCHLORIDE □ α-PROPYLAMINE-2-METHYL-PROPIONANILIDEHYDROCHLORIDE □ 2-(PROPYLAMINO)-o-PROPIONOTOLUIDIDE HYDROCHLORIDE

TOXICITY DATA WITH REFERENCE
ipr-rat LD50:148 mg/kg NIIRDN 6,692,82
scu-rat LD50:790 mg/kg NIIRDN 6,692,82
ivn-rat LD50:56,600 μg/kg NIIRDN 6,692,82
ipr-mus LD50:30 mg/kg 29ZVAB -,99,69
scu-mus LD50:632 mg/kg NIIRDN 6,692,82
ivn-mus LD50:55 mg/kg NIIRDN 6,692,82

SAFETY PROFILE: Poison by intraperitoneal and intravenous routes. Moderately toxic by subcutaneous route. When heated to decomposition it emits very toxic fumes of HCl and NO_x.

CMS320 CAS:498-23-7 HR: 2
CITRACONIC ACID
mf: $C_5H_6O_4$ mw: 130.11

PROP: Needles from Et_2O/pet ether. Obtained by carefully heating citric acid at about 175°. Hygroscopic, monoclinic crystals; characteristic odor. D: 1.62, mp: about 90° with decomp. Freely sol in water, alc, ether; sltly sol in chloroform; insol in benzene or petr ether.

SYNS: (Z)-2-METHYL-2-BUTENEDIOIC ACID (9CI) □ cis-METHYLBUTENEDIOIC ACID □ 2-METHYL-2-BUTENEDIOIC ACID □ METHYLMALEIC ACID

TOXICITY DATA WITH REFERENCE
orl-rat LD50:1320 mg/kg FCTXAV 2,327,64
orl-mus LD50:2260 mg/kg FCTXAV 2,327,64
orl-gpg LD50:1350 mg/kg FCTXAV 2,327,64

SAFETY PROFILE: Moderately toxic by ingestion. When heated to decomposition it emits acrid smoke and fumes.

CMS322 CAS:616-02-4 HR: 3
CITRACONIC ANHYDRIDE
mf: $C_5H_4O_3$ mw: 112.09

PROP: A liquid. Mp: 7°, bp: 213°.

SYNS: CITRACONIC ACID ANHYDRIDE □ 3-METHYL-2,5-FURANDIONE □ METHYLMALEIC ANHYDRIDE □ α-METHYLMALEIC ANHYDRIDE □ 2-METHYLMALEIC ANHYDRIDE □ 3-METHYLMALEIC ANHYDRIDE □ MONOMETHYLMALEIC ANHYDRIDE

TOXICITY DATA WITH REFERENCE
skn-rbt 10 mg/24H open JIHTAB 26,269,44
orl-rat LD50:2600 mg/kg JIHTAB 26,269,44
skn-rbt LD50:218 mg/kg SHELL*
skn-gpg LD50:1247 mg/kg JIHTAB 26,269,44

CONSENSUS REPORTS: Reported in EPA TSCA Inventory.

SAFETY PROFILE: Poison by skin contact. Moderately toxic by ingestion. A skin irritant. When heated to decomposition it emits acrid smoke and fumes. See also ANHYDRIDES.

CMS323 **CAS:7492-66-2** **HR: 1**
CITRAL DIETHYL ACETAL
mf: $C_{14}H_{26}O_2$ mw: 226.40

SYNS: 1,1-DIETHOXY-3,7-DIMETHYL-2,6-OCTADIENE □ 3,7-DIMETHYL-2,6-OCTADIENAL DIETHYL ACETAL □ 2,6-OCTADIENAL, 3,7-DIMETHYL-, DIETHYL ACETAL □ 2,6-OCTADIENE, 1,1-DIETHOXY-3,7-DIMETHYL-(9CI)

TOXICITY DATA WITH REFERENCE
skn-rbt 500 mg/24H MOD FCTOD7 21,667,83
mma-sat 3333 μg/plate EMMUEG 11(Suppl 12),1,88
orl-rat LD50:>5 g/kg FCTOD7 21,677,83
skn-rbt LD50:>5 g/kg FCTOD7 21,677,83

CONSENSUS REPORTS: Reported in EPA TSCA Inventory.

SAFETY PROFILE: Low toxicity by ingestion and skin contact. A skin irritant. Mutation data reported. When heated to decomposition it emits acrid smoke and irritating vapors.

CMS324 **CAS:66408-78-4** **HR: 1**
CITRAL ETHYLENE GLYCOL ACETAL
mf: $C_{12}H_{20}O_2$ mw: 196.32

SYNS: CITRACETAL □ 1,3-DIOXOLANE, 2-(2,6-DIMETHYL-1,5-HEPTADIENYL)-

TOXICITY DATA WITH REFERENCE
skn-rbt 500 mg/24H MOD FCTXAV 17,749,79

CONSENSUS REPORTS: Reported in EPA TSCA Inventory.

SAFETY PROFILE: A skin irritant. When heated to decomposition it emits acrid smoke and irritating fumes.

CMS325 **HR: 2**
CITRAL METHYLANTHRANILATE, SCHIFF'S BASE
mf: $C_{18}H_{23}NO_2$ mw: 285.42

SYN: N-(3,7-DIMETHYL-2,6-OCTADIENYLIDENE)ANTHRANILIC ACID METHYL ESTER

TOXICITY DATA WITH REFERENCE
skn-rbt 500 mg/24H MOD FCTOD7 20(Suppl),651,82
orl-rat LD50:3800 mg/kg FCTOD7 20(Suppl),651,82
skn-rbt LD50:2500 mg/kg FCTOD7 20(Suppl),651,82

SAFETY PROFILE: Moderately toxic by skin contact and ingestion. A skin irritant. When heated to decomposition it emits toxic fumes of NO_x.

CMS500 **CAS:25425-12-1** **HR: 3**
CITREOVIRIDIN
mf: $C_{23}H_{30}O_6$ mw: 402.53

PROP: Deep-yellow crystals from MeOH. Mp: 107–111°.

SYNS: CITREOVIRIDINE □ 4-METHOXY-5-METHYL-6-(7-METHYL-8-(TETRAHYDRO-3,4-DIHYDROXY-2,4,5-TRIMETHYL-2-FURANYL)-1,3,5,7-OCTATETRAENYL)-2H-PYRAN-2-ONE

TOXICITY DATA WITH REFERENCE
orl-rat TDLo:60 mg/kg (female 8-11D post):TER FCTOD7 24,1315,86
orl-rat TDLo:40 mg/kg (female 8-11D post):TER FCTOD7 24,1315,86
orl-rat TDLo:60 mg/kg (female 8-11D post):REP FCTOD7 24,1315,86
orl-rat TDLo:60 mg/kg (female 12-15D post):TER FCTOD7 24,1315,86
scu-rat LD50:3600 μg/kg JJEMAG 42,91,72
orl-mus LD50:29 mg/kg JJEMAG 42,91,72
ipr-mus LD50:7200 μg/kg JJEMAG 42,91,72
scu-mus LD50:9600 μg/kg RCOCB8 59,31,88

CONSENSUS REPORTS: EPA Genetic Toxicology Program.

SAFETY PROFILE: Poison by ingestion, subcutaneous, and intraperitoneal routes. When heated to decomposition it emits acrid smoke and irritating fumes.

CMS750 **CAS:77-92-9** **HR: 3**
CITRIC ACID
mf: $C_6H_8O_7$ mw: 192.14

PROP: Colorless, odorless crystals (crystals are monoclinic holohedra and crystallize from hot conc aq soln); acid taste. Mp: 135° (monohydrate), mp: 153° (anhydrous), bp: decomp; d: 1.665, flash p: 212°F. Very sol in H_2O and EtOH; mod sol in Et_2O.

SYNS: ACILETTEN □ CITRETTEN □ CITRIC ACID, anhydrous □ CITRO □ FEMA No. 2306 □ 2-HYDROXY-1,2,3-PROPANETRICARBOXYLIC ACID □ β-HYDROXYTRICARBALLYLIC ACID □ KYSELINA CITRONOVA (CZECH)

TOXICITY DATA WITH REFERENCE
skn-rbt 500 mg/24H MOD 28ZPAK -,105,72
eye-rbt 750 μg/24H SEV 28ZPAK -,105,72
orl-rat LD50:3 g/kg OYYAA2 43,561,92
ipr-rat LD50:883 mg/kg JPETAB 94,65,48
scu-rat LD50:5500 mg/kg TAKHAA 30,25,71
orl-mus LD50:5040 mg/kg TAKHAA 30,25,71
ipr-mus LD50:903 mg/kg TXCYAC 62,203,90
scu-mus LD50:2700 mg/kg TAKHAA 30,25,71
ivn-mus LD50:42 mg/kg JPETAB 94,65,48
orl-rbt LDLo:7000 mg/kg IECHAD 15,628,23
ivn-rbt LD50:330 mg/kg JPETAB 94,65,48

CONSENSUS REPORTS: Reported in EPA TSCA Inventory.

SAFETY PROFILE: Poison by intravenous route. Moderately toxic by subcutaneous and intraperitoneal routes. Mildly toxic by ingestion. A severe eye and moderate skin irritant. An irritating organic acid, some allergenic properties. Combustible liquid. Potentially explosive reaction with metal nitrates. When heated to decomposition it emits acrid smoke and fumes.

CMS775 **CAS:518-75-2** **HR: 3**
CITRININ
mf: $C_{13}H_{14}O_5$ mw: 250.27

PROP: Lemon-yellow needles from EtOH or MeOH. Mp: 178–179° (decomp).

SYNS: ANTIMYCIN □ 3H-2-BENZOPYRAN-7-CARBOXYLIC ACID, 4, 6-DIHYDRO-8-HYDROXY-3,4,5-TRIMETHYL-6-OXO-, (3R-trans)- □ (3R, 4S)-4,6-DIHYDRO-8-HYDROXY-3,4,5-TRIMETHYL-6-OXO-3H-2-BENZOPYRAN-7-CARBOXYLIC ACID

TOXICITY DATA WITH REFERENCE
skn-gpg 40 mg/24H SEV JANCA2 57,1121,74
dnd-esc 100 mg/L AEMIDF 52,1273,86
pic-esc 300 mg/L AEMIDF 52,1273,86
dnr-bcs 20 µg/disc CNREA8 36,445,76
ipr-mus TDLo:30 mg/kg (10D post):TER FCTXAV 14,175,76
scu-rat TDLo:30 mg/kg (8D post):TER JTEHD6 13,553,84
scu-rat TDLo:35 mg/kg (10D post):REP TXCYAC 25,151,82
scu-rat TDLo:35 mg/kg (10D post):TER TXCYAC 25,151,82
orl-rat TDLo:25,200 mg/kg/60W-C:NEO CALEDQ 17,281,83
orl-rat TD:13 g/kg/32W-C:ETA FGIGDO 5,77,81
ipr-rat LD50:67 mg/kg JPETAB 88,173,46
scu-rat LD50:67 mg/kg JPETAB 88,173,46
orl-mus LD50:112 mg/kg TXAPA9 37,139,76
ipr-mus LD50:35 mg/kg JPETAB 88,173,46
scu-mus LD50:73 mg/kg FCTXAV 15,29,77
orl-rbt LD50:134 mg/kg FCTOD7 21,487,83
ipr-rbt LD50:50 mg/kg FCTOD7 21,487,83
ivn-rbt LD50:19 mg/kg JPETAB 88,173,46
scu-gpg LD50:37 mg/kg JPETAB 88,173,46
orl-ham LD50:75 mg/kg FCTXAV 16,355,78
ipr-ham LD50:66 mg/kg FCTXAV 16,355,79

CONSENSUS REPORTS: IARC Cancer Review: Group 3 IMEMDT 7,56,87, Animal Limited Evidence IMEMDT 40,67,86

SAFETY PROFILE: Poison by ingestion and other routes. An experimental teratogen. Other experimental reproductive effects. A severe skin irritant. Questionable carcinogen with experimental neoplastigenic and tumorigenic data. Mutation data reported. When heated to decomposition it emits acrid smoke and irritating fumes.

CMS845 **CAS:106-23-0** **HR: 2**
CITRONELLAL
mf: $C_{10}H_{18}O$ mw: 154.25

PROP: Colorless to sltly yellow liquid; intense lemon-citronella-rose odor. D: 0.850–0.860, refr index: 1.446–1.456, flash p: 170°F. Sol in alc, most oils; sltly sol in propylene glycol; insol in glycerin and water.

SYNS: 3,7-DIMETHYL-6-OCTENAL □ FEMA No. 2307

SAFETY PROFILE: Combustible liquid. When heated to decomposition it emits acrid smoke and irritating fumes.

CMS850 **CAS:107-75-5** **HR: 1**
CITRONELLAL HYDRATE
mf: $C_{10}H_{20}O_2$ mw: 172.30

PROP: Colorless liquid; sweet, floral, lily odor. D: 0.918–0.923, refr index: 1.447–1.450, flash p: 212°F. Sol in fixed oils and propylene glycol; insol in glycerin.

SYNS: CYCLALIA □ CYCLOSIA □ 3,7-DIMETHYL-7-HYDROXYOCTANAL □ FEMA No. 2583 □ FIXOL □ HYDROXYCITRONELLAL (FCC) □ 7-HYDROXY-3,7-DIMETHYLOCTAN-1-AL □ 7-HYDROXYCITRONELLAL □ 7-HYDROXY-3,7-DIMETHYL OCTANAL □ LAURINE □ LILYL ALDEHYDE □ MUSUET SYNTHETIC □ MUSUETTINE PRINCIPLE □ PHIXIA

TOXICITY DATA WITH REFERENCE
skn-rbt 500 mg/24H FCTXAV 12,921,74

CONSENSUS REPORTS: Reported in EPA TSCA Inventory.

SAFETY PROFILE: A skin irritant. Combustible liquid. When heated to decomposition it emits acrid smoke and irritating fumes. See also ALDEHYDES.

CMT000 **CAS:8000-29-1** **HR: 2**
CITRONELLA OIL

SYNS: ESSENTIAL OIL of CYMBOPOGON NARDUS □ OIL of CITRONELLA □ OILS, CINTONELLA □ ZITRONELL OEL (GERMAN)

TOXICITY DATA WITH REFERENCE
skn-rbt 500 mg/24H FCTXAV 11,1067,73
eye-rbt 500 mg SEV AJOPAA 29,1363,46
dnr-bcs 5 L/disc TOFOD5 8,91,85
orl-rat LD50:7200 mg/kg PHARAT 14,435,59
ipr-rat LD50:713 mg/kg IJEBA6 9,515,71
orl-mus LD50:4600 mg/kg TOFOD5 8,91,85
skn-rbt LD50:4700 mg/kg FCTXAV 11,1067,73

CONSENSUS REPORTS: Reported in EPA TSCA Inventory.

SAFETY PROFILE: Moderately toxic by intraperitoneal route. Mildly toxic by ingestion and skin contact. Mutation data reported. A skin and severe eye irritant. When heated to decomposition it emits acrid smoke and fumes.

CMT050 **CAS:2436-90-0** **HR: 1**
CITRONELLENE
mf: $C_{10}H_{18}$ mw: 138.28

PROP: Bp: 154–155°, d: 0.757 @ 20°/4°.

SYNS: DIHYDROMYRCENE □ 3,7-DIMETHYL-1,6-OCTADIENE □ 1, 6-OCTADIENE, 3,7-DIMETHYL-

TOXICITY DATA WITH REFERENCE
skn-rbt 500 mg/24H MOD FCTOD7 21,845,83

CONSENSUS REPORTS: Reported in EPA TSCA Inventory.

SAFETY PROFILE: A skin irritant. When heated to decomposition it emits acrid smoke and irritating fumes.

C

CMT125 **CAS:502-47-6** **HR: 2**
CITRONELLIC ACID
mf: $C_{10}H_{18}O_2$ mw: 170.28

SYN: 3,7-DIMETHYL-6-OCTENOIC ACID

TOXICITY DATA WITH REFERENCE
skn-rbt 500 mg/24H MOD FCTOD7 20(Suppl),653,82
orl-rat LD50:2610 mg/kg FCTOD7 20(Suppl),653,82
skn-rbt LD50:450 mg/kg FCTOD7 20(Suppl),653,82

CONSENSUS REPORTS: Reported in EPA TSCA Inventory.

SAFETY PROFILE: Moderately toxic by ingestion and skin contact. A skin irritant. Reaction with ozone produces an explosive product. When heated to decomposition it emits acrid smoke and fumes.

CMT250 **CAS:106-22-9** **HR: 3**
CITRONELLOL
mf: $C_{10}H_{20}O$ mw: 156.30

PROP: Colorless oily liquid; rose odor. D: 0.850–0.860, refr index: 1.454–1.462, flash p: 215°F. Sol in fixed oils, propylene glycol; sltly sol in water; insol in glycerin @ 225°.

SYNS: CEPHROL □ 2,6-DIMETHYL-2-OCTEN-8-OL □ 3,7-DIMETHYL-6-OCTEN-1-OL □ FEMA No. 2309 □ FEMA No. 2980 □RHODI-NOL □ RODINOL

TOXICITY DATA WITH REFERENCE
skn-man 16 mg/48H MOD CTOIDG 94(8),41,79
skn-rbt 100 mg/24H SEV CTOIDG 94(8),41,79
skn-gpg 100 mg/24H SEV CTOIDG 94(8),41,79
orl-rat LD50:3450 mg/kg FCTXAV 13,757,75
ivn-mus LDLo:100 mg/kg CBCCT* 5,139,53
ims-mus LD50:4000 mg/kg J3ICA7 21,342,62
skn-rbt LD50:2650 mg/kg FCTXAV 13,757,75

CONSENSUS REPORTS: Reported in EPA TSCA Inventory.

SAFETY PROFILE: Poison by intravenous route. Moderately toxic by ingestion, skin contact, and intramuscular routes. A severe skin irritant. A combustible liquid. When heated to decomposition it emits acrid smoke and irritating fumes. See also ALCOHOLS.

CMT300 **CAS:7492-67-3** **HR: 1**
CITRONELLOXYACETALDEHYDE
mf: $C_{12}H_{22}O_2$ mw: 198.34

SYNS: ACETALDEHYDE, ((3,7-DIMETHYL-6-OCTENYL)OXY)- □ CITRONELLYLOXYACETALDEHYDE □ 6,10-DIMETHYL-3-OXA-9-UNDECENAL

TOXICITY DATA WITH REFERENCE
orl-rat LD50:>5 g/kg FCTXAV 12,861,74
skn-rbt LD50:>5 g/kg FCTXAV 12,861,74

CONSENSUS REPORTS: Reported in EPA TSCA Inventory.

SAFETY PROFILE: Low toxicity by ingestion and skin contact. When heated to decomposition it emits acrid smoke and irritating vapors.

CMT500 **CAS:68039-38-3** **HR: 1**
CITRONELLYL-2-BUTENOATE
mf: $C_{14}H_{24}O_2$ mw: 224.38

SYNS: 2-BUTENOIC ACID-3,7-DIMETHYL-6-OCTENYL ESTER □ CITRONELLYL-α-CROTONATE □ 3,7-DIMETHYL-6-OCTEN-1-OL CROTONATE

TOXICITY DATA WITH REFERENCE
skn-rbt 500 mg/24H MOD FCTXAV 14,725,76

CONSENSUS REPORTS: Reported in EPA TSCA Inventory.

SAFETY PROFILE: A skin irritant. See also ESTERS. When heated to decomposition it emits acrid smoke and irritating fumes.

CMT600 **HR: 2**
CITRONELLYL BUTYRATE
mf: $C_{14}H_{26}O_2$ mw: 226.36

PROP: Colorless liquid; strong, fruity-rosy odor. D: 0.873–0.883; refr index: 1.444–1.448, flash p: 212°F. Misc in alc, ether, chloroform, most oils; insol water @ 245°.

SYNS: 3,7-DIMETHYL-6-OCTEN-1-YL BUTYRATE □ FEMA No. 2312

SAFETY PROFILE: Combustible liquid. When heated to decomposition it emits acrid smoke and irritating fumes.

CMT750 **CAS:105-85-1** **HR: 1**
CITRONELLYL FORMATE
mf: $C_{11}H_{20}O_2$ mw: 184.31

PROP: Colorless liquid; strong, fruity odor. D: 0.890–0.93, refr index: 1.443–1.452, flash p: 197°F. Sol in alc, fixed oils; sltly sol in propylene glycol; insol in glycerin, water @ 235°.

SYNS: 3,7-DIMETHYL-6-OCTEN-1-OL FORMATE □ 2,6-DIMETHYL-2-OCTEN-8-YL FORMATE □ 3,7-DIMETHYL-6-OCTEN-1-YL FORMATE □ FEMA No. 2314 □ FORMIC ACID, CITRONELLYL ESTER □ FORMIC ACID-3,7-DIMETHYL-6-OCTEN-1-YL ESTER

TOXICITY DATA WITH REFERENCE
skn-hmn 20 mg/48H MLD FCTXAV 11,1073,73
skn-rbt 500 mg/24H FCTXAV 11,1073,73
orl-rat LD50:8400 mg/kg FCTXAV 11,1011,73

CONSENSUS REPORTS: Reported in EPA TSCA Inventory.

SAFETY PROFILE: Mildly toxic by ingestion. A human skin irritant. Combustible liquid. When heated to decomposition it emits acrid smoke and irritating fumes. See also ESTERS and FORMIC ACID.

CMT900 **HR: 2**
CITRONELLYL ISOBUTYRATE
mf: $C_{14}H_{26}O_2$ mw: 226.36

PROP: Colorless liquid; rosy-fruity odor. D: 0.870–0.880, refr index: 1.440–1.448, flash p: 212°F. Misc in alc, chloroform, ether, most oils; insol in water @ 249°.

SYNS: 3,7-DIMETHYL-6-OCTEN-1-YL ISOBUTYRATE □ FEMA No. 2313

SAFETY PROFILE: Combustible liquid. When heated to decomposition it emits acrid smoke and irritating fumes.

CMU000 CAS:51566-62-2 HR: 1
CITRONELLYL NITRILE

SYN: 3,7-DIMETHYL-6-OCTENENITRILE

TOXICITY DATA WITH REFERENCE
skn-rbt 500 mg/24H MLD FCTXAV 17,509,79
orl-rat LD50:5300 mg/kg FCTXAV 17,509,79

CONSENSUS REPORTS: Cyanide and its compounds are on the Community Right-To-Know List.

SAFETY PROFILE: Mildly toxic by ingestion. A skin irritant. See also NITRILES. When heated to decomposition it emits toxic fumes of NO_x and CN^-.

CMU050 CAS:139-70-8 HR: 1
CITRONELLYL PHENYLACETATE
mf: $C_{18}H_{26}O_2$ mw: 274.44

SYNS: ACETIC ACID, PHENYL-, 3,7-DIMETHYL-6-OCTENYL ESTER □ BENZENEACETIC ACID, 3,7-DIMETHYL-6-OCTENYL ESTER (9CI) □ 3,7-DIMETHYL-6-OCTEN-1-YL PHENYLACETATE

TOXICITY DATA WITH REFERENCE
skn-rbt 500 mg/24H MLD FCTOD7 20,657,82

CONSENSUS REPORTS: Reported in EPA TSCA Inventory.

SAFETY PROFILE: A skin irritant. When heated to decomposition it emits acrid smoke and irritating fumes.

CMU100 HR: 2
CITRONELLYL PROPIONATE
mf: $C_{13}H_{24}O_2$ mw: 212.33

PROP: Colorless liquid; fruity-rosy odor. D: 0.877–0.886, refr index: 1.443–1.449, flash p: 212°F. Misc in alc, most oils; insol in water @ 242°.

SYN: FEMA No. 2316

SAFETY PROFILE: Combustible liquid. When heated to decomposition it emits acrid smoke and irritating fumes.

CMU300 HR: 1
CITRUS HYSTRIX DC., fruit peel extract

PROP: Thailand plant belonging to the family Rutaceae JOETD7 13,105,85

TOXICITY DATA WITH REFERENCE
orl-rat TDLo:16 g/kg (female 15-22D post):TER JOETD7 13,105,85
orl-rat TDLo:25 g/kg (female 8-12D post):REP JOETD7 13,105,85
orl-rat TDLo:20 g/kg (female 2-5D post):REP JOETD7 13,105,85
orl-rat LDLo:100 g/kg JOETD7 13,105,85

SAFETY PROFILE: Slightly toxic by ingestion. An experimental teratogen. Other experimental reproductive effects. When heated to decomposition it emits acrid smoke and irritating fumes.

CMU320 CAS:3687-67-0 HR: 1
C.I. VAT BLACK 1
mf: $C_{20}H_9BrClNO_2S$ mw: 442.72

SYNS: C.I. 73670 □ DURINDONE PRINTING BLACK BL □ HELANTHRENE BLACK BL □ INDANTHRENE PRINTING BLACK BL □ 3-INDOLINONE, 5-BROMO-2-(9-CHLORO-3-OXONAPHTHO(1,2-b)THIEN-2(3H)-YLIDENE)- □ NAKO FAST GREY BL □ PARADONE PRINTING BLACK BLSF □ PARADONE PRINTING BLACK TLSF □ SOLANTHRENE GREY BL □ SOLANTHRENE PRINTING BLACK BL □ SOLANTHRENE PRINTING BLACK BLN □ SOLANTHRENE PRINTING BLACK TL □ SOLANTHRENE PRINTING BLACK TLN □ THIOINDIGO BLACK □ TINA GREY BL □ TINA PRINTING BLACK BL □ VAT BLACK 1

TOXICITY DATA WITH REFERENCE
ipr-rat LD50:4250 mg/kg GISAAA 50(8),91,85

CONSENSUS REPORTS: Reported in EPA TSCA Inventory.

SAFETY PROFILE: Low toxicity by intraperitoneal route. When heated to decomposition it emits toxic vapors of NO_x, SO_x, Br^-, and Cl^-.

CMU475 CAS:2278-50-4 HR: 1
C.I. VAT BLACK 8
mf: $C_{45}H_{19}N_3O_4$ mw: 665.67

SYNS: BENZADONE GREY M □ 1H-BENZ(6,7)INDAZOLO(2,3,4-fgh)NAPHTH(2″,3″:6′,7′)INDOLO(3′,2′:5,6)ANTHR A(2,1,9-mna) ACRIDINE-5,8,13,25-TETRAONE □ CALEDON GREY M □ CERN KYPOVA 8 □ C.I. 71000 □ INDANTHREN GREY M □ INDANTHREN GREY MG □ MIKETHRENE GREY M □ MIKETHRENE GREY MG □ NIHONTHRENE GREY M □ OSTANTHREN GREY M □ PARADONE GREY M □ PARADONE GREY MG □ SED OSTANTHRENOVA M □ VAT GRAY S □ VAT GREY S

TOXICITY DATA WITH REFERENCE
eye-rbt 500 mg/24H MLD 85JCAE-,1324,86
orl-rat LD50:11,200 mg/kg 85JCAE-,1324,86
ipr-mus LD50:3 g/kg GISAAA 51(1),78,86

CONSENSUS REPORTS: Reported in EPA TSCA Inventory.

SAFETY PROFILE: Mildly toxic by intraperitoneal and ingestion routes. An eye irritant.When heated to decomposition it emits toxic fumes of NO_x.

CMU500 CAS:6424-76-6 HR: 1
C.I. VAT BLUE 16
mf: $C_{36}H_{18}O_4$ mw: 514.54

SYNS: CALCOLOID NAVY BLUE 2GC □ CALEDON DARK BLUE G □ CALEDON PRINTING NAVY G □ CARBANTHRENE NAVY BLUE G □ C.I. 71200 □ 16,17-ETHYLENEDIOXYVIOLANTHRONE □ INDANTHRENE NAVY BLUE G □ MIKETHRENE MARINE BLUE G □ MODR NAMORNICKA OSTANTHRENOVA G (CZECH) □ NIHONTHRENE NAVY BLUE G □ PALANTHRENE NAVY BLUE G □ PARADONE NAVY BLUE G □ ROMANTRENE NAVY BLUE FG

TOXICITY DATA WITH REFERENCE
eye-rbt 500 mg/24H MLD 28ZPAK -,243,72

CONSENSUS REPORTS: Reported in EPA TSCA Inventory.

SAFETY PROFILE: An eye irritant. When heated to decomposition it emits acrid smoke and irritating fumes.

CMU750 **CAS:6373-20-2** **HR: 1**
C.I. VAT BLUE 22
mf: $C_{34}H_{12}Cl_4O_2$ mw: 594.26

SYNS: C.I. 59815 (CZECH) □ C.I. 59820 □ INDANTHREN NAVY BLUE TRR □ MODR NAMORNICKA OSTANTHRENOVA RA (CZECH) □ PALANTHRENE NAVY BLUE RB

TOXICITY DATA WITH REFERENCE
eye-rbt 500 mg/24H MLD 28ZPAK -,244,72

CONSENSUS REPORTS: Reported in EPA TSCA Inventory.

SAFETY PROFILE: An eye irritant. When heated to decomposition it emits toxic fumes of Cl^-.

CMU770 **CAS:2475-33-4** **HR: 1**
C.I. VAT BROWN 1
mf: $C_{42}H_{18}N_2O_6$ mw: 646.62

SYNS: AHCOVAT BROWN BR □ AMANTHRENE BROWN BR □ BENZADONE BROWN BR □ BROWN SK □ CALCOLOID BROWN BR □ CALEDON DARK BROWN 3R □ CARBANTHRENE BROWN BR □ CHEMITHRENE BROWN BR □ C.I. 70800 □ C.I. 70802 □ CIBANONE BROWN BR □ CIBANONE BROWN FBR □ C.I. VAT BROWN 44 □ DINAPHTHO(2,3-a:2′,3′-i)NAPHTH(2′,3′:6,7)INDOLO(2,3-c)CARBAZOLE-5,10,15,17,22,24-HEXAONE, 16,23-DIHYDRO- □ FENANTHREN BROWN BR □ HELANTHRENE BROWN GR □ HNED KYPOVA 1 □ INDANTHREN BRONZE BR □ INDANTHREN BROWN BR □ INDANTHREN BROWN GR □ INDANTHRENE BROWN BR □ HNED OSTANTHRENOVA BR □ MAYVAT BROWN BR □ MIKETHRENE BROWN BR □ MIKETHRENE BROWN GR □ NAPHTH(2′,3′:6,7)INDOLO(2,3-c)DINAPHTHO(2,3-a:2′,3′-i)CARBAZOLE-5,10,15, 17,22,24-HEXONE □ NIHONTHRENE BROWN BR □ NIHONTHRENE BROWN GR □ NYANTHRENE BROWN RB □ OSTANTHREN BROWN BR □ OSTANTHRENE BROWN BR □ PALANTHRENE BROWN BR □ PARADONE RED BROWN 2RD □ PONSOL BROWN RBT □ ROMANTRENE BROWN FBR □ ROMANTRENE BROWN FGR □ SANDOTHRENE BROWN NBR □ SOLANTHRENE BROWN BR □ SOLANTHRENE BROWN JR □ TINON BROWN BR □ TYRIAN BROWN I-BR

TOXICITY DATA WITH REFERENCE
eye-rbt 500 mg/24H MLD 85JCAE-,1325,86

CONSENSUS REPORTS: Reported in EPA TSCA Inventory.

SAFETY PROFILE: An eye irritant. When heated to decomposition it emits toxic vapors of NO_x.

CMU800 **CAS:6247-46-7** **HR: 1**
C.I. VAT BROWN 25
mf: $C_{43}H_{25}N_3O_7$ mw: 695.71

SYNS: CARBANTHRENE RED BROWN 5R □ C.I. 69020 □ 16H-DINAPHTHO(2,3-a:2′,3′-i)CARBAZOLE-5,10,15,17-TETRAONE, 6,9-DIBENZAMIDO-1-METHOXY- □ FENANTHRENE RED BROWN 5R □ HNED CERVENAVA OSTANTHRENOVA 5 RF □ HNED KYPOVA 25 □ INDANTHRENE RED BROWN 5RF □ INDANTHRENE REDDISH BROWN 5RF □ INDANTHREN RED BROWN 5RF □ MIKETHRENE RED BROWN 5RF □ OSTANTHREN REDDISH BROWN 5RF

TOXICITY DATA WITH REFERENCE
eye-rbt 500 mg/24H MLD 85JCAE-,1325,86

SAFETY PROFILE: An eye irritant.When heated to decomposition it emits toxic fumes of NO_x.

CMU810 **CAS:3271-76-9** **HR: 1**
C.I. VAT GREEN 3
mf: $C_{31}H_{15}NO_3$ mw: 449.47

SYNS: ANTHRA(2,1,9-mna)NAPHTH(2,3-h)ACRIDINE-5,10,15-TRIONE □ C.I. 69500 □ VAT GREEN 3 □ ZELEN KYPOVA 3 □ ZELEN OLIVOVA OSTANTHRENOVA B

TOXICITY DATA WITH REFERENCE
eye-rbt 500 mg/24H MLD 85JCAE-,1330,86

CONSENSUS REPORTS: Reported in EPA TSCA Inventory.

SAFETY PROFILE: An eye irritant. When heated to decomposition it emits toxic vapors of NO_x.

CMU815 **CAS:3263-31-8** **HR: 2**
C.I. VAT ORANGE 5
mf: $C_{20}H_{16}O_4S_2$ mw: 384.48

SYNS: AHCOVAT PRINTING ORANGE R □ ALGOL ORANGE RF □ AMANTHRENE ORANGE R □ (($\Delta^{2,2'}$(3H,3′H))-BIBENZO(b)THIOPHENE)-3,3′-DIONE, /S CALCOLOID PRINTING ORANGE RE □ CALCALOID PRINTING ORANGE RYW □ C.I. 73335 □ CIBA ORANGE R □ CIBA ORANGE RDL □ CIBA ORANGE RP □ 6,6′ DIETHOXYTHIOINDIGO □ DURINDONE ORANGE R □ DURINDONE ORANGE RP □ DURINDONE PRINTING ORANGE R □ HELIANE ORANGE RF □ HELINDON ORANGE R □ HOSTAVAT ORANGE R □ MIKETHRENE ORANGE R □ NIHONTHRENE FAST ORANGE R □ ORANZ KYPOVA 5 □ PALANTHRENE ORANGE R □ SANDOTHRENE ORANGE R □ SOLINDENE ORANGE R □ SULFANTHRENE ORANGE R □ SULFANTHRENE ORANGE RS □ THIOINDIGO ORANGE KKh □ TINA ORANGE R □ TYRIAN ORANGE A-RF □ VAT ORANGE R □ VAT ORANGE RF □ VAT PRINTING ORANGE R

TOXICITY DATA WITH REFERENCE
orl-rat LDLo:1500 mg/kg 85JCAE-,1337,86
ipr-rat LD50:2050 mg/kg GISAAA 50(8),91,85
ipr-mus LD50:1 g/kg GISAAA 50(8),91,85

CONSENSUS REPORTS: Reported in EPA TSCA Inventory.

SAFETY PROFILE: Moderately toxic by ingestion and intraperitoneal routes. When heated to decomposition it emits toxic vapors of SO_x.

CMU820 **CAS:4424-06-0** **HR: D**
C.I. VAT ORANGE 7
mf: $C_{26}H_{12}N_4O_2$ mw: 412.42

SYNS: BISBENZIMIDAZO(2,1-b:2′,1′ i)BENZO(lmn)(3,8)PHENANTHROLINE-8,17-DIONE □ BORDEAUX RRN □ BRILLIANT ORANGE GR □ C.I. 71105 □ CIBANONE BRILLIANT ORANGE GR □ C.I. PIGMENT ORANGE 43 □ FENANTHREN BRILLIANT ORANGE GR □ HOSTAPERM ORANGE GR □ HOSTAPERM VAT ORANGE GR □ HOSTAVAT

BRILLIANT ORANGE GR □ INDANTHREN BRILLIANT ORANGE GR □ INDANTHRENE BRILLIANT ORANGE GR □ INDANTHRENE BRILLIANT ORANGE GRP □ INDOFAST ORANGE OV 5983 □ MIKETHREN BRILLIANT ORANGE GR □ MIKETHRENE ORANGE GR □ OSTANTHRENE ORANGE GR □ OSTANTHREN ORANGE GR □ PALANTHRENE BRILLIANT ORANGE GR □ PARADONE BRILLIANT ORANGE GR □ PARADONE BRILLIANT ORANGE GR NEW □ trans-PERINONE □ PV FAST ORANGE GRL □ SANYO PERMANENT ORANGE D 213 □ SANYO PERMANENT ORANGE D 616 □ SOLANTHRENE BRILLIANT ORANGE JR □ SYMULER FAST ORANGE GRD □ THRENE BRILLIANT ORANGE GR □ TINON BRILLIANT ORANGE GR □ VAT BRILLIANT ORANGE □ VAT SCARLET 2Zh

TOXICITY DATA WITH REFERENCE
mmo-sat 33 μg/plate EMMUEG 11(Suppl 12),1,88

CONSENSUS REPORTS: Reported in EPA TSCA Inventory.

SAFETY PROFILE: Mutation data reported. When heated to decomposition it emits toxic vapors of NO_x.

CMU825 CAS:4203-77-4 HR: 1
C.I. VAT RED 13
mf: $C_{32}H_{22}N_4O_2$ mw: 494.58

SYNS: AHCOVAT RUBINE R □ (3,3′-BIANTHRA(1,9-cd)PYRAZOLE)-6,6′(1H,1′H)-DIONE, 1,1′-DIETHYL- □ CARBANTHRENE RED G 2B □ CARBANTHRENE RED G 2BP □ CERVEN KYPOVA 13 □ C.I. 70320 □ CIBANONE RED 6B □ CIBANONE RED F 6B □ C.I. PIGMENT RED 195 □ FENANTHREN RUBINE R □ INDANTHRENE RUBINE R □ INDANTHREN RUBINE R □ INDANTHREN RUBINE RS □ INDO MAROON LAKE RV 6666 □ NIHONTHRENE RED BB □ NYANTHRENE RED G 2B □ PALANTHRENE RED G 2B □ PONSOL RED 2B □ PONSOL RED 2BD □ SANDOTHRENE RED N 6B □ TINON RED 6B □ VAT RED 13

TOXICITY DATA WITH REFERENCE
skn-rbt 500 mg/24H MLD 85JCAE-,1325,86
eye-rbt 500 mg/24H MLD 85JCAE-,1325,86

CONSENSUS REPORTS: Reported in EPA TSCA Inventory.

SAFETY PROFILE: A skin and eye irritant. When heated to decomposition it emits toxic vapors of NO_x.

CMU850 CAS:542-46-1 HR: 1
cis-CIVETONE
mf: $C_{17}H_{30}O$ mw: 250.47

PROP: Crystals; strong musty odor. Mp: 31–32°, bp: 342°.

SYNS: CIVETONE □ 9-CYCLOHEPTADECEN-1-ONE □ 9-CYCLOHEPTADECEN-1-ONE, (Z)-(8CI,9CI)

TOXICITY DATA WITH REFERENCE
skn-rbt 500 mg/24H MOD FCTXAV 14,727,76

CONSENSUS REPORTS: Reported in EPA TSCA Inventory.

SAFETY PROFILE: A skin irritant. When heated to decomposition it emits acrid smoke and irritating fumes.

CMU875 CAS:76541-72-5 HR: 2
CLANICLOR
mf: $C_{11}H_{17}ClO_7P_2$ mw: 358.67

SYNS: (4-CHLOROPHENYL)(DIMETHOXYPHOSPHINYL)METHYL PHOSPHORIC ACID DIMETHYL ESTER □ DIMETHYL-α-(DIMETHOXYPHOSPHINYL)-p-CHLOROBENZYL PHOSPHATE □ SR-202

TOXICITY DATA WITH REFERENCE
orl-rat LD50:3200 mg/kg DRFUD4 7,271,82
ipr-rat LD50:580 mg/kg DRFUD4 7,271,82
orl-mus LD50:2 g/kg DRFUD4 7,271,82
ipr-mus LD50:500 mg/kg DRFUD4 7,271,82

SAFETY PROFILE: Moderately toxic by ingestion and intraperitoneal routes. When heated to decomposition it emits toxic fumes of Cl^- and PO_x.

CMU890 CAS:64741-62-4 HR: 1
CLARIFIED SLURRY OIL

SYNS: CATALYTIC CRACKED CLARIFIED OIL □ CAT CRACKED CLARIFIED OIL-DECANTED OIL □ CLARIFIED OILS (PETROLEUM), CATALYTIC CRACKED

TOXICITY DATA WITH REFERENCE
skn-rat TDLo:600 mg/kg (female 1-20D post):REP TXCYAC 5,587,89
orl-rat LD50:4300 mg/kg JACTDZ 1,136,90
skn-rbt LD:>2 g/kg JACTDZ 1,136,90

CONSENSUS REPORTS: Reported in EPA TSCA Inventory.

SAFETY PROFILE: Low toxicity by ingestion and skin contact. Experimental reproductive effects. When heated to decomposition it emits acrid smoke and irritating vapors.

CMV000 CAS:149-29-1 HR: 3
CLAVACIN
mf: $C_7H_6O_4$ mw: 154.13

PROP: Colorless crystals, prisms or plates from Et_2O or $CHCl_3$. Mp: 111°.

SYNS: CLAIRFORMIN □ 2,4-DIHYDROXY-2H-PYRAN-Δ-3(6H),α-ACETIC ACID-3,4-LACTONE □ (2,4-DIHYDROXY-2H-PYRAN-3(6H)-YLIDENE)ACETIC ACID-3,4-LACTONE □ EXPANSIN □ GIGANTIN □ 4-HYDROXY-4H-FURO(3,2-C)PYRAN-2(6H)-ONE □ LEUCOPIN □ MYCOIN □ PATULIN □ PENATIN □ PENICIDIN □ TERCININ □ TERININ

TOXICITY DATA WITH REFERENCE
dni-hmn:lym 50 mg/L FCTOD7 20,893,82
oms-hmn:hla 3200 mg/L FCTOD7 20,893,82
orl-rat TDLo:135 mg/kg (5W male/5W pre-20D preg):REP JTEHD6 2,713,77
orl-mus TDLo:24 mg/kg (female 14-19D post):REP FCTXAV 16,243,78
orl-rat TDLo:675 mg/kg (5W male/5W pre-20D preg):REP JTEHD6 2,713,77
orl-rat TDLo:135 mg/kg (MGN):TER JTEHD6 2,713,77
scu-rat TDLo:232 mg/kg/58W-I:NEO BJCAAI 15,85,61
orl-rat LD50:27,790 μg/kg ARCVBP 8,41,77
ipr-rat LD50:4590 μg/kg ARCVBP 8,41,77
ivn-rat LD50:8570 μg/kg ARCVBP 8,41,77
orl-mus LD50:17 mg/kg TXAPA9 45,275,78

ipr-mus LD50:5 mg/kg APTOA6 2,109,46
scu-mus LD50:10 mg/kg FCTXAV 18,181,80
ivn-mus LD50:5 mg/kg 85GDA2 5,439,81
ice-mus LD50:570 μg/kg TXCYAC 13,91,79

CONSENSUS REPORTS: IARC Cancer Review: Group 3 IMEMDT 7,56,87; Animal Limited Evidence IMEMDT 10,205,76; Animal Inadequate Evidence IMEMDT 40,83,86. EPA Genetic Toxicology Program.

SAFETY PROFILE: Poison by ingestion, subcutaneous, intracerebral, intraperitoneal, intravenous, and possibly other routes. An experimental teratogen. Other experimental reproductive effects. Human mutation data reported. Questionable carcinogen with experimental neoplastigenic and teratogenic data. An antimicrobial agent. When heated to decomposition it emits acrid smoke and irritating fumes.

CMV250 CAS:57943-81-4 HR: 2
CLAVULANIC ACID SODIUM SALT

PROP: Feathery needles. Mp: 68°.

SYNS: ANTIBIOTIC MM 14151 □ MM 14151

TOXICITY DATA WITH REFERENCE
scu-mus LD50:4500 mg/kg JANTAJ 30,77-27,77
ipr-mus LD50:4000 mg/kg JANTAJ 30,77-27,77

SAFETY PROFILE: Moderately toxic by intraperitoneal route. Mildly toxic by subcutaneous route. When heated to decomposition it emits toxic fumes of NO_x and Na_2O.

CMV325 CAS:57645-91-7 HR: 3
CLEBOPRIDE MALATE
mf: $C_{20}H_{24}ClN_3O_2 \cdot C_4H_6O_5$ mw: 508.02

SYNS: 4-AMINO-5-CHLORO-2-METHOXY-N-(1-BENZYL-4-PIPERIDYL)BENZAMIDE MALATE □ CLEBOPRIDE HYDROGEN MALATE □ LAS

TOXICITY DATA WITH REFERENCE
orl-rat TDLo:12,500 μg/kg (female 17-20D post):REP KSRNAM 16,5661,82
orl-rat TDLo:8 g/kg (9W male/7D pre/1-7D preg):REP KSRNAM 16,5670,82
orl-rat TDLo:219 mg/kg (female 35D pre):REP OYYAA2 25,845,83
orl-rat TDLo:228 mg/kg (female 91D pre):REP OYYAA2 25,803,83
orl-rat TDLo:9 g/kg (male 36D pre):REP OYYAA2 25,877,83
orl-rat LD50:2540 mg/kg OYYAA2 25,803,83
ipr-rat LD50:155 mg/kg OYYAA2 25,803,83
scu-rat LD50:4850 mg/kg OYYAA2 25,803,83
ivn-rat LD50:39 mg/kg OYYAA2 25,803,83
ims-rat LD50:1450 mg/kg OYYAA2 25,803,83
orl-mus LD50:490 mg/kg OYYAA2 25,803,83
ipr-mus LD50:145 mg/kg OYYAA2 25,803,83
scu-mus LD50:305 mg/kg OYYAA2 25,803,83
ivn-mus LD50:51 mg/kg OYYAA2 25,803,83
ims-mus LD50:260 mg/kg OYYAA2 25,803,83

SAFETY PROFILE: Poison by subcutaneous, intramuscular, intravenous and intraperitoneal routes. Moderately toxic by ingestion. Experimental reproductive effects. When heated to decomposition it emits toxic fumes of Cl^- and NO_x.

CMV375 CAS:25047-48-7 HR: 3
CLEISTANTHIN A
mf: $C_{28}H_{28}O_{11}$ mw: 540.52

PROP: A solid. Mp: 135–136°.

SYNS: CIBA GO.4350 □ DIPHYLLIN-3,4-o-DIMETHYL XYLOSIDE □ SHENG BAI XIN (CHINESE)

TOXICITY DATA WITH REFERENCE
orl-rat LD50:12,500 μg/kg PHMGBN 4,347,70
ipr-rat LD50:2630 μg/kg PHMGBN 4,347,70
ivn-rat LD50:2580 μg/kg PHMGBN 4,347,70
orl-mus LD50:38,110 μg/kg PHMGBN 4,347,70
orl-mky LDLo:40 mg/kg PHMGBN 4,347,70
ivn-mky LDLo:1 mg/kg PHMGBN 4,347,70
orl-cat LDLo:100 mg/kg PHMGBN 4,347,70
ipr-cat LDLo:5 mg/kg PHMGBN 4,347,70
orl-rbt LDLo:50 mg/kg PHMGBN 4,347,70
ipr-rbt LDLo:2500 μg/kg PHMGBN 4,347,70

SAFETY PROFILE: Poison by ingestion, intravenous and intraperitoneal routes. When heated to decomposition it emits acrid smoke and fumes.

CMV390 HR: 3
CLEMATIS

PROP: A perennial climbing herb. The leaves resemble those of poison ivy. It is native to Canada and the northern United States, and is available commercially throughout the United States.

SYNS: BLUEBELL □ BLUE JESSAMINE □ CABELLOS de ANGEL (CUBA, PUERTO RICO) □ CABEZA de VIEJO (MEXICO) □ CASCARITA □ CLEMATITE AUX GEAUX (CANADA) □ CURL FLOWER □ CURLY HEADS □ DEVILS HAIR □ DEVILS THREAD □ FLAMULA (CUBA) □ HEADACHE WEED □ HERBE AUX GEAUX (CANADA) □ LEATHER FLOWER □ LIANE BON GARCON (HAITI) □ PIPE STEM □ SUGARBOWLS □ TRAVELERS JOY □ VASE FLOWER □ VASE VINE □ VIRGIN'S BOWER □ YERBA de PORDIOSEROS (CUBA)

SAFETY PROFILE: The whole plant contains the poison protoanemonin, which is a direct irritant and drying agent on the skin and mucous membranes. Ingestion causes pain and blistering of the mouth possibly followed by vomiting, diarrhea, cramps, bloody and painful urination, dizziness, fainting, confusion, and convulsions.

CMV400 CAS:1163-36-6 HR: 3
CLEMIZOLE HDYROCHLORIDE
mf: $C_{19}H_{20}ClN_3 \cdot ClH$ mw: 362.33

PROP: A solid. Mp: 246°. Insol in Et_2O; sol in H_2O.

SYNS: ALLERCUR □ ALLERCURE HYDROCHLORIDE □ 1-(p-CHLOROBENZIL)-2-PIRROLIDIL-METIL-BENZIMIDAZOLO CLORIDATO (ITALIAN) □ 1-(p-CHLOROBENZYL)-2-(1-PYRROLIDINYLMETHYL) BENZIMIDAZOLE HYDROCHLORIDE □ 1-p-CHLOROBENZYL-PYRROLIDYL-METHYLENE-BENZIMIDAZOLE HYDROCHLORIDE □ P 48 □ REACTROL

C

TOXICITY DATA WITH REFERENCE
orl-rat LD50:1950 mg/kg JAPMA8 49,18,60
ivn-rat LD50:74 mg/kg JAPMA8 49,18,60
orl-mus LD50:837 mg/kg JAPMA8 49,18,60
ipr-mus LD50:290 mg/kg FRPSAX 14,194,59
ivn-mus LD50:75 mg/kg JAPMA8 49,18,60
ivn-gpg LD50:26 mg/kg AIPTAK 113,313,58

SAFETY PROFILE: Poison by intravenous and intraperitoneal routes. Moderately toxic by ingestion. When heated to decomposition it emits toxic fumes of NO_x and HCl.

CMV500 **CAS:34148-01-1** **HR: 3**
CLIDANAC
mf: $C_{16}H_{19}ClO_2$ mw: 278.80

SYNS: BRITAI □ 6-CHLORO-5-CYCLOHEXYL-2,3-DIHYDRO-1H-INDENE-1-CARBOXYLIC ACID (9CI) □ 6-CHLORO-5-CYCLOHEXYL-1-INDANCARBOXYLIC ACID □ INDANAL □ TAI 284

TOXICITY DATA WITH REFERENCE
orl-rat LD50:41 mg/kg JMCMAR 15,1297,72
scu-rat LD50:62,700 μg/kg IYKEDH 12,668,81
ivn-rat LD50:88,400 μg/kg IYKEDH 12,668,81
orl-mus LD50:825 mg/kg NIIRDN 6,217,82
scu-mus LD50:860 mg/kg IYKEDH 12,668,81
ivn-mus LD50:212 mg/kg IYKEDH 12,668,81
orl-rbt LD50:380 mg/kg NIIRDN 6,217,82

SAFETY PROFILE: Poison by ingestion, subcutaneous, and intravenous routes. An anti-inflammatory and antipyretic agent. When heated to decomposition it emits toxic fumes of Cl^-.

CMV675 **CAS:18323-44-9** **HR: 3**
CLINDAMYCIN
mf: $C_{18}H_{33}ClN_2O_5S$ mw: 425.04

PROP: Yellow, amorphous solid.

SYNS: 7(S)-CHLORO-7-DEOXYLINCOMYCIN □ CLEOCIN □ CLINDAMYCINE (FRENCH) □ DALACIN C □ SOBELIN □ U-21,251

TOXICITY DATA WITH REFERENCE
scu-rat LD50:2618 mg/kg TXAPA9 18,185,71
orl-rbt LDLo:1 mg/kg RMVEAG 156,915,80
orl-ham LDLo:1 mg/kg ARZNAD 34,794,84

SAFETY PROFILE: A poison by ingestion. Moderately toxic by subcutaneous route. When heated to decomposition it emits toxic fumes of Cl^-, SO_x, and NO_x.

CMV680 **HR: 2**
CLINDAMYCIN-2-PALMITATE MONOHYDROCHLORIDE
mf: $C_{34}H_{63}ClN_2O_6S{\cdot}ClH$ mw: 699.96

TOXICITY DATA WITH REFERENCE
orl-rat TDLo:2450 mg/kg (9-15D preg):TER KSRNAM 7,1724,73
orl-mus TDLo:2450 mg/kg (7-13D preg):REP KSRNAM 7,1724,73
scu-rat LD50:2 g/kg TXAPA9 21,516,72
orl-mus LD50:1956 mg/kg TXAPA9 21,516,72

SAFETY PROFILE: Moderately toxic by ingestion and subcutaneous routes. An experimental teratogen. Experimental reproductive effects. When heated to decomposition it emits toxic fumes of SO_x, NO_x, and Cl^-.

CMV690 **CAS:24729-96-2** **HR: 3**
CLINDAMYCIN-2-PHOSPHATE
mf: $C_{18}H_{34}ClN_2O_8PS$ mw: 505.02

PROP: A solid. Mp: 109–114°.

SYNS: 7(S)-CHLORO-7-DEOXYLINCOMYCIN-2-PHOSPHATE □ CLEOCIN PHOSPHATE □ CLINDAMYCIN PHOSPHATE □ SOBELIN □ U-28,508

TOXICITY DATA WITH REFERENCE
ipr-rat TDLo:700 mg/kg (9-15D preg):REP KSRNAM 7,1697,73
ipr-rat TDLo:700 mg/kg (9-15D preg):TER KSRNAM 7,1697,73
ivn-wmn TDLo:12 mg/kg:BPR SMJOAV 75,768,82
orl-rat LD50:1832 mg/kg TXAPA9 27,308,74
ipr-rat LD50:745 mg/kg IYKEDH 14,484,83
scu-rat LD50:3861 mg/kg IYKEDH 14,484,83
ivn-rat LD50:321 mg/kg IYKEDH 14,484,83
orl-mus LD50:2539 mg/kg IYKEDH 14,484,83
ipr-mus LD50:784 mg/kg DECRDP 3,79,77
scu-mus LD50:1036 mg/kg IYKEDH 14,484,83
ivn-mus LD50:820 mg/kg IYKEDH 14,484,83
ims-mus LD50:1100 mg/kg IYKEDH 14,484,83

SAFETY PROFILE: Poison by intravenous route. Moderately toxic by ingestion, intramuscular, and subcutaneous routes. Human systemic effects by intravenous route: pulse rate increase, blood pressure lowering. An experimental teratogen. Experimental reproductive effects. When heated to decomposition it emits toxic fumes of Cl^-, PO_x, SO_x, and NO_x.

CMV700 **CAS:30299-08-2** **HR: 3**
CLINOFIBRATE
mf: $C_{29}H_{36}O_6$ mw: 468.64

PROP: Off-white powder. Mp: 143–146° (decomp). Sol in methanol, ethanol, acetone, chloroform, glacial acetic acid. Sltly sol in CCl_4. Practically insol in water.

SYNS: 2,2′-CYCLOHEXANE-1,1-DIYLBIS(p-PHENYLENEOXY)BIS(2-METHYLBUTYRIC ACID) □ 2,2′-(CYCLOHEXYLIDENEBIS(4,1-PHENYLENEOXY)BIS(2-METHYLBUTANOIC ACID) □ 2,2′-(4,4′-CYCLOHEXYLIDENEDIPHENOXY)-2,2′-DIMETHYLDIBUTYRIC ACID □ LIPOCLIN □ S 8527

TOXICITY DATA WITH REFERENCE
ipr-rat LD50:205 mg/kg IYKEDH 12,933,81
scu-rat LD50:1930 mg/kg NIIRDN 6,219,82
orl-mus LD50:1600 mg/kg NIIRDN 6,219,82
ipr-mus LD50:255 mg/kg NIIRDN 6,219,82
scu-mus LD50:350 mg/kg NIIRDN 6,219,82
ivn-mus LD50:150 mg/kg IYKEDH 12,933,81

SAFETY PROFILE: Poison by subcutaneous, intravenous, and intraperitoneal routes. Moderately toxic by ingestion. When heated to decomposition it emits acrid smoke and fumes.

CMW000 **CAS:1622-61-3** **HR: 1**
CLOAZEPAM
mf: $C_{15}H_{10}ClN_3O_3$ mw: 315.73

PROP: Crystals from $EtOH/CH_2Cl_2$. Mp: 236.5–238.5°.

SYNS: 5-(o-CHLOROPHENYL)-1,3-DIHYDRO-7-NITRO-2H-1,4-BENZODIAZEPIN-2-ONE □ 5-(o-CHLOROPHENYL)-7-NITRO-1H-1,4-BENZODIAZEPIN-2(3H)-ONE □ CLOAZEPAM □ CLONAZEPAM □ RIVOTRIL □ RO 4-8180 □ RO 5-4023

TOXICITY DATA WITH REFERENCE
scu-mus TDLo:8 mg/kg (female 9D post):TER DGDFA5 22,61,80
orl-rbt TDLo:24 mg/kg (female 6-13D post):TER YACHDS 5,2457,77
orl-rbt TDLo:24 mg/kg (female 6-13D post):REP YACHDS 5,2457,77
orl-rat TDLo:3640 mg/kg (female 26W pre):REP YACHDS 4,1441,76
orl-rat TDLo:29,120 mg/kg (male 26W pre):REP YACHDS 4,1441,76
orl-rat TDLo:1820 mg/kg (26W male):REP YACHDS 4,1441,76
orl-mus TDLo:3300 µg/kg (female 6-16D post):TER TXAPA9 40,365,77
orl-man TDLo:21 mg/kg/26W-I:PNS PGMJAO 63,311,87
orl-man TDLo:1329 µg/kg/31D-I:SYS AJGAAR 83,576,88
orl-rat LD50:>15 g/kg DRUGAY 6,235,82
ipr-rat LD50:14,200 mg/kg IYKEDH 11,811,80
orl-mus LD50:2 g/kg PJPPAA 35,77,83
ipr-mus LD50:13,300 mg/kg IYKEDH 11,811,80

CONSENSUS REPORTS: EPA Genetic Toxicology Program.

SAFETY PROFILE: Mildly toxic by ingestion. Human systemic effects by ingestion: liver function tests impaired, fasciculations. Experimental teratogenic and reproductive effects. An anticonvulsant. When heated to decomposition it emits very toxic fumes of Cl^- and NO_x. See also DIAZEPAM.

CMW250 **CAS:2726-03-6** **HR: 3**
CLOBENZEPAM HYDROCHLORIDE
mf: $C_{17}H_{18}ClN_3O \cdot ClH$ mw: 352.29

PROP: Prisms from $EtOH/Et_2O$. Mp: 222–226°.

SYNS: 5,10-DIHYDRO-7-CHLOR-10-(2-(DIMETHYLAMINO)ETHYL)-11H-DIBENZO(b,e)(1,4)DIAZEPIN-11-ONE HCl □ TARPAN

TOXICITY DATA WITH REFERENCE
orl-rat LD50:242 mg/kg MEXPAG 6,205,62
ivn-rat LD50:26 mg/kg MEXPAG 6,205,62
orl-mus LD50:324 mg/kg 27ZQAG -,66,72
ipr-mus LD50:75 mg/kg JMCMAR 14,153,71
ivn-mus LD50:28,500 µg/kg MEXPAG 6,205,62
orl-gpg LD50:262 mg/kg MEXPAG 6,205,62

SAFETY PROFILE: Poison by ingestion, intravenous, and intraperitoneal routes. An antihistamine. When heated to decomposition it emits very toxic fumes of Cl^- and NO_x. See also DIAZEPAM.

CMW300 **CAS:25122-46-7** **HR: 3**
CLOBETASOL PROPIONATE
mf: $C_{25}H_{32}ClFO_5$ mw: 467.02

PROP: Creamy-white powder. Mp: 195.5–197°.

SYNS: 21-CHLORO-9-FLUORO-11-β,17-DIHYDROXY-16-β-METHYLPREGNA-1,4-DIENE-3,20-DIONE-17-PROPIONATE □ CLOBETASOL-17-PROPIONATE

TOXICITY DATA WITH REFERENCE
skn-rat TDLo:17,500 mg/kg (35D male):REP OYYAA2 27,1217,84
ipr-rat LD50:351 mg/kg NIIRDN 6,241,82
scu-rat LD50:366 mg/kg IYKEDH 9,1066,78
ipr-mus LD50:118 mg/kg NIIRDN 6,241,82
scu-mus LD50:81,700 µg/kg NIIRDN 6,241,82

SAFETY PROFILE: Poison by subcutaneous and intraperitoneal routes. Experimental reproductive effects. When heated to decomposition it emits toxic fumes of F^- and Cl^-.

CMW400 **CAS:25122-57-0** **HR: 2**
CLOBETASONE BUTYRATE
mf: $C_{26}H_{32}ClFO_5$ mw: 479.03

PROP: A solid. Mp: 90–100°.

SYNS: 21-CHLORO-9-FLUORO-17-HYDROXY-16-β-METHYLPREGNA-1,4-DIENE-3,11,20-TRIONE BUTYRATE □ CLOBETASONE-17-BUTYRATE □ EUMOVATE □ MOLIVATE □ SN 203

TOXICITY DATA WITH REFERENCE
scu-rat TDLo:3 mg/kg (female 17-22D post):REP KSRNAM 14,359,80
scu-rat TDLo:3300 µg/kg (female 7-17D post):TER KSRNAM 14,343,80
skn-rat TDLo:11 g/kg (7-17D preg):TER SKIZAB 36,91,80
scu-rbt TDLo:390 µg/kg (female 6-18D post):REP KSRNAM 14,373,80
skn-rat TDLo:11 g/kg (7-17D preg):REP SKIZAB 36,91,80
scu-rbt TDLo:780 µg/kg (female 6-18D post):REP KSRNAM 14,373,80
scu-rbt TDLo:97,500 ng/kg (female 6-18D post):TER KSRNAM 14,373,80
ipr-rat LD50:1510 mg/kg JTSCDR 5,45,80
ipr-mus LD50:500 mg/kg IYKEDH 14,838,83

SAFETY PROFILE: Moderately toxic by intraperitoneal route. An experimental teratogen. Other experimental reproductive effects. When heated to decomposition it emits toxic fumes of F^- and Cl^-.

CMW459 **CAS:14860-49-2** **HR: 3**
CLOBUTINOL
mf: $C_{14}H_{22}ClNO$ mw: 255.82

PROP: Bp: 179–180°.

SYNS: (p-CHLORO-α-(2-(DIMETHYLAMINO)-1-METHYLETHYL)-α-METHYL-PHENETHYL ALCOHOL □ 1-(4-CHLOROPHENYL)-2,3-DIMETHYL-4-DIMETHYLAMINO-2-BUTANOL □ p-CLORO-α-(2-DIMETILAMINO)-1-METILETIL)-α-METIL FENETIL ALCOOL (ITALIAN)

TOXICITY DATA WITH REFERENCE
orl-rat LD50:802 mg/kg IYKEDH 6,119,75

ipr-rat LD50:165 mg/kg IYKEDH 6,119,75
scu-rat TDLo:775 mg/kg IYKEDH 6,119,75
ivn-rat LD50:63 mg/kg IYKEDH 6,119,75
orl-mus LD50:334 mg/kg IYKEDH 6,119,75
ipr-mus LD50:128 mg/kg IYKEDH 6,119,75
scu-mus LD50:262 mg/kg OYYAA2 8,1067,74
ivn-mus LD50:53 mg/kg IYKEDH 6,119,75
ivn-dog LD50:45,300 µg/kg OYYAA2 8,1067,74

SAFETY PROFILE: Poison by ingestion, subcutaneous, intravenous, and intraperitoneal routes. When heated to decomposition it emits very toxic fumes of Cl^- and NO_x. See also ALCOHOLS.

CMW500 CAS:1215-83-4 HR: 3
CLOBUTINOL HYDROCHLORIDE
mf: $C_{14}H_{22}ClNO \cdot ClH$ mw: 292.28

PROP: Crystals. Mp: 169–170°.

SYNS: BIOTERTUSSIN □ (2-(p-CHLOROBENZYL)-3-DIMETHYLAMINOMETHYL-2-BUTANOL HYDROCHLORIDE □ 4-CHLORO-α-(2-(DIMETHYLAMINO)-1-METHYLETHYL)-α-METHYLBENZENEETHANOL HYDROCHLORIDE □ p-CHLORO-α-(2-DIMETHYLAMINO-1-METHYLETHYL)-α-METHYLPHENETHYL ALCOHOL HYDROCHLORIDE □ 1-p-CHLOROPHENYL-2,3-DIMETHYL-4-DIMETHYLAMINO-2-BUTANOL HYDROCHLORIDE □ p-CLORO-α-(2-DIMETILAMINO)-1-METILETIL)-α-METIL FENETIL ALCOOL CLORIDRATO (ITALIAN) □ KAT 256 □ PERTOXIL □ SILOMAT □ SILONIST

TOXICITY DATA WITH REFERENCE
orl-rat LD50:802 mg/kg NIIRDN 6,241,82
ipr-rat LD50:151 mg/kg NIIRDN 6,241,82
scu-rat LD50:702 mg/kg NIIRDN 6,241,82
ivn-rat LD50:63 mg/kg NIIRDN 6,241,82
orl-mus LD50:334 mg/kg NIIRDN 6,241,82
ipr-mus LD50:125 mg/kg NIIRDN 6,241,82
scu-mus LD50:262 mg/kg OYYAA2 4,961,70
ivn-mus LD50:40,900 µg/kg NIIRDN 6,241,82
ivn-dog LD50:45,300 µg/kg OYYAA2 4,961,70

SAFETY PROFILE: Poison by ingestion, subcutaneous, intravenous, and intraperitoneal routes. An antitussive. When heated to decomposition it emits toxic fumes of Cl^- and NO_x. See also ALCOHOLS.

CMW550 CAS:77174-66-4 HR: 2
CLOCONAZOLE HYDROCHLORIDE
mf: $C_{18}H_{15}ClN_2O \cdot HCl$ mw: 347.26

PROP: Crystals from EtOAc/MeCN. Mp: 148.5–150°.

SYNS: 1-(1-(2-((3-CHLOROBENZYL)OXY)PHENYL)VINYL)-1H-IMIDAZOLE HYDROCHLORIDE □ 1-(1-(2-((3-CHLOROPHENYL)METHOXY)PHENYL)ETHENYL)-1H-IMIDAZOLE HYDROCHLORIDE □ 1-(1-(2-((3-CHLOROPHENYL)METHOXY)PHENYL)ETHENYL)-1H-IMIDAZOLE MONOHYDROCHLORIDE (9CI) □ 710674-S

TOXICITY DATA WITH REFERENCE
scu-rat TDLo:1050 mg/kg (female 14D pre):REP KSRNAM 18,4917,84
scu-rat TDLo:1050 mg/kg (female 14D pre):TER KSRNAM 18,4917,84
scu-rat TDLo:5250 mg/kg (male 15W pre):REP KSRNAM 18,4917,84
scu-rat TDLo:42 mg/kg (female 14D pre):TER KSRNAM 18,4917,84
orl-rat LD50:2 g/kg DRFUD4 10,451,85
skn-rat LD50:>600 mg/kg KSRNAM 18,4811,84
ipr-rat LD50:204 mg/kg KSRNAM 18,4811,84
scu-rat LD50:4700 mg/kg KSRNAM 18,4811,84
orl-mus LD50:1350 mg/kg KSRNAM 18,4811,84
ipr-mus LD50:94 mg/kg KSRNAM 18,4811,84

SAFETY PROFILE: Moderately toxic by ingestion, skin contact, and subcutaneous routes. An experimental teratogen. Other experimental reproductive effects. When heated to decomposition it emits toxic fumes of Cl^- and NO_x.

CMW600 CAS:4755-59-3 HR: 3
CLODAZONE
mf: $C_{18}H_{20}ClN_3O$ mw: 329.86

PROP: Bp: 163° @ 0.03 mm.

SYN: AW-14-2446

TOXICITY DATA WITH REFERENCE
orl-rat LD50:2000 mg/kg INPHB6 1,214,68
ipr-rat LD50:142 mg/kg INPHB6 1,214,68
ivn-rat LD50:77 mg/kg INPHB6 1,214,68
orl-mus LD50:715 mg/kg INPHB6 1,214,68
ivn-mus LD50:90 mg/kg INPHB6 1,214,68
orl-gpg LD50:505 mg/kg INPHB6 1,214,68

SAFETY PROFILE: Poison by intravenous and intraperitoneal routes. Moderately toxic by ingestion. When heated to decomposition it emits toxic fumes of Cl^- and NO_x.

CMW700 CAS:511-13-7 HR: 3
CLOFEDANOL HYDROCHLORIDE
mf: $C_{17}H_{20}ClNO \cdot ClH$ mw: 326.29

SYNS: BAYER-186 □ BAYER B-186 □ CHLOPHEDIANOL HYDROCHLORIDE □ 2-CHLORO-α-(2-(DIMETHYLAMINO)-ETHYL)BENZHYDROL HYDROCHLORIDE □ 2-CHLORO-α-(2-(DIMETHYLAMINO)ETHYL)-α-PHENYL-BENZENEMETHANOL HYDROCHLORIDE (9CI) □ 1-o-CHLOROPHENYL-1-PHENYL-3-DIMETHYLAMINO-1-PROPANOL HYDROCHLORIDE □ CHLORPHEDIANOL HYDROCHLORIDE □ CLOPHEDIANOL HYDROCHLORIDE □ COLDRIN □ DETIGON □ DETIGON-BAYER □ α-(2-DIMETHYLAMINOETHYL)-o-CHLOROBENZHYDROL HYDROCHLORIDE □ REFUGAL □ SK 74 □ SL 501 □ ULO □ ULONE

TOXICITY DATA WITH REFERENCE
orl-mus TDLo:192 mg/kg (female 7-12D post):REP GNRIDX 7,177,73
orl-rat TDLo:192 mg/kg (female 9-14D post):TER GNRIDX 7,177,73
orl-rat TDLo:192 mg/kg (female 9-14D post):REP GNRIDX 7,177,73
orl-rat TDLo:6720 mg/kg (female 28D pre):REP OYYAA2 9,137,75
orl-rat TDLo:2520 mg/kg (28D male):REP OYYAA2 9,137,75
orl-rat LD50:350 mg/kg USXXAM #3031377
ipr-rat LD50:295 mg/kg IYKEDH 12,933,81
scu-rat LD50:440 mg/kg GNRIDX 7,177,73
ivn-rat LD50:53 mg/kg GNRIDX 7,177,73
ims-rat LD50:268 mg/kg GNRIDX 7,177,73
orl-mus LD50:284 mg/kg THERAP 15,93,60

ipr-mus LD50:130 mg/kg GNRIDX 7,177,73
scu-mus LD50:144 mg/kg THERAP 15,93,60
ivn-mus LD50:42 mg/kg THERAP 15,93,60
orl-dog LD50:84 mg/kg YKYUA6 32,1281,81
scu-mam LD50:186 mg/kg 27ZQAG -,369,72

SAFETY PROFILE: Poison by ingestion, subcutaneous, intramuscular, intravenous, and intraperitoneal routes. Experimental reproductive effects. An experimental teratogen. When heated to decomposition it emits toxic fumes of NO_x and HCl.

CMW750 CAS:17449-96-6 HR: 3
CLOFEXAMIDE PHENYLBUTAZONE
mf: $C_{14}H_{21}ClN_2O_2 \cdot C_{19}H_{20}N_2O_2$ mw: 593.23

SYNS: 2-(p-CHLOROPHENOXY)-N-(2-(DIETHYLAMINO)ETHYL) ACETAMIDE COMPOUND with 4-BUTYL-1,2-DIPHENYL-3,5-PYRAZOLIDINEDIONE (1:1) ☐ CLOFEXAMIDE-PHENYLBUTAZONE MIXTURE ☐ CLOFEZON ☐ CLOFEZONE ☐ PERCLUSON ☐ PERCLUSONE

TOXICITY DATA WITH REFERENCE
orl-mus TDLo:1600 mg/kg (female 7-14D post):TER OYYAA2 18,235,79
orl-rat TDLo:1280 mg/kg (female 8-15D post):TER OYYAA2 18,235,79
orl-rat TDLo:1920 mg/kg (8-15D preg):TER OYYAA2 18,235,79
orl-rat TDLo:1920 mg/kg (8-15D preg):REP OYYAA2 18,235,79
orl-rat TDLo:640 mg/kg (female 8-15D post):REP OYYAA2 18,235,79
orl-rat TDLo:9 g/kg (female 30D pre):REP OYYAA2 18,213,79
orl-rat TDLo:3 g/kg (30D male):REP OYYAA2 18,213,79
orl-rat LD50:1950 mg/kg OYYAA2 15,41,78
ipr-rat LD50:1500 mg/kg NIIRDN 6,237,82
orl-mus LD50:1700 mg/kg OYYAA2 17,935,79
ipr-mus LD50:1650 mg/kg NIIRDN 6,237,82
orl-rbt LD50:390 mg/kg CHTPBA 3,53,68
orl-gpg LD50:720 mg/kg CHTPBA 3,53,68

SAFETY PROFILE: Poison by ingestion. Moderately toxic by intraperitoneal route. An experimental teratogen. Other experimental reproductive effects. When heated to decomposition it emits very toxic fumes of Cl^- and NO_x.

CMX000 CAS:882-09-7 HR: 3
CLOFIBRIC ACID
mf: $C_{10}H_{11}ClO_3$ mw: 214.66

PROP: Crystals from MeOH or H_2O. Mp: 118–119°.

SYNS: (p-CHLOROPHENOXY)DIMETHYL-ACETIC ACID ☐ α-(p-CHLOROPHENOXY)ISOBUTYRIC ACID ☐ 2-(4-CHLOROPHENOXY)-2-METHYLPROPANOIC ACID ☐ 2-(p-CHLOROPHENOXY)-2-METHYLPROPIONIC ACID ☐ CHLOROPHIBRINIC ACID ☐ CLOFIBRINIC ACID ☐ CLOFIBRINSAEURE (GERMAN)

TOXICITY DATA WITH REFERENCE
dns-rat-orl 2100 mg/kg/1W-C CRNGDP 12,1557,91
unk-hmn TDLo:260 mg/kg (91D):LIV ATHSBL 36,159,80
orl-rat LD50:897 mg/kg ARZNAD 30,2023,80
scu-rat LD50:120 mg/kg PHARAT 22,167,67
orl-mus LD50:1170 mg/kg ARZNAD 27,1173,77
ipr-mus LD50:290 mg/kg PHARAT 22,167,67
scu-mus LD50:683 mg/kg RPTOAN 33,150,70

SAFETY PROFILE: Poison by intraperitoneal and subcutaneous routes. Moderately toxic by ingestion. Liver damage in humans by an unspecified route. Mutation data reported. When heated to decomposition it emits toxic fumes of Cl^-.

CMX500 CAS:911-45-5 HR: 3
CLOMIPHENE
mf: $C_{26}H_{28}ClNO$ mw: 406.00

PROP: A solid. Mp: 116.5–118°.

SYNS: CHLOMAPHENE ☐ CHLORAMIFENE ☐ CHLORAMIPHENE ☐ 2-(4-(2-CHLORO-1,2-DIPHENYLETHENYL)PHENOXY)-N,N-DIETHYLETHANAMINE ☐ 2-(p-(β-CHLORO-α-PHENYLSTYRYL)PHENOXY)-TRIETHYLAMINE ☐ CISCLOMIPHENE ☐ CLOMEPHENE B ☐ CLOMIFENE ☐ 1-(p-(β-DIETHYLAMINOETHOXY)PHENYL)-1,2-DIPHENYLCHLOROETHYLENE

TOXICITY DATA WITH REFERENCE
unr-wmn TDLo:5 mg/kg (5D pre):REP UMJOAJ 45,59,76
unr-wmn TDLo:15 mg/kg (female 15D pre):REP UMJOAJ 45,59,76
unr-wmn TDLo:120 mg/kg (15D pre):TER UMJOAJ 45,59,76
unr-wmn TDLo:10 mg/kg (5D pre):REP UMJOAJ 45,59,76
unr-wmn TDLo:15 mg/kg (15D pre):TER LANCAO 2,1262,75
unr-wmn TDLo:5 mg/kg (5D pre):TER UMJOAJ 45,59,76
orl-rat TDLo:4500 μg/kg (female 8-22D post):REP FESTAS 16,195,65
orl-rat TDLo:486 mg/kg (female 6-14D post):TER TXAPA9 10,565,67
orl-rat TDLo:2400 μg/kg (8D pre):REP FESTAS 16,195,65
orl-mus TDLo:3 mg/kg (female 2D post):REP JRPFA4 15,223,68
orl-rat TDLo:9900 μg/kg (female 6-14D post):REP TXAPA9 10,565,67
orl-rat TDLo:7480 μg/kg (55D male):REP JRPFA4 14,39,67
orl-rat TDLo:9900 μg/kg (female 6-14D post):TER TXAPA9 10,565,67
orl-rat TDLo:99 mg/kg (female 6-14D post):TER TXAPA9 10,565,67
orl-wmn TDLo:15 mg/kg/13W-I:CAR,SPN,SKN LANCAO 2,1176,77
orl-mus TDLo:552 mg/kg/69W-I:ETA PEXTAR 11,440,69
orl-wmn TDLo:5 mg/kg/5D-I:SKN BMJOAE 292,380,86
orl-mus LD50:1700 mg/kg FEPRA7 20,419,61
ipr-mus LD50:390 mg/kg FEPRA7 20,419,61

SAFETY PROFILE: Poison by intraperitoneal route. Moderately toxic by ingestion. Experimental teratogenic data. Human systemic effects by ingestion: dermatitis. Human reproductive effects by unspecified routes: death of fetus, stillbirth, and poor viability. Human teratogenic effects by unspecified routes include developmental abnormalities of the central nervous system, cardiovascular system, gastrointestinal system, and urogenital system. Other experimental teratogenic and reproductive effects. Questionable human carcinogen producing spinal cord and skin tumors. When heated to decomposition it emits very toxic fumes of Cl^- and NO_x.

CMX700 CAS:50-41-9 HR: 2
racemic-CLOMIPHENE CITRATE
mf: $C_{26}H_{28}ClNO \cdot C_6H_8O_7$ mw: 598.14

SYNS: CHLORAMIPHENE □ CHLORAMIPHENE CITRATE □ 2-CHLORO-1-(p-(β-DIETHYLAMINOETHOXY)PHENYL)-1,2-DIPHENYLETHYLENE □ 2-(p-(2-CHLORO-1,2-DIPHENYL VINYL)PHENOXY)TRIETHYLAMINE CITRATE (1:1) □ CLOMID □ CLOMIFEN CITRATE □ CLOMIFENO □ CLOMIPHENE CITRATE □ CLOMIPHENE DIHYDROGEN CITRATE □ CLOMIPHENE-R □ CLOMIPHINE □ CLOMIVID □ CLOMPHID □ 1-(p-(β-DIETHYLAMINO ETHOXY)PHENYL)-1,2-DIPHENYL-2-CHLOROETHYLENE CITRATE □ DYNERIC □ GENOZYM □ IKACLOMIN □ MER-41 □ MRL 41 □ NSC 35770 □ OMIFIN

TOXICITY DATA WITH REFERENCE
dnd-esc 25 mg/L MUREAV 165,57,86
dni-esc 50 mg/L MUREAV 165,57,86
orl-man TDLo:43 mg/kg (60D male):REP JCEMAZ 29,638,69
unr-wmn TDLo:15 mg/kg (15D pre):TER LANCAO 2,1107,81
par-rat TDLo:2 mg/kg (female 5D post):REP SCIEAS 204,629,79
orl-rat TDLo:3 mg/kg (female 9-14D post):REP OYYAA2 4,635,70
orl-rat TDLo:6 mg/kg (female 9-14D post):REP OYYAA2 4,635,70
scu-rat TDLo:2 mg/kg (female 8D post):TER PSDTAP 12,299,71
scu-rat TDLo:2 mg/kg (female 10D post):TER PSDTAP 12,299,71
orl-rat TDLo:6 mg/kg (female 9-14D post):TER OYYAA2 4,635,70
orl-rat TDLo:100 μg/kg (female 1D post):REP JRPFA4 13,373,67
orl-rat TDLo:150 mg/kg (30D male):REP FEPRA7 20,419,61
scu-mus TDLo:100 mg/kg (female 2-3D post):REP AJOGAH 147,633,83
scu-rbt TDLo:30 mg/kg (female 10-12D post):REP FESTAS 18,18,67
orl-rat TDLo:7 mg/kg (female 1-7D post):REP FESTAS 13,472,62
orl-rat TDLo:600 μg/kg (female 6D pre):REP PSEBAA 105,197,60
scu-mus TDLo:100 mg/kg (female 1-2D post):REP AJOGAH 147,633,83
scu-rat TDLo:2500 μg/kg (female 1D pre):REP JSTBBK 12,47,80
orl-dog TDLo:1325 mg/kg (female 53W pre):REP TXAPA9 9,44,66
orl-rat TDLo:3500 μg/kg (male 7D pre):REP TXAPA9 21,80,72
ims-rat TDLo:448 mg/kg (male 28D pre):REP INJFA3 24,170,79
scu-rat TDLo:2 mg/kg (female 12D post):TER JSTBBK 12,47,80
scu-rat TDLo:2500 μg/kg:ETA JSTBBK 12,47,80
par-rat TDLo:2 mg/kg:ETA OGSUA8 35,12,80
par-rat TD:50 mg/kg:ETA SCIEAS 197,164,77
orl-rat LD50:5750 mg/kg TXAPA9 9,44,66
ipr-rat LD50:530 mg/kg TXAPA9 9,44,66
orl-mus LD50:1700 mg/kg OYYAA2 3,187,69

CONSENSUS REPORTS: IARC Cancer Review: Group 3 IMEMDT 7,172,87; Human Inadequate Evidence IMEMDT 21,551,79; Animal Inadequate Evidence IMEMDT 21,551,79. EPA Genetic Toxicology Program.

SAFETY PROFILE: Moderately toxic by ingestion and intraperitoneal routes. Human reproductive effects by ingestion: changes in spermatogenesis and effects on testes, epididymis, and sperm duct. Human teratogenic effects by an unspecified route: developmental abnormalities of the eye and ear. Experimental reproductive effects. Questionable carcinogen with experimental tumorigenic data. Used as a drug to induce ovulation and for the treatment of oligospermia. When heated to decomposition it emits very toxic fumes of Cl^- and NO_x.

CMX760 CAS:4205-91-8 HR: 3
CLONIDINE HYDROCHLORIDE
mf: $C_9H_9Cl_2N_3 \cdot ClH$ mw: 266.57

PROP: A solid. Mp: 310–312°.

SYNS: CATAPRES □ CATAPRESAN □ 2-(2,6-DICHLOROANILINO)-2-IMIDAZOLINE HYDROCHLORIDE □ 2,6-DICHLORO-N-2-IMIDAZOLIDINYLIDENE-BENZENAMINE HYDROCHLORIDE □ 2-(2,6-DICHLOROPHENYLAMINO)-2-IMIDAZOLIN HYDROCHLORID (GERMAN) □ 2-((2,6-DICHLOROPHENYL)IMINO)IMIDAZOLIDINE MONOHYDROCHLORIDE □ DIXARIT □ HEMITON □ ISOGLAUCON □ ST-155

TOXICITY DATA WITH REFERENCE
scu-uns TDLo:2080 μg/kg (female 8-20D post):REP TJADAB 31,10B,85
ivn-dom TDLo:5 μg/kg (female 1D post):TER ANESAV 67,A449,87
ivn-dom TDLo:5 μg/kg (female 1D post):REP ANESAV 67,A449,87
orl-rat TDLo:6 mg/kg (30D pre):REP ZPPLBF 114,251,75
orl-inf TDLo:390 μg/kg PEDIAU 72,500,83
orl-wmn TDLo:180 μg/kg:BAH,CVS,PUL AJEMEN 7,343,89
orl-inf TDLo:390 μg/kg:NAH,CVS PEDIAU 72,500,83
orl-wmn TDLo:126 μg/kg/4W-I:GIT BMJOAE 292,174,86
orl-man TDLo:69 μg/kg:CNS,CVS LANCAO 2,694,76
orl-chd TDLo:70 μg/kg:CNS,CVS,PUL AJDCAI 137,171,83
orl-rat LD50:126 mg/kg IYKEDH 18,366,87
ipr-rat LD50:100 mg/kg IYKEDH 9,829,78
scu-rat LD50:77 mg/kg IYKEDH 18,366,87
ivn-rat LD50:29 mg/kg ARZNAD 16,1038,66
orl-mus LD50:135 mg/kg IYKEDH 18,366,87
ipr-mus LD50:100 mg/kg IYKEDH 9,829,78
scu-mus LD50:59 mg/kg YKKZAJ 95,966,75
ivn-mus LD50:17,600 μg/kg NIIRDN 6,235,82
orl-dog LD50:30 mg/kg PBPSDY 1,67,77
orl-rbt LD50:80 mg/kg ARZNAD 16,1038,66

SAFETY PROFILE: Poison by ingestion, subcutaneous, intravenous, and intraperitoneal routes. Human systemic effects by ingestion: coma, effects on sleep, fall in blood pressure, gastrointestinal system effects, increase in pulse rate, pulse rate decrease, reduced blood pressure, respiratory depression, respiratory system effects, somnolence. An experimental teratogen. Other experimental reproductive effects. When heated to decomposition it emits toxic fumes of NO_x and Cl^-.

C

CMX770 **CAS:17737-65-4** **HR: 3**
CLONIXIC ACID
mf: $C_{13}H_{11}ClN_2O_2$ mw: 262.71

PROP: A solid. Mp: 233–235°.

SYNS: CHLONIXIN ☐ CLONIXIN ☐ CLONIXINE ☐ 2-(2′-METHYL-3′-CHLORO)ANILINONICOTINIC ACID ☐ SCH 10304

TOXICITY DATA WITH REFERENCE
orl-rat TDLo:280 mg/kg (female 9-15D post):REP KSRNAM 9,453,75
orl-mus TDLo:560 mg/kg (female 7-13D post):TER KSRNAM 9,453,75
orl-rat TDLo:2240 mg/kg (28D male):REP OYYAA2 7,655,73
orl-rat LD50:335 mg/kg OYYAA2 7,655,73
ipr-rat LD50:148 mg/kg OYYAA2 7,655,73
scu-rat LD50:325 mg/kg OYYAA2 7,655,73
orl-mus LD50:400 mg/kg OYYAA2 7,655,73
ipr-mus LD50:198 mg/kg OYYAA2 7,655,73
scu-mus LD50:263 mg/kg OYYAA2 7,655,73

SAFETY PROFILE: Poison by ingestion, subcutaneous, and intraperitoneal routes. An experimental teratogen. Other experimental reproductive effects. When heated to decomposition it emits toxic fumes of Cl^- and NO_x.

CMX800 **CAS:3703-76-2** **HR: 3**
CLOPERASTINE
mf: $C_{20}H_{24}ClNO$ mw: 329.90

PROP: Bp: 172–174°.

SYN: 1-(2-((4-CHLOROPHENYL)PHENYLMETHOXY)ETHYL)PIPERIDINE (9CI)

TOXICITY DATA WITH REFERENCE
orl-rat LD50:2325 mg/kg NIIRDN 6,241,82
ipr-rat LD50:120 mg/kg NIIRDN 6,241,82
orl-mus LD50:600 mg/kg NIIRDN 6,241,82
ipr-mus LD50:96 mg/kg NIIRDN 6,241,82

SAFETY PROFILE: Poison by intraperitoneal route. Moderately toxic by ingestion. When heated to decomposition it emits toxic fumes of Cl^- and NO_x.

CMX820 **CAS:22199-30-0** **HR: 3**
CLOPERASTINE HYDROCHLORIDE
mf: $C_{20}H_{24}ClNO{\cdot}ClH$ mw: 366.36

SYNS: CLOPERASTINA CLORIDRATO (ITALIAN) ☐ 1-(2-(p-CLORO-α-FENILBENZILOSSI)ETIL)PIPERIDINA CLORIDRATO (ITALIAN)

TOXICITY DATA WITH REFERENCE
orl-rat LD50:1986 mg/kg BCFAAI 122,384,83
ipr-rat LD50:150 mg/kg BCFAAI 122,384,83
orl-mus LD50:553 mg/kg BCFAAI 122,384,83
ipr-mus LD50:140 mg/kg BCFAAI 122,384,83

SAFETY PROFILE: Poison by intraperitoneal route. Moderately toxic by ingestion. When heated to decomposition it emits toxic fumes of NO_x and HCl.

CMX840 **CAS:525-26-8** **HR: 2**
CLOPERIDONE HYDROCHLORIDE
mf: $C_{21}H_{23}ClN_4O_2{\cdot}ClH$ mw: 435.39

PROP: A solid. Mp: 240–241°.

SYNS: 3-(3-(4-(m-CHLOROPHENYL)-1-PIPERAZINYL)PROPYL)-2,4(1H,3H)-QUINAZOLINEDIONE HYDROCHLORIDE ☐ MA 1337

TOXICITY DATA WITH REFERENCE
orl-rat LD50:7650 mg/kg JPETAB 148,151,65
ipr-rat LD50:3290 mg/kg JPETAB 148,151,65
ipr-mus LD50:2610 mg/kg JPETAB 148,151,65

SAFETY PROFILE: Moderately toxic by intraperitoneal route. Mildly toxic by ingestion. When heated to decomposition it emits toxic fumes of NO_x and Cl^-.

CMX850 **CAS:2971-90-6** **HR: 1**
CLOPIDOL
mf: $C_7H_7Cl_2NO$ mw: 192.05

PROP: Insol in H_2O.

SYNS: COCCIDIOSTAT C ☐ COYDEN ☐ 3,5-DICHLORO-2,6-DIMETHYL-4-PYRIDINOL ☐ LERBEK ☐ METHYLCHLOROPINDOL ☐ METHYLCHLORPINDOL ☐ METILCLORPINDOL

TOXICITY DATA WITH REFERENCE
orl-rat LD50:18 g/kg MEIEDD 10,341,83

OSHA PEL: Total Dust: 15 mg/m³; Respirable Fraction: 5 mg/m³
ACGIH TLV: TWA 10 mg/m³

SAFETY PROFILE: A nuisance dust. When heated to decomposition it emits very toxic fumes of Cl^- and NO_x.

CMX860 **CAS:60086-22-8** **HR: 3**
CLOPIPAZAN MESYLATE
mf: $C_{19}H_{18}ClNO{\cdot}CH_4O_3S$ mw: 407.94

SYNS: 4(2-CHLORO-9H-XANTHEN-9-YLIDENE)-1-METHYLPIPERIDINE METHANESULFONATE ☐ SKF-69,634

TOXICITY DATA WITH REFERENCE
orl-hmn TDLo:9140 μg/kg/28D DRFUD4 6,15,81
orl-rat LD50:319 mg/kg DRFUD4 6,15,81
ipr-rat LD50:80 mg/kg DRFUD4 6,15,81
ipr-mus LD50:57 mg/kg DRFUD4 6,15,81

SAFETY PROFILE: Poison by ingestion and intraperitoneal routes. Unspecified human systemic effects by ingestion. When heated to decomposition it emits toxic fumes of Cl^-, NO_x, and SO_x.

CMX920 **CAS:34255-03-3** **HR: 3**
CLOQUINOZINE TARTRATE
mf: $C_{16}H_{22}ClNC_4H_6O_6$ mw: 413.94

SYNS: 3-(p-CHLOROBENZYL)OCTAHYDRO-QUINOLIZINE TARTRATE (1:1) ☐ 3-(p-CHLOROBENZYL)QUINOLIZIDINE TARTRATE ☐ QB-1

TOXICITY DATA WITH REFERENCE
ivn-rat LD50:45 mg/kg JJPAAZ 16,353,66
scu-mus LD50:430 mg/kg JJPAAZ 16,353,66
ivn-mus LD50:31,800 μg/kg JJPAAZ 16,353,66

SAFETY PROFILE: Poison by intravenous route. Moderately toxic by subcutaneous route. When heated to decomposition it emits toxic fumes of Cl^- and NO_x.

CMY000 **CAS:17780-75-5** **HR: 3**
CLORGYLINE HYDROCHLORIDE
mf: $C_{13}H_{15}Cl_2NO•ClH$ mw: 308.65

PROP: A solid. Mp: 98.5–100°. Sol in H_2O and $CHCl_3$.

SYN: N-METHYL-N-PROPARGYL-3-(2,4-DICHLOROPHENOXY)PROPYLAMINE HYDROCHLORIDE

TOXICITY DATA WITH REFERENCE
orl-rat LD50:210 mg/kg BCPCA6 17,1285,68
ivn-rat LD50:62 mg/kg BCPCA6 17,1285,68
orl-mus LD50:430 mg/kg BCPCA6 17,1285,68
ipr-mus LD50:350 mg/kg JMCMAR 21,56,78
scu-mus LD50:400 mg/kg BCPCA6 17,1285,68
ivn-mus LD50:94 mg/kg BCPCA6 17,1285,68

SAFETY PROFILE: Poison by ingestion, intraperitoneal, intravenous, and subcutaneous routes. When heated to decomposition it emits very toxic fumes of HCl and NO_x.

CMY030 **HR: 3**
CLOSTRIDIUM BOTULINUM TOXIN

SYNS: BOTULINUSTOXIN □ TOXIN, CLOSTRIDIUM BOTULINUM

TOXICITY DATA WITH REFERENCE
scu-mus LD50:30 pg/kg NATWAY 56,615,69
unr-mus LDLo:5 ng/kg 29QKAZ 2,631,72

SAFETY PROFILE: A deadly human poison. When heated to decomposition it emits acrid smoke and irritating vapors.

CMY050 **HR: 1**
CLOSTRIDIUM BUTYRICUM MIYAIRI, powder

TOXICITY DATA WITH REFERENCE
orl-rat LD50:>5 g/kg OYYAA2 34,215,87

SAFETY PROFILE: Low toxicity by ingestion. When heated to decomposition it emits acrid smoke and irritating vapors.

CMY070 **HR: 3**
CLOSTRIDIUM DIFFICILE TOXIN

SYN: TOXIN, CLOSTRIDIUM DIFFICILE

TOXICITY DATA WITH REFERENCE
ipr-mus LD50:13 μg/kg INFIBR 34,1036,81

SAFETY PROFILE: A poison by intraperitoneal route. When heated to decomposition it emits acrid smoke and irritating vapors.

CMY090 **HR: 3**
CLOSTRIDIUM DIFFICILE TOXIN B

SYNS: TOXB-DIF □ TOXIN B, CLOSTRIDIUM DIFFICILE

TOXICITY DATA WITH REFERENCE
scu-mus LD:>5 μg/kg TOXIA6 29,877,91
ivn-mus LD50:>200 mg/kg TOXIA6 29,877,91

SAFETY PROFILE: A poison by subcutaneous and intravenous routes. When heated to decomposition it emits acrid smoke and irritating vapors.

CMY130 **HR: 3**
CLOSTRIDIUM NOVYI α-TOXIN

SYNS: α-TOXIN, CLOSTRIDIUM NOVYI □ TOX-α-NOV

TOXICITY DATA WITH REFERENCE
scu-mus LD50:300 ng/kg TOXIA6 29,877,91
ivn-mus LD50:200 ng/kg TOXIA6 29,877,91

SAFETY PROFILE: A poison by subcutaneous and intravenous routes. When heated to decomposition it emits acrid smoke and irritating vapors.

CMY150 **HR: 3**
CLOSTRIDIUM OEDEMATIENS TYPE A TOXIN

SYN: TOXIN, CLOSTRIDIUM OEDEMATIENS, TYPE A

TOXICITY DATA WITH REFERENCE
ims-mus LD50:100 μg/kg JJMCAQ 36,67,83

SAFETY PROFILE: A poison by intramuscular route. When heated to decomposition it emits acrid smoke and irritating vapors.

CMY170 **HR: 1**
CLOSTRIDIUM PERFRINGENS EXOTOXIN

SYN: EXOTOXIN, CLOSTRIDIUM PERFRINGENS

TOXICITY DATA WITH REFERENCE
ivn-mus LDLo:25 g/kg PSEBAA 112,463,63

SAFETY PROFILE: Low toxicity by intravenous route. When heated to decomposition it emits acrid smoke and irritating vapors.

CMY190 **HR: 3**
CLOSTRIDIUM PERFRINGENS β-TOXIN

SYN: β-TOXIN, CLOSTRIDIUM PERFRINGENS

TOXICITY DATA WITH REFERENCE
ipr-mus LD50:4500 ng/kg TOXIA6 25,1301,87
ivn-mus LD50:310 ng/kg TOXIA6 25,1301,87

SAFETY PROFILE: A poison by intraperitoneal and intravenous routes. When heated to decomposition it emits acrid smoke and irritating vapors.

CMY220 **HR: 3**
CLOSTRIDIUM PERFRINGENS (Welchii) TYPE A ENTEROTOXIN

SYN: ENTEROTOXIN, CLOSTRIDIUM PERFRINGENS, TYPE A

TOXICITY DATA WITH REFERENCE
ivn-rat LDLo:50 μg/kg TOXIA6 29,751,91
ipr-mus LD50:39 μg/kg EJBCAI 149,287,85
ivn-mus LD50:81,160 ng/kg INFIBR 12,440,75
ivn-rbt LDLo:14,560 ng/kg JCVPAR 83,265,73

SAFETY PROFILE: A poison by intravenous route. When heated to decomposition it emits acrid smoke and irritating vapors.

CMY240 **HR: 3**
CLOSTRIDIUM SORDELLII TOXIN

SYNS: LT-SOR ☐ TOXIN, CLOSTRIDIUM SORDELLII

TOXICITY DATA WITH REFERENCE
scu-mus LD50:300 ng/kg TOXIA6 29,877,91
ivn-mus LD50:50 ng/kg TOXIA6 29,877,91

SAFETY PROFILE: A poison by subcutaneous and intravenous routes. When heated to decomposition it emits acrid smoke and irritating vapors.

CMY260 **HR: 3**
CLOSTRIDIUM TETANI TOXIN

TOXICITY DATA WITH REFERENCE
scu-mus LD50:100 pg/kg NATWAY 56,615,69

SAFETY PROFILE: A poison by subcutaneous route. When heated to decomposition it emits acrid smoke and irritating vapors.

CMY280 **HR: 3**
CLOSTRIDIUM TETANI, TYPE BE TOXIN

PROP: Bichainal extracellular toxin separated from Cl. TETANI NSAPCC 338,99,88

TOXICITY DATA WITH REFERENCE
scu-mus LD50:2 ng/kg NSAPCC 338,99,88

SAFETY PROFILE: A deadly poison by subcutaneous route. When heated to decomposition it emits acrid smoke and irritating vapors.

CMY300 **HR: 3**
CLOSTRIDIUM TETANI, TYPE S TOXIN

PROP: Single-chain intracellular toxin separated from Cl. TETANI NSAPCC 338,99,88

TOXICITY DATA WITH REFERENCE
scu-mus LD50:4 ng/kg NSAPCC 338,99,88

SAFETY PROFILE: A deadly poison by subcutaneous route. When heated to decomposition it emits acrid smoke and irritating vapors.

CMY475 **CAS:8000-34-8** **HR: 2**
CLOVE BUD OIL

SYNS: NELKEN OEL (GERMAN) ☐ OIL of CLOVE ☐ OILS, CLOVE

TOXICITY DATA WITH REFERENCE
dnr-bcs 30 μL/disc TOFOD5 8,91,85
orl-rat LD50:2650 mg/kg FCTXAV 13,761,75
skn-rbt LD50:5000 mg/kg FCTXAV 13,761,75

CONSENSUS REPORTS: EPA Genetic Toxicology Program. Reported in EPA TSCA Inventory.

SAFETY PROFILE: Moderately toxic by ingestion. Mildly toxic by skin contact. Mutation data reported. When heated to decomposition it emits acrid smoke and fumes.

CMY500 **CAS:8015-97-2** **HR: 2**
CLOVE LEAF OIL MADAGASCAR

PROP: From steam distillation of leaves of *Eugenis caryophyllata* Thunberg (*Eugenia aromatica* L. Baill.) (Fam. Myrtaceae). Pale-yellow liquid. Ref. index: 1.527–1.538 @ 20°. Sol in propylene glycol, fixed oils; insol in glycerin, mineral oil.

SYNS: CLOVE LEAF OIL ☐ OILS, CLOVE LEAF

TOXICITY DATA WITH REFERENCE
skn-mus 100% FCTXAV 16,695,78
skn-rbt 500 mg/24H SEV FCTXAV 16,695,78
skn-pig 100% FCTXAV 16,695,78
orl-rat LD50:1370 mg/kg FCTXAV 16,695,78
skn-rbt LD50:1200 mg/kg FCTXAV 16,695,78

CONSENSUS REPORTS: Reported in EPA TSCA Inventory.

SAFETY PROFILE: Moderately toxic by ingestion and skin contact. A severe skin irritant. When heated to decomposition it emits acrid smoke and fumes.

CMY510 **CAS:8000-34-8** **HR: 2**
CLOVE OIL, stem

SYN: OILS, CLOVE STEM

TOXICITY DATA WITH REFERENCE
orl-rat LD50:2020 mg/kg FCTXAV 13,765,75

CONSENSUS REPORTS: Reported in EPA TSCA Inventory.

SAFETY PROFILE: Moderately toxic by ingestion. When heated to decomposition it emits acrid smoke and irritating vapors.

CMY525 **CAS:24166-13-0** **HR: 2**
CLOXAZOLAZEPAM
mf: $C_{17}H_{14}Cl_2N_2O_2$ mw: 349.23

PROP: Crystals. Mp: 200° (decomp). Freely sol in glacial acetic acid; sparingly sol in chloroform; sltly sol in acetone, dehydrated ethanol, ethyl acetate, and benzene. Practically insol in water.

SYNS: CLOXAZOLAM ☐ CS 370 ☐ ENADEL ☐ LUBALIX ☐ MT 14-411 ☐ OLCADIL ☐ SEPAZON ☐ TOLESTAN

TOXICITY DATA WITH REFERENCE
orl-mus TDLo:1800 mg/kg (7-12D preg):TER SKKNAJ 23,180,71
orl-rat LD50:1780 mg/kg IYKEDH 5,106,74
orl-mus LD50:2630 mg/kg IYKEDH 5,106,74

SAFETY PROFILE: Moderately toxic by ingestion. An experimental teratogen. When heated to decomposition it emits toxic fumes of Cl^- and NO_x. A minor tranquilizer. See also DIAZEPAM.

CMY535 **CAS:15311-77-0** **HR: 3**
CLOXYPENDYL
mf: $C_{20}H_{25}ClN_4OS{\cdot}2ClH$ mw: 477.92

SYNS: 2-(4-(3-(3-CHLORO-10H-PYRIDO(3,2-b)-1,4-BENZOTHIAZINE-10-YL)PROPYL))-1-PIPERAZINYLETHANOL ☐ 4-(3-(3-CHLORO-10H-PYRIDO(3,2-b)(1,4)-BENZOTHIAZIN-10-YL)PROPYL)-1-PIPERAZINE ETHANOL ☐ 2-(4-(3-(3-CHLORO-10H-PYRIDO(3,2-b)(1,4)BENZOTHIAZIN-1-OYL)PROPYL)-1-PIPERAZINYL) ETHANOL

TOXICITY DATA WITH REFERENCE
orl-rat LD50:657 mg/kg ARZNAD 18,435,68
ipr-rat LD50:280 mg/kg ARZNAD 18,435,68
orl-mus LD50:610 mg/kg ARZNAD 18,435,68
ipr-mus LD50:166 mg/kg ARZNAD 18,435,68
scu-mus LD50:683 mg/kg ARZNAD 18,435,68
ivn-mus LD50:91 mg/kg ARZNAD 18,435,68

SAFETY PROFILE: Poison by intravenous and intraperitoneal routes. Moderately toxic by ingestion and subcutaneous routes. When heated to decomposition it emits toxic fumes of NO_x, SO_x, and HCl.

CMY650 **CAS:5786-21-0** **HR: 3**
CLOZAPINE
mf: $C_{18}H_{19}ClN_4$ mw: 326.86

PROP: Yellow crystals from Me_2CO/pet ether. Mp: 183–184°.

SYNS: 8-CHLORO-11-(4-METHYL-1-PIPERAZINYL)-5H-DIBENZO(b,e)(1,4)DIAZEPINE ☐ CLOZAPIN ☐ HF-1854 ☐ IPROX ☐ LEPONEX ☐ LEPOTEX ☐ W-801

TOXICITY DATA WITH REFERENCE
sln dmg-orl 2 mg/2D SOGEBZ 11,718,75
cyt-hmn:lym 10 mg/L HUGEDQ 38,77,77
orl-rat TDLo:1080 mg/kg (16-22D preg/20D post):REP FRPPAO 26,585,71
orl-rat TDLo:96 mg/kg (female 9-14D post):REP KSRNAM 7,696,73
orl-mus TDLo:48 mg/kg (female 7-12D post):TER KSRNAM 7,696,73
orl-rat TDLo:24 mg/kg (9-14D preg):REP KSRNAM 7,696,73
orl-rat TDLo:24 mg/kg (9-14D preg):TER KSRNAM 7,696,73
orl-hmn TDLo:6428 µg/kg/3D-I:CNS ARZNAD 22,919,72
orl-wmn TDLo:52 mg/kg/13D-I:GIT,SYS BMJOAE 305,810,92
orl-hmn TDLo:5 mg/kg/7D-I:CVS ARZNAD 22,919,72
orl-man TDLo:4286 µg/kg/11D-I:GIT JCPYDR 13,155,93
orl-rat LD50:251 mg/kg KSRNAM 7,667,73
scu-rat LD50:240 mg/kg KSRNAM 7,667,73
ivn-rat LD50:41,600 µg/kg KSRNAM 7,667,73
ims-rat LD50:210 mg/kg FRPPAO 26,585,71
orl-mus LD50:150 mg/kg JMCMAR 23,878,80
ipr-mus LD50:98 mg/kg ARZNAD 32,668,82
scu-mus LD50:194 mg/kg KSRNAM 7,667,73
ivn-mus LD50:36,500 µg/kg KSRNAM 7,667,73
orl-dog LD50:145 mg/kg FRPPAO 26,585,71
orl-gpg LD50:510 mg/kg FRPPAO 26,585,71

SAFETY PROFILE: Poison by ingestion, subcutaneous, intramuscular, intravenous, and intraperitoneal routes. Human systemic effects by ingestion: body temperature increase, body temperature increase, changes in blood cell count, changes in kidney tubules, diarrhea, hallucinations or distorted perceptions, hypermotility, pulse rate increase, somnolence. An experimental teratogen. Other experimental reproductive effects. Human mutation data reported. Used as a sedative. When heated to decomposition it emits toxic fumes of Cl^- and NO_x. See also DIAZEPAM.

CMY725 **CAS:9001-13-2** **HR: 1**
COAGULASE

SYNS: HEMOCOAGULASE ☐ HEPTOCOAGULASE ☐ PLASMA COAGULASE ☐ PLASMOCOAGULASE ☐ REPTILASE S ☐ RP-093 ☐ STAPHYLOCOAGULASE ☐ THROMBIN COAGULASE

TOXICITY DATA WITH REFERENCE
scu-rat TDLo:24,000 units/kg (9-14D preg):TER KSRNAM 9,2337,75
scu-rat TDLo:24,000 units/kg (9-14D preg):REP KSRNAM 9,2337,75
ivn-rat LD50:16,0200 units/kg KSRNAM 9,2304,75
scu-mus LD50:107 g/kg TOIZAG 15,383,68
ivn-mus LD50:53,700 mg/kg TOIZAG 15,383,68

SAFETY PROFILE: An experimental teratogen. Other experimental reproductive effects.

CMY750 **HR: 3**
COAL CONVERSION MATERIALS, SRC-II HEAVY DISTILLATE

SYN: SRC-II HEAVY DISTILLATE

TOXICITY DATA WITH REFERENCE
skn-mus TDLo:2870 mg/kg/2Y-I:CAR NTIS** CONF-801143
skn-mus TD:285 g/kg/2Y-I:CAR NTIS** CONF-801143
skn-mus TD:28,704 mg/kg/2Y-I:CAR NTIS** CONF-801143

SAFETY PROFILE: Suspected carcinogen with experimental carcinogenic data by skin contact.

CMY760 **HR: 3**
COAL DUST

PROP: Black powder or dust.

SYNS: ANTHRACITE PARTICLES ☐ COAL FACINGS ☐ COAL, GROUND BITUMINOUS (DOT) ☐ COAL-MILLED ☐ COAL SLAG-MILLED ☐ SEA COAL

TOXICITY DATA WITH REFERENCE
ihl-rat TCLo:6600 µg/m³/6H/86W-I:ETA AIHAAP 42,382,81
ihl-rat TC:14,900 µg/m³/6H/86W-I:ETA AIHAAP 42,382,81

OSHA PEL: Respirable Quartz Fraction less than 5% SiO_2: TWA 2 mg/m³; Respirable Quartz Fraction greater than or equal to 5% SiO_2: 0.1 mg/m³
ACGIH TLV: TWA 2 mg/m³ (fraction $<$ 5% quartz); TWA 0.1 mg/m³ (fraction $>$ 5% quartz)

SAFETY PROFILE: Questionable carcinogen with experimental tumorigenic data. Variable toxicity depending upon SiO_2 content. See also SILICA. Moderately flammable when exposed to heat, flame, or chemical reaction with oxidizers. Slightly explosive when exposed to flame.

CMY800 CAS:8007-45-2 HR: 3
COAL TAR

SYNS: CARBO-CORT □ COAL TAR, AEROSOL □ COAL TAR SOLUTION USP □ CRUDE COAL TAR □ ESTAR □ IMPERVOTAR □ LAV □ LAVATAR □ PICIS CARBONIS □ PIXALBOL □ PIX CARBONIS □ PIX LITHANTHRACIS □ POLYTAR BATH □ SUPERTAH □ SYNTAR □ TAR □ TAR, COAL □ ZETAR

TOXICITY DATA WITH REFERENCE
skn-hmn 15 μg/3D-I MLD 85DKA8 -,127,77
skn-rbt 5%/3H MLD SCPHA4 43,11,75
mmo-sat 5 μg/plate NTIS** PB84-138973
mma-esc 50 μg/plate NTIS** PB84-138973
orl-mus TDLo:12 g/kg/30W-C:ETA AJCAA7 26,552,36
ihl-mus TCLo:22 g/m^3/55W-I:CAR JJIND8 39,175,67
skn-mus TDLo:64 g/kg/36W-I:CAR AMIHBC 4,299,51
skn-mus TD:8400 mg/kg/64W-I:ETA ADMFAU 242,176,72

CONSENSUS REPORTS: NTP 7th Annual Report on Carcinogens. IARC Cancer Review: Group 1 IMEMDT 7,175,87; Animal Sufficient Evidence IMEMDT 34,65,84; IMEMDT 35,83,85; IMEMDT 3,22,73; Human Sufficient Evidence IMEMDT 34,65,84; IMEMDT 3,22,73; Human Limited Evidence IMEMDT 35,83,85. Reported in EPA TSCA Inventory.

OSHA PEL: TWA 0.2 mg/m^3; Carcinogen
DFG MAK: Human Carcinogen.
NIOSH REL: (Coal Tar Products) TWA 0.1 mg/m^3
DOT CLASSIFICATION: 3; *Label:* Flammable Liquid

SAFETY PROFILE: Confirmed human carcinogen with experimental carcinogenic and tumorigenic data. Mutation data reported. A human and experimental skin irritant. A flammable liquid. When heated to decomposition it emits acrid smoke and irritating fumes.

For occupational chemical analysis use NIOSH: Coal Tar Pitch Volatiles, 5023.

CMY825 CAS:8001-58-9 HR: 3
COAL TAR CREOSOTE

SYNS: AWPA #1 □ BRICK OIL □ COAL TAR OIL □ COAL TAR OIL (DOT) □ CREOSOTE □ CREOSOTE, from COAL TAR □ CREOSOTE OIL □ CREOSOTE P1 □ CREOSOTUM □ CRESYLIC CREOSOTE □ HEAVY OIL □ LIQUID PITCH OIL □ NAPHTHALENE OIL □ PRESERV-O-SOTE □ RCRA WASTE NUMBER U051 □ TAR OIL □ WASH OIL

TOXICITY DATA WITH REFERENCE
mma-sat 20 μg/plate MUREAV 119,21,83
bfa-rat/sat 250 mg/kg IAPUDO 59,279,84
orl-rat TDLo:52,416 mg/kg (female 91D pre):REP OYYAA2 21,899,81
orl-rat TDLo:14 g/kg (91D male):REP OYYAA2 21,899,81
skn-mus TDLo:99 g/kg/33W-I:CAR FAATDF 7,228,86
orl-rat LD50:725 mg/kg TXAPA9 6,378,64
orl-mus LD50:433 mg/kg OYYAA2 21,899,81

CONSENSUS REPORTS: NTP 7th Annual Report on Carcinogens. IARC Cancer Review: Group 2A IMEMDT 7,177,87; Animal Sufficient Evidence, Human Limited Evidence IMEMDT 35,83,85; Animal Sufficient Evicence IMEMDT 3,22,73. Reported in EPA TSCA Inventory.

NIOSH REL: (Coal Tar Products) TWA 0.1 mg/m^3 CHE fraction
DOT CLASSIFICATION: 3; *Label:* Flammable Liquid

SAFETY PROFILE: Confirmed carcinogen with experimental carcinogenic data. Poison by ingestion. Experimental reproductive effects. Mutation data reported. A flammable liquid. When heated to decomposition it emits acrid smoke and fumes.

CMY900 CAS:65996-92-1 HR: 3
COAL TAR DISTILLATE
DOT: UN 1136

SYNS: COAL TAR DISTILLATES, flammable (DOT) □ COAL TAR DISTILLATES □ DISTILLATES (COAL TAR)

CONSENSUS REPORTS: Reported in NTP 7th Annual Report on Carcinogens, 1992. Reported in EPA TSCA Inventory.

DOT CLASSIFICATION: 3; *Label:* Flammable Liquid

SAFETY PROFILE: A carcinogen. A flammable liquid. When heated to decomposition it emits acrid smoke and irritating vapors.

CMZ100 CAS:65996-93-2 HR: 3
COAL TAR PITCH VOLATILES

SYNS: PITCH □ PITCH, COAL TAR

TOXICITY DATA WITH REFERENCE
skn-mus TDLo:36 g/kg/18W-I:CAR AJIMD8 2,59,81
skn-mus TD:4200 mg/kg/31W-I:NEO TXAPA9 18,41,71
skn-mus TD:82 g/kg/52W-I:CAR HYSAAV 33(5),180,68

CONSENSUS REPORTS: IARC Cancer Review: Group 1 IMEMDT 7,174,87; Animal Sufficient Evidence, Human Sufficient Evidence IMEMDT 35,83,85; Human Sufficient Evidence IMEMDT 3,22,73. Reported in EPA TSCA Inventory.

OSHA PEL: TWA 0.2 mg/m^3; Carcinogen
ACGIH TLV: TWA 0.2 mg/m^3 (volatile), Confirmed Human Carcinogen
NIOSH REL: (Coal Tar Products) TWA 0.1 mg/m^3 CHE fraction
DOT CLASSIFICATION: 3; *Label:* Flammable Liquid

SAFETY PROFILE: Confirmed carcinogen with experimental carcinogenic and neoplastigenic data by skin contact. When heated to decomposition it emits acrid smoke and fumes.

For occupational chemical analysis use OSHA: #ID-58 or NIOSH: Coal Tar Pitch Volatiles 5023.

CNA250 CAS:7440-48-4 HR: 3
COBALT
mf: Co mw: 58.93

PROP: Gray, hard, magnetic, lustrous, ductile, somewhat malleable, silvery-blue metal. Mp: 1495°, bp: 28°

@ 3100 mm. Exists in two allotropic forms. At room temperature, the hexagonal form is more stable than the cubic form; both forms can exist at room temperature. Stable in air or toward water at ordinary temperatures. D 8.92, mp 1493°, bp about 3100°, Brinell hardness: 125, latent heat of fusion 62 cal/g, latent heat of vaporization 1500 cal/g, specific heat (15–100°): 0.1056 cal/g/°C. Readily sol in dil HNO_3; very slowly attacked by HCl or cold H_2SO_4. The hydrated salts of cobalt are red, and the sol salts form red solns that become blue on adding conc HCl.

SYNS: AQUACAT □ C.I. 77320 □ COBALT-59 □ KOBALT (GERMAN, POLISH) □ NCI-C60311 □ SUPER COBALT

TOXICITY DATA WITH REFERENCE
ims-rat TDLo:126 mg/kg:NEO NATUAS 173,822,54
imp-rbt TDLo:75 mg/kg:ETA ZEKBAI 52,425,42
ims-rat TD:126 mg/kg:NEO BJCAAI 10,668,56
orl-rat LD50:6171 mg/kg JACTDZ 1,686,92
ivn-rat LDLo:100 mg/kg EQSFAP 1,1,75
itr-rat LDLo:25 mg/kg NTIS** AEC-TR-6710
ipr-mus LDLo:100 mg/kg EQSFAP 1,1,75
orl-rbt LDLo:750 mg/kg AIPTAK 62,347,39
ivn-rbt LDLo:100 mg/kg EQSSDX 1,1,75

CONSENSUS REPORTS: Reported in EPA TSCA Inventory. Cobalt and its compounds are on the Community Right-To-Know List.

OSHA PEL: TWA 0.05 mg/m^3
ACGIH TLV: (metal, dust, and fume)TWA 0.02 mg(Co)/m^3; Animal Carcinogen
DFG TRK:0.5 mg/m^3 calculated as cobalt in that portion of dust that can possibly be inhaled in the production of cobalt powder and catalysts; hard metal (tungsten carbide) and magnet production (processing of powder, machine pressing, and mechanical processing of unsintered articles); other cobalt alloys and compounds: 0.1 mg/m^3 calculated as cobalt in that portion of dust that can possibly be inhaled. Animal Carcinogen, Suspected Human Carcinogen.
NIOSH REL: (Cobalt): Insufficient evidence for recommending limit

SAFETY PROFILE: Confirmed carcinogen with experimental neoplastigenic and tumorigenic data. Poison by intravenous, intratracheal, and intraperitoneal routes. Moderately toxic by ingestion. Inhalation of the dust may cause pulmonary damage. The powder may cause dermatitis. Ingestion of soluble salts produces nausea and vomiting by local irritation. Powdered cobalt ignites spontaneously in air. Flammable when exposed to heat or flame. Explosive reaction with hydrazinium nitrate, ammonium nitrate + heat, and 1,3,4,7-tetramethylisoindole (at 390°C). Ignites on contact with bromine pentafluoride. Incandescent reaction with acetylene or nitryl fluoride. See also COBALT COMPOUNDS.

For occupational chemical analysis use OSHA: #ID-125G or NIOSH: Cobalt, 7027; Elements, 7300.

CNA300 CAS:917-69-1 HR: 2
COBALT(III) ACETATE
mf: $C_6H_{12}O_6 \cdot Co$ mw: 239.11

SYNS: ACETIC ACID, COBALT(3+) SALT □ COBALT ACETATE □ COBALT(3+) ACETATE □ COBALTIC ACETATE □ COBALT TRIACETATE

CONSENSUS REPORTS: IARC Cancer Review: Group 2B IMEMDT 52,363,91; Animal Inadequate Evidence IMEMDT 52,363,91; Human Inadequate Evidence IMEMDT 52,363,91. Reported in EPA TSCA Inventory.

SAFETY PROFILE: A questionable carcinogen. When heated to decomposition it emits toxic vapors of Co.

CNA500 CAS:6147-53-1 HR: 3
COBALT ACETATE TETRAHYDRATE
mf: $C_4H_6O_4 \cdot Co \cdot 4H_2O$ mw: 249.11

SYNS: ACETIC ACID, COBALT(2+) SALT, TETRAHYDRATE □ COBALT DIACETATE TETRAHYDRATE □ COBALTOUS ACETATE TETRAHYDRATE □ OCTAN KOBALTNATY (CZECH)

TOXICITY DATA WITH REFERENCE
skn-rbt 500 mg/24H MOD 28ZPAK -,21,72
eye-rbt 500 mg/24H MLD 28ZPAK -,21,72
cyt-hmn:lym 600 µg/L CYGEDX 12(3),46,78
orl-rat LD50:708 mg/kg FCTOD7 20,311,82

CONSENSUS REPORTS: Cobalt and its compounds are on the Community Right-To-Know List. IARC Cancer Review: Group 2B IMEMDT 52,363,91; Human Inadequate Evidence IMEMDT 52,363,91.

SAFETY PROFILE: Moderately toxic by ingestion. Questionable carcinogen. A skin and eye irritant. Human mutation data reported. See also COBALT COMPOUNDS. When heated to decomposition it emits acrid smoke and irritating fumes.

CNA750 CAS:11114-92-4 HR: 3
COBALT ALLOY, Co,Cr

SYNS: COBALT-CHROMIUM ALLOY □ DIN 2.4602 □ DIN 2.4964 □ HASTELLOY C □ HAYNES 25 □ HAYNES ALLOY NUMBER 25 □ HAYNES STELLITE 21 □ HEV-4 □ HS 21 □ HS 25 □ Kh15N55M16 □ Kh15N55M16V □ S 816 □ STELLITE 8 □ STELLITE 21 □ STELLITE 23 □ STELLITE 25 □ STELLITE 27 □ STELLITE 30 □ STELLITE 31 □ STELLITE 36 □ STELLITE 8A □ STELLITE C □ VITALLIUM □ ZIMALLOY

TOXICITY DATA WITH REFERENCE
ims-rat TDLo:140 mg/kg:ETA LANCAO 1,564,71

CONSENSUS REPORTS: NTP 7th Annual Report on Carcinogens. IARC Cancer Review: Group 2B IMEMDT 52,363,91; Animal Limited Evidence IMEMDT 23,205,80; Animal Limited Evidence IMEMDT 52,363,91; Human Inadequate Evidence IMEMDT 52,363,91. Cobalt and its compounds, as well as chromium and its compounds, are on the Community Right-To-Know List.

OSHA PEL: TWA 1 mg(Cr)/m^3; 0.1 mg(Co)/m^3 (fume and dust)
ACGIH TLV: TWA 0.5 mg(Cr)/m^3; Not Classifiable as a Carcinogen.
NIOSH REL: (Cobalt) Insufficient evidence for recommending limit

SAFETY PROFILE: Confirmed carcinogen with experimental tumorigenic data. Violent reaction with molten

Li. See also COBALT COMPOUNDS and CHROMIUM COMPOUNDS.

CNB000 **HR: 3**
COBALT(III) AMIDE
mf: CoH_6N_3 mw: 107.00

$Co(NH_2)_3$

CONSENSUS REPORTS: Cobalt and its compounds are on the Community Right-To-Know List.

SAFETY PROFILE: Powdered material will spontaneously explode in air. When heated it converts to cobalt(III) nitride that ignites spontaneously in air. When heated to decomposition it emits toxic fumes of NO_x. See also COBALT COMPOUNDS and AMIDES.

CNB099 **HR: 3**
COBALT(II) AZIDE
mf: CoN_6 mw: 142.97

$Co(N_3)_2$

CONSENSUS REPORTS: Cobalt and its compounds are on the Community Right-To-Know List.

SAFETY PROFILE: Explodes when heated to 200°C. When heated to decomposition it emits toxic fumes of NO_x. See also COBALT COMPOUNDS and AZIDES.

CNB250 **CAS:7789-43-7** **HR: 3**
COBALT(II) BROMIDE
mf: Br_2Co mw: 218.75

PROP: Green crystals. Mp: 678°. Sol in H_2O, Me_2CO, and MeOH.

SYNS: COBALT DIBROMIDE □ COBALTOUS BROMIDE

TOXICITY DATA WITH REFERENCE
orl-rat LD50:406 mg/kg FCTOD7 20,311,82

CONSENSUS REPORTS: Reported in EPA TSCA Inventory. Cobalt and its compounds are on the Community Right-To-Know List.

SAFETY PROFILE: A poison by ingestion. Exothermic reaction when heated with sodium. When heated to decomposition it emits toxic fumes of Br^-. See also COBALT COMPOUNDS and BROMIDES.

CNB475 **CAS:513-79-1** **HR: 2**
COBALT(2+) CARBONATE
mf: $CO_3{\cdot}Co$ mw: 118.94

SYNS: CARBONIC ACID, COBALT(2+) SALT (1:1) □ C.I. 77353 □ COBALT CARBONATE □ COBALT CARBONATE (1:1) □ COBALT MONOCARBONATE □ COBALTOUS CARBONATE

TOXICITY DATA WITH REFERENCE
orl-rat LD50:640 mg/kg SchF## 16MAY86

CONSENSUS REPORTS: IARC Cancer Review: Group 2B; Human Inadequate Evidence IMEMDT 52,363,91. Reported in EPA TSCA Inventory.

SAFETY PROFILE: Moderately toxic by ingestion. A questionable carcinogen. When heated to decomposition it emits toxic fumes of Co.

CNB495 **CAS:12602-23-2** **HR: 2**
COBALT CARBONATE HYDROXIDE
mf: $C_2H_6Co_5O_{12}$ mw: 516.73

SYNS: COBALT, BIS(CARBONATO(2-))HEXAHYDROXYPENTA- □ COBALT CARBONATE, COBALT DIHYDROXIDE (2:3)

TOXICITY DATA WITH REFERENCE
dnr-bcs 50 mmol/L MUREAV 77,109,80

CONSENSUS REPORTS: IARC Cancer Review: Group 2B IMEMDT 52,363,91; Human Inadequate Evidence IMEMDT 52,363,91. Reported in EPA TSCA Inventory.

SAFETY PROFILE: Questionable carcinogen. Mutation data reported. When heated to decomposition it emits toxic vapors of Co.

CNB500 **CAS:10210-68-1** **HR: 3**
COBALT CARBONYL
mf: $C_8Co_2O_8$ mw: 341.94

$(OC)_3Co{:}(CO)_2{:}Co(CO)_3$

PROP: Air-sensitive orange-red platelets or crystals. D: 1.87, mp: 51°, decomp above 52°. Decomp on exposure to air. Insol in water; sol in organic solvents.

SYNS: COBALT OCTACARBONYL □ COBALT TETRACARBONYL □ COBALT TETRACARBONYL DIMER □ DI-mu-CARBONYLHEXACARBONYLDICOBALT □ DICOBALT CARBONYL □ DICOBALT OCTACARBONYL □ OCTACARBONYLDICOBALT

TOXICITY DATA WITH REFERENCE
ihl-rat LC50:165 mg/m³ VRDEA5 (2),101,69
ipr-rat LD50:754 mg/kg GISAAA 38(1),97,73
ihl-mus LC50:26,900 μg/m³/2H VRDEA5 (2),101,69
ipr-mus LD50:378 mg/kg GISAAA 38(1),97,73

CONSENSUS REPORTS: IARC Cancer Review: Group 2B IMEMDT 52,363,91; Human Inadequate Evidence IMEMDT 52,363,91. Reported in EPA TSCA Inventory. Cobalt and its compounds are on the Community Right-To-Know List. EPA Extremely Hazardous Substances List.

OSHA PEL: TWA 0.1 mg(Co)/m³
ACGIH TLV: TWA 0.1 mg(Co)/m³

SAFETY PROFILE: Poison by inhalation and intraperitoneal routes. Questionable carcinogen. Decomposes in air to form a product that ignites spontaneously in air. When heated to decomposition it emits acrid smoke and fumes. See also CARBONYLS and COBALT COMPOUNDS.

For occupational chemical analysis use OSHA: #ID-125g.

CNB599 **CAS:7646-79-9** **HR: 3**
COBALT(II) CHLORIDE
mf: Cl_2Co mw: 129.83

PROP: Pale blue hygroscopic powder or crystals. Mp: 735°, bp: 1049°, d: 3.348. Sol in H_2O and polar org solvs.

SYNS: COBALT DICHLORIDE ☐ COBALT MURIATE ☐ COBALTOUS CHLORIDE ☐ COBALTOUS DICHLORIDE ☐ KOBALT CHLORID (GERMAN)

TOXICITY DATA WITH REFERENCE
mrc-bcs 50 mmol/L MUREAV 77,109,80
mmo-smc 100 mmol/L CPBTAL 33,1571,85
mrc-smc:3 g/L MUREAV 155,117,85
dni-hmn:hla 1 mmol/L MUREAV 92,427,82
dns-ham:emb 200 µmol/L MUREAV 131,173,84
msc-ham:lng 200 µmol/L MUREAV 68,259,79
ipr-rat TDLo:30 g/kg (15-16D preg):TER TJADAB 29(3),23A,84
ipr-rat TDLo:30 g/kg (15-16D preg):REP TJADAB 29(3),23A,84
orl-rat TDLo:11 mg/kg (1-22D preg):REP GISAAA 45(2)6,80
ipr-mus TDLo:25 mg/kg (10D preg):TER JPMSAE 58,766,69
scu-rat TDLo:400 mg/kg:CAR LBANAX 11,43,77
orl-chd TDLo:48 mg/kg:CNS,END,MET JAMAAP 157,117,55
orl-chd LDLo:1500 mg/kg 34ZIAG -,182,69
orl-rat LD50:80 mg/kg HYSAAV 36,277,71
ivn-rat LD50:20 mg/kg AIPTAK 143,219,63
orl-mus LD50:80 mg/kg HYSAAV 36,277,71
ipr-mus LD50:49 mg/kg AEPPAE 244,17,62
scu-mus LDLo:100 mg/kg 27ZWAY 3.2,1444,-
ivn-dog LDLo:36 mg/kg HBAMAK 4,1289,35
orl-rbt LDLo:1272 mg/kg SMSJAR 26,131,1826
scu-rbt LDLo:200 mg/kg HBAMAK 4,1289,35
orl-gpg LD50:55 mg/kg HYSAAV 36,277,71
skn-gpg LDLo:165 mg/kg AEHLAU 11,201,65
ipr-gpg LDLo:165 mg/kg AEHLAU 11,201,65

CONSENSUS REPORTS: IARC Cancer Review: Group 2B IMEMDT 52,363,91; Animal Limited Evidence IMEMDT 52,363,91; Human Inadequate Evidence IMEMDT 52,363,91. Reported in EPA TSCA Inventory. EPA Genetic Toxicology Program. Cobalt and its compounds are on the Community Right-To-Know List.

SAFETY PROFILE: Suspected carcinogen with experimental carcinogenic data. Poison experimentally by ingestion, skin contact, intraperitoneal, intravenous, and subcutaneous routes. Moderately toxic to humans by ingestion. Human systemic effects by ingestion: anorexia, goiter (increased thyroid size), and weight loss. Experimental teratogenic and reproductive effects. Human mutation data reported. Incompatible with metals (e.g., sodium and potassium). See also COBALT. When heated to decomposition it emits toxic fumes of Cl^-.

CNB750 HR: 3
COBALT(III) CHLORIDE
mf: Cl_3Co mw: 165.29

CONSENSUS REPORTS: Cobalt and its compounds are on the Community Right-To-Know List.

SAFETY PROFILE: Ignites on contact with lithium. Incompatible with pentacarbonyl iron and zinc. When heated to decomposition it emits toxic fumes of Cl^-. See also COBALT COMPOUNDS and CHLORIDES.

CNB800 CAS:7791-13-1 HR: 3
COBALT(II) CHLORIDE HEXAHYDRATE
mf: $Cl_2Co \cdot 6H_2O$ mw: 237.95

PROP: Red monoclinic crystals. Mp: 86°. Sol in H_2O and Me_2CO.

SYNS: COBALT CHLORIDE, HEXAHYDRATE (8CI, 9CI) ☐ COBALT (2+) CHLORIDE HEXAHYDRATE ☐ COBALT DICHLORIDE HEXAHYDRATE ☐ COBALTOUS CHLORIDE, HEXAHYDRATE

TOXICITY DATA WITH REFERENCE
ivn-mus TDLo:47,590 mg/kg (8D preg):TER ENVRAL 33,47,84
orl-rat LD50:766 mg/kg FCTOD7 20,311,82
ipr-rat LD50:35 mg/kg JAPYAA 32,315,72
scu-rat LDLo:121 mg/kg EQSSDX 1,1,75
ipr-mus LD50:90 mg/kg AEPPAE 244,17,62
scu-mus LDLo:100 mg/kg EQSSDX 1,1,75
ivn-dog LDLo:30,300 µg/kg EQSSDX 1,1,75
scu-rbt LDLo:200 mg/kg EQSSDX 1,1,75
ivn-rbt LDLo:25,400 µg/kg EQSSDX 1,1,75

CONSENSUS REPORTS: Cobalt and its compounds are on the Community Right-To-Know List.

SAFETY PROFILE: Poison by subcutaneous, intravenous, and intraperitoneal routes. Moderately toxic by ingestion. An experimental teratogen. Experimental reproductive effects. When heated to decomposition it emits toxic fumes of Cl^-. See also COBALT and CHLORIDES.

CNB850 HR: 3
COBALT COMPOUNDS

CONSENSUS REPORTS: Cobalt and its compounds are on the Community Right-To-Know List.

DFG TRK: 0.5 mg/m^3 calculated as cobalt in that portion of dust that can possibly be inhaled in the production of cobalt powder and catalysts; hard metal (tungsten carbide) and magnet production (processing of powder, machine pressing, and mechanical processing of unsintered articles); other cobalt alloys and compounds: 0.1 mg/m^3 calculated as cobalt in that portion of dust that can possibly be inhaled. Animal Carcinogen, Suspected Human Carcinogen.

SAFETY PROFILE: Confirmed carcinogen with experimental neoplastigenic and tumorigenic data. Cobalt has a low toxicity by ingestion. Ingestion of soluble salts produces nausea and vomiting by local irritation. In animals, administration of cobalt salts produces an increase in the total red cell mass of the blood. In humans, a single case of poisoning with liver and kidney damage has been attributed to cobalt. Locally, cobalt has been shown to produce dermatitis and investigators have been able to demonstrate a hypersensitivity of the skin to cobalt. There have been reports of hematologic, digestive, and pulmonary changes in humans. See also specific compounds.

C

CNC000 **CAS:71-48-7** **HR: 3**
COBALT DIACETATE
mf: $C_4H_6O_4$•Co mw: 177.03

PROP: Light-pink crystals. Very sol in H_2O; sol in EtOH.

SYNS: ACETIC ACID, COBALT(2+) SALT ☐ COBALT ACETATE ☐ COBALT(2+) ACETATE ☐ COBALT(II) ACETATE ☐ COBALTOUS DIACETATE

TOXICITY DATA WITH REFERENCE
otr-ham:emb 200 μmol/L CNREA8 39,193,79
dnd-ham:emb 200 μmol/L CNREA8 39,193,79
orl-rat LD50:503 mg/kg FCTOD7 20,311,82
ivn-mus LD50:31 mg/kg BJPCAL 23,455,64
ivn-rbt LD50:25 mg/kg BJPCAL 23,455,64

CONSENSUS REPORTS: IARC Cancer Review: Group 2B IMEMDT 52,363,91; Human Inadequate Evidence IMEMDT 52,363,91. Cobalt and its compounds are on the Community Right-To-Know List. Reported in EPA TSCA Inventory. EPA Genetic Toxicology Program.

NIOSH REL: (Cobalt Metal, Dust and Fume) TWA 0.05 mg/m^3

SAFETY PROFILE: Poison by intravenous route. Moderately toxic by ingestion. Questionable carcinogen. Mutation data reported. See also COBALT COMPOUNDS. When heated to decomposition it emits acrid smoke and irritating fumes.

CNC050 **CAS:26490-63-1** **HR: 2**
COBALT(II)FLUOBORATE
mf: B_2CoF_8 mw: 232.55

SYNS: BORATE(1-), TETRAFLUORO-, COBALT(2+) (8CI,9CI) ☐ COBALT BOROFLUORIDE ☐ COBALT BORON TETRAFLUORIDE ☐ COBALTOUS TETRAFLUOROBORATE

TOXICITY DATA WITH REFERENCE
orl-rat LDLo:500 mg/kg NCNSA6 5,27,53

CONSENSUS REPORTS: Reported in EPA TSCA Inventory.

OSHA PEL: TWA 2.5 $mg(F)/m^3$
NIOSH REL: (Fluorides, Inorganic): TWA 2.5 $mg(F)/m^3$

SAFETY PROFILE: Moderately toxic by ingestion. When heated to decomposition it emits toxic vapors of B, Co, F^-.

CNC100 **CAS:10026-17-2** **HR: 3**
COBALT(II) FLUORIDE
mf: CoF_2 mw: 96.93

PROP: Small light-brown or reddish-pink, hexagonal crystals. Bp: 1400°. Reacts with water. Sltly sol in water.

SYNS: COBALT DIFLUORIDE ☐ COBALTOUS FLUORIDE

TOXICITY DATA WITH REFERENCE
orl-rat LD50:150 mg/kg FCTOD7 20,311,82

CONSENSUS REPORTS: Reported in EPA TSCA Inventory.

OSHA PEL: TWA 2.5 $mg(F)/m^3$
ACGIH TLV: TWA 2.5 $mg(F)/m^3$
NIOSH REL: (Fluorides, Inorganic): 10H TWA 2.5 $mg(F)/m^3$

SAFETY PROFILE: Poison by ingestion. When heated to decomposition it emits toxic fumes of Coand F^-.

CNC230 **CAS:16842-03-8** **HR: 3**
COBALT HYDROCARBONYL
mf: C_4HCoO_4 mw: 171.98

PROP: Light-yellow liquid or gas. Unstable, but distillable in current of CO. Mp: −33°, bp: 10°. Sol org solvs; spar sol in H_2O.

TOXICITY DATA WITH REFERENCE
ihl-rat LC50:165 mg/m^3/30M 34ZIAG -,182,69

CONSENSUS REPORTS: Cobalt and its compounds are on the Community Right-To-Know List.

OSHA PEL: TWA 0.1 $mg(Co)/m^3$
ACGIH TLV: TWA 0.1 $mg(Co)/m^3$

SAFETY PROFILE: Poison by inhalation. See also COBALT COMPOUNDS.

For occupational chemical analysis use OSHA: #ID-125g.

CNC231 **CAS:21041-93-0** **HR: 2**
COBALT(II) HYDROXIDE
mf: CoH_2O_2 mw: 92.95

SYNS: COBALT DIHYDROXIDE ☐ COBALT(2+) HYDROXIDE ☐ COBALTOUS HYDROXIDE

CONSENSUS REPORTS: IARC Cancer Review: Group 2B IMEMDT 51,363,91; Human Inadequate Evidence IMEMDT 51,363,91. Reported in EPA TSCA Inventory.

SAFETY PROFILE: Questionable carcinogen. When heated to decomposition it emits toxic vapors of Co.

CNC232 **CAS:12016-80-7** **HR: 2**
COBALT HYDROXIDE OXIDE
mf: $CoHO_2$ mw: 91.94

SYNS: COBALT HYDROXIDE OXIDE (CoO(OH)) ☐ COBALT OXIDE HYDROXIDE (CoOOH) ☐ COBALT OXYHYDROXIDE

CONSENSUS REPORTS: IARC Cancer Review: Group 2B IMEMDT 52,363,91; Human Inadequate Evidence IMEMDT 52,363,91. Reported in EPA TSCA Inventory.

SAFETY PROFILE: Questionable carcinogen. When heated to decomposition it emits toxic vapors of Co.

CNC500 **CAS:10141-05-6** **HR: 3**
COBALT(II) NITRATE
mf: CoN_2O_6 mw: 182.95

PROP: Mp: 55°, d: 1.87.

SYNS: COBALT DINITRATE ☐ COBALTOUS NITRATE ☐ NITRIC ACID, COBALT(2+) SALT

TOXICITY DATA WITH REFERENCE
itt-rat TDLo:14,636 μg/kg (1D male):REP JRPFA4 7,21,64

scu-rbt TDLo: 4530 μg/kg/5D-C: ETA COREAF 236,1387,53
orl-rat LD50: 434 mg/kg FCTOD7 20,311,82
orl-rbt LDLo: 250 mg/kg EQSSDX 1,1,75
scu-rbt LDLo: 75 mg/kg EQSSDX 1,1,75
ims-pgn LDLo: 50 mg/kg HBAMAK 4,1289,35
scu-frg LDLo: 150 mg/kg HBAMAK 4,1289,35

CONSENSUS REPORTS: Reported in EPA TSCA Inventory. Cobalt and its compounds are on the Community Right-To-Know List.

SAFETY PROFILE: Poison by ingestion, intramuscular, and subcutaneous routes. Experimental reproductive effects. Questionable carcinogen with experimental tumorigenic data. Used in animal feed. Explosive reaction with ammonium hexacyanoferrate(II) at 220°C. Potentially explosive reaction with carbon. When heated to decomposition it emits toxic fumes of NO_x. See also COBALT COMPOUNDS and NITRATES.

CNC750 CAS:12139-70-7 HR: 3
COBALT(II) NITRIDE
mf: CoN mw: 72.94

CONSENSUS REPORTS: Cobalt and its compounds are on the Community Right-To-Know List.

SAFETY PROFILE: Powder will spontaneously explode in air. When heated to decomposition it emits toxic fumes of NO_x. See also COBALT COMPOUNDS and NITRIDES.

CND000 CAS:63919-21-1 HR: 3
COBALT NITROPRUSSIDE

SYN: COBALT NITROSOPENTACYANOFERRATE(3)

TOXICITY DATA WITH REFERENCE
orl-rat LD50: 147 mg/kg AIPTAK 172,487,68
ipr-rat LD50: 15 mg/kg AIPTAK 172,487,68
orl-mus LD50: 74 mg/kg AIPTAK 172,487,68
ipr-mus LD50: 10,700 μg/kg AIPTAK 172,487,68
ivn-rbt LDLo: 9400 μg/kg AIPTAK 172,487,68

CONSENSUS REPORTS: Cobalt and its compounds, as well as cyanide and its compounds, are on the Community Right-To-Know List.

SAFETY PROFILE: Poison by ingestion, intravenous, and intraperitoneal routes. When heated to decomposition it emits very toxic fumes of CN^- and NO_x. See also COBALT COMPOUNDS and CYANIDE.

CND020 CAS:1308-06-1 HR: 2
COBALT OXIDE
mf: Co_3O_4 mw: 240.79

SYNS: COBALTIC-COBALTOUS OXIDE □ COBALTO-COBALTIC OXIDE □ COBALTO-COBALTIC TETROXIDE □ COBALTOSIC OXIDE □ COBALT TETRAOXIDE □ TRICOBALT TETRAOXIDE □ TRICOBALT TETROXIDE

TOXICITY DATA WITH REFERENCE
orl-rat LD50: >5 g/kg JACTDZ 1,696,92

CONSENSUS REPORTS: IARC Cancer Review: Group 2B IMEMDT 52,363,91; Animal Inadequate Evidence IMEMDT 52,363,91; Human Inadequate Evidence IMEMDT 52,363,91. Reported in EPA TSCA Inventory.

SAFETY PROFILE: Low toxicity by ingestion. A questionable carcinogen. When heated to decomposition it emits toxic vapors of Co.

CND125 CAS:1307-96-6 HR: 3
COBALT(II) OXIDE
mf: CoO mw: 74.93

PROP: Powder, cubic, hexagonal crystals or solid. Color varies from olive green to red, depending on the particle size, but the commercial material is usually dark gray and contains about 76% Co. The color of anhydrous CoO depends upon the degree of dispersion. It may be yellow, gray, brown , reddish, bluish or black. Mp: about 1935°, d: 5.7 to 6.7. Practically insol in water or alc; sol in acids or alkalies. Easily reduced to Co by C or CO. Reacts at high temperatures with silica, alumina, and zinc oxide to form pigments.

SYNS: C.I. 77322 □ C.I. PIGMENT BLACK 13 □ COBALT BLACK □ COBALT MONOOXIDE □ COBALT MONOXIDE □ COBALTOUS OXIDE □ COBALT OXIDE □ COBALT(2+) OXIDE □ MONOCOBALT OXIDE □ ZAFFRE

TOXICITY DATA WITH REFERENCE
ims-rat TDLo: 135 mg/kg: CAR CNREA8 22,152,62
ims-rat TD: 90 mg/kg: ETA CNREA8 22,158,62
orl-rat LD50: 202 mg/kg FCTOD7 20,311,82
itr-rat LDLo: 50 mg/kg NTIS** AEC-TR-6710
scu-mus LD50: 125 mg/kg ZVKOA6 19,186,74
ims-mus LDLo: 800 mg/kg CNREA8 22,152,62
orl-dog LDLo: 89 mg/kg EQSSDX 1,1,75

CONSENSUS REPORTS: IARC Cancer Review: Group 2B IMEMDT 52,363,91; Animal Sufficient Evidence; IMEMDT 52,363,91; Human Inadequate Evidence IMEMDT 52,363,91. Cobalt and its compounds are on the Community Right-To-Know List. Reported in EPA TSCA Inventory.

SAFETY PROFILE: Suspected carcinogen with experimental carcinogenic and tumorigenic data. Poison by ingestion, subcutaneous, and intratracheal routes. Moderately toxic by intramuscular route. Violent reaction with hydrogen peroxide. See also COBALT. Note: The commercial oxides are usually not definite chemical compounds but mixtures of the cobalt oxides.

For occupational chemical analysis use NIOSH: Cobalt, 7027; Elements, 7300.

CND825 CAS:1308-04-9 HR: 2
COBALT(III) OXIDE
mf: Co_2O_3 mw: 165.86

SYNS: C.I. 77323 □ COBALTIC OXIDE □ COBALT OXIDE (8CI, 9CI) □ COBALT(3+) OXIDE □ COBALT PEROXIDE □ COBALT SESQIOXIDE □ COBALT SESQUIOXIDE □ COBALT TRIOXIDE □ DICOBALT OXIDE □ DICOBALT TRIOXIDE

TOXICITY DATA WITH REFERENCE
ipr-rat LDLo: 703 mg/kg EQSFAP 1,1,75
scu-mus LD50: 2064 mg/kg ZVKOA6 19,186,74

CONSENSUS REPORTS: IARC Cancer Review: Group

2B IMEMDT 52,363,91; Human Inadequate Evidence IMEMDT 52,363,91. Reported in EPA TSCA Inventory. Cobalt compounds are on the Community Right-To-Know List.

SAFETY PROFILE: Moderately toxic by intraperitoneal and subcutaneous routes. Questionable carcinogen. Violent reaction with hydrogen peroxide. The oxide increases the sensitivity of nitroalkanes (e.g. nitromethane, nitroethane, and 1-nitropropane) to heat or detonation. See also COBALT COMPOUNDS.

CND900 CAS:13478-33-6 HR: 3
COBALT(II) PERCHLORATE, HEXAHYDRATE
mf: $Cl_2O_8•Co•6H_2O$ mw: 365.95

SYNS: COBALT DIPERCHLORATE HEXAHYDRATE ☐ COBALTOUS PERCHLORATE, HEXAHYDRATE ☐ COBALT PERCHLORATE HEXAHYDRATE ☐ PERCHLORIC ACID, COBALT(II) SALT, HEXAHYDRATE

TOXICITY DATA WITH REFERENCE
ipr-mus LD50:160 mg/kg JAFCAU 14,512,66

DOT CLASSIFICATION: 5.1; *Label:* Oxidizer

SAFETY PROFILE: A poison by intraperitoneal route. An oxidizer. When heated to decomposition it emits toxic vapors of Co and Cl^-.

CND920 CAS:13455-36-2 HR: 3
COBALT(II) PHOSPHATE
mf: $O_8P_2•3Co$ mw: 366.73

SYNS: COBALTOUS PHOSPHATE ☐ COBALT PHOSPHATE ☐ PHOSPHORIC ACID, COBALT(2+) SALT (2:3)

TOXICITY DATA WITH REFERENCE
orl-rat LD50:387 mg/kg FCTOD7 20,311,82

CONSENSUS REPORTS: Reported in EPA TSCA Inventory.

NIOSH REL: (Cobalt Metal, dust and fume): TWA 0.05 mg/m^3

SAFETY PROFILE: A poison by ingestion. When heated to decomposition it emits toxic vapors of PO_x and Co.

CNE000 CAS:68956-82-1 HR: 3
COBALT RESINATE, precipitated
DOT: UN 1318
mf: $Co(C_{44}H_{62}O_4)_2$ mw: 1368.81

PROP: Brown-red powder.

CONSENSUS REPORTS: Reported in EPA TSCA Inventory. Cobalt and its compounds are on the Community Right-To-Know List.

DOT CLASSIFICATION: 4.1; *Label:* Flammable Solid

SAFETY PROFILE: A dangerous fire hazard when exposed to heat, flame, oxidizers or air. Ignites spontaneously in air. See also COBALT COMPOUNDS. When heated to decomposition it emits acrid smoke and irritating fumes.

CNE125 CAS:10124-43-3 HR: 3
COBALT(II) SULFATE (1:1)
mf: $O_4S•Co$ mw: 154.99

PROP: Red to lavender or dark bluish solid or dimorphic, orthorhombic crystals. D: 3.71. Stable to 708°. Dissolves slowly in boiling water.

SYNS: COBALTOUS SULFATE ☐ COBALT SULFATE ☐ COBALT SULFATE (1:1) ☐ COBALT (2+) SULFATE ☐ COBALT(II) SULPHATE ☐ SULFURIC ACID, COBALT(2+) SALT (1:1)

TOXICITY DATA WITH REFERENCE
orl-rat LD50:424 mg/kg FCTOD7 20,311,82
ipr-mus LD50:143 mg/kg COREAF 256,1043,63
ivn-dog LDLo:20 mg/kg HBAMAK 4,1289,35
orl-rbt LDLo:1800 mg/kg HBAMAK 4,1289,35

CONSENSUS REPORTS: IARC Cancer Review: Group 2B IMEMDT 52,363,91; Human Inadequate Evidence IMEMDT 52,363,91. Cobalt and its compounds are on the Community Right-To-Know List. EPA Genetic Toxicology Program. Reported in EPA TSCA Inventory.

SAFETY PROFILE: Poison by intravenous and intraperitoneal routes. Moderately toxic by ingestion. Questionable carcinogen. When heated to decomposition it emits toxic fumes of SO_x. See also COBALT COMPOUNDS.

CNE200 CAS:1317-42-6 HR: 3
COBALT(II) SULFIDE
mf: CoS mw: 90.99

PROP: Exists in two forms. α-CoS: black, amorphous powder. Sol in HCl. β CoS: gray powder or reddish-silver octahedral crystals. Mp: above 1100°, d: 5.45. Practically insol in water; sol in acids.

SYNS: COBALT MONOSULFIDE ☐ COBALTOUS SULFIDE ☐ COBALT SULFIDE ☐ COBALT SULFIDE (amorphous)

TOXICITY DATA WITH REFERENCE
otr-ham:emb 1 mg/L CNREA8 42,2757,82
dnd-ham:ovr 10 mg/L CRNGDP 3,657,82
ims-rat TDLo:180 mg/kg:ETA CNREA8 22,158,62

CONSENSUS REPORTS: IARC Cancer Review: Group 2B IMEMDT 52,363,91; Animal Limited Evidence IMEMDT 52,363,91; Human Inadequate Evidence IMEMDT 52,363,91. Reported in EPA TSCA Inventory. Cobalt and its compounds are on the Community Right-To-Know List.

SAFETY PROFILE: Suspected carcinogen with experimental tumorigenic data. Mutation data reported. If dried at 300°C it ignites spontaneously in air. See also COBALT COMPOUNDS and SULFIDES.

CNE250 CAS:10026-18-3 HR: 3
COBALT TRIFLUORIDE
mf: CoF_3 mw: 226.47

PROP: Brown crystals. Sol in EtOH, Et_2O, and C_6H_6.

CONSENSUS REPORTS: Cobalt and its compounds are on the Community Right-To-Know List.

SAFETY PROFILE: A powerful fluorinating agent and

C

oxidizer. Violent reaction with hydrocarbons or water. Very exothermic reaction when warmed with silicon. When heated to decomposition it emits toxic fumes of F^-. See also COBALT COMPOUNDS and FLUORIDES.

CNE375 **CAS:24699-40-9** **HR: 3**
COBEN
mf: $C_{19}H_{25}N_3 \cdot ClH$ mw: 331.93

PROP: Crystals with sltly bitter taste. Mp: 183–185°.

SYNS: 1-(2-(PHENYL(2-PYRIDYLMETHYL)AMINO)ETHYL)PIPERIDINE HYDROCHLORIDE □ N-PHENYL-N-(2-PYRIDYLMETHYL)-2-PIPERIDINOETHYLAMINE HYDROCHLORIDE □ N-(2-PICOLYL)-N-PHENYL-N-(2-PIPERIDINOETHYL)AMINE HYDROCHLORIDE □ PICOPERIDAMINE HYDROCHLORIDE □ PICOPERINE HYDROCHLORIDE □ N-(2-PIPERIDINOETHYL)-N-(2-PYRIDYLMETHYL)ANILINE HYDROCHLORIDE □ 1-(2-(N-(2-PYRIDYLMETHYL)ANILINO)ETHYL)PIPERIDINE HYDROCHLORIDE □ N-(2-PYRIDYLMETHYL)-N-PHENYL-N-2-(PIPERIDINOETHYL)AMINE HYDROCHLORIDE □ TAT-3 HYDROCHLORIDE

TOXICITY DATA WITH REFERENCE
orl-mus TDLo:180 mg/kg (8-13D preg):TER TAKHAA 29,297,70
orl-rat LD50:740 mg/kg KSRNAM 4,403,70
ipr-rat LD50:69 mg/kg KSRNAM 4,403,70
scu-rat LD50:480 mg/kg KSRNAM 4,403,70
ivn-rat LD50:11,500 µg/kg KSRNAM 4,403,70
orl-mus LD50:210 mg/kg KSRNAM 4,403,70
ipr-mus LD50:55,500 µg/kg KSRNAM 4,403,70
scu-mus LD50:142 mg/kg KSRNAM 4,403,70
ivn-mus LD50:10 mg/kg KSRNAM 4,403,70
orl-dog LDLo:100 mg/kg ARZNAD 19,1916,69
ivn-dog LDLo:25 mg/kg ARZNAD 19,1916,69

SAFETY PROFILE: Poison by ingestion, subcutaneous, intravenous, and intraperitoneal routes. An experimental teratogen. When heated to decomposition it emits toxic fumes of NO_x and HCl.

CNE500 **CAS:29091-05-2** **HR: 2**
COBEXO
mf: $C_{11}H_{13}F_3N_4O_4$ mw: 322.28

PROP: Yellow crystals. Mp: 98–99°. Very sltly sol in H_2O; sol in EtOH; very sol in Me_2CO.

SYNS: COBEX □ N[3],N[3]-DIETHYL-2,4-DINITRO-6-(TRIFLUOROMETHYL)-1,3-BENZENEDIAMINE □ N[4],N[4]-DIETHYL-α,α,α-TRIFLUORO-3,5-DINITRO-TOLUENE-2,4-DIAMINE □ DINITRAMINE □ DINITROAMINE □ USB-3584

TOXICITY DATA WITH REFERENCE
orl-rat LD50:3000 mg/kg 28ZEAL 5,80,76
skn-rbt LD50:2000 mg/kg FMCHA2 -,D106,80

SAFETY PROFILE: Moderately toxic by ingestion and skin contact. An herbicide. When heated to decomposition it emits very toxic fumes of F^- and NO_x. See also FLUORIDES and NITRO COMPOUNDS of AROMATIC HYDROCARBONS.

CNE750 **CAS:50-36-2** **HR: 3**
COCAINE
mf: $C_{17}H_{21}NO_4$ mw: 303.39

PROP: Colorless to white crystals or prisms from alc. Mp: 98°, bp: 187–188°. Volatile, especially above 90°. Soluble in alcohol, chloroform, ether, oil turpentine, olive oil, liquid petrolatum, acetone, ethyl acetate, carbon disulfide. Sparingly soluble in water.

SYNS: BENZOYLMETHYLECGONINE □ BERNICE □ BERNIES □ BURESE □ 2-β-CARBOMETHOXY-3-β-BENZOXYTROPANE □ "C" CARRIE □ CECIL □ CHOLLY □ (−)-COCAINE □ β-COCAINE □ l-COCAINE □ COKE □ CORINE □ ECGONINE, METHYL ESTER, BENZOATE (ESTER) □ ERITROXILINA □ ERYTROXYLIN □ GIRL □ GOLD DUST □ HAPPY DUST □ 3-β-HYDROXY-1-α-H,5-α-H-TROPANE-2-β-CARBOXYLIC ACID METHYL ESTER, BENZOATE □ KOKAIN □ KOKAN □ KOKAYEEN □ METHYL-3-β-HYDROXY-1-α-H,5-α-H-TROPANE-2-β-CARBOXYLATE BENZOATE (ESTER) □ NEUROCAINE □ STAR DUST □ 2-β-TROPANECARBOXYLIC ACID, 3-β-HYDROXY-, METHYL ESTER, BENZOATE (ESTER) □ 3-TROPANYLBENZOATE-2-CARBOXYLIC ACID METHYL ESTER

TOXICITY DATA WITH REFERENCE
eye-rbt 16% MLD JAPMA8 42,685,53
orl-man LDLo:7353 µg/kg 85DCAI 2,73,70
orl-hmn TDLo:714 mg/kg:CNS JAMAAP 238,1391,77
unk-hmn LDLo:286 µg/kg 34ZIAG -,183,69
ipr-rat LD50:70 mg/kg JLCMAK 15,731,30
scu-rat LD50:250 mg/kg JLCMAK 15,731,30
ivn-rat LD50:17,500 µg/kg JLCMAK 15,731,30
orl-mus LD50:99 mg/kg ARZNAD 16,1275,66
ipr-mus LD50:75 mg/kg JATOD3 4,19,80
scu-mus LDLo:125 mg/kg APBOAI 12,189,66
ivn-mus LD50:30 mg/kg RPTOAN 35(3),114,72
scu-dog LDLo:3500 µg/kg BDHU** -,-,36
ivn-dog LD50:13 mg/kg AIPTAK 235,328,78
orl-rbt LDLo:126 mg/kg 27ZIAQ -,78,73
scu-rbt LDLo:50 mg/kg AEPPAE 160,53,31
ivn-rbt LD50:17 mg/kg JLCMAK 15,731,30

DOT CLASSIFICATION: Forbidden

SAFETY PROFILE: A human poison by ingestion and possibly other routes. Poison experimentally by ingestion, intraperitoneal, intravenous, subcutaneous, and parenteral routes. Human central nervous system effects by ingestion and possibly other routes: general anesthesia, hallucinations or distorted perceptions, and convulsions. An eye irritant. A widely abused, controlled substance. Abuse leads to habituation or addiction. In medicine, it is used as a local narcotic anesthetic applied topically to mucous membranes. The free base is soluble in fats and thus is used for ointments and oily solutions. For water-soluble applications, the sulfate or hydrochloride is used. See also ESTERS. When heated to decomposition it emits highly toxic fumes.

CNF000 **CAS:53-21-4** **HR: 3**
COCAINE HYDROCHLORIDE
mf: $C_{17}H_{21}NO_4 \cdot ClH$ mw: 339.85

PROP: A solid. Mp: 200–202°. Soluble in water, alc, chloroform, glycerol, acetone, ether, and oils. Decomposes when heated.

SYNS: COCAIN-CHLORHYDRAT (GERMAN) □ COCAINE CHLO-

RIDE □ (−)-COCAINE HYDROCHLORIDE □ l-COCAINE HYDROCHLORIDE □ COCAINE MURIATE □ 3-β-HYDROXY-1-α-H,5-α-H-TROPANE-2-β-CARBOXYLIC ACID METHYL ESTER, BENZOATE (ESTER), HYDROCHLORIDE □ SAL de MERCK

TOXICITY DATA WITH REFERENCE
eye-rbt 2% MLD ARZNAD 8,181,58
cyt-ham:lng 830 mg/L GMCRDC 27,95,81
scu-rat TDLo:1080 mg/kg (male 72D pre):REP JOAND3 10,17,89
scu-mus TDLo:60 mg/kg (female 9D post):REP JPMSAE 69,703,80
scu-rat TDLo:910 mg/kg (female 7-19D post):TER NRTXDN 10,51,88
ipr-rat TDLo:300 mg/kg (8-12D preg):TER TJADAB 21,37A,80
ipr-rat TDLo:10 mg/kg (1D male):REP PHBHA4 4,677,69
scu-rat TDLo:780 mg/kg (female 7-19D post):REP NRTXDN 10,51,88
scu-mus TDLo:60 mg/kg (female 9D post):TER JPMSAE 69,703,80
scu-mus TDLo:60 mg/kg (female 7D post):TER JPMSAE 69,703,80
ipr-rat LD50:78 mg/kg ARZNAD 8,181,58
orl-mus LD50:96 mg/kg RPOBAR 2,279,70
ipr-mus LD50:68 mg/kg AIPTAK 189,198,71
scu-mus LD50:30 mg/kg ARZNAD 8,181,58
ivn-mus LD50:15 mg/kg AIPTAK 105,221,56
ivn-dog LD50:21 mg/kg AIPTAK 235,328,78
scu-rbt LDLo:100 mg/kg JPETAB 47,255,33
scu-gpg LDLo:60 mg/kg JPETAB 47,255,33

SAFETY PROFILE: Poison by ingestion, intravenous, intraperitoneal, and subcutaneous routes. An experimental teratogen. Other experimental reproductive effects. An eye irritant. Mutation data reported. A widely abused, controlled substance. Abuse leads to habituation or addiction. In medicine, it is used as a local anesthetic and central nervous system stimulant. Incompatible with calomel; mercuric oxide; silver nitrate. When heated to decomposition it emits very toxic fumes of NO_x and HCl. See also COCAINE.

CNF050 CAS:1343-78-8 HR: D
COCHINEAL (dye)

SYNS: C.I. 75470 □ COCHENILLE DYE □ COCHINEAL □ COCHINEAL TINCTURE □ ENJI

TOXICITY DATA WITH REFERENCE
mmo-sat 100 mg/plate SKEZAP 23,86,82
mmo-sat 1 mg/plate AMONDS 3,253,80
cyt-ham:lng 5100 mg/L GANMAX 27,95,81

CONSENSUS REPORTS: Reported in EPA TSCA Inventory.

SAFETY PROFILE: Mutation data reported. When heated to decomposition it emits acrid smoke and irritating vapors.

CNF109 CAS:4611-05-6 HR: 3
COCHLIOBOLIN
mf: $C_{25}H_{36}O_4$ mw: 400.61

PROP: Crystals. Mp: 182°.

SYNS: COCHLIOBOLIN A □ OPHIOBOLIN □ OPHIOBOLIN A

TOXICITY DATA WITH REFERENCE
orl-mus LD50:238 mg/kg 85GDA2 6,143,81
ipr-mus LD50:21 mg/kg 85GDA2 6,143,81
scu-mus LD50:73 mg/kg 85GDA2 6,143,81
ivn-mus LD50:12 mg/kg 85GDA2 6,143,81

SAFETY PROFILE: Poison by ingestion, subcutaneous, intravenous, and intraperitoneal routes. When heated to decomposition it emits acrid smoke and fumes.

CNF159 CAS:11051-88-0 HR: 3
COCHLIODINOL
mf: $C_{32}H_{30}N_2O_4$ mw: 506.64

PROP: Purple crystals. Mp: 213–215°.

SYNS: 3,6-BIS(5-(3-METHYL-2-BUTENYL)INDOL-3-YL)-2,5-DIHYDROXY-p BENZOQUINONE □ 2,5-DIHYDROXY-3,6-BIS(5-(3-METHYL-2-BUTENYL)-1H-INDOL-3-YL)-2,5-CYCLOHEXADIENE-1,4-DIONE

TOXICITY DATA WITH REFERENCE
orl-mus LD50:221 mg/kg 85GDA2 3,310,80
ipr-mus LD50:41 mg/kg 85GDA2 3,310,80
scu-mus LD50:147 mg/kg 85GDA2 3,310,80

SAFETY PROFILE: Poison by ingestion, subcutaneous, and intraperitoneal routes. When heated to decomposition it emits toxic fumes of NO_x.

CNF250 CAS:104-61-0 HR: 2
COCONUT ALDEHYDE
mf: $C_9H_{16}O_2$ mw: 156.25

PROP: Colorless to sltly yellow liquid; coconut odor. D: 0.958–0.966, refr index: 1.446–1.450, flash p: 212°F. Sol in alc, fixed oils, propylene glycol; insol in water.

SYNS: ALDEHYDE C-18 □ γ-N AMYLBUTYROLACTONE □ FEMA No. 2781 □ 4-HYDROXYNONANOIC ACID, γ-LACTONE □ γ-NONALACTONE (FCC) □ 1,4-NONALOLIDE □ PRUNOLIDE

TOXICITY DATA WITH REFERENCE
skn-rbt 500 mg/24H MLD FCTXAV 13,681,75
dnr-bcs 20 mg/disc OIGZSE 34,267,85
orl-rat LD50:6600 mg/kg FCTXAV 13,681,75
orl-gpg LD50:3440 mg/kg FCTXAV 2,327,64

CONSENSUS REPORTS: Reported in EPA TSCA Inventory.

SAFETY PROFILE: Moderately toxic by ingestion. A skin irritant. Mutation data reported. Combustible liquid. When heated to decomposition it emits acrid smoke and irritating fumes. See also ALDEHYDES.

CNF325 CAS:61788-90-7 HR: 1
COCONUT DIMETHYL AMINE OXIDE

TOXICITY DATA WITH REFERENCE
skn-hmn 2500 μg/24H MOD AKEDAX 235,180,69
skn-rbt 10 mg MLD JSCCA5 22,411,71
skn-rbt 230 mg/5W open MLD JSCCA5 22,411,71
skn-gpg 115 mg/5W open MLD JSCCA5 22,411,71

CONSENSUS REPORTS: Reported in EPA TSCA Inventory.

SAFETY PROFILE: A human skin irritant. When heated to decomposition it emits toxic fumes of NO_x. See also AMINES.

CNF340 **CAS:68988-02-3** **HR: 1**
N-COCOPYRROLIDINONE

SYN: 2-PYRROLIDINONE, 1-COCO ALKYL derivs.

TOXICITY DATA WITH REFERENCE
skn-rbt 500 mg/24H SEV FCTOD7 26,475,88
eye-rbt 100 mg MOD FCTOD7 26,475,88
orl-rat LDLo:>10 g/kg FCTOD7 26,475,88

CONSENSUS REPORTS: Reported in EPA TSCA Inventory.

SAFETY PROFILE: Low toxicity by ingestion. A severe eye and skin irritant. When heated to decomposition it emits toxic vapors of NO_x.

CNF390 **CAS:53-84-9** **HR: 1**
CODEHYDROGENASE I
mf: $C_{21}H_{27}N_7O_{14}P_2$ mw: 663.49

SYNS: ADENINE-NICOTINAMIDE DINUCLEOTIDE □ CODEHYDRASE I □ COENZYME I □ COZYMASE □ COZYMASE I □ DIPHOSPHOPYRIDINE NUCLEOTIDE □ DPN □ ENZOPRIDE □ NAD □ NAD+ □ β-NAD □ NADIDE □ NICOTINAMIDE-ADENINE DINUCLEOTIDE □ NICOTINEAMIDE ADENINE DINUCLEOTIDE □ PYRIDINIUM, 3-CARBAMOYL-1-β-D-RIBOFURANOSYL-, HYDROXIDE, 5'-ESTER with ADENOSINE 5'-5'-(TRIHYDROGEN PYROPHOSPHATE), inner salt

TOXICITY DATA WITH REFERENCE
mmo-sat 100 µg/plate ABCHA6 45,327,81
ipr-mus LD50:4333 mg/kg PCJOAU 20,160,86

CONSENSUS REPORTS: Reported in EPA TSCA Inventory.

SAFETY PROFILE: Mildly toxic by intraperitoneal route. Mutation data reported. When heated to decomposition it emits toxic vapors of NO_x and PO_x.

CNF400 **CAS:53-59-8** **HR: 2**
CODEHYDROGENASE II
mf: $C_{21}H_{28}N_7O_{17}P_3$ mw: 743.47

SYNS: CODEHYDRASE II □ COENZYME II □ COZYMASE II □ NADP □ β-NADP □ NAD PHOSPHATE □ PYRIDINIUM, 3-CARBAMOYL-1-β-D-RIBOFURANOSYL-, HYDROXIDE, 5',5'-ESTER with ADENOSINE 2'-(DIHYDROGEN PHOSPHATE) 5'-(TRIHYDROGEN PYROPHOSPHATE), inner salt □ TPN □ β-TPN □ TPN (NUCLEOTIDE) □ TRIPHOSPHOPYRIDINE NUCLEOTIDE

TOXICITY DATA WITH REFERENCE
ipr-mus LD50:3166 mg/kg PCJOAU 20,160,86

CONSENSUS REPORTS: Reported in EPA TSCA Inventory.

SAFETY PROFILE: Moderately toxic by intraperitoneal route. When heated to decomposition it emits toxic vapors of NO_x and PO_x.

CNF500 **CAS:76-57-3** **HR: 3**
CODEINE
mf: $C_{18}H_{21}NO_3$ mw: 299.40

PROP: Prisms from Et_2O or $C_6H_6)H_2O$; octahedra or rhombic prisms. Mp: 155°, d: 1.32 @ 20°/4°. Sol in EtOH, $CHCl_3$, and Me_2CO; sltly sol in CCl_4 and H_2O.

SYNS: METHYLMORPHINE □ N-METHYL-NORDOCEINE □ MORPHINE-3-METHYL ETHER □ MORPHINE MONOMETHYL ETHER

TOXICITY DATA WITH REFERENCE
orl-rat TDLo:350 mg/kg (6-15D preg):TER ARZNAD 26,551,76
orl-rat TDLo:1200 mg/kg (6-15D preg):REP ARZNAD 26,551,76
scu-rat TDLo:8600 µg/kg (1D male):REP JPETAB 198,340,76
unr-man LDLo:12 mg/kg 85DCAI 2,73,70
orl-rat LD50:427 mg/kg JMCMAR 16,782,73
ipr-rat LD50:100 mg/kg ARZNAD 26,551,76
scu-rat LD50:229 mg/kg ARZNAD 24,600,74
ivn-rat LD50:75 mg/kg JPPMAB 25,929,73
orl-mus LD50:250 mg/kg MDCHAG 5,318,65
ipr-mus LD50:60 mg/kg APFRAD 8,261,50
scu-mus LD50:84,100 µg/kg ARZNAD 24,600,74
ivn-mus LD50:54 mg/kg ARZNAD 16,617,66
ims-mus LD50:290 mg/kg OYYAA2 13,97,77
ivn-dog LD50:69 mg/kg CPBTAL 7,372,59
ivn-rbt LD50:34 mg/kg EXPEAM 18,446,62

SAFETY PROFILE: A human poison by an unspecified route. An experimental poison by ingestion, intraperitoneal, intravenous, intramuscular, and subcutaneous routes. Human reproductive effects. An experimental teratogen. Other experimental reproductive effects. An addictive drug. Flammable when exposed to heat or flame. To fight fire, use alcohol foam. When heated to decomposition it emits toxic fumes of NO_x. See also MORPHINE and ETHERS.

CNF750 **CAS:1422-07-7** **HR: 3**
CODEINE HYDROCHLORIDE
mf: $C_{18}H_{21}NO_3 \cdot ClH$ mw: 335.86

PROP: Prisms. Mp: 287° (as dihydrate).

TOXICITY DATA WITH REFERENCE
orl-rat LD50:750 mg/kg ARZNAD 21,719,71
scu-rat LD50:215 mg/kg ARZNAD 21,727,71
orl-mus LD50:365 mg/kg BJPCAL 16,209,61
ipr-mus LD50:115 mg/kg ARZNAD 25,873,75
scu-mus LD50:270 mg/kg JMCMAR 19,1054,76
ivn-mus LD50:67 mg/kg BJPCAL 16,209,61
scu-rbt LDLo:36 mg/kg JPETAB 66,182,39
par-rbt LDLo:24 mg/kg JPETAB 66,182,39

SAFETY PROFILE: Poison by ingestion, subcutaneous, intravenous, intraperitoneal and parenteral routes. An addictive drug. See also CODEINE. When heated to decomposition it emits very toxic fumes of NO_x and HCl.

CNG250 **CAS:3688-66-2** **HR: 3**
CODEINE NICOTINATE (ESTER)
mf: $C_{24}H_{24}N_2O_4$ mw: 404.50

PROP: Crystals from EtOH. Mp: 134–135°.

SYNS: LYOPECT □ NICOCODINE □ NICOTINIC ACID-7,8-DIDEHYDRO-4,5-α-EPOXY-3-METHOXY-17-METHYLMORPHINAN-6-α-YL ESTER □ NICOTINIC ACID, ESTER with CODEINE □ RC 146

TOXICITY DATA WITH REFERENCE
ipr-rat LD50:37 mg/kg AIPTAK 143,466,63
orl-rat LD50:840 mg/kg AIPTAK 143,466,63
orl-mus LD50:280 mg/kg AIPTAK 143,466,63
ipr-mus LD50:48 mg/kg AIPTAK 143,466,63

SAFETY PROFILE: Poison by ingestion and intraperitoneal routes. See also CODEINE. An addictive drug. When heated to decomposition it emits toxic fumes of NO_x.

CNG500 **CAS:52-28-8** **HR: 3**
CODEINE PHOSPHATE
mf: $C_{18}H_{21}NO_3 \cdot H_3O_4P$ mw: 397.40

SYN: MORPHINAN-6-α-OL, 7,8 DIDEHYDRO-4,5-α-EPOXY-3-METHOXY-17-METHYL-, PHOSPHATE (1:1)

TOXICITY DATA WITH REFERENCE
scu-mus TDLo:110 mg/kg (9D preg):TER DGDFA5 22,61,80
orl-wmn TDLo:3600 μg/kg/2D-I:EYE,BPR CPHADV 5,15,86
orl-man TDLo:5143 μg/kg/5D-I CPHADV 5,15,86
orl-rat LD50:85 mg/kg USXXAM #3163649
ipr-rat LD50:104 mg/kg JPETAB 133,117,61
scu-rat LD50:312 mg/kg JPETAB 154,161,66
ivn-rat LD50:54 mg/kg AIPTAK 103,200,55
ims-rat LD50:208 mg/kg AIPTAK 190,124,71
orl-mus LD50:237 mg/kg JPETAB 128,384,60
ipr-mus LD50:110 mg/kg ARZNAD 19,1916,69
scu-mus LD50:80 mg/kg USXXAM #3031377
ivn-mus LD50:62 mg/kg JJPAAZ 17,538,67
ims-mus LD50:191 mg/kg AIPTAK 190,124,71
ivn-dog LD50:978,800 μg/kg OYYAA2 4,961,70
orl-rbt LDLo:100 mg/kg HBTXAC 1,72,55

CONSENSUS REPORTS: EPA Genetic Toxicology Program.

SAFETY PROFILE: Poison by ingestion, subcutaneous, intravenous, intramuscular, and intraperitoneal routes. Human systemic effects by ingestion: miosis (pupillary constriction), sleep disturbance, and blood pressure lowering. An experimental teratogen. Used as an analgesic. See also CODEINE. When heated to decomposition it emits very toxic fumes of NO_x and PO_x.

CNG675 **CAS:5913-76-8** **HR: 3**
CODEINE PHOSPHATE SESQUIHYDRATE
mf: $C_{18}H_{21}NO_3 \cdot H_3O_4P \cdot 3/2H_2O$ mw: 424.38

SYN: 7,8-DIDEHYDRO-4,5-α-EPOXY-3-METHOXY-17-METHYL-MORPHINAN-6-α-OL PHOSPHATE SESQUIHYDRATE (3:3:2)

TOXICITY DATA WITH REFERENCE
orl-rat LD50:209 mg/kg KSRNAM 5,1011,71
ipr-rat LD50:38 mg/kg KSRNAM 5,1011,71
scu-rat LD50:81 mg/kg KSRNAM 5,1011,71
orl-mus LD50:290 mg/kg KSRNAM 5,1787,71
ipr-mus LD50:83 mg/kg KSRNAM 5,1011,71
scu-mus LD50:191 mg/kg YKKZAJ 81,740,61
ivn-mus LD50:70 mg/kg KSRNAM 5,1787,71
ivn-dog LD50:97,800 μg/kg YKKZAJ 81,740,61
orl-rbt LDLo:100 mg/kg MEIEDD 10,350,83

SAFETY PROFILE: Poison by ingestion, subcutaneous, intravenous, and intraperitoneal routes. When heated to decomposition it emits toxic fumes of NO_x and PO_x. A narcotic analgesic and antitussive. See also CODEINE and CODEINE PHOSPHATE.

CNG750 **CAS:1420-53-7** **HR: 3**
CODEINE SULFATE
mf: $C_{36}H_{42}N_2O_6 \cdot O_4S$ mw: 694.86

SYN: MORPHINAN-6-α-OL, 7,8-DIDEHYDRO-4,5-α-EPOXY-3-METHOXY-17-METHYL-, SULFATE (2:1) (salt)

TOXICITY DATA WITH REFERENCE
cyt-ofs-ipr 100 mg/L BEXBBO 16,425,80
scu-mus TDLo:100 mg/kg (9D preg):REP JPMSAE 66,1727,77
scu-mus TDLo:100 mg/kg (9D preg):TER JPMSAE 66,1727,77
orl-rat LD50:430 mg/kg AIPTAK 123,48,59
ipr-rat LD50:107 mg/kg TXCYAC 14,217,79
scu-rat LD50:332 mg/kg JPETAB 134,332,61
ivn-rat LD50:55 mg/kg JPETAB 134,332,61
orl-mus LD50:395 mg/kg JPETAB 134,332,61
ipr-mus LD50:159 mg/kg AIPTAK 136,333,62
scu-mus LD50:183 mg/kg JPETAB 134,332,61
ivn-mus LD50:68 mg/kg JPETAB 134,332,61

SAFETY PROFILE: Poison by ingestion, intraperitoneal, intravenous, and subcutaneous routes. An experimental teratogen. Other experimental reproductive effects. Mutation data reported. An addictive, narcotic drug. When heated to decomposition it emits very toxic fumes of NO_x and SO_x. See also CODEINE.

CNG760 **CAS:33956-49-9** **HR: 2**
CODLELURE
mf: $C_{12}H_{22}O$ mw: 182.34

SYNS: AI3-34872 □ CODLEMONE □ (E,E)-8,10-DODECADIEN-1-OL □ 8,10-DODECADIEN-1-OL, (E,E)- □ 8E,10E-DODECADIEN-1-OL □ trans-8,trans-10-DODECADIEN-1-OL □ ENT 34872 □ PHEROCON CM

TOXICITY DATA WITH REFERENCE
orl-rat LD50:>3200 mg/kg SPEADM 78-1,49,78

CONSENSUS REPORTS: Reported in EPA TSCA Inventory.

SAFETY PROFILE: Moderately toxic by ingestion. When heated to decomposition it emits acrid smoke and irritating vapors.

C

CNG825 **HR: 3**
COFFEE SENNA

PROP: An annual herb (up to 3 feet tall) with smooth compound leaves and yellow flowers. The seed pod is about 4.5 inches long and thin with dark olive green seeds. It is common, particularly along highways and coastal areas, from Virginia to Texas, Hawaii, and Guam.

SYNS: 'AUKO'I (HAWAII) □ BICHE PRIETO (MEXICO) □ CASSIA OCCIDENTALIS □ DANDELION (JAMAICA) □ HEDIONDA (PUERTO RICO) □ MIKIPALAOA (HAWAII) □ PISS-A-BED (JAMAICA) □ POIS PUANTE (HAITI) □ STINKING WEED □ STYPTIC WEED □ WILD COFFEE □ YERBA HEDIONDA (CUBA)

SAFETY PROFILE: The whole plant contains the irritant chrysarobin, the cathartic emodin and toxalbumin. Reported poisonings are the result of ingesting raw seeds. Roasted seeds are used as a coffee substitute. The raw seeds have a strong laxative action in humans. Chronic ingestion may cause damage to the muscles, kidneys, liver and lungs, and may result in death. See also CHRYSAROBIN; 6-METHYL-1,3,8-TRIHYDROXYANTHRAQUINONE; and ABRIN.

CNG830 **CAS:64-86-8** **HR: 3**
COLCHICINE
mf: $C_{22}H_{25}NO_6$ mw: 399.48

PROP: Pale-yellow scales or powder. Mp: 155–157°. Crystals from ethyl acetate, pale-yellow needles. One gram dissolves in 22 mL water, 220 mL ether, 100 mL benzene; freely sol in alcohol or chloroform; practically insol in petr ether.

SYNS: 7-ACETAMIDO-6,7-DIHYDRO-1,2,3,10-TETRAMETHOXY-BENZO(a)HEPTALEN-9(5H)-ONE □ N-ACETYL TRIMETHYLCOLCHICINIC ACID METHYL ETHER □ COLCHICIN (GERMAN) □ COLCHICINA (ITALIAN) □ 7-α-H-COLCHICINE □ COLCHINEOS □ COLCHISOL □ COLCIN □ COLSALOID □ CONDYLON □ NSC 757 □ N-(5,6,7,9-TETRAHYDRO-1,2,3,10-TETRAMETHOXY-9-OXOBENZO(α)HEPTALEN-7-YL)-ACETAMIDE

TOXICITY DATA WITH REFERENCE
eye-rbt 1%/3D SEV AJOPAA 31,837,48
dnd-hmn:fbr 1 mg/L BBACAQ 824,117,85
sln-mus-ipr 200 μg/kg ENMUDM 8(Suppl 6),51,86
scu-rat TDLo:1200 μg/kg (18-20D preg):REP DEPBA5 9,119,76
orl-mus TDLo:80 mg/kg (female 8-12D post):REP TCMUD8 6,361,86
scu-rat TDLo:1200 μg/kg (18-20D preg):TER DEPBA5 9,119,76
ivn-ham TDLo:10 mg/kg (female 8D post):TER PSEBAA 112,775,63
scu-rat TDLo:5 mg/kg (female 8D post):REP JOMAAL 36,516,55
ipr-mus TDLo:2500 μg/kg (male 1D pre):REP TXAPA9 23,288,72
ipr-rat TDLo:250 μg/kg (1D male):REP HIKYAJ 12,1339,66
scu-rbt TDLo:22,500 μg/kg (male 30D pre):REP JPETAB 115,319,55
ipr-mus TDLo:500 μg/kg (female 7D post):TER TJADAB 18,31,78
scu-rat TDLo:1200 μg/kg (18-20D preg):TER DEPBA5 9,119,76
scu-mus TDLo:1250 μg/kg (female 9D post):TER CAJPBD 7,61,67
ipr-mus TDLo:500 μg/kg (female 8D post):TER TJADAB 18,31,78
orl-wmn TDLo:320 μg/kg:PNS LANCAO 2,1271,87
orl-hmn LDLo:86 μg/kg:PUL,GIT,MET 34ZIAG -,184,69
orl-man TDLo:12,514 μg/kg/2Y-I NEJMAG 316,1562,87
orl-man LDLo:11 mg/kg:PUL,SYS CMAJAX 138,335,88
ivn-wmn LDLo:360 μg/kg/6D:BLD,KID NPMDAD 9,1587,80
ivn-man LDLo:143 μg/kg/5D-I:BLD NEJMAG 309,310,83
par-hmn TDLo:710 μg/kg:CNS,CVS,GIT AIMDAP 137,394,77
scu-rat LDLo:4 mg/kg AIPTAK 84,257,50
ivn-rat LD50:1600 μg/kg TXAPA9 13,50,68
ice-rat LDLo:1 mg/kg AIPTAK 109,386,57
par-rat LDLo:1000 μg/kg FRPSAX 15,533,60
orl-mus LD50:5886 μg/kg NCISP* JAN86
ipr-mus LD50:2 mg/kg BAFEAG 42,308,55
scu-mus LD50:1200 μg/kg NIIRDN 6,280,82
ivn-mus LD50:1700 μg/kg NIIRDN 6,280,82
ims-mus LD50:1197 μg/kg JMCMAR 24,257,50
orl-dog LDLo:125 μg/kg 85DZAJ -,315,68
scu-dog LDLo:571 μg/kg HBAMAK 4,1337,35

CONSENSUS REPORTS: EPA Extremely Hazardous Substances List. EPA Genetic Toxicology Program. Reported in EPA TSCA Inventory.

SAFETY PROFILE: A human poison by ingestion and intravenous routes. Poison experimentally by most routes. Human systemic effects: aplastic anemia, blood pressure depression, body temperature decrease, changes in kidney tubules, dyspnea, flaccid paralysis without anesthesia, gastrointestinal effects, kidney damage and hemorrhaging, muscle contraction or spasticity, muscle weakness, nausea or vomiting, respiratory stimulation, and somnolence. An experimental teratogen. Experimental reproductive effects. A severe eye irritant. Human mutation data reported. Inhibits the formation of microtubules and thus impairs cell division. When heated to decomposition it emits toxic fumes of NO_x.

CNG835 **CAS:41826-92-0** **HR: 3**
COLIBIL
mf: $C_{16}H_{22}O_6$ mw: 310.38

PROP: Colorless needles from aq ethanol or plates from aq acetone. Mp: 150–151°. Stable to heat, humidity, indoor diffused sunlight.

SYNS: CHOLIBIL □ SUPACAL □ TREPIBUTONE □ 3-(2,4,5-TRIETHOXYBENZOYL)PROPIONIC ACID □ 2,4,5-TRIETHOXY-γ-OXOBENZENEBUTANOIC ACID

TOXICITY DATA WITH REFERENCE
orl-rat TDLo:9100 mg/kg (female 4W pre):REP TAKHAA 36,263,77
orl-rat LD50:2450 mg/kg IYKEDH 11,811,80
ipr-rat LD50:410 mg/kg IYKEDH 11,811,80
scu-rat LD50:570 mg/kg IYKEDH 11,811,80
ivn-rat LD50:350 mg/kg IYKEDH 11,811,80
orl-mus LD50:1340 mg/kg IYKEDH 11,811,80
ipr-mus LD50:510 mg/kg YAKUD5 22,1501,80
scu-mus LD50:625 mg/kg YAKUD5 22,1501,80
ivn-mus LD50:500 mg/kg IYKEDH 11,811,80

SAFETY PROFILE: Poison by intravenous route. Moderately toxic by ingestion, subcutaneous, and intraperitoneal routes. Experimental reproductive effects. When heated to decomposition it emits acrid fumes and smoke.

CNH125 **CAS:26591-12-8** **HR: 2**
COLORFIX
mf: $(C_2H_4N_4 \cdot CH_2O)_x$

SYNS: CARFLOC D 1000 □ CETRAMIN □ DICYANDIAMIDE-FORMALDEHYDE ADDUCT □ DICYANDIAMIDE-FORMALDEHYDE POLYMER □ DICYANDIAMIDE-FORMALDEHYDE RESIN □ DRASIL 507 □ FIXATIVE IS □ FIXER IS □ NEOFIX FP □ NIKAFLOC D 1000 □ NONISOLD □ PARAMEL DC □ PERMINAL FC-P □ RESIN T □ SATILAN PLZ □ SATILAN RL 2 □ SUMISET D □ SYNTEFIX □ US 2 □ WARCO F 71

TOXICITY DATA WITH REFERENCE
skn-rbt 500 mg/24H MLD 28ZPAK -,305,72
eye-rbt 100 mg/24H MOD 28ZPAK -,305,72
orl-rat LD50:1750 mg/kg 28ZPAK -,305,72

SAFETY PROFILE: Moderately toxic by ingestion. An eye and skin irritant. When heated to decomposition it emits toxic fumes of NO_x and CN^-. See also ALDEHYDES.

CNH275 **HR: 3**
COMBRETODENDRON AFRICANUM (Welw), extract

PROP: Plant belonging to the family *Lecythidaceae* (THERAP 22,325,67).

TOXICITY DATA WITH REFERENCE
scu-rat TDLo:1 g/kg (14D preg):TER THERAP 22,325,67
scu-mus TDLo:1 g/kg (female 1D post):REP THERAP 22,325,67
scu-rat TDLo:1 g/kg (14D preg):REP THERAP 22,325,67
ivn-rbt TDLo:50 mg/kg (female 1D pre):REP THERAP 22,325,67
scu-rat TDLo:1 g/kg (female 1D pre):REP THERAP 22,325,67
ipr-rat LD50:1 g/kg THERAP 22,325,67
scu-rat LD50:5 g/kg THERAP 22,325,67
ipr-mus LD50:950 mg/kg THERAP 22,325,67
ivn-rbt LD50:250 mg/kg THERAP 22,325,67

SAFETY PROFILE: Poison by intravenous route. Moderately toxic by intraperitoneal route. An experimental teratogen. Other experimental reproductive effects.

CNH300 **CAS:67110-84-3** **HR: 3**
COMPOUND 20-438
mf: $C_{22}H_{27}N \cdot ClH$ mw: 341.96

SYN: (4aRS,5RS,9bRS)-2,3,4,4a,5,9b-HEXAHYDRO-2-ETHYL-7-METHYL-5-p-TOLYL-1H-INDENO(1,2-c)PYRIDINE HYDROCHLORIDE

TOXICITY DATA WITH REFERENCE
orl-mus TDLo:10 mg/kg (male 1D pre):REP MUREAV 66,113,79
orl-rat TDLo:7500 μg/kg (male 1D pre):REP TJADAB 24(2),50A,81
orl-rat TDLo:30 mg/kg (1D male):REP TJADAB 24(2),50A,81
orl-rat TDLo:30 mg/kg (male 1D pre):REP ARTODN 7,171,84
orl-mus LDLo:400 mg/kg MUREAV 66,113,79

SAFETY PROFILE: Poison by ingestion. Experimental reproductive effects. When heated to decomposition it emits toxic fumes of NO_x.

C

CNH375 **CAS:4091-50-3** **HR: 3**
COMPOUND 48/80
mf: $C_{10}H_{15}NO$ mw: 165.26

TOXICITY DATA WITH REFERENCE
ipr-mus TDLo:1200 mg/kg (1-6D preg):REP JRPFA4 6,179,63
ivn-rat LD50:2400 μg/kg AIPTAK 120,353,59
ivn-mus LD50:1950 μg/kg AIPTAK 120,353,59
ivn-rbt LD50:1500 μg/kg AIPTAK 120,353,59
ivn-gpg LD50:1290 μg/kg AIPTAK 120,353,59

SAFETY PROFILE: Poison by intravenous route. Experimental reproductive effects. When heated to decomposition it emits toxic fumes of NO_x. See also AMINES.

CNH500 **CAS:27114-11-0** **HR: 3**
COMPOUND 69/183
mf: $C_{22}H_{25}FN_2O \cdot 2ClH$ mw: 425.41

SYNS: 3-(γ-(p-FLUOROBENZOYL)PROPYL)-2,3,4,4a,5,6-HEXAHYDRO-1(H)-PYRAZINO(1,2A)QUINOLINE HCl □ 4′-FLURO-4-(1,2,4,4a,5,6-HEXAHYDRO-3H-PYRANZINO(1,2-A)QUINOLIN-3-YL)-BUTYROPHENONE 2HCl

TOXICITY DATA WITH REFERENCE
orl-rat LD50:800 mg/kg DRFUD4 4,185,79
ipr-rat LD50:161 mg/kg ARZNAD 28,1641,78
orl-mus LD50:1 g/kg DRFUD4 4,185,79
ipr-mus LD50:300 mg/kg JMCMAR 13,516,70
ivn-mus LD50:95 mg/kg ARZNAD 28,1641,78

SAFETY PROFILE: Poison by intraperitoneal and intravenous routes. Moderately toxic by ingestion. When heated to decomposition it emits very toxic fumes of F^-, NO_x, and HCl.

CNH550 **HR: 3**
COMPOUND 14045 METHIODIDE
mf: $C_{20}H_{32}NO \cdot I$ mw: 429.43

SYN: 1-(3-CYCLOHEXYL-3-HYDROXY-3-PHENYLPROPYL)-1-METHYLPYRROLIDINIUM IODIDE

TOXICITY DATA WITH REFERENCE
orl-mus LD50:515 mg/kg JAPMA8 43,408,54
ipr-mus LD50:86 mg/kg JAPMA8 43,408,54
ivn-mus LD50:12,510 μg/kg JAPMA8 43,408,54
scu-dog LDLo:100 mg/kg JAPMA8 43,408,54

SAFETY PROFILE: Poison by subcutaneous, intravenous, and intraperitoneal routes. Moderately toxic by ingestion. When heated to decomposition it emits toxic fumes of I^- and NO_x. See also IODIDES.

CNH625 **CAS:11028-71-0** **HR: 3**
CONCANAVALIN A

PROP: Bp: 141–142° @ 19 mm. The most extensively investigated member of the lectin family of plant proteins. Unlike most lectins, it lacks covalently bound carbohydrate and therefore is not a glycoprotein.

SYNS: CON A □ NSC 143504 □ RICIN-TOXIN CON A

TOXICITY DATA WITH REFERENCE
cyt-oin:emb 100 mg/L DKBSAS 223,316,75
cyt-mnl:emb 25 mg/L EXPEAM 23,1568,76
cyt-nml:emb 25 mg/L EXPEAM 32,1568,76
dns-mus-ivn 400 µg/kg JJMCAQ 36,43,83
dns-mus:oth 1 mg/L JJIND8 65,1321,80
ivn-mus TDLo:5 mg/kg (female 7D post):TER NDKIA2 32,230,81
par-ham TDLo:800 µg/kg (female 3D post):REP CCPTAY 35,507,87
ivn-mus TDLo:5 mg/kg (female 7D post):TER SEIJBO 19,125,79
ivn-mus TDLo:2500 µg/kg (female 7D post):TER SEIJBO 19,125,79
ipr-mus LD50:41,500 µg/kg NCISP* JAN86
ivn-mus LD50:50 mg/kg ARZNAD 30,759,80

SAFETY PROFILE: Poison by intravenous and intraperitoneal routes. An experimental teratogen. Other experimental reproductive effects. Mutation data reported. When heated to decomposition it emits toxic fumes of NO_x.

CNH650 **CAS:66231-56-9** **HR: 2**
CONCTASE C

TOXICITY DATA WITH REFERENCE
ipr-rat LD50:1050 mg/kg OYYAA2 19,503,80
scu-rat LD50:2660 mg/kg OYYAA2 19,503,80
ipr-mus LD50:735 mg/kg OYYAA2 19,503,80
scu-mus LD50:850 mg/kg OYYAA2 19,503,80

SAFETY PROFILE: Moderately toxic by subcutaneous and intraperitoneal routes.

CNH660 **HR: 3**
CONESSINE HYDROCHLORIDE
mf: $C_{24}H_{40}N_2 \cdot ClH$ mw: 393.12

SYNS: CHLORHYDRATE de CONESSINE (FRENCH) □ 3-β-(DIMETHYLAMINO)CON-5-ENINE HYDROCHLORIDE

TOXICITY DATA WITH REFERENCE
ipr-mus LDLo:126 mg/kg APFRAD 7,549,49
scu-mus LDLo:150 mg/kg APFRAD 7,549,49
ivn-gpg LDLo:105 mg/kg APFRAD 7,549,49
ivn-frg LDLo:100 mg/kg

SAFETY PROFILE: Poison by subcutaneous, intravenous, and intraperitoneal routes. When heated to decomposition it emits toxic fumes of NO_x and HCl.

CNH730 **CAS:1604-01-9** **HR: 3**
γ-CONICEIN
mf: $C_8H_{15}N$ mw: 125.24

PROP: Alkaline liquid or oil; mousy odor. Bp: 171°, volatile with steam, d: 0.8753, n (16/D) 1.4661. Sltly sol in water; freely sol in alc, chloroform, and ether.

SYNS: γ-CONICEINE □ 2,3,4,5-TETRAHYDRO-6-PROPYL-PYRIDINE

TOXICITY DATA WITH REFERENCE
orl-pig TDLo:17 mg/kg (female 30-45D post):TER AJVRAH 46,1368,85
orl-mus LD50:12 mg/kg JPPMAB 15,1,63
scu-mus LD50:12 mg/kg JPPMAB 15,1,63
ivn-mus LD50:2600 µg/kg JPPMAB 15,1,63

SAFETY PROFILE: Poison by ingestion, subcutaneous, and intravenous routes. An experimental teratogen. When heated to decomposition it emits toxic fumes of NO_x.

CNH750 **HR: 2**
CONIUM MACULATUM

PROP: Colorless, oily liquid with mousy odor. Bp: 166.5°, fp: −2.5°, d: 0.844–0.848 @ 20°/4°. The toxic component of poison hemlock (CTOXAO 12,49,78).

SYN: KURDUMANA, root extract

TOXICITY DATA WITH REFERENCE
ipr-rat LD50:750 mg/kg IJEBA6 16,228,78

SAFETY PROFILE: Moderately toxic by intraperitoneal route. Combustible when exposed to heat or flame.

CNH775 **CAS:8001-64-7** **HR: 3**
CONVALLARIN

TOXICITY DATA WITH REFERENCE
scu-mus LDLo:70 mg/kg 27ZWAY E,1,78,-
scu-rbt LDLo:10 mg/kg HBAMAK 4,1289,35
ivn-rbt LDLo:4 mg/kg HBAMAK 4,1289,35
orl-pgn LDLo:100 mg/kg HBAMAK 4,1289,35
scu-pgn LDLo:3 mg/kg HBAMAK 4,1289,35
orl-frg LDLo:200 mg/kg HBAMAK 4,1289,35
scu-frg LDLo:15 mg/kg HBAMAK 4,1289,35

SAFETY PROFILE: Poison by ingestion, subcutaneous, and intravenous routes.

CNH780 **CAS:508-75-8** **HR: 3**
CONVALLATOXIN
mf: $C_{29}H_{42}O_{10}$ mw: 550.71

PROP: Crystals or prisms from methanol or ether. Mp: 235–242°. Sol in alc, acetone; sltly sol in chloroform, ethyl acetate, and water; practically insol in ether, petr ether.

SYNS: CONVALLAOTOXIN □ CONVALLATON □ CONVALLATOXOSIDE □ CONVALLOTOXIN □ CORGLYCON □ CORGLYCONE □ CORGLYKON □ KORGLYKON □ MANNOPYRANOSIDE, STROPHANTHIDIN-3 6-DEOXY-, α-l- □ RHAMNOSIDE, STROPHANTHIDIN-3, α-l- □ STROPHANTHIDIN-3-(6-DEOXY-α-l-MANNOPYRANOSIDE) □ STROPHANTHIDIN-α-l-RHAMNOSIDE

C

TOXICITY DATA WITH REFERENCE
ivn-rat LD50:15 mg/kg ARZNAD 11,848,61
ipr-mus LD50:10 mg/kg AIPTAK 155,165,65
scu-mus LD50:15 mg/kg FATOAO 44,342,81
ivn-mus LD50:1 mg/kg CSLNX* NX#00462
ivn-mky LDLo:90 µg/kg ARZNAD 13,412,63
ipr-cat LD50:200 µg/kg AIPTAK 155,165,65
ivn-cat LD50:76 mg/kg ARZNAD 13,220,63
idu-cat LDLo:1290 µg/kg ARZNAD 20,229,70
ivn-pig LDLo:50 µg/kg ARZNAD 20,229,70
idu-pig LDLo:3170 µg/kg ARZNAD 20,229,70

SAFETY PROFILE: Poison by subcutaneous, intravenous, intraduodenal, intraperitoneal, and possibly other routes. When heated to decomposition it emits acrid smoke and fumes.

CNH785 CAS:3253-62-1 HR: 3
CONVALLATOXOL

mf: $C_{29}H_{44}O_{10}$ mw: 552.73

PROP: Needles or prisms from MeOH/Et_2O. Mp: 170–172°.

SYNS: CONVALLOTOXOL □ CONVALOTOXOL □COVALLATOXOL □ α-l-STROPHANTHIDOL-3,6-DEOXY-MANNOPYRANOSIDE

TOXICITY DATA WITH REFERENCE
ivn-cat LD50:87 µg/kg JPETAB 111,365,54
ivn-rat LD50:56 mg/kg AIPTAK 155,165,65
ipr-mus LD50:30 mg/kg AIPTAK 155,165,65
ipr-cat LD50:130 µg/kg AIPTAK 155,165,65

SAFETY PROFILE: Poison by intravenous and intraperitoneal routes. When heated to decomposition it emits acrid smoke and fumes.

CNH789 HR: 3
COONTIE

PROP: The coontie looks like a low palm with a short trunk and a few 2-foot long pinnate leaves. It produces seed cones. It grows on the southeastern coast of Georgia, Florida, the Bahamas, the northwestern coast of Jamaica, the Dominican Republic, and Puerto Rico.

SYNS: BAY BUSH (BAHAMAS) □ COMPTIE □ FLORIDA ARROWROOT □ GUAYIGA (DOMINICAN REPUBLIC) □ MARUNGUEY (PUERTO RICO) □ PALMITA de JARDIN (PUERTO RICO) □ SAGO CYCAS □ SEMINOLE BREAD □ YUGILLA (CUBA) □ ZAMIA PUMILA

SAFETY PROFILE: The roots and trunk contain the poison cycasin. Ingestion of these plant parts may cause persistent vomiting, diarrhea, colic, depression, and muscular paralysis. Washing the grated root with water renders it edible. See also CYCASIN.

CNH792 CAS:8001-61-4 HR: 2
COPAIBA OIL

PROP: From steam distillation of South American *Copaifera* L. (Fam. Leguminosae) balsam. Colorless to yellow liquid; characteristic odor, aromatic, slightly bitter taste. D: 0.880–0.907; ref. index: 1.493–1.500 @ 20°. Sol in alc, fixed oils, mineral oil.

SYNS: BALSAM CAPTIVI □ BALSAMS, COPAIBA □ COPAIBA BALSAM □ COPAIBA OLEORESIN □ JESUIT'S BALSAM

TOXICITY DATA WITH REFERENCE
orl-rat LD50:5 g/kg FCTXAV 14,687,76

CONSENSUS REPORTS: Reported in EPA TSCA Inventory.

SAFETY PROFILE: Mildly toxic by ingestion. Large doses cause vomiting and diarrhea. Can also cause dermatitis and kidney damage. When heated to decomposition it emits acrid smoke and irritating fumes.

CNH800 CAS:11078-23-2 HR: 2
COPIAMYCIN

mf: $C_{55}H_{97}NO_{22}$ mw: 1124.53

PROP: Crystals from MeOH. Mp: 144° (decomp).

SYN: NSC-110326

TOXICITY DATA WITH REFERENCE
orl-mus LD50:1030 mg/kg 85GDA2 7,53,81
ipr-mus LD50:1200 mg/kg 85GDA2 7,53,81
scu-mus LD50:1050 mg/kg 85GDA2 7,53,81

SAFETY PROFILE: Moderately toxic by ingestion, subcutaneous, and intraperitoneal routes. When heated to decomposition it emits toxic fumes of NO_x.

CNI000 CAS:7440-50-8 HR: 2
COPPER

mf: Cu mw: 63.54

PROP: Reddish, malleable and ductile metal. Slowly weathers to green patina. Mp: 1083°, bp: 25° @ 2595 mm, d: 8.92, vap press: 1 mm @ 1628°.

SYNS: ALLBRI NATURAL COPPER □ ANAC 110 □ ARWOOD COPPER □ BRONZE POWDER □ CDA 101 □ CDA 102 □ CDA 110 □ CDA 122 □ C.I. 77400 □ C.I. PIGMENT METAL 2 □ COPPER-AIRBORNE □ COPPER BRONZE □ COPPER-MILLED □ COPPER SLAG-AIRBORNE □ COPPER SLAG-MILLED □ 1721 GOLD □ GOLD BRONZE □ KAFAR COPPER □ M1 (COPPER) □ M2 (COPPER) □ OFHC Cu □ RANEY COPPER

TOXICITY DATA WITH REFERENCE
orl-rat TDLo:152 mg/kg (22W pre):TER GISAAA 45(3),8,80
iut-rat TDLo:250 µg/kg (female 1D pre):REP IJEBA6 19,1124,81
orl-rat TDLo:1210 µg/kg (female 35W pre):REP GISAAA 42(8),30,77
orl-rat TDLo:1520 µg/kg (female 22W pre):TER GISAAA 45(3),8,80
ipl-rat TDLo:100 mg/kg:ETA AIHAAP 41,836,80
orl-hmn TDLo:120 µg/kg:GIT PHRPA6 73,910,58

CONSENSUS REPORTS: Reported in EPA TSCA Inventory. Copper and its compounds are on the Community Right-To-Know List.

OSHA PEL: TWA (dust, mist) 1 mg(Cu)/m³; (fume) 0.1 mg/m³
ACGIH TLV: TWA (dust, mist) 1 mg(Cu)/m³; (fume) 0.2 mg/m³
DFG MAK: (dust) 1 mg/m³; (fume) 0.1 mg/m³

SAFETY PROFILE: Questionable carcinogen with experimental tumorigenic data. Experimental teratogenic and reproductive effects. Human systemic effects by ingestion: nausea and vomiting. See also COPPER COMPOUNDS. Liquid copper explodes on contact with water. Potentially explosive reaction with actylenic compounds, 3-bromopropyne, ethylene oxide, lead azide, and ammonium nitrate. Ignites on contact with chlorine, chlorine trifluoride, fluorine (above 121°), and hydrazinium nitrate (above 70°). Reacts violently with C_2H_2, bromates, chlorates, iodates, ($Cl_2 + OF_2$), dimethyl sulfoxide + trichloroacetic acid, ethylene oxide, H_2O_2, hydrazine mononitrate, hydrazoic acid, H_2S + air; $Pb(N_3)_2$, K_2O_2, NaN_3, Na_2O_2 sulfuric acid. Incandescent reaction with potassium dioxide. Incompatible with 1-bromo-2-propyne.

For occupational chemical analysis use OSHA: #ID-125G or NIOSH: Copper, 7029; Welding and Brazing Fume, 7200; Elements, 7300.

CNI250 CAS:142-71-2 HR: 3
COPPER ACETATE
mf: $C_4H_6O_4 \cdot Cu$ mw: 181.64

PROP: Greenish-blue powder or small crystals.

SYNS: ACETATE de CUIVRE (FRENCH) □ ACETIC ACID, CUPRIC SALT □ COPPER(2+) ACETATE □ COPPER(II) ACETATE □ COPPER DIACETATE □ COPPER(2+) DIACETATE □ CRYSTALLIZED VERDIGRIS □ CRYSTALS of VENUS □ CUPRIC ACETATE □ CUPRIC DIACETATE □ NEUTRAL VERDIGRIS □ OCTAN MEDNATY (CZECH)

TOXICITY DATA WITH REFERENCE
scu-rat TDLo:40 mg/kg (7-10D preg):REP CRSBAW 166,1237,72
orl-rat LD50:595 mg/kg MarJV# 29MAR77
scu-rat LD50:350 mg/kg PMDCAY 15,211,77
ipr-mus LD50:2500 µg/kg BCPCA6 30,771,81

CONSENSUS REPORTS: Reported in EPA TSCA Inventory. Copper and its compounds are on the Community Right-To-Know List.

ACGIH TLV: TWA 1 mg(Cu)/m³

SAFETY PROFILE: Poison by subcutaneous and intraperitoneal routes. Moderately toxic by ingestion. Experimental reproductive effects. When heated to decomposition it emits acrid smoke and irritating fumes. See also COPPER COMPOUNDS.

CNI325 CAS:6046-93-1 HR: 3
COPPER(II) ACETATE MONOHYDRATE
mf: $C_4H_6O_4 \cdot Cu \cdot H_2O$ mw: 199.66

PROP: Green crystals. Mp: 115°. Sol in H_2O and EtOH; sltly sol in Et_2O.

SYNS: COPPER(2+) ACETATE, MONOHYDRATE □ COPPER DIACETATE MONOHYDRATE □ CUPRIC ACETATE MONOHYDRATE

TOXICITY DATA WITH REFERENCE
orl-rat LD50:710 mg/kg AIHAAP 30,470,69
orl-mus LDLo:1600 mg/kg AECTCV 14,111,85
ipr-mam LD50:5 mg/kg JANSAG 55,337,82

CONSENSUS REPORTS: Copper and its compounds are on the Community Right-To-Know List.

SAFETY PROFILE: Poison by intraperitoneal route. Moderately toxic by ingestion. See also COPPER COMPOUNDS

CNI500 CAS:12540-13-5 HR: 3
COPPER ACETYLIDE
mf: C_2Cu mw: 87.56

PROP: A black or brown solid.

SYN: COPPER CARBIDE

CONSENSUS REPORTS: Copper and its compounds are on the Community Right-To-Know List.

ACGIH TLV: TWA 1 mg(Cu)/m³
DOT CLASSIFICATION: Forbidden

SAFETY PROFILE: Ignites and then explodes when heated to 100°C. Much more sensitive to impact, friction, and heat than copper(I) acetylide (the red-brown form). See also COPPER COMPOUNDS and ACETYLIDES.

CNI600 CAS:11133-98-5 HR: 3
COPPER ALLOY, Cu, Be

SYN: COPPER-BERYLLIUM ALLOY

TOXICITY DATA WITH REFERENCE
ihl-hmn TCLo:300 ng(Be)/m³:PUL AEHLAU 9,473,64

CONSENSUS REPORTS: IARC Cancer Review: Group 1 IMEMDT 58,41,93; Human Sufficient Evidence IMEMDT 58,41,93; Animal Sufficient Evidence IMEMDT 1,17,72; Animal Sufficient Evidence IMEMDT 23,143,80; Animal Sufficient Evidence IMEMDT 58,41,93. Copper and its compounds, as well as beryllium and its compounds, are on the Community Right-To-Know List.

OSHA PEL: TWA 0.002 mg(Be)/m³; STEL 0.005 mg(Be)/m³/30M; CL 0.025 mg(Be)/m³
ACGIH TLV: TWA 0.002 mg(Be)/m³, Suspected Human Carcinogen.
NIOSH REL: (Beryllium) CL not to exceed 0.0005 mg(Be)/m³

SAFETY PROFILE: Confirmed carcinogen. Cases of berylliosis have been reported from exposure to so-called low beryllium alloys. Human systemic effects by inhalation: dyspnea, fibrosing alveolitis, weight loss or decreased weight gain. See also BERYLLIUM COMPOUNDS and COPPER COMPOUNDS. When heated to decomposition it emits very toxic fumes of BeO.

CNI900 HR: 3
COPPER ARSENATE HYDROXIDE
mf: $AsCu_2HO_5$ mw: 283.01

PROP: A green solid.

SYNS: COPPER ARSENATE (BASIC) □ CUPROUS ARSENATE, BASIC

CONSENSUS REPORTS: Copper and its compounds as well as arsenic and its compounds are on the Community Right-To-Know List.

NIOSH REL: CL 2 µg/m³/15M

SAFETY PROFILE: A poison by various routes. See also ARSENIC COMPOUNDS and COPPER COMPOUNDS. When heated to decomposition it emits toxic fumes of As.

CNJ750 CAS:12069-69-1 HR: 3
COPPER(II) CARBONATE HYDROXIDE (2:1:2)
mf: $CO_3 \cdot H_2O_2 \cdot 2Cu$ mw: 221.11

PROP: Green powder. Mp: decomp @ 200°, d: 4.0.

SYNS: BASIC COPPER CARBONATE □ BASIC CUPRIC CARBONATE □ (CARBONATO)DIHYDROXYDICOPPER □ CHESTNUT COMPOUND □ COPPER CARBONATE HYDROXIDE □ CUPRIC CARBONATE □ DICOPPER DIHYDROXYCARBONATE □ KOP KARB □ KUPFERCARBONAT (GERMAN) □ MALACHITE

TOXICITY DATA WITH REFERENCE
orl-rat LD50:1350 mg/kg SKEZAP 26,605,85
orl-rbt LD50:159 mg/kg GUCHAZ 6,128,73
orl-pgn LDLo:1000 mg/kg AUVJA2 16,147,40
orl-dck LDLo:900 mg/kg AUVJA2 16,147,40
orl-brd LD50:900 mg/kg PCOC** -,258,66

CONSENSUS REPORTS: Copper and its compounds are on the Community Right-To-Know List. Reported in EPA TSCA Inventory.

SAFETY PROFILE: Poison by ingestion. When heated to decomposition it emits acrid smoke and fumes. See also COPPER COMPOUNDS.

CNJ900 CAS:26500-47-8 HR: 2
COPPER CHLORATE
DOT: UN 2721
mf: $HClO_3 \cdot xCu$ mw: 529.24

SYNS: CHLORIC ACID, COPPER SALT □ COPPER CHLORATE (DOT)

ACGIH TLV: TWA 1 mg(Cu)/m³
DOT CLASSIFICATION: 5.1; *Label:* Oxidizer

SAFETY PROFILE: An oxidizer. When heated to decomposition it emits toxic fumes of Cl^-.

CNK250 CAS:7758-89-6 HR: 3
COPPER(I) CHLORIDE
mf: ClCu mw: 98.99

PROP: Cubic, white crystals. Stable in dry air; becomes green on exposure to moist air and brown on exposure to light. Undergoes a phase transition at 4°. Mp: 430°, bp: 1490°, d: 3.53, vap press: 1 mm @ 546°.

SYNS: CHLORID MEDNY (CZECH) □ COPPER MONOCHLORIDE □ CUPROUS CHLORIDE □ CUPROUS DICHLORIDE □ DICOPPER DICHLORIDE

TOXICITY DATA WITH REFERENCE
cyt-rat/ast 120 mg/kg GANNA2 54,155,63
orl-rat LD50:140 mg/kg EQSSDX 1,1,75
orl-mus LD50:347 mg/kg GTPZAB 35(2),42,91
scu-gpg LD50:100 mg/kg EQSSDX 1,1,75

CONSENSUS REPORTS: Copper and its compounds are on the Community Right-To-Know List. Reported in EPA TSCA Inventory. EPA Genetic Toxicology Program.

SAFETY PROFILE: Poison by ingestion and subcutaneous routes. Mutation data reported. Reacts violently with potassium or with lithium nitride + heat. When heated to decomposition it emits toxic fumes of Cl^-. See also CHLORIDES and COPPER COMPOUNDS.

CNK500 CAS:7447-39-4 HR: 3
COPPER(II) CHLORIDE (1:2)
DOT: UN 2802
mf: Cl_2Cu mw: 134.44

PROP: Yellowish-brown, hygroscopic powder. Mp: 498°, d: 3.054.

SYNS: COPPER BICHLORIDE □ COPPER(2+) CHLORIDE □ COPPER(II) CHLORIDE □ CUPRIC CHLORIDE □ CUPRIC DICHLORIDE

TOXICITY DATA WITH REFERENCE
dnd-mic-uns 2 mmol/L SCIEAS 198,513,77
mmo-smc 100 µmol/L CPBTAL 33,1571,85
ipr-mus TDLo:35 mg/kg (female 12-18D post):REP TXCYAC 64,19,90
ipr-mus LD50:7400 µg/kg AEPPAE 244,17,62
ivn-mus LD50:17,500 µg/kg EJMCA5 19,425,84
ihl-rat TCLo:20 µg/m³/26W-I GISAAA 52(7),74,87

CONSENSUS REPORTS: Copper and its compounds are on the Community Right-To-Know List. Reported in EPA TSCA Inventory.

DOT CLASSIFICATION: 8; *Label:* Corrosive

SAFETY PROFILE: Poison by intravenous and intraperitoneal routes. Experimental reproductive effects. Mutation data reported. See also COPPER COMPOUNDS and CHLORIDES. Can react violently with K and Na. When heated to decomposition it emits toxic fumes of Cl^-.

CNK559 CAS:1332-40-7 HR: 3
COPPER CHLORIDE, mixed with COPPER OXIDE, HYDRATE
mf: $Cl_2Cu_4H_6O_6$ mw: 427.12

SYNS: AGRIZAN □ BASF-GRUNKUPFER □ BASIC COPPER CHLORIDE □ BLITOX □ BLITOX 50 □ BLUE COPPER □ BLUE COPPER-50 □ CHEMOCIN □ CHEMPAR □ COBOX □ COBOX BLUE □ COLLOIDOX □ COPPER CHLORIDE, BASIC □ COPPER CHLORIDE OXIDE □ COPPER CHLORIDE OXIDE, HYDRATE (9CI) □ COPPER CHLOROXIDE □ COPPER OC FUNGICIDE □ COPPER OXYCHLORIDE □ COPPERSAN □ COPPESAN □ COPPESAN BLUE □ COPRANTOL □ COPREX □ COPROSAN BLUE □ COP-TOX □ COXYSAN □ CU-56 □ CUPRAL 45 □ CUPRAMAR □ CUPRAMER □ CUPRANTOL □ CUPRAVET □ CUPRAVIT □ CUPRAVIT FORTE □ CUPRAVIT GREEN □ CUPRICOL □ CUPRIC OXIDE CHLORIDE □ CUPRITOX □ CUPROKYLT □ CUPROL □ CUPROSAN □ CUPROSANA □ CUPROSAN BLUE □ CUPROVINOL □ CUPROX □ CUPROXOL □ DEVICOPPER □ FALIGRUEN □ FYCOL 8 □ FYTOLAN □ KAURITIL □ KILEX □ KT 35 □ KUPFEROXYCHLORID (GERMAN) □ KUPRICOL □ KUPRIKOL □ MICROCOP □ MIEDZIAN □ MIEDZIAN 50 □ NEORAM BLU □ OXICOB □ OXIVOR □ OXYCHLORUE DE CUIVRE □ OXYCLOR □ OXYCUR □ PARRYCOP □ PEPROSAN □ RECOP □ RHODIACUIVRE □ TAMRAGHOL □ TRICOP 50 □ VIRICUIVRE □ VITIGRAN □ VITIGRAN BLUE

TOXICITY DATA WITH REFERENCE
orl-hmn LDLo:200 mg/kg FAONAU 53A,43,74
orl-rat LD50:700 mg/kg PHJOAV 185,361,60

CONSENSUS REPORTS: Copper and its compounds are on the Community Right-To-Know List.

SAFETY PROFILE: A human poison by ingestion. A pesticide. When heated to decomposition it emits toxic fumes of Cl^-. See also COPPER COMPOUNDS and individual components.

CNK599 **HR: 3**
COPPER(I) CHLOROACETYLIDE
mf: $C_4Cl_2Cu_2$ mw: 246.04

CONSENSUS REPORTS: Copper and its compounds are on the Community Right-To-Know List.

SAFETY PROFILE: An explosive. When heated to decomposition it emits toxic fumes of Cl^-. See also COPPER COMPOUNDS, CHLORIDES, and ACETYLIDES.

CNK609 **HR: 3**
COPPER CHROMATE
mf: $CrCuO_4$ mw: 179.54

CONSENSUS REPORTS: Copper and its compounds, as well as chromium and its compounds, are on the Community Right-To-Know List.

SAFETY PROFILE: Ignites on contact with hydrogen sulfide gas. See also COPPER COMPOUNDS and CHROMIUM COMPOUNDS.

CNK625 **CAS:866-82-0** **HR: 2**
COPPER(I) CITRATE
mf: $C_6H_4O_7{\cdot}2Cu$ mw: 315.18

SYNS: CITRIC ACID, COPPER(2+) SALT (8CI) □ COPPER CITRATE □ COPPER SALT 2-HYDROXY-1,2,3-PROPANETRICARBOXYLIC ACID (1:2) □ CUPRIC CITRATE □ CUPROCITROL □ 2-HYDROXY-1,2,3-PROPANETRICARBOXYLIC ACID COPPER(2+) SALT (1:2) (9CI)

TOXICITY DATA WITH REFERENCE
ivn-ham TDLo:1800 µg/kg (8D preg):REP BIREBV 11,97,74
ipr-ham TDLo:2700 µg/kg (8D preg):TER TJADAB 21,89,80
ivn-ham TDLo:1800 µg/kg (8D preg):TER BIREBV 11,97,74
orl-rat LD50:1580 mg/kg MarJV# 29MAR77

CONSENSUS REPORTS: Copper and its compounds are on the Community Right-To-Know List. Reported in EPA TSCA Inventory.

SAFETY PROFILE: Moderately toxic by ingestion. Experimental reproductive effects. An experimental teratogen. See also COPPER COMPOUNDS.

CNK700 **CAS:55158-44-6** **HR: 3**
COPPER-COBALT-BERYLLIUM

SYNS: COPPER ALLOY, Cu, Be, Co □ BERYLLIUM-COPPER-COBALT ALLOY

CONSENSUS REPORTS: IARC Cancer Review: Group 1 IMEMDT 58,41,93; Human Sufficient Evidence IMEMDT 58,41,93; Animal Sufficient Evidence IMEMDT 1,17,72; Animal Sufficient Evidence IMEMDT 23,143,80; Animal Sufficient Evidence IMEMDT 58,41,93. Copper, cobalt, and beryllium and their compounds are on the Community Right-To-Know List.

OSHA PEL: TWA 0.002 mg(Be)/m^3; STEL 0.005 mg(Be)/m^3/30M; CL 0.025 mg(Be)/m^3
ACGIH TLV: TWA 0.002 mg(Be)/m^3, Suspected Human Carcinogen
NIOSH REL: (Beryllium) CL not to exceed 0.0005 mg (Be)/m^3; (Cobalt) Insufficient evidence for recommending limit

SAFETY PROFILE: Confirmed carcinogen. See COPPER, BERYLLIUM and COBALT COMPOUNDS. When heated to decomposition it emits very toxic fumes of BeO.

CNK750 **HR: 3**
COPPER COMPOUNDS

CONSENSUS REPORTS: Copper and its compounds are on the Community Right-To-Know List.

SAFETY PROFILE: As the sublimed oxide, copper may be responsible for one form of metal fume fever. In animals, inhalation of copper dust has caused hemolysis of the red blood cells, deposition of hemofuscin in the liver and pancreas, and injury to the lung cells; injection of the dust has caused cirrhosis of the liver and pancreas, and a condition closely resembling hemochromatosis, or bronzed diabetes. However, considerable trial exposure to copper compounds has not resulted in such disease. As regards local effect, copper chloride and sulfate have been reported as causing irritation of the skin and conjunctivae, possibly on an allergic basis. Cuprous oxide is irritating to the eyes and upper respiratory tract. Discoloration of the skin is often seen in persons handling copper, but this does not indicate any actual injury. There is an excess of cancer cases in the copper smelting industry. In humans the ingestion of a large quantity of copper sulfate has caused vomiting, gastric pain, dizziness, exhaustion, anemia, cramps, convulsions, shock, coma, and death. Symptoms attributed to damage to the nervous system and kidney have been recorded, jaundice has been observed, and, in some cases, the liver has been enlarged. Deaths have been reported to have occurred following the ingestion of as little as 27 g of the salt, while other victims have recovered after having taken up to 120 g. Many copper-containing compounds are used as fungicides. Many copper salts form highly unstable acetylides. Those formed in basic solutions from (Cu^+ salts + C_2H_2) are less stable than those formed from Cu^{++} salts. Copper salts + hydrazine react strongly, and with nitromethane these salts are explosive.

C

CNL000 **CAS:544-92-3** **HR: 3**
COPPER CYANIDE
DOT: UN 1587
mf: CCuN mw: 89.56

PROP: Monoclinic, white prisms; white-cream powder; dark-green orthorhombic crystals; dark red monoclinic crystals. Mp: 474° in N_2, bp: decomp, d: 2.92. Sol in NH_3 (aq); insol in H_2O and alcohols.

SYNS: COPPER(I) CYANIDE □ CUPRICIN □ CUPROUS CYANIDE □ RCRA WASTE NUMBER P029

CONSENSUS REPORTS: Reported in EPA TSCA Inventory. Cyanide and its compounds, as well as copper and its compounds, are on the Community Right-To-Know List.

ACGIH TLV: TWA 1 mg(Cu)/m^3
DOT CLASSIFICATION: 6.1; *Label:* Poison

SAFETY PROFILE: A poison. Reacts violently with magnesium. When heated to decomposition it emits very toxic CN^- and NO_x. See also CYANIDE and COPPER COMPOUNDS.

CNL250 **CAS:14763-77-0** **HR: 3**
COPPER(II) CYANIDE
mf: C_2CuN_2 mw: 115.58

PROP: Yellowish-green powder. Mp: decomp before melting.

SYNS: COPPER CYANIDE (DOT) □ COPPER CYANAMIDE □ CUPRIC CYANIDE (DOT) □ CYANURE de CUIVRE (FRENCH)

TOXICITY DATA WITH REFERENCE
ipr-rat LDLo:50 mg/kg NCNSA6 5,27,53

CONSENSUS REPORTS: Copper and its compounds, as well as cyanide and its compounds, are on the Community Right-To-Know List.

ACGIH TLV: TWA 1 mg(Cu)/m^3
DOT CLASSIFICATION: 6.1; *Label:* Poison, KEEP AWAY FROM FOOD

SAFETY PROFILE: Poison by intraperitoneal route. See also CYANIDE and COPPER COMPOUNDS. Incompatible with magnesium. When heated to decomposition it emits toxic fumes of NO_x and CN^-.

CNL500 **CAS:137-29-1** **HR: 3**
COPPER DIMETHYLDITHIOCARBAMATE
mf: $C_6H_{12}N_2S_4 \cdot Cu$ mw: 303.98

SYNS: COMPOUND-4018 □ CUMATE □ DIMETHYLDITHIOCARBAMIC ACID COPPER SALT □ WOLFEN

TOXICITY DATA WITH REFERENCE
mmo-sat 2 mg/plate MUREAV 68,313,79
mma-sat 1 mg/plate MUREAV 68,313,79
ihl-rat LCLo:210 mg/m^3/4H EPASR* 8EHQ-0786-0616
ipr-rat LDLo:25 mg/kg NCNSA6 5,14,53
ipr-mus LDLo:8 mg/kg CBCCT* 4,227,52

CONSENSUS REPORTS: Reported in EPA TSCA Inventory. Copper and its compounds are on the Community Right-To-Know List.

SAFETY PROFILE: Poison by inhalation and intraperitoneal route. Mutation data reported. See also COPPER COMPOUNDS and CARBAMATES. When heated to decomposition it emits very toxic fumes of NO_x and SO_x.

CNL625 **CAS:32061-49-7** **HR: 3**
COPPER(II)-1,3-DI(5-TETRAZOLYL)TRIAZENIDE
mf: $C_4H_4CuN_{22}$ mw: 423.77

Cu(-NN=NN= CN=NNHC=NN=NNH)$_2$

CONSENSUS REPORTS: Copper compounds are on the Community Right-To-Know List.

SAFETY PROFILE: A heat- and friction-sensitive explosive. Upon decomposition it emits toxic fumes of NO_x. See also COPPER COMPOUNDS and EXPLOSIVES.

CNL750 **CAS:54453-03-1** **HR: 3**
COPPER EDTA COMPLEX

SYN: (ETHYLENEDINITRILO)TETRAACETIC ACID COPPER(II) COMPLEX

TOXICITY DATA WITH REFERENCE
ipr-mus LD50:2090 μg(Cu)/kg PABIAQ 11,853,63

CONSENSUS REPORTS: Copper and its compounds are on the Community Right-To-Know List. Reported in EPA TSCA Inventory.

SAFETY PROFILE: Poison by intraperitoneal route. See also COPPER COMPOUNDS. When heated to decomposition it emits toxic fumes of NO_x.

CNM000 **HR: 3**
COPPER FUME
mf: Cu mw: 63.54

SYN: MIEDZ (POLISH)

TOXICITY DATA WITH REFERENCE
ihl-hmn TCLo:1 mg/m^3:IRR DTLVS* 3,59,71

CONSENSUS REPORTS: Copper and its compounds are on the Community Right-To-Know List.

OSHA PEL: Fume: TWA 0.1 mg/m^3
ACGIH TLV: Fume: TWA 0.2 mg/m^3
DFG MAK: Fume: 0.1 mg/m^3

SAFETY PROFILE: Human systemic irritant by inhalation. See also COPPER COMPOUNDS.

CNM250 **HR: 3**
COPPER(I) HYDRIDE
mf: Cu_2H_2 mw: 129.11

PROP: Red-brown crystals. Mp: decomp @ 60°, d: 6.38.

CONSENSUS REPORTS: Copper and its compounds are on the Community Right-To-Know List.

SAFETY PROFILE: Ignites spontaneously in air. Ignites on contact with fluorine, bromine or iodine. See also COPPER COMPOUNDS and HYDRIDES.

CNM500 **CAS:20427-59-2** **HR: 2**
COPPER HYDROXIDE
mf: $H_2O_2 \cdot Cu$ mw: 97.56

PROP: Blue, gelatinous or amorphous powder. D: 3.368.

SYNS: COMAC □ COPPER DIHYDROXIDE □ COPPER(2+) HYDROXIDE □ CUPRAVIT BLAU □ CUPRAVIT BLUE □ CUPRIC HYDROXIDE □ KOCIDE □ KUPRABLAU □ PARASOL

TOXICITY DATA WITH REFERENCE
orl-rat LD50:1 g/kg FMCHA2-,C81,91
orl-qal LD50:3400 mg/kg PEMNDP 9,184,91

CONSENSUS REPORTS: Copper and its compounds are on the Community Right-To-Know List. Reported in EPA TSCA Inventory.

SAFETY PROFILE: Moderately toxic by ingestion. See also COPPER COMPOUNDS.

CNM600 **CAS:1333-22-8** **HR: 2**
COPPER HYDROXIDE SULFATE
mf: $Cu_4H_6O_{10}S$ mw: 452.28

SYNS: COPPER OXYSULFATE □ CUPROXAT □ CUPROXAT FLOWABLE □ TOP COP TRI BASIC □ TRIBASIC COPPER SULFATE

TOXICITY DATA WITH REFERENCE
orl-rat LD50:2500 mg/kg FMCHA2-,C87,91
skn-rat LD50:>2 g/kg FMCHA2-,C87,91

CONSENSUS REPORTS: Reported in EPA TSCA Inventory.

SAFETY PROFILE: Moderately toxic by ingestion. Low toxicity by skin contact. When heated to decomposition it emits toxic vapors of SO_x and Cu.

CNM750 **CAS:3251-23-8** **HR: 2**
COPPER(II) NITRATE
mf: CuN_2O_6 mw: 187.55

PROP: Large, blue-green, deliquescent, orthorhombic crystals. D: 2.047, sublimes at 150–225°, mp: 255–256°. Sol in water, ethyl acetate, dioxane; dissolves in and reacts vigorously with ether.

SYNS: COPPER DINITRATE □ COPPER(2+) NITRATE □ CUPRIC DINITRATE □ CUPRIC NITRATE (DOT)

TOXICITY DATA WITH REFERENCE
skn-rbt 500 mg SEV FCTOD7 20,563,82
eye-rbt 100 mg SEV FCTOD7 20,573,82
eye-rbt 100 mg/4S rns SEV FCTOD7 20,573,82
orl-rat LD50:940 mg/kg EQSSDX 1,1,75

CONSENSUS REPORTS: Copper and its compounds are on the Community Right-To-Know List. Reported in EPA TSCA Inventory.

ACGIH TLV: TWA 1 mg(Cu)/m^3

SAFETY PROFILE: Moderately toxic by ingestion. A severe eye and skin irritant. Potentially explosive reaction above 220°C with ammonium or potassium hexacyanoferrate(II). Reaction with ammonia + potassium amide gives explosive product. Violent reaction with acetic anhydride. May ignite on prolonged contact with paper. Concentrated solutions may ignite in contact with tin or aluminum foil. Used as a fungicide, herbicide, and as a catalyst component in solid rocket fuel. When heated to decomposition it emits toxic fumes of NO_x. See also COPPER COMPOUNDS and NITRATES.

CNN000 **CAS:10031-43-3** **HR: 2**
COPPER(II) NITRATE, TRIHYDRATE (1:2:3)
mf: $N_2O_6 \cdot Cu \cdot 3H_2O$ mw: 241.62

PROP: Emerald green to dark green transparent blue, deliquescent crystals. Mp: 114.5°, d: 2.047. Sol in H_2O forming acidic soln; sol in EtOH; sltly sol in EtOAc.

SYNS: COPPER DINITRATE TRIHYDRATE □ CUPRIC NITRATE TRIHYDRATE □ GERHARDITE □ NITRIC ACID COPPER(2+) SALT TRIHYDRATE

TOXICITY DATA WITH REFERENCE
cyt-rat/ast 600 mg/kg GANNA2 54,155,63
orl-rat LD50:940 mg/kg AIHAAP 30,470,69

CONSENSUS REPORTS: Copper and its compounds are on the Community Right-To-Know List.

SAFETY PROFILE: Moderately toxic by ingestion. Mutation data reported. Can ignite on prolonged contact with paper. Can explode when finely mixed with potassium ferrocyanide. Concentrated solutions may ignite in contact with tin foil. See also COPPER COMPOUNDS. When heated to decomposition it emits toxic fumes of NO_x.

CNN250 **HR: 3**
COPPER(I) NITRIDE
mf: Cu_3N mw: 123.63

PROP: Dark-green powder. Mp: decomp @ 300°, d: 5.84 @ 25°/4°.

CONSENSUS REPORTS: Copper and its compounds are on the Community Right-To-Know List.

SAFETY PROFILE: Explodes on heating in air. When mixed with HNO_3 (conc) explodes violently. See also COPPER COMPOUNDS and NITRIDES.

CNN399 **HR: 3**
COPPER 1,3,5-OCTATRIEN-7-YNIDE
mf: C_8H_7Cu mw: 166.69

$H(CH{=}CH)_3C{\equiv}CCu$

CONSENSUS REPORTS: Copper and its compounds are on the Community Right-To-Know List.

SAFETY PROFILE: Explodes when heated in air. When heated to decomposition it emits acrid smoke and fumes. See also COPPER COMPOUNDS and ACETYLENE COMPOUNDS.

CNN500 **CAS:10290-12-7** **HR: 3**
COPPER ORTHOARSENITE
DOT: UN 1586
mf: $AsCuHO_3$ mw: 187.47

PROP: Yellowish-green powder. Mp: decomp.

SYNS: ACID COPPER ARSENITE □ AIR-FLO GREEN □ ARSONIC ACID, COPPER(2+) SALT (1:1) (9CI) □ COPPER ARSENITE, solid (DOT) □ CUPRIC ARSENITE □ CUPRIC GREEN □ SCHEELES GREEN □ SCHEELE'S MINERAL □ SWEDISH GREEN

CONSENSUS REPORTS: Arsenic and its compounds, as well as copper and its compounds, are on the Community Right-To-Know List.

OSHA PEL: Cancer Hazard
ACGIH TLV: TWA 0.2 mg(As)/m³ (Proposed: 0.01 mg (As)/m³; Human Carcinogen)
NIOSH REL: (Arsenic, Inorganic) CL 0.002 mg(As) /m³/15M
DOT CLASSIFICATION: 6.1; *Label:* Poison

SAFETY PROFILE: Poison. When heated to decomposition it emits toxic fumes of As. See also ARSENIC COMPOUNDS and COPPER COMPOUNDS.

CNN750 CAS:53421-36-6 HR: 2
COPPER(I) OXALATE
mf: $C_2Cu_2O_4$ mw: 183.10

PROP: Solid, light bluish-green powder.

SYN: CUPRIC OXALATE

CONSENSUS REPORTS: Copper and its compounds are on the Community Right-To-Know List.

SAFETY PROFILE: Explodes weakly when heated slightly. See also OXALATES and COPPER COMPOUNDS.

CNN755 CAS:814-91-5 HR: 2
COPPER(II) OXALATE
mf: $C_2H_2O_4{\cdot}Cu$ mw: 153.58

SYNS: COPPER OXALATE □ CUPRIC OXALATE □ ETHANEDIOIC ACID, COPPER(2+) SALT (1:1) □ OXALIC ACID, COPPER(2+) SALT (1:1)

DOT CLASSIFICATION: 6.1; *Label:* KEEP AWAY FROM FOOD

SAFETY PROFILE: Explodes weakly when heated slightly. See also OXALATES and COPPER COMPOUNDS.

CNO000 CAS:1317-39-1 HR: 2
COPPER(I) OXIDE
mf: Cu_2O mw: 143.08

PROP: Octahedral, cubic, red or yellow crystals or powder. Mp: 1235°, bp: loses O_2 @ 1800° (decomp), d: 6.0. Insol in water.

SYNS: BROWN COPPER OXIDE □ C.I. 77402 □ COPOX □ COPPER NORDOX □ COPPER SARDEX □ CUPPER OXIDE (RUSSIAN) □ CUPROUS OXIDE □ DICOPPER MONOXIDE □ FUNGIMAR □ KUPFEROXYDUL (GERMAN) □ OLEOCUIVRE □ OLEO NORDOX □ RED COPPER OXIDE □ YELLOW CUPROCIDE

TOXICITY DATA WITH REFERENCE
ihl-rat TCLo:11 μg/m³/24H (14W male):REP GISAAA 41(6),8,76
orl-rat LD50:470 mg/kg AIHAAP 30,470,69

CONSENSUS REPORTS: Reported in EPA TSCA Inventory. Copper and its compounds are on the Community Right-To-Know List.

SAFETY PROFILE: Moderately toxic by ingestion. Experimental reproductive effects. A fungicide. Violent, potentially explosive reaction with concentrated peroxyformic acid. Violent reaction when heated with aluminum. See also COPPER COMPOUNDS.

For occupational chemical analysis use NIOSH: Copper, 7029; Elements, 7300.

CNO250 CAS:1317-38-0 HR: 3
COPPER(II) OXIDE
mf: CuO mw: 79.54

PROP: Fine brownish-black powder or crystals. Bp: decomp @ 1026°, d: 6.4, mp: 1326°. Insol in water and alc.

SYNS: BLACK COPPER OXIDE □ C.I. 77403 □ C.I. PIGMENT BLACK 15 □ COPPER BROWN □ COPPER(II) OXIDE □ COPPER MONOOXIDE □ COPPER MONOXIDE □ COPPER(2+) OXIDE □ CUPRIC OXIDE

TOXICITY DATA WITH REFERENCE
itr-rat LDLo:278 mg/kg TPKVAL 11,47,69

CONSENSUS REPORTS: Copper and its compounds are on the Community Right-To-Know List.

SAFETY PROFILE: Poison by intratracheal route. Used as fungicide. Also a trace mineral added to animal feeds. Explodes when heated with powdered aluminum; anilinium perchlorate; hydrogen; magnesium; phthalic anhydride. Ignites on contact with dichloromethylsilane; hydrogen sulfide; hydrogen trisulfide. Incandescent reaction when heated with boron; rubidium acetylide (at 350°C); potassium; sodium; phospham. Reacts violently with $CsHC_2$; hydrazine; PN_2H; Ti; Zr. Incompatible with metals and reductants (e.g., hydroxylamine or hydrazine). See also COPPER COMPOUNDS.

For occupational chemical analysis use NIOSH: Copper, 7029.

CNO325 CAS:15061-57-1 HR: 3
COPPER(I) PERCHLORATE
mf: $Cu_2Cl_2O_8$ mw: 325.99

CONSENSUS REPORTS: Copper compounds are on the Community Right-To-Know List.

SAFETY PROFILE: Mixtures with 1,4-oxathiane, carbon monoxide, or alkenes (e.g. ethylene, allene, or 1,3-butadiene) form heat-sensitive explosive complexes. See also COPPER COMPOUNDS and PERCHLORATES.

CNO350 CAS:10294-46-9 HR: 3
COPPER(II) PERCHLORATE
mf: $CuCl_2O_8$ mw: 262.45

CONSENSUS REPORTS: Copper compounds are on the Community Right-To-Know List.

SAFETY PROFILE: Forms explosive complexes with

polyfunctional amines. See also COPPER COMPOUNDS and PERCHLORATES.

CNO500 CAS:17031-32-2 HR: 3
COPPER(II) PERCHLORATE, DIHYDRATE
mf: $Cl_2O_8{\cdot}Cu{\cdot}2H_2O$ mw: 298.48

SYNS: CUPRIC DIPERCHLORATE TETRAHYDRATE □ PERCHLORIC ACID, COPPER(II) SALT, DIHYDRATE (8CI, 9CI)

TOXICITY DATA WITH REFERENCE
ipr-rat LD50:29 mg/kg JAFCAU 14,512,66

CONSENSUS REPORTS: Copper and its compounds are on the Community Right-To-Know List.

DOT CLASSIFICATION: 5.1; *Label:* Oxidizer

SAFETY PROFILE: Poison by intraperitoneal route. See also COPPER COMPOUNDS and COPPER(II) PERCHLORATE. An oxidizer. When heated to decomposition it emits toxic fumes of Cl^-.

CNO750 CAS:34461-68-2 HR: 3
COPPER(II) PHOSPHINATE
mf: $CuH_2O_4P_2$ mw: 191.50

$Cu(OPHOH)_2$

CONSENSUS REPORTS: Copper and its compounds are on the Community Right-To-Know List.

SAFETY PROFILE: Explodes when heated to about 90°C or on impact. When heated to decomposition it emits toxic fumes of PO_x. See also COPPER COMPOUNDS.

CNP000 CAS:13991-87-2 HR: 3
COPPER SORBATE
mf: $C_{12}H_{14}O_4{\cdot}Cu$ mw: 285.80

SYN: SORBIC ACID, COPPER SALT

TOXICITY DATA WITH REFERENCE
orl-mus LD50:2800 mg/kg CHDDAT 266,1080,68
ipr-mus LDLo:15 mg/kg CHDDAT 266,1080,68

CONSENSUS REPORTS: Copper and its compounds are on the Community Right-To-Know List.

SAFETY PROFILE: Poison by intraperitoneal routes. Moderately toxic by ingestion. See also COPPER COMPOUNDS. When heated to decomposition it emits acrid smoke and irritating fumes.

CNP250 CAS:7758-98-7 HR: 3
COPPER(II) SULFATE (1:1)
mf: $O_4S{\cdot}Cu$ mw: 159.60

PROP: Blue or white rhombic crystals or crystalline granules or hygroscopic powder. D: 2.284, mp: 200°. Very sol in H_2O; sol in MeOH or glycerol; sltly sol in EtOH.

SYNS: BCS COPPER FUNGICIDE □ BLUE COPPER □ BLUE STONE □ BLUE VITRIOL □ COPPER MONOSULFATE □ COPPER SULFATE □ CP BASIC SULFATE □ CUPRIC SULFATE □ KUPFERSULFAT (GERMAN) □ ROMAN VITRIOL □ SULFATE de CUIVRE (FRENCH) □ SULFURIC ACID, COPPER(2+) SALT (1:1) □ TNCS 53 □ TRINAGLE

TOXICITY DATA WITH REFERENCE
dni-mus-ipr 20 g/kg ARGEAR 51,605,81
dns-ham:emb 200 μmol/L MUREAV 131,173,84
ivn-mus TDLo:3200 μg/kg (female 8D post):TER WRABDT 186,297,79
ipr-rat TDLo:7500 μg/kg (3D preg):REP BECTA6 25,702,80
ivn-mus TDLo:3200 μg/kg (female 7D post):REP WRABDT 186,297,79
itt-rat TDLo:3192 μg/kg (male 1D pre):REP JRPFA4 7,21,64
scu-mus TDLo:12,768 μg/kg (30D male):REP JRPFA4 7,21,64
ivn-ham TDLo:2130 μg/kg (female 8D post):TER BIREBV 11,97,74
par-ckn TDLo:10 mg/kg:ETA BEXBAN 9,519,40
orl-man LDLo:857 mg/kg:GIT ATXKA8 17,20,58
orl-cld TDLo:150 mg/kg:SYS,BLD AJDCAI 131,149,77
orl-hmn LDLo:50 mg/kg JAMAAP 235,801,76
orl-hmn TDLo:11 mg/kg:GIT LANCAO 2,700,60
orl-rat LD50:300 mg/kg 36SBA8 1,507,77
scu-rat LD50:43 mg/kg PESTD5 16,252,75
unr-rat LD50:520 mg/kg GTPZAB 26(6),21,82
ipr-mus LD50:18 mg/kg COREAF 256,1043,63
scu-mus LDLo:500 μg/kg TJIZAF 48,313,78
ivn-mus LDLo:50 mg/kg HBTXAC 1,76,56
ivn-rbt LD50:10 mg/kg JIDHAN 31,301,49

CONSENSUS REPORTS: Copper and its compounds are on the Community Right-To-Know List. Reported in EPA TSCA Inventory. EPA Genetic Toxicology Program.

ACGIH TLV: TWA 1 $mg(Cu)/m^3$

SAFETY PROFILE: A human poison by ingestion. An experimental poison by ingestion, subcutaneous, parenteral, intravenous, and intraperitoneal routes. Human systemic effects by ingestion: gastritis, diarrhea, nausea or vomiting, damage to kidney tubules, and hemolysis. Questionable carcinogen with experimental tumorigenic data. An experimental teratogen. Other experimental reproductive effects. Mutation data reported. Reacts violently with hydroxylamine; magnesium. See also COPPER COMPOUNDS and SULFATES. When heated to decomposition it emits toxic fumes of SO_x.

CNP500 CAS:7758-99-8 HR: 3
COPPER(II) SULFATE PENTAHYDRATE (1:1:5)
mf: $O_4S{\cdot}Cu{\cdot}5H_2O$ mw: 249.70

PROP: Blue crystals from water. Mp: loses $4H_2O$ @ 110°. Very sol in H_2O; sol in MeOH and glycerol; sltly sol in EtOH.

SYNS: BLUE COPPERRAS □ BLUESTONE □ BLUE VITRIOL □ COPPERFINE-ZINC □ CSP □ CUPRIC SULFATE PENTAHYDRATE □ KUPFERSULFAT-PENTAHYDRAT (GERMAN) □ KUPFERVITRIOL (GERMAN) □ ROMAN VITRIOL □ SALZBURG VITRIOL □ SULFURIC ACID, COPPER(2+) SALT, PENTAHYDRATE □ TRIANGLE

TOXICITY DATA WITH REFERENCE
dni-hmn:lym 76 μmol/L IAAAAM 79,83,86
cyt-rat/ast 300 mg/kg GANNA2 54,155,63

orl-hmn TDLo: 272 mg/kg: LIV,KID,BLD IPRAA8 18,807,65
orl-hmn LDLo: 1088 mg/kg IPRAA8 18,807,65
unk-man LDLo: 221 mg/kg 85DCAI 2,73,70
orl-rat LD50: 300 mg/kg 85ARAE 2,182,77
ipr-rat LD50: 18,700 μg/kg GISAAA 53(4),78,88
ipr-mus LD50: 33 mg/kg BCPCA6 14,289,65
orl-dog LDLo: 60 mg/kg HBAMAK 4,1289,35
scu-gpg LDLo: 62 mg/kg BMJOAE 2,217,13
orl-pgn LDLo: 1000 mg/kg AUVJA2 16,147,40
orl-dom LDLo: 5 mg/kg JCVPAR 82,47,72
orl-mam LD50: 470 mg/kg FMCHA2 -,C62,83
ipr-mam LD50: 7500 μg/kg JANSAG 55,337,82
orl-bwd LDLo: 300 mg/kg AUVJA2 16,147,40

CONSENSUS REPORTS: Copper and its compounds are on the Community Right-To-Know List.

SAFETY PROFILE: A human poison by an unspecified route. Moderately toxic to humans by ingestion. An experimental poison by ingestion, subcutaneous, and intraperitoneal routes. Human systemic effects by ingestion: jaundice, unspecified urinary system effects, and hemolysis. Human mutation data reported. Used widely as a fungicide. When heated to decomposition it emits toxic fumes of SO_x. See also COPPER COMPOUNDS and COPPER(II) SULFATE.

CNQ000 CAS:1317-40-4 HR: 3
COPPER (II) SULFIDE
mf: CuS mw: 95.60

PROP: Found in nature as the mineral covellite, or indigo copper (blue, hexagonal or monoclinic crystals). Black powder or crystals. Stable in air when dry. Oxidizes to $CuSO_4$ in moist air. Mp: transition @ 103°; bp: 200° (decomp @ 220)°; d: 4.6. Practically insol in water, alc, dil acids, alkalies; sol in KCN soln, NH_4OH, hot HNO_3.

SYNS: C.I. 77450 □ C.I. PIGMENT BLUE 34 □ COPPER BLUE □ COPPER MONOSULFIDE □ COPPER SULFIDE □ COPPER(2+) SULFIDE □ CUPRIC SULFIDE □ HORACE VERNET'S BLUE □ MONOCOPPER MONOSULFIDE □ OIL BLUE

TOXICITY DATA WITH REFERENCE
cyt-rat/ast 150 mg/kg GANNA2 54,155,63
otr-ham: emb 5 mg/L CNREA8 42,2757,82
dnd-ham: ovr 10 mg/L CRNGDP 3,657,82

CONSENSUS REPORTS: Copper and its compounds are on the Community Right-To-Know List. Reported in EPA TSCA Inventory.

SAFETY PROFILE: Mutation data reported. Explodes on contact with magnesium chlorate, zinc chlorate, cadmium chlorate, or concentrated solutions of chloric acid. Can react violently with H_2O_2, NH_4MgNO_3 + water, $Zn(ClO_3)_2$. See also COPPER COMPOUNDS and SULFIDES.

CNQ250 CAS:62126-20-9 HR: 3
COPPER(I) TETRAHYDROALUMINATE
mf: $AlCuH_4$ mw: 94.55

PROP: Unstable, decomp @ −70°C.

CONSENSUS REPORTS: Copper and its compounds are on the Community Right-To-Know List.

SAFETY PROFILE: Ignites spontaneously in air. Unstable. See also COPPER COMPOUNDS and ALUMINUM COMPOUNDS.

CNQ375 CAS:25267-55-4 HR: 3
COPPER-2,4,5-TRICHLOROPHENOLATE
mf: $C_{12}H_6Cl_6O_2$•Cu mw: 458.42

SYNS: COPPER TRICHLOROPHENOLATE □ CTCP □ TRIKHLORFENOLYAT MEDI (RUSSIAN)

TOXICITY DATA WITH REFERENCE
orl-rat LD50: 5500 mg/kg 85GMAT -,40,82
unr-rat LDLo: 100 mg/kg KSKZAN 18(5),55,80
orl-mus LD50: 3333 mg/kg 85GMAT-,40,82
orl-rbt LD50: 1537 mg/kg 85GMAT -,40,82
ihl-gpg LCLo: 200 mg/m³/1H 85GMAT -,40,82

CONSENSUS REPORTS: Copper and its compounds are on the Community Right-To-Know List.

SAFETY PROFILE: Poison by inhalation and possibly other routes. Moderately toxic by ingestion. When heated to decomposition it emits toxic fumes of Cl^-. See also COPPER COMPOUNDS and CHLOROPHENOLS.

CNQ500 HR: 2
COPPER-ZINC ALLOYS
mf: Cu-Zn mw: 128.91

CONSENSUS REPORTS: Copper and its compounds, as well as zinc and its compounds, are on the Community Right-To-Know List.

SAFETY PROFILE: Potentially explosive reaction with alkyl bromides. Violent reaction with diiodomethane + ether. See also COPPER COMPOUNDS and ZINC COMPOUNDS.

CNQ750 HR: 3
COPPER-ZINC CHROMATE COMPLEX

SYNS: CHROMIC ACID, COPPER-ZINC-COMPLEX □ CRAG FUNGICIDE 658 □ ZINC-COPPER CHROMATE COMPLEX

CONSENSUS REPORTS: Copper, zinc, chromium and their compounds are on the Community Right-To-Know List.

OSHA PEL: CL 0.1 mg(CrO_3)/m³
ACGIH TLV: TWA 0.01 mg(Cr)/m³; Confirmed Human Carcinogen
DFG MAK: Human Carcinogen.
NIOSH REL: (Chromium (VI)) TWA 0.001 mg(Cr(VI))/m³

SAFETY PROFILE: A poison. See also CHROMIUM COMPOUNDS, ZINC COMPOUNDS, and COPPER COMPOUNDS. A fungicide. When heated to decomposition it emits toxic fumes of ZnO.

CNR000 **CAS:8001-31-8** **HR: 3**
COPRA (OIL)
DOT: UN 1363

PROP: From the kernel of the fruit of the coconut palm *Cocos nucifera*. Fatty solid or liquid; sweet, nutty taste. Mp: 21–27°.

SYNS: COCONUT BUTTER □ COCONUT MEAL PELLETS, containing 6–13% moisture and no more than 10% residual fat (DOT) □ COCONUT OIL (FCC) □ COCONUT PALM OIL □ COPRA (DOT) □ COPRA PELLETS (DOT) □ FREE COCONUT OIL

CONSENSUS REPORTS: Reported in EPA TSCA Inventory.

DOT CLASSIFICATION: 4.2; *Label:* Spontaneously Combustible

SAFETY PROFILE: Flammable solid when exposed to heat or flame. May spontaneously heat and ignite if stored wet and hot.

CNR125 **CAS:314-35-2** **HR: 3**
CORAFIL
mf: $C_{13}H_{21}N_5O_2$ mw: 279.39

PROP: Waxy solid. Mp: 75°. Very sol in water, acetone; sltly sol in ethanol, ether.

SYNS: CAMPHOPHYLINE □ DIAETHYLAMINOAETHYL-THEOPHYLLIN (GERMAN) □ 7-(2-(DIETHYLAMINO)ETHYL)-3,7-DIHYDRO-1,3-DIMETHYL-1H-PURINE-2,6-DIONE (9CI) □ DIETHYLAMINOETHYL THEOPHYLLINE □ 7-(2-DIETHYLAMINO)ETHYLTHEOPHYLLINE □ 7-(DIETHYLAMINOETHYL)THEOPHYLLINE □ ETAMINOPHYLLINE □ ETAMIPHYLLIN □ ETAMIPHYLLINE □ MILLIPHYLLINE □ MILLOPHYLLINE □ PAREPHYLLIN □ QUERYL □ R-3588 □ SOLUFILINA □ SOLUPHYLINE

TOXICITY DATA WITH REFERENCE
orl-mus LD50:1237 mg/kg JPETAB 116,343,56
ipr-mus LD50:254 mg/kg JPETAB 116,343,56
scu-mus LD50:183 mg/kg ARZNAD 4,649,54

SAFETY PROFILE: Poison by subcutaneous and intraperitoneal routes. Moderately toxic by ingestion. When heated to decomposition it emits toxic fumes of NO_x.

CNR135 **HR: 3**
CORAL PLANT

PROP: The various species of Jatropha range in size from shrubs to small trees. Most are perennials. They bear a 3-sided seed capsule, and each section carries one seed. They are commonly cultivated, and most grow wild in tropical areas.

SYNS: BARBADOS NUT □ BELLYACHE BUSH □ CORAL VEGETAL (CUBA) □ CUIPU (MEXICO) □ FRAILECILLO (CUBA) □ GOUT STALK □ HIGUERETA CIMARRONA □ JATROPHA CATHARTICA □ JATROPHA CURCAS □ JATROPHA GOSSYPIIFOLIA □ JATROPHA INTEGERRIMA □ JATROPHA MACRORHIZA □ JATROPHA MULTIFIDA □ JATROPHA PODAGRICA □ JICAMILLA (MEXICO, TEXAS) □ MALA MUJER (MEXICO) □ PEREGRINA (PUERTO RICO, CUBA, US) □ PHYSIC NUT □ PINON (PUERTO RICO, DOMINICAN REPUBLIC) □ ROSE-FLOWERED JATROPHA □ SPICY JATROPHA □ TARTAGO (PUERTO RICO) □ TAUTUBA (PUERTO RICO) □ TINAJA (PUERTO RICO) □ TUATUA (PUERTO RICO)

SAFETY PROFILE: The seeds contain the poison jatrophin (curcin) a toxalbumin that inhibits protein synthesis in the intestinal wall. Ingestion of only one seed can rapidly cause severe nausea, vomiting, and diarrhea, with resulting fluid and electrolyte loss. See also ABRIN.

CNR150 **HR: 3**
CORAL SNAKE VENOM

SYNS: M.F. FULVIUS VENOM □ M. FULVIUS FULVIUS VENOM □ MICRURUS FULVIUS FULVIUS VENOM

TOXICITY DATA WITH REFERENCE
ipr-mus LD50:538 μg/kg TOXIA6 13,139,75
ivn-mus LD50:225 μg/kg TOXIA6 13,139,75
ims-mus LD50:200 μg/kg BIJOAK 193,899,81

SAFETY PROFILE: Deadly poison by intramuscular, intravenous, and intraperitoneal routes.

CNR250 **CAS:61503-59-1** **HR: 3**
CORALYNE SULFOACETATE
mf: $C_{22}H_{22}NO_4 \cdot C_2H_3O_5S$ mw: 503.56

SYNS: NSC 154890 □ 2,3,10,11-TETRAMETHOXY-8-METHYL-DIBENZO(a,g)QUINOLIZINIUM SALT with SULFOACETIC ACID (1:1)

TOXICITY DATA WITH REFERENCE
ipr-mus LD50:173 mg/kg NCISP* JAN86
ivn-mus LD50:221 mg/kg TXAPA9 37,165,76
ivn-dog LDLo:37 mg/kg TXAPA9 37,165,76

SAFETY PROFILE: Poison by intravenous and intraperitoneal routes. When heated to decomposition it emits very toxic fumes of SO_x and NO_x.

CNR500 **CAS:6452-71-7** **HR: 3**
CORETAL
mf: $C_{15}H_{23}NO_3$ mw: 265.39

SYNS: 1-(o-ALLYLOXY)PHENOXY)-3-(ISOPROPYLAMINO)-2-PROPANOL □ 1-(ISOPROPYLAMINO)-2-HYDROXY-3-(o-(ALLYLOXY)PHENOXY)PROPANE □ OXPRENOLOL □ 1-((1-METHYLETHYL)AMINO)-3-(2-(2-PROPENYLOXY)PHENOXY)-2-PROPANOL

TOXICITY DATA WITH REFERENCE
orl-hmn TDLo:50 mg/kg:CVS,PUL BMJOAE 1,776,77
orl-rat LD50:730 mg/kg ARZNAD 18,164,68
scu-rat LD50:940 mg/kg ARZNAD 18,164,68
ivn-rat LD50:33 mg/kg ARZNAD 18,164,68
ipr-mus LD50:170 mg/kg PJPPAA 25,145,73
scu-mus LD50:245 mg/kg ARZNAD 18,164,68
ivn-mus LD50:20 mg/kg ARZNAD 27,1022,77
orl-cat LD90:200 mg/kg ARZNAD 18,164,68
ivn-rbt LD90:20 mg/kg ARZNAD 18,164,68

SAFETY PROFILE: Poison by ingestion, subcutaneous, intravenous, and intraperitoneal routes. Human systemic effects by ingestion: decreased pulse rate and blood pressure, and unspecified changes in the cardiovascular and respiratory systems. When heated to decomposition it emits toxic fumes of NO_x. See also ALLYL COMPOUNDS.

CNR675 **CAS:42200-33-9** **HR: 3**
CORGARD
mf: $C_{17}H_{27}NO_4$ mw: 309.45

PROP: Crystalline powder. Mp: 124–136°. Freely sol in alc, propylene glycol; sltly sol in chloroform. Insol in acetone, benzene, ether, and hexane.

SYNS: ANABET □ 1-(tert-BUTYLAMINO)-3-((5,6,7,8-TETRAHYDRO-cis-6,7-DIHYDROXY-1-NAPHTHYL)OXY)-2-PROPANOL □ 5-(3-((1,1-DIMETHYLETHYL)AMINO)-2-HYDROXYPROPOXY)-1,2,3,4-TETRAHYDRO-2,3-NAPHTHALENEDIOL □ NADOLOL □ SOLGOL □ SQ 11725 □ 2,3-cis-1,2,3,4-TETRAHYDRO-5-((2-HYDROXY-3-tert-BUTYLAMINO)PROPOXY)-2,3-NAPHTHALENEDIOL

TOXICITY DATA WITH REFERENCE
orl-rat TDLo:550 mg/kg (7-17D preg):REP YACHDS 11,5119,83
orl-rat TDLo:1650 mg/kg (female 7-17D post):REP YACHDS 11,5119,83
orl-rat TDLo:42 g/kg (male 9W pre):REP YACHDS 11,5119,83
orl-rat TDLo:550 mg/kg (7-17D preg):TER YACHDS 11,5119,83
orl-rbt TDLo:1300 mg/kg (female 6-18D post):REP TXAPA9 44,379,78
orl-rat TDLo:42 g/kg (male 9W pre):REP YACHDS 11,5119,83
orl-man TDLo:571 μg/kg:EYE AIMEAS 97,454,82
orl-rat LD50:5300 mg/kg MEIEDD 10,909,83
ipr-rat LD50:322 mg/kg IYKEDH 16,1461,85
ivn-rat LD50:59,200 μg/kg IYKEDH 16,1461,85
orl-mus LD50:3800 mg/kg YACHDS 9,527,81
ipr-mus LD50:298 mg/kg IYKEDH 16,1461,85
ivn-mus LD50:47,100 μg/kg IYKEDH 16,1461,85

SAFETY PROFILE: Poison by intravenous and intraperitoneal routes. Moderately toxic by ingestion. Human systemic effects by ingestion: visual field changes. An experimental teratogen. Other experimental reproductive effects. When heated to decomposition it emits toxic fumes of NO_x. Used to reduce cardiac arrhythmias, lower blood pressure and treat angina pectoris.

CNR725 **CAS:2571-86-0** **HR: 3**
CORIAMYRTIN
mf: $C_{15}H_{18}O_5$ mw: 278.33

PROP: Bitter, monoclinic prisms or crystals. Mp: 229–230°. Sltly sol in water, cold alc; freely sol in hot alc and in ether.

SYNS: CORIAMYRTINE □ CORIAMYRTIONE

TOXICITY DATA WITH REFERENCE
scu-rat LDLo:1 mg/kg JPETAB 57,410,36
ivn-rat LDLo:700 μg/kg JPETAB 57,410,36
ipr-mus LD50:3 mg/kg JMCMAR 11,729,68
scu-mus LD50:3234 μg/kg JAPMA8 29,2,40
ivn-mus LDLo:1 mg/kg JPETAB 57,410,36
unr-mus LD50:1 mg/kg MEIEDD 10,361,83
unr-cat LDLo:237 μg/kg JPETAB 57,361,36
scu-rbt LD50:930 μg/kg JAPMA8 29,2,40
ivn-rbt LD50:371 μg/kg JAPMA8 29,2,40
scu-gpg LDLo:2439 μg/kg JPETAB 57,361,36
scu-frg LDLo:10 mg/kg JPETAB 57,410,36
unr-frg LD50:5800 μg/kg MEIEDD 10,361,83

SAFETY PROFILE: Poison by subcutaneous, intravenous, intraperitoneal, and possibly other routes. When heated to decomposition it emits acrid smoke and fumes.

CNR735 **CAS:8008-52-4** **HR: 2**
CORIANDER OIL

PROP: From steam distillation of ripe fruit of *Coriandrum sativum* L. (Fam. Umbelliferae). Colorless liquid; characteristic odor and taste. D: 0.863–0.875, refr index: 1.462 @ 20°.

SYNS: OIL of CORIANDER □ OILS, CORIANDER

TOXICITY DATA WITH REFERENCE
skn-rbt 500 mg/24H FCTXAV 11,1077,73
dnr-bcs 10 mg/disc TOFOD5 8,91,85
orl-rat LD50:4130 mg/kg FCTXAV 11,1077,73
orl-mus LD50:3520 mg/kg TOFOD5 8,91,85

CONSENSUS REPORTS: Reported in EPA TSCA Inventory.

SAFETY PROFILE: Moderately toxic by ingestion. Mutation data reported. A skin irritant. When heated to decomposition it emits acrid smoke and fumes.

CNR740 **HR: 3**
CORIARI MYRTIFOLIA

PROP: Small shrubs or trees that produce small green flowers and purple-black berries. They are grown as ornamentals in the southern U.S. and California.

SAFETY PROFILE: The fruit contains the poison coriamyrtin. Ingestion may cause convulsions similar to those produced by picrotoxin. See also PICROTOXIN.

CNR750 **CAS:20153-98-4** **HR: 3**
CORMELIAN
mf: $C_{31}H_{44}N_2O_{10}\cdot 2ClH$ mw: 677.69

PROP: Solid monohydrate. Mp: 194–198°.

SYNS: ASTA C □ N,N'-(BIS-ω-HYDROXYPROPYL)HOMOPIPERAZINE 3,4,5-TRIMETHOXYBENZOATE DIHYDROCHLORIDE □ N,N'-BIS(3-(3,4,5-TRIMETHOXYBENZOYLOXY)PROPYL)HOMOPIPERAZINE DIHYDROCHLORIDE □ 1,4-BIS(3-(3,4,5-TRIMETHOXYBENZOYLOXY)-PROPYL)PERHYDRO-1,4-DIAZEPINE DIHYDROCHLORIDE □ COMELIAN □ DILAZEP DIHYDROCHLORIDE

TOXICITY DATA WITH REFERENCE
orl-mus TDLo:6 g/kg (female 7-12D post):TER KSRNAM 8,2084,74
orl-rat TDLo:30 g/kg (female 30D pre):REP KSRNAM 8,2050,74
orl-rat TDLo:60 g/kg (30D male):REP KSRNAM 8,2050,74
orl-rat TDLo:1750 mg/kg (female 7-13D post):TER KSRNAM 8,4368,74
ipr-rat LD50:85 mg/kg ARZNAD 22,667,72
scu-rat LD50:369 mg/kg KSRNAM 8,2044,74
ivn-rat LD50:13,700 μg/kg KSRNAM 8,2044,74
ivn-rat LD50:2860 mg/kg KSRNAM 8,2044,74
ipr-mus LD50:161 mg/kg ARZNAD 22,667,72
scu-mus LD50:154 mg/kg ARZNAD 22,667,72
ivn-mus LD50:16,800 μg/kg KSRNAM 8,2044,74

ivn-dog LD50:11 mg/kg ARZNAD 22,667,72

SAFETY PROFILE: Poison by subcutaneous, intravenous, and intraperitoneal routes. Moderately toxic by ingestion. An experimental teratogen. Other experimental reproductive effects. Used as a coronary vasodilator and as an anti-anginal agent. When heated to decomposition it emits very toxic fumes of HCl and NO_x.

CNR825 **HR: 3**
CORMELIAN-DIGOTAB
mf: $C_{31}H_{44}N_2O_{10}•2ClH•C_{43}H_{66}O_{15}$ mw: 1500.78

SYNS: ASTA CD 072 □ DILAZEP/β-ACETYLDIGOXIN

TOXICITY DATA WITH REFERENCE
ipr-rat LD50:83,200 μg/kg ARZNAD 24,1914,74
orl-mus LD50:1980 mg/kg ARZNAD 24,1914,74
ipr-mus LD50:142 mg/kg ARZNAD 24,1914,74
orl-dog LD50:422 μg/kg ARZNAD 24,1914,74

SAFETY PROFILE: Poison by intraperitoneal route. Moderately toxic by ingestion. When heated to decomposition it emits toxic fumes of NO_x and HCl. See also CORMELIAN.

CNS000 **CAS:8001-30-7** **HR: 1**
CORN OIL

PROP: Light-yellow, clear, oily liquid; faint characteristic odor. Mp: −10°, flash p: 490°F (CC), d: 0.92, autoign temp: 740°F. From wet milling of *Zea mays* (85DIA2 2,70,77).

TOXICITY DATA WITH REFERENCE
skn-hmn 300 mg/3D-I MLD 85DKA8 -,127,77
orl-rat TDLo:12,500 mg/kg (15-19D preg):TER TJADAB 27,75A,83

CONSENSUS REPORTS: Reported in EPA TSCA Inventory.

SAFETY PROFILE: Human skin irritant. An experimental teratogen. May be an allergen. Combustible liquid when exposed to heat or flame. Dangerous spontaneous heating may occur during storage if leaks impregnate rags, waste, etc. To fight fire, use CO_2, dry chemical.

CNS200 **CAS:4503-12-2** **HR: 2**
CORONARIDINE HYDROCHLORIDE
mf: $C_{21}H_{26}N_2O_2•ClH$ mw: 374.95

SYNS: (−)-CORONARIDINE MONOHYDROCHLORIDE □ IBOGAMINE-18-CARBOXYLIC ACID, METHYL ESTER, HYDROCHLORIDE □ IBOGAMINE-18-CARBOXYLIC ACID, METHYL ESTER, MONOHYDROCHLORIDE (9CI) □ 6,9-METHANO-8H-PYRIDO(1′,2′:1,2)AZEPINO(4,5-b)INDOLE-6(6aH)-CARBOXYLIC ACID, 7,8,9,10,12,13- HEXAHYDRO-6-ETHYL-13a-HYDROXY-, METHYL ESTER, HYDROCHLORIDE

TOXICITY DATA WITH REFERENCE
orl-rat TDLo:50 mg/kg (female 10D pre):REP JPMSAE 62,1199,73
orl-mus LDLo:500 mg/kg JPMSAE 52,598,63
ipr-cat LDLo:20 mg/kg JPMSAE 52,598,63
ivn-cat LDLo:10 mg/kg JPMSAE 52,598,63

SAFETY PROFILE: Moderately toxic by ingestion, intraperitoneal, and intravenous routes. Experimental reproductive effects. When heated to decomposition it emits toxic fumes of NO_x and HCl.

CNS750 **CAS:50-23-7** **HR: 3**
CORTISOL
mf: $C_{21}H_{30}N_{40}O_5S$ mw: 954.97

PROP: Striated blocks from 2-propanol. Mp: 220° (decomp).

SYNS: AEROSEB-HC □ ANTI-INFLAMMATORY HORMONE □ BARSEB HC □ CETACORT □ COBADEX □ COMPOUND F □ CORTDOME □ CORTISOL ALCOHOL □ CORTISPRAY □ DERMACORT □ EF CORLIN □ GENACORT □ HC □ HEB-CORT □ HIDRO-COLISONA □ 11-β-HYDROCORTISONE □ HYDROCORTISONE FREE ALCOHOL □ HYDROCORTISYL □ HYDROCORTONE □ 17-HYDROXYCORTICOSTERONE □ 11-β-HYDROXYCORTISONE □ HYTONE LOTION □ KENDALL'S COMPOUND F □ NSC 10483 □ OPTEF □PERMI-CORT □ 4-PREGNENE-11-β,17-α,21-TRIOL 3,20-DIONE □ REICHSTEIN'S SUBSTANCE M □ SCHEROSON F □ 11-β,17,21-TRIHYDROXYPREGN-4-ENE-3,20-DIONE □ 11-β,17-α-21-TRIHYDROXY-4-PREGNENE-3,20-DIONE

TOXICITY DATA WITH REFERENCE
dns-rat-ipr 20 mg/kg BEXBAN 94,1511,82
dni-rat:mmr 1 mg/L AMBPBZ 85,57,77
cyt-mus-ipr 50 mg/kg PSEBAA 171,109,82
dni-gpg:lng 1 mg/L PSEBAA 171,109,82
dns-ckn:emb 10 mg/L ITCSAF 20,172,84
ipr-rat TDLo:80 mg/kg (female 14-15D post):REP BBRCA9 75,125,77
scu-rat TDLo:330 mg/kg (female 1-22D post):REP AMEBA7 40,297,62
scu-rbt TDLo:6 mg/kg (female 24-26D post):REP PEREBL 12,38,78
ims-mus TDLo:200 mg/kg (female 12D post):TER JCGBDF 7,341,87
ims-rat TDLo:500 mg/kg (female 13D post):TER TCMUD8 3,313,83
par-rat TDLo:50 mg/kg (female 16-18D post):TER PEREBL 11,282,77
ims-mus TDLo:400 mg/kg (female 11-14D post):REP JDREAF 50,609,71
orl-rat TDLo:210 mg/kg (14D pre):REP FESTAS 24,284,73
scu-rat TDLo:330 mg/kg (female 1-22D post):TER AMEBA7 40,297,62
scu-gpg TDLo:12 mg/kg (female 16-17D post):TER ANREAK 144,155,62
ipc-mus TDLo:4 mg/kg (female 11D post):TER 40YJAX -,207,79
ipr-rat LD50:150 mg/kg JJPAAZ 21,377,71
scu-rat LD50:449 mg/kg TXAPA9 8,250,66

CONSENSUS REPORTS: Reported in EPA TSCA Inventory.

SAFETY PROFILE: Poison by intraperitoneal route. Moderately toxic by subcutaneous route. An experimental teratogen. Other experimental reproductive effects. Mutation data reported. A steroid. When heated to decomposition it emits very toxic fumes of SO_x and NO_x. See also CORTISONE.

CNS800 **CAS:53-06-5** **HR: 3**
CORTISONE
mf: $C_{21}H_{28}O_5$ mw: 360.49

PROP: Rhombohedral platelets from 95% alc. Mp: 220–224° (some decomp). Fairly sol in cold methanol, ethanol, and acetone; much less sol in ether, benzene, and chloroform; sltly sol in water (28 mg/100 mL at 25°).

SYNS: ADRENALEX □ ADRESON □ COMPOUND E □ CORLIN □ CORTADREN □ CORTISAL □ CORTISATE □ CORTISTAL □ CORTIVITE □ CORTOGEN □ CORTONE □ 11-DEHYDRO-17-HYDROXYCORTICOSTERONE □ 17α,21-DIHYDROXY-4-PREGNENE-3,11,20-TRIONE □ 17-HYDROXY-11-DEHYDROCORTICOSTERONE □ 17-α-HYDROXY-11-DEHYDROCORTICOSTERONE □ 17α,21-HYDROXYPREGN-4-ENE-3,11,20-TRIONE □ INCORTIN □ KE □ KENDALL'S COMPOUND E □ PREGN-4-EN-17α,21-DIOL-3,11,20-TRIONE □ 4-PREGNENE-17α,21-DIOL-3,11,20-17,21-DIHYDROXY- □ Δ^4-PREGNENE-17α,21-DIOL-3,11,20-TRIONE □ REICHSTEIN'S SUBSTANCE FA □ SCHEROSON □ WINTERSTEINER'S COMPOUND F

TOXICITY DATA WITH REFERENCE

dns-rat-scu 168 mg/kg/3D-C ENDOAO 94,1637,74
dnd-mus:lvr 100 μmol/L ENZYAS 41,183,71
orl-wmn TDLo:160 mg/kg (16W pre/1-39W preg):REP OBGNAS 1,276,53
orl-wmn TDLo:90 mg/kg (90D preg):TER AJOGAH 65,237,53
unr-wmn TDLo:476 mg/kg (1-34W preg):TER LANCAO 2,730,56
ims-rat TDLo:82,500 μg/kg (female 11D post):REP ACENA7 60,36,69
ims-rat TDLo:110 mg/kg (female 1-22D post):REP JCINAO 41,710,62
ims-mus TDLo:300 mg/kg (female 12D post):REP JTEHD6 10,541,82
ims-mus TDLo:300 mg/kg (female 12D post):REP TCMUD8 4,403,84
par-rbt TDLo:32 mg/kg (female 20-27D post):TER JOENAK 14,284,56
ims-rat TDLo:82,500 μg/kg (female 11D post):TER ACENA7 60,36,69
unr-rat TDLo:144 mg/kg (male 9D pre):REP ANENAG 24(Suppl 3),1,63
unr-mus TDLo:120 mg/kg (female 1D pre):REP FESTAS 12,346,61
ipr-mus TDLo:200 mg/kg (female 11D post):REP JDREAF 45,1767,66
par-rat TDLo:35 mg/kg (female 7D pre):REP JEMEAV 102,347,55
ims-rat TDLo:225 mg/kg (female 13-22D post):TER ACENA7 50,104,65
scu-rat TDLo:500 mg/kg (8-17D preg):TER AGDJAI 18,34,70
ims-rat TDLo:110 mg/kg (female 1-22D post):TER ACENA7 53,547,66
orl-mus TDLo:1 g/kg (female 6-15D post):TER JJATDK 5,97,85

SAFETY PROFILE: Human teratogenic effects by ingestion and possibly other routes: developmental abnormalities of the eye and ear, craniofacial area, musculoskeletal system, cardiovascular system and other neonatal measures or effects. Other experimental teratogenic effects. Additional reproductive effects in experimental animals. Mutation data reported. When heated to decomposition it emits acrid smoke and fumes. See also CORTISOL.

CNS825 **CAS:50-04-4** **HR: 2**
CORTISONE-21-ACETATE
mf: $C_{23}H_{30}O_6$ mw: 402.53

PROP: Needles from Me_2CO. Mp: 239–241°.

SYNS: ACETATE CORTISONE □ 21-ACETOXY-17,α-HYDROXYPREGN-4-ENE-3,11,20-TRIONE □ 21-ACETOXY-17,α-HYDROXY-3,11,20-TRIKETOPREGNENE-4 □ 21-(ACETYLOXY)-17-HYDROXY-PREGN-4-ENE-3,11,20-TRIONE (9CI) □ ADRESON □ ARTRIONA □ BIOCORT ACETATE □ COMPOUND E ACETATE □ CORTADREN □ CORTELAN □ CORTISAL □ CORTISATE □ CORTISONE ACETATE □ CORTISONE MONOACETATE □ CORTISTAB □ CORTISYL □ CORTIVITE □ CORTOGEN □ CORTOGEN ACETATE □ CORTONE □ CORTONE ACETATE □ 11-DEHYDRO-17-HYDROXYCORTICOSTERONE ACETATE □ 11-DEHYDRO-17-HYDROXYCORTICOSTERONE-21-ACETATE □ 17,21-DIHYDROXYPREGN-4-ENE-3,11,20-TRIONE ACETATE □ 17,21-DIHYDROXY-PREGN-4-ENE-3,11,20-TRIONE 21-ACETATE □ INCORTIN □ IRISONE ACETATE □ 4-PREGNENE-17,α,21-DIOL-3,11,20-TRIONE 21-ACETATE □ RICORTEX □ SCHEROSON

TOXICITY DATA WITH REFERENCE

mmo-bcs 5 g/L MUREAV 42,19,77
dnr-bcs 5 g/L MUREAV 42,19,77
ims-rat TDLo:300 mg/kg (female 12-15D post):REP ACPAAN 8,217,67
par-rat TDLo:60 mg/kg (female 16-19D post):REP ENDOAO 54,384,54
ims-dog TDLo:37,500 μg/kg (female 20-34D post):TER JZKEDZ 6,37,80
scu-rat TDLo:500 mg/kg (female 12-13D post):TER DPTHDL 4,89,82
ims-rat TDLo:700 mg/kg (female 16-22D post):TER HMMRA2 10,425,78
ims-rat TDLo:3500 mg/kg (female 16-22D post):REP BNEOBV 32,211,77
ims-mus TDLo:200 mg/kg (female 11-12D post):REP ANREAK 142,479,62
scu-rat TDLo:400 mg/kg (female 1-8D post):REP FESTAS 28,464,77
orl-rat TDLo:300 mg/kg (20-21D preg):REP PRGLBA 4,93,73
orl-rat TDLo:56 mg/kg (male 14D pre):REP JPETAB 113,27,55
ims-dog TDLo:375 mg/kg (female 20-34D post):TER JZKEDZ 6,37,80
ocu-rbt TDLo:2880 μg/kg (female 6-18D post):TER TXAPA9 16,773,70
ipr-mus TDLo:182 mg/kg (female 13D post):TER TJADAB 24,79,81
par-rat TDLo:60 mg/kg (female 16-19D post):TER ENDOAO 54,384,54
scu-mus TDLo:50 mg/kg (female 14D post):TER TJADAB 30(1),11A,84
ims-rat TDLo:500 mg/kg (female 8-17D post):TER TJADAB 4,383,71
par-rat TDLo:600 mg/kg (female 7-14D post):TER ACENA7 22,23,56

ims-dog TDLo:150 mg/kg (female 20-34D post):TER JZKEDZ 6,37,80
ipr-mus TDLo:35 mg/kg:ETA EXPEAM 33,1640,77

SAFETY PROFILE: Questionable carcinogen with experimental tumorigenic data. An experimental teratogen. Experimental reproductive effects. Mutation data reported. When heated to decomposition it emits acrid smoke and fumes. See also CORTISON.

CNT250 HR: 3
CORUNDUM FUME

PROP: Half finely divided alumina, half silica (JPBAA7 69,81,55).

TOXICITY DATA WITH REFERENCE
itr-rat LDLo:90 mg/kg JPBAA7 69,81,55

SAFETY PROFILE: Poison by intratracheal route. See also ALUMINUM OXIDE (2:3) and SILICA.

CNT325 HR: 3
CORYDALOID

PROP: Extracted from *Corydalis bulbosa dc* (KSRNAM 5,1129,71).

TOXICITY DATA WITH REFERENCE
orl-rat TDLo:60 mg/kg (multi) :REP KSRNAM 5,1153,71
orl-rat TDLo:600 mg/kg (multi) :REP KSRNAM 5,1153,71
orl-rat TDLo:4550 mg/kg (26W male):REP KSRNAM 5,1129,71
orl-rat LD50:4100 mg/kg KSRNAM 5,1129,71
ipr-rat LD50:175 mg/kg KSRNAM 5,1129,71
scu-rat LD50:220 mg/kg KSRNAM 5,1129,71
orl-mus LD50:917 mg/kg KSRNAM 5,1129,71
ipr-mus LD50:205 mg/kg KSRNAM 5,1129,71
scu-mus LD50:155 mg/kg KSRNAM 5,1129,71

SAFETY PROFILE: Poison by subcutaneous and intraperitoneal routes. Moderately toxic by ingestion. Experimental reproductive effects.

CNT350 CAS:8023-88-9 HR: 2
COSTUS OIL

SYNS: CHANGALA □ CHOB-i-QUT □ COSTUS ROOT □ ESSENTIAL OIL □ GOSHTAM □ KASHMIRJA □ KASTAM □ KOOST □ KOOT □ KOST □ KUR □ KUSHTHA □ KUSTA □ KUTH □ OILS, COSTUS □ OUPLATE □ PACHAK □ PATCHUK □ PUTCHUK □ SEPPUDY □ SEPUDDY □ UPALET

TOXICITY DATA WITH REFERENCE
skn-mus 100% MLD FCTXAV 12,867,74
mmo-sat 10 µg/plate EMMUEG 10,141,87
orl-rat LD50:3400 mg/kg FCTXAV 12,867,74
skn-rbt LD50:>5 g/kg FCTXAV 12,867,74

CONSENSUS REPORTS: Reported in EPA TSCA Inventory.

SAFETY PROFILE: Moderately toxic by ingestion. Low toxicity by skin contact. A skin irritant. Mutation data reported. When heated to decomposition it emits acrid smoke and irritating vapors.

CNT625 CAS:82-54-2 HR: 3
COTARNIN
mf: $C_{12}H_{15}NO_4$ mw: 237.28

PROP: Small needles from benzene. Decomp @ 132–133°. Sol in alc, chloroform, ether, benzene; sltly sol in water; sol in dilute acids, in ammonia or sodium carbonate soln, but only sltly sol in potassium hydroxide soln. Aq or alcoholic solns are yellow.

SYNS: COTARNINE □ 5,6,7,8-TETRAHYDRO-4-METHOXY-6-METHYL-1,2-DIOXOLO-(4,5-g)ISOQUINOLIN-5-OL (9CI)

TOXICITY DATA WITH REFERENCE
unr-dog LDLo:200 mg/kg HBAMAK 4,1289,35
unr-rbt LDLo:180 mg/kg HBAMAK 4,1289,35
unr-frg LDLo:40 mg/kg HBAMAK 4,1289,35

SAFETY PROFILE: Poison by an unspecified route. When heated to decomposition it emits toxic fumes of NO_x.

CNT750 HR: 2
COTTON DUST

TOXICITY DATA WITH REFERENCE
ihl-hmn TCLo:10,000 mg/m³/10Y:PUL BJIMAG 17,1,60
ipr-rat LD50:2 g/kg JDGRAX 5(1),93,73
ipr-gpg LD50:4 g/kg JDGRAX 5(1),93,73

OSHA PEL: TWA 1 mg/m³ (raw dust); 0.2 mg/m³ (yarn manufacturing); 0.75 mg/m³ (slashing and weaving); 0.5 mg/m³ (other operations)
ACGIH TLV: TWA 0.2 mg/m³ (raw dust)
DFG MAK: 1.5 mg/m³ (raw cotton)
NIOSH REL: (Cotton Dust) CL 0.200 mg/m³ lint-free

SAFETY PROFILE: Human pulmonary effects. Causes a mild febrile condition of the lungs resembling metal fume fever. Coarser grades of cotton contain more dust than the finer varieties, and therefore constitute a greater hazard. It is considered an inert dust and indeed it is, within the meaning of the term. However, it can cause some illness, due to the allergens or fungi in the cotton or on the dust. Workers in processing rooms may develop conjunctivitis or blepharitis from the burned products of the gassing of the double yarn. It is a mild allergen. Inhalation may produce bronchial asthma, sneezing, and eczema in sensitized persons. Moderate fire and explosion hazard when exposed to heat or flame; can react with oxidizing materials.

CNU000 CAS:8001-29-4 HR: 2
COTTONSEED OIL (unhydrogenated)

PROP: Oily, pale-yellow, nearly odorless liquid from seeds of species of *Gossypium hirsutum.* Flash p: 486°F (CC), fp: 0–5°, d: 0.915–0.921 @ 25°/25°, autoign temp: 650°F.

SYNS: DEODORIZED WINTERIZED COTTONSEED OIL □ NCI-C50168

TOXICITY DATA WITH REFERENCE
ipr-rat TDLo:2256 mg/kg (5-15D preg):TER JDREAF 51,1632,72
orl-mus TDLo:2940 g/kg/35W-C:ETA LPDSAP 17,115,82

SAFETY PROFILE: Questionable carcinogen with experimental tumorigenic data. Experimental teratogenic effects. An allergen. Combustible liquid when exposed to heat or flame. However, if allowed to impregnate rags or oily waste, it can become a dangerous hazard due to spontaneous heating. To fight fire, use CO_2, dry chemical.

CNU750 CAS:56-72-4 HR: 3
COUMAPHOS
mf: $C_{14}H_{16}ClO_5PS$ mw: 362.78

PROP: A solid. Mp: 95°.

SYNS: AGRIDIP □ ASUNTHOL □ BAYER 21/199 □ BAYMIX 50 □ 3-CHLORO-7-HYDROXY-4-METHYL-COUMARIN-O,O-DIETHYL PHOSPHOROTHIOATE □ 3-CHLORO-7-HYDROXY-4-METHYL-COUMARIN-O-ESTER with O,O-DIETHYL PHOSPHOROTHIOATE □ O-3-CHLORO-4-METHYL-7-COUMARINYL-O,O-DIETHYL PHOSPHOROTHIOATE □ 3-CHLORO-4-METHYL-7-COUMARINYL DIETHYL PHOSPHOROTHIOATE □ 3-CHLORO-4-METHYL-7-HYDROXYCOUMARIN DIETHYL THIOPHOSPHORIC ACID ESTER □ 3-CHLORO-4-METHYLUMBELLIFERONE-O-ESTER with O,O-DIETHYL PHOSPHOROTHIOATE □ CUMAFOS (DUTCH) □ O,O-DIAETHYL-O-(3-CHLOR-4-METHYL-CUMARIN-7-YL)-MONOTHIOPHOSPHAT (GERMAN) □ O,O-DIETHYL-O-(3-CHLOOR-4-METHYL-CUMARIN-7-YL)MONOTHIOFOSFAAT (DUTCH) □ O,O-DIETHYL-O-(3-CHLORO-4-METHYL-7-COUMARINYL)PHOSPHOROTHIOATE □ O,O-DIETHYL-O-(3-CHLORO-4-METHYLCOUMARINYL-7) THIOPHOSPHATE □ O,O-DIETHYL-O-(3-CHLORO-4-METHYL-2-OXO-2H-BENZOPYRAN-7-YL)PHOSPHOROTHIOATE □ O,O-DIETHYL-3-CHLORO-4-METHYL-7-UMBELLIFERONE THIOPHOSPHATE □ O,O-DIETHYL-O-(3-CHLORO-4-METHYLUMBELLIFERYL)PHOSPHOROTHIOATE □ DIETHYL-3-CHLORO-4-METHYLUMBELLIFERYL THIONOPHOSPHATE □ DIETHYL THIOPHOSPHORIC ACIDESTER of 3-CHLORO-4-METHYL-7-HYDROXYCOUMARIN □ O,O-DIETIL-O-(3-CLORO-4-METIL-CUMARIN-7-IL-MONOTIOFOSFATO) (ITALIAN) □ DIOLICE □ ENT 17,956 □ MELDONE □ NCI-C08662 □ THIOPHOSPHATE de O,O-DIETHYLE et de O-(3-CHLORO-4-METHYL-7-COUMARINYLE) (FRENCH) □ UMBETHION

TOXICITY DATA WITH REFERENCE
otr-rat:emb 1400 ng/plate JJATDK 1,190,81
orl-rat LD50:13 mg/kg DOEAAH 35,25,79
ihl-rat LC50:303 mg/m3 VHTODE 24,87,82
skn-rat LD50:860 mg/kg WRPCA2 9,119,70
ipr-rat LD50:7500 μg/kg PSEBAA 129,699,68
orl-mus LD50:28 mg/kg HYSAAV 33(12),334,68
ipr-mus LD50:200 mg/kg PCOC** -,282,66
orl-rbt LD50:80 mg/kg HYSAAV 33(12),334,68
skn-rbt LD50:500 mg/kg VHTODE 24,87,82

CONSENSUS REPORTS: NCI Carcinogenesis Bioassay (feed); No Evidence: mouse, rat NCITR* NCI-CG-TR-96,79. EPA Extremely Hazardous Substances List.

SAFETY PROFILE: Poison by ingestion, skin contact, inhalation, and intraperitoneal routes. Mutation data reported. When heated to decomposition, it emits very toxic fumes of SO_x, PO_x, and Cl^-. See also COUMARIN.

CNU825 CAS:7400-08-0 HR: 2
4-COUMARIC ACID
mf: $C_9H_8O_3$ mw: 164.17

PROP: Needles. Mp: 210–213°. Crystallizes in anhydrous form from conc hot aq soln, but as monohydrate from dilute aq soln on slow cooling. Sltly sol in cold water; sol in hot water, alc, and ether. Practically insol in benzene and ligroin.

SYNS: p-COUMARIC ACID □ p-CUMARIC ACID □ p-HYDROXYCINNAMIC ACID □ 4-HYDROXYCINNAMIC ACID □ 4'-HYDROXYCINNAMIC ACID □ p-HYDROXYPHENYLACRYLIC ACID □ β-(4-HYDROXYPHENYL)ACRYLIC ACID □ 3-(4-HYDROXYPHENYL)-2-PROPENOIC ACID

TOXICITY DATA WITH REFERENCE
orl-mus TDLo:50 mg/kg (female 6D post):REP CCPTAY 20,49,79
orl-rat TDLo:2800 mg/kg (56D male):REP CCPTAY 23,677,81
orl-mus TDLo:35 mg/kg (6D preg):REP IJEBA6 16,1285,78
orl-uns TDLo:9600 mg/kg (female 12D pre):REP SCIEAS 195,575,77
ipr-mus LD50:657 mg/kg YKKZAJ 104,793,84

CONSENSUS REPORTS: Reported in EPA TSCA Inventory.

SAFETY PROFILE: Moderately toxic by intraperitoneal route. Experimental reproductive effects. When heated to decomposition it emits acrid smoke and fumes. See also COUMARIN.

CNU850 CAS:614-60-8 HR: 3
trans-o-COUMARIC ACID
mf: $C_9H_8O_3$ mw: 164.17

SYNS: CINNAMIC ACID, o-HYDROXY-, (E)- □ trans-o-HYDROXYCINNAMIC ACID □ trans-2-HYDROXYCINNAMIC ACID □ (E)-3-(2-HYDROXYPHENYL)-2-PROPENOIC ACID □ 2-PROPENOIC ACID, 3-(2-HYDROXYPHENYL)-, (E)-(9CI)

TOXICITY DATA WITH REFERENCE
ivn-mus LD50:180 mg/kg CSLNX* NX#02587

CONSENSUS REPORTS: Reported in EPA TSCA Inventory.

SAFETY PROFILE: Poison by intravenous route. When heated to decomposition it emits acrid smoke and irritating vapors.

CNU875 CAS:496-41-3 HR: 3
COUMARILIC ACID
mf: $C_9H_6O_3$ mw: 162.15

SYN: 2-BENZOFURANCARBOXYLIC ACID

TOXICITY DATA WITH REFERENCE
ivn-mus LD50:320 mg/kg CSLNX* NX#02495

CONSENSUS REPORTS: Reported in EPA TSCA Inventory.

SAFETY PROFILE: Poison by intravenous route. When heated to decomposition it emits acrid smoke and irritating vapors.

CNV000 CAS:91-64-5 HR: 3
COUMARIN
mf: $C_9H_6O_2$ mw: 146.15

PROP: Rhombic crystals; fragrant, pleasant odor; burn-

C

ing taste. Mp: 70°, bp: 291.0°, vap press: 1 mm @ 106.0°. Sol in EtOH, hot H_2O, and alkalis.

SYNS: 2H-1-BENZOPYRAN-2-ONE □ 1,2-BENZOPYRONE □ cis-o-COUMARINIC ACID LACTONE □ COUMARINIC ANHYDRIDE □ o-HYDROXYCINNAMIC ACID LACTONE □ o-HYDROXYZIMTSAEURE-LACTON (GERMAN) □ NCI-C60297 □ 2-OXO-1,2-BENZOPYRAN □ RAT-TEX □ TONKA BEAN CAMPHOR

TOXICITY DATA WITH REFERENCE
mma-sat 1 mg/plate ENMUDM 5(Suppl 1),3,83
dnd-mam:Lym 20 mmol/L PNASA6 48,686,62
orl-mus TDLo:3600 mg/kg (6-17D preg):TER ARZNAD 17,97,67
orl-rat TD:200 g/kg/2Y-C:ETA TXCYAC 1,93,73
orl-rat LD50:293 mg/kg FCTXAV 12,385,74
orl-mus LD50:196 mg/kg YKKZAJ 83,1124,63
ipr-mus LD50:220 mg/kg ARZNAD 15,897,65
scu-mus LD50:242 mg/kg YKKZAJ 83,1124,63
orl-gpg LD50:202 mg/kg FCTXAV 2,327,64

CONSENSUS REPORTS: IARC Cancer Review: Group 3 IMEMDT 7,56,87; Animal Limited Evidence IMEMDT 10,113,76. EPA Genetic Toxicology Program. Reported in EPA TSCA Inventory.

SAFETY PROFILE: Poison by ingestion, intraperitoneal, and subcutaneous routes. Questionable carcinogen with experimental tumorigenic data. Experimental teratogenic effects. Mutation data reported. Combustible when exposed to heat or flame. When heated to decomposition it emits acrid smoke and fumes. See also KETONES and ANHYDRIDES.

CNV500 CAS:4434-05-3 HR: 3
COUMERMYCIN AL
mf: $C_{55}H_{59}N_5O_{20}$ mw: 1110.19

PROP: A solid. Mp: 258–260° (decomp).

SYNS: COUMAMYCIN □ NOTOMYCIN A1

TOXICITY DATA WITH REFERENCE
mmo-esc 200 mg/L NATUAS 271,385,78
orl-mus LD50:2000 mg/kg 85ERAY 1,147,78
ipr-mus LD50:159 mg/kg 85ERAY 1,147,78
scu-mus LD50:250 mg/kg AACHAX -,786,65
ivn-mus LD50:25 mg/kg 85ERAY 1,147,78
ims-mus LD50:500 mg/kg 85ERAY 1,147,78

SAFETY PROFILE: Poison by subcutaneous, intramuscular, intravenous, and intraperitoneal routes. Moderately toxic by ingestion. Mutation data reported. When heated to decomposition it emits toxic fumes of NO_x.

CNW000 CAS:136-78-7 HR: 3
CRAG HERBICIDE
mf: $C_8H_7Cl_2O_5S{\cdot}Na$ mw: 309.10

PROP: Crystals. Mp: 170°. Very sol in water.

SYNS: CRAG HERBICIDE 1 □ CRAG SESONE □ 2,4-DES-Na □ 2,4-DES-NATRIUM (GERMAN) □ 2-(2,4-DICHLOROPHENOXY)ETHANOL HYDROGEN SULFATE SODIUM SALT □ 2,4-DICHLOROPHENOXYETHYL SULFATE, SODIUM SALT □ DISUL □ DISUL-Na □ DISUL-SODIUM □ NATRIUM-2,4-DICHLORPHENOXYATHYLSULFAT (GERMAN) □ SES □ SESONE (ACGIH) □ SODIUM-2-(2,4-DICHLOROPHENOXY)ETHYL SULFATE □ SODIUM-2,4-DICHLOROPHENOXYETHYL SULPHATE □ SODIUM-2,4-DICHLOROPHENYL CELLOSOLVE SULFATE

TOXICITY DATA WITH REFERENCE
orl-rat LD50:480 mg/kg JAFCAU 9,382,61
orl-mam LD50:1230 mg/kg FMCHA2 -,C212,83

OSHA PEL: TWA Total Dust: 10 mg/m^3; Respirable Fraction: 5 mg/m^3
ACGIH TLV: TWA 10 mg/m^3

SAFETY PROFILE: A poison by ingestion. Strong solutions are skin irritants. An herbicide. When heated to decomposition it emits very toxic fumes of Cl^-, Na_2O, and SO_x.

CNW100 CAS:11005-94-0 HR: 3
CRANOMYCIN

TOXICITY DATA WITH REFERENCE
orl-mus LD50:8600 µg/kg 85FZAT -,232,67
ipr-mus LD50:760 µg/kg 85FZAT -,232,67
ivn-mus LD50:840 µg/kg 85FZAT -,232,67

SAFETY PROFILE: Poison by ingestion, intravenous, and intraperitoneal routes.

CNW105 CAS:101516-88-5 HR: 3
CRANOMYCIN HYDROCHLORIDE

TOXICITY DATA WITH REFERENCE
orl-mus LD50:8600 mg/kg 85ERAY 1,453,78
ipr-mus LD50:760 µg/kg 85ERAY 1,453,78
ivn-mus LD50:840 µg/kg 85ERAY 1,453,78

SAFETY PROFILE: Poison by ingestion, intravenous, and intraperitoneal routes.

CNW125 CAS:55769-64-7 HR: 3
CRAVITEN
mf: $C_{32}H_{48}N_2O_{10}{\cdot}2ClH$ mw: 693.74

PROP: Crystals from alc. Mp: 83–113° (hydrates).

SYNS: BUTOBENDINE DIHYDROCHLORIDE □ CRAVITEN □ M-71

TOXICITY DATA WITH REFERENCE
orl-rat TDLo:1500 mg/kg (6-15D preg):TER PJPPAA 32,893,80
orl-ham TDLo:2250 mg/kg (6-10D preg):TER PJPPAA 32,893,80
orl-rat LD50:1500 mg/kg DRFUD4 5,610,80
ipr-rat LD50:142 mg/kg PJPPAA 32,823,80
ivn-rat LD50:15,800 µg/kg MEIEDD 10,212,83
orl-mus LD50:4500 mg/kg DRFUD4 5,610,80
ipr-mus LD50:550 mg/kg PJPPAA 32,823,80
ivn-mus LD50:18 mg/kg DRFUD4 5,610,80
ivn-rbt LD50:5100 µg/kg PJPPAA 32,823,80

SAFETY PROFILE: Poison by intravenous and intraperitoneal routes. Moderately toxic by ingestion. An experimental teratogen. When heated to decomposition it emits toxic fumes of NO_x and HCl. See also ESTERS.

CNW500 CAS:1319-77-3 HR: 3
CRESOL
DOT: UN 2022
mf: C_7H_8O mw: 108.15

PROP: Mixture of isomeric cresols obtained from coal tar, colorless or yellowish to brown-yellow or pinkish liquid, phenolic odor. Mp: 10.9–35.5°, bp: 191–203°, flash p: 178°F, d: 1.030–1.038 @ 25°/25°, vap press: 1 mm @ 38–53°, vap d: 3.72.

SYNS: ACIDE CRESYLIQUE (FRENCH) □ BACILLOL □ CRESOLI (ITALIAN) □ CRESYLIC ACID □ HYDROXYTOLUOLE (GERMAN) □ KRESOLE (GERMAN) □ KRESOLEN (DUTCH) □ KREZOL (POLISH) □ RCRA WASTE NUMBER U052 □ TEKRESOL □ ar-TOLUENOL □ TRICRESOL

TOXICITY DATA WITH REFERENCE
orl-rat LD50:1454 mg/kg NTIS** PB214-270
orl-mus LD50:760 mg/kg KSGZA3 36,932,82
skn-rbt LD50:2000 mg/kg TXAPA9 42,417,77

CONSENSUS REPORTS: Community Right-To-Know List.

OSHA PEL: TWA 5 ppm (skin)
ACGIH TLV: TWA 5 ppm
DFG MAK: (all isomers) 5 ppm (22 mg/m³)
NIOSH REL: (Cresol) TWA 10 mg/m³
DOT CLASSIFICATION: 6.1; *Label:* Poison

SAFETY PROFILE: A poison by ingestion. Moderately toxic by skin contact. Corrosive to skin and mucous membranes. Systemic poisoning has rarely been reported, but it is possible that absorption may result in damage to the kidneys, liver, and nervous system. The main hazard accompanying its use in industry lies in severe chemical burns and dermatitis. Flammable when exposed to heat or flame; can react vigorously with oxidizing materials. Slightly explosive in the form of vapor when exposed to heat or flame. Explosive Range: 1.35% @ 300°F. Reacts violently with HNO_3, oleum, or chlorosulfonic acid. When heated to decomposition it emits highly toxic and irritating fumes. To fight fire, use foam, CO_2, dry chemical. See also other cresol entries and PHENOL.

For occupational chemical analysis use OSHA: #32 or NIOSH: Cresols 2001.

CNW750 CAS:108-39-4 HR: 3
m-CRESOL
DOT: UN 2076
mf: C_7H_8O mw: 108.15

PROP: Colorless to yellowish liquid, phenolic odor. Mp: 10.9° bp: 202.8°, lel: 1.1% @ 302°F, flash p: 202°F, d: 1.04 @ 22°/4°, autoign temp: 1038°F, vap press: 1 mm @ 52.0°, vap d: 3.72.

SYNS: 3-CRESOL □ m-CRESYLIC ACID □ 1-HYDROXY-3-METHYLBENZENE □ m-HYDROXYTOLUENE □ m-KRESOL □ m-METHYLPHENOL □ 3-METHYLPHENOL □ m-OXYTOLUENE □ RCRA WASTE NUMBER U052 □ m-TOLUOL

TOXICITY DATA WITH REFERENCE
skn-rbt 517 mg/24H SEV BIOFX* 3-5/69
eye-rbt 103 mg SEV BIOFX* 3-5/69
dni-hmn:hla 10 µmol/L/4H BECTA6 32,220,84
scu-rbt TDLo:134 g/kg/(6-18D preg):TER OYYAA2 16,1191,78
skn-mus TDLo:2280 mg/kg/20W-I:NEO CNREA8 19,413,59
orl-rat LD50:242 mg/kg BIOFX* 3-5/69
skn-rat LD50:1100 mg/kg GTPZAB 18(2),58,74
scu-rat LDLo:900 mg/kg HBTXAC 5,56,59
orl-mus LD50:828 mg/kg GTPZAB 18,58,74
ipr-mus LD50:168 mg/kg HBTXAC 5,56,59
scu-mus LDLo:450 mg/kg HBAMAK 4,1361,35
ivn-dog LDLo:150 mg/kg HBTXAC 5,56,59
scu-cat LDLo:180 mg/kg JPETAB 80,233,44
orl-rbt LDLo:1400 mg/kg JPETAB 80,233,44
skn-rbt LD50:2050 mg/kg BIOFX* 3-5/69
ivn-rbt LDLo:280 mg/kg JPETAB 80,233,44
ipr-gpg LDLo:100 mg/kg HBAMAK 4,1361,35

CONSENSUS REPORTS: Community Right-To-Know List. Reported in EPA TSCA Inventory. EPA Genetic Toxicology Program.

OSHA PEL: TWA 5 ppm (skin)
ACGIH TLV: TWA 5 ppm
NIOSH REL: (Cresol) TWA 10 mg/m³
DOT CLASSIFICATION: 6.1; *Label:* Poison

SAFETY PROFILE: Poison by ingestion, intravenous, intraperitoneal, and subcutaneous routes. Moderately toxic by skin contact. Severe eye and skin irritant. An experimental teratogen. Human mutation data reported. Questionable carcinogen with experimental neoplastigenic data. Flammable when exposed to heat or flame. Moderately explosive in the form of vapor when exposed to heat or flame. See also other cresol entries and PHENOL.

For occupational chemical analysis use NIOSH: Cresols, 2001.

CNX000 CAS:95-48-7 HR: 3
o-CRESOL
DOT: UN 2076
mf: C_7H_8O mw: 108.15

PROP: Crystals or liquid darkening with exposure to air and light. Mp: 30°, bp: 191°, flash p: 178°F, d: 1.05 @ 20°/4°, autoign temp: 1110°F, vap press: 1 mm @ 38.2°, vap d: 3.72, lel: 1.4% @ 300°F.

SYNS: 2-CRESOL □ o-CRESYLIC ACID □ 1-HYDROXY-2-METHYLBENZENE □ o-HYDROXYTOLUENE □ o-KRESOL (GERMAN) □ o-METHYLPHENOL □ 2-METHYLPHENOL □ ORTHOCRESOL □ o-OXYTOLUENE □ RCRA WASTE NUMBER U052 □ o-TOLUOL

TOXICITY DATA WITH REFERENCE
skn-rbt 524 mg/24H SEV BIOFX* 4-5/69
eye-rbt 105 mg SEV BIOFX* 4-5/69
sce-hmn:fbr 8 mmol/L MUREAV 137,51,84
skn-mus TDLo:4800 mg/kg/12W-I:NEO CNREA8 19,413,59
orl-rat LD50:121 mg/kg BIOFX* 4-5/69
skn-rat LD50:620 mg/kg GTPZAB 18(2),58,74
scu-rat LDLo:65 mg/kg RMSRA6 15,561,1895
orl-mus LD50:344 mg/kg GTPZAB 18,58,74
ihl-mus LC50:179 mg/m³/2H 85GMAT -,40,82
skn-mus LD50:620 mg/kg 85GMAT -,40,82

ipr-mus LDLo:200 mg/kg RBPMAZ 22,1,52
scu-mus LDLo:410 mg/kg ZHINAV 64,113,1909
ivn-dog LDLo:80 mg/kg HBTXAC 5,56,59
scu-cat LDLo:55 mg/kg JPETAB 80,233,44

CONSENSUS REPORTS: EPA Extremely Hazardous Substances List. Community Right-To-Know List. EPA Genetic Toxicology Program. Reported in EPA TSCA Inventory.

OSHA PEL: TWA 5 ppm (skin)
ACGIH TLV: TWA 5 ppm
NIOSH REL: (Cresol) TWA 10 mg/m^3
DOT CLASSIFICATION: 6.1; *Label:* Poison

SAFETY PROFILE: Poison by ingestion, inhalation, subcutaneous, intravenous, and intraperitoneal routes. Moderately toxic by skin contact. A severe eye and skin irritant. Human mutation data reported. Questionable carcinogen with experimental neoplastigenic data. Flammable when exposed to heat, flame, or oxidants. To fight fire, water may be used to blanket fire; foam, fog, mist, dry chemical. See also other cresol entries and PHENOL.

For occupational chemical analysis use NIOSH: Cresols, 2001.

CNX250 **CAS:106-44-5** **HR: 3**
p-CRESOL
DOT: UN 2076
mf: C_7H_8O mw: 108.15

PROP: Found in a score of essential oils, including ylang-ylang and oil of jasmine (FCTXAV 12,385,74). Crystals, phenolic odor. Mp: 35.5°, bp: 201.8°, lel: 1.1% @ 302°F, flash p: 202°F, d: 1.0341 @ 20°/4°, autoign temp: 1038°F, vap press: 1 mm @ 53.0°, vap d: 3.72.

SYNS: 4-CRESOL ☐ p-CRESYLIC ACID ☐ 1-HYDROXY-4-METHYLBENZENE ☐ p-HYDROXYTOLUENE ☐ 4-HYDROXYTOLUENE ☐ p-KRESOL ☐ 1-METHYL-4-HYDROXYBENZENE ☐ p-METHYLPHENOL ☐ 4-METHYLPHENOL ☐ p-OXYTOLUENE ☐ PARAMETHYL PHENOL ☐ RCRA WASTE NUMBER U052 ☐ p-TOLUOL ☐ p-TOLYL ALCOHOL

TOXICITY DATA WITH REFERENCE
skn-rbt 517 mg/24H SEV BIOFX* 5-5/69
eye-rbt 103 mg SEV BIOFX* 5-5/69
skn-mus TDLo:2280 mg/kg/20W-I:NEO CNREA8 19,413,59
orl-rat LD50:207 mg/kg BIOFX* 5-5/69
skn-rat LD50:750 mg/kg GTPZAB 18,58,74
scu-rat LDLo:500 mg/kg HBTXAC 5,58,59
unr-rat LD50:1440 mg/kg GTPPAF 8,145,72
orl-mus LD50:344 mg/kg GTPZAB 18,58,74
ipr-mus LD50:25 mg/kg HBTXAC 5,58,59
scu-mus LDLo:150 mg/kg HBAMAK 4,1361,35
unk-mus LD50:160 mg/kg BJCAAI 6,160,52
scu-cat LDLo:80 mg/kg JPETAB 80,233,44
orl-rbt LDLo:620 mg/kg JPETAB 80,233,44
skn-rbt LD50:301 mg/kg BIOFX* 5-5/69
scu-rbt LDLo:300 mg/kg HBAMAK 4,1361,35
ivn-rbt LDLo:180 mg/kg JPETAB 80,233,44
scu-gpg LDLo:200 mg/kg HBTXAC 5,58,59
scu-frg LDLo:150 mg/kg HBAMAK 4,1361,35

CONSENSUS REPORTS: Community Right-To-Know List. Reported in EPA TSCA Inventory. EPA Genetic Toxicology Program.

OSHA PEL: TWA 5 ppm (skin)
ACGIH TLV: TWA 5 ppm
NIOSH REL: (Cresol) TWA 10 mg/m^3
DOT CLASSIFICATION: 6.1; *Label:* Poison

SAFETY PROFILE: Poison by ingestion, skin contact, subcutaneous, intravenous, and intraperitoneal routes. A severe skin and eye irritant. Questionable carcinogen with experimental neoplastigenic data by itself and with 7,12-dimethyl benz(a)anthracene. Combustible when exposed to heat or flame. Moderately explosive in the form of vapor when exposed to heat or flame. To fight fire, use CO_2, dry chemical, alcohol foam. See also other cresol entries and PHENOL.

For occupational chemical analysis use NIOSH: Cresols, 2001; Phenol and p-Cresol in Urine, 8305.

CNX400 **CAS:596-27-0** **HR: 3**
o-CRESOLPHTHALEIN
mf: $C_{22}H_{18}O_4$ mw: 346.40

SYNS: 3′,3″-DIMETHYLPHENOLPHTHALEIN ☐ PHENOLPHTHALEIN, 3′,3″-DIMETHYL-

TOXICITY DATA WITH REFERENCE
ivn-mus LD50:320 mg/kg CSLNX* NX#02167

CONSENSUS REPORTS: Reported in EPA TSCA Inventory.

SAFETY PROFILE: Poison by intravenous route. When heated to decomposition it emits acrid smoke and irritating vapors.

CNX625 **CAS:83-40-9** **HR: 3**
2,3-CRESOTIC ACID
mf: $C_8H_8O_3$ mw: 152.16

PROP: Needles from water or alc (aq). White to sltly reddish, odorless crystals. Mp: 169–170°, volatile with steam. Sltly sol in cold water, more sol in hot water; sol in chloroform, alc, ether, alkali hydroxides.

SYNS: ACIDO ORTOCRESOTINICO (ITALIAN) ☐ ACIDO 3-OSSI-5-METIL-BENZOICO (ITALIAN) ☐ CRESOTIC ACID ☐ o-CRESOTIC ACID ☐ CRESOTINIC ACID ☐ β-CRESOTINIC ACID ☐ o-CRESOTINIC ACID ☐ 2,3-CRESOTINIC ACID ☐ HOMOSALICYLIC ACID ☐ 2-HYDROXY-3-METHYL-BENZOIC ACID (9CI) ☐ 3-METHYLSALYCILIC ACID ☐ 3-MS

TOXICITY DATA WITH REFERENCE
dni-hmn:lym 1 mmol/L BCPCA6 29,1275,80
orl-rat LD50:445 mg/kg BIALAY 10,270,63
orl-mus LD50:1 g/kg APSXAS 7,289,70
ivn-mus LD50:345 mg/kg FRPSAX 20,506,65

CONSENSUS REPORTS: Reported in EPA TSCA Inventory.

SAFETY PROFILE: Poison by intravenous route. Moderately toxic by ingestion. Mutation data reported. When heated to decomposition it emits acrid smoke and fumes.

CNX700 **HR: 2**
CRESSA CRETICA Linn., extract

PROP: Indian plant belonging to the family Convolvulaceae IJEBA6 22,312,84

TOXICITY DATA WITH REFERENCE
orl-ham TDLo:500 mg/kg (female 1-5D post):REP IJEBA6 22,312,84
ipr-mus LD50:681 mg/kg IJEBA6 22,312,84

SAFETY PROFILE: Moderately toxic by intraperitoneal route. Experimental reproductive effects. When heated to decomposition it emits acrid smoke and irritating fumes.

CNX800 **HR: 3**
CROCUS

PROP: A type of lilly that grows from a bulb and produces tubular purple or white flowers. They are grown only as houseplants or outdoor ornamentals.

SYNS: AUTUMN CROCUS □ COLCHICUM AUTUMNALE □ COLCHICUM SPECIOSUM □ COLCHICUM VERNUM □ FALL CROCUS □ MEADOW SAFFRON □ MYSTERIA □ VELLORITA (CUBA) □ WONDER BULB

SAFETY PROFILE: The whole plant contains the poison colchicine. Ingestion of any part of the plant causes a burning pain in the mouth, intense thirst, nausea, vomiting, abdominal cramps, severe diarrhea, and sometimes kidney damage. Colchicine is excreted slowly so the effects may persist for some time. See also COLCHICINE.

CNX825 **CAS:15826-37-6** **HR: 2**
CROMOGLYCATE DISODIUM
mf: $C_{23}H_{14}O_{11}\cdot 2Na$ mw: 512.35

PROP: A solid. Mp: 241–242° (decomp).

SYNS: AARANE □ AARARRE □ CROMOGLYCATE □ CROMOLYN SODIUM □ CROMOLYN SODIUM SALT □ DISODIUM CHROMOGLYCATE □ DISODIUM CROMOGLICATE □ DISODIUM CROMOGLYCATE □ DISODIUM-5,5′-((2-HYDROXYTRIMETHYLENE)DIOXY)-BIS (4-OXO-4H-1-BENZOPYRAN-2-CARBOXYLATE) □ FPL 670 □ FRENASMA □ INOSTRAL □ INTAL □ LOMUDAL □ LOMUDAS □ NALCROM □ NASMIL □ RYNACROM □ SODIUM CROMOGLYCATE □ SODIUM CROMOLYN

TOXICITY DATA WITH REFERENCE
orl-wmn TDLo:96 mg/kg/6D-I:SKN BMJOAE 289,470,84
orl-hmn TDLo:34 mg/kg/4W:PUL BMJOAE 2,916,76
scu-rat LD50:6 g/kg NIIRDN 6,244,82
ipr-mus LD50:4100 mg/kg NIIRDN 6,244,82
scu-mus LD50:4400 mg/kg NIIRDN 6,244,82
ivn-mus LD50:3300 mg/kg NIIRDN 6,244,82
ivn-rbt LD50:2 g/kg KSRNAM 4,189,70

SAFETY PROFILE: Moderately toxic by intravenous route. Mildly toxic by subcutaneous and intraperitoneal routes. Human systemic effects by ingestion: allergic dermatitis, respiratory depression and other lung effects. When heated to decomposition it emits toxic fumes of Na_2O.

CNX830 **HR: 3**
CROTALUS CERASTES VENOM

SYN: VENOM, SNAKE, CROTALUS CERASTES

TOXICITY DATA WITH REFERENCE
ipr-mus LD50:2080 μg/kg 14FHAR -,409,63
ivn-mus LD50:2600 μg/kg 14FHAR -,409,63
ipr-mam LD50:4 mg/kg CLPTAT 8,849,67

SAFETY PROFILE: A deadly poison by intravenous and intraperitoneal routes.

CNY000 **HR: 3**
CROTALUS DURISSUS TERRIFICUS VENOM

SYN: VENOM, SNAKE, CROTALUS DURISSUS TERRIFICUS

TOXICITY DATA WITH REFERENCE
ipr-mus LD50:400 μg/kg TOXIA6 9,131,71
scu-mus LD50:600 μg/kg TOXIA6 23,825,85
ivn-mus LD50:47 μg/kg TOXIA6 23,361 85
ims-mus LD50:1400 μg/kg TOXIA6 15,129,77
unr-mus LD50:200 μg/kg TOXIA6 19,473,81
ims-dog LDLo:1500 μg/kg COREAF 260,5408,65
ipr-mam LD50:300 μg/kg CLPTAT 8,849,67

SAFETY PROFILE: A deadly poison by subcutaneous, intramuscular, intravenous, intraperitoneal, and possibly other routes.

CNY300 **HR: 3**
CROTALUS HORRIDUS VENOM

SYN: VENOM, SNAKE, CROTALUS HORRIDUS

TOXICITY DATA WITH REFERENCE
ipr-mus LD50:2940 μg/kg 14FHAR -,409,63
scu-mus LD50:24,900 μg/kg AJTMAQ 31,489,51
ivn-mus LD50:2630 μg/kg 14FHAR -,409,63
scu-dog LDLo:60 mg/kg AJPHAP 173,535,53
ivn-dog LDLo:500 μg/kg AJPHAP 173,535,53

SAFETY PROFILE: Poison by subcutaneous, intravenous, and intraperitoneal routes.

CNY325 **HR: 3**
CROTALUS RUBER RUBER VENOM

SYN: VENOM, SNAKE, CROTALUS RUBER RUBER

TOXICITY DATA WITH REFERENCE
ipr-mus LD50:4650 μg/kg 14FHAR -,409,63
ivn-mus LD50:3700 μg/kg 14FHAR -,409,63
ims-mus LD50:11,500 μg/kg TOXIA6 23,769,85
ipr-mam LD50:6690 μg/kg CLPTAT 8,849,67
ivn-mam LD50:3700 μg/kg CLPTAT 8,849,67

SAFETY PROFILE: Poison by intramuscular, intravenous, and intraperitoneal routes.

CNY350 **HR: 3**
CROTALUS SCUTULATUS SCUTULATUS VENOM

SYN: VENOM, SNAKE, CROTALUS SCUTULATUS SCUTULATUS

TOXICITY DATA WITH REFERENCE
ivn-rat LD50:1 mg/kg 85EGD4 -,211,78

ipr-mus LD50:110 μg/kg TOXIA6 16,81,78
scu-mus LD50:320 μg/kg TOXIA6 23,825,85
ivn-mus LD50:30 μg/kg TOXIA6 23,11,85
ims-mus LD50:700 μg/kg TOXIA6 15,129,77
unr-mus LD50:200 μg/kg TOXIA6 19,473,81

SAFETY PROFILE: A deadly poison by subcutaneous, intramuscular, intravenous, intraperitoneal and possibly other routes.

CNY375 **HR: 3**
CROTALUS SCUTULATUS VENOM

SYN: VENOM, SNAKE, CROTALUS SCUTULATUS SCUTULATUS

TOXICITY DATA WITH REFERENCE
ipr-mus LD50:178 μg/kg TOXIA6 9,131,71
ivn-mus LD50:178 μg/kg TOXIA6 9,131,71
ipr-mam LD50:230 μg/kg CLPTAT 8,849,67
ivn-mam LD50:210 μg/kg CLPTAT 8,849,67

SAFETY PROFILE: A deadly poison by intravenous and intraperitoneal routes.

CNY390 **HR: 3**
CROTALUS VIRIDIS CERERUS VENOM

SYN: VENOM, MIDGET FADED RATTLESNAKE, CROTALUS VIRIDUS CERBERUS

TOXICITY DATA WITH REFERENCE
ipr-mus LD50:2500 μg/kg TOXIA6 15,129,77
ims-mus LD50:6 mg/kg TOXIA6 15,129,77

SAFETY PROFILE: Poison by intramuscular and intraperitoneal routes.

CNY750 **HR: 3**
CROTALUS VIRIDIS CONCOLOR VENOM

SYN: VENOM, MIDGET FADED RATTLESNAKE, CROTALUS VIRIDIS CONCOLOR

TOXICITY DATA WITH REFERENCE
ipr-mus LD50:250 μg/kg TOXIA6 15,129,77
ivn-mus LD50:45 μg/kg TOXIA6 23,361,85
ims-mus LD50:1200 μg/kg TOXIA6 15,129,77
unr-mus LD50:200 μg/kg TOXIA6 19,473,81

SAFETY PROFILE: Deadly poison by intramuscular, intravenous, intraperitoneal, and possibly other routes.

COA000 **HR: 3**
CROTALUS VIRIDIS HELLERI VENOM

SYN: VENOM, MIDGET FADED RATTLESNAKE, CROTALUS VIRIDIS HELLERI

TOXICITY DATA WITH REFERENCE
ipr-mus LD50:1560 μg/kg PAASAH 44,145,56
scu-mus LD50:3560 μg/kg PAASAH 44,145,56
ivn-mus LD50:580 μg/kg TOXIA6 16,431,78
ims-mus LD50:5100 μg/kg TOXIA6 15,129,77
ivn-dog LD50:50 μg/kg PWPSA8 16,58,73
ivn-cat LD50:1 g/kg PWPSA8 16,58,73
ipr-mam LD50:1600 μg/kg CLPTAT 8,849,67
ivn-mam LD50:1290 μg/kg CLPTAT 8,849,67

SAFETY PROFILE: A deadly poison by subcutaneous, intramuscular, intravenous, and intraperitoneal routes.

COB000 **CAS:21284-11-7** **HR: 2**
CROTOCIN
mf: $C_{19}H_{24}O_5$ mw: 332.43

PROP: Crystals from MeOH. Mp: 126°.

SYNS: ANTIBIOTIC T □ 7-β,8-β:12,13-DIEPOXY-TRICHOTHEC-9-EN-6-β-OL □ (4-β(Z),7-β,8-β)-7,8:12,13-DIEPOXY-TRICHOTHEC-9-EN-4-OL 2-BUTENOATE

TOXICITY DATA WITH REFERENCE
skn-gpg 332 ng MLD FAATDF 4(2, Pt 2),5124,84
mma-sat 100 μg/plate CNREA8 38,536,78
orl-mus LD50:1000 mg/kg 85GDA2 6,183,81
ipr-mus LD50:810 mg/kg 85ERAY 3,2043,78
ivn-mus LD50:700 mg/kg 85GDA2 6,183,81

SAFETY PROFILE: Moderately toxic by ingestion, intravenous, and intraperitoneal routes. A skin irritant. Mutation data reported. When heated to decomposition it emits acrid smoke and fumes.

COB250 **CAS:4170-30-3** **HR: 3**
CROTONALDEHYDE
DOT: UN 1143
mf: C_4H_6O mw: 70.09

PROP: Water-white, mobile liquid; pungent suffocating odor. Bp: 104°, fp: −76.0°, lel: 2.1%, uel: 15.5%, flash p: 55°F, d: 0.853 @ 20°/20°, vap d: 2.41, autoign temp: 405°F.

SYNS: 2-BUTENAL □ CROTONIC ALDEHYDE □ CROTONALDEHYDE, stabilized (DOT) □ KROTONALDEHYD (CZECH) □ β-METHYLACROLEIN □ RCRA WASTE NUMBER U053

TOXICITY DATA WITH REFERENCE
mmo-sat 250 μmol/L ENMUDM 7(Suppl 3),56,85
sln-dmg-par 3500 ppm ENMUDM 7,677,85
trn-dmg-par 3500 ppm ENMUDM 7,677,85
dnd-mam:lym 21,500 mg/L/16H CNREA8 44,990,84
orl-rat TDLo:2664 mg/kg/2Y-C:CAR CNREA8 46,1285,86
orl-rat LD50:206 mg/kg GTPZAB 26(8),53,82
ihl-rat LC50:200 mg/m³/2H GTPZAB 26(8),53,82
orl-mus LD50:104 mg/kg GTPZAB 26(8),53,82
ihl-mus LD50:580 mg/m³/2H GTPZAB 26(8),53,82

CONSENSUS REPORTS: EPA Extremely Hazardous Substances List. Reported in EPA TSCA Inventory.

ACGIH TLV: TWA 2 ppm

OSHA PEL: TWA 2 ppm
DFG MAK: Suspected Carcinogen.
DOT CLASSIFICATION: 3; *Label:* Flammable Liquid, Poison

SAFETY PROFILE: Suspected carcinogen with experimental carcinogenic data. Poison by ingestion and inhalation. Mutation data reported. An eye, skin, and mucous membrane irritant. A lachrymating material that can cause corneal burns and is very dangerous to the eyes. Caution: Keep away from heat and open flame. Keep container closed. Use with adequate ventilation. Extremely irritating to eyes, skin, mucous membranes.

When necessary, the lacrimatory effect of the vapors may be counteracted by ammonia fumes. Dangerous fire hazard when exposed to heat or flame; can react with oxidizing materials. To fight fire, use alcohol foam, CO_2, dry chemical. Reacts violently with 1,3-butadiene. Violent hypergolic reaction with concentrated nitric acid. When heated to decomposition it emits acrid smoke and fumes. See also ALDEHYDES.

COB260 CAS:123-73-9 HR: 3
(E)-CROTONALDEHYDE
mf: C_4H_6O mw: 70.10

PROP: Water-white, mobile liquid; pungent, suffocating odor. Mp: −69°, bp: 102.2°, fp: −76.0°, lel: 2.1%, uel: 15.5%, flash p: 55°F, d: 0.853 @ 20°/20°, vap d: 2.41, autoign temp: 450°F. Moderately sol in H_2O.

SYNS: ALDEHYDE CROTONIQUE (FRENCH) □ trans-2-BUTENAL □ (E)-2-BUTENAL □ CROTONAL □ CROTONALDEHYDE □ CROTONIC ALDEHYDE □ 1,2-ETHANEDIOL DIPROPANOATE (9CI) □ ETHYLENE GLYCOL DIPROPIONATE (8CI) □ ETHYLENE PROPIONATE □ β-METHYL ACROLEIN □ NCI-C56279 □ PROPYLENE ALDEHYDE □ RCRA WASTE NUMBER U053 □ TOPANEL

TOXICITY DATA WITH REFERENCE
eye-hmn 45 ppm AIHAAP 28,561,67
skn-rbt 500 mg open MLD UCDS** 4/21/67
spm-mus-ipr 30 mg/kg MUREAV 39-317-77
ihl-hmn TCLo:12 mg/m³/10M:IRR JAMAAP 165,1908,57
ihl-rat LC50:4000 mg/m³/30M APTOA6 6,299,50
scu-rat LD50:140 mg/kg APTOA6 6,299,50
orl-mus LD50:240 mg/kg BIJOAK 34,1196,40
ipr-mus LD50:160 mg/kg ZolH## 23OCT75
scu-mus LD50:160 mg/kg APTOA6 6,299,50
scu-dog LDLo:568 mg/kg AEXPBL 18,218,1884
skn-rbt LD50:380 mg/kg UCDS** 4/21/67
skn-gpg LD50:1331 mg/kg AEXPBL 18,218,1884

CONSENSUS REPORTS: Reported in EPA TSCA Inventory.

OSHA PEL: TWA 2 ppm
ACGIH TLV: TWA 2 ppm
DFG MAK: Suspected Carcinogen.

SAFETY PROFILE: A poison by ingestion, subcutaneous, and intraperitoneal routes. Mutation data reported. A lachrymating material that is very dangerous to the eyes. Human respiratory system irritant by inhalation. Can cause corneal burns and is irritating to the skin. In case of contact, immediately flush the skin or eyes with water for at least 15 minutes and get medical attention. See also ALDEHYDES. Dangerous fire hazard when exposed to heat or flame. To fight fire, use alcohol foam, CO_2, dry chemical. Incompatible with 1,3-butadiene and oxidizing materials. When heated to decomposition it emits acrid smoke and fumes.

For occupational chemical analysis use NIOSH: Crotonaldehyde P&CAM 285.

COB500 CAS:3724-65-0 HR: 3
CROTONIC ACID
DOT: UN 2823
mf: $C_4H_6O_2$ mw: 86.10

PROP: Colorless, needle-like crystals. Bp: 185°, mp: 72°, flash p: 190°F (COC), d: 1.018 @ 15°/4°, vap press: 0.19 mm @ 20°, vap d: 2.97.

SYNS: α-BUTENOIC ACID □ 2-BUTENOIC ACID □ CROTONIC ACID, solid □ α-CROTONIC ACID □ 3-METHYLACRYLIC ACID □ β-METHYLACRYLIC ACID

TOXICITY DATA WITH REFERENCE
skn-rbt 10 mg/24H open JIHTAB 26,269,44
orl-rat LD50:1000 mg/kg JIHTAB 26,269,44
ipr-rat LD50:100 mg/kg 34ZIAG -,190,69
orl-mus LD50:4800 mg/kg BIJOAK 34,1196,40
ipr-mus LD50:25 mg/kg 38MKAJ 2C,4953,82
scu-mus LD50:3590 mg/kg JPPMAB 21,85,69
skn-rbt LD50:600 mg/kg 85JCAE-,309,86
skn-gpg LD50:200 mg/kg 38MKAJ 2C,4953,82 34ZIAG -,190,69

CONSENSUS REPORTS: Reported in EPA TSCA Inventory.

DOT CLASSIFICATION: 8; *Label:* Corrosive

SAFETY PROFILE: Poison by intraperitoneal route. Moderately toxic by ingestion, skin contact, and subcutaneous routes. A powerful corrosive and irritant. Flammable when exposed to heat or flame; can react with oxidizing materials. To fight fire, use alcohol foam, CO_2, dry chemical. When heated to decomposition it emits acrid smoke and irritating fumes.

COB750 CAS:623-70-1 HR: 3
α-CROTONIC ACID ETHYL ESTER
DOT: UN 1862
mf: $C_6H_{10}O_2$ mw: 114.16

PROP: Colorless, monoclinic prisms or water-white liquid; pungent odor. Mp: 45° (solid). Bp: 209° (solid), 137° (liquid), flash p: 36.0°F, d: 0.9207 @ 20°/20°, vap d: 3.93.

SYNS: 2-BUTENOIC ACID, ETHYL ESTER, (E)-(9CI) □ trans-2-BUTENOIC ACID ETHYL ESTER □ CROTONATE d'ETHYLE (FRENCH) □ ETHYL (E)-2-BUTENOATE □ ETHYLCROTONATE □ ETHYL (E)-CROTONATE □ ETHYL trans-CROTONATE □ ETHYL CROTONATE (DOT)

TOXICITY DATA WITH REFERENCE
skn-rbt 10 mg/24H open JIHTAB 26,269,44
eye-rbt 5 mg SEV AJOPAA 29,1363,46
orl-rat LD50:3000 mg/kg JIHTAB 26,269,44

CONSENSUS REPORTS: Reported in EPA TSCA Inventory.

DOT CLASSIFICATION: 3; *Label:* Flammable Liquid

SAFETY PROFILE: Moderately toxic by ingestion and probably by inhalation. A skin, mucous membrane, and severe eye irritant. Very dangerous fire hazard when exposed to heat or flame; can react vigorously with oxidizing materials. To fight fire, use foam, CO_2, or dry chemical. See also ESTERS. When heated to decomposition it emits acrid smoke and fumes.

COB825 CAS:623-43-8 HR: 2
(E)-CROTONIC ACID METHYL ESTER
mf: $C_5H_8O_2$ mw: 100.13

PROP: Liquid. Bp: 121°.

SYNS: trans-2-BUTENOIC ACID METHYL ESTER □ METHYL trans-2-BUTENOATE □ METHYL CROTONATE □ METHYL α-CROTONATE □ METHYL E-CROTONATE □ METHYL trans-CROTONATE

TOXICITY DATA WITH REFERENCE
skn-rbt 500 mg/24H MOD FCTXAV 17,865,79
orl-mus LD50:1600 mg/kg FCTXAV 17,865,79
skn-gpg LD50:10 g/kg FCTXAV 17,865,79

CONSENSUS REPORTS: Reported in EPA TSCA Inventory.

SAFETY PROFILE: Moderately toxic by ingestion. Mildly toxic by skin contact. A skin irritant. When heated to decomposition it emits acrid smoke and fumes. See also ESTERS.

COB900 CAS:623-68-7 HR: 2
CROTONIC ANHYDRIDE
mf: $C_8H_{10}O_3$ mw: 154.18

SYNS: ANHYDRID KYSELINY KROTONOVE □ 2-BUTENOIC ACID, ANHYDRIDE (9CI) □ CROTONIC ACID ANHYDRIDE

TOXICITY DATA WITH REFERENCE
skn-rbt 10 mg/24H open MLD AIHAAP 23,95,62
orl-rat LD50:2830 mg/kg AIHAAP 23,95,62

CONSENSUS REPORTS: Reported in EPA TSCA Inventory.

SAFETY PROFILE: Moderately toxic by ingestion. A skin irritant. When heated to decomposition it emits acrid smoke and irritating fumes.

COC250 CAS:8001-28-3 HR: 3
CROTON OIL

PROP: Oil from the seeds of *Croton tiglium* (BJCAAI 10,72,56). Brownish-yellow, viscid oil; sltly offensive odor. Composition: croton resin, glycerides of fatty acids, and crotin. D: 0.935 @ 25°/25°.

SYNS: CROTONOEL (GERMAN) □ CROTON RESIN □ CROTON TIGLIUM L. OIL □ OLEUM TIGLII □ OLIO DI CROTON (ITALIAN)

TOXICITY DATA WITH REFERENCE
skn-mus 190 ng MLD CNREA8 28,2338,68
skn-gpg 400 μg/20H JIDEAE 24,35,55
mma-sat 2500 μg/plate BJCAAI 37,873,78
dni-hmn:fbr 10 ppm CNREA8 35,1392,75
dni-hmn:lym 50 ppm BBRCA9 45,630,71
skn-mus TDLo:2 mg/kg/27W-I:NEO BJCAAI 7,482,53
skn-mus TD:285 mg/kg/42W-I:ETA BJCAAI 10,72,56
skn-mus TD:340 μg/kg/17W-I:ETA CNREA8 28,653,68
skn-mus TD:150 mg/kg/15W-I:NEO CNREA8 19,413,59
skn-mus TD:75 mg/kg/25W-I:ETA JNCIAM 39,1217,67
skn-mus TD:160 mg/kg/8W-I:ETA GANNA2 59,187,68
skn-mus TD:850 mg/kg/18W-I:ETA BJCAAI 7,472,53
skn-mus TD:120 mg/kg/3W-C:ETA ZEKBAI 65,325,63
skn-mus TD:5714 mg/kg/14W-I:ETA BJCAAI 11,206,57
skn-mus TD:540 mg/kg/45W-I:ETA IJCNAW 1,491,66
ipr-mus LDLo:1 mg/kg TXAPA9 23,288,72
ipr-frg LD50:60 mg/kg ZEKBAI 65,325,63
par-frg LD50:60 mg/kg ZEKBAI 65,325,63
unk-frg LD50:60 mg/kg ZEKBAI 65,325,63

CONSENSUS REPORTS: Reported in EPA TSCA Inventory.

SAFETY PROFILE: Poison by parenteral, intraperitoneal, and possibly other routes. A skin and eye irritant. An allergen. Human mutation data reported. Questionable carcinogen with experimental neoplastigenic and tumorigenic data by skin contact. When heated to decomposition it emits toxic fumes.

COC300 CAS:4786-20-3 HR: 3
CROTONONITRILE
mf: C_4H_5N mw: 67.10

SYNS: CROTONIC NITRILE □ CROTONIQUE NITRILE □ CROTONITRILE □ 1-CYANOPROPENE □ β-METHYLACRYLONITRILE □ 1-PROPENYL CYANIDE

TOXICITY DATA WITH REFERENCE
orl-rat LD50:501 mg/kg GISAAA 37(4),10,72
orl-mus LD50:396 mg/kg GISAAA 37(4),10,72
ivn-rbt LDLo:60 mg/kg COREAF 153,895,11
orl-gpg LD50:272 mg/kg GISAAA 37(4),10,72
scu-gpg LDLo:230 mg/kg COREAF 153,895,11

CONSENSUS REPORTS: Reported in EPA TSCA Inventory.

SAFETY PROFILE: A poison by ingestion, intravenous, and subcutaneous routes. When heated to decomposition it emits toxic vapors of NO_x and CN^-.

COC500 CAS:503-17-3 HR: 3
CROTONYLENE
DOT: UN 1144
mf: CH_3CCCH_3 mw: 54.09

PROP: Liquid. Mp: −32.8°, bp: 27°, flash p: <−4°F, lel: 1.4%, d: 0.688 @ 25°, vap d: 1.91.

SYNS: 2-BUTYNE □ DIMETHYLACETYLENE

CONSENSUS REPORTS: Reported in EPA TSCA Inventory.

DOT CLASSIFICATION: 3; *Label:* Flammable Liquid

SAFETY PROFILE: A simple asphyxiant. Very dangerous fire hazard when exposed to heat or flame; can react with oxidizing materials. Moderately explosive in the form of vapor when exposed to heat or flame. To fight fire, use foam, CO_2, dry chemicals. See also ACETYLENE COMPOUNDS and ARGON (for a description of simple asphyxiants).

COC750 HR: 2
α-(N-CROTONYL-N-ETHYL)AMINO-BUTYRIC ACID mixed with BUTRIC ACID, α-(N-CROTONYL-N-PROPYL)AMINO

PROP: Equal parts of dimethylamines of N-crotonyl-αethylaminobutyric acid and of N-crotonyl-α-propylaminobutyric acid. (JPETAB 128,176,60)

SYN: DCB

TOXICITY DATA WITH REFERENCE
ipr-mus LD50:698 mg/kg JPETAB 128,176,60
scu-mus LD50:800 mg/kg JPETAB 128,176,60

SAFETY PROFILE: Moderately by subcutaneous and intraperitoneal routes. When heated to decomposition it emits toxic fumes of NO_x.

COC875 CAS:9007-40-3 HR: 3
CROTOXIN

PROP: Crystals.

TOXICITY DATA WITH REFERENCE
ipr-mus LD50:50 mg/kg PNASA6 68,1560,71
scu-mus LD50:400 μg/kg BBACAQ 5,98,50
ivn-mus LD50:108 μg/kg NNAPBA 270,274,71

SAFETY PROFILE: Poison by subcutaneous, intravenous, and intraperitoneal routes.

COD000 CAS:7700-17-6 HR: 3
CROTOXYPHOS
mf: $C_{14}H_{19}O_6P$ mw: 314.30

PROP: Straw-colored liquid. D: 1.19 @ 25°, bp: 135°.

SYNS: CIODRIN □ CIODRIN VINYL PHOSPHATE □ CIOVAP □ CYODRIN □ DECROTOX □ (E)-3-((DIMETHOXYPHOSPHINYL) OXY)-2-BUTENOIC ACID 1-PHENYLETHYL ESTER (9CI) □ O,O-DIMETHYL-O-(1-METHYL-2-CARBOXY-α-PHENYLETHYL)VINYL PHOSPHATE □ DIMETHYL-cis-1-METHYL-2-(1-PHENYLETHOXYCARBONYL) VINYL PHOSPHATE □ DIMETHYL PHOSPHATE of α-METHYLBENZYL-3-HYDROXY-cis-CROTONATE □ DUO-KILL □ ENT 24,717 □ (E)-3-HYDROXY-CROTONIC ACID α-METHYLBENZYL ESTER, DIMETHYL PHOSPHATE □ 1-METHYLBENZYL-3-(DIMETHOXYPHOSPHINYLOXO) ISOCROTONATE □ α-METHYL BENZYL-3-(DIMETHOXY-PHOSPHINYLOXY)-cis-CROTONATE □ α-METHYLBENZYL-3-HYDROXY-CROTONATE DIMETHYL PHOSPHATE □ PANTOZOL 1 □ cis-2-(1-PHENYLETHOXY)CARBONYL-1-METHYLVINYL DIMETHYLPHOSPHATE □ SD 4294 □ SHELL SD 4294 □ VOLFAZOL

TOXICITY DATA WITH REFERENCE
mma-smc 5000 ppm NTIS** PB80-133226
msc-mus:lym 180 mg/L NTIS** PB84-138973
orl-rat LD50:38,400 μg/kg GISAAA 38(6),30,73
skn-rat LD50:202 mg/kg WRPCA2 9,119,70
scu-rat LD50:47 mg/kg BJPCBM 40,124,70
orl-mus LD50:39,800 μg/kg GISAAA 38(6),30,73
ipr-mus LD50:71 mg/kg JPPMAB 19,612,67
scu-mus LD50:15 mg/kg JPPMAB 19,612,67
ivn-mus LD50:4500 μg/kg JPPMAB 19,612,67
skn-rbt LD50:385 mg/kg PCOC** -,244,66

CONSENSUS REPORTS: Reported in EPA TSCA Inventory.

SAFETY PROFILE: Poison by ingestion, skin contact, subcutaneous, intravenous, and intraperitoneal routes. Mutation data reported. An insecticide. When heated to decomposition it emits highly toxic PO_x.

COD100 CAS:10141-07-8 HR: 2
CROTYLIDENE DICROTONATE
mf: $C_{12}H_{16}O_4$ mw: 224.28

SYNS: 2-BUTENOIC ACID, 2-BUTENYLIDENE ESTER, (E,E,E)-(8CI, 9CI) □ 2-BUTENYLIDENE CROTONATE □ CROTONIC ACID, 2-BUTENYLIDENE ESTER

TOXICITY DATA WITH REFERENCE
skn-rbt 10 mg/24H open MLD AIHAAP 23,95,62
orl-rat LD50:2590 mg/kg AIHAAP 23,95,62

SAFETY PROFILE: Moderately toxic by ingestion. A skin irritant. When heated to decomposition it emits acrid smoke and irritating fumes.

COD475 CAS:294-93-9 HR: 2
12-CROWN-4
mf: $C_8H_{16}O_4$ mw: 176.24

PROP: Liquid. D: 1.089, mp: 16°. Sol in $CHCl_3$, toluene, and dichloroethane.

SYNS: EOCT □ ETHYLENE OXIDE CYCLIC TETRAMER □ 1,4,7,10-TETRAOXACYCLODODECANE

TOXICITY DATA WITH REFERENCE
skn-rbt 100 mg/24H MLD DCTODJ 8,451,85
eye-rbt 50 mg MLD DCTODJ 8,451,85
ihl-rat TCLo:1 ppm/7H (15D male):REP TXAPA9 27,342,74
ihl-rat TCLo:500 ppb/7H (15D male):REP TXAPA9 27,342,74
orl-rat LD50:2830 mg/kg DCTODJ 1,339,78
ipr-rat LD50:1550 mg/kg DCTODJ 8,451,85
orl-mus LD50:3150 mg/kg TXAPA9 44,263,78
ipr-mus LD50:1290 mg/kg DCTODJ 8,451,85

SAFETY PROFILE: Moderately toxic by ingestion and intraperitoneal routes. Experimental reproductive effects. An eye and skin irritant.

COD500 CAS:17455-13-9 HR: 2
18-CROWN-6
mf: $C_{12}H_{24}O_6$ mw: 264.36

$OC_2H_4(OC_2H_4)_4OCH_2CH_2$

PROP: Crystals. Mp: 38–39.5°, bp: 116° @ 0.2 mm. Sol in water.

SYN: 1,4,7,10,13,16-HEXANOXACYCLOOCTADECANE

TOXICITY DATA WITH REFERENCE
skn-rbt 100 mg/24H MLD DCTODJ 8,451,85
eye-rbt 50 mg MOD DCTODJ 8,451,85
orl-rat LD50:525 mg/kg GISAAA 53(10),92,88
ipr-rat LD50:830 mg/kg DCTODJ 8,451,85
orl-mus LD50:540 mg/kg GISAAA 53(10),92,88
ipr-mus LD50:464 mg/kg DCTODJ 8,451,85

CONSENSUS REPORTS: Reported in EPA TSCA Inventory.

SAFETY PROFILE: Moderately toxic by ingestion and intraperitoneal routes. A skin and eye irritant. When heated to decomposition it emits acrid smoke and irritating fumes.

COD575 **CAS:14187-32-7** **HR: 2**
CROWN 18
mf: $C_{20}H_{24}O_6$ mw: 360.41

PROP: Fibrous needles from C_6H_6. Mp: 162.5–163.5°. Sol in C_6H_6, $CHCl_3$, and dioxan, Py; sltly sol in H_2O, and EtOH.

SYNS: DIBENZO-18-CROWN-6 □ 6,7,9,10,17,18,20,21-OCTAHYDRO-DIBENZO(b,k)(1,4,7,10,13,16)HEXAOXYCYCLOOCTADECIN

TOXICITY DATA WITH REFERENCE
skn-rbt 100 mg/24H MLD DCTODJ 8,451,85
eye-rbt 50 mg MOD DCTODJ 8,451,85
orl-rat LD50:2600 mg/kg DCTODJ 8,451,85
ipr-rat LD50:560 mg/kg GTPZAB 31(2),48,87
orl-mus LD50:4500 mg/kg GISAAA 52(11),72,87
ipr-mus LD50:430 mg/kg GTPZAB 31(2),48,87

SAFETY PROFILE: Moderately toxic by ingestion and intraperitoneal routes. An eye and skin irritant. When heated to decomposition it emits acrid smoke and fumes.

COD675 **HR: 3**
CROWN FLOWER

PROP: Large shrubs that produce seeds and pods similar to milkweed. The crown flower grows to 15 feet. The small crown flower grows to 6 feet. Crown flowers are cultivated in south Florida and Hawaii. The small crown flower is a weed in the West Indies.

SYNS: ALGODON de SEDA (CUBA, PUERTO RICO) □ CALOTROPIS (VARIOUS SPECIES) □ C. GIGANTEA □ C. PROCERA □ FRENCH JASMINE □ GIANT MILKWOOD □ MUDAR □ PUA KALAUNU (HAWAII) □ TULA (PUERTO RICO)

SAFETY PROFILE: The whole plant contains cardiac glycosides. The sap also contains calcium oxalate and an allergan and can produce severe inflammation on contact with the eye. Chewing any part of the plant results in burning pain in the lips, mouth, and throat, possibly followed by inflammation and blistering. Large amounts are seldom swallowed due to the bitter taste. Ingestion may cause inflammation of the stomach and intestines and cardiac arrythmias. See also DIGITALIS and OXALATES.

COD750 **CAS:68308-34-9** **HR: 3**
CRUDE SHALE OILS
DOT: UN 1288

SYNS: BLUE OIL □ GREEN OIL □ RAW SHALE OIL □ SHALE OIL (DOT) □ UNFINISHED LUBRICATING OIL

TOXICITY DATA WITH REFERENCE
skn-rbt 500 mg/72H AIHAAP 40,460,79
mmo-sat 200 μg/plate MUREAV 90,233,81
mma-sat 100 nL/plate ENVRAL 39,19,86
cyt-mus-ipr 1500 mg/kg ENMUDM 4,639,82
sce-mus-ipr 1500 mg/kg ENMUDM 4,408,82
ipr-mus TDLo:200 mg/kg (1D male):REP NTIS** BNL-51002
skn-mus TDLo:20,700 mg/kg/69W-I:CAR EVHPAZ 38,149,81
skn-mus TD:26,208 mg/kg/2Y-I:CAR NTIS** CONF-801143
skn-mus TD:265 g/kg/2Y-I:CAR NTIS** CONF-801143
skn-mus TD:31,200 mg/kg/2Y-I:CAR NTIS** CONF-790334-3
skn-mus TD:28,800 mg/kg/30W-I:ETA NTIS** CONF-790334-3
skn-mus TD:168 g/kg/14W-I:ETA BMJOAE 2,1104,22
skn-mus TD:216 g/kg/18W-I:ETA BMJOAE 2,1104,22
skn-mus TD:120 g/kg/30W-I:CAR NTIS** BNL-51002
skn-mus TD:86,400 mg/kg/22W-I:ETA NTIS** BNL-51002
orl-rat LD50:8 g/kg AIHAAP 40,460,79
orl-mus LD50:11 g/kg NTIS** Conf. 800680-1
ipr-mus LD50:4300 mg/kg NTIS** Conf. 800680-1
skn-rbt LD50:5 g/kg NTIS** BNL-51002-273,79

CONSENSUS REPORTS: IARC Cancer Review: Group 1 IMEMDT 7,339,87; Human Sufficient Evidence IMEMDT 35,161,85; Animal Limited Evidence IMEMDT 35,161,85; Animal Sufficient Evidence IMEMDT 3,22,73.

DOT CLASSIFICATION: 3; *Label:* Flammable Liquid

SAFETY PROFILE: Confirmed human carcinogen with experimental carcinogenic, neoplastigenic, and tumorigenic data. Mildly toxic by ingestion, skin contact, and intraperitoneal routes. A skin irritant. Experimental reproductive effects. Mutation data reported. Flammable when exposed to heat and flame. When heated to decomposition it emits acrid smoke and fumes.

COD850 **CAS:299-86-5** **HR: 3**
CRUFORMATE
mf: $C_{12}H_{19}ClNO_3P$ mw: 291.74

SYNS: AMIDOFOS □ AMIDOPHOS □ o-(4-terz.-BUTIL-2-CLORO-FENIL)-o-METIL-FOSFORAMMIDE (ITALIAN) □ o-(4-tert BUTYL-2-CHLOOR-FENYL)-o-METHYL-FOSFORZUUR-N-METHYL-AMIDE (DUTCH) □ 4-tert-BUTYL-2-CHLORO PHENYL METHYL METHYL PHOSPHORAMIDATE □ 4-tert.-BUTYL 2-CHLOROPHENYL METHYL-PHOSPHORAMIDATE de METHYLE (FRENCH) □ o-(4-tert-BUTYL-2-CHLOR-PHENYL)-o-METHYL-PHOSPHORSAEURE-N-METHYL AMID (GERMAN) □ CRUFOMATE □ CRUFOMATE A □ DOWCC 132 □ ENT 25,602-X □ o-METHYL-o-2-CHLORO-4-tert-BUTYLPHENYL-N-METHYLAMIDOPHOSPHATE □ MONTREL □ RUELENE □ RUELENE DRENCH □ RUELENE 25E □ RULENE

TOXICITY DATA WITH REFERENCE
skn-mus TDLo:100 mg/kg (female 2D pre):REP JESEDU 16,141,81
skn-mus TDLo:200 mg/kg (female 2D pre):REP JESEDU 16,141,81
skn-mus TDLo:100 mg/kg (2D pre):REP JPFCD2 13,169,78
orl-rat LD50:460 mg/kg TXAPA9 14,515,69
ihl-rat LCLo:12 mg/m³/4H 85GMAT -,29,82
unk-rat LD50:770 mg/kg 30ZDA9 -,317,71
orl-rbt LD50:400 mg/kg 28ZEAL 5,59,76
skn-rbt LD50:2000 mg/kg SPEADM 78-1,37,78
orl-gpg LD50:1000 mg/kg FCTXAV 6,185,68
orl-bwd LD50:100 mg/kg TXAPA9 21,315,72
orl-mam LD50:251 mg/kg VETNAL 54(12),94,78

OSHA PEL: TWA 5 mg/m³
ACGIH TLV: TWA 5 mg/m³

SAFETY PROFILE: A poison by ingestion and inhalation. Moderately toxic via skin contact and possibly other routes. Experimental reproductive effects. When heated to decomposition it emits very toxic fumes of PO_x, NO_x, and Cl^-.

COE000 **CAS:16919-19-0** **HR: 3**
CRYPTOHALITE
DOT: UN 2854
mf: $F_6Si•2H_4N$ mw: 178.19

PROP: Mp: subl, d: 2.01.

SYNS: AMMONIUM FLUOSILICATE □ AMMONIUM FLUOROSILICATE (DOT) □ AMMONIUM HEXAFLUOROSILICATE □ AMMONIUM SILICOFLUORIDE □ AMMONIUM SILICON FLUORIDE □ DIAMMONIUM FLUOSILICATE □ DIAMMONIUM HEXAFLUOROSILICATE(2-) □ DIAMMONIUM SILICON HEXAFLUORIDE □ FLUOSILICATE de AMMONIUM (FRENCH)

TOXICITY DATA WITH REFERENCE
orl-rat LDLo: 100 mg/kg NCNSA6 5,27,53
orl-mus LD50: 70 mg/kg GISAAA 53(11),80,88
orl-gpg LDLo: 150 mg/kg CRSBAW 124,133,37
scu-frg LDLo: 224 mg/kg CRSBAW 124,133,37

OSHA PEL: TWA 2.5 mg(F)/m^3
NIOSH REL: (Fluorides, Inorganic) TWA 2.5 mg(F)/m^3
DOT CLASSIFICATION: 6.1; *Label:* KEEP AWAY FROM FOOD

SAFETY PROFILE: Poison by ingestion and subcutaneous routes. See also HEXAFLUROSILICATE (2-) DIHYDROGEN and FLUORIDES. When heated to decomposition it emits very toxic fumes of F^-, NH_3, and NO_x.

COE100 **CAS:37226-23-6** **HR: 3**
CRYSTALLOMYCIN

TOXICITY DATA WITH REFERENCE
orl-mus LD50: 1500 mg/kg 85ERAY 1,393,78
ipr-mus LD50: 109 mg/kg ANTBAL 4(4),63,59
scu-mus LD50: 220 mg/kg ANTBAL 4(4),63,59
ivn-mus LD50: 124 mg/kg ANTBAL 4(4),63,59
scu-rbt LDLo: 200 mg/kg ANTBAL 4(4),63,59
ivn-rbt LDLo: 50 mg/kg ANTBAL 4(4),63,59
scu-gpg LDLo: 125 mg/kg ANTBAL 4(4),63,59

SAFETY PROFILE: Poison by subcutaneous, intravenous, and intraperitoneal routes. Moderately toxic by ingestion.

COE125 **CAS:10380-77-5** **HR: 3**
CT 3318
mf: $C_{31}H_{42}N_2O_3•2I$ mw: 744.55

SYNS: DIIODOMETHYLATE de la BIS(PIPERIDINOMETHYL-COUMARANYL-5)CETONE (FRENCH) □ DIIODOMETILATO del BISPIPERIDINOMETILCUMARANIL-5-CHETONE (ITALIAN)

TOXICITY DATA WITH REFERENCE
ivn-rat LD50: 1500 μg/kg AIPTAK 120,53,59
scu-mus LD50: 1100 μg/kg AIPTAK 106,395,56
ivn-dog LDLo: 400 μg/kg AIPTAK 106,395,56
scu-cat LDLo: 700 μg/kg AIPTAK 106,395,56
ivn-rbt LDLo: 250 μg/kg AIPTAK 106,395,56
scu-gpg LDLo: 400 μg/kg AIPTAK 106,395,56
ivn-ckn LD50: 1 mg/kg AIPTAK 120,53,59

SAFETY PROFILE: A deadly poison by subcutaneous and intravenous routes. When heated to decomposition it emits toxic fumes of NO_x and I^-.

C

COE175 **CAS:8007-87-2** **HR: 1**
CUBEB OIL

PROP: From steam distillation of mature, unripe fruit of *piper cubeba* L. (Fam. Piperaceae). Colorless to light green liquid; spicy odor, sltly acrid taste. D: 0.898–0.928, refr index: 1.492–1.502 @ 20°. Sol in fixed oils, mineral oil; insol in glycerin, propylene glycol.

SYNS: OIL OF CUBEB □ OILS, CUBEB

TOXICITY DATA WITH REFERENCE
skn-rbt 500 mg/24H FCTXAV 14,729,76

CONSENSUS REPORTS: Reported in EPA TSCA Inventory.

SAFETY PROFILE: A skin irritant. When heated to decomposition it emits acrid smoke and irritating fumes.

COE250 **CAS:18444-66-1** **HR: 3**
CUCURBITACIN E
mf: $C_{32}H_{44}O_8$ mw: 556.76

SYNS: CUCURBITACINE-E □ α-ELATERIN

TOXICITY DATA WITH REFERENCE
orl-mus LD50: 340 mg/kg CHTPBA 5,205,70

SAFETY PROFILE: Poison by ingestion. When heated to decomposition it emits acrid smoke and irritating fumes.

COE500 **CAS:122-03-2** **HR: 2**
CUMALDEHYDE
mf: $C_{10}H_{12}O$ mw: 148.22

PROP: Found in at least 50 essential oils, such as cumin, eucalyptus species, cinnamon, boldo, and rue, and as main constituent of oil of *Pectis papposa harn* and *gray* (FCTXAV 12,385,74). Colorless to pale-yellow liquid; pungent odor of cumin. D: 0.976–0.980, refr index: 1.529–1.534, flash p: 199°F. Sol in alc, ether; insol in water.

SYNS: p-CUMIC ALDEHYDE □ CUMINALDEHYDE □ CUMINIC ALDEHYDE (FCC) □ CUMINYL ALDEHYDE □ FEMA No. 2341 □ p-ISOPROPYLBENZALDEHYDE □ 4-ISOPROPYLBENZALDEHYDE □ p-ISOPROPYLBENZENECARBOXALDEHYDE □ 4-(1-METHYLETHYL)-BENZALDEHYDE (9CI)

TOXICITY DATA WITH REFERENCE
skn-rbt 500 mg/24H FCTXAV 12,395,74
orl-rat LD50: 1390 mg/kg FCTXAV 2,327,64
orl-mus LD50: 2400 mg/kg BIJOAK 34,1196,40
skn-rbt LD50: 2800 mg/kg FCTXAV 12,395,74

CONSENSUS REPORTS: Reported in EPA TSCA Inventory.

SAFETY PROFILE: Moderately toxic by ingestion and skin contact. A skin irritant. Combustible liquid. When heated to decomposition it emits acrid smoke and irritating fumes. See also ALDEHYDES.

COE750 CAS:98-82-8 HR: 3
CUMENE
DOT: UN 1918
mf: C_9H_{12} mw: 120.21

PROP: Colorless liquid. Mp: −96.0°, bp: 152°, flash p: 111°F, d: 0.864 @ 20°/4°, vap press: 10 mm @ 38.3°, autoign temp: 795°F, lel: 0.9%, uel: 6.5%, vap d: 4.1.

SYNS: BENZENE ISOPROPYL ☐ CUM ☐ CUMEEN (DUTCH) ☐ 2-FENILPROPANO (ITALIAN) ☐ 2-FENYL-PROPAAN (DUTCH) ☐ ISOPROPILBENZENE (ITALIAN) ☐ ISOPROPYLBENZEEN (DUTCH) ☐ ISOPROPYL BENZENE ☐ ISOPROPYLBENZOL ☐ ISOPROPYL-BENZOL (GERMAN) ☐ 2-PHENYLPROPANE ☐ RCRA WASTE NUMBER U055

TOXICITY DATA WITH REFERENCE
skn-rbt 10 mg/24H open MLD AMIHBC 4,119,51
skn-rbt 100 mg/24H MOD 85JCAE-,33,86
eye-rbt 86 mg MLD AMIHAB 14,387,56
eye-rbt 500 mg/24H MLD 85JCAE-,33,86
ihl-hmn TCLo:200 ppm:NOSE,CNS,PUL TGNCDL 2,39,61
orl-rat LD50:1400 mg/kg AMIHAB 14,387,56
ihl-rat LC50:8000 ppm/4H AMIHBC 4,119,51
ihl-mus LC50:24,700 mg/m³/2H 85GMAT -,78,82

CONSENSUS REPORTS: Community Right-To-Know List. Reported in EPA TSCA Inventory. EPA Genetic Toxicology Program.

OSHA PEL: TWA 50 ppm (skin)
ACGIH TLV: TWA 50 ppm (skin)
DFG MAK: 50 ppm (245 mg/m³)
DOT CLASSIFICATION: 3; *Label:* Flammable Liquid

SAFETY PROFILE: Moderately toxic by ingestion. Mildly toxic by inhalation and skin contact. Human systemic effects by inhalation: an antipsychotic, unspecified changes in the sense of smell and respiratory system. An eye and skin irritant. Potential narcotic action. Central nervous system depressant. There is no apparent difference between the toxicity of natural cumene or that derived from petroleum. See also BENZENE and TOLUENE. Flammable liquid when exposed to heat or flame; can react with oxidizing materials. Violent reaction with HNO_3; oleum; chlorosulfonic acid. To fight fire, use foam, CO_2, dry chemical.

For occupational chemical analysis use NIOSH: Hydrocarbons, Aromatic, 1501.

COF000 CAS:93-53-8 HR: 2
CUMENE ALDEHYDE
mf: $C_9H_{10}O$ mw: 134.19

PROP: Colorless liquid; floral odor. D: 0.998–1.006, refr index: 1.515–1.520, flash p: 156°F. Sol in fixed oils; sltly sol in propylene glycol; insol in glycerin.

SYNS: FEMA No. 2886 ☐ α-FORMYLETHYLBENZENE ☐ HYACINTHAL ☐ HYDRATROP ALDEHYDE ☐ HYDRATROPIC ALDEHYDE ☐ α-METHYL PHENYLACETALDEHYDE ☐ α-METHYL-α-TOLUIC ALDEHYDE ☐ 2-PHENYLPROPANAL ☐ α-PHENYLPROPIONALDEHYDE ☐ 2-PHENYLPROPIONALDEHYDE (FCC)

TOXICITY DATA WITH REFERENCE
orl-rat LD50:2800 mg/kg FCTXAV 2,327,64

CONSENSUS REPORTS: Reported in EPA TSCA Inventory.

SAFETY PROFILE: Moderately toxic by ingestion. Combustible liquid. When heated to decomposition it emits acrid smoke and irritating fumes. See also ALDEHYDES.

COF250 CAS:64-00-6 HR: 3
m-CUMENOL METHYLCARBAMATE
mf: $C_{11}H_{15}NO_2$ mw: 193.27

SYNS: COMPOUND 10854 ☐ m-CUMENYL METHYLCARBAMATE ☐ ENT 25,500 ☐ ENT 25,543 ☐ HERCULES 5727 ☐ HIP ☐ m-ISOPROPYLPHENOL-N-METHYLCARBAMATE ☐ m-ISOPROPYLPHENYL METHYLCARBAMATE ☐ m-ISOPROPYLPHENYL-N-METHYLCARBAMATE ☐ 3-ISOPROPYLPHENYL METHYLCARBAMATE ☐ N-METHYL-m-ISOPROPYLPHENYL CARBAMATE ☐ N-METHYL-3-ISOPROPYLPHENYL CARBAMATE ☐ OMS-15 ☐ UC 10854 ☐ UNION CARBIDE UC-10,854

TOXICITY DATA WITH REFERENCE
orl-rat LD50:29 mg/kg TXAPA9 21,315,72
skn-rat LD50:113 mg/kg 31ZOAD 1,263,68
ipr-rat LD50:14,200 μg/kg BWHOA6 44(1-3),241,71
ivn-rat LD50:3150 μg/kg BJIMAG 22,317,65
ims-rat LD50:14 mg/kg BJIMAG 22,317,65
unk-rat LD50:41 mg/kg 30ZDA9 -,193,71
orl-mus LD50:16 mg/kg JAFCAU 18,793,70
ipr-mus LDLo:6 mg/kg TXAPA9 6,402,64
ivn-mus LD50:1410 μg/kg CSLNX* NX#02085
ims-dog LDLo:13 mg/kg BJIMAG 22,317,65
skn-rbt LD50:40 mg/kg FMCHA2 -,D323,80
orl-gpg LD50:10 mg/kg 31ZOAD 1,263,68
orl-ckn LD50:12 mg/kg TXAPA9 11,49,67
orl-bwd LD50:3200 μg/kg TXAPA9 21,315,72

CONSENSUS REPORTS: EPA Extremely Hazardous Substances List.

SAFETY PROFILE: Poison by ingestion, skin contact, intraperitoneal, intravenous, intramuscular, and possibly other routes. A pesticide. See also CARBAMATES. When heated to decomposition it emits toxic fumes of NO_x.

COF325 CAS:8014-13-9 HR: 2
CUMIN OIL

PROP: From steam distillation of *Cuminum cyminum* L. Light-yellow to brown liquid; strong odor. D: 0.905–0.925, refr index: 1.501 @ 20°. Sol in fixed oils, mineral oil; very sol in glycerin, propylene glycol.

SYNS: CUMMIN ☐ OILS, CUMIN

TOXICITY DATA WITH REFERENCE
skn-rbt 500 mg/24H MOD FCTXAV 12,869,74
mmo-sat 100 μL/plate NUCADQ 1,10,79
dnr-bcs 10 mg/disc TOFOD5 8,91,85
bfa-rat/sat 2500 mg/kg NUCADQ 1,10,79
orl-rat LD50:2500 mg/kg FCTXAV 12,869,74
skn-rbt LD50:3560 mg/kg FCTXAV 12,869,74

CONSENSUS REPORTS: Reported in EPA TSCA Inventory.

SAFETY PROFILE: Moderately toxic by ingestion and skin contact. A skin irritant. Mutation data reported. When heated to decomposition it emits acrid smoke and irritating fumes.

COF400 CAS:599-64-4 HR: 3
p-CUMYLPHENOL
mf: $C_{15}H_{16}O$ mw: 212.31

SYNS: p-(α-CUMYL)PHENOL □ p-(α-α-DIMETHYLBENZYL)PHENOL □ 4-(DIMETHYLPHENYLMETHYL)PHENOL □ 4-HYDROXYDIPHENYLDIMETHYLMETHANE □ 4-(1-METHYL-1-PHENETHYL)PHENOL □ PHENOL, p-(α-α-DIMETHYLBENZYL)- □ PHENOL, 4-(1 METHYL-1-PHENETHYL)-(9CI)

TOXICITY DATA WITH REFERENCE
orl-frg LD50:335 mg/kg GTPZAB 12(12),44,68

CONSENSUS REPORTS: Reported in EPA TSCA Inventory.

SAFETY PROFILE: Poison by ingestion. When heated to decomposition it emits acrid smoke and irritating vapors.

COF500 CAS:12002-03-8 HR: 3
CUPRIC ACETOARSENITE
mf: $C_4H_6As_6Cu_4O_{16}$ mw: 1013.78

PROP: Emerald-green powder.

SYNS: (ACETATO)(TRIMETAARSENITO)DICOPPER □ ACETOARSENITE de CUIVRE (FRENCH) □ BASLE GREEN □ C.I. 77410 □ C.I. PIGMENT GREEN 21 (9CI) □ COPPER ACETOARSENITE (DOT) □ COPPER ACETOARSENITE, solid (DOT) □ EMERALD GREEN □ ENT 884 □ FRENCH GREEN □ GENUINE PARIS GREEN □ IMPERIAL GREEN □ KING'S GREEN □ MEADOW GREEN □ MINERAL GREEN □ MITIS GREEN □ MOSS GREEN □ MOUNTAIN GREEN □ NEUWIED GREEN □ NEW GREEN □ ORTHO P-G BAIT □ PARIS GREEN □ PARROT GREEN □ PATENT GREEN □ POWDER GREEN □ SCHWEINFURTERGRUEN □ SCHWEINFURT GREEN □ SOWBUG & CUTWORM BAIT □ SWEDISH GREEN □ VIENNA GREEN

TOXICITY DATA WITH REFERENCE
orl-rat LD50:22 mg/kg PCOC** -,254,66
unk-mam LD50:18 mg/kg 30ZDA9 -,388,71

CONSENSUS REPORTS: Arsenic and its compounds, as well as copper and its compounds, are on the Community Right-To-Know List.

OSHA PEL: TWA 0.5 mg(As)/m^3
DOT CLASSIFICATION: 6.1; *Label:* Poison

SAFETY PROFILE: Poison by ingestion and possibly other routes. An insecticide. When heated to decomposition it emits very toxic fumes of As. See also ARSENIC COMPOUNDS and COPPER COMPOUNDS.

COF750 CAS:8063-06-7 HR: 3
CURARE

PROP: Brown, brittle, resinous mass. Rendering of an Indian name given to the unstandardized extracts derived mainly from the bark of various species of *Strychnos* and *Chondodendron* (12VSA5 9,347,76).

SYNS: INTOCOSTRINE □ OURARI □ URARI □ WOORALI □ WOORARI □ WOURARA

TOXICITY DATA WITH REFERENCE
unr-man LDLo:735 μg/kg 85DCAI 2,73,70
ipr-mus LD50:3200 μg/kg AIPTAK 153,308,65
scu-mus LD50:500 μg/kg NATWAY 56,615,69
ivn-mus LD50:140 μg/kg PSEBAA 118,756,65
ivn-dog LD50:1200 μg/kg JPETAB 82,266,44
orl-rbt LDLo:270 mg/kg AEXPBL 61,283,09
scu-rbt LDLo:2700 μg/kg AEXPBL 61,283,09

SAFETY PROFILE: A deadly human poison by an unspecified route. A deadly experimental poison by ingestion, intraperitoneal, intravenous, and subcutaneous routes. When heated to decomposition it emits highly toxic fumes.

COF825 CAS:22260-42-0 HR: 3
CURARINE
mf: $C_{38}H_{44}N_2O_6$ mw: 624.84

SYNS: CURARIN □ (+)-CURARINE □ (1′-α)-7′,12′-DIHYDROXY-6,6′-DIMETHOXY-2,2,2′,2′-TETRAMETHYLTUBOCURARANIUM (9CI) □ 2,2′-DIMETHYL-BEBEERINIUM (8CI)

TOXICITY DATA WITH REFERENCE
scu-mus LDLo:380 μg/kg HBAMAK 4,1289,35
scu-dog LDLo:340 μg/kg FDWU** -,-,31
scu-cat LDLo:340 μg/kg HBAMAK 4,1289,35
orl-rbt LDLo:334 μg/kg AEXPBL 61,283,09
scu-rbt LDLo:340 μg/kg HBAMAK 4,1289,35
ivn-rbt LDLo:80 μg/kg HBAMAK 4,1289,35
scu-gpg LDLo:90 μg/kg HBAMAK 4,1289,35
scu-pgn LDLo:618 μg/kg FDWU** -,-,31
ims-pgn LDLo:618 μg/kg HBAMAK 4,1289,35
scu-frg LDLo:16,800 μg/kg FDWU** -,-,31

SAFETY PROFILE: A deadly poison by ingestion, subcutaneous, intramuscular and intravenous routes. When heated to decomposition it emits toxic fumes of NO_x.

COF850 HR: 2
CURCUMA LONGA Linn., rhizome extract

PROP: Indian plant belonging to the family Zingiberaceae (IJEBA6 6,232,68).

SYNS: HALDI RHIZOME EXTRACT □ TURMERIC rhizome extract

TOXICITY DATA WITH REFERENCE
orl-rat TDLo:700 mg/kg (1-7D preg):REP IJEBA6 16,1077,78
orl-rat LD50:12,200 mg/kg IJMRAQ 59,1289,71
ipr-mus LD50:430 mg/kg IJMRAQ 64,601,76

SAFETY PROFILE: Moderately toxic by intraperitoneal route. Mildly toxic by ingestion. Experimental reproductive effects.

COG000 **CAS:8024-37-1** **HR: 2**
CURCUMIN

SYNS: C.I. 75300 □ CURCUMA OIL □ CURCUMINE □ NCI-C60015 □ TURMERIC OIL □ TURMERIC OLEORESIN

TOXICITY DATA WITH REFERENCE
skn-rbt 500 mg/24H MLD FCTOD7 21,839,83
cyt-ham:lng 20 mg/L GMCRDC 27,95,81
ipr-mus LD50:1500 mg/kg IJMRAQ 64,601,76

CONSENSUS REPORTS: Reported in EPA TSCA Inventory.

SAFETY PROFILE: Moderately toxic by intraperitoneal route. A skin irritant. Mutation data reported. When heated to decomposition it emits acrid smoke and irritating fumes.

COG500 **CAS:62355-03-7** **HR: 3**
CUSCOHYGRINE BIS(METHYL BENZENESULFONATE)
mf: $C_{15}H_{30}N_2O•C_{12}H_{10}O_6S_2$ mw: 568.81

SYNS: 1,3-BIS(N,N-DIMETHYL-2-PYRROLIDINIUM)PROPANE DIBENZENESULFONATE □ 2,2'-(2-OXO-TRIMETHYLENE)BIS(1,1-DIMETHYLPYRROLIDINIUM) DIBENZENESUFLONATE

TOXICITY DATA WITH REFERENCE
ipr-mus LD50:117 mg/kg PCJOAU 11,478,77
ipr-cat LD50:117 mg/kg PCJOAU 11,135,77

SAFETY PROFILE: Poison by intraperitoneal route. See also SULFONATES. When heated to decomposition it emits very toxic fumes of NO_x and SO_x.

COH000 **HR: 3**
CUTTING OILS

SAFETY PROFILE: Often carcinogenic. The cause of "cutting oil" dermatitis is generally due to an insoluble oil. However it can occasionally be caused by a soluble oil. Many have looked for a causative factor other than the oil itself. Bacteria have frequently been blamed, although insoluble oils are usually sterile, while the soluble oils may contain bacteria. The metal slivers that occur in these oils after use have also been blamed as well as the sulfur, chlorine, and inhibitors that they contain. The oil itself can plug the pores and form boils. Combustible when exposed to heat or flame. See also MINERAL OIL.

COH250 **CAS:140-87-4** **HR: 3**
CYANACETIC ACID HYDRAZIDE
mf: $C_3H_5N_3O$ mw: 99.11

PROP: Mp: 115°, a solid.

SYNS: AB-42 □ ARMAZAL □ CIANAZIL □ CYACETACID □ CYACETACIDE □ CYACETAZID □ CYACETAZIDE □CYANACETHYDRAZIDE □ CYANACETIC ACID HYDRAZIDE □CYANACETOHYDRA-ZIDE □ CYANACETYLHYDRAZIDE □ CYANAZIDE □ CYANIZIDE □ CYANOACETHYDRAZIDE □ CYANOACETIC ACID HYDRAZIDE □ CYANOACETOHYDRAZIDE □ α-CYANOACETOHYDRAZIDE □ CYANOACETYLHYDRAZIDE □ CYANOETHYDRAZIDE □ CYAZID □ CYAZIDE □ DICTYCIDE □ DICTYZIDE □ HELMOX □ HIDACIAN □ HIDACIANN □ KYANACETHYDRAZID □ LEANDIN □ MACKREAZID □ MALONITRILE HYDRAZIDE □ MALONONITRILE HYDRAZIDE □ NEOHYDRAZID □ REACID □ REAZID □ REAZIDE □ TSIAZID □ USAF KF-18

TOXICITY DATA WITH REFERENCE
orl-mus LD50:250 mg/kg CLDND* AD277-689
scu-mus LD50:228 mg/kg ABMGAJ 21,635,68

CONSENSUS REPORTS: Cyanide and its compounds are on the Community Right-To-Know List.

SAFETY PROFILE: Poison by ingestion and intraperitoneal routes. When heated to decomposition it emits toxic fumes of NO_x and CN^-. See also NITRILES.

COH500 **CAS:420-04-2** **HR: 3**
CYANAMIDE
mf: CH_2N_2 mw: 42.05

PROP: Deliquescent crystals. Mp: 45°, bp: 260°, flash p: 285°F, d: 1.282, vap d: 1.45.

SYNS: AMIDOCYANOGEN □ CARBAMONITRILE □ CARBIMIDE □ CYANOAMINE □ CYANOGENAMIDE □ CYANOGEN NITRIDE □ HYDROGEN CYANAMIDE □ USAF EK-1995

TOXICITY DATA WITH REFERENCE
orl-rat TDLo:2600 mg/kg (male 70D pre):REP PHTXA6 61,20,87
orl-rat TDLo:1750 mg/kg (male 70D pre):REP PHTXA6 61,20,87
orl-rat TDLo:2450 mg/kg (male 70D pre):REP PHTXA6 61,20,87
orl-rat TDLo:208 mg/kg (male 70D pre):REP PHTXA6 61,20,87
orl-rat LD50:125 mg/kg MEIEDD 10,383,83
ihl-rat LCLo:86 mg/m³/4H 85GMAT -,40,82
skn-rat LD50:84 mg/kg 85GMAT -,40,82
ipr-rat LDLo:200 mg/kg PSEBAA 54,254,43
ipr-mus LD50:200 mg/kg NTIS** AD277-689
skn-rbt LD50:590 mg/kg 34ZIAG -,190,69

CONSENSUS REPORTS: Reported in EPA TSCA Inventory. Cyanide and its compounds are on the Community Right-To-Know List.

OSHA PEL: TWA 2 mg/m³
ACGIH TLV: TWA 2 mg/m³

SAFETY PROFILE: Poison by ingestion, inhalation, and intraperitoneal routes. Moderately toxic by skin contact. Experimental reproductive effects. Combustible when exposed to heat or flame. To fight fire, use CO_2, dry chemical. Thermally unstable. Contact with moisture (water), acids, or alkalies may cause a violent reaction above 40°. Concentrated aqueous solutions may undergo explosive polymerization. Mixture with 1,2-phenylenediamine salts may cause explosive polymerization. When heated to decomposition or on contact with acid or acid fumes, it emits toxic fumes of CN^- and NO_x. See also CYANIDE and AMIDES.

COH750 **HR: 3**
CYANATES

SAFETY PROFILE: Variable. See individual entry. When

heated to decomposition, or on contact with acid or acid fumes, they emit toxic fumes of CN^-.

COI000 **CAS:4027-17-2** **HR: 3**
CYANATOTRIBUTYLSTANNANE
mf: $C_{13}H_{27}NOSn$ mw: 332.10

SYN: TRIBUTYLTIN CYANATE

TOXICITY DATA WITH REFERENCE
ivn-mus LD50:6300 µg/kg CSLNX* NX#02224

CONSENSUS REPORTS: Cyanide and its compounds are on the Community Right-To-Know List.

OSHA PEL: TWA 0.1 mg(Sn)/m³ (skin)
ACGIH TLV: TWA 0.1 mg(Sn)/m³ (skin) (Proposed: TWA 0.1 mg(Sn)/m³; STEL 0.2 mg(Sn)/m³ (skin))
NIOSH REL: (Organotin Compounds) TWA 0.1 mg(Sn)/m³

SAFETY PROFILE: Poison by intravenous route. Tributyl tin compounds are very toxic to marine life. See also CYANATES and TIN COMPOUNDS. When heated to decomposition it emits toxic fumes of NO_x and CN^-.

For occupational chemical analysis use NIOSH: Organotin Compounds 5504.

COI125 **HR: 3**
CYANEA CAPILLATA TOXIN

SYN: TOXIN, JELLYFISH, SCYPHOZOAN, CYANEA CAPILLATA

TOXICITY DATA WITH REFERENCE
ivn-rat LD50:800 µg/kg TOXIA6 15,3,77
ipr-mus LD50:300 µg/kg TOXIA6 15,3,77
ivn-mus LD50:300 µg/kg TOXIA6 15,3,77
ivn-rbt LD50:300 µg/kg TOXIA6 15,3,77
ivn-ckn LD50:2 mg/kg TOXIA6 15,3,77

SAFETY PROFILE: A deadly poison by intravenous and intraperitoneal routes.

COI250 **CAS:917-61-3** **HR: 3**
CYANIC ACID, SODIUM SALT
DOT: UN 2207/UN 2478/UN 3080
mf: CNO•Na mw: 65.01

SYNS: CYANSAN ☐ SAN-CYAN ☐ SODIUM ISOCYANATE ☐ WEECON ☐ ZASSOL

TOXICITY DATA WITH REFERENCE
orl-hmn TDLo:5400 mg/kg/24W:EYE,MET AROPAW 94,927,76
orl-rat LD50:1500 mg/kg JPETAB 185,653,73
ims-rat LD50:310 mg/kg BJPCAL 1,186,46
orl-mus LDLo:4 mg/kg APFRAD 19,740,61
ipr-mus LD50:260 mg/kg JPETAB 185,653,73

CONSENSUS REPORTS: Cyanide and its compounds are on the Community Right-To-Know List. Reported in EPA TSCA Inventory.

DOT CLASSIFICATION: 6.1; *Label:* KEEP AWAY FROM FOOD (UN 2207); DOT Class: 6.1; *Label:* Poison (UN 2206); DOT Class: 6.1; *Label:* Poison, Flammable Liquid (UN 3080); DOT Class: 3; *Label:* Flammable Liquid, Poison (UN 2478)

SAFETY PROFILE: Poison by ingestion, intraperitoneal, and intramuscular routes. Human systemic effects by ingestion: weight loss, changes in the visual field, and other eye effects. See also CYANATES. When heated to decomposition it emits very toxic fumes of CN^- and Na_2O.

COI500 **CAS:57-12-5** **HR: 3**
CYANIDE
DOT: UN 1935
mf: CN^- mw: 26.02

SYNS: CARBON NITRIDE ION (CN¹⁻) ☐ CYANIDE(CN¹⁻) ☐ CYANIDE, dry (UN 1588) ☐ CYANIDE ANION ☐ CYANIDE (CN¹⁻) ☐ CYANIDE ION ☐ CYANIDE(CN¹⁻) ION ☐ CYANIDE SOLUTIONS (DOT) ☐ CYANURE ☐ HYDROCYANIC ACID, ION(CN¹⁻) ☐ ISOCYANIDE ☐ RCRA WASTE NUMBER P030

TOXICITY DATA WITH REFERENCE
ipr-mus LD50:3 mg/kg NATUAS 228,1315,70

CONSENSUS REPORTS: Cyanide and its compounds are on the Community Right-To-Know List.

OSHA PEL: TWA 5 mg(CN)/m³
ACGIH TLV: CL 5 mg/m³ (skin)
DFG MAK: 5 mg/m³
NIOSH REL: (Cyanide) TWA CL 5 mg/m³/10M
DOT CLASSIFICATION: 6.1; *Label:* Poison, KEEP AWAY FROM FOOD

SAFETY PROFILE: Very poisonous by most routes. Cyanide directly stimulates the chemoreceptors of the carotid and aortic bodies with a resultant hyperpnea (increase in the depth and rate of respiration). Cardiac irregularities are often noted, but the heart invariably outlasts the respirations. Death is due to respiratory arrest of central origin. It can occur within seconds or minutes of the inhalation of high concentrations of HCN gas. Because of slower absorption, death may be more delayed after the ingestion of cyanide salts, but the critical events still occur within the first hour. Two other sources of cyanide have been responsible for human poisoning: the naturally occurring amygdalin and the drug nitroprusside.

Amygdalin is a cyanogenic glycoside found in apricot, peach, and similar fruit pits and in sweet almonds (Sayre and Kaymakcalan, 1941). It is a chemical combination of glucose, benzaldehyde, and cyanide from which the latter can be released by the action of β-glucosidase or emulsion. Although these enzymes are not found in mammalian tissues, the human intestinal microflora appears to possess these or similar enzymes capable of effecting cyanide release resulting in human poisoning. For this reason, amygdalin may be as much as 40 times more toxic by the oral route as compared with intravenous injection. Amygdalin is the major ingredient of Laetrile, and this alleged anticancer drug has also been responsible for human cyanide poisoning.

An ethical drug that may also cause cyanide poisoning in overdose is the potent vascular smooth-muscle relaxant sodium nitroprusside. Although nitroprusside is related chemically to ferricyanide, unlike the latter it penetrates into erythrocytes and reacts with hemoglobin to release its cyanide (Smith and Kruszyna, 1974). Fortunately, the therapeutic margin for nitroprusside appears to be quite large.

Cyanide is commonly found in certain rat and pest poisons, silver and metal polishes, photographic solutions, and fumigating products. Compounds such as potassium cyanide can also be readily purchased from chemical stores. Cyanide is readily absorbed from all routes, including the skin, mucous membranes, and by inhalation, although alkali salts of cyanide are toxic only when ingested. Death may occur with ingestion of even small amounts of sodium or potassium cyanide and can occur within minutes or hours depending on route of exposure. Inhalation of toxic fumes represents a potentially rapidly fatal type of exposure. A blood cyanide level of greater than 0.2 μg/mL is considered toxic. Lethal cases have usually had levels above 1 μg/mL.

Clinically, cyanide poisoning is reported to produce a bitter, almond odor on the breath of the patient; however, only a small proportion of the population is genetically able to discern this characteristic odor. Typically, cyanide has a bitter, burning taste, and, following poisoning, symptoms of salivation, nausea without vomiting, anxiety, confusion, vertigo, giddiness, lower-jaw stiffness, convulsions, opisthotonos, paralysis, coma, cardiac arrhythmias, and transient respiratory stimulation followed by respiratory failure may occur. Bradycardia is a common finding, but in most cases heartbeat usually outlasts respiration (Wexler et al., 1947). A prolonged expiratory phase is considered to be characteristic of cyanide poisoning. (Casarett and Doull's, *Toxicology, The Basic Science of Poisons 2nd ed.* Doull, Klaassen and Amdur (eds), Macmillan Pub. Co. Inc. New York, NY).

The volatile cyanides resemble HCN physiologically, inhibiting tissue oxidation and causing death through asphyxia. Cyanogen is probably as toxic as HCN; the nitriles are generally considered somewhat less toxic, probably because of their lower volatility. The nonvolatile cyanide salts appear to be relatively nontoxic systemically, so long as they are not ingested and care is taken to prevent the formation of HCN. Workers, such as electroplaters and picklers, who are daily exposed to cyanide solutions may develop a "cyanide" rash, characterized by itching, and by macular, papular, and vesicular eruptions. Frequently there is secondary infection. Exposure to small amounts of cyanide compounds over long periods of time is reported to cause loss of appetite, headache, weakness, nausea, dizziness, and symptoms of irritation of the upper respiratory tract and eyes. See also specific cyanide compounds.

Flammable by chemical reaction with heat, moisture, acid. Many cyanides rather easily evolve HCN, a flammable gas that is highly toxic. Carbon dioxide from the air is sufficiently acidic to liberate HCN from cyanide solutions. Reaction with hypochlorite solutions may be violent at pH 10.0–10.3. Cyanides explode if melted with nitrites or chlorates at about 450°. Violent reaction with F_2, Mg, nitrates, HNO_3, nitrites. Metal cyanides are easily oxidized and may be thermally unstable. N-cyano derivatives may be reactive or unstable. Many organic nitriles can be very reactive under the right conditions. When heated to decomposition or on contact with acid, acid fumes, water, or steam, cyanides emit toxic and flammable vapors of CN^-. See also HYDROCYANIC ACID.

For occupational chemical analysis use OSHA: #ID-120 or NIOSH: Cyanides (Aerosol and Gas) 7904.

COI750 CAS:528-58-5 HR: 1
CYANIDOL
mf: $C_{15}H_{11}O_6$ mw: 287.26

SYNS: 1-BENZOPYRYLIUM, 2-(3,4-DIHYDROXYPHENYL)-3,5,7-TRIHYDROXY-, CHLORIDE ☐ CYANIDIN ☐ CYANIDIN CHLORIDE ☐ CYANIDOL CHLORIDE ☐ 2-(3,4-DIHYDROXYPHENYL)-3,5,7-TRIHYDROXY-1-BENZOPYRYLIUM CHLORIDE ☐ GASTROTELOS ☐ IdB-1027 ☐ 3,3′4′,5,7-PENTAHYDROXYFLAVYLIUM CHLORIDE

TOXICITY DATA WITH REFERENCE
mnt-hmn:lyms 200 mg/L MUREAV 246,205,91
sce-hmn:lyms 100 mg/L MUREAV 246,205,91
orl-dog LD:>3 g/kg DRFUD4 14,224,89
orl-uns LD:>6 g/kg DRFUD4 14,224,89

SAFETY PROFILE: Low toxicity by ingestion. Mutation data reported. When heated to decomposition it emits acrid smoke and irritating fumes.

COJ250 CAS:107-91-5 HR: 3
2-CYANOACETAMIDE
mf: $C_3H_4N_2O$ mw: 84.09

PROP: White powder. Mp: 119°, bp: decomp.

SYNS: CYANACETAMIDE ☐ CYANOACETAMIDE ☐ CYANOIMINOACETIC ACID ☐ MALONAMIDE NITRILE ☐ MALONAMONITRILE ☐ NITRILOMALONAMIDE ☐ USAF KF-14

TOXICITY DATA WITH REFERENCE
orl-mus LD50:1680 mg/kg KHZDAN 9,50,66
ipr-mus LD50:200 mg/kg NTIS** AD691-689
scu-gpg LD50:1155 mg/kg MELAAD 47,192,56

CONSENSUS REPORTS: Cyanide and its compounds are on the Community Right-To-Know List. Reported in EPA TSCA Inventory.

SAFETY PROFILE: Poison by intraperitoneal route. Moderately toxic by ingestion and subcutaneous routes. See also NITRILES. When heated to decomposition it emits toxic fumes of NO_x and CN^-.

COJ500 CAS:372-09-8 HR: 3
CYANOACETIC ACID
mf: $C_3H_3NO_2$ mw: 85.07

PROP: A solid. Mp: 66°, bp: 108° @ 15 mm.

SYNS: ACIDE CYANACETIQUE (FRENCH) ☐ CAA ☐ CYANESSIGSAEURE (GERMAN) ☐ MALONIC MONONITRILE ☐ MONOCYANOACETIC ACID ☐ USAF KF-17

TOXICITY DATA WITH REFERENCE
orl-rat LD50:1500 mg/kg LONZA# 12JAN81
ipr-mus LD50:200 mg/kg NTIS** AD691-490
scu-rbt LDLo:1900 mg/kg AIPTAK 5,161,1899
scu-frg LDLo:1300 mg/kg AIPTAK 5,161,1899

CONSENSUS REPORTS: Reported in EPA TSCA

Inventory. Cyanide and its compounds are on the Community Right-To-Know List.

SAFETY PROFILE: Poison by intraperitoneal route. Moderately toxic by ingestion and subcutaneous routes. See also NITRILES. Mixture with furfuryl alcohol explodes when heated. When heated to decomposition it emits toxic fumes of NO_x and CN^-.

COJ625 CAS:16130-58-8 HR: 3
CYANOACETYL CHLORIDE
mf: C_3H_2ClNO mw: 103.51

CONSENSUS REPORTS: Cyanide compounds are on the Community Right-To-Know List.

SAFETY PROFILE: A poison. A storage hazard. It may explode at room temperature. Upon decomposition it emits toxic fumes of Cl^-, NO_x, and CN^-. See also CYANIDE and CHLORIDES.

COK125 CAS:873-74-5 HR: 3
4-CYANOANILINE
mf: $C_7H_6N_2$ mw: 118.15

SYNS: p-AMINOBENZONITRILE (8CI) ☐ 4-AMINOBENZONITRILE (9CI) ☐ p-CYANOANILINE

TOXICITY DATA WITH REFERENCE
orl-mus LD50:650 mg/kg APFRAD 41,391,83
ipr-mus LD50:155 mg/kg JMCMAR 17,900,74
orl-qal LD50:23,700 µg/kg AECTCV 12,355,83
orl-bwd LD50:23,700 µg/kg AECTCV 12,355,83

CONSENSUS REPORTS: EPA Genetic Toxicology Program. Reported in EPA TSCA Inventory. Cyanide and its compounds are on the Community Right-To-Know List.

SAFETY PROFILE: Poison by ingestion and intraperitoneal routes. When heated to decomposition it emits toxic fumes of NO_x and CN^-. See also CYANIDE and ANILINE DYES.

COK250 CAS:105-07-7 HR: 3
p-CYANOBENZALDEHYDE
mf: C_8H_5NO mw: 131.14

SYNS: 4-CYANOBENZALDEHYDE ☐ p-CYANOBENZENECARBOXALDEHYDE ☐ p-FORMYLBENZONITRILE ☐ 4-FORMYLBENZONITRILE ☐ TEREPHTHALALDEHYDONITRILE ☐ USAF KF-1

TOXICITY DATA WITH REFERENCE
ipr-mus LD50:100 mg/kg NTIS** AD277-689

CONSENSUS REPORTS: Cyanide and its compounds are on the Community Right-To-Know List. Reported in EPA TSCA Inventory.

SAFETY PROFILE: Poison by intraperitoneal route. See also NITRILES and ALDEHYDES. When heated to decomposition it emits toxic fumes of NO_x and CN^-.

COK300 CAS:17174-98-0 HR: 3
2-CYANOBENZAMIDE
mf: $C_8H_6N_2O$ mw: 146.16

SYNS: BENZAMIDE, 2-CYANO- ☐ BENZAMIDE, o-CYANO-(8CI) ☐ o-CYANOBENZAMIDE

TOXICITY DATA WITH REFERENCE
ipr-rat LD50:391 mg/kg APFRAD 48,23,90
ipr-mus LD50:545 mg/kg APFRAD 48,23,90

CONSENSUS REPORTS: Reported in EPA TSCA Inventory.

SAFETY PROFILE: A poison by intraperitoneal route. When heated to decomposition it emits toxic vapors of NO_x.

COK659 CAS:60633-76-3 HR: 3
CYANOBORANE OLIGOMER
mf: $(CH_2BN)_n$

CONSENSUS REPORTS: Cyanide compounds are on the Community Right-To-Know List.

SAFETY PROFILE: A heat- and shock-sensitive explosive. When heated to decomposition it emits toxic fumes of NO_x and CN^-. See also CYANIDE; BORANES; and EXPLOSIVES.

COL000 HR: 3
N-CYANO-2-BROMOETHYLCYCLOHEXYLAMINE
mf: $C_9H_{15}BrN$ mw: 217.14

CONSENSUS REPORTS: Cyanide and its compounds are on the Community Right-To-Know List.

SAFETY PROFILE: A poison. May explode when heated to 160°C. When heated to decomposition it emits toxic fumes of Br^-, NO_x, and CN^-. See also CYANIDE, AMINES and BROMIDES.

COL125 CAS:82423-05-0 HR: 3
CYANOCYCLINE A
mf: $C_{22}H_{26}N_4O_5$ mw: 426.52

TOXICITY DATA WITH REFERENCE
dni-esc 2 mg/L JANTAJ 36,1228,83
oms-esc 2 mg/L JANTAJ 26,1228,83
dni-mus:ast 1 mg/L JANTAJ 36,1228,83
oms-mus:ast 100 µg/L JANTAJ 36,1228,83
ipr-mus LD50:10 mg/kg JANTAJ 35,771,82

CONSENSUS REPORTS: Cyanide and its compounds are on the Community Right-To-Know List.

SAFETY PROFILE: Poison by intraperitoneal route. Mutation data reported. When heated to decomposition it emits toxic fumes of NO_x and CN^-. See also CYANIDE.

COL250 CAS:1769-99-9 HR: 3
5-CYANO-10,11-DIHYDRO-5-(3-DIMETHYLAMINOPROPYL)-5H-DIBENZO (a,d)CYCLOHEPTENE HYDROCHLORIDE
mf: $C_{21}H_{24}N_2$•ClH mw: 340.93

SYNS: 10,11-DIHYDRO-5-CYANO-N,N-DIMETHYL-5H-DIBENZO(a,d) CYCLOHEPTENE-5-PROPYLAMINE HCl ☐ UCB 4208

TOXICITY DATA WITH REFERENCE
orl-rat LD50:1250 mg/kg 27ZQAG -,91,72
ivn-rat LD50:28 mg/kg 27ZQAG -,91,72
ims-rat LD50:620 mg/kg 27ZQAG -,91,72
ipr-mus LD50:120 mg/kg JMCMAR 6,251,63
ivn-mus LD50:56 mg/kg CSLNX* NX#01197

CONSENSUS REPORTS: Cyanide and its compounds are on the Community Right-To-Know List.

SAFETY PROFILE: Poison by intravenous and intraperitoneal routes. Moderately toxic by ingestion and intramuscular routes. When heated to decomposition it emits toxic fumes of NO_x, CN^-, and HCl. See also CYANIDE.

COL750 CAS:683-45-4 HR: 3
CYANODIMETHYLARSINE
mf: C_3H_6AsN mw: 131.02

$N{\equiv}CAs(CH_3)_2$

SYN: DIMETHYLCYANOARSINE

TOXICITY DATA WITH REFERENCE
orl-rat LD50:50 mg/kg NCNSA6 5,13,53
ihl-mus LCLo:400 mg/m³/15M ZGEMAZ 13,523,21
ihl-dog LCLo:900 mg/m³/8M ZGEMAZ 13,523,21
ihl-cat LCLo:100 mg/m³/20M ZGEMAZ 13,523,21

CONSENSUS REPORTS: Cyanide and its compounds, as well as arsenic and its compounds, are on the Community Right-To-Know List.

OSHA PEL: TWA 0.5 mg(As)/m³

SAFETY PROFILE: Poison by ingestion and inhalation. Ignites spontaneously in air. See also ARSENIC COMPOUNDS and CYANIDE. When heated to decomposition it emits very toxic fumes of As, CN^-, and NO_x.

COM075 HR: 3
α-CYANO-2,6-DIMETHYLERGOLINE-8-PROPIONAMIDE
mf: $C_{20}H_{24}N_4O$ mw: 336.48

SYN: ERGOLINE-8-PROPIONAMIDE, α-CYANO-2,6-DIMETHYL-

TOXICITY DATA WITH REFERENCE
orl-rat TDLo:4 mg/kg (female 5D post):REP ARZNAD 33,1094,83
orl-mus LD50:200 mg/kg ARZNAD 33,1094,83

SAFETY PROFILE: Poison by ingestion. Experimental reproductive effects. When heated to decomposition it emits toxic fumes of NO_x.

COM125 CAS:7790-03-6 HR: 2
2-(2-CYANOETHOXY)ETHYL ESTER ACRYLIC ACID
mf: $C_8H_{11}NO_3$ mw: 169.20

TOXICITY DATA WITH REFERENCE
skn-rbt 10 mg/24H open MLD AIHAAP 23,95,62
orl-rat LD50:1120 mg/kg AIHAAP 23,95,62
skn-rbt LD50:750 mg/kg AIHAAP 23,95,62

CONSENSUS REPORTS: Cyanide and its compounds are on the Community Right-To-Know List.

SAFETY PROFILE: Moderately toxic by ingestion and skin contact. A skin irritant. When heated to decomposition it emits toxic fumes of NO_x and CN^-. See also CYANIDE and ESTERS.

COM250 CAS:10141-15-8 HR: 2
4-CYANOETHOXY-2-METHYL-2-PENTANOL
mf: $C_9H_{17}NO_2$ mw: 171.27

TOXICITY DATA WITH REFERENCE
skn-rbt 10 mg/24H open MLD AIHAAP 23,95,62
orl-rat LDLo:3200 mg/kg AIHAAP 23,95,62
skn-rbt LDLo:1500 mg/kg AIHAAP 23,95,62

CONSENSUS REPORTS: Cyanide and its compounds are on the Community Right-To-Know List.

SAFETY PROFILE: Moderately toxic by ingestion and skin contact. A skin irritant. See also NITRILES. When heated to decomposition it emits toxic fumes of NO_x and CN^-.

COM300 CAS:57966-95-7 HR: 2
2-CYANO-N-((ETHYLAMINO)CARBONYL)-2-(METHOXYIMINO)-ACETAMIDE
mf: $C_7H_{10}N_4O_3$ mw: 198.21

SYNS: ACETAMIDE, 2-CYANO-N-((ETHYLAMINO)CARBONYL)-2-(METHOXYIMINO)- ☐ CURZATE ☐ CYMOXANIL ☐ DPX 3217 ☐ DPX 3217M

TOXICITY DATA WITH REFERENCE
orl-rat LD50:1100 mg/kg FMCHA2-,C90,91
skn-rbt LD50:>3 g/kg DOVEAA 35,179,81
orl-gpg LD50:1096 mg/kg PEMNDP 9,206,91

CONSENSUS REPORTS: Reported in EPA TSCA Inventory.

SAFETY PROFILE: Moderately toxic by ingestion. When heated to decomposition it emits toxic vapors of NO_x.

COM500 CAS:65216-94-6 HR: 3
N-(CYANOETHYL)DIETHYLENETRIAMINE

TOXICITY DATA WITH REFERENCE
skn-rbt 500 mg MLD SCCUR* -,7,61
orl-rat LD50:4550 mg/kg AMIHAB 17,129,58
ipr-rat LD50:261 mg/kg AMIHAB 17,129,58
ipr-mus LD50:222 mg/kg AMIHAB 17,129,58

CONSENSUS REPORTS: Cyanide and its compounds are on the Community Right-To-Know List.

SAFETY PROFILE: Poison by intraperitoneal route. Mildly toxic by ingestion. A skin irritant. When heated to decomposition it emits toxic fumes of NO_x and CN^-. See also CYANIDE.

COM750 CAS:1001-58-7 HR: 3
β-CYANOETHYLMERCAPTAN
mf: C_3H_5NS mw: 87.15

TOXICITY DATA WITH REFERENCE
ipr-mus LD50:100 mg/kg NTIS** AD691-490

CONSENSUS REPORTS: Reported in EPA TSCA Inventory. Cyanide and its compounds are on the Community Right-To-Know List.

SAFETY PROFILE: Poison by intraperitoneal route. When heated to decomposition it emits very toxic fumes of SO_x and NO_x.

CON000 CAS:1071-22-3 HR: 2
2-CYANOETHYLTRICHLOROSILANE
mf: $C_3H_4Cl_3NSi$ mw: 188.52

SYN: β-CYANOETHYLTRICHLOROSILANE

TOXICITY DATA WITH REFERENCE
skn-rbt 100 μg/24H open AIHAAP 23,95,62
orl-rat LD50:2000 mg/kg AIHAAP 23,95,62

CONSENSUS REPORTS: Cyanide and its compounds are on the Community Right-To-Know List. Reported in EPA TSCA Inventory.

SAFETY PROFILE: Moderately toxic by ingestion. A skin irritant. When heated to decomposition it emits very toxic fumes of NO_x and Cl^-. See also CYANIDE.

CON250 CAS:919-31-3 HR: 1
(2-CYANOETHYL)TRIETHOXYSILANE
mf: $C_9H_{19}NO_3Si$ mw: 217.38

SYNS: β-CYANOETHYLTRIETHOXYSILANE □ PROPIONITRILE, 3-(TRIETHOXYSILYL)- □ TRIETHOXY-2-KYANETHYLSILAN

TOXICITY DATA WITH REFERENCE
skn-rbt 10 mg/24H open MLD AIHAAP 23,95,62
orl-rat LD50:5630 mg/kg AIHAAP 23,95,62
skn-rbt LD50:5950 mg/kg AIHAAP 23,95,62

CONSENSUS REPORTS: Reported in EPA TSCA Inventory. Cyanide and its compounds are on the Community Right-To-Know List.

SAFETY PROFILE: Mildly toxic by ingestion and skin contact. A skin irritant. See also CYANIDE. When heated to decomposition it emits toxic fumes of NO_x and CN^-.

CON300 CAS:13067-93-1 HR: 3
CYANOFENPHOS
mf: $C_{15}H_{14}NO_2PS$ mw: 303.33

PROP: Crystalline solid. Mp: 83°, n (25/D) 1.5839. Solubility in water at 30°: 6 ppm. Moderately sol in ketones and aromatic solvents. Vap. press at 25°: 0.0000132 mm Hg.

SYNS: B-10094 □ CP 19699 □ CYANOPHENPHOS □ o-p-CYANOPHENYL-o-ETHYL PHENYLPHOSPHONOTHIOATE □ o-(4-CYANOPHENYL)-o-ETHYL PHENYLPHOSPHONOTHIOATE □ CYP □ ENT 25,832 □ ENT 25,832-a □ o-ETHYL-o-4-CYANOPHENYL PHENYLPHOSPHOROTHIOATE □ o-ETHYLPHENYLPHOSPHONOTHIOATE-o-ESTER with p-HYDROXYBENZONITRILE □ EXPERIMENTAL INSECTICIDE S-4087 □ MONSANTO CP-19699 □ PHENYLPHOSPHONOTHIOIC ACID-o-ETHYL ESTER-o-ESTER with p-HYDROXYBENZONITRILE □ PHOSPHOROTHIOIC ACID-o-(4-CYANOPHENYL)-o-ETHYL PHENYL ESTER □ S 4087 □ STAUFFER B-10094 □ SURECIDE □ UPJOHN U-32714

TOXICITY DATA WITH REFERENCE
orl-rat LD50:28,500 μg/kg ABCHA6 26,257,62
orl-mus LD50:43,700 μg/kg FMCHA2 -,C224,83
ipr-mus LD50:45 mg/kg ABCHA6 26,257,62
scu-mus LD50:122 mg/kg GUCHAZ 6,257,73
orl-gpg LDLo:50 mg/kg JEENAI 61,1261,68
scu-gpg LDLo:100 mg/kg JEENAI 61,1261,68
orl-ckn LD50:20 mg/kg TXAPA9 11,49,67

CONSENSUS REPORTS: Cyanide and its compounds are on the Community Right-To-Know List.

SAFETY PROFILE: Poison by ingestion, intraperitoneal, and subcutaneous routes. Toxic to fish and bees. When heated to decomposition it emits toxic fumes of CN^-, PO_x, and SO_x. See also CYANIDE.

CON500 CAS:353-18-4 HR: 3
2-CYANO-2′-FLUORODIETHYL ETHER
mf: C_5H_8FNO mw: 117.14

SYNS: 2-CYANOETHYL-2′-FLUOROETHYLETHER □ 2-FLUORO-2′-CYANODIETHYL ETHER

TOXICITY DATA WITH REFERENCE
ipr-mus LD50:10 mg/kg 30ZFAF -,140,59
scu-mus LD50:10 mg/kg JCSOA9 -,2774,49

CONSENSUS REPORTS: Cyanide and its compounds are on the Community Right-To-Know List.

SAFETY PROFILE: Poison by intraperitoneal and subcutaneous routes. See also CYANIDE and ETHERS. When heated to decomposition it emits toxic F^-, NO_x, and CN^-.

CON825 CAS:4474-17-3 HR: 3
CYANOFORMYL CHLORIDE
mf: C_2ClNO mw: 89.48

CONSENSUS REPORTS: Cyanide compounds are on the Community Right-To-Know List.

SAFETY PROFILE: A poison. Violent reaction with water. When heated to decomposition it emits toxic fumes of Cl^-, CN^-, and NO_x. See also CYANIDE and CHLORIDES.

COO000 CAS:460-19-5 HR: 3
CYANOGEN
DOT: UN 1026
mf: C_2N_2 mw: 52.04

N≡CC≡N

PROP: Colorless gas, pungent odor. Mp: −34.4°, bp: −21.0°, d: 0.866 @ 17°/4°, lel: 6.6%, uel: 32%, vap d: 1.8.

SYNS: CARBON NITRIDE □ CYANOGENE (FRENCH) □ CYANOGEN GAS (DOT) □ DICYANOGEN □ ETHANEDINITRILE □ NITRI-

LOACETONITRILE □ OXALIC ACID DINITRILE □ OXALONITRILE □ OXALYL CYANIDE □ PRUSSITE □ RCRA WASTE NUMBER P031

TOXICITY DATA WITH REFERENCE
eye-hmn 16 ppm/6M AIHAAP 21,121,60
ihl-hmn TCLo:16 ppm:EYE,NOSE AIHAAP 21,121,60
ihl-rat LC50:350 ppm/1H AIHAAP 21,121,60
unk-dog LDLo:15 mg/kg AIPTAK 3,77,1897
scu-rbt LDLo:13 mg/kg AIPTAK 3,77,1897
scu-pgn LDLo:9 mg/kg AIPTAK 3,77,1897
scu-frg LDLo:43 mg/kg AIPTAK 3,77,1897

CONSENSUS REPORTS: Reported in EPA TSCA Inventory. Cyanide and its compounds are on the Community Right-To-Know List.

OSHA PEL: TWA 10 ppm
ACGIH TLV: TWA 10 ppm
DFG MAK: 10 ppm (22 mg/m³)
DOT CLASSIFICATION: 2.3; *Label:* Poison Gas, Flammable Gas

SAFETY PROFILE: A poison by subcutaneous and possibly other routes. Moderately toxic by inhalation. Human systemic effects by inhalation: damage to the olfactory nerves, and irritation of the conjunctiva. A systemic irritant by inhalation and subcutaneous routes. A human eye irritant. Very dangerous fire hazard when exposed to heat, flames (sparks), or oxidizers. To fight fire, stop flow of gas. Potentially explosive reaction with powerful oxidants (e.g., dichlorine oxide; fluorine; oxygen; ozone). When heated to decomposition or on contact with acid, acid fumes, water, or steam, will react to produce highly toxic fumes of NO_x and CN^-. See also other cyanogen entries and CYANIDE.

COO250 CAS:764-05-6 HR: 3
CYANOGEN AZIDE
mf: CN_4 mw: 68.04

CONSENSUS REPORTS: Cyanide and its compounds are on the Community Right-To-Know List.

SAFETY PROFILE: A poison. Explodes violently with mild mechanical, thermal, or electrical shock. May spontaneously explode in storage even when cooled to −20 °C. Reacts with 10% sodium hydroxide to form the violently explosive 5-azidotetrazolide. See also other cyanogen entries; CYANIDE; and AZIDES.

COO500 CAS:506-68-3 HR: 3
CYANOGEN BROMIDE
DOT: UN 1889
mf: CBrN mw: 105.93

PROP: Colorless needles. Mp: 52°, bp: 61.6°, d: 2.015 @ 20°/4°, vap press: 100 mm @ 22.6°.

SYNS: BROMINE CYANIDE □ BROMOCYAN □BROMOCYANOGEN □ BROMURE de CYANOGEN (FRENCH) □ CAMPILIT □ CYANOBROMIDE □ CYANOGEN MONOBROMIDE □ RCRA WASTE NUMBER U246 □ TL 822

TOXICITY DATA WITH REFERENCE
ihl-hmn LCLo:92 ppm/10M NTIS** PB214-270
ihl-mus LCLo:500 mg/m³/10M NDRC** No.9-4-1-9,43

CONSENSUS REPORTS: Cyanide and its compounds are on the Community Right-To-Know List. EPA Extremely Hazardous Substances List. Reported in EPA TSCA Inventory.

DOT CLASSIFICATION: 6.1; *Label:* Poison, Corrosive

SAFETY PROFILE: A human and experimental poison by inhalation. Corrosive. When heated to decomposition it emits very toxic fumes of CN^- and Br^-. Possibly unstable. See also other cyanogen entries; CYANIDE; and BROMIDES.

COO750 CAS:506-77-4 HR: 3
CYANOGEN CHLORIDE
DOT: UN 1589
mf: CClN mw: 61.47

PROP: Colorless liquid or gas; lacrymatory and irritating odor. Mp: −6.5°, bp: 13.1°, d: 1.218 @ 4°/4°, vap press: 1010 mm @ 20°, vap d: 1.98.

SYNS: CHLORCYAN □ CHLORINE CYANIDE □ CHLOROCYAN □ CHLOROCYANIDE □ CHLOROCYANOGEN □ CHLORURE DE CYANOGENE □ CYANOGEN CHLORIDE (ACGIH,OSHA) □ CYANOGEN CHLORIDE, inhibited (DOT) □ RCRA WASTE NUMBER P033

TOXICITY DATA WITH REFERENCE
eye-hmn 100 mg/m³/2M SEV BJEPA5 33,241,46
ihl-hmn TCLo:10 mg/m³:EYE WHOTAC -,31,70
ihl-man TCLo:2 g/m³:SKN NTIS** PB158-508
ihl-rat LC50:5400 mg/m³/3M NTIS** PB158-508
ihl-mus LC50:3 g/m³/30S NTIS** PB158-508
scu-mus LDLo:39 mg/kg 27ZWAY 1.1,779,-
ihl-dog LC50:3800 mg/m³/1M NTIS** PB158-508
scu-dog LDLo:5 mg/kg HBAMAK 4,1341,35
ihl-mky LC50:4400 mg/m³/1M NTIS** PB158-508
orl-cat LD50:6 mg/kg NTIS** PB158-508
ihl-cat LC50:6 g/m³/1M NTIS** PB158-508
ihl-rbt LC50:6 g/m³/7M NTIS** PB158-508
scu-rbt LDLo:20 mg/kg HBAMAK 4,1341,35
ihl-gpg LC50:5500 mg/m³/2M NTIS** PB158-508
scu-pgn LDLo:8700 µg/kg HBAMAK 4,1341,35
ihl-dom LC50:3600 mg/m³/2M NTIS** PB158-508

CONSENSUS REPORTS: Cyanide and its compounds are on the Community Right-To-Know List. Reported in EPA TSCA Inventory.

OSHA PEL: CL 0.3 ppm
ACGIH TLV: CL 0.3 ppm
DOT CLASSIFICATION: 2.3; *Label:* Poison Gas, Flammable Gas

SAFETY PROFILE: Poison by ingestion, subcutaneous, and possibly other routes. Toxic by inhalation. Human systemic effects by inhalation: lacrimation, conjunctiva irritation, and chronic pulmonary edema or congestion. A primary irritant. A severe human eye irritant. An insecticide. Flammable when exposed to heat or flame. When heated to decomposition or on contact with water or steam, it will react to produce highly toxic and corrosive fumes of Cl^-, CN^-, and NO_x. See also other cyanogen entries, CYANIDE, and CHLORIDES.

COO825 **CAS:1495-50-7** **HR: 3**
CYANOGEN FLUORIDE
mf: CFN mw: 45.02

CONSENSUS REPORTS: Cyanide and its compounds are on the Community Right-To-Know List.

SAFETY PROFILE: A poison. Explosive polymerization is catalyzed by hydrogen fluoride. When heated to decomposition it emits toxic fumes of F^- and NO_x. See also other cyanogen entries, CYANIDE, and FLUORIDES.

COP000 **CAS:506-78-5** **HR: 3**
CYANOGEN IODIDE
mf: CIN mw: 152.92

PROP: Colorless solid. Mp: 146.5°, vap press: 1 mm @ 25.2°.

SYNS: IODINE CYANIDE □ JODCYAN

TOXICITY DATA WITH REFERENCE
scu-rat LDLo:44 mg/kg 27ZWAY 1.1,779,-
scu-mus LDLo:27 mg/kg HBAMAK 4,1289,35
scu-dog LDLo:19 mg/kg HBAMAK 4,1289,35
orl-cat LDLo:18 mg/kg HBAMAK 4,1289,35
scu-cat LDLo:20 mg/kg HBAMAK 4,1289,35
scu-rbt LDLo:360 mg/kg HBAMAK 4,1289,35
scu-frg LDLo:110 mg/kg HBAMAK 4,1289,35

CONSENSUS REPORTS: Cyanide and its compounds are on the Community Right-To-Know List. EPA Extremely Hazardous Substances List. Reported in EPA TSCA Inventory.

DOT CLASSIFICATION: 6.1; *Label:* Poison, KEEP AWAY FROM FOOD

SAFETY PROFILE: A poison by ingestion and subcutaneous routes. Violent reaction with P. See other cyanogen entries; CYANIDE and IODIDES. When heated to decomposition it emits very toxic fumes of NO_x, CN^-, and I^-.

COP125 **CAS:461-58-5** **HR: 3**
CYANOGUANIDINE
mf: $C_2H_4N_4$ mw: 84.08

SYNS: DICYANDIAMIN □ DICYANODIAMIDE

TOXICITY DATA WITH REFERENCE
orl-rat LD:>500 mg/kg NCNSA6 5,17,53
unr-rat LDLo:600 mg/kg BCPCA6 14,1325,65

CONSENSUS REPORTS: Cyanide compounds are on the Community Right-To-Know List.

SAFETY PROFILE: Moderately toxic by ingestion. Mixtures with ammonium nitrate, potassium chlorate, and related compounds are powerful explosives. When heated to decomposition it emits toxic fumes of CN^- and NO_x. See also CYANIDE and AMIDES.

COP400 **CAS:6071-81-4** **HR: 3**
S-1-CYANO-2-HYDROXY-3-BUTENE
mf: C_5H_7NO mw: 97.13

SYNS: S-3-HYDROXY-4-PENTENONITRILE □ 4-PENTENONITRILE, 3-HYDROXY-, S-

TOXICITY DATA WITH REFERENCE
scu-rat TDLo:150 mg/kg (female 8D post):TER FCTXAV 18,159,80
scu-rat LD50:200 mg/kg FCTXAV 18,159,80
unr-mus LD50:170 mg/kg JAFCAU 17,483,69

SAFETY PROFILE: Poison by subcutaneous route. An experimental teratogen. When heated to decomposition it emits toxic fumes of NO_x.

COP500 **CAS:31065-88-0** **HR: 3**
CYANOHYDROXYMERCURY
mf: CHHgNO mw: 243.62

TOXICITY DATA WITH REFERENCE
scu-mus LDLo:10 mg/kg MOLAAF 73,751,39

CONSENSUS REPORTS: Cyanide and its compounds, as well as mercury and its compounds, are on the Community Right-To-Know List.

OSHA PEL: CL 0.1 mg(Hg)/m³ (skin)
ACGIH TLV: TWA 0.1 mg(Hg)/m³ (skin)
NIOSH REL: (Organomercury): TWA 0.01 mg/m³; STEL 0.03 mg/m³ (skin)

SAFETY PROFILE: Poison by subcutaneous route. See also MERCURY COMPOUNDS and CYANIDE. When heated to decomposition it emits very toxic fumes of Hg, NO_x, and CN^-.

COP525 **CAS:17380-21-1** **HR: 3**
5-CYANO-3-INDOLYL ISOPROPYL KETONE
mf: $C_{13}H_{12}N_2O$ mw: 212.27

SYNS: INDOLE-5-CARBONITRILE, 3-(2-METHYLPROPIONYL)- □ 3-(2-METHYLPROPIONYL)-5-INDOLECARBONITRILE

TOXICITY DATA WITH REFERENCE
ivn-mus LD50:178 mg/kg CSLNX* NX#12039

DOT CLASSIFICATION: 3; *Label:* Flammable Liquid

SAFETY PROFILE: A poison by intravenous route. A flammable liquid. When heated to decomposition it emits toxic vapors of NO_x.

COP550 **CAS:17380-19-7** **HR: 3**
5-CYANO-3-INDOLYLMETHYL KETONE
mf: $C_{11}H_8N_2O$ mw: 184.21

SYNS: 3-ACETYLINDOLE-5-CARBONITRILE □ INDOLE-5-CARBONITRILE, 3-ACETYL-

TOXICITY DATA WITH REFERENCE
ivn-mus LD50:32 mg/kg CSLNX* NX#12036

DOT CLASSIFICATION: 3; *Label:* Flammable Liquid

SAFETY PROFILE: A poison by intravenous route. A flammable liquid. When heated to decomposition it emits toxic vapors of NO_x.

COP600 **HR: 3**
α-CYANO-6-ISOBUTYLERGOLINE-8-PROPIONAMIDE
mf: $C_{22}H_{28}N_4O$ mw: 364.54

SYN: ERGOLINE-8-PROPIONAMIDE, α-CYANO-6-ISOBUTYL-

TOXICITY DATA WITH REFERENCE
orl-rat TDLo:24 mg/kg (female 5D post):REP ARZNAD 33,1094,83
orl-mus LD50:200 mg/kg ARZNAD 33,1094,83

SAFETY PROFILE: Poison by ingestion. Experimental reproductive effects. When heated to decomposition it emits toxic fumes of NO_x.

COP700 **CAS:63278-33-1** **HR: 2**
α-((CYANOMETHOXY)IMINO)-BENZACETONITRILE
mf: $C_{10}H_7N_3O$ mw: 185.20

SYNS: BENZENEACETONITRILE, α-((CYANOMETHOXY)IMINO)- □ CGA-43089 □ CONCEP □ α-((CYANOMETHOXY)IMINO)BENZENEACETONITRILE □ CYOMETRINIL

TOXICITY DATA WITH REFERENCE
orl-rat LD50:2277 mg/kg 85JCAE-,926,86
skn-rat LD50:>3100 mg/kg 85JCAE-,926,86

CONSENSUS REPORTS: Reported in EPA TSCA Inventory.

SAFETY PROFILE: Moderately toxic by ingestion and skin contact. When heated to decomposition it emits toxic vapors of NO_x.

COP750 **CAS:1001-55-4** **HR: 3**
CYANOMETHYL ACETATE
mf: $C_4H_5NO_2$ mw: 99.10

PROP: Colorless liquid. Mp: −22.5°, bp: 200°, d: 1.123 @ 15°.

SYN: GLYCOLONITRILE ACETATE

TOXICITY DATA WITH REFERENCE
skn-rbt 10 mg/24H open JIHTAB 30,63,48
eye-rbt 20 mg open SEV JIHTAB 30,63,48
orl-rat LD50:32 mg/kg JIHTAB 30,63,48
ihl-rat LCLo:16 ppm/4H JIHTAB 31,343,49
skn-rbt LD50:43 mg/kg JIHTAB 30,63,48

CONSENSUS REPORTS: Cyanide and its compounds are on the Community Right-To-Know List.

SAFETY PROFILE: A poison by skin contact, ingestion, and inhalation. A skin and severe eye irritant. When heated to decomposition it emits very toxic fumes of NO_x and CN^-. See also NITRILES.

COP759 **HR: 3**
5-CYANO-5-METHYLTETRAZOLE
mf: $C_3H_3N_5$ mw: 109.09

N=NNCH₃N= CCN (ring: N…C)

CONSENSUS REPORTS: Cyanide compounds are on the Community Right-To-Know List.

SAFETY PROFILE: Explosive reaction with aluminum hydride. When heated to decomposition it emits toxic fumes of CN^- and NO_x. See also CYANIDE.

COP775 **CAS:1884-64-6** **HR: 3**
CYANONITRENE
mf: CN_2 mw: 40.02

CONSENSUS REPORTS: Cyanide compounds are on the Community Right-To-Know List.

SAFETY PROFILE: An explosive. When heated to decomposition it emits toxic fumes of CN^- and NO_x. See also CYANIDE.

COQ325 **CAS:68597-10-4** **HR: 3**
2-CYANO-4-NITROBENZENEDIAZONIUM HYDROGEN SULFATE
mf: $C_7H_4N_4O_6S$ mw: 272.19

CONSENSUS REPORTS: Cyanide compounds are on the Community Right-To-Know List.

SAFETY PROFILE: Reacts violently with sulfuric acid when heated. When heated to decomposition it emits toxic fumes of NO_x, CN^-, and SO_x. See also CYANIDE, SULFATES, and NITRO COMPOUNDS of AROMATIC HYDROCARBONS.

COQ375 **CAS:56092-91-2** **HR: 3**
3-(3-CYANO-1,2,4-OXADIAZOL-5-YL)-4-CYANO FURAZAN-2-(5-) OXIDE
mf: $C_6N_6O_3$ mw: 204.10

CONSENSUS REPORTS: Cyanide compounds are on the Community Right-To-Know List.

SAFETY PROFILE: Reacts explosively with hydrazines and other nitrogenous bases (e.g., mono- or dimethyl hydrazine, piperidine, piperazine, and diethylamine). When heated to decomposition it emits toxic fumes of CN^- and NO_x. See also CYANIDE.

COQ399 **CAS:2636-26-2** **HR: 3**
CYANOPHOS
mf: $C_9H_{10}NO_3PS$ mw: 243.23

PROP: Yellow to reddish-yellow transparent liquid. Bp: 119–120° (slt decomp), mp: 14–15°, n (32.5/D) 1.5404. Very sol in methanol, ethanol, acetone, chloroform. Sparingly sol in n-hexane, kerosene; sltly sol in water. Rapid decomp under alkaline conditions and upon exposure to light.

SYNS: BAY 34727 □ BAYER 34727 □ CIAFOS □ O-p-CYANOPHENYL O,O-DIMETHYL PHOSPHOROTHIOATE □ O-(4-CYANOPHENYL) O,O-DIMETHYL PHOSPHOROTHIOATE □ CYANOX □ CYAP □ O,O-DIMETHYL-O-(4-CYANO-PHENYL)-MONOTHIOPHOSPHAT (GERMAN) □ O,O-DIMETHYL-O-p-CYANOPHENYL PHOSPHOROTHIOATE □ O,O-DIMETHYL-O-4-CYANOPHENYL PHOSPHOROTHIOATE □ O,O-DIMETHYL-O-4-CYANOPHENYL THIOPHOSPHATE □ ENT 25,675 □ MAY & BAKER S-4084 □ PHOSPHOROTHIOIC ACID o-(4-CYANOPHENYL)-9,9-DIMETHYL ESTER □ S 4084 □ SUMITOMO S 4084 □ SUNITOMO S 4084

TOXICITY DATA WITH REFERENCE
orl-rat LD50:25 mg/kg SPEADM 78-1,26,78

skn-rat LD50:800 mg/kg 28ZEAL 5,62,76
orl-mus LD50:324 mg/kg GISAAA 48(9),76,83
ipr-mus LD50:880 mg/kg MEIEDD 10,322,83
orl-gpg LD50:324 mg/kg GISAAA 48(9),76,83
orl-ckn LD50:24 mg/kg TXAPA9 11,49,67
skn-mam LD50:2010 mg/kg GTPZAB 21(7),34,77

CONSENSUS REPORTS: Cyanide and its compounds are on the Community Right-To-Know List. On EPA Extremely Hazardous Substances List.

SAFETY PROFILE: Poison by ingestion. Moderately toxic by skin contact and intraperitoneal routes. An insecticide and cholinesterase inhibitor. See also PARATHION. When heated to decomposition it emits toxic fumes of NO_x, PO_x, CN^-, and SO_x.

COQ750 CAS:627-26-9 HR: 3
1-CYANOPROPENE
mf: C_4H_5N mw: 67.09

$N{\equiv}CCH{=}CHCH_3$

PROP: Bp: 118–119, flash p: 60.8°F.

SYN: 2-BUTENENITRILE

CONSENSUS REPORTS: Cyanide and its compounds are on the Community Right-To-Know List.

SAFETY PROFILE: A poison. Very reactive. A very dangerous fire hazard when exposed to heat or flame. When heated to decomposition it emits toxic fumes of CN^-. See also CYANIDE.

COR325 CAS:1190-16-5 HR: 3
3-CYANOPROPYLDICHLOROMETHYLSILANE
mf: $C_5H_9Cl_2NSi$ mw: 182.14

SYN: 4-(DICHLOROMETHYLSILYL)BUTYRONITRILE

TOXICITY DATA WITH REFERENCE
skn-rbt 100 µg/24H open AIHAAP 23,95,62
skn-rbt 5 mg/24H SEV 85JCAE-,1225,86
eye-rbt 750 µg/24H SEV 85JCAE-,1225,86
orl-rat LD50:2830 mg/kg AIHAAP 23,95,62
ihl-mus LCLo:770 mg/m³/2H 85JCAE-,1225,86
skn-rbt LD50:1490 mg/kg AIHAAP 23,95,62

CONSENSUS REPORTS: Cyanide and its compounds are on the Community Right-To-Know List. Reported in EPA TSCA Inventory.

SAFETY PROFILE: Moderately toxic by inhalation, ingestion, and skin contact. A severe skin and eye irritant. When heated to decomposition it emits toxic fumes of Cl^-, NO_x and CN^-. See also NITRILES.

COR500 CAS:1067-99-8 HR: 2
(3-CYANOPROPYL)DIETHOXY(METHYL) SILANE
mf: $C_9H_{19}NO_2Si$ mw: 201.38

SYNS: BUTYRONITRILE, 4-(DIETHOXYMETHYLSILYL)- □ DIETHOXY-3-KYANPROPYL-METHYLSILAN □ SILANE, (3-CYANOPROPYL)DIETHOXY(METHYL)-

TOXICITY DATA WITH REFERENCE
skn-rbt 10 mg/24H open MLD AIHAAP 23,95,62
orl-rat LD50:3730 mg/kg AIHAAP 23,95,62

CONSENSUS REPORTS: Reported in EPA TSCA Inventory.

SAFETY PROFILE: Moderately toxic by ingestion. A skin irritant. When heated to decomposition it emits toxic fumes of NO_x and Si.

COR600 CAS:40561-27-1 HR: 3
2-CYANO-2-PROPYL NITRATE
mf: $C_4H_6N_2O_3$ mw: 130.10

$(CH_3)_2C(CN)ONO_2$

CONSENSUS REPORTS: Cyanide compounds are on the Community Right-To-Know List.

SAFETY PROFILE: An impact-sensitive explosive. When heated to decomposition it emits toxic fumes of NO_x and CN^-. See also CYANIDE and NITRATES.

COR750 CAS:1071-27-8 HR: 2
(3-CYANOPROPYL) TRICHLOROSILANE
mf: $C_4H_6Cl_3NSi$ mw: 202.55

SYNS: BUTYRONITRILE, 4-(TRICHLOROSILYL)- □ SILANE, (3-CYANOPROPYL)TRICHLORO- □ SILANE, TRICHLORO(3-CYANOPROPYL)- □ TRICHLOR-3-KYANPROPYLSILAN

TOXICITY DATA WITH REFERENCE
orl-rat LD50:2830 mg/kg AIHAAP 23,95,62

CONSENSUS REPORTS: Reported in EPA TSCA Inventory.

SAFETY PROFILE: Moderately toxic by ingestion. When heated to decomposition it emits toxic vapors of NO_x and Cl^-.

COS500 CAS:60560-33-0 HR: 2
2-CYANO-3-(4-PYRIDYL)-1-(1,2,3,TRIMETHYLPROPYL)GUANIDINE
mf: $C_{13}H_{19}N_5$ mw: 245.37

SYNS: P 1134 □ PINACIDIL □ PND

TOXICITY DATA WITH REFERENCE
orl-rat LD50:570 mg/kg JMCMAR 21,773,78
orl-mus LD50:490 mg/kg NYKZAU 86,341,85

CONSENSUS REPORTS: Cyanide and its compounds are on the Community Right-To-Know List.

SAFETY PROFILE: Moderately toxic by ingestion. When heated to decomposition it emits very toxic fumes of NO_x and CN^-. See also CYANIDE.

COS800 CAS:1067-47-6 HR: 1
(3-CYANOPROPYL) TRIETHOXYSILANE
mf: $C_{10}H_{21}NO_3Si$ mw: 231.41

SYNS: BUTYRONITRILE, 4-(TRIETHOXYSILYL)- □ SILANE, (3-CYANOPROPYL)TRIETHOXY- □ SILANE, TRIETHOXY(3-CYANOPROPYL)- □ TRIETHOXY-3-KYANPROPYLSILAN

TOXICITY DATA WITH REFERENCE
skn-rbt 100 µg/24H open AIHAAP 23,95,62

orl-rat LD50:4920 mg/kg AIHAAP 23,95,62

CONSENSUS REPORTS: Reported in EPA TSCA Inventory.

SAFETY PROFILE: Mildly toxic by ingestion. A skin irritant. When heated to decomposition it emits toxic fumes of NO_x and Si.

COS825 CAS:70247-32-4 HR: 3
2-(5-CYANOTETRAZOLE)PENTAMMINECOBALT(III) PERCHLORATE
mf: $C_2H_{16}Cl_3CoN_{10}O_{12}$ mw: 537.50

CONSENSUS REPORTS: Cyanide compounds and cobalt compounds are on the Community Right-To-Know List.

SAFETY PROFILE: A relatively insensitive explosive. When heated to decomposition it emits toxic fumes of Cl^-, NO_x, and CN^-. See also CYANIDE, PERCHLORATES, and COBALT COMPOUNDS.

COS899 CAS:63833-98-7 HR: 3
CYANOTRIMEPRAZINE MALEATE
mf: $C_{19}H_{21}N_3S{\bullet}C_4H_4O_4$ mw: 439.57

SYNS: CIANATIL MALEATE □ CYAMEMAZINE MALEATE □ CYAMEPROMAZINE MALEATE □ CYANO-3-(DIMETHYLAMINO-3-METHYL-2-PROPYL)-10-PHENOTHIAZINE MALEATE □ 10-(3-(DIMETHYLAMINO)-2-METHYLPROPYL)PHENOTHIAZINE-2-CARBONITRILE MALEATE □ KYAMEPROMAZINE MALEATE □ 7204 RP

TOXICITY DATA WITH REFERENCE
orl-mus LD50:640 mg/kg CRSBAW 155,1029,61
ipr-mus LD50:210 mg/kg CRSBAW 155,1029,61
scu-mus LD50:690 mg/kg CRSBAW 155,1029,61
ivn-mus LD50:90 mg/kg CRSBAW 155,1029,61

CONSENSUS REPORTS: Cyanide and its compounds are on the Community Right-To-Know List.

SAFETY PROFILE: Poison by intravenous and intraperitoneal routes. Moderately toxic by ingestion and subcutaneous routes. See also NITRILES. When heated to decomposition it emits very toxic fumes of CN^-, NO_x, and SO_x.

COT000 CAS:16176-02-6 HR: 3
2-CYANO-1,2,3-TRIS(DIFLUOROAMINO)PROPANE
mf: $C_4H_4F_6N_4$ mw: 222.10

$F_2NC(CN)(CH_2NF_2)_2$

CONSENSUS REPORTS: Cyanide and its compounds are on the Community Right-To-Know List.

SAFETY PROFILE: A shock-sensitive explosive. When heated to decomposition it emits toxic fumes of F^-, NO_x, and CN^-. See also CYANIDE and FLUORIDES.

COU000 CAS:14901-08-7 HR: 3
CYCASIN
mf: $C_8H_{16}N_2O_7$ mw: 252.26

PROP: A solid. Mp: 154° (decomp)

SYNS: CYCAS REVOLUTA GLUCOSIDE □ CYKAZINE □ β-d-GLUCOSYLOXYAZOXYMETHANE □ METHYLAZOXYMETHANOL GLUCOSIDE □ METHYLAZOXYMETHANOL-β-d-GLUCOSIDE □ (METHYL-ONN-AZOXY)METHYL-β-d-GLUCOPYRANOSIDE

TOXICITY DATA WITH REFERENCE
mma-sat 10 μmol/plate CNREA8 39,3780,79
dnd-rat-orl 56 mg/kg MUREAV 54,39,78
orl-rat TDLo:100 mg/kg:ETA JCROD7 100,231,81
orl-rat TDLo:143 mg/kg (female 8-13D post):ETA JJIND8 38,233,67
orl-rat TDLo:230 mg/kg (female 2-5D post):ETA JJIND8 38,233,67
ipr-rat TDLo:90 mg/kg:ETA IGKEAO 44,211,74
scu-rat TDLo:2500 μg/kg:ETA RCOCB8 2,627,71
rec-rat TDLo:600 mg/kg/11W-I:ETA GANNA2 66,449,75
orl-mus TDLo:300 mg/kg:ETA CNREA8 29,1658,69
scu-mus TDLo:20 mg/kg/2W-I:ETA RCOCB8 2,627,71
orl-ham TDLo:150 mg/kg:CAR CNREA8 31,283,71
orl-rat TD:21 g/kg/3W-C:CAR FEPRA7 23,1386,64
orl-rat TD:480 mg/kg/2D-C:ETA APAVAY 340,151,65
scu-rat TD:234 mg/kg:ETA JJIND8 74,1275,85
orl-rat TD:1670 mg/kg/24W-C:ETA NRINA3 30,931,72
orl-rat LD50:270 mg/kg GANNA2 62,353,71
orl-mus LD50:500 mg/kg FEPRA7 31,1493,72
orl-rbt LDLo:30 mg/kg FEPRA7 31,1493,72
orl-gpg LDLo:20 mg/kg FEPRA7 31,1493,72
orl-ham LDLo:250 mg/kg FEPRA7 31,1493,72

CONSENSUS REPORTS: EPA Genetic Toxicology Program. IARC Cancer Review: Group 2B IMEMDT 7,56,87; Human Inadequate Evidence IMEMDT 10,121,76; Animal Sufficient Evidence IMEMDT 10,121,76; IMEMDT 1,157,72.

SAFETY PROFILE: Confirmed carcinogen with experimental carcinogenic and tumorigenic data. A poison by ingestion. An experimental teratogen. Mutation data reported. When heated to decomposition it emits toxic fumes of NO_x.

COU250 CAS:532-76-3 HR: 3
CYCLAINE HYDROCHLORIDE
mf: $C_{16}H_{23}NO_2{\bullet}ClH$ mw: 297.86

PROP: Bitter crystals. Mp: 182–184°

SYNS: CYCLAINE □ 1-(CYCLOHEXYLAMINO)-2-PROPANOL BENZOATE (ESTER) HYDROCHLORIDE □ D 109 □ HEXYLCAINE HYDROCHLORIDE

TOXICITY DATA WITH REFERENCE
orl-mus LD50:1080 mg/kg 29ZVAB -,57,69
scu-mus LD50:1080 mg/kg CLDND* 5,683,71
scu-rbt LD50:164 mg/kg CLDND*-,57,69
scu-gpg LD50:166 mg/kg JPETAB 93,388,48
ivn-rbt LD50:14 mg/kg 29ZVAB -,57,69

SAFETY PROFILE: Poison by intravenous and subcutaneous routes. Moderately toxic by ingestion. A local anesthetic. When heated to decomposition it emits very toxic fumes of HCl and NO_x.

COU500 CAS:103-95-7 HR: 2
CYCLAMEN ALDEHYDE
mf: $C_{13}H_{18}O$ mw: 190.31

PROP: Colorless liquid; strong, floral odor. D: 0.946–0.952, refr index: 1.503–1.508. Sol in fixed oils; insol in propylene glycol, glycerin.

SYNS: ALDEHYDE B ☐ CYCLAMAL ☐ FEMA No. 2743 ☐ p-ISOPROPYL-α-METHYLHYDROCINNAMIC ALDEHYDE ☐ p-ISOPROPYL-α-METHYLPHENYLPROPYL ALDEHYDE ☐ α-METHYL-p-ISOPROPYLHYDROCINNAMALDEHYDE ☐ 2-METHYL-3-(p-ISOPROPYLPHENYL) PROPIONALDEHYDE

TOXICITY DATA WITH REFERENCE
skn-hmn 15 mg/48H MLD FCTXAV 12,385,74
orl-rat LD50:3810 mg/kg FCTXAV 2,327,64

CONSENSUS REPORTS: Reported in EPA TSCA Inventory.

SAFETY PROFILE: Moderately toxic by ingestion. A human skin irritant. See also ALDEHYDES. When heated to decomposition it emits acrid smoke and irritating fumes.

COU510 CAS:7149-24-8 HR: 1
CYCLAMEN ALDEHYDE DIETHYL ACETAL
mf: $C_{17}H_{28}O_2$ mw: 264.45

SYNS: HYDROCINNAMALDEHYDE, p-ISOPROPYL-α-METHYL-, DIETHYL ACETAL ☐ α-METHYL-p-ISOPROPYL HYDROCINNAMIC ALDEHYDE DIETHYL ACETAL

TOXICITY DATA WITH REFERENCE
skn-rbt 500 mg/24H MOD FCTXAV 14,731,76

CONSENSUS REPORTS: Reported in EPA TSCA Inventory.

SAFETY PROFILE: A skin irritant. When heated to decomposition it emits acrid smoke and irritating fumes.

COU525 CAS:29886-96-2 HR: 1
CYCLAMEN ALDEHYDE DIMETHYL ACETAL
mf: $C_{15}H_{24}O_2$ mw: 236.39

SYNS: HYDROCINNAMALDEHYDE, p-ISOPROPYL-α-METHYL-, DIMETHYL ACETAL ☐ α-METHYL-p-ISOPROPYLHYDROCINNAMIC ALDEHYDE DIMETHYL ACETAL ☐ PROPIONALDEHYDE, 3-(p-ISOPROPYLBENZYL)-, DIMETHYL ACETAL

TOXICITY DATA WITH REFERENCE
skn-rbt 500 mg/24H MOD FCTOD7 20,659,82

SAFETY PROFILE: A skin irritant. When heated to decomposition it emits acrid smoke and irritating fumes.

COV125 HR: 3
CYCLAMIDOMYCIN
mf: $C_7H_{10}N_{20}$ mw: 374.37

SYNS: DESDANINE ☐ PYRACRYMYCIN-1 ☐ 1-PYRROLINE-2-ACRYLAMIDE

TOXICITY DATA WITH REFERENCE
orl-mus LD50:240 mg/kg 85GDA2 5,80,81
ipr-mus LD50:150 mg/kg 85GDA2 5,80,81
scu-mus LD50:150 mg/kg 85GDA2 5,80,81
ivn-mus LD50:125 mg/kg 85GDA2 5,80,81

SAFETY PROFILE: Poison by ingestion, subcutaneous, intravenous, and intraperitoneal routes. When heated to decomposition it emits toxic fumes of NO_x.

COV500 CAS:3572-80-3 HR: 3
CYCLAZOCINE
mf: $C_{18}H_{25}NO$ mw: 271.44

PROP: Crystals from MeOH. Mp: 201–204°

SYNS: 2-CYCLOPROPYLMETHYL-5,9-DIMETHYL-2′-HYDROXY-6,7-BENEOMORPHAN ☐ 3-CYCLOPROPYLMETHYL-6(eq),11(ax)-DIMETHYL-2,6-METHANO-3-BENZAZOCIN-8-OL ☐ 3-(CYCLOPROPYLMETHYL) 1-1,2,3,4,5,6-HEXAHYDRO-6,11-DIMETHYL-2,6-METHANO-3-BENZAZOCIN-8-OL ☐ 2-CYCLOPROPYLMETHYL-2′-HYDROXY-5,9-DIMETHYL-6,7-BENZOMORPHAN ☐ NIH 7981 ☐ NSC-107429 ☐ WIN 20740

TOXICITY DATA WITH REFERENCE
orl-rbt TDLo:13 mg/kg (6-18D preg):REP TXAPA9 31,534,75
scu-rat LD50:310 mg/kg JPETAB 143,141,64
ivn-rat LD50:32 mg/kg AIPTAK 165,112,67
ipr-mus LDLo:10 mg/kg RCOCB8 17,255,77
scu-mus LD50:153 mg/kg AIPTAK 241,79,79
ivn-mus LD50:28 mg/kg AIPTAK 165,112,67

SAFETY PROFILE: A poison by subcutaneous, intravenous, and intraperitoneal routes. Experimental reproductive effects. Used in the treatment of narcotic addiction. When heated to decomposition it emits toxic fumes of NO_x.

COV750 CAS:3741-38-6 HR: 3
CYCLIC ETHYLENE SULFITE
mf: $C_2H_4O_3S$ mw: 108.12

PROP: A liquid. Bp: 169–172°

SYNS: 1,3,2-DIOXATHIOLANE-2-OXIDE (9CI) ☐ ETHYLENE SULFITE ☐ 1,2-ETHYLENE SULFITE ☐ GLYCOL SULFITE

TOXICITY DATA WITH REFERENCE
ipr-mus TDLo:768 mg/kg/64W-I:ETA JNCIAM 53,695,74
ipr-mus LD50:250 mg/kg CBCCT* 5,341,53

CONSENSUS REPORTS: Reported in EPA TSCA Inventory.

SAFETY PROFILE: Poison by intraperitoneal route. Questionable carcinogen with experimental tumorigenic data. When heated to decomposition it emits toxic fumes of SO_x. See also SULFITES.

COV800 CAS:26741-53-7 HR: 1
CYCLIC NEOPENTANETETRAYL BIS(2,4-DI-tert-BUTYLPHENYL)ESTER PHOSPHOROUS ACID
mf: $C_{33}H_{50}O_6P_2$ mw: 604.77

SYNS: MARK PEP 24 ☐ 2,4,8,10-TETRAOXA-3,9-DIPHOSPHASPIRO (5.5)UNDECANE, 3,9-BIS(2,4-BIS(1,1-DIMETHYLETHYL)PHENOXY)- ☐ ULTRANOX 624 ☐ ULTRANOX 626 ☐ WESTON 626 ☐ WESTON MDW 626

TOXICITY DATA WITH REFERENCE
orl-rat LD50:5580 mg/kg EPASR* 8EHQ-1287-0706
ihl-rat LC50:>2 g/m^3 EPASR* 8EHQ-1287-0706
skn-rbt LD50:>200 mg/kg EPASR* 8EHQ-1287-0706

CONSENSUS REPORTS: Reported in EPA TSCA Inventory.

SAFETY PROFILE: Low toxicity by ingestion, inhalation, and skin contact. When heated to decomposition it emits toxic vapors of PO_x.

COV825 **HR: 3**
CYCLOBARBITAL-SALICYLAMIDE COMPLEX
mf: $C_{12}H_{16}N_2O_3•C_7H_7NO_2$ mw: 373.45

TOXICITY DATA WITH REFERENCE
ipr-rat LD50:300 mg/kg KSRNAM 4,2536,70
scu-rat LD50:1780 mg/kg KSRNAM 4,2536,70
ipr-mus LD50:580 mg/kg KSRNAM 4,2536,70
scu-mus LD50:2300 mg/kg KSRNAM 4,2536,70

SAFETY PROFILE: Poison by intraperitoneal route. Moderately toxic by subcutaneous route. When heated to decomposition it emits toxic fumes of NO_x. See also TETRAHYDROPHENOBARBITAL, BARBITURATES, and SALICYLAMIDE.

COW000 **CAS:287-23-0** **HR: 1**
CYCLOBUTANE
mf: C_4H_8 mw: 56.12

PROP: A gas. Mp: −50°, bp: 12.9°, flash p: <50°F(CC), d: 0.708 @ 11°, vap d: 1.93, lel: 1.8%. Sol in EtOH, Me_2CO; insol H_2O.

SYN: TETRAMETHYLENE

SAFETY PROFILE: May be a simple asphyxiant. See also CYCLOHEXANE. Very dangerous fire hazard when exposed to heat or flame; can react with oxidizing materials. To fight fire, stop flow of gas; CO_2, dry chemicals, or water spray. When heated to decomposition it emits acrid smoke and fumes.

COW250 **HR: 1**
CYCLOBUTENE
mf: C_4H_6 mw: 54.09

PROP: Gas. Bp: 2.4°; d: 0.733 @ 0°/4°; flash p: <15°F.

SYN: CYCLOBUTYLENE

SAFETY PROFILE: May be a simple asphyxiant. Dangerous fire hazard when exposed to heat or flame; can react with oxidizing materials. When heated to decomposition it emits acrid smoke and fumes.

COW675 **CAS:77287-90-2** **HR: 3**
17-CYCLOBUTYLMETHYL-3-HYDROXY-6-METHYLENE-8-β-METHYLMORPHINAN
mf: $C_{22}H_{29}NO•CH_4O_3S$ mw: 419.63

SYNS: (8-β)-17-(CYCLOBUTYLMETHYL)-6-METHYLENEMORPHINAN-3-OL METHANESULFONATE □ TR5379M

TOXICITY DATA WITH REFERENCE
orl-rat LD50:750 mg/kg FAATDF 3,478,83
ivn-rat LD50:22,300 μg/kg FAATDF 3,478,83
orl-mus LD50:365 mg/kg FAATDF 3,478,83
ivn-mus LD50:35 mg/kg FAATDF 3,478,83

SAFETY PROFILE: Poison by ingestion and intravenous routes. When heated to decomposition it emits toxic fumes of SO_x and NO_x. A narcotic antagonist.

COW700 **CAS:512-16-3** **HR: 2**
CYCLOBUTYROL
mf: $C_{10}H_{18}O_3$ mw: 186.28

PROP: Colorless crystals from ether-petr ether. Mp: 81–82°. Sltly sol in water, petr ether. Very sol in alcohols, acetone, dioxane, chloroform, ether.

SYNS: 1-CYCLOHEXANOL-α-BUTYRIC ACID □ α-ETHYL-1-HYDROXYCYCLOHEXANEACETIC ACID □ HEBUCOL □ α-(1-HYDROXYCYCLOHEXYL)BUTYRIC ACID □ 1-HYDROXY-α-ETHYLCYCLOHEXYLACETIC ACID □ JL 130

TOXICITY DATA WITH REFERENCE
orl-rat LD50:4820 mg/kg NIIRDN 6,313,82
scu-rat LD50:3230 mg/kg NIIRDN 6,313,82
ivn-rat LD50:1760 mg/kg NIIRDN 6,313,82
scu-mus LD50:4200 mg/kg NIIRDN 6,313,82
ivn-mus LD50:2900 mg/kg NIIRDN 6,313,82
ivn-rbt LD50:1920 mg/kg NIIRDN 6,313,82

CONSENSUS REPORTS: Reported in EPA TSCA Inventory.

SAFETY PROFILE: Moderately toxic by intravenous and subcutaneous routes. Mildly toxic by ingestion. When heated to decomposition it emits acrid smoke and fumes.

COW750 **CAS:12663-46-6** **HR: 3**
CYCLOCHLOROTINE
mf: $C_{24}H_{30}Cl_2N_5O_7$ mw: 571.49

PROP: White needles. Mp: 255°, decomp. Chlorine containing peptide produced by *P. islandicum* (85CVA2 5,177,70).

SYN: ISLANDITOXIN

TOXICITY DATA WITH REFERENCE
orl-mus TDLo:362 mg/kg/32W-C:ETA FCTXAV 10,193,72
orl-mus LD50:6550 μg/kg FCTXAV 10,193,72
ipr-mus LD50:330 μg/kg CTOXAO 17,45,80
scu-mus LD50:475 μg/kg FCTXAV 10,193,72
ivn-mus LD50:335 μg/kg FCTXAV 10,193,72

CONSENSUS REPORTS: IARC Cancer Review: Group 3 IMEMDT 7,56,87; Animal Limited Evidence IMEMDT 10,139,76.

SAFETY PROFILE: A deadly poison by ingestion, subcutaneous, intraperitoneal, and intravenous routes. Questionable carcinogen with experimental tumorigenic data. When heated to decomposition it emits very toxic fumes of Cl^- and NO_x.

C

COW780 CAS:72117-72-7 HR: 1
α-CYCLOCITRYLIDENE-4-METHYLBUTAN-3-ONE
mf: $C_{15}H_{24}O$ mw: 220.39

SYNS: DIMETHYLIONONE □ 1,3-DIMETHYL-α-IONONE □ 1-PENTEN-3-ONE, 1-(2,6,6-TRIMETHYL-2-CYCLOHEXEN-1-YL)-2-METHYL-

TOXICITY DATA WITH REFERENCE
skn-rbt 500 mg/24H MOD FCTXAV 16,717,78

CONSENSUS REPORTS: Reported in EPA TSCA Inventory.

SAFETY PROFILE: A skin irritant. When heated to decomposition it emits acrid smoke and irritating fumes.

COW825 CAS:51022-69-6 HR: 3
CYCLOCORT
mf: $C_{28}H_{35}FO_7$ mw: 502.63

PROP: Crystals from Me_2CO/hexane.

SYNS: (11-β,16-α)-21-(ACETYLOXY)-16,17-(CYCLOPENTYLIDENEBIS (OXY))-9-FLUORO-11-HYDROXYPREGNA-1,4-DIENE-3,20-DIONE □ AMCINONIDE □ CL-34699 □ PENTICORT

TOXICITY DATA WITH REFERENCE
ipr-rat LD50:243 mg/kg IYKEDH 13,637,82
scu-rat LD50:145 mg/kg IYKEDH 13,637,82
ipr-mus LD50:896 mg/kg IYKEDH 13,637,82
scu-mus LD50:143 mg/kg IYKEDH 13,637,82

SAFETY PROFILE: Poison by subcutaneous and intraperitoneal routes. A steroid. When heated to decomposition it emits toxic fumes of F^-.

COW875 CAS:31698-14-3 HR: 2
CYCLOCYTIDINE
mf: $C_9H_{11}N_3O_4$ mw: 225.23

PROP: A cytostatic agent and intermediate in the synthesis of cytarabine.

SYNS: ANCITABINE □ ANCYTABINE □ 2,2′-ANHYDROARABINOSYLCYTOSINE □ ANHYDROARA C □ ANHYDROCYTIDINE □ 2,2′-ANHYDROCYTIDINE □ o-2,2′-CYCLOCYTIDINE □ 2,2′-CYCLOCYTIDINE □ 2,2′-o-CYCLOCYTIDINE

TOXICITY DATA WITH REFERENCE
ipr-rat LD50:1700 mg/kg EKFMA7 9,31,80
ivn-rat LD50:820 mg/kg IYKEDH 7,108,76
orl-mus LD50:3400 mg/kg EKFMA7 9,31,80
ipr-mus LD50:1600 mg/kg EKFMA7 9,31,80
scu-mus LD50:4050 mg/kg IYKEDH 7,108,76
ivn-mus LD50:800 mg/kg IYKEDH 7,108,76

SAFETY PROFILE: Moderately toxic by ingestion, intravenous, and intraperitoneal routes. Mildly toxic by subcutaneous route. When heated to decomposition it emits toxic fumes of NO_x.

COW900 CAS:10212-25-6 HR: 3
CYCLOCYTIDINE HYDROCHLORIDE
mf: $C_9H_{11}N_3O_4 \cdot ClH$ mw: 261.69

PROP: Needles from MeOH/Me_2CO. A solid. Mp: 266–267° decomp.

SYNS: ALEXAN □ ANCITABINE HYDROCHLORIDE □ 2,2′-ANHYDROARABINOSYLCYTOSINE HYDROCHLORIDE □ 2,2′-ANHYDRO-1-β-d-ARABINOFURANOSYLCYTOSINE HYDROCHLORIDE □ 2,2′-ANHYDROCYTARABINE HYDROCHLORIDE □ 2,2′-ANHYDROCYTIDINE HYDROCHLORIDE □ 1-β-d-ARABINOFURANOSYL-2,2′-ANHYDRO-CYTOSINE HYDROCHLORIDE □ CYCLO-CMP HYDROCHLORIDE □ CYCLOCYTIDINE □ 2,2′-CYCLOCYTIDINE HYDROCHLORIDE □ 2,2′-o-CYCLOCYTIDINE HYDROCHLORIDE □ o-2,2′-CYCLOCYTIDINE MONOHYDROCHLORIDE □ NSC-145668

TOXICITY DATA WITH REFERENCE
sce-hmn:lym 100 ng/L MUREAV 53,215,78
dni-mus:leu 10 mg/L CPBTAL 20,2286,72
ipr-rat TDLo:800 mg/kg (female 8-15D post):REP OYYAA2 8,1681,74
ipr-rat TDLo:800 mg/kg (female 8-15D post):TER OYYAA2 8,1681,74
ipr-rat TDLo:400 mg/kg (8-15D preg):TER OYYAA2 8,1681,74
ipr-rat TDLo:12 g/kg (female 30D pre):REP OYYAA2 8,1469,74
ipr-rat TDLo:8250 mg/kg (male 2 year(s) pre):REP OYYAA2 8,1693,74
ivn-rbt TDLo:125 mg/kg (female 8-17D post):TER OYYAA2 8,1681,74
ivn-rbt TDLo:60 mg/kg (female 8-17D post):TER OYYAA2 8,1681,74
ipr-rat TDLo:400 mg/kg (8-15D preg):TER OYYAA2 8,1681,74
scu-wmn TDLo:120 mg/kg:CVS,GIT,BLD CTRRDO 62,455,78
ipr-rat LD50:3800 mg/kg YAKUD5 21,359,79
ivn-rat LD50:820 mg/kg YAKUD5 21,359,79
ipr-mus LD50:2528 mg/kg NCISP* JAN86
ivn-mus LD50:800 mg/kg NIIRDN 6,55,82
ivn-dog LD50:344 mg/kg OYYAA2 8,353,74
ivn-mky LD50:1045 mg/kg OYYAA2 8,353,74

SAFETY PROFILE: Poison by intravenous route. Moderately toxic by intraperitoneal route. Human systemic effects by subcutaneous route: blood pressure depression, nausea or vomiting, and changes in bone marrow. An experimental teratogen. Other experimental reproductive effects. Human mutation data reported. When heated to decomposition it emits toxic fumes of NO_x and HCl.

COW925 CAS:7585-39-9 HR: 3
β-CYCLODEXTRIN
mf: $C_{42}H_{70}O_{35}$ mw: 1135.12

SYNS: α-CYCLOAMYLOSE □ CYCLOHEPTAAMYLOSE □ β-CYCLOHEPTAAMYLOSE □ CYCLOHEPTAGLUCOSAN □ CYCLOMALTOHEPTAOSE □ CYCLOPHEPTAGLUCAN □ β-DEXTRIN □ SCHAARDINGER α-DEXTRIN

TOXICITY DATA WITH REFERENCE
orl-rat LD50:18,800 mg/kg OYYAA2 26,287,83
ipr-rat LD50:356 mg/kg 48THAM 1,109,82
scu-rat LD50:3700 mg/kg OYYAA2 26,287,83
ivn-rat LD50:1008 mg/kg AJPAA4 83,367,76
ipr-mus LD50:330 mg/kg 48THAM 1,109,82
scu-mus LD50:412 mg/kg 48THAM 1,109,82

SAFETY PROFILE: Poison by intraperitoneal route. Moderately toxic by subcutaneous and intravenous routes. Mildly toxic by ingestion. When heated to decomposition it emits acrid smoke and fumes. See also DEXTRINS.

COW930 **CAS:947-04-6** **HR: 2**
CYCLODODECALACTAM
mf: $C_{12}H_{23}NO$ mw: 197.36

SYNS: AZACYCLOTRIDECAN-2-ONE □ 2-AZACYCLOTRIDECANONE

TOXICITY DATA WITH REFERENCE
ipr-mus LD50:500 mg/kg JPMSAE 60,1058,71

CONSENSUS REPORTS: Reported in EPA TSCA Inventory.

SAFETY PROFILE: Moderately toxic by intraperitoneal route. When heated to decomposition it emits toxic vapors of NO_x.

COW935 **HR: 2**
cis,trans,trans-CYCLODODECA-1,5,9-TRIENE
mf: $C_{12}H_{18}$ mw: 162.30

SYNS: CDT □ 1,5,9-CYCLODODECATRIENE (Z,E,E)

TOXICITY DATA WITH REFERENCE
skn-mus 100%/12D open SEV BJIMAG 25,75,68
skn-rbt 2670 mg SEV BJIMAG 25,75,68
skn-rbt 20 g/31D-I open SEV BJIMAG 25,75,68
skn-gpg 10 g/31D-I open SEV BJIMAG 25,75,68
eye-rbt 89 mg MLD BJIMAG 25,75,68

SAFETY PROFILE: An eye and severe skin irritant. When heated to decomposition it emits acrid smoke and fumes.

COX000 **CAS:31717-87-0** **HR: 3**
CYCLODODECYL-2,6-DIMETHYLMORPHOLINE ACETATE
mf: $C_{18}H_{36}NO \cdot C_2H_4O_2$ mw: 342.61

PROP: D: 0.930. Misc in H_2O.

SYNS: N-CYCLODODECYL-2,6-DIMETHYLMORPHOLINIUM ACETATE □ CYCLOMORPH □ DODEMORFE (FRANCE)

TOXICITY DATA WITH REFERENCE
orl-rat LD50:2500 mg/kg DOVEAA 27,144,73
ipr-mus LD50:320 mg/kg GUCHAZ 6,244,73

SAFETY PROFILE: Poison by intraperitoneal route. Moderately toxic by ingestion. When heated to decomposition it emits toxic fumes of NO_x.

COX325 **CAS:152-53-4** **HR: 3**
CYCLOGUANIL HYDROCHLORIDE
mf: $C_{11}H_{14}ClN_5 \cdot ClH$ mw: 288.21

PROP: Prisms from H_2O. A solid. Mp: 210–215°.

SYNS: 1-(4-CHLOROPHENYL)-1,6-DIHYDRO-6,6-DIMETHYL-1,3,5-TRIAZINE-2,4-DIAMINE MONOHYDROCHLORIDE □ 4,6-DIAMINO-1-(p-CHLOROPHENYL)-1,2-DIHYDRO-2,2-DIMETHYL-s-TRIAZINE MONOHYDROCHLORIDE

TOXICITY DATA WITH REFERENCE
ipr-mus LD50:54 mg/kg TXAPA9 18,487,71
scu-mus LD50:220 mg/kg ATMPA2 74,393,80
ivn-dog LDLo:24 mg/kg TXAPA9 18,487,71

SAFETY PROFILE: Poison by subcutaneous, intravenous, and intraperitoneal routes. When heated to decomposition it emits toxic fumes of NO_x and Cl^-.

COX400 **CAS:516-21-2** **HR: 3**
CYCLOGUANYL
mf: $C_{11}H_{14}ClN_5$ mw: 251.75

PROP: Prisms from $CHCl_3/Et_2O$. A solid. Mp: 146

SYNS: 4-AMINO-6-p-CHLOROANILINO-1,2-DWUHYDRO-2,2-DWUMETHYLO-1,3,5-TROJAZYNA □ CHLORGUANIDE TRIAZINE □ CGT □ 1-(p-CHLOROPHENYL)-4,6-DIAMINO-2,2-DIMETHYL-1,2-DIHYDRO-s-TRIAZINE □ 1-p-CHLOROPHENYL-1,2-DIHYDRO-2,2-DIMETHYL-4,6-DIAMINO-s-TRIAZINE □ CYCLOGUANIL □ WR 5473 □ s-TRIAZINE, 1,2-DIHYDRO-1-(p-CHLOROPHENYL)-4,6-DIAMINO-2,2-DIMETHYL-

TOXICITY DATA WITH REFERENCE
mmo-omi 100 µg/plate AACHAX 12,84,77
orl-rat TDLo:50 mg/kg (female 1D post):TER BEBMAE 77(6),56,74
ipr-rat LD50:98 mg/kg DIPHAH 10,81,58

SAFETY PROFILE: Poison by intraperitoneal route. An experimental teratogen. Mutation data reported. When heated to decomposition it emits toxic fumes of NO_x.

COX500 **CAS:291-64-5** **HR: 3**
CYCLOHEPTANE
mf: C_7H_{14} mw: 98.19

PROP: An oil. Mp: −12°, bp: 118–120°, flash p: 59°F, d: 0.810 @ 20°/4°, vap d: 3.3.

SYN: SUBERANE

SAFETY PROFILE: Dangerous fire hazard when exposed to heat or flame; can react with oxidizing materials. To fight fire, use foam, CO_2, dry chemicals. See also CYCLOHEXANE.

COY000 **CAS:544-25-2** **HR: 3**
1,3,5-CYCLOHEPTATRIENE
DOT: UN 2603
mf: C_7H_8 mw: 92.14

CH=CH(CH=CH)$_2$ CH$_2$ (ring)

PROP: A liquid. D: 0.888 18.5°/4°, bp: 117° at 749 mm, flash p: 39.2°F.

SYNS: CYCLOHEPTATRIENE (DOT) □ TROPILIDENE □ TROPILIDIN

TOXICITY DATA WITH REFERENCE
cyt-rat:lvr 100 mg/L MUREAV 155,57,85
orl-rat LD50:57 mg/kg AOHYA3 10,123,67
skn-rat LD50:442 mg/kg AOHYA3 10,123,67
orl-mus LD50:171 mg/kg AOHYA3 10,123,67

DOT CLASSIFICATION: 3; *Label:* Flammable Liquid, Poison

SAFETY PROFILE: Poison by ingestion and skin contact. Mutation data reported. A very dangerous fire hazard when exposed to heat, flame, or oxidizers. Potentially violent reaction with nitrogen monoxide. When heated to decomposition it emits acrid smoke and fumes.

COY100 **CAS:12125-77-8** **HR: 3**
CYCLOHEPTATRIENE MOLYBDENUM TRICARBONYL
mf: $C_{10}H_8MoO_3$ mw: 272.12

PROP: Red hexagonal prisms from hexane. Mp: 95° decomp.

PROP: Mp: 100–101° decomp.

SYN: MOLYBDENUM, TRICARBONYL(1,3,5-CYCLOHEPTATRIENE)-

TOXICITY DATA WITH REFERENCE
ivn-mus LD50:32 mg/kg CSLNX* NX#04764

OSHA PEL: TWA Total Dust: 10 mg/m³; Respirable Fraction: 5 mg/m³
ACGIH TLV: TWA 10 mg(Mo)/m³

SAFETY PROFILE: Poison by intravenous route. A skin and eye irritant. When heated to decomposition it emits toxic fumes of Mo.

COY250 **CAS:628-92-2** **HR: 3**
CYCLOHEPTENE
mf: C_7H_{12} mw: 96.174

$CH{=}CH(CH_2)_4CH_2$ (ring)

PROP: Flash p: <73.4°F.

SAFETY PROFILE: A dangerous fire hazard when exposed to heat, flame, or oxidizers. When heated to decomposition it emits acrid smoke and fumes.

COY500 **CAS:509-86-4** **HR: 3**
CYCLOHEPTENYL ETHYLBARBITURIC ACID
mf: $C_{13}H_{18}N_2O_3$ mw: 250.33

SYNS: 5-(1-CYCLOHEPTEN-1-YL)-5-ETHYLBARBITURIC ACID □ CYCLOHEPTENYLETHYLMALONYLUREA □ 5-(1-CYCLOHEPTEN-1-YL)-5-ETHYL-2,4,6(1H,3H,5H)-PYRIMIDINETRIONE (9CI) □ 5-ETHYL-5-(1'-CYCLOHEPTENYL)-BARBITURIC ACID □ G 475 □ HEPTABARB □ HEPTABARBITAL □ HEPTABARBITONE □ HEPTABARBUM □ HEPTADORM □ HEPTAMAL □ MEDAPAN □ MEDOMIN □ MEDOMINE

TOXICITY DATA WITH REFERENCE
ipr-rat LD50:220 mg/kg JPETAB 93,101,48
ipr-mus LD50:210 mg/kg ARZNAD 9,360,59
ivn-dog LD50:105 mg/kg JPETAB 93,101,48
ivn-rbt LD50:119 mg/kg JPETAB 93,101,48

SAFETY PROFILE: A poison by intraperitoneal and intravenous routes. Psychotropic effects by ingestion. When heated to decomposition it emits toxic fumes of NO_x. See also BARBITURATES.

CPA500 **CAS:592-57-4** **HR: 3**
1,3-CYCLOHEXADIENE
mf: C_6H_8 mw: 80.13

$CH_2(CH{=}CH)_2CH_2$ (ring)

PROP: A liquid. Mp: −89°, bp: 80.5°, flash p: <73.4°F.

SAFETY PROFILE: A dangerous fire hazard when exposed to heat, flame, or oxidizers. It forms explosive polymeric oxides on exposure to air. When heated to decomposition it emits acrid smoke and fumes.

CPA775 **CAS:4998-76-9** **HR: 3**
CYCLOHEXANAMINE HYDROCHLORIDE
mf: $C_6H_{13}N{\cdot}ClH$ mw: 135.66

SYNS: AMINOCYCLOHEXANE HYDROCHLORIDE □ AMINOHEXAHYDROBENZENE HYDROCHLORIDE □ HEXAHYDROANILINE HYDROCHLORIDE

TOXICITY DATA WITH REFERENCE
spm-rat-orl 32,400 mg/kg/90D TXCYAC 8,143,77
orl-rat TDLo:1 g/kg (female 6-15D post):TER TOLED5 17,137,83
orl-mus TDLo:1 g/kg (female 6-15D post):REP TOLED5 17,137,83
orl-rat TDLo:73 g/kg (female 2 year(s) pre):REP FCTXAV 14,255,76
orl-rat TDLo:32,400 mg/kg (90D male):REP TXCYAC 8,143,77
orl-rat LD50:720 mg/kg GISAAA 51(9),80,86
ipr-rat LDLo:350 mg/kg LIFSAK 8,843,69
orl-mus LD50:760 mg/kg GISAAA 51(9),80,86
ipr-mus LD50:300 mg/kg TXAPA9 22,465,72

CONSENSUS REPORTS: EPA Genetic Toxicology Program. Reported in EPA TSCA Inventory.

SAFETY PROFILE: Poison by intraperitoneal route. Moderately toxic by ingestion. An experimental teratogen. Other experimental reproductive effects. Mutation data reported. When heated to decomposition it emits toxic fumes of NO_x and HCl.

CPB000 **CAS:110-82-7** **HR: 3**
CYCLOHEXANE
DOT: UN 1145
mf: C_6H_{12} mw: 84.18

PROP: Colorless, mobile liquid; pungent odor. Mp: 6.5°, bp: 81°, fp: 4.6°, flash: p: 1.4°F, ULC: 90–95, lel: 1.3%, uel: 8.4%, d: 0.7791 @ 20°/4°, autoign temp: 473°F, vap press: 100 mm @ 60.8°, vap d: 2.90. Prac insol in H_2O; sol in MeOH; misc most org solvs.

SYNS: CICLOESANO (ITALIAN) □ CYCLOHEXAAN (DUTCH) □ CYCLOHEXAN (GERMAN) □ CYKLOHEKSAN (POLISH) □ HEXAHYDROBENZENE □ HEXAMETHYLENE □ HEXANAPHTHENE □ RCRA WASTE NUMBER U056

TOXICITY DATA WITH REFERENCE
skn-rbt 1548 mg/2D-I JIHTAB 25,199,43
dnd-esc 10 μmol/L MUREAV 89,95,81
orl-rat LD50:29,820 mg/kg JIHTAB 25,415,43
orl-mus LD50:813 mg/kg NPIRI* 1,17,74
ihl-rbt TCLo:7444 ppm/6H/2W-I JIDHAN 25,323,43

orl-rbt LDLo:5500 mg/kg JIHTAB 25,199,43
ivn-rbt LDLo:77 mg/kg JPMRAB 3,1,28

CONSENSUS REPORTS: Community Right-To-Know List.

OSHA PEL: TWA 300 ppm
ACGIH TLV: TWA 300 ppm
DFG MAK: 300 ppm (1050 mg/m^3)
DOT CLASSIFICATION: 3; *Label:* Flammable Liquid

SAFETY PROFILE: Poison by intravenous route. Moderately toxic by ingestion. A systemic irritant by inhalation and ingestion. A skin irritant. Mutation data reported. Flammable liquid. Dangerous fire hazard when exposed to heat or flame; can react with oxidizing materials. Moderate explosion hazard in the form of vapor when exposed to flame. When mixed hot with liquid dinitrogen tetraoxide an explosion can result. To fight fire, use foam, CO_2, dry chemical, spray, fog. When heated to decomposition it emits acrid smoke and fumes.

For occupational chemical analysis use NIOSH: Hydrocarbons, Bp 36-126°C, 1500.

CPB050 CAS:1122-56-1 HR: 1
CYCLOHEXANECARBOXAMIDE
mf: $C_7H_{13}NO$ mw: 127.21

PROP: Prisms from H_2O. Mp: 186–188°. Hygroscopic. Very sol in EtOH, Et_2O.

SYNS: CYCLOHEXAMETHYLENE CARBAMIDE □ CYCLOHEXANAMIDE □ CYCLOHEXANEFORMAMIDE □ CYCLOHEXYLCARBOXAMIDE □ CYCLOHEXYL CARBOXYAMIDE □ HEXAHYDROBENZOIC ACID AMIDE

TOXICITY DATA WITH REFERENCE
skn-rbt 500 mg/24H MOD 33NFA8 -,2,75

CONSENSUS REPORTS: Reported in EPA TSCA Inventory.

SAFETY PROFILE: A skin irritant. When heated to decomposition it emits toxic fumes of NO_x.

CPB100 CAS:694-83-7 HR: 2
1,2-CYCLOHEXANEDIAMINE
mf: $C_6H_{14}N_2$ mw: 114.22

PROP: Bp: 92–93° @ 18 mm, d: 0.931, flash p: 167°F.

SYN: 1,2-DIAMINOCYCLOHEXANE

TOXICITY DATA WITH REFERENCE
skn-rbt 500 mg/24H MOD JACTDZ 1,8,90
orl-rat LDLo:1 g/kg JACTDZ 1,8,90
ihl-rat LCLo:3200 mg/m^3/4H TOXID9 12,357,92

SAFETY PROFILE: Slightly toxic by ingestion and inhalation. A skin irritant. A combustible liquid. When heated to decomposition it emits toxic fumes of NO_x.

CPB120 CAS:482-54-2 HR: 2
1,2-CYCLOHEXANEDIAMINETETRAACETIC ACID
mf: $C_{14}H_{22}N_2O_8$ mw: 346.38

SYNS: ACETIC ACID, (1,2-CYCLOHEXYLENEDINITRILO)TETRA- □ CDTA □ CGTA □ CHEL 600 □ COMPLEXON IV □ 1,2-CYCLOHEXANEDIAMINE-N,N,N′,N′-TETRAACETIC ACID □ 1,2-CYCLOHEXYLENEDIAMINETETRAACETIC ACID □ (1,2-CYCLOHEXYLENEDINITRILO)TETRAACETIC ACID □ CYDTA □ DCTA □ 1,2-DIAMINOCYCLOHEXANETETRAACETIC ACID □ 1,2-DIAMINOCYCLOHEXANE-N,N′-TETRAACETIC ACID □ GLYCINE, N,N′-1,2-CYCLOHEXANEDIYLBIS(N-(CARBOXYMETHYL))- (9CI) □ KOMPLEXON IV □ KYSELINA 1,2-CYKLOHEXYLENDIAMINTETRAOCTOVA □ OCTA

TOXICITY DATA WITH REFERENCE
ipr-rat LD50:413 mg/kg TOLED5 32,37,86
ipr-mus LD50:150 mg/kg NTIS** AD691-490

CONSENSUS REPORTS: Reported in EPA TSCA Inventory.

SAFETY PROFILE: Moderately toxic by intraperitoneal route. When heated to decomposition it emits toxic vapors of NO_x.

CPB625 CAS:1569-69-3 HR: 3
CYCLOHEXANETHIOL
DOT: UN 3054
mf: $C_6H_{12}S$ mw: 116.24

PROP: Oil. D: 0, 0.991, bp: 158–160°. Sol in EtOH, $CHCl_3$; insol in H_2O.

SYNS: CYCLOHEXYL MERCAPTAN (DOT) □ CYKLOHEXANTHIOL □ CYKLOHEXYLMERKAPTAN (CZECH)

TOXICITY DATA WITH REFERENCE
skn-rbt 2 mg/24H SEV 85JCAE-,984,86
eye-rbt 500 mg/24H MLD 85JCAE-,984,86

CONSENSUS REPORTS: Reported in EPA TSCA Inventory.

NIOSH REL: (Cyclohexanethiol) CL 0.5 ppm/15M
DOT CLASSIFICATION: 3; *Label:* Flammable Liquid

SAFETY PROFILE: An eye and severe skin irritant. When heated to decomposition it emits toxic fumes of SO_x. See also MERCAPTANS.

CPB650 CAS:3570-93-2 HR: 3
1,2,3-CYCLOHEXANETRIONE TRIOXIME
mf: $C_6H_9N_3O_3$ mw: 171.16

HON:CCH$_2$C(:NOH)CH$_2$C(:NOH) CH$_2$

PROP: Pale yellow crystals. Sol in Me_2CO, EtOH, MeOH.

SAFETY PROFILE: Explodes violently when heated to 155°C. Potentially explosive reaction with sulfinyl chloride. When heated to decomposition it emits toxic fumes of NO_x.

CPB750 CAS:108-93-0 HR: 3
CYCLOHEXANOL
mf: $C_6H_{12}O$ mw: 100.16

PROP: Colorless needles or viscous liquid; hygroscopic, camphor-like odor. Mp: 24°, bp: 161.5°, flash p: 154°F (CC), d: 0.9449 @ 25°/4°, vap press: 1 mm @ 21.0°, vap d: 3.45, autoign temp: 572°F. Sol in EtOH, Et_2O; mod sol in H_2O; misc in nonpolar solvents.

SYNS: ADRONAL □ ANOL □ CICLOESANOLO (ITALIAN) □ CYCLOHEXYL ALCOHOL □ CYKLOHEKSANOL (POLISH) □ HEXAHYDROPHENOL □ HEXALIN □ HYDRALIN □ HYDROPHENOL □ HYDROXYCYCLOHEXANE □ NAXOL

TOXICITY DATA WITH REFERENCE
eye-hmn 100 ppm JIHTAB 25,282,43
skn-rbt 14,600 μg/24H open MLD AIHAAP 23,95,62
eye-rbt 2 mg SEV AJOPAA 29,1363,46
cyt-hmn:leu 100 μmol/L DBTEAD 19,215,71
dnd-mam:lym 150 mmol/L PNASA6 48,686,62
scu-rat TDLo:315 mg/kg (21D male):REP IJEBA6 17,1305,79
ihl-hmn TCLo:75 ppm:NOSE,EYE,PUL JIHTAB 25,282,43
orl-rat LD50:2060 mg/kg MDZEAK 8,244,67
ipr-mus LD50:1352 mg/kg ARZNAD 19,1254,69
scu-mus LD50:2480 mg/kg REMBA8 5,7,67
ivn-mus LD50:272 mg/kg AIPTAK 135,342,62
ims-mus LD50:1000 mg/kg JSICAZ 21,342,62
orl-rbt LDLo:2200 mg/kg JIHTAB 25,199,43
skn-rbt LDLo:12 g/kg JIHTAB 25,199,43
ipr-rbt LDLo:1420 mg/kg JPMRAB 3,1,28

CONSENSUS REPORTS: EPA Genetic Toxicology Program. Reported in EPA TSCA Inventory.

OSHA PEL: TWA 50 ppm (skin)
ACGIH TLV: TWA 50 ppm (skin)
DFG MAK: 50 ppm (200 mg/m³)

SAFETY PROFILE: Poison by intravenous route. Moderately toxic by ingestion, subcutaneous, and intramuscular routes. Mildly toxic by skin contact. Human systemic effects by inhalation: conjunctiva irritation, and changes in the olfactory and respiratory systems. Has caused damage to kidneys, liver, and blood vessels in experimental animals. Experimental reproductive effects. Human mutation data reported. A severe eye irritant. Narcotic-like action. Has caused liver, kidney, vascular injury in experimental animals. Flammable when exposed to heat or flame; can react with oxidizing materials. Ignites on contact with chromium trioxide. Violent reaction with HNO_3. Incompatible with oxidants. To fight fire, use alcohol foam, foam, CO_2, dry chemical. When heated to decomposition it emits acrid smoke and fumes. See also ALCOHOLS.

For occupational chemical analysis use NIOSH: Alcohols III, 1402.

CPC000 CAS:108-94-1 HR: 3
CYCLOHEXANONE
DOT: UN 1915
mf: $C_6H_{10}O$ mw: 98.16

PROP: Colorless oily liquid; acetone-like odor. Mp: −45.0°, bp: 155°. ULC: 35–40, lel: 1.1% @ 100°, flash p: 111°F, d: 0.9478 @ 20°/4°, autoign temp: 788°F, vap press: 10 mm @ 38.7°, vap d: 3.4. Mod sol in H_2O.

SYNS: CICLOESANONE (ITALIAN) □ CYCLOHEXANON (DUTCH) □ CYKLOHEKSANON (POLISH) □ HEXANON □ KETOHEXAMETHYLENE □ NADONE □ NCI-C55005 □ PIMELIC KETONE □ RCRA WASTE NUMBER U057 □ SEXTONE

TOXICITY DATA WITH REFERENCE
eye-hmn 75 ppm JIHTAB 25,282,43
skn-rbt 500 mg open MLD UCDS**
eye-rbt 4740 μg SEV AJOPAA 29,1363,46
mma-sat 20 μL/L EJMBA2 18,213,83
mmo-bcs 200 μL/L EJMBA2 18,213,83
sce-ham:ovr 7500 μL/L ENMUDM 7(Suppl 3),60,85
orl-mus TDLo:11 g/kg (female 8-12D post):REP TCMUD8 6,361,86
ihl-rat TDLo:105 mg/m³/4H (1-20D preg):REP TPKVAL 14,26,75
ihl-hmn TCLo:75 ppm:NOSE,EYE,PUL JIHTAB 25,282,43
orl-rat LD50:1535 mg/kg AIHAAP 30,470,69
ihl-rat LC50:8000 ppm/4H NPIRI* 1,18,74
scu-rat LD50:2170 mg/kg JIHTAB 25,415,43
orl-mus LD50:1400 mg/kg NTIS** AD-A066-307
ipr-mus LD50:1350 mg/kg COREAF 254,2245,62
scu-mus LDLo:1300 mg/kg AEXPBL 50,199,1903
ivn-dog LDLo:630 mg/kg 14CYAT 2,1719,63
orl-rbt LDLo:1600 mg/kg JIHTAB 25,199,43
skn-rbt LD50:948 mg/kg AIHAAP 30,470,69

CONSENSUS REPORTS: Reported in EPA TSCA Inventory.

OSHA PEL: TWA 25 ppm (skin)
ACGIH TLV: TWA 25 ppm (skin)
DFG MAK: 50 ppm (200 mg/m³)
NIOSH REL: (Ketone (Cyclohexanone)) TWA 100 mg/m³
DOT CLASSIFICATION: 3; *Label:* Flammable Liquid

SAFETY PROFILE: Moderately toxic by ingestion, inhalation, subcutaneous, intravenous, and intraperitoneal routes. A skin and severe eye irritant. Human systemic effects by inhalation: changes in the sense of smell, conjunctiva irritation, and unspecified respiratory system changes. Human irritant by inhalation. Mild narcotic properties have also been ascribed to it. Human mutation data reported. Experimental reproductive effects. Flammable liquid when exposed to heat or flame; can react vigorously with oxidizing materials. Slight explosion hazard in its vapor form, when exposed to flame. Explosive reaction with nitric acid at 75°C. Reaction with hydrogen peroxide + nitric acid forms an explosive peroxide. To fight fire, use alcohol foam, dry chemical, or CO_2. When heated to decomposition it emits acrid smoke and irritating fumes. See also KETONES and CYCLOHEXANE.

For occupational chemical analysis use OSHA: #01 or NIOSH: Ketones I (Desorption in CS_2) 1300.

CPC250 HR: 2
CYCLOHEXANONE-Δ
mf: C_6H_8O mw: 96.12

PROP: Liquid. Bp: 155.5°, flash p: 93°F (CC), vap d: 3.31, vap press: 4 mm @ 20°.

SAFETY PROFILE: Skin contact can cause a dermatitis. Irritating to eyes, skin, and mucous membranes. Can damage the liver and kidneys. Dangerous fire hazard when exposed to flame and heat; can react with oxidizing materials. To fight fire, use CO_2, dry chemical.

CPC300 **CAS:78-18-2** **HR: 2**
CYCLOHEXANONE PEROXIDE
mf: $C_{12}H_{22}O_5$ mw: 246.34

SYNS: CYCLOHEXANOL, 1-((1-HYDROPEROXYCYCLOHEXYL)DIOXY)- □ 1-HYDROPEROXYCYCLOHEXYL-1-HYDROXYCYCLOHEXYL PEROXIDE □ 1-HYDROXY-1-HYDROPEROXYDICYCLOHEXYL PEROXIDE □ 1-HYDROXY-1'-HYDROPEROXYDICYCLOHEXYL PEROXIDE □ PEROXIDE, 1-HYDROPEROXYCYCLOHEXYL 1-HYDROXYCYCLOHEXYL

TOXICITY DATA WITH REFERENCE
eye-rbt 80 mg/1M RNS SEV ZAARAM 8,25,58
par-mus LD50:2 g/kg NCPBBY Jan/Feb,69

CONSENSUS REPORTS: Reported in EPA TSCA Inventory.

DFG MAK: Strong Skin Effects

SAFETY PROFILE: Slightly toxic by parenteral route. A severe eye and skin irritant. When heated to decomposition it emits acrid smoke and irritating vapors.

CPC579 **CAS:110-83-8** **HR: 3**
CYCLOHEXENE
DOT: UN 2256
mf: C_6H_{10} mw: 82.15

PROP: Colorless liquid. Bp: 83°, fp: −103.7°, flash p: <21.2°F, d: 0.8102 @ 20°/4°, vap press: 160 mm @ 38°, autoign temp: 590°F, vap d: 2.8, lel: 1.2%.

SYNS: BENZENETETRAHYDRIDE □ CYKLOHEKSEN (POLISH) □ 1,2,3,4-TETRAHYDROBENZENE

CONSENSUS REPORTS: Reported in EPA TSCA Inventory.

OSHA PEL: 300 ppm
ACGIH TLV: 300 ppm
DFG MAK: 300 ppm (1015 mg/m³)
DOT CLASSIFICATION: 3; *Label:* Flammable Liquid

SAFETY PROFILE: Moderately toxic by inhalation and ingestion. A very dangerous fire hazard when exposed to flame; can react with oxidizers. Dangerous; keep away from heat and open flame. To fight fire, use foam, CO_2, dry chemical.

For occupational chemical analysis use NIOSH: Hydrocarbons, Bp 36-126°C, 1500.

CPC625 **CAS:100-45-8** **HR: 2**
3-CYCLOHEXENE-1-CARBONITRILE
mf: C_7H_9N mw: 107.17

SYN: 3-CYCLOHENENYL CYANIDE

TOXICITY DATA WITH REFERENCE
eye-rbt 500 mg open AMIHBC 10,61,54
eye-rbt 500 mg/24H MLD 85JCAE-,903,86
orl-rat LD50:460 mg/kg AMIHBC 10,61,54
ihl-rat LCLo:124 ppm/4H AMIHBC 10,61,54
skn-rbt LD50:9460 mg/kg AMIHBC 10,61,54

CONSENSUS REPORTS: Cyanide and its compounds are on the Community Right-To-Know List.

SAFETY PROFILE: Moderately toxic by ingestion and inhalation. A skin and eye irritant. When heated to decomposition it emits toxic fumes of NO_x and CN^-. See also NITRILES.

CPC650 **CAS:4771-80-6** **HR: 2**
3-CYCLOHEXENE-1-CARBOXYLIC ACID
mf: $C_7H_{10}O_2$ mw: 126.17

PROP: A liquid. Mp: 17°, bp: 237° @ 745 mm.

SYN: KYSELINA 1,2,5,6-TETRAHYDROBENZOOVA

TOXICITY DATA WITH REFERENCE
skn-rbt 10 mg/24H open SEV AMIHBC 10,61,54
skn-rbt 5 mg/24H SEV 85JCAE-,315,86
eye-rbt 250 µg open SEV AMIHBC 10,61,54
orl-rat LD50:4260 mg/kg AMIHBC 10,61,54
skn-rbt LD50:1000 mg/kg AMIHBC 10,61,54

SAFETY PROFILE: Moderately toxic by skin contact. Mildly toxic by ingestion. A severe eye and skin irritant. When heated to decomposition it emits acrid smoke and fumes.

CPD000 **CAS:286-20-4** **HR: 3**
CYCLOHEXENE OXIDE
mf: $C_6H_{10}O$ mw: 98.16

PROP: Clear liquid. Bp: 130–131°, flash p: 81°F, d: 0.9678 @ 25°/4°, vap d: 3.5.

SYNS: CCHO □ CYCLOHEXANE OXIDE □ CYCLOHEXENE EPOXIDE □ CYCLOHEXENE-1-OXIDE □ 1,2-CYCLOHEXENE OXIDE □ CYCLOHEXYLENE OXIDE □ 1,2-EPOXYCYCLOHEXANE □ 7-OXABICYCLO(4.1.0)HEPTANE □ TETRAMETHYLENEOXIRANE

TOXICITY DATA WITH REFERENCE
mmo-klp 5 mmol/L MUREAV 89,269,81
mmo-sat 10 µmol/plate BCPCA6 29,1068,80
mma-sat 1 mg/plate MUREAV 58,217,78
msc-ham:lng 5 mmol/L CBINA8 51,77,84
unk-mus TDLo:79 mg/kg:ETA RARSAM 3,193,63
orl-rat LD50:1090 mg/kg AIHAAP 30,470,69
ihl-rat LCLo:2000 ppm/4H AIHAAP 30,470,69
ipr-rat LD50:549 mg/kg TXAPA9 52,422,80
ims-mus LD50:1000 mg/kg JSICAZ 21,342,62
skn-rbt LD50:630 mg/kg AIHAAP 30,470,69

CONSENSUS REPORTS: Reported in EPA TSCA Inventory. EPA Genetic Toxicology Program.

SAFETY PROFILE: Moderately toxic by ingestion, skin contact, intraperitoneal, and intramuscular routes. Mildly toxic by inhalation. Questionable carcinogen with experimental tumorigenic data. Mutation data reported. A flammable liquid and dangerous fire hazard when exposed to heat or flame. When heated to decomposition it emits acrid smoke and irritant fumes.

CPD250 **CAS:930-68-7** **HR: 3**
2-CYCLOHEXEN-1-ONE
mf: C_6H_8O mw: 96.14

PROP: A liquid. Bp: 169–171°.

SYN: CYCLOHEXENONE

TOXICITY DATA WITH REFERENCE
eye-rbt 98 mg AIHAAP 33,338,72
mmo-sat 15 mmol/L MUREAV 93,305,82
dns-rat:lvr 10 µmol/L MUREAV 221,263,89
orl-rat LD50:220 mg/kg AIHAAP 33,338,72
ihl-rat LC50:250 ppm/4H AIHAAP 33,338,72
ipr-mus LD50:170 mg/kg ZolH## 23OCT75
skn-rbt LD50:70 mg/kg AIHAAP 33,338,72

CONSENSUS REPORTS: Reported in EPA TSCA Inventory.

SAFETY PROFILE: A poison by ingestion, inhalation, intraperitoneal, and skin contact routes. Mutation data reported. When heated to decomposition it emits acrid smoke and irritant fumes. See also KETONES.

CPD500 CAS:19143-00-1 HR: 3
S-2-((4-CYCLOHEXEN-3-YLBUTYL)AMINO)ETHYL THIOSULFATE
mf: $C_{12}H_{23}NO_3S_2$ mw: 293.48

SYN: 2-((4-CYCLOHEXEN-3-YLBUTYL)AMINO)ETHANETHIOL HYDROGEN SULFATE (ESTER)

TOXICITY DATA WITH REFERENCE
orl-mus LD50:900 mg/kg JMCMAR 11,1190,68
ipr-mus LD50:75 mg/kg JMCMAR 11,1190,68

SAFETY PROFILE: Poison by intraperitoneal route. Moderately toxic by ingestion. See also THIOSULFATES. When heated to decomposition it emits very toxic fumes of NO_x and SO_x.

CPD625 CAS:77251-47-9 HR: 2
1-(2-CYCLOHEXEN-1-YLCARBONYL)-2-METHYLPIPERIDINE
mf: $C_{13}H_{21}NO$ mw: 207.35

SYN: AI3-37220

TOXICITY DATA WITH REFERENCE
skn-rbt 500 mg/24H MLD NTIS** AD-A087-646
eye-rbt 100 mg/24H MLD NTIS** AD-A087-646
orl-rat LDLo:1270 mg/kg NTIS** AD-A087-646

SAFETY PROFILE: Moderately toxic by ingestion. A skin and eye irritant. When heated to decomposition it emits toxic fumes of NO_x.

CPD650 CAS:1820-50-4 HR: 1
3-(3-CYCLOHEXENYL)-2,4-DIOXASPIRO(5.5)UNDEC-8-ENE
mf: $C_{15}H_{22}O_2$ mw: 234.37

SYN: 2,4-DIOXASPIRO(5.5)UNDEC-8-ENE, 3-(3-CYCLOHEXENYL)-

TOXICITY DATA WITH REFERENCE
orl-rat LD50:5190 mg/kg TXAPA9 28,313,74

CONSENSUS REPORTS: Reported in EPA TSCA Inventory.

SAFETY PROFILE: Low toxicity by ingestion. When heated to decomposition it emits acrid smoke and irritating vapors.

CPD750 CAS:100-40-3 HR: 3
CYCLOHEXENYLETHYLENE
mf: C_8H_{12} mw: 108.20

PROP: Liquid. Bp: 128°, fp: −109°, flash p: 60°F (TOC), d: 0.832 @ 20°/4°, autoign temp: 517°F, vap press: 25.8 mm @ 38°, vap d: 3.76.

SYNS: BUTADIENE DIMER □ 4-ETHENYL-1-CYCLOHEXENE □ NCI-C54999 □ 1,2,3,4-TETRAHYDROSTYRENE □ 1-VINYLCYCLOHEXENE-3 □ 1-VINYLCYCLOHEX-3-ENE □ 4-VINYLCYCLOHEXENE □ 4-VINYLCYCLOHEXENE-1 □ 4-VINYL-1-CYCLOHEXENE

TOXICITY DATA WITH REFERENCE
orl-mus TDLo:103 g/kg/2Y-I:CAR,REP JTEHD6 21,507,87
skn-mus TDLo:16 g/kg/54W-I:ETA JNCIAM 31,41,63
orl-mus TD:103 g/kg/2Y-I:NEO,REP NTPTR* NTP-TR-303,86
orl-mus TD:206 g/kg/2Y-I:CAR NTPTR* NTP-TR-303,86
orl-rat LD50:2563 mg/kg AIHAAP 30,470,69
ihl-rat LCLo:8000 ppm/4H AIHAAP 30,470,69
ihl-mus LC50:27,000 mg/m³ IARC** 11,-,76
skn-rbt LD50:16,640 mg/kg AIHAAP 30,470,69

CONSENSUS REPORTS: IARC Cancer Review: Group 3 IMEMDT 7,56,87; Animal Inadequate Evidence IMEMDT 11,277,76; Animal Limited Evidence IMEMDT 39,181,86. NTP Carcinogenesis Studies (gavage); Clear Evidence: mouse NTPTR* NTP-TR-303,86; Inadequate Studies: rat NTPTR* NTP-TR-303,86. Reported in EPA TSCA Inventory.

ACGIH TLV: TWA 0.1 ppm; Suspected Human Carcinogen (Proposed: TWA 0.1 ppm; Animal Carcinogen)

SAFETY PROFILE: Moderately toxic by ingestion and inhalation. Mildly toxic by skin contact. Questionable carcinogen with experimental carcinogenic, neoplastigenic, and tumorigenic data. Experimental reproductive effects. Dangerous fire hazard when exposed to heat, flame, or oxidizers. Can react with oxidizers. To fight fire, use foam, CO_2, dry chemical.

CPE125 CAS:4845-05-0 HR: 3
2-CYCLOHEXENYL HYDROPEROXIDE
mf: $C_6H_{10}O_2$ mw: 114.14

SAFETY PROFILE: A heat-sensitive explosive. When heated to decomposition it emits acrid smoke and fumes. See also PEROXIDES and EXPLOSIVES.

CPE500 CAS:10137-69-6 HR: 2
CYCLOHEXENYL TRICHLOROSILANE
DOT: UN 1762
mf: $C_6H_9Cl_3Si$ mw: 215.59

PROP: Colorless, fuming liquid; HCl odor. Bp: 202°; d: 1.263 @ 25°/25°; flash p: 200°F (COC).

SYNS: CYCLOHEXENE, 4-(TRICHLOROSILYL)- □ CYCLOHEXENYLTRICHLOROSILANE (DOT) □ TRICHLORO-3-CYCLOHEXENYLSILANE

TOXICITY DATA WITH REFERENCE
skn-rbt 5 mg/24H SEV 85JCAE-,1227,86
eye-rbt 250 µg/24H SEV 85JCAE-,1227,86
orl-rat LD50:2830 mg/kg AIHAAP 23,95,62
skn-rbt LD50:630 mg/kg AIHAAP 23,95,62

CONSENSUS REPORTS: Reported in EPA TSCA Inventory.

DOT CLASSIFICATION: 8; *Label:* Corrosive

SAFETY PROFILE: Moderately toxic by ingestion and skin contact. An eye and severe skin irritant. A corrosive material. It fumes in moist air, releasing HCl. Combustible when exposed to heat or flame. When heated to decomposition it emits toxic fumes of Cl^-. See also CHLOROSILANES.

CPE750 CAS:66-81-9 HR: 3
CYCLOHEXIMIDE
mf: $C_{15}H_{23}NO_4$ mw: 281.39

PROP: Crystals. Mp: 119–121°. Moderately sol in water; sol in chloroform, ether, and acetone.

SYNS: ACTI-AID □ ACTIDIONE □ ACTIDIONE TGF □ACTIDONE □ ACTISPRAY □ 3-(2-(3,5-DIMETHYL-2-OXOCYCLOHEXYL)-2-HYDROXYETHYL)GLUTARIMIDE □ HIZAROCIN □ KAKEN □ NARAMYCIN □ NEOCYCLOHEXIMIDE □ NSC-185 □ U-4527

TOXICITY DATA WITH REFERENCE
skn-rbt 5 mg/24H rns TXCYAC 14,117,49
skn-rbt 1%/24H MOD NTIS** PB-274-414
dni-hmn:oth 100 mg/L BBACAQ 696,15,82
dni-hmn:oth 300 nmol/L CNREA8 44,2421,84
ipr-rat TDLo:1 mg/kg (female 18D post):REP BJEPA5 51,361,70
ipr-mus TDLo:30 mg/kg (female 9D post):TER TJADAB 25,345,82
par-rat TDLo:750 µg/kg (female 18D post):TER PCBRD2 140,13,83
ipr-rat TDLo:300 µg/kg (13D preg):TER TJADAB 23,25,81
ipr-rat TDLo:300 µg/kg (13D preg):REP TJADAB 23,25,81
ipr-rat TDLo:1 mg/kg (4D preg):REP INJFA3 27,238,82
ipr-rat TDLo:250 µg/kg (female 10D post):TER FAATDF 4,352,84
ipr-rat TDLo:1 mg/kg (female 15D post):TER BJEPA5 51,361,70
ipr-mus TDLo:30 mg/kg (female 10D post):TER TJADAB 17,41A,78
orl-rat LD50:2 mg/kg UPJOH* 2(6),-,71
ipr-rat LD50:3700 µg/kg JPETAB 136,400,62
scu-rat LD50:2500 µg/kg ANTCAO 10,682,60
ivn-rat LD50:2 mg/kg ANTCAO 10,682,60
orl-mus LD50:133 mg/kg UPJOH* 2(6),-,71
ipr-mus LD50:100 mg/kg CNCRA6 30,9,63
scu-mus LD50:160 mg/kg UPJOH* 2(6),-,71
ivn-mus LD50:150 mg/kg JACSAT 69,474,47
orl-dog LD50:65 mg/kg PCOC** -,292,66
orl-mky LD50:60 mg/kg GUCHAZ 6,146,73

CONSENSUS REPORTS: EPA Extremely Hazardous Substances List. EPA Genetic Toxicology Program.

SAFETY PROFILE: A poison by ingestion, subcutaneous, intraperitoneal and intravenous routes. An experimental teratogen. Other experimental reproductive effects. Human mutation data reported. A skin irritant. A pesticide. When heated to decomposition it emits toxic fumes of NO_x.

CPF000 CAS:622-45-7 HR: 3
CYCLOHEXYL ACETATE
DOT: UN 2243
mf: $C_8H_{14}O_2$ mw: 142.22

PROP: Pale yellow liquid; fruity odor. Bp: 177°, d: 0.996, vap d: 4.9, flash p: 136°F, autoign temp: 633°F.

SYNS: CYCLOHEXANOL ACETATE □ CYCLOHEXANOLAZETAT (GERMAN) □ CYCLOHEXANYL ACETATE □ CYCLOHEXYLESTER KYSELINY OCTOVE

TOXICITY DATA WITH REFERENCE
skn-rbt 100 mg/24H MOD 85JCAE-,359,86
eye-rbt 500 mg/24H MLD 85JCAE-,359,86
ihl-hmn TCLo:3000 mg/m³/45M:IRR AHYGAJ 78,260,13
orl-rat LD50:6730 mg/kg TXAPA9 28,313,74
scu-cat LDLo:606 mg/kg AHYGAJ 78,260,13
skn-rbt LD50:10 g/kg TXAPA9 28,313,74

CONSENSUS REPORTS: Reported in EPA TSCA Inventory.

DOT CLASSIFICATION: 3; *Label:* Flammable Liquid

SAFETY PROFILE: Moderately toxic by subcutaneous route. Mildly toxic by ingestion and skin contact. Human systemic effects by inhalation: conjunctiva irritation and unspecified respiratory system changes. A systemic irritant to humans. A skin and eye irritant. Flammable liquid when exposed to heat or flame. When heated to decomposition it emits acrid smoke and irritating fumes.

CPF500 CAS:108-91-8 HR: 3
CYCLOHEXYLAMINE
DOT: UN 2357
mf: $C_6H_{13}N$ mw: 99.20

PROP: Liquid; strong, fishy odor. Mp: −17.7°, bp: 134.5°, flash p: 69.8°F, d: 0.865 @ 25°/25°, autoign temp: 560°F, vap d. 3.42. Misc in H_2O, organic solvents.

SYNS: AMINOCYCLOHEXANE □ AMINOHEXAHYDROBENZENE □ CHA □ CYCLOHEXANAMINE □ HEXAHYDROANILINE □ HEXAHYDROBENZENAMINE

TOXICITY DATA WITH REFERENCE
skn-hmn 125 mg/48H SEV AMIHBC 5,311,52
cyt-hmn:leu 10 µmol/L/5H MUREAV 39,1,76
cyt-ham:fbr 10 mg/L MUREAV 39,1,76
dni-hmn:hla 100 µg/L INHEAO 9,188,71
orl-mus TDLo:600 mg/kg (female 6-11D post):TER SEIJBO 11,51,71
orl-mus TDLo:120 mg/kg (female 6-11D post):TER SEIJBO 11,51,71
ipr-rat TDLo:300 mg/kg (male 1D pre):REP FCTXAV 10,29,72
orl-mus TDLo:600 mg/kg (female 6-11D post):REP SEIJBO 11,51,71
ipr-rat TDLo:100 mg/kg (male 1D pre):REP FCTXAV 10,29,72
orl-rat TDLo:5600 mg/kg (4W male):REP FCTXAV 19,291,81
orl-dog TDLo:15,750 mg/kg (male 9W pre):REP FCTXAV 19,291,81
orl-rat LD50:156 mg/kg SKEZAP 14,542,73
ihl-rat LC50:7500 mg/m³ GTPZAB 7(11),51,63

orl-mus LD50:224 mg/kg 85GMAT -,41,82
ihl-mus LC50:1070 mg/m³ GTPZAB 7(11),51,63
scu-mus LD50:1150 mg/kg VOONAW 4,659,58
ipr-mus LD50:129 mg/kg PCJOAU 22,469,88
skn-rbt LD50:277 mg/kg AIHAAP 30,470,69
par-rbt LDLo:500 mg/kg IECHAD 29,1247,37
ipr-mam LD50:200 mg/kg AMIHBC 5,311,52

CONSENSUS REPORTS: IARC Cancer Review: Group 3 IMEMDT 7,178,87; Animal Limited Evidence IMEMDT 7,178,87. EPA Extremely Hazardous Substances List. EPA Genetic Toxicology Program. Reported in EPA TSCA Inventory.

OSHA PEL: TWA 10 ppm
ACGIH TLV: TWA 10 ppm
DFG MAK: 10 ppm (40 mg/m³)
DOT CLASSIFICATION: 8; *Label:* Corrosive, Flammable Liquid

SAFETY PROFILE: A poison by ingestion, skin contact, and intraperitoneal routes. Experimental teratogenic and reproductive effects. A severe human skin irritant. Can cause dermatitis and convulsions. Human mutation data reported. Questionable carcinogen. Flammable liquid. Dangerous fire hazard when exposed to heat, flame, or oxidizers. To fight fire, use alcohol foam, CO_2, dry chemical. When heated to decomposition it emits toxic fumes of NO_x.

CPG000 CAS:58695-41-3 HR: 2
CYCLOHEXYLAMINO ACETIC ACID
mf: $C_8H_{15}NO_2$ mw: 157.24

SYNS: CYKLOHEXYLAMINACETAT (CZECH) □ OCTAN CYKLOHEXYLAMINU (CZECH)

TOXICITY DATA WITH REFERENCE
skn-rbt 500 mg/24H SEV 28ZPAK -,64,72
eye-rbt 500 mg/24H MLD 28ZPAK -,64,72
orl-rat LD50:2120 mg/kg 28ZPAK -,64,72

SAFETY PROFILE: Moderately toxic by ingestion. An eye and severe skin irritant. When heated to decomposition it emits toxic fumes of NO_x.

CPG125 CAS:2842-38-8 HR: 1
2-(CYCLOHEXYLAMINO)ETHANOL
mf: $C_8H_{17}NO$ mw: 143.26

SYNS: ABROMEEN E-25 □ N-(2-HYDROXYETHYL)CYCLOHEXYLAMINE

TOXICITY DATA WITH REFERENCE
scu-mus TDLo:900 mg/kg (6-14D preg):TER NTIS** PB223-160
orl-rat LD50:38,300 mg/kg 34ZIAG -,61,69

SAFETY PROFILE: Very mildly toxic by ingestion. An experimental teratogen. An eye irritant. When heated to decomposition it emits toxic fumes of NO_x.

CPG250 CAS:65210-28-8 HR: 3
2-(2-(CYCLOHEXYLAMINO))ETHYL-2-METHYL-1,3-BENZODIOXOLE HYDROCHLORIDE
mf: $C_{16}H_{23}NO_2$•ClH mw: 297.86

TOXICITY DATA WITH REFERENCE
ivn-rat LD50:20 mg/kg EJMCA5 12,413,77
ipr-mus LD50:79 mg/kg EJMCA5 12,413,77

SAFETY PROFILE: Poison by intravenous and intraperitoneal routes. When heated to decomposition it emits very toxic fumes of HCl and NO_x.

CPG500 CAS:57281-35-3 HR: 3
4-(CYCLOHEXYLAMINO)-1-(NAPHTHALENYLOXY)-2-BUTANOL
mf: $C_{20}H_{27}NO_2$ mw: 313.48

SYN: CHINOIN 103

TOXICITY DATA WITH REFERENCE
orl-mus LD50:178 mg/kg DRFUD4 4,12,79
ivn-mus LD50:50 mg/kg DRFUD4 4,12,79

SAFETY PROFILE: Poison by ingestion and intravenous routes. When heated to decomposition it emits toxic fumes of NO_x.

CPG625 CAS:64011-62-7 HR: 3
dl-1-CYCLOHEXYL-2-AMINOPROPANE HYDROCHLORIDE
mf: $C_9H_{19}N$•ClH mw: 177.75

SYN: (±)-α-METHYLCYCLOHEXANEETHYLAMINE HYDROCHLORIDE

TOXICITY DATA WITH REFERENCE
ipr-rat LD50:65 mg/kg JPETAB 100,267,50
ipr-rbt LDLo:100 mg/kg JPETAB 100,267,50
ipr-gpg LDLo:50 mg/kg JPETAB 100,267,50

SAFETY PROFILE: Poison by intraperitoneal route. When heated to decomposition it emits toxic fumes of NO_x and HCl.

CPG700 CAS:103-00-4 HR: 3
1-CYCLOHEXYLAMINO-2-PROPANOL
mf: $C_9H_{19}NO$ mw: 157.29

SYNS: 2-PROPANOL, 1-(CYCLOHEXYLAMINO)- □ USAF DO-19

TOXICITY DATA WITH REFERENCE
ipr-mus LD50:100 mg/kg NTIS** AD277-689

CONSENSUS REPORTS: Reported in EPA TSCA Inventory.

SAFETY PROFILE: Poison by intraperitoneal route. When heated to decomposition it emits toxic vapors of NO_x.

CPH500 CAS:15860-21-6 HR: 2
CYCLOHEXYLAMMONIUM STEARATE
mf: $C_{18}H_{36}O_2$•$C_6H_{13}N$ mw: 383.74

SYN: STEARIC ACID with CYCLOHEXYLAMINE (1:1)

TOXICITY DATA WITH REFERENCE
skn-hmn 500 mg/48H SEV AMIHBC 5,311,52

skn-rbt 500 mg MLD AMIHBC 5,311,52
ipr-mam LD50:4 g/kg AMIHBC 5,311,52

SAFETY PROFILE: Moderately toxic by intraperitoneal route. A severe human skin irritant. When heated to decomposition it emits toxic fumes of NO_x and NH_3. See also STEARIC ACID and CYCLOHEXYLAMINE.

CPI250 CAS:95-33-0 HR: 3
N-CYCLOHEXYL-2-BENZOTHIAZOLESULFENAMIDE
mf: $C_{13}H_{16}N_2S_2$ mw: 264.43

PROP: Light tan or buff powder. Mp: 103–104°, d: 1.27 @ 25°.

SYNS: ACCELERATOR CZ □ ACCICURE HBS □ BENZOTHIAZYL-2-CYCLOHEXYLSULFENAMIDE □ CBS □ CONAC A □ CONAC S □ CURAX □ N-CYCLOHEXYL-2-BENZOTHIAZOLESULFENAMIDE □ N-CYCLOHEXYL-2-BENZOTHIAZYLSULFENAMIDE □ DELAC S □ DURAX □ EKAGOM CBS □ NOCCELER CZ □ PENNAC CBS □ RHODIFAX 16 □ ROYAL CBTS □ SANCELER CM-PO □ SANTOCURE □ SANTOCURE VULCANIZATION ACCELERATOR □ SOXINOL CZ □ SULFENAMIDE TS □ SULFENAX □ SULFENAX CB □ SULFENAX CB 30 □ SULFENAX CB/K □ THIOHEXAM □ VULCAFOR CBS □ VULCAFOR HBS □ VULKACIT C □ VULKACIT CZ □ VULKACIT CZ/C □ VULKACIT CZ/K

TOXICITY DATA WITH REFERENCE
par-rat TDLo:400 mg/kg (4-11D preg):TER BEXBAN 93,107,82
par-rat TDLo:400 mg/kg (4-11D preg):REP BEXBAN 93,107,82
orl-mus TDLo:76 g/kg/78W-I:ETA NTIS** PB223-159
orl-rat LD50:5300 mg/kg JACTDZ 1,105,90
ivn-mus LD50:32 mg/kg CSLNX* NX#02243

CONSENSUS REPORTS: Reported in EPA TSCA Inventory.

SAFETY PROFILE: A poison by intravenous route. Questionable carcinogen with experimental tumorigenic data. An experimental teratogen. Experimental reproductive effects. When heated to decomposition it emits toxic fumes of SO_x and NO_x.

CPI300 CAS:1551-44-6 HR: 1
CYCLOHEXYL BUTYRATE
mf: $C_{10}H_{18}O_2$ mw: 170.28

SYNS: BUTANOIC ACID, CYCLOHEXYL ESTER (9CI) □ BUTYRIC ACID, CYCLOHEXYL ESTER □ CYCLOHEXYL BUTANOATE

TOXICITY DATA WITH REFERENCE
orl-rat LDLo:5 g/kg FCTOD7 26,293,88
skn-gpg LD50:>5 g/kg FCTOD7 26,293,88

CONSENSUS REPORTS: Reported in EPA TSCA Inventory.

SAFETY PROFILE: Low toxicity by ingestion and skin contact. When heated to decomposition it emits acrid smoke and irritating vapors.

CPI350 CAS:63441-20-3 HR: 1
1-(CYCLOHEXYLCARBONYL)-3-METHYLPIPERIDINE
mf: $C_{13}H_{23}NO$ mw: 209.37

SYNS: AI3-36537 □ PIPERIDINE, 1-(CYCLOHEXYLCARBONYL)-3-METHYL-

TOXICITY DATA WITH REFERENCE
eye-rbt 100 mg/24H MOD AEHA** 51-029-76

SAFETY PROFILE: An eye irritant. When heated to decomposition it emits toxic fumes of NO_x.

CPI375 CAS:32921-23-6 HR: 3
4-(CYCLOHEXYLCARBONYL)PYRIDINE
mf: $C_{12}H_{15}NO$ mw: 189.28

SYNS: CYCLOHEXYL 4-PYRIDYL KETONE □ KETONE, CYCLOHEXYL 4-PYRIDYL □ PYRIDINE, 4-(CYCLOHEXYLCARBONYL)-

TOXICITY DATA WITH REFERENCE
ipr-mus LD50:175 mg/kg JMCMAR 14,551,71

DOT CLASSIFICATION: 3; *Label:* Flammable Liquid

SAFETY PROFILE: A poison by intraperitoneal route. A flammable liquid. When heated to decomposition it emits toxic vapors of NO_x.

CPI400 CAS:542-18-7 HR: 1
CYCLOHEXYL CHLORIDE
mf: $C_6H_{11}Cl$ mw: 118.62

SYNS: CHLOROCYCLOHEXANE □ CYCLOHEXANE, CHLORO- □ MONOCHLOROCYCLOHEXANE

TOXICITY DATA WITH REFERENCE
ihl-rat LCLo:31 g/m³ GTPZAB 10(1),49,66
ihl-mus LC50:31 g/m³/2H 85GMAT-,36,82

CONSENSUS REPORTS: Reported in EPA TSCA Inventory.

SAFETY PROFILE: Slightly toxic by inhalation. When heated to decomposition it emits toxic vapors of Cl^-.

CPJ000 CAS:32808-51-8 HR: 3
4-(4-CYCLOHEXYL-3-CHLOROPHENYL)-4-OXOBUTYRIC ACID
mf: $C_{16}H_{19}ClO_3$ mw: 294.80

PROP: A solid. Mp: 90–92°

SYNS: l'ACIDE BUCLOXIQUE (FRENCH) □ BENZENEBUTANOIC ACID, 3-CHLORO-4-CYCLOHEXYL-α-OXO- □ BUCLOSINSAEURE (GERMAN) □ BUCLOXIC ACID □ BUCLOXONIC ACID □ 804 CB □ 3-(3-CHLORO-4-CYCLOHEXYLBENZOYL)PROPIONIC ACID □ 3-CHLORO-4-CYCLOHEXYL-α-OXOBENZENEBUTANOIC ACID □ 4-(3-CHLORO-4-CYCLOHEXYLPHENYL)-4-OXO-BUTYRIC ACID □ ESFAR

TOXICITY DATA WITH REFERENCE
orl-mus TDLo:975 mg/kg (6-18D preg):TER ARZNAD 24,1398,74
orl-rbt TDLo:550 mg/kg (7-28D preg):REP ARZNAD 24,1398,74
orl-rat TDLo:3600 mg/kg (72D pre):REP ARZNAD 24,1398,74
orl-rat LD50:120 mg/kg ARZNAD 24,1364,74
ipr-rat LD50:195 mg/kg ARZNAD 24,1398,74

orl-mus LD50:852 mg/kg ARZNAD 24,1398,74
ipr-mus LD50:1100 mg/kg ARZNAD 24,1364,74

SAFETY PROFILE: Poison by intraperitoneal and ingestion routes. Experimental teratogenic and reproductive effects. An anti-inflammatory agent. When heated to decomposition it emits toxic fumes of Cl^-.

CPJ250 CAS:32808-53-0 HR: 3
4-(4-CYCLOHEXYL-3-CHLOROPHENYL)-4-OXOBUTYRIC ACID CALCIUM SALT
mf: $C_{32}H_{36}Cl_2O_6$•Ca mw: 627.66

SYNS: ACIDE BUCLOXIQUE CALCIUM (FRENCH) □ l'ACIDE (CYCLOHEXYL-4, CHLORO-3, PHENYL)-4,OXO-4, BUTYRIQUE CALCIUM (FRENCH) □ BUCLOXIC ACID CALCIUM □ BUCLOXIC ACID CALCIUM SALT □ BUCLOXINSAEURE KALZIUM (FERMAN) □ BUCLOXONIC ACID CALCIUM SALT □ CALCIUM BUCLOXATE □ CALCIUM ESFAR □ CB 804 CALCIUM □ 4-(3-CHLOR-4-CYCLOHEXYL-PHENYL)-4-OXO-BUTTERSAEURE KALZIUM (GERMAN) □ 3-(3-CHLORO-4-CYCLOHEXYLBENZOYL)PROPIONIC ACID CALCIUM SALT □ 3-CHLORO-4-CYCLOHEXYL-α-OXO-BENZENEBUTANOIC ACID □ 3-CHLORO-4-CYCLOHEXYL-α-OXOBENZENEBUTANOIC ACID CALCIUM SALT □ 4-(3-CHLORO-4-CYCLOHEXYLPHENYL)-4-OXOBUTYRIC ACID CALCIUM SALT □ ESFAR CALCIUM

TOXICITY DATA WITH REFERENCE
orl-rat LD50:175 mg/kg ARZNAD 24,1364,74
ipr-rat LD50:200 mg/kg ARZNAD 24,1364,74
orl-mus LD50:1700 mg/kg ARZNAD 24,1360,74
ipr-mus LD50:1700 mg/kg ARZNAD 24,1364,74

SAFETY PROFILE: Poison by ingestion and intraperitoneal routes. Used as an anti-inflammatory agent. When heated to decomposition it emits toxic fumes of Cl^-.

CPJ500 CAS:92-64-8 HR: 3
CYCLOHEXYLCYANOETHYLETHANOLAMINE
mf: $C_{11}H_{14}N_2O$ mw: 190.27

SYNS: 2-(N-(2-CYANOETHYL)-N-CYCLOHEXYL)AMINO-ETHANOL □ N-(β-CYANOETHYL)-N-(β-HYDROXYETHY)-ANILINE □ N-β-HYDROXYETHYL-N-β-KYANETHYLANILIN (CZECH)

TOXICITY DATA WITH REFERENCE
eye-rbt 500 mg/24H MLD 28ZPAK -,162,73
orl-rat LD50:3210 mg/kg GISAAA 53(10),92,88
orl-mus LD50:1450 mg/kg GISAAA 53(10),92,88
ipr-mus LD50:200 mg/kg NTIS** AD691-490

CONSENSUS REPORTS: Cyanide and its compounds are on the Community Right-To-Know List.

SAFETY PROFILE: Poison by intraperitoneal route. Moderately toxic by ingestion. An eye irritant. When heated to decomposition it emits toxic fumes of NO_x and CN^-. See also NITRILES.

CPK000 HR: 2
N-CYCLOHEXYL-N-DIETHYLTHIOCARBONYL SULFONAMIDE

SYN: THIOPENTEX

TOXICITY DATA WITH REFERENCE
skn-hmn 250 mg/48H MOD AMIHBC 5,311,52
skn-rbt 500 mg MLD AMIHBC 5,311,52
ipr-mam LD50:1200 mg/kg AMIHBC 5,311,52

SAFETY PROFILE: Moderately toxic by intraperitoneal route. A human skin irritant. When heated to decomposition it emits very toxic fumes of SO_x.

C

CPK500 CAS:131-89-5 HR: 3
2-CYCLOHEXYL-4,6-DINITROPHENOL
mf: $C_{12}H_{14}N_2O_5$ mw: 266.28

PROP: Crystals. Mp: 104°

SYNS: 6-CICLOESIL-2,4-DINITRO-FENOLO (ITALIAN) □ 2-CYCLOHEXYL-4,6-DINITROFENOL (DUTCH) □ 6-CYCLOHEXYL-2,4-DINITROPHENOL □ DINEX □ DINITROCYCLOHEXYLPHENOL □ DINITRO-o-CYCLOHEXYLPHENOL □ 2,4-DINITRO-6-CYCLOHEXYLPHENOL □ 4,6-DINITRO-o-CYCLOHEXYLPHENOL □ DINITROCYCLOHEXYLPHENOL (DOT) □ DN DRY MIX No. 1 □ DN DUST No. 12 □ DNOCHP □ DOWSPRAY 17 □ DRY MIX No. 1 □ ENT 157 □ PEDINEX (FRENCH) □ RCRA WASTE NUMBER P034 □ SN 46

TOXICITY DATA WITH REFERENCE
skn-rbt 105 mg/9D-I MOD JIHTAB 30,10,48
orl-rat LD50:65 mg/kg ARSIM* 20,9,66
orl-mus LD50:50 mg/kg 85DPAN -,-,71/76
ipr-mus LD50:25 mg/kg BCPCA6 18,1389,69
scu-mus LDLo:30 mg/kg UCPHAQ 1,151,39
ivn-dog LDLo:8 mg/kg AIPTAK 50,20,35
orl-rbt LDLo:100 mg/kg UCPHAQ 1,151,39
scu-rbt LDLo:40 mg/kg UCPHAQ 1,151,39
orl-gpg LD50:50 mg/kg PCOC** -,417,66
skn-gpg LDLo:1000 mg/kg PCOC** -,417,66
scu-gpg LDLo:20 mg/kg PCOC** -,417,66

SAFETY PROFILE: A poison by ingestion, intraperitoneal, intravenous, subcutaneous, and possibly other routes. Moderately toxic by skin contact. A skin irritant. Fire hazard. See also NITRATES and PHENOLS. Can react with oxidizers. When heated to decomposition it emits toxic fumes of NO_x.

CPK625 CAS:52694-54-9 HR: 3
(+)-1-CYCLOHEXYL-4-(1,2-DIPHENYLETHYL)PIPERAZINE DIHYDROCHLORIDE
mf: $C_{24}H_{32}N_3$•2ClH mw: 421.50

SYN: (S)-1-CYCLOHEXYL-4-(1,2-DIPHENYLETHYL)-PIPERAZINE DIHYDROCHLORIDE

TOXICITY DATA WITH REFERENCE
ivn-rat LD50:8 mg/kg AIPTAK 221,105,76
orl-mus LD50:274 mg/kg AIPTAK 221,105,76
scu-mus LD50:320 mg/kg AIPTAK 221,105,76
ivn-mus LD50:18,500 μg/kg AIPTAK 221,105,76

SAFETY PROFILE: Poison by ingestion, subcutaneous, and intravenous routes. When heated to decomposition it emits toxic fumes of NO_x and HCl.

CPL100 CAS:10328-51-5 HR: 3
N,N′-(1,4-CYCLOHEXYLENEDIMETHYLENE)BIS(2-(1-AZIRIDINYL)ACETAMIDE)
mf: $C_{16}H_{28}N_4O_2$ mw: 308.48

SYNS: ACETAMIDE, N,N′-(1,4-CYCLOHEXYLENEDIMETHYLENE)BIS (2-(1-AZIRIDINYL)- □ 1-AZIRIDINEACETAMIDE, N,N′-(1,4-CYCLOHEX-

YLENEDIMETHYLENE)BIS- □ N,N'-BIS-AZIRIDINYLACETYL-1,4-CYCLOHEXYLDIMETHYLENEDIAMINE

TOXICITY DATA WITH REFERENCE
cyt-rat-orl 300 μg/kg MUREAV 31,115,75
ipr-mus TDLo:50 mg/kg (male 5D pre):REP EXPEAM 24,924,68
orl-mus LD50:71 mg/kg EXPEAM 24,924,68
ipr-mus LD50:45 mg/kg EXPEAM 24,924,68

SAFETY PROFILE: Poison by ingestion and intraperitoneal routes. Experimental reproductive effects. Mutation data reported. When heated to decomposition it emits toxic fumes of NO_x.

CPL250 CAS:4442-79-9 HR: 2
2-CYCLOHEXYLETHANOL
mf: $C_8H_{16}O$ mw: 128.24

SYNS: CYCLOHEXYLETHYL ALCOHOL □ HEXAHYDROPHENYLETHYL ALCOHOL

TOXICITY DATA WITH REFERENCE
orl-rat LD50:940 mg/kg FCTXAV 13,785,75
skn-rbt LD50:1220 mg/kg FCTXAV 13,785,75

CONSENSUS REPORTS: Reported in EPA TSCA Inventory.

SAFETY PROFILE: Moderately toxic by ingestion and skin contact. When heated to decomposition it emits acrid smoke and irritating fumes. See also ALCOHOLS.

CPL750 CAS:13908-93-5 HR: 3
CYCLOHEXYL FLUOROETHYL NITROSOUREA
mf: $C_9H_{16}FN_3O_2$ mw: 217.28

SYNS: CFNU □ 3-CYCLOHEXYL-1-(2-FLUOROETHYL)-1-NITROSOUREA □ N'-CYCLOHEXYL-N-(2-FLUOROETHYL)-N-NITROSOUREA □ FCNU □ 1-FLUOROETHYL-3-CYCLOHEXYL-1-NITROSOUREA □ NSC 87974 □ SRI 2619

TOXICITY DATA WITH REFERENCE
orl-rat LD50:18,500 μg/kg TXAPA9 10,397,67
ivn-rat LD50:12 mg/kg TXAPA9 10,397,67
orl-mus LD50:111 mg/kg TXAPA9 10,397,67
ipr-mus LD10:34 mg/kg CNREA8 34,194,74
scu-mus LD50:25,210 μg/kg NCISP* JAN86
ivn-mus LD50:51 mg/kg TXAPA9 10,397,67

SAFETY PROFILE: Poison by ingestion, subcutaneous, intravenous, and intraperitoneal routes. Many N-nitroso compounds are carcinogens. When heated to decomposition it emits very toxic fumes of F^- and NO_x. See also N-NITROSO COMPOUNDS.

CPM250 CAS:78128-81-1 HR: 3
3-CYCLOHEXYL-4-HYDROXY-2(5H)FURANONE
mf: $C_{10}H_{14}O_3$ mw: 182.24

SYN: α-CYCLOHEXYL-β-HYDROXY-$\Delta^{\alpha,\beta}$-BUTENOLID (GERMAN)

TOXICITY DATA WITH REFERENCE
scu-mus LD50:416 mg/kg ARZNAD 11,277,61
ivn-mus LD50:155 mg/kg ARZNAD 11,277,61

SAFETY PROFILE: Poison by intravenous route. Moderately toxic by subcutaneous route. When heated to decomposition it emits acrid smoke and irritating fumes. See also KETONES.

CPM750 CAS:6856-43-5 HR: 3
1-(3-CYCLOHEXYL-3-HYDROXY-3-PHENYLPROPYL)-1-METHYL-PIPERIDINIUM IODIDE
mf: $C_{20}H_{31}NO•CH_3I$ mw: 443.46

SYNS: 1-CYCLOHEXYL-1-PHENYL-3-PIPERIDINO-PROPANOL, METHYLIODIDE □ α-CYCLOHEXYL-α-(2-(PIPERIDINO)ETHYL)-BENZYLALCOHOL METHYLIODIDE □ WIN 1593

TOXICITY DATA WITH REFERENCE
orl-mus LD50:2520 mg/kg JPETAB 110,282,54
ivn-mus LD50:12 mg/kg JPETAB 110,282,54

SAFETY PROFILE: Poison by intravenous route. Moderately toxic by ingestion. See also IODIDES. When heated to decomposition it emits very toxic fumes of I^- and NO_x.

CPN500 CAS:3173-53-3 HR: 3
CYCLOHEXYL ISOCYANATE
DOT: UN 2488
mf: $C_7H_{11}NO$ mw: 125.19

PROP: Oil. Mp: 168–170°.

SYNS: CYCLOHEXANE, ISOCYANATO-(9CI) □ ISOCYANATOCYCLOHEXANE □ ISOCYANIC ACID, CYCLOHEXYL ESTER □ NSC 87419

TOXICITY DATA WITH REFERENCE
mmo-sat 150 μg/plate ABCHA6 44,3017,80
ipr-mus LD50:13 mg/kg NCISP* JAN86
ivn-mus LD50:18 mg/kg CSLNX* NX#04502

CONSENSUS REPORTS: Reported in EPA TSCA Inventory.

DOT CLASSIFICATION: 6.1; *Label:* Poison; DOT Class: 6.1; *Label:* Poison, Flammable Liquid; DOT Class: 3; *Label:* Flammable Liquid, Poison

SAFETY PROFILE: Poison by intravenous and intraperitoneal routes. Mutation data reported. A flammable liquid when exposed to heat or flame. When heated to decomposition it emits toxic fumes of NO_x. See also CYANATES and ESTERS.

CPN750 CAS:3687-61-4 HR: 3
2-(N-CYCLOHEXYL-N-ISOPROPYLAMINOMETHYL)-1,3,4-OXADIAZOLE
mf: $C_{15}H_{21}N_3O_2$ mw: 275.39

PROP: Bp: 148° @ 0.1.

SYNS: AF 594 □ 5-(2-(DIETHYLAMINO)ETHYL)-3-(p-METHOXYPHENYL)-1,2,4-OXADIAZOLE □ N,N-DIETHYL-3-(4-METHOXYPHENYL)-1,2,4-OXADIAZOLE-5-ETHANAMINE □ 3-p-METHOXYPHENYL-5-DIETHYLAMINOETHYL-1,2,4-OXADIAZOLE □ MEXOLAMINE □ R 1067

TOXICITY DATA WITH REFERENCE
orl-rat LD50:1899 mg/kg ARZNAD 12,539,62
ipr-rat LD50:288 mg/kg ARZNAD 12,539,62
orl-mus LD50:722 mg/kg ARZNAD 12,539,62
ipr-mus LD50:331 mg/kg ARZNAD 12,539,62
scu-mus LD50:691 mg/kg ARZNAD 12,539,62

ivn-mus LD50:83 mg/kg ARZNAD 12,539,62
orl-gpg LD50:700 mg/kg ARZNAD 12,539,62
ipr-gpg LD50:233 mg/kg ARZNAD 12,539,62

SAFETY PROFILE: Poison by intraperitoneal and intravenous routes. Moderately toxic by ingestion and subcutaneous routes. When heated to decomposition it emits toxic fumes of NO_x.

CPO500 CAS:4388-82-3 HR: 3
1,1-CYCLOHEXYL-2-METHYLAMINOPROPANE-5,5-PHENYLETHYLBARBITURATE
mf: $C_{12}H_{12}N_2O_3{\cdot}C_{10}H_{21}N$ mw: 387.58

SYNS: BARBEXACLONE □ BARBEXACLONUM □ BARBITURIC ACID, 1-(1-(1-CYCLOHEXYL-N-METHYL-2-PROPANAMINE)-5-ETHYL-5-PHENYL □ CHP-PHENOBARBITALAT (GERMAN) □ MALIASIN

TOXICITY DATA WITH REFERENCE
orl-rat LD50:306 mg/kg ARZNAD 13,613,63
orl-mus LD50:334 mg/kg ARZNAD 13,613,63

SAFETY PROFILE: Poison by ingestion. Human reproductive effects. When heated to decomposition it emits toxic fumes of NO_x. See also BARBITURATES.

CPP000 CAS:59182-63-7 HR: 3
N-(2-CYCLOHEXYL-1-METHYLETHYL)-3,3-DIPHENYLPROPYLAMINE HYDROCHLORIDE
mf: $C_{24}H_{33}N{\cdot}ClH$ mw: 372.04

PROP: Crystals from 2-propanol. Mp: 175–176°.

SYNS: DROPRENILAMINE HYDROCHLORIDE □ MG 8926 □ VALCOR

TOXICITY DATA WITH REFERENCE
orl-rat LD50:1550 mg/kg ARZNAD 26,2127,76
ipr-rat LD50:65 mg/kg ARZNAD 26,2127,76
orl-mus LD50:2850 mg/kg ARZNAD 26,2127,76
ipr-mus LD50:68 mg/kg ARZNAD 26,2127,76

SAFETY PROFILE: Poison by intraperitoneal route. Moderately toxic by ingestion. When heated to decomposition it emits very toxic fumes of HCl and NO_x.

CPP750 CAS:70907-61-8 HR: 3
1-(2-CYCLOHEXYLPHENOXY)-1-(2-IMIDAZOLINYL) ETHANE HYDROCHLORIDE
mf: $C_{17}H_{24}N_2O{\cdot}ClH$ mw: 308.89

SYN: MG 18512

TOXICITY DATA WITH REFERENCE
orl-mus LD50:1650 μg/kg ARZNAD 29,729,79
ipr-mus LD50:300 μg/kg ARZNAD 29,729,79

SAFETY PROFILE: A deadly poison by ingestion and intraperitoneal routes. When heated to decomposition it emits very toxic fumes of HCl and NO_x.

CPQ250 CAS:77-37-2 HR: 3
1-CYCLOHEXYL-1-PHENYL-3-PYRROLIDINO-1-PROPANOL
mf: $C_{19}H_{29}NO$ mw: 287.49

SYNS: 1-CYCLOHEXYL-1-PHENYL-3-(1-PYRROLIDINYL)-1-PROPANOL □ ELORINE □ KEMADRINE □ LERGINE □ METANIN □ OSNERVAN □ PROCIDLIDINA □ PROCYCLIDINE □ PROCYKLIDIN □ PROSYKLIDIN □ SPAMOL □ TRICICLIDINA □ TRICILOID □ TRICOLOID □ TRICYCLAMOL □ VAGOSIN

TOXICITY DATA WITH REFERENCE
ipr-mus LD50:131 mg/kg 27ZQAG -,291,72
ivn-mus LD50:60 mg/kg 27ZQAG -,291,72

SAFETY PROFILE: A poison by intraperitoneal and intravenous routes. When heated to decomposition it emits toxic fumes of NO_x.

CPQ275 CAS:6837-24-7 HR: 3
1-CYCLOHEXYL-2-PYRROLIDINONE
mf: $C_{10}H_{17}NO$ mw: 167.28

SYNS: N-CYCLOHEXYLPYRROLIDINONE □ N-CYCLOHEXYLPYRROLIDONE □ 2-PYRROLIDINONE, 1-CYCLOHEXYL-

TOXICITY DATA WITH REFERENCE
skn-rbt 500 mg/24H SEV FCTOD7 26,475,88
eye-rbt 100 mg SEV FCTOD7 26,475,88
orl-rat LD50:370 mg/kg FCTOD7 26,475,88
ihl-rat LC50:120 ppm/1H FAATDF 4,587,84
orl-rbt LD50:657 mg/kg FAATDF 4,587,84
skn-rbt LD50:1600 mg/kg FAATDF 4,587,84

CONSENSUS REPORTS: Reported in EPA TSCA Inventory.

SAFETY PROFILE: Poison by ingestion. Moderately toxic by inhalation and skin contact. A severe skin and eye irritant. When heated to decomposition it emits toxic fumes of NO_x.

CPQ625 CAS:100-88-9 HR: 3
N-CYCLOHEXYLSULPHAMIC ACID
mf: $C_6H_{13}NO_3S$ mw: 179.26

PROP: Crystals; sweet-sour taste. Mp: 169–170°. Fairly strong acid. Very sparingly soluble in water. Slowly hydrolyzed by hot water.

SYNS: CYCLAMATE □ CYCLAMIC ACID □ CYCLOHEXANESULPHAMIC ACID □ CYCLOHEXYLAMIDOSULPHURIC ACID □ CYCLOHEXYLAMINESULPHONIC ACID □ CYCLOHEXYLSULFAMIC ACID (9CI) □ CYCLOHEXYLSULPHAMIC ACID □ HEXAMIC ACID □ SUCARYL □ SUCARYL ACID

TOXICITY DATA WITH REFERENCE
orl-man TDLo:22 g/kg/77W-C:CAR,KID JOURAA 118,258,77
orl-man TD:131 g/kg/5Y-C:CAR,KID JOURAA 118,258,77
orl-man TD:164 g/kg/6Y-C:CAR,KID JOURAA 118,258,77
orl-rat LD50:12 g/kg AJMSA9 225,551,53
ivn-rat LD50:4 g/kg AJMSA9 225,551,53
orl-mus LD50:10 g/kg AJMSA9 225,551,53
ivn-mus LD50:180 mg/kg CSLNX* NX#01774

CONSENSUS REPORTS: Reported in EPA TSCA Inventory.

SAFETY PROFILE: Suspected human carcinogen producing bladder tumors. Poison by intravenous route. Mildly toxic by ingestion. When heated to decomposition it emits toxic fumes of SO_x and NO_x.

CPQ650 CAS:29396-39-2 HR: 3
3-CYCLOHEXYLSYDNONE IMINE MONOHYDRO-CHLORIDE
mf: $C_8H_{13}N_3O{\bullet}ClH$ mw: 203.70

SYN: N-(ZYKLOHEXYL)-SYDNONIMIN HYDROCHLORID (GERMAN)

TOXICITY DATA WITH REFERENCE
orl-mus LD50:206 mg/kg ABMGAJ 14,369,65
ipr-mus LD50:63 mg/kg OYYAA2 2,280,68
ivn-mus LD50:70 mg/kg JMCMAR 14,1013,71

SAFETY PROFILE: Poison by ingestion, intravenous and intraperitoneal routes. When heated to decomposition it emits toxic fumes of NO_x and HCl.

CPQ700 CAS:17796-82-6 HR: 2
N-(CYCLOHEXYLTHIO)PHTHALIMIDE
mf: $C_{14}H_{15}NO_2S$ mw: 261.36

SYNS: N-CYCLOHEXYLSULFENYLPHTHALIMIDE □ 1H-ISOINDOLE-1,3(2H)-DIONE, 2-(CYCLOHEXYLTHIO)- □ PHTHALIMIDE, N-(CYCLOHEXYLTHIO)- □ SANTOGARD PVI

TOXICITY DATA WITH REFERENCE
orl-rat TDLo:405 mg/kg (female 1-22D post):REP GISAAA 53(10),90,88
orl-rat TDLo:365 g/kg/2Y-C:NEO EPASR* 8EHQ-0786-0681
orl-rat LD50:2600 mg/kg EPASR* 8EHQ-0786-0681
orl-mus LD50:5100 mg/kg GISAAA 52(3),70,87
skn-rbt LD50:>5 g/kg EPASR* 8EHQ-0786-0681

CONSENSUS REPORTS: Reported in EPA TSCA Inventory.

SAFETY PROFILE: Moderately toxic by ingestion. Questionable carcinogen with experimental neoplastigenic data. Experimental reproductive effects. When heated to decomposition it emits toxic vapors of NO_x and SO_x

CPR000 CAS:664-95-9 HR: 2
1-CYCLOHEXYL-3-p-TOLYSULFONYLUREA
mf: $C_{14}H_{20}N_2O_3S$ mw: 296.42

PROP: Fine, white crystals. Mp: 174–176°.

SYNS: 1-CICLOESIL-3-p-TOLILSOLFONILUREA (ITALIAN) □ CYCHLORAL □ CYCLAMID □ CYCLAMIDE □ 1-CYCLOHEXYL-3-p-TOLUENESULFONYLUREA □ DIABORAL □ GLICOSIL □ GLYCYCLAMIDE □ N-(4-METHYLBENZENESULFONYL)-N'-CYCLOHEXYLUREA □ TOLCYCLAMIDE □ TOLHEXAMIDE □ 1-(p-TOLYLSULFONYL)-3-CYCLOHEXYLUREA

TOXICITY DATA WITH REFERENCE
orl-rat TDLo:800 mg/kg (9D preg):TER FATOAO 28,616,65
ipr-rat LD50:870 mg/kg FRPSAX 12,268,57
ipr-mus LD50:1150 mg/kg RPOBAR 2,280,70

SAFETY PROFILE: Moderately toxic by intraperitoneal route. An experimental teratogen. When heated to decomposition it emits toxic fumes of SO_x and NO_x.

CPR250 CAS:98-12-4 HR: 3
CYCLOHEXYLTRICHLOROSILANE
DOT: UN 1763
mf: $C_6H_{11}Cl_3Si$ mw: 217.61

PROP: A liquid. Bp: 198.6–200°.

SYNS: CYCLOHEXANE, 1-(TRICHLOROSILYL)- □ SILANE, TRICHLOROCYCLOHEXYL-

CONSENSUS REPORTS: Reported in EPA TSCA Inventory.

DOT CLASSIFICATION: 8; *Label:* Corrosive

SAFETY PROFILE: A highly toxic and corrosive material. When heated to decomposition it emits toxic fumes of Cl^-. See also CHLOROSILANES.

CPR500 CAS:16607-80-0 HR: 3
1-CYCLOHEXYLTRIMETHYLAMINE
mf: $C_9H_{19}N$ mw: 141.29

PROP: bp: 76° @ 29 mm.

SYN: N,N-DIMETHYL-N-CYCLOHEXYLMETHYLAMINE

TOXICITY DATA WITH REFERENCE
orl-rat LD50:1230 mg/kg AIHAAP 30,470,69
skn-rbt LD50:210 mg/kg AIHAAP 30,470,69

SAFETY PROFILE: Poison by skin contact. Moderately toxic by ingestion. See also AMINES. When heated to decomposition it emits toxic fumes of NO_x.

CPR750 CAS:742-20-1 HR: 3
CYCLOMETHIAZIDE
mf: $C_{13}H_{18}ClN_3O_4S_2$ mw: 379.91

SYNS: CYCLOPENTHIAZIDE □ 3-CYCLOPENTYLMETHYL HYDROCHLOROTHIAZIDE DERIV □ NAVIDREX □ NAVIDRIX □ SALIMED □ SALIMID □ SU 8341 □ TSIKLOMETIAZID

TOXICITY DATA WITH REFERENCE
ivn-rat LD50:142 mg/kg AIPTAK 131,325,61
ivn-mus LD50:232 mg/kg MEIEDD 10,394,83

SAFETY PROFILE: Poison by intravenous route. An antihypertensive agent. When heated to decomposition it emits very toxic fumes of SO_x, NO_x, and Cl^-.

CPR800 CAS:121-82-4 HR: 3
CYCLONITE
DOT: UN 0072/UN 0118/UN 0391/UN 0483
mf: $C_3H_6N_6O_6$ mw: 222.15

PROP: White, crystalline powder. Mp: 202°.

SYNS: CYCLONITE, desensitized (UN 0483) (DOT) □ CYCLONITE, wetted (UN 0072) (DOT) □ CYCLOTRIMETHYLENENITRAMINE □ CYCLOTRIMETHYLENETRINITRAMINE □ CYCLOTRIMETHYLENETRINITRAMINE, desensitized (UN 0483) (DOT) □ CYCLOTRIMETHYLENETRINITRAMINE, wetted (UN 0072) (DOT) □ CYKLONIT □ ESAIDRO-1,3,5-TRINITRO-1,3,5-TRIAZINA (ITALIAN) □ HEKSOGEN (POLISH) □ HEXAHYDRO-1,3,5-TRINITRO-1,3,5-TRIAZIN (GERMAN) □ HEXAHYDRO-1,3,5-TRINITRO-1,3,5-TRIAZIN □ HEXAHYDRO-1,3,5-TRINITRO-1,3,5-TRIAZINE □ HEXOGEEN (DUTCH) □ HEXOGEN □ HEXOGEN, desensitized (UN 0483) (DOT) □ HEXOGEN (Explosive) □ HEXOGEN 5W □ HEXOGEN, wetted (UN 0072) (DOT) □ HEXOLITE

☐ HEXOLITE, dry or wetted with <15% water, by weight (UN 0118) (DOT) ☐ PBX(AF) 108 ☐ PBXW 108(E) ☐ RDX ☐ RDX, desensitized (UN 0483) (DOT) ☐ RDX and HMX MIXTURES, desensitized with not <10% phleg matizer by weight (UN 0391) (DOT) ☐ RDX, wetted with not <15% water by weight (UN 0072) (DOT) ☐ RDX and HMX MIXTURES, wetted with not <15% water by weight (UN 0391) (DOT) ☐ T4 ☐ 1,3,5-TRIAZINE, HEXAHYDRO-1,3,5-TRINITRO-(9CI) ☐ TRIMETHYLEENTRINITRAMINE (DUTCH) ☐ TRIMETHYLENETRINITRAMINE ☐ sym-TRIMETHYLENETRINITRAMINE ☐ TRINITROCYCLOTRIMETHYLENE TRIAMINE ☐ 1,3,5-TRINITRO-1,3,5-TRIAZACYCLOHEXANE

TOXICITY DATA WITH REFERENCE
orl-rat TDLo:3 g/kg (13W male/13W pre-22D preg):REP NTIS** AD-A092-531
orl-rat TDLo:10 g/kg (13W male/13W pre-22D preg):REP NTIS** AD-A092-531
orl-rat TDLo:20 mg/kg (female 6-15D post):TER NTIS** AD-A166-249
orl-cld TDLo: 85 mg/kg JTCTDW 24,305,86
orl-rat LD50:100 mg/kg TXAPA9 39,531,77
ipr-rat LDLo:10 mg/kg EATR** EB-TR-73040
ivn-rat LDLo:18 mg/kg EATR** EB-TR-73040
orl-mus LD50:59 mg/kg NTIS** AD-A092-531
ivn-mus LD50:19 mg/kg EATR** EB-TR-73040
orl-cat LDLo:100 mg/kg FATOAO 7,43,44
orl-rbt LDLo:500 mg/kg FATOAO 7,43,44
ivn-gpg LD50:25 mg/kg EATR** EB-TR-73040

CONSENSUS REPORTS: Reported in EPA TSCA Inventory.

OSHA PEL: TWA 1.5 mg/m^3 (skin)
ACGIH TLV: TWA 1.5 mg/m^3 (skin)
DOT CLASSIFICATION: EXPLOSIVE 1.1D; *Label:* EXPLOSIVE 1.1D

SAFETY PROFILE: Poison by ingestion, intraperitoneal, and intravenous routes. An experimental teratogen. Other experimental reproductive effects. A corrosive irritant to skin, eyes, and mucous membranes. Cases of epileptiform convulsions have been reported from exposure. It is one of the most powerful high explosives in use today. Has more shattering power than TNT and is often mixed with TNT as a bursting charge for aerial bombs, mines, and torpedoes. It is easily initiated by mercury fulminate, which may be used as a booster. When heated to decomposition it emits toxic fumes of NO_x. See also AMINES, NITRATES, and EXPLOSIVES, HIGH.

CPR825 CAS:1552-12-1 HR: 2
cis,cis-CYCLOOCTA-1,5-DIENE
mf: C_8H_{12} mw: 108.20

PROP: Bp: 151–152°

SYNS: COD ☐ 1,5-CYCLOOCTADIENE (Z,Z)

TOXICITY DATA WITH REFERENCE
skn-mus 100%/12D open SEV BJIMAG 25,75,68
skn-rbt 2640 mg SEV BJIMAG 25,75,68
skn-rbt 20 g/31D-I open SEV BJIMAG 25,75,68
eye-rbt 88 mg MLD BJIMAG 25,75,68
skn-gpg 10 g/31D-I open SEV BJIMAG 25,75,68

SAFETY PROFILE: An eye and severe skin irritant. When heated to decomposition it emits acrid smoke and fumes.

CPR840 CAS:12245-39-5 HR: 3
(1,5-CYCLOOCTADIENE)(2,4-PENTANEDIONATO) RHODIUM
mf: $C_{13}H_{19}O_2Rh$ mw: 310.23

PROP: Yellow crystals from pet ether. Mp: 125–128° decomp. Sol in hexane, $CHCl_3$.

SYNS: ACETYLACETONATE-1,5-CYCLOOCTADIENE RHODIUM ☐ RHODIUM, ((1,2,5,6-eta)-1,5-CYCLOOCTADIENE)(2,4-PENTANEDIONATO-O,O′)- ☐ RHODIUM, (1,5-CYCLOOCTADIENE)(2,4-PENTANEDIONATO)-

TOXICITY DATA WITH REFERENCE
ipr-mus LD50:34 mg/kg CBINA8 45,1,83

OSHA PEL: TWA 0.1 mg(Rh)/m^3
ACGIH TLV: TWA 1 mg(Rh)/m^3

SAFETY PROFILE: Poison by intraperitoneal route. When heated to decomposition it emits toxic fumes of Rh.

CPS000 CAS:115-25-3 HR: 1
CYCLOOCTAFLUOROBUTANE
DOT: UN 1976
mf: C_4F_8 mw: 200.03

PROP: Colorless, odorless gas. Mp: −41.4°, bp: −6.04°, d (liquid): 1.513 @ −70°F.

SYNS: FC-C 318 ☐ FREON C-318 ☐ HALOCARBON C-138 ☐ OCTAFLUOROCYCLOBUTANE (DOT) ☐ PERFLUOROCYCLOBUTANE ☐ PROPELLANT C318 ☐ R-C 318

TOXICITY DATA WITH REFERENCE
sln-dmg-ihl 99 pph/10M ENVRAL 7,275,74

CONSENSUS REPORTS: EPA Genetic Toxicology Program. Reported in EPA TSCA Inventory.

DOT CLASSIFICATION: 2.2; *Label:* Nonflammable Gas

SAFETY PROFILE: Mildly toxic by ingestion and inhalation. Can cause slight transient effects at high concentrations. No anesthesia or central nervous system effects. Nonflammable gas. Mutation data reported. When heated to decomposition it emits highly toxic fumes of F^-.

CPS500 CAS:629-20-9 HR: 3
1,3,5,7-CYCLOOCTATETRAENE
mf: C_8H_8 mw: 104.15

CH=CH(CH=CH)$_2$CH=CH (ring)

PROP: A liquid. Mp: −7°, bp: 142–143°, fp: −4.7°, vap press: 7.9 mm @ 25°, flash p: <71.6°F, d: 0.921 20°/4°.

SAFETY PROFILE: May be a simple asphyxiant. A dangerous fire hazard when exposed to heat or flame; can react with oxidizing materials. To fight fire, use spray, mist, fog, foam, dry chemicals. Reaction with oxygen gives explosive peroxide by-products. When heated to decomposition it emits acrid smoke and fumes.

CPT000 **CAS:2163-69-1** **HR: 3**
3-CYCLOOCTYL-1,1-DIMETHYLUREA
mf: $C_{11}H_{22}N_2O$ mw: 198.35

PROP: Crystals. Mp: 138°. Sltly sol in H_2O; sol in C_6H_6 and Me_2CO; very sol in MeOH.

SYNS: ALIPUR-O □ 3-CYCLOOCTYL-1,1-DIMETHYLHARNSTOFF (GERMAN) □ N-CYCLOOCTYL-N',N'-DIMETHYLUREA □CYCLOURON □ CYCLURON

TOXICITY DATA WITH REFERENCE
orl-rat LD50:1500 mg/kg PCOC** -,294,66
ihl-rat LD50:1125 mg/kg EQSFAP 3,618,75
ipr-mus LD50:300 mg/kg 85DPAN -,-,71/76
unk-mus LD50:300 mg/kg 30ZDA9 -,229,71
orl-mam LD50:2600 mg/kg 85GYAZ -,81,71

SAFETY PROFILE: Poison by intraperitoneal and possibly other routes. Moderately toxic by ingestion and inhalation. A pesticide. When heated to decomposition it emits toxic fumes of NO_x.

CPT750 **CAS:4449-51-8** **HR: 3**
CYCLOPAMINE
mf: $C_{27}H_{41}NO_2$ mw: 411.69

PROP: Needles from EtOH. Mp: 237–238°. Derived from *Veratrum californicum* (TJADAB 3,175,70).

SYNS: ALKALOID V □ 11-DEOXOJERVINE

TOXICITY DATA WITH REFERENCE
orl-rbt TDLo:116 mg/kg (female 7-8D post):TER PSEBAA 136,1174,71
orl-ham TDLo:250 mg/kg (female 7D post):REP PSEBAA 149,302,75
orl-rat TDLo:120 mg/kg (6-9D preg):TER LANCAO 1,1187,73
unr-mus TDLo:300 mg/kg (10D preg):TER TJADAB 29(2),22A,84
orl-mus LDLo:180 mg/kg PSEBAA 149,302,75
orl-ham LDLo:170 mg/kg PSEBAA 149,302,75

SAFETY PROFILE: A poison by ingestion. An experimental teratogen. Other experimental reproductive effects. When heated to decomposition it emits toxic fumes of NO_x.

CPU250 **CAS:502-72-7** **HR: 1**
CYCLOPENTADECANONE
mf: $C_{15}H_{28}O$ mw: 224.43

PROP: A solid. Mp: 63°, bp: 120 @ 0.3 mm.

SYN: NORMUSCONE

TOXICITY DATA WITH REFERENCE
skn-rbt 500 mg/24H MLD FCTXAV 14,659,76

CONSENSUS REPORTS: Reported in EPA TSCA Inventory.

SAFETY PROFILE: A skin irritant. When heated to decomposition it emits acrid smoke and irritating fumes. See also KETONES.

CPU500 **CAS:542-92-7** **HR: 3**
1,3-CYCLOPENTADIENE
mf: C_5H_6 mw: 66.11

PROP: Colorless liquid. Mp: −85°, bp: 41–42°, d: 0.80475 @ 19°/4°. flash p: 77°F. Misc in EtOH and C_6H_6.

SYNS: CYCLOPENTADIENE □ PENTOLE □ PYROPENTYLENE □ R-PENTINE

TOXICITY DATA WITH REFERENCE
ihl-rat LC50:39 g/m³ GTPZAB 9(12),13,65
ihl-mus LC50:14 g/m³ GTPZAB 9(12),13,65

CONSENSUS REPORTS: Reported in EPA TSCA Inventory.

OSHA PEL: TWA 75 ppm
ACGIH TLV: TWA 75 ppm
DFG MAK: 75 ppm (200 mg/m³)

SAFETY PROFILE: Low toxicity by ingestion. A dangerous fire hazard when exposed to heat or flame; can react with oxidizing materials. Moderate explosion hazard in the form of gas when exposed to heat or by chemical reaction. It decomposes violently at high temperatures and pressures. Dimerization is highly exothermic. Explosive reaction with fuming nitric acid, dinitrogen tetroxide, sulfuric acid. Reaction with nitrogen oxide + oxygen forms an explosive product. Reaction with oxygen forms a flame-sensitive explosive product. Ignites on contact with oxygen + ozone. Reacts vigorously on contact with potassium hydroxide. Incompatible with oxides of nitrogen, sulfuric acid. When heated to decomposition it emits acrid smoke and fumes.

For occupational chemical analysis use NIOSH: 1,3-Cyclopentadiene, 2523.

CPU750 **CAS:21254-73-9** **HR: 3**
CYCLOPENTADIENYL GOLD(1)
mf: C_5H_5Au mw: 262.06

SAFETY PROFILE: Ignites with friction or low heat. Will burn easily. When heated to decomposition it emits acrid smoke and fumes. See also GOLD.

CPV000 **CAS:12079-65-1** **HR: 3**
CYCLOPENTADIENYLMANGANESE TRICARBONYL
mf: $C_8H_5MnO_3$ mw: 204.07

PROP: Pale-yellow crystals with camphoraceous odor. Mp: 76.8–77.1°.

SYNS: MANGANESE CYCLOPENTADIENYL TRICARBONYL □ MCT

TOXICITY DATA WITH REFERENCE
orl-rat LD50:22 mg/kg TXCYAC 34,341,85
ihl-rat LCLo:120 mg/m³/2H HYSAAV 30,40,65
ipr-rat LD50:14 mg/kg TXCYAC 34,341,85
orl-mus LD50:150 mg/kg GISAAA 28(4),29,63
ivn-mus LD50:710 μg/kg CSLNX* NX#11285

CONSENSUS REPORTS: Reported in EPA TSCA Inventory. Manganese and its compounds are on the Community Right-To-Know List.

OSHA PEL: TWA 0.1 mg(Mn)/m³ (skin)
ACGIH TLV: TWA 0.1 mg(Mn)/m³

SAFETY PROFILE: A poison by ingestion, inhalation, intraperitoneal, and intravenous routes. A mild narcotic which can damage kidneys. When heated to decomposition it emits acrid smoke and irritating fumes. See also MANGANESE COMPOUNDS and CARBON MONOXIDE.

CPV500 CAS:4984-82-1 HR: 3
CYCLOPENTADIENYL SODIUM
mf: C_5H_5Na mw: 88.08

CH=CHCHNaCH=CH

PROP: Colorless crystals. Sol in ethers.

SAFETY PROFILE: Ignites spontaneously in air. Evaporation of a solution leaves a pyrophoric residue. Mixture with lead(II)nitrate may be explosive above 100°C. When heated to decomposition it emits toxic fumes of Na_2O.

CPV609 CAS:538-02-3 HR: 3
CYCLOPENTAMINE HYDROCHLORIDE
mf: $C_9H_{19}N{\cdot}ClH$ mw: 177.75

SYNS: CLOPANE HYDROCHLORIDE □ N,α-DIMETHYLCYCLOPENTANEETHYLAMINE HYDROCHLORIDE

TOXICITY DATA WITH REFERENCE
orl-rat LD50:169 mg/kg 29ZVAB -,37,69
orl-mus LD50:169 mg/kg JPETAB 93,423,48
ivn-mus LD50:41,600 μg/kg JPETAB 93,423,48

SAFETY PROFILE: Poison by ingestion and intravenous routes. When heated to decomposition it emits toxic fumes of NO_x and HCl. See also AMINES.

CPV750 CAS:287-92-3 HR: 3
CYCLOPENTANE
DOT: UN 1146
mf: C_5H_{10} mw: 70.15

PROP: Colorless liquid. Bp: 49.3°, fp: −93.7°, flash p: 19.4°F, autoign temp: 716°F, d: 0.745 @ 20°/4°, vap press: 400 mm @ 31.0°, vap d: 2.42.

SYN: PENTAMETHYLENE

CONSENSUS REPORTS: Reported in EPA TSCA Inventory.

OSHA PEL: TWA 600 ppm
ACGIH TLV: TWA 600 ppm
DOT CLASSIFICATION: 3; *Label:* Flammable Liquid

SAFETY PROFILE: Mildly toxic by ingestion and inhalation. High concentrations have narcotic action. A very dangerous fire hazard when exposed to heat or flame; can react with oxidizers. To fight fire, use foam, CO_2, dry chemical. When heated to decomposition it emits acrid smoke and fumes.

CPW250 CAS:35944-73-1 HR: 3
1,3-CYCLOPENTANEDISULFONYL DIFLUORIDE
mf: $C_5H_8F_2O_4S_2$ mw: 234.25

SYN: PHILIPS 2133

TOXICITY DATA WITH REFERENCE
orl-rat LD50:7900 μg/kg TXAPA9 21,315,72
orl-mus LDLo:94 mg/kg AECTCV 14,111,85
orl-bwd LD50:1300 μg/kg TXAPA9 21,315,72

SAFETY PROFILE: A deadly poison by ingestion. See also FLUORIDES. When heated to decomposition it emits very toxic fumes of F^- and SO_x.

CPW300 CAS:1679-07-8 HR: 3
CYCLOPENTANETHIOL
DOT: UN 1228/UN 3071
mf: $C_5H_{10}S$ mw: 102.21

SYNS: CYCLOPENTYL MERCAPTAN □ MERCAPTOCYCLOPENTANE

TOXICITY DATA WITH REFERENCE
orl-mus LD50:2680 mg/kg DCTODJ 3,249,80

DOT CLASSIFICATION: 3; *Label:* Flammable Liquid, Poison (UN1228); DOT Class: 6.1; *Label:* Poison, Flammable Liquid (UN3071)

CONSENSUS REPORTS: Reported in EPA TSCA Inventory.

SAFETY PROFILE: A human poison. A flammable liquid. When heated to decomposition it emits toxic vapors of SO_x.

CPW325 CAS:54573-23-8 HR: 3
4,5-CYCLOPENTANOFURAZAN-N-OXIDE
mf: $C_5H_6N_2O_2$ mw: 126.11

SYN: TRIMETHYLENEFUROXAN

SAFETY PROFILE: Decomposes explosively at 150°C. Upon decomposition it emits toxic fumes of NO_x.

CPW500 CAS:120-92-3 HR: 3
CYCLOPENTANONE
DOT: UN 2245
mf: C_5H_8O mw: 84.13

O:C(CH$_2$)$_3$ CH$_2$

PROP: Liquid with a pleasant odor. Mp: −58.2°, bp: 130.6°, flash p: 79°F, d: 0.9509 @ 18°/4°, vap d: 2.3. Sparingly sol in H_2O.

SYNS: ADIPIC KETONE □ DUMASIN □ KETOCYCLOPENTANE □ KETOPENTAMETHYLENE

TOXICITY DATA WITH REFERENCE
skn-rbt 500 mg/24H FCTXAV 17,241,79
skn-rbt 500 mg MLD FCTOD7 20,573,82
eye-rbt 100 mg SEV FCTOD7 20,573,82
eye-rbt 100 mg/4S rns SEV FCTOD7 20,573,82
ipr-mus LD50:1950 mg/kg COREAF 254,2245,62
scu-mus LDLo:2600 mg/kg AEXPBL 50,199,1903
scu-frg LDLo:3000 mg/kg AEXPBL 50,199,1903

CONSENSUS REPORTS: Reported in EPA TSCA Inventory.

DOT CLASSIFICATION: 3; *Label:* Flammable Liquid

SAFETY PROFILE: Moderately toxic by intraperitoneal and subcutaneous routes. A skin and severe eye irritant. Dangerous fire hazard when exposed to heat or flame; can react with oxidizers. To fight fire, use alcohol foam, foam, CO_2, dry chemical. Potentially explosive reaction with hydrogen peroxide + nitric acid. When heated to decomposition it emits acrid smoke and fumes. See also KETONES.

CPW750 CAS:1192-28-5 HR: 2
CYCLOPENTANONE OXIME
mf: C_5H_9NO mw: 99.13

$CH_2(CH_2)_3C{=}NOH$

PROP: A solid. Mp: 57.5°, bp: 196°.

TOXICITY DATA WITH REFERENCE
unk-mus LD50:1200 mg/kg PCJOAU 12,227,78

CONSENSUS REPORTS: Reported in EPA TSCA Inventory.

SAFETY PROFILE: Moderately toxic by an unspecified route. Violent reaction when heated with 85% sulfuric acid. When heated to decomposition it emits toxic fumes of NO_x.

CPX750 CAS:142-29-0 HR: 3
CYCLOPENTENE
DOT: UN 2246
mf: C_5H_8 mw: 68.13

PROP: A liquid. Mp: −135.3°, bp: 44.242°, flash p: −20°F, d: 0.77199 @ 20°.

TOXICITY DATA WITH REFERENCE
orl-rat LD50:1656 mg/kg AIHAAP 30,470,69
skn-rbt LD50:1231 mg/kg AIHAAP 30,470,69

CONSENSUS REPORTS: Reported in EPA TSCA Inventory.

DOT CLASSIFICATION: 3; *Label:* Flammable Liquid

SAFETY PROFILE: Moderately toxic by ingestion and skin contact. A very dangerous fire hazard when exposed to flame or heat; can react with oxidizing materials. Keep away from heat and open flame. To fight fire, use foam, CO_2, dry chemical.

CPY800 CAS:936-52-7 HR: 3
N-(1-CYCLOPENTEN-1-YL)-MORPHOLINE
mf: $C_9H_{15}NO$ mw: 153.25

SYNS: MORPHOLINE, 4-(1-CYCLOPENTEN-1-YL)- □ (1-MORPHOLINOCYCLOPENTENE)

TOXICITY DATA WITH REFERENCE
ivn-mus LD50:320 mg/kg CSLNX* NX#02169

CONSENSUS REPORTS: Reported in EPA TSCA Inventory.

SAFETY PROFILE: Poison by intravenous route. When heated to decomposition it emits toxic vapors of NO_x.

CPZ125 CAS:5870-29-1 HR: 3
CYCLOPENTOLATE HYDROCHLORIDE
mf: $C_{17}H_{25}NO_3 \cdot ClH$ mw: 327.89

PROP: Crystals from EtOAc. Mp: 137–141°. Insol in Et_2O; sol in H_2O and EtOH.

SYNS: CYCLOGYL □ β-DIMETHYLAMINOETHYL (1-HYDROXYCYCLOPENTYL)PHENYLACETATE HYDROCHLORIDE □ 2-(DIMETHYLAMINO)ETHYL 1-HYDROXY-α-PHENYLCYCLOPENTANEACETATE HYDROCHLORIDE

TOXICITY DATA WITH REFERENCE
scu-chd TDLo:40 μg/kg:CNS AROPAW 87,634,72
ocu-cld TDLo:50 μg/kg/I:BAH,EYE AJOPAA 105,91,88
scu-rat LD50:2235 mg/kg DRUGAY-,444,90
orl-mus LD50:960 mg/kg NIIRDN 6,314,82
ipr-mus LD50:314 mg/kg JPETAB 106,141,52
ivn-mus LD50:84 mg/kg JPETAB 106,141,52

SAFETY PROFILE: Poison by intravenous and intraperitoneal routes. Moderately toxic by ingestion. Human systemic effects: convulsions, distorted perceptions, hallucinations, toxic psychosis. When heated to decomposition it emits toxic fumes of NO_x and HCl. See also ESTERS.

CQA000 CAS:1003-03-8 HR: 3
CYCLOPENTYLAMINE
mf: $C_5H_{11}N$ mw: 85.15

PROP: A liquid. D: 0.869 @ 20°/4°, fp −85.7°, bp: 107–108°, flash p: 55.4°F. Misc in H_2O.

SYNS: AMINOCYCLOPENTANE □ CB 1689 □ CYCLOPENTANAMINE

TOXICITY DATA WITH REFERENCE
ipr-rat LDLo:100 mg/kg BCPCA6 5,108,60

CONSENSUS REPORTS: Reported in EPA TSCA Inventory.

SAFETY PROFILE: A poison by intraperitoneal route. A dangerous fire hazard when exposed to heat or flame. When heated to decomposition it emits toxic fumes of NO_x. See also AMINES.

CQA100 CAS:67239-27-4 HR: 3
CYCLOPENTYL 3,4-DIHYDROXYPHENYL KETONE
mf: $C_{12}H_{14}O_3$ mw: 206.26

SYN: KETONE, CYCLOPENTYL 3,4-DIHYDROXYPHENYL

TOXICITY DATA WITH REFERENCE
ipr-mus LD50:650 mg/kg JMCMAR 7,178,64

DOT CLASSIFICATION: 3; *Label:* Flammable Liquid

SAFETY PROFILE: Moderately toxic by intraperitoneal route. A flammable liquid. When heated to decomposition it emits acrid smoke and irritating vapors.

CQB250 **CAS:40202-39-9** **HR: 3**
2-CYCLOPENTYL-4,6-DINITROPHENOL
mf: $C_{11}H_{12}N_2O_5$ mw: 252.25

SYN: DINITROCYCLOPENTYLPHENOL

TOXICITY DATA WITH REFERENCE
ivn-dog LDLo: 10 mg/kg AIPTAK 50,20,35
ivn-pgn LDLo: 5 mg/kg AIPTAK 50,20,35

SAFETY PROFILE: Poison by intravenous route. When heated to decomposition it emits toxic fumes of NO_x. See also NITRO COMPOUNDS of AROMATIC HYDROCARBONS and PHENOLS.

CQB275 **CAS:10137-73-2** **HR: 2**
CYCLOPENTYL ETHER
mf: $C_{10}H_{18}O$ mw: 154.28

PROP: A liquid. Bp: 13 80° @ 13 mm.

SYN: ETHER, DICYCLOPENTYL

TOXICITY DATA WITH REFERENCE
skn-rbt 10 mg/24H open MLD AIHAAP 23,95,62
orl-rat LD50: 470 mg/kg AIHAAP 23,95,62
ihl-rat LCLo: 250 ppm/4H AIHAAP 23,95,62
skn-rbt LD50: 1410 mg/kg AIHAAP 23,95,62

SAFETY PROFILE: Moderately toxic by ingestion, inhalation, and skin contact. A skin irritant. When heated to decomposition it emits acrid smoke and irritating fumes.

CQC250 **CAS:21208-99-1** **HR: 3**
S-2-((5-CYCLOPENTYLPENTYL)AMINO)ETHYL THIOSULFATE
mf: $C_{12}H_{25}NO_3S_2$ mw: 295.50

TOXICITY DATA WITH REFERENCE
orl-mus LD50: 800 mg/kg JMCMAR 11,1190,68
ipr-mus LD50: 25 mg/kg JMCMAR 11,1190,68

SAFETY PROFILE: Poison by intraperitoneal route. Moderately toxic by ingestion. When heated to decomposition it emits very toxic fumes of NO_x and SO_x. See also THIOSULFATES.

CQC500 **CAS:6055-19-2** **HR: 3**
CYCLOPHOSPHAMIDE HYDRATE
mf: $C_7H_{15}Cl_2N_2O_2P \cdot H_2O$ mw: 279.13

SYNS: 1-BIS(2-CHLOROETHYL)AMINO-1-OXA-2-AZA-5-OXAPHOSPHORIDINE MONOHYDRATE □ 2-(BIS(2-CHLOROETHYL)AMINO)-1-OXA-3-AZA-2-PHOSPHOCYCLOHEXANE 2-OXIDE MONOHYDRATE □ (BIS(CHLORO-2-ETHYL)AMINO)-2-TETRAHYDRO-3,4,5,6-OXAZAPHOSPHORINE-1,3,2-OXIDE-2-MONOHYDRATE □ BIS(2-CHLOROETHYL) PHOSPHORAMIDE CYCLIC PROPANOLAMIDE ESTER MONOHYDRATE □ N,N-BIS(β-CHLOROETHYL)-N′,O-PROPYLENEPHOSPHORIC ACID ESTER AMINE MONOHYDRATE □ N,N-BIS(2-CHLOROETHYL)TETRAHYDRO-2H-1,3,2-OXAPHOSPHORIN-2-AMINE-2-OXIDE MONOHYDRATE □ N,N-BIS(β-CHLOROETHYL)-N′,O-TRIMETHYLENEPHOSPHORIC ACID ESTER DIAMIDE MONOHYDRATE □ CB-4564 □ CLAFEN □ CYCLIC N′,O-PROPYLENE ESTER of N,N-BIS(2-CHLOROETHYL)PHOSPHORODIAMIDIC ACID MONOHYDRATE □ CYCLOPHOSPHAMIDE MONOHYDRATE □ CYCLOPHOSPHAMIDUM □ CYCLOPHOSPHAN □ CYCLOPHOSPHANE □ CYCLOPHOSPHANUM □ CYTOPHOSPHAN □ CYTOXAN □ 2-(DI(2-CHLOROETHYL)AMINO)-1-OXA-3-AZA-2-PHOSPHACYCLOHEXANE-2-OXIDE MONOHYDRATE □ N,N-DI(2-CHLOROETHYL)AMINO-N,O-PROPYLENE PHOSPHORIC ACID ESTER DIAMIDE MONOHYDRATE □ ENDOXANA □ ENDOXAN-ASTA □ ENDOXAN MONOHYDRATE □ ENDOXAN R □ ENDUXAN □ GENOXAL □ MITOXAN □ NSC 26271 □ PROCYTOX □ SEMDOXAN □ SENDOXAN □ SENDUXAN

TOXICITY DATA WITH REFERENCE
bfa-mus/smc 500 mg/kg EVSRBT 24,893,81
ivn-rat TDLo: 27,500 μg/kg (7-17D preg): REP KSRNAM 16,517,82
ivn-rat TDLo: 27,500 μg/kg (7-17D preg): TER KSRNAM 16,517,82
ivn-rat TDLo: 310 mg/kg (62D male): REP KSRNAM 16,508,82
orl-rat LD50: 94 mg/kg TXAPA9 4,324,62
ipr-rat LD50: 121 mg/kg FRMBAZ 18,409,70
orl-mus LD50: 350 mg/kg TXAPA9 4,324,62
ivn-mus LD50: 275 mg/kg TXAPA9 4,324,62
orl-dog LD50: 44 mg/kg TXAPA9 4,324,62

CONSENSUS REPORTS: IARC Cancer Review: Human Sufficient Evidence IMEMDT 26,165,81; Human Limited Evidence IMEMDT 9,135,75; Animal Sufficient Evidence IMEMDT 9,135,75; Animal Sufficient Evidence IMEMDT 26,165,81.

SAFETY PROFILE: Confirmed human carcinogen. Poison by ingestion and intravenous routes. Experimental reproductive effects. Mutation data reported. When heated to decomposition it emits toxic fumes of Cl^-, PO_x, and NO_x.

CQC600 **HR: 3**
CYCLOPHOSPHAMIDE and MNU (1:2)
mf: $C_7H_{15}Cl_2N_2O_2P \cdot C_4H_5N_6O_2$ mw: 430.26

PROP: A combination of these two drugs is used in chemotherapy to combat far advanced malignant tumors. (ZKKOBW 89,311,77)

SYN: MNU and CYCLOPHOSPHAMIDE (2:1)

TOXICITY DATA WITH REFERENCE
ivn-hmn TDLo: 48 mg/kg/28D-I: GIT,BLD ZKKOBW 89,311,77

SAFETY PROFILE: Human systemic effects by intravenous route: nausea or vomiting and bone marrow changes. When heated to decomposition it emits very toxic fumes of Cl^-, NO_x, and PO_x.

CQC650 **CAS:50-18-0** **HR: 3**
CYCLOPHOSPHORAMIDE
mf: $C_7H_{15}Cl_2N_2O_2P$ mw: 261.11

PROP: Crystals. Mp: 41–45°. Water-sol; sltly sol in organic solvents.

SYNS: ASTA □ ASTA B518 □ B 518 □ N,N-BIS-(β-CHLORAETHYL)-N′,O-PROPYLEN-PHOSPHORSAEURE-ESTER-DIAMID (GERMAN) □ 2-(BIS(2-CHLOROETHYL)AMINO)-2H-1,3,2-OXAAZAPHOSPHORINE 2-OXIDE □ N,N-BIS(2-CHLOROETHYL)-N′-(3-HYDROXYPROPYL)PHOSPHORODIAMIDIC ACID intramol. ESTER □ BIS(2-CHLOROETHYL) PHOSPHORAMIDE-CYCLIC PROPANOLAMIDE ESTER □ N,N-BIS(2-CHLOROETHYL)-N′,O-PROPYLENEPHOSPHORIC ACID ESTER DIAM-

IDE □ N,N-BIS(2-CHLOROETHYL)TETRAHYDRO-2H-1,3,2-OXAPHOSPHORIN-2-AMINE-2-OXIDE □ N,N-BIS(β-CHLOROETHYL)-N',O-TRIMETHYLENEPHOSPHORIC ACID ESTER DIAMIDE □ CB 4564 □ CLAFEN □ CLAPHENE □ CP □ CPA □ CTX □ CY □ CYCLOPHOSPHAMIDE □ CYCLOPHOSPHAMIDUM □ CYCLOPHOSPHAN □ CYCLOSTIN □ CYTOPHOSPHAN □ CYTOXAN □ N,N-DI(2-CHLOROETHYL)-N,o-PROPYLENE-PHOSPHORIC ACID ESTER DIAMIDE □ ENDOXAN □ ENDOXANAL □ GENOXAL □ HEXADRIN □MITOXAN □ NCI-C04900 □ NEOSAR □ NSC 26271 □ 2-H-1,3,2-OXAZAPHOSPHORINANE □ PROCYTOX □ RCRA WASTE NUMBER U058 □ SEMDOXAN □ SENDUXAN □ SK 20501 □ ZYKLOPHOSPHAMID (GERMAN)

TOXICITY DATA WITH REFERENCE

sce-hmn:oth 500 mg/L ENMUDM 7(Suppl 3),26,85
sce-hmn:fbr 10 μmol/L CNREA8 45,3626,85
mma-ham:lng 10 mg/L MUREAV 157,189,85
orl-wmn TDLo:60 mg/kg (60D preg):REP NEJMAG 289,2259,73
orl-wmn TDLo:107 mg/kg (21W male):REP LANCAO 2,156,75
unr-chd TDLo:180 mg/kg (90D male):REP LANCAO 2,426,72
mul-wmn TDLo:215 mg/kg (43-70D preg):TER AIMEAS 74,87,71
mul-wmn TDLo:382 mg/kg (1-39W preg):TER JAMAAP 188,423,64
par-rat TDLo:10 mg/kg (male 1D pre):REP TJADAB 24(1),52A,81
ipr-mus TDLo:20 mg/kg (female 12D post):REP TCMUD8 7,7,87
orl-rat TDLo:110 mg/kg (female 7-17D post):REP IYKEDH 7,367,76
ivn-mus TDLo:5 mg/kg (female 12D post):REP JTEHD6 18,25,86
scu-mus TDLo:4 mg/kg (female 3D post):TER TCMUD8 6,115,86
orl-mus TDLo:100 mg/kg (male 1D pre):TER FCTXAV 13,317,75
scu-rat TDLo:60 mg/kg (female 3D post):TER JEEMAF 41,65,77
orl-rat TDLo:91,800 μg/kg (male 18D pre):TER BIREBV 37,317,87
ipr-rat TDLo:250 mg/kg (female 5W pre):REP SCIEAS 211,80,81
orl-mus TDLo:20 mg/kg (female 10D post):REP TXAPA9 45,344,78
orl-rat TDLo:100 mg/kg (1D male):REP MUREAV 48,267,77
orl-dom TDLo:30 mg/kg (male 1D pre):REP JANSAG 34,903,72
orl-rat TDLo:90 mg/kg (female 1-20D post):REP TXAPA9 27,602,74
orl-mus TDLo:100 mg/kg (male 1D pre):REP FCTXAV 13,317,75
ipr-mus TDLo:3744 mg/kg (female 52D pre):REP PSEBAA 133,190,70
scu-rat TDLo:60 mg/kg (female 3D post):REP BIANA6 19,247,81
orl-rat TDLo:122 mg/kg (male 18D pre):REP BIREBV 37,317,87
orl-rat TDLo:91,800 μg/kg (male 18D pre):REP BIREBV 37,317,87
orl-rat TDLo:110 mg/kg (female 7-17D post):TER IYKEDH 7,367,76
orl-mus TDLo:20 mg/kg (female 10D post):TER JTEHD6 6,155,80
ipr-rat TDLo:20 mg/kg (female 13D post):TER TJADAB 24,1,81
orl-rat TDLo:5 mg/kg (female 11D post):TER TJADAB 28,11A,83
par-rat TDLo:20 mg/kg (female 14D post):TER IJEBA6 20,838,82
ipr-mus TDLo:1 mg/kg (female 12D post):TER JTEHD6 18,25,86
orl-rat TDLo:110 mg/kg (female 7-17D post):TER IYKEDH 7,367,76
orl-wmn TDLo:1890 mg/kg/3Y-I:CAR,BLD JHMJAX 142,211,78
orl-man TDLo:2310 mg/kg/4.5Y-C:CAR,GIT BMJOAE 280,524,80
unr-man TDLo:857 mg/kg/3Y-C:CAR,BLD JCPAAK 26,649,73
unr-wmn TDLo:1050 mg/kg/69W-C:CAR,BLD JCPAAK 26,649,73
orl-rat TDLo:475 mg/kg/100W-I:CAR IJCNAW 23,706,79
ivn-rat TDLo:676 mg/kg/1Y-I:CAR ARZNAD 20,1461,70
ipr-mus TDLo:1950 mg/kg/26W-I:NEO RRCRBU 52,1,75
scu-mus TDLo:1352 mg/kg/1Y-I:CAR,TER ARZNAD 20,1461,70
orl-wmn TD:2700 mg/kg/6Y-C:CAR,KID URGABW 17,105,78
orl-rat TD:1270 mg/kg/87W-I:CAR IJCNAW 23,706,79
orl-rat TD:698 mg/kg/89W-I:CAR IJCNAW 23,706,79
orl-man TD:1078 mg/kg/3Y-C:CAR,KID RIHYAC 32,1073,78
scu-mus TD:3600 mg/kg/76W-I:CAR ARHEAW 22,1338,79
orl-man TD:1800 mg/kg/6Y-C:CAR,KID JOURAA 126,544,81
orl-hmn TD:920 mg/kg/3Y-C:CAR,KID AIMEAS 91,221,79
orl-rat TD:1075 mg/kg/86W-I:CAR CANCAR 51,606,83
orl-man TD:1190 mg/kg/4Y-I:CAR,BLD MEDIAV 58,32,79
orl-wmn TD:1760 mg/kg/4Y-C:CAR,KID SJRHAT 12,73,83
orl-wmn TDLo:45 mg/kg:KID ARHEAW 15,530,72
orl-man TDLo:56 mg/kg/26D-I:BLD AJMEAZ 81,1059,86
orl-cld TDLo:2500 μg/kg AJDCAI 140,1094,86
orl-man TDLo:56 mg/kg/4W-I:SYS SMJOAV 78,222,85
orl-hmn TDLo:20 mg/kg:GIT,SYS,SKN ARHEAW 12,663,69
orl-wmn LDLo:16 mg/kg/4D-I:BLD AJMSA9 254,48,67
ivn-wmn TDLo:60 mg/kg/9W-I AIMDAP 145,548,85
ivn-wmn TDLo:13,500 μg/kg:EYE AIMEAS 116,92,92
mul-man LDLo:45 mg/kg/26W-I:BLD AJMSA9 254,48,67
orl-rat LD50:160 mg/kg JJATDK 9,235,89
ipr-rat LD50:40 mg/kg CPCHAO 18,307,62
scu-rat LD50:144 mg/kg KSRNAM 16,431,82
ivn-rat LD50:148 mg/kg KSRNAM 16,431,82
orl-mus LD50:137 mg/kg RPTOAN 36,240,73
scu-mus LD50:200 mg/kg ASBDD9 2,95,79
ivn-mus LD50:140 mg/kg 17TVAO-,97,69
par-mus LD50:315 mg/kg TRPLAU 13,316,72
ipr-dog LDLo:50 mg/kg KSRNAM 16,431,82
ivn-dog LDLo:11 mg/kg CCSUBJ 2,191,65
ivn-mky LDLo:45 mg/kg CCSUBJ 2,191,65

CONSENSUS REPORTS: NTP 7th Annual Report on Carcinogens. IARC Cancer Review: Group 1 IMEMDT

7,182,87; Human Sufficient Evidence IMEMDT 26,165,81; Animal Sufficient Evidence IMEMDT 26,165,81; IMEMDT 9,135,75; Human Limited Evidence IMEMDT 9,135,75. NCI Carcinogenesis Studies (ipr); Clear Evidence: mouse, rat RRCRBU 52,1,75. EPA Genetic Toxicology Program.

SAFETY PROFILE: Confirmed human carcinogen producing leukemia, Hodgkin's disease, gastrointestinal and bladder tumors. Experimental carcinogenic, neoplastigenic, and teratogenic data. A human poison by ingestion and many other routes. Human systemic effects: kidney changes (hepatic dysfunction), leukopenia (reduced white blood cell count), nausea and alopecia (loss of hair), liver changes, agranulocytosis. Human reproductive and teratogenic effects by multiple routes: spermatogenesis, testical changes, epididymis and sperm duct changes, menstrual cycle changes, fetal developmental abnormalities of the craniofacial area, musculoskeletal, and cardiovascular systems. Experimental reproductive effects. Human mutation data reported. A powerful skin irritant. Used as an immunosuppressive agent in nonmalignant diseases. When heated to decomposition it emits highly toxic fumes of PO_x, NO_x, and Cl^-.

CQD000 CAS:18172-33-3 HR: 3
α-CYCLOPIAZONIC ACID
mf: $C_{20}H_{20}N_2O_3$ mw: 336.42

PROP: A solid. Mp: 245–246°.

SYN: CYCLOPIAZONIC ACID

TOXICITY DATA WITH REFERENCE
mma-sat 1 µmol/plate AEMIDF 47,1355,84
orl-rat TDLo:20 mg/kg (8-11D preg):REP JTEHD6 14,585,84
orl-mus TDLo:30 mg/kg (female 4D post):REP RCOCB8 55,303,87
orl-mus TDLo:30 mg/kg (female 2D post):REP RCOCB8 55,303,87
orl-mus TDLo:30 mg/kg (female 3D post):REP RCOCB8 55,303,87
orl-rat TDLo:240 mg/kg (male 14D pre):REP FCTOD7 22,993,84
orl-rat TDLo:4 mg/kg (8-11D preg):TER JTEHD6 14,585,84
orl-rat LD50:36 mg/kg TXAPA9 18,114,71
ipr-rat LD50:2 mg/kg TXAPA9 18,114,71
orl-mus LD50:64 mg/kg RCOCB8 55,303,87
ipr-mus LD50:13 mg/kg FCTOD7 23,831,85

CONSENSUS REPORTS: EPA Genetic Toxicology Program.

SAFETY PROFILE: Poison by ingestion and intraperitoneal routes. An experimental teratogen. Experimental reproductive effects. Mutation data reported. When heated to decomposition it emits toxic fumes of NO_x.

CQD750 CAS:75-19-4 HR: 3
CYCLOPROPANE
DOT: UN 1027
mf: C_3H_6 mw: 42.09

$CH_2CH_2CH_2$ (ring)

PROP: Colorless gas with ethereal odor. Mp: −126.6°, bp: −33.5°, lel: 2.4%, uel: 10.4%, d: 1.879 g/L @ 0°, autoign temp: 932°F. Mod sol in H_2O; very sol in EtOH and Et_2O. A minor constituent of MAPP gas.

SYNS: CYCLOPROPANE, liquefied (DOT) □ TRIMETHYLENE

TOXICITY DATA WITH REFERENCE
cyt-ckn-ihl 20 pph/3H ANESAV 34,157,71

CONSENSUS REPORTS: IARC Cancer Review: Animal No Adequate Data IMEMDT 7,93,87. Reported in EPA TSCA Inventory.

DOT CLASSIFICATION: 2.1; *Label:* Flammable Gas

SAFETY PROFILE: Mutation data reported. Questionable carcinogen. High concentrations are narcotic. Human reproductive effects. Very dangerous fire hazard when exposed to heat or flame; can react with oxidizing materials. Explosion Hazard: Moderate in the form of vapor when exposed to heat or flame. To fight fire, stop flow of gas, then use CO_2, dry chemical, or water spray. When heated to decomposition it emits acrid smoke and fumes.

CQE250 CAS:765-30-0 HR: 2
CYCLOPROPYLAMINE
mf: C_3H_7N mw: 57.10

$CH_2CH_2CHNH_2$ (ring)

PROP: D: 0.824 @ 20°/4°, bp: 50°, flash p: 33.8°F. Misc in H_2O.

SAFETY PROFILE: A very dangerous fire hazard when exposed to heat or flame. When heated to decomposition it emits toxic fumes of NO_x. See also AMINES.

CQE750 CAS:540-47-6 HR: 3
CYCLOPROPYL METHYL ETHER
mf: C_4H_8O mw: 72.11

PROP: Liquid. Mp: −119°, bp: 44.7°, d: 0.786 @ 25°/4°, flash p: $<$50°F.

SYNS: CYCLOPROPANE, METHOXY-(9CI) □ CYCLOPROPYL METHYL ETHER □ CYPROME ETHER □ METHOXYCYCLOPROPANE □ 1-METHOXYCYCLOPROPANE □ METHYL CYCLOPROPYL ETHER

TOXICITY DATA WITH REFERENCE
ihl-mus LC50:126 g/m³/15M ANESAV 11,455,50

SAFETY PROFILE: Moderately toxic by inhalation. A very dangerous fire hazard when exposed to heat or flame. When heated to decomposition it emits acrid smoke and fumes. Can form unstable explosive peroxides. See also ETHERS and PEROXIDES.

CQF079 CAS:42281-59-4 HR: 3
(−)-17-CYCLOPROPYLMETHYLMORPHINAN-3,4-DIOL
mf: $C_{20}H_{27}NO_2$ mw: 313.48

PROP: A solid. Mp: 173–175°

SYNS: (L)-BC-2605 □ BRISTOL LABORATORIES BC 2605 □ 1-N-CYCLOPROPYLMETHYL-3,14-DIHYDROXYMORPHINAN □ (−)-3,14-DIHYDROXY-N-(CYCLOBUTYLMETHYL)MORPHINAN □ OXILORPHAN

TOXICITY DATA WITH REFERENCE
orl-hmn TDLo: 14 µg/kg: CNS,GIT,KID JCPCBR 16(4),183,76
scu-hmn TDLo: 26 µg/kg: CNS DRFUD4 2,746,77
scu-mus LD50: 315 mg/kg JPETAB 193,23,75
ivn-mus LD50: 32 mg/kg JPETAB 193,23,75

SAFETY PROFILE: Poison by subcutaneous and intravenous routes. Human systemic effects by ingestion and subcutaneous routes: somnolence, tremors, ataxia (loss of muscle coordination), hypermotility, diarrhea, and changes in the kidney, ureter, or bladder. When heated to decomposition it emits toxic fumes of NO_x.

CQF099 CAS:16590-41-3 HR: 2
N-CYCLOPROPYLMETHYLNOROXYMORPHONE
mf: $C_{20}H_{23}NO_4$ mw: 341.44

SYNS: (5-α)-17-(CYCLOPROPYLMETHYL-4,5-EPOXY-3,14-DIHYDROXY-MORPHINAN-6-ONE) (9CI) □ N-CYCLOPROPYLMETHYL-14-HYDROXYDIHYDROMORPHINONE □ EN 1639 □ EN 1939 □ NALTREXONE □ UM-792

TOXICITY DATA WITH REFERENCE
sce-hmn:lym 1 g/L ENMUDM 1,180,79
cyt-hmn:lym 1 g/L ENMUDM 1,180,79
sln-dmg-orl 10 g/L/24H MUREAV 66,129,79
sln-dmg-par 10 g/L MUREAV 66,129,79
scu-mus TDLo: 21 mg/kg (female 10-19D post): REP DEPBA5 21,283,88
scu-mus LD50: 551 mg/kg ANYAA9 281,321,76

CONSENSUS REPORTS: EPA Genetic Toxicology Program.

SAFETY PROFILE: Moderately toxic by subcutaneous route. Experimental reproductive effects. Human mutation data reported. When heated to decomposition it emits toxic fumes of NO_x. See also MORPHINE and KETONES.

CQF125 CAS:33453-19-9 HR: 2
1-CYCLOPROPYLMETHYL-4-PHENYL-6-CHLORO-2 (1H)-QUINAZOLINONE
mf: $C_{18}H_{15}ClN_2O$ mw: 310.80

SYNS: 6-CHLORO-1-(CYCLOPROPYLMETHYL)-4-PHENYL-2(1H)-QUINAZOLINONE □ SL-512

TOXICITY DATA WITH REFERENCE
orl-rat LD50: 2 g/kg ARZNAD 23,1266,73
ipr-rat LD50: 790 mg/kg ARZNAD 23,1266,73
orl-mus LD50: 1800 mg/kg ARZNAD 23,1266,73
ipr-mus LD50: 660 mg/kg ARZNAD 23,1266,73

SAFETY PROFILE: Moderately toxic by ingestion and intraperitoneal route. When heated to decomposition it emits toxic fumes of Cl^- and NO_x.

CQG250 CAS:26399-36-0 HR: 2
N-(CYCLOPROPYLMETHYL)-α,α,α-TRIFLUORO-2,6-DINITRO-N-PROPYL-p-TOLUIDINE
mf: $C_{14}H_{16}F_3N_3O_4$ mw: 347.33

PROP: Yellow-orange solid. Mp: 32°. Very sparingly sol in H_2O.

SYNS: CGA 10832 □ ER5461 □ GA-10832 □ PREGARD □ PROFLURALIN □ TOLBAN

TOXICITY DATA WITH REFERENCE
skn-rbt 218 mg open MLD CIGET* -,-,77
eye-rbt 44 mg SEV CIGET* -,-,77
orl-rat LD50: 1808 mg/kg FMCHA2 -,D310,80
ihl-rat LCLo: 3970 mg/m³ CIGET* -,-,77
skn-rbt LD50: 13,754 mg/kg CIGET* -,-,77

SAFETY PROFILE: Moderately toxic by ingestion and inhalation. Mildly toxic by skin contact. A skin and severe eye irritant. An herbicide. See also FLUORIDES. When heated to decomposition it emits very toxic fumes of F^- and NO_x.

CQG750 CAS:4163-15-9 HR: 3
CYCLORPHAN
mf: $C_{20}H_{27}NO$ mw: 297.48

SYNS: 17-(CYCLOPROPYLMETHYL)MORPHINAN-3-OL □ (−)-3-HYDROXY-N-CYCLOPROPYLMETHYLMORPHINAN

TOXICITY DATA WITH REFERENCE
ivn-rat LD50: 23 mg/kg AIPTAK 165,112,67
scu-mus LD50: 215 mg/kg AIPTAK 165,112,67
ivn-mus LD50: 24 mg/kg AIPTAK 165,112,67

SAFETY PROFILE: Poison by intravenous and subcutaneous routes. When heated to decomposition it emits toxic fumes of NO_x.

CQH000 CAS:68-41-7 HR: 3
CYCLOSERINE
mf: $C_3H_6N_2O_2$ mw: 102.11

PROP: Crystals. Mp: 155–156° decomp. Sol in H_2O and alkalis. Produced by *Streptomyces orchidaceus* (ANTCAO 6,360,56).

SYNS: d-R-AMINO-3-ISOSSAZOLIDONE (ITALIAN) □ d-4-AMINO-3-ISOXAZOLIDINONE □ d-4-AMINO-3-ISOXAZOLIDONE □ CICLOSERINA (ITALIAN) □ CYCLOMYCIN □ CYCLO-d-SERINE □ E-733-A □ FARMISERINE □ I-1431 □ JN-21 □ K-300 □ MIROSERINA □ NOVOSERIN □ ORIENTOMYCIN □ d-OXAMICINA (ITALIAN) □ d-OXAMYCIN □ OXYMYCIN □ PA 94 □ RO-1-9213 □ SEROMYCIN □ TISOMYCIN □ WASSERINA

TOXICITY DATA WITH REFERENCE
orl-hmn TDLo: 560 mg/kg/4W-I: CNS DICHAK 29,241,56
orl-wmn TDLo: 60 mg/kg: CNS BMJOAE 1,907,65
unr-wmn TDLo: 40 mg/kg/2D-I: CNS TUBEAS 38,297,57
unr-man TDLo: 64 mg/kg/4D-I: CNS ABANAE 3,148,55/56
orl-mus LD50: 5290 mg/kg 85ERAY 2,906,78
ipr-mus LD50: 180 mg/kg 85FZAT -,238,67
scu-mus LD50: 1400 mg/kg YKYUA6 31,1085,80
ivn-mus LD50: 560 mg/kg YKYUA6 31,1085,80
scu-dog LDLo: 2000 mg/kg ANTCAO 6,708,56
scu-mky LDLo: 4000 mg/kg ANTCAO 6,708,56

CONSENSUS REPORTS: EPA Genetic Toxicology Program.

SAFETY PROFILE: Poison by intraperitoneal route. Moderately toxic by intravenous and subcutaneous routes. Mildly toxic by ingestion. Human systemic effects by ingestion and possibly other routes: wakefulness, sleep, altered sleep time, hallucinations, distorted perceptions, tremors, convulsions, and coma. An antibiotic used in the treatment of human pulmonary tuberculosis. When heated to decomposition it emits toxic fumes of NO_x.

CQH100 **CAS:59865-13-3** **HR: 2**
CYCLOSPORIN A
mf: $C_{62}H_{111}N_{11}O_{12}$ mw: 1202.84

PROP: Needles from Me_2CO. Mp: 148–151°.

SYNS: ANTIBIOTIC S 7481F1 □ CICLOSPORIN □ CYCLOSPORIN □ CYCLOSPORINE □ CYCLOSPORINE A □ OL 27-400 □ S 7481F1 □ SANDIMMUN □ SANDIMMUNE

TOXICITY DATA WITH REFERENCE
sce-hmn:lyms 1 mg/L IGAYAY 134,403,85
ims-rbt TDLo:210 mg/kg (14D pre):REP INJFA3 29,218,84
orl-man TDLo:259 mg/kg/2W-C:CAR CEDEDE 8,159,83
orl-wmn TDLo:62,500 µg/kg/5D-I:SYS LANCAO 1,1221,86
orl-man TDLo:20 mg/kg/2D-I:BLD LANCAO 2,1092,86
unr-man TDLo:30 mg/kg/4D-I:SYS AIMEAS 107,786,87
orl-rat LD50:1489 mg/kg IYKEDH 17,365,86
ipr-rat LD50:147 mg/kg IYKEDH 17,365,86
scu-rat LD50:286 mg/kg IYKEDH 17,365,86
ivn-rat LD50:24 mg/kg IYKEDH 17,365,86
orl-mus LD50:2803 mg/kg IYKEDH 17,365,86
ivn-mus LD50:96 mg/kg IYKEDH 17,365,86
ivn-rbt LD50:10 mg/kg TOPADD 14,73,86

SAFETY PROFILE: Questionable human carcinogen producing Hodgkin's disease. Experimental carcinogenic data. Experimental reproductive effects. Poison by intraperitoneal and intravenous routes. Moderately toxic by ingestion. Human systemic effects by ingestion: increased body temperature, cyanosis. Mutation data reported. When heated to decomposition it emits toxic fumes of NO_x.

CQH250 **CAS:2691-41-0** **HR: 3**
CYCLOTETRAMETHYLENE TETRANITRAMINE
DOT: UN 0226/UN 0484
mf: $C_4H_8N_8O_8$ mw: 296.20

$O_2NN[CH_2N(NO_2)]_3CH_2$ (ring)

PROP: A solid. Mp: 286°.

SYNS: CYCLOTETRAMETHYLENETETRANITRAMINE, desensitized (UN 0483) (DOT) □ CYCLOTETRAMETHYLENETETRANITRAMINE (dry or unphlegmatized) (DOT) □ CYCLOTETRAMETHYLENETETRANITRAMINE, wetted (UN 0226) (DOT) □ HMX □ HMX (dry or unphlegmatized) (DOT) □ HMX, wetted (UN 0226) (DOT) □ beta HMY □ HW 4 □ LX 14-0 □ OCTOGEN □ OCTOGEN, desensitized (UN 0483) (DOT) □ OCTOGEN, wetted with not <15% water, by weight (UN 0226) (DOT) □ OKTOGEN □ TETRAMETHYLENETETRANITRAMINE

TOXICITY DATA WITH REFERENCE
orl-mus LD50:1500 mg/kg GISAAA 40(11),17,75
ivn-dog LDLo:40 mg/kg EATR** EB-TR-73040
orl-gpg LD50:300 mg/kg GISAAA 40(11),17,75
ivn-gpg LD50:28 mg/kg EATR** EB-TR-73040

CONSENSUS REPORTS: Reported in EPA TSCA Inventory.

DOT CLASSIFICATION: Forbidden (dry or unphlegmatized); DOT Class: EXPLOSIVE 1.1D; *Label:* EXPLOSIVE 1.1D

SAFETY PROFILE: A poison by ingestion and intravenous routes. An explosive. Decomposes violently at 279°C. When heated to decomposition it emits toxic fumes of NO_x. See also EXPLOSIVES, HIGH.

CQH325 **CAS:860-79-7** **HR: 3**
CYCLOVIROBUXINE D
mf: $C_{26}H_{46}N_2O$ mw: 402.74

SYNS: BEBUXINE □ CYCLOVIROBUXIN D □ CYCLOVIROBUXINE

TOXICITY DATA WITH REFERENCE
orl-mus LD50:293 mg/kg CYLPDN 3,101,82
ipr-mus LD50:9200 µg/kg CYLPDN 3,101,82
ivn-mus LD50:8900 µg/kg CYLPDN 3,101,82

SAFETY PROFILE: Poison by ingestion, intravenous, and intraperitoneal routes. When heated to decomposition it emits toxic fumes of NO_x.

CQH500 **CAS:126-02-3** **HR: 3**
CYCRIMINE HYDROCHLORIDE
mf: $C_{19}H_{29}NO \cdot ClH$ mw: 323.95

PROP: Crystals with bitter taste. Mp: 241–244°. Sol in H_2O, EtOH, and $CHCl_3$.

SYNS: COMPOUND 8958 □ α-CYCLOPENTYL-α-PHENYL-1-PIPERIDINEPROPANOL HYDROCHLORIDE □ PAGITANE HYDROCHLORIDE □ 1-PHENYL-1-CYCLOPENTYL-3-PIPERIDINO-1-PROPANOL HYDROCHLORIDE

TOXICITY DATA WITH REFERENCE
orl-rat LD50:628 mg/kg 27ZQAG -,218,72
orl-mus LD50:349 mg/kg 27ZQAG -,218,72
ipr-mus LD50:250 mg/kg NTIS** AD691-490
ivn-mus LD50:50 mg/kg 27ZQAG -,218,72

SAFETY PROFILE: Poison by ingestion, intraperitoneal, and intravenous routes. When heated to decomposition it emits very toxic fumes of HCl and NO_x.

CQH625 **CAS:7199-29-3** **HR: 2**
CYHEPTAMIDE
mf: $C_{16}H_{15}NO$ mw: 237.29

PROP: Long needles from acetonitrile. Mp: 193–194°. Sol in chloroform; sparingly sol in methanol, acetone; sltly sol in ethanol, ether. Practically insol in water.

SYNS: AY 8682 □ BS 7029 □ CYHEPTAMINE □ DIBENZO(a,d)CYCLOHEPTADIENE-5-CARBOXAMIDE □ DIBENZO(a,d)(1,4)-CYCLOHEPTADIENE-5-CARBOXAMIDE □ 10,11-DIHYDRO-5H-DIBENZO(a,d)CYCLOHEPTENE-5-CARBOXAMIDE □ ICI 51426

TOXICITY DATA WITH REFERENCE
orl-rat LD50:2400 mg/kg 27ZQAG -,68,72
ipr-rat LD50:2000 mg/kg 27ZQAG -,68,72
orl-mus LD50:1830 mg/kg 27ZQAG -,68,72
ipr-mus LD50:630 mg/kg JMCMAR 7,88,64

SAFETY PROFILE: Moderately toxic by ingestion and intraperitoneal routes. An anticonvulsant. When heated to decomposition it emits toxic fumes of NO_x.

CQH650 **CAS:13121-70-5** **HR: 3**
CYHEXATIN
mf: $C_{18}H_{34}OSn$ mw: 385.21

PROP: White solid. Very sparingly sol in H_2O; sparingly sol in Me_2CO and MeOH; sol in $CHCl_3$.

SYNS: DOWCO-213 □ ENT 27,395-X □ M 3180 □ PLICTRAN □ PLYCTRAN □ TCTH □ TRICYCLOHEXYLHYDROXYSTANNANE □ TRICYCLOHEXYLHYDROXYTIN □ TRICYCLOHEXYLTIN HYDROXIDE □ TRICYCLOHEXYLZINNHYDROXID (GERMAN)

TOXICITY DATA WITH REFERENCE
orl-rat TDLo:108 mg/kg (MGN):REP EQSFAP 4,80,75
orl-rat LD50:180 mg/kg KSKZAN 16(2),59,78
ihl-rat LC50:244 mg/m^3 GISAAA 49(2),74,84
skn-rat LD50:446 mg/kg FAATDF 7,299,86
ipr-rat LD50:13 mg/kg DOWCC* 47(7),80,82
orl-rbt LD50:458 mg/kg GISAAA 47(7),80,82
skn-rbt LD50:2422 mg/kg GISAAA 47(7),80,82
orl-gpg LD50:780 mg/kg TRIPA7 -,1,73
orl-ckn LD50:654 mg/kg TRIPA7 -,1,73
orl-dom LDLo:150 mg/kg TXAPA9 31,66,75

OSHA PEL: TWA 0.1 mg(Sn)/m^3; TWA 5 mg/m^3
ACGIH TLV: TWA 5 mg/m^3; TWA 0.1 mg(Sn)/m^3; STEL 0.2 mg/m^3 (skin)
NIOSH REL: (Organotin Compounds) TWA 0.1 mg(Sn)/m^3.

SAFETY PROFILE: Poison by ingestion, inhalation, and intraperitoneal routes. Moderately toxic by skin contact. Experimental reproductive effects. When heated to decomposition it emits acrid smoke and irritating fumes. See also TIN COMPOUNDS.

For occupational chemical analysis use NIOSH: Organotin Compounds, 5504.

CQH750 **CAS:508-77-0** **HR: 3**
CYMARIN
mf: $C_{30}H_{44}O_9$ mw: 548.74

PROP: Crystals from MeOH. Mp: 148°

SYNS: CYMARINE □ 3-β-(β-d-CYMAROSYLOXY)-5,14-DIHYDROXY-19-OXO-5-β-CARD-20(22)-ENOLIDE □ STROPHANTHIDIN-d-CYMAROSID (GERMAN) □ K-STROPHANTHIN-α

TOXICITY DATA WITH REFERENCE
ivn-rat LD50:20 mg/kg AIPTAK 155,165,65
ipr-mus LD50:12 mg/kg AIPTAK 155,165,65
ivn-cat LDLo:95 μg/kg MEIEDD 10,397,83
unr-cat LDLo:110 μg/kg AIPTAK 148,471,64

SAFETY PROFILE: Poison by intravenous, intraperitoneal, and possibly other routes. Used as a cardiotonic. When heated to decomposition it emits acrid smoke and fumes.

CQI000 **CAS:99-87-6** **HR: 3**
p-CYMENE
mf: $C_{10}H_{14}$ mw: 134.24

PROP: Colorless to pale-yellow liquid; odorless. Mp: −68.2°, bp: 176°, lel: 0.7%, @ 100°, ULC: 30–35, flash p: 117°F (CC), d: 0.853, refr index: 1.489, autoign temp: 817°F, vap d: 4.62, vap press: 1 mm @ 17.3°, flash p: (technical) 127°F, uel (technical): 5.6%. Found in nearly 100 volatile oils, including lemongrass, sage, thyme, coriander, star anise, and cinnamon (FCTXAV 12,385,74). Sol in alc, ether, acetone, benzene.

SYNS: CAMPHOGEN □ CYMENE □ CYMOL □ DOLCYMENE □ FEMA No. 2356 □ 4-ISOPROPYL-1-METHYLBENZENE □ p-ISOPROPYLTOLUENE □ p-METHYL-CUMENE □ p-METHYLISOPROPYL BENZENE □ 1-METHYL-4-ISOPROPYLBENZENE □ PARACYMENE □ PARACYMOL

TOXICITY DATA WITH REFERENCE
skn-rbt 500 mg/24H MOD FCTXAV 12,401,74
orl-rat LD50:4750 mg/kg FCTXAV 2,327,64

CONSENSUS REPORTS: Reported in EPA TSCA Inventory.

SAFETY PROFILE: Mildly toxic by ingestion. Humans sustain central nervous system effects at low doses. A skin irritant. Flammable liquid. Explosion Hazard: Slight in the form of vapor. To fight fire, use foam, CO_2, dry chemical. When heated to decomposition it emits acrid smoke and fumes.

CQI250 **CAS:536-60-7** **HR: 2**
p-CYMEN-7-OL
mf: $C_{10}H_{14}O$ mw: 150.24

PROP: Bp: 246.

SYNS: CUMIC ALCOHOL □ CUMINIC ALCOHOL □ CUMINOL □ CUMINYL ALCOHOL □ CUMYL ALCOHOL □ p-ISOPROPYLBENZYL ALCOHOL

TOXICITY DATA WITH REFERENCE
skn-rbt 500 mg/24H MOD FCTXAV 12,871,74
orl-rat LD50:1020 mg/kg FCTXAV 12,871,74
skn-rbt LD50:2500 mg/kg FCTXAV 12,871,74

CONSENSUS REPORTS: Reported in EPA TSCA Inventory.

SAFETY PROFILE: Moderately toxic by ingestion and skin contact. A skin irritant. When heated to decomposition it emits acrid smoke and irritating fumes. See also ALCOHOLS.

CQI500 **CAS:2631-37-0** **HR: 3**
m-CYM-5-YL METHYLCARBAMATE
mf: $C_{12}H_{17}NO_2$ mw: 207.30

PROP: Crystals. Mp: 87–88°. Very sparingly sol in H_2O; sol in org solvents.

SYNS: CARBAMULT □ ENT 27,300 □ ENT 27,300-A □ EP 316 □ METHYLCARBAMIC ACID-m-CYM-5-YL ESTER □ 3-METHYL-5-ISOPROPYLPHENYL-N-METHYLCARBAMATE □ (3-METHYL-5-ISOPROPYLPHE-

NYL)-N-METHYLCARBAMAT (GERMAN) □ 3-METHYL-5-(1-METHYLETHYL)PHENOLMETHYLCARBAMATE □ MINACIDE □ MORTON EP-316 □ PROMECARB □ SCHERING 34615 □ UC 9880 □ UNION CARBIDE UC-9880

TOXICITY DATA WITH REFERENCE
orl-rat LD50:60 mg/kg MEIEDD 10,1122,83
skn-rat LD50:450 mg/kg WRPCA2 9,119,70
ipr-rat LD50:27,200 μg/mg BWHOA6 44(1-3),241,71
ivn-rat LD50:5 mg/kg BJIMAG 22,317,65
ims-rat LD50:44 mg/kg BJIMAG 22,317,65
orl-mus LD50:16 mg/kg BESAAT 15,131,69
orl-gpg LDLo:25 mg/kg JEENAI 61(5),1261,68
scu-gpg LDLo:25 mg/kg JEENAI 61(5),1261,68

CONSENSUS REPORTS: EPA Extremely Hazardous Substances List.

SAFETY PROFILE: Poison by ingestion, intraperitoneal, intravenous, intramuscular, and subcutaneous routes. Moderately toxic by skin contact. An insecticide. When heated to decomposition it emits toxic fumes of NO_x. See also CARBAMATES.

CQI750 CAS:22936-86-3 HR: 2
CYPRAZINE
mf: $C_9H_{14}ClN_5$ mw: 227.73

SYNS: 6-CHLOR-N-CYCLOPROPYL-N′-(1-METHYLETHYL)-1,3,5-TRIZAINE-2,4-DIAMINE □ 2-CHLORO-4-CYCLOPROPYLAMINO-6-ISOPROPYLAMINO-sec-TRIAZINE □ 2-CHLORO-4-CYCLOPROPYLAMINO-6-ISOPROPYLAMINO-1,3,5-TRIAZINE □ OUTFOX □ S-6115 □ S-9115

TOXICITY DATA WITH REFERENCE
orl-rat LD50:1200 mg/kg FMCHA2 -,D227,80
skn-rbt LD50:7500 mg/kg GUCHAZ 6,147,73

SAFETY PROFILE: Moderately toxic by ingestion. Mildly toxic by skin contact. An herbicide. When heated to decomposition it emits very toxic fumes of Cl^- and NO_x.

CQJ000 CAS:8013-86-3 HR: 1
CYPRESS OIL

PROP: The constituents include furfural, d-α-pinene, d-camphene, cymene, d-terpineol, l-cadinene, sylvestrene, cypress camphor, and cedrol (FCTXAV 16,637,78).

TOXICITY DATA WITH REFERENCE
skn-rbt 500 mg/24H MOD FCTXAV 16,699,78

CONSENSUS REPORTS: Reported in EPA TSCA Inventory.

SAFETY PROFILE: A skin irritant. When heated to decomposition it emits acrid smoke and irritating fumes. See also individual components.

CQJ250 CAS:2759-71-9 HR: 3
CYPROMID
mf: $C_{10}H_9Cl_2NO$ mw: 230.10

SYNS: CIPROMID □ CLOBBER □ 3,4′-DICHLOROCYCLOPROPANECARBOXANILIDE □ N-(3,4-DICHLOROPHENYL)CYCLOPROPANECARBOXAMIDE

TOXICITY DATA WITH REFERENCE
orl-rat LD50:215 mg/kg WRPCA2 9,119,70
orl-rbt LD50:3028 mg/kg 28ZEAL 5,64,76
skn-rbt LD50:3038 mg/kg WRPCA2 7,135,68

SAFETY PROFILE: A poison by ingestion. Moderately toxic by skin contact. An herbicide. When heated to decomposition it emits very toxic fumes of HCl and NO_x.

C

CQJ500 CAS:427-51-0 HR: 2
CYPROSTERONE ACETATE
mf: $C_{24}H_{29}ClO_4$ mw: 416.98

PROP: Crystals from diisopropyl ether. Mp: 200–201°.

SYNS: 17-α-ACETOXY-6-CHLORO-1-α,2-α-METHYLENEPREGNA-4,6-DIENE-3,20-DIONE □ 6-CHLORO-1,2-α-METHYLENE-6-DEHYDRO-17-α-HYDROXYPROGESTERONE ACETATE □ 6-CHLORO-Δ^6-1,2-α-METHYLENE-17-α-HYDROXYPROGESTERONE ACETATE □ 6-CHLORO-1,2-α-METHYLENE-17-α-HYDROXY-Δ^6-PROGESTERONE ACETATE □ CPA □ CYPROTERONE ACETATE □ CYPROTERON-R ACETATE □ 1,2-α-METHYLENE-6-CHLORO-Δ^6-17-α-HYDROXYPROGESTERONE ACETATE □ 1,2-α-METHYLENE-6-CHLORO-PREGNA-4,6-DIENE-3,20-DIONE 17-α-ACETATE □ 1,2-α-METHYLENE-6-CHLORO-$\Delta^{4,6}$-PREGNADIENE-17-α-OL-3,20-DIONE 17-α-ACETATE □ 1,2-α-METHYLENE-6-CHLORO-$\Delta^{4,6}$-PREGNADIENE-17-α-OL-3,20-DIONE ACETATE □ NSC-81430 □ PREGNA-4,6-DIENE-3,20-DIONE, 6-CHLORO-17-HYDROXY-1-α,2-α-METHYLENE-, ACETATE □ SH 714

TOXICITY DATA WITH REFERENCE
dns-rat-orl 40 mg/kg CBINA8 31,287,80
orl-man TDLo:320 mg/kg (16W male):REP JRPFA4 32,365,78
orl-man TDLo:320 mg/kg (male 16W pre):REP JRPFA4 32,365,73
orl-man TDLo:1 mg/kg (14D male):REP CCPTAY 14,403,76
orl-man TDLo:98,570 μg/kg (4W male):REP INURAQ 6,638,69
ims-gpg TDLo:800 mg/kg (female 30-45D post):REP JRPFA4 37,315,74
scu-mus TDLo:12 mg/kg (female 12-17D post):REP PHBHA4 21,515,78
scu-mus TDLo:30 mg/kg (female 2D post):TER TJADAB 25,27,82
ims-mus TDLo:200 mg/kg (female 1-10D post):TER JRPFA4 32,299,73
orl-rat TDLo:110 mg/kg (female 11D pre):REP EJPHAZ 1,438,67
ims-mus TDLo:1200 mg/kg (female 10-15D post):REP JRPFA4 32,299,73
scu-rat TDLo:525 mg/kg (male 35D pre):REP JOENAK 35,209,66
ims-mky TDLo:233 mg/kg (male 84D pre):REP AVBIB9 10,197,73
orl-rat TDLo:4 mg/kg (female 1D pre):REP ARZNAD 35,459,85
ims-mus TDLo:120 mg/kg (female 10-15D post):REP JRPFA4 32,299,73
scu-rat TDLo:180 mg/kg (female 18D pre):REP AEFTAA 10,57,79
scu-rat TDLo:70 mg/kg (male 7D pre):REP IJEBA6 15,788,77
scu-rat TDLo:35 mg/kg (male 7D pre):REP JRPFA4 49,237,77

imp-rat TDLo:7840 μg/kg (male 28D pre):REP CCPTAY 12,517,75
scu-mus TDLo:30 mg/kg (female 5D post):TER ASBDD9 2,113,79
scu-mus TDLo:30 mg/kg (female 8D post):TER TJADAB 25,27,82
scu-rat TDLo:200 mg/kg (female 17-20D post):TER EXMDA4 (101),168,65
orl-rat TDLo:200 mg/kg (female 16-19D post):TER ACENA7 44,380,63
orl-rat TLDo:27 g/kg/78W:ETA LANCAO 2,688,76
ipr-rat LD50:565 mg/kg IYKEDH 13,349,82
ipr-mus LD50:3300 mg/kg NIIRDN 6,APP-4,82

CONSENSUS REPORTS: EPA Genetic Toxicology Program.

SAFETY PROFILE: Questionable carcinogen with experimental tumorigenic and teratogenic data. Moderately toxic by intraperitoneal route. Human reproductive effects by ingestion and possibly other routes: abnormal spermatogenesis, changes in the testes, epididymis, and sperm duct, impotence, and other paternal effects. Experimental reproductive effects. Mutation data reported. Used as a drug to arrest precocious puberty in children and hirsutism in women. A steroid. When heated to decomposition it emits toxic fumes of Cl^-.

CQJ750 CAS:56-17-7 HR: 3
CYSTAMINE DIHYDROCHLORIDE
mf: $C_4H_{12}N_2S_2 \cdot 2ClH$ mw: 225.22

PROP: Prisms from EtOH. Mp: 212°.

SYNS: AED □ 2-AMINOETHYL DISULFIDE DIHYDROCHLORIDE □ 2,2'-DITHIO-BIS-(ETHYLAMINE) DIHYDROCHLORIDE □ USAF CB-34

TOXICITY DATA WITH REFERENCE
orl-rat TDLo:1250 mg/kg (13-17D preg):REP JEEMAF 8,94,60
scu-rat LDLo:200 mg/kg AEPPAE 185,461,37
ipr-mus LD50:405 mg/kg ARZNAD 21,284,71
scu-cat LDLo:200 mg/kg AEPPAE 185,461,37
scu-gpg LDLo:300 mg/kg AEPPAE 185,461,37

CONSENSUS REPORTS: Reported in EPA TSCA Inventory.

SAFETY PROFILE: A poison by subcutaneous and intraperitoneal routes. Experimental reproductive effects. When heated to decomposition it emits very toxic fumes of HCl, SO_x, and NO_x. See also SULFIDES.

CQK000 CAS:52-90-4 HR: 2
l-CYSTEINE
mf: $C_3H_7NO_2S$ mw: 121.17

PROP: A solid. Mp: 178°. Sol in H_2O, AcOH, and NH_3. An amino acid derived from cystine, occurring naturally in the l-form, which will be considered here. Colorless crystals; sol in water, ammonium hydroxide, and acetic acid; insol in ether, acetone, benzene, carbon disulfide, and carbon tetrachloride.

SYNS: CYSTEIN □ CYSTEINE □ l-(+)-CYSTEINE □ HALF-CYSTEINE □ HALF-CYSTINE □ β-MERCAPTOALANINE □ THIOSERINE

TOXICITY DATA WITH REFERENCE
mmo-sat 60 μmol/plate BCPCA6 34,3725,85
dns-hmn:fbr 1 mmol/L CALEDQ 5,199,78
orl-mus TDLo:3600 mg/kg (female 7-12D post):REP TOIZAG 24,667,77
ipr-rat TDLo:100 mg/kg (female 12D post):REP PSEBAA 139,62,72
orl-mus TDLo:27,600 mg/kg (female 23D pre):REP ENDOAO 69,331,61
orl-mus TDLo:6 g/kg (male 30D pre):REP OYYAA2 7,1251,73
orl-rat LD50:1890 mg/kg AGACBH 4,125,74
ipr-rat LD50:1620 mg/kg OYYAA2 7,1251,73
scu-rat LD50:1550 mg/kg OYYAA2 7,1251,73
orl-mus LD50:660 mg/kg ARTODN 41,79,78
ipr-mus LD50:1400 mg/kg OYYAA2 7,1251,73
scu-mus LD50:1360 mg/kg OYYAA2 7,1251,73

CONSENSUS REPORTS: Reported in EPA TSCA Inventory. EPA Genetic Toxicology Program.

SAFETY PROFILE: Moderately toxic by ingestion, intraperitoneal, and subcutaneous routes. Experimental reproductive effects. Human mutation data reported. When heated to decomposition it emits very toxic fumes of SO_x and NO_x.

CQK100 CAS:51025-94-6 HR: 2
CYSTEINE-GERMANIC ACID
mf: $C_3H_{11}GeNO_6S$ mw: 261.80

SYNS: (l-CYSTEINE)TETRAHYDROXYGERMANIUM □ DB □ GERMANIUM, (l-CYSTEINE)TETRAHYDROXY-

TOXICITY DATA WITH REFERENCE
scu-rat TDLo:70 mg/kg (female 7-13D post):TER YIKUAO 22,107,73
orl-rat LD50:3400 mg/kg TOIZAG 20,180,73
ipr-rat LD50:1090 mg/kg TOIZAG 20,180,73
scu-rat LD50:1200 mg/kg TOIZAG 20,180,73
orl-mus LD50:3320 mg/kg TOIZAG 20,180,73
ipr-mus LD50:2350 mg/kg TOIZAG 20,180,73
scu-mus LD50:2160 mg/kg TOIZAG 20,180,73

SAFETY PROFILE: Moderately toxic by ingestion, intraperitoneal, and subcutaneous routes. An experimental teratogen. When heated to decomposition it emits toxic fumes of NO_x and SO_x.

CQK125 CAS:58100-26-8 HR: 2
CYSTEINE HYDRAZIDE
mf: $C_3H_9N_3OS$ mw: 135.18

SYN: l-HYDRAZIDE CYSTEINE

TOXICITY DATA WITH REFERENCE
orl-rat LD50:1050 mg/kg AITEAT 27,733,79
ipr-rat LD50:430 mg/kg AITEAT 27,733,79
orl-mus LD50:1010 mg/kg AITEAT 27,733,79
ipr-mus LD50:525 mg/kg AITEAT 27,733,79

SAFETY PROFILE: Moderately toxic by ingestion and intraperitoneal routes. When heated to decomposition it emits toxic fumes of SO_x and NO_x. See also l-CYSTEINE.

CQK250 **CAS:52-89-1** **HR: 2**
l-CYSTEINE HYDROCHLORIDE
mf: $C_3H_7NO_2S{\cdot}ClH$ mw: 157.63

PROP: White crystalline powder; characteristic acetic taste. Mp: 175° (decomp). Sol in water, alc.

SYNS: CYSTEINE CHLORHYDRATE □ CYSTEINE HYDROCHLORIDE □ l-CYSTEINE HYDROCHLORIDE □ l-CYSTEINE MONOHYDROCHLORIDE (FCC)

TOXICITY DATA WITH REFERENCE
mma-sat 20 mg/plate FCTOD7 22,623,84
cyt-ham:fbr 2 g/L FCTOD7 22,623,84
ipr-mus LD50:1250 mg/kg NTIS** AD691-490
ivn-mus LD50:771 mg/kg JJANAX 38,137,85
unk-mus LD50:3 g/kg BJCAAI 6,160,52

CONSENSUS REPORTS: Reported in EPA TSCA Inventory.

SAFETY PROFILE: Moderately toxic by intraperitoneal, intravenous, and possibly other routes. Mutation data reported. When heated to decomposition it emits very toxic fumes of NO_x, SO_x, and Cl^-.

CQK325 **CAS:56-89-3** **HR: 1**
l-CYSTINE
mf: $C_6H_{12}N_2O_4S_2$ mw: 240.30

PROP: Plates or prisms. Mp: 258–261° (decomp) (sealed tube). Sol in hot H_2O, mineral acids, and aq alkali. Naturally occurring levorotatory form. Colorless to white hexagonal tablets from water. Decomp 260–261°. d-Cystine: Crystals. Sltly sol in water. dl-Cystine, the synthetic racemic form: Crystals. Sltly sol in water. meso-Cystine, the internally compensated form: Crystals. Sltly sol in water.

SYNS: CYSTEINE DISULFIDE □ CYSTIN □ (−)-CYSTINE □ CYSTINE ACID □ DICYSTEINE □ β,β′-DITHIODIALANINE □ GELUCYSTINE □ OXIDIZED l-CYSTEINE

TOXICITY DATA WITH REFERENCE
orl-rat LDLo:25 g/kg OYYAA2 15,199,78

SAFETY PROFILE: Low toxicity by ingestion. When heated to decomposition it emits toxic fumes of PO_x and SO_x.

CQK500 **CAS:53317-25-2** **HR: 3**
l-CYSTINE-BIS(N,N-β-CHLOROETHYL)HYDRAZIDEHYDROBROMIDE
mf: $C_{14}H_{28}Cl_4N_6O_2S_2{\cdot}2BrH$ mw: 680.24

SYN: CYDRIN

TOXICITY DATA WITH REFERENCE
ipr-rat LD50:47 mg/kg NEOLA4 24,401,77
ipr-mus LD50:71 mg/kg NEOLA4 24,401,77
scu-mus LD50:76 mg/kg NEOLA4 24,401,77

SAFETY PROFILE: Poison by intraperitoneal and subcutaneous routes. When heated to decomposition it emits very toxic fumes of HBr, SO_x, NO_x, and Cl^-.

CQK600 **CAS:21739-91-3** **HR: 3**
CYTEMBENA
mf: $C_{11}H_8BrO_4{\cdot}Na$ mw: 307.09

SYNS: ACRYLIC ACID, 3-p-ANISOYL-3-BROMO-, SODIUM SALT, (E)- □ (E)-3-p-ANISOYL-3-BROMOACRYLIC ACID SODIUM SALT □ 2-BUTENOIC ACID, 3-BROMO-4-(4-METHOXYPHENYL)-4-OXO-, SODIUM SALT, (E)- (9CI) □ MBBA □ NCI-C50737 □ NSC-104801 □ SODNA SUL KYSELINY cis-β-4-METHOXYBENZOYL-β-BROMAKRYLOVE

TOXICITY DATA WITH REFERENCE
mmo-sat 100 μg/plate SCIEAS 236,933,87
msc-mus:lyms 25 mg/L SCIEAS 236,933,87
ipr-rat LD:14 mg/kg/2Y-I:CAR,REP NTPTR* NTP-TR-207,81
ipr-rat TDLo:7 mg/kg/2Y-I:CAR,REP NTPTR* NTP-TR-207,81
ipr-rat TDLo:7 mg/kg/2Y-I:CAR NTPTR* NTP-TR-207,81
ipr-rat TD:14 mg/kg/2Y-I:CAR NTPTR* NTP-TR-207,81
ipr-rat LD50:155 mg/kg CKFRAY 29,106,80
scu-rat LD50:155 mg/kg CKFRAY 29,106,80
ivn-rat LD50:245 mg/kg CKFRAY 29,106,80
ipr-mus LD50:50 mg/kg CKFRAY 29,106,80
scu-mus LD50:52 mg/kg CKFRAY 29,106,80
ivn-mus LD50:98 mg/kg CKFRAY 29,106,80

CONSENSUS REPORTS: NTP Carcinogenesis Bioassay (ipr): Clear Evidence: rat NTPTR* NTP-TR-207,81; NTP Carcinogenesis Bioassay (ipr): No Evidence: mouse NTPTR* NTP-TR-207,81

SAFETY PROFILE: Poison by intraperitoneal, subcutaneous, and intravenous routes. Questionable carcinogen with experimental carcinogenic data. Experimental reproductive effects. Mutation data reported. When heated to decomposition it emits toxic fumes of NaO_2 and Br^-.

CQL250 **CAS:115-93-5** **HR: 3**
CYTHIOATE
mf: $C_8H_{12}NO_5PS_2$ mw: 297.30

PROP: A solid. Insol in H_2O; sol in C_6H_6, Me_2CO, Et_2O, and EtOH.

SYNS: AC 26,691 □ AMERICAN CL-26691 □ AMERICAN CYANAMID CL-26,691 □ O-(4-(AMINOSULFONYL)PHENYL) O,O-DIMETHYL PHOSPHOROTHIOATE □ BENZENESULFONAMIDE, p-HYDROXY-, O-ESTER with O,O-DIMETHYL PHOSPHOROTHIOATE □ CL 26691 □ CYFLEE □ O,O-DIMETHYL O-p-SULFAMOYLPHENYL PHOSPHOROTHIOATE □ ENT 25,640 □ PHOSPHOROTHIOIC ACID, O-(4-(AMINOSULFONYL)PHENYL) O,O-DIMETHYL ESTER (9CI) □ PROBAN

TOXICITY DATA WITH REFERENCE
orl-rat LD50:160 mg/kg FMCHA2 -,D88,80
orl-mus LD50:38 mg/kg BESAAT 15,116,69

CONSENSUS REPORTS: Reported in EPA TSCA Inventory.

SAFETY PROFILE: Poison by ingestion. An insecticide. When heated to decomposition it emits very toxic fumes of NO_x, PO_x, and SO_x.

CQL500 **CAS:485-35-8** **HR: 3**
CYTISINE
mf: $C_{11}H_{14}N_2O$ mw: 190.27

PROP: A solid. Mp: 155°.

C

SYNS: BAPTITOXIN □ BAPTITOXINE □ CYSTISINE □CYTITONE □ 1,2,3,4,5,6-HEXAHYDRO-1,5-METHANO-8H-PYRIDO(1,2-A)(1,5)DIAZOCIN-8-ONE □ SOPHORINE □ ULEXINE

TOXICITY DATA WITH REFERENCE
scu-rat LDLo:20 mg/kg 85IXA4-,589,48
orl-mus LD50:101 mg/kg BJPCBM 35,161,69
ipr-mus LD50:9400 μg/kg BJPCBM 35,161,69
ivn-mus LD50:1730 μg/kg BJPCBM 35,161,69
inv-cat LD50:400 μg/kg ITOBAO (2),104,78

CONSENSUS REPORTS: Reported in EPA TSCA Inventory.

SAFETY PROFILE: Poison by ingestion, intravenous, and intraperitoneal routes. A toxin found in some plants. When heated to decomposition it emits toxic fumes of NO_x.

CQM125 CAS:14930-96-2 HR: 3
CYTOCHALASIN B
mf: $C_{29}H_{37}NO_5$ mw: 479.67

PROP: Needles from Me_2CO. Mp: 218–221°.

SYN: PHOMIN

TOXICITY DATA WITH REFERENCE
dni-hmn:oth 1 mg/L CNREA8 45,311,85
dni-hmn:hla 1 μmol/L MUREAV 92,427,82
cyt-hmn:fbr 1 mg/L JCLBA3 89,194,81
cyt-hmn:oth 1 mg/L JNCIAM 52,653,74
cyt-mus:mmr 1 mg/L ITCSAF 19,58,83
ipr-ham TDLo:5 mg/kg (female 8D post):TER TJADAB 22,59,80
orl-mus TDLo:1500 μg/kg (female 8D post):TER TJADAB 35,87,87
orl-mus TDLo:1500 μg/kg (female 8D post):TER TJADAB 35,87,87
ipr-rat LD50:11 mg/kg TOXIA6 17,137,79
ipr-mus LD50:30 mg/kg FEPRA7 38,438,79

CONSENSUS REPORTS: EPA Genetic Toxicology Program.

SAFETY PROFILE: Poison by intraperitoneal route. An experimental teratogen. Human mutation data reported. When heated to decomposition it emits toxic fumes of NO_x.

CQM250 CAS:36011-19-5 HR: 3
CYTOCHALASIN E
mf: $C_{28}H_{32}NO_7$ mw: 494.62

PROP: A solid. Mp: 206–208° decomp. Food storage mold metabolite of *Aspergillus clavatus* (TXAPA9 32,135,75).

TOXICITY DATA WITH REFERENCE
ipr-mus TDLo:3900 μg/kg (female 7-9D post):TER TJADAB 25,11,82
ipr-mus TDLo:3600 μg/kg (female 7-9D post):TER TJADAB 25,11,82
ipr-mus TDLo:8100 μg/kg (female 7-9D post):TER FEPRA7 38,438,79
ipr-mus LD50:2700 μg/kg FEPRA7 38,438,79
ipr-gpg LD50:500 μg/kg JJEMAG 48,105,78

SAFETY PROFILE: A poison by intraperitoneal route. An experimental teratogen. When heated to decomposition it emits toxic fumes of NO_x.

CQM500 CAS:65-46-3 HR: 2
CYTOSINE RIBOSIDE
mf: $C_9H_{13}N_3O_5$ mw: 243.25

PROP: Needles. Mp: 230° decomp.

SYNS: 4-AMINO-1-β-d-RIBOFURANOSYL-2(1H)-PYRIMIDINONE □ CYTIDINE □ 1-β-RIBOFURANOSYLCYTOSINE

TOXICITY DATA WITH REFERENCE
pic-esc 1 g/L ZAPOAK 12,583,72
oms-hmn:oth 100 μmol/L JIDEAE 65,52,75
dnd-mam:lym 150 μmol/L PNASA6 48,686,62
ipr-mus LD50:2700 mg/kg RPTOAN 40,66,77

CONSENSUS REPORTS: Reported in EPA TSCA Inventory.

SAFETY PROFILE: Moderately toxic by intraperitoneal route. Human mutation data reported. When heated to decomposition it emits toxic fumes of NO_x.

CQM600 CAS:71-30-7 HR: 2
CYTOSINIMINE
mf: $C_4H_5N_3O$ mw: 111.12

SYNS: 4-AMINO-2-HYDROXYPYRIMIDINE □ 4-AMINO-2(1H)-PYRIMIDINONE □ CYTOSINE (8CI) □ 2(1H)-PYRIMIDINONE, 4-AMINO-

TOXICITY DATA WITH REFERENCE
ipr-mus LD50:>2222 mg/kg JPETAB 207,504,78

CONSENSUS REPORTS: Reported in EPA TSCA Inventory.

SAFETY PROFILE: Moderately toxic by intraperitoneal route. When heated to decomposition it emits toxic vapors of NO_x.

CQM750 CAS:3543-75-7 HR: 3
CYTOSTASAN
mf: $C_{16}H_{21}Cl_2N_3O_2 \cdot ClH$ mw: 394.76

SYNS: IMET 3393 □ γ(1-METHYL-5-BIS(β-CHLORAETHYL)AMINOBENZIMIDAZOLYL)BUTTERSAEUREHYDROCHLORID(GERMAN) □ γ-(1-METHYL-5-BIS(β-CHLOROAETHYL) AMINOBENZIMIDAZOYL) BUTTERSAUERHYDROCHLORID(GERMAN)

TOXICITY DATA WITH REFERENCE
ipr-mus TDLo:70 mg/kg (7D preg):REP ZPPLBF 110,1067,71
ipr-mus TDLo:70 mg/kg (7D preg):TER ZPPLBF 110,1067,71
ipr-mus TDLo:70 mg/kg (7D preg):TER ZPPLBF 110,1067,71
orl-mus TDLo:250 mg/kg/4D-I:CAR ARGEAR 43,16,74
ipr-mus TDLo:50 mg/kg/4D-I:CAR ARGEAR 43,16,74
orl-rat LD50:200 mg/kg ATSUDG 8,504,85
ivn-rat LD50:40 mg/kg ATSUDG 8,504,85
orl-mus LD50:250 mg/kg ARGEAR 43,16,74
ipr-mus LD50:100 mg/kg ARGEAR 43,16,74
ivn-mus LD50:80 mg/kg ATSUDG 8,504,85

SAFETY PROFILE: A poison by ingestion, intravenous,

and intraperitoneal routes. Questionable carcinogen with experimental carcinogenic and teratogenic data. Experimental reproductive effects. When heated to decomposition it emits very toxic fumes of HCl and NO_x.

CQN000 CAS:4465-94-5 HR: 3
CYTOXAL ALCOHOL
mf: $C_7H_{17}Cl_2N_2O_3P{\bullet}C_6H_{13}N$ mw: 378.33

SYNS: 2-(BIS(2-CHLOROETHYL)AMINO)TETRAHYDROOXAZA-PHOSPHORINE CYCLOHEXYLAMINE SALT □ N,N-BIS(2-CHLOROETHYL)-N′-(3-HYDROXYPROPYL)PHOSPHORODIAMIDATE, CYCLOHEXYLAMMONIUM SALT □ N,N-BIS(2-CHLOROETHYL)-N′-3-PHOSPHORODIAMIDIC ACID HYDROXYLPROPYLCYCLOHEXYLAMINE SALT □ CYTOXYL ALCOHOL CYCLOHEXYLAMMONIUM SALT □ NCI-C04922 □ NSC-52695

TOXICITY DATA WITH REFERENCE
mmo-sat 100 µg/plate NTPTB* JAN82
mma-sat 100 µg/plate NTPTB* JAN82
dni-hmn:lym 500 µmol/L AGACBH 4,117,74
ipr-mus TDLo:29 mg/kg (11D preg):TER TJADAB 4,141,71
ipr-mus TDLo:290 mg/kg (11D preg):REP TJADAB 4,141,71
ipr-mus TDLo:290 mg/kg (11D preg):TER TJADAB 4,141,71
ipr-rat TDLo:2900 mg/kg/26W-I:CAR RRCRBU 52,1,75
ipr-mus TDLo:3900 mg/kg/26W-I:ETA CANCAR 40S,1935,77
orl-mus LD50:1618 mg/kg NCISP* JAN86
scu-mus LD50:966 mg/kg NCISP* JAN86
ivn-mus LD50:400 mg/kg NCISP* JAN86

CONSENSUS REPORTS: NCI Carcinogenesis Studies (ipr); Clear Evidence: mouse, rat RRCRBU 52,1,75.

SAFETY PROFILE: Suspected carcinogen with experimental carcinogenic and tumorigenic data. Poison by intravenous route. Moderately toxic by ingestion and subcutaneous routes. Experimental teratogenic and reproductive effects. Human mutation data reported. When heated to decomposition it emits very toxic fumes of NO_x, NH_3, PO_x, and Cl^-. See also ALCOHOLS.

D

DAA800 **CAS:94-75-7** **HR: 3**
2,4-D
mf: $C_8H_6Cl_2O_3$ mw: 221.04

PROP: White powder or crystals. Mp: 141°; bp: 160° @ 0.4 mm; vap d: 7.63.

SYNS: ACIDE-2,4-DICHLORO PHENOXYACETIQUE (FRENCH) □ ACIDO (2,4-DICLORO-FENOSSI)-ACETICO (ITALIAN) □ AGROTECT □ AMIDOX □ AMOXONE □ AQUA-KLEEN □ BARRAGE □ BH 2,4-D □ BRUSH-RHAP □ B-SELEKTONON □ CHIPCO TURF HERBICIDE "D" □ CHLOROXONE □ CITRUS FIX □ CROP RIDER □ CROTILIN □ D 50 □ DACAMINE □ 2,4-D ACID □ DEBROUSSAILLANT 600 □ DECAMINE □ DED-WEED □ DED-WEED LV-69 □ DEHERBAN □ DESORMONE □ (2,4-DICHLOOR-FENOXY)-AZIJNZUUR (DUTCH) □ DICHLOROPHENOXYACETIC ACID □ 2,4-DICHLOROPHENOXYACETIC ACID (DOT) □ 2,4-DICHLORPHENOXYACETIC ACID □ (2,4-DICHLOR-PHENOXY)-ESSIGSAEURE (GERMAN) □ DICOPUR □ DICOTOX □ DINOXOL □ DMA-4 □ DORMONE □ 2,4-DWUCHLOROFENOKSYOCTOSY KWAS (POLISH) □ EMULSAMINE BK □ EMULSAMINE E-3 □ ENT 8,538 □ ENVERT 171 □ ENVERT DT □ ESTERON □ ESTERON 99 □ ESTERON 76 BE □ ESTERON BRUSH KILLER □ ESTERON 99 CONCENTRATE □ ESTERONE FOUR □ ESTERON 44 WEED KILLER □ ESTONE □ FARMCO □ FERNESTA □ FERNIMINE □ FERNOXONE □ FOREDEX 75 □ FORMOLA 40 □ HEDONAL (The herbicide) □ HERBIDAL □ HIVOL-44 □ IPANER □ KROTILINE □ KWASU 2,4-DWUCHLOROFENOKSYOCTOWEGO □ KWAS 2,4-DWUCHLOROFENOKSYOCTOWY □ KYSELINA 2,4-DICHLORFENOXYOCTOVA □ LAWN-KEEP □ MACRONDRAY □ MIRACLE □ MONOSAN □ MOXONE □ NETAGRONE 600 □ NSC 423 □ PENNAMINE □ PENNAMINE D □ PHENOX □ PIELIK □ PLANOTOX □ PLANTGARD □ RCRA WASTE NUMBER U240 □ RHODIA □ SALVO □ SPRITZ-HORMIN/2,4-D □ SUPER D WEEDONE □ SUPERORMONE CONCENTRE □ TRANS AMINE □ TRIBUTON □ TRINOXOL □ U 46 □ U 46DP □ U-5043 □ VERGEMASTER □ VERTON D □ VERTON 2D □ VERTRON 2D □ VIDON 638 □ VISKO-RHAP □ VISKO-RHAP DRIFT HERBICIDES □ VISKO-RHAP LOW VOLATILE 4L □ WEED-AG-BAR □ WEEDAR-64 □ WEED-B-GON □ WEEDEZ WONDER BAR □ WEEDONE LV4 □ WEED TOX □ WEEDTROL

TOXICITY DATA WITH REFERENCE
skn-rbt 500 mg/24H MLD 28ZPAK -,279,72
eye-rbt 750 μg/24H SEV 28ZPAK -,279,72
sce-hmn:lym 10 mg/L JOHEA8 73,224,82
dni-ham:ovr 1 mmol/L TOLED5 29,137,85
orl-rat TDLo:500 mg/kg (female 6-15D post):REP FCTXAV 9,801,71
orl-mus TDLo:707 mg/kg (female 11-14D post):TER AECTCV 6,33,77
orl-mus TDLo:900 mg/kg (female 6-14D post):TER NTIS** PB223-160
orl-rat TDLo:1 g/kg (female 6-15D post):TER TXAPA9 22,14,72
orl-mus TDLo:900 mg/kg (female 6-14D post):REP NTIS** PB223-160
scu-mus TDLo:900 mg/kg (female 6-14D post):REP NTIS** PB223-160
orl-rat TDLo:220 μg/kg (1-22D preg):TER GISAAA 50(10),76,85
orl-rat TDLo:500 mg/kg (female 6-15D post):TER FCTXAV 9,801,71
orl-man TDLo:2 g/kg:BAH,PUL ARTODN 66,518,92
orl-man TDLo:5714 mg/kg:BAH,CVS ARTODN 66,518,92
orl-hmn LDLo:80 mg/kg:GIT,CNS ARPAAQ 94,270,72
orl-man LDLo:93 mg/kg:CNS PAREAQ 14,225,52
orl-rat LD50:370 mg/kg FMCHA2 -,C68,83
skn-rat LD50:1500 mg/kg WRPCA2 9,119,70
ipr-rat LDLo:666 mg/kg JIHTAB 29,85,47
orl-mus LD50:347 mg/kg RPZHAW 31,373,80
ipr-mus LDLo:125 mg/kg TXAPA9 23,288,72
orl-dog LD50:100 mg/kg AEHLAU 7,202,63
orl-rbt LDLo:800 mg/kg AMPMAR 12,26,51
skn-rbt LD50:1400 mg/kg AFDOAQ 16,3,52

CONSENSUS REPORTS: IARC Cancer Review: Group 2B IMEMDT 7,156,87; Human Limited Evidence IMEMDT 41,357,86; Animal Inadequate Evidence IMEMDT 15,111,77; Human Inadequate Evidence IMEMDT 15,111,77. EPA Genetic Toxicology Program. Reported in EPA TSCA Inventory. Community Right-To-Know List.

OSHA PEL: TWA 10 mg/m^3
ACGIH TLV: TWA 10 mg/m^3
DFG MAK: 10 mg/m^3

SAFETY PROFILE: Suspected human carcinogen. Experimental teratogenic and reproductive effects. Poison by ingestion, intravenous, and intraperitoneal routes. Moderately toxic by skin contact. Human systemic effects by ingestion: change in heart rate, coma, convulsions, nausea or vomiting, respiratory depression, somnolence. Can cause liver and kidney injury. A skin and severe eye irritant. Human mutation data reported. When heated to decomposition it emits toxic fumes of Cl^-.

For occupational chemical analysis use NIOSH: 2,4-D and 2,4,5-T, 5001.

DAB200 **CAS:33400-47-4** **HR: 3**
D-10,242
mf: $C_{22}H_{24}N_4O_2 \cdot ClH$ mw: 412.96

SYN: 2-AMINO-6-((1,2-DIPHENYLETHYL)AMINO)-3-PYRIDINECARBAMIC ACID ETHYL ESTER, MONOHYDROCHLORIDE

TOXICITY DATA WITH REFERENCE
orl-rat LD50:163 mg/kg DRFUD4 7,801,82
orl-mus LD50:295 mg/kg DRFUD4 7,801,82
orl-dog LD50:40 mg/kg DRFUD4 7,801,82

SAFETY PROFILE: Poison by ingestion. When heated to decomposition it emits toxic fumes of NO_x and HCl. See also CARBAMATES.

DAB400 CAS:39196-18-4 HR: 3
DACAMOX
mf: $C_9H_{18}N_2O_2S$ mw: 218.35

PROP: A solid. Mp: 56.5–57.5°.

SYNS: DIAMOND SHAMROCK DS-15647 □ 3,3-DIMETHYL-1-(METHYLTHIO)-2-BUTANONE-o-((METHYLAMINO)CARBONYL)OXIME □ DS-15647 □ ENT 27,851 □ RCRA WASTE NUMBER P045 □ THIOFANOX

TOXICITY DATA WITH REFERENCE
orl-rat LD50:8500 μg/kg 85ARAE 1,45,77
ihl-rat LC50:70 mg/m³ DOVEAA 31,158,77
skn-rbt LD50:39 mg/kg SPEADM 78-1,60,78
orl-qal LD50:1200 μg/kg JAFCAU 29,779,81
orl-dck LD50:109 mg/kg PEMNDP 9,818,91

CONSENSUS REPORTS: Reported in EPA TSCA Inventory. EPA Extremely Hazardous Substances List.

SAFETY PROFILE: Poison by ingestion, inhalation, and skin contact. When heated to decomposition it emits very toxic fumes of NO_x and SO_x.

DAB600 CAS:4342-03-4 HR: 3
DACARBAZINE
mf: $C_6H_{10}N_6O$ mw: 182.22

PROP: Ivory microcrystals. Mp: 250–255° (decomp).

SYNS: DETICENE □ DIC □ (DIMETHYLTRIAZENO)IMIDAZOLECARBOXAMIDE □ 4-(DIMETHYLTRIAZENO)IMIDAZOLE-5-CARBOXAMIDE □ 4-(3,3-DIMETHYL-1-TRIAZENO)IMIDAZOLE-5-CARBOXAMIDE □ 4-(5)-(3,3-DIMETHYL-1-TRIAZENO)IMIDAZOLE-5(4)-CARBOXAMIDE □ 5-(DIMETHYLTRIAZENO)IMIDAZOLE-4-CARBOXAMIDE □ 5-(3,3-DIMETHYLTRIAZENO)IMIDAZOLE-4-CARBOXAMIDE □ 5-(3,3-DIMETHYL-1-TRIAZENO)IMIDAZOLE-4-CARBOXAMIDE □ 5-(3,3-DIMETHYL-1-TRIAZENYL)-1H-IMIDAZOLE-4-CARBOXAMIDE □ DTIC □ DTIC-DOME □ NCI-C04717 □ NSC-45388

TOXICITY DATA WITH REFERENCE
mma-sat 100 μg/plate CRNGDP 3,467,82
sce-ham:ovr 200 mg/L CNREA8 43,577,83
ipr-rat TDLo:50 mg/kg (12D preg):TER CNREA8 33,2231,73
ipr-rat TDLo:400 mg/kg (12D preg):TER CNREA8 33,2231,73
ipr-rat TDLo:100 mg/kg (12D preg):TER CNREA8 33,2231,73
ipr-rat TDLo:200 mg/kg (12D preg):TER ANREAK 186,461,76
orl-rat TDLo:1730 mg/kg/15W-C:CAR JNCIAM 54,951,75
ipr-rat TDLo:25 mg/kg (20D preg):ETA,TER ARGEAR 50,3-06,80
ipr-rat TDLo:3900 mg/kg/26W-I:CAR RRCRBU 52,1,75
ipr-mus TDLo:1950 mg/kg/26W-I:CAR RRCRBU 52,1,75
orl-rat TD:3700 mg/kg/14W-C:CAR PAACA3 11,73,70
ivn-hmn TDLo:3500 μg/kg:GIT,BLD,BIO CCROBU 57,83,73
orl-rat LD50:2147 mg/kg YACHDS 9,3105,81
ipr-rat LD50:350 mg/kg ARGEAR 50,3-06,80
ivn-rat LD50:411 mg/kg YACHDS 9,3105,81
orl-mus LD50:2032 mg/kg YACHDS 9,3105,81
ipr-mus LD50:567 mg/kg CTRRDO 62,721,78
par-ham LD10:250 mg/kg JSONAU 15,355,80

CONSENSUS REPORTS: NTP 7th Annual Report on Carcinogens. IARC Cancer Review: Group 2B IMEMDT 7,184,87; Human Limited Evidence IMEMDT 26,203,81; Animal Sufficient Evidence IMEMDT 26,203,81. NCI Carcinogenesis Studies (ipr); Clear Evidence: mouse, rat RRCRBU 52,1,75. EPA Genetic Toxicology Program.

SAFETY PROFILE: Confirmed carcinogen with experimental carcinogenic and tumorigenic data. Poison by intraperitoneal and parenteral routes. Moderately toxic by ingestion and intravenous routes. Experimental teratogenic effects. Human systemic effects by intravenous route: nausea or vomiting, leukopenia (reduced white blood cell count), and changes in dehydrogenase enzymatic activity. Mutation data reported. When heated to decomposition it emits toxic fumes of NO_x.

DAB700 HR: 2
DAFFODIL

PROP: Bulb-producing plants with leaves that emerge directly from the bulb which looks much like an onion. A leafless stem carries one or more white or yellow flowers, which have a trumpet-shaped section growing from the center of a flat corona. They are native to Europe and northern Africa, and are cultivated as ornamentals in the United States.

SYNS: JONQUIL □ NARCISO (CUBA, MEXICO) □ NARCISSUS □ NARCISSUS POETICUS □ NARCISSUS PSEUDONARCISSUS □ PACIENCIA

SAFETY PROFILE: The bulbs contain the poisonous lycorine and related alkaloids. They are sometimes mistaken for onions. Ingestion of large amounts can cause nausea, persistent vomiting, and diarrhea.

DAB750 CAS:18067-13-5 HR: 3
DAIPIN
mf: $C_{18}H_{24}NO_4 \cdot CH_3O_4S$ mw: 429.53

SYNS: DD 234 □ ESPASMO GASIUM □ N-METHYLHYOSCINE METHYL SULFATE □ N-METHYLSCOPOLAMINE METHOSULFATE □ METHYLSCOPOLAMINE METHYL SULFATE □ N-METHYLSCOPOLAMINE METHYL SULFATE □ METHYLSCOPOLAMMONIUM METHYLSULFATE □ SANDRIX

TOXICITY DATA WITH REFERENCE
scu-rat TDLo:45,500 μg/kg (91D male):REP KSRNAM 7,192,73
orl-rat LD50:5590 mg/kg IYKEDH 4,90,73
ipr-rat LD50:212 mg/kg IYKEDH 4,90,73
scu-rat LD50:1340 mg/kg IYKEDH 4,90,73
ivn-rat LD50:56,800 μg/kg IYKEDH 4,90,73
orl-mus LD50:3010 mg/kg IYKEDH 4,90,73
ipr-mus LD50:116 mg/kg IYKEDH 4,90,73
scu-mus LD50:558 mg/kg IYKEDH 4,90,73
ivn-mus LD50:36,900 μg/kg IYKEDH 4,90,73

SAFETY PROFILE: Poison by intravenous and intraperitoneal routes. Moderately toxic by ingestion and subcutaneous routes. Experimental reproductive effects. When heated to decomposition it emits toxic fumes of SO_x and NO_x.

DAB800 CAS:1172-18-5 HR: 3
DALMANE
mf: $C_{21}H_{23}ClFN_3O \cdot 2ClH$ mw: 460.84

PROP: A solid. Mp: 208–218°.

SYNS: BENOZIL ☐ DALMADORM ☐ DALMADORM HYDROCHLORIDE ☐ DALMATE ☐ DORMODOR ☐ FELISON ☐ FLURAZEPAM DIHYDROCHLORIDE ☐ FLURAZEPAM HYDROCHLORIDE ☐ ID 480 DIHYDROCHLORIDE ☐ INSUMIN ☐ LUNIPAK ☐ NSC-78559 ☐ RO 5-6901 ☐ SOMLAN

TOXICITY DATA WITH REFERENCE
orl-rat LD50:978 mg/kg NIIRDN 6,7-06,82
ipr-rat LD50:179 mg/kg NIIRDN 6,7-06,82
scu-rat LD50:859 mg/kg NIIRDN 6,7-06,82
ivn-rat LD50:40,500 μg/kg NIIRDN 6,7-06,82
orl-mus LD50:660 mg/kg 26RAAN -,47,73
ipr-mus LD50:201 mg/kg NIIRDN 6,7-06,82
scu-mus LD50:440 mg/kg OYYAA2 14,637,77
ivn-mus LD50:66,900 μg/kg NIIRDN 6,7-06,82
orl-rbt LD50:568 mg/kg AIPTAK 178,216,69

SAFETY PROFILE: Poison by intravenous and intraperitoneal routes. Moderately toxic by ingestion and subcutaneous routes. Habituating and possibly addictive. An hypnotic and sedative. When heated to decomposition it emits very toxic fumes of F^-, NO_x, and Cl^-.

DAB807 CAS:5934-69-0 HR: 3
DAMANTOYLDIAZOMETHANE
mf: $C_{12}H_{16}N_2O$ mw: 204.30

SYN: KETONE, 1-ADAMANTYLDIAZO METHYL

TOXICITY DATA WITH REFERENCE
ipr-mus LD:>600 mg/kg PCJOAU 8,396,74

DOT CLASSIFICATION: 3; *Label:* Flammable Liquid

SAFETY PROFILE: Moderately toxic by intraperitoneal route. A flammable liquid. When heated to decomposition it emits toxic vapors of NO_x and Cl^-.

DAB815 CAS:2307-55-3 HR: 3
2,4-D AMMONIUM SALT
mf: $C_8H_6Cl_2O_3 \cdot H_3N$ mw: 238.08

TOXICITY DATA WITH REFERENCE
orl-rat TDLo:100 μg/kg (female 9D post):TER GISAAA 44(4),70,79
skn-mus TDLo:1300 mg/kg/86W-I:ETA VPITAR 33(5),83,74
ipr-mus LDLo:250 mg/kg JIDHAN 29,85,47

CONSENSUS REPORTS: IARC Cancer Review: Animal Inadequate Evidence IMEMDT 15,111,77.

SAFETY PROFILE: A poison by intraperitoneal route. Questionable carcinogen with experimental tumorigenic data. An experimental teratogen. When heated to decomposition it emits very toxic fumes of Cl^-, NO_x, and NH_3.

DAB820 CAS:101052-67-9 HR: 3
DAMPA D

PROP: Contains 3% dibenzthion, 0.3% dichlorothiocyanoaniline, 1% diphenylhydramine (NIIRDN 6,455,82).

TOXICITY DATA WITH REFERENCE
ipr-rat LD50:390 mg/kg NIIRDN 6,445,82
orl-mus LD50:1880 mg/kg NIIRDN 6,445,82
ipr-mus LD50:380 mg/kg NIIRDN 6,445,82
scu-mus LD50:940 mg/kg NIIRDN 6,445,82

SAFETY PROFILE: Poison by intraperitoneal route. Moderately toxic by ingestion and subcutaneous routes. A topical antibacterial agent. When heated to decomposition it emits toxic fumes of SO_x, NO_x, and CN^-. See also BENZHYDRYL, THIOCYANATES, and individual components.

DAB825 CAS:39515-41-8 HR: 3
DANITOL
mf: $C_{22}H_{23}NO_3$ mw: 349.43

PROP: Synthetic pyrethroid insecticide with repellent and contact activity. Pale yellow oil, n (26/D) 1.5283.

SYNS: α-CYANO-3-PHENOXYBENZYL 2,2,3,3-TETRAMETHYL-1-CYCLOPROPANECARBOXYLATE ☐ FENPROPANAGE ☐FENPROPATHRIN ☐ GENPROPATHRIN ☐ MEOTHRIN ☐ RODY ☐ S 32-06 ☐ SD 417-06 ☐ WL 417-06

TOXICITY DATA WITH REFERENCE
orl-rat LD50:18 mg/kg PSSCBG 8,579,77
ivn-rat LDLo:2500 μg/kg ARTODN 45,325,80
ipr-rat LD50:180 mg/kg JPIFAN (38),21,81
scu-rat LD50:900 mg/kg JPIFAN (38),21,81
orl-mus LD50:58 mg/kg JPIFAN (38),21,81
skn-mus LD50:740 mg/kg JPIFAN (38),21,81
ipr-mus LD50:210 mg/kg JPIFAN (38),21,81
skn-rbt LD50:2000 mg/kg FMCHA2 -,C70,83

CONSENSUS REPORTS: Cyanide and its compounds are on the Community Right-To-Know List.

SAFETY PROFILE: Poison by ingestion, intraperitoneal, and intravenous routes. Moderately toxic by skin contact. When heated to decomposition it emits toxic fumes of NO_x and CN^-. See also CYANIDE and ESTERS.

DAB830 CAS:17230-88-5 HR: 3
DANOCRINE
mf: $C_{22}H_{27}NO_2$ mw: 337.50

PROP: Crystals from acetone. Mp: 224.4–226.8°.

SYNS: CHRONOGYN ☐ CYCLOMEN ☐ DANAZOL ☐ DANOL ☐ LADOGAL ☐ 17-α-2,4-PREGNADIEN-20-YNO(2,3-d)ISOXAZOL-17-OL ☐ 17-α-PREGNA-2,4-DIEN-20-YNO(2,3-d)ISOXAZOL-17-OL ☐ 17-α-PREGN-4-EN-20-YNO(2,3-d)ISOXAZOL-17-OL ☐ WIN 17757 ☐ WINOBANIN

TOXICITY DATA WITH REFERENCE
orl-wmn TDLo:16 μg/kg (female 5D post):REP CCPTAY 27,39,83
orl-wmn TDLo:90 mg/kg (female 90D pre):REP JRPMAP 17,98,76
orl-wmn TDLo:120 mg/kg (female 30D pre):REP INJFA3 25,75,80

orl-man TDLo:1028 mg/kg (17W male):REP CCPTAY 7,357,73
orl-man TDLo:430 mg/kg (male 21W pre):REP JCEMAZ 32,522,71
orl-mky TDLo:1800 mg/kg (female 90D pre):REP JIMRBV 5(Suppl 3),1,77
orl-rat TDLo:600 mg/kg (female 6D pre):REP JIMRBV 5(Suppl 3),1,77
orl-mky TDLo:3600 mg/kg (female 90D pre):REP JIMRBV 5(Suppl 3),1,77
orl-rat TDLo:5600 mg/kg (female 14D pre):REP JIMRBV 5(Suppl 3),1,77
orl-rat TDLo:37,500 µg/kg (female 3D pre):REP FESTAS 25,367,74
scu-rat TDLo:93 mg/kg (male 7D pre):REP JRPFA4 77,233,86
orl-rat TDLo:462 mg/kg (male 14D pre):REP FESTAS 25,367,74
scu-rat TDLo:520 mg/kg (male 39D pre):REP JRPFA4 77,233,86
orl-rat TDLo:1400 mg/kg (male 14D pre):REP FESTAS 25,367,74
orl-wmn TDLo:2920 mg/kg/2Y-C:CAR JSONAU 28,114,85
orl-wmn TDLo:120 mg/kg/10D-I:SKN AIMDAP 145,2251,85
ipr-mus LD50:6770 mg/kg IYKEDH 14,484,83

SAFETY PROFILE: Human systemic effects by ingestion: allergic dermatitis. Human male reproductive effects by ingestion: changes in spermatogenesis, impotence, and other unspecified effects. Human female reproductive effects by ingestion: menstrual cycle changes or disorders, changes in fertility, and other unspecified effects. Other experimental reproductive effects. Questionable human carcinogen producing liver tumors. When heated to decomposition it emits toxic fumes of NO_x.

DAB840 CAS:14663-23-1 HR: 2
DANTRIUM HEMIHEPTAHYDRATE
mf: $C_{14}H_9N_4O_5 \cdot Na$ mw: 336.26

SYNS: DANTOROLENE SODIUM ◻ DANTRIUM ◻ DANTROLENE SODIUM ◻ F-400 ◻ 2,4-IMIDAZOLIDINEDIONE, 1-(((5-(4-NITROPHENYL)-2-FURANYL)METHYLENE)AMINO)-, SODIUM SALT ◻ 1-((5-(p-NITROPHENYL)FURFURYLIDENE)AMINO)HYDANTOIN SODIUM

TOXICITY DATA WITH REFERENCE
orl-rat TDLo:500 mg/kg (14D pre/7-17D preg):REP KSRNAM 11,2218,77
orl-rat TDLo:1820 mg/kg (female 26W pre):REP KSRNAM 11,2691,77
orl-rat TDLo:1050 mg/kg (35D male):REP KSRNAM 11,2691,77
orl-rat TDLo:500 mg/kg (14D pre/7-17D preg):TER KSRNAM 11,2218,77
orl-hmn TDLo:320 mg/kg:BAH,GIT JAMAAP 231,662,75
orl-wmn LDLo:600 mg/kg:GIT,SYS NYSJAM 77,1759,77
orl-rat LD50:7431 mg/kg IYKEDH 12,668,81
ipr-rat LD50:413 mg/kg IYKEDH 12,668,81
orl-mus LD50:1188 mg/kg JPMSAE 69,327,80
ipr-mus LD50:534 mg/kg IYKEDH 12,668,81

SAFETY PROFILE: Moderately toxic by intraperitoneal and ingestion routes. Human systemic effects by ingestion: anorexia (human), hepatitis, nausea or vomiting, somnolence. An experimental teratogen. Other experimental reproductive effects. When heated to decomposition it emits toxic fumes of NO_x and Na_2O.

DAB845 CAS:7261-97-4 HR: 2
DANTROLENE
mf: $C_{14}H_{10}N_4O_5$ mw: 314.28

PROP: Crystals from DMF (aq). Mp: 279–280°.

SYN: HYDANTOIN, 1-((5-(p-NITROPHENYL)FURFURYLIDENE)AMINO)-

TOXICITY DATA WITH REFERENCE
unr-man TDLo:9386 mg/kg/3Y-C:CAR PGMJAO 56,261,80

SAFETY PROFILE: Questionable human carcinogen producing Hodgkin's disease. When heated to decomposition it emits toxic fumes of NO_x.

DAB850 CAS:28164-88-7 HR: 3
DAPHNETOXIN
mf: $C_{27}H_{30}O_8$ mw: 482.57

PROP: Crystals. Mp: 194–196°.

SYN: ORTHOBENZOIC ACID, cyclic 7,8,10a-ESTER with 5,6-EPOXY-4,5,6,6a,7,8,9,10,10a,10b-DECAHYDRO-3a,4,7,8,10a-PENTAHYDROXY-5-(HYDROXYMETHYL)-8-ISOPROPENYL-2,10-DIMETHYLBEN Z(e)AZULEN-3(3aH)-ONE

TOXICITY DATA WITH REFERENCE
skn-hmn 5 µg/24H SEV TOXIA6 19,841,81
ipr-mus LD50:1100 µg/kg SCYYDZ 9(2),48,89

SAFETY PROFILE: Poison by intraperitoneal route. A severe skin irritant. When heated to decomposition it emits acrid smoke and irritating fumes.

DAB875 CAS:71-81-8 HR: 3
DARBID
mf: $C_{23}H_{33}N_2O \cdot I$ mw: 480.48

PROP: Crystals or amorphous powder. Mp: 198–201° (decomp). The methiodide is freely sol in boiling water, methanol, ethanol, chloroform; practically insol in ether.

SYNS: 2,2-DIPHENYL-4-DIISOPROPYLAMINOBUTYRAMIDE METHIODIDE ◻ DIPRAMID ◻ DIPRAMIDE ◻ ISAMID ◻ ISOPROPAMIDE ◻ ISOPROPAMIDE IODIDE ◻ MARYGIN-M ◻ 5579 MD ◻ PIACCAMIDE ◻ PRIAMIDE ◻ PRIAZIMIDE ◻ R 79 ◻ SANULCIN ◻ SKF 4740 ◻ TYRIMIDE

TOXICITY DATA WITH REFERENCE
orl-mus LD50:1600 mg/kg AIPTAK 103,120,55
ipr-mus LD50:120 mg/kg AIPTAK 103,120,55
ivn-mus LD50:12,779 mg/kg AIPTAK 103,100,55

SAFETY PROFILE: Poison by intraperitoneal route. Moderately toxic by ingestion. When heated to decomposition it emits toxic fumes of I^- and NO_x.

DAB879 CAS:469-62-5 HR: 3
DARVON
mf: $C_{22}H_{29}NO_2$ mw: 339.52

PROP: Crystals from pet ether. Mp: 75–76°.

SYNS: DEXTROPROPOXYPHENE ☐ α-(+)-4-DIMETHYLAMINO-1,2-DIPHENYL-3-METHYL-2-BUTANOL PROPIONATE ESTER ☐ DOLENE ☐ DOLOXENE ☐ (+)-PROPOXYPHENE ☐ d-PROPOXYPHENE ☐ PROXAGESIC ☐ SK 65

TOXICITY DATA WITH REFERENCE
orl-hmn TDLo:650 mg/kg:PUL,CNS,CVS AMSVAZ 200,241,76
orl-rat LD50:135 mg/kg AIPTAK 178,446,69
ipr-rat LD50:50 mg/kg AIPTAK 178,446,69
scu-rat LD50:79,100 μg/kg ARZNAD 24,600,74
orl-mus LD50:270 mg/kg ARZNAD 24,600,74
ipr-mus LD50:110 mg/kg AIPTAK 178,446,69
scu-mus LD50:113 mg/kg ARZNAD 24,600,74
ivn-mus LD50:25 mg/kg JPETAB 134,154,61

SAFETY PROFILE: Poison by ingestion, intraperitoneal, subcutaneous, and intravenous routes. Human systemic effects by ingestion: change in cardiac rate, respiratory depression and coma. When heated to decomposition it emits toxic fumes of NO_x.

DAB880 CAS:26570-10-5 HR: 3
DARVON-N
mf: $C_{10}H_8O_3S•C_{22}H_{29}NO_2•H_2O$ mw: 565.78

PROP: Crystals.

SYNS: d-4-DIMETHYLAMINO-3-METHYL-1,2-DIPHENYL-2-PROPIONOXYBUTANENAPHTHALENE-2-SULPHONATE HYDRATE ☐ d-PROPOXYPHENE NAPSYLATE HYDRATE ☐ S-9700

TOXICITY DATA WITH REFERENCE
orl-rat LD50:485 mg/kg KSRNAM 5,1011,71
ipr-rat LD50:56 mg/kg KSRNAM 5,1011,71
scu-rat LD50:169 mg/kg KSRNAM 5,1011,71
orl-mus LD50:973 mg/kg OYYAA2 5,359,71
ipr-mus LD50:171 mg/kg KSRNAM 5,2043,71
scu-mus LD50:749 mg/kg KSRNAM 5,1011,71

SAFETY PROFILE: Poison by ingestion, subcutaneous, and intraperitoneal routes. When heated to decomposition it emits toxic fumes of SO_x and NO_x. See also d-PROPOXYPHENE HYDROCHLORIDE and SULFONATES.

DAC000 CAS:20830-81-3 HR: 3
DAUNOMYCIN
mf: $C_{27}H_{29}NO_{10}$ mw: 527.57

PROP: Thin, red needles. Mp: 190° (decomp). Isolated from cultures of a *Streptomyces* (CNREA8 32,1029,72).

SYNS: ACETYLADRIAMYCIN ☐ CERUBIDIN ☐ DAUNAMYCIN ☐ DAUNORUBICIN ☐ DAUNORUBICINE ☐ DM ☐ FI6339 ☐ LEUKAEMOMYCIN C ☐ NCI-C04693 ☐ NSC-82151 ☐ RCRA WASTE NUMBER U059 ☐ RP 13057 ☐ 13,057 R.P. ☐ RUBIDOMYCIN ☐ RUBIDOMYCINE ☐ RUBOMYCIN C ☐ RUBOMYCIN C 1 ☐ STREPTOMYCES PEUCETIUS

TOXICITY DATA WITH REFERENCE
mmo-sat 1 μg/plate ENMUDM 7,129,85
pic-sat 800 ng/plate MUREAV 110,243,83
dni-hmn:oth 30 nmol/L CNREA8 44,2421,84
ipr-rat TDLo:8 mg/kg (7-14D preg):TER CRSBAW 163,1299,69
ipr-rat TDLo:4 mg/kg (6-9D preg):TER TJADAB 17,151,78
ipr-rat TDLo:16 mg/kg (7-14D preg):REP CRSBAW 163,1299,69
scu-mus TDLo:35 mg/kg (female 7D pre):REP ARZNAD 17,948,67
ipr-rat TDLo:2200 μg/kg/7W-I:ETA CANCAR 40,1935,77
ipr-rat TDLo:9 mg/kg (7-9D preg):TER CRSBAW 163,1299,69
ivn-rat TDLo:6250 μg/kg:CAR CNREA8 38,1444,78
ipr-mus TDLo:2340 μg/kg/26W-I:ETA CANCAR 40,1935,77
scu-mus TDLo:15 g/kg/12W-I:CAR RRCRBU 20,73,69
ivn-rat TD:5 mg/kg:NEO,REP CNREA8 32,1029,72
ivn-rat TD:13 mg/kg:CAR EXPEAM 27,1209,71
ivn-rat TD:10 mg/kg:CAR JJIND8 66,81,81
ivn-rat TD:40 mg/kg/4W-I:CAR AFCPDR 34,369,82
ivn-rat TD:10 mg/kg:CAR LAPPA5 38,71,78
ivn-rat TD:5 mg/kg:ETA CNREA8 36,2065,76
scu-mus TD:15 mg/kg/12W-I:NEO ARZNAD 17,948,67
orl-hmn LDLo:6 mg/kg 34ZIAG -,521,69
ipr-rat LD50:20 mg/kg ADTEAS 3,181,68
ivn-rat LD50:13 mg/kg NCINS* -,304,67
orl-mus LD50:205 mg/kg YKYUA6 25,573,74
ipr-mus LD50:2500 μg/kg NCINS* -,304,67
scu-mus LD50:16 mg/kg COREAF 257,1813,63
ivn-mus LD50:18 mg/kg 85FZAT -,243,67

CONSENSUS REPORTS: IARC Cancer Review: Group 2B IMEMDT 7,56,87; Animal Sufficient Evidence IMEMDT 10,145,76. NCI Carcinogenesis Studies (ipr); Clear Evidence: rat CANCAR 40,1935,77; No Evidence: mouse CANCAR 40,1935,77. EPA Genetic Toxicology Program.

SAFETY PROFILE: Confirmed carcinogen with experimental carcinogenic, neoplastigenic, and tumorigenic data. Human poison by ingestion. Experimental poison by subcutaneous, intravenous, and intraperitoneal routes. Experimental teratogenic and reproductive effects. Human mutation data reported. When heated to decomposition it emits toxic fumes of NO_x. See also DAUNOMYCIN HYDROCHLORIDE.

DAC200 CAS:23541-50-6 HR: 3
DAUNOMYCIN HYDROCHLORIDE
mf: $C_{27}H_{29}NO_{10}•ClH$ mw: 564.03

PROP: Thin, red needles. Mp: 188–190°. Sol in H_2O.

SYNS: CERUBIDINE ☐ DAUNOBLASTIN ☐ DAUNOBLASTINA ☐ DAUNOMYCIN CHLOROHYDRATE ☐ DAUNORUBICIN HYDROCHLORIDE ☐ NDC 0082-4155 ☐ NSC-82151 ☐ ONDENA ☐ RP 13057 HYDROCHLORIDE ☐ RUBIDOMYCIN ☐ RUBIDOMYCIN HYDROCHLORIDE ☐ RUBOMYCIN C

TOXICITY DATA WITH REFERENCE
mmo-sat 83 ng/plate CNREA8 38,2148,78
mmo-asn 250 mg/L MUREAV 97,293,82
orl-rat LD50:290 mg/kg NIIRDN 6,437,82
ipr-rat LD50:14,300 μg/kg NIIRDN 6,437,82
scu-rat LD50:33,200 μg/kg NIIRDN 6,437,82

ivn-rat LD50:14,300 μg/kg NIIRDN 6,437,82
orl-mus LD50:205 mg/kg NIIRDN 6,437,82
ipr-mus LD50:3050 μg/kg NIIRDN 6,437,82
scu-mus LD50:28,800 μg/kg NIIRDN 6,437,82

CONSENSUS REPORTS: EPA Genetic Toxicology Program.

SAFETY PROFILE: Poison by ingestion, intraperitoneal, subcutaneous, and intravenous routes. Mutation data reported. An antineoplastic agent. When heated to decomposition it emits very toxic fumes of NO_x and HCl. See also DAUNOMYCIN.

DAC300 **CAS:28008-55-1** **HR: 3**
DAUNOMYCINOL
mf: $C_{27}H_{31}NO_{10}$ mw: 529.59

SYNS: ANTIBIOTIC 20-798RP □ DAUNORUBICINOL □ DIHYDRODAUNOMYCIN □ 13-DIHYDRODAUNOMYCIN □ 13-DIHYDRODAUNORUBICIN □ DUBORIMYCIN □ 1-HYDROXY-13-DIHYDRODAUNOMYCIN □ LEUKAEMOMYCIN D

TOXICITY DATA WITH REFERENCE
pic-esc 1400 μg/L MUREAV 77,197,80
dni-mus:leu 7600 nmol/L JANTAJ 34,1596,81
oms-mus:leu 3 μmol/L JANTAJ 34,1596,81
ipr-mus LD50:6500 μg/kg 85GDA2 3,122,80

SAFETY PROFILE: Poison by intraperitoneal route. Mutation data reported. When heated to decomposition it emits toxic fumes of NO_x. See also DAUNOMYCIN.

DAC400 **CAS:8016-03-3** **HR: 1**
DAVANA OIL

PROP: Distilled from *Artemisia pallens* (FCTXAV 14,659,76).

TOXICITY DATA WITH REFERENCE
skn-rbt 500 mg/24H MOD FCTXAV 14,737,76

CONSENSUS REPORTS: Reported in EPA TSCA Inventory.

SAFETY PROFILE: A moderate skin irritant. When heated to decomposition it emits acrid smoke and irritating fumes.

DAC450 **CAS:12011-76-6** **HR: 2**
DAWSONITE
mf: CH_2AlO_5•Na mw: 144.00

SYN: CRYSTALLINE DEHYDROXY SODIUM ALUMINUM, CARBONATE

TOXICITY DATA WITH REFERENCE
imp-rat TDLo:200 mg/kg:NEO JJIND8 67,965,81

SAFETY PROFILE: Questionable carcinogen with experimental neoplastigenic data.

DAC500 **HR: 2**
DAY BLOOMING JESSAMINE

PROP: Evergreen shrubs with smooth-edged, oval leaves and clusters of white, tubular flowers. The day blooming jessamine has flowers that are fragrant during the day with black berries. The night blooming jessamine has flowers that are fragrant at night with white berries. They are native to the West Indies and grow wild and are under cultivation in the United States (Florida, Texas) and Guam.

SYNS: 'ALA-AUMOE (HAWAII) □ C. DIURNUM □ C. NOCTURNUM □ CESTRUM (VARIOUS SPECIES) □ CHINESE INKBERRY □ DAMA de DIA (PUERTO RICO) □ DAMA de NOCHE (PUERTO RICO) □ GALAN de DIA (CUBA) □ GALAN de NOCHE (CUBA) □ HUELE de NOCHE (MEXICO) □ JASMIN de NUIT (HAITI) □ KUPAOA (HAWAII) □ LILAS de NUIT (HAITI) □ MAKAHALA (HAWAII) □ NIGHT BLOOMING JESSAMINE □ ONAONA-IAPANA (HAWAII)

SAFETY PROFILE: The berries and sap contain toxic saponins and nicotine. Ingestion of these plant parts may result in inflammation of the stomach and intestines.

DAC800 **CAS:33857-26-0** **HR: 2**
DCDD
mf: $C_{12}H_6Cl_2O_2$ mw: 253.08

PROP: Colorless crystals. Mp: 201–202°.

SYNS: 2,7-DICHLORODIBENZODIOXIN □ 2,7-DICHLORODIBENZO-p-DIOXIN □ 2,7-DICHLORODIBENZO(b,e)(1,4)DIOXIN □ NCI-C03667

TOXICITY DATA WITH REFERENCE
eye-rbt 2 mg MLD EVHPAZ 5,87,73
orl-rat TDLo:5 mg/kg (6-15D preg):TER ADCSAJ 120,70,73
orl-mus TDLo:378 g/kg/90W-C:ETA NCITR* NCI-CG-TR-123,79
orl-mus TD:756 g/kg/90W-C:ETA NCITR* NCI-CG-TR-123,79

CONSENSUS REPORTS: IARC Cancer Review: Group 3 IMEMDT 7,56,87; Animal Inadequate Evidence IMEMDT 15,41,77; Human No Adequate Data IMEMDT 15,41,77. NCI Carcinogenesis Bioassay (feed); Clear Evidence: mouse NCITR* NCI-CG-TR-123,79; No Evidence: rat NCITR* NCI-CG-TR-123,79.

SAFETY PROFILE: An eye irritant. Experimental teratogenic data. Questionable carcinogen with experimental tumorigenic data. When heated to decomposition it emits toxic fumes of Cl^-.

DAC975 **CAS:66826-72-0** **HR: 2**
cis-DCPO
mf: $C_3H_4Cl_2O$ mw: 126.97

PROP: Bp: 78–80° @ 130 mm.

SYNS: cis-2-CHLORO-3-(CHLOROMETHYL)OXIRANE □ cis-1,3-DICHLORO-1,2-EPOXYPROPANE □ cis-1,3-DICHLOROPROPENE OXIDE

TOXICITY DATA WITH REFERENCE
otr-ham:emb 5 μmol/L JJIND8 69,531,82
skn-mus TDLo:400 mg/kg/53W-I:CAR CNREA8 43,159,83
scu-mus TDLo:20 mg/kg/71W-I:CAR CNREA8 43,159,83

SAFETY PROFILE: Questionable carcinogen with experimental carcinogenic data. Mutation data reported. When heated to decomposition it emits toxic fumes of Cl^-.

DAD040 **CAS:61848-70-2** **HR: 3**
cis-DDCP
mf: $C_6H_4Cl_2N_2Pt$ mw: 370.11

PROP: Pale yellow crystals. Sol in DMF.

SYNS: 1,2-DIAMINOCYCLOHEXANEPLATINUM(II) CHLORIDE □ DICHLORO(1,2-CYCLOHEXANEDIAMINE)PLATINUM □ DICHLORO(1,2-DIAMINOCYCLOHEXANE)PLATINUM □ DICHLORO(1,2-DIAMINO-CYCLOHEXANE)PLATINUM(II) □ cis-DICHLORO-1,2-DIAMINOCY-CLOHEXANE PLATINUM(II) □ NSC 194814 □ PT 155

TOXICITY DATA WITH REFERENCE
mmo-sat 20 nmol/plate CNREA8 41,4368,81
dnd-sat 10 mg/L/20H-C CNREA8 41,4368,81
oms-bcs 9900 nmol/L/3H-C CNREA8 41,4368,81
ipr-mus TDLo:20,530 μg/kg/10W-I:NEO CNREA8 41,4368,81
ipr-mus TD:41,060 μg/kg/10W-I:NEO CNREA8 41,4368,81
ipr-mus LD50:40,130 μg/kg NCISP* JAN86

SAFETY PROFILE: Poison by intraperitoneal route. Questionable carcinogen with experimental neoplastigenic data. Mutation data reported. When heated to decomposition it emits toxic fumes of Cl^- and NO_x. See also PLATINUM COMPOUNDS.

DAD050 **HR: 2**
trans(+)-DDCP
mf: $C_6H_{14}Cl_2N_2Pt$ mw: 380.176

SYNS: (SP-4-2)-trans(+)-DICHLORO(1,2-CYCLOHEXANEDIAMINE-N,N')-(9CI) □ XX 212

TOXICITY DATA WITH REFERENCE
mmo-sat 20 nmol/plate CNREA8 41,4368,81
dnd-sat 10 mg/L/20H-C CNREA8 41,4368,81
oms-bcs 20 μmol/L/3H-C CNREA8 41,4368,81
ipr-mus TDLo:20,530 μg/kg/10W-I:NEO CNREA8 41,4368,81
ipr-mus TD:41,060 μg/kg/10W-I:NEO CNREA8 41,4368,81

SAFETY PROFILE: Questionable carcinogen with experimental neoplastigenic data. Mutation data reported. When heated to decomposition it emits toxic fumes of Cl^- and NO_x. See also PLATINUM COMPOUNDS.

DAD075 **CAS:61848-66-6** **HR: 2**
trans(−)-DDCP
mf: $C_6H_{14}Cl_2N_2Pt$ mw: 380.21

PROP: Bright yellow solid. Sol in DMF.

SYN: trans(−)-DICHLORO-1,2-DIAMINOCYCLOHEXANEPLATINUM (II)

TOXICITY DATA WITH REFERENCE
mmo-sat 20 nmol/plate CNREA8 41,4368,81
dnd-sat 10 mg/L/20H-C CNREA8 41,4368,81
oms-bcs 13 μmol/L/3H-C CNREA8 41,4368,81
ipr-mus TDLo:20,530 μg/kg/10W-I:NEO CNREA8 41,4368,81
ipr-mus TD:41,060 μg/kg/10W-I:NEO CNREA8 41,4368,81

SAFETY PROFILE: Questionable carcinogen with experimental neoplastigenic data. Mutation data reported. When heated to decomposition it emits toxic fumes of Cl^- and NO_x. See also PLATINUM COMPOUNDS.

DAD200 **CAS:50-29-3** **HR: 3**
DDT
mf: $C_{14}H_9Cl_5$ mw: 354.48

PROP: Colorless crystals or white to sltly off-white powder. Odorless or with slight aromatic odor. Mp: 108.5–109°.

SYNS: AGRITAN □ ANOFEX □ ARKOTINE □ AZOTOX □ α,α-BIS (p-CHLOROPHENYL)-β,β,β-TRICHLOROETHANE □ 1,1-BIS-(p-CHLOROPHENYL)-2,2,2-TRICHLOROETHANE □ 2,2-BIS(p-CHLOROPHENYL)-1,1,1-TRICHLOROETHANE □ BOSAN SUPRA □ BOVIDERMOL □ CHLOROPHENOTHAN □ CHLOROPHENOTHANE □ CHLOROPHENOTOXUM □ CITOX □ CLOFENOTANE □ p,p'-DDT □ DEDELO □ DEOVAL □ DETOX □ DETOXAN □ DIBOVAN □ DICHLORODIPHENYLTRICHLOROETHANE □ DICHLORODIPHENYLTRICHLOROETHANE (DOT) □ p,p'-DICHLORODIPHENYLTRICHLOROETHANE □ 4,4'-DICHLORODIPHENYLTRICHLOROETHANE □ DICOPHANE □ DIDIGAM □ DIDIMAC □ DIPHENYLTRICHLOROETHANE □ DODAT □ DYKOL □ ENT 1,506 □ ESTONATE □ GENITOX □ GESAFID □ GESAPON □ GESAREX □ GESAROL □ GUESAPON □ GUESAROL □ GYRON □ HAVERO-EXTRA □ HILDIT □ IVORAN □ IXODEX □ KOPSOL □ MICRO DDT 75 □ MUTOXIN □ NCI-C00464 □ NEOCID □ PARACHLOROCIDUM □ PEB1 □ PENTACHLORIN □ PENTECH □ PPZEIDAN □ R50 □ RCRA WASTE NUMBER U061 □ RUKSEAM □ SANTOBANE □ TECH DDT □ 1,1,1-TRICHLOOR-2,2-BIS(4-CHLOOR FENYL)-ETHAAN (DUTCH) □ 1,1,1-TRICHLOR-2,2-BIS(4-CHLOR-PHENYL)-AETHAN (GERMAN) □ TRICHLOROBIS (4-CHLOROPHENYL) ETHANE □ 1,1,1-TRICHLORO-2,2-DI(4-CHLOROPHENYL)-ETHANE □ 1,1'-(2,2,2-TRICHLOROETHYLIDENE)BIS(4-CHLOROBENZENE) □ 1,1,1-TRICLORO-2,2-BIS(4-CLORO-FENIL)-ETANO (ITALIAN) □ ZEIDANE □ ZERDANE

TOXICITY DATA WITH REFERENCE
dni-dmg:oth 250 ppb EXPEAM 41,745,85
cyt-hmn:lym 200 μg/L/72H MUREAV 40,131,76
dnd-mus:ast 15 μmol/L MUREAV 89,95,81
orl-mus TDLo:504 mg/kg (lactating female 21D post):REP ENPBBC 4,189,74
orl-dog TDLo:3540 mg/kg (multi) :REP AECTCV 6,83,77
orl-rat TDLo:430 mg/kg (female 1-22D post):REP BECTA6 12,373,74
ipr-rat TDLo:21 mg/kg (lactating female 21D post):REP NATUAS 228,1222,70
scu-mus TDLo:418 mg/kg (female 6-14D post):TER NTIS** PB223-160
orl-rbt TDLo:150 mg/kg (female 7-9D post):TER AIPTAK 192,286,71
orl-rat TDLo:1890 mg/kg (female 36W pre):REP AECTCV 3,479,75/76
orl-rbt TDLo:150 mg/kg (female 7-9D post):REP AIPTAK 192,286,71
orl-rat TDLo:100 mg/kg (male 1D pre):REP FCTXAV 11,53,73
orl-mus TDLo:124 mg/kg (female 62D pre):REP ENPBBC 3,127,73
scu-mus TDLo:40 mg/kg (female 3D pre):REP IJMRAQ 65,576,77
orl-rbt TDLo:150 mg/kg (female 7-9D post):REP AIPTAK 192,286,71
ipr-rat TDLo:60 mg/kg (female 3D pre):REP TXAPA9 18,348,71

orl-rat TDLo: 112 mg/kg (56D male): REP IJEBA6 16,1002,78
orl-rat TDLo: 250 mg/kg (female 15-19D post): TER BNEOBV 26,283,75
orl-rat TDLo: 1225 mg/kg/7W-C: CAR TUMOAB 61,113,75
orl-mus TDLo: 24 mg/kg (MGN): NEO,TER JNCIAM 51,983,73
orl-mus TDLo: 73 mg/kg/26W-C: CAR FCTXAV 7,215,69
scu-mus TDLo: 370 mg/kg/80W-I: NEO IJCNAW 19,725,77
orl-ham TDLo: 21,280 mg/kg/38W-I: ETA,TER PSEBAA 134,113,70
orl-rat TD: 12,096 mg/kg/3Y-C: NEO TUMOAB 68,11,82
orl-mus TD: 7560 mg/kg/90W-C: NEO FCTXAV 11,433,73
orl-mus TD: 5600 mg/kg/80W-I: NEO IJCNAW 19,725,77
orl-rat TD: 8100 mg/kg/2Y-C: ETA TXAPA9 11,88,67
orl-mus TD: 3150 mg/kg/15W-C: ETA ZKKOBW 82,25,74
orl-mus TD: 3408 mg/kg (MGN): NEO,TER IJCNAW 11,688,73
orl-rat TD: 19 g/kg/2Y-C: NEO IJCNAW 19,179,77
orl-rat TD: 438 mg/kg/2Y-C: NEO REONBL 26,177,79
orl-rat TD: 17,976 mg/kg/2Y-C: NEO TUMOAB 68,11,82
orl-rat TD: 24,192 mg/kg/3Y-C: NEO TUMOAB 68,11,82
orl-inf LDLo: 150 mg/kg BMJOAE 2,845,45
orl-man TDLo: 6 mg/kg: CNS,GIT,SKN CMEP** -,1,56
orl-hmn TDLo: 16 mg/kg: CNS CMEP** -,1,56
orl-hmn LDLo: 500 mg/kg: CNS,CVS,PUL MEIEDD 10,409,83
orl-hmn TDLo: 5 mg/kg: CNS PHARAT 2,268,47
unr-man LDLo: 221 mg/kg 85DCAI 2,73,70
orl-rat LD50: 87 mg/kg DOEAAH 35,25,79
skn-rat LD50: 1931 mg/kg SPEADM 74-1,-,74
ipr-rat LD50: 9100 μg/kg PESTD5 17,351,76
scu-rat LD50: 1500 mg/kg BMJOAE 1,865,45
ivn-rat LD50: 68 mg/kg ANTBAL 14,316,69
orl-mus LD50: 135 mg/kg FEPRA7 12,368,53
ipr-mus LD50: 32 mg/kg PESTD5 17,351,76
ivn-mus LD50: 68,500 μg/kg ANTBAL 14,316,69
skn-rbt LD50: 300 mg/kg BMJOAE 1,865,45

CONSENSUS REPORTS: NTP 7th Annual Report on Carcinogens. IARC Cancer Review: Group 2B IMEMDT 7,186,87; Animal Sufficient Evidence IMEMDT 5,83,74; Human Inadequate Evidence IMEMDT 5,83,74. NCI Carcinogenesis Bioassay (feed); No Evidence: mouse, rat NCITR* NCI-CG-TR-131,78. Reported in EPA TSCA Inventory. EPA Genetic Toxicology Program.

OSHA PEL: TWA 1 mg/m³ (skin)
ACGIH TLV: TWA 1 mg/m³
NIOSH REL: (DDT) TWA 0.5 mg/m³; avoid skin contact
DFG MAK: 1 mg/m³

SAFETY PROFILE: Confirmed carcinogen with experimental carcinogenic, neoplastigenic, tumorigenic, and teratogenic data. Human poison by ingestion. Experimental poison by ingestion, skin contact, subcutaneous, intravenous, and intraperitoneal routes. Experimental reproductive effects. Human systemic effects by ingestion: anesthetic, convulsions, headache, analgesia, cardiac arrhythmias, nausea or vomiting, sweating, and unspecified pulmonary changes. Human mutation data reported. An insecticide. When heated to decomposition it emits toxic fumes of Cl^-. See also CHLORINATED HYDROCARBONS, AROMATIC.

A dose of 20 g has proved highly dangerous though not fatal to a human. This dose was taken by 5 persons who vomited an unknown portion of the material and even so recovered only incompletely after 5 weeks. Smaller doses produced less important symptoms with relatively rapid recovery. Experimental ingestion of 1.5 g resulted in great discomfort and moderate neurological changes including paresthesia, tremor, moderate ataxia, exaggeration of part of the reflexes, headache, and fatigue. Vomiting followed only after 11 hours. Recovery was complete on the following day. The fatal dose of DDT for humans is not known. Judging from the literature, no one has ever been killed by DDT in the absence of other insecticides and/or a variety of toxic solvents. However, these common solvent formulations are highly fatal when taken in small doses, partly because of the toxicity of the solvent, and perhaps because of the increased absorbability of the DDT; several fatal cases in humans have been reported. Little is known of the hazard of chronic DDT poisoning. Human volunteers have ingested up to 35 mg/day for 21 months with no ill effects.

DDT and some of its degradation products, particularly DDE, are stored in fat. This storage effect leads to a concentration of DDT at higher levels of the food chain. DDT stored in the fat is at least largely inactive since a greater total dose may be stored in an experimental animal than is sufficient as a lethal dose for that same animal if given at one time. A study based on 75 human cases reported an average of 5.3 ppm of DDT stored in the fat. A higher content of DDT and its derivatives (up to 434 ppm of DDE and 648 ppm of DDT) was found in workers who had very extensive exposure. Without exception, the samples were taken from persons who were either asymptomatic or suffering from some disease completely unrelated to DDT. Careful hospital examination of workers who had been very extensively exposed and who had volunteered for examination revealed no abnormality that could be attributed to DDT. Much higher levels have been found in humans than have been observed in the fat of experimental animals that were apparently asymptomatic. DDT stored in the fat is eliminated only very gradually when further dosage is discontinued. However, weight loss can speed the release of this stored DDT (and DDE) into the blood. After a single dose, the secretion of DDT in the milk and its excretion in the urine reach their height within a day or two and continue at a lower level thereafter.

DAD500 CAS:72732-50-4 HR: 3
DEACETYLDEMETHYLTHYMOXAMINE
mf: $C_{13}H_{21}NO_2$ mw: 223.35

SYNS: (2-(4-HYDROXY-2-ISOPROPYL-5-METHYLPHENOXY)ETHYL) METHYLAMINE □ 2-METHYL-4-(2-(METHYLAMINO)ETHOXY)-5-ISOPROPYL-PHENOL □ 2-METHYL-4-(2-(METHYLAMINO)ETHOXY)-5-(1-METHYLETHYL)-PHENOL

TOXICITY DATA WITH REFERENCE
orl-mus LD50: 340 mg/kg KSRNAM 16,1147,82
ipr-mus LD50: 84 mg/kg KSRNAM 16,1147,82
ivn-mus LD50: 28 mg/kg KSRNAM 16,1147,82

SAFETY PROFILE: Poison by ingestion, intravenous,

and intraperitoneal routes. When heated to decomposition it emits toxic fumes of NO_x.

DAD600 **CAS:34114-98-2** **HR: 3**
DEACETYL-HT-2 TOXIN
mf: $C_{19}H_{30}O_7$ mw: 370.49

SYNS: 12,13-EPOXY-3-α,4-β,8-α,15-TETRAHYDROXYTRICHOTHEC-9-ENE-8-ISOVALERATE ☐ TOXIN T-2 TRIOL ☐ 3-α,4-β,15-TRIHYDROXY-8,α-(3-METHYLBUTYRYLOXY)-12,13-EPOXYTRICHOTHEC-9-ENE

TOXICITY DATA WITH REFERENCE
ipr-mus LD50:108 mg/kg AEMIDF 46,120,83
orl-ckn LD50:30,180 μg/kg AEMIDF 35,636,78

SAFETY PROFILE: Poison by ingestion and intraperitoneal routes. When heated to decomposition it emits acrid smoke and fumes.

DAD650 **CAS:19855-39-1** **HR: 3**
DEACETYLLANATOSIDE B
mf: $C_{47}H_{74}O_{19}$ mw: 943.21

PROP: Prisms or needles from $CHCl_3$/MeOH/Et_2O. Mp: 240° (decomp).

SYNS: DEACETYL-LANATOSIDE B (8CI) ☐ DESACETYL-LANTOSID B (GERMAN) ☐ DESACETYLLANATOSIDE B ☐ GLUCOGITOXIN ☐ PURPUREA B ☐ PURPUREA GLYCOSIDE B ☐ PURPUREAGLYKOSID B (GERMAN)

TOXICITY DATA WITH REFERENCE
ivn-cat LD50:370 μg/kg 85ELDJ -,193,63
orl-frg LD50:8600 μg/kg JJPAAZ 9,91,60
scu-frg LD50:940 μg/kg JJPAAZ 9,91,60

SAFETY PROFILE: Poison by ingestion, subcutaneous, and intravenous routes. When heated to decomposition it emits acrid smoke and fumes.

DAD850 **CAS:35231-36-8** **HR: 3**
DEACETYLTHYMOXAMINE
mf: $C_{14}H_{23}NO$ mw: 221.38

SYNS: DESACETYLTHYMOXAMINE ☐ 4-(2-(DIMETHYLAMINO)ETHOXY)-5-ISOPROPYL-2-METHYLPHENOL ☐ 4-(2-(DIMETHYLAMINO)ETHOXY)-2-METHYL-5-(1-METHYLETHYL)PHENOL ☐ (2-(4-HYDROXY-2-ISOPROPYL-5-METHYLPHENOXY)ETHYL)DIMETHYLAMINE

TOXICITY DATA WITH REFERENCE
orl-mus LD50:335 mg/kg KSRNAM 16,1147,82
ipr-mus LD50:82 mg/kg KSRNAM 16,1147,82
scu-mus LD50:120 mg/kg KSRNAM 16,1147,82
ivn-mus LD50:29 mg/kg KSRNAM 16,1147,82

SAFETY PROFILE: Poison by ingestion, subcutaneous, intravenous, and intraperitoneal routes. When heated to decomposition it emits toxic fumes of NO_x.

DAD880 **HR: 3**
DEADLY NIGHTSHADE

PROP: A heavily branched perennial plant about 3-feet tall with 6-inch oval leaves and red sap. The flowers are about 1-inch long and grow from the leaf joints. The berries are about 0.5-inch in diameter and purple to black in color. It is native to Eurasia and North Africa, and is cultivated in the United States.

SYNS: ATROPA BELLADONNA ☐ BELLADONNA ☐ BLACK NIGHTSHADE ☐ NIGHTSHADE ☐ SLEEPING NIGHTSHADE

SAFETY PROFILE: The whole plant contains various belladonna alkaloids including atropine. Human systemic effects by ingestion: difficulty in speaking and swallowing, rapid heartbeat, fever, dilation of the pupil, blurred vision, hallucinations, urinary retention, and delirium. See also ATROPINE and BELLADONNA.

DAE100 **HR: 3**
DEATH CAMAS

PROP: Bulb-producing plants with 1.5-foot long grassy leaves. It produces yellow or green-white flowers at the end of a leafless stalk. They grow wild throughout Canada and the United States, except southern Florida and Hawaii.

SYNS: ALKALI GRASS ☐ HOG'S POTATO ☐ MYSTER GRASS ☐ POISON SEGO ☐ SAND CORN ☐ SOAP PLANT ☐ SQUIRREL FOOD ☐ WATER LILY ☐ WILD ONION ☐ ZIGADENUS (VARIOUS SPECIES) ☐ ZYGADENUS (VARIOUS SPECIES)

SAFETY PROFILE: All parts of the plant contain the poisons zygadenine, zygacine, germidines and protoveratridine. Human systemic effects by ingestion: mouth pain, persistent vomiting, headache, dizziness, slowed heartbeat, low blood pressure, and convulsions.

DAE200 **CAS:2862-16-0** **HR: 3**
7-DEAZAINOSINE
mf: $C_{11}H_{13}N_3O_5$ mw: 267.27

PROP: Needles from H_2O. Mp: 241–244°.

SYNS: DEAMINOHYDROXYTUBERCINDIN ☐ 7-β-d-RIBOFURANOSYL-7H-PYRROLO(2,3-d)PYRIMIDIN-4-OL

TOXICITY DATA WITH REFERENCE
orl-rat LD50:26 mg/kg CNREA8 29,116,69
ipr-rat LD50:25 mg/kg CNREA8 29,116,69
scu-rat LD50:24 mg/kg CNREA8 29,116,69
ipr-mus LD50:30 mg/kg CNREA8 29,116,69
orl-dog LDLo:48 mg/kg CNREA8 29,116,69
ivn-dog LDLo:48 mg/kg CNREA8 29,116,69

SAFETY PROFILE: Poison by ingestion, intraperitoneal, subcutaneous, and intravenous routes. When heated to decomposition it emits toxic fumes of NO_x.

DAE400 **CAS:17702-41-9** **HR: 3**
DECABORANE
DOT: UN 1868
mf: $B_{10}H_{14}$ mw: 122.24

PROP: Colorless needles or crystals. Solid by sublimation. Mp: 99.6°, d: 0.94. (solid), d: 0.78 (liquid @ 100°), bp: 213°, vap press: 19 mm @ 100°. Sol in CS_2.

SYN: DECABORANE(14)

TOXICITY DATA WITH REFERENCE
orl-rat LD50:64 mg/kg MLSR** #8,51
ihl-rat LC50:46 ppm/4H AMIHAB 17,362,58

skn-rat LD50:740 mg/kg AMIHAB 11,132,55
ipr-rat LD50:23 mg/kg AMIHBC 8,335,53
orl-mus LD50:41 mg/kg AMIHAB 11,132,55
ihl-mus LC50:12 ppm/4H NTIS** AD224-0-06
ipr-mus LD50:33 mg/kg AMIHAB 11,132,55
skn-rbt LD50:71 mg/kg AMIHAB 11,132,55
ipr-rbt LD50:28 mg/kg AMIHBC 8,335,53

CONSENSUS REPORTS: Reported in EPA TSCA Inventory. EPA Extremely Hazardous Substances List.

OSHA PEL: TWA 0.05 ppm; STEL 0.15 ppm (skin)
ACGIH TLV: TWA 0.05 ppm; STEL 0.15 ppm (skin)
DFG MAK: 0.05 ppm (0.3 mg/m^3)
DOT CLASSIFICATION: 4.1; *Label:* Flammable Solid, Poison

SAFETY PROFILE: Poison by inhalation, ingestion, skin contact, and intraperitoneal routes. Ignites in O_2 at 100°C. Forms impact-sensitive explosive mixtures with ethers (e.g., dioxane) and halocarbons (e.g., carbon tetrachloride). Incompatible with dimethyl sulfoxide. When heated to decomposition it emits toxic fumes of boron oxides. See also BORON COMPOUNDS and BORANES.

DAE425 CAS:2227-17-0 HR: 2
DECACHLOROBI-2,4-CYCLOPENTADIEN-1-YL
mf: $C_{10}Cl_{10}$ mw: 474.60

PROP: Tan crystals or yellow prisms from pet ether. Mp: 122–123°. Stable to alkali

SYNS: BIS(PENTACHLOR-2,4-CYCLOPENTADIEN-1-YL) □ BIS(PENTACHLOROCYCLOPENTADIENYL) □ BIS(PENTACHLORO-2,4-CYCLOPENTADIEN-1-YL) □ DECACHLOR □ 1,1′,2,2′,3,3′,4,4′,5,5′-DECACHLOROBI-2,4-CYCLOPENTADIEN-1-YL □ DIENOCHLOR □ ENT 25,718 □ HOOKER HRS-16 □ HOOKER HRS 1654 □ HRS 16 □ HRS 16A □ HRS 1654 □ PENTAC □ PENTAC WP

TOXICITY DATA WITH REFERENCE
mmo-sat 10 μg/plate MUREAV 116,185,83
mma-sat 500 μg/plate MUREAV 116,185,83
orl-rat LD50:1200 mg/kg ARSIM* 20,13,66

SAFETY PROFILE: Moderately toxic by ingestion. Mutation data reported. See also CHLORINATED HYDROCARBONS, ALIPHATIC.

DAE450 CAS:25152-84-5 HR: 2
trans,trans-2,4-DECADIENAL
mf: $C_{10}H_{16}O$ mw: 152.23

PROP: Yellow liquid; chicken fat odor. D: 0.806–0.876, refr index: 1.514–1.516, bp: 58–61° @ 0.05 mm, flash p: 212°F. Sol in alc, fixed oils; insol in water @ 104°.

SYNS: (E,E)-2,4-DECADIENAL □ (2E,4E) DECADIENAL □ (2E,4E)-2,4-DECADIENAL □ (E,E)-2,4-DECADIENAL □ FEMA No. 3135 □ HEPTENYL ACROLEIN

SAFETY PROFILE: Combustible liquid. When heated to decomposition it emits acrid smoke and irritating fumes.

DAE500 CAS:26660-76-4 HR: 3
DECADONIUM DIIODIDE
mf: $C_{34}H_{62}N_2$•2I mw: 752.78

SYNS: 1,10-BIS(N-METHYL-N-(1′-ADAMANTYL)AMINO)DECANE DIIODOMETHYLATE □ N,N′-DECAMETHYLENEBIS((1-ADAMANTYL)DIMETHYLAMMONIUM, DIIODIDE □ 1,10-(N-METHYL-N-(1′-ADAMANTYL)AMINO)DECANE DIIODOMETHYLATE

TOXICITY DATA WITH REFERENCE
ivn-mus LD50:780 μg/kg RPTOAN 33,185,70
ivn-cat LDLo:100 mg/kg RPTOAN 33,185,70
ivn-rbt LDLo:50 mg/kg RPTOAN 33,185,70

SAFETY PROFILE: Poison by intravenous route. When heated to decomposition it emits toxic fumes of I^- and NO_x.

DAE525 CAS:2392-39-4 HR: 3
DECADRON PHOSPHATE
mf: $C_{22}H_{28}FO_8P$•2Na mw: 516.45

PROP: Crystals. Mp: 233–235°.

SYNS: DEXACORT □ DEXADRESON □ DEXAGRO □ DEXAMETHASONE DISODIUM PHOSPHATE □ DEXAMETHASONE SODIUM PHOSPHATE □ DEXAMETHAZONE SODIUM PHOSPHATE □ DISODIUM DEXAMETHASONE PHOSPHATE □ 9-FLUORO-11-β,17,21-TRIHYDROXY-16-α-METHYLPREGNA-1,4-DIENE-3,20-DIONE-21-(DIHYDROGEN PHOSPHATE) DISODIUM SALT □ MEGACORT □ SODIUM DEXAMETHASONE PHOSPHATE □ SOLDESAM □ SOLU-DECADRON □ SPERSADOX □ TURBINAIRE

TOXICITY DATA WITH REFERENCE
ipr-rat TDLo:400 μg/kg (female 19-20D post):REP PEDIAU 65,287,80
ivn-rat TDLo:5500 μg/kg (female 7-17D post):REP KSRNAM 19,4521,85
orl-mus TDLo:10 mg/kg (female 6-15D post):TER BCFAAI 119,391,80
orl-rat TDLo:1 mg/kg (female 6-15D post):TER BCFAAI 119,391,80
par-rbt TDLo:4 mg/kg (female 11D post):REP ANIFAC 3,73,62
scu-mus TDLo:12,800 μg/kg (female 11-14D post):REP SEIJBO 13,245,73
ivn-rat TDLo:2500 μg/kg (multi) :REP KSRNAM 19,4539,85
ims-hor TDLo:2739 μg/kg (female 21D pre):REP JRFSAR 32,247,82
ivn-rat TDLo:3 mg/kg (female 30D pre):REP KSRNAM 19,4001,85
ims-rbt TDLo:9 mg/kg (female 21-26D post):REP JOENAK 64,363,75
ivn-rat TDLo:1800 μg/kg (male 26W pre):REP KSRNAM 19,4067,85
ivn-rbt TDLo:325 μg/kg (female 6-18D post):TER KSRNAM 19,4555,85
par-rbt TDLo:8 mg/kg (female 11-14D post):TER ANIFAC 3,73,62
ivn-rat TDLo:2500 μg/kg (multi) :TER KSRNAM 19,4539,85
ivn-rat TDLo:5500 μg/kg (female 7-17D post):TER KSRNAM 19,4521,85
ivn-inf TDLo:1500 μg/kg/3D-I:CNS LANCAO 1,632,87
ivn-wmn TDLo:320 mg/kg:GIT LANCAO 1,1035,86
ivn-cld TDLo:1 mg/kg:GIT AJEMEN 10,268,92

ivn-man TDLo: 357 μg/kg: GIT AJEMEN 10,268,92
orl-mus LD50: 1800 mg/kg ATSUDG 7,90,84
ipr-mus LD50: 550 mg/kg ATSUDG 7,90,84
ivn-mus LD50: 112 mg/kg KSRNAM 19,3961,85

CONSENSUS REPORTS: Reported in EPA TSCA Inventory.

SAFETY PROFILE: Poison by intravenous route. Moderately toxic by ingestion and intraperitoneal routes. Human systemic effects by intravenous route: peritonitis, central nervous system, and gastrointestinal changes. An experimental teratogen. Other experimental reproductive effects. When heated to decomposition it emits toxic fumes of F^-, PO_x, and Na_2O.

DAE600 CAS:15652-38-7 HR: 3
DECAFENTIN
mf: $C_{28}H_{36}P \cdot C_{18}H_{15}BrClSn$ mw: 868.99

SYNS: A-36 ☐ CELA A-36 ☐ DECYLTRIPHENYLPHOSPHONIUM BROMOCHLOROTRIPHENYLSTANNATE ☐ (DECYL-TRIPHENYL-PHOSPHONIUM)-TRIPHENYL-BROM-CHLOR-STANNAT (GERMAN) ☐ STANNOPLUS ☐ STANNORAM ☐ STANNPLOUS

TOXICITY DATA WITH REFERENCE
orl-rat LD50: 700 mg/kg FMCHA2 -,D287,80
orl-mus LD50: 550 mg/kg BESAAT 15,120,69
skn-rbt LD50: 310 mg/kg BESAAT 15,120,69

OSHA PEL: TWA 0.1 mg(Sn)/m³ (skin)
ACGIH TLV: TWA 0.1 mg(Sn)/m³ (skin) (Proposed: TWA 0.1 mg(Sn)/m³; STEL 0.2 mg(Sn)/m³ (skin))
NIOSH REL: (Organotin Compounds) TWA 0.1 mg(Sn)/m³

SAFETY PROFILE: Poison by skin contact. Moderately toxic by ingestion. When heated to decomposition it emits very toxic fumes of PO_x, Br^-, and Cl^-. A pesticide. See also TIN COMPOUNDS.

For occupational chemical analysis use NIOSH: Organotin Compounds 5504.

DAE625 CAS:41409-50-1 HR: 3
DECAFLUOROBUTYRAMIDINE
mf: $C_4F_{10}N_2$ mw: 266.04

$C_3F_7C(:NF)NF_2$

SAFETY PROFILE: A shock-sensitive explosive. Upon decomposition it emits toxic fumes of F^- and NO_x.

DAE695 CAS:19590-85-3 HR: 3
cis-N-(DECAHYDRO-2-METHYL-5-ISOQUINOLYL)-3,4,5-TRIMETHOXYBENZAMIDE
mf: $C_{20}H_{30}N_2O_4$ mw: 362.52

SYNS: M-30 ☐ cis-5,8,10-H-5-(3,4,5-TRIMETHOXYBENZAMIDO)-2-METHYL DECAHYDROISOQUINOLINE

TOXICITY DATA WITH REFERENCE
ipr-mus LD50: 278 mg/kg DECRDP 10,197,84
ivn-dog LDLo: 40 mg/kg DECRDP 10,197,84
ivn-rbt LDLo: 31,600 μg/kg DECRDP 10,197,84

SAFETY PROFILE: Poison by intravenous and intraperitoneal routes. When heated to decomposition it emits toxic fumes of NO_x.

DAE700 CAS:27460-73-7 HR: 3
trans-N-(DECAHYDRO-2-METHYL-5-ISOQUINOLYL)-3,4,5-TRIMETHOXYBENZAMIDE
mf: $C_{20}H_{30}N_2O_4$ mw: 362.52

SYNS: M-32 ☐ (E)-N-(2-METHYLDECAHYDROISOQUINOL-5-YL)-3,4,5-TRIMETHOXY-BENZAMIDE ☐ trans-N-(2-METHYLDECAHYDROISOQUINOL-5-YL)-3,4,5-TRIMETHOXYBENZAMIDE ☐ 5-(3,4,5-TRIMETHOXYBENZAMIDO)-2-METHYL-trans-DECAHYDROISOQUINOLINE ☐ trans-9,10-t-5-H-5-(3,4,5-TRIMETHOXYBENZAMIDO)-2-METHYL DECAHYDROISOQUINOLINE

TOXICITY DATA WITH REFERENCE
ipr-mus LD50: 221 mg/kg JMCMAR 23,2-06,80
ivn-dog LDLo: 40 mg/kg DECRDP 10,197,84
ivn-rbt LDLo: 17,800 μg/kg DECRDP 10,197,84

SAFETY PROFILE: Poison by intravenous and intraperitoneal routes. When heated to decomposition it emits toxic fumes of NO_x.

DAE800 CAS:91-17-8 HR: 3
DECAHYDRONAPHTHALENE
DOT: UN 1147
mf: $C_{10}H_{18}$ mw: 138.28

PROP: Water-white liquid with slight menthol odor. Mp (cis): −43.3°, mp (trans): −30.7°, bp (cis): 195.6°, bp (trans): 187.3°, flash p: 136°F, (CC), autoign temp: 482°F, vap press (cis): 1 mm @ 22.5°, vap press (trans): 10 mm @ 47.2°, d: (cis) 0.8963 @ 20°/4°, vap d: 4.76, lel: 0.7% @ 212°F, uel: 4.9% @ 212°F.

SYNS: BICYCLO(4.4.0)DECANE ☐ DEC ☐ DECALIN ☐ DECALIN (DOT) ☐ DECALIN SOLVENT ☐ DE-KALIN ☐ DEKALINA (POLISH) ☐ NAPHTHALANE ☐ NAPHTHANE ☐ PERHYDRONAPHTHALENE

TOXICITY DATA WITH REFERENCE
skn-rbt 10 mg/24H open MLD AMIHBC 4,119,51
eye-rbt 500 mg open AMIHBC 4,119,51
ihl-mus TCLo: 50 ppm/24H/90D-C: CAR FAATDF 5,785,85
ihl-hmn TCLo: 100 ppm: NOSE,EYE,PUL TGNCDL 2,40,61
orl-rat LD50: 4170 mg/kg AMIHBC 4,119,51
ihl-rat LCLo: 500 ppm/4H AMIHBC 4,119,51
ihl-mus LCLo: 993 ppm/4H NTIS** AD-A-062-138
skn-rbt LD50: 5900 mg/kg AMIHBC 4,119,51
ihl-gpg LCLo: 319 ppm/8H NTIS** AD-A086-341

CONSENSUS REPORTS: Reported in EPA TSCA Inventory.

DOT CLASSIFICATION: 3; *Label:* Flammable Liquid

SAFETY PROFILE: Moderately toxic by inhalation and ingestion. Questionable carcinogen with experimental carcinogenic and neoplastigenic data. Mildly toxic by skin contact. Human systemic effects by inhalation: conjunctiva irritation, unspecified olfactory and pulmonary system changes. Can cause kidney damage. Mutation data reported. A skin and eye irritant. Flammable liquid when exposed to heat or flame, can react with oxidizing materials. To fight fire, use foam, CO_2, dry chemical. When heated to decomposition it emits acrid smoke and fumes.

DAF000 **CAS:825-51-4** **HR: 2**
DECAHYDRO-2-NAPHTHALENOL
mf: $C_{10}H_{18}O$ mw: 154.28

PROP: D: 0.996, bp: 109° @ 14 mm, flash p: >112°.

SYNS: DECAHYDRONAPHTHALEN-2-OL ☐ DECAHYDRONAPTHOL-2 ☐ DECAHYDRO-β-NAPHTHOL ☐ trans-DECAHYDRO-β-NAPHTHOL ☐ 2-DECALINOL ☐ 2-DECALOL ☐ 2-HYDROXYDECALIN

TOXICITY DATA WITH REFERENCE
skn-rbt 500 mg/24H MOD FCTXAV 12,873,74
orl-rat LD50:>5 g/kg FCTXAV 12,873,74
skn-rbt LD50:>5 g/kg FCTXAV 12,873,74

CONSENSUS REPORTS: Reported in EPA TSCA Inventory.

SAFETY PROFILE: A moderate skin irritant. Low toxicity by ingestion and skin contact. Flammable or combustible liquid. When heated to decomposition it emits acrid smoke and irritating fumes.

DAF100 **CAS:10519-11-6** **HR: 1**
DECAHYDRO-β-NAPHTHYL ACETATE
mf: $C_{12}H_{20}O_2$ mw: 196.32

SYN: 2-NAPHTHOL, DECAHYDRO-, ACETATE

TOXICITY DATA WITH REFERENCE
skn-rbt 500 mg/24H MOD FCTXAV 17,755,79

CONSENSUS REPORTS: Reported in EPA TSCA Inventory.

SAFETY PROFILE: A skin irritant. When heated to decomposition it emits acrid smoke and irritating fumes.

DAF150 **CAS:10519-12-7** **HR: 1**
DECAHYDRO-β-NAPHTHYL FORMATE
mf: $C_{11}H_{18}O_2$ mw: 182.29

SYNS: DECALINYL FORMATE ☐ 2-NAPHTHALENOL, DECAHYDRO-, FORMATE ☐ 2-NAPHTHOL, DECAHYDRO-, FORMATE ☐ SANTALOZONE

TOXICITY DATA WITH REFERENCE
skn-rbt 500 mg/24H MOD FCTXAV 17,757,79

CONSENSUS REPORTS: Reported in EPA TSCA Inventory.

SAFETY PROFILE: A skin irritant. When heated to decomposition it emits acrid smoke and irritating fumes.

DAF200 **CAS:705-86-2** **HR: 1**
Δ-DECALACTONE
mf: $C_{10}H_{18}O_2$ mw: 170.28

PROP: Colorless liquid; coconut, fruity odor, butterlike on dilution. Refr index: 1.456–1.459. Very sol in alc and propylene glycol; insol in water @ 281°.

SYNS: AMYL-Δ-VALEROLACTONE ☐ DECANOLIDE-1,5 ☐ FEMA No. 2361

TOXICITY DATA WITH REFERENCE
skn-rbt 500 mg/24H MLD FCTXAV 14,659,76
eye-rbt 100 mg MLD NTIS** AD-A053-896

CONSENSUS REPORTS: Reported in EPA TSCA Inventory.

SAFETY PROFILE: A skin and eye irritant. When heated to decomposition it emits acrid smoke and irritating fumes.

D

DAF300 **CAS:52918-63-5** **HR: 3**
DECAMETHRINE
mf: $C_{22}H_{19}Br_2NO_3$ mw: 505.24

PROP: Crystals or powder. Mp: 100°. Sol in ethanol, acetone, and dioxane; insol in water.

SYNS: BUTOFLIN ☐ BUTOX ☐ DECAMETHRIN ☐ DECIS ☐ DEKAMETRIN (HUNGARIAN) ☐ DELTAMETHRIN ☐ ESBECYTHRIN ☐ JMC 45498 ☐ K-OTHRIN ☐ NRDC 161 ☐ RU 22974

TOXICITY DATA WITH REFERENCE
cyt-ofs-mul 100 nL/L JFIBA9 26,13,85
orl-rat TDLo:70 mg/kg (7-20D preg):REP JEPTDQ 2(3),751,79
orl-mus TDLo:50 mg/kg (female 8-12D post):REP TCMUD8 7,7,87
orl-mus TDLo:30 mg/kg (7-16D preg):TER JEPTDQ 2(3),751,79
orl-rat LD50:30 mg/kg FAATDF 7,299,86
ihl-rat LC50:785 mg/m³/2H JEPTDQ 2(3),751,79
ivn-rat LD50:2526 mg/kg PCBPBS 30,79,88
orl-mus LD50:3450 mg/kg IJTEDP 6,127,84
ice-mus LD50:26,100 μg/kg PCBPBS 24,200,85
ivn-dog LD50:3440 μg/kg IJTEDP 6,127,84

CONSENSUS REPORTS: EPA Genetic Toxicology Program.

SAFETY PROFILE: Poison by ingestion, inhalation, intravenous and intracerebral routes. An experimental teratogen. Other experimental reproductive effects. Mutation data reported. When heated to decomposition it emits toxic fumes of Br^-, CN^-, and NO_x. See also ESTERS.

DAF350 **CAS:541-02-6** **HR: 1**
DECAMETHYLCYCLOPENTASILOXANE
mf: $C_{10}H_{30}O_5Si_5$ mw: 370.85

SYNS: CYCLIC DIMETHYLSILOXANE PENTAMER ☐ DEKAMETHYLCYKLOPENTASILOXAN ☐ DIMETHYLSILOXANE PENTAMER ☐ DOW CORNING 345 ☐ DOW CORNING 345 FLUID ☐ KF 995 ☐ NUC SILICONE VS 7158 ☐ SF 1202 ☐ SILICON SF 1202 ☐ UNION CARBIDE 7158 SILICONE FLUID ☐ VS 7158

TOXICITY DATA WITH REFERENCE
skn-rbt 500 mg/24H MLD 85JCAE -,1234,86
eye-rbt 500 mg/24H MLD 85JCAE -,1234,86

CONSENSUS REPORTS: Reported in EPA TSCA Inventory.

SAFETY PROFILE: A skin and eye irritant. When heated to decomposition it emits acrid smoke and irritating vapors.

DAF450 CAS:63884-28-6 HR: 3
1,1′-DECAMETHYLENEBIS(1-METHYLPIPERIDINIUM IODIDE)
mf: $C_{22}H_{46}N_2•2I$ mw: 592.50

TOXICITY DATA WITH REFERENCE
scu-mus LD50:3 mg/kg YKKZAJ 74,1267,54
ivn-mus LD50:20 mg/kg BJPCAL 10,124,55
ivn-rbt LDLo:3 mg/kg

SAFETY PROFILE: Poison by subcutaneous and intravenous routes. When heated to decomposition it emits toxic fumes of NO_x and I^-. See also IODIDES.

DAF600 CAS:541-22-0 HR: 3
DECAMETHYLENEBIS(TRIMETHYLAMMONIUM BROMIDE)
mf: $C_{16}H_{38}N_2•2Br$ mw: 418.38

PROP: Crystals from MeOH/Me_2CO. Mp: 268–270° (decomp). Insol in Et_2O.

SYNS: DECACURAN ☐ DECAMETHONIUM ☐ DECAMETHONIUM BROMIDE ☐ DECAMETHONIUM DIBROMIDE ☐ DECAMETHYLENE-1, 10-BISTRIMETHYLAMMONIUM DIBROMIDE ☐ N,N,N,N′,N′,N′-HEXAMETHYL-1,10-DECANEDIAMINIUM DIBROMIDE ☐ SYNCURINE

TOXICITY DATA WITH REFERENCE
ipr-rat LD50:2900 µg/kg TXAPA9 14,67,69
orl-mus LD50:190 mg/kg JPETAB 118,395,56
ipr-mus LD50:900 µg/kg JAPMA8 45,792,56
scu-mus LD50:4 mg/kg YKKZAJ 74,911,54
ivn-mus LD50:630 µg/kg AIPTAK 105,221,56
ivn-cat LD50:35 µg/kg JPETAB 118,395,56
ivn-rbt LD50:125 µg/kg JPETAB 118,395,56

CONSENSUS REPORTS: Reported in EPA TSCA Inventory.

SAFETY PROFILE: Deadly poison by ingestion, intraperitoneal, subcutaneous, and intravenous routes. When heated to decomposition it emits very toxic fumes of NO_x, NH_3 , and Br^-. See also BROMIDES.

DAF800 CAS:1420-40-2 HR: 3
DECAMETHYLENEBIS(TRIMETHYLAMMONIUM DIIODIDE)
mf: $C_{16}H_{38}N_2•2I$ mw: 512.36

PROP: A solid. Mp: 243–246°.

SYNS: DECAMETHIONIUM IODIDE ☐ DECAMETHONIUM DIIODIDE ☐ DECAMETHONIUM IODIDE ☐ DECAMETHYLENEBIS(TRIMETHYLAMMONIUM IODIDE) ☐ EULISSIN A ☐ EULIXINE ☐ N,N,N, N¹,N¹,N¹-HEXAMETHYL-1,10-DECANEDIAMINIUM DIIODIDE ☐ PROCURAN

TOXICITY DATA WITH REFERENCE
orl-rat LD50:85 mg/kg PSEBAA 120,511,65
ipr-rat LD50:1550 µg/kg TXAPA9 36,585,76
ivn-rat LD50:3020 µg/kg BJPCAL 4,381,49
orl-mus LD50:115 mg/kg PSEBAA 120,511,65
scu-mus LD50:5 mg/kg YKKZAJ 74,911,54
ivn-mus LD50:838 µg/kg BJPCAL 4,381,49
ivn-rbt LD50:200 µg/kg RISSAF 12,158,49
ivn-ckn LD50:80 µg/kg AIPTAK 122,152,59

SAFETY PROFILE: Poison by ingestion, intraperitoneal, intravenous, and subcutaneous routes. When heated to decomposition it emits very toxic fumes of NO_x, NH_3, and I^-.

DAG000 CAS:112-31-2 HR: 2
1-DECANAL
mf: $C_{10}H_{20}O$ mw: 156.30

PROP: Colorless to light-yellow liquid; floral, fatty odor. D: 0.830 @ 15°/4°, bp: 208°, flash p: 185°F, mp: −5 (approx). Found in over 50 sources including citrus oils, citronella, and lemongrass. (FCTXAV 11,477,73). Sol in 80% alc, fixed oils, volatile oils, and mineral oils; insol in water and glycerol.

SYNS: ALDEHYDE C10 ☐ C-10 ALDEHYDE ☐ CAPRALDEHYDE ☐ CAPRIC ALDEHYDE ☐ CAPRINALDEHYDE ☐ CAPRINIC ALDEHYDE ☐ DECALDEHYDE ☐ n-DECALDEHYDE ☐ DECANAL ☐ n-DECANAL ☐ DECANALDEHYDE ☐ DECYL ALDEHYDE ☐ n-DECYL ALDEHYDE ☐ 1-DECYL ALDEHYDE ☐ DECYLIC ALDEHYDE ☐ FEMA No. 2362

TOXICITY DATA WITH REFERENCE
skn-rbt 14,372 µg/24H open SEV AIHAAP 23,95,62
skn-rbt 500 mg/24H MLD FCTXAV 11,1079,73
dnr-bcs 5 mg/disc OIGZSE 34,267,85
orl-rat LD50:3730 mg/kg FCTXAV 11,477,73
skn-rbt LD50:5040 mg/kg FCTXAV 11,477,73

CONSENSUS REPORTS: Reported in EPA TSCA Inventory.

SAFETY PROFILE: Moderately toxic by ingestion. Mildly toxic by skin contact. A severe skin irritant. Mutation data reported. Combustible liquid. When heated to decomposition it emits acrid smoke and irritating fumes. See also ALDEHYDES.

DAG200 CAS:112-31-2 HR: 2
1-DECANAL (mixed isomers)

SYN: FEMA No. 2362

TOXICITY DATA WITH REFERENCE
skn-rbt 14,372 µg/24H open SEV AIHAAP 23,95,62
orl-rat LD50:3730 mg/kg AIHAAP 23,95,62

CONSENSUS REPORTS: Reported in EPA TSCA Inventory.

SAFETY PROFILE: Moderately toxic by ingestion. A severe skin irritant. See also 1-DECANAL. When heated to decomposition it emits acrid smoke and fumes.

DAG400 CAS:124-18-5 HR: 2
DECANE
DOT: UN 2247
mf: $C_{10}H_{22}$ mw: 142.29

PROP: Liquid. Mp: −30°, bp: 174°, lel: 0.8%, uel: 5.4%, flash p: 115°F (CC), d: 0.730 @ 20°/4°, autoign temp: 410°F, vap press: 1 mm @ 16.5°, vap d: 4.90.

SYN: n-DECANE (DOT)

TOXICITY DATA WITH REFERENCE
skn-mus TDLo:25 g/kg/52W-I:ETA TXAPA9 9,70,66
ihl-mus LC50:72,300 mg/m³/2H GTPZAB 26(8),53,82

CONSENSUS REPORTS: Reported in EPA TSCA Inventory.

DOT CLASSIFICATION: 3; *Label:* Flammable Liquid

SAFETY PROFILE: Questionable carcinogen with experimental tumorigenic data. A simple asphyxiant. Narcotic in high concentrations. Flammable liquid when exposed to heat or flame. Can react with oxidizing materials. Moderately explosive in its vapor form. To fight fire, use foam, CO_2, dry chemical. Emitted from modern building materials (CENEAR 69,22,91). See also ARGON for discussion of asphyxiants.

DAG600 CAS:2016-57-1 HR: 3
1-DECANEAMINE
DOT: UN 2733/UN 2734
mf: $C_{10}H_{23}N$ mw: 157.34

PROP: Liquid. Mp: 15.0°, bp: 218.0–218.5° @ 747 mm, flash p: 210°F, d: 0.79 @ 20°, vap d: 5.5.

SYNS: 1-AMINODECANE □ DECYLAMINE

TOXICITY DATA WITH REFERENCE
skn-rbt 100 µg/24H open AIHAAP 23,95,62
orl-rat LD50:280 mg/kg AIHAAP 23,95,62
skn-rbt LD50:350 mg/kg AIHAAP 23,95,62
ipr-mus LD50:71 mg/kg EJMCA5 26,517,91

CONSENSUS REPORTS: Reported in EPA TSCA Inventory.

DOT CLASSIFICATION: 8; *Label:* Corrosive, Flammable Liquid (UN 2734); DOT Class: 3; *Label:* Flammable Liquid, Corrosive (UN 2733)

SAFETY PROFILE: Poison by ingestion and skin contact. A skin irritant. Flammable when exposed to heat or flame; can react with oxidizing materials. To fight fire, use alcohol foam, foam, dry chemical. When heated to decomposition it emits toxic fumes of NO_x. See AMINES; and AMINES, FATTY.

DAG650 CAS:646-25-3 HR: 3
1,10-DECANEDIAMINE
mf: $C_{10}H_{24}N_2$ mw: 172.36

TOXICITY DATA WITH REFERENCE
orl-rat LDLo:500 mg/kg JPETAB 90,260,47
ipr-mus LDLo:125 mg/kg CBCCT* 4,377,52

CONSENSUS REPORTS: Reported in EPA TSCA Inventory.

SAFETY PROFILE: Poison by intraperitoneal route. Moderately toxic by ingestion. When heated to decomposition it emits toxic vapors of NO_x.

DAH400 CAS:334-48-5 HR: 3
DECANOIC ACID
mf: $C_{10}H_{20}O_2$ mw: 172.30

PROP: White crystals or needles with an unpleasant odor. D: 0.8858 @ 40°/4°, bp: 270°, mp: 31.5°. Sol in EtOH, Et_2O, Me_2CO, C_6H_6, $CHCl_3$, and alkalis.

SYNS: CAPRIC ACID □ n-CAPRIC ACID □ CAPRINIC ACID □ CAPRYNIC ACID □ n-DECANOIC ACID □ n-DECOIC ACID □ DECYLIC ACID □ n-DECYLIC ACID □ HEXACID 1095 □ NEO-FAT 10 □ 1-NONANECARBOXYLIC ACID

TOXICITY DATA WITH REFERENCE
skn-rbt 500 mg/24H MOD FCTXAV 17,735,79
sln-smc 14,500 ppb ANYAA9 407,186,83
ivn-mus LD50:129 mg/kg APTOA6 18,141,61

CONSENSUS REPORTS: Reported in EPA TSCA Inventory.

SAFETY PROFILE: Poison by intravenous route. Mutation data reported. A moderate skin irritant. When heated to decomposition it emits acrid smoke and irritating fumes.

DAH450 CAS:10024-58-5 HR: 1
DECANOIC ACID, DIESTER with TRIETHYLENE GLYCOL (mixed isomers)
mf: $C_{26}H_{50}O_6$ mw: 458.76

SYN: DIDECANOYLTRIETHYLENE GLYCOL ESTER (mixed isomers)

TOXICITY DATA WITH REFERENCE
skn-rbt 10 mg/24H open MLD AIHAAP 23,95,62
orl-rat LD50:7460 mg/kg AIHAAP 23,95,62
skn-rbt LD50:11,200 mg/kg AIHAAP 23,95,62

CONSENSUS REPORTS: Reported in EPA TSCA Inventory.

SAFETY PROFILE: Mildly toxic by ingestion and skin contact. When heated to decomposition it emits acrid smoke and irritating fumes.

DAI000 CAS:26909-37-5 HR: 3
10-DECARBAMOYLMITOMYCIN C
mf: $C_{14}H_{17}N_3O_4$ mw: 291.34

SYNS: DCMC □ DECARBAMOYLMITOMYCIN C □ DECARBAMYLMITOMYCIN C

TOXICITY DATA WITH REFERENCE
mmo-sat 2500 ng/plate MUREAV 149,485,85
sce-hmn:lym 100 nmol/L MUREAV 149,485,85
sce-ham:lng 100 µg/L CNREA8 44,3270,84
ivn-mus LD50:33,800 µg/kg YKKZAJ 92,1218,72

SAFETY PROFILE: Poison by intravenous route. Human mutation data reported. When heated to decomposition it emits toxic fumes of NO_x.

DAI200 CAS:5053-08-7 HR: 3
DECASPIRIDE HYDROCHLORIDE
mf: $C_{15}H_{20}N_2O_2 \cdot ClH$ mw: 296.83

PROP: A solid. Mp: 232–233° (decomp).

SYNS: CHLORHYDRATE de PHENETHYL-8-OXA-1-DIAZA-3,8-SPIRO (4,5)DECANONE-2 (FRENCH) □ ESPIRAN □ 8-N-FENETIL-1-OXA-2-OXO-3,8-DIAZASPIRO-(4,5)-DECANO CLORIDRATO (ITALIAN) □ FENSPIRIDE □ FENSPIRIDE HYDROCHLORIDE □ JP 428 HYDROCHLORIDE □ NAT-333 HYDROCHLORIDE □ NDR-5998A HYDROCHLORIDE □ 8-(2-PHENYLETHYL)-1-OXA-3,8-DIAZASPIRO(4.5)DECAN-2-ONE HYDROCHLORIDE □ PHENETHYL-8-OXA-1-DIAZA-3,8-SPIRO(4,5)DECANONE-2-HYDROCHLORIDE □ PNEUMOREL □ RESPIRIDE □ TEGENCIA HYDROCHLORIDE □ VIARESPAN HYDROCHLORIDE

TOXICITY DATA WITH REFERENCE
orl-rat LD50:437 mg/kg ARZNAD 19,1263,69
ivn-rat LD50:122 mg/kg AIPTAK 193,111,71
orl-mus LD50:250 mg/kg AIPTAK 193,111,71
ipr-mus LD50:230 mg/kg ARZNAD 19,1263,69
ivn-mus LD50:106 mg/kg MEIEDD 10,575,83
ivn-dog LD50:74 mg/kg BCFAAI 117,343,78
orl-gpg LD50:260 mg/kg AIPTAK 193,111,71
ipr-gpg LD50:210 mg/kg AIPTAK 193,111,71

SAFETY PROFILE: Poison by ingestion, intraperitoneal, and intravenous routes. A bronchodilator and antiadrenergic agent. When heated to decomposition it emits very toxic fumes of NO_x and HCl.

DAI350 CAS:3913-71-1 HR: 2
2-DECENAL
mf: $C_{10}H_{18}O$ mw: 154.28

PROP: Sltly yellow liquid; orange odor. D: 0.836–0.846, refr index: 1.452–1.457. Sol in alc, fixed oils; insol in water.

SYNS: trans-2-DECEN-1-AL □ DECENALDEHYDE □ FEMA No. 2366

TOXICITY DATA WITH REFERENCE
skn-rbt 500 mg/24H SEV FCTXAV 17,761,79
orl-rat LD50:5000 mg/kg FCTXAV 17,761,79
skn-rbt LD50:3400 mg/kg FCTXAV 17,761,79

CONSENSUS REPORTS: Reported in EPA TSCA Inventory.

SAFETY PROFILE: Moderately toxic by skin contact. Mildly toxic by ingestion. A severe skin irritant. When heated to decomposition it emits acrid smoke and fumes. See also ALDEHYDES.

DAI360 CAS:21662-09-9 HR: 1
cis-4-DECENAL
mf: $C_{10}H_{18}O$ mw: 154.28

PROP: Colorless to sltly yellow liquid; fatty, orangelike odor. D: 0.847, refr index: 1.442–1.444, Sol in alc, fixed oils; insol in water.

SYNS: cis-4-DECEN-1-AL (FCC) □ FEMA No. 3264

TOXICITY DATA WITH REFERENCE
skn-gpg 100%/24H MOD FCTOD7 20,663,82
orl-mus LD50:>5 g/kg FCTOD7 20,663,82
skn-gpg LD50:>5 g/kg FCTOD7 20,663,82

CONSENSUS REPORTS: Reported in EPA TSCA Inventory.

SAFETY PROFILE: Low toxicity by ingestion and skin contact. A skin irritant. When heated to decomposition it emits acrid smoke and irritating fumes.

DAI400 CAS:13019-22-2 HR: 3
9-DECEN-1-OL
mf: $C_{10}H_{20}O$ mw: 156.30

PROP: D: 0.875, bp: 108° @ 8 mm, flash p: 104°F.

SYNS: ω-DECENOL □ 1-DECEN-10-OL □ DECYLENIC ALCOHOL

TOXICITY DATA WITH REFERENCE
skn-rbt 100% FCTXAV 12,405,74

CONSENSUS REPORTS: Reported in EPA TSCA Inventory.

SAFETY PROFILE: A skin irritant. A flammable liquid. When heated to decomposition it emits acrid smoke and irritating fumes. See also ALCOHOLS.

DAI450 CAS:50816-18-7 HR: 1
9-DECENYL ACETATE
mf: $C_{12}H_{22}O_2$ mw: 198.34

PROP: Bp: 95–97° @ 0.09 mm.

SYNS: ACETIC ACID, 9-DECENYL ESTER □ 9-DECEN-1-OL, ACETATE □ DECENYL ACETATE

TOXICITY DATA WITH REFERENCE
skn-rbt 500 mg/24H MLD FCTOD7 20,665,82
orl-rat LD50:>5 g/kg FCTOD7 20,665,82
skn-rbt LD50:>5 g/kg FCTOD7 20,665,82

CONSENSUS REPORTS: Reported in EPA TSCA Inventory.

SAFETY PROFILE: Low toxicity by ingestion and skin contact. A skin irritant. When heated to decomposition it emits acrid smoke and irritating fumes.

DAI460 CAS:13560-89-9 HR: 3
DECHLORANE PLUS
mf: $C_{18}H_{12}Cl_{12}$ mw: 653.70

PROP: Colorless crystals. Mp: >325°. Soluble in o-dichlorobenzene.

SYNS: DECHLORANE 605 □ DECHLORANE PLUS 515 □ DECHLORANE PLUS 2520

TOXICITY DATA WITH REFERENCE
orl-rat LD50:25 g/kg LitL## -06MAY85
ihl-rat LC50:2250 mg/m³ LitL## -06MAY85
skn-rbt LD50:8 g/kg LitL## -06MAY85

SAFETY PROFILE: Poison by inhalation. Mildly toxic by skin contact and ingestion. When heated to decomposition it emits toxic fumes of Cl^-. See also CHLORINATED HYDROCARBONS, AROMATIC.

DAI475 HR: 3
DECLINAX
mf: $C_{10}H_{13}N_3 \cdot BrH$ mw: 256.18

SYNS: DEBRISOQUIN HYDROBROMIDE □ 3,4-DIHYDRO-2(1H)-ISOQUINOLINECARBOXYAMIDINE HYDROBROMIDE □ RO 5-3307

TOXICITY DATA WITH REFERENCE
orl-mus LD50:242 mg/kg CTCEA9 6,299,64
ipr-mus LD50:150 mg/kg CTCEA9 6,299,64
scu-mus LD50:163 mg/kg CTCEA9 6,299,64
ivn-mus LD50:40,500 μg/kg CTCEA9 6,299,64

SAFETY PROFILE: Poison by ingestion, subcutaneous, intravenous, and intraperitoneal routes. When heated to decomposition it emits toxic fumes of NO_x and HBr.

DAI485 **CAS:64-73-3** **HR: 3**
DECLOMYCIN HYDROCHLORIDE
mf: $C_{21}H_{21}ClN_2O_8 \cdot ClH$ mw: 501.35

SYNS: 7-CHLORO-6-DEMETHYLTETRACYCLINE HYDROCHLORIDE ☐ CHLORTETRIN ☐ DEMECLOCYCLINE HYDROCHLORIDE ☐ DEMETHYLCHLOROTETRACYCLINE HYDROCHLORIDE ☐ DEMETHYLCHLORTETRACYCLINE HYDROCHLORIDE ☐ DEMETRACICLINA ☐ DETRAVIS ☐ LEDERMYCIN HYDROCHLORIDE ☐ MECICLIN ☐ MEXOCINE

TOXICITY DATA WITH REFERENCE
unr-wmn TDLo:48 μg/kg (28W preg):REP PEDIAU 34,423,64
iut-rat TDLo:50 mg/kg (1D pre):REP CCPTAY 29,553,84
orl-man TDLo:69 mg/kg/4D-I:SYS AJKDDP 5,270,85
orl-rat LD50:2372 mg/kg TXAPA9 18,185,71
ivn-rat LD50:94 mg/kg DRUGAY-,704,90
orl-mus LD50:2150 mg/kg DRUGAY-,704,90
ivn-mus LD50:275 mg/kg AISMAE 43,143,62

SAFETY PROFILE: Poison by intravenous route. Moderately toxic by ingestion. Human systemic effects by ingestion: depressed renal function tests, urine composition changes, weight loss or decreased weight gain. Human female reproductive effects by an unreported route: delayed effects on newborn. An antibacterial agent. Experimental reproductive effects. When heated to decomposition it emits toxic fumes of NO_x and Cl^-.

DAI500 **CAS:2156-96-9** **HR: 2**
DECYL ACRYLATE
mf: $C_{13}H_{24}O_2$ mw: 212.37

SYN: n-DECYL ACRYLATE

TOXICITY DATA WITH REFERENCE
skn-rbt 10 mg/24H open SEV AIHAAP 23,95,62
orl-rat LD50:6460 mg/kg AIHAAP 23,95,62
skn-rbt LD50:6300 mg/kg AIHAAP 23,95,62

CONSENSUS REPORTS: Reported in EPA TSCA Inventory.

SAFETY PROFILE: Mildly toxic by ingestion and skin contact. A severe skin irritant. When heated to decomposition it emits smoke and acrid fumes.

DAI600 **CAS:112-30-1** **HR: 2**
DECYL ALCOHOL
mf: $C_{10}H_{22}O$ mw: 158.32

PROP: Found in sweet orange and a few other essential oils (FCTXAV 11,95,73). Colorless, viscous, refractive liquid; floral fruity odor. Mp: 7°, fp: 7°, bp: 232–239° @ 700 mm, flash p: 180°F (OC), d: 0.8297 @ 20°/4°, refr index: 1.435–1.439, vap press: 1 mm @ 69.5°, vap d: 5.3. Sol in alc, ether, mineral oil, propylene glycol, fixed oils; insol in glycerin water @ 233°.

SYNS: AGENT 504 ☐ ALCOHOL C-10 ☐ ANTAK ☐ C 10 ALCOHOL ☐ CAPRIC ALCOHOL ☐ CAPRINIC ALCOHOL ☐ DECANAL DIMETHYL ACETAL ☐ DECANOL ☐ n-DECANOL ☐ 1-DECANOL (FCC) ☐ n-DECATYL ALCOHOL ☐ n-DECYL ALCOHOL ☐ DECYLIC ALCOHOL ☐ DYTOL S-91 ☐ EPAL 10 ☐ FEMA No. 2365 ☐ LOROL 22 ☐ NONYLCARBINOL ☐ PRIMARY DECYL ALCOHOL ☐ ROYALTAC ☐ SIPOL L10

TOXICITY DATA WITH REFERENCE
skn-hmn 75 mg/3D-I SEV 85DKA8 -,127,77
skn-rbt 2600 mg/kg/24H MOD AIHAAP 34,493,73
eye-rbt 83 mg SEV AIHAAP 34,493,73
orl-rat TDLo:35,295 mg/kg (female 1-15D post):REP ONGZAC 22(1),71,91
skn-mus TDLo:12 g/kg/25W-I:ETA TXAPA9 9,70,66
orl-rat LD50:4720 mg/kg AIHAAP 34,493,73
ipr-rat LD50:800 mg/kg 38MKAJ 2C,4631,82
orl-mus LD50:6500 mg/kg FMCHA2 -,C208,83
ihl-mus LC50:4 g/m³/2H 85GMAT -,42,82
skn-rbt LD50:3560 mg/kg FCTXAV 11,95,73

CONSENSUS REPORTS: Reported in EPA TSCA Inventory.

SAFETY PROFILE: Moderately toxic by skin contact. Mildly toxic by ingestion and inhalation. A severe human skin and eye irritant. Experimental reproductive effects. Questionable carcinogen with experimental tumorigenic data. Combustible when exposed to heat or flame; can react with oxidizing materials. To fight fire, use foam, CO_2, dry chemical. When heated to decomposition it emits acrid smoke and irritating fumes. See also ALCOHOLS.

DAI800 **CAS:66988-15-6** **HR: 2**
DECYL ALCOHOL (mixed isomers)
mf: $C_{10}H_{22}O$ mw: 158.32

SYNS: DECANOL (mixed isomers) ☐ FAIR 85 ☐ SPROUT-OFF ☐ TOBACCO SUCKER CONTROL AGENT 148

TOXICITY DATA WITH REFERENCE
skn-rbt 10 mg/24H open SEV AMIHBC 4,119,51
eye-rbt 500 mg open AMIHBC 4,119,51
orl-rat LD50:9800 mg/kg AMIHBC 4,119,51
skn-rbt LD50:3560 mg/kg AMIHBC 4,119,51

SAFETY PROFILE: Moderately toxic by skin contact. Mildly toxic by ingestion. An eye and severe skin irritant. When heated to decomposition it emits acrid smoke and irritating fumes. See also ALCOHOLS and DECYL ALCOHOL.

DAJ000 **CAS:1322-98-1** **HR: 3**
DECYL BENZENE SODIUM SULFONATE
mf: $C_{16}H_{25}O_3S \cdot Na$ mw: 320.46

SYNS: SODIUM DECYLBENZENESULFONAMIDE ☐ SODIUM DECYLBENZENESULFONATE

TOXICITY DATA WITH REFERENCE
eye-rbt 450 mg SEV AROPAW 40,668,48
eye-rbt 1% SEV JAPMA8 38,428,49
orl-mus LD50:2000 mg/kg PSTGAW 3,1,45
ivn-mus LD50:115 mg/kg JAPMA8 38,428,49

CONSENSUS REPORTS: Reported in EPA TSCA Inventory.

SAFETY PROFILE: Poison by intravenous route. Moderately toxic by ingestion. A severe eye irritant. When heated to decomposition it emits toxic fumes of SO_x. See also SULFONATES.

DAJ200 **CAS:28519-06-4** **HR: 1**
DECYL CHLORIDE (mixed isomers)
mf: $C_{10}H_{21}Cl$ mw: 176.76

SYN: CHLORODECANE

TOXICITY DATA WITH REFERENCE
skn-rbt 500 mg MLD 34ZIAG -,745,69
orl-rat LD50:45,300 mg/kg AIHAAP 23,95,62
skn-rbt LD50:5660 mg/kg AIHAAP 23,95,62

SAFETY PROFILE: Mildly toxic by skin contact and ingestion. A mild skin irritant. When heated to decomposition it emits toxic fumes of Cl^-. See also CHLORINATED HYDROCARBONS, ALIPHATIC.

DAJ300 **CAS:23489-03-4** **HR: 3**
1-DECYL-1-ETHYLPIPERIDINIUM BROMIDE
mf: $C_{17}H_{36}N \cdot Br$ mw: 334.45

TOXICITY DATA WITH REFERENCE
orl-mus LD50:202 mg/kg PSDTAP 15,331,74
ipr-mus LD50:34,964 μg/kg PSDTAP 15,331,74
ivn-mus LD50:4293 μg/kg PSDTAP 15,331,74

SAFETY PROFILE: Poison by ingestion, intravenous, and intraperitoneal routes. When heated to decomposition it emits toxic fumes of NO_x and Br^-.

DAJ400 **CAS:14817-09-5** **HR: 2**
4-n-DECYLOXY-3,5-DIMETHOXYBENZOIC ACID AMIDE
mf: $C_{19}H_{31}NO_4$ mw: 337.51

PROP: Crystals from MeOH. Mp: 72–74°.

SYNS: DECIMEMIDE □ 4-(DECYLOXY)-3,5-DIMETHOXYBENZAMIDE □ DENEGYT □ EGYT-1050

TOXICITY DATA WITH REFERENCE
orl-rat LD50:1650 mg/kg 27ZQAG -,377,72
orl-mus LD50:2950 mg/kg 27ZQAG -,377,72

SAFETY PROFILE: Moderately toxic by ingestion. An anticonvulsant. When heated to decomposition it emits toxic fumes of NO_x.

DAJ450 **CAS:6285-34-3** **HR: 3**
4-DECYLOXY-2-HYDROXYPHENYL 4-DECYLOXYPHENYL KETONE
mf: $C_{33}H_{50}O_4$ mw: 510.83

TOXICITY DATA WITH REFERENCE
ivn-mus LD50:100 mg/kg CSLNX* NX#03310

DOT CLASSIFICATION: 3; *Label:* Flammable Liquid

SAFETY PROFILE: A poison by intravenous route. A flammable liquid. When heated to decomposition it emits acrid smoke and irritating vapors.

DAJ500 **CAS:2082-84-0** **HR: 3**
DECYLTRIMETHYLAMMONIUM BROMIDE
mf: $C_{13}H_{30}N \cdot Br$ mw: 280.35

SYN: AMMONIUM, DECYLTRIMETHYL-, BROMIDE

TOXICITY DATA WITH REFERENCE
ivn-rat LD50:5500 μg/kg APTOA6 47,17,80
ivn-mus LD50:2800 μg/kg APTOA6 47,17,80

CONSENSUS REPORTS: Reported in EPA TSCA Inventory.

SAFETY PROFILE: Poison by intravenous route. When heated to decomposition it emits toxic vapors of NO_x and Br^-.

DAJ800 **CAS:8024-14-4** **HR: 2**
DEERTONGUE INCOLORE

PROP: Found in leaves of *Liatris odoratissima*; contains coumarin (12VSA5 9,5315,76).

SYNS: DEER'S TONGUE □ LIATRIS □ LIATRIX OLEORESIN □ VANILLA PLANT

TOXICITY DATA WITH REFERENCE
orl-rat LD50:730 mg/kg FCTXAV 14,743,76
skn-rbt LD50:3670 mg/kg FCTXAV 14,743,76

SAFETY PROFILE: Moderately toxic by ingestion and skin contact. When heated to decomposition it emits toxic fumes of NO_x. See also COUMARIN.

DAK000 **CAS:56283-74-0** **HR: 3**
16-DEETHYL-3-o-DEMETHYL-16-METHYL-3-o-(1-OXO-PROPYL)MONENSIN
mf: $C_{37}H_{62}O_{12}$ mw: 698.99

PROP: Prisms from EtOAc. Mp: 151–153°.

SYN: LAIDLOMYCIN

TOXICITY DATA WITH REFERENCE
ipr-mus LD50:5 mg/kg 85ERAY 1,814,78
scu-mus LD50:2500 μg/kg 85ERAY 1,814,78
ivn-mus LD50:1 mg/kg 85ERAY 1,814,78

SAFETY PROFILE: Poison by intraperitoneal, subcutaneous, and intravenous routes. When heated to decomposition it emits acrid smoke and irritating fumes.

DAK200 **CAS:70-51-9** **HR: 3**
DEFEROXAMINE
mf: $C_{25}H_{48}N_6O_8$ mw: 560.79

PROP: Crystals from EtOH (aq). Mp: 138–140°

SYNS: 30-AMINO-3,14,25-TRIHYDROXY-3,9,14,20,25-PENTAAZATRIACONTANE-2,10,13,21,24-PENTAONE □ N-BENZOYLFERRIOXAMINE B □ DEFEROXAMINUM □ DEFERRIOXAMINE □ DEFERRIOXAMINE B □ DESFERAL □ DESFERRAL □ DESFERRIN □ DESFERRIOXAMINE □ DESFERRIOXAMINE B □ Df B □ DFO □ DFOA □ DFOM □ NSC-52760

TOXICITY DATA WITH REFERENCE
dni-hmn:Cells-uns 50 μmol/L TXCYAC 5,427,91
dni-hmn:lvr 50 μmol/L TXCYAC 5,427,91
scu-cld TDLo:12 g/kg/17W-I:EYE NEJMAG 314,869,86
scu-hmn TDLo:37 g/kg/2Y-I:EYE NEJMAG 314,869,86
ivn-wmn TDLo:40 mg/kg:EYE NPRNAY 46,211,87
ivn-man TDLo:86 mg/kg/1H-C:BLD AJKDDP 6,254,85 ADCHAK 63,250,88

scu-rat LD50:12,240 mg/kg IYKEDH 6,119,75
ivn-rat LD50:329 mg/kg TXAPA9 18,185,71
orl-mus LD50:1340 mg/kg ARZNAD 17,748,67
ipr-mus LD50:1680 mg/kg KSRNAM 4,99,70
scu-mus LD50:1450 mg/kg IYKEDH 6,119,75

SAFETY PROFILE: Poison by intravenous route. Moderately toxic by ingestion, intraperitoneal, and subcutaneous routes. Human systemic effects: changes in hearing acuity, eye hemorrhage, optic nerve neuropathy, thrombocytopenia, visual field changes. Human mutation data reported. When heated to decomposition it emits toxic fumes of NO_x.

DAK300 CAS:138-14-7 HR: 3
DEFEROXAMINE MESYLATE
mf: $C_{25}H_{48}N_6O_8 \cdot CH_4O_3S$ mw: 656.90

PROP: Crystals from EtOH (aq). Mp: 148–149°.

SYNS: DEFEROXAMINE MESILATE □ DEFEROXAMINE METHANESULFONATE □ DESFERAL METHANESULFONATE □ DESFERRIOXAMINE B MESYLATE □ DESFERRIOXAMINE B METHANESULFONATE □ DESFERRIOXAMINE METHANESULFONATE

TOXICITY DATA WITH REFERENCE
scu-wmn TDLo:150 mg/kg/4D-I:GIT BMJOAE 291,448,85
scu-man TDLo:143 mg/kg/4D-I:EYE,GIT BMJOAE 291,448,85
par-man TDLo:16 g/kg/34W-I:CVS,GIT AJKDDP 10,71,87
unr-wmn TDLo:900 mg/kg/30W-I:CNS,CVS AJMEAZ 87,468,89
orl-rat LD50:17,300 mg/kg NIIRDN 6,496,82
scu-rat LD50:5500 mg/kg NIIRDN 6,496,82
ivn-rat LD50:330 mg/kg NIIRDN 6,496,82
orl-mus LD50:15,200 mg/kg NIIRDN 6,496,82
ipr-mus LD50:1240 mg/kg OYYAA2 7,1181,73
scu-mus LD50:1280 mg/kg OYYAA2 7,1181,73
ivn-mus LD50:273 mg/kg NIIRDN 6,496,82

SAFETY PROFILE: Poison by intravenous route. Moderately toxic by subcutaneous and intraperitoneal routes. Mildly toxic by ingestion. Human systemic effects: acute pulmonary edema, blood changes, cardiomyopathy, degenerative brain changes, diarrhea, hypermotility, nausea or vomiting, pericarditis, ulceration or bleeding from small intestine, visual field changes. When heated to decomposition it emits toxic fumes of NO_x and SO_x.

DAK400 CAS:1740-19-8 HR: 2
DEHYDROABIETIC ACID
mf: $C_{21}H_{30}O_2$ mw: 314.51

PROP: Crystals from EtOH (aq). Mp: 172–173°.

SYNS: DHA □ 13-ISOPROPYLPODOCARPA-8,11,13-TRIEN-15-OIC ACID

TOXICITY DATA WITH REFERENCE
orl-rat LD50:1710 mg/kg BECTA6 18,42,77

CONSENSUS REPORTS: Reported in EPA TSCA Inventory.

SAFETY PROFILE: Moderately toxic by ingestion. When heated to decomposition it emits acrid smoke and irritating fumes.

DAK600 CAS:434-16-2 HR: 2
7-DEHYDROCHOLESTEROL
mf: $C_{27}H_{44}O$ mw: 384.71

PROP: Plates from MeOH (aq) and Et_2O. Mp: 150–151° (anhyd).

SYNS: (3-β)CHOLESTA-5,7-DIEN-3-OL □ 5,7-CHOLESTADIEN-3-β-OL □ Δ^7-CHOLESTEROL □ $\Delta^{5,7}$-CHOLESTEROL □ DEHYDROCHOLESTERIN (GERMAN) □ 7-DEHYDROCHOLESTERIN □ DEHYDROCHOLESTEROL □ 7,8-DIDEHYDROCHOLESTEROL □ PROVITAMIN D_3

TOXICITY DATA WITH REFERENCE
scu-mus TDLo:800 mg/kg/4W-I:ETA NATWAY 60,525,73

CONSENSUS REPORTS: Reported in EPA TSCA Inventory.

SAFETY PROFILE: Questionable carcinogen with experimental tumorigenic data. When heated to decomposition it emits acrid smoke and irritating fumes.

DAK800 CAS:1059-86-5 HR: 2
7-DEHYDROCHOLESTEROL ACETATE
mf: $C_{29}H_{46}O_2$ mw: 426.75

PROP: Crystals from MeOH. Mp: 129–130°.

SYNS: CHOLESTA-5,7-DIEN-3-β-OL ACETATE □ 7-DEHYDROCHOLESTERYL ACETATE

TOXICITY DATA WITH REFERENCE
scu-mus TDLo:200 mg/kg/90D-I:ETA NATUAS 209,1026,66

SAFETY PROFILE: Questionable carcinogen with experimental tumorigenic data. When heated to decomposition it emits acrid smoke and irritating fumes.

DAL000 CAS:81-23-2 HR: 2
DEHYDROCHOLIC ACID
mf: $C_{24}H_{34}O_5$ mw: 402.58

PROP: Crystals from Me_2CO. Mp: 237°.

SYNS: ACIDE DEHYDROCHOLIQUE (FRENCH) □ ACOLEN □ BILIDREN □ BILOSTAT □ CHOLAGON □ CHOLAN DH □ CHOLIMED □ CHOLOGON □ CHOLOLIN □ DECHOLIN □ DEHYCHOL □ DEHYDROCHOLSAEURE (GERMAN) □ DEHYSTOLIN □ DEIDROCOLICO VITA □ DHC □ DIDOCOL □ DIDROCOLO □ DILABIL □ DILAHIL □ DRENOBYL □ EREBILE □ FELACRINOS □ HYKOLEX □ KETOCHOL □ KETOCHOLANIC ACID □ NOVOCOLIN □ OXYCHOLIN □ PROCHOLON □ SANOCHOLEN □ TRIKETOCHOLANIC ACID □ 3,7,12-TRIKETOCHOLANIC ACID □ 3,7,12-TRIOXOCHOLANIC ACID □ 3,7,12-TRIOXO-5-β-CHOLAN-24-OIC ACID

TOXICITY DATA WITH REFERENCE
orl-rat LD50:4 g/kg AIPTAK 116,154,58
ivn-rat LD50:750 mg/kg NIIRDN 6,495,82
ims-rat LD50:1500 mg/kg AEPPAE 222,244,54
orl-mus LD50:3100 mg/kg ARZNAD 12,857,62
scu-mus LD50:1620 mg/kg AIPTAK 116,154,58
ivn-mus LD50:1492 mg/kg JJPAAZ 22,235,72

CONSENSUS REPORTS: Reported in EPA TSCA Inventory.

SAFETY PROFILE: Moderately toxic by ingestion, intravenous, intramuscular, and subcutaneous routes. When

D

heated to decomposition it emits acrid smoke and fumes.

DAL030 **HR: 3**
DEHYDROEPIANDROSTERONE SODIUM SULFATE DIHYDRATE
mf: $C_{19}H_{27}O_5S{\cdot}Na{\cdot}2H_20$ mw: 426.55

SYNS: 3-β-HYDROXYANDROST-5-EN-17-ONE ESTER with SODIUM SULFATE DIHYDRATE ☐ PRASTERONE SODIUM SULFATE DIHYDRATE ☐ SODIUM ANDROST-5-EN-17-ONE-3-β-YL SULFATE DIHYDRATE

TOXICITY DATA WITH REFERENCE
ivn-wmn TDLo:4 mg/kg:GIT,SKN JMGZAI 18(5),10,81
ipr-rat LD50:523 mg/kg JMGZAI 18(5),10,81
scu-rat LD50:1005 mg/kg JMGZAI 18(5),10,81
ivn-rat LD50:468 mg/kg JMGZAI 18(5),10,81
ipr-mus LD50:460 mg/kg JMGZAI 18(5),10,81
scu-mus LD50:899 mg/kg JMGZAI 18(5),10,81
ivn-mus LD50:274 mg/kg JMGZAI 18(5),10,81

SAFETY PROFILE: Poison by intravenous route. Moderately toxic by intraperitoneal and subcutaneous routes. Human systemic effects by intravenous route: dermatitis, nausea, and vomiting. A steroid. When heated to decomposition it emits toxic fumes of SO_x and Na_2O.

DAL040 **HR: 3**
DEHYDROEPIANDROSTERONE SULFATE SODIUM
mf: $C_{19}H_{27}O_5S{\cdot}Na$ mw: 390.51

SYNS: DHA-S SODIUM ☐ MYLIS ☐ PRASTERONE SODIUM SULFATE ☐ SODIUM DEHYDROEPIANDROSTERONE SULFATE

TOXICITY DATA WITH REFERENCE
ipr-rat TDLo:500 mg/kg (8-17D preg):REP OYYAA2 12,201,76
ipr-mus TDLo:3 g/kg (female 7-16D post):REP OYYAA2 12,201,76
ipr-mus TDLo:100 mg/kg (female 7-16D post):TER OYYAA2 12,201,76
ipr-mus TDLo:1 g/kg (female 7-16D post):REP OYYAA2 12,201,76
ipr-rat TDLo:100 mg/kg (female 8-17D post):TER OYYAA2 12,201,76
ipr-rat TDLo:1 g/kg (female 8-17D post):TER OYYAA2 12,201,76
ivn-wmn TDLo:4 mg/kg:GIT,SKN JMGZAI 18(5),10,81
ipr-rat LD50:523 mg/kg KSRNAM 10,1852,76
scu-rat LD50:1005 mg/kg KSRNAM 10,1852,76
ivn-rat LD50:468 mg/kg KSRNAM 10,1852,76
ipr-mus LD50:460 mg/kg KSRNAM 10,1852,76
scu-mus LD50:899 mg/kg KSRNAM 10,1852,76
ivn-mus LD50:274 mg/kg KSRNAM 10,1852,76

SAFETY PROFILE: Poison by intravenous route. Moderately toxic by intraperitoneal and subcutaneous routes. Experimental teratogenic and reproductive effects. Human systemic effects by intravenous route: nausea or vomiting, dermatitis. A steroid. When heated to decomposition it emits toxic fumes of SO_x and Na_2O.

DAL060 **CAS:26400-24-8** **HR: 3**
DEHYDROHELIOTRIDINE
mf: $C_8H_{11}NO_2$ mw: 153.20

SYN: 3,8-DIDEHYDRO-HELIOTRIDINE

TOXICITY DATA WITH REFERENCE
dni-mus:kdy 5 mg/L CBINA8 4,421,71/72
dnd-dom:kdy 10 mg/L CBINA8 10,133,75
dns-dom:kdy 10 µmol/L CBINA8 13,243,76
dni-dom:kdy 40 mg/L CBINA8 10,133,75
oms-dom:kdy 10 mg/L CBINA8 10,133,75
ipr-rat TDLo:60 mg/kg (14D preg):TER JPTLAS 131,339,80
ipr-rat TDLo:30 mg/kg (female 14D post):TER JPTLAS 131,339,80
ipr-rat TDLo:40 mg/kg (female 14D post):TER JPTLAS 131,339,80
ipr-rat TDLo:40 mg/kg (14D preg):TER JPTLAS 131,339,80
ipr-rat LDLo:92 mg/kg CBINA8 12,299,76

SAFETY PROFILE: Poison by intraperitoneal route. An experimental teratogen. Mutation data reported. When heated to decomposition it emits toxic fumes of NO_x.

DAL100 **CAS:23107-11-1** **HR: 3**
DEHYDROHELIOTRINE
mf: $C_{16}H_{25}NO_5$ mw: 311.42

TOXICITY DATA WITH REFERENCE
ipr-rat TDLo:30 mg/kg (female 14D post):TER JPTLAS 131,339,80
ipr-rat TDLo:40 mg/kg (female 14D post):TER JPTLAS 131,339,80
ipr-rat TDLo:40 mg/kg (14D preg):TER JPTLAS 131,339,80
ipr-rat TDLo:60 mg/kg (14D preg):TER JPTLAS 131,339,80
ipr-rat LDLo:62 mg/kg CBINA8 12,299,76

SAFETY PROFILE: Poison by intraperitoneal route. An experimental teratogen. When heated to decomposition it emits toxic fumes of NO_x.

DAL200 **CAS:3343-10-0** **HR: 2**
1,2-DEHYDRO-3-METHYLCHOLANTHRENE
mf: $C_{21}H_{14}$ mw: 266.35

SYNS: DEHYDRO-3-METHYLCHOLANTHRENE ☐ 3-METHYLBENZ(j)-ACEANTHRYLENE ☐ 3-METHYLCHOLANTHRYLENE ☐ 20-METHYLCHOLANTHRYLENE

TOXICITY DATA WITH REFERENCE
skn-mus TDLo:85 mg/kg/20W-I:CAR CBINA8 22,69,78
scu-mus TDLo:120 mg/kg/6W-I:NEO IJCNAW 2,505,67

SAFETY PROFILE: Questionable carcinogen with experimental carcinogenic and neoplastigenic data. When heated to decomposition it emits acrid smoke and irritating fumes.

DAL300 CAS:72-63-9 HR: 2
1-DEHYDRO-17-α-METHYLTESTOSTERONE
mf: $C_{20}H_{28}O_2$ mw: 300.42

PROP: Crystals from acetone + ether. Mp: 163–164°.

SYNS: ABIROL □ ANABOLIN □ CIBA 17309 BA □ COMPOUND 17309 □ CREIN □ DANABOL □ DEHYDROMETHYLTESTERONE □ Δ1-DEHYDROMETHYLTESTOSTERONE □ DIANABOL □ DIANABOLE □ GEABOL □ 17-β-HYDROXY-17-α-METHYLANDROSTRA-1,4-DIEN-3-ONE □ MA □ METANABOL □ METANDIENON □ METANDIENONE □ METANDIENONUM □ METANDROSTENOLON □ METANDROSTENOLONE □ METASTENOL □ METHANDIENONE □METHANDROLONE □ METHANDROSTENOLONE □ 17-α-METHYL-17-β-HYDROXY-1,4-ANDROSTADIEN-3-ONE □ Δ'-17-METHYLTESTOSTERONE □ Δ^1-17-α-METHYLTESTOSTERONE □ NEROBOL □ NEROBOLETTES □ NSC-42722 □ PROTOBOLIN □ STENOLON □ STENOLONE

TOXICITY DATA WITH REFERENCE
dlt-mus-orl 200 mg/kg/10D-I VPITAR 38(4),63,80
orl-man TDLo:13 mg/kg (60D male):REP CCPTAY 15,151,77
scu-rat TDLo:7 mg/kg (female 14D pre):REP CCPTAY 5,489,72
orl-rat TDLo:200 mg/kg (female 10D pre):REP ARZNAD 14,330,64
orl-rat TDLo:200 mg/kg (male 10D pre):REP ARZNAD 14,330,64
orl-man TDLo:561 mg/kg/7Y-C:CAR LANCAO 2,1273,72
ipr-rat LD50:425 mg/kg PCJOAU 20,143,86

CONSENSUS REPORTS: EPA Genetic Toxicology Program.

SAFETY PROFILE: Moderately toxic by intraperitoneal route. Human reproductive effects by ingestion route: changes in spermatogenesis. Other experimental reproductive effects. Questionable human carcinogen producing liver tumors. Mutation data reported. When heated to decomposition it emits acrid smoke and irritating fumes. See also TESTOSTERONE.

DAL350 CAS:23291-96-5 HR: 2
DEHYDROMONOCROTALINE
mf: $C_{16}H_{21}NO_6$ mw: 323.38

SYNS: MONOCROTALINE, 3,8-DIDEHYDRO- □ 20-NORCROTALANAN-11,15-DIONE, 3,8-DIDEHYDRO-14,19-DIHYDRO-12,13-DIHYDROXY-, (13-α-14-α)-

TOXICITY DATA WITH REFERENCE
skn-mus TDLo:1504 mg/kg/47W-I:ETA CALEDQ 17,61,82
ipr-rat LDLo:16 mg/kg CBINA8 12,299,76

SAFETY PROFILE: Poison by intraperitoneal route. Questionable carcinogen with experimental tumorigenic data. When heated to decomposition it emits toxic fumes of NO_x.

DAL400 CAS:23107-12-2 HR: 3
DEHYDRORETRONECINE
mf: $C_8H_{11}NO_2$ mw: 153.20

SYNS: 3,8-DIDEHYDRORETRONECINE □ (R)-2,3-DIHYDRO-1-HYDROXY-1H-PYRROLIZINE-7-METHANOL

TOXICITY DATA WITH REFERENCE
mmo-sat 500 μg/plate MUREAV 149,485,85
sce-hmn:lym 1 μmol/L MUREAV 149,485,85
dni-rat:lvr 100 μmol/L CBINA8 30,325,80
oms-rat:lvr 20 μmol/L CBINA8 30,325,80
otr-ham:kdy 250 μg/L CRNGDP 1,161,80
scu-rat TDLo:350 mg/kg/1Y-I:NEO JNCIAM 56,787,76
skn-mus TDLo:120 mg/kg/28W-I:CAR JJIND8 61,85,78
scu-mus TDLo:80 mg/kg/4W-I:CAR JJIND8 61,85,78
skn-mus TD:1438 mg/kg/47W-I:CAR CALEDQ 17,61,82
ipr-rat LD50:122 mg/kg CBINA8 12,299,76

CONSENSUS REPORTS: IARC Cancer Review: Animal Sufficient Evidence IMEMDT 10,333,76.

SAFETY PROFILE: Confirmed carcinogen with experimental carcinogenic and neoplastigenic data. Poison by intraperitoneal route. Human mutation data reported. When heated to decomposition it emits toxic fumes of NO_x.

D

DAL600 CAS:84-17-3 HR: 3
DEHYDROSTILBESTROL
mf: $C_{18}H_{18}O_2$ mw: 266.36

PROP: A solid. Mp: 121–122°.

SYNS: 3,4-BIS(p-HYDROXYPHENYL)-2,4-HEXADIENE □ 3,4-BIS(4-HYDROXYPHENYL)-2,4-HEXADIENE □ CYCLADIENE □DIENESTROL □ DIENOESTROL □ β-DIENOESTROL □ DIENOL □ 4,4'-(1,2-DIETHYLIDENE-1,2-ETHANEDIYL)BISPHENOL □ p,p'-(DIETHYLIDENEETHYLENE)DIPHENOL □ 4,4'-(DIETHYLIDENEETHYLENE)DIPHENOL □ DINOVEX □ DI(p-OXYPHENYL)-2,4-HEXADIENE □ DV □ ESTRAGARD □ ESTRODIENOL □ ESTRORAL □ FOLLIDIENE □ FOLLORMON □ GYNEFOLLIN □ HORMOFEMIN □ 4,4'-HYDROXY-γ,Δ-DIPHENYL-β,Δ-HEXADIENE □ ISODIENESTROL □ OESTRASID □ OESTRODIENE □ OESTRODIENOL □ OESTRORAL □ PARA-DIEN □ RESTROL □ RETALON □ SEXADIEN □ SYNESTROL □ TESERENE □ WILLNESTROL

TOXICITY DATA WITH REFERENCE
sce-hmn:fbr 5 nmol/L NATUAS 281,392,79
dns-ham:emb 3 mg/L CNREA8 44,184,84
orl-rat TDLo:40 μg/kg (3D preg):REP JRPFA4 13,101,67

CONSENSUS REPORTS: IARC Cancer Review: Animal Inadequate Evidence IMEMDT 21,161,79; Human Limited Evidence IMEMDT 21,161,79.

SAFETY PROFILE: Suspected human carcinogen. Human mutation data reported. Experimental reproductive effects. Used as a drug for the treatment of postmenopausal symptoms. When heated to decomposition it emits acrid smoke and irritating fumes.

DAM000 CAS:60504-95-2 HR: 3
DEISOVALERYL BLASTMYCIN
mf: $C_{21}H_{28}N_2O_8$ mw: 436.51

PROP: A solid. Mp: 186–188°. Produced by *Streptomyces* sp. 5140-Al (JANTAJ 29,804,76).

SYNS: N-(7-BUTYL-4,9-DIMETHYL-2,6-DIOXO-8-HYDROXY-1,5-DIOXONAN-3-YL)-3-FORMAMIDOSALICYLAMIDE □ N-(7-BUTYL-8-HYDROXY-4,9-DIMETHYL-2,6-DIOXO-1,5-DIOXONAN-3-YL)-3-FORMAMIDOSALICYLAMIDE

TOXICITY DATA WITH REFERENCE
ipr-mus LD50:25 mg/kg JANTAJ 29,804,76
ivn-mus LD50:15 mg/kg JANTAJ 29,804,76

SAFETY PROFILE: Poison by intraperitoneal and intravenous routes. When heated to decomposition it emits toxic fumes of NO_x.

DAM315 **CAS:83435-67-0** **HR: 3**
DELAPRIL HYDROCHLORIDE
mf: $C_{26}H_{32}N_2O_5 \cdot ClH$ mw: 489.06

PROP: Plates from Me_2CO. Mp: 166–170° (decomp).

SYNS: CV 3317 □ N-(N-((S)-1-ETHOXYCARBONYL-3-PHENYLPROPYL)-l-ALANYL)-N-(INDAN-2-YL)GLYCINE HYDROCHLORIDE □ GLYCINE, N-(2,3-DIHYDRO-1H-INDEN-2-YL)-N-(N-(1-(ETHOXYCARBONYL)-3-PHENYLPROPYL)-l-A LANYL)-, MONOHYDROCHLORIDE, (S)-

TOXICITY DATA WITH REFERENCE
orl-rat TDLo:8700 mg/kg (female 15-22D post):REP YACHDS 15(Suppl 1),203,87
orl-rat LD50:8260 mg/kg YACHDS 15(Suppl 1),203,87
ipr-rat LD50:208 mg/kg YACHDS 15(Suppl 1),203,87
scu-rat LD50:5900 mg/kg YACHDS 15(Suppl 1),203,87
orl-mus LD50:3120 mg/kg YACHDS 15(Suppl 1),203,87
ipr-mus LD50:164 mg/kg YACHDS 15(Suppl 1),203,87
scu-mus LD50:2340 mg/kg YACHDS 15(Suppl 1),203,87

SAFETY PROFILE: Poison by intraperitoneal route. Moderately toxic by ingestion and subcutaneous routes. Experimental reproductive effects. When heated to decomposition it emits toxic fumes of NO_x and HCl.

DAM325 **HR: D**
DELATESTRYL and DEPO-PROVERA
mf: $C_{24}H_{34}O_4 \cdot C_{26}H_{41}O_3$ mw: 788.25

SYNS: DELATESTRYL and DEPO-MEDROXYPROGESTERONE ACETATE □ DEPO-MEDROXYPROGESTERONE ACETATE and DELATESTRYL □ DEPO-MEDROXYPROGESTERONE ACETATE and TESTOSTERONE ENANTHATE □ DEPO-PROVERA and DELATESTRYL □ DEPO-PROVERA and TESTOSTERONE ENANTHATE □ TESTOSTERONE ENANTHATE and DEPO-MEDROXYPROGESTERONE ACETATE □ TESTOSTERONE ENANTHATE and DEPO-PROVERA

TOXICITY DATA WITH REFERENCE
ims-man TDLo:15 mg/kg (3D male):REP CCPTAY 15,627,77
ims-man TDLo:23,570 μg/kg (2D male):REP CCPTAY 15,627,77

SAFETY PROFILE: Human male reproductive effects by intramuscular route: spermatogenesis, impotence. A steroid. When heated to decomposition it emits acrid smoke and fumes. See other testosterone entries.

DAM400 **CAS:528-53-0** **HR: 3**
DELPHINIDOL
mf: $C_{15}H_{11}O_7$ mw: 303.26

PROP: Dark-red-brown crystals.

SYNS: 3,3′,4′,5,5′,7-HEXAHYDROXYFLAVYLIUM ACID ANION □ 3,4,7-TRIHYDROXY-2-(3,4,5-TRIHYDROXYPHENYL)BENZOPYRYLIUM, ACID ANION

TOXICITY DATA WITH REFERENCE
ipr-rat LD50:2350 mg/kg CHTPBA 2,33,67
ivn-rat LD50:240 mg/kg CHTPBA 2,33,67
ipr-mus LD50:4110 mg/kg CHTPBA 2,33,67
ivn-mus LD50:840 mg/kg CHTPBA 2,33,67

SAFETY PROFILE: Poison by intravenous route. Moderately toxic by intraperitoneal route. When heated to decomposition it emits acrid smoke and irritating fumes.

DAM600 **CAS:57-42-1** **HR: 3**
DEMAROL
mf: $C_{15}H_{21}NO_2$ mw: 247.37

PROP: A solid. Mp: 30°, bp: 155° @ 5 mm.

SYNS: DEMEROL □ DOLCONTRAL □ DOLOSAL □ DOLSIN □ ETHYL-1-METHYL-4-PHENYLISONIPECOTATE □ ETHYL-1-METHYL-4-PHENYLPIPERIDINE-4-CARBOXYLATE □ ISONIPECAINE □ LIDOL □ MEPERIDINE □ N-METHYL-4-PHENYL-4-CARBETHOXYPIPERIDINE □ 1-METHYL-4-PHENYLISONIPECOTIC ACID, ETHYL ESTER □ 1-METHYL-4-PHENYL-PIPERIDIN-4-CARBON-SAEURE-AETHYLESTER-HYDROCHLORID (GERMAN) □ 1-METHYL-4-PHENYLPIPERIDINE-4-CARBOXYLIC ACID ETHYL ESTER □ NEMEROL □ PETHIDINETER □ PETHIDOINE □ PHETIDINE □ PIPERSAL □ PIRIDOSAL

TOXICITY DATA WITH REFERENCE
ims-wmn TDLo:2 mg/kg (39W preg):REP AJOGAH 64,1368,52
ipr-rat TDLo:1540 μg/kg (lactating female 1-7D post):REP NRTXDN 7,349,86
ipr-rat TDLo:3080 μg/kg (lactating female 1-7D post):REP NRTXDN 7,349,86
unr-man LDLo:15 mg/kg 85DCAI 2,73,70
unr-man TDLo:57 mg/kg:CNS NEJMAG 312,509,85
orl-rat LD50:162 mg/kg AIPTAK 180,155,69
ipr-rat LD50:87 mg/kg JAPMA8 47,323,58
scu-rat LD50:113 mg/kg ARZNAD 24,600,74
ivn-rat LD50:29 mg/kg DDREDK 1,83,81
idu-rat LD50:90 mg/kg AIPTAK 180,155,69
orl-mus LD50:200 mg/kg AIPTAK 135,376,62
ipr-mus LD50:150 mg/kg AIPTAK 135,376,62
par-mus LD50:178 mg/kg JMCMAR 11,889,68

SAFETY PROFILE: A human poison by an unspecified route. Poison experimentally by ingestion, subcutaneous, intravenous, intradermal, parenteral, and intraperitoneal routes. Human systemic effects by unspecified route: changes in sleep patterns and muscle weakness. Human reproductive effects by intramuscular route: effects on measurements and viability of newborn. A pharmaceutical pain killer. When heated to decomposition it emits toxic fumes of NO_x. See also ESTERS.

DAM700 **CAS:50-13-5** **HR: 3**
DEMEROL HYDROCHLORIDE
mf: $C_{15}H_{21}NO_2 \cdot ClH$ mw: 283.83

PROP: Minute crystals. Mp: 186–189°. Insol in C_6H_6 and Et_2O.

SYNS: ALGIL □ ALODAN (GEROT) □ ANTIDUROL □ CENTRALGIN □ CHLORBICYCLENE (FRENCH) □ DEMEROL □ DISPADOL □ DOLANTAL □ DOLANTIN □ DOLANTIN HYDROCHLORIDE □ DOLANTOL □ DOLAREN □ DOLARGAN □ DOLCONTRAL □ DOLENAL □ DOLENOL □ DOLESTINE □ DOLIN □ DOLOGAL □ DOLONEURINE □ DOLOPETHIN □ DOLOSAL □ DOLVANOL □ ENDOLAT □ ETHYL-1-METHYL-4-PHENYLISONIPECOTATE HYDROCHLORIDE □ ETHYL-1-METHYL-4-PHENYLPIPERIDINE-4-CARBOXYLATE HYDRO-

CHLORIDE □ ETHYL-1-METHYL-4-PHENYLPIPERIDYL-4-CARBOXYLATE HYDROCHLORIDE □ ISONIPECAINE HYDROCHLORIDE □ LIDOL □ LYDOL □ MEFEDINA □ MEPADIN □ MEPERIDINE HYDROCHLORIDE □ MEPHEDINE □ 1-METHYL-4-CARBETHOXY-4-PHENYLPIPERIDINE HYDROCHLORIDE □ N-METHYL-4-PHENYL-4-CARBETHOXYPIPERIDINE HYDROCHLORIDE □ 1-METHYL-4-PHENYL-4-CARBOETHOXYPIPERIDINE HYDROCHLORIDE □ 1-METHYL-4-PHENYLISONIPECOTIC ACID ETHYL ESTER HYDROCHLORIDE □ OPERIDINE □ PANTALGINE □ PENTANTIN □ PENTANTIN HYDROCHLORIDE □ PETHIDINE CHLORIDE □ PETIDIN □ PIRIDOSAL □ S 140 □ SAUTERALGYL □ SPASMEDAL □ SPASMODOLIN □ SYNELAUDINE □ WY 554

TOXICITY DATA WITH REFERENCE
mnt-mus-ipr 8 mg/kg IJMRAQ 75,112,82
ivn-wmn TDLo:735 μg/kg (female 39W post):TER JPEMAO 16,23,88
scu-ham TDLo:146 mg/kg (female 8D post):TER AJOGAH 123,705,75
scu-ham TDLo:250 mg/kg (female 8D post):TER AJOGAH 123,705,75
ims-wmn TDLo:43 mg/kg/2D-I AEMED3 14,1007,85
ims-wmn TDLo:13,500 μg/kg/3D-I AJPSAO 144,1062,87
orl-rat LD50:170 mg/kg JPETAB 103,147,51
ipr-rat LD50:65 mg/kg TXCYAC 14,217,79
scu-rat LD50:175 mg/kg ARZNAD 3,238,53
ivn-rat LD50:17 mg/kg AIPTAK 115,213,58
orl-mus LD50:178 mg/kg NIIRDN 6,755,82
ipr-mus LD50:120 mg/kg JPETAB 119,26,57
orl-mus LD50:178 mg/kg ARZNAD 28,164,78
ivn-dog LD50:68 mg/kg AIPTAK 149,571,64
par-frg LD50:515 mg/kg JPETAB 103,147,51

SAFETY PROFILE: Poison by ingestion, subcutaneous, intravenous, and intraperitoneal routes. Moderately toxic by parenteral route. Experimental teratogenic effects. Mutation data reported. An analgesic. When heated to decomposition it emits very toxic fumes of HCl and NO_x.

DAN000 CAS:58957-92-9 HR: 3
4-DEMETHOXYDAUNOMYCIN
mf: $C_{26}H_{27}NO_9$ mw: 497.54

SYNS: 4-DEMETHOXYDAUNORUBICIN □ NSC-256439

TOXICITY DATA WITH REFERENCE
mmo-sat 2 μg/plate ENMUDM 7,129,85
dns-rat:lvr 2 mg/L CNREA8 44,5599,84
dnd-mus:fbr 1800 nmol/L CNREA8 37,4523,77
dnd-mam:lym 200 nmol/L BBRCA9 69,744,76
orl-mus LD50:16 mg/kg CTRRDO 61,893,77
ipr-mus LD50:3 mg/kg CNREA8 48,926,88
ivn-mus LD50:4 mg/kg CTRRDO 61,893,77

SAFETY PROFILE: Poison by ingestion, intraperitoneal, and intravenous routes. Mutation data reported. When heated to decomposition it emits toxic fumes of NO_x. See also DAUNOMYCIN.

DAN200 HR: 3
N-DEMETHYLACLACINOMYCIN A
mf: $C_{41}H_{51}NO_{13}$ mw: 765.93

TOXICITY DATA WITH REFERENCE
mma-sat 1 nmol/plate CNREA8 38,1782,78
ipr-mus LD50:46 mg/kg JANTAJ 31,78-90,78

SAFETY PROFILE: Poison by intraperitoneal route. Mutation data reported. When heated to decomposition it emits toxic fumes of NO_x.

DAN300 CAS:7336-36-9 HR: 3
2-DEMETHYLCOLCHICINE
mf: $C_{21}H_{24}NO_6$ mw: 386.46

PROP: Crystals from $CHCl_3$. Mp: 110–140°, mp: 178–180° (double mp).

SYN: O^2-DEMETHYLCOLCHICINE

TOXICITY DATA WITH REFERENCE
ice-rat LDLo:75 μg/kg AIPTAK 109,386,57
orl-mus LD50:19 mg/kg AIPTAK 94,453,53
ipr-mus LD50:16 mg/kg AIPTAK 109,386,57
ims-mus LD50:16,577 μg/kg JMCMAR 24,257,81

SAFETY PROFILE: Poison by ingestion, intramuscular, intracerebral, and intraperitoneal routes. When heated to decomposition it emits toxic fumes of NO_x.

DAN375 CAS:477-29-2 HR: 3
3-DEMETHYLCOLCHICINE GLUCOSIDE
mf: $C_{27}H_{33}NO_{11}$ mw: 547.61

PROP: Cryst from EtOH. Mp: 192–195°, mp: 216–218° (block).

SYN: COLCHICOSIDE

TOXICITY DATA WITH REFERENCE
ice-rat LDLo:75 μg/kg AIPTAK 109,386,57
par-rat LDLo:75 μg/kg FRPSAX 15,533,60
orl-mus LD50:84 mg/kg AIPTAK 94,386,57
ipr-mus LD50:280 mg/kg AIPTAK 109,386,57
ivn-mus LD50:320 mg/kg AIPTAK 94,453,53

SAFETY PROFILE: Poison by ingestion, parenteral, intracerebral, intravenous, and intraperitoneal routes. When heated to decomposition it emits toxic fumes of NO_x. See also COLCHICINE.

DAO500 CAS:298-03-3 HR: 3
DEMETON
mf: $C_8H_{19}O_3PS_2$ mw: 258.36

PROP: Liquid. D: 1.119 @ 21°/4°, bp: 92–93° @ 0.15 mm.

SYNS: BAYER 8169 □ DEMETON-O □ DIAETHYLTHIOPHOSPHORSAEUREESTER des AETHYLTHIOGLYKOL (GERMAN) □ O,O-DIETHYL-O-(2-ETHTHIOETHYL)PHOSPHOROTHIOATE □ DIETHYL 2-ETHTHIOETHYL THIONOPHOSPHATE □ O,O-DIETHYL-O-2-(ETHYLTHIO)ETHYL PHOSPHOROTHIOATE □ O,O-DIETHYL-2-ETHYLTHIO ETHYL PHOSPHOROTHIOATE □ DIETHYL-2-(ETHYLTHIO(ETHYL PHOSPHOROTHIONATE)) □ DI-SEPTON □ E-1059 □ ETHYLTHIOMETON □ MERCAPTOFOS (RUSSIAN) □ THIODEMETON □ THIOLMECAPTOPHOS

TOXICITY DATA WITH REFERENCE
mmo-esc 200 μg/plate NTIS** PB80-133226
mma-esc 200 μg/plate NTIS** PB80-133226
dnr-bcs 5 μg/disc NTIS** PB80-133226
mma-smc 1000 ppm NTIS** PB80-133226
dns-hmn:fbr 100 mg/L NTIS** PB80-133226

cyt-ham-ipr 2 mg/kg ARTODN 58,152,85
orl-rat LD50:7500 μg/kg GUCHAZ 6,161,73
scu-mus LD50:15 mg/kg AEPPAE 217,144,53
skn-rbt LDLo:100 mg/kg GTPZAB 2(5),7,58
ipr-ham LD50:10 mg/kg ARTODN 58,152,85

SAFETY PROFILE: Poison by ingestion, skin contact, subcutaneous, and intraperitoneal routes. Human mutation data reported. An insecticide. When heated to decomposition it emits toxic fumes of PO_x and SO_x. See also DEMETON-O + DEMETON-S and other demeton entries.

For occupational chemical analysis use NIOSH: Demeton, 5514.

DAO600 CAS:8065-48-3 HR: 3
DEMETON-O + DEMETON-S
mf: $C_8H_{19}O_3PS_2 \cdot C_8H_{19}O_3PS_2$ mw: 516.72

PROP: A light-brown liquid; sulfur compound odor.

SYNS: BAY 10756 □ BAYER 8169 □ DEMETON □ DEMOX □ DIETHOXY THIOPHOSPHORIC ACID ESTER of 2-ETHYLMERCAPTOETHANOL □ O,O-DIETHYL 2-ETHYLMERCAPTOETHYL THIOPHOSPHATE □ O,O-DIETHYL O(and S)-2-(ETHYLTHIO)ETHYL PHOSPHOROTHIOATE MIXTURE □ E 1059 □ ENT 17,295 □ MERCAPTOPHOS □ PHOSPHOROTHIOIC ACID-O,O-DIETHYL-O-(2-(ETHYLTHIO)ETHYL) ESTER, mixed with O,O-DIETHYL S-(2-(ETHYLTHIO)ETHYL) ESTER (7:3) □ SYSTEMOX □ SYSTOX □ ULV

TOXICITY DATA WITH REFERENCE
sce-hmn:lym 10 mg/L MUREAV 88,307,81
sce-ham:lng 10 mg/L MUREAV 88,307,81
ipr-mus TDLo:10 mg/kg/(7-11D preg):TER TXAPA9 24,324,73
orl-hmn LDLo:171 μg/kg CMEP** -,1,56
orl-man TDLo:144 mg/kg/24D-I TXAPA9 14,603,69
orl-rat LD50:1700 μg/kg TXAPA9 21,315,72
ihl-rat LCLo:15 mg/m³/4H 85GMAT -,52,82
skn-rat LD50:8200 μg/kg TXAPA9 2,88,60
ipr-rat LD50:2500 μg/kg APCRAW 4,117,61
ivn-rat LD50:1750 μg/kg BLLIAX 38,151,58
ims-rat LDLo:3 mg/kg 13ZGAF -,220,62
orl-mus LD50:7850 μg/kg BLLIAX 38,151,58
ipr-mus LD50:4 mg/kg PSEBAA 129,699,68
ivn-mus LD50:3900 μg/kg BLLIAX 38,151,58
ims-dog LD50:3650 μg/kg 13ZGAF -,221,62
ihl-cat LCLo:15 mg/m³/4H 85GMAT -,52,82
ims-cat LD50:3900 μg/kg 13ZGAF -,220,62
skn-rbt LD50:24 mg/kg SPEADM 74-1,-,74

CONSENSUS REPORTS: EPA Genetic Toxicology Program. EPA Extremely Hazardous Substances List.

OSHA PEL: TWA 0.1 mg/m³ (skin)
ACGIH TLV: TWA 0.01 ppm (skin)
DFG MAK: 0.01 ppm (0.1 mg/m³)

SAFETY PROFILE: A deadly human poison by ingestion. Poison experimentally by ingestion, inhalation, skin contact, intramuscular, intravenous, subcutaneous, and intraperitoneal routes. An experimental teratogen. Human mutation data reported. An insecticide that inhibits cholinesterase in humans and animals and thus causes the buildup of acetylcholine. Doses are cumulative. If illness occurs, it is acute in nature, whether caused by a single large dose or by repeated exposure. Persons poisoned with demeton may be expected to show the following symptoms: headache, giddiness, blurred vision, weakness, nausea, diarrhea, and discomfort in the chest. When heated to decomposition it emits very toxic fumes of PO_x and SO_x. See also PARATHION and other DEMETON entries.

For occupational chemical analysis use NIOSH: Demeton, 5514.

DAO800 CAS:867-27-6 HR: 3
DEMETON-O-METHYL
mf: $C_6H_{15}O_3PS_2$ mw: 230.30

SYNS: BAY 15203 □ DEMETON-O-METILE (ITALIAN) □ O,O-DIMETHYL-O-(2-AETHYLTHIO-AETHYL) MONOTHIOPHOSPHAT (GERMAN) □ O,O-DIMETHYL-O-ETHYLMERCAPTOETHYL THIOPHOSPHATE □ O,O-DIMETHYL 2-ETHYLMERCAPTOETHYL THIOPHOSPHATE, THIONO ISOMER □ O,O-DIMETHYL-O-(2-ETHYL-THIO-ETHYL)-MONOTHIOFOSFAAT (DUTCH) □ O,O-DIMETHYL-O-2-(ETHYLTHIO)ETHYL PHOSPHOROTHIOATE □ O,O-DIMETIL-O-(2-ETILTIOETIL)-MONOTIOFOSFATO (ITALIAN) □ ENT 18,862 □ β-ETHYLMERCAPTOETHYL DIMETHYL THIONOPHOSPHATE □ O-(2-(ETHYLTHIO)ETHYL)-O,O-DIMETHYL PHOSPHOROTHIOATE □ 2-(ETHYLTHIO)ETHYL DIMETHYL PHOSPHOROTHIONATE □ METHYL-DEMETON-O □ O-METHYLDEMETON □ METHYLCISTOX □ METHYLMERCAPTOPHOS □ METHYLSYSTOX □ THIOPHOSPHATE de O,O-DIMETHYLE et de O-2-ETHYLTHIO-ETHYLE (FRENCH)

TOXICITY DATA WITH REFERENCE
ihl-rat TCLo:500 μg/m³/24H (female 1-22D post):REP HYSAAV 36(1),34,71
ihl-rat TCLo:200 ng/m³/(1-22D preg):TER HYSAAV 36(1),34,71
ihl-rat TCLo:100 μg/m³/24H (female 1-22D post):REP HYSAAV 36(1),34,71
ihl-rat TCLo:700 ng/m³/24H (1-22D preg):REP HYSAAV 36(1),34,71
ihl-rat TCLo:500 μg/m³/24H (female 1-22D post):TER HYSAAV 36(1),34,71
orl-rat LD50:75 mg/kg FATOAO 22,559,59
ivn-rat LD50:216 mg/kg BIJOAK 67,187,57
orl-mus LD50:46 mg/kg FATOAO 22,559,59
orl-cat LDLo:30 mg/kg FATOAO 22,559,59
ihl-cat LCLo:20 mg/m³ GISAAA 28(3),21,63
skn-rbt LDLo:75 mg/kg FATOAO 22,559,59

SAFETY PROFILE: Poison by ingestion, skin contact, inhalation, and intravenous routes. Experimental teratogenic and reproductive effects. When heated to decomposition it emits very toxic fumes of PO_x and SO_x. See also DEMETON-O + DEMETON-S and other DEMETON entries.

DAP000 CAS:301-12-2 HR: 3
DEMETON-O-METHYL SULFOXIDE
mf: $C_6H_{15}O_3PS_2$ mw: 230.30

PROP: Yellow liquid. D: 1.289 @ 20°/4°, bp: 106° @ 0.01 mm. Sol in H_2O, most org solvs; insol in pet ether.

SYNS: BAY 21097 □ DEMETON-S-METHYL-SULFOXID (GERMAN) □ DEMETON-S-METHYL SULFOXIDE □ DEMETON-METHYL SULPHOXIDE □ O,O-DIMETHYL-S-(2-AETHYLSULFINYL-AETHYL)-THIOLPHOSPHAT (GERMAN) □ O,O-DIMETHYL-S-(2-ETHTHIONYLETHYL)

PHOSPHOROTHIOATE ☐ DIMETHYL-S-(2-ETHTHIONYLETHYL) THIOPHOSPHATE ☐ O,O-DIMETHYL-S-(2-ETHYLSULFINYL-ETHYL)-MONOTHIOFOSFAAT (DUTCH) ☐ O,O-DIMETHYL-S-(2-(ETHYLSULFINYL)ETHYL) PHOSPHOROTHIOATE ☐ O,O-DIMETHYL-S-(2-ETHYLSULFINYL)ETHYL THIOPHOSPHATE ☐ O,O-DIMETHYL-S-ETHYLSULPHINYLETHYL PHOSPHOROTHIOLATE ☐ O,O-DIMETHYL-S-(3-OXO-3-THIA-PENTYL)-MONOTHIOPHOSPHAT (GERMAN) ☐ O,O-DIMETIL-S-(2-ETIL-SOLFINIL-ETIL)-MONOTIOFOSFATO (ITALIAN) ☐ ENT 24,964 ☐ S-(2-(ETHYLSULFINYL)ETHYL)-O,O-DIMETHYL PHOSPHOROTHIOATE ☐ ISOMETHYLSYSTOX SULFOXIDE ☐ METAISOSYSTOXSULFOXIDE ☐ METASYSTEMOX ☐ METASYSTOX-R ☐ METHYL DEMETON-O-SULFOXIDE ☐ METILMERCAPTOFOSOKSID ☐ OXYDEMETONMETHYL ☐ OXYDEMETON-METILE (ITALIAN) ☐ R 2170 ☐ THIOPHOSPHATE de O,O-DIMETHYLE et de S-2-ETHYLSULFINYLETHYLE (FRENCH)

TOXICITY DATA WITH REFERENCE
mmo-sat 50 μg/plate MUREAV 124,97,83
sce-hmn:lym 20 mg/L MUREAV 124,97,83
mnt-mus-ipr 10 mg/kg MUREAV 124,97,83
orl-rat LD50:30 mg/kg AEPPAE 234,352,58
ihl-rat LC50:1500 mg/m³/1H 85JFAN A309,83
skn-rat LD50:100 mg/kg WRPCA2 9,119,70
ipr-rat LD50:20 mg/kg GUCHAZ 6,385,73
ivn-rat LD50:47 mg/kg BIJOAK 67,187,57
orl-mus LD50:30 mg/kg SPEADM 78-1,29,78
ipr-mus LD50:8 mg/kg TXAPA9 4,621,62
orl-gpg LD50:120 mg/kg PCOC** -,721,66
orl-pgn LD50:15 mg/kg TXAPA9 20,57,71

CONSENSUS REPORTS: EPA Genetic Toxicology Program.

SAFETY PROFILE: Poison by ingestion, skin contact, intravenous, and intraperitoneal routes. Human mutation data reported. When heated to decomposition it emits very toxic fumes of PO_x and SO_x. See also other demeton entries.

DAP200 CAS:126-75-0 HR: 3
DEMETON-S
mf: $C_8H_{19}O_3PS_2$ mw: 258.36

SYNS: O,O-DIAETHYL-S-(2-AETHYLTHIO-AETHYL)-MONOTHIOPHOSPHAT (GERMAN) ☐ DIAETHYLTHIOPHOSPHORSAEUREESTER des AETHYLTHIOGLYKOL (GERMAN) ☐ DIETHYL-S-(2-ETHIOETHYL) THIOPHOSPHATE ☐ O,O-DIETHYL-S-(2-ETHTHIOETHYL)PHOSPHOROTHIOATE ☐ O,O-DIETHYL-S-ETHYL-2-ETHYLMERCAPTOPHOSPHOROTHIOLATE ☐ O,O-DIETHYL-S-(2-ETHYLTHIO-ETHYL)-MONOTHIOFOSFAAT (DUTCH) ☐ O,O-DIETHYL-S-2-(ETHYLTHIO) ETHYL PHOSPHOROTHIOATE ☐ O,O-DIETHYL-S-(2-(ETHYLTHIO) ETHYL) PHOSPHOROTHIOLATE (USDA) ☐ O,O-DIETIL-S-(2-ETILTIO-ETIL)-MONOTIOFOSFATO (ITALIAN) ☐ O,O-DIETYL-S-2-ETYLMERKAPTOETYLTIOFOSFAT (CZECH) ☐ 2-(ETHYLTHIO)-ETHANETHIOL S-ESTER with O,O-DIETHYL PHOSPHOROTHIOATE ☐ ISODEMETON ☐ IZOSYSTOX (CZECH) ☐ PO-SYSTOX ☐ THIOLDEMETON ☐ THIOL SYSTOX ☐ THIOPHOSPHATE de O,O-DIETHYLE et de S-(2-ETHYLTHIO-ETHYLE) (FRENCH)

TOXICITY DATA WITH REFERENCE
orl-rat LD50:1500 μg/kg AEPPAE 217,144,53
ipr-rat LD50:1500 μg/kg AMIHAB 13,606,56
ipr-mus LD50:1850 μg/kg BLLIAX 38,151,58
scu-mus LD50:6 mg/kg AEPPAE 217,144,53
ipr-gpg LD50:5500 μg/kg AMIHAB 13,606,56

SAFETY PROFILE: Poison by ingestion, intraperitoneal, and subcutaneous routes. When heated to decomposition it emits very toxic fumes of PO_x and SO_x. See also DEMETON-O + DEMETON-S and other demeton entries.

For occupational chemical analysis use NIOSH: Demeton, 5514.

D

DAP400 CAS:919-86-8 HR: 3
DEMETON-S-METHYL
mf: $C_6H_{15}O_3PS_2$ mw: 230.30

PROP: Pale-yellow oil. Bp: 89° @ 0.15 mm, d: 1.21 @ 20°/4°. Sltly sol in H_2O.

SYNS: BAY 18436 ☐ BAYER 25/154 ☐ DEMETON-S-METILE (ITALIAN) ☐ O,O-DIMETHYL-S-(2-AETHYLTHIO-AETHYL)-MONOTHIOPHOSPHAT (GERMAN) ☐ O,O-DIMETHYL-S-(2-ETHTHIOETHYL) PHOSPHOROTHIOATE ☐ DIMETHYL-S-(2-ETHTHIOETHYL)THIOPHOSPHATE ☐ O,O-DIMETHYL-S-ETHYLMERCAPTOETHYL THIOPHOSPHATE ☐ O,O-DIMETHYL-S-ETHYLMERCAPTOETHYL THIOPHOSPHATE, THIOLO ISOMER ☐ O,O-DIMETHYL-S-(2-ETHYLTHIO-ETHYL)-MONOTHIOFOSFAAT (DUTCH) ☐ O,O-DIMETHYL-S-(2-(ETHYLTHIO)ETHYL)PHOSPHOROTHIOATE ☐ O,O-DIMETHYL-S-(3-THIA-PENTYL)-MONOTHIOPHOSPHAT (GERMAN) ☐ O,O-DIMETIL-S-(2-ETILITIO-ETIL)-MONOTIOFOSFATO (ITALIAN) ☐ DURATOX ☐ S-(2-(ETHYLTHIO)ETHYL)-O,O-DIMETHYL PHOSPHOROTHIOATE ☐ S-(2-(ETHYLTHIO)ETHYL)DIMETHYL PHOSPHOROTHIOLATE ☐ S-(2-(ETHYLTHIO)ETHYL)-O,O-DIMETHYL THIOPHOSPHATE ☐ ISOMETASYSTOX ☐ ISOMETHYLSYSTOX ☐ METAISOSEPTOX ☐ METAISOSYSTOX ☐ METASYSTOX FORTE ☐ METHYL DEMETON THIOESTER ☐ METHYL ISOSYSTOX ☐ METHYL-MERCAPTOFOS TEOLOVY ☐ THIOPHOSPHATE de O,O-DIMETHYLE et de S-2-ETHYLTHIOETHYLE (FRENCH)

TOXICITY DATA WITH REFERENCE
mmo-sat 5 μL/plate MUREAV 28,405,75
sln-dmg-orl 80 ppm MUREAV 28,405,75
orl-rat LD50:30 mg/kg FMCHA2-,C197,91
ihl-rat LC50:500 mg/m³/4H 85DPAN -,-,71/76
skn-rat LD50:85 mg/kg PHJOAV 185,361,60
ipr-rat LD50:7500 μg/kg TXAPA9 4,621,62
ivn-rat LDLo:54 mg/kg BCPCA6 6,244,61
orl-gpg LD50:110 mg/kg TXAPA9 4,621,62
ipr-gpg LD50:12,500 μg/kg TXAPA9 4,621,62

CONSENSUS REPORTS: Reported in EPA TSCA Inventory. EPA Extremely Hazardous Substances List.

SAFETY PROFILE: Poison by ingestion, inhalation, skin contact, intraperitoneal, and intravenous routes. Mutation data reported. An insecticide. When heated to decomposition it emits very toxic fumes of PO_x and SO_x. See also DEMETON-O + DEMETON-S and other demeton entries.

DAP600 CAS:17040-19-6 HR: 3
DEMETON-S-METHYL-SULPHONE
mf: $C_6H_{15}O_5PS_2$ mw: 262.30

PROP: Crystals. Mp: 60°, bp: 120° @ 0.03 mm.

SYNS: BAYER 20315 ☐ DEMETON-S-METHYLSULFON (GERMAN) ☐ DEMETON-S-METHYLSULFONE ☐ O,O-DIMETHYL-S-(2-AETHYLSULFONYL-AETHYL)-THIOLPHOSPHAT (GERMAN) ☐ O,O-DIMETHYL-S-(2-ETHSULFONYLETHYL)PHOSPHOROTHIOATE ☐ DIMETHYL-S-(2-ETHSULFONYLETHYL)THIOPHOSPHATE ☐ O,O-DIMETHYL-S-ETHYL-

2-SULFONYLETHYL PHOSPHOROTHIOLATE □ O,O-DIMETHYL-S-E-THYLSULPHONYLETHYL PHOSPHOROTHIOLATE □ DIOXYDEME-TON-S-METHYL □ E 158 □ ISOMETASYSTOX SULFONE □ ISOME-THYLSYSTOX SULFONE □ M 3/158 □ METAISOSYSTOX-SOLFON 20 315

TOXICITY DATA WITH REFERENCE
mmo-esc 5 µL/plate MUREAV 28,405,75
orl-rat LD50:32,400 µg/kg BIJOAK 67,187,57
ihl-rat LC50:195 mg/m³/4H GUCHAZ 6,165,73
skn-rat LD50:500 mg/kg GUCHAZ 6,165,73
ipr-rat LD50:21 mg/kg GUCHAZ 6,165,73
ivn-rat LD50:22 mg/kg BIJOAK 67,187,57
orl-rbt LD50:50 mg/kg 85GYAZ -,20,71
orl-gpg LD50:120 mg/kg TXAPA9 4,621,62
ipr-gpg LD50:85 mg/kg TXAPA9 4,621,62

SAFETY PROFILE: Poison by ingestion, inhalation, intraperitoneal, and intravenous routes. Moderately toxic by skin contact. Mutation data reported. An insecticide. When heated to decomposition it emits very toxic fumes of PO_x and SO_x. See also DEMETON-O + DEMETON-S and other DEMETON entries.

DAP700 CAS:2955-38-6 HR: 2
DEMETRIN
mf: $C_{19}H_{17}ClN_2O$ mw: 324.83

PROP: Crystals from methanol. Mp: 144–147°.

SYNS: CENTRAX □ 7-CHLORO-1-(CYCLOPROPYLMETHYL)-1,3-DIHYDRO-5-PHENYL-2H-1,4-BENZODIAZEPIN-2-ONE □ 7-CHLORO-1-CYCLOPROPYLMETHYL-5-PHENYL-1H-1,4-BENZODIAZEPIN-2(3H)-ONE □ K-373 □ LYSANXIA □ PRAZEPAM □ SEDAPRAN □ SETTIMA □ TREPIDAN □ VERSTRAN

TOXICITY DATA WITH REFERENCE
orl-mus TDLo:504 mg/kg (female 1-21D post):REP PLRCAT 9,325,77
orl-rat TDLo:250 mg/kg (female 8-17D post):REP OYYAA2 15,797,78
orl-rat TDLo:2500 mg/kg (female 8-17D post):REP OYYAA2 15,797,78
orl-mus TDLo:1656 mg/kg (male 5D pre):REP PLRCAT 9,187,77
orl-rat TDLo:10 g/kg (female 8-17D post):TER OYYAA2 15,797,78
orl-rat TDLo:21 g/kg (female 14D pre):REP OYYAA2 15,813,78
orl-mus TDLo:800 mg/kg (male 5D pre):REP RCOCB8 13,601,76
orl-rat TDLo:10 g/kg (female 8-17D post):REP OYYAA2 15,797,78
orl-dog TDLo:910 mg/kg (female 26W pre):REP OYYAA2 15,777,78
orl-rat TDLo:28 g/kg (female 28D pre):REP OYYAA2 15,247,78
orl-mus TDLo:624 mg/kg (female 26D pre):REP RCOCB8 11,155,75
orl-dog TDLo:1200 mg/kg (female 30D pre):REP OYYAA2 15,759,78
orl-dog TDLo:1200 mg/kg (male 30D pre):REP OYYAA2 15,759,78
orl-dog TDLo:60 mg/kg (male 60D pre):REP OYYAA2 15,777,78
orl-rat TDLo:35 g/kg (35D male):REP OYYAA2 15,247,78
orl-rat TDLo:2500 mg/kg (female 8-17D post):TER OYYAA2 15,797,78
orl-rat TDLo:5284 mg/kg/2Y-C:ETA TXAPA9 57,39,81
orl-mus TDLo:14 g/kg/80W-C:ETA TXAPA9 57,39,81
orl-rat TD:55 g/kg/2Y-C TXAPA9 57,39,81
orl-mus LD50:2300 mg/kg JKXXAF #79-92631
ipr-mus LD50:1020 mg/kg OYYAA2 15,241,78

SAFETY PROFILE: Moderately toxic by ingestion and intraperitoneal routes. Experimental teratogenic and reproductive effects. Questionable carcinogen with experimental tumorigenic data. Note: This is a controlled substance (depressant) listed in the U.S. Code of Federal Regulations, Title 21 Part 1308.14 (1985). A tranquilizer and muscle relaxant. When heated to decomposition it emits toxic fumes of Cl^- and NO_x.

DAP800 CAS:62-97-5 HR: 3
DEMOTIL
mf: $C_{20}H_{24}N{\bullet}CH_3O_4S$ mw: 389.55

PROP: A solid. Mp: 194–195°. Sol in H_2O.

SYNS: N,N-DIMETHYL-4-PIPERIDYLIDENE-1,1-DIPHENYLMETHANE METHYLSULFATE □ DIPHEMANIL □ DIPHEMANIL METHYLSULFATE □ DIPHENATIL □ DIPHENMANIL METHYLSULFATE □ DIPHENMETHANIL □ DIPHENMETHANIL METHYLSULFATE □ 4-(DIPHENYLMETHYLENE)-1,1-DIMETHYLPIPERIDINIUM METHYLSULFATE □ NIVELONA □ p-(α-PHENYLBENZYLIDENE)-1,1-DIMETHYLPIPERIDINIUM METHYL SULFATE □ PRANTAL □ PRANTAL METHYLSULFATE □ VAGOPHEMANIL □ VAGOPHEMANIL METHYL SULFATE □ VARITON

TOXICITY DATA WITH REFERENCE
orl-rat LD50:1107 mg/kg PSEBAA 78,576,51
ivn-rat LD50:5 mg/kg OYYAA2 23,461,82
orl-mus LD50:317 mg/kg PSEBAA 78,576,51
ipr-mus LD50:47 mg/kg PSEBAA 78,576,51
ivn-mus LD50:4012 µg/kg AIPTAK 103,100,55
ivn-dog LD50:42 mg/kg PSEBAA 78,576,51
orl-gpg LD50:404 mg/kg PSEBAA 78,576,51

SAFETY PROFILE: Poison by ingestion, intraperitoneal, and intravenous routes. An anticholinergic agent. When heated to decomposition it emits very toxic fumes of SO_x and NO_x.

DAP812 CAS:3734-33-6 HR: 2
DENATONIUM BENZOATE
mf: $C_{21}H_{29}N_2O{\bullet}C_7H_5O_2$ mw: 446.64

SYNS: AMMONIUM, BENZYLDIETHYL((2,6-XYLYLCARBAMOYL) METHYL)-, BENZOATE □ BENZENEMETHANAMINIUM, N-(2-((2,6-DIMETHYLPHENYL)AMINO)-2-OXOETHYL)-N,N-DIETHYL-, BENZOATE □ BITREX □ THS-839 □ WIN 16568

TOXICITY DATA WITH REFERENCE
orl-rat LD50:584 mg/kg NTIS** PB81-139339
orl-rbt LD50:508 mg/kg

CONSENSUS REPORTS: Reported in EPA TSCA Inventory.

SAFETY PROFILE: Moderately toxic by ingestion. When heated to decomposition it emits toxic vapors of NO_x.

DAP815 **HR: 3**
DENDROASPIS ANGUSTICEPS VENOM

SYN: VENOM, SNAKE, DENDROASPIS ANGUSTICEPS

TOXICITY DATA WITH REFERENCE
ipr-mus LD50:2800 μg/kg 19DDA6 1,283,67
scu-mus LD50:3320 μg/kg 19DDA6 1,223,67
ivn-mus LD50:381 μg/kg TOXIA6 2,5,64
ivn-mam LD50:450 μg/kg CLPTAT 8,849,67

SAFETY PROFILE: Poison by subcutaneous, intravenous, and intraperitoneal routes.

DAP820 **HR: 3**
DENDROASPIS JAMESONI VENOM

SYNS: D. JAMESONI VENOM □ VENOM, SNAKE, DENDROASPIS JAMESONI

TOXICITY DATA WITH REFERENCE
ipr-mus LD50:440 μg/kg TOXIA6 13,295,75
scu-mus LD50:1020 μg/kg 19DDA6 1,223,67
ivn-mus LD50:840 μg/kg 19DDA6 1,223,67

SAFETY PROFILE: Poison by subcutaneous, intravenous, and intraperitoneal routes.

DAP825 **CAS:35306-34-4** **HR: 3**
DENDROBINE HYDROCHLORIDE
mf: $C_{16}H_{25}NO_2{\cdot}ClH$ mw: 299.88

SYN: DENDROBAN-12-ONE HYDROCHLORIDE

TOXICITY DATA WITH REFERENCE
ivn-rat LDLo:20 mg/kg JPETAB 55,319,35
ivn-mus LDLo:20 mg/kg JPETAB 55,319,35
ivn-rbt LDLo:17 mg/kg JPETAB 55,319,35
ivn-gpg LDLo:22 mg/kg JPETAB 55,319,35

SAFETY PROFILE: Poison by intravenous route. When heated to decomposition it emits toxic fumes of NO_x and HCl.

DAP840 **HR: 2**
DENDROCALAMUS MEMBRANACEUS Munro, extract excluding roots

PROP: Indian plant belonging to the family Poaceae (IJEBA6 22,312,84).

TOXICITY DATA WITH REFERENCE
orl-ham TDLo:500 mg/kg (female 1-5D post):REP IJEBA6 22,312,84
ipr-mus LD50:825 mg/kg IJEBA6 22,312,84

SAFETY PROFILE: Moderately toxic by intraperitoneal route. Experimental reproductive effects. When heated to decomposition it emits acrid smoke and irritating fumes.

DAP850 **CAS:71771-90-9** **HR: 2**
DENOPAMINE
mf: $C_{18}H_{23}NO_4$ mw: 317.42

PROP: A solid. Mp: 130–140°.

SYNS: BENZENEMETHANOL, α-(((2-(3,4-DIMETHOXYPHENYL)ETHYL)AMINO)METHYL)-4-HYDROXY-, (R)- □ (-)-(R)-1-(p-HYDROXYPHENYL)-2-((3,4-DIMETHOXYPHENETHYL)AMINO)ETHANOL □ TA 064

TOXICITY DATA WITH REFERENCE
orl-rat TDLo:810 mg/kg (female 17-21D post):REP OYYAA2 32,769,86
orl-rbt TDLo:1300 mg/kg (female 6-18D post):TER OYYAA2 32,769,86
orl-rat TDLo:6600 mg/kg (female 7-17D post):TER OYYAA2 32,769,86
orl-rat TDLo:6600 mg/kg (female 7-17D post):REP OYYAA2 32,769,86
orl-rat LD50:9369 mg/kg OYYAA2 32,751,86
ipr-rat LD50:1785 mg/kg OYYAA2 32,751,86

SAFETY PROFILE: Moderately toxic by intraperitoneal route. Experimental reproductive effects. When heated to decomposition it emits toxic fumes of NO_x.

DAP875 **CAS:26166-37-0** **HR: 3**
DENUDATINE
mf: $C_{22}H_{33}NO_2$ mw: 343.56

PROP: A solid. Mp: 248–249°.

SYN: 16,17-DIDEHYDRO-21-ETHYL-4-METHYL-7,20-CYCLOATIDANE-11-β,15-β-DIOL

TOXICITY DATA WITH REFERENCE
ivn-rat LD50:130 mg/kg CYLPDN 3,104,82
orl-mus LD50:290 mg/kg CYLPDN 3,104,82
ivn-mus LD50:128 mg/kg CYLPDN 3,104,82

SAFETY PROFILE: Poison by ingestion and intravenous routes. When heated to decomposition it emits toxic fumes of NO_x.

DAP880 **CAS:77234-90-3** **HR: 3**
DENZIMOL HYDROCHLORIDE
mf: $C_{19}H_{20}N_2O{\cdot}ClH$ mw: 328.87

PROP: Crystals from EtOH. Mp: 194–196°.

SYNS: α-(p-PHENETHYLPHENYL)-1-IMIDAZOLEETHANOL MONOHYDROCHLORIDE □ N-(β-(4-(β-PHENYLETHYL)PHENYL)-β-HYDROXYETHYL)IMIDAZOLE HYDROCHLORIDE □ REC 15-1533

TOXICITY DATA WITH REFERENCE
ipr-rat LD50:332 mg/kg ARZNAD 33,1155,83
ivn-rat LD50:22 mg/kg ARZNAD 33,1155,83
orl-mus LD50:434 mg/kg JMCMAR 24,727,81
ipr-mus LD50:246 mg/kg ARZNAD 33,1168,83
ivn-mus LD50:46 mg/kg ARZNAD 33,1168,83
ivn-rbt LD50:22,400 μg/kg ARZNAD 33,1168,83

SAFETY PROFILE: Poison by intravenous and intraperitoneal routes. Moderately toxic by ingestion. When heated to decomposition it emits toxic fumes of NO_x and HCl.

DAP900 **CAS:8044-51-7** **HR: 1**
DEOBASE

SYNS: DEODORIZED KEROSENE □ DEODORIZED KEROSINE

TOXICITY DATA WITH REFERENCE
skn-rbt 500 mg/24H MLD FCTOD7 20(Suppl),699,82

eye-rbt 100 mg/24H MLD FCTOD7 20(Suppl),699,82
orl-rat LD50:45 g/kg AMIHBC 2,420,50
orl-rbt LD50:8840 mg/kg AMIHBC 2,420,50

SAFETY PROFILE: Mildly toxic by ingestion. An eye and skin irritant. See also KEROSENE.

DAQ000 CAS:73825-59-9 HR: 3
11-DEOXO-12-β,13-α-DIHYDRO-11-α-HYDROXYJERVINE
mf: $C_{26}H_{40}NO_3$ mw: 414.67

SYN: 12-β,13-α-DIHYDROJERVINE-11-α-OL

TOXICITY DATA WITH REFERENCE
orl-ham TDLo:113 mg/kg (7D preg):REP JAFCAU 26,561,78
orl-ham TDLo:150 mg/kg (7D preg):TER JAFCAU 26,561,78
orl-ham LDLo:150 mg/kg JAFCAU 26,561,78

SAFETY PROFILE: Poison by ingestion. Experimental teratogenic and reproductive effects. When heated to decomposition it emits toxic fumes of NO_x. See also JERVINE.

DAQ002 CAS:51340-26-2 HR: 3
11-DEOXO-12-β,13-α-DIHYDRO-11-β-HYDROXYJERVINE
mf: $C_{26}H_{40}NO_3$ mw: 414.67

SYNS: 12-β,13-α-DIHYDRO-11-DEOXO-11-β-HYDROXYJERVINE ☐ 12-β,13-α-DIHYDROJERVINE-11-β-OL ☐ JERVINE, 11-DEOXO-12-β,13-α-DIHYDRO-11-β-HYDROXY-

TOXICITY DATA WITH REFERENCE
orl-ham TDLo:150 mg/kg (female 7D post):REP 41CIAR -,409,78
orl-ham LDLo:150 mg/kg JAFCAU 26,561,78

SAFETY PROFILE: Poison by ingestion. Experimental reproductive effects. When heated to decomposition it emits toxic fumes of NO_x.

DAQ100 HR: D
9-DEOXO-16,16-DIMETHYL-9-METHYLENE-PGE2
mf: $C_{23}H_{38}O_4$ mw: 378.61

SYNS: 9-DEOXO-16,16-DIMETHYL-9-METHYLENEPROSTAGLANDIN E2 ☐ 7-(5,5-DIMETHYL-3-HYDROXY-2-(3-HYDROXY-1-OCTENYL)CYCLOPENTYL)-5-HEPTENOIC ACID

TOXICITY DATA WITH REFERENCE
ivg-wmn TDLo:800 μg/kg (13W preg):REP CCPTAY 22,153,80
ivg-wmn TDLo:3 mg/kg (16W preg):REP CCPTAY 22,153,80

SAFETY PROFILE: Human reproductive effects by intravaginal route: changes in the uterus, cervix, and vagina; terminates pregnancy. A steroid. When heated to decomposition it emits acrid smoke and fumes. See various prostaglandins.

DAQ110 CAS:85559-57-5 HR: 3
11-DEOXOGLYCYRRHETINIC ACID HYDROGEN MALEATE SODIUM SALT
mf: $C_{34}H_{51}O_6$•7Na

TOXICITY DATA WITH REFERENCE
ipr-rat LD50:67,800 μg/kg EPXXDW #69380
scu-rat LD50:98,300 μg/kg EPXXDW #69380
ivn-rat LD50:54,200 μg/kg EPXXDW #69380

SAFETY PROFILE: Poison by subcutaneous, intravenous, and intraperitoneal routes. When heated to decomposition it emits toxic fumes of Na_2O.

DAQ400 CAS:83-44-3 HR: 3
DEOXYCHOLATIC ACID
mf: $C_{24}H_{40}O_4$ mw: 392.64

PROP: A white crystalline powder from EtOH. Mp: 187–189°. Sol in alc, acetone; sltly sol in ether and chloroform; insol in water.

SYNS: CHOLEIC ACID ☐ CHOLEREBIC ☐ CHOLOREBIC ☐ DEGALOL ☐ DEOXYCHOLIC ACID (FCC) ☐ 7-α-DEOXYCHOLIC ACID ☐ DESOXYCHOLIC ACID ☐ DESOXYCHOLSAEURE (GERMAN) ☐ 3,12-DIHYDROXYCHOLANIC ACID ☐ 3-α,12-α-DIHYDROXYCHOLANIC ACID ☐ 3-α,12-α-DIHYDROXY-5-β-CHOLANOIC ACID ☐ 3-α,12-α-DIHYDROXY-5-β-CHOLAN-24-OIC ACID ☐ 3-α,12-α-DIHYDROXYCHOLANSAEURE (GERMAN) ☐ DROXOLAN ☐ 17-β-(1-METHYL-3-CARBOXYPROPYL)-ETIOCHOLANE-3-α,12-α-DIOL ☐ PYROCHOL ☐ SEPTOCHOL

TOXICITY DATA WITH REFERENCE
mmo-sat 20 mg/L MUREAV 158,45,85
sln-smc 100 mg/L CRNGDP 5,447,84
otr-ham:emb 7250 μg/L TOLED5 9,177,81
ipr-rat TDLo:166 mg/kg (female 12-19D post):REP TJADAB 42,215,90
skn-mus TDLo:2700 mg/kg/10W-I:ETA BJCAAI 10,363,56
scu-mus TDLo:1120 mg/kg/22W-I:ETA NATUAS 145,627,40
scu-mus TD:1400 mg/kg/22W-I:ETA PRLBA4 129,439,40
orl-rat LD50:1 g/kg NAIZAM 33,71,82
orl-mus LD50:1 g/kg NAIZAM 33,71,82
ipr-mus LD50:130 mg/kg ARZNAD 20,323,70
ivn-rbt LDLo:1000 mg/kg ZGEMAZ 52,779,26

CONSENSUS REPORTS: Reported in EPA TSCA Inventory.

SAFETY PROFILE: Poison by intraperitoneal route. Moderately toxic by ingestion and intravenous routes. Questionable carcinogen with experimental tumorigenic data. Experimental reproductive effects. Mutation data reported. When heated to decomposition it emits acrid smoke and irritating fumes.

DAQ600 CAS:64-85-7 HR: 2
11-DEOXYCORTICOSTERONE
mf: $C_{21}H_{30}O_3$ mw: 330.51

SYNS: CORTEXONE ☐ DESOXYCORTICOSTERONE ☐ DESOXYCORTONE ☐ 21-HYDROXYPREGN-4-ENE-3,20-DIONE ☐ 21-HYDROXYPROGESTERONE ☐ 4-PREGNEN-21-OL-3,20-DIONE

TOXICITY DATA WITH REFERENCE
ims-rat TDLo: 14,500 μg/kg (7D pre/1-22D preg): REP JCINAO 41,710,62
par-rat TDLo: 8750 μg/kg (7D pre): REP JEMEAV 102,347,59
unr-mus LD50: 1000 mg/kg JMCMAR 7,673,64

SAFETY PROFILE: Moderately toxic by unspecified route. Experimental reproductive effects. A steroid. When heated to decomposition it emits acrid smoke and irritating fumes.

DAQ800 CAS:56-47-3 HR: 2
11-DEOXYCORTICOSTERONE ACETATE
mf: $C_{23}H_{32}O_4$ mw: 372.55

PROP: Crystals from Me_2CO. Mp: 159–161°.

SYNS: 21-ACETOXY-3,20-DIKETOPREGN-4-ENE □ CORTACET □ CORTATE □ CORTENIL □ CORTESAN □ CORTEXONE ACETATE □ CORTIFAR □ CORTIGEN □ CORTINAQ □ CORTIRON □COR-TIVIS □ CORTIXYL □ DCA □ DECORTIN □ DECORTON □ DECOSTERONE □ DECOSTRATE □ 11-DEOXYCORTICOSTERONE-21-ACETATE □ DEOXYCORTONE ACETATE □ DESCORTERONE □ DESCOTONE □ DESOXYCORTICOSTERONE ACETATE □ DESOXYCORTONE ACETATE □ DOCA □ DOCA ACETATE □ DOC-AC □ DOC ACETATE □ DORCOSTRIN □ DOXO □ 21-HYDROXYPREGN-4-ENE-3,20-DIONE-21-ACETATE □ KRINOCORTS □ OCRITEN □ ORGANON'S DOCA ACETATE □ PERCORTEN □ PERCOTOL □ 4-PREGNENE-3,20-DIONE-21-OL ACETATE □ PRIMOCORT □ PRIMOCORTAN □ SINCORTEX □ STERAQ □ SYNCORT □ SYNCORTA □ SYNCORTYL □ UNIDOCAN

TOXICITY DATA WITH REFERENCE
par-rat TDLo: 316 mg/kg (female 56D pre): REP PSEBAA 120,238,65
par-rat TDLo: 224 mg/kg (female 56D pre): REP PSEBAA 120,238,65
unr-rat TDLo: 16 mg/kg (male 2D pre): REP ANENAG 24(Suppl 3),1,63
scu-rat TDLo: 1258 mg/kg (20D male): REP ENDOAO 28,129,41
scu-mus TDLo: 520 mg/kg/13W-I: ETA PSEBAA 83,14,53

SAFETY PROFILE: Questionable carcinogen with experimental tumorigenic data. Experimental reproductive effects. A steroid. When heated to decomposition it emits acrid smoke and irritating fumes.

DAR000 CAS:55297-96-6 HR: 3
14-DEOXY-14-((2-DIETHYLAMINOETHYL)MERCAPTOACETOXY)-MUTILIN HYDROGEN FUMARATE
mf: $C_{28}H_{47}NO_4S{\bullet}C_4H_4O_4$ mw: 609.90

PROP: Crystals from Me_2CO. Mp: 147–148°.

SYNS: SQ 22947 □ TIAMUTIN

TOXICITY DATA WITH REFERENCE
orl-mus LD50: 841 mg/kg AMACCQ 7,507,75
scu-mus LD50: 521 mg/kg AMACCQ 7,507,75
orl-ckn LD50: 1860 mg/kg AMACCQ 7,507,75
ims-ckn LD50: 270 mg/kg AMACCQ 7,507,75
orl-trk LD50: 1345 mg/kg AMACCQ 7,507,75

SAFETY PROFILE: Poison by intramuscular route. Moderately toxic by ingestion and subcutaneous routes. When heated to decomposition it emits very toxic fumes of NO_x and SO_x.

DAR100 CAS:4298-16-2 HR: 3
dl-DEOXYEPHEDRINE HYDROCHLORIDE
mf: $C_{10}H_{15}N{\bullet}ClH$ mw: 185.72

SYNS: dl-DESOXYEPHEDRINE HYDROCHLORIDE □ dl-METHAMPHETAMINE HYDROCHLORIDE □ dl-N-METHYL-β-PHENYLISOPROPYLAMINE HYDROCHLORIDE

TOXICITY DATA WITH REFERENCE
ipr-rat LD50: 54,480 μg/kg JAPMA8 37,223,48
orl-mus LD50: 143 mg/kg JAPMA8 37,223,48
ipr-mus LD50: 61,160 μg/kg JAPMA8 37,223,48

SAFETY PROFILE: Poison by ingestion and intraperitoneal routes. A powerful central nervous system stimulant. When heated to decomposition it emits toxic fumes of NO_x and HCl. See also various amphetamines.

DAR150 CAS:447-25-6 HR: 3
6-DEOXY-6-FLUOROGLUCOSE
mf: $C_6H_{11}FO_5$ mw: 182.17

TOXICITY DATA WITH REFERENCE
orl-rat TDLo: 120 mg/kg (1D male): REP CCPTAY 25,535,82
orl-rat LDLo: 240 mg/kg CCPTAY 25,535,82
orl-mus LDLo: 480 mg/kg CCPTAY 25,535,82

SAFETY PROFILE: Poison by ingestion. Experimental reproductive effects. When heated to decomposition it emits toxic fumes of F^-.

DAR400 CAS:50-91-9 HR: 3
2′-DEOXY-5-FLUOROURIDINE
mf: $C_9H_{11}FN_2O_5$ mw: 246.22

PROP: Crystals from butyl acetate. Mp: 150–151°.

SYNS: DEOXYFLUOROURIDINE □ 1-β-d-2′-DEOXYRIBOFURANOSYL-5-FLUOROURACIL □ FDUR □ FLOXURIDIN □ FLOXURIDINE □ FLUORODEOXYURIDINE □ β-5-FLUORO-2′-DEOXYURIDINE □ 5-FLUORODEOXYURIDINE □ 5-FLUORO-2-DEOXYURIDINE □ 5-FLUORO-2′-DEOXYURIDINE □ 5-FLUOROURACIL DEOXYRIBOSIDE □ 5-FLUOROURACIL-2′-DEOXYRIBOSIDE □ FLUORURIDINE DEOXYRIBOSE □ FUDR □ 5-FUDR □ NSC-27640 □ RO 5-0360

TOXICITY DATA WITH REFERENCE
dnd-hmn: leu 500 nmol/L CNREA8 43,5145,83
dni-hmn: oth 1 μmol/L CNREA8 42,3005,82
oms-hmn: oth 1 μmol/L CNREA8 42,3005,82
ipr-mus TDLo: 10 mg/kg (female 16D post): REP TJADAB 29(2),5B,84
unr-mus TDLo: 45 mg/kg (12D preg): TER AJANA2 137,87,73
ipr-mus TDLo: 25 mg/kg (10D preg): TER TJADAB 23,241,81
unr-rat TDLo: 25 mg/kg (10D preg): REP TJADAB 19,43A,79
ipr-rat TDLo: 100 mg/kg (12D preg): REP JOANAY 126,37,78
ipr-rat TDLo: 25 mg/kg (10D preg): TER FAATDF 4,352,84
ipr-rat TDLo: 100 mg/kg (12D preg): TER JOANAY 126,37,78

ipr-rat TDLo: 25 mg/kg (12D preg): TER TJADAB 25,95,82
ipr-rat TDLo: 100 mg/kg (12D preg): TER JOANAY 124,494,77
ivn-hmn TDLo: 5 mg/kg/14D-C: GIT CANCAR 34,972,74
par-wmn TDLo: 173 mg/kg/82W-I: SKN JNMAAE 79,669,87
orl-rat LD50: 215 mg/kg NCILB* NIH-NCI-E-C-72-3252,73
ipr-rat LD50: 1600 mg/kg CPCHAO 18,307,62
orl-mus LD50: 147 mg/kg NCILB* NIH-NCI-E-C-72-3252,73

CONSENSUS REPORTS: EPA Genetic Toxicology Program.

SAFETY PROFILE: Poison by ingestion. Moderately toxic by intraperitoneal route. An experimental teratogen. Other experimental reproductive effects. Human systemic effects: hypermotility, diarrhea, nausea, vomiting and other gastrointestinal effects, allergic dermatitis, and bone marrow changes. Human mutation data reported. When heated to decomposition it emits very toxic fumes of F^- and NO_x.

DAR600 CAS:154-17-6 HR: 3
2-DEOXYGLUCOSE
mf: $C_6H_{12}O_5$ mw: 164.18

PROP: Crystals from acetone or butanone. Mp: 142–144°. α-Form: Crystals from isopropanol. Mp: 134–136°.

SYNS: 2-DEOXY-d-ARABINO-HEXOSE □ 2-DEOXY-3-ARABINO-HEXOSE □ d-2-DEOXYGLUCOSE □ 2-DEOXY-d-GLUCOSE □ 2-DE-OXY-d-GLUCOSE (FRENCH) □ 2-DG □ NSC 15193

TOXICITY DATA WITH REFERENCE
orl-rat TDLo: 4 g/kg (7-14D preg): TER 85DJA5 -,95,71
orl-rat TDLo: 2 g/kg (7-8D preg): REP 85DJA5 -,95,71
scu-rat LD50: 250 mg/kg APFRAD 39,327,81
ipr-rat LD50: 2000 mg/kg JPPMAB 17,814,65

CONSENSUS REPORTS: Reported in EPA TSCA Inventory.

SAFETY PROFILE: Poison by subcutaneous route. Moderately toxic by intraperitoneal route. An experimental teratogen. Other experimental reproductive effects. When heated to decomposition it emits acrid smoke and fumes.

DAR800 CAS:961-07-9 HR: 2
2′-DEOXYGUANOSINE
mf: $C_{10}H_{13}N_5O_4$ mw: 267.28

PROP: Crystals. Mp: 300–301° (also said to be indefinite).

SYNS: DEOXYGUANOSINE □ GUANINE DEOXYRIBOSIDE

TOXICITY DATA WITH REFERENCE
dni-ham: lng 1 mmol/L BICMBE 64,809,82
cyt-ham: fbr 500 μmol/L CYTOAN 49,667,84
sce-ham: fbr 500 μmol/L CYTOAN 49,667,84
dnd-mam: lym 50 mmol/L PNASA6, 48,686,62
ipr-rat LD50: >800 mg/kg ADTEAS 3,181,68

CONSENSUS REPORTS: Reported in EPA TSCA Inventory. EPA Genetic Toxicology Program.

SAFETY PROFILE: Moderately toxic by intraperitoneal route. Mutation data reported. When heated to decomposition it emits toxic fumes of NO_x.

DAS000 CAS:54-42-2 HR: 2
2′-DEOXY-5-IODOURIDINE
mf: $C_9H_{11}IN_2O_5$ mw: 354.12

PROP: Crystals from H_2O. Mp: 160°.

SYNS: ALLERGAN 211 □ DENDRID □ 1-(2-DEOXY-β-d-RIBOFURANOSYL)-5-IODOURACIL □ 1-β-d-2′-DEOXYRIBOFURANOSYL-5-IODOURACIL □ EMANIL □ HERPESIL □ HERPIDU □ HERPLEX □ HERPLEX LIQUIFILM □ IDEXUR □ IDOXENE □ IDOXURIDIN □ IDOXURIDINE □ IDU □ IDUCHER □ IDULEA □ IDUOCULOS □ IDUR □ IDURIDIN □ 5-IODODEOXYURIDINE □ 5-IODO-2′-DEOXYURIDINE □ 5-IODOURACIL DEOXYRIBOSIDE □ IUDR □ 5-IUDR □ JODDEOXIURIDIN □ KERECID □ NSC 39661 □ OPHTHALMADINE □ SK&F 14287 □ STOXIL □ SYNMIOL

TOXICITY DATA WITH REFERENCE
sce-hmn: lym 50 mg/L BMJOAE 283,817,81
sce-hmn: fbr 50 mg/L BMJOAE 283,817,81
msc-hmn: lym 100 μmol/L LIFSAK 19,563,76
msc-ham: lng 1 mg/L CRNGDP 6,1207,85
dni-rbt: kdy 1 mg/L JMCMAR 24,390,81
scu-mus TDLo: 600 mg/kg (female 16-18D post): REP TJADAB 11,103,75
ocu-rbt TDLo: 1350 mg/kg (female 6-18D post): TER AROPAW 93,46,75
ipr-mus TDLo: 100 mg/kg (7D preg): REP EXPEAM 29,198,73
ipr-mus TDLo: 300 mg/kg (7D preg): TER EXPEAM 29,198,73
ipr-mus TDLo: 100 mg/kg/8W-I: CAR EXPEAM 29,1132,73
ipr-rat LD50: 4000 mg/kg ADTEAS 3,181,68
ipr-mus LD50: 1000 mg/kg JJIND8 62,911,79

CONSENSUS REPORTS: Reported in EPA TSCA Inventory. EPA Genetic Toxicology Program.

SAFETY PROFILE: Moderately toxic by intraperitoneal route. Experimental teratogenic and reproductive effects. Questionable carcinogen with experimental carcinogenic data. Human mutation data reported. When heated to decomposition it emits very toxic fumes of I^- and NO_x.

DAS400 CAS:10356-92-0 HR: 2
1-DEOXY-1-(N-NITROSOMETHYLAMINO)-d-GLUCITOL
mf: $C_7H_{16}N_2O_6$ mw: 224.25

SYNS: 1-DEOXY-1-(METHYLNITROSAMINO)-d-GLUCITOL □ 1-N-METHYL-N-NITROSAMINO-1-DEOXY-d-GLUCITOLE □ 1-(N-METHYL-N-NITROSOAMINO)-1-DEOXY-d-GLUCITOL □ 1-N-METHYL-N-NITROSOAMINO-1-DESOXY-d-GLUCIT (GERMAN)

TOXICITY DATA WITH REFERENCE
orl-rat TDLo: 2500 mg/kg/25W-I: ETA ZEKBAI 75,296,71

SAFETY PROFILE: Questionable carcinogen with experimental tumorigenic data. Many N-nitroso compounds are carcinogens. When heated to decomposition it emits toxic fumes of NO_x. See also NITROSAMINES.

DAS600 **CAS:37636-51-4** **HR: 3**
3′-DEOXYPAROMOMYCIN I
mf: $C_{23}H_{45}N_5O_{13}$ mw: 599.73

PROP: Amorphous powder. Mp: 178–184° (decomp).

SYNS: LIVIDOMYCIN B □ QUINTOMYCIN D

TOXICITY DATA WITH REFERENCE
orl-mus LD50:10 g/kg YKYUA6 31,1085,80
scu-mus LD50:1245 µg/kg YKYUA6 31,1085,80
ivn-mus LD50:123 mg/kg 85ERAY 1,682,78
ims-mus LD50:1343 µg/kg YKYUA6 31,1085,80

SAFETY PROFILE: Poison by intravenous, intramuscular, and subcutaneous routes. Mildly toxic by ingestion. When heated to decomposition it emits toxic fumes of NO_x.

DAS800 **CAS:56530-49-5** **HR: 1**
12-DEOXY-PHORBOL-20-ACETATE-13-DODECANOATE
mf: $C_{34}H_{52}O_7$ mw: 572.86

SYN: 12-DEOXY-PHORBOL-13-DODECANOATE-20-ACETATE

TOXICITY DATA WITH REFERENCE
skn-mus 500 ng/4H APTOA6 37,250,75
skn-mus 2500 ng/24H APTOA6 37,250,75

SAFETY PROFILE: A skin irritant. When heated to decomposition it emits acrid smoke and irritating fumes.

DAT000 **CAS:25090-71-5** **HR: 1**
12-DEOXYPHORBOL-20-ACETATE-13-ISOBUTYRATE
mf: $C_{26}H_{36}O_7$ mw: 460.62

SYN: 12-DEOXYPHORBOL-13-ISOBUTYRATE-20-ACETATE

TOXICITY DATA WITH REFERENCE
skn-mus 500 ng/4H APTOA6 37,250,75
skn-mus 6400 ng/24H APTOA6 37,250,75

SAFETY PROFILE: A skin irritant. When heated to decomposition it emits acrid smoke and irritating fumes.

DAT200 **CAS:25090-73-7** **HR: 1**
12-DEOXY-PHORBOL-20-ACETATE-13-(2-METHYLBUTYRATE)
mf: $C_{27}H_{38}O_7$ mw: 474.65

SYN: 12-DEOXY-PHORBOL-13-α-METHYLBUTYRATE-20-ACETATE

TOXICITY DATA WITH REFERENCE
skn-mus 390 ng OPEN ARTODN 44,279,80
skn-mus 2800 ng/24H APTOA6 37,250,75

SAFETY PROFILE: A skin irritant. When heated to decomposition it emits acrid smoke and irritating fumes.

DAT400 **CAS:56602-09-6** **HR: 1**
12-DEOXY-PHORBOL-20-ACETATE-13-OCTENOATE
mf: $C_{30}H_{42}O_7$ mw: 514.72

SYN: 12-DEOXY-PHORBOL-13-OCTENOATE-20-ACETATE

TOXICITY DATA WITH REFERENCE
skn-mus 1800 ng/4H APTOA6 37,250,75
skn-mus 4000 ng/24H APTOA6 37,250,75

SAFETY PROFILE: A skin irritant. When heated to decomposition it emits acrid smoke and irritating fumes.

D

DAT600 **CAS:25090-72-6** **HR: 1**
12-DEOXY-PHORBOL-20-ACETATE-13-TIGLATE
mf: $C_{27}H_{36}O_7$ mw: 472.63

SYN: 12-DEOXYPHORBOL-13-TIGLATE-20-ACETATE

TOXICITY DATA WITH REFERENCE
skn-mus 1800 ng/4H APTOA6 37,250,75
skn-mus 7800 ng/24H APTOA6 37,250,75

SAFETY PROFILE: A skin irritant. When heated to decomposition it emits acrid smoke and irritating fumes.

DAT800 **CAS:69883-99-4** **HR: 1**
12-DEOXYPHORBOL-13-(4-ACETOXYPHENYLACETATE)-20-ACETATE
mf: $C_{32}H_{38}O_9$ mw: 566.4

TOXICITY DATA WITH REFERENCE
skn-mus 130 ng OPEN ARTODN 44,279,80

SAFETY PROFILE: A skin irritant. When heated to decomposition it emits acrid smoke and irritating fumes.

DAU000 **CAS:65700-60-9** **HR: 1**
12-DEOXYPHORBOL-13-ANGELATE
mf: $C_{25}H_{34}O_6$ mw: 430.59

TOXICITY DATA WITH REFERENCE
skn-mus 720 ng OPEN ARTODN 44,279,80

SAFETY PROFILE: A skin irritant. When heated to decomposition it emits acrid smoke and irritating fumes.

DAU200 **CAS:65700-59-6** **HR: 2**
12-DEOXYPHORBOL-13-ANGELATE-20-ACETATE
mf: $C_{27}H_{36}O_7$ mw: 472.63

TOXICITY DATA WITH REFERENCE
skn-mus 3 µg OPEN ARTODN 44,279,80
skn-mus TDLo:180 mg/kg/24W-I:ETA EXPEAM 30,1438,74

SAFETY PROFILE: Questionable carcinogen with experimental tumorigenic data. A skin irritant. When heated to decomposition it emits acrid smoke and irritating fumes.

DAU400 **CAS:56726-04-6** **HR: 1**
12-DEOXY-PHORBOL-13-DECDIENOATE-20-ACETATE
mf: $C_{32}H_{44}O_7$ mw: 540.76

SYN: 12-DEOXY-PHORBOL-20-ACETATE-13-DECDIENOATE

TOXICITY DATA WITH REFERENCE
skn-mus 1100 ng/4H APTOA6 37,250,75
skn-mus 3500 ng/24H APTOA6 37,250,75

SAFETY PROFILE: A skin irritant. When heated to decomposition it emits acrid smoke and irritating fumes.

DAU600 CAS:56530-47-3 HR: 1
12-DEOXY-PHORBOL-13-DODECANOATE
mf: $C_{32}H_{50}O_6$ mw: 530.82

TOXICITY DATA WITH REFERENCE
skn-mus 110 ng/4H APTOA6 37,250,75
skn-mus 280 ng/24H APTOA6 37,250,75

SAFETY PROFILE: A skin irritant. When heated to decomposition it emits acrid smoke and irritating fumes.

DAW200 CAS:28152-97-8 HR: 1
12-DEOXY-PHORBOL-13-α-METHYLBUTYRATE
mf: $C_{25}H_{36}O_6$ mw: 432.61

TOXICITY DATA WITH REFERENCE
skn-mus 400 ng/4H APTOA6 37,250,75
skn-mus 1600 ng/24H APTOA6 37,250,75

SAFETY PROFILE: A skin irritant. When heated to decomposition it emits acrid smoke and irritating fumes. See also ESTERS.

DAY600 CAS:28152-96-7 HR: 1
12-DEOXY-PHORBOL-13-TIGLATE
mf: $C_{25}H_{34}O_6$ mw: 430.59

TOXICITY DATA WITH REFERENCE
skn-mus 300 ng/4H APTOA6 37,250,75
skn-mus 600 ng/24H APTOA6 37,250,75

SAFETY PROFILE: A skin irritant. When heated to decomposition it emits acrid smoke and irritating fumes.

DAY800 CAS:61-67-6 HR: 3
4-DEOXYPYRIDOXAL
mf: $C_8H_{11}NO_2$ mw: 153.20

SYNS: DEOXYPYRIDOXINE ☐ 4-DEOXYPYRIDOXINE ☐ 4-DEOXYPYRIDOXOL ☐ DESOXYPYRIDOXINE ☐ 4,6-DIMETHYL-5-HYDROXY-3-PYRIDINEMETHANOL

TOXICITY DATA WITH REFERENCE
mmo-esc 500 mg/L/1H CRSUBM 3,69,55
orl-rat TDLo:400 mg/kg (16D pre):REP PSEBAA 68,274,48
orl-rat TDLo:300 mg/kg (12D pre):REP PSEBAA 68,274,48
ipr-mus LDLo:250 mg/kg JMCMAR 16,865,73

SAFETY PROFILE: A poison by intraperitoneal route. Experimental reproductive effects. Mutation data reported. When heated to decomposition it emits toxic fumes of NO_x.

DAY825 CAS:148-51-6 HR: 3
4-DEOXYPYRIDOXOL HYDROCHLORIDE
mf: $C_8H_{11}NO_2 \cdot ClH$ mw: 189.66

PROP: Crystals from alcohol + ether + acetone. Mp: 257°. Sol in water and alc.

SYNS: DESOXYPYRIDOXIME HYDROCHLORIDE ☐ 5-HYDROXY-4,6-DIMETHYL-3-PYRIDINEMETHANOL HYDROCHLORIDE ☐ NSC 3063

TOXICITY DATA WITH REFERENCE
orl-rat TDLo:11,500 µg/kg (1-20D preg):TER SCIEAS 169,1329,70
orl-rat TDLo:11,500 µg/kg (1-20D preg):REP SCIEAS 169,1329,70
ipr-mus LD50:150 mg/kg NTIS** AD691-490
orl-ckn LD50:1570 mg/kg JMCMAR 17,1235,74

SAFETY PROFILE: Poison by intraperitoneal route. Moderately toxic by ingestion. Experimental teratogenic and reproductive effects. When heated to decomposition it emits toxic fumes of NO_x and HCl.

DAY835 CAS:60504-57-6 HR: 3
1-DEOXYPYRROMYCIN
mf: $C_{30}H_{35}NO_{10}$ mw: 569.66

SYNS: ACLACINOMYCIN T ☐ AKLAVIN ☐ ANTIBIOTIC MA 144T1 ☐ DOXYPYRROMYCIN ☐ MA 144T1

TOXICITY DATA WITH REFERENCE
dni-mus:leu 630 nmol/L JANTAJ 34,1596,81
oms-mus:leu 280 nmol/L JANTAJ 34,1596,81
ipr-mus LD50:16,500 µg/kg JANTAJ 32,79-39,79
ivn-mus LD50:18,100 µg/kg JANTAJ 32,79-39,79

CONSENSUS REPORTS: EPA Genetic Toxicology Program.

SAFETY PROFILE: Poison by intraperitoneal and intravenous routes. Mutation data reported. When heated to decomposition it emits toxic fumes of NO_x.

DAZ117 CAS:53-36-1 HR: 3
DEPO-MEDRATE
mf: $C_{24}H_{32}O_6$ mw: 416.56

SYNS: DEPO-MEDROL ☐ DEPO-MEDRONE ☐ DEPO-METHYLPREDNISOLONE ☐ DEPO-METHYLPREDNISOLONE ACETATE ☐ DEPOT-MEDROL ☐ MEDROL ACETATE ☐ METHYLPREDNISOLONE ACETATE ☐ METHYLPREDNISOLONE 21-ACETATE ☐ 6-METHYLPREDNISOLONE ACETATE ☐ 6-α-METHYLPREDNISOLONE ACETATE ☐ MPA ☐ 11-β,17,21-TRIHYDROXY-6-α-METHYL-PREGNA-1,4-DIENE-3,20-DIONE 21-ACETATE ☐ U 8210 ☐ URBASON CRYSTAL SUSPENSION

TOXICITY DATA WITH REFERENCE
ims-rbt TDLo:1800 µg/kg (female 7-18D post):REP ANREAK 193,598,79
scu-rat TDLo:45,500 µg/kg (female 13W pre):REP JTSCDR 10(Suppl 1),11,85
scu-rat TDLo:45,500 µg/kg (13W male):REP JTSCDR 10(Suppl 1,),11,85
ims-rbt TDLo:1200 µg/kg (female 7-18D post):TER ANREAK 193,598,79
ims-mus TDLo:330 mg/kg (female 10D post):TER TXAPA9 56,23,80
scu-rat LD50:265 mg/kg JTSCDR 10(Suppl 1),1,85
ipr-mus LD50:2145 mg/kg NIIRDN 6,833,82

scu-mus LD50:1320 mg/kg JTSCDR 10(Suppl 1,),1,85

SAFETY PROFILE: Poison by subcutaneous route. Moderately toxic by intraperitoneal route. An experimental teratogen. Other experimental reproductive effects. A steroid. When heated to decomposition it yields acrid smoke and fumes.

DAZ118 HR: 3
(±)-DEPRENIL HYDROCHLORIDE
mf: $C_{13}H_{17}N \cdot ClH$ mw: 223.77

SYNS: (±)-N,α-DIMETHYL-N-2-PROPYNYLPHENETHYLAMINE HYDROCHLORIDE □ (±)-E-250 □ (±)-PHENYLISOPROPYLMETHYLPROPYNYLAMINE HYDROCHLORIDE

TOXICITY DATA WITH REFERENCE
scu-rat LD50:218 mg/kg APACAB 32,377,67
ivn-rat LD50:75 mg/kg APACAB 32,377,67

SAFETY PROFILE: Poison by subcutaneous and intravenous routes. When heated to decomposition it emits toxic fumes of NO_x and HCl. See also AMINES.

DAZ120 CAS:4528-52-3 HR: 3
(+)-DEPRENIL HYDROCHLORIDE
mf: $C_{13}H_{17}N \cdot ClH$ mw: 223.77

SYNS: (+)-DEPRENYL HYDROCHLORIDE □ (+)-N,α-DIMETHYL-N-2-PROPYNYLPHENETHYLAMINE HYDROCHLORIDE □ (+)-E-250 □ (+)-PHENYLISOPROPYLMETHYLPROPYNYLAMINE HYDROCHLORIDE

TOXICITY DATA WITH REFERENCE
scu-rat LD50:208 mg/kg APACAB 32,377,67
ivn-rat LD50:72,500 μg/kg APACAB 32,377,67

SAFETY PROFILE: Poison by subcutaneous and intravenous routes. When heated to decomposition it emits toxic fumes of NO_x and HCl. See also AMINES.

DAZ125 CAS:14611-52-0 HR: 3
(−)-DEPRENYL HYDROCHLORIDE
mf: $C_{13}H_{17}N \cdot ClH$ mw: 223.77

PROP: A solid. Mp: 170–171°.

SYNS: 1-DEPRENIL HYDROCHLORIDE □ (−)-N,α-DIMETHYL-N-2-PROPYNYLBENZENEETHANAMINE HYDROCHLORIDE □ (−)-E-250 □ ELDEPRYL □ JUMEX □ (−)-PHENYLISOPROPYLMETHYLPROPYNYLAMINE

TOXICITY DATA WITH REFERENCE
scu-rat LD50:280 mg/kg APACAB 32,377,67
ivn-rat LD50:81 mg/kg APACAB 32,377,67
unr-mus LD50:121 mg/kg APACAB 32,377,67

SAFETY PROFILE: Poison by subcutaneous, intravenous, and possibly other routes. When heated to decomposition it emits toxic fumes of NO_x and HCl.

DAZ135 CAS:17124-74-2 HR: 3
DEPT
mf: $C_{13}H_{21}N_3S \cdot 2BrH$ mw: 413.27

SYNS: 2-DIETHYLAMINOETHYL-N′-PHENYLISOTHIURONIUM BROMIDE HYDROBROMIDE □ 2-(2-(DIETHYLAMINO)ETHYL)-1-PHENYL-2-THIOPSEUDOUREA DIHYDROBROMIDE

TOXICITY DATA WITH REFERENCE
ipr-mus LD50:65,900 μg/kg YKKZAJ 88,156,68
scu-mus LD50:100 mg/kg YKKZAJ 88,156,68
ivn-mus LD50:3500 μg/kg YKKZAJ 88,156,68

SAFETY PROFILE: Poison by subcutaneous, intravenous, and intraperitoneal routes. When heated to decomposition it emits toxic fumes of NO_x, SO_x, and HBr.

D

DAZ140 CAS:10139-98-7 HR: 3
DEPTROPINE METHOBROMIDE
mf: $C_{24}H_{30}NO \cdot Br$ mw: 428.46

SYNS: BS 7020a □ 3-α-((10,11-DIHYDRO-5H-DIBENZO(a,d)CYCLOHEPTEN-5-YL)OXY)-8-METHYLTROPANIUM BROMIDE

TOXICITY DATA WITH REFERENCE
orl-rat LD50:800 mg/kg AIPTAK 192,105,71
ipr-rat LD50:7600 μg/kg AIPTAK 192,105,71
ivn-rat LD50:1200 μg/kg AIPTAK 102,105,71
orl-mus LD50:680 mg/kg AIPTAK 192,105,71
ivn-mus LD50:1150 μg/kg AIPTAK 192,105,71
orl-dog LD50:71 mg/kg AIPTAK 192,105,71
orl-rbt LD50:391 mg/kg AIPTAK 192,105,71

SAFETY PROFILE: Poison by ingestion, intravenous, and intraperitoneal routes. When heated to decomposition it emits toxic fumes of Br^- and NO_x.

DBA000 HR: 3
DERRIS ELLIPTICA, root

PROP: The constituents of Derris are rotenone, dequelin, tephrosin, and toxicarol.

SYNS: DEGUELIA ROOT □ DERRIS RESINS □ DERRIS ROOT □ TUBA ROOT

TOXICITY DATA WITH REFERENCE
orl-rat LDLo:100 mg/kg IECHAD 28,815,36
orl-mus LD50:350 mg/kg PSEBAA 54,140,43
orl-dog LDLo:100 mg/kg IECHAD 28,815,36
orl-rbt LDLo:200 mg/kg IECHAD 28,815,36
orl-gpg LDLo:75 mg/kg IECHAD 28,815,36

SAFETY PROFILE: Poison by ingestion. An insecticide. When heated to decomposition it emits acrid smoke and fumes. See also ROTENONE.

DBA175 CAS:49720-72-1 HR: 3
DESACETYLCOLCHICINE-d-TARTRATE
mf: $C_{20}H_{23}NO_5 \cdot C_4H_6O_6 \cdot H_2O$ mw: 525.56

SYNS: DEACETYLCOLCHICINE l-TARTRATE □ N-DEACETYLCOLCHICINE l-TARTRATE(1:1), HYDRATE □ METHYL ETHERTRIMETHYLCOLCHICINIC ACID-l-TARTRATE □ METHYL ETHER TRIMETHYLCOLCHICINIC ACID-d-TARTRATE, HYDRATE □ NCI 1136 □ NSC-36354 □ SKF 250 □ TMCA □ TMCA METHYL ESTER d-TARTRATE, HYDRATE

TOXICITY DATA WITH REFERENCE
orl-hmn TDLo:200 μg/kg:GIT,BLD NCISP* JAN86
par-rat LD50:3 mg/kg PMDCAY 9,1,73
ipr-mus LD50:70,520 μg/kg NCISP* JAN86
par-mus LD50:8 mg/kg PMDCAY 9,1,73
par-dog LD50:9500 μg/kg PMDCAY 9,1,73

SAFETY PROFILE: Poison by parenteral, and intraperitoneal routes. Human systemic effects by ingestion: nausea or vomiting, leukopenia and thrombocytopenia. When heated to decomposition it emits toxic fumes of NO_x.

DBA200 **CAS:2731-16-0** **HR: 3**
N-DESACETYLTHIOCOLCHICINE
mf: $C_{20}H_{23}NO_4S$ mw: 373.50

SYNS: BENZO(a)HEPTALEN-9(5H)-ONE, 7-AMINO-6,7-DIHYDRO-1, 2,3-TRIMETHOXY-10-(METHYLTHIO)-, (S)- □ COLCHICINE, N-DEACETYL-10-DEMETHOXY-10-METHYLTHIO- □ CORPS R. 261 □ N-DEACETYL-10-DEMETHOXY-10-METHYLTHIOCOLCHICINE □ N-DEACETYLMETHYLTHIOCOLCHICINE □ N-DEACETYLTHIOCOLCHICINE □ N-DESACETYLTHIOCOLCHICINE □ R 261 □ THIO-COLCIRAN

TOXICITY DATA WITH REFERENCE
ivn-mky TDLo:6 mg/kg (13-14W preg):TER ACEDAB 166,435,72
ivn-hmn TDLo:71 µg/kg/D:GIT,BLD BAFEAG 42,308,55
ipr-rat LD50:175 mg/kg PSEBAA 98,479,58
ims-rat LD50:37 mg/kg PSEBAA 98 479,58
ice-rat LDLo:2 mg/kg AIPTAK 109,386,57
par-rat LDLo:2000 µg/kg FRPSAX 15,533,60
orl-mus LD50:50 mg/kg BAFEAG 42,308,55
ipr-mus LD50:66 mg/kg JMCMAR 24,636,81
ivn-mus LD50:80 mg/kg AIPTAK 107,150,56

SAFETY PROFILE: Poison by ingestion, intraperitoneal, intravenous, intramuscular, parenteral, and intracerebral routes. An experimental teratogen. Human systemic effects by intravenous route: hypermotility, diarrhea, nausea, vomiting, and leukopenia (reduced white blood cell count). When heated to decomposition it emits very toxic fumes of NO_x and SO_x. See also COLCHICINE.

DBA250 **CAS:57645-49-5** **HR: 2**
DESBENZYL CLEBOPRIDE
mf: $C_{13}H_{18}ClN_3O_2$ mw: 283.79

SYN: 4-AMINO-5-CHLORO-2-METHOXY-N-(4-PIPERIDYL)BENZAMIDE

TOXICITY DATA WITH REFERENCE
orl-rat TDLo:3500 mg/kg (35D pre):REP OYYAA2 25,865,83
orl-rat LD50:2550 mg/kg OYYAA2 25,865,83
orl-mus LD50:1750 mg/kg OYYAA2 25,865,83

SAFETY PROFILE: Moderately toxic by ingestion. Experimental reproductive effects. When heated to decomposition it emits toxic fumes of Cl^- and NO_x.

DBA450 **HR: 3**
DESERT ROSE

PROP: A 6- to 10-foot tall shrub with a swollen trunk and branches. It has thick, dark green leaves and large pink or purple conical flowers. It is cultivated in hot and arid areas of the United States and Jamaica.

SYNS: ADENIUM (VARIOUS SPECIES) □ MOCK AZALEA

SAFETY PROFILE: The toxin is a cardiac glycoside similar to digitalis and found throughout the plant. No cases of poisoning have been reported in the United States; however, cardiac glycosides can cause death by their effect on heart function. See also DIGITALIS.

DBA475 **HR: 3**
DESETHYLAPRINDINE HYDROCHLORIDE
mf: $C_{20}H_{26}N_2$•ClH mw: 330.94

SYNS: AC 2197 HYDROCHLORIDE □ N-ETHYL-N′-2-INDANYL-N′-PHENYL-1,3-PROPANEDIAMINE HYDROCHLORIDE

TOXICITY DATA WITH REFERENCE
orl-rat LD50:525 mg/kg OYYAA2 27,353,84
ivn-rat LD50:22,100 µg/kg OYYAA2 27,353,84
orl-mus LD50:543 mg/kg OYYAA2 27,353,84
ivn-mus LD50:24,500 µg/kg OYYAA2 27,353,84

SAFETY PROFILE: Poison by intravenous route. Moderately toxic by ingestion. When heated to decomposition it emits toxic fumes of NO_x and HCl.

DBA500 **HR: 2**
DESGLUCODIGITONIN

TOXICITY DATA WITH REFERENCE
ipr-mus LD10: 20 mg/kg:ETA PCJOAU 11,749,77

SAFETY PROFILE: Questionable carcinogen with experimental tumorigenic data. When heated to decomposition it emits acrid smoke and irritating fumes.

DBA600 **CAS:5626-16-4** **HR: 3**
DESMETHYLDOXEPIN
mf: $C_{17}H_{17}NO$ mw: 251.35

SYNS: DIBENZ(b,e)OXEPIN-$\Delta^{11(6H)}$,γ-PROPYLAMINE □ KS 1675

TOXICITY DATA WITH REFERENCE
ipr-rat LD50:100 mg/kg ARZNAD 15,863,65
orl-mus LD50:118 mg/kg ARZNAD 15,863,65
ipr-mus LD50:92 mg/kg ARZNAD 15,863,65
ivn-mus LD50:29 mg/kg ARZNAD 15,863,65
ivn-rbt LD50:13,400 µg/kg ARZNAD 15,863,65

SAFETY PROFILE: Poison by ingestion, intravenous, and intraperitoneal routes. When heated to decomposition it emits toxic fumes of NO_x. See also DOXEPIN.

DBA700 **HR: 2**
DESMETHYLMISONIDAZOLE
mf: $C_7H_{11}N_3O_4$ mw: 201.21

SYNS: 3-METHOXY-2-(2-NITRO-1-IMIDAZOLYL)-1-PROPANOL □ NSC-261036

TOXICITY DATA WITH REFERENCE
orl-rat LD50:5076 mg/kg NTIS** PB81-121212
ivn-rat LD50:2532 mg/kg NTIS** PB81-121212
ivn-dog LDLo:1800 mg/kg NTIS** PB81-121212

SAFETY PROFILE: Moderately toxic by intravenous route. Mildly toxic by ingestion. When heated to decomposition it emits toxic fumes of NO_x.

DBA800 CAS:300-42-5 HR: 3
DESOXYEPHEDRINE HYDROCHLORIDE
mf: $C_{10}H_{15}N{\cdot}ClH$ mw: 185.72

PROP: Crystals from $EtOH/Et_2O$. Mp: 135–136°.

SYNS: A 884 □ AMDRAM □ AMEDRINE □ AMPHEDROXY □ AMPHEDROXYN □ APAMINE □ BOMBITA □ C 6379 □ CORVITIN □ DAROPERVAMIN □ DEA OXO-5 □ DEOFED □ DEOXYEPHEDRINE □ DEPOXIN □ DESAMINE □ DESFEDRIN □ DESOSSIEFEDRINA □ DES-OXA-D □ DESOXEDRINE □ DESOXIN □ DESOXO-5 □ DESOXYFED □ DESOXYN □ DESOXYPHED □ DESTIM □ DETREX □ DEXOPHRINE □ DEXOVAL □ N,α-DIMETHYLPHENETHYLAMINE HYDROCHLORIDE □ DOPIDRIN □ DOXEPHRIN □ DOXYFED □ DRINALFA □ EFFROXINE □ ESTIMULEX □ EUPHODRIN □ FENYPRIN □ GEROBIT □ GEROVIT □ HEROPON □ ISOPHEN □ KEMODRIN □ LANAZINE □ MADRINE □ METAMFETAMINA □METAMPHETAMIN □ METANFETAMINA □ METHAMPHETAMINE HYDROCHLORIDE □ METHEDRINE □ METHEDRINE HYDROCHLORIDE □ METHOXYN □ METHYLAMPHETAMINE HYDROCHLORIDE □ METHYLBENZEDRIN □ METHYLISOMIN □ N-METHYL-β-PHENYLISOPROPYLAMINHYDROCHLORID (GERMAN) □ METHYLPROPAMINE □ NEODRINE □ NEOPHARMEDRINE □ NORODIN □ OXYDRENE □ OXYFED □ PERVITIN □ PHILOPON □ PREMODRIN □ SEMOXYDRINE □ SPEED □ STIMULEX □ TONEDRIN □ VONEDRINE

TOXICITY DATA WITH REFERENCE
eye-rbt 2% MLD ARZNAD 8,708,58
scu-rat TDLo:210 mg/kg (female 1-21D post):REP DEPBA5 8,397,75
scu-rat TDLo:156 mg/kg (female 7-19D post):REP NYKZAU 85,79,85
scu-rat TDLo:42 mg/kg (1-21D preg):REP DEPBA5 8,397,75
ipr-rat TDLo:2250 μg/kg (1D male):REP PHBHA4 4,677,69
orl-hmn TDLo:17 mg/kg:CNS,CVS KLWOAZ 17,1580,38
orl-rat LD50:29 mg/kg ARZNAD 24,166,74
ipr-rat LD50:19 mg/kg ARZNAD 8,708,58
scu-rat LD50:79,700 μg/kg JJPAAZ 35,273,84
ivn-rat LD50:13 mg/kg HEPHD2 55,527,80
orl-mus LD50:15 mg/kg ARZNAD 24,166,74
ipr-mus LD50:4000 μg/kg ARZNAD 11,271,61
scu-mus LD50:9 mg/kg ARZNAD 11,271,61
ivn-mus LD50:15 mg/kg HEPHD2 55,527,80
orl-rbt LD50:10,500 μg/kg RCOCB8 3,215,72
scu-rbt LD50:205 mg/kg JPETAB 86,284,46
ivn-rat LD50:3 mg/kg RCOCB8 3,215,72
ims-rbt LD50:220 mg/kg JPETAB 86,284,46

SAFETY PROFILE: Poison by ingestion, intravenous, intraperitoneal, subcutaneous, and intramuscular routes. Human systemic effects by ingestion: altered sleep patterns, anorexia, and change in heart rate. Experimental reproductive effects. An eye irritant. A powerful central nervous system stimulant. When heated to decomposition it emits very toxic fumes of NO_x and HCl. See also BENZEDRINE.

DBB000 CAS:7632-10-2 HR: 3
DESOXYN
mf: $C_{10}H_{15}N$ mw: 149.26

SYNS: ANADREX □ DEOXYEPHEDRINE □ DESOXYEPHEDRINE □ METHAMPHETAMINE □ METHEDRINE □ METHYLAMPHETAMINE □ N-METHYLAMPHETAMINE □ N-METHYL-β-PHENYLISOPROPYLAMIN (GERMAN) □ N-METHYL-β-PHENYLISOPROPYLAMINE □ PERVERTIN □ 1-PHENYL-2-METHYLAMINO-PROPAN (GERMAN) □ 1-PHENYL-2-METHYLAMINOPROPANE □ α-PHENYL-β-METHYLAMINOPROPANE □ STIMULEX

TOXICITY DATA WITH REFERENCE
orl-rat LDLo:70 mg/kg AEPPAE 195,647,40
ipr-rat LDLo:32 mg/kg AEPPAE 195,647,40
scu-rat LD50:30 mg/kg AIPTAK 159,442,66
orl-mus LD50:34 mg/kg JPETAB 131,115,61
ipr-mus LD50:15 mg/kg CPBTAL 22,1459,74
scu-mus LD50:10 mg/kg AEPPAE 241,182,61
ivn-mus LD50:10 mg/kg 27ZIAQ -,-,65
scu-cat LDLo:50 mg/kg 27ZIAQ -,158,73

SAFETY PROFILE: Poison by ingestion, intraperitoneal, subcutaneous, and intravenous routes. A powerful central nervous system stimulant. When heated to decomposition it emits toxic fumes of NO_x. See also BENZEDRINE.

D

DBB200 CAS:125-33-7 HR: 3
2-DESOXYPHENOBARBITAL
mf: $C_{12}H_{14}N_2O_2$ mw: 218.28

SYNS: 5-AETHYL-5-PHENYL-HEXAHYDROPYRIMIDIN-4,6-DION (GERMAN) □ CYRAL □ 2-DEOXYPHENOBARBITAL □ DESOXYPHENOBARBITONE □ 5-ETHYLDIHYDRO-5-PHENYL-4,6(1H,5H)-PYRIMIDINEDIONE □ 5-ETHYLHEXAHYDRO-4,6-DIOXO-5-PHENYLPHYIMIDINE □ 5-ETHYLHEXAHYDRO-5-PHENYLPYRIMIDINE-4,6-DIONE □ 5-ETHYL-5-PHENYLHEXAHYDROPYRIMIDINE-4,6-DIONE □ HEXADIONA □ HEXAMIDINE □ HEXAMIDINE (the antispasmodic) □ LEPIMIDIN □ LEPSIRAL □ MAJSOLIN □ MIDONE □ MILEPSIN □ MISODINE □ MISOLYNE □ MIZODIN □ MIZOLIN □ MYLEPSIN □ MYLEPSINUM □ MYSEDON □ MYSOLINE □ NCI-C56360 □ 5-PHENYL-5-ETHYL-HEXAHYDROPYRIMIDINE-4,6-DIONE □ PRILEPSIN □ PRIMACIONE □ PRIMACLONE □ PRIMACONE □ PRIMAKTON □ PRIMIDON □ PRIMIDONE □ PRYSOLINE □ PYRIMIDONE MEDIPETS □ ROE 101 □ SERTAN

TOXICITY DATA WITH REFERENCE
mmo-sat 6666 μg/plate ENMUDM 8(Suppl 7),1,86
cyt-hmn:leu 1 mg/L AJOGAH 116,867,73
orl-wmn TDLo:2025 mg/kg (female 1-39W post):REP TJADAB 32,13,85
unr-wmn TDLo:10,640 mg/kg (female 1-38W post):REP TJADAB 26(1),36A,82
orl-wmn TDLo:10,350 mg/kg (1-39W preg):REP JOPDAB 94,835,79
orl-mus TDLo:875 mg/kg (male 5D pre):REP TOLED5 34,149,86
orl-mus TDLo:990 mg/kg (female 6-16D post):TER TXAPA9 40,365,77
orl-mus TDLo:275 mg/kg (female 6-16D post):TER EPILAK 18,1,77
orl-mus TDLo:330 mg/kg (female 6-16D post):TER TXAPA9 40,365,77
orl-wmn TDLo:38 g/kg/7Y-I DICPBB 17,551,83
orl-rat LD50:1500 mg/kg NIIRDN 6,691,82
ipr-rat LD50:240 mg/kg PCJOAU 25,305,91
orl-mus LD50:280 mg/kg TXAPA9 34,271,75
ipr-mus LD50:332 mg/kg PCJOAU 15,403,81

SAFETY PROFILE: Poison by ingestion and intraperitoneal routes. Human teratogenic effects include developmental abnormalities of the craniofacial area, skin and skin appendages, and cardiovascular system. Human reproductive effects: effects on newborn, including unusual growth statistics, drug dependence, physical and other neonatal changes. Experimental teratogenic and reproductive effects. Human mutation data reported. An addictive drug. When heated to decomposition it emits toxic fumes of NO_x. See also BARBITURATES.

DBB400 CAS:14918-35-5 HR: 3
DESTOMYCIN A
mf: $C_{20}H_{37}N_3O_{13}$ mw: 527.60

PROP: Powder. Mp: 180–190° (decomp). Sol in H_2O.

SYN: DESTONATE 20

TOXICITY DATA WITH REFERENCE
orl-mus LD50:50 mg/kg JAJAAA 18,38,65
ivn-mus LD50:5 mg/kg JAJAAA 18,38,65

SAFETY PROFILE: Poison by ingestion and intravenous routes. When heated to decomposition it emits toxic fumes of NO_x.

DBB500 CAS:37209-31-7 HR: 3
DETRALFATE

SYNS: APD ☐ DEXTRAN SULFATE SODIUM ALUMINIUM

TOXICITY DATA WITH REFERENCE
ipr-rat LD50:200 mg/kg IYKEDH 3,24,72
scu-rat LD50:1600 mg/kg IYKEDH 3,24,72
ipr-mus LD50:257 mg/kg IYKEDH 3,24,72
scu-mus LD50:2200 mg/kg IYKEDH 3,24,72

SAFETY PROFILE: Poison by intraperitoneal route. Moderately toxic by subcutaneous route. Used as an antipeptic ulcer agent. When heated to decomposition it emits toxic fumes of SO_x and Na_2O. See also DEXTRAN SULFATE SODIUM.

DBB600 CAS:67293-88-3 HR: 3
DEUTERIOMORPHINE
mf: $C_{17}H_{16}D_3NO_3$ mw: 288.37

TOXICITY DATA WITH REFERENCE
scu-mus LDLo:400 mg/kg SCIEAS 134,1078,61
ice-mus LD50:11,400 μg/kg SCIEAS 134,1078,61

SAFETY PROFILE: Poison by subcutaneous and intracerebral routes. When heated to decomposition it emits toxic fumes of NO_x. See also MORPHINE.

DBB800 CAS:7782-39-0 HR: 3
DEUTERIUM
mf: D_2 mw: 4.03

PROP: A gas. Chemically the same as hydrogen. Solid D_2 has hcp structure. Mp: −254.3°, bp: −249.3°. Lel: 5%, uel: 75%.

SYN: D_2

SAFETY PROFILE: Very dangerous fire and explosion hazard when exposed to heat, flame, sparks, and oxidizers. To fight fire, stop flow of gas. See also HYDROGEN.

DBC000 CAS:14333-26-7 HR: 3
DEUTERIUM FLUORIDE
mf: DF mw: 21.01

PROP: Chemically the same as hydrogen fluoride.

TOXICITY DATA WITH REFERENCE
ihl-rat LC50:1095 ppm/1H AMRL** TR-74-78,74
ihl-mus LC50:324 ppm/1H AMRL** TR-74-78,74

OSHA PEL: TWA 2.5 mg(F)/m³
ACGIH TLV: TWA 2.5 mg(F)/m³
NIOSH REL: TWA (Inorganic Fluorides) 2.5 mg(F)/m³

SAFETY PROFILE: Moderately toxic by inhalation. A dangerously reactive, powerful oxidant. When heated to decomposition it emits toxic fumes of F^-. See also FLUORIDES and HYDROFLUORIC ACID.

DBC500 CAS:55541-30-5 HR: 3
DEXAMETHASONE 17,21-DIPROPIONATE
mf: $C_{28}H_{37}FO_7$ mw: 504.65

SYNS: DEXAMETHASONE DIPROPIONATE ☐ 9-FLUORO-11-β,17,21-TRIHYDROXY-16-α-METHYLPREGNA-1,4-DIENE-3,20-DIONE-17,21-DIPROPIONATE ☐ THS-101

TOXICITY DATA WITH REFERENCE
scu-rat TDLo:1100 μg/kg (female 7-17D post):REP IYKEDH 17,310,86
scu-rat TDLo:11 mg/kg (female 7-17D post):TER TJADAB 28,12A,83
scu-rat TDLo:252 μg/kg (male 9W pre):TER IYKEDH 17,347,86
scu-rat TDLo:4550 μg/kg (female 91D pre):REP OYYAA2 28,775,84
scu-rbt TDLo:390 μg/kg (female 6-18D post):TER IYKEDH 17,327,86
ipr-rat LD50:36,500 μg/kg OYYAA2 28,687,84
scu-rat LD50:26 mg/kg OYYAA2 28,687,84
ipr-mus LD50:276 mg/kg IYKEDH 18,474,87
scu-mus LD50:183 mg/kg IYKEDH 18,474,87
scu-rbt LD50:7600 μg/kg OYYAA2 28,687,84

SAFETY PROFILE: Poison by subcutaneous and intraperitoneal routes. An experimental teratogen. Other experimental reproductive effects. A steroid. When heated to decomposition it emits toxic fumes of F^-.

DBC510 CAS:2265-64-7 HR: 2
DEXAMETHASONE ISONICOTINATE
mf: $C_{28}H_{32}FNO_6$ mw: 497.61

PROP: Crystals. Mp: 256°.

SYNS: AUXILSON ☐ DEXAMETHASONE-21-ISONICOTINATE ☐ 9-FLUORO-11-β,17,21-TRIHYDROXY-16-α-METHYLPREGNA-1,4-DIENE-3,20-DIONE, 21-ISONICOTINATE ☐ H3 111 ☐ PYRIDIN-4-CARBONSAEURE-(DEXAMETHASON-21')-ESTER (GERMAN) ☐ PYRIDINE-4-CARBOXYLIC ACID-(DEXAMETHASONE-21') ESTER ☐ VOREN

TOXICITY DATA WITH REFERENCE
ihl-rat TCLo:4178 mg/m³/10M (44D pre):REP OYYAA2 8,255,74

ihl-rat TCLo:4178 mg/m³/10M (44D male):REP OYYAA2 8,255,74
orl-rat LD50:3562 mg/kg TOIZAG 20,769,73
ipr-rat LD50:313 mg/kg TOIZAG 20,769,73
scu-rat LD50:3297 mg/kg TOIZAG 20,769,73
orl-mus LD50:3470 mg/kg TOIZAG 20,769,73

SAFETY PROFILE: Moderately toxic by intraperitoneal route. Experimental reproductive effects. A steroid. When heated to decomposition it emits toxic fumes of F^- and NO_x.

DBC550 CAS:466-11-5 HR: 2
DEXAMETHASONE SODIUM SULFATE
mf: $C_{22}H_{29}FO_8S{\cdot}Na$ mw: 495.56

PROP: A solid. Mp: 207–209°.

SYNS: DEXAMETHASONE SODIUM HEMISULFATE □ DEXA-SCHEROSON (INJECTABLE) □ 9-FLUORO-11-β,17,21-TRIHYDROXY-16-α-METHYLPREGNA-1,4-DIENE-3,20-DIONE-21-(HYDROGEN SULFATE), MONOSODIUM SALT

TOXICITY DATA WITH REFERENCE
ipr-rat LD50:2180 mg/kg NIIRDN 6,477,82
scu-rat LD50:3450 mg/kg NIIRDN 6,477,82
ipr-mus LD50:1660 mg/kg NIIRDN 6,477,82
scu-mus LD50:2020 mg/kg NIIRDN 6,477,82

SAFETY PROFILE: Moderately toxic by intraperitoneal and subcutaneous routes. A steroid. When heated to decomposition it emits toxic fumes of F^-, SO_x, and Na_2O.

DBC575 CAS:33755-46-3 HR: 3
DEXAMETHASONE VALERATE
mf: $C_{27}H_{37}FO_6$ mw: 476.64

SYNS: DEXAMETHASONE-17-VALERATE □ DV-17 □ 9-FLUORO-11-β,17,21-TRIHYDROXY-16-α-METHYLPREGNA-1,4-DIENE-3,20-DIONE-17-VALERATE

TOXICITY DATA WITH REFERENCE
skn-rat TDLo:18 μg/kg (17-22D preg):REP JZKEDZ 8,235,82
scu-rat TDLo:3 mg/kg (male 3D pre):REP OYYAA2 23,961,82
skn-rat TDLo:10,920 μg/kg (26W male):REP JZKEDZ 8,55,82
scu-rat LD50:117 mg/kg IYKEDH 17,1106,86
skn-mus LD50:1784 mg/kg JZKEDZ 8,23,82
scu-mus LD50:238 mg/kg JZKEDZ 8,23,82

SAFETY PROFILE: Poison by subcutaneous route. Moderately toxic by skin contact. Experimental reproductive effects. A steroid. When heated to decomposition it emits toxic fumes of F^-.

DBC800 CAS:9004-54-0 HR: 2
DEXTRAN 1

PROP: A linear water-sol polymer of average molecular weight 200,000 (ARPAAQ 67,589,59).

TOXICITY DATA WITH REFERENCE
ipr-rat TDLo:2500 mg/kg:NEO,REP AMPLAO 67,589,59
scu-rat TDLo:2500 mg/kg:NEO,REP AMPLAO 67,589,59
ipr-mus TDLo:8000 mg/kg:ETA AMPLAO 67,589,59
scu-mus TDLo:8000 mg/kg:ETA AMPLAO 67,589,59

CONSENSUS REPORTS: Reported in EPA TSCA Inventory.

SAFETY PROFILE: Questionable carcinogen with experimental neoplastigenic, tumorigenic, and teratogenic data. Other experimental reproductive effects. When heated to decomposition it emits acrid smoke and fumes. See also other DEXTRANS.

D

DBD000 CAS:9004-54-0 HR: 2
DEXTRAN 2

PROP: A linear, water-sol polymer of average molecular weight 100,000 (ARPAAQ 67,589,59).

TOXICITY DATA WITH REFERENCE
ipr-rat TDLo:2500 mg/kg:NEO,REP AMPLAO 67,589,59
scu-rat TDLo:2500 mg/kg:NEO AMPLAO 67,589,59
ipr-mus TDLo:8000 mg/kg:NEO AMPLAO 67,589,59

CONSENSUS REPORTS: Reported in EPA TSCA Inventory.

SAFETY PROFILE: Questionable carcinogen with experimental neoplastigenic data. When heated to decomposition it emits acrid smoke and fumes. See also other DEXTRANS.

DBD200 CAS:9004-54-0 HR: 2
DEXTRAN 5

PROP: A highly branched, water-sol polymer (ARPAAQ 67,589,59).

TOXICITY DATA WITH REFERENCE
scu-mus TDLo:8000 mg/kg:NEO AMPLAO 67,589,59

CONSENSUS REPORTS: Reported in EPA TSCA Inventory.

SAFETY PROFILE: Questionable carcinogen with experimental neoplastigenic data. When heated to decomposition it emits acrid smoke and fumes. See also other DEXTRANS.

DBD400 CAS:9004-54-0 HR: 3
DEXTRAN 10

PROP: A branched, water-sol polymer of average molecular weight 89,400 (ARPAAQ 67,589,59).

TOXICITY DATA WITH REFERENCE
scu-rat TDLo:2500 mg/kg:CAR AMPLAO 67,589,59
ipr-mus TDLo:8000 mg/kg:CAR AMPLAO 67,589,59
ivn-rbt TDLo:8750 mg/kg/I:CAR AMPLAO 67,589,59

CONSENSUS REPORTS: Reported in EPA TSCA Inventory.

SAFETY PROFILE: Suspected carcinogen with experimental carcinogenic data. When heated to decomposition it emits acrid smoke and fumes. See also other DEXTRANS.

DBD600 **CAS:9004-54-0** **HR: 2**
DEXTRAN 11

PROP: A highly branched, water-sol polymer of average molecular weight 71,400 (ARPAAQ 67,589,59).

TOXICITY DATA WITH REFERENCE
ipr-rat TDLo:2500 mg/kg:NEO AMPLAO 67,589,59
scu-mus TDLo:8000 mg/kg:ETA AMPLAO 67,589,59
ivn-rbt TDLo:8750 mg/kg:ETA AMPLAO 67,589,59

CONSENSUS REPORTS: Reported in EPA TSCA Inventory.

SAFETY PROFILE: Questionable carcinogen with experimental neoplastigenic and tumorigenic data. When heated to decomposition it emits acrid smoke and fumes. See also other DEXTRANS.

DBD700 **CAS:9004-54-0** **HR: 2**
DEXTRAN 70

PROP: Chemical and physical properties of the dextrans vary with the methods of production. Native dextrans usually have high molecular weight; lower-molecular-weight clinical dextrans are usually prepared by depolymerization of native dextrans or by synthesis. All dextrans are composed exclusively of α-d-glucopyranosyl units, differing only in degree of branching and chain length.

SYNS: DEXTRAN □ DEXTRAVEN □ EXPANDEX □ GENTRAN □ HEMODEX □ INTRADEX □ MACROSE □ ONKOTIN □ PLAVOLEX □ POLYGLUCIN □ PROMIT

TOXICITY DATA WITH REFERENCE
ivn-rbt TDLo:675 g/kg (8-16D preg):TER OYYAA2 6,1119,72
ivn-rbt TDLo:3640 g/kg (91D male):REP YACHDS 3,369,75
ivn-rbt TDLo:600 g/kg (30D male):REP YACHDS 3,369,75
ivn-wmn TDLo:6 mg/kg:CVS,SKN BMJOAE 2,1502,76
scu-mus LD50:13,900 mg/kg OYYAA2 6,1023,72
ivn-mus LD50:12,100 mg/kg OYYAA2 6,1023,72
ivn-rbt LD50:17,400 mg/kg OYYAA2 6,1023,72

CONSENSUS REPORTS: Reported in EPA TSCA Inventory.

SAFETY PROFILE: Human systemic effects by intravenous route: dermatitis and changes in the vascular and cardiac systems. Experimental reproductive effects. An experimental teratogen. See also various DEXTRANS.

DBD750 **CAS:9011-18-1** **HR: 2**
DEXTRAN SULFATE SODIUM

PROP: White powder from alcohol + ether. Freely sol in water.

SYNS: ASURO □ COLYONAL □ DEXTRARINE □ DEXULATE □ DS-M-1 □ MDS

TOXICITY DATA WITH REFERENCE
orl-rat TDLo:330 g/kg/94W-C:CAR CRNGDP 3,353,82
orl-rat TD:335 g/kg/19W-C:CAR JJIND8 66,579,81
orl-rat TD:600 g/kg/69W-C:CAR CALEDQ 18,29,83
orl-rat LD50:20,600 mg/kg KSRNAM 13,1318,79
ivn-rat LD50:473 mg/kg KSRNAM 13,1318,79
orl-mus LD50:21 g/kg NIIRDN 6,481,82
ivn-mus LD50:15,800 mg/kg NIIRDN 6,481,82
ivn-rbt LD50:19 g/kg NIIRDN 6,481,82

SAFETY PROFILE: Moderately toxic by intravenous route. Questionable carcinogen with experimental carcinogenic data. When heated to decomposition it emits toxic fumes of SO_x and Na_2O. See also other DEXTRAN entries.

DBD800 **CAS:9004-53-9** **HR: 1**
DEXTRINS
mf: $(C_6H_{10}O_5)_n \cdot xH_2O$

PROP: An intermediate product formed by the hydrolysis of starches. It describes a class of substances. Yellow or white powder or granules. Sol in boiling water to give a gummy solution; insol in alc and ether; forms colloids.

SYNS: ARTIFICIAL GUM □ DEXTRANS □ STARCH GUM □ TAPIOCA □ VEGETABLE GUM

SAFETY PROFILE: Mildly toxic by intravenous route. When heated to decomposition it emits acrid smoke and irritating fumes.

DBE000 **CAS:21888-96-0** **HR: 3**
DEXTROBENZETIMIDE HYDROCHLORIDE
mf: $C_{23}H_{26}N_2O_2 \cdot ClH$ mw: 398.97

SYNS: (+)-1-BENZYL-4-(2,6-DIOXO-3-PHENYL-3-PIPERIDYL)PIPERIDINE HYDROCHLORIDE □ (+)-2-(1-BENZYL-4-PIPERIDYL)-2-PHENYLGLUTARIMIDE HYDROCHLORIDE □ (+)-3-(1-BENZYL-4-PIPERIDYL)-3-PHENYLPIPERIDINE-2,6-DIONE HYDROCHLORIDE □ DEXBENZETIMIDE HYDROCHLORIDE □ DEXETIMIDE HYDROCHLORIDE □ (s)-3-PHENYL-1′-(PHENYLMETHYL)-(3,4′-BIPIPERIDINE)-2,6-DIONE HYDROCHLORIDE □ R 16470 □ TREMBLEX

TOXICITY DATA WITH REFERENCE
ivn-rat LD50:45 mg/kg 27ZQAG -,219,72
ivn-mus LD50:45 mg/kg ARZNAD 21,1365,71

SAFETY PROFILE: Poison by intravenous route. An anticholinergic agent used to treat Parkinson's disease. When heated to decomposition it emits very toxic fumes of HCl and NO_x.

DBE100 **CAS:5653-80-5** **HR: 3**
DEXTROMETHADONE
mf: $C_{21}H_{27}NO$ mw: 309.49

PROP: Crystals from 2-propanol. Mp: 100–101°.

SYNS: d-6-(DIMETHYLAMINO)-4,4-DIPHENYL-3-HEPTANONE □ (s)-6-(DIMETHYLAMINO)-4,4-DIPHENYL-3-HEPTANONE □ (+)-METHADONE □ d-METHADONE □ l-(+)-METHADONE □ s-(+)-METHADONE □ 6s-METHADONE

TOXICITY DATA WITH REFERENCE
dni-rat:tes 50 μmol/L BCPCA6 27,123,78
oms-rat:tes 50 μmol/L BCPCA6 27,123,78
scu-rat TDLo:17,800 μg/kg (1D male):REP JPETAB 198,340,76
orl-mus LD50:252 mg/kg JPETAB 110,135,54
ipr-mus LD50:74 mg/kg BJPCAL 9,280,54

scu-mus LD50:80 mg/kg BJPCAL 9,280,54

SAFETY PROFILE: Poison by ingestion, subcutaneous, and intraperitoneal routes. Experimental reproductive effects. Mutation data reported. Addictive. When heated to decomposition it emits toxic fumes of NO_x. See also METHADONE.

DBE150 CAS:125-71-3 HR: 3
DEXTROMETHORPHAN
mf: $C_{18}H_{25}NO$ mw: 271.44

SYNS: BA 2666 □ d-METHORPHAN □ Δ-METHORPHAN

TOXICITY DATA WITH REFERENCE
orl-mus LD50:210 mg/kg JPMSAE 60,1523,71
scu-mus LD50:112 mg/kg CPBTAL 7,372,59
ivn-dog LDLo:22 mg/kg CPBTAL 7,372,59

SAFETY PROFILE: Poison by ingestion, subcutaneous, and intravenous routes. When heated to decomposition it emits toxic fumes of NO_x. See also DEXTROMETHORPHAN HYDROBROMIDE.

DBE200 CAS:125-69-9 HR: 3
DEXTROMETHORPHAN HYDROBROMIDE
mf: $C_{18}H_{25}NO\cdot BrH$ mw: 352.36

PROP: A solid. Mp: 122–124°.

SYNS: ANTUSSAN □ DEMORPHAN □ DEXTROMETHORPHAN BROMIDE □ DEXTROMETORPHAN HYDROBROMIDE □ DORMETHAN □ MEDICON □ METHORATE HYDROBROMIDE □ d-METHORPHAN HYDROBROMIDE □ d-3-METHOXY-N-METHYLMORPHINAN HYDROBROMIDE □ 3-METHOXY-17-METHYL-9-α,13-α,14-α-MORPHINAN HYDROBROMIDE □ METRORAT □ RO 1-5470/5 □ ROMILAR □ ROMILAR HYDROBROMIDE □ TUSILAN □ TUSSADE

TOXICITY DATA WITH REFERENCE
orl-chd TDLo:30 mg/kg:CNS PEDIAU 59,117,77
orl-rat LD50:350 mg/kg JPETAB 109,189,53
scu-rat LD50:423 mg/kg OYYAA2 6,1207,72
orl-mus LD50:165 mg/kg JPETAB 109,189,53
scu-mus LD50:153 mg/kg ARZNAD 26,353,76
ivn-mus LD50:34 mg/kg OYYAA2 6,1207,72
ivn-dog LDLo:30 mg/kg ARZNAD 19,1916,69
ivn-cat LD50:19,800 μg/kg OYYAA2 6,1207,72
ivn-rbt LD50:15 mg/kg JPETAB 109,189,53
orl-gpg LD50:336 mg/kg OYYAA2 6,1207,72
scu-gpg LD50:150 mg/kg OYYAA2 6,1207,72

SAFETY PROFILE: Poison by ingestion, subcutaneous, and intravenous routes. Human systemic effects by ingestion: ataxia (loss of muscle coordination), excitement, and motor activity changes. When heated to decomposition it emits very toxic fumes of NO_x and HBr.

DBE600 CAS:101563-89-7 HR: 3
DEXTROMYCIN HYDROCHLORIDE
mf: $C_{23}H_{46}N_6O_{13}\cdot ClH$ mw: 651.21

TOXICITY DATA WITH REFERENCE
scu-mus LD50:750 mg/kg 85ERAY 1,743,78
ivn-mus LD50:50 mg/kg 85ERAY 1,743,78

SAFETY PROFILE: Poison by intravenous route. Moderately toxic by subcutaneous route. When heated to decomposition it emits very toxic fumes of HCl and NO_x.

DBE625 CAS:72050-78-3 HR: 1
DEXTROPROPOXYPHENE NAPSYLATE

TOXICITY DATA WITH REFERENCE
orl-mus TDLo:1920 mg/kg (7-12D preg):REP OYYAA2 4,1031,70
orl-mus TDLo:3600 mg/kg (7-12D preg):TER OYYAA2 4,1031,70
orl-hmn TDLo:28 mg/kg/1W:SKN BMJOAE 2,674,77

SAFETY PROFILE: Human systemic effects by ingestion: dermatitis. An experimental teratogen. Other experimental reproductive effects. See also d-PROPOXYPHENE HYDROCHLORIDE.

DBE800 CAS:125-73-5 HR: 3
DEXTRORPHAN
mf: $C_{17}H_{23}NO$ mw: 257.41

PROP: Crystals. Mp: 198–199°.

SYNS: (+)-cis-1,3,4,9,10,10A-HEXAHYDRO-11-METHYL-2H-10,4a-IMINOETHANOPHENANTHREN-6-OL □ d-3-HYDROXY-N-METHYLMORPHINAN

TOXICITY DATA WITH REFERENCE
scu-rat TDLo:25,750 μg/kg (female 5-12D post):REP JPETAB 200,255,77
scu-rat LD50:800 mg/kg CLDND* 22,1370,57
scu-mus LD50:131 mg/kg JOCEAH 22,1370,57
ivn-mus LD50:65 mg/kg CLDND* 22,1370,57

SAFETY PROFILE: Poison by subcutaneous and intravenous routes. Experimental reproductive effects. When heated to decomposition it emits toxic fumes of NO_x.

DBE825 CAS:143-98-6 HR: 3
DEXTRORPHAN TARTRATE
mf: $C_{17}H_{23}NO\cdot C_4H_6O_6$ mw: 407.51

SYNS: d-3-HYDROXY-N-METHYLMORPHINAN TARTRATE □ RO 1-6794

TOXICITY DATA WITH REFERENCE
orl-rat LD50:1100 mg/kg JPETAB 109,189,53
scu-rat LD50:800 mg/kg JPETAB 109,189,53
orl-mus LD50:385 mg/kg JPETAB 109,189,53
scu-mus LD50:350 mg/kg JPETAB 109,189,53
ivn-mus LD50:42 mg/kg 31ZPAG 2,77,66
ivn-rbt LD50:27,500 μg/kg JPETAB 109,189,53

SAFETY PROFILE: Poison by ingestion, subcutaneous, and intravenous routes. When heated to decomposition it emits toxic fumes of NO_x.

DBE835 CAS:70052-12-9 HR: 1
α-DFMO
mf: $C_6H_{12}F_2N_2O_2$ mw: 182.20

SYNS: α-DIFLUOROMETHYLORNITHINE □ 2-(DIFLUOROMETHYL)ORNITHINE □ RMI 71782

D

TOXICITY DATA WITH REFERENCE
dni-rat-ipr 2800 mg/kg BBACAQ 696,179,82
scu-rat TDLo:1 g/kg (16-20D preg):REP NTOTDY 7,57,85
scu-rat TDLo:2500 mg/kg (female 15-17D post):REP TJADAB 33,40C,86
scu-mus TDLo:800 mg/kg (female 8D post):REP BJPCBM 69,335P,80
iut-rat TDLo:50 μg/kg (female 4D post):REP CCPTAY 28,93,83
ipr-rat TDLo:1600 mg/kg (4-7D preg):REP CCPTAY 24,215,81
scu-rat TDLo:2500 mg/kg (female 15-17D post):TER TJADAB 33,40C,86
ivn-wmn LDLo:67 mg/kg/40M-C:CVS,BPR AIMEAS 105,141,86

SAFETY PROFILE: Human systemic effects by intravenous route: cardiomyopathy, pulse rate decrease, fall in blood pressure. An experimental teratogen. Other experimental reproductive effects. Mutation data reported. When heated to decomposition it emits toxic fumes of F^- and NO_x.

DBE875 **CAS:70711-41-0** **HR: 3**
DHAQ DIACETATE
mf: $C_{22}H_{28}N_4O_6 \cdot 2C_2H_4O_2$ mw: 564.66

SYNS: 5,8-BIS((2-((2-HYDROXYETHYL)AMINO)ETHYL)AMINO)-1,4-DIHYDROXY-9,10-ANTHRACENEDIONE DIACETATE □ 5,8-BIS((2-((2-HYDROXYETHYL)AMINO)ETHYL)AMINO)-1,4-DIHYDROXYANTHRAQUINONE 1,4-DIACETATE □ 1,4-DIHYDROXY-5,8-BIS(2-((2-HYDROXYETHYL)AMINO)ETHYLAMINO)-9,10-ANTHRACENEDIONE DIACETATE □ NSC 299195

TOXICITY DATA WITH REFERENCE
mmo-sat 6500 μmol/L CNREA8 41,376,81
mma-sat 6500 μmol/L CNREA8 41,376,81
cyt-ham:ovr 10 nmol/L CNREA8 41,376,81
sce-ham:ovr 10 nmol/L CNREA8 41,376,81
ipr-mus LD50:61,300 μg/kg NCISP* JAN86
ivn-dog LDLo:25 mg/kg NTIS** PB297-169

SAFETY PROFILE: Poison by intravenous and intraperitoneal routes. Mutation data reported. When heated to decomposition it emits toxic fumes of NO_x.

DBE885 **CAS:24477-37-0** **HR: 3**
DIABENOR
mf: $C_{20}H_{26}N_4O_5S$ mw: 434.56

PROP: A solid. Mp: 198°.

SYNS: 1-CYCLOHEXYL-3-((p-(2-(5-METHYL-3-ISOXAZOLECARBOXAMIDO)ETHYL)PHENYL)SULFONYL)UREA □ GLISOLAMIDE □ N-(4-(β-(5-METILOSSAZOL-3-CARBOSSAMIDO)-ETIL)-BENZENESOLFONIL)-N'-CICLOESIL-UREA □ P.M. 434,526

TOXICITY DATA WITH REFERENCE
orl-rat TDLo:1500 mg/kg (6-15D preg):TER BCFAAI 117,348,78
ipr-rat LD50:320 mg/kg BCFAAI 117,348,78
scu-rat LD50:900 mg/kg BCFAAI 117,348,78
ipr-mus LD50:230 mg/kg BCFAAI 117,348,78
scu-mus LD50:720 mg/kg BCFAAI 117,348,78

SAFETY PROFILE: Poison by intraperitoneal route. Moderately toxic by subcutaneous route. An experimental teratogen. When heated to decomposition it emits toxic fumes of SO_x and NO_x.

DBF000 **CAS:63019-65-8** **HR: 2**
N-1-DIACETAMIDOFLUORENE
mf: $C_{17}H_{15}NO_2$ mw: 265.33

SYNS: N-FLUOREN-1-YLDIACETAMIDE □ N-1-FLUORENYLDIACETAMIDE

TOXICITY DATA WITH REFERENCE
orl-rat TDLo:5200 mg/kg/52W-C:ETA JNCIAM 24,149,60

SAFETY PROFILE: Questionable carcinogen with experimental tumorigenic data. When heated to decomposition it emits toxic fumes of NO_x.

DBF200 **CAS:642-65-9** **HR: 3**
2-DIACETAMIDOFLUORENE
mf: $C_{17}H_{15}NO_2$ mw: 265.33

SYNS: N-ACETYL-N-9H-FLUOREN-2-YL-ACETAMIDE □ N-DIACETYL-2-AMINOFLUORENE □ N,N-DIACETYL-2-AMINOFLUORENE □ 2-DIACETYLAMINOFLUORENE □ N,N-DIACETYL-2-FLUORENAMINE □ F-diAA □ 2-FLUORENYLDIACETAMIDE □ N-FLUOREN-2-YLDIACETAMIDE □ N-2-FLUORENYLDIACETAMIDE

TOXICITY DATA WITH REFERENCE
orl-rat TDLo:970 mg/kg/23W-C:NEO JNCIAM 10,1201,50
skn-rat TDLo:290 mg/kg/77W-I:CAR JNCIAM 10,1201,50
orl-ham TDLo:6174 mg/kg/49W-C:CAR GANNA2 59,239,68
orl-rat TD:475 mg/kg/5W-C:ETA AICCA6 20,1364,64
orl-rat TD:4000 mg/kg/23W-C:ETA CNREA8 7,730,47
orl-rat TD:945 mg/kg/9W-I:ETA JNCIAM 34,697,65
orl-rat TD:1540 mg/kg/19W-I:ETA CNREA8 28,2177,68
orl-rat TD:613 mg/kg/8W-I:ETA JNCIAM 31,1407,63
orl-rat TD:1138 mg/kg/16W-I:ETA JNCIAM 29,933,62
orl-rat TD:1365 mg/kg/16W-I:ETA EJCAAH 12,137,76

SAFETY PROFILE: Suspected carcinogen with experimental carcinogenic, neoplastigenic, and tumorigenic data. When heated to decomposition it emits toxic fumes of NO_x.

DBF400 **CAS:51325-35-0** **HR: 2**
2,4-DIACETAMIDO-6-(5-NITRO-2-FURYL)-s-TRIAZINE
mf: $C_{11}H_{10}N_6O_5$ mw: 306.27

SYN: N,N'-(6-(5-NITRO-2-FURYL)-s-TRIAZINE-2,4-DIYL)BISACETAMIDE

TOXICITY DATA WITH REFERENCE
mma-sat 100 ng/plate MUREAV 48,295,77
orl-rat TDLo:57 g/kg/46W-C:CAR JNCIAM 51,403,73

CONSENSUS REPORTS: EPA Genetic Toxicology Program.

SAFETY PROFILE: Questionable carcinogen with experimental carcinogenic data. Mutation data reported. When heated to decomposition it emits toxic fumes of NO_x.

DBF600 **CAS:102-62-5** **HR: 2**
1,3-DIACETIN
mf: $C_7H_{12}O_5$ mw: 176.19

PROP: Crystals. D: 1.178 @ 15°/15°, bp: 280°, mp: 40°.

SYNS: 1,3-DIACETATE GLYCEROL ☐ 1,2-DIACETATE 1,2,3-PROPANETRIOL ☐ DIACETIN ☐ 1,2-DI-ACETIN ☐ 2,3-DIACETIN ☐ DIACETYL GLYCERINE ☐ DIGLYCERIDE ACETIC ACID ☐ GLYCEROL DIACETATE ☐ GLYCERYL-1,3-DIACETATE ☐ (HYDROXYMETHYL) ETHYLENE ACETATE

TOXICITY DATA WITH REFERENCE
orl-mus LD50:8500 mg/kg JPETAB 65,89,39
scu-mus LD50:3500 mg/kg JPETAB 65,89,39
ivn-mus LD50:2300 mg/kg JPETAB 65,89,39

CONSENSUS REPORTS: Reported in EPA TSCA Inventory.

SAFETY PROFILE: Moderately toxic by subcutaneous and intravenous routes. Mildly toxic by ingestion. When heated to decomposition it emits acrid smoke and irritating fumes.

DBF750 **CAS:123-42-2** **HR: 3**
DIACETONE ALCOHOL
DOT: UN 1148
mf: $C_6H_{12}O_2$ mw: 116.18

PROP: Liquid; oily; faint pleasant odor. Mp: −47 to −54°, bp: 164°, flash p: 148°F, d: 0.9306 @ 25°/4°, autoign temp: 1118°F, vap d: 4.00, vap press: 1.1 mm @ 20°, lel: 1.8%, uel: 6.9%, flash p: (acetone free): 136°F. Sol in water.

SYNS: DIACETONALCOHOL (DUTCH) ☐ DIACETONALCOOL (ITALIAN) ☐ DIACETONALKOHOL (GERMAN) ☐ DIACETONE ☐ DIACETONE-ALCOOL (FRENCH) ☐ DIKETONE ALCOHOL ☐ 4-HYDROXY-2-KETO-4-METHYLPENTANE ☐ 4-HYDROXY-4-METHYL-PENTAN-2-ON (GERMAN, DUTCH) ☐ 4-HYDROXY-4-METHYLPENTANONE-2 ☐ 4-HYDROXY-4-METHYL PENTAN-2-ONE ☐ 4-HYDROXY-4-METHYL-2-PENTANONE ☐ 4-IDROSSI-4-METIL-PENTAN-2-ONE (ITALIAN) ☐ 2-METHYL-2-PENTANOL-4-ONE ☐ PYRANTON ☐ TYRANTON

TOXICITY DATA WITH REFERENCE
eye-hmn 100 ppm/15M JIHTAB 28,262,46
skn-rbt 10 mg/24H open JIHTAB 30,63,48
skn-rbt 500 mg open MLD UCDS** 6/29/59
eye-rbt 5 mg SEV AJOPAA 29,1363,46
ihl-hmn TCLo:100 ppm:EYE,CNS,GIT JIHTAB 30,63,48
ihl-hmn TCLo:400 ppm:PUL NPIRI* 1,21,74
orl-rat LD50:4000 mg/kg JIHTAB 30,63,48
ipr-mus LD50:933 mg/kg SCCUR* -,3,61
skn-rbt LD50:13,500 mg/kg NPIRI* 1,21,74

CONSENSUS REPORTS: Reported in EPA TSCA Inventory.

OSHA PEL: TWA 50 ppm
ACGIH TLV: TWA 50 ppm
DFG MAK: 50 ppm (240 mg/m^3)
NIOSH REL: (Ketones) TWA 240 mg/m^3
DOT CLASSIFICATION: 3; *Label:* Flammable Liquid

SAFETY PROFILE: Moderately toxic by ingestion and intraperitoneal routes. Mildly toxic by skin contact. Human systemic effects by inhalation: headache, nausea or vomiting, eye and pulmonary changes. A skin, mucous membrane, and severe eye irritant. Can cause anemia and damage to liver and kidneys. Narcotic in high concentration. Flammable liquid when exposed to heat or flame; can react with oxidizing materials. Explosive in the form of vapor when exposed to heat or flame. To fight fire, use alcohol foam, foam, CO_2, dry chemical. When heated to decomposition it emits acrid smoke and irritating fumes. See also KETONES.

For occupational chemical analysis use NIOSH: Alcohols III, 1402.

DBF800 **CAS:1067-33-0** **HR: 3**
DIACETOXYDIBUTYL STANNANE
mf: $C_{12}H_{24}O_4Sn$ mw: 351.05

PROP: Clear, colorless oily liquid with a slight acetic acid odor. Bp: 144.5–145.5° @ 10 mm (decomp), mp: 8.5–10°, fp: 5–10°, flash p: 290°F (OC), d: 1.31 @ 25°, vap d: 12.1. Sol in org solvs.

SYNS: BA 2726 ☐ BIS(ACETYLOXY)DIBUTYLSTANNANE ☐ DIACETOXYBUTYLTIN ☐ DIACETOXYDIBUTYLTIN ☐ DIBUTYL TIN DIACETATE ☐ FOMREZ SUL-3 ☐ NCI-C02028 ☐ T 1 (Catalyst)

TOXICITY DATA WITH REFERENCE
orl-rat TDLo:110 mg/kg (female 7-17D post):REP AECTCV 23,216,92
orl-rat TDLo:15,200 µg/kg (female 8D post):REP BECTA6 49,715,92
orl-rat LD50:32 mg/kg NCILB* NIH-NCI-E-C-72-3252,73
orl-mus LD50:46 mg/kg NCILB* NIH-NCI-E-C-72-3252,73
ivn-mus LD50:18 mg/kg CSLNX* NX#02348

CONSENSUS REPORTS: NCI Carcinogenesis Bioassay (feed); Inadequate Studies: mouse, rat NCITR* NCI-CG-TR-183,79. Reported in EPA TSCA Inventory.

OSHA PEL: TWA 0.1 mg(Sn)/m^3 (skin)
ACGIH TLV: TWA 0.1 mg(Sn)/m^3 (skin) (Proposed: TWA 0.1 mg(Sn)/m^3; STEL 0.2 mg(Sn)/m^3 (skin))
NIOSH REL: (Organotin Compounds) TWA 0.1 mg(Sn)/m^3

SAFETY PROFILE: Poison by ingestion and intravenous routes. Experimental reproductive effects. Combustible when exposed to heat or flame; can react with oxidizing materials. To fight fire, use water, foam, CO_2, dry chemical. When heated to decomposition it emits acrid smoke and irritating fumes. See also TIN COMPOUNDS.

For occupational chemical analysis use NIOSH: Organotin Compounds 5504.

DBF875 **CAS:10140-75-7** **HR: 1**
1,1-DIACETOXY-2,3-DICHLOROPROPANE
mf: $C_7H_{10}Cl_2O_4$ mw: 229.07

SYN: 1,1-PROPANEDIOL, 2,3-DICHLORO-, DIACETATE

TOXICITY DATA WITH REFERENCE
skn-rbt 10 mg/24H open SEV AIHAAP 23,95,62
orl-rat LD50:320 mg/kg AIHAAP 23,95,62
skn-rbt LD50:1000 mg/kg AIHAAP 23,95,62

SAFETY PROFILE: A poison by ingestion. Mildly toxic

by skin contact. A severe skin irritant. When heated to decomposition it emits toxic fumes of Cl^-.

DBG200 **CAS:73785-34-9** **HR: 2**
trans-7,8-DIACETOXY-7,8-DIHYDROBENZO(a)PYRENE
mf: $C_{24}H_{16}O_4$ mw: 368.40

SYN: trans-BP-7,8-DIHYDRODIOL DIACETATE

TOXICITY DATA WITH REFERENCE
scu-mus TDLo: 13 mg/kg: NEO JJIND8 64,617,80

SAFETY PROFILE: Questionable carcinogen with experimental neoplastigenic data. When heated to decomposition it emits acrid smoke and irritating fumes.

DBG499 **HR: 3**
2-((2,4-DIACETYL-5-BENZOFURANYL)OXY)TRIETHYLAMINE HYDROCHLORIDE
mf: $C_{18}H_{23}NO_4 \cdot ClH$ mw: 353.88

TOXICITY DATA WITH REFERENCE
orl-mus LDLo: 150 mg/kg APFRAD 41,603,83

SAFETY PROFILE: A poison by ingestion. When heated to decomposition it emits toxic fumes of NO_x and HCl.

DBG900 **CAS:922-89-4** **HR: 3**
N,N′-DIACETYL-N,N′-DINITRO-1,2-DIAMINOETHANE
mf: $C_6H_{10}N_4O_6$ mw: 234.17

$CH_3CO(NO_2)C_2H_4N(NO_2)OCCH_3$

PROP: Plates from Me_2CO. Mp: 136° (decomp).

SAFETY PROFILE: A powerful oxidizer. Decomposes violently at 142°C. Upon decomposition it emits toxic fumes of NO_x. See also NITRO COMPOUNDS of AROMATIC HYDROCARBONS.

DBH000 **CAS:95-45-4** **HR: 3**
DIACETYL DIOXIME
mf: $C_4H_8N_2O_2$ mw: 116.14

PROP: Triclinic crystals from EtOH (aq). Mp: 238–240°. Sol in EtOH, Me_2CO, Et_2O, and alk soln; very sltly sol in H_2O and $CHCl_3$.

SYNS: 2,3-DIISONITROSOBUTANE □ DIMETHYLGLYOXIME

TOXICITY DATA WITH REFERENCE
otr-ham: emb 100 μg/L FCTXAV 18,289,80
orl-rat LDLo: 250 mg/kg NCNSA6 5,26,53

CONSENSUS REPORTS: Reported in EPA TSCA Inventory.

SAFETY PROFILE: Poison by ingestion. Mutation data reported. When heated to decomposition it emits toxic fumes of NO_x.

DBH200 **CAS:17598-65-1** **HR: 3**
DIACETYLLANATOSIDE
mf: $C_{47}H_{74}O_{19}$ mw: 943.11

PROP: Crystals from methanol, decomp 265–268°. One part dissolves in 5000 parts water, 200 parts methanol, and 2500 parts ethanol; very sltly sol in chloroform; practically insol in ether.

SYNS: DEACETYLLANATOSIDE C □ DESACE □ DESACETYLLANATOSIDE C □ DESACI □ DESCETYLDIGILANIDE C □ DESLANATOSIDE □ DESLANOSIDE □ DIACETYLLANATOSID C (GERMAN) □ SEDIRANIDO

TOXICITY DATA WITH REFERENCE
ipr-mus LD50: 10,360 μg/kg NIIRDN 6,489,82
ivn-mus LD50: 8100 μg/kg NIIRDN 6,489,82
ivn-cat LD50: 230 μg/kg 85ELDJ -,190,63

SAFETY PROFILE: Poison by intravenous and intraperitoneal routes. When heated to decomposition it emits acrid smoke and fumes. See also LANATOSIDE A and LANATOSIDE C.

DBH400 **CAS:1502-95-0** **HR: 3**
DIACETYLMORPHINE HYDROCHLORIDE
mf: $C_{21}H_{23}NO_5 \cdot ClH$ mw: 405.91

PROP: Prisms. Mp: 231–232° (decomp). Sol in H_2O and EtOH.

SYNS: DIAMORPHINE HYDROCHLORIDE □ HEROIN HYDROCHLORIDE

TOXICITY DATA WITH REFERENCE
scu-mus TDLo: 80 mg/kg (9D preg): TER DGDFA5 22,61,80
ipr-mus TDLo: 65 mg/kg (female 9D post): TER TJADAB 31,235,85
scu-ham TDLo: 148 mg/kg (female 8D post): TER AJOGAH 123,705,75
ipr-mus LD50: 240 mg/kg AIPTAK 122,434,59
ivn-mus LD50: 38 mg/kg TXAPA9 6,334,64

SAFETY PROFILE: Poison by intraperitoneal and intravenous routes. An experimental teratogen. A narcotic. Addictive. When heated to decomposition it emits very toxic fumes of NO_x and HCl. See also HEROIN and MORPHINE.

DBH800 **CAS:73622-67-0** **HR: 3**
3,4-DI(ACETYLTHIOMETHYL)-5-HYDROXY-6-METHYLPYRIDINE HYDROBROMIDE
mf: $C_{12}H_{15}NO_3S_2 \cdot BrH$ mw: 366.32

SYNS: 4,5-DI(MERCAPTOMETHYL)-2-METHYL-3-PYRIDINOL DITHIOACETATE HYDROBROMIDE □ 4,5-DIMERCAPTOPYRIDOXINDI-THIOACETAT HYDROBROMID (GERMAN)

TOXICITY DATA WITH REFERENCE
orl-rat LD50: 870 mg/kg ARZNAD 11,922,61
scu-rat LD50: 132 mg/kg ARZNAD 11,922,61
ivn-rat LD50: 41 mg/kg ARZNAD 11,922,61
orl-mus LD50: 409 mg/kg ARZNAD 11,922,61
scu-mus LD50: 121 mg/kg ARZNAD 11,922,61
ivn-mus LD50: 49 mg/kg ARZNAD 11,922,61

SAFETY PROFILE: Poison by subcutaneous and intravenous routes. Moderately toxic by ingestion. When heated to decomposition it emits very toxic fumes of NO_x, SO_x, and HBr. See also MERCAPTANS and BROMIDES.

DBI099 **CAS:10311-84-9** **HR: 3**
DIALIFOR
mf: $C_{14}H_{17}NO_4PS_2$ mw: 393.86

PROP: Crystals from toluene/hexane. Mp: 67–69°.

SYNS: S-(2-CHLORO-1-(1,3-DIHYDRO-1,3-DIOXO-2H-ISOINDOL-2-YL)ETHYL)-O,O-DIETHYL PHOSPHORODITHIOATE □ S-(2-CHLORO-1-PHTHALIMIDOETHYL)-O,O-DIETHYL PHOSPHORODITHIOATE □ O,O-DIETHYL-S-(2-CHLORO-1-PHTHALIMIDOETHYL)PHOSPHORODITHIOATE □ ENT 27,320 □ HERCULES 14503 □ PHOSPHORODITHIOIC ACID-S-(2-CHLORO-1-(1,3-DIHYDRO-1,3-DIOXO-2H-ISOINDOL-2-YL))ETHYL-O,O-DIETHYL ESTER □ PHOSPHORODITHIOIC ACID-S-((2-CHLORO-1-PHTHALIMIDOETHYL)-O,)-DIETHYL ESTER □ TORAK

TOXICITY DATA WITH REFERENCE
orl-ham TDLO:100 mg/kg (8D preg):TER,REP TXAPA9 16,24,70
orl-ham TDLO:200 mg/kg (7D preg):TER,REP TXAPA9 16,24,70
orl-rat LD50:5 mg/kg BESAAT 15,122,69
skn-rat LD50:28 mg/kg FAATDF 7,299,86
orl-mus LD50:39 mg/kg BESAAT 15,122,69
orl-dog LD50:94 mg/kg 85DPAN -,-,71/76
orl-rbt LD50:35 mg/kg BESAAT 15,122,69

SAFETY PROFILE: Poison by ingestion and skin contact. An experimental teratogen. Other experimental reproductive effects. When heated to decomposition it emits toxic fumes of SO_x, PO_x, and NO_x.

DBI159 **HR: 3**
DIALKYLZINCS
mf: R_2Zn

SAFETY PROFILE: Flammable when exposed to heat or flame. Some may explode on contact with water. Potentially explosive reaction with acyl halides (during the production of ketones); allyl chlorides; and alcohols. When heated to decomposition it emits toxic fumes of ZnO. See also ZINC COMPOUNDS.

DBI200 **CAS:2303-16-4** **HR: 3**
DIALLATE
mf: $C_{10}H_{17}Cl_2NOS$ mw: 270.24

PROP: Brown liquid. Bp: 150° @ 9 mm, mp: 25–30°. Sltly sol in water; sol in organic solvents.

SYNS: AVADEX □ CP 15,336 □ DATC □ 2,3-DCDT □ DIALLAAT (DUTCH) □ DIALLAT (GERMAN) □ 2,3-DICHLORALLYL-N,N-(DIISOPROPYL)-THIOCARBAMAT (GERMAN) □ S-(2,3-DICHLOR-ALLYL)-N,N-DIISOPROPYL-MONOTHIOCARBAMAAT (DUTCH) □ DICHLOROALLYL DIISOPROPYLTHIOCARBAMATE □ S-2,3-DICHLOROALLYL DIISOPROPYLTHIOCARBAMATE □ 2,3-DICHLOROALLYL-N,N-DIISOPROPYLTHIOLCARBAMATE □ 2,3-DICHLORO-2-PROPENE-1-THIOL DIISOPROPYLCARBAMATE □ S-(2,3-DICHLORO-2-PROPENYL)ESTER, BIS(1-METHYLETHYL) CARBAMOTHIOIC ACID □ S-(2,3-DICLORO-ALLIL)-N,N-DIISOPROPIL-MONOTIOCARBAMMATO (ITALIAN) □ DI-ISOPROPYLTHIOLOCARBAMATE de S-(2,3-DICHLOROALLYLE) (FRENCH) □ PYRADEX □ RCRA WASTE NUMBER U062

TOXICITY DATA WITH REFERENCE
mmo-sat 500 μg/plate MUREAV 85,45,81
cyt-ham:ovr 200 μmol/L MUREAV 85,45,81
orl-rat TDLo:4095 mg/kg/78W-C:ETA JJIND8 67,75,81
orl-mus TDLo:68 g/kg/84W-C:CAR JNCIAM 42,1101,69
unr-mus TDLo:1 mg/kg:ETA JPFCD2 15,929,80
orl-rat LD50:395 mg/kg RREVAH 10,97,65
skn-rat LD50:2124 mg/kg FAATDF 7,299,86
orl-dog LD50:510 mg/kg 28ZEAL 5,71,76
skn-rbt LD50:2000 mg/kg WRPCA2 9,119,70
orl-gpg LD50:420 mg/kg HYSAAV 34(10-12),356,69

CONSENSUS REPORTS: IARC Cancer Review: Group 3 IMEMDT 7,56,87; Animal Limited Evidence IMEMDT 30,235,83; Animal Sufficient Evidence IMEMDT 12,69,76. EPA Genetic Toxicology Program. Community Right-To-Know List.

SAFETY PROFILE: Poison by ingestion. Moderately toxic by skin contact. Questionable carcinogen with experimental carcinogenic and tumorigenic data. Mutation data reported. When heated to decomposition it emits very toxic fumes of Cl^-, NO_x, and SO_x. See also CARBAMATES and ALLYL COMPOUNDS.

DBI600 **CAS:124-02-7** **HR: 3**
DIALLYLAMINE
DOT: UN 2359
mf: $C_6H_{11}N$ mw: 97.18

PROP: Liquid, sol in water. D: 0.7889 @ 20°, bp: 112°, fp: −100°, flash p.: 69.8°F.

SYNS: DI-2-PROPENYLAMINE □ N-2-PROPENYL-2-PROPEN-1-AMINE

TOXICITY DATA WITH REFERENCE
skn-rbt 100 μg/24H open AIHAAP 23,95,62
eye-rbt 50 mg/20S rns SEV AEHLAU 1,343,60
ihl-man TCLo:5 ppm/5M:PUL,EYE AEHLAU 1,343,60
orl-rat LD50:650 mg/kg AIHAAP 23,95,62
ihl-rat LC50:795 ppm/8H AEHLAU 1,343,60
orl-mus LD50:355 mg/kg 85GMAT-,43,82
ipr-mus LD50:187 mg/kg AEHLAU 1,343,60
skn-rbt LD50:280 mg/kg AIHAAP 23,95,62
ihl-mam LC50:2100 mg/m³ TPKVAL 14,80,75

CONSENSUS REPORTS: Reported in EPA TSCA Inventory.

DOT CLASSIFICATION: 3; *Label:* Flammable Liquid

SAFETY PROFILE: Poison by ingestion, skin contact, and intraperitoneal routes. Moderately toxic by ingestion and inhalation. Human systemic effects by inhalation route: eye lacrimation, and changes in the trachea or bronchi. A skin and severe eye irritant. A dangerous fire hazard when exposed to heat or flame. When heated to decomposition it emits toxic fumes of NO_x. See also AMINES and ALLYL COMPOUNDS.

DBI800 **CAS:6392-46-7** **HR: 3**
4-DIALLYLAMINO-3,5-DIMETHYLPHENYL-N-METHYLCARBAMATE
mf: $C_{16}H_{22}N_2O_2$ mw: 274.40

PROP: Powder. Insol in water; sol in alc and benzene.

SYNS: ALLYLOXYCARB □ ALLYXYCARB □ APC □ BAY 50282 □ 4-(DIALLYLAMINO)-3,5-XYLENOL METHYLCARBAMATE (ester) □ 4-DIALLYL-AMINO-3,5-XYLYL N-METHYLCARBAMATE □ 3,5-DIMETHYL-4-

DIALLYLAMINOPHENYL-N-METHYLCARBAMATE □ ENZOSE □ HYDROL

TOXICITY DATA WITH REFERENCE
orl-rat LD50:89 mg/kg TXAPA9 21,315,72
orl-mus LDLo:62 mg/kg AECTCV 14,111,85
orl-bwd LD50:13 mg/kg JAFCAU 15,287,67

SAFETY PROFILE: Poison by ingestion. An insecticide. When heated to decomposition it emits toxic fumes of NO_x. See also CARBAMATES, ALLYL COMPOUNDS, and ESTERS.

DBJ100 CAS:13988-24-4 HR: 3
1,1-DIALLYL-3-(1,4-BENZODIOXAN-2-YLMETHYL)-3-METHYLUREA
mf: $C_{17}H_{22}N_2O_3$ mw: 302.41

SYNS: A 2275 □ 2-(N,N-DIALLYLCARBAMYLMETHYL)AMINOMETHYL-1,4-BENZODIOXAN □ N-((2,3-DIHYDRO-1,4-BENZODIOXIN-2-YL) METHYL)-N-METHYL-N',N'-DI-2-PROPENYL-UREA (9CI)

TOXICITY DATA WITH REFERENCE
orl-mus LD50:447 mg/kg JAPMA8 48,409,59
ipr-mus LD50:230 mg/kg JAPMA8 48,409,59
ivn-mus LD50:55 mg/kg JAPMA8 48,409,59

SAFETY PROFILE: Poison by intravenous and intraperitoneal routes. Moderately toxic by ingestion. When heated to decomposition it emits toxic fumes of NO_x. See also ALLYL COMPOUNDS.

DBJ200 CAS:538-08-9 HR: 3
DIALLYLCYANAMIDE
mf: $C_7H_{10}N_2$ mw: 122.19

PROP: Colorless, mobile liquid when pure. Mp: <−70°, bp: 105–110° @ 18 mm, (slight decomp), d: 0.9021, vap d: 4.1.

SYN: N-CYANODIALLYLAMINE

TOXICITY DATA WITH REFERENCE
ipr-mus LDLo:125 mg/kg CBCCT* 4,318,52

CONSENSUS REPORTS: Reported in EPA TSCA Inventory.

SAFETY PROFILE: Poison by intraperitoneal route. Possibly more toxic than cyanamide. When heated to decomposition or on contact with acid or acid fumes it emits highly toxic fumes of CN^-. See CYANAMIDE, ALLYL COMPOUNDS, and AMINES.

DBJ400 CAS:17381-88-3 HR: 3
DIALLYLDIBROMO STANNANE
mf: $C_6H_{10}Br_2Sn$ mw: 360.67

SYN: DIALLYLTIN DIBROMIDE

TOXICITY DATA WITH REFERENCE
ivn-mus LD50:18 mg/kg CSLNX* NX#03064

OSHA PEL: TWA 0.1 mg(Sn)/m³ (skin)
ACGIH TLV: TWA 0.1 mg(Sn)/m³ (skin) (Proposed: TWA 0.1 mg(Sn)/m³; STEL 0.2 mg(Sn)/m³ (skin))
NIOSH REL: (Organotin Compounds) TWA 0.1 mg(Sn)/m³

SAFETY PROFILE: Poison by intravenous route. When heated to decomposition it emits toxic fumes of Br^-. See also ALLYL COMPOUNDS, BROMIDES, and TIN COMPOUNDS.

For occupational chemical analysis use NIOSH: Organotin Compounds 5504.

DBJ600 CAS:37764-25-3 HR: 2
N,N-DIALLYDICHLOROACETAMIDE
mf: $C_8H_{11}Cl_2NO$ mw: 208.10

PROP: Clear viscous liquid. D: 1.202 @ 20°/20°. Sltly sol in H_2O.

SYNS: COMPOUND R-25788 □ N,N-DIALLYL-2,2-DICHLOROACETAMIDE □ 2,2-DICHLORO-N,N-DI-2-PROPENYLACETAMIDE □ R-25788 □ STAUFFER R-25788

TOXICITY DATA WITH REFERENCE
orl-rat LD50:2 g/kg 85ARAE 3,108,76/77
ihl-rat LC50:5500 mg/m³/1H PEMNDP 9,251,91

CONSENSUS REPORTS: Reported in EPA TSCA Inventory.

SAFETY PROFILE: Moderately toxic by ingestion. When heated to decomposition it emits very toxic fumes of Cl^- and NO_x. See also ALLYL COMPOUNDS.

DBK000 CAS:557-40-4 HR: 3
DIALLYL ETHER
DOT: UN 2360
mf: $C_6H_{10}O$ mw: 98.16

PROP: Liquid, odor of radishes. Bp: 94.3°, d: 0.805, vap d: 3.38, flash p: 20°F (OC).

SYNS: ALLYLETHER □ 3,3'-OXYBIS(1-PROPENE) □ PROPENYL ETHER

TOXICITY DATA WITH REFERENCE
skn-rbt 500 mg/24H MLD 85JCAE-,250,86
eye-rbt 100 mg/24H MOD 85JCAE-,250,86
eye-rbt 20 mg open JIHTAB 31,60,49
orl-rat LD50:320 mg/kg JPMSAE 63,1068,74
skn-rbt LD50:600 mg/kg JIHTAB 31,60,49

CONSENSUS REPORTS: Reported in EPA TSCA Inventory.

DOT CLASSIFICATION: 3; *Label:* Flammable Liquid, Poison

SAFETY PROFILE: Poison by ingestion. Moderately toxic by skin contact. A skin and eye irritant. A dangerous fire hazard when exposed to heat, flame, or oxidizing materials. To fight fire, use alcohol foam. Reacts with air to form explosive peroxides. Violent explosions have occurred during distillation. When heated to decomposition it emits acrid smoke and fumes. See also ALLYL COMPOUNDS and ETHERS.

DBK100 CAS:5164-11-4 HR: 2
1,1-DIALLYLHYDRAZINE
mf: $C_6H_{12}N_2$ mw: 112.20

SYNS: 1,1-DAH □ DIALLYLHYDRAZINE □ HYDRAZINE, 1,1-DI-2-PROPENYL-(9CI)

TOXICITY DATA WITH REFERENCE
orl-mus TDLo:56,000 mg/kg/2Y-C:CAR ANTRD4 1,259,81
orl-mus TD:59,220 mg/kg/90W-C:CAR ANTRD4 1,259,81

SAFETY PROFILE: Questionable carcinogen with experimental carcinogenic data. When heated to decomposition it emits toxic fumes of NO_x.

DBK120 CAS:26072-78-6 HR: 2
1,2-DIALLYLHYDRAZINE DIHYDROCHLORIDE
mf: $C_6H_{12}N_2 \cdot 2ClH$ mw: 185.12

SYN: 1,2-DAH HYDROCHLORIDE

TOXICITY DATA WITH REFERENCE
orl-mus TDLo:78,176 mg/kg/80W-C:CAR ONCOBS 39,104,82
orl-mus TD:89,152 mg/kg/80W-C:CAR ONCOBS 39,104,82

SAFETY PROFILE: Questionable carcinogen with experimental carcinogenic data. When heated to decomposition it emits toxic fumes of NO_x.

DBK200 CAS:999-21-3 HR: 3
DIALLYL MALEATE
mf: $C_{10}H_{12}O_4$ mw: 196.22

PROP: Liquid. Vap d: 6.6.

SYNS: MALEIC ACID, DIALLYL ESTER □ SIPOMER DAM

TOXICITY DATA WITH REFERENCE
skn-rbt 10 mg/24H open MLD JIHTAB 31,60,49
skn-rbt 500 mg MOD SCCUR* -,3,61
eye-rbt 100 mg AJOPAA 29,1363,46
orl-rat LD50:300 mg/kg JIHTAB 31,60,49
orl-mus LD50:493 mg/kg SCCUR* -,3,61
ipr-mus LD50:160 mg/kg SCCUR* -,3,61
skn-rbt LD50:1150 mg/kg JIHTAB 31,60,49

CONSENSUS REPORTS: Reported in EPA TSCA Inventory.

SAFETY PROFILE: Poison by ingestion and intraperitoneal routes. Moderately toxic by skin contact. A skin and eye irritant. When heated to decomposition it emits acrid smoke and irritating fumes. See also ALLYL COMPOUNDS and ESTERS.

DBK400 CAS:15180-03-7 HR: 3
N,N'-DIALLYLNORTOXIFERINIUM DICHLORIDE
mf: $C_{44}H_{50}N_4O_2 \cdot 2Cl$ mw: 737.88

SYNS: ALCUONIUM DICHLORIDE □ ALCURONIUM CHLORIDE □ ALLOFERIN □ DIALFERIN □ DIALLYLNORTOXIFERINE DICHLORIDE □ 4,4'-DIDEMETHYL-4,4'-DI-2-PROPENYLTOXIFERINE I DICHLORIDE □ RO 4-3816

TOXICITY DATA WITH REFERENCE
orl-rat LD50:27,600 μg/kg OYYAA2 3,390,69
ipr-rat LD50:270 μg/kg GNRIDX 1,349,67
scu-rat LD50:280 μg/kg OYYAA2 3,390,69
orl-mus LD50:38,500 μg/kg GNRIDX 1,349,67
ipr-mus LD50:610 μg/kg OYYAA2 3,390,69
scu-mus LD50:610 μg/kg GNRIDX 1,349,67
ivn-mus LD50:240 μg/kg OYYAA2 3,390,69

SAFETY PROFILE: Poison by ingestion, intraperitoneal, subcutaneous, and intravenous routes. Used as a skeletal muscle relaxant. When heated to decomposition it emits very toxic fumes of NO_x and Cl^-. See also ALLYL COMPOUNDS.

D

DBK800 CAS:3382-99-8 HR: 2
2,6-DIALLYLPHENOL
mf: $C_{12}H_{10}O$ mw: 170.22

TOXICITY DATA WITH REFERENCE
skn-mus TDLo:4600 mg/kg/12W-I:NEO CNREA8 19,413,59

SAFETY PROFILE: Questionable carcinogen with experimental neoplastigenic data by skin contact. When heated to decomposition it emits acrid smoke and irritating fumes. See also ALLYL COMPOUNDS and PHENOL.

DBL000 CAS:23679-20-1 HR: 3
DIALLYL PHOSPHITE
mf: $C_6H_{11}O_3P$ mw: 162.12

$(H_2C{=}CHCH_2O)_2P(O)H$

PROP: Oil. D: 1.08 @ 25°/25°, bp: 80° @ 2 mm.

SYN: DI-2-PROPENYL PHOSPHONITE

SAFETY PROFILE: A poison. May explode during distillation. When heated to decomposition it emits toxic fumes of PO_x. See also ALLYL COMPOUNDS.

DBL200 CAS:131-17-9 HR: 2
DIALLYL PHTHALATE
mf: $C_{14}H_{14}O_4$ mw: 246.28

PROP: Nearly colorless, oily liquid. Bp: 157°, flash p: 330°F, d: 1.120 @ 20°/20°, vap d: 8.3.

SYNS: DAPON 35 □ DAPON R □ DI-2-PROPENYL ESTER, 1,2-BENZENEDICARBOXYLIC ACID □ NCI-C50657 □ PHTHALIC ACID, DIALLYL ESTER □ o-PHTHALIC ACID, DIALLYL ESTER

TOXICITY DATA WITH REFERENCE
eye-rbt 500 mg AJOPAA 29,1363,46
mma-mus:lyms 67,200 μg/L SCIEAS 236,933,87
orl-rat LD50:770 mg/kg
cyt-ham:ovr 200 mg/L SCIEAS 236,933,87
orl-rat TDLo:52 g/kg/2Y-I:CAR EVHPAZ 65,271,86
orl-mus TDLo:156 g/kg/2Y-I:CAR EVHPAZ 65,271,86
orl-rat LD50:656 mg/kg NTPTR* NTP-TR-284,85
skn-rbt LD50:3400 mg/kg 14CYAT 2,1904,63
ipr-mus LD50:700 mg/kg 14CYAT 2,1904,63
skn-rbt LDLo:2800 mg/kg FEPRA7 5,191,46
scu-rbt LDLo:1000 mg/kg FEPRA7 5,191,46

CONSENSUS REPORTS: NTP Carcinogenesis Studies

(gavage); Equivocal Evidence: rat NTPTR* NTP-TR-284,85. Reported in EPA TSCA Inventory.

SAFETY PROFILE: Suspected carcinogen with experimental carcinogenic data. Moderately toxic by ingestion, skin contact, intraperitoneal, and subcutaneous routes. An eye irritant. Mutation data reported. Combustible when exposed to heat or flame; can react with oxidizing materials. To fight fire use CO_2 or dry chemical. When heated to decomposition it emits acrid smoke and irritating fumes. See also ALLYL COMPOUNDS and ESTERS.

DBL300 CAS:91297-11-9 HR: 3
DIALLYL SELENIDE
mf: $C_6H_{10}Se$ mw: 184.11

SYN: SELENIDE, DIALLYL-

TOXICITY DATA WITH REFERENCE
ivn-mus LD50:100 mg/kg CSLNX* NX#09268

OSHA PEL: TWA 0.2 mg(Se)/m^3
ACGIH TLV: TWA 0.2 mg(Se)/m^3

SAFETY PROFILE: Poison by intravenous route. When heated to decomposition it emits toxic fumes of Se.

DBL700 CAS:21187-98-4 HR: 3
DIAMICRON
mf: $C_{15}H_{21}N_3O_3S$ mw: 323.45

PROP: Crystals from anhydrous ethanol. Mp: 180–182°.

SYNS: 1-(3-AZABICYCLO(3.3.0)OCT-3-YL)-3-(p-TOLYLSULFONYL) UREA □ GLICLAZIDE □ N-(((HEXAHYDROCYCLOPENTA(c)PYRROL-2(1H)-YL)AMINO)CARBONYL)-4-METHYL-BENZENESULFONAMIDE □ 1-(HEXAHYDROCYCLOPENTA(c)PYRROL-2(1H)-YL)-3-(p-TOLYLSULFONYL)UREA □ N-(4-METHYLBENZENESULFONYL)-N'-(3-AZABICYCLO (3.3.0)OCT-3-YL)UREA □ NORDIALEX □ S 852 □ S 1702 □ SE 1702

TOXICITY DATA WITH REFERENCE
orl-rat TDLo:4400 mg/kg (7-17D preg):REP YACHDS 9,3551,81
orl-rat TDLo:1100 mg/kg (female 7-17D post):REP YACHDS 9,3551,81
orl-rat TDLo:1 g/kg (female 17-22D post):REP YACHDS 9,3551,81
orl-rat TDLo:2200 mg/kg (7-17D preg):TER YACHDS 9,3551,81
orl-rat TDLo:4400 mg/kg (7-17D preg):TER YACHDS 9,3551,81
orl-rat LD50:5 g/kg YACHDS 8,2661,80
ivn-rat LD50:382 mg/kg YACHDS 8,2661,80
scu-mus LD50:1034 mg/kg YACHDS 8,2661,80
ivn-mus LD50:295 mg/kg YACHDS 8,2661,80

SAFETY PROFILE: Poison by intravenous route. Moderately toxic by subcutaneous route. Mildly toxic by ingestion. Experimental teratogenic and reproductive effects. When heated to decomposition it emits toxic fumes of SO_x and NO_x.

DBL800 CAS:140-64-7 HR: 3
DIAMIDINE
mf: $C_{19}H_{24}N_4O_2 \cdot C_4H_{12}O_8S_2$ mw: 592.75

PROP: Hygroscopic crystals with slt butyric odor. Mp: 180°. Sol in H_2O and glycerol.

SYNS: 4,4'-DIAMIDINODIPHENOXYPENTANE DI(β-HYDROXYETHANESULFONATE □ 4,4'-DIAMIDINO-α,ω-DIPHENOXYPENTANE ISETHIONATE □ LOMIDIN □ LOMIDINE □ LOMIDINE ISOETHIONATE □ M & B 800 □ p,p'-(PENTAMETHYLENEDIOXY)DIBENZAMIDINE BIS (β-HYDROXYETHANESULFONATE) □ PENTAMIDINE DIISETHIONATE □ PENTAMIDINE ISETHIONATE □ 2512 R.P. □ R.P. 2512 □ USAF XR-10

TOXICITY DATA WITH REFERENCE
dni-mus:leu 29,500 µg/L INNDDK 1,103,83
ims-wmn TDLo:6 mg/kg/D:BLD,SKN CANCAR 34,441,74
ipr-mus LD50:63 mg/kg ANTCAO 2,581,52
scu-mus LD50:120 mg/kg ANTCAO 2,581,52
ivn-mus LD50:15,100 µg/kg ANTCAO 2,581,52

SAFETY PROFILE: Poison by intraperitoneal, subcutaneous, and intravenous routes. Human systemic effects by intramuscular route: hemorrhage and dermatitis. Mutation data reported. When heated to decomposition it emits very toxic fumes of NO_x and SO_x.

DBM000 CAS:100-33-4 HR: 3
4,4'-DIAMIDINODIPHENOXYPENTANE
mf: $C_{19}H_{24}N_4O_2$ mw: 340.47

PROP: Plates from H_2O. Mp: 186° (decomp).

SYNS: 4,4'-DIAMIDINO-α,ω-DIPHENOXYPENTANE □ p,p'-(PENTAMETHYLENE-DIOXY)BIS-BENZAMIDINE □ p,p'-(PENTAMETHYLENEDIOXY)DIBENZAMIDINE □ 4,4'-(PENTAMETHYLENEDIOXY)DIBENZAMIDINE □ PENTAMIDINE □ 4,4'-(1,5-PENTANEDIYLBIS(OXY))BIS-BENZENECARBOXIMIDAMIDE, (9CI)

TOXICITY DATA WITH REFERENCE
ims-man TDLo:28 mg/kg/1W-I:KID,MET AIMDAP 145,2247,85
par-man TDLo:4 mg/kg/3D:CNS,MET LANCAO 246,338,44
ipr-mus LD50:50 mg/kg JMCMAR 18,794,75

SAFETY PROFILE: Poison by intraperitoneal route. Human systemic effects by intramuscular and parenteral routes: anesthesia, changes in kidney tubules, fever, and convulsions. When heated to decomposition it emits toxic fumes of NO_x.

DBM400 CAS:6275-69-0 HR: 3
4,4'-DIAMIDINO-1,3-DIPHENOXYPROPANE DIHYDROCHLORIDE
mf: $C_{17}H_{20}N_4O_2 \cdot 2ClH$ mw: 385.33

PROP: A solid. Mp: 202–204° (decomp).

SYNS: 4,4'-DIAMIDINO-α,γ-DIPHENOXYPROPANE DIHYDROCHLORIDE □ M & B 782 DIHYDROCHLORIDE □ PANAMIDIN DIHYDROCHLORIDE □ PROPAMIDINE DIHYDROCHLORIDE □ 4,4'-(1,3-PROPANEDIYLBIS(OXY))BIS-BENZENECARBOXIMIDAMIDE, DIHYDROCHLORIDE □ 4,4'-(TRIMETHYLENEDIOXY)DIBENZAMIDINE DIHYDROCHLORIDE

TOXICITY DATA WITH REFERENCE
ipr-mus LD50:40 mg/kg ANTCAO 2,581,52
scu-mus LD50:55 mg/kg ATMPA2 37,1,43
ivn-mus LD50:35,100 μg/kg ANTCAO 2,581,52

SAFETY PROFILE: Poison by intraperitoneal, subcutaneous, and intravenous routes. When heated to decomposition it emits very toxic fumes of HCl and NO_x.

DBM800 CAS:59-33-6 HR: 3
DIAMINIDE MALEATE
mf: $C_{17}H_{23}N_3O \cdot C_4H_4O_4$ mw: 401.51

PROP: A solid. Mp: 100–101°. Very sol in H_2O.

SYNS: AH □ ANISOPYRADAMINE □ ANTHISAN MALEATE □ ANTIHIST □ N-DIMETHYLAMINOETHYL-N-p-METHOXY-α-AMINOPYRIDINE MALEATE □ 2-((2-(DIMETHYLAMINO)ETHYL)(p-METHOXYBENZYL)AMINO)PYRIDINE BIMALEATE □ 2-((2-(DIMETHYLAMINO)ETHYL)(p-METHOXYBENZYL)AMINO)PYRIDINE MALEATE □ N,N-DIMETHYL-N'-(4-METHOXYBENZYL)-N'-(2-PYRIDYL)ETHYLENEDIAMINE MALEATE □ HISTATEX □ MEPYRAMINE MALEATE □ N-p-METHOXYBENZYL-N'-N'-DIMETHYL-N-α-PYRIDYLETHYLENEDIAMINE MALEATE □ MINIHIST □ NEOANTERGAN MALEATE □ PARAMAL □ PARAMINYL MALEATE □ PYMAFED □ PYRA MALEATE □ PYRANILAMINE MALEATE □ PYRANINYL □ PYRANISAMINE MALEATE □ PYRILAMINE MALEATE □ RENSTAMIN □ 2786 R.P. MALEATE □ STANGEN MALEATE □ STATOMIN MALEATE □ THYLOGEN MALEATE

TOXICITY DATA WITH REFERENCE
dns-rat:lvr 5 μmol/L ENMUDM 3,11,81
orl-mus TDLo:4200 mg/kg (female 1-21D post):REP ARZNAD 18,188,68
orl-mus TDLo:420 mg/kg (female 1-21D post):REP ARZNAD 18,188,68
orl-rat TDLo:100 mg/kg (1D preg):REP LIFSAK 2,303,63
orl-rat TDLo:77 g/kg/2Y-C:ETA FCTOD7 22,27,84
orl-chd LDLo:42 mg/kg LANCAO 2,809,52
orl-wmn TDLo:200 mg/kg JAMAAP 257,660,87
orl-rat LD50:365 mg/kg TXAPA9 1,42,59
scu-rat LD50:150 mg/kg CRSBAW 144,887,50
orl-mus LD50:220 mg/kg AIPTAK 155,47,65
ipr-mus LD50:102 mg/kg JPETAB 90,224,47
ivn-mus LD50:23 mg/kg AIPTAK 155,47,65
ivn-gpg LD50:24,400 μg/kg AIPTAK 113,313,58

CONSENSUS REPORTS: Reported in EPA TSCA Inventory.

SAFETY PROFILE: A human poison by ingestion. An experimental poison by ingestion, subcutaneous, intravenous, and intraperitoneal routes. Experimental reproductive effects. Questionable carcinogen with experimental tumorigenic data. Mutation data reported. An antihistamine. When heated to decomposition it emits toxic fumes of NO_x.

DBN000 CAS:3407-94-1 HR: 3
2,6-DIAMINOACRIDINE
mf: $C_{13}H_{11}N_3$ mw: 209.27

PROP: Yellow crystals from H_2O. Mp: 355°.

SYNS: ACRAMINE RED □ 2,6-ACRIDINEDIAMINE □ 3,7-DIAMINOACRIDINE □ DIFLAVINE (ACRIDINE)

TOXICITY DATA WITH REFERENCE
ipr-mus LD50:300 mg/kg BJEPA5 28,1,47
scu-mus LD50:130 mg/kg BJEPA5 28,1,47

SAFETY PROFILE: Poison by intraperitoneal and subcutaneous routes. When heated to decomposition it emits toxic fumes of NO_x.

D

DBN200 HR: 2
3,6-DIAMINOACRIDINE HYDROCHLORIDE HEMIHYDRATE
mf: $Cl_3H_{11}N_3 \cdot ClH \cdot 1/2H_2O$ mw: 254.74

SYNS: NCI-C04137 □ PROFLAVINE MONOHYDROCHLORIDE HEMIHYDRATE

TOXICITY DATA WITH REFERENCE
orl-rat TDLo:27 g/kg/109W-C:ETA NCITR* NCI-CG-TR-5,77
orl-mus TDLo:35 g/kg/104W-C:ETA NCITR* NCI-CG-TR-5,77

CONSENSUS REPORTS: IARC Cancer Review: Animal Inadequate Evidence IMEMDT 24,195,80. NCI Carcinogenesis Bioassay (feed); Inadequate Studies: mouse NCITR* NCI-CG-TR-5,77; Clear Evidence: rat NCITR* NCI-CG-TR-5,77.

SAFETY PROFILE: Questionable carcinogen with experimental tumorigenic data. When heated to decomposition it emits very toxic fumes of NO_x and Cl^-.

DBN600 CAS:92-62-6 HR: 3
3,6-DIAMINOACRIDINIUM
mf: $C_{13}H_{11}N_3$ mw: 209.27

PROP: Yellow crystals. Mp: 284–286°

SYNS: 3,6-ACRIDINEDIAMINE □ 2,8-DIAMINOACRIDINE □ 3,6-DIAMINOACRIDINE □ 2,8-DIAMINOACRIDINIUM □ 3,7-DIAMINO-5-AZAANTHRACENE □ ISOFLAV BASE □ PROFLAVIN □ PROFLAVINE □ PROFOLIOL □ PROFORMIPHEN □ PROFUNDOL □ PROFURA □ PROGARMED □ PRO-GEN □ PROGESIC

TOXICITY DATA WITH REFERENCE
dnd-esc 10 mg/L MUREAV 107,1,83
otr-mus:emb 1 μmol/L MOPMA3 21,739,82
dni-mus:emb 1 μmol/L MOPMA3 21,739,82
oms-mus:emb 1 μmol/L MOPMA3 21,739,82
sce-ham:ovr 20 μg/L ENMUDM 4,65,82
ipr-mus LD50:50 mg/kg CHTHBK 16,371,71
scu-mus LD50:140 mg/kg BJEPA5 28,1,47
ivn-cat LDLo:11,111 μg/kg LANCAO 196,838,19

CONSENSUS REPORTS: IARC Cancer Review: Animal Inadequate Evidence IMEMDT 24,195,80. Reported in EPA TSCA Inventory. EPA Genetic Toxicology Program.

SAFETY PROFILE: Poison by intravenous, intraperitoneal, and subcutaneous routes. Questionable carcinogen. Mutation data reported. When heated to decomposition it emits toxic fumes of NO_x. See also other DIAMINOACRIDINE entries.

DBO000 CAS:615-05-4 HR: 3
2,4-DIAMINOANISOLE
mf: $C_7H_{10}N_2O$ mw: 138.19

PROP: Needles. Mp: 68°.

SYNS: C.I. 76050 □ C.I. OXIDATION BASE 12 □ 2,4-DAA □ 2,4-DIAMINEANISOLE □ 2,4-DIAMINOANISOL □ 2,4-DIAMINOANISOLE BASE □ m-DIAMINOANISOLE 1,3-DIAMINO-4-METHOXYBENZENE □ 2,4-DIAMINO-1-METHOXYBENZENE □ FURRO L □ 4-METHOXY-1,3-BENZENEDIAMINE □ p-METHOXY-m-PHENYLENEDIAMINE □ 4-METHOXY-m-PHENYLENEDIAMINE □ 4-MMPD □ PELAGOL DA □ PELAGOL GREY L □ PELAGOL L

TOXICITY DATA WITH REFERENCE
skn-rbt 12,500 µg/24H MLD FCTXAV 15,607,77
mma-sat 5 µg/plate MUREAV 79,289,80
dnd-hmn:fbr 50 µmol/L MUREAV 127,107,84
dnd-rat:lvr 500 µmol/L CBINA8 31,35,80
sce-mus-ipr 12 mg/kg MUREAV 108,225,83
scu-uns TDLo:65,100 µg/kg (female 10D post):REP JIDOAA 35,387,86
orl-rat LD50:460 mg/kg FCTXAV 15,607,77
ipr-rat LD50:116 mg/kg BCPCA6 30,2715,81

CONSENSUS REPORTS: IARC Cancer Review: Group 2B IMEMDT 7,56,87; Human Limited Evidence IMEMDT 27,103,82. EPA Genetic Toxicology Program. Reported in EPA TSCA Inventory.

DFG MAK: Animal Carcinogen, Suspected Human Carcinogen.
NIOSH REL: (2,4-diaminoanisole) Reduce to lowest feasible level

SAFETY PROFILE: Confirmed carcinogen. Poison by intraperitoneal route. Moderately toxic by ingestion. Experimental reproductive effects. Human mutation data reported. A skin irritant. When heated to decomposition it emits toxic fumes of NO_x. See also other DIAMINOANISOLE entries.

DBO400 CAS:39156-41-7 HR: 3
2,4-DIAMINOANISOLE SULPHATE
mf: $C_7H_{10}N_2O \cdot H_2O_4S$ mw: 236.27

SYNS: ANISOLE, 2,4-DIAMINO-, HYDROGEN SULFATE □ ANISOLE, 2,4-DIAMINO-, SULFATE □ 1,3-BENZENEDIAMINE, 4-METHOXY, SULFATE (1:1) (9CI) □ C.I. 76051 □ C.I. OXIDATION BASE 12A □ 2,4-DAA SULFATE □ 2,4-DIAMINOANISOLE SULPHATE □ 2,4-DIAMINOANISOL SULPHATE □ 2,4-DIAMINO-1-METHOXYBENZENE □ 1,3-DIAMINO-4-METHOXYBENZENE SULPHATE □ 2,4-DIAMINO-1-METHOXYBENZENE SULPHATE □ 2,4-DIAMINOSOLE SULPHATE □ DURAFUR BROWN MN □ FOURAMINE BA □ FOURRINE 76 □ FOURRINE SLA □ FURRO SLA □ 4-METHOXY-1,3-BENZENEDIAMINE SULFATE □ 4-METHOXY-1,3-BENZENEDIAMINE SULFATE (1:1) □ 4-METHOXY-1,3-BENZENEDIAMINE SULPHATE □ 4-METHOXY-m-PHENYLENEDIAMINE SULFATE □ p-METHOXY-m-PHENYLENEDIAMINE SULPHATE □ 4-METHOXY-m-PHENYLENEDIAMINE SULPHATE □ 4-MMPD SULPHATE □ NAKO TSA □ NCI-C01989 □ OXIDATION BASE 12A □ PELAGOL BA □ PELAGOL GREY □ PELAGOL GREY SLA □ PELAGOL SLA □ RENAL SLA □ URSOL SLA □ ZOBA SLE

TOXICITY DATA WITH REFERENCE
mma-sat 1 µg/plate ENMUDM 7(Suppl 5),1,85
sln-dmg-orl 15,100 µmol/L/3D MUREAV 48,181,77
sln-nsc 150 mg/L MUREAV 167,35,86
mrc-smc 500 mg/L MUREAV 78,243,80
orl-rat TDLo:18 g/kg/10W-C:CAR JJIND8 65,197,80
orl-mus TDLo:157 g/kg/78W-C:CAR NCITR* NCI-CG-TR-84,78
orl-rat TD:33 g/kg/78W-C:ETA NCITR* NCI-CG-TR-84,78
orl-rat TD:51 g/kg/72W-C:CAR NCITR* NCI-CG-TR-84,78
orl-mus TD:1310 g/kg/78W-C:NEO IARC** 27,103,82
orl-rat TD:689 g/kg/82W-C:CAR IARC** 27,103,82
orl-rat TD:1377 g/kg/82W-C:CAR IARC** 27,103,82
orl-rat TD:2730 g/kg/78W-C:CAR IARC** 27,103,82
orl-rat TD:2870 g/kg/82W-C:CAR IARC** 27,103,82
orl-rat TD:350 g/kg/10W-C:ETA IARC** 27,103,82
ipr-rat LD50:372 mg/kg JTEHD6 2,657,77

CONSENSUS REPORTS: NTP 7th Annual Report on Carcinogens. IARC Cancer Review: Animal Sufficient Evidence IMEMDT 27,103,82; Animal Inadequate Evidence IMEMDT 16,51,78. NCI Carcinogenesis Bioassay (feed); Clear Evidence: mouse, rat NCITR* NCI-CG-TR-84,78. Reported in EPA TSCA Inventory. EPA Genetic Toxicology Program. Community Right-To-Know List.

SAFETY PROFILE: Confirmed carcinogen with experimental carcinogenic, neoplastigenic, and tumorigenic data. Poison by intraperitoneal route. Mutation data reported. When heated to decomposition it emits very toxic fumes of NO_x and SO_x. See also other DIAMINOANISOLE entries.

DBO600 CAS:5327-72-0 HR: 1
1,4-DIAMINOANTHRACENE-9,10-DIOL
mf: $C_{14}H_{12}N_2O_2$ mw: 240.28

SYNS: 1,4-DIAMINO-9,10-ANTHRACENEDIOL □ 1,4-DIAMINO-9,10-DIHYDROXYANTHRACEN (CZECH) □ LEUKO-1,4-DIAMINOANTHRACHINON (CZECH) □ LEUCO-1,4-DIAMINOANTHRAQUINONE

TOXICITY DATA WITH REFERENCE
eye-rbt 100 mg/24H MOD 28ZPAK -,111,72

CONSENSUS REPORTS: Reported in EPA TSCA Inventory.

SAFETY PROFILE: An eye irritant. When heated to decomposition it emits toxic fumes of NO_x.

DBO800 CAS:1758-68-5 HR: 3
1,2-DIAMINOANTHRAQUINONE
mf: $C_{14}H_{10}N_2O_2$ mw: 238.26

PROP: Violet crystals with bronze lustre from $PhNO_2$.

SYN: 1,2-DAA (RUSSIAN)

TOXICITY DATA WITH REFERENCE
mmo-sat 100 µg/plate MUREAV 40,203,76
ipr-rat LD50:2700 mg/kg GTPZAB 21(12),27,77
ivn-mus LD50:320 mg/kg CSLNX* NX#01102

CONSENSUS REPORTS: Reported in EPA TSCA Inventory.

SAFETY PROFILE: Poison by intravenous route. Moderately toxic by intraperitoneal route. Mutation data reported. When heated to decomposition it emits toxic fumes of NO_x.

DBP000 CAS:128-95-0 HR: 3
1,4-DIAMINOANTHRAQUINONE
mf: $C_{14}H_{10}N_2O_2$ mw: 238.26

PROP: Dark-violet crystals with metallic cast from EtOH. Mp: 268°.

SYNS: ACETATE RED VIOLET R ☐ ACETOQUINONE LIGHT HELIOTROPE NL ☐ ACETYLON FAST RED VIOLET R ☐ AMACEL HELIOTROPE R ☐ AMAPLAST RED VIOLET P 2R ☐ 1,4-ANTHRAQUINONYLDIAMINE ☐ ARTISIL VIOLET 2RP ☐ CELANTHRENE RED VIOLET R ☐ CELLITON FAST RED VIOLET ☐ CELLUTATE RED VIOLET RH ☐ C.I. 61100 ☐ CIBACET VIOLET 2R ☐ C.I. DISPERSE VIOLET 1 ☐ CILLA FAST RED VIOLET RN ☐ C.I. SOLVENT VIOLET 11 ☐ DIACELLITON FAST VIOLET 5R ☐ 1,4-DIAMINO-9,10-ANTHRACENEDIONE ☐ 1,4-DIAMINOANTHRACHINON (CZECH) ☐ DISPERSE VIOLET K ☐ DURANOL VIOLET WR ☐ FENACET FAST VIOLET 5R ☐ GRACET VIOLET 2R ☐ GRASOL VIOLET R ☐ INTERCHEM ACETATE VIOLET R ☐ KRISOLAMINE ☐ MICROSETILE VIOLET 3R ☐ MIDETON FAST RED VIOLET R ☐ NACELAN VIOLET 4R ☐ NYLOQUINONE VIOLET R ☐ OIL VIOLET R ☐ ORACET VIOLET 2R ☐ PERLITON VIOLET 3R ☐ RESIREN VIOLET TR ☐ SEACYL VIOLET R ☐ SERISOL BRILLIANT VIOLET 2R ☐ SETACYL VIOLET R ☐ SETILE VIOLET 3R ☐ SUPRACET BRILLIANT VIOLET 3R ☐ TRANSETILE VIOLET P 3R

TOXICITY DATA WITH REFERENCE
eye-rbt 500 mg/24H MLD 28ZPAK -,121,72
mmo-sat 100 μg/plate MUREAV 40,203,76
mma-sat 100 μg/plate MUREAV 40,203,76
mmo-omi 12,800 μg/L HEREAY 99,209,83
orl-rat LDLo:5790 mg/kg 28ZPAK -,121,72
ipr-rat LD50:250 mg/kg GTPZAB 21(12),27,77

CONSENSUS REPORTS: Reported in EPA TSCA Inventory. EPA Genetic Toxicology Program.

SAFETY PROFILE: A poison by intraperitoneal route. Moderately toxic by intraperitoneal route. Mildly toxic by ingestion. Mutation data reported. An eye irritant. When heated to decomposition it emits toxic fumes of NO_x.

DBP200 CAS:129-44-2 HR: 2
1,5-DIAMINOANTHRAQUINONE
mf: $C_{14}H_{10}N_2O_2$ mw: 238.26

PROP: Red crystals from EtOH or AcOH. Sltly sol in EtOH, Et_2O, C_6H_6, $CHCl_3$, and Me_2CO.

SYNS: 1,5-ANTHRAQUINONYLDIAMINE ☐ 1,5-DIAMINOANTHRACHINON (CZECH) ☐ 1,5-DIAMINO-9,10-ANTHRAQUINONE

TOXICITY DATA WITH REFERENCE
eye-rbt 500 mg/24H MLD 28ZPAK -,122,72
mma-sat 500 μg/plate MUREAV 40,203,76
ipr-rat LD50:1300 mg/kg GTPZAB 21(12),27,77

CONSENSUS REPORTS: Reported in EPA TSCA Inventory.

SAFETY PROFILE: Moderately toxic by intraperitoneal route. Mutation data reported. An eye irritant. When heated to decomposition it emits toxic fumes of NO_x.

DBP400 CAS:129-42-0 HR: 1
1,8-DIAMINOANTHRAQUINONE
mf: $C_{14}H_{10}N_2O_2$ mw: 238.26

PROP: Red crystals from EtOH, AcOH, $PhNO_2$ or Py. Mp: 265°. Sltly sol in Et_2O.

SYNS: 1,5-ANTHRAQUINONYLDIAMINE ☐ 1,8-ANTHRAQUINONYLDIAMINE ☐ 1,8-DIAMINOANTHRACHINON (CZECH) ☐ 1,5-DIAMINOANTHRAQUINONE

TOXICITY DATA WITH REFERENCE
eye-rbt 500 mg/24H MLD 28ZPAK -,122,72

CONSENSUS REPORTS: Reported in EPA TSCA Inventory.

SAFETY PROFILE: An eye irritant. When heated to decomposition it emits toxic fumes of NO_x.

DBP909 CAS:145-49-3 HR: 3
1,5-DIAMINOANTHRARUFIN
mf: $C_{14}H_{10}N_2O_4$ mw: 270.26

PROP: Dark red crystals or powder. Sol concentration in H_2SO_4; sltly sol in H_2O and EtOH.

SYNS: 4,8-DIAMINOANTHRARUFIN ☐ 1,5-DIAMINO-4,8-DIHYDROXY-9,10-ANTHRACENEDIONE ☐ 1,5-DIAMINO-4,8-DIHYDROXYANTHRAQUINONE ☐ leuco-1,5-DIAMINO-4,8-DIHYDROXYANTHRAQUINONE ☐ 4,8-DIAMINO-1,5-DIHYDROXYANTHRAQUINONE ☐ 1,5-DIHYDROXY-4,8-DIAMINOANTHRACHINON (CZECH) ☐ 1,5-DIHYDROXY-4,8-DIAMINOANTHRAQUINONE

TOXICITY DATA WITH REFERENCE
eye-rbt 20 mg/24H MOD 28ZPAK -,103,72
mmo-sat 50 μg/plate MUREAV 40,203,76
mma-sat 50 μg/plate MUREAV 40,203,76
ivn-mus LD50:56 mg/kg CSLNX* NX#03923

CONSENSUS REPORTS: Reported in EPA TSCA Inventory.

SAFETY PROFILE: Poison by intravenous route. An eye irritant. Mutation data reported. When heated to decomposition it emits toxic fumes of NO_x.

DBP999 CAS:495-54-5 HR: 2
DIAMINOAZOBENZENE
mf: $C_{12}H_{12}N_4$ mw: 212.28

PROP: White or pale yellow crystals from water. Mp: 118–118.5°, bp: 286°, d: 1.139, vap press: 1 mm @ 99.8°.

SYNS: CHRYSOIDIN A ☐ C.I. 11270 ☐ 2,4-DIAMINOAZOBENZEN (CZECH) ☐ ORANZ ZASADITA 2 (CZECH) ☐ 4-(PHENYLAZO)-m-PHENYLENEDIAMINE

TOXICITY DATA WITH REFERENCE
eye-rbt 20 mg/24H MOD 28ZPAK -,237,72
mmo-sat 50 μg/plate MUREAV 44,9,77
dns-rat:lvr 500 ng/well MUREAV 240,227,90
orl-rat LD50:1650 mg/kg 28ZPAK -,237,72

CONSENSUS REPORTS: Reported in EPA TSCA Inventory.

SAFETY PROFILE: Moderately toxic by ingestion. An eye irritant. Mutation data reported. When heated to decomposition it emits toxic fumes of NO_x.

DBQ125 CAS:58338-59-3 HR: 3
2′,4-DIAMINOBENZANILIDE
mf: $C_{13}H_{13}N_3O$ mw: 227.26

PROP: Crystals in EtOH. Mp: 180–181°.

SYNS: 4-AMINO-N-(2′-AMINOPHENYL)BENZAMIDE ☐ GOE 1734

TOXICITY DATA WITH REFERENCE
mma-sat 5 mg/plate CTRRDO 69,1415,85
dnd-rat-orl 10 mg/kg CTRRDO 69,1415,85
orl-rat TDLo:6678 mg/kg/2Y-C:NEO CALEDQ 60,237,91
orl-rat LD50:95 mg/kg CTRRDO 69,1415,85
orl-mus LD50:625 mg/kg CTRRDO 69,1415,85

SAFETY PROFILE: Poison by ingestion. Mutation data reported. Questionable carcinogen with experimental tumorigenic data reported. When heated to decomposition it emits toxic fumes of NO_x.

DBQ200 CAS:618-56-4 HR: 3
3,5-DIAMINOBENZOIC ACID DIHYDROCHLORIDE
mf: $C_7H_8N_2O_2{\cdot}2ClH$ mw: 225.09

SYN: BENZOIC ACID, 3,5-DIAMINO-, DIHYDROCHLORIDE

TOXICITY DATA WITH REFERENCE
ivn-mus LD50:180 mg/kg CSLNX* NX#00464

CONSENSUS REPORTS: Reported in EPA TSCA Inventory.

SAFETY PROFILE: Poison by intravenous route. When heated to decomposition it emits toxic vapors of NO_x, HCl, and Cl^-.

DBQ250 CAS:128-94-9 HR: 1
4,5-DIAMINOCHRYSAZIN
mf: $C_{14}H_{10}N_2O_4$ mw: 270.26

SYNS: 9,10-ANTHRACENEDIONE, 1,8-DIAMINO-4,5-DIHYDROXY- ☐ ANTHRAQUINONE, 1,8-DIAMINO-4,5-DIHYDROXY- ☐ 1,8-DIAMINOCHRYSAZINE ☐ 1,8-DIAMINO-4,5-DIHYDROXYANTHRACHINON ☐ 1,8-DIAMINO-4,5-DIHYDROXYANTHRAQUINONE ☐ 1,8-DIAMINO-4,5-DIHYDROXY-9,10-ANTHRAQUINONE ☐ 4,5-DIAMINO-1,8-DIHYDROXYANTHRAQUINONE ☐ 1,8-DIHYDROXY-4,5-DIAMINOANTHRACHINON

TOXICITY DATA WITH REFERENCE
eye-rbt 100 mg/24H MOD 85JCAE -,653,86

CONSENSUS REPORTS: Reported in EPA TSCA Inventory.

SAFETY PROFILE: An eye irritant. When heated to decomposition it emits toxic vapors of NO_x.

DBQ800 CAS:3385-21-5 HR: 3
1,3-DIAMINOCYCLOHEXANE
mf: $C_6H_{14}N_2$ mw: 114.22

SYN: 1,3-CYCLOHEXANEDIAMINE

TOXICITY DATA WITH REFERENCE
orl-rat LD50:390 mg/kg AMIHAB 17,129,58
ihl-rat LCLo:2900 mg/m³/4H TOXID9 12,357,92
orl-mus LD50:543 mg/kg AMIHAB 17,129,58

CONSENSUS REPORTS: Reported in EPA TSCA Inventory.

SAFETY PROFILE: Poison by ingestion. When heated to decomposition it emits toxic fumes of NO_x. See also AMINES.

DBR000 CAS:68772-00-9 HR: 3
2,4-DIAMINO-5-(p-(p-((p-(2,4-DIAMINO-1-ETHYLPYRIMIDINIUM-5-YL)PHENYL)CARBAMOYLCINNAMAMIDO)PHENYL)-1-ETHYLPYRIMIDINIUM), DI-p-TOLUENE SULFONATE
mf: $C_{34}H_{36}N_{10}O_2{\cdot}2C_7H_7O_3S$ mw: 959.20

TOXICITY DATA WITH REFERENCE
dnd-mus:lym 400 nmol/L JMCMAR 22,134,79
ipr-mus LD10:27 mg/kg JMCMAR 22,134,79

SAFETY PROFILE: Poison by intraperitoneal route. Mutation data reported. When heated to decomposition it emits very toxic fumes of NO_x and SO_x. See also SULFONATES.

DBR200 CAS:68797-80-7 HR: 3
2,4-DIAMINO-5-(p-(p-((-(2,4-DIAMINO-1-METHYLPYRIMIDINIUM-5-YL)PHENYL)CARBAMOYL)CINNAMAMIDO)PHENYL-1-METHYLPYRIMIDINIUM)-DI-p-TOLUENE SULFONATE
mf: $C_{32}H_{32}N_{10}O_2{\cdot}2C_7H_7O_3S$ mw: 931.14

TOXICITY DATA WITH REFERENCE
dnd-mus:lym 400 nmol/L JMCMAR 22,134,79
ipr-mus LD10:50 mg/kg JMCMAR 22,134,79

SAFETY PROFILE: Poison by intraperitoneal route. Mutation data reported. When heated to decomposition it emits very toxic fumes of NO_x and SO_x. See also SULFONATES.

DBR400 CAS:609-20-1 HR: 2
1,4-DIAMINO-2,6-DICHLOROBENZENE
mf: $C_6H_6Cl_2N_2$ mw: 177.04

PROP: Needles. Mp: 124–125°.

SYNS: C.I. 37020 ☐ 2,6-DICHLORO-p-PHENYLENEDIAMINE ☐ NCI-C50260

TOXICITY DATA WITH REFERENCE
mma-sat 1 mg/plate ENMUDM 8(Suppl 7),1,86
orl-mus TDLo:87 g/kg/2Y-C:CAR NTPTR* NTP-TR-219,82
orl-mus TD:260 g/kg/2Y-C:CAR NTPTR* NTP-TR-219,82
orl-rat LD50:700 mg/kg NTPTR* NTP-TR-219,82F

CONSENSUS REPORTS: IARC Cancer Review: Group 3 IMEMDT 7,56,87; Animal Limited Evidence IMEMDT 39,325,86. NTP Carcinogenesis Bioassay (feed); Clear Evidence: mouse NTPTR* NTP-TR-219,82; No Evidence: rat NTPTR* NTP-TR-219,82. Reported in EPA TSCA Inventory.

SAFETY PROFILE: Questionable carcinogen with experimental carcinogenic data. Moderately toxic by ingestion. Mutation data reported. When heated to decomposition it emits very toxic fumes of Cl^- and NO_x.

DBT000 CAS:4702-64-1 HR: 1
1,5-DIAMINO-4,8-DIHYDROXY-3-(p-METHOXYPHENYL)ANTHRAQUINONE
mf: $C_{21}H_{16}N_2O_5$ mw: 376.39

SYN: MODR OSTACETOVA SE-LB (CZECH)

TOXICITY DATA WITH REFERENCE
eye-rbt 500 mg/24H MLD 28ZPAK -,245,72
orl-rat LD50:6380 mg/kg 28ZPAK -,245,72

CONSENSUS REPORTS: Reported in EPA TSCA Inventory.

SAFETY PROFILE: Mildly toxic by ingestion. An eye irritant. When heated to decomposition it emits toxic fumes of NO_x.

DBT200 CAS:92-26-2 HR: 3
3,6-DIAMINO-2,7-DIMETHYLACRIDINE
mf: $C_{15}H_{15}N_3$ mw: 237.33

PROP: Crystals from $PhNH_2$. Mp: 325°.

SYNS: ACRIDINE YELLOW BASE □ 2,8-DIAMINO-3,7-DIMETHYLACRIDINE

TOXICITY DATA WITH REFERENCE
mma-sat 400 nmol/L ENMUDM 3,11,81
dns-rat:lvr 5 μmol/L ENMUDM 3,11,81
scu-mus LD50:280 mg/kg BJEPA5 28,1,47

CONSENSUS REPORTS: Reported in EPA TSCA Inventory.

SAFETY PROFILE: Poison by subcutaneous route. Mutation data reported. When heated to decomposition it emits toxic fumes of NO_x.

DBU800 CAS:13426-91-0 HR: 2
1,2-DIAMINOETHANE COPPER COMPLEX
DOT: UN 1761
mf: $C_2H_{10}N_2$•xCu mw: 506.92

SYNS: COPPER-ETHYLENEDIAMINE COMPLEX □ CUPRIETHYLENE DIAMINE □ CUPRIETHYLENEDIAMINE, solution (DOT) □ ETHANE, 1, 2-DIAMINO-, COPPER COMPLEX □ KOMEEN □ KOPLEX AQUATIC HERBICIDE

TOXICITY DATA WITH REFERENCE
orl-rat LD50:750 mg/kg FMCHA2-,C179,91
skn-rbt LD50:>8 g/kg FMCHA2-,C179,91

CONSENSUS REPORTS: Copper and its compounds are on the Community Right-To-Know List.

DOT CLASSIFICATION: 8; *Label:* Corrosive, Poison

SAFETY PROFILE: Moderately toxic by ingestion. An irritating and corrosive material to the skin, eyes, and mucous membranes. When heated to decomposition it emits toxic fumes of NO_x. See also COPPER COMPOUNDS.

DBV200 CAS:97194-20-2 HR: 3
4,6-DIAMINO-1-(2-ETHYLPHENYL)-2-METHYL-2-PROPYL-s-TRIAZINE HYDROCHLORIDE
mf: $C_{15}H_{23}N_5$•ClH mw: 309.89

TOXICITY DATA WITH REFERENCE
orl-mus LD50:700 mg/kg JMCMAR 6,370,63
ipr-mus LD50:78 mg/kg JMCMAR 6,370,63

SAFETY PROFILE: Poison by intraperitoneal route. Moderately toxic by ingestion. When heated to decomposition it emits very toxic fumes of NO_x and HCl.

DBV400 CAS:1239-45-8 HR: 3
2,7-DIAMINO-10-ETHYL-9-PHENYLPHENANTHRIDINIUM BROMIDE
mf: $C_{21}H_{20}N_3$•Br mw: 394.35

PROP: Dark-red crystals from EtOH. Mp: 238–240°.

SYNS: 3,8-DIAMINO-5-ETHYL-6-PHENYLPHENANTHRIDINIUM BROMIDE □ 2,7-DIAMINO-9-PHENYL-10-ETHYLPHENANTHRIDINIUM BROMIDE □ 2,7-DIAMINO-9-PHENYLPHENANTHRIDINE ETHOBROMIDE □ DROMILAC □ ETHIDIUM BROMIDE □ HOMIDIUM BROMIDE □ RD 1572

TOXICITY DATA WITH REFERENCE
mma-sat 200 ng/plate MUREAV 127,31,84
sln-dmg-mul 3 mmol/L MUREAV 138,169,84
dni-hmn:hla 40 μmol/L MUREAV 92,427,82
ipr-mus LD50:20 mg/kg CNREA8 34,2699,74
scu-mus LD50:110 mg/kg ATMPA2 46,285,52

CONSENSUS REPORTS: EPA Genetic Toxicology Program.

SAFETY PROFILE: Poison by intraperitoneal and subcutaneous routes. Human mutation data reported. When heated to decomposition it emits very toxic fumes of NO_x and Br^-. See also BROMIDES.

DBW000 CAS:50309-02-9 HR: 3
4-(((4-(((4-(2,4-DIAMINO-1-ETHYLPYRIMIDINIUM-5-YL)PHENYL)AMINO)CARBONYL)PHENYL)AMINO)-1-ETHYLQUINOLINIUM) DIIODIDE
mf: $C_{30}H_{31}N_7O$•2I mw: 759.48

TOXICITY DATA WITH REFERENCE
dnd-mus:lym 1240 nmol/L JMCMAR 22,134,79
ipr-mus LD10:32 mg/kg JMCMAR 22,134,79

SAFETY PROFILE: Poison by intraperitoneal route. Mutation data reported. When heated to decomposition it emits very toxic fumes of NO_x and I^-.

DBW100 CAS:10308-83-5 HR: 3
DIAMINOGUANIDINIUM NITRATE
mf: $CH_8N_6O_3$ mw: 152.11

SAFETY PROFILE: Decomposes violently above 260°C. A powerful oxidizer. When heated to decomposition it emits toxic fumes of NO_x. See also NITRATES.

DBW200 **CAS:70548-53-7** **HR: 3**
(l)-2,6-DIAMINO-1-HEXANETHIOL DIHYDROCHLORIDE
mf: $C_6H_{16}N_2S{\bullet}2ClH$ mw: 221.22

TOXICITY DATA WITH REFERENCE
orl-mus LD50:450 mg/kg JMCMAR 22,631,79
ipr-mus LD50:88 mg/kg JMCMAR 22,631,79

SAFETY PROFILE: Poison by intraperitoneal route. Moderately toxic by ingestion. When heated to decomposition it emits very toxic fumes of Cl^-, SO_x, and NO_x.

DBW600 **CAS:70561-82-9** **HR: 2**
S-2,6-DIAMINOHEXYL DIHYDROGEN PHOSPHOROTHIOATE DIHYDRATE
mf: $C_6H_{17}N_2O_3PS{\bullet}2H_2O$ mw: 264.32

SYN: PHOSPHOROTHIOIC ACID-S-2,6-DIAMINOHEXYL ESTER, DIHYDRATE

TOXICITY DATA WITH REFERENCE
orl-mus LD50:900 mg/kg JMCMAR 22,631,79
ipr-mus LD50:450 mg/kg JMCMAR 22,631,79

SAFETY PROFILE: Moderately toxic by ingestion and intraperitoneal routes. When heated to decomposition it emits very toxic fumes of NO_x, PO_x, and SO_x. See also ESTERS.

DBX000 **CAS:2872-48-2** **HR: 2**
1,4-DIAMINO-2-METHOXYANTHRAQUINONE
mf: $C_{15}H_{12}N_2O_3$ mw: 268.29

SYNS: ACETATE FAST PINK 3B □ AMACEL CERISE B □ 9,10-ANTHRACENEDIONE, 1,4-DIAMINO-2-METHOXY-(9CI) □ ARTISIL BRILLIANT ROSE 5BP □ CELANTHRENE FAST PINK 3B □ CELLITON FAST PINK FF3B □ CELLITON FAST PINK FF3BA-CF □ CELLITON ROSE FF3B □ CERVEN DISPERZNI 11 □ C.I. 62015 □ CIBACET BRILLIANT PINK 4BN □ CIBACETE BRILLIANT PINK 4BN □ C.I. DISPERSE RED 11 □ CILLA FAST PINK FF3B □ C.I. SOLVENT VIOLET 26 □ 1,4-DA-2-MOA □ 1,4-DIAMINO-2-METHOXY-9,10-ANTHRACENEDIONE □ DISPERSE BRILLIANT PINK □ DISPERSE BRILLIANT ROSE □ DISPERSE RED 11 □ DISPERSOL RED B 3B □ DURANOL RED X3B □ FENACET FAST PINK 3BE □ INTERCHEM ACETATE PINK 3B □ MIKETON FAST PINK FF 3B □ NACELAN PINK 3B □ PALANIL VIOLET 6R □ PERLITON RED VIOLET FFB □ SAMARON RED VIOLET F3B □ SERIPLAS RED X3B □ SERISOL BLILLIANT RED X 3B □ SETACYL RED P-3B □ SOLVENT VIOLET 26 □ SUPRACET FAST PINK 3B □ TERASIL BRILLIANT PINK 4BN □ VIOLET ROZPOUSTEDLOVA 26

TOXICITY DATA WITH REFERENCE
skn-rbt 500 mg MOD NTIS** AD-A172-758
orl-rat LD50:708 mg/kg NTIS** AD-A172-758
ipr-rat LD50:700 mg/kg GTPZAB 21(12),27,77

CONSENSUS REPORTS: Reported in EPA TSCA Inventory.

SAFETY PROFILE: Moderately toxic by ingestion and intraperitoneal route. A skin irritant. When heated to decomposition it emits toxic fumes of NO_x.

DBX400 **CAS:8048-52-0** **HR: 3**
3,6-DIAMINO-10-METHYLACRIDINIUM CHLORIDE with 3,6-ACRIDINEDIAMINE
mf: $C_{14}H_{14}N_3{\bullet}Cl{\bullet}C_{13}H_{11}N_3$ mw: 469.03

SYNS: ACRIFLAVIN □ ACRIFLAVINE mixture with PROFLAVINE □ ACRIFLAVINIUM CHLORIDE □ ACRIFLAVINIUM CHLORIDUM □ ACRIFLAVON □ ANGIFLAN □ ASSIFLAVINE □ AVLON □ BIALFLAVINA □ BIOACRIDIN □ BOVOFLAVIN □ BURNOL □ BUROFLAVIN □ CHOLIFLAVIN □ CHROMOFLAVINE □ DIACRID □ 3,6-DIAMINOACRIDINE mixture with 3,6-DIAMINO-10-METHYLACRIDINIUM CHLORIDE □ 2,8-DIAMINO-10-METHYLACRIDINIUM CHLORIDE mixture with 2,8-DIAMINOACRIDINE □ EUFLAVINE □ FLAVACRIDINUM HYDROCHLORICUM □ FLAVINE □ FLAVIOFORM □ FLAVIPIN □ FLAVISEPT □ GLYCO-FLAVINE □ GONACRINE □ ISRAVIN □ MEDIFLAVIN □ NEUTRAL ACRIFLAVINE □ PANFLAVIN □ PANTONSILETTEN □ TRACHOSEPT □ TRIPLA-ETILO □ TRYPAFLAVINE □ VETAFLAVIN □ XANTHACRIDINUM □ ZORIFLAVIN

TOXICITY DATA WITH REFERENCE
mmo-esc 5 μg/plate MUREAV 131,193,84
sln-dmg-mul 5000 ppm MUREAV 138,169,84
dni-hmn:hla 10 μmol/L RAREAE 37,334,69
scu-rat TDLo:160 mg/kg/60W-I:ETA SEMEAS 159,778,52

CONSENSUS REPORTS: IARC Cancer Review: Group 3 IMEMDT 7,56,87; Animal Inadequate Evidence IMEMDT 13,31,77. EPA Genetic Toxicology Program.

SAFETY PROFILE: Poison by subcutaneous route. Questionable carcinogen with experimental tumorigenic data. Human mutation data reported. A topical antiseptic used in the treatment of gonorrhea. When heated to decomposition it emits very toxic fumes of NO_x and Cl^-. See also 3,6-DIAMINO-10-METHYLACRIDINIUM CHLORIDE.

DBX875 **CAS:7319-47-3** **HR: 3**
2,4-DIAMINO-5-METHYL-6-sec-BUTYLPYRIDO(2,3-d) PYRIMIDINE
mf: $C_{12}H_{17}N_5$ mw: 231.34

SYNS: BW 283U □ BW 58-283 □ BW 58-283b □ 2,4-DIAMINO-5-METHYL-6-(BUT-2-YL)PYRIDO(2,3-d)PYRIMIDINE

TOXICITY DATA WITH REFERENCE
oms-hmn:lym 11 nmol/L BCPCA6 25,1947,76
orl-rat LD50:100 mg/kg 14XBAV -,367,64
orl-mus LD50:57 mg/kg JMCMAR 11,711,68

SAFETY PROFILE: Poison by ingestion. Human mutation data reported. When heated to decomposition it emits toxic fumes of NO_x.

DBY000 **CAS:13897-55-7** **HR: 2**
2,4-DIAMINO-1-METHYLCYCLOHEXANE
mf: $C_7H_{16}N_2$ mw: 128.25

SYNS: 2,4-DIAMINOMETHYLCYCLOHEXANE □ 1-METHYL-2,4-CYCLOHEXANEDIAMINE □ 4-METHYL-1,3-CYCLOHEXANEDIAMINE

TOXICITY DATA WITH REFERENCE
orl-rat LD50:1410 mg/kg AIHAAP 30,470,69
skn-rbt LD50:500 mg/kg AIHAAP 30,470,69

SAFETY PROFILE: Moderately toxic by ingestion and skin contact. When heated to decomposition it emits toxic fumes of NO_x. See also AMINES.

DBY300 CAS:17168-83-1 HR: 3
1,2-DIAMINO-2-METHYLPROPANE AQUADIPEROXO CHROMIUM(IV)
mf: $C_4H_{14}CrN_2O_5$ mw: 222.16

$[CH_3CH(NH_2)CH_2NH_2Cr(H_{2O})(O_2)_2]\cdot H_2O$

CONSENSUS REPORTS: Chromium and its compounds are on the Community Right-To-Know List.

SAFETY PROFILE: Explodes at 83°C and is potentially explosive at 20°C. When heated to decomposition it emits toxic fumes of NO_x. See also CHROMIUM COMPOUNDS and PEROXIDES.

DBY500 CAS:18921-70-5 HR: 3
N-(4-(((2,4-DIAMINO-5-METHYL-6-QUINAZOLINYL) METHYL)AMINO)BENZOYL)-l-ASPARTIC ACID
mf: $C_{21}H_{22}N_6O_5$ mw: 438.49

PROP: A solid. Mp: 269–271° (decomp).

SYNS: N-(p-(((2,4-DIAMINO-5-METHYL-6-QUINAZOLINYL)METHYL) AMINO)BENZOYL)-l-ASPARTIC ACID □ METHASQUIN □ NSC 122870 □ SK 29836

TOXICITY DATA WITH REFERENCE
oms-hmn:lym 17 nmol/L BCPCA6 25,1947,76
ipr-mus LD50:50,110 µg/kg NCISP* JAN86
scu-mus LD50:158 mg/kg NCISP* JAN86

SAFETY PROFILE: Poison by subcutaneous and intraperitoneal routes. Human mutation data reported. When heated to decomposition it emits toxic fumes of NO_x.

DBY600 CAS:3545-88-8 HR: 3
1,7-DIAMINO-8-NAPHTHOL-3,6-DISULPHONIC ACID
mf: $C_{10}H_{10}N_2O_7S_2$ mw: 334.34

SYN: 3,6-NAPHTHALENEDISULFONIC ACID, 1,7-DIAMINO-8-HYDROXY-

TOXICITY DATA WITH REFERENCE
scu-rat TDLo:25 mg/kg (female 8D post):REP NATUAS 208,1219,65
scu-rat LD50:66 mg/kg NATUAS 208,1219,65

SAFETY PROFILE: Poison by subcutaneous route. Experimental reproductive effects. When heated to decomposition it emits toxic fumes of NO_x and SO_x.

DBY700 CAS:82-33-7 HR: 2
1,4-DIAMINO-5-NITRO ANTHRAQUINONE
mf: $C_{14}H_9N_3O_4$ mw: 283.26

PROP: Violet needles from EtOH.

SYNS: 9,10-ANTHRACENEDIONE, 1,4-DIAMINO-5-NITRO-(9CI) □ CELLITON FAST VIOLET B □ CELLITON FAST VIOLET BA-CF □ CELLITON VIOLET B □ C.I. 62030 □ CIBACET BRILLIANT VIOLET 3B □ C.I. DISPERSE VIOLET 8 □ CILLA FAST VIOLET B □ DIACELLITON FAST VIOLET B □ 1,4-DIAMINO-5-NITRO-9,10-ANTHRACENEDIONE □ DIANIX FAST VIOLET B □ DISPERSE VIOLET 2S □ DURANOL BRILLIANT BLUE VIOLET BR □ DURANOL BRILLIANT VIOLET BR □ FENACET FAST VIOLET B □ KAYALON FAST VIOLET BR □ MIKETON FAST VIOLET B □ NITROCRESOLAMINE □ PALANIL VIOLET 3B □ PERLITON VIOLET B □ SAMARON BRILLIANT VIOLET B □ SERISOL FAST VIOLET B □ SUPRACET FAST VIOLET B □ TERASIL BRILLIANT VIOLET 3B □ VIOLET 2S □ VONTERYL VIOLET 2B

TOXICITY DATA WITH REFERENCE
mma-sat 50 µg/plate MUREAV 40,203,76
skn-rat TDLo:22,500 mg/kg/65W-I:ETA VINIT* #1684-81
skn-rat TD:25,600 mg/kg/65W-I:ETA VINIT* #1684-81
ivn-mus LD50:56 mg/kg CSLNX* NX#01786

CONSENSUS REPORTS: Reported in EPA TSCA Inventory.

SAFETY PROFILE: Poison by intravenous route. Questionable carcinogen with experimental tumorigenic data. Mutation data reported. When heated to decomposition it emits toxic fumes of NO_x.

DBY800 CAS:720-69-4 HR: 2
4,6-DIAMINO-2-(5-NITRO-2-FURYL)-S-TRIAZINE
mf: $C_7H_6N_6O_3$ mw: 222.19

TOXICITY DATA WITH REFERENCE
mma-sat 100 ng/plate MUREAV 40,9,76
dnr-sat 500 nmol/well CNREA8 34,2266,74
mmo-esc 300 nmol/well CNREA8 34,2266,74
mrc-esc 500 nmol/well CNREA8 34,2266,74
orl-rat TDLo:13 g/kg/46W-C:CAR JNCIAM 51,403,73

CONSENSUS REPORTS: EPA Genetic Toxicology Program.

SAFETY PROFILE: Questionable carcinogen with experimental carcinogenic data. Mutation data reported. When heated to decomposition it emits toxic fumes of NO_x.

DCA200 CAS:95-86-3 HR: 3
2,4-DIAMINOPHENOL
mf: $C_6H_8N_2O$ mw: 124.16

PROP: Leaflets. Mp: 78–80° (decomp).

TOXICITY DATA WITH REFERENCE
mma-sat 10 µg/plate BCPCA6 26,729,77
ipr-mus LDLo:50 mg/kg RBPMAZ 22,1,52

SAFETY PROFILE: Poison by intraperitoneal route. Mutation data reported. When heated to decomposition it emits toxic fumes of NO_x. See also AMINES and PHENOL.

DCA300 CAS:27653-49-2 HR: 3
2,4-DIAMINO-5-PHENYL-6-ETHYLPYRIMIDINE
mf: $C_{12}H_{14}N_4$ mw: 214.30

SYN: 2,4-DIAMINO-6-ETHYL-5-PHENYLPYRIMIDINE

TOXICITY DATA WITH REFERENCE
orl-rat TDLo:50 mg/kg (14D preg):TER SJDBA9 7,45,76
orl-rat TDLo:1 mg/kg (female 9D post):REP AAGEAA 71(7),29,76
orl-rat TDLo:10 mg/kg (female 9D post):TER AAGEAA 71(7),29,76
orl-rat LD50:126 mg/kg AAGEAA 71,29,76

SAFETY PROFILE: Poison by ingestion. Experimental

teratogenic and reproductive effects. When heated to decomposition it emits toxic fumes of NO_x.

DCA600 **CAS:942-31-4** **HR: 3**
2,4-DIAMINO-5-PHENYLTHIAZOLE HYDROCHLORIDE
mf: $C_9H_9N_3S{\bullet}ClH$ mw: 227.73

PROP: A solid. Mp: 272°.

SYNS: AMIFENATSOL HYDROCHLORIDE □ AMIPHENAZOLE HYDROCHLORIDE □ AMPHENAZOLE HYDROCHLORIDE □ AMPHISOL HYDROCHLORIDE □ DAFTAZOL HYDROCHLORIDE □ DAPTAZILE HYDROCHLORIDE □ DAPTAZOLE HYDROCHLORIDE □ 2,4-DIAMINO 5-PHENYLTHIAZOL CHLORHYDRATE (FRENCH) □ 2,4-DIAMINO-5-PHENYLTHIAZOLE MONOHYDROCHLORIDE □ DPT □ FENAMIZOL HYDROCHLORIDE □ PHENAMIZOLE HYDROCHLORIDE □ 5-PHENYL-2,4-THIAZOLEDIAMINE MONOHYDROCHLORIDE (9CI)

TOXICITY DATA WITH REFERENCE
ipr-rat LD50:300 mg/kg 27ZQAG -,201,72
orl-mus LD50:372 mg/kg 27ZQAG -,201,72
ipr-mus LD50:200 mg/kg 27ZQAG -,201,72
scu-mus LD50:310 mg/kg JPPMAB 21,668,69

SAFETY PROFILE: Poison by ingestion, intraperitoneal, and subcutaneous routes. When heated to decomposition it emits very toxic fumes of NO_x, SO_x, and HCl.

DCB000 **CAS:38304-91-5** **HR: 3**
2,4-DIAMINO-6-PIPERIDINOPYRIMIDINE-3-OXIDE
mf: $C_9H_{15}N_5O$ mw: 209.29

PROP: A solid. Mp: 260° (decomp).

SYNS: 6-AMINO-1,2-DIHYDRO-1-HYDROXY-2-IMINO-4-PIPERIDINOPYRIMIDINE □ 2,4-DIAMINO-6-PIPERIDINILPIRIMIDINA-3-OSSIDO (ITALIAN) □ 2,3-DIHYDRO-3-HYDROXY-2-IMINO-6-(1-PIPERIDINYL)-4-PYRIMIDINAMINE □ LONITEN □ MINOSSIDILE (ITALIAN) □ MINOXIDIL □ 6-PIPERIDINO-2,4-DIAMINOPYRIMIDINE-3-OXIDE □ 6-(1-PIPERIDINYL)-2,4-PYRIMIDINEDIAMINE-3-OXIDE □ U-10,858

TOXICITY DATA WITH REFERENCE
scu-rat TDLo:1320 mg/kg (female 7-17D post):REP IYKEDH 23,756,92
scu-rat TDLo:2080 mg/kg (female 17-21D post):REP IYKEDH 23,774,92
orl-wmn TDLo:200 μg/kg:CVS,GIT AIMDAP 146,2075,86
orl-man TDLo:107 μg/kg/3D-I:BLD AIMEAS 92,874,80
orl-man LDLo:69 mg/kg/34W-I:CVS WJMDA2 143,527,85
orl-rat LD50:1321 mg/kg TXAPA9 39,1,77
ipr-rat LD50:759 mg/kg TXAPA9 39,1,77
ivn-rat LD50:29 mg/kg MEIEDD 10,888,83
ipr-mus LD50:560 mg/kg BCFAAI 121,16,82
ivn-mus LD50:51 mg/kg TXAPA9 39,1,77

SAFETY PROFILE: Poison by intravenous route. Moderately toxic by ingestion and intraperitoneal routes. Human systemic effects by ingestion: arrhythmias, blood pressure lowering, nausea or vomiting, pericarditis, thrombocytopenia. An antihypertensive agent. When heated to decomposition it emits toxic fumes of NO_x.

DCB100 **CAS:3148-72-9** **HR: 2**
DIAMINOPROPANOL TETRA ACETIC ACID
mf: $C_{11}H_{18}N_2O_9$ mw: 322.31

SYNS: ACETIC ACID, DIAMINOPROPANOLTETRA- □ ACETIC ACID, ((2-HYDROXYTRIMETHYLENE)DINITRILO)TETRA-(8CI) □ ACETIC ACID, ((2-HYDROXY-1,3-TRIMETHYLENE)DINITRILO)TETRA- □ DHPTA □ DPTA □ DTA □ GLYCINE, N,N'-(2-HYDROXY-1,3-PROPANEDIYL)BIS(N-(CARBOXYMETHYL)-(9CI) □ ((2-HYDROXYTRIMETHYLENE)DINITRILO)TETRAACETIC ACID

TOXICITY DATA WITH REFERENCE
ipr-rat LDLo:800 mg/kg KODAK• 21MAY71

CONSENSUS REPORTS: Reported in EPA TSCA Inventory.

SAFETY PROFILE: Moderately toxic by intraperitoneal route. When heated to decomposition it emits toxic vapors of NO_x.

DCB200 **CAS:2997-01-5** **HR: 2**
DI(3-AMINOPROPOXY)ETHANE
mf: $C_8H_{20}N_2O_2$ mw: 176.30

SYN: PROPYLAMINE, 3,3'-(ETHYLENEDIOXY)BIS-

TOXICITY DATA WITH REFERENCE
orl-rat LD50:2830 mg/kg AIHAAP 30,470,69
skn-rbt LD50:1250 mg/kg AIHAAP 30,470,69

CONSENSUS REPORTS: Reported in EPA TSCA Inventory.

SAFETY PROFILE: Moderately toxic by ingestion and skin contact. When heated to decomposition it emits toxic vapors of NO_x.

DCC100 **HR: 3**
m-DI-(2-AMINOPROPYL)BENZENE DIHYDROCHLORIDE
mf: $C_{12}H_{20}N_2{\bullet}2ClH$ mw: 265.26

SYN: α,α'-DIMETHYL-m-BENZENEBIS(ETHYLAMINE) DIHYDROCHLORIDE

TOXICITY DATA WITH REFERENCE
ipr-mus LD50:139 mg/kg AITEAT 11,441,63
scu-mus LD50:570 mg/kg AITEAT 11,441,63
ivn-mus LD50:18 mg/kg AITEAT 11,441,63

SAFETY PROFILE: Poison by intravenous and intraperitoneal routes. Moderately toxic by subcutaneous route. When heated to decomposition it emits toxic fumes of NO_x and HCl.

DCC125 **CAS:26076-87-9** **HR: 3**
p-DI-(2-AMINOPROPYL)BENZENE DIHYDROCHLORIDE
mf: $C_{12}H_{20}N_2{\bullet}2ClH$ mw: 265.26

SYN: α,α'-DIMETHYL-p-BENZENEBIS(ETHYLAMINE) DIHYDROCHLORIDE

TOXICITY DATA WITH REFERENCE
ipr-mus LD50:177 mg/kg AITEAT 11,441,63
scu-mus LD50:650 mg/kg AITEAT 11,441,63
ivn-mus LD50:23 mg/kg AITEAT 11,441,63

SAFETY PROFILE: Poison by intravenous and intraperitoneal routes. Moderately toxic by subcutaneous route. When heated to decomposition it emits toxic fumes of NO_x and HCl.

DCC200 **CAS:50309-03-0** **HR: 3**
4-(((4-(((4-(2,4-DIAMINO-1-PROPYLPYRIMIDINIUM-5-YL)PHENYL)AMINO)CARBONYL)PHENYL)AMINO)-1-PROPYLQUINOLINIUM) DIIODIDE
mf: $C_{32}H_{35}N_7O\cdot 2I$ mw: 787.54

TOXICITY DATA WITH REFERENCE
dnd-mus:lym 1240 nmol/L JMCMAR 22,134,79
ipr-mus LD10:40 mg/kg JMCMAR 22,134,79

SAFETY PROFILE: Poison by intraperitoneal route. Mutation data reported. When heated to decomposition it emits very toxic fumes of NO_x and I^-. See also IODIDES.

DCC400 **CAS:71-44-3** **HR: 3**
DIAMINOPROPYLTETRAMETHYLENEDIAMINE
mf: $C_{10}H_{26}N_4$ mw: 202.40

PROP: Deliquescent crystals. Bp: 141–142° @ 0.5 mm.

SYNS: N,N'-BIS(3-AMINOPROPYL)-1,4-BUTANEDIAMINE □ 1,4-BIS (AMINOPROPYL) BUTANEDIAMINE □ N,N'-BIS(3-AMINOPROPYL)-1,4-DIAMINOBUTANE □ GERONTINE □ MUSCULAMINE □ NEURIDINE □ SPERMINE □ SPERMINE, PURISS

TOXICITY DATA WITH REFERENCE
mmo-sat 100 μmol/L AMACCQ 9,77,76
cyt-hmn:hla 2 mmol/L JCLLAX 78,217,71
dni-rat:lvr 100 μmol/L BIJOAK 146,697,75
ipr-rat LD50:33 mg/kg TXAPA9 88,433,87
ivn-rat LD50:65 mg/kg AIPTAK 165,374,67
orl-mus LD30:650 mg/kg AGACBH 14,228,84
ipr-mus LDLo:8 mg/kg TXAPA9 23,288,72
scu-mus LD30:280 mg/kg AGACBH 14,228,84
ivn-mus LD50:56 mg/kg CSLNX* NX#00641

CONSENSUS REPORTS: Reported in EPA TSCA Inventory.

SAFETY PROFILE: Poison by intraperitoneal, subcutaneous, and intravenous routes. Moderately toxic by ingestion. Human mutation data reported. When heated to decomposition it emits toxic fumes of NO_x. See also AMINES.

DCC800 **CAS:141-86-6** **HR: 3**
2,6-DIAMINOPYRIDINE
mf: $C_5H_7N_3$ mw: 109.15

PROP: Crystals or leaflets. Mp: 121.5°, bp: 148–150° @ 5 mm.

TOXICITY DATA WITH REFERENCE
mma-sat 50 μg/plate ESKHA5 (94),28,76
ipr-mus LD50:100 mg/kg JMCMAR 8,296,65
ivn-mus LD50:56 mg/kg CSLNX* NX#00146

CONSENSUS REPORTS: Reported in EPA TSCA Inventory.

SAFETY PROFILE: Poison by intravenous and intraperitoneal routes. Mutation data reported. When heated to decomposition it emits toxic fumes of NO_x.

DCD000 **CAS:54-96-6** **HR: 3**
3,4-DIAMINOPYRIDINE
mf: $C_5H_7N_3$ mw: 109.15

PROP: A solid. Mp: 218–219°.

SYNS: DIAMINO-3,4-PYRIDINE □ SC10

TOXICITY DATA WITH REFERENCE
ipr-mus LD50:20 mg/kg JMCMAR 8,296,65
scu-mus LD50:35 mg/kg AIPTAK 150,413,64
ivn-mus LD50:13 mg/kg APFRAD 26,345,68
orl-bwd LD50:75 mg/kg AECTCV 12,355,83

SAFETY PROFILE: Poison by ingestion, intraperitoneal, intravenous, and subcutaneous routes. When heated to decomposition it emits toxic fumes of NO_x.

DCE000 **CAS:636-23-7** **HR: 3**
2,4-DIAMINOTOLUENE DIHYDROCHLORIDE
mf: $C_7H_{10}N_2\cdot 2ClH$ mw: 195.11

SYN: METATOLYLENEDIAMINE DIHYDROCHLORIDE

TOXICITY DATA WITH REFERENCE
orl-rat TDLo:9900 mg/kg/78W-C:NEO JEPTDQ 2,325,78
orl-mus TDLo:32 g/kg/78W-C:NEO JEPTDQ 2,325,78
orl-rat TD:20 g/kg/78W-C:NEO JEPTDQ 2,325,78
orl-mus TD:65 g/kg/78W-C:NEO JEPTDQ 2,325,78
orl-mam LDLo:3000 mg/kg JIDHAN 13,87,31
ipr-mus LD50:80 mg/kg NCIBR* NIH-NCI-E-68-1311,10,73

CONSENSUS REPORTS: Reported in EPA TSCA Inventory.

SAFETY PROFILE: Poison by intraperitoneal route. Moderately toxic by ingestion. Questionable carcinogen with experimental neoplastigenic data. When heated to decomposition it emits very toxic fumes of NO_x and HCl.

DCE200 **CAS:615-45-2** **HR: 3**
2,5-DIAMINOTOLUENE DIHYDROCHLORIDE
mf: $C_7H_{10}N_2\cdot 2ClH$ mw: 195.11

SYNS: 2-METHYL-1,4-BENZENEDIAMINE DIHYDROCHLORIDE □ p-TOLUENEDIAMINE DIHYDROCHLORIDE

TOXICITY DATA WITH REFERENCE
mmo-sat 10 μg/plate MUREAV 245,15,90
ipr-mus TDLo:50 mg/kg (8D preg):TER FCTXAV 15,447,77
orl-rat LDLo:100 mg/kg NCNSA6 5,11,53

CONSENSUS REPORTS: Reported in EPA TSCA Inventory.

SAFETY PROFILE: Poison by ingestion. Experimental teratogenic effects. Mutation data reported. When heated to decomposition it emits very toxic fumes of NO_x and HCl. See also CHLORIDES.

DCE400 **CAS:15481-70-6** **HR: 3**
2,6-DIAMINOTOLUENE DIHYDROCHLORIDE
mf: $C_7H_{10}N_2\cdot 2ClH$ mw: 195.11

SYN: NCI-C50317

TOXICITY DATA WITH REFERENCE
mmo-sat 10 μg/plate SCIEAS 236,933,87
msc-mus:lym 200 mg/L SCIEAS 236,933,87
cyt-ham:lng 250 mg/L MUREAV 241,175,90
orl-rat LDLo:1 g/kg NTPTR* NCI-TR-200,80
orl-mus LDLo:100 mg/kg NTPTR* NCI-TR-200,80

CONSENSUS REPORTS: Carcinogenesis Bioassay Completed; Results Negative NCITR* NCI-CG-TR-200,80. NCI Carcinogenesis Bioassay (feed); No Evidence: mouse, rat NCITR* NCI-CG-TR-200,80.

SAFETY PROFILE: A poison by ingestion. When heated to decomposition it emits very toxic fumes of NO_x and HCl.

DCE600 **CAS:615-50-9** **HR: 3**
2,5-DIAMINOTOLUENE SULFATE
mf: $C_7H_{10}N_2 \cdot H_2O_4S$ mw: 220.27

SYNS: C.I. 76043 ☐ p-DIAMINOTOLUENE SULFATE ☐ 2,5-DIAMINOTOLUENE SULPHATE ☐ 2-METHYL-1,4-BENZENEDIAMINE SULFATE ☐ 2-METHYL-p-PHENYLENEDIAMINE SULPHATE ☐ NCI-C01832 ☐ p-TOLUENEDIAMINE SULFATE ☐ 2,5-TOLUENEDIAMINE SULFATE ☐ TOLUENE-2,5-DIAMINE, SULFATE (1:1) (8CI) ☐ p-TOLUENEDIAMINE SULPHATE ☐ TOLUENE-2,5-DIAMINE SULPHATE ☐ p-TOLUYLENEDIAMINE SULPHATE ☐ TOLUYLENE-2,5-DIAMINE SULPHATE ☐ p-TOLYLENEDIAMINE SULPHATE

TOXICITY DATA WITH REFERENCE
mmo-sat 250 μg/plate JJIND8 71,293,83
orl-rat LD50:98 mg/kg JTEHD6 2,657,77
ipr-rat LD50:49 mg/kg JTEHD6 2,657,77

CONSENSUS REPORTS: IARC Cancer Review: Animal Indefinite Evidence IMEMDT 16,97,78. NCI Carcinogenesis Bioassay Completed; Results Indefinite: mouse, rat NCITR* NCI-CG-TR-126,78. Reported in EPA TSCA Inventory.

SAFETY PROFILE: Poison by ingestion and intraperitoneal routes. Mutation data reported. When heated to decomposition it emits very toxic fumes of NO_x and SO_x. See also SULFATES.

DCE800 **CAS:1326-22-3** **HR: 3**
(4-((4,6-DIAMINO-m-TOLYL)IMINO)-2,5-CYCLOHEXADIEN-1-YLIDENE) DIMETHYLAMMONIUM CHLORIDE
mf: $C_{15}H_{19}N_4 \cdot Cl$ mw: 290.83

SYNS: CHLORIDE of DIAMINOMETHYLPHENYLDIMETHYL-p-BENZOQUINONE-DIIMINE ☐ C.I. 50411 ☐ C.I. 50435 ☐ TOLUYLENE BLUE (BASIC DYE)

TOXICITY DATA WITH REFERENCE
ivn-mus LD50:56 mg/kg CSLNX* NX#04800

SAFETY PROFILE: Poison by intravenous route. When heated to decomposition it emits very toxic fumes of NO_x and Cl^-, and NH_3.

DCF000 **CAS:767-17-9** **HR: 3**
4,6-DIAMINO-1,3,5-TRIAZINE-2-THIONE
mf: $C_3H_5N_5S$ mw: 143.19

SYNS: 4,6-DIAMINO-s-TRIAZINE-2-THIONE ☐ 4,6-DIAMINO-1,3,5-TRIAZINE-2(1H)-THIONE ☐ TETRAHYDRO-4,6-DIIMINO-s-TRIAZINE-2(1H)-THIONE ☐ USAF B-45 ☐ USAF CY-14

TOXICITY DATA WITH REFERENCE
ipr-mus LD50:200 mg/kg NTIS** AD277-689

CONSENSUS REPORTS: Reported in EPA TSCA Inventory.

SAFETY PROFILE: Poison by intraperitoneal route. When heated to decomposition it emits very toxic fumes of NO_x and SO_x.

DCF200 **CAS:1455-77-2** **HR: 2**
3,5-DIAMINO-s-TRIAZOLE
mf: $C_2H_5N_5$ mw: 99.12

PROP: Prisms from H_2O. Mp: 206°.

SYNS: GUANAZOLE ☐ NCI-C04819 ☐ NSC 1895

TOXICITY DATA WITH REFERENCE
mmo-sat 252 μmol/plate MUREAV 231,251,90
oms-hmn:oth 500 μg/L CNREA8 32,2661,72
ipr-rat TDLo:2500 mg/kg/7W-I:ETA CANCAR 40(Suppl 4),1935,77
ipr-mus TDLo:9750 mg/kg/26W-I:ETA CANCAR 40(Suppl 4),1935,77
ivn-hmn TDLo:8620 mg/kg/5D:BLD CCROBU 59,1117,75

CONSENSUS REPORTS: NCI Carcinogenesis Studies (ipr); Equivocal Evidence: rat CANCAR 40,1935,77; No Evidence: mouse CANCAR 40,1935,77. Reported in EPA TSCA Inventory.

SAFETY PROFILE: Human systemic effects by intravenous route: leukopenia (reduced white blood cell count) and thrombocytopenia (reduced blood platelet count). Human mutation data reported. Questionable carcinogen with experimental tumorigenic data. When heated to decomposition it emits toxic fumes of NO_x.

DCF600 **CAS:6818-18-4** **HR: 2**
3,10-DIAMINOTRICYCLO(5.2.1.0^{2,6})DECANE
mf: $C_{10}H_{15}N_2O_4$ mw: 227.27

SYN: HEXAHYDRO-1,8-DIAMINO-4,7-METHANOINDAN

TOXICITY DATA WITH REFERENCE
orl-rat LD50:1070 mg/kg TXAPA9 28,313,74
skn-rbt LD50:530 mg/kg TXAPA9 28,313,74

SAFETY PROFILE: Moderately toxic by ingestion and skin contact. When heated to decomposition it emits toxic fumes of NO_x.

DCF700 **CAS:36039-40-4** **HR: 3**
2,5-DIAMINOTROPONE
mf: $C_7H_8N_2O$ mw: 136.17

SYN: 2,5-DIAMINO-2,4,6-CYCLOPHETATRIEN-1-ONE

TOXICITY DATA WITH REFERENCE
ipr-mus LD50:174 mg/kg CPBTAL 20,60,72
scu-mus LD50:361 mg/kg CPBTAL 20,60,72
ivn-mus LD50:192 mg/kg CPBTAL 20,60,72

SAFETY PROFILE: Poison by subcutaneous, intravenous, and intraperitoneal routes. When heated to decomposition it emits toxic fumes of NO_x.

DCF710 **CAS:34692-97-2** **HR: 3**
2,5-DIAMINOTROPONE HYDROCHLORIDE
mf: $C_7H_8N_2O•ClH$ mw: 172.63

SYN: 2,5-DIAMINO-2,4,6-CYCLOHEPTATRIEN-1-ONE HDYROCHLORIDE

TOXICITY DATA WITH REFERENCE
ipr-mus LD50:174 mg/kg YKKZAJ 91,1307,71
scu-mus LD50:361 mg/kg YKKZAJ 91,1307,71
ivn-mus LD50:192 mg/kg YKKZAJ 92,19,72

SAFETY PROFILE: Poison by subcutaneous, intravenous, and intraperitoneal routes. When heated to decomposition it emits toxic fumes of NO_x and HCl.

DCF725 **CAS:28965-70-0** **HR: 3**
DIAMMINEBORONIUM HEPTAHYDROTETRABORATE
mf: $B_5H_{15}N_2$ mw: 97.18

SYN: PENTABORANE(9)DIAMMONIATE

SAFETY PROFILE: Decomposes violently at room temperature. When heated to decomposition it emits toxic fumes of NH_3 and NO_x. See also BORON COMPOUNDS and BORANES.

DCF750 **CAS:23777-63-1** **HR: 3**
DIAMMINEBORONIUM TETRAHYDROBORATE
mf: $B_2H_{12}N_2$ mw: 61.73

SAFETY PROFILE: Ignites when heated in air. When heated to decomposition it emits toxic fumes of NO_x. See also BORON COMPOUNDS and BORANES.

DCF800 **CAS:41575-87-5** **HR: 3**
cis-DIAMMINEDINITRATO PLATINUM (II)
mf: $H_6N_4O_4Pt$ mw: 321.17

SYN: PLATINUM(II) DINITRODIAMMINE

TOXICITY DATA WITH REFERENCE
oth-esc 10 mmol/L NEOLA4 37,667,90
ipr-mus LDLo:7 mg/kg BICHBX 2,187,73

CONSENSUS REPORTS: Reported in EPA TSCA Inventory.

SAFETY PROFILE: Poison by intraperitoneal route. Mutation data reported. Explosive decomp @ 200°C. When heated to decomposition it emits toxic fumes of NO_x. See also PLATINUM COMPOUNDS and NITRATES.

DCG000 **HR: 3**
DIAMMINEMALONATO PLATINUM (II)
mf: $C_3H_8N_2O_4Pt$ mw: 331.22

TOXICITY DATA WITH REFERENCE
mmo-sat 2 μg/plate MUREAV 77,45,80
mma-sat 2 μg/plate MUREAV 77,45,80
ipr-mus LD50:225 mg/kg CBINA8 5,415,72

CONSENSUS REPORTS: Reported in EPA TSCA Inventory.

SAFETY PROFILE: Poison by intraperitoneal route. Mutation data reported. When heated to decomposition it emits toxic fumes of NO_x. See also PLATINUM COMPOUNDS.

DCG600 **CAS:28068-05-5** **HR: 3**
DIAMMINEPALLADIUM (II) NITRATE
mf: $H_6N_4O_6Pd$ mw: 264.48

$[(H_3N)_2Pd][NO_2]_2$

SAFETY PROFILE: The dry nitrate is a moderately impact-sensitive explosive. When heated to decomposition it emits toxic fumes of NO_x. See also NITRATES.

DCG800 **CAS:7784-44-3** **HR: 3**
DIAMMONIUM HYDROGEN ARSENATE
DOT: UN 1546
mf: $AsH_3O_4•2H_3N$ mw: 176.03

PROP: White powder or plate-like colorless crystals or prisms. Mp: decomp to yield NH_3.

SYNS: AMMONIUM ACID ARSENATE ☐ AMMONIUM ARSENATE, solid (DOT) ☐ DIAMMONIUM ARSENATE ☐ DIAMMONIUM MONOHYDROGEN ARSENATE ☐ DIBASIC AMMONIUM ARSENATE ☐ SECONDARY AMMONIUM ARSENATE

CONSENSUS REPORTS: Arsenic and its compounds are on the Community Right-To-Know List.

OSHA PEL: Cancer Hazard
ACGIH TLV: TWA 0.2 mg(As)/m³ (Proposed: 0.01 mg (As)/m³; Human Carcinogen)
NIOSH REL: (Inorganic Arsenic) CL 0.002 mg(As)/m³/15M
DOT CLASSIFICATION: 6.1; *Label:* Poison

SAFETY PROFILE: A poison. When heated to decomposition it emits very toxic fumes of As, NO_x, and NH_3. See also ARSENIC.

DCH000 **CAS:3164-29-2** **HR: 3**
DIAMMONIUM TARTRATE
mf: $C_4H_6O_6•2H_3N$ mw: 184.18

SYNS: AMMONIUM TARTRATE (DOT) ☐ AMMONIUM-d-TARTRATE ☐ 2,3-DIHYDROXYBUTANEDIOIC ACID, DIAMMONIUM SALT ☐ l-TARTARIC ACID, AMMONIUM SALT ☐ TARTARIC ACID, DIAMMONIUM SALT

TOXICITY DATA WITH REFERENCE
scu-rbt LD50:1130 mg/kg HBAMAK 4,1289,35
ivn-rbt LD50:113 mg/kg HBAMAK 4,1289,35

CONSENSUS REPORTS: Reported in EPA TSCA Inventory.

SAFETY PROFILE: Poison by intravenous route. Moderately toxic by subcutaneous route. When heated to decomposition it emits very toxic fumes of NH_3 and NO_x.

DCH200 **CAS:2050-92-2** **HR: 3**
DIAMYL AMINE
DOT: UN 2841
mf: $C_{10}H_{23}N$ mw: 157.34

PROP: Water-white liquid. Bp: 210–211°, flash p: 124°F, d: 0.777 @ 20°/20°, vap d: 5.42.

SYNS: DI-n-AMYLAMINE (DOT) ☐ DIPENTYLAMINE ☐ PENTYL PENTYLAMINE

TOXICITY DATA WITH REFERENCE
skn-rbt 500 mg open SEV UCDS** 8/9/68
orl-rat LD50:270 mg/kg UCDS** 8/9/68
ihl-rat LCLo:63 ppm/4H AIHAAP 23,95,62
skn-rbt LD50:350 mg/kg AIHAAP 23,95,62

CONSENSUS REPORTS: Reported in EPA TSCA Inventory.

DOT CLASSIFICATION: 6.1; *Label:* KEEP AWAY FROM FOOD

SAFETY PROFILE: Poison by inhalation, ingestion, and skin contact. A severe skin irritant. See also AMINES. Flammable liquid when exposed to heat or flame; can react with oxidizing materials. To fight fire, use alcohol foam, foam, CO_2, dry chemical. When heated to decomposition it emits toxic fumes of NO_x.

DCH400 **CAS:79-74-3** **HR: 3**
2,5-DI-tert-AMYLHYDROQUINONE
mf: $C_{16}H_{26}O_2$ mw: 250.42

SYNS: 2,5-BIS(1,1-DIMETHYLPROPYL)HYDROQUINONE ☐ 2,5-DI-tert-PENTYLHYDROQUINONE ☐ SANTOUAR A ☐ SANTOVAR A ☐ USAF B-21

TOXICITY DATA WITH REFERENCE
orl-rat LD50:2 g/kg IPSTB3 3,93,76
ipr-mus LD50:200 mg/kg NTIS** AD277-689
orl-rbt LD50:2 g/kg IPSTB3 3,93,76

CONSENSUS REPORTS: Reported in EPA TSCA Inventory.

SAFETY PROFILE: Poison by intraperitoneal route. Low toxicity by ingestion. When heated to decomposition it emits acrid smoke and irritating fumes.

DCH600 **CAS:13256-06-9** **HR: 2**
DI-n-AMYLNITROSAMINE
mf: $C_{10}H_{22}N_2O$ mw: 186.34

SYNS: DIAMYLNITROSAMIN (GERMAN) ☐ DIPENTYLNITROSAMINE ☐ DI-n-PENTYLNITROSAMINE ☐ N-NITROSODIPENTYLAMINE ☐ N-NITROSODI-n-PENTYLAMINE

TOXICITY DATA WITH REFERENCE
mma-sat 465 µg/plate PNASA6 72,5135,75
mma-ham:lng 500 µmol/L IAPUDO 27,179,80
orl-rat TDLo:7733 mg/kg/8W-C:CAR IJCNAW 27,249,81
scu-rat TDLo:11 g/kg/25W-I:ETA NATWAY 49,111,62
scu-mus TDLo:11 g/kg/17W-I:ETA BEXBAN 83,83,77
scu-rat TD:12 g/kg/49W-C:ETA ZEKBAI 69,103,67
orl-rat TD:48 g/kg/51W-C:ETA ZEKBAI 69,103,67
orl-rat LD50:1750 mg/kg NATWAY 48,134,61
scu-rat LD50:3000 mg/kg ZEKBAI 69,103,67

CONSENSUS REPORTS: EPA Genetic Toxicology Program.

SAFETY PROFILE: Moderately toxic by ingestion and subcutaneous routes. Questionable carcinogen with experimental carcinogenic and tumorigenic data. Mutation data reported. When heated to decomposition it emits toxic fumes of NO_x. See also NITROSAMINES.

For occupational chemical analysis use OSHA: #38.

DCH800 **CAS:28652-04-2** **HR: 2**
DIAMYLPHENOL
mf: $C_{16}H_{26}O$ mw: 234.42

PROP: Liquid. Bp: 278°, flash p: 260°F (OC), d: 0.93–0.94.

SYN: DIPENTYL PHENOL

TOXICITY DATA WITH REFERENCE
skn-hmn 250 mg/48H MOD AMIHBC 5,311,52
skn-rbt 500 mg SEV AMIHBC 5,311,52
ipr-rat LD50:620 mg/kg AMIHBC 5,311,52

SAFETY PROFILE: Moderately toxic by intraperitoneal route. A severe skin irritant experimentally. A human skin irritant. Combustible when exposed to heat or flame. Can react with oxidizing materials. To fight fire, use CO_2, dry chemical. When heated to decomposition it emits acrid smoke and fumes. See also PHENOL.

DCI000 **CAS:120-95-6** **HR: 3**
DI-tert-AMYLPHENOL
mf: $C_{16}H_{26}O$ mw: 234.42

SYNS: 2,4-DI-tert-AMYLPHENOL ☐ 2,4-DI-tert-PENTYLPHENOL ☐ PHENOL, 2,4-DI-tert-PENTYL- ☐ PRODOX 156

TOXICITY DATA WITH REFERENCE
eye-rbt 100 mg MOD IHFCAY 6,1,67
orl-rat LD50:330 mg/kg IHFCAY 6,1,67

CONSENSUS REPORTS: Reported in EPA TSCA Inventory.

DOT CLASSIFICATION: 6.1; *Label:* KEEP AWAY FROM FOOD

SAFETY PROFILE: Poison by ingestion. An eye irritant. When heated to decomposition it emits acrid smoke and irritating fumes. See also PHENOL.

DCI400 **CAS:35865-33-9** **HR: 3**
DIANEMYCIN
mf: $C_{47}H_{78}O_{14}$ mw: 867.25

PROP: Crystals. Mp: 72–74° from Me_2CO (aq), mp: 156–157° from EtOH (aq), vap d: 8.1.

TOXICITY DATA WITH REFERENCE
orl-mus LD50:150 mg/kg 37ASAA 3,47,78
ipr-mus LD50:9 mg/kg 37ASAA 3,47,78
scu-mus LD50:40 mg/kg 85ERAY 1,805,78

SAFETY PROFILE: Poison by ingestion, intraperitoneal, and subcutaneous routes. When heated to decomposition it emits acrid smoke and irritating fumes.

DCI600 **CAS:23261-20-3** **HR: 3**
DIANHYDROGALACTITOL
mf: $C_6H_{10}O_4$ mw: 146.16

SYNS: DAD □ DAG □ DIANHYDROCULCITOL □ 1,2:5,6-DIANHYDRODULCITOL □ 1,2:5,6-DIANHYDROGALACTITOL □ 1,2:5,6-DIEPOXYDULCITOL □ DULCITOLDIEPOXIDE □ NSC 132313

TOXICITY DATA WITH REFERENCE
mmo-sat 100 µg/plate CRNGDP 3,333,82
dni-mus-ipr 5 mg/kg NEOLA4 31,667,84
orl-rat LD50:14 mg/kg CCROBU 56,593,72
ipr-rat LD50:11 mg/kg CCROBU 56,593,72
ivn-rat LD50:16 mg/kg CCROBU 56,593,72
orl-mus LD50:7899 µg/kg NCISP* JAN86
ipr-mus LD50:15 mg/kg CCROBU 56,593,72
scu-mus LD50:16,500 µg/kg NCISP* JAN86
ivn-mus LD50:21 mg/kg CCROBU 56,593,72
ivn-dog LDLo:16 mg/kg CTRRDO 60,1585,76

SAFETY PROFILE: Poison by ingestion, intravenous, subcutaneous, and intraperitoneal routes. Mutation data reported. When heated to decomposition it emits acrid smoke and irritating fumes.

DCI800 **CAS:19895-66-0** **HR: 3**
DIANHYDROMANNITOL
mf: $C_6H_{10}O_4$ mw: 146.16

SYNS: 1:2:5:6-DIANHYDRO-d-MANNITOL □ NSC-133129

TOXICITY DATA WITH REFERENCE
dnd-ckn:leu 30 mmol/L TELEAY (29),2477,75
dni-rbt:bmr 274 µmol/L BCPCA6 25,1705,76
ipr-rat LD50:14 mg/kg EJCAAH 4,617,68
ipr-mus LD50:20 mg/kg EJCODS 18,573,82

SAFETY PROFILE: Poison by intraperitoneal route. Mutation data reported. When heated to decomposition it emits acrid smoke and irritating fumes.

DCJ000 **CAS:73928-11-7** **HR: 3**
DIANILINOMERCURY
mf: $C_{12}H_{12}HgN_2$ mw: 384.85

SYN: N,N'-MERCURIDIANILINE

TOXICITY DATA WITH REFERENCE
ivn-mus LD50:180 mg/kg CSLNX* NX#05148

CONSENSUS REPORTS: Mercury and its compounds are on the Community Right-To-Know List.

OSHA PEL: CL 0.1 mg(Hg)/m³ (skin)
ACGIH TLV: TWA 0.1 mg(Hg)/m³ (skin)
NIOSH REL: (Mercury, Aryl and Inorganic): CL 0.1 mg/m³ (skin)

SAFETY PROFILE: Poison by intravenous route. See also MERCURY COMPOUNDS. When heated to decomposition it emits very toxic fumes of NO_x and Hg.

DCJ200 **CAS:119-90-4** **HR: 3**
o-DIANISIDINE
mf: $C_{14}H_{16}N_2O_2$ mw: 244.32

PROP: Colorless leaflets or crystals. Mp: 137–138°, flash p: 403°F, vap d: 8.5. Sol in C_6H_6 and AcOH; sltly sol in H_2O.

SYNS: ACETAMINE DIAZO BLACK RD □ AMACEL DEVELOPED NAVY SD □ AZOENE FAST BLUE BASE □ AZOFIX BLUE B SALT □ AZOGNE FAST BLUE B □ BLUE BN BALSE □ BRENTAMINE FAST BLUE B BASE □ CELLITAZOL B □ C.I. 24110 □ C.I. AZOIC DIAZO COMPONENT 48 □ CIBACETE DIAZO NAVY BLUE 2B □ C.I. DISPERSE BLACK 6 □ DIACELLITON FAST GREY G □ DIACEL NAVY DC □ o-DIANISIDIN (CZECH, GERMAN) □ o-DIANISIDINA (ITALIAN) □ O,O'DIANISIDINE □ 3,3'-DIANISIDINE □ DIATO BLUE BASE B □ 3,3'-DIMETHOXYBENZIDIN (CZECH) □ 3,3'-DIMETHOXYBENZIDINE □ 3,3'-DIMETOSSIBENZODINA (ITALIAN) □ FAST BLUE B BASE □ HILTONIL FAST BLUE B BASE □ HILTOSAL FAST BLUE B SALT □ HINDASOL BLUE B SALT □ KAKO BLUE B SALT □ KAYAKU BLUE B BASE □ LAKE BLUE B BASE □ MEISEI TERYL DIAZO BLUE HR □ MITSUI BLUE B BASE □ NAPHTHANIL BLUE B BASE □ NEUTROSEL NAVY BN □ RCRA WASTE NUMBER U091 □ SANYO FAST BLUE SALT B □ SETACYL DIAZO NAVY R □ SPECTROLENE BLUE B

TOXICITY DATA WITH REFERENCE
mmo-sat 333 µg/plate ENMUDM 5(Suppl 1),3,83
mma-sat 1 µg/plate IGAYAY 123,18,82
sce-ham:ovr 500 µg/L ENMUDM 7,1,85
oms-dog:oth 100 µmol/L CNREA8 44,1893,84
dnd-dog:oth 100 µmol/L CNREA8 44,1893,84
orl-rat TDLo:12 g/kg/56W-I:ETA GTPZAB 9,18,65
orl-ham TDLo:588 g/kg/70W-C:ETA PAACA3 10,78,69
orl-rat LD50:1920 mg/kg 28ZPAK -,119,72
orl-dog LDLo:600 mg/kg AEXPBL 58,167,1907

CONSENSUS REPORTS: NTP 7th Annual Report On Carcinogens. IARC Cancer Review: Group 2B IMEMDT 7,198,87; Animal Sufficient Evidence IMEMDT 4,41,74. EPA Genetic Toxicology Program. Community Right-To-Know List. Reported in EPA TSCA Inventory.

DFG MAK: Animal Carcinogen, Suspected Human Carcinogen
NIOSH REL: (Benzidine-based dye) Reduce to lowest feasible level

SAFETY PROFILE: Confirmed carcinogen with experimental tumorigenic data. Moderately toxic by ingestion. Mutation data reported. Combustible when exposed to heat or flame. When heated to decomposition it emits toxic fumes of NO_x.

For occupational chemical analysis use OSHA: #ID-71.

DCJ400 **CAS:91-93-0** **HR: 3**
DIANISIDINE DIISOCYANATE
mf: $C_{16}H_{12}N_2O_4$ mw: 296.30

SYNS: 4,4'-DIISOCYANATO-3,3'-DIMETHOXY-1,1'-BIPHENYL □ 3,3'-DIMETHOXYBENZIDINE-4,4'-DIISOCYANATE □ 3,3'-DIMETHOXY-4,4'-BIPHENYLENE DIISOCYANATE □ NCI-C02175

TOXICITY DATA WITH REFERENCE
mmo-sat 3300 ng/plate ENMUDM 7(Suppl 5),1,85
mma-sat 3 µg/plate ENMUDM 7(Suppl 5),1,85
orl-rat TDLo:565 g/kg/78W-I:CAR NCITR* NCI-CG-TR-128,79
orl-rat TD:1200 g/kg/78W-I:CAR NCITR* NCI-CG-TR-128,79
ivn-mus LD50:180 mg/kg CSLNX* NX#02411

CONSENSUS REPORTS: IARC Cancer Review: Group 3

IMEMDT 7,56,87; Animal Limited Evidence IMEMDT 39,279,86. NCI Carcinogenesis Bioassay (feed); No Evidence: mouse NCITR* NCI-CG-TR-128,79; Clear Evidence: rat NCITR* NCI-CG-TR-128,79.

NIOSH REL: (Diisocyanates) TWA 0.005 ppm; CL 0.02 ppm/10M

SAFETY PROFILE: Poison by intravenous route. A strong sensitizer. Questionable carcinogen with experimental carcinogenic data. When heated to decomposition it emits toxic fumes of NO_x. See also CYANATES.

DCJ450 HR: 3
DIANTHUS SUPERBUS L., extract

PROP: A Southeast Asian carnation belonging to the family Caryophyllaceae (ZKPAK 2,366,69).

TOXICITY DATA WITH REFERENCE
scu-mus TDLo:20 g/kg (female 2D pre):REP MPHEAE 16,414,67
scu-mus TDLo:40 g/kg (female 4-8D post):REP IZKPAK 2,366,69
scu-mus TDLo:40 g/kg (female 1-5D post):REP IZKPAK 2,366,69
ipr-mus LD50:100 mg/kg IJEBA6 22,312,84

SAFETY PROFILE: Poison by intraperitoneal route. Experimental reproductive effects. When heated to decomposition it emits acrid smoke and irritating fumes.

DCJ600 CAS:13601-02-0 HR: 3
DIAQUODIAMMINEPLATINUM DINITRATE
mf: $H_{10}N_2O_2Pt{\cdot}N_2O_6$ mw: 389.23

SYN: cis-DIAQUODIAMMINEPLATINUM(II) DINITRATE

TOXICITY DATA WITH REFERENCE
mmo-sat 5 μg/plate MUREAV 48,139,77
idr-hmn TDLo:40 mg/kg:SKN CNREA8 35,2766,75
ipr-mus LDLo:5 mg/kg BICHBX 2,187,73

SAFETY PROFILE: Poison by intraperitoneal route. Human systemic skin effects by intradermal route. Mutagenic data reported. When heated to decomposition it emits toxic fumes of NO_x. See also NITRATES and PLATINUM COMPOUNDS.

DCJ800 CAS:61790-53-2 HR: 1
DIATOMACEOUS EARTH

PROP: Composed of skeletons of small aquatic plants related to algae and contains as much as 88% amorphous silica (DTLVS* 4,120,80). White to buff colored solid. Insol in water; sol in hydrofluoric acid.

SYNS: AMORPHOUS SILICA ☐ CELITE ☐ D.E. ☐ DIATOMACEOUS EARTH, NATURAL ☐ DIATOMACEOUS SILICA ☐ DIATOMITE ☐ INFUSORIAL EARTH ☐ KIESELGUHR ☐ SILICA, AMORPHOUS-DIATOMACEOUS EARTH (UNCALCINED) (ACGIH)

CONSENSUS REPORTS: IARC Cancer Review: Group 3 IMEMDT 7,341,87; Animal Inadequate Evidence IMEMDT 42,39,87; Human Inadequate Evidence IMEMDT 42,39,87. Reported in EPA TSCA Inventory.

OSHA PEL: TWA 6 mg/m^3
ACGIH TLV: TWA (nuisance particulate) 10 mg/m^3 of total dust (when toxic impurities are not present, e.g., quartz $<1\%$)
DFG MAK: 4 mg/m^3 as fine dust

SAFETY PROFILE: A nuisance dust that may cause fibrosis of the lungs. Roasting or calcining at high temperatures produces cristobalite and tridymite, thus increasing the fibrogenicity of the material. A questionable carcinogen.

DCJ850 CAS:7084-07-3 HR: 3
DIATRIN HYDROCHLORIDE
mf: $C_{15}H_{20}N_2S{\cdot}ClH$ mw: 296.89

PROP: A solid. Mp: 186–187°. Sol in H_2O.

SYNS: ENSTAMINE HYDROCHLORIDE ☐ METHAPHENILENE HYDROCHLORIDE ☐ NILHISTIN

TOXICITY DATA WITH REFERENCE
orl-mus LD50:550 mg/kg JPETAB 93,210,48
ipr-mus LD50:117 mg/kg MEIEDD 10,854,83
scu-mus LD50:160 mg/kg JPETAB 93,210,48
ivn-mus LD50:45 mg/kg JPETAB 93,210,48
ivn-rbt LD50:30 mg/kg JPETAB 93,210,48
orl-gpg LD50:900 mg/kg JPETAB 93,210,48
scu-gpg LD50:140 mg/kg JPETAB 93,210,48
ivn-gpg LD50:30 mg/kg JPETAB 93,210,48

SAFETY PROFILE: Poison by subcutaneous, intravenous, and intraperitoneal routes. Moderately toxic by ingestion. When heated to decomposition it emits toxic fumes of SO_x, NO_x, and HCl. See also AMINES.

DCK000 CAS:117-96-4 HR: 1
DIATRIZOIC ACID
mf: $C_{11}H_9I_3N_2O_4$ mw: 613.92

PROP: Crystals from EtOH (aq).

SYNS: AMIDOTRIZOIC ACID ☐ 3,5-BIS(ACETYLAMINO)-2,4,6-TRIIODOBENZOIC ACID ☐ 3,5-DIACETAMIDO-2,4,6-TRIIODOBENZOIC ACID ☐ DIAT (GERMAN) ☐ DIATRIZOESAURE (GERMAN) ☐ ODISTON ☐ UROGRAFIN ACID ☐ UROGRANOIC ACID ☐ UROTRAST

TOXICITY DATA WITH REFERENCE
ipr-rat LD50:14,300 mg/kg ARZNAD 15,222,65
ivn-rat LD50:11,300 mg/kg ARZNAD 15,222,65
ivn-mus LD50:8900 mg/kg KSRNAM 19,2411,85
ipr-gpg LD50:13 g/kg ARZNAD 15,222,65

SAFETY PROFILE: Mildly toxic by intraperitoneal and intravenous routes. When heated to decomposition it emits very toxic fumes of NO_x and I^-.

DCK200 CAS:34494-09-2 HR: 2
6,12-DIAZAANTHANTHRENE SULFATE
mf: $C_{20}H_{10}N_2{\cdot}H_2O_4S$ mw: 376.40

SYNS: ACRIDINO(2,1,9,8-klmna)ACRIDINE SULFATE ☐ 6,12-DIAZAANTHANTHRENE SULPHATE

TOXICITY DATA WITH REFERENCE
imp-rat TDLo:600 mg/kg:ETA NEOLA4 18,591,71

SAFETY PROFILE: Questionable carcinogen with ex-

perimental tumorigenic data. When heated to decomposition it emits very toxic fumes of NO_x and SO_x. See also SULFATES.

DCK400 **CAS:280-57-9** **HR: 2**
1,4-DIAZABICYCLO(2,2,2)OCTANE
mf: $C_6H_{12}N_2$ mw: 112.20

$C_2H_4NC_2H_4\ NCH_2CH_2$

PROP: Hygroscopic crystals. Mp: 158°, bp: 174°.

SYNS: BICYCLO(2,2,2)-1,4-DIAZAOCTANE □ DABCO □ DABCO CRYSTAL □ DABCO EG □ DABCO 33LV □ DABCO R-8020 □ DABCO S-25 □ D 33LV □ 1,4-ETHYLENEPIPERAZINE □ TRIETHYLENEDIAMINE

TOXICITY DATA WITH REFERENCE
skn-rbt 2500 μg open MLD TXAPA9 4,522,62
eye-rbt 25 mg MOD TXAPA9 4,522,62
orl-rat LD50:1700 mg/kg ZHYGAM 20,393,74
orl-rbt LD50:1100 mg/kg GISAAA 45(5),67,80
orl-gpg LD50:2250 mg/kg GISAAA 45(5),67,80

CONSENSUS REPORTS: Reported in EPA TSCA Inventory.

SAFETY PROFILE: Moderately toxic by ingestion. A skin and eye irritant, allergen and skin sensitizer. A powerful base. Forms an explosive complex with hydrogen peroxide. Mixtures with carbon auto-ignite at 230°C. Very exothermic reaction with cellulose nitrate. When heated to decomposition it emits toxic fumes of NO_x. See also AMINES.

DCK500 **HR: 3**
1,4-DIAZABICYCLO(2.2.2)OCTANE HYDROGEN PEROXIDATE
mf: $C_6H_{12}N_2 \cdot H_2O_2$ mw: 146.19

$C_2H_4NC_2H_4\ NCH_2CH_2 \cdot H_2O_2$

SAFETY PROFILE: Complex explodes when dried at room temperature. Upon decomposition it emits toxic fumes of NO_x. See also PEROXIDES.

DCK700 **CAS:283-66-9** **HR: 3**
1,6-DIAZA-3,4,8,9,12,13-HEXAOXABICYCLO(4.4.4) TETRADECANE
mf: $C_6H_{12}N_2O_6$ mw: 208.17

PROP: Crystals.

SYN: HEXAMETHYLENETRIPEROXYDIAMINE

DOT CLASSIFICATION: Forbidden

SAFETY PROFILE: The dry material is a powerful explosive that is heat- and shock-sensitive. Explodes on contact with bromine or sulfuric acid. When heated to decomposition it emits toxic fumes of NO_x. See also PEROXIDES.

DCK759 **CAS:439-14-5** **HR: 3**
DIAZEPAM
mf: $C_{16}H_{13}ClN_2O$ mw: 284.76

PROP: Plates or crystals from Me_2CO/pet ether. Mp: 125–126°.

SYNS: ALBORAL □ AMIPROL □ ANSIOLISINA □ APAURIN □ APOZEPAM □ ATENSINE □ ATILEN □ BIALZEPAM □ CALMOCITENE □ CERCINE □ 7-CHLORO-1,3-DIHYDRO-1-METHYL-5-PHENYL-2H-1,4-BENZODIAZEPIN-2-ONE □ 7-CHLORO-1-METHYL-5-3H-1,4-BENZODIAZEPIN-2(1H)-ONE □ 7-CHLORO-1-METHYL-2-OXO-5-PHENYL-3H-1,4-BENZODIAZEPINE □ 7-CHLORO-1-METHYL-5-PHENYL-2H-1,4-BENZODIAZEPIN-2-ONE □ 7-CHLORO-1-METHYL-5-PHENYL-1,3-DIHYDRO-2H-1,4-BENZODIAZEPIN-2-ONE □ CONDITION □ DIACEPAN □ DIAPAM □ DIAZETARD □ DIENPAX □ DIPAM □ DOMALIUM □ DUKSEN □ E-PAM □ ERIDAN □ FAUSTAN □ FRUSTAN □ GIHITAN □ KIATRIUM □ LEMBROL □ LEVIUM □ LIBERETAS □ METHYL DIAZEPINONE □ 1-METHYL-5-PHENYL-7-CHLORO-1,3-DIHYDRO-2H-1,4-BENZODIAZEPIN-2-ONE □ MOROSAN □ NSC-77518 □ PACITRAN □ PAXATE □ PLIDAN □ QUETINIL □ QUIATRIL □ RELAMINAL □ RELANIUM □ RENBORIN □ SAROMET □ SEDIPAM □ SEDUXEN □ SERENACK □ SERENZIN □ SETONIL □ SONACON □ STESOLID □ TENSOPAM □ TRANIMUL □ TRANQUIRIT □ UMBRIUM □ UNISEDIL □ VALEO □ VALITRAN □ VALIUM □ VATRAN □ VIVAL □ ZIPAN

TOXICITY DATA WITH REFERENCE
mma-sat 958 nmol/plate CNREA8 38,4478,78
cyt-wmn-unr 328 mg/kg/78W AJOGAH 107,456,70
cyt-hmn:leu 10 mg/L AJOGAH 103,836,69
orl-wmn TDLo:22,800 μg/kg (25-36W preg):REP JOPDAB 90,123,77
ivn-wmn TDLo:400 μg/kg (female 39W post):TER JOGBAS 79,635,72
orl-wmn TDLo:11,600 μg/kg (female 43D post):TER TJADAB 30,179,84
orl-mus TDLo:504 mg/kg (female 1-21D post):REP PLRCAT 9,325,77
scu-rat TDLo:1400 μg/kg (female 15-21D post):REP TOLED5 16,131,83
orl-rat TDLo:675 mg/kg (female 17-22D post):REP OYYAA2 17,787,79
scu-rat TDLo:20 mg/kg (female 13-20D post):REP PSCHDL 79,332,83
orl-mus TDLo:828 mg/kg (male 5D pre):REP PLRCAT 9,187,77
orl-rbt TDLo:1170 mg/kg (female 6-18D post):TER JZKEDZ 7,79,81
orl-rat TDLo:5 g/kg (female 8-17D post):TER OYYAA2 15,797,78
orl-rat TDLo:10,500 mg/kg (female 14D pre):REP OYYAA2 15,813,78
orl-mus TDLo:2 g/kg (male 5D pre):REP RCOCB8 13,601,76
orl-mus TDLo:140 mg/kg (female 13D post):REP TXAPA9 32,53,75
orl-dog TDLo:1820 mg/kg (female 26W pre):REP OYYAA2 15,777,78
orl-rat TDLo:27,300 mg/kg (female 91D pre):REP OYYAA2 20,1061,80
orl-mus TDLo:312 mg/kg (female 26D pre):REP RCOCB8 11,155,75
orl-rat TDLo:220 mg/kg (female 1-22D post):REP BEXBAN 78,1156,74

D

ipr-rat TDLo: 500 mg/kg (male 10D pre): REP ARANDR 3,31,79
orl-dog TDLo: 300 mg/kg (male 60D pre): REP OYYAA2 15,777,78
orl-rat TDLo: 14 g/kg (male 35D pre): REP KSRNAM 20,1445,86
scu-mus TDLo: 45 mg/kg (female 9D post): TER DGDFA5 22,61,80
orl-mus TDLo: 100 mg/kg (female 12D post): TER TXAPA9 32,53,75
ipr-ham TDLo: 280 mg/kg (female 8D post): TER LIFSAK 29,2141,81
orl-mus TDLo: 42 g/kg/80W-C: ETA TXAPA9 57,39,81
ivn-inf TDLo: 150 µg/kg: SKN,BIO BMJOAE 2,298,77
orl-man TDLo: 143 µg/kg: EYE BCPHBM 1,335,74
ims-wmn TDLo: 181 µg/kg: CNS,CVS BMJOAE 1,144,77
ivn-man TDLo: 143 µg/kg: PUL,CNS JAMAAP 238,1052,77
ivn-man TDLo: 71 µg/kg/1M-C: CVS DICPBB 17,125,83
orl-rat LD50: 352 mg/kg JTCTDW 20,271,83
ipr-rat LD50: 46,500 µg/kg IYKEDH 23,682,92
scu-rat LD50: 6350 µg/kg IYKEDH 23,682,92
orl-mus LD50: 48 mg/kg PCJOAU 17,30,83
skn-mus LD50: 800 mg/kg AREAD8 (4),57,80
ipr-mus LD50: 47 mg/kg AIPTAK 253,164,81
ivn-mus LD50: 25 mg/kg ARZNAD 31,2180,81
par-mus LD50: 150 mg/kg RPTOAN 33,70,70
ivn-rbt LD50: 9 mg/kg IJNEAQ 5,305,66

CONSENSUS REPORTS: IARC Cancer Review: Group 3 IMEMDT 7,189,87; Animal Inadequate Evidence IMEMDT 7,189,87; Human Inadequate Evidence IMEMDT 7,189,87. Reported in EPA TSCA Inventory. EPA Genetic Toxicology Program.

SAFETY PROFILE: Poison by ingestion, parenteral, subcutaneous, intravenous, and intraperitoneal routes. Moderately toxic by skin contact. Questionable carcinogen with experimental tumorigenic data. Human systemic effects: dermatitis, effect on inflammation or mediation of inflammation, change in cardiac rate, somnolence, respiratory depression, and other respiratory changes, visual field changes, diplopia (double vision), change in motor activity, muscle contraction or spasticity, ataxia (loss of muscle coordination), an antipsychotic and general anesthetic. Human reproductive effects by ingestion and intravenous routes causing developmental abnormalities of the fetal cardiovascular (circulatory) system and postnatal effects. Experimental teratogenic and reproductive effects. Human mutation data reported. An allergen. A drug for the treatment of anxiety. When heated to decomposition it emits very toxic fumes of Cl^- and NO_x.

DCL100 CAS:13556-50-8 HR: 3
1,3-DIAZIDOBENZENE
mf: $C_6H_4N_6$ mw: 160.14

PROP: Pale-yellow needles by steam distillation.

SAFETY PROFILE: Ignites and may explode weakly on contact with concentrated acids. When heated to decomposition it emits toxic fumes of NO_x. See also AZIDES.

DCL125 CAS:2294-47-5 HR: 3
1,4-DIAZIDOBENZENE
mf: $C_6H_4N_6$ mw: 160.16

PROP: Yellow crystals from Et_2O. Mp: 83°.

SYNS: BENZENE, 1,4-DIAZIDO- □ p-DIAZIDOBENZENE (DOT) □ 1,4-DIAZIDOBENZENE □ p-PHENYLENE DIAZIDE

DOT CLASSIFICATION: Forbidden

SAFETY PROFILE: Explodes violently when heated. When heated to decomposition it emits toxic fumes of NO_x. See also AZIDES.

DCL200 CAS:67880-17-5 HR: 3
1,2-DIAZIDOCARBONYL HYDRAZINE
mf: $C_2H_2N_8O_2$ mw: 170.12

SYN: HYDRAZINE DICARBONIC ACID DIAZIDE (DOT)

DOT CLASSIFICATION: Forbidden

SAFETY PROFILE: A heat- and impact-sensitive explosive. When heated to decomposition it emits toxic fumes of NO_x. See also AZIDES.

DCL300 CAS:26157-96-0 HR: 3
2,5-DIAZIDO-3,6-DICHLOROBENZOQUINONE
mf: $C_6Cl_2N_6O_2$ mw: 259.01

O:CC(N_3)=CClCO•C(N_3)= CCl

SAFETY PROFILE: A moderately impact-sensitive explosive. Upon decomposition it emits toxic fumes of NO_x. See also AZIDES.

DCL350 CAS:4774-73-6 HR: 3
DIAZIDODIMETHYLSILANE
mf: $C_2H_6N_6Si$ mw: 142.20

PROP: Liquid. Bp: 144.3°.

SAFETY PROFILE: A storage hazard. It may explode spontaneously. When heated to decomposition it emits toxic fumes of NO_x. See also AZIDES.

DCL400 CAS:67880-20-0 HR: 3
1,1-DIAZIDOETHANE
mf: $C_2H_4N_6$ mw: 112.09

PROP: Liquid with chloroform odor. Bp: 38° @ 14 mm.

SAFETY PROFILE: An extremely unstable explosive. Upon decomposition it emits toxic fumes of NO_x. See also AZIDES.

DCL600 CAS:629-13-0 HR: 3
1,2-DIAZIDOETHANE
mf: $C_2H_4N_6$ mw: 112.09

PROP: Bp: 54–55° @ 11 mm.

SYN: ETHANE, 1,2-DIAZIDO-

DOT CLASSIFICATION: Forbidden

SAFETY PROFILE: Explodes on heating or on contact

with sulfuric acid. Upon decomposition it emits toxic fumes of NO_x. See also AZIDES.

DCM000 CAS:67880-21-1 HR: 3
DIAZIDOMALONONITRILE
mf: C_3N_8 mw: 148.09

$(N_3)_2C(C{\equiv}N)_2$

PROP: Liquid.

SYN: DIAZIDODICYANOMETHANE

CONSENSUS REPORTS: Cyanide and its compounds are on the Community Right-To-Know List.

SAFETY PROFILE: The pure material is an unpredictable explosive. When heated to decomposition it emits toxic fumes of CN^- and NO_x. See also AZIDES and NITRILES.

DCM499 HR: 3
1,3-DIAZIDO-2-NITROAZAPROPANE
mf: $C_2H_4N_8O_2$ mw: 172.11

SYN: N,N-BIS(AZIDOMETHYL)NITRIC AMIDE

SAFETY PROFILE: An explosive. A high energy component of solid rocket propellants. When heated to decomposition it emits toxic fumes of NO_x. See also AZIDES.

DCM600 CAS:22750-69-2 HR: 3
1,3-DIAZIDOPROPENE
mf: $C_3H_4N_6$ mw: 124.10

$N_3CH{=}CHCH_2N_3$

PROP: Yellow volatile liquid with sltly fishy odor. Bp: 78–79° @ 26 mm.

SAFETY PROFILE: An unpredictable explosive. When heated to decomposition it emits toxic fumes of NO_x and CN^-. See also AZIDES.

DCM700 CAS:74273-75-9 HR: 3
2,6-DIAZIDOPYRAZINE
mf: $C_4H_2N_8$ mw: 162.11

$N_3C{=}NC(N_3){=}CHN{=}CH$ (ring)

SAFETY PROFILE: A heat- and impact-sensitive explosive. It may be detonated by heating to 200°C or by a hammer blow. Upon decomposition it emits toxic fumes of NO_x. See also AZIDES.

DCM750 CAS:333-41-5 HR: 3
DIAZINON
mf: $C_{12}H_{21}N_2O_3PS$ mw: 304.38

PROP: Liquid with faint ester-like odor. Bp: 84° @ 0.002 mm, d: 1.116 @ 20°/4°. Miscible in organic solvents.

SYNS: ALFA-TOX □ BASUDIN □ BASUDIN 10 G □ BAZUDEN □ DAZZEL □ O,O-DIAETHYL-O-(2-ISOPROPYL-4-METHYL-PYRIMIDIN-6-YL)-MONOTHIOPHOSPHAT (GERMAN) □ O,O-DIAETHYL-O-(2-ISOPROPYL-4-METHYL)-6-PYRIMIDYL-THIONOPHOSPHAT (GERMAN) □ DIANON □ DIATERR-FOS □ DIAZAJET □ DIAZATOL □ DIAZIDE □ DIAZINONE □ DIAZITOL □ DIAZOL □ O,O-DIETHYL-O-(2-ISOPROPYL-4-METHYL-PYRIMIDIN-6-YL)MONOTHIOFOSFAAT (DUTCH) □ O,O-DIETHYL-O-(2-ISOPROPYL-4-METHYL-6-PYRIMIDINYL)PHOSPHOROTHIOATE □ O,O-DIETHYL-O-(2-ISOPROPYL-6-METHYL-4-PYRIMIDINYL) PHOSPHOROTHIOATE □ DIETHYL 4-(2-ISOPROPYL-6-METHYLPYRIMIDINYL)PHOSPHOROTHIONATE □ O,O-DIETHYL-O-(2-ISOPROPYL-4-METHYL-6-PYRIMIDYL)PHOSPHOROTHIOATE □ O,O-DIETHYL-O-(2-ISOPROPYL-4-METHYL-6-PYRIMIDYL) THIONOPHOSPHATE □ O,O-DIETHYL-2-ISOPROPYL-4-METHYLPYRIMIDYL-6-THIOPHOSPHATE □ O,O-DIETHYL-O-6-METHYL-2-ISOPROPYL-4-PYRIMIDINYL PHOSPHOROTHIOATE □ O,O-DIETIL-O-(2-ISOPROPIL-4-METIL-PIRIMIDIN-6-IL)-MONOTIOFOSFATO (ITALIAN) □DIMPYLATE □ DIPOFENE □ DIZINON □ DYZOL □ ENT 19,507 □ G 301 □ G-24480 □ GARDENTOX □ GEIGY 24480 □ O-2-ISOPROPYL-4-METHYLPYRIMIDYL-O,O-DIETHYL PHOSPHOROTHIOATE □ ISOPROPYLMETHYLPYRIMIDYL DIETHYL THIOPHOSPHATE □ KAYAZINON □ KAYAZOL □ NCI-C08673 □ NEDCIDOL □ NEOCIDOL □ NIPSAN □ NUCIDOL □ SAROLEX □ SPECTRACIDE □ THIOPHOSPHATE de O,O-DIETHYLE et de o-2-ISOPROPYL-4-METHYL-6-PYRIMIDYLE (FRENCH)

TOXICITY DATA WITH REFERENCE
skn-rbt 500 mg open MOD CIGET* -,-,77
eye-rbt 100 mg SEV CIGET* -,-,77
cyt-hmn:lym 500 µg/L TSITAQ 18,1490,76
cyt-ham:lng 100 mg/L/27H MUREAV 66,277,79
orl-mus TDLo:3960 µg/kg (female 1-22D post):REP JTEHD6 3,989,77
orl-mus TDLo:3780 µg/kg (female 1-21D post):REP JEPTDQ 2(2),357,78
orl-mus TDLo:189 mg/kg (female 1-21D post):REP JEPTDQ 4(5-6),53,80
ipr-rat TDLo:200 mg/kg (female 11D post):TER AEHLAU 16,805,68
ipr-rat TDLo:150 mg/kg (female 11D post):TER AEHLAU 16,805,68
orl-rat TDLo:63,500 µg/kg (female 10D post):REP JFMAAQ 54,452,67
ipr-rat TDLo:150 mg/kg (female 11D post):REP AEHLAU 16,805,68
orl-pig TDLo:570 mg/kg (female 1-16W post):TER 32OAAP -,253,73
orl-mus TDLo:189 mg/kg (female 1-21D post):TER JEPTDQ 2(2),357,78
orl-mus TDLo:210 µg/kg (female 1-21D post):TER JEPTDQ 4(5-6),53,80
orl-rat TDLo:45 mg/kg (female 8-12D post):TER JFMAAQ 54,452,67
orl-rat TDLo:26,400 µg/kg (12-15D preg):TER JFMAAQ 54,452,67
orl-hmn TDLo:214 mg/kg:CNS,SKN CTOXAO 12,435,78
orl-rat LD50:66 mg/kg DOEAAH 35,25,79
ihl-rat LC50:3500 mg/m³/4H FMCHA2 -,C75,83
skn-rat LD50:180 mg/kg PMJMAQ -,156,57
ipr-rat LD50:65 mg/kg ARZNAD 5,436,55
orl-mus LD50:17 mg/kg SKEZAP 24,268,83
ihl-mus LC50:1600 mg/m³/4H PSDTAP 15,239,74
skn-mus LD50:2750 mg/kg JTEHD6 9,491,82
ipr-mus LD50:33 mg/kg TXAPA9 2,495,60
scu-mus LD50:58 mg/kg OIZAAV 71,6099,59
ivn-mus LD50:180 mg/kg CSLNX* NX#00023
orl-rbt LD50:143 mg/kg YKYUA6 31,459,80
skn-rbt LD50:180 mg/kg CMEP** -,1,56

CONSENSUS REPORTS: NCI Carcinogenesis Bioassay

D

(feed); No Evidence: mouse, rat NCITR* NCI-CG-TR-137,79. Reported in EPA TSCA Inventory. EPA Genetic Toxicology Program.

OSHA PEL: TWA 0.1 mg/m^3 (skin)
ACGIH TLV: TWA 0.1 mg/m^3
DFG MAK: 1 mg/m^3

SAFETY PROFILE: Poison by ingestion, skin contact, subcutaneous, intravenous, and intraperitoneal routes. Mildly toxic by inhalation. Human systemic effects by ingestion: changes in motor activity, muscle weakness, and sweating. Experimental teratogenic and reproductive effects. A skin and severe eye irritant. Human mutation data reported. When heated to decomposition it emits very toxic fumes of NO_x, PO_x, and SO_x.

For occupational chemical analysis use OSHA: #ID-62.

DCM875 CAS:76429-98-6 HR: 3
DIAZIRINE-3,3-DICARBOXYLIC ACID
mf: $C_3H_2N_2O_4$ mw: 130.06

PROP: A solid. Mp: 76°.

SAFETY PROFILE: The potassium salts of this acid are unstable explosives. When heated to decomposition it emits toxic fumes of NO_x.

DCN200 CAS:6832-13-9 HR: 3
DIAZOACETALDEHYDE
mf: $C_2H_2N_2O$ mw: 70.05

PROP: D: 1.159 @ 20°, bp: 40° @ 10 mm.

SAFETY PROFILE: A powerful, heat-sensitive explosive. When heated to decomposition it emits toxic fumes of NO_x. See also AZIDES and ALDEHYDES.

DCN600 CAS:38726-91-9 HR: 2
2-(DIAZOACETAMINO)-N-ETHYLACETAMIDE
mf: $C_6H_{10}N_4O_2$ mw: 170.20

SYNS: 2-((DIAZOACETYL)AMINO)-N-ETHYLACETAMIDE □ N-DIAZOACETYLGLYCINEETHYLAMIDE □ N-ETHYLDIAZOACETYLGLYCINE AMIDE

TOXICITY DATA WITH REFERENCE
dns-mus:fbr 2500 mmol/L JCROD7 94,7,79
ipr-mus LD50:3281 mg/kg ARZNAD 23,690,73

CONSENSUS REPORTS: EPA Genetic Toxicology Program.

SAFETY PROFILE: Moderately toxic by intraperitoneal route. Mutation data reported. When heated to decomposition it emits toxic fumes of NO_x.

DCN800 CAS:623-73-4 HR: 3
DIAZOACETIC ESTER
mf: $C_4H_6N_2O_2$ mw: 114.12

$N_2CHCO{\cdot}OC_2H_5$

PROP: Yellow oil with pungent odor. Mp: −22°, bp: 141° @ 720 mm.

SYNS: DAAE □ DIAZOACETIC ACID, ETHYL ESTER □ DIAZOESSIGSAEURE-AETHYLESTER (GERMAN) □ EDA □ ETHOXYCARBONYLDIAZOMETHANE □ ETHYL DIAZOACETATE

TOXICITY DATA WITH REFERENCE
orl-rat TDLo:2025 mg/kg/81W-I:ETA XENOBH 3,271,73
skn-rat TDLo:4167 mg/kg/48W-I:CAR ARGEAR 55,117,85
ipr-rat TDLo:30 mg/kg/20W-I:CAR ARGEAR 55,117,85
ivn-rat TDLo:75 mg/kg/30W-I:CAR ARGEAR 55,117,85
ivn-rat TD:1330 mg/kg/53W-I:ETA BTPGAZ 146,33,72
orl-rat TD:2400 mg/kg/68W-C:ETA NATWAY 50,99,63
ivn-rat TD:2850 mg/kg/57W-I:ETA XENOBH 3,271,73
ivn-rat TD:1000 mg/kg/40W-I:ETA PSEBAA 135,219,70
ivn-rat TD:2150 mg/kg 86W-I:ETA ADRCAC 19,51,65
ivn-rat TD:4300 mg/kg/86W-I:ETA ADRCAC 19,51,65
ivn-rat TD:100 mg/kg:CAR CRNGDP 3,785,82
orl-rat LD50:400 mg/kg XENOBH 3,271,73
ivn-rat LD50:280 mg/kg PSEBAA 135,219,70

SAFETY PROFILE: Poison by ingestion and intravenous routes. Questionable carcinogen with experimental carcinogenic and tumorigenic data. Can explode. Explodes on contact with tris(dimethylamino) antimony. When heated to decomposition it emits toxic fumes of NO_x. See also ESTERS.

DCN875 CAS:13138-21-1 HR: 3
DIAZOACETONITRILE
mf: C_2HN_3 mw: 67.05

CONSENSUS REPORTS: Cyanide and its compounds are on the Community Right-To-Know List.

SAFETY PROFILE: The precipitate from concentrated solutions is an extremely friction-sensitive explosive. When heated to decomposition it emits toxic fumes of NO_x and CN^-. See also NITRILES.

DCO200 CAS:38726-90-8 HR: 2
2-((DIAZOACETYL)AMINO)-N-METHYLACETAMIDE
mf: $C_5H_8N_4O_2$ mw: 156.17

SYNS: N-DIAZOACETYLGLYCINE METHYLAMIDE □ N-METHYLDIAZOACETYLGLYCINE AMIDE

TOXICITY DATA WITH REFERENCE
dns-mus:fbr 2500 mmol/L JCROD7 94,7,79
ipr-mus LD50:3672 mg/kg ARZNAD 23,690,73

CONSENSUS REPORTS: EPA Genetic Toxicology Program.

SAFETY PROFILE: Moderately toxic by intraperitoneal route. Mutation data reported. When heated to decomposition it emits toxic fumes of NO_x.

DCO509 CAS:19932-64-0 HR: 3
DIAZOACETYL AZIDE
mf: C_2HN_5O mw: 111.06

PROP: Mp: 7°C

SAFETY PROFILE: The solid and the liquid are powerful explosives sensitive to impact or friction. When heated to decomposition it emits toxic fumes of NO_x. See also AZIDES.

DCO600 CAS:999-29-1 HR: 2
DIAZOACETYLGLYCINE ETHYL ESTER
mf: $C_6H_9N_3O_3$ mw: 132.18

SYN: N-DIAZOACETYLGLYCINE ETHYL ESTER

TOXICITY DATA WITH REFERENCE
mmo-sat 10 µg/plate AMACCQ 6,655,74
dns-mus:fbr 2500 mmol/L JCROD7 94,7,79
ipr-mus LD50:1149 mg/kg ARZNAD 23,690,73

CONSENSUS REPORTS: EPA Genetic Toxicology Program.

SAFETY PROFILE: Moderately toxic by intraperitoneal route. Mutation data reported. When heated to decomposition it emits toxic fumes of NO_x. See also ESTERS.

DCO800 CAS:820-75-7 HR: 3
N-(DIAZOACETYL)GLYCINE HYDRAZINE
mf: $C_4H_7N_5O_2$ mw: 157.16

SYNS: N-DIAZOACETILGLICINA-IDRAZIDE (ITALIAN) □ DIAZOACETYLGLYCINE HYDRAZIDE □ N-DIAZOACETYL GLYCYLHYDRAZIDE □ NSC-58404

TOXICITY DATA WITH REFERENCE
mmo-sat 10 µg/plate AMACCQ 6,655,74
mma-sat 10 µg/plate PNASA6 72,5135,75
dns-mus:fbr 2500 µmol/L JCROD7 94,7,79
ipr-mus TDLo:1200 mg/kg/4D-I:CAR BSIBAC 45,227,69
ipr-mus TD:720 mg/kg/4D-I:NEO BSIBAC 45,227,69
ipr-rat LD50:1335 mg/kg CCROBU 53,13,69
ivn-rat LD50:1595 mg/kg CCROBU 53,13,69
orl-mus LD50:3852 mg/kg CCROBU 53,13,69
ipr-mus LD50:2575 mg/kg CCROBU 53,13,69
scu-mus LD50:2985 mg/kg CCROBU 53,13,69
ivn-dog LDLo:39 mg/kg CCROBU 53,13,69
ivn-rbt LDLo:833 mg/kg CCROBU 53,13,69

CONSENSUS REPORTS: EPA Genetic Toxicology Program.

SAFETY PROFILE: Poison by intravenous route. Moderately toxic by ingestion, intraperitoneal, and subcutaneous routes. Questionable carcinogen with experimental carcinogenic and neoplastigenic data. Mutation data reported. When heated to decomposition it emits toxic fumes including NO_x.

DCP300 CAS:6087-56-5 HR: 3
4-DIAZO-N,N-DIMETHYLANILIN CHLOROZINCATE
mf: $C_8H_{10}N_3 \cdot Cl_3Zn$ mw: 319.93

SYNS: BENZENEDIAZONIUM, 4-(DIMETHYLAMINO)-, TRICHLOROZINCATE(1-) □ BENZENEDIAZONIUM, p-(DIMETHYLAMINO)-, TRICHLOROZINCATE(1-) (8CI) □ DIAZO A □ DIAZO 4VZS □ p-DIMETHYLAMINOBENZENEDIAZONIUM CHLOROZINCATE (6CI) □ p-(DIMETHYLAMINO)BENZENEDIAZONIUM TRICHLOROZINCATE (7CI) □ 4-(DIMETHYLAMINO)BENZENEDIAZONIUM TRICHLOROZINCATE (1-)

TOXICITY DATA WITH REFERENCE
orl-rat LD50:93 mg/kg EPASR* 8EHQ-0391-1153

CONSENSUS REPORTS: Reported in EPA TSCA Inventory.

SAFETY PROFILE: A poison by ingestion. When heated to decomposition it emits toxic vapors of Zn, NO_x, and Cl^-.

DCP400 CAS:7008-85-7 HR: 3
5-DIAZOIMIDAZOLE-4-CARBOXAMIDE
mf: $C_4H_3N_5O$ mw: 137.12

SYNS: DIAZO-ICA □ DIAZOIMIDAZOLE-4-CARBOXAMIDE □ NSC 22420

TOXICITY DATA WITH REFERENCE
dni-esc 10 µg/L BCPCA6 18,1463,69
oms-esc 500 µg/L BCPCA6 18,1463,69
dni-rat:lvr 100 µmol/L CNREA8 2827,76
orl-rat TDLo:2175 mg/kg/14W-C:ETA JNCIAM 54,951,75
ipr-mus LD50:1002 µg/kg NCISP* JAN86

SAFETY PROFILE: A deadly poison by intraperitoneal route. Questionable carcinogen with experimental tumorigenic data. Mutation data reported. When heated to decomposition it emits toxic fumes of NO_x.

DCP600 CAS:64038-55-7 HR: 2
5-DIAZOIMIDAZOLE-4-CARBOXAMIDE HYDROCHLORIDE
mf: $C_4H_3N_5O \cdot HCl$ mw: 137.12

TOXICITY DATA WITH REFERENCE
orl-rat TDLo:105 mg/kg/10W-C:ETA JNCIAM 54,951,75

SAFETY PROFILE: Questionable carcinogen with experimental tumorigenic data. When heated to decomposition it emits very toxic fumes of NO_x and HCl.

DCP700 CAS:59348-62-8 HR: 3
DIAZOMALONIC ACID
mf: $C_3H_2N_2O_4$ mw: 130.06

SAFETY PROFILE: The impure acid and the diethyl ester are explosive. When heated to decomposition it emits toxic fumes of NO_x. See also AZIDES.

DCP775 CAS:1618-08-2 HR: 3
DIAZOMALONONITRILE
mf: C_3N_4 mw: 92.06

$N_2C(C{\equiv}N)_2$

PROP: Pale-yellow crystals solid.

SYN: DIAZODICYANOMETHANE

CONSENSUS REPORTS: Cyanide and its compounds are on the Community Right-To-Know List.

SAFETY PROFILE: An explosive sensitive to sparks or heating to 75°C. When heated to decomposition it emits toxic fumes of NO_x and CN^-. See also NITRILES and CYANIDE.

DCP800 **CAS:334-88-3** **HR: 3**
DIAZOMETHANE
mf: CH_2N_2 mw: 42.05

PROP: Yellow gas at ordinary temp which forms yellow solns in ethereal solvs. Mp: −145°, bp: −23°, d: 1.45.

SYNS: AZIMETHYLENE □ DIAZIRINE

TOXICITY DATA WITH REFERENCE
mmo-nsc 250 mmol/L HEREAY 35,521,49
ihl-rat TCLo: 272 mg/m³/26W-I: ETA BJCAAI 16,92,62
ihl-mus TCLo: 272 mg/m³/26W-I: ETA BJCAAI 16,92,62
scu-mus TDLo: 48 g/kg/52W-I: ETA BJCAAI 16,92,62

CONSENSUS REPORTS: IARC Cancer Review: Group 3 IMEMDT 7,56,87; Animal Sufficient Evidence IMEMDT 7,223,74. EPA Genetic Toxicology Program. Community Right-To-Know List.

OSHA PEL: TWA 0.2 ppm
ACGIH TLV: TWA 0.2 ppm
DFG MAK: Animal Carcinogen, Suspected Human Carcinogen

SAFETY PROFILE: Confirmed carcinogen with experimental tumorigenic data. A poisonous irritant by inhalation. A powerful allergen. It can cause pulmonary edema and frequently causes hypersensitivity leading to asthmatic symptoms. Mutation data reported. Highly explosive when shocked, exposed to heat or by chemical reaction. Undiluted liquid or gas may explode on contact with alkali metals, rough surfaces, heat (100°C), high-intensity light or shock. When heated to decomposition or on contact with acid or acid fumes it emits highly toxic fumes of NO_x. Incompatible with alkali metals; calcium sulfate.

For occupational chemical analysis use NIOSH: Diazomethane, 2515.

DCP880 **CAS:40953-35-3** **HR: 3**
2-DIAZONIO-4,5-DICYANOIMIDAZOLIDE
mf: C_5N_6 mw: 144.10

NC(CN)=C(CN)N= CN_2

PROP: Crystals from MeCN.

SYN: DIAZODICYANOIMIDAZOLE

CONSENSUS REPORTS: Cyanide and its compounds are on the Community Right-To-Know List.

SAFETY PROFILE: Explodes when heated above 150°C. The dry material is very shock sensitive. When heated to decomposition it emits toxic fumes of NO_x and CN^-. See also CYANIDE.

DCQ400 **CAS:157-03-9** **HR: 3**
6-DIAZO-5-OXONORLEUCINE
mf: $C_6H_9N_3O_3$ mw: 171.18

PROP: Yellow crystals from EtOH (aq). Mp: 142–150° (decomp).

SYNS: DIAZO-OXO-NORLEUCINE □ 6-DIAZO-5-OXO-l-NORLEUCINE □ DON □ NSC 7365

TOXICITY DATA WITH REFERENCE
pic-esc 5 µg/plate CNREA8 43,2819,83
ims-mus TDLo: 400 µg/kg (female 11D post): TER TJADAB 23,29A,81
ipr-rat TDLo: 100 µg/kg (female 7-8D post): TER PSEBAA 94,33,57
ipr-rat TDLo: 300 µg/kg (female 8D post): TER PSEBAA 94,33,57
ipr-rat TDLo: 100 µg/kg (female 7-8D post): REP PSEBAA 94,33,57
ipr-rat TDLo: 5 mg/kg (15D preg): TER TJADAB 15,281,77
ipr-rat LD50: 80 mg/kg CPCHAO 18,307,62
orl-mus LD50: 197 mg/kg NCISP* JAN86
ipr-mus LDLo: 300 mg/kg JOENAK 18,204,59
ivn-mus LD50: 74 mg/kg 85ERAY 2,1253,78

CONSENSUS REPORTS: EPA Genetic Toxicology Program.

SAFETY PROFILE: Poison by ingestion, intraperitoneal, and intravenous routes. Experimental teratogenic and reproductive effects. Mutation data reported. When heated to decomposition it emits toxic fumes of NO_x.

DCQ500 **CAS:64781-77-7** **HR: 3**
4-DIAZO-5-PHENYL-1,2,3-TRIAZOLE
mf: $C_8H_5N_5$ mw: 171.16

$N_2CN=N\text{-}N= CC_6H_5$

SAFETY PROFILE: A heat-sensitive explosive. Upon decomposition it emits toxic fumes of NO_x. See also AZIDES.

DCQ525 **CAS:5239-06-5** **HR: 3**
1,3-DIAZOPROPANE
mf: $C_3H_4N_4$ mw: 96.11

SYN: PROPANE, 1,3-BIS(DIAZO)-

DOT CLASSIFICATION: Forbidden

SAFETY PROFILE: An explosive forbidden for transport. When heated to decomposition it emits toxic vapors of NO_x.

DCQ550 **CAS:2032-04-4** **HR: 3**
3-DIAZOPROPENE
mf: $C_3H_4N_2$ mw: 68.08

$H_2C=CHCHN_2$

PROP: Unstable, used only in soln.

SYN: VINYLDIAZOMETHANE

SAFETY PROFILE: A storage hazard. It is potentially explosive and should be stored in the dark below 0°C. When heated to decomposition it emits toxic fumes of NO_x. See also AZIDES.

DCQ575 **HR: 2**
3-DIAZOTYRAMINE HYDROCHLORIDE
mf: $C_8H_9N_3O$ mw: 163.20

PROP: A nitrosated product of TYRAMINE

SYNS: 4-(2-AMINOETHYL)-6-DIAZO-2,4-CYCLOHEXADIENONE HYDROCHLORIDE □ TYRAMINE, 3-DIAZO-, HYDROCHLORIDE

TOXICITY DATA WITH REFERENCE
orl-rat TDLo:77,700 mg/kg/2Y-C:CAR CRNGDP 8,527,87

SAFETY PROFILE: Questionable carcinogen with experimental carcinogenic data. When heated to decomposition it emits toxic fumes of NO_x.

DCQ600 CAS:2435-76-9 HR: 3
DIAZOURACIL
mf: $C_4H_2N_4O_2$ mw: 138.10

PROP: Crystals from H_2O. Mp: 198° (decomp).

SYNS: 5-DIAZOPYRIMIDINE-2,4(3H)-DIONE □ 5-DIAZO-2,4(1H, 3H)-PYRIMIDINEDIONE □ 5-DIAZOURACIL □ 2,4-DIOSSI-5-DIAZO-PIRIMIDINA (ITALIAN) □ 2,6-DIOXO-5-DIAZOPYRIMIDINE □ DU □ NSC 23519 □ (1,2,3)OXADIAZOLO(5,4-d)PYRIMIDIN-5(4H)-ONE

TOXICITY DATA WITH REFERENCE
dnd-sat 10 mg/L MILEDM 1,169,76
dnd-esc 10 mg/L MILEDM 1,169,76
dns-rat:lvr 50 µmol/L CALEDQ 13,187,81
ipr-mus LD50:30,800 µg/kg NCISP* JAN86
scu-mus LD10:22 mg/kg EJCAAH 10,667,74

CONSENSUS REPORTS: Reported in EPA TSCA Inventory.

SAFETY PROFILE: Poison by subcutaneous and intraperitoneal routes. Mutation data reported. When heated to decomposition it emits toxic fumes of NO_x.

DCQ650 CAS:94362-44-4 HR: 2
DIAZO V

SYN: DIAZO RESIN V

TOXICITY DATA WITH REFERENCE
scu-rat TDLo:900 mg/kg/26W-I:NEO VINIT* #5689-83

SAFETY PROFILE: Questionable carcinogen with experimental neoplastigenic data. When heated to decomposition it emits acrid smoke and irritating fumes.

DCQ700 CAS:364-98-7 HR: 3
DIAZOXIDE
mf: $C_8H_7ClN_2O_2S$ mw: 230.68

PROP: Crystals from dilute alc. Mp: 330–331°. Sol in alc and alkaline solns; insol in water.

SYNS: 7-CHLORO-3-METHYL-2H-1,2,4-BENZOTHIADIAZINE-1,1-DIOXIDE □ 7-CLORO-3-METIL-2H-1,2,4-BENZOTIODIAZINE-1,1-DIOSSIDO (ITALIAN) □ DIAZOSSIDO (ITALIAN) □ DIZOXIDE □ EUDEMINE INJECTION □ HYPERSTAT □ HYPERTONALUM □ MUTABASE □ PROGLICEM □ SRG 95213

TOXICITY DATA WITH REFERENCE
ivn-dom TDLo:65 mg/kg (110-114D preg):TER JCNDBK 11,206,71
orl-rat LD50:980 mg/kg AIPTAK 143,446,63
ipr-rat LD50:510 mg/kg AIPTAK 143,446,63
orl-mus LD50:444 mg/kg JPETAB 136,344,62
ipr-mus LD50:326 mg/kg JPETAB 136,344,62
ivn-mus LD50:228 mg/kg JPETAB 136,344,62

SAFETY PROFILE: Poison by intravenous and intraperitoneal routes. Moderately toxic by ingestion. An experimental teratogen. When heated to decomposition it emits toxic fumes of Cl^-, SO_x, and NO_x.

DCQ800 CAS:34493-98-6 HR: 3
DIBEKACIN
mf: $C_{18}H_{37}N_5O_8$ mw: 451.60

PROP: A solid.

SYNS: DEBECACIN □ DIDEOXYKANAMYCIN B □ 3′,4′-DIDEOXYKANAMYCIN B □ DKB □ ORBICIN

TOXICITY DATA WITH REFERENCE
ims-rat TDLo:2400 mg/kg (female 10-21D post):REP JJANAX 26,40,73
ims-rat TDLo:900 mg/kg (10-15D preg):TER JJANAX 26,40,73
ims-mus TDLo:1500 mg/kg (female 8-13D post):TER JJANAX 26,40,73
ipr-rat LD50:16,760 µg/kg IYKEDH 6,119,75
scu-rat LD50:23,870 µg/kg IYKEDH 6,119,75
ivn-rat LD50:12,510 µg/kg IYKEDH 6,119,75
orl-mus LD50:763 mg/kg IYKEDH 6,119,75
ipr-mus LD50:11,960 µg/kg IYKEDH 6,119,75
scu-mus LD50:15,980 µg/kg IYKEDH 6,119,75
ivn-mus LD50:8950 µg/kg IYKEDH 6,119,75
ims-mus LD50:373 mg/kg JOPHDQ 4,356,81

CONSENSUS REPORTS: EPA Genetic Toxicology Program.

SAFETY PROFILE: Poison by intraperitoneal, subcutaneous, intramuscular, and intravenous routes. Moderately toxic by ingestion. Experimental teratogenic and reproductive effects. An antibacterial agent. When heated to decomposition it emits toxic fumes of NO_x.

DCR200 CAS:55-43-6 HR: 3
DIBENAMINE HYDROCHLORIDE
mf: $C_{16}H_{18}ClN \cdot ClH$ mw: 296.26

SYNS: N-(2-CHLOROETHYL)DIBENZYLAMINE HYDROCHLORIDE □ DIBENAMINE □ N,N-DIBENZYLAMINOETHYL CHLORIDE HYDROCHLORIDE □ DIBENZYLCHLORETHAMINE HYDROCHLORIDE □ DIBENZYLCHLORETHYLAMINE HYDROCHLORIDE □ N,N-DIBENZYL-β-CHLOROETHYLAMINE HYDROCHLORIDE □ N,N-DIBENZYL-2-CHLOROETHYLAMINE HYDROCHLORIDE □ SYMPATHOLYTIN

TOXICITY DATA WITH REFERENCE
ipr-mus LD50:75 mg/kg JPETAB 89,167,47
scu-mus LD50:400 mg/kg JPETAB 89,167,47
ivn-mus LD50:50 mg/kg JPETAB 89,167,47
ivn-cat LDLo:35 mg/kg JPETAB 89,167,47

SAFETY PROFILE: Poison by subcutaneous, intravenous, and intraperitoneal routes. An adrenergic blocker and diagnostic aid (pheochromocytoma). When heated to decomposition it emits very toxic fumes of Cl^- and NO_x.

D

DCR300 **CAS:5385-75-1** **HR: 2**
DIBENZ(a,e)ACEANTHRYLENE
mf: $C_{24}H_{14}$ mw: 302.38

PROP: Yellow needles from C_6H_6. Mp: 232°.

SYNS: DIBENZO(a,e)FLUORANTHENE □ 2,3,5,6-DIBENZOFLUORANTHENE

TOXICITY DATA WITH REFERENCE
mma-sat 500 nmol/L CRNGDP 5,1263,84
dns-mus:emb 1 μmol/L CRNGDP 5,379,84
skn-mus TDLo:2880 μg/kg/15W-I:CAR CRNGDP 8,461,87

CONSENSUS REPORTS: IARC Cancer Review: Group 3 IMEMDT 7,56,87; Animal Limited Evidence IMEMDT 32,321,83

SAFETY PROFILE: Questionable carcinogen with experimental carcinogenic data. Mutation data reported. When heated to decomposition it emits acrid smoke and fumes.

DCR400 **CAS:203-20-3** **HR: 3**
DIBENZ(a,j)ACEANTHRYLENE
mf: $C_{24}H_{14}$ mw: 302.38

PROP: Bright-yellow needles from C_6H_6. Mp: 181–181.3°.

SYN: 15,16-BENZDEHYDROCHOLANTHRENE

TOXICITY DATA WITH REFERENCE
scu-mus TDLo:800 mg/kg/9W-I:ETA AJCAA7 28,334,36
ivn-mus TDLo:20 mg/kg:ETA JNCIAM 1,225,40
ivn-mus LDLo:10 mg/kg JNCIAM 1,225,40

SAFETY PROFILE: Poison by intravenous route. Questionable carcinogen with experimental tumorigenic data. When heated to decomposition it emits acrid smoke and irritating fumes.

DCR600 **CAS:201-42-3** **HR: 2**
13H-DIBENZ(bc,j)ACEANTHRYLENE
mf: $C_{23}H_{14}$ mw: 290.37

PROP: Greenish-yellow prisms from C_6H_6/pet ether. Mp: 266–267°.

SYNS: 13H-ACENAPHTHO(1,8-ab)PHENANTHRENE □ 1′,9-METHYLENE-1,2:5,6-DIBENZANTHRACENE

TOXICITY DATA WITH REFERENCE
scu-mus TDLo:400 mg/kg:ETA AJCAA7 28,334,36

SAFETY PROFILE: Questionable carcinogen with experimental tumorigenic data. When heated to decomposition it emits acrid smoke and irritating fumes.

DCR800 **CAS:517-85-1** **HR: 2**
4H-DIBENZ(f,g,j)ACEANTHRYLENE, 5,5a,6,7-TETRAHYDRO-
mf: $C_{23}H_{18}$ mw: 294.41

SYNS: ANG.-STERANTHREN (GERMAN) □ ANG-STERANTHRENE

TOXICITY DATA WITH REFERENCE
skn-mus TDLo:1120 mg/kg/35W-I:ETA ZEKBAI 62,217,57
scu-mus TDLo:40 mg/kg:ETA ZEKBAI 62,217,57

SAFETY PROFILE: Questionable carcinogen with experimental tumorigenic data. When heated to decomposition it emits acrid smoke and irritating fumes.

DCS200 **CAS:1977-10-2** **HR: 3**
DIBENZACEPIN
mf: $C_{18}H_{18}ClN_3O$ mw: 327.84

PROP: Pale-yellow cryst from pet ether. Mp: 109–110°.

SYNS: 2-CHLORO-11-(4-METHYL-1-PIPERAZINYL)-DIBENZO(b,f)(1,4)OXAZEPINE □ 2-CHLORO-11-(4-METHYL-1-PIPERAZINYL)-DIBENZO(b,f)(1,4)OXOAZEPINE □ CL-62362 □ CL-71563 □ CLOXAZEPINE □ DIBENZOAZEPINE □ HF3170 □ LOXAPINE □ LW 3170 □ OXILAPINE □ S-805 □ SUM 3170

TOXICITY DATA WITH REFERENCE
orl-mus TDLo:6 mg/kg (female 7-12D post):REP OYYAA2 4,305,70
orl-rat TDLo:24 mg/kg (9-14D preg):REP OYYAA2 4,305,70
orl-rat TDLo:6 mg/kg (9-14D preg):REP OYYAA2 4,305,70
orl-rat TDLo:960 mg/kg (female 30D pre):REP OYYAA2 4,293,70
orl-rat TDLo:455 mg/kg (female 60D pre):REP OYYAA2 4,293,70
orl-rat TDLo:960 mg/kg (male 30D pre):REP OYYAA2 4,293,70
orl-rat LD50:151 mg/kg OYYAA2 4,293,70
ipr-rat LD50:35 mg/kg OYYAA2 4,293,70
scu-rat LD50:350 mg/kg OYYAA2 4,293,70
ivn-rat LD50:18 mg/kg OYYAA2 4,293,70
orl-mus LD50:40 mg/kg 27ZQAG -,79,72
ipr-mus LD50:27 mg/kg 27ZQAG -,79,72
scu-mus LD50:53 mg/kg OYYAA2 4,293,70
ivn-mus LD50:22 mg/kg OYYAA2 4,293,70

SAFETY PROFILE: Poison by ingestion, intraperitoneal, subcutaneous, and intravenous routes. Experimental teratogenic and reproductive effects. A tranquilizer. Many dibenz-azepine compounds have central nervous system effects. When heated to decomposition it emits very toxic fumes of Cl^- and NO_x.

DCS400 **CAS:226-36-8** **HR: 3**
DIBENZ(a,h)ACRIDINE
mf: $C_{21}H_{13}N$ mw: 279.35

PROP: Yellow crystals. Mp: 228°.

SYNS: 7-AZADIBENZ(a,h)ANTHRACENE □ DB(a,h)AC □ DIBENZ(a,d)ACRIDINE □ 1,2,5,6-DIBENZACRIDINE □ 1,2,5,6-DIBENZOACRIDINE □ 1,2,5,6-DINAPHTHACRIDINE

TOXICITY DATA WITH REFERENCE
otr-rat:emb 1 mg/L JJIND8 51,799,73
imp-rat TDLo:5 mg/kg:CAR CALEDQ 20,97,83
orl-mus TDLo:13 g/kg/63W-I:ETA PRLBA4 129,439,40
skn-mus TDLo:540 mg/kg/45W-I:ETA ACRSAJ 4,315,56
scu-mus TDLo:430 mg/kg/24W-I:ETA PRLBA4 123,343,37
ivn-mus TDLo:10 mg/kg:ETA JNCIAM 1,225,40
skn-mus TD:790 mg/kg/33W-I:ETA PRLBA4 117,318,35
scu-mus TD:1600 mg/kg/24W-I:ETA PRLBA4 129,439,40
imp-rat TD:1500 μg/kg:ETA CALEDQ 20,97,83

CONSENSUS REPORTS: NTP 7th Annual Report On Carcinogens. IARC Cancer Review: Group 2B IMEMDT 7,56,87; Animal Sufficient Evidence IMEMDT 32,277,83; IMEMDT 3,247,73. EPA Genetic Toxicology Program.

SAFETY PROFILE: Confirmed carcinogen with experimental carcinogenic and tumorigenic data. Mutation data reported. When heated to decomposition it emits toxic fumes of NO_x. See also ANTHRACENE.

DCS600 CAS:224-42-0 HR: 3
DIBENZ(a,j)ACRIDINE
mf: $C_{21}H_{13}N$ mw: 279.35

PROP: Yellow crystals. Mp: 216–217°.

SYNS: 7-AZADIBENZ(a,j)ANTHRACENE □ DB(a,j)AC □ DIBENZ(a,f)ACRIDINE □ 1,2,7,8-DIBENZACRIDINE □ 3,4,5,6-DIBENZACRIDINE □ DIBENZO(a,j)ACRIDINE □ 3,4,6,7-DINAPHTHACRIDINE

TOXICITY DATA WITH REFERENCE
mma-sat 4 μg/plate BJCAAI 37,873,78
dnd-esc 10 μmol/L PNCCA2 5,39,65
pic-esc 50 mg/L CNREA8 41,532,81
orl-mus TDLo:2520 mg/kg (21D pre):REP DABSAQ 29,4777,69
skn-mus TDLo:99 mg/kg/99W-I:CAR CALEDQ 37,337,87
scu-mus TDLo:40 mg/kg:ETA JJIND8 1,225,40
skn-mus TD:700 mg/kg/29W-I:ETA PRLBA4 117,318,35
skn-mus LD :590 mg/kg/25W-I:ETA BAFEAG 42,186,55

CONSENSUS REPORTS: NTP 7th Annual Report On Carcinogens. IARC Cancer Review: Group 2B IMEMDT 7,56,87; Animal Sufficient Evidence IMEMDT 32,283,83; IMEMDT 3,254,73. EPA Genetic Toxicology Program.

SAFETY PROFILE: Confirmed carcinogen with experimental carcinogenic and tumorigenic data. Experimental reproductive effects. Mutation data reported. When heated to decomposition it emits toxic fumes of NO_x. See also ANTHRACENE.

DCS800 CAS:224-53-3 HR: 2
DIBENZ(c,h)ACRIDINE
mf: $C_{21}H_{13}N$ mw: 279.35

PROP: Yellow crystals from EtOH. Mp: 189°.

SYNS: 14-AZADIBENZ(a,j)ANTHRACENE □ 1,2,7,8-DIBENZACRIDINE (FRENCH) □ 3,4:5,6-DIBENZACRIDINE

TOXICITY DATA WITH REFERENCE
mma-sat 4 μg/plate BJCAAI 37,873,78
skn-mus TDLo:2040 mg/kg/85W-I:ETA PRLBA4 129,439,40
scu-mus TDLo:4400 mg/kg/65W-I:ETA PRLBA4 129,439,40
skn-mus TD:300 mg/kg/25W-I:ETA ACRSAJ 4,315,56

SAFETY PROFILE: Questionable carcinogen with experimental tumorigenic data. Mutation data reported. When heated to decomposition it emits toxic fumes of NO_x. See also ANTHRACENE.

DCT000 CAS:63918-83-2 HR: 2
DIBENZ(a,j)ACRIDINE METHOSULFATE
mf: $C_{21}H_{13}N{\cdot}C_2H_6O_4S$ mw: 405.49

SYN: 3,4:5,6-DIBENZACRIDINE METHOSULFATE

TOXICITY DATA WITH REFERENCE
scu-mus TDLo:3750 mg/kg/56W-I:ETA PRLBA4 129,439,40

SAFETY PROFILE: Questionable carcinogen with experimental tumorigenic data. When heated to decomposition it emits very toxic fumes of SO_x and NO_x. See also SULFONATES.

DCT050 CAS:51-50-3 HR: 3
DIBENZAMINE
mf: $C_{16}H_{18}ClN$ mw: 259.80

SYNS: N-(2-CHLOROETHYL)DIBENZYLAMINE □ DIBENZYL CHLORETHYLAMINE □ N,N-DIBENZYL-β-CHLOROETHYLAMINE □ SYMPATHOLYTIN

TOXICITY DATA WITH REFERENCE
dni-mus ivg 2 pph JIDEAE 62,378,74
unr-mus TDLo:2 mg/kg (1D pre):REP FESTAS 12,346,61
orl-rat LD50:2400 mg/kg CLDND*
ipr-mus LD50:395 mg/kg AIPTAK 155,69,65
scu-mus LD50:800 mg/kg JPETAB 97,25,49
ivn-mus LD50:50 mg/kg AIPTAK 105,317,56

SAFETY PROFILE: Poison by intravenous and intraperitoneal routes. Moderately toxic by ingestion and subcutaneous routes. Experimental reproductive effects. Can cause leukopenia (reduced white blood cell count). Mutation data reported. When heated to decomposition it emits very toxic fumes of Cl^- and NO_x. See also AROMATIC AMINES.

DCT400 CAS:53-70-3 HR: 3
DIBENZ(a,h)ANTHRACENE
mf: $C_{22}H_{14}$ mw: 278.36

PROP: Silvery leaflets from AcOH. Mp: 266–267°.

SYNS: 1,2:5,6-BENZANTHRACENE □ DBA □ DB(a,h)A □ 1,2,5,6-DBA □ 1,2,5,6-DIBENZANTHRACEEN (DUTCH) □ 1,2:5,6-DIBENZANTHRACENE □ 1,2:5,6-DIBENZ(a)ANTHRACENE □ DIBENZO(a,h)ANTHRACENE □ 1,2:5,6-DIBENZOANTHRACENE □ RCRA WASTE NUMBER U063

TOXICITY DATA WITH REFERENCE
dnd-hmn:emb 360 nmol/L CBINA8 22,257,78
dnd-esc 10 μmol/L MUREAV 89,95,81
otr-rat-orl 200 mg/kg CNREA8 40,1157,80
msc-mus:lym 4250 μg/L MUREAV 106,101,82
msc-ham:lng 500 μg/L MUREAV 136,65,84
scu-rat TDLo:2400 μg/kg/50D-I:NEO 85DLAB -,-,75
orl-mus TDLo:4160 mg/kg/26W-I:CAR JPBAA7 49,21,39
skn-mus TDLo:1200 mg/kg/50W-I:CAR 14JTAF -,275,65
scu-mus TDLo:445 μg/kg:CAR CRNGDP 11,1721,90
ivn-mus TDLo:40 mg/kg:NEO PHRPA6 54,1158,39
imp-mus TDLo:80 mg/kg:CAR BJCAAI 11,212,57
mul-mus TDLo:40 mg/kg/12D-I:ETA PHRPA6 52,637,37
scu-gpg TDLo:250 mg/kg/24D-I:ETA AKBNAE 51,112,38
ivn-gpg TDLo:30 mg/kg:ETA JNCIAM 13,705,52
ims-pgn TDLo:6 mg/kg:CAR JNCIAM 32,905,64

irn-frg TDLo: 12 mg/kg: NEO CNREA8 24,1969,64
imp-mus TD: 14 mg/kg: NEO AJPAA4 16,287,40
scu-mus TD: 78 μg/kg: NEO JNCIAM 3,503,43
orl-mus TD: 4520 mg/kg/36W-C: CAR JNCIAM 1,17,40
imp-mus TD: 200 mg/kg: NEO AJCAA7 36,201,39
skn-mus TD: 6 μg/kg: NEO CNREA8 20,1179,60
scu-mus TD: 6 mg/kg: ETA IJCNAW 32,765,83
skn-mus TD: 400 mg/kg/40W-I: NEO CNREA8 22,78,62
imp-mus TD: 100 mg/kg: CAR BMBUAQ 14,147,58
scu-rat TD: 135 mg/kg/9W-I: NEO PSEBAA 68,330,48
scu-mus TD: 400 mg/kg/10W-I: NEO IJCNAW 2,500,67
ivn-mus LDLo: 10 mg/kg JNCIAM 1,225,40

CONSENSUS REPORTS: NTP 7th Annual Report On Carcinogens. IARC Cancer Review: Group 2A IMEMDT 7,56,87; Animal Sufficient Evidence IMEMDT 32,299,83; IMEMDT 3,178,73. EPA Genetic Toxicology Program. Reported in EPA TSCA Inventory.

SAFETY PROFILE: Confirmed carcinogen with experimental carcinogenic, tumorigenic, and neoplastigenic data. Poison by intravenous route. Human mutation data reported. When heated to decomposition it emits acrid smoke and irritating fumes.

For occupational chemical analysis use NIOSH: Polynuclear Aromatic Hydrocarbons (HPLC), 5506; (GC), 5515.

DCT600 CAS:224-41-9 HR: 2
DIBENZ(a,j)ANTHRACENE
mf: $C_{22}H_{14}$ mw: 278.36

PROP: Crystals from AcOH. Mp: 196°. Very sltly sol in EtOH and Et_2O.

SYN: 1,2:7,8-DIBENZANTHRACENE

TOXICITY DATA WITH REFERENCE
mma-sat 1 μg/plate MUREAV 51,311,78
skn-mus TDLo: 252 mg/kg/81W-I: ETA JNCIAM 44,641,70
scu-mus TDLo: 4 mg/kg: ETA JNCIAM 1,45,40
skn-mus TD: 1250 mg/kg/52W-I: ETA PRLBA4 117,318,35
scu-mus TD: 16 mg/kg: ETA JNCIAM 44,641,70
scu-mus TD: 4 mg/kg: ETA JNCIAM 1,45,40

CONSENSUS REPORTS: IARC Cancer Review: Group 3 IMEMDT 7,56,87; Animal Limited Evidence IMEMDT 32,309,83.

SAFETY PROFILE: Questionable carcinogen with experimental tumorigenic data. Mutation data reported. When heated to decomposition it emits acrid smoke and irritating fumes. See also ANTHRACENE.

DCT800 HR: 2
1,2,5,6-DIBENZANTHRACENECHOLEIC ACID
mf: $C_{96}H_{160}O_{16}•C_{22}H_{14}$ mw: 1848.92

SYN: 3-α-12-α-DIHYDROXY-5-β-CHOLAN-24-OIC ACID with DIBENZ (a,h)ANTHRACENE

TOXICITY DATA WITH REFERENCE
scu-mus TDLo: 800 mg/kg/9W-I: ETA JNCIAM 2,99,41

SAFETY PROFILE: Questionable carcinogen with experimental tumorigenic data. When heated to decomposition it emits acrid smoke and irritating fumes. See also ANTHRACENE.

DCU200 CAS:4665-48-9 HR: 2
1,2:5,6-DIBENZANTHRACENE-9,10-endo-α,β-SUCCINIC ACID
mf: $C_{26}H_{18}O_4$ mw: 394.44

SYN: 7,14-DIHYDRO-7,14-ETHANODIBENZ(a,b)ANTHRACENE-15,16-DICARBOXYLIC ACID

TOXICITY DATA WITH REFERENCE
mmo-esc 2040 mg/L/4H GENTAE 39,141,54
scu-rat TDLo: 665 mg/kg/50D-I: CAR,REP 85DLAB -,-,75

SAFETY PROFILE: Questionable carcinogen with experimental carcinogenic data. Experimental reproductive effects. Mutation data reported. When heated to decomposition it emits acrid smoke and irritating fumes. See also ANTHRACENE.

DCU600 CAS:63041-44-1 HR: 2
DIBENZANTHRANYL GLYCINE COMPLEX
mf: $C_{25}H_{18}N_2O_3$ mw: 394.45

SYNS: 1:2:5:6-DIBENZANTHRACENE-9-CARBAMIDO-ACETIC ACID ☐ N-(DIBENZ(a,h)ANTHRACEN-7-YLCARBAMOYL)GLYCINE

TOXICITY DATA WITH REFERENCE
scu-mus TDLo: 80 mg/kg/12W-I: ETA AJCAA7 35,203,39

SAFETY PROFILE: Questionable carcinogen with experimental tumorigenic data. When heated to decomposition it emits toxic fumes of NO_x.

DCU800 CAS:116-71-2 HR: 2
DIBENZANTHRONE
mf: $C_{34}H_{16}O_2$ mw: 456.50

PROP: Bluish-black powder from $PhNO_2$.

SYNS: DINAPHTHO(1,2,3-cd:3′,2′,1′-lm)PERYLENE-5,10-DIONE ☐ VIOLANTHRONE

TOXICITY DATA WITH REFERENCE
ipr-rat LD50: 5000 mg/kg RPTOAN 40,137,77
ipr-mus LD50: 2600 mg/kg RPTOAN 40,137,77

CONSENSUS REPORTS: Reported in EPA TSCA Inventory.

SAFETY PROFILE: Moderately toxic by intraperitoneal route. When heated to decomposition it emits acrid smoke and irritating fumes.

DCV000 CAS:116-90-5 HR: 2
4,4′-DIBENZANTHRONIL
mf: $C_{34}H_{18}O_2$ mw: 458.52

SYNS: 4,4′-DI-7H-BENZ(de)ANTHRACEN-7-ONE ☐ 4,4′-DIBENZANTHRONYL

TOXICITY DATA WITH REFERENCE
eye-rbt 500 mg/24H SEV 28ZPAK -,60,72
ipr-rat LD50: 2400 mg/kg RPTOAN 40,137,77
ipr-mus LD50: 1100 mg/kg RPTOAN 40,137,77

CONSENSUS REPORTS: Reported in EPA TSCA Inventory.

SAFETY PROFILE: Moderately toxic by intraperitoneal route. A severe eye irritant. When heated to decomposition it emits acrid smoke and irritating fumes.

DCV200 CAS:298-46-4 HR: 3
5H-DIBENZ(b,f)AZEPINE-5-CARBOXAMIDE
mf: $C_{15}H_{12}N_2O$ mw: 236.29

PROP: Crystals from EtOH or C_6H_6. Mp: 204–206°.

SYNS: BISTON □ CARBAMAZEPEN □ CARBAMAZEPINE □ CARBAMEZEPINE □ 5-CARBAMOYL-5H-DIBENZ(b,f)AZEPINE □ 5-CARBAMOYLDIBENZO(b,f)AZEPINE □ 5-CARBAMOYL-5H-DIBENZO(b,f)AZEPINE □ 5-CARBAMYLDIBENZO(b,f)AZEPINE □ 5-CARBAMYL-5H-DIBENZO(b,f)AZEPINE □ CARBAZEPINE □ FINLEPSIN □ G 32883 □ GEIGY 32883 □ STAZEPIN □ TEGRETAL □ TEGRETOL □ TELESMIN □ TIMONIL

TOXICITY DATA WITH REFERENCE
orl-rat TDLo:3600 mg/kg (14D pre/1-22D preg):REP TJADAB 29(3),33A,84
orl-rat TDLo:765 mg/kg (9-17D preg):TER EAMJAV 60,407,83
orl-rat TDLo:765 mg/kg (9-17D preg):REP EAMJAV 60,407,83
orl-mus TDLo:92 g/kg (male 2W pre):REP TCMUD8 6,393,86
orl-mus TDLo:5628 mg/kg (female 7-12D post):TER TJADAB 23,33A,81
orl-mus TDLo:440 mg/kg (6-16D preg):TER TXAPA9 40,365,77
orl-man TDLo:160 mg/kg/3W-I:SKN JCPYDR 5,185,85
orl-cld TDLo:1050 mg/kg/6W-I AJPSAO 143,1176,85
orl-man TDLo:253 mg/kg/6W-I:GIT,SYS JCPYDR 6,251,86
orl-cld TDLo:19 mg/kg/4W-I:PNS PEDIAU 73,841,84
orl-wmn TDLo:28 mg/kg/4D-I AJPSAO 143,1328,86
orl-hmn TDLo:43 mg/kg:CNS,GIT BMJOAE 1,754,77
orl-wmn TDLo:100 mg/kg/17D-I:BLD JCLPDE 45,315,84
orl-man LDLo:54 mg/kg/9D-I:CNS,KID,BLD CMAJAX 132,1040,85
orl-wmn LDLo:1920 mg/kg/17W-I:BLD AJPSAO 142,974,85
orl-rat LD50:1957 mg/kg JKXXAF #79-163823
ipr-rat LD50:293 mg/kg ARZNAD 30,477,80
orl-mus LD50:936 mg/kg RPTOAN 32,131,69
ipr-mus LD50:270 mg/kg JKXXAF #79-163823

CONSENSUS REPORTS: EPA Genetic Toxicology Program.

SAFETY PROFILE: A human poison by ingestion. Poison experimentally by intraperitoneal route. Human systemic effects by ingestion: aplastic anemia, sleep, hallucinations, distorted perceptions, nausea or vomiting, somnolence, dermatitis, ataxia (loss of muscle coordination), urine volume increase, and agranulocytosis, liver function tests impaired, fasciculations, and thrombocytopenia. Human reproductive effects. Experimental teratogenic and reproductive effects. An analgesic and anticonvulsant. When heated to decomposition it emits toxic fumes of NO_x. See also DIAZEPAM.

DCV400 CAS:28058-62-0 HR: 3
5H-DIBENZ(b,f)AZEPINE, 3-CHLORO-5-(3-(4-CARBAMOYL-4-PIPERIDINOPIPERIDINO)PROPYL)-10,11-DIHYDRO-, DIHYDROCHLORIDE, MONOHYDRATE
mf: $C_{28}H_{37}ClN_4O\cdot2ClH\cdot H_2O$ mw: 571.07

PROP: Bitter-tasting powder. Mp: 267°.

SYN: Y-4153

TOXICITY DATA WITH REFERENCE
orl-mus TDLo:60 mg/kg (7-12D preg):TER OYYAA2 5,663,71
orl-rat TDLo:7 g/kg (35D male):REP OYYAA2 5,643,71
orl-rat LD50:6800 mg/kg ARZNAD 21,391,71
ipr-rat LD50:125 mg/kg ARZNAD 21,391,71
orl-mus LD50:2550 mg/kg ARZNAD 21,391,71
ipr-mus LD50:160 mg/kg ARZNAD 21,391,71
scu-mus LD50:6500 mg/kg ARZNAD 21,391,71

SAFETY PROFILE: Poison by intraperitoneal route. Moderately toxic by ingestion. Mildly toxic by subcutaneous route. Experimental teratogenic and reproductive effects. A psychotropic drug. When heated to decomposition it emits very toxic fumes of Cl^- and NO_x.

DCV800 CAS:315-72-0 HR: 3
4-(3-(5H-DIBENZ(b,f)AZEPIN-5-YL)PROPYL)-1-PIPERAZINEETHANOL
mf: $C_{23}H_{29}N_3O$ mw: 363.55

PROP: A solid. Mp: 100–101°.

SYNS: ENDISON □ G 33040 □ GR 33040 □ 5-(3-(4-(2-HYDROXYETHYL)-1-PIPERAZINYL)PROPYL)-5H-DIBENZ(b,f)AZEPINE □ INSIDON □ NISIDANA □ OPIPRAMOL □ OPRAMIDOL

TOXICITY DATA WITH REFERENCE
orl-rat LD50:1110 mg/kg FRPPAO 25,519,70
ipr-rat LD50:95 mg/kg AIPTAK 148,560,64
scu-rat LD50:497 mg/kg AIPTAK 148,560,64
ivn-rat LD50:32 mg/kg AIPTAK 148,560,64
orl-mus LD50:443 mg/kg FRPPAO 25,519,70
ipr-mus LD50:120 mg/kg AIPTAK 148,560,64
scu-mus LD50:315 mg/kg AIPTAK 148,560,64
ivn-mus LD50:45 mg/kg AIPTAK 148,560,64
ivn-rbt LD50:11 mg/kg AIPTAK 148,560,64

SAFETY PROFILE: Poison by intraperitoneal, intravenous, and subcutaneous routes. Moderately toxic by ingestion. Many dibenz-azepine compounds have central nervous system effects. When heated to decomposition it emits toxic fumes of NO_x.

DCW000 CAS:15727-43-2 HR: 3
DI(BENZENEDIAZONIUM)ZINC TETRACHLORIDE
mf: $C_{12}H_{10}Cl_4N_4Zn$ mw: 417.42

PROP: After drying, it can explode.

SYN: BENZENEDIAZONIUM TETRACHLOROZINCATE

CONSENSUS REPORTS: Zinc and its compounds are on the Community Right-To-Know List.

SAFETY PROFILE: May be a light-, heat-, and shock-sensitive explosive. When heated to decomposition it

D

emits toxic fumes of NO_x, Cl^-, and ZnO. See also ZINC COMPOUNDS.

DCW400 CAS:29342-61-8 HR: 3
DIBENZENESULFONYL PEROXIDE
mf: $C_{12}H_{10}O_6S_2$ mw: 314.34

PROP: Waxy crystals. Mp: 66°.

SAFETY PROFILE: Explodes when heated to 53°C and when shocked. Decomposes violently when stored at room temperature. Explodes on contact with boiling water or fuming nitric acid. When heated to decomposition it emits toxic fumes of SO_x. See also PEROXIDES.

DCW600 CAS:4498-32-2 HR: 3
DIBENZEPIN
mf: $C_{18}H_{21}N_3O$ mw: 295.37

PROP: A solid. Mp: 116–117°.

SYNS: DIBENZEPINE □ 5,10-DIHYDRO-10-(2-(DIMETHYLAMINO) ETHYL)-5-METHYL-11H-DIBENZO(b,e)(1,4)DIAZEPIN-11-ONE □ 10-(2-(DIMETHYLAMINO)ETHYL)-5,10-DIHYDRO-5-METHYL-11H-DIBENZO(B,E)(1,4)DIAZEPIN-11-ONE □ 10-(2-(DIMETHYLAMINO)ETHYL)-5-METHYL-5H-DIBENZO(b,e)(1,4)DIAZEPIN-11(10H)-ONE □ HF 1927

TOXICITY DATA WITH REFERENCE
orl-rat LD50:220 mg/kg INPHB6 1,214,68
ipr-rat LD50:70 mg/kg INPHB6 1,214,68
scu-rat LD50:542 mg/kg IYKEDH 6,119,75
ivn-rat LD50:22 mg/kg INPHB6 1,214,68
orl-mus LD50:194 mg/kg IYKEDH 6,119,75
ipr-mus LD50:60 mg/kg ARZNAD 21,1727,71
scu-mus LD50:90 mg/kg ARZNAD 19,458,69
ivn-mus LD50:22 mg/kg INPHB6 1,214,68

SAFETY PROFILE: Poison by ingestion, intraperitoneal, intravenous, and subcutaneous routes. Many dibenz-azepine compounds have central nervous system effects. When heated to decomposition it emits toxic fumes of NO_x. See also DIAZEPAM.

DCW800 CAS:315-80-0 HR: 3
DIBENZEPINE HYDROCHLORIDE
mf: $C_{18}H_{21}N_3O{\cdot}ClH$ mw: 331.88

PROP: A solid. Mp: 234–240°.

SYNS: DIBENZEPIN HYDROCHLORIDE □ HF 1927 □ HYDROFLUORIDE-1927 WANDER □ 5-METHYL-10-β-DIMETHYLAMINOAETHYL-10, 11-DIHYDRO-11-OXO-5-DIBENZO(b,e)(1,4)DIAZEPIN □ NEODALIT □ NOVERIL □ NOVERYL

TOXICITY DATA WITH REFERENCE
orl-rat LD50:220 mg/kg 27ZQAG -,70,72
ipr-rat LD50:70 mg/kg 27ZQAG -,70,72
scu-rat LD50:520 mg/kg NIIRDN 6,341,82
ivn-rat LD50:22 mg/kg 27ZQAG -,70,72
orl-mus LD50:174 mg/kg NIIRDN 6,341,82
ipr-mus LD50:64 mg/kg FATOAO 35,274,72
scu-mus LD50:98 mg/kg FRPPAO 25,519,70

SAFETY PROFILE: Poison by ingestion, intravenous, subcutaneous, and intraperitoneal routes. An antidepressant. Many dibenz-azepine compounds have central nervous system effects. When heated to decomposition it emits very toxic fumes of HCl and NO_x. See also DIAZEPAM.

DCX000 CAS:201-65-0 HR: 2
1,2,3,4-DIBENZFLUORENE
mf: $C_{21}H_{14}$ mw: 266.35

SYN: 13H-INDENO(1,2-1)PHENANTHRENE

TOXICITY DATA WITH REFERENCE
skn-mus TDLo:1040 mg/kg/43W-I:ETA PRLBA4 129,439,40

SAFETY PROFILE: Questionable carcinogen with experimental tumorigenic data. When heated to decomposition it emits acrid smoke and irritating fumes.

DCX300 CAS:1210-35-1 HR: 1
DIBENZO(a,d)CYCLOHEPTADIEN-5-ONE
mf: $C_{15}H_{12}O$ mw: 208.27

SYNS: DIBENZOCYCLOHEPTENONE □ 5H-DIBENZO(a,d)CYCLOHEPTEN-5-ONE, 10,11-DIHYDRO- □ DIBENZOSUBERAN-5-ONE □ DIBENZOSUBERONE □ 2,3:6,7-DIBENZOSUBERONE □ DIENONE □ 10, 11-DIHYDRO-5H-DIBENZO(a,d)CYCLOHEPTEN-5-ONE

TOXICITY DATA WITH REFERENCE
eye-rbt 100 mg MOD JACTDZ 1,186,92

CONSENSUS REPORTS: Reported in EPA TSCA Inventory.

SAFETY PROFILE: An eye irritant. When heated to decomposition it emits acrid smoke and irritating vapors.

DCX400 CAS:193-40-8 HR: 2
DIBENZ(c,f)INDENO(1,2,3-ij)(2,7)NAPHTHYRIDINE
mf: $C_{22}H_{12}N_2$ mw: 304.36

PROP: Pale-yellow prisms from xylene. Mp: 270–272°.

TOXICITY DATA WITH REFERENCE
skn-mus TDLo:1200 mg/kg/52W-I:NEO BJCAAI 17,266,63

SAFETY PROFILE: Questionable carcinogen with experimental neoplastigenic data. When heated to decomposition it emits toxic fumes of NO_x.

DCX600 CAS:207-84-1 HR: 2
7H-DIBENZO(a,g)CARBAZOLE
mf: $C_{20}H_{13}N$ mw: 267.34

PROP: Crystals from Me_2CO or C_6H_6.

SYN: 1,2,5,6-DIBENZCARBAZOLE

TOXICITY DATA WITH REFERENCE
skn-mus TDLo:275 mg/kg/23W-I:ETA PRLBA4 122,429,37
skn-mus TD:900 mg/kg/38W-I:ETA PRLBA4 131,170,72

SAFETY PROFILE: Questionable carcinogen with experimental tumorigenic data. When heated to decomposition it emits toxic fumes of NO_x.

DCX800 CAS:239-64-5 HR: 2
7H-DIBENZO(a,i)CARBAZOLE
mf: $C_{20}H_{13}N$ mw: 267.34

PROP: Crystals from AcOH. Mp: 223.5–224°.

SYN: 1,2,7,8-DIBENZCARBAZOLE

TOXICITY DATA WITH REFERENCE
skn-mus TDLo:515 mg/kg/43W-I:ETA PRLBA4 122,429,37

SAFETY PROFILE: Questionable carcinogen with experimental tumorigenic data. When heated to decomposition it emits toxic fumes of NO_x.

DCY000 CAS:194-59-2 HR: 3
7H-DIBENZO(c,g)CARBAZOLE
mf: $C_{20}H_{13}N$ mw: 267.34

PROP: Needles or crystals from alc. Mp: 158°.

SYNS: 7-AZA-7H-DIBENZO(c,g)FLUORENE □ 7H-DB(c,g)C □ 3,4,5,6-DIBENZCARBAZOL □ 3,4,5,6-DIBENZCARBAZOLE □ 3,4,5,6-DIBENZOCARBAZOLE □ 3,4,5,6-DINAPHTHACARBAZOLE

TOXICITY DATA WITH REFERENCE
mma-sat 20 μg/plate MUREAV 198,15,88
dnd-mus-scu 44 μmol/kg CRNGDP 6,1271,85
scu-rat TDLo:150 mg/kg/17W-I:ETA PRLBA4 122,429,37
orl-mus TDLo:280 mg/kg/32W-I:ETA BJCAAI 4,203,50
skn-mus TDLo:99 mg/kg/99W-I:CAR CALEDQ 37,337,87
ipr-mus TDLo:13 mg/kg:ETA BIJOAK 32,1460,38
orl-mus TDLo:280 mg/kg/32W-I:ETA BJCAAI 4,203,50
skn-mus LD :90 mg/kg/23W-I:ETA PRLBA4 122,429,37
ivn-mus TDLo:10 mg/kg:ETA JNCIAM 1,225,40
imp-mus TDLo:40 mg/kg:ETA BJCAAI 6,412,52
imp-dog TDLo:65 mg/kg/52W-I:ETA JPBAA7 68,561,54
itr-ham TDLo:72 mg/kg/18W-I:NEO JTEHD6 3,935,77
ipr-mus LDLo:13 mg/kg BIJOAK 32,1460,38

CONSENSUS REPORTS: NTP 7th Annual Report On Carcinogens. IARC Cancer Review: Group 2B IMEMDT 7,56,87; Animal Sufficient Evidence IMEMDT 32,315,83; IMEMDT 3,260,73.

SAFETY PROFILE: Confirmed carcinogen with experimental carcinogenic, neoplastigenic, and tumorigenic data. Poison by intraperitoneal route. Mutation data reported. When heated to decomposition it emits toxic fumes of NO_x.

DCY200 CAS:189-64-0 HR: 3
DIBENZO(b,def)CHRYSENE
mf: $C_{24}H_{14}$ mw: 302.38

PROP: Golden-orange plates from trichlorobenzene. Mp: 315°.

SYNS: BD(a,h)P □ DIBENZO(a,h)PYRENE □ 1,2,6,7-DIBENZOPYRENE □ 3,4,8,9-DIBENZOPYRENE □ 3,4,8,9-DIBENZPYRENE

TOXICITY DATA WITH REFERENCE
mma-sat 12,500 pmol/plate CNREA8 41,2589,81
msc-ham:lng 30 μg/L CNREA8 42,1646,82
imp-rat TDLo:100 mg/kg:ETA NEOLA4 26,23,79
skn-mus TDLo:287 mg/kg/30W-I:CAR ZKKOBW 89,113,77
scu-mus TDLo:72 mg/kg/9W-I:ETA COREAF 246,1477,58
skn-mus TD:330 mg/kg/15W:ETA AVBNAN 56,39,39
skn-mus TD:700 mg/kg/42W-I:ETA PRLBA4 129,439,40
skn-mus TD:1060 mg/kg/44W-I:ETA PRLBA4 123,343,37

CONSENSUS REPORTS: NTP 7th Annual Report On Carcinogens. IARC Cancer Review: Group 2B IMEMDT 7,56,87; Animal Sufficient Evidence IMEMDT 32,331,83; IMEMDT 3,207,73.

SAFETY PROFILE: Confirmed carcinogen with experimental carcinogenic and tumorigenic data. When heated to decomposition it emits acrid smoke and irritating fumes. Mutation data reported.

DCY400 CAS:191-30-0 HR: 3
DIBENZO(def,p)CHRYSENE
mf: $C_{24}H_{14}$ mw: 302.38

PROP: Pale-yellow plates from C_6H_6/EtOH. Mp: 164–165°

SYNS: BA 51-090462 □ DB(a,l)P □ DIBENZO(a,d)PYRENE □ DIBENZO(a,l)PYRENE □ 1,2:3,4-DIBENZOPYRENE □ 1,2,9,10-DIBENZOPYRENE □ 2,3:4,5-DIBENZOPYRENE □ 1,2,3,4-DIBENZPYRENE □ 4,5,6,7-DIBENZPYRENE

TOXICITY DATA WITH REFERENCE
par-rat TDLo:96,761 μg/kg:CAR JCREA8 115,67,89
skn-mus TDLo:890 mg/kg/37W-I:ETA PRLBA4 123,343,37
scu-mus TDLo:48 mg/kg/4W-I:ETA NATWAY 55,43,68
scu-mus TD:72 mg/kg/9W-I:ETA COREAF 258,3387,64

CONSENSUS REPORTS: NTP 7th Annual Report On Carcinogens. IARC Cancer Review: Group 2B IMEMDT 7,56,87; Animal Sufficient Evidence IMEMDT 32,343,83; Animal Limited Evidence IMEMDT 3,224,73

SAFETY PROFILE: Confirmed carcinogen with experimental tumorigenic data. When heated to decomposition it emits acrid smoke and irritating fumes.

DCY600 CAS:63040-54-0 HR: 2
DIBENZO(b,def)CHRYSENE-7-CARBOXALDEHYDE
mf: $C_{25}H_{14}O$ mw: 330.39

SYN: 5-FORMYL-3,4:8,9-DIBENZOPYRENE

TOXICITY DATA WITH REFERENCE
scu-mus TDLo:72 mg/kg/9W-I:ETA COREAF 252,1711,61

SAFETY PROFILE: Questionable carcinogen with experimental tumorigenic data. When heated to decomposition it emits acrid smoke and irritating fumes. See also ALDEHYDES.

DCY800 CAS:2869-59-2 HR: 2
DIBENZO(def,p)CHRYSENE-10-CARBOXALDEHYDE
mf: $C_{25}H_{14}O$ mw: 330.39

SYN: 5-FORMYL-1,2:3,4-DIBENZOPYRENE

TOXICITY DATA WITH REFERENCE
scu-mus TDLo:72 mg/kg/9W-I:ETA COREAF 259,3899,64

SAFETY PROFILE: Questionable carcinogen with experimental tumorigenic data. When heated to decomposition it emits acrid smoke and irritating fumes.

DCZ000 **CAS:128-66-5** **HR: 2**
DIBENZO(b,def)CHRYSENE-7,14-DIONE
mf: $C_{24}H_{12}O_2$ mw: 332.36

PROP: C.I. vat yellow 4 tested in NCITR* NCI-CG-TR-134,79 consists of 18.2% dibenzo(b,def)chrysene-7,14-dione, 30.8% sorbitol, 5.5% lomar twc, 2.7% glycerin, and 42.8% water (NCITR* NCI-CG-TR-134,79).

SYNS: AHCOVAT PRINTING GOLDEN YELLOW □ AMANTHRENE GOLDEN YELLOW □ ANTHRAVAT GOLDEN YELLOW □ ARLANTHRENE GOLDEN YELLOW □ BENZADONE GOLDEN YELLOW □ CALCOLOID GOLDEN YELLOW □ CALEDON GOLDEN YELLOW □ CALEDON PRINTING YELLOW □ CARBANTHRENE GOLDEN YELLOW □ C.I. 59100 □ CIBANONE GOLDEN YELLOW □ C.I. VAT YELLOW □ DIBENZO(a,b)PYRENE-7,14-DIONE □ 2,3,7,8-DIBENZOPYRENE-1,6-QUINONE □ 1′,2′,6′,7′-DIBENZPYRENE-7,14-QUINONE □ FEMANTHREN GOLDEN YELLOW □ GOLDEN YELLOW □ HELANTHRENE YELLOW □ HOSTAVAT GOLDEN YELLOW □ INDANTHRENE GOLDEN YELLOW □ LEUCOSOL GOLDEN YELLOW □ MAYVAT GOLDEN YELLOW □ MIKETHRENE GOLD YELLOW □ NCI-C03565 □ NIHONTHRENE GOLDEN YELLOW □ NYANTHRENE GOLDEN YELLOW □ PALANTHRENE GOLDEN YELLOW □ PARADONE GOLDEN YELLOW □ PHARMANTHRENE GOLDEN YELLOW □ ROMANTRENE GOLDEN YELLOW □ SANDOTHRENE PRINTING YELLOW □ SOLANTHRENE BRILLIANT YELLOW □ TINON GOLDEN YELLOW □ TYRION YELLOW □ VAT GOLDEN YELLOW □ YELLOW

TOXICITY DATA WITH REFERENCE
orl-mus TDLo:7420 g/kg/2Y-C:CAR NCITR* NCI-CG-TR-134,79
orl-mus TD:2225 g/kg/106W-C:ETA NCITR* NCI-CG-TR-134,79

CONSENSUS REPORTS: NCI Carcinogenesis Bioassay Completed; Results Positive: Mouse NCITR* NCI-CG-TR-134,79; Negative: Rat NCITR* NCI-CG-TR-134,79. Reported in EPA TSCA Inventory. Community Right-To-Know List.

SAFETY PROFILE: Questionable carcinogen with experimental carcinogenic and tumorigenic data. When heated to decomposition it emits acrid smoke and irritating fumes.

DDA600 **CAS:438-60-8** **HR: 3**
N-3-(5H-DIBENZO(a,d)CYCLOHEPTEN-5-YL)PROPYL-N-METHYLAMINE
mf: $C_{19}H_{21}N$ mw: 263.41

SYNS: 5-(3-METHYLAMINOPROPYL)-5H-DIBENZO(a,d)CYCLOHEPTENE □ MK 240 □ PROTRIPTYLINE □ PROTRYPTYLINE □ TRIPTIL □ VIVACTIL

TOXICITY DATA WITH REFERENCE
orl-rat LD50:240 mg/kg FRPPAO 25,519,70
ipr-rat LD50:42 mg/kg FRPPAO 25,519,70
orl-mus LD50:269 mg/kg FRPPAO 25,519,70
ipr-mus LD50:67 mg/kg FRPPAO 25,519,70
scu-mus LD50:192 mg/kg FRPPAO 25,519,70
ivn-mus LD50:30 mg/kg JMCMAR 17,65,74
orl-rbt LD50:310 mg/kg FRPPAO 25,519,70
ivn-rbt LD50:8200 mg/kg FRPPAO 25,519,70

SAFETY PROFILE: Poison by ingestion, intraperitoneal, subcutaneous, and intravenous routes. When heated to decomposition it emits toxic fumes of NO_x.

DDA800 **CAS:262-12-4** **HR: 3**
DIBENZO-p-DIOXIN
mf: $C_{12}H_8O_2$ mw: 184.20

PROP: Crystals from MeOH. Mp: 119°.

SYNS: DIBENZODIOXIN □ DIBENZO(1,4)DIOXIN □ DIBENZO(b.e)(1,4)DIOXIN □ DIPHENYLENE DIOXIDE □ NCI-C03656 □ OXANTHRENE □ PHENODIOXIN

TOXICITY DATA WITH REFERENCE
skn-mus TDLo:110 g/kg/58W-I:ETA EVHPAZ 5,163,73
orl-rat LD50:1220 mg/kg CMSHAF 19,989,89
ipr-rat LD50:30 mg/kg PHBUA9 3,337,55
orl-mus LD50:866 mg/kg CMSHAF 19,989,89

CONSENSUS REPORTS: IARC Cancer Review: Group 3 IMEMDT 7,56,87; Animal Inadequate Evidence IMEMDT 15,41,77; Human No Adequate Data IMEMDT 15,41,77. NCI Carcinogenesis Bioassay Completed; Results Negative NCITR* NCI-CG-TR-122,79.

SAFETY PROFILE: A poison by intraperitoneal route. Moderately toxic by ingestion. Questionable carcinogen with experimental tumorigenic data. When heated to decomposition it emits acrid smoke and irritating fumes.

DDB000 **CAS:207-83-0** **HR: 2**
13H-DIBENZO(a,g)FLUORENE
mf: $C_{21}H_{14}$ mw: 266.35

PROP: Plates from AcOH or EtOAc. Mp: 174–175°, bp: 195–200° @ 0.5 mm.

SYN: 1,2,5,6-DIBENZOFLUORENE

TOXICITY DATA WITH REFERENCE
skn-mus TDLo:48 mg/kg/15W-I:ETA CNREA8 11,301,51
skn-mus TD:240 mg/kg/37W-I:ETA CNREA8 11,892,51
skn-mus TD:1220 mg/kg/51W-I:ETA PRLBA4 123,343,37

SAFETY PROFILE: Questionable carcinogen with experimental tumorigenic data. When heated to decomposition it emits acrid smoke and irritating fumes.

DDB200 **CAS:239-60-1** **HR: 2**
13H-DIBENZO(a,i)FLUORENE
mf: $C_{21}H_{14}$ mw: 266.35

PROP: Plates from EtOAc. Mp: 234°.

SYN: 1,2,7,8-DIBENZFLUORENE

TOXICITY DATA WITH REFERENCE
skn-mus TDLo:1340 mg/kg/56W-I:ETA PRLBA4 129,439,40

SAFETY PROFILE: Questionable carcinogen with experimental tumorigenic data. When heated to decomposition it emits acrid smoke and irritating fumes.

DDB400 **CAS:36115-09-0** **HR: 2**
1,1′-(2,8-DIBENZOFURADIYL)BIS(2-(DIMETHYLAMINOETHANONE)) DIHYDROCHLORIDE HYDRATE (2:5)
mf: $C_{20}H_{22}NO_3 \cdot 2ClH \cdot 5/2H_2O$ mw: 442.35

SYN: RMI 11567 DA

TOXICITY DATA WITH REFERENCE
orl-mus LD50:2700 mg/kg ALACBI 12,77,79
scu-mus LD50:1000 mg/kg ALACBI 12,77,79

SAFETY PROFILE: Moderately toxic by ingestion and subcutaneous routes. When heated to decomposition it emits very toxic fumes of HCl and NO_x.

DDB600 **CAS:3693-22-9** **HR: 2**
2-DIBENZOFURANAMINE
mf: $C_{12}H_9NO$ mw: 183.22

PROP: Crystals from EtOH. Mp: 128°.

SYNS: 2-ADO ☐ 3-AMINODIBENZOFURAN ☐ 2-AMINODIPHENYLENE OXIDE

TOXICITY DATA WITH REFERENCE
dns-mus-orl 80 mg/kg BIJOAK 111,12P,69
orl-rat TDLo:168 mg/kg/90W-I:CAR ZEKBAI 61,45,56
orl-mus TDLo:22 g/kg/52W-C:ETA BECCAN 46,271,68

SAFETY PROFILE: Questionable carcinogen with experimental carcinogenic and tumorigenic data. Mutation data reported. When heated to decomposition it emits toxic fumes of NO_x.

DDB800 **CAS:4106-66-5** **HR: 2**
3-DIBENZOFURANAMINE
mf: $C_{12}H_9NO$ mw: 183.22

PROP: Crystals from EtOH. Mp: 94°.

SYNS: 2-AMINODIPHENYLENOXYD (GERMAN) ☐ DIBENZOFURANYLAMINE

TOXICITY DATA WITH REFERENCE
orl-rat TDLo:1400 mg/kg/66W-C:CAR ZEKBAI 61,45,56

SAFETY PROFILE: Questionable carcinogen with experimental carcinogenic data. When heated to decomposition it emits toxic fumes of NO_x.

DDC000 **CAS:5834-25-3** **HR: 2**
N-3-DIBENZOFURANYLACETAMIDE
mf: $C_{14}H_{11}NO_2$ mw: 225.26

SYNS: 3-ACETAMIDODIBENZFURANE ☐ 3-ACETAMIDODIBENZOFURAN ☐ 3-ACETYLAMINODIBENZOFURAN ☐ 3-DIBENZOFURANYLACETAMIDE

TOXICITY DATA WITH REFERENCE
orl-rat TDLo:4496 mg/kg/35W-C:ETA CNREA8 9,504,49

SAFETY PROFILE: Questionable carcinogen with experimental tumorigenic data. When heated to decomposition it emits toxic fumes of NO_x.

DDC100 **CAS:6407-29-0** **HR: 3**
2-DIBENZOFURANYLPHENYL METHANONE
mf: $C_{19}H_{12}O_2$ mw: 272.31

SYN: KETONE, DIBENZOFURAN-2-YL PHENYL

TOXICITY DATA WITH REFERENCE
ipr-mus LD50:1 g/kg RPTOAN 48,143,85

DOT CLASSIFICATION: 3; *Label:* Flammable Liquid

SAFETY PROFILE: Low toxicity by intraperitoneal route. A flammable liquid. When heated to decomposition it emits acrid smoke and irritating vapors.

DDC200 **CAS:192-47-2** **HR: 3**
DIBENZO(h,rst)PENTAPHENE
mf: $C_{28}H_{16}$ mw: 352.44

PROP: Pale yellow needles. Mp: 321°.

SYNS: TRIBENZO(a,e,i)PYRENE ☐ (1,2,4,5,7,8)TRIBENZOPYRENE ☐ (1,2,4,5,8,9)TRIBENZOPYRENE ☐ 1,2:4,5:8,9-TRIBENZOPYRENE

TOXICITY DATA WITH REFERENCE
scu-mus TDLo:72 mg/kg/9W-I:ETA COREAF 259,3899,64

CONSENSUS REPORTS: IARC Cancer Review: Group 3 IMEMDT 7,56,87; Animal Limited Evidence IMEMDT 3,197,73

SAFETY PROFILE: Suspected carcinogen with experimental tumorigenic data. When heated to decomposition it emits acrid smoke and irritating fumes.

DDC400 **CAS:188-96-5** **HR: 2**
DIBENZO(cd,lm)PERYLENE
mf: $C_{26}H_{14}$ mw: 326.40

PROP: Golden-yellow rhombs from xylene. Mp: 380–381°, bp: 160° @ 12 mm (subl).

SYN: PEROPYRENE

TOXICITY DATA WITH REFERENCE
scu-mus TDLo:72 mg/kg/9W-I:ETA CHDDAT 266,301,68

SAFETY PROFILE: Questionable carcinogen with experimental tumorigenic data. When heated to decomposition it emits acrid smoke and irritating fumes.

DDC600 **CAS:215-64-5** **HR: 2**
DIBENZO(a,c)PHENAZINE
mf: $C_{20}H_{12}N_2$ mw: 280.34

PROP: Pale-yellow crystals from EtOH or EtOAc. Mp: 217°.

SYN: 1,2,3,4-DIBENZPHENAZINE

TOXICITY DATA WITH REFERENCE
imp-rat TDLo:7 mg/kg:ETA COREAF 240,1738,55

SAFETY PROFILE: Questionable carcinogen with experimental tumorigenic data. When heated to decomposition it emits toxic fumes of NO_x.

D

DDC800 **CAS:226-47-1** **HR: 2**
DIBENZO(a,h)PHENAZINE
mf: $C_{20}H_{12}N_2$ mw: 280.34

PROP: Yellow crystals from AcOH or C_6H_6. Mp: 284°. Sltly sol in most solvs.

SYNS: 7,14-DIAZADIBENZ(a,h)ANTHRACENE □ DIBENZ(a,h)PHENAZINE □ 1,2:5,6-DIBENZPHENAZINE

TOXICITY DATA WITH REFERENCE
imp-rat TDLo:7 mg/kg:ETA COREAF 240,1738,55
imp-rat TD:100 mg/kg:ETA NEOLA4 25,641,78

CONSENSUS REPORTS: EPA Genetic Toxicology Program.

SAFETY PROFILE: Questionable carcinogen with experimental tumorigenic data. When heated to decomposition it emits toxic fumes of NO_x.

DDD000 **CAS:1785-74-6** **HR: 3**
DIBENZOSUBERONE OXIME
mf: $C_{15}H_{13}NO$ mw: 223.29

SYN: 10,11-DIHYDRO-5H-DIBENZO(a,d)CYCLOHEPTEN-5-ONE OXIME

TOXICITY DATA WITH REFERENCE
ipr-mus LD50:350 mg/kg JMCMAR 7,88,64
ivn-mus LD50:100 mg/kg CSLNX* NX#01997

SAFETY PROFILE: Poison by intraperitoneal and intravenous routes. When heated to decomposition it emits toxic fumes of NO_x.

DDD400 **CAS:54818-88-1** **HR: 2**
N-2-DIBENZOTHIENYLACETAMIDE
mf: $C_{14}H_{11}NOS$ mw: 241.32

SYN: 2-ACETYLAMINODIBENZOTHIOPHENE

TOXICITY DATA WITH REFERENCE
orl-rat TDLo:4680 mg/kg/32W-C:CAR CNREA8 15,188,55

SAFETY PROFILE: Questionable carcinogen with experimental carcinogenic data. When heated to decomposition it emits very toxic fumes of SO_x and NO_x.

DDD600 **CAS:64057-52-9** **HR: 2**
N-3-DIBENZOTHIENYLACETAMIDE
mf: $C_{14}H_{11}NOS$ mw: 241.32

SYNS: 3-ACETAMIDODIBENZTHIOPHENE □ 3-ACETAMINODIBENZOTHIOPHENE □ 3-ACETYLAMINODIBENZOTHIOPHENE

TOXICITY DATA WITH REFERENCE
orl-rat TDLo:4739 mg/kg/35W-C:ETA CNREA8 9,504,49
orl-rat LD50:1195 mg/kg JPETAB 99,450,50

SAFETY PROFILE: Moderately toxic by ingestion. Questionable carcinogen with experimental tumorigenic data. When heated to decomposition it emits very toxic fumes of NO_x and SO_x.

DDD800 **CAS:63020-21-3** **HR: 2**
N-3-DIBENZOTHIENYLACETAMIDE-5-OXIDE
mf: $C_{14}H_{11}NO_2S$ mw: 257.32

SYNS: 3-ACETAMIDODIBENZTHIOPHENE OXIDE □ 3-ACETYLAMINODIBENZOTHIOPHENE-5-OXIDE

TOXICITY DATA WITH REFERENCE
orl-rat TDLo:5103 mg/kg/35W-C:ETA CNREA8 9,504,49

SAFETY PROFILE: Questionable carcinogen with experimental tumorigenic data. When heated to decomposition it emits very toxic fumes of SO_x and NO_x.

DDE000 **CAS:35556-06-0** **HR: 2**
1,1′-(2,8-DIBENZOTHIOPHENEDIYL)BIS(2-(DIMETHYLAMINO)ETHANONE) DIHYDROCHLORIDE TRIHYDRATE
mf: $C_{20}H_{22}N_2O_2S{\cdot}2ClH{\cdot}3H_2O$ mw: 481.48

SYN: RMI 11877 DA

TOXICITY DATA WITH REFERENCE
orl-mus LD50:2930 mg/kg ALACBI 12,77,79
scu-mus LD50:820 mg/kg ALACBI 12,77,79

SAFETY PROFILE: Moderately toxic by ingestion and subcutaneous routes. When heated to decomposition it emits very toxic fumes of HCl, SO_x and NO_x.

DDE200 **CAS:257-07-8** **HR: 3**
DIBENZ(b,f)(1,4)OXAZEPINE
mf: $C_{13}H_9NO$ mw: 195.23

SYNS: CR □ EA 3547

TOXICITY DATA WITH REFERENCE
skn-hmn 500 µg/1H MLD BJDEAZ 90,657,74
eye-man 17 ng APTOA6 35,412,74
eye-rbt 5 mg MLD ARTODN 34,183,75
eye-rbt 1540 ng APTOA6 35,412,74
eye-gpg 682 ng APTOA6 35,412,74
ihl-rat TCLo:2 mg/m³/5M (6-15D preg):TER TXAPA9 29,301,74
ivn-rbt TDLo:47,400 µg/kg (14-16D preg):REP TXAPA9 29,301,74
ihl-mus TCLo:236 mg/m³/18W-I:CAR TOLED5 17,13,83
ihl-mus TC:204 mg/m³/18W-I:ETA TOLED5 17,13,83
orl-rat LD50:563 mg/kg IAEC** 17JUN74
ipr-rat LD50:164 mg/kg IAEC** 17JUN74
ivn-rat LD50:26 mg/kg IAEC** 17JUN74
orl-mus LD50:770 mg/kg IAEC** 17JUN74
ihl-mus LCLo:1500 mg/m³/2H TXCYAC 8,347,77
ipr-mus LD50:242 mg/kg IAEC** 17JUN74
ivn-mus LD50:37,200 µg/kg IAEC** 17JUN74

CONSENSUS REPORTS: Reported in EPA TSCA Inventory.

SAFETY PROFILE: Poison by intraperitoneal and intravenous routes. Moderately toxic by ingestion and inhalation. Experimental teratogenic and reproductive effects. A human skin and eye irritant. Questionable carcinogen with experimental carcinogenic and tumorigenic data. When heated to decomposition it emits toxic fumes of NO_x.

DDE300 CAS:2743-38-6 HR: 1
DIBENZOYLTARTARIC ACID
mf: $C_{18}H_{14}O_8$ mw: 358.32

PROP: Crystals. Mp: 89–92°.

SYNS: BUTANEDIOIC ACID, 2,3-BIS(BENZOYLOXY)-, (R-(R*,R*))- □ TARTARIC ACID, DIBENZOATE

TOXICITY DATA WITH REFERENCE
eye-rbt 100 mg MOD FCTOD7 20,573,82
eye-rbt 100 mg/4S RNS MLD FCTOD7 20,573,82

CONSENSUS REPORTS: Reported in EPA TSCA Inventory.

SAFETY PROFILE: An eye irritant. When heated to decomposition it emits acrid smoke and irritating fumes.

DDF000 CAS:73926-80-4 HR: 3
DIBENZYLBUTYLSULFONIUM IODIDE MERCURIC IODIDE

SYN: DIBENZYLBUTYLSULFONIUM IODIDE with MERCURY IODIDE (1:1)

TOXICITY DATA WITH REFERENCE
ivn-mus LD50:56 mg/kg CSLNX* NX#01719

CONSENSUS REPORTS: Mercury and its compounds are on the Community Right-To-Know List.

NIOSH REL: (Mercury, Aryl and Inorganic): CL 0.1 mg/m³ (skin)

SAFETY PROFILE: Poison by intravenous route. See also IODIDES and MERCURY IODIDE. When heated to decomposition it emits very toxic fumes of Hg, I^-, and SO_x.

DDF200 CAS:101833-83-4 HR: 3
1-1,3-DIBENZYLDECAHYDRO-2-OXOIMIDAZO(4,5-c) THIENO(1,2-a)THIOLIUM-2-OXO-10-BORANESULFONATE
mf: $C_{22}H_{25}N_2OS \cdot C_{10}H_{15}O_4S$ mw: 596.86

SYNS: 1-3,4-(1′,3′-DIBENZYL-2′-KETO-IMIDAZOLIDO)-1,2-TRIMETHYLENE THIOPHANIUM CAMPHOR SULFONATE □ NU-2221

TOXICITY DATA WITH REFERENCE
ipr-mus LD50:135 mg/kg JPETAB 97,48,49
ivn-mus LD50:23 mg/kg JPETAB 97,48,49

SAFETY PROFILE: Poison by intraperitoneal and intravenous routes. When heated to decomposition it emits very toxic fumes of SO_x and NO_x.

DDF400 CAS:59766-02-8 HR: 2
7,14-DIBENZYLDIBENZ(a,h)ANTHRACENE
mf: $C_{36}H_{26}$ mw: 458.62

SYN: 9,10-DIBENZYL-1,2,5,6-DIBENZANTHRACENE

TOXICITY DATA WITH REFERENCE
skn-mus TDLo:1250 mg/kg/52W-I:ETA PRLBA4,111,485,32

SAFETY PROFILE: Questionable carcinogen with experimental tumorigenic data. When heated to decomposition it emits acrid smoke and irritating fumes.

DDF600 CAS:63957-48-2 HR: 3
DIBENZYL(5-DIBENZYLAMINO-2,4-PENTADIENYLIDENE)AMMONIUM CHLORIDE SESQUIHYDRATE
mf: $C_{33}H_{33}N_2 \cdot Cl \cdot 3/2H_2O$ mw: 520.16

TOXICITY DATA WITH REFERENCE
orl-mus LD50:500 mg/kg JMCMAR 12,806,69
ipr-mus LD50:200 mg/kg JMCMAR 12,806,69

SAFETY PROFILE: Poison by intraperitoneal route. Moderately toxic by ingestion. When heated to decomposition it emits very toxic fumes of NO_x, NH_3, and Cl^-.

DDF700 CAS:122-65-6 HR: 3
N,N′-DIBENZYLDITHIOOXAMIDE
mf: $C_{16}H_{16}N_2S_2$ mw: 300.46

SYNS: OXAMIDE, N,N′ DIBENZYLDITHIO- □ USAF MK-1

TOXICITY DATA WITH REFERENCE
ipr-mus LD50:50 mg/kg NTIS** AD277-689

CONSENSUS REPORTS: Reported in EPA TSCA Inventory.

SAFETY PROFILE: Poison by intraperitoneal route. When heated to decomposition it emits toxic vapors of NO_x and SO_x.

DDF800 CAS:122-75-8 HR: 3
N,N′-DIBENZYLETHYLENEDIAMINE DIACETATE
mf: $C_{16}H_{20}N_2 \cdot 2C_2H_4O_2$ mw: 360.50

SYNS: DBED DIACETATE □ ETHYLENEDIAMINE, N,N′-DIBENZYL-, DIACETATE

TOXICITY DATA WITH REFERENCE
ims-mus LD50:138 mg/kg ANTCAO 4,633,54

CONSENSUS REPORTS: Reported in EPA TSCA Inventory.

SAFETY PROFILE: Poison by intramuscular route. When heated to decomposition it emits toxic vapors of NO_x.

DDG400 CAS:3412-76-8 HR: 3
N,N′-DIBENZYLETHYLENEDIAMINE DIHYDROCHLORIDE
mf: $C_{16}H_{20}N_2 \cdot 2ClH$ mw: 313.30

SYN: DBED DIHYDROCHLORIDE

TOXICITY DATA WITH REFERENCE
orl-mus LD50:630 mg/kg ANTCAO 1,504,51
ipr-mus LD50:104 mg/kg ANTCAO 1,504,51
scu-mus LD50:200 mg/kg ARZNAD 9,628,59

SAFETY PROFILE: Poison by intraperitoneal and subcutaneous routes. Moderately toxic by ingestion. When heated to decomposition it emits very toxic fumes of HCl and NO_x.

DDG600 CAS:73926-81-5 HR: 3
DIBENZYLETHYLSULFONIUM IODIDE MERCURIC IODIDE

SYN: DIBENZYLETHYLSULFONIUM IODIDE with MERCURY IODIDE (1:1)

TOXICITY DATA WITH REFERENCE
ivn-mus LD50:56 mg/kg CSLNX* NX#01718

CONSENSUS REPORTS: Mercury and its compounds are on the Community Right-To-Know List.

NIOSH REL: TWA 0.05 mg(Hg)/m^3

SAFETY PROFILE: Poison by intravenous route. See also IODIDES and MERCURY IODIDE. When heated to decomposition it emits very toxic fumes of Hg, I^-, and SO_x.

DDG800 CAS:63-92-3 HR: 3
DIBENZYLINE HYDROCHLORIDE
mf: $C_{18}H_{22}ClNO{\cdot}ClH$ mw: 340.32

PROP: Cryst from EtOH/Et_2O. Mp: 137.5–140°.

SYNS: 688A □ BENSYLYT □ 2-(N-BENZYL-2-CHLOROETHYLAMINO)-1-PHENOXYPROPANE HYDROCHLORIDE □ BENZYL(2-CHLOROETHYL)(1-METHYL-2-PHENOXYETHYL)AMINE HYDROCHLORIDE □ N-BENZYL-N-PHENOXYISOPROPYL-β-CHLORETHYLAMINE HYDROCHLORIDE □ BENZYLYT □ BLOCADREN □ N-(2-CHLOROETHYL)-N-(1-METHYL-2-PHENOXYETHYL)BENZENEMETHANAMINE HYDROCHLORIDE □ N-(2-CHLOROETHYL)-N-(1-METHYL-2-PHENOXYETHYL)BENZYLAMINE HYDROCHLORIDE □ DIBENZYLENE □ DIBENZYLIN □ DIBENZYRAN □ FENOXYBENZAMIN □ NCI-C01661 □ PHENOXYBENZAMIDE HYDROCHLORIDE □ N-PHENOXYISOPROPYL-N-BENZYL-β-CHLOROETHYLAMINE HYDROCHLORIDE □ N-2-PHENOXYISOPROPYL-N-BENZYL-CHLOROETHYLAMINE HYDROCHLORIDE □ SKF 688A

TOXICITY DATA WITH REFERENCE
mmo-sat 3 µg/plate EMMUEG 11(Suppl 12),1,88
mma-sat 10 µg/plate EMMUEG 11(Suppl 12),1,88
orl-rat TDLo:12 mg/kg (5-12D preg):TER RCOCB8 7,701,74
par-rat TDLo:24,500 µg/kg (male 35D pre):REP CCPTAY 29,189,84
ipr-rat TDLo:780 mg/kg/1Y-I:CAR NCITR* NCI-CG-TR-72,78
ipr-mus TDLo:3900 mg/kg/52W-I:CAR NCITR* NCI-CG-TR-72,78
ipr-rat TD:1560 mg/kg/1Y-I:CAR NCITR* NCI-CG-TR-72,78
orl-man TDLo:7143 µg/kg/5D-I:SYS AIMEAS 107,119,87
orl-rat LDLo:800 mg/kg JPETAB 110,463,54
orl-mus LD50:900 mg/kg AIPTAK 108,102,56
ipr-mus LD50:228 mg/kg TXAPA9 28,227,74
scu-mus LD50:105 mg/kg ARZNAD 17,305,67
ivn-mus LD50:63,750 µg/kg EJPHAZ 9,289,70

CONSENSUS REPORTS: NTP 7th Annual Report on Carcinogens. IARC Cancer Review: Group 2B IMEMDT 7,56,87; Animal Sufficient Evidence IMEMDT 24,185,80. NCI Carcinogenesis Bioassay Completed; Results Positive: mouse, rat NCITR* NCI-CG-TR-72,78.

SAFETY PROFILE: Confirmed carcinogen with experimental carcinogenic and teratogenic data. Poison by intraperitoneal, intravenous, and subcutaneous routes. Human systemic effects by ingestion: changes in tubules, including acute renal failure, acute tubular necrosis. Moderately toxic by ingestion. Other experimental reproductive effects. Mutation data reported. A long-acting adrenergic blocker. When heated to decomposition it emits very toxic fumes of NO_x and Cl^-.

DDH000 CAS:780-24-5 HR: 3
DIBENZYLMERCURY
mf: $C_{14}H_{14}Hg$ mw: 382.87

PROP: Colorless crystals or needles from alc. Mp: 111°. Insol in Et_2O, pet ether; sltly sol in EtOH and C_6H_6; sol in organic solvents.

TOXICITY DATA WITH REFERENCE
ivn-mus LD50:56 mg/kg CSLNX* NX#03272

OSHA PEL: TWA 0.01 mg(Hg)/m^3; STEL 0.03 mg/m^3 (skin)
ACGIH TLV: TWA 0.01 mg(Hg)/m^3; STEL 0.03 mg(Hg)/m^3
NIOSH REL: (Mercury, Aryl and Inorganic): CL 0.1 mg/m^3 (skin)

SAFETY PROFILE: Poison by intravenous route. See also MERCURY COMPOUNDS, ORGANIC. When heated to decomposition it emits toxic fumes of Hg.

DDH200 CAS:2144-45-8 HR: 3
DIBENZYL PEROXYDICARBONATE
mf: $C_{16}H_{14}O_6$ mw: 302.30

SYNS: DIBENZYL PEROXYDICARBONATE, >87% with water (DOT) □ PEROXYDICARBONIC ACID, DIBENZYL ESTER

DOT CLASSIFICATION: Forbidden

SAFETY PROFILE: An unstable peroxide forbidden for transport. When heated to decomposition it emits acrid smoke and irritating vapors.

DDH400 CAS:17176-77-1 HR: 3
DIBENZYL PHOSPHITE
mf: $C_{14}H_{15}O_3P$ mw: 262.25

PROP: A liquid. Mp: 17°, bp: 110–120° @ 0.1 mm. Decomposes @ 160°.

SAFETY PROFILE: Potentially explosive decomposition when heated. When heated to decomposition it emits toxic fumes of PO_x.

DDH600 CAS:3666-67-9 HR: 3
2,2-DIBENZYL-4-(2-PIPERIDYL)-1,3-DIOXOLANE HYDROCHLORIDE
mf: $C_{22}H_{27}NO_2{\cdot}ClH$ mw: 373.96

TOXICITY DATA WITH REFERENCE
orl-mus LD50:200 mg/kg JMCMAR 9,127,66
ivn-mus LD50:25 mg/kg JMCMAR 9,127,66

SAFETY PROFILE: Poison by ingestion and intravenous routes. When heated to decomposition it emits very toxic fumes of HCl and NO_x.

DDH800 CAS:621-08-9 HR: 2
DIBENZYLSULFOXIDE
mf: $C_{14}H_{14}OS$ mw: 230.34

PROP: Leaflets from EtOH or H_2O. Mp: 133–135°. Sol in $CHCl_3$ and CH_2Cl_2.

SYNS: BENZYL SULFOXIDE □ DIBENZYL SULPHOXIDE

TOXICITY DATA WITH REFERENCE
ipr-mus LD50:600 mg/kg IJRBA3 3,41,61

CONSENSUS REPORTS: Reported in EPA TSCA Inventory.

SAFETY PROFILE: Moderately toxic by intraperitoneal route. When heated to decomposition it emits toxic fumes of SO_x.

DDH900 CAS:2964-06-9 HR: 3
DIBERAL
mf: $C_{12}H_{20}N_2O_3$ mw: 240.34

SYNS: 5-(1,3-DIMETHYLBUTYL)-5-ETHYLBARBITURIC ACID □ DMBEB □ 5-(1,3-DIMETHYLBUTYL)-5-ETHYL-2,4,6(1H,3H,5H)-PYRIMIDINETRIONE (9CI)

TOXICITY DATA WITH REFERENCE
orl-rat LD50:75 mg/kg JMPCAS 1,31,59
orl-mus LD50:70 mg/kg JMPCAS 1,31,59
ipr-mus LD50:17 mg/kg JMPCAS 1,31,59
orl-dog LD50:10 mg/kg JMPCAS 1,31,59
ivn-dog LD50:2500 µg/kg JMPCAS 1,31,59
ivn-rbt LDLo:7 mg/kg JACSAT 58,1354,36

SAFETY PROFILE: Poison by ingestion, intravenous, and intraperitoneal routes. When heated to decomposition it emits toxic fumes of NO_x. See also BARBITURATES.

DDI000 CAS:13084-46-3 HR: 3
DI-1,2-BIS(DIFLUOROAMINO)ETHYL ETHER
mf: $C_4H_6F_8N_4O$ mw: 278.10

$[F_2NCH_2CH(NF_2)]_2O$

SAFETY PROFILE: An impact-sensitive explosive. When heated to decomposition it emits toxic fumes of F^- and NO_x. See also ETHERS.

DDI200 CAS:1345-07-9 HR: 3
DIBISMUTH TRISULFIDE
mf: Bi_2S_3 mw: 514.15

PROP: Dark brown or black crystals with metallic lustre. Mp: 850°.

SYNS: BISMUTH SESQUISULFIDE □ BISMUTH(3+) SULFIDE □ C.I. 77172

TOXICITY DATA WITH REFERENCE
orl-rat LD50:5 g/kg GTPZAB 30(6),16,86
orl-mus LD50:10 g/kg GTPZAB 30(6),16,86

CONSENSUS REPORTS: Reported in EPA TSCA Inventory.

SAFETY PROFILE: Low toxicity by ingestion. Possibly explosive during preparation. When heated to decomposition it emits toxic fumes of Bi. See also BISMUTH COMPOUNDS and SULFIDES.

DDI450 CAS:19287-45-7 HR: 3
DIBORANE
DOT: UN 1911/NA 1911
mf: B_2H_6 mw: 27.68

PROP: Colorless air and moisture-sensitive gas; sickly-sweet odor. Mp: −165.5°, bp: −92.5°, d: 0.447 (liquid @ −112°), 0.577 (solid @ −183°), vap press: 224 mm @ −112°, autoign temp: 38–52°, lel: 0.9%, uel: 98%, flash p: −90°F. Sol in THF as BH_3-THF complex.

SYNS: BOROETHANE □ BORON HYDRIDE □ DIBORANE MIXTURES (NA 1911) □ DIBORON HEXAHYDRIDE

TOXICITY DATA WITH REFERENCE
ihl-rat LC50:40 ppm/4H 14KTAK -,693-64
ihl-mus LC50:29 ppm/4H TXAPA9 4,215,62
ihl-dog LCLo:125 ppm/2H AMIHAB 13,346,56
ihl-ham LCLo:50 ppm/8H AMIHAB 21,519,60

CONSENSUS REPORTS: Reported in EPA TSCA Inventory. EPA Extremely Hazardous Substances List.

OSHA PEL: TWA 0.1 ppm
ACGIH TLV: TWA 0.1 ppm
DFG MAK: 0.1 ppm (0.1 mg/m³)
DOT CLASSIFICATION: 2.1; *Label:* Flammable Gas (NA 1911); DOT Class: 2.3; *Label:* Poison Gas, Flammable Gas

SAFETY PROFILE: Poison by inhalation. An irritant to skin, eyes, and mucous membranes comparable to chlorine, fluorine, arsine, and phosgene. The liquid causes local inflammation, blisters, redness, and swelling. Injuries to central nervous system, liver, and kidneys have also been produced in experimental animals. Similar observations have been reported in humans, resulting at times in a reaction resembling metal fume fever. Human exposure to pentaborane has produced signs of severe central nervous system irritation such as drowsiness, dizziness, visual disturbances, muscle twitching, and in severe cases, painful muscle spasm. Dangerously flammable when exposed to heat or flame or by chemical reaction. On contact with moisture, hydrogen is usually evolved. Highly explosive when exposed to heat or flame. Explosive reaction with air, tetravinyllead, O_2 above 165°C, octanol oxime + sodium hydroxide, benzene vapor, HNO_3 Cl_2. Violent reaction with halocarbon liquids. Other boron hydrides evolve H_2 upon contact with moisture or can propagate a flame rapidly enough to cause an explosion. Heat can cause these materials to decompose violently or at least to evolve H_2. They also react with water or steam to evolve hydrogen. Reaction with Al or Li forms complex hydrides that may ignite spontaneously in air. Powerful oxidizing agents, such as chlorine gas, etc., can react violently with boron hydrides. Pentaborane (stable) is spontaneously flammable in air. See also BORANES and HYDRIDES.

For occupational chemical analysis use NIOSH: Diborane, 6006.

DDI500 CAS:12505-77-0 HR: 3
DIBORON OXIDE
mf: B_2O_2 mw: 53.62

PROP: Colorless solid. Sol in MeOH and EtOH.

SAFETY PROFILE: Violent reaction when heated to 400°C. When heated to decomposition it emits acrid smoke and fumes. See also BORON COMPOUNDS.

DDI600 CAS:13701-67-2 HR: 3
DIBORON TETRACHLORIDE
mf: B_2Cl_4 mw: 163.43

PROP: Colorless liquid; easily hydrolyzed. Mp: −93°, bp: 66.5°.

SAFETY PROFILE: May explode on contact with air or during reaction with dimethylmercury. When heated to decomposition it emits fumes of Cl^-. See also BORON COMPOUNDS and CHLORIDES.

DDI800 CAS:13965-73-6 HR: 3
DIBORON TETRAFLUORIDE
mf: B_2F_4 mw: 97.61

PROP: Colorless gas. Mp: −56°, bp: −34°.

SAFETY PROFILE: The gas explodes in the presence of oxygen. It ignites or reacts vigorously with mercury(II) oxide, manganese dioxide, and copper(II) oxide. When heated to decomposition it emits toxic fumes of F^-. See also BORON COMPOUNDS and FLUORIDES.

DDI900 CAS:77-48-5 HR: 3
DIBROMANTINE
mf: $C_5H_6Br_2N_2O_2$ mw: 285.95

SYNS: DIBROMANTIN □ N,N′-DIBROMODIMETHYLHYDANTOIN □ 1,3-DIBROMO-5,5-DIMETHYL-2,4-IMIDAZOLIDINEDIONE □ HYDANTOIN, 1,3-DIBROMO-5,5-DIMETHYL- □ 2,4-IMIDAZOLIDINEDIONE, 1,3-DIBROMO-5,5-DIMETHYL-(9CI)

TOXICITY DATA WITH REFERENCE
orl-rat LD50:250 mg/kg GISAAA 36(10),108,71
ihl-rat LCLo:29 g/m³/1H EPASR* 8EHQ-0281-0382
skn-rbt LDLo:20 g/kg EPASR* 8EHQ-0581-0382

CONSENSUS REPORTS: Reported in EPA TSCA Inventory.

SAFETY PROFILE: Poison by ingestion and inhalation routes. Slightly toxic by skin contact. When heated to decomposition it emits toxic vapors of NO_x and Br^-.

DDJ000 CAS:10318-26-0 HR: 3
DIBROMDULCITOL
mf: $C_6H_{12}Br_2O_4$ mw: 308.00

PROP: A solid. Mp: 187–188°.

SYNS: DBD □ 1,6-DIBROMODIDEOXYDULCITOL □ 1,6-DIBROMO-1,6-DIDEOXYDULCITOL □ 1,6-DIBROMO-1,6-DIDEOXYGALACTITOL □ 1,6-DIBROMO-1,6-DIDEOXY-d-GALACTITOL □ DIBROMODULCITOL □ 1,6-DIBROMODULCITOL □ ELOBROMOL □ GALACTICOL □ MITOLAC □ MITOLACTOL □ NCI-C04795 □ NSC-104800

TOXICITY DATA WITH REFERENCE
mmo-sat 100 µg/plate CRNGDP 3,333,82
dnd-rat-ipr 110 mg/kg CBINA8 47,133,83
bfa rat/sat 450 mg/kg CRNGDP 3,333,82
sce-ham:oth 5500 ng/L CNREA8 43,4530,83
dnd-mam:lym 150 mmol/L CBINA8 47,133,83
ipr-rat TDLo:5850 mg/kg/26W-I:NEO RRCRBU 52,1,75
ipr-mus TDLo:3500 mg/kg/26W-I:NEO RRCRBU 52,1,75
orl-hmn TDLo:72 mg/kg/D:ETA ANBCB3 23,50,78
orl-rat LD50:1000 mg/kg CCROBU 56,593,72
ipr-rat LD50:470 mg/kg CCROBU 56,593,72
orl-mus LD50:1238 mg/kg NCISP* JAN86
ipr-mus LD50:550 mg/kg ARZNAD 17,145,67
orl-rbt LD50:300 mg/kg CCROBU 56,593,72

CONSENSUS REPORTS: NCI Carcinogenesis Bioassay Completed; Results Positive: mouse, rat (RRCRBU 52,1,75).

SAFETY PROFILE: Poison by ingestion. Moderately toxic by intraperitoneal route. Questionable carcinogen with experimental carcinogenic, neoplastigenic, and tumorigenic data. Human mutation data reported. An anti-cancer agent taken orally. When heated to decomposition it emits very toxic fumes of Br^-.

DDJ400 CAS:3252-43-5 HR: 2
DIBROMOACETONITRILE
mf: C_2HBr_2N mw: 198.86

TOXICITY DATA WITH REFERENCE
mma-sat 16 µg/plate ENMUDM 8(Suppl 7),1,86
dnd-hmn:lym 50 µmol/L FAATDF 6,447,86
orl-rat TDLo:750 mg/kg (7-21D post):REP TXCYAC 46,83,87
skn-mus TDLo:2400 mg/kg/2W-I:CAR FAATDF 5,1065,85
orl-rat LD50:245 mg/kg EVHPAZ 69,183,86
orl-mus LD50:289 mg/kg EVHPAZ 69,183,86
ivn-mus LD50:56 mg/kg CSLNX* NX#05210

CONSENSUS REPORTS: Cyanide and its compounds are on the Community Right-To-Know List. Reported in EPA TSCA Inventory.

SAFETY PROFILE: Poison by intravenous route. Questionable carcinogen with experimental carcinogenic data. Experimental reproductive effects. Human mutation data reported. See also NITRILES and BROMIDES. When heated to decomposition it emits very toxic fumes of NO_x, Br^-, and CN^-.

DDJ600 CAS:99-73-0 HR: 3
2,4′-DIBROMOACETOPHENONE
mf: $C_8H_6Br_2O$ mw: 277.96

PROP: Needles from EtOH. Mp: 110–111°.

SYNS: p-BROMPHENACYL-8 □ p-BROMOPHENACYL BROMIDE □ 4-BROMOPHENACYL BROMIDE □ α,p-DIBROMOACETOPHENONE

TOXICITY DATA WITH REFERENCE
ivn-mus LD50:18 mg/kg CSLNX* NX#02407

CONSENSUS REPORTS: Reported in EPA TSCA Inventory.

SAFETY PROFILE: Poison by intravenous route. See

also BROMIDES. When heated to decomposition it emits toxic fumes of Br^-.

DDJ800 CAS:624-61-3 HR: 3
DIBROMOACETYLENE
mf: C_2Br_2 mw: 183.83

PROP: Heavy liquid with unpleasant odor. Mp: −25°, bp: explodes, d: 2 (approx), vap d: 6.35.

DOT CLASSIFICATION: Forbidden

SAFETY PROFILE: Ignites spontaneously in air. Explodes when heated. When heated to decomposition it emits toxic fumes of Br^-. See also ACETYLENE COMPOUNDS.

DDJ850 CAS:27695-54-1 HR: 3
2-((4-(DIBROMOACETYL)PHENYL)AMINO)-2-ETHOXY-1-(4-NITROPHENYL)ETHANONE
mf: $C_{18}H_{16}Br_2N_2O_5$ mw: 500.18

SYNS: ETHANONE, 2-((4-(DIBROMOACETYL)PHENYL)AMINO)-2-ETHOXY-1-(4-NITROPHENYL)- □ KETONE, 2-((4-(DIBROMOACETYL) PHENYL)AMINO)-2-ETHOXY-1-(4-NITROPHENYL)-

TOXICITY DATA WITH REFERENCE
ipr-mus LD50:1550 mg/kg ARZNAD 23,573,73

DOT CLASSIFICATION: 3; *Label:* Flammable Liquid

SAFETY PROFILE: Moderately toxic by intraperitoneal route. A flammable liquid. When heated to decomposition it emits toxic vapors of NO_x and Br^-.

DDJ875 CAS:81-98-1 HR: 2
3,9-DIBROMO-7H-BENZ(de)ANTHRACEN-7-ONE
mf: $C_{17}H_8Br_2O$ mw: 388.07

SYNS: 6-Bz-1-DIBROMBENZANTHRON (CZECH) □ 3,9-DIBROMBENZANTHRONE □ 2,7-DIBROMOMESOBENZANTHRONE

TOXICITY DATA WITH REFERENCE
eye-rbt 500 mg/24H MLD 28ZPAK -,89,72
ipr-rat LD50:4900 mg/kg RPTOAN 40,137,77
ipr-mus LD50:1410 mg/kg RPTOAN 40,137,77

CONSENSUS REPORTS: Reported in EPA TSCA Inventory.

SAFETY PROFILE: Moderately toxic by intraperitoneal route. An eye irritant. When heated to decomposition it emits toxic fumes of Br^-.

DDJ900 CAS:26249-12-7 HR: 3
DIBROMOBENZENE
mf: $C_6H_4Br_2$ mw: 235.92

SYNS: BENZENE, DIBROMO- □ UN2711 (DOT)

TOXICITY DATA WITH REFERENCE
ipr-mus LD50:780 mg/kg GTPZAB 20(12),52,76

DOT CLASSIFICATION: 3; *Label:* Flammable Liquid

CONSENSUS REPORTS: Reported in EPA TSCA Inventory.

SAFETY PROFILE: Moderately toxic by intraperitoneal route. A flammable liquid. When heated to decomposition it emits toxic vapors of Br^-.

DDK050 CAS:108-36-1 HR: 2
1,3-DIBROMOBENZENE
mf: $C_6H_4Br_2$ mw: 235.92

SYNS: BENZENE, m-DIBROMO- □ BENZENE, 1,3-DIBROMO-(9CI) □ m-DIBROMOBENZENE

TOXICITY DATA WITH REFERENCE
orl-mus LD50:2250 mg/kg GISAAA 44(12),19,79
ipr-mus LD50:900 mg/kg GISAAA 44(12),19,79

CONSENSUS REPORTS: Reported in EPA TSCA Inventory.

SAFETY PROFILE: Moderately toxic by ingestion and intraperitoneal routes. When heated to decomposition it emits toxic vapors of Br^-.

DDK600 CAS:6305-43-7 HR: 3
2,2′-DIBROMOBIACETYL
mf: $C_4H_4Br_2O_2$ mw: 243.90

PROP: Crystals from $CHCl_3$. Mp: 116–117°.

SYN: α,α′-DIBROMOBIACETYL

TOXICITY DATA WITH REFERENCE
ipr-mus LD50:9400 μg/kg JNCIAM 31,297,63
ivn-mus LD50:10 mg/kg CSLNX* NX#00598
ivn-dog LD50:21 mg/kg JNCIAM 31,297,63

CONSENSUS REPORTS: Reported in EPA TSCA Inventory.

SAFETY PROFILE: Poison by intravenous and intraperitoneal routes. When heated to decomposition it emits toxic fumes of Br^-. See also BROMIDES.

DDK800 CAS:26637-71-8 HR: 3
DIBROMOBICYCLOHEPTANE (mixed isomers)
mf: $C_7H_{10}Br_2$ mw: 253.99

SYNS: DIBROMOBICYCLOHEPTANE □ DIBROMONORBORNANE

TOXICITY DATA WITH REFERENCE
orl-rat LD50:210 mg/kg AIHAAP 30,470,69
skn-rbt LD50:250 mg/kg AIHAAP 30,470,69

SAFETY PROFILE: Poison by ingestion and skin contact. When heated to decomposition it emits toxic fumes of Br^-. See also BROMIDES.

DDK875 CAS:36333-41-2 HR: 3
1,4-DIBROMO-1,3-BUTADIYNE
mf: C_4Br_2 mw: 207.85

BrC≡CC≡CBr

SAFETY PROFILE: Explodes at room temperature. Upon decomposition it emits toxic fumes of Br^-. See also ACETYLENE COMPOUNDS and BROMIDES.

D

DDL000 **CAS:110-52-1** **HR: 3**
1,4-DIBROMOBUTANE
mf: $C_4H_8Br_2$ mw: 215.94

PROP: A liquid. Fp −20°, bp: 197–198°, d: 1.81 @ 20°/4°.

SYNS: DBB □ 1,4-DIBROMBUTAN (GERMAN)

TOXICITY DATA WITH REFERENCE
mmo-sat 10 μmol/plate MUREAV 141,11,84
ipr-mus LD50:300 mg/kg ARZNAD 14,668,64

CONSENSUS REPORTS: Reported in EPA TSCA Inventory.

SAFETY PROFILE: Poison by intraperitoneal route. Mutation data reported. When heated to decomposition it emits toxic fumes of Br^-. See also BROMIDES.

DDL400 **CAS:6974-12-5** **HR: 3**
1,4-DIBROMO-2-BUTENE
mf: $C_4H_6Br_2$ mw: 213.92

SYN: TL 80

TOXICITY DATA WITH REFERENCE
skn-rbt 1 mg/24H AMIHBC 10,61,54
eye-rbt 50 μg open SEV AMIHBC 10,61,54
orl-rat LD50:75 mg/kg AMIHBC 10,61,54
ihl-mus LCLo:1260 mg/m³/10M NDRC** NDCrc-132,Aug,42
ipr-mus LDLo:4 mg/kg CBCCT* 5,338,53

SAFETY PROFILE: Poison by ingestion and intraperitoneal routes. Moderately toxic by inhalation. A skin and severe eye irritant. When heated to decomposition it emits toxic fumes of Br^-. See also BROMIDES.

DDL600 **CAS:821-06-7** **HR: 3**
trans-1,4-DIBROMOBUT-2-ENE
mf: $C_4H_6Br_2$ mw: 213.92

PROP: Leaflets from pet ether. Mp: 54°, bp: 74–76° @ 1.8 mm.

SYNS: DIBROMOBUTENE □ 1,4-trans-DIBROMOBUTENE-2

TOXICITY DATA WITH REFERENCE
skn-rbt 500 mg SEV SCCUR* -,3,61
orl-rat LD50:62 mg/kg SCCUR* -,9,61
orl-mus LD50:29 mg/kg SCCUR* -,9,61

CONSENSUS REPORTS: Reported in EPA TSCA Inventory.

SAFETY PROFILE: Poison by ingestion. A severe skin irritant. When heated to decomposition it emits toxic fumes of Br^-. See also BROMIDES.

DDL800 **CAS:96-12-8** **HR: 3**
1,2-DIBROMO-3-CHLOROPROPANE
DOT: UN 2872
mf: $C_3H_5Br_2Cl$ mw: 236.35

PROP: Bp: 196°, flash p: 170°F (TOC).

SYNS: BBC 12 □ 1-CHLORO-2,3-DIBROMOPROPANE □ 3-CHLORO-1,2-DIBROMOPROPANE □ DBCP □ DIBROMCHLORPROPAN (GERMAN) □ 1,2-DIBROM-3-CHLOR-PROPAN (GERMAN) □ DIBROMOCHLOROPROPANE □ 1,2-DIBROMO-3-CLORO-PROPANO (ITALIAN) □ 1,2-DIBROOM-3-CHLOORPROPAAN (DUTCH) □ FUMAGON □ FUMAZONE □ NCI-C00500 □ NEMABROM □ NEMAFUME □ NEMAGON □ NEMAGONE □ NEMAGON SOIL FUMIGANT □ NEMANAX □ NEMAPAZ □ NEMASET □ NEMATOCIDE □ NEMATOX □ NEMAZON □ OS 1897 □ OXY DBCP □ RCRA WASTE NUMBER U066 □ SD 1897

TOXICITY DATA WITH REFERENCE
skn-rbt 10 g SEV TXAPA9 3,545,61
eye-rbt 1% MLD TXAPA9 3,545,61
dni-hmn:hla 10 mmol/L MUREAV 92,427,82
mma-sat 500 ng/plate ENMUDM 7(Suppl 3),15,85
spm-rbt-orl 375 mg/kg/10W-I FAATDF 6,628,86
orl-rat TDLo:50 mg/kg (male 5D pre):TER MUREAV 77,71,80
orl-rat TDLo:500 mg/kg (6-15D preg):TER BECTA6 21,483,79
orl-rat TDLo:250 mg/kg (male 5D pre):REP MUREAV 101,321,82
ihl-rbt TCLo:10 ppm/6H (male 40D pre):REP FAATDF 2,241,82
orl-rat TDLo:375 mg/kg (male 75D pre):REP HYSAAV 36(1-3),344,71
ihl-rat TCLo:10 ppm/6H (male 4W pre):REP FAATDF 3,104,83
ihl-rbt TCLo:1 ppm/6H (male 70D pre):REP FAATDF 2,241,82
orl-rat TDLo:200 mg/kg (1D male):REP TOXID9 4,135,84
scu-rat TDLo:19 mg/kg (male 19D pre):REP TXAPA9 90,299,87
orl-rat TDLo:5475 mg/kg/73W-I:CAR NCITR* NCI-CG-TR-28,78
ihl-rat TCLo:600 ppb/6H/2Y-I:CAR BJCAAI 42,772,80
scu-rat TDLo:240 mg/kg/12W-I:CAR TOLED5 31(Suppl),202,86
orl-mus TDLo:49 g/kg/47W-I:CAR NCITR* NCI-CG-TR-28,78
skn-mus TDLo:100 g/kg/74W-I:CAR JJIND8 63,1433,79
orl-rat TD:9280 mg/kg/64W-I:CAR NCITR* NCI-CG-TR-28,78
ihl-mus TC:3 ppm/6H/2Y-I:CAR JCROD7 98,75,80
ihl-rat TC:600 ppb/6H/76W-I:CAR EVHPAZ 47,365,83
ihl-rat TC:3 ppm/6H/2Y-I:CAR BJCAAI 42,772,80
ihl-mus TC:600 ppb/6H/76W-I:CAR EVHPAZ 47,365,83
ihl-rat TC:600 ppb/6H/2Y-I:CAR NTPTR* NTP-TR-206,82
ihl-rat TC:3 ppm/6H/84W-I:CAR NTPTR* NTP-TR-206,82
ihl-mus TC:600 ppm/6H/2Y-I:CAR NTPTR* NTP-TR-206,82
ihl-mus TC:3 ppm/6H/76W-I:CAR NTPTR* NTP-TR-206,82
scu-rat LD :240 mg/kg/12W-I:CAR TOLED5 31(Suppl),202,86
orl-rat LD50:170 mg/kg FMCHA2 -,C76,83
ihl-rat LC50:103 ppm/8H FEPRA7 15,448,56
scu-rat LD50:100 mg/kg TXCYAC 27,287,83
orl-mus LD50:257 mg/kg GUCHAZ 6,172,73
ipr-mus LD50:123 mg/kg MUREAV 68,169,79
orl-rbt LD50:180 mg/kg TXAPA9 3,545,61
skn-rbt LD50:1400 mg/kg TXAPA9 3,545,61
orl-ckn LD50:60 mg/kg TXAPA9 3,545,61

CONSENSUS REPORTS: NTP 7th Annual Report on Carcinogens. IARC Cancer Review: Group 2B IMEMDT 7,191,87; Animal Sufficient Evidence IMEMDT 15,139,77; Human Limited Evidence IMEMDT 20,83,79; Animal Sufficient Evidence IMEMDT 20,83,79. NCI

Carcinogenesis Bioassay Completed; Results Positive: mouse, rat NCITR* NCI-CG-TR-28,78. EPA Genetic Toxicology Program. Community Right-To-Know List. Reported in EPA TSCA Inventory.

OSHA PEL: TWA 0.001 ppm; Cancer Hazard
DFG MAK: Animal Carcinogen, Suspected Human Carcinogen
NIOSH REL: (Dibromochloropropane) CL 0.01 ppm/30M
DOT CLASSIFICATION: 6.1; *Label:* KEEP AWAY FROM FOOD

SAFETY PROFILE: Confirmed human carcinogen with experimental carcinogenic and teratogenic data. Poison by ingestion, inhalation, and subcutaneous routes. Moderately toxic by skin contact. An eye and severe skin irritant. Narcotic in high concentrations. Has been implicated in causing human sterility in male factory workers. Human mutation data reported. A soil fumigant. Combustible. When heated to decomposition it emits toxic fumes of Cl^- and Br^-. See also CHLORIDES and BROMIDES.

DDM000 CAS:10222-01-2 HR: 3
α,α-DIBROMO-α-CYANOACETAMIDE
mf: $C_3H_2Br_2N_2O$ mw: 241.89

SYNS: DBNPA □ DIBROMOCYANOACETAMIDE □ 2,2-DIBROMO-3-NITRILOPROPIONAMIDE

TOXICITY DATA WITH REFERENCE
skn-rbt 500 mg SEV PHMCAA 15,226,73
eye-rbt 100 mg SEV PHMCAA 15,226,73
ivn-mus LD50:10 mg/kg CSLNX* NX#07890
orl-mam LD50:118 mg/kg PHMCAA 15,226,73

CONSENSUS REPORTS: Cyanide and its compounds are on the Community Right-To-Know List. Reported in EPA TSCA Inventory.

SAFETY PROFILE: Poison by ingestion and intravenous routes. A severe skin and eye irritant. When heated to decomposition it emits very toxic fumes of Br^- and NO_x. See also NITRILES.

DDM200 CAS:1689-99-2 HR: 3
2,6-DIBROMO-4-CYANOPHENYL OCTANOATE
mf: $C_{15}H_{17}Br_2NO_2$ mw: 403.15

SYNS: BROMOXYNIL OCTANOATE □ 3,5-DIBROMO-4-OCTANOYLOXYBENZONITRILE □ NCR CE EE DOV7

TOXICITY DATA WITH REFERENCE
orl-rat LD50:250 mg/kg 28ZEAL 5,30,76
orl-mus LD50:245 mg/kg 28ZEAL 5,30,76
orl-rbt LD50:2 g/kg GUCHAZ 6,56,73

CONSENSUS REPORTS: Cyanide and its compounds are on the Community Right-To-Know List. Reported in EPA TSCA Inventory.

SAFETY PROFILE: Poison by ingestion. When heated to decomposition it emits very toxic fumes of NO_x and Br^-. See also NITRILES.

For occupational chemical analysis use NIOSH: Bromoxynil and Bromoxynil Octanoate, 5010.

DDM300 CAS:3322-93-8 HR: 2
1,2-DIBROMO-4-(1,2-DIBROMOETHYL)CYCLOHEXANE
mf: $C_8H_{12}Br_4$ mw: 427.84

SYNS: CITEX BCL 462 □ CYCLOHEXANE, 1,2-DIBROMO-4-(1,2-DIBROMOETHYL)- □ 1-(1,2-DIBROMOETHYL)-3,4-DIBROMOCYCLOHEXANE □ SAYTEX BCL 462

TOXICITY DATA WITH REFERENCE
slt-mus:lyms 40 mg/L EMMUEG 17,196,91
cyt-ham:lng 125 mg/L MUREAV 241,175,90
sce-ham:ovr 40 mg/L EMMUEG 13,60,89
orl-rat LD50:3220 mg/kg TSCAT* OTS0505764

CONSENSUS REPORTS: Reported in EPA TSCA Inventory.

SAFETY PROFILE: Moderately toxic by ingestion. Mutation data reported. When heated to decomposition it emits toxic vapors of Br^-.

DDM400 CAS:996-08-7 HR: 3
DIBROMODIBUTYLSTANNANE
mf: $C_8H_{18}Br_2Sn$ mw: 392.77

PROP: Mp: 20°.

SYNS: DIBROMODIBUTYLTIN □ DIBUTYL TIN DIBROMIDE

TOXICITY DATA WITH REFERENCE
orl-rbt LDLo:150 mg/kg SAIGBL 15,3,73
skn-rbt LDLo:1000 mg/kg SAIGBL 15,3,73

OSHA PEL: TWA 0.1 mg(Sn)/m³ (skin)
ACGIH TLV: TWA 0.1 mg(Sn)/m³ (skin) (Proposed: TWA 0.1 mg(Sn)/m³; STEL 0.2 mg(Sn)/m³ (skin))
NIOSH REL: (Organotin Compounds) TWA 0.1 mg(Sn)/m³

SAFETY PROFILE: Poison by ingestion. Moderately toxic by skin contact. See also TIN COMPOUNDS. When heated to decomposition it emits toxic fumes of Br^-.

For occupational chemical analysis use NIOSH: Organotin Compounds 5504.

DDM600 CAS:77966-70-2 HR: 3
2′,6′-DIBROMO-2-(DIETHYLAMINO)-p-ACETOTOLUIDIDE HYDROCHLORIDE
mf: $C_{13}H_{18}Br_2N_2O•ClH$ mw: 414.61

SYNS: C 3039 □ 2′,6′-DIBROMO-2-(DIETHYLAMINO)-4′-METHYLACETANILIDE HYDROCHLORIDE

TOXICITY DATA WITH REFERENCE
eye-rbt 2% MLD ARZNAD 8,270,58
ipr-rat LD50:234 mg/kg ARZNAD 8,270,58
scu-mus LD50:505 mg/kg ARZNAD 8,270,58

SAFETY PROFILE: Poison by intraperitoneal route. Moderately toxic by subcutaneous route. An eye irritant. When heated to decomposition it emits very toxic fumes of Br^-, NO_x, and HCl.

DDM800 **CAS:52400-80-3** **HR: 3**
5,6-DIBROMO-2-(2-(2-(DIETHYLAMINO)ETHYLAMINO) ETHYL)-2-METHYL-1,3-BENZODIOXOLE DIHYDROCHLORIDE
mf: $C_{16}H_{24}Br_2N_2O_2 \cdot 2ClH$ mw: 509.16

TOXICITY DATA WITH REFERENCE
ivn-rat LD50:40 mg/kg EJMCA5 12,413,77
ipr-mus LD50:150 mg/kg EJMCA5 12,413,77

SAFETY PROFILE: Poison by intravenous and intraperitoneal routes. When heated to decomposition it emits very toxic fumes of HCl, NO_x, and Br^-.

DDN100 **CAS:51877-12-4** **HR: 3**
1,2-DIBROMO-1,2-DIISOCYANATOETHANE POLYMERS
mf: $(C_4H_2Br_2N_2O_2)_{2\ OR\ 3}$

$(O{=}N{=}CCHBrCHBrC{=}N{=}O)_{2\ OR\ 3}$

CONSENSUS REPORTS: Cyanide and its compounds are on the Community Right-To-Know List.

SAFETY PROFILE: Vigorous or explosive reaction on heating with 2-phenyl-2-propyl hydroperoxide. When heated to decomposition it emits toxic fumes of CN^-, Br^-, and NO_x. See also CYANIDE.

DDN150 **CAS:72957-64-3** **HR: 2**
2,2-DIBROMO-1,3-DIMETHYLCYCLOPROPANOIC ACID
mf: $C_6H_8Br_2O_2$ mw: 271.94

$CH_3CHCBr_2C(CH_3)CO \cdot OH$

SAFETY PROFILE: Vigorous exothermic reaction on contact with tert-butylamine. When heated to decomposition it emits toxic fumes of Br^-.

DDN200 **CAS:4713-59-1** **HR: 3**
DIBROMODIPHENYLSTANNANE
mf: $C_{12}H_{10}Br_2Sn$ mw: 432.73

PROP: White or colorless crystals. Mp: 38°; bp: 230° @ 42 mm. Sol in alc and ether.

SYNS: DIPHENYLDIBROMOTIN □ DIPHENYLTIN DIBROMIDE

TOXICITY DATA WITH REFERENCE
ivn-mus LD50:71 mg/kg CSLNX* NX#05803

CONSENSUS REPORTS: Polybrominated biphenyl compounds are on the Community Right-To-Know List.

OSHA PEL: TWA 0.1 mg(Sn)/m³ (skin)
ACGIH TLV: TWA 0.1 mg(Sn)/m³ (skin) (Proposed: TWA 0.1 mg(Sn)/m³; STEL 0.2 mg(Sn)/m³ (skin))
NIOSH REL: (Organotin Compounds) TWA 0.1 mg(Sn)/m³

SAFETY PROFILE: Poison by intravenous route. See also TIN COMPOUNDS and BROMIDES. When heated to decomposition it emits toxic fumes of Br^-.

For occupational chemical analysis use NIOSH: Organotin Compounds 5504.

DDN700 **CAS:56411-66-6** **HR: 3**
2,3-DIBROMO-5,6-EPOXY-7,8-DIOXABICYCLO(2.2.2) OCTANE
mf: $C_6H_6Br_2O_3$ mw: 285.92

SAFETY PROFILE: Explodes on heating. When heated to decomposition it emits toxic fumes of Br^-. See also PEROXIDES.

DDN800 **CAS:557-91-5** **HR: 2**
1,1-DIBROMOETHANE
mf: $C_2H_4Br_2$ mw: 187.88

PROP: Liquid. Insol in water; sol in organic solvents. D: 2.06 @ 20.5°/4°, bp: 112.5° @ 755 mm.

SYNS: ETHYLIDENE BROMIDE □ ETHYLIDENE DIBROMIDE

TOXICITY DATA WITH REFERENCE
mma-sat 10 μmol/plate EVHPAZ 21,79,77
dnr-esc 10 μL/plate EVHPAZ 21,79,77
rec-rbt LDLo:1250 mg/kg JPETAB 34,223,28

CONSENSUS REPORTS: Reported in EPA TSCA Inventory.

SAFETY PROFILE: Moderately toxic by rectal route. Mutation data reported. Violent reaction with magnesium. When heated to decomposition it emits toxic fumes of Br^-. See also BROMIDES.

DDO400 **CAS:20404-94-8** **HR: 3**
1,2-DIBROMOHEPTAFLUOROISOBUTYL METHYL ETHER
mf: $C_5H_3Br_2F_7O$ mw: 371.90

TOXICITY DATA WITH REFERENCE
orl-mus LD50:1150 mg/kg TXAPA9 14,114,69
ipr-mus LD50:140 mg/kg TXAPA9 14,114,69

SAFETY PROFILE: Poison by intraperitoneal route. Moderately toxic by ingestion. See also ETHERS, BROMIDES, and FLUORIDES. When heated to decomposition it emits very toxic fumes of Br^- and F^-.

DDO450 **CAS:661-95-0** **HR: 1**
1,2-DIBROMOHEXAFLUOROPROPANE
mf: $C_3Br_2F_6$ mw: 309.85

SYNS: 1,2-DIBROMO-1,1,2,3,3,3-HEXAFLUOROPROPANE □ PROPANE, 1,2-DIBROMO-1,1,2,3,3,3-HEXAFLUORO-

TOXICITY DATA WITH REFERENCE
orl-rat LDLo:8624 mg/kg RADLAX 105,323,72

CONSENSUS REPORTS: Reported in EPA TSCA Inventory.

SAFETY PROFILE: Mildly toxic by ingestion. When heated to decomposition it emits toxic vapors of F^- and Br^-.

DDO800 CAS:629-03-8 HR: 3
1,6-DIBROMOHEXANE
mf: $C_6H_{12}Br_2$ mw: 244.00

PROP: A liquid. Mp: −2.3°, bp: 239–241° (slt decomp), d: 1.60 @ 15°.

SYNS: DBH □ 1,6-DIBROMOHEXAN (GERMAN)

TOXICITY DATA WITH REFERENCE
skn-rbt 500 mg MLD FCTOD7 20,563,82
eye-rbt 100 mg MLD FCTOD7 20,573,82
eye-rbt 100 mg/30S rns MLD FCTOD7 20,573,82
mmo-sat 10 μmol/plate MUREAV 141,11,84
ipr-mus LD50:270 mg/kg ARZNAD 14,668,64

CONSENSUS REPORTS: Reported in EPA TSCA Inventory.

SAFETY PROFILE: Poison by intraperitoneal route. A skin and eye irritant. Mutation data reported. When heated to decomposition it emits very toxic fumes of Br^-. See also BROMIDES.

DDP000 CAS:1689-84-5 HR: 3
3,5-DIBROMO-4-HYDROXYBENZONITRILE
mf: $C_7H_3Br_2NO$ mw: 276.93

PROP: Needles. Mp: 187°.

SYNS: BRITTOX □ BROMINAL □ BROMINEX □ BROMINIL □ BROMOXYNIL □ BROXYNIL □ BUCTRIL □ BUCTRIL INDUSTRIAL □ BUTILCHLOROFOS □ CHIPCO BUCTRIL □ CHIPCO CRAB-KLEEN □ 2,6-DIBROMO-4-CYANOPHENOL □ 3,5-DIBROMO-4-HYDROXYPHENYLCYANIDE □ ENT 20,852 □ 4-HYDROXY-3,5-DIBROMOBENZONITRILE □ MB 10064 □ ME4 BROMINAL □ NU-LAWN WEEDER □ OXYTRIL M

TOXICITY DATA WITH REFERENCE
orl-rat LD50:190 mg/kg WRPCA2 9,119,70
orl-mus LD50:110 mg/kg GUCHAZ 6,55,73
ivn-mus LD50:56 mg/kg CELNX* NX#02212
orl-rbt LD50:260 mg/kg 85DPAN -,-,71/76
orl-gpg LD50:63 mg/kg GUCHAZ 6,55,73
orl-dck LD50:200 mg/kg DOEAAH 35,25,79

CONSENSUS REPORTS: Cyanide and its compounds are on the Community Right-To-Know List.

SAFETY PROFILE: Poison by ingestion and intravenous routes. An herbicide. When heated to decomposition it emits highly toxic fumes of NO_x, CN^-, and Br^-. See also NITRILES.

For occupational chemical analysis use NIOSH: Bromoxynil and Bromoxynil Octanoate, 5010.

DDP200 CAS:3562-84-3 HR: 3
3,5-DIBROMO-4-HYDROXYPHENYL-2-ETHYL-3-BENZOFURANYL KETONE
mf: $C_{17}H_{12}Br_2O_3$ mw: 424.11

PROP: Yellow prisms. Mp: 151°.

SYNS: BENZBROMARON □ BENZBROMARONE □ DESURIC □ 3-(3,5-DIBROMO-4-HYDROXYBENZOYL-2-ETHYLBENZOFURAN □ (3,5-DIBROMO-4-HYDROXYPHENYL)(2-ETHYL-3-BENZOFURANYL)METHANONE □ EXURATE □ L2214 □ MINURIC □ MJ 10061 □ URICOVAC

TOXICITY DATA WITH REFERENCE
orl-rat TDLo:360 mg/kg (9-14D preg):TER SHNSAS 16,1521,79
orl-rat TDLo:480 mg/kg (9-14D preg):REP SHNSAS 16,1521,79
orl-rat TDLo:360 mg/kg (9-14D preg):REP SHNSAS 16,1521,79
orl-rat LD50:248 mg/kg IYKEDH 10,232,79
ipr-rat LD50:239 mg/kg IYKEDH 10,232,79
scu-rat LD50:1230 mg/kg IYKEDH 10,232,79
orl-mus LD50:618 mg/kg IYKEDH 10,232,79
ipr-mus LD50:146 mg/kg OYYAA2 6,341,72
scu-mus LD50:4120 mg/kg IYKEDH 10,232,79
ivn-mus LD50:77 mg/kg OYYAA2 6,341,72

CONSENSUS REPORTS: Reported in EPA TSCA Inventory.

DOT CLASSIFICATION: 3; *Label:* Flammable Liquid

SAFETY PROFILE: Poison by ingestion, intravenous and intraperitoneal routes. Moderately toxic by subcutaneous route. Experimental teratogenic and reproductive effects. A uricosuric agent which promotes the excretion of uric acid in the urine. A flammable liquid. When heated to decomposition it emits toxic fumes of Br^-. See also KETONES.

DDP300 CAS:73343-74-5 HR: 3
3,5-DIBROMO-4-HYDROXYPHENYL 2-MESITYL-3-BENZOFURANYL KETONE
mf: $C_{24}H_{18}Br_2O_3$ mw: 514.24

SYNS: BENZOFURAN, 3-(3,5-DIBROMO-4-HYDROXYBENZOYL)-2-MESITYL □ (DIBROMO-3,5 HYDROXY-4 BENZOYL)-3 MESITYL-2 BENZOFURANNE □ (3,5-DIBROMO-4-HYDROXYPHENYL)(2-(2,4,6-TRIMETHYLPHENYL)-3-BENZOFURANYL)M ETHANONE □ KETONE, 3,5-DIBROMO-4-HYDROXYPHENYL 2-MESITYL-3-BENZOFURANYL □ METHANONE, (3,5-DIBROMO-4-HYDROXYPHENYL)(2-(2,4,6-TRIMETHYLPHENYL)-3-BENZOFURANYL)-

TOXICITY DATA WITH REFERENCE
ipr-mus LDLo:2 g/kg EJMCA5 14,517,79

DOT CLASSIFICATION: 3; *Label:* Flammable Liquid

SAFETY PROFILE: Low toxicity by ingestion. A flammable liquid. When heated to decomposition it emits toxic vapors of Br^-.

DDP400 CAS:1122-10-7 HR: 3
DIBROMOMALEINIMIDE
mf: $C_4HBr_2NO_2$ mw: 254.88

TOXICITY DATA WITH REFERENCE
ipr-mus TDLo:3100 μg/kg (9D preg):REP ARTODN 37,15,76
ipr-mus LD50:11 mg/kg ARTODN 37,15,76

SAFETY PROFILE: Poison by intraperitoneal route. An experimental teratogen. Other experimental reproductive effects. When heated to decomposition it emits very toxic fumes of Br^- and NO_x.

DDP600 **CAS:488-41-5** **HR: 2**
1,6-DIBROMOMANNITOL
mf: $C_6H_{12}Br_2O_4$ mw: 308.00

PROP: Crystals from MeOH/dichloroethane. Mp: 176–178°.

SYNS: DBM □ DIBROMANNIT □ DIBROMANNITOL □ d-DIBROMANNITOL □ 1,6-DIBROMO-1,6-DIDEOXY-d-MANNITOL □ 1,6-DIBROMO-1,6-d-DIDESOXYMANNITOL □ MITOBROMOL □ MITOBRONITOL □ MYEBROL □ MYELOBROMOL □ NCI-C04762 □ NSC-94100 □ R 54

TOXICITY DATA WITH REFERENCE
mmo-sat 1 mg/plate CNREA8 43,4530,83
mma-sat 667 µg/plate ENMUDM 8(Suppl 7),1,86
sce-hmn:lym 10 nmol/L NGCJAK 15,1085,80
cyt-mus-ivn 90 mg/kg MUREAV 60,329,79
sce-ham:oth 1300 ng/L CNREA8 43,4530,83
orl-mus TDLo:150 mg/kg (female 11D post):TER KSRNAM 6,30,72
orl-rat TDLo:300 mg/kg (9-14D preg):TER KSRNAM 6,30,72
orl-rat TDLo:300 mg/kg (female 9D post):REP KSRNAM 6,30,72
orl-rat TDLo:3900 mg/kg (26D pre):REP OYYAA2 6,831,72
orl-rat TDLo:3900 mg/kg (male 26D pre):REP OYYAA2 6,831,72
orl-mus TDLo:300 mg/kg (female 12D post):TER KSRNAM 6,30,72
orl-rat TDLo:500 mg/kg (10D preg):TER HYDKAK 17,67,75
ipr-rat TDLo:9750 mg/kg/26W-I:NEO RRCRBU 52,1,75
ipr-mus TDLo:7000 mg/kg/26W-I:NEO RRCRBU 52,1,75
orl-rat LD50:1500 mg/kg NIIRDN 6,810,82
ipr-rat LD50:900 mg/kg EJCAAH 4,617,68
scu-rat LD50:1240 mg/kg NIIRDN 6,810,82
ivn-rat LD50:1370 mg/kg IYKEDH 8,680,77
orl-mus LD50:1380 mg/kg NIIRDN 6,810,82
ipr-mus LD50:900 mg/kg NIIRDN 6,810,82
scu-mus LD50:2200 mg/kg NIIRDN 6,810,82
ivn-mus LD50:2200 mg/kg IYKEDH 8,680,77
orl-rbt LD50:1080 mg/kg OYYAA2 6,831,72

CONSENSUS REPORTS: NCI Carcinogenesis Bioassay Completed; Results Positive: mouse, rat (RRCRBU 52,1,75).

SAFETY PROFILE: Moderately toxic by ingestion, intravenous, intraperitoneal, and subcutaneous routes. Questionable carcinogen with experimental carcinogenic and neoplastigenic data. Experimental teratogenic and reproductive effects. Human mutation data reported. When heated to decomposition it emits toxic fumes of Br^-.

DDP800 **CAS:74-95-3** **HR: 3**
DIBROMOMETHANE
DOT: UN 2664
mf: CH_2Br_2 mw: 173.85

PROP: Colorless, heavy liquid. Fp: −52.7°, bp: 95.6–97.4°, d: 2.485 @ 25°/25°, vap d: 6.05. Sltly sol in water.

SYNS: METHYLENE BROMIDE □ METHYLENE DIBROMIDE □ RCRA WASTE NUMBER U068

TOXICITY DATA WITH REFERENCE
mmo-sat 100 ng/plate BECTA6 24,590,80
ihl-rat LC50:40 g/m³/2H 85GMAT -,82,82
scu-mus LD50:3738 mg/kg TXAPA9 4,354,62
rec-rbt LDLo:5000 mg/kg JPETAB 34,223,28

CONSENSUS REPORTS: Community Right-To-Know List. Reported in EPA TSCA Inventory.

DOT CLASSIFICATION: 6.1; *Label:* KEEP AWAY FROM FOOD

SAFETY PROFILE: A poison. Moderately toxic by subcutaneous route. Mildly toxic by inhalation. Mutation data reported. Mixtures with potassium explode on light impact. When heated to decomposition it emits toxic fumes of Br^-. See also BROMIDES.

DDQ100 **CAS:10218-83-4** **HR: 3**
N,N-DIBROMOMETHYLAMINE
mf: CH_3Br_2N mw: 188.85

SAFETY PROFILE: An explosive very sensitive to impact or shock. Upon decomposition it emits toxic fumes of Br^- and NO_x. See also AMINES.

DDQ125 **HR: 3**
DIBROMOMETHYLBORANE
mf: CH_3BBr_2 mw: 185.65

SAFETY PROFILE: Ignites spontaneously in air. Explodes on mixing with sodium-potassium alloys. When heated to decomposition it emits toxic fumes of Br^-. See also BORANES.

DDQ400 **CAS:3296-90-0** **HR: 2**
DIBROMONEOPENTYL GLYCOL
mf: $C_5H_{10}Br_2O_2$ mw: 261.97

SYNS: 2,2-BIS(BROMOMETHYL)-1,3-PROPANEDIOL □ DIBROMONEOPENTYL GLYCOL □ DIBROMOPENTAERYTHRITOL □ FR 1138 □ NCI-C55516 □ PENTAERYTHRITOL DIBROMIDE □ PENTAERYTHRITOL DIBROMOHYDRIN

TOXICITY DATA WITH REFERENCE
cyt-ham:ovr 800 mg/L EMMUEG 10(Suppl 10),1,87
orl-mus TDLo:124 g/kg (male 15W pre):REP FAATDF 13,245,89
orl-mus TDLo:62 g/kg (female 15W pre):REP FAATDF 13,245,89
orl-mus TDLo:62 g/kg (male 15W pre):REP FAATDF 13,245,89
orl-rat LD50:3458 mg/kg JCTODH 7,77,80

CONSENSUS REPORTS: Reported in EPA TSCA Inventory.

SAFETY PROFILE: Moderately toxic by ingestion. Experimental reproductive effects. Mutation data reported. When heated to decomposition it emits toxic fumes of Br^-.

DDQ500 CAS:99-28-5 HR: 3
2,6-DIBROMO-4-NITROPHENOL
mf: $C_6H_3Br_2NO_3$ mw: 296.92

SYN: PHENOL, 2,6-DIBROMO-4-NITRO-

TOXICITY DATA WITH REFERENCE
ivn-mus LD50:56 mg/kg CSLNX* NX#03498

CONSENSUS REPORTS: Reported in EPA TSCA Inventory.

SAFETY PROFILE: Poison by intravenous route. When heated to decomposition it emits toxic vapors of NO_x and Br^-.

DDQ800 CAS:57541-73-8 HR: 2
3,4-DIBROMONITROSOPIPERIDINE
mf: $C_5H_8Br_2N_2O$ mw: 271.97

SYN: N-NITROSO-3,4-DIBROMOPIPERIDINE

TOXICITY DATA WITH REFERENCE
mmo-sat 200 µg/plate MUREAV 56,131,77
mma-sat 1 µmol/plate MUREAV 56,131,77
pic-esc 10 mg/L TCMUE9 1,91,84
sln-dmg-orl 1 mmol/L/24H MUREAV 67,27,79
orl-rat TDLo:1090 mg/kg/27W C:ETA CNREA8 35,3209,75

CONSENSUS REPORTS: EPA Genetic Toxicology Program.

SAFETY PROFILE: Mutation data reported. Questionable carcinogen with experimental tumorigenic data. Many N-nitroso compounds are carcinogens. When heated to decomposition it emits very toxic fumes of Br^- and NO_x. See also N-NITROSO COMPOUNDS.

DDR100 HR: 3
2-(3,5-DIBROMO-2-PENTYLOXYBENZYLOXY)TRIETHYLAMINE
mf: $C_{18}H_{29}Br_2NO_2$ mw: 451.30

TOXICITY DATA WITH REFERENCE
orl-rat LD50:350 mg/kg JPETAB 121,210,57
ipr-rat LD50:90 mg/kg JPETAB 121,210,57
orl-mus LD50:160 mg/kg JPETAB 121,210,57
ipr-mus LD50:90 mg/kg JPETAB 121,210,57

SAFETY PROFILE: Poison by ingestion and intraperitoneal routes. When heated to decomposition it emits toxic fumes of Br^- and NO_x. See also AMINES and BROMIDES.

DDR150 CAS:615-58-7 HR: 3
2,4-DIBROMOPHENOL
mf: $C_6H_4Br_2O$ mw: 251.92

SYN: PHENOL, 2,4-DIBROMO-

TOXICITY DATA WITH REFERENCE
orl-mus LD50:282 mg/kg GISAAA 44(12),19,79
ipr-mus LD50:160 mg/kg GISAAA 44(12),19,79

CONSENSUS REPORTS: Reported in EPA TSCA Inventory.

SAFETY PROFILE: Poison by ingestion and intraperitoneal routes. When heated to decomposition it emits toxic vapors of Br^-.

DDR200 CAS:696-24-2 HR: 3
DIBROMOPHENYLARSINE
mf: $C_6H_5AsBr_2$ mw: 311.85

PROP: Bp: 285° (decomp), d: 2.103.

SYNS: PHENYLARSONOUS DIBROMIDE □ PHENYLDIBROMOARSINE

TOXICITY DATA WITH REFERENCE
skn-rat LD50:15 mg/kg JPBAA7 58,411,46
skn-rbt LD50:4 mg/kg JPBAA7 58,411,46
ivn-rbt LD50:500 µg/kg JPBAA7 58,411,46
skn-gpg LD50:6 mg/kg JPBAA7 58,411,46

CONSENSUS REPORTS: Arsenic and its compounds are on the Community Right-To-Know List.

OSHA PEL: TWA 0.5 mg(As)/m^3

SAFETY PROFILE: Poison by skin contact and intravenous routes. See also ARSENIC COMPOUNDS. When heated to decomposition it emits very toxic fumes of As and Br^-.

DDR400 CAS:78-75-1 HR: 3
1,2-DIBROMOPROPANE
mf: $C_3H_6BR_2$ mw: 201.91

PROP: Colorless liquid. Mp: −55°, bp: 140–142°, n (20/D) 1.5203, d: 1.933. Sltly soluble in water; miscible with organic solvents.

SYN: PROPYLENE DIBROMIDE

TOXICITY DATA WITH REFERENCE
mmo-sat 10 µmol/plate ENMUDM 2,59,80
mma-sat 10 µmol/plate ENMUDM 2,59,80
sln-dmg-orl 5 mmol/L EXPEAM 30,621,74
orl-rat LD50:741 mg/kg GISAAA 41(3),105,76
ihl-rat LC50:15,344 mg/m^3 GTPZAB 19(9),36,75
orl-mus LD50:676 mg/kg GISAAA 41(3),105,76
ipr-mus LD50:75 mg/kg NTIS** AD691-490

CONSENSUS REPORTS: EPA Genetic Toxicology Program. Reported in EPA TSCA Inventory.

SAFETY PROFILE: Poison by intraperitoneal route. Moderately toxic by ingestion. Mildly toxic by inhalation. Mutation data reported. When heated to decomposition it emits toxic fumes of Br^-. See also PROPANE and BROMIDES.

DDR600 CAS:78-75-1 HR: 3
2,3-DIBROMOPROPANE
mf: $C_3H_6Br_2$ mw: 201.91

PROP: Colorless liquid. Mp: −55°, bp: 139.6–142.6°, fp: <−75°, d: 1.940 @ 25°/25°, vap d: 7.0.

SYN: PROPYLENE DIBROMIDE

TOXICITY DATA WITH REFERENCE
sln-dmg-orl 5 mmol/L EXPEAM 30,621,74
orl-rat LD50:741 mg/kg GISAAA 41(3),105,76

ihl-rat LC50: 12 g/m³/4H 85GMAT-,43,82
orl-mus LD50: 676 mg/kg GISAAA 41(3),105,76
ipr-mus LD50: 75 mg/kg NTIS** AD691-490

CONSENSUS REPORTS: Reported in EPA TSCA Inventory.

SAFETY PROFILE: Poison by intraperitoneal route. Moderately toxic by ingestion. Mutation data reported. When heated to decomposition it emits toxic fumes of Br^-. See also PROPANE and BROMIDES.

DDR800 CAS:96-21-9 HR: 3
1,3-DIBROMO-2-PROPANOL
mf: $C_3H_6Br_2O$ mw: 217.91

PROP: D: 2.12 @ 25°/4°, bp: 219° (part decomp).

SYN: GLYCEROL-α,γ-DIBROMOHYDRINE

TOXICITY DATA WITH REFERENCE
mmo-sat 100 nmol/plate ENMUDM 2,59,80
ipr-mus LD50: 150 mg/kg ARZNAD 17,145,67

CONSENSUS REPORTS: Reported in EPA TSCA Inventory.

SAFETY PROFILE: Poison by intraperitoneal route. Mutation data reported. When heated to decomposition it emits toxic fumes of Br^-.

DDS000 CAS:96-13-9 HR: 3
2,3-DIBROMOPROPANOL
mf: $C_3H_6Br_2O$ mw: 217.91

SYNS: 2,3-DIBROMO-1-PROPANOL □ NCI-C55436 □ USAF DO-42

TOXICITY DATA WITH REFERENCE
sln-dmg-orl 500 ppm ENMUDM 7,349,85
trn-dmg-orl 500 ppm ENMUDM 7,349,85
dnd-rat: oth 1 μmol/L CRNGDP 6,705,85
otr-ham: emb 500 nmol/L CRNGDP 6,705,85
msc-ham: lng 20 μmol/L MUREAV 124,213,83
ipr-mus LDLo: 125 mg/kg NTIS** AD277-689

CONSENSUS REPORTS: Reported in EPA TSCA Inventory. EPA Genetic Toxicology Program.

SAFETY PROFILE: Poison by intraperitoneal route. Mutation data reported. When heated to decomposition it emits toxic fumes of Br^-. See also BROMIDES.

DDS100 CAS:18791-02-1 HR: 2
2,3-DIBROMOPROPANOYL CHLORIDE
mf: $C_3H_3Br_2ClO$ mw: 250.33

PROP: Bp: 192–193° (part decomp).

SYNS: α,β-DIBROMOPROPIONYL CHLORIDE □ 2,3-DIBROMOPROPIONYL CHLORIDE

TOXICITY DATA WITH REFERENCE
orl-rat LD50: 1200 mg/kg GISAAA 49(4),90,84
ihl-rat LCLo: 1360 mg/m³ GISAAA 49(4),90,84
ihl-mus LC50: 19,200 mg/m³ GISAAA 49(4),90,84

SAFETY PROFILE: Moderately toxic by inhalation and ingestion. When heated to decomposition it emits toxic fumes of Cl^- and Br^-.

DDS200 CAS:513-31-5 HR: 3
2,3-DIBROMOPROPENE
mf: $C_3H_4Br_2$ mw: 199.89

PROP: D: 2.035 @ 25°/4°, bp: 139–140°.

TOXICITY DATA WITH REFERENCE
mmo-sat 100 nmol/plate ENMUDM 2,59,80
ivn-mus LD50: 100 mg/kg CSLNX* NX#03619

CONSENSUS REPORTS: Reported in EPA TSCA Inventory.

SAFETY PROFILE: Poison by intravenous route. Mutation data reported. When heated to decomposition it emits toxic fumes of Br^-.

DDS400 CAS:5221-17-0 HR: 3
2,3-DIBROMOPROPIONALDEHYDE
mf: $C_3H_4Br_2O$ mw: 215.89

SYN: DIBROMOPROPANAL

TOXICITY DATA WITH REFERENCE
mmo-sat 1 nmol/plate MUREAV 78,113,80
ipr-mus LD50: 5 mg/kg JAFCAU 30,627,82
ivn-mus LD50: 56 mg/kg CSLNX* NX#02408

SAFETY PROFILE: Poison by intravenous route. Mutagenic data reported. When heated to decomposition it emits toxic fumes of Br^-. See also ALDEHYDES and BROMIDES.

DDS600 CAS:521-74-4 HR: 3
5,7-DIBROMO-8-QUINOLINOL
mf: $C_9H_5Br_2NO$ mw: 302.97

PROP: Needles from EtOH. Mp: 196°. Insol in dil acids; sol in $CHCl_3$, AcOH, and EtOH.

SYNS: BRODIAR □ BROXYKINOLIN □ BROXYQUINOLINE □ COLEPUR □ COLIPAR □ 5,7-DIBROMO-8-HYDROXYQUINOLINE □ DIBROMOXYQUINOLINE □ FENILOR □ PARAMIBE

TOXICITY DATA WITH REFERENCE
orl-chd TDLo: 1000 mg/kg/27D: CNS,PUL LANCAO 1,922,68
orl-rat LDLo: 10 g/kg KSRNAM 4,27,70
ipr-rat LD50: 1140 mg/kg KSRNAM 4,27,70
orl-mus LD50: 7420 mg/kg KSRNAM 4,27,70
ipr-mus LD50: 325 mg/kg KSRNAM 4,27,70

CONSENSUS REPORTS: Reported in EPA TSCA Inventory. EPA Genetic Toxicology Program.

SAFETY PROFILE: Poison by intraperitoneal route. Mildly toxic by ingestion. Human systemic effects by ingestion: muscle weakness, ataxia (loss of muscle coordination), and gastritis. When heated to decomposition it emits very toxic fumes of Br^- and NO_x.

DDT000 CAS:15091-30-2 HR: 3
3,4-DIBROMOSULFOLANE
mf: $C_4H_6Br_2O_2S$ mw: 277.98

SYN: 3,4-DIBROMOTETRAHYDROTHIOPHENE-1,1-DIOXIDE

TOXICITY DATA WITH REFERENCE
ipr-mus LD50:9500 μg/kg RPTOAN 41,257,78
ivn-mus LD50:56 mg/kg CSLNX* NX#03181

CONSENSUS REPORTS: Reported in EPA TSCA Inventory.

SAFETY PROFILE: Poison by intraperitoneal and intravenous routes. See also BROMIDES. When heated to decomposition it emits very toxic fumes of Br^- and SO_x.

DDT200 CAS:85-79-0 HR: 3
DIBUCAINE
mf: $C_{20}H_{29}N_3O_2$ mw: 343.52

PROP: Hygroscopic crystals. Mp: 64°.

SYNS: 2-BUTOXY-N-(β-DIETHYLAMINOETHYL)CINCHONINAMIDE □ 2-BUTOXY-N-(2 (DIETHYLAMINO)ETHYL)CINCHONINAMIDE □ 2-BUTOXYQUINOLINE-4-CARBOXYLIC ACID DIETHYLAMINOETHYLAMIDE □ α-BUTYLOXYCINCHONINIC ACID DIETHYLETHYLENEDIAMIDE □ 2-BUTYOXY-N-(2-(DIETHYLAMINO)ETHYL)-4-QUINOLINECARBOXAMIDE □ CINCHOCAINE □ DERMACAINE □ N-(2-(DIETHYLAMINO)ETHYL)-2-BUTOXYCINCHONINAMIDE □ NUPERCAINAL □ NUPERCAINE □ SOVCAINE

TOXICITY DATA WITH REFERENCE
orl-chd LDLo:50 mg/kg 34ZIAG -,209,69
ipr-rat LDLo:7 mg/kg TXAPA9 1,156,59
ipr-mus LD50:24,500 μg/kg ARZNAD 10,925,60
scu-mus LD50:28,500 μg/kg ARZNAD 10,925,60
ivn-mus LDLo:6 mg/kg JAPMA8 39,4,50
scu-rbt LD50:8500 μg/kg PSEBAA 29,368,32
ivn-rbt LD50:2500 μg/kg PSEBAA 29,368,32
scu-gpg LDLo:112 mg/kg PHREA7 12,190,32

SAFETY PROFILE: A human poison by ingestion. Poison experimentally by subcutaneous, intravenous, and intraperitoneal routes. When heated to decomposition it emits toxic fumes of NO_x.

DDT250 CAS:37235-82-8 HR: 3
DIBUSMUTH DICHROMIUM NONAOXIDE
mf: $Bi_2Cr_2O_9$ mw: 665.95

SYN: BISMUTH CHROMATE

CONSENSUS REPORTS: Chromium compounds are on the Community Right-To-Know List.

SAFETY PROFILE: May ignite on contact with H_2S. When heated to decomposition it emits acrid smoke and fumes. See also CHROMIUM COMPOUNDS and BISMUTH COMPOUNDS.

DDT300 CAS:519-88-0 HR: 3
DIBUTAMIDE
mf: $C_{17}H_{28}N_2O_2$ mw: 292.47

PROP: Rods from ethanol + 10% ether. Mp: 134°. Practically insol in water; sol in ethanol, isopropanol, glacial acetic acid.

SYNS: AMBUCETAMID □ AMBUCETAMIDE □ BERSEN □ DIBUTAMID (GERMAN) □ α-DIBUTYL-AMINO-4-METHOXYBENZENEACETAMIDE (9CI) □ α-DIBUTYL-AMINO-p-METHOXYPHENYLACETAMIDE □ α-DIBUTYLAMINO-α-(p-METHOXYPHENYL)ACETAMIDE □ 2-DIBUTYLAMINO-2-(p-METHOXYPHENYL)ACETAMIDE □ α-p-METHOXYPHENYL-α-DI-n-BUTYLAMINOACETAMIDE □ MERITIN □ R 5

TOXICITY DATA WITH REFERENCE
ivn-rat LD50:61 mg/kg ARZNAD 11,929,61
orl-mus LD50:813 mg/kg JAPMA8 46,564,57
ipr-mus LD50:92 mg/kg JAPMA8 46,564,57
ivn-mus LD50:62,200 μg/kg JAPMA8 46,564,57

SAFETY PROFILE: Poison by intravenous and intraperitoneal routes. Moderately toxic by ingestion. When heated to decomposition it emits toxic fumes of NO_x.

DDT400 CAS:871-22-7 HR: 2
1,1-DIBUTOXYETHANE
mf: $C_{10}H_{22}O_2$ mw: 174.32

SYNS: ACETALDEHYDE DIBUTYL ACETAL □ DIBUTYL ACETAL □ 1,1'-(ETHYLIDENEBIS(OXY))BISBUTANE)

TOXICITY DATA WITH REFERENCE
skn-rbt 10 mg/24H open SEV AMIHBC 10,61,54
eye rbt 500 mg open AMIHBC 10,61,54
orl-rat LD50:8790 mg/kg AMIHBC 10,61,54

SAFETY PROFILE: Mildly toxic by ingestion. An eye and severe skin irritant. Combustible. To fight fire, use water, foam, CO_2, dry chemical. When heated to decomposition it emits acrid smoke and irritating fumes. See also ALDEHYDES.

DDT500 CAS:141-17-3 HR: 1
DIBUTOXYETHOXYETHYL ADIPATE
mf: $C_{22}H_{42}O_8$ mw: 434.64

SYNS: ADIPIC ACID, BIS(2-(2-BUTOXYETHOXY)ETHYL) ESTER □ HEXANEDIOIC ACID, BIS(2 (2-BUTOXYETHOXY)ETHYL) ESTER (9CI) □ TP-95 □ WAREFLEX

TOXICITY DATA WITH REFERENCE
orl-rat LD50:6 g/kg NPIRI* 2,16,75

CONSENSUS REPORTS: Reported in EPA TSCA Inventory.

SAFETY PROFILE: Slightly toxic by ingestion. When heated to decomposition it emits acrid smoke and irritating vapors.

DDT800 CAS:111-92-2 HR: 3
n-DIBUTYLAMINE
DOT: UN 2248
mf: $C_8H_{19}N$ mw: 129.28

PROP: A liquid. Mp: −61.9°; bp: 159°, flash p: 125°F (OC), d: 0.76, vap d: 4.46, vap press: 2 mm @ 20°. Sol in water and alcohol.

SYNS: N-BUTYL-1-BUTANAMINE □ DI-n-BUTYLAMINE □ DI(n-BUTYL)AMINE (DOT)

TOXICITY DATA WITH REFERENCE
skn-rbt 10 mg/24H open SEV AMIHBC 10,61,54
skn-rbt 500 mg open MOD UCDS** 3/25/70
eye-rbt 250 μg open SEV AMIHBC 10,61,54
cyt-ham:fbr 200 mg/L/48H MUREAV 48,337,77
orl-rat LD50:220 mg/kg ZHYGAM 20,393,74
ihl-rat LCLo:500 ppm/4H AEHLAU 1,343,60

scu-rat LDLo: 330 mg/kg JPETAB 20,435,23
orl-mus LD50: 290 mg/kg GISAAA 40(11),21,75
skn-rbt LD50: 1010 mg/kg AMIHBC 10,61,54
orl-gpg LD50: 230 mg/kg GISAAA 40(11),21,75

CONSENSUS REPORTS: Reported in EPA TSCA Inventory. EPA Genetic Toxicology Program.

DOT CLASSIFICATION: 8; *Label:* Corrosive, Flammable Liquid

SAFETY PROFILE: Poison by ingestion and subcutaneous routes. Moderately toxic by skin contact and inhalation. Corrosive. A severe skin and eye irritant. Mutation data reported. Flammable liquid when exposed to heat or flame; can react with oxidizing materials. To fight fire, use alcohol foam, foam, CO_2, dry chemical. Exothermic reaction with cellulose nitrate does not proceed to ignition. When heated to decomposition it emits toxic fumes of NO_x.

DDU200 CAS:77966-79-1 HR: 3
2-(DIBUTYLAMINO)-2′,6′-ACETOXYLIDIDE HYDROCHLORIDE
mf: $C_{18}H_{30}N_2O•ClH$ mw: 326.96

SYN: C 3103

TOXICITY DATA WITH REFERENCE
eye-rbt 2% SEV ARZNAD 8,407,58
ipr-rat LD50: 221 mg/kg ARZNAD 8,407,58
scu-mus LD50: 805 mg/kg ARZNAD 8,407,58

SAFETY PROFILE: Poison by intraperitoneal route. Moderately toxic by subcutaneous route. A severe eye irritant. When heated to decomposition it emits very toxic fumes of NO_x and HCl. See also AMINES.

DDU600 CAS:102-81-8 HR: 3
2-N-DIBUTYLAMINOETHANOL
DOT: UN 2873
mf: $C_{10}H_{23}NO$ mw: 173.34

PROP: Liquid. Bp: 222°, flash p: 220°F (OC), d: 0.85, vap d: 6.0.

SYNS: BU2AE ☐ DIBUTYLAMINOETHANOL ☐ 2-DIBUTYLAMINOETHANOL ☐ 2-DI-n-BUTYLAMINOETHANOL ☐ N,N-DI-n-BUTYLAMINOETHANOL (DOT) ☐ β-N-DIBUTYLAMINOETHYL ALCOHOL ☐ N,N-DIBUTYLETHANOLAMINE ☐ N,N-DIBUTYL-N-(2-HYDROXYETHYL) AMINE

TOXICITY DATA WITH REFERENCE
skn-rbt 10 mg/24H open AMIHBC 10,61,54
skn-rbt 5 mg/24H SEV 85JCAE-,695,86
eye-rbt 20 mg/24H open SEV AMIHBC 10,61,54
orl-rat LD50: 1070 mg/kg AMIHBC 10,61,54
ipr-rat LD50: 144 mg/kg TXAPA9 12,486,68
ipr-mus LD50: 52 mg/kg RCRVAB 38,975,69
skn-rbt LD50: 1680 mg/kg AMIHBC 10,61,54
ipr-mam LD50: 120 mg/kg TXAPA9 8,344,66

CONSENSUS REPORTS: Reported in EPA TSCA Inventory.

OSHA PEL: TWA 2 ppm
ACGIH TLV: TWA 0.5 ppm (skin)

DOT CLASSIFICATION: 6.1; *Label:* KEEP AWAY FROM FOOD

SAFETY PROFILE: Poison by intraperitoneal route. Moderately toxic by ingestion and skin contact. A severe eye and skin irritant. Combustible; can react with oxidizing materials. To fight fire, use CO_2, dry chemical. When heated to decomposition it emits toxic fumes of NO_x. See also AMINES and ALCOHOLS.

For occupational chemical analysis use NIOSH: Aminoethanol Compounds, 2007.

DDV200 CAS:102-83-0 HR: 3
3-(DIBUTYLAMINO)PROPYLAMINE
mf: $C_{11}H_{26}N_2$ mw: 186.39

TOXICITY DATA WITH REFERENCE
skn-rbt 100 μg/24H open AIHAAP 23,95,62
orl-rat LD50: 820 mg/kg AIHAAP 23,95,62
skn-rbt LD50: 270 mg/kg AIHAAP 23,95,62

CONSENSUS REPORTS: Reported in EPA TSCA Inventory.

SAFETY PROFILE: Poison by skin contact. Moderately toxic by ingestion. A skin irritant. Ignites on contact with cellulose nitrate. When heated to decomposition it emits toxic fumes of NO_x.

DDV225 CAS:7128-68-9 HR: 3
DI-N-BUTYLAMMONIUM HEXAFLUOROARSENATE
mf: $C_8H_{19}N•AsF_6H$ mw: 319.21

SYN: DIBUTYLAMINE, HEXAFLUOROARSENATE(1-)

TOXICITY DATA WITH REFERENCE
ivn-mus LD50: 180 mg/kg CSLNX* NX#04252

OSHA PEL: TWA 0.5 mg(As)/m^3

SAFETY PROFILE: Poison by intravenous route. When heated to decomposition it emits toxic fumes of NO_x, As, and F^-.

DDV250 CAS:2850-61-5 HR: 3
DIBUTYLARSINIC ACID
mf: $C_8H_{19}AsO_2$ mw: 222.19

PROP: Prisms from H_2O or crystals from Me_2CO. Mp: 138°. Sol in EtOH; sltly sol in H_2O.

SYNS: ARSINE OXIDE, DIBUTYLHYDROXY- ☐ ARSINIC ACID, DIBUTYL-(9CI)

TOXICITY DATA WITH REFERENCE
ivn-mus LD50: 18 mg/kg CSLNX* NX#01190

OSHA PEL: TWA 0.5 mg(As)/m^3

SAFETY PROFILE: Poison by intravenous route. When heated to decomposition it emits toxic fumes of As.

DDV400 CAS:17013-41-1 HR: 3
5,5-DIBUTYLBARBITURIC ACID
mf: $C_{12}H_{20}N_2O_3$ mw: 240.34

SYNS: DIBUTYLBARBITURIC ACID ☐ 5,5-DIBUTYL-2,4,6(1H,3H,5H)-PYRIMIDINETRIONE

TOXICITY DATA WITH REFERENCE
ipr-mus LD50:232 mg/kg JPETAB 89,356,47
orl-cat LDLo:350 mg/kg JPETAB 26,371,25
scu-rbt LDLo:500 mg/kg JACSAT 45,243,23

SAFETY PROFILE: Poison by ingestion and intraperitoneal routes. Moderately toxic by subcutaneous route. An hypnotic agent. When heated to decomposition it emits toxic fumes of NO_x. See also BARBITURATES.

DDV450 CAS:1012-72-2 HR: 3
p-DI-tert-BUTYLBENZENE
mf: $C_{14}H_{22}$ mw: 190.36

SYNS: BENZENE, BIS(1,1-DIMETHYLETHYL)- ☐ BENZENE, p-DI-tert-BUTYL-

DOT CLASSIFICATION: 3; *Label:* Flammable Liquid

CONSENSUS REPORTS: Reported in EPA TSCA Inventory.

SAFETY PROFILE: A flammable liquid. When heated to decomposition it emits acrid smoke and irritating vapors.

DDV500 CAS:719-22-2 HR: 1
2,6-DI-tert-BUTYL-p-BENZOQUINONE
mf: $C_{14}H_{20}O_2$ mw: 220.34

SYNS: p-BENZOQUINONE, 2,6-DI-tert-BUTYL- ☐ DBQ

TOXICITY DATA WITH REFERENCE
ipr-mus LD50:2270 mg/kg TOLED5 6,173,80

CONSENSUS REPORTS: Reported in EPA TSCA Inventory.

SAFETY PROFILE: Moderately toxic by intraperitoneal route. When heated to decomposition it emits acrid smoke and irritating vapors.

DDV600 CAS:77-58-7 HR: 3
DIBUTYLBIS(LAUROYLOXY)STANNANE
mf: $C_{32}H_{64}O_4Sn$ mw: 631.65

PROP: Pale yellow liquid to colorless solid (when pure). Mp: 23°, bp: non-distillable @ 10 mm, flash p: 455°F (OC), d: 1.066 @ 20°/20°, vap d: 21.8.

SYNS: BIS(DODECANOYLOXY)DI-n-BUTYLSTANNANE ☐ BIS(LAUROYLOXY)DIBUTYLSTANNANE ☐ BIS(LAUROYLOXY)DI(n-BUTYL)STANNANE ☐ BUTYNORATE ☐ DBTL ☐ DIBUTYLBIS(LAUROYLOXY)TIN ☐ DI-n-BUTYLTIN DI(DODECANOATE) ☐ DIBUTYLTIN DILAURATE (USDA) ☐ DIBUTYLTIN LAURATE ☐ DIBUTYL-ZINN-DILAURAT (GERMAN) ☐ FOMREZ SUL-4 ☐ LAUDRAN DI-n-BUTYLCINICITY (CZECH) ☐ LAURIC ACID, DIBUTYLSTANNYLENE derivative ☐ LAURIC ACID, DIBUTYLSTANNYLENE SALT ☐ STABILIZER D-22 ☐ THERM CHEK 820 ☐ TIN DIBUTYL DILAURATE ☐ TINOSTAT

TOXICITY DATA WITH REFERENCE
skn-rbt 500 mg/24H MLD 28ZPAK -,230,72
eye-rbt 100 mg/24H MOD 28ZPAK -,230,72
orl-rat LD50:175 mg/kg ARZNAD 10,44,60
ipr-rat LDLo:85 mg/kg BJPCAL 10,16,55
orl-mus LDLo:710 mg/kg AECTCV 14,111,85

CONSENSUS REPORTS: Reported in EPA TSCA Inventory.

OSHA PEL: TWA 0.1 mg(Sn)/m³ (skin)
ACGIH TLV: TWA 0.1 mg(Sn)/m³ (skin) (Proposed: TWA 0.1 mg(Sn)/m³; STEL 0.2 mg(Sn)/m³ (skin))
NIOSH REL: (Organotin Compounds) TWA 0.1 mg(Sn)/m³

SAFETY PROFILE: Poison by ingestion and intraperitoneal routes. A skin and eye irritant. Avoid the vapor produced by heating. Combustible when exposed to heat or flame; reacts with oxidizers. When heated to decomposition it emits acrid smoke and fumes. See also TIN COMPOUNDS.

For occupational chemical analysis use NIOSH: Organotin Compounds 5504.

DDV800 CAS:78-46-6 HR: 3
DIBUTYL BUTANEPHOSPHONATE
mf: $C_{12}H_{27}O_3P$ mw: 250.36

PROP: Colorless liquid with mild pleasant odor. Bp: 160–162° @ 20 mm, flash p: 311° (COC), d: 8.62.

SYNS: DIBUTYL BUTYLPHOSPHONATE ☐ NSC 2666

TOXICITY DATA WITH REFERENCE
ipr-mus LDLo:125 mg/kg CBCCT* 7,789,55
ivn-mus LD50:56 mg/kg CSLNX* NX#03463

CONSENSUS REPORTS: Reported in EPA TSCA Inventory.

SAFETY PROFILE: Poison by intraperitoneal and intravenous routes. Combustible when exposed to heat or flame. It can react vigorously with oxidizing materials. To fight fire, use foam, CO_2, or dry chemical. When heated to decomposition it emits toxic fumes of PO_x.

DDW000 CAS:532-49-0 HR: 3
DI-n-BUTYL-CARBAMYLCHOLINE SULPHATE
mf: $C_{30}H_{66}N_4O_4 \cdot O_4S$ mw: 643.06

PROP: Hygroscopic powder. Mp: 166° (decomp).

SYNS: DIBULINESULFAT ☐ DIBULINE SULFATE ☐ DIBUTOLINE ☐ DIBUTOLINE SULFATE ☐ 1-(((DIBUTYLAMINO)CARBONYL)OXY)-N-ETHYL-N,N-DIMETHYLETHANAMINIUM SULFATE (2:1) ☐ (2-DIBUTYLCARBAMYLOXYETHYL)-DIMETHYLETHYLAMMONIUM SULFATE ☐ DIMETHYL-ETHYL-β-HYDROXYETHYL-AMMONIUM-SULFATE-DI-n-BUTYLCARBAMATE ☐ DIMETHYLETHYL-β-HYDROXYETHYLAMMONIUM SULFATE DIBUTYLURETHAN ☐ ETHYL(2-HYDROXYETHYL)DIMETHYL-AMMONIUM SULFATE (SALT), BIS(DIBUTYLCARBAMATE)

TOXICITY DATA WITH REFERENCE
ipr-rat LD50:22 mg/kg JPETAB 84,105,45
scu-mus LD50:49 mg/kg CLDND*

SAFETY PROFILE: Poison by intraperitoneal and subcutaneous routes. See also CARBAMATES and SULFATES. When heated to decomposition it emits very toxic fumes of NO_x, NH_3, and SO_x.

DDW200 **CAS:112-73-2** **HR: 2**
DIBUTYL CARBITOL
mf: $C_{12}H_{26}O_3$ mw: 218.38

PROP: Practically colorless liquid, characteristic odor, sltly sol in water. D: 0.8853 @ 20°/20°, bp: 256°, fp: −60.2°, flash p: 245°F (OC).

SYNS: BIS(BUTOXYETHYL) ETHER □ BIS(2-BUTOXYETHYL) ETHER □ BUTYL DIGLYME □ 2,2′-DIBUTOXYETHYL ETHER □ DIETHYLENEGLYCOL DIBUTYL ETHER □ DIETHYLENEGLYCOL DI-n-BUTYL ETHER □ 1,1′-(OXYBIS(2,1-ETHANEDIYLOXY))BISBUTANE □ 5, 8,11-TRIOXAPENTADECANE

TOXICITY DATA WITH REFERENCE
skn-rbt 500 mg open MLD UCDS** 4/21/67
eye-rbt 500 mg open AMIHBC 10,61,54
orl-mus TDLo:16 g/kg (7-14D preg):REP EVHPAZ 57,141,84
orl-rat LD50:3900 mg/kg AMIHBC 10,61,54
skn-rbt LD50:4040 mg/kg AMIHBC 10,61,54

CONSENSUS REPORTS: Reported in EPA TSCA Inventory. Glycol ether compounds are on the Community Right-To-Know List.

SAFETY PROFILE: Moderately toxic by ingestion. Mildly toxic by skin contact. Experimental reproductive effects. A skin and eye irritant. See also GLYCOL ETHERS. Combustible when exposed to heat or flame. To fight fire, use foam or alcohol foam. When heated to decomposition it emits acrid smoke and irritating fumes.

DDW400 **CAS:112-48-1** **HR: 2**
DIBUTYL CELLOSOLVE
mf: $C_{10}H_{22}O_2$ mw: 174.32

SYNS: 1,2-DIBUTOXYETHANE □ 1,1′-(1,2-ETHANEDIYLBIS(OXY)) BIS-BUTANE □ ETHYLENE GLYCOL DIBUTYL

TOXICITY DATA WITH REFERENCE
skn-rbt 500 mg open MLD UCDS** 11/7/57
eye-rbt 500 mg open AMIHBC 10,61,54
orl-rat LD50:3250 mg/kg UCDS** 11/7/57
skn-rbt LD50:3560 mg/kg AMIHBC 10,61,54

SAFETY PROFILE: Moderately toxic by ingestion and skin contact. A skin and eye irritant. When heated to decomposition it emits acrid smoke and irritating fumes.

DDX200 **CAS:63041-48-5** **HR: 2**
9,10-DI-n-BUTYL-1,2,5,6-DIBENZANTHRACENE
mf: $C_{30}H_{30}$ mw: 390.60

TOXICITY DATA WITH REFERENCE
skn-mus TDLo:1250 mg/kg/52W-I:ETA PRLBA4 117,318,35

SAFETY PROFILE: Questionable carcinogen with experimental tumorigenic data. When heated to decomposition it emits acrid smoke and irritating fumes.

DDX600 **CAS:28660-63-1** **HR: 3**
DI-n-BUTYL(DIBUTYRYLOXY)STANNANE
mf: $C_{16}H_{32}O_4Sn$ mw: 407.17

SYNS: DI-n-BUTYLTIN DIBUTYRATE □ MASELNAN DI-n-BUTYLCINICITY (CZECH)

TOXICITY DATA WITH REFERENCE
skn-rbt 2 mg/24H SEV 85JCAE-,1252,86
eye-rbt 20 mg/24H MOD 85JCAE-,1252,86
orl-rat LD50:90,700 μg/kg 85JCAE-,1252,86

OSHA PEL: TWA 0.1 mg(Sn)/m³ (skin)
ACGIH TLV: TWA 0.1 mg(Sn)/m³ (skin) (Proposed: TWA 0.1 mg(Sn)/m³; STEL 0.2 mg(Sn)/m³ (skin))
NIOSH REL: (Organotin Compounds) TWA 0.1 mg(Sn)/m³

SAFETY PROFILE: Poison by ingestion. A severe eye and skin irritant. See also TIN COMPOUNDS. When heated to decomposition it emits acrid smoke and irritating fumes.

For occupational chemical analysis use NIOSH: Organotin Compounds 5504.

DDY000 **CAS:4593-81-1** **HR: 3**
DIBUTYLDICHLOROGERMANE
mf: $C_8H_{18}Cl_2Ge$ mw: 257.75

SYNS: DI-n-BUTYLGERMANEDICHLORIDE □ DICHLORODIBUTYLGERMANE

TOXICITY DATA WITH REFERENCE
msc-ham:ovr 100 mg/L TXAPA9 64,482,82
orl-rat LD50:100 mg/kg 85JCAE-,1241,86
ipr-rat LDLo:100 mg/kg CHDDAT 262,1302,66
orl-mus LD50:96 mg/kg 85JCAE-,1241,86
ipr-mus LDLo:96 mg/kg CHDDAT 262,1302,66

CONSENSUS REPORTS: Reported in EPA TSCA Inventory.

SAFETY PROFILE: Poison by ingestion and intraperitoneal routes. Mutation data reported. See also GERMANIUM COMPOUNDS. When heated to decomposition it emits very toxic fumes of Cl^-.

DDY200 **CAS:683-18-1** **HR: 3**
DIBUTYLDICHLOROSTANNANE
mf: $C_8H_{18}Cl_2Sn$ mw: 303.85

PROP: White, crystalline solid. Mp: 43°, bp: 135° @ 10 mm, flash p: 335°F (OC), d: 1.36 @ 50°, vap press: 2 mm @ 100°, vap d: 10.5.

SYNS: CHLORID DI-n-BUTYLCINICITY (CZECH) □ D.B.T.C. □ DIBUTYLDICHLOROTIN □ DIBUTYLTIN CHLORIDE □ DIBUTYLTIN DICHLORIDE □ DI-n-BUTYLTIN DICHLORIDE □ DI-n-BUTYL-ZINN-DICHLORID (GERMAN) □ DICHLORODIBUTYLSTANNANE □ DICHLORODIBUTYLTIN

TOXICITY DATA WITH REFERENCE
skn-rbt 500 mg/24H SEV 28ZPAK -,226,72
eye-rbt 50 μg/24H SEV 28ZPAK -,226,72
msc-ham:ovr 100 μg/L TXAPA9 64,482,82
dns-rbt:oth 10 μg/L JTEHD6 16,229,85
dni-rbt:oth 100 μg/L JTEHD6 16,229,85

orl-rat TDLo:45 mg/kg (female 7-15D post):REP TOLED5 58,347,91
orl-rat TDLo:20 mg/kg (female 8D post):REP TXCYAC 73,81,92
orl-rat LD50:100 mg/kg ARZNAD 10,44,60
ipr-rat LDLo:7500 µg/kg JOCMA7 2,183,60
ivn-rat LDLo:10 mg/kg BJIMAG 15,15,58
orl-mus LD50:70 mg/kg PHARAT 39,572,84
ivn-mus LD50:180 mg/kg CSLNX* NX#00182
orl-rbt LD50:50 µg/kg 85JCAE-,1248,86
skn-rbt LDLo:1360 mg/kg SAIGBL 15,3,73
ivn-rbt LDLo:5 mg/kg BJIMAG 15,15,58
ivn-gpg LDLo:5 mg/kg BJIMAG 15,15,58

CONSENSUS REPORTS: Reported in EPA TSCA Inventory.

OSHA PEL: TWA 0.1 mg(Sn)/m³ (skin)
ACGIH TLV: TWA 0.1 mg(Sn)/m³ (skin) (Proposed: TWA 0.1 mg(Sn)/m³; STEL 0.2 mg(Sn)/m³ (skin))
NIOSH REL: (Organotin Compounds) TWA 0.1 mg(Sn)/m³

SAFETY PROFILE: Poison by ingestion, intravenous, and intraperitoneal routes. Moderately toxic by skin contact. A severe skin and eye irritant. Experimental reproductive effects. Mutation data reported. See also TIN COMPOUNDS. Combustible when exposed to heat or flame. A dangerous material; emits highly toxic fumes of HCl; will react with water or steam to produce heat and toxic fumes; can react vigorously with oxidizing materials. To fight fire, use water, foam, CO_2, dry chemical.

For occupational chemical analysis use NIOSH: Organotin Compounds 5504.

DDY400 CAS:7483-25-2 HR: 1
DIBUTYL (DIETHYLENE GLYCOL BISPHTHALATE)

SYNS: DIETHYLENE GLYCOL, DIESTER with BUTYLPHTHALATE □ HOWFLEX GBP

TOXICITY DATA WITH REFERENCE
orl-rat LD50:11 g/kg EVHPAZ 4,3,73
ipr-rat LD50:11,200 mg/kg FCTXAV 4,383,66
orl-mus LD50:10,300 mg/kg FCTXAV 4,383,66
ipr-mus LD50:8400 mg/kg FCTXAV 4,383,66

SAFETY PROFILE: Mildly toxic by ingestion and intraperitoneal routes. When heated to decomposition it emits acrid smoke and irritating fumes.

DDY600 CAS:10584-98-2 HR: 2
DIBUTYLDI(2-ETHYLHEXYLOXYCARBONYLMETHYLTHIO)STANNANE
mf: $C_{28}H_{56}O_4S_2Sn$ mw: 639.65

SYNS: BIS(2-ETHYLHEXYLOXYCARBONYLMETHYLTHIO)DIBUTYLSTANNANE □ BIS(2-ETHYLHEXYLTHIOGLYCOLATE)DIBUTYLTIN □ DI-n-BUTYLTIN DI-2-ETHYLHEXYLTHIOGLYCOLATE □ DI-n-BUTYLZINN DI-2-AETHYLHEXYL THIOGLYKOLAT (GERMAN)

TOXICITY DATA WITH REFERENCE
orl-rat LD50:510 mg/kg ARZNAD 19,934,69

CONSENSUS REPORTS: Reported in EPA TSCA Inventory.

OSHA PEL: TWA 0.1 mg(Sn)/m³ (skin)
ACGIH TLV: TWA 0.1 mg(Sn)/m³ (skin) (Proposed: TWA 0.1 mg(Sn)/m³; STEL 0.2 mg(Sn)/m³ (skin))
NIOSH REL: (Organotin Compounds) TWA 0.1 mg(Sn)/m³

SAFETY PROFILE: Moderately toxic by ingestion. See also TIN COMPOUNDS. When heated to decomposition it emits toxic fumes of SO_x.

For occupational chemical analysis use NIOSH: Organotin Compounds 5504.

DDY800 CAS:563-25-7 HR: 3
DIBUTYLDIFLUOROSTANNANE
mf: $C_8H_{18}F_2Sn$ mw: 270.95

SYN: DIBUTYLTIN DIFLUORIDE

TOXICITY DATA WITH REFERENCE
orl-mus LDLo:470 mg/kg AECTCV 14,111,85
orl-rbt LDLo:200 mg/kg SAIGBL 15,3,73

CONSENSUS REPORTS: Reported in EPA TSCA Inventory.

OSHA PEL: TWA 0.1 mg(Sn)/m³ (skin)
ACGIH TLV: TWA 0.1 mg(Sn)/m³ (skin) (Proposed: TWA 0.1 mg(Sn)/m³; STEL 0.2 mg(Sn)/m³ (skin))
NIOSH REL: (Organotin Compounds) TWA 0.1 mg(Sn)/m³

SAFETY PROFILE: Poison by ingestion. See also TIN COMPOUNDS and FLUORIDES. When heated to decomposition it emits toxic fumes of F^-.

For occupational chemical analysis use NIOSH: Organotin Compounds 5504.

DDZ000 CAS:7392-96-3 HR: 3
DIBUTYL(DIFORMYLOXY)STANNANE
mf: $C_{10}H_{20}O_4Sn$ mw: 322.99

SYNS: DI-n-BUTYLTIN DIFORMATE □ MRAVENCAN DI-n-BUTYLCINICITY (CZECH)

TOXICITY DATA WITH REFERENCE
skn-rbt 2 mg/24H SEV 85JCAE-,1249,86
eye-rbt 5 mg/24H SEV 28ZPAK -,227,72
orl-rat LD50:60,900 µg/kg 28ZPAK -,227,72

OSHA PEL: TWA 0.1 mg(Sn)/m³ (skin)
ACGIH TLV: TWA 0.1 mg(Sn)/m³ (skin) (Proposed: TWA 0.1 mg(Sn)/m³; STEL 0.2 mg(Sn)/m³ (skin))
NIOSH REL: (Organotin Compounds) TWA 0.1 mg(Sn)/m³

SAFETY PROFILE: Poison by ingestion. A severe skin and eye irritant. When heated to decomposition it emits acrid and irritating fumes. See also TIN COMPOUNDS.

For occupational chemical analysis use NIOSH: Organotin Compounds 5504.

DEA000 CAS:2865-19-2 HR: 3
DIBUTYLDIIODOSTANNANE
mf: $C_8H_{18}I_2Sn$ mw: 486.75

SYN: DIBUTYLTIN DIIODIDE

TOXICITY DATA WITH REFERENCE
orl-rbt LDLo: 150 mg/kg SAIGBL 15,3,73
skn-rbt LDLo: 1000 mg/kg SAIGBL 15,3,73

OSHA PEL: TWA 0.1 mg(Sn)/m³ (skin)
ACGIH TLV: TWA 0.1 mg(Sn)/m³ (skin) (Proposed: TWA 0.1 mg(Sn)/m³; STEL 0.2 mg(Sn)/m³ (skin))
NIOSH REL: (Organotin Compounds) TWA 0.1 mg(Sn)/m³

SAFETY PROFILE: Poison by ingestion. Moderately toxic by skin contact. See also TIN COMPOUNDS and IODIDES. When heated to decomposition it emits toxic fumes of I^-.

For occupational chemical analysis use NIOSH: Organotin Compounds 5504.

DEA100 CAS:88-27-7 HR: 2
2,6-DI-tert-BUTYL-α-(DIMETHYLAMINO)-p-CRESOL
mf: $C_{17}H_{29}NO$ mw: 263.47

SYNS: AGIDOL 3 □ p-CRESOL, 2,6-DI-tert-BUTYL-α-(DIMETHYLAMINO)- □ ETHYL 703 □ ETHYL ANTIOXIDANT 703 □ F 1 □ F 1 (ANTIOXIDANT) □ OMI □ PHENOL, 4-((DIMETHYLAMINO)METHYL)-2,6-BIS(1,1-DIMETHYLETHYL)-(9CI)

TOXICITY DATA WITH REFERENCE
orl-rat LD50: 1030 mg/kg IPSTB3 3,93,76

CONSENSUS REPORTS: Reported in EPA TSCA Inventory.

SAFETY PROFILE: Moderately toxic by ingestion. When heated to decomposition it emits toxic vapors of NO_x.

DEA200 CAS:3231-93-4 HR: 3
2,2-DIBUTYL-1,3-DIOXA-2-STANNA-7,9-DITHIACYCLODODECAN-4,12-DIONE
mf: $C_{15}H_{28}O_4S_2Sn$ mw: 455.24

SYN: 2,2-DIBUTYL-1,3-DIOXA-7,9-DITHIA-2-STANNACYCLODODECAN

TOXICITY DATA WITH REFERENCE
ivn-mus LD50: 320 mg/kg CSLNX* NX#02853

OSHA PEL: TWA 0.1 mg(Sn)/m³ (skin)
ACGIH TLV: TWA 0.1 mg(Sn)/m³ (skin) (Proposed: TWA 0.1 mg(Sn)/m³; STEL 0.2 mg(Sn)/m³ (skin))
NIOSH REL: (Organotin Compounds) TWA 0.1 mg(Sn)/m³

SAFETY PROFILE: Poison by intravenous route. See also TIN COMPOUNDS. When heated to decomposition it emits toxic fumes of SO_x.

For occupational chemical analysis use NIOSH: Organotin Compounds 5504.

DEA400 CAS:4981-24-2 HR: 3
2,2-DIBUTYL-1,3-DIOXA-2-STANNA-7-THIACYCLODECAN-4,10-DIONE
mf: $C_{14}H_{26}O_4SSn$ mw: 409.15

SYNS: DIBUTYLDIHYDROXYSTANNANE-3,3′-THIODIPROPIONATE □ 2,2-DIBUTYL-1,3,7,2-DIOXATHIASTANNECANE-4,10-DIONE □ DIBUTYLTIN 3,3′-THIODIPROPIONATE

TOXICITY DATA WITH REFERENCE
ivn-mus LD50: 180 mg/kg CSLNX* NX#02851

OSHA PEL: TWA 0.1 mg(Sn)/m³ (skin)
ACGIH TLV: TWA 0.1 mg(Sn)/m³ (skin) (Proposed: TWA 0.1 mg(Sn)/m³; STEL 0.2 mg(Sn)/m³ (skin))
NIOSH REL: (Organotin Compounds) TWA 0.1 mg(Sn)/m³

SAFETY PROFILE: Poison by intravenous route. See also TIN COMPOUNDS. When heated to decomposition it emits toxic fumes of SO_x.

For occupational chemical analysis use NIOSH: Organotin Compounds 5504.

DEA600 CAS:3465-74-5 HR: 3
DIBUTYLDIPENTANOYLOXYSTANNANE
mf: $C_{18}H_{36}O_4Sn$ mw: 435.23

SYNS: DI-n-BUTYLTIN DIPENTANOATE □ DI(PENTANOYLOXY)DIBUTYLSTANNANE □ VALERAN DI-n-BUTYLCINICITY (CZECH)

TOXICITY DATA WITH REFERENCE
skn-rbt 500 mg/24H SEV 28ZPAK -,228,72
eye-rbt 5 mg/24H SEV 28ZPAK -,228,72
orl-rat LD50: 134 mg/kg 28ZPAK -,228,72

OSHA PEL: TWA 0.1 mg(Sn)/m³ (skin)
ACGIH TLV: TWA 0.1 mg(Sn)/m³ (skin) (Proposed: TWA 0.1 mg(Sn)/m³; STEL 0.2 mg(Sn)/m³ (skin))
NIOSH REL: (Organotin Compounds) TWA 0.1 mg(Sn)/m³

SAFETY PROFILE: Poison by ingestion. A severe skin and eye irritant. See also TIN COMPOUNDS. When heated to decomposition it emits acrid and irritating fumes.

For occupational chemical analysis use NIOSH: Organotin Compounds 5504.

DEB400 CAS:3465-73-4 HR: 3
DIBUTYLDIPROPIONYLOXYSTANNANE
mf: $C_{14}H_{28}O_4Sn$ mw: 379.11

SYNS: DI-n-BUTYLTIN DIPROPIONATE □ PROPINAN DI-n-BUTYLCINICITY (CZECH)

TOXICITY DATA WITH REFERENCE
skn-rbt 500 mg/24H SEV 28ZPAK -,228,72
eye-rbt 5 mg/24H SEV 28ZPAK -,228,72
orl-rat LD50: 70,900 μg/kg 28ZPAK -,228,72

OSHA PEL: TWA 0.1 mg(Sn)/m³ (skin)
ACGIH TLV: TWA 0.1 mg(Sn)/m³ (skin) (Proposed: TWA 0.1 mg(Sn)/m³; STEL 0.2 mg(Sn)/m³ (skin))
NIOSH REL: (Organotin Compounds) TWA 0.1 mg(Sn)/m³

SAFETY PROFILE: Poison by ingestion. A severe skin and eye irritant. See also TIN COMPOUNDS. When heated to decomposition it emits acrid smoke and irritating fumes.

For occupational chemical analysis use NIOSH: Organotin Compounds 5504.

DEB600 CAS:67057-34-5 HR: 3
DIBUTYLDITHIOCARBAMIC ACID-S-TRIBUTYLSTANNYL ESTER
mf: $C_{21}H_{45}NS_2Sn$ mw: 494.48

SYNS: ((DIBUTYLDITHIOCARBAMOYL)OXY)TRIBUTYLSTANNANE ☐ TRIBUTYLTIN-S,S'-DIBUTYLDITHIOCARBAMATE

TOXICITY DATA WITH REFERENCE
ivn-mus LD50:56 mg/kg CSLNX* NX#04817

OSHA PEL: TWA 0.1 mg(Sn)/m³ (skin)
ACGIH TLV: TWA 0.1 mg(Sn)/m³ (skin) (Proposed: TWA 0.1 mg(Sn)/m³; STEL 0.2 mg(Sn)/m³ (skin))
NIOSH REL: (Organotin Compounds) TWA 0.1 mg(Sn)/m³

SAFETY PROFILE: Poison by intravenous route. See also TIN COMPOUNDS, CARBAMATES, and ESTERS. When heated to decomposition it emits very toxic fumes of NO_x and SO_x.

For occupational chemical analysis use NIOSH: Organotin Compounds 5504.

DEB800 CAS:26818-53-1 HR: 3
N,N-DI-sec-BUTYL DITHIOOXAMIDE
mf: $C_{10}H_{20}N_2S_2$ mw: 232.44

TOXICITY DATA WITH REFERENCE
ivn-rat LDLo:5 mg/kg JPETAB 121,32,57
ivn-dog LDLo:5 mg/kg JPETAB 121,32,57
ivn-cat LDLo:5 mg/kg JPETAB 121,32,57
ivn-rbt LD50:2.3 mg/kg JPETAB 121,32,57
ivn-gpg LDLo:5 mg/kg JPETAB 121,32,57

SAFETY PROFILE: Poison by intravenous route. When heated to decomposition it emits very toxic fumes of NO_x and SO_x.

DEC000 CAS:625-22-9 HR: 3
DIBUTYL ESTER SULFURIC ACID
mf: $C_8H_{18}O_4S$ mw: 210.32

PROP: Liquid with sharp odor. Bp: 115–116° @ 6 mm.

SYNS: DI-n-BUTYLSULFAT (GERMAN) ☐ DIBUTYL SULFATE

TOXICITY DATA WITH REFERENCE
orl-rat TDLo:12 g/kg/24W-I:ETA ZEKBAI 74,241,70
scu-rat TDLo:9500 mg/kg/19W-I:ETA ZEKBAI 74,241,70
scu-rat LD50:5000 mg/kg ZEKBAI 74,241,70
orl-rbt LDLo:192 mg/kg AEXPBL 47,113,02

SAFETY PROFILE: Poison by ingestion. Mildly toxic by subcutaneous route. Questionable carcinogen with experimental tumorigenic data. See also ESTERS and SULFATES. When heated to decomposition it emits toxic fumes of SO_x.

DEC200 CAS:625-17-2 HR: 3
DI-sec-BUTYL FLUOROPHOSPHONATE
mf: $C_8H_{18}FO_3P$ mw: 212.23

SYNS: DI-sec-BUTYL ESTER PHOSPHOROFLUORIDIC ACID ☐ DI-sec-BUTYLFLUOROPHOSPHATE ☐ T-1835 ☐ TL 1266

TOXICITY DATA WITH REFERENCE
ihl-man TCLo:1 ppm/5M:EYE,CNS,PUL JCSOA9 -,635,49
ihl-rat LC50:4 g/kg/m³/10M NTIS** PB158-508
ihl-mus LC50:540 mg/m³/10M JCSOA9 -,635,49
ihl dog LC50:4 g/m³/10M NTIS** PB-158-508
ihl-mky LC50:100 mg/m³//2M NTIS** PB158-508
ihl-cat LC50:6 g/m³/10M NTIS** PB158-508
ihl-rbt LC50:5 g/m³/10M NTIS** PB158-508

SAFETY PROFILE: Poison by inhalation. Human systemic effects by inhalation including: miosis (pupillary constriction), somnolence, and respiratory changes. When heated to decomposition it emits very toxic fumes of F^- and PO_x.

DEC400 CAS:761-65-9 HR: 3
N,N-DI-n-BUTYLFORMAMIDE
mf: $C_9H_{19}NO$ mw: 157.29

PROP: Liquid. Bp: 235°.

SYNS: DBF ☐ DIBUTYLAMID KYSELINY MRAVENCI ☐ N,N-DI-n-BUTYLFORMAMIDE

TOXICITY DATA WITH REFERENCE
skn-rat TDLo:1200 mg/kg (10D preg):TER TXAPA9 41,35,77
ipr-rat LD50:390 mg/kg TXAPA9 26,596,73
ipr-mus LD50:300 mg/kg TXAPA9 26,596,73

CONSENSUS REPORTS: Reported in EPA TSCA Inventory.

SAFETY PROFILE: Poison by intraperitoneal route. An experimental teratogen. When heated to decomposition it emits toxic fumes of NO_x.

DEC600 CAS:105-75-9 HR: 3
DIBUTYL FUMARATE
mf: $C_{12}H_{20}O_4$ mw: 228.32

PROP: Colorless, clear, mobile liquid; typical odor. Bp: 285.1°, fp: −19°, flash p: 300°F (OC), d: 0.986 @ 20°/20°, vap d: 7.88.

SYNS: DIBUTYLESTER KYSELINY FUMAROVE ☐ FUMARIC ACID, DIBUTYL ESTER

TOXICITY DATA WITH REFERENCE
skn-rbt 10 mg/24H open MLD AMIHBC 4,119,51
eye-rbt 500 mg open AMIHBC 4,119,51
orl-rat LD50:8530 mg/kg AMIHBC 4,119,51
ipr-mus LD50:250 mg/kg NTIS** AD691-490
skn-rbt LD50:16 g/kg AMIHBC 4,119,51

CONSENSUS REPORTS: Reported in EPA TSCA Inventory.

SAFETY PROFILE: Poison by intraperitoneal route. Mildly toxic by ingestion and skin contact. An eye, skin, and mucous membrane irritant. Combustible when exposed to heat or flame; can react with oxidizing materials. To fight fire, use foam, CO_2, dry chemical. When heated to decomposition it emits acrid smoke and fumes.

DEC699 CAS:4835-11-4 HR: 3
N,N′-DIBUTYLHEXAMETHYLENEDIAMINE
mf: $C_{14}H_{32}N_2$ mw: 228.48

SYNS: DBHMD □ DIBUTYLHEXAMETHYLENEDIAMINE □ N,N′-DI-BUTYL-1,6-HEXANEDIAMINE □ 1,6-N,N′-DIBUTYLHEXANEDIAMINE

TOXICITY DATA WITH REFERENCE
ihl-rat LD50:220 mg/m³/4H FCTOD7 22,425,84

CONSENSUS REPORTS: EPA Extremely Hazardous Substances List. Reported in EPA TSCA Inventory.

SAFETY PROFILE: Poison by inhalation. A corrosive alkali. A severe eye, skin, and mucous membrane irritant. Strong alkalies are markedly corrosive and penetrating to the skin and mucous membranes. Human systemic effects by ingestion: acute circulatory shock; burns in the mouth, throat, and esophagus; suffocation due to glottal or laryngeal swelling; perforation and inflammation of the esophagus and the tracheobronchial tree; aspiration pneumonia. Scar formation can cause delayed problems with swallowing, and stomach filling and emptying. The immediate symptoms of ingestion are: visible burns in mouth, drooling, gagging, vomiting, chest and upper abdominal pain, difficulty in breathing, or apnea (respiratory arrest); collapse and cardiac arrest may occur. Flammable or poisonous gases may accumulate in tanks or hopper cars. This material may react violently with water. To fight small fires, use dry chemical, carbon dioxide, water spray, or foam. To fight large fires, use water spray, fog, or foam. When heated to decomposition it emits toxic fumes of NO_x. See also AMINES.

DEC725 CAS:7422-80-2 HR: 2
1,1-DIBUTYLHYDRAZINE
mf: $C_8H_{20}N_2$ mw: 144.30

SYNS: 1,1-DBH □ N,N-DIBUTYLHYDRAZINE □ 1,1-DI-n-BUTYLHYDRAZINE

TOXICITY DATA WITH REFERENCE
mmo-sat 16,640 nmol/plate MUREAV 278,215,92 (-S9)
orl-mus TDLo:49,280 mg/kg/2Y-C:CAR CRNGDP 2,651,81

SAFETY PROFILE: Mutation data reported. Questionable carcinogen with experimental carcinogenic data. When heated to decomposition it emits toxic fumes of NO_x.

DEC775 CAS:78776-28-0 HR: 2
1,2-DI-n-BUTYLHYDRAZINE DIHYDROCHLORIDE
mf: $C_8H_{20}N_2•2ClH$ mw: 217.22

TOXICITY DATA WITH REFERENCE
orl-mus TDLo:92 g/kg/90W-C:CAR EXPEAM 37,773,81
orl-mus TD:142 g/kg/90W-C:CAR EXPEAM 37,773,81

SAFETY PROFILE: Questionable carcinogen with experimental carcinogenic data. When heated to decomposition it emits toxic fumes of NO_x and HCl.

DEC800 CAS:88-58-4 HR: 2
2,5-DI-t-BUTYLHYDROQUINONE
mf: $C_{14}H_{22}O_2$ mw: 222.36

SYNS: 2,5-DI-tert-BUTYLBENZENE-1,4-DIOL □ HYDROQUINONE, 2,5-DI-tert-BUTYL-

TOXICITY DATA WITH REFERENCE
orl-ham TDLo:134 g/kg/24W-C:NEO CRNGDP 12,1341,91
orl-rat LDLo:800 mg/kg KODAK* 21MAY71

CONSENSUS REPORTS: Reported in EPA TSCA Inventory.

SAFETY PROFILE: Moderately toxic by ingestion. Questionable carcinogen with experimental neoplastigenic data. When heated to decomposition it emits acrid smoke and irritating vapors.

DED000 CAS:10537-47-0 HR: 3
(3,5-DI-tert-BUTYL-4-HYDROXYBENZYLIDENE)MALONONITRILE
mf: $C_{18}H_{22}N_2O$ mw: 282.42

PROP: Crystals or solid. Mp: 140–141°.

SYNS: ((3,5-BIS(1,1-DIMETHYLETHYL)-4-HYDROXYPHENYL)METHYLENE)PROPANEDINITRILE □ 2-((3,5-BIS(1,1-DIMETHYL)-4-HYDROXYPHENYL)METHYLENE)PROPANEDINITRILE □ ENT 27,910 □ GCP 5126 □ GULF S-15126 □ MALONOBEN □ S-15126

TOXICITY DATA WITH REFERENCE
orl-rat LD50:87 mg/kg SPEADM 78-1,21,78
skn-rbt LD50:226 mg/kg SPEADM 74-1,-,74

CONSENSUS REPORTS: Cyanide and its compounds are on the Community Right-To-Know List.

SAFETY PROFILE: Poison by ingestion and skin contact. See also NITRILES. When heated to decomposition it emits toxic fumes of NO_x and CN^-.

DED400 CAS:2587-84-0 HR: 3
DIBUTYL LEAD DIACETATE
mf: $C_{12}H_{24}O_4Pb$ mw: 439.55

SYN: DIACETOXYDIBUTYLPLUMBANE

TOXICITY DATA WITH REFERENCE
orl-rat LD50:34 mg/kg JJATDK 1,247,81
ipr-rat LDLo:10 mg/kg CRSBAW 164,209,70
orl-mus LD50:115 mg/kg CRSBAW 162,1456,68
ipr-mus LD50:6 mg/kg CRSBAW 164,209,70
ivn-mus LD50:6 mg/kg CRSBAW 164,209,70
orl-dom LDLo:30 mg/kg REMVAY 22,85,69

CONSENSUS REPORTS: Lead and its compounds are on the Community Right-To-Know List.

SAFETY PROFILE: Poison by ingestion, intraperitoneal, and intravenous routes. See also LEAD COMPOUNDS. When heated to decomposition it emits toxic fumes of Pb.

DED500 CAS:6280-99-5 HR: 2
dl-DIBUTYL MALATE
mf: $C_{12}H_{22}O_5$ mw: 246.34

SYNS: BUTANEDIOIC ACID, HYDROXY-, DIBUTYL ESTER, (+-)- □ DIBUTYL (+-)-HYDROXYBUTANEDIOATE □ ENT-337 □ MALIC ACID, DIBUTYL ESTER, (+-)-

TOXICITY DATA WITH REFERENCE
orl-rat LD50:>9699 mg/kg NTIS** AD-A010-337
ihl-rat LC:>4 g/m³/8H NTIS** AD-A010-337
ipr-rat LD50:>1272 mg/kg NTIS** AD-A010-337

CONSENSUS REPORTS: Reported in EPA TSCA Inventory.

SAFETY PROFILE: Moderately toxic by intraperitoneal route. Low toxicity by ingestion and inhalation. When heated to decomposition it emits acrid smoke and irritating vapors.

DED600 CAS:105-76-0 HR: 3
DIBUTYL MALEATE
mf: $C_{12}H_{20}O_4$ mw: 228.32

PROP: Liquid. Mp: −85° (sets to a glass), bp: 281°, flash p: 285°F (OC), d: 0.9964 @ 20°/20°, vap d: 7.9.

SYNS: 2-BUTENEDIOIC ACID, DIBUTYL ESTER □ DBM □ MALEIC ACID, DIBUTLY ESTER □ RC COMONOMER DBM □ STAFLEX DBM

TOXICITY DATA WITH REFERENCE
skn-rbt 500 mg open MLD UCDS** 11/27/63
eye-rbt 500 mg open AMIHBC 10,61,54
orl-rat LD50:3730 mg/kg UCDS** 11/27/63
orl-mus LD50:2400 mg/kg ARZNAD 14,670,64
ipr-mus LD50:150 mg/kg NTIS** AD691-490
skn-rbt LD50:10 g/kg NPIRI* 2,19,75

CONSENSUS REPORTS: Reported in EPA TSCA Inventory.

SAFETY PROFILE: Poison by intraperitoneal route. Moderately toxic by ingestion. Mildly toxic by skin contact. An eye and skin irritant. See also ESTERS and n-BUTYL ALCOHOL. Combustible when exposed to heat or flame; can react with oxidizing materials. To fight fire, use foam, CO_2, dry chemical, alcohol foam. When heated to decomposition it emits acrid smoke and irritating fumes.

DED800 CAS:15535-69-0 HR: 3
DIBUTYLMALOYLOXYSTANNANE
mf: $C_{12}H_{22}O_5Sn$ mw: 365.03

SYN: DIBUTYLTIN MALATE

TOXICITY DATA WITH REFERENCE
ivn-mus LD50:56 mg/kg CSLNX* NX#03637

OSHA PEL: TWA 0.1 mg(Sn)/m³ (skin)
ACGIH TLV: TWA 0.1 mg(Sn)/m³ (skin) (Proposed: TWA 0.1 mg(Sn)/m³; STEL 0.2 mg(Sn)/m³ (skin))
NIOSH REL: (Organotin Compounds) TWA 0.1 mg(Sn)/m³

SAFETY PROFILE: Poison by intravenous route. See also TIN COMPOUNDS. When heated to decomposition it emits acrid smoke and irritating fumes.
For occupational chemical analysis use NIOSH: Organotin Compounds 5504.

DEE000 CAS:629-35-6 HR: 3
DIBUTYLMERCURY
mf: $C_8H_{18}Hg$ mw: 314.85

PROP: Liquid. Bp: 105° @ 10 mm, d: 1.779, vap d: 10.8. Sol most org solvs; insol in H_2O.

SYN: DIBUTYLRTUT

TOXICITY DATA WITH REFERENCE
ipr-mus LDLo:7800 μg/kg CBCCT* 4,230,52

CONSENSUS REPORTS: Mercury and its compounds are on the Community Right-To-Know List. Reported in EPA TSCA Inventory.

OSHA PEL: TWA 0.01 mg(Hg)/m³; STEL 0.03 mg/m³ (skin)
ACGIH TLV: TWA 0.01 mg(Hg)/m³; STEL 0.03 mg(Hg)/m³
NIOSH REL: (Mercury, Organo): TWA 0.01 mg/m³; STEL 0.03 mg/m³ (skin)

SAFETY PROFILE: Poison by intraperitoneal route. See also MERCURY COMPOUNDS, ORGANIC. Flammable when exposed to heat or flame. Can react vigorously with oxidizing materials. When heated to decomposition or on contact with acid or acid fumes it emits highly toxic fumes of mercury.

DEE200 CAS:691-88-3 HR: 3
DI-sec-BUTYLMERCURY
mf: $C_8H_{18}Hg$ mw: 314.85

PROP: Liquid; unstable in air and light. Bp: 93–96° @ 18 mm.

TOXICITY DATA WITH REFERENCE
ipr-mus LDLo:31 mg/kg CBCCT* 4,230,52

CONSENSUS REPORTS: Mercury and its compounds are on the Community Right-To-Know List.

OSHA PEL: TWA 0.01 mg(Hg)/m³; STEL 0.03 mg/m³ (skin)
ACGIH TLV: TWA 0.01 mg(Hg)/m³; STEL 0.03 mg(Hg)/m³
NIOSH REL: (Mercury, Organo): TWA 0.01 mg/m³; STEL 0.03 mg/m³ (skin)

SAFETY PROFILE: Poison by intraperitoneal route. See also MERCURY COMPOUNDS, ORGANIC. When heated to decomposition it emits toxic fumes of Hg.

DEE400 CAS:3405-45-6 HR: 3
N,N-DIBUTYLMETHYLAMINE
mf: $C_9H_{21}N$ mw: 143.31

PROP: Colorless liquid, amine odor. Insol in water; sol in alcohol and ether, miscible with hydrocarbons. D: 0.7613 @ 20°/20°, bp: 159.6°, fp: −62°, flash p: 125°F (OC).

TOXICITY DATA WITH REFERENCE
skn-rbt 10 mg/24H open MLD AIHAAP 23,95,62

D

orl-rat LD50:540 mg/kg AIHAAP 23,95,62
ihl-rbt LCLo:250 ppm/4H AIHAAP 23,95,62
skn-rbt LDLo:880 mg/kg AIHAAP 23,95,62

CONSENSUS REPORTS: Reported in EPA TSCA Inventory.

SAFETY PROFILE: Moderately toxic by ingestion, inhalation, and skin contact. A skin irritant. Flammable liquid when exposed to heat or flame. To fight fire, use dry chemical, fog, mist, CO_2. When heated to decomposition it emits toxic fumes of NO_x. See also AMINES.

DEE600 CAS:1301-14-0 HR: 3
2,6-DI-tert-BUTYLNAPHTHALENESULFONIC ACID SODIUM SALT
mf: $C_{18}H_{23}O_3S \cdot Na$ mw: 342.46

SYNS: BECANTAL □ BECANTEX □ BECANTYL □ 2,6-DI-tert-BUTYL NAPHTALENE SULFONATE SODIQUE (FRENCH) □ KEUTEN □ L. 1633 □ LINCTUSSAL □ SODIUM-2,6-DI-tert-BUTYLNAPHTHALENESULFONATE

TOXICITY DATA WITH REFERENCE
orl-rat LDLo:5000 mg/kg CLDND*
scu-gpg LDLo:250 mg/kg AIPTAK 97,34,54

SAFETY PROFILE: Poison by subcutaneous route. Mildly toxic by ingestion. When heated to decomposition it emits toxic fumes of SO_x and Na_2O. See also SULFONATES.

DEE800 CAS:728-40-5 HR: 3
2,6-DI-tert-BUTYL-4-NITROPHENOL
mf: $C_{14}H_{21}NO_3$ mw: 251.33

$[(CH_3)_3C]_2O_2NC_6H_2OH$

PROP: Yellow plates or needles from EtOH or pet ether. Mp: 157.5° (decomp).

SAFETY PROFILE: Explodes when heated to 100°C. May explode spontaneously. See also NITRO COMPOUNDS of AROMATIC HYDROCARBONS and PHENOLS.

DEF090 CAS:2406-25-9 HR: 3
DI-tert-BUTYL NITROXIDE
mf: $C_8H_{18}NO$ mw: 144.27

SYNS: DTBN □ NITROXIDE, BIS(1,1-DIMETHYLETHYL) (9CI) □ NITROXIDE, DI-tert-BUTYL

TOXICITY DATA WITH REFERENCE
orl-mus LD50:505 mg/kg JPETAB 141,349,63
ivn-mus LD50:53,800 µg/kg JPETAB 141,349,63

CONSENSUS REPORTS: Reported in EPA TSCA Inventory.

SAFETY PROFILE: Poison by intravenous route. Moderately toxic by ingestion. When heated to decomposition it emits toxic vapors of NO_x.

DEF150 CAS:27371-95-5 HR: 3
2,2-DIBUTYL-1,3,2-OXATHIASTANNOLANE
mf: $C_{10}H_{22}OSSn$ mw: 309.07

SYN: 1,3,2-OXATHIASTANNOLANE, 2,2-DIBUTYL-

TOXICITY DATA WITH REFERENCE
orl-rat LD50:60 mg/kg GTPZAB 32(9),50,88
skn-rat LD50:2500 mg/kg GTPZAB 32(9),50,88
orl-mus LD50:13,700 µg/kg GTPZAB 32(9),50,88
ivn-mus LD50:180 mg/kg CSLNX* NX#02078

OSHA PEL: TWA 0.1 mg(Sn)/m³ (skin)
ACGIH TLV: TWA 0.1 mg(Sn)/m³; STEL 0.2 mg/m³ (skin)
NIOSH REL: (Organotin Compounds): 10H TWA 0.1 mg(Sn)/m³

SAFETY PROFILE: Poison by ingestion and intravenous routes. Moderately toxic by skin contact. When heated to decomposition it emits toxic fumes of SO_x and Sn.

For occupational chemical analysis use NIOSH: Organotin Compounds 5504.

DEF200 CAS:78-20-6 HR: 2
2,2-DIBUTYL-1,3,2-OXATHIASTANNOLANE-5-OXIDE
mf: $C_{10}H_{20}O_2SSn$ mw: 323.05

SYNS: DIBUTYL(THIOACETOXY)STANNANE □ DI-n-BUTYLZINN THIOGLYKOLAT (GERMAN)

TOXICITY DATA WITH REFERENCE
orl-rat LD50:510 mg/kg TRIPA7 -,1,73

CONSENSUS REPORTS: Reported in EPA TSCA Inventory.

OSHA PEL: TWA 0.1 mg(Sn)/m³ (skin)
ACGIH TLV: TWA 0.1 mg(Sn)/m³ (skin) (Proposed: TWA 0.1 mg(Sn)/m³; STEL 0.2 mg(Sn)/m³ (skin))
NIOSH REL: (Organotin Compounds) TWA 0.1 mg(Sn)/m³

SAFETY PROFILE: Moderately toxic by ingestion. See also TIN COMPOUNDS. When heated to decomposition it emits toxic fumes of SO_x.

For occupational chemical analysis use NIOSH: Organotin Compounds 5504.

DEF400 CAS:818-08-6 HR: 3
DIBUTYLOXOSTANNANE
mf: $C_8H_{18}OSn$ mw: 248.95

PROP: White, amorphous powder or polymeric infusible solid. Mp: decomp without melting, bulk density: 0.5, vap d: 8.6.

SYNS: DBOT □ DIBUTYLOXIDE of TIN □ DIBUTYLOXOTIN □ DIBUTYLSTANNANE OXIDE □ DIBUTYLTIN OXIDE □ DI-n-BUTYLTIN OXIDE □ DI-n-BUTYL-ZINN-OXYD (GERMAN) □ KYSLICNIK DI-n-BUTYLCINICITY (CZECH)

TOXICITY DATA WITH REFERENCE
skn-rbt 500 mg/24H MLD 28ZPAK -,226,72
eye-rbt 100 mg/24H MOD 28ZPAK -,226,72
orl-rat LD50:44,900 µg/kg 28ZPAK -,226,72
ipr-rat LD50:40 mg/kg FCTXAV 7,47,69

orl-rbt LDLo: 1500 mg/kg SAIGBL 15,3,73

CONSENSUS REPORTS: Reported in EPA TSCA Inventory.

OSHA PEL: TWA 0.1 mg(Sn)/m^3 (skin)
ACGIH TLV: TWA 0.1 mg(Sn)/m^3 (skin) (Proposed: TWA 0.1 mg(Sn)/m^3; STEL 0.2 mg(Sn)/m^3 (skin))
NIOSH REL: (Organotin Compounds) TWA 0.1 mg(Sn)/m^3

SAFETY PROFILE: Poison by ingestion and intraperitoneal routes. A skin and eye irritant. Flammable when exposed to flame; can react with oxidizing materials. To fight fire, use dry chemical, fog, CO_2. When heated to decomposition it emits acrid smoke and irritating fumes. See also TIN COMPOUNDS.

For occupational chemical analysis use NIOSH: Organotin Compounds 5504.

DEF800 CAS:5510-99-6 HR: 3
2,6-DI-sec-BUTYLPHENOL
mf: $C_{14}H_{22}O$ mw: 206.36

PROP: Amber liquid. Bp: 152°–165° @ 25 mm, fp: −50°, flash p: 280°F, d: 0.936 @ 25°/4°.

SYN: 2,6-DI-sec-BUTYLFENOL (CZECH)

TOXICITY DATA WITH REFERENCE
skn-rbt 2 mg/24H SEV 85JCAE-,228,86
eye-rbt 50 μg/24H SEV 28ZPAK -,56,72
ivn-mus LD50: 60 mg/kg JMCMAR 23,1350,80
ivn-rbt LDLo: 10 mg/kg JMCMAR 23,1350,80

SAFETY PROFILE: Poison by intravenous route. A severe skin and eye irritant. Combustible when exposed to heat or flame; can react with oxidizing materials. To fight fire, use foam, CO_2, dry chemical. When heated to decomposition it emits acrid and irritating fumes. See also PHENOL.

DEG000 CAS:96-76-4 HR: 3
2,4-DI-tert-BUTYLPHENOL
mf: $C_{14}H_{22}O$ mw: 206.36

PROP: Tan crystals. Mp: 56.5°, bp: 263.5°, flash p: 265°F, d: 0.907 @ 60°/4°, vap press: 1 mm @ 84.5°.

SYNS: ANTIOXIDANT No. 33 □ PRODOX 146 □ PRODOX 146A-85X

TOXICITY DATA WITH REFERENCE
ipr-mus LD50: 25 mg/kg NTIS** AD691-490
ivn-mus LD50: 100 mg/kg JMCMAR 23,1350,80

CONSENSUS REPORTS: Reported in EPA TSCA Inventory.

SAFETY PROFILE: Poison by intraperitoneal and intravenous routes. Combustible when exposed to heat or flame. Can react with oxidizing materials. Violent reaction with HNO_3. To fight fire, use foam, CO_2, dry chemical. When heated to decomposition it emits acrid smoke and fumes. See also PHENOL.

DEG100 CAS:128-39-2 HR: 3
2,6-DI-tert-BUTYLPHENOL
mf: $C_{14}H_{22}O$ mw: 206.36

SYNS: 2,6-BIS(tert-BUTYL)PHENOL □ ETHANOX 701 □ PHENOL, 2,6-DI-tert-BUTYL-

TOXICITY DATA WITH REFERENCE
ivn-mus LD50: 120 mg/kg JMCMAR 23,1350,80

CONSENSUS REPORTS: Reported in EPA TSCA Inventory.

SAFETY PROFILE: Poison by intravenous route. When heated to decomposition it emits acrid smoke and irritating vapors.

DEG150 CAS:3286-98-4 HR: 2
4,6-DI-tert-BUTYL-α-PHENYL-o-CRESOL
mf: $C_{21}H_{28}O$ mw: 296.49

SYNS: AI 3-29183 □ o-CRESOL, 4,6-DI-tert-BUTYL-α-PHENYL- □ PHENOL, 2,4-BIS(1,1-DIMETHYLETHYL)-6-(PHENYLMETHYL)-

TOXICITY DATA WITH REFERENCE
orl-mus LD50: 3430 mg/kg JAFCAU 27,1007,79

CONSENSUS REPORTS: Reported in EPA TSCA Inventory.

SAFETY PROFILE: Moderately toxic by ingestion. When heated to decomposition it emits acrid smoke and irritating vapors.

DEG200 CAS:101-96-2 HR: 3
N,N′-DI-sec-BUTYL-p-PHENYLENEDIAMINE
mf: $C_{14}H_{24}N_2$ mw: 220.40

PROP: Liquid. Mp: 17.8°, flash p: 285°F (OC), d: 0.94–0.95 @ 24°/24°.

SYN: TENAMENE 2

TOXICITY DATA WITH REFERENCE
orl-rat LDLo: 200 mg/kg KODAK* -,71
ihl-rat LCLo: 600 mg/m^3/6H KODAK* -,71
skn-gpg LD50: 5000 mg/kg RCTEA4 45(3),627,72

CONSENSUS REPORTS: Reported in EPA TSCA Inventory.

SAFETY PROFILE: Poison by ingestion. Moderately toxic by inhalation and skin contact. Corrosive to skin. A mild allergen. Symptoms of exposure are sweating, flushing, shortness of breath, and slow pulse. Combustible when exposed to heat or flame; can react with oxidizing materials. To fight fire, use foam, CO_2, dry chemical. When heated to decomposition it emits toxic fumes of NO_x. See also AMINES.

DEG400 CAS:2655-19-8 HR: 2
3,5-DI-tert-BUTYLPHENYLMETHYLCARBAMATE
mf: $C_{16}H_{25}NO_2$ mw: 263.42

SYNS: BUTACARB □ BUTACARBE (FRENCH)

TOXICITY DATA WITH REFERENCE
orl-rat LD50: 1800 mg/kg SPEADM 78-1,57,78
orl-mus LD50: 3200 mg/kg SPEADM 78-1,57,78

orl-dog LD50:1000 mg/kg SPEADM 78-1,57,78

SAFETY PROFILE: Moderately toxic by ingestion. When heated to decomposition it emits toxic fumes of NO_x. See also CARBAMATES.

DEG600 CAS:2528-36-1 HR: 2
DIBUTYL PHENYL PHOSPHATE
mf: $C_{14}H_{23}O_4P$ mw: 286.34

SYN: PHOSPHORIC ACID, DIBUTYL PHENYL ESTER

TOXICITY DATA WITH REFERENCE
orl-rat TDLo:48 g/kg (male 10W pre):REP FAATDF 16,117,91
orl-rat LD50:2140 mg/kg GTPZAB 25(4),46,81
orl-mus LD50:1790 mg/kg GTPZAB 25(4),46,81

CONSENSUS REPORTS: Reported in EPA TSCA Inventory.

ACGIH TLV: TWA 0.3 ppm (skin)

SAFETY PROFILE: Moderately toxic by ingestion. Experimental reproductive effects. When heated to decomposition it emits toxic fumes of PO_x.

DEG700 CAS:107-66-4 HR: 2
DIBUTYL PHOSPHATE
mf: $C_8H_{19}PO_4$ mw: 210.2

PROP: Pale-amber liquid or oil. Bp: 135–138° @ 0.05 mm, decomp >100°. Sol in butanol and CCl_4.

SYNS: DIBUTYL ACID PHOSPHATE □ DIBUTYL HYDROGEN PHOSPHATE □ DIBUTYL PHOSPHATE □ DI-n-BUTYL PHOSPHATE

TOXICITY DATA WITH REFERENCE
orl-rat LD50:3200 mg/kg 14CYAT -,1918

CONSENSUS REPORTS: Reported in EPA TSCA Inventory.

OSHA PEL: TWA 1 ppm; STEL 2 ppm
ACGIH TLV: TWA 1 ppm; STEL 2 ppm

SAFETY PROFILE: Moderately toxic by ingestion. When heated to decomposition it emits toxic fumes of PO_x. See also PHOSPHATES.

For occupational chemical analysis use NIOSH: Dibutyl Phosphate, 5017.

DEG800 CAS:1809-19-4 HR: 3
DIBUTYL PHOSPHITE
mf: $C_8H_{19}O_3P$ mw: 194.24

PROP: A liquid. Bp: 116–117° @ 8 mm, flash p: 120°F, d: 0.995 @ 20°/4°, vap press: <1 mm @ 20°, vap d: 6.7.

SYNS: BUTYL ALCOHOL HYDROGEN PHOSPHITE □ DIBUTYL HYDROGEN PHOSPHITE □ MOBIL DBHP

TOXICITY DATA WITH REFERENCE
skn-rbt 10 mg/24H open MLD JIHTAB 31,60,49
eye-rbt 250 μg open SEV JIHTAB 31,60,49
orl-rat LD50:3200 mg/kg ALBRW* #OPB-3,84
skn-rbt LD50:1990 mg/kg JIHTAB 31,60,49

CONSENSUS REPORTS: Reported in EPA TSCA Inventory.

SAFETY PROFILE: Moderately toxic by ingestion and skin contact. A skin and severe eye irritant. Flammable liquid when exposed to heat or flame or by chemical reaction. Many phosphites decompose to evolve phosphine when heated. Explosion Hazard: See PHOSPHINE. Can react vigorously with oxidizing materials. To fight fire, use foam, CO_2, dry chemical. Dangerous; when heated to decomposition or on contact with acid or acid fumes it emits highly toxic fumes of PO_x.

DEH200 CAS:84-74-2 HR: 3
DIBUTYL PHTHALATE
mf: $C_{16}H_{22}O_4$ mw: 278.38

PROP: Oily liquid, mild odor. Mp: −35°, bp: 340°, flash p: 315°F (CC), d: 1.047–1.049 @ 20°/20°, autoign temp: 757°F, vap d: 9.58.

SYNS: BENZENE-o-DICARBOXYLIC ACID DI-n-BUTYL ESTER □ o-BENZENEDICARBOXYLIC ACID, DIBUTYL ESTER □ n-BUTYL PHTHALATE (DOT) □ CELLUFLEX DPB □ DBP □ DIBUTYL-1,2-BENZENEDICARBOXYLATE □ DI-n-BUTYL PHTHALATE □ ELAOL □ HEXAPLAS M/B □ PALATINOL C □ POLYCIZER DBP □ PX 104 □ RCRA WASTE NUMBER U069 □ STAFLEX DBP □ WITCIZER 300

TOXICITY DATA WITH REFERENCE
mmo-sat 100 μg/plate JTEHD6 16,61,85
cyt-ham:fbr 30 mg/L/24H MUREAV 48,337,77
ipr-rat TDLo:6 g/kg (female 3-9D post):REP EVHPAZ 3,91,73
orl-rat TDLo:2520 mg/kg (1-21D preg):TER TXAPA9 26,253,73
orl-rat TDLo:12,600 mg/kg (female 1-21D post):TER TXAPA9 26,253,73
orl-mus TDLo:20 g/kg (female 6-13D post):REP TCMUD8 7,29,87
ipr-rat TDLo:1017 mg/kg (female 5-15D post):REP JPMSAE 61,51,72
orl-mus TDLo:7200 mg/kg (female 1-18D post):REP OEKSDJ 8,29,77
orl-mus TDLo:16,800 mg/kg (male 7D pre):REP TOLED5 5,413,80
orl-rat TDLo:16,800 mg/kg (male 7D pre):REP NFGZAD 28,159,82
orl-rat TDLo:8400 μg/kg (7D male):REP TXAPA9 53,35,80
orl-mus TDLo:7200 mg/kg (female 1-18D post):TER OEKSDJ 8,29,77
ipr-rat TDLo:305 mg/kg (female 5-15D post):TER JPMSAE 61,51,72
orl-hmn TDLo:140 mg/kg:CNS,GIT,KID SMWOAS 84,1243,54
orl-rat LD50:8000 mg/kg FMCHA2 -,C76,83
ihl-rat LC50:4250 mg/m^3 GTPZAB 17(8),26,73
skn-rat LDLo:6 g/kg 85GMAT -,44,82
ipr-rat LD50:3050 mg/kg JPMSAE 61,51,72
orl-mus LD50:5289 mg/kg GTPZAB 17(11),51,73
ihl-mus LC50:25 g/m^3/2H 85GMAT -,44,82
ivn-mus LD50:720 mg/kg KEKHB8 (3),19,73

CONSENSUS REPORTS: On EPA Extremely Hazardous Substances List by error. On the Community Right-To-

Know List. EPA Genetic Toxicology Program. Reported in EPA TSCA Inventory.

OSHA PEL: TWA 5 mg/m³
ACGIH TLV: TWA 5 mg/m³

SAFETY PROFILE: Moderately toxic by intraperitoneal and intravenous routes. Mildly toxic by ingestion. Human systemic eye effects by ingestion; hallucinations; distorted perceptions; nausea or vomiting; and kidney, ureter, or bladder changes. Experimental teratogenic and reproductive effects. Mutation data reported. Combustible when exposed to heat or flame; can react with oxidizing materials. Violent reaction with Cl_2. Incompatible with chlorine. To fight fire, use CO_2, dry chemical. When heated to decomposition it emits acrid smoke and fumes. See also ESTERS, PHTHALIC ACID, and n-BUTYL ALCOHOL.

For occupational chemical analysis use NIOSH: Dibutyl Phthalate, 5020.

DEH300 CAS:1187-33-3 HR: 3
N,N-DIBUTYLPROPIONAMIDE
mf: $C_{11}H_{23}NO$ mw: 185.35

SYN: PROPIONAMIDE, N,N-DIBUTYL-

TOXICITY DATA WITH REFERENCE
ipr-mus LDLo:125 mg/kg CBCCT* 5,288,53

CONSENSUS REPORTS: Reported in EPA TSCA Inventory.

SAFETY PROFILE: Poison by intraperitoneal route. When heated to decomposition it emits toxic vapors of NO_x.

DEH600 CAS:109-43-3 HR: 1
DIBUTYL SEBACATE
mf: $C_{18}H_{34}O_4$ mw: 314.52

PROP: Clear liquid. Bp: 180° @ 3 mm, fp: −11°, flash p: 353°F (COC), d: 0.936 @ 20°/20°, vap d: 10.8.

SYNS: BIS(n-BUTYL)SEBACATE □ DECANEDIOIC ACID, DIBUTYL ESTER □ DI-n-BUTYL SEBACATE □ KODAFLEX DBS □ MONOPLEX DBS □ POLYCIZER DBS □ PX 404 □ SEBACIC ACID, DIBUTYL ESTER □ STAFLEX DBS

TOXICITY DATA WITH REFERENCE
orl-rat TDLo:418 g/kg (10W male/10D pre):REP AMIHBC 7,310,53
orl-rat LD50:16 g/kg NPIRI* 2,22,75

CONSENSUS REPORTS: Reported in EPA TSCA Inventory.

SAFETY PROFILE: Mildly toxic by ingestion. Experimental reproductive effects. Combustible liquid when exposed to heat or flame; can react with oxidizing materials. To fight fire, use CO_2, dry chemical. When heated to decomposition it emits acrid smoke and fumes. See also ESTERS and n-BUTYL ALCOHOL.

DEH650 CAS:7399-02-2 HR: 3
2,2′-((DIBUTYLSTANNYLENE)BIS(THIO))BISACETIC ACID DINONYL ESTER
mf: $C_{30}H_{60}O_4S_2Sn$ mw: 667.71

SYNS: ACETIC ACID, 2,2′-((DIBUTYLSTANNYLENE)BIS(THIO))BIS-, DINONYL ESTER □ ACETIC ACID, ((DIBUTYLSTANNYLENE)DITHIO) DI-, DINONYL ESTER (8CI) □ MELLITE 131 □ 8-OXA-3,5-DITHIA-4-STANNAHEPTADECANOIC ACID, 4,4-DIBUTYL-7-OXO-, NONYLESTER (9CI)

TOXICITY DATA WITH REFERENCE
orl-mus LD50:150 mg/kg ERNFA7 11,424,66

OSHA PEL: 8H TWA 0.1 mg(Sn)/m³ (skin)
ACGIH TLV: TWA 0.1 mg(Sn)/m³; STEL 0.2 mg/m³ (skin)
NIOSH REL: (Organotin Compounds) TWA 0.1 mg(Sn)/m³

SAFETY PROFILE: Poison by ingestion. When heated to decomposition it emits toxic fumes of SO_x and Sn.

For occupational chemical analysis use NIOSH: Organotin Compounds 5504.

DEH700 CAS:1962-75-0 HR: 2
DIBUTYL TEREPHTHALATE
mf: $C_{16}H_{22}O_4$ mw: 278.38

SYNS: 1,4-BENZENEDICARBOXYLIC ACID, DIBUTYL ESTER □ TEREPHTHALIC ACID, DIBUTYL ESTER

TOXICITY DATA WITH REFERENCE
ipr-mus LDLo:1392 mg/kg JPMSAE 56,1446,67

SAFETY PROFILE: Moderately toxic by intraperitoneal route. When heated to decomposition it emits acrid smoke and irritating vapors.

DEH800 CAS:23535-89-9 HR: 3
DIBUTYL(TETRACHLOROPHTHALATO)STANNANE
mf: $C_{16}H_{18}Cl_4O_4Sn$ mw: 534.83

SYNS: 3,3-DIBUTYL-6,7,8,9-TETRACHLORO-2,4,3-BENZODIOXASTANNEPIN-1,5-DIONE □ DIBUTYLTIN TETRACHLOROPHTHALATE

TOXICITY DATA WITH REFERENCE
ivn-mus LD50:180 mg/kg CSLNX* NX#02077

OSHA PEL: TWA 0.1 mg(Sn)/m³ (skin)
ACGIH TLV: TWA 0.1 mg(Sn)/m³ (skin) (Proposed: TWA 0.1 mg(Sn)/m³; STEL 0.2 mg(Sn)/m³ (skin))
NIOSH REL: (Organotin Compounds) TWA 0.1 mg(Sn)/m³

SAFETY PROFILE: Poison by intravenous route. See also TIN COMPOUNDS. When heated to decomposition it emits toxic fumes of Cl^-.

For occupational chemical analysis use NIOSH: Organotin Compounds 5504.

DEI000 CAS:109-46-6 HR: 3
1,3-DIBUTYLTHIOUREA
mf: $C_9H_{20}N_2S$ mw: 188.37

PROP: White to light tan powder or needles from alc. Mp: 78°, vap d: 6.5.

SYNS: N,N'-DIBUTYLTHIOUREA ☐ 1,3-DI-n-BUTYL-2-THIOUREA ☐ 1,3-DIBUTYL-2-THIOUREA ☐ PENNZONE B ☐ THIATE U ☐ USAF EK-2138

TOXICITY DATA WITH REFERENCE
orl-rat TDLo: 225 mg/kg (female 6-20D post): REP FAATDF 17,399,91
orl-rat TDLo: 375 mg/kg (female 6-20D post): REP FAATDF 17,399,91
orl-rat LD50: 350 mg/kg JPETAB 90,260,47
ipr-mus LD50: 800 mg/kg NTIS** AD277-689

CONSENSUS REPORTS: Reported in EPA TSCA Inventory.

SAFETY PROFILE: Poison by ingestion. Moderately toxic by intraperitoneal route. When heated to decomposition it emits very toxic fumes of NO_x and SO_x.

DEI200 CAS:4253-22-9 HR: 3
DIBUTYLTHIOXOSTANNANE
mf: $C_8H_{18}SSn$ mw: 265.01

SYNS: DIBUTYLTIN SULFIDE ☐ TIN DIBUTYL MERCAPTIDE

TOXICITY DATA WITH REFERENCE
cyt-rat-unr 100 µg/kg GISAAA 38(8),10,73
orl-rat LD50: 145 mg/kg UBZHD4 50,695,78
orl-mus LD50: 145 mg/kg UBZHD4 50,695,78

CONSENSUS REPORTS: Reported in EPA TSCA Inventory.

OSHA PEL: TWA 0.1 mg(Sn)/m^3 (skin)
ACGIH TLV: TWA 0.1 mg(Sn)/m^3 (skin) (Proposed: TWA 0.1 mg(Sn)/m^3; STEL 0.2 mg(Sn)/m^3 (skin))
NIOSH REL: (Organotin Compounds) TWA 0.1 mg(Sn)/m^3

SAFETY PROFILE: Poison by ingestion. Mutation data reported. See also TIN COMPOUNDS and SULFIDES. When heated to decomposition it emits toxic fumes of SO_x.

For occupational chemical analysis use NIOSH: Organotin Compounds 5504.

DEI400 CAS:73927-86-3 HR: 3
DI-n-BUTYLTIN BISMETHANESULFONATE
mf: $C_{10}H_{24}O_6S_2Sn$ mw: 423.15

SYN: BIS((METHYLSULFONYL)OXY)DIBUTYLSTANNANE

TOXICITY DATA WITH REFERENCE
ivn-mus LD50: 10 mg/kg CSLNX* NX#02276

OSHA PEL: TWA 0.1 mg(Sn)/m^3 (skin)
ACGIH TLV: TWA 0.1 mg(Sn)/m^3 (skin) (Proposed: TWA 0.1 mg(Sn)/m^3; STEL 0.2 mg(Sn)/m^3 (skin))
NIOSH REL: (Organotin Compounds) TWA 0.1 mg(Sn)/m^3

SAFETY PROFILE: Poison by intravenous route. See also TIN COMPOUNDS and SULFONATES. When heated to decomposition it emits toxic SO_x.

For occupational chemical analysis use NIOSH: Organotin Compounds 5504.

DEI600 CAS:19706-58-2 HR: 3
DI-n-BUTYL TIN DI(HEXADECYLMALEATE)
mf: $C_{48}H_{88}O_8Sn$ mw: 912.05

SYN: HEXADECYLMALEINAN DI-n-BUTYLCINICITY (CZECH)

TOXICITY DATA WITH REFERENCE
eye-rbt 100 mg/24H MOD 28ZPAK -,231,72
orl-rat LD50: 386 mg/kg 28ZPAK -,231,72

OSHA PEL: TWA 0.1 mg(Sn)/m^3 (skin)
ACGIH TLV: TWA 0.1 mg(Sn)/m^3 (skin) (Proposed: TWA 0.1 mg(Sn)/m^3; STEL 0.2 mg(Sn)/m^3 (skin))
NIOSH REL: (Organotin Compounds) TWA 0.1 mg(Sn)/m^3

SAFETY PROFILE: Poison by ingestion. An eye irritant. See also TIN COMPOUNDS. When heated to decomposition it emits acrid smoke and irritating fumes.

For occupational chemical analysis use NIOSH: Organotin Compounds 5504.

DEI800 CAS:69239-37-8 HR: 3
DI-n-BUTYLTIN DI(MONONONYL)MALEATE
mf: $C_{42}H_{76}O_8Sn$ mw: 827.87

SYNS: BIS(NONYLOXYMALEOYLOXY)DIOCTYLSTANNANE ☐ DI-n-BUTYL-ZINN-DI(MONONONYL)MALEINAT (GERMAN) ☐ DIOCTYLBIS (NONYLOXYMALEOYLOXY)STANNANE

TOXICITY DATA WITH REFERENCE
orl-rat LD50: 170 mg/kg ARZNAD 19,934,69

OSHA PEL: TWA 0.1 mg(Sn)/m^3 (skin)
ACGIH TLV: TWA 0.1 mg(Sn)/m^3 (skin) (Proposed: TWA 0.1 mg(Sn)/m^3; STEL 0.2 mg(Sn)/m^3 (skin))
NIOSH REL: (Organotin Compounds) TWA 0.1 mg(Sn)/m^3

SAFETY PROFILE: Poison by ingestion. See also TIN COMPOUNDS. When heated to decomposition it emits acrid smoke and irritating fumes.

For occupational chemical analysis use NIOSH: Organotin Compounds 5504.

DEJ000 CAS:13323-62-1 HR: 3
DIBUTYLTIN DIOLEATE
mf: $C_{44}H_{54}O_4Sn$ mw: 765.67

SYNS: BIS(OLEOYLOXY)DIBUTYLSTANNANE ☐ CN 447 ☐ DIBUTYLBIS(OLEOYLOXY)STANNANE ☐ DIBUTYLBIS((1-OXO-9-OCTADECENYL)OXY)STANNANE (Z,Z)

TOXICITY DATA WITH REFERENCE
ivn-mus LD50: 32 mg/kg CSLNX* NX#03563

CONSENSUS REPORTS: Reported in EPA TSCA Inventory.

OSHA PEL: TWA 0.1 mg(Sn)/m^3 (skin)
ACGIH TLV: TWA 0.1 mg(Sn)/m^3 (skin) (Proposed: TWA 0.1 mg(Sn)/m^3; STEL 0.2 mg(Sn)/m^3 (skin))
NIOSH REL: (Organotin Compounds) TWA 0.1 mg(Sn)/m^3

SAFETY PROFILE: Poison by intravenous route. See also TIN COMPOUNDS. When heated to decomposition it emits acrid smoke and irritating fumes.

DEJ100 **CAS:78-04-6** **HR: 2**
DIBUTYLTIN MALEATE
mf: $C_{12}H_{20}O_4Sn$ mw: 347.01

SYNS: ADVASTAB DBTM □ ADVASTAB T290 □ ADVASTAB T340 □ BT 31 □ 2,2-DIBUTYL-1,3,2-DIOXASTANNEPIN-4,7-DIONE □ DIBUTYL(MALEOYLDIOXY)TIN □ DIBUTYLSTANNYLENE MALEATE □ 1,3,2-DIOXASTANNEPIN-4,7-DIONE, 2,2-DIBUTYL- □ IRGasTAB T 4 □ IRGasTAB T 150 □ IRGasTAB T 290 □ KS 4B □ MA300A □ MARKURE UL2 □ NUODEX V 1525 □ STANCLERET 157 □ STANN RC 40F □ STAVINOR 1300SN □ STAVINOR SN 1300 □ TN 3J □ TVS-MA 300 □ TVS-N 2000E

TOXICITY DATA WITH REFERENCE
orl-mus LDLo:470 mg/kg AECTCV 14,111,85

CONSENSUS REPORTS: Reported in EPA TSCA Inventory.

SAFETY PROFILE: Moderately toxic by ingestion. When heated to decomposition it emits toxic fumes of Sn.

DEJ200 **CAS:78-06-8** **HR: 3**
DIBUTYLTIN MERCAPTOPROPIONATE
mf: $C_{11}H_{22}O_2SSn$ mw: 337.08

SYNS: 2,2-DIBUTYLDIHYDRO-6H-1,3,2-OXATHIASTANNIN-6-ONE □ 2,2-DIBUTYL-1-OXA-2-STANNA-3-THIACYCLOHEXAN-6-ONE □ DIBUTYLTIN-S,O-3-MERCAPTOPROPIONATE □ DIBUTYLTIN-O,S-MERCAPTOPROPIONATE □ DIBUTYLTIN-S,O-β-MERCAPTOPROPIONATE □ DIBUTYL(3-MERCAPTOPROPIONATO(2-))TIN □ MERCAPTOPROPIONIC ACID, DIBUTYLTIN SALT

TOXICITY DATA WITH REFERENCE
ivn-mus LD50:100 mg/kg CSLNX* NX#02852

CONSENSUS REPORTS: Reported in EPA TSCA Inventory.

OSHA PEL: TWA 0.1 mg(Sn)/m³ (skin)
ACGIH TLV: TWA 0.1 mg(Sn)/m³ (skin) (Proposed: TWA 0.1 mg(Sn)/m³; STEL 0.2 mg(Sn)/m³ (skin))
NIOSH REL: (Organotin Compounds) TWA 0.1 mg(Sn)/m³

SAFETY PROFILE: Poison by intravenous route. See also TIN COMPOUNDS and MERCAPTANS. When heated to decomposition it emits toxic fumes of SO_x.

For occupational chemical analysis use NIOSH: Organotin Compounds 5504.

DEJ400 **CAS:31052-46-7** **HR: 3**
DICAESIUM SELENIDE
mf: Cs_2Se mw: 344.78

SYNS: CESIUM SELENIDE □ DICESIUM SELENIDE

CONSENSUS REPORTS: Selenium and its compounds are on the Community Right-To-Know List.

OSHA PEL: TWA 0.2 mg(Se)/m³
ACGIH TLV: TWA 0.2 mg(Se)/m³
DFG MAK: 0.1 mg(Se)/m³

SAFETY PROFILE: Ignites in air when warmed. When heated to decomposition it emits toxic fumes of Se. See also SELENIUM COMPOUNDS.

DEJ600 **HR: 3**
DICARBADODECABORANYLMETHYLETHYL SULFIDE
mf: $C_5H_{18}B_{10}S$ mw: 218.39

SYN: CARBORANYLMETHYLETHYL SULFIDE

TOXICITY DATA WITH REFERENCE
skn-rbt 1% SEV NTIS** AD-A041-973
orl-rat TDLo:660 mg/kg (6-16D preg):TER AEHA** 51-044-74/76
orl-rat LD50:2085 mg/kg AEHA** 51-044-74/76
skn-rbt LD50:3890 mg/kg AEHA** 51-044-74/76
ivn-rbt LDLo:320 mg/kg AEHA** 51-044-74/76

SAFETY PROFILE: Poison by intravenous route. Moderately toxic by ingestion and skin contact. An experimental teratogen. A severe skin irritant. See also BORON COMPOUNDS. When heated to decomposition it emits toxic fumes of SO_x.

DEJ800 **HR: 3**
DICARBADODECABORANYLMETHYLPROPYL SULFIDE
mf: $C_6H_{20}B_{10}S$ mw: 232.42

SYN: CARBORANYLMETHYLPROPYL SULFIDE

TOXICITY DATA WITH REFERENCE
skn-rbt 1% SEV NTIS** AD-A041-973
orl-rat TDLo:2090 mg/kg (6-16D preg):TER AEHA** 51-044-74/76
orl-rat TDLo:2090 mg/kg (6-16D preg):REP AEHA** 51-044-74/76
orl-rat LD50:3440 mg/kg AEHA** 51-044-74/76
skn-rbt LD50:3160 mg/kg AEHA** 51-044-74/76
ivn-rbt LDLo:320 mg/kg AEHA** 51-044-74/76

SAFETY PROFILE: Poison by intravenous route. Moderately toxic by ingestion and skin contact. An experimental teratogen. Other experimental reproductive effects. A severe skin irritant. See also BORON COMPOUNDS. When heated to decomposition it emits toxic fumes of SO_x.

DEJ849 **CAS:68348-85-6** **HR: 3**
DICARBONYL MOLYBDENUM DIAZIDE
mf: $C_2MoN_6O_2$ mw: 236.00

SAFETY PROFILE: An extremely sensitive explosive. It may be initiated by touch or on contact with traces of water. When heated to decomposition it emits toxic fumes of NO_x. See also MOLYBDENUM COMPOUNDS, CARBONYLS, and AZIDES.

DEJ859 **HR: 3**
DICARBONYLPYRAZINE RHODIUM(1) PERCHLORATE
mf: $C_6H_4ClN_2O_6Rh$

$[(OC)_2RhC_4H_4N_2]_n\ [ClO4]_n$

SAFETY PROFILE: The complex explodes violently when heated. When heated to decomposition it emits toxic fumes of Cl^- and NO_x. See also PERCHLORATES, CARBONYLS, and RHODIUM.

DEJ880 **CAS:68379-32-8** **HR: 3**
DICARBONYLTUNGSTEN DIAZIDE
mf: $C_2N_6O_2W$ mw: 323.91

SAFETY PROFILE: An extremely sensitive explosive. It may be initiated by touch or on contact with traces of water. When heated to decomposition it emits toxic fumes of NO_x. See also TUNGSTEN COMPOUNDS, CARBONYLS, and AZIDES.

DEK000 **CAS:56455-90-4** **HR: 2**
DICARBOXIDINE HYDROCHLORIDE
mf: $C_{20}H_{24}N_2O_6 \cdot 2ClH$ mw: 461.38

SYNS: 4,4'-((4,4'-DIAMINO-(1,1'-BIPHENYL)-3,3'-DIYL)BIS(OXY)) BISBUTANOIC ACID, DIHYDROCHLORIDE □ HYDROCHLORIC ACID DICARBOXIDE

TOXICITY DATA WITH REFERENCE
scu-rat TDLo:21,250 mg/kg/2Y-I:ETA JJIND8 62,301,79
scu-rat TD:37,500 mg/kg/2Y-I:ETA VOONAW 25(7),43,79

SAFETY PROFILE: Questionable carcinogen with experimental tumorigenic data. When heated to decomposition it emits very toxic fumes of NO_x and HCl.

DEK200 **CAS:6362-79-4** **HR: 2**
3,5-DICARBOXYBENZENESULFONIC ACID, SODIUM SALT
mf: $C_8H_5O_7S \cdot Na$ mw: 268.18

SYN: 3,5-DIKARBOXYBENZENSULFONAN SODNY (CZECH)

TOXICITY DATA WITH REFERENCE
eye-rbt 20 mg/24H SEV 28ZPAK -,185,72
orl-rat LD50:6450 mg/kg 28ZPAK -,185,72

CONSENSUS REPORTS: Reported in EPA TSCA Inventory.

SAFETY PROFILE: Mildly toxic by ingestion. A severe eye irritant. When heated to decomposition it emits toxic fumes of SO_x and Na_2O.

DEK400 **CAS:73758-56-2** **HR: 2**
DICARBOXYDINE
mf: $C_{20}H_{24}N_2O_6$ mw: 388.46

SYNS: γ,γ'-,3,3'-BENZIDINE DIOXYDIBUTYRIC ACID □ 3,3'-BENZIDINE-γ,γ'-DIOXYDIBUTYRIC ACID □ 4,4'-(3,3'-DIAMINO-p,p'-BIPHENYLENEDIOXY)DIBUTYRIC ACID

TOXICITY DATA WITH REFERENCE
scu-rat TDLo:19 g/kg/2Y-I:ETA JJIND8 62,301,79

SAFETY PROFILE: Questionable carcinogen with experimental tumorigenic data. When heated to decomposition it emits toxic fumes of NO_x.

DEK600 **CAS:12014-93-6** **HR: 3**
DICERIUM TRISULFIDE
mf: Ce_2S_2 mw: 344.37

PROP: Red-brown (α-form), blood red (β-form) or cinnabar-colored (γ-form). Solid. Mp: 2060 (18°) (γ-form).

SYN: CERIUM TRISULFIDE

SAFETY PROFILE: The powder explodes spontaneously in air. When heated to decomposition it emits toxic fumes of SO_x. See also SULFIDES and CERIUM COMPOUNDS.

DEL000 **CAS:79-43-6** **HR: 2**
DICHLORACETIC ACID
DOT: UN 1764
mf: $C_2H_2Cl_2O_2$ mw: 128.94

PROP: Colorless, corrosive liquid; pungent odor. Mp (a): 10°, (b): −4°, bp: 194°, d: 1.5634 @ 20°/4°, vap press: 1 mm @ 44.0°, vap d: 4.45.

SYNS: BICHLORACETIC ACID □ DCA □ DICHLORETHANOIC ACID □ 2,2-DICHLOROACETIC ACID □ DICHLOROETHANOIC ACID □ KYSELINA DICHLOROCTOVA □ URNER'S LIQUID

TOXICITY DATA WITH REFERENCE
skn-rbt 10 mg/24H open MOD AMIHBC 4,119,51
skn-rbt 2 mg/24H SEV 85JCAE-,570,86
eye-rbt 50 μg open SEV AMIHBC 4,119,51
orl-mus TDLo:427 g/kg/61W-C:CAR TXAPA9 90,183,87
orl-rat LD50:2820 mg/kg AMIHBC 4,119,51
skn-rbt LD50:510 mg/kg AMIHBC 4,119,51

CONSENSUS REPORTS: Reported in EPA TSCA Inventory.

DOT CLASSIFICATION: 8; *Label:* Corrosive

SAFETY PROFILE: Moderately toxic by skin contact and ingestion. It is corrosive to the skin, eyes, and mucous membranes. Questionable carcinogen with experimental tumorigenic data. Will react with water or steam to produce toxic and corrosive fumes. When heated to decomposition it emits toxic fumes of Cl^-. See also CHLORIDES.

DEL200 **CAS:50264-69-2** **HR: 2**
1-(2,4-DICHLORBENZYL)INDAZOLE-3-CARBOXYLIC ACID
mf: $C_{15}H_{10}Cl_2N_2O_2$ mw: 321.17

PROP: Crystals from EtOH. Mp: 207°.

SYNS: AF 1890 □ DICA □ 1-(2,4-DICHLOROBENZYL)-1H-INDAZOLE-3-CARBOXYLIC ACID □ 1-((2,4-DICHLOROPHENYL)METHYL)-1H-INDAZOLE-3-CARBOXYLIC ACID □ DICLONDAZOLIC ACID □ LONIDAMINE

TOXICITY DATA WITH REFERENCE
spm-rat-orl 250 mg/kg/5D-C EXMPA6 23,357,75
spm-mky-orl 250 mg/kg/5D JRPFA4 52,275,78
orl-rat TDLo:90 mg/kg (female 6-15D post):TER ARTODN 5,197,82
orl-mky TDLo:250 mg/kg (male 5D pre):REP JRPFA4 52,275,78
orl-rat TDLo:35 mg/kg (1D male):REP PSSYDG 14,453,78
orl-rat TDLo:175 mg/kg (6-15D preg):TER ATSUDG 5,197,82
orl-rat TDLo:350 mg/kg (6-15D preg):TER ATSUDG 5,197,82
orl-rat LD50:1700 mg/kg CHTHBK 27,91,81
ipr-rat LD50:525 mg/kg CHTHBK 27,91,81

orl-mus LD50:900 mg/kg CHTHBK 27,91,81
ipr-mus LD50:435 mg/kg CHTHBK 27,91,81

SAFETY PROFILE: Moderately toxic by ingestion and intraperitoneal routes. Experimental teratogenic and reproductive effects. Mutation data reported. When heated to decomposition it emits very toxic fumes of Cl^- and NO_x.

DEL300 CAS:773-76-2 HR: 1
5,7-DICHLOR-8-HYDROXYCHINOLIN
mf: $C_9H_5Cl_2NO$ mw: 214.05

SYNS: CHLOFUCID □ CHLOROXINE □ CHLOROXYQUINOLINE □ CHLORQUINOL □ CHQ □ CLOFUZID □ DICHLOROHYDROXYQUINOLINE □ 5,7-DICHLORO-8-HYDROXYQUINOLINE □ 5,7-DICHLOROOXINE □ DICHLOROQUINOLINOL □ DICHLOROXIN □ 5,7-DICHLOROXINE □ ENDIARON □ QUESYL □ 8-QUINOLINOL, 5,7-DICHLORO- □ QUINOLOR □ QUIXALIN

TOXICITY DATA WITH REFERENCE
uns-bac-esc 100 μmol/L MUREAV 307,141,94
orl-cat LDLo:2 g/kg ARZNAD 22,1307,72

CONSENSUS REPORTS: Reported in EPA TSCA Inventory.

SAFETY PROFILE: Slightly toxic by ingestion. Mutation data reported. When heated to decomposition it emits toxic vapors of NO_x and Cl^-.

DEL600 CAS:7791-21-1 HR: 3
DICHLORINE OXIDE
mf: Cl_2O mw: 86.906

PROP: Reddish-brown liquid; can be stored in CCl_4. Mp: −120.6°, bp: 2°. Very sol in CCl_4; sol in H_2O with slow decomp to HOCl.

SAFETY PROFILE: The liquid at 2°C is an unstable spark- and touch-sensitive explosive. The gas may explode when heated above 42°C. A powerful oxidizing agent. Explodes on contact with alcohols, ammonia, antimony, antimony sulfide, arsenic, barium sulfide, calcium phosphide, carbon, carbon disulfide vapor, charcoal, cork, dicyanogen, ethers, hydrogen sulfide, mercury sulfide, nitrogen oxide, paper, phosphine, phosphorus, potassium, rubber, sulfur, tin sulfide, turpentine, and other oxidizable materials. Self-explodes. Incompatible with carbon, dicyanogen, diphenylmercury, nitrogen oxide, oxidizable materials, and potassium. Explosive reaction when heated above 50°C with many hydrocarbons (e.g., butadiene, ethane, ethylene, methane, propane).

DEL800 CAS:17496-59-2 HR: 3
DICHLORINE TRIOXIDE
mf: Cl_2O_3 mw: 118.91

PROP: Brown crystals solid at certain temp, slowly decomp at higher temp to Cl_2O_6.

SAFETY PROFILE: An unstable explosive gas. When heated to decomposition it emits toxic fumes of Cl^-.

DEM000 CAS:5571-97-1 HR: 3
DICHLORMETHAZANONE
mf: $C_{11}H_{11}Cl_2NO_3S$ mw: 308.19

PROP: A solid. Mp: 122.5–126.1°.

SYNS: DICHLORMEZANONE □ 2-(3,4-DICHLOROPHENYL)-3-METHYL-4-METATHIAZANONE-1,1-DIOXIDE □ 2-(3,4-DICHLOROPHENYL)TETRAHYDRO-3-METHYL-4H-1,3-THIAZIN-4-ONE-1,1-DIOXIDE □ WIN 12267

TOXICITY DATA WITH REFERENCE
orl-rat LD50:1050 mg/kg TXAPA9 1,168,59
orl-mus LD50:840 mg/kg JPETAB 122,517,57
ipr-mus LD50:570 mg/kg TXAPA9 1,168,59
orl-cat LD50:300 mg/kg TXAPA9 1,168,59
ipr-cat LD50:400 mg/kg TXAPA9 1,168,59

SAFETY PROFILE: Poison by ingestion and intraperitoneal routes. When heated to decomposition it emits very toxic fumes of Cl^-, SO_x, and NO_x.

DEM200 CAS:79-02-7 HR: 3
2,2-DICHLOROACETALDEHYDE
mf: $C_2H_2Cl_2O$ mw: 112.94

PROP: Colorless liquid, polymerizes slowly to white solid. Bp: 90–91°, fp: −50°, flash p: 140°F (CC), d: 1.436 @ 25°/4°. Vap press: 50 mm @ 20°, vap d: 3.9.

SYNS: CHLORALDEHYDE □ DICHLOROACETALDEHYDE □ α,α-DICHLOROACETALDEHYDE

TOXICITY DATA WITH REFERENCE
mmo-sat 10 mg/plate CBINA8 30,9,80
mmo-omi 10 μL/plate CBINA8 30,9,80
sln-asn 10 mmol/L MUREAV 138,33,84

CONSENSUS REPORTS: Reported in EPA TSCA Inventory.

SAFETY PROFILE: Mutation data reported. Flammable liquid when exposed to heat or flame. To fight fire, use water, foam, CO_2, dry chemical. When heated to decomposition it emits toxic fumes of Cl^-. See also ACETALDEHYDE and CHLORIDES.

DEM300 CAS:683-72-7 HR: 2
2,2-DICHLOROACETAMIDE
mf: $C_2H_3Cl_2NO$ mw: 127.96

SYNS: ACETAMIDE, DICHLORO- □ ACETAMIDE, 2,2-DICHLORO- (8CI,9CI) □ DICHLOROACETAMIDE

TOXICITY DATA WITH REFERENCE
ipr-mus LDLo:1750 mg/kg JACSAT 63,1437,41

CONSENSUS REPORTS: Reported in EPA TSCA Inventory.

SAFETY PROFILE: Moderately toxic by intraperitoneal route. When heated to decomposition it emits toxic vapors of NO_x and Cl^-.

D

DEM800 **CAS:116-54-1** **HR: 3**
DICHLOROACETIC ACID METHYL ESTER
DOT: UN 2299
mf: $C_3H_4Cl_2O_2$ mw: 142.97

PROP: Colorless liquid, ethereal odor. Bp: 143.0°, d: 1.3809 @ 19.2°/19.2°, vap d: 4.93.

SYNS: METHYL DICHLOROACETATE (DOT) □ METHYL DICHLOROETHANOATE

TOXICITY DATA WITH REFERENCE
ihl-cat LCLo: 2000 ppm/30M TXAPA9 19,1,71

CONSENSUS REPORTS: Reported in EPA TSCA Inventory.

DOT CLASSIFICATION: 6.1; *Label:* KEEP AWAY FROM FOOD

SAFETY PROFILE: Poisonous irritant to the skin, eyes, and mucous membranes. Hydrolyzes upon contact with moisture to form a product corrosive to tissue. See also DICHLOROACETIC ACID and ESTERS. Dangerous; when heated to decomposition it emits highly toxic fumes of phosgene and Cl^-.

DEM825 **CAS:4124-30-5** **HR: 2**
DICHLOROACETIC ANHYDRIDE
mf: $C_4H_2Cl_4O_3$ mw: 239.86

PROP: Liquid. Bp: 215–216° (decomp).

TOXICITY DATA WITH REFERENCE
skn-rbt 10 mg/24H open MOD AMIHBC 4,119,51
eye-rbt 50 µg open SEV AMIHBC 4,119,51
orl-rat LD50: 2820 mg/kg AMIHBC 4,119,51
skn-rbt LD50: 470 mg/kg AMIHBC 4,119,51

CONSENSUS REPORTS: Reported in EPA TSCA Inventory.

SAFETY PROFILE: Moderately toxic by ingestion and skin contact. A skin and severe eye irritant. When heated to decomposition it emits toxic fumes of Cl^-. See also ANHYDRIDES and CHLORIDES.

DEN000 **CAS:3018-12-0** **HR: 3**
DICHLOROACETONITRILE
mf: C_2HCl_2N mw: 109.94

PROP: Liquid. D: 1.374 @ 11.5 mm, bp: 113°.

SYN: DICHLOROMETHYL CYANIDE

TOXICITY DATA WITH REFERENCE
mmo-sat 1 nmol/plate ENMUDM 5,447,83
dnd-hmn: lym 50 µmol/L FAATDF 6,447,86
orl-rat TDLo: 825 mg/kg (female 7-21D post): REP TXCYAC 46,83,87
orl-rat TDLo: 325 mg/kg (female 6-18D post): TER TJADAB 35,58A,87
orl-rat LD50: 330 mg/kg EVHPAZ 69,183,86
orl-mus LD50: 270 mg/kg EVHPAZ 69,183,86

CONSENSUS REPORTS: IARC Cancer Review: Group 3 IMEMDT 52,269,91; Animal Inadequate Evidence IMEMDT 52,269,91; Human No Available Data IMEMDT 52,269,91. Cyanide and its compounds are on the Community Right-To-Know List. EPA Genetic Toxicology Program.

SAFETY PROFILE: Poison by ingestion. An experimental teratogen. Other experimental reproductive effects. Human mutation data reported. When heated to decomposition it emits toxic fumes of Cl^-, CN^-, and NO_x. See also NITRILES and CHLORIDES.

DEN200 **CAS:2648-61-5** **HR: 3**
2,2-DICHLOROACETOPHENONE
mf: $C_8H_6Cl_2O$ mw: 189.04

PROP: Liquid or crystals. Mp: 21°, bp: 249° (decomp), d: 1.34 @ 15°, vap d: 6.5.

SYNS: α,α-DICHLOROACETOPHENONE □ ω,ω-DICHLOROACETOPHENONE □ PHENACYLIDENE CHLORIDE

TOXICITY DATA WITH REFERENCE
ihl-mus LCLo: 940 mg/m³/10M NDRC** NDCrc-132,Aug,42
ivn-mus LD50: 100 mg/kg CSLNX* NX#03021

CONSENSUS REPORTS: Reported in EPA TSCA Inventory.

SAFETY PROFILE: Poison by intravenous route. Moderately toxic by inhalation. When heated to decomposition it emits toxic fumes of Cl^-.

DEN400 **CAS:79-36-7** **HR: 2**
DICHLOROACETYL CHLORIDE
DOT: UN 1765
mf: C_2HCl_3O mw: 147.38

PROP: Fuming liquid, acrid odor, misc in ether. D: 1.5315 @ 16°/4°, bp: 108°, flash p: 151°F, vap d: 5.8.

SYNS: CHLORID KYSELINY DICHLOROCTOVE □ CHLORURE de DICHLORACETYLE (FRENCH) □ DICHLORACETYL CHLORIDE □ α,α-DICHLOROACETYL CHLORIDE □ 2,2-DICHLOROACETYL CHLORIDE □ DICHLOROACETYL CHLORIDE (DOT) □ DICHLOROETHANOYL CHLORIDE

TOXICITY DATA WITH REFERENCE
skn-rbt 2 mg/24H SEV 85JCAE-,571,86
eye-rbt 50 µg open SEV AMIHBC 4,119,51
scu-mus TDLo: 2 mg/kg/80W-I: ETA CNREA8 43,159,83
orl-rat LD50: 2460 mg/kg AMIHBC 4,119,51
ihl-rat LCLo: 2000 ppm/4H AMIHBC 4,119,51
skn-rbt LD50: 650 mg/kg AMIHBC 4,119,51

CONSENSUS REPORTS: Reported in EPA TSCA Inventory.

DOT CLASSIFICATION: 8; *Label:* Corrosive

SAFETY PROFILE: Questionable carcinogen with experimental tumorigenic data. Moderately toxic by ingestion, inhalation, and skin contact. Corrosive to the skin, eyes, and mucous membranes. Flammable when exposed to heat or flame. When heated to decomposition it emits toxic fumes of Cl^-. See also CHLORIDES.

DEN600 CAS:7572-29-4 HR: 3
DICHLOROACETYLENE
mf: C_2Cl_2 mw: 94.92

PROP: Volatile liquid. Mp: −66 to −64°, bp: 33°.

SYNS: DICHLOROETHYNE □ ETHYNE, DICHLORO-(9CI)

TOXICITY DATA WITH REFERENCE
mmo-sat 4000 ppm MUREAV 117,21,83
mma-sat 4000 ppm MUREAV 117,21,83
ihl-rat TCLo:14 ppm/6H/77W-I:CAR CRNGDP 5,1411,84
ihl-mus TCLo:2 ppm/24H/77W-I:CAR CRNGDP 5,1411,84
ihl-mus LC50:19 ppm/6H FCTXAV 13,511,75
ihl-rbt LCLo:307 ppm/1H FCTXAV 16,227,78

CONSENSUS REPORTS: IARC Cancer Review: Group 3 IMEMDT 7,56,87; Animal Limited Evidence IMEMDT 39,369,86.

OSHA PEL: CL 0.1 ppm
ACGIH TLV: CL 0.1 ppm
DFG MAK: Animal Carcinogen, Suspected Human Carcinogen.
DOT CLASSIFICATION: Forbidden

SAFETY PROFILE: Confirmed carcinogen with experimental carcinogenic data. Poison by inhalation. Central nervous system effects. Can be formed by thermal decomposition (>70°) from trichloroethylene. Symptoms include a disabling nausea and intense jaw pain. Strong explosive when shocked or exposed to heat or air. Can react vigorously with oxidizing materials. When heated to decomposition or on contact with acid or acid fumes it emits highly toxic fumes of Cl^-. See also ACETYLENE COMPOUNDS and CHLORINATED HYDROCARBONS, ALIPHATIC.

DEN820 CAS:24518-45-4 HR: 3
2-((4-(DICHLOROACETYL)PHENYL)AMINO)-2-ETHOXY-1-(4-NITROPHENYL)ETHANONE
mf: $C_{18}H_{16}Cl_2N_2O_5$ mw: 411.26

SYNS: ETHANONE, 2-((4-(DICHLOROACETYL)PHENYL)AMINO)-2-ETHOXY-1-(4-NITROPHENYL)- □ KETONE, 2-((4-(DICHLOROACETYL)PHENYL)AMINO)-2-ETHOXY-1-(4-NITROPHENYL)-

TOXICITY DATA WITH REFERENCE
ipr-mus LD50:3 g/kg ARZNAD 23,573,73

DOT CLASSIFICATION: 3; *Label:* Flammable Liquid

SAFETY PROFILE: Moderately toxic by intraperitoneal route. A flammable liquid. When heated to decomposition it emits toxic vapors of NO_x and Cl^-.

DEN840 CAS:27695-59-6 HR: 3
2-((4-(DICHLOROACETYL)PHENYL)AMINO)-2-HYDROXY-1-(4-METHOXYPHENYL)ETHANONE
mf: $C_{17}H_{15}Cl_2NO_4$ mw: 368.23

SYNS: ETHANONE, 2-((4-(DICHLOROACETYL)PHENYL)AMINO)-2-HYDROXY-1-(4-METHOXYPHENYL)- □ KETONE, 2-((4-(DICHLOROACETYL)PHENYL)AMINO)-2-HYDROXY-1-(4-METHOXYPHENYL)-

TOXICITY DATA WITH REFERENCE
ipr-mus LD50:560 mg/kg ARZNAD 23,573,73

DOT CLASSIFICATION: 3; *Label:* Flammable Liquid

SAFETY PROFILE: Moderately toxic by intraperitoneal route. A flammable liquid. When heated to decomposition it emits toxic vapors of NO_x and Cl^-.

DEN860 CAS:27695-58-5 HR: 3
2-((4-(DICHLOROACETYL)PHENYL)AMINO)-2-HYDROXY-1-(4-METHYLPHENYL)ETHANONE
mf: $C_{17}H_{15}Cl_2NO_3$ mw: 352.23

SYNS: ETHANONE, 2-((4-(DICHLOROACETYL)PHENYL)AMINO)-2-HYDROXY-1-(4-METHYLPHENYL)- □ KETONE, 2-((4-(DICHLOROACETYL)PHENYL)AMINO)-2-HYDROXY-1-(4-METHYLPHENYL)-

TOXICITY DATA WITH REFERENCE
ipr-mus LD50:2500 mg/kg ARZNAD 23,573,73

DOT CLASSIFICATION: 3; *Label:* Flammable Liquid

SAFETY PROFILE: Moderately toxic by intraperitoneal route. A flammable liquid. When heated to decomposition it emits toxic vapors of NO_x and Cl^-.

DEN880 CAS:27695-60-9 HR: 3
2-((4-(DICHLOROACETYL)PHENYL)AMINO)-2-HYDROXY-1-(4-PHENOXYPHENYL)ETHANONE
mf: $C_{22}H_{17}Cl_2NO_4$ mw: 430.30

SYNS: ETHANONE, 2-((4-(DICHLOROACETYL)PHENYL)AMINO)-2-HYDROXY-1-(4-PHENOXYPHENYL)- □ KETONE, 2-((4-(DICHLOROACETYL)PHENYL)AMINO)-2-HYDROXY-1-(4-PHENOXYPHENYL)-

TOXICITY DATA WITH REFERENCE
ipr-mus LD50:250 mg/kg ARZNAD 23,573,73

DOT CLASSIFICATION: 3; *Label:* Flammable Liquid

SAFETY PROFILE: A poison by intraperitoneal route. A flammable liquid. When heated to decomposition it emits toxic vapors of NO_x and Cl^-.

DEN900 CAS:27695-57-4 HR: 3
2-((4-(DICHLOROACETYL)PHENYL)AMINO)-2-HYDROXY-1-PHENYLETHANONE
mf: $C_{16}H_{13}Cl_2NO_3$ mw: 338.20

SYNS: ETHANONE, 2-((4-(DICHLOROACETYL)PHENYL)AMINO)-2-HYDROXY-1-PHENYL- □ KETONE, 2-((4-(DICHLOROACETYL)PHENYL)AMINO)-2-HYDROXY-1-PHENYL-

TOXICITY DATA WITH REFERENCE
ipr-mus LD50:1300 mg/kg ARZNAD 23,573,73

DOT CLASSIFICATION: 3; *Label:* Flammable Liquid

SAFETY PROFILE: Moderately toxic by intraperitoneal route. A flammable liquid. When heated to decomposition it emits toxic vapors of NO_x and Cl^-.

DEN910 CAS:27700-43-2 HR: 3
2-((4-(DICHLOROACETYL)PHENYL)AMINO)-2-HYDROXY-1-(4-(PHENYLTHIO)PHENYL)ET HANONE
mf: $C_{22}H_{17}Cl_2NO_3S$ mw: 446.36

SYNS: ETHANONE, 2-((4-(DICHLOROACETYL)PHENYL)AMINO)-2-HYDROXY-1-(4-(PHENYLTHIO)PHENYL)- □ KETONE, 2-((4-(DICHLOROACETYL)PHENYL)AMINO)-2-HYDROXY-1-(4-(PHENYLTHIO)PHENYL)-

D

TOXICITY DATA WITH REFERENCE
ipr-mus LD50:2500 mg/kg ARZNAD 23,573,73

DOT CLASSIFICATION: 3; *Label:* Flammable Liquid

SAFETY PROFILE: Moderately toxic by intraperitoneal route. A flammable liquid. When heated to decomposition it emits toxic vapors of NO_x, SO_x, and Cl^-.

DEO290 CAS:554-00-7 HR: 3
2,4-DICHLOROANILINE
mf: $C_6H_5Cl_2N$ mw: 162.02

SYNS: ANILINE, 2,4-DICHLORO- □ BENZENAMINE, 2,4-DICHLORO- (9CI) □ 2,4-DICHLORANILIN □ 2,4-DICHLOROBENZENAMINE

TOXICITY DATA WITH REFERENCE
orl-rat LD50:1600 mg/kg TSCAT* OTS 206512
ipr-rat LD50:400 mg/kg TSCAT* OTS 206512
orl-mus LD50:400 mg/kg TSCAT* OTS 206512
ipr-mus LD50:400 mg/kg TSCAT* OTS 206512
orl-cat LDLo:113 mg/kg AEXPBL 72,241,13

CONSENSUS REPORTS: Reported in EPA TSCA Inventory.

SAFETY PROFILE: Poison by ingestion and intraperitoneal routes. When heated to decomposition it emits toxic vapors of NO_x and Cl^-.

DEO295 CAS:95-82-9 HR: 3
2,5-DICHLOROANILINE
mf: $C_6H_5Cl_2N$ mw: 162.02

PROP: Needles from ligroin. Mp: 50°, bp: 251°.

SYNS: AMARTHOL FAST SCARLETT GG BASE □ AZOBASE DCA □ AZOEN FAST SCARLET 2G BASE □ AZOFIX SCARLET GG SALT □ C.I. 37010 □ C.I. AZOIC DIAZO COMPONENT 3 □ DEVOL SCARLET A (FREE BASE) □ 2,5-DICHLOROANILIN (CZECH) □ 2,5-DICHLOROBENZENEAMINE □ DURGASOL SCARLET GG SALT □ FAST RED SGG BASE □ HILTONIL FAST SCARLET 2G BASE □ HILTOSAL FAST SCARLET 2G SALT □ HINDAMINE SCARLET GG □ KAKO SCARLET GG SALT □ KAMBAMINE SCARLET GG BASE □ KAYAKU SCARLET GG BASE □ LAKE SCARLET GG BASE □ MEISEI SCARLET GG SALT □ MITUSI SCARLET GG BASE □ NAPHTHANIL SCARLET 2G BASE □ NAPHTOELAN MITSUI SCARLET GG SALT □ SANYO FAST SCARLET GG BASE □ SCARLET BASE CIBA I □ SPECTROLENE SCARLET 2G □ SYMULON SCARLET 2G SALT

TOXICITY DATA WITH REFERENCE
orl-rat LD50:1600 mg/kg TSCAT* OTS 206512
ipr-rat LD50:400 mg/kg TSCAT* OTS 206512
orl-mus LD50:1600 mg/kg TSCAT* OTS 206512
ipr-mus LD50:400 mg/kg TSCAT* OTS 206512
ivn-mus LD50:56 mg/kg CSLNX* NX#00202

CONSENSUS REPORTS: Reported in EPA TSCA Inventory.

SAFETY PROFILE: Poison by intraperitoneal and intravenous route. Moderately toxic by ingestion. Explodes spontaneously. When heated to decomposition it emits highly toxic fumes of Cl^- and NO_x. See also ANILINE and CHLORIDES.

DEO300 CAS:95-76-1 HR: 3
3,4-DICHLOROANILINE
mf: $C_6H_5Cl_2N$ mw: 162.02

PROP: Crystals or needles from ligroin. Mp: 71–72°, bp: 144–146° @ 15 mm. Practically insol in water; very sol in alcohol, ether; sltly sol in benzene.

SYNS: 1-AMINO-3,4-DICHLOROBENZENE □ DCA □ 3,4-DCA □ 3,4-DICHLORANILIN □ 3,4-DICHLORANILINE □ 4,5-DICHLOROANILINE □ 3,4-DICHLOROBENZENAMINE (9CI)

TOXICITY DATA WITH REFERENCE
skn-rbt 2 mg/24H SEV 85JCAE-,612,86
eye-rbt 250 μg/24H SEV 28ZPAK -,96,72
mmo-asn 200 mg/L CJMIAZ 16,369,70
orl-rat LD50:648 mg/kg 28ZPAK -,96,72
ihl-rat LCLo:65 mg/m³/4H TSCAT* OTS 215198
ipr-rat LD50:280 mg/kg LPPTAK 27,306,79
orl-mus LD50:740 mg/kg GTPZAB 13(5),29,69
ipr-mus LD50:310 mg/kg LPPTAK 27,306,79
skn-cat LD50:700 mg/kg GTPZAB 13(5),29,69
orl-bwd LD50:237 mg/kg AECTCV 12,355,83

CONSENSUS REPORTS: EPA Genetic Toxicology Program. Reported in EPA TSCA Inventory.

SAFETY PROFILE: Poison by ingestion and intraperitoneal routes. Moderately toxic by skin contact. A severe eye and skin irritant. Mutation data reported. When heated to decomposition it emits toxic fumes of Cl^- and NO_x. See also ANILINE DYES and CHLORIDES.

DEO500 CAS:70278-00-1 HR: 3
N,N-DICHLOROANILINE
mf: $C_6H_5CL_2N$ mw: 162.02

SAFETY PROFILE: A poison. An oil which explodes spontaneously at room temperature. When heated to decomposition it emits toxic fumes of Cl^- and NO_x. See also ANILINE and CHLORIDES.

DEO600 CAS:15307-79-6 HR: 3
(o-((2,6-DICHLOROANILINO)PHENYL))ACETIC ACID SODIUM SALT
mf: $C_{14}H_{10}Cl_2NO_2{\cdot}Na$ mw: 318.14

PROP: Crystals from H_2O. Mp: 283–285°.

SYNS: (o-(2,6-DICHLOROANILINO)PHENYL)ACETIC ACID MONOSODIUM SALT □ 2-((2,6-DICHLOROPHENYL)AMINO)BENZENEACETIC ACID MONOSODIUM SALT □ DICHRONIC □ DICLOFENAC SODIUM □ DICLOPHENAC SODIUM □ GP 45840 □ KRIPLEX □ NERIODIN □ PROPHENATIN □ SODIUM (o-(2,6-DICHLOROANILINO)PHENYL)ACETATE □ SODIUM (o-((2,6-DICHLOROPHENYL)AMINO)PHENYL)ACETATE □ TSUDOHMIN □ VALETAN □ VOLTAREN □ VOLTAROL

TOXICITY DATA WITH REFERENCE
orl-rat TDLo:24 mg/kg (female 9-14D post):REP KSRNAM 6,1673,72
orl-rat TDLo:6 mg/kg (9-14D preg):TER KSRNAM 6,1673,72
orl-rbt TDLo:10 mg/kg (female 1D pre):REP FESTAS 38,238,82
orl-rat TDLo:312 mg/kg (22W male):REP KSRNAM 6,1521,72

orl-rat TDLo: 1 mg/kg (21D preg): TER OYYAA2 27,117,84
orl-wmn TDLo: 180 mg/kg/13W-I BMJOAE 295,182,87
orl-wmn TDLo: 270 mg/kg/90D-I: GIT,BLD AIMDAP 152,625,92
orl-wmn TDLo: 112 mg/kg/8W-I: GIT,SKN AIMDAP 152,625,92
orl-rat LD50: 53 mg/kg TOIZAG 28,99,81
ipr-rat LD50: 25 mg/kg NIIRDN 6,311,82
scu-rat LD50: 83 mg/kg IYKEDH 5,106,74
ivn-rat LD50: 117 mg/kg IYKEDH 5,106,74
orl-mus LD50: 125 mg/kg ARZNAD 34,280,84
ipr-mus LD50: 130 mg/kg IYKEDH 5,106,74
scu-mus LD50: 390 mg/kg NIIRDN 6,311,82
ivn-mus LD50: 116 mg/kg IYKEDH 5,106,74
orl-dog LD50: 59 mg/kg KSRNAM 6,1521,72

SAFETY PROFILE: Poison by ingestion, intravenous, intraperitoneal, and subcutaneous routes. Experimental teratogenic and reproductive effects. Human systemic effects by ingestion: changes in erythrocyte (RBC) count, dermatitis, diarrhea, hypermotility, ulceration or bleeding from large intestine. An anti-inflammatory agent. When heated to decomposition it emits very toxic fumes of Cl^-, Na_2O, and NO_x.

DEO700 CAS:82-46-2 HR: 1
1,5-DICHLORO-9,10-ANTHRAQUINONE
mf: $C_{14}H_6Cl_2O_2$ mw: 277.10

PROP: Yellow needles from AcOH. Mp: 245°.

SYNS: 9,10-ANTHRACENEDIONE, 1,5-DICHLORO- □ 1,5-DICHLORANTHRACHINON □ 1,5-DICHLOROANTHRAQUINONE

TOXICITY DATA WITH REFERENCE
eye-rbt 500 mg/24H MLD 85JCAE-,566,86

CONSENSUS REPORTS: Reported in EPA TSCA Inventory.

SAFETY PROFILE: An eye irritant. When heated to decomposition it emits toxic fumes of Cl^-.

DEO750 CAS:82-43-9 HR: 1
1,8-DICHLORO-9,10-ANTHRAQUINONE
mf: $C_{14}H_6Cl_2O_2$ mw: 277.10

PROP: Pale-yellow needles from $PhNO_2$. Mp: 201–202°.

SYNS: 9,10-ANTHRACENEDIONE, 1,8-DICHLORO- □ 1,8-DICHLORANTHRACHINON □ 1,8-DICHLOROANTHRAQUINONE

TOXICITY DATA WITH REFERENCE
eye-rbt 500 mg/24H MLD 85JCAE-,566,86

CONSENSUS REPORTS: Reported in EPA TSCA Inventory.

SAFETY PROFILE: An eye irritant. When heated to decomposition it emits toxic fumes of Cl^-.

DEP400 CAS:63834-20-8 HR: 3
2-DICHLOROARSINOPHENOXATHIIN
mf: $C_{12}H_7AsCl_2OS$ mw: 345.07

SYN: TL 472

TOXICITY DATA WITH REFERENCE
orl-rat LDLo: 250 mg/kg NCNSA6 5,13,53
ihl-mus LCLo: 400 mg/m³/10M NDRC** NDCrc-132,Dec,42

CONSENSUS REPORTS: Arsenic and its compounds are on the Community Right-To-Know List.

OSHA PEL: TWA 0.5 mg(As)/m³

SAFETY PROFILE: Poison by ingestion and inhalation. See also ARSENIC COMPOUNDS. When heated to decomposition it emits very toxic fumes of As, Cl^-, and SO_x.

DEP599 CAS:541-73-1 HR: 2
m-DICHLOROBENZENE
mf: $C_6H_4Cl_2$ mw: 147.00

PROP: Liquid. D: 1.288 @ 20°/4°, fp −26.25°, bp: 173°.

SYN: 1,3-DICHLOROBENZENE

TOXICITY DATA WITH REFERENCE
mrc-smc 5 ppm NTIS** PB84-138973
ipr-mus LD50: 1062 mg/kg MUTAEX 2,111,87

CONSENSUS REPORTS: Reported in EPA TSCA Inventory. Community Right-To-Know List.

SAFETY PROFILE: Moderately toxic by intraperitoneal route. Mutation data reported. When heated to decomposition it emits toxic fumes of Cl^-. See also o-DICHLOROBENZENE and p-DICHLOROBENZENE.

DEP600 CAS:95-50-1 HR: 3
o-DICHLOROBENZENE
DOT: UN 1591
mf: $C_6H_4Cl_2$ mw: 147.00

PROP: Clear liquid. Mp: −17.5°, bp: 180.5°, fp: −22°, flash p: 151°F, d: 1.307 @ 20°/20°, vap d: 5.05, autoign temp: 1198°F, lel: 2.2%, uel: 9.2%.

SYNS: BENZENE, 1,2-DICHLORO- □ CHLOROBEN □ CHLORODEN □ CLOROBEN □ DCD □ o-DICHLOR BENZOL □ o-DICHLOROBENZENE □ 1,2-DICHLOROBENZENE □ o-DICHLOROBENZENE (ACGIH,OSHA) □ o-DICHLOROBENZENE (DOT) □ DILANTIN DB □ DILATIN DB □ DIZENE □ DOWTHERM E □ NCI-C54944 □ ODB □ ODCB □ ORTHODICHLOROBENZENE □ ORTHODICHLOROBENZOL □ SPECIAL TERMITE FLUID □ TERMITKIL

TOXICITY DATA WITH REFERENCE
eye-rbt 100 mg/30S rns MLD AMIHAB 17,180,58
spm-rat-ipr 250 mg/kg JACTDZ 4(2),224,85
ipr-rat TDLo: 50 mg/kg (1D male): REP JACTDZ 4(1),224,85
ihl-rat TCLo: 200 ppm/6H (6–15D preg): TER FAATDF 5,190,85
orl-rat LD50: 500 mg/kg WRPCA2 7,135,68
ihl-rat LCLo: 821 ppm/7H AMIHAB 17,180,58
ipr-rat LD50: 840 mg/kg MEPAAX 20,519,69
orl-mus LD50: 4386 g/kg YKYUA6 32,471,81
ivn-mus LDLo: 400 mg/kg JPBAA7 44,281,37
orl-rbt LD50: 500 mg/kg 85ARAE 3,32,76/77
ivn-rbt LDLo: 250 mg/kg JPBAA7 44,281,37
orl-gpg LDLo: 2000 mg/kg 14CYAT 2,1336,63
ihl-gpg LCLo: 800 ppm/24H JPBAA7 44,281,37

CONSENSUS REPORTS: IARC Cancer Review: Group 3 IMEMDT 7,192,87; Animal Inadequate Evidence IMEMDT 7,231,74, IMEMDT 29,213,82; Human Inadequate Evidence IMEMDT 7,231,74, IMEMDT 29,213,82. Reported in EPA TSCA Inventory. Community Right-To-Know List.

OSHA PEL: CL 50 ppm
ACGIH TLV: TWA 25 ppm, STEL 50 ppm
DFG MAK: 50 ppm (300 mg/m^3)
DOT CLASSIFICATION: 6.1; *Label:* KEEP AWAY FROM FOOD

SAFETY PROFILE: Poison by ingestion and intravenous routes. Moderately toxic by inhalation and intraperitoneal routes. An experimental teratogen. Other experimental reproductive effects. An eye, skin, and mucous membrane irritant. Causes liver and kidney injury. Questionable carcinogen. Mutation data reported. A pesticide. Flammable when exposed to heat or flame. Can react vigorously with oxidizing materials. To fight fire, use water, foam, CO_2, or dry chemical. Slow reaction with aluminum may lead to explosion during storage in a sealed aluminum container. When heated to decomposition it emits toxic fumes of Cl^-. See also CHLOROBENZENE; and CHLORINATED HYDROCARBONS, AROMATIC.

For occupational chemical analysis use NIOSH: Hydrocarbons, Halogenated, 1003.

DEP800 **CAS:106-46-7** **HR: 3**
p-DICHLOROBENZENE
DOT: UN 1592
mf: $C_6H_4Cl_2$ mw: 147.00

PROP: White crystals or leaflets with strong penetrating odor. Mp: 54°, bp: 174°, flash p: 150°F (CC), d: 1.4581 @ 20.5°/4°, vap press: 10 mm @ 54.8°, vap d: 5.08.

SYNS: p-CHLOROPHENYL CHLORIDE ☐ p-DICHLOORBENZEEN (DUTCH) ☐ 1,4-DICHLOORBENZEEN (DUTCH) ☐ p-DICHLORBENZOL (GERMAN) ☐ 1,4-DICHLOR-BENZOL (GERMAN) ☐ DI-CHLORICIDE ☐ 1,4-DICHLOROBENZENE (MAK) ☐ DICHLOROBENZENE, PARA, solid (DOT) ☐ p-DICHLOROBENZOL ☐ p-DICLOROBENZENE (ITALIAN) ☐ 1,4-DICLOROBENZENE (ITALIAN) ☐ EVOLA ☐ NCI-C54955 ☐ PARACIDE ☐ PARA CRYSTALS ☐ PARADI ☐ PARADI-CHLORBENZOL (GERMAN) ☐ PARADICHLOROBENZENE ☐ PARADI-CHLOROBENZOL ☐ PARADOW ☐ PARAMOTH ☐ PARANUGGETS ☐ PARAZENE ☐ PDB ☐ PDCB ☐ PERSIA-PERAZOL ☐ RCRA WASTE NUMBER U070 ☐ RCRA WASTE NUMBER U071 ☐ RCRA WASTE NUMBER U072 ☐ SANTOCHLOR

TOXICITY DATA WITH REFERENCE
eye-hmn 80 ppm AMIHAB 14,138,56
mmo-asn 200 mg/L CJMIAZ 16,369,70
orl-rat TDLo:10 g/kg (female 6–15D post):TER BECTA6 37,164,86
ihl-rbt TCLo:800 ppm/6H (female 6–18D post):TER FAATDF 5,190,85
orl-rat TDLo:7500 mg/kg (female 6–15D post):TER BECTA6 37,164,86
orl-rat TDLo:155 g/kg/2Y-I:CAR NTPTR* NTP-TR-319,87
orl-mus TDLo:155 g/kg/2Y-I:CAR NTPTR* NTP-TR-319,87
orl-hmn TDLo:300 mg/kg:EYE,PUL,GIT PCOC** -,851,66
orl-hmn LDLo:857 mg/kg 34ZIAG-,210,69
unr-hmn LDLo:357 mg/kg YKYUA6 31,1499,80
unr-man LDLo:221 mg/kg 85DCAI 2,73,70
orl-rat LD50:500 mg/kg WRPCA2 9,119,70
ipr-rat LD50:2562 mg/kg JAPMA8 38,124,49
orl-mus LD50:2950 mg/kg GUCHAZ 6,183,73
ipr-mus LD50:2 g/kg MUTAEX 2,111,87
scu-mus LD50:5145 mg/kg TOIZAG 20,772,73
orl-rbt LD50:2830 mg/kg YKYUA6 29,453,78
orl-gpg LDLo:2800 mg/kg 14CYAT 2,1338,63

CONSENSUS REPORTS: NTP 7th Annual Report on Carcinogens. IARC Cancer Review: Group 2B IMEMDT 7,192,87; Animal Inadequate Evidence IMEMDT 7,231,74; IMEMDT 29,213,82. Human Inadequate Evidence IMEMDT 7,231,74; Reported in EPA TSCA Inventory. EPA Genetic Toxicology Program. Community Right-To-Know List.

OSHA PEL: TWA 75 ppm; STEL 110 ppm
ACGIH TLV: TWA 75 ppm; STEL 110 ppm; (Proposed: 10 ppm; Animal Carcinogen)
DFG MAK: 50 ppm (300 mg/m^3)
DOT CLASSIFICATION: 6.1; *Label:* KEEP AWAY FROM FOOD

SAFETY PROFILE: Confirmed carcinogen with experimental carcinogenic data. An experimental teratogen. A human poison by an unspecified route. Moderately toxic to humans by ingestion. Moderately toxic experimentally by ingestion, subcutaneous, and intraperitoneal routes. Mildly toxic by subcutaneous route. Other experimental reproductive effects. Human systemic effects by ingestion: unspecified changes in the eyes, lungs, thorax and respiration, and decreased motility or constipation. Can cause liver injury in humans. A human eye irritant. Mutation data reported. A fumigant. Flammable liquid when exposed to heat, flame, or oxidizers. Dangerous; can react vigorously with oxidizing materials. To fight fire, use water, foam, CO_2, dry chemical. When heated to decomposition it emits toxic fumes of Cl^-. See also CHLORINATED HYDROCARBONS, AROMATIC.

For occupational chemical analysis use NIOSH: Hydrocarbons, Halogenated, 1003.

DEQ000 **CAS:5836-73-7** **HR: 3**
3,4-DICHLOROBENZENE DIAZOTHIOUREA
mf: $C_7H_6Cl_2N_4S$ mw: 249.13

SYNS: CHLOROPROMURITE ☐ (3,4-DICHLOOR-FENYL-AZO)-THIOUREUM (DUTCH) ☐ 1-(3′,4′-DICHLOROBENZENEDIAZOL)-2-THIOUREA ☐ 3,4-DICHLOROBENZENE DIAZOTHIOCARBAMID ☐ 3,4-DICHLOROPHENYLAZOTHIOUREA ☐ 3,4-DICHLOROPHENYL-AZO-THIOUREE (FRENCH) ☐ (3,4-DICHLOR-PHENYL-AZO)-THIOHARNSTOFF (GERMAN) ☐ (3,4-DICLORO-FENIL-AZO)-TIOUREA (ITALIAN) ☐ MURITAN ☐ PROMURIT ☐ PROMURITE

TOXICITY DATA WITH REFERENCE
orl-rat LD50:280 μg/kg FEPRA7 8,282,49
ipr-rat LD50:200 μg/kg FEPRA7 8,282,49
orl-mus LD50:1 mg/kg 28ZEAL 5,188,76
ipr-mus LD50:1350 μg/kg FEPRA7 8,282,49
orl-dog LD50:1 mg/kg 28ZEAL 5,188,76
ipr-rbt LD50:1750 μg/kg FEPRA7 8,282,49
ipr-gpg LD50:1900 μg/kg FEPRA7 8,282,49

SAFETY PROFILE: A deadly poison by ingestion and

intraperitoneal routes. When heated to decomposition it emits very toxic fumes of Cl^-, NO_x, and SO_x.

DEQ200 CAS:120-97-8 HR: 3
4,5-DICHLORO-m-BENZENEDISULFONAMIDE
mf: $C_6H_6Cl_2N_2O_4S_2$ mw: 305.16

PROP: Needles from DMSO (aq). Mp: 239–241°. Sol in alkalis.

SYNS: CB 8000 □ DARANIDE □ DASANIDE □DICHLOFENAM-IDE □ 4,5 DICHLORO-1,3-BENZENEDISULFONAMIDE □ 4,5-DICHLORO 1, 3-DISULFAMOYLBENZENE □ DICHLOROPHENAMIDE □ 3,4-DI-CHLORO-5-SULFAMYLBENZENESULFONAMIDE □ DICHLORPHENAM-IDE □ 1,3-DISULFAMYL-4,5-DICHLOROBENZENE □ ORATROL

TOXICITY DATA WITH REFERENCE
scu-mus TDLo:96 mg/kg (female 15D post):TER DE-BIAO 27,395,72
orl-rat TDLo:4642 mg/kg (1-22D preg):TER PSEBAA 126,6,67
orl-rat LD50:2600 mg/kg 29ZVAB -,41,69
orl-mus LD50:1710 mg/kg 29ZVAB -,41,69
ipr-mus LD50:304 mg/kg THERAP 19,1423,64
ivn-mus LD50:643 mg/kg 29ZVAB -,41,69
ivn-dog LD50:200 mg/kg 29ZVAB -,41,69

SAFETY PROFILE: Poison by intravenous and intraperitoneal routes. Moderately toxic by ingestion. An experimental teratogen. When heated to decomposition it emits very toxic fumes of Cl^-, NO_x, and SO_x.

DEQ600 CAS:91-94-1 HR: 3
3′,3′-DICHLOROBENZIDINE
mf: $C_{12}H_{10}Cl_2N_2$ mw: 253.14

PROP: Crystals or needles from alc. Mp: 133°. Insol in water; sol in alc, benzene, and glacial acetic acid.

SYNS: C.I. 23060 □ CURITHANE C126 □ DCB □ 4,4′-DIAMINO-3, 3′-DICHLOROBIPHENYL □ 4,4′-DIAMINO-3,3′-DICHLORODIPHENYL □ 3,3′-DICHLORBENZIDIN (CZECH) □ 3,3′-DICHLOROBENZIDINA (SPANISH) □ DICHLOROBENZIDINE □ o,o′-DICHLOROBENZIDINE □ 3,3′-DICHLOROBENZIDINE □ DICHLOROBENZIDINE BASE □ 3,3′-DICHLOROBIPHENYL-4,4′-DIAMINE □ 3,3′-DICHLORO-4,4′-BIPHE-NYLDIAMINE □ 3,3′-DICHLORO-4,4′-DIAMINOBIPHENYL □ 3,3′-DI-CHLORO-4,4′ DIAMINO(1,1-BIPHENYL) □ RCRA WASTE NUMBER U073

TOXICITY DATA WITH REFERENCE
mmo-sat 50 μg/plate CALEDQ 4,21,77
dns-hmn:hla 100 nmol/L CNREA8 38,2621,78
bfa-rat/sat 40 mg/kg SAIGBL 23,426,81
otr-ham:kdy 80 μg/L BJCAAI 37,873,78
dnd-mam:lym 25,500 nmol/L CBINA8 38,369,82
orl-rat TDLo:17 g/kg/50W-C:CAR TXAPA9 31,159,75
scu-rat TDLo:3600 mg/kg/61W-I:ETA VOONAW 21(6),110,75
orl-mus TDLo:5100 mg/kg/43W-I:ETA VOONAW 5(5),524,59
scu-mus TDLo:320 mg/kg (female 15-21D post):CAR BEXBAN 78,1402,74
scu-mus TDLo:5200 mg/kg/47W-I:ETA VOONAW 5(5),524,59
orl-dog TDLo:17 g/kg/7Y-I:ETA DUPON* HL623-74,74
orl-ham TDLo:176 g/kg/70W-C:ETA PAACA3 10,78,69
orl-rat LD :20 g/kg/52W-I:ETA VOONAW 5(5),524,59
orl-rat LD :21 g/kg/50W-C:CAR TXAPA9 31,159,75
scu-rat LD :7 g/kg/43W-I:ETA VOONAW 5(5),524,59

CONSENSUS REPORTS: NTP 7th Annual Report on Carcinogens. IARC Cancer Review: Group 2B IMEMDT 7,193,87; Human Inadequate Evidence IMEMDT 29,239,82; Animal Sufficient Evidence IMEMDT 29,239,82; IMEMDT 4,49,74. Reported in EPA TSCA Inventory. Community Right-To-Know List. EPA Genetic Toxicology Program.

OSHA PEL: Cancer Suspect Agent
ACGIH TLV: Suspected Human Carcinogen.
DFG TRK:0.1 mg/m³, Animal Carcinogen, Suspected Human Carcinogen.
NIOSH REL: (Benzidine-based Dye) Reduce to lowest feasible level.

SAFETY PROFILE: Confirmed carcinogen with experimental carcinogenic and tumorigenic data. Human mutation data reported. When heated to decomposition it emits very toxic fumes of Cl^- and NO_x.

For occupational chemical analysis use OSHA: #ID-65 or NIOSH: Benzidine and 3,3'-Dichlorobenzidine, 5509.

DEQ800 CAS:612-83-9 HR: 3
3,3′-DICHLOROBENZIDINE DIHYDROCHLORIDE
mf: $C_{12}H_{10}Cl_2N_2{\cdot}2ClH$ mw: 326.06

SYN: 3,3′-DICHLORO-(1,1′-BIPHENYL)-4,4′-DIAMINE DIHYDRO-CHLORIDE

TOXICITY DATA WITH REFERENCE
mmo-sat 10 μg/plate ENMUDM 5(Suppl 1),3,83
mma-sat 1 μg/plate ENMUDM 5(Suppl 1),3,83
orl-rat LD50:3820 mg/kg 34ZIAG -,211,69

CONSENSUS REPORTS: NTP 7th Annual Report on Carcinogens. Reported in EPA TSCA Inventory.

OSHA PEL: Cancer Suspect Agent

SAFETY PROFILE: Confirmed carcinogen. Moderately toxic by ingestion. Mutation data reported. When heated to decomposition it emits very toxic fumes of Cl^- and NO_x.

DER000 CAS:510-15-6 HR: 3
4,4′-DICHLOROBENZILIC ACID ETHYL ESTER
mf: $C_{16}H_{14}Cl_2O_3$ mw: 325.20

PROP: Viscous liquid, sometimes yellow, sltly sol in water. Bp: 146–148°, vap press: 2.2×10^{-6} mm @ 20°.

SYNS: ACAR □ ACARABEN 4E □ AKAR □ BENZILAN □ BENZ-o-CHLOR □ CHLORBENZILATE □ CHLOROBENZYLATE □ COM-POUND 338 □ 4,4′-DICHLORBENZILSAEUREAETHYLESTER (GERMAN) □ 4,4′-DICHLOROBENZILATE □ ENT 18,596 □ ETHYL 4-CHLORO-α-(4-CHLOROPHENYL)-α-HYDROXYBENZENEACETATE □ ETHYL-p,p′-DICHLOROBENZILATE □ ETHYL-4,4′-DICHLOROBENZILATE □ ETH-YL-4,4′-DICHLORODIPHENYL GLYCOLLATE □ ETHYL-4,4′-DICHLORO-PHENYL GLYCOLLATE □ ETHYL ESTER of 4,4′-DICHLOROBENZILIC ACID □ ETHYL-2-HYDROXY-2,2-BIS(4-CHLOROPHENYL)ACETATE □ FOLBEX □ FOLBEX SMOKE-STRIPS □ G 338 □ G 23992 □ GEIGY 338 □ KOP MITE □ NCI C00408 □ NCI-C60413 □ RCRA WASTE NUMBER U038

TOXICITY DATA WITH REFERENCE
skn-rbt 125 mg open MLD CIGET* -,-,77
eye-rbt 25 mg MOD CIGET* -,-,77
orl-rat TDLo:5475 mg/kg/2Y-C:CAR CTOXAO 16,67,80
orl-mus TDLo:71 g/kg/82W-C:CAR JNCIAM 42,1101,69
orl-mus TD:210 g/kg/78W-C:CAR NCITR* NCI-CG-TR-75,78
orl-mus TD:125 g/kg/83W-C:CAR DIGEBW 16,308,77
orl-rat TD:17,520 mg/kg/2Y-C:CAR CTOXAO 16,67,80
orl-rat TD:72 g/kg/78W-C:ETA NCITR* NCI-CG-TR-75,78
orl-rat TD:1752 mg/kg/2Y-C:NEO CTOXAO 16,67,80
orl-rat LD50:700 mg/kg WRPCA2 9,119,70
orl-mus LD50:729 mg/kg GUCHAZ 6,106,73
orl-ham LD50:700 mg/kg TXAPA9 48,A192,79

CONSENSUS REPORTS: IARC Cancer Review: Group 3 IMEMDT 7,56,87; Animal Limited Evidence IMEMDT 30,73,83; Animal Sufficient Evidence IMEMDT 5,75,74. NCI Carcinogenesis Bioassay Completed; Results Positive: mouse NCITR* NCI-CG-TR-75,78. NCI Carcinogenesis Bioassay Completed; Results Indefinite: rat NCITR* NCI-CG-TR-75,78. Community Right-To-Know List. Reported in EPA TSCA Inventory.

SAFETY PROFILE: Suspected carcinogen with experimental carcinogenic, neoplastigenic, and tumorigenic data. Moderately toxic by ingestion. A skin and eye irritant. A pesticide. When heated to decomposition it emits toxic fumes of Cl^-.

DER100 CAS:50-84-0 HR: 2
2,4-DICHLOROBENZOIC ACID
mf: $C_7H_4Cl_2O_2$ mw: 191.01

SYN: BENZOIC ACID, 2,4-DICHLORO-

TOXICITY DATA WITH REFERENCE
scu-mus LD50:1200 mg/kg BCPCA6 13,1538,64
orl-mus LD50:830 mg/kg SKEZAP 20,332,79

CONSENSUS REPORTS: Reported in EPA TSCA Inventory.

SAFETY PROFILE: Moderately toxic by ingestion and subcutaneous route. When heated to decomposition it emits toxic vapors of Cl^-.

DER400 CAS:50-79-3 HR: 2
2,5-DICHLOROBENZOIC ACID
mf: $C_7H_4Cl_2O_2$ mw: 191.01

PROP: A solid. Mp: 155.5°.

TOXICITY DATA WITH REFERENCE
uns-sat 1 g/L MUREAV 264,1,91
scu-mus LD50:1200 mg/kg BCPCA6 13,1538,64

CONSENSUS REPORTS: Reported in EPA TSCA Inventory.

SAFETY PROFILE: Moderately toxic by subcutaneous route. Mutation data reported. When heated to decomposition it emits toxic fumes of Cl^-.

DER600 CAS:51-44-5 HR: 3
3,4-DICHLOROBENZOIC ACID
mf: $C_7H_4Cl_2O_2$ mw: 191.01

PROP: Needles from EtOH (aq) or C_6H_6. Mp: 208–209°.

SYNS: SYNSTIGMINE □ SYNTOSTIGMIN □ VAGOSTIGMIN

TOXICITY DATA WITH REFERENCE
scu-mus LD50:400 mg/kg BCPCA6 13,1538,64

CONSENSUS REPORTS: Reported in EPA TSCA Inventory.

SAFETY PROFILE: Poison by subcutaneous route. When heated to decomposition it emits toxic fumes of Cl^-.

DER800 CAS:1194-65-6 HR: 2
2,6-DICHLOROBENZONITRILE
mf: $C_7H_3Cl_2N$ mw: 172.01

PROP: White solid. Mp: 142.5–143.5°. Almost insol in water; sol in organic solvents.

SYNS: CARSORON □ CASORON 133 □ CODE H 133 □ 2,6-DBN □ DBN (the herbicide) □ DCB □ DECABANE □ DICHLOBENIL (DOT) □ 2,6-DICHLORBENZONITRIL (GERMAN) □ DU-SPREX □ H 133 □ H 1313 □ NIA 5996 □ NIAGARA 5006 □ NIAGARA 5,996

TOXICITY DATA WITH REFERENCE
ipr-mus TDLo:260 μg/kg/39D-I:ETA AAATAP 122,107,79
scu-mus TDLo:260 μg/kg/39D-I:ETA AAATAP 122,107,79
orl-rat LD50:2710 mg/kg RREVAH 10,97,65
orl-mus LD50:2056 mg/kg 28ZEAL 5,73,76
skn-rbt LD50:1350 mg/kg GUCHAZ 6,177,73
orl-gpg LD50:681 mg/kg PCOC** -,337,66

CONSENSUS REPORTS: Cyanide and its compounds are on the Community Right-To-Know List. EPA Genetic Toxicology Program.

SAFETY PROFILE: Moderately toxic by ingestion and skin contact. Questionable carcinogen with experimental tumorigenic data. Does not hydrolyze to HCN in body. Less toxic than most aliphatic nitriles. When heated to decomposition it emits toxic fumes of Cl^-, CN^-, and NO_x. See also BENZONITRILE, CHLORIDES, and NITRILES.

DES000 CAS:90-98-2 HR: 3
p,p′-DICHLOROBENZOPHENONE
mf: $C_{13}H_8Cl_2O$ mw: 251.11

PROP: A solid. Mp: 147–148°, bp: 353°.

SYNS: DBP □ DCB □ 4,4′-DICHLOROBENZOPHENONE □ USAF DO-4

TOXICITY DATA WITH REFERENCE
ipr-mus LD50:200 mg/kg NTIS** AD277-689

CONSENSUS REPORTS: Reported in EPA TSCA Inventory.

SAFETY PROFILE: Poison by intraperitoneal route. When heated to decomposition it emits toxic fumes of Cl^-.

DES400 CAS:697-91-6 HR: 2
2,6-DICHLORO-p-BENZOQUINONE
mf: $C_6H_2Cl_2O_2$ mw: 176.98

PROP: Yellow prisms from EtOH. Mp: 121°. Sol in EtOH, $CHCl_3$, Me_2CO, C_6H_6, and Et_2O.

SYNS: 2,6-DICHLOQUINONE □ 2,6-DICHLORO-2,5-CYCLOHEXA-DIENE-1,4-DIONE □ 2,6-DICHLORO-p-QUINONE

TOXICITY DATA WITH REFERENCE
orl-rat LDLo:500 mg/kg NCNSA6 5,20,53

CONSENSUS REPORTS: Reported in EPA TSCA Inventory.

SAFETY PROFILE: Moderately toxic by ingestion. Mutation data reported. When heated to decomposition it emits very toxic fumes of Cl^-.

DET000 CAS:12041-76-8 HR: 3
DICHLOROBENZYL ALCOHOL
mf: $C_7H_6Cl_2O$ mw: 177.03

PROP: Crystals. Vap d: 6.1.

SYNS: BAYER 4245 □ RAPIDOSEPT

TOXICITY DATA WITH REFERENCE
orl-rat LD50:810 mg/kg AIHAAP 30,470,69
skn-rbt LD50:400 mg/kg AIHAAP 30,470,69

SAFETY PROFILE: Poison by skin contact. Moderately toxic by ingestion. An insecticide. When heated to decomposition it emits toxic fumes of Cl^-. See also ALCOHOLS and CHLORIDES.

DET400 CAS:1966-58-1 HR: 2
3,4-DICHLOROBENZYL METHYLCARBAMATE
mf: $C_9H_9Cl_2NO_2$ mw: 234.09

SYNS: 3,4-DICHLOROBENZENEMETHANOL METHYLCARBAMATE □ ROWMATE □ SIRMATE □ UC 22,463

TOXICITY DATA WITH REFERENCE
orl-rat LD50:1870 mg/kg WRPCA2 9,119,70
orl-mus LD50:1620 mg/kg 31ZOAD 1,141,68

SAFETY PROFILE: Moderately toxic by ingestion. When heated to decomposition it emits very toxic fumes of Cl^- and NO_x. See also CARBAMATES.

DET600 CAS:62046-37-1 HR: 2
3,4-DICHLOROBENZYL METHYLCARBAMATE with 2, 3-DICHLOROBENZYL METHYLCARBAMATE (80:20)
mf: $C_9H_9Cl_2NO_2$ mw: 234.09

SYNS: CHLORXYLAM □ 2,3(or 3,4)-DICHLOROBENZENEMETHANOL METHYL CARBAMATE □ ENT 25,736 □ ROWMATE □ SIRMATE □ U-17004 □ UC 22,463

TOXICITY DATA WITH REFERENCE
orl-rat LD50:1870 mg/kg 28ZEAL 4,144,69

SAFETY PROFILE: Moderately toxic by ingestion. When heated to decomposition it emits very toxic fumes of Cl^- and NO_x. See also individual components and CARBAMATES.

DEU000 CAS:6358-85-6 HR: 1
2,2′-((3,3′-DICHLORO(1,1′-BIPHENYL)-4,4′-DIYL)-BIS (AZO))BIS(3-OXO-N-PHENYL)BUTANAMIDE
mf: $C_{32}H_{26}Cl_2N_6O_4$ mw: 629.54

SYNS: AMAZON YELLOW X2485 □ BENZIDINE LACQUER YELLOW G □ BENZIDINE YELLOW □ BENZIDINE YELLOW TONER YT-378 □ BIS(ACETYL-N-PHENYLCARBAMYLMETHYL)-4,4′-DISAZO-3,3′-DICHLOROBIPHENYL □ BRILLIANT YELLOW SLURRY □ CARNELIO YELLOW GX □ C.I. 21090 □ C.I. PIGMENT YELLOW 12 □ DAINICHI BENZIDINE YELLOW GRT □ DAIRYLIDE YELLOW AAA □ DALTOLITE FAST YELLOW GT □ DIARYLANILIDE YELLOW □ 2,2′-((3,3′-DICHLORO(1,1′-DIPHENYL)-4,4′-DIYL)BIS(AZO)BIS(3-OXO-N)-PHENYLBUTANAMIDE □ ELJON YELLOW BG □ GRAPHTOL YELLOW A-HG □ HANCOCK YELLOW 10010 □ HELIC YELLOW GW □ IRGALITE YELLOW BO □ ISOL BENZIDINE YELLOW G □ KROMON YELLOW MTB □ LIGHT YELLOW JB □ LODESTONE YELLOW YB-57 □ MONOLITE YELLOW GT □ NCI-C03269 □ No. 49 CONCENTRATED BENZIDINE YELLOW □ PERMANENT YELLOW GHG □ PIGMENT YELLOW GT □ RANGOON YELLOW □ RECOLITE YELLOW GB □ SANYO BENZIDINE YELLOW-B □ SEGNALE LIGHT YELLOW 2GR □ SILOTON YELLOW GTX □ SYMULER FAST YELLOW GF □ VERONA YELLOW X-1791 □ VULCAFOR FAST YELLOW GTA

TOXICITY DATA WITH REFERENCE
orl-rat LD50:>10,800 mg/kg SCIEAS 36(1-4),10,89

CONSENSUS REPORTS: NCI Carcinogenesis Bioassay Completed; Results Negative NCITR* NCI-CG-Tr-30,78. Reported in EPA TSCA Inventory.

SAFETY PROFILE: Low toxicity by ingestion. When heated to decomposition it emits very toxic fumes of Cl^- and NO_x. See also CHLORIDES and AMIDES.

DEU100 CAS:70134-26-8 HR: 3
DICHLOROBIS(2-CHLOROCYCLOHEXYL)SELENIUM
mf: $C_{12}H_{20}Cl_4Se$ mw: 385.08

SYN: SELENIUM, DICHLOROBIS(2-CHLOROCYCLOHEXYL)-

TOXICITY DATA WITH REFERENCE
ivn-mus LD50:56 mg/kg CSLNX* NX#04576

OSHA PEL: TWA 0.2 mg(Se)/m³
ACGIH TLV: TWA 0.2 mg(Se)/m³

SAFETY PROFILE: Poison by intravenous route. When heated to decomposition it emits toxic fumes of Se and Cl^-.

DEU115 CAS:18252-65-8 HR: 3
cis-DICHLOROBIS(DIMETHYLSELENIDE)PLATINUM(II)
mf: $C_4H_{12}Cl_2PtSe_2$ mw: 484.07

SYNS: NSC 271675 □ PLATINUM(II), BIS(METHYL SELENIDE)DICHLORO-, cis- □ PLATINUM, DICHLOROBIS(METHYL SELENIDE)-, cis- □ PLATINUM, DICHLOROBIS(SELENOBIS(METHANE))-(SP-4-2)

TOXICITY DATA WITH REFERENCE
ivn-mus LD50:50 mg/kg CTRRDO 61,1519,77 85INA8 5,492,86

OSHA PEL: TWA 0.2 mg(Se)/m³
ACGIH TLV: TWA 0.2 mg(Se)/m³; TWA 0.002 mg(Pt)/m³

SAFETY PROFILE: Poison by intravenous route. When

D

heated to decomposition it emits toxic fumes of Se, Pt, and Cl^-.

DEU125 CAS:74037-18-6 HR: 3
DICHLOROBIS(2-ETHOXYCYCLOHEXYL)SELENIUM
mf: $C_{16}H_{30}Cl_2O_2Se$ mw: 404.32

SYN: SELENIUM, DICHLOROBIS(2-ETHOXYCYCLOHEXYL)-

TOXICITY DATA WITH REFERENCE
ivn-mus LD50:180 mg/kg CSLNX* NX#04578

OSHA PEL: TWA 0.2 mg(Se)/m^3
ACGIH TLV: TWA 0.2 mg(Se)/m^3

SAFETY PROFILE: Poison by intravenous route. When heated to decomposition it emits toxic fumes of Se and Cl^-.

DEU200 CAS:38780-42-6 HR: 3
cis-DICHLOROBIS(PYRROLIDINE)PLATINUM(II)
mf: $C_8H_{18}Cl_2N_2Pt$ mw: 408.27

SYN: cis-DIPYRROLIDINEDICHLOROPLATINUM(II)

TOXICITY DATA WITH REFERENCE
mmo-sat 100 nmol/plate CNREA8 39,913,79
scu-rat LD:89 mg/kg/6W-I:ETA CNREA8 39,913,79
ipr-mus LD50:240 mg/kg CBINA8 5,415,72

SAFETY PROFILE: Poison by intraperitoneal route. Questionable carcinogen with experimental tumorigenic data. Mutation data reported. See also PLATINUM COMPOUNDS. When heated to decomposition it emits very toxic fumes of Cl^- and NO_x.

DEU259 CAS:2899-02-7 HR: 3
N,N′-DICHLOROBIS(2,4,6-TRICHLOROPHENYL) UREA
mf: $C_{13}H_4Cl_8N_2O$ mw: 487.81

$O:C(NClC_6H_2Cl_3)_2$

SAFETY PROFILE: Violent or explosive reaction on mixing with dimethyl sulfoxide. Ignites on contact with ammonia, ammonium carbonate, or organic amines. A fabric treatment mixture of the urea with 1-(4-nitrophenylazo)-2-naphthol + zinc oxide may ignite spontaneously in storage, especially if heated. When heated to decomposition it emits toxic fumes of Cl^- and NO_x.

DEU300 CAS:10325-39-0 HR: 3
DICHLOROBORANE
mf: BCl_2H mw: 82.72

PROP: Gas, readily disproportionates; air and moisture-sensitive.

SAFETY PROFILE: A poison. Ignites spontaneously in air. When heated to decomposition it emits toxic fumes of Cl^-. See also BORON COMPOUNDS, BORANES, and CHLORIDES.

DEU400 CAS:1653-19-6 HR: 1
2,3-DICHLORO-1,3-BUTADIENE
mf: $C_4H_4Cl_2$ mw: 122.98

PROP: Liquid. D: 1.183 @ 20°/4°, bp: 98°.

SYN: 2,3-DICHLOR-1,3-BUTADIEN (CZECH)

TOXICITY DATA WITH REFERENCE
skn-rbt 500 mg/24H SEV 28ZPAK -,29,72
eye-rbt 500 mg/24H MLD 28ZPAK -,29,72

CONSENSUS REPORTS: Reported in EPA TSCA Inventory.

SAFETY PROFILE: An eye and severe skin irritant. When heated to decomposition it emits toxic fumes of Cl^-. See also CHLORINATED HYDROCARBONS, ALIPHATIC.

DEU509 CAS:51104-87-1 HR: 3
1,4-DICHLORO-1,3-BUTADIYNE
mf: C_4Cl_2 mw: 118.95

ClC≡CC≡CCl

PROP: Liquid. Mp: 11.0–11.2°.

SAFETY PROFILE: Explodes when heated above 70°C. When heated to decomposition it emits toxic fumes of Cl^-. See also CHLORINATED HYDROCARBONS, ALIPHATIC; and ACETYLENE COMPOUNDS.

DEU650 CAS:926-57-8 HR: 2
1,3-DICHLORO-2-BUTENE
mf: $C_4H_6Cl_2$ mw: 125.00

SYN: 2-BUTENE, 1,3-DICHLORO-

TOXICITY DATA WITH REFERENCE
ihl-rat LC50:3930 mg/m^3 ZKMAAX (6),66,69
ihl-mus LC50:4400 mg/m^3 ZKMAAX (6),66,69

CONSENSUS REPORTS: Reported in EPA TSCA Inventory.

SAFETY PROFILE: Moderately toxic by inhalation. When heated to decomposition it emits toxic vapors of Cl^-.

DEV000 CAS:764-41-0 HR: 3
1,4-DICHLORO-2-BUTENE
mf: $C_4H_6Cl_2$ mw: 125.00

PROP: Colorless liquid. Mp: 1–3°, bp: 156°, d: 1.183 @ 25°/4°.

SYNS: DCB □ 1,4-DCB □ 1,4-DICHLOROBUTENE-2 (MAK) □ RCRA WASTE NUMBER U074

TOXICITY DATA WITH REFERENCE
skn-rbt 10 mg/24H open SEV AMIHBC 4,119,51
eye-rbt 20 mg open SEV AMIHBC 4,119,51
mmo-sat 1 mmol/L ARTODN 41,249,79
mma-sat 1 mmol/L ARTODN 41,249,79
sln-dmg-orl 2 mmol/L/3D-I 35WYAM -,63,76
cyt-rat-ihl 1700 μg/m^3/30D-I ZKMAAX 25,335,85
orl-rat TDLo:750 μg/m^3(75D male pre):REP GISAAA 51(7),77,86

ihl-rat TCLo:5 ppm/6H (6-15D preg):TER TXAPA9 64,125,82
ihl-rat TCLo:1 ppm/6H/82W-I:CAR EPASR* 8EHQ-0985-0567
ihl-rat TC:100 ppb/6H/82W-I:NEO EPASR* 8EHQ-0985-0567
orl-rat LD50:89 mg/kg AMIHBC 4,119,51
ihl-rat LCLo:62 ppm/4H AMIHBC 4,119,51
orl-mus LD50:190 mg/kg GTPZAB 29(4),49,85
ihl-mus LC50:920 mg/m^3 GTPZAB 29(4),49,85
ivn-mus LD50:56 mg/kg CSLNX* NX#01103
skn-rbt LD50:620 mg/kg AMIHBC 4,119,51

CONSENSUS REPORTS: Reported in EPA TSCA Inventory. EPA Genetic Toxicology Program.

ACGIH TLV: (Proposed: TWA 0.005 ppm; Suspected Human Carcinogen)
DFG MAK: Animal Carcinogen, Suspected Human Carcinogen

SAFETY PROFILE: Confirmed carcinogen with experimental carcinogenic and neoplastigenic data. Poison by ingestion, inhalation, and intravenous routes. Moderately toxic by skin contact. An experimental teratogen. Other experimental reproductive effects. Mutation data reported. A severe skin and eye irritant. When heated to decomposition it emits toxic fumes of Cl^-. See also CHLORINATED HYDROCARBONS, ALIPHATIC.

DEV100 CAS:760-23-6 HR: 2
3,4-DICHLORO-1-BUTENE
mf: $C_4H_6Cl_2$ mw: 125.00

SYN: 1-BUTENE, 3,4-DICHLORO-

TOXICITY DATA WITH REFERENCE
cyt-rat-ihl 13,700 μg/m^3/30D-I ZKMAAX 25,335,85
orl-rat TDLo:75 μg/kg (male 75D pre):REP GISAAA 51(7),77,86
orl-mus LD50:724 mg/kg GISAAA 51(7),77,86

CONSENSUS REPORTS: Reported in EPA TSCA Inventory.

SAFETY PROFILE: Moderately toxic by ingestion. Experimental reproductive effects. Mutation data reported. When heated to decomposition it emits toxic fumes of Cl^-.

DEV200 CAS:11069-19-5 HR: 3
DICHLOROBUTENE
DOT: NA 2920
mf: $C_4H_6Cl_2$ mw: 125.00

SYNS: BUTENE, DICHLORO- ☐ DICHLOROBUTYLENE

DOT CLASSIFICATION: 8; *Label:* Corrosive, Flammable Liquid

SAFETY PROFILE: A flammable liquid. When heated to decomposition it emits toxic vapors of Cl^-.

DEV300 CAS:42520-97-8 HR: 3
2,2′-DICHLORO-N-BUTYLDIETHYLAMINE
mf: $C_8H_{17}Cl_2N$ mw: 198.16

SYNS: N-N-BIS(2-CHLOROETHYL)BUTYLAMINE ☐ N-BUTYL-2,2′-DICHLORODIETHYLAMINE ☐ TL 513

TOXICITY DATA WITH REFERENCE
dns-rat-ipr 10 mg/kg CRNGDP 1,621,80
dns-rat-orl 10 mg/kg CRNGDP 1,621,80
unr-rat LD50:1 mg/kg PHBUA9 1,297,53
ihl-mus LCLo:350 mg/m^3/10M NDRC** NDCrc-132,Dec,42

SAFETY PROFILE: Poison by inhalation and possibly other routes. Mutation data reported. When heated to decomposition it emits toxic fumes of Cl^- and NO_x.

DEV400 CAS:821-10-3 HR: 3
1,4-DICHLORO-2-BUTYNE
mf: $C_4H_4Cl_2$ mw: 122.98

PROP: D: 1.26° @ 20°/4°, bp: 165–168°.

SYN: 1,4-DICHLOROBUTYNE

TOXICITY DATA WITH REFERENCE
ivn-mus LD50:56 mg/kg CSLNX* NX#02969

CONSENSUS REPORTS: Reported in EPA TSCA Inventory.

SAFETY PROFILE: Poison by intravenous route. When heated to decomposition it emits toxic fumes of Cl^-. Probably a dangerous fire and explosion hazard. See also ACETYLENE COMPOUNDS; and CHLORINATED HYDROCARBONS, ALIPHATIC.

DEV600 CAS:1918-18-9 HR: 2
3,4-DICHLOROCARBANILIC ACID METHYL ESTER
mf: $C_8H_7Cl_2NO_2$ mw: 220.06

SYNS: (3,4-DICHLOROPHENYL)CARBAMIC ACID METHYL ESTER ☐ MCC ☐ METHYL-3,4-DICHLOROCARBANILATE ☐ METHYL-N-(3,4-DICHLOROPHENYL) CARBAMATE ☐ NIA 2,995 ☐ NIA 2995J ☐ SWEP

TOXICITY DATA WITH REFERENCE
orl-rat LD50:522 mg/kg GUCHAZ 6,477,73
skn-rbt LD50:2480 mg/kg WRPCA2 9,119,70

SAFETY PROFILE: Moderately toxic by ingestion and skin contact. A pesticide. See also CARBAMATES and ESTERS. When heated to decomposition it emits very toxic fumes of Cl^- and NO_x.

DEV800 CAS:101-05-3 HR: 3
2,4-DICHLORO-6-o-CHLORANILINO-s-TRIAZINE
mf: $C_9H_5Cl_3N_4$ mw: 275.53

PROP: White to tan crystals, insol in water. Mp: 160°.

SYNS: ANILAZIN ☐ ANILAZINE ☐ B-622 ☐ BORTRYSAN ☐ 2-(2-CHLORANILIN)-4,6-DICHLOR-1,3,5-TRIAZIN (GERMAN) ☐ (o-CHLOROANILINO)DICHLOROTRIAZINE ☐ 2,4-DICHLORO-6-(o-CHLOROANILINO)-s-TRIAZINE ☐ 2,4-DICHLORO-6-(2-CHLOROANILINO)-1,3,5-TRIAZINE ☐ 4,6-DICHLORO-N-(2-CHLOROPHENYL)-1,3,5-TRIAZIN-2-AMINE ☐ DIREZ ☐ DYRENE ☐ DYRENE 50W ☐ ENT 26,058 ☐ KEMATE ☐ NCI-C08684 ☐ TRIASYN ☐ TRIAZIN ☐ TRIAZINE (pesticide) ☐ ZINOCHLOR

TOXICITY DATA WITH REFERENCE
skn-man 0.1% MOD LANCAO 2,1252,80
skn-rbt 500 mg SEV 34ZIAG -,235,69
otr-rat:emb 990 ng/plate JJATDK 1,190,81
orl-rat LD50:2700 mg/kg ARSIM* 20,9,66
ipr-rat LD50:16 mg/kg JAFCAU 21,140,73
orl-mus LD50:6020 mg/kg YKYUA6 30,623,79
ipr-mus LD50:30 mg/kg JAFCAU 21,140,73
orl-rbt LD50:400 mg/kg 34ZIAG -,235,69

CONSENSUS REPORTS: NCI Carcinogenesis Bioassay Completed; Results Negative NCITR* NCI-CG-TR-104,78. EPA Genetic Toxicology Program.

SAFETY PROFILE: Poison by intraperitoneal routes. A human skin irritant. A severe skin irritant experimentally. Mutation data reported. A fungicide. When heated to decomposition it emits very toxic fumes of Cl^- and NO_x.

DEW000 CAS:333-25-5 HR: 3
DICHLORO(2-CHLOROVINYL)ARSINE OXIDE

SYN: LEWISITE I OXIDE

TOXICITY DATA WITH REFERENCE
orl-rat LD50:5 mg/kg JPBAA7 58,411,46
orl-rbt LD50:3 mg/kg JPBAA7 58,411,46
ivn-rbt LD50:1 mg/kg JPBAA7 58,411,46
orl-gpg LD50:2 mg/kg JPBAA7 58,411,46
scu-gpg LD50:200 μg/kg JPBAA7 58,411,46

CONSENSUS REPORTS: Arsenic and its compounds are on the Community Right-To-Know List.

OSHA PEL: TWA 0.5 mg(As)/m^3

SAFETY PROFILE: Poison by ingestion, intravenous, and subcutaneous routes. See also ARSENIC COMPOUNDS. When heated to decomposition it emits very toxic fumes of Cl^- and As. See also CHLOROVINYLARSINE DICHLORIDE.

DEW200 CAS:26270-58-6 HR: 3
5-(3,4-DICHLOROCINNAMOYL)-4,7-DIMETHOXY-6-(2-DIMETHYLAMINOETHOXY)BENZOFURAN MALEATE
mf: $C_{24}H_{23}Cl_2NO_5{\cdot}C_4H_4O_4$ mw: 592.46

TOXICITY DATA WITH REFERENCE
orl-mus LD50:320 mg/kg CHTPBA 8,479,73
ivn-mus LD50:24 mg/kg CHTPBA 8,479,73

SAFETY PROFILE: Poison by ingestion and intravenous routes. When heated to decomposition it emits very toxic fumes of Cl^- and NO_x.

DEW400 CAS:20373-56-2 HR: 3
2,6-DICHLORO-N-CYCLOPROPYL-N-ETHYL ISONICOTINAMIDE
mf: $C_{11}H_{12}Cl_2N_2O$ mw: 259.15

SYN: ABBOTT-28440

TOXICITY DATA WITH REFERENCE
orl-rat LD50:78 mg/kg 27ZQAG -,196,72
ipr-rat LD50:54 mg/kg 27ZQAG -,196,72
orl-mus LD50:123 mg/kg 27ZQAG -,196,72
ipr-mus LD50:129 mg/kg 27ZQAG -,196,72

SAFETY PROFILE: Poison by ingestion and intraperitoneal routes. When heated to decomposition it emits very toxic fumes of Cl^- and NO_x.

DEX000 CAS:14913-33-8 HR: 3
trans-DICHLORODIAMMINEPLATINUM(II)
mf: $C_{12}H_6N_2Pt$ mw: 300.07

PROP: Pale yellow crystals. Mp: 270° (decomp). Less sol in H_2O than *cis* -form; sol in DMF and DMSO.

SYNS: trans-DIAMMINEDICHLOROPLATINUM(II) □ trans-PLATINUM(II)DIAMMINEDICHLORIDE

TOXICITY DATA WITH REFERENCE
mma-sat 2 μg/plate MUREAV 77,45,80
dnd-hmn:fbr 50 μmol/L/4H CNREA8 42,145,82
dnd-hmn:lng 100 μmol/L CBINA8 36,345,81
dnd-hmn:oth 20 mg/L CNREA8 45,6232,85
msc-ham:lng 100 mg/L CNREA8 44,3270,84
ipr-mus TDLo:32,408 μg/kg/10W-I:ETA CNREA8 41,4368,81
ipr-mus LD50:27 mg/kg CBINA8 5,415,72

CONSENSUS REPORTS: EPA Genetic Toxicology Program.

SAFETY PROFILE: Poison by intraperitoneal route. Questionable carcinogen with experimental tumorigenic data. Human mutation data reported. See also PLATINUM COMPOUNDS. When heated to decomposition it emits toxic fumes of NO_x and Cl^-.

DEX400 CAS:84-58-2 HR: 3
2,3-DICHLORO-5,6-DICYANOBENZOQUINONE
mf: $C_8Cl_2N_2O_2$ mw: 227.00

PROP: Amber needles from $CHCl_3$. Mp: 215–217°.

SYN: 4,5-DICHLORO-3,6-DIOXO-1,4-CYCLOHEXADIENE-1,2-DICARBONITRILE

TOXICITY DATA WITH REFERENCE
ipr-mus LD50:31 mg/kg CHTHBK 16,371,71
ivn-mus LD50:13 mg/kg CSLNX* NX#07894

CONSENSUS REPORTS: Reported in EPA TSCA Inventory. Cyanide and its compounds are on the Community Right-To-Know List.

SAFETY PROFILE: Poison by intraperitoneal and intravenous route. When heated to decomposition it emits very toxic fumes of Cl^-, CN^-, and NO_x. See also NITRILES.

DEX600 CAS:17751-20-1 HR: 3
2′,6′-DICHLORO-2-(DIETHYLAMINO)ACETANILIDE HYDROCHLORIDE
mf: $C_{12}H_{16}Cl_2N_2O{\cdot}ClH$ mw: 311.66

SYN: C 3053

TOXICITY DATA WITH REFERENCE
eye-rbt 2% MLD ARZNAD 8,270,58
ipr-rat LD50:420 mg/kg ARZNAD 8,270,58
ipr-mus LD50:365 mg/kg ARZNAD 8,270,58

scu-mus LD50:775 mg/kg ARZNAD 8,270,58

SAFETY PROFILE: Poison by intraperitoneal route. Moderately toxic by subcutaneous route. An eye irritant. When heated to decomposition it emits very toxic fumes of Cl^- and NO_x.

DEX800 CAS:41572-59-2 HR: 3
7,8-DICHLORO-10-(2-(DIETHYLAMINO)ETHYL)ISOALLOXAZINE HYDROCHLORIDE
mf: $C_{16}H_{17}Cl_2N_5O_2{\cdot}ClH$ mw: 418.74

TOXICITY DATA WITH REFERENCE
scu-mus LD50:15 mg/kg CMTRAG 2,96,61
ivn-mus LD50:22 mg/kg CMTRAG 2,96,61

SAFETY PROFILE: Poison by subcutaneous and intravenous routes. When heated to decomposition it emits very toxic fumes of NO_x and Cl^-.

DEY000 CAS:77791-63-0 HR: 3
2′,6′-DICHLORO-2-(2-(DIETHYLAMINO)ETHYL)METHYLAMINOACETANILIDE DIHYDROCHLORIDE
mf: $C_{15}H_{23}Cl_2N_3O{\cdot}2ClH$ mw: 405.23

SYN: C 5365

TOXICITY DATA WITH REFERENCE
ipr-rat LD50:140 mg/kg ARZNAD 9,262,59
scu-mus LD50:585 mg/kg ARZNAD 9,262,59

SAFETY PROFILE: Poison by intraperitoneal route. Moderately toxic by subcutaneous route. When heated to decomposition it emits very toxic fumes of Cl^- and NO_x.

DEY200 CAS:101651-69-8 HR: 3
2′,6′-DICHLORO-2-(2-(DIETHYLAMINO)ETHYL)THIOACETANILIDE HYDROCHLORIDE
mf: $C_{14}H_{20}Cl_2N_2OS{\cdot}ClH$ mw: 371.78

SYN: C 4910

TOXICITY DATA WITH REFERENCE
ipr-rat LD50:172 mg/kg ARZNAD 9,683,59
scu-mus LD50:340 mg/kg ARZNAD 9,683,59

SAFETY PROFILE: Poison by intraperitoneal and subcutaneous routes. When heated to decomposition it emits very toxic fumes of Cl^-, NO_x, and SO_x.

DEY400 CAS:93405-68-6 HR: 3
7,8-DICHLORO-10-(3-(DIETHYLAMINO)-2-HYDROXYPROPYL)ISOALLOXAZINE SULFATE
mf: $C_{17}H_{19}Cl_2N_5O_3{\cdot}H_2O_4S$ mw: 510.39

TOXICITY DATA WITH REFERENCE
ipr-rat LD50:18 mg/kg CMTRAG 2,96,61
scu-mus LD50:74 mg/kg CMTRAG 2,96,61
ivn-mus LD50:75 mg/kg CMTRAG 2,96,61

SAFETY PROFILE: Poison by intraperitoneal, intravenous, and subcutaneous routes. When heated to decomposition it emits very toxic fumes of SO_x, Cl^-, and NO_x.

DEY600 CAS:101652-01-1 HR: 3
7,8-DICHLORO-10-(4-(DIETHYLAMINO)-1-METHYLBUTYL)ISOALLOXAZINE HYDROCHLORIDE
mf: $C_{19}H_{23}Cl_2N_5O_2{\cdot}ClH$ mw: 460.83

TOXICITY DATA WITH REFERENCE
scu-mus LD50:18 mg/kg CMTRAG 2,96,61
ivn-mus LD50:48 mg/kg CMTRAG 2,96,61

SAFETY PROFILE: Poison by subcutaneous and intravenous routes. When heated to decomposition it emits very toxic fumes of HCl and NO_x.

DEY800 CAS:1719-53-5 HR: 3
DICHLORODIETHYLSILANE
DOT: UN 1767
mf: $C_4H_{10}Cl_2Si$ mw: 157.13

PROP: Liquid. Mp: −96°, bp: 131.0°, d: 1.05, vap d: 5.41, flash p: 75.2°F.

SYN: DIETHYLDICHLOROSILANE (DOT)

TOXICITY DATA WITH REFERENCE
orl-rat LDLo:1000 mg/kg JIHTAB 30,332,48
ipr-rat LDLo:100 mg/kg JIHTAB 30,332,48

CONSENSUS REPORTS: Reported in EPA TSCA Inventory.

DOT CLASSIFICATION: 8; *Label:* Corrosive, Flammable Liquid

SAFETY PROFILE: Poison by intraperitoneal route. Moderately toxic by ingestion. Corrosive to tissue. Dangerous fire hazard when exposed to heat, flame, or oxidizers. Can react vigorously with oxidizing materials. To fight fire, use foam, CO_2, dry chemical. When heated to decomposition or in reaction with water or steam it emits toxic and corrosive fumes of Cl^-. See also CHLOROSILANES.

DEZ000 CAS:866-55-7 HR: 3
DICHLORODIETHYLSTANNANE
mf: $C_4H_{10}Cl_2Sn$ mw: 247.73

PROP: Water-white crystals. Mp: 85°, bp: 277°.

SYNS: DIAETHYLZINNDICHLORID (GERMAN) □ DICHLORODIETHYLTIN □ DIETHYLDICHLOROSTANNANE □ DIETHYLSTANNYL DICHLORIDE □ DIETHYLTIN CHLORIDE □ DIETHYLTIN DICHLORIDE

TOXICITY DATA WITH REFERENCE
orl-rat LDLo:160 mg/kg BJIMAG 15,15,58
ivn-rat LD50:20,600 µg/kg AEPPAE 242,370,61

OSHA PEL: TWA 0.1 mg(Sn)/m³ (skin)
ACGIH TLV: TWA 0.1 mg(Sn)/m³ (skin) (Proposed: TWA 0.1 mg(Sn)/m³; STEL 0.2 mg(Sn)/m³ (skin))
NIOSH REL: (Organotin Compounds) TWA 0.1 mg(Sn)/m³

SAFETY PROFILE: Poison by ingestion and intravenous routes. See also TIN COMPOUNDS and CHLORIDES. When heated to decomposition it emits toxic fumes of Cl^-.

For occupational chemical analysis use NIOSH: Organotin Compounds 5504.

DFA000 CAS:1649-08-7 HR: 1
1,2-DICHLORO-1,1-DIFLUOROETHANE
mf: $C_2H_2Cl_2F_2$ mw: 134.94

PROP: A liquid. Fp: −101°, bp: 45–47°, d: 1.416 @ 20°/4°.

TOXICITY DATA WITH REFERENCE
ihl-rat TCLo:2000 ppm/6H (male 14W pre):REP EPASR* 8EHQ-0587-0676
ihl-rat LCLo:20,000 ppm/4H TXAPA9 19,1,71

CONSENSUS REPORTS: Reported in EPA TSCA Inventory.

SAFETY PROFILE: Experimental reproductive effects. Mildly toxic by inhalation. When heated to decomposition it emits very toxic fumes of Cl^- and F^-.

DFA200 CAS:27156-03-2 HR: 2
DICHLORODIFLUOROETHYLENE
mf: $C_2Cl_2F_2$ mw: 132.92

PROP: Liquid. Vap d: 4.6.

SAFETY PROFILE: Moderately toxic by inhalation. A skin, eye, and mucous membrane irritant. Will react with water or steam to produce toxic and corrosive fumes. When heated to decomposition it emits toxic fumes of F^- and Cl^-.

DFA300 CAS:79-35-6 HR: 2
1,1-DICHLORO-2,2-DIFLUOROETHYLENE
mf: $C_2Cl_2F_2$ mw: 132.92

PROP: Volatile liquid. Fp: −127.1 to −126.7°, bp: 20.4° @ 764.3 mm.

SYNS: 1,1-DIFLUORO-2,2-DICHLOROETHYLENE ☐ GENETRON 1112A ☐ GENETRONE 1112A

TOXICITY DATA WITH REFERENCE
ihl-rat LC50:505 mg/m³/4H GTPZAB 21(5),36,77
ihl-mus LC50:610 mg/m³/4H GTPZAB 21(5),36,77
ihl-gpg LC50:700 mg/m³/4H GTPZAB 21(5),36,77

SAFETY PROFILE: Moderately toxic by inhalation. When heated to decomposition it emits toxic fumes of F^- and Cl^-.

DFA400 CAS:76-38-0 HR: 2
2,2-DICHLORO-1,1-DIFLUOROETHYL METHYL ETHER
mf: $C_3H_4Cl_2F_2O$ mw: 164.97

PROP: Liquid. D: 1.426 @ 20°/4°, mp: −35°, bp: 105°.

SYNS: ANALGIZER ☐ ANECOTAN ☐ 2,2-DICHLORO-1,1-DIFLUORO-1-METHOXYETHANE ☐ INGALAN ☐ INGALAN (RUSSIAN) ☐ INHALAN ☐ METHOFLURANE ☐ METHOXANE ☐ METHOXYFLUORAN ☐ METHOXYFLUORANE ☐ METHOXYFLURANE ☐ METOFANE ☐ METOXFLURAN ☐ METOXIFLURAN ☐ MOF ☐ NSC-110432 ☐ PENTHRANE ☐ PENTRAN ☐ PENTRANE

TOXICITY DATA WITH REFERENCE
eye-rbt 100 mg MOD FEPRA7 35,729,76
cyt-hmn:lym 200 ppm/24H ENVRAL 12,366,76
oms-ham:fbr 1 pph ANESAV 43,21,75
ihl-rat TCLo:100 ppm/8H (1-21D preg):TER ANESAV 48,11,78
ihl-rat TCLo:400 ppm/8H (female 1-21D post):TER ANESAV 48,11,78
ihl-mus TCLo:2 ppm/4H (female 6-15D post):TER AACRAT 59,421,80
ihl-hmn TCLo:3500 ppm/1H:KID CANJAE 21,294,74
orl-rat LD50:3600 mg/kg 85GMAT -,53,82
ihl-rat LC50:33,500 mg/m³/4H 85GMAT -,53,82
ihl-mus LC50:21,500 mg/m³ GTPZAB 22(7),55,78

CONSENSUS REPORTS: IARC Cancer Review: Animal Inadequate Evidence IMEMDT 7,93,87. EPA Genetic Toxicology Program.

NIOSH REL: (Waste Anesthetic Gases and Vapors) CL 2 ppm/1H

SAFETY PROFILE: Moderately toxic by ingestion. Mildly toxic by inhalation. Human systemic effects by inhalation: depressed renal function. An experimental teratogen. Human mutation data reported. An eye irritant. See also ETHERS. When heated to decomposition it emits very toxic fumes of Cl^- and F^-.

DFA600 CAS:75-71-8 HR: 1
DICHLORODIFLUOROMETHANE
DOT: UN 1028
mf: CCl_2F_2 mw: 120.91

PROP: Colorless, almost odorless gas. Mp: −158°, bp: −29°, vap press: 5 atm @ 16.1°.

SYNS: ALGOFRENE TYPE 2 ☐ ARCTON 6 ☐ DIFLUORODICHLOROMETHANE ☐ DWUCHLORODWUFLUOROMETAN (POLISH) ☐ ELECTRO-CF 12 ☐ ESKIMON 12 ☐ F 12 ☐ FC 12 ☐ FLUOROCARBON-12 ☐ FREON 12 ☐ FREON F-12 ☐ FRIGEN 12 ☐ GENETRON 12 ☐ HALON ☐ ISCEON 122 ☐ ISOTRON 12 ☐ KAISER CHEMICALS 12 ☐ LEDON 12 ☐ PROPELLANT 12 ☐ RCRA WASTE NUMBER U075 ☐ R12 (DOT) ☐ REFRIGERANT 12 ☐ UCON 12 ☐ UCON 12/HALOCARBON 12

TOXICITY DATA WITH REFERENCE
ihl-hmn TCLo:200,000 ppm/30M:EYE,PUL,LIV EJTXAZ 9,385,76
ihl-rat LC50:80 pph/30M EJTXAZ 9,385,76
ihl-mus LC50:76 pph/30M EJTXAZ 9,385,76
ihl-rbt LC50:80 pph/30M EJTXAZ 9,385,76
ihl-gpg LC50:80 pph/30M EJTXAZ 9,385,76

CONSENSUS REPORTS: Reported in EPA TSCA Inventory. EPA Genetic Toxicology Program.

OSHA PEL: TWA 1000 ppm
ACGIH TLV: TWA 1000 ppm
DFG MAK: 1000 ppm (5000 mg/m³)
DOT CLASSIFICATION: 2.2; *Label:* Nonflammable Gas

SAFETY PROFILE: Human systemic effects by inhalation: conjunctiva irritation, fibrosing alveolitis, and liver changes. Narcotic in high concentrations. Nonflammable gas. Can react violently with Al. When heated to decomposition it emits highly toxic fumes of phosgene, Cl^-, and F^-.

For occupational chemical analysis use NIOSH: see Dichlorodifluoromethane and 1,2-Dichlorotetrafluoroethane, 1018.

DFB400 CAS:56275-41-3 **HR: 1**
DICHLORODIFLUOROMETHANE with 1,1-DIFLUOROETHANE
DOT: UN 1954
mf: $C_2H_4F_2 \cdot CCl_2F_2$ mw: 186.97

SYNS: DICHLORODIFLUOROMETHANE and DIFLUOROETHANE AZEOTROPIC MIXTURE (DOT) □ FREON 500 □ R500 (DOT) □ UCON 500/HALOCARBON 500

DOT CLASSIFICATION: 2.2; *Label:* Nonflammable Gas

SAFETY PROFILE: A simple asphyxiant. See also components as listed. When heated to decomposition it emits very toxic fumes of Cl^- and F^-.

DFC000 **HR: 3**
DICHLORODIFLUOROMETHANE with TRICHLOROTRIFLUOROETHANE
mf: $CCl_3F \cdot CCl_2F_2$ mw: 258.27

SYN: DICHLORODIFLUOROMETHANE–TRICHLOROTRIFLUOROETHANE MIXTURE

TOXICITY DATA WITH REFERENCE
ihl-rat LC50:30 pph/30M EJTXAZ 9,385,76
ihl-mus LC50:22 pph/30M EJTXAZ 9,385,76
ihl-gpg LC50:50 pph/30M EJTXAZ 9,385,76

SAFETY PROFILE: Very mildly toxic by inhalation. See components as listed. When heated to decomposition it emits very toxic fumes of Cl^- and F^-.

DFC200 CAS:2767-41-1 **HR: 3**
DICHLORODIHEXYLSTANNANE
mf: $C_{12}H_{26}Cl_2Sn$ mw: 359.97

SYN: DIHEXYLTIN DICHLORIDE

TOXICITY DATA WITH REFERENCE
orl-rat LDLo:160 mg/kg BJIMAG 15,15,58
ivn-rat LDLo:10 mg/kg BJIMAG 15,15,58

OSHA PEL: TWA 0.1 mg(Sn)/m^3 (skin)
ACGIH TLV: TWA 0.1 mg(Sn)/m^3 (skin) (Proposed: TWA 0.1 mg(Sn)/m^3; STEL 0.2 mg(Sn)/m^3 (skin))
NIOSH REL: (Organotin Compounds) TWA 0.1 mg(Sn)/m^3

SAFETY PROFILE: Poison by ingestion and intravenous routes. See also TIN COMPOUNDS. When heated to decomposition it emits toxic fumes of Cl^-.

For occupational chemical analysis use NIOSH: Organotin Compounds 5504.

DFC800 CAS:33770-60-4 **HR: 3**
(2,5-DICHLORO-3,6-DIHYDROXY-p-BENZOQUINOLATO)MERCURY
mf: $C_6Cl_2HgO_4$ mw: 407.55

SYNS: 2,5-DICHLORO-3,6-DIHYDROXY-p-BENZOQUINONE, MERCURY SALT □ (2,5-DICHLORO-3,6-DIHYDROXY-p-BENZOQUINONE), MERCURY SALT

TOXICITY DATA WITH REFERENCE
ivn-mus LD50:10 mg/kg CSLNX* NX#04223

CONSENSUS REPORTS: Mercury and its compounds are on the Community Right-To-Know List.

OSHA PEL: CL 0.1 mg(Hg)/m^3 (skin)
ACGIH TLV: TWA 0.1 mg(Hg)/m^3 (skin)
NIOSH REL: (Mercury, Aryl and Inorganic): CL 0.1 mg/m^3 (skin)

SAFETY PROFILE: Poison by intravenous route. See also MERCURY COMPOUNDS. When heated to decomposition it emits very toxic fumes of Cl^- and Hg.

DFD000 CAS:10331-57-4 **HR: 3**
5,5′-DICHLORO-2,2′-DIHYDROXY-3,3′-DINITROBIPHENYL
mf: $C_{12}H_6Cl_2N_2O_6$ mw: 345.10

SYNS: BAY 9015 □ BAYER 9015 □ BILEVON M □ 3,3′-DICHLORO-5,5′-DINITRO-O,O′-BIPHENOL (FRENCH) □ 4,4′-DICHLORO-6,6′-DINITRO-O,O′-BIPHENOL □ 5,5′-DICHLORO-3,3′-DINITRO(1,1′-BIPHENYL)-2,2′-DIOL □ ME 3625 □ MENICHLOPHOLAN □ NICLOFOLAN

TOXICITY DATA WITH REFERENCE
orl-ham TDLo:10 mg/kg/(8D preg):TER JETOAS 4,525,71
orl-rat LD50:10 mg/kg TXAPA9 21,315,72
orl-ham LD50:50 mg/kg JETOAS 4,525,71
orl-dom LDLo:15 mg/kg FAZMAE 17,108,73
orl-bwd LD50:13 mg/kg TXAPA9 21,315,72

SAFETY PROFILE: A poison by ingestion. An experimental teratogen. When heated to decomposition it emits very toxic fumes of Cl^- and NO_x. See also NITRO COMPOUNDS of AROMATIC HYDROCARBONS and CHLORIDES.

DFD200 CAS:29202-04-8 **HR: 3**
3,4-DICHLORO-2,5-DILITHIOTHIOPHENE
mf: $C_4Cl_2Li_2S$ mw: 164.89

SAFETY PROFILE: The dry material is a slightly shock-sensitive explosive. When heated to decomposition it emits toxic fumes of Cl^- and SO_x. See also LITHIUM COMPOUNDS.

DFD400 CAS:17010-61-6 **HR: 2**
3′,4′-DICHLORO-4-DIMETHYLAMINOAZOBENZENE
mf: $C_{14}H_{13}Cl_2N_3$ mw: 294.20

SYNS: BENZENAMINE, 4-((3,4-DICHLOROPHENYL)AZO)-N,N-DIMETHYL-(9CI) □ 3′,4′-Cl2-DAB □ p-((3,4-DICHLOROPHENYL)AZO)-N,N-DIMETHYLANILINE

TOXICITY DATA WITH REFERENCE
orl-rat TDLo:15,120 mg/kg/36W-C:CAR CBINA8 53,107,85
orl-rat TD:11 g/kg/17W-I:ETA CNREA8 30,1520,70

SAFETY PROFILE: Questionable carcinogen with experimental carcinogenic and tumorigenic data. When heated to decomposition it emits toxic fumes of Cl^- and NO_x.

D

DFD600 CAS:101652-02-2 HR: 3
7,8-DICHLORO-10-(2-(DIMETHYLAMINO)ETHYL)I-SOALLOXAZINE SULFATE
mf: $C_{14}H_{13}Cl_2N_5O_2{\cdot}H_2O_4S$ mw: 452.30

TOXICITY DATA WITH REFERENCE
scu-mus LD50:30 mg/kg CMTRAG 2,96,61
ivn-mus LD50:18 mg/kg CMTRAG 2,96,61

SAFETY PROFILE: Poison by subcutaneous and intravenous routes. See also SULFATES and CHLORIDES. When heated to decomposition it emits very toxic fumes of SO_x, NO_x, and Cl^-.

DFE000 CAS:97864-38-5 HR: 3
7,8-DICHLORO-10-(3-(DIMETHYLAMINO)PROPYL)I-SOALLOXAZINE HYDROCHLORIDE
mf: $C_{15}H_{15}Cl_2N_5O_2{\cdot}ClH$ mw: 404.71

TOXICITY DATA WITH REFERENCE
ipr-rat LD50:22 mg/kg CMTRAG 2,96,61
scu-mus LD50:23 mg/kg CMTRAG 2,96,61
ivn-mus LD50:34 mg/kg CMTRAG 2,96,61

SAFETY PROFILE: Poison by intraperitoneal, subcutaneous, and intravenous routes. When heated to decomposition it emits very toxic fumes of NO_x and Cl^-.

DFE100 CAS:594-84-3 HR: 3
2,2-DICHLORO-3,3-DIMETHYLBUTANE
mf: $C_6H_{12}Cl_2$ mw: 155.07

$CH_3CCl_2C(CH_3)_3$

SAFETY PROFILE: Violent reaction with sodium hydroxide. When heated to decomposition it emits toxic fumes of Cl^-. See also CHLORINATED HYDROCARBONS, ALIPHATIC.

DFE200 CAS:118-52-5 HR: 2
1,3-DICHLORO-5,5-DIMETHYL HYDANTOIN
mf: $C_5H_6Cl_2N_2O_2$ mw: 197.03

$ClNCO{\cdot}NClCO{\cdot}C(CH_3)_2$ (ring)

PROP: Crystals, liberates chlorine on contact with hot water; prisms from $CHCl_3$. Mp: 132°. Subl @ 100°; conflagrates @ 212°; d: 1.5 @ 20°, vap d: 6.8. Sol in H_2O; mod in sol AcOH and EtOH.

SYNS: DACTIN ☐ DAKTIN ☐ DANTOIN ☐ DCA ☐ DICHLORANTIN ☐ DICHLORODIMETHYLHYDANTOIN ☐ 1,3-DICHLORO-5,5-DIMETHYL-2,4-IMIDAZOLIDINEDIONE ☐ 1,3-DICHLORO-5,5'-METHYLHYDANTOIN ☐ HALANE ☐ HYDAN ☐ HYDAN (antiseptic) ☐ NCI-C03054 ☐ OMCHLOR

TOXICITY DATA WITH REFERENCE
skn-rbt 500 mg/24H SEV EPASR* 8EHQ-0281-0382
skn-rbt 100 mg/24H SEV EPASR* 8EHQ-0281-0382
sln-dmg-par 250 ppm ENMUDM 7,677,85
otr-rat:emb 6300 ng/plate JJATDK 1,190,81
orl-rat LD50:542 mg/kg DTLVS* 4,129,80
ihl-rat LCLo:20 g/m³/1H EPASR* 8EHQ-0281-0382
orl-rbt LD50:1520 mg/kg GISAAA 47(6),76,82
orl-gpg LD50:1350 mg/kg GISAAA 47(6),76,82

CONSENSUS REPORTS: Reported in EPA TSCA Inventory.

OSHA PEL: TWA 0.2 mg/m³; STEL 0.4 mg/m³
ACGIH TLV: TWA 0.2 mg/m³; STEL 0.4 mg/m³

SAFETY PROFILE: Moderately toxic by ingestion. Mildly toxic by inhalation. A severe skin irritant. Mutation data reported. Avoid excessive contact because of effects of active chlorine on skin. Some of the hydantoins are central nervous system depressants. Mixtures with xylene may explode. Will react with water or steam to produce toxic and corrosive fumes. When heated to decomposition it emits toxic fumes of Cl^- and NO_x. See also CHLORIDES.

DFE229 CAS:40580-75-4 HR: 3
DICHLORO(4,5-DIMETHYL-o-PHENYLENEDIAMMINE) PLATINUM(II)
mf: $C_8H_{12}Cl_2N_2Pt$ mw: 402.21

SYN: cis-DICHLORO(4,5-DIMETHYL-O-PHENYLENEDIAMMINE)PLATINUM(II)

TOXICITY DATA WITH REFERENCE
mmo-sat 10,200 nmol/L JMCMAR 23,459,80
ipr-mus LD50:283 mg/kg CBINA8 11,145,75

CONSENSUS REPORTS: Reported in EPA TSCA Inventory.

SAFETY PROFILE: Poison by intraperitoneal route. Mutation data reported. See also TIN COMPOUNDS. When heated to decomposition it emits toxic fumes of Cl^-.

DFE259 CAS:75-78-5 HR: 3
DICHLORODIMETHYLSILANE
DOT: UN 1162
mf: $C_2H_6Cl_2Si$ mw: 129.06

PROP: Liquid. D: 1.06 @ 20°/4°, mp: −16°.

SYNS: DIMETHYLDICHLOROSILANE (DOT) ☐ DIMETHYL-DICHLORSILAN

TOXICITY DATA WITH REFERENCE
skn-rbt 20 mg/24H MOD 85JCAE-,1219,86
eye-rbt 5 mg/24H SEV 85JCAE-,1219,86
orl-rat LD50:5660 μL/kg JACTDZ 12,573,93
ihl-rat LC50:930 ppm/4H 85JCAE-,1219,86
ipr-rat LDLo:10 mg/kg JIDHAN 30,332,48
ihl-mus LC50:300 mg/m³/2H TPKVAL 3,23,61

CONSENSUS REPORTS: Reported in EPA TSCA Inventory.

DOT CLASSIFICATION: 3; *Label:* Flammable Liquid, Corrosive

SAFETY PROFILE: Poison by ingestion and intraperitoneal routes. Moderately toxic by inhalation. A skin and severe eye irritant. Violent reaction on contact with water. When heated to decomposition it emits toxic fumes of Cl^-. See also CHLOROSILANES.

DFE300 **CAS:59183-17-4** **HR: 3**
3,6-DICHLORO-3,6-DIMETHYLTETRAOXANE
mf: $C_4H_6Cl_2O_4$ mw: 189.00

$Cl(CH_3)COOCCl(CH_3)O\ O$

SAFETY PROFILE: An extremely shock- and heat-sensitive explosive. When heated to decomposition it emits toxic fumes of Cl^-. See also PEROXIDES.

DFE469 **CAS:58270-08-9** **HR: 3**
(trans-4)-DICHLORO(4,4-DIMETHYLZINC 5((((METHYLAMINO)CARBONYL)OXY)IMINO)PENTANENITRILE)
mf: $C_9H_{15}Cl_2N_3O_2Zn$ mw: 333.54

PROP: Powder. Mp: 120–125°. Sol in H_2O.

SYNS: AC 85258 □ ETHIENOCARB

TOXICITY DATA WITH REFERENCE
orl-rat LD50:9 mg/kg KSKZAN 16(2),65,78
skn-rat LD50:857 mg/kg KSKZAN 16(2),65,78

CONSENSUS REPORTS: EPA Extremely Hazardous Substances List. Reported in EPA TSCA Inventory. Zinc and its compounds, as well as cyanide and its compounds, are on the Community Right-To-Know List.

SAFETY PROFILE: Poison by ingestion. Moderately toxic by skin contact. When heated to decomposition it emits toxic fumes of Cl^-, NO_x, CN^-, and ZnO. See also ZINC COMPOUNDS and NITRILES.

DFE550 **CAS:1587-41-3** **HR: 3**
DICHLORODINITROMETHANE
mf: $CCl_2N_2O_4$ mw: 174.93

PROP: Liquid. Bp: 121–122.5°.

SAFETY PROFILE: Explodes when heated. Upon decomposition it emits toxic fumes of Cl^- and NO_x. See also NITROMETHANE.

DFE600 **CAS:3883-43-0** **HR: 2**
trans-2,3-DICHLORO-1,4-DIOXANE
mf: $C_4H_6Cl_2O_2$ mw: 157.00

PROP: Crystals. Mp: 28–30°, bp: 97–98° @ 20 mm.

SYN: trans-2,3-DICHLORO-p-DIOXANE

TOXICITY DATA WITH REFERENCE
skn-mus TDLo:4080 mg/kg/68W-I:ETA JNCIAM 53,695,74
scu-mus TDLo:1260 mg/kg/63W-I:NEO JNCIAM 53,695,74
orl-rat LD50:1410 mg/kg AIHAAP 30,470,69
skn-rbt LD50:440 mg/kg AIHAAP 30,470,69

SAFETY PROFILE: Moderately toxic by ingestion and skin contact. Questionable carcinogen with experimental neoplastigenic and tumorigenic data. When heated to decomposition it emits toxic fumes of Cl^-.

DFE800 **CAS:28675-08-3** **HR: 2**
DICHLORODIPHENYL OXIDE
mf: $C_{12}H_8Cl_2O$ mw: 239.10

PROP: Liquid. Vap d: 8.2.

SYNS: DICHLOROPHENYL ETHER □ PHENYL ETHER, DICHLORO

TOXICITY DATA WITH REFERENCE
orl-gpg LDLo:1000 mg/kg 14CYAT 2,1707,63

OSHA PEL: TWA 0.5 mg/m³

SAFETY PROFILE: Moderately toxic by ingestion. When heated to decomposition it emits toxic fumes of Cl^-. See also ETHERS and CHLORIDES.

DFF000 **CAS:80-10-4** **HR: 3**
DICHLORO DIPHENYLSILANE
DOT: UN 1769
mf: $C_{12}H_{10}Cl_2Si$ mw: 253.21

PROP: Colorless liquid. Mp: −22°, bp: 303°, d: 1.19 @ 20°, vap d: 8.45.

SYNS: DICHLOR-DIFENYLSILAN □ DIPHENYL DICHLOROSILANE (DOT)

TOXICITY DATA WITH REFERENCE
skn-rbt 500 mg/24H MOD 28ZPAK -,221,72
eye-rbt 5 mg/24H SEV 28ZPAK -,221,72

CONSENSUS REPORTS: Reported in EPA TSCA Inventory.

DOT CLASSIFICATION: 8; *Label:* Corrosive

SAFETY PROFILE: A poison irritant to skin, eyes, and mucous membranes. See also CHLOROSILANES. Can react vigorously with oxidizing materials. When heated to decomposition or on contact with acid or acid fumes it emits toxic fumes of Cl^-.

DFF200 **CAS:77791-64-1** **HR: 2**
2′,6′-DICHLORO-2-(DIPROPYLAMINO)ACETANILIDE HYDROCHLORIDE
mf: $C_{14}H_{20}Cl_2N_2O\cdot ClH$ mw: 339.72

SYN: C 3057

TOXICITY DATA WITH REFERENCE
eye-rbt 2% MOD ARZNAD 8,407,58
ipr-rat LD50:450 mg/kg ARZNAD 8,407,58
ipr-mus LD50:550 mg/kg ARZNAD 8,407,58
scu-mus LD50:2250 mg/kg ARZNAD 8,407,58

SAFETY PROFILE: Moderately toxic by intraperitoneal and subcutaneous routes. An eye irritant. When heated to decomposition it emits very toxic fumes of Cl^- and NO_x.

DFF400 **CAS:867-36-7** **HR: 3**
DICHLORODIPROPYLSTANNANE
mf: $C_6H_{14}Cl_2Sn$ mw: 275.79

PROP: Colorless crystals. Mp: 82.5–83°, bp: 118–121° @ 10 mm.

SYNS: DICHLORODIPROPYLTIN ☐ DIPROPYLTIN CHLORIDE ☐ DIPROPYLTIN DICHLORIDE ☐ DI-n-PROPYLTIN DICHLORIDE

TOXICITY DATA WITH REFERENCE
orl-rat LDLo:160 mg/kg BJIMAG 15,15,58

OSHA PEL: TWA 0.1 mg(Sn)/m^3 (skin)
ACGIH TLV: TWA 0.1 mg(Sn)/m^3 (skin) (Proposed: TWA 0.1 mg(Sn)/m^3; STEL 0.2 mg(Sn)/m^3 (skin))
NIOSH REL: (Organotin Compounds) TWA 0.1 mg(Sn)/m^3

SAFETY PROFILE: Poison by ingestion. See also TIN COMPOUNDS and CHLORIDES. When heated to decomposition it emits toxic fumes of Cl^-.

For occupational chemical analysis use NIOSH: Organotin Compounds 5504.

DFF500 CAS:15227-42-6 HR: 3
cis-DICHLORO(DIPYRIDINE)PLATINUM(II)
mf: $C_{10}H_{10}Cl_2N_2Pt$ mw: 424.21

PROP: Sulfur yellow solid. Mp: 224° (decomp). Sol in $CHCl_3$, Me_2CO, and DMF.

SYN: DICHLORODIPYRIDINEPLATINUM(II) (Z)

TOXICITY DATA WITH REFERENCE
mmo-sat 100 µg/plate MUREAV 95,79,82
ipr-mus LDLo:131 mg/kg JPMSAE 65,315,76

SAFETY PROFILE: Poison by intraperitoneal route. Mutation data reported. When heated to decomposition it emits toxic fumes of Cl^- and NO_x. See also PLATINUM COMPOUNDS.

DFF600 CAS:3583-47-9 HR: 2
1,4-DICHLORO-2,3-EPOXYBUTANE
mf: $C_4H_6Cl_2O$ mw: 141.00

TOXICITY DATA WITH REFERENCE
skn-rbt 10 mg/24H open MLD AIHAAP 23,95,62
skn-rbt 500 mg/24H MLD 85JCAE-,771,86
eye-rbt 500 mg/24H MLD 85JCAE-,771,86
mmo-klp 5 mmol/L MUREAV 89,269,81
mma-sat 1 mmol/L ARTODN 41,249,79
orl-rat LDLo:710 mg/kg AIHAAP 23,95,62
skn-rbt LDLo:2830 mg/kg AIHAAP 23,95,62

SAFETY PROFILE: Moderately toxic by ingestion and skin contact. A skin and eye irritant. Mutation data reported. When heated to decomposition it emits toxic fumes of Cl^-.

DFF800 CAS:1300-21-6 HR: 2
DICHLOROETHANE
mf: $C_2H_4Cl_2$ mw: 98.96

PROP: Lel: 5.6%, uel: 11.4%.

TOXICITY DATA WITH REFERENCE
ihl-rat TCLo:57 mg/m^3/4H (female 26W pre):REP GTPZAB 19(7),20,75
ihl-rat TCLo:15 mg/m^3/4H (female 96D pre):TER AKGIAO 53(2),57,77
ihl-rat TCLo:15 mg/m^3/4H (16W pre):REP GISAAA 41(6),100,76
ihl-rat TCLo:15 mg/m^3/4H (female 96D pre):REP AKGIAO 53(2),57,77
orl-rat LD50:1120 mg/kg HYSAAV 32,349,67
orl-mus LD50:625 mg/kg HYSAAV 32,349,67
ihl-mus LCLo:10 g/m^3 GISAAA 20(8),19,55
skn-rbt LD50:3890 mg/kg UCDS** 3/23/70

SAFETY PROFILE: Moderately toxic by ingestion and skin contact. Mildly toxic by inhalation. An experimental teratogen. Other experimental reproductive effects by inhalation. When heated to decomposition it emits very toxic fumes of Cl^-. See also ETHYLENE DICHLORIDE; and CHLORINATED HYDROCARBONS, ALIPHATIC.

DFF809 CAS:75-34-3 HR: 3
1,1-DICHLOROETHANE
DOT: UN 2362
mf: $C_2H_4Cl_2$ mw: 98.96

PROP: Colorless liquid; aromatic, ethereal odor; hot, saccharine taste. Mp: −97.7°, lel: 5.6%, fp: −98°, bp: 57.3°, flash p: 22°F (TOC), d: 1.174 @ 20°/4°, vap press: 230 mm @ 25°, vap d: 3.44, autoign temp: 856°F.

SYNS: AETHYLIDENCHLORID (GERMAN) ☐ CHLORINATED HYDROCHLORIC ETHER ☐ CHLORURE d'ETHYLIDENE (FRENCH) ☐ CLORURO di ETILIDENE (ITALIAN) ☐ 1,1-DICHLOORETHAAN (DUTCH) ☐ 1,1-DICHLORAETHAN (GERMAN) ☐ 1,1-DICLOROETANO (ITALIAN) ☐ ETHYLIDENE CHLORIDE ☐ ETHYLIDENE DICHLORIDE ☐ NCI-C04535 ☐ RCRA WASTE NUMBER U076

TOXICITY DATA WITH REFERENCE
ihl-rat TCLo:6000 ppm/7H (6–15D preg):TER TXAPA9 28,452,74
orl-mus TDLo:185 g/kg/78W-I:ETA,TER NCITR* NCI-CG-TR-66,78
orl-mus TD:1300 g/kg/78W-I:ETA,TER NCITR* NCI-CG-TR-66,78
orl-rat LD50:725 mg/kg HYSAAV 32,349,67
ihl-rat LCLo:16,000 ppm/4H JIDHAN 31,343,49

CONSENSUS REPORTS: NCI Carcinogenesis Bioassay (gavage); Inadequate Studies: mouse, rat NCITR* NCI-CG-TR-66,78. Reported in EPA TSCA Inventory.

OSHA PEL: TWA 100 ppm
ACGIH TLV: TWA 200 ppm; STEL 250 ppm
DFG MAK: 100 ppm (400 mg/m^3)
NIOSH REL: (1,1-Dichloroethane): Handle with caution.
DOT CLASSIFICATION: 3; *Label:* Flammable Liquid

SAFETY PROFILE: Moderately toxic by ingestion. Experimental teratogenic effects. Questionable carcinogen with experimental tumorigenic data. Liver damage reported in experimental animals. A very dangerous fire hazard and moderate explosion hazard when exposed to heat or flame; can react vigorously with oxidizing materials. To fight fire, use alcohol foam, water, foam, CO_2, dry chemical. When heated to decomposition it emits highly toxic fumes of phosgene and Cl^-.

For occupational chemical analysis use NIOSH: Hydrocarbons, Halogenated, 1003.

DFG159 CAS:10140-87-1 HR: 3
1,2-DICHLOROETHANOL ACETATE
mf: $C_4H_6Cl_2O_2$ mw: 157.00

PROP: Water-white liquid. Bp: 79–79.5° @ 33 mm; d: 1.296/20°C. Flash p: 307°F. Insoluble in water.

SYN: 1,2-DICHLOROETHYL ACETATE

TOXICITY DATA WITH REFERENCE
ihl-rat LCLo:16 ppm/4H JIHTAB 31,343,49

CONSENSUS REPORTS: EPA Extremely Hazardous Substances List.

SAFETY PROFILE: Poison by inhalation. Combustible when exposed to heat or flame. The vapor is potentially explosive. To fight small fires, use dry chemical, carbon dioxide, water spray, or foam. To fight large fires, use water spray, fog, or foam. May explode on heating with nitrates. When heated to decomposition it emits toxic fumes of Cl^- and phosgene.

DFG200 CAS:72-00-4 HR: 3
2,2-DICHLOROETHENYL DIETHYL PHOSPHATE
mf: $C_6H_{11}Cl_2O_4P$ mw: 249.04

SYNS: 2,2-DICHLOROVINYL DIETHYL PHOSPHATE □ O-(2,2-DI-CHLORVINYL)-O,O-DIETHYLPHOSPHAT (GERMAN) □ DICHLORVOS-ETHYL

TOXICITY DATA WITH REFERENCE
mmo-sat 5 μL/plate MUREAV 28,405,75
orl-rat LD50:2500 μg/kg EQSFAP 3,173,74
orl-mus LDLo:42 mg/kg AECTCV 14,111,85
ipr-mus LD50:12 mg/kg ARZNAD 5,746,55

SAFETY PROFILE: Poison by ingestion and intraperitoneal routes. Mutation data reported. See also PHOSPHATES and ESTERS. When heated to decomposition it emits very toxic fumes of Cl^- and PO_x.

DFG700 CAS:5960-88-3 HR: 3
2,2-DICHLOROETHYLAMINE
mf: $C_2H_5Cl_2N$ mw: 113.97

PROP: A liquid. Bp: 64° @ 58 mm.

SAFETY PROFILE: A poison. Solutions in ether are violently explosive at 80°C and 260 mbar. When heated to decomposition it emits toxic fumes of Cl^- and NO_x.

DFH000 CAS:10072-25-0 HR: 2
9-(2-(DI(2-CHLOROETHYL)AMINO)ETHYLAMINO)-6-CHLORO-2-METHOXYACRIDINE
mf: $C_{20}H_{22}Cl_3N_3O{\cdot}2ClH{\cdot}H_2O$ mw: 517.74

SYNS: 9-(2-(BIS(2-CHLOROETHYL)AMINO)ETHYLAMINO)-6-CHLORO-2-METHOXYACRIDINE DIHYDROCHLORIDE □ ICR-48b □ NSC-34372 □ QUINACRINE ETHYL MUSTARD

TOXICITY DATA WITH REFERENCE
ipr-mus TDLo:16 mg/kg/4W:CAR JNCIAM 36,915,66

SAFETY PROFILE: Questionable carcinogen with experimental carcinogenic data. When heated to decomposition it emits very toxic fumes of Cl^- and NO_x.

DFH100 HR: 3
o-(p-DI-(2-CHLOROETHYL)AMINOPHENYL)-dl-TYROSINE DIHYDROCHLORIDE
mf: $C_{19}H_{22}Cl_2N_2O_3{\cdot}2ClH$ mw: 470.25

SYN: o-(p-DI(2-CHLORAETHYL)-AMINOPHENYL)-dl-TYROSIN-DIHYDROCHLORID (GERMAN)

TOXICITY DATA WITH REFERENCE
orl-rat LD50:620 mg/kg GWXXBX #2644941
ipr-rat LD50:62 mg/kg GWXXBX #2644941
ivn-rat LD50:62 mg/kg GWXXBX #2644941
orl-mus LD50:360 mg/kg GWXXBX #2644941
ipr-mus LD50:110 mg/kg GWXXBX #2644941
ivn-mus LD50:30 mg/kg GWXXBX #2644941

SAFETY PROFILE: Poison by ingestion, intravenous, and intraperitoneal routes. When heated to decomposition it emits toxic fumes of NO_x and Cl^-.

DFH200 CAS:598-14-1 HR: 3
DICHLOROETHYLARSINE
DOT: UN 1892
mf: $C_2H_5AsCl_2$ mw: 174.89

PROP: Colorless liquid; fruity, biting, irritating odor. Mp: −65°, bp: 156° decomp, d: 1.742 @ 14°, vap press: 2.29 mm @ 21.5°, vap d: 6.03. Sol in H_2O; misc in EtOH and C_6H_6.

SYNS: ARSENIC DICHLOROETHANE □ ARSONOUS DICHLORIDE, ETHYL-(9CI) □ DICK (GERMAN) □ ED □ ETHYLARSONOUS DICHLORIDE □ ETHYLIDICHLORARSINE □ ETHYLIDICHLOROARSINE (DOT) □ TL 214

TOXICITY DATA WITH REFERENCE
ihl-hmn LCLo:14 ppm/30M NTIS** PB214-270
ihl-mus LC50:1555 mg/m³/10M NTIS** PB158-508
skn-mus LDLo:20 mg/kg NTIS** PB158-508
ihl-cat LCLo:12 ppm/40M ZGEMAZ 13,523,21
scu-cat LDLo:1 mg/kg ZGEMAZ 13,523,21

CONSENSUS REPORTS: Arsenic and its compounds are on the Community Right-To-Know List.

OSHA PEL: TWA 0.5 mg(As)/m³
DOT CLASSIFICATION: 6.1; *Label:* Poison

SAFETY PROFILE: A human poison by inhalation. Experimentally, a deadly poison by inhalation and subcutaneous routes, and probably by ingestion. A severe irritant. A military poison gas. Can react with oxidizing materials. Will react with water or steam to produce toxic and corrosive fumes. Dangerous; on contact with acid or acid fumes it emits highly toxic fumes of Cl^-, As and phosgene. See also ARSENIC COMPOUNDS.

DFH300 CAS:1739-53-3 HR: 3
DICHLOROETHYLBORANE
mf: $C_2H_5BCl_2$ mw: 110.78

PROP: Air- and moisture-sensitive liquid. Bp: 51.5°.

SAFETY PROFILE: Ignites spontaneously in air. When heated to decomposition it emits toxic fumes of Cl^-. See also BORANES and BORON COMPOUNDS.

D

DFH600 CAS:321-55-1 HR: 2
O,O-DI(2-CHLOROETHYL)-O-(3-CHLORO-4-METHYL-COUMARIN-7-YL) PHOSPHATE
mf: $C_{14}H_{14}Cl_3O_6P$ mw: 415.60

PROP: Crystals from EtOH. Mp: 91°.

SYNS: O,O-BIS(2-CHLOROETHYL)-O-(3-CHLORO-4-METHYL-7-COUMARINYL) PHOSPHATE □ 2-CHLOROETHANOL HYDROGEN PHOSPHATE ESTER with 3-CHLORO-7-HYDROXY-4-METHYLCOUMARIN □ 2-CHLOROETHANOL PHOSPHATE DIESTER ESTER with 3-CHLORO-7-HYDROXY-4-METHYLCOUMARIN □ 3-CHLORO-7-HYDROXY-4-METHYLCOUMARIN BIS(2-CHLOROETHYL)PHOSPHATE □ 3-CHLORO-4-METHYL-UMBELLIFERONE BIS(2-CHLOROETHYL)PHOSPHATE □ DI-(2-CHLOROETHYL)-3-CHLORO-4-METHYLCOUMARIN-7-YL PHOSPHATE □ DI-(2-CHLOROETHYL)-3-CHLORO-4-METHYL-7-COUMARINYL PHOSPHATE □ EUSTIDIL □ GALLOXON □ GALOXANE □ 96H60 □ HALOXON □ HELMIRANE □ HELMIRON □ HELMIRONE □ LOXON □ LUXON □ LXON

TOXICITY DATA WITH REFERENCE
dni-hmn:oth 10 mg/L JTEHD6 10,143,82
orl-rat LD50:900 mg/kg FAZMAE 17,108,73
ipr-ckn LD50:800 mg/kg BCPCA6 16,1183,67
orl-dom LD50:763 mg/kg AJVRAH 41,1857,80

SAFETY PROFILE: Moderately toxic by ingestion and intraperitoneal routes. Human mutation data reported. When heated to decomposition it emits very toxic fumes of PO_x and Cl^-. See also other coumarin entries.

DFH800 CAS:25323-30-2 HR: 3
DICHLOROETHYLENE
DOT: UN 1150
mf: $C_2H_2Cl_2$ mw: 96.94

TOXICITY DATA WITH REFERENCE
ihl-mus LCLo:76 g/m³/2H AEXPBL 83,235,18
ihl-gpg LCLo:155 g/m³/1H AEXPBL 83,235,18
orl-mam LDLo:2500 mg/kg UGLAAD 121,375,59

DOT CLASSIFICATION: 3; *Label:* Flammable Liquid

SAFETY PROFILE: Moderately toxic by ingestion. Mildly toxic by inhalation. Flammable when exposed to heat or flame. When heated to decomposition it emits toxic fumes of Cl^-. See also VINYLIDENE CHLORIDE.

DFI200 CAS:156-59-2 HR: 1
cis-DICHLOROETHYLENE
mf: $C_2H_2Cl_2$ mw: 96.94

HCCl=CHCl

PROP: Colorless liquid, pleasant odor. Mp: −80.5°, bp: 59°, lel: 9.7%, uel: 12.8%, flash p: 39°F, d: 1.291 @ 15°/4°, vap press: 400 mm @ 41.0°, vap d: 3.34.

SYN: 1,2-DICHLOROETHYLENE

TOXICITY DATA WITH REFERENCE
mmo-smc 100 mmol/L TCMUD8 4,365,84
mma-smc 40 mmol/L TCMUD8 4,365,84
mrc-smc 100 mmol/L TCMUD8 4,365,84
dns-rat:lvr 4300 µmol/L CRNGDP 5,1629,84
ihl-mus LCLo:65,000 mg/m³/2H AHBAAM 116,131,36
ihl-cat LCLo:20,000 mg/m³/6H AHBAAM 116,131,36

CONSENSUS REPORTS: Reported in EPA TSCA Inventory.

DFG MAK: 200 ppm (790 mg/m³)

SAFETY PROFILE: Mildly toxic by ingestion and inhalation. In high concentration it is irritating and narcotic. Has produced liver and kidney injury in experimental animals. Mutation data reported. Sometimes thought to be nonflammable, however, it is a dangerous fire hazard when exposed to heat or flame. Reaction with solid caustic alkalies or their concentrated solutions produces chloracetylene gas, which ignites spontaneously in air. Reacts violently with N_2O_4, KOH, Na, NaOH. Moderate explosion hazard in the form of vapor when exposed to flame. Can react vigorously with oxidizing materials. To fight fire, use water spray, foam, CO_2, dry chemical. When heated to decomposition it emits toxic fumes of Cl^-. See also VINYLIDENE CHLORIDE and CHLORINATED HYDROCARBONS, ALIPHATIC.

DFI210 CAS:540-59-0 HR: 3
1,2-DICHLOROETHYLENE
mf: $C_2H_2Cl_2$ mw: 96.94

PROP: Liquid with ethereal odor. Bp: 55°.

SYNS: ACETYLENE DICHLORIDE □ 1,2-DICHLOR-AETHEN (GERMAN) □ sym-DICHLOROETHYLENE □ DICHLORO-1,2-ETHYLENE (FRENCH) □ DIOFORM □ NCI-C56031

TOXICITY DATA WITH REFERENCE
skn-rbt 100 mg/24H MOD 85JCAE-,105,86
sln-Mold-asn 750 ppm MUREAV 266,117,92
orl-rat LD50:770 mg/kg ARSIM* 20,10,66
ipr-mus LD50:2 g/kg EJTXAZ 7,247,74
ihl-frg LCLo:117 mg/m³/1H AISFAR 15,1,37

CONSENSUS REPORTS: Reported in EPA TSCA Inventory. Community Right-To-Know List.

OSHA PEL: TWA 200 ppm
ACGIH TLV: TWA 200 ppm
DFG MAK: 200 ppm (790 mg/m³)

SAFETY PROFILE: Poison by inhalation. Moderately toxic by ingestion. A skin irritant. When heated to decomposition it emits highly toxic fumes of Cl^-. See also ACETYLENE COMPOUNDS; and CHLORINATED HYDROCARBONS, ALIPHATIC.

For occupational chemical analysis use NIOSH: Hydrocarbons, Halogenated, 1003.

DFI800 CAS:3967-55-3 HR: 2
1,2-DICHLOROETHYLENE CARBONATE
mf: $C_3H_2Cl_2O_3$ mw: 156.95

SYN: 4,5-DICHLORO-2-OXO-1,3-DIOXOLANE

TOXICITY DATA WITH REFERENCE
scu-mus TDLo:648 mg/kg/54W-I:ETA JNCIAM 48,1431,72

SAFETY PROFILE: Questionable carcinogen with experimental tumorigenic data. When heated to decomposition it emits toxic fumes of Cl^-.

DFJ000 **CAS:14096-51-6** **HR: 3**
DICHLORO(ETHYLENEDIAMMINE)PLATINUM(II)
mf: $C_2H_8Cl_2N_2Pt$ mw: 326.11

PROP: Yellow crystals. Sol in H_2O.

SYNS: ETHYLENEDIAMINEDICHLORIDE PLATINUM (II) □ PLATINUM ETHYLENEDIAMMINE DICHLORIDE □ Pt-05

TOXICITY DATA WITH REFERENCE
mmo-sat 2 μg/plate MUREAV 77,45,80
dni-hmn:oth 25 μmol/L IJCNAW 6,207,70
ipr-mus LDLo:14 mg/kg BCPCA6 2,187,73

SAFETY PROFILE: Poison by intraperitoneal route. Human mutation data reported. See also PLATINUM COMPOUNDS. When heated to decomposition it emits very toxic fumes of Cl^- and NO_x.

DFJ050 **CAS:111-44-4** **HR: 3**
DICHLOROETHYL ETHER
DOT: UN 1916
mf: $C_4H_8Cl_2O$ mw: 143.02

PROP: Colorless, stable liquid. Bp: 178.5°, fp: −51.9°, flash p: 131°F (CC), d: 1.2220 @ 20°/20°, autoign temp: 696°F, vap press: 0.7 mm @ 20°, vap d: 4.93. Misc in Et_2O, MeOH, and C_6H_6.

SYNS: BIS(β-CHLOROETHYL) ETHER □ BIS(2 CHLOROETHYL) ETHER □ CHLOREX □ 1-CHLORO-2-(β-CHLOROETHOXY)ETHANE □ CHLOROETHYL ETHER □ CLOREX □ DCEE □ 2,2′-DICHLOORETHYLETHER (DUTCH) □ 2,2′-DICHLOR-DIAETHYLAETHER (GERMAN) □ 2,2′-DICHLORETHYL ETHER □ β,β-DICHLORODIETHYL ETHER □ DICHLOROETHER □ DI(β-CHLOROETHYL)ETHER □ β,β′-DICHLOROETHYL ETHER □ sym-DICHLOROETHYL ETHER □ 2,2′-DICHLOROETHYL ETHER (MAK) □ DICHLOROETHYL OXIDE □ 2,2′-DICLOROETILETERE (ITALIAN) □ DWUCHLORODWUETYLOWY ETER (POLISH) □ ENT 4,504 □ ETHER DICHLORE (FRENCH) □ 1,1′-OXYBIS(2-CHLORO)ETHANE □ OXYDE de CHLORETHYLE (FRENCH) □ RCRA WASTE NUMBER U025

TOXICITY DATA WITH REFERENCE
skn-rbt 10 mg/24H open JIHTAB 30,63,48
skn-rbt 500 mg open MLD UCDS** 12/29/71
eye-rbt 20 mg AJOPAA 29,1363,46
mmo-sat 1 mL/plate/2H DHEFDK FDA-78-1046,78
mma-sat 1 mg/plate ENMUDM 8 (Suppl 7),1,86
orl-mus TDLo:33 g/kg/79W-C:CAR JNCIAM 42,1101,69
scu-mus TDLo:2400 mg/kg/60W-I:ETA JNCIAM 48,1431,72
orl-rat LD50:75 mg/kg JIHTAB 30,63,48
ihl-rat LC50:330 mg/m³/4H 85GMAT -,45,82
orl-mus LD50:112 mg/kg 85GMAT -,45,82
ihl-mus LC50:650 mg/m³/2H 85GMAT -,45,82
skn-rbt LD50:720 mg/kg UCDS** 12/29/71
skn-gpg LD50:300 mg/kg JIHTAB 30,63,48

CONSENSUS REPORTS: IARC Cancer Review: Group 3 IMEMDT 7,56,87; Animal Sufficient Evidence IMEMDT 9,117,75. Reported in EPA TSCA Inventory. On Community Right-To-Know List. On EPA Extremely Hazardous Substances List.

OSHA PEL: TWA 5 ppm; STEL 10 ppm (skin)
ACGIH TLV: TWA 5 ppm; STEL 10 ppm (skin)
DFG MAK: 10 ppm (60 mg/m³)

DOT CLASSIFICATION: 6.1; *Label:* Poison, Flammable Liquid

SAFETY PROFILE: A poison by ingestion, skin contact, and inhalation. A skin, eye, and mucous membrane irritant. Questionable carcinogen with experimental carcinogenic and tumorigenic data. Mutation data reported. Exposure to 1000 ppm for 30 to 60 minutes may result in death within days. The odor is easily detectable at 35 ppm which causes only slight irritation. Flammable liquid when exposed to heat, flame, or oxidants. Dangerous explosion hazard; reacts vigorously with oleum, chlorosulfonic acid. Reacts with water or steam to evolve toxic and corrosive fumes. Can react vigorously with oxidizing materials. To fight fire, use water, foam, mist, fog, spray, dry chemical. When heated to decomposition it emits toxic fumes of Cl^-. See also ETHERS.

For occupational chemical analysis use NIOSH: sym-Dichloroethyl Ether, 1004.

DFJ100 **CAS:90584-32-0** **HR: 3**
1,2-DICHLOROETHYL HYDROPEROXIDE
mf: $C_2H_4Cl_2O_2$ mw: 130.96

SAFETY PROFILE: Undergoes rapid exothermic decomposition at room temperature. When heated to decomposition it emits toxic fumes of Cl^-. See also PEROXIDES.

DFJ200 **CAS:63917-06-6** **HR: 3**
DI-2-CHLOROETHYL MALEATE
mf: $C_8H_{10}Cl_2O_4$ mw: 241.08

SYN: DI(2-CHLOROETHYL) ESTER, MALEIC ACID

TOXICITY DATA WITH REFERENCE
orl-rat LD50:71 mg/kg TXAPA9 28,313,74
skn-rbt LD50:140 mg/kg TXAPA9 28,313,74

SAFETY PROFILE: Poison by ingestion and skin contact. When heated to decomposition it emits toxic fumes of Cl^-.

DFJ400 **CAS:20198-77-0** **HR: 3**
2,3-DICHLORO-N-ETHYLMALEINIMIDE
mf: $C_6H_5Cl_2NO_2$ mw: 194.02

SYN: N-ETHYL-DICHLOROMALEINIMIDE

TOXICITY DATA WITH REFERENCE
ipr-mus TDLo:6200 μg/kg/(9D preg):REP ARTODN 37,15,76
ipr-mus TDLo:6200 μg/kg/(9D preg):TER ARTODN 37,15,76
ipr-mus LD50:15 mg/kg ARTODN 37,15,76
ivn-mus LD50:5600 μg/kg CSLNX* NX#03694

SAFETY PROFILE: Poison by intraperitoneal and intravenous routes. An experimental teratogen. Experimental reproductive effects. When heated to decomposition it emits very toxic fumes of Cl^- and NO_x.

DFJ500 **CAS:10232-90-3** **HR: 2**
2-(1,2-DICHLOROETHYL)-4-METHYL-1,3-DIOXOLANE
mf: $C_6H_{10}Cl_2O_2$ mw: 185.06

SYN: 1,3-DIOXOLANE, 2-(1,2-DICHLOROETHYL)-4-METHYL-

TOXICITY DATA WITH REFERENCE
skn-rbt 10 mg/24H open MLD AIHAAP 23,95,62
orl-rat LD50:620 mg/kg AIHAAP 23,95,62
skn-rbt LD50:1010 mg/kg AIHAAP 23,95,62

SAFETY PROFILE: Moderately toxic by ingestion and skin contact. A skin irritant. When heated to decomposition it emits toxic fumes of Cl^-.

DFJ800 **CAS:1125-27-5** **HR: 3**
DICHLOROETHYLPHENYLSILANE
DOT: UN 2435
mf: $C_8H_{10}Cl_2Si$ mw: 205.17

PROP: Liquid.

SYN: ETHYL PHENYL DICHLOROSILANE (DOT)

CONSENSUS REPORTS: Reported in EPA TSCA Inventory.

DOT CLASSIFICATION: 8; *Label:* Corrosive

SAFETY PROFILE: Poison by ingestion and inhalation. A poison irritant to skin, eyes, and mucous membranes. Corrosive. Will react with water or steam to produce toxic and corrosive fumes. Can react with oxidizing materials. When heated to decomposition it emits toxic fumes of Cl^- and phenol. See also CHLOROSILANES.

DFK000 **CAS:1789-58-8** **HR: 3**
DICHLOROETHYLSILANE
DOT: UN 1183
mf: $C_2H_6Cl_2Si$ mw: 129.07

PROP: Liquid. Vap d: 4.45, bp: 74–75°, flash p: <73.4°F.

SYN: ETHYL DICHLOROSILANE (DOT)

CONSENSUS REPORTS: Reported in EPA TSCA Inventory.

DOT CLASSIFICATION: 4.3; *Label:* Danger When Wet, Corrosive, Flammable Liquid

SAFETY PROFILE: Poison by ingestion and inhalation. A severe irritant to skin, eyes, and mucous membranes. Corrosive. Dangerous fire hazard if exposed to heat, open flames, or powerful oxidizers. Will react with water or steam to produce heat and toxic and corrosive fumes. To fight fire, use foam, dry chemical, mist, spray. When heated to decomposition it emits toxic fumes of Cl^- and phosgene. See also CHLOROSILANES.

DFK200 **CAS:63918-89-8** **HR: 3**
2-2′-DI(3-CHLOROETHYLTHIO)DIETHYL ETHER
mf: $C_8H_{16}Cl_2OS_2$ mw: 263.26

SYNS: BIS(β-CHLOROETHYLTHIOETHYL) ETHER □ BIS(2-CHLOROETHYLTHIOETHYL) ETHER □ 1,1′-OXYBIS(2-(2-CHLOROETHYL) THIOETHANE

TOXICITY DATA WITH REFERENCE
sln-dmg-ihl 100 pph/5M PREBA3 62B,284,46/47
ihl-hmn LCLo:400 mg/m³ SCJUAD 4,33,67
ihl-mus LC50:1650 mg/m³/10M NTIS** PB158-508

CONSENSUS REPORTS: EPA Genetic Toxicology Program.

SAFETY PROFILE: A human poison by inhalation. Mutation data reported. When heated to decomposition it emits very toxic fumes of SO_x and Cl^-. See also ETHERS and CHLORIDES.

DFK400 **CAS:10138-21-3** **HR: 2**
DICHLOROETHYLVINYLSILANE
mf: $C_4H_8Cl_2Si$ mw: 155.11

SYN: ETHYLVINYLDICHLOROSILANE

TOXICITY DATA WITH REFERENCE
skn-rbt 100 µg/24H open AIHAAP 23,95,62
skn-rbt 5 mg/24H SEV 85JCAE-,1223,86
eye-rbt 250 µg/24H SEV 85JCAE-,1223,86
orl-rat LDLo:2830 mg/kg AIHAAP 23,95,62
ihl-rat LCLo:8000 ppm/4H AIHAAP 23,95,62
skn-rbt LD50:750 mg/kg AIHAAP 23,95,62

CONSENSUS REPORTS: Reported in EPA TSCA Inventory.

SAFETY PROFILE: Moderately toxic by ingestion and skin contact. Mildly toxic by inhalation. A severe skin and eye irritant. When heated to decomposition it emits toxic fumes of Cl^-. See also CHLOROSILANES.

DFK600 **CAS:97-17-6** **HR: 3**
DICHLOROFENTHION
mf: $C_{10}H_{13}Cl_2O_3PS$ mw: 315.16

PROP: A liquid. A nonvolatile, residual organic phosphate nematocide and insecticide. Bp: 126–131° @ 0.2 mm, d: 1.3. Insol in water; sol in most organic solvents.

SYNS: BROMEX □ O,O-DIAETHYL-O-2,4-DICHLOR-PHENYL-MONOTHIOPHOSPHAT (GERMAN) □ O,O-DIAETHYL-O-2,4-DICHLOR-PHENYL-THIONOPHOSPHAT (GERMAN) □ DICHLOFENTHION □ DICHLOFENTION □ 2,4-DICHLORO-PHENOL-O-ESTER with O,O-DIETHYL PHOSPHOROTHIOATE □ O-2,4-DICHLOROPHENYL-O,O-DIETHYL PHOSPHOROTHIOATE □ 2,4-DICHLORO-PHENYL DIETHYL PHOSPHOROTHIONATE □ O,O-DIETHYL-O-(2,4-DICHLOOR-FENYL)-MONOTHIOFOSFAAT (DUTCH) □ O,O-DIETHYL-O-(2,4-DICHLORO-PHENYL) PHOSPHOROTHIOATE □ DIETHYL 2,4-DICHLOROPHENYL PHOSPHOROTHIONATE □ O,O-DIETHYL-O-2,4-DICHLOROPHENYL THIOPHOSPHATE □ O,O-DIETIL-O-(2,4-DICLORO-FENIL)-MONOTIOFOSFATO (ITALIAN) □ ECP □ ENT 17,470 □ HEXA-NEMA □ MOBILAWN □ NEMACIDE □ THIOPHOSPHATE de O-2,4-DICHLOROPHENYLE et de O,O-DIETHYLE (FRENCH) □ TRI-VC 13 □ VC13 NEMACIDE

TOXICITY DATA WITH REFERENCE
sce-hmn:lyms 2 mg/L MUREAV 102,89,82
orl-rat LD50:172 mg/kg FAATDF 7,299,86
skn-rat LD50:355 mg/kg FAATDF 7,299,86
skn-rbt LD50:6000 mg/kg 31ZOAD 1,136,68
orl-pgn LD50:75 mg/kg ASTTA8 (680),157,79
orl-ckn LD50:148 mg/kg TXAPA9 11,49,67
orl-qal LD50:316 mg/kg ASTTA8 (680),157,79

orl-mam LD50:270 mg/kg FMCHA2 -,C160,83
orl-bwd LD50:14 mg/kg TXAPA9 21,315,72

SAFETY PROFILE: Poison by ingestion and skin contact. A very toxic insecticide. Mutation data reported. See also ESTERS and PARATHION. When heated to decomposition it emits very toxic fumes of PO_x, SO_x, and Cl^-.

DFL000 CAS:75-43-4 HR: 1
DICHLOROFLUOROMETHANE
DOT: UN 1029
mf: $CHCl_2F$ mw: 102.92

PROP: Heavy, colorless gas. Mp: −135°, bp: 8.9°, d: 1.48, vap press: 2 atm @ 28.4°, vap d: 3.82.

SYNS: ALGOFRENE TYPE 5 ☐ ARCTON 7 ☐ DICHLOROMONOFLUOROMETHANE (OSHA, DOT) ☐ DWUCHLOROFLUOROMETAN (POLISH) ☐ FC 21 ☐ FLUORODICHLOROMETHANE ☐ FREON 21 ☐ GENETRON 21 ☐ R21 (DOT)

TOXICITY DATA WITH REFERENCE
ihl-rat TCLo:1 pph/6H (6-15D preg):REP TXAPA9 45,293,78
ihl-rat LC50:49,900 ppm/4H DTLVS* 4,132,80
ihl-gpg LCLo:10 pph/1H FLCRAP 1,197,67

CONSENSUS REPORTS: Reported in EPA TSCA Inventory.

OSHA PEL: TWA 10 ppm
ACGIH TLV: TWA 10 ppm
DFG MAK: 10 ppm (45 mg/m³)
DOT CLASSIFICATION: 2.2; *Label:* Nonflammable Gas

SAFETY PROFILE: Mildly toxic by inhalation. Experimental reproductive effects. When heated to decomposition it emits very toxic fumes of Cl^- and F^-.

For occupational chemical analysis use NIOSH: see Dichlorofluoromethane, 2516.

DFL100 CAS:498-67-9 HR: 3
(DICHLOROFLUOROMETHYL)BENZENE
mf: $C_7H_5Cl_2F$ mw: 179.02

SYNS: BENZENE, (DICHLOROFLUOROMETHYL)- ☐ α-α-DICHLORO-α-FLUOROTOLUENE ☐ TOLUENE, α-α-DICHLORO-α-FLUORO-

TOXICITY DATA WITH REFERENCE
orl-rat LD50:450 mg/kg GTPZAB 31(4),46,87
ihl-rat LC50:2 g/m³ GTPZAB 31(4),46,87
orl-mus LD50:1875 mg/kg GTPZAB 31(4),46,87
ihl-mus LC50:3100 mg/m³ GTPZAB 31(4),46,87
ihl-uns LC50:3200 mg/m³ GTPZAB 30(3),6,86

CONSENSUS REPORTS: Reported in EPA TSCA Inventory.

SAFETY PROFILE: Poison by inhalation route. Moderately toxic by ingestion. When heated to decomposition it emits toxic vapors of F^- and Cl^-.

DFL200 CAS:1085-98-9 HR: 2
N-(DICHLOROFLUOROMETHYLTHIO)-N′,N′-DIMETHYL-N-PHENYLSULFAMIDE
mf: $C_9H_{11}Cl_2FN_2O_2S_2$ mw: 333.24

PROP: Powder. Mp: 105–106°. Practically insol in H_2O; sltly sol in MeOH; sol in xylene.

SYNS: BAY 47531 ☐ BAYER 47531 ☐ DICHLOFLUANID ☐ DICHLOFLUANIDE ☐ N-DICHLORFLUORMETHYLTHIO-N′,N′-DIMETHYLAMINOSULFONSAEUREANILID (GERMAN) ☐ N-(DICHLORFLUOR-METHYL-THIO)-N′,N′-DIMETHYL-N-PHENYL-SCHWEFEL-SAEUREDIAMID (GERMAN) ☐ 1,1-DICHLORO-N-((DIMETHYLAMINO)SULFONYL)-1-FLUORO-N-PHENYLMETHANE SULFENAMIDE ☐ N-((DICHLOROFLUOROMETHYL)THIO)-N-((DIMETHYLAMINO)SULFONYL)ANILINE ☐ N-(DICHLOROFLUOROMETHYLTHIO)-N-(DIMETHYLSULFAMOYL)ANILINE ☐ N,N-DIMETHYL-N′-PHENYL-N′-FLUORODICHLOROMETHYLTHIOSULFAMIDE ☐ ELVARON ☐ EPAREN ☐ EUPAREN ☐ EUPARENE ☐ KU 13-032-C ☐ KUE 13032c

TOXICITY DATA WITH REFERENCE
mmo-esc 10 μg/plate MUREAV 116,185,83
mma-esc 100 μg/plate MUREAV 116,185,83
orl-rat LD50:500 mg/kg WRPCA2 9,119,70
ihl-rat LC50:300 mg/m³/4H 85JFAN A136,86
skn-rat LD50:1 g/kg GUCHAZ 6,179,73
orl-mus LD50:1250 mg/kg MEIEDD 10,442,83
orl-cat LD50:1 g/kg 85GYAZ -,97,71

CONSENSUS REPORTS: EPA Genetic Toxicology Program.

SAFETY PROFILE: Moderately toxic by ingestion and skin contact. Mutation data reported. A pesticide. When heated to decomposition it emits very toxic fumes of Cl^-, F^-, NO_x, and SO_x.

DFL400 CAS:731-27-1 HR: 3
N′-DICHLOROFLUOROMETHYLTHIO-N,N-DIMETHYL-N′-(4-TOLYL)SULFAMIDE
mf: $C_{10}H_{13}Cl_2FN_2O_2S_2$ mw: 347.27

PROP: Colorless to pale-yellow powder. Mp: 95–97°. Sol in C_6H_6, and xylene, sltly sol in H_2O and MeOH.

SYNS: N,N-DIMETHYL-N′-(4-TOLYL)-N′-(DICHLORFLUORMETHYLTHIO)SULFAMID (GERMAN) ☐ N,N-DIMETHYL-N-(4-TOLYL)-N-(DICHLOROFLUOR-METHYLTHIO)SULFAMIDE

TOXICITY DATA WITH REFERENCE
orl-rat LD50:1000 mg/kg GUCHAZ 6,505,73
skn-rat LD50:500 mg/kg GUCHAZ 6,505,73
orl-rbt LD50:500 mg/kg 28ZEAL 5,225,76
orl-gpg LD50:250 mg/kg 85DPAN -,-,71/76
orl-brd LD50:1000 mg/kg 28ZEAL 5,225,76

SAFETY PROFILE: Poison by ingestion. Moderately toxic by skin contact. When heated to decomposition it emits very toxic fumes of Cl^-, F^-, NO_x, and SO_x.

DFL600 CAS:15230-48-5 HR: 3
DICHLOROGERMANE
mf: Cl_2GeH_2 mw: 145.51

PROP: Colorless liquid. Mp: −68.0°, bp: 69.5°, d: 1.90 @ −68°, vap d: 5.0.

SAFETY PROFILE: Reaction with ammonia forms heat-sensitive explosive product. See also HYDROCHLORIC ACID and GERMANIUM COMPOUNDS. When heated to decomposition it emits toxic fumes of Cl^-.

DFL709 CAS:58941-14-3 HR: 3
N,N-DICHLOROGLYCINE
mf: $C_2H_3Cl_2NO_2$ mw: 143.96

SAFETY PROFILE: Explodes when heated to 65°C. When heated to decomposition it emits toxic fumes of Cl^- and NO_x.

DFL800 CAS:16260-59-6 HR: 3
1,6-DICHLORO-2,4-HEXADIYNE
mf: $C_6H_4Cl_2$ mw: 147.00

$ClCH_2(C{=}C)_2CH_2Cl$

PROP: Bp: 55–58° @ 0.5 mm.

SAFETY PROFILE: An extremely shock-sensitive explosive. Upon decomposition it emits toxic fumes of Cl^-. See also CHLORINATED HYDROCARBONS, ALIPHATIC.

DFM000 CAS:303-04-8 HR: 3
2,3-DICHLOROHEXAFLUOROBUTENE-2
mf: $C_4Cl_2F_6$ mw: 232.94

SYNS: DCHFB □ 2,3-DICHLOROHEXAFLUORO-2-BUTENE □ 2,3-DICHLORO-1,1,1,4,4,4-HEXAFLUOROBUTENE-2

TOXICITY DATA WITH REFERENCE
orl-rat LDLo:1000 mg/kg DOWCC* -,-,63
ihl-rat LC50:16 ppm/4H BJANAD 37,716,65
ihl-mus LC50:26 ppm/4H BJANAD 37,716,65
ihl-dog LC50:182 ppm/4H JETOAS 4,517,71
ihl-mky LC50:54 ppm/3H ANESAV 26,140,65

SAFETY PROFILE: Poison by inhalation. Moderately toxic by ingestion. When heated to decomposition it emits very toxic fumes of Cl^- and F^-.

DFM025 CAS:356-18-3 HR: 3
1,2-DICHLORO-1,2,3,3,4,4-HEXAFLUOROCYCLOBUTANE
mf: $C_4Cl_2F_6$ mw: 232.94

SYNS: CYCLOBUTANE, 1,2-DICHLOROHEXAFLUORO- □ CYCLOBUTANE, 1,2-DICHLORO-1,2,3,3,4,4-HEXAFLUORO-(9CI) □ 1,2-DICHLOROHEXAFLUOROCYCLOBUTANE □ 1,2-DICHLOROPERFLUOROCYCLOBUTANE

TOXICITY DATA WITH REFERENCE
ihl-mus LCLo:21 pph/30S ANASAB 16,3,61

SAFETY PROFILE: Poison by inhalation. When heated to decomposition it emits toxic vapors of NO_x and Cl^-.

DFM050 CAS:706-79-6 HR: 3
1,2-DICHLOROHEXAFLUOROCYCLOPENTENE
mf: $C_5Cl_2F_6$ mw: 244.95

SYNS: CYCLOPENTENE, 1,2-DICHLOROHEXAFLUORO- □ CYCLOPENTENE, 1,2-DICHLORO-3,3,4,4,5,5-HEXAFLUORO-(9CI) □ 1,2-DICHLORO-3,3,4,4,5,5-HEXAFLUOROCYCLOPENTENE □ 1,2-DICHLOROPERFLUOROCYCLOPENTENE

TOXICITY DATA WITH REFERENCE
orl-rat LD50:280 mg/kg GISAAA 30(11),6,65
orl-mus LD50:276 mg/kg GISAAA 30(11),6,65
ihl-mus LC50:2100 mg/m³/2H 85JCAE-,163,86
orl-rbt LD50:280 mg/kg GISAAA 30(11),6,65
orl-gpg LD50:280 mg/kg GISAAA 30(11),6,65

CONSENSUS REPORTS: Reported in EPA TSCA Inventory.

SAFETY PROFILE: Poison by ingestion. Moderately toxic by inhalation. When heated to decomposition it emits toxic vapors of F^- and Cl^-.

DFM099 HR: 3
4,5-DICHLORO-3,3,4,5,6,6-HEXAFLUORO-1,2-DIOXANE
mf: $C_4Cl_2F_6O_2$ mw: 264.94

$F_2C(CFCl)_2CF_2O\,O$ (ring)

SAFETY PROFILE: Explodes violently when heated. When heated to decomposition it emits toxic fumes of F^- and Cl^-. See also CHLORIDES and FLUORIDES.

DFM200 CAS:13442-13-2 HR: 2
6,7-DICHLORO-4-(HYDROXYAMINO)QUINOLINE-1-OXIDE
mf: $C_9H_6Cl_2N_2O_2$ mw: 245.07

TOXICITY DATA WITH REFERENCE
scu-mus TDLo:120 mg/kg/50D-I:ETA BCPCA6 16,631,67

SAFETY PROFILE: Questionable carcinogen with experimental tumorigenic data. When heated to decomposition it emits very toxic fumes of Cl^- and NO_x.

DFM600 CAS:101652-05-5 HR: 3
6,7-DICHLORO-10-(3-(N-(2-HYDROXYETHYL)ETHYLAMINO))ISOALLOXAZINE SULFATE
mf: $C_{17}H_{19}Cl_2N_5O_3{\cdot}H_2O_4S$ mw: 510.39

TOXICITY DATA WITH REFERENCE
ipr-rat LD50:30 mg/kg CMTRAG 2,96,61
scu-mus LD50:38 mg/kg CMTRAG 2,96,61
ivn-mus LD50:90 mg/kg CMTRAG 2,96,61

SAFETY PROFILE: Poison by intraperitoneal, subcutaneous, and intravenous routes. See also SULFATES and CHLORIDES. When heated to decomposition it emits very toxic fumes of SO_x, Cl^-, and NO_x.

DFM800 CAS:101652-07-7 HR: 3
6,7-DICHLORO-10-(3-(N-(2-HYDROXYETHYL)METHYLAMINO)PROPYL) ISOALLOXAZINE SULFATE
mf: $C_{16}H_{17}Cl_2N_5O_3{\cdot}H_2O_4S$ mw: 496.36

TOXICITY DATA WITH REFERENCE
ipr-rat LD50:45 mg/kg CMTRAG 2,96,61
ipr-mus LD50:75 mg/kg CMTRAG 2,96,61
scu-mus LD50:97 mg/kg CMTRAG 2,96,61
ivn-mus LD50:75 mg/kg CMTRAG 2,96,61

ims-mus LD50:40 mg/kg CMTRAG 2,96,61

SAFETY PROFILE: Poison by intraperitoneal, intravenous, intramuscular, and subcutaneous routes. When heated to decomposition it emits very toxic fumes of SO_x, NO_x, and Cl^-.

DFM875 CAS:90742-91-9 HR: 3
1-(2,5-DICHLORO-6-(1-(1H-IMIDAZOL-1-YL)VINYL) PHENOXY)-3-(ISOPROPYLAMINO)-2-PROPANOL HYDROCHLORIDE
mf: $C_{17}H_{21}Cl_2N_3O_2 \cdot ClH$ mw: 406.77

SYNS: 711389 S □ 1 (1 (2 (3 ISOPROPYLAMINO 2 HYDROXYPROPOXY)-3,6-DICHLOROPHENYL)VINYL)-1H-IMIDAZOLE HCl

TOXICITY DATA WITH REFERENCE
unr-rat LD50:245 mg/kg DRFUD4 10,472,85
ivn-mus LD50:19,800 µg/kg JMCMAR 27,1142,84
unr-mus LD50:171 mg/kg DRFUD4 10,472,85

SAFETY PROFILE: Poison by intravenous and possibly other routes. When heated to decomposition it emits toxic fumes of Cl^- and NO_x.

DFN400 CAS:59-61-0 HR: 3
3,4-DICHLORO-α-((ISOPROPYLAMINO)METHYL)BENZYL ALCOHOL
mf: $C_{11}H_{15}Cl_2NO$ mw: 248.17

SYNS: DCI □ DICHLORISOPRENALINE (GERMAN) □ DICHLORISOPROTERENOL □ 3,4-DICHLOR-ISOPROTERENOL (GERMAN) □ 3,4-DICHLORO-α-(((1-METHYLETHYL)AMINO)METHYL)BENZENEMETHANOL □ N-(β-(3,4-DICHLOROPHENYL)-β-HYDROXYETHYL)ISOPROPYLAMINE □ 1-(3,4-DICHLOROPHENYL)-2-ISOPROPYLAMINOETHANOL □ β HYDROXY N ISOPROPYL 3,4 DICHLOROPHENETHYLAMINE

TOXICITY DATA WITH REFERENCE
orl-mus LD50:165 mg/kg ARZNAD 18,48,68
ivn-mus LD50:39 mg/kg ARZNAD 18,48,68

SAFETY PROFILE: Poison by ingestion and intravenous routes. When heated to decomposition it emits very toxic fumes of Cl^- and NO_x.

DFN425 CAS:130-20-1 HR: 2
3,3′-DICHLOROINDANTHRONE
mf: $C_{28}H_{12}Cl_2N_2O_4$ mw: 511.32

SYNS: AHCOVAT BLUE BCF □ ALIZANTHRENE BLUE RC □ AMANTHRENE BLUE BCL □ 5,9,14,18-ANTHRAZINETETRONE, 7,16-DICHLORO-6,15-DIHYDRO- □ ATIC VAT BLUE BC □ BENZADONE BLUE RC □ BLUE K □ CALCOLOID BLUE BLC □ CALCOLOID BLUE BLD □ CALCOLOID BLUE BLFD □ CALCOLOID BLUE BLR □ CALEDON BLUE XRC □ CARBANTHRENE BLUE BCF □ CARBANTHRENE BLUE BCS □ CARBANTHRENE BLUE RBCF □ CARBANTHRENE BLUE RCS □ C.I. 69825 □ CIBANONE BLUE FG □ CIBANONE BLUE FGF □ CIBANONE BLUE FGL □ CIBANONE BLUE GF □ C.I. VAT BLUE 6 □ D and C BLUE No. 9 □ DICHLOROINDANTHRONE □ 7,16-DICHLOROINDANTHRONE □ FENAN BLUE BCS □ FENANTHREN BLUE BC □ FENANTHREN BLUE BD □ HARMONE B 79 □ HELANTHRENE BLUE BC □ INDANTHREN BLUE BC □ INDANTHREN BLUE BCA □ INDANTHREN BLUE BCS □ INDANTHRENE BLUE BC □ INDANTHRENE BLUE BCF □ INDO BLUE B-I □ INDO BLUE WD 279 □ INDOTONER BLUE B 79 □ INTRAVAT BLUE GF □ MIKETHRENE BLUE BC □ MIKETHRENE BLUE BCS □ MONOLITE FAST BLUE 2RV □ MONOLITE FAST BLUE 2RVSA □ NAVINON BLUE BC □ NAVINON BRILLIANT BLUE RCL □ NIHONTHRENE BLUE BC □ NIHONTHRENE BRILLIANT BLUE RCL □ NYANTHRENE BLUE BFP □ OSTANTHREN BLUE BCL □ OSTANTHREN BLUE BCS □ PALANTHRENE BLUE BC □ PALANTHRENE BLUE BCA □ PARADONE BLUE RC □ PERNITHRENE BLUE BC □ PONSOL BLUE BCS □ PONSOL BLUE BF □ PONSOL BLUE BFD □ PONSOL BLUE BFDP □ PONSOL BLUE BFN □ PONSOL BLUE BFND □ PONSOL BLUE BFP □ RESINATED INDO BLUE B 85 □ ROMANTRENE BLUE FBC □ SANDOTHRENE BLUE NG □ SANDOTHRENE BLUE NGR □ SANDOTHRENE BLUE NGW □ SOLANTHRENE BLUE B □ SOLANTHRENE BLUE F-SBA □ SOLANTHRENE BLUE SB □ TINON BLUE GF □ TINON BLUE GL □ VAT BLUE 6 □ VAT BLUE KD □ VAT FAST BLUE BCS □ VAT GREEN B □ VAT SKY BLUE K □ VAT SKY BLUE KD □ VAT SKY BLUE KP 2F

TOXICITY DATA WITH REFERENCE
orl-mus LD50:1800 mg/kg GNAMAP 14,152,75
skn-mus LD50:25 g/kg GNAMAP 14,152,75

SAFETY PROFILE: Moderately toxic by ingestion. Mildly toxic by skin contact. When heated to decomposition it emits toxic vapors of NO_x and Cl^-.

DFN450 CAS:1324-55-6 HR: 1
DICHLOROISOVIOLANTHRONE
mf: $C_{34}H_{14}Cl_2O_2$ mw: 525.38

SYNS: AHCOVAT BRILLIANT VIOLET 2R □ AHCOVAT BRILLIANT VIOLET 4R □ AMANTHRENE BRILLIANT VIOLET RR □ ARLANTHRENE VIOLET 4R □ ATIC VAT BRILLIANT PURPLE 4R □ BENZADONE BRILLIANT PURPLE 2R □ BENZADONE BRILLIANT PURPLE 4R □ BENZO(rst)PHENANTHRO(10,1,2-cde)PENTAPHENE-9,18-DIONE, DICHLORO- □ BRILLIANT VIOLET K □ CALCOLOID VIOLET 4RD □ CALCOLOID VIOLET 4RP □ CALEDON BRILLIANT PURPLE 4R □ CALEDON BRILLIANT PURPLE 4RP □ CALEDON PRINTING PURPLE 4R □ CARBANTHRENE BRILLIANT VIOLET 4R □ CARBANTHRENE VIOLET 2R □ CARBANTHRENE VIOLET 2RP □ C.I. 60010 □ CIBANONE VIOLET F 4R □ CIBANONE VIOLET F 2RB □ CIBANONE VIOLET 2R □ CIBANONE VIOLET 4R □ C.I. PIGMENT VIOLET 31 □ C.I. VAT VIOLET 1 (8CI) □ FENANTHREN BRILLIANT VIOLET 2R □ FENANTHREN BRILLIANT VIOLET 4R □ INDANTHREN BRILLIANT VIOLET 4R □ INDANTHREN BRILLIANT VIOLET RR □ INDANTHRENE BRILLIANT VIOLET 4R □ INDANTHRENE BRILLIANT VIOLET RR □ INDANTHREN PRINTING VIOLET F 4R □ INDOFAST VIOLET LAKE □ NIHONTHRENE BRILLIANT VIOLET 4R □ NIHONTHRENE BRILLIANT VIOLET RR □ NYANTHRENE BRILLIANT VIOLET 4R □ PONOLITH FAST VIOLET 4RN □ SANDOTHRENE VIOLET N 4R □ SANDOTHRENE VIOLET N 2RB □ SANDOTHRENE VIOLET 4R □ SOLANTHRENE BRILLIANT VIOLET F 2R □ SYMULER FAST VIOLET R □ TINON VIOLET B 4RP □ TINON VIOLET 4R □ TINON VIOLET 2RB □ VAT BRIGHT VIOLET K □ VAT BRILLIANT VIOLET K □ VAT BRILLIANT VIOLET KD □ VAT BRILLIANT VIOLET KP □ VIOLET KYPOVA 1 □ VIOLET PIGMENT 31

TOXICITY DATA WITH REFERENCE
orl-mus LD50:6700 mg/kg 85JCAE -,1329,86

CONSENSUS REPORTS: Reported in EPA TSCA Inventory.

SAFETY PROFILE: Mildly toxic by ingestion. When heated to decomposition it emits toxic vapors of Cl^-.

DFN500 CAS:36417-16-0 HR: 3
DICHLOROLAWSONE
mf: $C_{13}H_8Cl_2O_3$ mw: 283.11

SYNS: DCL □ DICHLOROALLYL LAWSONE □ 2-HYDROXY-3-(3,3-DICHLOROALLYL)-1,4-NAPHTHOQUINONE □ NSC 126771

TOXICITY DATA WITH REFERENCE
oms-mus:leu 6300 nmol/L CNREA8 39,4868,79
orl-rat LD50:281 mg/kg NCISP* JAN86
orl-mus LD50:192 mg/kg NCISP* JAN86
ipr-mus LD50:37,780 μg/kg NCISP* JAN86

SAFETY PROFILE: Poison by ingestion and intraperitoneal routes. Mutation data reported. When heated to decomposition it emits toxic fumes of Cl^-. See also ALLYL COMPOUNDS.

DFN700 CAS:1122-17-4 HR: 3
DICHLOROMALEIC ANHYDRIDE
mf: $C_4Cl_2O_3$ mw: 166.95

ClC=CClCO•OC O

SAFETY PROFILE: Mixtures with sodium chloride + urea undergo vigorous exothermic reaction above 118°C. When heated to decomposition it emits toxic fumes of Cl^-. See also ANHYDRIDES.

DFN800 CAS:1193-54-0 HR: 3
DICHLOROMALEIMIDE
mf: $C_4HCl_2NO_2$ mw: 165.96

CO•CCl=CClCO• NH

PROP: A solid. Mp: 175°.

SYNS: DICHLOROMALEINIMIDE □ 3,4-DICHLORO-2,5-PYRROLIDINEDIONE

TOXICITY DATA WITH REFERENCE
ipr-mus TDLo:25 mg/kg (9D preg):REP ARTODN 37,15,76
ipr-mus TDLo:25 mg/kg (9D preg):TER ARTODN 37,15,76
ipr-mus LD50:31 mg/kg ARTODN 37,15,76

SAFETY PROFILE: Poison by intraperitoneal route. Experimental teratogenic and reproductive effects. When heated to decomposition it emits very toxic fumes of Cl^- and NO_x.

DFO000 CAS:528-74-5 HR: 2
3'5'-DICHLOROMETHOTREXATE
mf: $C_{20}H_{20}Cl_2N_8O_5$ mw: 523.38

SYNS: DCM □ DICHLOROAMETHOPTERIN □ 3',5'-DICHLOROAMETHOPTERIN □ 3',5'-DICHLORO-4-AMINO-4-DEOXY-N_{10}-METHYLPTEROGLUTAMIC ACID □ N-(3,5-DICHLORO-4-((2,4-DIAMINO-6-PTERIDINYL METHYL)METHYLAMINO)BENZOYL)GLUTAMIC ACID □ DICHLOROMETHOTREXATE □ NCI-C04875 □ NSC-29630

TOXICITY DATA WITH REFERENCE
mma-sat 1 mg/plate ENMUDM 5(Suppl 1),3,83
ipr-mus TDLo:500 mg/kg (female 9D post):REP TCMUD8 7,7,87
ipr-rat TDLo:75 mg/kg/7W-I:ETA CANCAR 40(Suppl 4),1935,77
ipr-mus TDLo:5850 mg/kg/26W-I:ETA CANCAR 40(Suppl 4),1935,77
ipr-mus LD50:655 mg/kg NCISP* JAN86
ivn-mus LD50:1021 mg/kg NTIS** PB82-172644

CONSENSUS REPORTS: NCI Carcinogenesis Studies (ipr): Equivocal Evidence: rat; No Evidence: mouse CANCAR 40,1935,77.

SAFETY PROFILE: Moderately toxic by intraperitoneal and intravenous routes. Questionable carcinogen with experimental tumorigenic data. Experimental reproductive effects. Mutation data reported. When heated to decomposition it emits very toxic fumes of NO_x and Cl^-.

DFO600 CAS:56776-25-1 HR: 3
((2,3-DICHLORO-4-METHOXYPHENYL)-2-FURANYL-METHANONE)-O-(2-(DIETHYLAMINO)ETHYL) OXIME,MONOMETHANE SULFONATE
mf: $C_{18}H_{22}Cl_2N_2O_3 \cdot CH_4O_3S$ mw: 481.43

SYNS: ANP 4364 □ (DICHLORO-2,3-METHOXY-4) PHENYL FURYL-2-O-(DIETHYLAMINOETHYL)-CETONE-OXIME (FRENCH)

TOXICITY DATA WITH REFERENCE
ipr-mus LD50:110 mg/kg EJTXAZ 8,122,75
ivn-mus LD50:6 mg/kg EJTXAZ 8,188,75

SAFETY PROFILE: Poison by intraperitoneal and intravenous routes. When heated to decomposition it emits very toxic fumes of SO_x, NO_x, and Cl^-.

DFO800 CAS:2164-09-2 HR: 2
3',4'-DICHLORO-2-METHYLACRYLANILIDE
mf: $C_{10}H_9Cl_2NO$ mw: 230.10

PROP: Solid. Insol in water but sol in acetone, alcohol, isophorone, DMSO. Mp: 128°.

SYNS: CHLORANOCRYL □ DCM □ DCMA □ 3,4-DICHLOROANILIDE-α-METHYLACRYLIC ACID □ 3',4'-DICHLORO-2-METHACRYLANILIDE □ N-(3,4-DICHLOROPHENYL)METHACRYLAMIDE □ N-(3,4-DICHLOROPHENYL)-2-METHYL-2-PROPENAMIDE □ DICRYL □ METHACRYLIC ACID-3,4-DICHLOROANILIDE □ NIA 4556 □ NIAGARA 4556

TOXICITY DATA WITH REFERENCE
orl-rat LD50:1800 mg/kg WRPCA2 9,119,70
skn-rat LD50:1780 mg/kg 31ZOAD 1,155,68
orl-mus LD50:410 mg/kg GTPZAB 21(12),30,77
ipr-mus LD50:3000 mg/kg ARZNAD 14,668,64
skn-rbt LD50:10 g/kg PCOC** -,375,66

SAFETY PROFILE: Moderately toxic by ingestion, skin contact, and intraperitoneal routes. An herbicide. When heated to decomposition it emits very toxic fumes of Cl^- and NO_x.

DFO900 CAS:7651-91-4 HR: 3
N,N-DICHLOROMETHYLAMINE
mf: CH_3Cl_2N mw: 99.95

PROP: Yellow liquid. Bp: 59–60°.

SAFETY PROFILE: Explodes on contact with water, sodium sulfide, or calcium hypochlorite. When heated to decomposition it emits toxic fumes of $Cl^-$and NO_x.

DFP200 CAS:593-89-5 HR: 3
DICHLOROMETHYLARSINE
DOT: NA 1556
mf: CH_3AsCl_2 mw: 160.86

PROP: Colorless liquid. Bp: 89–91° @ 200 mm, fp: −59°, flash p: >221°F, d: 1.84 @ 20°/4°, vap press: 10 mm @ 24.3°, vap d: 5.40, mp: −42.5°.

SYNS: ARSONOUS DICHLORIDE, METHYL-(9CI) □ METHYLARSINE DICHLORIDE □ METHYLARSONOUS DICHLORIDE □ METHYLDICHLORARSINE □ METHYLDICHLOROARSINE (DOT) □ TL 294

TOXICITY DATA WITH REFERENCE
ihl-mus LC50:2700 mg/m³/10M NTIS** PB158-508

CONSENSUS REPORTS: Arsenic and its compounds are on the Community Right-To-Know List.

OSHA PEL: TWA 0.5 mg/(As)/m³
DOT CLASSIFICATION: 6.1; *Label:* Poison

SAFETY PROFILE: Poison irritant to skin, eyes, and mucous membranes and poison by ingestion and inhalation. A blistering type of military poison. It is rapidly detoxified in the body. A moderately persistent gas. Combustible when exposed to heat or flame. To fight fire, use water, foam, CO_2, dry chemical. Explosive reaction with chlorine. Can react vigorously with oxidizing materials. Dangerous; when heated to decomposition or on contact with acid or acid fumes it emits highly toxic fumes of Cl^- and As. See also CHLOROVINYLARSINE DICHLORIDE and ARSENIC COMPOUNDS.

DFP500 HR: 3
1-(2,4-DICHLORO-β-(p-METHYLBENZYLOXY)PHENETHYL)IMIDAZOLE NITRATE
mf: $C_{19}H_{18}Cl_2N_2O·HNO_3$ mw: 424.31

TOXICITY DATA WITH REFERENCE
orl-rat LD50:915 mg/kg IYKEDH 12,933,81
ipr-rat LD50:240 mg/kg IYKEDH 12,933,81
scu-rat LD50:1420 mg/kg IYKEDH 12,933,81
ivn-rat LD50:50 mg/kg IYKEDH 12,933,81
orl-mus LD50:720 mg/kg IYKEDH 12,933,81
ipr-mus LD50:180 mg/kg IYKEDH 12,933,81
scu-mus LD50:840 mg/kg IYKEDH 12,933,81
ivn-mus LD50:42 mg/kg IYKEDH 12,933,81

SAFETY PROFILE: Poison by intravenous and intraperitoneal routes. Moderately toxic by ingestion and subcutaneous routes. When heated to decomposition it emits toxic fumes of Cl^- and NO_x. See also NITRATES.

DFP600 CAS:58-54-8 HR: 3
2,3-DICHLORO-4-(2-METHYLENEBUTYRL)PHENOXY ACETIC ACID
mf: $C_{13}H_{12}Cl_2O_4$ mw: 303.15

PROP: A solid. Mp: 122°.

SYNS: CRINURYL □ (2,3-DICHLORO-4-(2-METHYLENEBUTYRYL) PHENOXY)ACETIC ACID □ (2,3-DICHLORO-4-(2-METHYLENE-1-OXOBUTYL)PHENOXY)ACETIC ACID □ EDECRIL □ EDECRIN □ EDECRINA □ ENDECRIL □ ETACRINIC ACID □ ETAKRINIC ACID □ ETHACRYNIC ACID □ HIDROMEDIN □ HYDROMEDIN □ (4-(2-METHYLENEBUTYRYL)-2,3-DICHLOROPHENOXY)ACETIC ACID □ METHYLENEBUTYRYL PHENOXYACETIC ACID □ MINGIT □ MK-595 □ OTACRIL □ REOMAX □ TALADREN □ UREGIT

TOXICITY DATA WITH REFERENCE
orl-wmn TDLo:4 mg/kg:EAR AIMDAP 117,715,66
orl-man TDLo:3 mg/kg:EAR,KID AIMDAP 117,715,66
ivn-wmn TDLo:3 mg/kg:EAR AIMDAP 117,715,66
orl-rat LD50:1 g/kg YAKUD5 21,775,79
ipr-rat LD50:43 mg/kg OYYAA2 2,411,68
orl-mus LD50:627 mg/kg MEIEDD 10,539,83
ivn-mus LD50:176 mg/kg MEIEDD 10,539,83

SAFETY PROFILE: Poison by intravenous route. Moderately toxic by ingestion. Human systemic effects by ingestion and intravenous routes: urine volume increase, impaired hearing, and tinnitus (ringing in the ears). A diuretic. When heated to decomposition it emits toxic fumes of Cl^-.

DFP800 CAS:1123-61-1 HR: 3
DICHLORO-N-METHYLMALEIMIDE
mf: $C_5H_3Cl_2NO_2$ mw: 179.99

SYNS: 2,3-DICHLORO-N-METHYLMALEIMIDE □ N-METHYLDICHLOROMALEINIMIDE

TOXICITY DATA WITH REFERENCE
ipr-mus TDLo:3100 µg/kg (9D preg):REP ARTODN 37,15,76
ipr-mus LD50:4 mg/kg ARTODN 37,15,76
ivn-mus LD50:10 mg/kg CSLNX* NX#03682

SAFETY PROFILE: Poison by intraperitoneal and intravenous routes. An experimental teratogen. Other experimental reproductive effects. When heated to decomposition it emits very toxic fumes of Cl^- and NO_x.

DFQ000 CAS:4885-02-3 HR: 2
α,α-DICHLOROMETHYL METHYL ETHER
mf: $C_2H_4Cl_2O$ mw: 114.96

PROP: A liquid. Bp: 82–84°.

SYNS: BIS(CHLOROPHENYL) ETHER □ α,α-DICHLOROMETHYL ETHER

TOXICITY DATA WITH REFERENCE
skn-mus TDLo:40 mg/kg:ETA ANYAA9 163,633,69

CONSENSUS REPORTS: Reported in EPA TSCA Inventory.

SAFETY PROFILE: Questionable carcinogen with experimental tumorigenic data. See also ETHERS and CHLORIDES. When heated to decomposition it emits toxic fumes of Cl^-.

DFQ100 CAS:76738-28-8 HR: 2
d-threo-2-(DICHLOROMETHYL)-α-(p-NITROPHENYL)-2-OXAZOLINE-4-METHANOL
mf: $C_{11}H_{10}Cl_2N_2O_4$ mw: 305.13

SYN: d-threo-2-DICLOROMETIL-4-((4′-NITROFENIL)-OSSIMETIL)-2-OSSAZOLINA (ITALIAN)

TOXICITY DATA WITH REFERENCE
orl-rat LD50:6 g/kg FRPSAX 10,3,55
orl-mus LD50:5700 mg/kg FRPSAX 10,3,55

ipr-mus LD50:4 g/kg FRPSAX 10,3,55
orl-gpg LD50:1000 mg/kg FRPSAX 10,3,55

SAFETY PROFILE: Moderately toxic by ingestion and intraperitoneal routes. When heated to decomposition it emits toxic fumes of Cl^- and NO_x.

DFQ200 CAS:84-57-1 HR: 2
2,5-DICHLORO-4-(3-METHYL-5-OXO-2-PYRAZOLIN-1-YL) BENZENESULFONIC ACID
mf: $C_{10}H_8Cl_2N_2O_4S$ mw: 323.16

SYNS: 2,5-DICHLORO-4-(4,5-DIHYDRO-3-METHYL-5-OXO-1H-PYRAZOL-1-YL)BENZENESULFONIC ACID □ DICHLORSULFOFENYL-METHYLPYRAZOLON (CZECH) □ KYSELINA 2,5-DICHLOR-4-(3'-METHYL-5'-PYRAZOLON-1'-YL)BENZENSULFONOVA (CZECH)

TOXICITY DATA WITH REFERENCE
eye-rbt 500 mg/24H SEV 28ZPAK -,186,72

CONSENSUS REPORTS: Reported in EPA TSCA Inventory.

SAFETY PROFILE: A severe eye irritant. See also SULFONATES. When heated to decomposition it emits very toxic fumes of SO_x, NO_x, and Cl^-.

DFQ400 CAS:57948-13-7 HR: 3
DICHLORO(4-METHYL-O-PHENYLENEDIAMMINE) PLATINUM(II)
mf: $C_7H_{10}Cl_2N_2Pt$ mw: 388.18

TOXICITY DATA WITH REFERENCE
mmo-sat 2500 nmoi/L JMCMAR 23,459,80
ipr-mus LD50:23 mg/kg RCRVAB 50,353,81

SAFETY PROFILE: Poison by intraperitoneal route. Mutation data reported. See also PLATINUM COMPOUNDS. When heated to decomposition it emits very toxic fumes of Cl^- and NO_x.

DFQ800 CAS:149-74-6 HR: 3
DICHLOROMETHYLPHENYLSILANE
DOT: UN 2437
mf: $C_7H_8Cl_2Si$ mw: 191.14

PROP: Colorless liquid. D: 1.18 @ 20°/4°, bp: 205°.

SYNS: METHYLPHENYLDICHLOROSILANE (DOT) □ PHENYLMETHYLDICHLOROSILANE

TOXICITY DATA WITH REFERENCE
ipr-rat LDLo:100 mg/kg 85GMAT -,99,82
ihl-mus LCLo:200 mg/m³/2H 85GMAT -,99,82
ipr-mus LDLo:100 mg/kg 85GMAT -,99,82
ihl-mus LCLo:170 mg/m³/2H 85JCAE-,1230,86
scu-mus LDLo:100 mg/kg 85GMAT -,99,82

CONSENSUS REPORTS: Reported in EPA TSCA Inventory. EPA Extremely Hazardous Substances List.

DOT CLASSIFICATION: 8; *Label:* Corrosive

SAFETY PROFILE: Poison by inhalation, subcutaneous, and intraperitoneal routes. Corrosive to eyes, skin, and mucous membranes. Flammable liquid. When heated to decomposition it emits toxic fumes of Cl^-. See also CHLOROSILANES.

DFS000 CAS:75-54-7 HR: 3
DICHLOROMETHYLSILANE
DOT: UN 1242
mf: CH_4Cl_2Si mw: 115.04

PROP: Colorless liquid; acrid hydrochloric acid-like odor. Bp: 41°, d: 1.10 @ 20/4°, mp: −93°, flash p: −26°F. Sol in benzene, ether, and heptane.

SYNS: METHYL DICHLOROSILANE (DOT) □ METHYL-DICHLORSILAN (CZECH)

TOXICITY DATA WITH REFERENCE
skn-rbt 2 mg/24H SEV 85JCAE-,1218,86
eye-rbt 20 mg/24H MOD 85JCAE-,1218,86
orl-rat LD50:2830 µL/kg JACTDZ 12,572,93
ihl-rat LC50:300 ppm/4H 85JCAE-,1218,86

CONSENSUS REPORTS: Reported in EPA TSCA Inventory.

DOT CLASSIFICATION: 4.3; *Label:* Danger When Wet, Corrosive, Flammable Liquid

SAFETY PROFILE: Moderately toxic by inhalation. Corrosive. A severe irritant to skin, eyes, and mucous membranes. Ignites spontaneously in air. A very dangerous fire hazard when exposed to heat or flame. Forms impact-sensitive explosive mixtures with potassium permanganate, lead(II) oxide, lead(IV) oxide, copper oxide, silver oxide. To fight fire, use water, foam, CO_2, mist. When heated to decomposition it emits toxic fumes of Cl^-. See also CHLOROSILANE.

DFS200 CAS:2700-89-2 HR: 3
1,2-DICHLORO-1-(METHYLSULFONYL)ETHYLENE
mf: $C_3H_4Cl_2O_2S$ mw: 175.03

SYN: CHEMAGRO D-113

TOXICITY DATA WITH REFERENCE
skn-rat 500 mg SEV 34ZIAG -,161,69
orl-rat LD50:61 mg/kg 34ZIAG -,161,69
skn-rat LD50:500 mg/kg 34ZIAG -,161,69
ipr-rat LD50:12,500 µg/kg 34ZIAG -,161,69
ipr-mus LD50:12,500 µg/kg 34ZIAG -,161,69
orl-gpg LD50:40 mg/kg 34ZIAG -,161,69
ipr-gpg LD50:12,500 µg/kg 34ZIAG -,161,69

SAFETY PROFILE: Poison by ingestion and intraperitoneal routes. Moderately toxic by skin contact. A severe skin irritant. When heated to decomposition it emits very toxic fumes of SO_x and Cl^-.

DFS600 CAS:31335-41-8 HR: 2
DICHLOROMETHYL TRICHLOROMETHYLTHIOSULFONE
mf: $C_2HCl_5O_2S_2$ mw: 298.40

SYNS: DICHLOROMETHANETHIOSULFONIC ACID-S-TRICHLOROMETHYL ESTER □ TRICHLORMETHYLESTER KYSELINY DICHLORMETHANTHIOSULFONOVE (CZECH)

TOXICITY DATA WITH REFERENCE
skn-rbt 500 mg/24H MLD 28ZPAK -,198,72
eye-rbt 20 mg/24H MOD 28ZPAK -,198,72
orl-rat LD50:3620 mg/kg 28ZPAK -,198,72

SAFETY PROFILE: Moderately toxic by ingestion. A skin and eye irritant. See also SULFONATES. When heated to decomposition it emits very toxic fumes of Cl^- and SO_x.

DFS700 CAS:675-62-7 HR: 3
DICHLOROMETHYL-3,3,3-TRIFLUOROPROPYLSILANE
mf: $C_4H_7Cl_2F_3Si$ mw: 211.10

SYN: SILANE, DICHLOROMETHYL(3,3,3-TRIFLUOROPROPYL)-

TOXICITY DATA WITH REFERENCE
ihl-mus LC50:300 mg/m³/2H 85JCAE -,1223,86

CONSENSUS REPORTS: Reported in EPA TSCA Inventory.

SAFETY PROFILE: Poison by inhalation. When heated to decomposition it emits toxic vapors of F^- and Cl^-.

DFS800 CAS:124-70-9 HR: 3
DICHLOROMETHYLVINYLSILANE
mf: $C_3H_6Cl_2Si$ mw: 141.08

PROP: A liquid. Flash p: −1°C, d: 1.08 @ 20°/4°, bp: 92°.

TOXICITY DATA WITH REFERENCE
ivn-mus LD50:56 mg/kg CSLNX* NX#03620

CONSENSUS REPORTS: Reported in EPA TSCA Inventory.

SAFETY PROFILE: Poison by intravenous route. A very dangerous fire hazard when exposed to heat, flame, or oxidizers. When heated to decomposition it emits toxic fumes of Cl^-. See also CHLOROSILANES.

DFT000 CAS:117-80-6 HR: 3
2,3-DICHLORO-1,4-NAPHTHOQUINONE
mf: $C_{10}H_4Cl_2O_2$ mw: 227.04

PROP: Golden-yellow crystals or needles from alc. Mp: 194–195°, vap d: 7.8. Insol in water; moderately sol in organic solvents.

SYNS: ALGISTAT □ COMPOUND 604 □ DICHLONE (DOT) □ 2,3-DICHLOR-1,4-NAPHTHOCHINON (GERMAN) □ 2,3-DICHLORO-1,4-NAPHTHALENEDIONE □ 2,3-DICHLORO-1,4-NAPHTHAQUINONE □ DICHLORONAPHTHOQUINONE □ 2,3-DICHLORONAPHTHOQUINONE □ 2,3-DICHLORO-α-NAPHTHOQUINONE □ 2,3-DICHLORONAPHTHOQUINONE-1,4 □ ENT 3,776 □ PHYGON □ PHYGON PASTE □ PHYGON SEED PROTECTANT □ PHYGON XL □ QUINTAR □ QUINTAR 540F □ SANQUINON □ UNIROYAL □ USR 604 □ U.S. RUBBER 604

TOXICITY DATA WITH REFERENCE
orl-mus TDLo:3300 mg/kg/78W-I:NEO NTIS** PB223-159
scu-mus TDLo:22 mg/kg:CAR NTIS** PB223-159
orl-rat LD50:160 mg/kg GTPZAB 16(5),52,72
orl-mus LD50:440 mg/kg 85GMAT-,47,82
ipr-mus LD50:30 mg/kg JMCMAR 26,570,83
skn-rbt LD50:5000 mg/kg FMCHA2 -,C77,83

CONSENSUS REPORTS: Reported in EPA TSCA Inventory.

SAFETY PROFILE: Poison by ingestion and intraperitoneal routes. Mildly toxic by skin contact. A skin, eye, and mucous membrane irritant. Large doses can cause central nervous system depression. Questionable carcinogen with experimental carcinogenic and neoplastigenic data. A fungicide and algaecide. When heated to decomposition it emits toxic fumes of Cl^-. See also CHLORIDES.

DFT400 CAS:89-61-2 HR: 2
1,4-DICHLORO-2-NITROBENZENE
mf: $C_6H_3Cl_2NO_2$ mw: 192.00

PROP: Prisms or plates from EtOH or EtOAc. D: 1.439 @ 75°/4°, mp: 56°, bp: 267°.

SYNS: 2,5-DICHLORNITROBENZEN (CZECH) □ 2,5-DICHLORONITROBENZENE □ NITRO-p-DICHLOROBENZENE

TOXICITY DATA WITH REFERENCE
skn-rbt 500 mg/24H MLD 28ZPAK -,94,72
eye-rbt 100 mg/24H MOD 28ZPAK -,94,72
mmo-sat 205 μg/plate MUREAV 116,217,83
orl-rat LD50:1210 mg/kg 28ZPAK -,94,72
orl-mus LD50:2850 mg/kg GTPZAB 25(8),50,81

CONSENSUS REPORTS: Reported in EPA TSCA Inventory.

SAFETY PROFILE: Moderately toxic by ingestion. A skin and eye irritant. Mutation data reported. See also CHLORINATED HYDROCARBONS, AROMATIC; and NITRO COMPOUNDS of AROMATIC HYDROCARBONS. When heated to decomposition it emits very toxic fumes of Cl^- and NO_x.

DFT600 CAS:99-54-7 HR: 2
1,2-DICHLORO-4-NITROBENZENE
mf: $C_6H_3Cl_2NO_2$ mw: 192.00

PROP: Liquid or needles from EtOH or CCl_4. Vap d: 6.6, mp: 42–43°, bp: 255–256°.

SYNS: DCNB □ 3,4-DICHLORNITROBENZEN (CZECH) □ 3,4-DICHLORONITROBENZENE

TOXICITY DATA WITH REFERENCE
skn-rbt 500 mg/24H MLD 28ZPAK -,94,72
eye-rbt 100 mg/24H MOD 28ZPAK -,94,72
mmo-sat 500 μg/plate AECTCV 9,533,80
mma-sat 500 μg/plate AECTCV 9,533,80
sln-dmg-par 200 ppm ENMUDM 7,677,85
orl-rat LD50:643 mg/kg 28ZPAK -,94,72
orl-mus LD50:1384 mg/kg 85GMAT-,47,82
skn-cat LD50:790 mg/kg 85GMAT-,47,82

CONSENSUS REPORTS: Reported in EPA TSCA Inventory.

SAFETY PROFILE: Moderately toxic by ingestion and skin contact. A skin and eye irritant. Mutation data reported. Potentially explosive reactions when heated with hydrogen + a catalyst. When heated to decomposition it emits very toxic fumes of NO_x and Cl^-. See also CHLORINATED HYDROCARBONS, AROMATIC; and NITRO COMPOUNDS of AROMATIC HYDROCARBONS.

DFT800 **CAS:1836-75-5** **HR: 3**
2,4-DICHLORO-4′-NITRODIPHENYL ETHER
mf: $C_{12}H_7Cl_2NO_3$ mw: 284.10

PROP: Crystals or solid. Mp: 70–71°. Very sltly sol in H_2O.

SYNS: 2′,4′-DICHLORO-4-NITROBIPHENYL ETHER □ 2,4-DICHLORO-1-(4-NITROPHENOXY)BENZENE □ 4-(2,4-DICHLOROPHENOXY)NITROBENZENE □ 2,4-DICHLOROPHENYL-p-NITROPHENYL ETHER □ 2,4-DICHLOROPHENYL-4-NITROPHENYL ETHER □ 2,4,-DICHLORPHENYL-4-NITROPHENYLAETHER (GERMAN) □ FW 925 □ MEZOTOX □ NCI-C00420 □ NICLOFEN □ NIP □ NITOFEN □ NITRAFEN □ NITRAPHEN □ NITROCHLOR □ 4′-NITRO-2,4-DICHLORODIPHENYL ETHER □ NITROFEN □ NITROFENE (FRENCH) □ NITROPHEN □ NITROPHENE □ PREPARATION 125 □ TOK □ TOK-2 □ TOK E □ TOK E-25 □ TOK E 40 □ TOKKORN □ TOK WP-50 □ TRIZILIN

TOXICITY DATA WITH REFERENCE
mmo-sat 33,300 ng/plate ENMUDM 7(Suppl 5),1,85
mma-sat 10 μg/plate ENMUDM 7(Suppl 5),1,85
otr-rat:emb 1500 ng/plate JJATDK 1,190,81
orl-mus TDLo:1100 mg/kg (female 7-17D post):REP SCIEAS 215,293,82
orl-rat TDLo:100 mg/kg (female 11D post):REP TXCYAC 20,209,81
orl-rat TDLo:37,530 μg/kg (female 6-18D post):REP TXCYAC 34,285,85
orl-rat TDLo:80 mg/kg (female 10-13D post):REP TXAPA9 86,22,86
orl-mus TDLo:660 mg/kg (female 5-15D post):REP TJADAB 24(1),11A,81
orl-rat TDLo:150 mg/kg (female 7-12D post):TER TCMUD8 6,339,86
orl-rat TDLo:62,500 μg/kg (female 6-15D post):TER TJADAB 34,213,86
orl-rat TDLo:62,500 μg/kg (6-15D preg):TER JACTDZ 4(1),221,85
orl-rat TDLo:80 mg/kg (female 10-13D post):TER TXAPA9 86,22,86
skn-rat TDLo:120 mg/kg (female 6-15D post):TER TXCYAC 28,37,83
skn-rat TDLo:12 mg/kg (female 6-15D post):TER TXCYAC 28,37,83
orl-mus TDLo:2520 mg/kg (female 1-21D post):TER TJADAB 32,45B,85
orl-mus TDLo:68,750 μg/kg (female 7-17D post):TER TXAPA9 67,1,83
orl-rat TDLo:42 g/kg/94W-C:CAR NCITR* NCI-CG-TR-26,78
orl-mus TDLo:24 g/kg/12W-C:CAR JJIND8 65,937,80
orl-mus TD:200 g/kg/78W-C:CAR NCITR* NCI-CG-TR-184,79
orl-mus TD:308 g/kg/78W-C:CAR NCITR* NCI-CG-TR-26,78
orl-mus TD:114 g/kg/58W-C:CAR NCITR* NCI-CG-TR-26,78
orl-mus TD:47 g/kg/12W-C:CAR JJIND8 65,937,80
orl-rat LD50:740 mg/kg HYSAAV 32,20,67
skn-rat LD50:5000 mg/kg AEHLAU 28,316,74
unk-rat LD50:3000 mg/kg 30ZDA9 -,109,71
orl-mus LD50:450 mg/kg HYSAAV 32,20,67
orl-cat LDLo:300 mg/kg HYSAAV 32,20,67
ihl-cat LCLo:620 mg/m³/4H HYSAAV 32,20,67
orl-rbt LD50:1620 mg/kg 28ZEAL 5,166,76

CONSENSUS REPORTS: NTP 7th Annual Report On Carcinogens. IARC Cancer Review: Group 2B IMEMDT 7,56,87; Animal Sufficient Evidence IMEMDT 30,271,83. NCI Carcinogenesis Bioassay (feed); No Evidence: rat NCITR* NCI-CG-TR-184,79; Clear Evidence: mouse, rat NCITR* NCI-CG-TR-26,78; Clear Evidence: mouse NCITR* NCI-CG-TR-184,79. EPA Genetic Toxicology Program. Community Right-To-Know List. Reported in EPA TSCA Inventory.

SAFETY PROFILE: Confirmed carcinogen with experimental carcinogenic data. Poison by ingestion. Moderately toxic by inhalation and possibly other routes. Experimental teratogenic and reproductive effects. A skin and severe eye irritant. Mutation data reported. A broad-spectrum herbicide. See also NITRO COMPOUNDS of AROMATIC HYDROCARBONS and ETHERS. When heated to decomposition it emits very toxic fumes of Cl^- and NO_x.

DFU000 **CAS:594-72-9** **HR: 3**
1,1-DICHLORO-1-NITROETHANE
DOT: UN 2650
mf: $C_2H_3Cl_2NO_2$ mw: 143.96

PROP: Liquid. Bp: 124°, flash p: 168°F(OC), d: 1.4153 @ 20°/20°, vap d: 4.97.

SYNS: 1,1-DICHLOOR-1-NITROETHAAN (DUTCH) □ 1,1-DICHLOR-1-NITROAETHAN (GERMAN) □ DICHLORONITROETHANE □ 1,1-DICLORO-1-NITROETANO (ITALIAN) □ ETHIDE

TOXICITY DATA WITH REFERENCE
orl-rat LD50:410 mg/kg BESAAT 12,161,66
ipr-mus LD50:240 mg/kg KHFZAN 10(6),53,76
orl-rbt LDLo:150 mg/kg JIHTAB 27,95,45
ihl-rbt LCLo:580 mg/m³/6H JIHTAB 27,95,45
ihl-gpg LCLo:580 mg/m³/6H JIHTAB 27,95,45

OSHA PEL: TWA 2 ppm
ACGIH TLV: TWA 2 ppm
DFG MAK: 10 ppm (60 mg/m³)
DOT CLASSIFICATION: 6.1; *Label:* Poison

SAFETY PROFILE: Poison by ingestion and intraperitoneal routes. Moderately toxic by inhalation. A strong irritant. Inhalation causes pulmonary edema. A fumigant for produce. Flammable when exposed to heat, flame, or oxidizers. Can react vigorously with oxidizing materials. To fight fire, use water, CO_2, dry chemical. When heated to decomposition it emits highly toxic fumes of Cl^- and NO_x.

For occupational chemical analysis use NIOSH: see 1,1-Dichloro-1-nitroethane, 1601.

DFU400 **CAS:6240-55-7** **HR: 2**
1,2-DICHLORO-3-NITRONAPHTHALENE
mf: $C_{10}H_5Cl_2NO_2$ mw: 242.06

PROP: Crystals from MeOH. Mp: 125–126°

TOXICITY DATA WITH REFERENCE
orl-rat TDLo:13 g/kg/52W-I:ETA,REP JNCIAM 41,985,68

SAFETY PROFILE: Questionable carcinogen with experimental tumorigenic data. Experimental reproductive effects. When heated to decomposition it emits very toxic fumes of Cl^- and NO_x. See also NITRO COMPOUNDS of AROMATIC HYDROCARBONS.

DFU600 CAS:609-89-2 HR: 3
2,4-DICHLORO-6-NITROPHENOL
mf: $C_6H_3Cl_2NO_3$ mw: 208.00

PROP: Yellow crystals from AcOH. Mp: 124–125°.

SYN: 2,4-DICHLOR-6-NITROFENOL (CZECH)

TOXICITY DATA WITH REFERENCE
eye-rbt 100 mg/24H SEV 28ZPAK -,80,72
orl-rat LD50:129 mg/kg 28ZPAK -,80,72

CONSENSUS REPORTS: Chlorophenol compounds are on the Community Right-To-Know List.

SAFETY PROFILE: Poison by ingestion. A severe eye irritant. When heated to decomposition it emits very toxic fumes of Cl^- and NO_x. See also CHLOROPHENOLS and NITRO COMPOUNDS of AROMATIC HYDROCARBONS.

DFU800 CAS:37169-10-1 HR: 3
2,4-DICHLORO-6-NITROPHENOL ACETATE
mf: $C_8H_5Cl_2NO_4$ mw: 250.04

SYN: 2,4-DICHLOR-6-NITROFENYLESTER KYSELINY OCTIVE (CZECH)

TOXICITY DATA WITH REFERENCE
skn-rbt 500 mg/24H MLD 28ZPAK -,93,72
eye-rbt 20 mg/24H MOD 28ZPAK -,93,72
orl-rat LD50:96 mg/kg 28ZPAK -,93,72

CONSENSUS REPORTS: Chlorophenol compounds are on the Community Right-To-Know List.

SAFETY PROFILE: Poison by ingestion. A skin and eye irritant. See also CHLOROPHENOLS and NITRO COMPOUNDS of AROMATIC HYDROCARBONS. When heated to decomposition it emits very toxic fumes of Cl^- and NO_x.

DFV200 CAS:14094-48-5 HR: 2
6,7-DICHLORO-4-NITROQUINOLINE-1-OXIDE
mf: $C_9H_4Cl_2N_2O_3$ mw: 259.05

TOXICITY DATA WITH REFERENCE
scu-mus TDLo:120 mg/kg/50D-I:ETA BCPCA6 16,631,67

SAFETY PROFILE: Questionable carcinogen with experimental tumorigenic data. When heated to decomposition it emits very toxic fumes of Cl^- and NO_x. See also NITRO COMPOUNDS of AROMATIC HYDROCARBONS.

DFV400 CAS:50-65-7 HR: 3
2′,5-DICHLORO-4′-NITROSALICYLANILIDE
mf: $C_{13}H_8Cl_2N_2O_4$ mw: 327.13

PROP: Pale-yellow crystals. Mp: 225–230°. Sltly sol in EtOH, $CHCl_3$, and Et_2O.

SYNS: BAY 2353 □ BAYER 73 □ BAYER 2353 □ BAYLUSCID □ CHEMAGRO 2353 □ 5-CHLORO-N-(2-CHLORO-4-NITROPHENYL)-2-HYDROXYBENZAMIDE □ 5-CHLORO-2′-CHLORO-4′-NITROSALICYLANILIDE □ 2-CHLORO-4-NITROPHENYLAMIDE-6-CHLOROSALICYLIC ACID □ N-(2-CHLORO-4-NITROPHENYL)-5-CHLOROSALICYLAMIDE □ CLONITRALID □ 2′,5-DICHLOR-4′-NITRO-SALIZYLSAEUREANILID (GERMAN) □ DICHLOSALE □ ENT 25,823 □ FENASAL □ HL 2447 □ 2-HYDROXY-5-CHLORO-N-(2-CHLORO-4-NITROPHENYL)BENZAMIDE □ IOMESAN □ IOMEZAN □ NICLOSAMIDE □ PHENASAL □ VERMITIN □ YOMESAN

TOXICITY DATA WITH REFERENCE
cyt-hmn:lym 6 mL/L MUREAV 173,81,86
sce-hmn:lym 8 mg/L MUREAV 173,81,86
orl-mus TDLo:300 mg/kg (male 5D pre):REP MUREAV 204,269,88
orl-rat LD50:2500 mg/kg 85JFAN A297,83
ipr-rat LD50:250 mg/kg ZTMPA5 13,1,62
orl-mus LD50:1000 mg/kg 85DPAN -,-,71/76
ivn-mus LD50:7500 μg/kg ARZNAD 10,884,60

SAFETY PROFILE: Poison by intravenous and intraperitoneal routes. Moderately toxic by ingestion. Experimental reproductive effects. Human mutation data reported. When heated to decomposition it emits very toxic fumes of Cl^- and NO_x.

DFV600 CAS:1420-04-8 HR: 3
2′,5-DICHLORO-4′-NITROSALICYLANILIDE-2-AMINOETHANOL SALT
mf: $C_{13}H_8Cl_2N_2O_4 \cdot C_2H_7NO$ mw: 388.23

PROP: A solid. Mp: 204°.

SYNS: BAYER 73 □ BAYER 25648 □ BAYLUSCID □ BAYLUSCIDE □ 5-CHLORO-N-(2-CHLORO-4-NITROPHENYL)-2-HYDROXYBENZAMIDE with 2-AMINOETHANOL (1:1) □ CLONITARLID □ 5,2′-DICHLORO-4′-NITROSALICYLANILIDE ETHANOLAMINE SALT □ 5,2-DICHLORO-4-NITROSALICYLIC ANILIDE-2-AMINOETHANOL SALT □ 2′,5-DICHLORO-4′-NITROSALICYLOYLANILIDE ETHANOLAMINE SALT □ ETHANOLAMINE SALT of 5,2′-DICHLORO-4′-NITROSALICYCLICANILIDE □ M 73 □ MOLLUSCICIDE BAYER 73 □ NCI-C00431 □ NICLOSAMIDE □ SR 73

TOXICITY DATA WITH REFERENCE
orl-rat LD50:500 mg/kg 85ARAE 3,103,76/77
ipr-rat LD50:250 mg/kg GUCHAZ 6,126,73

CONSENSUS REPORTS: NCI Carcinogenesis Bioassay (feed); Inadequate Studies: mouse, rat NCITR* NCI-CG-TR-91,78.

SAFETY PROFILE: Poison by intraperitoneal route. Moderately toxic by ingestion. Many N-nitroso compounds are carcinogens. A pesticide. When heated to decomposition it emits very toxic fumes of NO_x and Cl^-. See also N-NITROSO COMPOUNDS.

DFW000 CAS:69112-96-5 HR: 2
2,2′-DICHLORO-N-NITROSODIPROPYLAMINE
mf: $C_6H_{12}Cl_2N_2O$ mw: 199.10

SYN: NITROSOBIS(2-CHLOROPROPYL)AMINE

TOXICITY DATA WITH REFERENCE
mmo-sat 10 μg/plate MUREAV 66,1,79
mma-sat 10 μg/plate MUREAV 66,1,79
orl-rat TDLo:1360 mg/kg/20W-I:ETA EESADV 2,421,78

SAFETY PROFILE: Questionable carcinogen with experimental tumorigenic data. Many N-nitroso compounds are carcinogens. Mutation data reported. When heated to decomposition it emits very toxic fumes of Cl^- and NO_x. See also N-NITROSO COMPOUNDS.

DFW200 **CAS:57541-72-7** **HR: 2**
3,4-DICHLORONITROSOPIPERIDINE
mf: $C_5H_8Cl_2N_2O$ mw: 183.05

SYN: N-NITROSO-3,4-DICHLOROPIPERIDINE

TOXICITY DATA WITH REFERENCE
mmo-sat 200 μg/plate MUREAV 56,131,77
mma-sat 10 nmol/plate MUREAV 57,85,78
sln-dmg-orl 200 μmol/L/24H MUREAV 67,27,79
sce-hmn:lym 100 μmol/L TCMUE9 1,129,84
orl-rat TDLo:169 mg/kg/30W-I:ETA ZKKOBW 92,221,78
orl-rat TD:366 mg/kg/15W-C:ETA CNREA8 35,3209,75
orl-rat TD:260 mg/kg/21W-C:ETA CNREA8 40,3325,80

CONSENSUS REPORTS: EPA Genetic Toxicology Program.

SAFETY PROFILE: Questionable carcinogen with experimental tumorigenic data. Human mutation data reported. Many N-nitroso compounds are carcinogens. See also N-NITROSO COMPOUNDS. When heated to decomposition it emits very toxic fumes of Cl^- and NO_x.

DFW600 **CAS:59863-59-1** **HR: 2**
3,4-DICHLORO-N-NITROSOPYRROLIDINE
mf: $C_4H_6Cl_2N_2O$ mw: 169.02

TOXICITY DATA WITH REFERENCE
mma-sat 250 μg/plate MUREAV 89,35,81
orl-rat TDLo:1550 mg/kg/31W-I:ETA CNREA8 36,1988,76

SAFETY PROFILE: Questionable carcinogen with experimental tumorigenic data. Mutation data reported. Many N-nitroso compounds are carcinogens. When heated to decomposition it emits very toxic fumes of Cl^- and NO_x. See also N-NITROSO COMPOUNDS.

DFX000 **CAS:30586-10-8** **HR: 3**
DICHLOROPENTANE
DOT: UN 1152
mf: $C_5H_{10}Cl_2$ mw: 141.05

PROP: Clear, light-yellow liquid. Bp: 130°, flash p: 106°F (OC), vap d: 4.86, d: 1.06–1.08 @ 20°.

SYN: DICHLOROPENTANES (DOT)

DOT CLASSIFICATION: 3; *Label:* Flammable Liquid

SAFETY PROFILE: Flammable liquid when exposed to heat or flame. Can react vigorously with oxidizing materials. To fight fire, use water, foam, CO_2, dry chemical. When heated to decomposition it emits highly toxic fumes of Cl^- and phosgene. See also 1,5-DICHLOROPENTANE; and CHLORINATED HYDROCARBONS, ALIPHATIC.

DFX200 **CAS:628-76-2** **HR: 3**
1,5-DICHLOROPENTANE
mf: $C_5H_{10}Cl_2$ mw: 141.05

PROP: D: 1.1, vap d: 4.9, bp: 180°, flash p: >80°F (OC). Insol in water.

TOXICITY DATA WITH REFERENCE
ipr-mus LDLo:64 mg/kg CBCCT* 2,189,50

CONSENSUS REPORTS: Reported in EPA TSCA Inventory.

SAFETY PROFILE: Poison by intraperitoneal route. Dangerous fire hazard when exposed to heat or flame. To fight fire, use alcohol foam or spray. Use of water is ineffective except as a blanket. When heated to decomposition it emits toxic fumes of Cl^-. See also CHLORINATED HYDROCARBONS, ALIPHATIC.

DFX400 **CAS:536-29-8** **HR: 3**
DICHLOROPHENARSINE HYDROCHLORIDE
mf: $C_6H_6AsCl_2NO{\cdot}ClH$ mw: 290.41

PROP: White hygroscopic powder from EtOH. Mp: 146–148°. Sol in H_2O with hydrolysis.

SYNS: 2-AMINO-4-DICHLOROARSINOPHENOL HYDROCHLORIDE □ (3-AMINO-4-HYDROXYPHENYL)ARSONOUS DICHLORIDE MONOHYDROCHLORIDE □ 3-AMINO-4-HYDROXYPHENYL DICHLORARSINE HYDROCHLORIDE □ (3-AMINO-4-HYDROXYPHENYL)DICHLOROARSINE HYDROCHLORIDE □ ARSECLOR □ CHLORARSOL □ CHLORASEN □ CLORARSEN □ DICHLOROMAPHARSEN □ FILARSEN □ FONTARSOL □ HALARSOL □ R.P. 2591

TOXICITY DATA WITH REFERENCE
par-hmn TDLo:957 μg/kg:GIT JPETAB 73,412,41
orl-rat LDLo:500 mg/kg NCNSA6 5,12,53
ivn-rat LD50:24,500 μg/kg AMIUAG 8,196,54
ipr-mus LD50:41 mg/kg PSEBAA 78,392,51
unr-mus LD50:44 mg/kg CNREA8 9,626,49
ivn-rbt LDLo:15 mg/kg JPETAB 73,412,41

CONSENSUS REPORTS: Arsenic and its compounds, as well as chlorophenol compounds, are on the Community Right-To-Know List.

OSHA PEL: TWA 0.5 mg(As)/m^3
ACGIH TLV: TWA 0.2 mg(As)/m^3

SAFETY PROFILE: Poison by intravenous, intraperitoneal, and possibly other routes. Moderately toxic by ingestion. Human systemic effects by parenteral route: hypermotility, diarrhea, nausea, vomiting. See also ARSENIC COMPOUNDS and CHLOROPHENOLS. When heated to decomposition it emits very toxic fumes of As, NO_x, and Cl^-.

DFX500 **CAS:576-24-9** **HR: 2**
2,3-DICHLOROPHENOL
mf: $C_6H_4Cl_2O$ mw: 163.00

TOXICITY DATA WITH REFERENCE
orl-mus LD50:2376 mg/kg TOLED5 29,39,85

CONSENSUS REPORTS: Reported in EPA TSCA Inventory.

SAFETY PROFILE: Moderately toxic by ingestion. When heated to decomposition it emits toxic vapors of Cl^-.

DFX800 CAS:120-83-2 HR: 3
2,4-DICHLOROPHENOL
mf: $C_6H_4Cl_2O$ mw: 163.00

PROP: Colorless crystals or needles. Mp: 45°, bp: 210°, flash p: 237°F, d: 1.383 @ 60°/25°, vap d: 5.62, vap press: 1 mm @ 53.0°.

SYNS: DCP ☐ 2,4-DCP ☐ NCI-C55345 ☐ RCRA WASTE NUMBER U081

TOXICITY DATA WITH REFERENCE
sln-ham:lng 500 μmol/L MUREAV 182,135,87
orl-rat TDLo:7500 mg/kg (female 6-15D post):TER TOXID9 4,167,84
scu-mus TDLo:666 mg/kg (female 6-14D post):TER NTIS** PB223-160
orl-rat TDLo:20 mg/kg (1-20D preg):TER GISAAA 41(11),102,76
skn-mus TDLo:16 g/kg/39W-I:CAR CNREA8 19,413,59
orl-rat LD50:580 mg/kg FEPRA7 2,76,43
ipr-rat LD50:430 mg/kg BJPCAL 13,20,58
scu-rat LD50:1730 mg/kg FEPRA7 2,76,43
orl-mus LD50:1276 mg/kg FAATDF 5,478,85
ipr-mus LD50:153 mg/kg JMCMAR 18,868,75

CONSENSUS REPORTS: IARC Cancer Review: Human Limited Evidence IMEMDT 41,319,86. Reported in EPA TSCA Inventory. EPA Genetic Toxicology Program. Community Right-To-Know List.

SAFETY PROFILE: Suspected carcinogen with experimental carcinogenic and teratogenic data. Poison by intraperitoneal route. Moderately toxic by ingestion and subcutaneous routes. An experimental teratogen. Mutation data reported. Combustible when exposed to heat or flame. Can react vigorously with oxidizing materials. To fight fire, use alcohol foam, foam, CO_2, dry chemical. When heated to decomposition, or on contact with acid or acid fumes, it emits highly toxic fumes of Cl^-. See also CHLOROPHENOLS.

DFX850 CAS:583-78-8 HR: 2
2,5-DICHLOROPHENOL
mf: $C_6H_4Cl_2O$ mw: 163.00

SYN: PHENOL, 2,5-DICHLORO-

TOXICITY DATA WITH REFERENCE
sce-mus-ipr 210 mg/kg JACTDZ 2(2),249,83
orl-rat LD50:580 mg/kg NTIS** PB85-143766
orl-mus LD50:946 mg/kg TOLED5 29,39,85

CONSENSUS REPORTS: Reported in EPA TSCA Inventory.

SAFETY PROFILE: Moderately toxic by ingestion. Mutation data reported. When heated to decomposition it emits toxic vapors of Cl^-.

DFY000 CAS:87-65-0 HR: 3
2,6-DICHLOROPHENOL
mf: $C_6H_4Cl_2O$ mw: 163.00

PROP: Needles from pet ether. Mp: 66–67°, bp: 219–220°.

SYNS: 2,6-DICHLORFENOL (CZECH) ☐ RCRA WASTE NUMBER U082

TOXICITY DATA WITH REFERENCE
skn-rbt 500 mg/24H SEV 28ZPAK -,79,72
eye-rbt 250 μg/24H SEV 28ZPAK -,79,72
ipr-rat LD50:390 mg/kg BJPCAL 13,20,58
orl-mus LD50:2120 mg/kg TOLED5 29,39,85

CONSENSUS REPORTS: Reported in EPA TSCA Inventory. EPA Genetic Toxicology Program. Chlorophenol compounds are on the Community Right-To-Know List.

SAFETY PROFILE: Poison by intraperitoneal route. Moderately toxic by ingestion. A severe skin and eye irritant. When heated to decomposition it emits toxic fumes of Cl^-. See also CHLOROPHENOLS.

DFY400 CAS:97-16-5 HR: 3
2,4-DICHLOROPHENOL BENZENESULFONATE
mf: $C_{12}H_8Cl_2O_3S$ mw: 303.16

SYNS: COMPOUND 923 ☐ 2,4-DICHLOROPHENYL BENZENESULFONATE ☐ 2,4-DICHLOROPHENYL BENZENESULPHONATE ☐ 2,4-DICHLOROPHENYL ESTER of BENZENESULFONIC ACID ☐ 2,4-DICHLOROPHENYL ESTER BENZENESULPHONIC ACID ☐ DPBS ☐ EM 923 ☐ GENITE ☐ GENITOL

TOXICITY DATA WITH REFERENCE
orl-mus TDLo:260 g/kg/78W-I:ETA NTIS** PB223-159
scu-mus TDLo:1000 mg/kg:CAR NTIS** PB223-159
orl-rat LDLo:1000 mg/kg BESAAT 12,117,66
unk-rat LD50:1400 mg/kg 30ZDA9 -,274,71
orl-dog LDLo:620 mg/kg AIPTAK 121,306,59
orl-rbt LD50:700 mg/kg PCOC** ,556,66
ivn-rbt LD50:115 mg/kg AIPTAK 121,306,59

CONSENSUS REPORTS: Chlorophenol compounds are on the Community Right-To-Know List.

SAFETY PROFILE: Poison by intravenous route. Moderately toxic by ingestion and possibly other routes. Questionable carcinogen with experimental carcinogenic and tumorigenic data. An irritant. A pesticide. See also CHLOROPHENOLS. When heated to decomposition it emits very toxic fumes of Cl^- and SO_x.

DFY425 CAS:95-77-2 HR: 2
3,4-DICHLOROPHENOL
mf: $C_6H_4Cl_2O$ mw: 163.00

SYN: PHENOL, 3,4-DICHLORO-

TOXICITY DATA WITH REFERENCE
orl-mus LD50:1685 mg/kg TOLED5 29,39,85

CONSENSUS REPORTS: Reported in EPA TSCA Inventory.

SAFETY PROFILE: Moderately toxic by ingestion. When heated to decomposition it emits toxic vapors of Cl^-.

DFY450 **CAS:591-35-5** **HR: 2**
3,5-DICHLOROPHENOL
mf: $C_6H_4Cl_2O$ mw: 163.00

SYN: PHENOL, 3,5-DICHLORO-

TOXICITY DATA WITH REFERENCE
orl-mus LD50:2389 mg/kg TOLED5 29,39,85

CONSENSUS REPORTS: Reported in EPA TSCA Inventory.

SAFETY PROFILE: Moderately toxic by ingestion. When heated to decomposition it emits toxic vapors of Cl^-.

DFY709 **CAS:1929-73-3** **HR: 2**
(2,4-DICHLOROPHENOXY)ACETIC ACID BUTOXYETHYL ESTER
mf: $C_{14}H_{18}Cl_2O_4$ mw: 321.22

SYNS: BUTOXYETHYL-2,4-DICHLOROPHENOXYACETATE □ 2,4-D BUTOXYETHANOL ESTER □ 2,4-D BUTOXYETHYL ESTER □ 2,4-D 2-BUTOXYETHYL ESTER

TOXICITY DATA WITH REFERENCE
orl-rat TDLo:1500 mg/kg (6-15D preg):TER TXAPA9 22,14,72
orl-rat LD50:831 mg/kg FAATDF 9,423,87

SAFETY PROFILE: Moderately toxic by ingestion. An experimental teratogen. When heated to decomposition it emits toxic fumes of Cl^-. See also ESTERS.

DFY800 **CAS:2008-39-1** **HR: 3**
(2,4-DICHLOROPHENOXY)ACETIC ACID DIMETHYLAMINE
mf: $C_{10}H_{11}Cl_2NO_3$ mw: 264.12

SYNS: 2,4-D ACETATE □ 2,4-D AMINE SALT □ BLADEX G □ 2,4-D DIMETHYLAMINE SALT □ DEFY □ DEMISE □ (2,4-DICHLOROPHENOXY)ACETATE DIMETHYLAMINE □ DIMETHYLAMINE SALT of 2,4-D □ DIMETHYLAMMONIUM 2,4-DICHLOROPHENOXYACETATE □ FORMULA 40 □ HORMIN □ PHORDENE □ REED AMINE 400

TOXICITY DATA WITH REFERENCE
cyt-hmn:lyms 500 μmol/L MUTAEX 1,241,86
orl-rat TDLo:3 g/kg (6-15D preg):TER TXAPA9 22,14,72
orl-rat LD50:625 mg/kg GISAAA 36(11),33,71
unr-mus LD50:300 mg/kg HYSAAV 31(9),383,66
skn-rbt LD50:2115 mg/kg FMCHA2 -,C73,83

SAFETY PROFILE: Poison by unreported route. Moderately toxic by ingestion and skin contact. An experimental teratogen. Human mutation data reported. A weed killer. When heated to decomposition it emits very toxic fumes of Cl^-, NH_3, and NO_x.

DFZ000 **CAS:1928-45-6** **HR: 2**
2,4-DICHLOROPHENOXYACETIC ACID PROPYLENE GLYCOL BUTYL ETHER ESTER
mf: $C_{15}H_{20}Cl_2O_4$ mw: 335.25

SYNS: 2,4-D PGBE □ 2,4-D PROPYLENE GLYCOL BUTYL ETHER ESTER

TOXICITY DATA WITH REFERENCE
orl-rat TDLo:114 mg/kg (6-15D preg):REP FCTXAV 9,801,71
orl-rat TDLo:114 mg/kg (6-15D preg):TER FCTXAV 9,801,71
orl-mus TDLo:1341 mg/kg (female 12-15D post):TER AECTCV 6,33,77
orl-rat TDLo:758 mg/kg (female 6-15D post):TER FCTXAV 9,801,71
orl-rat TDLo:190 mg/kg (female 6-15D post):TER FCTXAV 9,801,71
orl-rat LD50:500 mg/kg NTIS** PB85-143766

CONSENSUS REPORTS: Glycol ether compounds are on the Community Right-To-Know List.

SAFETY PROFILE: Moderately toxic by ingestion. An experimental teratogen. Other experimental reproductive effects. A pesticide. When heated to decomposition it emits toxic fumes of Cl^-. See also GLYCOL ETHERS.

DGA000 **CAS:94-82-6** **HR: 2**
4-(2,4-DICHLOROPHENOXY)BUTYRIC ACID
mf: $C_{10}H_{10}Cl_2O_3$ mw: 249.10

PROP: Crystals from MeOH (aq). Mp: 118–119°.

SYNS: BUTOXON □ BUTOXONE □ BUTOXONE AMINE □ BUTOXONE ESTER □ BUTYRAC □ BUTYRAC ESTER □ 2,4-DB □ 2,4-D BUTYRIC □ γ-(2,4-DICHLOROPHENOXY)BUTYRIC ACID □ EMBUTOX □ EMBUTOX KLEAN-UP □ LEGUMEX D

TOXICITY DATA WITH REFERENCE
dnr-esc 5 mg/disc NTIS** PB80-133226
dnr-bcs 5 mg/disc NTIS** PB80-133226
orl-rat TDLo:17 mg/kg (1-7D preg):REP GISAAA 41(2),20,76
orl-rat TDLo:416 mg/kg (female 9D post):TER GISAAA 41(2),20,76
orl-rat TDLo:416 mg/kg (9D preg):TER GISAAA 41(2),20,76
orl-rat TDLo:17 mg/kg (1-7D preg):TER GISAAA 41(2),20,76
orl-rat LD50:700 mg/kg RREVAH 10,97,65
skn-rat LD50:800 mg/kg WRPCA2 9,119,70

CONSENSUS REPORTS: EPA Genetic Toxicology Program.

SAFETY PROFILE: Moderately toxic by ingestion and skin contact. An experimental teratogen. Other experimental reproductive effects. Mutation data reported. An herbicide. When heated to decomposition it emits toxic fumes of Cl^-.

DGA200 **CAS:14255-88-0** **HR: 3**
5,6-DICHLORO-1-PHENOXYCARBONYL-2-TRIFLUOROMETHYLBENZIMIDAZOLE
mf: $C_{15}H_7Cl_2F_3N_2O_2$ mw: 375.14

PROP: Yellow crystals. Mp: 103°. Sol in dioxan and Me_2CO.

SYNS: 5,6-DICHLORO-2-TRIFLUOROMETHYLBENZIMIDAZOLE-1-CARBOXYLATE □ 5,6-DICHLORO-2-(TRIFLUOROMETHYL)-1H-BENZIMIDAZOLE-1-CARBOXYLIC ACID PHENYL ESTER □ ENT 27,438 □ FENAZAFLOR □ FENOFLURAZOLE □ FENOZAFLOR □ FENZAFLOR □ FISONS NC 5016 □ LOVOZAL □ NC 5016 □ NSC 191025 □ PHENYL-5,6-DICHLORO-2-TRIFLUOROMETHYL-BENZIMIDAZOLE-1-CARBOXYLATE □ TARZOL

TOXICITY DATA WITH REFERENCE
orl-rat LD50:283 mg/kg FMCHA2 -,C103,83
skn-rat LD50:700 mg/kg WRPCA2 9,119,70
ipr-rat LD50:168 mg/kg GUCHAZ 6,275,73
orl-mus LD50:1600 mg/kg MRLAB3 33,839,68
ipr-mus LD50:42 mg/kg BCPCA6 18,1389,69
orl-dog LD50:50 mg/kg 85DPAN -,-,71/76
orl-rbt LD50:28 mg/kg GUCHAZ 6,275,73
orl-gpg LD50:59 mg/kg 31ZOAD 1,222,68
orl-ckn LD50:50 mg/kg MRLAB3 33,839,68
orl-mam LD50:3717 mg/kg NTIS** PB288-416

SAFETY PROFILE: Poison by ingestion and intraperitoneal routes. Moderately toxic by skin contact. When heated to decomposition it emits very toxic fumes of F^-, Cl^-, and NO_x.

DGA400 CAS:73986-95-5 HR: 2
2,4-DICHLOROPHENOXY ETHANEDIOL
mf: $C_8H_8Cl_2O_3$ mw: 223.06

SYN: 2,4-DICHLOROPHENOXY-1,2-ETHANEDIOL

TOXICITY DATA WITH REFERENCE
skn-rbt 10 mg/24H open MLD AMIHBC 4,119,51
eye-rbt 50 µg open SEV AMIHBC 4,119,51
orl-rat LD50:1070 mg/kg AMIHBC 4,119,51
skn-rbt LD50:420 mg/kg AMIHBC 4,119,51

SAFETY PROFILE: Moderately toxic by ingestion and skin contact. A skin and severe eye irritant. When heated to decomposition it emits very toxic fumes of Cl^-.

DGA425 CAS:73791-41-0 HR: 3
3-(2,4-DICHLOROPHENOXY)-2-HYDROXYPROPYL-o-CHLOROPHENYL ARSINIC ACID
mf: $C_{15}H_{14}AsCl_3O_4$ mw: 439.56

SYNS: ARSINE OXIDE, (o-CHLOROPHENYL)(3-(2,4-DICHLOROPHENOXY)-2-HYDROXYPROPYL)HYDROXY- □ (o-CHLOROPHENYL)(3-(2,4-DICHLOROPHENOXY)-2-HYDROXYPROPYL)HYDROXYARSINEOXIDE

TOXICITY DATA WITH REFERENCE
ivn-mus LD50:56 mg/kg CSLNX* NX#06928

OSHA PEL: TWA 0.5 mg(As)/m³

SAFETY PROFILE: Poison by intravenous route. When heated to decomposition it emits toxic fumes of As and Cl^-.

DGA800 CAS:23712-05-2 HR: 3
2-((3,4-DICHLOROPHENOXY)METHYL)-2-IMIDAZOLINE HYDROCHLORIDE
mf: $C_{10}H_{10}Cl_2N_2O$ mw: 245.12

PROP: A solid. Mp: 244–245°.

SYNS: DH-524 □ 2-((3,4-DICHLOROPHENOXY)METHYL)-2-IMIDAZOLINE MONOHYDROCHLORIDE

TOXICITY DATA WITH REFERENCE
orl-rat LD50:100 mg/kg 27ZQAG -,220,72
orl-mus LD50:111 mg/kg 27ZQAG -,220,72
ipr-mus LD50:41 mg/kg USXXAM #4020167

SAFETY PROFILE: Poison by ingestion and intraperitoneal routes. When heated to decomposition it emits very toxic fumes of Cl^- and NO_x.

DGB000 CAS:120-36-5 HR: 3
2-(2,4-DICHLOROPHENOXY) PROPIONIC ACID
mf: $C_9H_8Cl_2O_3$ mw: 235.07

SYNS: ACIDE-2-(2,4-DICHLORO-PHENOXY) PROPIONIQUE (FRENCH) □ ACIDO-2-(2,4-DICLORO-FENOSSI)-PROPIONICO (ITALIAN) □ CORNOX RD □ CORNOX RK □ DESORMONE □ 2-(2,4-DICHLOOR-FENOXY)-PROPIONZUUR (DUTCH) □ α-(2,4-DICHLOROPHENOXY) PROPIONIC ACID □ DICHLOROPROP □ 2-(2,4-DICHLOR-PHENOXY)-PROPIONSAEURE (GERMAN) □ DICHLORPROP □ 2,4-DP □ 2-(2,4-DP) □ HEDONAL □ HEDONAL DP □ HORMATOX □ KILDIP □ POLYCLENE □ POLYMONE □ POLYTOX □ RD 406 □ SERITOX 50 □ U46 □ U46 DP-FLUID □ VISKO-RHAP □ WEEDONE DP □ WEEDONE 170

TOXICITY DATA WITH REFERENCE
mmo-smc 700 mg/L ZAPOAK 9,483,69
orl-rat TDLo:20 mg/kg (female 4-18D post):REP TJADAB 33,11A,86
orl-mus TDLo:3 g/kg (female 6-15D post):TER ARZNAD 33,1479,83
orl-mus TDLo:5 g/kg (female 6-15D post):REP ARZNAD 33,1479,83
orl-mus TDLo:4 g/kg (female 6-15D post):TER ARZNAD 33,1479,83
orl-rat LD50:344 mg/kg JACTDZ 1,177,92
skn-rat LD50:1880 mg/kg GISAAA 50(9),22,85
orl-mus LD50:309 mg/kg RPZHAW 31,373,80
skn-mus LD50:1400 mg/kg 28ZEAL 5,76,76

CONSENSUS REPORTS: IARC Cancer Review: Group 2B IMEMDT 7,156,87; Human Limited Evidence IMEMDT 41,357,86. Reported in EPA TSCA Inventory.

SAFETY PROFILE: Suspected carcinogen. Poison by ingestion. Moderately toxic by skin contact. An experimental teratogen. Other experimental reproductive effects. Mutation data reported. A fumigant. When heated to decomposition it emits toxic fumes of Cl^-.

DGB100 HR: 1
(+)-2-(2,4-DICHLOROPHENOXY)PROPIONIC ACID
mf: $C_9H_8Cl_2O_3$ mw: 235.07

SYN: (+)-2-(2,4-DICHLORPHENOXY)PROPIONSAFEURE (GERMAN)

TOXICITY DATA WITH REFERENCE
orl-mus TDLo:3 g/kg (6-15D preg):TER ARZNAD 33,1479,83
orl-mus TDLo:4 g/kg (6-15D preg):REP ARZNAD 33,1479,83
orl-mus TDLo:5 g/kg (6-15D preg):TER ARZNAD 33,1479,83
orl-mus TDLo:4 g/kg (6-15D preg):TER ARZNAD 33,1479,83
orl-rat LD50:1 g/kg FMCHA2 -,C100,89

SAFETY PROFILE: Slightly toxic by ingestion. Experimental teratogenic and reproductive effects. When heated to decomposition it emits toxic fumes of Cl^-. See also CHLORINATED HYDROCARBONS, AROMATIC.

D

DGB200 **CAS:6965-71-5** **HR: 2**
2-(2,5-DICHLOROPHENOXY)PROPIONIC ACID
mf: $C_9H_8Cl_2O_3$ mw: 235.07

SYN: α-(2,5-DICHLOROPHENOXY)PROPIONIC ACID

TOXICITY DATA WITH REFERENCE
scu-mus TDLo:100 mg/kg:ETA NTIS** PB223-159

SAFETY PROFILE: Questionable carcinogen with experimental tumorigenic data. When heated to decomposition it emits toxic fumes of Cl^-.

DGB400 **CAS:39637-16-6** **HR: 3**
(2,4-DICHLOROPHENOXY)TRIBUTYLSTANNANE
mf: $C_{18}H_{30}Cl_2OSn$ mw: 452.07

SYN: TRI-n-BUTYL-2,4-DICHLOROPHENOXYTIN

TOXICITY DATA WITH REFERENCE
ivn-mus LD50:56 mg/kg CSLNX* NX#01883

OSHA PEL: TWA 0.1 mg(Sn)/m^3 (skin)
ACGIH TLV: TWA 0.1 mg(Sn)/m^3 (skin) (Proposed: TWA 0.1 mg(Sn)/m^3; STEL 0.2 mg(Sn)/m^3 (skin))
NIOSH REL: (Organotin Compounds) TWA 0.1 mg(Sn)/m^3

SAFETY PROFILE: Poison by intravenous route. Tributyl tin compounds are extremely toxic to marine life. See also TIN COMPOUNDS and CHLORIDES. When heated to decomposition it emits toxic fumes of Cl^-.

For occupational chemical analysis use NIOSH: Organotin Compounds 5504.

DGB480 **CAS:2621-62-7** **HR: 1**
N-(2,5-DICHLOROPHENYL)ACETAMIDE
mf: $C_8H_7Cl_2NO$ mw: 204.06

SYNS: ACETAMIDE, N-(2,5-DICHLOROPHENYL)-(9CI) □ ACETANILIDE, 2′,5′-DICHLORO- □ 2,5-DICHLORACETANILID □ 2′,5′-DICHLOROACETANILIDE

TOXICITY DATA WITH REFERENCE
orl-rat LD50:4100 mg/kg 85JCAE -,579,86

CONSENSUS REPORTS: Reported in EPA TSCA Inventory.

SAFETY PROFILE: Mildly toxic by ingestion. When heated to decomposition it emits toxic vapors of NO_x and Cl^-.

DGB500 **CAS:4205-90-7** **HR: 3**
2-(2,6-DICHLOROPHENYLAMINO)-2-IMIDAZOLINE
mf: $C_9H_9Cl_2N_3$ mw: 230.11

PROP: A solid. Mp: 136–138°.

SYNS: 734571A □ BENZENAMINE, 2,6-DICHLORO-N-2-IMIDAZOLIDINYLIDENE- (9CI) □ CLONIDIN □ CLONIDINE □ 2-IMIDAZOLINE, 2-(2,6-DICHLOROANILINO)-

TOXICITY DATA WITH REFERENCE
skn-rbt 500 mg MLD OYYAA2 45,257,93
scu-rat TDLo:2080 μg/kg (female 8-20D post):REP NRTXDN 9,559,88
orl-mus TDLo:20 mg/kg (female 11D post):TER TJADAB 32,19A,85
orl-mus TDLo:10 mg/kg (female 11D post):TER TJADAB 32,19A,85
orl-man TDLo:2857 ng/kg:BPR AIMDAP 143,2195,83
orl-rat LD50:157 mg/kg IYKEDH 18,366,87
scu-rat LD50:108 mg/kg IYKEDH 18,366,87
orl-mus LD50:108 mg/kg IYKEDH 18,366,87
scu-mus LD50:364 mg/kg IYKEDH 18,366,87

SAFETY PROFILE: Poison by ingestion and subcutaneous routes. Human systemic effects by ingestion: blood pressure lowering. An experimental teratogen. Other experimental reproductive effects. A skin irritant. When heated to decomposition it emits toxic fumes of NO_x and Cl^-.

DGB600 **CAS:696-28-6** **HR: 3**
DICHLOROPHENYLARSINE
mf: $C_6H_5AsCl_2$ mw: 222.93

PROP: Colorless gas or liquid, changes to yellow. Bp: 127–129° @ 13 mm, fp: −15.6°, d: 1.654 @ 20°, vap press: 0.021 mm @ 20°, vap d: 7.7.

SYNS: ARSONOUS DICHLORIDE, PHENYL-(9CI) □ DICHLOR-FENYLARSIN □ FDA □ FENILDICLOROARSINA (ITALIAN) □ FENYLDICHLORARSIN □ PHENYLARSINEDICHLORIDE □ PHENYLARSONOUS DICHLORIDE □ PHENYL DICHLORARSINE □ PHENYLDICHLOROARSINE □ RCRA WASTE NUMBER P036 □ TL 69

TOXICITY DATA WITH REFERENCE
ihl-rat LCLo:1400 mg/m^3/10M NDRC** NDCrc-132,Jan,42
skn-rat LD50:16 mg/kg JPBAA7 58,411,46
ihl-mus LC50:3300 mg/m^3/10M NTIS** PB158-508
skn-mus LD50:4 mg/kg NTIS** PB158-508
ivn-mus LD50:500 μg/kg JPBAA7 58,411,46
skn-rbt LD50:5 mg/kg JPBAA7 58,411,46
ivn-rbt LD50:500 mg/kg JPBAA7 58,411,46
skn-gpg LD50:4 mg/kg JPBAA7 58,411,46

CONSENSUS REPORTS: Arsenic and its compounds are on the Community Right-To-Know List. Reported in EPA TSCA Inventory. EPA Extremely Hazardous Substances List.

OSHA PEL: TWA 0.5 mg(As)/m^3

SAFETY PROFILE: Poison by inhalation, ingestion, skin contact, and intravenous routes. See also ARSENIC. A lachrymator type of military poison gas. When exposed to heat, water, or steam it reacts to produce corrosive fumes of Cl^-. When heated to decomposition it emits highly toxic fumes of arsenic.

DGB800 **CAS:15460-48-7** **HR: 2**
N-(3,4-DICHLOROPHENYL)-1-AZIRIDINECARBOXAMIDE
mf: $C_9H_8Cl_2N_2O$ mw: 231.09

SYN: 3,4-DICHLOROPHENYL-N-CARBAMOYLAZIRIDINE

TOXICITY DATA WITH REFERENCE
ipr-mus TDLo:20 mg/kg/4W-I:NEO CNREA8 29,2184,69

SAFETY PROFILE: Questionable carcinogen with ex-

perimental neoplastigenic data. When heated to decomposition it emits very toxic fumes of Cl^- and NO_x.

DGB875 CAS:873-51-8 HR: 3
DICHLOROPHENYLBORANE
mf: $C_6H_5BCl_2$ mw: 158.82

PROP: Moisture-sensitive fuming liquid. D: 1.194, mp: 7°, bp: 175°.

SAFETY PROFILE: The hot borane ignites in air. When heated to decomposition it emits toxic fumes of Cl^-. See also BORANES.

DGC100 CAS:76714-88-0 HR: 2
(E)-1-(2,4-DICHLOROPHENYL)-4,4-DIMETHYL-2-(1,2,4-TRIAZOL-1-YL)PENTEN-3-OL
mf: $C_{15}H_{17}Cl_2N_3O$ mw: 326.25

SYNS: S 3308 □ 1H-1,2,4-TRIAZOLE-1-ETHANOL, β-((2,4-DICHLOROPHENYL)METHYLENE)-α-(1,1-DIMETHYLETHYL)-, (E)-

TOXICITY DATA WITH REFERENCE
eye-rbt 100 mg MLD EPASR* 8EHQ-0485-0548
orl-rat LD50:474 mg/kg EPASR* 8EHQ-0485-0548

SAFETY PROFILE: Moderately toxic by ingestion. An eye irritant. When heated to decomposition it emits toxic fumes of NO_x and Cl^-.

DGC600 CAS:38780-39-1 HR: 3
cis-DICHLORO(o-PHENYLENEDIAMINE)PLATINUM(II)
mf: $C_6H_8Cl_2N_2Pt$ mw: 374.15

SYN: DICHLORO(1,2-PHENYLENEDIAMMINE)PLATINUM(II)

TOXICITY DATA WITH REFERENCE
mmo-sat 2300 nmol/L JMCMAR 23,459,80
dni-bac-esc 2500 nmol/L CBINA8 29,327,80
ipr-mus LD50:48 mg/kg CBINA8 5,415,72

SAFETY PROFILE: Poison by intraperitoneal route. Mutation data reported. See PLATINUM COMPOUNDS. When heated to decomposition it emits very toxic fumes of Cl^- and NO_x.

DGC800 CAS:34643-46-4 HR: 2
O-(2,4-DICHLOROPHENYL)-O-ETHYL-S-PROPYL-PHOSPHORODITHIOATE
mf: $C_{11}H_{15}Cl_2O_2PS_2$ mw: 345.25

PROP: Liquid. D: 1.3 @ 20°/4°, bp: 125–128° @ 0.1 mm. Very sltly sol in H_2O.

SYNS: BAY NTN 8629 □ BIDERON □ O-ETHYL-O-(2,4-DICHLOROPHENYL)-S-n-PROPYL-DITHIOPHOSPHATE □ NTN-8629 □ PROTHIOPHOS □ TOKUTHION

TOXICITY DATA WITH REFERENCE
orl-rat LD50:875 mg/kg GISAAA 53(5),92,88
skn-rat LD50:3900 mg/kg KONODE 20,94,76
orl-mus LD50:570 mg/kg GISAAA 53(5),92,88
skn-mus LD50:1600 mg/kg KONODE 20,94,76

SAFETY PROFILE: Moderately toxic by ingestion and skin contact. When heated to decomposition it emits very toxic fumes of Cl^-, PO_x, and SO_x.

DGC850 CAS:305-15-7 HR: 2
(2,5-DICHLOROPHENYL)HYDRAZINE
mf: $C_6H_6Cl_2N_2$ mw: 177.04

SYN: HYDRAZINE, (2,5-DICHLOROPHENYL)-

TOXICITY DATA WITH REFERENCE
orl-rat LDLo:500 mg/kg NCNSA6 5,20,53

CONSENSUS REPORTS: Reported in EPA TSCA Inventory.

SAFETY PROFILE: Moderately toxic by ingestion. When heated to decomposition it emits toxic vapors of NO_x and Cl^-.

DGD075 CAS:33175-34-7 HR: 3
3,4-DICHLOROPHENYL HYDROXYLAMINE
mf: $C_6H_5Cl_2NO$ mw: 178.02

$Cl_2C_6H_3NHOH$

SAFETY PROFILE: Decomposes exothermically at 80° C. It is a chemical intermediate in the production of aniline from 3,4-dichloronitrobenzene and its thermal decomposition has caused violent explosions in plant-scale reactors. When heated to decomposition it emits toxic fumes of Cl^- and NO_x. See also AROMATIC AMINES.

DGD085 CAS:31225-17-9 HR: 2
N-(3,4-DICHLOROPHENYL)-N′-HYDROXYUREA
mf: $C_7H_6Cl_2N_2O_2$ mw: 221.05

SYNS: DICHLORPHENYLOXYUREA □ UREA, N-(3,4-DICHLOROPHENYL)-N′-HYDROXY-

TOXICITY DATA WITH REFERENCE
orl-rat LD50:1850 mg/kg GISAAA 40(7),46,75
orl-mus LD50:2 g/kg GISAAA 40(7),46,75
orl-rbt LD50:1800 mg/kg GISAAA 40(7),46,75

CONSENSUS REPORTS: Reported in EPA TSCA Inventory.

SAFETY PROFILE: Moderately toxic by ingestion. When heated to decomposition it emits toxic vapors of NO_x and Cl^-.

DGD100 HR: 3
1-(3,4-DICHLOROPHENYL)-5-ISOPROPYLBIGUANIDE HYDROCHLORIDE
mf: $C_{11}H_{15}Cl_2N_5 \cdot ClH$ mw: 324.67

SYNS: N^1-3,4-DICHLOROPHENYL-N^5-ISOPROPYLDIGUANIDE HYDROCHLORIDE □ M5943

TOXICITY DATA WITH REFERENCE
orl-mus LD50:100 mg/kg BJPCAL 5,438,50
ipr-mus LD50:25 mg/kg BJPCAL 5,438,50
ivn-mus LD50:25 mg/kg BJPCAL 5,438,50

SAFETY PROFILE: Poison by ingestion, intravenous, and intraperitoneal routes. When heated to decomposition it emits toxic fumes of NO_x and HCl.

DGD400 **CAS:3687-13-6** **HR: 2**
2,4-DICHLOROPHENYLMETHANESULFONATE
mf: $C_7H_6Cl_2O_3S$ mw: 241.09

SYNS: SD 7727 ☐ SHELL SD 7,727

TOXICITY DATA WITH REFERENCE
orl-rat LD50:2793 mg/kg 28ZEAL 4,153,69
orl-mus LDLo:1070 mg/kg AECTCV 14,111,85
skn-rbt LD50:2500 mg/kg 28ZEAL 5,204,76

SAFETY PROFILE: Moderately toxic by ingestion and skin contact. When heated to decomposition it emits very toxic fumes of Cl^- and SO_x. See also SULFONATES.

DGD600 **CAS:330-55-2** **HR: 3**
3-(3,4-DICHLOROPHENYL)-1-METHOXYMETHYLUREA
mf: $C_9H_{10}Cl_2N_2O_2$ mw: 249.11

PROP: Solid. Mp: 93–94°. Sltly sol in water; partially sol in acetone and alc.

SYNS: 3-(3,4-DICHLOOR-FENYL)-1-METHOXY-1-METHYLUREUM (DUTCH) ☐ 3-(3,4-DICHLORO-FENIL)-1-METOSSI-1-METIL-UREA (ITALIAN) ☐ 3-(3,4-DICHLOROPHENYL)-1-METHOXY-1-METHYLUREA ☐ N'-(3,4-DICHLOROPHENYL)-N-METHOXY-N-METHYLUREA ☐ 1-(3,4-DICHLOROPHENYL)3-METHOXY-3-METHYLUREE (FRENCH) ☐ N-(3,4-DICHLOROPHENYL)-N'-METHYL-N'-METHOXYUREA ☐ 3-(3,4-DICHLOR-PHENYL)-1-METHOXY-1-METHYL-HARNSTOFF (GERMAN) ☐ 3-(4,5-DICHLORPHENYL)-1-METHOXY-1-METHYLHARNSTOFF (GERMAN) ☐ DU PONT 326 ☐ DUPONT HERBICIDE 326 ☐ GARNITAN ☐ HERBICIDE 326 ☐ HOE 2810 ☐ LINEX 4L ☐ LINOROX ☐ LINUREX ☐ LINURON ☐ LINURON (herbicide) ☐ LOREX ☐ LOROX ☐ LOROX LINURON WEED KILLER ☐ METHOXYDIURON ☐ 1-METHOXY-1-METHYL-3-(3,4-DICHLOROPHENYL)UREA ☐ PREMALIN ☐ SARCLEX ☐ SCARCLEX ☐ SINURON

TOXICITY DATA WITH REFERENCE
dni-mus-orl 500 mg/kg MUREAV 58,353,78
orl-rat LD50:1146 mg/kg FAATDF 7,299,86
ihl-rat LD50:48 mg/m³/4H 85GMAT -,48,82
orl-mus LD50:2400 mg/kg 85GMAT -,48,82
orl-dog LD50:500 mg/kg GUCHAZ 6,317,73
orl-ckn LD50:3765 mg/kg VETNAL 58(7),63,82

CONSENSUS REPORTS: Reported in EPA TSCA Inventory. EPA Genetic Toxicology Program.

SAFETY PROFILE: Poison by inhalation. Moderately toxic by ingestion. Mutation data reported. A selective herbicide used in farming. When heated to decomposition it emits very toxic fumes of Cl^- and NO_x. See also 3-(p-CHLOROPHENYL)-1,1-DIMETHYLUREA.

DGD800 **CAS:299-85-4** **HR: 3**
O-(2,4-DICHLOROPHENYL)-O-METHYLISOPROPYLPHOSPHORAMIDOTHIOATE
mf: $C_{10}H_{14}Cl_2NO_2PS$ mw: 314.18

SYNS: O-(2,4-DICHLOROPHENYL)-O-METHYL-N-ISOPROPYLPHOSPHORAMIDOTHIOATE ☐ DMPA ☐ DOW 1329 ☐ DOWCO 118 ☐ ENT 25,647 ☐ ISOPROPYLPHOSPHORAMIDOTHIOIC ACID-O-2,4-DICHLOROPHENYL-O-METHYL ESTER ☐ K 22023 ☐ (1-METHYLETHYL) PHOSPHORAMIDOTHIOIC ACID O-(2,4-DICHLOROPHENYL)-O-METHYL ESTER ☐ OMS 115 ☐ ZYTRON

TOXICITY DATA WITH REFERENCE
orl-rat LDLo:270 mg/kg FMCHA2 -,C260,83
unr-dog LD50:1000 mg/kg 30ZDA9 -,353,71
skn-rbt LD50:1680 mg/kg GUCHAZ 6,242,73
orl-gpg LD50:210 mg/kg 31ZOAD 1,191,68
orl-ckn LD50:1357 mg/kg TXAPA9 6,147,64
orl-bwd LD50:100 mg/kg TXAPA9 21,315,72

SAFETY PROFILE: Poison by ingestion. Moderately toxic by skin contact. An herbicide and plant growth regulator. When heated to decomposition it emits very toxic fumes of Cl^-, NO_x, PO_x, and SO_x.

DGE200 **CAS:13412-64-1** **HR: 2**
3-(2,6-DICHLOROPHENYL)-5-METHYL-4-ISOXAZOLYL PENICILLIN SODIUM MONOHYDRATE
mf: $C_{19}H_{16}Cl_2N_3O_5S{\cdot}Na{\cdot}H_2O$ mw: 510.35

PROP: Monohydrate. Mp: 222–225° (decomp). Sol in H_2O.

SYNS: BLP-1011 ☐ BRISPEN ☐ BRL-1702 ☐ CONSTAPHYL ☐ DICHLOR STAPENOR ☐ DICLOCIL ☐ DICLOXACILLIN SODIUM MONOHYDRATE ☐ DICLOXACILLIN SODIUM SALT ☐ DYCILL ☐ DYNAPEN ☐ MDI-PC ☐ NOXABEN ☐ P 1011 ☐ PATHOCIL ☐ PENSINT ☐ SODIUM DICLOXACILLIN ☐ SODIUM DICLOXACILLIN MONOHYDRATE ☐ STAMPEN ☐ STPHCILLIN A BANYU ☐ SYNTARPEN ☐ VERACILLIN

TOXICITY DATA WITH REFERENCE
orl-rat LD50:3579 mg/kg TXAPA9 18,185,71
ipr-rat LD50:630 mg/kg ARZNAD 15,322,65
ivn-rat LD50:520 mg/kg JJANAX 21,274,68
orl-mus LD50:4560 mg/kg JJANAX 21,274,68
ipr-mus LD50:1000 mg/kg JJANAX 21,274,68
scu-mus LD50:1100 mg/kg JJANAX 21,274,68
ivn-mus LD50:875 mg/kg JJANAX 21,274,68
ivn-rbt LD50:600 mg/kg JJANAX 21,274,68

SAFETY PROFILE: Moderately toxic by ingestion, intraperitoneal, intravenous, and subcutaneous routes. An antibacterial agent. When heated to decomposition it emits very toxic fumes of Na_2O, NO_x, SO_x, and Cl^-. See also other penicillin entries.

DGE400 **CAS:644-97-3** **HR: 3**
DICHLOROPHENYLPHOSPHINE
DOT: UN 2798
mf: $C_6H_5Cl_2P$ mw: 178.98

PROP: Pungent, odorous liquid. D: 1.33 @ 20°/4°, bp: 99–101° @ 5 mm.

SYNS: PHENYLDICHLOROPHOSPHINE ☐ PHENYLPHOSPHINE DICHLORIDE ☐ PHENYLPHOSPHONOUS ACID DICHLORIDE ☐ PHENYLPHOSPHONOUS DICHLORIDE ☐ PHENYL PHOSPHORUS DICHLORIDE ☐ PHENYL PHOSPHORUS DICHLORIDE (DOT)

CONSENSUS REPORTS: Reported in EPA TSCA Inventory.

DOT CLASSIFICATION: 8; *Label:* Corrosive

SAFETY PROFILE: A poison irritant to skin, eyes, and mucous membranes and poison by ingestion and inhalation. When heated to decomposition it emits very toxic fumes of Cl^- and PO_x. See also PHOSPHINE.

DGE800 **CAS:1698-53-9** **HR: 2**
4,5-DICHLORO-2-PHENYL-3(2H)-PYRIDAZINONE
mf: $C_{10}H_6Cl_2N_2O$ mw: 241.08

PROP: Prisms from EtOH (aq). Mp: 163–164°.

SYN: 1-FENYL-4,5-DICHLOR-6-PYRIDAZINON (CZECH)

TOXICITY DATA WITH REFERENCE
eye-rbt 20 mg/24H SEV 28ZPAK -,151,72
orl-rat LD50:2520 mg/kg 28ZPAK -,151,72

SAFETY PROFILE: Moderately toxic by ingestion. A severe eye irritant. When heated to decomposition it emits very toxic fumes of Cl^- and NO_x.

DGF000 **CAS:24096-53-5** **HR: 2**
N-(3,5-DICHLOROPHENYL)SUCCINIMIDE
mf: $C_{10}H_7Cl_2NO_2$ mw: 244.08

SYNS: 1-(3,5-DICHLOROPHENYL)-2,5-PYRROLIDINEDIONE □ DIMETHACHLON □ OHRIC

TOXICITY DATA WITH REFERENCE
orl-rat TDLo:17 g/kg/8W-C:ETA GANNA2 67,147,76
orl-mus LD50:1250 mg/kg FMCHA2 -,C172,83

SAFETY PROFILE: Moderately toxic by ingestion. Questionable carcinogen with experimental tumorigenic data. When heated to decomposition it emits very toxic fumes of Cl^- and NO_x.

DGF200 **CAS:27137-85-5** **HR: 3**
(DICHLOROPHENYL)TRICHLOROSILANE
DOT: UN 1766
mf: $C_6H_3Cl_5Si$ mw: 280.43

PROP: Straw-colored liquid, sol in benzene and perchloroethylene (mixture of isomers). D: 1.562, bp: 260°, flash p: 286°F.

SYNS: DICHLOROPHENYLTRICHLOROSILANE (DOT) □ TRICHLORO(DICHLOROPHENYL)SILANE

TOXICITY DATA WITH REFERENCE
ipr-rat LDLo:100 mg/kg 85GMAT -,48,82
orl-mus LDLo:100 mg/kg 85GMAT -,48,82
ihl-mus LCLo:80 mg/m³/2H 85GMAT -,48,82
ipr-mus LDLo:100 mg/kg 85GMAT -,48,82
scu-mus LDLo:100 mg/kg 85GMAT -,48,82
ihl-mam LCLo:80 mg/m³ CHABA8 57,12828a,62

CONSENSUS REPORTS: Reported in EPA TSCA Inventory. EPA Extremely Hazardous Substances List.

DOT CLASSIFICATION: 8; *Label:* Corrosive

SAFETY PROFILE: Poison by ingestion, inhalation, subcutaneous, and intraperitoneal routes. Corrosive to the eyes, skin, and mucous membranes. On contact with moisture it releases corrosive HCl. Combustible when exposed to heat or flame. When heated to decomposition it emits toxic fumes of Cl^-. See also CHLOROSILANES.

DGF300 **CAS:22591-21-5** **HR: 3**
DICHLOROPINACOLIN
mf: $C_6H_{10}Cl_2O$ mw: 169.06

SYNS: 2-BUTANONE, 1,1-DICHLORO-3,3-DIMETHYL- □ 1,1-DICHLORO-3,3-DIMETHYL-2-BUTANONE □ DICHLOROMETHYL tert-BUTYL KETONE □ α-α-DICHLOROPINACOLIN □DICHLOROPINAKOLIN □ ω,ω-DICHLORPINAKOLIN

TOXICITY DATA WITH REFERENCE
orl-rat LD50:3350 mg/kg GISAAA 52(12),92,87
orl-mus LD50:4150 mg/kg GISAAA 52(12),92,87
orl-rbt LD50:3550 mg/kg GISAAA 52(12),92,87

CONSENSUS REPORTS: Reported in EPA TSCA Inventory.

DOT CLASSIFICATION: 3; *Label:* Flammable Liquid

SAFETY PROFILE: Moderately toxic by ingestion. A flammable liquid. When heated to decomposition it emits toxic vapors of Cl^-.

DGF400 **CAS:78-99-9** **HR: 3**
1,1-DICHLOROPROPANE
mf: $C_3H_6Cl_2$ mw: 112.99

PROP: Flash p: 69.8°F, lel: 3.1%, d: 1.143 @ 10°, bp: 88.3°.

SYN: PROPYLIDENE CHLORIDE

TOXICITY DATA WITH REFERENCE
eye-rbt 500 mg AMIHBC 10,61,54
eye-rbt 500 mg open AMIHBC 10,61,54
orl-rat LD50:6500 mg/kg AMIHBC 10,61,54
ihl-rat LCLo:4000 ppm/4H AMIHBC 10,61,54
skn-rbt LD50:14 g/kg AMIHBC 10,61,54

SAFETY PROFILE: Mildly toxic by ingestion, inhalation, and skin contact. An eye irritant. A very dangerous fire hazard when exposed to heat, flame, or oxidizers. When heated to decomposition it emits toxic fumes of Cl^-. See also PROPYLENE DICHLORIDE; and CHLORINATED HYDROCARBONS, ALIPHATIC.

DGF800 **CAS:142-28-9** **HR: 2**
1,3-DICHLOROPROPANE
mf: $C_3H_6Cl_2$ mw: 112.99

PROP: Colorless liquid. Bp: 120.4°, d: 1.201 @ 15°, vap d: 3.90, flash p: 69.8°F.

SYN: TRIMETHYLENE DICHLORIDE

TOXICITY DATA WITH REFERENCE
mmo-sat 10 μmol/plate ENMUDM 2,59,80
mma-sat 10 μmol/plate ENMUDM 2,59,80
orl-dog LDLo:3000 mg/kg AJHYA2 16,325,32

CONSENSUS REPORTS: Reported in EPA TSCA Inventory.

SAFETY PROFILE: Moderately toxic by ingestion. Mutation data reported. A very dangerous fire hazard when exposed to heat or flame. When heated to decomposition it emits highly toxic fumes of Cl^- and phosgene. See also CHLORINATED HYDROCARBONS, ALIPHATIC; and PROPYLENE DICHLORIDE.

DGF900 **CAS:594-20-7** **HR: 3**
2,2-DICHLOROPROPANE
mf: $C_3H_6Cl_2$ mw: 112.99

$H_3CCCl_2CH_3$

PROP: D: 1.096, bp: 70.5°.

SAFETY PROFILE: Reacts explosively with dimethylzinc. When heated to decomposition it emits toxic fumes of Cl^-. See also PROPYLENE DICHLORIDE; and CHLORINATED HYDROCARBONS, ALIPHATIC.

DGG000 **CAS:8003-19-8** **HR: 3**
DICHLOROPROPANE-DICHLOROPROPENE MIXTURE
mf: $C_3H_6Cl_2 \cdot C_3H_4Cl_2$ mw: 223.96

PROP: D-D Soil fumigant consists of chlorinated C_3 hydrocarbons (100%), 1,3-dichloropropene, 3,3-dichloropropene, 1,2-dichloropropane, 2,3-dichloropropene, and related C_3 chlorinated hydrocarbons (SHELL*).

SYNS: D-D ☐ DD MIXTURE ☐ DD SOIL FUMIGANT ☐ 1,3-DICHLOROPROPENE and 1,2-DICHLOROPROPANE MIXTURE ☐ DICHLORPROPAN-DICHLORPROPENGEMISCH (GERMAN) ☐ DOWFUME N ☐ ENT 8,420 ☐ NEMAFENE ☐ TELONE ☐ VIDDEN D

TOXICITY DATA WITH REFERENCE
skn-rbt 500 mg/24H SEV AMIHBC 7,118,53
eye-rbt 5 mg SEV AMIHBC 7,118,53
mma-sat 500 μg/plate CNREA8 37,1915,77
orl-rat LD50:140 mg/kg ARSIM* 20,8,66
ihl-rat LC50:1000 ppm/4H AMIHBC 7,118,53
skn-rat LD50:779 mg/kg PEMNDP 8,263,87
orl-mus LD50:3 mg/kg ARSIM* 20,8,66
skn-rbt LD50:2100 mg/kg PCOC** -,371,66

SAFETY PROFILE: Poison by ingestion and inhalation. Moderately toxic by skin contact. Severe skin and eye irritant. Mutation data reported. A fumigant. When heated to decomposition it emits toxic fumes of Cl^-. See also PROPYLENE DICHLORIDE; and CHLORINATED HYDROCARBONS, ALIPHATIC.

DGG400 **CAS:96-23-1** **HR: 3**
1,3-DICHLORO-2-PROPANOL
DOT: UN 2750
mf: $C_3H_6Cl_2O$ mw: 128.99

PROP: Colorless liquid, ether-like odor. Bp: 174°, d: 1.367 @ 20°/4°, vap press: 1 mm @ 28.0°, vap d: 4.45, flash p: 165°F (OC), mp: −4°. Sol in H_2O and Et_2O.

SYNS: DICHLOROHYDRIN ☐ α-DICHLOROHYDRIN ☐ sym-DICHLOROISOPROPYL ALCOHOL ☐ 1,3-DICHLOROPROPANOL-2 (DOT) ☐ GLYCEROL α,γ-DICHLOROHYDRIN ☐ sym-GLYCEROL DICHLOROHYDRIN ☐ U 25,354

TOXICITY DATA WITH REFERENCE
skn-rbt 10 mg/24H open MLD AIHAAP 23,95,62
mmo-sat 1 μmol/plate ENMUDM 2,59,80
mma-sat 100 μg/plate SCIEAS 200,785,78
dni-hmn:hla 2500 μmol/L MUREAV 92,427,82
orl-rat LD50:110 mg/kg AIHAAP 23,95,62
ihl-rat LCLo:125 ppm/4H AIHAAP 23,95,62
orl-mus LD50:100 mg/kg 85GMAT -,46,82
skn-rbt LD50:800 mg/kg AIHAAP 23,95,62

CONSENSUS REPORTS: Reported in EPA TSCA Inventory. EPA Genetic Toxicology Program.

DOT CLASSIFICATION: 6.1; *Label:* Poison

SAFETY PROFILE: Poison by ingestion and inhalation. Moderately toxic by skin contact. Human mutation data reported. A skin irritant. Action may be similar to that of carbon tetrachloride, but more irritating to mucous membranes. Flammable when exposed to heat, flame, or oxidizers. To fight fire, use alcohol foam, dry chemical, fog, mist, or spray. Dangerous; when heated to decomposition it emits highly toxic fumes of Cl^- and phosgene.

DGG500 **CAS:513-88-2** **HR: 3**
1,1-DICHLOROPROPANONE
mf: $C_3H_4Cl_2O$ mw: 126.97

SYNS: DCP ☐ α-α-DICHLOROACETONE ☐ 1,1-DICHLOROACETONE ☐ DICHLOROMETHYL METHYL KETONE ☐ 2-PROPANONE, 1,1-DICHLORO-

TOXICITY DATA WITH REFERENCE
mmo-sat 1 μmol/plate TXAPA9 91,46,87
mma-sat 1 μmol/plate TXAPA9 91,46,87

DOT CLASSIFICATION: 3; *Label:* Flammable Liquid

SAFETY PROFILE: Mutation data reported. A flammable liquid. When heated to decomposition it emits toxic vapors of Cl^-.

DGG600 **CAS:616-23-9** **HR: 3**
2,3-DICHLOROPROPANOL
mf: $C_3H_6Cl_2O$ mw: 128.99

SYNS: 1,2-DICHLOROPROPANOL-3 ☐ 2,3-DICHLORO-1-PROPANOL ☐ 1,2-DICHLORO-3-PROPANOL ☐ GLYCEROL-α,β-DICHLOROHYDRIN

TOXICITY DATA WITH REFERENCE
skn-rbt 10 mg/24H JIHTAB 30,63,48
eye-rbt 6800 μg SEV AJOPAA 29,1363,46
mma-sat 262 μg/plate MUREAV 57,381,78
mma-esc 131 μg/plate MUREAV 57,381,78
orl-rat LD50:90 mg/kg JIHTAB 30,63,48
ihl-rat LCLo:500 ppm/4H JIHTAB 31,343,49
skn-rbt LD50:200 mg/kg JIHTAB 30,63,48

CONSENSUS REPORTS: Reported in EPA TSCA Inventory.

SAFETY PROFILE: Poison by ingestion and skin contact. Moderately toxic by inhalation. A skin and severe eye irritant. Mutation data reported. When heated to decomposition it emits toxic fumes of Cl^-. See also CHLORINATED HYDROCARBONS, AROMATIC.

DGG700 **CAS:26952-23-8** **HR: 3**
DICHLOROPROPENE
DOT: UN 2047
mf: $C_3H_4Cl_2$ mw: 110.97

SYNS: DICHLOROPROPYLENE ☐ 1-PROPENE, DICHLORO-

CONSENSUS REPORTS: Reported in EPA TSCA Inventory.

DOT CLASSIFICATION: 3; *Label:* Flammable Liquid

SAFETY PROFILE: A flammable liquid. When heated to decomposition it emits toxic vapors of Cl^-.

DGG800 CAS:563-54-2 HR: 2
1,2-DICHLOROPROPENE
mf: $C_3H_4Cl_2$ mw: 110.97

PROP: Liquid. Bp: 75°, vap d: 3.83.

SYNS: DICHLOR □ 1,2-DICHLOROPROPYLENE □ DICHLORPROPEN-GEMISCH (GERMAN) □ PDC □ PROPYLENE DICHLORIDE □ RCRA WASTE NUMBER U083

TOXICITY DATA WITH REFERENCE
mmo-sat 10 μL/plate JSFAAE 32,826,81
mma-sat 10 μL/plate JSFAAE 32,826,81
mmo-asn 10 μL/plate PMRSDJ 2,87,81
orl-rat LD50:2 g/kg 85ARAE 3,20,76/77
skn-rbt LD50:8750 mg/kg 34ZIAG -,744,69

SAFETY PROFILE: Moderately toxic by ingestion. Mildly toxic by skin contact. Mutation data reported. When heated to decomposition it emits toxic fumes of Cl^-. See also CHLORINATED HYDROCARBONS, ALIPHATIC.

DGG950 CAS:542-75-6 HR: 3
1,3-DICHLOROPROPENE
DOT: UN 2047
mf: $C_3H_4Cl_2$ mw: 110.97

PROP: Liquid. Bp: 103–110°, flash p: 95°F, d: 1.22, vap d: 3.8.

SYNS: α-CHLOROALLYL CHLORIDE □ γ-CHLOROALLYL CHLORIDE □ DICHLOROPROPENE (DOT) □ 1,3-DICHLOROPROPYENE-1 □ DICHLOROPROPYLENE □ α,γ-DICHLOROPROPYLENE □ 1,3-DICHLOROPROPYLENE □ NCI-C03985 □ RCRA WASTE NUMBER U084 □ TELONE □ TELONE II SOIL FUMIGANT □ VIDDEN D

TOXICITY DATA WITH REFERENCE
mmo-sat 33 μg/plate ENMUDM 5(Suppl 1),3,83
mma-sat 10 μmol/plate ENMUDM 2,59,80
sln-dmg-orl 5750 ppm ENMUDM 7,325,85
sce-ham:ovr 900 nmol/L CNJGA8 22,681,80
orl-rat TDLo:15,600 mg/kg/2Y-I:CAR NTPTR* NTP-TR-269,85
orl-mus TDLo:31,200 mg/kg/2Y-I:CAR NTPTR* NTP-TR-269,85
ihl-mus TCLo:60 ppm/6H/5D/2Y:NEO FAATDF 12,418,89
orl-rat LD50:470 mg/kg DOWCC* MSD-405
ihl-rat LC50:500 ppm 85JFAN A446,83
skn-rat LD50:775 mg/kg FMCHA2-,C103,91
ipr-rat LD50:175 mg/kg TOXID9 5,2,85
orl-mus LD50:640 mg/kg FAATDF 8,562,87
ihl-mus LC50:4650 mg/m³/2H 85GMAT-,48,82
skn-rbt LD50:504 mg/kg DOWCC* MSD-405
ihl-gpg LCLo:400 ppm/7H AIHAAP 38,217,77

CONSENSUS REPORTS: NTP 7th Annual Report On Carcinogens. IARC Cancer Review: Group 2B IMEMDT 7,195,87; Human Inadequate Evidence IMEMDT 41,113,86; Animal Sufficient Evidence IMEMDT 41,113,86. NTP Carcinogenesis Studies (gavage); Clear Evidence: mouse, rat NTPTR* NTP-TR-269,86. Reported in EPA TSCA Inventory. EPA Genetic Toxicology Program. Community Right-To-Know List.

OSHA PEL: TWA 1 ppm (skin)
ACGIH TLV: TWA 1 ppm (skin)
DFG MAK: Animal Carcinogen, Suspected Human Carcinogen
DOT CLASSIFICATION: 3; *Label:* Flammable Liquid

SAFETY PROFILE: Confirmed carcinogen with experimental carcinogenic data. Poison by ingestion. Poison by ingestion and intraperitoneal routes. Moderately toxic by skin contact. Mildly toxic by inhalation. A strong irritant. Mutation data reported. A pesticide. A flammable liquid and dangerous fire hazard when exposed to heat, flame, or oxidizers. Reacts vigorously with oxidizing materials. To fight fire, use water, foam, CO_2, dry chemical. When heated to decomposition it emits toxic fumes of Cl^-. See also ALLYL COMPOUNDS and CHLORIDES.

DGH200 CAS:10061-01-5 HR: 3
cis-1,3-DICHLOROPROPENE
mf: $C_3H_4Cl_2$ mw: 110.97

PROP: Flash p: 69.8°F (21°C). Bp: 104.3°.

SYNS: (Z)-1,3-DICHLOROPROPENE □ cis-1,3-DICHLOROPROPYLENE

TOXICITY DATA WITH REFERENCE
mmo-sat 20 μg/plate CNREA8 37,1915,77
mma-sat 20 μg/plate CNREA8 37,1915,77
dns-hmn:hla 100 μmol/L CALEDQ 20,263,83
scu-mus TDLo:9240 mg/kg/77W-I:NEO JJIND8 63,1433,79

CONSENSUS REPORTS: EPA Genetic Toxicology Program.

DFG MAK: Animal Carcinogen, Suspected Human Carcinogen

SAFETY PROFILE: Confirmed carcinogen with experimental neoplastigenic data. Human mutation data reported. A dangerous fire hazard when exposed to heat, flame, or oxidizers. When heated to decomposition it emits toxic fumes of Cl^-. See also CHLORINATED HYDROCARBONS, ALIPHATIC.

DGH225 CAS:10061-02-6 HR: 3
trans-1,3-DICHLOROPROPENE
mf: $C_3H_4Cl_2$ mw: 110.97

PROP: Liquid with chloroform odor. Bp: 112°, flash p: 69.8°F (21°C).

SYNS: (E)-1,3-DICHLOROPROPENE □ trans-1,3-DICHLOROPROPYLENE

TOXICITY DATA WITH REFERENCE
mmo-sat 20 μg/plate CNREA8 37,1915,77
mma-sat 20 μg/plate CNREA8 37,1915,77
dns-hmn:hla 100 μmol/L CALEDQ 20,263,83

CONSENSUS REPORTS: EPA Genetic Toxicology Program.

DFG MAK: Animal Carcinogen, Suspected Human Carcinogen

SAFETY PROFILE: Human mutation data reported. A dangerous fire hazard when exposed to heat, flame, or oxidizers. When heated to decomposition it emits toxic fumes of Cl^-. See also CHLORINATED HYDROCARBONS, ALIPHATIC.

DGH400 CAS:78-88-6 HR: 3
2,3-DICHLOROPROPENE
mf: $C_3H_4Cl_2$ mw: 110.97

PROP: Flash p: 50°F, bp: 94°.

SYNS: 2,3-DICHLORO-1-PROPENE □ 2,3-DICHLOROPROPYLENE □ NSC 60520

TOXICITY DATA WITH REFERENCE
skn-rbt 10 mg/24H open SEV AIHAAP 23,95,62
mmo-sat 20 µg/plate CNREA8 37,1915,77
mma-sat 20 µg/plate CNREA8 37,1915,77
dns-hmn:hla 100 µmol/L CALEDQ 20,263,83
orl-rat LD50:320 mg/kg AIHAAP 23,95,62
ihl-rat LCLo:500 ppm/4H AIHAAP 23,95,62
ihl-mus LC50:3100 mg/m³/2H 85GMAT -,48,82
skn-rbt LD50:1580 mg/kg AIHAAP 23,95,62

CONSENSUS REPORTS: Reported in EPA TSCA Inventory. EPA Genetic Toxicology Program.

SAFETY PROFILE: Poison by ingestion. Moderately toxic by inhalation and skin contact. Human mutation data reported. A severe skin irritant. A very dangerous fire hazard when exposed to heat, flame, or oxidizers. When heated to decomposition it emits toxic fumes of Cl^-. See also CHLORINATED HYDROCARBONS, ALIPHATIC.

DGH500 CAS:66826-73-1 HR: 2
trans-1,3-DICHLOROPROPENE OXIDE
mf: $C_3H_4Cl_2O$ mw: 126.97

PROP: Bp: 95–96° @ 132 mm.

SYNS: trans-2-CHLORO-3-(CHLOROMETHYL)OXIRANE □ trans-DCPO □ trans-1,3-DICHLORO-1,2-EPOXYPROPANE

TOXICITY DATA WITH REFERENCE
otr-ham:emb 10 µmol/L JJIND8 69,531,82
skn-mus TDLo:400 mg/kg/73W-I:CAR CNREA8 43,159,83
scu-mus TDLo:20 mg/kg/71W-I:CAR CNREA8 43,159,83

SAFETY PROFILE: Questionable carcinogen with experimental carcinogenic data. Mutation data reported. When heated to decomposition it emits toxic fumes of Cl^-.

DGH800 CAS:10140-89-3 HR: 3
2,3-DICHLORO PROPIONALDEHYDE
mf: $C_3H_4Cl_2O$ mw: 126.97

PROP: Liquid. Vap d: 4.4.

SYNS: 1,2-DICHLORO-3-PROPIONAL □ α,β-DICHLOROPROPIONALDEHYDE

TOXICITY DATA WITH REFERENCE
skn-rbt 10 mg/24H open SEV AMIHBC 4,119,51
skn-rbt 500 mg MOD SCCUR* -,3,61
eye-rbt 50 µg open SEV AMIHBC 4,119,51
mmo-sat 1 nmol/plate MUREAV 78,113,80
orl-rat LD50:160 mg/kg AMIHBC 4,119,51
ihl-rat LCLo:2500 ppb/4H SCCUR* -,3,61
orl-mus LD50:250 mg/kg SCCUR* -,3,61
ihl-mus LCLo:9300 ppm/15M SCCUR* -,3,61
skn-rbt LD50:78 mg/kg AMIHBC 4,119,51

SAFETY PROFILE: Poison by ingestion and skin contact. Mildly toxic by inhalation. A severe skin and eye irritant. Mutation data reported. When heated to decomposition it emits toxic fumes of Cl^-. See also ALDEHYDES and CHLORIDES.

DGI000 CAS:709-98-8 HR: 3
DICHLOROPROPIONANILIDE
mf: $C_9H_9Cl_2NO$ mw: 218.09

PROP: Light-brown solid (pure); liquid (technical grade). Mp (pure): 85–89°, bp (technical grade): 91–95°.

SYNS: BAY 30130 □ CHEM RICE □ CRYSTAL PROPANIL-4 □ DCPA □ N-(3,4-DICHLOROPHENYL)PROPANAMIDE □ N-(3,4-DICHLOROPHENYL)PROPIONAMIDE □ 3,4-DICHLOROPROPIONANILIDE □ 3′,4′-DICHLOROPROPIONANILIDE □ DIPRAM □ DPA □ FARMCO PROPANIL □ FW 734 □ GRASCIDE □ HERBAX TECHNICAL □ MONTROSE PROPANIL □ PROPANEX □ PROPANID □ PROPANIDE □ PROPANIL □ PROPIONIC ACID-3,4-DICHLOROANILIDE □ PROP-JOB □ RISELECT □ ROGUE □ ROSANIL □ S 10165 □ STAM □ STAM F 34 □ STAM LV 10 □ STAM M-4 □ STAMPEDE □ STAMPEDE 3E □ STAM SUPERNOX □ STREL □ SUPERNOX □ SURCOPUR □ SURPUR □ VERTAC

TOXICITY DATA WITH REFERENCE
dnr-bcs 100 µg/disc NTIS** PB80-133226
cyt-mus-unr 100 mg/kg TGANAK 14(6),41,80
cyt-mus-orl 100 mg/kg CYGEDX 14(6),38,80
orl-rat LD50:367 mg/kg KHZDAN 27,451,84
orl-mus LD50:360 mg/kg GTPZAB 21(12),30,77
orl-dog LD50:1217 mg/kg TXAPA9 23,650,72
skn-rbt LD50:4830 mg/kg FMCHA2 -,C197,83
orl-mam LD50:2527 mg/kg NTIS** PB288-416

CONSENSUS REPORTS: EPA Genetic Toxicology Program.

SAFETY PROFILE: Poison by ingestion. Moderately toxic by an unspecified route. Mildly toxic by skin contact. Mutation data reported. When heated to decomposition it emits very toxic fumes of Cl^- and NO_x.

DGI400 CAS:75-99-0 HR: 2
2,2-DICHLOROPROPIONIC ACID
mf: $C_3H_4Cl_2O_2$ mw: 142.97

PROP: White to tan powder. D: 1.39 @ 22.6°/4°, bp: 185–190°. Sol in water.

SYNS: BASFAPON □ BASFAPON B □ BASFAPON/BASFAPON N □ BASINEX □ BH DALAPON □ CRISAPON □ DALAPON (USDA) □ DALAPON 85 □ DED-WEED □ DEVIPON □ α-DICHLOROPROPIONIC ACID □ α,α-DICHLOROPROPIONIC ACID □ DOWPON □ DOWPON

M □ GRAMEVIN □ KENAPON □ LIROPON □ PROPROP □ RADAPON □ REVENGE □ UNIPON

TOXICITY DATA WITH REFERENCE
skn-rbt 100 μg/24H open AIHAAP 23,95,62
mmo-omi 500 ppm IJEBA6 11,114,73
skn-rat LD50:>5 g/kg FAATDF 7,299,86

CONSENSUS REPORTS: EPA Genetic Toxicology Program. Reported in EPA TSCA Inventory.

OSHA PEL: TWA 1 ppm
ACGIH TLV: TWA 1 ppm
DFG MAK: 1 ppm (6 mg/m^3)

SAFETY PROFILE: A corrosive with low toxicity by skin contact. A skin irritant. Mutation data reported. When heated to decomposition it emits toxic fumes of Cl^-.

DGI600 CAS:127-20-8 HR: 2
α,α-DICHLOROPROPIONIC ACID SODIUM SALT
mf: $C_3H_3Cl_2O_2{\cdot}Na$ mw: 164.95

SYNS: BASFAPON B □ DALAPON □ DALAPON SODIUM □ DALAPON SODIUM SALT □ 2,2-DICHLOROPROPIONIC ACID, SODIUM SALT □ DOWPON □ 2,2-DPA □ GRAMEVIN □ NATRIUMSALZ DER 2,2-DICHLORPROPIONSAEURE □ RADAPON □ SODIUM DALAPON □ SODIUM-α,α-DICHLOROPROPIONATE □ SODIUM-2,2-DICHLOROPROPIONATE □ UNIPON

TOXICITY DATA WITH REFERENCE
cyt-mus-unr 200 mg/kg TGANAK 16(1),45,82
orl-rat LD50:3860 mg/kg WRPCA2 9,119,70
orl-mus LD50:4600 mg/kg JPIFAN (11),42,72
orl-rbt LD50:3400 mg/kg 85DPAN ,-,71/76
orl-gpg LD50:3400 mg/kg 85DPAN -,-,71/76
orl-ckn LD50:5600 mg/kg DOEAAH 35,25,79
orl-mam LD50:4 g/kg 85GYAZ -,86,71

CONSENSUS REPORTS: Reported in EPA TSCA Inventory. EPA Genetic Toxicology Program.

DFG MAK: 1 ppm (6 mg/m^3)

SAFETY PROFILE: Moderately toxic by ingestion. Mutation data reported. When heated to decomposition it emits toxic fumes of Na_2O and Cl^-.

DGI700 CAS:26952-23-8 HR: 3
DICHLOROPROPYLENE
DOT: UN 2047
mf: $C_3H_4Cl_2$ mw: 110.97

SYN: 1-PROPENE, DICHLORO-

CONSENSUS REPORTS: Reported in EPA TSCA Inventory.

DOT CLASSIFICATION: 3; *Label:* Flammable Liquid

SAFETY PROFILE: A flammable liquid. When heated to decomposition it emits toxic vapors of Cl^-.

DGJ100 CAS:1702-17-6 HR: 2
3,6-DICHLORO-2-PYRIDINECARBOXYLIC ACID
mf: $C_6H_3Cl_2NO_2$ mw: 192.00

PROP: Crystals from C_6H_6. Mp: 152–153°.

SYNS: CLOPYRALID □ 3,6-DICHLOROPICOLINIC ACID □ DOWCO 290 □ KYSELINA 3,6-DICHLORPIKOLINOVA □ LONTREL □ LONTREL 3 □ MATRIGON □ PICOLINIC ACID, 3,6-DICHLORO- □ 2-PYRIDINECARBOXYLIC ACID, 3,6-DICHLORO-(9CI) □ XRM 3972

TOXICITY DATA WITH REFERENCE
orl-rat TDLo:150 mg/kg (female 6-15D post):TER FAATDF 4,91,84
orl-rat LD50:4300 mg/kg 85JFAN A433,85
ipr-rat LD50:900 mg/kg DOVEAA 31,387,77
orl-dck LD50:1465 mg/kg PEMNDP 8,189,87

SAFETY PROFILE: Moderately toxic by ingestion. An experimental teratogen. When heated to decomposition it emits toxic fumes of NO_x and Cl^-.

D

DGJ200 CAS:3428-24-8 HR: 3
4,5-DICHLOROPYROCATECHOL
mf: $C_6H_4Cl_2O_2$ mw: 179.00

PROP: Prisms from $CHCl_3$. Mp: 116–117°.

SYNS: 4,5-DICHLORO-1,2-BENZENEDIOL □ 4,5-DICHLOROCATECHOL

TOXICITY DATA WITH REFERENCE
mmo-smc 75 mg/L MUREAV 119,273,83
ivn-mus LD50:42 mg/kg CSLNX* NX#07864

SAFETY PROFILE: Poison by intravenous route. Mutation data reported. When heated to decomposition it emits toxic fumes of Cl^-.

DGJ250 CAS:86-98-6 HR: 3
4,7-DICHLOROQUINOLINE
mf: $C_9H_5Cl_2N$ mw: 198.05

SYNS: QUINOLINE, 4,7-DICHLORO- □ TL 1473

TOXICITY DATA WITH REFERENCE
mma-sat 500 nmol/plate MUREAV 42,335,77
scu-mus LDLo:80 mg/kg NDRC** 30101,8,45

CONSENSUS REPORTS: Reported in EPA TSCA Inventory.

SAFETY PROFILE: Poison by subcutaneous route. Mutation data reported. When heated to decomposition it emits toxic vapors of NO_x and Cl^-.

DGJ950 CAS:2213-63-0 HR: 3
2,3-DICHLOROQUINOXALINE
mf: $C_8H_4Cl_2N_2$ mw: 199.04

SYN: QUINOXALINE, 2,3-DICHLORO-

TOXICITY DATA WITH REFERENCE
ivn-mus LD50:5600 μg/kg CSLNX* NX#03256

CONSENSUS REPORTS: Reported in EPA TSCA Inventory.

SAFETY PROFILE: Poison by intravenous route. When heated to decomposition it emits toxic vapors of NO_x and Cl^-.

DGK000 **CAS:1919-43-3** **HR: 1**
2,3-DICHLOROQUINOXALINE-6-CARBONYLCHLORIDE
mf: $C_9H_3Cl_3N_2O$ mw: 261.49

SYN: 2,3-DICHLOROCHINOXALIN-6-KARBONYLCHLORID (CZECH)

TOXICITY DATA WITH REFERENCE
skn-rbt 20 mg/24H MOD 85JCAE-,868,86
eye-rbt 100 mg/24H SEV 28ZPAK -,150,72

CONSENSUS REPORTS: Reported in EPA TSCA Inventory.

SAFETY PROFILE: A skin and severe eye irritant. When heated to decomposition it emits very toxic fumes of Cl^- and NO_x.

DGK200 **CAS:320-72-9** **HR: 3**
3,5-DICHLOROSALICYLIC ACID
mf: $C_7H_4Cl_2O_3$ mw: 207.01

PROP: Crystals from EtOH (aq). Mp: 219–220° (subl).

SYN: USAF DO-68

TOXICITY DATA WITH REFERENCE
ipr-mus LD50:50 mg/kg NTIS** AD277-689

CONSENSUS REPORTS: Reported in EPA TSCA Inventory.

SAFETY PROFILE: Poison by intraperitoneal route. When heated to decomposition it emits toxic fumes of Cl^-.

DGK250 **CAS:3401-80-7** **HR: 3**
3,6-DICHLOROSALICYLIC ACID
mf: $C_7H_4Cl_2O_3$ mw: 207.01

SYNS: BENZOIC ACID, 3,6-DICHLORO-2-HYDROXY-(9CI) □ 3,6-DICHLORO-2-HYDROXYBENZOIC ACID □ SALICYLIC ACID, 3,6-DICHLORO-

TOXICITY DATA WITH REFERENCE
orl-rat LD50:1560 mg/kg GISAAA 43(10),95,78
orl-mus LD50:660 mg/kg GISAAA 43(10),95,78
ipr-mus LD50:50 mg/kg NTIS** PB85-143766

CONSENSUS REPORTS: Reported in EPA TSCA Inventory.

SAFETY PROFILE: A poison by intraperitoneal route. Moderately toxic by ingestion. When heated to decomposition it emits toxic vapors of Cl^-.

DGK300 **CAS:4109-96-0** **HR: 3**
DICHLOROSILANE
mf: Cl_2H_2Si mw: 101.01

CONSENSUS REPORTS: Reported in EPA TSCA Inventory.

PROP: A gas. Mp: −122°, bp: 8.3°.

SYNS: CHLOROSILANE □ DICHLOROSILANE □ SILICON CHLORIDE HYDRIDE

TOXICITY DATA WITH REFERENCE
ihl-rat LC50:215 ppm JTSCDR 18,394,93

SAFETY PROFILE: Moderately toxic by inhalation. Ignites spontaneously in air. Confined mixtures with air are spontaneously explosive. When heated to decomposition it emits toxic fumes of Cl^-. See also CHLOROSILANES.

DGK400 **CAS:73926-91-7** **HR: 2**
2,2′-DICHLORO-4,4′-STILBENEDIAMINE
mf: $C_{14}H_{12}Cl_2N_2$ mw: 279.18

SYNS: 4:4′-DIAMINO-2:2′-DICHLOROSTILBENE □ 2,2′-DICHLORO-4,4′-STILBENAMINE

TOXICITY DATA WITH REFERENCE
scu-rat TDLo:1400 mg/kg/W-I:ETA BMBUAQ 14,141,58

SAFETY PROFILE: Questionable carcinogen with experimental tumorigenic data. When heated to decomposition it emits very toxic fumes of Cl^- and NO_x.

DGK600 **CAS:73926-92-8** **HR: 2**
3,3′-DICHLORO-4,4′-STILBENEDIAMINE
mf: $C_{14}H_{12}Cl_2N_2$ mw: 279.18

SYN: 4:4′-DIAMINO-3:3′-DICHLOROSTILBENE

TOXICITY DATA WITH REFERENCE
scu-rat TDLo:200 mg/kg/W-I:ETA BMBUAQ 14,141,58

SAFETY PROFILE: Questionable carcinogen with experimental tumorigenic data. When heated to decomposition it emits very toxic fumes of Cl^- and NO_x.

DGK900 **CAS:2736-23-4** **HR: 2**
2,4-DICHLORO-5-SULFAMOYLBENZOIC ACID
mf: $C_7H_5Cl_2NO_4S$ mw: 270.09

SYNS: BENZOIC ACID, 5-(AMINOSULFONYL)-2,4-DICHLORO- □ BENZOIC ACID, 2,4-DICHLORO-5-SULFAMOYL- □ 2,4-DICHLORO-5-SULPHAMOYLBENZOIC ACID

TOXICITY DATA WITH REFERENCE
eye-rbt 100 mg SEV FCTOD7 20,573,82
eye-rbt 100 mg/4S RNS MLD FCTOD7 20,573,82
ipr-mus LD50:15 g/kg PCJOAU 19,697,85

SAFETY PROFILE: Slightly toxic by intraperitoneal route. A severe eye irritant. When heated to decomposition it emits toxic fumes of NO_x, SO_x, and Cl^-.

DGL200 **CAS:3001-57-8** **HR: 3**
3,4-DICHLOROSULFOLANE
mf: $C_4H_6Cl_2O_2S$ mw: 189.06

SYNS: DAC PRO □ 3,4-DICHLOROTETRAHYDROTHIOPHENE-1,1-DIOXIDE □ DICHLOROTHIOLANE DIOXIDE □ PRD EXPERIMENTAL NEMATOCIDE

TOXICITY DATA WITH REFERENCE
orl-rat LD50:482 mg/kg 28ZEAL 5,76,76
ipr-mus LD50:23 mg/kg RPTOAN 41,257,78
ivn-mus LD50:56 mg/kg CSLNX* NX#03183
skn-rbt LD50:1130 mg/kg TXAPA9 28,313,74

SAFETY PROFILE: Poison by intravenous and intraperitoneal routes. Moderately toxic by ingestion and skin contact. When heated to decomposition it emits very toxic fumes of Cl^- and SO_x.

DGL400 CAS:127-21-9 HR: 3
DICHLOROTETRAFLUOROACETONE
mf: $C_3Cl_2F_4O$ mw: 198.93

PROP: A colorless liquid. Miscible with water and many organic solvents. Bp: 44°, fp: <−100°, d: 1.52 @ 20°/4°.

SYNS: ACETONE-1,3-DICHLORO-1,1,3,3-TETRAFLUOROACETONE □ sym-DICHLOROTETRAFLUOROACETONE □ 1,3-DICHLORO-1,1,3,3-TETRAFLUORO-2-PROPANONE

TOXICITY DATA WITH REFERENCE
orl-rat LD50:61 mg/kg TXAPA9 7,592,65
ihl-rat LCLo:50 ppm/6H TXAPA9 7,592,65
skn-rat LD50:91 mg/kg TXAPA9 7,592,65
ivn-mus LD50:180 mg/kg CSLNX* NX#01749
skn-rbt LD50:146 mg/kg 34ZIAG -,213,69

SAFETY PROFILE: Poison by ingestion, skin contact, inhalation, and intravenous routes. When heated to decomposition it emits very toxic fumes of Cl^- and F^-.

DGL600 CAS:1320-37-2 HR: 1
DICHLOROTETRAFLUOROETHANE
DOT: UN 1958
mf: $C_2Cl_2F_4$ mw: 170.92

PROP: Colorless gas. Bp: 3.5°.

SYNS: DWUCHLOROCZTEROFLUOROETAN (POLISH) □ R114 (DOT) □ TETRAFLUORODICHLOROETHANE

TOXICITY DATA WITH REFERENCE
ihl-mus LCLo:700,000 ppm/30M AMPMAR 30,447,69

CONSENSUS REPORTS: Reported in EPA TSCA Inventory.

OSHA PEL: TWA 1000 ppm
ACGIH TLV: TWA 1000 ppm
DOT CLASSIFICATION: 2.2; *Label:* Nonflammable Gas

SAFETY PROFILE: A mildly toxic irritant; narcotic in high concentrations. An asphyxiant. Reacts violently with alcohol. When heated to decomposition it emits toxic fumes of F^- and Cl^-.

DGL800 CAS:3511-19-1 HR: 2
2,3-DICHLOROTETRAHYDROFURAN
mf: $C_4H_6Cl_2O$ mw: 141.00

TOXICITY DATA WITH REFERENCE
scu-mus TDLo:888 mg/kg/74W-I:ETA JNCIAM 48,1431,72

SAFETY PROFILE: Questionable carcinogen with experimental tumorigenic data. When heated to decomposition it emits toxic fumes of Cl^-. See also TETRAHYDROFURAN and CHLORIDES.

DGL875 CAS:6522-40-3 HR: 3
endo-2,5-DICHLORO-7-THIABICYCLO(2.2.1) HEPTANE
mf: $C_6H_8Cl_2S$ mw: 183.10

SAFETY PROFILE: Mixtures with dimethylformamide + sodium tetrahydroborate explode when heated. When heated to decomposition it emits toxic fumes of Cl^- and SO_x.

DGM600 CAS:1918-13-4 HR: 3
2,6-DICHLOROTHIOBENZAMIDE
mf: $C_7H_5Cl_2NS$ mw: 206.09

PROP: A solid. Mp: 152°. Sltly sol in H_2O.

SYNS: CHLOROTHIAMIDE □ DCBN □ 2,6-DICHLOROBENZENECARBOTHIOAMIDE □ SD 7961 □ WL-5792

TOXICITY DATA WITH REFERENCE
cyt-mus-unr 500 mg/kg TGANAK 14(6),41,80
cyt-mus-orl 500 mg/kg CYGEDX 14(6),38,80
orl-rat LD50:757 mg/kg WRPCA2 9,119,70
skn-rat LD50:1000 mg/kg WRPCA2 9,119,70
ipr-rat LD50:242 mg/kg ATXKA8 23,42,67
orl-mus LD50:500 mg/kg ATXKA8 23,42,67
orl-rbt LD50:300 mg/kg ATXKA8 23,42,67
orl-ckn LD50:500 mg/kg 28ZEAL 5,50,76
orl-dom LDLo:125 mg/kg ATXKA8 23,42,67

CONSENSUS REPORTS: EPA Genetic Toxicology Program.

SAFETY PROFILE: Poison by ingestion and intraperitoneal route. Moderately toxic by skin contact. Mutation data reported. An herbicide. When heated to decomposition it emits very toxic fumes of Cl^-, NO_x, and SO_x.

DGM700 CAS:95-73-8 HR: 2
2,4-DICHLOROTOLUENE
mf: $C_7H_6Cl_2$ mw: 161.03

SYNS: BENZENE, 2,4-DICHLORO-1-METHYL-(9CI) □ 2,4-DICHLORO-1-METHYLBENZENE □ TOLUENE, 2,4-DICHLORO-

TOXICITY DATA WITH REFERENCE
orl-rat LD50:4600 mg/kg GISAAA 53(2),80,88
orl-mus LD50:2900 mg/kg GISAAA 53(2),80,88
orl-gpg LD50:5 g/kg GISAAA 53(2),80,88

CONSENSUS REPORTS: Reported in EPA TSCA Inventory.

SAFETY PROFILE: Moderately toxic by ingestion. When heated to decomposition it emits toxic vapors of Cl^-.

DGM875 CAS:644-62-2 HR: 3
N-(2,6-DICHLORO-m-TOLYL)ANTHRANILIC ACID
mf: $C_{14}H_{11}Cl_2NO_2$ mw: 296.16

PROP: Powder or white crystals from acetone that lose water. Mp: 287–291°. Solubility (mg/mL): water 0.03; 0.1N NaOH 28, pH of saturated aq soln: approx. 6.9.

SYNS: ARQUEL □ 2-((2,6-DICHLORO-3-METHYLPHENYL)AMINO)-BENZOIC ACID (9CI) □ INF 4668 □ MECLOFENAMIC ACID □ MECLOPHENAMIC ACID

TOXICITY DATA WITH REFERENCE
ims-rbt TDLo:2750 μg/kg (female 1-3D post):REP BIREBV 14,451,76
orl-rbt TDLo:200 mg/kg (1D pre):REP FESTAS 38,238,82
orl-pig TDLo:15 mg/kg (female 16W post):REP JRPFA4 65,157,82
orl-rat LD50:100 mg/kg AGACBH 7,481,77
ipr-rat LD50:109 mg/kg JPETAB 148,422,65

SAFETY PROFILE: Poison by ingestion. Experimental reproductive effects. When heated to decomposition it emits toxic fumes of Cl^- and NO_x.

DGN000 CAS:29098-15-5 HR: 3
N-(2,6-DICHLORO-m-TOLYL)ANTHRANILIC ACID ETHOXYMETHYL ESTER
mf: $C_{17}H_{17}Cl_2NO_3$ mw: 354.25

PROP: Crystals from EtOH. Mp: 73–74°.

SYNS: 2-((2,6-DICHLORO-3-METHYLPHENYL)AMINO)BENZOIC ACID ETHOXYMETHYL ESTER ☐ ESTERE ETOSSIMETILICO dell' ACIDO N-(2,6-DICLORO-m-TOLIL)ANTRANILICO (ITALIAN) ☐ ETHOXYMETHYL-N-(2,6-DICHLORO-m-TOLYL)ANTHRANILATE ☐ ETOCLOFENE ☐ ETOFEN ☐ TEROFENAMATE

TOXICITY DATA WITH REFERENCE
orl-mus LD50:918 mg/kg DRFUD4 1,421,76
ipr-mus LD50:300 mg/kg DRFUD4 1,421,76

SAFETY PROFILE: Poison by intraperitoneal route. Moderately toxic by ingestion. An analgesic and anti-inflammatory agent. See also ESTERS. When heated to decomposition it emits very toxic fumes of Cl^- and NO_x.

DGN200 CAS:2782-57-2 HR: 2
1,3-DICHLORO-s-TRIAZINE-2,4,6(1H,3H,5H)-TRIONE
DOT: UN 2465
mf: $C_3H_2Cl_2N_3O_3$ mw: 198.98

PROP: White crystals, chlorine odor. Mp: 226–226.7°. Moderately sol in water.

SYNS: ACL 70 ☐ CDB 60 ☐ DICHLOROCYANURIC ACID ☐ DICHLOROISOCYANURATE ☐ DICHLOROISOCYANURIC ACID ☐ DICHLOROISOCYANURIC ACID, dry or dichloroisocyanuric acid salts (DOT) ☐ FI CLOR 71 ☐ HILITE 60 ☐ ISOCYANURIC ACID, DICHLORO- ☐ ISOCYANURIC DICHLORIDE ☐ ORCED ☐ KYSELINA DICHLORISOKYANUROVA (CZECH) ☐ TROCLOSENE

TOXICITY DATA WITH REFERENCE
skn-rbt 500 mg SEV 34ZIAG -,167,69
eye-rbt 100 MG SEV 34ZIAG -,167,69
orl-hmn LDLo:3570 mg/kg:GIT 34ZIAG -,167,69
orl-rat LD50:1173 mg/kg MarJV# 29MAR77

CONSENSUS REPORTS: Reported in EPA TSCA Inventory.

DOT CLASSIFICATION: 5.1; *Label:* Oxidizer

SAFETY PROFILE: Moderately toxic by ingestion. Human systemic effects by ingestion: ulceration or bleeding from stomach. Autopsy findings include gastrointestinal tract irritation, tissue edema, liver and kidney congestion. A severe eye and skin irritant. When heated to decomposition it emits chlorides and carbon monoxide.

DGN400 CAS:4499-01-8 HR: 2
2-(4,6-DICHLORO-s-TRIAZIN-2-YLAMINO)-4-(4-AMINO-3-SULFO-1-ANTHRAQUINONYLAMINO)BENZENESULFONIC ACID, DISODIUM SALT
mf: $C_{23}H_{12}Cl_2N_6O_8S_2 \cdot 2Na$ mw: 681.41

SYN: MODR BRILANTNI OSTAZINOVA S-R (CZECH)

TOXICITY DATA WITH REFERENCE
eye-rbt 100 mg/24H SEV 28ZPAK -,242,72
orl-rat LD50:8980 mg/kg 28ZPAK -,242,72

CONSENSUS REPORTS: Reported in EPA TSCA Inventory.

SAFETY PROFILE: Mildly toxic by ingestion. A severe eye irritant. When heated to decomposition it emits very toxic fumes of Cl^-, NO_x, Na_2O, and SO_x.

DGN600 CAS:73826-58-1 HR: 1
4-(4,6-DICHLORO-s-TRIAZIN-2-YLAMINO)-5-HYDROXY-6-(2-HYDROXY-5-NITROPHENYLAZO)-2,7-NAPHTHALENEDISULFONIC ACID
mf: $C_{19}H_{11}Cl_2N_7O_{10}S_2$ mw: 632.39

SYN: CERN OSTAZINOVA H-N (CZECH)

TOXICITY DATA WITH REFERENCE
skn-rbt 500 mg/24H MLD 28ZPAK -,234,72
eye-rbt 500 mg/24H MLD 28ZPAK -,234,72
orl-rat LD50:9120 mg/kg 28ZPAK -,234,72

SAFETY PROFILE: Mildly toxic by ingestion. A skin and eye irritant. When heated to decomposition it emits very toxic fumes of SO_x, NO_x, and Cl^-.

DGN800 CAS:6522-86-7 HR: 1
5-(3,5-DICHLORO-s-TRIAZINYLAMINO)-4-HYDROXY-3-PHENYLAZO-2,7-NAPHTHALENEDISULFONIC ACID
mf: $C_{19}H_{12}Cl_2N_6O_7S_2$ mw: 571.39

SYN: CERVEN BRILANTNI OSTAZINOVA S-5B (CZECH)

TOXICITY DATA WITH REFERENCE
eye-rbt 100 mg/24H MOD 28ZPAK -,235,72
orl-rat LD50:7460 mg/kg 28ZPAK -,235,72

SAFETY PROFILE: Mildly toxic by ingestion. An eye irritant. When heated to decomposition it emits very toxic fumes of Cl^-, NO_x, and SO_x.

DGO000 CAS:73816-75-8 HR: 1
2-(6-(4,6-DICHLORO-s-TRIAZINYL)METHYLAMINO-1-HYDROXY-3-SULFONAPHTHYLAZO)-1,5-NAPHTHALENEDISULFONIC ACID
mf: $C_{24}H_{16}Cl_2N_6O_{10}S_3$ mw: 715.54

SYN: ORANZ BRILANTNI OSTAZINOVA S-2R (CZECH)

TOXICITY DATA WITH REFERENCE
skn-rbt 500 mg/24H MLD 28ZPAK -,237,72
eye-rbt 500 mg/24H MLD 28ZPAK -,237,72
orl-rat LD50:8500 mg/kg 28ZPAK -,237,72

SAFETY PROFILE: Mildly toxic by ingestion. A skin and eye irritant. When heated to decomposition it emits very toxic fumes of Cl^-, NO_x, and SO_x.

DGO400 **CAS:3615-21-2** **HR: 3**
4,5-DICHLORO-2-TRIFLUOROMETHYLBENZIMIDAZOLE
mf: $C_8H_3Cl_2F_3N_2$ mw: 255.03

PROP: A solid. Mp: 212–214°.

SYNS: CHLORFLURAZOLE □ CHLOROFLURAZOLE □ NC 3363

TOXICITY DATA WITH REFERENCE
orl-rat LD50:13,080 µg/kg PSSCBG 15,31,84
ipr-mus LD50:14 mg/kg BCPCA6 18,1389,69
orl-ckn LD50:34 mg/kg GUCHAZ 6,101,73

CONSENSUS REPORTS: EPA Extremely Hazardous Substances List.

SAFETY PROFILE: Poison by ingestion and intraperitoneal routes. A pesticide. When heated to decomposition it emits very toxic fumes of Cl^-, NO_x, and F^-.

DGO600 **CAS:64048-90-4** **HR: 3**
DICHLORO(m-TRIFLUOROMETHYLPHENYL)ARSINE
mf: $C_7H_4AsCl_2F_3$ mw: 291.94

TOXICITY DATA WITH REFERENCE
ihl-mus LCLo:380 mg/m³ NDRC** -,12,43
ihl-hmn LCLo:28 ppm/10M NTIS** PB214-270

CONSENSUS REPORTS: Arsenic and its compounds are on the Community Right-To-Know List.

OSHA PEL: TWA 0.5 mg(As)/m³

SAFETY PROFILE: A human poison by inhalation. See also ARSENIC COMPOUNDS. When heated to decomposition it emits very toxic fumes of As, Cl^-, and F^-.

DGO800 **CAS:594-31-0** **HR: 3**
DICHLOROTRIPHENYLANTIMONY
mf: $C_{18}H_{15}Cl_2Sb$ mw: 423.98

PROP: Crystals from MeOH or EtOH/$CHCl_3$. Mp: 143°.

SYNS: ANTIMONY TRIPHENYLDICHLORIDE □ DICHLOROTRIPHENYLSTIBINE □ TRIPHENYLANTIMONY DICHLORIDE

TOXICITY DATA WITH REFERENCE
orl-rat LD50:195 mg/kg MarJV# 29MAR77

CONSENSUS REPORTS: Reported in EPA TSCA Inventory. Antimony and its compounds are on the Community Right-To-Know List.

OSHA PEL: TWA 0.5 mg(Sb)/m³
ACGIH TLV: TWA 0.5 mg(Sb)/m³
NIOSH REL: TWA 0.5 mg/m³

SAFETY PROFILE: Poison by ingestion. See also ANTIMONY COMPOUNDS. When heated to decomposition it emits very toxic fumes of Cl^- and Sb.

DGP000 **CAS:627-72-5** **HR: 3**
S-DICHLOROVINYL-l-CYSTEINE
mf: $C_5H_7Cl_2NO_2S$ mw: 216.09

SYNS: S-(1,2-DICHLOROETHYLENEYL)-l-CYSTEINE □ l-3-((1,2-DICHLOROVINYL)THIO)ALANINE

TOXICITY DATA WITH REFERENCE
mma-sat 1 µg/plate CBINA8 54,15,85
ipr-rat LDLo:50 mg/kg FCTXAV 3,67,65
ipr-mus LD50:45 mg/kg FCTXAV 3,67,65
ivn-rbt LDLo:10 mg/kg FCTXAV 3,67,65
ipr-gpg LDLo:20 mg/kg FCTXAV 3,67,65

SAFETY PROFILE: Poison by intravenous and intraperitoneal routes. Mutation data reported. When heated to decomposition it emits very toxic fumes of Cl^-, NO_x, and SO_x.

DGP200 **CAS:626-16-4** **HR: 3**
α,α′-DICHLORO-m-XYLENE
mf: $C_8H_8Cl_2$ mw: 175.06

PROP: Crystals from EtOH. Mp: 32–34°, bp: 254°.

SYNS: 1,3-BIS(CHLOROMETHYL)BENZENE □ m-XYLYLENE DICHLORIDE

TOXICITY DATA WITH REFERENCE
mmo sat 100 µg/plate MUREAV 191,79,87
ivn-mus LD50:100 mg/kg CSLNX* NX#03828

CONSENSUS REPORTS: Reported in EPA TSCA Inventory.

SAFETY PROFILE: Poison by intravenous route. Mutation data reported. See also CHLORINATED HYDROCARBONS, AROMATIC. When heated to decomposition it emits toxic fumes of Cl^-.

DGP400 **CAS:612-12-4** **HR: 3**
α,α′-DICHLORO-o-XYLENE
mf: $C_8H_8Cl_2$ mw: 175.06

PROP: Crystals from pet ether. D: 1.39 @ 0°, mp: 55°, bp: 239–241°.

SYNS: 1,2-BIS(CHLOROMETHYL)BENZENE □ o-XYLYLENE DICHLORIDE

TOXICITY DATA WITH REFERENCE
mmo-sat 100 µg/plate MUREAV 191,79,87
ivn-mus LD50:320 mg/kg CSLNX* NX#03225

CONSENSUS REPORTS: Reported in EPA TSCA Inventory.

SAFETY PROFILE: Poison by intravenous route. Mutation data reported. See also CHLORINATED HYDROCARBONS, AROMATIC. When heated to decomposition it emits toxic Cl^-.

DGP600 **CAS:623-25-6** **HR: 2**
α,α′-DICHLORO-p-XYLENE
mf: $C_8H_8Cl_2$ mw: 175.06

PROP: Platelets from hexane or EtOH. Mp: 104.1–104.3°, bp: 240–245° (decomp).

SYNS: 1,4-BIS(CHLOROMETHYL)BENZENE □ p-XYLYLENE DICHLORIDE

TOXICITY DATA WITH REFERENCE
skn-rbt 500 mg MOD 34ZIAG -,213,69
mmo-sat 100 µg/plate MUREAV 191,79,87

orl-rat LD50:1780 mg/kg 34ZIAG -,213,69

CONSENSUS REPORTS: Reported in EPA TSCA Inventory.

SAFETY PROFILE: Moderately toxic by ingestion. A skin and eye irritant. Mutation data reported. See also CHLORINATED HYDROCARBONS, AROMATIC. When heated to decomposition it emits toxic fumes of Cl^-.

DGP800 CAS:120-67-2 HR: 2
2,4-DICHLORPHENYL "CELLOSOLVE"
mf: $C_8H_8Cl_2O_2$ mw: 207.06

SYN: 2-(2,4-DICHLOROPHENOXY)ETHANOL

TOXICITY DATA WITH REFERENCE
orl-rat LD50:1410 mg/kg UCDS** 12/29/71
skn-rbt LD50:1250 mg/kg UCDS** 12/29/71

SAFETY PROFILE: Moderately toxic by skin contact and ingestion. When heated to decomposition it emits toxic fumes of Cl^-.

DGP900 CAS:62-73-7 HR: 3
DICHLORVOS
mf: $C_4H_7Cl_2O_4P$ mw: 220.98

PROP: Liquid with aromatic odor. Bp: 120° @ 14 mm, bp: 88° @ 3 mm. Sltly sol in water and glycerin; misc with aromatic and chlorinated hydrocarbon solvents and alc.

SYNS: APAVAP ☐ ASTROBOT ☐ ATGARD ☐ BAY 19149 ☐ BENFOS ☐ BIBESOL ☐ BREVINYL ☐ CANOGARD ☐ CEKUSAN ☐ CHLORVINPHOS ☐ CYANOPHOS ☐ CYPONA ☐ DDVF ☐ DDVP ☐ DEDEVAP ☐ DERIBAN ☐ DERRIBANTE ☐ DEVIKOL ☐ (2,2-DICHLOOR-VINYL)-DIMETHYL-FOSFAAT (DUTCH) ☐ DICHLOORVO (DUTCH) ☐ DICHLORFOS (POLISH) ☐ 2,2-DICHLOROETHENOL DIMETHYL PHOSPHATE ☐ 2,2-DICHLOROETHENYL DIMETHYL PHOSPHATE ☐ 2,2-DICHLOROETHENYL PHOSPHORIC ACID DIMETHYL ESTER ☐ DICHLOROPHOS ☐ DICHLOROVAS ☐ 2,2-DICHLOROVINYL ALCOHOL, DIMETHYL PHOSPHATE ☐ 2,2-DICHLOROVINYL DIMETHYL PHOSPHATE ☐ 2,2-DICHLOROVINYL DIMETHYL PHOSPHORIC ACID ESTER ☐ DICHLOROVOS ☐ DICHLORPHOS ☐ (2,2-DICHLOR-VINYL)-DIMETHYL-PHOSPHAT (GERMAN) ☐ O-(2,2-DICHLORVINYL)-O,O-DIMETHYLPHOSPHAT (GERMAN) ☐ (2,2-DICLORO-VINIL)DIMETILFOSFATO (ITALIAN) ☐ DIMETHYL-2,2-DICHLOROETHENYL PHOSPHATE ☐ DIMETHYL DICHLOROVINYL PHOSPHATE ☐ O,O-DIMETHYL DICHLOROVINYL PHOSPHATE ☐ DIMETHYL-2,2-DICHLOROVINYL PHOSPHATE ☐ O,O-DIMETHYL-O-2,2-DICHLOROVINYL PHOSPHATE ☐ O,O-DIMETHYL-O-(2,2-DICHLOR-VINYL)-PHOSPHAT (GERMAN) ☐ DIVIPAN ☐ DQUIGARD ☐ DUO-KILL ☐ DURAVOS ☐ ENT 20,738 ☐ EQUIGEL ☐ ESTROSEL ☐ ESTROSOL ☐ FECAMA ☐ FLY-DIE ☐ FLY FIGHTER ☐ HERKAL ☐ KRECALVIN ☐ LINDAN ☐ MAFU ☐ MARVEX ☐ MOPARI ☐ NCI-C00113 ☐ NERKOL ☐ NOGOS ☐ NO-PEST ☐ NO-PEST STRIP ☐ NSC-6738 ☐ NUVA ☐ OKO ☐ OMS 14 ☐ PHOSPHATE de DIMETHYLE et de 2,2-DICHLOROVINYLE (FRENCH) ☐ PHOSPHORIC ACID-2,2-DICHLOROETHENYL DIMETHYL ESTER ☐ PHOSVIT ☐ SD-1750 ☐ SZKLARNIAK ☐ TAP 9VP ☐ TASK ☐ TASK TABS ☐ TENAC ☐ TETRAVOS ☐ VAPONA ☐ VAPONITE ☐ VERDICAN ☐ VERDIPOR ☐ VINYLOFOS ☐ VINYLOPHOS

TOXICITY DATA WITH REFERENCE
dns-hmn:oth 65 mmol/L PSSCBG 15,439,84
dni-hmn:lym 62 mg/L TUMOAB 66,425,80
cyt-ham-ipr 3 mg/kg ARTODN 58,152,85
orl-rat TDLo:39,200 µg/kg (14-21D preg):REP NUPOBT 15,255,77
ihl-rbt TCLo:4 mg/m³/23H (female 1-28D post):TER ATXKA8 30,29,72
orl-rbt TDLo:65 mg/kg (female 6-13D post):REP TJADAB 20,383,79
unr-rat TDLo:40 mg/kg (male 2D pre):REP BECTA6 15,458,76
orl-pig TDLo:255 mg/kg (female 41-70D post):TER ZVRAAX 27,662,80
ipr-rat TDLo:15 mg/kg (11D preg):TER AEHLAU 16,805,68
orl-rat TDLo:4120 mg/kg/2Y-C:NEO NTPTR* NTP-TR-342,89
orl-rat TDLo:2060 mg/kg/2Y-C:CAR JJCREP 82,157,91
orl-mus TDLo:20,600 mg/kg/2Y-C:CAR NTPTR* NTP-TR-342,89
orl-rat LD50:17 mg/kg JPIFAN (13),36,72
ihl-rat LC50:15 mg/m³/4H GISAAA 33(12),35,68
skn-rat LD50:70,400 µg/kg APYPAY 32,507,81
ipr-rat LD50:23,300 µg/kg IJPPAZ 31,19,87
scu-rat LD50:10,800 µg/kg APYPAY 32,507,81
orl-mus LD50:61 mg/kg OSDIAF 15,553,66
ihl-mus LC50:13 mg/m³/4H GISAAA 33(12),35,68
skn-mus LD50:206 mg/kg ABCHA6 27,684,63
ipr-mus LD50:22 mg/kg JAFCAU 11,91,63
scu-mus LD50:24 mg/kg IMSUAI 31,170,62
ivn-mus LD50:18 mg/kg CSLNX* NX#00004
orl-dog LD50:100 mg/kg 85JFAN A141,84
orl-rbt LD50:10 mg/kg BEXBAN 83,32,77
skn-rbt LD50:107 mg/kg BESAAT 12,117,66

CONSENSUS REPORTS: IARC Cancer Review: Group 2B IMEMDT 53,267,91; Animal Sufficient Evidence IMEMDT 53,267,91; Animal Inadequate Evidence IMEMDT 20,97,79; Human No Adequate Data IMEMDT 20,97,79; Human Inadequate Evidence IMEMDT 53,267,91. NCI Carcinogenesis Bioassay (feed); No Evidence: mouse, rat NCITR* NCI-CG-TR-10,77. EPA Genetic Toxicology Program. Community Right-To-Know List. EPA Extremely Hazardous Substances List.

OSHA PEL: TWA 1 mg/m³ (skin)
ACGIH TLV: TWA 0.1 ppm (skin)
DFG MAK: 0.1 ppm (1 mg/m³)

SAFETY PROFILE: Confirmed carcinogen with carcinogenic and tumorigenic data. Poison by ingestion, inhalation, skin contact, subcutaneous, intravenous, and intraperitoneal routes. Experimental teratogenic and reproductive effects. Human mutation data reported. A cholinesterase inhibitor, it is used in flea (pest) collars for pets. No neurotoxicity has been observed. It is very rapidly metabolized and excreted. When heated to decomposition it emits very toxic fumes of Cl^- and PO_x. See also PARATHION.

For occupational chemical analysis use OSHA: #ID-62.

DGQ200 CAS:116-52-9 HR: 1
DICLORALUREA
mf: $C_5H_6Cl_6N_2O_3$ mw: 354.83

SYNS: 1,3-BIS(1-HYDROXY-2,2,2-TRICHLOROETHYL)UREA ☐ 1,3-BIS(2,2,2-TRICHLORO-1-HYDROXYETHYL)UREA ☐ CRAG DCU-73w ☐ CRAG EXPERIMENTAL HERBICIDE 2 ☐ CRAG HERBICIDE 2 ☐

DCM □ DCU □ DICHLORAL UREA □ DKhM □ EH2 □ EXPERIMENTAL HERBICIDE 2

TOXICITY DATA WITH REFERENCE
skn-rbt 50 mg open MLD UCDS** 4/25/68
cyt-mus-orl 5 g/kg CYGEDX 11(4),62,77
orl-rat LD50:6680 mg/kg 28ZEAL 5,74,76

SAFETY PROFILE: Mildly toxic by ingestion. A skin irritant. Mutation data reported. A pesticide. When heated to decomposition it emits very toxic fumes of Cl^- and NO_x.

DGQ300 CAS:12045-01-1 HR: 3
DICOBALT BORIDE
mf: BCo_2 mw: 128.68

CONSENSUS REPORTS: Cobalt compounds are on the Community Right-To-Know List.

SAFETY PROFILE: Ignites spontaneously in air when dry. See also COBALT COMPOUNDS and BORON COMPOUNDS.

DGQ400 CAS:36499-65-7 HR: 3
DICOBALT EDETATE
mf: $C_{10}H_{12}CoN_2O_8 \cdot Co$ mw: 406.10

SYNS: Ba 2724 □ COBALT(2)-EDATHAMIL □ DICOBALT EDTA □ ((ETHYLENEDINITRILO)TETRAACETATO(2−)) COBALTATE(2−) COBALT(2+) SALT □ KOBALT-EDTA (GERMAN)

TOXICITY DATA WITH REFERENCE
ipr-rat LD50:100 mg/kg AEPPAE 243,254,62
ivn-rat LD50:43 mg/kg AIPTAK 143,219,63

CONSENSUS REPORTS: Cobalt and its compounds are on the Community Right-To-Know List.

NIOSH REL: (Cobalt) TWA Insufficient evidence for recommending limit

SAFETY PROFILE: Poison by intravenous and intraperitoneal routes. See also COBALT COMPOUNDS. When heated to decomposition it emits toxic fumes of NO_x.

DGQ500 CAS:965-52-6 HR: 3
DICOFERIN
mf: $C_{12}H_9N_3O_5$ mw: 275.24

PROP: Crystals from pyridine. Mp: 298°. Practically insol in water.

SYNS: BACIFURANE □ DIARLIDAN □ ERCEFUROL □ERCEFU-RYL □ 4-HYDROXY-BENZOIC ACID ((5-NITRO-2-FURANYL)METHYLENE) HYDRAZIDE □ p-HYDROXYBENZOIC ACID (5-NITROFURFURYLIDENE)HYDRAZIDE □ NIFUROXAZID □ NIFUROXAZIDE □ (NITRO-5′ FURFURYLIDENE-2′) HYDROXY-4 BENZHYDRAZIDE (FRENCH) □ PENTOFURYL □ R.C. 27-109 □ RC 30-109

TOXICITY DATA WITH REFERENCE
mmo-sat 500 ng/plate MUREAV 157,1,85
mma-sat 500 ng/plate MUREAV 157,1,85
orl-mus LDLo:6000 mg/kg APTRAD 21,207,63
ipr-mus LD50:100 mg/kg JPPMAB 16,663,64

SAFETY PROFILE: Poison by intraperitoneal route. Mildly toxic by ingestion. Mutation data reported. When heated to decomposition it emits toxic fumes of NO_x.

DGQ600 CAS:1117-94-8 HR: 3
DICOPPER(I) ACETYLIDE
mf: C_2Cu_2 mw: 151.10

CuC≡CCu

CONSENSUS REPORTS: Copper and its compounds are on the Community Right-To-Know List.

SAFETY PROFILE: An unstable material. It explodes on impact or heating to 100°C. If warmed in air or oxygen it explodes on subsequent contact with acetylene. The sensitivity of the acetylide when precipitated from solution increases with acidity of the solution. It ignites on contact with chlorine; bromine vapor; or finely divided iodine. Reaction with silver nitrate solutions produces a sensitive, explosive mixture of silver acetylide and silver. When heated to decomposition it emits acrid smoke and fumes. See also COPPER COMPOUNDS and ACETYLIDES.

DGQ625 CAS:86425-12-9 HR: 3
DICOPPER(I)-1,5-HEXADIYNIDE
mf: $C_6H_4Cu_2$ mw: 203.19

$(-CH_2C{\equiv}CCu)_2$

CONSENSUS REPORTS: Copper and its compounds are on the Community Right-To-Know List.

SAFETY PROFILE: The dry material explodes at room temperature. When heated to decomposition it emits acrid smoke and fumes. See also COPPER COMPOUNDS and ACETYLIDES.

DGQ650 CAS:41084-90-6 HR: 3
DICOPPER(I) KETENIDE
mf: C_2Cu_2O mw: 167.11

CONSENSUS REPORTS: Copper and its compounds are on the Community Right-To-Know List.

SAFETY PROFILE: Mildly explosive when dry. See also COPPER COMPOUNDS.

DGQ700 CAS:27134-24-3 HR: 2
DICRESOL
mf: $C_{14}H_{14}O_2$ mw: 214.28

SYN: ar,ar′-DIMETHYL-(1,1′-BIPHENYL)-ar,ar′-DIOL (9CI)

TOXICITY DATA WITH REFERENCE
orl-rat LD50:1625 mg/kg GTPZAB 20(9),53,76
skn-rat LD50:825 mg/kg GTPZAB 20(9),53,76
orl-mus LD50:651 mg/kg GTPZAB 20(9),53,76

SAFETY PROFILE: Moderately toxic by ingestion and skin contact. When heated to decomposition it emits acrid smoke and fumes.

DGQ859 **HR: 3**
DICROTONYL PEROXIDE
mf: $C_8H_{10}O_4$ mw: 170.16

$(CH_3CH{=}CHCO{\cdot}O{-})_2$

SAFETY PROFILE: A very shock-sensitive explosive. When heated to decomposition it emits acrid smoke and fumes. See also PEROXIDES.

DGQ875 **CAS:141-66-2** **HR: 3**
DICROTOPHOS
mf: $C_8H_{16}NO_5P$ mw: 237.22

SYNS: BIDIRL □ BIDRIN □ C 709 □ CARBICRON □ CIBA 709 □ DIAPADRIN □ DICROTOFOS (DUTCH) □ 3-(DIMETHOXYPHOSPHINYLOXY)-N,N-DIMETHYL-cis-CROTONAMIDE □ 3-(DIMETHOXYPHOSPHINYLOXY)-N,N-DIMETHYLISOCROTONAMIDE □ 3-(DIMETHYLAMINO)-1-METHYL-3-OXO-1-PROPENYL DIMETHYL PHOSPHATE □ cis-2-DIMETHYLCARBAMOYL-1-METHYLVINYL DIMETHYLPHOSPHATE □ O,O-DIMETHYL-O-(2-DIMETHYL-CARBAMOYL-1-METHYL-VINYL) PHOSPHAT (GERMAN) □ O,O-DIMETHYL-O-(N,N-DIMETHYLCARBAMOYL-1-METHYLVINYL) PHOSPHATE □ O,O-DIMETHYL-O-(1,4-DIMETHYL-3-OXO-4-AZA-PENT-1-ENYL)FOSFAAT (DUTCH) □ O,O-DIMETHYL-O-(1,4-DIMETHYL-3-OXO-4-AZA-PENT-1-ENYL)PHOSPHATE □ DIMETHYLPHOSPHATE ESTER with 3-HYDROXY-N,N-DIMETHYL-cis-CROTONAMIDE □ DIMETHYL PHOSPHATE of 3-HYDROXY-N,N-DIMETHYL-cis-CROTONAMIDE □ O,O-DIMETIL-O-(1,4-DIMETIL-3-OXO-4-AZA-PENT-1-ENIL)-FOSFATO (ITALIAN) □ EKTAFOS □ ENT 24,482 □ 3-HYDROXYDIMETHYL CROTONAMIDE DIMETHYL PHOSPHATE □ 3-HYDROXY-N,N-DIMETHYL-cis-CROTONAMIDE DIMETHYL PHOSPHATE □ PHOSPHATE de DIMETHYLE et de 2-DIMETHYLCARBAMOYL-1-METHYL VINYLE (FRENCH) □ SD 3562 □ SHELL SD-3562

TOXICITY DATA WITH REFERENCE
mma-sat 500 μg/plate JTEHD6 16,403,85
mrc-smc 30 mmol/L/5H MUREAV 32,133,75
orl-rat LD50:13 mg/kg ARSIM* 20,6,66
ihl-rat LC50:90 mg/m³/4H PSDTAP 15,239,74
skn-rat LD50:42 mg/kg WRPCA2 9,119,70
scu-rat LD50:8137 μg/kg BJPCBM 40,124,70
orl-mus LD50:11 mg/kg GUCHAZ 6,196,73
ipr-mus LD50:9500 μg/kg TXAPA9 16,446,70
scu-mus LD50:11,500 μg/kg JPPMAB 19,612,67
ivn-mus LD50:9900 μg/kg JPPMAB 19,612,67
skn-rbt LD50:168 mg/kg GUCHAZ 6,196,73

CONSENSUS REPORTS: EPA Farm Worker Reentry (39 FR 16888,74). EPA Genetic Toxicology Program. EPA Extremely Hazardous Substances List.

OSHA PEL: TWA 0.25 mg/m³ (skin)
ACGIH TLV: TWA 0.25 mg/m³ (skin)

SAFETY PROFILE: Poison by ingestion, inhalation, skin contact, subcutaneous, intravenous, and intraperitoneal routes. Mutation data reported. Used to control the coffee borer and certain economically important pests of cotton. When heated to decomposition it emits very toxic fumes of NO_x and PO_x. See also ESTERS.

DGR200 **CAS:12001-89-7** **HR: 3**
DICUMENE CHROMIUM
mf: $C_{18}H_{24}{\cdot}Cr$ mw: 292.42

SYNS: BIS(CUMENE)CHROMIUM □ BIS(pi-CUMENE)CHROMIUM □ BIS(ISOPROPYLBENZENE)CHROMIUM □ DICUMENYLCHROMIUM

TOXICITY DATA WITH REFERENCE
skn-rbt 500 mg open MOD UCDS** 4/21/67
eye-rbt 15 mg MOD UCDS** 4/21/67
orl-rat LD50:810 mg/kg AIHAAP 30,470,69
ivn-mus LD50:2200 μg/kg CSLNX* NX#06778
skn-rbt LD50:22 mg/kg UCDS** 4/21/67

CONSENSUS REPORTS: Chromium and its compounds are on the Community Right-To-Know List.

OSHA PEL: TWA 1 mg(Cr)/m³
ACGIH TLV: TWA 0.05 mg(Cr)/m³; Confirmed Human Carcinogen
NIOSH REL: (Chromium(VI)) CL 1 μg(Cr(VI))/m³

SAFETY PROFILE: A confirmed carcinogen. Poison by skin contact and intravenous routes. Moderately toxic by ingestion. A skin and eye irritant. See also CHROMIUM COMPOUNDS. When heated to decomposition it emits acrid and irritating fumes.

For occupational chemical analysis use NIOSH: Chromium Hexavalent 7024.

DGR400 **CAS:25566-92-1** **HR: 3**
DICUMYLMETHANE
mf: $C_{19}H_{24}$ mw: 252.43

SYN: DKM (RUSSIAN)

TOXICITY DATA WITH REFERENCE
orl-mus LDLo:15 g/kg TPKVAL 6,73,64
ihl-mus LCLo:122 mg/m³ TPKVAL 6,73,64
scu-mus LDLo:19 g/kg TPKVAL 6,73,64

SAFETY PROFILE: Poison by inhalation. Mildly toxic by ingestion and subcutaneous routes. When heated to decomposition it emits acrid smoke and irritating fumes.

DGR600 **CAS:80-43-3** **HR: 1**
DI-α-CUMYL PEROXIDE
mf: $C_{18}H_{22}O_2$ mw: 270.40

SYNS: ACTIVE DICUMYL PEROXIDE □ BIS(α,α-DIMETHYLBENZYL) PEROXIDE □ CUMENE PEROXIDE □ CUMYL PEROXIDE □ DICUMYL PEROXIDE (DOT) □ DI-CUP □ DI-CUP 40 KF □ DI-CUPR □ DIISOPROPYLBENZENE PEROXIDE □ ISOPROPYLBENZENE PEROXIDE □ LUPERCO □ LUPEROX □ LUPEROX 500R □ LUPEROX 500T □ VAROX DCP-R □ VAROX DCP-T

TOXICITY DATA WITH REFERENCE
orl-rat LD50:4100 mg/kg BSPII* 1/75-19B

CONSENSUS REPORTS: Reported in EPA TSCA Inventory.

SAFETY PROFILE: Mildly toxic by ingestion. See also PEROXIDES. When heated to decomposition it emits acrid smoke and irritating fumes.

DGS000 **CAS:1071-98-3** **HR: 3**
DICYANOACETYLENE
mf: C_4N_2 mw: 76.09

$N{\equiv}CC{\equiv}C{\equiv}N$

PROP: A solid or liquid. Mp: 21°, bp: 76°.

SYN: 2-BUTYNEDINITRILE

CONSENSUS REPORTS: Cyanide and its compounds are on the Community Right-To-Know List.

SAFETY PROFILE: The pure material and concentrated solutions are potentially explosive. Ignites in air at 130°C. The flame temperature in oxygen can reach 4700°C. When heated to decomposition it emits toxic fumes of CN^- and NO_x. See also NITRILES and ACETYLENE COMPOUNDS.

DGS200 **CAS:1119-69-3** **HR: 3**
1,4-DICYANO-2-BUTENE
mf: $C_6H_6N_2$ mw: 106.13

$N{\equiv}CCH_2CHCHCH_2C{\equiv}N$

PROP: Accelerated polymerization decomp of dicyanobutene.

CONSENSUS REPORTS: Cyanide and its compounds are on the Community Right-To-Know List.

SAFETY PROFILE: Decomposes violently when heated. Upon decomposition it emits toxic fumes of NO_x and CN^-. See also CYANIDE.

DGS300 **CAS:1557-57-9** **HR: 3**
DICYANODIAZENE
mf: C_2N_4 mw: 80.05

SYN: AZOCARBONITRILE

CONSENSUS REPORTS: Cyanide and its compounds are on the Community Right-To-Know List.

SAFETY PROFILE: An explosive sensitive to shock or heating in a closed container. Upon decomposition it emits toxic fumes of CN^- and NO_x. See also CYANIDE.

DGS600 **CAS:111-97-7** **HR: 3**
DI(2-CYANOETHYL)SULFIDE
mf: $C_6H_8N_2S$ mw: 140.22

PROP: White crystals or needles from alc. Mp: (α) 28.65°,(β): 22.10°, d: 1.1095 @ 30°, bp: 112–113° @ 17 mm.

SYNS: β,β′-DICYANODIETHYL SULFIDE ☐ NITRIL KYSELINY β,β′-THIODIPROPIONOVE (CZECH) ☐ β,β′-THIODIPROPIONITRILE ☐ USAF HA-5

TOXICITY DATA WITH REFERENCE
skn-rbt 500 mg/24H MLD 28ZPAK -,172,72
eye-rbt 500 mg AJOPAA 29,1363,46
orl-cat LD50:4210 mg/kg JIHTAB 31,60,49
ipr-mus LD50:300 mg/kg NTIS** AD277-689
orl-rat LD50:4500 mg/kg JIHTAB 31,60,49

CONSENSUS REPORTS: Cyanide and its compounds are on the Community Right-To-Know List. Reported in EPA TSCA Inventory.

SAFETY PROFILE: Poison by intraperitoneal route. Moderately toxic by ingestion. A skin and eye irritant. When heated to decomposition it emits very toxic fumes of NO_x, SO_x, and CN^-. See also NITRILES.

DGS700 **CAS:55644-07-0** **HR: 3**
DICYANOFURAZAN
mf: C_4N_4O mw: 120.07

$N{\equiv}CC{=}NON{=}CC{\equiv}N$

CONSENSUS REPORTS: Cyanide and its compounds are on the Community Right-To-Know List.

SAFETY PROFILE: A powerful but insensitive explosive. Explodes on contact with nitrogenous bases (e.g., hydrazine, mono- or di-methylhydrazine, piperidine, piperazine, diethylamine). Upon decomposition it emits toxic fumes of NO_x and CN^-. See also CYANIDE.

DGS800 **CAS:55644-07-0** **HR: 3**
DICYANOFURAZAN-N-OXIDE
mf: $C_4N_4O_2$ mw: 136.07

$N{\equiv}CC{=}NON(O){=}CC{\equiv}N$

SYN: DICYANOFUROXAN

CONSENSUS REPORTS: Cyanide and its compounds are on the Community Right-To-Know List.

SAFETY PROFILE: Explodes on contact with nitrogenous bases (e.g., hydrazine, mono- or di-methylhydrazine, piperidine, piperazine, diethylamine). Upon decomposition it emits toxic fumes of NO_x and CN^-. See also CYANIDE.

DGT000 **CAS:4331-98-0** **HR: 3**
DICYANOGEN-N,N-DIOXIDE
mf: $C_2N_2O_2$ mw: 84.04

$({-}C{\equiv}NO)_2$

PROP: A liquid. Mp: −12.5° to −11°.

CONSENSUS REPORTS: Cyanide and its compounds are on the Community Right-To-Know List.

SAFETY PROFILE: Decomposes explosively at −45°C under vacuum. Upon decomposition it emits toxic fumes of NO_x and CN^-. See also CYANIDE.

DGT200 **CAS:38780-37-9** **HR: 3**
cis-DICYCLOBUTYLAMMINEDICHLOROPLATINUM(II)
mf: $C_8H_{18}Cl_2N_2Pt$ mw: 408.27

PROP: Pale yellow crystals. Sol in polar org solvs.

SYNS: cis-BIS(CYCLOBUTYLAMMINE)DICHLOROPLATINUM(II) ☐ cis-DICHLOROBIS(CYCLOBUTYLAMMINE)PLATINUM(II)

TOXICITY DATA WITH REFERENCE
mmo-sat 1 μmol/plate CBINA8 26,179,79
ipr-mus LD50:90 mg/kg CBINA8 5,415,72

SAFETY PROFILE: Poison by intraperitoneal route. Mutagenic data reported. See also PLATINUM COMPOUNDS. When heated to decomposition it emits very toxic fumes of Cl^- and NO_x.

DGT300 **CAS:17455-23-1** **HR: 3**
DICYCLOHEXANO-24-CROWN-8
mf: $C_{24}H_{44}O_8$ mw: 380.61

PROP: Pale yellow viscous liquid. D: 1.105 @ 4°/20°. Sol in EtOH and $CHCl_3$; sltly sol in H_2O.

SYN: TETRACOSAHYDRO DIBENZ(b,n)(1,4,7,10,13,16,19,22)OCTAOXACYCLOTETRACOSIN

TOXICITY DATA WITH REFERENCE
skn-rbt 100 mg/24H MLD DCTODJ 8,451,85
eye-rbt 50 mg MOD DCTODJ 8,451,85
orl-rat LD50:75 mg/kg DCTODJ 8,451,85
ipr-rat LD50:10 mg/kg DCTODJ 8,451,85
ipr-mus LD50:12 mg/kg DCTODJ 8,451,85

SAFETY PROFILE: Poison by ingestion and intraperitoneal routes. An eye and skin irritant. When heated to decomposition it emits acrid smoke and fumes.

DGT500 **CAS:849-99-0** **HR: 1**
DICYCLOHEXYL ADIPATE
mf: $C_{18}H_{30}O_4$ mw: 310.48

SYNS: ERGOPLAST ADC ☐ HEXANEDIOIC ACID, DICYCLOHEXYL ESTER (9CI)

TOXICITY DATA WITH REFERENCE
ipr-rat TDLo:1020 mg/kg (5-15D preg):TER JPMSAE 62,1596,73
ipr-rat TDLo:1700 mg/kg (female 5-15D post):TER JPMSAE 62,1596,73
ipr-rat TDLo:170 mg/kg (female 5-15D post):REP JPMSAE 62,1596,73
ipr-rat TDLo:510 mg/kg (female 5-15D post):TER JPMSAE 62,1596,73
orl-rat LD50:16 g/kg RPZHAW 18,283,67
ipr-rat LD50:5101 mg/kg JPMSAE 62,1596,73

CONSENSUS REPORTS: Reported in EPA TSCA Inventory.

SAFETY PROFILE: Mildly toxic by ingestion and intraperitoneal routes. Experimental teratogenic and reproductive effects. When heated to decomposition it emits acrid smoke and fumes.

DGT600 **CAS:101-83-7** **HR: 3**
N,N-DICYCLOHEXYLAMINE
DOT: UN 2565
mf: $C_{12}H_{23}N$ mw: 181.36

PROP: Liquid, fishy odor. Mp: 20°, fp: −0.1°, bp: 256°, flash p: >210°F (OC), d: 0.910, vap d: 6.27.

SYNS: DCHA ☐ DICHA ☐ N-CYCLOHEXYLCYCLOHEXANAMINE ☐ DICYCLOHEXYLAMINE (DOT) ☐ DICYKLOHEXYLAMIN (CZECH) ☐ DODECAHYDRODIPHENYLAMINE

TOXICITY DATA WITH REFERENCE
cyt-hmn:leu 200 µg/L INHEAO 9,188,71
skn-rbt 2 mg/24H SEV 85JCAE-,468,86
eye-rbt 750 µg/24H SEV 85JCAE-,468,86
orl-rat TDLo:40 g/kg/52W-I:ETA VOONAW 4,659,58
scu-mus TDLo:2404 mg/kg/48W-I:ETA VOONAW 4,659,58
orl-rat LD50:373 mg/kg 85JCAE-,468,86
scu-mus LD50:135 mg/kg VOONAW 4,659,58
scu-rbt LDLo:500 mg/kg IECHAD 29,1247,37

CONSENSUS REPORTS: IARC Cancer Review: Group 3 IMEMDT 7,178,87; Animal Inadequate Evidence IMEMDT 22,55,80. Reported in EPA TSCA Inventory.

DOT CLASSIFICATION: 8; *Label:* Corrosive

SAFETY PROFILE: Poison by ingestion and subcutaneous routes. Corrosive. A severe skin and eye irritant. Questionable carcinogen with experimental tumorigenic data. Human mutation data reported. Combustible when exposed to heat or flame; can react with oxidizing materials. To fight fire, use alcohol foam, CO_2, dry chemical. When heated to decomposition it emits toxic fumes of NO_x. See also CYCLOHEXYLAMINE.

DGU200 **CAS:3129-91-7** **HR: 3**
DICYCLOHEXYLAMINE NITRITE
DOT: UN 2687
mf: $C_{12}H_{23}N{\cdot}HNO_2$ mw: 228.38

SYNS: DECHAN ☐ DICHAN (CZECH) ☐ DICYCLOHEXYLAMINONITRITE ☐ DICYCLOHEXYLAMMONIUM NITRITE ☐ DICYKLOHEXYLAMIN NITRIT (CZECH) ☐ DICYNIT (CZECH) ☐ DODECAHYDROPHENYLAMINE NITRITE ☐ DUSITAN DICYKLOHEXYLAMINU (CZECH)

TOXICITY DATA WITH REFERENCE
scu-rat TDLo:2400 mg/kg/48W-I:ETA VOONAW 4,659,58
scu-mus TDLo:2040 mg/kg/52W-I:ETA VOONAW 4,659,58
orl-rat LD50:284 mg/kg 28ZPAK -,68,72
orl-mus LD50:80 mg/kg GISAAA 30(8),35,65
scu-mus LD50:155 mg/kg VOONAW 4,659,58
orl-gpg LD50:350 mg/kg UCPHAQ 2,231,49

CONSENSUS REPORTS: Reported in EPA TSCA Inventory.

DOT CLASSIFICATION: 4.1; *Label:* Flammable Solid

SAFETY PROFILE: Poison by ingestion and subcutaneous routes. Questionable carcinogen with experimental tumorigenic data. A flammable liquid. When heated to decomposition it emits very toxic fumes of HNO_2 and NO_x. See also NITRITES.

DGU400 **CAS:63915-52-6** **HR: 2**
DICYCLOHEXYLAMINE PENTANOATE
mf: $C_{12}H_{23}N{\cdot}C_5H_{10}O_2$ mw: 283.51

SYNS: DICYKLOHEXYLAMINKAPRONAT (CZECH) ☐ KAPRONAN DICYKLOHEXYLAMINU (CZECH)

TOXICITY DATA WITH REFERENCE
skn-rbt 500 mg/24H MOD 28ZPAK -,68,72
eye-rbt 250 µg/24H SEV 28ZPAK -,68,72
orl-rat LD50:3290 mg/kg 28ZPAK -,68,72

SAFETY PROFILE: Moderately toxic by ingestion. A skin and severe eye irritant. When heated to decomposition it emits toxic fumes of NO_x.

DGU709 CAS:38780-35-7 HR: 3
cis-DICYCLOHEXYLAMMINEDICHLOROPLATINUM(II)
mf: $C_{12}H_{26}Cl_2N_2Pt$ mw: 464.39

SYNS: cis-DICHLOROBIS(CYCLOHEXYLAMINE)PLATINUM(II) □ cis-HAD

TOXICITY DATA WITH REFERENCE
mma-sat 30 μg/plate MUREAV 95,79,82
dni-ham:ovr 40 mg/L CBINA8 35,189,81
cyt-ham:ovr 40 mg/L CBINA8 35,189,81
ipr-mus LD50:12 mg/kg BJCAAI 36,420,77

SAFETY PROFILE: Poison by intraperitoneal route. Mutation data reported. See also PLATINUM COMPOUNDS. When heated to decomposition it emits very toxic fumes of Cl^- and NO_x.

DGU800 CAS:4979-32-2 HR: 1
N,N-DICYCLOHEXYL-2-BENZOTHIAZOLESULFENAMIDE
mf: $C_{19}H_{26}N_2S_2$ mw: 346.59

SYN: N,N-DICYKLOHEXYLBENZTHIAZOLSULFENAMID (CZECH)

TOXICITY DATA WITH REFERENCE
skn-rbt 20 mg/24H MOD 85JCAE-,1101,86
eye-rbt 500 mg/24H MLD 85JCAE-,1101,86
orl-rat LD50:6420 mg/kg 85JCAE-,1101,86

CONSENSUS REPORTS: Reported in EPA TSCA Inventory.

SAFETY PROFILE: Mildly toxic by ingestion. An eye and skin irritant. When heated to decomposition it emits very toxic fumes of SO_x and NO_x.

DGV100 CAS:16069-36-6 HR: 3
DICYCLOHEXYL-18-CROWN-6
mf: $C_{20}H_{36}O_6$ mw: 372.56

PROP: Colorless or pale yellow wax. Mp: 38–54°, bp: 344°.

SYNS: DICYCLOHEXANO-18-CROWN-6 □ EICOSAHYDRO DIBENZO(b,k)(1,4,7,10,13,16)HEXAOXACYCLOOCTADECIN

TOXICITY DATA WITH REFERENCE
skn-rbt 100 mg/24H MLD DCTODJ 8,451,85
eye-rbt 50 mg MOD DCTODJ 8,451,85
orl-rat TDLo:105 mg/kg (female 1-21D post):TER GISAAA 52(11),72,87
orl-rat LD50:176 mg/kg DCTODJ 8,451,85
skn-rat LDLo:130 mg/kg MEIEDD 10,373,83
ipr-rat LD50:55 mg/kg DCTODJ 8,451,85
orl-mus LD50:192 mg/kg GISAAA 52(11),72,87
ipr-mus LD50:53 mg/kg DCTODJ 8,451,85

CONSENSUS REPORTS: Reported in EPA TSCA Inventory.

SAFETY PROFILE: Poison by ingestion, skin contact, and intraperitoneal routes. An experimental teratogen. An eye and skin irritant. When heated to decomposition it emits acrid smoke and fumes.

DGV200 CAS:587-15-5 HR: 3
DICYCLOHEXYL FLUOROPHOSPHONATE
mf: $C_{12}H_{22}FO_3P$ mw: 264.31

SYN: DICYCLOHEXYLFLUOROPHOSPHATE

TOXICITY DATA WITH REFERENCE
ihl-rat LD50:1200 mg/m³/10M NTIS** PB158-508
ihl-mus LD50:800 mg/m³/10M NTIS** PB158-508
ihl-dog LD50:1 g/m³/10M NTIS** PB158-508
ihl-rbt LD50:1200 mg/m³/10M NTIS** PB158-508
ihl gpg LD50:6 g/m³/10M NTIS** PB158-508

SAFETY PROFILE: Poison by inhalation. See also FLUORIDES. When heated to decomposition it emits very toxic fumes of F^- and PO_x.

DGV600 CAS:119-60-8 HR: 3
DICYCLOHEXYL KETONE
mf: $C_{13}H_{22}O$ mw: 194.35

PROP: Bp: 145° @ 15 mm.

TOXICITY DATA WITH REFERENCE
ivn-mus LD50:56 mg/kg CSLNX* NX#06751

CONSENSUS REPORTS: Reported in EPA TSCA Inventory.

DOT CLASSIFICATION: 3; *Label:* Flammable Liquid

SAFETY PROFILE: Poison by intravenous route. See also KETONES. A flammable liquid. When heated to decomposition it emits acrid and irritating fumes.

DGV650 CAS:1561-49-5 HR: 2
DICYCLOHEXYL PEROXIDE CARBONATE
mf: $C_{14}H_{22}O_6$ mw: 286.36

SYNS: DICYCLOHEXYL PEROXYDICARBONATE, not >91% with water (UN 2153) (DOT) □ DICYCLOHEXYL PEROXYDICARBONATE, technically pure (UN 2152) (DOT) □ PEROXYDICARBONIC ACID, DICYCLOHEXYL ESTER

CONSENSUS REPORTS: Reported in EPA TSCA Inventory.

SAFETY PROFILE: A peroxide. Handle carefully. When heated to decomposition it emits acrid smoke and irritating vapors.

DGV700 CAS:84-61-7 HR: 1
DICYCLOHEXYL PHTHALATE
mf: $C_{20}H_{26}O_4$ mw: 330.46

SYNS: 1,2-BENZENEDICARBOXYLIC ACID, DICYCLOHEXYL ESTER □ ERGOPLAST.FDC □ HF 191 □ KP 201 □ PHTHALIC ACID, DICYCLOHEXYL ESTER □ UNIMOLL 66

TOXICITY DATA WITH REFERENCE
orl-rat LD50:30 g/kg EVHPAZ 3,61,73
orl-rat TDLo:10,500 mg/kg/7D-C APTOA6 51,217,82

CONSENSUS REPORTS: Reported in EPA TSCA Inventory.

SAFETY PROFILE: Mildly toxic by ingestion. When heated to decomposition it emits acrid smoke and irritating vapors.

D

DGV900 **CAS:22771-17-1** **HR: 3**
DICYCLOHEXYLTIN OXIDE
mf: $C_{12}H_{22}OSn$ mw: 301.03

SYN: STANNANE, DICYCLOHEXYLOXO-

TOXICITY DATA WITH REFERENCE
orl-rat LD50:355 mg/kg GISAAA 48(3),55,83

OSHA PEL FINAL: TWA 0.1 mg(Sn)/m³ (skin)
ACGIH TLV: TWA 0.1 mg(Sn)/m³ (skin)

SAFETY PROFILE: Poison by ingestion. When heated to decomposition it emits toxic fumes of Sn.

DGW000 **CAS:77-73-6** **HR: 3**
DICYCLOPENTADIENE
DOT: UN 2048
mf: $C_{10}H_{12}$ mw: 132.22

PROP: Colorless crystals. Mp: 32.9°, bp: 166.6°, d: 0.976 @ 35°, vap press: 10 mm @ 47.6°, vap d: 4.55, flash p: 90°F (OC).

SYNS: BICYCLOPENTADIENE □ BISCYCLOPENTADIENE □ 1,3-CYCLOPENTADIENE, DIMER □ DICYKLOPENTADIEN (CZECH) □ DIMER CYKLOPENTADIENU (CZECH) □ 3a,4,7,7a-TETRAHYDRO-4,7-METHANOINDENE

TOXICITY DATA WITH REFERENCE
skn-rbt 10 mg/24H open SEV AMIHBC 10,61,54
skn-rbt 9300 µg/24H open SEV AIHAAP 23,95,62
skn-rbt 500 mg/24H MOD 28ZPAK -,27,72
eye-rbt 500 mg/24H MLD 85JCAE-,50,86
orl-rat LD50:353 mg/kg TXAPA9 20,552,71
ihl-rat LC50:372 ppm/4H 28ZPAK -,27,72
ipr-rat LD50:200 mg/kg NCIUS* PH 43-64-886,JAN,65
orl-mus LD50:190 mg/kg 40QBA3 -,448,78
ipr-mus LD50:200 mg/kg NCIUS* PH 43-64-886,JAN,65
skn-rbt LD50:5080 mg/kg TXAPA9 20,552,71
orl-brd LD50:1010 mg/kg NTIS** AD-A087-257

CONSENSUS REPORTS: Reported in EPA TSCA Inventory. EPA Genetic Toxicology Program.

OSHA PEL: TWA 5 ppm
ACGIH TLV: TWA 5 ppm
DFG MAK: 0.5 ppm (3 mg/m³)
DOT CLASSIFICATION: 3; *Label:* Flammable Liquid

SAFETY PROFILE: Poison by ingestion and intraperitoneal routes. Moderately toxic by inhalation. Mildly toxic by skin contact. A severe skin and moderate eye irritant. Dangerous fire hazard when exposed to heat or flame; can react with oxidizing materials. To fight fire, use alcohol foam. When heated to decomposition it emits acrid smoke and fumes.

DGW200 **CAS:1271-19-8** **HR: 3**
DICYCLOPENTADIENYLDICHLOROTITANIUM
mf: $C_{10}H_{10}Cl_2Ti$ mw: 249.00

PROP: Bright-red acicular crystals from toluene. Mp: 289°.

SYNS: DICHLOROBIS(ETA⁵-2,4-CYCLOPENTADIEN-1-YL-TITANIUM (9CI) □ DICHLORODICYCLOPENTADIENYLTITANIUM □ DICHLORODI-pi-CYCLOPENTADIENYLTITANIUM □ DICHLOROTITANOCENE □ DICYCLOPENTADIENYLTITANIUMDICHLORIDE □ NCI-C04502 □ TITANIUM FERROCENE □ TITANOCENE □ TITANOCENE, DICHLORIDE

TOXICITY DATA WITH REFERENCE
mmo-sat 100 µg/plate ENMUDM 5(Suppl 1),3,83
mma-sat 1 mg/plate ENMUDM 5(Suppl 1),3,83
otr-rat:emb 2960 µg/L JJIND8 67,1303,81
otr-mus:fbr 800 µg/L JJIND8 67,1303,81
otr-ham:emb 100 µg/L JJIND8 67,1303,81
ipr-mus TDLo:30 mg/kg (10D preg):TER TXCYAC 33,171,84
ipr-mus TDLo:60 mg/kg (10D preg):TER TXCYAC 33,171,84
orl-rat TDLo:26 g/kg/2Y-I:ETA NTPTR* NTP-TR-399,91
ims-rat TDLo:720 mg/kg/2Y-I:NEO NCIUS* PH 43-64-886,JUL,68
ims-mus TDLo:75 mg/kg:NEO NCIUS* PH 43-64-886,JUL,68
ims-rat TD:900 mg/kg/2Y-I:NEO PWPSA8 14,68,71
ims-rat TD:430 mg/kg/81W-I:ETA NCIUS* PH 43-64-886,AUG,69
ipr-rat LD50:25 mg/kg NCIUS* PH 43-64-886,JAN,65
ipr-mus LD50:60 mg/kg NCIUS* PH 43-64-886,AUG,64
ivn-mus LD50:180 mg/kg CSLNX* NX#00774

CONSENSUS REPORTS: Reported in EPA TSCA Inventory. EPA Genetic Toxicology Program.

SAFETY PROFILE: Poison by intravenous and intraperitoneal routes. Questionable carcinogen with experimental neoplastigenic, tumorigenic, and teratogenic data. Mutation data reported. See also TITANIUM COMPOUNDS. When heated to decomposition it emits toxic fumes of Cl^-.

DGW300 **HR: 2**
DICYCLOPENTA(c,lmn)PHENANTHREN-1(9H)-ONE, 2, 3-DIHYDRO-
mf: $C_{18}H_{12}O$ mw: 244.30

SYNS: 2,3-DIHYDRODICYCLOPENTA(c,lmn)PHENANTHREN-1(9H)-ONE □ 15,16-DIHYDRO-1,11-METHANOCYCLOPENTA(a)PHENANTHREN-17-ONE

TOXICITY DATA WITH REFERENCE
skn-mus TDLo:16 mg/kg:CAR CRNGDP 5,1485,84

SAFETY PROFILE: Questionable carcinogen with experimental carcinogenic data. When heated to decomposition it emits acrid smoke and irritating fumes.

DGW400 **CAS:50976-02-8** **HR: 1**
DICYCLOPENTENYL ACRYLATE
mf: $C_{13}H_{14}O_2$ mw: 202.27

SYNS: ACRYLIC ACID ((3a,4,7,7a-TETRAHYDRO)-4,7-METHANOINDENYL) ESTER □ DCPA □ DICYCLOPENTENYL ACRYLATE □ 2-PROPENOIC ACID, 3a,4,7,7a-TETRAHYDRO-4,7-METHANO-1H-INDENYL ESTER

TOXICITY DATA WITH REFERENCE
skn-rbt 500 mg open MOD UCDS** 3/28/72
orl-rat LD50:11 g/kg UCDS** 3/28/72
skn-rbt LD50:45,200 mg/kg TXAPA9 28,313,74

CONSENSUS REPORTS: EPA Extremely Hazardous Substances List. Reported in EPA TSCA Inventory.

SAFETY PROFILE: Mildly toxic by ingestion and skin contact. A skin irritant. See also ESTERS. When heated to decomposition it emits acrid smoke and irritating fumes.

DGW450 CAS:75662-22-5 HR: 1
DICYCLOPENTENYLOXYETHYL METHACRYLATE
mf: $C_{16}H_{22}O_3$ mw: 262.38

SYNS: 2-PROPENOIC ACID, 2-METHYL-, 2-((3a,4,5,6,7,7a-HEXAHYDRO-4,7-METHANO-1H-INDEN-5-YL)OXY) ETHYL ESTER ☐ QM 657

TOXICITY DATA WITH REFERENCE
skn-rbt 500 mg/24H MLD JTEHD6 16,39,85

SAFETY PROFILE: A skin irritant. When heated to decomposition it emits acrid smoke and irritating fumes.

DGW600 CAS:2001-81-2 HR: 3
(2-(DICYCLOPENTYLACETOXY)ETHYL)TRIETHYLAMMONIUM BROMIDE
mf: $C_{20}H_{38}NO_2$•Br mw: 404.50

PROP: A solid. Mp: 185–186°.

SYNS: AETHOBROMID DES α,α-DICYCLOPENTYLESSIGSAEURE-β'-DIAETHYLAMINO AETHYLESTER (GERMAN) ☐ DICYCLOPENTYLACETIC ACID-β-DIETHYLAMINOETHYL ESTER ETHOBROMIDE ☐ 2-((DICYCLOPENTYLACETYL)OXY)-N,N,N-TRIETHYL ETHANAMINIUM BROMIDE ☐ DIETHYLAMINOETHYL-α,α-DICYCLOPENTYLACETATE ETHOBROMIDE ☐ DIPENINBROMID (GERMAN) ☐ HL 267 ☐ Sa 267 ☐ TRIETHYL(2-HYDROXYETHYL)-AMMONIUM BROMIDE DICYCLOPENTYLACETATE ☐ UNOSPASTON

TOXICITY DATA WITH REFERENCE
orl-rat LD50:780 mg/kg ARZNAD 15,878,65
ipr-rat LD50:84,400 μg/kg OYYAA2 2,70,60
scu-rat LD50:299 mg/kg OYYAA2 2,70,68
ivn-rat LD50:6600 μg/kg ARZNAD 15,878,65
orl-mus LD50:570 mg/kg NIIRDN 6,350,82
ipr-mus LD50:88 mg/kg ARZNAD 10,911,60
scu-mus LD50:80 mg/kg NIIRDN 6,350,82
ivn-mus LD50:6200 μg/kg NIIRDN 6,350,82

SAFETY PROFILE: Poison by subcutaneous, intravenous, and intraperitoneal routes. Moderately toxic by ingestion. See also BROMIDES. An antispasmodic agent. When heated to decomposition it emits very toxic fumes of Br^- and NO_x.

DGW875 CAS:16102-24-2 HR: 3
DICYCLOPROPYLDIAZOMETHANE
mf: $C_7H_{10}N_2$ mw: 122.17

$(C_3H_5)_2CN_2$

SAFETY PROFILE: Decomposes violently above −15°C. When heated to decomposition it emits toxic fumes of NO_x.

DGX000 CAS:5232-99-5 HR: 3
3,3-DICYCLOPROPYL-2-(ETHOXYCARBONYL)ACRYLONITRILE
mf: $C_{18}H_{15}NO_2$ mw: 277.34

PROP: A solid. Mp: 95–97°.

SYNS: α-CARBETHOXY-β,β-BISCYCLOPROPYL ACRYLONITRILE ☐ 2-CYANO-3,3-DIPHENYLACRYLIC ACID, ETHYL ESTER ☐ 2-CYANO-3,3-DIPHENYL-2-PROPENOIC ACID, ETHYL ESTER ☐ USAF A -15972 ☐ UV ABSORBER-2

TOXICITY DATA WITH REFERENCE
ipr-mus LD50:100 mg/kg NTIS** AD277-689

CONSENSUS REPORTS: Cyanide and its compounds are on the Community Right-To-Know List. Reported in EPA TSCA Inventory.

SAFETY PROFILE: Poison by intraperitoneal route. See also ESTERS and NITRILES. When heated to decomposition it emits toxic fumes of NO_x.

DGX200 CAS:7173-51-5 HR: 3
DIDECYL DIMETHYL AMMONIUM CHLORIDE
mf: $C_{22}H_{48}N$•Cl mw: 362.16

SYNS: ALIQUAT 203 ☐ BARDAC 22 ☐ BIO-DAC 50-22 ☐ BTC 1010 ☐ N-DECYL-N,N-DIMETHYL-1-DECANAMINIUM CHLORIDE (CI) ☐ DIMETHYLDIDECYLAMMONIUM CHLORIDE ☐ QUATERNIUM-12

TOXICITY DATA WITH REFERENCE
skn-rbt 500 mg SEV NTIS** AD867-663
orl-rat LD50:84 mg/kg NTIS** AD867-663
ipr-rat LD50:45 mg/kg NTIS** AD867-663
orl-mus LD50:268 mg/kg NTIS** AD867-663
ipr-mus LD50:11 mg/kg NTIS** AD867-663
ipr-gpg LDLo:7 mg/kg NTIS** AD867-663

CONSENSUS REPORTS: Reported in EPA TSCA Inventory.

SAFETY PROFILE: Poison by ingestion and intraperitoneal routes. A severe skin irritant. A fungicide. When heated to decomposition it emits very toxic fumes of NO_x, NH_3, and Cl^-.

DGX600 CAS:84-77-5 HR: 2
DIDECYL PHTHALATE
mf: $C_{28}H_{46}O_4$ mw: 446.74

PROP: A clear liquid. Bp: 252° @ 4 mm, fp: −53°, flash p: 450°F (COC), d: 0.964–0.968 @ 20°/20°.

SYN: DI-N-DECYL PHTHALATE

TOXICITY DATA WITH REFERENCE
skn-rbt 10 mg/24H open MLD AIHAAP 23,95,62
ipr-mus LDLo:2233 mg/kg JPMSAE 56,1446,67
skn-rbt LD50:17 g/kg AIHAAP 23,95,62

CONSENSUS REPORTS: Reported in EPA TSCA Inventory.

SAFETY PROFILE: Moderately toxic by intraperitoneal route. Mildly toxic by skin contact. See also ESTERS and PHTHALIC ACID. A skin irritant. Combustible when exposed to heat or flame; can react with oxidizing materials. To fight fire, use foam, CO_2, dry chemical.

DGY000 **CAS:19763-77-0** **HR: 3**
7,8-DIDEHYDRO-4,5-α-EPOXY-14-HYDROXY-3-METHOXY-17-METHYLMORPHINAN-6-ONE-N-OXIDE
mf: $C_{18}H_{21}NO_5$ mw: 331.40

SYN: DIHYDROOXYCODEINON-N-OXYD (GERMAN)

TOXICITY DATA WITH REFERENCE
ivn-mus LD50:800 mg/kg ARZNAD 7,594,57
ivn-rbt LD50:75 mg/kg ARZNAD 7,594,57

SAFETY PROFILE: Poison by intravenous route. When heated to decomposition it emits toxic fumes of NO_x.

DHA300 **CAS:77327-05-0** **HR: 3**
DIDEMNIN B
mf: $C_{57}H_{89}N_7O_{15}$ mw: 1112.53

SYN: NSC-325319

TOXICITY DATA WITH REFERENCE
dni-mus:leu 2500 ng/L CALEDQ 23,279,84
dni-mus:oth 400 µg/L CNREA8 44,1796,84
oms-mus:oth 400 µg/L CNREA8 44,1796,84
oms-mus:leu 2500 ng/L CALEDQ 23,279,84
ivn-rat LD50:860 µg/kg NTIS** PB84-192251
ivn-mus LD50:1530 µg/kg NTIS** PB84-192251
ivn-dog LDLo:418 µg/kg NTIS** PB84-192251

SAFETY PROFILE: A deadly poison by intravenous route. Mutation data reported. When heated to decomposition it emits toxic fumes of NO_x.

DHA325 **CAS:4097-22-7** **HR: 2**
2′,3′-DIDEOXYADENOSINE
mf: $C_{10}H_{13}N_5O_2$ mw: 235.28

PROP: A solid. Mp: 184–186°.

TOXICITY DATA WITH REFERENCE
dni-esc 37 mg/L AMACCQ 19,424,81
dni-mus 144 mg/L AMACCQ 19,424,81
dni-omi 19 mg/L AMACCQ 19,424,81
orl-mus LD50:5 g/kg AMACCQ 19,424,81
scu-mus LD50:1320 mg/kg AMACCQ 19,424,81

SAFETY PROFILE: Moderately toxic by subcutaneous route. Mildly toxic by ingestion. Mutation data reported. When heated to decomposition it emits toxic fumes of NO_x.

DHA400 **CAS:64070-13-9** **HR: 3**
3′,4′-DIDEOXYKANAMYCIN B SULFATE
mf: $C_{18}H_{37}N_5O_8 \cdot O_4S$ mw: 547.66

SYN: DKB SULFATE

TOXICITY DATA WITH REFERENCE
ipr-rat LD50:799 mg/kg JJANAX 26,221,73
scu-rat LD50:1376 mg/kg JJANAX 26,221,73
ivn-rat LD50:140 mg/kg JJANAX 26,221,73
ims-rat LD50:560 mg/kg JJANAX 26,221,73
ipr-mus LD50:431 mg/kg JJANAX 26,221,73
scu-mus LD50:521 mg/kg JJANAX 26,221,73
ivn-mus LD50:63 mg/kg JJANAX 26,221,73
ims-mus LD50:396 mg/kg JJANAX 26,221,73

SAFETY PROFILE: Poison by intramuscular and intravenous routes. Moderately toxic by intraperitoneal and subcutaneous routes. When heated to decomposition it emits very toxic fumes of NO_x and SO_x.

DHA425 **CAS:53866-33-4** **HR: 2**
2,4-DIDEUTERIOESTRADIOL
mf: $C_{18}H_{22}D_2O_2$ mw: 274.42

SYN: ESTRA-1,3,5(10)-TRIENE-2,4-D2-3,17-DIOL, (17-β)-

TOXICITY DATA WITH REFERENCE
imp-ham TDLo:360 mg/kg/15W-I:CAR MOPMA3 23,278,83

SAFETY PROFILE: Questionable carcinogen with experimental carcinogenic data. When heated to decomposition it emits acrid smoke and irritating fumes.

DHA450 **CAS:14621-84-2** **HR: 3**
DIDEUTERODIAZOMETHANE
mf: CD_2N_2 mw: 44.06

SAFETY PROFILE: An explosive. When heated to decomposition it emits toxic fumes of NO_x. See also DIAZOMETHANE.

DHB309 **HR: 1**
DIEFFENBACHIA (VARIOUS SPECIES)

PROP: Tall, unbranched ornamentals with long, ivory-spotted leaves. They are extremely popular as indoor plants, and are grown outdoors in southern Florida and Hawaii.

SYNS: CAMILICHIGUI (MEXICO) ☐ CANNE-A-GRATTER (HAITI) ☐ CANNE-MADERE (HAITI) ☐ DICHA (CUBA) ☐ DIEFFENBACHIA MACUALTA ☐ DIEFFENBACHIA SEQUINE ☐ DUMBCANE ☐ DUMB PLANT ☐ MOTHER-IN-LAW'S TONGUE PLANT ☐ PELA PUERCO (DOMINICAN REPUBLIC) ☐ RABANO (PUERTO RICO) ☐ TUFT ROOT

SAFETY PROFILE: The leaf contains calcium oxalate crystals and chewing them results in burning pain in the lips, mouth, and throat, possibly followed by inflammation and blistering. The cut or crushed leaves may also cause contact dermatitis or conjunctivitis. Systemic effects are usually not seen because of the insolubility of calcium oxalate. See also OXALATES.

DHB400 **CAS:60-57-1** **HR: 3**
DIELDRIN
DOT: UN 2761
mf: $C_{12}H_8Cl_6O$ mw: 380.90

PROP: White crystals; odorless. Mp: 176–177°, vap d: 13.2. Insol in water; sol in common organic solvents.

SYNS: ALVIT ☐ COMPOUND 497 ☐ DIELDREX ☐ DIELDRINE (FRENCH) ☐ DIELDRITE ☐ ENT 16,225 ☐ HEOD ☐ HEXACHLOROEPOXYOCTAHYDRO-endo,exo-DIMETHANONAPHTHALENE ☐ 3,4,5,6,9,9-HEXACHLORO-1a,2,2a,3,6,6a,7,7a-OCTAHYDRO-2,7:3,6-DIMETHANONAPHTH(2,3-b)OXIRENE ☐ ILLOXOL ☐ INSECTICIDE No. 497 ☐ NCI-C00124 ☐ OCTALOX ☐ PANORAM D-31 ☐ QUINTOX ☐ RCRA WASTE NUMBER P037

TOXICITY DATA WITH REFERENCE
mmo-sat 1 mg/L JOHEA8 68,184,77
mma-hmn:fbr 1 μmol/L MUREAV 42,161,77
dns-hmn:fbr 1 μmol/L MUREAV 42,161,77
dni-hmn:hla 400 μmol/L MUREAV 92,427,82
otr-rat-orl 5 mg/kg CNREA8 40,1157,80
orl-rat TDLo:14 μg/kg (multi) :REP PCBPBS 13,20,80
orl-dog TDLo:219 mg/kg (female 44W pre):REP JAVMA4 123,28,53
orl-ham TDLo:30 mg/kg (female 8D post):TER TJADAB 9,11,74
orl-mus TDLo:2250 μg/kg (female 6-14D post):TER TJADAB 16,57,77
orl-ham TDLo:30 mg/kg (female 8D post):REP TJADAB 9,11,74
orl-mus TDLo:12,500 μg/kg (male 1D pre):REP FCTXAV 13,317,75
orl-mus TDLo:6250 μg/kg (male 5D pre):REP ENVRAL 9,26,75
orl-rat TDLo:336 μg/kg (56D male):REP 32OAAP -,189,73
orl-uns TDLo:21 mg/kg (male 30W pre):REP DABBBA 35,1129,74
orl-mus TDLo:30,600 μg/kg (6-14D preg):TER THERAP 25,907,70
orl-mus TDLo:15 mg/kg (female 9D post):TER TJADAB 9,11,74
orl-mus TDLo:4500 μg/kg (female 6-14D post):TER TJADAB 16,57,77
orl-rat TDLo:200 mg/kg/2Y-C:ETA FCTXAV 2,551,64
orl-mus TDLo:546 mg/kg/65W-C:CAR ARTODN Suppl.2,197,79
orl-mus TD:11 g/kg/3Y-C:NEO FCTXAV 11,415,73
orl-mus TD:610 mg/kg/73W-C:NEO FCTXAV 11,433,73
orl-mus TD:714 mg/kg/85W-C:CAR LAINAW 44,392,81
orl-mus TD:8 mg/kg/2Y-C:ETA CRNGDP 3,941,82
orl-mus TD:4550 mg/kg/65W-C:CAR CNREA8 41,3615,81
orl-man LDLo:65 mg/kg 34ZIAG -,215,69
unk-hmn LDLo:28 mg/kg ATXKA8 22,115,66
orl-rat LD50:38,300 μg/kg JAFCAU 3,402,55
ihl-rat LC50:13 mg/m³/4H 85GMAT -,73,82
skn-rat LD50:56 mg/kg RPZHAW 18,161,67
ipr-rat LD50:35 mg/kg CBPCBB 85,437,86
ivn-rat LD50:9 mg/kg BJIMAG 21,269,64
orl-mus LD50:38 mg/kg SPEADM 78-1,13,78
ipr-mus LDLo:26 mg/kg TXAPA9 23,288,72
ivn-mus LD50:10.5 mg/kg TXAPA9 23,408,72
orl-dog LD50:65 mg/kg GUCHAZ 6,198,73
orl-mky LD50:3 mg/kg 32ZDAL -,79,70
ihl-cat LC50:80 mg/m³/4H GTPZAB 8(4),30,64

CONSENSUS REPORTS: IARC Cancer Review: Group 3 IMEMDT 7,196,87; Human Inadequate Evidence IMEMDT 5,125,74; Animal Sufficient Evidence IMEMDT 5,125,74. NCI Carcinogenesis Bioassay (feed); Clear Evidence: mouse NCITR* NCI-CG-TR-21,78; No Evidence: rat NCITR* NCI-CG-TR-22,78; Inadequate Studies: rat NCITR* NCI-CG-TR-21,78.

OSHA PEL: TWA 0.25 mg/m³ (skin)
ACGIH TLV: TWA 0.25 mg/m³ (skin)
DFG MAK: 0.25 mg/m³
NIOSH REL: (Dieldrin) Lowest reliable detectable level
DOT CLASSIFICATION: 6.1; *Label:* Poison

SAFETY PROFILE: A human poison by ingestion and possibly other routes. Poison experimentally by inhalation, ingestion, skin contact, intravenous, and intraperitoneal routes. Experimental teratogenic and reproductive data. Absorbed readily through the skin and by other routes. It is a central nervous system stimulant. Questionable carcinogen with experimental carcinogenic, neoplastigenic, and tumorigenic data. Human mutation data reported. An insecticide. Dieldrin is considerably more toxic than DDT by ingestion and skin contact. Dieldrin or its derivatives may accumulate in the body from chronic low dosages. When heated to decomposition it emits toxic fumes of Cl^-. See also ALDRIN.

D

DHB550 CAS:13029-44-2 HR: 2
(E,E)-DIENESTROL
mf: $C_{18}H_{18}O_2$ mw: 266.36

PROP: A solid. Mp: 227–228°.

SYNS: α-DIENESTROLPHENOL, 4,4'-(DIETHYLIDENEETHYLENE)DI-, trans-, (E,E)- □ PHENOL, 4,4'-(1,2-DIETHYLIDENE-1,2-ETHANEDIYL) BIS-, (E,E)-(9CI)

TOXICITY DATA WITH REFERENCE
mnt-ham:emb 10 mg/L TOLED5 31(Suppl),204,86
imp-ham TDLo:640 mg/kg/38W-I:ETA CNREA8 43,5200,83

SAFETY PROFILE: Questionable carcinogen with experimental tumorigenic data. Mutation data reported. When heated to decomposition it emits acrid smoke and irritating fumes.

DHB600 CAS:298-18-0 HR: 3
dl-DIEPOXYBUTANE
mf: $C_4H_6O_2$ mw: 86.10

PROP: Colorless liquid. Bp: 144.5°, mp: 4°, d: 1.112 @ 18°/4°.

SYNS: dl-BUTADIENE DIOXIDE □ 1,2:3,4-DIANHYDRO-dl-THREITOL □ (±)-1,2:3,4-DIEPOXYBUTANE □ dl-1,2:3,4-DIEPOXYBUTANE

TOXICITY DATA WITH REFERENCE
mmo-ssp 20 mmol/L ADWMAX -,193,62
dns-ham:lvr 1 μmol/L ENMUDM 6,1,84
scu-rat TDLo:335 mg/kg/67W-I:ETA JNCIAM 37,825,66
skn-mus TDLo:132 mg/kg/66W-I:CAR CNREA8 43,159,83
scu-mus TDLo:13,200 μg/kg/68W-I:CAR CNREA8 43,159,83
skn-mus TD:13 g/kg/11W-I:ETA JNCIAM 31,41,63
skn-mus TD:24 g/kg/68W-I:CAR 14JTAF -,275,74
scu-mus TD:260 mg/kg/65W-I:NEO JNCIAM 37,825,66
orl-rat LD50:210 mg/kg IHFCAY 6,1,67
ihl-rat LC50:56 ppm/4H IHFCAY 6,1,67
skn-mus LDLo:400 mg/kg JNCIAM 31,41,63
skn-rbt LD50:800 mg/kg IHFCAY 6,1,67

CONSENSUS REPORTS: IARC Cancer Review: Group 2B IMEMDT 7,56,87; Animal Sufficient Evidence IMEMDT 11,115,76. EPA Genetic Toxicology Program.

SAFETY PROFILE: Confirmed carcinogen with experimental carcinogenic, neoplastigenic, and tumorigenic data. Poison by ingestion, inhalation, and skin contact.

Mutation data reported. When heated to decomposition it emits acrid smoke and irritating fumes.

DHB800 **CAS:564-00-1** **HR: 3**
meso-1,2,3,4-DIEPOXYBUTANE
mf: $C_4H_6O_2$ mw: 86.10

PROP: A liquid. Mp: −19°, bp: 140°.

SYNS: (R*,S*)-2,2′-BIOXIRANE □ 1,2:3,4-DIANHYDROERYTHRITOL □ meso-DIEPOXYBUTANE □ (R*,S*)-DIEPOXYBUTANE □ ERYTHRITOL ANHYDRIDE

TOXICITY DATA WITH REFERENCE
mmo-sat 100 μg/plate ENMUDM 6(Suppl 2),1,84
mmo-ssp 51 mmol/L ADWMAX -,193,62
skn-mus TDLo: 26 g/kg/22W-I: NEO 14JTAF -,275,64
skn-mus TD: 26 g/kg/22W-I: ETA JNCIAM 31,41,63
skn-mus LDLo: 400 mg/kg JNCIAM 31,41,63

CONSENSUS REPORTS: IARC Cancer Review: Group 2B IMEMDT 7,56,87; Animal Sufficient Evidence IMEMDT 11,115,76.

SAFETY PROFILE: Confirmed carcinogen with experimental carcinogenic, neoplastigenic, and tumorigenic data. Poison by skin contact. Mutation data reported. When heated to decomposition it emits acrid smoke and irritating fumes.

DHC000 **CAS:24854-67-9** **HR: 2**
1,2,9,10-DIEPOXYDECANE
mf: $C_{10}H_{18}O_2$ mw: 170.28

TOXICITY DATA WITH REFERENCE
unk-mus TDLo: 510 mg/kg: ETA RARSAM 3,193,63

SAFETY PROFILE: Questionable carcinogen with experimental tumorigenic data. When heated to decomposition it emits acrid smoke and irritating fumes.

DHC200 **CAS:63869-17-0** **HR: 2**
DIEPOXYDIHYDROMYRCENE
mf: $C_{10}H_{18}O_2$ mw: 170.28

SYN: DIEPOXYDIHYDRO-7-METHYL-3-METHYLENE-1,6-OCTADIENE

TOXICITY DATA WITH REFERENCE
unk-mus TDLo: 5100 mg/kg: ETA RARSAM 3,193,63

SAFETY PROFILE: Questionable carcinogen with experimental tumorigenic data. When heated to decomposition it emits acrid smoke and irritating fumes.

DHC309 **CAS:56411-67-7** **HR: 3**
2,3:5,6-DIEPOXY-7,8-DIOXABICYCLO[2.2.2]OCTANE
mf: $C_6H_6O_4$ mw: 142.11

PROP: Crystals.

SAFETY PROFILE: Explodes when heated. When heated to decomposition it emits acrid smoke and fumes. See also PEROXIDES.

DHC600 **CAS:4247-19-2** **HR: 2**
1,2,6,7-DIEPOXYHEPTANE
mf: $C_7H_{12}O_2$ mw: 128.19

TOXICITY DATA WITH REFERENCE
skn-mus TDLo: 2400 mg/kg/20W-I: NEO JNCIAM 35,707,65
skn-mus TD: 7560 mg/kg/63W-I: NEO 14JTAF -,275,64

SAFETY PROFILE: Questionable carcinogen with experimental neoplastigenic data. When heated to decomposition it emits acrid smoke.

DHC800 **CAS:1888-89-7** **HR: 2**
1,2:5,6-DIEPOXYHEXANE
mf: $C_6H_{10}O_2$ mw: 114.16

TOXICITY DATA WITH REFERENCE
mmo-ssp 116 mmol/L ADWMAX -,193,62
skn-mus TDLo: 6960 mg/kg/29W-I: ETA JNCIAM 39,1217,67
scu-mus TDLo: 2068 mg/kg/47W-I: NEO JNCIAM 37,825,66
unk-mus TDLo: 4600 mg/kg: ETA RARSAM 3,193,63
skn-mus TD: 28 g/kg/46W-I: ETA 14JTAF -,275,64

CONSENSUS REPORTS: EPA Genetic Toxicology Program.

SAFETY PROFILE: Questionable carcinogen with experimental neoplastigenic and tumorigenic data. Mutation data reported. When heated to decomposition it emits acrid smoke and irritating fumes.

DHD200 **CAS:6341-85-1** **HR: 2**
1,2,3,4-DIEPOXY-2-METHYLBUTANE
mf: $C_5H_8O_2$ mw: 100.13

SYN: 2-METHYL-2,2′-BIOXIRANE (9CI)

TOXICITY DATA WITH REFERENCE
mmo-sat 7500 μmol/L MUREAV 156,77,85
unk-mus TDLo: 800 mg/kg: ETA RARSAM 3,193,63

SAFETY PROFILE: Questionable carcinogen with experimental tumorigenic data. Mutation data reported. When heated to decomposition it emits acrid smoke and irritating fumes.

DHD400 **CAS:24829-11-6** **HR: 2**
1,2:8,9-DIEPOXYNONANE
mf: $C_9H_{16}O_2$ mw: 156.25

TOXICITY DATA WITH REFERENCE
unr-mus TDLo: 3750 mg/kg: ETA RARSAM 3,193,63

SAFETY PROFILE: Questionable carcinogen with experimental tumorigenic data. When heated to decomposition it emits acrid smoke and irritating fumes.

DHD600 **CAS:3012-69-9** **HR: 2**
9,10:12,13-DIEPOXYOCTADECANOIC ACID
mf: $C_{18}H_{32}O_4$ mw: 312.50

SYN: 9,10:12,13-DIEPOXYSTEARIC ACID

TOXICITY DATA WITH REFERENCE
skn-mus TDLo:3360 mg/kg/28W-I:ETA JNCIAM 31,41,63

SAFETY PROFILE: Questionable carcinogen with experimental tumorigenic data. When heated to decomposition it emits acrid smoke and irritating fumes.

DHD800 CAS:2426-07-5 HR: 3
1,2:7,8-DIEPOXYOCTANE
mf: $C_8H_{14}O_2$ mw: 142.22

PROP: A liquid. Bp: 105–108° @ 14 mm.

SYNS: 2,2′-(1,4-BUTANEDIYL)BISOXIRANE ☐ 1,2-EPOXY-7,8-EPOXYOCTANE ☐ OXIRANE, 2,2′-(1,4-BUTANEDIYL)BIS-(9CI)

TOXICITY DATA WITH REFERENCE
skn-rbt 100 µg/24H open AIHAAP 23,95,62
mmo-klp 500 µmol/L MUREAV 89,269,81
mma-sat 990 µg/plate PNASA6 72,5135,75
mmo-nsc 75 mmol/L CNREA8 32,1890,72
cyt-ham:lng 6 µmol/L JEPTDQ 2(2),587,79
msc-ham:lng 6 µmol/L JEPTDQ 2(2),587,79
skn-mus TDLo:6600 mg/kg/55W-I:ETA JNCIAM 39,1217,67
orl-rat LD50:1070 mg/kg AIHAAP 23,95,62
skn-rbt LD50:320 mg/kg AIHAAP 23,95,62

CONSENSUS REPORTS: EPA Genetic Toxicology Program.

SAFETY PROFILE: Poison by skin contact. Moderately toxic by ingestion. A skin irritant. Questionable carcinogen with experimental tumorigenic data. Mutation data reported. When heated to decomposition it emits acrid and irritating fumes.

DHE000 CAS:4051-27-8 HR: 2
1,2,4,5-DIEPOXYPENTANE
mf: $C_5H_8O_2$ mw: 100.13

SYNS: 2,2′-METHYLENEBIS OXIRANE (9CI) ☐ 1:4-PENTADIENE DIOXIDE

TOXICITY DATA WITH REFERENCE
mmo-nsc 75 mmol/L CNREA8 32,1890,72
cyt-rat-ipr 2500 mg/kg BJPCAL 6,235,51
skn-mus TDLo:70 g/kg/58W-I:CAR JNCIAM 35,707,65
skn-mus TD:76 g/kg/63W-I:NEO 14JTAF -,275,64

SAFETY PROFILE: Questionable carcinogen with experimental carcinogenic and neoplastigenic data. Mutation data reported. When heated to decomposition it emits acrid smoke and irritating fumes.

DHE100 HR: 3
DIEPOXYPIPERAZINE
mf: $C_{10}H_{18}N_2O_2$ mw: 198.30

SYNS: N,N′-BIS(2,3-EPOXYPROPYL)PIPERAZINE ☐ 1,4-BIS-(2,3-ETHOXYPROPYL)PIPERAZINE ☐ DIGLYCIDYL PIPERAZINE ☐ EPOXYPIPERAZINE ☐ NSC 74437

TOXICITY DATA WITH REFERENCE
orl-mus LD50:91,380 µg/kg NCISP* JAN86
ipr-mus LD50:40,770 µg/kg NCISP* JAN86
scu-mus LD50:199 mg/kg NCISP* JAN86

SAFETY PROFILE: Poison by ingestion, subcutaneous, and intraperitoneal routes. When heated to decomposition it emits toxic fumes of NO_x.

DHE485 HR: 3
DIESEL EXHAUST

TOXICITY DATA WITH REFERENCE
ihl-rat TCLo:4900 µg/m³/8H/2Y-C:CAR DTESD7 13,349,86
ihl-rat TC:7 mg/m³/7H/2Y-I:CAR FAATDF 9,208,87

ACGIH TLV: (Proposed: TWA 0.15 mg/m³; Suspected Human Carcinogen)

SAFETY PROFILE: Suspected carcinogen with experimental carcinogenic data. When heated to decomposition it emits acrid smoke and irritating fumes.

DHE700 HR: 2
DIESEL EXHAUST PARTICLES

PROP: Particulate samples collected from the exhaust of a 1979 2.3 L diesel powered automobile running on No. 2 diesel fuel TXAPA9 56,110,80

TOXICITY DATA WITH REFERENCE
mmo-sat 200 µg/plate TXAPA9 56,110,80
mma-sat 200 µg/plate TXAPA9 56,110,80
msc-hmn:lyms 100 mg/L DTESD7 10,277,82
sce-mus-uns 300 mg/kg DTESD7 10,265,82
ihl-rat TCLo:2200 µg/m³/16H/2Y-I:NEO DTESD7 13,471,86
ihl-rat TC:8300 µg/kg/6H/86W-I:ETA AIHAAP 42,382,81
ihl-rat TC:8300 µg/m³/6H/86W-I:ETA AIHAAP 42,382,81
ihl-rat TC:7 mg/m³/7H/2Y-I:ETA ACGHD2 13,3,85

NIOSH REL: (Diesel Exhaust) TWA reduce to lowest feasible level

SAFETY PROFILE: Questionable carcinogen with experimental neoplastigenic and tumorigenic data. Human mutation data reported. When heated to decomposition it emits acrid smoke and irritating fumes.

DHE800 CAS:68476-30-2 HR: 3
DIESEL FUEL MARINE
DOT: NA 1993

PROP: Brown, sltly viscous liquid. Flash p: 100°F; d: <1; autoign temp: 494°F.

SYNS: API No.2 FUEL OIL ☐ FUEL OIL NO.2. (DOT) ☐ GAS OIL ☐ GAS OIL (DOT) ☐ HOME HEATING OIL No.2 ☐ #2 HOME HEATING OILS ☐ NUMBER 2 BURNER FUEL ☐ NUMBER 2 FUEL OIL

TOXICITY DATA WITH REFERENCE
skn-rbt 500 mg/24H MOD 52MLA2 1,1,83
eye-rbt 100 mg/30S MLD 52MLA2 1,1,83
skn-mus TDLo:243 g/kg/97W-I:CAR FAATDF 9,297,87
orl-rat LD50:14,500 mg/kg 52MLA2 1,1,83

CONSENSUS REPORTS: Reported in EPA TSCA Inventory. IARC Cancer Review: Group 3 IMEMDT 45,239,89; IARC Cancer Review: Animal Limited Evidence IMEMDT 45,239,89

DOT CLASSIFICATION: 3; *Label:* Flammable Liquid

SAFETY PROFILE: Mildly toxic by ingestion. A moderate skin and eye irritant. Questionable carcinogen with experimental carcinogenic data. A flammable liquid and dangerous fire hazard when exposed to heat, flame, or oxidizers. To fight fire, use foam, CO_2, dry chemical. When heated to decomposition it emits acrid smoke and irritating fumes. See also MINERAL OIL and KEROSENE.

DHE900 CAS:68334-30-5 HR: 2
DIESEL FUELS
DOT: NA 1993

SYNS: AUTOMOTIVE DIESEL OIL □ DIESEL FUEL (DOT) □ DIESEL OIL (PETROLEUM) □ DIESEL OILS □ DIESEL TEST FUEL □ FUELS, DIESEL □ OLEJ NAPEDOWY III

TOXICITY DATA WITH REFERENCE
skn-rbt 500 mg MOD NTIS** AD-A172-198
orl-rat LD50:9 g/kg 52MLA2 1,1,83

CONSENSUS REPORTS: IARC Cancer Review: Group 3 IMEMDT 45,219,89; Human Inadequate Evidence IMEMDT 45,219,89. Reported in EPA TSCA Inventory.

DOT CLASSIFICATION: 3; *Label:* None

SAFETY PROFILE: Low toxicity by ingestion. A skin irritant. Questionable carcinogen. When heated to decomposition it emits acrid smoke and irritating vapors.

DHF000 CAS:111-42-2 HR: 3
DIETHANOLAMINE
mf: $C_4H_{11}NO_2$ mw: 105.16

PROP: A faintly colored, viscous liquid or deliquescent prisms. Mp: 28°, bp: 270° (decomp), flash p: 305°F (OC), d: 1.0919 @ 30°/20°, autoign temp: 1224°F, vap press: 5 mm @ 138°, vap d: 3.65. Very sol in water.

SYNS: BIS(2-HYDROXYETHYL)AMINE □ DEA □ DIAETHANOLAMIN (GERMAN) □ DIETHANOLAMIN (CZECH) □ DIETHYLOLAMINE □ 2,2′-DIHYDROXYDIETHYLAMINE □ DI(2-HYDROXYETHYL) AMINE □ DIOLAMINE □ 2,2′-IMINOBISETHANOL □ 2,2′-IMINODIETHANOL □ NCI-C55174

TOXICITY DATA WITH REFERENCE
skn-rbt 50 mg open MLD UCDS** 12/29/71
skn-rbt 500 mg/24H MLD 28ZPAK -,109,72
eye-rbt 5500 mg SEV AJOPAA 29,1363,46
eye-rbt 750 µg/24H SEV 28ZPAK -,109,72
orl-rat TDLo:18,382 mg/kg (male 14D pre):REP NTPTR* NIH-92-3343
orl-rat LD50:710 mg/kg TXAPA9 17,498,70
ipr-rat LD50:120 mg/kg EVSSAV 2,289,68
scu-rat LD50:2200 mg/kg EVSSAV 2,289,68
ivn-rat LD50:778 mg/kg EVSSAV 2,289,68
ims-rat LD50:1500 mg/kg EVSSAV 2,289,68
orl-mus LD50:3300 mg/kg GISAAA 29(11),25,64
ipr-mus LD50:2300 mg/kg TXAPA9 22,175,72
scu-mus LD50:3553 mg/kg ARZNAD 4,649,54
skn-rbt LD50:12,200 mg/kg NPIRI* 1,24,74
orl-gpg LD50:2 g/kg DTLVS* 4,140,80

CONSENSUS REPORTS: Community Right-To-Know List. Reported in EPA TSCA Inventory.

OSHA PEL: TWA 3 ppm
ACGIH TLV: TWA 0.46 ppm (skin)

SAFETY PROFILE: Poison by intraperitoneal route. Moderately toxic by ingestion and subcutaneous routes. Mildly toxic by skin contact. A severe eye and mild skin irritant. Experimental reproductive effects. Combustible when exposed to heat or flame; can react with oxidizing materials. To fight fire, use alcohol foam, water, CO_2, dry chemical. When heated to decomposition it emits toxic fumes such as NO_x. See also AMINES.

For occupational chemical analysis use NIOSH: Aminoethanol Compounds II 3509.

DHF200 CAS:5716-15-4 HR: 2
DIETHANOLAMMONIUM MALEIC HYDRAZIDE
mf: $C_4H_{11}NO_2 \cdot C_4H_4N_2O_2$ mw: 217.26

SYNS: 6-HYDROXY-3-(2H)-PYRIDAZINONE DIETHANOLAMINE □ 2,2′-IMINODI-ETHANOL with 1,2-DIHYDRO-3,6-PYRIDAZINEDIONE (1:1) □ MALEIC HYDRAZIDE DIETHANOLAMINE SALT □ MAZIDE 30 □ NCI-C54660 □ SLO-GRO

TOXICITY DATA WITH REFERENCE
mma-sat 50 µL/plate MUREAV 66,247,79
scu-rat TDLo:1300 mg/kg/65W-I:ETA TXCYAC 1,301,73
scu-rat TD:1350 mg/kg/60W-I:ETA NATUAS 180,62,57
orl-rat LD50:3900 mg/kg FMCHA2-,C188,91

CONSENSUS REPORTS: Reported in EPA TSCA Inventory.

SAFETY PROFILE: Moderately toxic by ingestion. Questionable carcinogen with experimental tumorigenic data. Mutation data reported. When heated to decomposition it emits toxic fumes of NO_x and NH_3.

DHF400 CAS:91-99-6 HR: 2
DIETHANOL-m-TOLUIDINE
mf: $C_{11}H_{17}NO_2$ mw: 195.29

SYNS: N,N-BIS(β-HYDROXYETHYL)-3-METHYLANILINE □ N,N-BIS (2-HYDROXYETHYL)-3-METHYLANILINE □ N,N-BIS(2-HYDROXYETHYL)-m-TOLUIDINE □ N,N-DIHYDROXYETHYL-m-TOLUIDINE □ EMERY 5709 □ 2,2′-((3-METHYLPHENYL)IMINO)BISETHANOL □ m-TOLYLDIETHANOLAMINE □ 2,2′-(m-TOLYLIMINO)DIETHANOL

TOXICITY DATA WITH REFERENCE
orl-rat LD50:3100 mg/kg LONZA# 02JUN80

CONSENSUS REPORTS: Reported in EPA TSCA Inventory.

SAFETY PROFILE: Moderately toxic by ingestion. When heated to decomposition it emits toxic fumes of NO_x.

DHF600 CAS:341-70-8 HR: 3
DIETHAZINE HYDROCHLORIDE
mf: $C_{18}H_{22}N_2S \cdot ClH$ mw: 334.94

PROP: A solid. Mp: 184–186°. Insol in Et_2O. Sol in H_2O.

SYNS: ANTIPAR □ APARKAZIN □ CASANTIN □ DEPARKIN □ 10-

(2-DIETHYLAMINO)ETHYLPHENOTHIAZINE HYDROCHLORIDE □ DIPARCOL □ LABITON □ THIANTAN □ THIONTAN

TOXICITY DATA WITH REFERENCE
dnd-esc 20 µmol/L MUREAV 89,95,81
scu-rat LD50:200 mg/kg PSEBAA 78,708,51
orl-mus LD50:450 mg/kg 27ZQAG-,20,72
ipr-mus LD50:225 mg/kg 27ZQAG-,20,72
scu-mus LD50:450 mg/kg THERAP 2,115,47
ivn-mus LD50:5 mg/kg THERAP 2,115,47
scu-rbt LD50:150 mg/kg 27ZQAG -,20,72
ivn-rbt LD50:2500 µg/kg THERAP 2,115,47
scu-gpg LD50:450 mg/kg AIPTAK 137,375,62

SAFETY PROFILE: Poison by intraperitoneal, subcutaneous, and intravenous routes. Moderately toxic by ingestion. Mutation data reported. When heated to decomposition it emits very toxic fumes of NO_x, SO_x, and HCl.

DHF800 CAS:6485-91-2 HR: 2
DIETHOXYCHLOROSILANE
mf: $C_4H_{11}ClO_2Si$ mw: 154.69

PROP: Liquid. Vap d: 5.33.

SYN: CHLORODIETHOXYSILANE

TOXICITY DATA WITH REFERENCE
skn-rbt 10 mg/24H open MLD JIHTAB 31,60,49
eye-rbt 250 µg open SEV JIHTAB 31,60,49
skn-rbt 500 mg/24H MLD 85JCAE-,1222,86
orl-rat LD50:6300 mg/kg JIHTAB 31,60,49
ihl-rat LCLo:1000 ppm/4H JIHTAB 31,343,49

SAFETY PROFILE: A severe eye and mild skin irritant. Mildly toxic by inhalation and ingestion. Reacts with water or steam to produce heat and toxic and corrosive fumes of HCl. When heated to decomposition it emits toxic fumes of Cl^-. See also CHLOROSILANES.

DHG000 CAS:78-62-6 HR: 3
DIETHOXYDIMETHYLSILANE
DOT: UN 2380
mf: $C_6H_{16}O_2Si$ mw: 148.31

$(CH_3CH_2O)_2Si(CH_3)_2$

PROP: A liquid. Bp: 114°, d: 0.86, vap press: 10 mm @ 13.3°, vap d: 5.1, flash p.: <73.4°F.

SYNS: DIMETHYL-DIETHOXYSILAN (CZECH) □ DIMETHYLDIETHOXYSILANE (DOT)

TOXICITY DATA WITH REFERENCE
skn-rbt 500 mg/24H MLD 85JCAE-,1226,86
eye-rbt 500 mg/24H MLD 85JCAE-,1226,86
eye-rbt 83 mg JIDHAN 30,332,48
orl-rat LD50:9280 mg/kg 85JCAE-,1226,86
ihl-rat LC50:20 g/m³/8H 85JCAE-,1226,86

CONSENSUS REPORTS: Reported in EPA TSCA Inventory.

DOT CLASSIFICATION: 3; *Label:* Flammable Liquid

SAFETY PROFILE: Mildly toxic by inhalation and ingestion. A skin and eye irritant. A dangerous fire hazard when exposed to heat, flame, or oxidizers. When heated to decomposition it emits acrid smoke and irritating fumes.

DHH200 CAS:21548-32-3 HR: 3
(DIETHOXYPHOSPHINYLIMINO)-1,3-DITHIETANE
mf: $C_6H_{12}NO_3PS_2$ mw: 241.28

PROP: Yellow liquid with mercaptan odor. D: 1.3 @ 25°. Sol in Me_2CO, $CHCl_3$, MeOH, and toluene; mod sol in H_2O.

SYNS: AC 64475 □ ACCONEM □ CL 64475 □ DIETHOXYPHOSPHINYLIMINO-2-DITHIETANNE-1,3 (FRENCH) □ 1,3-DITHIETAN-2-YLIDENE PHOSPHORAMIDIC ACID DIETHYL ESTER □ FOSTHIETAN □ GEOFOS □ NEM-A-TAK

TOXICITY DATA WITH REFERENCE
orl-rat LD50:4700 µg/kg FMCHA2 -,C167,83
orl-mus LD50:18 mg/kg MEIEDD 10,607,83
skn-rbt LD50:27,400 µg/kg FMCHA2 -,C167,83

CONSENSUS REPORTS: Reported in EPA TSCA Inventory. EPA Extremely Hazardous Substances List.

SAFETY PROFILE: Poison by ingestion and skin contact. When heated to decomposition it emits very toxic fumes of NO_x, PO_x, and SO_x.

DHH400 CAS:950-10-7 HR: 3
2-(DIETHOXYPHOSPHINYLIMINO)-4-METHYL-1,3-DITHIOLANE
mf: $C_8H_{16}NO_3PS_2$ mw: 269.34

SYNS: AC 47470 □ AMERICAN CYANAMID CL-47470 □ CL-47,470 □ CYCLIC PROPYLENE (DIETHOXYPHOSPHINYL)DITHIOIMIDOCARBONATE □ CYTROLANE □ p,p-DIETHYL CYCLIC PROPYLENE ESTER of PHOSPHONODITHIOIMIDOCARBONIC ACID □ DIETHYL (4-METHYL-1,3-DITHIOLAN-2-YLIDENE)PHOSPHOROAMIDATE □ EI-47470 □ ENT 25,991 □ MEPHOSFOLAN □ (4-METHYL-1,3-DITHIOLAN-2-YLIDENE)PHOSPHORAMIDIC ACID, DIETHYL ESTER

TOXICITY DATA WITH REFERENCE
orl-rat LD50:9 mg/kg BESAAT 15,122,69
orl-mus LD50:11 mg/kg BESAAT 15,122,69
skn-rbt LD50:28,700 ug/kg FMCHA2 -,C68,83
orl-ckn LD50:2800 ug/kg EXPEAM 30,63,74

CONSENSUS REPORTS: EPA Extremely Hazardous Substances List. Reported in EPA TSCA Inventory. EPA Genetic Toxicology Program.

SAFETY PROFILE: Poison by ingestion and skin contact. When heated to decomposition it emits very toxic fumes of NO_x, PO_x, and SO_x.

DHH600 CAS:74038-45-2 HR: 3
(α-(DIETHOXYPHOSPHINYL)-p-METHOXYBENZYL)BIS (2-CHLOROPROPYL)ANTIMONITE
mf: $C_{18}H_{30}Cl_2O_7PSb$ mw: 582.10

SYN: (α-HYDROXY-p-METHOXYBENZYL)-PHOSPHONIC ACID DIETHYL ESTER, ESTER with BIS(2-CHLOROPROPYL) ANTIMONATE(III)

TOXICITY DATA WITH REFERENCE
ivn-mus LD50:180 mg/kg CSLNX* NX#01812

CONSENSUS REPORTS: Antimony and its compounds are on the Community Right-To-Know List.

OSHA PEL: TWA 0.5 mg(Sb)/m^3
ACGIH TLV: TWA 0.5 mg(Sb)/m^3
NIOSH REL: (Antimony) TWA 0.5 mg/m^3

SAFETY PROFILE: Poison by intravenous route. See also ANTIMONY COMPOUNDS. When heated to decomposition it emits fumes of Sb, PO_x, and Cl^-.

DHH800 CAS:3054-95-3 HR: 3
3,3-DIETHOXYPROPENE
DOT: UN 2374 (DOT)
mf: $C_7H_{14}O_2$ mw: 130.21

$H_2C{=}CHCH(OCH_2CH_3)_2$

PROP: Flash p: <73.4°.

SYNS: ACROLEIN ACETAL □ ACRYLALDEHYDE DIETHYL ACETAL □ 3,3-DIETHOXY-1-PROPENE □ PROPENAL DIETHYL ACETAL □ 1-PROPENE, 3,3-DIETHOXY-(9CI)

TOXICITY DATA WITH REFERENCE
mma-sat 5 μg/plate TCMUD8 1,259,80

DOT CLASSIFICATION: 3; *Label:* Flammable Liquid

SAFETY PROFILE: Mutation data reported. A flammable liquid. A dangerous fire hazard when exposed to heat, flame, or oxidizers. When heated to decomposition it emits acrid smoke and fumes.

DHI000 CAS:97-96-1 HR: 3
DIETHYL ACETALDEHYDE
DOT: UN 1178
mf: $C_6H_{12}O$ mw: 100.18

PROP: Colorless liquid; pungent odor. Fp: −89°, bp: 117°, flash p: 70°F (OC), d: 0.808–0.814, vap press: 13.7 mm @ 20°, vap d: 3.45, lel: 1.2%, uel: 7.7%. Misc in alc, ether; sltly sol in water.

SYNS: ALDEHYDE-2-ETHYLBUTYRIQUE (FRENCH) □ 2-ETHYLBUTANAL □ ETHYL BUTYRALDEHYDE □ α-ETHYLBUTYRALDEHYDE □ ETHYL BUTYRALDEHYDE (DOT) □ 2-ETHYLBUTYRALDEHYDE (DOT, FCC) □ 2-ETHYLBUTYRIC ALDEHYDE □ FEMA No. 2426

TOXICITY DATA WITH REFERENCE
skn-rbt 500 mg open MLD UCDS** 12/14/71
orl-rat LD50:3980 mg/kg AMIHBC 4,119,51
ihl-rat LCLo:8000 ppm/4H AMIHBC 4,119,51

CONSENSUS REPORTS: Reported in EPA TSCA Inventory.

DOT CLASSIFICATION: 3; *Label:* Flammable Liquid

SAFETY PROFILE: Moderately toxic by ingestion. Mildly toxic by inhalation. A skin irritant. Flammable liquid. Can react vigorously with oxidizing materials. To fight fire, use alcohol foam, CO_2, dry chemical. When heated to decomposition it emits acrid smoke and fumes. See also ALDEHYDES.

DHI200 CAS:685-91-6 HR: 3
N,N-DIETHYLACETAMIDE
mf: $C_6H_{13}NO$ mw: 115.20

PROP: Liquid. Mp: <65°, bp: 180°, flash p: 170°F, d: 0.92, vap d: 4.0.

TOXICITY DATA WITH REFERENCE
orl-rat TDLo:910 mg/kg/73W-I:ETA JNCIAM 35,949,65
orl-rat LD50:1500 mg/kg ARZNAD 19,1073,69
ivn-rat LD50:1 g/kg ARZNAD 28,1571,78
ipr-mus LD50:1690 mg/kg ARZNAD 20,1242,70
ivn-dog LD50:1 g/kg ARZNAD 28,1571,78
ivn-rbt LDLo:1920 mg/kg ARZNAD 20,1242,70
ivn-ckn LDLo:3900 mg/kg ARZNAD 20,1242,70

CONSENSUS REPORTS: Reported in EPA TSCA Inventory.

SAFETY PROFILE: Moderately toxic by ingestion and intraperitoneal routes. Questionable carcinogen with experimental tumorigenic data. Flammable when exposed to heat or flame. To fight fire, use foam, mist, CO_2, dry chemical. When heated to decomposition it emits toxic fumes of NO_x.

DHI400 CAS:88-09-5 HR: 3
DIETHYLACETIC ACID
mf: $C_6H_{12}O_2$ mw: 116.18

PROP: Colorless, volatile liquid; rancid odor. Fp −15°, mp: −93°, bp: 190°, flash p: 78°F (CC), d: 0.917, vap press: 10 mm @ 15.3°, vap d: 4.0, autoign temp: 865°F. Misc in alc, ether, water.

SYNS: 2-ETHYL BUTANOIC ACID □ α-ETHYLBUTYRIC ACID □ 2-ETHYLBUTYRIC ACID (FCC) □ FEMA No. 2429 □ 3-PENTANECARBOXYLIC ACID

TOXICITY DATA WITH REFERENCE
skn-rbt 10 mg/24H open MLD AMIHBC 10,61,54
eye-rbt 250 μg open SEV AMIHBC 10,61,54
orl-rat LD50:2200 mg/kg AMIHBC 10,61,54
skn-rbt LD50:520 mg/kg AMIHBC 10,61,54

CONSENSUS REPORTS: Reported in EPA TSCA Inventory.

SAFETY PROFILE: Moderately toxic by ingestion and skin contact. An irritant to skin and mucous membranes. A severe eye irritant. See also ESTERS. Narcotic in high concentrations. Flammable liquid. To fight fire, use CO_2, dry chemical, alcohol foam. When heated to decomposition it emits acrid smoke and fumes.

DHI600 CAS:2235-46-3 HR: 1
DIETHYLACETOACETAMIDE
mf: $C_8H_{15}NO_2$ mw: 157.24

PROP: Misc liquid. D: 0.995 @ 20°/20°, bp: decomp, fp: −70°, flash p: 250°F (COC).

SYNS: N,N-DIETHYLACETOACETAMIDE □ N,N-DIETHYL-3-OXOBUTANAMIDE (9CI)

TOXICITY DATA WITH REFERENCE
skn-rbt 500 mg/24H MLD 85JCAE-,729,86
orl-rat LD50:4760 mg/kg AIHAAP 23,95,62

CONSENSUS REPORTS: Reported in EPA TSCA Inventory.

SAFETY PROFILE: Mildly toxic by ingestion. A skin irritant. Combustible when exposed to heat or flame.

When heated to decomposition it emits toxic fumes of NO_x.

DHI800 CAS:63019-57-8 HR: 2
1-DIETHYLACETYLAZIRIDINE
mf: $C_8H_{15}NO$ mw: 141.24

SYN: DIETHYLACETYLETHYLENEIMINE

TOXICITY DATA WITH REFERENCE
cyt-rat-ipr 50 mg/kg BJPCAL 9,306,54
scu-rat TDLo:400 mg/kg/35W-I:NEO BJPCAL 9,306,54
scu-mus TDLo:488 mg/kg/20W-I:NEO BJPCAL 9,306,54
scu-rat TD:420 mg/kg/28W-I:NEO BJPCAL 9,306,54

SAFETY PROFILE: Questionable carcinogen with experimental neoplastigenic data. Mutation data reported. When heated to decomposition it emits toxic fumes of NO_x.

DHI850 CAS:762-21-0 HR: 3
DIETHYL ACETYLENE DICARBOXYLATE
mf: $C_8H_{10}O_4$ mw: 170.16

$CH_3CH_2OCO{\cdot}OC{\equiv}CCO{\cdot}OCH_2CH_3$

SAFETY PROFILE: Mixture with 1,3,5-cyclooctatriene explodes when heated to 60°C. When heated to decomposition it emits acrid smoke and fumes. See also ACETYLENE COMPOUNDS.

DHI875 CAS:2014-30-4 HR: 2
l-N,N-DIETHYLALANINE-6-CHLORO-o-TOLYL ESTER HYDROCHLORIDE
mf: $C_{14}H_{20}ClNO_2{\cdot}ClH$ mw: 306.26

SYN: FC 676

TOXICITY DATA WITH REFERENCE
skn-rbt 200 mg MOD BCFAAI 107,310,68
eye-rbt 1 g MOD BCFAAI 107,310,68
scu-mus LD50:2000 mg/kg BCFAAI 107,310,68

SAFETY PROFILE: Moderately toxic by subcutaneous route. An eye and skin irritant. When heated to decomposition it emits toxic fumes of NO_x and HCl. See also ESTERS.

DHI880 CAS:760-19-0 HR: 3
DIETHYLALUMINUM BROMIDE
mf: $C_4H_{10}AlBr$ mw: 165.01

$(CH_3CH_2)_2AlBr$

SAFETY PROFILE: Ignites on contact with nitromethane. When heated to decomposition it emits toxic fumes of Br^-. See also ALUMINUM COMPOUNDS and BROMIDES.

DHI885 CAS:96-10-6 HR: 3
DIETHYLALUMINUM CHLORIDE
mf: $C_4H_{10}AlCl$ mw: 110.56

$(CH_3CH_2)_2AlCl$

SYNS: CHLORODIETHYLALUMINUM ☐ DIETHYLALUMINUM MONOCHLORIDE ☐ DIETHYLCHLOROALUMINUM

TOXICITY DATA WITH REFERENCE
ihl-rat LC50:11 g/m^3 85JCAE-,1215,86

CONSENSUS REPORTS: Reported in EPA TSCA Inventory.

ACGIH TLV: TWA 2 $mg(Al)/m^3$

SAFETY PROFILE: Reaction with chlorine azide may form an explosive product. When heated to decomposition it emits toxic fumes of Cl^-. See also ALUMINUM COMPOUNDS and CHLORIDES.

DHJ200 CAS:109-89-7 HR: 3
DIETHYLAMINE
DOT: UN 1154
mf: $C_4H_{11}N$ mw: 73.16

PROP: Colorless liquid, ammonia-like odor. Mp: −38.9°, bp: 55.5°, flash p: −0.4°F, d: 0.711 @ 18°/4°, fp −50, autoign temp: 594°F, vap press: 400 mm @ 38.0°, vap d: 2.53, lel: 1.8%, uel: 10.1%.

SYNS: 2-AMINOPENTANE ☐ DIAETHYLAMIN (GERMAN) ☐ N,N-DIETHYLAMINE ☐ DIETILAMINA (ITALIAN) ☐ DWUETYLOAMINA (POLISH) ☐ N-ETHYL-ETHANAMINE

TOXICITY DATA WITH REFERENCE
skn-rbt 10 mg/24H MLD AMIHBC 4,119,51
skn-rbt 500 mg open MLD UCDS** 5/21/71
eye-rbt 50 µg open SEV AMIHBC 4,119,51
ihl-mam LC50:5000 mg/m^3 TPKVAL 14,80,75
orl-rat LD50:540 mg/kg AEHLAU 1,343,60
ihl-rat LC50:4000 ppm/4H AEHLAU 1,343,60
skn-rbt LD50:820 mg/kg UCDS** 5/21/71

CONSENSUS REPORTS: Reported in EPA TSCA Inventory.

OSHA PEL: TWA 10 ppm; STEL 25 ppm
ACGIH TLV: TWA 5 ppm; STEL 15; Not Classifiable As Carcinogen
DFG MAK: 10 ppm (30 mg/m^3)
DOT CLASSIFICATION: 3; *Label:* Flammable Liquid

SAFETY PROFILE: Moderately toxic by ingestion, inhalation, and skin contact. A skin and severe eye irritant. Exposure to strong vapor can cause severe cough and chest pains. Contact with liquid can damage eyes, possibly permanently; contact with skin causes necrosis and vesiculation. A very dangerous fire hazard when exposed to heat, flame, or oxidizers. To fight fire, use alcohol foam, CO_2, dry chemical. Explodes on contact with dicyanofurazan. Violent reaction with sulfuric acid. Ignites on contact with cellulose nitrate of sufficiently high surface area. When heated to decomposition it emits toxic fumes of NO_x. See also AMINES.

For occupational chemical analysis use OSHA: #41 or NIOSH: Amines, Aliphatic 2010.

DHJ400 **CAS:3213-15-8** **HR: 3**
2-(DIETHYLAMINO)ACETANILIDE
mf: $C_{12}H_{18}N_2O$ mw: 206.32

SYNS: (DIETHYLAMINO)ACYLANILIDE □ 2-(DIETHYLAMINO)-N-PHENYLACETAMIDE □ V 343

TOXICITY DATA WITH REFERENCE
eye-rbt 2% MLD ARZNAD 8,270,58
ipr-rat LD50:465 mg/kg ARZNAD 8,270,58
ipr-mus LD50:235 mg/kg ARZNAD 8,270,58
scu-mus LD50:800 mg/kg ARZNAD 8,270,58

SAFETY PROFILE: Poison by intraperitoneal route. Moderately toxic by subcutaneous route. An eye irritant. When heated to decomposition it emits toxic fumes of NO_x.

DHJ600 **CAS:3010-02-4** **HR: 3**
N,N-DIETHYLAMINOACETONITRILE
mf: $C_6H_{12}N_2$ mw: 112.20

SYNS: (DIETHYLAMINO)ACETONITRILE □ N,N-DIETHYLGLYCINONITRILE □ NITRIL KISELINY DIETHYLAMINOOCTOVE (CZECH)

TOXICITY DATA WITH REFERENCE
skn-rbt 10 mg/24H open MLD AMIHBC 4,119,51
skn-rbt 10 mg/24H open MLD AIHAAP 23,95,62
eye-rbt 750 µg open SEV AMIHBC 4,119,51
ihl-rat LC50:125 ppm/4H AMIHBC 4,119,51
skn-rbt LD50:360 mg/kg AMIHBC 4,119,51

CONSENSUS REPORTS: Cyanide and its compounds are on the Community Right-To-Know List. Reported in EPA TSCA Inventory.

SAFETY PROFILE: Poison by skin contact. Moderately toxic by inhalation. A skin and severe eye irritant. When heated to decomposition it emits toxic fumes of NO_x. See also NITRILES.

DHJ800 **CAS:77966-26-8** **HR: 2**
2-(DIETHYLAMINO)-o-ACETOPHENETIDIDE, HYDROCHLORIDE
mf: $C_{14}H_{22}N_2O_2 \cdot ClH$ mw: 286.84

SYNS: C 3095 □ 2-(DIETHYLAMINO)-2'-ETHOXYACETANILIDE, HYDROCHLORIDE

TOXICITY DATA WITH REFERENCE
eye-rbt 2% MLD ARZNAD 8,270,58
ipr-rat LD50:443 mg/kg ARZNAD 8,270,58
scu-mus LD50:1295 mg/kg ARZNAD 8,270,58

SAFETY PROFILE: Moderately toxic by intraperitoneal and subcutaneous routes. An eye irritant. When heated to decomposition it emits very toxic fumes of NO_x and HCl.

DHK000 **CAS:77966-27-9** **HR: 3**
2-(DIETHYLAMINO)-p-ACETOPHENETIDIDE, HYDROCHLORIDE
mf: $C_{14}H_{22}N_2O_2 \cdot ClH$ mw: 286.84

SYN: C 3094

TOXICITY DATA WITH REFERENCE
eye-rbt 2% MLD ARZNAD 8,270,58
ipr-rat LD50:349 mg/kg ARZNAD 8,270,58
scu-mus LD50:720 mg/kg ARZNAD 8,270,58

SAFETY PROFILE: Poison by intraperitoneal route. Moderately toxic by subcutaneous route. An eye irritant. When heated to decomposition it emits very toxic fumes of NO_x and HCl.

DHK200 **CAS:6304-07-0** **HR: 3**
2-(DIETHYLAMINO)-o-ACETOTOLUIDIDE HYDROCHLORIDE
mf: $C_{13}H_{20}N_2O \cdot ClH$ mw: 256.81

SYNS: C 3080 □ 2-(DIMETHYLAMINO)-2'-METHYLACETANILIDE HYDROCHLORIDE

TOXICITY DATA WITH REFERENCE
ipr-rat LD50:430 mg/kg ARZNAD 8,270,58
ipr-mus LD50:280 mg/kg ARZNAD 8,270,58
scu-mus LD50:825 mg/kg ARZNAD 8,270,58

SAFETY PROFILE: Poison by intraperitoneal route. Moderately toxic by subcutaneous route. When heated to decomposition it emits very toxic fumes of NO_x and HCl.

DHK400 **CAS:137-58-6** **HR: 3**
2-(DIETHYLAMINO)-2′,6′-ACETOXYLIDIDE
mf: $C_{14}H_{22}N_2O$ mw: 234.38

PROP: Needles from C_6H_6 or EtOH. Mp: 68–69°, bp: 180–182° @ 4 mm.

SYNS: ANESTACON □ DIETHYLAMINOACETO-2,6-XYLIDIDE □ α-DIETHYLAMINOACETO-2,6-XYLIDIDE □ α-DIETHYLAMINO-2,6-ACETOXYLIDIDE □ DIETHYLAMINOACET-2,6-XYLIDIDE □ α-DIETHYLAMINO-2,6-DIMETHYLACETANILIDE □ ω-DIETHYLAMINO-2,6-DIMETHYLACETANILIDE □ α-DIETILAMINO-2,6-DIMETILACETANILIDE (ITALIAN) □ DUNCAINE □ GRAVOCAIN □ ISICAINA □ LEOSTESIN □ LIDA-MANTLE □ LIDOCAINE □ LIGNOCAINE □ MARICAINE □ RUCAINA □ SOLCAIN □ XILOCAINA (ITALIAN) □ XYCIANE □ XYLESTESIN □ XYLOCAIN □ XYLOCITIN □ XYLOTOX

TOXICITY DATA WITH REFERENCE
mma-sat 50 µmol/plate JAFCAU 34,157,86
ims-rat TDLo:6 mg/kg (female 11D post):REP NETOD7 8,61,86
imp-rat TDLo:7500 mg/kg (female 3-17D post):TER TJADAB 33,73C,86
orl-cld TDLo:21 mg/kg:BAH,BLD,PUL JTCTDW 30,413,92
orl-wmn TDLo:39 mg/kg:CNS,CVS NEJMAG 306,381,82
ivn-wmn TDLo:16 mg/kg:CVS,PUL EJCPAS 22,129,82
ivn-man TDLo:8643 µg/kg/4H-C:BAH AIMEAS 97,149,82
ivn-hmn TDLo:23 mg/kg:BAH,PUL ATXKA8 28,72,71
ivn-man TDLo:1700 µg/kg/2M-C:BAH,CPR,PUL AJEMEN 4,143,86
orl-rat LD50:317 mg/kg BCFAAI 110,330,71
ipr-rat LD50:133 mg/kg AIPTAK 243,97,80
scu-rat LD50:335 mg/kg EJMCA5 9,188,74
ivn-rat LDLo:25 mg/kg BJANAD 23,153,51
orl-mus LD50:220 mg/kg ARZNAD 16,1275,66
ipr-mus LD50:102 mg/kg JMCMAR 24,1059,81
scu-mus LD50:238 mg/kg JMCMAR 28,714,85

CONSENSUS REPORTS: Reported in EPA TSCA Inventory.

SAFETY PROFILE: Poison by ingestion, intravenous, intraperitoneal, and subcutaneous routes. Human systemic effects: blood pressure lowering, changes in heart rate, coma, convulsions, distorted perceptions, dyspnea, excitement, hallucinations, muscle contraction or spasticity, pulse rate, respiratory depression, toxic psychosis. An experimental teratogen. Other experimental reproductive effects. A local anesthetic. Mutation data reported. When heated to decomposition it emits toxic fumes of NO_x.

DHK600 CAS:73-78-9 HR: 3
2-DIETHYLAMINO-2′,6′-ACETOXYLIDIDE HYDROCHLORIDE
mf: $C_{14}H_{22}N_2O{\bullet}ClH$ mw: 270.84

PROP: A solid. Mp: 127–129°. Sol in water.

SYNS: ANESTACON HYDROCHLORIDE □ 2-(DIETHYLAMINO)-2′,6′-ACETOXYLIDIDE MONOHYDROCHLORIDE □ α-DIETHYLAMINO-2,5-ACETOXYLIDINE HYDROCHLORIDE □ ω-DIETHYLAMINO-2,6-DIMETHYLACETANILIDE HYDROCHLORIDE □ 2-(DIETHYLAMINO)-N-(2,6-DIMETHYLPHENYL)ACETAMIDE MONOHYDROCHLORIDE □ DUNCAINE HYDROCHLORIDE □ GRAVOCAIN HYDROCHLORIDE □ ISICAINE HYDROCHLORIDE □ LEOSTESIN HYDROCHLORIDE □ LIDOCAINE HYDROCHLORIDE □ LIDOTHESIN HYDROCHLORIDE □ LIGNOCAINE HYDROCHLORIDE □ RUCAINA HYDROCHLORIDE □ S 202 □ XYCAINE HYDROCHLORIDE □ XYLESTESIN HYDROCHLORIDE □ XYLOCAINE HYDROCHLORIDE □ XYLOCARD □ XYLOCITIN HYDROCHLORIDE □ XYLONEURAL □ XYLOTOX HYDROCHLORIDE

TOXICITY DATA WITH REFERENCE
skn-rbt 3% MLD AIPTAK 137,410,62
eye-rbt 3% MLD AIPTAK 137,410,62
imp-rat TDLo:7500 mg/kg (female 3-17D post):TER ANESAV 65,626,86
imp-rat TDLo:7500 mg/kg (female 3-17D post):REP ANESAV 65,626,86
ivn-cld TDLo:60 mg/kg/1H JTCTDW 24,51,86
ivn-man TDLo:9 mg/kg/4H-C:CVS DICPBB 19,669,85
ivn-man TDLo:7143 μg/kg:BPR CHETBF 61,682,72
imp-man TDLo:5714 μg/kg CMAJAX 137,219,87
ipr-rat LD50:122 mg/kg JPETAB 111,224,54
scu-rat LD50:570 mg/kg RPOBAR 2,299,70
ivn-rat LD50:21 mg/kg RPOBAR 2,299,70
orl-mus LD50:220 mg/kg RPOBAR 2,298,70
ipr-mus LD50:63 mg/kg AIPTAK 274,253,85
scu-mus LD50:163 mg/kg PSEBAA 103,353,60
ivn-mus LD50:15 mg/kg JPPMAB 14(Suppl),48T,62
ims-mus LD50:260 mg/kg RPOBAR 2,298,70
itr-rbt LD50:28 mg/kg AIPTAK 200,359,72

CONSENSUS REPORTS: Reported in EPA TSCA Inventory. EPA Genetic Toxicology Program.

SAFETY PROFILE: Poison by ingestion, intraperitoneal, intravenous, subcutaneous, intramuscular, and intratracheal routes. Human systemic effects: somnolence, respiratory depression, low blood pressure, cardiomyopathy including infarction, pulse rate increase. An experimental teratogen. Other experimental reproductive effects. A skin and eye irritant. An anesthetic. When heated to decomposition it emits very toxic fumes of NO_x and HCl.

DHK800 CAS:77966-82-6 HR: 3
2-(DIETHYLAMINO)-3′,5′-ACETOXYLIDIDE HYDROCHLORIDE
mf: $C_{14}H_{22}N_2O{\bullet}ClH$ mw: 270.84

SYNS: C 3065 □ 2-(DIETHYLAMINO)-3′,5′-DIMETHYLACETANILIDE HYDROCHLORIDE

TOXICITY DATA WITH REFERENCE
eye-rbt 2% MOD ARZNAD 8,270,58
ipr-rat LD50:260 mg/kg ARZNAD 8,270,58
scu-mus LD50:550 mg/kg ARZNAD 8,270,58

SAFETY PROFILE: Poison by intraperitoneal route. Moderately toxic by subcutaneous route. An eye irritant. When heated to decomposition it emits very toxic fumes of NO_x and HCl.

DHL200 CAS:77967-24-9 HR: 2
N-(2-DIETHYLAMINO)ACETYLANTHRANILIC ACID, ETHYL ESTER HYDROCHLORIDE
mf: $C_{15}H_{22}N_2O_3{\bullet}ClH$ mw: 314.85

SYNS: C 3102 □ N-((DIETHYLAMINO)ACETYL)ANTHRANILIC ACID, ETHYL ESTER, HYDROCHLORIDE

TOXICITY DATA WITH REFERENCE
eye-rbt 2% MOD ARZNAD 8,270,58
ipr-rat LD50:447 mg/kg ARZNAD 8,270,58
scu-mus LD50:2150 mg/kg ARZNAD 8,270,58

SAFETY PROFILE: Moderately toxic by intraperitoneal and subcutaneous routes. An eye irritant. See also ESTERS. When heated to decomposition it emits very toxic fumes of NO_x and HCl.

DHL400 CAS:77967-25-0 HR: 2
N-(2-DIETHYLAMINO)ACETYLANTHRANILIC ACID, METHYL ESTER, HYDROCHLORIDE
mf: $C_{14}H_{20}N_2O_3{\bullet}ClH$ mw: 300.82

SYNS: C 3089 □ N-((DIETHYLAMINO)ACETYL)ANTHRANILIC ACID, METHYL ESTER, HYDROCHLORIDE

TOXICITY DATA WITH REFERENCE
eye-rbt 2% MLD ARZNAD 8,270,58
ipr-rat LD50:615 mg/kg ARZNAD 8,270,58
scu-mus LD50:2400 mg/kg ARZNAD 8,270,58

SAFETY PROFILE: Moderately toxic by intraperitoneal and subcutaneous routes. An eye irritant. See also ESTERS. When heated to decomposition it emits very toxic fumes of HCl and NO_x.

DHL600 HR: 3
β-4-(1-DIETHYLAMINOACETYL-2-PIPERIDYL)-2,2-DIPHENYL-1,3-DIOXOLANE HYDROCHLORIDE
mf: $C_{26}H_{34}N_2O_3{\bullet}ClH$ mw: 459.08

TOXICITY DATA WITH REFERENCE
orl-mus LD50:150 mg/kg JMCMAR 9,127,66
ivn-mus LD50:12 mg/kg JMCMAR 9,127,66

SAFETY PROFILE: Poison by ingestion and intravenous routes. When heated to decomposition it emits very toxic fumes of NO_x and HCl.

D

DHL800 **CAS:1027-14-1** **HR: 3**
DIETHYLAMINOACETYL-2,4,6-TRIMETHYLANILINE HYDROCHLORIDE
mf: $C_{15}H_{24}N_2O{\cdot}ClH$ mw: 284.87

PROP: A solid. Mp: 140°.

SYNS: 2-DIETHYLAMINO-2′,4′,6′-TRIMETHYLACETANILIDE HYDROCHLORIDE □ 2-(DIETHYLAMINO)-2′,4′,6′-TRIMETHYLACETANILIDE MONOHYDROCHLORIDE □ 2-(DIETHYLAMINO)-N-(2,4,6-TRIMETHYLPHENYL)ACETAMIDE MONOHYDROCHLORIDE □ MESIDICAINE HYDROCHLORIDE □ MESOCAINE HYDROCHLORIDE □ MESOKAIN HYDROCHLORIDE □ TRIMECAINE □ TRIMECAINE HYDROCHLORIDE □ TRIMEKAIN HYDROCHLORIDE □ N-sym-TRIMETHYLPHENYLDIETHYLAMINOACETAMIDE HYDROCHLORIDE

TOXICITY DATA WITH REFERENCE
ipr-mus LD50:172 mg/kg AFPCAG 29,81,76
scu-mus LD50:295 mg/kg MEIEDD 10,1386,83

SAFETY PROFILE: Poison by intraperitoneal and subcutaneous routes. When heated to decomposition it emits very toxic fumes of NO_x and HCl.

DHM000 **CAS:5123-63-7** **HR: 1**
3-(DIETHYLAMINO)BENZENESULFONIC ACID, SODIUM SALT
mf: $C_{10}H_{14}NO_3S{\cdot}Na$ mw: 251.30

SYN: N,N-DIETHYLMETANILAN SODNY (CZECH)

TOXICITY DATA WITH REFERENCE
skn-rbt 500 mg/24H MLD 28ZPAK -,185,72
eye-rbt 20 mg/24H MOD 28ZPAK -,185,72
orl-rat LD50:7470 mg/kg 28ZPAK -,185,72

CONSENSUS REPORTS: Reported in EPA TSCA Inventory.

SAFETY PROFILE: Mildly toxic by ingestion. A skin and eye irritant. When heated to decomposition it emits very toxic fumes of NO_x, Na_2O, and SO_x. See also SULFONATES.

DHM309 **CAS:1877-24-3** **HR: 3**
2-(DIETHYLAMINO)BUTYRIC ACID-2,6-XYLYL ESTER HYDROCHLORIDE
mf: $C_{16}H_{25}NO_2{\cdot}ClH$ mw: 299.88

SYN: FC 455

TOXICITY DATA WITH REFERENCE
skn-rbt 200 mg MOD BCFAAI 107,310,68
eye-rbt 1 g MOD BCFAAI 107,310,68
scu-mus LD50:230 mg/kg BCFAAI 107,310,68
ivn-mus LD50:11 mg/kg BCFAAI 107,310,68

SAFETY PROFILE: Poison by subcutaneous and intravenous routes. An eye and skin irritant. When heated to decomposition it emits toxic fumes of NO_x and HCl. See also ESTERS.

DHM400 **CAS:77985-21-8** **HR: 3**
3-(DIETHYLAMINO)-2′,6′-BUTYROXYLIDIDE HYDROCHLORIDE
mf: $C_{16}H_{26}N_2O{\cdot}ClH$ mw: 298.90

SYN: C 5125

TOXICITY DATA WITH REFERENCE
ipr-rat LD50:40 mg/kg ARZNAD 8,544,58
ipr-mus LD50:30 mg/kg ARZNAD 8,544,58
scu-mus LD50:86 mg/kg ARZNAD 8,544,58

SAFETY PROFILE: Poison by intraperitoneal and subcutaneous routes. When heated to decomposition it emits very toxic fumes of HCl and NO_x.

DHM500 **CAS:2869-83-2** **HR: 2**
3-(DIETHYLAMINO)-7-((p-(DIMETHYLAMINO)PHENYL) AZO)-5-PHENYLPHENAZINIUM CHLORIDE
mf: $C_{30}H_{31}N_6{\cdot}Cl$ mw: 511.12

PROP: Dark green crystals or powder.

SYNS: C.I. 11050 □ JANUS GREEN B □ JANUS GREEN V

TOXICITY DATA WITH REFERENCE
cyt-ham:ovr 20 μmol/L/5H-C ENMUDM 1,27,79
scu-rat TDLo:960 mg/kg/13W-I:ETA GANNA2 44,293,53

CONSENSUS REPORTS: Reported in EPA TSCA Inventory.

SAFETY PROFILE: Questionable carcinogen with experimental tumorigenic data. Mutation data reported. When heated to decomposition it emits toxic fumes of NO_x.

DHN800 **CAS:77791-20-9** **HR: 3**
2-(DIETHYLAMINO)-N,N-DIPHENYLACETAMIDE HYDROCHLORIDE
mf: $C_{18}H_{22}N_2O{\cdot}ClH$ mw: 318.88

SYN: C 3135

TOXICITY DATA WITH REFERENCE
ipr-rat LD50:210 mg/kg ARZNAD 8,609,58
scu-mus LD50:222 mg/kg ARZNAD 8,609,58

SAFETY PROFILE: Poison by intraperitoneal and subcutaneous routes. When heated to decomposition it emits very toxic fumes of NO_x and HCl.

DHO000 **CAS:59960-90-6** **HR: 3**
2-(DIETHYLAMINO)-N-(DIPHENYLMETHYL)ACETAMIDE HYDROCHLORIDE
mf: $C_{19}H_{24}N_2O{\cdot}ClH$ mw: 332.91

SYN: C 3209

TOXICITY DATA WITH REFERENCE
eye-rbt 2% MOD ARZNAD 8,609,58
ipr-rat LD50:165 mg/kg ARZNAD 8,609,58
scu-mus LD50:399 mg/kg APSXAS 5,429,68
ivn-mus LD50:26 mg/kg APSXAS 5,429,68

SAFETY PROFILE: Poison by intraperitoneal, intravenous, and subcutaneous routes. An eye irritant. When heated to decomposition it emits very toxic fumes of NO_x and HCl.

DHO400 **CAS:1942-52-5** **HR: 3**
DIETHYLAMINOETHANETHIOL HYDROCHLORIDE
mf: $C_6H_{15}NS{\bullet}ClH$ mw: 169.74

PROP: Solid (as hydrochloride). Sol in H_2O and EtOH.

SYN: USAF E-4

TOXICITY DATA WITH REFERENCE
ipr-mus LD50:100 mg/kg NTIS** AD277-689

CONSENSUS REPORTS: Reported in EPA TSCA Inventory.

SAFETY PROFILE: Poison by intraperitoneal route. When heated to decomposition it emits very toxic fumes of Cl^-, SO_x, and NO_x.

DHO500 **CAS:100-37-8** **HR: 3**
2-DIETHYLAMINOETHANOL
DOT: UN 2686
mf: $C_6H_{15}NO$ mw: 117.22

PROP: Colorless, hygroscopic liquid. Bp: 162°, flash p: 140°F (OC), d: 0.8851 @ 20°/20°, vap press: 1.4 mm @ 20°, vap d: 4.03. Sol in water.

SYNS: DEAE ☐ DIAETHYLAMINOAETHANOL (GERMAN) ☐ DIETHYLAMINOETHANOL ☐ β-DIETHYLAMINOETHANOL ☐ N-DIETHYLAMINOETHANOL ☐ 2-(DIETHYLAMINO)ETHANOL ☐ 2-N-DIETHYLAMINOETHANOL ☐ DIETHYLAMINOETHANOL (DOT) ☐ β-DIETHYLAMINOETHYL ALCOHOL ☐ DIETHYLETHANOLAMINE ☐ N,N-DIETHYLETHANOLAMINE ☐ N,N-DIETHYL-N-(β-HYDROXYETHYL) AMINE ☐ 2-HYDROXYTRIETHYLAMINE

TOXICITY DATA WITH REFERENCE
skn-rbt 10 mg/24H open JIHTAB 26,269,44
skn-rbt 500 mg open MLD UCDS** 6/11/63
eye-rbt 5 mg SEV UCDS** 6/11/63
ihl-hmn TCLo:200 ppm:GIT 34ZIAG -,216,69
orl-rat LD50:1300 mg/kg JIHTAB 26,269,44
ihl-rat LCLo:4500 mg/m³/4H GTPZAB 14(11),52,70
ipr-rat LD50:1220 mg/kg TXAPA9 12,486,68
ihl-mus LC50:5000 mg/m³ GTPZAB 14(11),52,70
ipr-mus LD50:192 mg/kg JPETAB 94,249,48
scu-mus LD50:1561 mg/kg ARZNAD 4,649,54
ivn-mus LD50:188 mg/kg ARZNAD 9,31,59
ims-mus LD50:416 mg/kg ARZNAD 9,31,59
skn-gpg LD50:884 mg/kg JIHTAB 26,269,44

CONSENSUS REPORTS: Reported in EPA TSCA Inventory.

OSHA PEL: TWA 10 ppm (skin)
ACGIH TLV: TWA 10 ppm (skin)
DFG MAK: 10 ppm (50 mg/m³)
DOT CLASSIFICATION: 3; *Label:* Flammable Liquid

SAFETY PROFILE: Poison by intraperitoneal and intravenous routes. Moderately toxic by ingestion, skin contact, subcutaneous, and intramuscular routes. Human systemic effects by inhalation: nausea or vomiting. A skin and severe eye skin irritant. Flammable liquid when exposed to heat or flame; can react with oxidizing materials. To fight fire, use alcohol foam, CO_2, dry chemical. When heated to decomposition it emits toxic fumes of NO_x. See also AMINES.

For occupational chemical analysis use NIOSH: Aminoethanol Compounds, 2007.

DHO600 **CAS:487-53-6** **HR: 3**
DIETHYLAMINOETHANOL-p-AMINOSALICYLATE
mf: $C_{13}H_{20}N_2O_3$ mw: 252.35

PROP: Oily liquid. Sol in $CHCl_3$.

SYNS: DIETHYLAMINOETHYL-p-AMINOSALICYLATE ☐ 2-DIETHYLAMINOETHYL-p-AMINOSALICYLATE ☐ 2-DIETHYLAMINOETHYL-4-AMINOSALICYLATE ☐ DIETHYLAMINOETHYL-3-HYDROXY-4-AMINOBENZOATE ☐ HYDROXYPROCAINE ☐ m-HYDROXYPROCAINE ☐ METAHYDROXYPROCAINE ☐ OXYCAINE ☐ OXYPROCAIN ☐ OXYPROCAINE

TOXICITY DATA WITH REFERENCE
ipr-rat LD50:190 mg/kg KLWOAZ 31,97,53
ivn-rat LD50:36 mg/kg KLWOAZ 31,97,53
ipr-mus LD50:125 mg/kg ARZNAD 2,112,52
scu-mus LD50:280 mg/kg KLWOAZ 31,97,53
ivn-mus LD50:47 mg/kg KLWOAZ 31,97,53

SAFETY PROFILE: Poison by intravenous and intraperitoneal routes. When heated to decomposition it emits toxic fumes of NO_x.

DHO700 **CAS:14426-20-1** **HR: 2**
2-DIETHYLAMINOETHANOL HYDROCHLORIDE
mf: $C_6H_{15}NO{\bullet}ClH$ mw: 153.68

TOXICITY DATA WITH REFERENCE
ipr-mus LD50:1162 mg/kg APBDAJ 290,131,57
scu-mus LD50:1260 mg/kg AIPTAK 112,36,57
ivn-mus LD50:458 mg/kg APBDAJ 290,131,57

CONSENSUS REPORTS: Reported in EPA TSCA Inventory.

SAFETY PROFILE: Moderately toxic by intraperitoneal, intravenous, and subcutaneous routes. When heated to decomposition it emits toxic fumes of NO_x and HCl.

DHO800 **CAS:102207-84-1** **HR: 3**
2-(2-(DIETHYLAMINO)ETHOXY)-2′,6′-ACETOXYLIDIDE HYDROCHLORIDE
mf: $C_{16}H_{26}N_2O_2{\bullet}ClH$ mw: 314.90

SYN: C 3054

TOXICITY DATA WITH REFERENCE
eye-rbt 2% MLD ARZNAD 9,113,59
ipr-rat LD50:245 mg/kg ARZNAD 9,113,59
scu-mus LD50:500 mg/kg ARZNAD 9,113,59

SAFETY PROFILE: Poison by intraperitoneal route. Moderately toxic by subcutaneous route. An eye irritant. When heated to decomposition it emits very toxic fumes of NO_x and HCl.

DHP000 **CAS:54099-23-9** **HR: 3**
2-(2-(DIETHYLAMINO)ETHOXY)ADAMANTANE ETHYL IODIDE
mf: $C_{18}H_{34}NO{\bullet}I$ mw: 407.43

SYN: (2-(2-ADAMANTYLOXY)ETHYL)TRIETHYL-AMMONIUMIODIDE

D

TOXICITY DATA WITH REFERENCE
orl-mus LD50:400 mg/kg FRPSAX 32,129,77
ipr-mus LD50:30 mg/kg FRPSAX 32,129,77

SAFETY PROFILE: Poison by ingestion and intraperitoneal routes. When heated to decomposition it emits very toxic fumes of NO_x, NH_3, and I^-. See also IODIDES.

DHP200 CAS:6376-26-7 HR: 3
o-(DIETHYLAMINOETHOXY)BENZANILIDE
mf: $C_{19}H_{24}N_2O_2$ mw: 312.45

PROP: A solid. Mp: 44°.

SYNS: o-DIAETHYLAMINOAETHOXY-BENZANILID (GERMAN) □ 2-(2-(DIETHYLAMINO)ETHOXY)BENZANILIDE

TOXICITY DATA WITH REFERENCE
scu-mus TDLo:1425 mg/kg (female 1-19D post): REP ARZNAD 18,658,68
scu-mus TDLo:950 mg/kg (female 1-19D post):TER ARZNAD 18,658,68
orl-mus LD50:335 mg/kg ARZNAD 16,1127,66
scu-mus LD50:225 mg/kg ARZNAD 18,658,68
ivn-mus LD50:29 mg/kg ARZNAD 16,1127,66

SAFETY PROFILE: Poison by ingestion, subcutaneous, and intravenous routes. An experimental teratogen. Other experimental reproductive effects. When heated to decomposition it emits toxic fumes of NO_x.

DHP400 CAS:5014-35-7 HR: 3
2-(2-(DIETHYLAMINO)ETHOXY)-5-BROMOBENZANILIDE
mf: $C_{19}H_{23}BrN_2O_2$ mw: 391.35

SYN: 5-BROMO-2-(2-(DIETHYLAMINO)ETHOXY)BENZANILIDE

TOXICITY DATA WITH REFERENCE
scu-mus TDLo:950 mg/kg (female 1-19D post):TER ARZNAD 18,658,68
scu-mus TDLo:285 mg/kg (1-19D preg):REP ARZNAD 18,658,68
scu-mus TDLo:285 mg/kg (1-19D preg):TER ARZNAD 18,658,68
ipr-mus LD50:250 mg/kg BCFAAI 101,785,62
scu-mus LD50:830 mg/kg ARZNAD 18,658,68

SAFETY PROFILE: Poison by intraperitoneal route. Moderately toxic by subcutaneous route. Experimental teratogenic and reproductive effects. When heated to decomposition it emits toxic fumes of Br^- and NO_x.

DHP450 CAS:17822-72-9 HR: 3
2-(2-(DIETHYLAMINO)ETHOXY)-2′-CHLORO-BENZANILIDE
mf: $C_{19}H_{23}ClN_2O_2$ mw: 346.89

SYN: BENZANILIDE, 2′-CHLORO-2-(2-(DIETHYLAMINO)ETHOXY)-

TOXICITY DATA WITH REFERENCE
scu-mus TDLo:950 mg/kg (female 1-19D post):TER ARZNAD 18,658,68
scu-mus LD50:215 mg/kg ARZNAD 18,658,68

SAFETY PROFILE: Poison by subcutaneous route. An experimental teratogen. When heated to decomposition it emits toxic fumes of NO_x and Cl^-.

DHP500 CAS:17822-73-0 HR: 3
2-(2-(DIETHYLAMINO)ETHOXY)-3′-CHLORO-BENZANILIDE
mf: $C_{19}H_{23}ClN_2O_2$ mw: 346.89

TOXICITY DATA WITH REFERENCE
scu-mus TDLo:570 mg/kg (1-19D preg):REP ARZNAD 18,658,68
scu-mus TDLo:950 mg/kg (1-19D preg):TER ARZNAD 18,658,68
scu-mus LD50:380 mg/kg ARZNAD 18,658,68

SAFETY PROFILE: Poison by subcutaneous route. Experimental teratogenic and reproductive effects. When heated to decomposition it emits toxic fumes of Cl^- and NO_x.

DHP550 CAS:17822-71-8 HR: 3
2-(2-(DIETHYLAMINO)ETHOXY)-4′-CHLORO-BENZANILIDE
mf: $C_{19}H_{23}ClN_2O_2$ mw: 346.89

SYN: BENZANILIDE, 4′-CHLORO-2-(2-(DIETHYLAMINO)ETHOXY)-

TOXICITY DATA WITH REFERENCE
scu-mus TDLo:570 mg/kg (female 1-19D post):REP ARZNAD 18,658,68
scu-mus LD50:260 mg/kg ARZNAD 18,658,68

SAFETY PROFILE: Poison by subcutaneous route. Experimental reproductive effects. When heated to decomposition it emits toxic fumes of NO_x and Cl^-.

DHP600 CAS:26270-62-2 HR: 3
6-(2-(DIETHYLAMINOETHOXY))-4,7-DIMETHOXY-5-CINNAMOYLBENZOFURANMALEATE
mf: $C_{25}H_{29}NO_5 \cdot C_4H_4O_4$ mw: 539.63

TOXICITY DATA WITH REFERENCE
orl-mus LD50:150 mg/kg CHTPBA 8,479,73
ivn-mus LD50:11 mg/kg CHTPBA 8,479,73

SAFETY PROFILE: Poison by ingestion and intravenous routes. When heated to decomposition it emits toxic fumes of NO_x.

DHQ000 CAS:41226-18-0 HR: 3
6-(2-DIETHYLAMINOETHOXY)-4,7-DIMETHOXY-5-(p-METHOXYCINNAMOYL)BENZOFURAN OXALATE
mf: $C_{26}H_{31}NO_6 \cdot C_2H_2O_4$ mw: 543.62

TOXICITY DATA WITH REFERENCE
orl-mus LD50:240 mg/kg CHTPBA 8,479,73
ivn-mus LD50:15 mg/kg CHTPBA 8,479,73

SAFETY PROFILE: Poison by ingestion and intravenous routes. When heated to decomposition it emits toxic fumes of NO_x. See also OXALATES.

DHQ200 **CAS:468-61-1** **HR: 3**
2-(2-DIETHYLAMINOETHOXY)ETHYL-2-ETHYL-2-PHENYLBUTYRATE
mf: $C_{20}H_{33}NO_3$ mw: 335.54

PROP: Yellow oil. Bp: 150° @ 0.5 mm.

SYNS: 2-(2-(DIETHYLAMINO)ETHOXY)ETHANOL-2-ETHYL-2-PHENYLBUTYRATE □ 2-(2-DIETHYLAMINOETHOXY)ETHYL α,α-DIETHYLPHENYLACETATE □ α,α-DIETHYLBENZENEACETIC ACID 2-(2-(DIETHYLAMINO)ETHOXY)ETHYL ESTER □ 2-ETHYL-2-PHENYLBUTYRIC ACID 2-(2-DIETHYLAMINOETHOXY)ETHYL ESTER □ OXELADIN □ PECTAMOL

TOXICITY DATA WITH REFERENCE
orl-rat LD50:183 mg/kg JPPMAB 9,446,57
orl-mus LD50:130 mg/kg JPPMAB 9,446,57
ivn-mus LD50:13 mg/kg JPPMAB 9,446,57

SAFETY PROFILE: Poison by ingestion and intravenous routes. When heated to decomposition it emits toxic fumes of NO_x.

DHQ500 **CAS:1045-21-2** **HR: 3**
DIETHYLAMINOETHOXYETHYL-1-PHENYL-1-CYCLOPENTANE CARBOXYLATE HYDROCHLORIDE
mf: $C_{20}H_{31}NO_3 \cdot ClH$ mw: 369.98

PROP: A solid. Mp: 71–72°.

SYN: 1-PHENYLCYCLOPENTANECARBOXYLIC ACID (2-(2-(DIETHYLAMINO)ETHOXY)ETHYL) ESTER HYDROCHLORIDE

TOXICITY DATA WITH REFERENCE
orl-rat LD50:830 mg/kg AIPTAK 103,200,55
ivn-rat LD50:25 mg/kg AIPTAK 103,200,55
orl-mus LD50:230 mg/kg AIPTAK 103,200,55
ivn-mus LD50:26,500 μg/kg AIPTAK 103,200,55

SAFETY PROFILE: Poison by ingestion and intravenous routes. When heated to decomposition it emits toxic fumes of NO_x and HCl. See also ESTERS.

DHQ600 **CAS:75348-40-2** **HR: 3**
6-(2-DIETHYLAMINOETHOXY)-N-(o-METHOXYPHENYL)NICOTINAMIDE HYDROCHLORIDE
mf: $C_{19}H_{25}N_3O_3 \cdot ClH$ mw: 379.93

TOXICITY DATA WITH REFERENCE
orl-mus LD50:500 mg/kg CHTPBA 8,226,73
ivn-mus LDLo:100 mg/kg CHTPBA 8,226,73

SAFETY PROFILE: Poison by intravenous route. Moderately toxic by ingestion. When heated to decomposition it emits very toxic fumes of HCl and NO_x.

DHQ800 **CAS:17822-74-1** **HR: 3**
2-(2-(DIETHYLAMINO)ETHOXY)-3-METHYLBENZANILIDE
mf: $C_{20}H_{25}N_2O_2$ mw: 325.47

TOXICITY DATA WITH REFERENCE
scu-mus TDLo:570 mg/kg (1-19D preg):REP ARZNAD 10,650,60
scu-mus TDLo:570 mg/kg (1-19D preg):TER ARZNAD 18,658,68
scu-mus LD50:285 mg/kg ARZNAD 18,658,68

SAFETY PROFILE: Poison by subcutaneous route. An experimental teratogen. When heated to decomposition it emits toxic fumes of NO_x.

DHR000 **CAS:5372-13-4** **HR: 3**
5-(2-(DIETHYLAMINO)ETHOXY)-3-METHYL-1-PHENYLPYRAZOLE
mf: $C_{16}H_{23}N_3O$ mw: 273.42

SYN: P-314

TOXICITY DATA WITH REFERENCE
orl-mus LD50:416 mg/kg ARZNAD 17,214,67
scu-mus LD50:306 mg/kg ARZNAD 17,214,67

SAFETY PROFILE: Poison by ingestion and subcutaneous routes. When heated to decomposition it emits toxic fumes of NO_x.

DHR800 **CAS:101692-44-8** **HR: 3**
1-DIETHYLAMINO-3-(p-ETHOXYPHENYL)INDAN CITRATE
mf: $C_{21}H_{27}NO \cdot C_6H_8O_7$ mw: 501.63

SYNS: 1-p-AETHOXYPHENYL-3-DIAETHYLAMINO-INDAN CITRAT (GERMAN) □ LABOR-NR 2683

TOXICITY DATA WITH REFERENCE
orl-rat LD50:860 mg/kg ARZNAD 11,915,61
ivn-rat LD50:97 mg/kg ARZNAD 11,915,61
orl-mus LD50:347 mg/kg ARZNAD 11,915,61
ivn-mus LD50:137 mg/kg ARZNAD 11,915,61

SAFETY PROFILE: Poison by ingestion and intravenous routes. When heated to decomposition it emits toxic fumes of NO_x.

DHR900 **CAS:73343-71-2** **HR: 3**
p-(2-(DIETHYLAMINO)ETHOXY)PHENYL 2-MESITYL-3-BENZOFURANYL KETONE HYDROCHLORIDE
mf: $C_{30}H_{33}NO_3 \cdot ClH$ mw: 492.10

SYNS: BENZOFURAN, 3-(p-(2-(DIETHYLAMINO)ETHOXY)BENZOYL)-2-MESITYL-, HYDROCHLORIDE □ KETONE, p-(2-(DIETHYLAMINO)ETHOXY)PHENYL 2-MESITYL-3-BENZOFURANYL □ METHANONE, (4-(2-(DIETHYLAMINO)ETHOXY)PHENYL)(2-(2,4,6-TRIMETHYLPHENYL)-3-BENZOFURA NYL)-, HCl

TOXICITY DATA WITH REFERENCE
ipr-mus LD50:740 mg/kg EJMCA5 14,517,79

DOT CLASSIFICATION: 3; *Label:* Flammable Liquid

SAFETY PROFILE: Moderately toxic by intraperitoneal route. A flammable liquid. When heated to decomposition it emits toxic vapors of NO_x and Cl^-.

DHS000 **CAS:67-98-1** **HR: 2**
(p-2-DIETHYLAMINOETHOXYPHENYL)-1-PHENYL-2-p-ANISYLETHANOL
mf: $C_{27}H_{33}NO_3$ mw: 419.61

SYNS: 1-(p-2-DIETHYLAMINOETHOXYPHENYL)-1-PHENYL-2-p-ANISYLETHANOL □ 1-(4-(2-DIETHYLAMINOETHOXY)PHENYL)-1-PHENYL 2 (p-ANISYL)ETHANOL □ 1-(p-(2-(DIETHYLAMINO)ETHOXY)PHENYL)-1-PHENYL-2-(p-METHOXYPHENYL)ETHANOL □ ETHAMOXYTRIPHETOL □ ETHANOXYTRIPHETOL □ MER 25

D

TOXICITY DATA WITH REFERENCE
unr-rbt TDLo:150 mg/kg (female 21-26D post):REP JRPMAP 4,137,70
orl-rat TDLo:550 mg/kg (22D post):REP ENDOAO 63,295,58
ims-rbt TDLo:150 mg/kg (female 21-26D post):REP TXAPA9 15,185,69
scu-mus TDLo:200 mg/kg (female 1-5D post):REP BJPCBM 57,487,76
orl-rat TDLo:250 mg/kg (female 13-22D post):REP ENDOAO 63,295,58
orl-rbt TDLo:25 mg/kg (female 1D post):TER ENDOAO 65,339,59
scu-mus TDLo:120 mg/kg (female 1-3D post):REP JRPFA4 9,277,65
par-mus TDLo:120 mg/kg (female 1-3D post):REP JOENAK 20,299,60
ims-rat TDLo:200 mg/kg (female 1D pre):REP ENDOAO 82,959,68
orl-rbt TDLo:225 mg/kg (female 3D pre):REP ENDOAO 69,396,61
orl-rat TDLo:45 mg/kg (female 1-9D post):REP CCPTAY 3,347,71
orl-rat TDLo:495 mg/kg (female 31D pre):REP ENDOAO 63,295,58
ims-dom TDLo:575 mg/kg (female 18-21W post):REP FESTAS 17,637,66
orl-rat TDLo:7 mg/kg (female 7D pre):REP CCPTAY 3,347,71
orl-rat TDLo:480 mg/kg (female 6-14D post):TER TXAPA9 10,565,67
orl-rat TDLo:15,400 μg/kg (female 1-14D post):TER TXAPA9 10,565,67
scu-mus TDLo:120 mg/kg (female 12-15D post): TER IRLCDZ 4,379,76
scu-mus TDLo:30 mg/kg/5D-I:ETA IRLCDZ 4,379,76
orl-mus LD50:1700 mg/kg ENDOAO 63,295,58

SAFETY PROFILE: Moderately toxic by ingestion. Questionable carcinogen with experimental tumorigenic data. Experimental teratogenic and reproductive effects. When heated to decomposition it emits toxic fumes of NO_x.

DHS200 **CAS:2192-21-4** **HR: 3**
1-(2-(2-(DIETHYLAMINO)ETHOXY)PHENYL)-3-PHENYL-1-PROPANONE HYDROCHLORIDE
mf: $C_{21}H_{27}NO_2•ClH$ mw: 361.95

PROP: A solid. Mp: 129–130°.

SYNS: ASAMEDOL □ BAXACOR □ CORODILAN □ DIALICOR □ (o-β-DIETHYLAMINOETHOXY)-PHENYL PROPIOPHENONE HYDROCHLORIDE □ ETAFENONE HYDROCHLORIDE □ HETAPHENONE □ L.G. 11,457 HYDROCHLORIDE □ PAGANO-COR □ β-PHENYL-o-(DIETHYLAMINOETHOXY)PROPIOPHENONE HYDROCHLORIDE □ RELICOR □ RELICOR HYDROCHLORIDE

TOXICITY DATA WITH REFERENCE
orl-rat LD50:716 mg/kg ARZNAD 19,1664,69
ivn-rat LD50:20,800 μg/kg ARZNAD 19,1664,69
orl-mus LD50:352 mg/kg NIIRDN 6,109,82
orl-dog LD50:50 mg/kg ARZNAD 19,1664,69

SAFETY PROFILE: Poison by ingestion and intravenous routes. When heated to decomposition it emits very toxic fumes of NO_x and HCl.

DHS400 **CAS:17066-89-6** **HR: 3**
2-(2-(DIETHYLAMINO)ETHOXY)-N-(2,6-XYLYL)CINCHONINAMIDE, HYDROCHLORIDE
mf: $C_{24}H_{29}N_3O_2•ClH$ mw: 428.02

SYNS: C 3223 □ 2-(2-(DIETHYLAMINO)ETHOXY)-N-(2,6-XYLYL)-4-QUINOLINECARBOXAMIDE, HYDROCHLORIDE

TOXICITY DATA WITH REFERENCE
eye-rbt 2% SEV ARZNAD 8,708,58
ipr-rat LD50:175 mg/kg ARZNAD 8,708,58
scu-mus LD50:465 mg/kg ARZNAD 8,708,58

SAFETY PROFILE: Poison by intraperitoneal route. Moderately toxic by subcutaneous route. A severe eye irritant. When heated to decomposition it emits very toxic fumes of NO_x and HCl.

DHS600 **CAS:102207-85-2** **HR: 3**
2-(N′-(2-(DIETHYLAMINO)ETHYL)ACETAMIDO)-2′,6′-ACETOXYLIDIDE HYDROCHLORIDE
mf: $C_{18}H_{29}N_3O_2•ClH$ mw: 355.96

SYN: C 3150

TOXICITY DATA WITH REFERENCE
ipr-rat LD50:92 mg/kg ARZNAD 9,167,59
scu-mus LD50:212 mg/kg ARZNAD 9,167,59

SAFETY PROFILE: Poison by intraperitoneal and subcutaneous routes. When heated to decomposition it emits very toxic fumes of NO_x and HCl.

DHS800 **CAS:77966-71-3** **HR: 3**
2-(DIETHYLAMINO)-N-ETHYL-o-ACETOTOLUIDIDE HYDROCHLORIDE
mf: $C_{15}H_{24}N_2O•ClH$ mw: 284.87

SYN: C 5123

TOXICITY DATA WITH REFERENCE
ipr-rat LD50:140 mg/kg ARZNAD 8,708,58
scu-mus LD50:170 mg/kg ARZNAD 8,708,58

SAFETY PROFILE: Poison by intraperitoneal and subcutaneous routes. When heated to decomposition it emits very toxic fumes of NO_x and HCl.

DHT000 **CAS:77966-80-4** **HR: 3**
N-(2-(DIETHYLAMINO)ETHYL)-2′,6′-ACETOXYLIDIDE HYDROCHLORIDE
mf: $C_{16}H_{26}N_2O•ClH$ mw: 298.90

SYN: C 3137

TOXICITY DATA WITH REFERENCE
ipr-rat LD50:125 mg/kg ARZNAD 8,708,58
scu-mus LD50:190 mg/kg ARZNAD 8,708,58

SAFETY PROFILE: Poison by intraperitoneal and subcutaneous routes. When heated to decomposition it emits very toxic fumes of NO_x and HCl.

DHT125 **CAS:2426-54-2** **HR: 3**
DIETHYLAMINOETHYL ACRYLATE
mf: $C_9H_{17}NO_2$ mw: 171.27

SYNS: ACRYLIC ACID-N,N-DIETHYLAMINOETHYL ESTER □ AGEFLEX FA-2 □ N,N-DIETHYLAMINOETHYL ACRYLATE □ β-DIETHYLAMINOETHYL ACRYLATE □ 2-(DIETHYLAMINO)ETHYL ACRYLATE

TOXICITY DATA WITH REFERENCE
skn-rbt 5 mg/24H SEV 85JCAE-,748,86
eye-rbt 250 μg/24H SEV 85JCAE-,748,86
orl-rat LD50:770 mg/kg UCDS** 9/15/64
ipr-mus LDLo:31 mg/kg CBCCT* 4,226,52
skn-rbt LD50:200 mg/kg UCDS** 12/30/71

CONSENSUS REPORTS: Reported in EPA TSCA Inventory.

SAFETY PROFILE: Poison by skin contact and intraperitoneal routes. Moderately toxic by ingestion. A severe skin and eye irritant. When heated to decomposition it emits toxic fumes of NO_x.

DHT200 **CAS:54099-14-8** **HR: 3**
N-(2-(DIETHYLAMINO)ETHYL)-1-ADAMANTANEACETAMIDE ETHYL IODIDE
mf: $C_{18}H_{32}N_2O \cdot C_2H_5I$ mw: 448.49

TOXICITY DATA WITH REFERENCE
orl-mus LD50:600 mg/kg FRPSAX 32,129,77
ipr-mus LD50:35 mg/kg FRPSAX 32,129,77

SAFETY PROFILE: Poison by intraperitoneal route. Moderately toxic by ingestion. See also IODIDES. When heated to decomposition it emits very toxic fumes of NO_x and I^-.

DHT300 **CAS:133-16-4** **HR: 3**
2-(DIETHYLAMINO)ETHYL-4-AMINO-2-CHLOROBENZOATE
mf: $C_{13}H_{19}ClN_2O_2$ mw: 270.79

PROP: A solid. Mp: 41.5–42.5°.

SYN: CHLOROPROCAINE

TOXICITY DATA WITH REFERENCE
ipr-mus LD50:266 mg/kg TXAPA9 54,501,80
scu-mus LD50:1069 mg/kg ANESAV 54,177,81
ivn-gpg LDLo:64 mg/kg JMPCAS 3,525,61

SAFETY PROFILE: Poison by intravenous and intraperitoneal routes. Moderately toxic by subcutaneous route. When heated to decomposition it emits toxic fumes of NO_x and Cl^-.

DHT400 **CAS:52401-04-4** **HR: 3**
2-(2-(2-DIETHYLAMINO)ETHYLAMINO)ETHYL-2,5-DIMETHYL-1,3-BENZODIOXOLE DIMALEATE
mf: $C_{17}H_{28}N_2O_2 \cdot 2C_4H_4O_4$ mw: 524.63

TOXICITY DATA WITH REFERENCE
ivn-rat LD50:40 mg/kg EJMCA5 12,413,77
ipr-mus LD50:150 mg/kg EJMCA5 12,413,77

SAFETY PROFILE: Poison by intravenous and intraperitoneal routes. When heated to decomposition it emits toxic fumes of NO_x.

DHT600 **CAS:52400-99-4** **HR: 3**
2-(2-(2-(DIETHYLAMINO)ETHYLAMINO)ETHYL)-2-METHYL-1,3-BENZODIOXOLE, DIMALEATE
mf: $C_{16}H_{26}N_2O_2 \cdot 2C_4H_4O_4$ mw: 510.60

TOXICITY DATA WITH REFERENCE
ivn-rat LD50:30 mg/kg EJMCA5 12,413,77
ipr-mus LD50:98 mg/kg EJMCA5 12,413,77

SAFETY PROFILE: Poison by intravenous and intraperitoneal routes. When heated to decomposition it emits toxic fumes of NO_x.

DHT800 **CAS:52400-82-5** **HR: 3**
2-(2-(DIETHYLAMINO)ETHYLAMINOMETHYL)-2-METHYL-1,3-BENZODIOXOLE DIMALEATE
mf: $C_{15}H_{24}N_2O_2 \cdot 2C_4H_4O_4$ mw: 496.57

TOXICITY DATA WITH REFERENCE
ivn-rat LD50:40 mg/kg EJMCA5 12,413,77
ipr-mus LD50:150 mg/kg EJMCA5 12,413,77

SAFETY PROFILE: Poison by intravenous and intraperitoneal routes. When heated to decomposition it emits toxic fumes of NO_x.

DHU000 **CAS:479-50-5** **HR: 3**
1-(2′-DIETHYLAMINO)ETHYLAMINO-4-METHYLTHIOXANTHENONE
mf: $C_{20}H_{24}N_2OS$ mw: 340.52

PROP: Yellow crystals from EtOH (aq). Mp: 64–65°.

SYNS: 1-((2-(DIETHYLAMINO)ETHYL)AMINO)-4-METHYL-9H-THIOXANTHEN-9-ONE □ LUCANTHON □ LUCANTHONE □ MIRACIL D □ NILODIN

TOXICITY DATA WITH REFERENCE
mmo-sat 500 μmol/plate SCIEAS 186,647,74
mma-sat 50 μg/plate TXAPA9 52,237,80
dnd-hmn:hla 3 mg/L/15M ECREAL 103,175,76
cyt-hmn:lym 4 mg/L/24H MUREAV 55,43,78
cyt-hmn:leu 340 μg/L/24H MUREAV 55,43,78
ivn-mus LD50:56 mg/kg CSLNX* NX#01110
ims-mus LD50:400 mg/kg CSHCAL 4,445,77

CONSENSUS REPORTS: EPA Genetic Toxicology Program.

SAFETY PROFILE: Poison by intravenous route. Human mutation data reported. When heated to decomposition it emits very toxic fumes of NO_x and SO_x.

DHU400 **CAS:65268-91-9** **HR: 3**
2-(3-(2-(DIETHYLAMINO)ETHYLAMINO)PROPYL)-2-METHYL-1,3-BENZODIOXOLE DIMALEATE
mf: $C_{17}H_{28}N_2O_2 \cdot 2C_4H_4O_4$ mw: 524.63

TOXICITY DATA WITH REFERENCE
ivn-rat LD50:30 mg/kg EJMCA5 12,413,77
ipr-mus LD50:88 mg/kg EJMCA5 12,413,77

SAFETY PROFILE: Poison by intravenous and intraperi-

D

toneal routes. When heated to decomposition it emits toxic fumes of NO_x.

DHU600 **CAS:52479-15-9** **HR: 3**
2-(3-(2-(DIETHYLAMINO)ETHYLAMINO)PROPYL)-2-METHYL-1,3-NAPHTHOL(2,3-d)DIOXOLE DIHYDROCHLORIDE
mf: $C_{21}H_{30}N_2O_2$•2ClH mw: 415.45

TOXICITY DATA WITH REFERENCE
ivn-rat LD50:30 mg/kg EJMCA5 12,413,77
ipr-mus LD50:150 mg/kg EJMCA5 12,413,77

SAFETY PROFILE: Poison by intravenous and intraperitoneal routes. When heated to decomposition it emits very toxic fumes of NO_x and HCl.

DHU900 **CAS:302-40-9** **HR: 3**
DIETHYLAMINOETHYL BENZILATE
mf: $C_{20}H_{25}NO_3$ mw: 327.46

PROP: Crystals. Mp: 51°, mp: 177–178°. Solubility in water (25°): 14.9/100 mL. Practically insol in ether.

SYNS: BENACTIZINA (ITALIAN) □ BENACTYZIN □ BENACTYZINE □ BENZILIC ACID-β-DIETHYLAMINOETHYL ESTER □ DIAZIL □ β-DIETHYLAMINOETHYL BENZILATE □ 2-(DIETHYLAMINO)ETHYL BENZILATE □ 2-(DIETHYLAMINO)ETHYL DIPHENYLGLYCOLATE □ DIPHENYLGLYCOLIC ACID 2-(DIETHYLAMINO)ETHYL ESTER □ α-HYDROXY-α-PHENYLBENZENEACETIC ACID-2-(DIETHYLAMINO)ETHYL ESTER

TOXICITY DATA WITH REFERENCE
cyt-ham-ipr 10 mg/kg ACNSAX 17,253,75
ims-rat LD50:135 mg/kg NTIS** AD-A099-745
ipr-mus LD50:100 mg/kg BCFAAI 111,293,72
scu-mus LD50:159 mg/kg RPTOAN 32,311,69

SAFETY PROFILE: Poison by subcutaneous, intramuscular, and intraperitoneal routes. Mutation data reported. When heated to decomposition it emits toxic fumes of NO_x.

DHV200 **CAS:77985-23-0** **HR: 3**
N-((2-DIETHYLAMINO)ETHYL)CARBAMIC ACID-6-CHLORO-o-TOLYL ESTER HYDROCHLORIDE
mf: $C_{14}H_{21}ClN_2O_2$•ClH mw: 321.28

SYN: C 5307

TOXICITY DATA WITH REFERENCE
eye-rbt 2% MLD ARZNAD 8,708,58
ipr-rat LD50:52 mg/kg ARZNAD 8,708,58
scu-mus LD50:84 mg/kg ARZNAD 8,708,58

SAFETY PROFILE: Poison by intraperitoneal and subcutaneous routes. An eye irritant. When heated to decomposition it emits very toxic fumes of $Cl^-$and NO_x. See also CARBAMATES.

DHV400 **CAS:77985-24-1** **HR: 3**
N-(2-(DIETHYLAMINO)ETHYLCARBAMIC ACID) MESITYL ESTER HYDROCHLORIDE
mf: $C_{16}H_{26}N_2O_2$•ClH mw: 314.90

SYN: C 5308

TOXICITY DATA WITH REFERENCE
ipr-rat LD50:62 mg/kg ARZNAD 8,708,58
scu-mus LD50:178 mg/kg ARZNAD 8,708,58

SAFETY PROFILE: Poison by intraperitoneal and subcutaneous routes. See also CARBAMATES. When heated to decomposition it emits very toxic fumes of NO_x and HCl.

DHV600 **CAS:77985-25-2** **HR: 3**
N-(2-(DIETHYLAMINO)ETHYL)CARBAMIC ACID-2,6-XYLYL ESTER HYDROCHLORIDE
mf: $C_{15}H_{24}N_2O_2$•ClH mw: 300.87

SYN: C 3221

TOXICITY DATA WITH REFERENCE
ipr-rat LD50:40 mg/kg ARZNAD 8,708,58
scu-mus LD50:136 mg/kg ARZNAD 8,708,58

SAFETY PROFILE: Poison by intraperitoneal and subcutaneous routes. See also CARBAMATES. When heated to decomposition it emits very toxic fumes of NO_x and HCl.

DHW200 **CAS:902-83-0** **HR: 3**
2-(DIETHYLAMINO)ETHYLCHLORODIPHENYLACETATE HYDROCHLORIDE
mf: $C_{20}H_{24}ClNO_2$•ClH mw: 382.36

SYNS: 2-CHLORO-2,2-DIPHENYLACETIC ACID-2-(DIETHYLAMINO) ETHYL ESTER HYDROCHLORIDE □ DIAMINOPHEN □ DIAPHEN (NEUROPLEGIC)

TOXICITY DATA WITH REFERENCE
ipr-mus LD50:76 mg/kg JPETAB 74,274,42

SAFETY PROFILE: Poison by intraperitoneal route. When heated to decomposition it emits very toxic fumes of Cl^- and NO_x.

DHW400 **CAS:3737-35-7** **HR: 3**
2-DIETHYLAMINOETHYL CYCLOPENTYL(2-THIENYL) GLYCOLATE HYDROCHLORIDE
mf: $C_{17}H_{27}NO_3S$•ClH mw: 361.97

SYNS: α-CYCLOPENTYL-2-THIOPHENEGLYCOLIC ACID-2-(DIETHYLAMINO)ETHYL ESTER HYDROCHLORIDE □ WIN-2299 HYDROCHLORIDE

TOXICITY DATA WITH REFERENCE
ivn-hmn TDLo:29 μg/kg:EYE,CNS AJPSAO 113,887,57
orl-mus LD50:770 mg/kg JPETAB 110,282,54
ivn-mus LD50:80 mg/kg JPETAB 110,282,54

SAFETY PROFILE: Poison by intravenous route. Moderately toxic by ingestion. Human systemic effects by intravenous route: dilation of the pupils, somnolence, and hallucinations or distorted perceptions. When heated to decomposition it emits very toxic fumes of HCl, SO_x, and NO_x.

DHW600 **CAS:9015-73-0** **HR: 2**
DIETHYLAMINOETHYL-DEXTRAN

SYNS: DEAE-D □ DIETHYLAMINOETHYLDEXTRAN POLYMER

TOXICITY DATA WITH REFERENCE
scu-mus TDLo: 1200 mg/kg/30W-I:NEO JNCIAM 50,387,73

SAFETY PROFILE: Questionable carcinogen with experimental neoplastigenic data. When heated to decomposition it emits toxic fumes of NO_x. See also various dextrans.

DHX600 **CAS:77945-09-6** **HR: 3**
N-(2-(DIETHYLAMINO)ETHYL)-2′,6′-DIMETHYLCROTONANILIDE HYDROCHLORIDE
mf: $C_{18}H_{28}N_2O{\cdot}ClH$ mw: 324.94

SYNS: C 3141 □ N-(2-(DIETHYLAMINO)ETHYL)-2′,6′-CROTONXYLILIDE HYDROCHLORIDE

TOXICITY DATA WITH REFERENCE
eye-rbt 2% MLD ARZNAD 8,708,58
ipr-rat LD50:58 mg/kg ARZNAD 8,708,58
scu-mus LD50:70 mg/kg ARZNAD 8,708,58

SAFETY PROFILE: Poison by intraperitoneal and subcutaneous routes. An eye irritant. When heated to decomposition it emits very toxic fumes of HCl and NO_x.

DHX800 **CAS:64-95-9** **HR: 3**
2-DIETHYLAMINOETHYL DIPHENYLACETATE
mf: $C_{20}H_{25}NO_2$ mw: 311.46

SYNS: ADIPHENIN □ ADIPHENINE □ BENZENEACETIC ACID, α-PHENYL-, 2-(DIETHYLAMINO)ETHYL ESTER, (9CI) □ 2-DIETHYLAMINOETHYLESTER KYSELINY DIFENYLOCTOVE □ DIFACIL □ DIPHACIL □ DIPHACYL □ DIPHENYLACETIC ACID DIETHYLAMINOETHYL ESTER □ DIPHENYLACETIC ACID, 2-(DIETHYLAMINO)ETHYL ESTER □ DIPHENYLACETYLDIETHYLAMINOETHANOL □ ESTER DWUETYLOAMINOETYLOWY KWASU DWUFENYLOOCTOWEGO □PATROVINE □ SPASMOLYTIN □ TRANSENTINE □ TRANZETIL □ TRASENTIN □ TRASENTINE □ TRAZENTYNA □ VEGANTINE □ WEGANTYNA

TOXICITY DATA WITH REFERENCE
orl-rat LDLo:1600 mg/kg APPNAH 1,4,50
scu-rat LDLo:1600 mg/kg APPNAH 1,4,50
ivn-rat LD50:27 mg/kg AJDDAL 18,241,51
orl-mus LD50:600 mg/kg CLDND* 2,201,68
scu-mus LD50:400 mg/kg BJPCAL 14,559,59
ivn-mus LD50:21,500 µg/kg JPETAB 104,269,52
ivn-dog LD50:35 mg/kg JPETAB 89,131,47
ivn-rbt LD50:30 mg/kg CLDND*

SAFETY PROFILE: Poison by intravenous and intraperitoneal routes. Moderately toxic by ingestion and subcutaneous routes. See also ESTERS. When heated to decomposition it emits toxic fumes of NO_x.

DHY200 **CAS:668-37-1** **HR: 3**
2-DIETHYLAMINOETHYLDIPHENYLCARBAMATE HYDROCHLORIDE
mf: $C_{19}H_{24}N_2O_2{\cdot}ClH$ mw: 348.91

PROP: Crystals from Me_2CO/EtOH. Mp: 182°.

SYNS: DIAETHYLAMINOAETHYL DIPHENYLCARBAMAT HYDROCHLORID (GERMAN) □ DIAMFEN HYDROCHLORIDE □ DIETHYLAMINOETHYLDIPHENYL CARBAMATE HYDROCHLORIDE □ DIETHYL(2-HYDROXYETHYL)AMMONIUM CHLORIDE DIPHENYLCARBAMATE □ DIPHENYLCARBAMIC ACID-2-(DIETHYLAMINO)ETHYL ESTER HYDROCHLORIDE □ 1352 HC □ SD 25

TOXICITY DATA WITH REFERENCE
orl-rat LD50:710 mg/kg ARZNAD 15,534,65
ivn-rat LD50:27 mg/kg ARZNAD 15,534,65
orl-mus LD50:270 mg/kg ARZNAD 15,534,65
ipr-mus LD50:76 mg/kg ARZNAD 15,534,65
scu-mus LD50:180 mg/kg AIPTAK 125,311,60
ivn-mus LD50:41 mg/kg ARZNAD 15,534,65

SAFETY PROFILE: Poison by ingestion, subcutaneous, intravenous, and intraperitoneal routes. When heated to decomposition it emits very toxic fumes of NO_x, NH_3, and HCl. See also CARBAMATES.

D

DHY300 **HR: 3**
3-(2-(DIETHYLAMINO)ETHYL)-5,5-DIPHENYLHYDANTOIN
mf: $C_{21}H_{25}N_2O_2$ mw: 337.44

SYN: DIETHYLAMINOETHYLDIPHENYLHYDANTOINE (FRENCH)

TOXICITY DATA WITH REFERENCE
orl-rat LD50:1300 mg/kg THERAP 11,1159,56
scu-mus LD50:1300 mg/kg THERAP 11,1159,56
ivn-mus LD50:37 mg/kg THERAP 11,1159,56

SAFETY PROFILE: Poison by intravenous route. Moderately toxic by ingestion and subcutaneous routes. When heated to decomposition it emits toxic fumes of NO_x.

DHY400 **CAS:548-68-5** **HR: 3**
S-(2-(DIETHYLAMINO)ETHYL)DIPHENYLTHIOACETIC ACID HYDROCHLORIDE
mf: $C_{20}H_{25}NOS{\cdot}ClH$ mw: 363.98

PROP: Rosettes of small needles from C_6H_6/pet ether; large prisms from EtOH/EtOAc. Mp: 129–130°.

SYNS: 23B □ S-(2-(DIETHYLAMINO)ETHYL)DIPHENYLTHIOACETIC ACID HYDROCHLORIDE □ β-DIETHYLAMINOETHYL DIPHENYLTHIOACETATE HYDROCHLORIDE □ 2-(DIETHYLAMINO)ETHYL DIPHENYLTHIOACETATE HYDROCHLORIDE □ 2-DIETHYLAMINOETHYL DIPHENYLTHIOLACETATE HYDROCHLORIDE □ DIPHENYLTHIOLACETIC ACID-2-DIETHYLAMINOETHYL ESTER HYDROCHLORIDE □ THIPHEN □ THIPHENAMIL HYDROCHLORIDE □ TIFEN □ TIPHEN □ TROCINAT □ TROCINATE □ TROCINATE HYDROCHLORIDE

TOXICITY DATA WITH REFERENCE
orl-rat LD50:2720 mg/kg 29ZVAB -,117,69
ipr-rat LD50:1500 mg/kg JPETAB 89,131,47
imp-rat LD50:1500 mg/kg 29ZVAB -,117,69
orl-mus LD50:443 mg/kg JPETAB 89,131,47
ipr-mus LD50:187 mg/kg JPETAB 89,131,47
ivn-mus LD50:30 mg/kg JPETAB 89,131,47
orl-dog LD50:1500 mg/kg 29ZVAB -,117,69
ivn-dog LD50:30 mg/kg 29ZVAB -,117,69
ivn-rbt LD50:19 mg/kg 29ZVAB -,117,69

SAFETY PROFILE: Poison by intravenous and intraperitoneal routes. Moderately toxic by ingestion and implant. See also ESTERS. When heated to decomposition it emits very toxic fumes of NO_x, SO_x, and HCl.

DHZ000 CAS:41542-50-1 HR: 2
S-DIETHYLAMINOETHYL ESTER-p-BROMOTHIOBENZO HYDROXIMIC ACID HYDROCHLORIDE
mf: $C_{13}H_{19}BrNOS\cdot ClH$ mw: 353.76

SYNS: p-BROMOBENZOYLTHIOHYDROXIMIC ACID-5-DIETHYLAMINOETHYL ESTER HYDROCHLORIDE □ p-BROMOTHIOBENZOHYDROXIMIC ACID-S-DIETHYLAMINOETHYL ESTER HYDROCHLORIDE □ DIETHYXIME

TOXICITY DATA WITH REFERENCE
ims-rat LD50:920 mg/kg RPTOAN 38,105,75
ims-mus LD50:810 mg/kg RPTOAN 38,105,75
ims-cat LD50:500 mg/kg RPTOAN 38,105,75
ims-rbt LD50:1 g/kg BEXBAN 83,32,77

SAFETY PROFILE: Moderately toxic by intramuscular route. When heated to decomposition it emits toxic fumes of Br^-, SO_x, NO_x, and HCl. See also ESTERS.

DHZ050 CAS:13406-60-5 HR: 3
1-(2-(DIETHYLAMINO)ETHYL)-2-(p-ETHOXYBENZYL)-5-BENZIMIDAZOLYL METHYL KETONE
mf: $C_{24}H_{31}N_3O_2$ mw: 393.58

SYN: KETONE, 1-(2-(DIETHYLAMINO)ETHYL)-2-(p-ETHOXYBENZYL)-5-BENZIMIDAZOLYL METHYL

TOXICITY DATA WITH REFERENCE
ipr-mus LD50:13 mg/kg CHTPBA 2,16,67

DOT CLASSIFICATION: 3; *Label:* Flammable Liquid

SAFETY PROFILE: A poison by intraperitoneal route. A flammable liquid. When heated to decomposition it emits toxic vapors of NO_x.

DHZ100 CAS:4929-77-5 HR: 3
1-(2-(DIETHYLAMINO)ETHYL)-2-((p-ETHOXYPHENYL)THIO)BENZIMIDAZOLE HYDROCHLORIDE
mf: $C_{21}H_{27}N_3OS\cdot ClH$ mw: 406.03

TOXICITY DATA WITH REFERENCE
orl-mus LD50:230 mg/kg YKKZAJ 87,296,67
ipr-mus LD50:84 mg/kg YKKZAJ 85,962,65
scu-mus LD50:190 mg/kg YKKZAJ 87,296,67
ivn-mus LD50:24 mg/kg YKKZAJ 87,296,67

SAFETY PROFILE: Poison by ingestion, subcutaneous, intravenous, and intraperitoneal routes. When heated to decomposition it emits toxic fumes of SO_x, NO_x, and HCl.

DIA000 CAS:77945-03-0 HR: 3
N-(2-(DIETHYLAMINO)ETHYL)-2-ETHOXY-N-(2,6-XYLYL)CINCHONINAMIDE HYDROCHLORIDE
mf: $C_{26}H_{33}N_3O_2\cdot ClH$ mw: 456.08

SYNS: C 3222 □ N-(2-(DIETHYLAMINO)ETHYL)-2-ETHOXY-N-(2,6-XYLYL)-4-QUINOLINECARBOXAMIDE HYDROCHLORIDE

TOXICITY DATA WITH REFERENCE
eye-rbt 2% SEV ARZNAD 8,708,58
ipr-rat LD50:150 mg/kg ARZNAD 8,708,58
scu-mus LD50:250 mg/kg ARZNAD 8,708,58

SAFETY PROFILE: Poison by intraperitoneal and subcutaneous routes. A severe eye irritant. When heated to decomposition it emits very toxic fumes of NO_x and HCl.

DIA400 CAS:102129-21-5 HR: 3
3-(2-(DIETHYLAMINO)ETHYL)-5-(2-FURYL)-1-PHENYL-1H-PYRAZOLINE HYDROCHLORIDE
mf: $C_{19}H_{25}N_3O\cdot ClH$ mw: 347.93

SYN: 1-PHENYL-3-(β-DIAETHYLAMINO-AETHYL)-5-FURYL-PYRAZOLIN-HYDROCHLORID (GERMAN)

TOXICITY DATA WITH REFERENCE
ipr-mus LD50:95 mg/kg ARZNAD 10,925,60
scu-mus LD50:180 mg/kg ARZNAD 10,925,60

SAFETY PROFILE: Poison by intraperitoneal and subcutaneous routes. When heated to decomposition it emits very toxic fumes of NO_x and HCl.

DIA800 CAS:39367-89-0 HR: 3
N-(2-(DIETHYLAMINO)ETHYL)-N-MESITYL-β-ISODURYLAMIDE HYDROCHLORIDE
mf: $C_{25}H_{36}N_2O\cdot ClH$ mw: 417.09

SYNS: C 3230 □ N-(2-(DIETHYLAMINO)ETHYL)-2,2',4,4',6,6'-HEXAMETHYLBENZANILIDE HYDROCHLORIDE

TOXICITY DATA WITH REFERENCE
eye-rbt 2% MLD ARZNAD 8,708,58
ipr-rat LD50:210 mg/kg ARZNAD 8,708,58
scu-mus LD50:700 mg/kg ARZNAD 8,708,58

SAFETY PROFILE: Poison by intraperitoneal route. Moderately toxic by subcutaneous route. An eye irritant. When heated to decomposition it emits very toxic fumes of HCl and NO_x.

DIB000 CAS:78372-03-9 HR: 3
1-(2-(DIETHYLAMINO)ETHYL)-3-MESITYL-1-METHYLUREA HYDROCHLORIDE
mf: $C_{17}H_{29}N_3O\cdot ClH$ mw: 327.95

SYN: C 3234

TOXICITY DATA WITH REFERENCE
eye-rbt 2% MLD ARZNAD 8,664,58
ipr-rat LD50:160 mg/kg ARZNAD 8,664,58
scu-mus LD50:287 mg/kg ARZNAD 8,664,58

SAFETY PROFILE: Poison by intraperitoneal and subcutaneous routes. An eye irritant. When heated to decomposition it emits very toxic fumes of NO_x and HCl.

DIB200 CAS:97702-94-8 HR: 3
2-(DIETHYLAMINO)-N-ETHYL-N-(1-MESITYLOXY-2-PROPYL)ACETAMIDE HYDROCHLORIDE
mf: $C_{20}H_{34}N_2O_2\cdot ClH$ mw: 371.02

SYN: C 2046

TOXICITY DATA WITH REFERENCE
eye-rbt 2% MLD ARZNAD 9,70,59
scu-mus LD50:140 mg/kg ARZNAD 9,70,59

SAFETY PROFILE: Poison by subcutaneous route. An eye irritant. When heated to decomposition it emits very toxic fumes of NO_x and HCl.

DIB300 CAS:105-16-8 HR: 1
2-(DIETHYLAMINO)ETHYL METHACRYLATE
mf: $C_{10}H_{19}NO_2$ mw: 185.30

SYNS: DAKTOSE B □ 2-DIETHYLAMINOETHYLESTER KYSELINY METHAKRYLOVE □ DIETHYLAMINOETHYL METHACRYLATE □ β-(DIETHYLAMINO)ETHYL METHACRYLATE □ 2-(N,N-DIETHYLAMINO) ETHYL METHACRYLATE □ METHACRYLIC ACID, 2-(DIETHYLAMINO) ETHYL ESTER □ 2-PROPENOIC ACID, 2-METHYL-, 2-(DIETHYLAMINO)ETHYL ESTER (9CI)

TOXICITY DATA WITH REFERENCE
orl-rat LD50:4696 mg/kg 85GMAT -,51,82
ihl-rat LC50:11 g/m³/4H 85GMAT -,51,82
ihl-mus LC50:12,100 mg/m³/2H 85GMAT -,51,82

CONSENSUS REPORTS: Reported in EPA TSCA Inventory.

SAFETY PROFILE: Mildly toxic by ingestion and inhalation. When heated to decomposition it emits toxic vapors of NO_x.

DIB600 CAS:1227-61-8 HR: 3
N-(2-(DIETHYLAMINO)ETHYL)-2-(p-METHOXYPHENOXY)ACETAMIDE
mf: $C_{15}H_{24}N_2O_3$ mw: 280.41

SYNS: MEPHEXAMIDE □ 2-(p-METHOXYPHENOXY) N (2 (DIETHYLAMINO)ETHYL)ACETAMIDE □ MEXEPHENAMIDE

TOXICITY DATA WITH REFERENCE
orl-mus LD50:1500 µg/kg CHTPBA 1,444,66
ivn-mus LD50:168 mg/kg 27ZQAG -,396,72
ivn-rbt LD50:135 mg/kg AMPYAT 123,141,65

SAFETY PROFILE: Poison by ingestion and intravenous routes. When heated to decomposition it emits toxic fumes of NO_x.

DIB800 CAS:97702-95-9 HR: 3
2-(DIETHYLAMINO)-N-ETHYL-N-(1-(p-METHOXYPHENOXY)-2-PROPYL)ACETAMIDE HYDROCHLORIDE
mf: $C_{18}H_{30}N_2O_3{\cdot}ClH$ mw: 358.96

SYN: C 2098

TOXICITY DATA WITH REFERENCE
eye-rbt 2% MLD ARZNAD 9,70,59
scu-mus LD50:355 mg/kg ARZNAD 9,70,59

SAFETY PROFILE: Poison by subcutaneous route. An eye irritant. When heated to decomposition it emits very toxic fumes of NO_x and HCl.

DIC000 CAS:77791-67-4 HR: 3
2-(2-(DIETHYLAMINO)ETHYL)METHYLAMINO-o-ACETANISIDIDE DIHYDROCHLORIDE
mf: $C_{16}H_{27}N_3O_2{\cdot}2ClH$ mw: 366.38

SYN: C 5416

TOXICITY DATA WITH REFERENCE
ipr-rat LD50:182 mg/kg ARZNAD 9,262,59
scu-mus LD50:850 mg/kg ARZNAD 9,262,59

SAFETY PROFILE: Poison by intraperitoneal route. Moderately toxic by subcutaneous route. When heated to decomposition it emits very toxic fumes of NO_x and HCl.

DIC200 CAS:77966-72-4 HR: 3
2-(2-(DIETHYLAMINO)ETHYL)METHYLAMINO-o-ACETOTOLUIDIDE DIHYDROCHLORIDE
mf: $C_{16}H_{27}N_3O{\cdot}2ClH$ mw: 350.38

SYN: C 5346

TOXICITY DATA WITH REFERENCE
ipr-rat LD50:150 mg/kg ARZNAD 9,262,59
scu-mus LD50:800 mg/kg ARZNAD 9,262,59

SAFETY PROFILE: Poison by intraperitoneal route. Moderately toxic by subcutaneous route. When heated to decomposition it emits very toxic fumes of NO_x and HCl.

DIC400 CAS:77966-81-5 HR: 3
2-(2-(DIETHYLAMINO)ETHYL)METHYLAMINO-2′,6′-ACETOXYLIDIDE DIHYDROCHLORIDE
mf: $C_{17}H_{29}N_3O{\cdot}2ClH$ mw: 364.41

SYN: C 5342

TOXICITY DATA WITH REFERENCE
ipr-rat LD50:125 mg/kg ARZNAD 9,262,59
scu-mus LD50:500 mg/kg ARZNAD 9,262,59

SAFETY PROFILE: Poison by intraperitoneal route. Moderately toxic by subcutaneous route. When heated to decomposition it emits very toxic fumes of NO_x and HCl.

DIC600 CAS:52479-14-8 HR: 3
2-((2-(DIETHYLAMINO)ETHYL)METHYLAMINO)ETHYL-2 METHYL 1,3 BENZODIOXOLE DIMALEATE
mf: $C_{17}H_{28}N_2O_2{\cdot}2C_4H_4O_4$ mw: 524.63

TOXICITY DATA WITH REFERENCE
ivn-rat LD50:8500 µg/kg EJMCA5 12,413,77
ipr-mus LD50:75 mg/kg EJMCA5 12,413,77

SAFETY PROFILE: Poison by intravenous and intraperitoneal routes. When heated to decomposition it emits toxic fumes of NO_x.

DIC800 CAS:65210-31-3 HR: 3
2-(2-(DIETHYLAMINO)ETHYL)-2-METHYL-1,3-BENZODIOXOLE HYDROCHLORIDE
mf: $C_{14}H_{21}NO_2{\cdot}ClH$ mw: 271.82

TOXICITY DATA WITH REFERENCE
ivn-rat LD50:45 mg/kg EJMCA5 12,413,77
ipr-mus LD50:100 mg/kg EJMCA5 12,413,77

SAFETY PROFILE: Poison by intravenous and intraperitoneal routes. When heated to decomposition it emits very toxic fumes of HCl and NO_x.

DID800 CAS:17617-13-9 HR: 3
1-DIETHYLAMINOETHYL-2-METHYL-3-PHENYL-1,2,3,4-TETRAHYDRO-4-QUINAZOLINONE
mf: $C_{21}H_{27}N_3O \cdot C_2H_2O_4$ mw: 427.55

SYN: 2,3-DIHYDRO-1-DIETHYLAMINOETHYL-2-METHYL-3-PHENYL-4(3H)-QUINAZOLINONE OXALATE

TOXICITY DATA WITH REFERENCE
orl-mus LD50:678 mg/kg JMCMAR 11,788,68
ipr-mus LD50:180 mg/kg JMCMAR 11,788,68

SAFETY PROFILE: Poison by intraperitoneal route. Moderately toxic by ingestion. When heated to decomposition it emits toxic fumes of NO_x.

DIE200 CAS:34963-48-9 HR: 3
1-DIETHYLAMINOETHYL-2-METHYL-1,2,3,4-TETRAHYDRO-3-(o-TOLYL)-4-QUINAZOLINONE OXALATE
mf: $C_{22}H_{29}N_3O \cdot C_2H_2O_4$ mw: 441.58

SYN: 2,3-DIHYDRO-1-DIETHYLAMINOETHYL-2-METHYL-3-(o-TOLYL)-4(3H)-QUINAZOLINONE OXALATE

TOXICITY DATA WITH REFERENCE
orl-mus LD50:620 mg/kg JMCMAR 11,788,68
ipr-mus LD50:156 mg/kg JMCMAR 11,788,68

SAFETY PROFILE: Poison by intraperitoneal route. Moderately toxic by ingestion. When heated to decomposition it emits toxic fumes of NO_x. See also OXALATES.

DIE300 HR: 3
DIETHYLAMINOETHYLMORPHINE
mf: $C_{23}H_{32}N_2O_3$ mw: 384.57

SYN: 7,8-DIDEHYDRO-3-(2-(DIETHYLAMINO)ETHOXY)-4,5-α-EPOXY-17-METHYLMORPHINAN-6-α-OL

TOXICITY DATA WITH REFERENCE
ipr-mus LD50:40 mg/kg APFRAD 8,261,50
scu-mus LD50:110 mg/kg THERAP 7,21,52
ivn-mus LD50:30 mg/kg THERAP 7,21,52

SAFETY PROFILE: Poison by subcutaneous, intravenous, and intraperitoneal routes. When heated to decomposition it emits toxic fumes of NO_x. See also MORPHINE.

DIE350 CAS:15451-93-1 HR: 3
1-(2-(DIETHYLAMINO)ETHYL)-2-p-PHENETIDINO-5-BENZIMIDAZOLYL METHYL KETONE
mf: $C_{23}H_{30}N_4O_2$ mw: 394.57

SYN: KETONE, 1-(2-(DIETHYLAMINO)ETHYL)-2-p-PHENETIDINO-5-BENZIMIDAZOLYL METHYL

TOXICITY DATA WITH REFERENCE
scu-mus LDLo:200 mg/kg CHTPBA 2,16,67

DOT CLASSIFICATION: 3; *Label:* Flammable Liquid

SAFETY PROFILE: A poison by subcutaneous route. A flammable liquid. When heated to decomposition it emits toxic vapors of NO_x.

DIE400 CAS:97702-97-1 HR: 3
2-(DIETHYLAMINO)-N-ETHYL-N-(1-PHENOXY-2-PROPYL)ACETAMIDE HYDROCHLORIDE
mf: $C_{17}H_{28}N_2O_2 \cdot ClH$ mw: 328.93

SYN: C 2053

TOXICITY DATA WITH REFERENCE
eye-rbt 2% MLD ARZNAD 9,70,59
scu-mus LD50:315 mg/kg ARZNAD 9,70,59

SAFETY PROFILE: Poison by subcutaneous route. An eye irritant. When heated to decomposition it emits very toxic fumes of NO_x and HCl.

DIE600 CAS:102128-91-6 HR: 3
3-(DIETHYLAMINO)-N-ETHYL-N-(1-PHENOXY-2-PROPYL)PROPIONAMIDE HYDROCHLORIDE
mf: $C_{18}H_{30}N_2O_2 \cdot ClH$ mw: 342.96

SYN: C 2052

TOXICITY DATA WITH REFERENCE
eye-rbt 2% HLD ARZNAD 9,70,59
scu-mus LD50:387 mg/kg ARZNAD 9,70,59

SAFETY PROFILE: Poison by subcutaneous route. An eye irritant. When heated to decomposition it emits very toxic fumes of HCl and NO_x.

DIF200 CAS:6009-67-2 HR: 3
3-(β-DIETHYLAMINOETHYL)-3-PHENYL-2-BENZOFURANONE HYDROCHLORIDE
mf: $C_{20}H_{23}NO_2 \cdot ClH$ mw: 345.90

PROP: A solid. Mp: 152–153°.

SYNS: ABBOTT'S A.P. 43 ☐ AMETHONE HYDROCHLORIDE ☐ AMOLANONE HYDROCHLORIDE ☐ AP 43 ☐ 3-PHENYL-3-DIETHYLAMINOETHYLBENZOFURANONE-2-HYDROCHLORIDE

TOXICITY DATA WITH REFERENCE
orl-rat LD50:600 mg/kg JPETAB 84,387,45
orl-mus LD50:660 mg/kg JPETAB 84,387,45
ipr-mus LD50:205 mg/kg JPETAB 84,387,45
scu-mus LD50:395 mg/kg JPETAB 84,387,45
scu-rbt LD50:250 mg/kg JPETAB 84,387,45
ivn-rbt LD50:23 mg/kg JPETAB 84,387,45

SAFETY PROFILE: Poison by intraperitoneal, subcutaneous, and intravenous routes. Moderately toxic by ingestion. An anesthetic. When heated to decomposition it emits very toxic fumes of NO_x and HCl.

DIF600 CAS:14557-50-7 HR: 3
2-(2-(DIETHYLAMINO)ETHYL)-2-PHENYL-4-PENTENOIC ACID ETHYL ESTER
mf: $C_{19}H_{29}NO_2$ mw: 303.49

SYN: UCB 6249

TOXICITY DATA WITH REFERENCE
orl-rat LD50:200 mg/kg 27ZQAG -,367,72
ipr-rat LD50:90 mg/kg 27ZQAG -,367,72
ivn-rat LD50:41 mg/kg 27ZQAG -,367,72
orl-mus LD50:200 mg/kg 27ZQAG -,367,72
ipr-mus LD50:60 mg/kg 27ZQAG -,367,72

ivn-mus LD50:28 mg/kg 27ZQAG -,367,72

SAFETY PROFILE: Poison by ingestion, intraperitoneal, and intravenous routes. See also ESTERS. When heated to decomposition it emits toxic fumes of NO_x.

DIH000 CAS:1111-44-0 HR: 3
1-(2-(DIETHYLAMINO)ETHYL)RESERPINE BITARTRATE
mf: $C_{39}H_{53}N_3O_9 \cdot 2C_4H_6O_6$ mw: 1008.15

PROP: A solid. Mp: 145–150° (decomp).

SYNS: BIETASERPINE BITARTRATE □ D. L. 152 □ TENSIBAR

TOXICITY DATA WITH REFERENCE
orl-rat TDLo:24 g/kg (30D pre):REP KSRNAM 5,81,71
orl-rat TDLo:3840 mg/kg (30D male):REP KSRNAM 5,81,71
orl-rat LD50:4885 mg/kg KSRNAM 5,81,71
ipr-rat LD50:496 mg/kg KSRNAM 5,81,71
scu-rat LD50:888 mg/kg KSRNAM 5,81,71
orl-mus LD50:620 mg/kg ARZNAD 14,1040,64
ipr-mus LD50:430 mg/kg ARZNAD 14,1040,64
scu-mus LD50:1061 mg/kg KSRNAM 5,81,71
ivn-mus LD50:215 mg/kg ARZNAD 14,1040,64

SAFETY PROFILE: Poison by intravenous route. Moderately toxic by ingestion, subcutaneous, and intraperitoneal routes. An experimental teratogen. Experimental reproductive effects. An antihypertensive agent. When heated to decomposition it emits toxic fumes of NO_x.

DIH600 CAS:17140-68-0 HR: 3
7-(2-(DIETHYLAMINO)ETHYL)THEOPHYLLINE HYDROCHLORIDE
mf: $C_{13}H_{21}N_5O_2 \cdot ClH$ mw: 315.85

PROP: A solid. Mp: 239–241°.

SYNS: CHLORHYDRATE de DIETHYLAMINOETHYLTHEOPHYLLINE (FRENCH) □ 7-(β-DIAETHYLAMINO-AETHYL)-THEOPHYLLIN-HYDROCHLORID (GERMAN) □ 7-(2-(DIETHYLAMINO)ETHYL)-3,7-DIHYDRO-1,3-DIMETHYL-1H-PURINE-2,6-DIONE HYDROCHLORIDE □ 7-(2-(N,N-DIETHYLAMINO)ETHYL)THEOPHYLLINE HYDROCHLORIDE □ ETAMIPHYLLINE HYDROCHLORIDE □ ETAMIPHYLLIN HYDROCHLORIDE □ SOLUFILINA

TOXICITY DATA WITH REFERENCE
ipr-mus LD50:182 mg/kg ARZNAD 8,190,58
scu-mus LD50:182 mg/kg ARZNAD 4,649,54
ivn-mus LD50:127 mg/kg AIPTAK 103,146,55

SAFETY PROFILE: Poison by intravenous, intraperitoneal, and subcutaneous routes. When heated to decomposition it emits very toxic fumes of HCl and NO_x. See also THEOPHYLLIN.

DIH800 CAS:101670-51-3 HR: 3
2-((2-DIETHYLAMINO)ETHYL)THIOACETANILIDE HYDROCHLORIDE
mf: $C_{14}H_{22}N_2OS \cdot ClH$ mw: 302.90

SYN: C 4928

TOXICITY DATA WITH REFERENCE
ipr-rat LD50:220 mg/kg ARZNAD 9,683,59
scu-mus LD50:1500 mg/kg ARZNAD 9,683,59

SAFETY PROFILE: Poison by intraperitoneal route. Moderately toxic by subcutaneous route. When heated to decomposition it emits very toxic fumes of NO_x, SO_x, and HCl.

DII000 CAS:102489-61-2 HR: 3
2-(2-(DIETHYLAMINO)ETHYL)THIO-o-ACETOTOLUIDIDE HYDROCHLORIDE
mf: $C_{15}H_{24}N_2OS \cdot ClH$ mw: 316.93

SYN: C 4924

TOXICITY DATA WITH REFERENCE
ipr-rat LD50:310 mg/kg ARZNAD 9,683,59
scu-mus LD50:1350 mg/kg ARZNAD 9,683,59

SAFETY PROFILE: Poison by intraperitoneal route. Moderately toxic by subcutaneous route. When heated to decomposition it emits very toxic fumes of NO_x, SO_x and HCl.

DII200 CAS:60-91-3 HR: 3
N-(DIETHYLAMINOETHYL)THIODIPHENYLAMINE
mf: $C_{18}H_{22}N_2S$ mw: 298.48

PROP: Oily liquid. Bp: 195–208°. Sol in EtOH, C_6H_6, $CHCl_3$, and acids; insol in H_2O.

SYNS: ANTIPAR □ CASANTIN □ DIETHAZIN □ DIETHAZINE □ N-(2′-DIETHYLAMINOETHYL)DIBENZOPARATHIAZINE □ 10-(2-DIETHYLAMINOETHYL)PHENOTHIAZINE □ DINEZIN □ DIPARCOL □ DOLISINA □ DOLSIMA □ EAZAMINE □ FOURNEAU 2987 □ LATIBON □ LODIBON □ 2987 R.P. □ THIANTAN

TOXICITY DATA WITH REFERENCE
ivn-rat LD50:28,500 μg/kg AIPTAK 103,371,55
orl-mus LD50:450 mg/kg AIPTAK 134,255,61
ipr-mus LD50:225 mg/kg ARZNAD 4,171,54
scu-mus LD50:450 mg/kg PSCBAY 2,17,63
ivn-mus LD50:65 mg/kg AIPTAK 120,450,59
ivn-rbt LD50:20 mg/kg AIPTAK 120,450,59

SAFETY PROFILE: Poison by intravenous and intraperitoneal routes. Moderately toxic by ingestion and subcutaneous routes. When heated to decomposition it emits very toxic fumes of NO_x and SO_x.

DII400 CAS:78109-87-2 HR: 3
N-(2-(DIETHYLAMINO)ETHYL)-2,4,6-TRIMETHYLBENZAMIDE HYDROCHLORIDE
mf: $C_{16}H_{26}N_2O \cdot ClH$ mw: 298.90

SYNS: C 3235 □ N-(2-(DIETHYLAMINO)ETHYL)-β-ISODURYLAMIDE HYDROCHLORIDE

TOXICITY DATA WITH REFERENCE
eye-rbt 2% MLD ARZNAD 8,708,58
ipr-rat LD50:171 mg/kg ARZNAD 8,708,58
scu-mus LD50:390 mg/kg ARZNAD 8,708,58

SAFETY PROFILE: Poison by intraperitoneal and subcutaneous routes. An eye irritant. When heated to decomposition it emits very toxic fumes of HCl and NO_x.

DII600 CAS:78280-29-2 HR: 3
N-(2-(DIETHYLAMINO)ETHYL)-2,6-XYLIDINE DIHYDROCHLORIDE
mf: $C_{14}H_{24}N_2$•2ClH mw: 293.32

SYN: C 3144

TOXICITY DATA WITH REFERENCE
eye-rbt 2% MLD ARZNAD 8,708,58
ipr-rat LD50:120 mg/kg ARZNAD 8,708,58
scu-mus LD50:400 mg/kg ARZNAD 8,708,58

SAFETY PROFILE: Poison by intraperitoneal and subcutaneous routes. An eye irritant. When heated to decomposition it emits very toxic fumes of NO_x and HCl.

DII800 CAS:97702-98-2 HR: 3
2-(DIETHYLAMINO)-N-ETHYL-N-(1-(2,4-XYLYLOXY)-2-PROPYL)ACETAMIDE HYDROCHLORIDE
mf: $C_{19}H_{32}N_2O_2$•ClH mw: 356.99

SYN: C 2039

TOXICITY DATA WITH REFERENCE
eye-rbt 2% MLD ARZNAD 9,70,59
scu-mus LD50:185 mg/kg ARZNAD 9,70,59

SAFETY PROFILE: Poison by subcutaneous route. An eye irritant. When heated to decomposition it emits very toxic fumes of NO_x and HCl.

DIJ000 CAS:97702-99-3 HR: 3
2-(DIETHYLAMINO)-N-ETHYL-N-(1-(3,5-XYLYLOXY)-2-PROPYL) ACETAMIDE HYDROCHLORIDE
mf: $C_{19}H_{32}N_2O_2$•ClH mw: 356.99

SYN: C 2057

TOXICITY DATA WITH REFERENCE
eye-rbt 2% MLD ARZNAD 9,70,59
scu-mus LD50:112 mg/kg ARZNAD 9,70,59

SAFETY PROFILE: Poison by subcutaneous route. An eye irritant. When heated to decomposition it emits very toxic fumes of NO_x and HCl.

DIJ200 CAS:78372-04-0 HR: 3
1-(2-(DIETHYLAMINO)ETHYL)-3-(2,6-XYLYL)UREA HYDROCHLORIDE
mf: $C_{15}H_{25}N_3O$•ClH mw: 299.89

SYN: C 3145

TOXICITY DATA WITH REFERENCE
ipr-rat LD50:260 mg/kg ARZNAD 8,664,58
scu-mus LD50:445 mg/kg ARZNAD 8,664,58

SAFETY PROFILE: Poison by intraperitoneal route. Moderately toxic by subcutaneous route. When heated to decomposition it emits very toxic fumes of NO_x and HCl.

DIJ300 CAS:1563-01-5 HR: 3
3-(4-DIETHYLAMINO-2-HYDROXYPHENYLAZO)-4-HYDROXYBENZENESULFONIC ACID
mf: $C_{16}H_{19}N_3O_5S$ mw: 365.44

SYN: BENZENESULFONIC ACID, 3-(4-DIETHYLAMINO-2-HYDROXYPHENYLAZO)-4-HYDROXY-

TOXICITY DATA WITH REFERENCE
ivn-mus LD50:320 mg/kg CSLNX* NX#03409

CONSENSUS REPORTS: Reported in EPA TSCA Inventory.

SAFETY PROFILE: Poison by intravenous route. When heated to decomposition it emits toxic vapors of NO_x and SO_x.

DIK000 CAS:77-51-0 HR: 3
4-(DIETHYLAMINO)-2-ISOPROPYL-2-PHENYLVALERONITRILE
mf: $C_{16}H_{24}N_2$ mw: 244.42

PROP: A liquid. Bp: 138–146° @ 3 mm.

SYN: TAT-1

TOXICITY DATA WITH REFERENCE
ipr-mus LD50:77 mg/kg OYYAA2 2,323,68
scu-mus LD50:127 mg/kg OYYAA2 2,323,68

CONSENSUS REPORTS: Cyanide and its compounds are on the Community Right-To-Know List.

SAFETY PROFILE: Poison by intraperitoneal and subcutaneous routes. See also NITRILES. When heated to decomposition it emits very toxic fumes of NO_x and CN^-.

DIK200 CAS:97703-02-1 HR: 3
2-(DIETHYLAMINO)-N-(1-MESITYLOXY-2-PROPYL)-N-METHYLACETAMIDE HYDROCHLORIDE
mf: $C_{19}H_{32}N_2O_2$•ClH mw: 356.99

SYN: C 2047

TOXICITY DATA WITH REFERENCE
eye-rbt 2% MLD ARZNAD 9,70,59
scu-mus LD50:180 mg/kg ARZNAD 9,70,59

SAFETY PROFILE: Poison by subcutaneous route. An eye irritant. When heated to decomposition it emits very toxic fumes of NO_x and HCl.

DIK400 CAS:97703-04-3 HR: 2
2-(DIETHYLAMINO)-N-(1-(p-METHOXYPHENOXY)-2-PROPYL)-N-METHYLACETAMIDE HYDROCHLORIDE
mf: $C_{17}H_{28}N_2O_3$•ClH mw: 344.93

SYN: C 2097

TOXICITY DATA WITH REFERENCE
eye-rbt 2% MLD ARZNAD 9,70,59
scu-mus LD50:590 mg/kg ARZNAD 9,70,59

SAFETY PROFILE: Moderately toxic by subcutaneous route. An eye irritant. When heated to decomposition it emits very toxic fumes of NO_x and HCl.

DIK600 CAS:77966-73-5 HR: 3
2-(DIETHYLAMINO)-N-METHYL-o-ACETOTOLUIDIDE HYDROCHLORIDE
mf: $C_{14}H_{22}N_2O{\cdot}ClH$ mw: 270.84

SYN: V 346

TOXICITY DATA WITH REFERENCE
ipr-rat LD50:260 mg/kg ARZNAD 8,609,58
ipr-mus LD50:200 mg/kg ARZNAD 8,609,58
scu-mus LD50:670 mg/kg ARZNAD 8,609,58

SAFETY PROFILE: Poison by intraperitoneal route. Moderately toxic by subcutaneous route. When heated to decomposition it emits very toxic fumes of NO_x and HCl.

DIK800 CAS:77966-83-7 HR: 3
2-(DIETHYLAMINO)-N-METHYL-2′,6′-ACETOXYLIDIDE HYDROCHLORIDE
mf: $C_{15}H_{24}N_2O{\cdot}ClH$ mw: 284.87

SYN: V 317

TOXICITY DATA WITH REFERENCE
eye-rbt 2% MLD ARZNAD 8,609,58
ipr-rat LD50:138 mg/kg ARZNAD 8,609,58
scu-mus LD50:280 mg/kg ARZNAD 8,609,58

SAFETY PROFILE: Poison by intraperitoneal and subcutaneous routes. An eye irritant. When heated to decomposition it emits very toxic fumes of NO_x and HCl.

DIL000 CAS:102259-67-6 HR: 3
6-((4-(DIETHYLAMINO)-1-METHYLBUTYL)AMINO)-2-METHYL-4,5,8-TRIMETHOXYQUINOLINE-1,5-NAPHTHALENE DISULFONATE
mf: $C_{22}H_{35}N_3O_3{\cdot}C_{10}H_8O_6S_2$ mw: 677.90

TOXICITY DATA WITH REFERENCE
ipr-mus LD50:60 mg/kg ARZNAD 20,1775,70
ivn-brd LD50:51 mg/kg ARZNAD 20,1775,70

SAFETY PROFILE: Poison by intraperitoneal and intravenous routes. When heated to decomposition it emits very toxic fumes of NO_x and SO_x.

DIL200 CAS:72820-33-8 HR: 3
6((-4-(DIETHYLAMINO)-1-METHYLBUTYL))-5,8-DIMETHOXY-2,4-DIMETHYLQUINOLINE-1,5-NAPHTHALENE DISULFONATE
mf: $C_{22}H_{35}N_3O_2{\cdot}C_{10}H_8O_6S_2$ mw: 661.90

TOXICITY DATA WITH REFERENCE
ipr-mus LD50:45 mg/kg ARZNAD 20,1775,70
ivn-brd LD50:47 mg/kg ARZNAD 20,1775,70

SAFETY PROFILE: Poison by intraperitoneal and intravenous routes. When heated to decomposition it emits very toxic fumes of NO_x and SO_x.

DIL400 CAS:91-44-1 HR: 3
7-DIETHYLAMINO-4-METHYLCOUMARIN
mf: $C_{14}H_{17}NO_2$ mw: 231.32

SYNS: COUMARIN 1 □ 7-(DIETHYLAMINO)-4-METHYL-2H-1-BENZOPYRAN-2-ONE

TOXICITY DATA WITH REFERENCE
orl-rat LD50:5 g/kg MVCRB3 2,193,73
orl-mus LD50:1780 mg/kg SCIEAS 36(1-4),10,89
ivn-mus LD50:180 mg/kg CSLNX* NX#03230

CONSENSUS REPORTS: Reported in EPA TSCA Inventory.

SAFETY PROFILE: Poison by intravenous route. Moderately toxic by ingestion. When heated to decomposition it emits toxic fumes of NO_x.

DIL600 CAS:67210-66-6 HR: 2
7-DIETHYLAMINO-4-METHYLCOUMARIN, HYDROGEN SULFATE
mf: $C_{14}H_{17}NO_2{\cdot}H_2O_4S$ mw: 329.40

TOXICITY DATA WITH REFERENCE
orl-rat LD50:1600 mg/kg MarJV# 29MAR77
orl-mus LD50:1780 mg/kg ESKHA5 (104),45,86

CONSENSUS REPORTS: Reported in EPA TSCA Inventory.

SAFETY PROFILE: Moderately toxic by ingestion. When heated to decomposition it emits very toxic fumes of SO_x and NO_x.

DIM000 HR: 3
2-(DIETHYLAMINO)-N-METHYL-N-(2-MESITYLOXYETHYL)ACETAMIDE HYDROCHLORIDE
mf: $C_{18}H_{30}N_2O_2{\cdot}ClH$ mw: 342.96

SYN: C 2060

TOXICITY DATA WITH REFERENCE
eye-rbt 2% MLD ARZNAD 8,761,58
scu-mus LD50:280 mg/kg ARZNAD 8,761,58

SAFETY PROFILE: Poison by subcutaneous route. An eye irritant. When heated to decomposition it emits very toxic fumes of NO_x and HCl.

DIM200 CAS:30185-90-1 HR: 2
5-(DIETHYLAMINO)METHYL-3-(1-METHYL-5-NITROIMIDAZOL-2-YLMETHYLENE AMINO)-2-OXAZOLIDINONE HYDROCHLORIDE
mf: $C_{13}H_{20}N_6O_4{\cdot}ClH$ mw: 360.85

TOXICITY DATA WITH REFERENCE
orl-mus LD50:3400 mg/kg JMCMAR 14,94,71
ipr-mus LD50:970 mg/kg JMCMAR 14,94,71

SAFETY PROFILE: Moderately toxic by ingestion and intraperitoneal routes. When heated to decomposition it emits very toxic fumes of NO_x and HCl.

D

DIM400 CAS:16417-75-7 HR: 3
2-(DIETHYLAMINO)-N-(2-METHYL-1-NAPHTHYL) ACETAMIDE HYDROCHLORIDE
mf: $C_{17}H_{22}N_2O•ClH$ mw: 306.87

SYN: V331

TOXICITY DATA WITH REFERENCE
eye-rbt 2% MLD ARZNAD 8,609,58
ipr-rat LD50:167 mg/kg ARZNAD 8,609,58
scu-mus LD50:300 mg/kg ARZNAD 8,609,58

SAFETY PROFILE: Poison by intraperitoneal and subcutaneous routes. An eye irritant. When heated to decomposition it emits very toxic fumes of NO_x and HCl.

DIM600 CAS:30096-82-3 HR: 3
4-DIETHYLAMINOMETHYL-2-(5-NITRO-2-THIENYL) THIAZOLE HYDROCHLORIDE
mf: $C_{12}H_{15}N_3O_2S_2•ClH$ mw: 333.88

TOXICITY DATA WITH REFERENCE
orl-mus LD50:300 mg/kg JMCMAR 18,794,75
ipr-mus LD50:100 mg/kg JMCMAR 18,794,75

SAFETY PROFILE: Poison by ingestion and intraperitoneal routes. When heated to decomposition it emits very toxic fumes of NO_x, SO_x, and HCl.

DIM800 CAS:97703-08-7 HR: 3
2-(DIETHYLAMINO)-N-METHYL-N-(2-PHENETHYL) ACETAMIDE HYDROCHLORIDE
mf: $C_{15}H_{24}N_2O•CClH$ mw: 284.87

SYN: C 6608

TOXICITY DATA WITH REFERENCE
ipr-rat LD50:232 mg/kg ARZNAD 9,113,59
scu-mus LD50:630 mg/kg ARZNAD 9,113,59

SAFETY PROFILE: Poison by intraperitoneal route. Moderately toxic by subcutaneous route. When heated to decomposition it emits very toxic fumes of NO_x and HCl.

DIN000 CAS:102128-92-7 HR: 2
3-(DIETHYLAMINO)-N-METHYL-N-(1-PHENOXY-2-PROPYL)PROPIONAMIDE HYDROCHLORIDE
mf: $C_{17}H_{28}N_2O_2•ClH$ mw: 328.93

SYN: C 1863

TOXICITY DATA WITH REFERENCE
eye-rbt 2% MLD ARZNAD 9,70,59
scu-mus LD50:480 mg/kg ARZNAD 9,70,59

SAFETY PROFILE: Moderately toxic by subcutaneous route. An eye irritant. When heated to decomposition it emits very toxic fumes of HCl and NO_x.

DIN200 CAS:97703-11-2 HR: 3
2-(DIETHYLAMINO)-N-METHYL-N-(3-PHENYLPROPYL) ACETAMIDE HYDROCHLORIDE
mf: $C_{16}H_{26}N_2O•ClH$ mw: 298.90

SYN: C 6610

TOXICITY DATA WITH REFERENCE
eye-rbt 2% MLD ARZNAD 9,113,59
ipr-rat LD50:100 mg/kg ARZNAD 9,113,59
scu-mus LD50:350 mg/kg ARZNAD 9,113,59

SAFETY PROFILE: Poison by intraperitoneal and subcutaneous routes. An eye irritant. When heated to decomposition it emits very toxic fumes of NO_x and HCl.

DIN400 CAS:77791-27-6 HR: 3
2-(DIETHYLAMINO)-N-(4-METHYL-2-PYRIDYL)ACETAMIDE DIHYDROCHLORIDE
mf: $C_{12}H_{19}N_3O•2ClH$ mw: 294.26

SYN: C 5334

TOXICITY DATA WITH REFERENCE
ipr-rat LD50:183 mg/kg ARZNAD 8,609,58
scu-mus LD50:370 mg/kg ARZNAD 8,609,58

SAFETY PROFILE: Poison by intraperitoneal and subcutaneous routes. When heated to decomposition it emits very toxic fumes of NO_x and HCl.

DIN600 CAS:23505-41-1 HR: 3
2-DIETHYLAMINO-6-METHYLPYRIMIDIN-4-YL DIETHYLPHOSPHOROTHIONATE
mf: $C_{13}H_{24}N_3O_3PS$ mw: 333.43

PROP: A liquid. D: 1.14 @ 20°.

SYNS: O-(2-(DIETHYLAMINO)-6-METHYL-4-PYRIMIDINYL)-O,O-DIETHYL PHOSPHOROTHIOATE ☐ O,O-DIETHYL O-(2-DIETHYLAMINO-6-METHYL-4-PYRIMIDINYL)PHOSPHOROTHIOATE ☐ FERNEX ☐ PIRIMIFOSETHYL ☐ PIRIMIPHOS-ETHYL ☐ PP211 ☐ PRIMICID ☐ PRIMOTEC ☐ PRINICID

TOXICITY DATA WITH REFERENCE
mmo-sat 5 μL/plate MUREAV 28,405,75
mmo-esc 5 μL/plate MUREAV 28,405,75
orl-rat LD50:140 mg/kg GUCHAZ 6,421,73
skn-rat LD50:1000 mg/kg 28ZEAL 5,184,76
orl-mus LD50:105 mg/kg 28ZEAL 5,184,76
orl-cat LD50:25 mg/kg 28ZEAL 5,184,76
orl-gpg LD50:50 mg/kg 28ZEAL 5,184,76
orl-bwd LD50:7500 μg/kg AECTCV 12,355,83

CONSENSUS REPORTS: EPA Extremely Hazardous Substances List.

SAFETY PROFILE: Poison by ingestion. Moderately toxic by skin contact. Mutation data reported. When heated to decomposition it emits very toxic fumes of NO_x, PO_x, and SO_x.

DIN800 CAS:29232-93-7 HR: 2
2-DIETHYLAMINO-6-METHYLPYRIMIDIN-4-YL DIMETHYL PHOSPHOROTHIONATE
mf: $C_{11}H_{20}N_3O_3PS$ mw: 305.37

PROP: Straw-colored oil. D: 1.157 @ 30 mm. Very sltly sol in H_2O. Misc in most org solvs.

SYNS: ACTELIC ☐ ACTELLIC ☐ ACTELLIFOG ☐ BLEX ☐ O-(2-DIETHYLAMINO-6-METHYLPYRIMIDIN-4-YL)-O,O-DIMETHYL PHOSPHOROTHIOATE ☐ O-(2-(DIETHYLAMINO)-6-METHYL-4-PYRIMIDINYL)-O,O-DIMETHYL PHOSPHOROTHIOATE ☐ ENT 27,699GC ☐ METHYL

PIRIMIPHOS □ PIRIMIFOS-METHYL □ PLANT PROTECTION PP511 □ PP511 □ PYRIMIDINE PHOSPHATE □ PYRIMIPHOS METHYL □ SILOSAN

TOXICITY DATA WITH REFERENCE
mmo-sat 5 μL/plate MUREAV 28,405,75
cyt-mus-unr 500 mg/kg TGANAK 16(1),45,82
orl-rat LD50:1250 mg/kg SSCMBX 20,33,83
orl-mus LD50:1180 mg/kg 28ZEAL 5,184,76
orl-rbt LD50:1150 mg/kg 28ZEAL 5,184,76
orl-gpg LD50:1000 mg/kg 28ZEAL 5,184,76

SAFETY PROFILE: Moderately toxic by ingestion. Mutation data reported. When heated to decomposition it emits very toxic fumes of NO_x, PO_x, and SO_x.

DIO200 CAS:15421-84-8 HR: 3
7-DIETHYLAMINO-5-METHYL-s-TRIAZOLO(1,5-a)PYRIMIDINE
mf: $N_5C_{10}H_{15}$ mw: 205.30

PROP: Bitter-tasting powder. Mp: 98–99.4°.

SYNS: AR 12008 □ N,N-DIETHYL-5-METHYL-(1,2,4)TRIAZOLO(1,5-a)PYRIMIDINE-7-AMINE □ 5-METHYL-7-DIETHYLAMINO-s-TRIAZOLO-(1,5-a)PYRIMIDINE □ ROCORNAL □ TRAPIDIL □ TRAPYMIN

TOXICITY DATA WITH REFERENCE
orl-mus TDLo:240 mg/kg (female 7-13D post):REP IYKEDH 6,418,75
orl-mus TDLo:504 mg/kg (female 15-21D post):REP IYKEDH 7,187,76
orl-rat TDLo:420 mg/kg (8-14D preg):REP IYKEDH 6,418,75
orl-rat TDLo:420 mg/kg (8-14D preg):TER IYKEDH 6,418,75
orl-mus TDLo:480 mg/kg (female 1-8D post):REP IYKEDH 7,187,76
orl-rat TDLo:43,680 mg/kg (male 26W pre):REP IYKEDH 7,170,76
orl-rbt TDLo:390 mg/kg (female 6-18D post):TER IYKEDH 7,195,76
orl-rat LD50:235 mg/kg MEIEDD 10,1369,83
ipr-rat LD50:100 mg/kg MEIEDD 10,1369,83
scu-rat LD50:100 mg/kg MEIEDD 10,1369,83
ivn-rat LD50:76 mg/kg MEIEDD 10,1369,83
orl-mus LD50:380 mg/kg MEIEDD 10,1369,83
ipr-mus LD50:155 mg/kg MEIEDD 10,1369,83
scu-mus LD50:132 mg/kg MEIEDD 10,1369,83
ivn-mus LD50:101 mg/kg IYKEDH 10,232,79

SAFETY PROFILE: Poison by ingestion, intraperitoneal, subcutaneous, and intravenous routes. An experimental teratogen. Other experimental reproductive effects. A coronary vasodilator. When heated to decomposition it emits toxic fumes of NO_x. See also AMINES.

DIO300 CAS:25953-06-4 HR: 3
5-(DIETHYLAMINO)-2-NITROSOPHENOL HYDROCHLORIDE
mf: $C_{10}H_{14}N_2O_2$•ClH mw: 230.72

TOXICITY DATA WITH REFERENCE
orl-rat LD50:315 mg/kg JMCMAR 13,370,70
scu-rat LD50:84 mg/kg JMCMAR 13,370,70
ivn-rat LD50:52 mg/kg JMCMAR 13,370,70
orl-mus LD50:122 mg/kg JMCMAR 13,370,70
ipr-mus LD50:36,800 μg/kg JMCMAR 13,370,70
ivn-mus LD50:34,500 μg/kg JMCMAR 13,370,70

SAFETY PROFILE: Poison by ingestion, intraperitoneal, subcutaneous, and intravenous routes. When heated to decomposition it emits toxic fumes of NO_x and HCl.

D

DIP000 CAS:7347-49-1 HR: 2
4-((4-(DIETHYLAMINO)PHENYL)AZO)PYRIDINE-1-OXIDE
mf: $C_{15}H_{18}N_4O$ mw: 270.37

SYN: N,N-DIETHYL-4-(4′-(PYRIDYL-1′-OXIDE)AZO)ANILINE

TOXICITY DATA WITH REFERENCE
orl-rat TDLo:6426 mg/kg/52W-C:NEO JNCIAM 37,365,66

SAFETY PROFILE: Questionable carcinogen with experimental neoplastigenic data. When heated to decomposition it emits toxic fumes of NO_x.

DIP100 CAS:5185-78-4 HR: 3
2-(p-(DIETHYLAMINOPHENYL))-1,3,2-DITHIARSENOLANE
mf: $C_{12}H_{18}AsNS_2$ mw: 315.35

SYN: 1,3,2-DITHIARSENOLANE, 2-(p-(DIETHYLAMINO)PHENYL)-

TOXICITY DATA WITH REFERENCE
ipr-mus LD50:16 mg/kg JMCMAR 9,221,66

OSHA PEL: TWA 0.5 mg(As)/m^3

SAFETY PROFILE: Poison by intraperitoneal route. When heated to decomposition it emits toxic fumes of NO_x, SO_x, and As.

DIP600 CAS:134-80-5 HR: 3
2-DIETHYLAMINOPROPIOPHENONE HYDROCHLORIDE
mf: $C_{13}H_{19}NO$•ClH mw: 241.79

PROP: Crystals. Mp: 168° (decomp).

SYNS: AMPHEPRAMONUM HYDROCHLORIDE □ α-BENZOYLTRIETHYLAMINE HYDROCHLORIDE □ α-BENZOYLTRIETHYLAMMONIUM CHLORIDE □ 2-(DIETHYLAMINO)-1-PHENYL-1-PROPANONE HYDROCHLORIDE □ DIETHYLPROPIONE HYDROCHLORIDE □ DIETHYLPROPION HYDROCHLORIDE □ 1-PHENYL-2-DIETHYLAMINOPROPANONE-1-HYDROCHLORIDE □ 1-PHENYL-2-DIETHYLAMINO-1-PROPANONE HYDROCHLORIDE □ REGENON □ REGONON HYDROCHLORIDE □ TENUATE □ TENUATE HYDROCHLORIDE □ TEPANIL

TOXICITY DATA WITH REFERENCE
orl-rat LD50:500 mg/kg JPETAB 137,365,62
ipr-rat LD50:139 mg/kg APTOA6 17,121,60
orl-mus LD50:450 mg/kg JPETAB 137,365,62
ipr-mus LD50:190 mg/kg AEPPAE 237,171,59
scu-mus LD50:450 mg/kg APSXAS 4,37,67
ivn-mus LD50:50 mg/kg APTOA6 17,182,60

SAFETY PROFILE: Poison by intraperitoneal and intravenous routes. Moderately toxic by ingestion and subcutaneous routes. When heated to decomposition it emits very toxic fumes of HCl, NH_3, and NO_x.

DIP800 **CAS:41240-93-1** **HR: 3**
2-(DIETHYLAMINO)-2′-PROPOXYACETANILIDE HYDROCHLORIDE
mf: $C_{15}H_{24}N_2O_2$•ClH mw: 300.87

SYN: C 3127

TOXICITY DATA WITH REFERENCE
eye-rbt 2% MOD ARZNAD 8,270,58
ipr-rat LD50:350 mg/kg ARZNAD 8,270,58
scu-mus LD50:1950 mg/kg ARZNAD 8,270,58

SAFETY PROFILE: Poison by intraperitoneal route. Moderately toxic by subcutaneous route. An eye irritant. When heated to decomposition it emits very toxic fumes of NO_x and HCl.

DIQ100 **CAS:14642-66-1** **HR: 3**
3-DIETHYLAMINOPROPYLAMINE
mf: $C_7H_{18}N_2$ mw: 130.23

$(CH_3CH_2)_2NN(CH_2)_3NH_2$

SAFETY PROFILE: Ignites spontaneously on cellulose nitrate of high surface area. When heated to decomposition it emits toxic fumes of NO_x. See also AMINES.

DIR000 **CAS:522-00-9** **HR: 3**
10-(2-DIETHYLAMINOPROPYL)PHENOTHIAZINE
mf: $C_{19}H_{24}N_2S$ mw: 312.51

PROP: Crystals. Mp: 53–55°.

SYNS: AETHOPROPROPAZIN □ ATHAPROPAZINE □ ATHOPROPAZIN □ DIBUTIL □ 10-(2-DIETHYLAMINO-2-METHYLETHYL)PHENOTHIAZINE □ 2-DIETHYLAMINO-1-PROPYL-N-DIBENZOPARATHIAZINE □ N,N-DIETHYL-α-METHYL-10H-PHENOTHIAZINE-10-ETHANAMINE □ ETHOPROPAZINE □ ETOPROPEZINA □ FEMPROPAZINE □ FENPROPAZINA □ ISOTAZIN □ ISOTHAZINE □ ISOTHIAZINE □ LYSIVANE □ PARCIDOL □ PARDIDOL □ PARDISOL □ PARFEZINE □ PARKIN □ PARKISOL □ PARPHEZEIN □ PARSIDOL □ PARSITAN □ PHENOPROPAZINE □ PHENOPROZINE □ PRODICTAZIN □ PRODIERAZINE □ PROFENAMINA (ITALIAN) □ PROFENAMINUM □ ROCHIPEL □ ROCIPEL □ RODIPAL □ RP 3356 □ SC 2538 □ SKF 2538 □ TOMIL □ W 483

TOXICITY DATA WITH REFERENCE
scu-rat LD50:200 mg/kg PSCBAY 2,17,63
ivn-rat LD50:15 mg/kg 27ZQAG -,41,72
orl-mus LD50:300 mg/kg BCFAAI 111,293,72
scu-mus LD50:500 mg/kg 27ZQAG -,41,72
ivn-mus LD50:50 mg/kg 27ZQAG -,41,72
scu-rbt LD50:200 mg/kg 27ZQAG -,41,72
ivn-rbt LD50:15 mg/kg 27ZQAG -,41,72

SAFETY PROFILE: Poison by ingestion, subcutaneous, and intravenous routes. An anticholinergic agent used to treat Parkinson's disease. When heated to decomposition it emits very toxic fumes of NO_x and SO_x.

DIR875 **CAS:38078-09-0** **HR: 3**
DIETHYLAMINOSULFUR TRIFLUORIDE
mf: $C_4H_{10}F_3NS$ mw: 161.18

$(CH_3CH_2)_2NSF_3$

PROP: Pale-yellow liquid. Bp: 46–47° @ 10 mm.

SAFETY PROFILE: Decomposes violently above 90°C. Decomposes explosively on contact with water. When heated to decomposition it emits toxic fumes of F^-, SO_x, and NO_x. See also FLUORIDES.

DIS000 **CAS:77791-66-3** **HR: 2**
2-(DIETHYLAMINO)-2′,4′,6′-TRICHLOROACETANILIDE HYDROCHLORIDE
mf: $C_{12}H_{15}Cl_3N_2O$•ClH mw: 346.10

SYN: V 340

TOXICITY DATA WITH REFERENCE
eye-rbt 2% MLD ARZNAD 8,270,58
ipr-rat LD50:487 mg/kg ARZNAD 8,270,58
ipr-mus LD50:490 mg/kg ARZNAD 8,270,58
scu-mus LD50:1165 mg/kg ARZNAD 8,270,58

SAFETY PROFILE: Moderately toxic by intraperitoneal and subcutaneous routes. An eye irritant. When heated to decomposition it emits very toxic fumes of Cl^- and NO_x.

DIS200 **CAS:2150-48-3** **HR: 3**
(6-(DIETHYLAMINO)-3H-XANTEN-3-YLIDENE)DIETHYLAMMONIUM CHLORIDE
mf: $C_{21}H_{27}N2O$•Cl mw: 358.95

SYNS: C.I. 45010 □ N-(6-(DIETHYLAMINO)-3H-XANTHEN-3-YLIDINE)-N-ETHYLETHANAMINIUM CHLORIDE □ ETHANIMINIUM-N-(6-(DIETHYLAMINO)-3H-XANTHEN-3-YLIDENE)-N-ETHYL CHLORIDE □ NSC 44690 □ PYRONIN B □ E TETRAETHYLPYRONIN

TOXICITY DATA WITH REFERENCE
sln-dmg-orl 1000 ppm AMNTA4 87,295,53
dnd-esc 20 µmol/L MUREAV 89,95,81
ipr-mus LD50:15,100 µg/kg NCISP* Jan 86

CONSENSUS REPORTS: Reported in EPA TSCA Inventory.

SAFETY PROFILE: Poison by intraperitoneal route. Mutation data reported. When heated to decomposition it emits toxic fumes of Cl^-, NO_x, and NH_3.

DIS500 **CAS:660-68-4** **HR: 3**
DIETHYLAMMONIUM CHLORIDE
mf: $C_4H_{11}N$•ClH mw: 109.62

SYNS: N-ETHYL-ETHANAMINE HYDROCHLORIDE (9CI) □ HYDROCHLORIDE DIETHYLAMINE

TOXICITY DATA WITH REFERENCE
orl-rat LD50:9900 mg/kg GISAAA 49(1),71,84
orl-mus LD50:4860 mg/kg GTPZAB 31(1),49,87
ipr-mus LD50:960 mg/kg JJPAAZ 17,475,67
scu-mus LD50:1130 mg/kg AIPTAK 112,36,57
ivn-mus LD50:320 mg/kg AIPTAK 112,36,57

CONSENSUS REPORTS: EPA Genetic Toxicology Program. Reported in EPA TSCA Inventory.

SAFETY PROFILE: Poison by intravenous route. Moderately toxic by subcutaneous and intraperitoneal routes. When heated to decomposition it emits toxic fumes of NO_x, NH_3, and HCl.

DIS600 CAS:2624-44-4 HR: 2
DIETHYLAMMONIUM-2,5-DIHYDROXYBENZENE SULFONATE
mf: $C_{10}H_{17}NO_5S$ mw: 263.34

PROP: Crystals from alc. Mp: 125°.

SYNS: AGLUMIN □ ALTODOR □ CYCLOHEXADIENOL-4-ONE-1-SULFONATE de DIETHYLAMINE (FRENCH) □ CYCLONAMINE □ DICYNENE □ DICYNONE □ DIETHYLAMMONIUM CYCLOHEXADIEN-4-OL-1-ONE-4-SULFONATE □ 2,5-DIHYDROXYBENZENESULFONIC ACID with N-ETHYLETHANAMINE □ DIIDROXI-1,4-BENZENESULFONATO-3-DI-ETILAMMONIUM (ITALIAN) □ E 141 □ ESELIN □ ETAMSYLATE □ ETHAMSYLATE □ 1-HYDROXY-4-OXO-2,5-CYCLOHEXADIENE-1-SULFONIC ACID compound with DIETHYLAMINE □ MD 141

TOXICITY DATA WITH REFERENCE
sln-asn 1 mg/L MUREAV 26,159,74
orl-rat LD50:7500 mg/kg DRUGAY-,180,90
scu-rat LD50:5250 mg/kg DRUGAY-,180,90
ivn-rat LD50:1350 mg/kg THERAP 15,110,60
ims-rat LD50:>4 mg/kg DRUGAY-,180,90
ivn-cat LDLo:1 g/kg THERAP 15,110,60

CONSENSUS REPORTS: Reported in EPA TSCA Inventory.

SAFETY PROFILE: Moderately toxic by intravenous route. Mutation data reported. When heated to decomposition it emits very toxic fumes of SO_x, NH_3, and NO_x. See also SULFONATES and AMINES.

DIS650 CAS:579-66-8 HR: 2
2,6-DIETHYLANILINE
mf: $C_{10}H_{15}N$ mw: 149.26

SYNS: ANILINE, 2,6-DIETHYL- □ BENZENAMINE, 2,6-DIETHYL-(9CI) □ 2,6-DIETHYLBENZENAMINE

TOXICITY DATA WITH REFERENCE
mma-sat 250 ng/plate BECTA6 35,696,85
orl-rat LD50:1800 mg/kg FAATDF 3,285,83
orl-rat TDLo:10,080 mg/kg/20D-C FAATDF 3,285,83

CONSENSUS REPORTS: Reported in EPA TSCA Inventory.

SAFETY PROFILE: Moderately toxic by ingestion. Mutation data reported. When heated to decomposition it emits toxic vapors of NO_x.

DIS700 CAS:91-66-7 HR: 2
N,N-DIETHYLANILINE
DOT: UN 2432
mf: $C_{10}H_{15}N$ mw: 149.26

PROP: Colorless to yellow liquid. D: (25/4) 0.9302, bp: 215–216°, mp: −38°, n (24/D) 1.5394. Volatile with steam. Sltly sol in alc, chloroform, ether. One gram dissolves in 70 mL water at 12°. Sol in acids; sltly sol in EtOH, Et_2O, $CHCl_3$.

SYNS: BENZENAMINE, N,N-DIETHYL-(9CI) □ DEA □ DIAETHYLANILIN (GERMAN) □ N,N-DIETHYLAMINOBENZENE □ N,N-DIETHYLAMINOBENZENE □ N,N-DIETHYLANILIN (CZECH) □ N,N-DIETHYLBENZENAMINE □ DIETHYLANILINE □ DIETHYLPHENYLAMINE

TOXICITY DATA WITH REFERENCE
orl-rat LD50:782 mg/kg MarJV# 29MAR77
ipr-rat LD50:420 mg/kg AGGHAR 15,447,57

CONSENSUS REPORTS: Reported in EPA TSCA Inventory.

DOT CLASSIFICATION: 6.1; *Label:* KEEP AWAY FROM FOOD

SAFETY PROFILE: Moderately toxic by ingestion and intraperitoneal routes. When heated to decomposition it emits toxic fumes of NO_x. See also ANILINE DYES.

DIS775 CAS:5185-76-2 HR: 3
N,N-DIETHYL-p-ARSANILIC ACID
mf: $C_{10}H_{16}AsNO_3$ mw: 273.19

SYN: p-ARSANILIC ACID, N,N-DIETHYL-

TOXICITY DATA WITH REFERENCE
ipr-mus LD50:7102 μg/kg JMCMAR 9,221,66

OSHA PEL: TWA 0.5 mg(As)/m^3

SAFETY PROFILE: Poison by intraperitoneal route. When heated to decomposition it emits toxic fumes of NO_x and As.

DIS800 CAS:692-42-2 HR: 3
DIETHYL ARSINE
mf: $C_4H_{11}As$ mw: 134.05

$(CH_3CH_2)_2AsH$

PROP: D: 1.13 @ 23.7°, bp: 105°.

CONSENSUS REPORTS: Arsenic and its compounds are on the Community Right-To-Know List.

SAFETY PROFILE: A poison. A dangerous fire hazard. It spontaneously ignites in air above 0°C. Upon decomposition it emits toxic fumes of As. See also ARSENIC COMPOUNDS.

DIS850 CAS:4964-27-6 HR: 3
DIETHYL ARSINIC ACID
mf: $C_4H_{11}AsO_2$ mw: 166.07

SYNS: ARSINE OXIDE, DIETHYLHYDROXY- □ DIETHYLHYDROXY ARSINE OXIDE

TOXICITY DATA WITH REFERENCE
ivn-mus LD50:180 mg/kg CSLNX* NX#03818

OSHA PEL: TWA 0.5 mg(As)/m^3

SAFETY PROFILE: Poison by intravenous route. When heated to decomposition it emits toxic fumes of As.

DIT300 CAS:1972-28-7 HR: 3
DIETHYL AZOFORMATE
mf: $C_6H_{10}N_2O_4$ mw: 174.16

$CH_3CH_2OCO{\cdot}N{=}NCO{\cdot}OCH_2CH_3$

PROP: A liquid. Fp: 6°, bp: 90–95° @ 5 mm.

SYN: 1,2-DIETHOXYCARBONYLDIAZENE

SAFETY PROFILE: A shock-sensitive explosive that also

burns explosively. When heated to decomposition it emits toxic fumes of NO_x. See also AZIDES.

DIT350 CAS:5256-74-6 HR: 3
DIETHYL AZOMALONATE
mf: $C_7H_{10}N_2O_4$ mw: 186.17

$N_2C(CO \cdot OCH_2CH_3)_2$

PROP: Yellow oil solidifying in ice to needles. Bp: 105° @ 12 mm.

SAFETY PROFILE: A poison. Potentially explosive when heated. When heated to decomposition it emits toxic fumes of NO_x.

DIT400 CAS:36911-94-1 HR: 2
6,8-DIETHYLBENZ(a)ANTHRACENE
mf: $C_{22}H_{20}$ mw: 284.42

TOXICITY DATA WITH REFERENCE
ims-rat TDLo:50 mg/kg:ETA JMCMAR 15,905,72

SAFETY PROFILE: Questionable carcinogen with experimental tumorigenic data. When heated to decomposition it emits acrid smoke and irritating fumes.

DIT600 CAS:36911-95-2 HR: 2
8,12-DIETHYLBENZ(a)ANTHRACENE
mf: $C_{22}H_{20}$ mw: 284.42

TOXICITY DATA WITH REFERENCE
ims-rat TDLo:50 mg/kg:ETA JMCMAR 15,905,72

SAFETY PROFILE: Questionable carcinogen with experimental tumorigenic data. When heated to decomposition it emits acrid smoke and irritating fumes.

DIT800 CAS:16354-52-2 HR: 2
9,10-DIETHYL-1,2-BENZANTHRACENE
mf: $C_{22}H_{20}$ mw: 284.42

SYN: 7,12-DIETHYLBENZ(a)ANTHRACENE

TOXICITY DATA WITH REFERENCE
skn-mus TDLo:380 mg/kg/16W-I:ETA PRLBA4 129,439,40

SAFETY PROFILE: Questionable carcinogen with experimental tumorigenic data. When heated to decomposition it emits acrid smoke and irritating fumes.

DIU000 CAS:25340-17-4 HR: 3
DIETHYL BENZENE
DOT: UN 2049
mf: $C_{10}H_{14}$ mw: 134.24

PROP: Colorless, mobile liquid. Bp: 183.8°, flash p: 134°F, d: 0.868 @ 25°/25°, autoign temp: 743–842°F, vap press: 1 mm @ 20.7°, vap d: 4.62.

SYNS: DIETHYLBENZENE (DOT) □ DIETHYLBENZOL

TOXICITY DATA WITH REFERENCE
skn-rbt 100% MOD AMIHAB 14,387,56
eye-rbt 88 mg MLD AMIHAB 14,387,56
orl-rat LDLo:5000 mg/kg 28ZRAQ -,57,60

CONSENSUS REPORTS: Reported in EPA TSCA Inventory.

DOT CLASSIFICATION: 3; *Label:* Flammable Liquid

SAFETY PROFILE: Mildly toxic by ingestion. A skin and eye irritant. Flammable liquid when exposed to heat or flame; can react with oxidizing materials. To fight fire, use CO_2, dry chemical. When heated to decomposition it emits acrid smoke and fumes. See also ETHYL BENZENE.

DIU200 CAS:141-93-5 HR: 2
m-DIETHYLBENZENE
mf: $C_{10}H_{14}$ mw: 134.24

PROP: A solid. Bp: 181–182°.

TOXICITY DATA WITH REFERENCE
orl-rat LDLo:5000 mg/kg 28ZRAQ -,57,60

CONSENSUS REPORTS: Reported in EPA TSCA Inventory.

SAFETY PROFILE: Moderately toxic by ingestion. When heated to decomposition it emits acrid and irritating fumes. See also DIETHYLBENZENE.

DIU400 CAS:1709-50-8 HR: 2
N,N-DIETHYLBENZENESULFONAMIDE
mf: $C_{10}H_{15}NO_2S$ mw: 213.32

TOXICITY DATA WITH REFERENCE
orl-rat TDLo:300 mg/kg (female 9D post):REP TXAPA9 23,376,72
orl-rat TDLo:300 mg/kg (female 9D post):TER TXAPA9 23,376,72
orl-rat LD50:890 mg/kg TXAPA9 23,376,72

SAFETY PROFILE: Moderately toxic by ingestion. An experimental teratogen. Other experimental reproductive effects. When heated to decomposition it emits very toxic fumes of SO_x and NO_x.

DIU500 CAS:75889-62-2 HR: 3
DIETHYL-4-(BENZOTHIAZOL-2-YL)BENZYLPHOSPHONATE
mf: $C_{18}H_{20}NO_3PS$ mw: 361.42

PROP: Needles from *n*-hexane. Mp: 96–97°.

SYNS: (p-(2-BENZOTHIAZOLYL)BENZYL)PHOSPHONIC ACID DIETHYL ESTER □ KB-944

TOXICITY DATA WITH REFERENCE
orl-rat TDLo:3 g/kg (30D pre):REP ARZNAD 32,1071,82
orl-rat LD50:1555 mg/kg ARZNAD 32,1068,82
ipr-rat LD50:479 mg/kg ARZNAD 32,1068,82
scu-rat LD50:2216 mg/kg ARZNAD 32,1068,82
orl-mus LD50:2807 mg/kg ARZNAD 32,1068,82
ipr-mus LD50:777 mg/kg ARZNAD 32,1068,82
ivn-dog LD50:76 mg/kg ARZNAD 32,1068,82

SAFETY PROFILE: Poison by intravenous route. Moderately toxic by ingestion, subcutaneous, and intraperitoneal routes. Experimental reproductive effects. A calcium antagonist. When heated to decomposition it emits toxic fumes of PO_x, SO_x, and NO_x. See also ESTERS.

DIU600 CAS:1080-32-6 HR: 3
DIETHYL BENZYLPHOSPHONATE
mf: $C_{11}H_{17}O_3P$ mw: 228.25

PROP: Oily liquid. Bp: 172° @ 25 mm.

TOXICITY DATA WITH REFERENCE
ivn-mus LD50:180 mg/kg CSLNX* NX#01630

CONSENSUS REPORTS: Reported in EPA TSCA Inventory.

SAFETY PROFILE: Poison by intravenous route. When heated to decomposition it emits toxic fumes of PO_x.

DIU800 CAS:13286-32-3 HR: 3
O,O-DIETHYL-S-BENZYL THIOPHOSPHATE
mf: $C_{11}H_{17}O_3PS$ mw: 260.31

PROP: D: 1.16 @ 20°/4°, bp: 110–114° @ 0.02 mm.

SYNS: IBP □ KITAZIN □ PHOSPHOROTHIOIC ACID-S-BENZYL-O, O-DIETHYL ESTER □ RICID

TOXICITY DATA WITH REFERENCE
orl-rat LD50:660 mg/kg 85ARAE 4,102,76/77
orl-mus LD50:230 mg/kg CHYCDW 14,197,80
ipr-mus LD50:96 mg/kg CHYCDW 14,197,80
ivn-mus LD50:570 mg/kg CHYCDW 14,197,80

SAFETY PROFILE: Poison by ingestion and intraperitoneal routes. Moderately toxic by intravenous route. When heated to decomposition it emits very toxic fumes of PO_x and SO_x.

DIV000 CAS:542-63-2 HR: 3
DIETHYLBERYLLIUM
mf: $C_4H_{10}Be$ mw: 67.13

PROP: Colorless liquid. Mp: −13°, bp: 63° @ 3 mm; vap d: 2.3.

CONSENSUS REPORTS: IARC Cancer Review: Group 1 IMEMDT 58,41,93; Human Sufficient Evidence IMEMDT 58,41,93; Animal Sufficient Evidence IMEMDT 1,17,72; Animal Sufficient Evidence IMEMDT 23,143,80; Animal Sufficient Evidence IMEMDT 58,41,93. Beryllium and its compounds are on the Community Right-To-Know List.

OSHA PEL: TWA 0.002 mg(Be)/m^3; STEL 0.005 mg(Be)/m^3/30M; CL 0.025 mg(Be)/m^3
ACGIH TLV: TWA 0.002 mg/m^3, Suspected Human Carcinogen

SAFETY PROFILE: Confirmed human carcinogen. Very poisonous. Dangerous fire hazard when exposed to heat or flame. Spontaneously flammable in air. Can react vigorously with oxidizing materials. To fight fire, use special extinguishing agents, dry chemical. Explodes on contact with water. Upon decomposition it emits poisonous fumes of BeO. See also BERYLLIUM COMPOUNDS.

DIV200 CAS:28616-48-0 HR: 3
DIETHYL BIS-DIMETHYLPYROPHOSPHORADIAMIDE (symmetrical)
mf: $C_8H_{22}N_2O_5P_2$ mw: 288.26

SYNS: sym-DIETHYL BIS(DIMETHYLAMIDO)PYROPHOSPHATE □ DIETHYL DI(DIMETHYLAMIDO)PYROPHOSPHATE (symmetrical) □ N, N,N',N'-TETRAMETHYL-p,p'-DIAMIDODIPHOSPHORIC ACID DIETHYL ESTER

TOXICITY DATA WITH REFERENCE
orl-rat LD50:12,400 µg/kg JPETAB 107,464,53
skn-rat LD50:10 mg/kg JPETAB 107,464,53
ipr-rat LD50:10 mg/kg JPETAB 107,464,53
ipr-mus LD50:16,400 µg/kg JPETAB 107,464,53
ivn-dog LD50:20 mg/kg JPETAB 107,464,53
ipr-gpg LD50:13 mg/kg JPETAB 107,464,53

SAFETY PROFILE: Poison by ingestion, skin contact, intravenous, and intraperitoneal routes. When heated to decomposition it emits very toxic fumes of NO_x and PO_x.

DIV400 CAS:65313-34-0 HR: 3
DIETHYLBISMUTH CHLORIDE
mf: $C_4H_{10}BiCl$ mw: 302.45

$(CH_3CH_2)_2BiCl$

SAFETY PROFILE: A dangerous fire hazard; ignites spontaneously in air. When heated to decomposition it emits toxic fumes of Bi. See also BISMUTH COMPOUNDS and CHLORIDES.

DIV600 CAS:2641-56-7 HR: 3
DIETHYLBIS(OCTANOYLOXY)STANNANE
mf: $C_{20}H_{40}O_4Sn$ mw: 463.29

SYNS: DIETHYLBIS(1-OXOOCTYL)OXY)STANNANE □ DIETHYLTIN DICAPRYLATE □ DIETHYLTIN DIOCTANOATE

TOXICITY DATA WITH REFERENCE
orl-rat LD50:330 mg/kg UBZHAZ 50,695,78
orl-mus LD50:330 mg/kg UBZHAZ 50,695,78

OSHA PEL: TWA 0.1 mg(Sn)/m^3 (skin)
ACGIH TLV: TWA 0.1 mg(Sn)/m^3 (skin) (Proposed: TWA 0.1 mg(Sn)/m^3; STEL 0.2 mg(Sn)/m^3 (skin))
NIOSH REL: (Organotin Compounds) TWA 0.1 mg(Sn)/m^3

SAFETY PROFILE: Poison by ingestion. See also TIN COMPOUNDS. When heated to decomposition it emits acrid and irritating fumes.

For occupational chemical analysis use NIOSH: Organotin Compounds 5504.

DIV800 HR: 3
DIETHYLCADMIUM
mf: $C_4H_{10}Cd$ mw: 170.5

PROP: An oil; decomp by moisture. D: 1.6562, mp: −21°, bp: 64°.

CONSENSUS REPORTS: Cadmium and its compounds are on the Community Right-To-Know List.

OSHA PEL: TWA 5 µg(Cd)/m^3

ACGIH TLV: TWA 0.05 mg(Cd)/m^3 (Proposed: TWA 0.01 mg(Cd)/m^3 (dust), Suspected Human Carcinogen; 0.002 mg(Cd)/m^3 (respirable dust), Suspected Human Carcinogen); BEI: 10 μg/g creatinine in urine; 10 μg/L in blood
DFG BAT: Blood 1.5 μg/dL; Urine 15 μg/dL, Suspected Carcinogen
NIOSH REL: (Cadmium) Reduce to lowest feasible level

SAFETY PROFILE: Confirmed human carcinogen. A poison. A dangerous fire and explosion hazard. Explodes when heated rapidly to 130°C. On exposure to air it forms white fumes that turn brown and explode. The vapor explodes when heated to 180°C. When heated to decomposition it emits highly toxic fumes of cadmium. See also CADMIUM COMPOUNDS.

DIW000 CAS:90-89-1 HR: 3
DIETHYLCARBAMAZINE
mf: $C_{10}H_{21}N_3O$ mw: 199.34

PROP: Crystals. Mp: 47–49°, bp: 108.5–111°.

SYNS: BITIRAZINE □ CARBAMAZINE □ CARBILAZINE □ CARICIDE □ CATACIDE □ CYPIP □ 1-DIETHYLCARBAMOYL-4-METHYLPIPERAZINE □ 1-DIETHYLCARBAMYL-4-METHYLPIPERZINE □ N,N-DIETHYL-4-METHYL-1-PIPERAZINECARBOXAMIDE □ DITRAZINE BASE □ ETHODRYL □ 84L □ 1-METHYL-4-DIETHYLCARBAMYLPIPERAZINE □ NOTEZINE □ RP 3799 □ SPATONIN

TOXICITY DATA WITH REFERENCE
oms-nml 10 mmol/L CUSCAM 52,303,83
orl-rat TDLo:630 mg/kg (7-11D preg):TER CYLPDN 4,201,83
orl-wmn TDLo:7300 mg/kg/1Y-I:SKN BMJOAE 291,632,85
ipr-mus LD50:240 mg/kg JPPMAB 22,306,70

CONSENSUS REPORTS: Reported in EPA TSCA Inventory.

SAFETY PROFILE: Poison by intraperitoneal route. Human systemic effects by ingestion: allergic dermatitis. An experimental teratogen. Mutation data reported. When heated to decomposition it emits toxic fumes of NO_x and HCl. An additive permitted in the food and drinking water of animals and/or for the treatment of food-producing animals.

DIW200 CAS:1642-54-2 HR: 3
DIETHYLCARBAMAZINE ACID CITRATE
mf: $C_{10}H_{21}N_3O \cdot C_6H_8O_7$ mw: 391.48

PROP: A solid. Mp: 137–138.5°.

SYNS: BANOCIDE □ CARICIDE □ CARITROL □ DICAROCIDE □ DIETHYLCARBAMAZANE CITRATE □ DIETHYLCARBAMAZINE CITRATE □ DIETHYLCARBAMAZINE HYDROGEN CITRATE □ 1-DIETHYLCARBAMOYL-4-METHYLPIPERAZINE DIHYDROGEN CITRATE □ N,N-DIETHYL-4-METHYL-1-PIPERAZINE CARBOXAMIDE CITRATE □ N,N-DIETHYL-4-METHYL-1-PIPERAZINECARBOXAMIDE DIHYDROGEN CITRATE □ N,N-DIETHYL-4-METHYL-1-PIPERAZINECARBOXAMIDE-2-HYDROXY-1,2,3-PROPANETRICARBOXYLATE □ DITRAZIN □ DITRAZIN CITRATE □ DITRAZINE □ DITRAZINE CITRATE □ ETHODRYL CITRATE □ FRANOCIDE □ FRANOZAN □ HETRAZAN □ LOXURAN □ 1-METHYL-4-DIETHYLCARBAMOYLPIPERAZINE CITRATE

TOXICITY DATA WITH REFERENCE
orl-rat TDLo:1800 mg/kg (8-16D preg):TER IJMRAQ 60,1529,72
orl-rat LD50:1400 mg/kg FAZMAE 17,108,73
ihl-rat LC50:309 mg/m^3/4H 85GMAT -,63,82
scu-rat LD50:1136 mg/kg NIIRDN 6,306,82
orl-mus LD50:660 mg/kg 29ZVAB -,41,69
ipr-mus LD50:425 mg/kg IJOCAP 26,748,87
scu-mus LD50:608 mg/kg NIIRDN 6,306,82
ivn-mus LD50:180 mg/kg CSLNX* NX#00466

CONSENSUS REPORTS: Reported in EPA TSCA Inventory.

SAFETY PROFILE: Poison by inhalation and intravenous routes. Moderately toxic by ingestion, subcutaneous, and intraperitoneal routes. An experimental teratogen. When heated to decomposition it emits toxic fumes of NO_x.

DIW300 CAS:5348-97-0 HR: 3
DIETHYLCARBAMAZINE HYDROCHLORIDE
mf: $C_{10}H_{21}N_{3O} \cdot ClH$ mw: 235.80

SYN: 1-DIETHYLCARBAMYL-4-METHYLPIPERAZINE HYDROCHLORIDE

TOXICITY DATA WITH REFERENCE
orl-rat LD50:1380 mg/kg ANYAA9 50,141,48
ipr-rat LD50:285 mg/kg CLDND*
ivn-rat LD50:150 mg/kg ANYAA9 50,141,48
orl-mus LD50:660 mg/kg ANYAA9 50,141,48
ipr-mus LD50:248 mg/kg ANYAA9 50,141,48
ivn-mus LD50:82 mg/kg ANYAA9 50,141,48

CONSENSUS REPORTS: Reported in EPA TSCA Inventory.

SAFETY PROFILE: A poison by intraperitoneal and intravenous routes. Moderately toxic by ingestion. When heated to decomposition it emits very toxic fumes of $NO_2$and HCl.

DIW400 CAS:88-10-8 HR: 2
DIETHYLCARBAMOYL CHLORIDE
mf: $C_5H_{10}ClNO$ mw: 135.61

PROP: Liquid. Mp: −44°, bp: 186°, vap d: 4.1.

SYNS: DIETHYLCARBAMIC CHLORIDE □ DIETHYLCARBAMIDOYL CHLORIDE □ N,N-DIETHYLCARBAMOYL CHLORIDE □ DIETHYLCARBAMYL CHLORIDE

TOXICITY DATA WITH REFERENCE
skn-mus TDLo:43,200 mg/kg/72W-I:CAR JACTDZ 6(4),479,87
ipr-mus LD50:750 mg/kg NTIS** AD691-490

CONSENSUS REPORTS: Reported in EPA TSCA Inventory.

DFG MAK: Suspected Carcinogen

SAFETY PROFILE: Suspected carcinogen with experimental carcinogenic data. Moderately toxic by intraperitoneal route. Mutation data reported. Reacts with water or steam to produce toxic and corrosive fumes. When heated to decomposition it emits highly toxic fumes of Cl^- and NO_x. See also CARBAMATES and CHLORIDES.

DIW600 **CAS:2425-25-4** **HR: 3**
O,O-DIETHYL-S-(CARBETHOXY)METHYL PHOSPHOROTHIOLATE
mf: $C_8H_{17}O_5PS$ mw: 256.28

SYNS: ACETOPHOS □ ACETOXON □ O,O-DIETHYL-S-CARBOETHOXYMETHYL PHOSPHOROTHIOATE □ O,O-DIETHYL-S-CARBOETHOXYMETHYL THIOPHOSPHATE □ PHOSPHOROTHIOIC ACID-O,O-DIETHYL ESTER-S-ESTER with ETHYL MERCAPTOACETATE

TOXICITY DATA WITH REFERENCE
orl-rat LD50:45 mg/kg HYSAAV 31,18,66
ipr-mus LD50:214 mg/kg JEENAI 51,714,58

SAFETY PROFILE: Poison by ingestion and intraperitoneal routes. When heated to decomposition it emits very toxic fumes of PO_x and SO_x.

DIW800 **CAS:112-36-7** **HR: 3**
DIETHYL CARBITOL
mf: $C_8H_{18}O_3$ mw: 162.26

PROP: Colorless liquid; sol in water and hydrocarbons. D: 0.9082 @ 20°/20°, bp: 189°, fp: −44°, flash p: 180°F (OC), vap d: 5.6.

SYNS: BIS(2-ETHOXYETHYL)ETHER □ DIETHYLENE GLYCOL DIETHYL ETHER □ 1-ETHOXY-2-(β-ETHOXYETHOXY)ETHANE □ ETHYL DIGLYME □ 3,6,9-TRIOXAUNDECANE

TOXICITY DATA WITH REFERENCE
eye-rbt 50 mg MOD UCDS** 4/25/58
orl-mus TDLo:24 g/kg (7-14D preg):REP EVHPAZ 57,141,84
orl-mus TDLo:3 g/kg (female 6-15D post):TER NTIS** PB88-168257
orl-rat LD50:4970 mg/kg UCDS** 4/25/58
orl-gpg LD50:1850 mg/kg JIHTAB 23,259,41

CONSENSUS REPORTS: Reported in EPA TSCA Inventory. Glycol ether compounds are on the Community Right-To-Know List.

SAFETY PROFILE: Moderately toxic by ingestion. An experimental teratogen. Other experimental reproductive effects. An eye irritant. Flammable when exposed to heat or flame. When heated to decomposition it emits acrid smoke and irritating fumes. See also GLYCOL ETHERS.

DIX000 **CAS:919-54-0** **HR: 2**
O,O-DIETHYL-S-CARBOETHOXYMETHYL DITHIOPHOSPHATE
mf: $C_8H_{17}O_4PS_2$ mw: 272.34

PROP: A liquid. D: 1.18 @ 20°/4°, bp: 137° @ 1.5 mm.

SYNS: ACETHION □ ACETHIONE □ ((DIETHOXYPHOSPHINOTHIOYL)THIO)ACETIC ACID, ETHYL ESTER □ O,O-DIETHYL-S-CARBOETHOXYMETHYL PHOSPHORODITHIOATE □ ENT 25,650 □ ETHOXYPHAS □ ETHOXYPHOS □ ETHYL ((DIETHOXYPHOSPHINOTHIOYL)THIO)ACETATE □ HERCULES 4580 □ PHOSPHORODITHIOIC ACID-O,O-DIETHYL ESTER-S-ESTER with ETHYL MERCAPTOACETATE

TOXICITY DATA WITH REFERENCE
orl-mus LD50:1200 mg/kg ARSIM* 20,12,66
ipr-mus LD50:1280 mg/kg JEENAI 51,714,58

SAFETY PROFILE: Moderately toxic by ingestion and intraperitoneal routes. When heated to decomposition it emits very toxic fumes of PO_x and SO_x. See also MERCAPTANS and ESTERS.

DIX200 **CAS:105-58-8** **HR: 3**
DIETHYL CARBONATE
DOT: UN 2366
mf: $C_5H_{10}O_3$ mw: 118.15

PROP: Colorless liquid, mild odor. Mp: 43°, bp: 125.8°, flash p: 77°F (OC), d: 0.975 @ 20°/4°, vap press: 10 mm @ 23.8°, vap d: 4.07.

SYNS: DEC □ DIAETHYLCARBONAT (GERMAN) □ DIETHYL CARBONATE (DOT) □ ETHOXYFORMIC ANHYDRIDE □ ETHYL CARBONATE □ EUFIN □ NCI-C60899

TOXICITY DATA WITH REFERENCE
ipr-ham TDLo:496 mg/kg (female 8D post):TER CNREA8 27,1696,67
ipr-ham TDLo:1004 mg/kg (female 8D post):TER CNREA8 27,1696,67
ipr-ham TDLo:496 mg/kg (8D preg):TER CNREA8 27,1696,67
orl-mus TDLo:500 mg/kg:ETA BCPCA6 2,168,59
ipr-mus TDLo:456 mg/kg:ETA BCPCA6 2,168,59
orl-rat LDLo:15 g/kg FAONAU 53A,52,74
scu-rat LD50:8500 mg/kg CLDND* 49,172.101,92

CONSENSUS REPORTS: Reported in EPA TSCA Inventory.

DOT CLASSIFICATION: 3; *Label:* Flammable Liquid

SAFETY PROFILE: Mildly toxic by subcutaneous route. Questionable carcinogen with experimental tumorigenic and teratogenic data. A dangerous fire hazard when exposed to heat or flame; can react with oxidizing materials. To fight fire, use foam, CO_2, dry chemical. When heated to decomposition it emits acrid smoke and fumes. See also ANHYDRIDES.

DIX400 **CAS:2315-36-8** **HR: 2**
N,N-DIETHYLCHLORACETAMIDE
mf: $C_6H_{12}ClNO$ mw: 149.64

SYN: N-CHLOROACETYLDIETHYLAMINE

TOXICITY DATA WITH REFERENCE
orl-rat LD50:500 mg/kg RREVAH 10,97,65
ihl-rat LC50:315 mg/m³/4H GTPZAB 35(6),36,91
skn-rat LD50:980 mg/kg GTPZAB 33(2),51,89
orl-mus LD50:226 mg/kg GTPZAB 35(6),36,91
ihl-mus LC50:270 mg/m³/2H GTPZAB 35(6),36,91
ihl-gpg LC50:175 mg/m³/4H GTPZAB 35(6),36,91

CONSENSUS REPORTS: Reported in EPA TSCA Inventory.

SAFETY PROFILE: Moderately toxic by ingestion, skin contact, and inhalation. When heated to decomposition it emits very toxic fumes of Cl^- and NO_x.

DIX600 **CAS:1757-18-2** **HR: 3**
O,O-DIETHYL-O-(2-CHLORO-1,2,5-DICHLOROPHENYLVINYL) PHOSPHOROTHIOATE
mf: $C_{12}H_{14}Cl_3O_3PS$ mw: 375.64

SYNS: AKTON □ AXIOM □ O-(2-CHLORO-1-(2,5-DICHLOROPHENYL))-O,O-DIETHYL ESTER PHOSPHOROTHIOIC ACID □ O-(2-CHLORO-1-(2,5-DICHLOROPHENYL)VINYL)-O,O-DIETHYL PHOSPHOROTHIOATE □ ENT 27,102 □ SD 9098 □ SHELL SD-9098

TOXICITY DATA WITH REFERENCE
orl-rat LD50:146 mg/kg WRPCA2 9,119,70
orl-mus LD50:89 mg/kg SPEADM 78-1,41,78
skn-rbt LD50:177 mg/kg SPEADM 78-1,41,78
orl-bwd LD50:75 mg/kg AECTCV 12,355,83

SAFETY PROFILE: Poison by ingestion and skin contact. An insecticide. When heated to decomposition it emits very toxic fumes of Cl^-, PO_x, and SO_x.

DIX800 **CAS:7173-84-4** **HR: 3**
O,O-DIETHYL-S-p-CHLOROPHENYL THIOMETHYLPHOSPHOROTHIOATE
mf: $C_{11}H_{16}ClO_3PS_2$ mw: 326.81

SYNS: S-((p-CHLOROPHENYLTHIO)METHYL)-O,O-DIETHYL PHOSPHORODITHIOATE □ DANIFOS

TOXICITY DATA WITH REFERENCE
orl-mus LD50:165 mg/kg BESAAT 15,118,69
skn-mus LD50:220 mg/kg BESAAT 15,118,69

SAFETY PROFILE: Poison by ingestion and skin contact. When heated to decomposition it emits very toxic fumes of Cl^-, PO_x, and SO_x.

DIY000 **CAS:814-49-3** **HR: 3**
DIETHYL CHLOROPHOSPHATE
mf: $C_4H_{10}ClO_3P$ mw: 172.56

PROP: Water-white liquid with irritating, unpleasant odor. Bp: 88–90° @ 15 mm, d: 1.21 @ 20°/4°, vap d: 5.94.

SYNS: CHLOROPHOSPHORIC ACID DIETHYL ESTER □ DIETHOXYPHOSPHORUS OXYCHLORIDE

TOXICITY DATA WITH REFERENCE
skn-rbt 10 mg/24H open MLD AIHAAP 23,95,62
orl-rat LD50:11 mg/kg AIHAAP 23,95,62
skn-rbt LD50:7900 µg/kg AIHAAP 23,95,62

CONSENSUS REPORTS: Reported in EPA TSCA Inventory.

SAFETY PROFILE: Deadly poison by skin contact. Poison by ingestion. A cholinesterase inhibitor. See also PARATHION. Trace HCl catalyzes a hazardous reaction during the preparation of diethyl phosphate from diethyl chlorophosphate. When heated to decomposition it emits very toxic fumes of Cl^- and PO_x.

DIY200 **CAS:1331-43-7** **HR: 3**
DIETHYLCYCLOHEXANE (mixed isomers)
mf: $C_{10}H_{20}$ mw: 140.30

PROP: Liquid, insol in water. D: 0.8037 @ 20°/20°, bp: 174°, fp: −100°, flash p: 120°F (OC), autoign temp: 465°F, lel: 0.8% @ 140°F, uel: 6.0% @ 230°F.

TOXICITY DATA WITH REFERENCE
skn-rbt 10 mg/24H open MLD AIHAAP 23,95,62
ihl-rat LCLo:2000 ppm/4H AIHAAP 23,95,62
orl-rat LD50:64 g/kg AIHAAP 23,95,62
skn-rbt LDLo:2500 mg/kg AIHAAP 23,95,62

SAFETY PROFILE: Moderately toxic by inhalation and skin contact. Mildly toxic by ingestion. A skin irritant. Flammable liquid when exposed to heat, flame, or oxidizers. To fight fire, use foam, mist, dry chemical. When heated to decomposition it emits acrid smoke and fumes.

DIY600 **CAS:100-38-9** **HR: 3**
N,N-DIETHYL CYSTEAMINE
mf: $C_6H_{15}NS$ mw: 133.28

SYNS: N-DIAETHYL CYSTEAMIN (GERMAN) □ DIETHYLAMINOETHANETHIOL □ 2-(DIETHYLAMINO)ETHANETHIOL □ 2-(DIETHYLAMINO)ETHYLMERCAPTAN □ β-DIETHYLAMINOETHYL MERCAPTAN □ DIETHYLCYSTEAMIN □ DIETHYLCYSTEAMINE □ N-DIETHYL CYSTEAMINE □ DIETHYL(2-MERCAPTOETHYL)AMINE

TOXICITY DATA WITH REFERENCE
orl-mus LD50:231 mg/kg GTPZAB 13(7),26,69
ihl-mus LD50:42,500 mg/m³ GTPZAB 13(7),26,69
ipr-mus LD50:96 mg/kg BCPCA6 14,289,65
scu-mus LD50:120 mg/kg AIPTAK 109,108,57

SAFETY PROFILE: Poison by ingestion, intraperitoneal, and subcutaneous routes. Mildly toxic by inhalation. When heated to decomposition it emits very toxic fumes of NO_x and SO_x. See also MERCAPTANS and AMINES.

DIY800 **CAS:104-78-9** **HR: 3**
N,N-DIETHYL-1,3-DIAMINOPROPANE
DOT: UN 2684
mf: $C_7H_{18}N_2$ mw: 130.27

PROP: Liquid. Bp: 165–170°, flash p: 138°F (OC), d: 0.82, vap d: 4.48.

SYNS: 1-AMINO-3-(DIETHYLAMINO)PROPANE □ N,N-DIETHYLAMINOPROPYLAMINE □ N-(3-DIETHYLAMINOPROPYL)AMINE □ 3-(DIETHYLAMINO)PROPYLAMINE (DOT) □ DIETHYLAMINOTRIMETHYLENAMINE

TOXICITY DATA WITH REFERENCE
skn-rbt 100 µg/24H open AIHAAP 23,95,62
orl-rat LD50:1410 mg/kg UCDS** 2/27/67
skn-rbt LD50:750 mg/kg AIHAAP 23,95,62

CONSENSUS REPORTS: Reported in EPA TSCA Inventory.

DOT CLASSIFICATION: 8; *Label:* Corrosive, Flammable Liquid

SAFETY PROFILE: Moderately toxic by ingestion and skin contact. Corrosive to the eyes, skin, and mucous membranes. A sensitizer. See also AMINES. Flammable liquid when exposed to heat or flame; can react with oxidizing materials. To fight fire, use foam, CO_2, dry chemical. When heated to decomposition it emits toxic fumes of NO_x.

DIZ100 **CAS:1609-47-8** **HR: 3**
DIETHYL DICARBONATE
mf: $C_6H_{10}O_5$ mw: 162.16

PROP: Viscous liquid; fruity odor. Bp: 58.5–62° @ 3 mm, d: 1.12, visc (20°): 1.97 cp. Soluble in alc, esters, ketones, and hydrocarbons.

SYNS: BAYCOVIN □ DEPC □ DICARBONIC ACID DIETHYL ESTER □ DIETHYL ESTER of PYROCARBONIC ACID □ DIETHYL OXYDIFORMATE □ DIETHYL PYROCARBONATE □ DIETHYL PYROCARBONIC ACID □ DKD □ ETHYL PYROCARBONATE □ OXYDIFORMIC ACID DIETHYL ESTER □ PIREF □ PYROCARBONATE d'ETHYLE (FRENCH) □ PYROCARBONIC ACID, DIETHYL ESTER □ PYROKOHLENSAEURE DIAETHYL ESTER (GERMAN)

TOXICITY DATA WITH REFERENCE
orl-rat LD50:850 mg/kg FAONAU 51A,69,72
ihl-rat LCLo:10 g/m³/1H ZLUFAR 114,292,61
ipr-rat LD50:47 mg/kg FAONAU 53A,52,74
orl-mus LD50:1558 mg/kg FAONAU 53A,52,74
orl-mus LD50:2027 mg/kg ZLUFAR 139,287,69
ihl-mus LCLo:10 ppm/1H FAONAU 51A,67,72
orl-dog LD50:500 mg/kg FAONAU 53A,52,74
orl-cat LDLo:100 mg/kg FAONAU 51A,69,72
orl-rbt LDLo:500 mg/kg FAONAU 51A,69,72

CONSENSUS REPORTS: Reported in EPA TSCA Inventory.

SAFETY PROFILE: Poison by ingestion, inhalation, and intraperitoneal routes. Concentrated DEPC is irritating to eyes, mucous membranes, and skin. When heated to decomposition it emits acrid smoke and fumes. See also ESTERS.

DJA200 **CAS:3152-41-8** **HR: 3**
O,O-DIETHYL-S-(3,4-DICHLOROPHENYL-THIO)METHYL PHOSPHOROTHIOATE
mf: $C_{11}H_{15}Cl_2O_2PS_3$ mw: 377.31

SYNS: S-((3,4-DICHLOROPHENYLTHIO)METHYL)-O,O-DIETHYL PHOSPHORODITHIOATE □ ENT 25,555-X □ G 27365 □ GEIGY G-27365

TOXICITY DATA WITH REFERENCE
orl-rat LD50:89 mg/kg ARSIM* 20,10,66
orl-mus LD50:175 mg/kg ARSIM* 20,10,66
orl-ckn LD50:143 mg/kg TXAPA9 11,49,67

SAFETY PROFILE: Poison by ingestion. When heated to decomposition it emits very toxic fumes of Cl^-, PO_x, and SO_x.

DJA300 **CAS:1474-80-2** **HR: 3**
DIETHYL DI(DIMETHYLAMIDO)PYROPHOSPHATE (unsymmetrical)
mf: $C_8H_{22}N_2O_5P_2$ mw: 288.26

SYNS: COMPOUND 6515 □ DIETHYL BIS-DIMETHYL PYROPHOSPHORDIAMIDE asym □ unsym-DIETHYL BIS(DIMETHYLAMIDO)PYROPHOSPHATE

TOXICITY DATA WITH REFERENCE
orl-rat LD50:3800 μg/kg JPETAB 107,464,53
skn-rat LD50:5 mg/kg JPETAB 107,464,53
ipr-rat LD50:2400 μg/kg JPETAB 107,464,53
ipr-mus LD50:4700 μg/kg JPETAB 107,464,53
ivn-dog LD50:10 mg/kg JPETAB 107,464,53
ipr-gpg LD50:5 mg/kg JPETAB 107,464,53

SAFETY PROFILE: Poison by ingestion, skin contact, intravenous, and intraperitoneal routes. When heated to decomposition it emits toxic fumes of PO_x and NO_x. See also ESTERS.

D

DJA325 **CAS:10161-85-0** **HR: 3**
O,O-DIETHYL Se-(2-DIETHYLAMINOETHYL)PHOSPHOROSELENOATE
mf: $C_{10}H_{24}NO_3PSe$ mw: 316.28

SYN: PHOSPHOROSELENOIC ACID, Se-(2-(DIETHYLAMINO)ETHYL) O,O-DIETHYL ESTER

TOXICITY DATA WITH REFERENCE
scu-mus LD50:60 μg/kg JMCMAR 10,115,67

OSHA PEL: TWA 0.2 mg(Se)/m³
ACGIH TLV: TWA 0.2 mg(Se)/m³

SAFETY PROFILE: Poison by subcutaneous route. When heated to decomposition it emits toxic fumes of NO_x, PO_x, and Se.

DJA400 **CAS:78-53-5** **HR: 3**
O,O-DIETHYL-S-(2-DIETHYLAMINOETHYL) THIOPHOSPHATE
mf: $C_{10}H_{24}NO_3PS$ mw: 269.38

PROP: A liquid. Bp: 110 @ 0.2 mm.

SYNS: AMITON □ CHIPMAN 6200 □ CITRAM □ S-(2-(DIETHYLAMINO)ETHYL)PHOSPHOROTHIOIC ACID-O,O-DIETHYL ESTER □ S-(DIETHYLAMINOETHYL)-O,O-DIETHYL PHOSPHOROTHIOATE □ DIETHYL S 2 DIETHYLAMINOETHYL PHOSPHOROTHIOATE □ (2-DIETHYLAMINO)ETHYLPHOSPHOROTHIOIC ACID-O,O-DIETHYL ESTER □ O,O-DIETHYL-S-2-DIETHYLAMINOETHYL PHOSPHOROTHIOATE □ O,O-DIETHYL-S-(β-DIETHYLAMINO)ETHYL PHOSPHOROTHIOLATE □ O,O-DIETHYL-S-DIETHYLAMINOETHYL PHOSPHOROTHIOLATE □ O,O-DIETHYL-S-2-DIETHYLAMINOETHYL PHOSPHOROTHIOLATE □ DSDP □ ENT 24,980-X □ INFERNO □ METRAMAC □ METRAMAK □ R-5,158 □ RHODIA-6200 □ TETRAM

TOXICITY DATA WITH REFERENCE
orl-rat LD50:3300 μg/kg ARSIM* 20,7,66
scu-rat LD50:150 μg/kg CJPPA3 44,745,66
scu-mus LD50:190 μg/kg CJPPA3 46,109,68
scu-rbt LD50:125 μg/kg CJPPA3 46,109,68
scu-gpg LD50:80 μg/kg CJPPA3 46,109,68
scu-ham LD50:210 μg/kg CJPPA3 46,109,68

CONSENSUS REPORTS: EPA Extremely Hazardous Substances List.

SAFETY PROFILE: A deadly poison by ingestion and subcutaneous routes. A cholinesterase inhibitor. An insecticide. See also PARATHION. When heated to decomposition it emits very toxic fumes of NO_x, PO_x, and SO_x.

DJA800 **CAS:106-20-7** **HR: 2**
2,2′-DIETHYLDIHEXYLAMINE
mf: $C_{16}H_{35}N$ mw: 241.52

PROP: Water-white liquid with sltly ammonia-like odor.

Flash p: 270°F (OC), d: 0.8062 @ 20°/20°, vap d: 8.35, bp: 281.1°. Sol in EtOH and Et_2O; sltly sol in H_2O.

SYNS: BIS-2-ETHYLHEXYLAMIN □ DI(2-ETHYLHEXYL)AMINE □ 2-ETHYL-N-(2-ETHYLHEXYL)-1-HEXANAMINE

TOXICITY DATA WITH REFERENCE
skn-rbt 10 mg/24H open SEV JIHTAB 31,60,49
skn-rbt 500 mg open MLD UCDS** 8/9/68
skn-rbt 500 mg/24H SEV 28ZPAK -,62,72
eye-rbt 750 µg SEV AJOPAA 29,1363,46
eye-rbt 50 µg/24H SEV 28ZPAK -,62,72
orl-rat LD50:1640 mg/kg JIHTAB 31,60,49
ipr-mus LD50:800 mg/kg CBCCT* 2,184,50
skn-rbt LD50:1190 mg/kg UCDS** 8/9/68

CONSENSUS REPORTS: Reported in EPA TSCA Inventory.

SAFETY PROFILE: Moderately toxic by ingestion, skin contact, and intraperitoneal routes. A severe skin and eye irritant. Combustible when exposed to heat or flame; can react with oxidizing materials. To fight fire, use alcohol foam, foam, CO_2, dry chemical. When heated to decomposition it emits toxic fumes of NO_x. See also AMINES.

DJB000 CAS:2767-55-7 HR: 3
DIETHYLDIIODOSTANNANE
mf: $C_4H_{10}I_2Sn$ mw: 430.63

PROP: Very sltly sol white crystals or white needles. Mp: 44°, bp: 240–245° (decomp).

SYNS: DIETHYLSTANNIUM DIIODIDE □ DIETHYLSTANNIUMDIJODID (GERMAN) □ DIETHYLTIN DIIODIDE □ TIN, DIETHYL-, DIIODIDE

TOXICITY DATA WITH REFERENCE
orl-rat LDLo:100 mg/kg BJPCAL 10,16,55
ipr-rat LDLo:26 mg/kg BJPCAL 10,16,55

OSHA PEL: TWA 0.1 $mg(Sn)/m^3$ (skin)
ACGIH TLV: TWA 0.1 $mg(Sn)/m^3$ (skin) (Proposed: TWA 0.1 $mg(Sn)/m^3$; STEL 0.2 $mg(Sn)/m^3$ (skin))
NIOSH REL: (Organotin Compounds) TWA 0.1 $mg(Sn)/m^3$

SAFETY PROFILE: Poison by ingestion and intraperitoneal routes. See also TIN COMPOUNDS and IODIDES. When heated to decomposition it emits toxic fumes of I^-.

For occupational chemical analysis use NIOSH: Organotin Compounds 5504.

DJB200 CAS:7773-34-4 HR: 2
α,α′-DIETHYL-4,4′-DIMETHOXYSTILBENE
mf: $C_{20}H_{24}O_2$ mw: 296.44

SYNS: 3,4-BIS(p-METHOXYPHENYL)-3-HEXENE □ DEPOT-OESTROMENINE □ DEPOT-OESTROMON □ 3,4-DIANISYL-3-HEXENE □ trans-α,α′-DIETHYL-4,4′-DIMETHOXYSTILBENE □ (E)-1,1′-(1,2-DIETHYL-1,2-ETHENE-DIYL)BIS(4-METHOXYBENZENE) □ DIETHYLSTILBESTROL DIMETHYL ETHER □ DIMESTROL □ 4,4′-DIMETHOXY-α,β-DIETHYLSTILBENE □ STILBESTROL DIMETHYL ETHER □ SYNTHILA

TOXICITY DATA WITH REFERENCE
scu-ham TDLo:560 g/kg/35W-I:ETA CNREA8 31,1251,71

SAFETY PROFILE: Questionable carcinogen with experimental tumorigenic data. When heated to decomposition it emits acrid and irritating fumes. See also DIETHYLSTILBESTEROL.

DJB400 CAS:17010-64-9 HR: 2
3′,4′-DIETHYL-4-DIMETHYLAMINOAZOBENZENE
mf: $C_{18}H_{23}N_3$ mw: 281.44

SYNS: BENZENAMINE, 4-((3,4-DIETHYLPHENYL)AZO)-N,N-DIMETHYL-(9CI) □ p-((3,4-DIETHYLPHENYL)AZO)-N,N-DIMETHYLANILINE □ N,N-DIMETHYL-p-((3,4-DIETHYLPHENYL)AZO)ANILINE □ 3′,4′-Et2-DAB

TOXICITY DATA WITH REFERENCE
orl-rat TDLo:8467 mg/kg/36W-C:CAR CBINA8 53,107,85
orl-rat TD:2511 mg/kg/17W-I:ETA CNREA8 30,1520,70

SAFETY PROFILE: Questionable carcinogen with experimental carcinogenic and tumorigenic data. When heated to decomposition it emits toxic fumes of NO_x.

DJB600 CAS:64048-13-1 HR: 3
p-DIETHYL-p′-DIMETHYLTHIOPYROPHOSPHATE
mf: $C_6H_{16}O_5P_2S_2$ mw: 294.28

SYN: p′-DIETHYL-p-DIMETHYL THIOPYROPHOSPHATE

TOXICITY DATA WITH REFERENCE
ims-rat LD50:1 mg/kg CJCHAG 34,1819,56
ipr-mus LD50:1800 µg/kg CJCHAG 34,1819,56
ims-mus LD50:1 mg/kg CJCHAG 34,1819,56

SAFETY PROFILE: A deadly poison by intramuscular and intraperitoneal routes. When heated to decomposition it emits very toxic fumes of SO_x and PO_x.

DJB800 CAS:7346-14-7 HR: 2
N,N′-DIETHYL-N,N′-DINITROSOETHYLENEDIAMINE
mf: $C_6H_{14}N_4O_2$ mw: 174.24

SYNS: N,N′-DINITROSO-N,N′-DIETHYLETHYLENEDIAMINE □ NSC 62579

TOXICITY DATA WITH REFERENCE
orl-rat TDLo:12 mg/kg/1Y-I:CAR JNCIAM 41,985,68
orl-rat TD:34 mg/kg/50W-I:CAR JNCIAM 41,985,68

SAFETY PROFILE: Questionable carcinogen with experimental carcinogenic data. When heated to decomposition it emits toxic fumes of NO_x. See also N-NITROSO COMPOUNDS.

DJC000 CAS:72-56-0 HR: 3
DIETHYLDIPHENYL DICHLOROETHANE
mf: $C_{18}H_{20}Cl_2$ mw: 307.28

PROP: Crystals. Mp: 60–61°.

SYNS: 1,1-BIS(p-ETHYLPHENYL)-2,2-DICHLOROETHANE □ 2,2-BIS(p-ETHYLPHENYL)-1,1-DICHLOROETHANE □ α,α-DICHLORO-2,2-BIS(p-ETHYLPHENYL)ETHANE □ 1,1-DICHLORO-2,2-BIS(p-ETHYLPHENYL)ETHANE □ 1,1-DICHLORO-2,2-BIS(4-ETHYLPHENYL)ETHANE □ 2,2-DICHLORO-1,1-BIS(p-ETHYLPHENYL)ETHANE □ DI(p-ETHYLPHE-

NYL)DICHLOROETHANE □ ETHYLAN □ p,p-ETHYL DDD □ p,p'-ETHYL-DDD □ NCI-C02868 □ PERTHANE □ Q-137

TOXICITY DATA WITH REFERENCE
mma-sat 333 µg/plate NTPTB* APR 82
scu-mus TDLo:900 mg/kg (6-14D preg):TER NTIS** PB223-160
orl-mus TDLo:210 g/kg/2Y-C:CAR TUMOAB 66,277,80
orl-mus TD:547 g/kg/2Y-C:ETA NCITR* NCI-CG-TR-156,79
orl-rat LD50:6600 mg/kg SPEADM 78-1,16,78
ivn-rat LD50:73 mg/kg AIPTAK 103,404,55
orl-mus LD50:6600 mg/kg 31ZOAD 1,142,68
ivn-mus LD50:173 mg/kg AIPTAK 103,404,55
orl-bwd LD50:9000 mg/kg DOEAAH 35,25,79

CONSENSUS REPORTS: NCI Carcinogenesis Bioassay (feed); Clear Evidence: mouse NCITR* NCI-CG-TR-156,79; No Evidence: rat NCITR* NCI-CG-TR-156,79.

SAFETY PROFILE: Poison by intravenous route. Mildly toxic by ingestion. Questionable carcinogen with experimental carcinogenic and tumorigenic data. Experimental teratogenic and reproductive effects. Mutation data reported. A pesticide. When heated to decomposition it emits toxic fumes of Cl^-. See also CHLORINATED HYDROCARBONS, ALIPHATIC.

DJC400 CAS:85-98-3 HR: 3
1,3-DIETHYL-1,3-DIPHENYLUREA
mf: $C_{17}H_{20}N_2O$ mw: 268.39

PROP: Colorless crystals. Mp: 73°, d: 1.12, bp: 326°, flash p: 302°F (CC), vap d: 9.3.

SYNS: BIS(N-ETHYL-N-PHENYL)UREA □ N,N-DIETHYLCARBANILIDE □ N,N'-DIETHYL-N,N'-DIPHENYLUREA □ sym-DIETHYLDIPHENYLUREA □ USAF EK-1047

TOXICITY DATA WITH REFERENCE
orl-rat LD50:2750 mg/kg GISAAA 41(5),21,76
orl-mus LD50:2500 mg/kg GISAAA 41(5),21,76
ipr-mus LD50:200 mg/kg NTIS** AD277-689

CONSENSUS REPORTS: Reported in EPA TSCA Inventory.

SAFETY PROFILE: Poison by intraperitoneal route. Moderately toxic by ingestion. Combustible when exposed to heat or flame. Probably a slight explosion hazard, although it is a component of smokeless explosive mixtures. When heated to decomposition it burns and emits very toxic fumes of NO_x. To fight fire, use dry chemical, CO_2, spray or mist. An explosion regulator.

DJC600 CAS:110-81-6 HR: 3
DIETHYLDISULFIDE
mf: $C_4H_{10}S_2$ mw: 122.26

PROP: Liquid or oil with garlic odor. Fp: −101.5°, bp: 154°, d: 0.99267 @ 20°/4°, vap d: 4.22, vap press: 4.28 mm @ 25°. Sltly sol in water.

SYN: DIETHYLDISULFID (CZECH)

TOXICITY DATA WITH REFERENCE
skn-rbt 500 mg/24H MLD 28ZPAK -,171,72
eye-rbt 100 mg/24H MOD 28ZPAK -,171,72
orl-rat LD50:2030 mg/kg 28ZPAK -,171,72

CONSENSUS REPORTS: Reported in EPA TSCA Inventory.

SAFETY PROFILE: Moderately toxic by ingestion. A skin and eye irritant. Flammable when exposed to heat or flame; can react vigorously with oxidizing materials. To fight fire, use foam or dry chemical. See also SULFIDES and SULFATES.

DJC800 CAS:147-84-2 HR: 2
DIETHYLDITHIOCARBAMIC ACID
mf: $C_5H_{11}NS_2$ mw: 149.29

SYNS: CARBAMODITHIOIC ACID, DIETHYL-(9CI) □ DIECA □ DIETHYL DITHIOCARBAMATE □ DIETHYLDITHIOCARBAMINIC ACID □ DIETHYLDITHIONE

TOXICITY DATA WITH REFERENCE
sce-hmn:lyms 10 nmol/L CNREA8 49,5239,89
ivn-rat LD50:1250 mg/kg PHMCAA 3,62,61
orl-rat LD50:1 g/kg CRTXB2 20,83,89
ivn-mus LD50:1750 mg/kg PHMCAA 3,62,61
ivn-dog LD50:1000 mg/kg PHMCAA 3,62,61

CONSENSUS REPORTS: Reported in EPA TSCA Inventory.

SAFETY PROFILE: Moderately toxic by intravenous route. Mutation data reported. See also CARBAMATES. When heated to decomposition it emits very toxic fumes of NO_x and SO_x.

DJC875 CAS:69654-93-9 HR: 3
DIETHYLDITHIOCARBAMIC ACID ANHYDROSULFIDE with DIMETHYLTHIOCARBAMIC ACID
mf: $C_8H_{16}N_2OS_3$ mw: 220.38

SYNS: DDS □ DIMETHYLCARBAMYL DIETHYLTHIOCARBAMYL SULFIDE

TOXICITY DATA WITH REFERENCE
ipr-rat LD50:900 µg/kg AIPTAK 108,27,56
ipr-mus LD50:800 µg/kg AIPTAK 108,27,56
ivn-dog LD50:200 µg/kg AIPTAK 108,27,56

SAFETY PROFILE: A deadly poison by intravenous and intraperitoneal routes. When heated to decomposition it emits toxic fumes of SO_x and NO_x. See also CARBAMATES and SULFIDES.

DJD000 CAS:1518-58-7 HR: 2
DIETHYLDITHIOCARBAMIC ACID DIETHYLAMINE SALT
mf: $C_9H_{21}N_2S_2$ mw: 221.44

PROP: Crystals from Et_2O. Mp: 83–84°.

SYNS: CARBAMODITHIOIC ACID, DIETHYL-, compd. with N-ETHYLETHANAMINE (1:1) (9CI) □ CONTRAMINE □ DIAETHYLAMMONIUM-DIAETHYLDITHIOCARBAMAT □ DIETHYLAMMONIUM DIETHYLDITHIOCARBAMATE □ USAF EK-2635

TOXICITY DATA WITH REFERENCE
ipr-mus LD50:500 mg/kg NTIS** AD277-689

CONSENSUS REPORTS: Reported in EPA TSCA Inventory.

D

SAFETY PROFILE: Moderately toxic by intraperitoneal route. When heated to decomposition it emits very toxic fumes of NO_x and SO_x. See also CARBAMATES.

DJD200 **CAS:17549-30-3** **HR: 2**
DIETHYLDITHIOCARBAMIC ACID LEAD(II) SALT
mf: $C_{10}H_{20}N_2S_4 \cdot Pb$ mw: 503.75

PROP: Prisms from Et_2O.

TOXICITY DATA WITH REFERENCE
ipr-rat LDLo:500 mg/kg NCNSA6 5,30,53

NIOSH REL: (Inorganic Lead) TWA 0.10 mg(Pb)/m³

CONSENSUS REPORTS: Lead and its compounds are on the Community Right-To-Know List.

SAFETY PROFILE: Moderately toxic by intraperitoneal route. See also LEAD COMPOUNDS and CARBAMATES. When heated to decomposition it emits very toxic fumes of Pb, SO_x, and NO_x.

DJD400 **CAS:136-92-5** **HR: 3**
DIETHYLDITHIOCARBAMIC ACID SELENIUM(II) SALT
mf: $C_{10}H_{20}N_2S_4 \cdot Se$ mw: 375.52

SYNS: ETHYL SELENAC □ SELENIUM DIETHYLDITHIOCARBAMATE

TOXICITY DATA WITH REFERENCE
orl-rat LDLo:250 mg/kg NCNSA6 5,40,53
ipr-rat LDLo:50 mg/kg NCNSA6 5,40,53

CONSENSUS REPORTS: Selenium and its compounds are on the Community Right-To-Know List.

OSHA PEL: TWA 0.2 mg(Se)/m³
ACGIH TLV: TWA 0.2 mg(Se)/m³
DFG MAK: 0.1 mg(Se)/m³

SAFETY PROFILE: Poison by ingestion and intraperitoneal routes. See also SELENIUM COMPOUNDS and CARBAMATES. When heated to decomposition it emits very toxic fumes of NO_x, SO_x, and Se.

DJD600 **CAS:111-46-6** **HR: 2**
DIETHYLENE GLYCOL
mf: $C_4H_{10}O_3$ mw: 106.14

$(HOC_2H_4)_2O$

PROP: Clear, colorless, practically odorless, syrupy liquid. Fp: −8°, mp: −6.5°, bp: 133° @ 14 mm, flash p: 255°F, d: 1.1184 @ 20°/20°, autoign temp: 444°F, vap press: 1 mm @ 91.8°, vap d: 3.66. Sol in water.

SYNS: BIS(2-HYDROXYETHYL) ETHER □ BRECOLANE NDG □ CARBITOL □ DEACTIVATOR E □ DEACTIVATOR H □ DEG □ DICOL □ DIGLYCOL □ DIHYDROXYDIETHYL ETHER □ β,β′-DIHYDROXYDIETHYL ETHER □ 2,2′-DIHYDROXYETHYL ETHER □ DISSOLVANT APV □ ETHYLENE DIGLYCOL □ GLYCOL ETHER □ GLYCOL ETHYL ETHER □ 3-OXAPENTANE-1,5-DIOL □ 3-OXA-1,5-PENTANEDIOL □ 2,2′-OXYBISETHANOL □ 2,2′-OXYDIETHANOL □ TL4N

TOXICITY DATA WITH REFERENCE
skn-hmn 112 mg/3D-I MLD 85DKA8 -,127,77
skn-rbt 500 mg MLD 34ZIAG -,731,69
eye-rbt 50 mg MLD JPETAB 42,355,31
orl-mus TDLo:343 g/kg multi:REP FAATDF 14,622,90
orl-rat TDLo:50 g/kg (1-20D preg): TER OYYAA2 27,801,84
scu-rat TDLo:2500 mg/kg/82W-I:NEO VINIT* #6801-83
orl-mus TDLo:420 mg/kg/22W-I:NEO VINIT* #6801-83
ihl-mus TCLo:4 mg/m³/2H/30W-I:CAR GISAAA 33(2),36,68
scu-mus TDLo:1250 mg/kg/66W-I:NEO VINIT* #6801-83
orl-rat TD:1752 g/kg/2Y-C:ETA IMSUAI 36,55,67
orl-rat TD:584 g/kg/2Y-C:ETA FEPRA7 4,149,45
orl-rat TD:840 mg/kg/81W-I:NEO VINIT* #6801-83
orl-hmn LD50:1000 mg/kg JIHTAB 21,173,39
orl-cld TDLo:2400 mg/kg JOPDAB 109,731,86
orl-rat LD50:12,565 mg/kg NPIRI* 1,25,74
ipr-rat LD50:7700 mg/kg 38MKAJ 2C,3836,82
scu-rat LD50:18,800 mg/kg 38MKAJ 2C,3836,82
orl-mus LD50:23,700 mg/kg FEPRA7 4,142,45
ihl-mus LCLo:130 mg/m³/2H GTPZAB 10(12),30,66
ipr-mus LD50:9719 mg/kg FEPRA7 6,342,47
scu-mus LDLo:5 g/kg JPETAB 42,355,31
orl-dog LD50:9000 mg/kg JPETAB 67,101,39
orl-cat LD50:3300 mg/kg JIHTAB 21,173,39
skn-rbt LD50:11,890 mg/kg NPIRI* 1,25,74
ivn-rbt LD50:2000 mg/kg JPETAB 59,93,37

CONSENSUS REPORTS: Reported in EPA TSCA Inventory. Glycol ether compounds are on the Community Right-To-Know List.

SAFETY PROFILE: Moderately toxic to humans by ingestion. Poison experimentally by inhalation. Moderately toxic by ingestion and intravenous routes. Questionable carcinogen with experimental carcinogenic, tumorigenic, and teratogenic data. An eye and human skin irritant. Combustible when exposed to heat or flame; can react with oxidizing materials. To fight fire, use alcohol foam, water, CO_2, dry chemical. Mixtures with sodium hydroxide decompose exothermically when heated to 230°C and release explosive hydrogen gas. When heated to decomposition it emits acrid smoke and irritating fumes. See also GLYCOL ETHERS.

DJD700 **CAS:13988-26-6** **HR: 2**
DIETHYLENE GLYCOL BISPHTHALATE
mf: $C_{12}H_{12}O_5$ mw: 236.24

SYNS: 2,5,8-BENZOTRIOXACYCLOUNDECIN-1,9-DIONE, 3,4,6,7-TETRAHYDRO-(9CI) □ HOWFLEX GBP

TOXICITY DATA WITH REFERENCE
orl-rat TDLo:45 g/kg/13W-C:ETA EJCAAH 5,415,69

CONSENSUS REPORTS: Reported in EPA TSCA Inventory.

SAFETY PROFILE: Questionable carcinogen with experimental tumorigenic data. When heated to decomposition it emits acrid smoke and irritating fumes.

DJD750 **CAS:628-68-2** **HR: 1**
DIETHYLENE GLYCOL DIACETATE
mf: $C_8H_{14}O_5$ mw: 190.22

SYN: ACETIC ACID, OXYDIETHYLENE ESTER

TOXICITY DATA WITH REFERENCE
eye-rbt 500 mg MOD AJOPAA 29,1363,46

CONSENSUS REPORTS: Reported in EPA TSCA Inventory.

SAFETY PROFILE: An eye irritant. When heated to decomposition it emits acrid smoke and irritating vapors.

DJD800 CAS:4246-51-9 HR: 2
DIETHYLENE GLYCOL DI(3-AMINOPROPYL) ETHER
mf: $C_{10}H_{24}N_2O_3$ mw: 220.36

SYNS: DI(3-AMINOPROPYL) ETHER of DIETHYLENE GLYCOL □ 3, 3'-(OXYBIS(2,1-ETHANEDIYLOXY))BIS-1-PROPANAMINE

TOXICITY DATA WITH REFERENCE
orl-rat LD50:4290 mg/kg AIHAAP 30,470,69
skn-rbt LD50:2500 mg/kg AIHAAP 30,470,69

CONSENSUS REPORTS: Reported in EPA TSCA Inventory. Glycol ether compounds are on the Community Right-To-Know List.

SAFETY PROFILE: Moderately toxic by skin contact. Mildly toxic by ingestion. When heated to decomposition it emits toxic fumes of NO_x. See also GLYCOL ETHERS.

DJE000 CAS:120-55-8 HR: 2
DIETHYLENE GLYCOL DIBENZOATE
mf: $C_{18}H_{18}O_5$ mw: 314.36

PROP: Crystals. Mp: 70°, bp: 210°, flash p: 365°F, vap d: 9.38.

SYNS: BENZOIC ACID, DIESTER with DIETHYLENE GLYCOL □ DIBENZOYLDIETHYLENEGLYCOL ESTER

TOXICITY DATA WITH REFERENCE
skn-rbt 500 mg/24H MLD 85JCAE-,716,86
eye-rbt 500 mg/24H MLD 85JCAE-,716,86
orl-rat LD50:2830 mg/kg AIHAAP 23,95,62
skn-rbt LD50:20 g/kg AIHAAP 23,95,62

CONSENSUS REPORTS: Reported in EPA TSCA Inventory. Glycol ether compounds are on the Community Right-To-Know List.

SAFETY PROFILE: Moderately toxic by ingestion. A skin and eye irritant. Combustible when exposed to heat or flame; can react with oxidizing materials. To fight fire, use water, foam, CO_2, or dry chemical. When heated to decomposition it emits acrid smoke and fumes. See also GLYCOL ETHERS.

DJE400 CAS:693-21-0 HR: 3
DIETHYLENE GLYCOL DINITRATE
DOT: UN 0075
mf: $C_4H_8N_2O_7$ mw: 196.14

$O(C_2H_4ONO_2)_2$

PROP: Liquid. Vap d: 6.76.

SYNS: BIS(HYDROXYAETHYL)-AETHER-DINITRAT (GERMAN) □ DIETHYLENEGLYCOL DINITRATE, containing at least 25% phlegmatizer (DOT) □ DIETHYLENGLYKOLDINITRATE (CZECH) □ DIGLYCOLDINITRAAT (DUTCH) □ DIGLYCOL (DINITRATE de) (FRENCH) □ DIGLYKOLDINITRAT (GERMAN) □ DI(HYDROXYETHYL) ETHER DINITRATE □ DINITRATE de DIETHYLENE-GLYCOL (FRENCH) □ DINITRODIGLICOL (ITALIAN) □ DINITRODIGLYKOL (CZECH)

TOXICITY DATA WITH REFERENCE
orl-rat LD50:753 mg/kg JACTDZ 12,602,93
orl-mus LD50:1250 mg/kg JOHYAY 17,114,73

CONSENSUS REPORTS: Reported in EPA TSCA Inventory. Glycol ether compounds are on the Community Right-To-Know List.

DOT CLASSIFICATION: Forbidden; DOT Class: EXPLOSIVE 1.1D; *Label:* EXPLOSIVE 1.1D (UN 0075)

SAFETY PROFILE: Moderately toxic by ingestion. Ingestion of this compound can cause a drop in blood pressure and cardiac disturbances. A dangerous fire hazard when exposed to heat or flame; can react vigorously with oxidizing or reducing materials. A dangerous explosive sensitive to heat, shock, and vibration. Used in low-freezing dynamites and some permissible explosives. Upon decomposition it emits toxic fumes of NO_x. See also GLYCOL ETHERS; NITRATES and EXPLOSIVES, HIGH.

DJE600 CAS:764-99-8 HR: 2
DIETHYLENE GLYCOL DIVINYL ETHER
mf: $C_8H_{14}O_3$ mw: 158.22

SYNS: BIS(2-VINYLOXYETHYL)ETHER □ DVEDEG (RUSSIAN)

TOXICITY DATA WITH REFERENCE
skn-rbt 10 mg/24H open MLD AIHAAP 23,95,62
orl-rat LD50:3730 mg/kg AIHAAP 23,95,62
orl-mus LD50:2570 mg/kg GISAAA 42(3),12,77
skn-rbt LD50:14,100 mg/kg AIHAAP 23,95,62

CONSENSUS REPORTS: Reported in EPA TSCA Inventory. Glycol ether compounds are on the Community Right-To-Know List.

SAFETY PROFILE: Moderately toxic by ingestion. Mildly toxic by skin contact. A skin irritant. When heated to decomposition it emits acrid smoke and fumes. See also GLYCOL ETHERS.

DJE800 CAS:1002-67-1 HR: 1
DIETHYLENE GLYCOL ETHYL METHYL ETHER
mf: $C_7H_{16}O_3$ mw: 148.23

SYN: 2-ETHOXYETHYL-2-METHOXYETHYLETHER

TOXICITY DATA WITH REFERENCE
skn-rbt 10 mg/24H open MLD AIHAAP 23,95,62
orl-rat LD50:6500 mg/kg AIHAAP 23,95,62
skn-rbt LD50:7070 mg/kg AIHAAP 23,95,62

CONSENSUS REPORTS: Glycol ether compounds are on the Community Right-To-Know List.

SAFETY PROFILE: Mildly toxic by ingestion and skin contact. A skin irritant. When heated to decomposition it emits acrid smoke and irritating fumes. See also GLYCOL ETHERS.

DJF000 **CAS:10143-53-0** **HR: 1**
DIETHYLENE GLYCOL ETHYLVINYL ETHER
mf: $C_8H_{16}O_3$ mw: 160.24

SYN: 2-ETHOXYETHYL-2-(VINYLOXY)ETHYL ETHER

TOXICITY DATA WITH REFERENCE
skn-rbt 10 mg/24H open MLD AIHAAP 23,95,62
orl-rat LD50:11,300 mg/kg AIHAAP 23,95,62
skn-rbt LD50:8410 mg/kg AIHAAP 23,95,62

CONSENSUS REPORTS: Reported in EPA TSCA Inventory. Glycol ether compounds are on the Community Right-To-Know List.

SAFETY PROFILE: Mildly toxic by ingestion and skin contact. A skin irritant. When heated to decomposition it emits acrid smoke and fumes. See also GLYCOL ETHERS.

DJF200 **CAS:112-34-5** **HR: 2**
DIETHYLENE GLYCOL MONOBUTYL ETHER
mf: $C_8H_{18}O_3$ mw: 162.26

PROP: Colorless liquid. Mp: −68.1°, bp: 230.6°, flash p: 172°F, d: 0.9553 @ 20°/4°, autoign temp: 442°F, vap press: 0.02 mm @ 20°, vap d: 5.58.

SYNS: BUCB □ BUTOXYDIETHYLENE GLYCOL □ BUTOXYDIGLYCOL □ 2-(2-BUTOXYETHOXY)ETHANOL □ BUTYL CARBITOL □ o-BUTYL DIETHYLENE GLYCOL □ BUTYL DIOXITOL □ DIETHYLENE GLYCOL-n-BUTYL ETHER □ DIGLYCOL MONOBUTYL ETHER □ DOWANOL DB □ EKTASOLVE DB □ GLYCOL ETHER DB □ JEFFERSOL DB □ POLY-SOLV DB

TOXICITY DATA WITH REFERENCE
eye-rbt 5 mg SEV AJOPAA 29,1363,46
orl-rat LD50:6560 mg/kg UCDS** 1/31/66
orl-rat LD50:5660 mg/kg DOWCC* MSD-41
orl-mus LD50:2400 mg/kg JACTDZ 12,139,93
ipr-mus LD50:850 mg/kg FEPRA7 6,342,47
skn-rbt LD50:4120 mg/kg UCDS** 1/31/66
orl-gpg LD50:2000 mg/kg JIHTAB 23,259,41

CONSENSUS REPORTS: Reported in EPA TSCA Inventory. Glycol ether compounds are on the Community Right-To-Know List.

DFG MAK: 100 mg/m³

SAFETY PROFILE: Moderately toxic by ingestion and intraperitoneal routes. Mildly toxic by skin contact. A severe eye irritant. Combustible when exposed to heat or flame; can react with oxidizing materials. To fight fire, use alcohol foam, CO_2, or dry chemical. When heated to decomposition it emits acrid smoke and irritating fumes. See also GLYCOL ETHERS.

DJF400 **CAS:16672-39-2** **HR: 1**
DI(ETHYLENE GLYCOL MONOBUTYL ETHER) PHTHALATE
mf: $C_{24}H_{38}O_8$ mw: 454.62

TOXICITY DATA WITH REFERENCE
orl-rat LD50:9700 mg/kg JIHTAB 23,259,41

CONSENSUS REPORTS: Reported in EPA TSCA Inventory. Glycol ether compounds are on the Community Right-To-Know List.

SAFETY PROFILE: Mildly toxic by ingestion. When heated to decomposition it emits acrid smoke and irritating fumes. See also GLYCOL ETHERS.

DJF600 **CAS:10143-54-1** **HR: 2**
DIETHYLENE GLYCOL MONO-2-CYANOETHYL ETHER
mf: $C_7H_{13}NO_3$ mw: 159.21

TOXICITY DATA WITH REFERENCE
skn-rbt 10 mg/24H open MLD AIHAAP 23,95,62
orl-rat LD50:13,400 mg/kg AIHAAP 23,95,62

CONSENSUS REPORTS: Cyanide and its compounds, as well as glycol ether compounds, are on the Community Right-To-Know List.

SAFETY PROFILE: Mildly toxic by ingestion. A skin irritant. When heated to decomposition it emits toxic fumes of NO_x and CN^-. See also GLYCOL ETHERS and NITRILES.

DJF800 **CAS:18912-80-6** **HR: 2**
DIETHYLENE GLYCOL MONOISOBUTYL ETHER
mf: $C_8H_{18}O_3$ mw: 162.26

SYNS: EKTASOLVE DIB □ ISOBUTOXY-2-ETHOXY-2-ETHANOL □ 2-(2-ISOBUTOXYETHOXY)ETHANOL

TOXICITY DATA WITH REFERENCE
skn-rbt 500 mg/24H MLD 85JCAE-,635,86
eye-rbt 20 mg/24H MOD 85JCAE-,635,86
orl-rat LD50:4920 mg/kg TXAPA9 28,313,74
ihl-rat LClo:6700 ppm/6H KODAK* 21MAY71
skn-rbt LD50:3560 mg/kg TXAPA9 28,313,74

CONSENSUS REPORTS: Reported in EPA TSCA Inventory. Glycol ether compounds are on the Community Right-To-Know List.

SAFETY PROFILE: Moderately toxic by skin contact. Mildly toxic by ingestion and inhalation. A skin and eye irritant. When heated to decomposition it emits acrid smoke and irritating fumes. See also GLYCOL ETHERS.

DJG000 **CAS:111-77-3** **HR: 2**
DIETHYLENE GLYCOL MONOMETHYL ETHER
mf: $C_5H_{12}O_3$ mw: 120.17

PROP: Hygroscopic, water-white liquid. Mp: −70°, bp: 194.2°, flash p: 200°F (OC), d: 1.0354 @ 20°/4°, vap press: 0.2 mm @ 20°, vap d: 4.14.

SYNS: DIETHYLENE GLYCOL METHYL ETHER □ DIGLYCOL MONOMETHYL ETHER □ DOWANOL DM □ ETHYLENE DIGLYCOL MONOMETHYL ETHER □ MECB □ METHOXYDIGLYCOL □ 2-(2-METHOXYETHOXY)ETHANOL □ β-METHOXY-β′-HYDROXYDIETHYL ETHER □ METHYL CARBITOL □ POLY-SOLV DM

TOXICITY DATA WITH REFERENCE
eye-rbt 500 mg MOD UCDS** 4/21/67
eye-rbt 500 mg/24H MLD 85JCAE -,628,86
orl-mus TDLo:32 g/kg (female 7-14D post):REP EVHPAZ 57,141,84

orl-rat TDLo: 21,650 mg/kg (female 7-16D post): TER FAATDF 6,430,86
orl-rat TDLo: 21,650 mg/kg (female 7-16D post): REP FAATDF 6,430,86
orl-rat TDLo: 21,650 mg/kg (7-16D preg): TER TJADAB 31,54A,85
orl-rat LD50: 5500 mg/kg 38MKAJ 2C,3957,82
ipr-rat LD50: 2722 mg/kg GNAMAP 29,37,90
ipr-mus LD50: 2611 mg/kg GNAMAP 29,37,90
orl-rbt LD50: 7190 mg/kg 38MKAJ 2C,3957,82
skn-rbt LD50: 650 mg/kg UCDS** 4/21/67
orl-gpg LD50: 4160 mg/kg JIHTAB 23,259,41

CONSENSUS REPORTS: Reported in EPA TSCA Inventory. Glycol ether compounds are on the Community Right-To-Know List.

SAFETY PROFILE: Moderately toxic by skin contact and intraperitoneal routes. Mildly toxic by ingestion. An experimental teratogen. Other experimental reproductive effects. An eye irritant. Combustible when exposed to heat or flame; can react with oxidizing materials. Reacts violently with $Ca(OCl)_2$, chlorosulfonic acid, and oleum. To fight fire, use dry chemical, alcohol foam, water spray or mist, CO_2. When heated to decomposition it emits acrid smoke and irritating fumes. See also GLYCOL ETHERS.

DJG200 CAS:10143-56-3 HR: 2
DIETHYLENEGLYCOL-MONO-2-METHYLPENTYL ETHER
mf: $C_{10}H_{22}O_3$ mw: 190.32

SYNS: DIETHYLENE GLYCOL MONOMETHYLPENTYL ETHER ☐ 2-METHYLPENTYL CARBITOL ☐ 2-(2-((2-METHYLPENTYL)OXY)ETHOXY)ETHANOL

TOXICITY DATA WITH REFERENCE
skn-rbt 10 mg/24H open MLD AIHAAP 23,95,62
eye-rbt 100 mg SEV 34ZIAG -,730,69
orl-rat LD50: 5660 mg/kg AIHAAP 23,95,62
skn-rbt LD50: 1580 mg/kg AIHAAP 23,95,62

CONSENSUS REPORTS: Glycol ether compounds are on the Community Right-To-Know List.

SAFETY PROFILE: Moderately toxic by skin contact. Mildly toxic by ingestion. A skin and severe eye irritant. When heated to decomposition it emits acrid smoke and irritating fumes. See also GLYCOL ETHERS.

DJG400 CAS:929-37-3 HR: 1
DIETHYLENE GLYCOL MONOVINYL ETHER
mf: $C_6H_{12}O_3$ mw: 132.18

SYNS: DEGMVE (RUSSIAN) ☐ DIETHYLENEGLYCOL VINYL ETHER ☐ DIETHYLEN-GLYCOL MONOVINYL ESTER ☐ 2-(2-(ETHENYLOXY)ETHOXY)ETHANOL

TOXICITY DATA WITH REFERENCE
orl-rat LD50: 4930 mg/kg GISAAA 42(3),12,77
orl-mus LD50: 4450 mg/kg GISAAA 39(11),94,74

CONSENSUS REPORTS: Glycol ether compounds are on the Community Right-To-Know List.

SAFETY PROFILE: Mildly toxic by ingestion. When heated to decomposition it emits acrid smoke and irritating fumes. See also GLYCOL ETHERS.

DJG600 CAS:111-40-0 HR: 3
DIETHYLENETRIAMINE
DOT: UN 2079
mf: $C_4H_{13}N_3$ mw: 103.20

$HN(C_2H_4NH_2)_2$

PROP: Yellow, viscous liquid; mild ammonia-like odor. Mp: −39°, bp: 207°, flash p: 215°F (OC), d: 0.9586 @ 20°/20°, autoign temp: 750°F, vap press: 0.22 mm @ 20°, vap d: 3.48. Misc in H_2O and EtOH.

SYNS: AMINOETHYLETHANEDIAMINE ☐ N-(2-AMINOETHYL)ETHYLENEDIAMINE ☐ 3-AZAPENTANE-1,5-DIAMINE ☐ BIS(β-AMINOETHYL)AMINE ☐ BIS(2-AMINOETHYL)AMINE ☐ D.E.H. 20 ☐ DETA ☐ 2,2′-DIAMINODIETHYLAMINE ☐ 2,2′-IMINOBISETHYLAMINE

TOXICITY DATA WITH REFERENCE
skn-rbt 10 mg/24H open SEV JIHTAB 31,60,49
skn-rbt 500 mg open MOD UCDS** 12/30/71
skn-rbt 500 mg IYKEDH 6,170,75
eye-rbt 750 μg open SEV JIHTAB 31,60,49
orl-rat LD50: 1080 mg/kg AMIHAB 17,129,58
ipr-rat LD50: 74 mg/kg AMIHAB 17,129,58
ipr-mus LD50: 71 mg/kg AMIHAB 17,129,58
skn-rbt LD50: 1090 mg/kg JIHTAB 31,60,49
skn-gpg LD50: 162 mg/kg JIHTAB 26,269,44

CONSENSUS REPORTS: Reported in EPA TSCA Inventory.

OSHA PEL: TWA 1 ppm
ACGIH TLV: TWA 1 ppm (skin)
DOT CLASSIFICATION: 8; *Label:* Corrosive

SAFETY PROFILE: Poison by skin contact and intraperitoneal routes. Moderately toxic by ingestion. Corrosive. A severe skin and eye irritant. High concentration of vapors causes irritation of respiratory tract, nausea, and vomiting. Repeated exposures can cause asthma and sensitization of skin. Combustible when exposed to heat or flame; can react with oxidizing materials. Mixture with nitromethane is a shock-sensitive explosive. Ignites on contact with cellulose nitrate of high surface area. To fight fire, use alcohol foam. When heated to decomposition it emits toxic fumes of NO_x. See also AMINES.

For occupational chemical analysis use OSHA: #ID-60 or NIOSH: Diethylenetriamine 2540.

DJG700 CAS:22042-96-2 HR: 1
DIETHYLENETRIAMINEPENTA(METHYLENEPHOSPHONIC ACID), SODIUM SALT
mf: $C_9H_{28}N_3O_{15}P_5 \cdot xNa$ mw: 734.18

SYNS: BRIQUEST 543-33S ☐ DEQUEST 2066 ☐ DEQUEST 2066 DEFLOCCULANT and SEQUESTRANT ☐ PHOSPHONIC ACID, (((PHOSPHONOMETHYL)IMINO)BIS(2,1-ETHANEDIYLNITRILOBIS(METHYLENE)))TETRAKIS-, SODIUM SAL ☐ PHOSPHONIC ACID, ((BIS(2-(BIS(PHOSPHONOMETHYL)AMINO)ETHYL)AMINO)METHYL)-, SODIUM SALT (8CI) ☐ SEQUION 40Na32 ☐ WAYPLEX 55S

TOXICITY DATA WITH REFERENCE
orl-rat LD50: >5 g/kg EPASR* 8EHQ-0790-0950/0951

skn-rbt LD50:>5 g/kg EPASR* 8EHQ-0790-0950/0951
orl-qal LD50:>2510 mg/kg EPASR* 8EHQ-0790-0950/0951

CONSENSUS REPORTS: Reported in EPA TSCA Inventory.

SAFETY PROFILE: Low toxicity by ingestion and skin contact. When heated to decomposition it emits toxic vapors of NO_x, NaO, and PO_x.

DJG800 CAS:67-43-6 **HR: 2**
(DIETHYLENETRINITRILO)PENTAACETIC ACID
mf: $C_{14}H_{23}N_3O_{10}$ mw: 393.40

PROP: Crystals from H_2O. Mp: 219–220°. Sol in alkalis and H_2O.

SYNS: ((CARBOXYMETHYLIMINO)BIS(ETHYLENENITRILO))TETRAACETIC ACID □ CHEL 330 □ CHEL 330 ACID □ CHEL DTPA □ DIETHYLENETRIAMINEPENTAACETIC ACID □ 1,1,4,7,7-DIETHYLENETRIAMINEPENTAACETIC ACID □ DTPA □ HAMP-EX ACID □ MONAQUEST □ PENTHAMIL □ PERMA KLEER □ 3,6,9-TRIS(CARBOXYMETHYL)-3,6,9-TRIAZAUNDECANEDIOIC ACID

TOXICITY DATA WITH REFERENCE
ipr-rat LDLo:665 mg/kg AHRTAN 13,295,62
ipr-mus LD50:543 mg/kg ARTODN 57,212,85

CONSENSUS REPORTS: Reported in EPA TSCA Inventory.

SAFETY PROFILE: Moderately toxic by intraperitoneal route. When heated to decomposition it emits toxic fumes of NO_x.

DJH200 CAS:7316-37-2 **HR: 2**
DIETHYL-β,γ-EPOXYPROPYLPHOSPHONATE
mf: $C_7H_{15}O_4P$ mw: 194.19

TOXICITY DATA WITH REFERENCE
ipr-mus TDLo:13 g/kg/64W-I:ETA JNCIAM 53,695,74
scu-mus TDLo:13 g/kg/63W-I:NEO JNCIAM 53,695,74

SAFETY PROFILE: Questionable carcinogen with experimental neoplastigenic and tumorigenic data. When heated to decomposition it emits toxic fumes of PO_x.

DJH500 CAS:2651-85-6 **HR: 3**
DIETHYL ETHANE PHOSPHONITE
mf: $C_6H_{15}O_2P$ mw: 150.16

$(CH_3CH_2O)_2PCH_2CH_3$

PROP: A liquid. Bp: 137–139°.

SAFETY PROFILE: Ignites spontaneously in air when a large surface area is exposed (e.g., on filter paper). When heated to decomposition it emits toxic fumes of PO_x.

DJH800 **HR: 3**
DIETHYLETHEROXODIPEROXOCHROMIUM(VI)
mf: $C_2H_6CrO_6$ mw: 178.07

PROP: Blue solid explodes powerfully @ −30°.

CONSENSUS REPORTS: Chromium and its compounds are on the Community Right-To-Know List.

SAFETY PROFILE: A very unstable explosive. See also CHROMIUM COMPOUNDS and PEROXIDES.

DJI000 CAS:2595-54-2 **HR: 3**
O,O-DIETHYL-S-(N-ETHOXYCARBONYL-N-METHYLCARBAMOYLMETHYL) PHOSPHORODITHIOATE
mf: $C_{10}H_{20}NO_5PS_2$ mw: 329.40

PROP: Pale-yellow oil. D: 1.223 @ 20°/20°, bp: 144° @ 0.02 mm. Sltly sol in H_2O.

SYNS: AFOS □ O,O-DIAETHYL-S-(3-METHYL-2,4-DIOXO-5-OXA-3-AZA-HEPTYL)-DITHIOPHOSPHAT (GERMAN) □ O,O-DIETHYL S-(N-ETHOXYCARBONYL-N-METHYLCARBAMOYLMETHYL) PHOSPHOROTHIOLOTHIONATE □ O,O-DIETHYL S-(N-METHYL-N-CARBOETHOXYCARBAMOYLMETHYL) DITHIOPHOSPHATE □ O,O-DIETHYL-S-(3-METHYL-2,4-DIOXO-5-OXA-3-AZA-HEPTYL)-DITHIOFOSFAAT (DUTCH) □ O,O-DIETIL-S-(N-ETOSSI-CARBONIL-N-METIL-CARBAMOIL-METIL)-DITIOFOSFATO (ITALIAN) □ DITHIOPHOSPHATE de O,O-DIETHYLE et de S-N-METHYL-N-CARBOETHOXY CARBAMOYLMETHYLE (FRENCH) □ N-ETHOXYCARBONYL-N-METHYLCARBAMOYLMETHYL-O,O-DIETHYL PHOSPHORODITHIOATE □ S-((ETHOXYCARBONYL)METHYLCARBAMOYL)METHYL-O,O-DIETHYL PHOSPHORODITHIOATE □ S-(N-ETHOXYCARBONYL-N-METHYLCARBAMOYLMETHYL)-DIETHYL PHOSPHORODITHIOATE □MARFOTOKS □ MC 474 □ MECARBAM □ MS 1053 □ MS 1143 □MURATOX □ MURFOTOX □ MUROTOX □ MURPHOTOX □ MURUTOX □ PENNSALT TD-72 □ PESTAN

TOXICITY DATA WITH REFERENCE
orl-rat LD50:36 mg/kg 28ZEAL 5,144,76
ihl-rat LC50:700 mg/m³/6H PEMNDP 9,538,91
skn-rat LD50:380 mg/kg WRPCA2 9,119,70
orl-mus LD50:106 mg/kg GUCHAZ 6,323,73
orl-gpg LDLo:25 mg/kg JEENAI 62,934,69
scu-gpg LDLo:50 mg/kg JEENAI 62,934,69

CONSENSUS REPORTS: EPA Genetic Toxicology Program.

SAFETY PROFILE: Poison by ingestion, skin contact, and subcutaneous routes. An insecticide. When heated to decomposition it emits very toxic fumes of SO_x, PO_x, and NO_x.

DJI200 CAS:52400-60-9 **HR: 3**
N,N-DIETHYL-N′-(2-(2-ETHYL-1,3-BENZODIOXOL-2-YL)ETHYL) ETHYLENEDIAMINE DIMALEATE
mf: $C_{17}H_{28}N_2O_2•2C_4H_4O_4$ mw: 524.63

SYN: 2-(2-(2-(DIETHYLAMINO)ETHYLAMINO)ETHYL)-2-ETHYL-1,3-BENZODIOXOLE DIMALEATE

TOXICITY DATA WITH REFERENCE
ivn-rat LD50:55 mg/kg EJMCA5 12,413,77
ipr-mus LD50:175 mg/kg EJMCA5 12,413,77

SAFETY PROFILE: Poison by intravenous and intraperitoneal routes. When heated to decomposition it emits toxic fumes of NO_x.

D

DJI400 **CAS:100-36-7** **HR: 3**
N,N-DIETHYLETHYLENEDIAMINE
DOT: UN 2685
mf: $C_6H_{16}N_2$ mw: 116.24

PROP: Liquid. Bp: 149–150°, flash p: 115°F (OC), d: 0.82 @ 20°/20°, vap d: 4.00.

SYNS: N,N-DIETHYL-1,2-ETHANEDIAMINE ☐ USAF AM-1

TOXICITY DATA WITH REFERENCE
skn-rbt 10 mg/24H open AMIHBC 10,61,54
eye-rbt 50 μg open SEV AMIHBC 10,61,54
orl-rat LD50:2830 mg/kg AMIHBC 10,61,54
ipr-mus LD50:300 mg/kg NTIS** AD277-689
skn-rbt LD50:820 mg/kg AMIHBC 10,61,54

CONSENSUS REPORTS: Reported in EPA TSCA Inventory.

DOT CLASSIFICATION: 8; *Label:* Corrosive, Flammable Liquid

SAFETY PROFILE: Poison by intraperitoneal route. Moderately toxic by ingestion and skin contact. A skin and severe eye irritant. Flammable liquid when exposed to heat or flame; can react with oxidizing materials. To fight fire, use alcohol foam, CO_2, dry chemical. When heated to decomposition it emits toxic fumes of NO_x. See also AMINES.

DJJ400 **CAS:358-74-7** **HR: 3**
DIETHYL FLUOROPHOSPHATE
mf: $C_4H_{10}FO_3P$ mw: 156.11

PROP: A liquid with a sweet or fruity odor. Bp: 80–80.5 @ 32 mm°, d: 1.14 @ 27°/4°, vap d: 5.38.

SYNS: FLUOPHOSPHORIC ACID, DIETHYL ESTER ☐ PHOSPHOROFLUORIDIC ACID, DIETHYL ESTER ☐ T-1036 ☐ TL 345

TOXICITY DATA WITH REFERENCE
ihl-rat LD50:7 g/m³/10M NTIS** PB158-508
ihl-mus LC50:500 mg/m³/10M JIHTAB 30,307,48
skn-mus LD50:35 mg/kg NTIS** PB158-508
ihl-gpg LC50:7 g/m³/10M NTIS** PB158-508

SAFETY PROFILE: Poison by inhalation and skin contact. See also FLUORIDES. When heated to decomposition or on contact with acid or acid fumes it emits highly toxic fumes of F^- and PO_x.

DJJ600 **CAS:617-84-5** **HR: 2**
DIETHYL FORMAMIDE
mf: $C_5H_{11}NO$ mw: 101.17

PROP: A liquid. D: 0.908 @ 19°, bp: 177–178°. Sol in H_2O, EtOH, and Et_2O.

TOXICITY DATA WITH REFERENCE
ipr-rat LD50:1740 mg/kg BIJOAK 85,72,62
ipr-mus LDLo:3200 mg/kg THERAP 26,409,71

CONSENSUS REPORTS: Reported in EPA TSCA Inventory.

SAFETY PROFILE: Moderately toxic by intraperitoneal route. When heated to decomposition it emits toxic fumes of NO_x.

DJJ800 **CAS:623-91-6** **HR: 2**
DIETHYL FUMARATE
mf: $C_8H_{12}O_4$ mw: 172.20

PROP: White crystals or liquid. Mp: 1–2°, bp: 218.5°, flash p: 220°F, d: 1.0529 @ 20°/20°, vap press: 1 mm @ 53.2°, vap d: 5.93.

SYN: ETHYL FUMARATE

TOXICITY DATA WITH REFERENCE
orl-rat LD50:1780 mg/kg AIHAAP 23,95,62

CONSENSUS REPORTS: Reported in EPA TSCA Inventory.

SAFETY PROFILE: Moderately toxic by ingestion. Combustible when exposed to heat or flame; can react with oxidizers. To fight fire, use alcohol foam, foam, CO_2, dry chemical. When heated to decomposition it emits acrid smoke and fumes. See also ESTERS, FUMARIC ACID, and ETHYL ALCOHOL.

DJJ829 **HR: 3**
DIETHYL GALLIUM HYDRIDE
mf: $C_4H_{11}Ga$ mw: 128.85

SAFETY PROFILE: Ignites spontaneously in air. Reacts violently with water. When heated to decomposition it emits acrid smoke and fumes. See also GALLIUM and HYDRIDES.

DJJ850 **CAS:26645-10-3** **HR: 3**
DIETHYL GOLD BROMIDE
mf: $C_4H_{10}AuBr$ mw: 335.02

SYNS: BROMODIETHYLGOLD ☐ DIETHYLGOLD BROMIDE (DOT)

DOT CLASSIFICATION: Forbidden

SAFETY PROFILE: Explodes at 70°C. When heated to decomposition it emits toxic fumes of Br^-. See also GOLD COMPOUNDS and BROMIDES.

DJJ875 **HR: 2**
DIETHYLGUANIDINE HYDROCHLORIDE DIHYDRATE
mf: $C_5H_{13}N_3{\cdot}ClH{\cdot}2H_2O$ mw: 187.71

TOXICITY DATA WITH REFERENCE
orl-rat LD50:1580 mg/kg GISAAA 49(1),71,84
orl-mus LD50:1600 mg/kg GISAAA 49(1),71,84
orl-gpg LD50:890 mg/kg GISAAA 49(1),71,84

SAFETY PROFILE: Moderately toxic by ingestion. When heated to decomposition it emits toxic fumes of NO_x and HCl.

DJK000 **CAS:3071-70-3** **HR: 3**
3,3′-DIETHYLHEPTAMETHINETHIACYANINE IODIDE
mf: $C_{25}H_{25}N_2S_2{\cdot}I$ mw: 544.54

TOXICITY DATA WITH REFERENCE
ipr-mus LD50:100 μg/kg JMCMAR 10,897,67

CONSENSUS REPORTS: Reported in EPA TSCA Inventory.

SAFETY PROFILE: Poison by intraperitoneal route.

When heated to decomposition it emits very toxic fumes of I^-, NO_x, and SO_x. See also IODIDES.

DJK100 **CAS:424-40-8** **HR: 2**
DIETHYL HEXAFLUOROGLUTARATE
mf: $C_9H_{10}F_6O_4$ mw: 296.19

PROP: A liquid. D: 1.34 @ 25°/4°, bp: 75–80° @ 3 mm.

SYNS: DIETHYL PERFLUOROGLUTARATE □ HEXAFLUOROGLUTARIC ACID DIETHYL ESTER □ HEXAFLUOROPENTANEDIOIC ACID DIETHYL ESTER

TOXICITY DATA WITH REFERENCE
orl-rat LD50:5 g/kg 85GMAT -,53,82
ihl-rat LC50:1300 mg/m³/4H 85GMAT -,53,82
orl-mus LD50:4200 mg/kg 85GMAT -,53,82
ihl-mus LCLo:10 g/m³/2H 85GMAT -,53,82

SAFETY PROFILE: Moderately toxic by inhalation. Mildly toxic by ingestion. When heated to decomposition it emits toxic fumes of F^-. See also ESTERS.

DJK200 **CAS:101-07-5** **HR: 2**
2-DI-(2-ETHYLHEXYL)AMINOETHANOL
mf: $C_{18}H_{39}NO$ mw: 285.58

SYN: 2-(BIS(2-ETHYLHEXYL)AMINO)ETHANOL

TOXICITY DATA WITH REFERENCE
skn-rbt 10 mg/24H open MLD AMIHBC 10,61,54
eye-rbt 500 mg open AMIHBC 10,61,54
orl-rat LD50:4920 mg/kg AMIHBC 10,61,54
skn-rbt LD50:2520 mg/kg AMIHBC 10,61,54

SAFETY PROFILE: Moderately toxic by skin contact. Mildly toxic by ingestion. A skin and eye irritant. When heated to decomposition it emits toxic fumes of NO_x. See also PARATHION.

DJK400 **CAS:5810-88-8** **HR: 2**
O,O′-DI(2-ETHYLHEXYL) DITHIOPHOSPHORIC ACID
mf: $C_{16}H_{35}O_2PS_2$ mw: 354.60

PROP: Crystals. Sol in $CHCl_3$, C_6H_6, and heptane.

TOXICITY DATA WITH REFERENCE
skn-rbt 100 µg/24H open AIHAAP 23,95,62
skn-rbt 2 mg/24H SEV 85JCAE-,1187,86
eye-rbt 750 µg/24H SEV 85JCAE-,1187,86
orl-rat LD50:2140 mg/kg AIHAAP 23,95,62
skn-rbt LD50:1250 mg/kg AIHAAP 23,95,62

CONSENSUS REPORTS: Reported in EPA TSCA Inventory.

SAFETY PROFILE: Moderately toxic by ingestion and skin contact. A skin and severe eye irritant. When heated to decomposition it emits very toxic fumes of PO_x and SO_x.

DJK600 **CAS:10143-60-9** **HR: 1**
DI-(2-ETHYLHEXYL) ETHER
mf: $C_{16}H_{34}O$ mw: 242.50

SYNS: BIS(2-ETHYLHEXYL)ETHER □ 1,1′-OXYBIS(2-ETHYLHEXANE)

TOXICITY DATA WITH REFERENCE
skn-rbt 10 mg/24H open MLD AMIHBC 10,61,54
eye-rbt 500 mg open AMIHBC 10,61,54
orl-rat LD50:34 g/kg AMIHBC 10,61,54

CONSENSUS REPORTS: Reported in EPA TSCA Inventory.

SAFETY PROFILE: Mildly toxic by ingestion. A skin and eye irritant. When heated to decomposition it emits acrid smoke and irritating fumes. See also ETHERS.

DJK800 **CAS:16111-62-9** **HR: 2**
DI(2-ETHYLHEXYL) PEROXYDICARBONATE
mf: $C_{18}H_{34}O_6$ mw: 346.52

SYNS: PEROXYDICARBONIC ACID, BIS(2-ETHYLHEXYL) ESTER □ PEROXYDICARBONIC ACID, DI(2-ETHYLHEXYL) ESTER

TOXICITY DATA WITH REFERENCE
orl-rat LD50:1020 mg/kg BSPII* 1/75-19B

CONSENSUS REPORTS: Reported in EPA TSCA Inventory.

SAFETY PROFILE: Moderately toxic by ingestion. When heated to decomposition it emits acrid smoke and irritating fumes. See also PEROXIDES, ORGANIC.

DJL000 **CAS:577-11-7** **HR: 3**
DI-(2-ETHYLHEXYL) SODIUM SULFOSUCCINATE
mf: $C_{20}H_{38}O_7S{\cdot}Na$ mw: 445.63

PROP: White, waxlike, plastic solid; octyl alcohol odor. Sol in hexane, glycerin, alc; sltly sol in water and org solvs.

SYNS: AEROSOL GPG □ ALCOPOL O □ ALPHASOL OT □ BEROL 478 □ BIS(ETHYLHEXYL) ESTER of SODIUM SULFOSUCCINIC ACID □ BIS(2-ETHYLHEXYL)SODIUM SULFOSUCCINATE □ BIS(2-ETHYLHEXYL)-S-SODIUM SULFOSUCCINATE □ 1,4-BIS(2-ETHYLHEXYL) SODIUM SULFOSUCCINATE □ 1,4-BIS(2-ETHYLHEXYL)SULFOBUTANEDIOIC ACID ESTER, SODIUM SALT □ CELANOL DOS 75 □ CLESTOL □ COLACE □ COMPLEMIX □ CONSTONATE □ COPROL □ DEFILIN □ DIOCTLYN □ DIOCTYLAL □ DIOCTYL ESTER of SODIUM SULFOSUCCINATE □ DIOCTYL ESTER of SODIUM SULFOSUCCINIC ACID □ DIOCTYL-MEDO FORTE □ DIOCTYL SODIUM SULFOSUCCINATE (FCC) □ DIOCTYL SULFOSUCCINATE SODIUM SALT □ DIOMEDICONE □ DIOSUCCIN □ DIOTILAN □ DIOVAC □ DOCUSATE SODIUM □ DOXINATE □ DOXOL □ DSS □ DULSIVAC □ DUOSOL □ 2-ETHYLHEXYL SULFOSUCCINATE SODIUM □ HUMIFEN WT 27G □ KONLAX □ KOSATE □ LAXINATE □ MANOXAL OT □ MERVAMINE □ MODANE SOFT □ MOLATOC □ MOLCER □ MOLOFAC □ MONAWET MD 70E □ NEKAL WT-27 □ NEVAX □ NIKKOL OTP 70 □ NORVAL □ OBSTON □ RAPISOL □ REGUTOL □ REQUTOL □ REVAC □ SANMORIN OT 70 □ SBO □ SOBITAL □ SODIUM BIS(2-ETHYLHEXYL) SULFOSUCCINATE □ SODIUM DI-(2-ETHYLHEXYL) SULFOSUCCINATE □ SODIUM DIOCTYL SULFOSUCCINATE □ SODIUM DIOCTYL SULPHOSUCCINATE □ SODIUM-2-ETHYLHEXYLSULFOSUCCINATE □ SODIUM SULFODI-(2-ETHYLHEXYL)SULFOSUCCINATE □ SOFTIL □ SOLIWAX □ SOLUSOL-75% □ SOLUSOL-100% □ SULFIMEL DOS □ TEX WET 1001 □ TRITON GR-5 □ VATSOL OT □ VELMOL □ WAXSOL □ WETAID SR

TOXICITY DATA WITH REFERENCE
skn-rbt 10 mg/24H MOD JPETAB 82,377,44
eye-rbt 250 µg MLD AROPAW 34,99,45
eye-rbt 1% SEV JAPMA8 38,428,49

orl-rat LD50:1900 mg/kg JSCCA5 13,469,62
ipr-rat LD50:590 mg/kg BCTKAG 7,161,74
orl-mus LD50:2643 mg/kg DCTODJ 1,89,77/78
ivn-mus LD50:60 mg/kg JAPMA8 38,428,49

CONSENSUS REPORTS: Reported in EPA TSCA Inventory.

SAFETY PROFILE: Poison by intravenous route. Moderately toxic by ingestion and intraperitoneal routes. A skin and severe eye irritant. See also ESTERS. When heated to decomposition it emits toxic fumes of SO_x and Na_2O.

DJL200 CAS:25430-97-1 HR: 3
DI-2-ETHYLHEXYLTIN DICHLORIDE
mf: $C_{16}H_{34}Cl_2Sn$ mw: 416.09

PROP: Crystals.

SYNS: DICHLORODI(2-ETHYLHEXYL)STANNANE □ DI(2-ETHYLHEXYL)TIN DICHLORIDE

TOXICITY DATA WITH REFERENCE
ivn-rat LD50:5 mg/kg JOCMA7 2,183,60

OSHA PEL: TWA 0.1 mg(Sn)/m³ (skin)
ACGIH TLV: TWA 0.1 mg(Sn)/m³ (skin) (Proposed: TWA 0.1 mg(Sn)/m³; STEL 0.2 mg(Sn)/m³ (skin))
NIOSH REL: (Organotin Compounds) TWA 0.1 mg(Sn)/m³

SAFETY PROFILE: Poison by intravenous route. See also TIN COMPOUNDS. When heated to decomposition it emits highly toxic fumes of Cl^-.

For occupational chemical analysis use NIOSH: Organotin Compounds 5504.

DJL400 CAS:1615-80-1 HR: 3
1,2-DIETHYLHYDRAZINE
mf: $C_4H_{12}N_2$ mw: 88.18

PROP: Bp: 86°, d: 0.797 @ 26°. Sol in alc and ether.

SYNS: 1,2-DIAETHYLHYDRAZINE (GERMAN) □ N-N'-DIETHYLHYDRAZINE □ sym-DIETHYLHYDRAZINE □ HYDRAZOETHANE □ HYDROAZOETHANE □ RCRA WASTE NUMBER U086 □ SDEH

TOXICITY DATA WITH REFERENCE
mmo-sat 83,200 nmol/plate MUREAV 278,215,92
ivn-rat TDLo:500 mg/kg (15D preg):TER IARCCD 4,45,73
ivn-rat TDLo:10 mg/kg (15D preg):TER XENOBH 3,271,73
scu-rat TDLo:700 mg/kg/28W-I:ETA NATWAY 53,557,66
ivn-rat TDLo:50 mg/kg (15D preg):CAR,TER IARCCD 4,45,73
ivn-rat TDLo:1850 mg/kg/37W-I:ETA XENOBH 3,271,73
ivn-rat TD:50 mg/kg (15D preg):ETA,TER FCTXAV 6,584,68

CONSENSUS REPORTS: IARC Cancer Review: Group 2B IMEMDT 7,56,87; Animal Sufficient Evidence IMEMDT 4,153,74.

SAFETY PROFILE: Confirmed carcinogen with experimental carcinogenic, tumorigenic, and teratogenic data. It is also a transplacental carcinogen. Mutation data reported. When heated to decomposition it emits toxic fumes of NO_x. See also HYDRAZINE.

DJL600 CAS:7699-31-2 HR: 2
1,2-DIETHYLHYDRAZINE DIHYDROCHLORIDE
mf: $C_4H_{12}N_2{\cdot}2ClH$ mw: 161.10

TOXICITY DATA WITH REFERENCE
ivn-rat TDLo:50 mg/kg (female 15D post):ETA,TER EXPEAM 24,561,68

SAFETY PROFILE: Questionable carcinogen with experimental tumorigenic and teratogenic data. When heated to decomposition it emits very toxic fumes of HCl and NO_x. See also 1,2-DIETHYLHYDRAZINE.

DJM800 CAS:53-46-3 HR: 3
DIETHYL(2-HYDROXYETHYL)METHYLAMMONIUM BROMIDE XANTHENE-9-CARBOXYLATE
mf: $C_{21}H_{26}NO_3{\cdot}Br$ mw: 420.39

PROP: Crystals from 2-propanol. Mp: 175–176°. Insol in Et_2O; sol in H_2O.

SYNS: ASABAINE □ AVAGAL □ BANTHIN □ BANTHINE BROMIDE □ β-DIETHYLAMINOETHYL XANTHENE-9-CARBOXYLATE METHOBROMIDE □ β-DIETHYLAMINOETHYL-9-XANTHENECARBOXYLATE METHOBROMIDE □ N,N-DIETHYL-N-METHYL-2-((9H-XANTHEN-9-YLCARBONYL)OXY)ETHANAMINIUM BROMIDE □ DOLADENE □ FRENOGASTRICO □ GASTRON □ GASTROSEDAN □ MANTHELINE □ METANTYL □ METAXAN □ METHANIDE □ METHANTHELINE BROMIDE □ METHELINA □ MTB 51 □ RESOBANTIN □ SC 2910 □ ULCINE □ ULCUDEXTER □ VAGAMIN □ VAGANTIN □ XANTELINE □ XANTHENE-9-CARBOXYLIC ACID ESTER with DIETHYL(2-HYDROXYETHYL) METHYLAMMONIUM BROMIDE

TOXICITY DATA WITH REFERENCE
orl-rat LD50:1660 mg/kg NIIRDN 6,357,82
orl-mus LD50:460 mg/kg NIIRDN 6,357,82
ipr-mus LD50:46 mg/kg JPETAB 106,141,52
scu-mus LDLo:600 mg/kg ARZNAD 8,107,58
ivn-mus LD50:4300 μg/kg AIPTAK 105,221,56
ivn-dog LDLo:23 mg/kg PSEBAA 78,576,51

SAFETY PROFILE: Poison by intraperitoneal and intravenous routes. Moderately toxic by ingestion and subcutaneous routes. Unspecified human reproductive effects. When heated to decomposition it emits very toxic fumes of NO_x, NH_3, and Br^-.

DJN000 CAS:3710-84-7 HR: 3
DIETHYLHYDROXYLAMINE
mf: $C_4H_{11}NO$ mw: 89.16

PROP: A liquid. D: 0.867 @ 20°/0°, mp: −8°, bp: 133°. Sol in water.

SYNS: DEHA □ N,N-DIETHYLHYDROXYLAMINE

TOXICITY DATA WITH REFERENCE
sln-dmg-orl 200 ppm CAES** 492-78,78
dns-hmn:leu 4000 ppm ENVRAL 20,99,79
dlt-rat-par 180 mg/kg CAES** 492-78,78
par-rat TDLo:18 mg/kg (1D male):REP ENVRAL 20,99,79
par-rat TDLo:180 mg/kg (1D male):REP ENVRAL 20,99,79
orl-rat LDLo:1600 mg/kg KODAK* 21MAY71

skn-rat LDLo: 100 mg/kg KODAK* 21MAY71
orl-mus LD80: 2150 mg/kg 34ZIAG -,217,69
ipr-mus LDLo: 1750 mg/kg 34ZIAG -,217,69
skn-rbt LDLo: 2000 mg/kg 34ZIAG -,217,69

CONSENSUS REPORTS: Reported in EPA TSCA Inventory. EPA Genetic Toxicology Program.

SAFETY PROFILE: Poison by skin contact. Moderately toxic by ingestion and intraperitoneal routes. Experimental reproductive effects. Human mutation data reported. When heated to decomposition it emits toxic fumes of NO_x. See also AMINES.

DJN400 **CAS:64048-99-3** **HR: 3**
DIETHYL(m-HYDROXYPHENYL)ARSINE METHIODIDE

SYNS: DIETHYL(m-HYDROXYPHENYL)METHYLARSONIUM IODIDE ☐ TL 1503

TOXICITY DATA WITH REFERENCE
scu-mus LDLo: 20 mg/kg NDRC** 30101,9,45

CONSENSUS REPORTS: Arsenic and its compounds are on the Community Right-To-Know List.

OSHA PEL: TWA 0.5 mg(As)/m^3

SAFETY PROFILE: Poison by subcutaneous route. See also ARSENIC COMPOUNDS and IODIDES. When heated to decomposition it emits very toxic fumes of As and I^-.

DJN489 **HR: 3**
DIETHYL HYDROXYTIN HYDROPEROXIDE
mf: $C_4H_{12}O_3Sn$ mw: 226.85

$(CH_3CH_2)Sn(OH)OOH$

SAFETY PROFILE: An explosive. When heated to decomposition it emits acrid smoke and fumes. See also TIN COMPOUNDS and PEROXIDES.

DJN600 **CAS:78-52-4** **HR: 3**
O,O-DIETHYL-S-2-ISOPROPYLMERCAPTOMETHYLDITHIOPHOSPHATE
mf: $C_8H_{19}O_2PS_3$ mw: 274.42

SYNS: AMERICAN CYANAMID 12,008 ☐ O,O-DIETHYL-S-(ISOPROPYLMERCAPTOMETHYL) PHOSPHORODITHIOATE ☐ O,O-DIETHYL-S-(ISOPROPYLTHIOMETHYL) PHOSPHORODITHIOATE ☐ ENT 22,865 ☐ EXPERIMENTAL INSECTICIDE 12008 ☐ (ISOPROPYLTHIO)-METHANETHIOL-S-ESTER with O,O-DIETHYL PHOSPHORODITHIOATE ☐ TM 12008

TOXICITY DATA WITH REFERENCE
orl-rat LD50: 1100 μg/kg ARSIM* 20,1,66
scu-rat LD50: 2 mg/kg JEENAI 50,356,57

SAFETY PROFILE: Poison by ingestion and subcutaneous routes. When heated to decomposition it emits very toxic fumes of PO_x and SO_x. See also MERCAPTANS.

DJN700 **CAS:24264-08-2** **HR: 3**
DIETHYLKETENE
mf: $C_6H_{10}O$ mw: 98.14

$(CH_3CH_2)_2C{=}C{=}O$

PROP: A liquid, giving pale-yellow crystals on chilling. D: 0.831 @ 20°/4°, bp: 88–89.5°.

SYN: 2-ETHYL-1-BUTENE-1-ONE

SAFETY PROFILE: Reacts with air to form explosive peroxides. When heated to decomposition it emits acrid smoke and fumes. See also PEROXIDES.

DJN750 **CAS:96-22-0** **HR: 3**
DIETHYL KETONE
DOT: UN 1156
mf: $C_5H_{10}O$ mw: 86.15

$(CH_3CH_2)_2C{=}O$

PROP: Colorless, mobile liquid; acetone-like odor. Mp: −42°, bp: 101°, flash p: 55°F, d: 0.8159 @ 19°/4°, vap d: 2.96, autoign temp: 842°F, lel: 1.6%. Mod sol in water; misc in alc and ether.

SYNS: DEK ☐ DIETHYLCETONE (FRENCH) ☐ DIMETHYLACETONE ☐ METACETONE ☐ METHACETONE ☐ PENTANONE-3 ☐ 3-PENTANONE ☐ PROPIONE

TOXICITY DATA WITH REFERENCE
skn-rbt 410 mg open MLD UCDS** 4/25/58
eye-rbt 50 mg MOD UCDS** 4/25/58
mrc-smc 14,800 ppm MUREAV 149,339,85
sln-smc 14,800 ppm MUREAV 149,339,85
eye-rbt 100 mg/24H MOD 85JCAE-,282,86
orl-rat LD50: 2140 mg/kg UCDS** 4/25/58
ihl-rat LCLo: 8000 ppm/4H AMIHBC 10,61,54
ipr-rat LDLo: 1250 mg/kg JIHTAB 27,1,45
ivn-mus LD50: 513 mg/kg JPMSAE 67,566,78
skn-rbt LD50: 20 g/kg AMIHBC 10,61,54

CONSENSUS REPORTS: Reported in EPA TSCA Inventory.

OSHA PEL: TWA 200 ppm
ACGIH TLV: TWA 200 ppm
DOT CLASSIFICATION: 3; *Label:* Flammable Liquid

SAFETY PROFILE: Moderately toxic by ingestion, intraperitoneal, and intravenous routes. A skin and eye irritant. Mutation data reported. Dangerous fire hazard when exposed to heat or flame; can react vigorously with oxidizing materials. To fight fire, use alcohol foam, foam, CO_2, dry chemical. Reacts with hydrogen peroxide + nitric acid to form a shock- and heat-sensitive explosive peroxide. When heated to decomposition it emits acrid smoke and irritating fumes. See also KETONES.

DJN800 **CAS:15773-47-4** **HR: 3**
DIETHYL LEAD DIACETATE
mf: $C_8H_{16}O_4Pb$ mw: 383.43

PROP: White crystals from cyclohexane. Mp: 130°.

TOXICITY DATA WITH REFERENCE
orl-mus LD50: 130 mg/kg CRSBAW 162,1456,68

CONSENSUS REPORTS: Lead and its compounds are on the Community Right-To-Know List.

SAFETY PROFILE: Poison by ingestion. See also LEAD COMPOUNDS. When heated to decomposition it emits toxic fumes of Pb.

DJN875 CAS:17498-10-1 HR: 3
DIETHYL LEAD DINITRATE
mf: $C_4H_{10}N_2O_6Pb$ mw: 389.33

CONSENSUS REPORTS: Lead and its compounds are on the Community Right-To-Know List.

PROP: White solid from $CHCl_3$.

SAFETY PROFILE: A poison. Unstable above 0°C and explodes when heated. When heated to decomposition it emits toxic fumes of NO_x. See also LEAD COMPOUNDS and NITRATES.

DJO000 CAS:50-37-3 HR: 3
N,N-DIETHYLLYSERGAMIDE
mf: $C_{20}H_{25}N_3O$ mw: 323.48

SYNS: ACID ☐ CUBES ☐ DELYSID ☐ 9,10-DIDEHYDRO-N,N-DIETHYL-6-METHYL-ERGOLINE-8-β-CARBOXAMIDE ☐ HEAVENLY BLUE ☐ LSD ☐ d-LSD ☐ LSD-25 ☐ LYSERGAMID ☐ LYSERGAURE DIETHYLAMID ☐ d-LYSERGIC ACID DIETHYLAMIDE ☐ LYSERGIC ACID DIETHYLAMIDE-25 ☐ LYSERGIDE ☐ LYSERGSAEUREDIAETHYLAMID ☐ PEARLY GATES ☐ ROYAL BLUE ☐ WEDDING BELLS

TOXICITY DATA WITH REFERENCE
mmo-sat 100 mg/L MUREAV 10,269,70
dlt-dmg-ipr 1 mg/kg ACNSAX 13,212,73
cyt-mus-orl 7500 ng/kg/5W-I NULSAK 19,153,76
cyt-ham-orl 375 μg/kg/5W NULSAK 21,206,78
cyt-ham:lng 1180 mg/L GMCRDC 27,95,81
orl-wmn TDLo:2500 mg/kg (1D pre):REP JAMAAP 212,1483,70
orl-rat TDLo:300 μg/kg (13-15D preg):REP PHMCAA 12,296,70
scu-rat TDLo:5 μg/kg (female 4D post):REP SCIEAS 157,459,67
scu-ham TDLo:84 ng/kg (female 8D post):TER SCIEAS 158,265,67
orl-rat TDLo:150 μg/kg (6-15D preg):TER AUCMBJ 61,221,73
scu-rat TDLo:10 μg/kg (male 1D pre):REP PSYPAG 10,44,66
scu-ham TDLo:84 ng/kg (female 8D post):REP SCIEAS 158,265,67
scu-mus TDLo:10 mg/kg (male 10D pre):REP JRPFA4 22,141,70
ipr-mus TDLo:20 μg/kg (female 8D post):TER FOMOAJ 31,502,72
ipr-mus TDLo:5 μg/kg (female 12D post):TER GEPHDP 7,395,76
ipr-mus TDLo:40 μg/kg (female 7D post):TER NATUAS 220,490,68
ipr-mus TDLo:200 μg/kg (female 6D post):TER SCIEAS 164,574,69
orl-hmn TDLo:700 ng/kg:CNS JPETAB 120,340,57
orl-hmn TDLo:2857 ng/kg:CNS 34ZIAG -,356,69
orl-hmn TDLo:857 ng/kg:CNS,GIT ARZNAD 16,220,66
ims-hmn TDLo:750 ng/kg:CNS PSYPAG 3,219,62
ivn-rat LD50:16 mg/kg IRMEA9 172,702,59
ipr-mus LD50:50 mg/kg NTIS** AD277-689
ivn-mus LD50:46 mg/kg IRMEA9 172,702,59
ivn-rbt LD50:300 μg/kg ANYAA9 66,668,57
scu-gpg LD50:16 mg/kg AIPTAK 137,375,62
orl-bwd LD50:1800 μg/kg TXAPA9 21,315,72

CONSENSUS REPORTS: EPA Genetic Toxicology Program.

SAFETY PROFILE: Poison by ingestion, subcutaneous, intraperitoneal, and intravenous routes. Mutation data reported. Human systemic effects by ingestion and intramuscular routes: euphoria, hallucinations, distorted perceptions, excitement, anorexia, nausea, and vomiting. An experimental teratogen. Other experimental reproductive effects. Mutation data reported. A much-abused hallucinogen. A federally regulated substance. When heated to decomposition it emits toxic fumes of NO_x.

DJO100 CAS:557-18-6 HR: 3
DIETHYL MAGNESIUM
mf: $C_4H_{10}Mg$ mw: 82.43

PROP: A solid. Mp: 176° (decomp).

CONSENSUS REPORTS: Reported in EPA TSCA Inventory.

SAFETY PROFILE: Ignites on contact with moist air, water, or carbon dioxide. See also MAGNESIUM COMPOUNDS.

DJO200 CAS:141-05-9 HR: 2
DIETHYL MALEATE
mf: $C_8H_{12}O_4$ mw: 172.20

PROP: Water-white liquid. Mp: −10, bp: 225.0°, flash p: 250°F (OC), d: 1.064, vap press: 1 mm @ 57.3°, vap d: 5.93.

SYNS: (Z)-2-BUTENEDIOIC ACID DIETHYL ESTER ☐ ETHYL MALEATE ☐ MALEIC ACID, DIETHYL ESTER

TOXICITY DATA WITH REFERENCE
skn-rbt 10 mg/24H open MLD JIHTAB 31,60,49
skn-rbt 530 mg open MLD UCDS** 10/29/57
eye-rbt 500 mg open JIHTAB 31,60,49
orl-rat LD50:3200 mg/kg JIHTAB 31,60,49
ipr-rat LD50:3070 mg/kg TXAPA9 52,422,80
skn-rbt LD50:4000 mg/kg UCDS** 10/29/57

CONSENSUS REPORTS: Reported in EPA TSCA Inventory.

SAFETY PROFILE: Moderately toxic by ingestion, skin contact, and intraperitoneal routes. A skin and eye irritant. Combustible when exposed to heat or flame; can react with oxidizing materials. To fight fire, use CO_2, dry chemical. When heated to decomposition it emits acrid smoke and irritating fumes.

DJO400 CAS:627-44-1 HR: 3
DIETHYL MERCURY
mf: $C_4H_{10}Hg$ mw: 258.73

PROP: Colorless liquid, hazel-like odor. Bp: 159°, d: 2.43 @ 20°. Sol in Et_2O; less sol in EtOH.

TOXICITY DATA WITH REFERENCE
dlt-rat-ihl 6 μg/m³/24H GISAAA 47(5),8,82
ihl-rat TCLo:6 μg/m³/24H (16W pre):TER GISAAA 47(5),8,82
ihl-hmn LCLo:1040 μg/m³/14W CJPEA4 34,158,43
orl-rat LD50:51 mg/kg 85GMAT -,52,82
ihl-rat LC50:258 mg/m³ GISAAA 38(1),100,73
orl-mus LD50:44 mg/kg 85GMAT -,52,82
ihl-mus LC50:91 mg/m³ GISAAA 38(1),100,73
ipr-mus LD50:45 mg/kg YKKZAJ 79,579,59

CONSENSUS REPORTS: Reported in EPA TSCA Inventory. Mercury and its compounds are on the Community Right-To-Know List.

OSHA PEL: TWA 0.01 mg(Hg)/m³; STEL 0.03 mg/m³ (skin)
ACGIH TLV: TWA 0.01 mg(Hg)/m³; STEL 0.03 mg(Hg)/m³

SAFETY PROFILE: A deadly human poison by inhalation. Poison by ingestion and intraperitoneal route. An experimental teratogen. See also MERCURY COMPOUNDS, ORGANIC. Flammable when exposed to heat or flame; can react with oxidizing materials. When heated to decomposition or on contact with acid or acid fumes it emits highly toxic fumes of Hg.

DJO800 CAS:50-11-3 HR: 2
5,5-DIETHYL-1-METHYLBARBITURIC ACID
mf: $C_9H_{14}N_2O_3$ mw: 198.25

PROP: Crystals from C_6H_6/pet ether; needles from H_2O. Mp: 154.5°. Sltly sol in hot H_2O.

SYNS: AN 23 □ 5,5-DIETHYL-1-METHYL-2,4,6(1H,3H,5H)-PYRIMIDINETRIONE □ ENDIEMALUM □ GEMONIL □ GEMONIT □ METABARBITAL □ METHARBITAL □ METHARBITONE □ METHARBUTAL □ METHYLBARBITAL □ N-METHYLBARBITAL □ 1-METHYLBARBITAL □ SCH 412

TOXICITY DATA WITH REFERENCE
orl-mus LD50:500 mg/kg 27ZIAQ -,-,65
ipr-mus LD50:500 mg/kg 27ZIAQ -,159,73

SAFETY PROFILE: Moderately toxic by ingestion and intraperitoneal routes. When heated to decomposition it emits toxic fumes of NO_x. See also BARBITURATES.

DJP500 CAS:132-19-4 HR: 3
N,N-DIETHYL-1-METHYL-3,3-DI-2-THIENYLALLYLAMINE HYDROCHLORIDE
mf: $C_{16}H_{21}NS_2$•ClH mw: 327.96

SYNS: 191C49 HYDROCHLORIDE □ DIETHIBUTIN HYDROCHLORIDE □ 3-DIETHYLAMINO-1,1-DI(2′-THIENYL)BUT-1-ENE HYDROCHLORIDE □ N,N-DIETHYL-4,4-DI-2-THIENYL-3-BUTEN-2-AMINE HYDROCHLORIDE □ N,N-DIETHYL-3,3-DI-2-THIENYL-1-METHYLALLYLAMINE HYDROCHLORIDE □ DIETHYLTHIAMBUTENE HYDROCHLORIDE □ NIH-4185 HYDROCHLORIDE □ THEMALON HYDROCHLORIDE □ THIAMBUIENE HYDROCHLORIDE

TOXICITY DATA WITH REFERENCE
scu-rat LD50:45 mg/kg BJPCAL 8,2,53
orl-mus LD50:204 mg/kg MEIEDD 10,1330,83
scu-mus LD50:81 mg/kg BJPCAL 8,2,53
ivn-mus LD50:16 mg/kg BJPCAL 8,2,53

SAFETY PROFILE: Poison by ingestion, subcutaneous, and intravenous routes. An analgesic and narcotic used in veterinary medicine. When heated to decomposition it emits very toxic fumes of NO_x, SO_x, and HCl. See also ALLYL COMPOUNDS.

DJP600 CAS:50285-72-8 HR: 3
1,1-DIETHYL-3-METHYL-3-NITROSOUREA
mf: $C_6H_{13}N_3O_2$ mw: 159.22

SYNS: NITROSO-1,1-DIETHYL-3-METHYLUREA □ NITROSOMETHYLDIAETHYLHARNSTOFF □ NITROSOMETHYLDIETHYLUREA □ 1-NITROSO-1-METHYL-3,3-DIETHYLUREA

TOXICITY DATA WITH REFERENCE
mma-sat 250 μg/plate JJIND8 67,1117,81
sce-ham:lng 500 μmol/L MUREAV 126,259,84
orl-rat TDLo:1400 mg/kg/50W-I:ETA ZKKOBW 83,315,75
orl-gpg TDLo:1200 mg/kg/32W-I:ETA CNREA8 40,1879,80
orl-ham TDLo:1592 mg/kg/W-I:ETA PAACA3 24,92,83
scu-ham TDLo:585 mg/kg/24W-I:ETA EXPADD 20,153,81
orl-rat TD:2475 mg/kg/33W-C:ETA JJIND8 65,451,80
scu-ham TD:623 mg/kg/22W-I:CAR CALEDQ 23,177,84
orl-ham TD:1592 mg/kg/W-I:ETA PAACA3 24,92,83
orl-rat TD:1911 mg/kg/32W-I:ETA IAPUDO 57,617,84
orl-ham TD:1592 mg/kg/29W-I:ETA IAPUDO 57,617,84
scu-ham LD50:283 mg/kg CALEDQ 23,177,84

SAFETY PROFILE: Poison by subcutaneous route. Questionable carcinogen with experimental carcinogenic and tumorigenic data. Mutation data reported. When heated to decomposition it emits toxic fumes of NO_x. See also N-NITROSO COMPOUNDS.

DJQ200 CAS:1605-58-9 HR: 3
DIETHYLMETHYLPHOSPHINE
mf: $C_5H_{13}P$ mw: 104.14

SAFETY PROFILE: May ignite spontaneously after long exposure to air. When heated to decomposition it emits toxic fumes of PO_x. See also PHOSPHINE.

DJQ300 CAS:7310-87-4 HR: 3
3,3′-DIETHYL-9-METHYLSELENOCARBOCYANINE IODIDE
mf: $C_{22}H_{23}N_2Se_2$•I mw: 600.29

SYN: BENZOSELENAZOLIUM, 3-ETHYL-2-(3-(3-ETHYL-2-BENZOSELENAZOLINYLIDENE)-2-METHYLPROPENYL)-, IODIDE

TOXICITY DATA WITH REFERENCE
ivn-mus LD50:18 mg/kg CSLNX* NX#02907

OSHA PEL: TWA 0.2 mg(Se)/m³
ACGIH TLV: TWA 0.2 mg(Se)/m³

SAFETY PROFILE: Poison by intravenous route. When heated to decomposition it emits toxic fumes of NO_x, Se, and I^-.

DJQ800 CAS:19481-39-1 HR: 3
DIETHYLMETHYLSULFONIUM IODIDEMERCURIC IODIDE (ADDITION COMPOUND)

SYN: DIETHYLMETHYL SULFONIUM IODINE with MERCURY IODIDE (1:1)

TOXICITY DATA WITH REFERENCE
ivn-mus LD50:18 mg/kg CSLNX* NX#01853

CONSENSUS REPORTS: Mercury and its compounds are on the Community Right-To-Know List.

NIOSH REL: (Mercury, Aryl and Inorganic): CL 0.1 mg/m^3 (skin)

SAFETY PROFILE: Poison by intravenous route. See also MERCURY COMPOUNDS, IODIDES, and SULFONATES. When heated to decomposition it emits very toxic fumes of SO_x, I^-, and Hg.

DJQ850 CAS:140-80-7 HR: 3
N,N-DIETHYL-4-METHYLTETRAMETHYLENEDIAMINE
mf: $C_9H_{22}N_2$ mw: 158.33

SYN: TETRAMETHYLENEDIAMINE, N,N-DIETHYL-4-METHYL-

TOXICITY DATA WITH REFERENCE
ivn-mus LD50:180 mg/kg CSLNX* NX#05228

CONSENSUS REPORTS: Reported in EPA TSCA Inventory.

SAFETY PROFILE: Poison by intravenous route. When heated to decomposition it emits toxic vapors of NO_x.

DJR700 CAS:2511-10-6 HR: 3
O,S-DIETHYL METHYLTHIOPHOSPHONATE
mf: $C_5H_{13}O_2PS$ mw: 168.21

SYNS: O,S-DIETHYL METHYLPHOSPHONOTHIOATE ☐ LG 61 ☐ METHYLPHOSPHONOTHIOIC ACID-O,S-DIETHYL ESTER ☐ OSDMP

TOXICITY DATA WITH REFERENCE
orl-rat LD50:6 mg/kg JAFCAU 32,774,84
ivn-mus LD50:1 mg/kg IAEC** 17JUN74
ivn-dog LD50:5620 μg/kg IAEC** 17JUN74
ivn-rbt LD50:2480 μg/kg IAEC** 17JUN74

SAFETY PROFILE: Poison by ingestion and intravenous routes. When heated to decomposition it emits toxic fumes of PO_x and SO_x.

DJR800 CAS:52-60-8 HR: 3
O,O-DIETHYL-O-(4-(METHYLTHIO)-3,5-XYLYL)PHOSPHOROTHIOATE
mf: $C_{13}H_{21}O_3PS_2$ mw: 320.43

SYNS: BAY 9017 ☐ BAY 37341 ☐ BAYER 37341 ☐ O-(3,5-DIMETHYL-4-(METHYLTHIO)PHENYL)-O,O-DIETHYL ESTER PHOSPHOROTHIOIC ACID ☐ O-(3,5-DIMETHYL-4-(METHYLTHIO)PHENYL)-O,O-DIETHYL PHOSPHOROTHIOATE ☐ ENT 25,673 ☐ 4-(METHYLTHIO)-3,5-XYLENOL-O-ESTER with O,O-DIETHYL PHOSPHOROTHIOATE

TOXICITY DATA WITH REFERENCE
orl-rat LD50:375 mg/kg TXAPA9 21,315,72
ipr-mus LDLo:500 mg/kg CBCCT* 5,337,53
orl-ckn LD50:18 mg/kg TXAPA9 11,49,67
orl-bwd LD50:4200 μg/kg TXAPA9 21,315,72

SAFETY PROFILE: Poison by ingestion. Moderately toxic by intraperitoneal route. When heated to decomposition it emits very toxic fumes of PO_x and SO_x.

DJS100 CAS:3060-37-5 HR: 3
α,α-DIETHYL-1-NAPHTHALENEACETIC ACID SODIUM SALT
mf: $C_{16}H_{17}O_2$•Na mw: 264.32

TOXICITY DATA WITH REFERENCE
ipr-rat LD50:466 mg/kg AIPTAK 154,297,65
scu-rat LD50:809 mg/kg AIPTAK 154,297,65
ivn-rat LD50:433 mg/kg AIPTAK 154,297,65
ipr-mus LD50:470 mg/kg AIPTAK 154,297,65
scu-mus LD50:964 mg/kg AIPTAK 154,297,65
ivn-mus LD50:324 mg/kg AIPTAK 154,297,65

SAFETY PROFILE: Poison by intravenous route. Moderately toxic by subcutaneous and intraperitoneal routes. When heated to decomposition it emits toxic fumes of Na_2O.

DJS200 CAS:59-26-7 HR: 3
N,N-DIETHYLNICOTINAMIDE
mf: $C_{10}H_{14}N_2O$ mw: 178.26

PROP: Pale yellow oil or crystals with sltly bitter taste. Mp: 24–26°, bp: 296–300° (decomp).

SYNS: ANACARDONE ☐ ANACORDONE ☐ ASTROCAR ☐ BETAPYRIMIDUM ☐ CAMPHOZONE ☐ CARBAMIDAL ☐ CARDAMINE ☐ CARDIAGEN ☐ CARDIAMID ☐ CARDIAMINA ☐ CARDIAMINE ☐ CARDIMON ☐ CITOCOR ☐ CORACON ☐ CORAETHAMIDE ☐ CORAETHAMIDUM ☐ CORALEPT ☐ CORAMINE ☐ CORAVITA ☐ CORAZONE ☐ CORDIAMID ☐ CORDIAMIN ☐ CORDIAMINE ☐ CORDITON ☐ CORDYNIL ☐ COREDIOL ☐ CORESPIN ☐ CORETHAMIDE ☐ CORETONE ☐ CORMED ☐ CORMID ☐ CORMOTYL ☐ CORNOTONE ☐ COROTONIN ☐ COROVIT ☐ CORVITAN ☐ CORVITOL ☐ CORVITONE ☐ CORYWAS ☐ DANAMINE ☐ DIAETHYL-NICOTINAMID (GERMAN) ☐ DIETHYL-NICOTAMIDE ☐ N,N-DIETHYL-3-PYRIDINECARBOXAMIDE ☐ DIETILAMIDE-CARBOPIRIDINA ☐DINACO-RYL ☐ DYNACORYL ☐ DYNAMICARDE ☐ ELITONE ☐ EUCORAN ☐ HANSACOR ☐ INICARDIO ☐ KARDIAMID ☐ KARDONYL ☐ KORDIAMIN ☐ LEPTAMIN ☐ MEDIAMID ☐ NIAMINE ☐ NICAMIDE ☐ NICETAMIDE ☐ NICETHAMIDE ☐ NICOR ☐ NICORDAMIN ☐ NICORINE ☐ NICORYL ☐ NICOTINIC ACID DIETHYLAMIDE ☐ NIKARDIN ☐ NIKETAMID ☐ NIKETHAROL ☐ NIKETHYL ☐ NIKETILAMID ☐ NIKORIN ☐ NIQUETAMIDA ☐ NISETAMIDE ☐PERCORAL ☐ PROCARDINE ☐ PROCORMAN ☐ PYRICAROYL ☐ PYRIDINE-3-CARBOXYDIETHYLAMIDE ☐ PYRIDINE-3-CARBOXYLIC ACID DIETHYLAMIDE ☐ REFORMIN ☐ REHORMIN ☐ SALVACARD ☐ SALVACORIN ☐ SOLYACORD ☐ STELLAMINE ☐ STIMINOL ☐ STIMULIN ☐ TONOCARD ☐ TONOCOR ☐ VASAZOL ☐ VENTRAMINE

TOXICITY DATA WITH REFERENCE
ipr-rat LD50:272 mg/kg PSDTAP 4,132,64
scu-rat LD50:240 mg/kg PSEBAA 62,19,46
ivn-rat LD50:191 mg/kg 27ZIAQ -,173,73
orl-mus LD50:188 mg/kg RPOBAR 1,423,64
ipr-mus LD50:174 mg/kg JPETAB 128,176,60
scu-mus LD50:200 mg/kg BCFAAI 111,293,72
ivn-mus LD50:180 mg/kg SMWOAS 85,305,55
scu-dog LDLo:175 mg/kg 27ZWAY -,-,37
ivn-dog LDLo:175 mg/kg 27ZWAY -,-,37
ipr-rbt LD50:225 mg/kg 27ZWAY -,-,37
scu-rbt LDLo:300 mg/kg KLWOAZ 12,1860,33

ivn-rbt LDLo: 150 mg/kg JPETAB 66,260,39
ipr-gpg LDLo: 250 mg/kg 27ZWAY -,-,37
scu-gpg LDLo: 300 mg/kg 27ZWAY -,-,37

SAFETY PROFILE: Poison by ingestion, intravenous, intraperitoneal, and subcutaneous routes. When heated to decomposition it emits toxic fumes of NO_x.

DJS500 CAS:7119-92-8 HR: 2
DIETHYLNITRAMINE
mf: $C_4H_{10}N_2O_2$ mw: 118.16

SYNS: N-ETHYL-N-NITROETHANAMINE (9CI) □ N-NITRODIETHYLAMINE

TOXICITY DATA WITH REFERENCE
mmo-esc 20 µmol/plate IAPUDO 57,485,84
hma-rat/sat 200 mg/kg CNREA8 41,3205,81
orl-rat TDLo: 18,200 mg/kg/2Y-C: ETA ARGEAR 52,629,82
ipr-mus LD50: 730 mg/kg PCJOAU 10,1504,76

SAFETY PROFILE: Moderately toxic by intraperitoneal route. Questionable carcinogen with experimental tumorigenic data. Mutation data reported. When heated to decomposition it emits toxic fumes of NO_x.

DJS800 CAS:3270-86-8 HR: 3
O,O-DIETHYL-S-(4-NITROPHENYL)THIOPHOSPHATE
mf: $C_{10}H_{14}NO_5PS$ mw: 291.28

SYNS: O,O-DIETHYL-S-p-NITROFENYLESTER KYSELINY THIOFOSFORECNE (CZECH) □ O,O-DIETHYL-S-(4-NITROPHENYL) PHOSPHOROTHIOATE □ O,O-DIETHYL-S-(4-NITROPHENYL)PHOSPHOROTHIOIC ACID ESTER □ PARATHION S □ S-PHENYL PARATHION

TOXICITY DATA WITH REFERENCE
orl-rat LD50: 4410 µg/kg 28ZPAK -,208,72
orl-mus LD50: 11,900 µg/kg 28ZPAK -,208,72
ipr-mus LD50: 107 µg/kg PHARAT 35,806,80
scu-mus LD50: 1250 µg/kg AMIHAB 11,487,55

SAFETY PROFILE: A deadly poison by ingestion, subcutaneous, and intraperitoneal routes. When heated to decomposition it emits very toxic fumes of NO_x, PO_x, and SO_x.

DJT000 CAS:597-88-6 HR: 3
O,S-DIETHYL-O-(4-NITROPHENYL)THIOPHOSPHATE
mf: $C_{10}H_{14}NO_5PS$ mw: 291.28

SYNS: O,S-DIETHYL-O-(p-NITROPHENYL) PHOSPHOROTHIOATE □ O,S-DIETHYL-O-(4-NITROPHENYL)PHOSPHOROTHIOATE □ O,S-DIETHYL-O-(4-NITROPHENYL)PHOSPHOROTHIOIC ACID ESTER □ O,S-DIETHYL-O-(p-NITROPHENYL)PHOSPHOROTHIOIC ACID ESTER □ S-ETHYL PARATHION □ ISOPARATHION

TOXICITY DATA WITH REFERENCE
orl-rat LD50: 17,900 µg/kg 28ZPAK -,209,72
skn-rat LD50: 75 mg/kg TXAPA9 2,523,60
ipr-rat LD50: 5500 µg/kg AMIHBC 6,9,52
ims-rat LD50: 10 mg/kg AIHAAP 19,190,58
orl-mus LD50: 25 mg/kg JPETAB 105,156,52
ipr-mus LD50: 50 µg/kg PHARAT 35,806,80
scu-mus LD50: 20 mg/kg PAREAQ 11,636,59
orl-gpg LD50: 32 mg/kg JPETAB 105,156,52

SAFETY PROFILE: Poison by ingestion, skin contact, intraperitoneal, subcutaneous, and intramuscular routes. When heated to decomposition it emits very toxic fumes of NO_x. See also PARATHION.

DJT100 CAS:376-50-1 HR: 3
DIETHYL OCTAFLUOROADIPATE
mf: $C_{10}H_{10}F_8O_4$ mw: 346.20

SYNS: DIETHYL OCTAFLUOROHEXANEDIOATE □ DIETHYL PERFLUOROADIPATE □ HEXANEDIOIC ACID, OCTAFLUORO-, DIETHYL ESTER

TOXICITY DATA WITH REFERENCE
orl-rat LD: >8400 mg/kg TPKVAL 12,142,71
ihl-rat LC: >200 mg/m³/4H TPKVAL 12,142,71

CONSENSUS REPORTS: Reported in EPA TSCA Inventory.

SAFETY PROFILE: May be a poison by inhalation. Low toxicity by ingestion. When heated to decomposition it emits toxic vapors of F^-.

DJT200 CAS:95-92-1 HR: 3
DIETHYL OXALATE
DOT: UN 2525
mf: $C_6H_{10}O_4$ mw: 146.16

PROP: Colorless, oily, aromatic liquid; decomp in water. Mp: −40.6°, bp: 185.4°, flash p: 168°F (OC), d: 1.079 @ 20°/4°, vap d: 5.04. Sltly sol in H_2O.

SYNS: DIETHYL ETHANEDIOATE □ ETHYL OXALATE □ ETHYL OXALATE (DOT) □ OXALIC ACID, DIETHYL ESTER

TOXICITY DATA WITH REFERENCE
orl-rat LD50: 400 mg/kg 14CYAT 2,1882,63
orl-mus LD50: 2000 mg/kg GISAAA 46(5),87,81

CONSENSUS REPORTS: Reported in EPA TSCA Inventory.

DOT CLASSIFICATION: 6.1; *Label:* KEEP AWAY FROM FOOD

SAFETY PROFILE: Poison by ingestion. Flammable liquid when exposed to heat or flame; can react with oxidizing materials. To fight fire, use foam, CO_2, dry chemical. When heated to decomposition it emits acrid smoke and fumes. See also OXALATES and ESTERS.

DJT400 CAS:702-54-5 HR: 3
5,5-DIETHYL-1,3-OXAZIN-2,4-DIONE
mf: $C_8H_{13}NO_3$ mw: 171.22

PROP: Crystals from Et_2O. Mp: 97–98°.

SYNS: DIETADIONE (ITALIAN) □ DIETHADION □DIETHADIONE □ 5,5-DIETHYLDIHYDRO-2H-1,3-OXAZINE-2,4(3H)-DIONE □ 5,5-DIETHYL-1,3-OXAZINE-2,4-DIONE □ 5,5-DIETHYLTETRAHYDRO-2H-1,3-OXAZINE-2,4(3H)-DIONE □ 5,5-DIETILDIIDRO-1,3-OSSAZIN-2,4-DIONE (ITALIAN) □ DIETROXINE □ DIHYDRO-5,5-DIETHYL-2H-1,3-OXAZINE-2,4(3H)-DIONE □ DIIDRO-5,5-DIETIL-2H-1,3-OSSAZIN-2,4(3H)-DIONE (ITALIAN) □ DIOXONE □ L 1811 □ LEDOSTEN □ LEPTON □ PERSISTEN □ TOCE □ TOCEN

TOXICITY DATA WITH REFERENCE
orl-rat LD50: 71 mg/kg JPPMAB 13,244,61

ipr-rat LD50:32 mg/kg JPPMAB 13,244,61
scu-rat LD50:39 mg/kg JPPMAB 13,244,61
orl-mus LD50:81 mg/kg RPOBAR 2,282,70
ipr-mus LD50:49 mg/kg RPOBAR 2,281,70
scu-mus LD50:61 mg/kg RPOBAR 2,281,70
ivn-mus LD50:32 mg/kg JPPMAB 13,244,61
ims-mus LD50:45 mg/kg RPOBAR 2,281,70

CONSENSUS REPORTS: Reported in EPA TSCA Inventory.

SAFETY PROFILE: Poison by ingestion, intravenous, intraperitoneal, subcutaneous, and intramuscular routes. An analeptic (central nervous system stimulant). When heated to decomposition it emits toxic fumes of NO_x.

DJT800 CAS:514-73-8 HR: 3
3,3′-DIETHYLPENTAMETHINETHIACYANINE IODIDE
mf: $C_{23}H_{24}N_2S_2•I$ mw: 519.51

PROP: Green needles from MeOH. Mp: 248° (decomp).

SYNS: ABMINTHIC ☐ ANELMID ☐ ANGUIFUGAN ☐ COMPOUND 01748 ☐ DEJO ☐ DELVEX ☐ DIETHYLTHIADICARBOCYANINE IODIDE ☐ 3,3′-DIETHYLTHIADICARBOCYANINE IODIDE ☐ DILOMBRIN ☐ DITHIAZANINE IODIDE /S DITHIAZANIN IODIDE ☐ DITHIAZININE ☐ EASTMAN 7663 ☐ 3-ETHYL-2-(5-(3-ETHYL-2-BENZOTHIAZOLINYLIDENE)-1,3-PENTADIENYL)BENZOTHIAZOLIUM IODIDE ☐ L-01748 ☐ NETOCYD ☐ NK 136 ☐ OMNI-PASSIN ☐ PARTEL ☐ TELMICID ☐ TELMID ☐ TELMIDE ☐ VERCIDON

TOXICITY DATA WITH REFERENCE
orl-mus LD50:20 mg/kg BSIBAC 44,1032,68
ipr-mus LD50:3 mg/kg JMCMAR 10,897,67
ivn-mus LD50:1 mg/kg CSLNX* NX#02015

CONSENSUS REPORTS: Reported in EPA TSCA Inventory.

SAFETY PROFILE: Poison by ingestion, intraperitoneal, and intravenous routes. When heated to decomposition it emits very toxic fumes of I^-, SO_x, and NO_x. See also IODIDES.

DJU200 CAS:512-48-1 HR: 3
2,2-DIETHYL-4-PENTENAMIDE
mf: $C_9H_{17}NO$ mw: 155.27

PROP: A solid. Mp: 75–76°.

SYNS: DIAETHYLALLYLACETAMIDE (GERMAN) ☐ EPINOVAL ☐ NOVONAL

TOXICITY DATA WITH REFERENCE
orl-hmn LDLo:300 mg/kg DMWOAX 102,1591,77
orl-hmn TDLo:100 mg/kg:CVS,PUL DMWOAX 102,1591,77
orl-rat LD50:400 mg/kg DMWOAX 102,1591,77
ipr-rat LD50:217 mg/kg ITMZBJ 17,305,80
orl-mus LDLo:300 mg/kg LDTU** -,-,31
orl-dog LD50:300 mg/kg DMWOAX 102,1591,77
rec-gpg LDLo:150 mg/kg LDTU** -,-,31

SAFETY PROFILE: Poison to humans by ingestion. An experimental poison by ingestion, rectal, and intraperitoneal routes. Human systemic effects by ingestion: muscle spasms, cardiac arrhythmias, and respiratory depression. When heated to decomposition it emits toxic fumes of NO_x.

DJU400 CAS:628-37-5 HR: 3
DIETHYL PEROXIDE
mf: $C_4H_{10}O_2$ mw: 90.12

$CH_3CH_2OOCH_2CH_3$

PROP: Lel: 2.3%, d: 0.8° @ 20°/4°, vap d: 7.7, bp: 62–63°. Sltly sol in H_2O.

SAFETY PROFILE: A shock- and heat-sensitive explosive. Reacts violently with O_2. When heated to decomposition it emits acrid smoke and fumes. See also PEROXIDES, ORGANIC.

DJU600 CAS:14666-78-5 HR: 3
DIETHYL PEROXYDICARBONATE
mf: $C_6H_{10}O_6$ mw: 178.14

$CH_3CH_2OCO•OOCO•OCH_2CH_3$

SYNS: DIETHYL PEROXYDICARBONATE, >27% in solution (DOT) ☐ DIETHYL PEROXYDIFORMATE ☐ ETHYL PEROXYCARBONATE ☐ PEROXYDICARBONIC ACID, DIETHYL ESTER

DOT CLASSIFICATION: Forbidden

SAFETY PROFILE: The impure material is a powerful explosive extremely sensitive to heat or impact. When heated to decomposition it emits acrid smoke and fumes. See also PEROXIDES.

DJU700 CAS:1006-59-3 HR: 3
2,6-DIETHYLPHENOL
mf: $C_{10}H_{14}O$ mw: 150.24

SYN: PHENOL, 2,6-DIETHYL-

TOXICITY DATA WITH REFERENCE
ipr-mus LD50:230 mg/kg JMPCAS 2,201,60
ivn-mus LD50:100 mg/kg JMCMAR 23,1350,80

CONSENSUS REPORTS: Reported in EPA TSCA Inventory.

SAFETY PROFILE: Poison by intravenous and intraperitoneal routes. When heated to decomposition it emits acrid smoke and irritating vapors.

DJU800 CAS:52400-58-5 HR: 3
N,N-DIETHYL-N′-(2-(2-PHENYL-1,3-BENZODIOXOL-2-YL)ETHYL)ETHYLENEDIAMINE DIMALEATE
mf: $C_{21}H_{28}N_2O_2•2C_4H_4O_4$ mw: 572.67

SYN: 2-(2-(2-(DIETHYLAMINO)ETHYLAMINO)ETHYL)-1,3-BENZODIOXOLE, DIMALEATE

TOXICITY DATA WITH REFERENCE
ivn-rat LD50:10 mg/kg EJMCA5 12,413,77
ipr-mus LD50:90 mg/kg EJMCA5 12,413,77

SAFETY PROFILE: Poison by intravenous and intraperitoneal routes. When heated to decomposition it emits toxic fumes of NO_x.

D

DJV000 CAS:23564-06-9 HR: 2
DIETHYL-4,4′-o-PHENYLENEBIS(3-THIOALLOPHANATE)
mf: $C_{14}H_{18}N_4O_4S_2$ mw: 370.48

PROP: Plates. Mp: 195°.

SYNS: BAS 3220 □ 1,2-BIS-(3-ETHOXYCARBONYLTHIOUREIDO) BENZENE □ 1,2-BIS(3-ETHOXYCARBONYL-2-THIOUREIDO) BENZENE □ CERCOBIN □ CLEARY 3336 □ ENOVIT □ ETHYL THIOPHANATE □ NF 35 (fungicide) □ PELT SOL □ (1,2-PHENYLENEBIS(IMINOCARBONOTHIOYL))BISCARBAMIC ACID DIETHYL ESTER □ THIOFANATE □ THIOPHANAT (GERMAN) □ THIOPHANATE ETHYL □ THIOPHENITE □ TIOFANATE ETILE (ITALIAN) □ TOPSIN □ 3336 TURF FUNGICIDE

TOXICITY DATA WITH REFERENCE
mmo-smc 5 ppm RSTUDV 6,161,76
sln-asn 100 mg/L EVHPAZ 31,81,79
orl-mus TDLo:336 g/kg (MGN):REP OYYAA2 4,23,70
orl-mus TDLo:15 g/kg (1-15D preg):REP OYYAA2 4,23,70
ipr-rat LD50:2400 mg/kg OYYAA2 4,5,70
ipr-mus LD50:3750 mg/kg OYYAA2 4,5,70

CONSENSUS REPORTS: EPA Genetic Toxicology Program.

SAFETY PROFILE: Moderately toxic by intraperitoneal route. Experimental reproductive effects. Mutation data reported. A fungicide. When heated to decomposition it emits very toxic fumes of NO_x and SO_x. See also CARBAMATES.

DJV200 CAS:93-05-0 HR: 3
DIETHYL-p-PHENYLENEDIAMINE
mf: $C_{10}H_{16}N_2$ mw: 164.28

PROP: A liquid. Bp: 260–262°.

SYNS: p-AMINODIETHYLANILINE □ p-(DIETHYLAMINO)ANILINE □ 4-(DIETHYLAMINO)ANILINE □ N,N′-DIETHYL-p-FENYLENDIAMIN □ N,N′-DIETHYL-p-PHENYLENEDIAMINE □ DIETHYL-PARA-PHENYLENEDIAMINE □ DPD

TOXICITY DATA WITH REFERENCE
mmo-sat 666 μg/plate EMMUEG 11(Suppl 12),1,88
cyt-ham:lng 2500 μg/L MUREAV 241,175,90
skn-hmn TDLo:73 μg/kg:SKN,BLD JIDHAN 4,386,23
scu-rat LDLo:100 mg/kg JIDHAN 4,386,23
ivn-dog LDLo:70 mg/kg JIDHAN 4,386,23
orl-cat LDLo:300 mg/kg JIDHAN 4,386,23
orl-rbt LDLo:450 mg/kg JIDHAN 4,386,23
skn-rbt LDLo:125 mg/kg JIDHAN 4,386,23
scu-rbt LDLo:250 mg/kg JIDHAN 4,386,23

CONSENSUS REPORTS: Reported in EPA TSCA Inventory.

SAFETY PROFILE: Poison by ingestion, skin contact, subcutaneous, and intravenous routes. Human systemic skin effects by skin contact: hemorrhage, allergic dermatitis, and primary irritation. Mutation data reported. When heated to decomposition it emits toxic fumes of NO_x. See also AMINES.

DJV250 CAS:6283-63-2 HR: 3
N,N′-DIETHYL-p-PHENYLENEDIAMINE SULFATE
mf: $C_{10}H_{16}N_2 \cdot H_2O_4S$ mw: 262.36

SYNS: p-PHENYLENEDIAMINE, N,N-DIETHYL-, SULFATE (1:1) □ SIRAN N,N-DIETHYL-p-FENYLENDIAMINU

TOXICITY DATA WITH REFERENCE
eye-rbt 500 mg/24H MLD 85JCAE -,479,86
mmo-sat 8 mg/plate MUREAV 238,1,90
orl-uns TDLo:36 mg/kg (male 60D pre):REP GISAAA 51(9),71,86
orl-rat LDLo:100 mg/kg GISAAA 22(5),41,57
scu-rat LDLo:100 mg/kg GISAAA 22(5),41,57
orl-mus LDLo:300 mg/kg GISAAA 22(5),41,57
scu-mus LDLo:80 mg/kg GISAAA 22(5),41,57

SAFETY PROFILE: Poison by ingestion and subcutaneous routes. Experimental reproductive effects. An eye irritant. Mutation data reported. When heated to decomposition it emits toxic fumes of SO_x.

DJV300 CAS:1665-59-4 HR: 3
N,N-DIETHYL-N′-PHENYLETHYLENEDIAMINE
mf: $C_{12}H_{20}N_2$ mw: 192.34

SYNS: ETHYLENEDIAMINE, N,N-DIETHYL-N′-PHENYL- □ 1167 F

TOXICITY DATA WITH REFERENCE
ivn-rbt LDLo:50 mg/kg AIPAAV 63,400,39

CONSENSUS REPORTS: Reported in EPA TSCA Inventory.

SAFETY PROFILE: Poison by intravenous route. When heated to decomposition it emits toxic vapors of NO_x.

DJV600 CAS:18854-01-8 HR: 3
O,O-DIETHYL-O-(5-PHENYL-3-ISOXAZOLYL) PHOSPHOROTHIOATE
mf: $C_{13}H_{16}NO_4PS$ mw: 313.33

PROP: Yellowish liquid. Bp: 160° @ 0.2 mm.

SYNS: O,O-DIETHYL-O-(3-(5-PHENYL)-1,2-ISOXAZOLYL)PHOSPHOROTHIOATE □ O,O-DIETHYL-O-(5-PHENYL-3-ISOXAZOLYL)PHOSPHOROTHIOIC ACID ESTER □ E-48 □ ISOXATHION □ KARPHOS □ SI-6711

TOXICITY DATA WITH REFERENCE
orl-rat LD50:112 mg/kg BESAAT 15,121,69
skn-rat LD50:450 mg/kg BESAAT 15,121,69
orl-mus LD50:79,100 μg/kg SKKNAJ 29,1,77
skn-mus LD50:193 mg/kg BESAAT 15,121,69
ipr-mus LD50:105 mg/kg SKKNAJ 29,1,77
scu-mus LD50:720 mg/kg SKKNAJ 29,1,77
orl-ckn LD50:21,600 μg/kg SKKNAJ 29,1,77

SAFETY PROFILE: Poison by ingestion, skin contact, and intraperitoneal routes. Moderately toxic by subcutaneous route. When heated to decomposition it emits very toxic fumes of NO_x, PO_x, and SO_x.

DJV800 CAS:64036-46-0 HR: 3
DIETHYL PHENYLTIN ACETATE
mf: $C_{12}H_{18}O_2Sn$ mw: 312.99

SYN: ACETOXYDIETHYLPHENYLSTANNANE

TOXICITY DATA WITH REFERENCE
orl-rat LDLo:50 mg/kg BJPCAL 10,16,55

OSHA PEL: TWA 0.1 mg(Sn)/m^3 (skin)
ACGIH TLV: TWA 0.1 mg(Sn)/m^3 (skin) (Proposed: TWA 0.1 mg(Sn)/m^3; STEL 0.2 mg(Sn)/m^3 (skin))
NIOSH REL: (Organotin Compounds) TWA 0.1 mg(Sn)/m^3

SAFETY PROFILE: Poison by ingestion. See also TIN COMPOUNDS. When heated to decomposition it emits acrid and irritating fumes.

For occupational chemical analysis use NIOSH: Organotin Compounds 5504.

DJW000 CAS:627-49-6 HR: 3
DIETHYL PHOSPHINE
mf: $C_4H_{11}P$ mw: 90.11

$(CH_3CH_2)_2PH$

PROP: A liquid. Bp: 85°, d: 1, vap d: 3.11.

SAFETY PROFILE: Poison by ingestion and inhalation. Flammable when exposed to heat or flame; spontaneously flammable in air. Can react vigorously with oxidizing materials. To fight fire, use foam, CO_2, dry chemical. When heated to decomposition it emits toxic fumes of PO_x. See also PHOSPHINE.

DJW200 CAS:7531-39-7 HR: 3
DIETHYLPHOSPHINIC ACID-p-NITROPHENYL ESTER
mf: $C_{10}H_{14}NO_4P$ mw: 243.22

SYN: p-NITROPHENYL ESTER of DIETHYLPHOSPHINIC ACID

TOXICITY DATA WITH REFERENCE
scu-rat LD50:3400 µg/kg FATOAO 42(3),299,79
ivn-rat LD50:3350 µg/kg FATOAO 42(3),299,79
scu-mus LD50:3400 µg/kg RPTOAN 42,106,79
ivn-mus LD50:3350 µg/kg RPTOAN 42,106,79

SAFETY PROFILE: Poison by subcutaneous and intravenous routes. See also ESTERS. When heated to decomposition it emits very toxic fumes of NO_x and PO_x.

DJW400 CAS:762-04-9 HR: 2
DIETHYL PHOSPHITE
mf: $C_4H_{11}O_3P$ mw: 138.12

$(CH_3CH_2O)_2P(:O)H$

PROP: Colorless liquid. D: 1.074 @ 20°/0°, bp: 187–188°.

SYNS: DIETHYL HYDROGEN PHOSPHITE □ PHOSPHOROUS ACID, DIETHYL ESTER

TOXICITY DATA WITH REFERENCE
orl-rat LD50:3900 mg/kg ALBRW* #OPB-3,84
skn-rbt LD50:2165 mg/kg AIHAAP 30,470,69

CONSENSUS REPORTS: Reported in EPA TSCA Inventory.

SAFETY PROFILE: Moderately toxic by ingestion and skin contact. Mixtures with 4-nitrophenol may explode if heated. When heated to decomposition it emits toxic fumes of PO_x.

DJW600 CAS:2524-04-1 HR: 3
O,O-DIETHYLPHOSPHOROCHLORIDOTHIOATE
DOT: UN 2751
mf: $C_4H_{10}ClO_2PS$ mw: 188.62

PROP: A liquid. D: 1.202 @ 20°/4°, bp: 94–96° @ 20 mm.

SYNS: CHLORO-PHOSPHONOTHIOIC ACID-O,O-DIETHYL ESTER □ DIETHYLCHLOROTHIOPHOSPHATE □ DIETHYLCHLORTHIOFOSFAT (CZECH) □ DIETHYLTHIOPHOSPHORYL CHLORIDE (DOT)

TOXICITY DATA WITH REFERENCE
orl-rat LDLo:1000 mg/kg 34ZIAG -,393,69
ihl-rat LC50:20 ppm/4H 85JCAE-,1172,86
orl-mus LD50:800 mg/kg 85GMAT-,51,82
unr-mus LD50:750 mg/kg GISAAA 56(5),6,91
ihl-mus LC50:725 mg/m^3/2H 85GMAT -,51,82
orl-rbt LD50:900 mg/kg HYSAAV 33(12),334,68
skn-rbt LDLo:250 mg/kg 34ZIAG -,393,69
orl-gpg LD50:810 mg/kg HYSAAV 33(12),334,68

CONSENSUS REPORTS: Reported in EPA TSCA Inventory.

DOT CLASSIFICATION: 8; *Label:* Corrosive

SAFETY PROFILE: Poison by inhalation and skin contact. Moderately toxic by ingestion. Corrosive. Probably a severe eye and skin irritant. See also ESTERS. When heated to decomposition it emits very toxic fumes of Cl^-, PO_x, and SO_x.

DJW800 CAS:2942-58-7 HR: 3
DIETHYL PHOSPHOROCYANIDATE
mf: $C_5H_{10}NO_3P$ mw: 163.13

PROP: Oil. Bp: 103–104° @ 20 mm.

SYNS: DIETHOXYPHOSPHORYL CYANIDE □ DIETHYLCYANOPHOSPHATE □ DIETHYL CYANOPHOSPHONATE

TOXICITY DATA WITH REFERENCE
ipr-mus LD50:1400 µg/kg PAREAQ 11,636,59
scu-mus LD50:25 mg/kg JCSOA9 -,699,48
ivn-rbt LD50:4 mg/kg JCSOA9 -,699,48

SAFETY PROFILE: Poison by intravenous, intraperitoneal, and subcutaneous routes. When heated to decomposition it emits very toxic fumes of NO_x and PO_x.

DJW875 CAS:1068-22-0 HR: 1
O,O-DIETHYL PHOSPHORODITHIOATE AMMONIUM
mf: $H_4N{\cdot}C_4H_{10}O_2PS_2$ mw: 203.28

PROP: Crystals. Mp: 164–166°.

TOXICITY DATA WITH REFERENCE
skn-rbt 500 mg MLD 34ZIAG -,97,69
eye-rbt 100 mg MLD 34ZIAG -,97,69
orl-rat LD50:7900 mg/kg 34ZIAG -,97,69

SAFETY PROFILE: Mildly toxic by ingestion. An eye and skin irritant. When heated to decomposition it emits toxic fumes of PO_x, SO_x, NH_3, and NO_x.

D

DJX000 CAS:84-66-2 HR: 3
DIETHYL PHTHALATE
mf: $C_{12}H_{14}O_4$ mw: 222.26

PROP: Clear, colorless liquid. Mp: −0.3°, bp: 298°, flash p: 325°F (OC), d: 1.110, vap d: 7.66.

SYNS: ANOZOL □ 1,2-BENZENEDICARBOXYLIC ACID, DIETHYL ESTER □ DIETHYL-o-PHTHALATE □ ESTOL 1550 □ ETHYL PHTHALATE □ NCI-C60048 □ NEANTINE □ PALATINOL A □ PHTHALIC ACID, DIETHYL ESTER □ PHTHALOL □ PHTHALSAEUREDIAETHYLESTER (GERMAN) □ PLACIDOL E □ RCRA WASTE NUMBER U088 □ SOLVANOL

TOXICITY DATA WITH REFERENCE
eye-rbt 112 mg JPETAB 82,377,44
mmo-sat 200 µg/plate JTEHD6 16,61,85
orl-mus TDLo:171 g/kg (male 7D pre):REP TXAPA9 88,255,87
ipr-rat TDLo:506 mg/kg (5-15D preg):TER JPMSAE 61,51,72
ipr-rat TDLo:506 mg/kg (5-15D preg):REP JPMSAE 61,51,72
orl-rat TDLo:53,480 mg/kg (14D male):REP FCTXAV 16,415,78
ihl-hmn TCLo:1000 mg/m³:EYE,PUL AGGHAR 5,1,33
orl-rat LD50:8600 mg/kg GTPZAB 24(3),25,80
ipr-rat LD50:5058 mg/kg JPMSAE 61,51,72
orl-mus LD50:6172 mg/kg GTPZAB 17(11),51,73
ipr-mus LD50:2749 mg/kg FEPRA7 6,342,47
orl-rbt LDLo:1000 mg/kg 14CYAT 2,1904,63
ivn-rbt LDLo:100 mg/kg AGGHAR 5,1,33
orl-gpg LD50:8600 mg/kg GTPZAB 24(3),25,80
scu-gpg LDLo:3000 mg/kg AGGHAR 5,1,33

CONSENSUS REPORTS: Reported in EPA TSCA Inventory.

OSHA PEL: TWA 5 mg/m³
ACGIH TLV: TWA 5 mg/m³

SAFETY PROFILE: Poison by intravenous route. Moderately toxic by ingestion, subcutaneous, and intraperitoneal routes. Human systemic effects by inhalation: lacrimation, respiratory obstruction, and other unspecified respiratory system effects. An eye irritant and systemic irritant by inhalation. An experimental teratogen. Other experimental reproductive effects. Narcotic in high concentrations. Combustible when exposed to heat or flame. To fight fire, use water spray, mist, foam. When heated to decomposition it emits acrid smoke and irritating fumes.

DJX200 CAS:5131-24-8 HR: 2
O,O-DIETHYLPHTHALIMIDOPHOSPHONOTHIOATE
mf: $C_{12}H_{14}NO_4PS$ mw: 299.30

PROP: A solid. Mp: 81.5–83.5°.

SYNS: O,O-DIAETHYL-N-PHTALIMIDOTHIOPHOSPHAT (GERMAN) □ O,O-DIETHYL-(1,2-DIHYDRO-1,3-DIOXO-2H-ISOINDOL-2-YL)PHOSPHONOTHIOATE □ O,O-DIETHYL PHTHALIMIDOTHIOPHOSPHATE

TOXICITY DATA WITH REFERENCE
orl-rat LD50:5660 mg/kg 85ARAE 4,104,76
unk-rat LD50:5000 mg/kg 30ZDA9 -,353,71
orl-rbt LD50:1000 mg/kg 85DPAN -,-,71/76
orl-gpg LD50:5660 mg/kg 28ZEAL 5,86,76
orl-ckn LD50:4500 mg/kg 31ZOAD 1,160,68

SAFETY PROFILE: Moderately toxic by ingestion. When heated to decomposition it emits very toxic fumes of NO_x, PO_x, and SO_x.

DJX250 CAS:1114-51-8 HR: 2
N,N-DIETHYLPROPANAMIDE
mf: $C_7H_{15}NO$ mw: 129.23

SYNS: DIETHYLAMIDE of PROPIONIC ACID □ N,N-DIETHYLPROPIONAMIDE □ PROPANAMIDE, N,N-DIETHYL-(9CI) □ PROPIONAMIDE, N,N-DIETHYL-

TOXICITY DATA WITH REFERENCE
ipr-mus LD50:770 mg/kg DIPHAH 18,245,66

CONSENSUS REPORTS: Reported in EPA TSCA Inventory.

SAFETY PROFILE: Moderately toxic by intraperitoneal route. When heated to decomposition it emits toxic vapors of NO_x.

DJX300 CAS:77650-95-4 HR: 3
N,N-DIETHYL-N′-((8-α)-6-PROPYLERGOLIN-8-YL) UREA
mf: $C_{22}H_{32}N_4O$ mw: 368.58

PROP: Crystals from EtOH. Mp: 132–134°.

SYNS: 1-((5R,8S,10R)-6-PROPYL-8-ERGOLINYL)-3,3-DIETHYLUREA □ PROTERGURIDE □ UREA, N,N-DIETHYL-N′-((8-α)-6-PROPYLERGOLIN-8-YL)-

TOXICITY DATA WITH REFERENCE
orl-rat TDLo:10 µg/kg (female 5D post):REP CCCCAK 52,2983,87
orl-rat TDLo:40 µg/kg (lactating female 4D post):REP CCCCAK 52,2983,87
ivn-mus LD50:37 mg/kg CCCCAK 52,2983,87

SAFETY PROFILE: Poison by intravenous route. Experimental reproductive effects. When heated to decomposition it emits toxic fumes of NO_x.

DJX350 CAS:96860-89-8 HR: 3
N,N-DIETHYL-N′-((8-α)-6-PROPYLERGOLIN-8-YL) UREA (Z)-2-BUTENEDIOATE
mf: $C_{22}H_{32}N_4O \cdot C_4H_4O_4$ mw: 484.66

SYN: UREA, N,N-DIETHYL-N′-((8-α)-6-PROPYLERGOLIN-8-YL)-, (Z)-2-BUTENEDIOATE (1:1)

TOXICITY DATA WITH REFERENCE
orl-rat TDLo:52,600 ng/kg (lactating female 4D post):REP CCCCAK 49,2828,84
orl-rat TDLo:13 µg/kg (female 5D post):REP CCCCAK 49,2828,84
ivn-mus LD50:48,650 µg/kg CCCCAK 49,2828,84

SAFETY PROFILE: Poison by intravenous route. Experimental reproductive effects. When heated to decomposition it emits toxic fumes of NO_x.

DJX400 CAS:5826-91-5 HR: 3
DIETHYL PROPYLMETHYLPYRIMIDYL THIOPHOSPHATE
mf: $C_{12}H_{21}N_2O_3PS$ mw: 304.38

SYNS: O,O-DIETHYL-O-(2-PROPYL-4-METHYLPYRIMIDINYL-6) PHOSPHOROTHIOATE ☐ O,O-DIETHYL-O-(2-N-PROPYL-4-METHYL-PYRIMIDYL-6)PHOSPHOROTHIOATE ☐ O,O-DIETHYL-O-(2-PROPYL-4-METHYL-6-PYRIMIDYL)PHOSPHOROTHIOIC ACID ESTER ☐ G-24622 ☐ PIRAZINON ☐ RCRA WASTE NUMBER P040

TOXICITY DATA WITH REFERENCE
orl-rat LD50:261 mg/kg PCOC** -,913,66
orl-mus LD50:50 mg/kg PCOC** -,913,66
skn-rbt LD50:3500 mg/kg 27ZTAP 3,53,69

SAFETY PROFILE: Poison by ingestion. Moderately toxic by skin contact. When heated to decomposition it emits very toxic fumes of NO_x, PO_x, and SO_x.

DJX800 CAS:4079-68-9 HR: 2
N,N-DIETHYL-2-PROPYNYLAMINE
mf: $C_7H_{13}N$ mw: 111.21

PROP: Bp: 120°.

TOXICITY DATA WITH REFERENCE
orl-rat LD50:1540 mg/kg TXAPA9 28,313,74
ihl-rat LCLo:1000 ppm/4H TXAPA9 28,313,74
skn-rbt LD50:570 mg/kg TXAPA9 28,313,74

CONSENSUS REPORTS: Reported in EPA TSCA Inventory.

SAFETY PROFILE: Moderately toxic by ingestion and skin contact. Mildly toxic by inhalation. When heated to decomposition it emits toxic fumes of NO_x.

DJY000 CAS:21600-43-1 HR: 3
3,3-DIETHYL-1-(m-PYRIDYL)TRIAZENE
mf: $C_9H_{14}N_4$ mw: 178.27

SYNS: PYDT ☐ 1-(PYRIDYL-3-)-3,3-DIAETHYL-TRIAZEN (GERMAN) ☐ m-PYRIDYL-DIETHYL-TRIAZENE ☐ 1-PYRIDYL-3,3-DIETHYLTRIAZENE ☐ 1-(PYRIDYL-3)-3,3-DIETHYLTRIAZENE ☐ 1-(3-PYRIDYL)-3,3-DIETHYLTRIAZENE

TOXICITY DATA WITH REFERENCE
sln-dmg-orl 2 mmol/L/3D-I ARTODN 43,201,80
mrc-smc 21 mmol/L MUREAV 21,123,73
cyt-hmn:leu 25 µmol/L MUREAV 21,123,73
hma-mus/smc 1600 µmol/kg MUREAV 21,123,73
cyt-ham:lng 10 mg/L MUREAV 88,197,81
ivn-rat TDLo:167 mg/kg (15D preg):TER IARCCD 4,45,73
scu-rat TDLo:60 mg/kg (10D preg):TER IARCCD 4,45,73
orl-rat TDLo:660 mg/kg/73W-I:ETA ZKKOBW 77,217,72
scu-rat TDLo:500 mg/kg/50W-I:NEO ZKKOBW 81,285,74
ivn-rat TDLo:55 mg/kg (15D preg):CAR,TER IARCCD 4,45,73
scu-rat TD:50 mg/kg:ETA ZKKOBW 81,285,74
orl-rat LD50:210 mg/kg ZKKOBW 77,217,72
scu-rat LD50:210 mg/kg ZKKOBW 81,285,74

SAFETY PROFILE: Poison by ingestion and subcutaneous routes. Questionable carcinogen with experimental carcinogenic, neoplastigenic, tumorigenic, and teratogenic data. Human mutation data reported. A transplacental carcinogen. When heated to decomposition it emits toxic fumes of NO_x.

DJY050 HR: 2
DIETHYL PYROCARBONATE mixed with AMMONIA
mf: $C_6H_{10}O_5 \cdot 2H_3N$ mw: 196.24

SYN: DEPC and AMMONIA

TOXICITY DATA WITH REFERENCE
orl-mus TDLo:1936 mg/kg/4W-I:ETA JCREA8 97,205,80

SAFETY PROFILE: Questionable carcinogen with experimental tumorigenic data. When heated to decomposition it emits toxic fumes of ammonia.

DJY100 CAS:60842-44-6 HR: 3
3-(2-(DIETHYLPYRROLIDINO)ETHOXY)-6-METHOXY-2-PHENYLBENZOFURAN HYDROCHLORIDE
mf: $C_{25}H_{31}NO_2 \cdot ClH$ mw: 414.03

SYNS: DBF ☐ 2-PHENYL-3-DIETHYLPYRROLIDINOETHOXY-6-METHOXYBENZOFURAN HYDROCHLORIDE

TOXICITY DATA WITH REFERENCE
orl-rat TDLo:20 mg/kg (8D preg):REP IJEBA6 5,80,67
ivg-rat TDLo:20 mg/kg (female 1D post):REP IJEBA6 12,370,74
ipr-rat LD50:388 mg/kg IJEBA6 5,80,67

SAFETY PROFILE: Poison by intraperitoneal route. Experimental reproductive effects. When heated to decomposition it emits toxic fumes of NO_x and HCl.

DJY200 CAS:13593-03-8 HR: 3
O,O-DIETHYL-O-2-QUINOXALYLTHIOPHOSPHATE
mf: $C_{12}H_{15}N_2O_3PS$ mw: 298.32

PROP: Crystals. Mp: 31–32°. Bp: 142° (decomp) @ 0.0003 mm.

SYNS: BAY 5821 ☐ BAY 77049 ☐ BAYRUSIL ☐ CHINALPHOS ☐ O,O-DIAETHYL-O-(CHINOXALYL-(2))-MONOTHIOPHOSPHAT (GERMAN) ☐ DIETHQUINALPHION ☐ DIETHQUINALPHIONE ☐ O,O-DIETHYL-O-(2-CHINOXALYL)PHOSPHOROTHIOATE ☐ O,O-DIETHYL-O-QUINOXALIN-2-YL PHOSPHOROTHIOATE ☐ O,O-DIETHYL-O-(2-QUINOXALINYL) PHOSPHOROTHIOATE ☐ O,O-DIETHYL-O-(2-QUINOXALYL) PHOSPHOROTHIOATE ☐ EKALUX ☐ ENT 27,394 ☐ NSC 190986 ☐ QUINALPHOS ☐ SAN 6538 I ☐ SANDOZ 6538 ☐ SPENCER S-6538 ☐ SRA 7312 ☐ WIE OBEN

TOXICITY DATA WITH REFERENCE
mnt-mus-orl 5 mg/kg FCTOD7 29,115,91
cyt-mus-orl 5 mg/kg FCTOD7 29,115,91
orl-gpg TDLo:30 mg/kg (female 15D post):REP BECTA6 24,739,80
ipr-rat TDLo:6500 µg/kg (male 26D pre):REP ANDRDQ 20,163,88
orl-rat LD50:26 mg/kg ARSIM* 20,21,66
ihl-rat LC50:175 mg/m³ 85DPAN -,-,71/76
skn-rat LD50:300 mg/kg BESAAT 15,124,69
orl-mus LD50:107 mg/kg 85JCAE -,1166,86
orl-dog LD50:100 mg/kg 85JCAE -,1166,86
par-ckn LD50:10,250 µg/kg IVEJAC 62,86,85

SAFETY PROFILE: Poison by ingestion, inhalation, skin

D

contact, parenteral, and intraperitoneal routes. Experimental reproductive effects. Mutation data reported. An insecticide. When heated to decomposition it emits very toxic fumes of NO_x, PO_x, and SO_x.

DJY400 **CAS:19311-91-2** **HR: 3**
N,N-DIETHYLSALICYLAMIDE
mf: $C_{11}H_{15}NO_2$ mw: 193.27

SYNS: N,N-DIETHYL-2-HYDROXYBENZAMIDE □ o-HYDROXY-N,N-DIETHYLBENZAMIDE □ SALICYLDIETHYLAMIDE

TOXICITY DATA WITH REFERENCE
orl-rat LD50:580 mg/kg JPETAB 108,450,53
ipr-rat LD50:350 mg/kg JPETAB 108,450,53
orl-mus LD50:850 mg/kg THERAP 8,237,53
ipr-mus LD50:200 mg/kg NTIS** AD691-490

SAFETY PROFILE: Poison by intraperitoneal route. Moderately toxic by ingestion. When heated to decomposition it emits toxic fumes of NO_x.

DJY600 **CAS:110-40-7** **HR: 1**
DIETHYL SEBACATE
mf: $C_{14}H_{26}O_4$ mw: 258.40

PROP: Colorless to sltly yellow liquid; faint fruity odor. D: 0.960–0.965, refr index: 1.435. Misc with alc, ether, other org sols, fixed oils; insol in water @ 302°.

SYNS: DIETHYL DECANEDIOATE □ DIETHYL-1,10-DECANEDIOATE □ ETHYL SEBACATE □ FEMA No. 2376 □ SEBACIC ACID, DIETHYL ESTER

TOXICITY DATA WITH REFERENCE
skn-rbt 500 mg/24H MLD FCTXAV 16,637,78
orl-rat LD50:14,470 mg/kg FCTXAV 2,327,64
orl-gpg LD50:7280 mg/kg FCTXAV 2,327,64

CONSENSUS REPORTS: Reported in EPA TSCA Inventory.

SAFETY PROFILE: Mildly toxic by ingestion. A skin irritant. See also ESTERS. When heated to decomposition it emits acrid smoke and irritating fumes.

DJY800 **CAS:5117-17-9** **HR: 3**
N,N-DIETHYLSELENOUREA
mf: $C_5H_{12}N_2Se$ mw: 179.15

SYNS: 1,1-DIETHYL-2-SELENOUREA □ USAF B-100

TOXICITY DATA WITH REFERENCE
ipr-mus LD50:10 mg/kg NTIS** AD277-689

CONSENSUS REPORTS: Reported in EPA TSCA Inventory. Community Right-To-Know List.

OSHA PEL: TWA 0.2 mg(Se)/m^3
ACGIH TLV: TWA 0.2 mg(Se)/m^3
DFG MAK: 0.1 mg(Se)/m^3

SAFETY PROFILE: Poison by intraperitoneal route. See also SELENIUM COMPOUNDS. When heated to decomposition it emits very toxic fumes of NO_x and Se.

DKA000 **CAS:40193-47-3** **HR: 2**
N,N-DIETHYL-4-STILBENAMINE
mf: $C_{18}H_{21}N$ mw: 251.40

SYNS: DIETHYLAMINO STILBENE □ 4-STILBENYL-N,N-DIETHYLAMINE

TOXICITY DATA WITH REFERENCE
scu-rat TDLo:160 mg/kg/9W-I:NEO XPHPAW 149,328,57
scu-rat TD:180 mg/kg/8W-I:ETA PTRMAD 241,147,48

SAFETY PROFILE: Questionable carcinogen with experimental neoplastigenic and tumorigenic data. When heated to decomposition it emits toxic fumes of NO_x.

DKA200 **CAS:522-40-7** **HR: 2**
α,α′-DIETHYL-(E)-4,4′-STILBENEDIOL BIS(DIHYDROGEN PHOSPHATE)
mf: $C_{18}H_{22}O_8P_2$ mw: 428.34

PROP: Crystals or powder. Mp: 249–252° (204–2°).

SYNS: DESdp □ 4,4′-(1,2-DIETHYL-1,2-ETHENEDIYL)BISPHENOL-(E)-BIS(DIHYDROGEN PHOSPHATE) □ DIETHYLSTILBESTEROL DIPHOSPHATE □ DIETHYLSTILBESTROL DIPHOSPHATE □ DIETHYLSTILBESTROL PHOSPHATE □ DIETHYLSTILBESTRYL DIPHOSPHATE □ FOSFESTROL □ HONVAN □ PHOSPHESTROL □ ST52-ASTA □ STILBESTROL DIPHOSPHATE □ STILPHOSTROL

TOXICITY DATA WITH REFERENCE
cyt-mus-ivn 10 mg/kg ENMUDM 1,184,79
dnd-ham:ovr 400 nmol/L ENMUDM 1,163,79
ipr-mus TDLo:400 mg/kg (6D preg):TER TJADAB 9,229,74
orl-rat TDLo:48 mg/kg (6D male):REP OYYAA2 20,1141,80
ivn-man TDLo:11 mg/kg/3D-I:BPR,SKN SMJOAV 75,248,82
ivn-mus LD50:630 mg/kg ARZNAD 18,666,68

CONSENSUS REPORTS: EPA Genetic Toxicology Program.

SAFETY PROFILE: Moderately toxic by intravenous route. An experimental teratogen. Other experimental reproductive effects. Human systemic effects by intravenous routes: blood pressure lowering, blood clotting factor change, sweating. Mutation data reported. When heated to decomposition it emits toxic fumes of PO_x. See also DIETHYLSTILBESTEROL.

DKA400 **CAS:63528-82-5** **HR: 2**
α,α′-DIETHYL-4,4′-STILBENEDIOL DISODIUM SALT
mf: $C_{18}H_{18}O_2{\cdot}2Na$ mw: 312.34

SYNS: DES DISODIUM SALT □ DIETHYLSTILBESTEROL DISODIUM SALT

TOXICITY DATA WITH REFERENCE
scu-mus TDLo:10 mg/kg (17D preg):REP CNREA8 37,1099,77
scu-mus TDLo:10 mg/kg/(15D preg):NEO,REP CNREA8 37,1099,77
scu-mus TDLo:10 mg/kg (female 15D post):NEO,TER CNREA8 37,1099,77

SAFETY PROFILE: Experimental teratogenic and reproductive effects. Questionable carcinogen with experimental neoplastigenic data. When heated to decomposition it emits toxic fumes of Na_2O. See also DIETHYLSTILBESTEROL.

DKA600 **CAS:56-53-1** **HR: 3**
DIETHYLSTILBESTEROL
mf: $C_{18}H_{20}O_2$ mw: 268.38

PROP: Small crystals or plates from EtOAc or C_6H_6. Mp: 171–172°.

SYNS: ACNESTROL □ AGOSTILBEN □ ANTIGESTIL □ BIO-DES □ 3,4-BIS(p-HYDROXYPHENYL) 3 HEXENE □ BUFON □ CLIMATERINE □ COMESTROL □ COMESTROL ESTROBENE □ CYREN □ DAWE'S DESTROL □ DEB □ DES (synthetic estrogen) □ DESMA □ DESTROL □ DIASTYL □ DIBESTROL □ DICORVIN □ DI-ESTRYL □ trans-4,4'-(1,2-DIETHYL-1,2-ETHENEDIYL)BISPHENOL □ 4,4'-(1,2-DIETHYL-1,2-ETHENEDIYL)BIS-PHENOL □ α,α'-DIETHYLSTILBENEDIOL □ α,α'-DIETHYL-(E)-4,4'-STILBENEDIOL □ α,α'-DIETHYL-4,4'-STILBENEDIOL □ trans-α,α'-DIETHYL-4,4'-STILBENEDIOL □ 2,2'-DIETHYL-4,4'-STILBENEDIOL □ trans-DIETHYLSTILBESTEROL □ DIETHYLSTILBESTROL □ trans-DIETHYLSTILBESTROL □ DIETHYLSTILBOESTEROL □ trans-DIETHYLSTILBOESTEROL □ DIETILESTILBESTROL (SPANISH) □ 4,4'-DIHYDROXYDIETHYLSTILBENE □ 4,4'-DIHYDROXY-α,β-DIETHYLSTILBENE □ 3,4'(4,4'-DIHYDROXYPHENYL)HEX-3-ENE □ DISTILBENE □ DOMESTROL □ DYESTROL □ ESTILBEN □ ESTRIL □ ESTROBENE □ ESTROGEN □ ESTROMENIN □ ESTROSYN □ FOLLIDIENE □ FONATOL □ GRAFESTROL □ GYNOPHARM □ HIBESTROL □ IDROESTRIL □ ISCOVESCO □ MAKAROL □ MENOSTILBEEN □ MICREST □ MICROEST □MILES-TROL □ NEO-OESTRANOL 1 □ NSC-3070 □ OEKOLP □ OESTROGENINE □ OESTROL VETAG □ OESTROMENIN □ OESTROMENSIL □ OESTROMENSYL □ OESTROMIENIN □ OESTROMON □ PABESTROL □ PALESTROL □ PERCUTATRINE OESTROGENIQUE ISCOVESCO □ PROTECTONA □ RCRA WASTE NUMBER U089 □ RUMESTROL 1 □ RUMESTROL 2 □ SEDESTRAN □ SERRAL □ SEXOCRETIN □ SIBOL □ SINTESTROL □ STIBILIUM □ STIL □ STILBESTROL □ STILBESTRONE □ STILBETIN □ STILBOEFRAL □ STILBOESTROFORM □ STILBOESTROL □ STILBOFOLLIN □ STILBOL □ STILKAP □ STILROL □ SYNESTRIN □ SYNTHOESTRIN □ SYNTHOFOLIN □ SYNTOFOLIN □ TAMPOVAGAN STILBOESTROL □ TYLOSTERONE □ VAGESTROL

TOXICITY DATA WITH REFERENCE
mma-esc 50 mg/L MUREAV 130,97,84
dnd-hmn:fbr 300 µmol/L ENMUDM 7,267,85
cyt-hmn:lym 100 µg/L PMRSDJ 5,457,85
orl-wmn TDLo:1730 mg/kg (1-39W preg):REP NEJMAG 292,334,75
unr-wmn TDLo:150 mg/kg (female 7-34W post):TER JOURAA 117,477,77
orl-wmn TDLo:5 mg/kg (5D pre):REP CCPTAY 14,375,76
orl-wmn TDLo:24 mg/kg (34W pre):REP ACEDAB 105,7,66
orl-wmn TDLo:2500 µg/kg (5D pre):REP FESTAS 24,95,73
orl-man TDLo:1065 mg/kg (30W male):REP ASUPAZ 73,199,68
orl-wmn TDLo:1730 mg/kg (1-39W preg):TER NEJMAG 292,334,75
orl-rat TDLo:100 µg/kg (female 15D post):REP JSTBBK 9,595,78
orl-mky TDLo:7200 µg/kg (female 18-23W post):REP JRPMAP 26,309,81
orl-rat TDLo:473 µg/kg (female 1-22D post):REP JONUAI 66,321,58
orl-uns TDLo:1038 µg/kg (female 30D pre):REP FEPRA7 15,575,56
scu-rat TDLo:1 mg/kg (female 19D post):REP CALEDQ 6,107,79
orl-rat TDLo:130 µg/kg (female 6-18D post):REP TJADAB 25,37,82
orl-rat TDLo:585 µg/kg (female 6-18D post):REP TJADAB 25,37,82
scu-rbt TDLo:6 µg/kg (female 1-3D post):TER FESTAS 20,211,69
orl-mky TDLo:1 mg/kg (female 1-6D post):REP AJOGAH 115,101,73
orl-ham TDLo:2250 µg/kg (female 1-9D post):REP CCPTAY 3,347,71
orl-rat TDLo:100 µg/kg (female 5D post):REP CCPTAY 14,487,76
scu-rat TDLo:250 µg/kg (female 1D pre):REP ACENA7 49,193,65
orl-rat TDLo:6300 ng/kg (female 7D pre):REP CCPTAY 3,347,71
orl-rat TDLo:100 µg/kg (male 10D pre):REP PSEBAA 120,725,65
par-rat TDLo:6 µg/kg (female 16-19D post):TER BEXBAN 82,1561,76
scu-mus TDLo:100 µg/kg (female 16D post):TER TXAPA9 47,279,79
scu-rat TDLo:3 mg/kg (female 19-21D post):TER BEXBAN 90,1597,80
orl-wmn TDLo:21 mg/kg (female 13-15W post):CAR,TER AJOGAH 137,220,80
orl-wmn TDLo:7655 µg/kg/4Y-C:CAR,REP BJOGAS 82,417,75
orl-wmn TDLo:7655 µg/kg/4Y-C:CAR BJOGAS 82,417,75
orl-wmn TDLo:21 mg/kg (female 13-15W post):CAR AJOGAH 137,220,80
unr-man TDLo:18,250 µg/kg/7Y-I:CAR JOURAA 112,160,74
unr-man TDLo:184 mg/kg/12Y-C:CAR JAMAAP 240,1510,78
mul-man TDLo:25 mg/kg/2Y-C:CAR UROTAQ 19,180,52
orl-rat TDLo:103 g/kg/2Y-C:CAR AEHLAU 19,489,69
scu-rat TDLo:6 µg/kg (female 15-18D post):NEO JTEHD6 5,1059,79
imp-rat LD :100 mg/kg/5W-C:NEO CNREA8 48,4158,88
imp-rat TDLo:11,500 µg/kg:CAR JJIND8 67,455,81
orl-mus TDLo:263 µg/kg/50W-C:CAR JJIND8 33,971,64
ipr-mus TDLo:1 mg/kg (female 17D post):CAR JJIND8 73,133,84
scu-mus TDLo:800 µg/kg (female 9-16D post):CAR CNREA8 45,5145,85
par-mus TDLo:340 mg/kg/34W-I:ETA CNREA8 2,759,42
imp-mus TDLo:8 mg/kg:ETA JJIND8 2,65,41
ivg-mus TDLo:5280 µg/kg/44W-I:ETA ANYAA9 75,543,59
scu-dog TDLo:17 mg/kg/29W-I:ETA AJEBAK 37,549,59
par-dog TDLo:12 mg/kg/23W-I:ETA AJEBAK 40,139,62
imp-mky TDLo:48 mg/kg:CAR LBASAE 23,493,73
orl-gpg TDLo:144 mg/kg/12W-I:ETA RSABAC 25,215,49
imp-gpg TDLo:1080 µg/kg:ETA BSBSAS 8,142,51
orl-ham TDLo:40 mg/kg (female 15D post):ETA CALEDQ 1,139,76
scu-ham TDLo:66 mg/kg/44W-I:CAR CALEDQ 31,181,86
scu-ham TDLo:816 mg/kg/1Y-I:NEO PSEBAA 66,195,47
imp-ham TDLo:160 mg/kg:CAR BJCAAI 8,451,54
imp-ham LD :160 mg/kg:CAR ZEKBAI 61,1,56

orl-man LD :69 mg/kg/5Y-C:NEO JOURAA 128,1044,82
scu-rat LD :1 mg/kg (female 19D post):NEO CALEDQ 6,107,79
imp-rat LD :25 mg/kg:CAR CNREA8 39,773,79
scu-mus LD :230 mg/kg/39W-I:NEO ZEKBAI 56,482,49
imp-ham LD :360 mg/kg/12W-I:CAR CNREA8 43,2678,83
imp-ham LD :864 mg/kg/12W-I:CAR CNREA8 43,4638,83
imp-rat LD :20 mg/kg:CAR CNREA8 43,4781,83
orl-mus LD :1075 µg/kg/4W-C:CAR FCTOD7 25,229,87
orl-man LD :45,990 µg/kg/3Y-C:CAR BJURAN 51,6,79
skn-hmn TDLo:60 µg/kg/14D AIHAAP 20,469,59
orl-rat LD50:>3 g/kg DRFUD4 8,413,83
ipr-rat LD50:34 mg/kg TXAPA9 24,142,73
orl-mus LD50:>3 g/kg DRFUD4 8,413,83
ipr-mus LD50:538 mg/kg DRFUD4 8,413,83
scu-mus LDLo:500 mg/kg KLWOAZ 18,156,39
ivn-mus LD50:300 mg/kg ARZNAD 18,666,68

CONSENSUS REPORTS: NTP 7th Annual Report on Carcinogens. IARC Cancer Review: Group 1 IMEMDT 7,273,87; Human Limited Evidence IMEMDT 6,55,74; IMEMDT 21,173,79; Animal Sufficient Evidence IMEMDT 21,173,79; IMEMDT 6,55,74. EPA Genetic Toxicology Program. Reported in EPA TSCA Inventory.

SAFETY PROFILE: Confirmed carcinogen producing skin, liver, and lung tumors in exposed humans as well as uterine and other reproductive system tumors in the female offspring of exposed women. Experimental carcinogenic, neoplastigenic, tumorigenic, and teratogenic data. A transplacental carcinogen. A human teratogen by many routes. Poison by intraperitoneal and subcutaneous routes. It causes glandular system effects by skin contact. Human reproductive effects by ingestion: abnormal spermatogenesis; changes in testes, epididymis, and sperm duct; menstrual cycle changes or disorders; changes in female fertility; unspecified maternal effects; developmental abnormalities of the fetal urogenital system; germ cell effects in offspring; and delayed effects in newborn. Implicated in male impotence and enlargement of male breasts. Other experimental reproductive effects. Mutation data reported. When heated to decomposition it emits acrid smoke and fumes. See also ETHINYL ESTRADIOL.

DKA800 CAS:63019-08-9 HR: 2
DIETHYLSTILBESTROL DIPALMITATE
mf: $C_{50}H_{80}O_4$ mw: 745.30

SYNS: α,α'-DIETHYL-4,4'-STILBENEDIOL DIPALMITATE □ 4,4'-DIHYDROXY-α,β-DIETHYLSTILBENE PALMITATE

TOXICITY DATA WITH REFERENCE
scu-rbt TDLo:535 µg/kg (5D male):REP ENDOAO 60,519,57
ims-rbt TDLo:38 mg/kg/45W-I:ETA,TER CANCAR 10,500,57

SAFETY PROFILE: Experimental teratogenic and reproductive effects. Questionable carcinogen with experimental tumorigenic data. When heated to decomposition it emits acrid smoke and irritating fumes. See also DIETHYLSTILBESTEROL.

DKB000 CAS:130-80-3 HR: 3
DIETHYLSTILBESTROL DIPROPIONATE
mf: $C_{24}H_{28}O_4$ mw: 380.52

PROP: Crystals or plates from MeOH. Mp: 104°.

SYNS: CLINESTROL □ CYREN B □ DESD □ DIBESTIL □ trans-4,4'-(1,2-DIETHYL-1,2-ETHENEDIYL)BISPHENOL DIPROPIONATE □ α,α'-DIETHYL-4,4'-STILBENEDIOL, DIPROPIONATE □ α,α'-DIETHYL-4,4'-STILBENEDIOL trans-DIPROPIONATE □ trans-α,α'-DIETHYL-4,4'-STILBENEDIOL DIPROPIONATE □ α,α'-DIETHYL-4,4'-STILBENEDIOL DIPROPIONYL ESTER □ DIETHYLSTILBENE DIPROPIONATE □ DIETHYLSTILBESTEROL DIPROPIONATE □ DIETHYLSTILBESTROL PROPIONATE □ DIHYDROXYDIETHYLSTILBENE DIPROPIONATE □ 4,4'-DIHYDROXY-α,β-DIETHYLSTILBENE DIPROPIONATE □ DIPROPIONATO de ESTILBENE (SPANISH) □ p,p'-DIPROPIONOXY-trans-α,β-DIETHYLSTILBENE □ DISTILBENE □ ESTILBEN □ ESTILBIN □ ESTROBEN □ ESTROBENE □ ESTROGENIN □ ESTROSTILBEN □ EUVESTIN □ GYNOLETT □ HORFEMINE □ NEO-OESTRANOL II □ OESTROGYNAEDRON □ ORESTOL □ PABESTROL □ SINCICLAN □ STILBESTROL DIETHYL DIPROPIONATE □ STILBESTROL DIPROPIONATE □ STILBESTROL PROPIONATE □ STILBESTRONATE □ STILBOESTROL DIPROPIONATE □ STILBOFAX □ STILRONATE □ SYNESTRIN □ SYNOESTRON □ SYNTESTRIN □ SYNTESTRINE □ WILLESTROL

TOXICITY DATA WITH REFERENCE
cyt-hmn:lym 1300 nmol/L TGANAK 16(2),24,82
sln-mus-ipr 50 mg/kg MUREAV 144,27,85
cyt-ham:fbr 5 mg/L CRNGDP 3,499,82
sce-ham:fbr 5 mg/L CRNGDP 3,499,82
orl-rat TDLo:60 µg/kg (3-4D preg):REP APTSAI 26(Suppl 1),1,68
scu-rat TDLo:70 µg/kg (female 28D pre):REP APTOA6 26(1),1,68
scu-ctl TDLo:200 ng/kg (female 1D pre):REP JRPFA4 18,359,69
ipr-rat TDLo:500 µg/kg (female 5D pre):REP IJMRAQ 60,287,72
unr-mus TDLo:200 µg/kg (female 13D post):TER CRSBAW 164,2372,70
scu-rat TDLo:80 mg/kg:ETA HMMRA2 7,228,75
imp-rat TDLo:113 mg/kg:ETA EJCAAH 5,99,69
imp-gpg TDLo:5 mg/kg:ETA,REP RSABAC 25,215,49
scu-ham TDLo:160 mg/kg:ETA BECCAN 30,60,52
scu-frg TDLo:133 mg/kg/32W-I:NEO BEXBAN 81,898,76

CONSENSUS REPORTS: IARC Cancer Review: Animal Sufficient Evidence IMEMDT 21,173,79. EPA Genetic Toxicology Program.

SAFETY PROFILE: Confirmed carcinogen with experimental tumorigenic data. An experimental teratogen. Other experimental reproductive effects. Human mutagenic data. When heated to decomposition it emits acrid smoke and irritating fumes. See also DIETHYLSTILBESTEROL.

DKB100 CAS:6052-82-0 HR: 2
DIETHYLSTILBOESTROL-3,4-OXIDE
mf: $C_{18}H_{20}O_3$ mw: 284.38

SYNS: DES-α,β-OXIDE □ DES-3,4-OXIDE □ α,α'-DIETHYL-α,α'-EPOXYBIBENZYL-4,4'-DIOL □ DIETHYLSTILBOESTROL-α,β-OXIDE

TOXICITY DATA WITH REFERENCE
dnd-hmn:leu 300 µmol/L BCPCA6 34,3251,85

sce-hmn:fbr 40 nmol/L NATUAS 281,392,79
imp-ham TDLo:640 mg/kg/38W-I:ETA CNREA8 43,5200,83

SAFETY PROFILE: Questionable carcinogen with experimental tumorigenic data. Human mutation data reported. When heated to decomposition it emits acrid smoke and fumes. See also DIETHYLSTILBESTEROL.

DKB110 CAS:64-67-5 HR: 3
DIETHYL SULFATE
DOT: UN 1594
mf: $C_4H_{10}O_4S$ mw: 154.20

PROP: Colorless, oily liquid; faint ethereal odor. Mp: −25°, bp: 209.5° (decomp to ethyl ether), flash p: 220°F (CC), d: 1.18 @ 18°/0°, autoign temp: 817°F, vap press: 1 mm @ 47.0°, vap d: 5.31. Insol in water; decomp by hot water; misc with alc and ether. Insol in water.

SYNS: DIAETHYLSULFAT (GERMAN) □ DIETHYLESTER KYSELINY SIROVE □ DIETHYL ESTER SULFURIC ACID □ DIETHYL TETRAOXOSULFATE □ DS □ ETHYL SULFATE

TOXICITY DATA WITH REFERENCE
skn-rbt 10 mg/24H SEV JIHTAB 31,60,49
skn-rbt 500 mg open MLD UCDS** 12/30/71
eye-rbt 2 mg SEV JIHTAB 31,60,49
mmo-sat 5 mg/plate MUREAV 57,141,78
mma-sat 5 mg/plate MUREAV 57,141,78
slt-mus-ipr 100 mg/kg/10W MUREAV 75,63,80
cyt-mus-ipr 150 mg/kg MUREAV 75,63,80
msc-ham:ovr 1 mmol/L MUREAV 57,217,78
ivn-rat TDLo:340 mg/kg (female 15D post):TER IARCCD 4,45,73
orl-rat TDLo:3700 mg/kg/81W-I:ETA ZEKBAI 74,241,70
scu-rat TDLo:800 mg/kg/32W-I:ETA ZEKBAI 74,241,70
ivn-rat TDLo:85 mg/kg (15D preg):CAR IARCCD 4,45,73
orl-rat LD50:880 mg/kg JIDHAN 31,60,49
ihl-rat LCLo:250 ppm/4H JIHTAB 31,343,49
scu-rat LD50:350 mg/kg ZEKBAI 74,241,70
orl-mus LD50:647 mg/kg NTIS** PB214-270
skn-rbt LD50:600 mg/kg UCDS** 12/30/71

CONSENSUS REPORTS: NTP 7th Annual Report on Carcinogens. IARC Cancer Review: Group 2A IMEMDT 7,198,87; Animal Sufficient Evidence IMEMDT 4,277,74. EPA Genetic Toxicology Program. Community Right-To-Know List. Reported in EPA TSCA Inventory.

DFG TRK:0.03 ppm; Animal Carcinogen, Human Suspected Carcinogen.
DOT CLASSIFICATION: 6.1; *Label:* Poison

SAFETY PROFILE: Confirmed with experimental carcinogenic and tumorigenic data. Poison by inhalation and subcutaneous routes. Moderately toxic by ingestion and skin contact. A severe skin irritant. An experimental teratogen. Mutation data reported. Combustible when exposed to heat or flame; can react with oxidizing materials. Moisture causes liberation of H_2SO_4. Violent reaction with potassium tert-butoxide. Reacts violently with 3,8-dinitro-6-phenylphenanthridine + water. Reaction with iron + water forms explosive hydrogen gas. To fight fire, use alcohol foam, H_2O foam, CO_2, dry chemicals. When heated to decomposition it emits toxic fumes of SO_x. See also SULFATES.

DKB119 CAS:623-81-4 HR: 3
DIETHYL SULFITE
mf: $C_4H_{10}O_3S$ mw: 138.18

PROP: D: 1.105 @ 0°/0, bp: 158°.

SYNS: DIETHYL SULFITE □ ETHYL SULFITE □ TL 158

TOXICITY DATA WITH REFERENCE
ihl-mus LC:>2700 mg/m³/10M NDRC** NDCrc-132,Apr42

CONSENSUS REPORTS: Reported in EPA TSCA Inventory.

SAFETY PROFILE: Low toxicity by inhalation. Reaction with chlorine fluoride gives the violently explosive ethyl fluorosulfate. When heated to decomposition it emits toxic fumes of SO_x. See also SULFITES.

DKB150 CAS:627-54-3 HR: 3
DIETHYL TELLURIDE
mf: $C_4H_{10}Te$ mw: 185.72

SYNS: DIETHYL TELLURIDE □ DIETHYLTELLURIUM □ ETHYL TELLURIDE (8CI) □ 1,1′-TELLUROBISETHANE

TOXICITY DATA WITH REFERENCE
orl-rat LD50:54,100 µg/kg GTPZAB 25(10),51,81
ihl-rat LC50:24 mg/m³ GTPZAB 25(10),51,81
orl-mus LD50:154 mg/kg GTPZAB 25(10),51,81
ihl-mus LC50:117 mg/m³ GTPZAB 25(10),51,81

SAFETY PROFILE: Poison by ingestion and inhalation. Ignites spontaneously in air. See also TELLURIUM COMPOUNDS.

DKB160 CAS:636-09-9 HR: 2
DIETHYL TEREPHTHALATE
mf: $C_{12}H_{14}O_4$ mw: 222.26

SYNS: 1,4-BENZENEDICARBOXYLIC ACID, DIETHYL ESTER □ TEREPHTHALIC ACID, DIETHYL ESTER

TOXICITY DATA WITH REFERENCE
ipr-mus LDLo:1111 mg/kg JPMSAE 56,1446,67

CONSENSUS REPORTS: Reported in EPA TSCA Inventory.

SAFETY PROFILE: Moderately toxic by intraperitoneal route. When heated to decomposition it emits acrid smoke and irritating vapors.

DKB165 CAS:41239-48-9 HR: 2
2,5-DIETHYLTETRAHYDROFURAN
mf: $C_8H_{16}O$ mw: 128.24

SYN: FURAN, 2,5-DIETHYLTETRAHYDRO-

TOXICITY DATA WITH REFERENCE
orl-rat LD50:3400 mg/kg JACTDZ 1,5,90

CONSENSUS REPORTS: Reported in EPA TSCA Inventory.

SAFETY PROFILE: Moderately toxic by ingestion. When heated to decomposition it emits acrid smoke and irritating vapors.

DKB175 CAS:22392-07-0 HR: 3
DIETHYL THALLIUM PERCHLORATE
mf: $C_4H_{10}ClO_4Tl$ mw: 361.96

CONSENSUS REPORTS: Thallium and its compounds are on the Community Right-To-Know List.

SAFETY PROFILE: A poison. Explodes at its mp 250°C. When heated to decomposition it emits toxic fumes of Cl^-. See also THALLIUM COMPOUNDS and PERCHLORATES.

DKB600 CAS:5827-03-2 HR: 3
N,N-DIETHYLTHIOCARBAMYL-O,O-DIISOPROPYLDITHIOPHOSPHATE
mf: $C_{11}H_{24}NO_2PS_3$ mw: 329.51

SYNS: DIETHYLDITHIOCARBAMIC ANHYDRIDE of O,O-DIISOPROPYL THIONOPHOSPHORIC ACID □ DIETHYLDITHIOCARBAMIC ANHYDROSULFIDE □ O,O-DIISOPROPYL-S-DIETHYLDITHIOCARBAMOYLPHOSPHORODITHIOATE □ O,O-DIISOPROPYL DITHIOPHOSPHORIC ACID ESTER of-N,N-S-DIETHYLTHIOCARBAMOYL-O,O-DIISOPROPYL PHOSPHOROTHIOATE □ DIISOPROPYL ESTER of DITHIOCARBAMYL PHOSPHOROTHIOIC ACID □ ENT 24,725

TOXICITY DATA WITH REFERENCE
orl-rat LD60:320 mg/kg 28ZEAL 4,175,69
orl-mus LD50:290 mg/kg ARSIM* 20,13,66

SAFETY PROFILE: Poison by ingestion. See also CARBAMATES and ANHYDRIDES. When heated to decomposition it emits very toxic fumes of PO_x, NO_x, and SO_x.

DKC200 CAS:69226-06-8 HR: 3
2,2-DIETHYL-3-THIOMORPHOLINONE
mf: $C_8H_{15}NOS$ mw: 173.30

TOXICITY DATA WITH REFERENCE
orl-mus LD50:2250 mg/kg JMCMAR 6,136,63
ipr-mus LD50:652 mg/kg JMCMAR 6,136,63
ivn-mus LD50:137 mg/kg JMCMAR 6,136,63
orl-rbt LD50:1500 mg/kg JMCMAR 6,136,63
ivn-rbt LD50:185 mg/kg JMCMAR 6,136,63

SAFETY PROFILE: Poison by intravenous route. Moderately toxic by ingestion and intraperitoneal routes. When heated to decomposition it emits very toxic fumes of NO_x and SO_x.

DKC400 CAS:105-55-5 HR: 3
1,3-DIETHYLTHIOUREA
mf: $C_5H_{12}N_2S$ mw: 132.25

PROP: Crystals. Mp: 77°. Sol in H_2O and EtOH.

SYNS: N,N′-DIETHYLTHIOCARBAMIDE □ N,N′-DIETHYLTHIOUREA □ 1,3-DIETHYL-2-THIOUREA □ NCI-C03816 □ PENNZONE E □ THIATE H □ U 15030 □ USAF EK-1803

TOXICITY DATA WITH REFERENCE
msc-mus:lym 1500 mg/L EMMUEG 12,85,88
orl-rat TDLo:11 g/kg/2Y-C:CAR NCITR* NCI-CG-TR-149,79
orl-rat LD50:316 mg/kg NCILB* NCI-E-C-72-3252,73
orl-mus LDLo:62 mg/kg AECTCV 14,111,85
ipr-mus LD50:500 mg/kg NTIS** AD277-689

CONSENSUS REPORTS: NCI Carcinogenesis Bioassay (feed); Clear Evidence: rat NCITR* NCI-CG-TR-149,79; No Evidence: mouse NCITR* NCI-CG-TR-149,79. Reported in EPA TSCA Inventory. EPA Genetic Toxicology Program.

SAFETY PROFILE: Poison by ingestion. Moderately toxic by intraperitoneal route. Questionable carcinogen with experimental carcinogenic data. Mutation data reported. When heated to decomposition it emits very toxic fumes of NO_x and SO_x.

DKC600 CAS:73940-85-9 HR: 3
DIETHYLTIN DI(10-CAMPHORSULFONATE)
mf: $C_{24}H_{40}O_8S_2Sn$ mw: 639.45

SYNS: BIS(10-CAMPHORSULFONATO)DIETHYLSTANNANE □ BIS(2-OXO-9-BORNANESULFONIC ACID) DIETHYLSTANNYL ESTER

TOXICITY DATA WITH REFERENCE
ivn-mus LD50:89 mg/kg CSLNX* NX#03150

OSHA PEL: TWA 0.1 mg(Sn)/m³ (skin)
ACGIH TLV: TWA 0.1 mg(Sn)/m³ (skin) (Proposed: TWA 0.1 mg(Sn)/m³; STEL 0.2 mg(Sn)/m³ (skin))
NIOSH REL: (Organotin Compounds) TWA 0.1 mg(Sn)/m³

SAFETY PROFILE: Poison by intravenous route. See also TIN COMPOUNDS; SULFONATES; and CAMPHOR. When heated to decomposition it emits toxic fumes of SO_x.

For occupational chemical analysis use NIOSH: Organotin Compounds 5504.

DKC800 CAS:134-62-3 HR: 3
DIETHYL-m-TOLUAMIDE
mf: $C_{12}H_{17}NO$ mw: 191.30

PROP: A liquid, sol in water, alc, and ether. Bp: 160° @ 19 mm, d: 0.996 @ 20°/4°.

SYNS: AI 3-22542 □ AUTAN □ BAKER'S ANTIFOL □ CHEMFORM □ DEET □ DELPHENE □ m-DELPHENE □ DET □ m-DET □ m-DETA □ DETAMIDE □ DIELTAMID □ N,N-DIETHYL-3-METHYLBENZAMIDE □ DIETHYLTOLUAMIDE □ N,N-DIETHYL-m-TOLUAMIDE □ ENT 20,218 □ ENT 22,542 □ FLYPEL □ METADELPHENE □ 3-METHYL-N,N-DIETHYLBENZAMIDE □ MGK DIETHYLTOLUAMIDE □ NAUGATUCK DET □ OFF □ REPEL □ REPPER-DET □ REPUDIN-SPECIAL □ m-TOLUIC ACID DIETHYLAMIDE

TOXICITY DATA WITH REFERENCE
skn-rbt 500 mg MOD NTIS** AD-A106-944
eye-rbt 100 mg NTIS** AD-A107-736
spm-rat-ihl 1500 mg/m³ AEHA** 75-51-0034-80
skn-rat TDLo:19 g/kg (1-19D preg):REP FATOAO 38,202,75
orl-wmn LDLo:950 mg/kg JAMAAP 258,1509,87
orl-wmn LDLo:950 mg/kg:BAH,PUL JAMAAP 258,1509,87
orl-man LDLo:679 mg/kg JAMAAP 258,1509,87
orl-cld TDLo:4750 mg/kg:BAH JAMAAP 258,1509,87
orl-wmn TDLo:950 mg/kg:EYE,BAH,PUL JAMAAP 258,1509,87
skn-hmn TDLo:35 mg/kg/5D:SKIN TXAPA9 1,97,59
orl-rat LD50:1950 mg/kg SPEADM 78-1,53,78
ihl-rat LC50:5950 mg/m³ AEHA** 75-51-0034-80

skn-rat LD50:5000 mg/kg MPPBAB 45,65,76
skn-mus LD50:3170 mg/kg YKYUA6 31,309,80
orl-rbt LD50:1584 mg/kg NTIS** AD-A082-131
skn-rbt LD50:3180 mg/kg TXAPA9 28,313,74
ivn-rbt LDLo:75 mg/kg TXAPA9 1,97,59

CONSENSUS REPORTS: Reported in EPA TSCA Inventory.

SAFETY PROFILE: Poison by intravenous route. Moderately toxic by ingestion and skin contact. Human systemic effects: coma, convulsions, dermatitis, mydriasis (pupillary dilation), nausea or vomiting, stiffness. An eye and skin irritant. Experimental reproductive effects by skin contact. Mutation data reported. Can cause central nervous system disturbances. A pesticide. DEET is the active ingredient in most commercial insect repellents. When heated to decomposition it emits toxic fumes of NO_x.

DKD000 CAS:2728-04-3 HR: 2
N,N-DIETHYL-o-TOLUAMIDE
mf: $C_{12}H_{17}NO$ mw: 191.30

SYN: 2-METHYL-N,N-DIETHYLBENZAMIDE

TOXICITY DATA WITH REFERENCE
skn-hmn 2500 µg/5D MLD TXAPA9 1,97,59
skn-rbt 2 mg/kg TXAPA9 1,97,59
orl-rat LD50:1210 mg/kg TXAPA9 7,772,65

SAFETY PROFILE: Moderately toxic by ingestion. A human skin irritant. When heated to decomposition it emits toxic fumes of NO_x.

DKD200 CAS:63980-20-1 HR: 1
DIETHYL TRIAZENE
mf: $C_4H_{11}N_3$ mw: 101.18

PROP: A liquid. Bp: 84° @ 145 mm.

SYNS: DIETHYL-TRIAZENE □ N,N'-DIETHYLTRIAZENE □ 1,3-DIETHYLTRIAZENE □ 1,3-DIETHYL-1-TRIAZENE □ 1,3-DIETHYLTRIAZINE □ 1-TRIAZENE, 1,3-DIETHYL-(9CI)

TOXICITY DATA WITH REFERENCE
mmo-esc 5 µmol/L CRNGDP 12,1161,91
orl-rat TDLo:400 mg/kg/20W-I:CAR CALEDQ 35,129,87
scu-rat TDLo:110 mg/kg (15D post):NEO,TER XENOBH 3,271,73
scu-rat LD50:220 mg/kg 85JCAE-,498,86

SAFETY PROFILE: Poison by subcutaneous route. Questionable carcinogen with experimental carcinogenic and tumorigenic data. An experimental teratogen. Mutation data reported. When heated to decomposition it emits toxic fumes of NO_x.

DKD500 CAS:757-44-8 HR: 1
DIETHYL (2-(TRIETHOXYSILYL)ETHYL)PHOSPHONIC ACID
mf: $C_{12}H_{29}O_6PSi$ mw: 328.47

SYNS: 2-(DIETHOXYPHOSPHINYL)ETHYLTRIETHOXYSILANE □ PHOSPHONIC ACID, (2-(TRIETHOXYSILYL)ETHYL)-, DIETHYL ESTER

TOXICITY DATA WITH REFERENCE
orl-rat LD50:17,200 mg/kg TXAPA9 28,313,74

CONSENSUS REPORTS: Reported in EPA TSCA Inventory.

SAFETY PROFILE: Mildly toxic by ingestion. When heated to decomposition it emits toxic vapors of PO_x.

DKD600 CAS:3218-37-9 HR: 3
N,N-DIETHYL-4-(α-(α,α,α-TRIFLUORO-o-TOLYL) BENZYLOXY)PENTYLAMINE CITRATE
mf: $C_{23}H_{30}F_3NO•C_6H_8O_7$ mw: 585.68

TOXICITY DATA WITH REFERENCE
orl-mus LDLo:300 mg/kg ARZNAD 14,964,64
ivn-mus LDLo:20 mg/kg ARZNAD 14,964,64

SAFETY PROFILE: Poison by ingestion and intravenous routes. See also FLUORIDES. When heated to decomposition it emits very toxic fumes of F^- and NO_x.

DKD650 CAS:634-95-7 HR: 2
1,1-DIETHYLUREA
mf: $C_5H_{12}N_2O$ mw: 116.19

SYNS: asym-DIETHYLUREA □ N,N-DIETHYLUREA □ UREA, 1,1-DIETHYL- □ UREA, N,N-DIETHYL-(9CI)

TOXICITY DATA WITH REFERENCE
ipr-mus LDLo:2905 mg/kg JPETAB 54,188,35

CONSENSUS REPORTS: Reported in EPA TSCA Inventory.

SAFETY PROFILE: Moderately toxic by intraperitoneal route. When heated to decomposition it emits toxic vapors of NO_x.

DKE200 CAS:304-84-7 HR: 3
N,N-DIETHYLVANILLAMIDE
mf: $C_{12}H_{17}NO_3$ mw: 223.30

PROP: Needles pet ether. Mp: 95–96°.

SYNS: DIETHYLAMIDE de VANILLIQUE □ 3-METHOXY-4-HYDROXYBENZOIC ACID DIETHYLAMIDE □ VANILLIC ACID DIETHYLAMIDE □ VANILLIC ACID-N,N-DIETHYLAMIDE □ VANILLINSAEURE-DIAETHYLAMID (GERMAN)

TOXICITY DATA WITH REFERENCE
ipr-rat LD50:28 mg/kg COREAF 243,609,56
orl-mus LD50:67 mg/kg RPOBAR 1,423,64
ipr-mus LD50:38 mg/kg RPOBAR 1,423,64
scu-mus LD50:57 mg/kg RPOBAR 1,423,64
ivn-mus LD50:15 mg/kg 27ZQAG -,377,72
orl-dog LD50:300 mg/kg 27ZQAG -,377,72
ivn-dog LD50:30 mg/kg 27ZQAG -,377,72
orl-rbt LDLo:50 mg/kg ARZNAD 20,367,70

SAFETY PROFILE: Poison by ingestion, intravenous, subcutaneous, and intraperitoneal routes. When heated to decomposition it emits toxic fumes of NO_x.

DKE400 **CAS:1851-77-0** **HR: 3**
DI(ETHYLXANTHOGEN)TRISULFIDE
mf: $C_6H_{10}O_2S_5$ mw: 274.46

SYNS: BEXT □ BIS(ETHOXYTHIOCARBONYL)TRISULFIDE □ BIS (ETHYLXANTHOGEN) TRISULFIDE □ DEFOLIANT 713 □ DI-ETHOXYTHIOKARBONYL-TRISULFID □ TRISULFIDE, BIS(ETHOXYTHIOCARBONYL)-

TOXICITY DATA WITH REFERENCE
orl-rat LD50:235 mg/kg 28ZEAL 5,69,76
orl-mus LD50:140 mg/kg 28ZEAL 5,69,76
ihl-mus LC50:300 mg/m³/3H 85JCAE-,1018,86

SAFETY PROFILE: Poison by ingestion. Moderately toxic by inhalation. When heated to decomposition it emits toxic fumes of SO_x.

DKE600 **CAS:557-20-0** **HR: 3**
DIETHYLZINC
DOT: UN 1366
mf: $C_4H_{10}Zn$ mw: 123.51

PROP: Liquid. Mp: −28°, bp: 118°, d: 1.187 @ 18°.

SYNS: ZINC ETHIDE □ ZINC ETHYL (DOT)

CONSENSUS REPORTS: Reported in EPA TSCA Inventory. Zinc and its compounds are on the Community Right-To-Know List.

DOT CLASSIFICATION: 4.2; *Label:* Spontaneously Combustible

SAFETY PROFILE: Presumed to be a poison. Ignites spontaneously in air. Dangerously flammable by spontaneous chemical reaction in air, or with oxidizing materials. A dangerous explosion hazard. Explosive reaction with alkenes + diiodomethane, sulfur dioxide. Reacts violently with bromine, water, nitro compounds. Ignites on contact with air, ozone, methanol, or hydrazine. Reacts violently with nonmetal halides (e.g., arsenic trichloride or phosphorus trichloride to produce pyrophoric triethyl arsine or triethyl phosphine. To fight fire, do not use water, foam, or halogenated extinguishing agents. Use dry materials, such as graphite, sand, etc. When heated to decomposition it emits toxic fumes of ZnO. See also ZINC COMPOUNDS.

DKE800 **CAS:63868-62-2** **HR: 3**
DIETROL
mf: $C_{12}H_{17}NO{\cdot}2C_4H_6O_6$ mw: 491.50

SYNS: ADPHEN □ BACARATE □ 3,4-DIMETHYL-2-PHENYLMORPHOLINE BITARTRATE □ HOURBESE □ LIMIT □ MINUS □ NEO-NILOREX □ OBEPAR □ PHENAZINE □ PHENDIMETRAZINE BITARTRATE □ PLEGINE □ REDUCTO □ STATOBEX □ SYMETRA □ TRIMSTAT □ TRIMTABS

TOXICITY DATA WITH REFERENCE
orl-mus LD50:285 mg/kg 27ZQAG -,285,72
ipr-mus LD50:410 mg/kg 27ZQAG -,285,72
ivn-mus LD50:170 mg/kg 27ZQAG -,285,72

SAFETY PROFILE: Poison by ingestion, intraperitoneal, and intravenous routes. When heated to decomposition it emits toxic fumes of NO_x.

DKF130 **CAS:59198-70-8** **HR: 3**
DIFLUCORTOLONE VALERATE
mf: $C_{27}H_{36}F_2O_5$ mw: 478.63

PROP: Crystals. Mp: 195–195.5°.

SYNS: DFV □ DIFLUCORTOLONE 21-VALERATE □ DIFLUCORTOLONVALERIANAT (GERMAN) □ 6-α,9-DIFLUORO-11-β-HYDROXY-16-α-METHYL-21-VALERYLOXY-1,4-PREGNADIENE-3,20-DIONE □ 6-α,2-DIFLUORO-11-β-HYDROXY-21-VALERYLOXY-16-α-METHYL-1,4-PREGNADIENE-3,20-DIONE □ NERISONA □ NERISONE □ TEMETEX

TOXICITY DATA WITH REFERENCE
scu-rat TDLo:700 μg/kg (female 7-13D post):REP YKRYAH 10,1357,77
scu-mus TDLo:700 μg/kg (female 7-13D post):REP YKRYAH 11,237,78
scu-mus TDLo:7 μg/kg (female 7-13D post):REP YKRYAH 11,237,78
scu-rat TDLo:700 μg/kg (female 7-13D post):TER YKRYAH 10,1357,77
skn-rat TDLo:25 g/kg (female 6-15D post):TER ARZNAD 26,1476,76
skn-rbt TDLo:650 mg/kg (female 6-18D post):REP ARZNAD 26,1476,76
skn-rat TDLo:35 g/kg (female 35D pre):REP IYKEDH 9,64,78
skn-rat TDLo:10,500 mg/kg (female 35D pre):REP IYKEDH 9,64,78
skn-rat TDLo:10,500 mg/kg (male 35D pre):REP IYKEDH 9,64,78
skn-rat TDLo:5 g/kg (6-15D preg):TER ARZNAD 26,634,76
skn-rat TDLo:25 g/kg (female 6-15D post):TER ARZNAD 26,1476,76
scu-rat TDLo:400 μg/kg (female 16-19D post):TER YKRYAH 10,1357,77
orl-rat LD50:3100 mg/kg ARZNAD 26,1476,76
ipr-rat LD50:19,500 μg/kg IYKEDH 8,165,77
scu-rat LD50:13 mg/kg ARZNAD 26,634,76
orl-mus LD50:4750 mg/kg IYKEDH 11,811,80
ipr-mus LD50:432 mg/kg ARZNAD 26,634,76
scu-mus LD50:140 mg/kg IYKEDH 11,811,80
orl-dog LD50:1 g/kg ARZNAD 26,1476,76
scu-rbt TDLo:26,300 μg/kg NIIRDN 6,339,82

SAFETY PROFILE: Poison by subcutaneous route. Moderately toxic by ingestion and intraperitoneal routes. An experimental teratogen. Experimental reproductive effects. When heated to decomposition it emits toxic fumes of F^-.

DKF170 **CAS:404-42-2** **HR: 3**
2,4′-DIFLUOROACETANILIDE
mf: $C_8H_7F_2NO$ mw: 171.16

TOXICITY DATA WITH REFERENCE
orl-rat LD50:2 mg/kg JSFAAE 8,400,57
ipr-rat LD50:5 mg/kg JSFAAE 8,400,57
orl-mus LD50:25 mg/kg YKKZAJ 88,1620,68

SAFETY PROFILE: Poison by ingestion and intraperitoneal routes. When heated to decomposition it emits toxic fumes of F^- and NO_x.

DKF200 CAS:381-73-7 HR: 3
DIFLUOROACETIC ACID
mf: $C_2H_2F_2O_2$ mw: 96.04

PROP: A liquid. Fp −0.35°, bp: 134°.

TOXICITY DATA WITH REFERENCE
ivn-mus LD50:180 mg/kg CSLNX* NX#03015

CONSENSUS REPORTS: Reported in EPA TSCA Inventory.

SAFETY PROFILE: Poison by intravenous route. Explosive reaction with fluorine + cesium fluoride. When heated to decomposition it emits toxic fumes of F^-. See also FLUORIDES.

DKF400 CAS:10405-27-3 HR: 3
DIFLUOROAMINE
mf: F_2HN mw: 53.01

PROP: Colorless gas, oxidizes HI to I_2. Decomp in rubber or plastic tubing and by many metals. Readily oxidized and reduced. Mp: −116°, bp: −23°.

SYN: FLUORIMIDE

SAFETY PROFILE: A shock-sensitive explosive. Tends to explode in the solid state or upon freezing. Explosive reaction with cesium fluoride + Benzenediazonium tetrafluoroborate. Violent reaction with ClO_2. When heated to decomposition it emits toxic fumes of F^- and NO_x. See also FLUORIDES and AMINES.

DKF600 CAS:17224-08-7 HR: 3
3-DIFLUOROAMINO-1,2,3-TRIFLUORODIAZIRIDINE
mf: CF_5N_3 mw: 149.02

SAFETY PROFILE: A strong oxidant. An explosive sensitive to shock and phase change. When heated to decomposition it emits toxic fumes of F^- and NO_x. See also FLUORIDES.

DKF620 CAS:56533-30-3 HR: 3
DIFLUOROAMMONIUM HEXAFLUOROARSENATE
mf: AsF_8H_2N mw: 242.93

CONSENSUS REPORTS: Arsenic and its compounds are on the Community Right-To-Know List.

SAFETY PROFILE: Arsenic compounds are generally poisons. A storage hazard; decomposes at room temperature after a delay period. When heated to decomposition it emits toxic fumes of F^-, NO_x, NH_3, and As. See also ARSENIC COMPOUNDS.

DKF800 CAS:372-18-9 HR: 3
m-DIFLUOROBENZENE
mf: $C_6H_4F_2$ mw: 114.09

PROP: A solid. Mp: −59°, bp: 83°, flash p: <32°C, d: 1.155 @ 18°/4°.

SAFETY PROFILE: A very dangerous fire hazard when exposed to heat or flame; can react vigorously with oxidizing materials. When heated to decomposition it emits toxic fumes of F^-.

DKG000 CAS:540-36-3 HR: 3
p-DIFLUOROBENZENE
mf: $C_6H_4F_2$ mw: 114.09

PROP: Mp: −13°, bp: 89°, flash p: 23°F, d: 1.168 @ 18°/4°.

SAFETY PROFILE: A very dangerous fire hazard when exposed to heat or flame; can react vigorously with oxidizing materials. When heated to decomposition it emits toxic fumes of F^-.

D

DKG100 CAS:368-68-3 HR: 3
3,4-DIFLUOROBENZENEARSONIC ACID
mf: $C_6H_5AsF_2O_3$ mw: 238.03

SYN: BENZENEARSONIC ACID, 3,4-DIFLUORO-

TOXICITY DATA WITH REFERENCE
ipr-mus LDLo:31,300 μg/kg CBCCT* 6,53,54

OSHA PEL: TWA 0.5 mg(As)/m³

SAFETY PROFILE: Poison by intraperitoneal route. When heated to decomposition it emits toxic fumes of As and F^-.

DKG400 CAS:61735-78-2 HR: 2
2,10-DIFLUOROBENZO(rst)PENTAPHENE
mf: $C_{24}H_{12}F_2$ mw: 338.36

SYN: 2,10-DIFLUORODIBENZO(a,i)PYRENE

TOXICITY DATA WITH REFERENCE
mma-sat 12,500 pmol/plate CNREA8 41,2589,81
scu-mus TDLo:20 mg/kg:ETA PAACA3 13,37,72

SAFETY PROFILE: Questionable carcinogen with experimental tumorigenic data. Mutation data reported. When heated to decomposition it emits toxic fumes of F^-.

DKG600 CAS:10578-16-2 HR: 3
DIFLUORODIAZENE
mf: F_2N_2 mw: 66.01

FN=NF

PROP: Colorless gas. Bp: −105.7°.

SAFETY PROFILE: Very unstable explosive. Explodes when frozen or condensed from gas. Explosive reaction with hydrogen above 90°C. When heated to decomposition it emits toxic fumes of F^- and NO_x.

DKG700 CAS:693-85-6 HR: 3
DIFLUORODIAZIRINE
mf: CF_2N_2 mw: 78.02

PROP: Gas. Bp: −91.3°.

SAFETY PROFILE: An explosive sensitive to heat or exposure to light. When heated to decomposition it emits toxic fumes of F^- and NO_x.

DKG850 **CAS:75-61-6** **HR: 1**
DIFLUORODIBROMOMETHANE
DOT: UN 1941
mf: CBr_2F_2 mw: 209.83

PROP: Colorless, heavy liquid. Fp: −141°, bp: 23° @ 24.5 mm, d: 2.288 @ 15°/4°, mp: −110° (−1°). Insol in water.

SYNS: DIBROMODIFLUOROMETHANE □ FREON 12-B2 □ HALON 1202 □ R12B2 (DOT)

TOXICITY DATA WITH REFERENCE
ihl-mus LCLo:67 g/m³/15M MRLR** #107,52

CONSENSUS REPORTS: Reported in EPA TSCA Inventory.

OSHA PEL: TWA 100 ppm
ACGIH TLV: TWA 100 ppm
DFG MAK: 100 ppm (860 mg/m³)
DOT CLASSIFICATION: 9; *Label:* None

SAFETY PROFILE: Mildly toxic by inhalation. A nonflammable liquid. When heated to decomposition it emits very toxic fumes of Br^- and F^-.

For occupational chemical analysis use NIOSH: see Dibromodifluoromethane, 1012.

DKG980 **CAS:351-63-3** **HR: 2**
2′,4′-DIFLUORO-4-DIMETHYLAMINOAZOBENZENE
mf: $C_{14}H_{13}F_2N_3$ mw: 261.30

TOXICITY DATA WITH REFERENCE
orl-rat TDLo:2727 mg/kg/14W-I:ETA CNREA8 18,469,58
orl-rat TD:3400 mg/kg/12W-C:ETA CNREA8 13,93,53

SAFETY PROFILE: Questionable carcinogen with experimental tumorigenic data. When heated to decomposition it emits toxic fumes of F^- and NO_x.

DKH000 **CAS:349-37-1** **HR: 2**
2′,5′-DIFLUORO-4-DIMETHYLAMINOAZOBENZENE
mf: $C_{14}H_{13}F_2N_3$ mw: 261.30

SYN: N,N-DIMETHYL-p-(2,5-DIFLUOROPHENYLAZO)ANILINE

TOXICITY DATA WITH REFERENCE
orl-rat TDLo:3400 mg/kg/12W-C:ETA CNREA8 13,93,53

SAFETY PROFILE: Questionable carcinogen with experimental tumorigenic data. When heated to decomposition it emits very toxic fumes of F^- and NO_x.

DKH100 **CAS:350-87-8** **HR: 2**
3′,5′-DIFLUORO-4-DIMETHYLAMINOAZOBENZENE
mf: $C_{14}H_{13}F_2N_3$ mw: 261.30

SYNS: ANILINE, N,N-DIMETHYL-p-(3,5-DIFLUOROPHENYLAZO)- □ p-((3,5-DIFLUOROPHENYL)AZO)-N,N-DIMETHYLANILINE □ N,N-DIMETHYL-p-(3,5-DIFLUOROPHENYLAZO)ANILINE

TOXICITY DATA WITH REFERENCE
orl-rat TDLo:3400 mg/kg/12W-C:ETA CNREA8 13,93,53

SAFETY PROFILE: Questionable carcinogen with experimental tumorigenic data. When heated to decomposition it emits toxic fumes of F^- and NO_x.

DKH200 **CAS:3582-17-0** **HR: 3**
DIFLUORODIMETHYLSTANNANE
mf: $C_2H_6F_2Sn$ mw: 186.77

PROP: White crystals. Bp: decomp <360°. Sol in water.

SYNS: DIMETHYLTIN DIFLUORIDE □ DIMETHYLTIN FLUORIDE

TOXICITY DATA WITH REFERENCE
ivn-mus LD50:56 mg/kg CSLNX* NX#00014

CONSENSUS REPORTS: Reported in EPA TSCA Inventory.

OSHA PEL: TWA 0.1 mg(Sn)/m³ (skin)
ACGIH TLV: TWA 0.1 mg(Sn)/m³ (skin) (Proposed: TWA 0.1 mg(Sn)/m³; STEL 0.2 mg(Sn)/m³ (skin))
NIOSH REL: (Organotin Compounds) TWA 0.1 mg(Sn)/m³

SAFETY PROFILE: Poison by intravenous route. See also TIN COMPOUNDS and FLUORIDES. When heated to decomposition it emits toxic fumes of F^-.

For occupational chemical analysis use NIOSH: Organotin Compounds 5504.

DKH250 **CAS:312-30-1** **HR: 3**
4,4′-DIFLUORO-3,3-DINITRODIPHENYL SULFONE
mf: $C_{12}H_6F_2N_2O_6S$ mw: 344.26

SYNS: BENZENE, 1,1′-SULFONYLBIS(4-FLUORO-3-NITRO)- □ BIS(4-FLUORO-3-NITROPHENYL)SULFONE □ p,p′-DIFLUORO-m,m′-DINITRODIPHENYL SULFONE □ 3,3′-DINITRO-4,4′-DIFLUORODIPHENYL SULFONE □ NDS □ SULFONE, BIS(4-FLUORO-3-NITROPHENYL)

TOXICITY DATA WITH REFERENCE
ivn-mus LD50:100 mg/kg CSLNX* NX#03495

CONSENSUS REPORTS: Reported in EPA TSCA Inventory.

SAFETY PROFILE: Poison by intravenous route. When heated to decomposition it emits toxic vapors of NO_x, SO_x, and F^-.

DKH825 **CAS:338-66-9** **HR: 3**
DIFLUORO-N-FLUOROMETHANIMINE
mf: CF_3N mw: 83.01

PROP: Gas with odor resembling HCN. Bp: −60°.

SAFETY PROFILE: Explodes when exposed to flame. When heated to decomposition it emits toxic fumes of F^- and NO_x.

DKH830 **CAS:16282-67-0** **HR: 3**
DIFLUOROMETHYLENE DIHYPOFLUORITE
mf: CF_4O_2 mw: 120.00

PROP: Gas or pale yellow liquid; colorless liquid at −1°; glassy material (very viscous) below −1°, bp: −62°, d: 1.20 @ 25°/4°.

SAFETY PROFILE: A strong oxidant. Potentially hazardous reactions with organic or easily oxidized materials. Potentially explosive reactions with trans-dichloroethylene, tetrafluoroethylene, and possibly other haloalkanes. When heated to decomposition it emits toxic fumes of F^-.

DKH875 **CAS:67037-37-0** **HR: 3**
dl-α-DIFLUOROMETHYLORNITHINE
mf: $C_6H_{12}F_2N_2O_2$ mw: 182.20

SYNS: α-DFMO □ EFLORNITHINE □ RMI 71782

TOXICITY DATA WITH REFERENCE
msc-hmn:lym 1250 μmol/L CNREA8 44,4272,84
cyt-ham:ovr 500 μmol/L CNREA8 44,4535,84
scu-ham TDLo:150 mg/kg (female 7D post):TER CCPTAY 28,159,83
orl-mus TDLo:3952 mg/kg (female 5-8D post):TER SCIEAS 208,505,80
orl-mus TDLo:3952 mg/kg (female 5-8D post):REP SCIEAS 208,505,80
scu-ham TDLo:300 mg/kg (female 7D post):REP CCPTAY 28,159,83
orl-rat TDLo:10 g/kg (5-9D preg):REP SCIEAS 208,505,80
ice-rat LD50:1364 μg/kg APTOA6 56,250,85

SAFETY PROFILE: Poison by intracerebral route. An experimental teratogen. Experimental reproductive effects. Human mutation data reported. When heated to decomposition it emits toxic fumes of F^- and NO_x. An ornithine-decarboxylase inhibitor.

DKH900 **CAS:446-35-5** **HR: 3**
2,4-DIFLUORONITROBENZENE
mf: $C_6H_3F_2NO_2$ mw: 159.10

SYN: BENZENE, 2,4-DIFLUORO-1-NITRO-

TOXICITY DATA WITH REFERENCE
mmo-sat 80 nL/plate MUREAV 116,217,83
orl-rat LD50:200 mg/kg JACTDZ 1,185,92

CONSENSUS REPORTS: Reported in EPA TSCA Inventory.

SAFETY PROFILE: A poison by ingestion. Mutation data reported. When heated to decomposition it emits toxic vapors of NO_x and F^-.

DKI400 **CAS:368-97-8** **HR: 3**
DIFLUOROPHENYLARSINE
mf: $C_6H_5AsF_2$ mw: 190.03

SYN: PHENYLDIFLUOROARSINE

TOXICITY DATA WITH REFERENCE
skn-rat LD50:15 mg/kg JPBAA7 58,411,46
skn-rbt LD50:4 mg/kg JPBAA7 58,411,46
ivn-rbt LD50:500 μg/kg JPBAA7 58,411,46
skn-gpg LD50:10 mg/kg JPBAA7 58,411,46

CONSENSUS REPORTS: Arsenic and its compounds are on the Community Right-To-Know List.

OSHA PEL: TWA 0.5 mg(As)/m^3

SAFETY PROFILE: Poison by skin contact and intravenous routes. See also FLUORIDES and ARSENIC COMPOUNDS. When heated to decomposition it emits very toxic fumes of As and F^-.

DKI600 **CAS:22494-42-4** **HR: 3**
5-(2,4-DIFLUOROPHENYL)SALICYLIC ACID
mf: $C_{13}H_8F_2O_3$ mw: 250.21

PROP: Crystals from C_6H_6/pet ether. Mp: 210–211°.

SYNS: DIFLUNISAL □ 2′,4′-DIFLUORO-4-HYDROXY-(1,1′-BIPHENYL)-3-CARBOXYLIC ACID □ 2′,4′-DIFLUORO-4-HYDROXY-3-BIPHENYLCARBOXYLIC ACID □ 2′,4′-DIFLUORO-4-HYDROXY-(1′,1-DIPHENYL)-3-CARBOXYLIC ACID □ DOLOBID □ DOLOBIL □ DOLOBIS □ FLOVACIL □ FLUNIGET □ 2-(HYDROXY)-5-(2,4-DIFLUOROPHENYL)BENZOIC ACID □ MK 647

TOXICITY DATA WITH REFERENCE
orl-rbt TDLo:520 mg/kg (female 6-18D post):TER TJADAB 30,319,84
orl-rbt TDLo:780 mg/kg (female 6-18D post):REP TJADAB 30,319,84
orl-hmn TDLo:71 mg/kg/5D-I:CNS,LIV BMJOAE 2,736,78
orl-man TDLo:429 mg/kg/17W-I:BLD ANZJB8 16,811,86
orl-rat LD50:392 mg/kg IYKEDH 15,688,84
ipr-rat LD50:159 mg/kg IYKEDH 15,688,84
scu-rat LD50:185 mg/kg IYKEDH 15,688,84
orl-mus LD50:439 mg/kg BCPHBM 4,19S,77
ipr-mus LD50:124 mg/kg IYKEDH 15,688,84
scu-mus LD50:220 mg/kg IYKEDH 15,688,84
orl-rbt LD50:603 mg/kg BCPHBM 4,19S,77

SAFETY PROFILE: Poison by ingestion, subcutaneous, and intraperitoneal routes. Human systemic effects by ingestion: tolerance, and cholestatic jaundice (due to the stoppage of the flow of bile), agranulocytosis, increased body temperature. An experimental teratogen. Other experimental reproductive effects. An analgesic and anti-inflammatory agent. When heated to decomposition it emits toxic fumes of F^-. See also FLUORIDES.

DKI800 **CAS:453-13-4** **HR: 3**
1,3-DIFLUORO-2-PROPANOL
mf: $C_3H_6F_2O$ mw: 96.09

PROP: A liquid. D: 1.24 @ 20°/4°, bp: 54–55° @ 34 mm.

TOXICITY DATA WITH REFERENCE
ivn-mus LD50:178 mg/kg CSLNX* NX#00467

CONSENSUS REPORTS: Reported in EPA TSCA Inventory.

SAFETY PROFILE: Poison by intravenous route. When heated to decomposition it emits toxic fumes of F^-.

DKJ200 **CAS:314-04-5** **HR: 2**
3,8-DIFLUOROTRICYCLOQUINAZOLINE
mf: $C_{21}H_{10}F_2N_4$ mw: 356.35

TOXICITY DATA WITH REFERENCE
skn-mus TDLo:1200 mg/kg/50W-I:NEO BCPCA6 14,323,65

SAFETY PROFILE: Questionable carcinogen with experimental neoplastigenic data. See also FLUORIDES. When heated to decomposition it emits very toxic fumes of F^- and NO_x.

D

DKJ225 **CAS:1510-31-2** **HR: 3**
1,1-DIFLUOROUREA
mf: $CH_2F_2N_2O$ mw: 96.04

PROP: Non-hygroscopic needles from CH_2Cl_2/CCl_4; hygroscopic platelets by sublimation. Mp: 41–41.5°.

SAFETY PROFILE: Concentrated aqueous solutions decompose above −20°C and evolve the explosive gases tetrafluorohydrazine and difluoramine. When heated to decomposition it emits toxic fumes of F^- and NO_x.

DKJ300 **CAS:23674-86-4** **HR: 3**
DIFLUPREDNATE
mf: $C_{27}H_{34}F_2O_7$ mw: 508.61

PROP: Crystals from methylene chloride/ether/petr ether. Mp: 191–194°.

SYNS: 6-α,9-α-DIFLUOROPREDNISOLONE 17-BUTYRATE 21-ACETATE □ 6-α,9-DIFLUORO-11-β,17,21-TRIHYDROXYPREGNA-1,4-DIENE-3,20-DIONE-21-ACETATE-17-BUTYRATE □ W 6309 □ EPITOPIC

TOXICITY DATA WITH REFERENCE
scu-rat TDLo:1100 ng/kg (female 7-17D post):REP IYKEDH 15,1035,84
scu-rat TDLo:1500 ng/kg (female 17-22D post):REP IYKEDH 15,1046,84
scu-rbt TDLo:1300 ng/kg (female 6-18D post):TER IYKEDH 15,1055,84
scu-rat TDLo:1100 μg/kg (female 7-17D post):TER IYKEDH 15,1035,84
scu-rat TDLo:840 μg/kg (9W male/2W pre-7D preg):TER IYKEDH 15,1028,84
skn-dog TDLo:21 mg/kg (female 24W pre):REP JTSCDR 10,107,85
scu-rbt TDLo:130 μg/kg (female 6-18D post):TER IYKEDH 15,1055,84
ipr-rat LD50:1 g/kg IYKEDH 15,1066,84
scu-rat LD50:2315 mg/kg IYKEDH 15,1066,84
ipr-mus LD50:146 mg/kg IYKEDH 15,1066,84
scu-mus LD50:105 mg/kg IYKEDH 15,1066,84

SAFETY PROFILE: Poison by subcutaneous and intraperitoneal routes. An experimental teratogen. Other experimental reproductive effects. When heated to decomposition it emits toxic fumes of F^-.

DKJ600 **CAS:628-36-4** **HR: 2**
1,2-DIFORMYLHYDRAZINE
mf: $C_2H_4N_2O_2$ mw: 88.08

PROP: Prisms. Mp: 162°. Sol in H_2O; sltly sol in EtOH.

SYNS: 1,2-DIFORMYLHYDRAZIN (GERMAN) □ HYDRAZODIFORMIC ACID

TOXICITY DATA WITH REFERENCE
orl-mus TDLo:5800 mg/kg/66W-C:CAR ZKKOBW 92,11,78

CONSENSUS REPORTS: Reported in EPA TSCA Inventory.

SAFETY PROFILE: Questionable carcinogen with experimental carcinogenic data. When heated to decomposition it emits toxic fumes of NO_x. See also HYDRAZINE.

DKK200 **CAS:19247-68-8** **HR: 3**
N,N-DIFURFURAL-n-PHENYLENEDIAMINE
mf: $C_{14}H_{12}N_2O_2$ mw: 240.28

TOXICITY DATA WITH REFERENCE
orl-rat LD50:1220 mg/kg 85GMAT -,54,82
ihl-rat LCLo:70 mg/m³/4H 85GMAT -,54,82
orl-mus LD50:400 mg/kg 85GMAT -,54,82

SAFETY PROFILE: Poison by inhalation and ingestion. When heated to decomposition it emits toxic fumes of NO_x.

DKK800 **CAS:63906-88-7** **HR: 3**
DIGAMMACAINE
mf: $C_{21}H_{26}N_2O\cdot ClH$ mw: 358.95

SYNS: 1-BENZAMIDO-1-PHENYL-3-PIPERIDINOPROPANE HYDROCHLORIDE □ N-(3-BENZAMIDO-3-PHENYL)PROPYL PIPERIDINE HYDROCHLORIDE

TOXICITY DATA WITH REFERENCE
scu-rat LDLo:112 mg/kg CLDND*
ivn-rat LDLo:16 mg/kg CLDND*
orl-mus LDLo:99 mg/kg CLDND*
scu-mus LDLo:105 mg/kg CLDND*
ivn-mus LD50:17 mg/kg CLDND*
ims-mus LDLo:96 mg/kg CLDND*

SAFETY PROFILE: Poison by ingestion, intramuscular, subcutaneous, and intravenous routes. When heated to decomposition it emits very toxic fumes of HCl and NO_x.

DKL200 **CAS:8031-42-3** **HR: 3**
DIGITALIS

PROP: Dried whole leaf of *Digitalis purpurea* (27ZTAP 3,54,69). Composition: digitoxin (0.2–0.4%), etc.

SYNS: DIGITALIS PURPUREA, LEAF □ DIGITANNOID □ FOXGLOVE

TOXICITY DATA WITH REFERENCE
ivn-man TDLo:12 μg/kg JAMAAP 241,164,79
unr-man LDLo:29 mg/kg 85DCAI 2,73,70
ipr-rat LDLo:50 mg/kg TXAPA9 1,156,59
ivn-mus LDLo:200 mg/kg AEXPBL 82,1,17
orl-cat LD50:244 mg/kg JAPMA8 44,607,55
ivn-cat LDLo:62 mg/kg JAPMA8 44,607,55
ipr-pgn LD50:220 mg/kg JPPMAB 14,96,62
ivn-pgn LDLo:670 mg/kg JPPMAB 14,96,62

SAFETY PROFILE: A deadly human poison by intravenous route. An experimental poison by ingestion, intravenous, and intraperitoneal routes. 2.5 g or 30 cc of the tincture is a toxic dose. An overdose can be fatal. It has been implicated in aplastic anemia. It contains digitalin, digitalein, digitonin, and digitoxin (the most toxic component). See also individual components.

DKL300 CAS:90028-00-5 HR: 3
DIGITALIS LANATA STANDARD

TOXICITY DATA WITH REFERENCE
orl-cat LD50:103 mg/kg JAPMA8 44,607,55
ivn-cat LDLo:29,900 μg/kg JAPMA8 44,607,55
ivn-gpg LDLo:81,200 μg/kg ARZNAD 17,1237,67
ivn-pgn LDLo:300 μg/kg JPHAA3 25,611,36

SAFETY PROFILE: Poison by ingestion and intravenous routes. See also DIGITALIS.

DKL400 CAS:11024-24-1 HR: 3
DIGITONIN
mf: $C_{56}H_{92}O_{29}$ mw: 1229.48

PROP: Crystals from EtOH. Mp: 235–240°.

SYN: DIGITIN

TOXICITY DATA WITH REFERENCE
sln-smc 250 ppb ANYAA9 407,186,83
ivn-rat LD50:4 mg/kg ARZNAD 12,815,62
orl-mus LDLo:90 mg/kg HBAMAK 4,1289,35
ipr-mus LD10:20 mg/kg PCJOAU 11,749,77
scu-mus LDLo:200 mg/kg HBAMAK 4,1289,35
ivn-mus LDLo:10 mg/kg HBAMAK 4,1289,35

CONSENSUS REPORTS: Reported in EPA TSCA Inventory.

SAFETY PROFILE: Poison by ingestion, intravenous, intraperitoneal, and subcutaneous routes. Mutation data reported. See also DIGITALIS. When heated to decomposition it emits acrid smoke and irritating fumes.

DKL800 CAS:71-63-6 HR: 3
DIGITOXIN
mf: $C_{41}H_{64}O_{13}$ mw: 765.05

PROP: A solid. Mp: 256–257°.

SYNS: ACEDOXIN ☐ ASTHENTHILO ☐ CARDIDIGIN ☐ CARDIGIN ☐ CARDITOXIN ☐ CRISTAPURAT ☐ CRYSTALLINE DIGITALIN ☐ CRYSTODIGIN ☐ DIGILONG ☐ DIGIMED ☐ DIGIMERCK ☐ DIGISIDIN ☐ DIGITALIN ☐ DIGITALINE (FRENCH) ☐ DIGITALINE CRISTALLISEE ☐ DIGITALINE NATIVELLE ☐ DIGITALINUM VERUM ☐ DIGITOPHYLLIN ☐ DIGITOXIGENIN-TRIDIGITOXOSID (GERMAN) ☐ DIGITOXIGENIN TRIDIGITOXOSIDE ☐ DITAVEN ☐ GLUCODIGIN ☐ LANATOXIN ☐ MONO-GLYCOCOARD ☐ MYODIGIN ☐ PURODIGIN ☐ PURPURID ☐ TARDIGAL ☐ TRI-DIGITOXOSIDE (GERMAN) ☐ UNIDIGIN

TOXICITY DATA WITH REFERENCE
eye-dog 1%/1D AJOPAA 56,889,63
orl-wmn TDLo:200 μg/kg (27-30W preg):REP AJCDAG 6,834,60
orl-wmn TDLo:400 μg/kg:CVS,GIT CTOXAO 18,679,81
orl-man LDLo:286 μg/kg:PNS,CVS CTOXAO 18,679,81
orl-inf TDLo:150 μg/kg:CVS,GIT AJDCAI 130,425,76
orl-wmn TDLo:300 μg/kg:CVS AJEMEN 2,504,84
orl-man TDLo:71 μg/kg:CVS AJEMEN 2,504,84
unk-man LDLo:44 μg/kg 85DCAI 2,73,70
orl-rat LD50:56 mg/kg TXAPA9 18,185,71
ivr-rat LD50:3900 μg/kg ARZNAD 11,848,61
orl-mus LD50:4950 μg/kg ARZNAD 8,557,58
scu-mus LD50:22,180 μg/kg AIPTAK 153,436,65
ivn-mus LD50:4100 μg/kg ARZNAD 8,557,58
ivn-dog LDLo:500 μg/kg HBAMAK 4,1343,35
orl-cat LD50:180 μg/kg AIPTAK 159,1,66
ipr-cat LD50:170 μg/kg AIPTAK 155,165,65
ivn-cat LDLo:180 μg/kg AEPPAE 184,181,37
idu-cat LDLo:373 μg/kg ARZNAD 19,687,69
iat-cat LDLo:440 μg/kg AEPPAE 141,329,29
ivn-rbt LDLo:1 mg/kg HBAMAK 4,1343,35
ivn-pig LDLo:400 μg/kg ARZNAD 20,229,70
idu-pig LDLo:550 μg/kg ARZNAD 20,229,70

CONSENSUS REPORTS: Reported in EPA TSCA Inventory. EPA Extremely Hazardous Substances List.

SAFETY PROFILE: A deadly poison by most routes. Human systemic effects: arrhythmias, cardiomyopathy, EKG changes, nausea or vomiting, paresthesia, pulse rate increase, thrombocytopenia. Human reproductive effects by ingestion: reduced viability of newborn. An eye irritant. When heated to decomposition it emits acrid smoke and irritating fumes. See also DIGITALIS.

DKL875 CAS:1339-93-1 HR: 3
DIGITOXOSIDE
mf: $C_{41}H_{64}O_{17}$ mw: 765.05

SYN: MONO-DIGITOXID (GERMAN)

TOXICITY DATA WITH REFERENCE
ivn-cat LDLo:270 μg/kg AEPPAE 237,222,59
ivn-gpg LDLo:310 μg/kg AEPPAE 237,222,59
ivn-pgn LDLo:400 μg/kg APFRAD 9,730,51

SAFETY PROFILE: A deadly poison by intravenous route. When heated to decomposition it emits acrid smoke and fumes. See also DIGITALIS.

DKM120 CAS:2095-06-9 HR: 2
N-N-DIGLYCIDYLANILINE
mf: $C_{12}H_{15}NO_2$ mw: 205.28

SYNS: BIS(2,3-EPOXYPROPYL)ANILINE ☐ N,N-BIS(2,3-EPOXYPROPYL)ANILINE ☐ BIS(EPOXYPROPYL)PHENYLAMINE ☐ N,N-DIGLYCIDYLANILIN (CZECH) ☐ N-N-DIGLYCIDYLPHENYLAMINE ☐ N-(OXIRANYLMETHYL)-N-PHENYL-OXIRANEMETHANAMINE (9CI)

TOXICITY DATA WITH REFERENCE
skn-rbt 500 mg/24H SEV 28ZPAK -,136,72
eye-rbt 500 mg/24H MLD 28ZPAK -,136,72
orl-rat LD50:1620 mg/kg AIHAAP 30,470,69
skn-rbt LD50:3560 mg/kg AIHAAP 30,470,69

CONSENSUS REPORTS: Reported in EPA TSCA Inventory.

SAFETY PROFILE: Moderately toxic by ingestion and skin contact. An eye and severe skin irritant. When heated to decomposition it emits toxic fumes of NO_x. See also ANILINE DYES.

DKM130 CAS:15336-81-9 HR: 2
N,N′-DIGLYCIDYL-5,5-DIMETHYLHYDANTOIN
mf: $C_{11}H_{16}N_2O_4$ mw: 240.29

SYNS: 1,3-BIS(2,3-EPOXYPROPYL)-5,5-DIMETHYLHYDANTOIN ☐ 5,5-DIMETHYL-1,3-BIS(OXIRANYLMETHYL)-2,4-IMIDAZOLIDINEDIONE ☐ 5,5-DIMETHYL-1,3-BIS(OXIRANYLMETHYL)-2,4-IMIDAZOLIDINEDIONE (9CI) ☐ XB 2793

D

TOXICITY DATA WITH REFERENCE
skn-mus TDLo:3513 mg/kg/93W-I:ETA NTIS** ORNL-5762
skn-mus TD:3792 mg/kg/2Y-I:ETA NTIS** ORNL-5762
skn-mus TD:5263 mg/kg/84W-I:ETA NTIS** ORNL-5762
skn-mus TD:6903 mg/kg/92W-I:ETA NTIS** ORNL-5762
skn-mus TD:12 g/kg/77W-I:ETA NTIS** ORNL-5762
skn-mus TD:13 g/kg/85W-I:ETA NTIS** ORNL-5762

CONSENSUS REPORTS: Reported in EPA TSCA Inventory.

SAFETY PROFILE: Questionable carcinogen with experimental tumorigenic data by skin contact. When heated to decomposition it emits toxic fumes of NO_x.

DKM200 CAS:2238-07-5 HR: 3
DIGLYCIDYL ETHER
mf: $C_6H_{10}O_3$ mw: 130.16

PROP: Liquid. D: 1.126 @ 25°/4°, bp: 98–99° @ 11 mm.

SYNS: BIS(2,3-EPOXYPROPYL)ETHER □ DGE □ DI(2,3-EPOXYPROPYL) ETHER

TOXICITY DATA WITH REFERENCE
skn-rbt 563 mg/3D SEV AMIHAB 14,250,56
eye-rbt 113 mg SEV AMIHAB 14,250,56
mmo-sat 50 µg/plate MUREAV 66,367,79
mma-sat 50 µg/plate MUREAV 66,367,79
unr-mus TDLo:1300 mg/kg:ETA RARSAM 3,193,63
orl-rat LD50:450 mg/kg AMIHAB 14,250,56
orl-mus LD50:170 mg/kg AMIHAB 14,250,56
ihl-mus LC50:30 ppm/4H AMIHAB 14,250,56
ivn-mus LD50:100 mg/kg CSLNX* NX#00108
skn-rbt LD50:1500 mg/kg AMIHAB 14,250,56
ivn-rbt LDLo:200 mg/kg AEHLAU 2,31,61

CONSENSUS REPORTS: EPA Extremely Hazardous Substances List. Reported in EPA TSCA Inventory. EPA Genetic Toxicology Program.

OSHA PEL: TWA 0.1 ppm
ACGIH TLV: TWA 0.1 ppm
DFG MAK: 0.1 ppm (0.6 mg/m³); Suspected Carcinogen
NIOSH REL: (Glycidyl Ethers) CL 1 mg/m³/15M

SAFETY PROFILE: Suspected carcinogen with experimental tumorigenic data. Poison by ingestion, inhalation, and intravenous routes. Moderately toxic by skin contact. A severe eye and skin irritant. Mutation data reported. Chronic exposure can cause bone marrow depression. When heated to decomposition it emits acrid smoke and fumes. See also ETHERS.

DKM400 CAS:63041-01-0 HR: 2
DIGLYCIDYL ETHER of N,N-BIS(2-HYDROXYPROPYL)-tert-BUTYLAMINE
mf: $C_{16}H_{31}NO_4$ mw: 301.48

SYN: 2,2'-BIS(2,3-EPOXYPROPOXY)-N-tert-BUTYLDIPROPYLAMINE

TOXICITY DATA WITH REFERENCE
scu-mus TDLo:3600 mg/kg/9W-I:ETA FCTXAV 4,365,66

SAFETY PROFILE: Questionable carcinogen with experimental tumorigenic data. See also ETHERS. When heated to decomposition it emits toxic fumes of NO_x.

DKM500 CAS:5493-45-8 HR: D
DIGLYCIDYL HEXAHYDROPHTHALATE
mf: $C_{14}H_{20}O_6$ mw: 284.34

SYNS: 1,2-CYCLOHEXANEDICARBOXYLIC ACID, BIS(OXIRANYLMETHYL)ESTER (9CI) □ 1,2-CYCLOHEXANEDICARBOXYLIC ACID, BIS(2,3-EPOXYPROPYL) ESTER □ DIGLYCIDYLESTER KYSELINY HEXAHYDROFTALOVE (CZECH) □ LEKUTHERM 2159 □ LEKUTHERM X 100 □ PHTHALIC ACID, HEXAHYDRO-, BIS(2,3-EPOXYPROPYL) ESTER □ PHTHALIC ACID, HEXAHYDRO-, DIGLYCIDYL ESTER

TOXICITY DATA WITH REFERENCE
mmo-sat 333 µg/plate MUREAV 172,105,86

CONSENSUS REPORTS: Reported in EPA TSCA Inventory.

SAFETY PROFILE: Mutation data reported. When heated to decomposition it emits acrid smoke and irritating vapors.

DKM800 CAS:63040-98-2 HR: 2
N,N-DIGLYCIDYL-p-TOLUENESULPHONAMIDE
mf: $C_{13}H_{17}NO_4S$ mw: 283.37

SYNS: N,N-BIS(2,3-EPOXYPROPYL)-p-TOLUENESULFONAMIDE □ N,N-DIGLYCIDYL-p-TOLUENESULFONAMIDE

TOXICITY DATA WITH REFERENCE
scu-mus TDLo:17 g/kg/43W-I:ETA FCTXAV 4,365,66

SAFETY PROFILE: Questionable carcinogen with experimental tumorigenic data. When heated to decomposition it emits very toxic fumes of NO_x and SO_x.

DKN000 CAS:628-89-7 HR: 2
DIGLYCOL CHLORHYDRIN
mf: $C_4H_9ClO_2$ mw: 124.58

PROP: Colorless liquid. Bp: 75–76° @ 5 mm, flash p: 225°F (OC), d: 1.18, 20°/4°, vap press: 0.17 mm @ 20°.

SYN: 2-(2-CHLOROETHOXY)ETHANOL

TOXICITY DATA WITH REFERENCE
skn-rbt 10 mg/24H open JIHTAB 26,269,44
eye-rbt 5 mg SEV AJOPAA 29,1363,46
orl-rat LD50:6300 mg/kg JIHTAB 26,269,44
skn-gpg LD50:3000 mg/kg JIHTAB 26,269,44

CONSENSUS REPORTS: Reported in EPA TSCA Inventory.

SAFETY PROFILE: Moderately toxic by skin contact. Mildly toxic by ingestion. A skin and severe eye irritant. Combustible when exposed to heat or flame; can react with oxidizing materials. To fight fire, use alcohol foam, CO_2, dry chemical. Will react with water or steam to produce toxic and corrosive fumes. When heated to decomposition it emits toxic fumes of Cl^-.

DKN250 CAS:54086-41-8 HR: 3
DIGOLD(I) KETENIDE
mf: C_2Au_2O mw: 433.96

SAFETY PROFILE: The dry material is a shock-sensitive explosive. Forms heat-sensitive explosive complexes with tertiary heterocyclic bases (e.g., pyridine, methyl pyridines, 2,6-dimethylpyridine, quinoline). See other gold compounds.

DKN300 **CAS:1672-46-4** **HR: 3**
DIGOXIGENINE
mf: $C_{23}H_{34}O_5$ mw: 390.57

PROP: Dihydrate: Prismatic rods from dilute alc. Anhydrous: As stout prisms from ethyl acetate. Mp: 220°.

SYNS: CARDOGENEN-(20:22)-TRIOL-(3-β,12,14) (GERMAN) □ DIGOXIGENIN □ LANADIGENIN □ $\Delta^{20:22}$-3-β,12-β,14,21-TETRAHYDROXYNORCHOLENIC ACID LACTONE □ 3 β,12-β,14-TRIHYDROXY-CARD-20(22)-ENOLIDE □ 3-β,12,14-TRIOXY-CARDEN-(20:22)-OLID (GERMAN) □ 3-β,12,14-TRIOXY-DIGEN-(20:22)-OLID (GERMAN)

TOXICITY DATA WITH REFERENCE
ivn-gpg LDLo:3545 µg/kg AIPTAK 153,536,65
par-pgn LDLo:1620 µg/kg CPBTAL 8,18,60

SAFETY PROFILE: A deadly poison by intravenous and parenteral routes. When heated to decomposition it emits acrid smoke and fumes. See also DIGITALIS.

DKN400 **CAS:20830-75-5** **HR: 3**
DIGOXIN
mf: $C_{41}H_{64}O_{14}$ mw: 781.05

PROP: White, crystalline powder. Crystals from alc (aq). Mp: 265° (decomp). Glycoside isolated from *Digitalis lanata* (JPETAB 52,1,34).

SYNS: CHLOROFORMIC DIGITALIN □ DIGACIN □ DIGITALIS GLYCOSIDE □ DIGOXIGENIN-TRIDIGITOXOSID (GERMAN) □ DIGOXINE □ HOMOLLE'S DIGITALIN □ LANICOR □ LANOXIN □ ROUGOXIN □ SK-DIGOXIN

TOXICITY DATA WITH REFERENCE
ims-dom TDLo:810 µg/kg (1-18W preg):TER AJOGAH 121,1100,75
orl-wmn TDLo:100 µg/kg:CNS,CVS,GIT NZMJAX 84,443,76
orl-man TDLo:333 µg/kg:EYE,CVS BMJOAE 287,392,83
orl-cld TDLo:127 µg/kg:BPR AFPYAE 34(1),137,86
ivn-inf LDLo:200 µg/kg:BPR PEDIAU 77,848,86
ivn-inf TDLo:50 µg/kg:BPR AFPYAE 34(1),137,86
scu-rat LD50:30 mg/kg PPHAD4 1,97,80
ivn-rat LD50:25 mg/kg AIPTAK 155,165,65
orl-mus LD50:17,780 µg/kg AIPTAK 153,436,65
ipr-mus LD50:3964 µg/kg CHTHBK 16,371,71
scu-mus LD50:12,880 µg/kg AIPTAK 153,436,65
ivn-mus LD50:7670 µg/kg PLRCAT 6,417,74
par-mus LDLo:2947 µg/kg SIHSD8 4(2),32,81
orl-dog LDLo:300 µg/kg CRSBAW 109,279,32

CONSENSUS REPORTS: EPA Extremely Hazardous Substances List. Reported in EPA TSCA Inventory. EPA Genetic Toxicology Program.

SAFETY PROFILE: A deadly poison by most routes. Human systemic effects by ingestion: anorexia, cardiac arrhythmias, nausea and vomiting, visual field changes, pulse rate decrease, fall in blood pressure. An experimental teratogen. When heated to decomposition it emits acrid and irritating fumes. See also DIGITALIS.

DKN600 **HR: 3**
ε-DIGOXIN ACETATE
mf: $C_{43}H_{66}O_{15}$ mw: 823.09

SYN: ε-ACETYLDIGOXIN (GERMAN)

TOXICITY DATA WITH REFERENCE
orl-gpg LD50:2500 µg/kg ARZNAD 15,481,65
ivn-gpg LD50:2900 µg/kg ARZNAD 15,481,65

SAFETY PROFILE: Poison by ingestion and intravenous routes. See also DIGOXIN. When heated to decomposition it emits acrid and irritating fumes.

DKN875 **HR: 3**
DIGOXIN PENTAFORMATE
mf: $C_{46}H_{64}O_{19}$ mw: 921.10

SYN: PENTAFORMYL DIGOXIN

TOXICITY DATA WITH REFERENCE
orl-mus LD50:9660 µg/kg AIPTAK 153,436,65
scu-mus LD50:9220 µg/kg AIPTAK 153,436,65
orl-gpg LDLo:8788 µg/kg AIPTAK 153,436,65
ivn-gpg LDLo:3515 µg/kg AIPTAK 153,436,65

SAFETY PROFILE: Poison by ingestion, subcutaneous, and intravenous routes. When heated to decomposition it emits acrid smoke and fumes. See also DIGOXIN.

DKO000 **CAS:51622-02-7** **HR: 3**
DIHEPTYLMERCURY
mf: $C_{14}H_{30}Hg$ mw: 399.03

TOXICITY DATA WITH REFERENCE
ipr-mus LDLo:15,600 µg/kg CBCCT* 4,230,52

CONSENSUS REPORTS: Mercury and its compounds are on the Community Right-To-Know List.

OSHA PEL: TWA 0.01 mg(Hg)/m³; STEL 0.03 mg/m³ (skin)
ACGIH TLV: TWA 0.01 mg(Hg)/m³; STEL 0.03 mg(Hg)/m³
NIOSH REL: (Mercury, Organo): TWA 0.01 mg/m³; STEL 0.03 mg/m³ (skin)

SAFETY PROFILE: Poison by intraperitoneal route. See also MERCURY COMPOUNDS, ORGANIC. When heated to decomposition it emits toxic fumes of Hg.

DKO600 **CAS:143-16-8** **HR: 3**
DIHEXYLAMINE
mf: $C_{12}H_{27}N$ mw: 185.40

PROP: Liquid or crystals. Mp: 193–195°, flash p: 220°F (OC), d: 0.78, vap d: 6.38.

SYN: DI-N-HEXYLAMINE

TOXICITY DATA WITH REFERENCE
skn-rbt 500 mg open MLD UCDS** 11/29/63
eye-rbt 750 µg/24H SEV 85JCAE ,438,86
orl-rat LD50:380 mg/kg UCDS** 11/29/63
ivn-mus LD50:10 mg/kg CSLNX* NX#05177
skn-rbt LD50:170 mg/kg AEHLAU 1,343,60

CONSENSUS REPORTS: Reported in EPA TSCA Inventory.

SAFETY PROFILE: Poison by ingestion, skin contact, and intravenous routes. A skin and severe eye irritant. Flammable when exposed to heat or flame; can react with oxidizing materials. To fight fire, use CO_2, dry chemical. When heated to decomposition it emits toxic fumes of NO_x. See also AMINES.

DKO800 CAS:112-58-3 HR: 2
DIHEXYL ETHER
mf: $C_{12}H_{26}O$ mw: 186.38

PROP: Mp: −43.0°, bp: 227°, flash p: 170°F (OC), d: 0.794, autoign temp: 365°F, vap d: 6.4.

SYNS: HEXYL ETHER □ N-HEXYL ETHER □ 1,1′-OXYBISHEXANE

TOXICITY DATA WITH REFERENCE
skn-rbt 10 mg/24H open SEV AMIHBC 10,61,54
skn-rbt 500 mg open MLD UCDS** 7/13/71
eye-rbt 500 mg open AMIHBC 10,61,54
orl-rat LD50:30,900 mg/kg AMIHBC 10,61,54
skn-rbt LD50:6900 mg/kg AMIHBC 10,61,54

CONSENSUS REPORTS: Reported in EPA TSCA Inventory.

SAFETY PROFILE: Mildly toxic by ingestion and skin contact. An eye and severe skin irritant. Combustible when exposed to heat or flame; can react with oxidizing materials. To fight fire, use foam, CO_2, dry chemical. See also ETHERS for explosion hazard.

DKP000 CAS:19139-31-2 HR: 1
DIHEXYL FUMARATE
mf: $C_{16}H_{28}O_4$ mw: 284.44

SYNS: DIHEXYL trans-BUTENEDIOATE □ FUMARIC ACID DIHEXYL ESTER □ HEXYL FUMARATE

TOXICITY DATA WITH REFERENCE
skn-rbt 500 mg/24H MLD FCTXAV 16,709,78

CONSENSUS REPORTS: Reported in EPA TSCA Inventory.

SAFETY PROFILE: A skin irritant. When heated to decomposition it emits acrid smoke and irritating fumes.

DKP200 CAS:18279-21-5 HR: 3
DIHEXYL LEAD DIACETATE
mf: $C_{16}H_{32}O_4Pb$ mw: 495.67

SYN: PLUMBANE, BIS(ACETYLOXY)DIHEXYL-

TOXICITY DATA WITH REFERENCE
orl-mus LD50:215 mg/kg CRSBAW 162,1456,68

CONSENSUS REPORTS: Lead and its compounds are on the Community Right-To-Know List.

SAFETY PROFILE: Poison by ingestion. See also LEAD COMPOUNDS. When heated to decomposition it emits toxic fumes of Pb.

DKP400 CAS:105-52-2 HR: 1
DIHEXYL MALEATE
mf: $C_{16}H_{28}O_4$ mw: 284.44

PROP: Liquid, sol in water <0.01% by weight @ 20°. D: 0.9602 @ 20°/20°, bp: 179° @ 10 mm, vap press: <0.01 mm @ 20°, fp: −70°, flash p: 290°F (OC).

SYNS: 2-BUTENEDIOIC ACID BIS(1,3-DIMETHYLBUTYL) ESTER □ DI(4-METHYL-2-AMYL) MALEATE □ DI(4-METHYL-2-PENTYL) MALEATE □ DMAM □ MALEIC ACID DI(1,3-DIMETHYLBUTYL) ESTER □ MALEIC ACID DIHEXYL ESTER

TOXICITY DATA WITH REFERENCE
skn-rbt 10 mg/24H open MLD AIHAAP 23,95,62
eye-rbt 500 mg open AMIHBC 10,61,54
orl-rat LD50:7340 mg/kg AIHAAP 23,95,62
skn-rbt LD50:12 g/kg AMIHBC 10,61,54

CONSENSUS REPORTS: Reported in EPA TSCA Inventory.

SAFETY PROFILE: Mildly toxic by ingestion and skin contact. A skin and eye irritant. Combustible material when exposed to heat of flame. See also ESTERS and MALEIC ACID. To fight fire, use dry chemical, CO_2, water, fog, mist. When heated to decomposition it emits acrid smoke and irritating fumes.

DKP600 CAS:84-75-3 HR: 1
DI-n-HEXYL PHTHALATE
mf: $C_{20}H_{30}O_4$ mw: 334.50

PROP: Liquid. Mp: −58°, bp: 210° @ 5 mm, flash p: 350°F, d: 0.995 @ 20°/20°, vap d: 11.5.

SYNS: 1,2-BENZENEDICARBOXYLIC ACID DIHEXYL ESTER □ DIHEXYL PHTHALATE □ PHTHALIC ACID DIHEXYL ESTER

TOXICITY DATA WITH REFERENCE
eye-rbt 500 mg open AMIHBC 10,61,54
eye-rbt 500 mg/24H MLD 85JCAE-,388,86
orl-mus TDLo:13 g/kg (male 7D pre):REP TXAPA9 88,255,87
orl-mus TDLo:30 g/kg (female 7D pre):REP TXAPA9 88,255,87
orl-mus TDLo:7920 mg/kg (female 6-13D post):REP TCMUD8 7,29,87
orl-mus TDLo:30 g/kg (male 7D pre):REP TXAPA9 88,255,87
orl-mus TDLo:13 g/kg (male 7D pre):REP TXAPA9 88,255,87
orl-mus TDLo:181 g/kg (male 18W pre):REP NTIS** PB85-249332
orl-rat TDLo:9600 mg/kg (4D male):REP EVHPAZ 45,77,82
orl-rat LD50:29,600 mg/kg EVHPAZ 3,131,73
skn-rbt LD50:20 g/kg EVHPAZ 4,3,73

CONSENSUS REPORTS: Reported in EPA TSCA Inventory.

SAFETY PROFILE: Very mildly toxic by ingestion and skin contact. Experimental reproductive effects. A skin and eye irritant. Combustible when exposed to heat or flame; can react with oxidizing materials. To fight fire, use foam, CO_2, dry chemical. See also PHTHALIC ACID and ESTERS.

DKP800 CAS:3006-15-3 HR: 2
DIHEXYL SODIUM SULFOSUCCINATE
mf: $C_{16}H_{30}O_7S{\cdot}Na$ mw: 389.51

PROP: Clear, viscous liquid.

SYNS: AEROSOL MA80 ☐ BUTANEDIOIC ACID, SULFO-, 1,4-DIHEXYL ESTER, SODIUM SALT (9CI) ☐ DIHEXYL SODIOSULFOSUCCINATE ☐ DIHEXYL SULFOSUCCINATE SODIUM SALT ☐ MONAWET MM 80 ☐ SODIUM DIHEXYL SULFOSUCCINATE ☐ SUCCINIC ACID, SULFO-, 1,4-DIHEXYL ESTER, SODIUM SALT (8CI) ☐ SV 1017

TOXICITY DATA WITH REFERENCE
orl-rat LD50:1750 mg/kg JACTDZ 1,109,90

CONSENSUS REPORTS: Reported in EPA TSCA Inventory.

SAFETY PROFILE: Moderately toxic by ingestion. Flammable when exposed to heat, flame, oxidizers. To fight fire, use water, spray, foam, mist, dry chemical. When heated to decomposition it emits toxic fumes of SO_x and Na_2O.

DKQ000 CAS:57-41-0 HR: 3
DIHYDANTOIN
mf: $C_{15}H_{12}N_2O_2$ mw: 252.29

PROP: A solid. Mp: 295–298°.

SYNS: ALEVIATIN ☐ ANTISACER ☐ AURANILE ☐ CAUSOIN ☐ CITRULLAMON ☐ COMITAL ☐ CONVUL ☐ DANTEN ☐ DANTINAL ☐ DANTOINAL KLINOS ☐ DANTOINE ☐ DENYL ☐ DIDAN-TDC-250 ☐ DIFENILHIDANTOINA (SPANISH) ☐ DIFENIN ☐ DIFHYDAN ☐ DIHYCON ☐ DI-HYDAN ☐ DILANTIN ☐ DILANTINE ☐ DINTOIN ☐ DIPHANTOIN ☐ DIPHEDAL ☐ DIPHENINE ☐ DIPHENTOIN ☐ DIPHENYLAN ☐ DIPHENYLHYDANTOIN ☐ 5,5-DIPHENYLHYDANTOIN ☐ DIPHENYLHYDANTOINE (FRENCH) ☐ 5,5-DIPHENYLIMIDAZOLIDIN-2,4-DIONE ☐ 5,5-DIPHENYL-2,4-IMIDAZOLIDINEDIONE ☐ DI-PHETINE ☐ DITOINATE ☐ DPH ☐ EKKO CAPSULES ☐ ELEPSINDON ☐ ENKELFEL ☐ EPAMIN ☐ EPANUTIN ☐ EPASMIR '5' ☐ EPDANTOINE SIMPLE ☐ EPELIN ☐ EPIFENYL ☐ EPIHYDAN ☐ EPILAN ☐ EPILANTIN ☐ EPINAT ☐ EPISED ☐ EPTAL ☐ EPTOIN ☐ FENANTOIN ☐ FENIDANTOIN 'S' ☐ FENYLEPSIN ☐ FENYTOINE ☐ GEROT-EPILAN-D ☐ HIDAN ☐ HIDANTILO ☐ HIDANTINA SENOSIAN ☐ HIDANTINA VITORIA ☐ HIDANTOMIN ☐ HYDANTAL ☐ HYDANTOIN ☐ ICTALIS SIMPLE ☐ IDANTOIN ☐ KESSODANTEN ☐ LABOPAL ☐ LEHYDAN ☐ LEPITOIN ☐ LEPSIN ☐ MINETOIN ☐ NCI-C55765 ☐ NEOSHIDANTOINA ☐ NOVANTOINA ☐ OM-HYDANTOINE ☐ OXYLAN ☐ PHANANTIN ☐ PHENATOINE ☐ RITMENAL ☐ SACERIL ☐ SANEPIL ☐ SILANTIN ☐ SODANTON ☐ SOLANTIN ☐ SYLANTOIC ☐ TACOSAL ☐ THILOPHENYL ☐ TOIN UNICELLES ☐ ZENTRONAL ☐ ZENTROPIL

TOXICITY DATA WITH REFERENCE
dnd-esc 50 μmol/L MUREAV 89,95,81
dni-hmn:lym 360 μmol/L TXAPA9 33,38,75
mnt-mus-ivn 500 μg/kg MUREAV 141,183,84
orl-wmn TDLo:1365 mg/kg (female 1-39W post):REP JOPDAB 89,662,76
unr-wmn TDLo:540 mg/kg (female 1-39W post):REP JOPDAB 89,662,76
unr-wmn TDLo:1620 mg/kg (female 1-39W post):REP JOPDAB 89,662,76
orl-wmn TDLo:1620 mg/kg (female 1-39W post):TER JAMAAP 244,1464,80
orl-wmn TDLo:1080 mg/kg (female 1-39W post):TER LANCAO 2,702,71
orl-wmn TDLo:540 mg/kg (female 1-39W post):TER JOPDAB 87,285,75
orl-rat TDLo:1300 mg/kg (female 7-19D post):REP TJADAB 24,115,81
orl-rat TDLo:50 mg/kg (female 7-16D post):REP BCPCA6 28,2585,79
orl-mus TDLo:630 mg/kg (female 8-12D post):REP TCMUD8 6,361,86
scu-rat TDLo:1050 mg/kg (female 1-21D post):REP PJABDW 58,78,82
orl-rat TDLo:2400 mg/kg (female 7-18D post):REP TJADAB 35,287,87
ipr-mus TDLo:58 mg/kg (female 13D post):TER TOLED5 18(Suppl 1),109,83
orl-rat TDLo:540 mg/kg (female 9-17D post):TER EAMJAV 60,407,83
scu-rat TDLo:900 mg/kg (female 10-15D post):TER TJADAB 12,291,75
orl-rat TDLo:270 mg/kg (female 9-17D post):TER EAMJAV 60,407,83
ipr-mus TDLo:75 mg/kg (female 10D post):TER TJADAB 13,32A,76
orl-mus TDLo:85 mg/kg (female 10D post):REP JDREAF 57,362,78
orl-rat TDLo:540 mg/kg (female 9-17D post):REP EAMJAV 60,407,83
par-rat TDLo:4800 mg/kg (male 60D pre):REP IJMDAI 15,615,79
orl-uns TDLo:330 mg/kg (female 8-18D post):REP JIDOAA 35,387,86
ipr-mus TDLo:150 mg/kg (female 11-13D post):REP JDREAF 58,1918,79
ipr-rat TDLo:50 mg/kg (female 1D pre):REP PSEBAA 119,982,65
ipr-rat TDLo:338 mg/kg (female 12-14D post):TER DABBBA 36,2145,75
orl-mus TDLo:640 mg/kg (female 14D pre):TER SCIEAS 211,483,81
orl-rat TDLo:1300 mg/kg (female 7-19D post):TER BJPCBM 59,494,77
scu-rat TDLo:600 mg/kg (female 13-15D post):TER TXAPA9 22,193,72
orl-mus TDLo:400 mg/kg (female 9-18D post):TER TJADAB 25(2),33A,82
orl-rat TDLo:1300 mg/kg (female 7-19D post):TER TJADAB 24,115,81
orl-wmn TDLo:1620 mg/kg (female 1-39W post):CAR,TER JAMAAP 244,1464,80
orl-wmn TDLo:730 mg/kg/1Y-C:CAR,BLD NKGZAE 46,1,83
unr-wmn TDLo:1620 mg/kg (female 1-39W post):CAR,TER JAMAAP 244,1464,80
orl-rat TDLo:1500 mg/kg:ETA CNREA8 26,619,66
orl-mus TDLo:5284 mg/kg/17W-C:ETA ZKKOBW 78,290,72
orl-rat TD:1500 mg/kg/27D-I:ETA CNREA8 28,924,68
orl-wmn TD:730 mg/kg/1Y-C:CAR,BLD NKGZAE 43,711,80
orl-chd TD:6023 mg/kg/1Y-C:NEO,BLD,SKN AJDEBP 22,28,81
orl-chd TDLo:3 mg/kg/3W:SKN CLPTAT 20,48,76
orl-chd LDLo:67 mg/kg NEURAI 11,138,61
orl-chd TDLo:11 mg/kg/D:CNS AJDCAI 130,75,76
orl-man TDLo:31 mg/kg/4D-I:CNS,PSY JCGADC 9,337,87

orl-wmn TDLo:106 mg/kg/2W-I:CNS,PSY JCGADC 9,337,87
orl-man TDLo:1300 mg/kg:BLD AIMEAS 69,557,68
orl-chd TDLo:140 mg/kg/:LIV,BLD,SKN PEDIAU 28,943,61
ivn-cld TDLo:15 mg/kg CPEDAM 24,467,85
ivn-cld TDLo:15 mg/kg PEDIAU 72,831,83
unr-cld TDLo:18 mg/kg/3D-I PEDIAU 72,831,83
orl-rat LD50:1635 mg/kg ARZNAD 33,1168,83
ivn-rat LD50:101 mg/kg JKXXAF #79-163823
orl-mus LD50:150 mg/kg JMCMAR 18,383,75
ipr-mus LD50:100 mg/kg JMCMAR 24,465,81
scu-mus LD50:110 mg/kg ARZNAD 4,723,54

CONSENSUS REPORTS: NTP 7th Annual Report On Carcinogens. IARC Cancer Review: Group 2B IMEMDT 7,319,87; Human Limited Evidence IMEMDT 13,201,77; Animal Sufficient Evidence IMEMDT 13,201,77. EPA Genetic Toxicology Program.

SAFETY PROFILE: Confirmed carcinogen producing lymphoma, Hodgkin's disease, tumors of the skin and appendages. Experimental carcinogenic and tumorigenic data. A human poison by ingestion. Poison experimentally by ingestion, subcutaneous, intravenous, and intraperitoneal routes. Moderately toxic by an unspecified route. Experimental teratogenic and reproductive effects. Human systemic effects by ingestion: dermatitis, change in motor activity (specific assay), ataxia (loss of muscle coordination), degenerative brain changes, encephalitis, hallucinations, distorted perceptions, irritability, and jaundice. Human teratogenic effects by ingestion: developmental abnormalities of the central nervous system, cardiovascular (circulatory) system, musculoskeletal system, craniofacial area, skin and skin appendages, eye, ear, other developmental abnormalities. Effects on newborn include abnormal growth statistics (e.g., reduced weight gain), physical abnormalities, other postnatal measures or effects, and delayed effects. Human mutation data reported. A drug for the treatment of grand mal and psychomotor seizures. When heated to decomposition it emits toxic fumes of NO_x.

DKQ200 CAS:63868-75-7 HR: 3
DIHYDRALAZINE HYDROCHLORIDE
mf: $C_8H_{10}N_6•ClH$ mw: 226.70

SYNS: C-7441 □ 1,4-DIHYDRAZINOPHTHALAZINE HYDROCHLORIDE

TOXICITY DATA WITH REFERENCE
orl-rat LD50:350 mg/kg RPOBAR 2,283,70
ipr-rat LD50:270 mg/kg RPOBAR 2,283,70
ivn-rat LD50:167 mg/kg RPOBAR 2,283,70
ipr-mus LD50:290 mg/kg JPETAB 101,368,51

SAFETY PROFILE: Poison by ingestion, intraperitoneal, and intravenous routes. When heated to decomposition it emits very toxic fumes of NO_x and HCl.

DKQ400 HR: 3
DIHYDRAZINECOBALT(II) CHLORATE
mf: $Cl_2CoH_8N_4O_6$ mw: 298.0

CONSENSUS REPORTS: Cobalt and its compounds are on the Community Right-To-Know List.

SAFETY PROFILE: A powerful explosive extremely sensitive to light impact, friction, or heating to 90°C. When heated to decomposition it emits toxic fumes of Cl^- and NO_x. See also COBALT COMPOUNDS and CHLORATES.

DKQ600 CAS:7327-87-9 HR: 3
1,4-DIHYDRAZINOPHTHALAZINE SULFATE
mf: $C_8H_{10}N_6•H_2O_4S$ mw: 288.32

PROP: Needles or crystals. Mp: 233° (decomp).

SYN: DIHYDRALAZINE SULFATE

TOXICITY DATA WITH REFERENCE
mmo-sat 100 μg/plate TCMUD8 5,339,85
mma-sat 100 μg/plate TCMUD8 5,339,85
dnr-rat:lvr 500 μmol/L SCIEAS 210,329,80
dnd-mus-ipr 159 mg/kg ENMUDM 4,605,82
sce-mus-ipr 79,500 μg/kg ENMUDM 4,605,82
orl-rat LD50:820 mg/kg RPOBAR 2,284,70
orl-mus LD50:400 mg/kg RPOBAR 2,284,70
ipr-mus LD50:159 mg/kg RPOBAR 2,283,70
scu-mus LD50:223 mg/kg RPOBAR 2,283,70

SAFETY PROFILE: Poison by ingestion, intraperitoneal, and subcutaneous routes. Mutation data reported. When heated to decomposition it emits very toxic fumes of NO_x and SO_x.

DKQ650 CAS:75034-93-4 HR: 2
DIHYDREL

SYN: DIMETHYLHYDRAZINIUM derivative of 2-CHLOROETHYLPHOSPHONIC ACID

TOXICITY DATA WITH REFERENCE
eye-rbt 40% GISAAA 51(10),85,86
orl-rat LD50:3500 mg/kg AIPTAK 123,48,59
orl-mus LD50:2333 mg/kg GISAAA 51(10),85,86

SAFETY PROFILE: Moderately toxic by ingestion. An eye irritant. When heated to decomposition it emits toxic fumes of NO_x, PO_x, and Cl^-.

DKR000 CAS:6258-06-6 HR: 1
9,10-DIHYDRO-1-AMINO-4-BROMO-9,10-DIOXO-2-ANTHRACENE SULFONIC ACID SODIUM SALT
mf: $C_{14}H_7BrNO_5S•Na$ mw: 404.18

SYNS: 1-AMINO-4-BROMANTHRACHINON-2-SULFONAN SODNY (CZECH) □ BROMANMINAN SODNY (CZECH)

TOXICITY DATA WITH REFERENCE
eye-rbt 500 mg/24H MLD 28ZPAK -,193,72

CONSENSUS REPORTS: Reported in EPA TSCA Inventory.

SAFETY PROFILE: An eye irritant. When heated to

decomposition it emits very toxic fumes of Br^-, Na_2O, NO_x, and SO_x.

DKR200 **CAS:3567-76-8** **HR: 2**
2,3-DIHYDRO-6-AMINO-2-(2-CHLOROETHYL)-4H-1,3-BENZOXAZIN-4-ONE
mf: $C_{10}H_{11}ClN_2O_2$ mw: 226.68

SYNS: A 350 □ 6-AMINO-2-(2-CHLOROETHYL)-2,3-DIHYDRO-4H-1,3-BENZOXAZIN-4-ONE □ AMINOCHLORTHENOXAZINE □ 2-(β-CHLOROETHYL)-2,3-DIHYDRO-4-OXO-6-AMINO-1,3-BENZOXAZINE

TOXICITY DATA WITH REFERENCE
orl-rat LD50:1958 mg/kg JPPMAB 16,502,64
ipr-rat LD50:1500 mg/kg ARZNAD 13,884,63
orl-mus LD50:10 g/kg ARZNAD 13,884,63
ipr-mus LD50:1941 mg/kg ARZNAD 13,884,63

SAFETY PROFILE: Moderately toxic by ingestion and intraperitoneal routes. An antipyretic and analgesic. When heated to decomposition it emits very toxic fumes of Cl^- and NO_x.

DKR400 **CAS:40847-64-1** **HR: 1**
9,10-DIHYDRO-1-AMINO-4-(3-(2-HYDROXYETHYL)AMINOSULFONYL-4-METHYLPHENYLAMINO)-9,10-DIOXO-2-ANTHRACENE SULFONIC ACID SODIUM SALT
mf: $C_{23}H_{20}N_3O_8S_2 \cdot Na$ mw: 553.57

SYN: MODR BRILANTNI ALIZARINOVA BRL (CZECH)

TOXICITY DATA WITH REFERENCE
eye-rbt 500 mg/24H MLD 28ZPAK -,241,72

CONSENSUS REPORTS: Reported in EPA TSCA Inventory.

SAFETY PROFILE: An eye irritant. When heated to decomposition it emits very toxic fumes of NO_x, Na_2O, and SO_x.

DKS400 **CAS:60968-08-3** **HR: 2**
1,2-DIHYDROBENZO(a)ANTHRACENE
mf: $C_{18}H_{14}$ mw: 230.32

SYN: 1,2-DIHYDROBENZ(a)ANTHRACENE

TOXICITY DATA WITH REFERENCE
skn-mus TDLo:18 mg/kg:NEO CNREA8 38,1705,78

SAFETY PROFILE: Questionable carcinogen with experimental neoplastigenic data by skin contact. When heated to decomposition it emits acrid smoke and irritating fumes.

DKS600 **CAS:60968-01-6** **HR: 2**
3,4-DIHYDROBENZO(a)ANTHRACENE
mf: $C_{18}H_{14}$ mw: 230.32

SYN: 3,4-DIHYDROBENZ(a)ANTHRACENE

TOXICITY DATA WITH REFERENCE
skn-mus TDLo:3700 μg/kg:NEO CNREA8 38,1705,78

SAFETY PROFILE: Questionable carcinogen with experimental neoplastigenic data by skin contact. When heated to decomposition it emits acrid smoke and irritating fumes.

DKS800 **CAS:10023-25-3** **HR: 2**
6,13-DIHYDROBENZO(e)(1)BENZOTHIOPYRANO(4,3-b)INDOLE
mf: $C_{19}H_{13}NS$ mw: 287.39

TOXICITY DATA WITH REFERENCE
mma-sat 100 μg/plate MUREAV 66,307,79
scu-mus TDLo:72 mg/kg/9W-I:NEO MUREAV 66,307,79
scu-mus TD:270 mg/kg/20W-I:NEO JNCIAM 46,1257,71

SAFETY PROFILE: Questionable carcinogen with experimental neoplastigenic data. Mutation data reported. When heated to decomposition it emits very toxic fumes of SO_x and NO_x.

DKT400 **CAS:100466-04-4** **HR: 2**
2,3-DIHYDRO-1H-BENZO(h,i)CHRYSENE
mf: $C_{21}H_{16}$ mw: 268.37

SYNS: 5:10-TRIMETHYLENE-1:2-BENZANTHRACENE □ 1:12-TRIMETHYLENECHRYSENE

TOXICITY DATA WITH REFERENCE
skn-mus TDLo:3000 mg/kg/43W-I:ETA AKBNAE 62(2),30,41
scu-mus TDLo:240 mg/kg/20D-I:ETA AKBNAE 62(2),30,41

SAFETY PROFILE: Questionable carcinogen with experimental tumorigenic data. When heated to decomposition it emits acrid smoke and fumes. See also CHRYSENE.

DKT500 **CAS:101018-97-7** **HR: 3**
(2,3-DIHYDRO-1,4-BENZODIOXIN-6-YL)(4-FLUOROPHENYL)METHANONE
mf: $C_{15}H_{11}FO_3$ mw: 258.26

SYNS: 1,4-BENZODIOXINE, 2,3-DIHYDRO-, 6-(4-FLUOROPHENYL)- □ KETONE, 2,3-DIHYDRO-1,4-BENZODIOXIN-6-YL 4-FLUOROPHENYL □ METHANONE, (2,3-DIHYDRO-1,4-BENZODIOXIN-6-YL)(4-FLUOROPHENYL)-

TOXICITY DATA WITH REFERENCE
orl-mus LD50:>800 mg/kg PCJOAU 19,611,85

DOT CLASSIFICATION: 3; *Label:* Flammable Liquid

SAFETY PROFILE: Moderately toxic by ingestion. A flammable liquid. When heated to decomposition it emits toxic vapors of F^-.

DKU000 **CAS:17573-23-8** **HR: 2**
7,8-DIHYDROBENZO(a)PYRENE
mf: $C_{20}H_{14}$ mw: 254.34

PROP: A solid. Mp: 128°.

TOXICITY DATA WITH REFERENCE
mma-sat 10 μg/plate PNASA6 72,5135,75
scu-mus TDLo:9 mg/kg:NEO JJIND8 64,617,80

CONSENSUS REPORTS: EPA Genetic Toxicology Program.

SAFETY PROFILE: Questionable carcinogen with experimental neoplastigenic data. Mutation data reported. When heated to decomposition it emits acrid smoke and irritating fumes.

DKU400 CAS:66788-01-0 HR: 2
9,10-DIHYDROBENZO(e)PYRENE
mf: $C_{20}H_{14}$ mw: 254.34

SYN: 9,10-H2 B(e)P

TOXICITY DATA WITH REFERENCE
mmo-sat nmol/plate CNREA8 40,1985,80
skn-mus TDLo:25 mg/kg:NEO CNREA8 40,203,80
scu-mus TDLo:9 mg/kg:ETA JJIND8 64,617,80

SAFETY PROFILE: Questionable carcinogen with experimental neoplastigenic and tumorigenic data. Mutation data reported. When heated to decomposition it emits acrid smoke and irritating fumes.

DKU875 HR: 3
3,4-DIHYDRO-2H-1,4-BENZOTHIAZINE HYDROCHLORIDE
mf: $C_8H_9NS{\cdot}ClH$ mw: 187.70

SYNS: BTZ □ DIIDROBENZO(1-4)TIAZINA CLORIDRATO (ITALIAN)

TOXICITY DATA WITH REFERENCE
orl-mus LD50:800 mg/kg AIPTAK 89,55,52
ipr-mus LD50:300 mg/kg AIPTAK 89,55,52
ivn-rbt LDLo:200 mg/kg AIPTAK 89,55,52

SAFETY PROFILE: Poison by intraperitoneal and intravenous routes. Moderately toxic by ingestion. When heated to decomposition it emits toxic fumes of SO_x, NO_x, and HCl.

DKV125 CAS:102366-79-0 HR: 2
7,8-DIHYDRO-N-BENZYLADENINE
mf: $C_{12}H_{13}N_5$ mw: 227.30

TOXICITY DATA WITH REFERENCE
orl-rat LD50:2100 mg/kg SHBOAO 32,530,78
scu-rat LD50:480 mg/kg SHBOAO 32,530,78
orl-mus LD50:2000 mg/kg SHBOAO 32,530,78
scu-mus LD50:1100 mg/kg SHBOAO 32,530,78

SAFETY PROFILE: Moderately toxic by ingestion and subcutaneous routes. When heated to decomposition it emits toxic fumes of NO_x.

DKV150 CAS:619-01-2 HR: 2
DIHYDROCARVEOL
mf: $C_{10}H_{18}O$ mw: 154.28

PROP: Colorless, oily liquid; spearmint odor. D: 0.921–0.926, refr index: 1.477–1.481, flash p: 153°F. Sol in alc, fixed oils; insol in water.

SYNS: 1,6-DIHYDROCARVEOL □ FEMA No. 2379 □ 8-p-MENTHEN-2-OL □ 6-METHYL-3-ISOPROPYLCYCLOHEXANOL

TOXICITY DATA WITH REFERENCE
skn-rbt 500 mg/24H MOD FCTXAV 17,771,79

CONSENSUS REPORTS: Reported in EPA TSCA Inventory.

SAFETY PROFILE: A moderate skin and eye irritant. A combustible liquid. When heated to decomposition it emits acrid smoke and irritating fumes.

DKV160 CAS:20777-49-5 HR: 1
DIHYDROCARVEYL ACETATE
mf: $C_{12}H_{20}O_2$ mw: 196.32

SYNS: CYCLOHEXANOL, 2-METHYL-5-(1-METHYLETHENYL)-, ACETATE,(1-α-2-β,5α- (9CI) □ DIHYDROCARVEOL ACETATE □ DIHYDROCARVYL ACETATE □ p-MENTH-8-EN-2-OL, ACETATE □ p-MENTH-8-EN-2-YL ACETATE □ 2-METHYL-5-(1-METHYLETHENYL)CYCLOHEXYL ACETATE

TOXICITY DATA WITH REFERENCE
skn-rbt 500 mg/24H MLD FCTOD7 21,843,83

CONSENSUS REPORTS: Reported in EPA TSCA Inventory.

SAFETY PROFILE: A skin irritant. When heated to decomposition it emits acrid smoke and irritating fumes.

DKV175 CAS:7764-50-3 HR: 2
d-DIHYDROCARVONE
mf: $C_{10}H_{16}O$ mw: 152.26

PROP: Colorless liquid; spearmint-like odor. D: 0.923–0.928, refr index: 1.470–1.474. Sol in alc, fixed oils; insol in water.

SYNS: FEMA No. 3565 □ p-MENTH-8-EN-2-ONE □ 8-p-MENTHEN-2-ONE □ d-2-METHYL-5-(1-METHYLENENYL)-CYCLOHEXANONE

TOXICITY DATA WITH REFERENCE
skn-rbt 500 mg/24H MOD FCTXAV 18,665,80
orl-rat LD50:>5 g/kg FCTXAV 18,665,80
scu-mus LD50:2900 mg/kg FCTXAV 18,665,80

CONSENSUS REPORTS: Reported in EPA TSCA Inventory.

SAFETY PROFILE: Moderately toxic by subcutaneous route. A skin irritant. When heated to decomposition it emits acrid smoke and irritating fumes.

DKV200 CAS:1016-75-7 HR: 2
2,3-DIHYDRO-6-CHLORO-2-(2-CHLOROETHYL)-4H-1,3-BENZOXAZIN-4-ONE
mf: $C_{10}H_9Cl_2NO_2$ mw: 246.10

TOXICITY DATA WITH REFERENCE
ipr-rat LD50:2000 mg/kg ARZNAD 13,884,63
orl-mus LD50:11,765 mg/kg ARZNAD 13,884,63
ipr-mus LD50:3143 mg/kg ARZNAD 13,884,63

SAFETY PROFILE: Moderately toxic by intraperitoneal route. Mildly toxic by ingestion. When heated to decomposition it emits very toxic fumes of Cl^- and NO_x.

DKV309 **CAS:28058-62-0** **HR: 3**
1′-(3-(10,11-DIHYDRO-3-CHLORO-5H-DIBENZ(b,f) AZEPIN-5-YL)PROPYL)-(1,4′-BIPIPERIDINE)-4′-CARBOXAMIDE DIHYDROCHLORIDE
mf: $C_{28}H_{37}ClN_4O{\cdot}2ClH$ mw: 552.99

SYNS: 3-CHLOROCARPIPRAMINE DIHYDROCHLORIDE □ CLOCAPRAMINE DIHYDROCHLORIDE □ CLOCARPRAMINE DIHYDROCHLORIDE

TOXICITY DATA WITH REFERENCE
orl-rat LD50:6200 mg/kg NIIRDN 6,231,82
ipr-rat LD50:105 mg/kg NIIRDN 6,231,82
orl-mus LD50:2550 mg/kg NIIRDN 6,231,82
ipr-mus LD50:160 mg/kg NIIRDN 6,231,82
scu-mus LD50:6300 mg/kg NIIRDN 6,231,82

SAFETY PROFILE: Poison by intraperitoneal route. Moderately toxic by ingestion. When heated to decomposition it emits toxic fumes of NO_x and HCl.

DKV400 **CAS:40762-15-0** **HR: 2**
1,3-DIHYDRO-7-CHLORO-5-(o-FLUOROPHENYL)-3-HYDROXY-1-(2-HYDROXYETHYL)-2H-1,4-BENZODIAZEPIN-2-ONE
mf: $C_{17}H_{14}ClFN_2O_3$ mw: 348.78

PROP: Crystals from CH_2Cl_2/pet ether. Mp: 138–140°

SYN: SAS 643

TOXICITY DATA WITH REFERENCE
unr-hmn TDLo:71 μg/kg:CNS DRFUD4 3,145,78
orl-rat LD50:2550 mg/kg ARZNAD 25,1294,75
ipr-rat LD50:586 mg/kg ARZNAD 25,1294,75
ipr-mus LD50:760 mg/kg JPPMAB 26,566,74

SAFETY PROFILE: Moderately toxic by ingestion and intraperitoneal routes. Human systemic effects by an unspecified route: somnolence. When heated to decomposition it emits very toxic fumes of Cl^-, F^-, and NO_x.

DKV700 **CAS:5991-71-9** **HR: 3**
2,3-DIHYDRO-7-CHLORO-2-OXO-5-PHENYL-1H-1,4-BENZODIAZEPINE-3-CARBOXYLIC ACID MONOPOTASSIUM SALT
mf: $C_{16}H_{10}ClN_2O_3{\cdot}K$ mw: 352.80

SYN: 4311 CB

TOXICITY DATA WITH REFERENCE
orl-mam LDLo:870 mg/kg CHTPBA 4,239,69
ipr-mam LDLo:300 mg/kg CHTPBA 4,239,69
scu-mam LDLo:450 mg/kg CHTPBA 4,239,69
ivn-mam LDLo:220 mg/kg CHTPBA 4,239,69

SAFETY PROFILE: Poison by intravenous and intraperitoneal routes. Moderately toxic by ingestion and subcutaneous routes. When heated to decomposition it emits toxic fumes of Cl^-, NO_x, and K_2O.

DKV800 **CAS:63041-49-6** **HR: 2**
meso-DIHYDROCHOLANTHRENE
mf: $C_{20}H_{16}$ mw: 256.36

SYN: 6,12,b-DIHYDROCHOLANTHRENE

TOXICITY DATA WITH REFERENCE
scu-mus TDLo:40 mg/kg:ETA CNREA8 1,695,41
scu-mus TD:100 mg/kg:ETA CNREA8 1,685,41

SAFETY PROFILE: Questionable carcinogen with experimental tumorigenic data. A cholinergic agent. When heated to decomposition it emits acrid smoke and irritating fumes.

D

DKW000 **CAS:360-68-9** **HR: 2**
DIHYDROCHOLESTEROL
mf: $C_{27}H_{48}O$ mw: 388.75

SYNS: (3-β,5-β)-CHOLESTAN-3-OL □ 3-β-CHOLESTANOL □ COPROSTANOL □ COPROSTAN-3-β-OL □ COPROSTEROL □ 3-β-HYDROXYCHOLESTANE □ KOPROSTERIN (GERMAN) □ STERCORIN □ XYMOSTANOL

TOXICITY DATA WITH REFERENCE
scu-mus TDLo:60 g/kg/2W-I:CAR NATWAY 60,525,73

SAFETY PROFILE: Questionable carcinogen with experimental neoplastigenic data. When heated to decomposition it emits acrid smoke and irritating fumes.

DKW100 **CAS:929-73-7** **HR: 2**
DODECYLAMINE, HYDROCHLORIDE
mf: $C_{12}H_{27}N{\cdot}ClH$ mw: 221.86

SYNS: DODECANAMINE HYDROCHLORIDE □ 1-DODECANAMINE, HYDROCHLORIDE (9CI) □ n-DODECYLAMINE HYDROCHLORIDE □ DODECYLAMMONIUM CHLORIDE □ n-DODECYLAMMONIUM CHLORIDE □ LAURYLAMINE HYDROCHLORIDE □ LAURYLAMMONIUM HYDROCHLORIDE

TOXICITY DATA WITH REFERENCE
orl-mus LD50:755 mg/kg PCJOAU 15,383,81

CONSENSUS REPORTS: Reported in EPA TSCA Inventory.

SAFETY PROFILE: Moderately toxic by ingestion. When heated to decomposition it emits toxic vapors of HCl and Cl^-.

DKW200 **CAS:41593-31-1** **HR: 2**
1,2-DIHYDROCHRYSENE
mf: $C_{18}H_{14}$ mw: 230.32

SYN: DIHYDROCHRYSENE

TOXICITY DATA WITH REFERENCE
ipr-mus TDLo:59 mg/kg/15D-I:CAR CNREA8 39,5063,79

SAFETY PROFILE: Questionable carcinogen with experimental carcinogenic data. When heated to decomposition it emits acrid smoke and irritating fumes. See also CHRYSENE.

DKW400 **CAS:71435-43-3** **HR: 2**
3,4-DIHYDROCHRYSENE
mf: $C_{18}H_{14}$ mw: 230.32

TOXICITY DATA WITH REFERENCE
ipr-mus TDLo:59 mg/kg/15D-I:ETA CNREA8 39,5063,79

SAFETY PROFILE: Questionable carcinogen with ex-

perimental tumorigenic data. When heated to decomposition it emits acrid smoke and irritating fumes. See also CHRYSENE.

DKW800 **CAS:125-28-0** **HR: 3**
DIHYDROCODEINE
mf: $C_{18}H_{23}NO_3$ mw: 301.42

PROP: A solid. Mp: 55°.

SYNS: CODHYDRINE □ COHYDRIN □ DEHACODIN □ DF 118 □ DIDRATE □ DIHYDRIN □ 7,8-DIHYDROCODEINE □ DIHYDRONEOPINE □ DROCODE □ HYDROCODIN □ 6-HYDROXY-3-METHOXY-N-METHYL-4,5-EPOXYMORPHINAN □ NADEINE □ NOVICODIN □ PARACODIN □ PARACODINE □ PARZONE □ RAPACODIN

TOXICITY DATA WITH REFERENCE
orl-hmn TDLo:6500 mg/kg/26W:EYE,CNS BMJOAE 1,1594,78
scu-wmn TDLo:6 mg/kg/36H-I:PUL BMJOAE 1,211,59
orl-rat LD50:240 mg/kg TXAPA9 1,42,59
scu-mus LD50:135 mg/kg THERAP 6,146,51
ivn-mus LD50:80 mg/kg THERAP 6,146,51
orl-rbt LDLo:400 mg/kg HBAMAK 4,1289,35
scu-gpg LDLo:80 mg/kg HBAMAK 4,1289,35

SAFETY PROFILE: Poison by ingestion, intravenous, and subcutaneous routes. Human systemic effects by ingestion and subcutaneous routes: somnolence, miosis (pupillary constriction), and respiratory depression. An analgesic. Can cause drug dependency with repeated doses. When heated to decomposition it emits toxic fumes of NO_x. See also CODEINE.

DKX000 **CAS:5965-13-9** **HR: 3**
DIHYDROCODEINE BITARTRATE
mf: $C_{18}H_{23}NO_3 \cdot C_4H_6O_6$ mw: 451.52

PROP: Crystals from MeOH. Mp: 192–193°, mp: 186–190° (commercial product).

SYNS: DF 118 □ DIHYDROCODEINE ACID TARTRATE □ DIHYDROCODEINE TARTRATE □ DIHYDROCODEINE TARTRATE (1:1)

TOXICITY DATA WITH REFERENCE
ivn-hmn TDLo:357 μg/kg/5M-C:CNS LANCAO 1,1425,82
orl-rat LD50:240 mg/kg TXAPA9 1,42,59
ipr-mus LD50:252 mg/kg AIPTAK 136,333,62
scu-mus LDLo:350 mg/kg JPETAB 51,35,34
scu-rbt LDLo:39 mg/kg JPETAB 66,182,39
par-rbt LDLo:116 mg/kg JPETAB 66,182,39

SAFETY PROFILE: Poison by ingestion, intraperitoneal, subcutaneous, and parenteral routes. Human systemic effects by intravenous route: irritability. When heated to decomposition it emits toxic fumes of NO_x. See also CODEINE.

DKX050 **CAS:34195-34-1** **HR: 3**
DIHYDROCODEINONE BITARTRATE
mf: $C_{18}H_{21}NO_3 \cdot C_4H_6O_6$ mw: 449.50

SYNS: CODEINONE, DIHYDRO-, TARTRATE □ HYDROCODONE BITARTRATE □ MORPHINAN-6-ONE, 4,5-α-EPOXY-3-METHOXY-17-METHYL-, TARTRATE (1:1)

TOXICITY DATA WITH REFERENCE
scu-ham TDLo:153 mg/kg (female 8D post):TER AJOGAH 123,705,75
orl-rat LD50:375 mg/kg TXAPA9 1,42,59

SAFETY PROFILE: Poison by ingestion. An experimental teratogen. When heated to decomposition it emits toxic fumes of NO_x.

DKX100 **CAS:2111-75-3** **HR: 2**
DIHYDROCUMINYL ALDEHYDE
mf: $C_{10}H_{14}O$ mw: 150.24

PROP: d-Form: Liquid. Bp: (745) 237°, d: (20/4) 0.953, n (20/D) 1.5058. l-Form: Liquid. Bp: 104–105° @ 10 mm, d: (20/4) 0.9645, n (20/D) 1.5069.

SYNS: 4-ISOPROPENYL-1-CYCLOHEXENE-1-CARBOXALDEHYDE □ p-MENTHA-1,8-DIEN-7-AL □ 4-(1-METHYLETHENYL)-1-CYCLOHEXENE-1-CARBOXALDEHYDE (9CI) □ PERILLA ALDEHYDE □ PERILLAL □ PERILLALDEHYDE □ PERILLYL ALDEHYDE

TOXICITY DATA WITH REFERENCE
skn-gpg 100%/24H MOD FCTOD7 20(Suppl),799,82
cyt-ham:fbr 50 mg/L FCTOD7 22,623,84
orl-mus LD50:1720 mg/kg FCTOD7 20(Suppl),799,82

CONSENSUS REPORTS: Reported in EPA TSCA Inventory.

SAFETY PROFILE: Moderately toxic by ingestion. A skin irritant. Mutation data reported. When heated to decomposition it emits acrid smoke and fumes. See also ALDEHYDES.

DKX600 **CAS:427-00-9** **HR: 3**
DIHYDRODEOXYMORPHINE
mf: $C_{17}H_{21}NO_2$ mw: 271.39

PROP: Crystals from EtOAc. Mp: 162–164°.

SYNS: 6-DEOXY-7,8-DIHYDROMORPHINE □ DESOMORPHINE □ DIHYDRODESOXYMORPHINE-D □ 4,5-EPOXY-3-HYDROXY-N-METHYLMORPHINAN □ 4,5-α-EPOXY-17-METHYLMORPHINAN-3-OL □ PERMONID

TOXICITY DATA WITH REFERENCE
scu-mus LDLo:104 mg/kg JPETAB 55,257,35
ivn-mus LD50:27 mg/kg YKKZAJ 84,268,64

SAFETY PROFILE: Poison by subcutaneous and intravenous routes. See also MORPHINE. When heated to decomposition it emits toxic fumes of NO_x.

DKX800 **CAS:73651-49-7** **HR: 1**
9,10-DIHYDRO-4,5-DIAMINO-1-HYDROXY-2,7-ANTHRACENE DISULFONIC ACID DISODIUM SALT
mf: $C_{14}H_8N_2O_9S_2 \cdot 2Na$ mw: 458.34

SYN: AZUROL ALIZARINOVY SW (CZECH)

TOXICITY DATA WITH REFERENCE
skn-rbt 500 mg/24H MLD 28ZPAK -,239,72
eye-rbt 500 mg/24H MLD 28ZPAK -,239,72

SAFETY PROFILE: A skin and eye irritant. When heated to decomposition it emits very toxic fumes of NO_x, Na_2O, and SO_x.

DKX875 CAS:74339-98-3 HR: 2
trans-1,2-DIHYDRODIBENZ(a,e)ACEANTHRYLENE-1,2-DIOL
mf: $C_{24}H_{16}O_2$ mw: 336.40

SYN: trans-12,13-DIHYDRO-12,13-DIHYDROXYDIBENZO(a,e)FLUORANTHENE

TOXICITY DATA WITH REFERENCE
skn-mus TDLo: 1200 μg/kg: ETA CRNGDP 8,461,87

SAFETY PROFILE: Questionable carcinogen with experimental tumorigenic data. When heated to decomposition it emits acrid smoke and irritating fumes.

DKX900 CAS:74340-04-8 HR: 2
trans-10,11-DIHYDRODIBENZ(a,e)ACEANTHRYLENE-10,11-DIOL
mf: $C_{24}H_{16}O_2$ mw: 336.40

SYN: trans-3,4-DIHYDRO-3,4-DIHYDROXYDIBENZO(a,e)FLUORANTHENE

TOXICITY DATA WITH REFERENCE
mma-sat 1200 nmol/L CRNGDP 5,1263,84
dnd-man: lyms 208 nmol CRNGDP 4,27,83
skn-mus TDLo: 1200 μg/kg: ETA CRNGDP 8,461,87

SAFETY PROFILE: Questionable carcinogen with experimental tumorigenic data. Mutation data reported. When heated to decomposition it emits acrid smoke and irritating fumes.

DKY000 CAS:153-34-4 HR: 2
5,6-DIHYDRODIBENZ(a,h)ANTHRACENE
mf: $C_{22}H_{16}$ mw: 280.38

PROP: Crystals from EtOH/C_6H_6. Mp: 194–195°.

TOXICITY DATA WITH REFERENCE
mma-sat 1 μg/plate MUREAV 51,311,78
skn-mus TDLo: 130 mg/kg/44W-I: NEO JNCIAM 34,1,65
scu-mus TDLo: 16 mg/kg: ETA JNCIAM 44,641,70

SAFETY PROFILE: Questionable carcinogen with experimental neoplastigenic and tumorigenic data. Mutation data reported. When heated to decomposition it emits acrid smoke and irritating fumes.

DKY200 CAS:16361-01-6 HR: 2
5,6-DIHYDRODIBENZ(a,j)ANTHRACENE
mf: $C_{22}H_{16}$ mw: 280.38

PROP: Crystals from C_6H_6/pet ether. Mp: 155.5–156.5°.

TOXICITY DATA WITH REFERENCE
mma-sat 1 μg/plate MUREAV 51,311,78
skn-mus TDLo: 268 mg/kg/85W-I: ETA JNCIAM 44,641,70

SAFETY PROFILE: Questionable carcinogen with experimental tumorigenic data. Mutation data reported. When heated to decomposition it emits acrid smoke and irritating fumes.

DKY400 CAS:57816-08-7 HR: 2
7,14-DIHYDRODIBENZ(a,h)ANTHRACENE
mf: $C_{22}H_{16}$ mw: 280.38

SYN: 9,10-DIHYDRO-1,2,5,6-DIBENZANTHRACENE

TOXICITY DATA WITH REFERENCE
mma-sat 1 μg/plate MUREAV 51,311,78
skn-mus TDLo: 1150 mg/kg/48W-I: ETA PRLBA4 129,439,40
skn-mus TD: 1250 mg/kg/52W-I: ETA PRLBA4 117,318,35

CONSENSUS REPORTS: EPA Genetic Toxicology Program.

SAFETY PROFILE: Questionable carcinogen with experimental tumorigenic data. Mutation data reported. When heated to decomposition it emits acrid smoke and irritating fumes.

DKY800 CAS:494-19-9 HR: 3
10,11-DIHYDRO-5-DIBENZ(b,f)AZEPINE
mf: $C_{14}H_{13}N$ mw: 195.28

PROP: Pale-yellow crystals by sublimation; prisms from pet ether. Mp: 110°.

SYN: IMINODIBENZYL

TOXICITY DATA WITH REFERENCE
eye-rbt 100 mg MLD FCTOD7 20,573,82
ivn-mus LD50: 320 mg/kg CSLNX* NX#01352

CONSENSUS REPORTS: Reported in EPA TSCA Inventory.

SAFETY PROFILE: Poison by intravenous route. An eye irritant. When heated to decomposition it emits toxic fumes of NO_x.

DLA000 CAS:63077-00-9 HR: 2
3,4-DIHYDRO-1,2,5,6-DIBENZCARBAZOLE
mf: $C_{20}H_{15}N$ mw: 269.36

SYN: 12,13-DIHYDRO-7H-DIBENZO(a,g)CARBAZOLE

TOXICITY DATA WITH REFERENCE
scu-mus TDLo: 120 mg/kg/9W-I: ETA BAFEAG 42,3,55

SAFETY PROFILE: Questionable carcinogen with experimental tumorigenic data. When heated to decomposition it emits toxic fumes of NO_x.

DLA100 HR: 2
5,8-DIHYDRODIBENZO(a,def)CHRYSENE
mf: $C_{24}H_{16}$ mw: 304.40

SYN: 5,8-DIHYDRO-3,4:9,10-DIBENZOPYRENE

TOXICITY DATA WITH REFERENCE
scu-mus TDLo: 72 mg/kg/9W-I: ETA COREAF 251,1322,60

SAFETY PROFILE: Questionable carcinogen with experimental tumorigenic data. When heated to decomposition it emits acrid smoke and irritating fumes.

D

DLA120 CAS:7350-86-9 HR: 2
7,14-DIHYDRODIBENZO(b,def)CHRYSENE
mf: $C_{24}H_{16}$ mw: 304.40

SYN: 5,10-DIHYDRO-3,4:8,9-DIBENZOPYRENE

TOXICITY DATA WITH REFERENCE
scu-mus TDLo:72 mg/kg/9W-I:ETA COREAF 251,1322,60

SAFETY PROFILE: Questionable carcinogen with experimental tumorigenic data. When heated to decomposition it emits acrid smoke and irritating fumes.

DLB400 CAS:84-16-2 HR: 3
DIHYDRODIETHYLSTILBESTROL
mf: $C_{18}H_{22}O_2$ mw: 270.40

PROP: Needles from benzene, thin plates from dilute alc. Mp: 185–188°. Freely sol in ether; sol in acetone, alc, methanol; sltly sol in benzene, chloroform. Sol in dilute solns of alkali hydroxides. Practically insol in water and in dil mineral acids.

SYNS: meso-3,4-BIS(p-HYDROXYPHENYL)-n-HEXANE □ 3,4-BIS(p-HYDROXYPHENYL)HEXANE □ CYCLOESTROL □ 4,4'-(1,2-DIETHYLETHYLENE)DIPHENOL □ DIHYDROSTILBESTROL □ 4,4'-DIHYDROXY-α,β-DIETHYLDIPHENYLETHANE □ 4,4'-DIHYDROXY-γ,Δ-DIPHENYLHEXANE □ γ,Δ-DI(p-HYDROXYPHENYL)-HEXANE □ meso-3,4-DI(p-HYDROXYPHENYL)-n-HEXANE □ EXTRA-PLEX □ HEXANOESTROL □ HEXESTROL □ meso-HEXESTROL □ HEXOESTROL □ HORMOESTROL □ SINESTROL □ SYNESTROL □ SYNTHOVO □ SYNTROGENE □ VITESTROL

TOXICITY DATA WITH REFERENCE
dns-ham:emb 1 mg/L CNREA8 44,184,84
scu-rat TDLo:3 mg/kg (female 19-21D post):REP BEXBAN 90,1597,80
orl-rat TDLo:250 μg/kg (5D pre):REP CHTPBA 4,1,69
orl-rat TDLo:546 mg/kg (female 91D pre):REP YACHDS 7,3355,79
orl-rat TDLo:2700 μg/kg (male 27D pre):REP YACHDS 7,3340,79
scu-rat TDLo:3 mg/kg (female 19-21D post):TER BEXBAN 90,1597,80
scu-mus TDLo:74 mg/kg/56W-I:CAR VRDEA5 (6),46,62
ivg-mus TDLo:18 mg/kg/17W-I:NEO VOONAW 22(3),68,76
scu-gpg TDLo:74 mg/kg/69W-I:NEO VRDEA5 (6),46,62
imp-gpg TDLo:540 μg/kg:ETA BSBSAS 8,142,51
scu-ham TDLo:360 mg/kg:ETA CBINA8 55,157,85
imp-ham TDLo:640 mg/kg/38W-I:ETA CNREA8 43,5200,83
scu-ham LD:800 mg/kg/26W-I:ETA VOONAW 7(7),35,61
ipr-rat LD50:200 mg/kg NIIRDN 6,743,82
scu-rat LD50:1000 mg/kg NIIRDN 6,743,82
orl-mus LD50:1000 mg/kg NIIRDN 6,743,82
ipr-mus LD50:200 mg/kg NIIRDN 6,743,82

SAFETY PROFILE: Poison by intraperitoneal route. Moderately toxic by ingestion and subcutaneous routes. Questionable carcinogen with experimental carcinogenic and neoplastigenic data. Experimental teratogenic and reproductive effects. Mutation data reported. See also DIETHYLSTILBESTROLS.

DLB800 CAS:28622-84-6 HR: 2
4,5-DIHYDRO-4,5-DIHYDROXYBENZO(a)PYRENE
mf: $C_{20}H_{14}O_2$ mw: 286.34

SYNS: BENZO(a)PYRENE, 4,5-DIHYDROXY-4,5-DIHYDRO- □ BP-4,5-DIHYDRODIOL

TOXICITY DATA WITH REFERENCE
dnd-hmn:fbr 30 μmol/L CBINA8 41,155,82
skn-mus TDLo:4580 μg/kg:NEO CRNGDP 3,371,78

SAFETY PROFILE: Questionable carcinogen with experimental neoplastigenic data. Mutation data reported. When heated to decomposition it emits acrid smoke and irritating fumes.

DLC000 CAS:24909-09-9 HR: 2
9,10-DIHYDRO-9,10-DIHYDROXYBENZO(a)PYRENE
mf: $C_{20}H_{14}O_2$ mw: 286.34

SYN: 9,10-DIHYDROBENZO(a)PYRENE-9,10-DIOL

TOXICITY DATA WITH REFERENCE
dnd-hmn:fbr 30 μmol/L CBINA8 41,155,82
skn-mus TDLo:1 mg/kg:NEO BJCAAI 34,523,76
skn-mus TD:4580 μg/kg:ETA CCSUDL 3,371,78
skn-mus TD:4580 μg/kg:NEO CNREA8 40,1981,80

CONSENSUS REPORTS: EPA Genetic Toxicology Program.

SAFETY PROFILE: Questionable carcinogen with experimental neoplastigenic and tumorigenic data by skin contact. Human mutation data reported. When heated to decomposition it emits acrid smoke and irritating fumes.

DLC400 CAS:58030-91-4 HR: 2
(±)-trans-9,10-DIHYDRO-9,10-DIHYDROXYBENZO(a) PYRENE
mf: $C_{20}H_{14}O_2$ mw: 286.34

SYN: BP-9,10-DIHYDRODIOL

TOXICITY DATA WITH REFERENCE
mma-sat 50 μg/plate CNREA8 41,270,81
skn-mus TDLo:4600 μg/kg:ETA CALEDQ 3,23,77

SAFETY PROFILE: Questionable carcinogen with experimental tumorigenic data. Mutation data reported. When heated to decomposition it emits acrid smoke and irritating fumes.

DLC600 CAS:37571-88-3 HR: 2
trans-4,5-DIHYDRO-4,5-DIHYDROXYBENZO(a)PYRENE
mf: $C_{20}H_{14}O_2$ mw: 286.34

SYNS: (E)-BENZO(a)PYRENE-4,5-DIHYDRODIOL □ trans-4,5-DIHYDROBENZO(a)PYRENE-4,5-DIOL □ trans-4,5-DIHYDROXY-4,5-DIHYDROBENZO(a)PYRENE

TOXICITY DATA WITH REFERENCE
mma-sat 30 mg/L ENMUDM 7,839,85
dni-omi 2 mg/L PNASA6 74,1378,77
otr-ham:emb 1 mg/L IJCNAW 19,814,77
sce-ham:ovr 8 mg/L MUREAV 50,367,78
msc-ham:lng 25 mg/L CNREA8 36,3350,76

skn-mus TDLo:1 mg/kg:NEO BJCAAI 34,523,76

CONSENSUS REPORTS: EPA Genetic Toxicology Program.

SAFETY PROFILE: Questionable carcinogen with experimental neoplastigenic data by skin contact. Mutation data reported. When heated to decomposition it emits acrid smoke and irritating fumes.

DLD200 CAS:64920-31-6 HR: 2
trans-1,2-DIHYDRO-1,2-DIHYDROXYCHRYSENE
mf: $C_{18}H_{14}O_2$ mw: 262.32

SYNS: (E)-1,2-DIHYDRO-1,2-CHRYSENEDIOL □ trans-1,2-DIHYDROCHRYSENE-1,2-DIOL □ trans-1,2-DIHYDROXY-1,2-DIHYDROCHRYSENE

TOXICITY DATA WITH REFERENCE
mmo-sat 5 μg/plate CNREA8 44,3408,84
mma-sat 37,500 pmol/plate BBRCA9 78,847,77
ipr-mus TDLo:67 mg/kg/15D-I:NEO CNREA8 39,5063,79

SAFETY PROFILE: Questionable carcinogen with neoplastigenic data. Mutation data reported. When heated to decomposition it emits acrid smoke and irritating fumes.

DLD400 CAS:66267-19-4 HR: 2
trans-3,4-DIHYDRO-3,4-DIHYDROXYDIBENZ(a,h)ANTHRACENE
mf: $C_{22}H_{16}O_2$ mw: 312.38

SYNS: trans-DBA-3,4-DIHYDRODIOL □ trans-3,4-DIHYDRO-3,4-DIHYDROXYDIBENZO(a,h)ANTHRACENE

TOXICITY DATA WITH REFERENCE
skn-mus TDLo:500 μg/kg:NEO CNREA8 39,1310,79

SAFETY PROFILE: Questionable carcinogen with experimental neoplastigenic data by skin contact. When heated to decomposition it emits acrid smoke and irritating fumes.

DLD600 CAS:68162-13-0 HR: 2
trans-3,4-DIHYDRO-3,4-DIHYDROXY-7,12-DIMETHYLBENZ(a)ANTHRACENE
mf: $C_{20}H_{18}O_2$ mw: 290.38

SYN: trans-3,4-DIHYDRO-3,4-DIHYDROXY DMBA

TOXICITY DATA WITH REFERENCE
mma-sat 2500 nmol/L CBINA8 32,257,80
mma-sat 5 μmol/L BBRCA9 83,1468,78
otr-mus:fbr 120 μg/L BBRCA9 85,357,78
sce-ham:ovr 2 mg/L CALEDQ 7,45,79
msc-ham:lng 120 μg/L/3H BJCAAI 39,540,79
skn-mus TDLo:105 μg/kg:NEO CNREA8 39,1934,79
skn-mus TD:34,846 ng/kg:ETA CNREA8 40,3661,80

SAFETY PROFILE: Questionable carcinogen with experimental neoplastigenic and tumorigenic data by skin contact. Mutation data reported. When heated to decomposition it emits acrid smoke and irritating fumes.

DLD800 CAS:65763-32-8 HR: 2
trans-8,9-DIHYDRO-8,9-DIHYDROXY-7,12-DIMETHYLBENZ(a)ANTHRACENE
mf: $C_{20}H_{18}O_2$ mw: 290.38

SYNS: (E)-8,9-DIHYDRO-8,9-DIHYDROXY-7,12-DIMETHYLBENZ(a)ANTHRACENE □ trans-8,9-DIHYDRO-8,9-DIHYDROXY DMBA

TOXICITY DATA WITH REFERENCE
mma-sat 5 μmol/L BBRCA9 83,1468,78
otr-mus:fbr 250 mg/L BBRCA9 85,357,78
sce-ham:ovr 8 mg/L CALEDQ 7,45,79
msc-ham:lng 250 μg/L/3H BJCAAI 39,540,79
skn-mus TDLo:1050 μg/kg:ETA CNREA8 39,1934,79

SAFETY PROFILE: Questionable carcinogen with experimental tumorigenic data by skin contact. Mutation data reported. When heated to decomposition it emits acrid smoke and irritating fumes.

DLE000 CAS:64598-80-7 HR: 2
(±)-(1R,2S,3R,4R)-3,4-DIHYDRO-3,4-DIHYDROXY-1,2-EPOXYBENZ(a)ANTHRACENE
mf: $C_{18}H_{14}O_3$ mw: 278.32

SYNS: BA-3,4-DIOL-1,2-EPOXIDE-1 □ BA-3,4-DIOL-1,2-EPOXIDE-2 □ BENZ(a)ANTHRACENE 3,4-DIOL-1,2-EPOXIDE-2 □ (±)-3-α,4-β-DIHYDROXY-1-α,2-α-EPOXY-1,2,3,4-TETRAHYDROBENZ(a)ANTHRACENE □ (E)-1,2,3,4-TETRAHYDRO-3-α,4-β-DIHYDROXY-1-α,2-α-EPOXYBENZ(a)ANTHRACENE

TOXICITY DATA WITH REFERENCE
mmo-sat 100 pmol/plate CNREA8 43,1656,83
mma-sat 150 pmol/plate CRNGDP 4,1631,83
msc-ham:lng 20 μmol/L CNREA8 43,1656,83
skn-mus TDLo:22 mg/kg:ETA CNREA8 38,1699,78

CONSENSUS REPORTS: EPA Genetic Toxicology Program.

SAFETY PROFILE: Questionable carcinogen with experimental tumorigenic data by skin contact. Mutation data reported. When heated to decomposition it emits acrid smoke and irritating fumes.

DLE200 CAS:64598-81-8 HR: 2
(±)-(1S,2R,3R,4R)-3,4-DIHYDRO-3,4-DIHYDROXY-1,2-EPOXYBENZ(a)ANTHRACENE
mf: $C_{18}H_{14}O_3$ mw: 278.32

SYN: BA-3,4-DIOL-1,2-EPOXIDE-1

TOXICITY DATA WITH REFERENCE
ipr-mus TDLo:3100 μg/kg/15D-I:NEO JJIND8 63,201,79

SAFETY PROFILE: Questionable carcinogen with experimental neoplastigenic data. When heated to decomposition it emits acrid smoke and irritating fumes.

DLE400 CAS:102420-56-4 HR: 2
trans-1,2-DIHYDRO-1,2-DIHYDROXYINDENO(1,2,3-cd)PYRENE
mf: $C_{22}H_{14}O_2$ mw: 310.36

SYN: IP-1,2-DIOL

TOXICITY DATA WITH REFERENCE
skn-mus TDLo:40 mg/kg/20D-I:ETA CRNGDP 7,1761,86

SAFETY PROFILE: Questionable carcinogen with experimental tumorigenic data. When heated to decomposition it emits acrid smoke and irritating fumes.

DLE500 CAS:83876-50-0 HR: 2
cis-5,6-DIHYDRO-5,6-DIHYDROXY-12-METHYLBENZ (a)ACRIDINE
mf: $C_{18}H_{15}NO_2$ mw: 277.34

SYN: BENZ(a)ACRIDINE-5,6-DIOL, 5,6-DIHYDRO-12-METHYL-, (Z)-

TOXICITY DATA WITH REFERENCE
scu-mus TDLo:72 mg/kg/12W-I:ETA JMCMAR 26,303,83

SAFETY PROFILE: Questionable carcinogen with experimental tumorigenic data. When heated to decomposition it emits toxic fumes of NO_x.

DLF200 CAS:64521-15-9 HR: 2
trans-8,9-DIHYDRO-8,9-DIHYDROXY-7-METHYLBENZ (a)ANTHRACENE
mf: $C_{18}H_{16}O_2$ mw: 264.34

TOXICITY DATA WITH REFERENCE
mma-sat 30 μg/L BBRCA9 75,427,77
otr-mus:fbr 1 mg/L IJCNAW 19,828,77
sce-ham:ovr 8 mg/L MUREAV 50,367,78
msc-ham:lng 1 mg/L IJCNAW 19,828,77
skn-mus TDLo:1000 μg/kg:ETA CALEDQ 3,247,77

CONSENSUS REPORTS: EPA Genetic Toxicology Program.

SAFETY PROFILE: Questionable carcinogen with experimental tumorigenic data by skin contact. Mutation data reported. When heated to decomposition it emits acrid smoke and irritating fumes.

DLF400 CAS:67411-81-8 HR: 2
1,2-DIHYDRO-1,2-DIHYDROXY-5-METHYLCHRYSENE
mf: $C_{19}H_{16}O_2$ mw: 276.35

SYN: 1,2-DIHYDRO-5-METHYL-1,2-CHRYSENEDIOL

TOXICITY DATA WITH REFERENCE
mma-sat 7200 pmol/plate CNREA8 38,2191,78
skn-mus TDLo:36 μg/kg:CAR CNREA8 45,6406,79
skn-mus TDLo:1200 μg/kg/18D-I:NEO CNREA8 40,1396,80

SAFETY PROFILE: Questionable carcinogen with experimental carcinogenic and neoplastigenic data. Mutation data reported. When heated to decomposition it emits acrid smoke and irritating fumes.

DLF600 CAS:67523-22-2 HR: 2
7,8-DIHYDRO-7,8-DIHYDROXY-5-METHYLCHRYSENE
mf: $C_{19}H_{16}O_2$ mw: 276.35

SYN: 7,8-DIHYDRO-5-METHYL-7,8-CHRYSENEDIOL

TOXICITY DATA WITH REFERENCE
mma-sat 2700 pmol/plate CNREA8 38,2191,78
skn-mus TDLo:1200 μg/kg/18D-I:NEO CNREA8 40,1396,80

SAFETY PROFILE: Questionable carcinogen with experimental neoplastigenic data by skin contact. Mutation data reported. When heated to decomposition it emits acrid and irritating fumes.

DLF700 CAS:81840-15-5 HR: 3
3,4-DIHYDRO-6-(4-(3,4-DIMETHOXYBENZOYL)-1-PIPERAZINYL)-2(1H)-QUINOLINONE
mf: $C_{22}H_{25}N_3O_4$ mw: 395.50

PROP: Crystals from EtOH/$CHCl_3$. Mp: 238–239.5°.

SYNS: 1-(3,4-DIMETHOXYBENZOYL)-4-(1,2,3,4-TETRAHYDRO-2-OXO-6-QUINOLINYL)PIPERA ZINE ☐ OPC-8212 ☐ PIPERAZINE, 1-(3,4-DIMETHOXYBENZOYL)-4-(1,2,3,4-TETRAHYDRO-2-OXO-6-QUINOLINYL)-

TOXICITY DATA WITH REFERENCE
orl-rat TDLo:1100 mg/kg (female 7-17D post):REP IYKEDH 18,875,87
orl-rbt TDLo:3900 mg/kg (female 6-18D post):TER IYKEDH 18,898,87
orl-rat TDLo:11 g/kg (female 7-17D post):TER IYKEDH 18,875,87
orl-rat TDLo:1100 mg/kg (female 7-17D post):TER IYKEDH 18,875,87
orl-rat TDLo:330 mg/kg (female 7-17D post):TER IYKEDH 18,875,87
ivn-rat LD50:79,300 μg/kg IYKEDH 18,922,87
ivn-mus LD50:56,300 μg/kg IYKEDH 18,922,87
ivn-dog LD50:63,300 μg/kg IYKEDH 18,922,87

SAFETY PROFILE: Poison by intravenous route. An experimental teratogen. Experimental reproductive effects. When heated to decomposition it emits toxic fumes of NO_x.

DLG000 CAS:22797-20-2 HR: 3
5,10-DIHYDRO-10-(2-(DIMETHYLAMINO)ETHYL)-8-ETHYLSULFONYL-5-METHYL-11H-DIBENZO(b,e) (1,4)DIAZEPIN-11-ONE
mf: $C_{20}H_{25}N_3O_3S$ mw: 387.54

SYN: SM-307

TOXICITY DATA WITH REFERENCE
orl-mus LD50:857 mg/kg JJPAAZ 21,47,71
ipr-mus LD50:354 mg/kg JJPAAZ 21,47,71
ivn-mus LD50:74 mg/kg JJPAAZ 21,47,71

SAFETY PROFILE: Poison by intraperitoneal and intravenous routes. Moderately toxic by ingestion. When heated to decomposition it emits very toxic fumes of NO_x and SO_x.

DLH200 CAS:739-71-9 HR: 3
10,11-DIHYDRO-5-(3-DIMETHYLAMINO-2-METHYLPROPYL)-5H-DIBENZ (b,f)AZEPINE
mf: $C_{20}H_{26}N_2$ mw: 294.48

SYNS: 10,11-DIHYDRO-N,N,β-TRIMETHYL-5H-DIBENZ(b,f)AZEPINE-5-PROPANAMINE ☐ 1-(3-DIMETHYLAMINO-2-METHYLPROPYL)-4,5-DIHYDRO-2,3:6,7-DIBENZAZEPINE ☐ 5-(3-(DIMETHYLAMINO)-2-METHYLPROPYL)-10,11-DIHYDRO-5H-DIBENZ(b,f)AZEPINE ☐ 5-(γ-DIMETHYLAMINO-β-METHYLPROPYL)-10,11-DIHYDRO-5H-DIBENZO(b,f) AZEPINE ☐ FI 6120 ☐ IL 6001 ☐ 2′-METIL-3′-DIMETILAMINO-PROPIL-5-IMINODIBENZILE (ITALIAN) ☐ 7162 RP ☐ SAPILENT ☐ SUR-

MONTIL □ TRIMEPRIMINA (ITALIAN) □ TRIMEPROPIMINE □ TRIMIPRAMINE

TOXICITY DATA WITH REFERENCE
orl-hmn TDLo:18 mg/kg:CVS PSDTAP 6,171,65
orl-mus LD50:250 mg/kg BCFAAI 102,753,63
ipr-mus LD50:145 mg/kg CRSBAW 155,307,61
scu-mus LD50:200 mg/kg BCFAAI 102,753,63
ivn-mus LD50:42 mg/kg CRSBAW 155,307,61

CONSENSUS REPORTS: EPA Genetic Toxicology Program.

SAFETY PROFILE: Poison by ingestion, intraperitoneal, subcutaneous, and intravenous routes. Human systemic effects by ingestion: unspecified heart effects. When heated to decomposition it emits toxic fumes of NO_x.

DLH600 CAS:50-49-7 HR: 3
5,6-DIHYDRO-N-(3-(DIMETHYLAMINO)PROPYL)-11H-DIBENZ(b,e)AZEPINE
mf: $C_{19}H_{24}N_2$ mw: 280.45

PROP: A liquid. Bp: 174–175° @ 0.1 mm.

SYNS: ANTIDEPRIN □ BERKOMINE □ CENSTIN □ 10,11-DIHYDRO-5-(3-(DIMETHYLAMINO)PROPYL)-5H-DIBENZ(b,f)AZEPINE □ 2, 2′-(3-DIMETHYLAMINOPROPYLAMINO)BIBENZYL □ 1-(3-DIMETHYLAMINOPROPYL)-4,5-DIHYDRO-2,3,6,7-DIBENZAZEPINE □ 5-(3-DIMETHYLAMINOPROPYL)-10,11-DIHYDRO-5H-DIBENZO(b,f)AZEPINE □ N-(γ-DIMETHYLAMINOPROPYL)IMINODIBENZYL □ 2,2′-(3-DIMETHYLAMINOPROPYLIMINO)DIBENZYL □ DIMIPRESSIN □ DPID □ DYNAPRIN □ DYNA-ZINA □ EUPRAMIN □ G 22355 □ IM □ IMIDOBENZYLE □ IMIPRAMINA (ITALIAN) □ IMIPRAMINE □ IMIPRIN □ IMIZIN □ IMIZINUM □ IMPRAMINE □ INTALPRAM □ IRAMIL □ IRMIN □ MELIPRAMIN □ MELIPRAMINE □ NELIPRAMIN □ PRAZEPINE □ PROMIBEN □ SURPLIX □ TIMOLET □ TOFRANIL

TOXICITY DATA WITH REFERENCE
dni-oin-unr 10 g/L JCLBA3 47,182a,70
cyt-oin-unr 10 g/L JCLBA3 47,182a,70
orl-rat TDLo:378 mg/kg (lactating female 21D post):REP PSCHDL 56,93,78
scu-rat TDLo:40 mg/kg (female 18-21D post):REP ADBBBW 8,171,72
orl-rat TDLo:165 mg/kg (female 14D pre):REP PBBHAU 3,799,75
orl-mus TDLo:87,500 µg/kg (female 7-13D post):REP OYYAA2 4,855,70
orl-rat TDLo:600 mg/kg (60D male/60D pre):REP ARZNAD 15,1218,65
scu-rbt TDLo:55 mg/kg (female 6-16D post):TER ARZNAD 15,1218,65
orl-rbt TDLo:390 mg/kg (female 6-18D post):REP IYKEDH 7,195,76
ipr-mus TDLo:188 mg/kg (male 5D pre):REP IMSCE2 14,1129,86
scu-rat TDLo:130 mg/kg (female 8-20D post):TER TJADAB 31,46A,85
scu-rbt TDLo:255 mg/kg (female 1-17D post):TER LANCAO 1,638,63
orl-rat TDLo:75 mg/kg (9-14D preg):TER OYYAA2 4,855,70
ipc-mus TDLo:80 µg/kg (female 15D post):TER 40YJAX -,207,79
orl-wmn LDLo:2 mg/kg/1D:PUL JPPMAB 16,265,64
orl-man LD50:30 mg/kg HEPHD2 55,527,80
orl-cld LD50:40 mg/kg HEPHD2 55,527,80
orl-wmn TDLo:3 mg/kg/32H-I:SKN JAMAAP 254,357,85
orl-chd TDLo:30 mg/kg:CNS AJDCAI 130,507,76
orl-cld LDLo:35 mg/kg SMWOAS 99,1157,69
orl-hmn LDLo:450 mg/kg:SKN BMJOAE 1,722,79
orl-hmn LD50:40 mg/kg PSDTAP 6,171,65
orl-man TDLo:8 mg/kg/3D-I:CNS LANCAO 2,568,59
orl-rat LD50:250 mg/kg PCJOAU 14,773,80
ipr-rat LD50:79 mg/kg AIPTAK 148,560,64
scu-rat LD50:250 mg/kg AIPTAK 148,560,64
ivn-rat LD50:15,900 µg/kg ARZNAD 29,193,79
orl-mus LD50:188 mg/kg PCJOAU 14,773,80
ipr-mus LD50:51,600 µg/kg BRXXAA #1460700
scu-mus LD50:195 µg/kg PCJOAU 15,412,81
ivn-mus LD50:21 mg/kg AIPTAK 245,283,80
orl-dog LDLo:100 mg/kg 27ZQAG -,78,72

CONSENSUS REPORTS: EPA Genetic Toxicology Program.

SAFETY PROFILE: A human poison by ingestion. An experimental poison by ingestion, subcutaneous, intravenous, and intraperitoneal routes. Human systemic effects by ingestion: somnolence, hallucinations, distorted perceptions, changes in motor activity, ataxia (loss of muscle coordination), coma, nausea and vomiting, irritative dermatitis. An experimental teratogen by ingestion. Other experimental reproductive effects. Mutation data reported. When heated to decomposition it emits toxic fumes of NO_x.

DLH630 CAS:113-52-0 HR: 3
10,11-DIHYDRO-5-(3-(DIMETHYLAMINO)PROPYL)-5H-DIBENZ(b,f)AZEPINE HYDROCHLORIDE
mf: $C_{19}H_{24}N_2$•ClH mw: 316.91

PROP: Crystals from Me_2CO. Mp: 174–175°.

SYNS: ANTIDEPRIN HYDROCHLORIDE □ BERKOMINE □ CENSTIM □ CENSTIN □ CHIMOREPTIN □ CHRYTEMIN □ CO CAP IMIPRAMINE 25 □ DEPRINOL □ 10,11-DIHYDRO-N,N-DIMETHYL-5H-DIBENZ(b,f)AZEPINE 5 PROPANAMINE MONOHYDROCHLORIDE □ 5-(3-DIMETHYLAMINOPROPYL)-10,11-DIHYDRO-5H-DIBENZ(b,f)AZEPINE HYDROCHLORIDE □ N-(3-DIMETHYLAMINOPROPYL)IMINODIBENZYL HYDROCHLORIDE □ N-(γ-DIMETILAMINOPROPIL)-IMINODIBENZILE CLORIDRATO (ITALIAN) □ DIMIPRESSIN □ DYNA-ZINA □ EFURANOL □ EUPRAMIN □ FEINALMIN □ G 22150 □ G 22355 □ IA-PRAM □ IMAVATE □ IMIDOBENZYLE /S IMIDOL □IMILANYLE □ IMIPRAMINA (ITALIAN) □ IMIPRAMINE □ IMIPRAMINE HYDROCHLORIDE □ IMIPRAMINE MONOHYDROCHLORIDE □ IMIPRIN □ IMP HYDROCHLORIDE □ INTALPRAM □ IPROGEN □ IRAMIL □ JANIMINE □ LOFEPRAMINE □ MELIPRAMIN □ MELIPRAMINE □ MELIPRAMINE HYDROCHLORIDE □ MELIPRAMIN HYDROCHLORIDE □ NSC 114900 □ PERSAMINE □ PERTOFRAM □ PRESAMINE □ PROMIBEN □ PYRLEUGAN □ SK-PRAMINE □ SK-PRAMINE HYDROCHLORIDE □ SURPLIX □ TEPERINE □ TIMOLET □ TOFRANIL □ TOFRANILE

TOXICITY DATA WITH REFERENCE
dni-oin-unr 10 g/L JCLBA3 47,182a,70
cyt-oin-unr 10 g/L JCLBA3 47,182a,70
orl-rat TDLo:285 mg/kg (female 14D pre-21D post):REP PSYPAG 41,237,75
scu-rat TDLo:65 mg/kg (female 8-20D post):REP TCMUD8 6,173,86
ipr-rat TDLo:210 mg/kg (female 15-21D post):REP DPTHDL 8,17,85

orl-rat TDLo:180 mg/kg (female 14D pre):REP PSYPAG 41,237,75
scu-rbt TDLo:165 mg/kg (female 3-18D post):REP APTOA6 20,186,63
orl-mus TDLo:9700 mg/kg (female 75D pre):REP RCOCB8 19,311,78
ipr-rat TDLo:225 mg/kg (female 23D pre):REP PSYPAG 32,337,73
scu-mus TDLo:125 mg/kg (female 9D post):TER DGDFA5 22,61,80
orl-rat TDLo:800 mg/kg (female 1-16D post):TER ARZNAD 15,1222,65
orl-rat TDLo:1 g/kg (female 7-16D post):REP ARZNAD 15,1222,65
orl-rat TDLo:200 mg/kg (female 7-16D post):REP ARZNAD 15,1222,65
ipr-ham TDLo:35 mg/kg (female 8D post):TER RCPBDC 5,275,80
scu-rat TDLo:130 mg/kg (female 8-20D post):TER TCMUD8 6,173,86
scu-rat TDLo:65 mg/kg (female 8-20D post):TER TCMUD8 6,173,86
orl-man TDLo:2143 µg/kg/2D-I JCLPDE 44,225,83
orl-chd TDLo:25 mg/kg:CNS,PUL JAMAAP 179,456,62
orl-chd LDLo:15 mg/kg BMJOAE 1,261,74
orl-chd TDLo:27 mg/kg:CNS,KID JAMAAP 230,1405,74
orl-wmn TDLo:107 mg/kg:CNS,CVS NEJMAG 268,33,63
orl-wmn TDLo:30 mg/kg:PNS,CNS BMJOAE 2,1458,59
orl-rat LD50:305 mg/kg TXAPA9 18,185,71
ipr-rat LD50:72 mg/kg ARZNAD 21,391,71
scu-rat LD50:217 mg/kg ARZNAD 21,391,71
ivn-rat LD50:18 mg/kg ATXKA8 21,30,65
orl-mus LD50:275 mg/kg THERAP 20,67,65
ipr-mus LD50:104 mg/kg ARZNAD 24,166,74
scu-mus LD50:189 mg/kg FRPPAO 25,519,70
ivn-mus LD50:27 mg/kg ARZNAD 31,75,81
ivn-mky LDLo:25 mg/kg IJEBA6 22,539,84
ipr-gpg LD50:85 mg/kg PHARAT 38,749,83
scu-gpg LD50:190 mg/kg AIPTAK 137,375,62

CONSENSUS REPORTS: Reported in EPA TSCA Inventory.

SAFETY PROFILE: Human poison by ingestion. An experimental poison by ingestion, intravenous, subcutaneous, and intraperitoneal routes. An experimental teratogen. Human systemic effects by ingestion: sleep, somnolence, convulsions, muscle contraction or spasticity, coma, blood pressure decrease, dyspnea (difficulty in breathing), paresthesia (abnormal sensations), and kidney changes. Experimental reproductive effects. Mutation data reported. Used in the treatment of depression. When heated to decomposition it emits very toxic fumes of NO_x and HCl. See also DIAZEPAM.

DLH800 CAS:35281-29-9 HR: 2
5,6-DIHYDRO-7,12-DIMETHYLBENZ(a)ANTHRACENE
mf: $C_{20}H_{18}$ mw: 258.38

TOXICITY DATA WITH REFERENCE
skn-mus TDLo:128 mg/kg/50W-I:CAR ZKKOBW 77,226,72

SAFETY PROFILE: Questionable carcinogen with experimental carcinogenic data by skin contact. When heated to decomposition it emits acrid smoke and irritating fumes.

DLI000 CAS:52171-93-4 HR: 2
3,4-DIHYDRO-1,11-DIMETHYLCHRYSENE
mf: $C_{20}H_{18}$ mw: 258.38

TOXICITY DATA WITH REFERENCE
mma-sat 20 µg/plate CNREA8 36,4525,76
skn-mus TDLo:120 mg/kg/50W-I:NEO CNREA8 34,1315,74

CONSENSUS REPORTS: EPA Genetic Toxicology Program.

SAFETY PROFILE: Questionable carcinogen with experimental neoplastigenic data by skin contact. Mutation data reported. When heated to decomposition it emits acrid smoke and irritating fumes. See also CHRYSENE.

DLI200 CAS:5831-16-3 HR: 2
16,17-DIHYDRO-11,17-DIMETHYLCYCLOPENTA(a) PHENANTHRENE
mf: $C_{19}H_{17}$ mw: 245.36

SYN: 11,17-DIMETHYL-16,17-DIHYDRO-15H-CYCLOPENTA(a)PHENANTHRENE

TOXICITY DATA WITH REFERENCE
mmo-sat 20 µg/plate CNREA8 36,4525,76
skn-mus TDLo:108 mg/kg/1Y-I:ETA PEXTAR 11,69,69

CONSENSUS REPORTS: EPA Genetic Toxicology Program.

SAFETY PROFILE: Questionable carcinogen with experimental tumorigenic data by skin contact. Mutation data reported. When heated to decomposition it emits acrid smoke and irritating fumes.

DLI300 CAS:85616-56-4 HR: 2
15,16-DIHYDRO-7,11-DIMETHYL-17H-CYCLOPENTA (a)PHENANTHREN-17-ONE
mf: $C_{19}H_{16}O$ mw: 260.35

SYN: 7,11-DIMETHYL-15,16-DIHYDROCYCLOPENTA(a)PHENANTHREN-17-ONE

TOXICITY DATA WITH REFERENCE
skn-mus TDLo:40 mg/kg/10W-I:CAR CNREA8 46,1817,86

SAFETY PROFILE: Questionable carcinogen with experimental carcinogenic data. When heated to decomposition it emits acrid smoke and irritating fumes.

DLI400 CAS:894-52-0 HR: 2
15,16-DIHYDRO-11,12-DIMETHYLCYCLOPENTA(a) PHENANTHREN-17-ONE
mf: $C_{19}H_{16}O$ mw: 260.35

TOXICITY DATA WITH REFERENCE
mma-sat 50 µg/plate CNREA8 36,4525,76
skn-mus TDLo:108 mg/kg/1Y-I:CAR PEXTAR 11,69,69
scu-mus TDLo:360 mg/kg:CAR PEXTAR 11,69,69

CONSENSUS REPORTS: EPA Genetic Toxicology Program.

SAFETY PROFILE: Questionable carcinogen with experimental carcinogenic data. Mutation data reported. When heated to decomposition it emits acrid smoke and irritating fumes.

DLI600 CAS:1920-21-4 HR: 2
2,3-DIHYDRO-2,5-DIMETHYL-2-FORMYL-1,4-PYRAN
mf: $C_8H_{12}O_2$ mw: 140.20

SYNS: 3,4-DIHYDRO-2,5-DIMETHYL-2H-PYRAN-2-CARBOXALDEHYDE □ METHACROLEIN DIMER □ METHACRYLALDEHYDE DIMER

TOXICITY DATA WITH REFERENCE
skn-rbt 10 mg/24H open MLD AMIHBC 10,61,54
eye-rbt 20 mg open AMIHBC 10,61,54
orl-rat LD50:2460 mg/kg AMIHBC 10,61,54

SAFETY PROFILE: Moderately toxic by ingestion. A skin and eye irritant. When heated to decomposition it emits acrid smoke and irritating fumes.

DLI650 CAS:78499-27-1 HR: 3
10,11-DIHYDRO-α-8-DIMETHYL-11-OXO-DIBENZ(b,f) OXEPIN-2-ACETIC ACID
mf: $C_{18}H_{16}O_4$ mw: 296.34

SYNS: AD-1590 □ DIBENZ(b,f)OXEPIN-2-ACETIC ACID, 10,11-DIHYDRO-α-8-DIMETHYL-11-OXO- □ 2-(8-METHYL-10,11-DIHYDRO-11-OXODIBENZ(b,f)OXEPIN-2-YL)PROPIONIC □ PROPIONIC ACID, 2-(8-METHYL-10,11-DIHYDRO-11-OXODIBENZ(b,f)OXEPIN-2-YL)-

TOXICITY DATA WITH REFERENCE
orl-rat TDLo:33 mg/kg (female 7-17D post):REP YACHDS 16,2797,88
orl-rat TDLo:1600 μg/kg (female 17-20D post):REP YACHDS 16,2815,88
orl-rat LD50:147 mg/kg JMCMAR 25,1065,82
ipr-rat LD50:391 mg/kg YACHDS 16,2701,88
scu-rat LD50:483 mg/kg YACHDS 16,2701,88
ivn-rat LDLo:160 mg/kg YKKZAJ 108,788,88
orl-mus LD50:212 mg/kg YACHDS 16,2701,88
ipr-mus LD50:191 mg/kg YACHDS 16,2701,88
scu-mus LD50:387 mg/kg YACHDS 16,2701,88

SAFETY PROFILE: Poison by ingestion, intraperitoneal, and subcutaneous routes. Experimental reproductive effects. When heated to decomposition it emits acrid smoke and irritating fumes.

DLJ500 CAS:66289-74-5 HR: 3
endo,endo-DIHYDRODI(NORBORNADIENE)
mf: $C_{14}H_{18}$ mw: 186.32

SYNS: 4,7-METHANO-2,3,8-METHENOCYCLOPENT(a)INDENE, DODECAHYDRO-, stereoisomer □ RJ 5 □ SHELLOYNE H

TOXICITY DATA WITH REFERENCE
ihl-rat TCLo:150 mg/m³/6H/1Y-I:ETA NTIS** AD-A134-150
ihl-rat TC:150 mg/m³/6H/1Y-I:ETA AETODY 7,133,84
orl-mus LDLo:250 mg/kg AMRL** TR-74-78,74

SAFETY PROFILE: Poison by ingestion. Questionable carcinogen with experimental tumorigenic data. When heated to decomposition it emits acrid smoke and irritating fumes.

DLJ600 CAS:14938-42-2 HR: 2
9,10-DIHYDRO-9,10-DIOXO-1,8-ANTHRACENEDISULFONIC ACID DIPOTASSIUM SALT
mf: $C_{14}H_6O_8S_2$•2K mw: 444.52

PROP: Yellow needles from H_2O. Sltly sol in H_2O.

SYNS: ANTHRACHINON-1,8-DISULFONAN DRASELNY (CZECH) □ ANTHRAQUINONEDISULFONIC ACID, DIPOTASSIUM SALT

TOXICITY DATA WITH REFERENCE
skn-rbt 500 mg/24H MLD 28ZPAK -,193,72
eye-rbt 100 mg/24H MOD 28ZPAK -,193,72
orl-rat LD50:15 g/kg 28ZPAK -,193,72

SAFETY PROFILE: Mildly toxic by ingestion. A skin and eye irritant. When heated to decomposition it emits toxic fumes of SO_x and K_2O.

DLJ700 CAS:853-35-0 HR: 1
9,10-DIHYDRO-9,10-DIOXO-1,5-ANTHRACENE DISULFONIC ACID DISODIUM SALT
mf: $C_{14}H_6O_8S_2$•2Na mw: 412.30

PROP: Yellow leaflets from H_2O. Sol in H_2O.

SYNS: ANTHRACHINON-1,5-DISULFONAN SODNY (CZECH) □ DISODIUM ANTHRAQUINONE-1,5-DISULFONATE □ SODIUM ANTHRAQUINONE-1,5-DISULFONATE

TOXICITY DATA WITH REFERENCE
eye-rbt 500 mg/24H MLD 28ZPAK ,193,72

CONSENSUS REPORTS: Reported in EPA TSCA Inventory.

SAFETY PROFILE: An eye irritant. When heated to decomposition it emits toxic fumes of SO_x and Na_2O. See also SULFONATES.

DLJ800 CAS:128-56-3 HR: 1
9,10-DIHYDRO-9,10-DIOXO-1-ANTHRACENE SULFONIC ACID SODIUM SALT
mf: $C_{14}H_7O_5S$•Na mw: 310.26

PROP: Yellow leaflets. Sltly sol H_2O.

SYNS: ANTHRACHINON-1-SULFONAN SODNY (CZECH) □ SODIUM ANTHRAQUINONE-1-SULFONATE

TOXICITY DATA WITH REFERENCE
eye-rbt 500 mg/24H MLD 28ZPAK -,192,72
orl-rat LD50:20 g/kg GISAAA 45(3),73,80
orl-mus LD50:32 g/kg GISAAA 45(3),73,80
orl-rbt LD50:14 g/kg GISAAA 45(3),73,80
orl-gpg LD50:32 g/kg GISAAA 45(3),73,80

CONSENSUS REPORTS: Reported in EPA TSCA Inventory.

SAFETY PROFILE: Very mildly toxic by ingestion. When heated to decomposition it emits toxic fumes of SO_x and Na_2O.

DLK000 CAS:521-24-4 HR: 2
3,4-DIHYDRO-3,4-DIOXO-1-NAPHTHALENE SULFONIC ACID SODIUM SALT
mf: $C_{10}H_6O_5S{\cdot}Na$ mw: 261.21

PROP: Golden-yellow needles from EtOH (aq) or orange powder. Sol in H_2O; mod sol in Me_2CO; insol in C_6H_6.

SYNS: β-NAPHTHOQUINONE-4-SULFONATE SODIUM SALT □ SODIUM-β-NAPHTHOQUINONE-4-SULFONATE □ SODIUM-1,2-NAPHTHOQUINONE-4-SULFONATE

TOXICITY DATA WITH REFERENCE
ipr-mus LD50:625 mg/kg NTIS** AD691-490

CONSENSUS REPORTS: Reported in EPA TSCA Inventory.

SAFETY PROFILE: Moderately toxic by intraperitoneal route. When heated to decomposition it emits toxic fumes of SO_x and Na_2O.

DLK200 CAS:3347-22-6 HR: 3
5,10-DIHYDRO-5,10-DIOXONAPHTHO(2,3-b)-p-DITHIIN-2,3-DICARBONITRILE
mf: $C_{14}H_4N_2O_2S_2$ mw: 296.32

PROP: Brown crystals. Mp: 225°.

SYNS: DELAN □ DELAN-COL □ 2,3-DICARBONITRILO-1,4-DIATHIAANTHRACHINON (GERMAN) □ 2,3-DICYANO-1,4-DITHIA-ANTHRAQUINONE □ 2,3-DINITRILO-1,4-DITHIA-ANTHRAQUINONE □ 2,3-DINITRILO-1,4-DITHIOANTHRACHINON (GERMAN) □ 1,4-DITHIAANTHRAQUINONE-2,3-DICARBONITRILE □ 1,4-DITHIAANTHRAQUINONE-2,3-DINITRILE □ DITHIANON □ DITHIANONE □ DTA □ IT 931 □ MV 119A □ STAUFFER MV-119A □ THYNON

TOXICITY DATA WITH REFERENCE
mor-mus:fbr 25 mg/L EMMUEG 21,81,93
orl-rat LD50:638 mg/kg FMCHA2 -,C86,83
unk-rat LD50:1015 mg/kg 30ZDA9 -,425,71
orl-mus LD50:1140 mg/kg 31ZOAD 1,189,68
orl-gpg LD50:110 mg/kg 28ZEAL 5,86,76

CONSENSUS REPORTS: Cyanide and its compounds are on the Community Right-To-Know List.

SAFETY PROFILE: Poison by ingestion. Moderately toxic by an unspecified route. Mutation data reported. A fungicide. See also NITRILES. When heated to decomposition it emits very toxic fumes of NO_x, SO_x, and CN^-.

DLK600 CAS:63041-56-5 HR: 2
7,14-DIHYDRO-7,14-DIPROPYLDIBENZ(a,h)ANTHRACENE-7,14-DIOL
mf: $C_{28}H_{28}O_2$ mw: 396.56

SYNS: 9,10-DIHYDRO-9,10-DIHYDROXY-9,10-DI-n-PROPYL-1,2:5,6-DIBENZANTHRACENE □ 9,10-DIHYDROXY-9,10-DI-n-PROPYL-9,10-DIHYDRO-1,2:5,6-DIBENZANTHRACENE □ 9,10-DI-n-PROPYL-9-10-DIHYDROXY-9,10-DIHYDRO-1,2,5,6-DIBENZANTHRACENE

TOXICITY DATA WITH REFERENCE
scu-rat TDLo:8 mg/kg:ETA JOCEAH 2,175,37

SAFETY PROFILE: Questionable carcinogen with experimental tumorigenic data. When heated to decomposition it emits acrid smoke and irritating fumes.

DLK700 CAS:24519-85-5 HR: 3
5,6-DIHYDRO-p-DITHIIN-2,3-DICARBOXIMIDE
mf: $C_6H_5NO_2S_2$ mw: 187.24

SYN: 3,6-DITHIA-3,4,5,6-TETRAHYDROPHTHALIMIDE

TOXICITY DATA WITH REFERENCE
ipr-mus TDLo:25 mg/kg (9D preg):REP MOPMA3 13,133,77
ipr-mus TDLo:50 mg/kg (9D preg):TER MOPMA3 13,133,77
ipr-mus LD50:64 mg/kg DPHFAK 23,113,71

SAFETY PROFILE: Poison by intraperitoneal route. An experimental teratogen. Experimental reproductive effects. When heated to decomposition it emits toxic fumes of SO_x and NO_x.

DLK750 HR: 2
1,2-DIHYDRO-1,2-EPOXYINDENO(1,2,3-cd)PYRENE
mf: $C_{22}H_{12}O$ mw: 292.34

SYN: INDENO(1,2,3-cd)PYRENE-1,2-OXIDE

TOXICITY DATA WITH REFERENCE
mmo-sat 1 μg/plate CNREA8 45,5421,85
skn-mus TDLo:40 mg/kg/20D-I:ETA CRNGDP 7,1761,86

SAFETY PROFILE: Questionable carcinogen with experimental tumorigenic data. Mutation data reported. When heated to decomposition it emits acrid smoke and irritating fumes.

DLK800 CAS:511-12-6 HR: 3
DIHYDROERGOTAMINE
mf: $C_{33}H_{37}N_5O_5$ mw: 583.65

PROP: Prisms from Me_2CO. Mp: 239°.

SYN: DEHYDROERGOTAMINE

TOXICITY DATA WITH REFERENCE
ivn-rat LD50:110 mg/kg HPPAAL 2,48,44
scu-mus LD50:80 mg/kg RPTOAN 31,53,68

SAFETY PROFILE: Poison by intravenous and subcutaneous routes. When heated to decomposition it emits toxic fumes of NO_x.

DLL000 CAS:5989-77-5 HR: 3
DIHYDROERGOTAMINE TARTRATE (2:1)
mf: $C_{66}H_{74}N_{10}O_{10}{\cdot}C_4H_6O_6$ mw: 1317.60

TOXICITY DATA WITH REFERENCE
ivn-rat LD50:110 mg/kg BSAMA5 2,1,46
ipr-mus LD50:210 mg/kg RPOBAR 2,284,70
ivn-mus LD50:118 mg/kg BSAMA5 2,1,46
scu-cat LD50:68 mg/kg BSAMA5 2,1,46
ivn-rbt LD50:25 mg/kg BSAMA5 2,1,46

SAFETY PROFILE: Poison by intravenous, intraperitoneal, and subcutaneous routes. When heated to decomposition it emits toxic fumes of NO_x.

DLL400 CAS:8067-24-1 HR: 3
DIHYDROERGOTOXINE METHANE SULFONATE

SYNS: CCK 179 ☐ CIRCANOL ☐ CO-DERGOCRINE MESYLATE ☐ DIHYDROERGOTOXINE MESYLATE ☐ DIHYDROERGOTOXINE METHANESULPHONATE ☐ DIHYDROERGOTOXINE MONOMETHANE-SULFONATE (SALT) ☐ DIHYDROERGOTOXIN MESYLATE ☐ DIHY-DROERGOTOXIN METHANESULFONATE ☐ HYDERGIN ☐ HYDER-GINE ☐ ISCHELIUM ☐ REDERGIN

TOXICITY DATA WITH REFERENCE
dlt-mus-ipr 100 mg/kg MUREAV 50,317,78
ipr-rat TDLo:10 mg/kg (female 12-21D post):REP AJPHAP 180,296,55
orl-rat LD50:>1 g/kg DRUGAY-,469,90
ipr-rat LD50:500 mg/kg DRUGAY-,469,90
scu-rat LD50:>2 g/kg DRUGAY-,469,90
ivn-rat LD50:86 mg/kg DRUGAY-,469,90

CONSENSUS REPORTS: EPA Genetic Toxicology Program.

SAFETY PROFILE: Poison by intravenous route. Experimental reproductive effects. Mutation data reported. When heated to decomposition it emits very toxic fumes of SO_x and NO_x.

DLL600 CAS:29734-68-7 HR: 3
DIHYDRO-β-ERYTHROIDINE HYDROBROMIDE
mf: $C_{16}H_{21}NO_3{\cdot}BrH$ mw: 356.30

TOXICITY DATA WITH REFERENCE
ipr-rat LDLo:30 mg/kg JPETAB 75,270,42
ivn-rat LD50:8900 μg/kg JPETAB 82,266,44
ivn-dog LD50:1100 μg/kg JPETAB 82,266,44
ivn-rbt LD50:2100 μg/kg JPETAB 82,266,44

SAFETY PROFILE: A deadly poison by intravenous and intraperitoneal routes. When heated to decomposition it emits very toxic fumes of Br^- and NO_x.

DLM000 CAS:42028-27-3 HR: 2
15,16-DIHYDRO-11-ETHYLCYCLOPENTA(a)PHENAN-THREN-17-ONE
mf: $C_{19}H_{16}O$ mw: 260.35

TOXICITY DATA WITH REFERENCE
mma-sat 50 μg/plate CNREA8 36,4525,76
skn-mus TDLo:120 mg/kg/50W-I:NEO CNREA8 33,832,73

CONSENSUS REPORTS: EPA Genetic Toxicology Program.

SAFETY PROFILE: Questionable carcinogen with experimental neoplastigenic data. Mutation data reported. When heated to decomposition it emits acrid smoke and irritating fumes.

DLM600 CAS:5096-24-2 HR: 2
2,3-DIHYDRO-3-ETHYL-6-METHYL-1H-CYCLOPENTA (a)ANTHRACENE
mf: $C_{20}H_{20}$ mw: 260.40

SYN: 3-ETHYL-2,3-DIHYDRO-6-METHYL-1H-CYCLOPENT(a)ANTHRA-CENE

TOXICITY DATA WITH REFERENCE
orl-rat TDLo:1000 mg/kg:ETA CNREA8 26,619,66
orl-rat LDLo:1000 mg/kg CNREA8 26,619,66

SAFETY PROFILE: Moderately toxic by ingestion. Questionable carcinogen with experimental tumorigenic data. When heated to decomposition it emits acrid smoke and irritating fumes.

D

DLN000 CAS:52831-41-1 HR: 2
6,13-DIHYDRO-2-FLUOROBENZO(g)(1)BENZOTHIO-PYRANO(4,3-b)INDOLE
mf: $C_{19}H_{12}FNS$ mw: 305.38

TOXICITY DATA WITH REFERENCE
mma-sat 30 μg/plate MUREAV 66,307,79
scu-mus TDLo:78 mg/kg/9W-I:NEO MUREAV 66,307,79

SAFETY PROFILE: Questionable carcinogen with experimental neoplastigenic data. Mutation data reported. When heated to decomposition it emits very toxic fumes of F^-, NO_x, and SO_x.

DLN200 CAS:52831-55-7 HR: 2
6,13-DIHYDRO-3-FLUOROBENZO(e)(1)BENZOTHIOPY-RANO(4,3-b)INDOLE
mf: $C_{19}H_{12}FNS$ mw: 305.38

TOXICITY DATA WITH REFERENCE
mma-sat 30 μg/plate MUREAV 66,307,79
scu-mus TDLo:72 mg/kg/9W-I:NEO MUREAV 66,307,79

SAFETY PROFILE: Questionable carcinogen with experimental neoplastigenic data. Mutation data reported. When heated to decomposition it emits very toxic fumes of F^-, NO_x, and SO_x.

DLN400 CAS:52831-67-1 HR: 2
6,13-DIHYDRO-4-FLUOROBENZO(e)(1)BENZOTHIOPY-RANO(4,3-b)INDOLE
mf: $C_{19}H_{12}FNS$ mw: 305.38

TOXICITY DATA WITH REFERENCE
mma-sat 30 μg/plate MUREAV 66,307,79
scu-mus TDLo:72 mg/kg/9W-I:NEO MUREAV 66,307,79

SAFETY PROFILE: Questionable carcinogen with experimental neoplastigenic data. Mutation data reported. When heated to decomposition it emits very toxic fumes of F^-, NO_x, and SO_x.

DLO000 CAS:22298-04-0 HR: 2
6,11-DIHYDRO-2-FLUORO(1)BENZOTHIOPYRANO(4,3-b)INDOLE
mf: $C_{15}H_{10}FNS$ mw: 255.32

SYN: 6,11-DIHYDRO-2-FLUORO-THIOPYRANO(4,3-b)BENZ(e)IN-DOLE

TOXICITY DATA WITH REFERENCE
mma-sat 100 μg/plate MUREAV 66,307,79
scu-mus TDLo:72 mg/kg/9W-I:NEO MUREAV 66,307,79

SAFETY PROFILE: Questionable carcinogen with experimental neoplastigenic data. Mutation data reported.

When heated to decomposition it emits very toxic fumes of F^-, SO_x, and NO_x.

DLO200 CAS:21243-26-5 HR: 2
6,11-DIHYDRO-4-FLUORO(1)BENZOTHIOPYRANO(4,3-b)INDOLE
mf: $C_{15}H_{10}FNS$ mw: 255.32

TOXICITY DATA WITH REFERENCE
mma-sat 90 μg/plate MUREAV 66,307,79
scu-mus TDLo:78 mg/kg/9W-I:NEO MUREAV 66,307,79

SAFETY PROFILE: Questionable carcinogen with experimental neoplastigenic data. Mutation data reported. When heated to decomposition it emits very toxic fumes of F^-, SO_x, and NO_x.

DLO400 CAS:18497-13-7 HR: 3
DIHYDROGEN HEXACHLOROPLATINATE HEXAHYDRATE
mf: $Cl_6Pt•2H•6H_2O$ mw: 521.97

SYN: PLATINATE(2-), HEXACHLORO-, DIHYDROGEN, HEXAHYDRATE

TOXICITY DATA WITH REFERENCE
ipr-mus LD50:82 mg/kg TXAPA9 49,41,79

OSHA PEL: TWA 0.002 mg(Pt)/m^3
ACGIH TLV: TWA 0.002 mg(Pt)/m^3

SAFETY PROFILE: Poison by intraperitoneal route. When heated to decomposition it emits toxic fumes of Pt and Cl^-.

DLO875 CAS:34257-95-9 HR: 3
DIHYDROHELENALIN
mf: $C_{15}H_{20}O_4$ mw: 264.35

PROP: Crystals. Mp: 223–226°.

SYNS: 11,13-DIHYDROHELENALIN ☐ PLENOLIN

TOXICITY DATA WITH REFERENCE
dni-mus:ast 2143 μmol/L JPMSAE 67,1235,78
orl-mus LD50:123 mg/kg PLMEAA 45,131,82
ipr-mus LD50:31 mg/kg PLMEAA 45,131,82

SAFETY PROFILE: Poison by ingestion and intraperitoneal routes. Mutation data reported.

DLO880 CAS:14168-01-5 HR: 2
β-DIHYDROHEPTACHLOR
mf: $C_{10}H_7Cl_7$ mw: 375.32

SYNS: β-DHC ☐ DILOR ☐ GL 2487

TOXICITY DATA WITH REFERENCE
orl-rat LD50:5 g/kg GISAAA 52(2),93,87
orl-mus LD50:1890 mg/kg VETNAL 52(7),100,76
orl-ckn LD50:2000 mg/kg VETNAL 52(7),100,76

SAFETY PROFILE: Moderately toxic by ingestion. When heated to decomposition it emits toxic fumes of Cl^-. See also HEPTACHLOR.

DLO950 CAS:83053-63-8 HR: 2
15,16-DIHYDRO-11-HYDROXYCYCLOPENTA(a)PHENANTHREN-17-ONE
mf: $C_{17}H_{12}O_2$ mw: 248.29

TOXICITY DATA WITH REFERENCE
skn-mus TDLo:400 mg/kg:ETA CRNGDP 3,677,82

SAFETY PROFILE: Questionable carcinogen with experimental tumorigenic data. When heated to decomposition it emits acrid smoke and irritating fumes.

DLP000 CAS:63918-74-1 HR: 3
6,7-DIHYDRO-6-(2-HYDROXYETHYL)-5H-DIBENZ(c,e)AZEPINE
mf: $C_{16}H_{17}NO$ mw: 239.34

SYN: RO 2-3599

TOXICITY DATA WITH REFERENCE
ipr-mus LD50:135 mg/kg JPETAB 103,10,51
ivn-mus LD50:32 mg/kg JPETAB 103,10,51

SAFETY PROFILE: Poison by intravenous and intraperitoneal routes. When heated to decomposition it emits toxic fumes of NO_x.

DLP200 CAS:55651-36-0 HR: 2
15,16-DIHYDRO-11-HYDROXYMETHYL-17H-CYCLOPENTA(a)PHENANTHREN-17-ONE

TOXICITY DATA WITH REFERENCE
mma-sat 1 μg/plate CNREA8 40,882,80
skn-mus TDLo:16 mg/kg:ETA CNREA8 40,882,80

SAFETY PROFILE: Questionable carcinogen with experimental tumorigenic data. Mutation data reported. When heated to decomposition it emits acrid smoke and fumes.

DLP400 CAS:55651-31-5 HR: 2
15,16-DIHYDRO-15-HYDROXY-11-METHYL-17H-CYCLOPENTA(a)PHENANTHREN-17-ONE

TOXICITY DATA WITH REFERENCE
mma-sat 1 μg/plate CNREA8 40,882,80
skn-mus TDLo:16 mg/kg:ETA CNREA8 40,882,80

SAFETY PROFILE: Questionable carcinogen with experimental tumorigenic data. Mutation data reported. When heated to decomposition it emits acrid smoke and fumes.

DLP600 CAS:24684-56-8 HR: 2
15,16-DIHYDRO-16-HYDROXY-11-METHYLCYCLO PENTA(a)PHENANTHREN-17-ONE
mf: $C_{18}H_{14}O_2$ mw: 262.32

SYN: 15,16-DIHYDRO-16-HYDROXY-11-METHYL-17H-CYCLOPENTA(a)PHENANTHREN-17-ONE

TOXICITY DATA WITH REFERENCE
mma-sat 1 μg/plate CNREA8 40,882,80
skn-mus TDLo:115 mg/kg/48W-I:NEO CNREA8 33,832,73

CONSENSUS REPORTS: EPA Genetic Toxicology Program.

SAFETY PROFILE: Questionable carcinogen with experimental neoplastigenic data. Mutation data reported. When heated to decomposition it emits acrid smoke and irritating fumes.

DLP800 CAS:31499-72-6 HR: 1
DIHYDRO-α-IONONE
mf: $C_{13}H_{22}O$ mw: 194.35

SYN: 4-(2,6,6-TRIMETHYL-2-CYCLOHEXEN-1-YL)-2-BUTANONE

TOXICITY DATA WITH REFERENCE
skn-rbt 500 mg/24H MOD FCTXAV 16,711,78

CONSENSUS REPORTS: Reported in EPA TSCA Inventory.

SAFETY PROFILE: A skin irritant. When heated to decomposition it emits acrid smoke and irritating fumes. See also KETONES.

DLQ000 CAS:6414-38-6 HR: 3
DIHYDROISOCODEINE ACID TARTRATE
mf: $C_{18}H_{23}NO_3 \cdot C_4H_6O_6$ mw: 451.52

SYNS: DIHYDROISOCODEINE TARTRATE □ DIHYDRO ISOCODEINE TARTRATE (1:1)

TOXICITY DATA WITH REFERENCE
scu-mus LDLo:1500 mg/kg JPETAB 51,35,34
scu-rbt LDLo:196 mg/kg JPETAB 66,182,39

SAFETY PROFILE: Poison by subcutaneous route. See also CODEINE. When heated to decomposition it emits toxic fumes of NO_x.

DLQ400 CAS:37795-69-0 HR: 3
2,3-DIHYDRO-9H-ISOXAZOLO(3,2-b)QUINAZOLIN-9-ONE
mf: $C_{10}H_8N_2O_2$ mw: 188.20

SYN: W-2429

TOXICITY DATA WITH REFERENCE
orl-rat LD50:395 mg/kg ARZNAD 27,770,77
ivn-rat LD50:278 mg/kg ARZNAD 27,770,77
orl-mus LD50:420 mg/kg ARZNAD 27,770,77
ivn-mus LD50:212 mg/kg ARZNAD 27,770,77
orl-dog LD50:700 mg/kg ARZNAD 27,793,77

SAFETY PROFILE: Poison by ingestion and intravenous routes. When heated to decomposition it emits toxic fumes of NO_x. See also KETONES.

DLQ600 CAS:1128-08-1 HR: 1
DIHYDROJASMONE
mf: $C_{11}H_{18}O$ mw: 166.29

PROP: Oil. D: 0.917 @ 18°/4°, bp: 140–147° @ 22 mm.

SYNS: 3-METHYL-2-n-PENTANYL-2-CYCLOPENTEN-1-ONE □ 3-METHYL-2-PENTYL-2-CYCLOPENTEN-1-ONE □ 2-PENTYL-3-METHYL-2-CYCLOPENTEN-1-ONE

TOXICITY DATA WITH REFERENCE
skn-rbt 500 mg/24H FCTXAV 12,517,74
orl-rat LD50:2500 mg/kg FCTXAV 12,523,74
skn-rbt LD50:>5 g/kg FCTXAV 12,523,74

CONSENSUS REPORTS: Reported in EPA TSCA Inventory.

SAFETY PROFILE: Moderately toxic by ingestion. A skin irritant. When heated to decomposition it emits acrid smoke and irritating fumes. See also KETONES.

D

DLQ800 CAS:21842-58-0 HR: 3
12,β,13,α-DIHYDROJERVINE
mf: $C_{26}H_{39}NO_3$ mw: 413.66

TOXICITY DATA WITH REFERENCE
orl-ham TDLo:113 mg/kg (7D preg):REP 41CIAR -,409,78
orl-ham TDLo:75 mg/kg (7D preg):TER JAFCAU 26,561,78
orl-ham LDLo:75 mg/kg JAFCAU 26,561,78

SAFETY PROFILE: Poison by ingestion. Experimental teratogenic and reproductive effects. When heated to decomposition it emits toxic fumes of NO_x.

DLR000 CAS:587-63-3 HR: 3
DIHYDROKAVAIN
mf: $C_{14}H_{16}O_3$ mw: 232.30

PROP: Prisms from Et_2O/pet ether. Mp: 56–60°. Extracted from the roots of *Piper methysticum forst* (AIPTAK 138,505,62).

SYNS: 7,8-DIHYDROKAWAIN □ 5,6-DIHYDRO-4-METHOXY-6-PHENETHYL-2H-PYRAN-2-ONE □ (S)-5,6-DIHYDRO-4-METHOXY-6-(2-PHENYLETHYL)-2H-PYRAN-2-ONE (9CI) □ DHK □ MARINDININ

TOXICITY DATA WITH REFERENCE
orl-mus LD50:920 mg/kg AIPTAK 138,505,62
ipr-mus LD50:325 mg/kg AIPTAK 138,505,62
ivn-mus LD50:53 mg/kg AIPTAK 177,261,69
ipr-rbt LD50:350 mg/kg AIPTAK 138,505,62

SAFETY PROFILE: Poison by intravenous and intraperitoneal routes. Moderately toxic by ingestion. When heated to decomposition it emits acrid smoke and irritating fumes.

DLR100 CAS:37686-84-3 HR: 3
9,10-α-DIHYDROLISURIDE
mf: $C_{20}H_{28}N_4O$ mw: 340.52

PROP: A solid. Mp: 202–204°.

SYNS: trans-DIHYDROLISURIDE □ 1-((5R,8S,10R)-6-METHYL-8-ERGOLINYL)-3,3-DIETHYLUREA □ TERGURID □ TERGURIDE □ UREA, N,N-DIETHYL-N′-((8-α)-6-METHYLERGOLIN-8-YL)- □ ZK 31224

TOXICITY DATA WITH REFERENCE
orl-wmn TDLo:30 μg/kg (lactating female 3D post):REP GOBIDS 26,33,88
orl-rat TDLo:100 μg/kg (female 5D post):REP CCCCAK 52,2983,87
orl-rat TDLo:880 μg/kg (lactating female 4D post):REP CCCCAK 52,2983,87
orl-rat LD50:100 mg/kg YACHDS 21,329,93
ipr-rat LD50:2 g/kg YACHDS 21,329,93
scu-rat LD50:>2 g/kg YACHDS 21,329,93

orl-mus LD50:160 mg/kg YACHDS 21,329,93
ipr-mus LD50:272 mg/kg YACHDS 21,329,93
scu-mus LD50:>2 g/kg YACHDS 21,329,93
ivn-mus LD50:75 mg/kg CCCCAK 52,2983,87

SAFETY PROFILE: Poison by ingestion and intravenous routes. Experimental reproductive effects. When heated to decomposition it emits toxic fumes of NO_x.

DLR150 CAS:37686-85-4 HR: 3
trans-9,10-DIHYDROLISURIDE HYDROGEN MALEATE
mf: $C_{20}H_{28}N_4O•C_4H_4O_4$ mw: 456.60

SYNS: N,N-DIETHYL-N′-((8-α)-6-METHYLERGOLIN-8-YL)UREA (Z)-2-BUTENEDIOATE □ DIRONYL □ 1-((5R,8S,10R)-6-METHYL-8-ERGOLINYL)-3,3-DIETHYLUREA HYDROGEN MALEATE □ TERGURIDE HYDROGEN MALEATE □ VUFB 6638 □ UREA, N,N-DIETHYL-N′-((8-α)-6-METHYLERGOLIN-8-YL)-, (Z)-2-BUTENEDIOATE (1:1)

TOXICITY DATA WITH REFERENCE
orl-rat TDLo:1180 μg/kg (lactating female 4D post):REP CCCCAK 49,2828,84
orl-rat TDLo:402 μg/kg (female 5D post):REP CCCCAK 49,2828,84
ivn-mus LD50:100 mg/kg CCCCAK 49,2828,84

SAFETY PROFILE: Poison by intravenous route. Experimental reproductive effects. When heated to decomposition it emits toxic fumes of NO_x.

DLR200 CAS:5836-85-1 HR: 2
15,16-DIHYDRO-11-METHOXYCYCLOPENTA(a)PHENANTHREN-17-ONE
mf: $C_{18}H_{14}O_2$ mw: 262.32

SYN: 11-METHOXY-15,16-DIHDYROCYCLOPENTA(a)PHENANTHREN-17-ONE

TOXICITY DATA WITH REFERENCE
skn-mus TDLo:108 mg/kg/1Y-I:CAR PEXTAR 11,69,69
skn-mus TD:1600 μg/kg:ETA CRNGDP 3,677,82

SAFETY PROFILE: Questionable carcinogen with experimental carcinogenic and tumorigenic data. When heated to decomposition it emits acrid smoke and irritating fumes. See also KETONES.

DLR300 CAS:41890-92-0 HR: 1
DIHYDROMETHOXYELGENOL
mf: $C_{11}H_{24}O_2$ mw: 188.35

SYNS: 3,7-DIMETHYL-7-METHOXY-2-OCTANOL □ ELESANT □ 2-OCTANOL, 3,7-DIMETHYL-7-METHOXY- □ OSIROL □ OSYROL

TOXICITY DATA WITH REFERENCE
skn-rbt 1% MLD FCTOD7 30,25S,92
orl-rat LD50:4490 mg/kg FCTOD7 30,25S,92

CONSENSUS REPORTS: Reported in EPA TSCA Inventory.

SAFETY PROFILE: Low toxicity by ingestion. A skin irritant. When heated to decomposition it emits acrid smoke and irritating vapors.

DLR600 CAS:30835-61-1 HR: 2
15,16-DIHYDRO-11-METHOXY-7-METHYLCYCLOPENTA(a)PHENANTHREN-17-ONE
mf: $C_{19}H_{16}O_2$ mw: 276.35

SYN: 15,16-DIHYDRO-11-METHOXY-7-METHYL-17H-CYCLOPENTA(a)PHENANTHREN-17-ONE

TOXICITY DATA WITH REFERENCE
mma-sat 20 μg/plate CNREA8 36,4525,76
skn-mus TDLo:96 mg/kg/40W-I:NEO CNREA8 33,832,73

CONSENSUS REPORTS: EPA Genetic Toxicology Program.

SAFETY PROFILE: Questionable carcinogen with experimental neoplastigenic data. Mutation data reported. When heated to decomposition it emits acrid smoke and irritating fumes. See also KETONES.

DLS000 CAS:70301-64-3 HR: 3
10,11-DIHYDRO-11-(p-METHOXYPHENYL)-2-(4-METHYL-1-PIPERAZINYL)PYRIDAZINO(3,4-b)(1,4)BENZOXAZEPINE
mf: $C_{23}H_{25}N_5O_2$ mw: 403.53

TOXICITY DATA WITH REFERENCE
scu-rat LD50:255 mg/kg PCJOAU 13,256,79
scu-mus LD50:255 mg/kg PCJOAU 13,256,79

SAFETY PROFILE: Poison by subcutaneous route. When heated to decomposition it emits toxic fumes of NO_x.

DLS600 CAS:58-28-6 HR: 3
10,11-DIHYDRO-5-(3-(METHYLAMINO)PROPYL)-5H-DIBENZ(b,f)AZEPINE HYDROCHLORIDE
mf: $C_{18}H_{22}N_2•ClH$ mw: 302.88

PROP: A solid. Mp: 214–218°.

SYNS: DESIPRAMINE HYDROCHLORIDE □ DESMETHYLIMIPRAMINE HYDROCHLORIDE □ DIMETHYLIMIPRAMINE HYDROCHLORIDE □ DMI HYDROCHLORIDE □ EX 4355 □ G 35020 □ GMI □ IMIPRAMINEDEMETHYL HYDROCHLORIDE □ IRENE □ JB 8181 □ N-(γ-METHYLAMINOPROPYL)IMINODIBENZYL HYDROCHLORIDE □ NORPRAMIN □ NORTIMIL □ NSC-114901 □ PERTOFRAN □ PERTOFRANE □ RMI9,384A

TOXICITY DATA WITH REFERENCE
cyt-oin-unr 10 g/L JCLBA3 47,182a,70
scu-rat TDLo:130 mg/kg (female 10-22D post):REP NETOD7 7,493,85
scu-mus TDLo:90 mg/kg (female 9D post):TER DGDFA5 22,61,80
orl-man TDLo:5 mg/kg/5D-I AJPSAO 142,386,85
orl-wmn TDLo:70 mg/kg/4W-I AJPSAO 142,386,85
orl-man TDLo:490 μg/kg JCLPDE 47,210,86
orl-wmn TDLo:14 mg/kg/1W-I:KID,SYS JCLPDE 44,153,83
orl-rat LD50:871 mg/kg ARZNAD 33,1411,83
ipr-rat LD50:55 mg/kg 27ZQAG -,70,72
scu-rat LD50:250 mg/kg WMWOA4 112,558,62
ivn-rat LD50:19 mg/kg 27ZQAG -,70,72
orl-mus LD50:315 mg/kg 27ZQAG -,70,72
ipr-mus LD50:88 mg/kg 27ZQAG -,70,72
scu-mus LD50:180 mg/kg WMWOA4 112,558,62

SAFETY PROFILE: Poison by ingestion, intraperitoneal, subcutaneous, and intravenous routes. Human systemic effects by ingestion: decreased urine volume, sodium level changes, chlorine level changes. An experimental teratogen. Other experimental reproductive effects. Mutation data reported. When heated to decomposition it emits very toxic fumes of NO_x and HCl.

DLS800 CAS:1563-67-3 HR: 3
2,3-DIHYDRO-2-METHYLBENZOPYRANYL-7,N-METHYLCARBAMATE
mf: $C_{11}H_{13}NO_3$ mw: 207.25

SYNS: A 468 □ BAY 48130 □ BAY 62863 □ BAYER 62863 □ C 1120 □ DECARBOFURAN □ ENT 27,324

TOXICITY DATA WITH REFERENCE
orl-rat LD50:43 mg/kg JAFCAU 20,923,72
orl-gpg LDLo:25 mg/kg JEENAI 61,1261,68
scu-gpg LDLo:25 mg/kg JEENAI 61,1261,68

SAFETY PROFILE: Poison by ingestion and subcutaneous routes. When heated to decomposition it emits toxic fumes of NO_x. See also CARBAMATES.

DLT000 CAS:7499-32-3 HR: 2
9,10-DIHYDRO-7-METHYLBENZO(a)PYRENE
mf: $C_{21}H_{16}$ mw: 268.37

SYN: 1′:2′-DIHYDRO-4′-METHYL-3:4-BENZPYRENE

TOXICITY DATA WITH REFERENCE
imp-mus TDLo:520 mg/kg/10W-I:ETA AJCAA7 36,211,39

SAFETY PROFILE: Questionable carcinogen with experimental tumorigenic data. When heated to decomposition it emits acrid smoke and irritating fumes.

DLT200 CAS:63041-50-9 HR: 2
meso-DIHYDRO-3-METHYLCHOLANTHRENE
mf: $C_{21}H_{18}$ mw: 270.39

SYN: 6,12b-DIHYDRO-3-METHYLCHOLANTHRENE

TOXICITY DATA WITH REFERENCE
scu-mus TDLo:40 mg/kg:ETA CNREA8 1,695,41
scu-mus TD:200 mg/kg:ETA CNREA8 1,685,41

SAFETY PROFILE: Questionable carcinogen with experimental tumorigenic data. When heated to decomposition it emits acrid smoke and irritating fumes.

DLT400 CAS:25486-92-4 HR: 2
11,12-DIHYDRO-3-METHYLCHOLANTHRENE
mf: $C_{21}H_{18}$ mw: 270.39

TOXICITY DATA WITH REFERENCE
mma-sat 1 μg/plate MUREAV 51,311,78
skn-mus TDLo:168 mg/kg/57W-I:ETA JNCIAM 44,641,70

SAFETY PROFILE: Questionable carcinogen with experimental tumorigenic data. Mutation data reported. When heated to decomposition it emits acrid smoke and irritating fumes.

DLT600 HR: 2
9,10-DIHYDRO-3-METHYLCHOLANTHRENE-1,9,10-TRIOL
mf: $C_{21}H_{18}O_3$ mw: 318.39

SYNS: (E)-1-HYDROXY-MC-9,10-DIHYDRODIOL □ (E)-1-HYDROXY-3-METHYLCHOLANTHRENE 9,10-DIHYDRODIOL

TOXICITY DATA WITH REFERENCE
skn-mus TDLo:127 μg/kg:NEO CNREA8 39,3549,79

SAFETY PROFILE: Questionable carcinogen with experimental neoplastigenic data. When heated to decomposition it emits acrid smoke and irritating fumes.

DLT800 CAS:40951-13-1 HR: 2
15,16-DIHYDRO-11-METHYL-17H-CYCLOPENTA(a)PHENANTHREN-17-OL
mf: $C_{18}H_{16}O$ mw: 248.34

TOXICITY DATA WITH REFERENCE
skn-mus TDLo:91 mg/kg/38W-I:NEO CNREA8 33,832,73

SAFETY PROFILE: Questionable carcinogen with experimental neoplastigenic data. When heated to decomposition it emits acrid smoke and irritating fumes.

DLU200 CAS:24684-42-2 HR: 2
16,17-DIHYDRO-11-METHYLCYCLOPENTA(a)PHENANTHREN-15-ONE
mf: $C_{18}H_{14}O$ mw: 246.32

TOXICITY DATA WITH REFERENCE
mma-sat 50 μg/plate CNREA8 36,4525,76
skn-mus TDLo:120 mg/kg/50W-I:ETA CNREA8 33,832,73

CONSENSUS REPORTS: EPA Genetic Toxicology Program.

SAFETY PROFILE: Questionable carcinogen with experimental tumorigenic data. Mutation data reported. When heated to decomposition it emits acrid smoke and irritating fumes. See also KETONES.

DLU400 CAS:30835-65-5 HR: 2
15,16-DIHYDRO-7-METHYLCYCLOPENTA(a)PHENANTHREN-17-ONE
mf: $C_{18}H_{14}O$ mw: 246.32

SYN: 15,16-DIHYDRO-7-METHYL-17H-CYCLOPENTA(a)PHENANTHREN-17-ONE

TOXICITY DATA WITH REFERENCE
mmo-sat 20 μg/plate CNREA8 36,4525,76
skn-mus TDLo:72 mg/kg/30W-I:ETA CNREA8 33,832,73

CONSENSUS REPORTS: EPA Genetic Toxicology Program.

SAFETY PROFILE: Questionable carcinogen with experimental tumorigenic data. Mutation data reported. When heated to decomposition it emits acrid smoke and irritating fumes. See also KETONES.

DLU600 CAS:5837-17-2 HR: 2
16,17-DIHYDRO-17-METHYLENE-15H-CYCLOPENTA (a)PHENANTHRENE
mf: $C_{18}H_{14}$ mw: 230.32

TOXICITY DATA WITH REFERENCE
skn-mus TDLo: 125 mg/kg/52W-I: ETA NATUAS 210,1281,66

SAFETY PROFILE: Questionable carcinogen with experimental tumorigenic data. When heated to decomposition it emits acrid smoke and irritating fumes.

DLU700 CAS:83053-62-7 HR: 2
15,16-DIHYDRO-11-METHYL-15-METHOXYCYCLOPENTA(a)PHENANTHREN-17-ONE
mf: $C_{19}H_{16}O_2$ mw: 276.35

TOXICITY DATA WITH REFERENCE
skn-mus TDLo: 1600 µg/kg: ETA CRNGDP 3,677,82

SAFETY PROFILE: Questionable carcinogen with experimental tumorigenic data. When heated to decomposition it emits acrid smoke and irritating fumes.

DLU800 CAS:29676-95-7 HR: 2
1,4-DIHYDRO-1-METHYL-7-(2-(5-NITRO-2-FURYL)VINYL)-4-OXO-1,8-NAPHTHYRIDINE-3-CARBOXYLIC ACID, POTASSIUM SALT
mf: $C_{16}H_{10}N_3O_6 \cdot K$ mw: 379.39

SYN: NFN

TOXICITY DATA WITH REFERENCE
orl-mus TDLo: 1411 mg/kg/14W-C: CAR JJIND8 69,1317,82
orl-mus TDLo: 3150 mg/kg/25W-C: CAR CIZAAZ 50,249,74
orl-mus TD: 3276 mg/kg/26W-C: CAR CIGZAF 50,249,74

SAFETY PROFILE: Questionable carcinogen with experimental carcinogenic data. When heated to decomposition it emits toxic fumes of NO_x and K_2O.

DLU900 CAS:27016-91-7 HR: 3
2,3-DIHYDRO-N-METHYL-7-NITRO-2-OXO-5-PHENYL-1H-1,4-BENZODIAZEPINE-1-CARBOXAMIDE
mf: $C_{17}H_{14}N_4O_4$ mw: 338.35

SYN: D 58SI

TOXICITY DATA WITH REFERENCE
orl-rat TDLo: 32,500 mg/kg (25D pre): REP TAKHAA 30,85,71
orl-rat TDLo: 7500 mg/kg (male 25D pre): REP TAKHAA 30,85,71
orl-rat TDLo: 2500 mg/kg (male 25D pre): REP TAKHAA 30,85,71
orl-rat LD50: 2280 mg/kg TAKHAA 29,153,70
ipr-rat LD50: 345 mg/kg TAKHAA 29,153,70
orl-mus LD50: 1230 mg/kg TAKHAA 29,134,70
ipr-mus LD50: 370 mg/kg TAKHAA 29,134,70

SAFETY PROFILE: Poison by intraperitoneal route. Moderately toxic by ingestion. Experimental reproductive effects. When heated to decomposition it emits toxic fumes of NO_x.

DLV000 CAS:2011-67-8 HR: 2
1,3-DIHYDRO-1-METHYL-7-NITRO-5-PHENYL-2H-1,4-BENZODIAZEPIN-2-ONE
mf: $C_{16}H_{13}N_3O_3$ mw: 295.32

PROP: Pale-yellow plates from EtOH. Mp: 156.5–157.5°.

SYNS: ELIMIN □ HYPNON □ 1-METHYLNITRAZEPAM □ 1-METHYL-7-NITRO-5-PHENYL-1,3-DIHYDRO-2H-1,4-BENZODIAZEPIN-2-ONE □ 1-METHYL-5-PHENYL-7-NITRO-1,3-DIHYDRO-2H-1,4-BENZODIAZEPIN-2-ONE □ NIMETAZEPAM □ S 1530

TOXICITY DATA WITH REFERENCE
orl-rat TDLo: 700 mg/kg (8-14D preg): REP RCOCB8 46,437,84
orl-rat LD50: 970 mg/kg ARZNAD 22,534,72
ipr-rat LD50: 970 mg/kg ARZNAD 22,534,72
orl-mus LD50: 750 mg/kg ARZNAD 22,534,72
ipr-mus LD50: 510 mg/kg OYYAA2 7,705,73

SAFETY PROFILE: Moderately toxic by ingestion and intraperitoneal routes. Experimental reproductive effects. An anticonvulsant and muscle relaxant. Related to diazepam. When heated to decomposition it emits toxic fumes of NO_x.

DLV200 CAS:5259-88-1 HR: 2
5,6-DIHYDRO-2-METHYL-1,4-OXATHIIN-3-CARBOXANILIDE-4,4-DIOXIDE
mf: $C_{12}H_{13}NO_4S$ mw: 267.32

PROP: Prisms from EtOH. Sltly sol in H_2O; sol in C_6H_6, MeOH, and EtOH; very sol in Me_2CO.

SYNS: DCMOD □ 2,3-DIHYDRO-5-CARBOXANILIDO-6-METHYL-1,4-OXATHIIN-4,4-DIOXIDE □ 5,6-DIHYDRO-2-METHYL-3-CARBOXANILIDO-1,4-OXATHIIN-4,4-DIOXID (GERMAN) □ 5,6-DIHYDRO-2-METHYL-N-PHENYL-1,4-OXATHIIN-3-CARBOXAMIDE-4,4-DIOXIDE □ DIOXIDE of VITAVAX □ F461 □ OXYCARBOXIN □ OXYCARBOXINE □ PLANTVAX □ PLANT WAX □ VITAVEX

TOXICITY DATA WITH REFERENCE
mrc-asn 8000 ppm ENMUDM 2,359,80
orl-rat LD50: 2000 mg/kg WRPCA2 9,119,70

SAFETY PROFILE: Moderately toxic by ingestion. Mutation data reported. A pesticide. When heated to decomposition it emits very toxic fumes of NO_x and SO_x.

DLV400 CAS:70301-54-1 HR: 3
10,11-DIHYDRO-2-(4-METHYL-1-PIPERAZINYL)-11-(2-ATHIAZOLYL)-PYRIDAZINO(3,4-b)(1,4)BENZOXAZEPINE
mf: $C_{19}H_{20}N_6OS$ mw: 380.51

TOXICITY DATA WITH REFERENCE
scu-rat LD50: 217 mg/kg PCJOAU 13,256,79
scu-mus LD50: 217 mg/kg PCJOAU 13,256,79

SAFETY PROFILE: Poison by subcutaneous route. When heated to decomposition it emits very toxic fumes of SO_x and NO_x.

DLV600 **CAS:70301-68-7** **HR: 3**
10,11-DIHYDRO-2-(4-METHYL-1-PIPERAZINYL)-11-(3,4-XYLYL)PYRIDAZINO(3,4-b)(1,4)BENZOXAZEPINE MALEATE
mf: $C_{24}H_{27}N_5O \cdot C_4H_4O_4$ mw: 517.64

TOXICITY DATA WITH REFERENCE
scu-rat LD50:155 mg/kg PCJOAU 13,256,79
scu-mus LD50:155 mg/kg PCJOAU 13,256,79

SAFETY PROFILE: Poison by subcutaneous route. When heated to decomposition it emits toxic fumes of NO_x.

DLV800 **CAS:3978-86-7** **HR: 3**
6,11-DIHYDRO-11-(1-METHYL-4-PIPERIDYLIDENE)-5H-BENZO(5,6)CYCLOHEPTA (1,2-b) PYRIDINE DIMALEATE
mf: $C_{20}H_{22}N_2 \cdot 2C_4H_4O_4$ mw: 522.60

PROP: Crystals from EtOAc/MeOH. Mp: 152–154°.

SYNS: AZATADINE DIMALEATE □ AZATADINE MELEATE □ IDULIAN □ OPTIMINE □ SCH 10649 □ ZADINE

TOXICITY DATA WITH REFERENCE
orl-rat LD50:440 mg/kg TXAPA9 18,185,71
ipr-rat LD50:166 mg/kg JZKEDZ 1,173,75
scu-rat LD50:1000 mg/kg JZKEDZ 1,173,75
orl-mus LD50:184 mg/kg JZKEDZ 1,173,75
ipr-mus LD50:105 mg/kg JZKEDZ 1,173,75
scu-mus LD50:178 mg/kg JZKEDZ 1,173,75
scu-mus LDLo:50 mg/kg JZKEDZ 1,79,75

SAFETY PROFILE: Poison by ingestion, subcutaneous, and intraperitoneal routes. An antihistamine. When heated to decomposition it emits toxic fumes of NO_x.

DLV900 **CAS:2346-00-1** **HR: 2**
4,5-DIHYDRO-2-METHYLTHIAZOLE
mf: C_4H_7NS mw: 101.18

SYNS: METHYL-2 Δ-2 THIAZOLINE □ THIAZOLE, 4,5-DIHYDRO-2-METHYL- □ 2-THIAZOLINE, 2-METHYL-

TOXICITY DATA WITH REFERENCE
ipr-mus LD50:600 mg/kg EJMCA5 20,16,85

CONSENSUS REPORTS: Reported in EPA TSCA Inventory.

SAFETY PROFILE: Moderately toxic by intraperitoneal route. When heated to decomposition it emits toxic vapors of NO_x and SO_x.

DLW600 **CAS:466-99-9** **HR: 3**
DIHYDROMORPHINONE
mf: $C_{17}H_{19}NO_3$ mw: 285.37

PROP: A solid. Mp: 266–267°.

SYNS: DIMO □ HYDROMORPHONE □ HYMORPHAN □ LAUDICON □ PARAMORPHAN

TOXICITY DATA WITH REFERENCE
orl-hmn LDLo:1428 μg/kg 34ZIAG -,223,69
scu-mus LD50:84 mg/kg JPETAB 52,468,34
ivn-mus LD50:104 mg/kg YKKZAJ 84,268,64

SAFETY PROFILE: A deadly human poison by ingestion. An experimental poison by ingestion and subcutaneous routes. When heated to decomposition it emits toxic fumes of NO_x. See also (−)MORPHINE.

DLX000 **CAS:18479-58-8** **HR: 2**
DIHYDROMYRCENOL
mf: $C_{10}H_{20}O$ mw: 156.30

SYN: 2,6-DIMETHYL-7-OCTEN-2-OL

TOXICITY DATA WITH REFERENCE
skn-rbt 500 mg/24H MLD FCTXAV 12,525,74
orl-rat LD50:3600 mg/kg FCTXAV 12,525,74

CONSENSUS REPORTS: Reported in EPA TSCA Inventory.

SAFETY PROFILE: Moderately toxic by ingestion. A skin irritant. When heated to decomposition it emits acrid smoke and irritating fumes.

DLX100 **CAS:88969-41-9** **HR: 2**
DIHYDROMYRCENYL ACETATE
mf: $C_{12}H_{22}O_2$ mw: 198.34

SYNS: 3-METHYLENE-7-METHYLOCTAN-7-YL ACETATE □ 2-METHYL-6-METHYLENE-2-OCTANOL ACETATE (ESTER)

TOXICITY DATA WITH REFERENCE
skn-rbt 500 mg/24H MLD FCTOD7 21,847,83
eye-rbt 10% MLD FCTOD7 21,847,83
skn-rbt LD50:2800 mg/kg FCTOD7 21,847,83

SAFETY PROFILE: Moderately toxic by skin contact. A skin and eye irritant. When heated to decomposition it emits acrid smoke and fumes.

DLX200 **CAS:529-34-0** **HR: 2**
3,4-DIHYDRO-1(2H)-NAPHTHALENONE
mf: $C_{10}H_{10}O$ mw: 146.20

PROP: A liquid. D: 1.099 @ 15.6°/4°, mp: 8°.

SYNS: α-TETRALONE □ 1-TETRALONE

TOXICITY DATA WITH REFERENCE
orl-rat LD50:810 mg/kg AIHAAP 30,470,69

CONSENSUS REPORTS: Reported in EPA TSCA Inventory.

SAFETY PROFILE: Moderately toxic by ingestion. When heated to decomposition it emits acrid smoke and irritating fumes.

DLX300 **CAS:31785-60-1** **HR: 2**
2,3-DIHYDRO-2-(1-NAPHTHYL)-4(1H)-QUINAZOLINONE
mf: $C_{18}H_{14}N_2O$ mw: 274.34

SYNS: 2,3-DIHYDRO-2-(1-NAPHTHALENYL)-4(1H)-QUINAZOLINONE □ NSC-145669 □ U-29,409

TOXICITY DATA WITH REFERENCE
orl-rat TDLo:480 mg/kg (8D male):REP PSEBAA 137,532,71
ipr-mus LD50:998 mg/kg NCISP* JAN86

SAFETY PROFILE: Moderately toxic by intraperitoneal route. Experimental reproductive effects. When heated to decomposition it emits toxic fumes of NO_x.

DLX400 CAS:124-90-3 HR: 3
DIHYDRONE HYDROCHLORIDE
mf: $C_{18}H_{21}NO_4$•ClH mw: 351.86

PROP: Long rods from H_2O. Mp: 270–272° (decomp).

SYNS: DIHYDROOXYCODEINONE HYDROCHLORIDE ☐ DIHYDROXYCODEINONE HYDROCHLORIDE ☐ DINARKON ☐ EUBINE ☐ EUCODAL ☐ EUKODAL ☐ EUTAGEN ☐ 14-HYDROXYDIHYDROCODEINONE HYDROCHLORIDE ☐ OXIKON ☐ OXYCODONE HYDROCHLORIDE ☐ OXYCODON HYDROCHLORIDE ☐ OXYCON ☐ OXYKODAL ☐ OXYKON ☐ PANCODINE ☐ PERCODAN HYDROCHLORIDE ☐ STUPENONE ☐ TECODIN ☐ TECODINE ☐TEKO-DIN ☐ THECODIN ☐ THECODINE ☐ THEKODIN

TOXICITY DATA WITH REFERENCE
scu-mus LDLo:350 μg/kg AEPPAE 194,296,40
ivn-cat LDLo:2500 μg/kg AEPPAE 194,296,40
scu-rbt LDLo:80 mg/kg HBAMAK 4,1289,35
ivn-rbt LDLo:45 mg/kg HBAMAK 4,1289,35
scu-frg LDLo:500 mg/kg HBAMAK 4,1289,35

SAFETY PROFILE: Poison by intravenous and subcutaneous routes. When heated to decomposition it emits very toxic fumes of NO_x and HCl. See also (–)MORPHINE.

DLX800 CAS:17247-77-7 HR: 2
1,2-DIHYDRO-2-(5′-NITROFURYL)-4-HYDROXYQUINAZOLINE-3-OXIDE
mf: $C_{12}H_9N_3O_5$ mw: 275.24

SYN: 1,2-DIHYDRO-2-(5′-NITROFURYL)-4-HYDROXY-CHINAZOLIN-3-OXID (GERMAN)

TOXICITY DATA WITH REFERENCE
orl-rat TDLo:40 g/kg/26W-C:ETA ZKKOBW 79,165,73

SAFETY PROFILE: Questionable carcinogen with experimental tumorigenic data. When heated to decomposition it emits toxic fumes of NO_x.

DLY000 CAS:146-22-5 HR: 3
1,3-DIHYDRO-7-NITRO-5-PHENYL-2H-1,4-BENZODIAZEPIN-2-ONE
mf: $C_{15}H_{11}N_3O_3$ mw: 281.29

PROP: Yellow crystals from EtOH. Mp: 224–226°.

SYNS: BENZALIN ☐ CALSMIN ☐ EATAN ☐ EPIBENZALIN ☐ EPINELBON ☐ EUNOCTIN ☐ HIPNAX ☐ HIPSAL ☐ LA 1 ☐ MOGADAN ☐ NELBON ☐ NEOZEPAM ☐ NEUCHLONIC ☐ NITRADOS ☐ NITRAZEPAM ☐ NITRENPAX ☐ 7-NITRO-5-PHENYL-2,3-DIHYDRO-1H-1,4-BENZODIAZEPIN-2-ONE ☐ NSC-58775 ☐ PAXISYN ☐ PELSON ☐ RADEDORM ☐ RELACT ☐ RO 4-5360 ☐ RO 5-3059 ☐ SOMNASED ☐ SOMNIBEL ☐ SOMNITE ☐ SONEBON ☐ SONNOLIN ☐ SUREM ☐ UNISOMNIA

TOXICITY DATA WITH REFERENCE
sln-dmg-orl 2 mg/9D SOGEBZ 11,718,75
spm-mus-orl 300 mg/kg/15D-C CYTBAI 36,45,83
orl-mus TDLo:126 mg/kg (female 1-21D post):REP PLRCAT 9,325,77
orl-mus TDLo:414 mg/kg (male 5D pre):REP PLRCAT 9,187,77
orl-mus TDLo:1 g/kg (male 5D pre):REP RCOCB8 13,601,76
orl-rat TDLo:100 mg/kg (8D preg):REP RCOCB8 46,437,84
orl-mus TDLo:156 mg/kg (female 26D pre):REP RCOCB8 11,155,75
orl-rat LD50:825 mg/kg TXAPA9 18,185,71
ipr-rat LD50:733 mg/kg JMCMAR 20,952,77
orl-mus LD50:550 mg/kg VINIT* #3206-79
ipr-mus LD50:275 mg/kg 27ZQAG -,165,72
ivn-mus LD50:130 mg/kg CSLNX* NX#01434
ivn-rbt LD50:520 mg/kg 27ZQAG -,165,72

CONSENSUS REPORTS: EPA Genetic Toxicology Program.

SAFETY PROFILE: Poison by intraperitoneal and intravenous routes. Moderately toxic by ingestion. Experimental reproductive effects. Mutation data reported. An anticonvulsant and hypnotic agent. When heated to decomposition it emits toxic fumes of NO_x. See also DIAZEPAM.

DLY200 CAS:33389-33-2 HR: 2
1,2-DIHYDRO-2-(5-NITRO-2-THIENYL)QUINAZOLIN-4(3H)-ONE
mf: $C_{12}H_9N_3O_3S$ mw: 275.30

SYN: 1,2-DIHYDRO-2-(5-NITRO-2-THIENYL)-4(3H)-QUINAZOLINONE

TOXICITY DATA WITH REFERENCE
mmo-sat 5 μg/plate CNREA8 35,3611,75
orl-rat TDLo:13 g/kg/49W-C:CAR JNCIAM 57,277,76

CONSENSUS REPORTS: EPA Genetic Toxicology Program.

SAFETY PROFILE: Questionable carcinogen with experimental carcinogenic data. Mutation data reported. When heated to decomposition it emits very toxic fumes of SO_x and NO_x.

DLY400 CAS:5413-60-5 HR: 1
DIHYDRONORDICYCLOPENTADIENYL ACETATE
mf: $C_{12}H_{16}O_2$ mw: 192.28

PROP: Oil with anise odor. Bp: 79–80° @ 3 mm.

SYNS: 3a,4,5,6,7,7a-HEXAHYDRO-4,7-METHANO-1H-INDEN-6-OL ACETATE ☐ TRICYCLODECEN-4-YL-8-ACETATE ☐ VERDYL ACETATE

TOXICITY DATA WITH REFERENCE
skn-rbt 500 mg/24H MOD FCTXAV 14,889,76

CONSENSUS REPORTS: Reported in EPA TSCA Inventory.

SAFETY PROFILE: A skin irritant. When heated to decomposition it emits acrid smoke and irritating fumes.

DLY700 CAS:3686-43-9 HR: 3
3,6-DIHYDRO-1,2,2H-OXAZINE
mf: C_4H_7NO mw: 85.11

ONHCH$_2$CH=CH CH$_2$

PROP: A liquid. Bp: 47–48° @ 8 mm.

SAFETY PROFILE: Reaction with nitric acid forms an explosive product. When heated to decomposition it emits toxic fumes of NO_x.

DLY800 CAS:7374-66-5 HR: 2
5,13-DIHYDRO-5-OXOBENZO(e)(2)BENZOPYRANO(4,3-b)INDOLE
mf: $C_{19}H_{11}NO_2$ mw: 285.31

SYNS: 5-OXO-5H-BENZO(E)ISOCHROMENO(4,3-b)INDOLE □ 5-OXO-5,13-DIHYDROBENZO(E)(2)BENZOPYRANO(4,3-b)INDOLE

TOXICITY DATA WITH REFERENCE
scu-mus TDLo:72 mg/kg/9W-I:ETA SCIEAS 158,387,67

SAFETY PROFILE: Questionable carcinogen with experimental tumorigenic data. When heated to decomposition it emits toxic fumes of NO_x.

DLZ000 CAS:56179-83-0 HR: 2
1,2-DIHYDROPHENANTHRENE
mf: $C_{14}H_{12}$ mw: 180.26

PROP: A liquid. Bp: 110–115° @ 0.3 mm.

TOXICITY DATA WITH REFERENCE
mmo-sat 25 nmol/plate CNREA8 39,4069,79
skn-mus TDLo:72 mg/kg:ETA CNREA8 39,4069,79

SAFETY PROFILE: Questionable carcinogen with experimental tumorigenic data. Mutation data reported. When heated to decomposition it emits acrid smoke and irritating fumes.

DMA000 CAS:28622-66-4 HR: 2
1,2-DIHYDRO-1,2-PHENANTHRENEDIOL
mf: $C_{14}H_{12}O_2$ mw: 212.26

SYN: PHENANTHRENE-1,2-DIHYDRODIOL

TOXICITY DATA WITH REFERENCE
mmo-sat 20 μg/plate CNREA8 49,20,89
skn-mus TDLo:85 mg/kg:ETA CNREA8 39,4069,79

SAFETY PROFILE: Questionable carcinogen with experimental tumorigenic data. Mutation data reported. When heated to decomposition it emits acrid smoke and irritating fumes.

DMA400 CAS:18264-88-5 HR: 2
N-(9,10-DIHYDRO-2-PHENANTHRYL)ACETAMIDE
mf: $C_{16}H_{15}NO$ mw: 237.32

SYN: 2-ACETYLAMINO-9,10-DIHYDROPHENANTHRENE

TOXICITY DATA WITH REFERENCE
orl-rat TDLo:4608 mg/kg/32W-C:CAR CNREA8 15,188,55

SAFETY PROFILE: Questionable carcinogen with experimental carcinogenic data. When heated to decomposition it emits toxic fumes of NO_x.

DMB000 CAS:21820-82-6 HR: 3
5-(2-(3,6-DIHYDRO-4-PHENYL-1(2H)-PYRIDYL)ETHYL)-3-METHYL-2-OXAZOL IDINONE
mf: $C_{17}H_{22}N_2O_2$ mw: 286.41

SYN: AHR-1680

TOXICITY DATA WITH REFERENCE
orl-rat LD50:340 mg/kg 27ZQAG -,200,72
ipr-rat LD50:140 mg/kg 27ZQAG -,200,72
ivn-rat LD50:71 mg/kg 27ZQAG -,200,72
orl-mus LD50:349 mg/kg 27ZQAG -,200,72
ipr-mus LD50:180 mg/kg 27ZQAG -,200,72
ivn-mus LD50:91 mg/kg 27ZQAG -,200,72
orl-dog LD50:300 mg/kg 27ZQAG -,200,72
ipr-gpg LD50:189 mg/kg 27ZQAG -,200,72

SAFETY PROFILE: Poison by ingestion, intraperitoneal, and intravenous routes. When heated to decomposition it emits toxic fumes of NO_x.

DMB200 CAS:66731-42-8 HR: 2
2,3-DIHYDROPHORBOL MYRISTATE ACETATE
mf: $C_{36}H_{58}O_8$ mw: 618.94

SYNS: 2,3-DIHYDROPHORBOL ACETATE MYRISTATE □ DPMA

TOXICITY DATA WITH REFERENCE
skn-mus TDLo:37 mg/kg/31W-I:NEO CNREA8 38,921,78

SAFETY PROFILE: Questionable carcinogen with experimental neoplastigenic data. When heated to decomposition it emits acrid smoke and irritating fumes.

DMC000 CAS:68-94-0 HR: 2
1,7-DIHYDRO-6H-PURIN-6-ONE
mf: $C_5H_4N_4O$ mw: 136.13

PROP: Needles.

SYNS: HYPOXANTHINE □ 9H-PURIN-6-OL □ PURIN-6(3H)-ONE □ 6(1H)-PURINONE

TOXICITY DATA WITH REFERENCE
ipr-mus TDLo:600 mg/kg (female 13D post):TER JJPAAZ 22,201,72
ipr-mus TDLo:1 g/kg (female 13D post):TER JJPAAZ 22,201,72
ipr-mus LD50:750 mg/kg NTIS** AD691-490

CONSENSUS REPORTS: Reported in EPA TSCA Inventory.

SAFETY PROFILE: Moderately toxic by intraperitoneal route. An experimental teratogen. When heated to decomposition it emits toxic fumes of NO_x.

D

DMC200 **CAS:110-87-2** **HR: 3**
DIHYDROPYRAN
mf: C_5H_8O mw: 84.13

$O(CH_2)_3CH{=}CH$ (ring)

PROP: Colorless, mobile liquid; ethereal odor. Bp: 86–87°, flash p: 0°F, d: 0.922 @ 19°/15, vap d: 2.90.

SYNS: Δ^2-DIHYDROPYRAN □ 3,4-DIHYDROPYRAN □ 2H-3,4-DIHYDROPYRAN

CONSENSUS REPORTS: Reported in EPA TSCA Inventory.

SAFETY PROFILE: A flammable and very dangerous fire hazard when exposed to heat or flame; can react vigorously with oxidizing materials. Keep away from heat and open flame. To fight fire, use alcohol foam, CO_2, or dry chemical. When heated to decomposition it emits acrid smoke and irritating fumes.

DMC600 **CAS:123-33-1** **HR: 2**
1,2-DIHYDROPYRIDAZINE-3,6-DIONE
mf: $C_4H_4N_2O_2$ mw: 112.10

PROP: Crystals. Mp: >300°. Sol in water and alc.

SYNS: BURTOLIN □ CHEMFORM □ DE-CUT □ DE-SPROUT □ 1,2-DIHYDRO-3,6-PYRADIZINEDIONE □ 1,2-DIHYDRO-3,6-PYRIDAZINEDIONE □ DREXEL-SUPER P □ ENT 18,870 □ FAIR 30 □ FAIR PS □ HYDRAZID KYSELINY MALEINOVE □ 6-HYDROXY-3(2H)-PYRIDAZINONE □ KMH □ MAH □ MAINTAIN 3 □ MALAZIDE □ MALEIC ACID HYDRAZIDE □ MALEIC HYDRAZIDE □ MALEIC HYDRAZIDE 30% □ MALEIC HYDRAZINE □ MALEIN 30 □ MALEINSAEUREHYDRAZID □ N,N-MALEOYLHYDRAZINE □ MALZID □ MH □ MH 30 □ MH-40 □ MH 36 BAYER □ RCRA WASTE NUMBER U148 □ REGULOX □ REGULOX W □ REGULOX 50 W □ RETARD □ ROYAL MH-30 □ ROYAL SLO-GRO □ SLO-GRO □ SPROUT/OFF □ SPROUT-STOP □ STUNTMAN □ SUCKER-STUFF □ SUPER-DE-SPROUT □ SUPER SPROUT STOP □ SUPER SUCKER-STUFF □ SUPER SUCKER-STUFF HC □ 1,2,3,6-TETRAHYDRO-3,6-DIOXOPYRIDAZINE □ VONDALHYDE □ VONDRAX

TOXICITY DATA WITH REFERENCE
cyt-grh-orl 5 mg CYTOAN 37,345,72
mma-sat 50 μL/plate MUREAV 66,247,79
dns-esc 30 μmol/L ZKKOBW 92,177,78
sln-dmg-orl 4000 ppm MUREAV 55,15,78
sln-dmg-par 4000 ppm NATUAS 207,439,65
cyt-mus-ipr 5000 ppm CISCB7 20,28,76
scu-rat TDLo:2600 mg/kg/65W-I:ETA BJCAAI 19,392,65
orl-rat LD50:3800 mg/kg WRPCA2 9,119,70

CONSENSUS REPORTS: IARC Cancer Review: Group 3 IMEMDT 7,56,87; Animal Inadequate Evidence IMEMDT 4,173,74. Reported in EPA TSCA Inventory.

SAFETY PROFILE: Moderately toxic by ingestion. Questionable carcinogen with experimental tumorigenic data. Mutation data reported. Can cause chronic liver damage and acute central nervous system effects. When heated to decomposition it emits highly toxic fumes of NO_x. See also HYDRAZINE.

DMC800 **HR: 3**
1,2-DIHYDROPYRIDO(2,1,e)TETRAZOLE
mf: $C_5H_4N_4$ mw: 120.12

$CH{=}CHCH{=}CHN\,C{=}NN{=}N$ (ring)

PROP: Explodes on touching with a hot rod.

SAFETY PROFILE: A friction- and heat-sensitive explosive. When heated to decomposition it emits toxic fumes of NO_x.

DMD000 **CAS:480-18-2** **HR: 2**
2,3-DIHYDROQUERCETIN
mf: $C_{15}H_{12}O_7$ mw: 304.27

SYNS: CATECHIN HYDRATE □ DIHYDROQUERCETIN □ (+)-DIHYDROQUERCETIN □ (2R,3R)-DIHYDROQUERCETIN □ 2,3-DIHYDRO-3,3′,4′,5,7-PENTAHYDROXYFLAVONE □ 2-(3,4-DIHYDROXYPHENYL)-2,3-DIHYDRO-3,5,7-TRIHYDROXY-4H-1-BENZOPYRAN-4-ONE □ (2R-trans)-2-(3,4-DIHYDROXYPHENYL)-2,3-DIHYDRO-3,5,7-TRIHYDROXY-4H-1-BENZOPYRAN-4-ONE □ DISTYLIN □ 3,3′,4′,5,7-PENTAHYDROXYFLAVANONE □ TAXIFOLIN □ TAXIFOLIOL

TOXICITY DATA WITH REFERENCE
mmo-sat 100 μg/plate ENMUDM 3,401,81
mma-sat 1660 nmol/plate MUREAV 54,297,78
cyt-ham:fbr 1 g/L/48H MUREAV 48,337,77
ipr-rat LD50:1200 mg/kg JJPAAZ 21,377,71
ipr-mus LD50:985 mg/kg RPTOAN 38,213,75

CONSENSUS REPORTS: Reported in EPA TSCA Inventory.

SAFETY PROFILE: Moderately toxic by intraperitoneal route. Mutation data reported. When heated to decomposition it emits acrid smoke and irritating fumes.

DMD200 **CAS:102338-88-5** **HR: 3**
4a,5-DIHYDRO-RIBOFLAVIN-5′-PHOSPHATE SODIUM SALT
mf: $C_{17}H_{22}N_4O_6P{\bullet}xNa$ mw: 570.33

TOXICITY DATA WITH REFERENCE
scu-mus LD50:375 mg/kg CMTRAG 2,96,61
ivn-mus LD50:420 mg/kg CMTRAG 2,96,61

SAFETY PROFILE: Poison by subcutaneous and intravenous routes. When heated to decomposition it emits very toxic fumes of NO_x, PO_x, and Na_2O.

DMD600 **CAS:94-58-6** **HR: 3**
DIHYDROSAFROLE
mf: $C_{10}H_{12}O_2$ mw: 164.22

PROP: An oily liquid. Bp: 228°, d: 1.0695 @ 20°.

SYNS: 1,2-(METHYLENEDIOXY)-4-PROPYLBENZENE □ 5-PROPYL-1,3-BENZODIOXOLE □ 4-PROPYL-1,2-METHYLENEDIOXYBENZENE □ RCRA WASTE NUMBER U090

TOXICITY DATA WITH REFERENCE
skn-rbt 500 mg/24H MLD FCTXAV 12,527,74
orl-mus TDLo:101 g/kg/81W-C:CAR FCTXAV 19,130,81
orl-mus TDLo:163 g/kg/81W-C:CAR JNCIAM 42,1101,69
orl-mus TD:101 g/kg/81W-C:CAR DIGEBW 19,42,79

orl-rat LD50:2260 mg/kg TXAPA9 7,18,65
orl-mus LD50:3700 mg/kg TXAPA9 7,18,65
ipr-mus LD50:2830 mg/kg COREAF 250,1148,60

CONSENSUS REPORTS: IARC Cancer Review: Group 2B IMEMDT 7,56,87; Animal Sufficient Evidence IMEMDT 10,231,76; Animal Limited Evidence IMEMDT 1,169,72. Reported in EPA TSCA Inventory. EPA Genetic Toxicology Program.

SAFETY PROFILE: Confirmed carcinogen with experimental carcinogenic data. Moderately toxic by ingestion and intraperitoneal routes. A skin irritant. When heated to decomposition it emits acrid smoke and irritating fumes.

DME000 CAS:128-46-1 HR: 3
DIHYDROSTREPTOMYCIN
mf: $C_{21}H_{41}N_7O_{12}$ mw: 583.69

SYNS: DHMS □ DST

TOXICITY DATA WITH REFERENCE
cyt-mus-par 100 mg/kg NULSAK 2,161,71
unr-wmn TDLo:260 mg/kg (19-22W preg):TER SJRDAH 50,61,69
ipr-mus TDLo:12 g/kg (female 1-20D post):REP KJMEA9 10,31,61
scu-rat TDLo:600 mg/kg (6-10D preg):TER OSDIAF 14,107,65
scu-rat TDLo:600 mg/kg (6-10D preg):REP OSDIAF 14,107,65
ivn-rat LD50:200 mg/kg JOBAAY 53,205,47
ipr-mus LD50:533 mg/kg UPJOH* 2(6),-,71
scu-rat LD50:1100 mg/kg ARZNAD 12,597,62
ivn-mus LD50:200 mg/kg 85GDA2 1,96,80
scu-mus LD50:1180 mg/kg ACHTA6 11,2,63
ivn-mus TDLo:200 mg/kg 85GDA2 1,96,80
ims-mus LD50:350 mg/kg AIMDAP 119,493,67

CONSENSUS REPORTS: EPA Genetic Toxicology Program.

SAFETY PROFILE: Poison by intravenous and intramuscular routes. Moderately toxic by subcutaneous and intraperitoneal routes. Human teratogenic effects by unspecified route: developmental abnormalities of the eye and ear. An experimental teratogen. Other experimental reproductive effects. Mutation data reported. A derivative of streptomycin; has anesthetic properties. When heated to decomposition it emits toxic fumes of NO_x.

DME300 CAS:67-96-9 HR: 3
DIHYDROTACHYSTEROL
mf: $C_{28}H_{46}O$ mw: 398.74

PROP: Needles from 90% methanol. Crystals from MeOH. Mp: 131–132° (125–1°). Insol in water. Easily sol in organic solvents.

SYNS: ANTITANIL □ ANTI-TETANY SUBSTANCE 10 □ A.T. 10 □ CALCAMINE □ DHT, □ DICHYSTROLUM □ DIHYDROTACHYSTEROL, □ DYGRATYL □ HYTAKEROL □ PARTEROL □ (E-β,5E,7E,10-α,22E)-9,10-SECOERGOSTA-5,7,22-TRIEN-3-OL (9CI)

TOXICITY DATA WITH REFERENCE
orl-rat TDLo:150 mg/kg (1-5D preg):REP CUSCAM 41,181,72
orl-mus LD50:288 mg/kg NIIRDN 6,330,82
ipr-mus LD50:104 mg/kg NIIRDN 6,330,82

SAFETY PROFILE: Poison by ingestion and intraperitoneal routes. Experimental reproductive effects. When heated to decomposition it emits acrid smoke and irritating fumes.

D

DME400 CAS:80-25-1 HR: 1
DIHYDROTERPINYL ACETATE
mf: $C_{12}H_{22}O_2$ mw: 198.34

SYNS: ACETIC ACID DIHYDROTERPINYL ESTER □ ACETIC ACID-p-MENTHAN-8-OL ESTER □ p-MENTHAN-8-OL ACETATE

TOXICITY DATA WITH REFERENCE
skn-rbt 500 mg/24H MLD FCTXAV 12,807,74

CONSENSUS REPORTS: Reported in EPA TSCA Inventory.

SAFETY PROFILE: A skin irritant. When heated to decomposition it emits acrid smoke and irritating fumes. See also ESTERS.

DME600 CAS:63681-01-6 HR: 3
1,2-DIHYDRO-2,2,4,6-TETRAMETHYLPYRIDINE
mf: $C_9H_{15}N$ mw: 137.25

TOXICITY DATA WITH REFERENCE
skn-rbt 500 mg SEV SCCUR* -,4,61
orl-rat LDLo:600 mg/kg SCCUR* -,4,61
orl-mus LD50:640 mg/kg SCCUR* -,4,61
skn-rbt LDLo:140 mg/kg SCCUR* -,4,61

SAFETY PROFILE: Poison by skin contact. Moderately toxic by ingestion. A severe skin irritant. When heated to decomposition it emits toxic fumes of NO_x.

DME700 CAS:21457-22-7 HR: 3
6,7-DIHYDRO-3,5,5,7-TETRAMETHYL-5H-THIAZOLO (3,2-a)PYRIMIDIN-7-OL HYDROCHLORIDE
mf: $C_{10}H_{16}N_2OS{\cdot}ClH$ mw: 248.80

TOXICITY DATA WITH REFERENCE
orl-mus LD50:820 mg/kg PHARAT 24,572,69
ipr-mus LD50:360 mg/kg PHARAT 24,572,69
scu-mus LD50:795 mg/kg PHARAT 24,572,69

SAFETY PROFILE: Poison by intraperitoneal route. Moderately toxic by ingestion and subcutaneous routes. When heated to decomposition it emits toxic fumes of SO_x, NO_x, and HCl.

DMF000 CAS:77-79-2 HR: 2
2,5-DIHYDROTHIOPHENE DIOXIDE
mf: $C_4H_6O_2S$ mw: 118.16

PROP: Crystals. Mp: 64.7°. Sol in H_2O.

SYNS: BUTADIENE SULFONE □ 2,5-DIHYDROTHIOPHENE-1,1-DIOXIDE □ 2,5-DIHYDROTHIOPHENE SULFONE □ NCI-C04557 □ SULFOL-3-ENE □ β-SULFOLENE □ 3-SULFOLENE

TOXICITY DATA WITH REFERENCE
orl-rat LD50:2830 mg/kg TXAPA9 28,313,74
ipr-mus LD50:1700 mg/kg PCJOAU 12,1568,78

CONSENSUS REPORTS: NCI Carcinogenesis Bioassay (gavage); No Evidence: mouse, rat NCITR* NCI-CG-TR-102,78. Reported in EPA TSCA Inventory.

SAFETY PROFILE: Moderately toxic by ingestion and intraperitoneal routes. When heated to decomposition it emits toxic fumes of SO_x.

DMF600 CAS:5831-17-4 HR: 2
16,17-DIHYDRO-11,12,17-TRIMETHYLCYCLOPENTA (a)PHENANTHRENE
mf: $C_{20}H_{19}$ mw: 259.39

SYN: 11,12,17-TRIMETHYL-16,17-DIHYDRO-15H-CYCLOPENTA(a) PHENANTHRENE

TOXICITY DATA WITH REFERENCE
skn-mus TDLo:108 mg/kg/1Y-I:CAR PEXTAR 11,69,69

SAFETY PROFILE: Questionable carcinogen with experimental carcinogenic data. When heated to decomposition it emits acrid smoke and irritating fumes.

DMF800 CAS:35764-73-9 HR: 3
cis-(±)-9,10-DIHYDRO-N,N,10-TRIMETHYL-2-(TRIFLUOROMETHYL)-9-ANTHRACENE PROPANAMINE
mf: $C_{21}H_{24}F_3N$ mw: 347.46

SYNS: (+−)-9,10-DIHYDRO-N,N,10-TRIMETHYL-2-(TRIFLUORMETHYL)-9-ANTHRACENPROPANAMIN (GERMAN) □ FLUOTRACEN □ SKF 28175

TOXICITY DATA WITH REFERENCE
orl-rat LD50:487 mg/kg ARZNAD 27,1589,77
orl-mus LD50:353 mg/kg ARZNAD 27,1589,77

SAFETY PROFILE: Poison by ingestion. An antipsychotic agent. When heated to decomposition it emits very toxic fumes of F^- and NO_x.

DMG400 CAS:89-84-9 HR: 2
2′,4′-DIHYDROXYACETOPHENONE
mf: $C_8H_8O_3$ mw: 152.16

PROP: Leaflets or needles. Mp: 147°.

SYNS: 4-ACETYLRESORCINOL □ 2,4-DIHYDROXYACETOPHENONE □ 1-(2,4-DIHYDROXYPHENYL)ETHANONE □ RESACETOPHENONE □ β-RESACETOPHENONE □ RESOACETOPHENONE

TOXICITY DATA WITH REFERENCE
eye-rbt 500 mg SEV IHFCAY 6,1,67
scu-rat TDLo:20 mg/kg (4D pre):REP JSICAZ 19,264,60
orl-rat LD50:2830 mg/kg IHFCAY 6,1,67

CONSENSUS REPORTS: Reported in EPA TSCA Inventory.

SAFETY PROFILE: Moderately toxic by ingestion. Experimental reproductive effects. A severe eye irritant. When heated to decomposition it emits acrid smoke and irritating fumes.

DMG600 CAS:490-78-8 HR: 2
2′,5′-DIHYDROXYACETOPHENONE
mf: $C_8H_8O_3$ mw: 152.16

PROP: Green needles from H_2O. Mp: 202–203°.

SYNS: ACETYLHYDROQUINONE □ 2-ACETYLHYDROQUINONE □ 2,5-DIHYDROXYACETOPHENONE □ QUINACETOPHENONE

TOXICITY DATA WITH REFERENCE
ipr-mus LDLo:500 mg/kg CBCCT* 5,140,53

CONSENSUS REPORTS: Reported in EPA TSCA Inventory.

SAFETY PROFILE: Moderately toxic by intraperitoneal route. When heated to decomposition it emits acrid smoke and irritating fumes.

DMG800 CAS:72-48-0 HR: 3
1,2-DIHYDROXY-9,10-ANTHRAQUINONE
mf: $C_{14}H_8O_4$ mw: 240.22

PROP: Orange-red crystals needles or prisms from alc or by sublimation. Bp: 430° (sublimes); mp: 289°. Very sltly sol in water.

SYNS: ALIZARIN □ ALIZARINA □ ALIZARIN B □ ALIZARINE □ ALIZARINE B □ ALIZARINE 3B □ ALIZARINE INDICATOR □ ALIZARINE LAKE RED 2P □ ALIZARINE LAKE RED 3P □ ALIZARINE LAKE RED IPX □ ALIZARINE L PASTE □ ALIZARINE NAC □ ALIZARINE PASTE 20% BLUISH □ ALIZARINE RED □ ALIZARINE RED B □ ALIZARINE RED B2 □ ALIZARINE RED IP □ ALIZARINE RED IPP □ ALIZARINE RED L □ ALIZARINPRIMEVEROSIDE □ ALIZARIN RED □ 9,10-ANTHRACENEDIONE, 1,2-DIHYDROXY- □ 1,2-ANTHRAQUINONEDIOL □ CERTIQUAL ALIZARINE □ C.I. 58000 □ C.I. MORDANT RED 11 □ C.I. PIGMENT RED 83 □ D and C ORANGE NUMBER 15 □ DEEP CRIMSON MADDER 10821 □ 1,2-DIHYDROXYANTHRACHINON □ 1,2-DIHYDROXYANTHRAQUINONE □ ELJON MADDER □ MITSUI ALIZARINE B □ SANYO CARMINE L2B □ TURKEY RED

TOXICITY DATA WITH REFERENCE
eye-rbt 500 mg/24H MLD 28ZPAK -,101,72
mmo-sat 100 μg/plate MUREAV 40,203,76
mma-sat 100 μg/plate MUREAV 40,203,76
dnr-bcs 2 mg/disc TRENAF 27,153,76
orl-bwd LD50:316 mg/kg AECTCV 12,355,83

CONSENSUS REPORTS: Reported in EPA TSCA Inventory. EPA Genetic Toxicology Program.

SAFETY PROFILE: Poison by ingestion. Mutation data reported. An eye irritant. Flammable when exposed to oxidizers and heat. When heated to decomposition it emits acrid smoke and irritating fumes.

DMH000 CAS:81-64-1 HR: 3
1,4-DIHYDROXYANTHRAQUINONE
mf: $C_{14}H_8O_4$ mw: 240.22

PROP: Red crystals from alc. Mp: 194°, bp: 200–202°, vap press: 1 mm @ 196.7°, vap d: 8.3.

SYNS: 1,4-DIHYDROXYANTHRACHINON (CZECH) □ 1,4-DIHYDROXY-9,10-ANTHRAQUINONE □ 1,4-DIOXYANTHRAQUINONE (RUSSIAN) □ QUINIZARIN

TOXICITY DATA WITH REFERENCE
eye-rbt 500 mg/24H MLD 28ZPAK -,102,72

mmo-sat 100 μg/plate MUREAV 40,203,76
ipr-rat LD50:2100 mg/kg GTPZAB 21(12),27,77
ivn-mus LD50:320 mg/kg CSLNX* NX#03274

CONSENSUS REPORTS: Reported in EPA TSCA Inventory.

SAFETY PROFILE: Poison by intravenous route. Moderately toxic by intraperitoneal route. Mutation data reported. An eye irritant. A weak allergen. When heated to decomposition it emits acrid smoke and irritating fumes.

DMH200 CAS:117-12-4 HR: 1
1,5-DIHYDROXYANTHRAQUINONE
mf: $C_{14}H_8O_4$ mw: 240.22

PROP: Green to yellow crystals from alc. Mp: 280°, bp: subl, vap d: 8.3.

SYNS: ANTHRARUFIN □ 1,5-DIHYDROXYANTHRACHINON (CZECH) □ 1,5-DIHYDROXY-9,10-ANTHRAQUINONE

TOXICITY DATA WITH REFERENCE
eye-rbt 500 mg/24H MLD 28ZPAK -,102,72
mmo-sat 50 μg/plate MUREAV 40,203,76

CONSENSUS REPORTS: Reported in EPA TSCA Inventory.

SAFETY PROFILE: An eye irritant. Mutation data reported. When heated to decomposition it emits acrid smoke and irritating fumes.

DMH400 CAS:117-10-2 HR: 2
1,8-DIHYDROXYANTHRAQUINONE
mf: $C_{14}H_8O_4$ mw: 240.22

PROP: Reddish-yellow needles or leaflets. Mp: 193°, vap d: 8.3. Sol in alc, alkalis.

SYNS: ALTAN □ ANTRAPUROL □ CHRYSAZIN □ DANTHRON □ DANTRON □ DIAQUONE □ 1,8-DIHYDROXY-9,10-ANTHRACENEDIONE □ 1,8-DIHYDROXYANTHRACHINON (CZECH) □ DIONONE □ DORBANE □ DORBANEX □ DUOLAX □ ISTIN □ LAXANORM □ LAXANTHREEN □ LAXIPUR □ LAXIPURIN □ LTAN □ MODANE □ USAF ND-59 □ ZWITSALAX

TOXICITY DATA WITH REFERENCE
eye-rbt 500 mg/24H MLD 28ZPAK -,102,72
mmo-sat 100 μg/plate MUREAV 40,203,76
dns-mus:lvr 20 μmol/L CNREA8 44,2918,84
orl-rat TDLo:292 g/kg/70W-C:CAR BJCAAI 52,781,85
orl-mus TDLo:129 g/kg/77W-C:CAR JJCREP 77,871,86
orl-mus TD:130 g/kg/77W-C:NEO TOLED5 31(Suppl),206,86
ipr-mus LD50:500 mg/kg NTIS** AD277-689

CONSENSUS REPORTS: Reported in EPA TSCA Inventory.

SAFETY PROFILE: Moderately toxic by intraperitoneal route. An eye irritant. Questionable carcinogen with experimental carcinogenic and neoplastigenic data. Human mutation data reported. A laxative. When heated to decomposition it emits acrid smoke and irritating fumes.

DMH600 CAS:84-60-6 HR: 3
2,6-DIHYDROXYANTHRAQUINONE
mf: $C_{14}H_8O_4$ mw: 240.22

PROP: Yellow needles from EtOH.

SYNS: ANTHRAFLAVIC ACID □ ANTHRAFLAVIN □ NSC-33531

TOXICITY DATA WITH REFERENCE
mma-sat 100 μg/plate MUREAV 40,203,76
ivn-mus LD50:180 mg/kg CSLNX* NX#06773

CONSENSUS REPORTS: Reported in EPA TSCA Inventory.

SAFETY PROFILE: Poison by intravenous route. Mutation data reported. When heated to decomposition it emits acrid smoke and irritating fumes. See other dihydroxyanthraquinone entries.

DMI300 CAS:20123-80-2 HR: 2
2,5-DIHYDROXYBENZENESULFONIC ACID CALCIUM SALT
mf: $C_{12}H_{10}O_{10}S_2 \cdot Ca$ mw: 418.42

PROP: White, powdery crystals from water. Mp: >300° (decomp). Color deepens to pink upon exposure to air. Very soluble in water and alc; practically insol in ether, benzene, chloroform.

SYNS: CALCIUM DOBESILATE □ DEXIUM □ DOBESILATE CALCIUM □ DOXIUM □ HYDROQUINONE CALCIUM SULFONATE

TOXICITY DATA WITH REFERENCE
orl-rat LD50:9400 mg/kg APFRAD 30,415,72
orl-mus LD50:7700 mg/kg APFRAD 20,415,72
ivn-mus LD50:775 mg/kg APFRAD 30,415,72

SAFETY PROFILE: Moderately toxic by intravenous route. Mildly toxic by ingestion. When heated to decomposition it emits toxic fumes of SO_x. See also SULFONATES and CALCIUM COMPOUNDS.

DMI400 CAS:2373-98-0 HR: 3
3,3'-DIHYDROXYBENZIDINE
mf: $C_{12}H_{12}N_2O_2$ mw: 216.26

PROP: Plates from Me_2CO. Mp: 160°.

SYNS: 6,6'-DIAMINO-m,m'-BIPHENOL □ 4,4'-DIAMINO-3,3'-BIPHENYLDIOL □ 3,3'-DIOXYBENZIDINE □ 3,3'-DWUOKSYBENZYDYNA (POLISH)

TOXICITY DATA WITH REFERENCE
pic-esc 100 mmol/L MDMIAZ 31,11,79
orl-rat TDLo:9950 mg/kg/52W-I:NEO VOONAW 7(2),33,61
scu-rat TDLo:5900 mg/kg/43W-I:CAR VOONAW 7(2),33,61
mul-rat TDLo:6900 mg/kg/43W-I:CAR VOONAW 7(2),33,61
orl-mus TDLo:11 g/kg/47W-I:ETA VOONAW 7(2),33,61
skn-mus TDLo:5040 mg/kg/52W-I:ETA VOONAW 7(2),33,61
scu-mus TDLo:1620 g/kg/45W-I:CAR AICCA6 7,46,50
scu-mus TD:14 g/kg/43W-I:ETA VOONAW 7(2),33,61

SAFETY PROFILE: Suspected carcinogen with experimental carcinogenic, neoplastigenic, and tumorigenic data. Mutation data reported. When heated to decomposition it emits toxic fumes of NO_x.

DMI600 CAS:131-56-6 HR: 3
2,4-DIHYDROXYBENZOPHENONE
mf: $C_{13}H_{10}O_3$ mw: 214.23

PROP: Needles from H_2O. Mp: 142.6–144.6°. Sol in concentrations of H_2SO_4.

SYNS: 2,4-DIHYDROXYBENZOFENON (CZECH) □ EASTMAN INHIBITOR DHPB □ QUINSORB 010 □ SYNTASE 100 □ UF 1 □ USAF DO-28 □ USAF ND-54 □ UVINUL 400

TOXICITY DATA WITH REFERENCE
eye-rbt 100 mg/24H MOD 28ZPAK -,101,72
orl-rat LD50:8600 mg/kg RPZHAW 19,179,68
ipr-mus LD50:100 mg/kg NTIS** AD277-689
ivn-mus LD50:85 mg/kg BJPCAL 22,221,64

CONSENSUS REPORTS: Reported in EPA TSCA Inventory.

SAFETY PROFILE: Poison by intravenous and intraperitoneal routes. Mildly toxic by ingestion. An eye irritant. When heated to decomposition it emits acrid smoke and irritating fumes.

DMJ000 CAS:143-62-4 HR: 3
3,β,14-DIHYDROXY-5,β-CARD-20(22)ENOLIDE
mf: $C_{23}H_{34}O_4$ mw: 374.57

PROP: Crystals from EtOH (aq). Mp: 253°.

SYNS: CARDOGENEN-(20:22)-DIOL-(3-β,14) (GERMAN) □ CERBERIGENIN □ DIGITOXIGENIN □ DIGITOXIGENINE □ (3-β,5-β)-3,14-DIHYDROXY-CARD-20(22)-ENOLIDE □ 3-β,14-DIOXY-CARDEN-(20:22)-OLID (GERMAN) □ 3-β,14-DIOXY-DIGEN-(20:22)-OLID (GERMAN) □ ECHUJETIN □ EVONOGENIN □ THEVETIGENIN □ $\Delta^{-20:22}$-3,14,21-TRIHYDROXYNORCHOLENIC ACID LACTONE

TOXICITY DATA WITH REFERENCE
ivn-rat LD50:1600 μg/kg ARZNAD 11,848,61
orl-mus LD50:26,170 μg/kg AIPTAK 153,436,65
scu-mus LD50:11,820 μg/kg AIPTAK 153,436,65
ivn-mus LD50:1131 μg/kg LIFSAK 37,775,85
ivn-cat LDLo:420 mg/kg AEPPAE 184,181,37
ivn-gpg LDLo:1419 μg/kg AIPTAK 153,436,65
par-pgn LDLo:600 μg/kg CPBTAL 8,18,60

SAFETY PROFILE: Poison by ingestion, subcutaneous, intravenous, and parenteral routes. When heated to decomposition it emits acrid smoke and irritating fumes.

DMJ200 CAS:128-13-2 HR: 3
3-α,7-β-DIHYDROXY-6-β-CHOLAN-24-OIC ACID
mf: $C_{24}H_{40}O_4$ mw: 392.64

PROP: Plates from EtOH. Mp: 203°.

SYNS: CHOLIT-URSAN □ DELURSAN □ DESTOLIT □ DEURSIL □ 3-α,7-β-DIHYDROXYCHOLANIC ACID □ 3,7-DIHYDROXYCHOLAN-24-OIC ACID □ 3-α,7-β-DIHYDROXY-5-β-CHOLANOIC ACID □ (3-α,5-β,7-β)-3,7-DIHYDROXYCHOLAN-24-OIC ACID □ 3-α,7-α-DIHYDROXY-CHOLANSAEURE (GERMAN) □ 3-α,7-β-DIOXYCHOLANIC ACID □ 17-β-(1-METHYL-3-CARBOXYPROPYL)ETIOCHOLANE-3-α,7-β-DIOL □ UDCA □ URSACOL □ URSO □ URSOCHOL □ URSODEOXYCHOL □ URSODEOXYCHOLIC ACID □ URSODESOXYCHOLIC ACID □ URSOFALK □ URSOLVAN

TOXICITY DATA WITH REFERENCE
mmo-sat 40 mg/L MUREAV 158,45,85
orl-rat TDLo:11 g/kg (female 7-17D post):REP OYYAA2 15,931,78
orl-rbt TDLo:65 mg/kg (female 6-18D post):REP OYYAA2 15,1133,78
orl-rbt TDLo:130 mg/kg (female 6-18D post):REP OYYAA2 15,1133,78
orl-rat TDLo:11,900 mg/kg (4-20D preg):TER AIPTAK 246,149,80
orl-rat TDLo:11,900 mg/kg (4-20D preg):REP AIPTAK 246,149,80
orl-rat TDLo:2975 mg/kg (female 4-20D post):TER AIPTAK 246,149,80
orl-rat TDLo:22 g/kg (male 9W pre):TER OYYAA2 15,923,78
orl-rat LD50:4600 mg/kg BCFAAI 126,282,87
ipr-rat LD50:890 mg/kg NIIRDN 6,95,82
ivn-rat LD50:310 mg/kg NIIRDN 6,95,82
ivn-mus LD50:240 mg/kg NIIRDN 6,95,82

SAFETY PROFILE: Poison by intraperitoneal and intravenous routes. Experimental teratogenic and reproductive effects. Mutation data reported. Stimulates the flow of bile to the duodenum (a cholagogic). When heated to decomposition it emits acrid smoke and irritating fumes.

DMJ400 CAS:32222-06-3 HR: 3
1a,25-DIHYDROXYCHOLECALCIFEROL
mf: $C_{27}H_{44}O_3$ mw: 416.71

PROP: Crystals from methyl formate. Mp: 118–119°.

SYNS: CALCITRIOL □ 1,25-DIHYDROXYCHOLECALCIFEROL □ 1-α,25-DIHYDROXYCHOLECALCIFEROL □ DIHYDROXYVITAMIN D3 □ 1-α,25-DIHYDROXYVITAMIN D3 □ Ro 215535 □ ROCALTROL □ (5Z,7E)-9,10-SECOCHESTA-5.7.10(19)-TRIENE-1-α,3-β,25-TRIOL □ (1-α,3-β,5Z,7E)-9,10-SECOCHOLESTA-5,7,10(19)-TRIENE-1,3,25-TRIOL □ SOLTRIOL

TOXICITY DATA WITH REFERENCE
dns-mus:fbr 2 μg/L CNREA8 46,604,86
scu-rat TDLo:15 μg/kg (female 16-21D post):REP ACATA5 111,343,81
orl-rbt TDLo:3600 ng/kg (female 7-18D post):REP TXAPA9 45,242,75
orl-rat TDLo:55 μg/kg (7-17D preg):TER YACHDS 11,4221,83
orl-rat TDLo:175 μg/kg (35D pre):REP YACHDS 11,4189,83
orl-rat TDLo:7 μg/kg (male 35D pre):REP YACHDS 11,4189,83
orl-rat TDLo:1400 ng/kg (male 35D pre):REP YACHDS 11,4189,83
scu-rat TDLo:15 μg/kg (female 16-21D post):TER ACATA5 111,343,81
scu-rat LD50:66 μg/kg YACHDS 11,4175,83
orl-mus LD50:1350 μg/kg YACHDS 11,4175,83
ipr-mus LD50:1900 μg/kg YACHDS 11,4175,83
scu-mus LD50:145 μg/kg YACHDS 11,4175,83

SAFETY PROFILE: A deadly poison by ingestion, intraperitoneal, and subcutaneous routes. Experimental teratogenic and reproductive effects. Mutation data reported. Enhances intestinal calcium transport and bone mineral mobilization. When heated to decomposition it emits acrid smoke and irritating fumes.

DMJ600 **CAS:2892-51-5** **HR: 2**
3,4-DIHYDROXY-3-CYCLOBUTENE-1,2-DIONE
mf: $C_4H_2O_4$ mw: 114.06

PROP: Crystals from water. Mp: 293° (decomp approx). Sol in water.

SYNS: DIHYDROXYCYCLOBUTENEDIONE □ 3,4-DIHYDROXYCYCLOBUTENE-1,2-DIONE □ QUADRATIC ACID □ SQUARIC ACID

TOXICITY DATA WITH REFERENCE
scu-mus TDLo:368 mg/kg/92W-I:ETA JNCIAM 46,143,71

CONSENSUS REPORTS: Reported in EPA TSCA Inventory.

SAFETY PROFILE: Questionable carcinogen with experimental tumorigenic data. When heated to decomposition it emits acrid smoke and fumes.

DMJ800 **CAS:128-59-6** **HR: 1**
DIHYDROXYDIBENZANTHRONE
mf: $C_{34}H_{16}O_4$ mw: 488.50

SYNS: 16,17-DIHYDROXYDIBENZANTHRONE □ DIHYDROXYVIOLANTHRON (CZECH) □ 16,17-DIHYDROXYVIOLANTHRONE

TOXICITY DATA WITH REFERENCE
eye-rbt 500 mg/24H MLD 28ZPAK -,104,72

CONSENSUS REPORTS: Reported in EPA TSCA Inventory.

SAFETY PROFILE: An eye irritant. When heated to decomposition it emits acrid smoke and fumes.

DMK200 **CAS:66267-18-3** **HR: 2**
trans-1,2-DIHYDROXY-1,2-DIHYDROBENZO(a,h)ANTHRACENE
mf: $C_{22}H_{16}O_2$ mw: 312.38

SYNS: DBA-1,2-DIHYDRODIOL □ (E)-1,2-DIHYDRO-1,2-DIHYDROXYDIBENZ(a,h)ANTHRACENE □ trans-1,2-DIHYDROXY-1,2-DIHYDROBENZ(a,h)ANTHRACENE

TOXICITY DATA WITH REFERENCE
skn-mus TDLo:2000 μg/kg:ETA CNREA8 39,1310,79

SAFETY PROFILE: Questionable carcinogen with experimental tumorigenic data. When heated to decomposition it emits acrid smoke and irritating fumes.

DMK400 **CAS:24961-49-7** **HR: 2**
trans-4,5-DIHYDROXY-4,5-DIHYDROBENZO(e)PYRENE
mf: $C_{20}H_{14}O_2$ mw: 286.34

SYNS: BENZO(e)PYRENE-4,5-DIHDYRODIOL □ 4,5-DIHYDRO-4,5-DIHYDROXYBENZO(e)PYRENE □ B(e)P-4,5-DIHYDRODIOL

TOXICITY DATA WITH REFERENCE
mma-sat 10 nmol/plate JBCHA3 254,4408,79
ipr-mus TDLo:32 mg/kg/15D-I:ETA CNREA8 40,203,80

SAFETY PROFILE: Questionable carcinogen with experimental tumorigenic data. Mutation data reported. When heated to decomposition it emits acrid smoke and irritating fumes.

DMK600 **CAS:66788-06-5** **HR: 2**
trans-9,10-DIHYDROXY-9,10-DIHYDROBENZO(e)PYRENE
mf: $C_{20}H_{14}O_2$ mw: 286.34

SYN: B(E)P 9,10-DIHYDRODIOL

TOXICITY DATA WITH REFERENCE
mma-sat 10 nmol/plate JBCHA3 254,4408,79
ipr-mus TDLo:32 mg/kg/15D-I:NEO CNREA8 40,203,80

SAFETY PROFILE: Questionable carcinogen with experimental neoplastigenic data. Mutation data reported. When heated to decomposition it emits acrid smoke and fumes.

DML000 **CAS:61443-57-0** **HR: 2**
(+,–)-trans-7,8-DIHYDROXY-7,8-DIHYDROBENZO(a)PYRENE
mf: $C_{20}H_{14}O_2$ mw: 286.34

SYN: BP-7,8-DIHYDRODIOL

TOXICITY DATA WITH REFERENCE
dnd-rat:lvr 20 μmol/L CRNGDP 3,861,82
msc-mus:fbr 200 nmol/L CNREA8 42,1866,82
orl-mus TDLo:206 mg/kg/6W-I:CAR JJIND8 62,1103,79
skn-mus TDLo:34 mg/kg/60W-I:NEO CNREA8 37,3356,77
scu-mus TDLo:10 mg/kg:NEO JJIND8 64,617,80
skn-mus TD:22 mg/kg/25W-I:ETA PNASA6 73,3867,76
skn-mus TD:2160 μg/kg:ETA CNREA8 37,4130,77

CONSENSUS REPORTS: EPA Genetic Toxicology Program.

SAFETY PROFILE: Questionable carcinogen with experimental carcinogenic, neoplastigenic, and tumorigenic data by skin contact. Mutation data reported. When heated to decomposition it emits toxic fumes of NO_x.

DML200 **CAS:60864-95-1** **HR: 2**
(–)-trans-7,8-DIHYDROXY-7,8-DIHYDROBENZO(a)PYRENE
mf: $C_{20}H_{14}O_2$ mw: 286.34

SYN: BP-7,8-DIHYDRODIOL

TOXICITY DATA WITH REFERENCE
mmo-sat 8 μg/plate MUREAV 58,361,78
otr-rat:lvr 10 mg/L CNREA8 40,1281,80
msc-ham:lng 40 nmol/L/2D CALEDQ 4,35,77
skn-mus TDLo:573 μg/kg:NEO CNREA8 37,2721,77
skn-mus TD:1144 μg/kg:NEO CCSUDL 3,371,78
skn-mus TD:1 mg/kg:NEO BJCAAI 34,523,76

CONSENSUS REPORTS: EPA Genetic Toxicology Program.

SAFETY PROFILE: Questionable carcinogen with ex-

perimental neoplastigenic data. Mutation data reported. When heated to decomposition it emits acrid smoke and fumes.

DML400 CAS:62314-67-4 HR: 2
(+)-trans-7,8-DIHYDROXY-7,8-DIHYDROBENZO(a)PYRENE
mf: $C_{20}H_{14}O_2$ mw: 286.34

TOXICITY DATA WITH REFERENCE
mma-sat 4 µg/plate MUREAV 58,361,78
otr-rat:lvr 10 mg/L CNREA8 40,1281,80
msc-ham:lng 1200 nmol/L/2D CALEDQ 4,35,77
skn-mus TDLo:573 µg/kg:NEO CNREA8 37,2721,77
skn-mus TD:1144 µg/kg:NEO CCSUDL 3,371,78

CONSENSUS REPORTS: EPA Genetic Toxicology Program.

SAFETY PROFILE: Questionable carcinogen with experimental neoplastigenic data. Mutation data reported. When heated to decomposition it emits acrid smoke and fumes.

DML775 CAS:69260-85-1 HR: 2
trans-3,4-DIHYDROXY-3,4-DIHYDRO-7,12-DIHYDROXYMETHYLBENZ(a)ANTHRACENE
mf: $C_{20}H_{18}O_4$ mw: 322.38

SYN: (E)-3,4-DIHYDROXY-3,4-DIHYDROBENZ(a)ANTHRACENE-7,12-DIMETHANOL

TOXICITY DATA WITH REFERENCE
mma-sat 50 nmol/plate CNREA8 40,3661,80
skn-mus TDLo:116 µg/kg:ETA CNREA8 40,3661,80

SAFETY PROFILE: Questionable carcinogen with experimental tumorigenic data. Mutation data reported. When heated to decomposition it emits acrid smoke and irritating fumes.

DML800 CAS:3343-12-2 HR: 2
11,12-DIHYDROXY-11,12-DIHYDRO-3-METHYLCHOLANTHRENE (E)
mf: $C_{21}H_{18}O_2$ mw: 302.39

SYNS: trans-11,12-DIHYDRO-11,12-DIHYDROXY-3-METHYLCHOLANTHRENE □ (E)-MC 11,12-DIHYDRODIOL □ (E)-11,12-DIHYDRO-3-METHYLCHOLANTHRENE-11,12-DIOL □ (E)-3-METHYLCHOLANTHRENE-11,12-DIHYDRODIOL □ trans-3-METHYL-11,12-DIHYDROCHOLANTHRENE-11,12-DIOL

TOXICITY DATA WITH REFERENCE
mma-sat 20 µmol/L BBRCA9 85,1568,78
sce-ham:ovr 1 mg/L CALEDQ 7,45,79
msc-ham:lng 1 mg/L BBRCA9 85,1568,78
skn-mus TDLo:121 µg/kg:ETA CNREA8 39,3549,79

CONSENSUS REPORTS: EPA Genetic Toxicology Program.

SAFETY PROFILE: Questionable carcinogen with experimental tumorigenic data. Mutation data reported. When heated to decomposition it emits acrid smoke and fumes.

DMM400 CAS:3179-90-6 HR: 2
5,8-DIHYDROXY-1,4-DIHYDROXYETHYLAMINOANTHRAQUINONE
mf: $C_{18}H_{18}N_2O_6$ mw: 358.38

SYNS: ACETATE TURQUOISE BLUE B □ ACETOQUINONE LIGHT GREEN BLUE JL □ AMACEL GREEN BLUE B □ AMACEL GREEN BLUE G □ ANTHRAQUINONE, 1,4-BIS((2-HYDROXYETHYL)AMINO)-5,8-DIHYDROXY- □ ANTHRAQUINONE, 1,4-DIHYDROXY-5,8-BIS((2-HYDROXYETHYL)AMINO)- □ ARTISIL BLUE GREEN GP □ CELANTHRENE FAST BLUE 2G □ CELLITON BLUE GREEN B □ CELLITON FAST BLUE GREEN B □ CELLITON FAST BLUE GREEN BA-CF □ CELUTATE GREEN BLUE BGH □ CIBACET BLUE GREEN C □ CIBACET BLUE GREEN CB □ CIBACET TURQUOISE BLUE G □ CIBACET TURQUOISE BLUE 2G □ CIBACET TURQUOISE BLUE 4G □ CILLA FAST BLUE GREEN B □ C.I. 62500 □ C.I. DISPERSE BLUE 7 □ C.I. SOLVENT BLUE 69 □ DIACELLITON FAST BLUE GREEN B □ 1,4-DIHYDROXY-5,8-BIS((2-HYDROXYETHYL)AMINO)-9,10-ANTHRACENEDIONE □ 1,4-DIOXYETHYLAMINO-5,8-DIOXYANTHRAQUINONE □ DISPERSE BLUE 7 □ DISPERSE BLUE GREEN □ DISPERSIVE blue-green □ 1,4-DOEA-5,8-DAPFA (RUSSIAN) □ DURANOL BLUE GREEN B □ DURANOL PRINTING BLUE GREEN B □ ESTEROQUINONE LIGHT BLUE 4JL □ FENACET FAST TURQUOISE B □ INTERCHEM ACETATE GREEN BLUE ALF □ INTERCHEM HISPERSE GREEN BLUE ALFH □ MIKETON FAST TURQUOISE BLUE G □ NACELAN BLUE CBG □ NYLOQUINONE BLUE 4J □ PALANIL BLUE 7G □ PERLITON BLUE GREEN B □ SAMARON BLUE 5G □ SERIPLAS BLUE GREEN BW □ SERISOL FAST BLUE GREEN B □ SERISOL FAST BLUE GREEN BW □ SETACYL BLUE 6GN □ SETACYL BLUE GREEN P-BS □ SETACYL TURQUOISE BLUE G □ SETACYL TURQUOISE BLUE 2G □ SETACYL TURQUOISE BLUE 4G □ SETACYL TURQUOISE BLUE GD □ SUPRACET BLUE GREEN B □ SUPRACET FAST GREEN BLUE B □ TERASIL BLUE GREEN CB □ TERASIL TURQUOISE BLUE G

TOXICITY DATA WITH REFERENCE
msc-mus:lym 10 mg/L EPASR* 8EHQ-1179-0321
hma-mus:sat 8 g/kg EPASR* 8EHQ-1179-0321
ipr-rat LD50:700 mg/kg GTPZAB 21(12),27,77

CONSENSUS REPORTS: Reported in EPA TSCA Inventory.

SAFETY PROFILE: Moderately toxic by intraperitoneal route. Mutation data reported. When heated to decomposition it emits toxic fumes of NO_x.

DMM600 CAS:10232-92-5 HR: 3
2,4-DIHYDROXY-3,3-DIMETHYLBUTYRONITRILE
mf: $C_6H_{11}NO_2$ mw: 129.18

TOXICITY DATA WITH REFERENCE
skn-rbt 500 mg/24H MLD 85JCAE -,917,86
eye-rbt 2 mg/24H SEV 85JCAE -,917,86
ihl-man TCLo:13 mg/m^3 (5Y male):REP GTPZAB 24(5),28,80
orl-rat LD50:310 mg/kg AIHAAP 23,95,62
skn-rbt LD50:130 mg/kg AIHAAP 23,95,62

CONSENSUS REPORTS: Cyanide and its compounds are on the Community Right-To-Know List.

SAFETY PROFILE: Poison by ingestion and skin contact. Human reproductive effects by inhalation: impaired spermatogenesis. When heated to decomposition it emits toxic fumes of NO_x and CN^-. See also NITRILES.

DMN400 CAS:81-55-0 HR: 3
1,8-DIHYDROXY-4,5-DINITROANTHRAQUINONE
mf: $C_{14}H_6N_2O_8$ mw: 330.22

SYNS: 9,10-ANTHRACENEDIONE, 1,8-DIHYDROXY-4,5-DINITRO-(9CI) □ 4,5-DINITROCHRYSAZIN □ NCI-C60742

TOXICITY DATA WITH REFERENCE
mmo-sat 100 µg/plate MUREAV 40,203,76
ivn-mus LD50:180 mg/kg CSLNX* NX#01788

CONSENSUS REPORTS: Reported in EPA TSCA Inventory.

SAFETY PROFILE: Poison by intravenous route. Mutation data reported. When heated to decomposition it emits toxic fumes of NO_x.

DMN450 CAS:947-42-2 HR: 3
DIHYDROXYDIPHENYLSILANE
mf: $C_{12}H_{12}O_2Si$ mw: 216.33

SYNS: DIFENYL-DIHYDROXYSILAN □ SILANE, DIHYDROXYDIPHENYL-

TOXICITY DATA WITH REFERENCE
orl-mus LD50:2150 mg/kg 85JCAE -,1237,86
ivn-mus LD50:180 mg/kg CSLNX* NX#04052

CONSENSUS REPORTS: Reported in EPA TSCA Inventory.

SAFETY PROFILE: Poison by intravenous route. Moderately toxic by ingestion. When heated to decomposition it emits acrid smoke and irritating vapors.

DMO500 CAS:64551-89-9 HR: 2
(±)-cis-3,4-DIHYDROXY-1,2-EPOXY-1,2,3,4-TETRAHYDROBENZ(a)ANTHRACENE
mf: $C_{18}H_{10}O_3$ mw: 274.28

SYNS: BENZ(a)ANTHRACENE, 3,4-DIHYDROXY-1,2-EPOXY-1,2,3,4-TETRAHYDRO-, (Z), (+)- □ (±)-cis-3,4-DIHYDROXY-1,2-EPOXY-1,2,3,4-TETRAHYDROBENZO(a)ANTHRACENE □ DIOL-EPOXIDE-1

TOXICITY DATA WITH REFERENCE
skn-mus TDLo:4400 µg/kg:NEO CNREA8 38,1705,78

SAFETY PROFILE: Questionable carcinogen with experimental neoplastigenic data. When heated to decomposition it emits acrid smoke and irritating fumes.

DMO600 CAS:64598-83-0 HR: 2
(±)trans-8-β,9-α-DIHYDROXY-10-α,11-α-EPOXY-8,9,10,11-TETRAHYDROBENZ(a)ANTHRACENE
mf: $C_{18}H_{14}O_3$ mw: 278.32

SYNS: BA-8,9-DIOL-10,11-EPOXIDE-1 □ (E)-8,9,10,11-TETRAHYDRO-8-β,9-α-DIHYDROXY-10-α,11-α-BENZ(a)ANTHRACENE

TOXICITY DATA WITH REFERENCE
mmo-sat 50 nmol/plate MUREAV 44,313,77
msc-ham:lng 600 µg/L BJCAAI 39,540,79
skn-mus TDLo:22 mg/kg:ETA CNREA8 38,1699,78

CONSENSUS REPORTS: EPA Genetic Toxicology Program.

SAFETY PROFILE: Questionable carcinogen with experimental tumorigenic data. Mutation data reported. When heated to decomposition it emits acrid smoke and fumes.

DMO800 CAS:63438-26-6 HR: 2
(+)-trans-3,4-DIHYDROXY-1,2-EPOXY-1,2,3,4-TETRAHYDROBENZ(a) ANTHRACENE
mf: $C_{18}H_{10}O_3$ mw: 274.28

SYNS: (E)-(+)-3,4-DIHYDROXY-1,2-EPOXY-1,2,3,4-TETRAHYDROBENZ(a)ANTHRACENE □ (+)-trans-3,4-DIHYDROXY-1,2-EPOXY-1,2,3,4-TETRAHYDROBENZO(a)ANTHRACENE □ DIOL-EPOXIDE 2

TOXICITY DATA WITH REFERENCE
skn-mus TDLo:4400 µg/kg:NEO CNREA8 38,1705,78
skn-mus TD:4390 µg/kg:NEO CNREA8 40,1981,80

SAFETY PROFILE: Questionable carcinogen with experimental neoplastigenic data. When heated to decomposition it emits acrid smoke and fumes.

DMP000 CAS:64838-75-1 HR: 2
(±)-trans-1,β,2,α-DIHYDROXY-3,α,4,α-EPOXY-1,2,3,4-TETRAHYDROBENZ(a)ANTHRACENE
mf: $C_{18}H_{14}O_3$ mw: 278.32

SYN: BA-1,2-DIOL-3,4-EPOXIDE-1

TOXICITY DATA WITH REFERENCE
skn-mus TDLo:22 mg/kg:ETA CNREA8 38,1699,78

SAFETY PROFILE: Questionable carcinogen with experimental tumorigenic data. When heated to decomposition it emits acrid smoke and fumes.

DMP200 CAS:64598-82-9 HR: 2
(±)-trans-8-β,9-α-DIHYDROXY-10-β,11-β-EPOXY-8,9,10,11-TETRAHYDROBENZ(a)ANTHRACENE
mf: $C_{18}H_{14}O_3$ mw: 278.32

SYN: (E)-(±)-8,9,10,11-TETRAHYDRO-8-β,9-α-DIHYDROXY 10 β,11 β EPOXYBENZ(a)ANTHRACENE

TOXICITY DATA WITH REFERENCE
mmo-sat 50 nmol/plate MUREAV 44,313,77
msc-ham:lng 600 µg/L BJCAAI 39,540,79
skn-mus TDLo:22 mg/kg:ETA CNREA8 38,1699,78

CONSENSUS REPORTS: EPA Genetic Toxicology Program.

SAFETY PROFILE: Questionable carcinogen with experimental tumorigenic data. When heated to decomposition it emits acrid smoke and fumes.

DMP600 CAS:63323-29-5 HR: 2
(+)cis-7,α,8,β-DIHYDROXY-9,α,10,α-EPOXY-7,8,9,10-TETRAHYDROBENZO(a)PYRENE
mf: $C_{20}H_{14}O_3$ mw: 302.34

SYNS: (+)-BP-7,α,8-β-DIOL-9,α,10,α-EPOXIDE 1 □ (+)-Z-7,8,9,10-TETRAHYDRO-7-α,8-β-DIHDYROXY-9-α,10-α EPOXYBENZO(a)PYRENE □ (+)-cis-7,8,9,10 TETRAHYDRO-7-β,8-α-DIHYDROXY-9-β,10-β-EPOXYBENZO(a)PYRENE

TOXICITY DATA WITH REFERENCE
mmo-sat 100 pmol/plate BBRCA9 77,1389,77

D

skn-mus TDLo: 1200 μg/kg: ETA CNREA8 39,67,79

SAFETY PROFILE: Questionable carcinogen with experimental tumorigenic data. Mutation data reported. When heated to decomposition it emits acrid smoke and fumes.

DMP800 **CAS:63357-09-5** **HR: 2**
(−)-cis-7,β,8,α-DIHYDROXY-9,β,10,β-EPOXY-7,8,9,10-TETRAHYDROBENZO(a)PYRENE
mf: $C_{20}H_{14}O_3$ mw: 302.34

PROP: Crystals from EtOAc. Mp: 214°.

SYNS: (−)BP-7,β,8,α-DIOL-9,β,10,β-EPOXIDE 1 □ (−)-Z-7,8,9,10-TETRAHYDRO-7-α,8-β-DIHDYROXY-9-α,10-α-EPOXYBENZO(a)PYRENE □ (−)-Z-7,8,9,10-TETRAHYDRO-7-β,8-α-DIHYDROXY-9-β,10-β-EPOXY-BENZO(a)PYRENE

TOXICITY DATA WITH REFERENCE
mmo-sat 100 pmol/plate BBRCA9 77,1389,77
dnd-mus-skn 8 μmol/kg CNREA8 44,1081,84
msc-ham: lng 1 μmol/L MUREAV 44,313,77
skn-mus TDLo: 1200 μg/kg: ETA CNREA8 39,67,79

SAFETY PROFILE: Questionable carcinogen with experimental tumorigenic data. Mutation data reported. When heated to decomposition it emits acrid smoke and irritating fumes.

DMP900 **CAS:58917-67-2** **HR: 2**
(±)-(E)-7,8-DIHYDROXY-9,10-EPOXY-7,8,9,10-TETRAHYDROBENZO(a)PYRENE
mf: $C_{20}H_{14}O_3$ mw: 302.34

SYN: BP 7,8-DIOL-9,10-EPOXIDE 2

TOXICITY DATA WITH REFERENCE
skn-mus TDLo: 2400 μg/kg: ETA CNREA8 39,67,79

SAFETY PROFILE: Questionable carcinogen with experimental tumorigenic data. When heated to decomposition it emits acrid smoke and irritating fumes.

DMQ000 **CAS:58917-67-2** **HR: 3**
(±)-trans-7,8-DIHYDROXY-9,10-EPOXY-7,8,9,10-TETRAHYDRO-BENZO(a)PYRENE
mf: $C_{20}H_{14}O_3$ mw: 302.34

SYNS: anti-BENZO(a)PYRENE-DIOLEPOXIDE □ anti-BP-DIOLEPOXIDE

TOXICITY DATA WITH REFERENCE
dni-omi 200 μg/L PNASA6 74,1378,77
dni-omi 100 μg/L PNASA6 74,1378,77
msc-ham: ovr 100 nmol/L MUREAV 112,329,83

SAFETY PROFILE: Mutation data reported. When heated to decomposition it emits acrid smoke and fumes.

DMR000 **CAS:58917-91-2** **HR: 2**
(±)-7,β,8,α-DIHYDROXY-9,β,10,β-EPOXY-7,8,9,10-TETRAHYDROBENZO(a)PYRENE
mf: $C_{20}H_{14}O_3$ mw: 302.34

SYNS: BPDE-syn □ B(a)P EPOXIDE I □ (±)-7-α,8-β-DIHYDROXY-9-α,10-α-EPOXY-7,8,9,10-TETRAHYDROBENZO(a)PYRENE □ (±)-7,8,8a,9a-TETRAHYDROBENZO(10,11)CHYRSENO(3,4-b)OXIRENE-7,8-DIOL □ (±)-7,8,9,10-TETRAHYDRO-7-α,8-β-DIHYDROXY-9-α,10-α-EPOXY-BENZO(a)PYRENE

TOXICITY DATA WITH REFERENCE
mmo-sat 300 pmol/plate CNREA8 36,3358,76
mma-sat 300 pmol/plate CNREA8 36,3358,76
dnr-hmn: fbr 1 μmol/L CBINA8 20,279,78
dnd-mus-skn 20 μmol/kg CRNGDP 3,1135,82
oms-mus-skn 20 μmol/kg CRNGDP 3,1135,82
dnd-mam: lym 600 nmol CRNGDP 3,267,82
skn-mus TDLo: 2420 μg/kg: ETA CCSUDL 3,371,78
skn-mus TD: 89 mg/kg/37W-I: ETA CNREA8 37,3356,77

SAFETY PROFILE: Questionable carcinogen with experimental tumorigenic data. Human mutation data reported. When heated to decomposition it emits acrid smoke and fumes.

DMR150 **HR: 2**
(±)-9-α-10-β-DIHYDROXY-11-β,12-β-EPOXY-9,10,11,12-TETRAHYDROBENZO(e)PYRENE
mf: $C_{20}H_{14}O_3$ mw: 302.34

SYNS: BENZO(e)PYRENE, 9,10-DIOL-11,12-EPOXIDE 1 (cis) □ B(e)P DIOL EPOXIDE-1 □ B(e)P 9,10-DIOL-11,12-EPOXIDE-1

TOXICITY DATA WITH REFERENCE
ipr-mus TDLo: 476 mg/kg: CAR CNREA8 41,915,81

SAFETY PROFILE: Questionable carcinogen with experimental carcinogenic data. When heated to decomposition it emits acrid smoke and irritating fumes.

DMR200 **HR: 2**
(±)-9,β,10,α-DIHYDROXY-11,α,12,α-EPOXY-9,10,11,12-TETRAHYDROBENZO(e)PYRENE
mf: $C_{20}H_{14}O_3$ mw: 302.3

SYN: B(E)P DIOL EPOXIDE-2

TOXICITY DATA WITH REFERENCE
mmo-sat 1 nmol/plate CNREA8 40,1985,80
mma-sat 1 nmol/plate CNREA8 40,1985,80
msc-ham: lng 1 nmol/L CNREA8 40,1985,80
ipr-mus TDLo: 476 mg/kg: CAR CNREA8 41,915,81

SAFETY PROFILE: Questionable carcinogen with experimental carcinogenic data. Mutation data reported. When heated to decomposition it emits acrid smoke and fumes.

DMS000 **HR: 2**
trans-1,2-DIHYDROXY-anti-3,4-EPOXY-1,2,3,4-TETRAHYDROCHRYSENE
mf: $C_{18}H_{14}O$ mw: 278.32

SYN: (+)-(E)-3,4-EPOXY-1,2,3,4-TETRAHYDRO-CHRYSENEDIOL

TOXICITY DATA WITH REFERENCE
skn-mus TDLo: 20 mg/kg: NEO CNREA8 40,1981,80

SAFETY PROFILE: Questionable carcinogen with experimental neoplastigenic data. When heated to decomposition it emits acrid smoke and fumes.

DMS200 CAS:72074-67-0 **HR: 2**
(±)-1,β,2,α-DIHYDROXY-3,α,4,α-EPOXY-1,2,3,4-TETRAHYDROCHRYSENE
mf: $C_{18}H_{14}O_3$ mw: 278.32

SYN: (±)-1,2,3,4-TETRAHYDRO-3,α,4,α-EPOXY-1,β,2,α-CHRYSENEDIOL

TOXICITY DATA WITH REFERENCE
mmo-sat 1 nmol/L CRNGDP 6,237,85
mma-sat 1 nmol/plate CNREA8 39,4069,79
ipr-mus TDLo:72 mg/kg/15D-I:CAR CNREA8 39,5063,79

SAFETY PROFILE: Questionable carcinogen with experimental carcinogenic data. Mutation data reported. When heated to decomposition it emits acrid smoke and fumes.

DMS400 CAS:72074-66-9 **HR: 2**
(±)-1,β,2,α-DIHYDROXY-3,β,4,β-EPOXY-1,2,3,4-TETRAHYDROCHRYSENE
mf: $C_{18}H_{14}O_3$ mw: 278.32

SYN: (±)-1,2,3,4-TETRAHYDRO-3,β,4,β-EPOXY-1-β,2-α-CHRYSENEDIOL

TOXICITY DATA WITH REFERENCE
mma-sat 1 nmol/plate CNREA8 39,4069,79
msc-ham:lng 1 nmol/plate CNREA8 39,4069,79
ipr-mus TDLo:72 mg/kg/15D-I:ETA CNREA8 39,5063,79

SAFETY PROFILE: Questionable carcinogen with experimental tumorigenic data. Mutation data reported. When heated to decomposition it emits acrid smoke and fumes.

DMT400 CAS:4500-29-2 **HR: 2**
N,N-DI(2-HYDROXYETHYL)CYCLOHEXYLAMINE
mf: $C_{10}H_{21}NO_2$ mw: 187.32

SYNS: ABBOMEEN E-2 □ ABBOMEEN E-2 AEROSOL □ 2,2'-CYCLOHEXYLIMINODIETHANOL

TOXICITY DATA WITH REFERENCE
orl-rat LD50:2600 mg/kg 34ZIAG -,61,69

CONSENSUS REPORTS: Reported in EPA TSCA Inventory.

SAFETY PROFILE: Moderately toxic by ingestion. When heated to decomposition it emits toxic fumes of NO_x.

DMT500 CAS:150-25-4 **HR: 2**
N,N-DIHYDROXYETHYL GLYCINE
mf: $C_6H_{13}NO_4$ mw: 163.20

SYN: GLYCINE, N,N-DIHYDROXYETHYL-

TOXICITY DATA WITH REFERENCE
ipr-mus LD50:1540 mg/kg REPMBN 10,391,62

CONSENSUS REPORTS: Reported in EPA TSCA Inventory.

SAFETY PROFILE: Moderately toxic by intraperitoneal route. When heated to decomposition it emits toxic vapors of NO_x.

DMT800 CAS:28005-74-5 **HR: 2**
DI-(HYDROXYETHYL)-o-TOLYLAMINE
mf: $C_{11}H_{17}NO_2$ mw: 195.29

SYNS: EMERY 5712 □ 2,2'-((2-METHYLPHENYL)IMINO)BISETHANOL □ 2,2'-(o-TOYLYIMINO)DIETHANOL □ o-TOLYLDIETHANOLAMINE

TOXICITY DATA WITH REFERENCE
skn-rbt 500 mg/24H MLD 85JCAE-,697,86
eye-rbt 750 μg/24H SEV 85JCAE-,697,86
orl-rat LDLo:2200 mg/kg AIHAAP 23,95,62
skn-rbt LDLo:1000 mg/kg AIHAAP 23,95,62

CONSENSUS REPORTS: Reported in EPA TSCA Inventory.

SAFETY PROFILE: Moderately toxic by ingestion and skin contact. A skin and severe eye irritant. When heated to decomposition it emits toxic fumes of NO_x. See also AMINES.

DMU000 CAS:1069-23-4 **HR: 3**
3,4-DIHYDROXY-1,5-HEXADIENE
mf: $C_6H_{10}O_2$ mw: 114.16

TOXICITY DATA WITH REFERENCE
orl-rat LD50:1620 mg/kg AIHAAP 30,470,69
skn-rbt LD50:400 mg/kg AIHAAP 30,470,69

SAFETY PROFILE: Poison by skin contact. Moderately toxic by ingestion. When heated to decomposition it emits acrid smoke and irritating fumes.

DMV200 CAS:485-47-2 **HR: 3**
2,2-DIHYDROXY-1,3-INDANDIONE
mf: $C_9H_6O_4$ mw: 178.15

PROP: Crystals. Pale yellow prisms. Turns reddish @ 125°, swells @ 139°, decomp @ 240°. Mp: 241–243° (becomes anhyd with reddening at 125–1°).

SYNS: 2,2-DIHYDROXY-1H-INDENE-1,3(2H)-DIONE □ 1,2,3-INDANTRIONE-2-HYDRATE □ 1,2,3-INDANTRIONE MONOHYDRATE □ NINHYDRIN □ NINHYDRIN HYDRATE □ TRIKETOHYDRINDENE HYDRATE

TOXICITY DATA WITH REFERENCE
orl-rat LDLo:250 mg/kg NCNSA6 5,28,53
ipr-mus LD50:78 mg/kg CRSBAW 151,719,57

CONSENSUS REPORTS: Reported in EPA TSCA Inventory.

SAFETY PROFILE: Poison by ingestion and intraperitoneal routes. When heated to decomposition it emits acrid smoke and fumes.

DMV400 CAS:99-11-6 **HR: 2**
2,6-DIHYDROXYISONICOTINIC ACID
mf: $C_6H_5NO_4$ mw: 155.12

PROP: Buff to gray powder. Microscopic plates from $CHCl_3$; crystals from H_2O. Sol in H_2O.

SYNS: CITRAZINIC ACID □ 2,6-DIHYDROXY-4-CARBOXYPYRIDINE □ KYSELINA CITRAZINOVA

TOXICITY DATA WITH REFERENCE
ipr-rat LDLo:800 mg/kg KODAK* -,-,71

CONSENSUS REPORTS: Reported in EPA TSCA Inventory.

SAFETY PROFILE: Moderately toxic by intraperitoneal route. A moderately irritating organic acid with some allergenic properties. When heated to decomposition it emits toxic fumes of NO_x.

DMV600 CAS:7683-59-2 HR: 3
3,4-DIHYDROXY-α-((ISOPROPYLAMINO)METHYL) BENZYL ALCOHOL
mf: $C_{11}H_{17}NO_3$ mw: 211.29

SYNS: A 21 □ ALEUDRIN □ ALUDRINE □ ASIPRENOL □ ASMALAR □ ASSIPRENOL □ BELLASTHMAN □ BRONKEPHRINE □ DIHYDROXYPHENYLETHANOLISOPROPYLAMINE □ 1-(3,4-DIHYDROXYPHENYL)-2-ISOPROPYLAMINOETHANOL □ EPINEPHRINE ISOPROPYL HOMOLOG □ 4-(1-HYDROXY-2-((1-METHYLETHYL)AMINO) ETHYL)-1,2-BENZENEDIOL □ IPA □ ISONORENE □ ISOPRENALINE □ ISOPROPYDRIN □ ISOPROPYLADRENALINE □ ISOPROPYLAMINOMETHYL-3,4-DIHYDROXYPHENYL CARBINOL □ α-(ISOPROPYLAMINOMETHYL)PROTOCATECHUYL ALCOHOL □ISOPROPYLARTERENOL □ N-ISOPROPYL-β-DIHYDROXYPHENYL-β-HYDROXYETHYLAMINE □ ISOPROPYL NORADRENALINE □ N-ISOPROPYLNORADRENALINE □ 1-ISOPROPYLNORADRENALINE □ ISOPROTERENOL □ l-ISOPROTERENOL □ ISORENIN □ ISUPREL □ ISUPREN □ LOMUPREN □ NEODRENAL □ NEO-EPININE □ NORISODRINE □ NOVODRIN □ PROTERNOL □ RESPIFRAL □ SAVENTRINE □ VAPO-N-ISO □ WIN 5162

TOXICITY DATA WITH REFERENCE
oms-rat-par 100 mg/kg BEXBAN 94,1458,82
dns-mus-ipr 341 mg/kg JCLBA3 56,605,73
scu-ham TDLo:3 µg/kg (female 8D post):TER PSEBAA 130,1168,69
scu-ham TDLo:3 µg/kg (female 8D post):REP PSEBAA 130,1168,69
ivn-rat TDLo:10 µg/kg (8D preg):REP DPTHDL 7(Suppl 1),72,84
ipr-rat TDLo:210 mg/kg (7D male):REP PSEBAA 153,170,76
ivn-rat TDLo:20 µg/kg (20D preg):TER TJADAB 24(1),24A,81
scu-ham TDLo:3 µg/kg (female 8D post):TER PSEBAA 130,1168,69
ims-hmn TDLo:14 µg/kg:CVS KLWOAZ 19,1303,40
orl-rat LD50:355 mg/kg USXXAM #4026897
ipr-rat LDLo:100 mg/kg FCTXAV 3,597,65
scu-rat LD50:600 µg/kg TOXID9 4,77,84
ivn-rat LD50:57 mg/kg TXAPA9 16,303,70
orl-mus LD50:450 mg/kg 27ZIAQ -,139,73
ipr-mus LD50:440 mg/kg JPETAB 164,290,68
scu-mus LD50:400 mg/kg ARZNAD 26,1404,76
ivn-mus LD50:83 mg/kg JPETAB 97,14,49
orl-dog LD50:600 mg/kg TXAPA9 8,353,66
ivn-dog LD50:50 mg/kg JPETAB 164,290,68
orl-rbt LD50:3070 mg/kg TXAPA9 8,353,66
ivn-rbt LD50:27 mg/kg TXAPA9 8,353,66
orl-gpg LD50:270 µg/kg JPETAB 164,290,68
scu-gpg LD50:320 µg/kg JPETAB 164,290,68

SAFETY PROFILE: Poison by ingestion, subcutaneous, intravenous, and intraperitoneal routes. An experimental teratogen. Other experimental reproductive effects. Human systemic effects by intramuscular route: increased pulse and cardiac rate. A bronchodilator. Mutation data reported. When heated to decomposition it emits toxic fumes of NO_x.

DMV800 CAS:586-06-1 HR: 3
3,5-DIHYDROXY-α-((ISOPROPYLAMINO)METHYL) BENZYL ALCOHOL
mf: $C_{11}H_{17}NO_3$ mw: 211.29

SYNS: METAPROTERENOL □ ORCIPRENALINE

TOXICITY DATA WITH REFERENCE
orl-wmn TDLo:800 µg/kg (1D pre):REP RDCNBM 5,31,81
orl-man TDLo:286 µg/kg AJMSA9 291,168,86
orl-rat LD50:3370 mg/kg TXAPA9 8,353,66
ivn-rat LD50:67,200 µg/kg TXAPA9 8,353,66
ipr-mus LDLo:240 mg/kg APTOA6 31,33,72
scu-mus LD50:406 mg/kg ARZNAD 26,1404,76
ivn-mus LD50:86 mg/kg APTOA6 38,474,76
orl-dog LD50:125 mg/kg TXAPA9 8,353,66
ivn-dog LD50:30 mg/kg TXAPA9 8,353,66
orl-rbt LD50:3110 mg/kg TXAPA9 8,353,66
ivn-rbt LD50:81,300 µg/kg TXAPA9 8,353,66

SAFETY PROFILE: Poison by ingestion, intraperitoneal, and intravenous routes. Moderately toxic by subcutaneous route. Human reproductive effects by ingestion: changes in the uterus, cervix, and vagina. When heated to decomposition it emits toxic fumes of NO_x.

DMW000 CAS:7361-61-7 HR: 3
5,6-DIHYDRO-2-(2,6-XYLIDINO)-4H-1,3-THIAZINE
mf: $C_{12}H_{16}N_2S$ mw: 220.36

PROP: Crystals from C_6H_6/pet ether. Sol in dil acids, C_6H_6, Me_2CO, and $CHCl_3$; insol in H_2O.

SYNS: BAY 1470 □ BAY VA 1470 □ N-(5,6-DIHYDRO-4H-1,3-THIAZINYL)-2,6-XYLIDINE □ 2-(2,6-DIMETHYLANILINO)-5,6-DIHYDRO-4H-1,3-THIAZINE □ 2-(2,6-DIMETHYLPHENYLAMINO)-4H-5,6-DIHYDRO-1,3-THIAZINE □ N-(2,6-DIMETHYLPHENYL)-5,6-DIHYDRO-4H-1,3-THIAZIN-2-AMINE □ N-(2,6-DIMETHYLPHENYL)-5,6-DIHYDRO-4H-1,3-THIAZINE-2-AMINE (9CI) □ ROMPUN □ WH 7286 □ XYLAZINE (USDA) □ XYLZIN

TOXICITY DATA WITH REFERENCE
orl-wmn TDLo:8 mg/kg:BAH,CVS CTOXAO 18,663,81
ims-wmn TDLo:734 µg/kg:EYE,CVS AJEMEN 4,222,86
ims-wmn TDLo:22 mg/kg:BAH,CVS AJEMEN 4,222,86
orl-rat LD50:130 mg/kg DTTIAF 75,565,68
orl-mus LD50:240 mg/kg DTTIAF 75,565,68
scu-mus LD50:121 mg/kg DTTIAF 75,565,68
ivn-mus LD50:18 mg/kg CSLNX* NX#10054

SAFETY PROFILE: Poison by ingestion, subcutaneous, and intravenous routes. Human systemic effects: change in motor activity, fall in blood pressure, miosis, pleural thickening, pulse rate decrease, somnolence. When heated to decomposition it emits very toxic fumes of NO_x and SO_x.

DMW200 CAS:526-84-1 HR: 3
DIHYDROXYMALEIC ACID
mf: $C_4H_4O_6$ mw: 148.07

$(:C(OH)CO \cdot OH)_2$

PROP: Plates.

SYN: DIHYDROXYBUTENEDIOIC ACID

SAFETY PROFILE: A storage hazard. It may explode in a sealed container. Slowly decomposes to release carbon dioxide. When heated to decomposition it emits acrid smoke and fumes.

DMX000 CAS:69260-83-9 HR: 2
(E)-3,4-DIHYDROXY-7-METHYL-3,4-DIHYDROBENZ(a) ANTHRACENE-12-METHANOL
mf: $C_{20}H_{18}O_3$ mw: 306.38

SYNS: trans-3,4-DIHYDRO-12-(HYDROXYMETHYL)-7-METHYL-BENZ(a)ANTHRACENE-3,4-DIOL □ trans-3,4-DIHYDROXY-3,4-DIHYDRO-7-METHYL-12-HYDROXYMETHYLBENZ(a)ANTHRACENE

TOXICITY DATA WITH REFERENCE
mma-sat 35 nmol/plate CNREA8 40,3661,80
mma-ham:lng 400 nmol/L PNASA6 76,862,79
skn-mus TDLo:110 μg/kg:ETA CNREA8 40,3661,80

SAFETY PROFILE: Questionable carcinogen with experimental tumorigenic data. Mutation data reported. When heated to decomposition it emits acrid smoke and fumes.

DMX200 CAS:2318-18-5 HR: 3
2,12-DIHYDROXY-4-METHYL-11,16-DIOXOSENECIONANIUM
mf: $C_{19}H_{28}NO_6$ mw: 366.48

PROP: Bevelled plates from EtOAc or Me_2CO. Mp: 196.5–197.5°.

SYNS: trans-15-ETHYLIDENE-12-β-HYDROXY-4,12-α,13-β-TRIMETHYL 8-OXO-4,8 SECOSENEC-1-ENINE □ 12-HYDROXY-4-METHYL-4,8-SECO-SENECIONAN-8,11,16-TRIONE □ NSC-89945 □ RENARDIN □ RENARDINE □ SENKIRKIN □ SENKIRKINE

TOXICITY DATA WITH REFERENCE
mma-sat 1 mg/plate MUREAV 68,211,79
sln-dmg-orl 10 μmol/L/3D-I FCTOD7 22,223,84
dns-rat:lvr 2 μmol/L CNREA8 45,3125,85
dns-mus:lvr 20 μmol/L CNREA8 45,3125,85
dns-ham:lvr 2 μmol/L CNREA8 45,3125,85
sce-ham:lng 60 μg/L MUREAV 142,209,85
ipr-rat TDLo:1320 mg/kg/56W-I:NEO JJIND8 63,469,79
orl-rat LDLo:200 mg/kg NATUAS 227,401,70
ipr-rat LD50:220 mg/kg JJIND8 63,469,79

CONSENSUS REPORTS: IARC Cancer Review: Group 3 IMEMDT 7,56,87; Animal Limited Evidence IMEMDT 31,231,83; Animal Inadequate Evidence IMEMDT 10,327,76.

SAFETY PROFILE: Poison by ingestion and intraperitoneal routes. Questionable carcinogen with experimental neoplastigenic data. Mutation data reported. When heated to decomposition it emits toxic fumes of NO_x.

DMX800 CAS:2033-94-5 HR: 3
3,4-DIHYDROXY-3-METHYL-4-PHENYL-1-BUTYNE
mf: $C_{11}H_{12}O_2$ mw: 176.23

SYNS: 3-METHYL-3,4-DIHYDROXY-4-PHENYL-BUTIN-1 (GERMAN) □ 3-METHYL-3,4-DIHYDROXY-4-PHENYL-1-BUTYNE □ 2-METHYL-1-PHENYL-3-BUTYNE-1,2-DIOL

TOXICITY DATA WITH REFERENCE
orl-rat LD50:730 mg/kg NYKZAU 64,351,68
scu-rat LD50:610 mg/kg NYKZAU 64,351,68
orl-mus LD50:710 mg/kg ARZNAD 13,728,63
ivn-mus LD50:240 mg/kg ARZNAD 13,728,63

SAFETY PROFILE: Poison by intravenous route. Moderately toxic by subcutaneous and ingestion routes. When heated to decomposition it emits acrid smoke and fumes. See also ACETYLENE COMPOUNDS.

DMZ000 CAS:2277-92-1 HR: 3
2,2′-DIHYDROXY-3,3′,5,5′,6-PENTACHLOROBENZANILIDE
mf: $C_{13}H_6Cl_5NO_3$ mw: 401.45

PROP: Cream-colored powder. Mp: 209–211°.

SYNS: DIPLIN □ ICI 46638 □ OXYCLOZANID □OXYCLOZANIDE □ 3,5,6,3′,5′-PENTACHLORO-2,2′-DIHYDROXYBENZANILIDE □ 3,3′,5,5′,6-PENTACHLORO-2,2′ DIHYDROXYBENZANILIDE □ 3,3′,5,5′,6-PENTACHLORO-2′-HYDROXYSALICYLANILIDE □ 2,3,5-TRICHLORO-N-(3,5-DICHLORO-2-HYDROXYPHENYL)-6-HYDROXYBENZAMIDE □ ZANIL □ ZANILOX

TOXICITY DATA WITH REFERENCE
orl-rat LD50:1000 mg/kg NATUAS 210,744,66
orl-dom LDLo:60 mg/kg VETRAX 78,267,66
orl-ctl LDLo:60 mg/kg VETRAX 78,267,66
ivn-dom LDLo:10 mg/kg VETRAX 78,267,66

SAFETY PROFILE: Poison by ingestion and intravenous routes. When heated to decomposition it emits very toxic fumes of Cl^- and NO_x.

DNA200 CAS:59-92-7 HR: 3
l-DIHYDROXYPHENYL-l-ALANINE
mf: $C_9H_{11}NO_4$ mw: 197.21

PROP: Prisms or needles from $H_2O + SO_2$; plates from EtOH (aq). Mp: 285.5° (decomp).

SYNS: 2-AMINO-3-(3,4-DIHYDROXYPHENYL)PROPANOIC ACID □ BENDOPA □ BIODOPA □ BROCADOPA □ CEREPAP □CIDANDOPA □ DA □ DEADOPA □ DIHYDROXY-l-PHENYLALANINE □ (−)-3-(3,4-DIHYDROXYPHENYL)-l-ALANINE □ β-(3,4-DIHYDROXYPHENYL)-α-ALANINE □ l-α-DIHYDROXYPHENYLALANINE □ l-β-(3,4-DIHYDROXYPHENYL)ALANINE □ l-3,4-DIHYDROXYPHENYL-α-ALANINE □ β-(3,4-DIHYDROXYPHENYL)-l-ALANINE □ 3-(3,4-DIHYDROXYPHENYL)-l-ALANINE □ 3,4-DIHYDROXYPHENYLALANINE □ (−)-3,4-DIHYDROXYPHENYLALANINE □ 3,4-DIHYDROXYPHENYL-l-ALANINE □ 3,4-DIHYDROXY-l-PHENYLALANINE □ l-3,4-DIHYDROXYPHENYLALANINE □ (−)-DOPA □ l-DOPA □ DOPAFLEX □ DOPAL □ DOPARKINE □ DOPASOL □ DOPRIN □ ELDOPAL □EURODO-PA □ HELFO DOPA □ l-o-HYDROXYTYROSINE □ 3-HYDROXY-l-TYROSINE □ INSULAMINA □ LARODOPA □ MAIPEDOPA □ PARDA □ RO 4-6316 □ SOBIODOPA □ VELDOPA

TOXICITY DATA WITH REFERENCE
dnr-bcs 500 μg/disc MUREAV 137,17,84

dni-hmn:fbr 3 mmol/L CNREA8 42,3783,82
orl-rat TDLo:3600 mg/kg (female 9-14D post):REP KSRNAM 4,2877,70
orl-mus TDLo:900 mg/kg (female 7-12D post):REP SKKNAJ 22,165,70
orl-rat TDLo:22 g/kg (female 1-22D post):REP NATUAS 239,285,72
orl-mus TDLo:2400 mg/kg (female 7-12D post):REP IYKEDH 3,53,72
orl-mus TDLo:4800 mg/kg (female 7-12D post):TER IYKEDH 3,53,72
orl-rat TDLo:8100 mg/kg (female 9-14D post):TER KSRNAM 4,2877,70
orl-mus TDLo:5 g/kg (female 6-15D post):REP AUCMBJ 61,251,73
ipr-rat TDLo:480 mg/kg (female 9-14D post):REP SKKNAJ 22,165,70
scu-rat TDLo:5 mg/kg (female 1D pre):REP ENDOAO 57,466,55
ipr-rat TDLo:4500 mg/kg (male 15D pre):REP IJMRAQ 71,46,80
orl-rbt TDLo:2250 mg/kg (female 7-15D post):TER AUCMBJ 61,251,73
orl-mus TDLo:1200 mg/kg (female 7-12D post):TER IYKEDH 3,53,72
orl-rat TDLo:500 mg/kg (1-5D preg):TER TXAPA9 38,251,76
orl-man TDLo:87,520 mg/kg/1.5Y-C:CAR,SKN NEURAI 24,340,74
orl-wmn TDLo:320 mg/kg/4D-I AHJOA2 110,488,85
orl-hmn TDLo:156 g/kg/10Y:CNS JNNPAU 34,502,71
orl-hmn TDLo:13 g/kg/1Y:CNS,PUL JNNPAU 34,668,71
orl-rat LD50:1780 mg/kg TXAPA9 28,1,74
ipr-rat LD50:624 mg/kg TXAPA9 28,1,74
orl-mus LD50:2363 mg/kg TXAPA9 28,1,74
ipr-mus LD50:588 mg/kg TXAPA9 28,1,74
scu-mus LD50:4449 mg/kg IYKEDH 3,186,72
ivn-mus LD50:450 mg/kg TXAPA9 28,1,74
orl-rbt LD50:609 mg/kg TXAPA9 28,1,74
orl-bwd LD50:100 mg/kg AECTCV 12,355,83

CONSENSUS REPORTS: Reported in EPA TSCA Inventory.

SAFETY PROFILE: Poison by ingestion. Moderately toxic by intravenous and intraperitoneal routes. Human systemic effects by ingestion: somnolence, hallucinations and distorted perceptions, toxic psychosis, motor activity changes, ataxia, dyspnea. Experimental teratogenic and reproductive effects. Questionable human carcinogen producing skin tumors. Human mutation data reported. An anticholinergic agent used as an anti-Parkinsonian drug. When heated to decomposition it emits toxic fumes of NO_x.

DNA600 CAS:13055-82-8 HR: 3
7-(3-(2-(3,5-DIHYDROXYPHENYL-2-HYDROXY-ETHYLAMINO)PROPYL))THEOPHYLLINE HYDROCHLORIDE

mf: $C_{18}H_{23}N_5O_5 \cdot ClH$ mw: 425.92

PROP: Crystals. Mp: 249–250°.

SYNS: BRONCHODIL □ BRONCHOSPASMIN □ REPROTEROL HYDROCHLORIDE □ 7-(3-((β,3,5-TRIHYDROXYPHENETHYL)AMINO)PROPYL)THEOPHYLLINE MONOHYDROCHLORIDE □ W-2946M

TOXICITY DATA WITH REFERENCE
orl-rat TDLo:35 g/kg (16-22D preg/28D post):REP ARZNAD 27,45,77
ivn-rbt TDLo:390 mg/kg (6-18D preg):REP ARZNAD 27,45,77
ivn-rat LD50:142 mg/kg ARZNAD 27,45,77
ivn-mus LD50:148 mg/kg ARZNAD 27,45,77
ivn-dog LD50:160 mg/kg ARZNAD 27,45,77

SAFETY PROFILE: Poison by intravenous route. Experimental reproductive effects. When heated to decomposition it emits very toxic fumes of HCl and NO_x. See also THEOPHYLLINE.

DNA800 CAS:555-30-6 HR: 3
l-(−)-3-(3,4-DIHYDROXYPHENYL)-2-METHYLALANINE

mf: $C_{10}H_{13}NO_4$ mw: 211.24

PROP: Crystals. Sol in isopropanol, EtOH, and H_2O.

SYNS: ALDOMET □ ALDOMETIL □ ALDOMIN □ ALPHA MEDOPA □ AMD □ BAYER 1440 L □ BAYPRESOL □ l(−)-β-(3,4-DIHYDROXYPHENYL)-α-METHYLALANINE □ DOPAMET □ DOPEGYT □ DOPTAEC □ 3-HYDROXY-α-METHYL-l-TYROSINE □ HYPERPAX □ l-(α-MD) □ MEDOMET □ MEDOPREN □ METHOPLAIN □ α-METHYL-l-3,4-DIHYDROXYPHENYLALANINE □ α-METHYL-β-(3,4-DIHYDROXYPHENYL)-l-ALANINE □ l-α-METHYL-3,4-DIHYDROXYPHENYLALANINE □ l-(−)-α-METHYL-β-(3,4-DIHYDROXYPHENYL)ALANINE □ METHYLDOPA □ α-METHYL-l-DOPA □ l-α-METHYLDOPA □ MK. B51 □ MK 351 □ NCI-C55721 □ NR.C 2294 □ PRESINOL □ PRESOLISIN □ SEDOMETIL □ SEMBRINA

TOXICITY DATA WITH REFERENCE
dlt-mus-orl 960 mg/kg CYTBAI 41,151,84
cyt-ham:lng 37 mg/L GMCRDC 27,95,81
unr-wmn TDLo:1680 mg/kg (28-36W preg):REP LANCAO 2,498,82
unr-wmn TDLo:10,950 mg/kg (2Y pre/1-12W preg):REP BMJOAE 283,194,81
orl-rat TDLo:7500 mg/kg (female 6-20D post):TER NTIS** PB86-245321
orl-mus TDLo:960 mg/kg (male 1D pre):REP CYTBAI 41,151,84
orl-rat TDLo:13 g/kg (male 65D pre):REP TXCYAC 41,305,86
orl-rat TDLo:7500 mg/kg (female 6-20D post):REP NTIS** PB86-245321
orl-mus TDLo:480 mg/kg (male 1D pre):REP CYTBAI 41,151,84
orl-rat TDLo:2100 mg/kg (1-12D preg):TER TXAPA9 38,251,76
orl-wmn TDLo:900 mg/kg/13W-I NEURAI 35,1668,85
orl-wmn TDLo:1830 mg/kg/17W-I:PNS SAMJAF 65,194,84
orl-man TDLo:1071 mg/kg/22W-I:SKN CUTIBC 38,187,86
orl-wmn TDLo:44 g/kg/3Y-I:GIT AHJOA2 105,1037,83
orl-rat LD50:5000 mg/kg 27ZQAG -,348,72
ipr-rat LD50:300 mg/kg 27ZQAG -,348,72
ipr-mus LD50:150 mg/kg JMCMAR 20,1378,77
ivn-mus LD50:1700 mg/kg NYKZAU 56,1103,60
orl-rbt LD50:713 mg/kg 27ZIAQ -,162,73
ivn-rbt LD50:713 mg/kg 27ZQAG -,348,72

CONSENSUS REPORTS: Reported in EPA TSCA Inventory. EPA Genetic Toxicology Program.

SAFETY PROFILE: Poison by intraperitoneal route. Moderately toxic by ingestion and intravenous routes. Human systemic effects by ingestion: fasciculations, hallucinations, distorted perceptions, tremors, allergic dermatitis, necrotic gastrointestinal changes. An experimental teratogen. Human reproductive effects: menstrual cycle changes or disorders, effects on newborn including abnormal neonatal measures and growth statistics, biochemical and metabolic changes. Experimental reproductive effects. Mutation data reported. When heated to decomposition it emits toxic fumes of NO_x.

DNB000 CAS:2589-47-1 HR: 3
17R,21-α-DIHYDROXY-4-PROPYLAJMALANIUM HYDROGEN TARTRATE
mf: $C_{23}H_{32}N_2O_2 \cdot C_4H_6O_6$ mw: 518.67

PROP: Crystals from EtOH/Et_2O. Mp: 149–152°.

SYNS: GT-1012 □ NEO-GILURYTMAL □ NPA □ PRAJMALINE BITARTRATE □ PRAJMALINE HYDROGEN TARTRATE □ N PROPYLAJMALINE BITARTRATE □ N-PROPYLAJMALINE HYDROGEN TARTRATE □ N-PROPYLAJMALINIUM BITARTRATE □ N-PROPYLAJMALINIUMHYDROGENTARTRAT (GERMAN) □ N⁴-PROPYLAJMALINIUM HYDROGEN TARTRATE

TOXICITY DATA WITH REFERENCE
orl-rbt TDLo:65 mg/kg (6-18D preg):TER ARZNAD 22,2085,72
orl-rat TDLo:120 mg/kg (6-15D preg):TER ARZNAD 22,2085,72
orl-wmn LDLo:22 mg/kg ARTODN 37,135,77
orl-man TDLo:1400 μg/kg:CNS BMJOAE 2,675,77
orl-rat LD50:54 mg/kg ARZNAD 22,2085,72
ivn-rat LD50:3400 μg/kg ARZNAD 22,2085,72
orl-mus LD50:45 mg/kg MEIEDD 10,1107,83
ivn-mus LD50:1700 μg/kg MEIEDD 10,1107,83

SAFETY PROFILE: Poison by ingestion and intravenous routes. An experimental teratogen. Human systemic effects by ingestion: hallucinations and distorted perceptions. Experimental reproductive effects. An antiarrhythmic agent. When heated to decomposition it emits toxic fumes of NO_x.

DNB200 CAS:53609-64-6 HR: 3
DI(2-HYDROXY-n-PROPYL)AMINE
mf: $C_6H_{14}N_2O_3$ mw: 162.22

SYNS: BHP □ N-BIS(2-HYDROXYPROPYL)NITROSAMINE □ 2,2′-BISHYDROXYPROPYLNITROSAMINE □ DHPN □ 2,2′-DIHYDROXY-DI-n-PROPYLNITROSOAMINE □ N,N-DI-(2-HYDROXYPROPYL)NITROSAMINE □ DIISOPROPANOLNITROSAMINE □ DIPN □ N-NITROSOBIS(2-HYDROXYPROPYL)AMINE □ N-NITROSO-N,N-DI(2-HYDROXYPROPYL)AMINE □ N-NITROSO-1,1′-IMINODI-2-PROPANOL □ 1,1′-NITROSOIMINODI-2-PROPANOL

TOXICITY DATA WITH REFERENCE
mma-sat 250 μg/plate MUREAV 111,135,83
otr-hmn:oth 5 mg/L BANRDU 12,15,82
dns-rat:lvr 5 mmol/L MUREAV 144,197,85
msc-ham:lng 700 μmol/L CNREA8 40,3463,80
orl-rat TDLo:4600 mg/kg/42W-C:CAR CRNGDP 5,167,84
ipr-rat TDLo:3000 mg/kg:CAR CALEDQ 6,115,79
scu-rat TDLo:1400 mg/kg/2W-I:CAR CTRRDO 63,1181,79
orl-mus TDLo:28 g/kg/16W-I:NEO CALEDQ 3,255,77
ipr-mus TDLo:125 mg/kg:CAR JTSCDR 10,315,85
scu-mus TDLo:6049 mg/kg/23W-I:CAR CALEDQ 9,257,80
ims-mus TDLo:125 mg/kg:ETA IGSBAL 109,99,84
orl-rbt TDLo:29 g/kg/25W-C:CAR CALEDQ 5,339,78
scu-gpg TDLo:7500 mg/kg/30W-I:CAR JNCIAM 58,387,77
orl-ham TDLo:1875 mg/kg/15W-I:CAR CNREA8 41,4715,81
skn-ham TDLo:688 mg/kg/43W-I:CAR CALEDQ 10,365,80
scu-ham TDLo:2375 mg/kg/19W-I:CAR CANCAR 36,379,75
scu-ham TDLo:100 mg/kg (14D post):NEO,TER ZEKBAI 90,119,77
scu-uns TDLo:20 g/kg/80W-I:CAR JJIND8 65,835,80
scu-ham TD:4125 mg/kg/33W-I:CAR CANCAR 36,379,75
scu-ham TD:2750 mg/kg/34W-I:CAR ZEKBAI 90,141,77
scu-ham TD:5250 mg/kg/15W-I:CAR CRNGDP 7,801,86
scu-rat TD:3000 mg/kg:CAR IGAYAY 107,248,78
ipr-rat TD:2800 mg/kg:NEO JJIND8 72,471,84
scu-rat TD:14,000 mg/kg/20W-I:CAR NAIZAM 32,670,81
scu-rat TD:3560 mg/kg/20W-I:CAR JNCIAM 58,361,77
ipr-rat TD:2000 mg/kg:CAR IAPUDO 41,611,82
scu-ham TD:500 mg/kg:CAR JCREA8 102,265,82
scu-ham TD:4250 mg/kg/17W-I:CAR CRNGDP 3,1021,82
scu-rat LD50:5000 mg/kg JJIND8 63,181,79
scu-mus LD50:5160 mg/kg CALEDQ 9,257,80
scu-gpg LD50:4900 mg/kg JNCIAM 58,387,77

SAFETY PROFILE: Suspected carcinogen with experimental carcinogenic, neoplastigenic, tumorigenic, and teratogenic data. Moderately toxic by subcutaneous route. Human mutation data reported. When heated to decomposition it emits toxic fumes of NO_x. See also NITROSAMINES.

DNB600 CAS:33372-40-6 HR: 2
4-(2,3-DIHYDROXYPROPYLAMINO)-2-(5-NITRO-2-THIENYL)QUINAZOLINE
mf: $C_{15}H_{14}N_4O_4S$ mw: 346.39

TOXICITY DATA WITH REFERENCE
mmo-sat 1 μg/plate CNREA8 35,3611,75
orl-rat TDLo:8313 mg/kg/47W-C:CAR JNCIAM 57,277,76

SAFETY PROFILE: Questionable carcinogen with experimental carcinogenic data. Mutation data reported. When heated to decomposition it emits very toxic fumes of NO_x and SO_x.

DNC000 CAS:479-18-5 HR: 3
7-(2,3-DIHYDROXYPROPYL)THEOPHYLLINE
mf: $C_{10}H_{14}N_4O_4$ mw: 254.28

SYNS: AFI-PHYLLIN □ ARISTOPHYLLIN □ ASTMAMASIT □ ASTROPHYLLIN □ CIRCAIN □ CIRCAIR □ CORONAL □ CORONARIN □ CORPHYLLIN □ COR-THEOPHYLLINE □ 7-(2,3-DIHYDROXYPROPYL)-3,7-DIHYDRO-1,3-DIMETHYL-1H-PURINE-2,5-DIONE □ DIHYDROXYPROPYL THEOPYLIN (GERMAN) □ DIHYDROXYPROPYL THEOPHYLLINE □ (1,2-DIHYDROXY-3-PROPYL)THIOPHYLLIN □ DILOR □ 1,3-DIMETHYL-7-(2,3-DIHYDROXYPROPYL)XANTHINE □ 7-(2,3-DIOXYPROPYL)THEOPHYLLINE □ DIPHYLLIN □ DIPROFILLIN □ DIPROFILLINE □ DIPROPHYLLIN □ DIPROPHYLLINE □ DT □ DYPHYLLINE □ GLYFYLLIN □ GLYPHYLLIN □ GLYPHYLLINE □ HIDROXITEOFILLINA □ HIPHYLLIN □ HYPHYLLINE □ LUFYLLIN □ NEOPHYLLIN □ NEOPHYLLINE □ NEOPHYLLIN M □ NEOSTENOVA-

SAN □ NEOTHYLLINE □ NEOTILINA □ NEO-VASOPHYLINE □ NEUFIL □ NEUTRAFIL □ NEUTRAFILLINA □ NEUTRAPHYLLIN □ NEUTRAPHYLLINE □ NEUTROXANTINA □ PROPYLPHYLLIN □ PROTHEOPHYLLINE □ PURIFILIN □ SIBEPHYLLIN □SIBEPHYLLINE □ SOLUFILIN □ SOLUFYLLIN □ SYNTHOPHYLLINE □ TEFILAN □ THEAL □ THEAL AMPULES □ THEFYLAN

TOXICITY DATA WITH REFERENCE
scu-rat LD50:1253 mg/kg AEPPAE 230,194,57
ivn-rat LD50:860 mg/kg AEPPAE 230,194,57
orl-mus LD50:1954 mg/kg JPETAB 116,343,56
ipr-mus LD50:1052 mg/kg ARZNAD 8,190,58
scu-mus LD50:120 mg/kg ARZNAD 6,601,56
ivn-mus LD50:1080 mg/kg RPOBAR 2,288,70

CONSENSUS REPORTS: Reported in EPA TSCA Inventory.

SAFETY PROFILE: Poison by subcutaneous route. Moderately toxic by ingestion, intraperitoneal, and intravenous routes. A smooth muscle relaxant. When heated to decomposition it emits toxic fumes of NO_x. See also THEOPHYLLINE.

DNC200 CAS:59-00-7 HR: 2
4,8-DIHYDROXYQUINALDIC ACID
mf: $C_{10}H_7NO_4$ mw: 205.18

PROP: Sulfur-yellow crystals. Mp: 286°. Insol in water; sol in aqueous alkali, hydroxides, and hot dil HCl.

SYNS: 4,8-DIHYDROXYQUINALDINIC ACID □ 4,8-DIHYDROXYQUINOLINE-2-CARBOXYLIC ACID □ XANTHURENIC ACID

TOXICITY DATA WITH REFERENCE
imp-mus TDLo:160 mg/kg:NEO ANYAA9 108,924,63

SAFETY PROFILE: Questionable carcinogen with experimental neoplastigenic data. When heated to decomposition it emits toxic fumes of NO_x.

DNC400 CAS:66788-03-2 HR: 2
trans-9,10-DIHYDROXY-9,10,11,12-TETRAHYDROBENZO(e)PYRENE
mf: $C_{20}H_{16}O_2$ mw: 288.36

SYN: B(E)P H4-9,10-DIOL

TOXICITY DATA WITH REFERENCE
mma-sat 10 nmol/plate JBCHA3 254,4408,79
skn-mus TDLo:69 mg/kg:ETA CNREA8 40,203,80

SAFETY PROFILE: Questionable carcinogen with experimental tumorigenic data. Mutation data reported. When heated to decomposition it emits acrid smoke and fumes.

DNC600 CAS:73771-79-6 HR: 2
trans-1,2-DIHYDROXY-1,2,3,4-TETRAHYDROCHRYSENE
mf: $C_{18}H_{16}O_2$ mw: 264.34

SYN: trans-1,2,3,4-TETRAHYDROCHRYSENE-1,2,-DIOL

TOXICITY DATA WITH REFERENCE
skn-mus TDLo:42 mg/kg:NEO CNREA8 38,1831,78

SAFETY PROFILE: Questionable carcinogen with experimental neoplastigenic data. When heated to decomposition it emits acrid smoke and irritating fumes.

DNC800 CAS:70443-38-8 HR: 2
trans-3,4-DIHYDROXY-1,2,3,4-TETRAHYDRODIBENZ(a,h)ANTHRACENE
mf: $C_{22}H_{18}O_2$ mw: 314.40

SYN: trans-3,4-DIHYDROXY-1,2,3,4-TETRAHYDRODIBENZO(a,h)ANTHRACENE

TOXICITY DATA WITH REFERENCE
mma-sat 60 µg/plate MUREAV 96,1,82
skn-mus TDLo:2010 mg/kg:ETA CNREA8 39,1310,79

SAFETY PROFILE: Questionable carcinogen with experimental tumorigenic data. Mutation data reported. When heated to decomposition it emits acrid smoke and fumes.

DND900 HR: 3
5,7-DIHYDROXYTETRAZOLO(1,5-a)PYRIDINE-6-CARBONITRILE
mf: $C_6H_3N_5O_2$ mw: 177.14

SYNS: 5,7-DIHYDROXY-PYRIDOTETRAZOLE-6-CARBONITRILE □ 1,3-DIOXY-2-NICOTINSAEURENITRIL-TETRAZOL (GERMAN)

TOXICITY DATA WITH REFERENCE
scu-mus LDLo:500 mg/kg BDVU** -,-,37
scu-gpg LDLo:346 mg/kg BDVU** -,-,37
scu-frg LDLo:3800 mg/kg BDVU** -,-,37

SAFETY PROFILE: Poison by subcutaneous route. When heated to decomposition it emits toxic fumes of NO_x.

DNE000 CAS:488-17-5 HR: 3
2,3-DIHYDROXYTOLUENE
mf: $C_7H_8O_2$ mw: 124.15

SYNS: 3-METHYL-1,2-BENZENEDIOL □ 3-METHYLCATECHOL □ 3-METHYLPYROCATECHOL □ 2,3-TOLUENEDIOL

TOXICITY DATA WITH REFERENCE
ivn-mus LD50:56 mg/kg CSLNX* NX#07878

CONSENSUS REPORTS: Reported in EPA TSCA Inventory.

SAFETY PROFILE: Poison by intravenous route. When heated to decomposition it emits acrid smoke and fumes.

DNE400 CAS:3468-11-9 HR: 2
1,3-DIIMINOISOINDOLINE
mf: $C_8H_7N_3$ mw: 145.18

PROP: Yellow crystals. Mp: 199° (decomp). Sol in alcohols, acids; sltly sol in H_2O.

SYNS: AFASTOGEN BLUE 5040 □ 1,3-DIIMINOISOINDOLIN (CZECH) □ FASTOGEN BLUE FP-3100 □ FASTOGEN BLUE SH-100 □ MODR FRALOSTANOVA 3G (CZECH) □ PHTHALIMIDIMIDE □ PHTHALOCYANINE BLUE 01206 □ PHTHALOGEN

TOXICITY DATA WITH REFERENCE
skn-rbt 500 mg/24H SEV 28ZPAK -,143,72

eye-rbt 250 μg/24H SEV 28ZPAK -,143,72
scu-rat TDLo:990 mg/kg/44W-I:ETA VOONAW 21(11),75,75
scu-mus TDLo:140 mg/kg/51W-I:CAR VOONAW 21(11),75,75

CONSENSUS REPORTS: Reported in EPA TSCA Inventory.

SAFETY PROFILE: A severe eye and skin irritant. Questionable carcinogen with experimental carcinogenic and tumorigenic data. When heated to decomposition it emits toxic fumes of NO_x.

DNE500 CAS:624-74-8 HR: 3
DIIODOACETYLENE
mf: C_2I_2 mw: 277.83

PROP: Needles from ligroin. Mp: 76.0–76.5°

SYNS: DIIODOETHYNE □ ETHYNE, DIIODO-

DOT CLASSIFICATION: Forbidden

SAFETY PROFILE: An explosive sensitive to impact, crushing, or heating to 84°C. When heated to decomposition it emits toxic fumes of I^-. See also IODIDES and ACETYLENE COMPOUNDS.

DNE700 CAS:615-42-9 HR: 3
1,2-DIIODOBENZENE
mf: $C_6H_4I_2$ mw: 329.91

PROP: Plates or prisms from pet ether. Mp: 27°, bp: 286–287°.

SAFETY PROFILE: Explodes violently when heated to 181°C in a sealed container. When heated to decomposition it emits toxic fumes of I^-. See also IODIDES.

DNE800 CAS:4460-32-6 HR: 2
N-2,5-DIIODOBENZOYL-N′,N′,N″,N″-DIETHYLENEPHOSPHORTRIAMIDE
mf: $C_{11}H_{12}I_2N_3O_2P$ mw: 503.03

SYNS: p,p-BIS(1-AZIRIDINYL)-2,5-DIIODOBENZOYLPHOSPINIC AMIDE □ N-(BIS(1-AZIRIDINYL)PHOSPHINYL)-2,5-DIIODOBENZAMIDE □ DIIODBENZOTEPH □ DIIODOBENZOTEF

TOXICITY DATA WITH REFERENCE
orl-rat LD50:500 mg/kg PCJOAU 12,689,78
ipr-rat LD50:500 mg/kg RPTOAN 41,135,78

SAFETY PROFILE: Moderately toxic by ingestion and intraperitoneal routes. When heated to decomposition it emits very toxic fumes of PO_x, NO_x, and I^-.

DNE875 CAS:53214-97-4 HR: 3
1,4-DIIODO-1,3-BUTADIYNE
mf: C_4I_2 mw: 174.95

IC≡CC≡CI

SAFETY PROFILE: A solid. Mp: 94–95°. Explodes at 100°C. When heated to decomposition it emits toxic fumes of I^-. See also IODIDES.

DNF000 CAS:15978-93-5 HR: 2
cis-DIIODODIAMMINEPLATINUM (II)
mf: $H_6I_2N_2Pt$ mw: 482.97

TOXICITY DATA WITH REFERENCE
pic-esc 1 μg/plate BBRCA9 90,209,79
idr-hmn TDLo:40 mg/kg:SKN CNREA8 35,2766,75

CONSENSUS REPORTS: Reported in EPA TSCA Inventory.

SAFETY PROFILE: Human systemic effects by intradermal route: unspecified effects on the skin. Mutation data reported. When heated to decomposition it emits very toxic fumes of I^- and NO_x. See also IODIDES and PLATINUM COMPOUNDS.

DNF200 CAS:1955-21-1 HR: 3
2,6-DIIODOHYDROQUINONE
mf: $C_6H_4I_2O_2$ mw: 361.90

PROP: Needles from H_2O. Mp: 144–145°.

SYNS: 2,6-DIIODO-1,4-BENZENEDIOL □ 2,6-DIIODOQUINOL

TOXICITY DATA WITH REFERENCE
ipr-mus LD50:237 mg/kg BCPCA6 12,885,63

CONSENSUS REPORTS: Reported in EPA TSCA Inventory.

SAFETY PROFILE: Poison by intraperitoneal route. When heated to decomposition it emits toxic fumes of I^-.

DNF300 CAS:618-76-8 HR: 2
3,5-DIIODO-4-HYDROXYBENZOIC ACID
mf: $C_7H_4I_2O_3$ mw: 389.91

SYN: BENZOIC ACID, 3,5-DIIODO-4-HYDROXY-

TOXICITY DATA WITH REFERENCE
orl-rat LD:>500 mg/kg NCNSA6 5,20,53
orl-mus LD50:4 g/kg JAPMA8 43,495,54
ipr-mus LD50:1000 mg/kg JMPCAS 2,213,60

CONSENSUS REPORTS: Reported in EPA TSCA Inventory.

SAFETY PROFILE: Moderately toxic by ingestion and intraperitoneal routes. When heated to decomposition it emits toxic vapors of I^-.

DNF400 CAS:2961-61-7 HR: 3
3,5-DIIODO-4-HYDROXYBENZONITRILE, LITHIUM SALT
mf: $C_7H_2I_2NO{\cdot}Li$ mw: 376.84

SYNS: BENTROL □ CERTOL □ 4-CYANO-2,6-DIJODPHENOL LITHIUMSALZ (GERMAN) □ 3,5-DIJOD-4-HYDROXY-BENZONITRILE LITHIUMSALZ (GERMAN) □ IOXYNIL, LITHIUM SALT

TOXICITY DATA WITH REFERENCE
orl-rat LD50:71 mg/kg RREVAH 10,97,65
skn-rat LD50:87 mg/kg 85GYAZ -,93,71
orl-mus LD50:190 mg/kg 85GYAZ -,93,71
orl-dog LD50:140 mg/kg PCOC** -,616,66
orl-ckn LD50:120 mg/kg 85GYAZ -,93,71

CONSENSUS REPORTS: Cyanide and its compounds are on the Community Right-To-Know List.

SAFETY PROFILE: Poison by ingestion and skin contact. When heated to decomposition it emits very toxic fumes of I^-, NO_x, CN^-, and Li_2O. See also NITRILES and LITHIUM COMPOUNDS.

DNF450 CAS:4662-17-3 HR: 3
3,5-DIIODO-4-HYDROXYPHENYL 2,5-DIMETHYL-3-FURYL KETONE
mf: $C_{13}H_{10}I_2O_3$ mw: 468.03

SYNS: DB 136 □ KETONE, 3,5-DIIODO-4-HYDROXYPHENYL 2,5-DIMETHYL-3-FURYL

TOXICITY DATA WITH REFERENCE
ipr-mus LD50:315 mg/kg AIPTAK 147,497,64
ivn-gpg LDLo:164 mg/kg AIPTAK 147,497,64

DOT CLASSIFICATION: 3; *Label:* Flammable Liquid

SAFETY PROFILE: A poison by intraperitoneal and intravenous routes. A flammable liquid. When heated to decomposition it emits toxic vapors of I^-.

DNF500 CAS:4568-82-5 HR: 3
3,5-DIIODO-4-HYDROXYPHENYL 2-FURYL KETONE
mf: $C_{11}H_6I_2O_3$ mw: 439.97

SYNS: DB 134 □ DIIODO-3,3 HYDROXY-4 BENZOYL 2 FURANNE □ KETONE, 3,5-DIIODO-4-HYDROXYPHENYL 2-FURYL

TOXICITY DATA WITH REFERENCE
ipr-mus LD50:132 mg/kg AIPTAK 147,497,64

DOT CLASSIFICATION: 3; *Label:* Flammable Liquid

SAFETY PROFILE: A poison by intraperitoneal route. A flammable liquid. When heated to decomposition it emits toxic vapors of F^-.

DNF550 CAS:73343-72-3 HR: 3
3,5-DIIODO-4-HYDROXYPHENYL 2-MESITYL-3-BENZOFURANYL KETONE
mf: $C_{24}H_{18}I_2O_3$ mw: 608.22

SYNS: BENZOFURAN, 3-(3,5-DIIODO-4-HYDROXYBENZOYL)-2-MESITYL- □ (DIIODO-3,5 HYDROXY-4 BENZOYL)-3 MESITYL-2 BENZOFURANNE □ KETONE, 3,5-DIIODO-4-HYDROXYPHENYL 2-MESITYL-3-BENZOFURANYL □ METHANONE, (4-HYDROXY-3,5-DIIODOPHENYL)(2-(2,4,6-TRIMETHYLPHENYL)-3-BENZOFURANYL)-

TOXICITY DATA WITH REFERENCE
ipr-mus LD50:840 mg/kg EJMCA5 14,517,79

DOT CLASSIFICATION: 3; *Label:* Flammable Liquid

SAFETY PROFILE: Moderately toxic by intraperitoneal route. A flammable liquid. When heated to decomposition it emits toxic vapors of F^-.

DNF600 CAS:83-73-8 HR: 3
DIIODOHYDROXYQUIN
mf: $C_9H_5I_2NO$ mw: 396.95

PROP: Crystals from xylene.

SYNS: DIIODOHYDROXYQUIN □ DIIODOHYDROXYQUINOLINE □ 5,7-DIIODO-8-HYDROXYQUINOLINE □ 5,7-DIIODO-OXINE □ DIIODOQUIN □ 5,7-DIIODO-8-QUINOLINOL □ DINOLEINE □ DIODOQUIN □ DIODOXYLIN □ DI-QUINOL □ DIREXIODE □ DISOQUIN □ DYODIN □ EMBEQUIN □ ENTEROSEPT □ FLORAQUIN □ FLUORAQUIN □ 8-HYDROXY-5,7-DIIODOQUINOLINE □ IODOQUINOL □ IOQUIN SUSPENSION □ LANODOXIN □ MOEBIQUIN □ QUINADOME □ SEARLEQUIN □ SEBAQUIN □ SS 578 □ YODOXIN □ ZOAQUIN

TOXICITY DATA WITH REFERENCE
dnr-esc 260 nmol/plate MUREAV 188,111,87
mnt-mus-orl 80 mg/kg MUREAV 222,219,89
orl-chd TDLo:120 g/kg/2Y-I:EYE LANCAO 1,261,66
ivn-mus LD50:56 mg/kg CSLNX* NX#03304
orl-cat LDLo:300 mg/kg AJTMAQ 24,29,44
orl-gpg LDLo:50 mg/kg AJTMAQ 24,29,44

CONSENSUS REPORTS: Reported in EPA TSCA Inventory.

SAFETY PROFILE: Poison by ingestion and intravenous routes. Human systemic effects by ingestion: eye effects. Mutation data reported. When heated to decomposition it emits very toxic fumes of I^- and NO_x.

DNF800 CAS:75-11-6 HR: 3
DIIODOMETHANE
mf: CH_2I_2 mw: 267.83

PROP: Light straw-colored to clear, heavy, refractive liquid. Mp: 5–6°, bp: 181° (part decomp), d: 3.33 @ 15°/15°, vap d: 9.25. Sltly sol in water.

SYNS: METHYLENE DIIODIDE □ METHYLENE IODIDE □ MI-GEE

TOXICITY DATA WITH REFERENCE
orl-cld LDLo:2778 µL/kg:BAH,PUL AEMED3 19,1171,90
ipr-rat LD50:403 mg/kg 34ZIAG -,756,69
ipr-mus LD50:467 mg/kg 34ZIAG -,756,69
scu-mus LD50:830 mg/kg TXAPA9 4,354,62

CONSENSUS REPORTS: Reported in EPA TSCA Inventory.

SAFETY PROFILE: Moderately toxic by intraperitoneal and subcutaneous routes. Probably an irritant and narcotic in high concentration. Human systemic effects: acute pulmonary edema, somnolence. Potentially explosive reaction with diethyl zinc + alkenes. Violent reaction with copper-zinc alloys + ether. Forms very shock-sensitive explosive mixtures with potassium, potassium-sodium alloys, and lithium. When heated to decomposition it emits toxic fumes of I^-. See also IODIDES.

DNF850 CAS:20018-09-1 HR: D
DIIODOMETHYL p-TOLYL SULFONE
mf: $C_8H_8I_2O_2S$ mw: 422.02

SYNS: AMICAL 48 □ BENZENE, 1-((DIIODOMETHYL)SULFONYL)-4-METHYL- □ 1-((DIIODOMETHYL)SULFONYL)-4-METHYLBENZENE

TOXICITY DATA WITH REFERENCE
orl-rat TDLo:5 g/kg (female 6-15D post):REP TOLED5 62,45,92
orl-rat TDLo:1250 mg/kg (female 6-15D post):REP TOLED5 62,45,92

CONSENSUS REPORTS: Reported in EPA TSCA Inventory.

SAFETY PROFILE: Experimental reproductive effects. When heated to decomposition it emits toxic vapors of SO_x and I^-.

DNG000 CAS:305-85-1 HR: 3
2,6-DIIODO-4-NITROPHENOL
mf: $C_6H_3I_2NO_3$ mw: 390.90

PROP: Light-yellow feathery crystals from AcOH. Mp: 157°.

SYNS: ANCYLOL □ DIISOPHENOL □ DISOFEN □ DISOPHENOL □ DNP

TOXICITY DATA WITH REFERENCE
orl-rat LD50:170 mg/kg TXAPA9 6,232,64
ipr-rat LD50:105 mg/kg TXAPA9 6,232,64
scu-rat LD50:122 mg/kg TXAPA9 6,232,64
ivn-rat LD50:105 mg/kg TXAPA9 6,232,64
orl-mus LD50:212 mg/kg TXAPA9 6,232,64
ipr-mus LD50:107 mg/kg TXAPA9 6,232,64
scu-mus LD50:110 mg/kg TXAPA9 6,232,64
ivn-mus LD50:88 mg/kg TXAPA9 6,232,64
par-uns LDLo:36 mg/kg FAZMAE 17,108,73

CONSENSUS REPORTS: Reported in EPA TSCA Inventory.

SAFETY PROFILE: Poison by ingestion, intraperitoneal, subcutaneous, intravenous, and parenteral routes. An anthelmintic. When heated to decomposition it emits very toxic fumes of I^- and NO_x. See also NITRO COMPOUNDS of AROMATIC HYDROCARBONS.

DNG200 CAS:3861-47-0 HR: 3
3,5-DIIODO-4-OCTANOYLOXYBENZONITRILE
mf: $C_{15}H_{17}I_2NO_2$ mw: 497.13

PROP: Wax. Mp: 59–60°.

SYNS: 4-CYANO-2,6-DIJODPHENOL CAPRYSAEUREESTER (GERMAN) □ 3,5-DIIODO-4-HYDROXYBENZONITRILE OCTANOATE □ 3,5-DIJOD-4-HYDROXY-BENZONITRIL CAPRYSAEUREESTER (GERMAN) □ IOXYNIL OCTANOATE □ M&B 11,461 □ RIP-15830 □ TOTRIL

TOXICITY DATA WITH REFERENCE
mrc-bcs 20 µg/disc/24H MUREAV 40,19,76
orl-rat LD50:190 mg/kg GUCHAZ 6,305,73
skn-rat LD50:500 mg/kg 85JFAN A233,83
orl-mus LD50:240 mg/kg GUCHAZ 6,305,73
skn-mus LD50:1240 mg/kg OYYAA2 1,78,67

CONSENSUS REPORTS: EPA Genetic Toxicology Program. Cyanide and its compounds are on the Community Right-To-Know List.

SAFETY PROFILE: Poison by ingestion. Moderately toxic by skin contact. Mutation data reported. An herbicide. When heated to decomposition it emits very toxic fumes of I^-, NO_x, and CN^-. See also NITRILES.

DNG400 CAS:300-37-8 HR: 2
3,5-DIIODO-4-OXO-1(4H)PYRIDINEACETIC ACID-2,2′-IMINODIETHANOL SALT
mf: $C_7H_5I_2O_3 \cdot C_4H_{11}NO_2$ mw: 510.09

PROP: Powder. Mp: 155–157° (decomp).

SYNS: CARDIOTRAST □ DIAETHANOLAMIN-3,5-DIJODPYRIDON-(4)-ESSIGSAEURE (GERMAN) □ DIATRAST □ DIETHANOLAMINE-3,5-DIIODO-4-PYRIDONE-N-ACETATE □ 3,5-DIIODO-4-PYRIDONE-N-ACETATE BIS(HYDROXYETHYL)AMMONIUM □ 3,5-DIIODO-4-PYRIDONE-N-ACETIC ACID, DIETHANOLAMINE SALT □ DIODON □ DIODONE □ DIODRAST □ ETHANOL,2,2′-IMINODI-,3,5-DIIODO-4-OXO-1(4H)-PYRIDINEACETATE (salt) □ ETHANOL,2,2′-IMINODI- with 3,5-DIIODO-4-OXO-1(4H)-PYRIDINEACETIC ACID (1:1) □ IODOPYRACET □ IODURON B □ IOPYRACIL □ METHYLGLUCAMINE-3,5-DIIODO-4-PYRIDONE-N-ACETATE □ MOSYLAN □ NEOMETHIODAL □ NEO-SKIODAN □ NEO-TENEBRYL □ NOSYDRAST □ OPARENOL □ PELVIRAN □ PER-ABRODIL □ PER-RADIOGRAPHOL □ PYELOSIL □ PYLUMBRIN □ PYRACETON □ RP 3203 □ SAVAC □ UMBRADIL □ URIODONE □ VASIODONE □ XUMBRADIL

TOXICITY DATA WITH REFERENCE
ivn-rat LD50:5400 mg/kg AEPPAE 222,584,54
ivn-mus LD50:6400 mg/kg JPETAB 116,394,56
ivn-dog LD50:2 g/kg BJRAAP 6,304,33
ivn-cat LD50:2800 mg/kg MECHAN 6,344,63
ivn-rbt LD50:4700 mg/kg MECHAN 6,344,63

SAFETY PROFILE: Moderately toxic by intravenous route. When heated to decomposition it emits very toxic fumes of NO_x and I^-.

DNG800 CAS:20389-01-9 HR: 3
DIIODOQUINONE
mf: $C_6H_2I_2O_2$ mw: 359.88

PROP: Yellow plates from pet ether; yellow leaflets from Et_2O. Mp: 177–178°.

SYN: 2,6-DIIODO-p-BENZOQUINONE

TOXICITY DATA WITH REFERENCE
ipr-mus LD50:84 mg/kg BCPCA6 12,885,63

CONSENSUS REPORTS: Reported in EPA TSCA Inventory.

SAFETY PROFILE: Poison by intraperitoneal route. When heated to decomposition it emits toxic fumes of I^-.

DNH000 CAS:133-91-5 HR: 2
3,5-DIIODOSALICYLIC ACID
mf: $C_7H_4I_2O_3$ mw: 389.91

PROP: White to pale pink, crystalline powder; needles from alc. Mp: 228–230° (decomp). Sltly sol in water.

TOXICITY DATA WITH REFERENCE
orl-rat LDLo:500 mg/kg NCNSA6 5,8,53
orl-mus LD50:450 mg/kg QJPPAL 19,483,46

CONSENSUS REPORTS: Reported in EPA TSCA Inventory.

SAFETY PROFILE: Moderately toxic by ingestion. A trace mineral added to animal feeds. When heated to decomposition it emits toxic fumes of I^-.

DNH125 **CAS:141-04-8** **HR: 2**
DIISOBUTYL ADIPATE
mf: $C_{14}H_{26}O_4$ mw: 258.40

SYNS: DIBA □ FTAFLEX DIBA □ ISOBUTYL ADIPATE

TOXICITY DATA WITH REFERENCE
ipr-rat TDLo:1190 mg/kg (5-15D preg):TER JPMSAE 62,1596,73
ipr-rat TDLo:595 mg/kg (5-15D preg):TER JPMSAE 62,1596,73
ipr-rat LD50:5950 mg/kg JPMSAE 62,1596,73
orl-gpg LD50:12,300 mg/kg GWXXBX #2703360

CONSENSUS REPORTS: Reported in EPA TSCA Inventory.

SAFETY PROFILE: Moderately toxic by intraperitoneal route. Mildly toxic by ingestion. Experimental teratogenic effects. When heated to decomposition it emits acrid smoke and fumes.

DNH400 **CAS:110-96-3** **HR: 3**
DIISOBUTYLAMINE
DOT: UN 2361
mf: $C_8H_{19}N$ mw: 129.28

PROP: Water-white liquid, amine odor. Fp −77°, mp: −70°, bp: 139°, flash p: 69.8°F, d: 0.745 @ 20°/4°, vap press: 10 mm @ 30.6°, vap d: 4.46. Sltly sol in water.

SYN: 2-METHYL-N-(2-METHYLPROPYL)-1-PROPANAMINE

TOXICITY DATA WITH REFERENCE
orl-rat LD50:258 mg/kg HYSAAV 34(7-9),426,69
orl-mus LD50:629 mg/kg HYSAAV 34(7-9),426,69
orl-gpg LD50:620 mg/kg HYSAAV 34(7-9),426,69

CONSENSUS REPORTS: Reported in EPA TSCA Inventory.

DOT CLASSIFICATION: 3; *Label:* Flammable Liquid

SAFETY PROFILE: Poison by ingestion. A dangerous fire hazard when exposed to heat or flame; can react vigorously with oxidizing materials. To fight fire, use alcohol foam, CO_2, dry chemical. When heated to decomposition it emits toxic fumes of NO_x.

DNH500 **CAS:102367-57-7** **HR: 3**
DIISOBUTYLAMINOBENZOYLOXYPROPYL THEOPHYLLINE
mf: $C_{25}H_{35}N_5O_4$ mw: 469.65

SYN: α-((DIISOBUTYLAMINO)METHYL)THEOPHYLLINE-8-ETHANOL BENZOATE (ester)

TOXICITY DATA WITH REFERENCE
orl-mus LD50:2567 mg/kg NIIRDN 6,306,82
ipr-mus LD50:1835 mg/kg NIIRDN 6,306,82
ivn-mus LD50:273 mg/kg NIIRDN 6,306,82

SAFETY PROFILE: Poison by intravenous route. Moderately toxic by ingestion and intraperitoneal routes. When heated to decomposition it emits toxic fumes of NO_x.

DNH800 **CAS:108-82-7** **HR: 2**
DIISOBUTYL CARBINOL
mf: $C_9H_{20}O$ mw: 144.29

PROP: Colorless liquid. Fp: −65°, bp: 179°, flash p: 165°F, d: 0.8121 @ 20°/20°, vap press: 0.3 mm @ 20°, vap d: 4.98, lel: 0.8% @ 212°F, uel: 6.1% @ 212°F.

SYNS: 2,6-DIMETHYL HEPTANOL-4 □ 2,6-DIMETHYL-4-HEPTANOL □ sec-NONYL ALCOHOL

TOXICITY DATA WITH REFERENCE
skn-rbt 10 mg/24H open MLD JIHTAB 31,60,49
skn-rbt 500 mg open MLD UCDS** 12/30/71
eye-rbt 500 mg open JIHTAB 31,60,49
orl-rat LD50:3560 mg/kg JIDHAN 31,60,49
ipr-rat LD50:800 mg/kg NPIRI* 1,22,74
orl-mus LD50:3530 mg/kg SCCUR* -,4,61
skn-rbt LD50:4600 mg/kg NPIRI* 1,22,74

CONSENSUS REPORTS: Reported in EPA TSCA Inventory.

SAFETY PROFILE: Moderately toxic by ingestion and intraperitoneal routes. Mildly toxic by skin contact. A powerful systemic irritant by inhalation. A skin and eye irritant. Can cause central nervous system and liver damage when ingested. Combustible when exposed to heat or flame; can react with oxidizing materials. To fight fire, use alcohol foam, foam, CO_2, dry chemical. When heated to decomposition it emits acrid smoke and fumes.

DNI200 **CAS:63919-00-6** **HR: 2**
DIISOBUTYLENE OXIDE
mf: $C_8H_{16}O$ mw: 128.24

SYN: EP-185

TOXICITY DATA WITH REFERENCE
skn-rbt 500 mg open MLD UCDS** 4/10/68
eye-rbt 500 mg open AMIHBC 10,61,54
orl-rat LD50:4920 mg/kg AMIHBC 10,61,54
ihl-rat LCLo:4000 ppm/4H AMIHBC 10,61,54
skn-rbt LD50:14 g/kg AMIHAB 14,250,56

SAFETY PROFILE: Moderately toxic by inhalation. Mildly toxic by ingestion and skin contact. A skin and eye irritant. When heated to decomposition it emits acrid smoke and fumes.

DNI400 **CAS:7283-69-4** **HR: 1**
DIISOBUTYL FUMARATE
mf: $C_{12}H_{20}O_4$ mw: 228.32

TOXICITY DATA WITH REFERENCE
skn-rbt 10 mg/24H open MLD AIHAAP 23,95,62
orl-rat LD50:8120 mg/kg AIHAAP 23,95,62
skn-rbt LD50:7490 mg/kg AIHAAP 23,95,62

SAFETY PROFILE: Mildly toxic by ingestion and skin contact. A skin irritant. When heated to decomposition it emits acrid smoke and fumes. See also ESTERS.

DNI600 CAS:1191-15-7 HR: 3
DIISOBUTYLHYDROALUMINUM
mf: $C_8H_{19}Al$ mw: 142.25

$[(CH_3)_2CHCH_2]_2AlH$

PROP: Colorless pyrophoric liquid. Fp: −80°, bp: 140° @ 4 mm, d: 0.798. Misc in hydrocarbon solvents. Sol in Et_2O, C_6H_6, toluene, and cyclohexane.

SYNS: AL-ALCHILI (ITALIAN) □ AL-DIISOBUTYL □ BIS(ISOBUTYL)HYDROALUMINUM □ DIISOBUTYLALUMINIUM HYDRIDE □ DIISOBUTYLALUMINUM HYDRIDE □ HYDROBIS(2-METHYLPROPYL) ALUMINUM □ HYDRODIISOBUTYLALUMINUM

TOXICITY DATA WITH REFERENCE
ihl-gpg LCLo: 70 g/m³/1H MELAAD 57,188,66

CONSENSUS REPORTS: Reported in EPA TSCA Inventory.

ACGIH TLV: TWA 2 mg(Al)/m³

SAFETY PROFILE: Mildly toxic by inhalation. Dangerous fire hazard; ignites spontaneously in air. To fight fire, do not use water, foam, or halogenated extinguishing agents. See also HYDRIDES and ALUMINUM COMPOUNDS.

DNI800 CAS:108-83-8 HR: 3
DIISOBUTYL KETONE
DOT: UN 1157
mf: $C_9H_{18}O$ mw: 142.27

PROP: Liquid. Bp: 166°, flash p: 140°F, d: 0.81, vap d: 4.9, lel: 0.8% @ 212°F, uel: 6.2% @ 212°F.

SYNS: DIISOBUTILCHETONE (ITALIAN) □ DI-ISOBUTYLCETONE (FRENCH) □ DIISOBUTYLKETON (DUTCH, GERMAN) □ *s* DIISOPROPYLACETONE □ 2,6-DIMETHYL-HEPTAN-4-ON (DUTCH, GERMAN) □ 2,6-DIMETHYLHEPTAN-4-ONE □ 2,6-DIMETHYL-4-HEPTANONE □ 2,6-DIMETIL-EPTAN-4-ONE (ITALIAN) □ ISOBUTYL KETONE □ ISOVALERONE □ VALERONE

TOXICITY DATA WITH REFERENCE
eye-hmn 25 ppm/15M MLD JIHTAB 28,262,46
skn-rbt 10 mg/24H open MLD JIHTAB 31,60,49
skn-rbt 500 mg open MLD UCDS** 12/15/71
eye-rbt 500 mg AJOPAA 29,1363,46
ihl-hmn TCLo: 50 ppm: EYE,CNS,GIT JIHTAB 30,63,48
orl-rat LD50: 5750 mg/kg NPIRI* 1,23,74
ihl-rat LCLo: 2000 ppm/4H JIHTAB 31,343,49
orl-mus LD50: 1416 mg/kg SCCUR* -,4,61
skn-rbt LD50: 16 g/kg NPIRI* 1,23,74

CONSENSUS REPORTS: Reported in EPA TSCA Inventory.

OSHA PEL: TWA 25 ppm
ACGIH TLV: TWA 25 ppm
DFG MAK: 50 ppm (290 mg/m³)
NIOSH REL: (Ketones) TWA 140 mg/m³
DOT CLASSIFICATION: 3; *Label:* Flammable Liquid

SAFETY PROFILE: Moderately toxic by ingestion and inhalation. Mildly toxic by skin contact. Human systemic effects by inhalation: headache, nausea or vomiting, and unspecified eye effects. An eye and skin irritant. Narcotic in high concentrations. Flammable liquid when exposed to heat or flame; can react with oxidizing materials. To fight fire, use CO_2, dry chemical, water spray, mist or fog. When heated to decomposition it emits acrid smoke and fumes. See also KETONES.

For occupational chemical analysis use NIOSH: Ketones I (desorption in CS_2) 1300.

DNJ000 CAS:61947-30-6 HR: 3
DIISOBUTYLOXOSTANNANE
mf: $C_8H_{18}OSn$ mw: 248.95

SYNS: DIISOBUTYLTIN OXIDE □ KYSLICNIK DIISOBUTYLCINICITY (CZECH)

TOXICITY DATA WITH REFERENCE
skn-rbt 500 mg/24H SEV 28ZPAK -,226,72
eye-rbt 100 mg/24H MOD 28ZPAK -,226,72
orl-rat LD50: 53,200 μg/kg 28ZPAK -,226,72

OSHA PEL: TWA 0.1 mg(Sn)/m³ (skin)
ACGIH TLV: TWA 0.1 mg(Sn)/m³ (skin) (Proposed: TWA 0.1 mg(Sn)/m³; STEL 0.2 mg(Sn)/m³ (skin))
NIOSH REL: (Organotin Compounds) TWA 0.1 mg(Sn)/m³

SAFETY PROFILE: Poison by ingestion. An eye and severe skin irritant. When heated to decomposition it emits acrid smoke and fumes. See also TIN COMPOUNDS.

For occupational chemical analysis use NIOSH: Organotin Compounds 5504.

DNJ400 CAS:84-69-5 HR: 2
DIISOBUTYL PHTHALATE
mf: $C_{16}H_{22}O_4$ mw: 278.38

PROP: Liquid. Mp: −64°, flash p: 385°F, d: 1.039–1.043, vap d: 9.59.

SYNS: DIBP □ DIISOBUTYLESTER KYSELINY FTALOVE □ HATCOL DIBP □ HEXAPLAS M/1B □ KODAFLEX DIBP □ PALATINOL IC

TOXICITY DATA WITH REFERENCE
ipr-rat TDLo: 375 mg/kg (female 5-15D post): TER JPMSAE 61,51,72
orl-mus TDLo: 32 g/kg (female 6-13D post): REP TCMUD8 7,29,87
ipr-rat TDLo: 1250 mg/kg (female 5-15D post): REP JPMSAE 61,51,72
orl-mus TDLo: 16,800 mg/kg (male 7D pre): REP TOLED5 5,413,80
orl-rat TDLo: 8400 μg/kg (7D male): REP TXAPA9 53,35,80
orl-rat LD50: 15 g/kg EVHPAZ 3,131,73
ipr-rat LD50: 3749 mg/kg JPMSAE 61,51,72
orl-mus LD50: 10 g/kg GTPZAB 29(12),39,85
ipr-mus LD50: 3990 mg/kg JSCCA5 28,667,77
skn-gpg LD50: 10 g/kg EVHPAZ 4,3,73

CONSENSUS REPORTS: Reported in EPA TSCA Inventory.

SAFETY PROFILE: Moderately toxic by intraperitoneal route. Mildly toxic by ingestion and skin contact. Experimental teratogenic and reproductive effects. Combustible when exposed to heat or flame. To fight fire, use foam, CO_2, dry chemical. When heated to decomposition it emits acrid smoke and fumes.

DNJ600 CAS:3437-84-1 HR: 3
DIISOBUTYRYL PEROXIDE
mf: $C_8H_{14}O_4$ mw: 174.20

$[(CH_3)_2CHCO{\bullet}O\text{-}]_2$

CONSENSUS REPORTS: Reported in EPA TSCA Inventory.

SAFETY PROFILE: May explode when dried at room temperature. When heated to decomposition it emits acrid smoke and fumes. See also PEROXIDES.

DNJ800 CAS:822-06-0 HR: 3
1,6-DIISOCYANATOHEXANE
DOT: UN 2281
mf: $C_8H_{12}N_2O_2$ mw: 168.22

$O{:}N{:}C(CH_2)_6C{:}N{:}O$

PROP: Oil. D: 1.053 @ 20°/4°, bp: 121–122° @ 9 mm.

SYNS: DESMODUR H □ DESMODUR N □ HEXAMETHYLENDIISOKYANAT □ HEXAMETHYLENE DIISOCYANATE □ HEXAMETHYLENE DIISOCYANATE (DOT) □ HEXAMETHYLENE-1,6-DIISOCYANATE □ 1,6-HEXAMETHYLENE DIISOCYANATE □ 1,6-HEXANEDIOL DIISOCYANATE □ HMDI □ ISOCYANIC ACID, DIESTER with 1,6-HEXANEDIOL □ ISOCYANIC ACID, HEXAMETHYLENE ESTER □ METYLENO-BIS-FENYLOIZOCYJANIAN □ SZESCIOMETYLENODWUIZOCYJANIAN □ TL 78

TOXICITY DATA WITH REFERENCE
orl-rat LD50:738 mg/kg AIHAAP 30,470,69
ihl-rat LCLo:60 mg/m³/4H GTPZAB 12(10),40,68
orl-mus LD50:350 mg/kg TAKHAA 39,202,80
ihl-mus LC50:30 mg/m³ 85GMAT-,74,82
ivn-mus LD50:5600 µg/kg CSLNX* NX#07805
skn-rbt LD50:593 mg/kg AIHAAP 30,470,69

CONSENSUS REPORTS: Reported in EPA TSCA Inventory.

ACGIH TLV: TWA 0.005 ppm
DFG MAK: 0.01 ppm (0.07 mg/m³)
NIOSH REL: (Diisocyanates) TWA 0.005 ppm; CL 0.02 ppm/10M
DOT CLASSIFICATION: 6.1; *Label:* Poison

SAFETY PROFILE: Poison by inhalation and intravenous routes. Moderately toxic by ingestion and skin contact. Potentially explosive reaction with alcohols + base. When heated to decomposition it emits toxic fumes of NO_x. See also CYANATES.

For occupational chemical analysis use OSHA: #42.

DNK100 CAS:4747-90-4 HR: 3
DIISOCYANATOMETHANE
mf: $C_3H_2N_2O_2$ mw: 98.06

SYN: METHYLENE DIISOCYANATE

SAFETY PROFILE: Polymerizes violently on contact with dimethyl formamide (DMF). When heated to decomposition it emits toxic fumes of NO_x.

DNK800 CAS:27215-10-7 HR: 2
DIISOOCTYL ACID PHOSPHATE
DOT: UN 1902
mf: $C_{16}H_{35}O_4P$ mw: 322.48

PROP: A corrosive liquid.

SYN: DIISOOCTYL PHOSPHATE (DOT)

CONSENSUS REPORTS: Reported in EPA TSCA Inventory.

DOT CLASSIFICATION: 8; *Label:* Corrosive

SAFETY PROFILE: Moderately toxic by irritation to skin, eyes, and mucous membranes. A corrosive compound. When heated to decomposition it emits toxic fumes of PO_x. See also PHOSPHATES.

DNL200 CAS:24423-68-5 HR: 3
DIISOPENTYLMERCURY
mf: $C_{10}H_{22}Hg$ mw: 342.91

SYNS: DIISOAMYLMERCURY □ DIISOPENTYLRTUT □ MERCURY, DIISOAMYL-

TOXICITY DATA WITH REFERENCE
ipr-mus LDLo:16 mg/kg CBCCT* 4,230,52

CONSENSUS REPORTS: Mercury and its compounds are on the Community Right-To-Know List.

OSHA PEL: TWA 0.01 mg(Hg)/m³; STEL 0.03 mg/m³ (skin)
ACGIH TLV: TWA 0.01 mg(Hg)/m³; STEL 0.03 mg(Hg)/m³
NIOSH REL: (Mercury, Organo): TWA 0.01 mg/m³; STEL 0.03 mg/m³ (skin)

SAFETY PROFILE: Poison by intraperitoneal route. Violent reaction on contact with iodine. When heated to decomposition it emits toxic fumes of Hg. See also MERCURY COMPOUNDS.

DNL400 CAS:63979-62-4 HR: 3
DIISOPENTYLOXOSTANNANE
mf: $C_{10}H_{22}OSn$ mw: 277.01

SYNS: DIISOPENTYLTIN OXIDE □ KYSLICNIK DIISOAMYLCINICITY (CZECH)

TOXICITY DATA WITH REFERENCE
skn-rbt 500 mg/24H SEV 28ZPAK -,227,72
eye-rbt 20 mg/24H MOD 28ZPAK -,227,72
orl-rat LD50:64,500 µg/kg 28ZPAK -,227,72

OSHA PEL: TWA 0.1 mg(Sn)/m³ (skin)
ACGIH TLV: TWA 0.1 mg(Sn)/m³ (skin) (Proposed: TWA 0.1 mg(Sn)/m³; STEL 0.2 mg(Sn)/m³ (skin))
NIOSH REL: (Organotin Compounds) TWA 0.1 mg(Sn)/m³

SAFETY PROFILE: Poison by ingestion. An eye and severe skin irritant. When heated to decomposition it emits acrid smoke and fumes. See also TIN COMPOUNDS.

For occupational chemical analysis use NIOSH: Organotin Compounds 5504.

DNL600 **CAS:110-97-4** **HR: 3**
DIISOPROPANOLAMINE
mf: $C_6H_{15}NO_2$ mw: 133.22

PROP: Mp: 42°, bp: 249°, flash p: 260°F (OC), d: 0.9890 @ 45°/20°, vap d: 4.59.

SYNS: BIS(2-HYDROXYPROPYL)AMINE □ BIS(2-PROPANOL) AMINE □ DIPA □ DIPROPYL-2,2′-DIHYDROXYAMINE □ 1,1′-IMINO-DI-2-PROPANOL

TOXICITY DATA WITH REFERENCE
skn-rbt 500 mg open MLD UCDS** 5/21/71
eye-rbt 50 mg SEV UCDS** 5/21/71
orl-rat LD50:4765 mg/kg GTPZAB 30(7),46,86
ipr-mus LD50:96 mg/kg AIMJA9 30,23,79

CONSENSUS REPORTS: Reported in EPA TSCA Inventory.

SAFETY PROFILE: Poison by intraperitoneal route. Mildly toxic by ingestion. A skin and severe eye irritant. Combustible when exposed to heat or flame; can react with oxidizing materials. To fight fire, use alcohol foam, CO_2, dry chemical. When heated to decomposition it emits toxic fumes of NO_x.

DNL800 **CAS:6938-94-9** **HR: 2**
DIISOPROPYL ADIPATE
mf: $C_{12}H_{22}O_4$ mw: 230.34

SYNS: ADIPIC ACID DIISOPROPYL ESTER □ CERAPHYL 230 □ HEXANEDIOIC ACID, BIS(1-METHYLETHYL) ESTER □ ISOPROPYL ADIPATE □ STANDAMUL DIPA □ WICKENOL 116

TOXICITY DATA WITH REFERENCE
skn-rbt 100 mg/24H MLD JACTDZ 3(3),101,84
ivn-rat LD50:640 mg/kg MRLR** No.256,54

CONSENSUS REPORTS: Reported in EPA TSCA Inventory.

SAFETY PROFILE: Moderately toxic by intravenous route. A skin irritant. When heated to decomposition it emits acrid smoke and fumes. See also ESTERS.

DNM200 **CAS:108-18-9** **HR: 3**
DIISOPROPYLAMINE
DOT: UN 1158
mf: $C_6H_{15}N$ mw: 101.22

PROP: Colorless liquid. Bp: 83–84°, flash p: 19.4°F. D: 0.722 @ 220.0°, vap d: 3.5.

SYNS: DIPA □ N-(1-METHYLETHYL)-2-PROPANAMINE □ 2-PROPANAMINE, N-(1-METHYLETHYL)-

TOXICITY DATA WITH REFERENCE
eye-rbt 750 μg open SEV AMIHBC 10,61,54
skn-rbt 500 mg/24H MLD 85JCAE-,433,86
mma-sat 1 μg/plate NUCADQ 3,129,82
orl-rat LD50:770 mg/kg AEHLAU 1,343,60
ihl-rat LC50:4800 mg/m³/2H 85GMAT -,54,82
orl-mus LD50:2120 mg/kg GISAAA 45(3),79,80
ihl-mus LC50:4200 mg/m³/2H 85GMAT -,54,82
ihl-cat LCLo:2207 ppm/72M JIHTAB 31,142,49
orl-rbt LD50:4700 mg/kg GISAAA 45(3),79,80
ihl-rbt LCLo:2207 ppm/150M JIHTAB 31,142,49
orl-gpg LD50:2800 mg/kg GISAAA 45(3),79,80
ihl-gpg LCLo:2207 ppm/82M JIHTAB 31,142,49
scu-gpg LDLo:1400 mg/kg JIHTAB 31,142,49
ihl-mam LC50:4200 mg/m³ TPKVAL 14,80,75

CONSENSUS REPORTS: Reported in EPA TSCA Inventory.

OSHA PEL: TWA 5 ppm (skin)
ACGIH TLV: TWA 5 ppm (skin)
DOT CLASSIFICATION: 3; *Label:* Flammable Liquid

SAFETY PROFILE: Moderately toxic by ingestion and subcutaneous routes. Mildly toxic by inhalation. Mutation data reported. A skin and severe eye irritant. Inhalation of fumes can cause pulmonary edema. A very dangerous fire hazard when exposed to heat or flame; can react vigorously with oxidizing materials. To fight fire, use alcohol foam, foam, CO_2, dry chemical. When heated to decomposition it emits toxic fumes of NO_x. See also AMINES.

D

DNM400 **CAS:660-27-5** **HR: 2**
DIISOPROPYLAMINE DICHLORACETATE
mf: $C_6H_{15}N \cdot C_2H_2Cl_2O_2$ mw: 230.16

SYNS: β-ANOXIN □ CUBISOL □ DADA □ DAPA □ DAPOCEL □ DEDYL □ DICHLOROACETATO di DIISOPROPILAMMONIO □ DICHLOROACETIC ACID, DIISOPROPYLAMINE SALT □ DIEDI □ DIISOPROPYLAMINE, compd. with DICHLOROACETIC ACID (1:1) □ DIISOPROPYLAMINE DICHLOROETHANOATE □ DIISOPROPYLAMMONIUM DICHLOROACETATE □ DIISOPROPYLAMMONIUM DICHLOROETHANOATE □ DIPA □ DISOTAT □ IS 401 □ KALODIL □ KRINO B 15 □ OXYPANGAM □ TENSICOR □ VASCULOPATINA

TOXICITY DATA WITH REFERENCE
mma-sat 10 μg/plate NUCADQ 3,129,82
ipr-rat LD50:840 mg/kg THERAP 16,136,61
orl-mus LD50:1700 mg/kg ARZNAD 13,109,63
ipr-mus LD50:750 mg/kg BCFAAI 97,608,58
scu-mus LD50:1330 mg/kg ARZNAD 13,109,63

CONSENSUS REPORTS: Reported in EPA TSCA Inventory.

SAFETY PROFILE: Moderately toxic by intraperitoneal, subcutaneous, and ingestion routes. Mutation data reported. When heated to decomposition it emits very toxic fumes of Cl^-, NH_3, and NO_x.

DNM600 **CAS:77966-84-8** **HR: 3**
2-(DIISOPROPYLAMINO)-2′,6′-ACETOXYLIDIDE HYDROCHLORIDE
mf: $C_{16}H_{26}N_2O \cdot ClH$ mw: 298.90

SYN: V 377

TOXICITY DATA WITH REFERENCE
eye-rbt 2% MLD ARZNAD 8,407,58
ipr-rat LD50:90 mg/kg ARZNAD 8,407,58
scu-mus LD50:197 mg/kg ARZNAD 8,407,58

SAFETY PROFILE: Poison by subcutaneous and intraperitoneal routes. An eye irritant. When heated to decomposition it emits very toxic fumes of NO_x and HCl.

DNN000 CAS:14549-32-7 HR: 3
2-(2-(DIISOPROPYLAMINO)ETHOXY)BUTYROPHENONE HYDROCHLORIDE
mf: $C_{18}H_{29}NO_2 \cdot ClH$ mw: 327.94

SYNS: 2-BUTYRYL-β-(N,N-DIISOPROPYL)PHENOXYETHYLAMINE HYDROCHLORIDE ☐ KETOCAINE HYDROCHLORIDE ☐ REC 7-0518

TOXICITY DATA WITH REFERENCE
orl-rat LD50:446 mg/kg ARZNAD 16,1275,66
scu-rat LD50:935 mg/kg ARZNAD 16,1275,66
orl-mus LD50:147 mg/kg ARZNAD 16,1275,66
ipr-mus LD50:102 mg/kg ARZNAD 16,1275,66
scu-mus LD50:217 mg/kg ARZNAD 16,1275,66
ivn-mus LD50:14 mg/kg ARZNAD 16,1275,66
ivn-cat LD50:6 mg/kg ARZNAD 16,1275,66

SAFETY PROFILE: Poison by ingestion, intravenous, subcutaneous, and intraperitoneal routes. When heated to decomposition it emits very toxic fumes of NO_x and HCl.

DNN600 CAS:3737-09-5 HR: 3
α-(2-(DIISOPROPYLAMINO)ETHYL)-α-PHENYL-2-PYRIDINEACETAMIDE
mf: $C_{21}H_{29}N_3O$ mw: 339.53

SYNS: DICORANTIL ☐ γ-DIISOPROPYLAMINO-α-PHENYL-α-(2-PYRIDYL)BUTYRAMIDE ☐ DISOPYRAMIDE ☐ H 3292 ☐ RITMODAN ☐ SC 7031 ☐ SEARLE 703

TOXICITY DATA WITH REFERENCE
orl-rat TDLo:420 mg/kg (female 9-15D post):REP KSRNAM 5,1641,71
orl-rat TDLo:1260 mg/kg (9-15D preg):REP KSRNAM 5,1641,71
orl-rat TDLo:11,200 mg/kg (28D male):REP KSRNAM 5,1628,71
orl-rat TDLo:1260 mg/kg (female 9-15D post):TER KSRNAM 5,1641,71
orl-hmn TDLo:4286 μg/kg:CVS,PUL NEJMAG 302,614,80
orl-rat LD50:333 mg/kg ARZNAD 38,1398,88
ipr-rat LD50:170 mg/kg NIIRDN 6,319,82
scu-rat LD50:800 mg/kg IYKEDH 9,829,78
orl-mus LD50:409 mg/kg NIIRDN 6,319,82
ipr-mus LD50:114 mg/kg KSRNAM 5,1628,71
scu-mus LD50:305 mg/kg NIIRDN 6,319,82
ivn-mus LD50:30 mg/kg JMCMAR 27,1142,84
ivn-dog LDLo:36 mg/kg JPETAB 136,114,62

SAFETY PROFILE: Poison by ingestion, intraperitoneal, intravenous, and subcutaneous routes. An experimental teratogen. Other experimental reproductive effects. Human systemic effects by ingestion: dyspnea, cardiac, and pulmonary changes. Experimental reproductive effects. When heated to decomposition it emits toxic fumes of NO_x.

DNN630 CAS:24544-04-5 HR: 2
2,6-DIISOPROPYL ANILINE
mf: $C_{12}H_{19}N$ mw: 177.32

SYNS: ANILINE, 2,6-DIISOPROPYL- ☐ BENZENAMINE, 2,6-BIS(1-METHYLETHYL)-

TOXICITY DATA WITH REFERENCE
orl-rat LD50:3204 mg/kg FAATDF 3,285,83

CONSENSUS REPORTS: Reported in EPA TSCA Inventory.

SAFETY PROFILE: Moderately toxic by ingestion. When heated to decomposition it emits toxic vapors of NO_x.

DNN709 CAS:25321-09-9 HR: 1
DIISOPROPYLBENZENE
mf: $C_{12}H_{18}$ mw: 162.30

TOXICITY DATA WITH REFERENCE
skn-rbt 100 mg/24H MOD 85JCAE-,39,86
eye-rbt 500 mg/24H MLD 85JCAE-,39,86
orl-rat LD50:6500 mg/kg TXAPA9 28,313,74
ihl-rat LCLo:5300 mg/m³/4H 85GMAT -,55,82
ihl-mus LCLo:5300 mg/m³/2H 85GMAT -,55,82
skn-rbt LD50:16 g/kg TXAPA9 28,313,74

CONSENSUS REPORTS: Reported in EPA TSCA Inventory.

SAFETY PROFILE: Mildly toxic by ingestion and skin contact. A skin and eye irritant. When heated to decomposition it emits acrid smoke and fumes.

DNN800 CAS:577-55-9 HR: 2
o-DIISOPROPYLBENZENE
mf: $C_{12}H_{18}$ mw: 162.30

PROP: Clear, colorless liquid. Mp: $<-55°$, bp: 205°, flash p: 170°F (OC), d: 0.863–0.867 @ 25°/25°, autoign temp: 840°F, vap d: 5.6.

TOXICITY DATA WITH REFERENCE
orl-rat LDLo:5000 mg/kg 28ZRAQ -,57,60

CONSENSUS REPORTS: Reported in EPA TSCA Inventory.

SAFETY PROFILE: Mildly toxic by ingestion. Combustible when exposed to heat or flame; can react with oxidizing materials. To fight fire, use foam, CO_2, dry chemical, water spray or mist. When heated to decomposition it emits acrid smoke and irritating fumes. See also CUMENE.

DNN829 CAS:99-62-7 HR: 2
1,3-DIISOPROPYLBENZENE
mf: $C_{12}H_{18}$ mw: 162.30

SYN: m-DIISOPROPYLBENZENE

TOXICITY DATA WITH REFERENCE
orl-rat LD50:7400 mg/kg 85GMAT -,54,82
orl-mus LD50:3100 mg/kg 85GMAT -,54,82
ipr-mus LD50:1650 mg/kg 85GMAT -,54,82

CONSENSUS REPORTS: Reported in EPA TSCA Inventory.

SAFETY PROFILE: Moderately toxic by ingestion and intraperitoneal routes. When heated to decomposition it emits acrid smoke and fumes.

DNN830 CAS:100-18-5 HR: 2
1,4-DIISOPROPYLBENZENE
mf: $C_{12}H_{18}$ mw: 162.30

SYNS: BENZENE, 1,4-BIS(1-METHYLETHYL)-(9CI) □ BENZENE, p-DIISOPROPYL- □ 1,4-BIS(1-METHYLETHYL)BENZENE □ p-DIISOPROPYLBENZENE □ p-DIISOPROPYLBENZOL

TOXICITY DATA WITH REFERENCE
orl-mus LD50:3400 mg/kg 85GMAT -,54,82
ipr-mus LD50:1650 mg/kg GTPZAB 14(2),41,70

CONSENSUS REPORTS: Reported in EPA TSCA Inventory.

SAFETY PROFILE: Moderately toxic by ingestion and intraperitoneal routes. When heated to decomposition it emits acrid smoke and irritating vapors.

DNN840 HR: 3
1,3-DIISOPROPYLBENZENE SODIUM SALT, DIHYDROPEROXIDE
mf: $C_{12}H_{18}•Na•2H_2O_2$ mw: 253.33

SYN: SODIUM-m-DIISOPROPYLBENZOL (Na-m) DIHYDROPEROXIDE (RUSSIAN)

TOXICITY DATA WITH REFERENCE
unr-rat LD50:1050 mg/kg GISAAA 42(4),11,77
unr-mus LD50:415 mg/kg GISAAA 42(4),11,77
unr-rbt LD50:320 mg/kg GISAAA 42(4),11,77

SAFETY PROFILE: Poison by an unspecified route. When heated to decomposition it emits toxic fumes of Na_2O. See also PEROXIDES, ORGANIC.

DNN850 HR: 2
1,4-DIISOPROPYLBENZENE SODIUM SALT, DIISOPEROXIDE
mf: $C_{12}H_{18}•Na•2H_2O_2$ mw: 253.33

SYN: SODIUM-p-DIISOPROPYLBENZOL (Na-p) DIHYDROPEROXIDE (RUSSIAN)

TOXICITY DATA WITH REFERENCE
unr-rat LD50:1250 mg/kg GISAAA 42(4),11,77
unr-mus LD50:660 mg/kg GISAAA 42(4),11,77
unr-rbt LD50:450 mg/kg GISAAA 42(4),11,77

SAFETY PROFILE: Moderately toxic by unspecified route. When heated to decomposition it emits toxic fumes of Na_2O. See also PEROXIDES, ORGANIC.

DNN900 CAS:95-29-4 HR: 2
N,N-DIISOPROPYL-2-BENZOTHIAZOLESULFENAMIDE
mf: $C_{13}H_{18}N_2S_2$ mw: 266.45

SYNS: 2-BENZOTHIAZOLESULFENAMIDE, N,N-DIISOPROPYL- □ DIPAC

TOXICITY DATA WITH REFERENCE
orl-mus LD50:3892 mg/kg GTPZAB 8(7),39,64

CONSENSUS REPORTS: Reported in EPA TSCA Inventory.

SAFETY PROFILE: Moderately toxic by ingestion. When heated to decomposition it emits toxic vapors of NO_x and SO_x.

DNO200 CAS:15721-33-2 HR: 3
DIISOPROPYLBERYLLIUM
mf: $C_6H_{14}Be$ mw: 95.19

$((CH_3)_2CH)_2Be$

CONSENSUS REPORTS: IARC Cancer Review: Group 1 IMEMDT 58,41,93; Human Sufficient Evidence IMEMDT 58,41,93; Animal Sufficient Evidence IMEMDT 1,17,72; Animal Sufficient Evidence IMEMDT 23,143,80; Animal Sufficient Evidence IMEMDT 58,41,93. Beryllium and its compounds are on the Community Right-To-Know List.

OSHA PEL: TWA 0.002 mg(Be)/m³; STEL 0.005 mg(Be)/m³/30M; CL 0.025 mg(Be)/m³
ACGIH TLV: TWA 0.002 mg(Be)/m³, Suspected Human Carcinogen
DFG TRK: 0.002 mg(Be)/m³. Animal Carcinogen, Suspected Human Carcinogen

SAFETY PROFILE: Confirmed human carcinogen. Explosive reaction on contact with water. When heated to decomposition it emits toxic fumes of BeO. See also BERYLLIUM COMPOUNDS.

DNO400 CAS:693-13-0 HR: 3
DIISOPROPYLCARBODIIMIDE
mf: $C_7H_{14}N_2$ mw: 126.23

PROP: Moisture-sensitive liquid. D: 0.806, bp: 145–148°.

SYNS: N,N'-METHANETETRAYLBIS-2-PROPANAMINE □ 2-PROPANAMINE, N,N'-METHANETETRAYLBIS-(9CI)

TOXICITY DATA WITH REFERENCE
ivn-mus LD50:36 mg/kg CSLNX* NX#05886

CONSENSUS REPORTS: Reported in EPA TSCA Inventory.

SAFETY PROFILE: Poison by intravenous route. When heated to decomposition it emits toxic fumes of NO_x.

DNO800 CAS:741-58-2 HR: 3
N-(2-(O,O-DIISOPROPYLDITHIOPHOSPHORYL)ETHYL) BENZENESULFONAMIDE
mf: $C_{14}H_{24}NO_4PS_3$ mw: 397.54

PROP: Viscous amber liquid or crystals. D: 1.25 @ 22°, mp: 34.4°.

SYNS: BENSULIDE □ BENZULFIDE □ BETAMEC □ BETASAN □ O,O-BIS(1-METHYLETHYL)-S-(2-((PHENYLSULFONYL) AMINO)ETHYL) PHEOSPHORODITHIOATE □ N-(β-O,O-DIISOPROPYLDITHIOPHOSPHORYLETHYL)BEZENESULFONAMIDE □ S-(O,O-DIISOPROPYL PHOSPHORODITHIOATE) ESTER of N-(2-MERCAPTOETHYL)BENZENESULFONAMIDE □ DISAN □ EXPORSAN □ N-(2-MERCAPTOETHYLBENZENESULFONAMIDE)-S-(O,O-DIISOPROPYL PHOSPHORODITHIOATE) □ PHOSPHORODITHIOIC ACID-O,O-BIS (1-METHYLETHYL)-S-((2-((PHENYLSULFONYL)AMINO)ETHYL ESTER □ PREFAR □ PRE-SAN □ R-4461

TOXICITY DATA WITH REFERENCE
orl-rat LD50:271 mg/kg FMCHA2 -,C31,83
skn-rat LD50:3950 mg/kg 31ZOAD 1,34,68
orl-mus LD50:1540 mg/kg JPIFAN (27),11,76
ipr-mus LD50:630 mg/kg JPIFAN (27),11,76
skn-rbt LD50:2000 mg/kg WRPCA2 9,119,70

SAFETY PROFILE: Poison by ingestion. Moderately toxic by skin contact. An herbicide. When heated to decomposition it emits very toxic fumes of NO_x, SO_x, and PO_x. See also ESTERS.

DNO900 **CAS:2973-10-6** **HR: 2**
DIISOPROPYL ESTER SULFURIC ACID
mf: $C_6H_{14}O_4S$ mw: 182.26

SYNS: DI-ISOPROPYLSULFAT (GERMAN) □ DI-ISOPROPYLSULFATE □ ISOPROPYL SULFATE

TOXICITY DATA WITH REFERENCE
scu-rat TDLo:300 mg/kg:ETA ZKKOBW 79,135,73
orl-rat LD50:1090 mg/kg AIHAAP 30,470,69
skn-rbt LD50:1410 mg/kg AIHAAP 30,470,69

SAFETY PROFILE: Moderately toxic by ingestion and skin contact. Questionable carcinogen with experimental tumorigenic data. When heated to decomposition it emits toxic fumes of SO_x. See also ESTERS and SULFATES.

DNP000 **CAS:96-80-0** **HR: 2**
N,N-DIISOPROPYL ETHANOLAMINE
mf: $C_8H_{19}NO$ mw: 145.28

SYNS: 2-DIISOPROPYLAMINOETHANOL □ DIISOPROPYL ETHANOLAMINE

TOXICITY DATA WITH REFERENCE
skn-rbt 500 mg open MLD UCDS** 6/6/69
eye-rbt 750 µg open SEV AMIHBC 10,61,54
orl-rat LD50:1070 mg/kg UCDS** 6/6/69
skn-rbt LD50:450 mg/kg AMIHBC 10,61,54

CONSENSUS REPORTS: Reported in EPA TSCA Inventory.

SAFETY PROFILE: Moderately toxic by ingestion and skin contact. A skin and severe eye irritant.

DNP600 **CAS:20652-39-5** **HR: 2**
N,N-DIISOPROPYL ETHYL CARBAMATE
mf: $C_9H_{19}NO_2$ mw: 173.29

SYNS: DIISOPROPYLCARBAMIC ACID, ETHYL ESTER □ DIISOPROPYL ETHYL CARBAMATE

TOXICITY DATA WITH REFERENCE
ipr-mus TDLo:6500 mg/kg/13W-I:ETA JNCIAM 9,35,48

SAFETY PROFILE: Questionable carcinogen with experimental tumorigenic data. When heated to decomposition it emits toxic fumes of NO_x. See also CARBAMATES.

DNP700 **CAS:121-05-1** **HR: 3**
N,N-DIISOPROPYL ETHYLENEDIAMINE
mf: $C_8H_{20}N_2$ mw: 144.30

SYNS: ETHYLENEDIAMINE, N,N-DIISOPROPYL- □ USAF AM-2

TOXICITY DATA WITH REFERENCE
ipr-mus LD50:200 mg/kg NTIS** AD277-689

CONSENSUS REPORTS: Reported in EPA TSCA Inventory.

SAFETY PROFILE: Poison by intraperitoneal route. When heated to decomposition it emits toxic vapors of NO_x.

DNQ200 **CAS:7283-70-7** **HR: 2**
DIISOPROPYL FUMARATE
mf: $C_{10}H_{16}O_4$ mw: 200.26

SYN: FUMARIC ACID, DIISOPROPYL ESTER

TOXICITY DATA WITH REFERENCE
skn-rbt 100 mg/24H MOD 85JCAE-,375,86
eye-rbt 500 mg open AMIHBC 10,61,54
orl-rat LD50:3250 mg/kg AMIHBC 10,61,54
skn-rbt LD50:10 g/kg AMIHBC 10,61,54

SAFETY PROFILE: Moderately toxic by ingestion. Mildly toxic by skin contact. A skin and eye irritant. When heated to decomposition it emits acrid smoke and fumes.

DNQ600 **CAS:1809-20-7** **HR: 2**
DIISOPROPYL HYDROGEN PHOSPHITE
mf: $C_6H_{15}O_3P$ mw: 166.18

PROP: A liquid. D: 1.00 @ 20°/4°, bp: 106–108° @ 53 mm.

SYNS: DIISOPROPYL PHOSPHITE □ DIISOPROPYLPHOSPHONATE □ O,O-DIISOPROPYL PHOSPHONATE □ ISOPROPYL PHOSPHONATE □ PHOSPHONIC ACID, BIS(1-METHYLETHYL) ESTER

TOXICITY DATA WITH REFERENCE
mmo-sat 5 µL/plate MUREAV 28,405,75
orl-rat LD50:1700 mg/kg GTPZAB 29(11),51,85
skn-rbt LD50:5700 mg/kg ALBRW* #OPB-3,84

CONSENSUS REPORTS: Reported in EPA TSCA Inventory.

SAFETY PROFILE: Moderately toxic by ingestion. Mildly toxic by skin contact. Mutation data reported. When heated to decomposition it emits toxic fumes of PO_x. See also ESTERS.

DNQ700 **CAS:86886-16-0** **HR: 3**
DIISOPROPYL HYPONITRITE
mf: $C_6H_{14}N_2O_2$ mw: 146.19

$(CH_3)_2CHON{=}NOCH(CH_3)_2$

SYN: BIS(2-PROPYLOXY)DIAZENE

SAFETY PROFILE: An impact-sensitive explosive. When heated to decomposition it emits toxic fumes of NO_x. See also NITRITES.

DNQ800 **CAS:1071-39-2** **HR: 3**
DIISOPROPYLMERCURY
mf: $C_6H_{14}Hg$ mw: 286.79

PROP: A liquid. Bp: 63° @ 10 mm, d: 2.00 @ 20°/4°, vap d: 9.9.

TOXICITY DATA WITH REFERENCE
ipr-mus LDLo:7800 µg/kg CBCCT* 4,320,52

CONSENSUS REPORTS: Mercury and its compounds are on the Community Right-To-Know List.

OSHA PEL: TWA 0.01 mg(Hg)/m^3; STEL 0.03 mg/m^3 (skin)
ACGIH TLV: TWA 0.01 mg(Hg)/m^3; STEL 0.03 mg(Hg)/m^3
NIOSH REL: (Mercury, Organo): TWA 0.01 mg/m^3; STEL 0.03 mg/m^3 (skin)

SAFETY PROFILE: Poison by intraperitoneal route. Mercury compounds are poisons. When heated to decomposition it emits toxic fumes of Hg. See also MERCURY COMPOUNDS, ORGANIC.

DNQ875 CAS:1445-75-6 HR: 2
DIISOPROPYL METHYLPHOSPHONATE
mf: $C_7H_{17}O_3P$ mw: 180.21

PROP: Bp: 66° @ 3 mm.

SYNS: DIISOPROPYL METHANEPHOSPHONATE □ DIMP

TOXICITY DATA WITH REFERENCE
orl-rat LD50:826 mg/kg 40QBA3 -,450,78
orl-mus LD50:1041 mg/kg 40QBA3 -,450,78
orl-dck LD50:1490 mg/kg NTIS** AD A087 257
orl-ctl LD50:750 mg/kg NTIS** AD-A093-673
orl-mam LD50:503 mg/kg NTIS** AD-A087-257
orl-brd LD50:1000 mg/kg NTIS** AD-A087-257

CONSENSUS REPORTS: Reported in EPA TSCA Inventory.

SAFETY PROFILE: Moderately toxic by ingestion. When heated to decomposition it emits toxic fumes of PO_x. See also ESTERS and PHOSPHONIC ACID.

DNR200 CAS:23668-76-0 HR: 3
DIISOPROPYLOXOSTANNANE
mf: $C_6H_{14}OSn$ mw: 220.89

PROP: Solid. Insol in water.

SYNS: DIISOPROPYLTIN OXIDE □ KYSLICNIK DIISOPROPYLCINICITY (CZECH)

TOXICITY DATA WITH REFERENCE
skn-rbt 500 mg/24H SEV 28ZPAK -,225,72
eye-rbt 20 mg/24H MOD 28ZPAK -,225,72
orl-rat LD50:57,700 µg/kg 28ZPAK -,225,72

OSHA PEL: TWA 0.1 mg(Sn)/m^3 (skin)
ACGIH TLV: TWA 0.1 mg(Sn)/m^3 (skin) (Proposed: TWA 0.1 mg(Sn)/m^3; STEL 0.2 mg(Sn)/m^3 (skin))
NIOSH REL: (Organotin Compounds) TWA 0.1 mg(Sn)/m^3

SAFETY PROFILE: Poison by ingestion. An eye and severe skin irritant. When heated to decomposition it emits acrid smoke and irritating fumes. See also TIN COMPOUNDS.

For occupational chemical analysis use NIOSH: Organotin Compounds 5504.

DNR309 CAS:3254-66-8 HR: 3
DIISOPROPYL PARAOXON
mf: $C_{12}H_{18}NO_6P$ mw: 303.28

SYNS: DIISOPROPYL-p-NITROPHENYL PHOSPHATE □ O,O-DIISOPROPYL-o,p-NITROPHENYL PHOSPHATE □ MIOTICOL □ PROPICOL

TOXICITY DATA WITH REFERENCE
orl-mus LD50:143 mg/kg JAFCAU 17,243,69
ipr-mus LD50:33 mg/kg JAFCAU 12,318,64

SAFETY PROFILE: Poison by ingestion and intraperitoneal routes. When heated to decomposition it emits very toxic fumes of NO_x and PO_x.

DNR400 CAS:105-64-6 HR: 3
DIISOPROPYL PERDICARBONATE
mf: $C_8H_{14}O_6$ mw: 206.22

$[(CH_3)_2CHOCO{\bullet}O\text{-})]_2$

PROP: Colorless, crystalline solid. Rapid decomp @ 63°F, mp: 8°–10°, d: 1.080 @ 15.5°/4°. Almost insol in water; miscible with aliphatic and aromatic hydrocarbons, esters, ethers, and chlorinated hydrocarbons.

SYNS: DIISOPROPYL PEROXYDICARBONATE □ ISOPROPYL PERCARBONATE □ ISOPROPYL PEROXYDICARBONATE □ PEROXYDICARBONATE d'ISOPROPYLE □ PEROXYDICARBONIC ACID, BIS(1-METHYLETHYL) ESTER

TOXICITY DATA WITH REFERENCE
eye-rbt 500 mg SEV IHFCAY 6,1,67
orl-rat LD50:2140 mg/kg IHFCAY 6,1,67
skn-rbt LD50:2025 mg/kg BSPII* 1/75-19B

CONSENSUS REPORTS: Reported in EPA TSCA Inventory.

SAFETY PROFILE: Moderately toxic by ingestion and skin contact. A severe eye irritant. Very dangerous fire hazard. Dangerously unstable above 10°C. An impact- and heat-sensitive explosive. Solutions may spontaneously explode (the hazard increases with concentration). Storage in sealed containers may be dangerous. Explodes on contact with amines or potassium iodide. May explode on contact with organic matter. When heated to decomposition it emits acrid smoke and fumes. See also PEROXIDES, ORGANIC.

DNR800 CAS:2078-54-8 HR: 3
2,6-DIISOPROPYLPHENOL
mf: $C_{12}H_{18}O$ mw: 178.30

PROP: A colorless liquid or solid. Fp: 17.9°, mp: 19°, bp: 136° @ 30 mm, flash p: 235°F (CC), d: 0.955 @ 20°/4°.

SYNS: 2,6-BIS(1-METHYLETHYL)PHENOL □ DIPRIVAN □ ICI 35868 □ PHENOL, 2,6-BIS(1-METHYLETHYL)-(9CI) □ PROPOFOL

TOXICITY DATA WITH REFERENCE
unr-wmn TDLo:2800 µg/kg (female 39W post):REP BJANAD 62,649,89
unr-man TDLo:2857 µg/kg/1D-I ANASAB 43,170,88
ivn-rat LD50:42 mg/kg YACHDS 21,11,93
ipr-mus LD50:170 mg/kg JMPCAS 2,201,60
ivn-mus LD50:50 mg/kg JMCMAR 23,1350,80
ivn-rbt LDLo:20 mg/kg JMCMAR 23,1350,80

CONSENSUS REPORTS: Reported in EPA TSCA Inventory.

SAFETY PROFILE: Poison by intravenous and intraperitoneal routes. Experimental reproductive effects. Combustible when exposed to heat or flame; can react with oxidizing materials. To fight fire, use foam, CO_2, dry chemical. When heated to decomposition it emits acrid smoke and fumes. See also PHENOL.

DNS000 CAS:26762-93-6 HR: 2
DIISOPROPYLPHENYLHYDROPEROXIDE (solution)
DOT: UN 2171
mf: $C_{12}H_{19}O_2$ mw: 195.30

PROP: Colorless to pale-yellow liquid.

SYN: DIISOPROPYLBENZENE HYDROPEROXIDE, not more than 72% in solution (DOT)

TOXICITY DATA WITH REFERENCE
unr-mus TDLo:391 mg/kg:ETA RARSAM 3,193,63

CONSENSUS REPORTS: Reported in EPA TSCA Inventory.

DOT CLASSIFICATION: Forbidden

SAFETY PROFILE: Questionable carcinogen with experimental tumorigenic data. A powerful oxidizer. When heated to decomposition it emits acrid smoke and fumes. See also PEROXIDES, ORGANIC.

DNS200 CAS:330-64-3 HR: 3
3,5-DIISOPROPYLPHENYL-N-METHYLCARBAMATE
mf: $C_{14}H_{21}NO_2$ mw: 235.36

SYNS: 3,5-BIS(1-METHYLETHYL)PHENOL METHYLCARBAMATE □ 3,5-BIS(1-METHYLETHYL)PHENYL ESTER METHYL CARBAMIC ACID □ 3,5-BIS(1-METHYLETHYL)PHENYL METHYLCARBAMATE □ 3,5-DIISOPROPYLPHENOL METHYLCARBAMATE □ 3,5-DIISOPROPYLPHENYL METHYLCARBAMATE □ DIP □ ENT 25,780 □ HOOKER HRS-1422

TOXICITY DATA WITH REFERENCE
orl-rat LD50:200 mg/kg TXAPA9 14,515,69
ipr-rat LD50:267 mg/kg BWHOA6 44(1-3),241,71
ivn-rat LD50:29,700 μg/kg BWHOA6 44(1-3),241,71
ipr-mus LD50:31 mg/kg BECTA6 2,163,67
orl-bwd LD50:10 mg/kg TXAPA9 21,315,72

SAFETY PROFILE: Poison by ingestion, intraperitoneal, and intravenous routes. When heated to decomposition it emits toxic fumes of NO_x. See also CARBAMATES.

DNS600 CAS:2303-17-5 HR: 3
N-DIISOPROPYLTHIOCARBAMIC ACID S-2,3,3-TRICHLORO-2-PROPENYL ESTER
mf: $C_{10}H_{16}Cl_3NOS$ mw: 304.68

PROP: Oil. Mp: 29–30°.

SYNS: AVADEX BW □ CP 23426 □ N-DIISOPROPYLTHIOCARBAMIC ACID-S-2,3,3-TRICHLOROALLYL ESTER □ N,N-DIISOPROPYL-2,3,3-TRICHLORALLYL-THIOLCARBAMAT (GERMAN) □ DIISOPROPYLTRICHLOROALLYLTHIOCARBAMATE □ DIPTHAL □ FAR-GO □ 2,3,3-THICHLORO-2-PROPENE-1-THIOL, DIISOPROPYLCARBAMATE □ TRIALLAT (GERMAN) □ TRIALLATE □ 2,3,3-TRICHLORALLYL-N,N-(DIISOPROPYL)-THIOCARBAMAT (GERMAN) □ 2,3,3-TRICHLOROALLYL DIISOPROPYLTHIOCARBAMATE □ S-2,3,3-TRICHLOROALLYL-N,N-DIISOPROPYLTHIOCARBAMATE

TOXICITY DATA WITH REFERENCE
mma-sat 100 nmol/L BCPCA6 32,3739,83
mma-bcs 50 μg/plate JAFCAU 29,268,81
cyt-ham:ovr 100 μmol/L MUREAV 85,45,81
orl-rat LD50:800 mg/kg 85JFAN A403,84
skn-rat LDLo:3500 mg/kg GISAAA 33(7)37,68
orl-mus LD50:930 mg/kg GISAAA 33(7),37,68
ihl-cat LCLo:400 mg/m³/4H GISAAA 33(7),37,68
skn-rbt LD50:2225 mg/kg 28ZEAL 5,226,76

CONSENSUS REPORTS: EPA Genetic Toxicology Program.

SAFETY PROFILE: Poison by inhalation. Moderately toxic by ingestion and skin contact. Mutation data reported. An herbicide. When heated to decomposition it emits very toxic fumes of Cl^-, NO_x, and SO_x. See also CARBAMATES and ESTERS.

DNS800 CAS:2986-17-6 HR: 2
DIISOPROPYL THIOUREA
mf: $C_7H_{16}N_2S$ mw: 160.31

SYNS: N,N′-BIS(1-METHYLETHYL)THIOUREA □ N,N′-DIISOPROPYLTHIOUREA □ 1,3-DIISOPROPYLTHIOUREA □ THIOUREA, N,N′-BIS(1-METHYLETHYL)-(9CI)

TOXICITY DATA WITH REFERENCE
orl-rat LD50:450 mg/kg JPETAB 90,260,47
orl-mus LDLo:1070 mg/kg AECTCV 14,111,85

CONSENSUS REPORTS: Reported in EPA TSCA Inventory.

SAFETY PROFILE: Moderately toxic by ingestion. When heated to decomposition it emits very toxic fumes of NO_x and SO_x.

DNT000 CAS:38802-82-3 HR: 3
DIISOPROPYLTIN DICHLORIDE
mf: $C_6H_{14}Cl_2Sn$ mw: 275.79

PROP: Colorless crystals. Sol in water. Mp: 84°

SYN: DICHLORODIISOPROPYLSTANNANE

TOXICITY DATA WITH REFERENCE
ivn-rat LD50:15 mg/kg JOCMA7 2,183,60

OSHA PEL: TWA 0.1 mg(Sn)/m³ (skin)
ACGIH TLV: TWA 0.1 mg(Sn)/m³ (skin) (Proposed: TWA 0.1 mg(Sn)/m³; STEL 0.2 mg(Sn)/m³ (skin))
NIOSH REL: (Organotin Compounds) TWA 0.1 mg(Sn)/m³

SAFETY PROFILE: Poison by intravenous route. When heated to decomposition it emits toxic fumes of Cl^-. See also TIN COMPOUNDS and CHLORIDES.

For occupational chemical analysis use NIOSH: Organotin Compounds 5504.

DNT200 CAS:49538-98-9 HR: 2
O,O-DIISOPROPYL-S-TRICYCLOHEXYLTIN PHOSPHORODITHIOATE
mf: $C_{24}H_{47}O_2PS_2Sn$ mw: 581.49

SYNS: ((DIISOPROPROXYPHOSPHINOTHIOYL)THIO)TRICYCLOHEXYL STANNANE □ R-28627

TOXICITY DATA WITH REFERENCE
orl-rat LD50:860 mg/kg SPEADM 74-1,-,74

OSHA PEL: TWA 0.1 mg(Sn)/m³ (skin)
ACGIH TLV: TWA 0.1 mg(Sn)/m³ (skin) (Proposed: TWA 0.1 mg(Sn)/m³; STEL 0.2 mg(Sn)/m³ (skin))
NIOSH REL: (Organotin Compounds) TWA 0.1 mg(Sn)/m³

SAFETY PROFILE: Moderately toxic by ingestion. When heated to decomposition it emits very toxic fumes of SO_x and PO_x. See also TIN COMPOUNDS.

For occupational chemical analysis use NIOSH: Organotin Compounds 5504.

DNT300 CAS:522-75-8 HR: 1
4,4′-DIISOTHIOINDIGO
mf: $C_{16}H_8O_2S_2$ mw: 296.36

SYNS: ANTINOLO RED B □ ($\Delta^{2,2'}$(3H,3′H)-BIBENZO(b)THIOPHENE)-3,3′-DIONE □ C.I. 73300 □ CIBA PINK B □ C.I. VAT RED 41 □ DURINDONE PRINTING RED B □ DURINDONE RED B □ DURINDONE RED BP □ HELIANE RED 5B □ HELINDON RED BB □ ISOTHIOINDIGO □ TETRA PINK B □ THIOINDIGO □ THIOINDIGO RED B □ THIOINDIGO RED S □ TINA PINK B □ TYRIAN RED A-5B □ VAT RED 5B

TOXICITY DATA WITH REFERENCE
ipr-rat LD50:4170 mg/kg GISAAA 50(8),91,85

CONSENSUS REPORTS: Reported in EPA TSCA Inventory.

SAFETY PROFILE: Mildly toxic by intraperitoneal route. When heated to decomposition it emits toxic vapors of SO_x.

DNU000 CAS:630-93-3 HR: 3
DILANTIN
mf: $C_{15}H_{11}N_2O_2$•Na mw: 274.27

PROP: Hygroscopic crystals with bitter, soapy taste.

SYNS: ALEPSIN □ ANTILEPSIN □ ANTISACER □ AURANILE □ CITRULLAMON □ DANTEN □ DANTOIN □ DENYL □ DENYLSODIUM □ DERIZENE □ DIFENIN □ DIFETOIN □ DIFHYDAN □ DI-HYDAN □ DIHYDANTOIN □ DILANTIN SODIUM □ DI-LEN □ DINTOINA □ DIPHANTOINE SODIUM □ DIPHEDAN □ DIPHENATE □ DIPHENIN □ DIPHENINE SODIUM □ DIPHENTOIN □ DIPHENYLAN SODIUM □ DIPHENYLHYDANTOIN SODIUM □ 5,5-DIPHENYLHYDANTOIN SODIUM □ 5,5-DIPHENYL-2,4-IMIDAZOLIDINE-DIONE, MONOSODIUM SALT □ DI-PHETINE □ DITOIN □ DIVULSAN □ DPH □ ENKEFAL □ EPAMIN □ EPANUTIN □ EPELIN □ EPIFENYL □ EPIHYDAN □ EPILAN-D □ EPILANTIN □ EPINAT □ EPTOIN □ FENANTOIN □ FENITOIN □ FENYTOINE □ HYDANTIN SODIUM □ HYDANTOIN SODIUM □ IDANTOIL □ IDANTOINAL □ LEPITOIN □ LEPITOIN SODIUM □ MINETOIN □ NOVANTOINA □ NOVODIPHENYL □ OM-HYDANTOINE SODIUM □ PHENYTOIN SODIUM □ SACERIL □ SDPH □ SODANTON □ SODIUM DIPHENYLHYDANTOIN □ SODIUM DIPHENYL HYDANTOINATE □ SODIUM-5,5-DIPHENYLHYDANTOINATE □ SODIUM-5,5-DIPHENYL-2,4-IMIDAZOLIDINEDIONE □ SOLANTOIN □ SOLANTYL □ SOLUBLE PHENYTOIN □ SYLANTOIC □ TACOSAL □ THILOPHENYT □ ZENTROPIL

TOXICITY DATA WITH REFERENCE
dnd-esc 50 µmol/L MUREAV 89,95,81
pic-esc 100 mg/L VIRLAX 99,257,79
orl-rat TDLo:3 g/kg (9-13D preg):REP TJADAB 27,149,83
orl-rat TDLo:640 mg/kg (female 5-20D post):REP TOLED5 34,107,86
scu-mus TDLo:50 mg/kg (female 12D post):TER PSEBAA 168,175,81
ipr-rat TDLo:1300 mg/kg (female 17-22D post):TER TOLED5 39,165,87
orl-rat TDLo:3 g/kg (9-13D preg):TER TJADAB 27,149,83
ipr-mus TDLo:263 mg/kg (female 11-13D post):TER JDREAF 58,1740,79
orl-rbt TDLo:1200 mg/kg (female 7-18D post):REP TJADAB 21,371,80
orl-cat TDLo:26 mg/kg (female 10-22D post):REP TJADAB 20,447,79
ipr-mus TDLo:189 mg/kg (female 8-10D post):TER TXAPA9 64,271,82
orl-rat TDLo:1280 mg/kg (female 5-20D post):TER NRTXDN 8,45,87
orl-mus TDLo:1500 mg/kg (female 6-15D post):TER ChaEB# 19JAN78
ipr-mus TDLo:165 mg/kg (female 8-10D post):TER TXAPA9 64,271,82
orl-rbt TDLo:900 mg/kg (female 7-18D post):TER TJADAB 21,371,80
ipr-rat TDLo:650 mg/kg (female 17-22D post):TER TOLED5 39,165,87
orl-man TDLo:70 mg/kg/17D-I:CNS,GIT,SKN NEJMAG 242,897,50
orl-wmn TDLo:4 mg/kg/D:END,SKN JAMAAP 176(6),491,61
orl-wmn LDLo:78 mg/kg:SKN,BLD,PUL ADSYAF 46,856,42
orl-man LDLo:647 mg/kg/21W-I:SKN AIMDAP 81,605,48
unr-man LDLo:29 mg/kg 85DCAI 2,73,70
orl-rat LD50:1530 mg/kg JPETAB 138,224,62
ipr-rat LD50:138 mg/kg IJPPAZ 10,5,66
ivn-rat LD50:104 mg/kg ARZNAD 33,1155,83
scu-rat LD50:230 mg/kg NYKZAU 56,377,60
orl-mus LD50:165 mg/kg IJEBA6 19,1047,81
ipr-mus LD50:103 mg/kg ARZNAD 30,12,80
scu-mus LD50:400 mg/kg BCPCA6 17,369,68
ivn-mus LD50:110 mg/kg ARZNAD 30,477,80
ivn-dog LDLo:90 mg/kg ARPAAQ 28,761,39

CONSENSUS REPORTS: IARC Cancer Review: Animal Sufficient Evidence IMEMDT 13,201,77. Reported in EPA TSCA Inventory.

SAFETY PROFILE: Confirmed carcinogen. Experimental teratogen. Other experimental reproductive effects. Poison by ingestion, subcutaneous, intravenous, and intraperitoneal routes. Human systemic effects by ingestion: anorexia, respiratory depression, nausea or vomiting, hemorrhage, dermatitis, and endocrine effects. Mutation data reported. An anticonvulsant and cardiac depressant used for the treatment of grand mal and psychomotor seizures. When heated to decomposition it emits very toxic fumes of NO_x and Na_2O.

DNU100 **CAS:456-59-7** **HR: 2**
DILATIN
mf: $C_{17}H_{24}O_3$ mw: 276.41

PROP: Crystals. Mp: 50–53°, bp: 192–194°. Practically insol in water; sol in lipoids and their solvents.

SYNS: ARTO-ESPASMOL □ BS 572 □ CAPILAN □ CICLOSPASMOL □ CLANDILON □ CYCLANDELATE □ CYCLERGINE □ CYCLOBRAL □ CYCLOLYT □ CYCLOMANDOL □ CYCLOSPASMOL □ α-HYDROXYBENZENEACETIC ACID 3,3,5-TRIMETHYLCYCLOHEXYL ESTER (9CI) □ NATIL □ NOVODIL □ PEREBRAL □ SAICLATE □ SANCYCLAN □ SEPYRON □ SPASMIONE □ SPASMOCYCLON □ SPASMOCYCLONE □ 3,3,5-TRIMETHYLCYCLOHEXANOL-α-PHENYL-α-HYDROXYACETATE □ 3,5,5-TRIMETHYLCYCLOHEXYL AMYGDALATE □ 3,3,5-TRIMETHYLCYCLOHEXYL MANDELATE

TOXICITY DATA WITH REFERENCE
orl-rat LD50:5 g/kg NIIRDN 6,310,82
ipr-rat LD50:2570 mg/kg AIPTAK 105,145,56
ipr-mus LD50:3780 mg/kg AIPTAK 105,145,56
ipr-dog LD50:2000 mg/kg AIPTAK 105,145,56
orl-gpg LD50:3950 mg/kg AIPTAK 105,145,56
ipr-gpg LD50:2480 mg/kg AIPTAK 105,145,56

CONSENSUS REPORTS: Reported in EPA TSCA Inventory.

SAFETY PROFILE: Moderately toxic by ingestion and intraperitoneal routes. When heated to decomposition it emits acrid smoke and fumes. See also ESTERS.

DNU200 **CAS:849-55-8** **HR: 3**
DILATOL HYDROCHLORIDE
mf: $C_{19}H_{25}NO_2$•ClH mw: 335.91

SYNS: ARLIDIN HYDROCHLORIDE □ BUPHENINE HYDROCHLORIDE □ DILATYL □ p-HYDROXY-α-(1-((1-METHYL-3-PHENYLPROPYL)AMINO)ETHYL)BENZYL ALCOHOL HYDROCHLORIDE □ 1-p-HYDROXYPHENYL-2-(1′-METHYL-3′-PHENYLPROPYLAMINO)-1-PROPANOL HYDROCHLORIDE □ NYLIDRIN HYDROCHLORIDE □ SUPRIFEN PSB HYDROCHLORIDE □ VERINA

TOXICITY DATA WITH REFERENCE
ipr-rat LD50:380 mg/kg 27ZQAG -,351,72
orl-mus LD50:250 mg/kg 27ZQAG -,351,72
ipr-mus LD50:136 mg/kg YKYUA6 24,431,73
ivn-mus LD50:40 mg/kg 27ZQAG -,351,72

SAFETY PROFILE: Poison by ingestion, intraperitoneal, and intravenous routes. When heated to decomposition it emits very toxic fumes of Cl^- and NO_x.

DNU300 **CAS:71-68-1** **HR: 3**
DILAUDID
mf: $C_{17}H_{19}NO_3$•ClH mw: 321.83

PROP: A solid. Mp: 305–315° (decomp).

SYNS: DIHYDROMORPHINONE HYDROCHLORIDE □ DILAUDID HYDROCHLORIDE □ 4,5-α-EPOXY-3-HYDROXY-17-METHYLMORPHINAN-6-ONE HYDROCHLORIDE □ HYDROMORPHONE HYDROCHLORIDE □ HYMORPHAN

TOXICITY DATA WITH REFERENCE
scu-ham TDLo:21,500 µg/kg (8D preg):TER AJOGAH 123,705,75
scu-ham TDLo:44 mg/kg (8D preg):TER AJOGAH 123,705,75
scu-rat LD50:51 mg/kg ARZNAD 3,238,53
scu-mus LD50:120 mg/kg ARZNAD 3,238,53
ivn-mus LD50:55 mg/kg TXAPA9 6,334,64
ivn-cat LDLo:3 mg/kg AEPPAE 194,296,40
ivn-rbt LDLo:2500 µg/kg AEPPAE 194,296,40

SAFETY PROFILE: Poison by subcutaneous and intravenous routes. Experimental teratogenic effects. A powerful analgesic. When heated to decomposition it emits very toxic fumes of NO_x and HCl. See also MORPHINE.

DNU310 **CAS:1421-28-9** **HR: 3**
DILAUDID HYDROCHLORIDE
mf: $C_{17}H_{21}NO_3$•ClH mw: 323.85

SYNS: DIHYDROMORPHINE HYDROCHLORIDE □ PARAMORFAN

TOXICITY DATA WITH REFERENCE
scu-mus LDLo:149 mg/kg JPETAB 52,468,34
ivn-mus LD50:55 mg/kg TXAPA9 6,334,64
orl-rbt LDLo:800 mg/kg HBAMAK 4,1289,35
scu-rbt LD50:50 mg/kg JPETAB 66,182,39
par-rbt LDLo:142 mg/kg JPETAB 66,182,39
scu-gpg LDLo:500 mg/kg HBAMAK 4,1289,35

SAFETY PROFILE: Poison by subcutaneous, intravenous, and parenteral routes. Moderately toxic by ingestion. When heated to decomposition it emits very toxic fumes of NO_x and HCl. See also MORPHINE.

DNU325 **CAS:2592-85-0** **HR: 3**
1,3-DILITHIOBENZENE
mf: $C_6H_4Li_2$ mw: 89.98

SAFETY PROFILE: An unstable explosive. See also LITHIUM COMPOUNDS and EXPLOSIVES.

DNU350 **CAS:15114-92-8** **HR: 3**
DILITHIUM-1,1-BIS(TRIMETHYLSILYL)HYDRAZIDE
mf: $C_6H_{18}Li_2N_2Si_2$ mw: 188.27

SAFETY PROFILE: Ignites spontaneously in air. Ignites or explodes on contact with nitric acid, fluorine gas, or liquid ozone + oxygen. When heated to decomposition it emits toxic fumes of NO_x. See also LITHIUM COMPOUNDS.

DNU400 **CAS:8006-75-5** **HR: 1**
DILL SEED OIL, EUROPEAN TYPE

PROP: From steam distillation of the dried ripe fruit of *Anethum graveolens* L. (Fam. Umbelliferae). Yellowish liquid; caraway odor and taste. D: 0.890–0.915, refr index: 1.4836 @ 20°. Sol in fixed oils, mineral oil, and propylene glycol; insol in glycerin.

SYNS: DILL FRUIT OIL □ DILL HERB OIL □ DILL OIL □ DILL SEED OIL □ DILL WEED OIL

TOXICITY DATA WITH REFERENCE
skn-rbt 500 mg/24H MOD FCTOD7 20 (Suppl),673,82
mma-sat 1 mg/plate JOPHDQ 3,236,80
orl-rat LD50:4040 mg/kg FCTXAV 14,659,76

CONSENSUS REPORTS: Reported in EPA TSCA Inventory.

SAFETY PROFILE: Mildly toxic by ingestion. A skin irritant. Mutation data reported. When heated to decomposition it emits acrid smoke and fumes.

DNU600 **CAS:33286-22-5** **HR: 3**
DILTIAZEM HYDROCHLORIDE
mf: $C_{22}H_{26}N_2O_4S{\cdot}ClH$ mw: 451.02

PROP: A solid. Mp: 187–188°. Insol in C_6H_6; sol in H_2O, MeOH, and $CHCl_3$.

SYNS: ANGINYL □ CADIZEM □ CARDIEM □ CRD-401 □ CRP-401 □ DILZEM □ HERBESSER □ TILDIEM

TOXICITY DATA WITH REFERENCE
orl-rat TDLo:700 mg/kg (female 15-21D post):REP KSRNAM 8,3401,74
orl-rat TDLo:60 mg/kg (9-14D preg):REP KSRNAM 8,3401,74
orl-mus TDLo:100 mg/kg (female 9D post):REP KSRNAM 8,3401,74
orl-mus TDLo:25 mg/kg (female 10D post):TER KSRNAM 8,3401,74
orl-rat TDLo:600 mg/kg (female 11D post):TER KSRNAM 8,3401,74
orl-mus TDLo:60 mg/kg (female 7-12D post):REP KSRNAM 8,3401,74
ivn-rat TDLo:378 mg/kg (female 14D pre):REP KSRNAM 21,4857,87
ipr-rat TDLo:80 mg/kg (female 13D post):REP OFAJAE 52,103,75
orl-rat TDLo:1200 mg/kg (9-14D preg):REP KSRNAM 8,3401,74
ipr-rbt TDLo:125 mg/kg (female 7-16D post):TER OFAJAE 52,103,75
ipr-rat TDLo:480 mg/kg (female 9-14D post):TER OFAJAE 52,103,75
orl-man TDLo:1286 μg/kg/1D AIMEAS 99,794,83
orl-wmn TDLo:19 mg/kg:SKN PGMJAO 64,467,88
orl-man TDLo:36 mg/kg/13D-I:SYS GASTAB 88,1260,85
orl-rat LD50:560 mg/kg JJPAAZ 22,467,72
scu-rat LD50:520 mg/kg JJPAAZ 22,467,72
ivn-rat LD50:38 mg/kg JMGZAI 11(1),12,74
orl-mus LD50:640 mg/kg JJPAAZ 22,467,72
ipr-mus LD50:177 mg/kg JMCMAR 29,820,86
scu-mus LD50:260 mg/kg JJPAAZ 22,467,72
ivn-mus LD50:58 mg/kg JJPAAZ 22,467,72

SAFETY PROFILE: Poison by subcutaneous, intravenous, and intraperitoneal routes. Moderately toxic by ingestion. Human systemic effects by ingestion: fall in blood pressure, pulse rate decrease, gastrointestinal effects, dermatitis, fibrous hepatitis. Experimental reproductive effects. An experimental teratogen. When heated to decomposition it emits toxic fumes of SO_x, NO_x, and HCl.

DNU850 **CAS:14465-96-4** **HR: 3**
DIMATIF
mf: $C_4H_{10}N_3PS$ mw: 163.20

SYNS: BIS(1-AZIRIDINYL)AMINOPHOSPINE SULFIDE □ p,p-BIS(1-AZIRIDINYL)PHOSPHINOTHIOIC AMIDE □ DIETHYLENEIMINEAMIDOTHIOPHOSPHORIC ACID □ ENT 61,969

TOXICITY DATA WITH REFERENCE
mmo-smc 1 mg/L TGANAK 18,455,84
cyt-hmn:lym 400 mg/L TGANAK 18,455,84
sce-hmn:lym 2 mg/L TGANAK 16(2),34,82
cyt-mus-unr 1 mg/kg TGANAK 18,455,84
orl-rat TDLo:50 mg/kg (6-15D preg):TER GISAAA 48(5),75,83
orl-rat TDLo:10 mg/kg (1-20D preg):REP GISAAA 48(5),75,83
orl-rat LD50:66 mg/kg GISAAA 48(5),75,83
orl-qal LD50:100 mg/kg JRPFA4 48,371,76

SAFETY PROFILE: Poison by ingestion. An experimental teratogen. Other experimental reproductive effects. Human mutation data reported. When heated to decomposition it emits toxic fumes of PO_x, NO_x, and SO_x.

DNU860 **CAS:4076-02-2** **HR: 2**
DIMAVAL
mf: $C_3H_7O_3S_3{\cdot}Na$ mw: 210.27

PROP: A solid. Mp: 229° (decomp).

SYNS: DIMAYAL □ 2,3-DIMERCAPTOPROPANE SODIUM SULPHONATE □ 2,3-DIMERCAPTOPROPANESULFONIC ACID SODIUM SALT □ 2,3-DIMERCAPTO-1-PROPANESULFONIC ACID SODIUM SALT □ meso-DIMERCAPTOSUCCINIC ACID SODIUM SALT □ DMPS □ SODIUM-2,3-DIMERCAPTOPROPANE-1-SULFONATE □ SODIUM-2,3-DITHIOLPROPANESULFONATE □ UNITHIOL □ UNITIOL □ UNITOL

TOXICITY DATA WITH REFERENCE
cyt-hmn:lym 1 mmol/L CYGEDX 8(4),31,74
orl-rat TDLo:28 g/kg (26W pre-21D post):REP ARZNAD 30,1291,80
ipr-rat LD50:1055 mg/kg ARZNAD 30,1291,80
ipr-mus LD50:1098 mg/kg TXAPA9 61,385,81

SAFETY PROFILE: Moderately toxic by intraperitoneal route. Experimental reproductive effects. Human mutation data reported. When heated to decomposition it emits toxic fumes of SO_x and Na_2O.

DNU875 **HR: 3**
DIMEBON DIHYDROCHLORIDE
mf: $C_{22}H_{22}N_3{\cdot}2ClH$ mw: 401.39

SYN: 9-(2-(2-METHYLPYRIDYL-5)ETHYL)-3,6-DIMETHYL-1,2,3,4-TETRAHYDRO-γ-CARBOLINE 2HCl

TOXICITY DATA WITH REFERENCE
orl-rat LD50:1132 mg/kg RPTOAN 48,103,85
ipr-rat LD50:160 mg/kg RPTOAN 48,103,85
ivn-rat LD50:59 mg/kg RPTOAN 48,103,85
orl-mus LD50:486 mg/kg FATOAO 47(3),75,84
ipr-mus LD50:145 mg/kg RPTOAN 48,103,85
scu-mus LD50:465 mg/kg FATOAO 47(3),75,84
ivn-mus LD50:90,500 μg/kg FATOAO 47(3),75,84

SAFETY PROFILE: Poison by intravenous and intraperitoneal routes. Moderately toxic by ingestion and subcutaneous routes. When heated to decomposition it emits toxic fumes of NO_x and HCl.

D

DNV000 **CAS:1165-48-6** **HR: 3**
DIMEFLINE
mf: $C_{20}H_{21}NO_3$ mw: 323.42

SYNS: 8-(DIMETHYLAMINOMETHYL)-7-METHOXY-3-METHYLFLAVONE □ 8-((DIMETHYLAMINO)METHYL)-7-METHOXY-3-METHYL-2-PHENYLFLAVONE □ DW 62 □ MALIVAN □ N-(7-METHOXY-3-METHYL-4-OXO-2-PHENYL-4H-CHROMEN-8-YL)METHYL-N,N-DIMETHYLAMINE □ REANIMIL □ REC 7/0267 □ REMEFLIN

TOXICITY DATA WITH REFERENCE
orl-rat LD50:40 mg/kg RPOBAR 1,423,64
ipr-rat LD50:6 mg/kg TXAPA9 18,185,71
ivn-rat LD50:1800 µg/kg RPOBAR 1,423,64
rec-rat LD50:10 mg/kg RPOBAR 1,423,64
orl-mus LD50:12 mg/kg RPOBAR 1,423,64
ipr-mus LD50:4800 µg/kg JMPCAS 3,471,61
scu-mus LD50:4 mg/kg RPOBAR 1,423,64
ivn-dog LDLo:1 mg/kg RPOBAR 1,423,64

SAFETY PROFILE: Poison by ingestion, intraperitoneal, intravenous, rectal, and subcutaneous routes. When heated to decomposition it emits very toxic fumes of NO_x.

DNV200 **CAS:2740-04-7** **HR: 3**
DIMEFLINE HYDROCHLORIDE
mf: $C_{20}H_{21}NO_3•ClH$ mw: 359.88

SYNS: DEMEFLINE □ 8-((DIMETHYLAMINO)METHYL)-7-METHOXY-3-METHYLFLAVONE HYDROCHLORIDE □ 8-((DIMETHYLAMINO)METHYL)-7-METHOXY-3-METHYL-2-PHENYL-4H-1-BENZOPYRAN-4-ONE HYDROCHLORIDE □ DW 62 □ 3-METHYL-7-METHOXY-8-(DIMETHYLAMINO-METHYL)-FLAVONE HYDROCHLORIDE □ NSC-114650 □ REC 7/0267 □ REMEFLIN

TOXICITY DATA WITH REFERENCE
orl-rat LD50:14 mg/kg NIIRDN 6,345,82
orl-mus LD50:12 mg/kg JPETAB 128,176,60
ipr-mus LD50:5 mg/kg JPETAB 128,176,60
scu-mus LD50:4 mg/kg JPETAB 128,176,60

SAFETY PROFILE: Poison by ingestion, intraperitoneal, and subcutaneous routes. When heated to decomposition it emits very toxic fumes of HCl and NO_x.

DNV600 **CAS:27292-46-2** **HR: 2**
2,3-DIMERCAPTOPROPYL-p-TOLYSULFIDE
mf: $C_{10}H_{14}S_3$ mw: 230.42

SYN: ANTARSIN

TOXICITY DATA WITH REFERENCE
orl-rat LDLo:2000 mg/kg FATOAO 30(2),226,67
ipr-rat LDLo:1000 mg/kg FATOAO 30(2),226,67
scu-rat LDLo:7000 mg/kg FATOAO 30(2),226,67

SAFETY PROFILE: Moderately toxic by ingestion and intraperitoneal routes. Mildly toxic by subcutaneous route. When heated to decomposition it emits toxic fumes of SO_x. See also MERCAPTANS.

DNV610 **CAS:2418-14-6** **HR: 2**
2,3-DIMERCAPTOSUCCINIC ACID
mf: $C_4H_6O_4S_2$ mw: 182.22

SYNS: BUTANEDIOIC ACID, 2,3-DIMERCAPTO-(9CI) □ 2,3-DIMERCAPTOBUTANEDIOIC ACID □ DIMERCAPTOSUCCINIC ACID □ α-β-DIMERCAPTOSUCCINIC ACID □ SUCCINIC ACID, 2,3-DIMERCAPTO- □ SUXIMER

TOXICITY DATA WITH REFERENCE
orl-rat LD50:4 g/kg YHHPAL 15,335,80
orl-mus LD50:6 g/kg YHHPAL 15,335,80
ipr-mus LD50:2478 mg/kg ARTODN 61,321,88

CONSENSUS REPORTS: Reported in EPA TSCA Inventory.

SAFETY PROFILE: Moderately toxic by intraperitoneal route. Slightly toxic by ingestion. When heated to decomposition it emits toxic vapors of SO_x.

DNV800 **CAS:304-55-2** **HR: 2**
meso-2,3-DIMERCAPTOSUCCINIC ACID
mf: $C_4H_6O_4S_2$ mw: 182.22

PROP: Crystals from EtOAc. Mp: 210–211° (decomp); dependent on rate of heating.

SYNS: (R*,S*)-2,3-DIMERCAPTOBUTANEDIOIC ACID □ meso-DIMERCAPTOSUCCINIC ACID □ DIM-SA □ DMS □ DMSA □ DTS □ Ro 1-7977 □ SUCCIMER

TOXICITY DATA WITH REFERENCE
orl-rat TDLo:1 g/kg (female 6-15D post):REP JTEHD6 30,191,90
orl-mus TDLo:17,600 mg/kg (female 14-21D post):REP LIFSAK 47,1745,90
scu-mus TDLo:16,400 mg/kg (female 6-15D post):REP FAATDF 11,715,88
orl-rat TDLo:10 g/kg (female 6-15D post):REP JTEHD6 30,181,90
orl-mus LD50:>5011 mg/kg CRTXB2 20,83,89
ipr-mus LD50:500 mg/kg NTIS** AD691-490
scu-mus LD50:1725 mg/kg AIPTAK 131,283,61
ivn-rbt LDLo:2700 mg/kg AIPTAK 131,283,61

CONSENSUS REPORTS: Reported in EPA TSCA Inventory.

SAFETY PROFILE: Moderately toxic by intraperitoneal, intravenous, and subcutaneous routes. Experimental reproductive effects. When heated to decomposition it emits toxic fumes of SO_x. See also MERCAPTANS.

DNW000 **CAS:63869-15-8** **HR: 3**
DIMERCUROUS METHANE ARSONATE
mf: $CH_3AsO_3•2Hg$ mw: 539.14

SYN: METHANEARSONIC ACID DIMERCURY SALT

TOXICITY DATA WITH REFERENCE
ipr-mus LD50:75 mg/kg NTIS** AD691-490

CONSENSUS REPORTS: Arsenic and its compounds, as well as mercury and its compounds, are on the Community Right-To-Know List.

OSHA PEL: TWA 0.5 mg(As)/m³; CL 0.1 mg(Hg)/m³ (skin)
ACGIH TLV: TWA 0.2 mg(As)/m³; 0.1 mg(Hg)/m³ (skin)
NIOSH REL: (Mercury, Organo): TWA 0.01 mg/m³; STEL 0.03 mg/m³ (skin)

SAFETY PROFILE: Poison by intraperitoneal route.

When heated to decomposition it emits very toxic fumes of As and Hg. See also MERCURY COMPOUNDS and ARSENIC COMPOUNDS.

DNW200 **CAS:12529-66-7** **HR: 3**
DIMERCURY IMIDE OXIDE
mf: $(HHg_2NO)_n$

$(Hg:N^+:HgOH^-)_n$

PROP: Yellow crystals.

SYNS: MILLON'S BASE ANHYDRIDE □ POLY(DIMERCURYIMMONIUM HYDOXIDE)

CONSENSUS REPORTS: Mercury and its compounds are on the Community Right-To-Know List.

SAFETY PROFILE: A severe explosion hazard if touched or heated. When heated to decomposition it emits toxic fumes of Hg and NO_x. See also MERCURY COMPOUNDS.

DNW400 **CAS:125-64-4** **HR: 3**
DIMERIN
mf: $C_{10}H_{17}NO_2$ mw: 183.28

SYNS: 3,3-DIETHYL-2,4-DIOXO-5-METHYLPIPERIDINE □ 3,3-DIETHYL-5-METHYL-2,4-PIPERIDINEDIONE □ 3,3-DIETHYL-5-METHYLPIPERIDINE-2,4-DIONE □ 2,4-DIOXY 3,3-DIETHYL-5-METHYLPIPERIDINE □ METHYPROLON □ METHYPRYLON □ METIPRILONE □ NOCTAN □ NOLUDAR □ RO 1-6463

TOXICITY DATA WITH REFERENCE
orl-wmn TDLo:600 mg/kg:CNS,PUL JAMAAP 198,1213,66
orl-hmn TDLo:26 mg/kg:CNS CTOXAO 6,563,73
orl-rat LD50:860 mg/kg 27ZQAG -,264,72
scu-rat LD50:400 mg/kg 27ZQAG -,264,72
ivn-rat LD50:380 mg/kg 27ZQAG -,264,72
orl-mus LD50:890 mg/kg JPETAB 118,139,56
ipr-mus LD50:1000 mg/kg 27ZQAG -,264,72
ivn-mus LD50:275 mg/kg 27ZQAG -,264,72
orl-dog LD50:300 mg/kg CLDND* ,261,72
scu-rbt LD50:500 mg/kg 27ZQAG -,264,72
ivn-rbt LD50:315 mg/kg 27ZQAG -,264,72

SAFETY PROFILE: Poison by ingestion, subcutaneous, and intravenous routes. Moderately toxic by intraperitoneal routes. Human systemic effects by ingestion: general anesthesia, sleep disorder, motor activity and pulmonary changes. When heated to decomposition it emits toxic fumes of NO_x.

DNW700 **CAS:4757-55-5** **HR: 3**
DIMETACRINE
mf: $C_{20}H_{26}N_2$ mw: 294.48

PROP: Free base. Bp: (1) 200°.

SYNS: DIMETHACIN □ DIMETHACINE □ 9,9-DIMETHYL-10-(3-(DIMETHYLAMINO)PROPYL)ACRIDAN

TOXICITY DATA WITH REFERENCE
ipr-rat TDLo:91 mg/kg (91D male):REP OYYAA2 8,949,74
orl-rat LD50:1850 mg/kg IYKEDH 6,530,75
ipr-rat LD50:181 mg/kg IYKEDH 6,530,75
scu-rat LD50:1076 mg/kg IYKEDH 6,530,75
orl-mus LD50:1293 mg/kg IYKEDH 6,530,75
ipr-mus LD50:206 mg/kg IYKEDH 6,530,75
scu-mus LD50:676 mg/kg IYKEDH 6,530,75
ivn-mus LD50:39,600 µg/kg IYKEDH 6,530,75

SAFETY PROFILE: Poison by intravenous and intraperitoneal routes. Moderately toxic by ingestion and subcutaneous routes. Experimental reproductive effects. When heated to decomposition it emits toxic fumes of NO_x.

DNW759 **CAS:32865-01-3** **HR: 3**
dl-DIMETANE MALEATE
mf: $C_{16}H_{19}BrN_2{\bullet}C_4H_4O_4$ mw: 435.36

PROP: Crystals. Mp: 133–134.5°.

SYNS: (±)-2-(p-BROMO-α-(2-(DIMETHYLAMINO)ETHYL)BENZYL) PYRIDINE MALEATE □ dl-BROMOPHENIRAMINE MALEATE □ (±)-BROMPHENIRAMINE MALEATE □ dl-BROMPHENIRAMINE MALEATE □ (±)-(Z)-γ-(4-BROMOPHENYL) N,N-DIMETHYL-2-PYRIDINEPROPANAMINE 2-BUTENEDIOATE (1:1)

TOXICITY DATA WITH REFERENCE
orl-rat LD50:161 mg/kg CMTRAG 3,120,61
ipr-rat LD50:113 mg/kg CMTRAG 3,120,61
orl-mus LD50:147 mg/kg CMTRAG 3,120,61
ipr-mus LD50:109 mg/kg CMTRAG 3,120,61
ivn-mus LD50:26 mg/kg CMTRAG 3,120,61
orl-gpg LD50:245 mg/kg CMTRAG 3,120,61

SAFETY PROFILE: Poison by ingestion, intravenous, and intraperitoneal routes. When heated to decomposition it emits toxic fumes of Br^- and NO_x.

DNW800 **CAS:2303-47-1** **HR: 2**
cis-1,4-DIMETHANE SULFONOXY-2-BUTENE
mf: $C_6H_{12}O_6S_2$ mw: 244.30

TOXICITY DATA WITH REFERENCE
sln-dmg-par 10 mmol/L JOGNAU 54,146,56
skn-mus TDLo:480 mg/kg/10W-I:NEO CNREA8 17,64,57

SAFETY PROFILE: Questionable carcinogen with experimental neoplastigenic data. Mutation data reported. When heated to decomposition it emits toxic fumes of SO_x.

DNX000 **CAS:1953-56-6** **HR: 2**
trans-1,4-DIMETHANE SULFONOXY-2-BUTENE
mf: $C_6H_{12}O_6S_2$ mw: 244.30

SYNS: CB 2095 □ 2-BUTENE-1,4-DIOL, DIMETHANESULFONATE, (E)-

TOXICITY DATA WITH REFERENCE
sln-dmg-par 10 mmol/L JOGNAU 54,146,56
skn-mus TDLo:480 mg/kg/10W-I:NEO CNREA8 17,64,57

SAFETY PROFILE: Questionable carcinogen with experimental neoplastigenic data. Mutation data reported. When heated to decomposition it emits toxic fumes of SO_x.

DNX200 CAS:2917-96-6 HR: 2
1,4-DIMETHANESULFONOXY-2-BUTYNE
mf: $C_6H_{10}O_6S_2$ mw: 242.28

SYN: CB2058

TOXICITY DATA WITH REFERENCE
sln-dmg-par 10 mmol/L JOGNAU 54,146,56
sln-dmg-unk 10 mmol/L ANYAA9 160,228,69
skn-mus TDLo:320 mg/kg/5W-I:NEO CNREA8 17,64,57

SAFETY PROFILE: Questionable carcinogen with experimental neoplastigenic data by skin contact. Mutation data reported. When heated to decomposition it emits toxic fumes of SO_x. See also ACETYLENE COMPOUNDS.

DNX300 CAS:1001-62-3 HR: 3
DIMETHANESULFONYL PEROXIDE
mf: $C_2H_6O_6S_2$ mw: 190.19

PROP: Crystals.

SAFETY PROFILE: Decomposes explosively after melting at 79°C. When heated to decomposition it emits toxic fumes of SO_x. See also PEROXIDES.

DNX400 CAS:2773-92-4 HR: 3
DIMETHISOQUIN HYDROCHLORIDE
mf: $C_{17}H_{24}N_2O{\cdot}ClH$ mw: 308.89

PROP: Off white powder. Mp: 144–148°. Insol in Et_2O; sol in H_2O, EtOH, and $CHCl_3$.

SYNS: 3-BUTYL-1-(2-(DIMETHYLAMINO)ETHOXY)ISOQUINOLINE HYDROCHLORIDE ☐ 2-((3-BUTYL-1-ISOQUINOLINYL)OXY)-N,N-DIMETHYLETHANAMINE MONOHYDROCHLORIDE ☐ 1-(β-DIMETHYLAMINOETHOXY)-3-N-BUTYLISOQUINOLINE HYDROCHLORIDE ☐ 1-(β-DIMETHYLAMINOETHOXY)-3-N-BUTYLISOQUINOLINE MONOHYDROCHLORIDE ☐ ISOCHINOL ☐ PRURALGAN ☐ PRURALGIN ☐ QUOTANE ☐ QUOTANE HYDROCHLORIDE

TOXICITY DATA WITH REFERENCE
ipr-rat LD50:45 mg/kg JPETAB 103,306,51
ivn-mus LD50:8 mg/kg ARZNAD 18,729,68
ivn-rbt LD50:5 mg/kg JPETAB 103,306,51

SAFETY PROFILE: Poison by intraperitoneal and intravenous routes. A topical anesthetic. When heated to decomposition it emits very toxic fumes of HCl and NO_x.

DNX500 CAS:8015-19-8 HR: 3
DIMETHISTERONE and ETHINYL ESTRADIOL
mf: $C_{23}H_{32}O_2{\cdot}C_{20}H_{24}O_2$ mw: 636.99

SYNS: ETHINYL ESTRADIOL and DIMETHISTERONE ☐ ORACON ☐ OVIN ☐ SECROVIN

TOXICITY DATA WITH REFERENCE
orl-wmn TDLo:273 mg/kg (77W pre):REP AJOGAH 107,717,70
orl-rbt TDLo:1250 mg/kg (female 5D pre):REP JIMAAD 37,322,61
orl-wmn TDLo:244 mg/kg/8Y-I:CAR AJOGAH 123,299,75
orl-wmn TDLo:92 mg/kg/3Y-I:CAR OBGNAS 47,639,76

SAFETY PROFILE: Suspected human carcinogen producing uterine tumors. Human reproductive effects by ingestion: abnormalities of the uterus, cervix, and vagina. A steroid. When heated to decomposition it emits acrid smoke and irritating fumes.

DNX600 CAS:116-01-8 HR: 3
DIMETHOATE-ETHYL
mf: $C_6H_{14}NO_3PS_2$ mw: 243.30

SYNS: AMERICAN CYANAMID 18706 ☐ B/77 ☐ O,O-DIMETHYL-S-(N-ETHYLCARBAMOYLMETHYL) DITHIOPHOSPHATE ☐ O,O-DIMETHYL-S-(N-ETHYLCARBAMOYLMETHYL) PHOSPHORODITHIOATE ☐ EI-18706 ☐ ENT 25,506 ☐ ETHOATE METHYL ☐ S-(2-(ETHYLAMINO-2-OXOETHYL))-O,O-DIMETHYL PHOSPHORODITHIOATE ☐ S-(N-ETHYLCARBAMOYLMETHYL) DIMETHYL PHOSPHORODITHIOATE ☐ FITIOS ☐ FITIOS B/77 ☐ N-MONOETHYLAMIDE of O,O-DIMETHYLDITHIOPHOSPHORYLACETIC ACID ☐ PHOSHOROTHIOIC ACID-S-(2-(ETHYLAMINO)-2-OXOETHYL)-O,O-DIMETHYL ESTER

TOXICITY DATA WITH REFERENCE
orl-rat LD50:125 mg/kg WRPCA2 9,119,70
skn-rat LD50:2000 mg/kg WRPCA2 9,119,70
ims-rat LD50:250 mg/kg FRPSAX 21,443,66
orl-mus LD50:350 mg/kg SPEADM 74-1,-,74

SAFETY PROFILE: Poison by ingestion and intramuscular routes. Moderately toxic by skin contact. A pesticide. When heated to decomposition it emits very toxic fumes of NO_x, PO_x, and SO_x.

DNX800 CAS:1113-02-6 HR: 3
DIMETHOATE OXYGEN ANALOG
mf: $C_5H_{12}NO_4PS$ mw: 213.21

PROP: Oil. D: 1.32 @ 20°/4°. Misc in H_2O; almost insol in hexane.

SYNS: O-ANALOG of DIMETHOATE ☐ BAY 45432 ☐ BAYER 45,432 ☐ DIMETHOATE O-ANALOG ☐ DIMETHOATE PO ISOLOGUE ☐ DIMETHOXON ☐ O,O-DIMETHYL-S-((N-METHYL-CARBAMOYL)-METHYL)MONOTHIOFOSFAAT (DUTCH) ☐ O,O-DIMETHYL-S-(N-METHYL-CARBAMOYL)-METHYL-MONOTHIOPHOSPHAT (GERMAN) ☐ O,O-DIMETHYL-S-((METHYLCARBAMOYL)METHYL)PHOSPHOROTHIOATE ☐ O,O-DIMETHYL-S-(N-METHYLCARBAMOYLMETHYL)PHOSPHOROTHIOATE ☐ O,O-DIMETHYL-S-(N-METHYLCARBAMOYLMETHYL) PHOSPHOROTHIOLATE ☐ DIMETHYL-S-(N-METHYL-CARBAMOYL-METHYL)PHOSPHOROTHIOLATE ☐ O,O-DIMETHYL-S-(N-METHYL-CARBAMOYLMETHYL) THIOPHOSPHATE ☐ O,O-DIMETHYL-S-(2-OXO-3-AZABUTYL)-MONOTHIOPHOSPHATE ☐ O,O-DIMETIL-S-(N-METIL-CARBAMOIL)-METIL-MONOTIOFOSFATO (ITALIAN) ☐ ENT 25,776 ☐ FOLIMAT ☐ OMETHOAT ☐ OMETHOATE ☐ PHOSPHOROTHIOIC ACID, O,O-DIMETHYL S-(2-(METHYLAMINO)-2-OXOETHYL) ESTER ☐ PO-DIMETHOATE ☐ THIOPHOSPHATE de O,O-DIETHYLE et de S-(N-METHYLCARBAMOYL) METHYLE (FRENCH)

TOXICITY DATA WITH REFERENCE
mmo-esc 5 μL/plate MUREAV 28,405,75
orl-rat LD50:30 mg/kg FMCHA2-,C144,91
skn-rat LD50:700 mg/kg WRPCA2 9,119,70
orl-mus LD50:24 mg/kg PCBPBS 1,248,71
ipr-mus LD50:180 mg/kg ACPMAP 16,7,63
orl-cat LD50:50 mg/kg 85DPAN -,-,71/76
orl-rbt LD50:50 mg/kg 85DPAN -,-,71/76

SAFETY PROFILE: Poison by ingestion and intraperitoneal routes. Moderately toxic by skin contact. Mutation data reported. An insecticide. When heated to decomposition it emits very toxic fumes of NO_x, PO_x, and SO_x.

DNY000 **CAS:94-15-5** **HR: 3**
DIMETHOCAINE
mf: $C_{16}H_{26}N_2O_2$ mw: 278.44

SYNS: 3-(DIETHYLAMINO)-2,2-DIMETHYL-1-PROPANOL-p-AMINO-BENZOATE □ LAROCAINE

TOXICITY DATA WITH REFERENCE
scu-mus LDLo:380 mg/kg AEPPAE 168,447,32
ivn-mus LDLo:40 mg/kg PHREA7 12,190,32
scu-rbt LDLo:150 mg/kg PHREA7 12,190,32
ivn-rbt LDLo:150 mg/kg PHREA7 12,190,32
scu-gpg LDLo:200 mg/kg PHREA7 12,190,32
par-frg LDLo:200 mg/kg AEPPAE 168,447,32

SAFETY PROFILE: Poison by subcutaneous, intravenous, and parenteral routes. When heated to decomposition it emits toxic fumes of NO_x.

DNY400 **CAS:17210-48-9** **HR: 2**
3,4′-DIMETHOXY-4-AMINOAZOBENZENE
mf: $C_{14}H_{15}N_3O_2$ mw: 257.32

SYN: 4-((p-METHOXYPHENYL)AZO)-o-ANISIDINE

TOXICITY DATA WITH REFERENCE
orl-rat TDLo:10 g/kg/24W-C:ETA GANNA2 59,131,68
orl-man TDLo:642 µg/kg:BLD,BIO IARC** 27,39,82

CONSENSUS REPORTS: IARC Cancer Review: Human Inadequate Evidence IMEMDT 27,39,82; Animal Inadequate Evidence IMEMDT 27,39,82.

SAFETY PROFILE: Questionable carcinogen with experimental tumorigenic data. Human systemic effects by ingestion: methemoglobinemia-carboxhemoglobinemia, and changes in porphyrin metabolism. When heated to decomposition it emits toxic fumes of NO_x.

DNY500 **CAS:2735-04-8** **HR: 2**
2,4-DIMETHOXYANILINE
mf: $C_8H_{11}NO_2$ mw: 153.20

PROP: Plates from pet ether or oil. Mp: 32.5–33.5°, bp: 75–80° @ 0.0006 mm.

TOXICITY DATA WITH REFERENCE
mma-sat 10 µg/plate ENMUDM 7(Suppl 5),1,85
otr-rat:emb 55 µg/plate JJATDK 1,190,81
orl-rat LD50:464 mg/kg NCILB* NCI-E-C-72-3252,73
orl-mus LD50:1 g/kg NCILB* NCI-E-C-72-3252,73

CONSENSUS REPORTS: Reported in EPA TSCA Inventory.

SAFETY PROFILE: Moderately toxic by ingestion. Mutation data reported. When heated to decomposition it emits toxic fumes of NO_x. See also ANILINE DYES.

DNY800 **CAS:6448-90-4** **HR: 1**
1,5-DIMETHOXYANTHRAQUINONE
mf: $C_{16}H_{12}O_4$ mw: 268.28

PROP: Pale-yellow needles from EtOH. Mp: 236°.

SYNS: 1,5-DIMETHOXY-9,10-ANTHRACENEDIONE □ 1,5-DIMETHOXYANTHRACHINON (CZECH)

TOXICITY DATA WITH REFERENCE
eye-rbt 500 mg/24H MLD 28ZPAK -,113,72

CONSENSUS REPORTS: Reported in EPA TSCA Inventory.

SAFETY PROFILE: An eye irritant. When heated to decomposition it emits acrid smoke and irritating fumes.

D

DNZ100 **CAS:476-70-0** **HR: 3**
1,10-DIMETHOXY-6a-α-APORPHINE-2,9-DIOL
mf: $C_{19}H_{21}NO_4$ mw: 327.41

PROP: d-Form: Crystals from ether. Mp: 162–164°. Very sltly sol in water or ether; sol in alc, chloroform, dilute acids. dl-Form: Mp: 159–162°.

SYNS: BOLDIN □ BOLDINE □ (+)-BOLDINE □ (S)-BOLDINE □ (+)-(S)-BOLDINE □ (S)-5,6,6a,7-TETRAHYDRO-1,10-DIMETHOXY-6-METHYL-4H-DIBENZO(de,g)QUINOLINE-2,9-DIOL □ UNIBOLDINA

TOXICITY DATA WITH REFERENCE
mmo-smc 50 mg/L MUREAV 260,145,91
mrc-smc 100 mg/L MUREAV 260,145,91
orl-mus LD50:450 mg/kg APFRAD 38,537,80
ipr-mus LD50:170 mg/kg APFRAD 38,537,80
ivn-mus LD50:90 mg/kg APFRAD 38,537,80

CONSENSUS REPORTS: Reported in EPA TSCA Inventory.

SAFETY PROFILE: Poison by intravenous and intraperitoneal routes. Moderately toxic by ingestion. Mutation data reported. When heated to decomposition it emits toxic fumes of NO_x.

DOA000 **CAS:16354-53-3** **HR: 2**
7,12-DIMETHOXYBENZ(a)ANTHRACENE
mf: $C_{20}H_{16}O_2$ mw: 288.36

TOXICITY DATA WITH REFERENCE
ims-rat TDLO:50 mg/kg:NEO CNREA8 29,506,69
ims-rat TD:50 mg/kg:NEO PNASA6 58,2253,67

SAFETY PROFILE: Questionable carcinogen with experimental neoplastigenic data. When heated to decomposition it emits acrid smoke and irritating fumes.

DOA200 **CAS:91-16-7** **HR: 2**
o-DIMETHOXYBENZENE
mf: $C_8H_{10}O_2$ mw: 138.18

PROP: Crystals from pet ether. Mp: 22.5°, bp: 206° @ 759 mm.

SYNS: 1,2-DIMETHOXYBENZENE □ PYROCATECHOL DIMETHYL ETHER □ VERATROL □ VERATROLE

TOXICITY DATA WITH REFERENCE
orl-rat LD50:890 mg/kg GTPZAB 26(2),54,82
orl-mus LD50:700 mg/kg GTPZAB 26(2),54,82

CONSENSUS REPORTS: Reported in EPA TSCA Inventory.

SAFETY PROFILE: Moderately toxic by ingestion. When

heated to decomposition it emits acrid smoke and irritating fumes.

DOA400 CAS:150-78-7 HR: 3
p-DIMETHOXYBENZENE
mf: $C_8H_{10}O_2$ mw: 138.18

PROP: Colorless leaflets or plates, odor of sweet clover. Mp: 55–56°, bp: 212.6°, d: 1.053 @ 55°/55°.

SYNS: DIMETHYL ETHER HYDROQUINONE ☐ DIMETHYLHYDROQUINONE ☐ DIMETHYLHYDROQUINONE ETHER ☐ DMB ☐ QUINOL DIMETHYL ETHER ☐ USAF AN-9 ☐ USAF UCTL-1791

TOXICITY DATA WITH REFERENCE
skn-rbt 6 g/12D-I MLD JIHTAB 31,79,49
skn-rbt 500 mg/24H MOD FCTXAV 16,715,78
skn-gpg 40%/24H MOD FCTXAV 16,715,78
orl-rat LD50:3600 mg/kg FCTXAV 16,715,78
ipr-rat LD50:1100 mg/kg JIHTAB 31,79,49
ipr-mus LD50:100 mg/kg NTIS** AD277-689

CONSENSUS REPORTS: Reported in EPA TSCA Inventory.

SAFETY PROFILE: Poison by intraperitoneal route. Moderately toxic by ingestion. A skin irritant. Flammable when exposed to heat or flame; can react with oxidizing materials. See also ETHERS.

DOA800 CAS:20325-40-0 HR: 3
3,3′-DIMETHOXYBENZIDINE DIHYDROCHLORIDE
mf: $C_{14}H_{16}N_2O_2 \cdot 2ClH$ mw: 317.24

SYNS: C.I. DISPERSE BLACK 6 DIHYDROCHLORIDE ☐ o-DIANISIDINE DIHYDROCHLORIDE ☐ 3,3-DIMETHOXY-(1,1′-BIPHENYL)-4,4′-DIAMINE DIHYDROCHLORIDE

TOXICITY DATA WITH REFERENCE
mmo-sat 100 nmol/plate MUREAV 136,33,84
mma-sat 10 nmol/plate EMMUEG 10,263,87
orl-rat TDLo:1040 mg/kg/1Y-I:CAR,REP JNCIAM 41,985,68
orl-mus TDLo:5760 mg/kg/2Y-I:ETA VOONAW 25(7),43,79
scu-mus TDLo:1152 mg/kg/2Y-I:ETA,REP VOONAW 25(7),43,79
orl-rat TD:11 g/kg/51W-I:CAR JNCIAM 41,985,68
orl-rat LD:6497 mg/kg/91W-C:CAR NTPTR* NTP-TR-372,90

CONSENSUS REPORTS: NTP 7th Annual Report On Carcinogens. NTP Carcinogenesis Studies (Gavage); Clear Evidence: Rat NCITR* NTP-TR-372,90. Reported in EPA TSCA Inventory.

NIOSH REL: (Benzidine-Based Dye) Reduce to lowest feasible level

SAFETY PROFILE: Confirmed carcinogen with experimental carcinogenic and tumorigenic data. Experimental reproductive data. Mutation data reported. When heated to decomposition it emits very toxic fumes of NO_x and HCl.

DOA810 CAS:64466-47-3 HR: 3
4,7-DIMETHOXY-2-BENZOFURANYL METHYL KETONE
mf: $C_{12}H_{12}O_4$ mw: 220.24

SYNS: 1-(4,7-DIMETHOXY-2-BENZOFURANYL)ETHANONE ☐ ETHANONE, 1-(4,7-DIMETHOXY-2-BENZOFURANYL)-(9CI) ☐ KETONE, 4,7-DIMETHOXY-2-BENZOFURANYL METHYL

TOXICITY DATA WITH REFERENCE
ipr-mus LD50:900 mg/kg EJMCA5 12,383,77

DOT CLASSIFICATION: 3; *Label:* Flammable Liquid

SAFETY PROFILE: Moderately toxic by intraperitoneal route. A flammable liquid. When heated to decomposition it emits acrid smoke and irritating vapors.

DOA815 CAS:64466-48-4 HR: 3
6,7-DIMETHOXY-2-BENZOFURANYL METHYL KETONE
mf: $C_{12}H_{12}O_4$ mw: 220.24

SYNS: 1-(6,7-DIMETHOXY-2-BENZOFURANYL)ETHANONE ☐ ETHANONE, 1-(6,7-DIMETHOXY-2-BENZOFURANYL)-(9CI) ☐ KETONE, 6,7-DIMETHOXY-2-BENZOFURANYL METHYL

TOXICITY DATA WITH REFERENCE
ipr-mus LD50:1200 mg/kg EJMCA5 12,383,77

DOT CLASSIFICATION: 3; *Label:* Flammable Liquid

SAFETY PROFILE: Moderately toxic by intraperitoneal route. A flammable liquid. When heated to decomposition it emits acrid smoke and irritating vapors.

DOA820 CAS:64466-49-5 HR: 3
1-(5,8-DIMETHOXY-2H-1-BENZOPYRAN-3-YL)ETHANONE
mf: $C_{13}H_{14}O_4$ mw: 234.27

SYNS: 5,8-DIMETHOXY-2H-1-BENZOPYRAN-3-YL METHYL KETONE ☐ ETHANONE, 1-(5,8-DIMETHOXY-2H-1-BENZOPYRAN-3-YL)-(9CI) ☐ KETONE, 5,8-DIMETHOXY-2H-1-BENZOPYRAN-3-YL METHYL

TOXICITY DATA WITH REFERENCE
ipr-mus LD50:1500 mg/kg EJMCA5 12,383,77

DOT CLASSIFICATION: 3; *Label:* Flammable Liquid

SAFETY PROFILE: Moderately toxic by intraperitoneal route. A flammable liquid. When heated to decomposition it emits acrid smoke and irritating vapors.

DOA830 CAS:64466-50-8 HR: 3
1-(7,8-DIMETHOXY-2H-1-BENZOPYRAN-3-YL)ETHANONE
mf: $C_{13}H_{14}O_4$ mw: 234.27

SYNS: 7,8-DIMETHOXY-2H-1-BENZOPYRAN-3-YL METHYL KETONE ☐ ETHANONE, 1-(7,8-DIMETHOXY-2H-1-BENZOPYRAN-3-YL)-(9CI) ☐ KETONE, 7,8-DIMETHOXY-2H-1-BENZOPYRAN-3-YL METHYL

TOXICITY DATA WITH REFERENCE
ipr-mus LD50:1200 mg/kg EJMCA5 12,383,77

DOT CLASSIFICATION: 3; *Label:* Flammable Liquid

SAFETY PROFILE: Moderately toxic by intraperitoneal

route. A flammable liquid. When heated to decomposition it emits acrid smoke and irritating vapors.

DOA875 **HR: 3**
β-(2,4-DIMETHOXY-5-BENZYLBENZOYL)PROPIONIC ACID SODIUM SALT
mf: $C_{19}H_{19}O_5 \cdot Na$ mw: 350.34

SYNS: 3-(4,6-DIMETHOXY-α-PHENYL-m-TOLUOYL)-PROPIONIC ACID SODIUM SALT □ SC-2657

TOXICITY DATA WITH REFERENCE
orl-mus LD50:1140 mg/kg JPETAB 100,421,50
ipr-mus LD50:225 mg/kg JPETAB 100,421,50
ivn-dog LDLo:113 mg/kg JPETAB 100,421,50

SAFETY PROFILE: Poison by intravenous and intraperitoneal routes. Moderately toxic by ingestion. When heated to decomposition it emits toxic fumes of Na_2O.

DOB200 **CAS:10143-66-5** **HR: 2**
1,3-DIMETHOXYBUTANE
mf: $C_6H_{14}O_2$ mw: 118.20

TOXICITY DATA WITH REFERENCE
skn-rbt 10 mg/24H open MLD AIHAAP 23,95,62
orl-rat LD50:3730 mg/kg AIHAAP 23,95,62
ihl-rat LCLo:8000 ppm AIHAAP 23,95,62
skn-rbt LD50:10 g/kg AIHAAP 23,95,62

SAFETY PROFILE: Moderately toxic by ingestion. Mildly toxic by skin contact. A skin irritant. When heated to decomposition it emits acrid smoke and irritating fumes.

DOB275 **HR: 3**
β-(2,4-DIMETHOXY-5-CYCLOHEXYLBENZOYL)PROPIONIC ACID
mf: $C_{18}H_{24}O_5$ mw: 320.42

TOXICITY DATA WITH REFERENCE
ivn-rat LD50:300 mg/kg AIPTAK 116,154,58
orl-mus LD50:400 mg/kg AIPTAK 116,154,58
scu-mus LD50:300 mg/kg AIPTAK 116,154,58

SAFETY PROFILE: Poison by ingestion, subcutaneous, and intravenous routes. When heated to decomposition it emits acrid smoke and fumes.

DOB300 **HR: 3**
β-(2,4-DIMETHOXY-5-CYCLOHEXYLBENZOYL)PROPIONIC ACID SODIUM SALT
mf: $C_{18}H_{23}O_5 \cdot Na$ mw: 342.36

SYN: SC-2644

TOXICITY DATA WITH REFERENCE
orl-mus LD50:1020 mg/kg JPETAB 100,421,50
ipr-mus LD50:214 mg/kg JPETAB 100,421,50
ivn-dog LDLo:107 mg/kg JPETAB 100,421,50

SAFETY PROFILE: Poison by intravenous and intraperitoneal routes. Moderately toxic by ingestion. When heated to decomposition it emits toxic fumes of Na_2O.

DOB325 **HR: 3**
β-(2,4-DIMETHOXY-5-CYCLOPENTYLMETHYLBENZOYL)PROPIONIC ACID SODIUM SALT
mf: $C_{18}H_{23}O_5 \cdot Na$ mw: 342.36

SYNS: 3-(α-CYCLOPENTYL-4,6-DIMETHOXY-m-TOLUOYL)-PROPIONIC ACID SODIUM SALT □ SC-2798

TOXICITY DATA WITH REFERENCE
orl-mus LD50:1060 mg/kg JPETAB 100,421,50
ipr-mus LD50:148 mg/kg JPETAB 100,421,50
ivn-dog LDLo:74 mg/kg JPETAB 100,421,50

SAFETY PROFILE: Poison by intravenous and intraperitoneal routes. Moderately toxic by ingestion. When heated to decomposition it emits toxic fumes of Na_2O.

DOB600 **CAS:63040-49-3** **HR: 2**
5,6-DIMETHOXYDIBENZ(a,h)ANTHRACENE
mf: $C_{24}H_{18}O_2$ mw: 338.42

SYNS: 3,4-DIMETHOXY-DBA □ 3,4-DIMETHOXY-1,2:5,6-DIBENZANTHRACENE

TOXICITY DATA WITH REFERENCE
skn-mus TDLo:600 mg/kg/60W-I:NEO CNREA8 22,78,62
scu-mus TDLo:80 mg/kg/4W-I:ETA CNREA8 22,78,62

SAFETY PROFILE: Questionable carcinogen with experimental neoplastigenic and tumorigenic data. When heated to decomposition it emits acrid smoke and irritating fumes.

DOC000 **CAS:41226-20-4** **HR: 3**
4,7-DIMETHOXY-6-(2-DIISOPROPYLAMINOETHOXY)-5-(p-METHOXYCINNAMOYL) BENZOFURAN OXALATE
mf: $C_{28}H_{35}NO_6 \cdot C_2H_2O_4$ mw: 571.68

TOXICITY DATA WITH REFERENCE
orl-mus LD50:185 mg/kg CHTPBA 8,479,73
ivn-mus LD50:13 mg/kg CHTPBA 8,479,73

SAFETY PROFILE: Poison by ingestion and intravenous routes. When heated to decomposition it emits toxic fumes of NO_x.

DOC800 **CAS:26270-60-0** **HR: 3**
4,7-DIMETHOXY-6-(2-DIMETHYLAMINOETHOXY)-5-(p-FLUOROCINNAMOYL)BENZOFURAN MALEATE
mf: $C_{23}H_{25}FNO_5 \cdot C_4H_4O_4$ mw: 530.57

TOXICITY DATA WITH REFERENCE
orl-mus LD50:430 mg/kg CHTPBA 8,479,73
ivn-mus LD50:14 mg/kg CHTPBA 8,479,73

SAFETY PROFILE: Poison by intravenous route. Moderately toxic by ingestion. When heated to decomposition it emits very toxic fumes of NO_x and F^-.

DOD200 CAS:52171-41-2 HR: 3
4-,7-DIMETHOXY-6-(2-DIMETHYLAMINOETHOXY)-5-(p-HYDROXYCINNAMOYL)BENZOFURAN OXALATE
mf: $C_{23}H_{25}NO_6 \cdot C_2H_2O_4$ mw: 501.53

TOXICITY DATA WITH REFERENCE
orl-mus LD50:640 mg/kg CHTPBA 8,479,73
ivn-mus LD50:35 mg/kg CHTPBA 8,479,73

SAFETY PROFILE: Poison by intravenous route. Moderately toxic by ingestion. When heated to decomposition it emits toxic fumes of NO_x.

DOD400 CAS:52171-37-6 HR: 3
4,7-DIMETHOXY-6-(2-DIMETHYLAMINOETHOXY)-5-(p-ISOPROPOXYCINNAMOYL)BENZOFURAN MALEATE
mf: $C_{26}H_{31}NO_6 \cdot C_4H_4O_4$ mw: 569.66

TOXICITY DATA WITH REFERENCE
orl-mus LD50:300 mg/kg CHTPBA 8,479,73
ivn-mus LD50:27 mg/kg CHTPBA 8,479,73

SAFETY PROFILE: Poison by ingestion and intravenous routes. When heated to decomposition it emits toxic fumes of NO_x.

DOD600 CAS:26270-59-7 HR: 3
4,7-DIMETHOXY-6-(2-DIMETHYLAMINOETHOXY)-5-(p-METHOXYCINNAMOYL)BENZOFURAN MALEATE
mf: $C_{24}H_{27}NO_6 \cdot C_4H_4O_4$ mw: 541.60

PROP: A solid. Mp: 128–129°.

TOXICITY DATA WITH REFERENCE
orl-mus LD50:255 mg/kg CHTPBA 8,479,73
ivn-mus LD50:25 mg/kg CHTPBA 8,479,73

SAFETY PROFILE: Poison by ingestion and intravenous routes. When heated to decomposition it emits toxic fumes of NO_x.

DOE000 CAS:7549-37-3 HR: 1
1,1-DIMETHOXY-3,7-DIMETHYL-2,6-OCTADIENE (cis and trans)
mf: $C_{12}H_{22}O_2$ mw: 198.34

SYNS: CITRAL DIMETHYL ACETAL □ 1,1-DIMETHOXY-3,7-DIMETHYL-2,6-OCTADIENE

TOXICITY DATA WITH REFERENCE
skn-rbt 500 mg/24H FCTXAV 11,1065,73

CONSENSUS REPORTS: Reported in EPA TSCA Inventory.

SAFETY PROFILE: A skin irritant. When heated to decomposition it emits acrid smoke and irritating fumes.

DOE200 CAS:120-20-7 HR: 3
3,4-DIMETHOXYDOPAMINE
mf: $C_{10}H_{15}NO_2$ mw: 181.26

PROP: Crystals from C_6H_6/pet ether. Mp: 124°, bp: 188° @ 15 mm, d: 1.08 @ 28°/4°, vap d: 6.25.

SYNS: DIMETHYOXYDOPAMINE □ 3,4-DIMETHOXYPHENETHYLAMINE □ 3,4-DIMETHOXY-β-PHENETHYLAMINE □ DIMETHOXYPHENYLETHYLAMINE □ 3,4-DIMETHOXYPHENYLETHYLAMINE □ 3,4-DIMETHOXY-β-PHENYLETHYLAMINE □ β-(3,4-DIMETHOXYPHENYL) ETHYLAMINE □ 2-(3,4-DIMETHOXYPHENYL)ETHYLAMINE □ 3,4-DIMETHOXYPHENYLETHYLAMINE (base) □ DIMETHYLMESCALINE □ DIMPEA □ DMPE □ DMPEA □ HOMOVERATRYLAMINE

TOXICITY DATA WITH REFERENCE
ipr-mus LD50:181 mg/kg YKKZAJ 97,1117,77
ivn-mus LD50:56 mg/kg CSLNX* NX#04483

CONSENSUS REPORTS: Reported in EPA TSCA Inventory.

SAFETY PROFILE: Poison by intravenous and intraperitoneal routes. Moderately toxic by intraperitoneal route. When heated to decomposition it emits toxic fumes of NO_x.

DOE600 CAS:110-71-4 HR: 3
1,2-DIMETHOXYETHANE
DOT: UN 2252
mf: $C_4H_{10}O_2$ mw: 90.14

$CH_3OC_2H_4OCH_3$

PROP: Liquid; sharp, ethereal odor. D: 0.86877, mp: −58°, bp: 82–83°, n (24/D) 1.3739, flash p: 4.5°C (40°F). Miscible with water and alc; sol in hydrocarbon solvents.

SYNS: DIMETHOXYETHANE □ α,β-DIMETHOXYETHANE □ 1,2-DIMETHOXYETHANE (DOT) □ DIMETHYLCELLOSOLVE □ 2,5-DIOXAHEXANE □ EGDME □ ETHYLENE DIMETHYL ETHER □ ETHYLENE GLYCOL DIMETHYL ETHER □ GLYCOL DIMETHYL ETHER □ GLYME □ MONOETHYLENE GLYCOL DIMETHYL ETHER □ MONOGLYME

TOXICITY DATA WITH REFERENCE
orl-mus TDLo:1960 mg/kg (7-10D preg):TER NISFAY 32,113,80
orl-mus TDLo:361 mg/kg (female 11D post):TER TJADAB 35,321,87
orl-mus TDLo:16 g/kg (female 7-14D post):REP EVHPAZ 57,141,84
orl-mus TDLo:1400 mg/kg (female 7-10D post):TER NISFAY 32,113,80

CONSENSUS REPORTS: Reported in EPA TSCA Inventory. Glycol ether compounds are on the Community Right-To-Know List.

DOT CLASSIFICATION: 3; *Label:* Flammable Liquid

SAFETY PROFILE: An experimental teratogen. Other experimental reproductive effects. Readily forms an explosive peroxide. A very dangerous fire hazard when exposed to heat, flame, or oxidizers. Mixture with lithium tetrahydroaluminate may ignite or explode if heated. When heated to decomposition it emits acrid smoke and fumes. See also GLYCOL ETHERS.

DOF000 CAS:10232-93-6 HR: 2
DI(2-METHOXYETHYL) MALEATE
mf: $C_{10}H_{16}O_6$ mw: 232.26

SYN: BIS(2-METHOXYETHYL)ESTER MALEIC ACID

TOXICITY DATA WITH REFERENCE
skn-rbt 10 mg/24H open MLD AMIHBC 4,119,51
eye-rbt 100 mg open AMIHBC 4,119,51
orl-rat LD50:3340 mg/kg AMIHBC 4,119,51
skn-rbt LD50:1940 mg/kg AMIHBC 4,119,51

SAFETY PROFILE: Moderately toxic by ingestion and skin contact. A skin and eye irritant. When heated to decomposition it emits acrid smoke and irritating fumes.

DOF400 CAS:117-82-8 HR: 3
DIMETHOXY ETHYL PHTHALATE
mf: $C_{14}H_{18}O_6$ mw: 282.32

PROP: Light-colored, clear liquid; mild aromatic odor. Mp: −40° (forms gel), bp: 190–210° @ 4 mm, flash p: 360°F, d: 1.171 @ 20°/20°, vap press: 0.3 mm @ 150°, vap d: 9.75.

SYNS: 1,2-BENZENEDICARBOXYLIC ACID BI(2-METHOXYETHYL) ESTER (9CI) □ BIS(METHOXYETHYL) PHTHALATE □ BIS(2-METHOXYETHYL) PHTHALATE □ DI(2-METHOXYETHYL)PHTHALATE □ DMEP □ KESSCOFLEX MCP □ 2-METHOXYETHYL PHTHALATE □ PHTHALIC ACID BIS(2-METHOXYETHYL) ESTER

TOXICITY DATA WITH REFERENCE
eye-rbt 100 mg MLD KODAK* #902511
skn-gpg 500 mg MLD KODAK* #902511
spm-rat-orl 1500 mg/kg ARTODN 53,71,83
dlt-mus-ipr 1190 mg/kg TXAPA9 29,35,74
ipr-mus TDLo:2380 mg/kg (male 1D pre):TER EVHPAZ 3,81,73
ipr-rat TDLo:374 mg/kg (female 5-15D post):TER JPMSAE 61,51,72
ipr-mus TDLo:2380 mg/kg (male 1D pre):REP EVHPAZ 3,81,73
ipr-mus TDLo:2380 µg/kg (male 1D pre):REP TXAPA9 29,35,74
ipr-rat TDLo:374 mg/kg (female 5-15D post):REP JPMSAE 61,51,72
orl-rat TDLo:1500 mg/kg (1D male):REP TJADAB 26(3),14A,82
orl-rat TDLo:593 mg/kg (female 10D post):TER TJADAB 33,93C,86
orl-rat TDLo:593 mg/kg (female 13D post):TER TJADAB 33,93C,86
ipr-rat TDLo:747 mg/kg (female 5-15D post):TER JPMSAE 61,51,72
orl-rat LDLo:2750 mg/kg 29ZWAE -,356,68
ihl-rat LCLo:1595 ppm/16H 14CYAT 2,1904,63
ipr-rat LD50:3735 mg/kg JPMSAE 61,51,72
ipr-mus LD50:2510 mg/kg 34ZIAG -,691,69
orl-gpg LD50:1600 mg/kg 14CYAT 2,1904,63

CONSENSUS REPORTS: EPA Genetic Toxicology Program. Reported in EPA TSCA Inventory.

SAFETY PROFILE: Moderately toxic by ingestion and intraperitoneal routes. Mildly toxic by inhalation. Experimental teratogenic and reproductive effects. A skin and eye irritant. Mutation data reported. A pesticide. Combustible when exposed to heat or flame; can react with oxidizing materials. To fight fire, use water, foam, CO_2, dry chemical. When heated to decomposition it emits acrid smoke and irritating fumes.

DOF600 CAS:134-96-3 HR: 2
3,5-DIMETHOXY-4-HYDROXYBENZALDEHYDE
mf: $C_9H_{10}O_4$ mw: 182.19

PROP: A solid. Mp: 113–114°, bp: 192–193° @ 14 mm.

SYNS: GALLALDEHYDE-3,5-DIMETHYL ETHER □ SYRINGALDEHYDE □ SYRINGEALDEHYDE □ SYRINGIC ALDEHYDE □ SYRINGYLALDEHYDE

TOXICITY DATA WITH REFERENCE
ipr-mus LD50:1000 mg/kg JMCMAR 7,178,64

CONSENSUS REPORTS: Reported in EPA TSCA Inventory.

SAFETY PROFILE: Moderately toxic by intraperitoneal route. When heated to decomposition it emits acrid smoke and irritating fumes. See also ALDEHYDES.

DOF800 CAS:52171-42-3 HR: 3
4,7-DIMETHOXY-5-(p-HYDROXYCINNAMOYL)-6-(2-PYRROLIDINYLETHOXY)BENZOFURAN MALEATE
mf: $C_{25}H_{27}NO_6•C_4H_4O_4$ mw: 553.61

TOXICITY DATA WITH REFERENCE
orl-mus LD50:1700 mg/kg CHTPBA 8,479,73
ivn-mus LD50:40 mg/kg CHTPBA 8,479,73

SAFETY PROFILE: Poison by intravenous route. Moderately toxic by ingestion. When heated to decomposition it emits toxic fumes of NO_x.

DOG600 CAS:15589-00-1 HR: 3
2,5-DIMETHOXY-4-METHYLAMPHETAMINE HYDROCHLORIDE
mf: $C_{12}H_{19}NO_2•ClH$ mw: 245.78

SYNS: 2,5-DIMETHOXY-α,4-DIMETHYLPHENETHYLAMINE HYDROCHLORIDE □ 1-(2,5-DIMETHOXY-4-METHYLPHENYL)-2-AMINOPROPANE

TOXICITY DATA WITH REFERENCE
ipr-rat LD50:32,500 µg/kg TXAPA9 45(1),49,78
orl-mus LD50:330 mg/kg TXAPA9 45(1),49,78
ipr-mus LD50:89 mg/kg JMCMAR 13,26,70
ivn-mus LD50:36 mg/kg TXAPA9 45(1),49,78
ivn-dog LD50:7200 µg/kg TXAPA9 45(1),49,78

SAFETY PROFILE: Poison by ingestion, intraperitoneal, and intravenous routes. A central nervous system stimulant. When heated to decomposition it emits very toxic fumes of NO_x and HCl. See also BENZEDRINE.

DOG700 CAS:1125-88-8 HR: 2
DIMETHOXYMETHYLBENZENE
mf: $C_9H_{12}O_2$ mw: 152.21

SYNS: BENZALDEHYDE, DIMETHYL ACETAL □ DIMETHOXYPHENYLMETHANE

TOXICITY DATA WITH REFERENCE
skn-rbt 500 mg/24H MLD FCTXAV 17,711,79
orl-rat LD50:1220 mg/kg FCTXAV 17,711,79

CONSENSUS REPORTS: Reported in EPA TSCA Inventory.

SAFETY PROFILE: Moderately toxic by ingestion. A skin

irritant. When heated to decomposition it emits acrid smoke and irritating fumes.

DOH400 CAS:3027-21-2 HR: 2
DIMETHOXYMETHYLPHENYLSILANE
mf: $C_9H_{14}O_2Si$ mw: 182.3

PROP: Bp: 96° @ 21 mm.

SYNS: DIMETHOXYPHENYLMETHYLSILANE ☐ METHYLPHENYLDIMETHOXYSILANE ☐ PHENYLMETHYLDIMETHOXYSILANE

TOXICITY DATA WITH REFERENCE
orl-rat LD50:892 mg/kg GTPZAB 22(2),50,78

CONSENSUS REPORTS: Reported in EPA TSCA Inventory.

SAFETY PROFILE: Moderately toxic by ingestion. When heated to decomposition it emits acrid smoke and irritating fumes.

DOI400 CAS:635-85-8 HR: 3
3,4-DIMETHOXYPHENETHYLAMINE HYDROCHLORIDE
mf: $C_{10}H_{15}NO_2 \cdot ClH$ mw: 217.72

PROP: A solid. Mp: 154–155°.

SYN: 3,4-DIMETHOXY-β-PHENYLETHYLAMINE HYDROCHLORIDE

TOXICITY DATA WITH REFERENCE
ipr-rat LD50:146 mg/kg TXAPA9 25,299,73
ipr-mus LD50:363 mg/kg TXAPA9 25,299,73
ivn-dog LD50:122 mg/kg TXAPA9 25,299,73
ivn-mky LD50:220 mg/kg TXAPA9 25,299,73
ipr-gpg LD50:375 mg/kg TXAPA9 25,299,73

SAFETY PROFILE: Poison by intraperitoneal and intravenous routes. When heated to decomposition it emits very toxic fumes of NO_x and HCl.

DOJ200 CAS:91-10-1 HR: 3
2,6-DIMETHOXYPHENOL
mf: $C_8H_{10}O_3$ mw: 154.18

PROP: A solid. Mp: 55–56°, bp: 262–267°. Sltly sol in H_2O.

SYNS: ALDRICH ☐ 1,3-DIMETHYL PYROGALLATE ☐ PYROGALLOL DIMETHYLETHER ☐ PYROGALLOL-1,3-DIMETHYL ETHER ☐ SYRINGOL

TOXICITY DATA WITH REFERENCE
orl-rat LD50:550 mg/kg FOMDAK 32,309,91
orl-mus LD50:2500 mg/kg BCTKAG 14,301,84
ivn-cat LDLo:100 mg/kg BJPCBM 53,93,75

CONSENSUS REPORTS: Reported in EPA TSCA Inventory.

SAFETY PROFILE: Poison by intravenous route. Moderately toxic by ingestion. When heated to decomposition it emits acrid smoke and irritating fumes. See also ETHERS.

DOJ400 CAS:69782-26-9 HR: 3
(2-(2,5-DIMETHOXYPHENOXY)ETHYL)HYDRAZINE HYDROCHLORIDE
mf: $C_{10}H_{16}N_2O_3 \cdot ClH$ mw: 248.74

TOXICITY DATA WITH REFERENCE
orl-mus LD50:125 mg/kg JMCMAR 6,63,63
ipr-mus LD50:125 mg/kg JMCMAR 6,63,63

SAFETY PROFILE: Poison by ingestion and intraperitoneal routes. When heated to decomposition it emits very toxic fumes of Cl^- and NO_x. See also HYDRAZINE.

DOJ600 CAS:69782-18-9 HR: 3
(2-(3,4-DIMETHOXYPHENOXY)ETHYL)HYDRAZINE HYDROCHLORIDE
mf: $C_{10}H_{16}N_2O_3 \cdot ClH$ mw: 248.74

TOXICITY DATA WITH REFERENCE
orl-mus LD50:90 mg/kg JMCMAR 6,63,63
ipr-mus LD50:90 mg/kg JMCMAR 6,63,63

SAFETY PROFILE: Poison by ingestion and intraperitoneal routes. When heated to decomposition it emits very toxic fumes of Cl^- and NO_x. See also HYDRAZINE.

DOJ700 CAS:27318-87-2 HR: 2
3-(3,5-DIMETHOXYPHENOXY)-1,2-PROPANEDIOL
mf: $C_{11}H_{16}O_5$ mw: 228.27

SYN: 3-(3′,5′-DIMETHOXYPHENOXY)PROPANEDIOL-(1,2)

TOXICITY DATA WITH REFERENCE
orl-rat LD50:2100 mg/kg ARZNAD 24,111,74
ipr-rat LD50:955 mg/kg ARZNAD 24,111,74
orl-mus LD50:2070 mg/kg ARZNAD 24,111,74
ipr-mus LD50:780 mg/kg ARZNAD 24,111,74

SAFETY PROFILE: Moderately toxic by ingestion and intraperitoneal routes. When heated to decomposition it emits acrid smoke and fumes.

DOJ800 CAS:24973-25-9 HR: 3
1-(2,5-DIMETHOXYPHENYL)-2-AMINOPROPANE
mf: $C_{11}H_{17}NO_2 \cdot ClH$ mw: 231.75

SYNS: 2,5-DIMETHOXYAMPHETAMINE HYDROCHLORIDE ☐ 2,5-DIMETHOXY-α-METHYLBENZENEETHANAMINE HYDROCHLORIDE ☐ 2,5-DIMETHOXY-α-METHYLPHENETHYLAMINE HYDROCHLORIDE ☐ 2,5-DIMETHOXY-α-METHYL-β-PHENYLETHYLAMINE HYDROCHLORIDE ☐ β-(2,5-DIMETHOXYPHENYL)ISOPROPYLAMINE HYDROCHLORIDE

TOXICITY DATA WITH REFERENCE
mmo-sat 10 mg/plate MUREAV 56,199,77
ipr-rat LD50:63 mg/kg TXAPA9 45,49,78
ipr-mus LD50:135 mg/kg JMCMAR 13,26,70
ivn-mus LD50:39 mg/kg TXAPA9 45,49,78
ivn-dog LD50:26 mg/kg TXAPA9 45,49,78

SAFETY PROFILE: Poison by intraperitoneal and intravenous routes. Mutation data reported. When heated to decomposition it emits very toxic fumes of NO_x and HCl.

DOK000 CAS:13078-75-6 **HR: 3**
1-(3,4-DIMETHOXYPHENYL)-2-AMINOPROPANE
mf: $C_{11}H_{17}NO_2$•ClH mw: 231.75

SYNS: 3,4-DIMETHOXYAMPHETAMINE HYDROCHLORIDE ☐ 3,4-DIMETHOXY-α-METHYL-β-PHENYLETHYLAMINEHYDROCHLORIDE

TOXICITY DATA WITH REFERENCE
ipr-rat LD50:48 mg/kg TXAPA9 25,299,73
ipr-mus LD50:168 mg/kg TXAPA9 25,299,73
ivn-dog LD50:59 mg/kg TXAPA9 25,299,73
ivn-mky LD50:53 mg/kg TXAPA9 25,299,73
ipr-gpg LD50:195 mg/kg TXAPA9 25,299,73

SAFETY PROFILE: Poison by intraperitoneal and intravenous routes. When heated to decomposition it emits very toxic fumes of NO_x and HCl.

DOK200 CAS:6358-53-8 **HR: 3**
1-((2,5-DIMETHOXYPHENYL)AZO)-2-NAPHTHOL
mf: $C_{18}H_{16}N_2O_3$ mw: 308.36

PROP: Mp: 156°. Sltly water-sol; mod sol in alc.

SYNS: C.I. 12156 ☐ C.I. SOLVENT RED 80 ☐ CITRUS RED No. 2 ☐ 2,5-DIMETHOXYBENZENEAZO-β-NAPHTHOL ☐ 1-((2,5-DIMETHOXYPHENYL)AZO)-2-NAPHTHALENOL ☐ 2,5-DIMETHOXY-1-(PHENYLAZO)-2-NAPHTHOL ☐ 1-(1-(2,5-DIMETHOXYPHENYL)AZO)-2-NAPHTHOL ☐ 1-(2,5-DIMETHYLOXYPHENYLAZO)-2-NAPHTHOL

TOXICITY DATA WITH REFERENCE
mmo-sat 500 µg/plate MUREAV 56,249,78
scu-mus TDLo:20 g/kg/80W-C:CAR FCTXAV 4,493,66
imp-mus TDLo:80 mg/kg:CAR BJCAAI 22,825,68

CONSENSUS REPORTS: IARC Cancer Review: Group 2B IMEMDT 7,56,87; Animal Sufficient Evidence IMEMDT 8,101,75. EPA Genetic Toxicology Program.

SAFETY PROFILE: Confirmed carcinogen with experimental carcinogenic data. Mutation data reported. When heated to decomposition it emits toxic fumes of NO_x.

DOK400 CAS:26011-83-6 **HR: 3**
3-(2,4-DIMETHOXYPHENYL)CROTONIC ACID MAGNESIUM SALT
mf: $C_{24}H_{26}O_8$•Mg mw: 466.81

SYNS: DIMECROTIC ACID MAGNESIUM SALT ☐ 2,4-DIMETHOXY-β-METHYLCINNAMIC ACID MAGNESIUM SALT ☐ HEPADIAL

TOXICITY DATA WITH REFERENCE
ipr-rat LD50:1000 µg/kg MEIEDD 10,466,83
ipr-mus LD50:1300 µg/kg MEIEDD 10,466,83

SAFETY PROFILE: A deadly poison by intraperitoneal route. When heated to decomposition it emits acrid smoke and irritating fumes. Stimulates the production of bile by the liver. See also MAGNESIUM COMPOUNDS.

DOK600 CAS:2801-68-5 **HR: 3**
2-(2,5-DIMETHOXYPHENYL)ISOPROPYLAMINE
mf: $C_{11}H_{17}NO_2$ mw: 211.29

SYN: C 1739

TOXICITY DATA WITH REFERENCE
ipr-rat LD50:170 mg/kg ARZNAD 8,708,58
scu-mus LD50:375 mg/kg ARZNAD 8,708,58

SAFETY PROFILE: Poison by intraperitoneal and subcutaneous routes. When heated to decomposition it emits toxic fumes of NO_x.

DOL400 CAS:64050-54-0 **HR: 3**
2-((DIMETHOXYPHOSPHINYL)OXY)-1H-BENZ(d,e)ISOQUINOLINE-1,3(2H)-DIONE
mf: $C_{14}H_{12}NO_6P$ mw: 321.24

SYN: PHOSPHORIC ACID, DIMETHYL ESTER, ester with N-HYDROXYNAPHTHALIMIDE

TOXICITY DATA WITH REFERENCE
orl-rat LD50:70 mg/kg TXAPA9 21,315,72
orl-bwd LD50:2400 µg/kg TXAPA9 21,315,72

SAFETY PROFILE: Poison by ingestion. When heated to decomposition it emits very toxic fumes of PO_x and NO_x. See also ESTERS.

DOL800 CAS:25601-84-7 **HR: 3**
3-(DIMETHOXYPHOSPHINYLOXY)-N-METHYL-N-METHOXY-cis-CROTONAMIDE
mf: $C_8H_{16}NO_6P$ mw: 253.22

SYNS: CIBA C-2307 ☐ ENT 27,625 ☐ 3-HYDROXY N METHOXY-N-METHYL-cis-CROTONAMIDE, DIMETHYL PHOSPHATE ☐ METHOCROTOPHOS ☐ (E)-(3-(METHOXYMETHYLAMINO)-1-METHYL-3-OXO-1-PROPENYL)DIMETHYL PHOSPHATE ☐ NSC 195154

TOXICITY DATA WITH REFERENCE
orl-rat LD50:2 mg/kg BESAAT 15,107,69
orl-mus LD50:2 mg/kg BESAAT 15,107,69
orl-dog LD50:7 mg/kg BESAAT 15,107,69
orl-rbt LD50:11 mg/kg BESAAT 15,107,69
skn-rbt LD50:107 mg/kg BESAAT 15,107,69

SAFETY PROFILE: A poison by ingestion and skin contact. When heated to decomposition it emits very toxic fumes of NO_x and PO_x.

DOM200 CAS:4744-10-9 **HR: 3**
1,1-DIMETHOXYPROPANE
mf: $C_5H_{12}O_2$ mw: 104.15

$(CH_3O)_2CHCH_2CH_3$

PROP: A liquid. Flash p: 50°F, bp: 89°.

SAFETY PROFILE: A very dangerous fire hazard when exposed to heat, flame, or oxidizers. When heated to decomposition it emits acrid smoke and fumes.

DOM400 CAS:77-76-9 **HR: 3**
2,2-DIMETHOXYPROPANE
mf: $C_5H_{12}O_2$ mw: 104.15

$(CH_3O)_2C(CH_3)_2$

PROP: A liquid. Flash p: 19.4°F, bp: 83° @ 20 mm.

SAFETY PROFILE: A very dangerous fire hazard when exposed to heat, flame, or oxidizers. When heated to 210°C it burns with a cool flame and then explodes. Explosive reaction with metal perchlorates (e.g., manga-

D

nese(II) perchlorate; nickel(II) perchlorate) above 65°C. When heated to decomposition it emits acrid smoke and fumes.

DOM600 CAS:6044-68-4 HR: 6
3,3-DIMETHOXYPROPENE
mf: $C_5H_{10}O_2$ mw: 102.14

$H_2C{=}CHCH(OCH_3)_2$

PROP: A liquid. Flash p: 66.2°F, bp: 40° @ 120 mm.

SAFETY PROFILE: A very dangerous fire hazard when exposed to heat, flame oxidizers. May form dangerous peroxides upon exposure to air. When heated to decomposition it emits acrid smoke and fumes. See also ALLYL COMPOUNDS.

DON000 CAS:26270-61-1 HR: 3
4,7-DIMETHOXY-6-(2-PYRROLIDINYLETHOXY)-5-CINNAMOYLBENZOFURAN MALEATE
mf: $C_{25}H_{27}NO_5 \cdot C_4H_4O_4$ mw: 537.61

TOXICITY DATA WITH REFERENCE
orl-mus LD50:250 mg/kg CHTPBA 8,479,73
ivn-mus LD50:10 mg/kg CHTPBA 8,479,73

SAFETY PROFILE: Poison by ingestion and intravenous routes. When heated to decomposition it emits toxic fumes of NO_x.

DON200 CAS:15233-65-5 HR: 3
2,6-DIMETHOXYQUINOL
mf: $C_8H_{10}O_4$ mw: 170.18

SYNS: 2,6-DIMETHOXY-1,4-BENZENEDIOL □ 2,6-DIMETHOXYHYDROQUINONE □ 3,5-DIMETHOXYHYDROQUINONE

TOXICITY DATA WITH REFERENCE
ivn-mus LD50:35 mg/kg BJPCBM 53,93,75
ivn-cat LDLo:30 mg/kg BJPCBM 53,93,75

SAFETY PROFILE: Poison by intravenous route. When heated to decomposition it emits acrid smoke and irritating fumes.

DON400 CAS:23435-31-6 HR: 3
2′,5′-DIMETHOXYSTILBENAMINE
mf: $C_{16}H_{17}NO_2$ mw: 255.34

SYNS: (trans)-2,5-DIMETHOXY-4′-AMINOSTILBENE □ 4-(2,5-DIMETHOXYPHENETHYL)ANILINE □ 4-(2-(2,5-DIMETHOXYPHENYL)ETHYL)BENZENAMINE □ 4-(2,5-DIMETHOXY)STILBENAMINE □ 2,5-DIMETHOXY-4′-STILBENAMINE

TOXICITY DATA WITH REFERENCE
orl-rat TDLo:2360 mg/kg/78W-C:CAR JEPTDQ 2,325,78
orl-mus TDLo:130 g/kg/78W-C:CAR JEPTDQ 2,325,78
orl-rat TD:4725 mg/kg/78W-C:CAR JEPTDQ 2,325,78

SAFETY PROFILE: Suspected carcinogen with experimental carcinogenic data. When heated to decomposition it emits toxic fumes of NO_x.

DON700 CAS:1230-33-7 HR: 2
3,6-DIMETHOXY-4-SULFANILAMIDOPYRIDAZINE
mf: $C_{12}H_{14}N_4O_4S$ mw: 310.36

SYNS: CS-61 □ N-(3,6-DIMETHOXY-4-PYRIDAZINYL)SULFANILAMIDE □ 4-SULFANILAMIDO-3,6-DIMETHOXYPYRIDAZINE

TOXICITY DATA WITH REFERENCE
orl-rat TDLo:3 g/kg (9-14D preg):TER TXAPA9 27,20,74
orl-rat TDLo:4500 mg/kg (9-14D preg):REP TXAPA9 27,20,74
orl-mus TDLo:6 g/kg (7-12D preg):TER TXAPA9 27,20,74
orl-mus LD50:2050 mg/kg ARZNAD 15,1441,65

SAFETY PROFILE: Moderately toxic by ingestion. Experimental teratogenic and reproductive effects. When heated to decomposition it emits toxic fumes of SO_x and NO_x.

DON800 CAS:696-59-3 HR: 3
2,5-DIMETHOXYTETRAHYDROFURAN
mf: $C_6H_{12}O_3$ mw: 131.16

$OCH(OCH_3)C_2H_4CHOCH_3$ (ring)

PROP: Colorless liquid. Bp: 145–147, vap d: 4.56, flash p: <50°F, d: 1.023 @ 20°. Misc in EtOH, C_6H_6, Me_2CO, Et_2O.

SAFETY PROFILE: A very dangerous fire hazard when exposed to heat, flame, or oxidizers. When heated to decomposition it emits acrid smoke and fumes. See also TETRAHYDROFURAN.

DOO400 CAS:6483-64-3 HR: 2
3,3′-DIMETHOXYTRIPHENYLMETHANE-4,4′-BIS(1″-AZO-2″-NAPHTHOL)
mf: $C_{41}H_{32}N_4O_4$ mw: 644.77

SYN: 1,1′-(BENZYLIDENEBIS((2-METHOXY-p-PHENYLENE))(AZO))DI-2-NAPHTHOL

TOXICITY DATA WITH REFERENCE
orl-rat TDLo:40 g/kg/83W-C:ETA ZEKBAI 57,530,51

CONSENSUS REPORTS: Reported in EPA TSCA Inventory.

SAFETY PROFILE: Questionable carcinogen with experimental tumorigenic data. When heated to decomposition it emits toxic fumes of NO_x.

DOO600 CAS:534-15-6 HR: 3
DIMETHYLACETAL
DOT: UN 2377
mf: $C_4H_{10}O_2$ mw: 90.14

PROP: Colorless liquid; strong aromatic odor. Bp: 64.5°, flash p: 34°F, d: 0.848 @ 25°, vap d: 3.1.

SYNS: ACETALDEHYDE DIMETHYL ACETAL □ 1,1-DIMETHOXYETHANE (DOT) □ DIMETHYL ALDEHYDE □ ETHYLIDENE DIMETHYL ETHER □ METHYL FORMYL

TOXICITY DATA WITH REFERENCE
skn-rbt 10 mg/24H open MLD JIHTAB 31,60,49

eye-rbt 20 mg open JIHTAB 31,60,49
orl-rat LD50:6500 mg/kg JIHTAB 31,60,49
ihl-rat LC50:3000 ppm/4H AMIHAB 12,623,55
orl-rbt LD50:4507 mg/kg PSEBAA 29,730,32
skn-rbt LD50:20 g/kg JIHTAB 31,60,49

CONSENSUS REPORTS: Glycol ether compounds are on the Community Right-To-Know List. Reported in EPA TSCA Inventory.

DOT CLASSIFICATION: 3; *Label:* Flammable Liquid

SAFETY PROFILE: Mildly toxic by inhalation, ingestion, and skin contact. A skin and eye irritant. A very dangerous fire hazard when exposed to heat, flame, or oxidizers. When exposed to heat or flame it can react vigorously with oxidizing materials. To fight fire, use foam, CO_2, dry chemical. When heated to decomposition it emits acrid smoke and irritating fumes. See also GLYCOL ETHERS.

DOO800 CAS:127-19-5 HR: 2
N,N-DIMETHYLACETAMIDE
mf: C_4H_9NO mw: 87.14

PROP: Colorless oily liquid; weak fishy odor. Mp: −20°, bp: 165°, d: 0.943 @ 20°/4°, vap d: 3.01, vap press: 1.3 mm @ 25°, flash p: 171°F (TOC), lel: 1.8%, uel: 11.5% @ 740 mm and 160°. Misc in water.

SYNS: ACETDIMETHYLAMIDE □ ACETIC ACID DIMETHYLAMIDE □ DIMETHYLACETAMIDE □ DIMETHYLACETONE AMIDE □ DIMETHYLAMIDE ACETATE □ DMA □ DMAC □ NSC 3138 □ U-5954

TOXICITY DATA WITH REFERENCE
skn-rbt 10 mg/24H open MLD AIHAAP 23,95,62
eye-rbt 100 mg MLD DCTODJ 9,147,86
dni-mus-unr 4400 mg/kg APHGAO 56,97,86
ihl-rat TCLo:300 ppm/6H (female 10W pre-3W post):REP FAATDF 7,132,86
skn-rat TDLo:2400 mg/kg (female 10-11D post):TER TXAPA9 41,35,77
orl-rat TDLo:5600 mg/kg (female 6-19D post):TER FAATDF 9,550,87
orl-rat TDLo:5600 mg/kg (female 6-19D post):REP FAATDF 9,550,87
ihl-rat TCLo:288 ppm/6H (male 10D pre):REP TOXID9 4,65,84
skn-rat TDLo:2400 mg/kg (female 10-11D post):TER TXAPA9 41,35,77
orl-rbt TDLo:3900 mg/kg (female 6-18D post):TER ARZNAD 30,1557,80
orl-rat LD50:4930 mg/kg DCTODJ 9,147,86
ihl-rat LC50:2475 ppm/1H DCTODJ 9,147,86
ipr-rat LD50:2750 mg/kg JRPFA4 4,219,62
ivn-rat LD50:2640 mg/kg ARZNAD 26,1581,76
orl-mus LD50:4620 mg/kg ARZNAD 26,1581,76
ihl-mus LC50:7200 mg/m³ CHYCDW 13,29,79
skn-mus LD50:9600 mg/kg CHYCDW 13,29,79
ipr-mus LD50:2800 mg/kg YKIGAK 31,327,80
ivn-mus LD50:3020 mg/kg ARZNAD 26,1581,76
skn-rbt LD50:2240 mg/kg AIHAAP 23,95,62

CONSENSUS REPORTS: Reported in EPA TSCA Inventory.

OSHA PEL: TWA 10 ppm (skin)
ACGIH TLV: TWA 10 ppm (skin)
DFG MAK: 10 ppm (35 mg/m³)

SAFETY PROFILE: Moderately toxic by skin contact, inhalation, intravenous, and intraperitoneal routes. Mildly toxic by ingestion. Experimental teratogenic and reproductive effects. A skin and eye irritant. Less toxic than dimethylformamide. Mutation data reported. Combustible when exposed to heat and flame. A moderate explosion hazard. Violent reaction with halogenated compounds (e.g., carbon tetrachloride, hexachlorocyclohexane) when heated above 90°C. Iron powder catalyzes the reaction so that it initiates at 71°C. When heated to decomposition it emits toxic fumes of NO_x.

For occupational chemical analysis use NIOSH: Dimethylacetamide 2004.

D

DOP000 CAS:2044-64-6 HR: 1
N,N-DIMETHYLACETOACETAMIDE
mf: $C_6H_{11}NO_2$ mw: 129.18

PROP: Liquid, misc in water and organic solvents. Bp: 220°, d: 1.049–1.052 @ 20°/20°, flash p: 252°F (COC).

SYN: N,N-DIMETHYL-3-OXOBUTANAMIDE

TOXICITY DATA WITH REFERENCE
skn-rbt 500 mg/24H MLD 85JCAE-,728,86
eye-rbt 500 mg/24H MLD 85JCAE-,728,86
orl-rat LD50:22,600 mg/kg AIHAAP 23,95,62
skn-rbt LD50:14,100 mg/kg AIHAAP 23,95,62

CONSENSUS REPORTS: Reported in EPA TSCA Inventory.

SAFETY PROFILE: Mildly toxic by ingestion and skin contact. A skin and eye irritant. Combustible when exposed to heat or flame. To fight fire, use water, fog, mist, CO_2, foam. When heated to decomposition it emits toxic fumes of NO_x.

DOP200 CAS:13265-60-6 HR: 3
O,O-DIMETHYL-S-(2-(ACETYLAMINO)ETHYL) DITHIOPHOSPHATE
mf: $C_6H_{14}NO_3PS_2$ mw: 243.30

SYNS: S-(2-(ACETYLAMINO)ETHYL)-O,O-DIMETHYL PHOSPHORODITHIOATE □ AMIPHOS □ CP 49674 □ DAEP □ O,O-DIMETHYL-S-(2-ACETAMIDOETHYL) ESTER PHOSPHORODITHIOIC ACID □ O,O-DIMETHYL-S-(2-ACETYLAMINOETHYL) PHOSPHORODITHIOATE □ N-((O,O-DIMETHYLPHOSPHORODITHIOYL)ETHYL)ACETAMIDE □ ENT 27,346 □ MONSANTO CP-49674 □ NSC 190945 □ PHOSPHORODITHIOIC ACID, O,O-DIMETHYL ESTER, S-ESTER with N-(2-MERCAPTOETHYL)ACETAMIDE

TOXICITY DATA WITH REFERENCE
orl-mus TDLo:600 mg/kg (1-15D preg):TER OYYAA2 6,621,72
orl-mus TDLo:600 mg/kg (1-15D preg):REP OYYAA2 6,621,72
orl-rat LD50:220 mg/kg ARSIM* 20,16,66
ihl-rat LCLo:40 mg/m³/4H 85GMAT -,55,82
skn-rat LD50:375 mg/kg 85GMAT -,55,82
orl-mus LD50:146 mg/kg 85GMAT -,55,82
skn-mus LD50:472 mg/kg OYYAA2 1,57,67
ipr-mus LD50:117 mg/kg OYYAA2 1,57,67

scu-mus LD50:245 mg/kg OYYAA2 1,57,67
skn-mky LD50:400 mg/kg OYYAA2 1,57,67

SAFETY PROFILE: Poison by ingestion, skin contact, intraperitoneal, and subcutaneous routes. Experimental teratogenic and reproductive effects. When heated to decomposition it emits very toxic NO_x, PO_x, and SO_x. See also MERCAPTANS.

DOP400 CAS:762-42-5 HR: 3
DIMETHYL ACETYLENEDICARBOXYLIC ACID
mf: $C_6H_6O_4$ mw: 142.12

$CH_3OCO{\cdot}C{\equiv}CCO{\cdot}OCH_3$

PROP: Bp: 98° @ 20 mm.

SYNS: ACETYLENEDICARBOXYLIC ACID, DIMETHYL ESTER ☐ 1,2-BIS(METHOXYCARBONYL)ETHYNE ☐ DI(CARBOMETHOXY)ACETYLENE ☐ DIMETHYL ETHYNEDICARBOXYLATE ☐ METHYL ACETYLENEDICARBOXYLATE

TOXICITY DATA WITH REFERENCE
orl-rat LDLo:50 mg/kg NCNSA6 5,13,53

CONSENSUS REPORTS: Reported in EPA TSCA Inventory.

SAFETY PROFILE: Poison by ingestion. A dienophile. Explosive reaction with 1-methylsilacyclopenta-2,4-diene at 150°C. Octakis(trifluorophosphine)dirhodium catalyzes explosive polymerization of the acid above 20°C. When heated to decomposition it emits acrid smoke and irritating fumes. See also ESTERS.

DOP600 CAS:30560-19-1 HR: 3
O,S-DIMETHYLACETYLPHOSPHOROAMIDOTHIOATE
mf: $C_4H_{10}NO_3PS$ mw: 183.18

SYNS: ACEPHAT (GERMAN) ☐ ACEPHATE ☐ ACETYLPHOSPHORAMIDOTHIOIC ACID-O,S-DIMETHYL ESTER ☐ CHEVRON RE 12,420 ☐ ENT 27,822 ☐ ORTHENE ☐ ORTHENE-755 ☐ ORTHO 12420 ☐ ORTRAN ☐ ORTRIL ☐ RE 12420 ☐ 75 SP

TOXICITY DATA WITH REFERENCE
mmo-sat 3 mg/plate NTIS** PB80-133226
mrc-smc 50,000 ppm NTIS** PB80-133226
mmo-esc 5 μL/plate MUREAV 28,405,75
dns-hmn:fbr 1 g/L NTIS** PB80-133226
msc-mus:lym 1 g/L NTIS** PB84-138973
orl-rat LD50:700 mg/kg MEIEDD 10,5,83
orl-mus LD50:233 mg/kg CHYCDW 14,226,80
ihl-mus LCLo:2200 mg/m³/5H TXAPA9 45,232,78
orl-dog LDLo:681 mg/kg GUCHAZ 6,1,73
skn-rbt LD50:2000 mg/kg 85DPAN -,-,71/76
orl-ckn LD50:852 mg/kg 28ZEAL 5,4,76
orl-dck LD50:350 mg/kg 28ZEAL 5,4,76
orl-mam LD50:321 mg/kg AECTCV 13,483,84
orl-brd LD50:106 mg/kg AECTCV 10,185,81

CONSENSUS REPORTS: EPA Genetic Toxicology Program.

SAFETY PROFILE: Poison by ingestion. Moderately toxic by skin contact and inhalation. Human mutation data reported. When heated to decomposition it emits very toxic fumes of NO_x, PO_x, and SO_x. See also ESTERS.

DOP800 CAS:2680-03-7 HR: 3
N,N-DIMETHYLACRYLAMIDE
mf: C_5H_9NO mw: 99.15

SYN: N,N-DIMETHYL-2-PROPENAMIDE

TOXICITY DATA WITH REFERENCE
orl-mus LD50:460 mg/kg ESKGA2 20,317,74
scu-mus LD50:580 mg/kg ESKGA2 20,317,74

CONSENSUS REPORTS: Reported in EPA TSCA Inventory.

SAFETY PROFILE: A poison by ingestion. When heated to decomposition it emits toxic fumes of NO_x.

DOQ300 CAS:627-93-0 HR: 2
DIMETHYL ADIPATE
mf: $C_8H_{14}O_4$ mw: 174.22

PROP: A liquid. Fp: 0°, mp: 8°.

SYNS: DIMETHYL HEXANEDIOATE ☐ METHYL ADIPATE

TOXICITY DATA WITH REFERENCE
ipr-rat TDLo:181 mg/kg (5-15D preg):REP JPMSAE 62,1596,73
ipr-rat TDLo:362 mg/kg (5-15D preg):TER JPMSAE 62,1596,73
ipr-rat LD50:1809 mg/kg JPMSAE 62,1596,73

CONSENSUS REPORTS: Reported in EPA TSCA Inventory.

SAFETY PROFILE: Moderately toxic by intraperitoneal route. Experimental teratogenic and reproductive effects. When heated to decomposition it emits acrid smoke and irritating fumes.

DOQ350 CAS:1191-16-8 HR: 1
3,3-DIMETHYLALLYL ACETATE
mf: $C_7H_{12}O_2$ mw: 128.19

PROP: Oil. Bp: 54–56° @ 32 mm.

SYNS: 2-BUTEN-1-OL, 3-METHYL-, ACETATE ☐ DIMETHYLALLYL ACETATE ☐ γ,γ-DIMETHYLALLYL ACETATE ☐ ISOPENT-2-ENYL ACETATE ☐ 3-METHYL-2-BUTENYL ACETATE ☐ PRENYL ACETATE

TOXICITY DATA WITH REFERENCE
skn-rbt 500 mg/24H MLD FCTOD7 20,817,82
orl-rat LD50:3 g/kg FCTOD7 20,817,82

CONSENSUS REPORTS: Reported in EPA TSCA Inventory.

SAFETY PROFILE: Mildly toxic by ingestion. A skin irritant. When heated to decomposition it emits acrid smoke and irritating fumes.

DOQ400 CAS:359-83-1 HR: 3
2-(3,3-DIMETHYLALLYL)CYCLAZOCINE
mf: $C_{19}H_{27}NO$ mw: 285.47

SYNS: 2-DIMETHYLALLYL-5,9-DIMETHYL-2′-HYDROXYBENZOMORPHAN ☐ 2-(3,3-DIMETHYLALLYL)-2′,2′-HYDROXY-5,9-DIMETHYL-6,7-BENZOMORPHAN ☐ FORTALGESIC ☐ FORTALIN ☐ FORTRAL ☐ 1,2,3,4,5,6-HEXAHYDRO-6,11-DIMETHYL-3-(3-METHYL-2-BUTENYL)-2,6-METHANO-3-BENZAZOCINE ☐ 2′-HYDROXY-5,9-DIMETHYL-2-(3,3-DI-

METHYLALLYL)-6,7-BENZOMORPHAN □ dl-2'-HYDROXY-5,9-DIMETHYL-2-(3,3-DIMETHYLALLYL)-6,7-BENZOMORPHAN □ II-C-2 □ KF-1820 □ LITICON □ 3-(3-METHYL-2-BUTENYL)-1,2,3,4,5,6-HEXAHYDRO-6,11-DIMETHYL-2,6-METHANO-3-BENZAZOCIN-8-OL □ NIH 7958 □ NSC-107430 □ PENTAGIN □ PENTAZOCINE □ SOSIGON □ TALWAN □ TALWIN □ WIN 20228

TOXICITY DATA WITH REFERENCE
ims-rat TDLo: 1500 mg/kg (female 30D pre): REP IYKEDH 13,198,82
scu-rat TDLo: 10,800 μg/kg (male 1D pre): REP JPETAB 198,340,76
ipr-rat TDLo: 1820 mg/kg (male 26W pre): REP IYKEDH 13,261,82
orl-hmn LDLo: 18 mg/kg CTOXAO 10,327,77
ims-wmn TDLo: 1400 mg/kg/3Y: CNS,SKN JAMAAP 231,271,75
ivn-hmn TDLo: 300 mg/kg/D: CNS BMJOAE 2,21,78
ivn-man TDLo: 3 mg/kg/2D-I: CNS JPETAB 143,149,64
ims-man TDLo: 83 mg/kg/4Y-I: MUS AIMDAP 143,2203,83
ims-hmn TDLo: 571 μg/kg: CNS,GIT JPETAB 143,149,64
orl-rat LD50: 1110 mg/kg KSRNAM 4,2145,70
scu-rat LD50: 61 mg/kg AMOKAG 35,179,81
ivn-rat LD50: 21 mg/kg 31ZPAG 2,174,66
ims-rat LD50: 175 mg/kg AIPTAK 190,124,71
orl-mus LD50: 205 mg/kg AMOKAG 35,179,81
ipr-mus LD50: 85 mg/kg CPBTAL 24,2912,76
scu-mus LD50: 80 mg/kg AMOKAG 35,179,81
ivn-mus LD50: 19,800 μg/kg NIIRDN 6,777,82
ims-mus LD50: 98 mg/kg AIPTAK 190,124,71

SAFETY PROFILE: Poison by ingestion, subcutaneous, intramuscular, intraperitoneal, and intravenous routes. Experimental reproductive effects. Human systemic effects by intramuscular and intravenous routes: wakefulness, euphoria, hallucinations or distorted perceptions, tremors, convulsions, excitement, motor activity changes, muscle weakness, analgesia, withdrawal, parasympathomimetic effects, nausea or vomiting, and dermatitis. Can cause drug dependency and other central nervous system effects. An analgesic. When heated to decomposition it emits toxic fumes of NO_x. See also ALLYL COMPOUNDS.

DOQ600 CAS:3639-66-5 HR: 3
2-(3,3-DIMETHYLALLYL)-5-ETHYL-2'-HYDROXY-9-METHYL-6,7-BENZOMORPHAN
mf: $C_{20}H_{29}NO$ mw: 299.50

SYN: 5-ETHYL-2'-HYDROXY-2(N)-(3-METHYL-2-BUTENYL)-9-METHYL-6,7-BENZOMORPHAN

TOXICITY DATA WITH REFERENCE
scu-rat LD50: 128 mg/kg JPETAB 143,141,64
ivn-rat LD50: 16 mg/kg JPETAB 143,141,64
scu-mus LD50: 116 mg/kg JPETAB 143,141,64
ivn-mus LD50: 16 mg/kg 31ZPAG 2,175,66

SAFETY PROFILE: Poison by subcutaneous and intravenous routes. When heated to decomposition it emits toxic fumes of NO_x. See also ALLYL COMPOUNDS.

DOQ700 CAS:1184-58-3 HR: 3
DIMETHYLALUMINUM CHLORIDE
mf: C_2H_6AlCl mw: 92.50

SAFETY PROFILE: Ignites spontaneously in air. Violent reaction on contact with water. When heated to decomposition it emits toxic fumes of Cl^-. See also ALUMINUM COMPOUNDS and CHLORIDES.

D

DOQ750 CAS:865-37-2 HR: 3
DIMETHYLALUMINUM HYDRIDE
mf: C_2H_7Al mw: 58.06

SAFETY PROFILE: Ignites on contact with traces of air or moisture. See also ALUMINUM COMPOUNDS and HYDRIDES.

DOQ800 CAS:124-40-3 HR: 3
DIMETHYLAMINE
DOT: UN 1032/UN 1160
mf: C_2H_7N mw: 45.10

PROP: Gas. D: 0.680 @ 0°/4°, mp: −96°, bp: 7°. Very sol in water.

SYNS: DIMETHYLAMINE, anhydrous (DOT) □ DIMETHYLAMINE, aqueous solution (DOT) □ DIMETHYLAMINE, solution (DOT) □ DMA □ N-METHYLMETHANAMINE □ RCRA WASTE NUMBER U092

TOXICITY DATA WITH REFERENCE
eye-rbt 50 mg/5M BJIMAG 23,153,66
cyt-rat-ihl 50 μg/m³ GISAAA 36(11),9,71
orl-rat LD50: 698 mg/kg HYSAAV 32,329,67
ihl-rat LC50: 4540 ppm/6H AIHAAP 43,411,82
orl-mus LD50: 316 mg/kg HYSAAV 32,329,67
ihl-mus LC50: 7650 ppm/2H AIHAAP 43,411 82
orl-rbt LD50: 240 mg/kg HYSAAV 32,329,67
ivn-rbt LDLo: 4 g/kg 85ESA3 11,509,89
orl-gpg LD50: 240 mg/kg HYSAAV 32,329,67
ihl-mam LC50: 3700 mg/m³ TPKVAL 14,80,75

CONSENSUS REPORTS: EPA Genetic Toxicology Program. Reported in EPA TSCA Inventory.

OSHA PEL: TWA 10 ppm
ACGIH TLV: TWA 5 ppm, STEL 15 ppm
DFG MAK: 2 ppm (4 mg/m³)
DOT CLASSIFICATION: 2.1; *Label:* Flammable Gas (UN 1032); DOT Class: 3; *Label:* Flammable Liquid (UN 1160)

SAFETY PROFILE: Poison by ingestion. Moderately toxic by inhalation and intravenous routes. Mutation data reported. An eye irritant. Corrosive to the eyes, skin, and mucous membranes. A flammable gas. When heated to decomposition it emits toxic fumes of NO_x. Incompatible with acrylaldehyde, fluorine, and maleic anhydride

DOR200 CAS:74-94-2 HR: 3
DIMETHYLAMINE BORANE
mf: $C_2H_7N{\cdot}BH_3$ mw: 58.94

PROP: Solid. Mp: 37°, bp: 49° @ 0.01 mm.

SYNS: BORANE with DIMETHYLAMINE (1:1) □ DMAB □ N-METHYLMETHANAMINE with BORANE (1:1)

TOXICITY DATA WITH REFERENCE
skn-rbt 50 mg MLD JOCMA7 1,46,59
eye-rbt 10 mg JOCMA7 1,46,59
dni-mus:ast 100 µmol/L JPMSAE 74,755,85
uns-mus:ast 100 µmol/L JPMSAE 74,755,85
orl-rat LD50:59 mg/kg AIHQA5 16,280,55
ipr-rat LD50:39 mg/kg 14KTAK -,693,64
ipr-mus LD50:200 mg/kg JPMSAE 69,1025,80
ivn-mus LD50:56 mg/kg CSLNX* NX#05150
ipr-rbt LD50:35,100 µg/kg 14KTAK -,693,64
ipr-gpg LD50:55,900 µg/kg 14KTAK -,693,64

CONSENSUS REPORTS: Reported in EPA TSCA Inventory.

SAFETY PROFILE: Poison by ingestion, intraperitoneal, and intravenous routes. A skin and eye irritant. Mutation data reported. When heated to decomposition it emits toxic fumes of NO_x. See also DIMETHYLAMINE and BORANE.

DOR400 CAS:2032-59-9 HR: 3
4-DIMETHYLAMINE m-CRESYL METHYLCARBAMATE
mf: $C_{11}H_{16}N_2O_2$ mw: 208.29

SYNS: A 363 □ AMINOCARB □ AMINOCARBE (FRENCH) □ BAY 44646 □ BAYER 5080 □ BAYER 44646 □ 4-DIMETHYLAMINO-3-CRESYL METHYLCARBAMATE □ 4-(DIMETHYLAMINO)-3-METHYLPHENOL METHYL CARBAMATE (ester) □ (4-DIMETHYLAMINO-3-METHYL-PHENYL)N-METHYL-CARBAMAAT (DUTCH) □ (4-DIMETHYLAMINO-3-METHYL-PHENYL)N-METHYL-CARBAMAT (GERMAN) □ (4-DIMETHYLAMINO-3-METHYL-PHENYL)N-METHYL-CARBAMATE □ 4-(DIMETHYLAMINO)-m-TOLYL METHYLCARBAMATE □ (4-DIMETILAMINO-3-METIL-FENIL)-N-METIL-CARBAMMATO (ITALIAN) □ ENT 25,784 □ MATACIL □ N-METHYLCARBAMATE de 4-DIMETHYLAMINO-3-METHYL PHENYLE (FRENCH) □ MITACIL

TOXICITY DATA WITH REFERENCE
mma-sat 5 mmol/L ENMUDM 5,384,83
cyt-ham:ovr 5 mmol/L ENMUDM 5,384,83
orl-rat LD50:30 mg/kg TXAPA9 21,315,72
skn-rat LD50:275 mg/kg WRPCA2 9,119,70
orl-mus LDLo:94 mg/kg AECTCV 14,111,85
ipr-mus LD50:7 mg/kg TXAPA9 6,402,64
orl-gpg LDLo:50 mg/kg JEENAI 60(3),733,67
scu-gpg LDLo:50 mg/kg JEENAI 60(3),733,67

SAFETY PROFILE: Poison by ingestion, skin contact, intraperitoneal, and subcutaneous routes. Mutation data reported. An insecticide used for forest insect control. When heated to decomposition it emits toxic fumes of NO_x. See also CARBAMATES.

DOR600 CAS:506-59-2 HR: 2
DIMETHYLAMINE HYDROCHLORIDE
mf: $C_2H_7N{\cdot}ClH$ mw: 81.56

SYNS: DIMETHYLAMMONIUM CHLORIDE □ HYDROCHLORIC ACID DIMETHYLAMINE □ N-METHYLMETHANAMINE HYDROCHLORIDE

TOXICITY DATA WITH REFERENCE
ipr-mus TDLo:3825 mg/kg (female 1-17D post):REP JTEHD6 32,319,91
orl-mus TDLo:12 g/kg/Y-C:NEO GISAAA 44(8),15,79
orl-rat LD50:1070 mg/kg GISAAA 32(6),12,67
orl-mus LD50:8100 mg/kg GISAAA 32(6),12,67
ipr-mus LD50:1570 mg/kg JJPAAZ 17,475,67
scu-mus LD50:2000 mg/kg AIPTAK 112,36,57
ivn-mus LD50:1210 mg/kg AIPTAK 112,36,57
orl-rbt LD50:1600 mg/kg GISAAA 32(6),12,67
orl-gpg LD50:1600 mg/kg GISAAA 32(6),12,67

CONSENSUS REPORTS: Reported in EPA TSCA Inventory. EPA Genetic Toxicology Program.

SAFETY PROFILE: Moderately toxic by ingestion, intravenous, subcutaneous, and intraperitoneal routes. Experimental reproductive effects. Questionable carcinogen with experimental neoplastigenic data. When heated to decomposition it emits very toxic fumes of NO_x, NH_3, and HCl.

DOR800 CAS:3426-62-8 HR: 2
DIMETHYLAMINE-2,3,6-TRICHLOROBENZOATE
mf: $C_7H_3Cl_3O_2{\cdot}C_2H_7N$ mw: 270.55

SYNS: BENZAC 1281 □ 2KF □ KF 2 (HERBICIDE) □ KYSELINA 2,3,6-TRICHLORBENZOOVA DIMETHYLAMONNA SUL □ METHANAMINE, N-METHYL-, 2,3,6-TRICHLOROBENZOATE □ POLIDIM □ 2,3,6-TRICHLOROBENZOIC ACID, DIMETHYLAMINE SALT □ TRYSBEN 200

TOXICITY DATA WITH REFERENCE
skn-gpg 500 mg open MOD DUPON*
orl-rat LD50:1644 mg/kg DUPON*

SAFETY PROFILE: Moderately toxic by ingestion. A skin irritant. When heated to decomposition it emits very toxic fumes of Cl^- and NO_x.

DOS000 CAS:315-18-4 HR: 3
4-(DIMETHYLAMINE)-3,5-XYLYL-N-METHYLCARBAMATE
mf: $C_{12}H_{18}N_2O_2$ mw: 222.32

PROP: Crystals. Mp: 85°, vap press: <0.1 mm @ 139°.

SYNS: 4-(DIMETHYLAMINO)-3,5-DIMETHYLPHENOL METHYLCARBAMATE (ESTER) □ 4-(DIMETHYLAMINO)-3,5-DIMETHYLPHENYL ESTER, METHYLCARBAMIC ACID □ 4-(DIMETHYLAMINO)-3,5-DIMETHYLPHENYL-N-METHYLCARBAMATE □ 4-(DIMETHYLAMINO)-3,5-XYLENOL METHYLCARBAMATE (ESTER) □ 4-(DIMETHYLAMINO)-3,5-XYLYL ESTER METHYLCARBAMIC ACID □ 4-DIMETHYLAMINO-3,5-XYLYL METHYLCARBAMATE □ 4-DIMETHYLAMINO-3,5-XYLYL-N-METHYLCARBAMATE □ 4-(N,N-DIMETHYLAMINO)-3,5-XYLYL N-METHYLCARBAMATE □ DOWCO 139 □ ENT 25,766 □ METHYL-4-DIMETHYLAMINO-3,5-XYLYL CARBAMATE □ METHYL-4-DIMETHYLAMINO-3,5-XYLYL ESTER of CARBAMIC ACID □ MEXACARBATE (DOT) □ NCI-C00544 □ OMS-47 □ ZACTRAN □ ZECTANE □ ZECTRAN □ ZEXTRAN

TOXICITY DATA WITH REFERENCE
scu-mus TDLo:90 mg/kg (6-14D preg):TER NTIS** PB223-160
orl-mus TDLo:1200 mg/kg/78W-I:NEO NTIS** PB223-159
orl-rat LD50:14 mg/kg JEENAI 62,1307,69
orl-mus LD50:12 mg/kg PSSCBG 2,10,71
skn-mus LD50:107 mg/kg JAFCAU 15,479,67
ipr-mus LD50:7800 µg/kg JAFCAU 16,561,68
orl-dog LD50:22 mg/kg FMCHA2 -,C196,89
orl-rbt LD50:37 mg/kg SPEADM 78-1,59,78

orl-gpg LD50:15 mg/kg PCOC** -,1232,66
orl-pgn LD50:5620 μg/kg ASTTA8 (680),157,79 JEENAI 62,1307,69

CONSENSUS REPORTS: IARC Cancer Review: Group 3 IMEMDT 7,56,87; Animal Inadequate Evidence IMEMDT 12,237,76. NCI Carcinogenesis Bioassay (feed); No Evidence: mouse, rat NCITR* NCI-CG-TR-147,78. EPA Extremely Hazardous Substances List.

SAFETY PROFILE: Poison by ingestion, skin contact, and intraperitoneal routes. Experimental teratogenic effects. Questionable carcinogen with experimental neoplastigenic data. When heated to decomposition it emits toxic fumes of NO_x. See also ESTERS and CARBAMATES.

DOS200 CAS:926-64-7 HR: 3
DIMETHYLAMINOACETONITRILE
DOT: UN 2378
mf: $C_4H_8N_2$ mw: 84.14

PROP: Flash p: <73.4°F.

SYNS: N-(CYANOMETHYL)DIMETHYLAMINE □ 2-DIMETHYLAMINOACETONITRILE (DOT) □ N,N-DIMETHYLGLYCINONITRILE

TOXICITY DATA WITH REFERENCE
skn-rbt 500 mg/24H MLD 85JCAE-,923,86
eye-rbt 20 mg/24H MOD 85JCAE-,923,86
orl-rat LD50:50 mg/kg AIHAAP 23,95,62
ihl-rat LCLo:250 ppm/4H AIHAAP 23,95,62
skn-rbt LD50:170 mg/kg EPASR* FYI-OTS-0483,0238
ocu-rbt LDLo:100 mg/kg EPASR* FYI-OTS-0482,0238

CONSENSUS REPORTS: Cyanide and its compounds are on the Community Right-To-Know List.

DOT CLASSIFICATION: 3; *Label:* Flammable Liquid, Poison

SAFETY PROFILE: Poison by ingestion, skin contact, and ocular routes. Moderately toxic by inhalation. A skin and eye irritant. A dangerous fire hazard when exposed to heat or flame. When heated to decomposition it emits toxic fumes of NO_x and CN^-. See also NITRILES.

DOS300 CAS:24869-88-3 HR: 3
(DIMETHYLAMINO)ACETYLENE
mf: C_4H_7N mw: 69.11

$(CH_3)_2NC{\equiv}CH$

PROP: Liquid. Bp: 53°.

SAFETY PROFILE: Reacts vigorously with water. When heated to decomposition it emits toxic fumes of NO_x. See also ACETYLENE COMPOUNDS.

DOS800 CAS:13365-38-3 HR: 3
9-(p-DIMETHYLAMINOANILINO)ACRIDINE
mf. $C_{21}H_{19}N_3$ mw: 313.43

SYNS: N'-9-ACRIDINYL-N,N-DIMETHYL-1,4-BENZENEDIAMINE □ 9-((p-DIMETHYLAMINO)PHENYL)AMINO))ACRIDINE □ NSC 13002 □ WIN 1701

TOXICITY DATA WITH REFERENCE
dnd-mam:lym 2 μmol/L JMCMAR 24,170,81
ivn-rat LD50:15 mg/kg NCIAL* -,326,67
ipr-mus LD50:80,970 μg/kg NCISP* JAN86
ivn-dog LDLo:5 mg/kg NCIAL* -,326,67
ivn-mky LDLo:5 mg/kg NCIAL* -,326,67

SAFETY PROFILE: Poison by intravenous and intraperitoneal routes. Mutation data reported. When heated to decomposition it emits toxic fumes of NO_x.

D

DOT000 CAS:58-15-1 HR: 3
DIMETHYLAMINOANTIPYRINE
mf: $C_{13}H_{17}N_3O$ mw: 231.33

PROP: Colorless leaflets, somewhat water-sol; crystals from toluene. Mp: 107–109°. Sol in H_2O, EtOH, $CHCl_3$, Et_2O, toluene, and acids.

SYNS: AMIDAZOPHEN □ AMIDOFEBRIN □ AMIDOPHEN □ AMIDOPHENAZONE □ AMIDOPYRAZOLINE □ AMIDOPYRIN □ AMINOFENAZONE (ITALIAN) □ AMINOPHENAZONE □ AMINOPYRINE □ ANAFEBRINA □ BRUFANEUXOL □ DAP □ DEREUMA □DIMAPYRIN □ DIMETHYLAMINO-ANALGESINE □ 4-(DIMETHYLAMINO)ANTIPYRINE □ DIMETHYLAMINOAZOPHENE □ 4-(DIMETHYLAMINO)-1,2-DIHYDRO-1,5-DIMETHYL-2-PHENYL-3H-PYRAZOL-3-ONE □ 4-DIMETHYLAMINO-2,3-DIMETHYL-1-PHENYL-3-PYRAZOLIN-5-ONE □ 4-DIMETHYLAMINO-2,3-DIMETHYL-1-PHENYL-5-PYRAZOLONE □ DIMETHYLAMINOPHENAZON (GERMAN) □DIMETHYLAMINOPHENAZONE □ 4-DIMETHYLAMINOPHENAZONE □ DIMETHYLAMINOPHENYLDIMETHYLPYRAZOLIN □ 4-DIMETHYLAMINO-1-PHENYL-2,3-DIMETHYLPYRAZOLONE □ 3-keto-1,5-DIMETHYL-4-DIMETHYLAMINO-2-PHENYL-2,3-DIHYDROPYRAZOLE □ 1,5-DIMETHYL-4-DIMETHYLAMINO-2-PHENYL-3-PYRAZOLONE □ 2,3-DIMETHYL-4-DIMETHYLAMINO 1 PHENYL-5-PYRAZOLONE □ DIPIRIN □ DIPYRIN □ FEBRININA □ FEBRON □ ITAMIDONE □ MAMALLET-A □ NETSUSARIN □ NOVAMIDON □ 1-PHENYL-2,3-DIMETHYL-4-DIMETHYLAMINOPYRAZOLONE-5 □ 1-PHENYL-2,3-DIMETHYL-4-DIMETHYLAMINOPYRAZOL-5-ONE □ PIRAMIDON □ PIRIDOL □PIROMIDINA □ POLINALIN □ PYRADONE □ PYRAMIDON □ PYRAMIDONE

TOXICITY DATA WITH REFERENCE
mma-sat 31 μmol/plate MUREAV 66,33,79
dni-mus:oth 100 mg/L ONCODU 19,183,80
otr-ham-orl 100 mg/kg IAPUDO 41,585,82
cyt-ham:fbr 3 mmol/L HDSKEK 10,63,85
msc-ham-orl 100 mg/kg IAPUDO 41,585,82
orl-rat TDLo:60 mg/kg (9-14D preg):REP SKNEA7 22,109,72
scu-mus TDLo:630 mg/kg (female 9-11D post):TER TJADAB 16,118,77
orl-rat TDLo:900 mg/kg (female 9-14D post):TER SKNEA7 22,109,72
scu-mus TDLo:600 mg/kg (female 7-9D post):REP RCOCB8 34,141,81
scu-mus TDLo:630 mg/kg (female 9-11D post):TER YKKZAJ 101,470,81
scu-mus TDLo:630 mg/kg (female 9-11D post):TER TXCYAC 29,281,84
orl-rat TDLo:60 mg/kg (9-14D preg):TER SKNEA7 22,109,72
orl-rat TDLo:4200 mg/kg/2.5Y-C:NEO ARGEAR 49,220,79
unr-man LDLo:220 mg/kg 85DCAI 2,73,70
orl-rat LD50:285 mg/kg RPTOAN 51,183,88
ipr-rat LD50:190 mg/kg NYKZAU 68,442,72
scu-rat LD50:295 mg/kg ARZNAD 8,229,58

ivn-rat LD50:98 mg/kg OYYAA2 16,1011,78
ims-rat LD50:340 mg/kg ARZNAD 10,665,60
orl-mus LD50:350 mg/kg PCJOAU 18,46,84
ipr-mus LD50:169 mg/kg RPTOAN 36,293,72
scu-mus LD50:248 mg/kg IYKEDH 8,494,77
ivn-mus LD50:78 mg/kg OYYAA2 16,1011,78
ims-mus LD50:306 mg/kg OYYAA2 13,109,77

CONSENSUS REPORTS: Reported in EPA TSCA Inventory. EPA Genetic Toxicology Program.

SAFETY PROFILE: Human poison by unspecified route. Experimental poison by ingestion, subcutaneous, intramuscular, intravenous, and intraperitoneal routes. Moderately toxic by parenteral route. Experimental teratogenic and reproductive effects. Questionable carcinogen when mixed with $NaNO_2$ (1:1). Mutation data reported. Can cause bone marrow depression resulting in leucopenia. Has been implicated in development of aplastic anemia. A tranquilizer. When heated to decomposition it emits toxic fumes of NO_x.

DOT200 HR: 2
4-(DIMETHYLAMINO)ANTIPYRINE mixed with SODIUM NITRITE (1:1)
mf: $C_{13}H_{17}N_3O{\bullet}NNaO_2$ mw: 300.33

SYNS: AMINOPHENAZONE mixed with SODIUM NITRITE (1:1) □ AMINOPYRINE mixed with SODIUM NITRITE (1:1) □ SODIUM NITRITE mixed with AMINOPYRINE (1:1) □ SODIUM NITRITE mixed with 4-(DIMETHYLAMINO)ANTIPYRINE (1:1)

TOXICITY DATA WITH REFERENCE
mmo-sat 1 mg/plate TOLED5 12,281,82
mma-sat 1 mg/plate TOLED5 12,281,82
cyt-rat-orl 600 mg/kg MFEPDX 1,225,79
hma-mus/sat 2 mmol/L/kg ATSUDG (4),49,80
hma-mus/esc 10 mg/kg CBINA8 35,199,81
orl-rat TDLo:3438 mg/kg/50W-I:CAR NATUAS 244,176,73
orl-rat TD:10 g/kg/40W-I:CAR IARCCD 14,461,76
orl-rat TD:8000 mg/kg/40W-I:CAR IARCCD 14,461,76

SAFETY PROFILE: Questionable carcinogen with experimental carcinogenic data. Mutation data reported. When heated to decomposition it emits toxic fumes of NO_x and Na_2O. See also SODIUM NITRITE.

DOT300 CAS:60-11-7 HR: 3
4-DIMETHYLAMINOAZOBENZENE
mf: $C_{14}H_{15}N_3$ mw: 225.32

PROP: Yellow, crystalline tablets; yellow leaflets from EtOH. Mp: 115°. Sol in EtOH, Me_2CO, and C_6H_6; insol in H_2O.

SYNS: ATUL FAST YELLOW R □ BENZENEAZODIMETHYLANILINE □ BRILLIANT FAST YELLOW □ BUTTER YELLOW □ CERASINE YELLOW GG □ C.I. 11020 □ C.I. SOLVENT YELLOW 2 □ DAB □ p-DIMETHYLAMINOAZOBENZEN (CZECH) □ DIMETHYLAMINOAZOBENZENE □ N,N-DIMETHYL-p-AMINOAZOBENZENE □ p-DIMETHYLAMINOAZOBENZENE □ N,N-DIMETHYL-4-AMINOAZOBENZENE □ 4-(N,N-DIMETHYLAMINO)AZOBENZENE □ DIMETHYLAMINOAZOBENZOL □ p-DIMETHYLAMINO-AZOBENZOL (GERMAN) □ 4-DIMETHYLAMINOAZOBENZOL □ 4-DIMETHYLAMINOPHENYLAZOBENZENE □ N,N-DIMETHYL-p-AZOANILINE □ N,N-DIMETHYL-p-PHENYLAZOANILINE □ N,N-DIMETHYL-4-(PHENYLAZO)BENZAMINE □ N,N-DIMETHYL-4-(PHENYLAZO)BENZENAMINE □ DIMETHYL YELLOW □ DIMETHYL YELLOW-N,N-DIMETHYLANILINE □ DMAB □ ENIAL YELLOW 2G □ FAST OIL YELLOW B □ FAT YELLOW □ GRASAL BRILLIANT YELLOW □ JAUNE de BEURRE (FRENCH) □ METHYL YELLOW □ OIL YELLOW □ OLEAL YELLOW 2G □ ORGANOL YELLOW ADM □ ORIENT OIL YELLOW GG □ P.D.A.B. □ PETROL YELLOW WT □ RCRA WASTE NUMBER U093 □ RESINOL YELLOW GR □ RESOFORM YELLOW GGA □ SILOTRAS YELLOW T2G □ SOMALIA YELLOW A □ STEAR YELLOW JB □ SUDAN YELLOW □ TOYO OIL YELLOW G □ USAF EK-338 □ WAXOLINE YELLOW AD □ YELLOW G SOLUBLE in GREASE □ ZLUT MASELNA (CZECH)

TOXICITY DATA WITH REFERENCE
dnr-esc 80 mg/L MUREAV 119,135,83
dni-hmn:hla 100 µmol/L MUREAV 92,427,82
ipr-mus TDLo:3 g/kg (5D male):REP PMRSDJ 1,712,81
scu-mus TDLo:200 mg/kg/(10D preg):TER OFAJAE 36,195,60
orl-rat TDLo:5426 mg/kg/17W-C:CAR CBINA8 53,107,85
skn-rat TDLo:1440 mg/kg/90W-I:NEO CNREA8 26,2406,66
ipr-mus TDLo:3830 µg/kg:NEO CNREA8 44,2540,84
scu-mus TDLo:4000 mg/kg (15-21D preg):CAR,TER BEXBAN 78,1402,75
mul-mus TDLo:400 mg/kg/I:ETA CNREA8 1,397,41
orl-dog TDLo:9600 mg/kg/69W-C:ETA JNCIAM 13,1497,53
orl-ham TDLo:9600 mg/kg/42W-I:ETA ARPAAQ 71,566,61
orl-rat TD:2600 mg/kg/13W-C:NEO CNREA8 17,387,57
orl-rat TD:13 g/kg/53W-C:NEO MEXPAG 4,1,61
orl-rat TD:1920 mg/kg/14W-C:ETA CNREA8 5,235,45
orl-rat TD:1800 mg/kg/14W-C:ETA CNREA8 8,141,48
orl-rat TD:2331 mg/kg/7W-I:ETA KTUNAA 32,229,79
orl-rat TD:17,200 mg/kg/17W-C:NEO ARPAAQ 81,162,66
orl-rat TD:8316 mg/kg/33W-C:NEO JPBAA7 59,1,47
orl-rat TD:3990 mg/kg/19W-C:NEO GANNA2 63,131,72
ipr-mus TD:11 mg/kg:NEO CNREA8 44,2540,84
orl-rat TD:800 mg/kg/64D-C:ETA ZEKBAI 61,327,56
orl-rat LD50:200 mg/kg ZEKBAI 69,103,67
ipr-rat LD50:230 mg/kg CNREA8 34,2274,74
orl-mus LD50:300 mg/kg GANNA2 54,455,63
ipr-mus LD50:230 mg/kg CNREA8 34,2274,74

CONSENSUS REPORTS: NTP 7th Annual Report On Carcinogens. IARC Cancer Review: Group 2B IMEMDT 7,56,87; Animal Sufficient Evidence IMEMDT 8,125,75. EPA Genetic Toxicology Program. Community Right-To-Know List. Reported in EPA TSCA Inventory.

OSHA PEL: Cancer Suspect Agent
NIOSH REL: (4-Dimethylaminoazobenzene): TWA use 29 CFR 1910.1015

SAFETY PROFILE: Confirmed carcinogen with experimental carcinogenic, neoplastigenic, and tumorigenic data. Poison by ingestion and intraperitoneal routes. Experimental teratogenic and reproductive effects. Human mutation data reported. When heated to decomposition it emits toxic fumes of NO_x.

DOT400 CAS:100-10-7 HR: 2
p-(DIMETHYLAMINO)BENZALDEHYDE
mf: $C_9H_{11}NO$ mw: 149.21

PROP: Small, granular, lemon-colored crystals (may turn pink upon exposure to light). Mp: 74°, bp:

176–177° @ 17 mm. Sltly water-sol; sol in alc, ether, chloroform, acetic acid, and many other organic solvents.

SYNS: 4-(DIMETHYLAMINO) BENZALDEHYDE □ 4-DIMETHYLAMINOBENZENECARBONAL □ EHRLICH'S REAGENT □ p-FORMYLDIMETHYLANILINE

TOXICITY DATA WITH REFERENCE
orl-rat LDLo:500 mg/kg JPETAB 90,260,47
ipr-rat LD50:620 mg/kg HINEL* AF33(657)-11756,64

CONSENSUS REPORTS: Reported in EPA TSCA Inventory.

SAFETY PROFILE: Moderately toxic by ingestion and intraperitoneal routes. When heated to decomposition it emits toxic fumes of NO_x. See also ALDEHYDES.

DOT600 CAS:443-30-1 HR: 2
1-(4-DIMETHYLAMINOBENZAL)INDENE
mf: $C_{18}H_{17}N$ mw: 247.36

SYNS: DABI □ (4-DIMETHYLAMINOBENZYLIDENE)INDENE □ N,N-DIMETHYL-α-INDOLYLIDENE-p-TOLUIDINE □ 4-(1H-INDEN-1-YLIDENEMETHYL)-N,N-DIMETHYLBENZENAMINE □ NSC-80087

TOXICITY DATA WITH REFERENCE
orl-rat TDLo:270 mg/kg/15D-I:ETA,REP NATUAS 222,383,69
orl-mus TDLo:4000 mg/kg/26W-C:ETA PTEUA6 7,229,72
ipr-rat LDLo:2000 mg/kg JMCMAR 13,770,70

SAFETY PROFILE: Moderately toxic by intraperitoneal route. Experimental reproductive effects. Questionable carcinogen with experimental tumorigenic data. When heated to decomposition it emits toxic fumes of NO_x.

DOT800 CAS:536-17-4 HR: 3
p-DIMETHYLAMINOBENZALRHODANINE
mf: $C_{12}H_{12}N_2OS_2$ mw: 264.38

PROP: Red crystals or powder. Mp: 285–288° (dec comp). Sltly sol in Me_2CO, EtOH, and $CHCl_3$; insol in H_2O.

SYNS: p-(DIMETHYLAMINO)BENZAL-5-RHODANINE □ 5-(p-DIMETHYLAMINOBENZAL)RHODANINE □ 5-(p DIMETHYLAMINOBENZOYLIDENE)RHODANINE □ p-DIMETHYLAMINOBENZYLIDENE RHODAMINE □ USAF PD-20

TOXICITY DATA WITH REFERENCE
ipr-mus LD50:150 mg/kg NTIS** AD603-561

CONSENSUS REPORTS: Reported in EPA TSCA Inventory.

SAFETY PROFILE: Poison by intraperitoneal route. When heated to decomposition it emits very toxic fumes of NO_x and SO_x.

DOU000 CAS:53004-03-8 HR: 3
5(4-DIMETHYLAMINOBENZENEAZO)TETRAZOLE
mf: $C_9H_{11}N_7$ mw: 217.24

$HNN{=}NN{=}CN{=}NC_6H_4N(CH_3)_2$

SAFETY PROFILE: Explodes when heated to 155°C. When heated to decomposition it emits toxic fumes of NO_x.

DOU600 CAS:140-56-7 HR: 3
p-DIMETHYLAMINOBENZENEDIAZOSODIUM SULPHONATE
mf: $C_8H_{10}N_3O_3S{\cdot}Na$ mw: 251.26

PROP: Yellow-brown crystals. Mod sol in water; sol in DMF.

SYNS: BAYER 5072 □ DAPA □ DAS □ DEKSONAL □ DEXON □ p-DIMETHYLAMINOBENZENE DIAZO SODIUM SULFONATE □ p-(DIMETHYLAMINO)BENZENEDIAZOSULFONATE □ p-DIMETHYLAMINOBENZENEDIAZOSULFONIC ACID, SODIUM SALT □ 4-DIMETHYLAMINOBENZENEDIAZOSULFONIC ACID, SODIUM SALT □ p-(DIMETHYLAMINO)BENZENEDIAZOSULPHONATE □ p-(DIMETHYLAMINO)BENZENEDIAZOSULPHONIC ACID, SODIUM SALT □ 4-DIMETHYLAMINOBENZENEDIAZOSULPHONIC ACID, SODIUM SALT □ p-DIMETHYLAMINOBENZOLDIAZOSULFONAT (NATRIUMSALZ) (GERMAN) □ (4-(DIMETHYLAMINO)PHENYL)DIAZENESULFONIC ACID, SODIUM SALT □ 4-((DIMETHYLAMINO)PHENYL)DIAZENESULFONIC ACID, SODIUM SALT □ p-(DIMETHYLAMINO)-PHENYLDIAZONATRIUMSULFONAT (GERMAN) □ N,N-DIMETHYL-p-ANILINEDIAZOSULFONIC ACID SODIUM SALT □ FENAMINOSULF □ GOLD ORANGE MP □ LESAN □ NCI-C03010 □ SODIUM-p-(DIMETHYLAMINO)BENZENEDIAZOSULFONATE □ SODIUM-4-(DIMETHYLAMINO)BENZENEDIAZOSULFONATE □ SODIUM-p-(DIMETHYLAMINO)BENZENEDIAZOSULPHONATE □ SODIUM-4-(DIMETHYLAMINO) BENZENEDIAZOSULPHONATE □ SODIUM-(4-(DIMETHYLAMINO) PHENYL)DIAZENESULFONATE □ TROPAEOLIN D

TOXICITY DATA WITH REFERENCE
mmo-sat 25 μg/plate YACHDS 13,4923,85
mmo-esc 25 μg/plate YACHDS 13,4923,85
orl-rat TDLo:25 mg/kg (7-11D preg):TER GYLPDN 4,201,83
orl-rat LD50:60 mg/kg WRPCA2 9,119,70
ipr-rat LD50:10,300 μg/kg 34ZIAG -,202,69
orl-mus LDLo:140 mg/kg AECTCV 14,111,85
ipr-mus LD50:60 mg/kg 34ZIAG -,202,69
ivn-mus LD50:56 mg/kg CSLNX* NX#00143
ipr-dog LDLo:5 mg/kg JPETAB 95,262,49
orl-rbt LD50:150 mg/kg 85DPAN -,-,71/76
ipr-rbt LDLo:10 mg/kg JPETAB 95,262,49

CONSENSUS REPORTS: IARC Cancer Review: Group 3 IMEMDT 7,56,87; Animal Inadequate Evidence IMEMDT 8,147,75. NCI Carcinogenesis Bioassay (feed); No Evidence: mouse, rat NCITR* NCI-CG-TR-101,78. EPA Genetic Toxicology Program.

SAFETY PROFILE: Poison by ingestion, intravenous, and intraperitoneal routes. Experimental teratogenic effects. Human mutation data reported. Questionable carcinogen. A fungicide. When heated to decomposition it emits very toxic fumes of NO_x, Na_2O, and SO_x.

DOU650 CAS:619-84-1 HR: 3
p-DIMETHYLAMINO BENZOIC ACID
mf: $C_9H_{11}NO_2$ mw: 165.21

SYN: BENZOIC ACID, p-(DIMETHYLAMINO)-

TOXICITY DATA WITH REFERENCE
ivn-mus LD50:180 mg/kg CSLNX* NX#04362

CONSENSUS REPORTS: Reported in EPA TSCA Inventory.

SAFETY PROFILE: Poison by intravenous route. When heated to decomposition it emits toxic vapors of NO_x.

DOU700 CAS:6843-30-7 HR: 3
5-DIMETHYLAMINO-3-BENZOYLINDOLE
mf: $C_{17}H_{16}N_2O$ mw: 264.35

SYN: KETONE, 5-DIMETHYLAMINO-3-INDOLYL PHENYL

TOXICITY DATA WITH REFERENCE
ivn-mus LD50:28 mg/kg CSLNX* NX#12193

DOT CLASSIFICATION: 3; *Label:* Flammable Liquid

SAFETY PROFILE: A poison by intravenous route. A flammable liquid. When heated to decomposition it emits toxic vapors of NO_x.

DOV000 CAS:63918-82-1 HR: 2
p-DIMETHYLAMINOBENZYLIDENE-3,4,5,6-DIBENZ-9-METHYLACRIDINE
mf: $C_{31}H_{24}N_2$ mw: 424.57

SYN: 14-(p-(DIMETHYLAMINO)STYRYL)DIBENZ(a,j)ACRIDINE

TOXICITY DATA WITH REFERENCE
scu-mus TDLo:200 mg/kg:ETA VOONAW 1,52,55

SAFETY PROFILE: Questionable carcinogen with experimental tumorigenic data. When heated to decomposition it emits toxic fumes of NO_x.

DOV200 CAS:13629-82-8 HR: 2
3,3′-DIMETHYL-4-AMINOBIPHENYL
mf: $C_{14}H_{15}N$ mw: 197.30

SYNS: 3,3′-DIMETHYL-4-AMINODIPHENYL □ 3,3′-DIMETHYL-4-BIPHENYLAMINE

TOXICITY DATA WITH REFERENCE
scu-rat TDLo:960 mg/kg/21W-I:CAR ANZJA7 29,38,59
scu-rat TD:2400 mg/kg/W-I:ETA BMBUAQ 14,141,58

SAFETY PROFILE: Questionable carcinogen with experimental carcinogenic and tumorigenic data. When heated to decomposition it emits toxic fumes of NO_x.

DOV400 CAS:63019-93-2 HR: 2
4-(DIMETHYLAMINO)-3-BIPHENYLOL
mf: $C_{14}H_{15}NO$ mw: 213.30

SYN: 4-DIMETHYLAMINO-3-HYDROXYDIPHENYL

TOXICITY DATA WITH REFERENCE
imp-mus TDLo:80 mg/kg:ETA BJCAAI 11,212,57

SAFETY PROFILE: Questionable carcinogen with experimental tumorigenic data. When heated to decomposition it emits toxic fumes of NO_x.

DOV800 CAS:64246-07-7 HR: 3
4-DIMETHYLAMINO-1,1-BIS((3,4-(METHYLENEDIOXY)PHENOXY)METHYL)-1-BUTANOL, METHYLCARBAMATE (ester), CITRATE
mf: $C_{24}H_{30}N_2O_8 \cdot C_6H_8O_7$ mw: 666.70

TOXICITY DATA WITH REFERENCE
orl-mus LD50:525 mg/kg FRPSAX 32,502,77
ivn-mus LD50:72 mg/kg FRPSAX 32,502,77

SAFETY PROFILE: Poison by intravenous route. Moderately toxic by ingestion. When heated to decomposition it emits toxic fumes of NO_x. See also CARBAMATES and ESTERS.

DOW875 HR: 3
p-DIMETHYLAMINO-CARVACROLDIMETHYLURETHANE METHIODIDE
mf: $C_{16}H_{27}N_2O_2 \cdot I$ mw: 406.35

SYN: (CARBOXYMETHYL)TRIMETHYLAMMONIUM IODIDE-5-(DIMETHYLAMINO)-4-ISOPROPYL-o-TOLYL ESTER

TOXICITY DATA WITH REFERENCE
orl-mus LDLo:95 mg/kg FEPRA7 5,184,46
scu-mus LDLo:4 µg/kg FEPRA7 5,184,46
scu-dog LDLo:345 µg/kg FEPRA7 5,184,46

SAFETY PROFILE: Poison by ingestion and subcutaneous routes. When heated to decomposition it emits toxic fumes of I^-, NH_3, and NO_x. See also IODIDES.

DOX000 CAS:5913-82-6 HR: 3
3-β-(DIMETHYLAMINO)CON-5-ENINE-DIHYDROBROMIDE
mf: $C_{24}H_{40}N_2 \cdot 2BrH$ mw: 356.66

PROP: A solid. Mp: 340° (decomp).

SYNS: CONESSINE DIHYDROBROMIDE □ KONESSIN DIHYDROBROMIDE □ NERIINE DIHYDRBROMIDE □ ROQUESSINE DIHYDROBROMIDE □ WRIGHTINE DIHYDROBROMIDE

TOXICITY DATA WITH REFERENCE
orl-mus LD50:390 mg/kg CHTPBA 5,129,70
ipr-mus LD50:85 mg/kg CHTPBA 5,129,70
ivn-mus LD50:27 mg/kg CHTPBA 5,129,70

SAFETY PROFILE: Poison by ingestion, intraperitoneal, and intravenous routes. When heated to decomposition it emits toxic fumes of NO_x and HBr.

DOX100 HR: 3
1-DIMETHYLAMINO-3-CYANO-3-PHENYL-4-METHYLHEXANE HYDROCHLORIDE
mf: $C_{16}H_{24}N_2 \cdot ClH$ mw: 280.88

SYNS: 2-(2-(DIMETHYLAMINO)ETHYL)-3-METHYL-2-PHENYLVALERONITRILE HYDROCHLORIDE □ Z-4

TOXICITY DATA WITH REFERENCE
orl-rat LD50:407 mg/kg JPETAB 117,451,56
orl-mus LD50:382 mg/kg JPETAB 117,451,56
ipr-mus LD50:186 mg/kg JPETAB 117,451,56

CONSENSUS REPORTS: Cyanide and its compounds are on the Community Right-To-Know List.

SAFETY PROFILE: Poison by ingestion and intraperitoneal routes. When heated to decomposition it emits toxic fumes of NO_x, CN^-, and HCl. See also NITRILES.

DOX200 CAS:23273-02-1 HR: 3
DIMETHYLAMINODIBORANE
mf: $C_2H_7N{\bullet}B_2H_6$ mw: 72.78

PROP: Volatile liquid, sensitive to air and moisture; can be stored for months in sealed tube *in vacuo.* Sol in ethers and aromatic hydrocarbons.

SYN: DIMETHYLAMINE with DIBORANE (1:1)

TOXICITY DATA WITH REFERENCE
ihl-rat LC50:248 mg/m³/4H 14KTAK -,693,64
ihl-mus LC50:182 mg/m³/4H 14KTAK -,693,64

SAFETY PROFILE: Poison by inhalation. Ignites spontaneously in air. When heated to decomposition it emits toxic fumes of NO_x. See also BORANES and AMINES.

DOX600 CAS:23103-98-2 HR: 3
2-(DIMETHYLAMINO)-5,6-DIMETHYL-4-PYRIMIDINYL-DIMETHYLCARBAMATE
mf: $C_{11}H_{18}N_4O_2$ mw: 238.33

PROP: A solid. Mp: 90.5°.

SYNS: ABOL ☐ AFICIDA ☐ APHOX ☐ DIMETHYLCARBAMIC ACID 2-(DIMETHYLAMINO) 5,6 DIMETHYL-4-PYRIMIDINYL ESTER ☐ 5,6-DIMETHYL-2-DIMETHYLAMINO-4-PYRIMIDINYLDIMETHYLCARBAMATE ☐ ENT 27,766 ☐ FERNOS ☐ PIRIMICARB ☐ PIRIMOR ☐ PP 062 ☐ PYRIMOR ☐ RAPID

TOXICITY DATA WITH REFERENCE
cyt-hmn:lym 10 mg/L CYGEDX 15(2),74,81
cyt-mus unr 2 mg/kg IGANAK 14(6),41,80
orl-rat LD50:147 mg/kg CHINAG 30,1018,69
orl-mus LD50:107 mg/kg 28ZEAL 5,184,76
orl-dog LD50:100 mg/kg 28ZEAL 5,184,76

CONSENSUS REPORTS: EPA Genetic Toxicology Program.

SAFETY PROFILE: Poison by ingestion. Human mutation data reported. An insecticide. When heated to decomposition it emits toxic fumes of NO_x. See also CARBAMATES.

DOY400 CAS:60-46-8 HR: 3
4-(DIMETHYLAMINO)-2,2-DIPHENYLVALERAMIDE
mf: $C_{19}H_{24}N_2O$ mw: 296.45

SYNS: AMINOPENTAMIDE ☐ BL 139 ☐ CENTRINE ☐ α-(2-(DIMETHYLAMINO)PROPYL)-α-PHENYLBENZENEACETAMIDE ☐ DIMEVAMIDE ☐ α,α-DIPHENYL-γ-DIMETHYLAMINOVALERAMIDE ☐ 3-METHYL-4-DIMETHYLAMINO-2,2-DIPHENYLBUTYRAMIDE ☐ VALERAMIDE-OM

TOXICITY DATA WITH REFERENCE
orl-mus LDLo:441 mg/kg CLDND*
ipr-mus LDLo:121 mg/kg CLDND*
ivn-mus LDLo:46 mg/kg CLDND*

SAFETY PROFILE: Poison by intraperitoneal, and intravenous routes. Moderately toxic by ingestion. An anticholinergic. Used in veterinary medicine as an anticonvulsant and anti-emetic. When heated to decomposition it emits toxic fumes of NO_x.

DOY600 CAS:13242-44-9 HR: 3
2-DIMETHYLAMINO ETHANETHIOL HYDROCHLORIDE
mf: $C_4H_{11}NS{\bullet}ClH$ mw: 141.68

PROP: A solid. Mp: 156–157°.

SYNS: CAPTAMINE HYDROCHLORIDE ☐ N-DIMETHYLCYSTEAMINE HYDROCHLORIDE ☐ N-(2-MERCAPTOETHYL)DIMETHYLAMINE HYDROCHLORIDE ☐ NSC 45463

TOXICITY DATA WITH REFERENCE
ipr-mus LD50:280 mg/kg YKKZAJ 93,25,73

CONSENSUS REPORTS: Reported in EPA TSCA Inventory.

SAFETY PROFILE: Poison by intraperitoneal route. When heated to decomposition it emits very toxic fumes of Cl^-, SO_x, and NO_x.

DOY800 CAS:108-01-0 HR: 3
N-DIMETHYLAMINOETHANOL
DOT: UN 2051
mf: $C_4H_{11}NO$ mw: 89.16

$HOC_2H_4N(CH_3)_2$

PROP: A liquid. Bp: 135°, flash p: 105°F (OC), d: 0.8866 @ 20°/4°, vap d: 3.03.

SYNS: DEANOL ☐ DIMETHYLAETHANOLAMIN (GERMAN) ☐ DIMETHYLAMINOAETHANOL (GERMAN) ☐DIMETHYLAMINOETHANOL ☐ β-DIMETHYLAMINOETHANOL ☐ N,N-DIMETHYLAMINOETHANOL ☐ 2-(DIMETHYLAMINO)ETHANOL ☐ β-DIMETHYLAMINOETHYL ALCOHOL ☐ DIMETHYLETHANOLAMINE ☐ N,N-DIMETHYLETHANOLAMINE ☐ DIMETHYLETHANOLAMINE (DOT) ☐ N,N-DIMETHYL-2-HYDROXYETHYLAMINE ☐ N,N-DIMETHYL-N-(2-HYDROXYETHYL) AMINE ☐ DMAE ☐ β-HYDROXYETHYLDIMETHYLAMINE

TOXICITY DATA WITH REFERENCE
skn-rbt 445 mg open MLD UCDS** 12/15/71
eye-rbt 750 μg open SEV AMIHBC 4,119,51
orl-rat LD50:2 g/kg ZHYGAM 20,393,74
ihl-rat LCLo:4500 mg/m³/4H GTPZAB 14(11),52,70
ipr-rat LD50:1080 mg/kg TXAPA9 12,486,68
ihl-mus LC50:3250 mg/m³ GTPZAB 14(11),52,70
ipr-mus LD50:234 mg/kg JPETAB 94,249,48
scu-mus LD50:961 mg/kg AEPPAE 225,428,55
skn-rbt LD50:1370 mg/kg AMIHBC 4,119,51

CONSENSUS REPORTS: Reported in EPA TSCA Inventory.

DOT CLASSIFICATION: 3; *Label:* Flammable Liquid

SAFETY PROFILE: Moderately toxic by ingestion, inhalation, skin contact, intraperitoneal, and subcutaneous routes. A skin and severe eye irritant. Used medically as a central nervous system stimulant. Flammable liquid when exposed to heat or flame; can react vigorously with oxidizing materials. Ignites spontaneously in contact with cellulose nitrate of high surface area. To fight fire, use alcohol foam, foam, CO_2, dry chemical. When heated to decomposition it emits toxic fumes of NO_x.

D

DOZ000 CAS:3635-74-3 HR: 2
2-DIMETHYLAMINOETHANOL-p-ACETAMIDOBENZOATE
mf: $C_{13}H_{18}N_2O_2$ mw: 234.33

PROP: Crystals from EtOH/EtOAc. Mp: 159–161.5°.

SYNS: 4-(ACETYLAMINO)BENZOIC ACID with 2-(DIMETHYLAMINO)ETHANOL (1:1) □ CERVOXAN □ DAYFEN □ DEANER □ DEANOL ACETAMIDOBENZOATE □ DEANOL-p-ACETAMIDOBENZOATE □ DIFORENE □ DMAE p-ACETAMIDOBENZOATE □ ELEVAN □ NERVOTON

TOXICITY DATA WITH REFERENCE
ipr-rat LD50:800 mg/kg 27ZQAG -,419,72
orl-mus LD50:3918 mg/kg 27ZQAG -,419,72
ipr-mus LD50:1020 mg/kg 27ZQAG -,419,72

SAFETY PROFILE: Moderately toxic by ingestion and intraperitoneal routes. An antidepressant. When heated to decomposition it emits toxic fumes of NO_x.

DOZ100 CAS:1421-89-2 HR: 3
DIMETHYLAMINOETHANOL ACETATE
mf: $C_6H_{13}NO_2$ mw: 131.20

SYNS: ACETIC ACID, 2-(DIMETHYLAMINO)ETHYL ESTER □ 2-DIMETHYLAMINOETHANOL ACETATE □ DIMETHYLAMINOETHYL ACETATE □ 2-(DIMETHYLAMINO)ETHYL ACETATE

TOXICITY DATA WITH REFERENCE
ipr-mus LD50:104 mg/kg IJNEAQ 8,131,69

CONSENSUS REPORTS: Reported in EPA TSCA Inventory.

SAFETY PROFILE: Poison by intraperitoneal route. When heated to decomposition it emits toxic vapors of NO_x.

DPA000 CAS:63980-59-6 HR: 2
2-(DIMETHYLAMINO)ETHANOL BITARTRATE
mf: $C_4H_{11}NO \cdot 2C_4H_4O_6$ mw: 385.32

SYNS: ATROL □ DIMETHAEN □ LIPARON □ RECREIN

TOXICITY DATA WITH REFERENCE
orl-rat LD50:2590 mg/kg 27ZQAG -,419,72
ipr-rat LD50:459 mg/kg 27ZQAG -,419,72
scu-rat LD50:1098 mg/kg 27ZQAG -,419,72
orl-mus LD50:3100 mg/kg SCIEAS 126,610,57

SAFETY PROFILE: Moderately toxic by ingestion, intraperitoneal, and subcutaneous routes. When heated to decomposition it emits toxic fumes of NO_x.

DPA500 HR: 3
2-(2-DIMETHYLAMINOETHOXY)CHALCONE CITRATE
mf: $C_{19}H_{21}NO_2 \cdot C_6H_8O_7$ mw: 487.55

TOXICITY DATA WITH REFERENCE
orl-mus LD50:603 mg/kg JAPMA8 47,640,58
ipr-mus LD50:158 mg/kg JAPMA8 47,640,58
ivn-mus LD50:40,800 µg/kg JAPMA8 47,640,58
ivn-dog LDLo:41 mg/kg JAPMA8 47,640,58

SAFETY PROFILE: Poison by intravenous and intraperitoneal routes. Moderately toxic by ingestion. When heated to decomposition it emits toxic fumes of NO_x.

DPA600 CAS:1704-62-7 HR: 3
2-(2-DIMETHYLAMINOETHOXY)ETHANOL
mf: $C_6H_{15}NO_2$ mw: 133.22

TOXICITY DATA WITH REFERENCE
skn-rbt 500 mg/24H MLD 85JCAE-,629,86
eye-rbt 750 µg/24H SEV 85JCAE-,629,86
orl-rat LD50:2460 mg/kg TXAPA9 28,313,74
skn-rbt LD50:1410 mg/kg TXAPA9 28,313,74

CONSENSUS REPORTS: Reported in EPA TSCA Inventory.

SAFETY PROFILE: Moderately toxic by ingestion and skin contact. A skin and severe eye irritant. When heated to decomposition it emits toxic fumes of NO_x.

DPB300 CAS:2439-35-2 HR: 3
DIMETHYLAMINOETHYL ACRYLATE
mf: $C_7H_{13}NO_2$ mw: 143.21

SYNS: ACRYLIC ACID, 2-(DIMETHYLAMINO)ETHYL ESTER □ ADAME □ 2-PROPENOIC ACID, 2-(DIMETHYLAMINO)ETHYL ESTER (9CI)

TOXICITY DATA WITH REFERENCE
orl-rat LD50:455 mg/kg EPASR* 8EHQ-1190-1119
ihl-rat LC50:66 mg/m³/4H EPASR* 8EHQ-0391-1119

SAFETY PROFILE: Poison by inhalation route. Moderately toxic by ingestion. When heated to decomposition it emits toxic vapors of NO_x.

DPB400 CAS:54099-13-7 HR: 3
N-(2-(DIMETHYLAMINO)ETHYL)-1-ADAMANTANEACETAMIDE ETHYL IODIDE
mf: $C_{16}H_{28}N_2O \cdot C_2H_5I$ mw: 420.43

TOXICITY DATA WITH REFERENCE
orl-mus LD50:600 mg/kg FRPSAX 32,129,77
ipr-mus LD50:75 mg/kg FRPSAX 32,129,77

SAFETY PROFILE: Poison by intraperitoneal route. Moderately toxic by ingestion. When heated to decomposition it emits very toxic fumes of NO_x and I^-. See also IODIDES.

DPC000 CAS:108-00-9 HR: 3
2-DIMETHYLAMINOETHYLAMINE
mf: $C_4H_{12}N_2$ mw: 88.15

PROP: Flash p: 51.8°F, bp: 107°.

SAFETY PROFILE: A very dangerous fire hazard when exposed to heat, flame, or oxidizers. When heated to decomposition it emits toxic fumes of NO_x.

DPC200 CAS:17599-02-9 HR: 3
2-(DIMETHYLAMINO)ETHYL-p-AMINOBENZOATE HYDROCHLORIDE
mf: $C_{11}H_{16}N_2O_2 \cdot ClH$ mw: 244.75

SYNS: 2-(DIMETHYLAMINO)ETHYL ESTER-p-AMINOBENZOIC ACID

HYDROCHLORIDE □ HCL SALZ des p-AMINO-BENZOESAEURE-DI-METHYLAMINO-AETHYL-ESTER (GERMAN)

TOXICITY DATA WITH REFERENCE
ipr-rat LDLo:370 mg/kg ARZNAD 1,154,51
ivn-rat LDLo:72 mg/kg ARZNAD 1,154,51

SAFETY PROFILE: Poison by intraperitoneal and intravenous routes. When heated to decomposition it emits very toxic fumes of HCl and NO_x.

DPC400 CAS:52400-61-0 HR: 3
2-(2-(2-(DIMETHYLAMINO)ETHYLAMINO)ETHYL)-2-METHYL-1,3-BENZODIOXOLEDI HYDROCHLORIDE
mf: $C_{14}H_{22}N_2O_2$•2ClH mw: 323.30

TOXICITY DATA WITH REFERENCE
ivn-rat LD50:51 mg/kg EJMCA5 12,413,77
ipr-mus LD50:175 mg/kg EJMCA5 12,413,77

SAFETY PROFILE: Poison by intravenous and intraperitoneal routes. When heated to decomposition it emits very toxic fumes of HCl and NO_x.

DPD200 CAS:942-46-1 HR: 3
α-(1-(DIMETHYLAMINO)ETHYL)BENZYL ALCOHOL HYDROCHLORIDE
mf: $C_{11}H_{17}NO$•ClH mw: 215.75

SYNS: METHYLEPHEDRINE HYDROCHLORIDE □ l-METHYLEPHEDRINE HYDROCHLORIDE □ N-METHYLEPHEDRINE HYDROCHLORIDE

TOXICITY DATA WITH REFERENCE
ipr-mus LD50:185 mg/kg AIPTAK 138,209,62
scu-mus LD50:699 mg/kg NIIRDN 6,827,82
ivn-rbt LDLo:55 mg/kg JPETAB 36,363,29
ivn-gpg LDLo:139 mg/kg AIPTAK 125,236,60

SAFETY PROFILE: Poison by intravenous and intraperitoneal routes. Moderately toxic by subcutaneous route. When heated to decomposition it emits very toxic fumes of NO_x and Cl^-.

DPD400 CAS:6152-43-8 HR: 3
(2-(DIMETHYLAMINO)ETHYL)(o-BENZYLPHENOXY) ETHERHYDROCHLORIDE
mf: $C_{17}H_{21}NO$•ClH mw: 291.85

SYNS: N-(2'-DIMETHYLAMINOAETHYL)-(o-BENZYLPHENOL)-AETHER HYDROCHLORID (GERMAN) □ HL 2153

TOXICITY DATA WITH REFERENCE
orl-mus LD50:305 mg/kg ARZNAD 8,219,58
ipr-mus LD50:164 mg/kg ARZNAD 8,219,58
ivn-mus LD50:60 mg/kg ARZNAD 8,219,58

SAFETY PROFILE: Poison by ingestion, intraperitoneal, and intravenous routes. When heated to decomposition it emits very toxic fumes of HCl and NO_x.

DPE000 CAS:51-68-3 HR: 3
DIMETHYLAMINOETHYL-4-CHLOROPHENOXYACETIC ACID
mf: $C_{12}H_{16}ClNO_3$ mw: 257.74

SYNS: ACEPHENE □ ANALUX □ ANP 235 □ AT SEFEN □ CENTROFENOXINA □ CEREBON □ CLOFENOXIN □ CLOPHENOXATE □ p-CHLOROPHENOXYACETIC ACID-β-DIMETHYLAMINOETHYL ESTER □ DEANOL-p-CHLOROPHENOXYACETATE □ DEANOLESTERE □ DIMETHYLAMINOETHYL-p-CHLOROPHENOXYACETATE □ EN 1627 □ HELFERGIN □ LICIDRIL □ LUCIDRYL □ MECLOFENOXANE □ MECLOPHENOXATE □ MUCIDRIL □ PROSERYL

TOXICITY DATA WITH REFERENCE
orl-rat LD50:2600 mg/kg 27ZQAG -,386,72
orl-mus LD50:1750 mg/kg 27ZQAG -,386,72
ipr-mus LD50:800 mg/kg BCFAAI 111,293,72
ivn-rbt LD50:150 mg/kg 27ZQAG -,386,72

SAFETY PROFILE: Poison by intravenous route. Moderately toxic by ingestion and intraperitoneal routes. When heated to decomposition it emits very toxic fumes of Cl^- and NO_x.

DPE100 CAS:33232-39-2 HR: 3
5-(2-(DIMETHYLAMINO)ETHYL)-2,3-DIHYDRO-3-HYDROXY-2-(p-METHOXYPHENYL)1,5-BENZOTHIAZEPIN-4(5H)-ONE-ACETATE (ESTER) HYDROCHLORIDE
mf: $C_{22}H_{26}N_2O_4S$•ClH mw: 451.02

TOXICITY DATA WITH REFERENCE
orl-rat LD50:560 mg/kg IYKEDH 5,106,74
scu-rat LD50:520 mg/kg IYKEDH 5,106,74
ivn-rat LD50:38 mg/kg IYKEDH 5,106,74
orl-mus LD50:640 mg/kg IYKEDH 5,106,74
scu-mus LD50:280 mg/kg IYKEDH 5,106,74
ivn-mus LD50:58 mg/kg IYKEDH 5,106,74

SAFETY PROFILE: Poison by subcutaneous and intravenous routes. Moderately toxic by ingestion. When heated to decomposition it emits toxic fumes of SO_x, NO_x, and HCl. See also ESTERS.

DPE200 CAS:2424-75-1 HR: 3
DIMETHYLAMINOETHYLDIPHENYLETHOXY ACETATE HYDROCHLORIDE
mf: $C_{20}H_{25}NO_3$•ClH mw: 363.92

PROP: Needles. Mp: 170–172°.

SYNS: AESTOCIN □ DIMENOXADOL HYDROCHLORIDE □ β'-DIMETHYLAMINOETHYL-α,α-DIPHENYL-α-ETHOXYACETATEHYDROCHLORIDE □ 2-(DIMETHYLAMINO)ETHYL ESTER HYDROCHLORIDE ETHOXYDIPHENYLACETIC ACID □ 2,2-DIPHENYL-2-ETHOXYACETIC ACID (2-(DIMETHYLAMINO)ETHYL) ESTER HYDROCHLORIDE □ ESTOCINE □ ESTOTSIN □ LOKARIN □ PROPALGYL

TOXICITY DATA WITH REFERENCE
ivn-rat LD50:66 mg/kg PCJOAU 8,189,74
orl-mus LD50:700 mg/kg MEIEDD 10,467,84
ipr-mus LD50:175 mg/kg PCJOAU 4,7,70
scu-mus LD50:179 mg/kg RPTOAN 32,317,69
ivn-mus LDLo:40 mg/kg JMCMAR 8,571,65

SAFETY PROFILE: Poison by intravenous, intraperitoneal, and subcutaneous routes. Moderately toxic by ingestion. An analgesic and anticonvulsant. When heated to decomposition it emits very toxic fumes of NO_x and HCl.

DPE800 CAS:78372-05-1 HR: 3
1-(2-(DIMETHYLAMINO)ETHYL)-1-ETHYL-3-MESITYLUREA HYDROCHLORIDE
mf: $C_{16}H_{27}N_3O{\bullet}ClH$ mw: 313.92

TOXICITY DATA WITH REFERENCE
ipr-rat LD50:123 mg/kg ARZNAD 8,664,58
scu-mus LD50:230 mg/kg ARZNAD 8,664,58

SAFETY PROFILE: Poison by intraperitoneal and subcutaneous routes. When heated to decomposition it emits very toxic fumes of NO_x and HCl.

DPF200 CAS:69884-15-7 HR: 2
N-(4-(1-(DIMETHYLAMINO)ETHYLIDENE)AMINO)PHENYL)-2-METHOXYACETAMIDE HYDROCHLORIDE
mf: $C_{13}H_{19}N_3O_2{\bullet}ClH$ mw: 285.81

PROP: A solid. Mp: 206–207°.

SYNS: AMIDANTEL ☐ BAY d8815 ☐ N-(4-((1-(DIMETHYLAMINO)-AETHYLIDEN)AMINO)PHENYL)-2-METHOXYACETAMID-HYDROCHLORID (GERMAN)

TOXICITY DATA WITH REFERENCE
orl-rat LD50:4693 mg/kg ARZNAD 29,31,79
orl-mus LD50:1207 mg/kg ARZNAD 29,31,79
scu-mus LD50:569 mg/kg ARZNAD 29,31,79
orl-dog LD50:500 mg/kg ARZNAD 29,31,79
orl-cat LD50:750 mg/kg ARZNAD 29,31,79
orl-rbt LD50:500 mg/kg ARZNAD 29,31,79

SAFETY PROFILE: Moderately toxic by ingestion and subcutaneous routes. When heated to decomposition it emits very toxic fumes of NO_x and HCl.

DPF600 CAS:61-50-7 HR: 3
3-(2-(DIMETHYLAMINO)ETHYL)INDOLE
mf: $C_{12}H_{16}N_2$ mw: 188.30

PROP: A solid. Mp: 48–49°.

SYNS: N,N-DIMETHYLTRYPTAMINE ☐ DMT

TOXICITY DATA WITH REFERENCE
ims-man TDLo:1 mg/kg:EYE,CNS,CVS PSYPAG 4,39,63
ipr-mus LD50:47 mg/kg YKKZAJ 94,1620,74
ivn-mus LD50:32 mg/kg CSLNX* NX#00740

SAFETY PROFILE: Poison by intravenous and intraperitoneal routes. Human systemic effects by intramuscular route: pupil dilation, hallucinations and distorted perceptions, blood pressure increase. When heated to decomposition it emits toxic fumes of NO_x.

DPG000 CAS:101831-88-3 HR: 3
3-(2-(DIMETHYLAMINO)ETHYL)INDOLESULFOSALICYLATE

SYN: N,N-DIMETHYL-β-3-INDOLYLETHYLAMINE SULFOSALICYLATE

TOXICITY DATA WITH REFERENCE
ipr-mus LD50:153 mg/kg RPTOAN 33,180,70
ivn-mus LD50:69 mg/kg RPTOAN 33,180,70

SAFETY PROFILE: Poison by intraperitoneal and intravenous routes. When heated to decomposition it emits very toxic fumes of NO_x and SO_x.

DPG109 CAS:487-93-4 HR: 3
3-(2-DIMETHYLAMINOETHYL)-5-INDOLOL
mf: $C_{12}H_{16}N_2O$ mw: 204.30

PROP: A solid. Mp: 146–147°, bp: 320° @ 0.1 mm.

SYNS: BUFOTENIN ☐ 3-(β-DIMETHYLAMINOETHYL)-5-HYDROXYINDOLE ☐ N,N-DIMETHYL-5-HYDROXYTRYPTAMINE ☐ N,N-DIMETHYLSEROTONIN ☐ 5-HYDROXY-N,N-DIMETHYLTRYPTAMINE

TOXICITY DATA WITH REFERENCE
ivn-hmn TDLo:57 μg/kg:PSY SCIEAS 123,886,56
ipr-mus LD50:290 mg/kg PSYPAG 16,385,70

SAFETY PROFILE: Poison by intraperitoneal route. Human systemic effects with very small amounts taken by intravenous route: psychotropic effects. A modified natural neurotransmitter. When heated to decomposition it emits toxic fumes of NO_x.

DPG200 CAS:101652-11-3 HR: 3
10-(2-(DIMETHYLAMINO)ETHYL)ISOALLOXAZINE SULFATE
mf: $C_{14}H_{15}N_5O_2{\bullet}H_2O_4S$ mw: 383.42

TOXICITY DATA WITH REFERENCE
scu-mus LD50:90 mg/kg CMTRAG 2,96,61
ivn-mus LD50:75 mg/kg CMTRAG 2,96,61

SAFETY PROFILE: Poison by subcutaneous and intravenous routes. When heated to decomposition it emits very toxic fumes of SO_x and NO_x.

DPG400 CAS:78372-06-2 HR: 3
1-(2-(DIMETHYLAMINO)ETHYL)-1-ISOPROPYL-3-(2,6-XYLYL)UREA HYDROCHLORIDE
mf: $C_{16}H_{27}N_3O{\bullet}ClH$ mw: 313.92

SYN: C 3215

TOXICITY DATA WITH REFERENCE
eye-rbt 2% MLD ARZNAD 8,664,58
ipr-rat LD50:41 mg/kg ARZNAD 8,664,58
scu-mus LD50:57 mg/kg ARZNAD 8,664,58

SAFETY PROFILE: Poison by intraperitoneal and subcutaneous routes. An eye irritant. When heated to decomposition it emits very toxic fumes of NO_x and HCl.

DPG600 CAS:2867-47-2 HR: 3
DIMETHYLAMINOETHYL METHACRYLATE
DOT: UN 2522
mf: $C_8H_{15}NO_2$ mw: 157.24

PROP: Liquid, sol in water and organic solvents. D: 0.933 @ 25°, bp: 182–190°, flash p: 165°F (TOC), vap d: 5.4.

SYNS: AGEFLEX FM-1 ☐ 2-(DIMETHYLAMINO)ETHANOL METHACRYLATE ☐ 2-(DIMETHYLAMINO)ETHYL ESTER METHACRYLIC ACID ☐ N,N-DIMETHYLAMINOETHYL METHACRYLATE ☐ β-DIMETHYLAMINOETHYL METHACRYLATE ☐ 2-(DIMETHYLAMINO)ETHYL METHACRYLATE ☐ USAF RH-3

TOXICITY DATA WITH REFERENCE
orl-rat LD50:1751 mg/kg 85GMAT -,55,82
ihl-rat LC50:620 mg/m³/4H 85GMAT -,55,82
ihl-mus LC50:1800 mg/m³/2H 85GMAT -,55,82

ipr-mus LD50:25 mg/kg NTIS** AD277-689

CONSENSUS REPORTS: Reported in EPA TSCA Inventory.

DOT CLASSIFICATION: 6.1; *Label:* Poison

SAFETY PROFILE: Poison by intraperitoneal route. Moderately toxic by ingestion and inhalation. A skin, eye, and mucous membrane irritant. A powerful lachrymator. Flammable when exposed to sparks, heat, open flame, or oxidizers. To fight fire, use alcohol foam, dry chemical, spray. When heated to decomposition it emits toxic fumes of NO_x. See also ESTERS.

DPH000 CAS:4724-58-7 HR: 3
2-DIMETHYLAMINOETHYL-2-METHYL-BENZHYDRYL ETHER CITRATE
mf: $C_{18}H_{23}NO\cdot C_6H_8O_7$ mw: 461.56

SYNS: BENHEXAL □ N,N-DIMETHYL-2-((o-METHYL-α-PHENYL-BENZYL)OXY)-ETHYLAMINE CITRATE □ NORFLEX □ ORFENADRINA □ ORPHENADRINE CITRATE □ R-528

TOXICITY DATA WITH REFERENCE
dnd esc 50 μmol/L MUREAV 89,95,81
ivn-rat LD50:26 mg/kg 27ZQAG -,373,72
ims-rat LD50:208 mg/kg 27ZQAG -,373,72
orl-mus LD50:150 mg/kg 29ZVAB -,83,69
ivn-mus LD50:37 mg/kg 27ZQAG -,373,72
ivn-rbt LD50:22 mg/kg IJNEAQ 5,305,66

SAFETY PROFILE: Poison by ingestion, intramuscular, and intravenous routes. Mutation data reported. When heated to decomposition it emits toxic fumes of NO_x. See also ETHERS.

DPH200 CAS:65210-30-2 HR: 3
2-(2-(DIMETHYLAMINO)ETHYL)-2-METHYL-1,3-BENZODIOXOLE HYDROCHLORIDE
mf: $C_{12}H_{17}NO_2\cdot ClH$ mw: 243.76

TOXICITY DATA WITH REFERENCE
ivn-rat LD50:60 mg/kg EJMCA5 12,413,77
ipr-mus LD50:200 mg/kg EJMCA5 12,413,77

SAFETY PROFILE: Poison by intravenous and intraperitoneal routes. When heated to decomposition it emits very toxic fumes of HCl and NO_x.

DPH400 CAS:104-19-8 HR: 3
1-(2-(DIMETHYLAMINO)ETHYL)-4-METHYLPIPERAZINE
mf: $C_9H_{21}N_3$ mw: 171.33

TOXICITY DATA WITH REFERENCE
skn-rbt 100 μg/24H open AIHAAP 23,95,62
skn-rbt 5 mg/24H SEV 85JCAE-,867,86
eye-rbt 750 μg/24H SEV 85JCAE-,867,86
orl-rat LD50:1420 mg/kg AIHAAP 23,95,62
skn-rbt LD50:390 mg/kg AIHAAP 23,95,62

CONSENSUS REPORTS: Reported in EPA TSCA Inventory.

SAFETY PROFILE: Poison by skin contact. Moderately toxic by ingestion. A severe skin and eye irritant. When heated to decomposition it emits toxic fumes of NO_x.

DPH600 CAS:4985-15-3 HR: 3
5-DIMETHYLAMINOETHYLOXYIMINO-5H-DIBENZO(a,d)CYCLOHEPTA-1,4-DIENE HYDROCHLORIDE
mf: $C_{19}H_{22}N_2O\cdot ClH$ mw: 330.89

PROP: A solid. Mp: 185–187°.

SYNS: AGEDAL □ BAY 1521 □ 5-(DIMETHYLAMINOAETHYL-OXYIMINO)-5H-DIBENZO(a,d)CYCLOHEPTA-1,4-DIENHYDROCHLORID (GERMAN) □ 5-(DIMETHYLAMINOOXYIMINO)-5H-DIBENZO(a,b)CYCLOHEPTA-1,4-DIENE HYDROCHLORIDE □ 5-(DIMETILAMINOETILOSIMINO-5H-DIBENZO(a,d)CICLOEPTA-1,4-DIENE) CLORIDRATO (ITALIAN) □ NOGEDAL □ NOXIPTILINE HYDROCHLORIDE □ NOXIPTILIN HYDROCHLORID (GERMAN) □ NOXIPTYLINE HYDROCHLORIDE

TOXICITY DATA WITH REFERENCE
orl-rat TDLo:16,800 mg/kg (24W male):REP KSRNAM 6,1897,72
orl-rat LD50:607 mg/kg FRPPAO 25,519,70
ipr-rat LD50:149 mg/kg FRPPAO 25,519,70
scu-rat LD50:985 mg/kg 27ZQAG -,84,72
ivn-rat LD50:12 mg/kg KSRNAM 6,1897,72
ims-rat LD50:209 mg/kg FRPPAO 25,519,70
orl-mus LD50:275 mg/kg KSRNAM 6,1897,72
ipr-mus LD50:93 mg/kg FRPPAO 25,519,70
scu-mus LD50:212 mg/kg KSRNAM 6,1897,72
ivn-mus LD50:21,300 μg/kg KSRNAM 6,1897,72
ims-mus LD50:144 mg/kg FRPPAO 25,519,70
orl-dog LD50:800 mg/kg ARZNAD 19,846,69
scu-dog LD50:100 mg/kg ARZNAD 19,846,69

SAFETY PROFILE: Poison by ingestion, subcutaneous, intravenous, intramuscular, and intraperitoneal routes. Experimental reproductive effects. When heated to decomposition it emits very toxic fumes of NO_x and HCl.

DPI000 CAS:5934-20-3 HR: 3
N-DIMETHYLAMINOETHYLPHENOTHIAZINE HYDROCHLORIDE
mf: $C_{16}H_{18}N_2S\cdot ClH$ mw: 306.88

PROP: A solid. Mp: 201–201.5°.

SYNS: N-(β-DIMETHYLAMINOETHYL)-PHENOTHIAZINEHYDROCHLORIDE □ FENETHAZINE HYDROCHLORIDE □ LISERGAN HYDROCHLORIDE □ RUTERGAN HYDROCHLORIDE

TOXICITY DATA WITH REFERENCE
ipr-mus LD50:115 mg/kg MEIEDD 10,572,83
scu-mus LD50:210 mg/kg 27ZQAG -,38,72

SAFETY PROFILE: Poison by intraperitoneal and subcutaneous routes. An antihistamine. When heated to decomposition it emits very toxic fumes of NO_x, SO_x, and HCl.

DPI400 CAS:52401-02-2 HR: 3
2-(2-(DIMETHYLAMINO)ETHYL)-2-PHENYL-1,3-BENZODIOXOLE HYDROCHLORIDE
mf: $C_{17}H_{19}NO_2$•ClH mw: 305.83

TOXICITY DATA WITH REFERENCE
ivn-rat LD50:23 mg/kg EJMCA5 12,413,77
ipr-mus LD50:83 mg/kg EJMCA5 12,413,77

SAFETY PROFILE: Poison by intravenous and intraperitoneal routes. When heated to decomposition it emits very toxic fumes of HCl and NO_x.

DPI600 CAS:74758-13-7 HR: 2
N-(2-(DIMETHYLAMINO)ETHYL)-N-(3-PHENYL-1-INDOLYL)ACETAMIDE HYDROCHLORIDE
mf: $C_{20}H_{23}N_3O$•ClH mw: 357.92

TOXICITY DATA WITH REFERENCE
orl-rat LD50:1100 mg/kg ARZNAD 30,919,80
orl-mus LD50:1000 mg/kg ARZNAD 30,919,80

SAFETY PROFILE: Moderately toxic by ingestion. When heated to decomposition it emits very toxic fumes of NO_x and HCl.

DPI700 HR: 3
β-DIMETHYLAMINOETHYL-2-PHENYLTETRAHYDROBENZOATE HYDROCHLORIDE
mf: $C_{17}H_{23}NO_2$•ClH mw: 309.87

SYNS: 2-PHENYL-3-CYCLOHEXENE-1-CARBOXYLIC ACID 2-DIMETHYLAMINOETHYL ESTER HYDROCHLORIDE □ S 187

TOXICITY DATA WITH REFERENCE
scu-rat LDLo:800 mg/kg APPNAH 1,4,50
ivn-rat LDLo:80 mg/kg APPNAH 1,4,50
ivn-rbt LDLo:40 mg/kg APPNAH 1,4,50
scu-gpg LDLo:267 mg/kg APPNAH 1,4,50
ivn-gpg LDLo:27 mg/kg APPNAH 1,4,50

SAFETY PROFILE: Poison by intravenous and subcutaneous routes. When heated to decomposition it emits toxic fumes of NO_x and HCl.

DPI750 CAS:96811-96-0 HR: 3
2-((2-(DIMETHYLAMINO)ETHYL)(SELENOPHENE-2-YLMETHYL)AMINO)PYRIDINE
mf: $C_{14}H_{19}N_3Se$ mw: 308.32

SYN: PYRIDINE, 2-((2-(DIMETHYLAMINO)ETHYL)(SELENOPHENE-2-YLMETHYL)AMINO)-

TOXICITY DATA WITH REFERENCE
par-mus LD50:90 mg/kg 43FLAV 4(3),559,80

OSHA PEL: TWA 0.2 mg(Se)/m^3
ACGIH TLV: TWA 0.2 mg(Se)/m^3

SAFETY PROFILE: Poison by parenteral route. When heated to decomposition it emits toxic fumes of NO_x and Se.

DPJ200 CAS:91-79-2 HR: 3
2-((2-DIMETHYLAMINOETHYL)-3-THENYLAMINO)PYRIDINE
mf: $C_{14}H_{19}N_3S$ mw: 261.42

PROP: Bp: 169–172° @ 1 mm.

SYNS: DIETHYLENDIAMINE □ N-(2-DIMETHYLAMINOETHYL)-N-2-PYRIDYL-3-THENYLAMINE □ METHAPYRILENE □ NCI-C60640 □ N-(α-PYRIDYL)-N-(β-THENYL)-N′,N′-DIMETHYLETHYLENEDIAMINE □ TENFIDIL □ THEFANIL □ THENFADIL □ THENYLDIAMINE □ WIN-2848

TOXICITY DATA WITH REFERENCE
dnd-esc 30 μmol/L MUREAV 89,95,81
orl-hmn TDLo:50 mg/kg:CNS,GIT CTOXAO 11,287,77
ipr-mus LD50:77 mg/kg CLDND•

SAFETY PROFILE: Poison by intraperitoneal route. Human systemic effects by ingestion: hallucinations and distorted perceptions, gastrointestinal changes. Mutation data reported. An antihistamine. When heated to decomposition it emits very toxic fumes of NO_x and SO_x.

DPJ400 CAS:135-23-9 HR: 3
2-((2-(DIMETHYLAMINO)ETHYL)-2-THENYL-AMINO)PYRIDINE HYDROCHLORIDE
mf: $C_{14}H_{19}N_3S$•ClH mw: 297.88

PROP: A solid. Mp: 162°. Very sol in water.

SYNS: BARHIST □ CAPATHYN □ CORYZOL □ N,N-DIMETHYL-N′-2-PYRIDINYL-N′-(2-THIENYLMETHYL)-1,2-ETHANEDIAMINE MONOHYDROCHLORIDE □ N,N-DIMETHYL-N′-(2-PYRIDYL)-N′-THENYLETHYLENEDIAMINE HYDROCHLORIDE □ N,N-DIMETHYL-N′-(2-THENYL)-N′-(2-PYRIDYL-ETHYLENE-DIAMINE HYDROCHLORIDE) □ DOZAR □ HISTADYL HYDROCHLORIDE □ HISTAFED □ HISTIDYL □ LULLAMIN □ METHACON □ METHAPYRILENE HYDROCHLORIDE □ METHAPYRILENE HYDROCHLORIDE (L.A.) □ METHAPYRILENE HYDROCHLORIDE (S.A.) □ METHOXYLENE □ PYRATHYN □ N-(2-PYRIDYL)-N-(2-THIENYL)-N,N′-DIMETHYL-ETHYLENEDIAMINE HYDROCHLORIDE □ SEMIKON □ SEMIKON HYDROCHLORIDE □ SOMNICAPS □ TEM-HISTINE □ TERALIN □ THENYLENE □ THENYLENE HYDROCHLORIDE □ THENYLPYRAMINE HYDROCHLORIDE □ W-53 HYDROCHLORIDE □ WIN 2848 HYDROCHLORIDE SALT

TOXICITY DATA WITH REFERENCE
mmo-smc 500 mg/L IAPUDO 57,721,84
mrc-smc 1250 mg/L IAPUDO 57,721,84
dns-rat :lvr 1 μmol/L CNREA8 42,3010,82
oms-rat:lvr 100 μmol/L MUREAV 135,131,84
orl-rat TDLo:9100 mg/kg/26/W-C:CAR TXAPA9 66,252,82
orl-rat TD:18,200 mg/kg/26W-C:CAR TXAPA9 66,252,82
orl-rat TD:22 g/kg/64W-C:CAR SCIEAS 209,817,80
orl-rat TD:9625 mg/kg/2Y-C:CAR FCTOD7 22,27,84
orl-rat TD:4813 mg/kg/2Y-C:ETA FCTOD7 22,27,84
orl-rat TD:10,080 mg/kg/24W-C:ETA TXAPA9 74,63,84
orl-man TDLo:429 μg/kg:GIT JPETAB 90,83,47
orl-rat LD50:521 mg/kg 29ZVAB -,73,69
scu-rat LD50:150 mg/kg 29ZVAB -,73,69
orl-mus LD50:182 mg/kg JPETAB 90,83,47
scu-mus LD50:75 mg/kg JPETAB 93,210,48
ivn-mus LD50:17,500 μg/kg JPETAB 93,210,48
ivn-gpg LD50:14,600 μg/kg AIPTAK 113,313,58

CONSENSUS REPORTS: NCI Carcinogenesis Studies (feed): Clear Evidence: rat FCTOD7 22,27,84; Clear Evidence: rat SCIEAS 209,817,80; (gavage); No

Evidence: guinea pig, ham JTEHD6 12,653,83. Reported in EPA TSCA Inventory.

SAFETY PROFILE: Poison by ingestion, intravenous, and subcutaneous routes. Human systemic effects by ingestion: gastritis. Questionable carcinogen with experimental carcinogenic and tumorigenic data. Mutation data reported. An antihistamine. When heated to decomposition it emits very toxic fumes of Cl^-, SO_x, and NO_x.

DPJ600 CAS:13261-62-6 HR: 2
2-DIMETHYLAMINOFLUORENE
mf: $C_{15}H_{15}N$ mw: 209.31

PROP: Crystals from EtOH. Mp: 180°.

SYNS: 2-DIMETHYLAMINO-FLUOREN (GERMAN) □ N,N-DIMETHYL-2-AMINOFLUORENE □ 2-FLUORENYLDIMETHYLAMINE

TOXICITY DATA WITH REFERENCE
orl-rat TDLo:1370 mg/kg/47W-I:ETA BJCAAI 6,89,52
orl-rat TD:4250 mg/kg/47W-I:ETA JBCHA3 221,845,56
orl-rat TD:3960 mg/kg/13W-C:ETA ONCOAR 8,233,55

SAFETY PROFILE: Questionable carcinogen with experimental tumorigenic data. When heated to decomposition it emits toxic fumes of NO_x.

DPJ800 CAS:87-01-4 HR: 2
7-DIMETHYLAMINO-4-METHYLCOUMARIN
mf: $C_{12}H_{13}NO_2$ mw: 203.26

SYNS: 2H-1-BENZOPYRAN-2-ONE, 7-(DIMETHYLAMINO)-4-METHYL-(9CI) □ COUMARIN 311 □ DAMC □ 7-(DIMETHYLAMINO)-4-METHYL-2H-1-BENZOPYRAN-2-ONE □ FBA 52

TOXICITY DATA WITH REFERENCE
orl-rat LDLo:1500 mg/kg CNREA8 26,619,66

CONSENSUS REPORTS: Reported in EPA TSCA Inventory.

SAFETY PROFILE: Moderately toxic by ingestion. When heated to decomposition it emits toxic fumes of NO_x.

DPK000 HR: 3
anti-8-(N,N-DIMETHYLAMINOMETHYL)DIBENZOBICYCLO(3.2.1)OCTADIENE HYDROCHLORIDE

SYN: 10,11-DIHYDRO-N,N-DIMETHYL-5,10-METHANO-5H-DIBENZO(a,d)CYCLOHEPTENE-12-METHANAMINE HCl

TOXICITY DATA WITH REFERENCE
orl-mus LD50:151 mg/kg DRFUD4 3,142,78
ivn-mus LD50:29 mg/kg DRFUD4 3,142,78

SAFETY PROFILE: Poison by ingestion and intravenous routes. When heated to decomposition it emits very toxic fumes of NO_x and HCl.

DPL000 CAS:55738-54-0 HR: 3
trans-2-((DIMETHYLAMINO)METHYLIMINO)-5-(2-(5-NITRO-2-FURYL)VINYL)-1,3,4-OXADIAZOLE
mf: $C_{11}H_{12}N_5O_4$ mw: 277.27

TOXICITY DATA WITH REFERENCE
mma-sat 100 ng/plate MUREAV 40,9,76
mmo-esc 300 nmol/well CNREA8 34,2266,74
orl-rat TDLo:42 g/kg/46W-I:CAR JNCIAM 51,403,73
orl-rat TD:42 g/kg/46W-C:CAR JNCIAM 54,841,75

CONSENSUS REPORTS: IARC Cancer Review: Group 2B IMEMDT 7,56,87; Animal Limited Evidence IMEMDT 7,147,74. EPA Genetic Toxicology Program.

SAFETY PROFILE: Confirmed carcinogen with experimental carcinogenic data. Mutation data reported. When heated to decomposition it emits toxic fumes of NO_x.

D

DPL200 CAS:2914-77-4 HR: 3
2-DIMETHYLAMINOMETHYL-1-(m-METHOXYPHENYL) CYCLOHEXANOL
mf: $C_{16}H_{25}NO_2$ mw: 263.42

TOXICITY DATA WITH REFERENCE
orl-mus LD50:395 mg/kg ARZNAD 28,107,78

SAFETY PROFILE: Poison by ingestion route. When heated to decomposition it emits toxic fumes of NO_x.

DPL900 CAS:63982-47-8 HR: 3
3-DIMETHYLAMINO-4-METHYLPHENYL ESTER-N-METHYLCARBAMIC ACID HYDROCHLORIDE
mf: $C_{11}H_{16}N_2O_2 \cdot ClH$ mw: 244.75

SYNS: N-METHYLURETHANE of HYDROCHLORIDE of 2-DIMETHYLAMINO-p-CRESOL □ T-1768

TOXICITY DATA WITH REFERENCE
orl-mus LD50:60 mg/kg JCSOA9 -,182,47
scu-mus LD50:10 mg/kg JCSOA9 -,182,47

SAFETY PROFILE: Poison by ingestion and subcutaneous routes. When heated to decomposition it emits toxic fumes of NO_x and HCl. See also CARBAMATES and ESTERS.

DPL950 CAS:56464-05-2 HR: 3
1-(4-(DIMETHYLAMINO)-2-METHYL-5-PHENYL-1H-PYRROL-3-YL)ETHANONE
mf: $C_{15}H_{18}N_2O$ mw: 242.35

SYNS: ETHANONE, 1-(4-(DIMETHYLAMINO)-2-METHYL-5-PHENYL-1H-PYRROL-3-YL)- □ KETONE, (4-(DIMETHYLAMINO)-2-METHYL-5-PHENYLPYRROL-3-YL) METHYL

TOXICITY DATA WITH REFERENCE
orl-mus LD50:1 g/kg FRPSAX 39,538,84

DOT CLASSIFICATION: 3; *Label:* Flammable Liquid

SAFETY PROFILE: Low toxicity by ingestion. A flammable liquid. When heated to decomposition it emits toxic vapors of NO_x.

DPM200 CAS:1477-79-8 HR: 3
3-DIMETHYLAMINO-2-METHYL-1-PHENYL-o-TOLYPROPANOL HYDROCHLORIDE
mf: $C_{19}H_{25}NO \cdot ClH$ mw: 319.91

SYNS: α-(2-(DIMETHYLAMINO)-1-METHYLETHYL)-2-METHYL-α-PHENYLBENZENEETHANOL, HCl □ SKF 70643-A

TOXICITY DATA WITH REFERENCE
orl-rat LD50:185 mg/kg JPPMAB 17,509,65
orl-mus LD50:245 mg/kg JPPMAB 17,509,65
orl-mam LD50:250 mg/kg JMCMAR 8,836,65

SAFETY PROFILE: Poison by ingestion. When heated to decomposition it emits very toxic fumes of HCl and NO_x.

DPM400 CAS:7005-47-2 HR: 3
2-DIMETHYLAMINO-2-METHYL-1-PROPANOL
mf: $C_6H_{15}NO$ mw: 117.22

SYNS: DMAMP □ USAF CS-1

TOXICITY DATA WITH REFERENCE
ipr-mus LDLo:25 mg/kg NTIS** AD277-689

CONSENSUS REPORTS: Reported in EPA TSCA Inventory.

SAFETY PROFILE: Poison by intraperitoneal route. When heated to decomposition it emits toxic fumes of NO_x.

DPN200 CAS:605-65-2 HR: 3
5-(DIMETHYLAMINO)-1-NAPHTHALENESULFONYL CHLORIDE
mf: $C_{12}H_{12}ClNO_2$ mw: 237.70

PROP: Crystals from Me_2CO (aq). Mp: 69°.

SYNS: 1-CHLOROSULFONYL-5-DIMETHYLAMINONAPHTHALENE □ DANSYL □ DANSYL CHLORIDE □ DIMETHYLAMINONAPHTHALENESULFONYL CHLORIDE □ 1-DIMETHYLAMINONAPHTHALENE-5-SULFONYL CHLORIDE □ 1-(DIMETHYLAMINO)-5-NAPHTHALENESULFONYLCHLORIDE □ 5-DIMETHYLAMINONAPHTHYL-5-SULFONYL CHLORIDE

TOXICITY DATA WITH REFERENCE
ivn-mus LD50:56 mg/kg CSLNX* NX#00262

CONSENSUS REPORTS: Reported in EPA TSCA Inventory.

SAFETY PROFILE: Poison by intravenous route. When heated to decomposition it emits very toxic fumes of Cl^- and NO_x.

DPN400 CAS:6632-68-4 HR: 2
1,3-DIMETHYL-4-AMINO-5-NITROSOURACIL
mf: $C_6H_8N_4O_3$ mw: 184.18

PROP: Violet crystals. Sol in alkalis; sltly sol in H_2O and Me_2CO; insol in EtOH, $CHCl_3$, dioxan, and Et_2O.

SYN: DANU

TOXICITY DATA WITH REFERENCE
scu-rat TDLo:16 g/kg/34W-I:ETA ZKKOBW 80,297,73
ipr-mus LD50:2000 mg/kg ZKKOBW 80,297,73

CONSENSUS REPORTS: Reported in EPA TSCA Inventory.

SAFETY PROFILE: Moderately toxic by intraperitoneal route. Questionable carcinogen with experimental tumorigenic data. When heated to decomposition it emits toxic fumes of NO_x.

DPN800 CAS:5882-48-4 HR: 3
4-DIMETHYLAMINOPHENOL HYDROCHLORIDE
mf: $C_8H_{11}NO{\cdot}ClH$ mw: 173.66

SYN: p-DIMETHYLAMINOPHENOL HYDROCHLORIDE

TOXICITY DATA WITH REFERENCE
orl-rat LD50:689 mg/kg TXCYAC 31,165,84
ipr-rat LD50:90 mg/kg ARTODN 34,333,75
ivn-rat LD50:57 mg/kg ARTODN 34,337,75
orl-mus LD50:946 mg/kg TXCYAC 31,165,84
ipr-mus LD50:83 mg/kg ARTODN 49,191,82
ivn-mus LD50:70 mg/kg TXCYAC 31,165,84
orl-gpg LD50:1032 mg/kg TXCYAC 31,165,84

SAFETY PROFILE: Poison by intraperitoneal and intravenous routes. Moderately toxic by ingestion. When heated to decomposition it emits very toxic fumes of NO_x and HCl.

DPO100 CAS:93407-11-5 HR: 3
3-DIMETHYLAMINO-2-PHENOXYPROPIOPHENONE HYDROCHLORIDE
mf: $C_{17}H_{19}NO_2{\cdot}ClH$ mw: 305.83

SYN: U-0172

TOXICITY DATA WITH REFERENCE
skn-rbt 2500 ppm MLD AIPTAK 137,410,62
eye-rbt 5000 ppm MLD AIPTAK 137,410,62
ipr-mus LD50:178 mg/kg AIPTAK 137,410,62

SAFETY PROFILE: Poison by intraperitoneal route. A skin and eye irritant. When heated to decomposition it emits toxic fumes of NO_x and HCl.

DPO200 CAS:539-17-3 HR: 3
4-(p-DIMETHYLAMINOPHENYLAZO)ANILINE
mf: $C_{14}H_{16}N_4$ mw: 240.34

SYNS: ACETILE DIAZO BLACK N □ ADAB □ p-AMINOBENZENEAZODIMETHYLANILINE □ 4-AMINO-DAB □ 4-AMINO-4′-DIMETHYLAMINOAZOBENZENE □ 4′-AMINO-N,N-DIMETHYL-4-AMINOAZOBENZENE □ 4-((4-AMINOPHENYL)AZO)-N,N-DIMETHYLBENZENAMINE □ C.I. 11025 □ C.I. DISPERSE BLACK 3 □ DIAZO NERO MICROSETILE G □ INTERCHEM ACETATE DEVELOPED BLACK □ MEISEI TERYL DIAZO BLACK CR □ MICROSETILE DIAZO BLACK G □ SUPRACET DIAZO BLACK A

TOXICITY DATA WITH REFERENCE
mma-sat 20 µg/plate CALEDQ 17,263,83
otr-rat:lvr 50 µmol/L CNREA8 43,5087,83
otr-ham:kdy 2500 µg/L BJCAAI 38,34,78
ipr-rat LD50:350 mg/kg CNREA8 34,2274,74
ipr-mus LD50:350 mg/kg CNREA8 34,2274,74

CONSENSUS REPORTS: Reported in EPA TSCA Inventory.

SAFETY PROFILE: Poison by intraperitoneal route. Mutation data reported. When heated to decomposition it emits toxic fumes of NO_x.

DPO275 **CAS:73688-85-4** **HR: 3**
4-(p-DIMETHYLAMINOPHENYLAZO)-BENZENEARSONIC ACID HYDROCHLORIDE
mf: $C_{14}H_{16}AsN_3O_3{\cdot}ClH$ mw: 385.71

PROP: Mp: 203° decomposes.

SYN: BENZENEARSONIC ACID, 4-(p-DIMETHYLAMINOPHENYLAZO)-, HYDROCHLORIDE

TOXICITY DATA WITH REFERENCE
ivn-mus LD50:25 mg/kg CSLNX* NX#05710

OSHA PEL: TWA 0.5 mg(As)/m³

SAFETY PROFILE: Poison by intravenous route. When heated to decomposition it emits toxic fumes of NO_x, As, and HCl.

DPO400 **CAS:18463-85-9** **HR: 2**
6-((p-(DIMETHYLAMINO)PHENYL)AZO)BENZOTHIAZOLE
mf: $C_{15}H_{14}N_4S$ mw: 282.39

SYNS: 6-DIMETHYLAMINOPHENYLAZOBENZOTHIAZOLE □ 6-DIMETHYLAMINOPHENYLAZOBENZTHIAZOLE □ N,N-DIMETHYL-p-(6-BENZTHIAZOLYLAZO)ANILINE □ N,N-DIMETHYL-4-(6′-BENZTHIAZOLYLAZO)ANILINE

TOXICITY DATA WITH REFERENCE
dns-rat-unr 10 mg/kg CRNGDP 6,611,85
dns-rat-orl 10 mg/kg MUREAV 156,1,85
orl-rat TDLo:540 mg/kg/4W-C:ETA JMCMAR 11,1074,68
orl-rat TD:600 mg/kg/60D-C:ETA CALEDQ 21,69,83

SAFETY PROFILE: Questionable carcinogen with experimental tumorigenic data. Mutation data reported. When heated to decomposition it emits very toxic fumes of SO_x and NO_x.

DPO600 **CAS:18559-92-7** **HR: 2**
7-((p-(DIMETHYLAMINO)PHENYL)AZO)BENZOTHIAZOLE
mf: $C_{15}H_{14}N_4S$ mw: 282.39

SYNS: N,N-DIMETHYL-p-(7-BENZTHIAZOLYLAZO)ANILINE □ N,N-DIMETHYL-4-(7′-BENZTHIAZOLYLAZO)ANILINE

TOXICITY DATA WITH REFERENCE
mma-sat 4 μg/plate MUREAV 93,67,82
orl-rat TDLo:1620 mg/kg/13W-C:ETA JMCMAR 11,1074,68

SAFETY PROFILE: Questionable carcinogen with experimental tumorigenic data. Mutation data reported. When heated to decomposition it emits very toxic fumes of NO_x and SO_x.

DPO800 **CAS:63040-63-1** **HR: 2**
4-((p-(DIMETHYLAMINO)PHENYL)AZO)ISOQUINOLINE
mf: $C_{17}H_{16}N_4$ mw: 276.37

SYN: N,N-DIMETHYL-4-(4′-ISOQUINOLINYLAZO)ANILINE

TOXICITY DATA WITH REFERENCE
orl-rat TDLo:3276 mg/kg/26W-C:ETA AICCA6 19,531,63

SAFETY PROFILE: Questionable carcinogen with experimental tumorigenic data. When heated to decomposition it emits toxic fumes of NO_x.

DPP000 **CAS:63040-64-2** **HR: 2**
5-((p-(DIMETHYLAMINO)PHENYL)AZO)ISOQUINOLINE
mf: $C_{17}H_{16}N_4$ mw: 276.37

SYN: N,N′-DIMETHYL-4-(5′-ISOQUINOLINYLAZO)ANILINE

TOXICITY DATA WITH REFERENCE
orl-rat TDLo:1092 mg/kg/26W-C:ETA AICCA6 19,531,63
orl-rat TD:2196 mg/kg/17W-C:ETA AICCA6 19,531,63

SAFETY PROFILE: Questionable carcinogen with experimental tumorigenic data. When heated to decomposition it emits toxic fumes of NO_x.

DPP200 **CAS:63040-65-3** **HR: 2**
7-((p-(DIMETHYLAMINO)PHENYL)AZO)ISOQUINOLINE
mf: $C_{17}H_{16}N_4$ mw: 276.37

SYN: N,N-DIMETHYL-4-(7′-ISOQUINOLINYLAZO)ANILINE

TOXICITY DATA WITH REFERENCE
orl-rat TDLo:3276 mg/kg/26W-C:ETA AICCA6 19,531,63

SAFETY PROFILE: Questionable carcinogen with experimental tumorigenic data. When heated to decomposition it emits toxic fumes of NO_x.

DPP400 **CAS:10318-23-7** **HR: 2**
5-((p-(DIMETHYLAMINO)PHENYL)AZO)ISOQUINOLINE-2-OXIDE
mf: $C_{17}H_{16}N_4O$ mw: 292.37

SYN: N,N-DIMETHYL-4-(5′-ISOQUINOLYL-2′-OXIDE)AZOANILINE

TOXICITY DATA WITH REFERENCE
orl-rat TDLo:720 mg/kg/17W-C:ETA AICCA6 19,531,66

SAFETY PROFILE: Questionable carcinogen with experimental tumorigenic data. When heated to decomposition it emits toxic fumes of NO_x.

DPP600 **CAS:19471-27-3** **HR: 2**
4-((p-(DIMETHYLAMINO)PHENYL)AZO)-2,5-LUTIDINE 1-OXIDE
mf: $C_{15}H_{18}N_4O$ mw: 270.37

SYN: N,N-DIMETHYL-4-(4′-(2′,5′-DIMETHYLPYRIDYL-1′-OXIDE)AZO)ANILINE

TOXICITY DATA WITH REFERENCE
orl-rat TDLo:2646 mg/kg/21W-C:ETA JNCIAM 41,855,68

SAFETY PROFILE: Questionable carcinogen with experimental tumorigenic data. When heated to decomposition it emits toxic fumes of NO_x.

DPP709 **CAS:19456-77-0** **HR: 2**
4-((p-(DIMETHYLAMINO)PHENYL)AZO)-3,5-LUTIDINE-1-OXIDE
mf: $C_{15}H_{18}N_4O$ mw: 270.37

SYNS: N,N-DIMETHYL-4-(4′-(3′,5′-DIMETHYLPYRIDYL 1′ OXIDE)AZO)ANILINE □ N,N DIMETHYL-4-((3′,5′-LUTIDYL-1′-OXIDE)AZO)ANILINE

TOXICITY DATA WITH REFERENCE
orl-rat TDLo:5292 mg/kg/21W-C:ETA JNCIAM 41,855,68

D

SAFETY PROFILE: Questionable carcinogen with experimental tumorigenic data. When heated to decomposition it emits toxic fumes of NO_x.

DPP800 CAS:7349-99-7 HR: 2
4-((4-(DIMETHYLAMINO)PHENYL)AZO)-2,6-LUTIDINE-1-OXIDE
mf: $C_{15}H_{18}N_4O$ mw: 270.37

SYNS: N,N-DIMETHYL-4-(4'-(2',6'-DIMETHYLPYRIDYL-1'-OXIDE) AZO)ANILINE □ 2,6-DIMETHYLPYRIDINE-1-OXIDE-4-AZO-p-DIMETHYLANILINE

TOXICITY DATA WITH REFERENCE
orl-rat TDLo:714 mg/kg/17W-C:NEO JNCIAM 37,365,66
orl-rat TD:4300 mg/kg/17W-C:ETA CNREA8 14,715,54

SAFETY PROFILE: Questionable carcinogen with experimental neoplastigenic and tumorigenic data. When heated to decomposition it emits toxic fumes of NO_x.

DPQ200 CAS:33804-48-7 HR: 3
4-((p-(DIMETHYLAMINO)PHENYL)AZO)-N-METHYLACETANILIDE
mf: $C_{17}H_{20}N_4O$ mw: 296.41

SYNS: N'-ACETYL-N'-METHYL-4'-AMINO-N,N-DIMETHYL-4-AMINOAZOBENZENE □ 4-(N-ACETYL-N-METHYL)AMINO-4'-(N',N'-DIMETHYLAMINO)AZOBENZENE □ N',N'-DIMETHYL-4'-AMINO-N-ACETYL-N-MONOMETHYL-4-AMINOAZOBENZENE □ N-(4-((4-(DIMETHYLAMINO) PHENYL)AZO)PHENYL)-N-METHYLACETAMIDE

TOXICITY DATA WITH REFERENCE
mma-sat 250 nmol/plate CNREA8 46,1654,86
dns-rat:lvr 1 μmol/L CNREA8 46,1654,86
orl-rat TDLo:1630 mg/kg/21W-C:ETA CNREA8 34,2274,74
ipr-rat LD50:370 mg/kg CNREA8 34,2274,74

SAFETY PROFILE: Poison by intraperitoneal route. Questionable carcinogen with experimental tumorigenic data. Mutation data reported. When heated to decomposition it emits toxic fumes of NO_x.

DPQ400 CAS:17400-65-6 HR: 2
5-((p-(DIMETHYLAMINO)PHENYL)AZO)-7-METHYLQUINOLINE
mf: $C_{18}H_{18}N_4$ mw: 290.40

SYNS: N,N-DIMETHYL-4-(5'-(7'-METHYLQUINOLYL)AZO)ANILINE □ 7'-METHYL-5'-(p-DIMETHYLAMINOPHENYLAZO)QUINOLINE

TOXICITY DATA WITH REFERENCE
orl-rat TDLo:540 mg/kg/30D-C:CAR JNCIAM 40,891,68

SAFETY PROFILE: Questionable carcinogen with experimental carcinogenic data. When heated to decomposition it emits toxic fumes of NO_x.

DPQ600 CAS:17416-18-1 HR: 2
5-((p-(DIMETHYLAMINO)PHENYL)AZO)QUINALDINE
mf: $C_{18}H_{18}N_4$ mw: 290.40

SYN: 2'-METHYL-5'-(p-DIMETHYLAMINOPHENYLAZO)QUINOLINE

TOXICITY DATA WITH REFERENCE
orl-rat TDLo:540 mg/kg/30D-C:CAR JNCIAM 40,891,68

SAFETY PROFILE: Questionable carcinogen with experimental carcinogenic data. When heated to decomposition it emits toxic fumes of NO_x.

DPQ800 CAS:17416-17-0 HR: 2
5-((p-(DIMETHYLAMINO)PHENYL)AZO)QUINOLINE
mf: $C_{17}H_{16}N_4$ mw: 276.37

SYNS: N,N-DIMETHYL-p-(5'-QUINOLYLAZO)ANILINE □ N,N-DIMETHYL-4-(5'-QUINOLYLAZO)ANILINE

TOXICITY DATA WITH REFERENCE
orl-rat TDLo:714 mg/kg/17W-C:CAR JNCIAM 40,891,68
orl-rat TD:720 mg/kg/17W-C:ETA AICCA6 19,531,63
orl-rat TD:714 mg/kg/17W-C:ETA JNCIAM 26,1461,61

SAFETY PROFILE: Questionable carcinogen with experimental carcinogenic and tumorigenic data. When heated to decomposition it emits toxic fumes of NO_x.

DPR000 CAS:30041-69-1 HR: 2
6-((p-(DIMETHYLAMINO)PHENYL)AZO)QUINOLINE
mf: $C_{17}H_{16}N_4$ mw: 276.37

SYNS: N,N-DIMETHYL-4-(6'-QUINOLYLAZO)ANILINE □ QUINOLINE-6-AZO-p-DIMETHYLANILINE

TOXICITY DATA WITH REFERENCE
dns-rat-orl 40 mg/kg CALEDQ 27,115,85
orl-rat TDLo:714 mg/kg/17W-C:ETA AICCA6 19,531,63

CONSENSUS REPORTS: EPA Genetic Toxicology Program.

SAFETY PROFILE: Questionable carcinogen with experimental tumorigenic data. Mutation data reported. When heated to decomposition it emits toxic fumes of NO_x.

DPR200 CAS:22750-85-2 HR: 2
5-((p-(DIMETHYLAMINO)PHENYL)AZO)QUINOLINE-1-OXIDE
mf: $C_{17}H_{16}N_4O$ mw: 292.37

SYN: N,N-DIMETHYL-4-((5'-QUINOLYL-l'-OXIDE)AZO)ANILINE

TOXICITY DATA WITH REFERENCE
orl-rat TDLo:714 mg/kg/17W-C:ETA AICCA6 19,531,63

SAFETY PROFILE: Questionable carcinogen with experimental tumorigenic data. When heated to decomposition it emits toxic fumes of NO_x.

DPR400 CAS:22750-86-3 HR: 2
6-((p-(DIMETHYLAMINO)PHENYL)AZO)QUINOLINE-1-OXIDE
mf: $C_{17}H_{16}N_4O$ mw: 292.37

SYN: N,N'-DIMETHYL-4-((6'-QUINOLYL-1'-OXIDE)AZO)ANILINE

TOXICITY DATA WITH REFERENCE
orl-rat TDLo:714 mg/kg/17W-C:ETA AICCA6 19,531,63

SAFETY PROFILE: Questionable carcinogen with experimental tumorigenic data. When heated to decomposition it emits toxic fumes of NO_x.

DPS200 CAS:24220-18-6 HR: 3
2-(p-DIMETHYLAMINOPHENYL)-1,6-DIMETHYLQUINOLINIUM CHLORIDE
mf: $C_{19}H_{21}N_2{\cdot}Cl$ mw: 312.87

TOXICITY DATA WITH REFERENCE
orl-mus LD50:50 mg/kg JMCMAR 13,122,70
ipr-mus LD50:10 mg/kg JMCMAR 13,122,70

SAFETY PROFILE: Poison by ingestion and intraperitoneal routes. When heated to decomposition it emits very toxic fumes of NO_x and Cl^-.

DPS600 CAS:2150-58-5 HR: 3
4-(p-DIMETHYLAMINOPHENYL)IMINO-2,5-CYCLOHEXADIENE-1-ONE
mf: $C_{14}H_{14}N_2O$ mw: 226.30

PROP: A solid. Mp: 133–134°. Sol in water.

TOXICITY DATA WITH REFERENCE
ipr-mus LD50:80 mg/kg JMCMAR 21,11,78

CONSENSUS REPORTS: Reported in EPA TSCA Inventory.

SAFETY PROFILE: Poison by intraperitoneal route. When heated to decomposition it emits toxic fumes of NO_x.

DPS700 CAS:102504-71-2 HR: 3
4-DIMETHYLAMINOPHENYL-2-((4-PHENYL-1,2,5,6-TETRAHYDRO-1-PYRIDYL)ETHYL) KETONE
mf: $C_{22}H_{26}N_2O$ mw: 334.50

SYNS: 4'-(DIMETHYLAMINO)-3-(4-PHENYL-1,2,5,6-TETRAHYDRO-1-PYRIDYL)PROPIOPHENONE □ PROPIOPHENONE, 4'-(DIMETHYLAMINO)-3-(4-PHENYL-1,2,3,6-TETRAHYDRO-1-PYRIDYL)-

TOXICITY DATA WITH REFERENCE
ivn-mus LD50:42 mg/kg CSLNX* NX#11073

DOT CLASSIFICATION: 3; *Label:* Flammable Liquid

SAFETY PROFILE: A poison by intravenous route. A flammable liquid. When heated to decomposition it emits toxic vapors of NO_x.

DPT800 CAS:108-16-7 HR: 3
1,1-DIMETHYLAMINOPROPANOL-2
mf: $C_5H_{14}N_2O$ mw: 118.21

PROP: A liquid. Fp: <−20°, bp: 122.5–126.2°, flash p: 90°F, d: 0.850 @ 25°/25°, vap d: 3.52.

SYNS: 1,1-DIMETHYLAMINOPROPAN-2-OL □ DIMETHYL(2-HYDROXYPROPYL)AMINE □ DIMETHYLISOPROPANOLAMINE

TOXICITY DATA WITH REFERENCE
skn-rbt 10 mg/24H MLD AMIHBC 10,61,54
skn-rbt 500 mg MOD FCTOD7 20 563,82
eye-rbt 250 µg SEV AMIHBC 10,61,54
eye-rbt 100 mg/4S rns SEV FCTOD7 20,573,82
orl-rat LD50:1890 mg/kg AMIHBC 10,61,54

CONSENSUS REPORTS: Reported in EPA TSCA Inventory.

SAFETY PROFILE: Moderately toxic by ingestion. A skin and severe eye irritant. Used in boiler water condensate corrosion control. Dangerous fire hazard when exposed to heat or flame. Can react vigorously with oxidizers. To fight fire, use foam, CO_2, dry chemical. When heated to decomposition it emits toxic fumes of NO_x. See also AMINES.

DPU000 CAS:1738-25-6 HR: 3
3-(DIMETHYLAMINO)PROPIONITRILE
mf: $C_5H_{10}N_2$ mw: 98.17

PROP: Liquid. Mp: −43°, bp: 170°, d: 0.8617, vap d: 3.35, flash p: 145°F.

SYN: β-DIMETHYLAMINOPROPIONITRILE

TOXICITY DATA WITH REFERENCE
skn-rbt 500 mg/24H MLD 85JCAE-,923,86
eye-rbt 20 mg/24H MOD 85JCAE-,923,86
orl-rat LD50:2600 mg/kg DCTODJ 2,223,79
ivn-mus LD50:180 mg/kg CSLNX* NX#00201
skn-rbt LD50:1410 mg/kg AIHAAP 23,95,62

CONSENSUS REPORTS: Reported in EPA TSCA Inventory. Cyanide and its compounds are on the Community Right-To-Know List.

SAFETY PROFILE: Poison by intravenous route. Moderately toxic by ingestion and skin contact. A skin and eye irritant. Flammable liquid when exposed to heat, flame, or oxidizers; can react with oxidizing materials. To fight fire, use foam, CO_2, dry chemical. When heated to decomposition it emits highly toxic fumes of NO_x and CN^-. See also NITRILES.

DPU400 CAS:879-72-1 HR: 3
3-(DIMETHYLAMINO)PROPIOPHENONE HYDROCHLORIDE
mf: $C_{11}H_{15}NO{\cdot}ClH$ mw: 213.73

SYN: β-DIMETHYLAMINOPROPIOPHENONE HYDROCHLORIDE

TOXICITY DATA WITH REFERENCE
scu-rat LDLo:200 mg/kg FATOAO 25,437,62
orl-mus LD50:100 mg/kg AITEAT 15,249,67
ipr-mus LD50:70 mg/kg AITEAT 15,249,67
scu-mus LD50:223 mg/kg PCJOAU 24,833,90

CONSENSUS REPORTS: Reported in EPA TSCA Inventory.

SAFETY PROFILE: Poison by ingestion, subcutaneous, and intraperitoneal routes. When heated to decomposition it emits very toxic fumes of HCl and NO_x.

DPU600 CAS:102571-36-8 HR: 3
1-(3-(DIMETHYLAMINO)PROPOXY)ADAMANTANE ETHYL IODIDE
mf: $C_{17}H_{32}NO{\cdot}I$ mw: 393.40

SYN: (2-(1-ADAMANTYLOXY)PROPYL)DIMETHYLETHYLAMMONIUM IODIDE

TOXICITY DATA WITH REFERENCE
orl-mus LD50:400 mg/kg FRPSAX 32,129,77
ipr-mus LD50:100 mg/kg FRPSAX 32,129,77

SAFETY PROFILE: Poison by ingestion and intraperitoneal routes. When heated to decomposition it emits very toxic fumes of I^-, NH_3, and NO_x. See also IODIDES.

DPU800 CAS:15083-53-1 HR: 3
5-(3-(DIMETHYLAMINO)PROPOXY)-3-METHYL-1-PHENYLPYRAZOLE
mf: $C_{15}H_{21}N_3O$ mw: 259.39

SYN: P-329

TOXICITY DATA WITH REFERENCE
orl-mus LD50:636 mg/kg ARZNAD 17,214,67
scu-mus LD50:350 mg/kg ARZNAD 17,214,67

SAFETY PROFILE: Poison by subcutaneous route. Moderately toxic by ingestion. When heated to decomposition it emits toxic fumes of NO_x.

DPW600 CAS:303-54-8 HR: 3
5-(3-(DIMETHYLAMINO)PROPYL)-5H-DIBENZ(b,f)AZEPINE
mf: $C_{19}H_{22}N_2$ mw: 278.43

PROP: Orange-red oil. Bp: 160–170° @ 0.4 mm.

TOXICITY DATA WITH REFERENCE
ivn-rat LD50:23 mg/kg AIPTAK 120,450,59
ivn-mus LD50:41 mg/kg AIPTAK 120,450,59
ivn-rbt LD50:6 mg/kg AIPTAK 120,450,59

SAFETY PROFILE: Poison by intravenous route. When heated to decomposition it emits toxic fumes of NO_x.

DPX200 CAS:5560-72-5 HR: 3
5-(3-(DIMETHYLAMINO)PROPYL)-6,7,8,9,10,11-HEXAHYDRO-5H-CYCLOOCT(b)INDOLE
mf: $C_{19}H_{28}N_2$ mw: 284.49

SYNS: GALATUR □ IPRINDOLE □ PRAMINDOLE □ PRONDOL □ WY-3263

TOXICITY DATA WITH REFERENCE
orl-rat LD50:484 mg/kg TXAPA9 18,185,71
ipr-rat LD50:187 mg/kg 27ZQAG -,130,72
orl-mus LD50:759 mg/kg FRPPAO 25,519,70
ipr-mus LD50:195 mg/kg 27ZQAG -,130,72
ivn-mus LD50:32 mg/kg CSLNX* NX#01206

SAFETY PROFILE: Poison by intravenous and intraperitoneal routes. Moderately toxic by ingestion. When heated to decomposition it emits toxic fumes of NO_x.

DPX400 CAS:303-70-8 HR: 3
5-(3-(DIMETHYLAMINO)PROPYL)-2-HYDROXY-10,11-DIHYDRO-5H-DIBENZ(b,f)AZEPINE
mf: $C_{19}H_{24}N_2O$ mw: 296.45

SYNS: GP 33679 □ 2-HYDROXYINIPRAMINE

TOXICITY DATA WITH REFERENCE
scu-rat TDLo:300 mg/kg (female 8-12D post):TER ARZNAD 19,1617,69
orl-rat LD50:2980 mg/kg ARZNAD 19,1617,69
ivn-rat LD50:19 mg/kg ARZNAD 19,1617,69
orl-mus LD50:733 mg/kg ARZNAD 19,1617,69
ivn-mus LD50:29 mg/kg ARZNAD 19,1617,69
ivn-rbt LD50:12,500 µg/kg ARZNAD 19,1617,69

SAFETY PROFILE: Poison by intravenous route. Moderately toxic by ingestion. Experimental teratogenic effects. When heated to decomposition it emits toxic fumes of NO_x.

DPX800 CAS:6202-23-9 HR: 3
5-(3-DIMETHYLAMINOPROPYLIDENE)-5H-DIBENZO-(a,d)CYCLOHEPTENE HYDROCHLORIDE
mf: $C_{20}H_{21}N•ClH$ mw: 311.88

PROP: A solid. Mp: 216–218°.

SYNS: CYCLOBENZAPRINE HYDROCHLORIDE □ 3-(5H-DIBENZO(a,d)CYCLOHEPTEN-5-YLIDENE)-N,N-DIMETHYL-1-PROPANAMINE HYDROCHLORIDE □ N,N-DIMETHYL-5H-DIBENZO(a,d)CYCLOHEPTENE-$\Delta^{5,\gamma}$-PROPYLAMINE HYDROCHLORIDE □ FLEXERIL □ FLEXIBAN □ PROHEPTATRIENE HYDROCHLORIDE □ PROHEPTATRIEN MONOHYDROCHLORIDE

TOXICITY DATA WITH REFERENCE
orl-man TDLo:5714 µg/kg/2D-I:BAH JCLPDE 54,39,93
orl-wmn TDLo:1 mg/kg/3D-I:BAH JCLPDE 54,39,93
orl-wmn TDLo:12 mg/kg:BAH JTCTDW 20,281,83
orl-wmn TDLo:8400 µg/kg/2W-I:BAH JCLPDE 44,151,83
orl-mus LD50:250 mg/kg 27ZQAG -,87,72
ivn-mus LD50:36 mg/kg AIPTAK 144,481,63

SAFETY PROFILE: Poison by ingestion and intravenous routes. Human systemic effects: distorted perceptions, euphoria, general anesthetic, hallucinations, pulse rate increase, somnolence, wakefulness. A muscle relaxant. When heated to decomposition it emits very toxic fumes of NO_x and HCl.

DPY200 CAS:897-15-4 HR: 3
11-(3-DIMETHYLAMINOPROPYLIDENE)-6,11-DIHYDRODIBENZO(b,e)THIEPINE HYDROCHLORIDE
mf: $C_{19}H_{21}NS•ClH$ mw: 331.93

SYNS: 3-DIBENZO(b,e)THIEPIN-11(6H)-YLIDENE-N,N-DIMETHYL-1-PROPANAMINE, HYDROCHLORIDE □ N,N-DIMETHYLDIBENZO(b,e)THIEPIN-$\Delta^{11(6H),\gamma}$-PROPYLAMINE HYDROCHLORIDE □ DOSULEPIN CHLORIDE □ DOSULEPIN HYDROCHLORIDE □ DOTHEIPIN HYDROCHLORIDE □ PROTHIADENE HYDROCHLORIDE □ PROTHIADEN HYDROCHLORIDE

TOXICITY DATA WITH REFERENCE
orl-rat TDLo:440 mg/kg (MGN):REP OYYAA2 27,1103,84
orl-rat TDLo:1080 mg/kg (female 17-22D post):REP IYKEDH 14,582,83
orl-rat TDLo:110 mg/kg (female 7-17D post):REP OYYAA2 27,1103,84
orl-rat TDLo:440 mg/kg (MGN):TER OYYAA2 27,1103,84
orl-wmn LDLo:40 mg/kg PGMJAO 60,442,84
orl-cld TDLo:90 mg/kg:CVS PGMJAO 60,442,84
orl-rat LD50:260 mg/kg 27ZQAG -,88,72
ipr-rat LD50:105 mg/kg IYKEDH 14,192,83
scu-rat LD50:760 mg/kg 27ZQAG -,88,72
ivn-rat LD50:24 mg/kg 27ZQAG -,88,72
orl-mus LD50:209 mg/kg 27ZQAG -,88,72
ipr-mus LD50:116 mg/kg IYKEDH 14,192,83
scu-mus LD50:620 mg/kg IYKEDH 14,192,83
ivn-mus LD50:29,200 µg/kg IYKEDH 14,192,83

SAFETY PROFILE: Poison by ingestion, intraperitoneal, and intravenous routes. Moderately toxic by subcutaneous route. Human systemic effects by ingestion: cardiomyopathy. Experimental teratogenic and reproductive effects. An antidepressant. When heated to decomposition it emits very toxic fumes of NO_x, SO_x, and HCl.

DPY600 CAS:51003-81-7 HR: 3
5-DIMETHYLAMINO-6-PROPYL-5H-INDENO(5,6-d)-1,3-DIOXOLE HYDROCHLORIDE
mf: $C_{15}H_{19}NO_2•ClH$ mw: 281.81

SYNS: pr-MDI □ 2-PROPYL-3-DIMETHYLAMINO-5,6-METHYLENEDIOXYINDENE HYDROCHLORIDE □ 2-N-PROPYL-3-DIMETHYLAMINO-5,6-METHYLENEDIOXYINDENE HYDROCHLORIDE

TOXICITY DATA WITH REFERENCE
ipr-rat LD50:175 mg/kg RCOCB8 26,85,79
ipr-mus LD50:185 mg/kg RCOCB8 26,85,79
ivn-mus LD50:40 mg/kg RCOCB8 26,85,79

SAFETY PROFILE: Poison by intraperitoneal and intravenous routes. When heated to decomposition it emits very toxic fumes of NO_x and HCl.

DQA400 CAS:60-87-7 HR: 3
10-(2-(DIMETHYLAMINO)PROPYL)PHENOTHIAZINE
mf: $C_{17}H_{20}N_2S$ mw: 284.45

SYNS: A-91033 □ APROBIT □ ATOSIL □ AVOMINE □ DIMAPP □ DIMETHYLAMINO-ISOPROPYL-PHENTHIAZIN (GERMAN) □ (2-DIMETHYLAMINO-2-METHYL)ETHYL-N-DIBENZOPARATHIAZINE □ N-(2'-DIMETHYLAMINO-2'-METHYL)ETHYLPHENOTHIAZINE □ 10-(2-(DIMETHYLAMINO)-2-METHYLETHYL)PHENOTHIAZINE □ N-DIMETHYLAMINO-2-METHYLETHYL THIODIPHENYLAMINE □ (DIMETHYLAMINO-2-PROPYL-10-PHENOTHIAZINE HYDROCHLORIDE (FRENCH) □ DIPRAZINE □ DIPROZIN □ FARGAN □ FENAZIL □ FENERGAN □ FENETAZINA □ HIBERNA □ HISTARGAN □ IERGIGAN □ ISOPHENERGAN □ ISOPROMETHAZINE □ LERCIGAN □ LERGIGAN □ LILLY 1516 □ LILLY 01516 □ NCI-C60673 □ PHARGAN □ PHENERGAN □ PHENSEDYL □ PILPOPHEN □ PIPOLPHEN □ PROAZAIMINE □ PROAZAMINE □ PROCIT □ PROMAZINAMIDE □ PROMETASIN □ PROMETAZIN □ PROMETHIAZINE □ PROMEZATHINE □ PROREX □ PROTAZINE □ PROTHAZIN □ PROVIGAN □ PYRETHIA □ PYRETHIAZINE □ ROMERGAN □ 3277 RP □ 3389 R.P. □ 4182 R.P. □ SKF 1498 □ SYNALGOS □ TANIDIL □ THIERGAN □ VALLERGINE □ WY 509

TOXICITY DATA WITH REFERENCE
eye-rbt 100 mg SEV FCTOD7 20 573,82
eye-rbt 100 mg/4S rns MLD FCTOD7 20573,82
dni-hmn:fbr 80 nmol/L DNSYAG 29,829,68
otr-ham:emb 10 mg/L ENMUDM 8(Suppl 6),4,86
msc-ham:lng 10 mg/L SHIGAZ 70,943,83
ipr-rat TDLo:350 mg/kg (10-16D preg):TER JPHYA7 164,138,62
par-rat TDLo:350 mg/kg (female 4D post):TER FATOAO 48(6),89,85
par-rat TDLo:350 mg/kg (female 4D post):REP FATOAO 48(6),89,85
skn-cld TDLo:13 mg/kg:EYE,PSY CMAJAX 130,1460,84
ipr-rat LDLo:110 mg/kg TXAPA9 1,156,59
scu-rat LD50:700 mg/kg CRSBAW 144,887,50
ivn-rat LD50:45 mg/kg AIPTAK 120,450,59
ims-rat LD50:169 mg/kg TXAPA9 18,185,71
orl-mus LD50:326 mg/kg ARZNAD 7,237,57
ipr-mus LD50:124 mg/kg JPETAB 108,201,53
scu-mus LD50:225 mg/kg PRPHA8 2,53,47
ivn-mus LD50:40 mg/kg TXAPA9 18,185,71
ims-mus LD50:175 mg/kg TXAPA9 18,185,71
orl-rbt LD50:580 mg/kg PHARAT 25,91,70
ivn-rbt LD50:19 mg/kg AIPTAK 120,450,59

SAFETY PROFILE: Poison by ingestion, intravenous, intramuscular, intraperitoneal, and subcutaneous routes. Human systemic effects by ingestion: pupillary dilation, wakefulness, hallucinations, and distorted perceptions. An experimental teratogen. Other experimental reproductive effects. Human mutation data reported. A severe eye irritant. When heated to decomposition it emits very toxic fumes of NO_x and SO_x.

D

DQA600 CAS:58-40-2 HR: 3
10-(3-(DIMETHYLAMINO)PROPYL)PHENOTHIAZINE
mf: $C_{17}H_{20}N_2S$ mw: 284.45

PROP: Oily liquid. Bp: 203–210° @ 0.3 mm.

SYNS: AMPAZINE □ N,N-DIMETHYL-10H-PHENOTHIAZINE-10-PROPANAMINE □ ESPARIN □ LIRANOL □ NEO-HIBERNEX □ PROMAZINE □ PROTACTYL □ SPARINE □ VEROPHEN □ WY 1094

TOXICITY DATA WITH REFERENCE
scu-rat TDLo:80 mg/kg (1-18D preg):REP JNEUAY 3,295,62
unr-mus TDLo:4 mg/kg (1D pre):REP FESTAS 12,346,61
orl-rat LD50:350 mg/kg ARZNAD 8,507,58
ipr-rat LDLo:210 mg/kg TXAPA9 1,156,59
scu-rat LD50:192 mg/kg ARZNAD 24,1798,74
ivn-rat LD50:14,500 μg/kg AIPTAK 119,311,59
orl-mus LD50:401 mg/kg ARZNAD 8,489,58
ipr-mus LD50:140 mg/kg AIPTAK 155,69,65
scu-mus LD50:110 mg/kg ARZNAD 24,1798,74
ivn-mus LD50:45 mg/kg APTOA6 19,87,62
ivn-rbt LD50:21 mg/kg AIPTAK 120,450,59

CONSENSUS REPORTS: EPA Genetic Toxicology Program.

SAFETY PROFILE: Human poison by unspecified route. An experimental poison by ingestion, subcutaneous, intravenous, and intraperitoneal routes. Experimental reproductive effects. When heated to decomposition it emits very toxic fumes of NO_x and SO_x.

DQA700 CAS:13082-24-1 HR: 3
10-(2-(DIMETHYLAMINO)PROPYL)PHENOTHIAZIN-2-YL MORPHOLINOMETHYL KETONE
mf: $C_{23}H_{29}N_3O_2S$ mw: 411.61

SYN: PHENOTHIAZINE, 10-(2-(DIMETHYLAMINO)PROPYL)-2-(MORPHOLINOACETYL)-

TOXICITY DATA WITH REFERENCE
ipr-mus LD50:410 mg/kg CHTPBA 1,397,66

DOT CLASSIFICATION: 3; *Label:* Flammable Liquid

SAFETY PROFILE: Moderately toxic by intraperitoneal route. A flammable liquid. When heated to decomposition it emits toxic vapors of SO_x and NO_x.

DQA710 CAS:13065-64-0 HR: 3
10-(3-(DIMETHYLAMINO)PROPYL)PHENOTHIAZIN-2-YL MORPHOLINOMETHYL KETONE
mf: $C_{23}H_{29}N_3O_2S$ mw: 411.61

SYN: PHENOTHIAZINE, 10-(3-(DIMETHYLAMINO)PROPYL)-2-(MORPHOLINOACETYL)-

TOXICITY DATA WITH REFERENCE
ipr-mus LD50:250 mg/kg CHTPBA 1,397,66

DOT CLASSIFICATION: 3; *Label:* Flammable Liquid

SAFETY PROFILE: A poison by intraperitoneal route. A flammable liquid. When heated to decomposition it emits toxic vapors of SO_x and NO_x.

DQB309 CAS:13713-13-8 HR: 3
2-(3-DIMETHYLAMINOPROPYL)-3a,4,7,7a-TETRAHYDRO-4,7-ETHANOISOINDOLINE DIMETHIODIDE
mf: $C_{19}H_{36}N_2 \cdot 2I$ mw: 546.37

TOXICITY DATA WITH REFERENCE
orl-rat LD50:1730 mg/kg SKNEA7 10,15,60
ivn-rat LD50:137 mg/kg SKNEA7 10,15,60
orl-mus LD50:2334 mg/kg SKNEA7 10,15,60
ivn-mus LD50:156 mg/kg SKNEA7 10,15,60
ivn-rbt LD50:82,200 μg/kg SKNEA7 10,15,60

SAFETY PROFILE: Poison by intravenous route. Moderately toxic by ingestion. When heated to decomposition it emits toxic fumes of I^- and NO_x. See also IODIDES.

DQB600 CAS:1122-58-3 HR: 3
4-DIMETHYLAMINOPYRIDINE
mf: $C_7H_{10}N_2$ mw: 122.19

PROP: Plates from EtOH. Mp: 114°. Very sol in H_2O.

SYNS: 4-DIMETHYLAMINEPYRIDINE □ γ-(DIMETHYLAMINO)PYRIDINE □ p-DIMETHYLAMINOPYRIDINE

TOXICITY DATA WITH REFERENCE
orl-mus LDLo:470 mg/kg AECTCV 14,111,85
ivn-mus LD50:56 mg/kg CSLNX* NX#04228

CONSENSUS REPORTS: Reported in EPA TSCA Inventory.

SAFETY PROFILE: Poison by intravenous route. Moderately toxic by ingestion. An acylation catalyst. When heated to decomposition it emits toxic fumes of NO_x.

DQB800 CAS:5585-67-1 HR: 3
2-(DIMETHYLAMINO) RESERPILINATE
mf: $C_{26}H_{35}N_3O_5$ mw: 469.64

SYNS: ANTIPRESSINE DIHYDROCHLORIDE □ 2-(DIMETHYLAMINO) RESERPILIN-24-OIC ACID ETHYL ESTER

TOXICITY DATA WITH REFERENCE
orl-rat LD50:2350 mg/kg OYYAA2 3,390,69
ipr-rat LD50:330 mg/kg OYYAA2 3,390,69
scu-rat LD50:1000 mg/kg OYYAA2 3,390,69
orl-mus LD50:2100 mg/kg OYYAA2 3,390,69
ipr-mus LD50:410 mg/kg OYYAA2 3,390,69
scu-mus LD50:980 mg/kg OYYAA2 3,390,69

SAFETY PROFILE: Poison by intraperitoneal route. Moderately toxic by ingestion and subcutaneous routes. When heated to decomposition it emits toxic fumes of NO_x.

DQC000 CAS:63019-60-3 HR: 2
7-(p-(DIMETHYLAMINO)STYRYL)BENZ(c)ACRIDINE
mf: $C_{27}H_{22}N_2$ mw: 374.51

SYN: p-DIMETHYLAMINOBENZYLIDEN-3,4-BENZ-9-METHYLACRIDINE

TOXICITY DATA WITH REFERENCE
scu-mus TDLo:200 mg/kg:ETA VOONAW 1,52,55

SAFETY PROFILE: Questionable carcinogen with experimental tumorigenic data. When heated to decomposition it emits toxic fumes of NO_x.

DQC200 CAS:63019-59-0 HR: 2
12-(p-DIMETHYLAMINO)STYRYLBENZ(a)ACRIDINE
mf: $C_{27}H_{22}N_2$ mw: 374.51

SYN: p-DIMETHYLAMINOBENZYLIDEN-1,2-BENZ-9-METHYL-ACRIDINE

TOXICITY DATA WITH REFERENCE
scu-mus TDLo:200 mg/kg:ETA VOONAW 1,52,55

SAFETY PROFILE: Questionable carcinogen with experimental tumorigenic data. When heated to decomposition it emits toxic fumes of NO_x.

DQC400 CAS:1628-58-6 HR: 2
2-(p-(DIMETHYLAMINO)STYRYL)BENZOTHIAZOLE
mf: $C_{17}H_{16}N_2S$ mw: 280.41

SYN: 2-(4-DIMETHYLAMINOSTYRYL)BENZOTHIAZOLE

TOXICITY DATA WITH REFERENCE
orl-rat TDLo:35 g/kg/1Y-I:NEO,REP JNCIAM 41,985,68

CONSENSUS REPORTS: Reported in EPA TSCA Inventory.

SAFETY PROFILE: Experimental reproductive effects. Questionable carcinogen with experimental neoplastigenic data. When heated to decomposition it emits very toxic fumes of NO_x and SO_x.

DQC600 CAS:19716-21-3 HR: 2
4-(p-(DIMETHYLAMINO)STYRYL)-6,8-DIMETHYLQUINOLINE
mf: $C_{21}H_{22}N_2$ mw: 302.45

SYN: 6,8-DIMETHYL-(4-p-(DIMETHYLAMINO)STYRYL)QUINOLINE

TOXICITY DATA WITH REFERENCE
orl-rat TDLo:38 mg/kg/51W-I:ETA,REP JNCIAM 41,985,68

SAFETY PROFILE: Experimental reproductive effects. Questionable carcinogen with experimental tumorigenic data. When heated to decomposition it emits toxic fumes of NO_x.

DQD000 CAS:897-55-2 HR: 3
4-(4-DIMETHYLAMINOSTYRYL)QUINOLINE
mf: $C_{19}H_{18}N_2$ mw: 274.39

SYNS: 2-(4-N,N-DIMETHYLAMINOSTYRYL)QUINOLINE □ 4-(p-(DIMETHYLAMINO)STYRYL)QUINOLINE

TOXICITY DATA WITH REFERENCE
ivn-mus TDLo: 100 mg/kg: NEO,REP CNREA8 25,938,65
ivn-mus LDLo: 160 mg/kg CNREA8 25,938,65

CONSENSUS REPORTS: Reported in EPA TSCA Inventory.

SAFETY PROFILE: Poison by intravenous route. Experimental reproductive effects. Questionable carcinogen with experimental neoplastigenic data. When heated to decomposition it emits toxic fumes of NO_x.

DQD200 CAS:21970-53-6 HR: 2
4-(p-(DIMETHYLAMINO)STYRYL)QUINOLINE MONOHYDROCHLORIDE
mf: $C_{19}H_{18}N_2 \cdot ClH$ mw: 310.85

SYN: NSC 63346

TOXICITY DATA WITH REFERENCE
orl-rat TDLo: 115 mg/kg/1Y-I: NEO,REP JNCIAM 41,985,68

SAFETY PROFILE: Experimental reproductive effects. Questionable carcinogen with experimental neoplastigenic data. When heated to decomposition it emits very toxic fumes of NO_x and HCl.

DQD400 CAS:1596-84-5 HR: 3
DIMETHYLAMINOSUCCINAMIC ACID
mf: $C_6H_{12}N_2O_3$ mw: 160.20

PROP: A solid. Mp: 154–155°.

SYNS: ALAR □ ALAR-85 □ AMINOZIDE □ B 995 □ BERNSTEINSAEURE-2,2-DIMETHYLHYDRAZID (GERMAN) □ B-NINE □ BUTANEDIOIC ACID MONO(2,2-DIMETHYLHYDRAZIDE) □ DAMINOZIDE (USDA) □ DIMAS □ N-DIMETHYL AMINO-β-CARBAMYL PROPIONIC ACID □ N-(DIMETHYLAMINO)SUCCINAMIC ACID □ N-DIMETHYLAMINO-SUCCINAMIDSAEURE (GERMAN) □ DMASA □ DMSA □ KYLAR □ NCI-C03827 □ SADH □ SUCCINIC ACID-2,2-DIMETHYLHYDRAZIDE □ SUCCINIC-1,1-DIMETHYL HYDRAZIDE

TOXICITY DATA WITH REFERENCE
mma-mus: lyms 1650 mg/L EMMUEG 12,85,88
msc-mus: lyms 156 mg/L EMMUEG 12,85,88
orl-rat TDLo: 182 g/kg/2Y-C: CAR NCITR* NCI-CG-TR-83,78
orl-mus TDLo: 2600 g/kg/62W-C: CAR CNREA8 37,3497,77
orl-mus TD: 873 g/kg/2Y-C: ETA NCITR* NCI-CG-TR-83,78
orl-rat LD50: 8400 mg/kg FMCHA2 -,D10,80
orl-mus LD50: 6300 mg/kg CHABA8 84,100483n,76
ipr-mus LD50: 1325 mg/kg CHABA8 22,126,73

CONSENSUS REPORTS: EPA Genetic Toxicology Program. NCI Carcinogenesis Bioassay (feed); Clear Evidence: mouse, rat NCITR* NCI-CG-TR-83,78.

SAFETY PROFILE: Suspected carcinogen with experimental carcinogenic and tumorigenic data. Moderately toxic by ingestion and intraperitoneal routes. When heated to decomposition it emits toxic fumes of NO_x.

DQD500 HR: 3
p-DIMETHYLAMINOTHYMOLDIMETHYLURETHANE METHIODIDE
mf: $C_{16}H_{27}N_2O_2 \cdot I$ mw: 406.35

SYN: (CARBOXYMETHYL)TRIMETHYLAMMONIUM IODIDE-6-(DIMETHYLAMINO)-4-ISOPROPYL-m-TOLYL ESTER

TOXICITY DATA WITH REFERENCE
orl-mus LDLo: 77,500 μg/kg FEPRA7 5,184,46
scu-mus LDLo: 15 μg/kg FEPRA7 5,184,46
scu-dog LDLo: 4600 μg/kg FEPRA7 5,184,46

SAFETY PROFILE: Poison by ingestion and subcutaneous routes. When heated to decomposition it emits toxic fumes of I^-, NH_3, and NO_x. See also IODIDES.

DQD600 CAS:7347-47-9 HR: 2
4-((4-(DIMETHYLAMINO)-m-TOLYL)AZO)-2-PICOLINE-1-OXIDE
mf: $C_{14}H_{16}N_4O$ mw: 256.34

SYN: N,N-DIMETHYL 2 METHYL-4-(4′-(2′-METHYLPYRIDYL-1′-OXIDE)AZO)ANILINE

TOXICITY DATA WITH REFERENCE
orl-rat TDLo: 2142 mg/kg/17W-C: NEO JNCIAM 37,365,66

SAFETY PROFILE: Questionable carcinogen with experimental neoplastigenic data. When heated to decomposition it emits toxic fumes of NO_x.

DQD800 CAS:7347-48-0 HR: 2
4-((4-(DIMETHYLAMINO)-o-TOLYL)AZO)-2-PICOLINE-1-OXIDE
mf: $C_{15}H_{18}N_4O$ mw: 270.37

SYNS: N,N′DIMETHYL-3-METHYL-4-(4′-(2′-METHYLPYRIDYL-1′OXIDE)AZO)ANILINE □ N,N′ DIMETHYL-4-(4′-(2′-METHYLPYRIDYL-1-OXIDE)AZO)-o-TOLUIDINE

TOXICITY DATA WITH REFERENCE
orl-rat TDLo: 2142 mg/kg/17W-C: NEO JNCIAM 37,365,66

SAFETY PROFILE: Questionable carcinogen with experimental neoplastigenic data. When heated to decomposition it emits toxic fumes of NO_x.

DQE000 CAS:19456-74-7 HR: 2
4-((4-(DIMETHYLAMINO)-m-TOLYL)AZO)-3-PICOLINE-1-OXIDE
mf: $C_{15}H_{18}N_4O$ mw: 270.37

SYN: N,N,2-TRIMETHYL-4-(4′-(3′-METHYLPYRIDYL-1′-OXIDE)AZO)ANILINE

TOXICITY DATA WITH REFERENCE
orl-rat TDLo: 4284 mg/kg/17W-C: ETA JNCIAM 41,85,68

SAFETY PROFILE: Questionable carcinogen with experimental tumorigenic data. When heated to decomposition it emits toxic fumes of NO_x.

D

DQE200 **CAS:19471-28-4** **HR: 2**
4-((4-(DIMETHYLAMINO)-o-TOLYL)AZO)-3-PICOLINE-1-OXIDE
mf: $C_{15}H_{18}N_4O$ mw: 270.37

SYN: N,N,3-TRIMETHYL-4-(4'-(3'-METHYLPYRIDYL-1'-OXIDE)AZO) ANILINE

TOXICITY DATA WITH REFERENCE
orl-rat TDLo:4284 mg/kg/17W-C:ETA JNCIAM 41,855,68

SAFETY PROFILE: Questionable carcinogen with experimental tumorigenic data. When heated to decomposition it emits toxic fumes of NO_x.

DQE400 **CAS:17400-68-9** **HR: 2**
5-((4-(DIMETHYLAMINO)-m-TOLYL)AZO)QUINOLINE
mf: $C_{18}H_{18}N_4$ mw: 290.40

SYNS: N,N-DIMETHYL-4-(5'-QUINOLYLAZO)-m-TOLUIDINE □ 3-METHYL-5'-(p-DIMETHYLAMINOPHENYLAZO)QUINOLINE

TOXICITY DATA WITH REFERENCE
orl-rat TDLo:2142 mg/kg/17W-C:CAR JNCIAM 40,891,68

SAFETY PROFILE: Questionable carcinogen with experimental carcinogenic data. When heated to decomposition it emits toxic fumes of NO_x.

DQE600 **CAS:17416-21-6** **HR: 2**
5-((4-(DIMETHYLAMINO)-o-TOLYL)AZO)QUINOLINE
mf: $C_{18}H_{18}N_4$ mw: 290.40

SYN: 2-METHYL-5'-(p-DIMETHYLAMINOPHENYLAZO)QUINOLINE

TOXICITY DATA WITH REFERENCE
orl-rat TDLo:2142 mg/kg/17W-C:CAR JNCIAM 40,891,68

SAFETY PROFILE: Questionable carcinogen with experimental carcinogenic data. When heated to decomposition it emits toxic fumes of NO_x.

DQE800 **CAS:14144-91-3** **HR: 3**
5-DIMETHYLAMINO-4-TOLYL METHYLCARBAMATE
mf: $C_{11}H_{16}N_2O_2$ mw: 208.29

SYNS: BAY 42696 □ 4-METHYL-3-DIMETHYLAMINOPHENYL ESTER-N-METHYLCARBAMIC ACID

TOXICITY DATA WITH REFERENCE
orl-rat LD50:46 mg/kg ATXKA8 27,311,71
scu-mus LDLo:20 mg/kg NTIS** PB158-508

SAFETY PROFILE: Poison by ingestion and subcutaneous routes. A strong oxidizing agent. When heated to decomposition it emits toxic fumes of NO_x. See also CARBAMATES and ESTERS.

DQE900 **CAS:2083-91-2** **HR: 3**
DIMETHYLAMINOTRIMETHYLSILANE
mf: $C_5H_{15}NSi$ mw: 117.27

PROP: Liquid. D: 0.74 @ 20°/4°, bp: 85°.

SAFETY PROFILE: Explosive reaction with xenon difluoride below 0°C. When heated to decomposition it emits toxic fumes of NO_x.

DQF000 **CAS:6120-10-1** **HR: 2**
4-DIMETHYLAMINO-3,5-XYLENOL
mf: $C_{10}H_{15}NO$ mw: 165.26

TOXICITY DATA WITH REFERENCE
orl-mus TDLo:33 g/kg/78W-I:ETA NTIS** PB223-159

SAFETY PROFILE: Questionable carcinogen with experimental tumorigenic data. When heated to decomposition it emits toxic fumes of NO_x.

DQF200 **CAS:19456-73-6** **HR: 2**
4-((4-(DIMETHYLAMINO)-2,3-XYLYL)AZO)PYRIDINE-1-OXIDE
mf: $C_{15}H_{18}N_4O$ mw: 270.37

SYN: N,N,2,3-TETRAMETHYL-4-(4'-(PYRIDYL-1'-OXIDE)AZO)ANILINE

TOXICITY DATA WITH REFERENCE
orl-rat TDLo:6552 mg/kg/26W-C:ETA JNCIAM 41,855,68

SAFETY PROFILE: Questionable carcinogen with experimental tumorigenic data. When heated to decomposition it emits toxic fumes of NO_x.

DQF400 **CAS:19456-75-8** **HR: 2**
4-((4-(DIMETHYLAMINO)-2,5-XYLYL)AZO)PYRIDINE-1-OXIDE
mf: $C_{15}H_{18}N_4O$ mw: 270.37

SYN: N,N,2,5-TETRAMETHYL-4-(4'-(PYRIDYL-1'-OXIDE)AZO)ANILINE

TOXICITY DATA WITH REFERENCE
orl-rat TDLo:3276 mg/kg/26W-C:ETA JNCIAM 41,855,68

SAFETY PROFILE: Questionable carcinogen with experimental tumorigenic data. When heated to decomposition it emits toxic fumes of NO_x.

DQF600 **CAS:19595-66-5** **HR: 2**
4-((4-(DIMETHYLAMINO)-3,5-XYLYL)AZO)PYRIDINE-1-OXIDE
mf: $C_{15}H_{18}N_4O$ mw: 270.37

SYN: N,N,2,6-TETRAMETHYL-4-(4'-(PYRIDYL-1'-OXIDE)AZO)ANILINE

TOXICITY DATA WITH REFERENCE
orl-rat TDLo:6552 mg/kg/26W-C:ETA JNCIAM 41,855,68

SAFETY PROFILE: Questionable carcinogen with experimental tumorigenic data. When heated to decomposition it emits toxic fumes of NO_x.

DQF650 **CAS:14488-49-4** **HR: 3**
DIMETHYLAMMONIUM PERCHLORATE
mf: $C_2H_8ClNO_4$ mw: 145.54

SAFETY PROFILE: An explosive salt. When heated to decomposition it emits toxic fumes of Cl^-, NH_3, and NO_x. See also PERCHLORATES.

DQF800 **CAS:121-69-7** **HR: 3**
N,N-DIMETHYLANILINE
DOT: UN 2253
mf: $C_8H_{11}N$ mw: 121.20

PROP: Yellowish-brown oily liquid. Mp: 2.5°, bp: 193.1°, flash p: 145°F (CC), d: 0.9557 @ 20°/4°, ULC: 20–25, autoign temp: 700°F, vap press: 1 mm @ 29.5°, vap d: 4.17.

SYNS: BENZENAMINE, N,N,-DIMETHYL-(9CI) □ (DIMETHYLAMINO)BENZENE □ N,N-DIMETHYLBENZENEAMINE □ DIMETHYLPHENYLAMINE □ N-DIMETHYL-ANILINE (OSHA) □ DIMETHYLPHENYLAMINE □ N,N-DIMETHYLPHENYLAMINE □ DWUMETYLOANILINA (POLISH) □ NCI-C56428 □ VERSNELLER NL 63/10

TOXICITY DATA WITH REFERENCE
skn-rbt 10 mg/24H open MLD AIHAAP 23,95,62
orl-hmn LDLo:50 mg/kg NCPBBY Jan/Feb,69
dnd-rat-ipr 485 mg/kg EMMUEG 21,349,93
dnd-mus-ipr 485 mg/kg EMMUEG 21,349,93
sce-ham:ovr 30 mg/L EMMUEG 13,60,89
orl-rat TDLo:15,450 mg/kg/2Y-C:ETA NTPTR* NTP-TR-360,89
orl-hmn LDLo:50 mg/kg:GIT NCPBBY Jan/Feb,69
orl-rat LD50:1410 mg/kg AIHAAP 23,95,62
ihl-rat LCLo:250 mg/m³/4H GISAAA 37(4),35,72
scu-rat LDLo:100 mg/kg 85GMAT-,55,82
orl-mus LDLo:350 mg/kg NTPTR* NTP-TR-360,89
skn-rbt LD50:1770 mg/kg AIHAAP 23,95,62
orl-rat TDLo-32,500 mg/kg/13W-I JTEHD6 29,77,90
orl-rat TDLo-16,250 mg/kg/13W-I NTPTR* NTP-TR-360,89
ihl-rat TCLo:10,700 µg/m³/5H/17W-I GISAAA 37(4),35,72
ihl-rat TCLo:300 µg/m³/24H/14W-C GISAAA 34(3),7,69
orl-mus TDLo:32,500 mg/kg/13W-I JTEHD6 29,77,90
orl-rat LD50:1410 mg/kg AIHAAP 23,95,62
ihl-rat LCLo:250 mg/m³/4H GISAAA 37(4),35,72

CONSENSUS REPORTS: Reported in EPA TSCA Inventory. Community Right-To-Know List.

OSHA PEL: TWA 5 ppm; STEL 10 ppm (skin)
ACGIH TLV: TWA 5 ppm, STEL 10 ppm (skin)
DFG MAK: 5 ppm (25 mg/m³); Suspected Carcinogen
DOT CLASSIFICATION: 6.1; *Label:* Poison

SAFETY PROFILE: Suspected carcinogen with equivocal tumorigenic data. Human poison by ingestion. Moderately toxic by inhalation and skin contact. A skin irritant. Human systemic effects by ingestion: nausea or vomiting. Physiological action is similar to, but less toxic than aniline. A central nervous system depressant. Mutation data reported. Flammable liquid when exposed to heat, flame, or oxidizers. Explodes on contact with benzoyl peroxide or diisopropyl peroxydicarbonate. To fight fire, use foam, CO_2, dry chemical. When heated to decomposition it emits highly toxic fumes of aniline and NO_x. See also ANILINE.

DQG000 **CAS:41217-05-4** **HR: 2**
6,12-DIMETHYLANTHANTHRENE
mf: $C_{24}H_{16}$ mw: 304.40

SYN: 6,12-DIMETHYL-BIBENZO(def,mno)CHRYSENE

TOXICITY DATA WITH REFERENCE
dnd-mam:lym 30 µmol/L CBINA8 47,87,83
scu-mus TDLo:72 mg/kg/9W-I:ETA COREAF 246,1477,58

SAFETY PROFILE: Questionable carcinogen with experimental tumorigenic data. Mutation data reported. When heated to decomposition it emits acrid smoke and irritating fumes.

DQG200 **CAS:781-43-1** **HR: 2**
9,10-DIMETHYLANTHRACENE
mf: $C_{16}H_{14}$ mw: 206.30

PROP: Crystals from EtOH. Mp: 180–181°, bp: 140–142° @ 0.5–1 mm.

TOXICITY DATA WITH REFERENCE
mmo-sat 20 µg/plate CRNGDP 6,1483,85
mma-esc 10 µg/plate PMRSDJ 1,387,81
dnr-esc 500 mg/L PMRSDJ 1,195,81
sln-dmg-par 5 mmol/L MUREAV 125,243,84
mrc-smc 200 ppm PMRSDJ 1,481,81
skn-mus TDLo:40 mg/kg/20D-I:CAR CRNGDP 6,1483,85
skn-mus TDLo:1100 mg/kg/46W-I:ETA CNREA8 2,157,42

CONSENSUS REPORTS: EPA Genetic Toxicology Program.

SAFETY PROFILE: Questionable carcinogen with experimental carcinogenic and tumorigenic data. Mutation data reported. When heated to decomposition it emits acrid smoke and irritating fumes.

DQG400 **CAS:18380-68-2** **HR: 3**
DIMETHYLANTIMONY CHLORIDE
mf: C_2H_6ClSb mw: 187.27

PROP: Thermally unstable yellowish liquid. Bp: 155–160°.

CONSENSUS REPORTS: Antimony and its compounds are on the Community Right-To-Know List.

SAFETY PROFILE: A poison. Ignites at 40°C in air. When heated to decomposition it emits toxic fumes of Cl^-. See also ANTIMONY COMPOUNDS and CHLORIDES.

DQG600 **CAS:593-57-7** **HR: 3**
DIMETHYLARSINE
mf: C_2H_7As mw: 106.07

PROP: Colorless liquid. Mp: −78°, bp: 36°, d: 1.213 @ 29°/4°, vap d: 3.65.

SYN: CACODYL HYDRIDE

CONSENSUS REPORTS: Arsenic and its compounds are on the Community Right-To-Know List.

SAFETY PROFILE: Arsenic compounds are generally poisons. Ignites spontaneously in air. It is more toxic than its oxidation products; reacts vigorously with oxidizing agents. To fight fire, exclude O_2, allow fire to burn, or apply water, foam, dry chemical, water spray, or CO_2. When heated to decomposition it emits toxic fumes of As. See also ARSINE and ARSENIC COMPOUNDS.

DQG700 CAS:13367-92-5 HR: 3
DIMETHYL ARSINIC SULFIDE
mf: $C_2H_6As_2S_2$ mw: 244.04

SYNS: ARSINE SULFIDE, DIMETHYLDI- □ DIMETHYLDIARSINE SULFIDE

TOXICITY DATA WITH REFERENCE
ivn-mus LD50:10 mg/kg CSLNX* NX#03919

OSHA PEL: TWA 0.5 mg(As)/m^3

SAFETY PROFILE: Poison by intravenous route. When heated to decomposition it emits toxic fumes of SO_x and As.

DQH000 CAS:28842-05-9 HR: 2
3,6′-DIMETHYLAZOBENZENE
mf: $C_{14}H_{14}N_2$ mw: 210.30

SYNS: 2:3′-AZOTOLUENE □ 2,3′-DIMETHYLAZOBENZENE

TOXICITY DATA WITH REFERENCE
orl-rat TDLo:26 g/kg/36W-C:NEO JPBAA7 58,275,46
mul-mus TDLo:400 mg/kg/I:ETA CNREA8 1,397,41
orl-rat TD:321 g/kg/36W-C:ETA JPBAA7 58,275,46

SAFETY PROFILE: Questionable carcinogen with experimental neoplastigenic and tumorigenic data. When heated to decomposition it emits toxic fumes of NO_x.

DQH200 CAS:35077-51-1 HR: 2
N,N′-DIMETHYL-4,4′-AZODIACETANILIDE
mf: $C_{18}H_{20}N_4O_2$ mw: 324.42

SYNS: N′-ACETYL-N′-MONOMETHYL-4′-AMINO-N-ACETYL-N-MONOMETHYL-4-AMINOAZOBENZENE □ N,N′-(AZODI-4,1-PHENYLENE)BIS (N-METHYLACETAMIDE) □ 4,4′-BIS(N-ACETYL-N-METHYLAMINO)AZOBENZENE

TOXICITY DATA WITH REFERENCE
dns-rat:lvr 1 µmol/L CNREA8 46,1654,86
orl-rat TDLo:2200 mg/kg/13W-C:ETA CNREA8 34,2274,74
ipr-rat LD50:480 mg/kg CNREA8 34,2274,74

SAFETY PROFILE: Moderately toxic by intraperitoneal route. Questionable carcinogen with experimental tumorigenic data. Mutation data reported. When heated to decomposition it emits toxic fumes of NO_x.

DQH509 CAS:2446-84-6 HR: 3
DIMETHYL AZODIFORMATE
mf: $C_4H_6N_2O_4$ mw: 146.11

$CH_3OCO{\cdot}N{=}NCO{\cdot}OCH_3$

PROP: Oil. Shock-sensitive, burns explosively. Fp: 10°, bp: 96° @ 25 mm.

SYNS: DIMETHYL AZOFORMATE □ DIMETHYL DIZENEDICARBOXYLATE

SAFETY PROFILE: A shock-sensitive explosive. It burns explosively when ignited. When heated to decomposition it emits toxic fumes of NO_x.

DQH550 HR: 2
1,3-DIMETHYLBENZ(e)ACEPHENANTHRYLENE
mf: $C_{22}H_{18}$ mw: 282.40

SYN: 1,3-DIMETHYLBENZO(b)FLUORANTHENE

TOXICITY DATA WITH REFERENCE
mma-sat 63 nmol/plate CRNGDP 6,1023,85
skn-mus TDLo:452 µg/kg/20D-I:ETA CRNGDP 6,1023,85

SAFETY PROFILE: Questionable carcinogen with experimental tumorigenic data. Mutation data reported. When heated to decomposition it emits acrid smoke and irritating fumes.

DQH600 CAS:3518-05-6 HR: 2
1,10-DIMETHYL-5,6-BENZACRIDINE
mf: $C_{19}H_{15}N$ mw: 257.35

SYN: 8,12-DIMETHYLBENZ(a)ACRIDINE

TOXICITY DATA WITH REFERENCE
scu-mus TDLo:200 mg/kg/4W-I:ETA ACRSAJ 4,315,56
scu-mus TDLo:250 mg/kg/10D-I:ETA BAFEAG 34,22,47

SAFETY PROFILE: Questionable carcinogen with experimental tumorigenic data. When heated to decomposition it emits toxic fumes of NO_x.

DQH800 CAS:17401-48-8 HR: 2
2,10-DIMETHYL-5,6-BENZACRIDINE
mf: $C_{19}H_{15}N$ mw: 257.35

SYN: 9,12-DIMETHYLBENZ(a)ACRIDINE

TOXICITY DATA WITH REFERENCE
skn-mus TDLo:540 mg/kg/45W-I:ETA ACRSAJ 4,315,56

SAFETY PROFILE: Questionable carcinogen with experimental tumorigenic data. When heated to decomposition it emits toxic fumes of NO_x.

DQI200 CAS:963-89-3 HR: 2
7,9-DIMETHYLBENZ(c)ACRIDINE
mf: $C_{19}H_{15}N$ mw: 257.35

SYN: 3,10-DIMETHYL-7,8-BENZACRIDINE (FRENCH)

TOXICITY DATA WITH REFERENCE
mma-sat 10 µg/plate ENMUDM 6(Suppl 2),1,84
otr-rat:emb 40,800 µg/L JJIND8 67,1303,81
otr-ham:emb 100 µg/L JJIND8 67,1303,81
skn-mus TDLo:180 mg/kg/15W-I:ETA ACRSAJ 4,315,56
scu-mus TDLo:200 mg/kg/4W-I:ETA ACRSAJ 4,315,56
scu-mus TD:250 mg/kg/10D-I:ETA BAFEAG 34,22,47

CONSENSUS REPORTS: EPA Genetic Toxicology Program.

SAFETY PROFILE: Questionable carcinogen with experimental tumorigenic data. Mutation data reported. When heated to decomposition it emits toxic fumes of NO_x.

DQI400 CAS:32740-01-5 HR: 2
7,11-DIMETHYLBENZ(c)ACRIDINE
mf: $C_{19}H_{15}N$ mw: 257.35

SYN: 1,10-DIMETHYL-7,8-BENZACRIDINE (FRENCH)

TOXICITY DATA WITH REFERENCE
mma-sat 1 nmol/plate GANNA2 70,749,79
skn-mus TDLo:360 mg/kg/30W-I:ETA ACRSAJ 4,315,56
scu-mus TDLo:200 mg/kg/4W-I:ETA ACRSAJ 4,315,56
scu-mus TD:250 mg/kg/10D-I:ETA BAFEAG 34,22,47

SAFETY PROFILE: Questionable carcinogen with experimental tumorigenic data. Mutation data reported. When heated to decomposition it emits toxic fumes of NO_x.

DQI600 CAS:53-69-0 HR: 3
5,7-DIMETHYL-1,2-BENZACRIDINE
mf: $C_{19}H_{15}N$ mw: 257.35

SYN: 8,10-DIMETHYL-BENZ(a)ACRIDINE

TOXICITY DATA WITH REFERENCE
mma-sat 500 nmol/L ENMUDM 3,11,81
dns-rat:lvr 50 µmol/L ENMUDM 3,11,81
irn-frg LDLo:11 mg/kg CNREA8 24,1969,64

SAFETY PROFILE: Poison by intrarenal route. Mutation data reported. When heated to decomposition it emits toxic fumes of NO_x.

DQI800 CAS:2381-40-0 HR: 2
6,9-DIMETHYL-1,2-BENZACRIDINE
mf: $C_{19}H_{15}N$ mw: 257.35

SYNS: 7,10-DIMETHYLBENZ(c)ACRIDINE □ 2,10-DIMETHYL-7,8-BENZACRIDINE (FRENCH)

TOXICITY DATA WITH REFERENCE
mma-sat 50 µg/plate PNASA6 72,5135,75
otr-ham:emb 2 mg/L EJCAAH 17,179,81
otr-ham:kdy 80 µg/L BJCAAI 37,873,78
skn-mus TDLo:190 mg/kg/16W-I:ETA ACRSAJ 4,315,56
scu-mus TDLo:200 mg/kg/4W-I:ETA ACRSAJ 4,315,56
scu-mus TD:250 mg/kg/10D-I:ETA BAFEAG 34,22,47

CONSENSUS REPORTS: EPA Genetic Toxicology Program.

SAFETY PROFILE: Questionable carcinogen with experimental tumorigenic data. Mutation data reported. When heated to decomposition it emits toxic fumes of NO_x.

DQJ000 CAS:611-74-5 HR: 2
N,N-DIMETHYLBENZAMIDE
mf: $C_9H_{11}NO$ mw: 149.21

PROP: Liquid or crystals. Mp: 40–41°, bp: 157–158° @ 35 mm.

SYN: DIMETHYL BENZMIDE

TOXICITY DATA WITH REFERENCE
orl-mus LD50:960 mg/kg TXAPA9 19,20,71

CONSENSUS REPORTS: Reported in EPA TSCA Inventory.

SAFETY PROFILE: Moderately toxic by ingestion. When heated to decomposition it emits toxic fumes of NO_x.

DQJ200 CAS:57-97-6 HR: 3
DIMETHYLBENZANTHRACENE
mf: $C_{20}H_{16}$ mw: 256.36

PROP: Leaflets from Me_2CO/EtOH. Mp: 122–123°.

SYNS: DBA □ DIMETHYLBENZ(a)ANTHRACENE □ 7,12-DIMETHYLBENZANTHRACENE □ 7,12-DIMETHYLBENZ(a)ANTHRACENE □ 9,10-DIMETHYL-BENZANTHRACENE □ 9,10-DIMETHYLBENZ(a)ANTHRACENE □ 9,10-DIMETHYL-1,2-BENZANTHRACENE □ 9,10-DIMETHYL-1,2-BENZANTHRAZEN (GERMAN) □ DIMETHYLBENZANTHRENE □ 7,12-DIMETHYLBENZO(a)ANTHRACENE □ 1,4-DIMETHYL-2,3-BENZPHENANTHRENE □ DMBA □ 7,12-DMBA □ NCI-C03918 □ RCRA WASTE NUMBER U094

TOXICITY DATA WITH REFERENCE
skn-mus 64 µg MLD CALEDQ 4,333,78
dnd-hmn:emb 220 nmol/L MUREAV 89,95,81
dni-hmn:lvr 1 mmol/L VOONAW 28(11),53,82
otr-mus:emb 300 µg/L PMRSDJ 5,659,85
ipr-rat TDLo:6060 µg/kg (female 8-12D post):REP IRLCDZ 7,358,79
orl-mus TDLo:50 mg/kg (female 7-16D post):REP TJADAB 17,33A,78
orl-rat TDLo:672 mg/kg (15D pre-5D post):TER DOESD6 54,410,81
ipr-rat TDLo:30,300 µg/kg (female 8-12D post):REP PPASAK 52,34,78
orl-rat TDLo:672 mg/kg (15D pre-5D post):REP DOESD6 54,410,81
orl-mus TDLo:330 mg/kg (female 7-16D post):REP TJADAB 17,33A,78
ivn-rat TDLo:20 mg/kg (female 5D post):TER IARCCD 4,112,73
orl-mus TDLo:150 mg/kg (female 14-18D post):REP VAPHDQ 372,29,76
ipr-mus TDLo:25 mg/kg (female 12D post):REP TJADAB 16,147,77
ipr-rat TDLo:25 mg/kg (female 8-12D post):REP DCTODJ 9,15,86
orl-gpg TDLo:1 g/kg (female 27D pre-5D post):REP DOESD6 54,410,81
ivn-rat TDLo:31 mg/kg (male 1D pre):REP JEMEAV 118,27,63
ipr-rat TDLo:25 mg/kg (female 8-12D post):TER FEPRA7 37,853,78
ivn-rat TDLo:20 mg/kg (female 9D post):TER IARCCD 4,112,73
ivn-rat TDLo:20 mg/kg (female 13D post):TER IARCCD 4,112,73
orl-mus TDLo:12,500 µg/kg (female 10D post):TER MUREAV 141,105,84
orl-rat TDLo:37,500 µg/kg (female 14-20D post):ETA,TER CRNGDP 3,413,78
orl-rat TDLo:15 mg/kg:CAR LIFSAK 7,259,60
skn-rat TDLo:60 mg/kg/30W-I:ETA JCUPBN 7,277,80
ipr-rat TDLo:24 mg/kg (20D preg):ETA,TER CCSUDL 3,413,78
ipr-rat TDLo:25 mg/kg:ETA CNREA8 39,3968,79

scu-rat TDLo: 500 μg/kg: CAR CRNGDP 6,769,85
ivn-rat TDLo: 15 mg/kg (21D preg): ETA,TER JNCIAM 52,1365,74
ivn-rat TDLo: 30 mg/kg: NEO KIDZAK 23(Suppl 1),34,71
ims-rat TDLo: 2400 μg/kg: CAR NTIS** DOE/EV/03140-5
ivn-rat TDLo: 15 mg/kg (21D preg): ETA,TER NEOLA4 23,285,76
par-rat TDLo: 1 mg/kg: CAR CMBID4 33,469,87
imp-rat TDLo: 11 μg/kg: ETA,TER NISFAY 34,1853,82
ipc-rat TDLo: 1250 μg/kg: ETA,TER GANNA2 62,55,71
orl-mus TDLo: 640 mg/kg/75D-I: NEO,REP CNREA8 48,425,88
skn-mus TDLo: 103 μg/kg: NEO CNREA8 39,1934,79
ipr-mus TDLo: 112 mg/kg (female 14-21D post): NEO,TER IJCNAW 4,219,69
scu-mus TDLo: 60 mg/kg (13-17D preg): NEO,TER LIFSAK 26,1955,80
scu-mus TDLo: 833 μg/kg: NEO BJCAAI 20,148,66
ivn-mus TDLo: 120 mg/kg (18-20D preg): NEO,REP VOONAW 20(8),65,74
ivn-mus TDLo: 4800 mg/kg/24W-I: CAR APJAAG 31,799,81
ivn-mus TDLo: 120 mg/kg (female 18-20D post): NEO,TER VOONAW 20(8),65,74
itr-mus TDLo: 4 mg/kg: CAR AJCAA7 35,538,39
imp-mus TDLo: 40 mg/kg: ETA OSOMAE 48,47,79
ivg-mus TDLo: 744 mg/kg/31W-I: CAR,REP BJCAAI 20,184,66
ivg-mus TDLo: 40 mg/kg (19D preg): CAR,TER VOONAW 22(6),44,76
par-dog TDLo: 1235 mg/kg/52W-I: ETA JJIND8 65,921,80
itr-dog TDLo: 11 mg/kg/53W-I: ETA JTCSAQ 49,364,65
imp-dog TDLo: 378 mg/kg/52W-C: ETA JJIND8 65,921,80
skn-mky TDLo: 1600 mg/kg/65W-I: CAR JNCIAM 57,1269,76
skn-rbt TDLo: 7 mg/kg/8W-I: ETA ONCOAR 15,98,62
ivn-rbt TDLO: 20 mg/kg (25D preg): NEO,TER BEXBAN 85,369,78
par-rbt TDLo: 40 mg/kg/16W-I: ETA AOUNAZ 102,111,83
skn-gpg TDLo: 240 mg/kg/20W-I: ETA PGTCA4 5,120,79
scu-gpg TDLo: 20 mg/kg/9W-I: ETA COREAF 252,1236,61
imp-gpg TDLo: 100 mg/kg: ETA COREAF 252,1236,61
orl-ham TDLo: 80 mg/kg: NEO PEXTAR 26,128,83
orl-ham TDLo: 25 mg/kg/(15D preg): NEO,TER PAACA3 18,1,77
skn-ham TDLo: 1600 μg/kg: CAR PEXTAR 26,128,83
scu-ham TDLo: 400 μg/kg: CAR BJCAAI 21,184,67
ivn-ham TDLo: 24 mg/kg/13W-I: ETA,REP INURAQ 15,42,77
ims-ham TDLo: 12,500 μg/kg: ETA CMBID4 24,127,79
itr-ham TDLo: 5200 μg/kg/13W-I: ETA EJCAAH 10,483,74
mul-ham TDLo: 272 g/kg/18W-I: NEO AEMBAP 91,57,77
ims-ckn TDLo: 125 mg/kg/2W-I: NEO BJCAAI 41,130,80
skn-grb TDLo: 70 mg/kg/7W-I: ETA CNREA8 26,844,66
ims-grb TDLo: 350 μg/kg/7W-I: ETA CNREA8 26,844,66
scu-rat TD: 20 mg/kg/39D-I: NEO CNREA8 31,1951,71
scu-rat TD: 13 mg/kg: NEO EJCAAH 6,417,70
ivn-rat TD: 30 mg/kg/7D-I: NEO ACLSCP 13,289,83
orl-rat TD: 150 mg/kg: CAR IJCNAW 26,349,80
imp-rat TD: 5 mg/kg: CAR NGGZAK 85,555,84
orl-rat TD: 10 mg/kg: CAR CNREA8 45,4827,85
orl-mus TD: 60 mg/kg/2W-I: CAR JTEHD6 10,131,82
orl-rat TD: 37,500 μg/kg: CAR CRNGDP 5,1539,84
scu-mus TD: 1880 μg/kg/47W-I: NEO JNCIAM 37,825,66
orl-rat TD: 15 mg/kg: CAR BCTRD6 4,129,84
orl-rat LD50: 327 mg/kg GANNA2 68,237,77
ivn-rat LD50: 54 mg/kg SCIEAS 147,1153,65
orl-mus LD50: 340 mg/kg SCIEAS 147,1153,65
ipr-mus LD50: 54 mg/kg PWPSA8 24,177,81
itr-mus LD50: 22,500 μg/kg PWPSA8 24,177,81
scu-gpg LDLo: 20 mg/kg COREAF 252,1236,61

CONSENSUS REPORTS: Reported in EPA TSCA Inventory. EPA Genetic Toxicology Program.

SAFETY PROFILE: Suspected carcinogen with experimental carcinogenic, neoplastigenic, tumorigenic, and teratogenic data. A transplacental carcinogen. Poison by ingestion, intravenous, subcutaneous, intraperitoneal, and intratracheal routes. Other experimental reproductive effects. Human mutation data reported. A skin irritant. When heated to decomposition it emits acrid smoke and irritating fumes.

DQJ400 CAS:313-74-6 HR: 2
1,12-DIMETHYLBENZ(a)ANTHRACENE
mf: $C_{20}H_{16}$ mw: 256.36

PROP: Yellow needles. Mp: 132°.

SYN: 1′,9-DIMETHYL-1,2-BENZANTHRACENE

TOXICITY DATA WITH REFERENCE
scu-mus TDLo: 72 mg/kg/9W-I: ETA BAFEAG 49,312,62

SAFETY PROFILE: Questionable carcinogen with experimental tumorigenic data. When heated to decomposition it emits acrid smoke and irritating fumes. See also DIMETHYLBENZANTHRACENE.

DQJ600 CAS:18429-70-4 HR: 2
4,5-DIMETHYLBENZ(a)ANTHRACENE
mf: $C_{20}H_{16}$ mw: 256.36

PROP: Needles from C_6H_6/EtOH. Mp: 138–139°.

SYN: 3,4′-DIMETHYL-1,2-BENZANTHRACENE

TOXICITY DATA WITH REFERENCE
scu-rat TDLo: 18 mg/kg: ETA PSEBAA 128,720,68

SAFETY PROFILE: Questionable carcinogen with experimental tumorigenic data. When heated to decomposition it emits acrid smoke and irritating fumes. See also DIMETHYLBENZANTHRACENE.

DQJ800 CAS:20627-28-5 HR: 2
6,7-DIMETHYLBENZ(a)ANTHRACENE
mf: $C_{20}H_{16}$ mw: 256.36

PROP: Pale-yellow needles from MeOH. Mp: 114°.

SYN: 4,10-DIMETHYL-1,2-BENZANTHRACENE

TOXICITY DATA WITH REFERENCE
scu-mus TDLo: 40 mg/kg: ETA JNCIAM 1,303,40

SAFETY PROFILE: Questionable carcinogen with experimental tumorigenic data. When heated to decomposition it emits acrid smoke and fumes. See also DIMETHYLBENZANTHRACENE.

DQK000 **CAS:317-64-6** **HR: 2**
6,8-DIMETHYLBENZ(a)ANTHRACENE
mf: $C_{20}H_{16}$ mw: 256.36

PROP: Needles from C_6H_6/EtOH. Mp: 138–139°.

SYN: 6,8-DIMETHYL-1,2-BENZANTHRACENE

TOXICITY DATA WITH REFERENCE
ims-rat TDLo:50 mg/kg:NEO CNREA8 29,506,69
skn-mus TDLo:240 mg/kg/37W-I:ETA CNREA8 11,892,51

SAFETY PROFILE: Questionable carcinogen with experimental neoplastigenic and tumorigenic data. When heated to decomposition it emits acrid smoke and irritating fumes. See also DIMETHYLBENZANTHRACENE.

DQK200 **CAS:568-81-0** **HR: 2**
6,12-DIMETHYLBENZ(a)ANTHRACENE
mf: $C_{20}H_{16}$ mw: 256.36

PROP: Needles from MeOH. Mp: 75°.

SYN: 4,9-DIMETHYL-1,2-BENZANTHRACENE

TOXICITY DATA WITH REFERENCE
imp-mus TDLo:80 mg/kg:ETA JNCIAM 2,241,41

SAFETY PROFILE: Questionable carcinogen with experimental tumorigenic data. When heated to decomposition it emits acrid smoke and irritating fumes. See also DIMETHYLBENZANTHRACENE.

DQK400 **CAS:35187-28-1** **HR: 2**
7,11-DIMETHYLBENZ(a)ANTHRACENE
mf: $C_{20}H_{16}$ mw: 256.36

PROP: Needles from C_6H_6/EtOH. Mp: 146°.

SYN: 8,10-DIMETHYL-1,2-BENZANTHRACENE

TOXICITY DATA WITH REFERENCE
scu-mus TDLo:40 mg/kg:ETA CNREA8 6,454,46

SAFETY PROFILE: Questionable carcinogen with experimental tumorigenic data. When heated to decomposition it emits acrid smoke and irritating fumes. See also DIMETHYLBENZANTHRACENE.

DQK600 **CAS:58430-00-5** **HR: 2**
5,6-DIMETHYL-1,2-BENZANTHRACENE
mf: $C_{20}H_{16}$ mw: 256.36

PROP: Plates from EtOH. Mp: 187–188°.

SYN: 8,9-DIMETHYLBENZ(a)ANTHRACENE

TOXICITY DATA WITH REFERENCE
skn-mus TDLo:500 mg/kg/21W-I:ETA PRLBA4 117,318,35

SAFETY PROFILE: Questionable carcinogen with experimental tumorigenic data. When heated to decomposition it emits acrid smoke and irritating fumes. See also DIMETHYLBENZANTHRACENE.

DQK800 **CAS:20627-31-0** **HR: 2**
5,9-DIMETHYL-1,2-BENZANTHRACENE
mf: $C_{20}H_{16}$ mw: 256.36

PROP: Plates from C_6H_6/EtOH. Mp: 135°.

SYNS: 5:9-DIMETHYL-1:2-BENZANTHRACENE □ 8,12-DIMETHYLBENZ(a)ANTHRACENE

TOXICITY DATA WITH REFERENCE
scu-mus TDLo:40 mg/kg:ETA AJCAA7 33,499,38

SAFETY PROFILE: Questionable carcinogen with experimental tumorigenic data. When heated to decomposition it emits acrid smoke and irritating fumes. See also DIMETHYLBENZANTHRACENE.

DQK900 **CAS:71964-72-2** **HR: 2**
7,12-DIMETHYLBENZ(a)ANTHRACENE-3,4-DIOL
mf: $C_{20}H_{16}O_2$ mw: 288.36

TOXICITY DATA WITH REFERENCE
mma-sat 10 nmol/plate 46OJAN-,675,81
skn-mus TDLo:35 µg/kg:ETA CNREA8 40,3661,80

SAFETY PROFILE: Questionable carcinogen with experimental tumorigenic data. Mutation data reported. When heated to decomposition it emits acrid smoke and irritating fumes.

DQL000 **CAS:604-81-9** **HR: 2**
5,10-DIMETHYL-1,2-BENZANTHRACENE
mf: $C_{20}H_{16}$ mw: 256.36

PROP: Plates from C_6H_6/EtOH. Mp: 147°.

SYN: 7,8-DIMETHYLBENZ(a)ANTHRACENE

TOXICITY DATA WITH REFERENCE
scu-mus TDLo:40 mg/kg:ETA AJCAA7 33,499,38
scu-mus TD:400 mg/kg:ETA JACSAT 58,2376,36

SAFETY PROFILE: Questionable carcinogen with experimental tumorigenic data. When heated to decomposition it emits acrid smoke and irritating fumes. See also DIMETHYLBENZANTHRACENE.

DQL200 **CAS:58429-99-5** **HR: 2**
6,7-DIMETHYL-1,2-BENZANTHRACENE
mf: $C_{20}H_{16}$ mw: 256.36

PROP: Crystals from EtOAc. Mp: 174°.

SYN: 9,10-DIMETHYLBENZ(a)ANTHRACENE

TOXICITY DATA WITH REFERENCE
mma-sat 2500 µg/plate BJCAAI 37,873,78
dns-rat:lvr 50 µmol/L ENMUDM 3,11,81
dnd-mus-skn 110 mg/L CNREA8 32,643,72
dni-mus-ipr 100 mg/kg MUREAV 46,305,77
otr-ham:kdy 80 µg/L BJCAAI 37,873,78
skn-mus TDLo:1250 mg/kg/52W-I:ETA PRLBA4 117,318,35

SAFETY PROFILE: Questionable carcinogen with experimental tumorigenic data. Mutation data reported. When heated to decomposition it emits acrid smoke and irritating fumes. See also DIMETHYLBENZANTHRACENE.

DQL400 **CAS:32976-87-7** **HR: 2**
7,12-DIMETHYLBENZ(a)ANTHRACENE, DEUTERATED
mf: $C_{20}D_{16}$ mw: 272.36

SYN: 7,12-DIMETHYLBENZ(a)ANTHRACENE-D16

TOXICITY DATA WITH REFERENCE
scu-mus TDLo:60 mg/kg/9W-I:ETA NATWAY 58,371,71

SAFETY PROFILE: Questionable carcinogen with experimental tumorigenic data. When heated to decomposition it emits acrid smoke and irritating fumes. See also DIMETHYLBENZANTHRACENE.

DQL800 **CAS:63019-25-0** **HR: 2**
9:10-DIMETHYL-1:2-BENZANTHRACENE-9:10-OXIDE
mf: $C_{20}H_{16}O$ mw: 272.36

SYN: 9:10-DIMETHYL-9-10-DIHYDRO-1,2-BENZANTHRACENE-9,10-OXIDE

TOXICITY DATA WITH REFERENCE
skn-mus TDLo:860 mg/kg/36W-I:ETA PRLBA4, 129,439,40

SAFETY PROFILE: Questionable carcinogen with experimental tumorigenic data. When heated to decomposition it emits acrid smoke and irritating fumes.

DQL820 **CAS:68908-87-2** **HR: 2**
1,3-DIMETHYLBENZENE, BENZYLATED

SYNS: BENZENE, 1,3-DIMETHYL-, BENZYLATED □ SANTOSOL 150

TOXICITY DATA WITH REFERENCE
orl-rat LD50:2333 mg/kg EPASR* 8EHQ-1090-0941
skn-rbt LD50:>5 g/kg EPASR* 8EHQ-1090-0941

CONSENSUS REPORTS: Reported in EPA TSCA Inventory.

SAFETY PROFILE: Moderately toxic by ingestion. Low toxicity by skin contact. When heated to decomposition it emits acrid smoke and irritating vapors.

DQL899 **CAS:68596-88-3** **HR: 3**
3,5-DIMETHYLBENZENEDIAZONIUM-2-CARBOXYLATE
mf: $C_9H_8N_2O_2$ mw: 176.18

PROP: Crystals. Explodes on melting.

SAFETY PROFILE: A powerful, heat-sensitive explosive. When heated to decomposition it emits toxic fumes of NO_x.

DQM000 **CAS:612-82-8** **HR: 3**
3,3′-DIMETHYLBENZIDINE DIHYDROCHLORIDE
mf: $C_{14}H_{16}N_2•2ClH$ mw: 285.24

SYNS: 4,4′-DIAMINO-3,3′-DIMETHYLBIPHENYL DIHYDROCHLORIDE □ 3,3′-DIMETHYLBIPHENYL-4,4′-BIPHENYLDIAMINE DIHYDROCHLORIDE □ 2,3′-DIMETHYLBIPHENYL-4,4′-DIAMINE DIHYDROCHLORIDE □ o-TOLIDINE DIHYDROCHLORIDE

TOXICITY DATA WITH REFERENCE
mma-sat 50 μg/plate CALEDQ 4,21,77
sln-dmg-orl 14 pph ENMUDM 7,325,85
sln-dmg-par 2750 ppm ENMUDM 7,325,85
orl-rat TDLo:1820 mg/kg/65W-C:CAR JACTDZ 10(2),255,91
orl-mus TDLo:15,288 mg/kg/78W-C:CAR FCTOD7 27,801,89
orl-rat TDLo:1120 mg/kg/14D-C NTPTR* NTP-TR-390,91
orl-rat TDLo:18,200 mg/kg/13W-C NTPTR* NTP-TR-390,91

CONSENSUS REPORTS: Reported in NTP Carcinogenesis Studies (Drinking); Clear Evidence: RAT NTPTR* NTP-TR-390,91. Reported in EPA TSCA Inventory.

SAFETY PROFILE: Suspected carcinogen with experimental carcinogenic data. Mutation data reported. When heated to decomposition it emits very toxic fumes of HCl and NO_x.

DQM100 **CAS:582-60-5** **HR: 3**
5,6-DIMETHYLBENZIMIDAZOLE
mf: $C_9H_{10}N_2$ mw: 146.21

SYN: BENZIMIDAZOLE, 5,6-DIMETHYL-

TOXICITY DATA WITH REFERENCE
ipr-mus LD50:400 mg/kg RPTOAN 41,249,78

CONSENSUS REPORTS: Reported in EPA TSCA Inventory.

SAFETY PROFILE: Poison by intraperitoneal route. When heated to decomposition it emits toxic vapors of NO_x.

DQM200 **CAS:18463-86-0** **HR: 2**
N,N-DIMETHYL-p-(4-BENZIMIDAZOLYAZO)ANILINE
mf: $C_{15}H_{15}N_5$ mw: 265.35

SYNS: 4-((p-(DIMETHYLAMINO)PHENYL)AZO)BENZIMIDAZOLE □ N,N-DIMETHYL-4(4′-BENZIMIDAZOLYLAZO)ANILINE

TOXICITY DATA WITH REFERENCE
orl-rat TDLo:1080 mg/kg/9W-C:ETA JMCMAR 11,1074,68

SAFETY PROFILE: Questionable carcinogen with experimental tumorigenic data. When heated to decomposition it emits toxic fumes of NO_x.

DQM400 **CAS:4699-26-7** **HR: 2**
6,12-DIMETHYLBENZO(1,2-b:5,4-b′)BIS(1)BENZOTHIOPHENE
mf: $C_{20}H_{14}S_2$ mw: 318.46

TOXICITY DATA WITH REFERENCE
scu-mus TDLo:80 mg/kg:ETA JNCIAM 18,555,57

SAFETY PROFILE: Questionable carcinogen with experimental tumorigenic data. When heated to decomposition it emits toxic fumes of SO_x.

DQM600 CAS:22781-23-3 HR: 3
2,2-DIMETHYL-1,3-BENZODIOX-4-OL METHYLCARBAMATE
mf: $C_{11}H_{13}NO_4$ mw: 223.25

PROP: Crystals. Mp: 129–130°. Very sltly sol in H_2O.

SYNS: BENCARBATE □ BENDIOCARB □ BICAM ULV □ 2,2-DIMETHYL-1,3-BENZDIOXOL-4-YL-N-METHYLCARBAMATE □ 2,2-DIMETHYLBENZO-1,3-DIOXOL-4-YL METHYLCARBAMATE □ 2,2-DIMETHYL-4-(N-METHYLAMINOCARBOXYLATO)-1,3-BENZODIOXOLE □ 2,2-DIMETHYL-4-(N-METHYLCARBAMATO)-1,3-BENZODIOXOLE □DY-CARB □ FICAM □ GARVOX □ 2,3-ISOPROPYLIDENEDIOXYPHENYL METHYLCARBAMATE □ MC6897 □ METHYLCARBAMIC ACID-2,3-(ISOPROPYLIDENEDIOXY)PHENYL ESTER □ MULTAMAT □ NIOMIL □ ROTATE □ TATTOO □ TURCAM

TOXICITY DATA WITH REFERENCE
orl-rat LD50:40 mg/kg FMCHA2 -,C29,83
skn-rat LD50:566 mg/kg PEMNDP 9,54,91
orl-mus LD50:45 mg/kg PSSCBG 3,735,72
orl-rbt LD50:35 mg/kg 85JFAN A029,84
orl-gpg LD50:35 mg/kg 85JFAN A029,84

CONSENSUS REPORTS: EPA Genetic Toxicology Program.

SAFETY PROFILE: Poison by ingestion. Moderately toxic by skin contact. When heated to decomposition it emits toxic fumes of NO_x. See also CARBAMATES.

DQM800 CAS:37750-86-0 HR: 2
6,12-DIMETHYLBENZO(1,2-b:4,5-b′)DITHIONAPHTHENE
mf: $C_{20}H_{14}S_2$ mw: 318.46

TOXICITY DATA WITH REFERENCE
mnt-mus:fbr 1500 µg/L NULSAK 6,17,63
cyt-mus:fbr 1500 µg/L NULSAK 6,17,63
scu-mus TDLo:80 mg/kg:ETA JNCIAM 18,555,57

SAFETY PROFILE: Questionable carcinogen with experimental tumorigenic data. Mutation data reported. When heated to decomposition it emits toxic fumes of SO_x.

DQM850 CAS:619-04-5 HR: 3
3,4-DIMETHYLBENZOIC ACID
mf: $C_9H_{10}O_2$ mw: 150.19

SYN: BENZOIC ACID, 3,4-DIMETHYL-

TOXICITY DATA WITH REFERENCE
ipr-mus LD50:316 mg/kg JMCMAR 11,1020,68

CONSENSUS REPORTS: Reported in EPA TSCA Inventory.

SAFETY PROFILE: Poison by intraperitoneal route. When heated to decomposition it emits acrid smoke and irritating vapors.

DQN000 CAS:16757-85-0 HR: 2
1,2-DIMETHYLBENZO(a)PYRENE
mf: $C_{22}H_{16}$ mw: 280.38

TOXICITY DATA WITH REFERENCE
cyt-ckn:leu 1 pph/30M BBRCA9 93,954,80
scu-mus TDLo:72 mg/kg/13W-I:ETA IJCNAW 3,238,68

SAFETY PROFILE: Questionable carcinogen with experimental tumorigenic data. Mutation data reported. When heated to decomposition it emits acrid smoke and irritating fumes.

DQN200 CAS:16757-86-1 HR: 2
1,3-DIMETHYLBENZO(a)PYRENE
mf: $C_{22}H_{16}$ mw: 280.38

TOXICITY DATA WITH REFERENCE
scu-mus TDLo:72 mg/kg/13W-I:ETA IJCNAW 3,238,68

SAFETY PROFILE: Questionable carcinogen with experimental tumorigenic data. When heated to decomposition it emits acrid smoke and irritating fumes.

DQN400 CAS:16757-88-3 HR: 2
1,4-DIMETHYLBENZO(a)PYRENE
mf: $C_{22}H_{16}$ mw: 280.38

TOXICITY DATA WITH REFERENCE
scu-mus TDLo:72 mg/kg/13W-I:ETA IJCNAW 3,238,68

SAFETY PROFILE: Questionable carcinogen with experimental tumorigenic data. When heated to decomposition it emits acrid smoke and irritating fumes.

DQN600 CAS:16757-90-7 HR: 2
1,6-DIMETHYLBENZO(a)PYRENE
mf: $C_{22}H_{16}$ mw: 280.38

TOXICITY DATA WITH REFERENCE
scu-mus TDLo:72 mg/kg/13W-I:ETA IJCNAW 3,238,68

SAFETY PROFILE: Questionable carcinogen with experimental tumorigenic data. When heated to decomposition it emits acrid smoke and irritating fumes.

DQN800 CAS:16757-87-2 HR: 2
2,3-DIMETHYLBENZO(a)PYRENE
mf: $C_{22}H_{16}$ mw: 280.38

TOXICITY DATA WITH REFERENCE
scu-mus TDLo:72 mg/kg/13W-I:ETA IJCNAW 3,238,68

SAFETY PROFILE: Questionable carcinogen with experimental tumorigenic data. When heated to decomposition it emits acrid smoke and irritating fumes.

DQO000 CAS:16757-91-8 HR: 2
3,6-DIMETHYLBENZO(a)PYRENE
mf: $C_{22}H_{16}$ mw: 280.38

TOXICITY DATA WITH REFERENCE
scu-mus TDLo:72 mg/kg/13W-I:ETA IJCNAW 3,238,68

SAFETY PROFILE: Questionable carcinogen with experimental tumorigenic data. When heated to decomposition it emits acrid smoke and irritating fumes.

D

DQO200 CAS:16757-84-9 HR: 2
3,12-DIMETHYLBENZO(a)PYRENE
mf: $C_{22}H_{16}$ mw: 280.38

TOXICITY DATA WITH REFERENCE
scu-mus TDLo:72 mg/kg/13W-I:ETA IJCNAW 3,238,68

SAFETY PROFILE: Questionable carcinogen with experimental tumorigenic data. When heated to decomposition it emits acrid smoke and irritating fumes.

DQO400 CAS:16757-89-4 HR: 2
4,5-DIMETHYLBENZO(a)PYRENE
mf: $C_{22}H_{16}$ mw: 280.38

TOXICITY DATA WITH REFERENCE
scu-mus TDLo:72 mg/kg/13W-I:ETA IJCNAW 3,238,68

SAFETY PROFILE: Questionable carcinogen with experimental tumorigenic data. When heated to decomposition it emits acrid smoke and irritating fumes.

DQO600 CAS:2818-89-5 HR: 2
2,5-DIMETHYLBENZOSELENAZOLE
mf: C_9H_9NSe mw: 210.15

SYN: 2,5-DIMETHYLBENZSELENAZOL (CZECH)

TOXICITY DATA WITH REFERENCE
skn-rbt 500 mg/24H MLD 28ZPAK -,222,72
eye-rbt 500 mg/24H MLD 28ZPAK -,222,72
orl-rat LD50:1060 mg/kg 28ZPAK -,222,72

CONSENSUS REPORTS: Community Right-To-Know List.

OSHA PEL: TWA 0.2 mg(Se)/m^3
ACGIH TLV: TWA 0.2 mg(Se)/m^3
DFG MAK: 0.1 mg(Se)/m^3

SAFETY PROFILE: Moderately toxic by ingestion. A skin and eye irritant. When heated to decomposition it emits very toxic fumes of NO_x and Se. See also SELENIUM COMPOUNDS.

DQO650 CAS:2626-34-8 HR: 3
5,6-DIMETHYL-2,1,3-BENZOSELENODIAZOLE
mf: $C_8H_8N_2Se$ mw: 211.14

SYN: 2,1,3-BENZOSELENADIAZOLE, 5,6-DIMETHYL-

TOXICITY DATA WITH REFERENCE
ivn-mus LD50:56 mg/kg CSLNX* NX#02249

OSHA PEL: TWA 0.2 mg(Se)/m^3
ACGIH TLV: TWA 0.2 mg(Se)/m^3

SAFETY PROFILE: Poison by intravenous route. When heated to decomposition it emits toxic fumes of NO_x and Se.

DQO800 CAS:95-26-1 HR: 2
2,5-DIMETHYLBENZOTHIAZOLE
mf: C_9H_9NS mw: 163.25

SYN: 2,5-DIMETHYLBENZTHIAZOL (CZECH)

TOXICITY DATA WITH REFERENCE
skn-rbt 500 mg/24H MOD 28ZPAK -,202,72
eye-rbt 20 mg/24H MOD 28ZPAK -,202,72
orl-rat LD50:957 mg/kg 28ZPAK -,202,72

CONSENSUS REPORTS: Reported in EPA TSCA Inventory.

SAFETY PROFILE: Moderately toxic by ingestion. A skin and eye irritant. When heated to decomposition it emits very toxic fumes of SO_x and NO_x.

DQP100 HR: 3
1,2-DIMETHYL-1H-BENZOTRIAZOLIUM IODIDE
mf: $C_8H_{10}N_3{\cdot}I$ mw: 275.11

SYN: 1,2-DIMETHYLBENZOTRIAZOLIUM JODID (GERMAN)

TOXICITY DATA WITH REFERENCE
scu-mus LDLo:436 mg/kg SDMU** -,-,36
ivn-mus LDLo:105 mg/kg SDMU** -,-,36
scu-frg LDLo:380 mg/kg SDMU** -,-,36

SAFETY PROFILE: Poison by intravenous route. Moderately toxic by subcutaneous route. When heated to decomposition it emits toxic fumes of I^- and NO_x. See also IODIDES.

DQP125 CAS:22713-35-5 HR: 3
1,3-DIMETHYL-3H-BENZOTRIAZOLIUM IODIDE

SYN: 1,3-DIMETHYLBENZOTRIAZOLIUM JODID (GERMAN)

TOXICITY DATA WITH REFERENCE
scu-mus LDLo:316 mg/kg SDMU** -,-,36
ivn-mus LDLo:127 mg/kg SDMU** -,-,36
scu-frg LDLo:250 mg/kg SDMU** -,-,36

SAFETY PROFILE: Poison by intravenous and subcutaneous routes. When heated to decomposition it emits toxic fumes of I^- and NO_x. See also IODIDES.

DQP400 CAS:32362-68-8 HR: 2
4,9-DIMETHYL-2,3-BENZTHIOPHANTHRENE
mf: $C_{18}H_{14}S$ mw: 262.38

TOXICITY DATA WITH REFERENCE
skn-mus TDLo:440 mg/kg/11W-I:ETA XPHPAW 149,477,51

SAFETY PROFILE: Questionable carcinogen with experimental tumorigenic data. When heated to decomposition it emits toxic fumes of SO_x.

DQP500 CAS:62346-96-7 HR: 1
2,4-DIMETHYLBENZYL ACETATE
mf: $C_{11}H_{14}O_2$ mw: 178.25

SYNS: BENZENEMETHANOL, 2,4-DIMETHYL-, ACETATE ☐ BENZYL ALCOHOL, 2,4-DIMETHYL-, ACETATE ☐ 2,4-DIMETHYLBENZENEMETHANOL ACETATE

TOXICITY DATA WITH REFERENCE
orl-mus LDLo:5 g/kg FCTOD7 30,21S,92
skn-gpg LD50:>5 g/kg FCTOD7 30,21S,92

CONSENSUS REPORTS: Reported in EPA TSCA Inventory.

SAFETY PROFILE: Low toxicity by ingestion and skin contact. When heated to decomposition it emits acrid smoke and irritating vapors.

DQP800 CAS:103-83-3 HR: 3
N,N-DIMETHYLBENZYLAMINE
DOT: UN 2619
mf: $C_9H_{13}N$ mw: 135.23

PROP: Corrosive liquid. Bp: 181°.

SYNS: ARALDITE ACCELERATOR 062 ☐ BDMA ☐ BENZYLDIMETHYLAMINE ☐ BENZYL-N,N-DIMETHYLAMINE ☐ N-BENZYLDIMETHYLAMINE ☐ N,N-DIMETHYLBENZENEMETHANAMINE ☐ N-(PHENYLMETHYL)DIMETHYLAMINE ☐ SUMINE 2015

TOXICITY DATA WITH REFERENCE
skn-rbt 500 mg/4H SEV DCTODJ 8,43,85
eye-rbt 5 mg SEV DCTODJ 8,43,85
orl-rat LD50:265 mg/kg KorCJ# 22AUG74
ihl-rat LC50:2062 mg/m³/4H DCTODJ 8,43,85
ihl-mus LC50:1800 mg/m³/2H 85GMAT -,56,82
skn-rbt LD50:1660 mg/kg DCTODJ 8,43,85

CONSENSUS REPORTS: Reported in EPA TSCA Inventory.

DOT CLASSIFICATION: 8; *Label:* Corrosive

SAFETY PROFILE: Poison by ingestion. Moderately toxic by inhalation and skin contact. Corrosive. A severe eye and skin irritant. Flammable when exposed to heat or flame. When heated to decomposition it emits toxic fumes of NO_x.

DQQ000 CAS:1875-92-9 HR: 3
DIMETHYLBENZYLAMINE HYDROCHLORIDE
mf: $C_9H_{13}N•ClH$ mw: 171.69

SYNS: DIMETHYLBENZYLAMMONIUM CHLORIDE ☐ USAF EL-78

TOXICITY DATA WITH REFERENCE
ipr-mus LD50:200 mg/kg NTIS** AD277-689

SAFETY PROFILE: Poison by intraperitoneal route. When heated to decomposition it emits toxic fumes of NO_x, NH_3, and HCl.

DQQ200 CAS:100-86-7 HR: 2
DIMETHYL BENZYL CARBINOL
mf: $C_{10}H_{14}O$ mw: 150.24

PROP: Needles or white crystalline solid; floral odor. D: 0.972–0.977, mp: 24°, bp: 214–216°, flash p: 198°F. Sol in fixed oils, mineral oil, propylene glycol; insol in glycerin.

SYNS: BENZYL DIMETHYL CARBINOL ☐ α,α-DIMETHYLPHENETHYL ALCOHOL ☐ 1,1-DIMETHYL-2-PHENYLETHANOL ☐ DMBC ☐ FEMA No. 2393

TOXICITY DATA WITH REFERENCE
orl-rat LD50:1280 mg/kg FCTXAV 2,327,64
orl-gpg LD50:988 mg/kg FCTXAV 2,327,64

CONSENSUS REPORTS: Reported in EPA TSCA Inventory.

SAFETY PROFILE: Moderately toxic by ingestion. Combustible liquid. When heated to decomposition it emits acrid smoke and irritating fumes.

DQQ375 HR: 2
DIMETHYL BENZYL CARBINYL ACETATE
mf: $C_{12}H_{16}O_2$ mw: 192.26

PROP: Colorless liquid to solid at room temp; floral, fruity odor. D: 0.995–1.002, refr index: 1.490–1.495, flash p: 212°F. Sol in fixed oils; sltly sol in propylene glycol; insol in water.

SYNS: α,α-DIMETHYLPHENETHYL ACETATE ☐ FEMA No. 2392

SAFETY PROFILE: Combustible liquid. When heated to decomposition it emits acrid smoke and irritating fumes.

DQQ380 HR: 2
DIMETHYL BENZYL CARBINYL BUTYRATE
mf: $C_{14}H_{20}O_2$ mw: 220.31

PROP: Colorless liquid; prunelike odor. D: 0.960–0.981, refr index: 1.473–1.493 @ 25°, flash p: 151°F. Sol in alc, fixed oils; insol in water, propylene glycol.

SYNS: α,α-DIMETHYLPHENRTHYL BUTYRATE ☐ FEMA No. 2394

SAFETY PROFILE: Combustible liquid. Use in accordance with good manufacturing practice.

DQQ400 CAS:67785-77-7 HR: 1
DIMETHYL BENZYL CARBINYL PROPIONATE
mf: $C_{13}H_{18}O_2$ mw: 206.31

SYNS: BENZYLISOPROPYL PROPIONATE ☐ α,α-DIMETHYLPHENETHYL ALCOHOL PROPIONATE

TOXICITY DATA WITH REFERENCE
skn-rbt 500 mg/24H MOD FCTXAV 18,669,80
orl-rat LD50:>5 g/kg FCTXAV 18,669,80
skn-rbt LD50:>5 g/kg FCTXAV 18,669,80

CONSENSUS REPORTS: Reported in EPA TSCA Inventory.

SAFETY PROFILE: Low toxicity by ingestion and skin contact. A skin irritant. When heated to decomposition it emits acrid smoke and irritating fumes.

DQQ600 CAS:6280-75-7 HR: 3
N,N′-DI(α-METHYLBENZYL)ETHYLENEDIAMINE
mf: $C_{18}H_{22}N_2$ mw: 266.42

SYN: N,N′-BIS(α-METHYLBENZYL)ETHYLENEDIAMINE

TOXICITY DATA WITH REFERENCE
skn-rbt 10 mg/24H open MLD AMIHBC 10,61,54
eye-rbt 20 mg open SEV AMIHBC 10,61,54
orl-rat LD50:1290 mg/kg AMIHBC 10,61,54
scu-mus LD50:200 mg/kg ARZNAD 9,628,59
skn-rbt LD50:530 mg/kg AMIHBC 10,61,54

SAFETY PROFILE: Poison by subcutaneous route. Moderately toxic by ingestion and skin contact. A skin and severe eye irritant. When heated to decomposition it emits toxic fumes of NO_x.

DQQ700 CAS:42609-52-9 HR: 1
1-(α,α-DIMETHYLBENZYL)-3-METHYL-3-PHENYLUREA
mf: $C_{17}H_{20}N_2O$ mw: 268.39

TOXICITY DATA WITH REFERENCE
orl-rat LD50:6130 mg/kg SHBOAO 32,488,78
scu-rat LD50:7810 mg/kg SHBOAO 32,488,78
orl-mus LD50:6830 mg/kg SHBOAO 32,488,78
scu-mus LD50:7600 mg/kg SHBOAO 32,488,78

SAFETY PROFILE: Mildly toxic by ingestion and subcutaneous routes. When heated to decomposition it emits toxic fumes of NO_x.

DQR200 CAS:506-63-8 HR: 3
DIMETHYL BERYLLIUM
mf: C_2H_6Be mw: 39.09

PROP: White needles or crystals. Bp: sublimes @ 200°.

CONSENSUS REPORTS: IARC Cancer Review: Group 1 IMEMDT 58,41,93; Human Sufficient Evidence IMEMDT 58,41,93; Animal Sufficient Evidence IMEMDT 1,17,72; Animal Sufficient Evidence IMEMDT 23,143,80; Animal Sufficient Evidence IMEMDT 58,41,93. Beryllium and its compounds are on the Community Right-To-Know List.

OSHA PEL: TWA 0.002 mg(Be)/m³; STEL 0.005 mg(Be)/m³/30M; CL 0.025 mg(Be)/m³
ACGIH TLV: TWA 0.002 mg(Be)/m³, Suspected Human Carcinogen

SAFETY PROFILE: Confirmed human carcinogen. A poison. Flammable when exposed to heat or flame; can react with oxidizing materials. Explosive reaction on contact with water. Ignites on contact with moist air or carbon dioxide. Upon decomposition it emits highly toxic fumes of BeO. See also BERYLLIUM COMPOUNDS.

DQR289 HR: 3
DIMETHYLBERYLLIUM-1,2-DIMETHOXYETHANE
mf: $C_2H_6Be \cdot C_4H_{10}O_2$ mw: 129.21

CONSENSUS REPORTS: IARC Cancer Review: Group 1 IMEMDT 58,41,93; Human Sufficient Evidence IMEMDT 58,41,93; Animal Sufficient Evidence IMEMDT 1,17,72; Animal Sufficient Evidence IMEMDT 23,143,80; Animal Sufficient Evidence IMEMDT 58,41,93. Beryllium and its compounds are on the Community Right-To-Know List.

SAFETY PROFILE: Confirmed human carcinogen. Ignites spontaneously in air. Upon decomposition it emits highly toxic fumes of BeO. See also BERYLLIUM COMPOUNDS.

DQR350 CAS:81-26-5 HR: 3
2,2′-DIMETHYL-1,1′-BIANTHRAQUINONE
mf: $C_{30}H_{18}O_4$ mw: 442.48

SYNS: 1,1′-BIANTHRACENE-9,9′,10,10′-TETRAONE, 2,2′-DIMETHYL- ☐ 2,2′-DIMETHYL-1,1′-BIANTHRACENE-9,9′,10,10′-TETRONE

TOXICITY DATA WITH REFERENCE
ivn-mus LD50:180 mg/kg CSLNX* NX#04855

CONSENSUS REPORTS: Reported in EPA TSCA Inventory.

SAFETY PROFILE: Poison by intravenous route. When heated to decomposition it emits acrid smoke and irritating vapors.

DQR600 CAS:657-24-9 HR: 3
1,1-DIMETHYLBIGUANIDE
mf: $C_4H_{11}N_5$ mw: 129.20

SYNS: N,N-DIMETHYLBIGUANIDE ☐ N,N-DIMETHYLDIGUANIDE ☐ FLUMAMINE ☐ GLUCOPHAGE ☐ GLUCOPHAGE LA 6023 ☐ GLUEOPHOGE ☐ LA 6023 ☐ MELBIN ☐ METFORMIN ☐ NNDG

TOXICITY DATA WITH REFERENCE
cyt-ham:lng 2 g/L/48H GMCRDC 27,95,81
orl-rat TDLo:6 g/kg (1-12D preg):TER COREAF 253,321,61
scu-mus LD50:230 mg/kg AITDAQ 2,1,54
ipr-mus LD50:247 mg/kg JMCMAR 10,521,67
scu-gpg LD50:146 mg/kg MEXPAG 8,237,63
par-frg LD50:5000 mg/kg AITDAQ 2,1,54

SAFETY PROFILE: Poison by subcutaneous and intraperitoneal routes. Mildly toxic by parenteral route. Experimental teratogenic effects. Mutation data reported. When heated to decomposition it emits toxic fumes of NO_x.

DQR800 CAS:1115-70-4 HR: 3
1,1-DIMETHYLBIGUANIDE HYDROCHLORIDE
mf: $C_4H_{11}N_5 \cdot ClH$ mw: 165.66

PROP: Prisms from H_2O. Mp: 232°. Sol in H_2O and EtOH; insol in Et_2O.

SYNS: DIABEFAGOS ☐ DIMETHYLBIGUANIDE HYDROCHLORIDE ☐ N,N-DIMETHYLIMIDODICARBONIMIDIC DIAMIDE MONOHYDROCHLORIDE ☐ GLUCOPHAGE ☐ HAURYMELLIN ☐ MEGUAN ☐ METFORMIN HYDROCHLORIDE ☐ METIGUANIDE

TOXICITY DATA WITH REFERENCE
skn-rbt 500 mg MLD FCTOD7 20,563,82
eye-rbt 100 mg MLD FCTOD7 20 573,82
eye-rbt 100 mg/4S rns MLD FCTOD7 20,573,82
orl-rat LD50:1 g/kg MEIEDD 10,849,83
scu-rat LD50:300 mg/kg MEIEDD 10,849,83
orl-mus LD50:1450 mg/kg NIIRDN 6,841,82
ipr-mus LD50:420 mg/kg NIIRDN 6,841,82
scu-mus LD50:620 mg/kg NIIRDN 6,841,82
ivn-mus LD50:180 mg/kg CSLNX* NX#04012

SAFETY PROFILE: Poison by intravenous and subcutaneous routes. Moderately toxic by ingestion. An eye and skin irritant. When heated to decomposition it emits toxic fumes of NO_x and Cl^-.

DQS000 CAS:91-97-4 HR: 3
3,3′-DIMETHYL-4,4′-BIPHENYLENE DIISOCYANATE
mf: $C_{16}H_{12}N_2O_2$ mw: 264.30

PROP: Crystals from chlorobenzene. Mp: 70°.

SYNS: 4,4′-DIISOCYANATO-3,3′-DIMETHYL-1,1′-BIPHENYL □ ISOCYANIC ACID, 3,3′-DIMETHYL-4,4′-BIPHENYLENE ESTER

TOXICITY DATA WITH REFERENCE
ivn-mus LD50:56 mg/kg CSLNX* NX#02412

CONSENSUS REPORTS: Reported in EPA TSCA Inventory.

NIOSH REL: TWA (Diisocyanates) 0.005 ppm; CL 0.02 ppm/10M

SAFETY PROFILE: Poison by intravenous route. When heated to decomposition it emits toxic fumes of NO_x. See also CYANATES and ESTERS.

DQS100 CAS:1134-35-6 HR: 3
4,4′-DIMETHYL-2,2′-BIPYRIDINE
mf: $C_{12}H_{12}N_2$ mw: 184.26

SYN: 2,2′-BIPYRIDINE, 4,4′-DIMETHYL-

TOXICITY DATA WITH REFERENCE
ipr-mus LD50:78,700 µg/kg TOXIA6 23,815,85

CONSENSUS REPORTS: Reported in EPA TSCA Inventory.

SAFETY PROFILE: Poison by intraperitoneal route. When heated to decomposition it emits toxic vapors of NO_x.

DQS600 CAS:63977-49-1 HR: 3
DIMETHYL-BIS(β-CHLOROETHYL)AMMONIUM CHLORIDE
mf: $C_6H_{14}Cl_2N•Cl$ mw: 206.56

SYNS: 2-CHLORO-N-(2-CHLOROETHYL)-N,N-DIMETHYLETHANAMINIUM CHLORIDE □ TL 379

TOXICITY DATA WITH REFERENCE
ihl-mus LCLo:1700 mg/m³/10M NDRC** NDCrc-132,Sept,42
ipr-mus LD50:67 mg/kg CANCAR 2,1055,49
scu-mus LDLo:100 mg/kg JPETAB 91,224,47

SAFETY PROFILE: Poison by intraperitoneal and subcutaneous routes. Moderately toxic by inhalation. When heated to decomposition it emits very toxic fumes of Cl^-, NO_x, and NH_3.

DQT100 HR: 3
N,N-DIMETHYL-(2-BROMOETHYL)HYDRAZINIUM BROMIDE
mf: $C_4H_{12}BrN_2•Br$ mw: 248.00

SYN: N,N-DIMETHYL-(2-BROMAETHYL)-HYDRAZINIUMBROMID (GERMAN)

TOXICITY DATA WITH REFERENCE
orl-mus LD50:230 mg/kg ABMGAJ 27,663,71
ipr-mus LD50:315 mg/kg ABMGAJ 27,663,71
ivn-mus LD50:72 mg/kg ABMGAJ 27,663,71

SAFETY PROFILE: Poison by ingestion, intravenous, and intraperitoneal routes. When heated to decomposition it emits toxic fumes of Br^- and NO_x.

DQT150 CAS:513-81-5 HR: 3
2,3-DIMETHYL-1,3-BUTADIENE
mf: C_6H_{10} mw: 82.14

$H_2C{=}C(CH_3)C(CH_3){=}CH_2$

PROP: Liquid. D: 0.726 @ 20°/4°, bp: 69–70°.

SAFETY PROFILE: Forms explosive polymeric peroxides on exposure to air. Explodes on contact with thiazyl fluoride. Ignites on contact with oxygen + ozone above −78°C. When heated to decomposition it emits acrid smoke and fumes.

DQT200 CAS:75-83-2 HR: 3
2,2-DIMETHYLBUTANE
mf: C_6H_{14} mw: 86.20

PROP: Liquid. Bp: 49.7°, mp: −98.2°, flash p: −54°F, fp: −101.9°, d: 0.649, autoign temp: 797°F, vap press: 400 mm @ 31.0°, vap d: 3.00, lel: 1.2%, uel: 7.0%.

SYN: NEOHEXANE (DOT)

CONSENSUS REPORTS: Reported in EPA TSCA Inventory.

OSHA PEL: TWA 500 ppm; STEL 1000 ppm
ACGIH TLV: TWA 500 ppm; STEL 1000 ppm
DFG MAK: 200 ppm (700 mg/m³)
NIOSH REL: (Alkanes) TWA 350 mg/m³

SAFETY PROFILE: Probably an irritant and narcotic in high concentration. A very dangerous fire and explosion hazard when exposed to heat or flame; can react vigorously with oxidizing materials. Keep away from heat or open flame. To fight fire, use foam, CO_2, dry chemical. When heated to decomposition it emits acrid smoke and irritating fumes.

DQT400 CAS:79-29-8 HR: 3
2,3-DIMETHYLBUTANE
DOT: UN 2457
mf: C_6H_{14} mw: 86.20

PROP: Liquid. Mp: −135°, bp: 58.0°, flash p: −20°F, d: 0.662 @ 20°/4°, autoign temp: 788°F, vap press: 400 mm @ 39.0°, vap d: 3.0, lel: 1.2%, uel: 7.0%.

CONSENSUS REPORTS: Reported in EPA TSCA Inventory.

OSHA PEL: TWA 500 ppm; STEL 1000 ppm
ACGIH TLV: TWA 500 ppm; STEL 1000 ppm
DFG MAK: 200 ppm (700 mg/m³)
NIOSH REL: TWA (Alkanes) 350 mg/m³
DOT CLASSIFICATION: 3; *Label:* Flammable Liquid

SAFETY PROFILE: Probably an irritant and narcotic in high concentration. A very dangerous fire and explosion hazard when exposed to heat or flame; can react vigorously with oxidizing materials. Keep away from heat and open flame. To fight fire, use foam, CO_2, dry chemical. When heated to decomposition it emits acrid smoke and irritating fumes.

DQU000 CAS:75-97-8 HR: 2
3,3-DIMETHYL-2-BUTANONE
mf: $C_6H_{12}O$ mw: 100.18

$CH_3CO{\cdot}C(CH_3)_3$

PROP: Liquid with camphoraceous odor. Bp: 106.0–106.1°, flash p: 53.6°F.

SYNS: tert-BUTYL METHYL KETONE □ KETONE, t-BUTYL METHYL □ METHYL tert-BUTYL KETONE □ METHYLTERT-BUTYL KETONE □ PINACOLIN □ PINACOLINE □ PINACOLONE □ PINAKOLIN □ PINAKOLIN (GERMAN)

TOXICITY DATA WITH REFERENCE
orl-rat LD50:610 mg/kg GISAAA 52(12),91,87
orl-mus LD50:1625 mg/kg GISAAA 52(12),91,87
ihl-mus LC50:5700 mg/m³ GISAAA 52(12),91,87
orl-rbt LD50:900 mg/kg GISAAA 52(12),91,87
scu-gpg LDLo:700 mg/kg MEIEDD 10,1072,83

CONSENSUS REPORTS: Reported in EPA TSCA Inventory.

DOT CLASSIFICATION: 3; *Label:* Flammable Liquid

SAFETY PROFILE: Moderately toxic by ingestion and subcutaneous routes. Slightly toxic by inhalation. A dangerous fire hazard when exposed to heat, flame, or oxidizers. When heated to decomposition it emits acrid smoke and irritating fumes. See also KETONES.

DQU200 CAS:3625-18-1 HR: 3
5-(1,3-DIMETHYL-2-BUTENYL)-5-ETHYL BARBITURIC ACID
mf: $C_{12}H_{18}N_2O_3$ mw: 238.32

SYNS: 5-(1,3-DIMETHYL-2-BUTENYL)-5-ETHYL-2,4,6(1H,3H,5H)PYRIMIDINETRIONE □ MCNEIL 481

TOXICITY DATA WITH REFERENCE
orl-mus LD50:18 mg/kg 27ZQAG -,177,72
ipr-mus LD50:3500 µg/kg PMDCAY 8,61,71
orl-dog LD50:2 mg/kg 27ZQAG -,177,72
ivn-dog LDLo:250 µg/kg 27ZQAG -,177,72
ivn-rbt LDLo:250 µg/kg 27ZQAG -,177,72

SAFETY PROFILE: Poison by ingestion, intravenous, and intraperitoneal routes. When heated to decomposition it emits toxic fumes of NO_x. See also BARBITURATES.

DQU400 CAS:36798-79-5 HR: 2
1-(2-(1,3-DIMETHYL-2-BUTENYLIDENE)HYDRAZINO) PHTHALAZINE
mf: $C_{14}H_{16}N_4$ mw: 240.34

PROP: A solid. Mp: 132–133°.

SYNS: BUDRALAZINE □ BUTERAZINE □ DJ-1461 □ MESITYL OXIDE (1-PHTHALAZINYL)HYDRAZONE □ 4-METHYL-3-PENTEN-2-ONE (1-PHTHALAZINYL)HYDRAZONE □ 1(2H)-PHTHALAZINONE (1,3-DIMETHYL-2-BUTENYLIDENE)HYDRAZONE

TOXICITY DATA WITH REFERENCE
orl-rat TDLo:55 mg/kg (7-17D preg):REP OYYAA2 21,321,81
orl-rat TDLo:1350 mg/kg (female 17-22D post):REP OYYAA2 21,331,81
orl-rbt TDLo:65 mg/kg (female 6-18D post):TER OYYAA2 21,343,81
orl-rat TDLo:1100 mg/kg (7-17D preg):TER OYYAA2 21,321,81
orl-rat LD50:620 mg/kg TXAPA9 44,431,78
ipr-rat LD50:3570 mg/kg TXAPA9 44,431,78
orl-mus LD50:1820 mg/kg TXAPA9 44,431,78
ipr-mus LD50:4020 mg/kg TXAPA9 44,431,78

SAFETY PROFILE: Moderately toxic by ingestion and intraperitoneal routes. Experimental teratogenic and reproductive effects. An antihypertensive agent. When heated to decomposition it emits toxic fumes of NO_x.

DQU600 CAS:108-09-8 HR: 3
1,3-DIMETHYL BUTYLAMINE
DOT: UN 2379
mf: $C_6H_{15}N$ mw: 101.22

PROP: A liquid. Bp: 106–109°, flash p: 55°F (OC), d: 0.750 @ 20°/20°.

SYN: 1,3-DIMETHYLBUTYLAMINE (DOT)

TOXICITY DATA WITH REFERENCE
orl-rat LDLo:600 mg/kg SCCUR* -,4,61
orl-mus LD50:470 mg/kg SCCUR* -,4,61
ihl-mus LCLo:1278 ppm/15M SCCUR* -,4,61
ivn-mus LD50:80 mg/kg CSLNX* NX#03558
skn-rbt LDLo:600 mg/kg SCCUR* -,4,61

CONSENSUS REPORTS: Reported in EPA TSCA Inventory.

DOT CLASSIFICATION: 3; *Label:* Flammable Liquid

SAFETY PROFILE: Poison by intravenous route. Moderately toxic by ingestion and skin contact. Mildly toxic by inhalation. A dangerous fire and explosion hazard when exposed to heat or flame; can react vigorously with oxidizing materials. To fight fire, use foam, CO_2, dry chemical. When heated to decomposition it emits toxic fumes of NO_x. See also AMINES.

DQU800 CAS:98-19-1 HR: 1
1,3-DIMETHYL-5-tert-BUTYLBENZENE
mf: $C_{12}H_{18}$ mw: 162.30

PROP: A liquid. D: 0.866 @ 20°, mp: −17, bp: 200–202°.

SYN: 5-tert-BUTYL-m-XYLENE

TOXICITY DATA WITH REFERENCE
orl-rat LDLo:5000 mg/kg 28ZRAQ -,58,60

CONSENSUS REPORTS: Reported in EPA TSCA Inventory.

SAFETY PROFILE: Mildly toxic by ingestion. When heated to decomposition it emits acrid smoke and irritating fumes.

DQV000 CAS:17874-34-9 HR: 1
4,6-DIMETHYL-8-tert-BUTYLCOUMARIN

SYN: 8-tert-BUTYL-4,6-DIMETHYLCOUMARIN

TOXICITY DATA WITH REFERENCE
skn-gpg 100% MLD FCTXAV 18,671,80
orl-mus LD50:>5 g/kg FCTXAV 18,671,80
skn-gpg LD50:>5 g/kg FCTXAV 18,671,80

CONSENSUS REPORTS: Reported in EPA TSCA Inventory.

SAFETY PROFILE: Low toxicity by ingestion and skin contact. A skin irritant. When heated to decomposition it emits toxic fumes of NO_x.

DQV200 CAS:6592-90-1 HR: 3
5-(1,3-DIMETHYLBUTYL)-5-ETHYL BARBITURIC ACID, SODIUM SALT
mf: $C_{12}H_{19}N_2O_3$•Na mw: 262.32

TOXICITY DATA WITH REFERENCE
ipr-rat LDLo:20 mg/kg JAPMA8 29,509,40
scu-rat LD50:22 mg/kg QJPPAL 12,657,39
par-rat LDLo:10 mg/kg JACSAT 58,585,36
ipr-mus LD50:24 mg/kg QJPPAL 12,657,39
scu-mus LD50:27 mg/kg QJPPAL 12,657,39
ivn-rbt LD50:20 mg/kg QJPPAL 12,657,39
ivn-gpg LD50:22 mg/kg QJPPAL 12,657,39

SAFETY PROFILE: Poison by intraperitoneal, subcutaneous, intravenous, and parenteral routes. When heated to decomposition it emits toxic fumes of NO_x and Na_2O. See also BARBITURATES.

DQV250 CAS:793-24-8 HR: 2
N-(1,3-DIMETHYLBUTYL)-N′-PHENYL-p-PHENYLENE-DIAMINE
mf: $C_{18}H_{24}N_2$ mw: 268.44

SYNS: ANTOZITE 67 □ ANTOZITE 67E □ NCI-C56315 □ p-PHENYLENEDIAMINE, N-(1,3-DIMETHYLBUTYL)-N′-PHENYL- □ SANTOFLEX 13 □ VULKANOX 4020

TOXICITY DATA WITH REFERENCE
orl-rat LD50:3580 mg/kg JACTDZ 1,67,90
skn-rbt LD50:>7940 mg/kg JACTDZ 1,67,90

CONSENSUS REPORTS: Reported in EPA TSCA Inventory.

SAFETY PROFILE: Moderately toxic by ingestion. Low toxicity by skin contact. When heated to decomposition it emits toxic vapors of NO_x.

DQV300 CAS:760-79-2 HR: 2
N,N-DIMETHYLBUTYRAMIDE
mf: $C_6H_{13}NO$ mw: 115.20

SYN: BUTYRAMIDE, N,N-DIMETHYL-

TOXICITY DATA WITH REFERENCE
ipr-mus LD50:2110 mg/kg AIHAAP 32,539,71
ivn-mus LD50:1620 mg/kg AIHAAP 32,539,71
ivn-rbt LD50:790 mg/kg AIHAAP 32,539,71

CONSENSUS REPORTS: Reported in EPA TSCA Inventory.

SAFETY PROFILE: Moderately toxic by intraperitoneal and intravenous routes. When heated to decomposition it emits toxic vapors of NO_x.

DQW600 CAS:1607-30-3 HR: 3
DI-2-METHYLBUTYRYL PEROXIDE
mf: $C_{10}H_{18}O_4$ mw: 202.25

$(CH_3CH_2CH(CH_3)CO•O-)_2$

SAFETY PROFILE: The pure material is unstable and explodes at room temperature. When heated to decomposition it emits acrid smoke and fumes. See also PEROXIDES.

DQW800 CAS:506-82-1 HR: 3
DIMETHYLCADMIUM
mf: C_2H_6Cd mw: 142.47

PROP: Oil or liquid; decomp by water; foul odor. D: 1.984; mp: −2.4°; bp: 106°.

CONSENSUS REPORTS: Cadmium and its compounds are on the Community Right-To-Know List.

OSHA PEL: TWA 5 µg(Cd)/m³
ACGIH TLV: TWA 0.05 mg(Cd)/m³ (Proposed: TWA 0.01 mg(Cd)/m³ (dust), Suspected Human Carcinogen; 0.002 mg(Cd)/m³ (respirable dust), Suspected Human Carcinogen); BEI: 10 µg/g creatinine in urine; 10 µg/L in blood
DFG BAT: Blood 1.5 µg/dL; Urine 15 µg/dL, Suspected Carcinogen
NIOSH REL: (Cadmium) Reduce to lowest feasible level

SAFETY PROFILE: Confirmed human carcinogen. Contact with air produces the friction-sensitive explosive dimethyl cadmium peroxide. Explodes when heated above 150°C. Ignition may occur on contact with air if the surface area is large. See also CADMIUM COMPOUNDS.

DQX300 CAS:3938-45-2 HR: 3
N,N-DIMETHYLCARBAMIC ACID, m-ISOPROPYL PHENYL ESTER
mf: $C_{12}H_{17}NO_2$ mw: 207.30

SYN: CARBAMIC ACID, N,N-DIMETHYL-, m-ISOPROPYLPHENYL ESTER

TOXICITY DATA WITH REFERENCE
skn-rbt 10 mg/24H open MLD AIHAAP 23,95,62
orl-rat LD50:160 mg/kg AIHAAP 23,95,62
skn-rbt LD50:280 mg/kg AIHAAP 23,95,62

SAFETY PROFILE: Poison by ingestion and skin contact. When heated to decomposition it emits toxic fumes of NO_x.

DQX800 **CAS:63884-71-9** **HR: 3**
DIMETHYLCARBAMIC ESTER of HORDENINE HYDROCHLORIDE
mf: $C_{13}H_{20}N_2O_2$•ClH mw: 272.81

SYNS: AR-41 ☐ N,N-DIMETHYLCARBAMIC ACID-4-(β-DIMETHYLAMINOETHYL)PHENYL ESTER, HYDROCHLORIDE ☐ N,N-DIMETHYLCARBAMIC ACID-p-(β-DIMETHYLAMINOETHYL)PHENYL ESTER, HYDROCHLORIDE ☐ N,N-DIMETHYL-p-(N',N'-DIMETHYLCARBAMOYLOXY)PHENETHYLAMINE, HYDROCHLORIDE

TOXICITY DATA WITH REFERENCE
orl-mus LDLo:75 mg/kg JPETAB 43,413,31
ivn-mus LD80:15 mg/kg NTIS** PB158-508

SAFETY PROFILE: Poison by ingestion and intravenous routes. When heated to decomposition it emits very toxic fumes of NO_x and HCl. See also ESTERS and CARBAMATES.

DQY000 **CAS:63884-67-3** **HR: 3**
DIMETHYLCARBAMIC ESTER of 2-OXYBENZYLDIETHYLAMINE HYDROCHLORIDE
mf: $C_{14}H_{22}N_2O_2$•ClH mw: 286.84

SYN: DIMETHYLCARBAMIC ACID-(α-(DIETHYLAMINO))-o-TOLYL ESTER, HYDROCHLORIDE

TOXICITY DATA WITH REFERENCE
orl-mus LDLo:5 mg/kg JPETAB 43,413,31
ivn-mus LDLo:1500 μg/kg NTIS** PB158-508

SAFETY PROFILE: Poison by ingestion and intravenous routes. When heated to decomposition it emits very toxic fumes of NO_x and HCl. See also ESTERS and CARBAMATES.

DQY400 **CAS:63680-76-2** **HR: 3**
DIMETHYLCARBAMIC ESTER of 8-OXYMETHYLQUINOLINIUM METHYLSULFATE
mf: $C_{13}H_{15}N_2O_2$•CH_3O_4S mw: 342.40

SYNS: N,N-DIMETHYLCARBAMIC ACID-8-QUINOLINYL ESTER METHOSULFATE ☐ 8-HYDROXY-1-METHYLQUINOLINIUM METHYLSULFATE DIMETHYLCARBAMATE

TOXICITY DATA WITH REFERENCE
orl-mus LDLo:200 mg/kg JPETAB 43,413,31
ivn-mus LDLo:500 μg/kg JPETAB 43,413,31

SAFETY PROFILE: Poison by ingestion and intravenous routes. When heated to decomposition it emits very toxic fumes of NO_x and SO_x. See also ESTERS and CARBAMATES.

DQY909 **CAS:51-60-5** **HR: 3**
3-(DIMETHYLCARBAMOXY)PHENYL TRIMETHYLAMMONIUM METHYL SULFATE
mf: $C_{12}H_{19}N_2O_2$•CH_3O_4S mw: 334.43

PROP: Crystals from EtOH. Mp: 142–145°.

SYNS: AR-32 ☐ N,N-DIMETHYLCARBAMIC ACID-3-DIMETHYLAMINOPHENYL ESTER METHOSULFATE ☐ DIMETHYLCARBAMIC ACID ESTER with (m-HYDROXYPHENYL)TRIMETHYLAMMONIUM METHYL SULFATE ☐ N,N-DIMETHYLCARBAMIC ACID-3-(TRIMETHYLAMMONIO)PHENYL ESTER METHYLSULFATE ☐ DIMETHYLCARBAMIC ESTER of 3-OXYPHENYLTRIMETHYLAMMONIUM METHYLSULFATE ☐ (3-(DIMETHYLCARBAMOYLOXY)PHENYL)TRIMETHYLAMMONIUM METHYLSULFATE ☐ EUSTIGMIN METHYLSULFATE ☐ HODOSTIN ☐ (m-HYDROXYPHENYL)TRIMETHYLAMMONIUM METHYL SULFATE DIMETHYLCARBAMATE ☐ (3-HYDROXYPHENYL)TRIMETHYLAMMONIUM METHYL SULFATE DIMETHYLCARBAMIC ESTER ☐ KIRKSTIGMINE METHYL SULFATE ☐ LEOSTIGMINE METHYL SULFATE ☐ NEOESERINE METHYL SULFATE ☐ NEOSTIGMETH ☐ NEOSTIGMINE METHOSULFATE ☐ NEOSTIGMINE METHYL SULFATE ☐ NEOSTIGMINE MONOMETHYLSULFATE ☐ NORMASTIGMIN ☐ PHILOSTIGMIN METHYL SULFATE ☐ POLSTIGMINE ☐ PROSERIN ☐ PROSERINE METHYL SULFATE ☐ PROSTIGMINE METHYLSULFATE ☐ SB-23 ☐ STIGMANOL METHYL SULFATE ☐ STIGMOSAN METHYL SULFATE ☐ SYNTHOSTIGMINE METHYL SULFATE ☐ TL-1394 ☐ VAGOSTIGMINE METHYL SULFATE

TOXICITY DATA WITH REFERENCE
scu-mus TDLo:1400 μg/kg (7-13D preg):REP KSRNAM 8,1986,74
scu-rat LD50:334 μg/kg TXAPA9 25,569,73
orl-mus LD50:7500 μg/kg JPETAB 99,16,50
ipr-mus LD50:230 μg/kg ATXKA8 29,39,72
scu-mus LD50:420 μg/kg JPETAB 99,16,50
ivn-mus LD50:160 μg/kg JPETAB 99,16,50
scu-dog LD50:50 μg/kg NTIS** PB158-508
scu-cat LDLo:50 μg/kg NTIS** PB158-508
ims-rbt LD50:310 μg/kg AIPTAK 81,276,50

SAFETY PROFILE: A deadly poison by ingestion, intravenous, subcutaneous, intraperitoneal, and intramuscular routes. Experimental reproductive effects. When heated to decomposition it emits very toxic fumes of SO_x, NH_3, and NO_x. See also CARBAMATES.

DQY950 **CAS:79-44-7** **HR: 3**
DIMETHYLCARBAMOYL CHLORIDE
DOT: UN 2262
mf: C_3H_6ClNO mw: 107.55

PROP: Liquid. Mp: −33°, bp: 165–167°, d: 1.678 @ 20°/4°, vap d: 3.73.

SYNS: CARBAMIC ACID, DIMETHYL-(9CI) ☐ CARBAMYL CHLORIDE, N,N-DIMETHYL- ☐ CHLORID KYSELINY DIMETHYLKARBAMINOVE ☐ CHLOROFORMIC ACID DIMETHYLAMIDE ☐ DDC ☐ DIMETHYLAMID KYSELINY CHLORMRAVENCI ☐ (DIMETHYLAMINO)CARBONYL CHLORIDE ☐ N,N-DIMETHYLAMINOCARBONYL CHLORIDE ☐ DIMETHYLCARBAMIC ACID CHLORIDE ☐ N,N-DIMETHYLCARBAMIC ACID CHLORIDE ☐ DIMETHYLCARBAMIC CHLORIDE ☐ DIMETHYLCARBAMIDOYL CHLORIDE ☐ N,N-DIMETHYLCARBAMIDOYL CHLORIDE ☐ N,N-DIMETHYLCARBAMOYL CHLORIDE ☐ DIMETHYL CARBAMOYL CHLORIDE (ACGIH,DOT) ☐ DIMETHYLCARBAMYL CHLORIDE ☐ N,N-DIMETHYLCARBAMYL CHLORIDE ☐ DIMETHYLCHLOROFORMAMIDE ☐ DIMETHYLKARBAMOYLCHLORID ☐ DMCC ☐ RCRA WASTE NUMBER U097 ☐ TL 389

TOXICITY DATA WITH REFERENCE
mma-sat 300 ng/plate ENMUDM 6(Suppl 2),1,84
mmo-esc 100 μg/plate ENMUDM 6(Suppl 2),1,84
ihl-rat TCLo:1 ppm/6H/6W-I:CAR CALEDQ 33,175,86
skn-mus TDLo:17,280 mg/kg/72W-I:CAR JACTDZ 6(4),479,87
ipr-mus TDLo:2560 mg/kg/64W-I:NEO JJIND8 53,695,74
scu-mus TDLo:894 mg/kg/52W-I:CAR JACTDZ 6(4),479,87
ihl-ham TCLo:1 ppm/6H/89W-I:CAR JEPTDQ 4(1),107,80
skn-mus LD :13 g/kg/55W-I:NEO JJIND8 48,1539,72
scu-mus LD :8 g/kg/40W-I:NEO JJIND8 53,695,74

ihl-rat 111 ppm/6H/6W-I:ETA PAACA3 21,106,80
ihl-rat 111 ppm:ETA OHSLAM 6,5,76
scu-mus LD :5200 mg/kg/26W-I:NEO JJIND8 48,1539,72
skn-mus LD :13 g/kg/55W-I:NEO JJIND8 53,695,74
ihl-ham 111 ppm/6H/2Y-I:ETA JEPTDQ 4,107,80
orl-rat LD50:1 g/kg ZAARAM 24,71,74
ihl-rat LC50:180 ppm/6H JEPTDQ 4(1),107,80
ihl-mus LCLo:1000 mg/m^3/10M NDRC** NDCrc-132,Oct42
ipr-mus LD50:300 mg/kg

CONSENSUS REPORTS: NTP 7th Annual Report On Carcinogens. IARC Cancer Review: Group 2A IMEMDT 7,199,87; Animal Sufficient Evidence IMEMDT 12,77,76; Human Inadequate Evidence IMEMDT 12,77,76. EPA Genetic Toxicology Program. Community Right-To-Know List. Reported in EPA TSCA Inventory.

ACGIH TLV: Suspected Human Carcinogen
DFG MAK: Animal Carcinogen, Suspected Human Carcinogen.
DOT CLASSIFICATION: 8; *Label:* Corrosive

SAFETY PROFILE: Confirmed carcinogen with experimental carcinogenic, neoplastigenic, and tumorigenic data. Poison by intraperitoneal route. Moderately toxic by inhalation and ingestion. Human mutation data reported. Can cause skin and papillary tumors by skin contact, and squamous cell carcinoma by inhalation. Will react with water or steam to produce toxic and corrosive fumes. A powerful lachrymator. When heated to decomposition it emits very toxic fumes of Cl^- and NO_x. See also CHLORIDES.

DQZ000 CAS:644-64-4 HR: 3
1-DIMETHYLCARBAMOYL-5-METHYL-3-PYRAZOLYL DIMETHYLCARBAMATE
mf: $C_{10}H_{16}N_4O_3$ mw: 240.30

PROP: A solid. Mp: 68–71°. Very sol in H_2O.

SYNS: DIMETHYLCARBAMIC ACID-1 ((DIMETHYLAMINO)CARBONYL)-5-METHYL-1H-PYRAZOL-3-YL ESTER □ DIMETHYLCARBAMIC ACID ESTER with 3-HYDROXY-N,N,5-TRIMETHYLPYRAZOLE-1-CARBOXAMIDE □ DIMETHYLCARBAMIC ACID-5-METHYL-1H-PYRAZOL-3-YL ESTER □ 2-DIMETHYLCARBAMOYL-3-METHYLPYRAZOLYL-(5)-N,N-DIMETHYLCARBAMAT □ 2-DIMETHYLCARBAMOYL-3-METHYL-5-PYRAZOLYL DIMETHYLCARBAMATE □ 2-(N,N-DIMETHYLCARBAMYL)-3-METHYLPYRAZOLYL-5 N,N-DIMETHYLCARBAMATE □ DIMETHYL 2-CARBAMYL-3-METHYLPYRAZOLYLDIMETHYLCARBAMATE (GERMAN) □ DIMETILAN □ DIMETILANE □ ENT 25595-X □ ENT 25,922 □ GEIGY 22870 □ 3-HYDROXY-N,N,5-TRIMETHYLPYRAZOLE-1-CARBOXAMIDE DIMETHYLCARBAMATE (ESTER) □ 5-METHYL-1H-PYRAZOL-3-YL DIMETHYLCARBAMATE □ SNIP □ SNIP FLY □ SNIP FLY BANDS

TOXICITY DATA WITH REFERENCE
orl-rat LD50:25 mg/kg PHJOAV 185,361,60
skn-rat LD50:600 mg/kg PHJOAV 185,361,60
orl-mus LD50:60 mg/kg GUCHAZ 6,222,73
ipr-mus LD50:12 mg/kg BECTA6 2,163,67
skn-rbt LD50:2000 mg/kg PCOC** -,393,66
orl-gpg LD50:63 mg/kg 85DPAN -,-,71/76

CONSENSUS REPORTS: EPA Extremely Hazardous Substances List.

SAFETY PROFILE: Poison by ingestion and intraperitoneal routes. Moderately toxic by skin contact. An insecticide. When heated to decomposition it emits toxic fumes of NO_x. See also CARBAMATES.

DRB200 CAS:611-92-7 HR: 2
N,N′-DIMETHYL CARBANILIDE
mf: $C_{15}H_{16}N_2O$ mw: 240.33

PROP: A solid. Mp: 122°.

SYNS: CENTRALITE II □ N,N′-DIMETHYL-N,N′-DIPHENYLUREA □ METHYL CENTRALITE □ UREA, N,N′-DIMETHYL-N,N′-DIPHENYL-(9CI)

TOXICITY DATA WITH REFERENCE
orl-rat LDLo:500 mg/kg JPETAB 90,260,47

CONSENSUS REPORTS: Reported in EPA TSCA Inventory.

SAFETY PROFILE: Moderately toxic by ingestion. When heated to decomposition it emits toxic fumes of NO_x.

DRB400 CAS:5826-73-3 HR: 2
DIMETHYL CARBATE
mf: $C_{11}H_{14}O_4$ mw: 210.25

PROP: Crystals when very pure; usually a syrup. Mp: 38°, bp: 139° @ 12.5 mm.

SYNS: cis-BICYCLO(2.2.1)HEPT-5-ENE-2,3-DICARBOXYLIC ACID, DIMETHYL ESTER □ (endo,endo)-BICYCLO(2.2.1)HEPT-5-ENE-2,3-DICARBOXYLIC ACID DIMETHYL ESTER □ cis-BICYCLO(2,2,1-HEPTENE-2,3-DICARBOXYLIC ACID) METHYL ESTER □ COMPOUND-3916 □ DIMALONE □ DIMELONE □ DIMETHYL cis-BICYCLO(2,2,1)-5-HEPTENE-2,3-DICARBOXYLATE □ cis-3,6-ENDOMETHYLENE-Δ^4-TETRAHYDROPHTHALIC ACID DIMETHYL ESTER □ NISY □ cis-5-NORBORNENE-2,3-DICARBOXYLIC ACID DIMETHYL ESTER

TOXICITY DATA WITH REFERENCE
eye-rbt 500 mg AJOPAA 29,1363,46
orl-rat LD50:1000 mg/kg GUCHAZ 6,211,73
orl-mus LD50:1400 mg/kg JPETAB 93,26,48

SAFETY PROFILE: Moderately toxic by ingestion. An eye irritant. An insect repellent. Combustible. When heated to decomposition it emits acrid smoke and irritating fumes.

DRB600 CAS:2088-72-4 HR: 3
O,O-DIMETHYL-S-CARBOETHOXYMETHYL THIOPHOSPHATE
mf: $C_6H_{13}O_5PS$ mw: 228.22

PROP: A liquid. Bp: 76–80° @ 0.01 mm.

SYNS: ((DIMETHOXYPHOSPHINYL)THIO)ACETIC ACID ETHYL ESTER □ O,O-DIMETHYL-S-(CARBETHOXY)METHYL PHOSPHOROTHIOLATE □ O,O-DIMETHYL ESTER PHOSPHOROTHIOIC ACID-S-ESTER with ETHYL MERCAPTOACETATE □ METHYLACETAPHOS □ METHYL ACETOPHOS □ METHYL ACETOXON

TOXICITY DATA WITH REFERENCE
orl-rat LD50:385 mg/kg HYSAAV 31,18,66
skn-rat LD50:220 mg/kg GISAAA 33(8),107,68
unr-rat LD50:1000 mg/kg 30ZDA9 -,351,71
orl-mus LD50:314 mg/kg GISAAA 33(8),107,68

SAFETY PROFILE: Poison by ingestion and skin contact. Moderately toxic by unspecified route. When heated to

D

decomposition it emits very toxic fumes of PO_x and SO_x. See also ESTERS.

DRB800 **CAS:64038-38-6** **HR: 2**
7,11-DIMETHYL-10-CHLOROBENZ(c)ACRIDINE
mf: $C_{19}H_{14}ClN$ mw: 291.79

SYNS: 2-CHLORO-1,10-DIMETHYL-7,8-BENZACRIDINE (FRENCH) □ 1,10-DIMETHYL-2-CHLORO-7,8-BENZACRIDINE (FRENCH)

TOXICITY DATA WITH REFERENCE
skn-mus TDLo:250 mg/kg/21W-I:ETA ACRSAJ 4,315,56
skn-mus TD:380 mg/kg/31W-I:ETA AICCA6 11,736,55

SAFETY PROFILE: Questionable carcinogen with experimental tumorigenic data. When heated to decomposition it emits very toxic fumes of NO_x and Cl^-.

DRC000 **CAS:4584-46-7** **HR: 3**
DIMETHYL(2-CHLOROETHYL)AMINE HYDROCHLORIDE
mf: $C_4H_{10}ClN{\cdot}ClH$ mw: 144.06

SYNS: 2-CHLORO-N,N-DIMETHYLETHYLAMINE HYDROCHLORIDE □ DIMETHYL-β-CHLOROETHYLAMINE HYDROCHLORIDE

TOXICITY DATA WITH REFERENCE
mmo-sat 1 mg/L ENMUDM 3,33,81
mmo-esc 1 μmol/L JPPMAB 31,67P,79
sln-dmg-orl 1700 mmol/L MUREAV 95,237,82
dns-rat:lvr 5 μmol/L ENMUDM 3,33,81
ipr-mus TDLo:720 mg/kg/8W-I:NEO CNREA8 39,391,79
ipr-mus LD50:280 mg/kg CANCAR 2,1055,49
scu-mus LD50:250 mg/kg JPETAB 97,25,49

CONSENSUS REPORTS: Reported in EPA TSCA Inventory.

SAFETY PROFILE: Poison by intraperitoneal and subcutaneous routes. Questionable carcinogen with experimental neoplastigenic data. Mutation data reported. When heated to decomposition it emits very toxic fumes of Cl^- and NO_x.

DRC400 **CAS:13508-53-7** **HR: 2**
DIMETHYLCHLOROMETHYLETHOXYSILANE
mf: $C_5H_{13}ClOSi$ mw: 152.72

SYN: DIMETHYL-CHLORMETHYL-ETHOXYSILAN (CZECH)

TOXICITY DATA WITH REFERENCE
skn-rbt 500 mg/24H MOD 28ZPAK -,217,72
eye-rbt 500 mg/24H MOD 28ZPAK -,217,72
orl-rat LD50:1550 mg/kg 28ZPAK -,217,72
ihl-rat LCLo:2560 ppm/4H 28ZPAK -,217,72

CONSENSUS REPORTS: Reported in EPA TSCA Inventory.

SAFETY PROFILE: Moderately toxic by ingestion and inhalation. A skin and eye irritant. When heated to decomposition it emits toxic fumes of Cl^-.

DRC500 **HR: 3**
DIMETHYL-2-CHLORO-4-NITROPHENYLTHIONOPHOSPHATE
mf: $C_8H_9ClNO_5PS$ mw: 297.65

$(CH_3O)_2P(S)OC_6H_3(Cl)NO_2$

SAFETY PROFILE: Ignites during thermal decomposition when heated to 270°C. When heated to decomposition it emits toxic fumes of Cl^-, NO_x, PO_x, and SO_x. See also NITRO COMPOUNDS of AROMATIC HYDROCARBONS.

DRC600 **CAS:10389-72-7** **HR: 3**
α,α-DIMETHYL-o-CHLOROPHENETHYLAMINE HYDROCHLORIDE
mf: $C_{10}H_{14}ClN{\cdot}ClH$ mw: 220.16

PROP: A solid. Mp: 245–246°.

SYNS: 2-CHLORO-α,α-DIMETHYLBENZENEETHANIAMINE HYDROCHLORIDE □ O-CHLORO-α,α-DIMETHYLPHENETHYLAMINE HYDROCHLORIDE □ CLORTERMINE HYDROCHLORIDE □ S 77 □ SU-10568 □ VORANIL

TOXICITY DATA WITH REFERENCE
orl-rat LD50:332 mg/kg TXAPA9 18,185,71
ipr-rat LD50:92 mg/kg APTOA6 17,121,60

SAFETY PROFILE: Poison by ingestion and intraperitoneal routes. When heated to decomposition it emits very toxic fumes of Cl^- and NO_x.

DRC800 **CAS:3789-77-3** **HR: 2**
N,N-DIMETHYL-p-((m-CHLOROPHENYL)AZO)ANILINE
mf: $C_{14}H_{14}ClN_3$ mw: 259.76

SYN: 3′-CHLORO-4-DIMETHYLAMINOAZOBENZENE

TOXICITY DATA WITH REFERENCE
orl-rat TDLo:4900 mg/kg/17W-C:NEO JEMEAV 87,139,48

SAFETY PROFILE: Questionable carcinogen with experimental neoplastigenic data. When heated to decomposition it emits very toxic fumes of Cl^- and NO_x.

DRD000 **CAS:3010-47-7** **HR: 2**
N,N-DIMETHYL-p-((o-CHLOROPHENYL)AZO)ANILINE
mf: $C_{14}H_{14}ClN_3$ mw: 259.76

SYN: 2′-CHLORO-4-DIMETHYLAMINOAZOBENZENE

TOXICITY DATA WITH REFERENCE
orl-rat TDLo:4900 mg/kg/17W-C:NEO JEMEAV 87,139,48

SAFETY PROFILE: Questionable carcinogen with experimental neoplastigenic data. When heated to decomposition it emits very toxic fumes of Cl^- and NO_x.

DRD800 **CAS:63041-62-3** **HR: 2**
2,3-DIMETHYLCHOLANTHRENE
mf: $C_{22}H_{18}$ mw: 282.40

SYN: 16:20-DIMETHYLCHOLANTHRENE

TOXICITY DATA WITH REFERENCE
scu-mus TDLo:400 mg/kg:ETA AJCAA7 28,334,36

SAFETY PROFILE: Questionable carcinogen with experimental tumorigenic data. When heated to decomposition it emits acrid smoke and irritating fumes.

DRD850 **CAS:85923-37-1** **HR: 2**
3,6-DIMETHYLCHOLANTHRENE
mf: $C_{22}H_{18}$ mw: 282.40

SYN: BENZ(j)ACEANTHRYLENE, 1,2-DIHYDRO-3,6-DIMETHYL-(9CI)

TOXICITY DATA WITH REFERENCE
mma-ham:lng 50 µg/L CALEDQ 28,223,85
msc-ham:lng 100 µg/L PAACA3 24,94,83
skn-mus TDLo:706 ng/kg:ETA CALEDQ 28,223,85

SAFETY PROFILE: Questionable carcinogen with experimental tumorigenic data. Mutation data reported. When heated to decomposition it emits acrid smoke and irritating fumes.

DRE000 **CAS:63041-61-2** **HR: 2**
15,20-DIMETHYLCHOLANTHRENE
mf: $C_{22}H_{18}$ mw: 282.40

SYN: 1,3-DIMETHYLCHOLANTHRENE

TOXICITY DATA WITH REFERENCE
scu-mus TDLo:200 µg/kg:ETA JNCIAM 2,99,41

SAFETY PROFILE: Questionable carcinogen with experimental tumorigenic data. When heated to decomposition it emits acrid smoke and irritating fumes.

DRE200 **CAS:15914-23-5** **HR: 2**
1,2-DIMETHYLCHRYSENE
mf: $C_{20}H_{16}$ mw: 256.36

PROP: Plates from C_6H_6. Mp: 263–264°.

TOXICITY DATA WITH REFERENCE
skn-mus TDLo:800 mg/kg/33W-I:ETA PRLBA4 129,439,40
scu-mus TDLo:3200 mg/kg/48W-I:ETA PRLBA4 129,439,40

SAFETY PROFILE: Questionable carcinogen with experimental tumorigenic data. When heated to decomposition it emits acrid smoke and irritating fumes. See also CHRYSENE.

DRE400 **CAS:52171-92-3** **HR: 2**
1,11-DIMETHYLCHRYSENE
mf: $C_{20}H_{16}$ mw: 256.36

SYN: 5,7-DIMETHYLCHRYSENE

TOXICITY DATA WITH REFERENCE
mma-sat 20 µg/plate CNREA8 36,4525,76
skn-mus TDLo:120 mg/kg/50W-I:NEO CNREA8 34,1315,74

CONSENSUS REPORTS: EPA Genetic Toxicology Program.

SAFETY PROFILE: Mutation data reported. Questionable carcinogen with experimental neoplastigenic data. When heated to decomposition it emits acrid smoke and irritating fumes. See also CHRYSENE.

DRE600 **CAS:63019-23-8** **HR: 2**
4,5-DIMETHYLCHRYSENE
mf: $C_{20}H_{16}$ mw: 256.36

TOXICITY DATA WITH REFERENCE
scu-mus TDLo:80 mg/kg:ETA CNREA8 3,606,43

SAFETY PROFILE: Questionable carcinogen with experimental tumorigenic data. When heated to decomposition it emits acrid smoke and irritating fumes. See also CHRYSENE.

D

DRE800 **CAS:3697-27-6** **HR: 2**
5,6-DIMETHYLCHRYSENE
mf: $C_{20}H_{16}$ mw: 256.36

PROP: Plates from C_6H_6/EtOH. Mp: 127–128°.

TOXICITY DATA WITH REFERENCE
scu-mus TDLo:80 mg/kg:ETA CNREA8 3,606,43

SAFETY PROFILE: Questionable carcinogen with experimental tumorigenic data. When heated to decomposition it emits acrid smoke and irritating fumes. See also CHRYSENE.

DRF000 **CAS:14207-78-4** **HR: 2**
5,11-DIMETHYLCHRYSENE
mf: $C_{20}H_{16}$ mw: 256.36

TOXICITY DATA WITH REFERENCE
skn-mus TDLo:400 µg/kg/20D-I:ETA CALEDQ 8,65,79

SAFETY PROFILE: Questionable carcinogen with experimental tumorigenic data. When heated to decomposition it emits acrid smoke and irritating fumes. See also CHRYSENE.

DRF200 **CAS:617-54-9** **HR: 1**
DIMETHYL CITRACONATE
mf: $C_7H_{10}O_4$ mw: 158.17

PROP: Bp: 210–211°.

SYNS: DIMETHYL METHYL MALEATE ☐ cis-2-METHYL-2-BUTENEDIOIC ACID, DIMETHYL ESTER ☐ METHYLMALEIC ACID, DIMETHYL ESTER

TOXICITY DATA WITH REFERENCE
skn-rbt 500 mg/24H MLD FCTXAV 14,659,76

CONSENSUS REPORTS: Reported in EPA TSCA Inventory.

SAFETY PROFILE: A skin irritant. When heated to decomposition it emits acrid smoke and irritating fumes. See also ESTERS.

DRF400 **CAS:675-09-2** **HR: 2**
4,6-DIMETHYLCOUMARIN
mf: $C_8H_{12}O_2$ mw: 140.20

PROP: A solid. Mp: 50–51°, bp: 140–142° @ 35 mm.

TOXICITY DATA WITH REFERENCE
ipr-mus LD50:750 mg/kg APTOA6 2,109,46

CONSENSUS REPORTS: Reported in EPA TSCA Inventory.

SAFETY PROFILE: Moderately toxic by intraperitoneal route. When heated to decomposition it emits acrid smoke and irritating fumes. See also COUMARIN.

DRF600 CAS:1467-79-4 HR: 3
DIMETHYLCYANAMIDE
mf: $C_3H_6N_2$ mw: 70.11

PROP: Colorless, mobile liquid. Mp: −41.0°, bp: 162–163°, flash p: 160°F (TCC), d: 0.8767 @ 30°, vap press: 40 mm @ 80°, vap d: 2.55.

TOXICITY DATA WITH REFERENCE
orl-rat LD50:146 mg/kg GTPZAB 19(11),23,75
ihl-rat LC50:2500 mg/m^3 GTPZAB 19(11),23,75
orl-mus LD50:73 mg/kg GTPZAB 19(11),23,75
ihl-mus LC50:2800 mg/m^3 GTPZAB 19(11),23,75
skn-mus LD50:125 mg/kg GTPZAB 19(11),23,75
ipr-mus LD50:40 mg/kg NTIS** AD691-490
orl-gpg LD50:146 mg/kg GTPZAB 19,23,75

CONSENSUS REPORTS: Reported in EPA TSCA Inventory.

SAFETY PROFILE: Poison by ingestion, skin contact, and intraperitoneal routes. Moderately toxic by inhalation. Flammable when exposed to heat, flame, or oxidizers. Can react with oxidizing materials. To fight fire, use foam, CO_2, or dry chemical. When heated to decomposition or in reaction with water or steam it produces toxic fumes of NO_x and CN^- and flammable vapors. See also CYANIDE.

DRF709 CAS:98-94-2 HR: 3
N,N-DIMETHYLCYCLOHEXANAMINE
DOT: UN 2264
mf: $C_8H_{17}N$ mw: 127.26

SYNS: CYCLOHEXYLDIMETHYLAMINE □ N-CYCLOHEXYLDIMETHYLAMINE □ (DIMETHYLAMINO)CYCLOHEXANE □ N,N-DIMETHYLAMINOCYCLOHEXANE □ DIMETHYLCYCLOHEXYLAMINE □ N,N-DIMETHYLCYCLOHEXYLAMINE (DOT) □ POLYCAT 8

TOXICITY DATA WITH REFERENCE
orl-rat LD50:348 mg/kg ZHYGAM 20,393,74
ihl-rat LC50:1889 mg/m^3/2H GTPZAB 28(5),54,84
orl-mus LD50:320 mg/kg GTPZAB 28(5),54,84
ihl-mus LC50:1100 mg/m^3/2H GTPZAB 28(5),54,84
orl-rbt LD50:620 mg/kg ZHYGAM 20,393,74
orl-gpg LD50:520 mg/kg ZHYGAM 20,393,74

CONSENSUS REPORTS: Reported in EPA TSCA Inventory.

DOT CLASSIFICATION: 8; *Label:* Corrosive

SAFETY PROFILE: Poison by ingestion. Moderately toxic by inhalation. When heated to decomposition it emits toxic fumes of NO_x.

DRF800 CAS:583-57-3 HR: 3
cis-1,2-DIMETHYLCYCLOHEXANE
mf: C_8H_{16} mw: 112.22

PROP: Flash p: 61.8°F.

SYNS: o-DIMETHYLCYCLOHEXANE □ 1,2-DIMETHYLCYCLOHEXANE (DOT)

SAFETY PROFILE: A very dangerous fire hazard when exposed to heat, flame, or oxidizers. When heated to decomposition it emits acrid smoke and fumes.

DRG000 CAS:591-21-9 HR: 3
1,3-DIMETHYLCYCLOHEXANE
mf: C_8H_{16} mw: 112.22

PROP: Flash p: 42.8°F.

SYN: m-DIMETHYLCYCLOHEXANE

SAFETY PROFILE: A very dangerous fire hazard when exposed to heat, flame, or oxidizers. When heated to decomposition it emits acrid smoke and fumes.

DRG200 CAS:589-90-2 HR: 3
1,4-DIMETHYLCYCLOHEXANE
mf: C_8H_{16} mw: 112.24

PROP: Liquid. Mp: 86°, bp: 119.5°, flash p: 50°F (CC), d: 0.77, vap press: 10 mm @ 10.2°, vap d: 3.86.

SAFETY PROFILE: Dangerous fire hazard when exposed to heat or flame; can react vigorously with oxidizing materials. Keep away from heat and open flame. To fight fire, use foam, CO_2, dry chemical.

DRG400 HR: 3
trans-1,2-DIMETHYLCYCLOHEXANE
mf: C_8H_{16} mw: 112.22

PROP: Flash p: 41.6°F.

SAFETY PROFILE: A very dangerous fire hazard when exposed to heat, flame, or oxidizers. When heated to decomposition it emits acrid smoke and fumes.

DRH200 CAS:5831-10-7 HR: 2
11,17-DIMETHYL-15H-CYCLOPENTA(a)PHENANTHRENE
mf: $C_{19}H_{16}$ mw: 244.35

TOXICITY DATA WITH REFERENCE
mma-sat 50 µg/plate CNREA8 36,4525,76
skn-mus TDLo:108 mg/kg/1Y-I:CAR PEXTAR 11,69,69

CONSENSUS REPORTS: EPA Genetic Toxicology Program.

SAFETY PROFILE: Questionable carcinogen with experimental carcinogenic data. Mutation data reported. When heated to decomposition it emits acrid smoke and irritating fumes.

DRH400 CAS:5831-09-4 HR: 2
12,17-DIMETHYL-15H-CYCLOPENTA(a)PHENANTHRENE
mf: $C_{19}H_{16}$ mw: 244.35

TOXICITY DATA WITH REFERENCE
mma-sat 50 μg/plate CNREA8 36,4525,76
skn-mus TDLo: 108 mg/kg/1Y-I: ETA PEXTAR 11,69,69

CONSENSUS REPORTS: EPA Genetic Toxicology Program.

SAFETY PROFILE: Questionable carcinogen with experimental tumorigenic data. Mutation data reported. When heated to decomposition it emits acrid smoke and irritating fumes.

DRH800 CAS:63020-69-9 HR: 2
3,4-DIMETHYL-1,2-CYCLOPENTENOPHENANTHRENE
mf: $C_{19}H_{18}$ mw: 246.37

SYN: 16,17-DIHYDRO-11,12-DIMETHYL-15H-CYCLOPENTA(a)PHENANTHRENE

TOXICITY DATA WITH REFERENCE
skn-mus TDLo: 1260 mg/kg/39W-I: ETA ARGEAR 6,1,53

SAFETY PROFILE: Questionable carcinogen with experimental tumorigenic data. When heated to decomposition it emits acrid smoke and irritating fumes.

DRI400 CAS:3546-11-0 HR: 2
3,3′-DIMETHYL-N,N′-DIACETYLBENZIDINE
mf: $C_{18}H_{20}N_2O_2$ mw: 296.40

SYNS: N,N′-DIACETYL-3,3′-DIMETHYLBENZIDINE □ 3′,3‴-DIMETHYL-4′,4‴-BIACETANILIDE

TOXICITY DATA WITH REFERENCE
mma-sat 10 μg/plate SAIGBL 23,168,81
orl-rat TDLo: 7900 mg/kg/43W-C: CAR CNREA8 16,525,56

SAFETY PROFILE: Questionable carcinogen with experimental carcinogenic data. Mutation data reported. When heated to decomposition it emits toxic fumes of NO_x.

DRI600 CAS:110-70-3 HR: 3
N,N′-DIMETHYLDIAMINOETHANE
mf: $C_4H_{12}N_2$ mw: 88.18

SYNS: 1,2-BIS(METHYLAMINO)ETHANE □ 2,5-DIAZAHEXANE □ N,N′-DIMETHYLETHANEDIAMINE □ N,N′-DIMETHYL-1,2-ETHANEDIAMINE □ N,N′-DIMETHYLETHYLENEDIAMINE □ sym-DIMETHYLETHYLENEDIAMINE □ ETHYLENEDIAMINE, N,N′-DIMETHYL- □ 1,2-ETHANEDIAMINE, N,N′-DIMETHYL-(9CI)

TOXICITY DATA WITH REFERENCE
ipr-mus LD50: 200 mg/kg EJMCA5 17,235,82

CONSENSUS REPORTS: Reported in EPA TSCA Inventory.

SAFETY PROFILE: Poison by intravenous route. When heated to decomposition it emits toxic vapors of NO_x.

DRI700 CAS:53534-20-6 HR: 3
1,1-DIMETHYLDIAZENIUM PERCHLORATE
mf: $C_2H_8ClN_2O_4$ mw: 159.55

SAFETY PROFILE: An impact-sensitive salt. When heated to decomposition it emits toxic fumes of Cl^- and NO_x. See also PERCHLORATES.

D

DRI800 CAS:35335-07-0 HR: 2
9,10-DIMETHYL-1,2,5,6-DIBENZANTHRACENE
mf: $C_{24}H_{18}$ mw: 306.42

PROP: Crystals from C_6H_6/EtOH. Mp: 204–205°.

SYNS: 9,10-DIMETHYL-DBA □ 7,14-DIMETHYLDIBENZ(a,h)ANTHRACENE

TOXICITY DATA WITH REFERENCE
msc-ham: lng 25 μg/L MUREAV 136,65,84
skn-mus TDLo: 200 mg/kg/20W-I: NEO CNREA8 22,78,62
scu-mus TDLo: 20 mg/kg: ETA CNREA8 22,78,62

SAFETY PROFILE: Questionable carcinogen with experimental neoplastigenic and tumorigenic data. Mutation data reported. When heated to decomposition it emits acrid smoke and irritating fumes.

DRJ000 CAS:63042-50-2 HR: 2
4,9-DIMETHYL-2,3,5,6-DIBENZOTHIOPHENTHRENE
mf: $C_{22}H_{16}S$ mw: 312.44

SYN: 7,13-DIMETHYLBENZO(b)PHENANTHRO(3,2-d)THIOPHENE

TOXICITY DATA WITH REFERENCE
scu-mus TDLo: 80 mg/kg: ETA JNCIAM 18,555,57

SAFETY PROFILE: Questionable carcinogen with experimental tumorigenic data. When heated to decomposition it emits toxic fumes of SO_x.

DRJ200 CAS:16924-32-6 HR: 3
1,1-DIMETHYLDIBORANE
mf: $C_2H_{10}B_2$ mw: 55.724

$(CH_3)_2B:H_2:BH_2$

PROP: Easily liquefied gas; sensitive to air and moisture. Readily disproportionates to other methylboranes. Bp: −1°, mp: −150°, flash p: <14°F. Sol in ethers and hydrocarbons.

SAFETY PROFILE: A very dangerous fire and explosion hazard when exposed to heat, flame, or oxidizers. See also BORON COMPOUNDS and BORANES.

DRJ400 CAS:17156-88-6 HR: 3
1,2-DIMETHYLDIBORANE
mf: $C_2H_{10}B_2$ mw: 55.724

$CH_3HB:H_2:BHCH_3$

PROP: Colorless gas, decomp by water. Mp: −125°; bp: 49°, flash p: <−67°.

SAFETY PROFILE: A very dangerous fire and explosion hazard when exposed to heat, flame, or oxidizers. See also BORON COMPOUNDS and BORANES.

DRJ800 CAS:78-63-7 HR: 2
2,5-DIMETHYL-2,5-DI(tert-BUTYLPEROXY)HEXANE
mf: $C_{16}H_{34}O_4$ mw: 290.50

PROP: Colorless to light-yellow liquid. D: 0.85, fp: 8°, flash p: >180°F (MOC), bp: 250°. Insol in water; sol in many organic solvents.

SYNS: 2,5-DIMETHYL-2,5-DI(t-BUTYLPEROXY)HEXANE □ PEROXIDE, (1,1,4,4-TETRAMETHYL-1,4-BUTANEDIYL)BIS((1,1-DIMETHYLETHYL) □ PEROXIDE, (1,1,4,4-TETRAMETHYLTETRAMETHYLENE)BIS (tert-BUTYL □ TRIGONOX 101-101/45 □ VAROX

TOXICITY DATA WITH REFERENCE
ipr-mus LDLo:1700 mg/kg BSPII* 1/75-19B

CONSENSUS REPORTS: Reported in EPA TSCA Inventory.

SAFETY PROFILE: Moderately toxic by intraperitoneal route. Combustible when exposed to heat, flames, or reducing agents. To fight fire, use water spray, foam, dry chemical. When heated to decomposition it emits acrid smoke and irritating fumes. Used in the polymerization of styrene and in crosslinking of various grades of polyethylene. See also PEROXIDES, ORGANIC.

DRJ825 CAS:1068-27-5 HR: 2
2,5-DIMETHYL-2,5-DI(tert-BUTYLPEROXY)HEXYNE-3
mf: $C_{16}H_{30}O_4$ mw: 286.46

SYN: 3-HEXYNE, 2,5-DIMETHYL-2,5-DI(t-BUTYLPEROXY)-

TOXICITY DATA WITH REFERENCE
ipr-mus LD50:1850 mg/kg BSPII* 1/75-19B

CONSENSUS REPORTS: Reported in EPA TSCA Inventory.

SAFETY PROFILE: Moderately toxic by intraperitoneal route. A peroxide. Handle carefully. When heated to decomposition it emits acrid smoke and irritating vapors.

DRK400 CAS:42149-31-5 HR: 2
2,5-DIMETHYL-1,2,5,6-DIEPOXYHEX-3-YNE
mf: $C_8H_{10}O_2$ mw: 138.18

TOXICITY DATA WITH REFERENCE
scu-mus TDLo:1040 mg/kg/26W-I:NEO JNCIAM 53,695,74

SAFETY PROFILE: Questionable carcinogen with experimental neoplastigenic data. When heated to decomposition it emits acrid smoke and irritating fumes. See also ACETYLENE COMPOUNDS.

DRK500 CAS:34983-45-4 HR: 2
trans-4,4′-DIMETHYL-α-α′-DIETHYLSTILBENE
mf: $C_{20}H_{24}$ mw: 264.44

SYNS: DMES □ STILBENE, α-α′-DIETHYL-4,4′-DIMETHYL-, (E)-

TOXICITY DATA WITH REFERENCE
orl-dog TDLo:6 g/kg/30D-C:ETA TXAPA9 21,582,72

SAFETY PROFILE: Questionable carcinogen with experimental tumorigenic data. When heated to decomposition it emits acrid smoke and irritating fumes.

DRK600 CAS:19072-57-2 HR: 3
2,6-DIMETHYL-1,1-DIETHYLPIPERIDINIUM BROMIDE
mf: $C_{11}H_{24}BrN$ mw: 250.27

SYNS: AGILENE □ SC-1950

TOXICITY DATA WITH REFERENCE
orl-rat LD50:2000 mg/kg JPETAB 99,435,50
orl-mus LD50:365 mg/kg JPETAB 99,435,50
ipr-mus LD50:40 mg/kg JPETAB 99,435,50
ivn-dog LDLo:25 mg/kg JPETAB 99,435,50
ivn-rbt LDLo:25 mg/kg JPETAB 99,435,50

SAFETY PROFILE: Poison by ingestion, intraperitoneal, and intravenous routes. When heated to decomposition it emits very toxic fumes of Br^- and NO_x.

DRK800 CAS:578-32-5 HR: 2
N,N-DIMETHYL-2,5-DIFLUORO-p-(2,5-DIFLUOROPHENYLAZO)ANILINE
mf: $C_{14}H_{11}F_4N_3$ mw: 297.28

SYN: 2,5,2′,5′-TETRAFLUORO-4-DIMETHYLAMINOAZOBENZENE

TOXICITY DATA WITH REFERENCE
orl-rat TDLo:6400 mg/kg/21W-C:NEO CNREA8 17,387,57

SAFETY PROFILE: Questionable carcinogen with experimental neoplastigenic data. When heated to decomposition it emits very toxic fumes of F^- and NO_x.

DRL000 CAS:351-65-5 HR: 2
N,N-DIMETHYL-p-(3,4-DIFLUOROPHENYLAZO)ANILINE
mf: $C_{14}H_{13}F_2N_3$ mw: 261.30

SYNS: 3′,4′-DIFLUORO-4-DIMETHYLAMINOAZOBENZENE □ N,N-DIMETHYL-3′,4′-DIFLUORO-4-(PHENYLAZO)BENZENEAMINE

TOXICITY DATA WITH REFERENCE
orl-rat TDLo:2356 mg/kg/17W-C:CAR CBINA8 53,107,85
orl-rat TD:3400 mg/kg/13W-C:NEO CNREA8 17,387,57

SAFETY PROFILE: Questionable carcinogen with experimental carcinogenic and neoplastigenic data. When heated to decomposition it emits very toxic fumes of F^- and NO_x.

DRL200 CAS:122-15-6 HR: 3
5,5-DIMETHYLDIHYDRORESORCINOL DIMETHYLCARBAMATE
mf: $C_{11}H_{17}NO_3$ mw: 211.29

SYNS: DIMETAN □ DIMETHYLCARBAMATE de 5,5-DIMETHYL DIHYDRORESORCINOL (FRENCH) □ DIMETHYLCARBAMIC ACID ester with 3-HYDROXY-5,5-DIMETHYL-2-CYCLOHEXEN-1-ONE □ 5,5-DIMETHYL-DIHYDRORESORCINOL-N,N-DIMETHYLCARBAMAT (GERMAN) □ 5,5-DIMETHYL-4,5-DIHYDRO-3-RESORCYL-DIMETHYL-CARBAMAT (GERMAN) □ (5,5-DIMETHYL-3-OXO-CYCLOHEX-1-EN-YL)-N,N-DIMETHYL-CARBAMAAT (DUTCH) □ 5,5-DIMETHYL-3-OXO-1-CYCLOHEXEN-1-YL DIMETHYLCARBAMATE □ 5,5-DIMETHYL-3-OXOCYCLOHEX-1-ENYL DIMETHYLCARBAMATE □ (5,5-DIMETHYL-3-OXO-CYCLOHEX-1-EN-YL)-N,N-DIMETHYL-CARBAMAT (GERMAN) □ (5,5-

DIMETIL-3-OXO-CICLOES-1-EN-IL)-N,N-DIMETIL-CARBAMMATO (ITALIAN) □ ENT 24,738 □ GEIGY 19258 □ 3-HYDROXY-5,5-DIMETHYL-2-CYCLOHEXEN-1-ONE DIMETHYLCARBAMATE

TOXICITY DATA WITH REFERENCE
orl-rat LD50:120 mg/kg WRPCA2 9,119,70
orl-mus LD50:90 mg/kg 85DPAN -,-,71/76
orl-dog LD50:50 mg/kg 85GYAZ -,66,71

SAFETY PROFILE: Poison by ingestion. When heated to decomposition it emits toxic fumes of NO_x. See also CARBAMATES.

DRL400 CAS:35653-70-4 HR: 2
2,4′-DIMETHYL-4-DIMETHYLAMINOAZOBENZENE
mf: $C_{16}H_{19}N_3$ mw: 253.38

TOXICITY DATA WITH REFERENCE
orl-rat TDLo:8940 mg/kg/35W-C:ETA ARZNAD 12,270,62

SAFETY PROFILE: Questionable carcinogen with experimental tumorigenic data. When heated to decomposition it emits toxic fumes of NO_x.

DRL450 CAS:3215-85-8 HR: 3
3,3-DIMETHYL-4-(DIMETHYLAMINO)-4-(o-METHOXYPHENYL)BUTYL o-METHOXYPHENYL KETONE
mf: $C_{23}H_{31}NO_3$ mw: 369.55

SYN: KETONE, 3,3-DIMETHYL-4-(DIMETHYLAMINO)-4-(o-METHOXYPHENYL)BUTYL o-METHOXYPHENYL

TOXICITY DATA WITH REFERENCE
orl-mus LD50:500 mg/kg JMCMAR 9,187,66

DOT CLASSIFICATION: 3; *Label:* Flammable Liquid

SAFETY PROFILE: Moderately toxic by ingestion. A flammable liquid. When heated to decomposition it emits toxic vapors of NO_x.

DRL460 CAS:3215-84-7 HR: 3
3,3-DIMETHYL-4-(DIMETHYLAMINO)-4-(p-METHOXYPHENYL)BUTYL p-METHOXYPHENYL KETONE
mf: $C_{23}H_{31}NO_3$ mw: 369.55

SYN: KETONE, 3,3-DIMETHYL-4-(DIMETHYLAMINO)-4-(p-METHOXYPHENYL)BUTYL p-METHOXYPHENYL

TOXICITY DATA WITH REFERENCE
orl-mus LD50:505 mg/kg JMCMAR 9,187,66

DOT CLASSIFICATION: 3; *Label:* Flammable Liquid

SAFETY PROFILE: Moderately toxic by ingestion. A flammable liquid. When heated to decomposition it emits toxic vapors of NO_x.

DRL600 CAS:38035-28-8 HR: 3
2,3-DIMETHYL-8-(DIMETHYLAMINOMETHYL)-7-METHOXYCHROMONE HYDROCHLORIDE
mf: $C_{15}H_{19}NO_3$•ClH mw: 297.81

SYNS: 4H-1-BENZOPYRAN-4-ONE, 8-((DIMETHYLAMINO)METHYL)-7-METHOXY-2,3-DIMETHYL-, HYDROCHLORIDE □ REC 7/0268

TOXICITY DATA WITH REFERENCE
orl-rat LD50:7800 μg/kg 27ZQAG -,156,72
ipr-rat LD50:3300 μg/kg 27ZQAG -,157,72
scu-rat LD50:2200 μg/kg 27ZQAG -,157,72
ipr-mus LD50:3300 μg/kg JMPCAS 3,471,61

SAFETY PROFILE: Poison by ingestion, intraperitoneal, and subcutaneous routes. When heated to decomposition it emits very toxic fumes of NO_x and HCl.

D

DRM000 CAS:3759-07-7 HR: 3
9,9-DIMETHYL-10-DIMETHYLAMINOPROPYLACRIDAN HYDROGEN TARTRATE
mf: $C_{20}H_{26}N_2$•$C_4H_4O_6$ mw: 442.56

PROP: A solid. Mp: 155–156°.

SYNS: DIMETACRINE BITARTRATE □ DIMETACRIN HYDROGENTARTRATE □ DIMETHACRINE TARTRATE □ 10-(3-(DIMETHYLAMINO)PROPYL)-9,9-DIMETHYLACRIDAN TARTRATE (1:1) □ 9,9-DIMETHYL-10-(3-DIMETHYLAMINO)PROPYLACRIDINE TARTRATE □ ISOTONIL □ ISTONYL □ MIROISTONIL □ MO 709 □ SD 709 □ ((R-R*,R*))-N,N,9,9-TETRAMETHYL-10(9H)-ACRIDINEPROPANAMINE-2,3-DIHYDROXYBUTANEDIOATE (1:1)

TOXICITY DATA WITH REFERENCE
orl-mus TDLo:175 mg/kg (female 7-13D post):REP OYYAA2 4,855,70
orl-mus TDLo:525 mg/kg (female 7-13D post):REP OYYAA2 4,855,70
ipr-rat TDLo:315 mg/kg (female 6-12D post):TER OYYAA2 5,129,71
orl-rat TDLo:450 mg/kg (9-14D preg):TER OYYAA2 4,855,70
ipr-rat TDLo:315 mg/kg (female 6-12D post):REP OYYAA2 5,129,71
orl-rat TDLo:300 mg/kg (female 9-14D post):TER OYYAA2 4,855,70
orl-rat LD50:1671 mg/kg ARZNAD 24,1098,74
ipr-rat LD50:203 mg/kg OYYAA2 4,855,70
scu-rat LD50:1214 mg/kg OYYAA2 4,855,70
ivn-rat LD50:38 mg/kg WKWOAO 78,21,66
orl-mus LD50:860 mg/kg WKWOAO 78,21,66
ipr-mus LD50:175 mg/kg WKWOAO 78,21,66
scu-mus LD50:798 mg/kg OYYAA2 4,855,70
ivn-mus LD50:40,900 μg/kg OYYAA2 4,855,70
orl-cat LD50:150 mg/kg WKWOAO 78,21,66
ivn-cat LD50:40 mg/kg WKWOAO 78,21,66

SAFETY PROFILE: Poison by ingestion, intravenous, and intraperitoneal routes. Moderately toxic by subcutaneous route. Experimental teratogenic and reproductive effects. When heated to decomposition it emits toxic fumes of NO_x.

DRM100 CAS:3215-88-1 HR: 3
3,3-DIMETHYL-4-(DIMETHYLAMINO)-4-(m-TOLYL)BUTYL m-TOLYL KETONE
mf: $C_{23}H_{31}NO$ mw: 337.55

SYN: KETONE, 3,3-DIMETHYL-4-(DIMETHYLAMINO)-4-(m-TOLYL)BUTYL m-TOLYL

TOXICITY DATA WITH REFERENCE
orl-mus LD50:1250 mg/kg JMCMAR 9,187,66

DOT CLASSIFICATION: 3; *Label:* Flammable Liquid

SAFETY PROFILE: Moderately toxic by ingestion. A

flammable liquid. When heated to decomposition it emits toxic vapors of NO_x.

DRM110 **CAS:3215-89-2** **HR: 3**
3,3-DIMETHYL-4-(DIMETHYLAMINO)-4-(o-TOLYL)BUTYL o-TOLYL KETONE
mf: $C_{23}H_{31}NO$ mw: 337.55

SYN: KETONE, 3,3-DIMETHYL-4-(DIMETHYLAMINO)-4-(o-TOLYL) BUTYL o-TOLYL

TOXICITY DATA WITH REFERENCE
orl-mus LD50:125 mg/kg JMCMAR 9,187,66

DOT CLASSIFICATION: 3; *Label:* Flammable Liquid

SAFETY PROFILE: A poison by ingestion. A flammable liquid. When heated to decomposition it emits toxic vapors of NO_x.

DRM120 **CAS:3215-87-0** **HR: 3**
3,3-DIMETHYL-4-(DIMETHYLAMINO)-4-(p-TOLYL)BUTYL p-TOLYL KETONE
mf: $C_{23}H_{31}NO$ mw: 337.55

SYN: KETONE, 3,3-DIMETHYL-4-(DIMETHYLAMINO)-4-(p-TOLYL) BUTYL p-TOLYL

TOXICITY DATA WITH REFERENCE
orl-mus LD50:755 mg/kg JMCMAR 9,187,66

DOT CLASSIFICATION: 3; *Label:* Flammable Liquid

SAFETY PROFILE: Moderately toxic by ingestion. A flammable liquid. When heated to decomposition it emits toxic vapors of NO_x.

DRM600 **CAS:4100-38-3** **HR: 3**
3,4-DIMETHYL-4-(3,4-DIMETHYL-5-ISOXAZOLYAZO)-ISOXAZOLIN-5-ONE
mf: $C_{10}H_{12}N_4O_3$ mw: 236.26

$O{:}CON{=}CCH_3\,C(CH_3)N{=}NC{=}CCH_3C(CH_3){=}N\,O$

SAFETY PROFILE: Explodes if heated rapidly to 100°C but is stable to impact or friction. When heated to decomposition it emits toxic fumes of NO_x.

DRM800 **CAS:63886-45-3** **HR: 3**
2,5-DIMETHYL-1-(5-(2,5-DIMETHYLPYRROLIDINO)-2,4-PENTADIENYLIDENE) PYRROLIDINIUM CHLORIDE SESQUIHYDRATE
mf: $C_{17}H_{29}N_2{\cdot}Cl{\cdot}3/2H_2O$ mw: 323.96

TOXICITY DATA WITH REFERENCE
orl-mus LD50:50 mg/kg JMCMAR 12,806,69
ipr-mus LD50:10 mg/kg JMCMAR 12,806,69

SAFETY PROFILE: Poison by ingestion and intraperitoneal routes. When heated to decomposition it emits very toxic fumes of NO_x and Cl^-.

DRN200 **CAS:40487-42-1** **HR: 3**
3,4-DIMETHYL-2,6-DINITRO-N-(1-ETHYLPROPYL)ANILINE
mf: $C_{13}H_{19}N_3O_4$ mw: 281.35

PROP: Orange-yellow crystals from MeOH. Mp: 56–57°. Very sltly sol in H_2O; sol in $CHCl_3$ and C_6H_6.

SYNS: AC 92553 □ N-(1-AETHYLPROPYL)-3,4-DIMETHYL-2,6-DINITROANILIN (GERMAN) □ N-(1-AETHYLPROPYL)-2,6-DINITRO-3,4-XYLIDIN (GERMAN) □ 2,5-DINITRO-N-(1-ETHYLPROPYL)-3,4-XYLIDINE □ N-(1-ETHYLPROPYL)-3,4-DIMETHYL-2,6-DINITROBENZENAMINE □ HERBADOX □ HORBADOX □ PAY-OFF □ PENDIMETHALIN □ PENOXALINE □ PHENOXALIN □ PROWL □ STOMP □ TENDIMETHALIN

TOXICITY DATA WITH REFERENCE
orl-rat TDLo:219 g/kg/2Y-C:ETA JACTDZ 12,107,93
orl-rat LD50:1050 mg/kg 85JFAN A314,84
ipr-rat LD50:500 mg/kg IJEBA6 25,463,87
orl-mus LD50:1340 mg/kg PEMNDP 9,656,91
ipr-mus LD50:220 mg/kg IJEBA6 25,463,87

SAFETY PROFILE: Poison by intraperitoneal route. Moderately toxic by ingestion. Questionable carcinogen with experimental tumorigenic data. An herbicide. When heated to decomposition it emits toxic fumes of NO_x.

DRN300 **CAS:14760-99-7** **HR: 3**
N,N′-DIMETHYL-N,N′-DINITROOXAMIDE
mf: $C_4H_6N_4O_6$ mw: 206.11

SAFETY PROFILE: An explosive. When heated to decomposition it emits toxic fumes of NO_x. See also EXPLOSIVES.

DRN400 **CAS:3844-60-8** **HR: 3**
1,6-DIMETHYL-1,6-DINITROSOBIUREA
mf: $C_4H_8N_6O_4$ mw: 204.18

SYNS: N,N′-DIMETHYL-N,N′-DINITROSO-1,2-HYDRAZINEDICARBOXAMIDE □ HYDRAZODICARBONSAEUREABIS(METHYLNITROSAMID) (GERMAN) □ HYDRAZODICARBOXYLIC ACID BIS(METHYLNITROSAMIDE) □ HYDROAZODICARBOXYBIS(METHYLNITROSAMIDE) □ NSC 409425 □ SRI 1666

TOXICITY DATA WITH REFERENCE
scu-rat TDLo:420 mg/kg/28W-I:ETA ZEKBAI 69,103,67
scu-rat LD50:200 mg/kg ZEKBAI 69,103,67
ipr-mus LD50:56,570 μg/kg NCISP* JAN86

SAFETY PROFILE: Poison by subcutaneous and intraperitoneal routes. Questionable carcinogen with experimental tumorigenic data. Many N-nitroso compounds are carcinogens. When heated to decomposition it emits toxic fumes of NO_x. See also N-NITROSO COMPOUNDS.

DRN600 **CAS:7601-87-8** **HR: 3**
N,N′-DIMETHYL-N,N′-DINITROSOOXAMIDE
mf: $C_4H_6N_4O_4$ mw: 174.14

$[CH_3N(N{:}O)CO{\cdot}{-}]_2$

SYNS: DIMETHYLDINITROSOOXAMID (GERMAN) □ N,N′-DINITROSO-N,N′-DIMETHYLOXAMID (GERMAN)

TOXICITY DATA WITH REFERENCE
mmo-smc 1 mmol/L/10M ZEVBA5 95,82,64
orl-rat LD50:96 mg/kg ZEKBAI 69,103,676

SAFETY PROFILE: Poison by ingestion. Mutation data reported. Many N-nitroso compounds are carcinogens. See also NITRATES for fire hazard. A heat- and shock-sensitive explosive. Can react vigorously with reducing materials. When heated to decomposition it emits highly toxic fumes of NO_x. See also N-NITROSO COMPOUNDS and EXPLOSIVES, HIGH.

DRN800 CAS:55556-88-2 HR: 2
2,5-DIMETHYLDINITROSOPIPERAZINE
mf: $C_6H_{14}N_4O_2$ mw: 174.24

PROP: Mixture approximately 25% cis and 75% trans conformers (CNREA8 35,1270,75).

SYNS: 2,5-DIMETHYL 1,4-DINITROSOPIPERAZINE □ 2,5-DIMETHYL-DNPZ □ DINITROSO-2,5-DIMETHYLPIPERAZINE

TOXICITY DATA WITH REFERENCE
mma-sat 25 μg/plate TCMUE9 1,13,84
mma-smc 50 μmol/plate MUREAV 77,143,80
orl-rat TDLo:2740 mg/kg/50W-I:ETA CNREA8 35,1270,75

CONSENSUS REPORTS: EPA Genetic Toxicology Program.

SAFETY PROFILE: Many N-nitroso compounds are carcinogens. Questionable carcinogen with experimental tumorigenic data. Mutation data reported. When heated to decomposition it emits toxic fumes of NO_x. See also N-NITROSO COMPOUNDS.

DRO000 CAS:55380-34-2 HR: 2
2,6-DIMETHYLDINITROSOPIPERAZINE
mf: $C_6H_{14}N_4O_2$ mw: 174.24

SYNS: 2,6-DIMETHYL-DNPZ □ DINITROSO-2,6-DIMETHYLPIPERAZINE □ N,N'-DINITROSO-2,6-DIMETHYLPIPERAZINE □ 1,4-DINITROSO-2,6-DIMETHYLPIPERAZINE □ DNDMP

TOXICITY DATA WITH REFERENCE
mma-smc 50 μmol/plate TCMUE9 1,13,84
mma-sat 50 μg/plate MUREAV 77,143,80
orl-rat TDLo:240 mg/kg/20W-I:CAR CRNGDP 4,1165,83
orl-gpg TDLo:4800 mg/kg/50W-I:ETA CNREA8 40,1879,80
orl-ham TDLo:1960 mg/kg35W-I:NEO CRNGDP 4,1165,83
orl-rat TD:1800 mg/kg/33W-I:ETA CNREA8 35,1270,75
orl-rat TD:1200 mg/kg/20W-I:CAR CRNGDP 4,1165,83
orl-rat TD:63 mg/kg/25W-I:ETA FCTOD7 21,601,83
orl-rat TD:958 mg/kg/27W-I:ETA IAPUDO 57,617,84
orl-ham TD:2091 mg/kg/67W-I:ETA IAPUDO 57,617,84

CONSENSUS REPORTS: EPA Genetic Toxicology Program.

SAFETY PROFILE: Questionable carcinogen with experimental carcinogenic, neoplastigenic, and tumorigenic data. Mutation data reported. A model carcinogen and carcinogenic metabolite. When heated to decomposition it emits toxic fumes of NO_x. See also N-NITROSO COMPOUNDS.

DRO200 CAS:6972-76-5 HR: 2
N,N'-DIMETHYL-N,N'-DINITROSO-1,3-PROPANEDIAMINE
mf: $C_5H_{12}N_4O_2$ mw: 160.21

SYNS: DINITROSODIMETHYLPROPANEDIAMINE □ N,N'-DINITROSO-N,N'-DIMETHYL-1,3-PROPANEDIAMINE □ NSC 62580

TOXICITY DATA WITH REFERENCE
orl-rat TDLo:360 mg/kg/48W-I:NEO JNCIAM 41,985,68
orl-rat TD:640 mg/kg/32W-I:NEO JNCIAM 41,985,68

SAFETY PROFILE: Questionable carcinogen with experimental neoplastigenic data. Many N-nitroso compounds are carcinogens. When heated to decomposition it emits toxic fumes of NO_x. See also N-NITROSO COMPOUNDS.

DRO800 CAS:25136-55-4 HR: 3
DIMETHYL DIOXANE
mf: $C_6H_{12}O_2$ mw: 116.18

PROP: Water-white liquid. Bp: 117.5°, flash p: 75°F, d: 0.9268, vap press: 15.4 mm @ 20°, vap d: 4.0.

SYNS: DIMETHYL-p-DIOXANE (DOT) □ DIMETHYLDIOXANES (DOT)

TOXICITY DATA WITH REFERENCE
skn-rbt 10 mg/24H open JIHTAB 30,63,48
eye-rbt 20 mg/24H MOD 85JCAE-,811,86
ihl-rat TCLo:500 mg/m³/24H (30D pre):REP TPKVAL 12,64,71
orl-rat LD50:3000 mg/kg JIHTAB 30,63,48
ihl-rat LCLo:8000 ppm/4H JIHTAB 30,63,48

SAFETY PROFILE: Moderately toxic by ingestion. Mildly toxic by inhalation. Experimental reproductive effects. A skin and eye irritant. A very dangerous fire hazard when exposed to heat or flame; can react vigorously with oxidizing materials. To fight fire, use foam, CO_2, dry chemical. When heated to decomposition it emits acrid smoke and irritating fumes.

DRP200 CAS:2033-24-1 HR: 3
2,2-DIMETHYL-m-DIOXANE-4,6-DIONE
mf: $C_6H_8O_4$ mw: 144.14

PROP: Pale yellow, crystalline solid. Mp: 96–104°.

SYNS: 2,2-DIMETHYL-1,3-DIOXANE-4,6-DIONE □ 2,2-DIMETHYL-4,6-DIOXO-m-DIOXANE

TOXICITY DATA WITH REFERENCE
ivn-mus LD50:180 mg/kg CSLNX* NX#04102

CONSENSUS REPORTS: Reported in EPA TSCA Inventory.

SAFETY PROFILE: Poison by intravenous route. When heated to decomposition it emits acrid smoke and irritating fumes.

DRP400 **CAS:2916-31-6** **HR: 3**
2,2-DIMETHYL-1,3-DIOXOLAN
mf: $C_5H_{10}O_2$ mw: 102.14

$OC(CH_3)_2OCH_2CH_2$

PROP: A liquid. Flash p: 30.2°F, bp: 91.5–93.0°.

SAFETY PROFILE: A very dangerous fire hazard when exposed to heat, flame, or oxidizers. When heated to decomposition it emits acrid smoke and fumes.

DRP600 **CAS:7122-04-5** **HR: 3**
2-(4,5-DIMETHYL-1,3-DIOXOLAN-2-YL)PHENYL-N-METHYLCARBAMATE
mf: $C_{13}H_{17}NO_4$ mw: 251.31

SYNS: C-10015 □ CIBA-GEIGY C-10015 □ ENT 27,410 □ FONDAREN □ NSC 191000 □ SAPRECON C

TOXICITY DATA WITH REFERENCE
orl-rat LD50:110 mg/kg FMCHA2 -,C210,83
orl-dog LD50:300 mg/kg 28ZEAL 5,118,76

SAFETY PROFILE: Poison by ingestion. When heated to decomposition it emits toxic fumes of NO_x. See also CARBAMATES.

DRP800 **CAS:957-51-7** **HR: 2**
N,N-DIMETHYL-2,2-DIPHENYLACETAMIDE
mf: $C_{16}H_{17}NO$ mw: 239.34

PROP: White solid or crystals. Mp: 134.5–135.5°. Very sltly sol in water; mod sol in acetone, dimethyl formamide, and phenyl cellosolve.

SYNS: DIAMIDE □ DIF 4 □ N,N-DIMETHYLDIPHENYLACETAMIDE □ N,N-DIMETHYL-α,α-DIPHENYLACETAMIDE □ N,N-DIMETHYL-α-PHENYLBENZENEACETAMIDE □ DIMID □ DIPHENAMID □ DIPHENAMIDE □ DIPHENYLAMIDE □ 2,2-DIPHENYL-N,N-DIMETHYLACETAMIDE □ DYMID □ ENIDE □ FDN □ FENAM □ LILLY 34,314 □ U 4513

TOXICITY DATA WITH REFERENCE
cyt-mus-unr 10 mg/kg TGANAK 16(1),45,82
orl-rat LD50:685 mg/kg JDGRAX 12(1-2),155,80
orl-mus LD50:600 mg/kg PCOC** -,431,66
ipr-mus LD50:500 mg/kg GUCHAZ 6,233,73
scu-mus LD50:800 mg/kg GUCHAZ 6,233,73
orl-dog LD50:1000 mg/kg 28ZEAL 5,84,76
orl-mky LD50:1000 mg/kg 28ZEAL 5,84,76
orl-rbt LD50:1500 mg/kg 28ZEAL 5,84,76

SAFETY PROFILE: Moderately toxic by ingestion, intraperitoneal, and subcutaneous routes. Mutation data reported. A pesticide. When heated to decomposition it emits toxic fumes of NO_x.

DRP875 **HR: 3**
N,N-DIMETHYL-2-(p-(1,2-DIPHENYL-1-BUTENYL)PHENOXY)ETHYLAMINE CITRATE
mf: $C_{26}H_{29}NO{\cdot}C_6H_8O_7$ mw: 563.70

TOXICITY DATA WITH REFERENCE
orl-rat LD50:1550 mg/kg IYKEDH 12,933,81
ipr-rat LD50:660 mg/kg IYKEDH 12,933,81
ivn-rat LD50:76 mg/kg IYKEDH 12,933,81
orl-mus LD50:6500 mg/kg IYKEDH 12,933,81
ipr-mus LD50:218 mg/kg IYKEDH 12,933,81
ivn-mus LD50:95 mg/kg IYKEDH 12,933,81

SAFETY PROFILE: Poison by intravenous and intraperitoneal routes. Moderately toxic by ingestion. When heated to decomposition it emits toxic fumes of NO_x. See also AMINES.

DRQ000 **CAS:13865-57-1** **HR: 2**
N,N-DIMETHYL-4-(DIPHENYLMETHYL)ANILINE
mf: $C_{21}H_{21}N$ mw: 287.43

SYNS: 4-DIMETHYLAMINOTRIPHENYLMETHAN (GERMAN) □ 4-DIMETHYLAMINOTRIPHENYLMETHANE

TOXICITY DATA WITH REFERENCE
scu-rat TDLo:1620 mg/kg/12W-I:ETA NATWAY 42,215,55

SAFETY PROFILE: Questionable carcinogen with experimental tumorigenic data. When heated to decomposition it emits toxic fumes of NO_x.

DRQ200 **CAS:997-95-5** **HR: 2**
2,2′-DIMETHYLDIPROPYLINITROSOAMINE
mf: $C_8H_{18}N_2O$ mw: 158.28

SYNS: DI-ISO-BUTYLNITROSAMINE □ DMDPN □ NITROSODIISOBUTYLAMINE □ N-NITROSODIISOBUTYLAMINE □ N-NITROSODI-ISO-BUTYLAMINE □ N-NITROSO-2,2′-DIMETHYLDI-n-PROPYLAMINE

TOXICITY DATA WITH REFERENCE
mma-sat 25 µg/plate TCMUE9 1,13,84
orl-rat TDLo:1750 mg/kg/30W-I:ETA JJIND8 62,407,79
scu-ham TDLo:3063 mg/kg/49W-I:ETA JNCIAM 55,1209,75
orl-rat TD:11 g/kg/50W-C:ETA CALEDQ 14,297,81
scu-ham LD50:5600 mg/kg JNCIAM 55,1209,75

SAFETY PROFILE: Mildly toxic by subcutaneous route. Questionable carcinogen with experimental neoplastigenic and tumorigenic data. Mutation data reported. Many nitrosamine compounds are carcinogens. When heated to decomposition it emits toxic fumes of NO_x. See also NITROSAMINES.

DRQ400 **CAS:624-92-0** **HR: 3**
DIMETHYLDISULFIDE
DOT: UN 2381
mf: $C_2H_6S_2$ mw: 94.20

PROP: A liquid. Flash p: 44.6°F, bp: 109.7°, d: 1.057 @ 16°/4°, vap press: 28.6 mm @ 25°, vap d: 3.24.

TOXICITY DATA WITH REFERENCE
ihl-rat LC50:15,850 µg/m³/2H GTPZAB 16(6),46,72
ihl-mus LC50:12,300 µg/m³/2H GTPZAB 16(6),46,72

CONSENSUS REPORTS: Reported in EPA TSCA Inventory. EPA Extremely Hazardous Substances List.

DOT CLASSIFICATION: 3; *Label:* Flammable Liquid

SAFETY PROFILE: Poison by inhalation. A very dangerous fire hazard when exposed to heat, flame, or oxidizers. Can react vigorously with oxidizing materials. See also SULFIDES.

DRQ600 CAS:598-64-1 HR: 2
DIMETHYLDITHIOCARBAMIC ACID with DIMETHYLAMINE (1:1)
mf: $C_5H_{12}N_2S_2$ mw: 164.31

SYNS: DIMETHYLDITHIOCARBAMIC ACID DIMETHYL AMINE SALT ☐ DIMETHYLDITHIOCARBAMIC ACID DIMETHYLAMMONIUM SALT

TOXICITY DATA WITH REFERENCE
orl-mus TDLo:29 g/kg/78W-I:ETA NTIS** PB223-159
scu-mus TDLo:464 mg/kg:ETA NTIS** PB223-159

CONSENSUS REPORTS: Reported in EPA TSCA Inventory.

SAFETY PROFILE: Questionable carcinogen with experimental tumorigenic data. When heated to decomposition it emits very toxic fumes of NO_x, NH_3, and SO_x. See also CARBAMATES.

DRQ650 CAS:51-82-1 HR: 3
N,N-DIMETHYLDITHIOCARBAMIC ACID DIMETHYLAMINOMETHYL ESTER
mf: $C_6H_{14}N_2S_2$ mw: 178.34

SYNS: CARBAMIC ACID, DITHIO-, N,N-DIMETHYL-, DIMETHYLAMINOMETHYL ESTER ☐ N,N-DIMETHYL-DITHIOCARBAMINSAEURE-DIMETHYLAMINOMETHYL-ESTER

TOXICITY DATA WITH REFERENCE
ipr-rat LD50:230 mg/kg ARZNAD 16,734,66
ipr-mus LD50:410 mg/kg ARZNAD 16,734,66

CONSENSUS REPORTS: Reported in EPA TSCA Inventory.

SAFETY PROFILE: Poison by intraperitoneal route. When heated to decomposition it emits toxic vapors of NO_x and SO_x.

DRR000 CAS:26419-73-8 HR: 3
2,4-DIMETHYL-1,3-DITHIOLANE-2-CARBOXALDEHYDE O-(METHYLCARBAMOYL)OXIME
mf: $C_8H_{14}N_2O_2S_2$ mw: 234.36

SYNS: 2,4-DIMETHYL-1,3-DITHIOLANE-2-CARBOXALDEHYDE O-((METHYLAMINO)CARBONYL)OXIME ☐ 2,4-DIMETHYL-2-FORMYL-1,3-DITHIOLANE OXIME METHYLCARBAMATE ☐ ENT 27,696 ☐ MBR 6168 ☐ 3M MBR 6168 ☐ TIRPATE

TOXICITY DATA WITH REFERENCE
orl-rat LD50:1 mg/kg WRPCA2 9,119,70
skn-rat LD50:300 mg/kg GUCHAZ 6,213,73

CONSENSUS REPORTS: EPA Extremely Hazardous Substances List.

SAFETY PROFILE: Poison by ingestion and skin contact. A pesticide. When heated to decomposition it emits very toxic fumes of NO_x and SO_x. See also CARBAMATES and ALDEHYDES.

DRR200 CAS:2540-82-1 HR: 3
O,O-DIMETHYL DITHIOPHOSPHORYLACETIC ACID-N-METHYL-N-FORMYLAMIDE
mf: $C_6H_{12}NO_4PS_2$ mw: 257.28

PROP: Yellow viscous oil or crystal mass. D: 1.361 @ 20°/4°, mp: 25–26°. Sltly sol in H_2O; misc in most org solvs.

SYNS: AFLIX ☐ ANTHIO ☐ ANTIO ☐ CP 53926 ☐ O,O-DIMETHYL-S-(N-FORMYL-N-METHYLCARBAMOYLMETHYL) PHOSPHORODITHIOATE ☐ O,O-DIMETHYL-S-(3-METHYL-2,4-DIOXO-3-AZA-BUTYL)-DITHIOFOSFAAT (DUTCH) ☐ O,O-DIMETHYL-S-(3-METHYL-2,4-DIOXO-3-AZA-BUTYL)-DITHIOPHOSPHAT (GERMAN) ☐ O,O-DIMETHYL-S-(N-METHYL-N-FORMYL-CARBAMOYLMETHYL)-DITHIOPHOSPHAT ☐ O,O-DIMETHYL-S-(N-METHYL-N-FORMYLCARBAMOYLMETHYL)PHOSPHORODITHIOATE ☐ O,O-DIMETHYL PHOSPHORODITHIOATE N-FORMYL-2-MERCAPTO-N-METHYLACETAMIDE-S-ESTER ☐ O,O-DIMETIL-S-(N-FORMIL-N-METIL-CARBAMOIL-METIL)-DITIOFOSFATO (ITALIAN) ☐ ENT 27,257 ☐ FORMOTHION ☐ S-(2-(FORMYLMETHYLAMINO)-2-OXOETHYL)-O,O-DIMETHYLPHOSPHORODITHIOATE ☐ N-FORMYL-N-METHYLCARBAMOYLMETHYL-O,O-DIMETHYL PHOSPHORODITHIOATE ☐ S-(N-FORMYL-N-METHYLCARBAMOYLMETHYL)-O,O-DIMETHYL PHOSPHORODITHIOATE ☐ S-(N-FORMYL-N-METHYLCARBAMOYLMETHYL) DIMETHYL PHOSPHOROTHIOLOTHIONATE ☐ S 6900 ☐ SAN 244 I ☐ SAN 6913 I ☐ SAN 7107 I ☐ SPENCER S-6900 ☐ VEL 4284

TOXICITY DATA WITH REFERENCE
mmo-sat 5 mg/plate MUREAV 116,185,83
mma-sat 5 mg/plate MUREAV 116,185,83
orl-rat LD50:250 mg/kg IRGGAJ 21,92,64
skn-rat LD50:353 mg/kg BJIMAG 26,59,69
ivn-rat LD50:35 mg/kg IRGGAJ 22,246,66
orl-mus LD50:190 mg/kg SPEADM 78-1,31,78
ihl-mus LC50:27 mg/m³ GISAAA 40(4),110,75
orl-cat LD50:210 mg/kg 85DPAN -,-,71/76
orl-rbt LD50:420 mg/kg SPEADM 78-1,31,78

CONSENSUS REPORTS: EPA Extremely Hazardous Substances List.

SAFETY PROFILE: Poison by ingestion, inhalation, skin contact, and intravenous routes. Mutation data reported. When heated to decomposition it emits very toxic fumes of NO_x, PO_x, and SO_x. See also ESTERS.

DRR400 CAS:2597-03-7 HR: 3
(O,O-DIMETHYLDITHIOPHOSPHORYLPHENYL)ACETIC ACID ETHYL ESTER
mf: $C_{12}H_{17}O_4PS_2$ mw: 320.38

SYNS: AIMSAN ☐ BAY 33051 ☐ BAYER 18510 ☐ CIDEMUL ☐ CIDIAL ☐ DIMEPHENTHIOATE ☐ DIMEPHENTHOATE ☐ O,O-DIMETHYL-S-(1-CARBOETHOXYBENZYL) DITHIOPHOSPHATE ☐ O,O-DIMETHYL-S-α-ETHOXY-CARBONYLBENZYL PHOSPHORODITHIOATE ☐ O,O-DIMETHYL-S-(PHENYLACETIC ACID ETHYL ESTER) PHOSPHORODITHIOATE ☐ O,O-DIMETHYL-S-(PHENYL)(CARBOETHOXY)METHYL PHOSPHORODITHIOATE ☐ (DIMETHYL-S-(PHENYLETHOXYCARBONYLMETHYL)PHOSPHOROTHIOLOTHIONATE) ☐ ELSAN ☐ ENT 23,438 ☐ ENT 27,386GC ☐ S-α-ETHOXYCARBONYLBENZYL-O,O-DIMETHYL PHOSPHORODITHIOATE ☐ S-α-ETHOXYCARBONYLBENZYL DIMETHYL PHOSPHOROTHIOLOTHIONATE ☐ ETHYL-α-((DIMETHOXYPHOSPHENOTHIOYL)THIO)BENZENEACETATE ☐ ETHYL-O,O-DIMETHYL PHOSPHORODITHIOYLPHENYL ACETATE ☐ ETHYL ESTER of O,O-DIMETHYLDITHIOPHOSPHORYL α-PHENYL ACETATE ACID ☐ ETHYL MERCAPTOPHENYLACETATE-O,O-DIMETHYL PHOSPHOROCITHIOATE ☐ FENTHOATE ☐ L-561 ☐

MONTECATINI L-561 □ NSC 190978 □ OMS 1075 □ PAP □ PAPTHION □ PHENDAL □ PHENTHOATE □ ROGODIAL □ S 2940 □ TANONE □ TH 346-1 □ TSIDIAL

TOXICITY DATA WITH REFERENCE
orl-rat LD50:200 mg/kg WRPCA2 9,119,70
skn-rat LD50:700 mg/kg WRPCA2 9,119,70
orl-mus LD50:150 mg/kg GUCHAZ 6,207,73
skn-mus LD50:2620 mg/kg GUCHAZ 6,217,73
orl-dog LD50:500 mg/kg SPEADM 78-1,40,78
orl-rbt LD50:72 mg/kg GUCHAZ 6,207,73

SAFETY PROFILE: Poison by ingestion. Moderately toxic by skin contact. An insecticide used for control of crop pests and mosquitoes. When heated to decomposition it emits very toxic fumes of PO_x and SO_x. See also ESTERS.

DRR500 CAS:18539-34-9 HR: 3
N,N-DIMETHYL-2-(DI-2,6-XYLYLMETHOXY)ETHYLAMINE HYDROCHLORIDE
mf: $C_{21}H_{29}NO \cdot ClH$ mw: 347.97

SYNS: BS 5933 □ β-DIMETHYLAMINOETHYL-2,6,2′,6′-TETRAMETHYLBENZHYDRYL ETHER HYDROCHLORIDE

TOXICITY DATA WITH REFERENCE
orl-mus LD50:250 mg/kg AIPTAK 135,442,62
ipr-mus LD50:80 mg/kg AIPTAK 135,442,62
scu-mus LD50:140 mg/kg AIPTAK 135,442,62
ivn-mus LD50:35 mg/kg AIPTAK 135,442,62
ivn-cat LD50:15 mg/kg AIPTAK 135,442,62
orl-gpg LD50:100 mg/kg AIPTAK 135,442,62

SAFETY PROFILE: Poison by ingestion, subcutaneous, intravenous, and intraperitoneal routes. When heated to decomposition it emits toxic fumes of NO_x and HCl. See also AMINES.

DRR700 CAS:26651-96-7 HR: 1
2,6-DIMETHYLDODECA-2,6,8-TRIEN-10-ONE
mf: $C_{14}H_{22}O$ mw: 206.36

SYNS: 7,11-DIMETHYL-4,6,10-DODECATRIEN-3-ONE □ 4,6,10-DODECATRIEN-3-ONE, 7,11-DIMETHYL- □ PSEUDOMETHYLIONONE

TOXICITY DATA WITH REFERENCE
orl-rat LD50:>5 g/kg FCTOD7 26,305,88
skn-rbt LDLo:5 g/kg FCTOD7 26,305,88

CONSENSUS REPORTS: Reported in EPA TSCA Inventory.

SAFETY PROFILE: Low toxicity by ingestion and skin contact. When heated to decomposition it emits acrid smoke and irritating vapors.

DRR800 CAS:112-18-5 HR: 2
N,N-DIMETHYLDODECYLAMINE
mf: $C_{14}H_{31}N$ mw: 213.46

SYNS: ADMA 2 □ ARMEEN DM-12D □ BARLENE 125 □ DDA □ N,N-DIMETHYL-1-DODECANAMINE □ N,N-DIMETHYLLAURYLAMINE □ DODECYLDIMETHYLAMINE □ N-DODECYLDIMETHYLAMINE □ LAURYLDIMETHYLAMINE □ N-LAURYLDIMETHYLAMINE □ MONOLAURYL DIMETHYLAMINE □ RC 5629

TOXICITY DATA WITH REFERENCE
skn-rbt 500 mg/24H SEV 28ZPAK -,63,72
eye-rbt 50 μg/24H SEV 28ZPAK -,63,72
orl-rat LD50:740 mg/kg CMEP** -,1,56

CONSENSUS REPORTS: Reported in EPA TSCA Inventory.

SAFETY PROFILE: Moderately toxic by ingestion. A severe skin and eye irritant. When heated to decomposition it emits toxic fumes of NO_x.

DRS000 CAS:1920-05-4 HR: 2
DIMETHYLDODECYLAMINE ACETATE
mf: $C_{14}H_{31}N \cdot C_2H_4O_2$ mw: 273.52

SYNS: N,N-DIMETHYLDODECYLAMINE ACETATE □ PENAR □ TRIPENAR

TOXICITY DATA WITH REFERENCE
orl-rat LD50:800 mg/kg 28ZEAL 4,186,69

CONSENSUS REPORTS: Reported in EPA TSCA Inventory.

SAFETY PROFILE: Moderately toxic by ingestion. When heated to decomposition it emits toxic fumes of NO_x.

DRS200 CAS:1643-20-5 HR: 2
DIMETHYLDODECYLAMINE-N-OXIDE
mf: $C_{14}H_{31}NO$ mw: 229.46

PROP: Very hygroscopic needles from dry toluene. Mp: 130–131°.

SYNS: AMMONYX LO □ AMONYX AO □ AROMOX DMMC-W □ CONCO XAL □ DDNO □ N,N-DIMETHYLDODECYLAMINE OXIDE □ N,N-DIMETHYL-DODECYLAMINOXID (CZECH) □ DODECYLDIMETHYLAMINE OXIDE □ N-DODECYLDIMETHYLAMINE OXIDE □ LAURYLDIMETHYLAMINE OXIDE □ NCI-C55129

TOXICITY DATA WITH REFERENCE
skn-rbt 500 mg/24H SEV 28ZPAK -,76,72
eye-rbt 50 μg/24H SEV 28ZPAK -,76,72

CONSENSUS REPORTS: Reported in EPA TSCA Inventory.

SAFETY PROFILE: A severe skin and eye irritant. When heated to decomposition it emits toxic fumes of NO_x.

DRS400 CAS:41892-01-7 HR: 3
N,N-DIMETHYL-n-DODECYL(2-HYDROXY-3-CHLOROPROPYL)AMMONIUM CHLORIDE
mf: $C_{17}H_{37}ClNO \cdot Cl$ mw: 342.45

TOXICITY DATA WITH REFERENCE
orl-rat LD50:1070 mg/kg TXAPA9 28,313,74
skn-rbt LD50:200 mg/kg TXAPA9 28,313,74

SAFETY PROFILE: Poison by skin contact. Moderately toxic by ingestion. When heated to decomposition it emits very toxic fumes of Cl^-, NH_3, and NO_x.

DRS600 CAS:38094-02-9 HR: 3
N,N-DIMETHYL-n-DODECYL(3-HYDROXYPROPENYL) AMMONIUM CHLORIDE
mf: $C_{17}H_{35}NO \cdot Cl$ mw: 304.98

TOXICITY DATA WITH REFERENCE
orl-rat LD50:1070 mg/kg TXAPA9 28,313,74
skn-rbt LD50:89 mg/kg TXAPA9 28,313,74

SAFETY PROFILE: Poison by skin contact. Moderately toxic by ingestion. When heated to decomposition it emits very toxic fumes of NO_x, NH_3, and Cl^-.

DRS800 CAS:120-08-1 HR: 3
6,7-DIMETHYLESCULETIN
mf: $C_{11}H_{10}O_4$ mw: 206.21

PROP: Needles from H_2O. Mp: 144°.

SYNS: AESCULETIN DIMETHYL ETHER □ 6,7-DIMETHOXYBENZOPYRAN-2-ONE □ 6,7-DIMETHOXYCOUMARIN □ ESCOPARONE □ ESCULETIN DIMETHYL ETHER □ SCOPARON □ SCOPARONE

TOXICITY DATA WITH REFERENCE
orl-rat TDLo:1180 mg/kg (8W male/2W pre-3W post):REP IJEBA6 17,740,79
orl-rat TDLo:725 mg/kg (15-22D preg/21D post):REP IJEBA6 17,740,79
orl-rat LD50:292 mg/kg DRFUD4 3,550,78
ipr-rat LD50:190 mg/kg DRFUD4 3,550,78
orl-mus LD50:280 mg/kg DRFUD4 3,550,78
ipr-mus LD50:180 mg/kg IJMRAQ 60,763,72

SAFETY PROFILE: Poison by ingestion and intraperitoneal routes. Experimental reproductive effects. An antihypertensive agent. When heated to decomposition it emits acrid smoke and irritating fumes.

DRT000 HR: 3
7,14-DIMETHYL-7,14-ETHANODIBENZ(a,b)ANTHRACENE-15,16-DICARBOXYLIC ACID
mf: $C_{24}H_{20}O_4$ mw: 372.44

SYN: 7,12-DIMETHYLBENZANTHRACENE-7,12-endo-α,β-SUCCINIC ACID

TOXICITY DATA WITH REFERENCE
scu-rat TDLo:600 mg/kg/50D-I:CAR,REP 85DLAB -,-,75
ipr-rat LDLo:297 mg/kg 85DLAB -,-,75
ipr-mus LDLo:247 mg/kg 85DLAB -,-,75

SAFETY PROFILE: Poison by intraperitoneal route. Experimental reproductive effects. Questionable carcinogen with experimental carcinogenic data. When heated to decomposition it emits acrid smoke and irritating fumes.

DRT089 HR: 3
(DIMETHYL ETHER)OXODIPEROXO CHROMIUM(VI)
mf: $C_2H_6CrO_6$ mw: 178.06

CONSENSUS REPORTS: Chromium and its compounds are on the Community Right-To-Know List.

SAFETY PROFILE: The solid material explodes violently above −30°C. See also CHROMIUM COMPOUNDS; PEROXIDES and ETHERS.

DRT200 CAS:79-64-1 HR: 1
6-α,21-DIMETHYLETHISTERONE
mf: $C_{23}H_{32}O_2$ mw: 340.55

PROP: Crystals. Mp: 102°.

SYNS: DIMETHESTERONE □ DIMETHISTERON □ DIMETHISTERONE □ 6-α,21-DIMETHYL-17-β-HYDROXY-17-α-PREG-4-EN-20-YN-3-ONE □ 6-α,21-DIMETHYL-17-β-HYDROXY-17-α-PREGN-4-EN-20-YN-3-ONE □ 17-α-ETHYNYL-6-α,21-DIMETHYLTESTOSTERONE □ 17-α-ETHYNYL-17-HYDROXY-6-α,21-DIMETHYLANDROST-4-EN-3-ONE □ (6-α,17-β)-17-HYDROXY-6-METHYL-17-(1-PROPYNYL)-ANDROST-4-EN-3-ONE □ 17-β-HYDROXY-6-α-METHYL-17-(1-PROPYNYL)ANDROST-4-EN-3-ONE □ LUTOGAN □ LUTOSAN □ 6-α-METHYL-17-α-PROPYNYLTESTOSTERONE □ 6-α-METHYL-17-(1-PROPYNYL)TESTOSTERONE □ P-5048 □ SECROSTERON

TOXICITY DATA WITH REFERENCE
scu-rat TDLo:70 mg/kg (female 14D pre):REP CCPTAY 5,57,72
scu-rat TDLo:28 mg/kg (female 14D pre):REP CCPTAY 5,57,72
orl-rat TDLo:100 mg/kg (17-20D preg):TER ECJPAE 24,77,77
orl-mus LD50:7650 mg/kg MEIEDD 10,469,83

CONSENSUS REPORTS: IARC Cancer Review: Animal Inadequate Evidence IMEMDT 21,377,79.

SAFETY PROFILE: Mildly toxic by ingestion. Questionable carcinogen. Experimental teratogenic and reproductive effects. A steroid used as a progestin and in the treatment of menstrual disorders. When heated to decomposition it emits acrid smoke and irritating fumes.

DRT400 CAS:67262-78-6 HR: 3
2′,6′-DIMETHYL-2-(2-ETHOXYETHYLAMINO)ACETANILIDE
mf: $C_{14}H_{22}N_2O_2$ mw: 250.38

SYN: 2-(2-ETHOXYETHYLAMINO)-2′,6′-ACETOXYLIDIDE

TOXICITY DATA WITH REFERENCE
ipr-mus LD50:130 mg/kg JPMSAE 67,595,78
ivn-mus LD50:35 mg/kg JPMSAE 67,595,78

SAFETY PROFILE: Poison by intraperitoneal and intravenous routes. When heated to decomposition it emits toxic fumes of NO_x.

DRT600 CAS:102207-86-3 HR: 3
2′,6′-DIMETHYL-2-(2-ETHOXYETHYLAMINO)ACETANILIDE HYDROCHLORIDE
mf: $C_{14}H_{22}N_2O_2 \cdot ClH$ mw: 286.84

SYN: 2-(2-ETHOXYETHYLAMINO)-2′,6′-ACETOXYLIDIDE HYDROCHLORIDE

TOXICITY DATA WITH REFERENCE
ipr-mus LD50:150 mg/kg JPMSAE 67,595,78
ivn-mus LD50:35 mg/kg JPMSAE 67,595,78

SAFETY PROFILE: Poison by intraperitoneal and intravenous routes. When heated to decomposition it emits very toxic fumes of NO_x and HCl.

D

DRU000 **CAS:3837-54-5** **HR: 2**
N,N-DIMETHYL-p-((3-ETHOXYPHENYL)AZO)ANILINE
mf: $C_{16}H_{19}N_3O$ mw: 269.38

SYN: 3′-ETHOXY-4-DIMETHYLAMINOAZOBENZENE

TOXICITY DATA WITH REFERENCE
orl-rat TDLo:9202 mg/kg/30W-C:NEO JEMEAV 87,139,48

SAFETY PROFILE: Questionable carcinogen with experimental neoplastigenic data. When heated to decomposition it emits toxic fumes of NO_x.

DRU200 **CAS:1825-58-7** **HR: 2**
DIMETHYLETHOXYPHENYLSILANE
mf: $C_{10}H_{16}OSi$ mw: 180.35

SYN: DIMETHYL-FENYL-ETHOXYSILAN (CZECH)

TOXICITY DATA WITH REFERENCE
skn-rbt 500 mg/24H MOD 28ZPAK -,221,72
eye-rbt 500 mg/24H MLD 28ZPAK -,221,72
orl-rat LD50:2460 mg/kg 28ZPAK -,221,72

CONSENSUS REPORTS: Reported in EPA TSCA Inventory.

SAFETY PROFILE: Moderately toxic by ingestion. A skin and eye irritant. When heated to decomposition it emits acrid smoke and irritating fumes.

DRU400 **CAS:2669-32-1** **HR: 3**
O,O-DIMETHYL-S-(5-ETHOXY-1,3,4-THIADIAZOLINYL-3-METHYL)DITHIOPHOSPHATE
mf: $C_7H_{13}N_2O_4PS_3$ mw: 316.37

SYNS: O,O-DIMETHYL-S-(5-ETHOXY-1,3,4-THIADIAZOL-2(3H)-ONYL-(3)-METHYL)DITHIOPHOSPHATE □ O,O-DIMETHYL-S-(5-ETHOXY-1,3,4-THIADIAZOL-2(3H)-ONYL-(3)-METHYL)PHOSPHORODITHIOATE □ ENT 27,238 □ GEIGY 12968 □ LYTHIDATHION □ NC-2962

TOXICITY DATA WITH REFERENCE
orl-rat LD50:268 mg/kg 28ZEAL 4,186,69

SAFETY PROFILE: Poison by ingestion. A pesticide. When heated to decomposition it emits very toxic fumes of NO_x, PO_x, and SO_x.

DRU600 **CAS:63021-00-1** **HR: 2**
DIMETHYL ETHYL ALLENOLIC ACID METHYL ETHER
mf: $C_{16}H_{18}O_3$ mw: 258.34

SYNS: ACIDE DIMETHYL-ETHYL-ALLENOLIQUE ETHER METHYLIQUE (FRENCH) □ α,α-DIMETHYL-2-(6-METHOXYNAPHTHYL)PROPIONIC ACID

TOXICITY DATA WITH REFERENCE
orl-mus TDLo:139 mg/kg/24W-I:ETA CRSBAW 146,916,52

SAFETY PROFILE: Questionable carcinogen with experimental tumorigenic data. When heated to decomposition it emits acrid smoke and irritating fumes. See also ETHERS.

DRV000 **CAS:529-05-5** **HR: 2**
1,4-DIMETHYL-7-ETHYLAZULENE
mf: $C_{14}H_{16}$ mw: 184.30

PROP: Blue oil. Bp: 161° @ 12 mm.

SYNS: BA 2784 □ CAMUZULENE □ CHAMAZULEN □ CHAMAZULENE □ DIMETHULENE □ DIMETHWLEN □ 7-ETHYL-1,4-DIMETHYLAZULENE □ KAMILLENOEL (GERMAN)

TOXICITY DATA WITH REFERENCE
orl-rat LD50:10 g/kg ARZNAD 19,615,69
ims-mus LD50:3 g/kg MEIEDD 10,283,83

SAFETY PROFILE: Moderately toxic by intramuscular route. Mildly toxic by ingestion. An anti-inflammatory and antipyretic agent. When heated to decomposition it emits acrid smoke and irritating fumes.

DRV200 **CAS:1420-07-1** **HR: 3**
2-(1,1-DIMETHYLETHYL)-4,6-DINITROPHENOL
mf: $C_{10}H_{12}N_2O_5$ mw: 240.24

PROP: Yellow solid. Mp: 125.5–126.5°.

SYNS: o-tert-BUTYL-4,6-DINITROPHENOL □ 2,4-DINITRO-6-tert-BUTYLPHENOL □ DINOTERB □ DNTBP □ HERBOGIL

TOXICITY DATA WITH REFERENCE
orl-rat LD50:62 mg/kg FMCHA2 -,C84,83
orl-mus LD50:25 mg/kg 28ZEAL 5,82,76
skn-gpg LD50:150 mg/kg 28ZEAL 5,82,76

CONSENSUS REPORTS: EPA Extremely Hazardous Substances List.

SAFETY PROFILE: Poison by ingestion and skin contact. A pesticide. When heated to decomposition it emits toxic fumes of NO_x. See also NITRO COMPOUNDS of AROMATIC HYDROCARBONS.

DRV500 **HR: 3**
N,N-DIMETHYL-N′-ETHYL-N′-1-NAPHTHYLETHYLENEDIAMINE
mf: $C_{16}H_{22}N_2$ mw: 242.40

TOXICITY DATA WITH REFERENCE
ipr-rat LDLo:80 mg/kg BJPCAL 11,1,56
ipr-mus LD50:121 mg/kg BJPCAL 11,1,56
scu-mus LD50:443 mg/kg BJPCAL 11,1,56

SAFETY PROFILE: Poison by intraperitoneal route. Moderately toxic by subcutaneous route. When heated to decomposition it emits toxic fumes of NO_x. See also AMINES.

DRV550 **HR: 3**
N,N-DIMETHYL-N′-ETHYL-N′-2-NAPHTHYLETHYLENEDIAMINE
mf: $C_{16}H_{22}N_2$ mw: 242.40

TOXICITY DATA WITH REFERENCE
ipr-rat LDLo:50 mg/kg BJPCAL 11,1,56
ipr-mus LD50:53 mg/kg BJPCAL 11,1,56
scu-mus LD50:287 mg/kg BJPCAL 11,1,56

SAFETY PROFILE: Poison by subcutaneous and intraperitoneal routes. When heated to decomposition it emits toxic fumes of NO_x. See also AMINES.

DRV600 CAS:50285-71-7 HR: 2
1,1-DIMETHYL-3-ETHYL-3-NITROSOUREA
mf: $C_5H_{11}N_3O_2$ mw: 145.19

SYNS: NITROSOAETHYLDIMETHYLHARNSTOFF □ NITROSO-1,1-DIMETHYL-3-ETHYLUREA □ NITROSOETHYLDIMETHYLUREA □ 1-NITROSO-1-ETHYL-3,3-DIMETHYLUREA

TOXICITY DATA WITH REFERENCE
mma-sat 250 μg/plate JJIND8 67,1117,81
orl-rat TDLo:1230 mg/kg/50W-I:ETA ZKKOBW 83,315,75

SAFETY PROFILE: Questionable carcinogen with experimental tumorigenic data. Mutation data reported. Many N-nitroso compounds are carcinogens. When heated to decomposition it emits toxic fumes of NO_x. See also N-NITROSO COMPOUNDS.

DRV850 CAS:27692-91-7 HR: 3
N,N-DIMETHYL-N′-ETHYL-N′-PHENYLETHYLENEDIAMINE
mf: $C_{19}H_{20}N_2$ mw: 192.34

SYN: 2325 RP

TOXICITY DATA WITH REFERENCE
ipr-rat LDLo:350 mg/kg BJPCAL 11,1,56
ipr-mus LD50:500 mg/kg BJPCAL 11,1,56
scu-mus LD50:1150 mg/kg BJPCAL 11,1,56

SAFETY PROFILE: Poison by intraperitoneal route. Moderately toxic by subcutaneous route. When heated to decomposition it emits toxic fumes of NO_x. See also AMINES.

DRW000 CAS:66967-65-5 HR: 3
DIMETHYLETHYL(3-(10H-PYRIDO(3,2-b))(1,4)BENZOTHIAZIN 10-YL)PROPYLAMMONIUM ETHYL SULFATE
mf: $C_{18}H_{24}N_3S•C_2H_5O_4S$ mw: 439.64

SYN: D 268

TOXICITY DATA WITH REFERENCE
orl-mus LD50:494 mg/kg ARZNAD 8,489,58
ipr-mus LD50:58 mg/kg ARZNAD 8,489,58

SAFETY PROFILE: Poison by intraperitoneal route. Moderately toxic by ingestion. When heated to decomposition it emits very toxic fumes of NO_x, NH_3, and SO_x.

DRX400 CAS:5581-40-8 HR: 3
DIMETHYL FANDANE
mf: $C_{17}H_{19}N$ mw: 237.37

SYNS: 2,3-DIHYDRO-N,N-DIMETHYL-3-PHENYL-1H-INDEN-1-AMINE □ DIMEFADANE □ N,N-DIMETHYL-3-PHENYL-1-INDANAMINE □ SK+F 1340

TOXICITY DATA WITH REFERENCE
orl-rat LD50:176 mg/kg TXAPA9 21,315,72
orl-bwd LD50:75 mg/kg TXAPA9 21,315,72

SAFETY PROFILE: Poison by ingestion. When heated to decomposition it emits toxic fumes of NO_x.

DRX600 CAS:23339-04-0 HR: 2
2,3-DIMETHYLFLUORANTHENE
mf: $C_{18}H_{14}$ mw: 230.32

TOXICITY DATA WITH REFERENCE
skn-mus TDLo:40 mg/kg/20D:NEO JNCIAM 49,1165,72

CONSENSUS REPORTS: IARC Cancer Review: Animal No Evidence IMEMDT 32,355,83.

SAFETY PROFILE: Questionable carcinogen with experimental neoplastigenic data. When heated to decomposition it emits acrid smoke and irritating fumes.

DRX800 CAS:38048-87-2 HR: 2
7,8-DIMETHYLFLUORANTHENE
mf: $C_{18}H_{14}$ mw: 230.32

TOXICITY DATA WITH REFERENCE
skn-mus TDLo:40 mg/kg/20D:NEO JNCIAM 49,1165,72

SAFETY PROFILE: Questionable carcinogen with experimental neoplastigenic data. An initiator. When heated to decomposition it emits acrid smoke and irritating fumes.

DRY000 CAS:25889-63-8 HR: 2
8,9-DIMETHYLFLUORANTHENE
mf: $C_{18}H_{14}$ mw: 230.32

TOXICITY DATA WITH REFERENCE
skn-mus TDLo:40 mg/kg/20D-I:ETA JNCIAM 49,1165,72

SAFETY PROFILE: Questionable carcinogen with experimental tumorigenic data. An initiator. When heated to decomposition it emits acrid smoke and irritating fumes.

DRY100 CAS:17057-98-6 HR: 2
1,9-DIMETHYLFLUORENE
mf: $C_{15}H_{14}$ mw: 194.29

SYN: 9H-FLUORENE, 1,9-DIMETHYL-

TOXICITY DATA WITH REFERENCE
mma-sat 10 μg/plate MUREAV 91,167,81
ipr-mus TDLo:42 mg/kg/3D-I:NEO JTEHD6 21,525,87

SAFETY PROFILE: Questionable carcinogen with experimental neoplastigenic data. Mutation data reported. When heated to decomposition it emits acrid smoke and irritating fumes.

DRY289 CAS:420-23-5 HR: 3
DIMETHYLFLUOROARSINE
mf: C_2H_6AsF mw: 123.99

CONSENSUS REPORTS: Arsenic and its compounds are on the Community Right-To-Know List.

SAFETY PROFILE: Arsenic compounds are poisons by many routes. Ignites spontaneously in air. When heated to decomposition it emits toxic fumes of F^- and As. See also ARSENIC COMPOUNDS.

DRY400 CAS:737-22-4 HR: 2
7,12-DIMETHYL-4-FLUOROBENZ(a)ANTHRACENE
mf: $C_{20}H_{15}F$ mw: 274.35

SYN: 4-FLUORO-7,12-DIMETHYLBENZ(a)ANTHRACENE

TOXICITY DATA WITH REFERENCE
ims-rat TDLo:10 mg/kg:NEO NATUAS 273,566,78

SAFETY PROFILE: Questionable carcinogen with experimental neoplastigenic data. When heated to decomposition it emits toxic fumes of F^-.

DRY600 CAS:794-00-3 HR: 2
7,12-DIMETHYL-5-FLUOROBENZ(a)ANTHRACENE
mf: $C_{20}H_{15}F$ mw: 274.35

SYN: 5-FLUORO-7,12-DIMETHYLBENZ(a)ANTHRACENE

TOXICITY DATA WITH REFERENCE
dni-hmn:hla 70 μmol/L MUREAV 92,427,82
scu-rat TDLo:823 mg/kg/10W-I:ETA JMCMAR 21,1076,78
skn-mus TDLo:110 μg/kg:ETA CNREA8 39,411,79

CONSENSUS REPORTS: EPA Genetic Toxicology Program.

SAFETY PROFILE: Questionable carcinogen with experimental tumorigenic data. Human mutation data reported. An initiator. When heated to decomposition it emits toxic fumes of F^-.

DRY800 CAS:2023-60-1 HR: 2
7,12-DIMETHYL-8-FLUOROBENZ(a)ANTHRACENE
mf: $C_{20}H_{15}F$ mw: 274.35

SYN: 8-FLUORO-7,12-DIMETHYLBENZ(a)ANTHRACENE

TOXICITY DATA WITH REFERENCE
ims-rat TDLo:10 mg/kg:NEO NATUAS 273,566,78

SAFETY PROFILE: Questionable carcinogen with experimental neoplastigenic data. When heated to decomposition it emits toxic fumes of F^-.

DRZ000 CAS:2023-61-2 HR: 2
7,12-DIMETHYL-11-FLUOROBENZ(a)ANTHRACENE
mf: $C_{20}H_{15}F$ mw: 274.35

SYN: 11-FLUORO-7,12-DIMETHYLBENZ(a)ANTHRACENE

TOXICITY DATA WITH REFERENCE
ims-rat TDLo:10 mg/kg:NEO NATUAS 273,566,78
skn-mus TDLo:110 μg/kg:ETA CNREA8 39,411,79

CONSENSUS REPORTS: EPA Genetic Toxicology Program.

SAFETY PROFILE: Questionable carcinogen with experimental neoplastigenic and tumorigenic data. An initiator. When heated to decomposition it emits toxic fumes of F^-.

DSA000 CAS:150-74-3 HR: 2
N,N-DIMETHYL-p-((p-FLUOROPHENYL)AZO)ANILINE
mf: $C_{14}H_{14}FN_3$ mw: 243.31

SYNS: 4-(DIMETHYLAMINO)-4′-FLUOROAZOBENZENE □ 4′-FLUORO-N,N-DIMETHYL-4-AMINOAZOBENZENE □ 4′-FLUORO-p-DIMETHYLAMINOAZOBENZENE □ 4′-FLUORO-4-DIMETHYLAMINOAZOBENZENE □ 4′-FLUORO-N,N-DIMETHYL-p-PHENYLAZOANILINE □ p-((p-FLUOROPHENYL)AZO)-N,N-DIMETHYLANILINE □ 4-((4-FLUOROPHENYL)AZO)-N,N-DIMETHYLBENZENAMINE

TOXICITY DATA WITH REFERENCE
dns-rat-orl 2520 mg/kg/12W-I CNREA8 29,2039,69
ipr-mus TDLo:400 mg/kg (8-9D preg):TER KAIZAN 37,179,62
ipr-mus TDLo:400 mg/kg (8-9D preg):REP KAIZAN 37,179,62
orl-rat TDLo:2720 mg/kg/14W-I:ETA CNREA8 18,469,58
orl-rat LD:3970 mg/kg/19W-C:ETA ARZNAD 12,270,62
orl-rat TD:3200 mg/kg/12W-C:ETA CNREA8 13,93,53
orl-rat TD:3150 mg/kg/13W-C:ETA CNREA8 9,652,49

SAFETY PROFILE: Experimental teratogenic and reproductive effects. Questionable carcinogen with experimental tumorigenic data. Mutation data reported. When heated to decomposition it emits very toxic fumes of F^- and NO_x.

DSA800 CAS:5954-50-7 HR: 3
DIMETHYL FLUOROPHOSPHATE
mf: $C_2H_6FO_3P$ mw: 128.05

PROP: Liquid. Mp: low, bp: 149°, d: 1.28, vap d: 4.42.

SYNS: FLUOPHOSPHORIC ACID, DIMETHYL ESTER □ PHOSPHOROFLUORIDIC ACID, DIMETHYL ESTER □ PF-1 □ T-1035 □ TL 311

TOXICITY DATA WITH REFERENCE
ihl-rat LC50:1800 mg/m³/1M NTIS** PB158-508
ihl-mus LC50:290 mg/m³/10M JIHTAB 30,307,48
skn-mus LD50:36 mg/kg NTIS** PB158-508
ipr-mus LD50:3 mg/kg NTIS** PB158-508
ivn-mus LD50:450 μg/kg NTIS** PB158-508
ihl-dog LC50:6 g/m³/1M NTIS** PB158-508
ivn-dog LD50:1 mg/kg NTIS** PB158-508
ihl-cat LC50:6 g/m³/1M NTIS** PB158-508

SAFETY PROFILE: Poison by inhalation, skin contact, and intravenous routes. When heated to decomposition it emits toxic fumes of F^- and PO_x. See also ESTERS, FLUORIDES, and PHOSPHATES.

DSB000 CAS:68-12-2 HR: 3
DIMETHYLFORMAMIDE
DOT: UN 2265
mf: C_3H_7NO mw: 73.11

$(CH_3)_2NCO{\cdot}H$

PROP: Colorless, mobile liquid; fishy or faint aminr odor. Mp: −61°, bp: 152.8°, lel: 2.2% @ 100°, uel: 15.2% @ 100°, flash p: 136°, d: 0.945 @ 22.4°/4°, autoign temp: 833°F, vap press: 3.7 mm @ 25°, vap d: 2.51. Misc in H_2O, EtOH, Et_2O, C_6H_6, and $CHCl_3$.

SYNS: DIMETHYLFORMAMID (GERMAN) □ N,N-DIMETHYL FORMAMIDE □ N,N-DIMETHYLFORMAMIDE (DOT) □ DIMETILFORMAMIDE (ITALIAN) □ DIMETYLFORMAMIDU (CZECH) □ DMF □ DMFA

□ DWUMETHYLOFORMAMID (POLISH) □ N-FORMYLDIMETHYLAMINE □ NCI-C60913 □ NSC 5356 □ U-4224

TOXICITY DATA WITH REFERENCE
skn-hmn 100%/24H MLD BJIMAG 13,51,56
skn-rbt 10 mg/24H open JIHTAB 30,63,48
eye-rbt 100 mg RNS SEV DCTODJ 9,147,86
mma-sat 600 µg/plate PMRSDJ 1,343,81
cyt-hmn:lym 100 nmol/L CHPUA4 31,548,81
ihl-rat TDLo:600 mg/m³/24H (1-19D preg):REP TPKVAL 13,75,73
ihl-rat TCLo:4 mg/m³/4H (1-19D preg):TER TPKVAL 14,32,75
skn-rat TDLo:3600 mg/kg (11-13D preg):TER TXAPA9 41,35,77
ipr-mus TDLo:2100 mg/kg (female 11D post):REP DEGEA3 30,455,75
ihl-rat TCLo:4 mg/m³/4H (1-19D preg):REP TPKVAL 14,32,75
orl-rbt TDLo:2600 mg/kg (female 6-18D post):TER ARZNAD 30,1557,80
ipr-mus TDLo:15,120 mg/kg (female 1-14D post):TER DEGEA3 30,455,75
orl-rat LD50:2800 mg/kg ZEKBAI 69,103,67
ipr-rat LD50:1400 mg/kg BJIMAG 13,51,56
scu-rat LD50:3800 mg/kg ARZNAD 15,618,65
ivn-rat LD50:2000 mg/kg ZEKBAI 69,103,67
orl-mus LD50:3750 mg/kg TPKVAL 1,54,61
ihl-mus LC50:9400 mg/m³/2H TPKVAL 1,54,61
ipr-mus LD50:650 mg/kg CNCRA6 30,9,63
scu-mus LD50:4500 mg/kg ARZNAD 15,618,65
ivn-mus LD50:2500 mg/kg ARZNAD 15,618,65
ims-mus LD50:3800 mg/kg ARZNAD 15,618,65
ivn-dog LD50:470 mg/kg ARZNAD 15,618,65
ipr-cat LD50:500 mg/kg BJIMAG 13,51,56
skn-rbt LD50:4720 mg/kg AIHAAP 30,470,69

CONSENSUS REPORTS: IARC Cancer Review: Group 2B IMEMDT 47,171,89; Human Limited Evidence IMEMDT 47,171,89; Animal Inadequate Evidence IMEMDT 47,171,89. EPA Genetic Toxicology Program. Reported in EPA TSCA Inventory.

OSHA PEL: TWA 10 ppm (skin)
ACGIH TLV: TWA 10 ppm (skin); BEI: 40 mg(N-methylformamide)/g creatinine at end of shift
DFG MAK: 10 ppm (30 mg/m³)
DOT CLASSIFICATION: 3; *Label:* Flammable Liquid

SAFETY PROFILE: Suspected carcinogen. Moderately toxic by ingestion, intravenous, subcutaneous, intramuscular, and intraperitoneal routes. Mildly toxic by skin contact and inhalation. Experimental teratogenic and reproductive effects. A skin and severe eye irritant. Human mutation data reported. Flammable liquid when exposed to heat or flame; can react with oxidizing materials. Explosion hazard when exposed to flame. Explosive reaction with bromine, potassium permanganate; triethylaluminum + heat. Forms explosive mixtures with lithium azide (shock sensitive above 200°C); uranium perchlorate. Ignition on contact with chromium trioxide. Violent reaction with chlorine; sodium hydroborate + heat; diisocyanatomethane; carbon tetrachloride + iron; 1,2,3,4,5,6-hexachlorocyclohexane + iron. Vigorous exothermic reaction with magnesium nitrate; sodium + heat; sodium hydride + heat; sulfinyl chloride + traces of iron or zinc; 2,4,6-trichloro-1,3,5-triazine (with gas evolution); and many other materials. Avoid contact with halogenated hydrocarbons; inorganic and organic nitrates; (2,5-dimethyl pyrrole + $P(OCl)_3$); C_6Cl_6; methylene diisocyanates; P_2O_3. To fight fire, use foam, CO_2, dry chemical. When heated to decomposition it emits toxic fumes of NO_x.

For occupational chemical analysis use OSHA: #ID-66 or NIOSH: Dimethylformamide, 2004.

D

DSB200 CAS:533-74-4 HR: 3
DIMETHYLFORMOCARBOTHIALDINE
mf: $C_5H_{10}N_2S_2$ mw: 162.29

PROP: Crystals from Me_2CO/hexane. Mp: 106°. Sol in alc.

SYNS: BASAMID □ BASAMID G □ BASAMID-GRANULAR □ BASAMID P □ BASAMID-PUDER □ CARBOTHIALDIN □ CARBOTHIALDINE □ CRAG 974 □ CRAG FUNGICIDE 974 □ CRAG NEMACIDE □ CRAG 85W □ DAZOMET □ 3,5-DIMETHYLPERHYDRO-1,3,5-THIADIAZIN-2-THION (CZECH, GERMAN) □ 3,5-DIMETHYLTETRAHYDRO-1,3,5-THIADIAZINE-2-THIONE □ 3,5-DIMETHYLTETRAHYDRO-1,3,5-2H THIADIAZINE-2-THIONE □ 3,5-DIMETHYL-1,2,3,5-TETRAHYDRO-1,3,5-THIADIAZINETHIONE-2 □ 3,5-DIMETHYLTETRAHYDRO-2H-1,3,5-THIADIAZINE-2-THIONE □ 3,5-DIMETHYL-1,3,5-2H-TETRAHYDROTHIADIAZINE-2-THIONE □ 3,5 DIMETHYL-2-THIONOTETRAHYDRO-1,3,5-THIADIAZINE □ 3,5-DIMETIL-PERIDRO-1,3,5-THIADIAZIN-2-TIONE (ITALIAN) □ DMTT □ FENNOSAN B 100 □ MICOFUME □ MYLON (CZECH) □ MYLONE □ MYLONE 85 □ N 521 □ NALCON 243 □ NEFUSAN □ PREZERVIT □ STAUFFER N 521 □ TETRAHYDRO-2H-3,5-DIMETHYL-1,3,5-THIADIAZINE-2-THIONE □ TETRAHYDRO-3,5-DIMETHYL-2H-1,3,5-THIADIAZINE-2-THIONE □ THIAZON □ THIAZONE □ 2-THIO-3,5-DIMETHYLTETRAHYDRO-1,3,5-THIADIAZINE □ TIAZON □ TROYSAN 142 □ UCC 974

TOXICITY DATA WITH REFERENCE
eye-rbt 500 mg/24H SEV 28ZPAK ,204,72
orl-rat LD50:320 mg/kg TXAPA9 9,521,66
ihl-rat LC50:8400 mg/m³/4H PEMNDP 9,225,91
skn-rat LD50:2260 mg/kg NNGADV 17,S327,92
ipr-rat LD50:87 mg/kg TXAPA9 9,521,66
orl-mus LD50:180 mg/kg TXAPA9 9,521,66
skn-mus LD50:2400 mg/kg NNGADV 17,S327,92
ipr-mus LDLo:50 mg/kg ARZNAD 21,121,71
scu-mus LDLo:500 mg/kg AIPTAK 12,447,04
ipr-dog LD50:47 mg/kg TXAPA9 9,521,66
orl-rbt LD50:120 mg/kg TXAPA9 9,521,66
ipr-rbt LD50:127 mg/kg TXAPA9 9,521,66
orl-gpg LD50:160 mg/kg TXAPA9 9,521,66

CONSENSUS REPORTS: Reported in EPA TSCA Inventory.

SAFETY PROFILE: Poison by ingestion and intraperitoneal routes. Moderately toxic by skin contact and subcutaneous routes. A severe eye irritant. A mild primary skin irritant and sensitizer. When heated to decomposition it emits very toxic fumes of NO_x and SO_x.

DSB400 CAS:2175-91-9 HR: 3
6,6-DIMETHYLFULVENE
mf: C_8H_{10} mw: 106.17

PROP: Bp: 118–119°.

SAFETY PROFILE: Peroxidizes in air to form a heat-sensitive explosive, insoluble peroxide. The peroxide ignites on contact with ether. When heated to decomposition it emits acrid smoke and fumes.

DSB600 **CAS:624-49-7** **HR: 2**
DIMETHYL FUMARATE
mf: $C_6H_8O_4$ mw: 144.14

PROP: Crystals. Mp: 102°, bp: 88.5° @ 12 mm.

SYNS: ALLOMALEIC ACID DIMETHYL ESTER ☐ BOLETIC ACID DIMETHYL ESTER ☐ trans-BUTENEDIOIC ACID DIMETHYL ESTER ☐ trans-1,2-ETHYLENEDICARBOXYLIC ACID DIMETHYL ESTER ☐ FUMARIC ACID, DIMETHYL ESTER

TOXICITY DATA WITH REFERENCE
skn-rbt 20 mg/24H MOD 85JCAE-,373,86
eye-rbt 250 µg/24H SEV 85JCAE-,373,86
orl-rat LD50:2240 mg/kg AIHAAP 30,470,69
skn-rbt LD50:1250 mg/kg AIHAAP 30,470,69

CONSENSUS REPORTS: Reported in EPA TSCA Inventory.

SAFETY PROFILE: Moderately toxic by ingestion and skin contact. A skin and severe eye irritant. When heated to decomposition it emits acrid smoke and irritating fumes. See also ESTERS.

DSB800 **CAS:28802-49-5** **HR: 2**
DIMETHYL FURANE
mf: C_6H_8O mw: 96.14

SYN: DIMETHYL FURAN

TOXICITY DATA WITH REFERENCE
skn-rbt 10 mg/24H open JIHTAB 26,269,44
eye-rbt 20 mg SEV AJOPAA 29,1363,46
orl-rat LD50:300 mg/kg JIDHAN 26,269,44
skn-gpg LD50:1000 mg/kg JIHTAB 26,269,44

SAFETY PROFILE: A poison by ingestion. Moderately toxic by skin contact. A skin and severe eye irritant. When heated to decomposition it emits acrid smoke and irritating fumes.

DSC000 **CAS:625-86-5** **HR: 2**
2,5-DIMETHYL FURANE
mf: C_6H_8O mw: 96.14

PROP: Colorless liquid. Mp: −63°, bp: 94°, flash p: 60.8°F. D: 0.9026 @ 17.7°/4°, vap d: 3.31.

TOXICITY DATA WITH REFERENCE
dnr-bcs 190 µg/disc DFSCDX 13,353,86
cyt-ham:ovr 8 mmol/L CALEDQ 13,89,81
ihl-rat LCLo:500 ppm/4H JIHTAB 31,343,49

CONSENSUS REPORTS: Reported in EPA TSCA Inventory.

SAFETY PROFILE: Moderately toxic by inhalation. Mutation data reported. A very dangerous fire hazard when exposed to heat or flame; can react vigorously with oxidizing materials. Keep away from heat and open flame. To fight fire, use alcohol foam, foam, CO_2, dry chemical. When heated to decomposition it emits acrid smoke and irritating fumes.

DSC100 **CAS:4568-81-4** **HR: 3**
2,5-DIMETHYL-3-FURYL p-HYDROXYPHENYL KETONE
mf: $C_{13}H_{12}O_3$ mw: 216.25

SYNS: DB 135 ☐ DIMETHYL-2,5 (HYDROXY 4 BENZOYL) 3 FURANNE ☐ KETONE, 2,5-DIMETHYL-3-FURYL p-HYDROXYPHENYL

TOXICITY DATA WITH REFERENCE
ipr-mus LD50:605 mg/kg AIPTAK 147,497,64
ivn-gpg LDLo:118 mg/kg AIPTAK 147,497,64

DOT CLASSIFICATION: 3; *Label:* Flammable Liquid

SAFETY PROFILE: A poison by intravenous route. Moderately toxic by intraperitoneal route. A flammable liquid. When heated to decomposition it emits acrid smoke and irritating vapors.

DSC400 **HR: 3**
DIMETHYLGOLD SELENOCYANATE
mf: C_3H_6AuNSe mw: 332.02

$(CH_3)_2AuSeC{\equiv}N$

CONSENSUS REPORTS: Cyanide and its compounds, as well as selenium and its compounds, are on the Community Right-To-Know List.

OSHA PEL: TWA 0.2 mg(Se)/m³
ACGIH TLV: TWA 0.2 mg(Se)/m³
DFG MAK: 0.1 mg(Se)/m³

SAFETY PROFILE: A very shock-sensitive explosive. It explodes when precipitated from aqueous solutions. When heated to decomposition it emits toxic fumes of Se, CN^-, and NO_x. See also SELENIUM COMPOUNDS and CYANIDE.

DSD400 **CAS:926-82-9** **HR: 3**
3,5-DIMETHYLHEPTANE
mf: C_9H_{20} mw: 128.26

$CH_3CH_2CH_3CHCH_2CHCH_3CH_2CH_3$

PROP: Liquid. Bp: 136°, flash p: 73.5°F, d: 0.723 @ 20°, vap press: 9.5 mm @ 25°, vap d: 4.42.

SAFETY PROFILE: No toxicity information. A probable irritant and narcotic in high concentration. A very dangerous fire hazard when exposed to heat or flame; can react vigorously with oxidizing materials. To fight fire, use CO_2, dry chemical. When heated to decomposition it emits acrid smoke and irritating fumes.

DSD600 **HR: 3**
4,4-DIMETHYLHEPTANE
mf: C_9H_{20} mw: 128.26

$(CH_3)_2C(CH_2CH_2CH_3)_2$

PROP: Liquid. Bp: 135.2°, flash p: 69.8°F, d: 0.72 @ 25°/4°, vap press: 10.4 mm @ 25°, vap d: 4.42.

SAFETY PROFILE: No toxicity information. A probable

irritant and narcotic in high concentration. A very dangerous fire hazard when exposed to heat or flame; can react vigorously with oxidizing materials. To fight fire, use CO_2, dry chemical. When heated to decomposition it emits acrid smoke and irritating fumes.

DSD775 CAS:106-72-9 HR: 1
2,6-DIMETHYL-5-HEPTENAL
mf: $C_9H_{16}O$ mw: 140.23

PROP: Pale-yellow liquid or oil; melon odor. D: 0.852–0.858, refr index: 1.443–1.448

SYN: FEMA No. 2497

SAFETY PROFILE: Skin and eye irritant. When heated to decomposition it emits acrid smoke and irritating fumes.

DSD800 CAS:2738-18-3 HR: 2
2,6-DIMETHYL-3-HEPTENE
mf: C_9H_{18} mw: 126.23

$(CH_3)_2CHCH{=}CHCH_2CH(CH_3)_2$

PROP: Clear liquid. Bp: 128.5–129°, flash p: 59.8°F, d: 0.722 @ 15.5°/15.5°, vap press: 28.4 mm @ 38°, vap d: 4.38.

SAFETY PROFILE: A probable irritant and narcotic in high concentration. A very dangerous fire hazard when exposed to heat or flame; can react vigorously with oxidizing materials. To fight fire, use foam, CO_2, dry chemical. When heated to decomposition it emits acrid smoke and irritating fumes.

DSE489 CAS:7226-23-5 HR: 3
1,3-DIMETHYLHEXAHYDROPYRIMIDONE
mf: $C_6H_{12}N_2O$ mw: 128.17

$CH_3NCO{\cdot}N(CH_3)C_2H_4\,CH_2$

PROP: Bp: 146° @ 44 mm.

SYN: DIMETHYLPROPYLENEUREA

SAFETY PROFILE: Explodes on contact with chromium trioxide. When heated to decomposition it emits toxic fumes of NO_x.

DSE509 CAS:584-94-1 HR: 3
2,3-DIMETHYLHEXANE
mf: C_8H_{18} mw: 115.67

$(CH_3)_2CHCHCH_3CH_2CH_2CH_3$

PROP: A clear liquid. Bp: 116°, flash p: 41.6°F, d: 0.716 @ 15.5°/15.5°, vap d: 4.1, autoign temp: 820°F.

SAFETY PROFILE: A probable irritant and narcotic in high concentration. A very dangerous fire hazard when exposed to heat or flame; can react vigorously with oxidizing materials. To fight fire, use foam, CO_2, dry chemical. When heated to decomposition it emits acrid smoke and irritating fumes.

DSE600 CAS:589-43-5 HR: 3
2,4-DIMETHYLHEXANE
mf: C_8H_{18} mw: 115.67

$(CH_3)_2CHCH_2CHCH_3CH_2CH_3$

PROP: A liquid. Bp: 109°, flash p: 50°F(OC), d: 0.705 @ 15.5°/15.5°, vap d: 3.9.

SAFETY PROFILE: A probable irritant and narcotic in high concentration. A very dangerous fire hazard when exposed to heat or flame; can react vigorously with oxidizing materials. To fight fire, use foam, CO_2, dry chemical. When heated to decomposition it emits acrid smoke and irritating fumes.

DSE800 HR: 3
DIMETHYLHEXANE DIHYDROPEROXIDE (dry)

PROP: Fine, white crystals; insol in hydrocarbons; sltly sol in water, esters, and glycerin; sol in other organic solvents. Mp: 104°.

SYNS: 2,5-DIMETHYL-2,5-DIHYDROPEROXYHEXANE, >82% with water (DOT) □ HEXANE, 2,5-DIMETHYL-, 2,5-DIHYDROPEROXIDE

CONSENSUS REPORTS: Reported in EPA TSCA Inventory.

DOT CLASSIFICATION: Forbidden

SAFETY PROFILE: A reactive peroxide. When heated to decomposition it emits acrid smoke and fumes. See also PEROXIDES, ORGANIC.

DSF200 CAS:53306-53-9 HR: 1
DI(3-METHYLHEXYL)PHTHALATE
mf: $C_{22}H_{34}O_4$ mw: 362.56

SYN: BIS(3-METHYLHEXYL)PHTHALIC ACID ESTER

TOXICITY DATA WITH REFERENCE
orl-rat LD50:35,500 mg/kg GTPZAB 24(3),25,80
orl-mus LD50:35,500 mg/kg GTPZAB 24(3),25,80
orl-gpg LD50:35,500 mg/kg GTPZAB 24(3),25,80

SAFETY PROFILE: Mildly toxic by ingestion. When heated to decomposition it emits acrid smoke and irritating fumes.

DSF300 CAS:77-71-4 HR: 2
5,5-DIMETHYLHYDANTOIN
mf: $C_5H_8N_2O_2$ mw: 128.15

PROP: Prisms from EtOH. Mp: 175°. Sol in H_2O.

SYNS: 5,5-DIMETHYL-2,4-IMIDAZOLIDINEDIONE □ DMH □ 2,4-IMIDAZOLIDINEDIONE, 5,5-DIMETHYL- □ NSC 8652 □ T10

TOXICITY DATA WITH REFERENCE
orl-rat LD50:7800 mg/kg GISAAA 47(6),76,82
scu-mus LD50:2800 mg/kg ARZNAD 4,723,54
unr-mus LD50:10 g/kg GISAAA 46(5),69,81
orl-rbt LD50:12,660 mg/kg GISAAA 47(6),76,82
orl-gpg LD50:8430 mg/kg GISAAA 47(6),76,82
unr-gpg LD50:5600 mg/kg GISAAA 46(5),69,81
unr-mam LD50:15,950 mg/kg GISAAA 46(5),69,81

CONSENSUS REPORTS: Reported in EPA TSCA Inventory.

SAFETY PROFILE: Moderately toxic by subcutaneous route. Mildly toxic by ingestion. When heated to decomposition it emits toxic fumes of NO_x.

DSF400 CAS:57-14-7 HR: 3
1,1-DIMETHYLHYDRAZINE
DOT: UN 1163
mf: $C_2H_8N_2$ mw: 60.12

PROP: Colorless liquid, ammonia-like odor. Hygroscopic, water-misc. Mp: −58°, bp: 63.3°, flash p: 5°F, d: 0.791 @ 22°, vap press: 157 mm @ 25°, vap d: 1.94, autoign temp: 480°F, lel: 2%, uel: 95%. Sol in H_2O and EtOH.

SYNS: DIMAZINE □ DIMETHYLHYDRAZINE □ asym-DIMETHYLHYDRAZINE □ N,N-DIMETHYLHYDRAZINE □ uns-DIMETHYLHYDRAZINE □ unsym-DIMETHYLHYDRAZINE □ 1,1-DIMETHYLHYDRAZINE (GERMAN) □ DIMETHYLHYDRAZINE, unsymmetrical (DOT) □ DMH □ NIESYMETRYCZNA DWU METYLOHYDRAZYNA (POLISH) □ RCRA WASTE NUMBER U098 □ UDMH (DOT)

TOXICITY DATA WITH REFERENCE
otr-hmn:fbr 167 µmol/L PNASA6 80,7219,83
dnd-hmn:fbr 300 µmol/L ENMUDM 7,267,85
ipr-rat TDLo:600 mg/kg (6-15D preg):TER JTEHD6 13,125,84
ipr-rat TDLo:600 mg/kg (6-15D preg):REP JTEHD6 13,125,84
orl-rat TDLo:150 mg/kg/7W-I:ETA NATUAS 246,491,73
scu-rat TDLo:21 mg/kg:ETA SUFOAX 31,413,80
orl-mus TDLo:5880 mg/kg/42W-C:CAR JNCIAM 50,181,73
ipr-mus TDLo:144 mg/kg/8W-I:ETA JNCIAM 42,337,69
scu-mus TDLo:420 mg/kg/21W-I:NEO CALEDQ 39,69,88
orl-ham TDLo:228 g/kg/48W-I:CAR CANCAR 40,2427,77
scu-ham TDLo:2686 mg/kg/72W-I:CAR CALEDQ 35,303,87
orl-mus TD:288 mg/kg/8W-I:ETA JNCIAM 42,337,69
scu-rat TD:400 mg/kg/20W-I:ETA SUFOAX 31,413,80
orl-rat TD:300 mg/kg/14W-I:ETA NATUAS 246,491,73
scu-mus LD :200 mg/kg/25W-I:ETA BEXBAN 92,1681,81
orl-rat LD50:122 mg/kg MEPAAX 24,71,73
ihl-rat LC50:252 ppm/4H AMIHAB 12,609,55
ipr-rat LD50:102 mg/kg TXAPA9 6,371,64
ivn-rat LD50:119 mg/kg MEPAAX 24,71,73
ice-rat LDLo:27 mg/kg BCPCA6 14,1901,65
orl-mus LD50:265 mg/kg MEPAAX 24,71,73
ihl-mus LC50:172 ppm/4H AMIHAB 12,609,55
ipr-mus LD50:113 mg/kg PSEBAA 124,172,67
ivn-mus LD50:250 mg/kg MEPAAX 24,71,73
ihl-dog LC50:3580 ppm/15M AIHAAP 24,137,63

CONSENSUS REPORTS: NTP 7th Annual Report On Carcinogens. IARC Cancer Review: Group 2B IMEMDT 7,56,87; Animal Sufficient Evidence IMEMDT 4,137,74. EPA Genetic Toxicology Program. Community Right-To-Know List. EPA Extremely Hazardous Substances List. Reported in EPA TSCA Inventory.

OSHA PEL: TWA 0.5 ppm (skin)
ACGIH TLV: TWA 0.5 ppm (skin); Suspected Human Carcinogen; (Proposed: TWA 0.01 ppm (skin); Suspected Human Carcinogen)
DFG MAK: Animal Carcinogen, Suspected Human Carcinogen
NIOSH REL: (Hydrazines) CL 0.15 mg/m³/2H
DOT CLASSIFICATION: 6.1; *Label:* Poison, Flammable Liquid, Corrosive

SAFETY PROFILE: Confirmed carcinogen with experimental carcinogenic, tumorigenic, and teratogenic data. Other experimental reproductive effects. Poison by ingestion, intraperitoneal, intravenous, and intracerebral routes. Moderately toxic by inhalation and skin contact. Human mutation data reported. A plant growth control agent. Corrosive. A powerful reducing agent. A dangerous fire hazard. It is hypergolic with many oxidants (e.g., dinitrogen tetroxide, hydrogen peroxide, and nitric acid). Dangerous when exposed to heat, flame, or oxidizers; can react vigorously with oxidizing materials such as air, fuming HNO_3, (HNO_3 + N_2O_4), NO. A high-energy propellant for liquid-fueled rockets. To fight fire, use alcohol foam, CO_2, dry chemical. When heated to decomposition it emits highly toxic fumes of NO_x. See also HYDRAZINE.

DSF600 CAS:540-73-8 HR: 3
1,2-DIMETHYLHYDRAZINE
DOT: UN 2382
mf: $C_2H_8N_2$ mw: 60.12

PROP: Clear, colorless, flammable, hygroscopic, fuming liquid; fishy ammonia odor. Flash p: <73.4°F, bp: 81°, mp: −9°, d: 0.8274 @ 20°/4°. Sol in H_2O, EtOH, etc.

SYNS: 1,2-DIMETHYLHYDRAZIN (GERMAN) □ DIMETHYLHYDRAZINE, symmetrical (DOT) □ N,N'-DIMETHYLHYDRAZINE □ sym-DIMETHYLHYDRAZINE □ 1,2-DIMETHYL-HYDRAZINE □ DMH □ HYDRAZOMETHANE □ RCRA WASTE NUMBER U099 □ SDMH □ SYMETRYCZNA DWUMETYLOHYDRAZYNA (POLISH)

TOXICITY DATA WITH REFERENCE
otr-hmn:fbr 230 µmol/L CALEDQ 29,265,85
dns-hmn:lng 100 µL/L NTIS** AD-A041-973
sce-mus-rec 20 mg/kg ENMUDM 8(Suppl 6),41,86
hma-mus/esc 50 µmol/kg MUREAV 148,1,85
dnd-ham:lng 2 mmol/L MUREAV 173,157,86
ipr-rat TDLo:100 mg/kg (6-15D preg):TER JTEHD6 13,125,84
orl-rat TDLo:120 mg/kg/4W-I:ETA AJCNAC 30,176,77
ipr-rat TDLo:260 mg/kg/13W-I:ETA CALEDQ 19,1,83
sce-mus-ipr 270 µmol/kg TCMUD8 9,219,89
scu-rat TDLo:90 mg/kg/28W-I:CAR BIMDB3 32,41,80
par-rat TDLo:160 mg/kg/16W-I:ETA PAACA3 21,67,80
orl-mus TDLo:12,500 µg/kg/24W-I:NEO TXAPA9 72,313,84
ipr-mus TDLo:25 mg/kg/24W-I:NEO TXAPA9 72,313,84
scu-mus TDLo:200 mg/kg/10W-I:CAR JJIND8 63,1081,79
rec-mus TDLo:360 mg/kg/18W-I:ETA FEPRA7 34,827,75
imp-rbt TDLo: 2 mg/kg:ETA CNREA8 35,2292,75
scu-ham TDLo:340 mg/kg/17W-I:CAR EXMPA6 27,19,77
scu-rat TD:120 mg/kg/12W-I:CAR NASDA6 32,270,81
scu-rat TD:20 mg/kg:CAR GANNA2 74,493,83
scu-rat TD:200 mg/kg/10W-I:CAR CRNGDP 3,1097,82
scu-mus TD:300 mg/kg/20W-I:CAR JNCIAM 52,999,74
scu-rat TD:80 mg/kg/4W-I:ETA CRNGDP 6,637,85
scu-rat TD:300 mg/kg/11W-I:NEO GASTAB 81,475,81
scu-rat TD:42 mg/kg 28W-I:NEO BIMDB3 32,41,80

scu-rat TD:420 mg/kg/28W-I:CAR CRNGDP 4,1175,83
scu-rat TD:200 mg/kg/20W-I:ETA DICRAG 23,137,80
scu-rat TD:400 mg/kg/10W-I:CAR LANCAO 2,1030,80
orl-rat LD50:100 mg/kg NATWAY 54,285,67
ihl-rat LCLo:280 ppm/4H AMIHAB 12,609,55
ipr-rat LD50:163 mg/kg MEPAAX 24,71,73
scu-rat LD50:220 mg/kg XENOBH 3,271,73
ivn-rat LD50:176 mg/kg MEPAAX 24,71,73
orl-mus LD50:36 mg/kg MEPAAX 24,71,73
ipr-mus LD50:35 mg/kg MEPAAX 24,71,73
scu-mus LD50:24 mg/kg TOLED5 8,87,81
ivn-mus LD50:29 mg/kg MEPAAX 24,71,73
ivn-dog LD50:100 mg/kg MEPAAX 24,71,73
ims-ham LD50:95 mg/kg ARZNAD 19,1891,69

CONSENSUS REPORTS: IARC Cancer Review: Group 2B IMEMDT 7,56,87; Animal Sufficient Evidence IMEMDT 4,145,74. EPA Genetic Toxicology Program.

DFG MAK: Animal Carcinogen, Suspected Human Carcinogen

DOT CLASSIFICATION: 3; *Label:* Flammable Liquid, Poison

SAFETY PROFILE: Confirmed carcinogen with experimental carcinogenic, neoplastigenic, tumorigenic, and teratogenic data. Poison by ingestion, intraperitoneal, intravenous, subcutaneous, and intramuscular routes. Moderately toxic by inhalation. Human mutation data reported. A very dangerous fire hazard when exposed to heat, flame, or oxidizers. A high-energy propellant for liquid-fueled rockets. When heated to decomposition it emits toxic fumes of NO_x.

DSF800 CAS:306-37-6 HR: 3
1,2-DIMETHYLHYDRAZINE DIHYDROCHLORIDE
mf: $C_2H_8N_2 \cdot 2ClH$ mw: 133.04

PROP: A solid. Mp: 170°.

SYNS: N,N'-DIMETHYLHYDRAZINE DIHYDROCHLORIDE ☐ sym-DIMETHYLHYDRAZINE DIHYDROCHLORIDE ☐ DMH

TOXICITY DATA WITH REFERENCE
otr-rat-ipr 100 mg/kg CALEDQ 26,191,82
dnd-rat-orl 1700 µg/kg/2D-C CRNGDP 4,529,83
dns-rat-orl 1700 µg/kg/4D-C CRNGDP 4,529,83
dni-rat-orl 1700 µg/kg/2D-C CRNGDP 4,529,83
dni-mus:oth 500 µmol/L JJIND8 68,1015,82
ims-ham TDLo:200 mg/kg (12D preg):REP APTOD9 19,A71,80
orl-rat TDLo:35 mg/kg:CAR CALEDQ 14,47,81
ipr-rat TDLo:160 mg/kg/16W-I:CAR NUCADQ 4,146,82
scu-rat TDLo:200 mg/kg/20W-I:CAR PSEBAA 151,237,76
orl-mus TDLo:23 mg/kg/2Y-C:CAR ANTRD4 2,365,82
ipr-mus TDLo:212 mg/kg/8W-I:ETA JNCIAM 42,337,69
scu-mus TDLo:150 mg/kg/10W-I:CAR CALEDQ 9,111,80
par-mus TDLo:133 mg/kg/6W-I:ETA CALEDQ 8,23,79
orl-gpg TDLo:2190 mg/kg/73W-I:CAR TXAPA9 38,647,76
scu-gpg TDL0:2040 mg/kg/34W-I:ETA TXAPA9 38,647,76
orl-ham TDLo:446 mg/kg/51W-C:NEO CNREA8 32,804,72
scu-rat TD:400 mg/kg/22W-I:CAR JJIND8 64,263,80
scu-rat TD:500 mg/kg/20W-I:CAR APLMAS 105,29,81
orl-mus TD:672 mg/kg/48W-C:CAR VAPHDQ 384,263,79
orl-rat TD:75 mg/kg/5W-I:CAR CNREA8 43,4083,83
orl-mus TD:720 mg/kg/24W-I:CAR VAPHDQ 384,263,79
scu-mus TD:480 mg/kg/16W-I:CAR VAPHDQ 384,263,79
orl-mus TD:284 mg/kg/40W-C:CAR TJXMAH 27,5,80
scu-mus TD:400 mg/kg/20W-I:CAR FCTXAV 19,281,81
scu-rat TD:240 mg/kg/16W-I:CAR JSGRA2 29,363,80
orl-rat TDLo:150 mg/kg/10W-I:CAR TXAPA9 55,417,80
orl-rat LD50:100 mg/kg 23HZAR -,267,70
scu-rat LD50:122 mg/kg NTIS** AD-A062-138
scu-mus LD50:25,400 µg/kg CRNGDP 3,603,82
scu-ham LD50:50 mg/kg NTIS** AD-A062-138

CONSENSUS REPORTS: Reported in EPA TSCA Inventory. EPA Genetic Toxicology Program.

SAFETY PROFILE: Suspected carcinogen with experimental carcinogenic, neoplastigenic, and tumorigenic data. Poison by ingestion and subcutaneous routes. Experimental reproductive effects. Mutation data reported. A rocket fuel. When heated to decomposition it emits very toxic fumes of HCl and NO_x.

DSG000 CAS:593-82-8 HR: 3
1,1-DIMETHYLHYDRAZINE HYDROCHLORIDE
mf: $C_2H_8N_2 \cdot ClH$ mw: 96.58

PROP: Hygroscopic crystals from EtOH. Mp: 83°.

TOXICITY DATA WITH REFERENCE
mmo-sat 20 mg/plate MUREAV 66,247,79
dni-hmn:hlas 6 mmol/L CRNGDP 13,2389,92
orl-rat TDLo:35 g/kg/73W-C:ETA ZEKBAI 69,103,67
orl-rat LD50:196 mg/kg AMIHAB 13,34,56
ipr-rat LD50:210 mg/kg AMIHAB 13,34,56
ivn-rat LD50:191 mg/kg AMIHAB 13,34,56
orl-mus LD50:426 mg/kg AMIHAB 13,34,56
ipr-mus LD50:400 mg/kg AMIHAB 13,34,56
ivn-mus LD50:402 mg/kg AMIHAB 13,34,56
ivn-dog LD50:96 mg/kg AMIHAB 13,34,56

NIOSH REL: (Hydrazines) CL 0.15 mg/m³/2H

SAFETY PROFILE: Poison by ingestion, intraperitoneal, and intravenous routes. Questionable carcinogen with experimental tumorigenic data. Mutation data reported. When heated to decomposition it emits very toxic fumes of HCl and NO_x.

DSG200 CAS:56400-60-3 HR: 3
1,2-DIMETHYLHYDRAZINE HYDROCHLORIDE
mf: $C_2H_8N_2 \cdot ClH$ mw: 96.58

SYNS: sym-DIMETHYLHYDRAZINE HYDROCHLORIDE ☐ DMH

TOXICITY DATA WITH REFERENCE
dns-mus-orl 20 mg/kg FEPRA7 33,596,74
orl-rat TDLo:150 mg/kg/5W-I:NEO JJIND8 63,1089,79
scu-rat TDLo:300 mg/kg/20W-I:CAR CANCAR 40,2502,77
scu-rat TDLo:120 mg/kg/6W-I:ETA CNREA8 33,940,73
scu-rat TD:20 mg/kg:ETA CNREA8 41,1240,81
scu-rat TD:80 mg/kg/4W-I:ETA CNREA8 41,1240,81
scu-rat TD:160 mg/kg/8W-I:ETA CNREA8 41,1240,81
scu-rat TD:320 mg/kg/16W-I:ETA CNREA8 41,1240,81
scu-rat TD:532 mg/kg/20W-I:ETA LAINAW 30,505,74
orl-rat TD:150 mg/kg/3W-I:ETA CALEDQ 25,311,85
orl-rat LD50:257 mg/kg AMIHAB 13,34,56
ipr-rat LD50:262 mg/kg AMIHAB 13,34,56

ivn-rat LD50:281 mg/kg AMIHAB 13,34,56
orl-mus LD50:58 mg/kg AMIHAB 13,34,56
ipr-mus LD50:56 mg/kg AMIHAB 13,34,56
scu-mus LD50:12 mg/kg BIJOAK 122,121,71
ivn-mus LD50:47 mg/kg AMIHAB 13,34,56
ivn-dog LD50:161 mg/kg AMIHAB 13,34,56

SAFETY PROFILE: Poison by ingestion, intraperitoneal, subcutaneous, and intravenous routes. Questionable carcinogen with experimental carcinogenic, neoplastigenic, and tumorigenic data. Mutation data reported. When heated to decomposition it emits very toxic fumes of HCl and NO_x. See also 1,1-DIMETHYL HYDRAZINE.

DSG400 CAS:26049-69-4 HR: 3
2-(2,2-DIMETHYLHYDRAZINO)-4-(5-NITRO-2-FURYL) THIAZOLE
mf: $C_9H_{10}N_4O_3S$ mw: 254.29

SYN: DMNT

TOXICITY DATA WITH REFERENCE
mma-sat 100 ng/plate MUREAV 40,9,76
dnr-sat 500 nmol/well CNREA8 34,2266,74
mmo-esc 300 nmol/well CNREA8 34,2266,74
mrc-esc 500 nmol/well CNREA8 34,2266,74
pic-esc 1 mg/L MUREAV 26,3,74
orl-rat TDLo:4800 mg/kg/46W-I:CAR CNREA8 30,897,70
orl-mus TDLo:15 g/kg/17W-C:CAR CNREA8 33,1593,73
orl-rat TD:31 g/kg/44W-C:CAR PAACA3 10,15,69

CONSENSUS REPORTS: EPA Genetic Toxicology Program.

SAFETY PROFILE: Suspected carcinogen with experimental carcinogenic data. Mutation data reported. When heated to decomposition it emits very toxic fumes of NO_x and SO_x.

DSG600 CAS:868-85-9 HR: 3
DIMETHYL HYDROGEN PHOSPHITE
mf: $C_2H_7O_3P$ mw: 110.06

PROP: D: 1.20 20°/4°, bp: 56.5° @ 8 mm.

SYNS: BIS(HYDROXYMETHYL)PHOSPHINE OXIDE ☐ DIMETHOXYPHOSPHINE OXIDE ☐ DIMETHYL ACID PHOSPHITE ☐ DIMETHYLESTER KYSELINY FOSFORITE (CZECH) ☐ DIMETHYLFOSFIT ☐ DIMETHYLFOSFONAT ☐ DIMETHYLHYDROGENPHOSPHITE ☐ DIMETHYL PHOSPHITE ☐ DIMETHYL PHOSPHONATE ☐ DIMETHYL PHOSPHOROUS ACID ☐ HYDROGEN DIMETHYL PHOSPHITE ☐ METHYL PHOSPHONATE ☐ NCI-C54773 ☐ PHOSPHOROUS ACID DIMETHYL ESTER

TOXICITY DATA WITH REFERENCE
skn-rbt 500 mg/24H MLD 28ZPAK -,215,72
eye-rbt 20 mg/24H MOD 28ZPAK -,215,72
mma-sat 7500 μg/plate ENMUDM 8(Suppl 7),1,86
orl-rat TDLo:103 g/kg/2Y-I:CAR NTPTR* NTP-TR-287,85
orl-rat LD50:3050 mg/kg ALBRW* #OPB-3,84
skn-rbt LD50:2400 mg/kg ALBRW* #OPB-3,84

CONSENSUS REPORTS: IARC Cancer Review: Group 3 IMEMDT 48,85,90; Animal Limited Evidence IMEMDT 48,85,90. NTP Carcinogenesis Studies (gavage); No Evidence: mouse NTPTR* NTP-TR-287,85; Clear Evidence: rat NTPTR* NTP-TR-287,85. Reported in EPA TSCA Inventory.

DFG MAK: Suspected Carcinogen

SAFETY PROFILE: Suspected carcinogen with experimental carcinogenic data. Moderately toxic by ingestion and skin contact. A skin and eye irritant. Mutation data reported. When heated to decomposition it emits toxic fumes of PO_x.

DSG700 CAS:654-42-2 HR: 3
2,6-DIMETHYLHYDROQUINONE
mf: $C_8H_{10}O_2$ mw: 138.18

PROP: Crystals from xylene. Mp: 149–151°.

SYNS: 2,6-DIMETHYL-1,4-BENZENEDIOL (9CI) ☐ DMHQ ☐ m-XHQ ☐ m-XYLOHYDROQUINONE ☐ 2,6-XYLOHYDROQUINONE ☐ 2,6-XYLOQUINOL

TOXICITY DATA WITH REFERENCE
orl-wmn TDLo:84 mg/kg (30W pre):REP ACEDAB 28,83,56
orl-man TDLo:7143 μg/kg (1D male):REP JMIABM 22,19,58
orl-mus TDLo:40 mg/kg (female 1D pre):REP JOENAK 14,228,56
orl-mus TDLo:40 mg/kg (female 1D pre):TER JOENAK 14,228,56
orl-rat TDLo:100 mg/kg (4D pre):REP ACEDAB 28,83,56
ipr-rat TDLo:950 mg/kg (female 19D pre):REP ENDOAO 57,466,55
scu-rat TDLo:3 mg/kg (female 1D pre):REP ENDOAO 57,466,55
unr-rat TDLo:50 mg/kg (female 5D pre):REP CMJRAY 49,354,52
orl-mus LD50:186 mg/kg IJEBA6 2,23,64
ipr-mus LD50:117 mg/kg IJEBA6 2,23,64
ipr-cat LDLo:20 mg/kg IJEBA6 2,23,64

SAFETY PROFILE: Poison by ingestion and intraperitoneal routes. An experimental teratogen. Human reproductive effects by ingestion: impaired spermatogenesis in men and changes in fertility in women. Experimental reproductive effects. When heated to decomposition it emits acrid smoke and irritating fumes.

DSH000 CAS:2019-14-9 HR: 3
DIMETHYL(2-HYDROXYETHYL)OCTYLAMMONIUM BROMIDE BENZILATE
mf: $C_{26}H_{38}NO_3\cdot Br$ mw: 492.56

PROP: Crystals from Me_2CO. Mp: 115°.

SYNS: AD-205 ☐ (2-BENZILOXYETHYL)DIMETHYLOCTYLAMMONIUM BROMIDE ☐ BENZILSAEURE-DIMETHYL-OCTYL-AMMONIUM-AETHYLESTER BROMIDE (GERMAN)

TOXICITY DATA WITH REFERENCE
orl-mus LD50:1150 mg/kg TXAPA9 5,225,63
ipr-mus LD50:104 mg/kg TXAPA9 5,225,63
scu-mus LD50:850 mg/kg TXAPA9 5,225,63
ivn-mus LD50:17,500 μg/kg TXAPA9 5,225,63

SAFETY PROFILE: Poison by intraperitoneal and intravenous routes. Moderately toxic by ingestion and subcutaneous routes. When heated to decomposition it emits very toxic fumes of NO_x, NH_3, and Br^-.

DSH200 CAS:56927-39-0 HR: 3
DIMETHYL(2-HYDROXYETHYL)PENTYLAMMONIUM BROMIDE BENZILATE
mf: $C_{23}H_{32}NO_3 \cdot Br$ mw: 450.47

SYN: BENZILSAEURE-DIMETHYL-PENTYL-AMMONIUM-AETHYLESTER BROMIDE (GERMAN)

TOXICITY DATA WITH REFERENCE
scu-mus LD50:230 mg/kg ARZNAD 10,763,60
ivn-mus LD50:12 mg/kg ARZNAD 10,763,60

SAFETY PROFILE: Poison by subcutaneous and intravenous routes. When heated to decomposition it emits very toxic fumes of NO_x, NH_3, and Br^-.

DSH400 CAS:69928-30-9 HR: 2
3,5-DIMETHYL-3-HYDROXYHEXANE-4-CARBOXYLIC ACID-β-LACTONE
mf: $C_9H_{16}O_2$ mw: 156.25

SYN: 4-ETHYL-3-ISOPROPYL-4-METHYL-1-OXACYCLOBUTAN-2-ONE

TOXICITY DATA WITH REFERENCE
eye-rbt 20 mg open SEV AMIHBC 10,61,54
orl-rat LD50:2700 mg/kg AMIHBC 10,61,54

SAFETY PROFILE: Moderately toxic by ingestion. A severe eye irritant. When heated to decomposition it emits acrid smoke and irritating fumes.

DSH700 CAS:66634-53-5 HR: 3
N,N-DIMETHYL-β-HYDROXYPHENETHYLAMINE
mf: $C_{10}H_{15}NO$ mw: 165.26

SYNS: α-(DIMETHYLAMINOMETHYL)BENZYL ALCOHOL ☐ β-HYDROXY-β-PHENYLETHYL DIMETHYLAMINE

TOXICITY DATA WITH REFERENCE
scu-mus LDLo:826 mg/kg AIPTAK 47,96,34
ivn-mus LD50:56 mg/kg CSLNX* NX#04125
ivn-rbt LDLo:132 mg/kg AIPTAK 47,96,34

SAFETY PROFILE: Poison by intravenous route. Moderately toxic by subcutaneous route. When heated to decomposition it emits toxic fumes of NO_x. See also AMINES.

DSH800 CAS:27945-43-3 HR: 3
3-(3,5-DIMETHYL-4-HYDROXYPHENYL)-2-METHYL-4(3H)-QUINAZOLINONE
mf: $C_{17}H_{16}N_2O_2$ mw: 280.35

SYNS: 3-(4-HYDROXY-3,5-XYLYL)-2-METHYL-4(3H)-QUINAZOLINONE ☐ 2-METHYL-3-(3,5-DIMETHYL-4-HYDROXYPHENYL)-3,4-DIHYDROQUINAZOLIN-4-ONE ☐ SRC-226

TOXICITY DATA WITH REFERENCE
orl-rat LD50:230 mg/kg IJPAAO 37,109,75
orl-mus LD50:775 mg/kg IJPAAO 37,109,75

SAFETY PROFILE: Poison by ingestion. When heated to decomposition it emits toxic fumes of NO_x.

DSI000 CAS:1505-26-6 HR: 3
1,2-DIMETHYL-3-(m-HYDROXYPHENYL)-3-PROPYLPYRROLIDINE
mf: $C_{15}H_{23}N$ mw: 217.39

SYN: m-(1,2-DIMETHYL-3-PROPYL-3-PYRROLIDINYL)PHENOL

TOXICITY DATA WITH REFERENCE
ipr-rat LDLo:97 mg/kg JMCMAR 8,316,65
orl-mus LDLo:300 mg/kg CHTPBA 7,450,72
ipr-mus LDLo:300 mg/kg CHTPBA 7,450,72

SAFETY PROFILE: Poison by ingestion and intraperitoneal routes. When heated to decomposition it emits toxic fumes of NO_x.

DSI200 CAS:66941-43-3 HR: 3
(2,2-DIMETHYL-3-HYDROXYPROPYL)TRIETHYLAMMONIUM BROMIDE TROPATE (ESTER)
mf: $C_{20}H_{34}NO_3 \cdot Br$ mw: 416.46

SYN: TROPASAEUREESTER DES 3-TRIAETHYLAMMONIUM 2,2-DIMETHYL-1-PROPANOLBROMID (GERMAN)

TOXICITY DATA WITH REFERENCE
ivn-rat LDLo:80 mg/kg AEPPAE 173,86,33
ivn-mus LDLo:70 mg/kg AEPPAE 173,86,33
unr-frg LDLo:3000 mg/kg AEPPAE 173,86,33

SAFETY PROFILE: Poison by intravenous route. Moderately toxic by an unspecified route. When heated to decomposition it emits very toxic fumes of NO_x, NH_3, and Br^-. See also ESTERS.

DSI489 CAS:29128-41-4 HR: 3
DIMETHYL HYPONITRILE
mf: $C_2H_6N_2O_2$ mw: 90.08

PROP: Fragrant liquid.

SYN: DIMETHOXYDIAZENE

CONSENSUS REPORTS: Cyanide and its compounds are on the Community Right-To-Know List.

SAFETY PROFILE: A dangerously unpredictable explosive. When heated to decomposition it emits toxic fumes of CN^- and NO_x. See also NITRILES.

DSI709 CAS:50-47-5 HR: 3
DIMETHYLIMIPRAMINE
mf: $C_{18}H_{22}N_2$ mw: 266.42

PROP: Bp: 172–174° @ 0.02 mm.

SYNS: DEMETHYLIMIPRAMINE ☐ DESIMIPRAMINE ☐ DESIPRAMIN ☐ DESIPRAMINE (D4) ☐ DESMETHYLIMIPRAMINE ☐ DMI ☐ DMI 50475 ☐ METHYLAMINOPROPYLIMINODIBENZYL ☐ MONODEMETHYLIMIPRAMINE ☐ NORIMIPRAMINE ☐ PENTOFRAN ☐ PERTOFRAN ☐ PERTOFRANE ☐ SERTOFRAN

TOXICITY DATA WITH REFERENCE
dnd-esc 20 μmol/L MUREAV 89,95,81
cyt-oin-unr 10 g/L JCLBA3 47,182a,70
scu-rat TDLo:16,250 μg/kg (female 8-20D post):REP ARTODN 7,504,84
orl-rat TDLo:500 mg/kg (female 7-16D post):TER ARZNAD 19,1617,69

orl-wmn LDLo:30 mg/kg:BRN,CNS 34ZIAG -,201,69
orl-chd LDLo:125 mg/kg:CNS,PUL PSYPAG 10,431,67
orl-hmn LDLo:30 mg/kg:CNS,BRN,PUL DMWOAX 93,117,68
orl-rat LD50:375 mg/kg ARZNAD 19,1617,69
ipr-rat LD50:48 mg/kg ARZNAD 20,1561,70
scu-rat LD50:183 mg/kg ARZNAD 20,1561,70
ivn-rat LD50:29 mg/kg AIPTAK 148,560,64
orl-mus LD50:448 mg/kg JJPAAZ 21,47,71
ipr-mus LD50:85 mg/kg ARZNAD 21,1727,71
scu-mus LD50:214 mg/kg FRPPAO 25,519,70
ivn-mus LD50:22 mg/kg APSXAS 12,173,75

SAFETY PROFILE: Human poison by ingestion. Experimental poison by ingestion, intraperitoneal, subcutaneous, and intravenous routes. Human systemic effects by ingestion: degenerative brain changes, tremors, coma, and cyanosis. An experimental teratogen. Other experimental reproductive effects. Mutation data reported. An antidepressant. Related to diazepam. When heated to decomposition it emits toxic fumes of NO_x.

DSI800 CAS:17309-87-4 HR: 2
N,N-DIMETHYL-p-(6-INDAZYLAZO)ANILINE
mf: $C_{15}H_{15}N_5$ mw: 265.35

SYNS: 6-((p-(DIMETHYLAMINO)PHENYL)AZO)-1H-INDAZOLE □ N,N-DIMETHYL-4-(6'-1H-INDAZYLAZO)ANILINE

TOXICITY DATA WITH REFERENCE
mma-sat 20 µg/plate MUREAV 93,67,82
orl-rat TDLo:2700 mg/kg/21W-C:CAR JMCMAR 12,1113,69

SAFETY PROFILE: Questionable carcinogen with experimental carcinogenic data. Mutation data reported. When heated to decomposition it emits toxic fumes of NO_x.

DSI889 HR: 3
DIMETHYLIODOARSINE
mf: C_2H_6AsI mw: 231.90

CONSENSUS REPORTS: Arsenic and its compounds are on the Community Right-To-Know List.

SAFETY PROFILE: Ignites when heated in air. When heated to decomposition it emits toxic fumes of I^- and As. See also ARSENIC COMPOUNDS and IODIDES.

DSJ200 CAS:10143-20-1 HR: 1
2,8-DIMETHYL-6-ISOBUTYLNONANOL-4
mf: $C_{15}H_{32}O$ mw: 228.47

TOXICITY DATA WITH REFERENCE
skn-rbt 10 mg/24H open MLD AMIHBC 10,61,54
eye-rbt 500 mg open AMIHBC 10,61,54
orl-rat LD50:16 g/kg AMIHBC 10,61,54

SAFETY PROFILE: Mildly toxic by ingestion. A skin and eye irritant. When heated to decomposition it emits acrid smoke and irritating fumes.

DSJ800 CAS:489-84-9 HR: 2
1,4-DIMETHYL-7-ISOPROPYLAZULENE
mf: $C_{15}H_{18}$ mw: 198.33

PROP: Blue-violet plates from EtOH or blue oil. Mp: 31.5°, bp: 167–168°.

SYNS: AZULON □ s-GUAIAZULENE

TOXICITY DATA WITH REFERENCE
orl-rat LD50:1550 mg/kg ARZNAD 19,615,69
orl-mus LD50:1300 mg/kg ARZNAD 19,615,69

CONSENSUS REPORTS: Reported in EPA TSCA Inventory.

SAFETY PROFILE: Moderately toxic by ingestion. When heated to decomposition it emits acrid smoke and irritating fumes.

DSK200 CAS:119-38-0 HR: 3
DIMETHYL-5-(1-ISOPROPYL-3-METHYLPYRAZOLYL) CARBAMATE
mf: $C_{10}H_{17}N_3O_2$ mw: 211.30

SYNS: DIMETHYLCARBAMATE-d'1-ISOPROPYL-3-METHYL-5-PYRAZOLYLE (FRENCH) □ DIMETHYLCARBAMIC ACID 3-METHYL-1-(1-METHYLETHYL)-1H-PYRAZOL-5-YL ESTER □ ENT 19,060 □ GEIGY G-23611 □ ISOLAN □ ISOLANE (FRENCH) □ (1-ISOPROPIL-3-METIL-1H-PIRAZOL-5-IL)-N,N-DIMETIL-CARBAMMATO (ITALIAN) □ (1-ISOPROPYL-3-METHYL-1H-PYRAZOL-5-YL)-N,N-DIMETHYLCARBAMAAT (DUTCH) □ (1-ISOPROPYL-3-METHYL-1H-PYRAZOL-5-YL)-N,N-DIMETHYL-CARBAMAT (GERMAN) □ ISOPROPYLMETHYLPYRAZOLYL DIMETHYLCARBAMATE □ 1-ISOPROPYL-3-METHYL-5-PYRAZOLYL DIMETHYLCARBAMATE □ 1-ISOPROPYL-3-METHYLPYRAZOLYL-(5)-DIMETHYLCARBAMATE □ 5-METHYL-2-ISOPROPYL-3-PYRAZOLYL DIMETHYLCARBAMATE □ PRIMIN □ SAOLAN

TOXICITY DATA WITH REFERENCE
mmo-smc 5 ppm RSTUDV 6,161,76
orl-mus TDLo:6600 mg/kg/78W-I:ETA NTIS** PB223-159
orl-rat LD50:10,800 µg/kg PESTD5 17,351,76
skn-rat LD50:5600 µg/kg 85DPAN -,-,71/76
ipr-rat LD50:2150 µg/kg PESTD5 17,351,76
orl-mus LD50:9800 µg/kg PESTD5 17,351,76
ipr-mus LD50:1 mg/kg TXAPA9 6,402,64
orl-bwd LD50:8600 µg/kg TXAPA9 21,315,72

CONSENSUS REPORTS: EPA Extremely Hazardous Substances List.

SAFETY PROFILE: Poison by ingestion, skin contact, and intraperitoneal routes. Questionable carcinogen with experimental tumorigenic data. Mutation data reported. An insecticide. When heated to decomposition it emits toxic fumes of NO_x. See also CARBAMATES.

DSK600 CAS:2674-91-1 HR: 3
O,O-DIMETHYL-S-ISOPROPYL-2-SULFINYLETHYL-PHOSPHOROTHIOATE
mf: $C_7H_{17}O_4PS_2$ mw: 260.33

SYNS: S-2-AETHYLSULFINYL-1-METHYL AETHYL-O,O DIMETHYL-MONOTHIOPHOSPHAT □ BAY 23655 □ BAYER 23655 □ ENT 25,674 □ ESP □ ESTON □ ESTOX □ S-2-ETHYL-SULFINYL-1-METHYL-ETHYL-O,O-DIMETHYL-MONOTHIOFOSFAAT □ S-2-ETHYL-SULPHINYL-1-METHYL-ETHYL-O,O-DIMETHYL PHOSPHOROTHIOLATE □ S-2-ETIL-SULFINIL-1-METIL-ETIL-O,O-DIMETIL-MONOTIOFOSFATO □ META-

SYSTOX-S ☐ OXYDEPROFOS ☐ OXYPHIONFOS ☐ PHOSPHOROTHIOIC ACID, O,O-DIMETHYL S-(ETHYLSULFINYL-(2-ISOPROPYL)) ESTER ☐ S410 ☐ THIOMETAN ☐ THIOPHOSPHATE de O,O-DIMETHYLE ET DE S-2-(ISOPROPYLSULFINYL)-ETHYLE

TOXICITY DATA WITH REFERENCE
mmo-sat 50 mg/plate MUREAV 116,185,83
orl-rat LD50:89 mg/kg NNGADV 17,S309,92
skn-rat LD50:820 mg/kg NNGADV 17,S309,92
ipr-rat LD50:27 mg/kg NNGADV 17,S309,92
scu-rat LD50:29 mg/kg NNGADV 17,S309,92
orl-mus LD50:58,700 μg/kg YKYUA6 30,623,79
ipr-mus LD50:30 mg/kg TXAPA9 4,621,62

SAFETY PROFILE: Poison by ingestion and intraperitoneal routes. Moderately toxic by skin contact. Mutation data reported. When heated to decomposition it emits very toxic fumes of PO_x and SO_x.

DSK800 CAS:36614-38-7 HR: 3
O,O-DIMETHYL-S-2-(ISOPROPYLTHIO)ETHYLPHOSPHORODITHIOATE
mf: $C_7H_{17}O_2PS_3$ mw: 260.39

PROP: Light yellow brown liquid with sltly aromatic odor. D: 1.18 @ 20°/4°, bp: 53 56° @ 0.01 mm.

SYNS: HODSON ☐ HOSALON ☐ HOSDON GRANULE ☐ S-2-ISOPROPYLTHIOETHYL-O,O-DIMETHYL PHOSPHORODITHIOATE ☐ ISOTHIOATE ☐ PHOSPHORODITHIOIC ACID-O,O-DIMETHYL-S-(2-((1-METHYLETHYL)THIO)ETHYL) ESTER

TOXICITY DATA WITH REFERENCE
orl-rat LD50:150 mg/kg 28ZEAL 5,134,76
orl-mus LD50:50 mg/kg FMCHA2 -,C128,83
skn-mus LD50:240 mg/kg FMCHA2 -,C128,83

SAFETY PROFILE: Poison by ingestion and skin contact. When heated to decomposition it emits very toxic fumes of PO_x and SO_x.

DSK900 CAS:534-13-4 HR: 2
1,3-DIMETHYLISOTHIOUREA
mf: $C_3H_8N_2S$ mw: 104.19

PROP: Colorless, exceedingly deliquescent crystals. Mp: 52°. Very sol in water, alc, acetone; sparingly sol in benzene, ether, carbon disulfide; very sltly sol in pet ether.

SYNS: DIMETHYLTHIOCARBAMIDE ☐ N,N'-DIMETHYLTHIOCARBAMIDE ☐ sym-DIMETHYLTHIOUREA ☐ 1,3-DIMETHYLTHIOUREA

TOXICITY DATA WITH REFERENCE
orl-rat TDLo:2 g/kg (14D preg):TER TJADAB 23,335,81
orl-rat TDLo:2 g/kg (14D preg):REP TJADAB 23,335,81
orl-mus LDLo:500 mg/kg TJADAB 23,335,81

CONSENSUS REPORTS: Reported in EPA TSCA Inventory.

SAFETY PROFILE: Moderately toxic by ingestion. Experimental teratogenic and reproductive effects. When heated to decomposition it emits toxic fumes of SO_x and NO_x.

DSK950 CAS:300-87-8 HR: 2
3,5-DIMETHYLISOXAZOLE
mf: C_5H_7NO mw: 97.13

SYNS: DMI ☐ 3,5-DWUMETYLOIZOKSAZOLU ☐ ISOXAZOLE, 3,5-DIMETHYL- ☐ U 21221

TOXICITY DATA WITH REFERENCE
ipr-mus LD50:880 mg/kg DIPHAH 18,19,66

CONSENSUS REPORTS: Reported in EPA TSCA Inventory.

SAFETY PROFILE: Moderately toxic by intraperitoneal route. When heated to decomposition it emits toxic vapors of NO_x.

DSL000 CAS:6155-81-3 HR: 2
N'-(3,4-DIMETHYL-5-ISOXAZOLYL)SULFANILAMIDE LITHIUM SALT
mf: $C_{11}H_{13}N_3O_3S{\bullet}Li$ mw: 274.27

TOXICITY DATA WITH REFERENCE
orl-mus LD50:1 g/kg JPETAB 88,47,46
scu-mus LD50:5000 mg/kg JPETAB 88,47,46
ivn-mus LD50:2500 mg/kg JPETAB 88,47,46

SAFETY PROFILE: Moderately toxic by ingestion and intravenous routes. Mildly toxic by subcutaneous route. When heated to decomposition it emits very toxic fumes of NO_x and SO_x. See also LITHIUM COMPOUNDS.

DSL200 CAS:2200-44-4 HR: 2
N'-(3,4-DIMETHYL-5-ISOXAZOLYL)SULFANILAMIDE SODIUM SALT
mf: $C_{11}H_{13}N_3O_3S{\bullet}Na$ mw: 290.32

TOXICITY DATA WITH REFERENCE
orl-rat LD50:1 g/kg JPETAB 88,47,46
ipr-rat LD50:3200 mg/kg JPETAB 88,47,46
orl-mus LD50:1 g/kg JPETAB 88,47,46
ivn-mus LD50:2300 mg/kg JPETAB 88,47,46

SAFETY PROFILE: Moderately toxic by ingestion, intraperitoneal, and intravenous routes. When heated to decomposition it emits very toxic fumes of NO_x, Na_2O, and SO_x.

DSL289 CAS:598-26-5 HR: 3
DIMETHYLKETENE
mf: C_4H_6O mw: 70.09

$(CH_3)_2C{=}C{=}O$

PROP: Pale-yellow liquid. Fp −97.5°, bp: 34°.

SYN: 2-METHYL-1-PROPENE-1-ONE

SAFETY PROFILE: Upon exposure to air it forms the very unstable explosive peroxide poly(peroxyisobutyrolactone). The peroxide is heat- and friction-sensitive and will also explode upon evaporation. When heated to decomposition it emits acrid smoke and fumes. See also PEROXIDES.

DSL400 **CAS:20917-34-4** **HR: 3**
DIMETHYL LEAD DIACETATE
mf: $C_6H_{12}O_4Pb$ mw: 355.37

SYN: DIACETOXYDIMETHYLPLUMBANE

TOXICITY DATA WITH REFERENCE
orl-mus LD50:120 mg/kg CRSBAW 162,1456,68

CONSENSUS REPORTS: Lead and its compounds are on the Community Right-To-Know List.

SAFETY PROFILE: Poison by ingestion. When heated to decomposition it emits toxic fumes of Pb. See also LEAD COMPOUNDS.

DSL600 **CAS:2999-74-8** **HR: 3**
DIMETHYLMAGNESIUM
mf: C_2H_6Mg mw: 54.38

PROP: A solid. Stable to 2°.

CONSENSUS REPORTS: Reported in EPA TSCA Inventory.

SAFETY PROFILE: The solid and its solution in ether ignite on contact with water. The powder ignites on contact with moist air. When heated to decomposition it emits irritating fumes of MgO. See also MAGNESIUM COMPOUNDS.

DSL800 **CAS:624-48-6** **HR: 2**
DIMETHYL MALEATE
mf: $C_6H_8O_4$ mw: 144.14

PROP: Liquid. Mp: −17.5°, bp: 205.0°, flash p: 235°F (OC), d: 1.15 @ 14°/4°, vap press: 1 mm @ 45.7°, vap d: 4.97.

SYNS: DIMETHYLESTER KYSELINY MALEINOVE □ MALEIC ACID, DIMETHYL ESTER □ METHYL MALEATE □ SIPOMER DMM

TOXICITY DATA WITH REFERENCE
eye-rbt 100 mg AJOPAA 29,1363,46
orl-rat LDLo:1410 mg/kg AIHAAP 23,95,62
skn-rbt LD50:530 mg/kg AIHAAP 23,95,62

CONSENSUS REPORTS: Reported in EPA TSCA Inventory.

SAFETY PROFILE: Moderately toxic by ingestion and skin contact. An eye irritant. Combustible when exposed to heat or flame; can react with oxidizing materials. To fight fire, use CO_2, dry chemical. See also ESTERS and MALEIC ACID.

DSM000 **CAS:766-39-2** **HR: 2**
α,β-DIMETHYLMALEIC ANHYDRIDE
mf: $C_6H_6O_3$ mw: 126.12

PROP: Pearly plates or leaflets. Mp: 96°, bp: 223°.

SYN: DIMETHYLMALEIC ANHYDRIDE

TOXICITY DATA WITH REFERENCE
scu-rat TDLo:2600 mg/kg/65W-I:ETA BJCAAI 19,392,65

CONSENSUS REPORTS: Reported in EPA TSCA Inventory.

SAFETY PROFILE: Questionable carcinogen with experimental tumorigenic data. When heated to decomposition it emits acrid smoke and irritating fumes. See also ANHYDRIDES.

DSM200 **CAS:108-59-8** **HR: 1**
DIMETHYL MALONATE
mf: $C_5H_8O_4$ mw: 132.13

PROP: Bp: 181°.

SYNS: DIMETHYL PROPANEDIOATE □ MALONIC ACID DIMETHYL ESTER □ METHYL MALONATE □ PROPANEDIOIC ACID DIMETHYL ESTER (9CI)

TOXICITY DATA WITH REFERENCE
skn-rbt 500 mg/24H FCTXAV 17,363,79
orl-rat LD50:5331 mg/kg FCTXAV 17,363,79

CONSENSUS REPORTS: Reported in EPA TSCA Inventory.

SAFETY PROFILE: Mildly toxic by ingestion. A skin irritant. Violent reaction with CH_3N_3 occurred with $NaOCH_3$ present. When heated to decomposition it emits acrid smoke and irritating fumes. See also ESTERS.

DSM289 **CAS:33212-68-9** **HR: 3**
DIMETHYL MANGANESE
mf: C_2H_6Mn mw: 85.01

CONSENSUS REPORTS: Manganese and its compounds are on the Community Right-To-Know List.

SAFETY PROFILE: An unstable explosive. Ignites spontaneously in air. See also MANGANESE COMPOUNDS.

DSM450 **CAS:593-74-8** **HR: 3**
DIMETHYL MERCURY
mf: C_2H_6Hg mw: 230.67

PROP: Volatile, colorless liquid with faint sweet odor. D: 3.1874 @ 20°/4°, bp: 92° @ 761 mm. Insoluble in water; very sol in alc and ether.

SYN: MERCURY, DIMETHYL

TOXICITY DATA WITH REFERENCE
dnd-mmo-omi 600 mg/L NATUAS 257,422,75
oth-mus:oth 25 mg/L MUREAV 17,93,73

CONSENSUS REPORTS: IARC Cancer Review: Group 2B IMEMDT 58,239,93; Human Inadequate Evidence IMEMDT 58,239,93. Reported in EPA TSCA Inventory.

OSHA PEL: TWA 0.01 mg(Hg)/m^3; CL 0.03 mg(Hg)/m^3 (skin)
ACGIH TLV: TWA 0.01 mg(Hg)/m^3; STEL 0.03 mg(Hg)/m^3

SAFETY PROFILE: Highly toxic. Mutation data reported. Easily flammable. When heated to decomposition it emits toxic fumes of Hg.

DSM500 CAS:50563-36-5 HR: 3
2,6-DIMETHYL-N-(2-METHOXYETHYL)CHLOROACETANILIDE
mf: $C_{13}H_{18}ClNO_2$ mw: 255.77

SYNS: A 4766 ☐ A 5089 ☐ ACETAMIDE, 2-CHLORO-N-(2,6-DIMETHYLPHENYL)-N-(2-METHOXYETHYL)- ☐ 2,6-ACETOXYLIDIDE, 2-CHLORO-N-(2-METHOXYETHYL)- ☐ 2-CHLORO-N-(2,6-DIMETHYLPHENYL)-N-(2-METHOXYETHYL)ACETAMIDE ☐ 2-CHLORO-N-(2-METHOXYETHYL)ACET-2′,6′-XYLIDIDE ☐ DIMETHACHLOR ☐ DIMETHACHLORE ☐ TERIDOX

TOXICITY DATA WITH REFERENCE
orl-rat LD50:1600 mg/kg PEMNDP 9,290,91
ihl-rat LC50:>750 mg/m³/2H 85JFAN A151,83
skn-rat LD50:>3170 mg/kg PEMNDP 9,290,91

CONSENSUS REPORTS: Reported in EPA TSCA Inventory.

SAFETY PROFILE: A poison by inhalation. Moderately toxic by ingestion and skin contact. When heated to decomposition it emits toxic vapors of NO_x and Cl^-.

DSM600 CAS:63938-21-6 HR: 3
2,3-DIMETHYL-7-METHOXY-8-(MORPHOLINOMETHYL)CHROMONE HYDROCHLORIDE
mf: $C_{17}H_{21}NO_4$•ClH mw: 339.85

SYN: 4H-1-BENZOPYRAN-4-ONE, 7-METHOXY-2,3-DIMETHYL-8-(4-MORPHOLINYLMETHYL)-, HYDROCHLORIDE

TOXICITY DATA WITH REFERENCE
orl-rat LD50:44 mg/kg 27ZQAG -,156,72
ipr-rat LD50:17 mg/kg 27ZQAG -,156,72
scu-rat LD50:19 mg/kg 27ZQAG -,156,72
ipr-mus LD50:17,400 μg/kg JMPCAS 3,471,61

SAFETY PROFILE: Poison by ingestion, intraperitoneal, and subcutaneous routes. When heated to decomposition it emits very toxic fumes of NO_x and HCl.

DSM800 CAS:3613-30-7 HR: 1
3,7-DIMETHYL-7-METHOXY-1-OCTANAL
mf: $C_{11}H_{22}O_2$ mw: 186.33

SYNS: HYDROXYCITRONELLA METHYL ETHER ☐ METHOXYCITRONELLAL METHYL ETHER

TOXICITY DATA WITH REFERENCE
skn-rbt 500 mg/24H MOD FCTXAV 14,807,76

CONSENSUS REPORTS: Reported in EPA TSCA Inventory.

SAFETY PROFILE: A skin irritant. When heated to decomposition it emits acrid smoke and irritating fumes. See also ETHERS.

DSN000 CAS:3009-55-0 HR: 2
N,N-DIMETHYL-p-(2-METHOXYPHENYLAZO)ANILINE
mf: $C_{15}H_{17}N_3O$ mw: 255.35

SYN: 2′-METHOXY-4-DIMETHYLAMINOAZOBENZENE

TOXICITY DATA WITH REFERENCE
orl-rat TDLo:6600 mg/kg/26W-C:NEO CNREA8 17,387,57

SAFETY PROFILE: Questionable carcinogen with experimental neoplastigenic data. When heated to decomposition it emits toxic fumes of NO_x.

DSN200 CAS:20691-83-2 HR: 2
N,N-DIMETHYL-p-(3-METHOXYPHENYLAZO)ANILINE
mf: $C_{15}H_{17}N_3O$ mw: 255.35

SYN: 3′-METHOXY-4-DIMETHYLAMINOAZOBENZENE

TOXICITY DATA WITH REFERENCE
orl-rat TDLo:2800 mg/kg/13W-C:NEO CNREA8 17,387,57

SAFETY PROFILE: Questionable carcinogen with experimental neoplastigenic data. When heated to decomposition it emits toxic fumes of NO_x.

DSN400 CAS:3009-50-5 HR: 2
N,N-DIMETHYL-p-(4-METHOXYPHENYLAZO)ANILINE
mf: $C_{15}H_{17}N_3O$ mw: 255.35

SYN: 4′-METHOXY-4-DIMETHYLAMINOAZOBENZENE

TOXICITY DATA WITH REFERENCE
orl-rat TDLo:6600 mg/kg/26W-C:NEO CNREA8 17,387,57
orl-rat TD:10,204 mg/kg/48W-C:ETA ARZNAD 12,270,62

SAFETY PROFILE: Questionable carcinogen with experimental neoplastigenic and tumorigenic data. When heated to decomposition it emits toxic fumes of NO_x.

DSN600 CAS:7203-92-1 HR: 3
3,3-DIMETHYL-1-p-METHOXYPHENYLTRIAZENE
mf: $C_9H_{13}N_3O$ mw: 179.25

SYNS: 1-p-METHOXYFENYL-3,3-DIMETHYLTRIAZEN (CZECH) ☐ 1-(p-METHOXYPHENYL)-3,3-DIMETHYLTRIAZENE ☐ 1-(4-METHYLOXYPHENYL)-3,3-DIMETHYLTRIAZINE

TOXICITY DATA WITH REFERENCE
sln-dmg-orl 1 mmol/L CBINA8 9,365,74
mrc-smc 1 mmol/L/1H CBINA8 9,365,74
hma-mus/smc 10 mmol/L CBINA8 9,365,74
scu-rat TDLo:1700 mg/kg/45W-I:CAR ZKKOBW 81,285,74
orl-rat LD50:347 mg/kg 28ZPAK -,119,72
scu-rat LD50:450 mg/kg ZKKOBW 81,285,74

CONSENSUS REPORTS: EPA Genetic Toxicology Program.

SAFETY PROFILE: Poison by ingestion. Moderately toxic by subcutaneous route. Questionable carcinogen with experimental carcinogenic data. Mutation data reported. When heated to decomposition it emits toxic fumes of NO_x.

DSN800 CAS:67262-79-7 HR: 3
2′,6′-DIMETHYL-2-(2-METHOXYPROPYLAMINO)ACETANILIDE HYDROCHLORIDE
mf: $C_{14}H_{22}N_2O_2$ mw: 250.38

SYN: 2-(2-METHOXYPROPYLAMINO)-2′,6′-ACETOXYLIDIDE HYDROCHLORIDE

TOXICITY DATA WITH REFERENCE
ipr-mus LD50:120 mg/kg JPMSAE 67,595,78
ivn-mus LD50:40 mg/kg JPMSAE 67,595,78

SAFETY PROFILE: Poison by intraperitoneal and intravenous routes. When heated to decomposition it emits toxic fumes of NO_x and HCl.

DSO000 CAS:950-37-8 HR: 3
O,O-DIMETHYL-S-(5-METHOXY-1,3,4-THIADIAZOLINYL-3-METHYL) DITHIOPHOSPHATE
mf: $C_6H_{11}N_2O_4PS_3$ mw: 302.34

PROP: Crystals from MeOH. Mp: 39–40°. Very sltly sol in H_2O.

SYNS: CIBA-GEIGY GS 13005 □ S-(2,3-DIHYDRO-5-METHOXY-2-OXO-1,3,4-THIADIAZOL-3-METHYL) □ (O,O-DIMETHYL)-S-(-2-METHOXY-Δ^2-1,3,4-THIADIAZOLIN-5-ON-4-YLMETHYL)DITHIOPHOSPHATE DIMETHYL PHOSPHOROTHIOLOTHIONATE □ O,O-DIMETHYL-S-(2-METHOXY-1,3,4-THIADIAZOL-5-(4H)-ONYL-(4)-METHYL)-DITHIOPHOSPHAT (GERMAN) □ O,O-DIMETHYL-S-(2-METHOXY-1,3,4-THIADIAZOL-5(4H)-ONYL-(4)-METHYL) PHOSPHORODITHIOATE □ O,O-DIMETHYL-S-((2-METHOXY-1,3,4 (4H)-THIODIAZOL-5-ON-4-YL)-METHYL)DITHIOFOSFAAT (DUTCH) □ O,O-DIMETIL-S-((2-METOSSI-1,3,4-(4H)-TIADIZAOL-5-ON-4-IL)-METIL)-DITIFOSFATO (ITALIAN) □ DMTP (JAPAN) □ ENT 27,193 □ FISONS NC 2964 □ GEIGY 13005 □ METHIDATHION □ S-((5-METHOXY-2-OXO-1,3,4-THIADIAZOL-3(2H)-YL)METHYL)-O,O-DIMETHYL PHOSPHORODITHIOATE □ SOMONIL □ SURPRACIDE □ ULTRACIDE

TOXICITY DATA WITH REFERENCE
eye-rbt 34 mg SEV CIGET* 8/1/73
orl-man TDLo:93 mg/kg:EYE,CNS HETOEA 9,415,90
orl-rat LD50:20 mg/kg WRPCA2 9,119,70
ihl-rat LC50:3600 mg/m³/4H FMCHA2 -,C224,83
orl-mus LD50:25 mg/kg BESAAT 15,122,69
orl-rbt LD50:63 mg/kg 31ZOAD 1,293,68
skn-rbt LD50:200 mg/kg FMCHA2 -,C224,83
orl-gpg LD50:25 mg/kg 31ZOAD 1,293,68
orl-ham LD50:30 mg/kg 31ZOAD 1,293,68
orl-ckn LD50:80 mg/kg 31ZOAD 1,293,68

CONSENSUS REPORTS: EPA Extremely Hazardous Substances List.

SAFETY PROFILE: Poison by ingestion and skin contact. Moderately toxic by inhalation. Human mutation data reported. Human systemic effects: coma, lacrimation, miosis. A severe eye irritant. An insecticide. When heated to decomposition it emits very toxic fumes of NO_x, PO_x, and SO_x.

DSO200 CAS:23422-53-9 HR: 3
N,N-DIMETHYL-N′-(((METHYLAMINO)CARBONYL) OXY)PHENYLMETHANIMIDAMIDE MONOHYDROCHLORIDE
mf: $C_{11}H_{15}N_3O_2$•ClH mw: 257.75

PROP: Powder. Very sol in H_2O; sol in MeOH; sltly sol in Me_2CO, $CHCl_3$, and hexane.

SYNS: CARZOL SP □ DICARZOL □ m-(((DIMETHYLAMINO)METHYLENE)AMINO)PHENYLMETHYL CARBAMATE,HYDROCHLORIDE □ 3-DIMETHYLAMINOMETHYLENEIMINOPHENYL-N-METHYLCARBAMATE, HYDROCHLORIDE □ ENT 27,566 □ EP-332 □ FORMETANATE HYDROCHLORIDE □ MORTON EP332 □ NOR-AM EP 332 □ SCHERING 36056 □ SN 36056

TOXICITY DATA WITH REFERENCE
orl-rat LD50:20 mg/kg FMCHA2-,C63,91
orl-dog LD50:19 mg/kg 28ZEAL 5,118,76
ipr-rat LD50:4700 µg/kg PCBPBS 1,445,71
orl-mus LD50:18 mg/kg 28ZEAL 5,118,76
skn-rbt LD50:10,200 mg/kg 28ZEAL 5,118,76
orl-ckn LD50:21,500 µg/kg 28ZEAL 5,118,76

CONSENSUS REPORTS: Reported in EPA TSCA Inventory. EPA Extremely Hazardous Substances List.

SAFETY PROFILE: Poison by ingestion and intraperitoneal routes. Mildly toxic by skin contact. When heated to decomposition it emits very toxic fumes of NO_x and HCl.

DSO500 CAS:56877-15-7 HR: 3
1,3-DIMETHYL-5-(METHYLAMINO)-4-PYRAZOLYL o-FLUOROPHENYL KETONE
mf: $C_{13}H_{14}FN_3O$ mw: 247.30

SYNS: (1,3-DIMETHYL-5-(METHYLAMINO)-1H-PYRAZOL-4-YL)(2-FLUOROPHENYL)METHANONE □ KETONE, 1,3-DIMETHYL-5-(METHYLAMINO)-4-PYRAZOLYL o-FLUOROPHENYL □ METHANONE, (1,3-DIMETHYL-5-(METHYLAMINO)-1H-PYRAZOL-4-YL)(2-FLUOROPHENYL)- □ PD 73093

TOXICITY DATA WITH REFERENCE
mma-sat 1 µmol/plate CRNGDP 7,2019,86

DOT CLASSIFICATION: 3; *Label:* Flammable Liquid

SAFETY PROFILE: Mutation data reported. A flammable liquid. When heated to decomposition it emits toxic vapors of NO_x and F^-.

DSO800 CAS:2449-49-2 HR: 2
N,N-DIMETHYL-α-METHYLBENZYLAMINE
mf: $C_{10}H_{15}N$ mw: 149.26

PROP: Bp: 194–195°.

SYN: N,N,α-TRIMETHYLBENZYLAMINE

TOXICITY DATA WITH REFERENCE
skn-rbt 10 mg/24H open AMIHBC 10,61,54
skn-rbt 5 mg/24H SEV 85JCAE-,451,86
eye-rbt 20 mg open AMIHBC 10,61,54
orl-rat LD50:420 mg/kg AMIHBC 10,61,54
ihl-rat LCLo:125 ppm/4H AMIHBC 10,61,54
skn-rbt LD50:890 mg/kg AMIHBC 10,61,54

SAFETY PROFILE: Moderately toxic by ingestion, skin contact, and inhalation. A severe skin and eye irritant. When heated to decomposition it emits toxic fumes of NO_x.

DSP400 CAS:60-51-5 HR: 3
O,O-DIMETHYL METHYLCARBAMOYLMETHYL PHOSPHORODITHIOATE
mf: $C_5H_{12}NO_3PS_2$ mw: 229.27

PROP: Crystals from Et_2O or toluene/hexane. Mp: 51–52°. Sol in H_2O, alcohols, $CHCl_3$, C_6H_6, and ketones.

SYNS: AC-12682 □ AMERICAN CYANAMID 12880 □ BI-58 □ CEKUTHOATE □ CL 12880 □ CYGON □ CYGON INSECTICIDE □ DAPHENE □ DE-FEND □ DEMOS-L40 □ DEVIGON □ DIMATE 267 □ DIMETATE □ DIMETHOAAT (DUTCH) □ DIMETHOAT (GERMAN) □ DIMETHOATE (USDA) □ DIMETHOAT TECHNISCH 95% □ DI-

METHOGEN □ O,O-DIMETHYLDITHIOPHOSPHORYLACETIC ACID-N-MONOMETHYLAMIDE SALT □ O,O-DIMETHYL-DITHIOPHOSPHORYL-ESSIGSAEURE MONOMETHYLAMID (GERMAN) □ O,O-DIMETHYL-S-(2-(METHYLAMINO)-2-OXOETHYL) PHOSPHORODITHIOATE □ O,O-DIMETHYL-S-(N-METHYLCARBAMOYL)-METHYLDITHIOFOSFAAT (DUTCH) □ (O,O-DIMETHYL-S-(N-METHYL-CARBAMOYLMETHYL)-DITHIOPHOSPHAT) (GERMAN) □ O,O-DIMETHYL-S-(N-METHYLCARBAMOYLMETHYL) DITHIOPHOSPHATE □ O,O-DIMETHYL-S-(N-METHYLCARBAMOYLMETHYL) PHOSPHORODITHIOATE □ O,O-DIMETHYL-S-(N-METHYLCARBAMYLMETHYL) THIOTHIONO-PHOSPHATE □ O,O-DIMETHYL-S-(N-MONOMETHYL)-CARBAMYL METHYLDITHIOPHOSPHATE □ O,O-DIMETHYL-S-(2-OXO-3-AZA-BUTYL)-DITHIOPHOSPHAT (GERMAN) □ O,O-DIMETIL-S-(N-METIL-CARBAMOIL-METIL)-DITIOFOSFATO (ITALIAN) □ DIMETON □ DIMEVUR □ DITHIOPHOSPHATE de O,O-DIMETHYLE et de S(-N-METHYLCARBAMOYL-METHYLE) (FRENCH) □ EI-12880 □ ENT 24,650 □ EXPERIMENTAL INSECTICIDE 12,880 □ FERKETHION □ FORTION NM □ FOSFAMID □ FOSFOTOX □ FOSTION MM □ L-395 □ LURGO □ S-METHYLCARBAMOYLMETHYL-O,O-DIMETHYL PHOSPHORODITHIOATE □ N-MONOMETHYLAMIDE of O,O-DIMETHYL DITHIOPHOSPHORYLACETIC ACID □ NC-262 □ NCI-C00135 □ PERFECTHION □ PHOSPHAMID □ PHOSPHORODITHIOIC ACID-O,O-DIMETHYL-S-(2-(METHYLAMINO)-2-OXOETHYL) ESTER □ RACUSAN □ RCRA WASTE NUMBER P044 □ REBELATE □ ROGODIAL □ ROGOR □ ROXION U.A. □ SINORATOX □ TRIMETION

TOXICITY DATA WITH REFERENCE
mma-sat 500 μg/plate JTEHD6 16,403,85
mma-hmn:fbr 100 μmol/L MUREAV 42,161,77
sce-hmn:lym 20 mg/L MUREAV 88,307,81
orl-mus TDLo:1050 mg/kg (MGN):REP TXAPA9 26,29,73
ipr-mus TDLo:40 mg/kg (female 1D post):TER BIRUAA 13,238,75
orl-mus TDLo:220 mg/kg (female 6-16D post):TER JESEDU 20,373,85
orl-rat TDLo:120 mg/kg (6-15D preg):TER BECTA6 22,522,79
orl-rat TDLo:256 mg/kg/4W-I:CAR ARGEAR 41,311,73
ims-rat TDLo:176 mg/kg/6W-I:CAR ARGEAR 41,311,73
orl-man TDLo:286 mg/kg:PNS,BAH ICMED9 12,110,86
orl-hmn LD50:30 mg/kg GUCHAZ 6,209,73
orl-man TDLo:300 mg/kg JTCTDW 24,69,86
orl-rat LD50:60 mg/kg YKYUA6 30,623,79
skn-rat LD50:353 mg/kg BJIMAG 26,59,69
ipr-rat LD50:100 mg/kg BJIMAG 21,52,64
scu-rat LD50:350 mg/kg BJIMAG 21,52,64
ivn-rat LD50:450 mg/kg BJIMAG 21,52,64
orl-mus LD50:60 mg/kg BJIMAG 21,52,64
ipr-mus LD50:45 mg/kg BJIMAG 21,52,64
orl-dog LD50:400 mg/kg SPEADM 78-1,31,78

CONSENSUS REPORTS: NCI Carcinogenesis Bioassay (feed); No Evidence: mouse, rat NCITR* NCI-CG-TR-4,77. Reported in EPA TSCA Inventory. EPA Genetic Toxicology Program. EPA Extremely Hazardous Substances List.

SAFETY PROFILE: A deadly human poison. Poison by ingestion, skin contact, intraperitoneal, and subcutaneous routes. Moderately toxic by intravenous route. Human systemic effects: coma, dyspnea, fasciculations. Questionable carcinogen with experimental carcinogenic data. Experimental teratogenic and reproductive effects. Human mutation data reported. When heated to decomposition it emits very toxic fumes of NO_x, PO_x, and SO_x. See also ESTERS.

DSP600 CAS:23135-22-0 HR: 3
N',N'-DIMETHYL-N-((METHYLCARBAMOYL)OXY)-1-METHYLTHIOOXAMIMIDIC ACID
mf: $C_7H_{13}N_3O_3S$ mw: 219.29

PROP: Solid in H_2O. Mp: 100–102°. Sol in H_2O, Me_2CO, EtOH, and MeOH.

SYNS: D-1410 □ 2-(DIMETHYLAMINO)-N-(((METHYLAMINO)CARBONYL)OXY)-2-OXOETHANIMIDOTHIOIC ACID METHYL ESTER □ 2-DIMETHYLAMINO-1-(METHYLTHIO)GLYOXAL-o-METHYLCARBAMOYLMONOXIME □ N,N-DIMETHYL-α-METHYLCARBAMOYLOXYIMINO-α-(METHYLTHIO)ACETAMIDE □ N',N'-DIMETHYL-N-((METHYLCARBAMOYL)OXY)-1-THIOOXAMIMIDIC ACID METHYL ESTER □ DPX 1410 □ INSECTICIDE-NEMATICIDE 1410 □ METHYL-2-(DIMETHYLAMINO)-N-(((METHYLAMINO)CARBONYL)OXY)-2-OXOETHANIMIDOTHIOATE □ METHYL-1-(DIMETHYLCARBAMOYL)-N-(METHYLCARBAMOYLOXY)THIOFORMIMIDATE □ S-METHYL-1-(DIMETHYLCARBAMOYL)-N-((METHYLCARBAMOYL)OXY)THIOFORMIMIDATE □ METHYL-N',N'-DIMETHYL-N-((METHYLCARBAMOYL)OXY)-1-THIOOXAMIMIDATE □ OXAMYL □ THIOXAMYL □ VYDATE □ VYDATE L INSECTICIDE/NEMATICIDE □ VYDATE L OXAMYL INSECTICIDE/NEMATOCIDE

TOXICITY DATA WITH REFERENCE
orl-rat TDLo:945 mg/kg (male 12W pre):REP FAATDF 7,106,86
orl-rat TDLo:1890 mg/kg (multi):REP FAATDF 7,106,86
orl-rat LD50:2500 μg/kg FAATDF 6,423,86
ihl-rat LC50:170 mg/m³/1H 85DPAN -,-,71/76
skn-rat LDLo:300 mg/kg FAATDF 6,423,86
orl-mus LD50:2300 μg/kg FAATDF 6,423,86
skn-rbt LD50:740 mg/kg SPEADM 78-1,61,78
orl-qal LD50:4180 μg/kg 85DPAN -,-,71/76

CONSENSUS REPORTS: EPA Extremely Hazardous Substances List.

SAFETY PROFILE: Poison by ingestion, skin contact, and inhalation. Experimental reproductive effects. Moderately toxic by skin contact. When heated to decomposition it emits very toxic fumes of NO_x and SO_x.

DSP650 CAS:69462-47-1 HR: 1
2,6-DIMETHYL-1-((2-METHYLCYCLOHEXYL)CARBONYL)PIPERIDINE
mf: $C_{15}H_{27}NO$ mw: 237.43

SYNS: AI3-36561 □ PIPERIDINE, 2,6-DIMETHYL-1-((2-METHYLCYCLOHEXYL)CARBONYL)-

TOXICITY DATA WITH REFERENCE
skn-rbt 500 mg/24H MLD AEHA** 51-029-76
eye-rbt 100 mg/24H MOD AEHA** 51-029-76

SAFETY PROFILE: A skin and eye irritant. When heated to decomposition it emits toxic fumes of NO_x.

DSQ000 CAS:122-14-5 HR: 3
DIMETHYL-3-METHYL-4-NITROPHENYLPHOSPHOROTHIONATE
mf: $C_9H_{12}NO_5PS$ mw: 277.25

PROP: Yellow oil. D: 1.323 @ 25°/4°, bp: 140–145° @ 0.1 mm. Insol in H_2O; sltly sol in ligroin.

SYNS: ACCOTHION □ ACEOTHION □ AGRIA 1050 □ AGRIYA 1050 □ AGROTHION □ AMERICAN CYANAMID CL-47,300 □ ARBO-

GAL ☐ BAY 41831 ☐ BAYER 41831 ☐ BAYER S 5660 ☐ CEKUTROTHION ☐ CL 47300 ☐ CP 47114 ☐ CYFEN ☐ CYTEL ☐ CYTEN ☐ O,O-DIMETHYL-O-(3-METHYL-4-NITROFENYL)-MONOTHIOFOSFAAT (DUTCH) ☐ O,O-DIMETHYL-O-(3-METHYL-4-NITRO-PHENYL)-MONOTHIOPHOSPHAT (GERMAN) ☐ O,O-DIMETHYL-O-(3-METHYL-4-NITROPHENYL) PHOSPHOROTHIOATE ☐ O,O-DIMETHYL-O-(3-METHYL-4-NITROPHENYL) THIOPHOSPHATE ☐ O,O-DIMETHYL-O-(3-METHYL) PHOSPHOROTHIOATE ☐ O,O-DIMETHYL-O-(4-NITRO-3-METHYLPHENYL)THIOPHOSPHATE ☐ O,O-DIMETHYL-O-4-NITRO-m-TOLYL PHOSPHOROTHIOATE ☐ O,O-DIMETIL-O-(3-METIL-4-NITROFENIL)-MONOTIOFOSFATO (ITALIAN) ☐ EI 47300 ☐ ENT 25,715 ☐ FALITHION ☐ FENITOX ☐ FENITROTHION ☐ FENITROTION (HUNGARIAN) ☐ FOLETHION ☐ H-35-F 87 (BVM) ☐ 8057HC ☐ KOTION ☐ MEP (Pesticide) ☐ METATHIONE ☐ METATION ☐ METHYLNITROPHOS ☐ MONSANTO CP 47114 ☐ NITROPHOS ☐ NOVATHION ☐ NUVANOL ☐ OLEOSUMIFENE ☐ OMS 43 ☐ OVADOFOS ☐ PENNWALT C-4852 ☐ PHENITROTHION ☐ S 112A ☐ S 5660 ☐ SUMITHIAN ☐ THIOPHOSPHATE de O,O-DIMETHYLE et de O-(3-METHYL-4-NITROPHENYLE) (FRENCH) ☐ VERTHION

TOXICITY DATA WITH REFERENCE
mmo-sat 500 μg/plate MUREAV 116,185,83
dni-omi 100 ppm NNGADV 9,325,84
orl-rat TDLo:90 mg/kg (female 7-15D post):REP NRTXDN 11,321,89
orl-wmn TDLo:800 mg/kg:GIT,PUL ARTODN 56,136,84
orl-man LDLo:429 μL/kg:BAH,PUL,GIT HUTODJ 6,403,87
orl-rat LD50:250 mg/kg TXAPA9 21,315,72
ihl-rat LC50:378 mg/m³/4H EGESAQ 24,173,80
skn-rat LD50:1250 mg/kg GISAAA 31(10),12,66
ipr-rat LD50:300 mg/kg TXAPA9 63,91,82
ivn-rat LD50:33 mg/kg ABCHA6 27,669,63
itr-rat LD50:950 mg/kg TXAPA9 63,91,82
orl-mus LD50:715 mg/kg HYSAAV 31,13,66
skn-mus LD50:2500 mg/kg ABCHA6 25,605,61
scu-mus LD50:1000 mg/kg ABCHA6 25,605,61

CONSENSUS REPORTS: EPA Genetic Toxicology Program. EPA Extremely Hazardous Substances List.

SAFETY PROFILE: Poison by ingestion, inhalation, intravenous, and intraperitoneal routes. Moderately toxic by skin contact, intratracheal, and subcutaneous routes. Human systemic effects: coma, diarrhea, dyspnea, gastrointestinal changes, hypermotility, nausea or vomiting, respiratory depression. Mutation data reported. When heated to decomposition it emits very toxic fumes of NO_x, PO_x, and SO_x.

DSQ600 CAS:3572-74-5 HR: 3
N,N-DIMETHYL-2-(α-METHYL-α-PHENYLBENZYLOXY) ETHYLAMINE
mf: $C_{18}H_{23}NO$ mw: 269.42

PROP: Bp: 125–135° @ 0.5 mm.

SYNS: N,N-DIMETHYL-2-((α-METHYL-α-PHENYLBENZYL)OXY)ETHYLAMINE ☐ SUBSTANZ NR. 1934 (GERMAN)

TOXICITY DATA WITH REFERENCE
ivn-mus LD50:43 mg/kg ARZNAD 4,189,54
scu-gpg LD50:54 mg/kg ARZNAD 4,189,54

SAFETY PROFILE: Poison by intravenous and subcutaneous routes. When heated to decomposition it emits toxic fumes of NO_x.

DSQ800 CAS:66-79-5 HR: 3
3,3-DIMETHYL-6-(((5-METHYL-3-PHENYL)-4-ISOXAZOLECARBOXAMIDE)-7-OXO)-4-THIA-1-AZABICYCLO(3.2.0)HEPTANE-2-CARBOXYLIC ACID

SYNS: BRL 1400 ☐ 5-METHYL-3-PHENYL-4-ISOXAZOLYL-PENICILLIN ☐ MPI-PC ☐ MPI-PENICILLIN ☐ OXACILLIN ☐ OXAZOCILLIN ☐ PENICILLIN P-12 ☐ PROSTAPHLYN ☐ STAPENOR

TOXICITY DATA WITH REFERENCE
oth-mic-omi 4 mg/L JOBAAY 170,1831,88
ivn-wmn TDLo:5560 mg/kg/20D-I:BLD SMJOAV 70,1245,77
ivn-man TDLo:3800 mg/kg/19D-I:BLD JOPDAB 89,769,76
ivn-cld TDLo:2550 mg/kg/17D-I:BLD JOPDAB 90,668,77
ivn-inf TDLo:3800 mg/kg/19D-I:BLD JOPDAB 89,769,76
orl-mus LD50:6500 mg/kg 85GMAT -,95,82
ivn-mus LD50:1500 mg/kg ARZNAD 15,322,65
orl-cat LDLo:750 mg/kg ARZNAD 15,322,65

SAFETY PROFILE: Poison by an unspecified route. Moderately toxic by ingestion and intravenous routes. Human systemic effects: agranulocytosis. Mutation data reported. When heated to decomposition it emits very toxic fumes of NO_x and SO_x. See also other penicillin entries.

DSR200 CAS:20241-03-6 HR: 3
3,3-DIMETHYL-1-(m-METHYLPHENYL)TRIAZENE
mf: $C_9H_{13}N_3$ mw: 163.25

SYNS: 3,3-DIMETHYL-1-(m-TOLYL)TRIAZENE ☐ 1-(m-METHYLPHENYL)-3,3-DIMETHYLTRIAZENE ☐ 1-(3-METHYLPHENYL)-3,3-DIMETHYLTRIAZENE

TOXICITY DATA WITH REFERENCE
mma-sat 400 nmol/L JMCMAR 22,473,79
orl-rat TDLo:250 mg/kg:CAR ZKKOBW 81,285,74
scu-rat TDLo:500 mg/kg:CAR ZKKOBW 81,285,74
scu-rat TD:2700 mg/kg/56W-I:CAR ZKKOBW 81,285,74
orl-rat LD50:300 mg/kg ZKKOBW 81,285,74
scu-rat LD50:500 mg/kg ZKKOBW 81,285,74
ipr-mus LD50:201 mg/kg JMCMAR 19,1299,76

SAFETY PROFILE: Poison by ingestion and intraperitoneal routes. Moderately toxic by subcutaneous route. Questionable carcinogen with experimental carcinogenic data. Mutation data reported. When heated to decomposition it emits toxic fumes of NO_x.

DSR400 CAS:756-79-6 HR: 2
DIMETHYL METHYLPHOSPHONATE
mf: $C_3H_9O_3P$ mw: 124.09

PROP: Pleasant-smelling liquid. Bp: 66–68° @ 10 mm.

SYNS: DMMP ☐ METHYLPHOSPHONIC ACID DIMETHYL ESTER ☐ NCI-C56762

TOXICITY DATA WITH REFERENCE
dlt-mus-orl 65 g/kg/13W-C MUREAV 138,213,84
cyt-ham:ovr 250 mg/L NTIS** AD-A124-785
orl-mus TDLo:33 g/kg (female 7-14D post):REP NTIS** PB85-220143
orl-rat TDLo:63 g/kg (63D male):REP TXAPA9 72,379,84

orl-rat TDLo:15,750 mg/kg (male 63D pre):REP TXAPA9 72,379,84
orl-rat TDLo:126 g/kg (male 63D pre):REP TXAPA9 72,379,84
orl-rat TDLo:515 g/kg/2Y-C:CAR FAATDF 11,91,88
orl-rat LD50:8210 mg/kg TSCAT* FYI-OTS-0483-0242
ivn-rat LD50:1050 mg/kg TSCAT* FYI-OTS-0483-0242
orl-mus LD50:>6810 mg/kg NTPTR* NTP-TR-323,87
ivn-mus LD50:912 mg/kg TSCAT* FYI-OTS-0483-0242

CONSENSUS REPORTS: Reported in EPA TSCA Inventory.

SAFETY PROFILE: Moderately toxic by intravenous route. Experimental reproductive effects. Questionable carcinogen with experimental carcinogenic data. Mutation data reported. An experimental nerve gas stimulant. A flame retardant. When heated to decomposition it emits toxic fumes of PO_x.

DSR600 CAS:50308-86-6 HR: 3
1,3-DIMETHYL-4-(p-((p-((1-METHYLPYRIDINIUM-4-YL) AMINO)PHENYL)CARBAMOYL)ANILINOQUINOLINIUM), DIBROMIDE
mf: $C_{30}H_{29}N_5O{\cdot}2Br$ mw: 635.46

TOXICITY DATA WITH REFERENCE
dnd-mus:lym 4900 nmol/L JMCMAR 22,134,79
ipr-mus LD10:5 mg/kg JMCMAR 22,134,79

SAFETY PROFILE: Poison by intraperitoneal route. Mutation data reported. When heated to decomposition it emits very toxic fumes of NO_x and Br^-.

DSR800 CAS:50308-87-7 HR: 3
1,6-DIMETHYL-4-((p-((p-((1-METHYLPYRIDINIUM-4-YL)AMINO)PHENYL) CARBAMOYL)ANILINO)QUINOLINIUM) DI-p-TOLUENESULFONATE
mf: $C_{30}H_{29}N_5O{\cdot}2C_7H_7O_3S$ mw: 818.04

TOXICITY DATA WITH REFERENCE
dnd-mus:lym 1050 nmol/L JMCMAR 22,134,79
ipr-mus LD10:15 mg/kg JMCMAR 22,134,79

SAFETY PROFILE: Poison by intraperitoneal route. Mutation data reported. When heated to decomposition it emits very toxic fumes of NO_x and SO_x.

DSS000 CAS:50425-34-8 HR: 3
1,8-DIMETHYL-4-((p-((p-((1-METHYLPYRIDINIUM-4-YL)AMINO)PHENYL)CARBAMOYL)ANILINO)QUINOLINIUM)DI-p-TOLUENESULFONATE
mf: $C_{30}H_{29}N_5O{\cdot}2C_7H_7O_3S$ mw: 818.04

TOXICITY DATA WITH REFERENCE
dnd-mus:lym 870 nmol/L JMCMAR 22,134,79
ipr-mus LD10:60 mg/kg JMCMAR 22,134,79

SAFETY PROFILE: Poison by intraperitoneal route. Mutation data reported. When heated to decomposition it emits very toxic fumes of NO_x and SO_x.

DSS200 CAS:7347-46-8 HR: 2
N,N-DIMETHYL-4-(2-METHYL-4-PYRIDYLAZO)ANILINE-N-OXIDE
mf: $C_{14}H_{16}N_4O$ mw: 256.34

SYNS: 4-((4-(DIMETHYLAMINO)PHENYL)AZO)-2-PICOLINE-1-OXIDE □ N,N-DIMETHYL-4-((2-METHYL-4-PYRIDINYL)AZO)BENZENAMINE-N-OXIDE □ N,N-DIMETHYL-4-(4'-(2'-METHYLPYRIDYL-1'-OXIDE)AZO)ANILINE □ 2'-MePO$_4$' □ 2-METHYLPYRIDINE-1-OXIDE-4-AZO-p-DIMETHYLANILINE

TOXICITY DATA WITH REFERENCE
orl-rat TDLo:714 mg/kg/17W-C:NEO JNCIAM 37,365,66
orl-rat TD:2142 mg/kg/17W-C:ETA JNCIAM 41,855,68

SAFETY PROFILE: Questionable carcinogen with experimental neoplastigenic and tumorigenic data. When heated to decomposition it emits toxic fumes of NO_x.

DSS800 CAS:3761-42-0 HR: 3
O,O-DIMETHYL-o-(4-(METHYLSULFONYL)-m-TOLYL) PHOSPHOROTHIOATE
mf: $C_{10}H_{15}O_5PS_2$ mw: 310.34

SYNS: O,O-DIMETHYL-o-((4-METHYLTHIO)-m-TOLYL) PHOSPHOROTHIOATE SULFONE □ FENTHIONE SULFONE

TOXICITY DATA WITH REFERENCE
orl-rat LD50:125 mg/kg GUCHAZ 6,279,73
ipr-rat LDLo:250 mg/kg TXAPA9 6,86,64
orl-mus LD50:210 mg/kg JOPHDQ 9,697,86

SAFETY PROFILE: Poison by ingestion and intraperitoneal routes. When heated to decomposition it emits very toxic fumes of PO_x and SO_x.

DSS900 CAS:39195-82-9 HR: 2
3,3-DIMETHYL-1-(METHYLTHIO)-2-BUTANONE OXIME
mf: $C_7H_{15}NOS$ mw: 161.29

SYN: 2-BUTANONE, 3,3-DIMETHYL-1-(METHYLTHIO)-, OXIME

TOXICITY DATA WITH REFERENCE
orl-rat LD50:813 mg/kg JAFCAU 23,963,75

CONSENSUS REPORTS: Reported in EPA TSCA Inventory.

SAFETY PROFILE: Moderately toxic by ingestion. When heated to decomposition it emits toxic vapors of NO_x and SO_x.

DST000 CAS:2032-65-7 HR: 3
3,5-DIMETHYL-4-METHYLTHIOPHENYL-N-METHYLCARBAMATE
mf: $C_{11}H_{15}NO_2S$ mw: 225.33

PROP: Crystals or powder. Mp: 117–118°. Sol in most org solvs; pract insol in H_2O.

SYNS: BAY 9026 □ BAYER 37344 □ 3,5-DIMETHYL-4-(METHYLTHIO)PHENOL METHYLCARBAMATE □ 3,5 DIMETHYL-4-METHYLTHIOPHENYL-N-CARBAMAT (GERMAN) □ DRAZA □ ENT 25,726 □ H 321 □ MERCAPTODIMETHUR (DOT) □ MESUROL □ METHIOCARB □ METHYL CARBAMIC ACID-4-(METHYLTHIO)-3,5-XYLYL ESTER □ 4-METHYLMERCAPTO-3,5-DIMETHYLPHENYL N-METHYLCARBAMATE □ 4-METHYLMERCAPTO-3,5-XYLYL METHYLCARBAMATE □ 4-METHYL-

D

THIO-3,5-DIMETHYLPHENYL METHYLCARBAMATE □ 4-(METHYLTHIO)-3,5-XYLENOL METHYLCARBAMATE □ 4-(METHYLTHIO)-3,5-XYLYL METHYLCARBAMATE □ METMERCAPTURON □ OMS-93

TOXICITY DATA WITH REFERENCE
orl-rat LD50:15 mg/kg FMCHA2 -,C150,83
skn-rat LD50:350 mg/kg PCOC** -,105,66
orl-mus LD50:25,200 μg/kg TOIZAG 17,60,70
ipr-mus LD50:16 mg/kg TXAPA9 6,402,64
orl-gpg LD50:40 mg/kg 85DPAN -,-,71/76

CONSENSUS REPORTS: EPA Extremely Hazardous Substances List.

SAFETY PROFILE: Poison by ingestion, skin contact, and intraperitoneal routes. Used as an insecticide, molluscicide, and bird repellent. When heated to decomposition it emits very toxic fumes of NO_x and SO_x. See also ESTERS and CARBAMATES.

DST200 CAS:55-37-8 HR: 3
O,O-DIMETHYL-O-4-(METHYLTHIO)-3,5-XYLYL PHOSPHOROTHIOATE
mf: $C_{11}H_{17}O_3PS_2$ mw: 292.37

SYNS: BAY 37342 □ BAYER 9013 □ BAYER 37342 □ O,O-DIMETHYL-O-(3,5-DIMETHYL-4-METHYLTHIOPHENYL) PHOSPHOROTHIOATE □ O-(3,5-DIMETHYL-4-(METHYLTHIO)PHENYL)-O,O-DIMETHYL PHOSPHOROTHIOATE □ ENT 25,684 □ G 347

TOXICITY DATA WITH REFERENCE
orl-rat LD50:1000 mg/kg TXAPA9 21,315,72
orl-mus LDLo:1070 mg/kg AECTCV 14,111,85
orl-ckn LD50:103 mg/kg TXAPA9 11,49,67
orl-bwd LD50:10 mg/kg TXAPA9 21,315,72

SAFETY PROFILE: Poison by ingestion. When heated to decomposition it emits very toxic fumes of PO_x and SO_x.

DST600 CAS:141-91-3 HR: 3
2,6-DIMETHYLMORPHOLINE
mf: $C_6H_{13}NO$ mw: 115.20

PROP: Liquid. D: 0.9346, bp: 146.6°, fp: −85°, flash p: 112°F (OC), vap d: 4.0.

SYN: 2,6-DIMETHYL-2,3,5,6-TETRAHYDRO-4H-1,4-OXAZINE

TOXICITY DATA WITH REFERENCE
skn-rbt 10 mg/24H open MLD AIHAAP 23,95,62
eye-rbt 2 mg/24H SEV 85JCAE-,888,86
mmo-sat 6666 μg/plate ENMUDM 5(Suppl 1),3,83
orl-rat LD50:2830 mg/kg UCDS** 11/13/61
skn-rbt LD50:710 mg/kg AIHAAP 23,95,62

CONSENSUS REPORTS: Reported in EPA TSCA Inventory.

SAFETY PROFILE: Moderately toxic by ingestion and skin contact. A skin and eye irritant. Mutation data reported. Flammable liquid when exposed to heat, flame, or oxidizers. To fight fire, use alcohol foam. When heated to decomposition it emits toxic fumes of NO_x.

DST800 CAS:597-25-1 HR: 3
DIMETHYLMORPHOLINOPHOSPHONATE
mf: $C_6H_{14}NO_4P$ mw: 195.18

PROP: A liquid. Bp: 96° @ 1 mm.

SYNS: DIMETHYL MORPHOLINOPHOSPHORAMIDATE □ DMMPA □ MORPHOLINOPHOSPHONIC ACID DIMETHYL ESTER □ 4-MORPHOLINYLPHOSPHONIC ACID DIMETHYL ESTER □ NCI-C54740

TOXICITY DATA WITH REFERENCE
msc-mus:lym 2200 mg/L NTPTR* NTP-TR-298,86
cyt-ham:ovr 3 g/L NTPTR* NTP-TR-298,86
sce-ham:ovr 3 g/L NTPTR* NTP-TR-298,86
orl-rat TDLo:309 g/kg/2Y-I:CAR NTPTR* NTP-TR-298,86
orl-rat LD50:6 g/kg NTPTR* NTP-TR-298,86
ipr-rat LD50:2400 mg/kg NTPTR* NTP-TR-298,86
ims-rat LD50:5200 mg/kg NTPTR* NTP-TR-298,86
orl-mus LD50:3300 mg/kg NTPTR* NTP-TR-298,86
ipr-mus LD50:5 g/kg NTPTR* NTP-TR-298,86
ivn-mus LD50:400 mg/kg NTPTR* NTP-TR-298,86
ims-mus LD50:4800 mg/kg NTPTR* NTP-TR-298,86
ivn-rbt LD50:350 mg/kg NTPTR* NTP-TR-298,86

CONSENSUS REPORTS: NTP Carcinogenesis Studies (gavage); Some Evidence: rat NTPTR* NTP-TR-298,86; No Evidence: mouse NTPTR* NTP-TR-298,86.

SAFETY PROFILE: Poison by intravenous route. Moderately toxic by ingestion and intraperitoneal routes. Questionable carcinogen with experimental carcinogenic data. Mutation data reported. When heated to decomposition it emits very toxic fumes of NO_x and PO_x. See also ESTERS.

DSU000 CAS:55-93-6 HR: 3
DIMETHYLMYLERAN
mf: $C_8H_{18}O_6S_2$ mw: 272.36

SYNS: DDM □ 2,5-DIMETHANESULFOMYLOXYHEXANE □ 1,4-DIMETHANESULFONOXY-1,4-DIMETHYLBUTANE □ 2,5-HEXANEDIOL DIMETHYLSULFONATE □ NSC-23890

TOXICITY DATA WITH REFERENCE
sln-dmg-orl 1 pph ZEVBA5 90,457,59
dlt-ofs-ipr 4 mg/kg MUREAV 58,263,78
spm-mus-ipr 4 mg/kg EXPEAM 30,178,74
dlt-mus-ipr 8 mg/kg IRLCDZ 5,341,77
ipr-mus LD50:16 mg/kg JNCIAM 56,609,76
ivn-dog LDLo:1 mg/kg CCSUBJ 2,203,65
ivn-mky LDLo:1 mg/kg CCSUBJ 2,203,65

CONSENSUS REPORTS: EPA Genetic Toxicology Program.

SAFETY PROFILE: Poison by intravenous and intraperitoneal routes. Mutation data reported. Used for treatment of chronic granulocytic leukemia. When heated to decomposition it emits very toxic fumes of SO_x.

DSU400 CAS:86-56-6 HR: 3
N,N-DIMETHYL-1-NAPHTHYLAMINE
mf: $C_{12}H_{13}N$ mw: 171.26

PROP: A liquid. D: 1.052 @ 4°/4°, bp: 272–274°.

SYNS: 1-DIMETHYLAMINONAPHTHALENE □ DIMETHYL-α-NAPH-

THYLAMINE □ α-DIMETHYLNAPHTHYLAMINE □ N,N-DIMETHYL-α-NAPHTHYLAMINE

TOXICITY DATA WITH REFERENCE
dnd-mus-ipr 50 mg/kg CRNGDP 2,265,81
orl-rat LDLo:500 mg/kg JPETAB 90,260,47
ipr-mus LD50:75 mg/kg NTIS** AD691-490

CONSENSUS REPORTS: Reported in EPA TSCA Inventory.

SAFETY PROFILE: Poison by intraperitoneal route. Moderately toxic by ingestion. Mutation data reported. When heated to decomposition it emits toxic fumes of NO_x.

DSU600 CAS:607-59-0 HR: 2
N,N-DIMETHYL-p-(1-NAPHTHYLAZO)ANILINE
mf: $C_{18}H_{17}N_3$ mw: 275.38

SYNS: DAN □ p-DIMETHYLAMINOBENZENEAZO-1-NAPHTHALENE □ p-DIMETHYLAMINOBENZENE-1-AZO-1-NAPHTHALENE

TOXICITY DATA WITH REFERENCE
dns-rat:lvr 100 μmol/L MUREAV 136,255,84
dns-rat-orl 100 mg/kg ENMUDM 7,101,85
dns-ham:lvr 10 μmol/L MUREAV 136,255,84
orl-rat TDLo:25 g/kg/79W-C:ETA JNCIAM 13,57,52
scu-rat TDLo:90 mg/kg/2W-I:CAR JNCIAM 18,843,57

SAFETY PROFILE: Questionable carcinogen with experimental carcinogenic and tumorigenic data. Mutation data reported. When heated to decomposition it emits toxic fumes of NO_x.

DSU800 CAS:613-65-0 HR: 2
N,N-DIMETHYL-4(2′-NAPHTHYLAZO)ANILINE
mf: $C_{18}H_{17}N_2$ mw: 261.37

SYNS: DA-2-N □ p-DIMETHYLAMINOBENZENE-1-AZO-2-NAPHTHALENE □ 2-(4-DIMETHYLAMINOPHENYLAZO)NAPHTHALENE

TOXICITY DATA WITH REFERENCE
dns-rat:lvr 10 μmol/L MUREAV 136,255,84
dns-rat-orl 100 mg/kg ENMUDM 7,101,85
dns-ham:lvr 2 μmol/L MUREAV 136,255,84
orl-rat TDLo:8630 mg/kg/230D-C:CAR JNCIAM 14,571,53
skn-mus TDLo:4800 mg/kg:ETA JNCIAM 13,1259,53

SAFETY PROFILE: Questionable carcinogen with experimental carcinogenic and tumorigenic data. Mutation data reported. When heated to decomposition it emits toxic fumes of NO_x.

DSV000 CAS:63019-14-7 HR: 2
N,N-DIMETHYL-p-(2-(1-NAPHTHYL)VINYL)ANILINE
mf: $C_{20}H_{19}N$ mw: 273.40

SYN: 1-(4′-DIMETHYLAMINOPHENYL)-2-(1′-NAPHTHYL)ETHYLENE

TOXICITY DATA WITH REFERENCE
scu-rat TDLo:215 mg/kg/W-I:ETA PTRMAD 241,147,48
scu-mus TDLo:320 mg/kg/W-I:ETA,REP PTRMAD 241,147,48

SAFETY PROFILE: Experimental reproductive effects. Questionable carcinogen with experimental tumorigenic data. When heated to decomposition it emits toxic fumes of NO_x.

DSV200 CAS:4164-28-7 HR: 3
DIMETHYLNITRAMINE
mf: $C_2H_6N_2O_2$ mw: 90.10

PROP: Needles from ligroin. Mp: 58°, bp: 187°. Sol in H_2O.

SYNS: DIMETHYLNITRAMIN (GERMAN) □ DIMETHYLNITROAMINE □ DMNM □ DMNO □ N-NITRODIMETHYLAMINE □ N-NITRODMA

TOXICITY DATA WITH REFERENCE
mma-sat 250 μmol/plate CRNGDP 5,809,84
hma-rat/sat 200 mg/kg CNREA8 41,3205,81
orl-rat TDLo:90 mg/kg/2Y-C:CAR CRNGDP 10,1977,89
orl-rat TD:34 g/kg/82W-C:ETA ZEKBAI 69,103,67
orl-rat TDLo:20 g/kg/1Y-C:ETA JJIND8 64,1435,80
orl-rat LD50:1095 mg/kg TXAPA9 33,185,75
ipr-rat LD50:897 mg/kg TXAPA9 33,185,75
ipr-mus LD50:399 mg/kg DCTODJ 1,363,78

SAFETY PROFILE: Poison by intraperitoneal route. Moderately toxic by ingestion. Questionable carcinogen with experimental tumorigenic data. Mutation data reported. When heated to decomposition it emits toxic fumes of NO_x.

DSV289 CAS:22691-91-4 HR: 3
3,3-DIMETHYL-1-NITRO-1-BUTYNE
mf: $C_6H_9NO_2$ mw: 127.14

$O_2NC{\equiv}CC(CH_3)_3$

PROP: Yellow-green liquid. Fp: −3°, bp: 55° @ 15 mm.

SYN: tert-BUTYLNITROACETATE

SAFETY PROFILE: Ignites and then explodes on contact with primary, secondary, and tertiary amines. When heated to decomposition it emits toxic fumes of NO_x. See also ACETYLENE COMPOUNDS and NITRO COMPOUNDS.

DSV400 CAS:59-35-8 HR: 2
4,6-DIMETHYL-2-(5-NITRO-2-FURYL)PYRIMIDINE
mf: $C_{10}H_9N_3O_3$ mw: 219.22

TOXICITY DATA WITH REFERENCE
orl-rat TDLo:8988 mg/kg/49W-C:CAR JNCIAM 57,277,76

SAFETY PROFILE: Questionable carcinogen with experimental carcinogenic data. When heated to decomposition it emits toxic fumes of NO_x.

DSV800 CAS:551-92-8 HR: 2
1,2-DIMETHYL-5-NITROIMIDAZOLE
mf: $C_5H_7N_3O_2$ mw: 141.15

PROP: Needles from H_2O. Mp: 138–139°.

SYNS: 1,2-DIMETHYL-5-NITRO-1H-IMIDAZOLE □ DIMETRIDAZOLE □ EMTRYL □ EMTRYLVET □ EMTRYMIX □ 8595 R.P.

D

TOXICITY DATA WITH REFERENCE
mmo-sat 25 μg/plate MUREAV 38,203,76
bfa-rat/sat 800 mg/kg MUREAV 97,171,82
orl-rat TDLo: 50 g/kg/46W-C:NEO JNCIAM 51,403,73

CONSENSUS REPORTS: EPA Genetic Toxicology Program.

SAFETY PROFILE: Questionable carcinogen with experimental neoplastigenic data. Mutation data reported. When heated to decomposition it emits toxic fumes of NO_x.

DSW500 CAS:5213-47-8 HR: 3
4,5-DIMETHYL-2-NITROIMIDAZOLE
mf: $C_5H_7N_3O_2$ mw: 141.15

TOXICITY DATA WITH REFERENCE
orl-mus LD50: 330 mg/kg AACHAX -,478,65
ipr-mus LD50: 158 mg/kg AACHAX -,478,65
scu-mus LD50: 297 mg/kg AACHAX -,478,65

SAFETY PROFILE: Poison by ingestion, subcutaneous, and intraperitoneal routes. When heated to decomposition it emits toxic fumes of NO_x.

DSW600 CAS:3837-55-6 HR: 2
N,N-DIMETHYL-p-((m-NITROPHENYL)AZO)ANILINE
mf: $C_{14}H_{14}N_4O_2$ mw: 270.32

SYN: 3′-NITRO-4-DIMETHYLAMINOAZOBENZENE

TOXICITY DATA WITH REFERENCE
orl-rat TDLo: 5184 mg/kg/17W-C:NEO JEMEAV 87,139,48

SAFETY PROFILE: Questionable carcinogen with experimental neoplastigenic data. When heated to decomposition it emits toxic fumes of NO_x.

DSW800 CAS:3010-38-6 HR: 2
N,N-DIMETHYL-p-((o-NITROPHENYL)AZO)ANILINE
mf: $C_{14}H_{14}N_4O_2$ mw: 270.32

SYN: 2′-NITRO-4-DIMETHYLAMINOAZOBENZENE

TOXICITY DATA WITH REFERENCE
orl-rat TDLo: 5184 mg/kg/17W-C:NEO JEMEAV 87,139,48

SAFETY PROFILE: Questionable carcinogen with experimental neoplastigenic data. When heated to decomposition it emits toxic fumes of NO_x.

DSX400 CAS:7227-92-1 HR: 3
3,3-DIMETHYL-1-(p-NITROPHENYL)TRIAZENE
mf: $C_8H_{10}N_4O_2$ mw: 194.22

SYNS: 1-p-NITROFENYL-3,3-DIMETHYLTRIAZEN (CZECH) □ 1-(p-NITROPHENYL-3,3-DIMETHYL-TRIAZEN (GERMAN) □ 1-(p-NITROPHENYL)-3,3-DIMETHYL-TRIAZENE □ 1-(4-NITROPHENYL)-3,3-DIMETHYLTRIAZENE

TOXICITY DATA WITH REFERENCE
mmo-sat 300 nmol/plate MUREAV 190,177,87
cyt-hmn:lym 1 μmol/L MUREAV 190,183,87
scu-rat TDLo: 3250 mg/kg/72W-I:NEO ZKKOBW 81,285,74
scu-rat TD: 330 mg/kg: ETA ZKKOBW 81,285,74
orl-rat LD50: 1660 mg/kg 28ZPAK -,133,72
scu-rat LD50: 350 mg/kg ZKKOBW 81,285,74

SAFETY PROFILE: Poison by subcutaneous route. Moderately toxic by ingestion. Questionable carcinogen with experimental neoplastigenic and tumorigenic data. Human mutation data reported. When heated to decomposition it emits toxic fumes of NO_x.

DSX800 CAS:37699-43-7 HR: 2
2,3-DIMETHYL-4-NITROPYRIDINE-1-OXIDE
mf: $C_7H_8N_2O_3$ mw: 168.17

TOXICITY DATA WITH REFERENCE
mmo-sat 100 nmol/plate GANNA2 70,799,79
dnr-esc 500 μg/well CNREA8 32,2369,72
scu-mus TDLo: 1760 mg/kg/15W-I:ETA GANNA2 70,799,79

SAFETY PROFILE: Questionable carcinogen with experimental tumorigenic data. Mutation data reported. When heated to decomposition it emits toxic fumes of NO_x.

DSY000 CAS:21816-42-2 HR: 2
2,5-DIMETHYL-4-NITROPYRIDINE-1-OXIDE
mf: $C_7H_8N_2O_3$ mw: 168.17

TOXICITY DATA WITH REFERENCE
mmo-sat 100 nmol/plate GANNA2 70,799,79
mmo-esc 500 μmol/L GANNA2 70,799,79
dnd-mus:fbr 500 μmol/L CNREA8 35,521,75
scu-mus TDLo: 1800 mg/kg/15W-I:ETA GANNA2 70,799,79

SAFETY PROFILE: Questionable carcinogen with experimental tumorigenic data. Mutation data reported. When heated to decomposition it emits toxic fumes of NO_x.

DSY600 CAS:138-89-6 HR: 3
N,N-DIMETHYL-p-NITROSOANILINE
DOT: UN 1369
mf: $C_8H_{10}N_2O$ mw: 150.20

$(CH_3)_2NC_6H_4N{:}O$

PROP: Green plates from Et_2O. Mp: 92.5–93.5°. Sol in EtOH, Et_2O; sltly sol in H_2O.

SYNS: ACCELERINE □ p-(DIMETHYLAMINO)NITROSOBENZENE □ 4-(DIMETHYLAMINO)NITROSOBENZENE □ DIMETHYL-p-NITROSOANILINE (DOT) □ N,N-DIMETHYL-4-NITROSOBENZENAMINE □ DIMETHYL(p-NITROSOPHENYL)AMINE □ NCI-C01821 □ NDMA □ p-NITROSO-N,N-DIMETHYLANILINE □ 4-NITROSODIMETHYLANILINE □ p-NITROSODIMETHYLANILINE (DOT) □ PARANITROSODIMETHYLANILIDE □ ULTRA BRILLIANT BLUE P

TOXICITY DATA WITH REFERENCE
mmo-sat 10 μg/plate ENMUDM 8(Suppl 7),1,86
mma-sat 33 μg/plate ENMUDM 8(Suppl 7),1,86
orl-rat TDLo: 7300 mg/kg/1Y-C:ETA PUOMA5 46,68,68
orl-rat LD50: 65 mg/kg NCIMR* NIH-71-E-2144
orl-gpg LDLo: 650 mg/kg JIDHAN 13,87,31
orl-mam LDLo: 650 mg/kg JIDHAN 13,87,31

CONSENSUS REPORTS: Reported in EPA TSCA Inventory.

DOT CLASSIFICATION: 4.2; *Label:* Spontaneously Combustible

SAFETY PROFILE: Poison by ingestion. Mutation data reported. Questionable carcinogen with experimental tumorigenic data. Flammable when exposed to heat, flame, or oxidizers. Violent reaction with acetic anhydride + acetic acid. When heated to decomposition it emits toxic fumes of NO_x.

DSY800 CAS:70786-64-0 HR: 2
3,2′-DIMETHYL-4-NITROSOBIPHENYL
mf: $C_{14}H_{13}NO$ mw: 211.28

TOXICITY DATA WITH REFERENCE
mmo-sat 120 µmol/plate JMCMAR 22,981,79
scu-ham TDLo:1173 mg/kg/37W-I:CAR CALEDQ 22,981,79

SAFETY PROFILE: Questionable carcinogen with experimental carcinogenic data. Mutation data reported. When heated to decomposition it emits toxic fumes of NO_x.

DSY889 HR: 3
1,2-DIMETHYLNITROSOHYDRAZINE
mf: $C_2H_7N_3$ mw: 73.10

SAFETY PROFILE: The liquid deflagrates on heating. When heated to decomposition it emits toxic fumes of NO_x. See also HYDRAZINE.

DSZ000 CAS:16339-12-1 HR: 3
N,O-DIMETHYL-N-NITROSOHYDROXYLAMINE
mf: $C_2H_6N_2O_2$ mw: 90.10

SYNS: N-METHOXY-N-NITROSOMETHYLAMINE □ N-NITROSOMETHOXYMETHYLAMINE □ N-NITROSOMETHYLMETHOXYAMINE □ N-NITROSO-N-METHYL-o-METHYLHYDROXYLAMIN (GERMAN) □ N-NITROSO-N-METHYL-o-METHYL-HYDROXYLAMINE

TOXICITY DATA WITH REFERENCE
mmo-sat 1 µg/plate MUREAV 51,319,78
mma-sat 1 µg/plate MUREAV 51,319,78
mmo-omi 1 pph/72H-C SOGEBZ 10,522,74
orl-rat TDLo:6000 mg/kg/50W-I:ETA ZKKOBW 89,31,77
ivn-rat LD50:130 mg/kg ZEKBAI 69,103,67

SAFETY PROFILE: Poison by intravenous route. Questionable carcinogen with experimental tumorigenic data. Mutation data reported. Many N-nitroso compounds are carcinogens. When heated to decomposition it emits toxic fumes of NO_x. See also N-NITROSO COMPOUNDS.

DTA000 CAS:1456-28-6 HR: 3
2,6-DIMETHYLNITROSOMORPHOLINE
mf: $C_6H_{12}N_2O_2$ mw: 144.20

SYNS: DIMETHYLNITROSOMORPHOLINE □ 2,6-DIMETHYL-N-NITROSOMORPHOLINE □ DMNM □ Me_2NMOR □ NITROSO-2,6-DIMETHYLMORPHOLINE □ N-NITROSO-2,6-DIMETHYLMORPHOLINE

TOXICITY DATA WITH REFERENCE
mmo-sat 1 mg/plate TCMUD8 1,295,80
mma-sat 50 nmol/plate MUREAV 57,1,78
dns-rat:lvr 1 mmol/L MUREAV 144,197,85
orl-rat TDLo:135 mg/kg/30W-I:ETA ZKKOBW 92,221,78
scu-rat TDLo:1684 mg/kg/18W-I:CAR CALEDQ 13,159,81
orl-gpg TDLo:960 mg/kg/12W-I:CAR CNREA8 40,1879,80
orl-ham TDLo:819 mg/kg/45W-I:CAR JNCIAM 60,371,78
scu-ham TDLo:2040 mg/kg/24W-I:CAR JCROD7 109,183,85
orl-ham TD:937 mg/kg/51W-I:CAR JNCIAM 58,429,77
orl-rat TD:2025 mg/kg/30W-I:ETA CNREA8 35,2123,75
orl-gpg TD:400 mg/kg/23W-I:CAR JJIND8 64,529,80
orl-rat TD:300 mg/kg/30W-I:ETA CRNGDP 1,501,80
orl-rat TD:300 mg/kg/30W-I:ETA CALEDQ 10,325,80
scu-rat TD:1198 mg/kg/20W-I:NEO CALEDQ 13,159,81
orl-rat TD:400 mg/kg/5W-I:ETA CALEDQ 16,281,82
orl-ham TD:644 mg/kg/35W-I:ETA VTPHAK 17,352,80
orl-ham TD:1073 mg/kg/29W-I:NEO JCROD7 109,183,85
scu-ham TD:1200 mg/kg/15W-I:NEO PAACA3 24,62,83
scu-rat LD50:387 mg/kg CALEDQ 13,159,81
orl-gpg LD50:280 mg/kg JJIND8 64,529,80
orl-ham LD50:367 mg/kg JNCIAM 58,429,77
scu-ham LD50:320 mg/kg JNCIAM 60,197,78

CONSENSUS REPORTS: EPA Genetic Toxicology Program.

SAFETY PROFILE: Suspected carcinogen with experimental carcinogenic, tumorigenic, and neoplastigenic data. Poison by ingestion and subcutaneous routes. Mutation data reported. Used as a model carcinogenic and carcinogenic metabolite. When heated to decomposition it emits toxic fumes of NO_x. See also N-NITROSO COMPOUNDS.

DTA050 HR: 2
2,6-DIMETHYL-4-NITROSOMORPHOLINE cis and trans mixture (2:1)
mf: $C_6H_{12}N_2O_2$ mw: 144.20

TOXICITY DATA WITH REFERENCE
orl-rat TDLo:75 g/kg/50W-I:CAR CRNGDP 3,911,82
orl-rat TD:138 g/kg/50W-I:CAR CRNGDP 3,911,82

SAFETY PROFILE: Questionable carcinogen with experimental carcinogenic data. When heated to decomposition it emits toxic fumes of NO_x.

DTA400 CAS:17721-95-8 HR: 2
2,6-DIMETHYLNITROSOPIPERIDINE
mf: $C_7H_{14}N_2O$ mw: 142.23

SYN: N-NITROSO-2,6-DIMETHYLPIPERIDINE

TOXICITY DATA WITH REFERENCE
orl-rat TDLo:2813 mg/kg/50W-I:ETA IJCNAW 16,318,75

CONSENSUS REPORTS: EPA Genetic Toxicology Program.

SAFETY PROFILE: Questionable carcinogen with experimental tumorigenic data. Many N-nitroso compounds are carcinogens. When heated to decomposition it emits toxic fumes of NO_x. See also N-NITROSO COMPOUNDS.

D

DTA600 CAS:65445-59-2 HR: 2
3,5-DIMETHYLNITROSOPIPERIDINE
mf: $C_7H_{14}N_2O$ mw: 142.23

SYNS: 3,5-DIMETHYL-1-NITROSOPIPERIDINE □ N-NITROSO-3,5-DIMETHYLPIPERIDINE

TOXICITY DATA WITH REFERENCE
mma-sat 1 μg/plate MUREAV 56,131,77
sln-dmg-orl 5 mmol/L/24H MUREAV 67,27,79
mma-smc 50 mmol/L/24H MUREAV 57,155,78
orl-rat TDLo:3100 mg/kg/50W-I:ETA JJIND8 68,989,82

CONSENSUS REPORTS: EPA Genetic Toxicology Program.

SAFETY PROFILE: Questionable carcinogen with experimental tumorigenic data. Mutation data reported. Many N-nitroso compounds are carcinogens. When heated to decomposition it emits toxic fumes of NO_x. See also N-NITROSO COMPOUNDS.

DTA690 CAS:78338-31-5 HR: 2
cis-3,5-DIMETHYL-1-NITROSOPIPERIDINE
mf: $C_7H_{14}N_2O$ mw: 142.23

SYNS: NITROSO-3,5-DIMETHYLPIPERIDINE cis-isomer □ PIPERIDINE, 3,5-DIMETHYL-1-NITROSO-, (Z)-

TOXICITY DATA WITH REFERENCE
mma-sat 100 μg/plate TCMUD8 1,295,80
orl-rat TDLo:2550 mg/kg/50W-I:ETA JJIND8 68,989,82

SAFETY PROFILE: Questionable carcinogen with experimental tumorigenic data. Mutation data reported. When heated to decomposition it emits toxic fumes of NO_x.

DTA700 CAS:78338-32-6 HR: 2
trans-3,5-DIMETHYL-1-NITROSOPIPERIDINE
mf: $C_7H_{14}N_2O$ mw: 142.23

SYNS: NITROSO-3,5-DIMETHYLPIPERIDINE trans-isomer □ PIPERIDINE, 3,5-DIMETHYL-1-NITROSO-, (E)-

TOXICITY DATA WITH REFERENCE
mma-sat 50 μg/plate TCMUE9 1,129,84
orl-rat TDLo:500 mg/kg/50W-I:ETA JJIND8 68,989,82

SAFETY PROFILE: Questionable carcinogen with experimental tumorigenic data. Mutation data reported. When heated to decomposition it emits toxic fumes of NO_x.

DTA800 CAS:55556-86-0 HR: 2
2,5-DIMETHYL-N-NITROSOPYRROLIDINE
mf: $C_6H_{12}N_2O$ mw: 128.20

TOXICITY DATA WITH REFERENCE
orl-rat TDLo:5625 mg/kg/50W-I:ETA CNREA8 36,1988,76

SAFETY PROFILE: Questionable carcinogen with experimental tumorigenic data. Many N-nitroso compounds are carcinogens. See also N-NITROSO COMPOUNDS. When heated to decomposition it emits toxic fumes of NO_x.

DTB200 CAS:13256-32-1 HR: 3
1,3-DIMETHYLNITROSOUREA
mf: $C_3H_7N_3O_2$ mw: 117.13

SYNS: DIMETHYLNITROSOHARNSTOFF (GERMAN) □ N,N'-DIMETHYLNITROSOUREA □ 1,3-DIMETHYL-N-NITROSOUREA □ NITROSODIMETHYLUREA □ N-NITROSODIMETHYLUREA

TOXICITY DATA WITH REFERENCE
mmo-omi 1 pph ANTBAL 27,738,82
dni-mus-ipr 80 mg/kg INSSDM 19,85,81
par-rat TDLo:50 mg/kg (female 9D post):TER IARCCD 4,112,73
par-rat TDLo:10 mg/kg (female 13D post):TER IARCCD 4,112,73
ivn-rat TDLo:20 mg/kg/(10D preg):TER APEPA2 257,296,67
orl-rat TDLo:1300 mg/kg/65W-C:ETA ZEKBAI 69,103,67
scu-rat TDLo:836 mg/kg/20W-I:NEO AMOKAG 32,119,78
ivn-rat TDLo:660 mg/kg/66W-I:ETA ZEKBAI 69,103,67
scu-mus TDLo:720 mg/kg/9W-I:CAR GANNA2 62,135,71
scu-ham TDLo:680 mg/kg/17W-I:CAR AMOKAG 28,333,74
orl-rat LD50:280 mg/kg ZEKBAI 69,103,67
ivn-rat LD50:280 mg/kg ZEKBAI 69,103,67

CONSENSUS REPORTS: EPA Genetic Toxicology Program.

SAFETY PROFILE: Suspected carcinogen with experimental carcinogenic, neoplastigenic, tumorigenic, and teratogenic data. Poison by ingestion and intravenous routes. Mutation data reported. When heated to decomposition it emits toxic fumes of NO_x. See also N-NITROSO COMPOUNDS.

DTB800 CAS:128-50-7 HR: 2
6,6-DIMETHYL-2-NORPINENE-2-ETHANOL
mf: $C_{11}H_{18}O$ mw: 166.29

SYNS: 6,6-DIMETHYLBICYCLO-(3.1.1)-2-HEPTENE-2-ETHANOL □ HOMOMYRETENOL □ NOPOL □ NOPOL (TERPENE)

TOXICITY DATA WITH REFERENCE
skn-rbt 500 mg/24H MOD FCTXAV 17,879,79
orl-rat LD50:890 mg/kg FCTXAV 17,879,79
ims-mus LD50:500 mg/kg JSICAZ 21,342,62

CONSENSUS REPORTS: Reported in EPA TSCA Inventory.

SAFETY PROFILE: Moderately toxic by ingestion and intramuscular routes. A skin irritant. When heated to decomposition it emits acrid smoke and irritating fumes.

DTC000 CAS:128-51-8 HR: 2
6,6-DIMETHYL-2-NORPINENE-2-ETHANOL ACETATE
mf: $C_{13}H_{20}O_2$ mw: 208.33

SYNS: CITROVIOL □ 6,6-DIMETHYLBICYCLO(3.1.1)-2-HEPTENE-2-ETHYL ACETATE □ LIGNYL ACETATE □ NOPOL ACETATE □ NOPYL ACETATE □ 2-PINENE-10-METHYL ACETATE

TOXICITY DATA WITH REFERENCE
skn-rbt 500 mg/24H MLD FCTXAV 12,943,74
orl-rat LD50:3000 mg/kg FCTXAV 12,943,74

CONSENSUS REPORTS: Reported in EPA TSCA Inventory.

SAFETY PROFILE: Moderately toxic by ingestion. A skin irritant. When heated to decomposition it emits acrid smoke and irritating fumes.

DTC400 CAS:124-28-7 HR: 3
N,N-DIMETHYLOCTADECYLAMINE
mf: $C_{20}H_{43}N$ mw: 297.64

PROP: A solid or liquid. Fp: 22.9°.

SYNS: ARMEEN DM 18D □ DIMANTINE □ N,N-DIMETHYLOKTADECYLAMIN (CZECH) □ DIMETHYLSTEARAMINE □ DYMANTHINE □ KEMAMINE 9902D □ STEARYLDIMETHYLAMINE

TOXICITY DATA WITH REFERENCE
skn-rbt 20 mg/24H MOD 85JCAE-,440,86
eye-rbt 20 mg/24H MOD 85JCAE-,440,86
ipr-rat LDLo:100 mg/kg NCNSA6 5,11,53
ipr-mus LD50:315 mg/kg EJMCA5 26,517,91

CONSENSUS REPORTS: Reported in EPA TSCA Inventory.

SAFETY PROFILE: Poison by intraperitoneal route. A skin and eye irritant. When heated to decomposition it emits toxic fumes of NO_x.

DTC600 CAS:122-19-0 HR: 3
DIMETHYLOCTADECYLBENZYLAMMONIUM CHLORIDE
mf: $C_{27}H_{50}N \cdot Cl$ mw: 424.23

SYNS: AMMONYX 4 □ AMMONYX CA SPECIAL □ ARQUAD DM18B-90 □ BARQUAT SB-25 □ BENZYLDIMETHYLSTEARYLAMMONIUM CHLORIDE □ BENZYLSTEARYLDIMETHYLAMMONIUM CHLORIDE □ CARSOQUAT SDQ-25 □ DEHYQUART STC-25 □ DIMETHYLBENZYLOCTADECYLAMMONIUM CHLORIDE □ INTEXAN SB-85 □ J SOFT C 4 □ KATAMINE AB □ NISSAN CATION S2-100 □ N-OCTADECYL-N-BENZYL-N,N-DIMETHYLAMMONIUMCHLORIDE □ OCTADECYLDIMETHYLBENZYLAMMONIUM CHLORIDE □ ORTHOSAN MB □ QUATERNOL 1 □ STEARALKONIUM CHLORIDE □ STEARYLDIMETHYLBENZYLAMMONIUM CHLORIDE □ STEBAC □ TALLOW BENZYL DIMETHYLAMMONIUM CHLORIDE □ TRITON X-40 □ VARISOFT SDC

TOXICITY DATA WITH REFERENCE
skn-hmn 3 mg/3D-I MLD 85DKA8 -,127,77
skn-man 125 mg/2D MLD PSTGAW 20,16,53
skn-rbt 1 mg/24H OYYAA2 6,329,72
eye-rbt 200 μg SEV PSTGAW 20,16,53
orl-rat LD50:1250 mg/kg JACTDZ 1(2),57,82
ipr-rat LD50:280 mg/kg KHFZAN 12(12),61,78
orl-mus LD50:760 mg/kg JACTDZ 1(2),57,82
ipr-mus LD50:175 mg/kg KHFZAN 12(12),61,78
orl-gpg LD50:500 mg/kg GISAAA 49(8),90,84

CONSENSUS REPORTS: Reported in EPA TSCA Inventory.

SAFETY PROFILE: Poison by intraperitoneal route. Moderately toxic by ingestion. A human skin irritant and severe experimental eye irritant. When heated to decomposition it emits very toxic fumes of NO_x, NH_3, and Cl^-.

DTC800 CAS:5392-40-5 HR: 2
3,7-DIMETHYL-2,6-OCTADIENAL
mf: $C_{10}H_{16}O$ mw: 152.26

PROP: Mobile, pale-yellow liquid; strong lemon odor. D: 0.891–0.897 @ 15°, refr index: 1.486–1.490, flash p: 198°F. Sol in 5 volumes of 60% alc; sol in all proportions of benzyl benzoate, diethyl phthalate, glycerin, propylene glycol, mineral oil, fixed oils, and 95% alc; insol in water.

SYNS: BUTOBEN □ BUTYL p-HYDROXYBENZOATE □ CITRAL (FCC) □ FEMA No. 2203 □ NCI-C56348 □ NERAL

TOXICITY DATA WITH REFERENCE
skn-hmn 40 mg/24H MLD FCTXAV 17,259,79
skn-man 16 mg/48H SEV CTOIDG 94(8),41,79
skn-rbt 100 mg/24H SEV CTOIDG 94(8),41,79
skn-gpg 1%/48H MOD JSCCA5 28,357,77
skn-gpg 100 mg/24H SEV CTOIDG 94(8),41,79
dnr-bcs 2222 μg/disc OIGZSE 34,267,85
skn-rat TDLo:27,600 mg/kg (female 60D pre):REP JRPFA4 55,347,79
skn-rat TDLo:46 g/kg (female 14W pre):REP FCTXAV 18,547,80
skn-rat TDLo:46 g/kg (female 14W pre):REP JRPFA4 55,347,79
skn-rat TDLo:28 g/kg (60D pre):REP FCTXAV 18,547,80
orl-rat LD50:4960 mg/kg FCTXAV 2,327,64
ipr-rat LD50:460 mg/kg JRPFA4 55,347,79
orl-mus LD50:6000 mg/kg BIJOAK 34,1196,40

CONSENSUS REPORTS: Reported in EPA TSCA Inventory.

SAFETY PROFILE: Moderately toxic by intraperitoneal route. Mildly toxic by ingestion. Experimental reproductive effects. A severe human and experimental skin irritant. Mutation data reported. Combustible liquid. When heated to decomposition it emits acrid smoke and irritating fumes.

DTD000 CAS:106-24-1 HR: 3
3,7-DIMETHYL-(E)-2,6-OCTADIEN-1-OL
mf: $C_{10}H_{18}O$ mw: 154.28

PROP: Colorless to pale-yellow, oily liquid; pleasant floral odor. D: 0.870–0.890 @ 15°, refr index: 1.469–1.478, mp: 15°, bp: 230°, flash p: 214°F. Sol in fixed oils, propylene glycol; sltly sol in water; insol in glycerin @ 230°.

SYNS: 2,6-DIMETHYL-trans-2,6-OCTADIEN-8-OL □ 3,7-DIMETHYL-trans-2,6-OCTADIEN-1-OL □ FEMA No. 2507 □ GERANIOL (FCC) □ GERANIOL ALCOHOL □ GERANIOL EXTRA □ GERANYL ALCOHOL □ GUANIOL □ LEMONOL

TOXICITY DATA WITH REFERENCE
skn-man 16 mg/24H SEV CTOIDG 94(8),41,79
skn-rbt 100 mg/24H SEV CTOIDG 94(8),41,79
skn-gpg 100 mg/24H SEV CTOIDG 94(8),41,79
orl-rat LD50:3600 mg/kg FCTXAV 2,327,64
scu-mus LD50:1090 mg/kg SIZEAR 3,73,52
ims-mus LD50:4000 mg/kg JSICAZ 21,342,62
ivn-rbt LDLo:50 mg/kg NYKZAU 58,394,62

CONSENSUS REPORTS: Reported in EPA TSCA Inventory.

SAFETY PROFILE: Poison by intravenous route. Moderately toxic by ingestion, subcutaneous, and intramuscular routes. A severe human skin irritant. Combustible liquid. When heated to decomposition it emits acrid smoke and irritating fumes.

DTD200 CAS:106-25-2 HR: 2
2-cis-3,7-DIMETHYL-2,6-OCTADIEN-1-OL
mf: $C_{10}H_{18}O$ mw: 154.28

PROP: Colorless oily liquid; sweet, rose odor. D: 0.875–0.880, refr index: 1.467–1.478, bp: 225–226°. Sol in alc, chloroform, ether, water @ 227°.

SYNS: 3,7-DIMETHYL-(Z)-2,6-OCTADIEN-1-OL □ FEMA No. 2770 □ NEROL (FCC)

TOXICITY DATA WITH REFERENCE
skn-rbt 500 mg/24H MOD FCTXAV 14,623,76
orl-rat LD50:4500 mg/kg FCTXAV 14,623,76
ims-mus LD50:3000 mg/kg JSICAZ 21,342,62

CONSENSUS REPORTS: Reported in EPA TSCA Inventory.

SAFETY PROFILE: Moderately toxic by intramuscular route. Mildly toxic by ingestion. A skin irritant. When heated to decomposition it emits acrid smoke and irritating fumes.

DTD400 CAS:5986-38-9 HR: 2
2,6-DIMETHYL-5,7-OCTADIEN-2-OL
mf: $C_{10}H_{18}O$ mw: 154.28

SYN: OCIMENOL

TOXICITY DATA WITH REFERENCE
orl-rat LD50:1700 mg/kg FCTXAV 14,817,76

CONSENSUS REPORTS: Reported in EPA TSCA Inventory.

SAFETY PROFILE: Moderately toxic by ingestion. When heated to decomposition it emits acrid smoke and irritating fumes.

DTD800 CAS:105-87-3 HR: 1
trans-3,7-DIMETHYL-2,6-OCTADIEN-1-OL ACETATE
mf: $C_{12}H_{20}O_2$ mw: 196.32

PROP: Colorless, sweet, clear, oily, liquid; odor of lavender. D: 0.907–0.918 @ 15°, refr index: 1.458–1.464, bp: 130–132° @ 16 mm, flash p: 219°F. Sol in alc, fixed oils, ether; sltly sol in propylene glycol; insol in water and glycerol.

SYNS: ACETIC ACID GERANIOL ESTER □ 3,7-DIMETHYL-2-trans-6-OCTADIENYL ACETATE □ trans-3,7-DIMETHYL-2,6-OCTADIEN-1-YL ACETATE □ trans-2,6-DIMETHYL-2,6-OCTADIEN-8-YL ETHANOATE □ FEMA No. 2509 □ GERANIOL ACETATE □ GERANYL ACETATE (FCC) □ NCI-C54728

TOXICITY DATA WITH REFERENCE
skn-man 16 mg/48H MLD CTOIDG 94(8),41,79
skn-rbt 100 mg/24H SEV CTOIDG 94(8),41,79
skn-gpg 100 mg/24H MOD CTOIDG 94(8),41,79
pic-esc 25 µg/well MUREAV 260,349,91
mma-mus:lyms 18 mg/L MUREAV 196,61,88
sce-ham:ovr 70 mg/L EMMUEG 10(Suppl 10),1,87
orl-rat LD50:6330 mg/kg FCTXAV 2,327,64

CONSENSUS REPORTS: NTP Carcinogenesis Studies (gavage): No Evidence: mouse, rat NTPTR* NTP-TR-252,87. Reported in EPA TSCA Inventory.

SAFETY PROFILE: Mildly toxic by ingestion. A human skin irritant. Mutation data reported. Combustible liquid. When heated to decomposition it emits acrid smoke and irritating fumes. See also ESTERS.

DTE000 CAS:56172-46-4 HR: 1
3,7-DIMETHYL-2-trans-6-OCTADIENYL CROTONATE
mf: $C_{14}H_{22}O_2$ mw: 222.36

SYNS: CROTONIC ACID GERNAIOL ESTER □ trans-3,7-DIMETHYL-2,6-OCTADIEN-1-OL-2-BUTENOATE □ 3-7-DIMETHYL-2,6-OCTADIENYL ESTER-2-BUTENOIC ACID □ GERANIOL CROTONATE □ GERANYL-2-BUTENOATE □ GERANYL CROTONATE

TOXICITY DATA WITH REFERENCE
skn-rbt 500 mg/24H MOD FCTXAV 12,891,74
orl-rat LD50:>5 g/kg FCTXAV 12,891,74
skn-rbt LD50:>5 g/kg FCTXAV 12,891,74

CONSENSUS REPORTS: Reported in EPA TSCA Inventory.

SAFETY PROFILE: Low toxicity by ingestion and skin contact. A skin irritant. When heated to decomposition it emits acrid smoke and irritating fumes. See also ESTERS.

DTE400 CAS:107-74-4 HR: 1
3,7-DIMETHYL-1,2-OCTANEDIOL
mf: $C_{10}H_{22}O_2$ mw: 174.32

SYNS: 3,7-DIMETHYL-7-HYDROXY-1-OCTANOL □ HYDROXYCITRONELLOL □ 7-HYDROXY-3,7-DIMETHYLOCTAN-1-OL

TOXICITY DATA WITH REFERENCE
skn-rbt 500 mg/24H MLD FCTXAV 12,923,74
orl-rat LD50:>5 g/kg FCTXAV 12,923,74
skn-rbt LD50:>5 g/kg FCTXAV 12,923,74

CONSENSUS REPORTS: Reported in EPA TSCA Inventory.

SAFETY PROFILE: Low toxicity by ingestion and skin contact. A skin irritant. When heated to decomposition it emits acrid smoke and irritating fumes.

DTE600 CAS:106-21-8 HR: 2
DIMETHYLOCTANOL
mf: $C_{10}H_{22}O$ mw: 158.32

PROP: Colorless liquid; sweet, rose odor. D: 0.26–0.842, refr index: 1.435. Sol in fixed oils, propylene glycol; insol in glycerin.

SYNS: DIHYDROCITRONELLOL □ 2,6-DIMETHYL-8-OCTANOL □ 3,7-DIMETHYL-1-OCTANOL (FCC) □ FEMA No. 2391 □ GERANIOL TETRAHYDRIDE □ PELARGOL □ PERHYDROGERANIOL □ TETRAHYDROGERANIOL

TOXICITY DATA WITH REFERENCE
skn-rbt 500 mg/24H FCTXAV 13,517,75

orl-rat LD50:>5 g/kg FCTXAV 12,535,74
skn-rbt LD50:2400 mg/kg FCTXAV 12,535,74

CONSENSUS REPORTS: Reported in EPA TSCA Inventory.

SAFETY PROFILE: Moderately toxic by skin contact. A skin irritant. When heated to decomposition it emits acrid smoke and irritating fumes.

DTE800 CAS:20780-49-8 HR: 1
3,7-DIMETHYLOCTANYL ACETATE
mf: $C_{12}H_{24}O_2$ mw: 200.36

PROP: Bp: 109–110° @ 12 mm.

SYNS: DIHYDROCITRONELLYL ACETATE □ 3,7-DIMETHYLOCTYL ACETATE □ TETRAHYDROGERANYL ACETATE

TOXICITY DATA WITH REFERENCE
skn-rbt 500 mg/24H MOD FCTXAV 18,673,80
skn-rbt LD50:5000 mg/kg FCTXAV 18,673,80

CONSENSUS REPORTS: Reported in EPA TSCA Inventory.

SAFETY PROFILE: Mildly toxic by skin contact. A skin irritant. When heated to decomposition it emits acrid smoke and irritating fumes.

DTF000 CAS:67874-80-0 HR: 1
3,7-DIMETHYLOCTANYL BUTYRATE
mf: $C_{14}H_{28}O_2$ mw: 228.42

SYNS: 3,7-DIMETHYLOCTYL ESTER BUTANOIC ACID □ TETRAHYDROGERANYL BUTYRATE

TOXICITY DATA WITH REFERENCE
skn-rbt 500 mg/24H MOD FCTXAV 18,649,80
orl-rat LD50:>5 g/kg FCTXAV 18,675,80
skn-rbt LD50:>5 g/kg FCTXAV 18,675,80

CONSENSUS REPORTS: Reported in EPA TSCA Inventory.

SAFETY PROFILE: Low toxicity by ingestion and skin contact. A skin irritant. When heated to decomposition it emits acrid smoke and irritating fumes.

DTF200 CAS:29714-87-2 HR: 1
DIMETHYLOCTATRIENE
mf: $C_{10}H_{16}$ mw: 136.26

SYNS: DIMETHYLOCTATRIENE (mixed isomer) □ OCIMENE

TOXICITY DATA WITH REFERENCE
skn-rbt 500 mg/24H MOD FCTXAV 16,829,78

SAFETY PROFILE: A skin irritant. When heated to decomposition it emits acrid smoke and irritating fumes.

DTF400 CAS:141-25-3 HR: 2
2,6-DIMETHYL-1-OCTEN-8-OL
mf: $C_{10}H_{20}O$ mw: 156.30

PROP: Flash p: 212°F.

SYNS: α-CITRONELLOL □ 3,7-DIMETHYL-7-OCTEN-1-OL □ FEMA No. 2981 □ RHODINOL (FCC)

TOXICITY DATA WITH REFERENCE
ims-mus LD50:4000 mg/kg JSICAZ 21,342,62

CONSENSUS REPORTS: Reported in EPA TSCA Inventory.

SAFETY PROFILE: Moderately toxic by intramuscular route. Combustible liquid. When heated to decomposition it emits acrid smoke and irritating fumes.

DTF410 CAS:106-22-9 HR: 2
2,6-DIMETHYL-2-OCTEN-8-OL
mf: $C_{10}H_{20}O$ mw: 156.30

SYNS: CEPHROL □ CITRONELLOL □ 3,7-DIMETHYL-6-OCTEN-1-OL □ 6-OCTEN-1-OL, 3,7-DIMETHYL- □ RHODINOL □ RODINOL

TOXICITY DATA WITH REFERENCE
skn-man 16 mg/48H MOD CTOIDG 94(8),41,79
skn-rbt 100 mg/24H SEV CTOIDG 94(8),41,79
skn-gpg 100 mg/24H SEV CTOIDG 94(8),41,79
orl-rat LD50:3450 mg/kg FCTXAV 13,757,75
scu-mus LD50:880 mg/kg SIZSAR 3,73,52
ims-mus LD50:4 g/kg JSICAZ 21,342,62
skn-rbt LD50:2650 mg/kg FCTXAV 13,757,75

CONSENSUS REPORTS: Reported in EPA TSCA Inventory.

SAFETY PROFILE: Moderately toxic by ingestion, subcutaneous, and skin contact routes. A human skin irritant. When heated to decomposition it emits acrid smoke and irritating vapors.

DTF800 CAS:141-16-2 HR: 1
2,6-DIMETHYL-2-OCTEN-8-YL BUTYRATE
mf: $C_{14}H_{26}O_2$ mw: 226.40

SYNS: BUTYRIC ACID-3,7-DIMETHYL-6-OCTENYL ESTER □ 3,7-DIMETHYL-6-OCTEN-1-OL BUTYRATE □ 2,6-DIMETHYL-2-OCTEN-8-OL-BUTYRATE □ RHODINYL BUTYRATE

TOXICITY DATA WITH REFERENCE
skn-rbt 500 mg MLD FCTXAV 14,849,76

CONSENSUS REPORTS: Reported in EPA TSCA Inventory.

SAFETY PROFILE: A skin irritant. When heated to decomposition it emits acrid smoke and irritating fumes.

DTF850 CAS:78-66-0 HR: 2
3,6-DIMETHYL-OCTYN-4-DIOL-(3,6)
mf: $C_{10}H_{18}O_2$ mw: 170.28

SYN: 4-OCTYN-3,6-DIOL, 3,6-DIMETHYL-

TOXICITY DATA WITH REFERENCE
orl-mus LD50:825 mg/kg ARZNAD 4,477,54

CONSENSUS REPORTS: Reported in EPA TSCA Inventory.

SAFETY PROFILE: Moderately toxic by ingestion. When

D

heated to decomposition it emits acrid smoke and irritating vapors.

DTG000 CAS:1854-26-8 HR: 2
DIMETHYLOL DIHYDROXYETHYLENE UREA
mf: $C_4H_{10}N_2O_5$ mw: 178.17

PROP: Hygroscopic crystals.

SYNS: ARKOFIX NG ☐ CASSURIT LR ☐ DEPREMOL G ☐ (4,5-DIHYDROXY-1,3-BIS(HYDROXYMETHYL))-2-IMIDAZOLIDINONE ☐ DIMETHYLOLGLYOXALUREA ☐ DMDHEU ☐ FIRMATEX RK ☐ FIXAPRET CP ☐ HYLITE LF ☐ KNITTEX LE ☐ NCI-C60322 ☐ NEUPERM GFN ☐ NS 11 ☐ PERMAFRESH 183 ☐ PROTOCOL C ☐ PROX DW ☐ READPRET KPN ☐ SARCOSET GM ☐ SUMITEX FSK ☐ SUMITEX NS ☐ VERAPRET DH ☐ WNM

TOXICITY DATA WITH REFERENCE
skn-rbt 500 mg/24H SEV 28ZPAK -,269,72
eye-rbt 500 mg/24H MLD 28ZPAK -,269,72
mmo-sat 3333 μg/plate ENMUDM 9(Suppl 9),1,87
slt-dmg-orl 60 ppb EMMUEG 23,51,94

CONSENSUS REPORTS: Reported in EPA TSCA Inventory.

SAFETY PROFILE: An eye and severe skin irritant. Mutation data reported. When heated to decomposition it emits toxic fumes of NO_x.

DTG200 CAS:10143-22-3 HR: 2
N,N-DIMETHYLOL-2-METHOXYETHYL CARBAMATE
mf: $C_6H_{13}NO_5$ mw: 179.20

TOXICITY DATA WITH REFERENCE
skn-rbt 500 mg open MLD UCDS** 5/19/66
eye-rbt 15 mg SEV UCDS** 3/7/66
orl-rat LD50:11 g/kg UCDS** 5/19/66

CONSENSUS REPORTS: Reported in EPA TSCA Inventory.

SAFETY PROFILE: Mildly toxic by ingestion. A skin and severe eye irritant. When heated to decomposition it emits toxic fumes of NO_x. See also CARBAMATES.

DTG400 CAS:126-30-7 HR: 2
DIMETHYLOLPROPANE
mf: $C_5H_{12}O_2$ mw: 104.17

PROP: White, crystalline solid or needles from C_6H_6. Mp: 129°, bp: 206° @ 747 mm.

SYNS: 2,2-DIMETHYL-1,3-PROPANEDIOL ☐ DIMETHYLTRIMETHYLENE GLYCOL ☐ NEOL ☐ NEOPENTYLENE GLYCOL ☐ NEOPENTYL GLYCOL ☐ NPG

TOXICITY DATA WITH REFERENCE
orl-rat LDLo:3200 mg/kg KODAK* -,-,71

CONSENSUS REPORTS: Reported in EPA TSCA Inventory.

SAFETY PROFILE: Moderately toxic by ingestion. Used in polyester manufacture. An insect repellent. Combustible when exposed to heat or flame; can react with oxidizing materials. When heated to decomposition it emits acrid smoke and irritating fumes. See also GLYCOLS.

DTG700 CAS:140-95-4 HR: 2
1,3-DIMETHYLOLUREA
mf: $C_3H_8N_2O_3$ mw: 120.13

PROP: Crystals from alc. Mp: 132.5°. Very sol in cold water, hot ethanol, and methanol.

SYNS: N,N′-BIS(HYDROXYMETHYL)UREA ☐ 1,3-BIS(HYDROXYMETHYL)UREA ☐ CAURITE ☐ CSI PASTE ☐ N,N′-DIHYDROXYMETHYLUREA ☐ DMU ☐ FINISH EN ☐ KAURIT S ☐ KNITTEX ASL ☐ METHURAL ☐ METHURIN (RUSSIAN) ☐ METURAL ☐ OXYMETHUREA ☐ PERMAFRESH 477 ☐ PROTESINE DMU ☐ UREOL P

TOXICITY DATA WITH REFERENCE
orl-rat LD50:3400 mg/kg GISAAA 44(3),68,79
orl-mus LD50:1795 mg/kg GISAAA 44(3),68,79
orl-rbt LD50:3200 mg/kg GISAAA 44(3),68,79

CONSENSUS REPORTS: Reported in EPA TSCA Inventory.

SAFETY PROFILE: Moderately toxic by ingestion. When heated to decomposition it emits toxic fumes of NO_x.

DTG750 CAS:51200-87-4 HR: 2
DIMETHYL OXAZOLIDINE
mf: $C_5H_{11}NO$ mw: 101.17

SYNS: 4,4-DIMETHYLOXAZOLIDINE ☐ OXAZOLIDINE A

TOXICITY DATA WITH REFERENCE
cyt-ham:ovr 500 ng/L EPASR* 8EHQ-0283-0470
orl-rat LD50:950 mg/kg CTOIDG 96(3),79,81
ihl-rat LC50:11,700 mg/m³ CTOIDG 96(3),79,81
skn-rbt LD50:1400 mg/kg CTOIDG 96(3),79,81

CONSENSUS REPORTS: Reported in EPA TSCA Inventory.

SAFETY PROFILE: Moderately toxic by ingestion and skin contact. Mildly toxic by inhalation. Mutation data reported. When heated to decomposition it emits toxic fumes of NO_x.

DTH000 CAS:1955-45-9 HR: 3
3,3-DIMETHYL-2-OXETHANONE
mf: $C_5H_8O_2$ mw: 100.13

SYNS: 3,3-DIMETHYL-2-OXETANONE ☐ DIMETHYL PROPIOLACTONE ☐ 3,3-DIMETHYL-β-PROPIOLACTONE ☐ NCI-C04126 ☐ PIVALIC ACID LACTONE ☐ PIVALOLACTONE

TOXICITY DATA WITH REFERENCE
mma-sat 333 μg/plate ENMUDM 7(Suppl 5),1,85
mma-esc 333 μg/plate ENMUDM 7(Suppl 5),1,85
orl-rat TDLo:216 g/kg/2Y-I:CAR NCITR* NCI-CG-TR-140,78
orl-rat TD:72 g/kg/69W:ETA NCITR* NCI-CG-TR-140,78
orl-rat LD50:1470 mg/kg NCILB* NIH-NCI-E-C-72-3252,73
orl-mus LD50:316 mg/kg NCILB* NIH-NCI-E-C-72-3252,73

CONSENSUS REPORTS: NCI Carcinogenesis Bioassay (gavage); No Evidence: mouse NCITR* NCI-CG-TR-140,78; Clear Evidence: rat NCITR* NCI-CG-TR-140,78. Reported in EPA TSCA Inventory.

SAFETY PROFILE: Poison by ingestion. Questionable carcinogen with experimental carcinogenic and tumorigenic data. Mutation data reported. When heated to decomposition it emits acrid smoke and irritating fumes.

DTH100 CAS:32568-89-1 HR: 2
5,5-DIMETHYL-3-(2-(OXIRANYLMETHOXY)PROPYL)-1-(OXIRANYLMETHYL)-2,4-IMIDAZ OLIDINEDIONE
mf: $C_{14}H_{22}N_2O_5$ mw: 298.38

SYN: 2,4-IMIDAZOLIDINEDIONE, 5,5-DIMETHYL-3-(2-(OXIRANYLMETHOXY)PROPYL)-1-(OXIRANYLMETHYL)-

TOXICITY DATA WITH REFERENCE
mmo-sat 1 mg/plate TSCAT* OTS 206476
mmo-smc 5 mg/plate TSCAT* OTS 206476
orl-rat LD50:1800 mg/kg TSCAT* OTS 206476
orl-mus LD50:1878 mg/kg TSCAT* OTS 206476

CONSENSUS REPORTS: Reported in EPA TSCA Inventory.

SAFETY PROFILE: Moderately toxic by ingestion. Mutation data reported. When heated to decomposition it emits toxic vapors of NO_x.

DTH200 CAS:2873-97-4 HR: 2
N-(1,1-DIMETHYL-3-OXOBUTYL)ACRYLAMIDE
mf: $C_9H_{15}NO_2$ mw: 169.25

PROP: Crystals or solid. Mp: 57–58°, bp: 120° @ 8 mm.

SYNS: DIACETONE ACRYLAMIDE □ N-(1,1-DIMETHYL-3-OXOBUTYL)-2-PROPENAMIDE □ N-(2-(2-METHYL-4-OXOPENTYL))ACRYLAMIDE

TOXICITY DATA WITH REFERENCE
orl-rat LD50:1770 mg/kg JACTDZ 1,113,90
orl-mus LD50:1303 mg/kg ARTODN 47,179,81

CONSENSUS REPORTS: Reported in EPA TSCA Inventory.

SAFETY PROFILE: Moderately toxic by ingestion. When heated to decomposition it emits toxic fumes of NO_x.

DTH400 CAS:2273-45-2 HR: 3
DIMETHYLOXOSTANNANE
mf: C_2H_6OSn mw: 164.77

PROP: White powder. Insol in water.

SYN: DIMETHYLTIN OXIDE

TOXICITY DATA WITH REFERENCE
ivn-mus LD50:100 mg/kg CSLNX* NX#03809

CONSENSUS REPORTS: Reported in EPA TSCA Inventory.

OSHA PEL: TWA 0.1 mg(Sn)/m³ (skin)
ACGIH TLV: TWA 0.1 mg(Sn)/m³ (skin) (Proposed: TWA 0.1 mg(Sn)/m³; STEL 0.2 mg(Sn)/m³ (skin))
NIOSH REL: (Organotin Compounds) TWA 0.1 mg(Sn)/m³

SAFETY PROFILE: Poison by intravenous route. When heated to decomposition it emits acrid smoke and irritating fumes. See also TIN COMPOUNDS.

For occupational chemical analysis use NIOSH: Organotin Compounds 5504.

DTH600 CAS:63951-48-4 HR: 2
α,γ-DIMETHYL-α-OXYMETHYL GLUTARALDEHYDE
mf: $C_8H_{14}O_3$ mw: 158.22

SYN: 2-(HYDROXYMETHYL)-2,4-DIMETHYLPENTANEDIAL

TOXICITY DATA WITH REFERENCE
skn-rbt 500 mg SEV SCCUR* -,4,61
orl-rat LD50:2040 mg/kg SCCUR* -,4,61
orl-mus LD50:570 mg/kg SCCUR* -,4,61

SAFETY PROFILE: Moderately toxic by ingestion. A severe skin irritant. When heated to decomposition it emits acrid smoke and irritating fumes. See also ALDEHYDES.

DTH700 CAS:3886-91-7 HR: 3
N,N-DIMETHYLPALMITAMIDE
mf: $C_{18}H_{37}NO$ mw: 283.56

SYNS: N,N-DIMETHYLHEXADECANAMIDE □ HEXADECANAMIDE, N,N-DIMETHYL-

TOXICITY DATA WITH REFERENCE
ipr-mus LD50:4800 mg/kg AIHAAP 32,539,71
ivn-mus LD50:220 mg/kg AIHAAP 32,539,71

CONSENSUS REPORTS: Reported in EPA TSCA Inventory.

SAFETY PROFILE: A poison by intravenous route. When heated to decomposition it emits toxic vapors of NO_x.

DTH800 CAS:3820-53-9 HR: 3
DIMETHYL PARANITROPHENYL THIONOPHOSPHATE
mf: $C_8H_{10}NO_5PS$ mw: 263.22

PROP: Crystals or solid. Vap d: 9.1, mp: 55–56°, d: 1.235 @ 20°/4°.

SYNS: O,O-DIMETHYL-S-p-NITROFENYL ESTER KYSELINY THIOFOSFORECEN (CZECH) □ O,O-DIMETHYL-S-(p-NITROPHENYL) PHOSPHOROTHIOATE □ O,O-DIMETHYL-S-(4-NITROPHENYL)THIOPHOSPHATE

TOXICITY DATA WITH REFERENCE
orl-rat LD50:43 mg/kg 28ZPAK -,208,72
scu-mus LD50:8 mg/kg AMIHAB 11,487,55

SAFETY PROFILE: Poison by ingestion and subcutaneous routes. When heated to decomposition it emits very toxic fumes of NO_x, PO_x, and SO_x.

DTI400 CAS:10143-23-4 HR: 2
2,3-DIMETHYL-1-PENTANOL
mf: $C_7H_{16}O$ mw: 116.23

SYN: 2,3-DIMETHYLPENTANOL

TOXICITY DATA WITH REFERENCE
skn-rbt 10 mg/24H open MLD AIHAAP 23,95,62
orl-rat LD50:2380 mg/kg AIHAAP 23,95,62

skn-rbt LD50:2500 mg/kg AIHAAP 23,95,62

SAFETY PROFILE: Moderately toxic by ingestion and skin contact. A skin irritant. When heated to decomposition it emits acrid smoke and irritating fumes.

DTI600 CAS:565-80-0 HR: 3
2,4-DIMETHYL-3-PENTANONE
mf: $C_7H_{14}O$ mw: 114.21

PROP: A liquid. Flash p: 59°C, d: 0.811 @ 20°/4°, bp: 124–125°.

SYNS: DIISOPROPYL KETONE □ ISOBUTYRONE □ ISOPROPYL KETONE □ PM 2763

TOXICITY DATA WITH REFERENCE
orl-rat LD50:3536 mg/kg EPASR* 8EHQ-0990-1062

DOT CLASSIFICATION: 3; *Label:* Flammable Liquid

SAFETY PROFILE: Moderately toxic by ingestion. A flammable liquid and very dangerous fire hazard when exposed to heat, flame, or oxidizers. When heated to decomposition it emits acrid smoke and fumes.

DTI709 CAS:71901-54-7 HR: 3
S,S-DIMETHYLPENTASULFUR HEXANITRIDE
mf: $C_2H_6N_6S_5$ mw: 274.41

SAFETY PROFILE: A powerful explosive. Upon decomposition it emits toxic fumes of SO_x and NO_x. See also NITRIDES.

DTJ000 CAS:690-02-8 HR: 3
DIMETHYL PEROXIDE
mf: $C_2H_6O_2$ mw: 62.07

PROP: Gas or liquid. Bp: 10°.

SAFETY PROFILE: Both the liquid and the vapor are powerful explosives extremely sensitive to heat or shock. Rough handling may cause ignition. When heated to decomposition it emits acrid smoke and fumes. See also PEROXIDES.

DTJ159 CAS:15411-45-7 HR: 3
DIMETHYLPEROXYCARBONATE
mf: $C_4H_6O_6$ mw: 150.09

$CH_3OCO•OOCO•OCH_3$

SAFETY PROFILE: Explodes when heated above 55°C or on impact. When heated to decomposition it emits acrid smoke and fumes. See also PEROXIDES.

DTJ200 CAS:22349-59-3 HR: 2
1,4-DIMETHYLPHENANTHRENE
mf: $C_{16}H_{14}$ mw: 206.30

PROP: Needles from MeOH. Mp: 50–51°, bp: 182–186° @ 6 mm.

TOXICITY DATA WITH REFERENCE
mma-sat 50 μg/plate MUREAV 116,91,83
skn-mus TDLo:40 mg/kg/20D-I:CAR CNREA8 41,3441,81

CONSENSUS REPORTS: IARC Cancer Review: Group 3 IMEMDT 7,56,87; Animal Inadequate Evidence IMEMDT 32,349,83; Human No Adequate Data IMEMDT 32,349,83.

SAFETY PROFILE: Questionable carcinogen with experimental carcinogenic data. Mutation data reported. When heated to decomposition it emits acrid smoke and irritating fumes.

DTJ400 CAS:122-09-8 HR: 3
α,α-DIMETHYLPHENETHYLAMINE
mf: $C_{10}H_{15}N$ mw: 149.26

SYNS: α,α-DIMETHYLBENZEETHANAMINE □ 1,1-DIMETHYL-2-PHENYLETHANAMINE □ α,α-DIMETHYL-β-PHENYLETHYLAMINE □ DUROMINE □ LIPOPILL □ LONAMIN □ MG 18370 □ MG 18570 □ MIRAPRONT □ PHENTERMINE □ 2-PHENYL-tert-BUTYLAMINE □ RCRA WASTE NUMBER P046 □ WILPO

TOXICITY DATA WITH REFERENCE
sln-asn 1 mg/L MUREAV 26,159,74
orl-man TDLo:1429 μg/kg:ANS THERAP 34,205,79
orl-mus LD50:105 mg/kg AIPTAK 178,62,69
ipr-mus LD50:71 mg/kg RCOCB8 14,677,76
ivn-mus LD50:14 mg/kg CSLNX* NX#03232

CONSENSUS REPORTS: Reported in EPA TSCA Inventory.

SAFETY PROFILE: Poison by ingestion, intravenous, and intraperitoneal routes. Human systemic effects by ingestion: sympathomimetic. Mutation data reported. When heated to decomposition it emits toxic fumes of NO_x.

DTK300 CAS:27691-62-9 HR: 1
N,N-DIMETHYL-3-PHENOTHIAZINESULFONAMIDE
mf: $C_{14}H_{14}N_2O_2S_2$ mw: 306.42

SYNS: 3-DIMETHYLSULPHAMIDOPHENOTHIAZINE □ 3-PHENOTHIAZINESULFONAMIDE, N,N-DIMETHYL- □ 10H-PHENOTHIAZINE-3-SULFONAMIDE, N,N-DIMETHYL-

TOXICITY DATA WITH REFERENCE
eye-rbt 100 mg MLD FCTOD7 20,573,82

SAFETY PROFILE: An eye irritant. When heated to decomposition it emits toxic fumes of NO_x and SO_x.

DTK600 CAS:2747-31-1 HR: 3
N,N-DIMETHYL-p-PHENYLAZOANILINE-N-OXIDE
mf: $C_{14}H_{15}N_3O$ mw: 241.32

SYNS: DAB-N-OXIDE □ 4-DIMETHYLAMINOAZOBENZENE AMINE-N-OXIDE □ N,N-DIMETHYLAMINOAZOBENZENE-N-OXIDE

TOXICITY DATA WITH REFERENCE
orl-rat TDLo:6300 mg/kg/30W-C:ETA GANNA2 54,455,63
scu-rat TDLo:278 mg/kg/12W-I:ETA CNREA8 27,1600,67
orl-mus TDLo:11 g/kg/26W-C:ETA GANNA2 54,455,63
orl-rat LD50:2200 mg/kg GANNA2 54,455,63
ipr-rat LD50:155 mg/kg GANNA2 54,455,63
orl-mus LD50:760 mg/kg GANNA2 54,455,63
ipr-mus LD50:175 mg/kg GANNA2 54,455,63

SAFETY PROFILE: Poison by intraperitoneal route.

Moderately toxic by ingestion. Questionable carcinogen with experimental tumorigenic data. When heated to decomposition it emits toxic fumes of NO_x.

DTK800 CAS:2438-49-5 HR: 2
N,N-DIMETHYL-4-PHENYLAZO-o-ANISIDINE
mf: $C_{15}H_{17}N_3O$ mw: 255.35

SYN: 3-METHOXY-4-DIMETHYLAMINOAZOBENZENE

TOXICITY DATA WITH REFERENCE
orl-rat TDLo:9800 mg/kg/34W-C:ETA CNREA8 21,1068,61
skn-rat TDLo:640 mg/kg/40W-I:ETA CNREA8 25,1784,65

SAFETY PROFILE: Questionable carcinogen with experimental tumorigenic data. When heated to decomposition it emits toxic fumes of NO_x.

DTL000 CAS:36576-23-5 HR: 2
2,3-DIMETHYL-4-(PHENYLAZO)BENZENAMINE
mf: $C_{14}H_{15}N_3$ mw: 225.32

PROP: Orange crystals from C_6H_6/ligroin. Mp: 98°.

SYN: 2,3 DIMETHYL-4-PHENYLAZOANILINE

TOXICITY DATA WITH REFERENCE
orl-mus TDLo:6000 mg/kg/26W-C:NEO FCTXAV 11,415,73

SAFETY PROFILE: Questionable carcinogen with experimental neoplastigenic data. When heated to decomposition it emits toxic fumes of NO_x.

DTL200 CAS:126-27-2 HR: 3
2-DI(N-METHYL-N-PHENYL-tert-BUTYL-CARBAMOYL-METHYL)AMINOETHANOL
mf: $C_{28}H_{41}N_3O_3$ mw: 467.72

PROP: Crystals from C_6H_6/hexane. Mp: 104–104.5°.

SYNS: BETALGIL □ N,N-DIS(N-METHYL-N-PHENYL-tert-BUTYLACETAMIDO)-β-HYDROXYETHYLAMINE □ EMOREN □ FH 099 □ H4 099 □ 2,2′-((2-HYDROXYETHYL)IMINO BIS(N-(α,α-DIMETHYLPHENETHYL))-N-METHYL-ACETAMIDE □ 2,2′-(2-HYDROXYETHYL)IMINO) BIS(N-(1,1-DIMETHYL-2-PHENYLETHYL))-N-METHYLACETAMIDE) □ MUCAINE □ MUCOXIN □ MUTHESA □ OXAINE □ OXETACAINE □ OXETHACAINA (ITALIAN) □ OXETHAZINE □ STOMACAIN □ TEPILTA □ TOPICAIN □ WY 806

TOXICITY DATA WITH REFERENCE
sln-asn 1 mg/L MUREAV 26,159,74
ipr-rat LD50:30 mg/kg GMITAB 134,642,75
ipr-mus LD50:27 mg/kg GMITAB 134,642,75
scu-mus LD50:58 mg/kg GMITAB 134,642,75

SAFETY PROFILE: Poison by intraperitoneal and subcutaneous routes. Mutation data reported. When heated to decomposition it emits toxic fumes of NO_x.

DTL600 CAS:99-98-9 HR: 3
N,N-DIMETHYL-p-PHENYLENEDIAMINE
mf: $C_8H_{12}N_2$ mw: 136.22

PROP: Reddish-violet crystals or needles. Mp: 41°, bp: 262°. Sol in EtOH, Me_2CO, C_6H_6, Et_2O, (aq) HCl, and $CHCl_3$; insol in H_2O.

SYN: DIMETHYL-p-PHENYLENEDIAMINE

TOXICITY DATA WITH REFERENCE
cyt-ham:lng 10 mg/L MUREAV 241,175,90
skn-hmn TDLo:14 µg/kg:SKN JIDHAN 4,386,23
orl-rat LDLo:50 mg/kg NCNSA6 5,11,53
scu-rat LDLo:50 mg/kg JIDHAN 4,386,23
ipr-mus LDLo:50 mg/kg RBPMAZ 22,1,52
skn-dog LDLo:84 mg/kg JIDHAN 4,386,23
ivn-dog LDLo:51 mg/kg JIDHAN 4,386,23
orl-cat LDLo:20 mg/kg JIDHAN 4,386,23
orl-rbt LDLo:150 mg/kg JIDHAN 4,386,23
ihl-rbt LCLo:500 ppb JIDHAN 4,386,23
skn-rbt LDLo:60 mg/kg JIDHAN 4,386,23
scu-rbt LDLo:60 mg/kg JIDHAN 4,386,23
ihl-gpg LCLo:240 ppb JIDHAN 4,386,23
scu-gpg LDLo:100 mg/kg JIDHAN 4,386,23

CONSENSUS REPORTS: Reported in EPA TSCA Inventory. EPA Extremely Hazardous Substances List.

SAFETY PROFILE: Poison by ingestion, inhalation, skin contact, subcutaneous, intraperitoneal, and intravenous routes. Human systemic effects by skin contact: primary skin irritation, allergic dermatitis, and hemorrhage. Mutation data reported. When heated to decomposition it emits toxic fumes of NO_x. See also AROMATIC AMINES.

DTL800 CAS:105-10-2 HR: 3
N,N-DIMETHYL-p-PHENYLENEDIAMINE
mf: $C_8H_{12}N_2$ mw: 136.22

PROP: Crystals from pet ether. Mp: 53°, bp: 149–150° @ 17 mm.

SYNS: p-AMINODIMETHYLANILINE □ C.I. 76075 □ p-DIMETHYLAMINOPHENYLAMINE □ N,N-DIMETHYL-1,4-BENZENEDIAMINE □ DIMETHYL-p-PHENYLENEDIAMINE □ DMPD

TOXICITY DATA WITH REFERENCE
mma-sat 5 µg/plate AEMIDF 42,641,81
dns-rat:lvr 100 µmol/L MUREAV 135,255,84
dns-ham:lvr 100 µmol/L MUREAV 135,255,84
ipr-rat LD50:21 mg/kg JPETAB 95,262,49
ipr-mus LD50:25 mg/kg JPETAB 95,262,49
ipr-dog LDLo:10 mg/kg JPETAB 95,262,49
ipr-rbt LD50:100 mg/kg JPETAB 95,262,49
ipr-gpg LD50:45 mg/kg JPETAB 95,262,49

SAFETY PROFILE: Poison by intraperitoneal route. Mutation data reported. When heated to decomposition it emits toxic fumes of NO_x.

DTM200 CAS:60160-75-0 HR: 3
N,N-DIMETHYL-p-PHENYLENEDIAMINE HEMISULFATE

TOXICITY DATA WITH REFERENCE
orl-rat LDLo:75 mg/kg NCNSA6 5,11,53

CONSENSUS REPORTS: Reported in EPA TSCA Inventory.

SAFETY PROFILE: Poison by ingestion. When heated to decomposition it emits very toxic fumes of NO_x and SO_x.

DTM400 **CAS:2052-46-2** **HR: 3**
N,N-DIMETHYL-p-PHENYLENEDIAMINE MONOHYDROCHLORIDE
mf: $C_8H_{12}N_2 \cdot ClH$ mw: 172.68

TOXICITY DATA WITH REFERENCE
orl-rat LDLo:100 mg/kg NCNSA6 5,11,53

CONSENSUS REPORTS: Reported in EPA TSCA Inventory.

SAFETY PROFILE: Poison by ingestion. When heated to decomposition it emits very toxic fumes of NO_x and HCl.

DTM600 **CAS:154-99-4** **HR: 3**
o,p-DIMETHYL-β-PHENYLETHYLHYDRAZINE DIHYDROGEN SULFATE

SYNS: β-(2,4-DIMETHYLPHENYL)ETHYLHYDRAZINE DIHYDROGEN SULPHATE □ LON 41

TOXICITY DATA WITH REFERENCE
scu-mus TDLo:160 mg/kg (female 7-10D post):REP JOENAK 49,635,71
orl-mus TDLo:240 mg/kg (1-6D preg):REP JOENAK 30,205,64
orl-mus LD50:250 mg/kg JOENAK 30,205,64
scu-mus LD50:250 mg/kg JOENAK 30,205,64

SAFETY PROFILE: Poison by ingestion and subcutaneous routes. Experimental reproductive effects. When heated to decomposition it emits very toxic fumes of SO_x and NO_x.

DTM800 **CAS:10158-43-7** **HR: 3**
DIMETHYLPHENYLETHYNYLTHALLIUM
mf: $C_{10}H_{11}Tl$ mw: 335.57

$(CH_3)_2TlC{\equiv}CPh$

CONSENSUS REPORTS: Thallium and its compounds are on the Community Right-To-Know List.

SAFETY PROFILE: An explosive sensitive to heating, stirring or impact. See also THALLIUM COMPOUNDS and ACETYLENE COMPOUNDS.

DTN000 **HR: 2**
2,4-DIMETHYLPHENYLMALEIMIDE
mf: $C_{12}H_{11}NO_2$ mw: 201.24

SYN: 2,4-DIMETHYL-N-PHENYLMALEIMIDE

TOXICITY DATA WITH REFERENCE
skn-rbt 100 µg/24H open AIHAAP 23,95,62
orl-rat LD50:710 mg/kg AIHAAP 23,95,62

SAFETY PROFILE: Moderately toxic by ingestion. A skin irritant. When heated to decomposition it emits toxic fumes of NO_x.

DTN100 **CAS:617-94-7** **HR: 2**
DIMETHYLPHENYLMETHANOL
mf: $C_9H_{12}O$ mw: 136.21

PROP: Prisms. Mp: 35–37°, bp: 202°.

SYNS: α-CUMYL ALCOHOL □ α,α-DIMETHYLBENZENEMETHANOL □ α,α-DIMETHYLBENZYL ALCOHOL □ DIMETHYLPHENYLCARBINOL □ 1-HYDROXYCUMENE □ PHENYLDIMETHYLCARBINOL □ 2-PHENYLISOPROPANOL

TOXICITY DATA WITH REFERENCE
skn-rbt 500 mg/24H SEV FCTOD7 20(Suppl), 675,82
orl-rat LD50:1300 mg/kg FCTOD7 20(Suppl),675,82
orl-mus LD50:1400 mg/kg ESKGA2 24,115,78
skn-rbt LD50:4300 mg/kg FCTOD7 20(Suppl),675,82

CONSENSUS REPORTS: Reported in EPA TSCA Inventory.

SAFETY PROFILE: Moderately toxic by ingestion. Mildly toxic by skin contact. A severe skin irritant. When heated to decomposition it emits acrid smoke and irritating fumes.

DTN200 **CAS:2655-14-3** **HR: 3**
3,5-DIMETHYLPHENYL-N-METHYLCARBAMATE
mf: $C_{10}H_{13}NO_2$ mw: 179.24

PROP: Crystals. Mp: 99°. Sol in most org solvs; very sltly sol in H_2O.

SYNS: DRC 3340 □ H-69 □ MACBAL □ MAQBARL □ 3,5-XMC □ 3,5-XYLENOL METHYLCARBAMATE □ 3,5-XYLENYL-N-METHYLCARBAMATE □ 3,5-XYLYL-N-METHYLCARBAMATE

TOXICITY DATA WITH REFERENCE
orl-rat LD50:542 mg/kg 85ARAE 1,44,77
orl-mus LD50:280 mg/kg OYYAA2 3,74,69
orl-rbt LD50:445 mg/kg SPEADM 78-1,56,78

CONSENSUS REPORTS: EPA Genetic Toxicology Program.

SAFETY PROFILE: Poison by ingestion. When heated to decomposition it emits toxic fumes of NO_x. See also CARBAMATES.

DTN775 **CAS:42013-48-9** **HR: 3**
5,5-DIMETHYL-2-PHENYLMORPHOLINE
mf: $C_{12}H_{17}NO$ mw: 191.30

SYNS: G 130 □ GP 130 □ 2-PHENYL-5-DIMETHYLTETRAHYDRO-1,4-OXAZINE

TOXICITY DATA WITH REFERENCE
orl-rat LD50:480 mg/kg ARZNAD 23,810,73
orl-mus LD50:380 mg/kg ARZNAD 23,810,73
ipr-mus LD50:100 mg/kg ARZNAD 23,810,73

SAFETY PROFILE: Poison by ingestion and intraperitoneal routes. When heated to decomposition it emits toxic fumes of NO_x.

DTN800 **CAS:7635-51-0** **HR: 3**
3,4-DIMETHYL-2-PHENYLMORPHOLINEHYDROCHLORIDE
mf: $C_{12}H_{17}NO \cdot ClH$ mw: 227.76

PROP: A solid. Mp: 191°.

SYNS: PHENDIMETRAZINE HYDROCHLORIDE □ d-2-PHENYL-3,4-DIMETHYLMORPHOLINE HYDROCHLORIDE

TOXICITY DATA WITH REFERENCE
orl-rat LD50:455 mg/kg 27ZQAG -,285,72

ipr-rat LD50: 245 mg/kg TXAPA9 2,589,60
scu-rat LD50: 435 mg/kg TXAPA9 2,589,60
orl-mus LD50: 340 mg/kg TXAPA9 2,589,60
ipr-mus LD50: 195 mg/kg TXAPA9 2,589,60
scu-mus LD50: 270 mg/kg 27ZQAG -,285,72
ivn-mus LD50: 92 mg/kg TXAPA9 2,589,60

SAFETY PROFILE: Poison by ingestion, intraperitoneal, subcutaneous, and intravenous routes. When heated to decomposition it emits very toxic fumes of NO_x and HCl.

DTN875 CAS:72586-68-6 HR: 2
1,3-DIMETHYL-3-PHENYL-1-NITROSOUREA
mf: $C_9H_{11}N_3O_2$ mw: 193.23

SYN: N,N'-DIMETHYL-N-NITROSO-N'-PHENYLUREA

TOXICITY DATA WITH REFERENCE
mmo-sat 1 µmol/plate CRNGDP 4,409,83
sce-ham:lng 100 nmol/L MUREAV 126,259,84
orl-rat TDLo: 58 mg/kg/20W-I:ETA CRNGDP 8,237,87

SAFETY PROFILE: Questionable carcinogen with experimental tumorigenic data. Mutation data reported. Many N-nitroso compounds are carcinogens. When heated to decomposition it emits toxic fumes of NO_x. See also N-NITROSO COMPOUNDS.

DTN896 CAS:672-66-2 HR: 3
DIMETHYLPHENYLPHOSPHINE
mf: $C_8H_{11}P$ mw: 150.09

$CH_3OCO{\cdot}OOCO{\cdot}OCH_3$

PROP: A liquid. Bp: 192°.

SAFETY PROFILE: Explodes when heated above 55°C or on impact. When heated to decomposition it emits acrid smoke and fumes. See also PEROXIDES.

DTO000 CAS:54-77-3 HR: 3
1,1-DIMETHYL-4-PHENYLPIPERAZINE IODIDE
mf: $C_{12}H_{19}N_2{\cdot}I$ mw: 318.23

SYNS: 1,1-DIMETHYL-4-PHENYLPIPERAZINIUM IODIDE □ DMPP □ DMPP IODIDE

TOXICITY DATA WITH REFERENCE
ipr-mus LD50: 18,500 µg/kg AIPTAK 97,186,54
ivn-mus LD50: 1600 µg/kg EJPHAZ 11,75,70
ims-mus LD50: 28 mg/kg JPETAB 103,330,51
ivn-rbt LD50: 1 mg/kg JPETAB 103,330,51

SAFETY PROFILE: Poison by intravenous, intraperitoneal, and intramuscular routes. When heated to decomposition it emits very toxic fumes of NO_x and I^-.

DTO100 CAS:10125-85-6 HR: 3
1,1-DIMETHYL-4-PHENYLPIPERIDINIUM IODIDE
mf: $C_{13}H_{20}N{\cdot}I$ mw: 317.24

TOXICITY DATA WITH REFERENCE
ipr-mus LD50: 17 mg/kg AIPTAK 97,186,54
ivn-mus LD50: 1333 µg/kg JMPCAS 2,449,60
ims-mus LD50: 28 mg/kg AIPTAK 97,186,54

SAFETY PROFILE: Poison by intramuscular, intravenous, and intraperitoneal routes. When heated to decomposition it emits toxic fumes of NO_x and I^-.

DTO200 CAS:3734-17-6 HR: 3
1,2-DIMETHYL-3-PHENYL-3-PYRROLIDYL PROPIONATE
mf: $C_{15}H_{21}NO_2$ mw: 247.37

PROP: Bp: 126–128° @ 1.1 mm.

SYNS: A-1981 □ COGESIC □ 1,2-DIMETHYL-3-PHENYL-3-PYRROLIDINOL PROPIONATE (ester) □ PRODILIDINE

TOXICITY DATA WITH REFERENCE
orl-rat LD50: 253 mg/kg JPETAB 134,332,61
ipr-rat LDLo: 133 mg/kg JMPCAS 5,441,62
scu-rat LD50: 188 mg/kg JPETAB 134,332,61
ivn-rat LD50: 74 mg/kg JPETAB 134,332,61
orl-mus LD50: 318 mg/kg JPETAB 134,332,61
scu-mus LD50: 194 mg/kg JPETAB 134,332,61
ivn-mus LD50: 91 mg/kg JPETAB 134,332,61

SAFETY PROFILE: Poison by ingestion, intravenous, intraperitoneal, and subcutaneous routes. When heated to decomposition it emits toxic fumes of NO_x.

DTO300 CAS:91481-04-8 HR: 3
1-(2,4-DIMETHYL-5-PHENYL-1H-PYRROL-3-YL)ETHANONE
mf: $C_{14}H_{15}NO$ mw: 213.30

SYNS: ETHANONE, 1-(2,4-DIMETHYL-5-PHENYL-1H-PYRROL-3-YL)- □ KETONE, (2,4-DIMETHYL-5-PHENYLPYRROL-3-YL) METHYL

TOXICITY DATA WITH REFERENCE
orl-mus LD50: 500 mg/kg FRPSAX 39,538,84

DOT CLASSIFICATION: 3; *Label.* Flammable Liquid

SAFETY PROFILE: Moderately toxic by ingestion. A flammable liquid. When heated to decomposition it emits toxic vapors of NO_x.

DTO600 CAS:1176-08-5 HR: 3
N,N-DIMETHYL-2-(α-PHENYL-o-TOLOXY)ETHYLAMINE DIHYDROGEN CITRATE
mf: $C_{17}H_{21}NO{\cdot}C_6H_8O_7$ mw: 447.53

PROP: Crystals from MeOH or H_2O. Mp: 138–140°.

SYNS: PHENYLTOLOXAMINE DIHYDROGEN CITRATE □ PRN

TOXICITY DATA WITH REFERENCE
orl-rat LD50: 1472 mg/kg TXAPA9 1,42,59
ipr-mus LD50: 246 mg/kg JAPMA8 42,587,53

CONSENSUS REPORTS: Reported in EPA TSCA Inventory.

SAFETY PROFILE: Poison by intraperitoneal route. Moderately toxic by ingestion. When heated to decomposition it emits toxic fumes of NO_x.

D

DTO800 CAS:6152-43-8 HR: 3
N,N-DIMETHYL-2-(α-PHENYL-o-TOLOXY)ETHYLAMINE HYDROCHLORIDE
mf: $C_{17}H_{21}NO•ClH$ mw: 291.85

SYNS: BRISTAMIN HYDROCHLORIDE □ PHENYLTOLOXAMINE HYDROCHLORIDE

TOXICITY DATA WITH REFERENCE
orl-mus LD50:305 mg/kg ARZNAD 8,219,58
ipr-mus LD50:163 mg/kg JAPMA8 42,587,53
ivn-mus LD50:33 mg/kg JAPMA8 42,587,53

SAFETY PROFILE: Poison by ingestion, intravenous, and intraperitoneal routes. Moderately toxic by ingestion. When heated to decomposition it emits very toxic fumes of HCl and NO_x.

DTP000 CAS:7227-91-0 HR: 3
3,3-DIMETHYL-1-PHENYLTRIAZENE
mf: $C_8H_{11}N_3$ mw: 149.22

$C_6H_5N{=}NN(CH_3)_2$

SYNS: 3,3-DIMETHYL-1-PHENYL-1-TRIAZENE □ DMPT □ 1-FENYL-3,3-DIMETHYLTRIAZIN □ NSC 3094 □ PDMT □ PDT □ 1-PHENYL-3,3-DIMETHYLTRIAZENE □ PHENYLDIMETHYLTRIAZINE □ X 119

TOXICITY DATA WITH REFERENCE
mmo-sat 1 μg/plate JNCIAM 62,873,79
mrc-smc 900 ppm JNCIAM 62,901,79
cyt-hmn:leu 25 μmol/L MUREAV 77,123,73
otr-ham:emb 100 μg/L NCIMAV 58,243,81
cyt-ham:lng 10 mg/L MUREAV 88,197,81
scu-rat TDLo:125 mg/kg (female 15D post):TER IARCCD 4,45,73
ipr-rat TDLo:30 mg/kg (female 12D post):TER TJADAB 39,53,89
orl-rat TDLo:310 mg/kg:CAR ZKKOBW 81,285,74
scu-rat TDLo:1250 mg/kg/59W-I:CAR ZKKOBW 81,285,74
scu-rat TDLo:75 mg/kg (23D preg):CAR IARCCD 4,45,73
ivn-rat TDLo:30,500 μg/kg/16W-I:ETA ZAPPAN 115,8,72
ivn-rat TDLo:10 mg/kg (22D preg):ETA ZAPPAN 115,8,72
scu-rat TD:1166 mg/kg/47W-I:ETA FCTXAV 6,579,68
orl-rat TD:1600 mg/kg/79W-C:CAR ZKKOBW 81,285,74
orl-rat TD:2040 mg/kg/47W-I:ETA FCTXAV 6,579,68
orl-rat TD:1650 mg/kg/33W-I:ETA XENOBH 3,271,73
scu-rat TD:1250 mg/kg/59W-I:ETA NATWAY 54,171,67
orl-rat LD50:310 mg/kg ZKKOBW 81,285,74
ipr-rat LD50:180 mg/kg CPCHAO 18,307,62
orl-mus LD50:200 mg/kg NCISP* JAN86
ipr-mus LD50:190 mg/kg JMCMAR 19,1299,76

CONSENSUS REPORTS: EPA Genetic Toxicology Program.

SAFETY PROFILE: Poison by ingestion and intraperitoneal routes. Questionable carcinogen with experimental carcinogenic and tumorigenic data. Experimental teratogenic effects. Human mutation data reported. Decomposes explosively on attempted distillation at atmospheric pressure. When heated to decomposition it emits toxic fumes of NO_x.

DTP400 CAS:101-42-8 HR: 2
1,1-DIMETHYL-3-PHENYLUREA
mf: $C_9H_{12}N_2O$ mw: 164.23

PROP: White crystals. Mp: 131–133°. Insol in water; sltly sol in hydrocarbons.

SYNS: BEET-KLEEN □ DIBAR □ N,N-DIMETHYL-N'-PHENYLUREA □ DYBAR □ FENIDIN □ FENULON □ FENURON □ N-PHENYL-N',N'-DIMETHYLUREA □ 1-PHENYL-3,3-DIMETHYLUREA □ 3-PHENYL-1,1-DIMETHYLUREA □ PDU □ PUD (HERBICIDE)

TOXICITY DATA WITH REFERENCE
dni-mus-orl 500 mg/kg MUREAV 58,353,78
orl-rat LD50:6400 mg/kg FMCHA2 -,D137,80
orl-mus LD50:4700 mg/kg GISAAA 47(3),82,82
orl-rbt LD50:4700 mg/kg GISAAA 47(3),82,82
orl-gpg LD50:3200 mg/kg GISAAA 40(10),22,75

CONSENSUS REPORTS: Reported in EPA TSCA Inventory.

SAFETY PROFILE: Moderately toxic by ingestion. Mutation data reported. When heated to decomposition it emits toxic fumes of NO_x.

DTP600 CAS:13171-22-7 HR: 3
DIMETHYL PHOSPHATE ESTER with 2-CHLORO-N-ETHYL-3-HYDROXYCROTONAMIDE
mf: $C_8H_{15}ClNO_5P$ mw: 271.66

SYNS: C-776 □ 2-CHLORO-3-(ETHYLAMINO)-1-METHYL-3-OXO-1-PROPENYL DIMETHYL ESTER PHOSPHORIC ACID □ 2-CHLORO-3-(ETHYLAMINO)-1-METHYL-3-OXO-1-PROPENYL DIMETHYL PHOSPHATE □ CIBA C-776 □ ENT 27,358 □ NSC 190956

TOXICITY DATA WITH REFERENCE
orl-rat LD50:37 mg/kg ARSIM* 20,7,66
ipr-mus LD50:7800 μg/kg TXAPA9 13,37,68
scu-gpg LDLo:100 mg/kg JEENAI 62(4),934,69

SAFETY PROFILE: Poison by ingestion, intraperitoneal, and subcutaneous routes. When heated to decomposition it emits very toxic fumes of Cl^-, NO_x, and PO_x.

DTP800 CAS:34491-04-8 HR: 3
DIMETHYL PHOSPHATE ESTER with 2-CHLORO-N-METHYL-3-HYDROXYCROTONAMIDE
mf: $C_7H_{13}ClNO_5P$ mw: 257.63

SYNS: CIBA C-768 □ ENT 27,357 □ NSC 190955

TOXICITY DATA WITH REFERENCE
orl-rat LD50:33 mg/kg ARSIM* 20,7,66
orl-gpg LDLo:100 mg/kg JEENAI 62(4),934,69
scu-gpg LDLo:50 mg/kg JEENAI 62(4),934,69

SAFETY PROFILE: Poison by ingestion and subcutaneous routes. When heated to decomposition it emits very toxic Cl^-, NO_x, and PO_x. See also ESTERS.

DTQ089 CAS:676-59-5 HR: 3
DIMETHYL PHOSPHINE
mf: C_2H_7P mw: 64.05

PROP: Liquid with a disgusting odor. Bp: 25°.

SAFETY PROFILE: Ignites spontaneously in air. When

heated to decomposition it emits toxic fumes of PO_x and phosphine. See also PHOSPHINE.

DTQ400 CAS:10265-92-6 HR: 3
O,S-DIMETHYL PHOSPHORAMIDOTHIOATE
mf: $C_2H_8NO_2PS$ mw: 141.14

PROP: Crystals. Mp: 40°. Sltly water-sol; sol in alc.

SYNS: ACEPHATE-MET □ BAY 71628 □ BAYER 71628 □ CHEVRON 9006 □ CHEVRON ORTHO 9006 □ O,S-DIMETHYL ESTER AMIDE of AMIDOTHIOATE □ ENT 27,396 □ HAMIDOP □ METAMIDOFOS ESTRELLA □ METHAMIDOPHOS □ MONITOR □ MTD □ NSC 190987 □ ORTHO 9006 □ PILLARON □ SRA 5172 □ TAHMABON □ TAMARON □ THIOPHOSPHORSAEURE-O,S-DIMETHYLESTERAMID (GERMAN)

TOXICITY DATA WITH REFERENCE
orl-rat TDLo: 10 mg/kg (female 6-15D post): TER VEMJA8 34,357,86
orl-man TDLo: 257 mg/kg: PNS,EYE,SKN NEJMAG 306,125,82
orl-wmn TDLo: 360 mg/kg: PNS,EYE,SKN NEJMAG 306,125,82
orl-rat LD50: 7500 μg/kg ARSIM* 20,7,66
ihl-rat LD50: 9 mg/kg TXAPA9 45,232,78
skn-rat LD50: 50 mg/kg 28ZEAL 5,149,76
ipr-rat LD50: 15 mg/kg PCBPBS 13,267,80
orl-mus LD50: 14 mg/kg PCBPBS 7,83,77
ihl-mus LD50: 19 mg/kg TXAPA9 45,232,78
orl-rbt LD50: 10 mg/kg 28ZEAL 5,149,76
skn-rbt LD50: 118 mg/kg GUCHAZ 6,333,73

CONSENSUS REPORTS: EPA Extremely Hazardous Substances List.

SAFETY PROFILE: Poison by ingestion, inhalation, skin contact, subcutaneous, and intraperitoneal routes. Human systemic effects by ingestion: fasciculations, pupillary constriction, and sweating. A cholinesterase inhibitor type of insecticide. When heated to decomposition it emits very toxic fumes of NO_x, PO_x, and SO_x. See also PARATHION.

DTQ600 CAS:2524-03-0 HR: 3
O,O-DIMETHYLPHOSPHOROCHLORIDOTHIOATE
mf: $C_2H_6ClO_2PS$ mw: 160.56

PROP: A liquid. D: 1.326, bp: 68° @ 12 mm.

SYNS: DIMETHYL CHLOROTHIOPHOSPHATE (DOT) □ DIMETHYLCHLORTHIOFOSAT (CZECH) □ O,O-DIMETHYLESTER KYSELINY CHLORTHIOFOSFORECNE (CZECH) □ DIMETHYL PHOSPHOROCHLORIDOTHIOATE (DOT) □ METHYL PCT □ PHOSPHOROCHLORIDOTHIOIC ACID-O,O-DIMETHYL ESTER

TOXICITY DATA WITH REFERENCE
orl-rat LDLo: 1000 mg/kg 34ZIAG -,393,69
ihl-rat LC50: 340 mg/m³/4H 85GMAT -,56,82
orl-mus LD50: 1800 mg/kg 85GMAT -,56,82
ihl-mus LC50: 320 mg/m³/2 85GMAT -,56,82
skn-rbt LDLo: 750 mg/kg 34ZIAG -,393,69

CONSENSUS REPORTS: Reported in EPA TSCA Inventory. EPA Extremely Hazardous Substances List.

SAFETY PROFILE: Poison by inhalation. Moderately toxic by ingestion and skin contact. Corrosive. When heated to decomposition it emits very toxic fumes of Cl^-, PO_x, and SO_x.

DTQ800 CAS:3581-11-1 HR: 3
O,O-DIMETHYL PHOSPHOROTHIOATE-O-ESTER with 4-HYDROXY-m-ANISONITRILE
mf: $C_{10}H_{12}NO_4PS$ mw: 273.26

SYNS: B 11163 □ O-(4-CYANO-2-METHOXYPHENYL)-O,O-DIMETHYL PHOSPHOROTHIOATE □ ENT 27,230 □ PHOSPHOROTHIOIC ACID-O,O-DIMETHYL-O-(4-CYANO-2-METHOXYPHENYL) ESTER □ PHOSPHOROTHIOIC ACID-O,O-DIMETHYL ESTER-O-ESTER with VANNILLONITRILE □ STAUFFER B-11163 □ TP540

TOXICITY DATA WITH REFERENCE
orl-rat LD50: 2710 mg/kg ARSIM* 20,21,66
orl-mus LD50: 4200 mg/kg TDKNAF 24,221,65
scu-gpg LDLo: 100 mg/kg JEENAI 61,1261,68

CONSENSUS REPORTS: Cyanide and its compounds are on the Community Right-To-Know List.

SAFETY PROFILE: Poison by subcutaneous route. Moderately toxic by ingestion. When heated to decomposition it emits very toxic fumes of NO_x, PO_x, CN^-, and SO_x. See also ESTERS and NITRILES.

DTR200 CAS:131-11-3 HR: 2
DIMETHYL PHTHALATE
mf: $C_{10}H_{10}O_4$ mw: 194.20

PROP: Colorless, odorless liquid. Mp: 0°, bp: 282.4°, flash p: 295°F (CC), d: 1.189 @ 25°/25°, autoign temp: 1032°F, vap d: 6.69, vap press: 1 mm @ 100.3°.

SYNS: AVOLIN □ 1,2-BENZENEDICARBOXYLIC ACID DIMETHYL ESTER □ DIMETHYL-1,2-BENZENEDICARBOXYLATE □ DIMETHYL BENZENEORTHODICARBOXYLATE □ DMP □ ENT 262 □ FERMINE □ METHYL PHTHALATE □ MIPAX □ NTM □ PALATINOL M □ PHTHALIC ACID METHYL ESTER □ PHTHALSAEUREDIMETHYLESTER (GERMAN) □ RCRA WASTE NUMBER U102 □ SOLVANOM □ SOLVARONE

TOXICITY DATA WITH REFERENCE
eye-rbt 119 mg JPETAB 82,377,44
mmo-sat 200 μg/plate JTEHD6 16,61,85
cyt-rat-skn 25 g/kg/4W-I FATOAO 40,454,77
ipr-rat TDLo: 1125 mg/kg (5-15D preg): TER JPMSAE 61,51,72
ipr-rat TDLo: 338 mg/kg (5-15D preg): TER JPMSAE 61,51,72
ipr-rat TDLo: 338 mg/kg (5-15D preg): REP JPMSAE 61,51,72
orl-rat LD50: 6800 mg/kg GTPZAB 24(3),25,80
ipr-rat LD50: 3375 mg/kg JPMSAE 61,51,72
orl-mus LD50: 6800 mg/kg GTPZAB 24(3),25,80
ipr-mus LD50: 1380 mg/kg IPSTB3 3,93,76
scu-mus LDLo: 6500 mg/kg EDWU** -,-,37
ihl-cat LCLo: 9630 mg/m³/6H EDWU** -,-,37
orl-rbt LD50: 4400 mg/kg JPETAB 93,26,48
orl-gpg LD50: 2400 mg/kg JPETAB 93,26,48
orl-ckn LD50: 8500 mg/kg JPETAB 93,26,48

CONSENSUS REPORTS: On EPA Extremely Hazardous Substances List by error. Reported in EPA TSCA Inventory. Community Right-To-Know List.

OSHA PEL: TWA 5 mg/m³
ACGIH TLV: TWA 5 mg/m³

SAFETY PROFILE: Moderately toxic by ingestion and intraperitoneal routes. Mildly toxic by inhalation. Experimental teratogenic and reproductive effects. Mutation data reported. An eye irritant. A pesticide and insect repellent. Combustible when exposed to heat or flame; can react with oxidizing materials. To fight fire, use CO_2, dry chemical. When heated to decomposition it emits acrid smoke and irritating fumes. See also ESTERS.

DTR400 CAS:106-55-8 HR: 2
2,5-DIMETHYLPIPERAZINE
mf: $C_6H_{14}N_2$ mw: 114.22

TOXICITY DATA WITH REFERENCE
eye-rbt 750 µg SEV AMIHBC 4,119,51
eye-rbt 250 µg/24H SEV 85JCAE-,865,86
orl-rat LD50:3160 mg/kg AMIHBC 4,119,51
skn-rbt LD50:800 mg/kg AMIHBC 4,119,51

CONSENSUS REPORTS: Reported in EPA TSCA Inventory.

SAFETY PROFILE: Moderately toxic by ingestion and skin contact. A severe eye irritant. When heated to decomposition it emits toxic fumes of NO_x.

DTR800 CAS:77966-85-9 HR: 3
2-(2,6-DIMETHYLPIPERIDINO)-2′,6′-ACETOXYLIDIDE HYDROCHLORIDE
mf: $C_{17}H_{26}N_2O•ClH$ mw: 310.91

SYN: V 374

TOXICITY DATA WITH REFERENCE
eye-rbt 2% MLD ARZNAD 8,407,58
ipr-rat LD50:53 mg/kg ARZNAD 8,407,58
scu-mus LD50:125 mg/kg ARZNAD 8,407,58

SAFETY PROFILE: Poison by intraperitoneal and subcutaneous routes. An eye irritant. When heated to decomposition it emits very toxic fumes of NO_x and HCl.

DTR850 HR: 2
DIMETHYLPOLYSILOXANE
mf: $[(CH_3)_2SiO—]$

PROP: Clear, colorless viscous liquid. D: 0.96, refr index: 1.400. Sol in hydrocarbon solvents; insol in water.

SYNS: DIMETHYL SILICONE □ POLYDIMETHYLSILOXANE

SAFETY PROFILE: Combustible liquid. When heated to decomposition it emits acrid smoke and irritating fumes.

DTS400 CAS:3282-30-2 HR: 3
2,2-DIMETHYLPROPANOYL CHLORIDE
DOT: UN 2438
mf: C_5H_9ClO mw: 120.59

PROP: Bp: 105–106°.

SYNS: 2,2-DIMETHYLPROPIONYL CHLORIDE □ NEOPANTANOYL CHLORIDE □ PIVALIC ACID CHLORIDE □ PIVALOLYL CHLORIDE □ PIVALOYL CHLORIDE □ PIVALYL CHLORIDE □ TRIMETHYL ACETYL CHLORIDE (DOT)

CONSENSUS REPORTS: Reported in EPA TSCA Inventory.

DOT CLASSIFICATION: 8; *Label:* Corrosive, Flammable Liquid, Poison

SAFETY PROFILE: A corrosive irritant to skin, eyes, and mucous membranes. The liquid is flammable when exposed to heat, flame, or oxidizers. When heated to decomposition it emits toxic fumes of Cl^-.

DTS500 CAS:26062-79-3 HR: 2
N,N-DIMETHYL-N-2-PROPENYL-2-PROPEN-1-AMINIUM CHLORIDE HOMOPOLYMER (9CI)
mf: $(C_8H_{16}N•Cl)x$

SYNS: AGEFLOC WT 20 □ AMMONIUM, DIALLYLDIMETHYL-, CHLORIDE, POLYMERS □ CALGON 261 □ CALGON 261LV □ CALGON POLYMER 261 □ CAT-FLOC □ CONDUCTIVE POLYMER 261 □ CP 261 □ CP 261LV □ E 261 □ LECTRAPEL □ 261LV □ MERCK 261 □ MERQUAT 100 □ PAS-H 10 □ PBK 1 □ PERCOL 1697 □ POLYMER 261 □ POLYMER 261LV □ POLYQUATERNIUM 6 □ QUATERNIUM 40 □ VPK 402

TOXICITY DATA WITH REFERENCE
orl-rat LD50:3 g/kg GISAAA 53(3),66,88
orl-mus LD50:1720 mg/kg GISAAA 53(3),66,88
orl-gpg LD50:3250 mg/kg GISAAA 53(3),66,88

CONSENSUS REPORTS: Reported in EPA TSCA Inventory.

SAFETY PROFILE: Moderately toxic by ingestion. When heated to decomposition it emits toxic vapors of NO_x and Cl^-.

DTS600 CAS:758-96-3 HR: 2
N,N-DIMETHYLPROPIONAMIDE
mf: $C_5H_{11}NO$ mw: 101.17

PROP: Bp: 165–178°.

TOXICITY DATA WITH REFERENCE
ipr-mus LD50:875 mg/kg AIHAAP 32,539,71
ivn-mus LD50:820 mg/kg AIHAAP 32,539,71

CONSENSUS REPORTS: Reported in EPA TSCA Inventory.

SAFETY PROFILE: Moderately toxic by intraperitoneal and intravenous routes. When heated to decomposition it emits toxic fumes of NO_x.

DTS625 CAS:95619-40-2 HR: 1
2,6-DIMETHYL-4-PROPOXY-BENZOIC ACID 2-METHYL-2-(1-PYRROLIDINYL)PROPYLESTER
mf: $C_{20}H_{31}NO_3•ClH$ mw: 369.98

SYNS: BENZOIC ACID, 2,6-DIMETHYL-4-PROPOXY-, 2-METHYL-2-(1-PYRROLIDINYL)PROPYL ESTER, HYDROCHLORIDE □ U-2363

TOXICITY DATA WITH REFERENCE
skn-rbt 5 pph MLD AIPTAK 137,410,62
eye-rbt 5000 ppm MLD AIPTAK 137,410,62

ipr-mus LD50:55,700 mg/kg AIPTAK 137,410,62

SAFETY PROFILE: Slightly toxic by intraperitoneal route. A skin and eye irritant. When heated to decomposition it emits toxic fumes of NO_x and HCl.

DTT400 CAS:24690-46-8 HR: 2
N,N-DIMETHYL-p-((p-PROPYLPHENYL)AZO)ANILINE
mf: $C_{17}H_{21}N_3$ mw: 267.41

SYN: 4'-N-PROPYL-4-DIMETHYLAMINOAZOBENZENE

TOXICITY DATA WITH REFERENCE
orl-rat TDLo:4284 mg/kg/17W-C:ETA JNCIAM 27,663,61

SAFETY PROFILE: Questionable carcinogen with experimental tumorigenic data. When heated to decomposition it emits toxic fumes of NO_x.

DTT600 CAS:23950-58-5 HR: 2
N-(1,1-DIMETHYLPROPYNYL)-3,5-DICHLOROBENZAMIDE
mf: $C_{12}H_{11}Cl_2NO$ mw: 256.14

PROP: A solid. Mp: 155–156°.

SYNS: 3,5-DICHLORO-N-(1,1-DIMETHYL-2-PROPYNYL)BENZAMIDE □ KERB □ PROMAMIDE □ PRONAMIDE □ PROPYZAMIDE □ RCRA WASTE NUMBER U192 □ RH 315

TOXICITY DATA WITH REFERENCE
orl-rat TDLo:1092 mg/kg/2Y-C:ETA ENVRAL 23,1,80
orl-mus TDLo:65,520 mg/kg/78W-C:CAR ENVRAL 23,1,80
orl-rat LD50:5620 mg/kg 85ARAE 2,217,77

SAFETY PROFILE: Mildly toxic by ingestion. Questionable carcinogen with experimental carcinogenic and tumorigenic data. An herbicide. When heated to decomposition it emits very toxic fumes of Cl^- and NO_x.

DTU200 CAS:2825-00-5 HR: 3
3,5-DIMETHYL-4H-PYRAN-4-ONE-2-METHOXY-6-(TETRAHYDRO-4-(β-METHYL-p-NITROCINNAMYLIDENE)-2-FURYL)
mf: $C_{22}H_{23}NO_6$ mw: 397.46

PROP: Yellow prisms. Mp: 158°.

SYNS: AUREOTHIN □ MYCOLUTEIN

TOXICITY DATA WITH REFERENCE
orl-mus LD50:3 mg/kg 85GDA2 5,390,81
ipr-mus LD50:1 mg/kg 85GDA2 5,388,81
scu-mus LD50:2 mg/kg 85GDA2 5,388,81
ivn-mus LD50:1260 μg/kg CSLNX* NX#02084

SAFETY PROFILE: Poison by ingestion, intravenous, intraperitoneal, and subcutaneous routes. When heated to decomposition it emits toxic fumes of NO_x.

DTU400 CAS:5910-89-4 HR: 3
2,3-DIMETHYLPYRAZINE
mf: $C_6H_8N_2$ mw: 108.16

PROP: Colorless liquid; nutty cocoa odor. D: 1.000–1.022 @ 20°, refr index: 1.506–1.509, flash p: 147°F (OC), d: 0.99, vap d: 3.72, bp: 156–158°. Misc with water, organic solvents. Sol in water and org solvs.

SYNS: 2,3-DIMETHYL-1,4-DIAZINE □ FEMA No. 3271

TOXICITY DATA WITH REFERENCE
orl-rat LD50:613 mg/kg DCTODJ 3,249,80
ipr-mus LD50:1390 mg/kg TXAPA9 17,244,70

CONSENSUS REPORTS: Reported in EPA TSCA Inventory.

SAFETY PROFILE: Moderately toxic by ingestion and intraperitoneal routes. Flammable liquid when exposed to heat, sparks, or flame. When heated to decomposition it emits toxic fumes of NO_x.

DTU600 CAS:123-32-0 HR: 3
2,5-DIMETHYLPYRAZINE
mf: $C_6H_8N_2$ mw: 108.16

PROP: Colorless liquid; potato taste. D: 0.980–1.000, refr index: 1.497–1.501, flash p: 147°F (OC), d: 0.99, vap d: 3.72, bp: 155°, mp: 15°. Misc with water, organic solvents. Sol in H_2O, EtOH, and Et_2O.

SYNS: 2,5-DIMETHYL-1,4-DIAZINE □ FEMA No. 3272

TOXICITY DATA WITH REFERENCE
mmo-smc 3300 μg/L FCTXAV 18,581,80
cyt-ham:ovr 2500 μg/L FCTXAV 18,581,80
orl-rat LD50:1020 mg/kg DCTODJ 3,249,80
ipr-mus LD50:1350 mg/kg TXAPA9 17,244,70

CONSENSUS REPORTS: Reported in EPA TSCA Inventory.

SAFETY PROFILE: Moderately toxic by ingestion and intraperitoneal routes. Mutation data reported. Flammable liquid when exposed to heat, open flame, spark, oxidizers. To fight fire, use water spray, mist, dry chemical, CO_2, foam. When heated to decomposition it emits toxic fumes of NO_x.

DTU800 CAS:108-50-9 HR: 2
2,6-DIMETHYLPYRAZINE
mf: $C_6H_8N_2$ mw: 108.16

PROP: Prisms or white to yellow crystals; nutty, coffee odor. Mp: 48°, d: 0.965 @ 50°, bp: 155.6°. Sol in H_2O, EtOH, and Et_2O.

SYN: FEMA No. 3273

TOXICITY DATA WITH REFERENCE
mmo-smc 3300 mg/L FCTXAV 18,581,80
cyt-ham:ovr 2500 mg/L FCTXAV 18,581,80
orl-rat LD50:880 mg/kg DCTODJ 3,249,80
ipr-mus LD50:1080 mg/kg TXAPA9 17,244,70

CONSENSUS REPORTS: Reported in EPA TSCA Inventory.

SAFETY PROFILE: Moderately toxic by ingestion and intraperitoneal route. Mutation data reported. When heated to decomposition it emits toxic fumes of NO_x.

DTU850 **CAS:67-51-6** **HR: 2**
3,5-DIMETHYLPYRAZOLE
mf: $C_5H_8N_2$ mw: 96.15

SYNS: DMP □ 3,5-DWUMETYLOPIRAZOLU □ PYRAZOLE, 3,5-DI-METHYL- □ TH 564 □ U 6245

TOXICITY DATA WITH REFERENCE
ipr-mus LD50:570 mg/kg DIPHAH 18,19,66

CONSENSUS REPORTS: Reported in EPA TSCA Inventory.

SAFETY PROFILE: Moderately toxic by intraperitoneal route. When heated to decomposition it emits toxic vapors of NO_x.

DTV089 **CAS:1073-23-0** **HR: 3**
2,6-DIMETHYLPYRIDINE-N-OXIDE
mf: C_7H_9NO mw: 123.15

$CH_3CH=CHCH=CHCCH_3=N:O$

SAFETY PROFILE: Explosive reaction with phosphoryl chloride. When heated to decomposition it emits toxic fumes of NO_x.

DTV200 **CAS:21600-42-0** **HR: 3**
(3,3-DIMETHYL-1-(m-PYRIDYL-N-OXIDE))TRIAZENE
mf: $C_7H_{10}N_4O$ mw: 166.21

SYNS: 3-(3′,3′-DIMETHYLTRIAZENO)-PYRIDIN-N-OXID (GERMAN) □ 3-(3′,3′-DIMETHYLTRIAZENO)PYRIDINE-N-OXIDE □ PYNDT □ 1-(PYRIDYL-3-N-OXID)-3,3-DIMETHYL-TRIAZEN (GERMAN) □ 1-(PYRIDYL-3-N-OXIDE)-3,3-DIMETHYLTRIAZENE

TOXICITY DATA WITH REFERENCE
sln-dmg-orl 700 µmol/L CBINA8 9,365,74
cyt-hmn:leu 25 µmol/L MUREAV 77,123,73
hma-mus/smc 400 µmol/L/Kg AGACBH 3,99,73
ivn-rat TDLo:490 mg/kg/38W-I:CAR ZKKOBW 81,285,74
ivn-rat TD:540 mg/kg/36W-I:ETA XENOBH 3,271,73
scu-rat LD50:200 mg/kg ZKKOBW 81,285,74
ivn-rat LD50:230 mg/kg ZKKOBW 81,285,74

CONSENSUS REPORTS: EPA Genetic Toxicology Program.

SAFETY PROFILE: Poison by intravenous and subcutaneous routes. Questionable carcinogen with experimental carcinogenic and tumorigenic data. Human mutation data reported. When heated to decomposition it emits toxic fumes of NO_x.

DTV400 **CAS:333-40-4** **HR: 3**
S-(4,6-DIMETHYL-2-PYRIMIDINYL)-O,O-DIETHYL PHOSPHORODITHIOATE
mf: $C_{10}H_{17}N_2O_2PS_2$ mw: 292.38

SYNS: ENT 25,737 □ STAUFFER R-3413

TOXICITY DATA WITH REFERENCE
orl-rat LD50:59 mg/kg ARSIM* 20,23,66
orl-ckn LD50:41 mg/kg TXAPA9 7,606,65

SAFETY PROFILE: Poison by ingestion. When heated to decomposition it emits very toxic fumes of NO_x, PO_x, and SO_x.

DTY200 **CAS:17025-30-8** **HR: 2**
N,N-DIMETHYL-4-(4′-QUINOLYLAZO)ANILINE
mf: $C_{17}H_{16}N_4$ mw: 276.37

SYN: 4-((p-(DIMETHYLAMINO)PHENYL)AZO)QUINOLINE

TOXICITY DATA WITH REFERENCE
orl-rat TDLo:2142 mg/kg/17W-C:ETA JNCIAM 26,1461,61

SAFETY PROFILE: Questionable carcinogen with experimental tumorigenic data. When heated to decomposition it emits toxic fumes of NO_x.

DTY400 **CAS:63042-68-2** **HR: 2**
N,N-DIMETHYL-4-((4′-QUINOLYL-1′-OXIDE)AZO)ANILINE
mf: $C_{17}H_{16}N_4O$ mw: 292.37

SYN: 4-((p-(DIMETHYLAMINO)PHENYL)AZO)QUINOLINE-1-OXIDE

TOXICITY DATA WITH REFERENCE
orl-rat TDLo:2142 mg/kg/17W-C:ETA JNCIAM 26,1461,61

SAFETY PROFILE: Questionable carcinogen with experimental tumorigenic data. When heated to decomposition it emits toxic fumes of NO_x.

DTY600 **CAS:70324-23-1** **HR: 3**
3,3-DIMETHYL-1(3-QUINOLYL)TRIAZENE
mf: $C_{11}H_{12}N_4$ mw: 144.22

$(CH_3)_2NN=NC_9H_6N$

SAFETY PROFILE: Crude material decomposes violently when dried. The pure material explodes at 131°C. When heated to decomposition it emits toxic fumes of NO_x.

DTY700 **CAS:2379-55-7** **HR: 3**
2,3-DIMETHYLQUINOXALINE
mf: $C_{10}H_{10}N_2$ mw: 158.22

SYN: QUINOXALINE, 2,3-DIMETHYL-

TOXICITY DATA WITH REFERENCE
ivn-mus LD50:180 mg/kg CSLNX* NX#00809

CONSENSUS REPORTS: Reported in EPA TSCA Inventory.

SAFETY PROFILE: Poison by intravenous route. When heated to decomposition it emits toxic vapors of NO_x.

DUA200 **CAS:23521-13-3** **HR: 2**
N,N-DIMETHYL-p-(5-QUINOXALYLAZO)ANILINE
mf: $C_{16}H_{15}N_5$ mw: 277.36

SYN: 5-((p-(DIMETHYLAMINO)PHENYL)AZO)QUINOXALINE

TOXICITY DATA WITH REFERENCE
orl-rat TDLo:2200 mg/kg/17W-C:CAR JMCMAR 12,1113,69

SAFETY PROFILE: Questionable carcinogen with ex-

perimental carcinogenic data. When heated to decomposition it emits toxic fumes of NO_x.

DUA400 **CAS:23521-14-4** **HR: 2**
N,N-DIMETHYL-p-(6-QUINOXALYAZO)ANILINE
mf: $C_{16}H_{15}N_5$ mw: 277.36

SYNS: 6-((p-(DIMETHYLAMINO)PHENYL)AZO)QUINOXALINE □ N,N-DIMETHYL-p-(6-QUINOXALINYLAZO)ANILINE

TOXICITY DATA WITH REFERENCE
orl-rat TDLo:1100 mg/kg/60D-C:CAR JMCMAR 12,1113,69

SAFETY PROFILE: Questionable carcinogen with experimental carcinogenic data. When heated to decomposition it emits toxic fumes of NO_x.

DUA600 **CAS:101652-10-2** **HR: 2**
7,8-DIMETHYL-10-(d-RIBO-2,3,4,5-TETRAHYDROXYPENTYL)-4a,5-DIHYDROISOALLOXAZINE
mf: $C_{17}H_{22}N_4O_6$ mw: 378.43

TOXICITY DATA WITH REFERENCE
ipr-rat LD50:965 mg/kg CMTRAG 2,96,61
ipr-mus LD50:800 mg/kg CMTRAG 2,96,61

SAFETY PROFILE: Moderately toxic by intraperitoneal route. When heated to decomposition it emits toxic fumes of NO_x.

DUA800 **CAS:1778-08-1** **HR: 2**
N,N-DIMETHYLSALICYLAMIDE
mf: $C_9H_{11}NO_2$ mw: 165.21

SYNS: SALICYLDIMETHYLAMIDE □ SAM

TOXICITY DATA WITH REFERENCE
orl-rat LD50:2300 mg/kg JPETAB 108,450,53
ipr-rat LD50:2000 mg/kg JPETAB 108,450,53
ipr-mus LD50:1100 mg/kg YKKZAJ 86,120,66

SAFETY PROFILE: Moderately toxic by ingestion and intraperitoneal routes. When heated to decomposition it emits toxic fumes of NO_x.

DUB000 **CAS:6918-51-0** **HR: 3**
DIMETHYL SELENATE
mf: $C_2H_6O_4Se$ mw: 173.03

CONSENSUS REPORTS: Selenium and its compounds are on the Community Right-To-Know List.

OSHA PEL: TWA 0.2 mg(Se)/m^3
ACGIH TLV: TWA 0.2 mg(Se)/m^3
DFG MAK: 0.1 mg(Se)/m^3

SAFETY PROFILE: Explodes when heated to 150°C. When heated to decomposition it emits toxic fumes of Se. See also SELENIUM COMPOUNDS.

DUB200 **CAS:593-79-3** **HR: 2**
DIMETHYL SELENIDE
mf: C_2H_6Se mw: 109.04

PROP: A liquid. Bp: 58°, d: 1.43 @ 25°, vap d: 3.75, mp: −87.2°.

SYNS: DIMETHYLSELENIUM □ METHYL SELENIDE □ METHYL SELENIUM

TOXICITY DATA WITH REFERENCE
ipr-rat LD50:2200 mg/kg PSEBAA 79,230,52
scu-rat LDLo:2180 mg/kg ARTODN 45,207,80
ipr-mus LD50:1800 mg/kg PSEBAA 79,230,52

CONSENSUS REPORTS: Selenium and its compounds are on the Community Right-To-Know List.

OSHA PEL: TWA 0.2 mg(Se)/m^3
ACGIH TLV: TWA 0.2 mg(Se)/m^3
DFG MAK: 0.1 mg(Se)/m^3

SAFETY PROFILE: Moderately toxic by intraperitoneal and subcutaneous routes. When heated to decomposition it emits toxic fumes of Se. See also SELENIUM COMPOUNDS.

DUB600 **CAS:63148-62-9** **HR: 2**
DIMETHYL SILOXANE

PROP: Viscosity 100 at 25 degrees (ISMJAV 22,15,63).

SYN: DOW-CORNING 200 FLUID-LOT No. AA-4163

TOXICITY DATA WITH REFERENCE
scu-mus TDLo:120 g/kg:ETA ISMJAV 22,15,63

CONSENSUS REPORTS: Reported in EPA TSCA Inventory.

SAFETY PROFILE: Questionable carcinogen with experimental tumorigenic data.

DUB600 **HR: 3**
(DIMETHYL SILYLMETHYL)TRIMETHYL LEAD
mf: $C_6H_{18}PbSi$ mw: 325.49

$(CH_3)_2SiHCH_2Pb(CH_3)_3$

CONSENSUS REPORTS: Lead and its compounds are on the Community Right-To-Know List.

SAFETY PROFILE: Lead compounds are generally poisons. Decomposes violently when heated above 100°C in the presence of oxygen. When heated to decomposition it emits toxic fumes of Pb. See also LEAD COMPOUNDS.

DUB800 **CAS:1145-73-9** **HR: 3**
N,N-DIMETHYL-4-STILBENAMINE
mf: $C_{16}H_{17}N$ mw: 223.34

SYNS: 4-DIMETHYLAMINOSTILBEN (GERMAN) □ N,N-DIMETHYL-4-AMINOSTILBENE □ N,N-DIMETHYL-p-STYRYLANILINE □ STILBENYL-N,N-DIMETHYLAMINE

TOXICITY DATA WITH REFERENCE
mma-sat 10 µg/plate JJIND8 71,293,83
mmo-bcs 5 g/L MUREAV 42,19,77
dnr-bcs 5 g/L MUREAV 42,19,77
ipr-rat TDLo:180 mg/kg/9W-I:ETA BJCAAI 6,392,52
orl-rat TD:520 mg/kg/49W-C:ETA ABMGAJ 9,87,62
orl-rat TD:383 mg/kg/46W-C:ETA ZEKBAI 67,135,65
orl-rat TD:305 mg/kg/43W-C:ETA ZEKBAI 61,230,56
orl-rat TD:370 mg/kg/26W-C:ETA ZEKBAI 65,272,63

orl-rat TD:4200 mg/kg/20W-C:CAR GANNA2 61,367,70
orl-rat LDLo:50 mg/kg CNREA8 26,619,66
ipr-rat LD50:70 mg/kg ZEKBAI 65,272,63

SAFETY PROFILE: Poison by ingestion and intraperitoneal routes. Questionable carcinogen with experimental carcinogenic and tumorigenic data. Mutation data reported. When heated to decomposition it emits toxic fumes of NO_x.

DUC000 CAS:838-95-9 HR: 3
(E)-N,N-DIMETHYL-4-STILBENAMINE
mf: $C_{16}H_{17}N$ mw: 223.34

SYNS: trans-p-(DIMETHYLAMINO)STILBENE □ trans-4-DIMETHYLAMINOSTILBENE □ 4-DIMETHYLAMINO-trans-STILBENE □ (E)-N,N,-DIMETHYL-4-(2-PHENYLETHENYL)BENZENAMINE □ trans-N,N-DIMETHYL-4-STILBENAMINE

TOXICITY DATA WITH REFERENCE
mma-sat 10 µg/plate PNASA6 72,5135,75
sln-dmg-orl 1500 µmol/L BIZNAT 102,271,83
oms-rat-orl 25 µmol/kg CBINA8 24,355,79
dns-rat:lvr 1 µmol/L ENMUDM 7,101,85
dns-rat-orl 40 mg/kg ENMUDM 7,101,85
orl-rat TDLo:240 mg/kg/20W-I:CAR ZEKBAI 74,200,70
scu-rat TDLo:135 mg/kg/W-I:ETA PTRMAD 241,147,48
orl-rat TD:275 mg/kg/26W-I:ETA PTRMAD 241,147,48
scu-rat TD:180 mg/kg/8W-I:ETA PTRMAD 241,147,48
orl-rat TD:580 mg/kg/32W-C:ETA BJCAAI 22,133,68
orl-rat TD:550 mg/kg/26W-C:ETA EXPEAM 12,185,56
orl-rat TD:305 mg/kg/44W-C:ETA EXPEAM 12,185,56
orl-rat LD50:50 mg/kg ARTODN 56,151,85

CONSENSUS REPORTS: EPA Genetic Toxicology Program.

SAFETY PROFILE: Poison by ingestion and subcutaneous routes. Questionable carcinogen with experimental carcinogenic and tumorigenic data. Mutation data reported. When heated to decomposition it emits toxic fumes of NO_x.

DUC200 CAS:14301-11-2 HR: 2
(Z)-N,N-DIMETHYL-4-STILBENAMINE
mf: $C_{16}H_{17}N$ mw: 223.34

SYNS: cis-4-DIMETHYLAMINOSTILBENE □ cis-N,N-DIMETHYL-4-STILBENAMINE

TOXICITY DATA WITH REFERENCE
orl-rat TDLo:310 mg/kg/20W-I:NEO ZEKBAI 74,200,70

SAFETY PROFILE: Questionable carcinogen with experimental neoplastigenic data. When heated to decomposition it emits toxic fumes of NO_x.

DUC400 CAS:7456-24-8 HR: 3
DIMETHYLSULFAMIDO-3-(DIMETHYLAMINO-2-PROPYL)-10-PHENOTHIAZINE
mf: $C_{19}H_{25}N_3O_2S_2$ mw: 391.59

SYNS: DIMETHOTHIAZINE □ 10-(2-(DIMETHYLAMINO)PROPYL)-N,N-DIMETHYLPHENOTHIAZINE-2-SULFONAMIDE □ 3-DIMETHYLSULFONAMIDO-10-(2-DIMETHYLAMINOPROPYL)PHENOTHIAZINE

TOXICITY DATA WITH REFERENCE
orl-rat TDLo:600 mg/kg (female 9-14D post):REP OYYAA2 4,381,70
orl-mus LD50:740 mg/kg AIPTAK 159,70,66
ipr-mus LD50:190 mg/kg AIPTAK 159,70,66
scu-mus LD50:475 mg/kg AIPTAK 159,70,66
ivn-mus LD50:100 mg/kg AIPTAK 159,70,66

SAFETY PROFILE: Poison by intraperitoneal and intravenous routes. Moderately toxic by ingestion and subcutaneous routes. Experimental reproductive effects. When heated to decomposition it emits very toxic fumes of NO_x and SO_x.

DUC600 CAS:15020-57-2 HR: 2
p-(N,N-DIMETHYLSULFAMOYL)PHENOL
mf: $C_8H_{11}NO_3S$ mw: 201.26

SYN: N,N-DIMETHYL-HYDROXYBENZENESULFONAMIDE

TOXICITY DATA WITH REFERENCE
orl-mus LD50:2290 mg/kg JAFCAU 15,845,67

CONSENSUS REPORTS: Reported in EPA TSCA Inventory.

SAFETY PROFILE: Moderately toxic by ingestion. When heated to decomposition it emits very toxic fumes of SO_x and NO_x.

DUD000 CAS:121-58-4 HR: 2
N,N-DIMETHYLSULFANILIC ACID
mf: $C_8H_{11}NO_3S$ mw: 201.26

PROP: Crystals from H_2O. Mp: 270–271°.

TOXICITY DATA WITH REFERENCE
par-mus LDLo:4000 mg/kg CBCCT* 7,695,55

CONSENSUS REPORTS: Reported in EPA TSCA Inventory.

SAFETY PROFILE: Moderately toxic by parenteral route. When heated to decomposition it emits very toxic fumes of NO_x and SO_x.

DUD100 CAS:77-78-1 HR: 3
DIMETHYL SULFATE
DOT: UN 1595
mf: $C_2H_6O_4S$ mw: 126.14

PROP: Colorless, odorless liquid. Mp: −31.8°, fp −27, bp: 188° (decomp), flash p: 182°F (OC), d: 1.332 @ 15°, vap d: 4.35, autoign temp: 370°F. Sltly sol in H_2O, hexane, EtOH, C_6H_6; sol in Et_2O and Me_2CO.

SYNS: DIMETHYLESTER KYSELINY SIROVE (CZECH) □ DIMETHYL MONOSULFATE □ DIMETHYLSULFAAT (DUTCH) □ DIMETHYLSULFAT (CZECH) □ DIMETILSOLFATO (ITALIAN) □ DMS □ DMS (METHYL SULFATE) □ DWUMETYLOWY SIARCZAN (POLISH) □ METHYLE (SULFATE de) (FRENCH) □ METHYL SULFATE (DOT) □ RCRA WASTE NUMBER U103 □ SULFATE de METHYLE (FRENCH) □ SULFATE DIMETHYLIQUE (FRENCH) □ SULFURIC ACID, DIMETHYL ESTER

TOXICITY DATA WITH REFERENCE
skn-rbt 10 mg/24H open SEV AMIHBC 4,119,51
eye-rbt 100 mg/4S rns SEV FCTOD7 20,573,82

eye-rbt 50 µg/24H SEV 28ZPAK -,177,72
mma-sat 4300 nmol/L/1H PNASA6 75,4465,78
dnr-omi 640 µg/plate BIZNAT 95,463,76
dnd-hmn:lym 1 mmol/L JACTDZ 1(3),125,82
ivn-rat TDLo:100 mg/kg (15D preg):TER IARCCD 4,45,73
ihl-rat TCLo:17 mg/m³/19W-I:ETA ZEKBAI 74,241,70
scu-rat TDLo:50 mg/kg:ETA ZEKBAI 74,241,70
ivn-rat TDLo:20 mg/kg (15D preg):CAR,REP IARCCD 4,45,73
ihl-hmn LCLo:97 ppm/10M 34ZIAG -,226,69
orl-rat LD50:205 mg/kg GTPZAB 23(3),28,79
ihl-rat LC50:45 mg/m³/4H GTPZAB 24(11),55,80
scu-rat LD50:100 mg/kg ZEKBAI 74,241,70
orl-mus LD50:140 mg/kg GTPZAB 23(3),28,79
ihl-mus LC50:280 mg/m³ GTPZAB 23(3),28,79
orl-rbt LDLo:45 mg/kg AEXPBL 47,113,02
scu-rbt LDLo:53 mg/kg AEXPBL 47,113,02
ivn-rbt LDLo:50 mg/kg AEXPBL 47,113,02
ihl-gpg LC50:32 ppm/1H 85JCAE-,1079,86

CONSENSUS REPORTS: NTP 7th Annual Report On Carcinogens. IARC Cancer Review: Group 2A IMEMDT 7,200,87; Animal Sufficient Evidence IMEMDT 4,271,74; Human Inadequate Evidence IMEMDT 4,271,74. EPA Genetic Toxicology Program. Community Right-To-Know List. EPA Extremely Hazardous Substances List. Reported in EPA TSCA Inventory.

OSHA PEL: TWA 0.1 ppm (skin)
ACGIH TLV: TWA 0.1 ppm (skin); Suspected Human Carcinogen
DFG TRK: Production: 0.02 ppm; Use: 0.04 ppm; Animal Carcinogen, Suspected Human Carcinogen
DOT CLASSIFICATION: 3; *Label:* Poison, Corrosive

SAFETY PROFILE: Confirmed carcinogen with experimental carcinogenic, tumorigenic, and teratogenic data. Human poison by inhalation. Experimental poison by ingestion, inhalation, intravenous, and subcutaneous routes. Other experimental reproductive effects. Human mutation data reported. A corrosive irritant to skin, eyes, and mucous membranes. There is no odor or initial irritation to give warning of exposure. On brief, mild exposures, conjunctivitis; catarrhal inflammation of the mucous membranes of the nose, throat, larynx, and trachea; and possibly some reddening of the skin develop after the latent period. With longer, heavier exposures, the cornea shows clouding, the irritation changes to the nasopharynx are more marked, and after 6 to 8 hours pulmonary edema may develop. Death may occur in 3 or 4 days. The liver and kidneys are frequently damaged. Spilling of the liquid on the skin can cause ulceration and local necrosis. In patients surviving severe exposure, there may be serious injury of the liver and kidneys, with suppression of urine, jaundice, albuminuria, and hematuria appearing. Death, resulting from the kidney or liver damage, may be delayed for several weeks. Flammable when exposed to heat, flame, or oxidizers. Can react with oxidizing materials. Violent reaction with NH_4OH and NaN_3. To fight fire, use water, foam, CO_2, dry chemical. When heated to decomposition it emits toxic fumes of SO_x. See also SULFATES.

For occupational chemical analysis use NIOSH: Dimethyl Sulfate, 2524.

DUD400 CAS:1003-78-7 HR: 3
2,4-DIMETHYL SULFOLANE
mf: $C_6H_{12}O_2S$ mw: 148.24

PROP: A liquid. Mp: −3°, bp: 280°, flash p: 280°F (OC), d: 1.14 @ 20°/4°, vap press: 0.006 mm @ 20°.

SYNS: DMS □ TETRAHYDRO-2,4-DIMETHYLTHIOPHENE-1,1-DIOXIDE

TOXICITY DATA WITH REFERENCE
orl-rat LDLo:100 mg/kg SCCUR* -,4,61
orl-mus LD50:140 mg/kg SCCUR* -,4-61
ipr-mus LD50:72 mg/kg AIHAAP 32,539,71
ivn-mus LD50:61 mg/kg AIHAAP 32,539,71
orl-rbt LD50:115 mg/kg SCCUR* -,4,61
skn-rbt LDLo:3600 mg/kg SCCUR* -,4,61
ivn-rbt LD50:36 mg/kg AIHAAP 32,539,71

SAFETY PROFILE: Poison by ingestion, intraperitoneal, and intravenous routes. Moderately toxic by skin contact. Combustible when exposed to heat or flame; can react with oxidizing materials. To fight fire, use water, foam, CO_2, dry chemical. When heated to decomposition it emits toxic fumes of SO_x. See also SULFATES.

DUD800 CAS:67-68-5 HR: 2
DIMETHYL SULFOXIDE
mf: C_2H_6OS mw: 78.14

PROP: Clear, water-white, hygroscopic liquid; garlic-onion-oyster odor. Mp: 18.5°, bp: 189°, flash p: 203°F (OC), d: 1.100 @ 20°, vap press: 0.37 mm @ 20°, lel: 3.0%, uel: 43%, autoign temp: 574°F (301°C). Misc in H_2O and org solvs.

SYNS: A 10846 □ DELTAN □ DEMASORB □ DEMAVET □ DEMESO □ DEMSODROX □ DERMASORB □ DIMETHYL SULPHOXIDE □ DIMEXIDE □ DIPIRARTRIL-TROPICO □ DMS-70 □ DMS-90 □ DMSO □ DOLICUR □ DOLIGUR □ DOMOSO □ DROMISOL □ DURASORB □ GAMASOL 90 □ HYADUR □ INFILTRINA □ M 176 □ METHYLSULFINYLMETHANE □ METHYL SULFOXIDE □ NSC-763 □ RIMSO-50 □ SOMIPRONT □ SQ 9453 □ SULFINYLBIS(METHANE) □ SYNTEXAN □ TOPSYM

TOXICITY DATA WITH REFERENCE
skn-rbt 10 mg/24H open MLD AIHAAP 23,95,62
skn-rbt 500 mg/24H MLD 28ZPAK -,177,72
eye-rbt 500 mg/24H MLD 28ZPAK -,177,72
mmo-esc 551 g/L MUREAV 130,97,84
oms-hmn:lym 140 mmol/L PNASA6 79,1171,82
ipr-ham TDLo:4400 mg/kg (female 8D post):TER LANCAO 1,208,66
ipr-mus TDLo:5500 mg/kg (female 10D post):TER AUCMBJ 61,131,73
ipr-rat TDLo:56 g/kg (6-12D preg):REP ANYAA9 141,110,67
scu-rat TDLo:30,750 mg/kg (female 8-10D post):REP PSEBAA 125,567,67
ivn-mus TDLo:240 g/kg (female 1-20D post):REP COREAF 260,327,65
ipr-mus TDLo:210 g/kg (female 6-12D post):TER ANYAA9 141,110,67

ipr-ham TDLo:5500 mg/kg (female 8D post):TER JEEMAF 16,49,66
orl-rat TDLo:59 g/kg/81W-I:ETA GTPZAB 28(5),39,84
scu-rat TDLo:220 g/kg/82W-I:ETA GTPZAB 28(5),39,84
orl-mus TDLo:65,340 mg/kg/66W-I:ETA GTPZAB 28(5),39,84
scu-mus TDLo:66 g/kg/66W-I:ETA GTPZAB 28(5),39,84
ivn-man TDLo:606 mg/kg:GIT,LIV LANCAO 2,1004,80
orl-rat LD50:14,500 mg/kg TXAPA9 15,74,69
ipr-rat LD50:8200 mg/kg FCTOD7 22,665,84
scu-rat LD50:12 g/kg ARZNAD 14,1050,64
ivn-rat LD50:5360 mg/kg TXAPA9 7,104,65
orl-mus LD50:7920 mg/kg CHTPBA 3,10,68
ipr-mus LD50:2500 mg/kg RPTOAN 35,300,72
ivn-mus LD50:3800 mg/kg 34ZIAG -,656,69
ivn-dog LD50:2500 mg/kg CNCRA6 31,7,63

CONSENSUS REPORTS: Reported in EPA TSCA Inventory. EPA Genetic Toxicology Program.

SAFETY PROFILE: Slightly toxic by ingestion. Moderately toxic by intravenous and intraperitoneal routes. Human systemic effects by intravenous route: nausea or vomiting and jaundice. Experimental teratogenic and reproductive effects. A skin and eye irritant. Questionable carcinogen with experimental tumorigenic data. Human mutation data reported. Can cause an anaphylactic reaction. Corneal opacity reported only in rabbits, dogs, and pigs. It freely penetrates the skin and may carry dissolved chemicals with it into the body. Combustible when exposed to heat or flame; can react with oxidizing materials. To fight fire, use water, foam, alcohol foam, CO_2, dry chemical. Violent or explosive reaction with many acyl, aryl, and nonmetal halides (e.g., acetyl chloride, benzenesulfonyl chloride, bromobenzoyl acetanilide, cyanuric chloride, iodine pentafluoride, $Mg(ClO_4)_2$, CH_3Br, NIO_4, oxalyl chloride, P_2O_3, phosphorus trichloride, phosphoryl chloride, silver fluoride, silver difluoride, sodium hydride, sulfur dichloride, disulfur dichloride, sulfuryl chloride, tetrachlorosilane, thionyl chloride). Violent or explosive reaction with boron compounds [e.g., borane, nonahydrononaborate(2-) ion], 4(4'-bromobenzoyl)acetanilide, carbonyl diisothiocyanate, dinitrogen tetraoxide, hexachlorocyclotriphosphazine, copper + trichloroacetic acid, metal alkoxides (e.g., potassium tert-butoxide, sodium isopropoxide), trifluoroacetic acid anhydride. Incompatible with magnesium perchlorate, metal oxosalts, perchloric acid, periodic acid, sulfur trioxide. Forms powerfully explosive mixtures with metal salts of oxoacids (e.g., aluminum perchlorate, sodium perchlorate, iron(III) nitrate). When heated to decomposition it emits toxic fumes of SO_x.

DUE000 CAS:120-61-6 HR: 2
DIMETHYL TEREPHTHALATE
mf: $C_{10}H_{10}O_4$ mw: 194.20

PROP: Crystals from Et_2O. Mp: 141–142°, bp: 284°.

SYNS: 1,4-BENZENE DICARBOXYLIC ACID DIMETHYL ESTER (9CI) □ DIMETHYL-1,4-BENZENE DICARBOXYLATE □ METHYL-4-CARBOMETHOXY BENZOATE □ NCI-C50055 □ TEREPHTHALIC ACID METHYL ESTER

TOXICITY DATA WITH REFERENCE
eye-rbt 500 mg/24H MOD 28ZPAK -,47,72
mnt-mus-ipr 200 μmol/kg MUREAV 204,703,88
orl-rat LD50:4390 mg/kg 28ZPAK -,47,72
ipr-rat LD50:3900 mg/kg AIHAAP 34,455,73

CONSENSUS REPORTS: NTP Carcinogenesis Bioassay (feed): Equivocal Evidence: Mouse NCITR* NCI-TR-121,79; (feed): No Evidence: Rat NCITR* NCI-TR-121,79. Reported in EPA TSCA Inventory.

SAFETY PROFILE: Moderately toxic by intraperitoneal route. Mildly toxic by ingestion. An eye irritant. Mutation data reported. When heated to decomposition it emits acrid smoke and irritating fumes.

DUF000 CAS:25486-91-3 HR: 2
7,12-DIMETHYL-8,9,10,11-TETRAHYDROBENZ(a)ANTHRACENE
mf: $C_{20}H_{20}$ mw: 260.40

SYN: 8,9,10,11-TETRAHYDRO-7,12-DIMETHYLBENZ(a) ANTHRACENE

TOXICITY DATA WITH REFERENCE
skn-mus TDLo:212 mg/kg/65W-I:ETA JNCIAM 44,641,70

SAFETY PROFILE: Questionable carcinogen with experimental tumorigenic data. When heated to decomposition it emits acrid smoke and irritating fumes.

DUF200 CAS:52171-94-5 HR: 2
1,11-DIMETHYL-1,2,3,4-TETRAHYDROCHRYSENE
mf: $C_{20}H_{20}$ mw: 260.40

TOXICITY DATA WITH REFERENCE
mma-sat 20 μg/plate CNREA8 36,4525,76
skn-mus TDLo:120 mg/kg/50W-I:ETA CNREA8 34,1315,74

CONSENSUS REPORTS: EPA Genetic Toxicology Program.

SAFETY PROFILE: Questionable carcinogen with experimental tumorigenic data. Mutation data reported. When heated to decomposition it emits acrid smoke and irritating fumes.

DUF400 CAS:4336-19-0 HR: 2
DIMETHYL TETRAHYDROPHTHALATE
mf: $C_{10}H_{14}O_4$ mw: 198.24

PROP: Crystals. Vap d: 6.83.

SYN: 1-CYCLOHEXENE-1,2-DICARBOXYLIC ACID DIMETHYL ESTER

TOXICITY DATA WITH REFERENCE
eye-rbt 500 mg AJOPAA 29,1363,46
orl-rat LD50:700 mg/kg JIHTAB 31,60,49

SAFETY PROFILE: Moderately toxic by ingestion. An eye irritant. Combustible when exposed to heat or flame; can react with oxidizing materials. When heated to decomposition it emits acrid smoke and irritating fumes. See also ESTERS.

DUF800 **HR: 3**
3,6-DIMETHYL-1,2,4,5-TETRAOXANE
mf: $C_4H_8O_4$ mw: 120.11

$CH_3CHOOCHCH_3O\ O$

SAFETY PROFILE: An extremely shock-sensitive explosive. May explode if touched. When heated to decomposition it emits acrid smoke and fumes.

DUG000 **HR: 3**
DIMETHYLTHALLIUM FULMINATE
mf: C_3H_6NOTl mw: 276.46

$(CH_3)_2TlC{\equiv}N{\cdot}O$

CONSENSUS REPORTS: Thallium and its compounds are on the Community Right-To-Know List.

SAFETY PROFILE: Highly explosive. See also THALLIUM COMPOUNDS and FULMINATES.

DUG089 **HR: 3**
DIMETHYLTHALLIUM-N-METHYLACETOHYDROXAMATE
mf: $C_5H_{12}NO_2Tl$ mw: 276.46

$(CH_3)_2TlON(CH_3)CO{\cdot}CH_3$

CONSENSUS REPORTS: Thallium and its compounds are on the Community Right-To-Know List.

SAFETY PROFILE: Highly explosive. It may explode below 160°C. See also THALLIUM COMPOUNDS and FULMINATES.

DUG200 **CAS:541-58-2** **HR: 3**
2,4-DIMETHYLTHIAZOLE
mf: C_5H_7NS mw: 113.19

PROP: Bp: 70–73° @ 50 mm, d: 1.506 @ 15°/4°. Sol in cold H_2O; less sol in hot H_2O; sol in EtOH and Et_2O.

TOXICITY DATA WITH REFERENCE
ipr-mus LD50:250 mg/kg NTIS** AD691-490

CONSENSUS REPORTS: Reported in EPA TSCA Inventory.

SAFETY PROFILE: Poison by intraperitoneal route. When heated to decomposition it emits very toxic fumes of NO_x and SO_x.

DUG425 **CAS:2530-10-1** **HR: 3**
DIMETHYLTHIENYLCETONE
mf: $C_8H_{10}OS$ mw: 154.24

SYNS: ETHANONE, 1-(2,5-DIMETHYL-3-THIENYL)- □ KETONE, 2,5-DIMETHYL-3-THIENYL METHYL

TOXICITY DATA WITH REFERENCE
ipr-mus LD50:260 mg/kg APFRAD 5,16,47

DOT CLASSIFICATION: 3; *Label:* Flammable Liquid

SAFETY PROFILE: A poison by intraperitoneal route. A flammable liquid. When heated to decomposition it emits toxic vapors of SO_x.

DUG450 **CAS:631-67-4** **HR: 2**
N,N-DIMETHYLTHIOACETAMIDE
mf: C_4H_9NS mw: 103.20

SYNS: ACETAMIDE, N,N-DIMETHYLTHIO- □ N,N-DIMETHYLETHANETHIOAMIDE □ DIMETHYLTHIOACETAMID □ DIMETHYLTHIOACETAMIDE □ ETHANETHIOAMIDE, N,N-DIMETHYL-(9CI)

TOXICITY DATA WITH REFERENCE
ipr-mus LD50:500 mg/kg AEPPAE 233,376,58

CONSENSUS REPORTS: Reported in EPA TSCA Inventory.

SAFETY PROFILE: Moderately toxic by intraperitoneal route. When heated to decomposition it emits toxic vapors of NO_x and SO_x.

DUG500 **CAS:152-20-5** **HR: 3**
DIMETHYLTHIOMETHYLPHOSPHATE
mf: $C_3H_9O_3PS$ mw: 156.15

$((CH_3O)_2(CH_3S)PO)$

PROP: A liquid. D: 1.25 @ 20°/4°, bp: 98–101° @ 11 mm.

SYNS: HC7901 □ METHYLPHOSPHOROTHIOATE □ O,O,S-TRIMETHYL PHOSPHOROTHIOATE

TOXICITY DATA WITH REFERENCE
orl-rat TDLo:40 mg/kg (female 20D post):REP ARTODN 61,378,88
orl-rat TDLo:2500 µg/kg (female 20D post):REP ARTODN 61,378,88
orl-rat TDLo:40 mg/kg (female 8-10D post):TER TOLED5 32,185,86
orl-rat TDLo:40 mg/kg (female 8-10D post):REP TOLED5 32,185,86
orl-rat TDLo:10 mg/kg (female 20D post):TER ARTODN 61,378,88
orl-rat LD50:15 mg/kg JAFCAU 27,463,79
ipr-rat LD50:51 mg/kg ARTODN 51,221,82
ivn-rat LD50:45 mg/kg DTESD7 8,631,80
orl-mus LD50:38 mg/kg PCBPBS 24,251,85
ipr-mus LD50:5 mg/kg ACPMAP 16,7,63
ivn-mus LD50:123 mg/kg DTESD7 8,631,80

SAFETY PROFILE: Poison by ingestion, intravenous, and intraperitoneal routes. An experimental teratogen. Other experimental reproductive effects. When heated to decomposition it emits toxic fumes of PO_x and SO_x.

DUG600 **CAS:50847-92-2** **HR: 3**
2,2-DIMETHYL-3-THIOMORPHOLINONE
mf: $C_6H_{11}NOS$ mw: 145.24

SYN: 2,2-DIMETHYL-3-THIOMORPHOLONE

TOXICITY DATA WITH REFERENCE
orl-mus LD50:3423 mg/kg JMCMAR 6,136,63
ipr-mus LD50:400 mg/kg NTIS** AD691-490

SAFETY PROFILE: Poison by intraperitoneal route. Moderately toxic by ingestion. When heated to decomposition it emits very toxic fumes of NO_x and SO_x.

D

DUG700 **CAS:531-53-3** **HR: 3**
DIMETHYLTHIONINE
mf: $C_{14}H_{14}N_3S{\cdot}Cl$ mw: 291.82

SYNS: 3-AMINO-7-(DIMETHYLAMINO)PHENOTHIAZIN-5-IUM CHLORIDE □ AZURE A □ C.I. 52005 □ asym-DIMETHYL-3,7-DIAMINOPHENAZATHIONIUM CHLORIDE □ PHENOTHIAZIN-5-IUM, 3-AMINO-7-(DIMETHYLAMINO)-, CHLORIDE

TOXICITY DATA WITH REFERENCE
cyt-ham:ovr 20 μmol/L/5H-C ENMUDM 1,27,79
ivn-rat LD50:37,720 μg/kg SMBUA9 9,96,51
ivn-mus LD50:59,040 μg/kg SMBUA9 9,96,51
ivn-rbt LD50:19,340 μg/kg SMBUA9 9,96,51

CONSENSUS REPORTS: Reported in EPA TSCA Inventory.

SAFETY PROFILE: Poison by intravenous route. Mutation data reported. When heated to decomposition it emits toxic vapors of NO_x, SO_x, and Cl^-.

DUG800 **CAS:2767-47-7** **HR: 3**
DIMETHYLTIN DIBROMIDE
mf: $C_2H_6Br_2Sn$ mw: 308.59

PROP: Colorless or white crystals. Mp: 76°, bp: 208–213°. Sol in water and organic solvents.

SYN: DIBROMODIMETHYL STANNANE

TOXICITY DATA WITH REFERENCE
ivn-mus LD50:56,200 μg/kg CSLNX* NX#02289

OSHA PEL: TWA 0.1 mg(Sn)/m³ (skin)
ACGIH TLV: TWA 0.1 mg(Sn)/m³ (skin) (Proposed: TWA 0.1 mg(Sn)/m³; STEL 0.2 mg(Sn)/m³ (skin))
NIOSH REL: (Organotin Compounds) TWA 0.1 mg(Sn)/m³

SAFETY PROFILE: Poison by intravenous route. When heated to decomposition it emits toxic fumes of Br^-. See also TIN COMPOUNDS and BROMIDES.

For occupational chemical analysis use NIOSH: Organotin Compounds 5504.

DUG825 **CAS:753-73-1** **HR: 3**
DIMETHYLTIN DICHLORIDE
mf: $C_2H_6Cl_2Sn$ mw: 219.67

SYNS: DICHLORID DIMETHYLCINICITY □ DICHLORODIMETHYLSTANNANE □ DICHLORODIMETHYLTIN □ DIMETHYLDICHLOROSTANNANE □ DIMETHYLDICHLOROTIN □ STANNANE, DICHLORODIMETHYL- □ TIN, DIMETHYL-, DICHLORIDE

TOXICITY DATA WITH REFERENCE
mmo-sat 10 μg/tube MUREAV 300,265,93
dnd-bcs 100 μg/disk MUREAV 280,195,92
orl-rat LD50:73,900 μg/kg TRIPA7-,1,73
ivn-rat LDLo:40 mg/kg BJIMAG 15,15,58
ivn-mus LD50:56 mg/kg CSLNX* NX#02187
orl-rbt LDLo:50 mg/kg SAIGBL 15,3,73

OSHA PEL: TWA 0.1 mg(Sn)/m³
ACGIH TLV: TWA 0.1 mg(Sn)/m³; STEL 0.2 mg/m³ (skin)
NIOSH REL: (organotin compounds): TWA 0.1 mg(Sn)/m³

CONSENSUS REPORTS: Reported in EPA TSCA Inventory.

SAFETY PROFILE: Poison by ingestion and intravenous routes. Mutation data reported. When heated to decomposition it emits toxic vapors of Sn and Cl^-.

For occupational chemical analysis use NIOSH: Organotin Compounds 5504.

DUG889 **CAS:40237-34-1** **HR: 3**
DIMETHYLTIN DINITRATE
mf: $C_2H_6N_6O_6Sn$ mw: 272.79

PROP: Colorless crystals.

SAFETY PROFILE: Explodes when heated. When heated to decomposition it emits toxic fumes of NO_x. See also TIN COMPOUNDS and NITRATES.

DUH200 **CAS:609-72-3** **HR: 3**
N,N-DIMETHYL-o-TOLUIDINE
mf: $C_9H_{13}N$ mw: 135.23

PROP: A liquid. Bp: 184.8°.

SYNS: DIMETHYL-o-TOLUIDINE □ o-METHYLDIMETHYLANILINE □ N,N,2-TRIMETHYLANILINE

TOXICITY DATA WITH REFERENCE
orl-rat LDLo:500 mg/kg JPETAB 90,260,47
ipr-mus LD50:338 mg/kg AISFAR 1,284,51

CONSENSUS REPORTS: Reported in EPA TSCA Inventory.

SAFETY PROFILE: Poison by intraperitoneal route. Moderately toxic by ingestion. When heated to decomposition it emits toxic fumes of NO_x.

For occupational chemical analysis use NIOSH: Amines, Aromatic 2002.

DUH400 **CAS:3010-57-9** **HR: 2**
N,N-DIMETHYL-4-(p-TOLYLAZO)ANILINE
mf: $C_{15}H_{17}N_3$ mw: 239.35

SYNS: N,N-DIMETHYL-4-((4-METHYLPHENYL)AZO)BENZENAMINE □ p′-METHYL-p-DIMETHYLAMINOAZOBENZENE □ 4′-METHYL-4-DIMETHYLAMINOAZOBENZENE

TOXICITY DATA WITH REFERENCE
mma-sat 500 nmol/plate MUREAV 121,95,83
dns-rat:lvr 10 μmol/L CNREA8 46,1654,86
ipr-mus TDLo:1 g/kg (8-9D preg):TER KAIZAN 37,179,62
ipr-mus TDLo:1 g/kg (8-9D preg):REP KAIZAN 37,179,62
orl-rat TDLo:7776 mg/kg/35W-C:ETA CNREA8 5,227,45

CONSENSUS REPORTS: EPA Genetic Toxicology Program.

SAFETY PROFILE: Questionable carcinogen with experimental tumorigenic and teratogenic data. Experimental reproductive effects. Mutation data reported. When heated to decomposition it emits toxic fumes of NO_x.

DUH600 CAS:55-80-1 HR: 2
N,N-DIMETHYL-p-(m-TOLYLAZO)ANILINE
mf: $C_{15}H_{17}N_3$ mw: 239.35

SYNS: 4-(N,N-DIMETHYLAMINO)-3′-METHYLAZOBENZENE □ N,N-DIMETHYL-p-(3′-METHYLPHENYLAZO)ANILINE □ N,N-DIMETHYL-4-((3-METHYLPHENYL)AZO)BENZENAMINE □ MDAB □ 3′-MDAB □ 3′-METHYLBUTTERGELB (GERMAN) □ 3′-METHYL-DAB □ 3′-METHYL-4-DIMETHYLAMINOAZOBENZEN (CZECH) □ M′-METHYL-p-DIMETHYLAMINOAZOBENZENE □ 3′-METHYL-4-DIMETHYLAMINOAZOBENZENE □ 3′-METHYL-N,N-DIMETHYL-4-AMINOAZOBENZENE □ 3′-METHYLDIMETHYLAMINOAZOBENZOL (GERMAN)

TOXICITY DATA WITH REFERENCE
otr-rat:lvr 240 μmol/L AMOKAG 39,231,85
dns-rat:lvr 1 μmol/L CNREA8 46,1654,86
orl-rat TDLo:1800 mg/kg/7W-C:CAR JJIND8 71,855,83
scu-rat TDLo:2750 mg/kg/1Y-I:ETA CNREA8 35,3798,75
imp-rat TDLo:2000 mg/kg/9W-C:ETA EXPEAM 34,788,78
scu-mus TDLo:7179 mg/kg (15-19D preg):NEO,TER CALEDQ 17,321,83
orl-ham TDLo:12 g/kg/38W-C:ETA ARPAAQ 71,566,61
orl-rat TD:2419 mg/kg/12W-C:CAR CBINA8 53,107,85
orl-rat TD:2520 mg/kg/12W-C:ETA GANNA2 63,31,72
orl-rat TD:2160 mg/kg/17W-C:ETA JMCMAR 11,1074,68
orl-rat TD:2600 mg/kg/13W-C:ETA CNREA8 17,387,57
orl-rat LD:5244 mg/kg/24W-C:CAR CRNGDP 8,719,87
orl-rat LD:2142 mg/kg/17W-C:ETA JJIND8 26,1461,61
orl-rat TD:1890 mg/kg/9W-C:ETA CALEDQ 9,299,80
orl-rat TDLo:2229 mg/kg/12W-C:CAR AJPAA4 87,189,77
orl-rat TD:1656 mg/kg/8W-I:ETA CNREA8 33,1119,73
orl-rat LDLo:1500 mg/kg 28ZPAK -,236,72
orl-mus LD50:17,700 mg/kg JKXXAF #80-157517
ipr-mus LD50:1530 mg/kg JJIND8 62,911,79

CONSENSUS REPORTS: Reported in EPA TSCA Inventory. EPA Genetic Toxicology Program.

SAFETY PROFILE: Moderately toxic by ingestion. An experimental teratogen. Questionable carcinogen with experimental carcinogenic, neoplastigenic, and tumorigenic data. Mutation data reported. When heated to decomposition it emits toxic fumes of NO_x.

DUH800 CAS:3731-39-3 HR: 2
N,N-DIMETHYL-p-((o-TOLYL)AZO)ANILINE
mf: $C_{15}H_{17}N_3$ mw: 239.35

SYNS: N,N-DIMETHYL-p-(2′-METHYLPHENYLAZO)ANILINE □ N,N-DIMETHYL-4-((2-METHYLPHENYL)AZO)BENZENAMINE □ o′-METHYL-p-DIMETHYLAMINOAZOBENZENE □ 2′-METHYL-4-DIMETHYLAMINOAZOBENZENE □ 2-METHYL-N,N-DIMETHYL-4-AMINOAZOBENZENE

TOXICITY DATA WITH REFERENCE
mma-sat 100 nmol/plate CALEDQ 1,91,75
dns-rat:lvr 10 μmol/L CNREA8 46,1654,86
orl-rat TDLo:14,414 mg/kg/52W-C:CAR CBINA8 53,107,85
orl-rat TD:5856 mg/kg/26W-C:ETA CNREA8 5,227,45

SAFETY PROFILE: Questionable carcinogen with experimental carcinogenic and tumorigenic data. Mutation data reported. When heated to decomposition it emits toxic fumes of NO_x.

DUI000 CAS:1933-50-2 HR: 3
4′-(3,3-DIMETHYL-1-TRIAZENO)ACETANILIDE
mf: $C_{10}H_{14}N_4O$ mw: 206.28

SYNS: AC 24055 □ 1-(p-ACETAMIDOPHENYL)-3,3-DIMETHYLTRIAZENE □ 1-(4-ACETAMINOPHENYL)-3,3-DIMETHYLTRIAZENE □ AMERICAN CYANAMID CL-24055 □ AMERICAN CYANIMID 24,055 □ ANTIFEEDANT 24005 □ ANTIFEEDING COMPOUND 24,055 □ CL 24055 □ CYANAMID 24055 □ 1,1-DIMETHYL-3-(p-ACETAMIDOPHENYL)TRIAZENE □ 4′-DIMETHYLTRIAZENOACETANILIDE □ N-(4-(3,3-DIMETHYL-1-TRIAZENYL)PHENYL)ACETAMIDE □ ENT 25,651

TOXICITY DATA WITH REFERENCE
mma-sat 1 μmol/L JMCMAR 22,473,79
sln-dmg-orl 1 μmol/L CBINA8 9,365,74
mrc-smc 10 mmol/L CBINA8 9,365,74
hma-mus/smc 1 mmol/L CBINA8 9,365,74
orl-rat LD50:510 mg/kg TXAPA9 21,315,72
skn-rbt LD50:1400 mg/kg 28ZEAL 4,195,69
orl-bwd LD50:56 mg/kg TXAPA9 21,315,72

CONSENSUS REPORTS: EPA Genetic Toxicology Program.

SAFETY PROFILE: Poison by ingestion. Moderately toxic by skin contact. Mutation data reported. When heated to decomposition it emits toxic fumes of NO_x.

DUI200 CAS:80266-48-4 HR: 3
4′-(3-(3,3-DIMETHYL-1-TRIAZENO)-9-ACRIDINYLAMINO)METHANESULFONANILIDE
mf: $C_{22}H_{22}N_6O_2S$ mw: 434.56

TOXICITY DATA WITH REFERENCE
mmo-sat 32 μmol/L JMCMAR 23,269,80

SAFETY PROFILE: Mutation data reported. When heated to decomposition it emits very toxic fumes of NO_x and SO_x.

DUI709 CAS:3585-32-8 HR: 3
1,3-DIMETHYL-1-TRIAZINE
mf: $C_2H_7N_3$ mw: 73.10

PROP: Liquid. Bp: 92°.

SYN: N,N′-DIMETHYLTRIAZENE

TOXICITY DATA WITH REFERENCE
mmo-esc 25 μmol/L CRNGDP 12,1161,91

SAFETY PROFILE: Mutation data reported. Explodes violently on contact with flame. Upon decomposition it emits toxic fumes of NO_x.

DUJ000 CAS:25724-50-9 HR: 3
3,5-DIMETHYL-1-(TRICHLOROMETHYLMERCAPTO)PYRAZOLE
mf: $C_6H_7Cl_3N_2S$ mw: 245.56

TOXICITY DATA WITH REFERENCE
orl-rat LD50:570 mg/kg AIHAAP 30,470,69
skn-rbt LD50:200 mg/kg AIHAAP 30,470,69

SAFETY PROFILE: Poison by skin contact. Moderately toxic by ingestion. When heated to decomposition it emits very toxic fumes of Cl^-, NO_x, SO_x. See also MERCAPTANS.

DUJ400 CAS:24602-86-6 HR: 2
2,6-DIMETHYL-4-TRIDECYLMORPHOLINE
mf: $C_{19}H_{39}NO$ mw: 297.59

PROP: Oil. Bp: 134° @ 0.5 mm. Misc in H_2O and org solvs.

SYNS: BAS 2205-F □ E-236 □ N-TRIDECYL-2,6-DIMETHYLMORPHOLIN (GERMAN) □ N-TRIDECYL-2,6-DIMETHYLMORPHOLINE □ 4-TRIDECYL-2,6-DIMETHYLMORPHOLINE

TOXICITY DATA WITH REFERENCE
orl-rat TDLo:5400 μg/kg (female 7-15D post):TER VPITAR 39(6),55,81
orl-rat TDLo:29,250 μg/kg (female 7-15D post):TER VPITAR 39(6),55,81
orl-rat TDLo:585 mg/kg (female 7-15D post):TER VPITAR 39(6),55,81
orl-rat TDLo:1300 mg/kg (female 1-20D post):TER VPITAR 39(6),55,81
orl-rat LD50:650 mg/kg GUCHAZ 6,522,73
orl-mus LD50:1560 mg/kg VPITAR 39(6),55,81
orl-cat LD50:540 mg/kg VPITAR 39(6),55,81
orl-rbt LD50:750 mg/kg 28ZEAL 5,229,76
orl-gpg LD50:1 g/kg VPITAR 39(6),55,81

SAFETY PROFILE: Moderately toxic by ingestion. Experimental teratogenic effects. When heated to decomposition it emits toxic fumes of NO_x.

DUK000 CAS:53780-34-0 HR: 2
2′,4′-DIMETHYL-5-((TRIFLUOROMETHYL)SULFONAMIDO)ACETANILIDE
mf: $C_{11}H_{13}F_3N_2O_3S$ mw: 310.32

PROP: Crystals. Mp: 183–185°. Sol in Me_2CO and MeOH; sltly sol in H_2O and C_6H_6.

SYNS: N-(2,4-DIMETHYL-5-(((TRIFLUOROMETHYL)SULFONYL)AMINO)PHENYL)ACETAMIDE □ EMBARK □ EMBARK PLANT GROWTH REGULATOR □ MBR 12325 □ MEFLUIDIDE □ VEL 3973 □ VISTAR □ VISTAR HERBICIDE

TOXICITY DATA WITH REFERENCE
orl-rat LD50:4 g/kg 85ARAE 3,66,76/77
orl-mus LD50:>1920 mg/kg PEMNDP 9,545,91

SAFETY PROFILE: Moderately toxic by ingestion. An herbicide and plant growth regulator. When heated to decomposition it emits very toxic fumes of F^-, NO_x, and SO_x.

DUK200 CAS:343-75-9 HR: 2
N,N-DIMETHYL-p-(2,4,6-TRIFLUOROPHENYLAZO)ANILINE
mf: $C_{14}H_{12}F_3N_3$ mw: 279.29

SYN: 2′,4′,6′-TRIFLUORO-4-DIMETHYLAMINOAZOBENZENE

TOXICITY DATA WITH REFERENCE
orl-rat TDLo:2700 mg/kg/13W-C:ETA CNREA8 13,93,53

SAFETY PROFILE: Questionable carcinogen with experimental tumorigenic data. When heated to decomposition it emits very toxic fumes of F^- and NO_x.

DUK800 CAS:2164-17-2 HR: 2
1,1-DIMETHYL-3-(α,α,α-TRIFLUORO-m-TOLYL) UREA
mf: $C_{10}H_{11}F_3N_2O$ mw: 232.23

PROP: Crystals. Mp: 163–164.5°. Sol in most org solvs; very sltly sol in H_2O.

SYNS: C 2059 □ CIBA 2059 □ COTORAN □ COTORAN MULTI 50WP □ COTTONEX □ N,N-DIMETHYL-N′-(3-TRIFLUOROMETHYLPHENYL)UREA □ 1,1-DIMETHYL-3-(3-TRIFLUOROMETHYLPHENYL) UREA □ FLUOMETURON □ HERBICIDE C-2059 □ LANEX □ NCI-C08695 □ PAKHTARAN □ 3-(5-TRIFLUORMETHYLPHENYL)-,1-DIMETHYLHARNSTOFF (GERMAN) □ N-(m-TRIFLUOROMETHYLPHENYL)-N′,N′-DIMETHYLUREA □ N-(3-TRIFLUOROMETHYLPHENYL)-N′-N′-DIMETHYLUREA □ 3-(m-TRIFLUOROMETHYLPHENYL)-1,1-DIMETHYLUREA

TOXICITY DATA WITH REFERENCE
mma-sat 1 μg/plate MUREAV 58,353,78
otr-rat:emb 56 μg/plate JJATDK 1,190,81
dni-mus-orl 1 g/kg MUREAV 58,353,78
orl-mus TDLo:87 g/kg/2Y-C:CAR NCITR* NCI-CG-TR-195,80
orl-rat LD50:6416 mg/kg PESTD5 17,351,76
ipr-rat LD50:685 mg/kg PESTD5 17,351,76
orl-mus LD50:900 mg/kg CIGET* -,-,77
ipr-mus LD50:552 mg/kg PESTD5 17,351,76
orl-rbt LD50:2500 mg/kg 85GMAT -,116,82
orl-gpg LD50:810 mg/kg 85GMAT -,116,82

CONSENSUS REPORTS: EPA Genetic Toxicology Program. IARC Cancer Review: Group 3 IMEMDT 7,56,87; Animal Inadequate Evidence IMEMDT 30,245,83. NCI Carcinogenesis Bioassay (feed); No Evidence: rat NCITR* NCI-CG-TR-195,80; Equivocal Evidence: mouse NCITR* NCI-CG-TR-195,80. Reported in EPA TSCA Inventory.

SAFETY PROFILE: Moderately toxic by ingestion and intraperitoneal routes. Questionable carcinogen with experimental carcinogenic data. Mutation data reported. When heated to decomposition it emits very toxic fumes of F^- and NO_x.

DUL200 CAS:2223-82-7 HR: 3
2,2-DIMETHYLTRIMETHYLENE ACRYLATE
mf: $C_{11}H_{16}O_4$ mw: 212.27

SYNS: DIMETHYLOLPROPANE DIACRYLATE □ 2,2-DIMETHYL-1,3-PROPANEDIOL DIACRYLATE □ 2,2-DIMETHYLTRIMETHYLENE ESTER ACRYLIC ACID □ NEOPENTYL GLYCOL DIACRYLATE □ 2-PROPENOIC ACID-2,2-DIMETHYL-1,3-PROPANEDIYL ESTER □ SR 247

TOXICITY DATA WITH REFERENCE
skn-rbt 500 mg open SEV UCDS** 11/30/71
skn-mus TDLo:46,800 mg/kg/28W-I:CAR JTEHD6 16,55,85
orl-rat LD50:5190 μL/kg UCDS** 10/5/77
skn-rbt LD50:283 μL/kg UCDS** 10/5/77

CONSENSUS REPORTS: Reported in EPA TSCA Inventory.

SAFETY PROFILE: Poison by skin contact. Mildly toxic by ingestion. A severe skin irritant. Questionable carcinogen with experimental carcinogenic data. When heated to decomposition it emits acrid smoke and irritating fumes. See also ESTERS.

DUL400 CAS:34522-40-2 HR: 2
N,N-DIMETHYL-4-(3,4,5-TRIMETHYLPHENYL)AZOANILINE
mf: $C_{17}H_{21}N_3$ mw: 267.41

SYNS: N,N-DIMETHYL-4-((3,4,5-TRIMETHYLPHENYL)AZO)BENZENAMINE ☐ N,N-3′,4′,5′-PENTAMETHYLAMINOAZOBENZENE

TOXICITY DATA WITH REFERENCE
orl-rat TDLo:4320 mg/kg/17W-C:ETA JMCMAR 15,212,72

SAFETY PROFILE: Questionable carcinogen with experimental tumorigenic data. When heated to decomposition it emits toxic fumes of NO_x.

DUL500 CAS:13271-93-7 HR: 3
1,2-DIMETHYL-2-TRIMETHYLSILYLHYDRAZINE
mf: $C_5H_{16}N_2Si$ mw: 132.28

$CH_3NHN(CH_3)SI(CH_3)_3$

SAFETY PROFILE: Explosive or violent reaction on contact with: 50/50 mixture of nitric + sulfuric acids; fuming nitric acid; fluorine; ozone + oxygen. When heated to decomposition it emits toxic fumes of NO_x.

DUL550 CAS:26464-99-3 HR: 3
DIMETHYLTRIMETHYLSILYLPHOSPHINE
mf: $C_5H_{15}PSi$ mw: 134.23

PROP: A liquid. Bp: 130°.

SAFETY PROFILE: Ignites spontaneously in air. Reaction with water forms the spontaneously flammable dimethylphosphine. When heated to decomposition it emits toxic fumes of PO_x. See also PHOSPHINE.

DUL589 HR: 3
1,4-DIMETHYL-2,3,7-TRIOXABICYCLO[2.2.1]HEPT-5-ENE
mf: $C_6H_8O_3$ mw: 128.13

SYN: 2,5-DIMETHYL-2,5-DIHYDROFURAN-2,5-ENDO PEROXIDE

SAFETY PROFILE: A very unstable explosive. When heated to decomposition it emits acrid smoke and fumes. See also PEROXIDES.

DUL800 CAS:5152-30-7 HR: 3
o,o′-DIMETHYLTUBOCURARINE
mf: $C_{40}H_{48}N_2O_6$ mw: 652.90

SYNS: DIMETHYL TUBOCURARINE ☐ o,o-DIMETHYLTUBOCURARINE ☐ N,N′,o,o-TETRAMETHYL-(+)-TUBOCURINE

TOXICITY DATA WITH REFERENCE
ivn-dog LD50:120 µg/kg RISSAF 13,339,50
ivn-rbt LD50:35 µg/kg RISSAF 13,339,50

SAFETY PROFILE: Poison by intravenous route. When heated to decomposition it emits toxic fumes of NO_x.

DUM000 CAS:7601-55-0 HR: 3
DIMETHYL TUBOCURARINE IODIDE
mf: $C_{40}H_{48}N_2O_6 \cdot 2I$ mw: 906.70

PROP: A solid. Mp: 266°.

SYNS: (+)-o,o′-DIMETHYLCHONDROCURARINE DIIODIDE ☐ DIMETHYLETHER of d-TUBOCURARINE IODIDE ☐ 6,6′,7′,12′-TETRAMETHOXY-2,2,2′,2′-TETRAMETHYLTUBOCURARANIUM DIIODIDE ☐ TUBOCURARINE DIMETHYL ETHER IODIDE ☐ d-TUBOCURARINE IODIDE DIMETHYL ETHER

TOXICITY DATA WITH REFERENCE
ivn-rat LD50:35 µg/kg JLCMAK 34,516,49
ivn-mus LD50:238 µg/kg JLCMAK 34,516,49
ivn-rbt LD50:32 µg/kg JLCMAK 34,516,49
ivn-gpg LD50:50 mg/kg JLCMAK 34,516,49

SAFETY PROFILE: Poison by intravenous route. When heated to decomposition it emits very toxic fumes of I^- and NO_x. See also IODIDES and ETHER.

DUM100 CAS:13265-01-5 HR: 3
α,3-DIMETHYLTYROSINE METHYL ESTER HYDROCHLORIDE
mf: $C_{12}H_{17}NO_3 \cdot ClH$ mw: 259.76

SYN: H 59/64

TOXICITY DATA WITH REFERENCE
orl-rat LD50:1400 mg/kg PSDTAP 10,206,69
orl-mus LD50:700 mg/kg PSDTAP 10,206,69
ivn-mus LD50:140 mg/kg PSDTAP 10,206,69

SAFETY PROFILE: Poison by intravenous route. Moderately toxic by ingestion. When heated to decomposition it emits toxic fumes of NO_x and HCl.

DUM150 CAS:598-94-7 HR: 1
1,1-DIMETHYLUREA
mf: $C_3H_8N_2O$ mw: 88.13

SYN: UREA, 1,1-DIMETHYL-

TOXICITY DATA WITH REFERENCE
ipr-mus LDLo:6610 mg/kg JPETAB 54,188,35

CONSENSUS REPORTS: Reported in EPA TSCA Inventory.

SAFETY PROFILE: Mildly toxic by intraperitoneal route. When heated to decomposition it emits toxic vapors of NO_x.

DUM200 CAS:96-31-1 HR: 2
1,3-DIMETHYLUREA
mf: $C_3H_8N_2O$ mw: 88.13

PROP: Colorless rhombic crystals. D: 1.14, mp: 106°, bp: 270°. Sol in water and alc.

SYNS: N,N′-DIMETHYLHARNSTOFF (GERMAN) ☐ N,N′-DIMETHYLUREA ☐ sym-DIMETHYLUREA ☐ SYMMETRIC DIMETHYLUREA

TOXICITY DATA WITH REFERENCE
mmo-clr 400 mmol/L FOMIAZ 20,452,75
dni-hmn:lym 40 mmol/L PNASA6 79,1171,82
orl-rat TDLo:2 g/kg (12D preg):TER TJADAB 23,335,81
orl-rat TDLo:2 g/kg (12D preg):REP TJADAB 23,335,81
ipr-mus LDLo:4962 mg/kg JPETAB 54,188,35

CONSENSUS REPORTS: Reported in EPA TSCA Inventory.

SAFETY PROFILE: Moderately toxic by intraperitoneal

route. Experimental teratogenic and reproductive effects. Human mutation data reported. When heated to decomposition it emits toxic fumes of NO_x.

DUM400 HR: 2
DIMETHYLUREA and SODIUM NITRITE

SYNS: DIMETHYLHARNSTOFF and NATRIUMNITRIT (GERMAN) □ SODIUM NITRITE and DIMETHYLUREA

TOXICITY DATA WITH REFERENCE
orl-rat TDLo: 14 g/kg/56D-C: ETA ARZNAD 21,1707,71

SAFETY PROFILE: Questionable carcinogen with experimental tumorigenic data. When heated to decomposition it emits toxic fumes of NO_x and Na_2O. See also SODIUM NITRITE.

DUM600 CAS:63019-76-1 HR: 2
p-N,N-DIMETHYLUREIDOAZOBENZENE
mf: $C_{15}H_{16}N_4O$ mw: 268.35

TOXICITY DATA WITH REFERENCE
scu-rat TDLo: 2600 mg/kg/83D-I: NEO BJPCAL 9,306,54

SAFETY PROFILE: Questionable carcinogen with experimental neoplastigenic data. When heated to decomposition it emits toxic fumes of NO_x.

DUM800 CAS:4849-32-5 HR: 3
m-(3,3-DIMETHYLUREIDO)PHENYL-tert-BUTYL CARBAMATE
mf: $C_{14}H_{21}N_3O_3$ mw: 279.38

SYNS: tert-BUTYLCARBAMIC ACID ESTER with 3-(m-HYDROXYPHENYL)-1,1-DIMETHYLUREA □ 3-((((DIMETHYLAMINO)CARBONYL)AMINO)PHENYL-1,1-DIMETHYLETHYL)CARBAMATE □ 1,1-DIMETHYL-3-((3-N-tert-BUTYLCARBAMYLOXY)-PHENYL)UREA □ m-(3,3-DIMETHYLHARNSTOFF)-PHENYL-tert-BUTYLCARBAMAT (GERMAN) □ FMC 11092 □ KARBUTILATE □ NIA 11092 □ TANDEX

TOXICITY DATA WITH REFERENCE
orl-rat LD50: 3000 mg/kg GUCHAZ 6,310,72
ivn-mus LD50: 320 mg/kg CSLNX* NX#03896

SAFETY PROFILE: Poison by intravenous route. Moderately toxic by ingestion. An herbicide. When heated to decomposition it emits toxic fumes of NO_x. See also CARBAMATES.

DUN400 CAS:63141-79-7 HR: 3
α-(2,2-DIMETHYLVINYL)-α-ETHYNYL-p-CRESOL
mf: $C_{13}H_{13}O$ mw: 185.26

SYN: DIMETHYLVINYLETHINYL-p-HYDROXYPHENYLMETHANE

TOXICITY DATA WITH REFERENCE
ihl-rat TCLo: 8900 μg/m³/24H (1-18D preg): REP GISAAA 41(1),95,76
ihl-mus LC50: 220 mg/m³ GISAAA 41(1),95,76

SAFETY PROFILE: Poison by inhalation. Experimental reproductive effects. When heated to decomposition it emits acrid smoke and irritating fumes.

DUN600 CAS:1468-37-7 HR: 3
DIMETHYLXANTHOGEN DISULFIDE
mf: $C_4H_6O_2S_4$ mw: 214.34

PROP: Crystals from C_6H_6. Mp: 23°.

SYNS: BIS(METHYLXANTHOGEN) DISULFIDE □ DI(METHOXYTHIOCARBONYL) DISULFIDE □ o,o-DIMETHYL DITHIOBIS(THIOFORMATE) □ DIMETHYL DIXANTHOGEN □ DIMETHYL XANTHIC DISULFIDE □ DIMEXAN □ DIMEXANO □ DI(THIONOCARBOMETHOXY) DISULFIDE □ THIOPEROXYDICARBONIC ACID DIMETHYL ESTER □ TRIDEX □ TRI-PE

TOXICITY DATA WITH REFERENCE
orl-rat LD50: 240 mg/kg WRPCA2 9,119,70

SAFETY PROFILE: Poison by ingestion. A pesticide. When heated to decomposition it emits toxic fumes of SO_x.

DUN800 CAS:18997-62-1 HR: 2
N,N-DIMETHYL-p-(2,3,XYLYLAZO)ANILINE
mf: $C_{16}H_{19}N_3$ mw: 253.38

SYNS: 2′,3′-DIMETHYL-4-DIMETHYLAMINOAZOBENZENE □ N,N-DIMETHYL-p-(2′,3′-DIMETHYLPHENYLAZO)ANILINE

TOXICITY DATA WITH REFERENCE
mma-sat 4 μg/plate MUREAV 93,67,82
orl-rat TDLo: 1075 mg/kg/8W-C: CAR CBINA8 53,107,85
orl-rat TDLo: 1080 mg/kg/30D-I: ETA JMCMAR 11,1234,68

SAFETY PROFILE: Questionable carcinogen with experimental carcinogenic and tumorigenic data. Mutation data reported. When heated to decomposition it emits toxic fumes of NO_x.

DUO000 CAS:3025-73-8 HR: 2
N,N-DIMETHYL-p-(3,4-XYLYLAZO)ANILINE
mf: $C_{16}H_{19}N_3$ mw: 253.38

SYNS: 3′,4′-DIMETHYL-4-DIMETHYLAMINOZOBENZENE □ N,N-DIMETHYL-p-(3′,4′-DIMETHYLPHENYLAZO)ANILINE

TOXICITY DATA WITH REFERENCE
orl-rat TDLo: 9030 mg/kg/25W-C: CAR CBINA8 53,107,85
orl-rat TD: 6600 mg/kg/26W-C: ETA CNREA8 17,387,57
orl-rat TD: 4523 mg/kg/17W-I: ETA CNREA8 30,1520,70
orl-rat TD: 6480 mg/kg/26W-C: NEO JMCMAR 15,212,72

SAFETY PROFILE: Questionable carcinogen with experimental carcinogenic, neoplastigenic, and tumorigenic data. When heated to decomposition it emits toxic fumes of NO_x.

DUO200 CAS:544-97-8 HR: 3
DIMETHYLZINC
mf: C_2H_6Zn mw: 95.44

PROP: A liquid. D: 1.386 @ 10.5°/4°, mp: −42.5°, bp: 46°.

CONSENSUS REPORTS: Zinc and its compounds are on the Community Right-To-Know List.

SAFETY PROFILE: A poison. Ignites spontaneously in air. Explodes in an oxygen atmosphere. Explosive reaction on contact with water; 2,2-dichloropropane.

When heated to decomposition it emits toxic fumes of ZnO. See also ZINC COMPOUNDS.

DUO300 CAS:89591-51-5 HR: 3
DIMETPRAMIDE
mf: $C_{16}H_{26}N_4O_4 \cdot ClH$ mw: 374.87

SYNS: N-(2-(DIETHYLAMINO)ETHYL)-4-(DIMETHYLAMINO)-2-METHOXY-5-NITROBENZAMIDE MONOHYDROCHLORIDE □ DIMETHPRAMIDE □ 4-DIMETHYLAMINO-5-NITRO-2-METHOXY-N-(2-DIETHYLAMINOETHYL)BENZAMIDE HYDROCHLORIDE

TOXICITY DATA WITH REFERENCE
eye-rbt 500 mg MLD RPTOAN 48,201,85
orl-rat LD50:856 mg/kg RPTOAN 48,201,85
orl-mus LD50:230 mg/kg RPTOAN 48,201,85
ipr-mus LD50:159 mg/kg RPTOAN 48,201,85
orl-gpg LD50:636 mg/kg RPTOAN 48,201,85

SAFETY PROFILE: Poison by ingestion and intraperitoneal routes. An eye irritant. When heated to decomposition it emits toxic fumes of NO_x and HCl.

DUO350 CAS:642-15-9 HR: 3
DIMIDIN
mf: $C_{28}H_{40}N_2O_9$ mw: 548.62

PROP: Crystals from MeOH, MeOH (aq), or Skellysolve B. Mp: 149–150°. Freely sol in alc, ether, acetone, chloroform. Very sltly sol in pet ether, benzene, carbon tetrachloride. Practically insol in water and in 5% aq solns of hydrochloric acid, sodium carbonate, and sodium bicarbonate.

SYNS: ANTIMYCIN A1 □ DIHYDROSAMIDIN □ ISOVALERIC ACID-8-ESTER with 3-FORMAMIDO-N-(7-HEXYL-8-HYDROXY-4,9-DIMETHYL-2,6-DIOXO-1,5-DIOXONAN-3-YL)SALICYLAMIDE

TOXICITY DATA WITH REFERENCE
orl-rat LD50:1469 mg/kg JDGRAX 7(2),1,75
ipr-mus LD50:7600 μg/kg 85FZAT -,144,67
scu-mus LD50:25 mg/kg 85FZAT -,144,67
ivn-mus LD50:900 μg/kg 85FZAT -,144,67
ims-mus LD50:1000 mg/kg JDGRAX 7(2),1,75

SAFETY PROFILE: Poison by subcutaneous, intravenous, and intraperitoneal routes. Moderately toxic by ingestion and intramuscular routes. When heated to decomposition it emits toxic fumes of NO_x. See also ESTERS.

DUO400 CAS:119-48-2 HR: 3
DIMORPHOLAMINE
mf: $C_{20}H_{38}N_4O_4$ mw: 398.62

SYNS: AMIPAN T □ N,N′-DIBUTYL-N,N′-DICARBOXYETHYLENE DIAMINEMORPHOLIDE □ N,N′-DIBUTYL-N,N′-DICARBOXYMORPHOLIDE-ETHYLENEDIAMINE □ N,N′-DI-n-BUTYLETHYLENEDIAMINE-N,N′-DICARBOXYBISMORPHOLIDE □ N,N′-1,2-ETHANEDIYLBIS(N-BUTYL-4-MORPHOLINECARBOXAMIDE) □ N,N′-ETHYLENEBIS(N-BUTYL-4-MORPHOLINECARBOXAMIDE) □ PRONTODIN □ 1064 TH □ THERALEPTIQUE □ THERAPTIQUE

TOXICITY DATA WITH REFERENCE
orl-rat LD50:270 mg/kg NIIRDN 6,347,82
scu-rat LD50:190 mg/kg NIIRDN 6,347,82
ivn-rat LD50:24 mg/kg NIIRDN 6,347,82
ims-rat LD50:122 mg/kg NIIRDN 6,347,82
orl-mus LD50:150 mg/kg MEIEDD 10,476,83
ipr-mus LD50:80 mg/kg AIPTAK 163,133,66
scu-mus LD50:104 mg/kg AIPTAK 163,133,66
ivn-mus LD50:42,200 μg/kg NIIRDN 6,347,82

SAFETY PROFILE: Poison by ingestion, intraperitoneal, intravenous, intramuscular, and subcutaneous routes. An analeptic agent (stimulant). When heated to decomposition it emits toxic fumes of NO_x. See also AMINES.

DUO500 CAS:69853-15-2 HR: 3
DIMORPHOLINIUM HEXACHLOROSTANNATE
mf: $C_8H_{10}Cl_6N_2O_2Sn$ mw: 497.59

SYN: MORPHOLINIUM, HEXACHLOROSTANNATE(2-) (2:1)

TOXICITY DATA WITH REFERENCE
ivn-mus LD50:18 mg/kg CSLNX* NX#02489

OSHA PEL: TWA 2 mg(Sn)/m^3
ACGIH TLV: TWA 2 mg(Sn)/m^3

SAFETY PROFILE: Poison by intravenous route. When heated to decomposition it emits toxic fumes of NO_x, Sn, and Cl^-.

DUO600 CAS:13071-27-7 HR: 3
1,5-DIMORPHOLINO-3-(1-NAPHTHYL)-PENTANE
mf: $C_{23}H_{32}N_2O_2$ mw: 368.57

SYNS: 4,4′-(3-(1-NAPHTHALENYL)-1,5-PENTANEDIYL)BISMORPHOLINE □ 4,4′-(3-(1-NAPHTHYL)-1,5-PENTAMETHYLENE)DIMORPHOLINE

TOXICITY DATA WITH REFERENCE
orl-rat LD50:708 mg/kg ARZNAD 18,1127,68
scu-rat LD50:830 mg/kg ARZNAD 18,1127,68
ivn-rat LD50:26 mg/kg ARZNAD 18,1127,68
orl-mus LD50:1700 mg/kg ARZNAD 18,1127,68
ipr-mus LD50:452 mg/kg JMCMAR 13,418,70
ivn-mus LD50:98 mg/kg ARZNAD 18,1127,68

SAFETY PROFILE: Poison by intravenous route. Moderately toxic by ingestion, subcutaneous, and intraperitoneal routes. When heated to decomposition it emits toxic fumes of NO_x.

DUO800 CAS:74749-73-8 HR: 1
DIMYRCETOL

PROP: A mixture of dihydromyrcenol and dihydromyrcenyl formate (FCTXAV 18,649,80).

TOXICITY DATA WITH REFERENCE
skn-rbt 500 mg/24H MOD FCTXAV 18,679,80
orl-rat LD50:4100 mg/kg FCTXAV 18,649,80

SAFETY PROFILE: Mildly toxic by ingestion. A skin irritant.

DUP000 CAS:258-76-4 HR: 2
DINAPHTHAZINE
mf: $C_{20}H_{12}N_2$ mw: 280.34

PROP: Shiny black platelets from Py.

SYN: DINAPHTAZIN (GERMAN)

TOXICITY DATA WITH REFERENCE
scu-mus TDLo: 200 mg/kg: ETA ZEKBAI 58,56,51

SAFETY PROFILE: Questionable carcinogen with experimental tumorigenic data. When heated to decomposition it emits toxic fumes of NO_x.

DUP100 CAS:2379-81-9 HR: 1
16H-DINAPHTHO(2,3-a:2′,3′-i)CARBAZOLE-5,10,15, 17-TETRAONE, 6,9-DIBENZAMIDO-
mf: $C_{42}H_{23}N_3O_6$ mw: 665.68

SYNS: AHCOVAT OLIVE ARN ☐ AHCOVAT OLIVE R ☐ AMANTHRENE OLIVE R ☐ ATIC VAT OLIVE R ☐ BENZADONE OLIVE R ☐ CALCOLOID OLIVE R ☐ CALCOLOID OLIVE RC ☐ CALCOLOID OLIVE RL ☐ CALEDONE OLIVE RP ☐ CALEDON OLIVE R ☐ CARBANTHRENE OLIVE R ☐ CERN KYPOVA 27 ☐ C.I. 69005 ☐ CIBANONE OLIVE F2R ☐ CIBANONE OLIVE 2R ☐ C.I. VAT BLACK 27 ☐ FENANTHREN OLIVE R ☐ INDANTHRENE OLIVE R ☐ INDANTHREN OLIVE R ☐ MAYVAT OLIVE AR ☐ MIKETHRENE OLIVE R ☐ NIHONTHRENE OLIVE R ☐ NYANTHRENE OLIVE R ☐ OLIV OSTANTHRENOVY R ☐ OSTANTHREN OLIVE R ☐ PALANTHRENE OLIVE R ☐ PARADONE OLIVE R ☐ PERNITHRENE OLIVE R ☐ PONSOL OLIVE AR ☐ PONSOL OLIVE ARD ☐ ROMANTRENE OLIVE FR ☐ SANDOTHRENE OLIVE N2R ☐ SOLANTHRENE OLIVE R ☐ TINON OLIVE 2R ☐ TYRIAN OLIVE I-R

TOXICITY DATA WITH REFERENCE
eye-rbt 500 mg/24H MLD 85JCAE -,1324,86

CONSENSUS REPORTS: Reported in EPA TSCA Inventory.

SAFETY PROFILE: An eye irritant. When heated to decomposition it emits toxic vapors of NO_x.

DUP200 CAS:29903-04-6 HR: 3
DI-(1-NAPHTHOYL)PEROXIDE
mf: $C_{22}H_{14}O_4$ mw: 342.35

PROP: White or pale-yellow solid from dioxan (aq). Mp: 98° (decomp).

SAFETY PROFILE: A friction-sensitive explosive. When heated to decomposition it emits acrid smoke and fumes. See also PEROXIDES.

DUP300 CAS:148-01-6 HR: 3
DINITOLMIDE
mf: $C_8H_7N_3O_5$ mw: 225.18

PROP: Yellowish solid or needles from EtOH (aq). Mp: 181°. Very sltly sol in water; sol in acetone, acetonitrile, and dimethyl formamide.

SYNS: COCCIDINE A ☐ COCCIDOT ☐ DINITOLMID ☐ 3,5-DINITRO-o-TOLUAMIDE ☐ D.O.T. ☐ 2-METHYL-3,5-DINITROBENZAMIDE ☐ ZOALENE ☐ ZOAMIX

TOXICITY DATA WITH REFERENCE
mmo-esc 500 μg/plate MUREAV 77,21,80
mrc-bcs 1 mg/disc MUREAV 77,21,80
orl-rat LD50: 600 mg/kg 29ZVAB -,537,69
ivn-dog LD50: 75 mg/kg PCOC** -,1252,66

OSHA PEL: TWA 5 mg/m³

ACGIH TLV: TWA 5 mg/m³

SAFETY PROFILE: Poison by intravenous route. Moderately toxic by ingestion. Mutation data reported. A strong exothermic reaction above 248°C has caused industrial explosions. When heated to decomposition it emits toxic fumes of NO_x. See also NITRO COMPOUNDS of AROMATIC HYDROCARBONS.

DUP400 CAS:96-91-3 HR: 3
4,6-DINITRO-2-AMINOPHENOL
mf: $C_6H_5N_3O_5$ mw: 199.14

PROP: Dark red crystals or needles from EtOH or prisms from $CHCl_3$. Mp: 169°; flash p: 410°F. Soluble in alc, benzene, glacial acetic acid, aniline, and ether; sparingly sol in water.

SYNS: ACIDE PICRAMIQUE (FRENCH) ☐ 2-AMINO-4,6-DINITROPHENOL ☐ C.I. OXIDATION BASE 21 ☐ FOURRINE 93 ☐ FOURRINE 4R ☐ FURRO 4R ☐ PICRAMIC ACID ☐ ZOBA 4R

TOXICITY DATA WITH REFERENCE
mmo-sat 5 μmol/plate NEZAAQ 38,533,83
mma-sat 5 μmol/plate NEZAAQ 38,533,83
ivn-dog LDLo: 150 mg/kg AIPTAK 50,20,35
ipr-pgn LDLo: 140 mg/kg AIPTAK 50,20,35

SAFETY PROFILE: Poison by intravenous and intraperitoneal routes. Mutation data reported. Combustible when exposed to heat, flame, or oxidizers. A powerful explosive when dry. May explode when shocked or heated. When heated to decomposition it emits toxic fumes of NO_x. See also EXPLOSIVES, HIGH.

DUP600 CAS:97-02-9 HR: 3
2,4-DINITROANILINE
mf: $C_6H_5N_3O_4$ mw: 183.14

PROP: Yellow, needle-like crystals. Mp: 188°, flash p: 435°F (CC), d: 1.615, vap d: 6.31. Insol in water.

SYNS: 2,4-DINITRANILINE ☐ 2,4-DINITROANILIN (GERMAN) ☐ 2,4-DINITROANILINA (ITALIAN) ☐ 2,4-DINITROBENZENAMIME ☐ DNA ☐ NCI-C60753

TOXICITY DATA WITH REFERENCE
eye-rbt 500 mg/24H MLD 28ZPAK -,132,72
mmo-sat 10 μg/plate ENMUDM 5(Suppl 1),3,83
mma-sat 2 μg/plate MUREAV 67,1,79
ihl-rat TCLo: 17 mg/m³/4H (1-22D preg): TER GTPZAB 26(4),47,82
ihl-rat TCLo: 1100 μg/m³/4H (1-7D preg): TER GTPZAB 26(4),47,82
ihl-rat TCLo: 17 mg/m³/4H (1-7D preg): REP GTPZAB 26(4),47,82
orl-rat LD50: 285 mg/kg TSCAT* OTS 206512
ipr-rat LDLo: 250 mg/kg NCNSA6 5,32,53
orl-mus LD50: 370 mg/kg GTPZAB 25(8),50,81
ipr-mus LDLo: 400 mg/kg JAPMA8 48,419,59
orl-gpg LD50: 1050 mg/kg GISAAA 47(10),15,82

CONSENSUS REPORTS: Reported in EPA TSCA Inventory.

SAFETY PROFILE: Poison by ingestion and intraperitoneal routes. Experimental teratogenic and reproductive effects. Mutation data reported. An eye irritant. Combustible and explosive when exposed to heat or flame; can react with oxidizing materials. To fight fire, use CO_2, dry chemical. Mixtures with charcoal ignite at 350°C. Vigorous reaction with chlorine + hydrochloric acid evolves gases. When heated to decomposition it emits highly toxic fumes of NO_x.

DUP800 CAS:119-27-7 HR: 3
2,4-DINITROANISOL
mf: $C_7H_6N_2O_5$ mw: 198.15

PROP: Colorless to yellow crystals from alc. Mp: 83°, bp: sublimes, d: 1.341 @ 20°/4°, vap d: 6.83.

SYNS: α-DINITROANISOLE □ 2,4-DINITROANISOLE □ 2,4-DINITROPHENYLMETHYL ETHER □ 1-METHOXY-2,4-DINITROBENZENE

TOXICITY DATA WITH REFERENCE
mmo-sat 10 μg/plate BCPCA6 26,729,77
orl-rat LDLo:100 mg/kg NCNSA6 5,16,53

CONSENSUS REPORTS: Reported in EPA TSCA Inventory.

SAFETY PROFILE: Poison by ingestion. Mutation data reported. When heated to decomposition it emits toxic fumes of NO_x. See also NITRO COMPOUNDS of AROMATIC HYDROCARBONS and NITRATES.

DUQ000 CAS:82-35-9 HR: 2
1,5-DINITROANTHRAQUINONE
mf: $C_{14}H_6N_2O_6$ mw: 298.22

PROP: Yellow needles from $PhNO_2$ or xylene. Mp: 384–385°. Sltly sol in AcOH; sltly sol in EtOH, Et_2O, and C_6H_6.

SYNS: 1,5-DINITRO-9,10-ANTHRACENEDIONE □ 1,5-DINITROANTHRACHINON (CZECH)

TOXICITY DATA WITH REFERENCE
eye-rbt 500 mg/24H MLD 28ZPAK -,121,72
ipr-rat LD50:3130 mg/kg GISAAA 49(4),90,84
orl-mus LD50:4750 mg/kg GISAAA 49(4),90,84

CONSENSUS REPORTS: Reported in EPA TSCA Inventory.

SAFETY PROFILE: Moderately toxic by intraperitoneal route. Mildly toxic by ingestion. An eye irritant. When heated to decomposition it emits toxic fumes of NO_x.

DUQ180 CAS:25154-54-5 HR: 3
DINITROBENZENE
DOT: UN 1597
mf: $C_6H_4N_2O_4$ mw: 168.12

SYNS: DINITROBENZENE, solution (DOT) □ DINITROBENZOL, solid (DOT)

OSHA PEL: TWA 1 mg/m³ (skin)
ACGIH TLV: TWA 0.15 ppm (skin)
DFG MAK: 0.15 ppm (1 mg/m³); Suspected Carcinogen
DOT CLASSIFICATION: 6.1; *Label:* Poison

SAFETY PROFILE: Suspected carcinogen. A poison. When heated to decomposition it emits toxic fumes of NO_x. See also o-DINITROBENZENE.

DUQ200 CAS:99-65-0 HR: 3
m-DINITROBENZENE
DOT: UN 1597
mf: $C_6H_4N_2O_4$ mw: 168.12

PROP: Yellowish crystals from alc. Mp: 89°, bp: 291°.

SYNS: BINITROBENZENE □ 1,3-DINITROBENZENE □ 2,4-DINITROBENZENE □ 1,3-DINITROBENZOL □ DWUNITROBENZEN (POLISH)

TOXICITY DATA WITH REFERENCE
mmo-sat 3300 ng/plate ENMUDM 2,531,80
mma-sat 100 nmol/plate MUREAV 58,11,78
eye-rbt 100 mg JACTDZ 1,168,92
orl-rat TDLo:150 mg/kg (male 10W pre):REP JTEHD6 19,477,86
orl-rat TDLo:33,600 μg/kg (female 16W pre):REP JTEHD6 7,829,81
orl-rat TDLo:90 mg/kg (male 12W pre):REP JTEHD6 19,477,86
orl-rat TDLo:50 mg/kg (male 1D pre):REP TXAPA9 92,54,88
orl-hmn LDLo:28 mg/kg 34ZIAG -226,69
skn-man TDLo:4 mg/kg/2D-I:CNS,PUL LANCAO 2,582,01
orl-rat LD50:83 mg/kg NTIS** AD-A066-307
ipr-rat LD50:28 mg/kg AEPPAE 207,446,49
orl-dog LDLo:600 mg/kg NTIS** AD-A066-307
ivn-dog LD50:10 mg/kg NTIS** AD-A066-307
orl-cat LDLo:27 mg/kg LANCAO 2,582,01
orl-rbt LDLo:400 mg/kg NTIS** AD-A066-307
orl-bwd LD50:42 mg/kg TXAPA9 21,315,72

CONSENSUS REPORTS: Reported in EPA TSCA Inventory. EPA Genetic Toxicology Program.

OSHA PEL: TWA 1 mg/m³ (skin)
ACGIH TLV: TWA 0.15 ppm (skin)
DFG MAK: 0.15 ppm (1 mg/m³); Suspected Carcinogen
DOT CLASSIFICATION: 6.1; *Label:* Poison

SAFETY PROFILE: Suspected carcinogen. Human poison by ingestion. Experimental poison by ingestion, intraperitoneal, and intravenous routes. Human systemic effects by skin contact: cyanosis and motor activity changes. Experimental reproductive effects. An eye irritant. Mutation data reported. Mixture with nitric acid is a high explosive. Mixture with tetranitromethane is a high explosive very sensitive to sparks. When heated to decomposition it emits toxic fumes of NO_x. See also o- and p-DINITROBENZENE.

DUQ400 CAS:528-29-0 HR: 3
o-DINITROBENZENE
DOT: UN 1597
mf: $C_6H_4N_2O_4$ mw: 168.12

PROP: Colorless needles or plates from alc. Mp: 118°, bp: 319°, flash p: 302°F (CC), d: 1.571 @ 0°/4°, vap d: 5.79. Sol in EtOH, and $CHCl_3$; sltly sol in H_2O.

SYN: 1,2-DINITROBENZENE

OSHA PEL: TWA 1 mg/m³ (skin)

D

ACGIH TLV: TWA 0.15 ppm (skin)
DFG MAK: 0.15 ppm (1 mg/m³); Suspected Carcinogen
DOT CLASSIFICATION: 6.1; *Label:* Poison

SAFETY PROFILE: Suspected carcinogen. Poison by inhalation and ingestion. Moderately toxic by skin contact. Can cause liver, kidney, and central nervous system injury. Combustible when exposed to heat or flame; can react vigorously with oxidizing materials. A severe explosion hazard when shocked or exposed to heat or flame. It is used in bursting charges and to fill artillery shells. Mixtures with nitric acid are highly explosive. To fight fire, use water, CO_2, dry chemical. Dangerous; when heated to decomposition it emits highly toxic fumes of NO_x and explodes. See also m- and p-DINITROBENZENE and NITRO COMPOUNDS of AROMATIC HYDROCARBONS.

DUQ600 CAS:100-25-4 HR: 3
p-DINITROBENZENE
DOT: UN 1597
mf: $C_6H_4N_2O_4$ mw: 168.12

PROP: White crystals, needles or prisms from alc. Mp: 173°, bp: 299°. Volatile with steam.

SYN: DITHANE A-4

TOXICITY DATA WITH REFERENCE
mmo-sat 5 µg/plate CRNGDP 6,727,85
mma-sat 25 µg/plate CRNGDP 6,727,85
orl-cat LDLo: 29 mg/kg 85ESA3 11,516,89

CONSENSUS REPORTS: Reported in EPA TSCA Inventory.

OSHA PEL: TWA 1 mg/m³ (skin)
ACGIH TLV: TWA 0.15 ppm (skin)
DFG MAK: 0.15 ppm (1 mg/m³); Suspected Carcinogen
DOT CLASSIFICATION: 6.1; *Label:* Poison

SAFETY PROFILE: Suspected carcinogen. Poison by ingestion. Mutation data reported. Mixture with nitric acid is a high explosive. When heated to decomposition it emits toxic fumes of NO_x. See also o- and m-DINITROBENZENE.

DUR200 CAS:528-76-7 HR: 3
2,4-DINITROBENZENESULFENYL CHLORIDE
mf: $C_6H_3ClN_2O_4S$ mw: 234.62

PROP: A solid. Mp: 95–96°.

SAFETY PROFILE: A heat-sensitive explosive. When heated to decomposition it emits toxic fumes of Cl^-, SO_x, and NO_x. See also NITRO COMPOUNDS of AROMATIC HYDROCARBONS.

DUR400 CAS:89-02-1 HR: 3
2,4-DINITROBENZENESULFONIC ACID
mf: $C_6H_4N_2O_7S$ mw: 248.18

PROP: Crystals. Mp: 106–108° (hydrate), mp: 130° (anhyd).

SYN: KYSELINA-2,4-DINITROBENZENSULFONOVA (CZECH)

TOXICITY DATA WITH REFERENCE
skn-rbt 500 mg/24H MOD 28ZPAK -,180,72
eye-rbt 500 mg/24H MLD 28ZPAK -,180,72
ivn-mus LD50: 320 mg/kg CSLNX* NX#01550

CONSENSUS REPORTS: Reported in EPA TSCA Inventory.

SAFETY PROFILE: Poison by intravenous route. A skin and eye irritant. When heated to decomposition it emits very toxic fumes of NO_x and SO_x.

DUR500 CAS:5128-28-9 HR: 3
4,6-DINITROBENZOFURAZAN-N-OXIDE
mf: $C_6H_2N_4O_6$ mw: 226.10

PROP: Yellow needles from AcOH. Mp: 172°.

SAFETY PROFILE: A powerful, high explosive as sensitive as picric acid. Forms impact-, friction-, or electric shock-sensitive explosive complexes with cysteine and nucleophiles (e.g., potassium hydrogen carbonate in water or methanol, potassium hydroxide in methanol, ammonia, hydroxylamine, hydrazine hydrate). Forms explosive adducts with furan, N-methylindole, and N-methylpyrrole. Forms adducts with ketones (e.g., acetone, cyclopentanone, cyclopentanedione, 2,4-pentanedione, 3-methyl-2,4-pentanedione), which can produce shock-sensitive, high explosive potassium salts. Upon decomposition it emits toxic fumes of NO_x. See also EXPLOSIVES, HIGH.

DUR800 CAS:87-31-0 HR: 3
5,7-DINITRO-1,2,3-BENZOXADIAZOLE
DOT: UN 0074
mf: $C_6H_2N_4O_5$ mw: 210.12

SYNS: DDNP □ DIAZODINITROPHENOL, wetted with not <40% H_2O or mixture of alcohol & H_2O (UN 0074) (DOT) □ DIAZO □ 2-DIAZO-4,6-DINITROBENZENE-1-OXIDE □ DIAZODINITROPHENOL (dry) (DOT) □ INITIATING EXPLOSIVE DIAZODINITROPHENOL (DOT)

DOT CLASSIFICATION: Forbidden; DOT Class: EXPLOSIVE 1.1A; *Label:* EXPLOSIVE 1.1A (UN 0074)

SAFETY PROFILE: An explosive. When heated to decomposition it emits toxic fumes of NO_x. See also NITRO COMPOUNDS of AROMATIC HYDROCARBONS; and EXPLOSIVES, HIGH.

DUS000 CAS:1528-74-1 HR: 2
4,4′-DINITROBIPHENYL
mf: $C_{12}H_8N_2O_4$ mw: 244.22

PROP: Needles from EtOH or toluene. Mp: 239–239.5°.

SYN: 4,4′-DINITROBIFENYL (CZECH)

TOXICITY DATA WITH REFERENCE
eye-rbt 500 mg/24H MLD 28ZPAK -,61,72
mmo-sat 2500 µg/plate MUREAV 91,321,81
orl-rat TDLo: 950 mg/kg/W-I: ETA TXAPA9 6,352,64

CONSENSUS REPORTS: Reported in EPA TSCA Inventory.

SAFETY PROFILE: An eye irritant. Questionable carcinogen with experimental tumorigenic data. Mutation data reported. When heated to decomposition it emits toxic fumes of NO_x.

DUS200 CAS:1817-73-8 HR: 2
2,4-DINITRO-6-BROMOANILINE
mf: $C_6H_4BrN_3O_4$ mw: 262.04

PROP: Yellow needles from AcOH or EtOH. Mp: 153–154°.

SYNS: 2,4-DINITRO-6-BROMANILIN (CZECH) ☐ 2-BROMO-4,6-DINITROANILINE ☐ 6-BROMO-2,4-DINITROANILINE ☐ BROMO DNA ☐ NCI-C60844

TOXICITY DATA WITH REFERENCE
eye-rbt 500 mg/24H SEV 28ZPAK -,94,72
mmo-sat 10 μg/plate ENMUDM 9(Suppl 9),1,87
orl-rat LD50:4100 mg/kg TSCAT* OTS 206481

CONSENSUS REPORTS: Reported in EPA TSCA Inventory.

SAFETY PROFILE: Mildly toxic by ingestion. A severe eye irritant. Mutation data reported. When heated to decomposition it emits very toxic fumes of Br^- and NO_x. See also 2,4-DINITROANILINE.

DUS400 CAS:28103-68-6 HR: 3
2,3-DINITRO-2-BUTENE
mf: $C_4H_6N_2O_4$ mw: 146.11

$CH_3C(NO_2)=C(NO_2)CH_3$

SAFETY PROFILE: Potentially explosive at 135°C/14 mbar. When heated to decomposition it emits toxic fumes of NO_x.

DUS500 CAS:29110-68-7 HR: 2
2,4-DINITRO-6-tert-BUTYLPHENYL METHANESULFONATE
mf: $C_{11}H_{14}N_2O_7S$ mw: 318.33

SYNS: HE 166 ☐ PREPARATION HE 166

TOXICITY DATA WITH REFERENCE
orl-rat TDLo:1820 mg/kg (52W pre):REP JTSCDR 9,161,84
orl-rat TDLo:2184 mg/kg (52W male):REP JTSCDR 9,161,84
orl-rat TDLo:280 mg/kg/32W-C:ETA JTSCDR 9,161,84
orl-rat TD:1120 mg/kg/32W-C:ETA JTSCDR 9,161,84

SAFETY PROFILE: Questionable carcinogen with experimental tumorigenic data. Experimental reproductive effects. An herbicide. When heated to decomposition it emits toxic fumes of SO_x and NO_x. See also SULFONATES.

DUS600 CAS:2401-85-6 HR: 3
2,4-DINITRO-1-CHLORO-NAPHTHALENE
mf: $C_{10}H_5ClN_2O_4$ mw: 252.62

PROP: Yellow needles from C_6H_6 or AcOH (aq). Mp: 146.5°.

SYN: 1-CHLORO-2,4-DINITRONAPHTHALENE

TOXICITY DATA WITH REFERENCE
orl-rat TDLo:5000 mg/kg:CAR CNREA8 26,619,66
orl-rat TD:5000 mg/kg:NEO CNREA8 28,924,68
unr-mam LD50:250 mg/kg 30ZDA9 -,81,71

CONSENSUS REPORTS: Reported in EPA TSCA Inventory.

SAFETY PROFILE: Poison by unspecified route. Questionable carcinogen with experimental carcinogenic and neoplastigenic data. When heated to decomposition it emits very toxic fumes of Cl^- and NO_x. See also 2,4-DINITROANILINE.

DUS700 CAS:534-52-1 HR: 3
DINITRO-o-CRESOL
mf: $C_7H_6N_2O_5$ mw: 198.15

PROP: Yellow, prismatic crystals from alc. Mp: 85.8°, vap d: 6.82.

SYNS: ANTINONIN ☐ ARBOROL ☐ CAPSINE ☐ CHEMSECT DNOC ☐ DEGRASSAN ☐ DEKRYSIL ☐ DETAL ☐ DINITROCRESOL ☐ 2,4-DINITRO-o-CRESOL ☐ 4,6-DINITRO-o-CRESOL ☐ 4,6-DINITRO-o-CRESOLO (ITALIAN) ☐ DINITRODENDTROXAL ☐ 3,5-DINITRO-2-HYDROXYTOLUENE ☐ 4,6-DINITRO-o-KRESOL (CZECH) ☐ 4,6-DINITROKRESOL (DUTCH) ☐ DINITROL ☐ DINITROMETHYL CYCLOHEXYLTRIENOL ☐ 2,4-DINITRO-6-METHYLPHENOL ☐ DINOC ☐ DINURANIA ☐ DITROSOL ☐ DN-DRY MIX No.2 ☐ DNOK (CZECH) ☐ DWUNITRO-o-KREZOL (POLISH) ☐ EFFUSAN ☐ ELGETOL ☐ELIPOL ☐ ENT 154 ☐ EXTRAR ☐ HEDOLIT ☐ K III ☐ KRENITE (OBS.) ☐ KRESAMONE ☐ KREZOTOL 50 ☐ LE DINITROCRESOL-4,6 (FRENCH) ☐ LIPAN ☐ 2-METHYL-4,6-DINITROPHENOL ☐ NITRADOR ☐ NITROFAN ☐ PROKARBOL ☐ RAFEX ☐ RAPHATOX ☐ RCRA WASTE NUMBER P047 ☐ SANDOLIN ☐ SELINON ☐ SINOX ☐ TRIFOCIDE ☐ TRIFRINA ☐ WINTERWASH ☐ ZAHLREICHE BEZEICHNUNGEN (GERMAN)

TOXICITY DATA WITH REFERENCE
skn-rbt 105 mg/9D-I MLD JIHTAB 30,10,48
eye-rbt 20 mg/24H MOD 85JCAE-,678,86
mma-sat 1 μmol/plate AIDZAC 10,305,82
sln-dmg-orl 250 μmol/L ARTODN Suppl. 4,59,80
orl-man TDLo:7500 μg/kg/7D:CNS CMEP** -,1,56
ihl-hmn TCLo:1 mg/m³:BRN,CVS,GIT HYSAAV 30,197,65
unr-man LDLo:29 mg/kg 85DCAI 2,73,70
orl-rat LD50:10 mg/kg 85ARAE 3,54,76
skn-rat LD50:200 mg/kg WRPCA2 9,119,70
ipr-rat LDLo:28 mg/kg TXAPA9 1,156,59
scu-rat LD50:25,600 μg/kg JPPMAB 4,1062,52
orl-mus LD50:47 mg/kg HYSAAV 30,197,65
ipr-mus LD50:19 mg/kg BCPCA6 18,1389,69
ivn-dog LDLo:15 mg/kg AIPTAK 50,20,35
ihl-cat LCLo:40 mg/m³ HYSAAV 30,197,65

CONSENSUS REPORTS: Reported in EPA TSCA Inventory. EPA Genetic Toxicology Program. Community Right-To-Know List. EPA Extremely Hazardous Substances List.

OSHA PEL: TWA 0.2 mg/m³ (skin)
ACGIH TLV: TWA 0.2 mg/m³ (skin)
DFG MAK: 0.2 mg/m³
NIOSH REL: (Dinitro-Ortho-Cresol) TWA 0.2 mg/m³
DOT CLASSIFICATION: 6.1; *Label:* Poison

SAFETY PROFILE: Human poison by unspecified route. Experimental poison by ingestion, inhalation, skin contact, intraperitoneal, and intravenous routes. Human systemic effects by ingestion and inhalation: somnolence, headache, abnormal brain recordings from specific areas of the central nervous system, cardiac and gastrointestinal changes. Mutation data reported. An eye and skin irritant. Less toxic than the para form, but is still highly toxic. A pesticide. See also NITRO COMPOUNDS of AROMATIC HYDROCARBONS and other dinitrocresol entries.

For occupational chemical analysis use NIOSH: Dinitro-o-cresol S166.

DUT000 CAS:497-56-3 HR: 3
3,5-DINITRO-o-CRESOL
mf: $C_7H_6N_2O_5$ mw: 198.15

PROP: Yellow prisms from EtOH. Mp: 85°.

TOXICITY DATA WITH REFERENCE
scu-rat LDLo:40 mg/kg XPHBAO 271,146,41

NIOSH REL: (Dinitro-Ortho-Cresol) TWA 0.2 mg/m^3

SAFETY PROFILE: Poison by subcutaneous route. When heated to decomposition it emits toxic fumes of NO_x. See also NITRO COMPOUNDS of AROMATIC HYDROCARBONS and other dinitrocresol entries.

DUT200 CAS:63989-82-2 HR: 3
3,5-DINITRO-p-CRESOL
mf: $C_7H_6N_2O_5$ mw: 198.15

PROP: Crystals or pale-yellow needles from toluene. Mp: 154–155°.

TOXICITY DATA WITH REFERENCE
ipr-pgn LDLo:20 mg/kg AIPTAK 50,20,35

SAFETY PROFILE: Poison by intraperitoneal route. Strong irritant to eyes, skin, and mucous membranes. Can cause brain, liver, and kidney damage by various routes. When heated to decomposition it emits toxic fumes of NO_x. See also 4,6-DINITRO-o-CRESOL.

DUT600 CAS:609-93-8 HR: 3
2,6-DINITRO-p-CRESOL
mf: $C_7H_6N_2O_5$ mw: 198.15

PROP: Yellow needles from pet ether. Mp: 84°.

SYNS: DINITRO-p-CRESOL ☐ DNPC ☐ VICTORIA ORANGE ☐ VICTORIA YELLOW

TOXICITY DATA WITH REFERENCE
mmo-sat 1 nmol/plate MUREAV 58,1,78
ipr-mus LD50:24.8 mg/kg JPPMAB 5,497,53

CONSENSUS REPORTS: Reported in EPA TSCA Inventory.

SAFETY PROFILE: Poison by intraperitoneal route. Mutation data reported. When heated to decomposition it emits toxic fumes of NO_x. See also other dinitrocresol entries.

DUT800 CAS:2980-64-5 HR: 3
4,6-DINITRO-o-CRESOL AMMONIUM SALT
mf: $C_7H_6N_2O_5 \cdot H_3N$ mw: 215.19

SYNS: AMMONIUM DNOC ☐ DINOZOL ☐ DINOZOL 50 ☐ DNOC AMMONIUM SALT ☐ ERBITOX ☐ KRESONIT E ☐ KREZAMON ☐ KREZONIT E ☐ 2-METHYL-4,6-DINITROPHENOL, AMMONIUM SALT ☐ SUPERELGETOL

TOXICITY DATA WITH REFERENCE
cyt-hmn:leu 20 µg/L EESADV 2,243,78
cyt-mus-orl 20 mg/kg EESADV 5,38,81
cyt-mus-ipr 10 mg/kg PHABDI 18,77,78
dlt-mus-ipr 10 mg/kg EESADV 2,401,78
spm-mus-ipr 10 mg/kg EESADV 2,243,78
orl-mus TDLo:20 mg/kg (9-12D preg):TER EESADV 5,38,81
orl-rat LDLo:50 mg/kg JPPMAB 4,1062,52
scu-rat LD50:27,500 µg/kg JPPMAB 4,1062,52

SAFETY PROFILE: Poison by ingestion and subcutaneous routes. An experimental teratogen. Human mutation data reported. An herbicide. When heated to decomposition it emits toxic fumes of NO_x and NH_3. See also other dinitrocresol entries.

DUU000 HR: 3
4,6-DINITRO-o-CRESOL DIETHYLAMINE SALT
mf: $C_7H_6N_2O_4 \cdot C_4H_{11}N$ mw: 255.31

TOXICITY DATA WITH REFERENCE
orl-rat LDLo:50 mg/kg JPPMAB 4,1062,52
scu-rat LD50:36,500 µg/kg JPPMAB 4,1062,52

SAFETY PROFILE: Poison by ingestion and subcutaneous routes. When heated to decomposition it emits toxic fumes of NO_x. See also other dinitrocresol entries.

DUU200 CAS:63989-84-4 HR: 3
4,6-DINITRO-o-CRESOL METHYLAMINE (1:1)

TOXICITY DATA WITH REFERENCE
ipr-mus LDLo:31 mg/kg CBCCT* 6,146,54

NIOSH REL: (Dinitro-Ortho-Cresol) TWA 0.2 mg/m^3

SAFETY PROFILE: Poison by intraperitoneal route. When heated to decomposition it emits toxic fumes of NO_x. See also NITRO COMPOUNDS of AROMATIC HYDROCARBONS and other dinitrocresol entries.

DUU400 CAS:63989-85-5 HR: 3
4,6-DINITRO-o-CRESOL MORPHOLINE (1:1)
mf: $C_7H_6N_2O_5 \cdot C_4H_9NO$ mw: 285.29

TOXICITY DATA WITH REFERENCE
ipr-mus LDLo:25 mg/kg CBCCT* 6,146,54

NIOSH REL: (Dinitro-Ortho-Cresol) TWA 0.2 mg/m^3

SAFETY PROFILE: Poison by intraperitoneal route. When heated to decomposition it emits toxic fumes of NO_x. See also NITRO COMPOUNDS of AROMATIC HYDROCARBONS and other dinitrocresol entries.

DUU600 CAS:2312-76-7 HR: 3
4,6-DINITRO-o-CRESOL SODIUM SALT
mf: $C_7H_5N_2O_5 \cdot Na$ mw: 220.13

PROP: Brilliant, orange-yellow dye.

SYNS: CORODINOC □ CRESOTOL □ DINITRO-o-CRESOL SODIUM SALT □ 3,5-DINITRO-o-CRESOL SODIUM SALT □ 2,4-DINITRO-6-METHYLPHENOL SODIUM SALT □ DINOC □ DNOC SODIUM SALT □ DYNOSOL □ EK 54 □ ELGETOL □ KRENITE (OBS.) □ KREZONITE □ 2-METHYL-4,6-DINITROPHENOL SODIUM SALT □ SINOX □ SODIUM-4,6-DINITRO-o-CRESOXIDE □ SODIUM SALT of 4,6-DINITRO-o-CRESOL

TOXICITY DATA WITH REFERENCE
orl-rat LD50:26 mg/kg SPEADM 74-1,-,74
skn-rat LD50:200 mg/kg SPEADM 74-1,-,74
scu-rat LDLo:20 mg/kg JPETAB 76,245,42
orl-dom LD50:200 mg/kg 85GYAZ -,75,71
orl-mam LD50:200 mg/kg GUCHAZ 6,243,73

CONSENSUS REPORTS: Reported in EPA TSCA Inventory.

NIOSH REL: (Dinitro-Ortho-Cresol) TWA 0.2 mg/m^3

SAFETY PROFILE: Poison by ingestion, skin contact, and subcutaneous routes. Flammable. A pesticide. When heated to decomposition it emits toxic fumes of Na_2O. See also other dinitrocresol entries.

DUU800 CAS:505-71-5 HR: 2
N,N′-DINITRO-1,2-DIAMINOETHANE
mf: $C_2H_6N_4O_4$ mw: 150.10

$(O_2NNHCH_2-)_2$

PROP: Crystals from H_2O. Mp: 174.5–176°.

SYNS: N,N′-DINITROETHYLENEDIAMINE □ N,N′-DINITROETHANEDIAMINE □ ETHANEDIAMINE, N,N′-DINITRO-(9CI) □ ETHYLENEDINITRAMINE □ ETHYLENEDINITROAMINE □ HALEITE

TOXICITY DATA WITH REFERENCE
ipr-mus LD50:540 mg/kg PCJOAU 10,1504,76

CONSENSUS REPORTS: Reported in EPA TSCA Inventory.

SAFETY PROFILE: Moderately toxic by intraperitoneal route. A relatively insensitive explosive. Decomposes violently at 202°C. Forms very impact-sensitive lead and silver salts. When heated to decomposition it emits toxic fumes of NO_x. See also NITRO COMPOUNDS.

DUV100 CAS:961-68-2 HR: 3
2,4-DINITRODIPHENYLAMINE
mf: $C_{12}H_9N_3O_4$ mw: 259.24

SYN: DIPHENYLAMINE, 2,4-DINITRO-

TOXICITY DATA WITH REFERENCE
ivn-mus LD50:180 mg/kg CSLNX* NX#06394

CONSENSUS REPORTS: Reported in EPA TSCA Inventory.

SAFETY PROFILE: Poison by intravenous route. When heated to decomposition it emits toxic vapors of NO_x.

DUV400 CAS:52129-71-2 HR: 2
3′,5′-DINITRO-4′-(DI-n-PROPYLAMINO)ACETOPHENONE
mf: $C_{14}H_{19}N_3O_5$ mw: 309.36

SYN: BUBAN 37

TOXICITY DATA WITH REFERENCE
orl-rat LD50:750 mg/kg FMCHA2 -,C38,83

CONSENSUS REPORTS: Reported in EPA TSCA Inventory.

SAFETY PROFILE: Moderately toxic by ingestion. When heated to decomposition it emits toxic fumes of NO_x.

DUV600 CAS:1582-09-8 HR: 2
2,6-DINITRO-N,N-DIPROPYL-4-(TRIFLUOROMETHYL) BENZENAMINE
mf: $C_{13}H_{16}F_3N_3O_4$ mw: 335.32

PROP: A solid. Mp: 48.5–49°. Very sol in Me_2CO and xylene; very sltly sol in H_2O. Technical product contains 84–88 ppm dipropylnitrosoamine NCITR* NCI-CG-TR-34,78.

SYNS: AGREFLAN □ AGRIFLAN 24 □ CRISALIN □ DIGERMIN □ 2,6-DINITRO-N,N-DI-N-PROPYL-α,α,α-TRIFLURO-p-TOLUIDINE □ 2,6-DINITRO-4-TRIFLUORMETHYL-N,N-DIPROPYLANILIN (GERMAN) □ 4-(DI-N-PROPYLAMINO)-3,5-DINITRO-1-TRIFLUOROMETHYLBENZENE □ N,N-DI-N-PROPYL-2,6-DINITRO-4-TRIFLUOROMETHYLANILINE □ N,N-DIPROPYL-4-TRIFLUOROMETHYL-2,6-DINITROANILINE □ ELANCOLAN □ L-36352 □ LILLY 36,352 □ M.T.F. □ NCI-C00442 □ NITRAN □ OLITREF □ SUPER-TREFLAN □ SU SEGURO CARPIDOR □ SYNFLORAN □ TREFANOCIDE □ TREFICON □ TREFLAM □ TREFLAN □ TREFLANOCIDE ELANCOLAN □ TRI-4 □ TRIFLORAN □ TRIFLUORALIN (USDA) □ α,α,α-TRIFLUORO-2,6-DINITRO-N,N-DIPROPYL-p-TOLUIDINE □ 4-(TRIFLUOROMETHYL)-2,6-DINITRO-N,N-DIPROPYLANILINE □ TRIFLURALIN □ TRIFLURALINA 600 □ TRIFLURALINE □ TRIFUREX □ TRIKEPIN □ TRIM □ TRISTAR

TOXICITY DATA WITH REFERENCE
mma-sat 1 mg/plate ENMUDM 8(Suppl 7),1,86
mrc-asn 100 μg/plate AISSAW 18,123,82
cyt-hmn:lym 2 ppm PATHAB 73,707,81
sce-hmn:lym 1 mg/L BSIBAC 60,2149,84
cyt-mus-ipr 200 mg/kg EESADV 4,263,80
orl-mus TDLo:10 mg/kg (6-15D preg):TER TJADAB 15,15A,77
ipr-mus TDLo:200 mg/kg (1D male):TER EESADV 4,263,80
ipr-mus TDLo:200 mg/kg (1D male):REP EESADV 4,263,80
orl-mus TDLo:180 g/kg/78W-C:CAR NCITR* NCI-CG-TR-34,78
ipr-mus TDLo:2600 μg/kg/39D-I:ETA PATHAB 73,707,81
scu-mus TDLo:2600 μg/kg/39D-I:ETA PATHAB 73,707,81
orl-mus TD:340 g/kg/78W-C:CAR NCITR* NCI-CG-TR-34,78
orl-rat LD50:1930 mg/kg FCTOD7 30,1031,92
ihl-rat LC50:2800 mg/m^3/1H NNGADV 16,557,91
skn-rat LD50:>5 g/kg WRPCA2 9,119,70
orl-mus LD50:3197 mg/kg NNGADV 16,557,91
ipr-mus LDLo:1500 mg/kg BECTA6 20,554,78
orl-dog LD50:>2 g/kg PEMNDP 9,851,91
orl-rbt LD50:>2 g/kg PEMNDP 9,851,91

CONSENSUS REPORTS: NCI Carcinogenesis Bioassay

(feed); Clear Evidence: mouse NCITR* NCI-CG-TR-34,78; No Evidence: rat NCITR* NCI-CG-TR-34,78. EPA Genetic Toxicology Program. Community Right-To-Know List.

SAFETY PROFILE: Moderately toxic by ingestion and intraperitoneal routes. Experimental teratogenic and reproductive effects. Questionable carcinogen with experimental carcinogenic and tumorigenic data. Human mutation data reported. When heated to decomposition it emits very toxic fumes of F^- and NO_x. See also FLUORIDES.

DUV700 CAS:69-78-3 HR: 2
2,2′-DINITRO-5,5′-DITHIODIBENZOIC ACID
mf: $C_{14}H_8N_2O_8S_2$ mw: 396.36

SYNS: BENZOIC ACID, 3,3′-DITHIOBIS(6-NITRO- □ 2,2′-DINITRO-5,5′-DITHIODIBENZOESAEURE □ 3,3′-DITHIOBIS(6-NITROBENZOIC ACID)

TOXICITY DATA WITH REFERENCE
ipr-mus LD50:2080 mg/kg ARZNAD 21,284,71

CONSENSUS REPORTS: Reported in EPA TSCA Inventory.

SAFETY PROFILE: Moderately toxic by intraperitoneal route. When heated to decomposition it emits toxic vapors of NO_x and SO_x.

DUV710 CAS:600-40-8 HR: 3
1,1-DINITROETHANE
mf: $C_2H_4N_2O_4$ mw: 120.08

SYNS: 1,1-DINITROETHANE (dry) (DOT) □ ETHANE, 1,1-DINITRO-

TOXICITY DATA WITH REFERENCE
ipr-mus LD50:250 mg/kg KHFZAN 10(6),53,76

DOT CLASSIFICATION: Forbidden

SAFETY PROFILE: A poison by intraperitoneal route. An unstable solid forbidden for transport. When heated to decomposition it emits toxic vapors of NO_x.

DUV720 CAS:7570-26-5 HR: 3
1,2-DINITROETHANE
mf: $C_2H_4N_2O_4$ mw: 120.08

SYN: ETHANE, 1,2-DINITRO-

DOT CLASSIFICATION: Forbidden

SAFETY PROFILE: An unstable solid forbidden for transport. When heated to decomposition it emits toxic vapors of NO_x.

DUW100 CAS:105735-71-5 HR: 2
3,7-DINITROFLUORANTHENE
mf: $C_{16}H_8N_2O_4$ mw: 292.26

TOXICITY DATA WITH REFERENCE
mmo-sat 250 pg/plate MUREAV 191,85,87
mma-sat 1 μg/plate MUREAV 191,85,87
dnr-bcs 10 ng/disc MUREAV 191,85,87
scu-rat TDLo:5 mg/kg/10W-I:CAR CRNGDP 8,1919,87
imp-rat TDLo:800 μg/kg:CAR CRNGDP 12,1003,91

CONSENSUS REPORTS: IARC Cancer Review: Group 3 IMEMDT 46,189,89; Animal Limited Evidence IMEMDT 46,189,89; Human No Adequate Data IMEMDT 46,189,89;

SAFETY PROFILE: Questionable carcinogen with experimental carcinogenic data. Mutation data reported. When heated to decomposition it emits toxic fumes of NO_x.

DUW120 CAS:22506-53-2 HR: 2
3,9-DINITROFLUORANTHENE
mf: $C_{16}H_8N_2O_4$ mw: 292.26

SYN: 4,12-DINITROFLUORANTHENE

TOXICITY DATA WITH REFERENCE
mmo-sat 250 pg/plate MUREAV 191,85,87
mma-sat 1 μg/plate MUREAV 191,85,87
dnr-bcs 10 ng/disc MUREAV 191,85,87
scu-rat TDLo:5 mg/kg/10W-I:CAR CRNGDP 8,1919,87
imp-rat TDLo:800 μg/kg:CAR CRNGDP 12,1003,91

CONSENSUS REPORTS: IARC Cancer Review: Group 3 IMEMDT 46,195,89; Animal Limited Evidence IMEMDT 46,195,89; Human No Adequate Data IMEMDT 46,195,89.

SAFETY PROFILE: Questionable carcinogen with experimental carcinogenic data. Mutation data reported. When heated to decomposition it emits toxic fumes of NO_x.

DUW200 CAS:5405-53-8 HR: 3
2,7-DINITROFLUORENE
mf: $C_{13}H_8N_2O_4$ mw: 256.23

PROP: Needles from $PhNO_2$. Mp: 334°. Sol in hot AcOH; almost insol in EtOH.

TOXICITY DATA WITH REFERENCE
mmo-sat 10 ng/plate MUREAV 143,213,85
uns-bac-esc 1250 ng/tube EMMUEG 18,41,91
dns-rat:lvr 250 μg/L MUREAV 190,159,87
dns-rat:lvr 100 μmol/L ENMUDM 3,11,81
dns-mus:lvr 2500 μg/L MUREAV 190,159,87
orl-rat TDLo:2700 mg/kg/17W-C:CAR CNREA8 22,1002,62

SAFETY PROFILE: A questionable carcinogen with experimental carcinogenic data. Mutation data reported. A very shock-sensitive explosive which also explodes above its melting point of 152°C. Upon decomposition it emits toxic fumes of NO_x. See also NITRO COMPOUNDS of AROMATIC HYDROCARBONS.

DUW400 CAS:70-34-8 HR: 3
2,4-DINITRO-1-FLUOROBENZENE
mf: $C_6H_3FN_2O_4$ mw: 186.11

PROP: Crystals or oil. Crystals mp: 27°; oil mp: 12°, bp: 137° @ 20 mm. Sol in ether, benzene, and propylene glycol.

SYNS: 2,4-DINITROFLUOROBENZENE □ 2,4-DNFB □ 1-FLUORO-2,4-DINITROBENZENE □ 1,2,4-FLUORODINITROBENZENE

TOXICITY DATA WITH REFERENCE
mmo-sat 500 nmol/L ENMUDM 3,11,81
mma-sat 33 μg/plate ENMUDM 5(Suppl 1),3,83
mmo-esc 5 μmol/L CRNGDP 3,139,82
mrc-smc 320 μmol/L MGGEAE 168,125,79
otr-ham:kdy 80 μg/L BJCAAI 37,873,78
orl-rat LDLo:50 mg/kg NCNSA6 5,17,53
skn-mus LDLo:100 mg/kg CNREA8 29,179,69
scu-mus LDLo:100 mg/kg BIJOAK 41,558,47

CONSENSUS REPORTS: Reported in EPA TSCA Inventory. EPA Genetic Toxicology Program.

SAFETY PROFILE: Poison by ingestion, skin contact, and subcutaneous routes. A powerful irritant and vesicant. Mutation data reported. Solutions in ether may explode when evaporated. When heated to decomposition it emits highly toxic fumes of NO_x and F^-. See also NITRO COMPOUNDS of AROMATIC HYDROCARBONS and FLUORIDES.

DUW500 CAS:119-15-3 HR: 2
2,4-DINITRO-p-HYDROXYDIPHENYLAMINE
mf: $C_{12}H_9N_3O_5$ mw: 275.24

SYNS: ACETAMINE YELLOW 2R □ ACETOQUINONE LIGHT YELLOW 2RZ □ AMACEL YELLOW RR □ CELLITON FAST YELLOW RR □ C.I. 10345 □ C.I. DISPERSE YELLOW 1 □ CILLA FAST YELLOW RR □ C.I. SOLVENT YELLOW 52 □ DISPERSE FAST YELLOW 2K □ DISPERSE YELLOW R □ DISPERSE YELLOW STABLE 2K □ DISPERSOL FAST YELLOW A □ DISPERSOL PRINTING YELLOW A □ DISPERSOL YELLOW B-A □ FAST DISPERSE YELLOW 2K □ FENACET FAST YELLOW 2R □ KAYALON FAST YELLOW RR □ MICROSETILE YELLOW 2R □ NYLOQUINONE YELLOW 2R □ PERLITON YELLOW RR □ PERMANENT YELLOW 2K □ PHENOL, p-(2,4-DINITROANILINO)- □ RELITON YELLOW R □ SERISOL FAST YELLOW A □ SETACYL YELLOW P-BS □ SRA GOLDEN YELLOW VIII □ SUPRACET FAST YELLOW 2R □ SUPRACET YELLOW RR □ SYNTEN YELLOW P 2R

TOXICITY DATA WITH REFERENCE
mmo-sat 33 μg/plate EMMUEG 11(Suppl 12),1,88
orl-rat LD50:>5 g/kg JSCCA5 23,259,72
ipr-rat LD50:5230 mg/kg GISAAA 52(11),94,87
orl-mus LD50:6550 mg/kg GISAAA 52(11),94,87
ipr-uns LD50:2500 mg/kg GISAAA 40(10),114,75

CONSENSUS REPORTS: Reported in EPA TSCA Inventory.

SAFETY PROFILE: Moderately toxic by intraperitoneal route. Mildly toxic by ingestion. Mutation data reported. When heated to decomposition it emits toxic vapors of NO_x.

DUW503 CAS:2536-18-7 HR: 3
1,3-DINITRO-2-IMIDAZOLIDONE
mf: $C_3H_4N_4O_5$ mw: 176.11

SYNS: 1,3-DINITRO-2-IMIDAZOLIDINONE □ 2-IMIDAZOLIDINONE, 1,3-DINITRO-

TOXICITY DATA WITH REFERENCE
ivn-mus LD50:18 mg/kg CSLNX* NX#04408

CONSENSUS REPORTS: Reported in EPA TSCA Inventory.

SAFETY PROFILE: Poison by intravenous route. When heated to decomposition it emits toxic vapors of NO_x.

DUW505 CAS:608-50-4 HR: 3
2,4-DINITROMESITYLENE
mf: $C_9H_{10}N_2O_4$ mw: 210.21

SYNS: BENZENE, 1,3,5-TRIMETHYL-2,4-DINITRO- □ 2,4-DINITRO-1,3,5-TRIMETHYLBENZENE (DOT) □ MESITYLENE, 2,4-DINITRO-

DOT CLASSIFICATION: Forbidden

SAFETY PROFILE: An unstable solid forbidden for transport. When heated to decomposition it emits toxic vapors of NO_x.

DUW507 CAS:625-76-3 HR: 3
DINITROMETHANE
mf: $CH_2N_2O_4$ mw: 106.05

SYN: METHANE, DINITRO-

DOT CLASSIFICATION: Forbidden

SAFETY PROFILE: An unstable substance forbidden for transport. When heated to decomposition it emits toxic vapors of NO_x and Cl^-.

DUX509 HR: 3
3,5-DINITRO-2-METHYLBENZENEDIAZONIUM-4-OXIDE
mf: $C_7H_4N_4O_5$ mw: 224.13

SAFETY PROFILE: A very shock-sensitive explosive. When heated to decomposition it emits toxic fumes of NO_x. See also EXPLOSIVES, HIGH.

DUX560 CAS:70343-15-6 HR: 3
2,5-DINITRO-3-METHYLBENZOIC ACID
mf: $C_8H_6N_2O_6$ mw: 226.15

PROP: Straw colored prisms from MeOH (aq). Mp: 180–181°.

SAFETY PROFILE: Mixtures with oleum + sodium azide are potentially explosive. When heated to decomposition it emits toxic fumes of NO_x. See also NITRO COMPOUNDS of AROMATIC HYDROCARBONS.

DUX600 CAS:10308-90-4 HR: 3
N,N′-DINITRO-N-METHYL-1,2-DIAMINOETHANE
mf: $C_3H_8N_4O_4$ mw: 164.14

$CH_3N(NO_2)C_2H_4NHNO_2$

PROP: Crystals. Mp: 120.5–121.8°.

SAFETY PROFILE: Decomposes violently at 210°C. When heated to decomposition it emits toxic fumes of NO_x.

D

DUX700 **CAS:605-71-0** **HR: 3**
1,5-DINITRONAPHTHALENE
mf: $C_{10}H_6N_2O_4$ mw: 218.17

PROP: Needles from AcOH or Me_2CO. Mp: 218°. Sol in hot C_6H_6 and Py.

SYN: 1,5-DINITRONAPHTHALENE

TOXICITY DATA WITH REFERENCE
mmo-sat 50 μg/plate MUREAV 91,321,81
uns-bac-esc 5 μg/tube EMMUEG 18,41,91
pic-esc 15,500 ng/well MUREAV 260,349,91

DFG MAK: Suspected Carcinogen (all isomers)

SAFETY PROFILE: A suspected carcinogen. Mutation data reported. Mixtures with sulfur or sulfuric acid (used in commercial reactions) may explode if heated to 120°C. Initiation temperature depends on the quality of the dinitronaphthalene. When heated to decomposition it emits toxic fumes of NO_x. See also NITRO COMPOUNDS of AROMATIC HYDROCARBONS.

DUX800 **CAS:605-69-6** **HR: 3**
2,4-DINITRO-1-NAPHTHOL
mf: $C_{10}H_6N_2O_5$ mw: 234.18

PROP: Yellow needles or leaflets from EtOH or $CHCl_3$. Mp: 140°, vap d: 8.08. Sltly sol in Et_2O, EtOH, C_6H_6, and H_2O.

SYNS: C.I. 10315 □ 2,4-DINITRO-1-NAFTOL □ 2-4 DINITRO-α-NAPHTOL □ 2-4 DINITRO-α-NAPHTOL (FRENCH) □ GOLDEN YELLOW □ MANCHESTER YELLOW □ MARITUS YELLOW □ NAPHTHOL YELLOW □ NAPHTHYLENE YELLOW □ SAFFRON YELLOW □ ZLUT MARCIOVA □ ZLUT NAFTOLOVA

TOXICITY DATA WITH REFERENCE
mmo-sat 3300 ng/plate ENMUDM 3,499,81
mmo-sat 50 μg/plate GDIKAN 29,278,81
skn-hmn TDLo:50 mg/kg:SKN XPHBAO 271,187,41
ivn-mus LD50:180 mg/kg CSLNX* NX#03278
ivn-dog LDLo:13,300 μg/kg AIPTAK 35,63,28
scu-gpg LDLo:80 mg/kg HBTXAC 1,118,56
ipr-pgn LDLo:15 mg/kg HBTXAC 1,118,56
ivn-pgn LDLo:15 mg/kg AIPTAK 50,20,35
ims-pgn LD50:1850 μg/kg HBTXAC 1,118,56
scu-frg LDLo:60 mg/kg HBTXAC 1,118,56

CONSENSUS REPORTS: Reported in EPA TSCA Inventory.

SAFETY PROFILE: Poison by subcutaneous, intramuscular, intravenous, and intraperitoneal routes. Human reproductive effects by skin contact: toxic to the skin. Mutation data reported. For fire, disaster, and explosion hazards, see NITRATES.

DUY200 **HR: 3**
2,6-DINITRO-4-PERCHLORYLPHENOL
mf: $C_6H_3ClN_2O_8$ mw: 266.56

SAFETY PROFILE: A very shock-sensitive explosive. An analog of picric acid. When heated to decomposition it emits toxic fumes of Cl^- and NO_x. See also PICRIC ACID and PERCHLORATES.

DUY400 **CAS:610-54-8** **HR: 3**
2,4-DINITROPHENETOLE
mf: $C_8H_8N_2O_5$ mw: 212.18

PROP: Crystals. Mp: 87°. Vap d: 7.32.

TOXICITY DATA WITH REFERENCE
mmo-sat 500 nmol/L ENMUDM 3,11,81
orl-rat LDLo:250 mg/kg NCNSA6 5,16,53

CONSENSUS REPORTS: Reported in EPA TSCA Inventory.

SAFETY PROFILE: Poison by ingestion. Mutation data reported. When heated to decomposition it emits toxic fumes of NO_x.

DUY600 **CAS:25550-58-7** **HR: 3**
DINITROPHENOL
DOT: UN 0076/UN 1320/UN 1599
mf: $C_6H_4N_2O_5$ mw: 184.12

SYNS: DINITROPHENOL □ DINITROPHENOL, dry or wetted with <15% water, by weight (UN 0076) (DOT) □ DINITROPHENOL, wetted with not <15% water, by weight (UN 1320) (DOT) □ DINITROPHENOL SOLUTIONS (UN 1599) (DOT)

TOXICITY DATA WITH REFERENCE
orl-rat LDLo:30 mg/kg 28ZEAL 4,198,69
orl-dog LDLo:30 mg/kg JPETAB 49,187,33
scu-rbt LDLo:30 mg/kg JPETAB 49,187,33

DOT CLASSIFICATION: EXPLOSIVE 1.1D; *Label:* EXPLOSIVE 1.1D, Poison (UN 076); DOT Class: 4.1; *Label:* Flammable Solid, Poison (UN 1320); DOT Class: 6.1; *Label:* Poison (UN 1599)

SAFETY PROFILE: Poison by ingestion and subcutaneous routes. An explosive and flammable solid. When heated to decomposition it emits toxic fumes of NO_x. See also NITRO COMPOUNDS of AROMATIC HYDROCARBONS.

DUY900 **CAS:66-56-8** **HR: 3**
2,3-DINITROPHENOL
mf: $C_6H_4N_2O_5$ mw: 184.12

PROP: Yellow needles or crystals. Mp: 144°, d: 1.681 @ 20°, vap d: 6.35. Sol in EtOH, Et_2O; sltly sol in H_2O.

SYNS: 2,3-DINITROFENOL □ 2,3-DINITROPHENOL

TOXICITY DATA WITH REFERENCE
mmo-sat 250 μg/plate SAIGBL 29,34,87
ipr-rat LD50:190 mg/kg JPPMAB 11,462,59
ipr-mus LD50:200 mg/kg JPPMAB 11,462,59
ipr-dog LDLo:1 g/kg JPPMAB 11,462,59

SAFETY PROFILE: Poison by intraperitoneal route. Inhalation of dust can be fatal. A skin irritant and an allergen. Mutation data reported. A powerful stimulant of the metabolism by excessive oxidation. For fire hazard, see NITRATES. Highly explosive when exposed to heat. It is used as a component of some shell and bomb charges. See also NITRO COMPOUNDS of AROMATIC HYDROCARBONS; and EXPLOSIVES, HIGH.

DUZ000 **CAS:51-28-5** **HR: 3**
2,4-DINITROPHENOL
mf: $C_6H_4N_2O_5$ mw: 184.12

PROP: Yellow crystals or plates from water. Mp: 113°, d: 1.683 @ 24°, vap d: 6.35. Sol in EtOH, Me_2CO, and C_6H_6; sltly sol in H_2O.

SYNS: ALDIFEN □ CHEMOX PE □ 2,4-DINITROFENOL (DUTCH) □ DINITROFENOLO (ITALIAN) □ α-DINITROPHENOL □ 2,4-DNP □ FENOXYL CARBON N □ 1-HYDROXY-2,4-DINITROBENZENE □ MAROXOL-50 □ NITRO KLEENUP □ NSC 1532 □ RCRA WASTE NUMBER P048 □ SOLFO BLACK B □ SOLFO BLACK BB □ SOLFO BLACK 2B SUPRA □ SOLFO BLACK G □ SOLFO BLACK SB □ TERTROSULPHUR BLACK PB □ TERTROSULPHUR PBR

TOXICITY DATA WITH REFERENCE
skn-rbt 300 mg/4W-I MLD JIHTAB 30,10,48
oms-ofs:oth 100 μmol/L AEEXAH (3),279,72
cyt-mus-ipr 10 g/kg IJMRAQ 59,1442,71
orl-rat TDLo:2040 mg/kg (8D pre-21D post):REP PSEBAA 32,678,35
ipr-mus TDLo:40,800 μg/kg (10-12D preg):TER FCTXAV 11,31,73
orl-hmn LDLo:36 mg/kg:BAH,CVS JAMAAP 101,1333,33
orl-rat LD50:30 mg/kg TXAPA9 21,315,72
ipr-rat LD50:20 mg/kg JPPMAB 17,814,65
scu-rat LD50:25 mg/kg JPETAB 49,187,33
orl-mus LD50:45 mg/kg FATOAO 28,493,65
ipr-mus LD50:26 mg/kg BCPCA6 18,1389,69
orl-dog LDLo:30 mg/kg JPETAB 49,187,33
ihl-dog LCLo:300 mg/m³/30M 85GMAT -,62,82
scu-gpg LDLo:25 mg/kg AEPPAE 192,331,39

CONSENSUS REPORTS: Reported in EPA TSCA Inventory. EPA Genetic Toxicology Program.

SAFETY PROFILE: A deadly human poison by ingestion. An experimental poison by ingestion, inhalation, intravenous, intraperitoneal, subcutaneous, and intramuscular routes. Moderately toxic by skin contact. Experimental teratogenic and reproductive effects. Human systemic effects: body temperature increase, change in heart rate, coma. A skin irritant. Mutation data reported. Phytotoxic. A pesticide. An explosive. Forms explosive salts with alkalies and ammonia. When heated to decomposition it emits toxic fumes of NO_x. See also NITRO COMPOUNDS of AROMATIC HYDROCARBONS.

DVA000 **CAS:329-71-5** **HR: 3**
2,5-DINITROPHENOL
mf: $C_6H_4N_2O_5$ mw: 184.12

PROP: Yellow crystals or needles. Mp: 108°. Sltly sol in cold water and alc; sol in hot alc, ether, and alkali hydroxides.

SYNS: γ-DINITROPHENOL □ 2,5-DNP

TOXICITY DATA WITH REFERENCE
mmo-sat 100 μg/plate SAIGBL 29,34,87
ipr-rat LD50:150 mg/kg JPPMAB 11,462,59
ipr-mus LD50:273 mg/kg JPPMAB 11,462,59
ipr-dog LDLo:100 mg/kg JPPMAB 11,462,59

SAFETY PROFILE: A poison by intraperitoneal route. Mutation data reported. When heated to decomposition it emits toxic fumes of NO_x. See also 2,4-DINITROPHENOL.

DVA200 **CAS:573-56-8** **HR: 3**
2,6-DINITROPHENOL
mf: $C_6H_4N_2O_5$ mw: 184.12

PROP: Yellow crystals from water. Mp: 63°, vap d: 6.35. Sltly sol in cold water, alc. Very sol in chloroform, ether, or boiling alc; also sol in fixed alkali solns.

SYNS: β-DINITROPHENOL □ 2,6-DINITROFENOL

TOXICITY DATA WITH REFERENCE
ipr-rat LD50:38 mg/kg JPPMAB 11,462,59
ipr-mus LD50:45 mg/kg JPPMAB 11,462,59
ipr-dog LDLo:50 mg/kg JPPMAB 11,462,59
ims-pgn LDLo:40 mg/kg JPETAB 49,187,33

SAFETY PROFILE: Poison by intramuscular route. Moderately explosive when exposed to heat. See also 2,4-DINITROPHENOL.

DVA400 **CAS:577-71-9** **HR: 3**
3,4-DINITROPHENOL
mf: $C_6H_4N_2O_5$ mw: 184.12

PROP: Yellowish needles from H_2O. Mp: 134°. Sol in EtOH and Et_2O; sltly sol in H_2O.

SYN: 3,4-DINITROFENOL

TOXICITY DATA WITH REFERENCE
mmo-sat 100 μg/plate SAIGBL 29,34,87
ipr-rat LD50:98 mg/kg JPPMAB 11,462,59
ipr-mus LD50:112 mg/kg JPPMAB 11,462,59
ipr-dog LDLo:500 mg/kg JPPMAB 11,462,59

SAFETY PROFILE: A poison by intraperitoneal route. Mutation data reported. When heated to decomposition it emits toxic fumes of NO_x. See also 2,4-DINITROPHENOL.

DVA600 **CAS:586-11-8** **HR: 3**
3,5-DINITROPHENOL
mf: $C_6H_4N_2O_5$ mw: 184.12

PROP: A solid. Mp: 126°.

TOXICITY DATA WITH REFERENCE
unk-rat LD50:45 mg/kg JPPMAB 11,462,59
unk-mus LD50:50 mg/kg JPPMAB 11,462,59
unk-dog LDLo:500 mg/kg JPPMAB 11,462,59

SAFETY PROFILE: A poison by unspecified routes. When heated to decomposition it emits toxic fumes of NO_x. See also 2,4-DINITROPHENOL.

DVA800 **CAS:1011-73-0** **HR: 3**
2,4-DINITROPHENOL SODIUM SALT
mf: $C_6H_3N_2O_5$•Na mw: 206.10

SYNS: SODIUM-2,4-DINITROPHENOL □ SODIUM-2,4-DINITROPHENOLATE □ SODIUM DNP

TOXICITY DATA WITH REFERENCE
scu-rat TDLo: 8 mg/kg (female 9D post): TER AEHLAU 8,648,64
scu-rat TDLo: 30 mg/kg (female 10D post): REP AEHLAU 8,648,64
scu-rat LDLo: 10 mg/kg JPETAB 48,410,33
scu-mus LD50: 50 mg/kg NYKZAU 56,23,60
scu-dog LDLo: 25 mg/kg JPETAB 48,410,33
ivn-dog LDLo: 20 mg/kg AIPTAK 50,20,35
ivn-pgn LDLo: 15 mg/kg AIPTAK 50,20,35

SAFETY PROFILE: Poison by subcutaneous and intravenous routes. An experimental teratogen. Other experimental reproductive effects. When heated to decomposition it emits toxic fumes of NO_x and Na_2O. See also 2,4-DINITROPHENOL.

DVB200 **HR: 2**
2,4-DINITROPHENYLACETYL CHLORIDE
mf: $C_8H_5ClN_2O_5$ mw: 244.60

SAFETY PROFILE: Potentially explosive when heated. When heated to decomposition it emits toxic fumes of Cl^- and NO_x. See also NITRO COMPOUNDS of AROMATIC HYDROCARBONS.

DVB850 **CAS:89-37-2** **HR: 3**
2,4-DINITROPHENYL-DIMETHYL-DITHIOCARBAMATE
mf: $C_9H_9N_3O_4S_2$ mw: 287.33

SYNS: CARBAMIC ACID, DIMETHYLDITHIO-, 2,4-DINITROPHENYL ESTER □ USAF SN-31

TOXICITY DATA WITH REFERENCE
ipr-mus LD50: 200 mg/kg NTIS** AD277-689

CONSENSUS REPORTS: Reported in EPA TSCA Inventory.

SAFETY PROFILE: Poison by intraperitoneal route. When heated to decomposition it emits toxic vapors of NO_x and SO_x.

DVC200 **CAS:2600-55-7** **HR: 3**
2,4-DINITROPHENYL-2,4-DINITRO-6-sec-BUTYLPHENYL CARBONATE
mf: $C_{17}H_{14}N_4O_{11}$ mw: 450.35

SYNS: B 377 □ CARBONIC ACID-2-sec-BUTYL-4,6-DINITROPHENYL-2,4-DINITROPHENYL ESTER (8CI) □ TRIBONATE

TOXICITY DATA WITH REFERENCE
orl-rat LD50: 108 mg/kg WRPCA2 7,135,68

SAFETY PROFILE: Poison by ingestion. When heated to decomposition it emits toxic fumes of NO_x. See also NITRO COMPOUNDS of AROMATIC HYDROCARBONS.

DVC400 **CAS:119-26-6** **HR: 3**
2,4-DINITROPHENYLHYDRAZINE
mf: $C_6H_6N_4O_4$ mw: 198.16

PROP: Blue-red crystalline powder with violet fluorescence. Mp: 198° (decomp). Sltly sol in water and alcohol.

SYNS: 2,4-DINITROFENYLHYDRAZIN (CZECH) □ 2,4-DNPH

TOXICITY DATA WITH REFERENCE
eye-rbt 500 mg/24H MOD 28ZPAK -,132,72
mmo-sat 5 μmol/L ENMUDM 3,11,81
mma-sat 1 μmol/plate MUREAV 58,11,78
mmo-omi 6 mg/L MUREAV 173,233,86
dnd-mus-ipr 1900 μmol/kg CNREA8 41,1469,81
orl-rat LD50: 654 mg/kg 28ZPAK -,132,72
ipr-mus LD50: 450 mg/kg CNREA8 41,1469,81

CONSENSUS REPORTS: Reported in EPA TSCA Inventory. EPA Genetic Toxicology Program.

SAFETY PROFILE: Moderately toxic by ingestion and intraperitoneal routes. An eye irritant. Mutation data reported. When heated to decomposition it emits toxic fumes of NO_x. See also HYDRAZINE. A dangerous explosive.

DVC600 **HR: 3**
2,4-DINITROPHENYLHYDRAZINIUMPERCHLORATE
mf: $C_6H_7ClN_4O_8$ mw: 219.48

SAFETY PROFILE: Explosive decomposition may occur during concentration by evaporation. When heated to decomposition it emits toxic fumes of Cl^- and NO_x. See also NITRO COMPOUNDS of AROMATIC HYDROCARBONS.

DVC700 **CAS:17508-17-7** **HR: 3**
o-(2,4-DINITROPHENYL)HYDROXYLAMINE
mf: $C_6H_5N_3O_5$ mw: 199.12

PROP: Pale-yellow needles from EtOH. Mp: 112–113°.

SAFETY PROFILE: Potentially explosive reaction with potassium hydride in THF solution. When heated to decomposition it emits toxic fumes of NO_x. See also NITRO COMPOUNDS of AROMATIC HYDROCARBONS.

DVC800 **CAS:63732-56-9** **HR: 3**
2,4-DINITROPHENYLMORPHINE HYDROCHLORIDE
mf: $C_{23}H_{21}N_3O_7•ClH$ mw: 487.93

SYN: 2,4-DINITROPHENYL ETHER of MORPHINE

TOXICITY DATA WITH REFERENCE
ipr-rat LDLo: 40 mg/kg UCPHAQ 1,59,38
ipr-mus LDLo: 300 mg/kg UCPHAQ 1,59,38
scu-mus LD50: 700 mg/kg JPETAB 67,127,39
scu-rbt LDLo: 100 mg/kg UCPHAQ 1,59,38
ivn-rbt LDLo: 3 mg/kg UCPHAQ 1,59,38
par-frg LDLo: 2000 mg/kg UCPHAQ 1,59,38

SAFETY PROFILE: Poison by intravenous, intraperitoneal, and subcutaneous routes. Moderately toxic by parenteral route. When heated to decomposition it emits very toxic fumes of HCl and NO_x. See also MORPHINE and ETHERS.

DVD000 CAS:2736-80-3 HR: 3
2,2-DINITRO-1,3-PROPANEDIOL
mf: $C_3H_6N_2O_6$ mw: 166.11

PROP: A solid. Mp: 140–142°.

TOXICITY DATA WITH REFERENCE
ipr-mus LD50:76 mg/kg KHFZAN 11(1),73,77

CONSENSUS REPORTS: Reported in EPA TSCA Inventory.

SAFETY PROFILE: Poison by intraperitoneal route. When heated to decomposition it emits toxic fumes of NO_x.

DVD200 CAS:918-52-5 HR: 3
2,2-DINITROPROPANOL
mf: $C_3H_6N_2O_5$ mw: 150.11

SYNS: 2,2-DINITRO-1-PROPANOL □ DNPOH □ NPOH

TOXICITY DATA WITH REFERENCE
ipr-mus LD50:280 mg/kg KHFZAN 11(1),73,77

CONSENSUS REPORTS: Reported in EPA TSCA Inventory.

SAFETY PROFILE: Poison by intraperitoneal route. When heated to decomposition it emits toxic fumes of NO_x.

DVD400 CAS:75321-20-9 HR: 3
1,3-DINITROPYRENE
mf: $C_{16}H_8N_2O_4$ mw: 292.26

PROP: Light-brown needles from C_6H_6/MeOH. Mp: 274–276°.

SYN: DINITROPYRENE

TOXICITY DATA WITH REFERENCE
mmo-esc 80 ng/plate MUREAV 142,163,85
msc-ham:lng 2500 µg/L CRNGDP 3,917,82
msc-ham:ovr 500 µg/L MUREAV 119,387,83
ipr-rat TDLo:23 mg/kg/4W-I:CAR DTESD7 13,279,86
scu-rat TDLo:16 mg/kg/10W-I:CAR CRNGDP 5,583,84
ipr-rat TD:35,071 µg/kg/4W-I:ETA CRNGDP 12,1187,91

CONSENSUS REPORTS: IARC Cancer Review: Group 3 IMEMDT 46,201,89; Animal Limited Evidence IMEMDT 46,201,89; Human No Adequate Data IMEMDT 46,201,89.

DFG MAK: Suspected Carcinogen

SAFETY PROFILE: Suspected carcinogen with experimental carcinogenic data. Mutation data reported. When heated to decomposition it emits toxic fumes of NO_x.

DVD600 CAS:42397-64-8 HR: 3
1,6-DINITROPYRENE
mf: $C_{16}H_8N_2O_4$ mw: 292.26

PROP: Light-brown needles from C_6H_6/MeOH.

SYN: DINITROPYRENE

TOXICITY DATA WITH REFERENCE
mmo-sat 600 pg/plate JJIND8 73,1359,84
dns-hmn:oth 500 nmol/L TXAPA9 79,28,85
dns-hmn:lvr 80 nmol/L ENMUDM 5,488,83
ipr-rat TDLo:23 mg/kg/4W-I:CAR DTESD7 13,279,86
scu-rat TDLo:9206 µg/kg/8W-I:CAR DTESD7 13,279,86
ipl-rat TDLo:600 µg/kg:CAR JJIND8 76,693,86
scu-mus TDLo:80 mg/kg/20W-I:CAR DTESD7 13,253,86
itr-ham TDLo:104 mg/kg/26W-I:CAR JJCREP 76,457,85
scu-rat TD:16 mg/kg/10W-I:CAR CALEDQ 25,239,85

CONSENSUS REPORTS: IARC Cancer Review: Group 2B IMEMDT 46,215,89; Animal Sufficient Evidence IMEMDT 46,215,89; Human No Adequate Data IMEMDT 46,215,89.

DFG MAK: Suspected Carcinogen

SAFETY PROFILE: Confirmed carcinogen with experimental carcinogenic data. Human mutation data reported. When heated to decomposition it emits toxic fumes of NO_x.

DVD800 CAS:42397-65-9 HR: 3
1,8-DINITROPYRENE
mf: $C_{16}H_8N_2O_4$ mw: 292.26

PROP: Light-brown needles from C_6H_6/MeOH.

SYN: DINITROPYRENE

TOXICITY DATA WITH REFERENCE
mmo-sat 1 nmol/plate SCIEAS 209,1039,80
mma-sat 1 µg/plate MUREAV 91,321,81
msc-hmn:lym 100 µg/L ENMUDM 5,457,83
msc-mus:lym 500 µg/L EVSRBT 25,397,82
ipr-rat TDLo:23 mg/kg/4W-I:CAR DTESD7 13,279,86
scu-rat TDLo:16 mg/kg/10W-I:CAR CRNGDP 5,583,84
scu-mus TDLo:40 mg/kg/20W-I:CAR DTESD7 13,253,86
scu-rat TD:160 µg/kg/10W-I:CAR CALEDQ 25,239,85
scu-mus TD:40 mg/kg/20W-I:CAR JJIND8 79,185,87
ipr-rat TD:35,071 µg/kg/4W-I:CAR CRNGDP 12,1187,91

CONSENSUS REPORTS: IARC Cancer Review: Group 2B IMEMDT 46,231,89; Animal Sufficient Evidence IMEMDT 46,231,89; Human No Adequate Data IMEMDT 46,231,89.

DFG MAK: Suspected Carcinogen

SAFETY PROFILE: Confirmed carcinogen with experimental carcinogenic data. Human mutation data reported. When heated to decomposition it emits toxic fumes of NO_x.

DVE000 CAS:1596-52-7 HR: 2
4,6-DINITROQUINOLINE-1-OXIDE
mf: $C_9H_5N_3O_5$ mw: 235.17

TOXICITY DATA WITH REFERENCE
cyt-omi 170 µmol/L GANNA2 60,155,69
mmo-smc 100 mg/L IGSBAL 85,127,72
scu-mus TDLo:560 mg/kg/I:ETA CPBTAL 17,544,69

SAFETY PROFILE: Questionable carcinogen with experimental carcinogenic data. Mutation data reported. When heated to decomposition it emits toxic fumes of NO_x.

D

DVE200 **CAS:13442-17-6** **HR: 2**
4,7-DINITROQUINOLINE-1-OXIDE
mf: $C_9H_5N_3O_5$ mw: 235.17

TOXICITY DATA WITH REFERENCE
skn-mus TDLo:300 mg/kg/25W-I:NEO GANNA2 60,523,69

SAFETY PROFILE: Questionable carcinogen with experimental neoplastigenic data. When heated to decomposition it emits toxic fumes of NO_x.

DVE260 **CAS:105-12-4** **HR: 3**
p-DINITROSOBENZENE
mf: $C_6H_4N_2O_2$ mw: 136.12

SYNS: BENZENE, p-DINITROSO- □ BENZENE, 1,4-DINITROSO-(9CI) □ 1,4-DINITROSOBENZENE

TOXICITY DATA WITH REFERENCE
orl-rat LD50:1020 mg/kg KCRZAE (11),38,85
ihl-rat LCLo:200 mg/m³/2H KCRZAE (11),38,85
orl-mus LD50:1230 mg/kg KCRZAE (11),38,85
ihl-mus LCLo:200 mg/m³/2H KCRZAE (11),38,85

CONSENSUS REPORTS: Reported in EPA TSCA Inventory.

SAFETY PROFILE: Poison by inhalation route. Moderately toxic by ingestion. When heated to decomposition it emits toxic vapors of NO_x.

DVE400 **CAS:13256-12-7** **HR: 3**
N,N′-DINITROSO-N,N′-DIMETHYLETHYLENEDIAMINE
mf: $C_4H_{10}N_4O_2$ mw: 146.18

SYNS: DIMETHYL-DI-NITROSO-AETHYLENDIAMIN (GERMAN) □ DIMETHYLDINITROSOETHYLENEDIAMINE □ N,N′-DIMETHYL-N,N′-DINITROSOETHYLENEDIAMINE

TOXICITY DATA WITH REFERENCE
mmo-omi 5000 ppm/24H-C SOGEBZ 10,522,74
orl-rat TDLo:500 mg/kg/17W-I:ETA GISAAA 39(9),80,74
orl-rat TDLo:570 mg/kg/43W-C:ETA ARZNAD 19,1077,69
orl-rat LD50:125 mg/kg GISAAA 39(9),80,74
orl-mus LD50:250 mg/kg GISAAA 39(9),80,74

SAFETY PROFILE: Poison by ingestion. Questionable carcinogen with experimental tumorigenic data. Mutation data reported. Many N-nitroso compounds are carcinogens. When heated to decomposition it emits toxic fumes of NO_x. See also N-NITROSO COMPOUNDS.

DVE600 **CAS:55557-00-1** **HR: 2**
DINITROSOHOMOPIPERAZINE
mf: $C_5H_{10}N_4O_2$ mw: 158.19

SYN: HEXAHYDRO-1,4-DINITROSO-1H-1,4-DIAZEPINE

TOXICITY DATA WITH REFERENCE
mmo-sat 1 µg/plate MUREAV 51,319,78
mmo-esc 16,700 µmol/L CNREA8 36,4099,76
msc-ham:ovr 5 µmol/L TCMUE9 1,129,84
orl-rat TDLo:65 mg/kg/2Y-I:CAR EESADV 6,513,82
orl-rat TD:105 mg/kg/30W-I:CAR EESADV 6,513,82
orl-rat TD:1530 mg/kg/31W-I:ETA CNREA8 35,1270,75

CONSENSUS REPORTS: EPA Genetic Toxicology Program.

SAFETY PROFILE: Questionable carcinogen with experimental carcinogenic and tumorigenic data. Mutation data reported. When heated to decomposition it emits toxic fumes of NO_x. See also N-NITROSO COMPOUNDS.

DVF000 **CAS:15973-99-6** **HR: 3**
DI(N-NITROSO)-PERHYDROPYRIMIDINE
mf: $C_4H_8N_4O_2$ mw: 144.16

PROP: A solid. Mp: 61–63°.

TOXICITY DATA WITH REFERENCE
ipr-rat TDLo:13 mg/kg/66W-I:CAR JCROD7 105,191,83
ipr-rat TDLo:525 mg/kg/35W-I:NEO CALEDQ 6,57,79
ipr-rat LD50:300 mg/kg CALEDQ 6,57,79

SAFETY PROFILE: Poison by intraperitoneal route. Questionable carcinogen with experimental carcinogenic and neoplastigenic data. When heated to decomposition it emits toxic fumes of NO_x. See also N-NITROSO COMPOUNDS.

DVF200 **CAS:140-79-4** **HR: 3**
DINITROSOPIPERAZINE
mf: $C_4H_8N_4O_2$ mw: 144.16

PROP: White or cream colored crystals. Mp: 158°, vap d: 4.97.

SYNS: DINITROSOPIPERAZIN (GERMAN) □ N,N′-DINITROSOPIPERAZINE □ 1,4-DINITROSOPIPERAZINE □ DNPZ □ NSC 339 □ USAF DO-36

TOXICITY DATA WITH REFERENCE
mma-smc 50 µmol/plate MUREAV 77,143,80
sce-hmn:lym 10 mmol/L TCMUE9 1,129,84
orl-mus TDLo:140 mg/kg (15-21D preg):REP CNREA8 40,2925,80
orl-mus TDLo:400 mg/kg (20D post):REP CNREA8 40,2925,80
orl-rat TDLo:1040 mg/kg/1Y-I:CAR JNCIAM 41,985,68
scu-rat TDLo:1070 mg/kg/53W-I:ETA ZEKBAI 69,103,67
orl-mus TDLo:140 mg/kg (15-21D preg):CAR,TER CNREA8 40,2925,80
orl-mus TDLo:1568 mg/kg/28W-C:NEO JNCIAM 46,1029,71
scu-mus TDLo:720 mg/kg/72W-I:ETA VOONAW 15(6),104,69
orl-rat TD:1800 mg/kg/64W-C:ETA ARZNAD 19,1077,69
scu-rat TD:1100 mg/kg/110W-I:ETA ARZNAD 19,1077,69
orl-rat TD:2250 mg/kg/50W-I:ETA ZKKOBW 77,257,72
orl-mus TD:7300 mg/kg/52W-C:NEO CNREA8 40,2925,80
orl-rat TD:560 mg/kg/10W-C:ETA CNREA8 42,4236,82
orl-rat TD:300 mg/kg/50W-I:ETA CNREA8 26,619,66
scu-rat TD:1086 mg/kg/69W-I:ETA ZEKBAI 66,138,64
orl-mus TD:5280 mg/kg/44W-I:ETA VOONAW 15(6),104,69
orl-rat TD:1120 mg/kg/20W-C:ETA CNREA8 42,4236,82
orl-rat TD:1680 mg/kg/30W-C:ETA CNREA8 42,4236,82
orl-rat LD50:160 mg/kg ZEKBAI 69,103,67
scu-rat LD50:160 mg/kg ZEKBAI 69,103,67
ipr-mus LD50:100 mg/kg NTIS** AD277-689

CONSENSUS REPORTS: EPA Genetic Toxicology Program. Reported in EPA TSCA Inventory.

SAFETY PROFILE: Suspected carcinogen with experimental carcinogenic, neoplastigenic, tumorigenic, and teratogenic data. Poison by ingestion, subcutaneous, and intraperitoneal routes. Experimental reproductive effects. Human mutation data reported. When heated to decomposition it emits toxic fumes of NO_x. See also N-NITROSO COMPOUNDS.

DVF300 **CAS:118-02-5** **HR: 3**
2,4-DINITROSO-m-RESORCINOL
mf: $C_6H_4N_2O_4$ mw: 168.12

SYNS: BENZENE-1,3-DIOL, 2,4-DINITROSO- □ 2,4-DINITRORESORCINOL (heavy metal salts of) (dry) (DOT) □ RESORCINOL, 2,4-DINITROSO-

TOXICITY DATA WITH REFERENCE
ipr-mus LD50:300 mg/kg JPETAB 119,522,57

DOT CLASSIFICATION: Forbidden

SAFETY PROFILE: A poison by intraperitoneal route. An unstable substance forbidden for transport. When heated to decomposition it emits toxic vapors of NO_x.

DVF400 **CAS:101-25-7** **HR: 3**
3,7-DINITROSO-1,3,5,7-TETRAAZABICYCLO[3.3.1]NONANE
DOT: UN 2972
mf: $C_5H_{10}N_6O_2$ mw: 186.18

SYNS: ACETO DNPT 40 □ ACETO DNPT 80 □ ACETO DNPT 100 □ CHKHZ 18 □ DINITROSOPENTAMETHYLENETETRAMINE □ N,N-DINITROSOPENTAMETHYLENETETRAMINE □ N[1],N[3]-DINITROSOPENTAMETHYLENETETRAMINE □ 3,4-DI-N-NITROSOPENTAMETHYLENETETRAMINE □ 3,7-DI-N-NITROSOPENTAMETHYLENETETRAMINE □ DNPMT □ DNPT □ 1,5-METHYLENE-3,7-DINITROSO-1,3,5,7-TETRAAZACYCLOOCTAINE □ 1,5-METHYLENE-3,7-DINITROSO-1,3,5,7-TETRAAZACYCLOOCTANE □ POROFOR CHKHC-18 □ POROPHOR B □ UNICEL-ND □ UNICEL NDX □ VULCACEL B-40 □ VULCACEL BN

TOXICITY DATA WITH REFERENCE
mmo-sat 500 µg/plate PMRSDJ 1,302,81
pic-esc 100 mg/L PMRSDJ 1,224,81
dnd-bcs 2 mg/disc PMRSDJ 1,175,81
sce-ham:ovr 80 mg/L PMRSDJ 1,538,81
otr-ham:kdy 73,500 µg/L PMRSDJ 1,626,81
orl-rat LD50:940 mg/kg MELAAD 58,22,67
ipr-rat LD50:220 mg/kg MELAAD 58,22,67
scu-rat LD50:220 mg/kg APACAB 28,209,65
ipr-mus LD50:130 mg/kg APACAB 28,209,65
scu-mus LD50:140 mg/kg APACAB 28,209,65
ivn-mus LD50:120 mg/kg APACAB 28,209,65
ivn-rbt LD50:130 mg/kg APACAB 28,209,65

CONSENSUS REPORTS: IARC Cancer Review: Group 3 IMEMDT 7,56,87; Animal No Evidence IMEMDT 11,241,76. Reported in EPA TSCA Inventory. EPA Genetic Toxicology Program.

DOT CLASSIFICATION: 4.1; *Label:* Flammable Solid, EXPLOSIVE

SAFETY PROFILE: Poison by intravenous, intraperitoneal, and subcutaneous routes. Moderately toxic by ingestion. Questionable carcinogen. Mutation data reported. Can ignite when handled and burns very rapidly. Many N-nitroso compounds are carcinogens. A blowing agent. When heated to decomposition it emits toxic fumes of NO_x. See also N-NITROSO COMPOUNDS.

DVF600 **CAS:128-42-7** **HR: 2**
4,4′-DINITRO-2,2′-STILBENEDISULFONIC ACID
mf: $C_{14}H_{10}N_2O_{10}S_2$ mw: 430.38

PROP: Yellow paste or brownish crystals.

SYNS: DINITROSTILBENEDISULFONIC ACID □ KYSELINA-4,4′-DINITROSTILBEN-2,2′-DISULFONOVA (CZECH)

TOXICITY DATA WITH REFERENCE
skn-rbt 500 mg/24H MOD 28ZPAK -,194,72
eye-rbt 500 mg/24H SEV 28ZPAK -,194,72
orl-rat LD50:12,600 mg/kg 85JCAE-,1062,86
orl-mus LD50:47 g/kg GISAAA 45(3),73,80
orl-rbt LD50:30 g/kg GISAAA 45(3),73,80
orl-gpg LD50:71 g/kg GISAAA 45(3),73,80

CONSENSUS REPORTS: Reported in EPA TSCA Inventory.

SAFETY PROFILE: Very low toxicity by ingestion. A skin and severe eye irritant. Can react vigorously with reducing materials. When heated to decomposition it emits very toxic fumes of NO_x and SO_x.

DVF800 **CAS:1594-56-5** **HR: 3**
2,4-DINITRO-1-THIOCYANOBENZENE
mf: $C_7H_3N_3O_4S$ mw: 225.19

SYNS: 2,4-DINITROPHENYL THIOCYANATE □ 2,4-DINITRO-RHODANBENZOL (GERMAN) □ 2,4-DINITROTHIOCYANATOBENZENE □ 2,4-DINITROTHIOCYANOBENZENE □ DNRB □ DNTB □ DRB □ GRYZBOL □ GRZYBOL □ NBT □ NIRIT □ NITRITE □ RHODANDINITROBENZOL □ RODATOX 60 □ TRIRODAZEEN □ TRI-RODAZENE □ 2317-W

TOXICITY DATA WITH REFERENCE
mma-sat 50 µg/plate MUREAV 40,19,76
hma-ofs/sat 450 µg/L CALEDQ 19,147,83
ipr-rat LDLo:30 mg/kg ARZNAD 16,870,66
orl-mus LD50:2750 mg/kg FMCHA2 -,D219,80
ipr-mus LDLo:30 mg/kg ARZNAD 21,121,71
orl-gpg LD50:1650 mg/kg 85GMAT -,62,82

CONSENSUS REPORTS: EPA Genetic Toxicology Program. Community Right-To-Know List.

SAFETY PROFILE: Poison by intraperitoneal route. Moderately toxic by ingestion. Mutation data reported. When heated to decomposition it emits very toxic fumes of NO_x, CN^-, and SO_x. See also CYANIDE.

DVG000 **CAS:5347-12-6** **HR: 3**
2,4-DINITROTHIOPHENE
mf: $C_4H_2N_2O_4S$ mw: 174.14

PROP: Leaflets from EtOH. Mp: 56°.

TOXICITY DATA WITH REFERENCE
mmo-sat 50 µmol/L MUREAV 118,153,83

mmo-klp 500 µmol/L MUREAV 118,153,83
ipr-mus LDLo: 1 mg/kg HBTXAC 5,171,59
ivn-mus LD50: 32 mg/kg CBCCT* 6,143,54
par-mus LDLo: 40 mg/kg CBCCT* 7,695,55

SAFETY PROFILE: Poison by intraperitoneal, intravenous, and parenteral routes. Mutation data reported. When heated to decomposition it emits very toxic fumes of NO_x and SO_x. See also NITRO COMPOUNDS.

DVG200 CAS:303-21-9 HR: 3
2,6-DINITROTHYMOL
mf: $C_{10}H_{12}N_2O_5$ mw: 240.24

PROP: Yellow prisms from pet ether. Mp: 55°. Very sol in Me_2CO and EtOH; sol in alkalis; sltly sol in H_2O.

SYNS: 2,4-DINITRO-6-ISOBROPYL-m-CRESOL □ DINITROTHYMOL 1-2-4 (FRENCH)

TOXICITY DATA WITH REFERENCE
orl-rat LDLo: 100 mg/kg NCNSA6 5,36,53
orl-mus TDLo: 140 mg/kg AECTCV 14,111,85
ivn-dog LD50: 15 mg/kg AIPTAK 50,20,35
ipl-pgn LD50: 10 mg/kg AIPTAK 50,20,35

CONSENSUS REPORTS: Reported in EPA TSCA Inventory.

SAFETY PROFILE: Poison by ingestion, intravenous, and implant routes. When heated to decomposition it emits toxic fumes of NO_x. See also NITRO COMPOUNDS of AROMATIC HYDROCARBONS.

DVG600 CAS:25321-14-6 HR: 3
DINITROTOLUENE
DOT: UN 2038
mf: $C_7H_6N_2O_4$ mw: 182.15

SYNS: BENZENE, METHYLDINITRO- □ DINITROPHENYLMETHANE □ DINITROTOLUENES, liquid or solid (DOT) □ METHYLDINITROBENZENE □ TOLUENE, ar,ar-DINITRO-

TOXICITY DATA WITH REFERENCE
dns-rat-orl 100 mg/kg CRNGDP 3,241,82
orl-rat TDLo: 196 mg/kg (7-20D preg): REP CIIT** DOCKET #10992/82
orl-rat TDLo: 1050 mg/kg (7-20D preg): REP CIIT** DOCKET #1099/82
orl-rat TDLo: 2100 mg/kg (female 7–20D post): REP FAATDF 5,948,85
orl-rat TDLo: 1050 mg/kg (7–20D preg): TER CIIT** DOCKET #1099/82
orl-rat TDLo: 12,775 mg/kg/Y-C: ETA PAACA3 24,91,83

CONSENSUS REPORTS: Reported in EPA TSCA Inventory. EPA Genetic Toxicology Program.

OSHA PEL: TWA 1.5 mg/m³ (skin)
ACGIH TLV: TWA 0.15 mg/m³ (skin); Suspected Human Carcinogen
DFG MAK: Animal Carcinogen, Suspected Human Carcinogen
NIOSH REL: (Dinitrotoluene): Reduce to lowest level
DOT CLASSIFICATION: 6.1; *Label:* Poison

SAFETY PROFILE: Confirmed carcinogen with experimental tumorigenic and teratogenic data. A poison. Experimental reproductive effects. Mutation data reported. Flammable. When heated to decomposition it emits toxic fumes of NO_x. See also 2,4-DINITROTOLUENE.

DVG800 CAS:602-01-7 HR: 2
2,3-DINITROTOLUENE
mf: $C_7H_6N_2O_4$ mw: 182.15

PROP: Needles from pet ether. Mp: 63°.

SYNS: 1-METHYL-2,3-DINITRO-BENZENE (9CI) □ 2,3-DNT

TOXICITY DATA WITH REFERENCE
skn-rbt 500 mg/24H MLD NTIS** AD-B011-150
mmo-sat 50 µg/plate ENMUDM 4,163,82
mma-sat 1 mg/plate NTIS** AD-A080-146
dnd-rat: lvr 300 µmol/L SinJF# 26OCT82
orl-rat LD50: 911 mg/kg NTIS** AD-A080-146
orl-mus LD50: 1072 mg/kg NTIS** PB214-270

CONSENSUS REPORTS: Reported in EPA TSCA Inventory.

OSHA PEL: TWA 1.5 mg/m³ (skin)
NIOSH REL: (Dinitrotoluene): Reduce to lowest level

SAFETY PROFILE: Moderately toxic by ingestion. Mutation data reported. A skin irritant. When heated to decomposition it emits toxic fumes of NO_x. See also 2,4-DINITROTOLUENE.

DVH000 CAS:121-14-2 HR: 3
2,4-DINITROTOLUENE
mf: $C_7H_6N_2O_4$ mw: 182.15

PROP: Yellow needles from CS_2. Mp: 69.5°, bp: 300°, d: 1.521 @ 15°, vap d: 6.27, flash p: 404°F.

SYNS: 2,4-DINITROTOLUOL □ DNT □ 2,4-DNT □ 1-METHYL-2,4-DINITROBENZENE □ NCI-C01865 □ RCRA WASTE NUMBER U105

TOXICITY DATA WITH REFERENCE
skn-rbt 500 mg/24H MLD NTIS** AD-B011-150
mma-sat 125 µg/plate ENMUDM 7(Suppl 5),1,85
dnd-rat: lvr 3 mmol/L SinJF# 26OCT82
oms-rat-orl 10 mg/kg JTEHD6 11,555,83
mmo-sat 10 µg/plate NTIS** AD-A080-146
cyt-mus-orl 840 µg/kg MUREAV 38,387,76
dlt-mus-orl 2 mg/kg MUREAV 38,387,76
orl-rat TDLo: 3094 mg/kg (13W male): REP NTIS** AD-A077-692
orl-rat TDLo: 16,425 mg/kg (female 1 year(s) pre): REP NTIS** AD-A077-692
orl-rat TDLo: 8463 mg/kg (male 13W pre): REP JACTDZ 4(4),243,85
orl-rat TDLo: 12,410 mg/kg (male 1 year(s) pre): REP NTIS** AD-A077-692
orl-rat TDLo: 2620 mg/kg/78W-C: NEO NCITR* NCI-CG-TR-54,78
orl-mus TDLo: 10,080 mg/kg/2Y-C: CAR JACTDZ 4(4),257,85
orl-rat TD: 5460 mg/kg/78W-C: NEO NCITR* NCI-CG-TR-54,78
orl-rat TD: 28 g/kg/2Y-C: CAR NTIS** AD-A080-146
orl-rat TD: 12,775 mg/kg/2Y-C: ETA NCITR* NCI-CG-TR-54,78

orl-mus TDLo:8760 mg/kg/2Y-C:NEO NTIS** AD-A080-146
orl-rat LD50:268 mg/kg NTIS** PB214-270
orl-mus LD50:790 mg/kg GTPZAB 25(8),50,81
scu-cat LDLo:25 mg/kg XPHBAO 271,110,41
orl-gpg LD50:1300 mg/kg GISAAA 42(10),12,77

CONSENSUS REPORTS: NCI Carcinogenesis Bioassay (feed); No Evidence: mouse NCITR* NCI-CG-TR-54,78; Some Evidence: rat NCITR* NCI-CG-TR-54,78. Reported in EPA TSCA Inventory.

OSHA PEL: TWA 1.5 mg/m³ (skin)
NIOSH REL: (Dinitrotoluene): Reduce to lowest level

SAFETY PROFILE: Suspected carcinogen with experimental carcinogenic and neoplastigenic data. Poison by ingestion and subcutaneous routes. Experimental reproductive effects. A skin irritant. Mutation data reported. An irritant and an allergen. Can cause anemia, methemoglobinemia, cyanosis, and liver damage. Combustible when exposed to heat or flame; can react with oxidizing materials. To fight fire, use water spray or mist, dry chemical. Decomposes when heated to 250°C. There are instances of explosion during manufacture or storage. Mixture with nitric acid is a high explosive. Mixture with sodium carbonate can decompose with significant pressure increase at 210°C. Mixtures with other alkalies may have the same effect. Ignites on contact with sodium oxide. When heated to decomposition it emits toxic fumes of NO_x.

For occupational chemical analysis use OSHA: #44.

DVH200 CAS:619-15-8 HR: 2
2,5-DINITROTOLUENE
mf: $C_7H_6N_2O_4$ mw: 182.15

PROP: Needles from EtOH. Mp: 52.5°.

SYNS: 2,5-DNT □ 2-METHYL-1,4-DINITROBENZENE

TOXICITY DATA WITH REFERENCE
skn-rbt 500 mg/24H MOD NTIS** AD-B011-150
mmo-sat 10 µg/plate NTIS** AD-A080-146
orl-rat LD50:517 mg/kg NTIS** AD-A080-146
orl-mus LD50:652 mg/kg NTIS** AD-A080-146

CONSENSUS REPORTS: Reported in EPA TSCA Inventory.

OSHA PEL: TWA 1.5 mg/m³ (skin)
NIOSH REL: (Dinitrotoluene): Reduce to lowest level

SAFETY PROFILE: Moderately toxic by ingestion. Mutation data reported. A skin irritant. When heated to decomposition it emits toxic fumes of NO_x. See also 2,4-DINITROTOLUENE.

DVH400 CAS:606-20-2 HR: 3
2,6-DINITROTOLUENE
mf: $C_7H_6N_2O_4$ mw: 182.15

PROP: Needles from EtOH. Mp: 66°.

SYNS: 2,6-DNT □ 2-METHYL-1,3-DINITROBENZENE □ RCRA WASTE NUMBER U106

TOXICITY DATA WITH REFERENCE
skn-rbt 500 mg/24H MLD NTIS** AD-B011-150
dnd-rat-orl 10 mg/kg JTEHD6 11,555,83
dns-rat-orl 5 mg/kg CRNGDP 3,241,82
orl-rat TD:2555 mg/kg/Y-C:ETA PAACA3 24,91,83
orl-rat TDLo:5110 mg/kg/Y-C:ETA PAACA3 24,91,83
orl-rat LD50:177 mg/kg NTIS** PB214-270
orl-mus LD50:621 mg/kg NTIS** AD-A080-146

CONSENSUS REPORTS: Reported in EPA TSCA Inventory.

OSHA PEL: TWA 1.5 mg/m³ (skin)
NIOSH REL: (Dinitrotoluene): Reduce to lowest level

SAFETY PROFILE: Poison by ingestion. A skin irritant. Questionable carcinogen with experimental tumorigenic data. Mutation data reported. When heated to decomposition it emits toxic fumes of NO_x. See also 2,4-DINITROTOLUENE.

DVH600 CAS:610-39-9 HR: 2
3,4-DINITROTOLUENE
mf: $C_7H_6N_2O_4$ mw: 182.15

PROP: A solid. Mp: 61°.

SYNS: 3,4-DNT □ 4-METHYL-1,2-DINITROBENZENE

TOXICITY DATA WITH REFERENCE
skn-rbt 500 mg/24H MLD NTIS** AD-B011-150
mmo-sat 1 mg/plate NTIS** AD-A080-146
mma-sat 10 µg/plate ENMUDM 4,163,82
dnd-rat:lvr 300 µmol/L SinJF# 26OCT82
orl-rat LD50:807 mg/kg NTIS** AD-A080-146
orl-mus LD50:747 mg/kg NTIS** AD-A080-146

CONSENSUS REPORTS: Reported in EPA TSCA Inventory.

OSHA PEL: TWA 1.5 mg/m³ (skin)
NIOSH REL: (Dinitrotoluene). Reduce to lowest level

SAFETY PROFILE: Moderately toxic by ingestion. Mutation data reported. A skin irritant. When heated to decomposition it emits toxic fumes of NO_x. See also 2,4-DINITROTOLUENE.

DVH800 CAS:618-85-9 HR: 3
3,5-DINITROTOLUENE
mf: $C_7H_6N_2O_4$ mw: 182.15

PROP: Needles from AcOH. Mp: 93°.

SYNS: 3,5-DNT □ 1-METHYL-3,5-DINITRO-BENZENE

TOXICITY DATA WITH REFERENCE
mmo-sat 100 µg/plate NTIS** AD-A080-146
orl-rat LD50:216 mg/kg NTIS** AD-A080-146
orl-mus LD50:607 mg/kg NTIS** AD-A080-146

OSHA PEL: TWA 1.5 mg/m³ (skin)
NIOSH REL: (Dinitrotoluene): Reduce to lowest level

SAFETY PROFILE: Poison by ingestion. Mutation data reported. Flammable when exposed to heat or flame; can react with oxidizing materials. A moderate explosion hazard when exposed to heat. To fight fire, use water, CO_2, dry chemical. When heated to decomposition it emits toxic fumes of NO_x. See also 2,4-DINITROTOLUENE; EXPLOSIVES, HIGH; and NITRATES.

DVI100 CAS:6393-42-6 HR: 3
2,6-DINITRO-p-TOLUIDINE
mf: $C_7H_7N_3O_4$ mw: 197.17

PROP: Yellow crystals from EtOH. Mp: 171–172°.

SYNS: 4-AMINO-3,5-DINITROTOLUENE □ 2,6-DINITRO-4-METHYLANILINE □ 4-METHYL-2,6-DINITROANILINE □ 4-METHYL-2,6-DINITROBENZENAMINE

TOXICITY DATA WITH REFERENCE
mmo-sat 1 μg/plate ENMUDM 4,163,82
ivn-mus LD50:320 mg/kg CSLNX* NX#03362

SAFETY PROFILE: Poison by intravenous route. Mutation data reported. When heated to decomposition it emits toxic fumes of NO_x. See also NITRO COMPOUNDS of AROMATIC HYDROCARBONS.

DVI600 CAS:6379-46-0 HR: 2
4,6-DINITRO-1,2,3-TRICHLOROBENZENE
mf: $C_6HCl_3N_2O_4$ mw: 271.44

PROP: Greenish-yellow needles from EtOH. Mp: 98°

SYNS: 1,2,3-TRICHLORO-4,6-DINITROBENZENE □ VANCIDE PB

TOXICITY DATA WITH REFERENCE
orl-mus TDLo:13 g/kg/78W-I:CAR NTIS** PB223-159
scu-mus TDLo:10 mg/kg:CAR NTIS** PB223-159

SAFETY PROFILE: Questionable carcinogen with experimental carcinogenic data. When heated to decomposition it emits very toxic fumes of NO_x and Cl^-. See also NITRO COMPOUNDS of AROMATIC HYDROCARBONS.

DVI800 CAS:8069-76-9 HR: 2
DINOCTON-O
mf: $C_{16}H_{22}N_2O_7$ mw: 354.40

PROP: A mixture of methyl-2,4-dinitro-6-(1-ethylhexyl)phenyl carbonate and methyl-2,4-dinitro-6-(1-propylpentyl)phenyl carbonate (30ZDA9 -,100,71).

SYNS: DINOCTON-6 □ MC 1945

TOXICITY DATA WITH REFERENCE
orl-rat LD50:1250 mg/kg 28ZEAL 5,82,76
skn-rat LD50:3000 mg/kg 31ZOAD 1,176,68

SAFETY PROFILE: Moderately toxic by ingestion and skin contact. When heated to decomposition it emits toxic fumes of NO_x.

DVJ000 CAS:84-76-4 HR: 2
DI-n-NONYL PHTHALATE
mf: $C_{26}H_{42}O_4$ mw: 418.68

PROP: Oil.

SYNS: BISOFLEX 91 □ DINONYL-1,2-BENZENEDICARBOXYLATE □ DIONONYL PHTHALATE

TOXICITY DATA WITH REFERENCE
orl-mus LD50:21,500 mg/kg GTPZAB 24(3),25,80
orl-gpg LD50:21,500 mg/kg GTPZAB 24(3),25,80

CONSENSUS REPORTS: Reported in EPA TSCA Inventory.

SAFETY PROFILE: Moderately toxic by ingestion. Used as a plasticizer. When heated to decomposition it emits acrid smoke and irritating fumes. See also ESTERS.

DVJ100 CAS:33854-16-9 HR: 3
DINOPROST METHYL ESTER
mf: $C_{21}H_{36}O_5$ mw: 368.57

SYNS: PGF2 METHYL ESTER □ PGF2-α METHYL ESTER □ PROSTAGLANDIN F2-α METHYL ESTER □ (5Z,9-α,11-α,13E,15S)-9,11,15-TRIHYDROXY-PROSTA-5,13-DIEN-1-OIC ACID, METHYL ESTER

TOXICITY DATA WITH REFERENCE
ivg-wmn TDLo:80 μg/kg (7W preg):TER JOPDAB 102,620,83
scu-ham TDLo:480 μg/kg (4D preg):REP JPETAB 186,67,73
ivg-wmn TDLo:80 μg/kg:GIT JOPDAB 102,620,83

SAFETY PROFILE: A human teratogen by intravaginal route with developmental abnormalities of the central nervous system and musculoskeletal system. Human systemic effects by intravaginal route: nausea or vomiting. Experimental reproductive effects. See also ESTERS.

DVJ200 CAS:363-24-6 HR: 3
DINOPROSTONE
mf: $C_{20}H_{32}O_5$ mw: 352.52

PROP: Crystals. Mp: 66–68°.

SYNS: (5Z,11-α,13E,15S)-11,15-DIHYDROXY-9-OXOPROSTA-5,13-DIEN-1-OIC ACID □ 7-(3-HYDROXY-2-(3-HYDROXY-1-OCTENYL)-5-OXOCYCLOPENTYL)-5-HEPTENOIC ACID □ PGE2 □ PROSTAGLANDIN E2 □ (−)-PROSTAGLANDIN E2 □ (15S)-PROSTAGLANDIN E2 □ PROSTIN E2 □ U-12062

TOXICITY DATA WITH REFERENCE
oms-mus:oth 100 nmol/L JIDEAE 66,313,76
spm-mus-ipr 3 mg/kg INJFA3 21,82,76
dns-gpg:lng 1 mg/L PSEBAA 171,109,82
ivg-wmn TDLo:2400 μg/kg (female 14W post):REP JOGBAS 78,294,71
ivn-wmn TDLo:100 ng/kg (female 23W post):REP BMJOAE 3,196,70
ivg-wmn TDLo:800 μg/kg (female 2W post):REP CCPTAY 3,173,71
orl-wmn TDLo:200 μg/kg (lactating female 3D post):REP JRPMAP 33,630,88
ivn-wmn TDLo:400 μg/kg (female 16W post):REP AJOGAH 102,317,68
orl-man TDLo:1 mg/kg (male 1D pre):REP FESTAS 50,789,88
orl-mus TDLo:1440 mg/kg (female 7-12D post):REP OYYAA2 8,787,74
ivg-wmn TDLo:60 μg/kg (female 41W post):REP BJOGAS 91,598,84
orl-mus TDLo:36 mg/kg (female 7-12D post):TER OYYAA2 8,787,74

ipr-rat TDLo:1200 μg/kg (female 9-14D post):TER AKGIAO 56(8),53,80
ipr-rat TDLo:2 mg/kg (female 15D post):REP NYKZAU 79,15,82
orl-ham TDLo:48 mg/kg (female 4-6D post):REP ACEDAB 170,3,72
ipr-mus TDLo:30 mg/kg (female 9-14D post):REP THERAP 33,671,78
orl-rat TDLo:1440 mg/kg (female 9-14D post):REP OYYAA2 8,787,74
scu-rat TDLo:17,500 μg/kg (female 1-7D post):REP NATUAS 222,287,69
scu-ham TDLo:3600 μg/kg (female 4-6D post):REP ACEDAB 170,3,72
ipr-rat TDLo:3600 μg/kg (female 20D post):REP NATUAS 250,330,74
scu-rat TDLo:14 mg/kg (male 14D pre):REP IJEBA6 19,112,81
scu-rat TDLo:35 mg/kg (male 35D pre):REP IJEBA6 19,112,81
scu-ham TDLo:800 μg/kg (female 8D post):TER PLIRDW 20,241,81
ipr-rat TDLo:75 mg/kg (female 6-10D post):TER TOLED5 1,3,77
ipr-rat TDLo:6 mg/kg (female 12-15D post):TER PRGLBA 3,299,73
orl-mus TDLo:720 mg/kg (female 7-12D post):TER OYYAA2 8,787,74
scu-mus TDLo:746 μg/kg (female 12D post):TER TXCYAC 5,97,75
orl-rat LD50:500 mg/kg OYYAA2 8,787,74
scu-rat LD50:31,600 μg/kg OYYAA2 8,787,74
ivn-rat LD50:59,500 μg/kg OYYAA2 8,787,74
orl-mus LD50:750 mg/kg OYYAA2 8,787,74
scu-mus LD50:19,700 μg/kg OYYAA2 8,787,74
ivn-mus LD50:23,200 μg/kg OYYAA2 8,787,74
ipr-ham LD50:1 mg/kg PLMEDD 17,309,85

SAFETY PROFILE: Poison by subcutaneous and intravenous routes. Moderately toxic by ingestion and intraperitoneal routes. An experimental teratogen. Human reproductive effects by intravenous, intraplacental, and intravaginal routes: changes in the uterus, cervix, and vagina; termination of pregnancy; and changes in fertility. Experimental reproductive effects. Mutation data reported. When heated to decomposition it emits acrid smoke and irritating fumes.

DVJ400 CAS:3204-27-1 HR: 3
DINOTERB ACETATE
mf: $C_{12}H_{14}N_2O_6$ mw: 282.28

SYNS: 2-tert-BUTYL-4,6-DINITROPHENYL ACETATE □ 2-(1,1-DIMETHYLETHYL)-4,6-DINITROPHENOL ACETATE □ MC 1108 □ P-1108

TOXICITY DATA WITH REFERENCE
orl-rat LD50:62 mg/kg WRPCA2 9,119,70
orl-rbt LD50:100 mg/kg 28ZEAL 5,83,76

SAFETY PROFILE: Poison by ingestion. When heated to decomposition it emits toxic fumes of NO_x. See also 2-(1,1-DIMETHYLETHYL)-4,6-DINITROPHENOL (dinoterb).

DVJ500 CAS:8015-43-8 HR: 3
DIOCIDE
mf: $C_{21}H_{38}N{\cdot}C_2H_5ClHgO{\cdot}Br$ mw: 665.62

SYNS: DIOCID □ 1-HEXADECYL-PYRIDINIUM BROMIDE mixture with CHLORO(2-HYDROXYETHYL)MERCURY

TOXICITY DATA WITH REFERENCE
orl-rat LD50:172 mg/kg PCJOAU 11,918,77
ipr-rat LD50:21,700 mg/kg PCJOAU 11,918,77
orl-mus LD50:54 mg/kg PCJOAU 11,918,77

CONSENSUS REPORTS: Mercury and its compounds are on the Community Right-To-Know List.

OSHA PEL: TWA 0.01 mg(Hg)/m³; STEL 0.03 mg/m³ (skin)
ACGIH TLV: TWA 0.01 mg(Hg)/m³; STEL 0.03 mg(Hg)/m³
NIOSH REL: (Mercury, Aryl and Inorganic): CL 0.1 mg/m³ (skin)

SAFETY PROFILE: Poison by ingestion. Mildly toxic by intraperitoneal route. When heated to decomposition it emits toxic fumes of Br^-, Cl^-, NO_x, and Hg. See also MERCURY COMPOUNDS.

DVJ600 CAS:1120-48-5 HR: 3
DIOCTYLAMINE
mf: $C_{16}H_{35}N$ mw: 241.52

PROP: A liquid. Mp: 14–15°, bp: 297–298°.

TOXICITY DATA WITH REFERENCE
ipr-mus LDLo:4 mg/kg CBCCT* 2,133,50

CONSENSUS REPORTS: Reported in EPA TSCA Inventory.

SAFETY PROFILE: Poison by intraperitoneal route. When heated to decomposition it emits toxic fumes of NO_x. See also AMINES.

DVJ800 CAS:3648-18-8 HR: 3
DIOCTYLDI(LAUROYLOXY)STANNANE
mf: $C_{40}H_{80}O_4Sn$ mw: 743.89

SYNS: BIS(DODECANOLOXY)DIOCTYLSTANNANE □ BIS(LAUROYLOXY)DIOCTYLSTANNANE □ DIDODECANOYLOXYDIOCTYLSTANNANE □ DIOCTYLBIS(LAUROYLOXY)STANNANE □ DIOCTYLDIDODECANOYLOXYSTANNANE □ DIOCTYLTIN DILAURATE □ DI-n-OCTYLTIN DILAURATE □ DI-n-OCTYL-ZINN DILAURAT (GERMAN)

TOXICITY DATA WITH REFERENCE
orl-rat LD50:6450 mg/kg ARZNAD 19,934,69
ipr-rat LD50:95 mg/kg ARZNAD 19,934,69

CONSENSUS REPORTS: Reported in EPA TSCA Inventory.

OSHA PEL: TWA 0.1 mg(Sn)/m³ (skin)
ACGIH TLV: TWA 0.1 mg(Sn)/m³ (skin) (Proposed: TWA 0.1 mg(Sn)/m³; STEL 0.2 mg(Sn)/m³ (skin))
NIOSH REL: (Organotin Compounds) TWA 0.1 mg(Sn)/m³

SAFETY PROFILE: Poison by intraperitoneal route. Mildly toxic by ingestion. When heated to decomposition it emits acrid smoke and irritating fumes. See also TIN COMPOUNDS.

For occupational chemical analysis use NIOSH: Organotin Compounds 5504.

DVK200 **CAS:16091-18-2** **HR: 2**
2,2-DIOCTYL-1,3,2-DIOXASTANNEPIN-4,7-DIONE
mf: $C_{20}H_{36}O_4Sn$ mw: 459.25

SYNS: DIOCTYLSTANNYLENE MALEATE ☐ DIOCTYLTIN MALEATE ☐ DI-n-OCTYLTIN MALEATE ☐ DI-n-OCTYLZINN MALEINAT ☐ ESTABEX U 18 ☐ LIV 1176 ☐ MELLITE 825 ☐ STANN OMF ☐ THERMOLITE 813 ☐ TVS 8105

TOXICITY DATA WITH REFERENCE
orl-rat LD50:4500 mg/kg ARZNAD 19,934,69
orl-mus LD50:775 mg/kg ERNFA7 11,424,66

CONSENSUS REPORTS: Reported in EPA TSCA Inventory.

OSHA PEL: TWA 0.1 mg(Sn)/m^3 (skin)
ACGIH TLV: TWA 0.1 mg(Sn)/m^3 (skin) (Proposed: TWA 0.1 mg(Sn)/m^3; STEL 0.2 mg(Sn)/m^3 (skin))
DFG MAK: 0.1 mg(Sn)/m^3 calculated as total dust
NIOSH REL: (Organotin Compounds) TWA 0.1 mg(Sn)/m^3

SAFETY PROFILE: Moderately toxic by ingestion. When heated to decomposition it emits acrid smoke and irritating fumes. See also TIN COMPOUNDS.

For occupational chemical analysis use NIOSH: Organotin Compounds 5504.

DVK400 **CAS:101-67-7** **HR: 1**
4,4′-DIOCTYLDIPHENYLAMINE
mf: $C_{28}H_{43}N$ mw: 393.72

TOXICITY DATA WITH REFERENCE
orl-rat LD50:8000 mg/kg TXAPA9 42,417,77

CONSENSUS REPORTS: Reported in EPA TSCA Inventory.

SAFETY PROFILE: Mildly toxic by ingestion. When heated to decomposition it emits toxic fumes of NO_x. See also AMINES.

DVK600 **CAS:141-02-6** **HR: 3**
DIOCTYL FUMARATE
mf: $C_{20}H_{36}O_4$ mw: 340.56

PROP: Clear, mobile liquid; mild odor. Bp: 211–220°, flash p: 365°F (COC), d: 0.942 @ 20°/20°.

SYNS: BIS(2-ETHYLHEXYL) FUMARATE ☐ 2-BUTENEDIOIC ACID BIS(2-ETHYLHEXYL) ESTER ☐ DI(2-ETHYLHEXYL) FUMARATE ☐ DOF ☐ 2-ETHYLHEXYL FUMARATE ☐ RC COMONOMER DOF

TOXICITY DATA WITH REFERENCE
skn-rbt 10 mg/24H open SEV AMIHBC 10,61,54
eye-rbt 500 mg open AMIHBC 10,61,54
orl-rat LD50:29,200 mg/kg AMIHBC 10,61,54
ipr-mus LD50:250 mg/kg NTIS** AD691-490

CONSENSUS REPORTS: Reported in EPA TSCA Inventory.

SAFETY PROFILE: Poison by intraperitoneal route. An eye and severe skin irritant. Combustible when exposed to heat or flame; can react with oxidizing materials. To fight fire, use foam, CO_2, dry chemical. See also ESTERS and FUMARIC ACID.

DVK709 **HR: 1**
DIOCTYLISOPENTYLPHOSPHINE OXIDE
mf: $C_{21}H_{45}OP$ mw: 344.63

TOXICITY DATA WITH REFERENCE
orl-rat LD50:8369 mg/kg GISAAA 47(8),27,82
ihl-rat LC50:3311 g/m^3 GISAAA 47(8),27,82
orl-mus LD50:9500 mg/kg GISAAA 47(8),27,82
ihl-mus LC50:1288 g/m^3 GISAAA 47(8),27,82

SAFETY PROFILE: Mildly toxic by inhalation and ingestion. When heated to decomposition it emits toxic fumes of PO_x and phosphine. See also PHOSPHINE.

DVK800 **CAS:2915-53-9** **HR: 1**
DIOCTYL MALEATE
mf: $C_{20}H_{36}O_4$ mw: 340.56

SYNS: BIS(1-OCTYL) MALEATE ☐ 2-BUTENEDIOIC ACID (Z)-, DIOCTYL ESTER (9CI) ☐ DI-N-OCTYL MALEATE ☐ DIOCTYL MALEATE ☐ PX-538

TOXICITY DATA WITH REFERENCE
orl-rat LD50:14,200 mg/kg NPIRI* 2,43,75
skn-rbt LD50:14 g/kg NPIRI* 2,43,75

CONSENSUS REPORTS: Reported in EPA TSCA Inventory.

SAFETY PROFILE: Mildly toxic by ingestion. When heated to decomposition it emits acrid smoke and fumes. See also ESTERS.

DVL000 **CAS:1116-76-3** **HR: 3**
N,N-DIOCTYL-1-OCTANAMINE
mf: $C_{24}H_{51}N$ mw: 353.76

PROP: Viscous oily liquid. D: 0.809, bp: 365–367°. Sol in common org solvs; insol in H_2O.

SYNS: ALAMINE 308 ☐ ALAMINE 336 ☐ TRICAPRYLYLAMINE ☐ TRI-n-OCTYLAMINE

TOXICITY DATA WITH REFERENCE
ipr-rat LD50:1 g/kg HYDRDA 3,201,78
ipr-mus LDLo:63 mg/kg CBCCT* 4,323,52

CONSENSUS REPORTS: Reported in EPA TSCA Inventory.

SAFETY PROFILE: Poison by intraperitoneal route. When heated to decomposition it emits toxic fumes of NO_x. See also AMINES.

DVL200 **CAS:15535-79-2** **HR: 2**
2,2-DIOCTYL-1,3,2-OXATHIASTANNOLANE-5-OXIDE
mf: $C_{18}H_{36}O_2SSn$ mw: 435.29

SYNS: DIOCTYLTHIOACETOXYSTANNANE ☐ DIOCTYLTIN THIOGLYCOLATE ☐ DI-n-OCTYLTIN THIOGLYCOLATE ☐ DI-n-OCTYLZINN THIOGLYKOLAT (GERMAN)

TOXICITY DATA WITH REFERENCE
orl-rat LD50:945 mg/kg ARZNAD 19,943,69

CONSENSUS REPORTS: Reported in EPA TSCA Inventory.

OSHA PEL: TWA 0.1 mg(Sn)/m³ (skin)
ACGIH TLV: TWA 0.1 mg(Sn)/m³ (skin) (Proposed: TWA 0.1 mg(Sn)/m³; STEL 0.2 mg(Sn)/m³ (skin))
NIOSH REL: (Organotin Compounds) TWA 0.1 mg(Sn)/m³

SAFETY PROFILE: Moderately toxic by ingestion. When heated to decomposition it emits toxic fumes of SO_x. See also TIN COMPOUNDS.

For occupational chemical analysis use NIOSH: Organotin Compounds 5504.

DVL400 **CAS:870-08-6** **HR: 2**
DIOCTYLOXOSTANNANE
mf: $C_{16}H_{34}OSn$ mw: 361.19

SYNS: DIOCTYLTIN OXIDE ☐ DI-n-OCTYLTIN OXIDE ☐ DI-n-OCTYL-ZINN OXYD (GERMAN) ☐ OXODIOCTYLSTANNANE

TOXICITY DATA WITH REFERENCE
orl-rat LD50:2500 mg/kg ARZNAD 19,934,69

CONSENSUS REPORTS: Reported in EPA TSCA Inventory.

OSHA PEL: TWA 0.1 mg(Sn)/m³ (skin)
ACGIH TLV: TWA 0.1 mg(Sn)/m³ (skin) (Proposed: TWA 0.1 mg(Sn)/m³; STEL 0.2 mg(Sn)/m³ (skin))
DFG MAK: 0.1 mg(Sn)/m³ calculated as total dust
NIOSH REL: (Organotin Compounds) TWA 0.1 mg(Sn)/m³

SAFETY PROFILE: Moderately toxic by ingestion. When heated to decomposition it emits acrid smoke and irritating fumes. See also TIN COMPOUNDS.

For occupational chemical analysis use NIOSH: Organotin Compounds 5504.

DVL600 **CAS:117-84-0** **HR: 2**
n-DIOCTYL PHTHALATE
mf: $C_{24}H_{38}O_4$ mw: 390.62

SYNS: o-BENZENEDICARBOXYLIC ACID DIOCTYL ESTER ☐ 1,2-BENZENEDICARBOXYLIC ACID DIOCTYL ESTER ☐ CELLUFLEX DOP ☐ DINOPOL NOP ☐ DIOCTYL-o-BENZENEDICARBOXYLATE ☐ DI-OCTYL PHTHALATE ☐ DNOP ☐ OCTYL PHTHALATE ☐ n-OCTYL PHTHALATE ☐ PX-138 ☐ RCRA WASTE NUMBER U107 ☐ VINICIZER 85

TOXICITY DATA WITH REFERENCE
skn-rbt 500 mg/24H MLD 28ZPAK -,48,72
eye-rbt 5 mg SEV AJOPAA 29,1363,46
eye-rbt 500 mg/24H MLD 28ZPAK -,48,72
orl-mus TDLo:78 g/kg (7-14D preg):REP NTIS** PB85-220143
ipr-rat TDLo:5 g/kg (5-15D preg):TER JPMSAE 61,51,72
orl-mus LD50:6513 mg/kg GTPZAB 17(10),51,73
ipr-mus LD50:65 g/kg JSCCA5 28,667,77

CONSENSUS REPORTS: On EPA Extremely Hazardous Substances List by error. Reported in EPA TSCA Inventory.

SAFETY PROFILE: Mildly toxic by ingestion. Experimental teratogenic and reproductive effects. A skin and severe eye irritant. Used as a plasticizer. When heated to decomposition it emits acrid smoke and irritating fumes. See also ESTERS.

DVL700 **CAS:117-81-7** **HR: 3**
DI-sec-OCTYL PHTHALATE
mf: $C_{24}H_{38}O_4$ mw: 390.62

PROP: A liquid. D: 0.986 @ 20°, mp: −46°, bp: 231° @ 5 mm.

SYNS: BEHP ☐ BIS(2-ETHYLHEXYL)-1,2-BENZENEDICARBOXYLATE ☐ BIS(2-ETHYLHEXYL)PHTHALATE ☐ BISOFLEX 81 ☐ BISOFLEX DOP ☐ COMPOUND 889 ☐ DAF 68 ☐ DEHP ☐ DI(2-ETHYLHEXYL)ORTHOPHTHALATE ☐ DI(2-ETHYLHEXYL)PHTHALATE ☐ DIOCTYL PHTHALATE ☐ DOP ☐ ETHYLHEXYL PHTHALATE ☐ ERGOPLAST FDO ☐ 2-ETHYLHEXYL PHTHALATE ☐ EVIPLAST 80 ☐ EVIPLAST 81 ☐ FLEXIMEL ☐ FLEXOL DOP ☐ FLEXOL PLASTICIZER DOP ☐ GOOD-RITE GP 264 ☐ HATCOL DOP ☐ HERCOFLEX 260 ☐ KODAFLEX DOP ☐ MOLLAN O ☐ NCI-C52733 ☐ NUOPLAZ DOP ☐ OCTOIL ☐ OCTYL PHTHALATE ☐ PALATINOL AH ☐ PHTHALIC ACID DIOCTYL ESTER ☐ PITTSBURGH PX-138 ☐ PLATINOL AH ☐ PLATINOL DOP ☐ RC PLASTICIZER DOP ☐ RCRA WASTE NUMBER U028 ☐ REOMOL DOP ☐ REOMOL D 79P ☐ SICOL 150 ☐ STAFLEX DOP ☐ TRUFLEX DOP ☐ VESTINOL AH ☐ VINICIZER 80 ☐ WITCIZER 312

TOXICITY DATA WITH REFERENCE
skn-rbt 500 mg/24H MLD 28ZPAK -,48,72
eye-rbt 500 mg AJOPAA 29,1363,46
eye-rbt 500 mg/24H MLD 28ZPAK -,48,72
dns-rat:lvr 500 µmol/L PMRSDJ 5,371,85
sln-ham:lvr 50 mg/L PMRSDJ 5,397,85
orl-mus TDLo:1 g/kg (female 7D post):TER EVHPAZ 45,71,82
orl-rat TDLo:7140 mg/kg (female 1-21D post):TER TXAPA9 26,253,73
ipr-mus TDLo:24 g/kg (female 7-9D post):REP ARTODN 56,263,85
ivn-mus TDLo:50 mg/kg (female 1D pre):REP NTIS** PB250-102
orl-mus TDLo:2040 mg/kg (female 1-17D post):REP NTIS** PB85-105674
ipr-mus TDLo:12,780 µg/kg (male 1D pre):REP TXAPA9 29,35,74
orl-rat TDLo:17,200 mg/kg (multi) :REP NEZAAQ 31,507,76
scu-mus TDLo:2970 mg/kg (male 3D pre):REP JTEHD6 16,71,85
orl-rat TDLo:35 mg/kg (female 14D pre):REP FCTXAV 15,389,77
ipr-rat TDLo:6 g/kg (female 3-9D post):REP EVHPAZ 3,91,73

ipr-rat TDLo:1 g/kg (male 10D pre):REP TXAPA9 62,121,82
orl-rat TDLo:6 g/kg (male 3D pre):REP ARTODN 59,290,86
orl-mus TDLo:4200 mg/kg (male 21D pre):REP OKEHDW 15,129,77
orl-rat TDLo:9766 mg/kg (female 12D post):TER TJADAB 35,41,87
orl-mus TDLo:2040 mg/kg (female 1-17D post):TER TJADAB 27,84A,83
ipr-rat TDLo:10 g/kg (female 5-15D post):TER JPMSAE 61,51,72
orl-rat TDLo:216 g/kg/2Y-C:CAR NTPTR* NTP-TR-217,82
orl-mus TDLo:260 g/kg/2Y-C:CAR NTPTR* NTP-TR-217,82
orl-rat TD:433 g/kg/2Y-C:CAR NTPTR* NTP-TR-217,82
orl-mus TD:519 g/kg/2Y-C:CAR NTPTR* NTP-TR-217,82
orl-mus TD:120 g/kg/24W-C:ETA CRNGDP 4,1021,83
orl-rat TD:524 g/kg/2Y-C:CAR EVHPAZ 65,271,86
orl-mus TD:262 g/kg/2Y-C:CAR EVHPAZ 65,271,86
orl-rat TD:438 g/kg/2Y-C:CAR CALEDQ 38,15,87
orl-man TDLo:143 mg/kg:GIT JIHTAB 27,130,45
orl-rat LD50:30,600 mg/kg EVHPAZ 3,131,73
skn-rat LDLo:4 g/kg GISAAA 45(6),35,80
ipr-rat LD50:30,700 mg/kg JIHTAB 27,130,45
ivn-rat LD50:250 mg/kg TXAPA9 45,230,78
orl-mus LD50:30 g/kg TJADAB 14,259,76
ipr-mus LD50:14 g/kg JPMSAE 55,158,66
ivn-mus LD50:1060 mg/kg NTIS** PB250-102
orl-rbt LD50:34 g/kg EVHPAZ 4,3,73
skn-rbt LD50:25 g/kg JIHTAB 27,130,45
orl-gpg LD50:26 g/kg IMEMDT 29,269,82
skn-gpg LD50:10 g/kg EVHPAZ 4,3,73

CONSENSUS REPORTS: NTP 7th Annual Report on Carcinogens. IARC Cancer Review: Group 2B IMEMDT 7,56,87; Human Inadequate Evidence IMEMDT 29,269,82; Animal Sufficient Evidence IMEMDT 29,269,82. NTP Carcinogenesis Bioassay (feed); Clear Evidence: mouse, rat NTPTR* NTP-TR-217,82. EPA Genetic Toxicology Program. Reported in EPA TSCA Inventory. Community Right-To-Know List.

OSHA PEL: TWA 5 mg/m^3; STEL 10 mg/m^3
ACGIH TLV: TWA 5 mg/m^3; STEL 10 mg/m^3
DFG MAK: 10 mg/m^3
NIOSH REL: (DEHP) Reduce to lowest feasible level

SAFETY PROFILE: Confirmed carcinogen with experimental carcinogenic and tumorigenic data. Experimental teratogenic data. Other experimental reproductive effects. Poison by intravenous route. Human systemic effects by ingestion: gastrointestinal tract effects. A mild skin and eye irritant. When heated to decomposition it emits acrid smoke.

For occupational chemical analysis use NIOSH: Di(2-ethylhexyl) phthalate, 5020.

DVL800 CAS:69226-45-5 HR: 3
DIOCTYL(1,2-PROPYLENEDIOXYBIS(MALEOYLDI-OXY))STANNANE
mf: $C_{27}H_{42}O_8Sn$ mw: 613.38

SYNS: DI-n-OCTYLTIN DI(1,2-PROPYLENEGLYCOLMALEATE) □ DI-n-OCTYL-ZINN-DI-(1,2-PROPYLENGLYKOLMALEINAT)(GERMAN)

TOXICITY DATA WITH REFERENCE
orl-rat LD50:4775 mg/kg ARZNAD 19,934,69
ipr-rat LD50:30 mg/kg ARZNAD 19,934,69

OSHA PEL: TWA 0.1 mg(Sn)/m^3 (skin)
ACGIH TLV: TWA 0.1 mg(Sn)/m^3 (skin) (Proposed: TWA 0.1 mg(Sn)/m^3; STEL 0.2 mg(Sn)/m^3 (skin))
NIOSH REL: (Organotin Compounds) TWA 0.1 mg(Sn)/m^3

SAFETY PROFILE: Poison by intraperitoneal route. Mildly toxic by ingestion. When heated to decomposition it emits acrid smoke and irritating fumes. See also TIN COMPOUNDS.

For occupational chemical analysis use NIOSH: Organotin Compounds 5504.

DVM000 CAS:3572-47-2 HR: 3
DIOCTYLTHIOXOSTANNANE
mf: $C_{16}H_{34}SSn$ mw: 377.25

SYN: DI-n-OCTYLTIN SULFIDE

TOXICITY DATA WITH REFERENCE
orl-rat LD50:1900 mg/kg ARZNAD 10,44,60
ivn-mus LD50:180 mg/kg CSLNX* NX#01771

CONSENSUS REPORTS: Reported in EPA TSCA Inventory.

OSHA PEL: TWA 0.1 mg(Sn)/m^3 (skin)
ACGIH TLV: TWA 0.1 mg(Sn)/m^3 (skin) (Proposed: TWA 0.1 mg(Sn)/m^3; STEL 0.2 mg(Sn)/m^3 (skin))
NIOSH REL: (Organotin Compounds) TWA 0.1 mg(Sn)/m^3

SAFETY PROFILE: Poison by intravenous route. When heated to decomposition it emits toxic fumes of SO_x. See also SULFIDES and TIN COMPOUNDS.

For occupational chemical analysis use NIOSH: Organotin Compounds 5504.

DVM200 CAS:27107-88-6 HR: 2
DI-n-OCTYLTIN BIS(BUTYL MERCAPTOACETATE)
mf: $C_{28}H_{56}O_4S_2Sn$ mw: 639.65

SYN: BIS(MERCAPTOACETATE)DIOCTYLTIN BIS(BUTYL) ESTER

TOXICITY DATA WITH REFERENCE
orl-mus LD50:1140 mg/kg ATXKA8 26,196,70

OSHA PEL: TWA 0.1 mg(Sn)/m^3 (skin)
ACGIH TLV: TWA 0.1 mg(Sn)/m^3 (skin) (Proposed: TWA 0.1 mg(Sn)/m^3; STEL 0.2 mg(Sn)/m^3 (skin))
NIOSH REL: (Organotin Compounds) TWA 0.1 mg(Sn)/m^3

SAFETY PROFILE: Moderately toxic by ingestion. When heated to decomposition it emits toxic fumes of SO_x. See also TIN COMPOUNDS.

For occupational chemical analysis use NIOSH: Organotin Compounds 5504.

DVM400 CAS:22205-30-7 HR: 2
DI-n-OCTYLTIN BIS(DODECYL MERCAPTIDE)
mf: $C_{44}H_{88}O_4S_2Sn$ mw: 864.13

SYN: BIS(MERCAPTO)DIOCTYLTIN BIS(DODECYL) ESTER

TOXICITY DATA WITH REFERENCE
orl-mus LD50:4000 mg/kg ATXKA8 26,196,70

CONSENSUS REPORTS: Reported in EPA TSCA Inventory.

OSHA PEL: TWA 0.1 mg(Sn)/m³ (skin)
ACGIH TLV: TWA 0.1 mg(Sn)/m³ (skin) (Proposed: TWA 0.1 mg(Sn)/m³; STEL 0.2 mg(Sn)/m³ (skin))
NIOSH REL: (Organotin Compounds) TWA 0.1 mg(Sn)/m³

SAFETY PROFILE: Moderately toxic by ingestion. When heated to decomposition it emits toxic fumes of SO_x. See also TIN COMPOUNDS and SULFIDES.

For occupational chemical analysis use NIOSH: Organotin Compounds 5504.

DVM600 CAS:10039-33-5 HR: 2
DI-n-OCTYLTIN BIS(2-ETHYLHEXYL MALEATE)
mf: $C_{40}H_{72}O_8Sn$ mw: 799.81

SYNS: BIS(HYDROGEN MALEATO)DIOCTYLTIN BIS(2-ETHYLHEXYL) ESTER ☐ DI-n-OCTYL-ZINN-BIS(2-AETHYLHEXYLMALEINAT) (GERMAN)

TOXICITY DATA WITH REFERENCE
orl-rat LD50:2760 mg/kg ARZNAD 19,934,69
orl-mus LD50:2700 mg/kg FCTXAV 8,655,70

CONSENSUS REPORTS: Reported in EPA TSCA Inventory.

OSHA PEL: TWA 0.1 mg(Sn)/m³ (skin)
ACGIH TLV: TWA 0.1 mg(Sn)/m³ (skin) (Proposed: TWA 0.1 mg(Sn)/m³, STEL 0.2 mg(Sn)/m³ (skin))
NIOSH REL: (Organotin Compounds) TWA 0.1 mg(Sn)/m³

SAFETY PROFILE: Moderately toxic by ingestion. When heated to decomposition it emits acrid smoke and irritating fumes. See also TIN COMPOUNDS.

For occupational chemical analysis use NIOSH: Organotin Compounds 5504.

DVM800 CAS:15571-58-1 HR: 2
DI-n-OCTYLTIN BIS(2-ETHYLHEXYL) MERCAPTOACETATE
mf: $C_{36}H_{72}O_4S_2Sn$ mw: 751.89

SYNS: BIS(2-ETHYLHEXYLTHIOGLYCOLATE)DIOCTYLTIN ☐ BIS(MERCAPTOACETATE)DIOCTYLTIN BIS(2-ETHYLHEXYL) ESTER ☐ 10-ETHYL-4,4-DIOCTYL-7-OXO-8-OXA-3,5-DITHIA-4-STANNATETRADECANOIC ACID-2-ETHYLHEXYL ESTER ☐ DI-N-OCTYLTIN-ITHIOGLYCOLIC ACID 2 ETHYLHEXYL ESTER ☐ DI N OCTYLTIN 2 ETHYLHEXYLDIMERCAPTOETHANOATE ☐ OTS 11

TOXICITY DATA WITH REFERENCE
orl-rat LD50:2100 mg/kg NAHRAR 13,343,69
orl-mus LD50:2010 mg/kg ATXKA8 26,196,70

CONSENSUS REPORTS: Reported in EPA TSCA Inventory.

OSHA PEL: TWA 0.1 mg(Sn)/m³ (skin)
ACGIH TLV: TWA 0.1 mg(Sn)/m³ (skin) (Proposed: TWA 0.1 mg(Sn)/m³; STEL 0.2 mg(Sn)/m³ (skin))
DFG MAK: 0.1 mg(Sn)/m³ calculated as total dust
NIOSH REL: (Organotin Compounds) TWA 0.1 mg(Sn)/m³

SAFETY PROFILE: Moderately toxic by ingestion. When heated to decomposition it emits toxic fumes of SO_x. See also TIN COMPOUNDS and ESTERS.

For occupational chemical analysis use NIOSH: Organotin Compounds 5504.

DVN000 CAS:69226-43-3 HR: 2
DI-n-OCTYLTIN BIS(LAURYLTHIOGLYCOLATE)
mf: $C_{44}H_{88}O_4S_2Sn$ mw: 864.13

SYNS: BIS(LAUROYLOXYCARBONYLMETHYLTHIO)DIOCTYLSTANNANE ☐ DI-n-OCTYL-ZINN-BIS(LAURYL-THIOGLYKOLAT) (GERMAN)

TOXICITY DATA WITH REFERENCE
orl-rat LD50:3700 mg/kg ARZNAD 19,934,69

OSHA PEL: TWA 0.1 mg(Sn)/m³ (skin)
ACGIH TLV: TWA 0.1 mg(Sn)/m³ (skin) (Proposed: TWA 0.1 mg(Sn)/m³; STEL 0.2 mg(Sn)/m³ (skin))
NIOSH REL: (Organotin Compounds) TWA 0.1 mg(Sn)/m³

SAFETY PROFILE: Moderately toxic by ingestion. When heated to decomposition it emits toxic fumes of SO_x. See also TIN COMPOUNDS.

For occupational chemical analysis use NIOSH: Organotin Compounds 5504.

DVN200 CAS:69226-46-6 HR: 2
DI-n-OCTYLTIN-1,4-BUTANEDIOL-BIS-MERCAPTOACETATE
mf: $C_{24}H_{46}O_4S_2Sn$ mw: 581.51

SYN: DI-n-OCTYL-ZINN-1,4-BUTANDIOL-BIS-MERCAPTOACETAT (GERMAN)

TOXICITY DATA WITH REFERENCE
orl-rat LD50:2950 mg/kg ARZNAD 19,934,69

OSHA PEL: TWA 0.1 mg(Sn)/m³ (skin)
ACGIH TLV: TWA 0.1 mg(Sn)/m³ (skin) (Proposed: TWA 0.1 mg(Sn)/m³; STEL 0.2 mg(Sn)/m³ (skin))
NIOSH REL: (Organotin Compounds) TWA 0.1 mg(Sn)/m³

SAFETY PROFILE: Moderately toxic by ingestion. When heated to decomposition it emits toxic fumes of SO_x. See also TIN COMPOUNDS and MERCAPTANS.

For occupational chemical analysis use NIOSH: Organotin Compounds 5504.

DVN300 **CAS:3542-36-7** **HR: 3**
DI-n-OCTYLTINDICHLORIDE
mf: $C_{16}H_{34}Cl_2Sn$ mw: 416.09

SYNS: DICHLORODIOCTYLSTANNANE □ DIOCTYLSTANNIUM DICHLORIDE □ DIOCTYLTIN DICHLORIDE □ DOTC □ DI-n-OCTYLZINN DICHLORID □ STANNANE, DICHLORODIOCTYL- □ STANNANE, DIOCTYLDICHLORO- □ TIN, DIOCTYL-, DICHLORIDE

TOXICITY DATA WITH REFERENCE
msc-ham:lng 1250 μg/L ARZNAD 36,1263,86
dns-rbt:Cells-uns 1 μg/L JTEHD6 16,229,85
orl-rat LD50:5500 mg/kg ARZNAD 19,934,69
ivn-rat LDLo:10 mg/kg BJIMAG 15,15,58
ivn-mus LD50:18 mg/kg CSLNX* NX#02188
orl-rbt LDLo:250 mg/kg SAIGBL 15,3,73

CONSENSUS REPORTS: Reported in EPA TSCA Inventory.

OSHA PEL: TWA 0.1 mg(Sn)/m³ (skin)
ACGIH TLV: TWA 0.1 mg(Sn)/m³; STEL 0.2 mg/m³ (skin)
NIOSH REL: (Organotin Compound): TWA 0.1 mg(Sn)/m³

SAFETY PROFILE: A poison by ingestion and intravenous routes. Mutation data reported. When heated to decomposition it emits toxic vapors of Sn and Cl^-.

For occupational chemical analysis use NIOSH: Organotin compounds 5504.

DVN400 **CAS:69226-44-4** **HR: 2**
DI-n-OCTYLTIN ETHYLENEGLYCOL DITHIOGLYCOLATE
mf: $C_{22}H_{42}O_4S_2Sn$ mw: 553.45

SYNS: DIOCTYL(ETHYLENEDIOXYBIS(CARBONYLMETHYLTHIO))STANNANE □ DI-n-OCTYL-ZINN AETHYLENGLYKOL-DITHIOGLYKOLAT (GERMAN) □ OTS 15

TOXICITY DATA WITH REFERENCE
orl-rat LD50:880 mg/kg ARZNAD 19,934,69

OSHA PEL: TWA 0.1 mg(Sn)/m³ (skin)
ACGIH TLV: TWA 0.1 mg(Sn)/m³ (skin) (Proposed: TWA 0.1 mg(Sn)/m³; STEL 0.2 mg(Sn)/m³ (skin))
NIOSH REL: (Organotin Compounds) TWA 0.1 mg(Sn)/m³

SAFETY PROFILE: Moderately toxic by ingestion. When heated to decomposition it emits toxic fumes of SO_x. See also TIN COMPOUNDS.

For occupational chemical analysis use NIOSH: Organotin Compounds 5504.

DVN600 **CAS:58229-88-2** **HR: 2**
DI-n-OCTYLTIN MERCAPTIDE

SYNS: DIOCTYLTIN MERCAPTIDE □ ERGOTERM OTGO

TOXICITY DATA WITH REFERENCE
orl-rat LDLo:750 mg/kg RPZHAW 19,329,68

OSHA PEL: TWA 0.1 mg(Sn)/m³ (skin)
ACGIH TLV: TWA 0.1 mg(Sn)/m³ (skin) (Proposed: TWA 0.1 mg(Sn)/m³; STEL 0.2 mg(Sn)/m³ (skin))
NIOSH REL: (Organotin Compounds) TWA 0.1 mg(Sn)/m³

SAFETY PROFILE: Moderately toxic by ingestion. When heated to decomposition it emits toxic fumes of SO_x. See also TIN COMPOUNDS and MERCAPTANS.

For occupational chemical analysis use NIOSH: Organotin Compounds 5504.

DVN800 **CAS:3033-29-2** **HR: 3**
DI-n-OCTYLTIN β-MERCAPTOPROPIONATE
mf: $C_{19}H_{38}O_2SSn$ mw: 449.32

SYNS: DIHYDRO-2,2-DIOCTYL-6H-1,3,2-OXATHIASTANNIN-6-ONE □ DIOCTYLTIN-β-MERCAPTOPROPIONATE □ DI-n-OCTYL-ZINN β-MERCAPTOPROPIONAT (GERMAN)

TOXICITY DATA WITH REFERENCE
orl-rat LD50:1850 mg/kg ARZNAD 19,934,69
ipr-rat LD50:6600 μg/kg TXAPA9 35,63,76

CONSENSUS REPORTS: Reported in EPA TSCA Inventory.

OSHA PEL: TWA 0.1 mg(Sn)/m³ (skin)
ACGIH TLV: TWA 0.1 mg(Sn)/m³ (skin) (Proposed: TWA 0.1 mg(Sn)/m³; STEL 0.2 mg(Sn)/m³ (skin))
NIOSH REL: (Organotin Compounds) TWA 0.1 mg(Sn)/m³

SAFETY PROFILE: Poison by intraperitoneal route. Moderately toxic by ingestion. When heated to decomposition it emits toxic fumes of SO_x. See also TIN COMPOUNDS and MERCAPTANS.

For occupational chemical analysis use NIOSH: Organotin Compounds 5504.

DVN909 **CAS:3594-15-8** **HR: 3**
DIOCTYLTIN-3,3′-THIODIPROPIONATE
mf: $C_{22}H_{42}O_4SSn$ mw: 521.39

SYN: 2,2-DIOCTYL-1,3-DIOXA-2-STANNA-7-THIADECAN-4,10-DIONE

TOXICITY DATA WITH REFERENCE
ivn-mus LD50:320 mg/kg CSLNX* NX#02854

OSHA PEL: TWA 0.1 mg(Sn)/m³ (skin)
ACGIH TLV: TWA 0.1 mg(Sn)/m³ (skin) (Proposed: TWA 0.1 mg(Sn)/m³; STEL 0.2 mg(Sn)/m³ (skin))
NIOSH REL: (Organotin Compounds) TWA 0.1 mg(Sn)/m³

SAFETY PROFILE: Poison by intravenous route. When heated to decomposition it emits toxic fumes of SO_x. See also TIN COMPOUNDS.

For occupational chemical analysis use NIOSH: Organotin Compounds 5504.

DVO000 **HR: 3**
DIOCYDE

TOXICITY DATA WITH REFERENCE
orl-rat LD50:172 mg/kg KHFZAN 11(7),44,77
orl-mus LD50:54 mg/kg KHFZAN 11(7),44,77
ipr-mus LD50:21,700 μg/kg KHFZAN 11(7),44,77

SAFETY PROFILE: Poison by ingestion and intraperitoneal routes.

DVO100 CAS:582-52-5 HR: 1
1:2,5:6-DI-O-ISOPROPYLIDENE-α-D-GLUCOFURANOSE
mf: $C_{12}H_{20}O_6$ mw: 260.32

SYNS: 1:2,5:6-DI-O-ISOPROPYLIDEN-α-D-GLUCOFURANOSE ☐ GLUCOFURANOSE, 1:2,5:6-DI-O-ISOPROPYLIDENE-, α-D-

TOXICITY DATA WITH REFERENCE
orl-mus LD50:4 g/kg ARZNAD 29,986,79

CONSENSUS REPORTS: Reported in EPA TSCA Inventory.

SAFETY PROFILE: Mildly toxic by ingestion. When heated to decomposition it emits acrid smoke and irritating vapors.

DVO175 CAS:63323-30-8 HR: 2
anti-DIOLEPOXIDE
mf: $C_{20}H_{12}O_3$ mw: 300.32

SYNS: (−)-7-α,8-β-DIHYDROXY-9-β,10-β-EPOXY-7,8,9,10-TETRAHYDROBENZO(a)PYRENE ☐ (−)-BP 7-α,8-β-DIOL-9-β,10-β-EPOXIDE 2 ☐ (−)-7,8,9,10-TETRAHYDRO-7-β,8-α-DIHYDROXY-9-α,10-α-EPOXYBENZO(a)PYRENE ☐ anti-DIOLEPOXIDE

TOXICITY DATA WITH REFERENCE
mmo-sat 100 pmol/plate BBRCA9 77,1389,77
cyt-ham:lng 300 µg/L IJCNAW 24,485,79
sce-ham:lng 600 µg/L IJCNAW 24,485,79
msc-ham:lng 10 µmol CRNGDP 3,1223,82
skn-mus TDLo:2400 µg/kg:ETA CNREA8 39,67,79

SAFETY PROFILE: Questionable carcinogen with experimental tumorigenic data by skin contact. Mutation data reported. When heated to decomposition it emits toxic fumes of NO_x.

DVO600 CAS:702-62-5 HR: 2
2-4-DIONE-1,3-DIAZASPIRO(4.5)DECANE
mf: $C_8H_{12}N_2O_2$ mw: 168.22

PROP: Crystals from EtOH (aq). Mp: 218–220°.

SYNS: CYCLOHEXANESPIRO-5′-HYDANTOIN ☐ SPIRO(CYCLOHEXANE-1,5′-HYDANTOIN)

TOXICITY DATA WITH REFERENCE
orl-mus LD50:420 mg/kg JMCMAR 8,239,65

CONSENSUS REPORTS: Reported in EPA TSCA Inventory.

SAFETY PROFILE: Moderately toxic by ingestion. When heated to decomposition it emits toxic fumes of NO_x.

DVO700 CAS:125-30-4 HR: 3
DIONIN HYDROCHLORIDE
mf: $C_{19}H_{23}NO_3 \cdot ClH$ mw: 349.89

PROP: Bitter-tasting powder. Mp: 123° (decomp).

SYNS: CODETHYLINE HYDROCHLORIDE ☐ DIONINE HYDROCHLORIDE ☐ ETHYLMORPHINE HYDROCHLORIDE ☐ o-ETHYLMORPHINE HYDROCHLORIDE

TOXICITY DATA WITH REFERENCE
orl-rat LD50:950 mg/kg ARZNAD 21,719,71
scu-rat LD50:200 mg/kg ARZNAD 21,727,71
orl-mus LD50:520 mg/kg ARZNAD 21,719,71
scu-mus LD50:265 mg/kg APFRAD 15,640,57

SAFETY PROFILE: Poison by subcutaneous route. Moderately toxic by ingestion. When heated to decomposition it emits toxic fumes of NO_x and HCl.

DVO809 CAS:17667-23-1 HR: 3
DIOSPYROL
mf: $C_{22}H_{18}O_4$ mw: 346.40

SYN: 6,6′-DIMETHYL-(2,2′-BINAPHTHALENE)-1,1′,8,8′-TETROL

TOXICITY DATA WITH REFERENCE
orl-rat LD50:3000 mg/kg DRFUD4 5,438,80
orl-mus LD50:3000 mg/kg DRFUD4 5,438,80
orl-dog LD50:2000 mg/kg DRFUD4 5,438,80
ipr-mky LD50:100 mg/kg DRFUD4 5,438,80
orl-ham LD50:3000 mg/kg DRFUD4 5,438,80

SAFETY PROFILE: Poison by intraperitoneal route. Moderately toxic by ingestion. When heated to decomposition it emits acrid smoke and fumes.

DVO819 CAS:101-08-6 HR: 3
DIOTHANE
mf: $C_{22}H_{27}N_3O_4$ mw: 397.52

SYNS: 3-PIPERIDINO-1,2-PROPANEDIOL DICARBANILATE ☐ 3-(1-PIPERIDYL)-1,2-PROPANE DICARBANILATE

TOXICITY DATA WITH REFERENCE
scu-mus LDLo:1200 mg/kg JPETAB 47,255,33
scu-rbt LDLo:300 mg/kg JPETAB 47,255,33
ivn rbt LDLo:15 mg/kg JPETAB 47,255,33
scu-gpg LDLo:400 mg/kg JPETAB 47,255,33

SAFETY PROFILE: Poison by subcutaneous and intravenous routes. When heated to decomposition it emits toxic fumes of NO_x.

DVP400 CAS:631-06-1 HR: 3
d-DIOXADROL HYDROCHLORIDE
mf: $C_{20}H_{23}NO_2 \cdot ClH$ mw: 345.90

PROP: A solid. Mp: 256–260°.

SYNS: CL-911C ☐ DEXOXADROL HYDROCHLORIDE ☐ d-2-(2,2-DIPHENYL-1,3-DIOXOLAN-4-YL)PIPERIDINE HYDROCHLORIDE ☐ d-2,2-DIPHENYL-4-(2-PIPERIDYL)-1,3-DIOXOLANE HYDROCHLORIDE ☐ NSC-526062 ☐ RELANE ☐ U-22,559A

TOXICITY DATA WITH REFERENCE
orl-hmn TDLo:400 µg/kg:CNS PSEBAA 118,352,65
orl-rat LD50:280 mg/kg AIPTAK 153,105,65
orl-mus LD50:380 mg/kg AIPTAK 153,105,65
ivn-rbt LD50:33 mg/kg AIPTAK 153,105,65

SAFETY PROFILE: Poison by ingestion and intravenous routes. Human systemic effects by ingestion: somnolence, hallucinations or distorted perceptions, and analgesia. When heated to decomposition it emits very toxic fumes of HCl and NO_x.

DVP600 **CAS:505-22-6** **HR: 3**
m-DIOXAN
mf: $C_4H_8O_2$ mw: 88.12

$O(CH_2)_3O\ CH_2$

PROP: A liquid. Bp: 105° @ 755 mm, flash p: 33.8°F. lel: 2%, uel: 22%. Sol in water.

SYNS: 1,3-DIOXACYCLOHEXANE □ 1,3-DIOXANE □ 1,3-PROPANEDIOL FORMAL

TOXICITY DATA WITH REFERENCE
mmo-sat 3333 µg/plate ENMUDM 5(Suppl 1),3,83
uns-sat 70 mg/L MUREAV 192,239,87
cyt-ham:ovr 1300 mg/L EMMUEG 10(Suppl 10),1,87
sce-ham:ovr 2080 mg/L EMMUEG 10(Suppl 10),1,87

SAFETY PROFILE: Mutation data reported. A very dangerous fire and explosion hazard when exposed to heat or flame; can react with oxidizing materials. Can form dangerous peroxides when exposed to air. When heated to decomposition it emits acrid smoke and fumes.

DVQ000 **CAS:123-91-1** **HR: 3**
DIOXANE
DOT: UN 1165
mf: $C_4H_8O_2$ mw: 88.11

$OC_2H_4OCH_2\ CH_2$

PROP: Colorless liquid with pleasant odor. Mp: 12°, fp: 11°, bp: 101.1°, lel: 2.0%, uel: 22.2%, flash p: 54°F (CC), d: 1.0353 @ 20°/4°, autoign temp: 356°F, vap press: 40 mm @ 25.2°, vap d: 3.03. Sol in EtOH and C_6H_6.

SYNS: DIETHYLENE DIOXIDE □ 1,4-DIETHYLENE DIOXIDE □ DIETHYLENE ETHER □ DI(ETHYLENE OXIDE) □ DIOKAN □ DIOKSAN (POLISH) □ DIOSSANO-1,4 (ITALIAN) □ DIOXAAN-1,4 (DUTCH) □ p-DIOXAN (CZECH) □ DIOXAN-1,4 (GERMAN) □ p-DIOXANE □ 1,4-DIOXANE (MAK) □ DIOXANNE (FRENCH) □ DIOXYETHYLENE ETHER □ GLYCOL ETHYLENE ETHER □ NCI-C03689 □ RCRA WASTE NUMBER U108 □ TETRAHYDRO-p-DIOXIN □ TETRAHYDRO-1,4-DIOXIN

TOXICITY DATA WITH REFERENCE
eye-hmn 300 ppm/15M JIHTAB 28,262,46
skn-rbt 515 mg open MLD UCDS** 12/17/71
eye-rbt 21 mg AJOPAA 29,1363,46
eye-gpg 10 µg MOD JPPMAB 11,150,59
dnd-rat:lvr 300 µmol/L SinJF# 26OCT82
oms-rat-ivn 50 mg/kg ARTODN 49,29,81
orl-rat TDLo:10 g/kg (6-15D preg):TER TOLED5 26,85,85
orl-rat TDLo:185 g/kg/2Y-C:CAR NCITR* NCI-CG-TR-80,78
ihl-rat TCLo:111 ppm/7H/2Y-C:ETA TXAPA9 30,287,74
orl-mus TDLo:239 g/kg/90W-C:CAR NCITR* NCI-CG-TR-80,78
skn-mus TDLo:14 g/kg/60W-I:ETA EVHPAZ 5,163,73
ipr-mus TDLo:12 g/kg/8W-I:NEO TXAPA9 82,19,86
orl-rat TD:416 g/kg/57W-C:ETA BJCAAI 24,164,70
orl-rat TD:408 g/kg/2Y-C:CAR NCITR* NCI-CG-TR-80,78
orl-mus TD:523 g/kg/90W-C:CAR NCITR* NCI-CG-TR-80,78
orl-rat TD:416 g/kg/57W-C:CAR BJCAAI 24,164,70
orl-rat TD:528 g/kg/63W-I:ETA JNCIAM 35,949,65
ihl-hmn TCLo:470 ppm:CNS,CVS,GIT AMIHAB 20,445,59
ihl-hmn TCLo:5500 ppm/1M:EYE,PUL PHRPA6 45,2023,30
ihl-hmn LCLo:470 ppm/3D PLENBW 7,22,75
ihl-rat LC50:46 g/m³/2H KBAMAJ 11(6),53,77
ipr-rat LD50:799 mg/kg ENVRAL 40,411,86
orl-mus LD50:5700 mg/kg JIHTAB 21,173,39
ihl-mus LC50:37 g/m³/2H 85GMAT -,63,82
ipr-mus LD50:790 mg/kg FEPRA7 6,342,47
orl-cat LD50:2000 mg/kg JIHTAB 21,173,39
ihl-cat LCLo:44 g/m³/7H KDPU** -,-,37
orl-rbt LD50:2000 mg/kg JIHTAB 21,173,39
skn-rbt LD50:7600 mg/kg UCDS** 12/17/71
ivn-rbt LDLo:1500 mg/kg JOHYAY 35,540,35
orl-gpg LD50:3150 mg/kg JIHTAB 23,259,41

CONSENSUS REPORTS: NTP 7th Annual Report on Carcinogens. IARC Cancer Review: Group 2B IMEMDT 7,201,87; Animal Sufficient Evidence IMEMDT 11,247,76. NCI Carcinogenesis Bioassay (oral); Clear Evidence: mouse, rat NCITR* NCI-CG-TR-80,78. EPA Genetic Toxicology Program. Glycol ether compounds are on the Community Right-To-Know List. Reported in EPA TSCA Inventory.

OSHA PEL: TWA 25 ppm (skin)
ACGIH TLV: TWA 25 ppm (skin)
DFG MAK: 50 ppm (180 mg/m³); Suspected Carcinogen (Proposed: 25 ppm (skin); Animal Carcinogen)
NIOSH REL: CL (Dioxane) 1 ppm/30M
DOT CLASSIFICATION: 3; *Label:* Flammable Liquid

SAFETY PROFILE: Confirmed carcinogen with experimental carcinogenic, neoplastigenic, tumorigenic, and teratogenic data. Poison by intraperitoneal route. Moderately toxic by ingestion and inhalation. Mildly toxic by skin contact. Human systemic effects by inhalation: lacrimation, conjunctiva irritation, convulsions, high blood pressure, unspecified respiratory and gastrointestinal system effects. Mutation data reported. An eye and skin irritant. The irritant effects probably provide sufficient warning, in acute exposures, to enable a worker to leave exposure before being seriously affected. Repeated exposure to low concentrations has resulted in human fatalities, the organs chiefly affected being the liver and kidneys.

A very dangerous fire and explosion hazard when exposed to heat or flame; can react vigorously with oxidizing materials. Violent reaction with (H_2 + Raney Ni), $AgClO_4$. Can form dangerous peroxides when exposed to air. Potentially explosive reaction with nitric acid + perchloric acid; Raney nickel catalyst (above 210°C). Forms explosive mixtures with decaborane (impact-sensitive); triethynylaluminum (sensitive to heating or drying). Violent reaction with sulfur trioxide. Incompatible with sulfur trioxide. To fight fire, use alcohol foam, CO_2, dry chemical. When heated to decomposition it emits acrid smoke and irritating fumes. See also GLYCOL ETHERS.

For occupational chemical analysis use NIOSH: Dioxane, 1602.

DVQ400 **CAS:766-15-4** **HR: 2**
m-DIOXANE-4,4-DIMETHYL
mf: $C_6H_{12}O_2$ mw: 116.18

PROP: Bp: 132°.

SYN: 4,4-DIMETHYLDIOXANE-1,3

TOXICITY DATA WITH REFERENCE
skn-rbt 10 mg/24H open MLD AIHAAP 23,95,62
skn-rbt 500 mg/24H MLD 85JCAE-,812,86
eye-rbt 2 mg/24H SEV 85JCAE-,812,86
ihl-rat TCLo:10 μg/m³/24H (16W pre):TER GISAAA 43(9),16,78
ihl-rat TCLo:10 μg/m³/24H (16W pre):REP GISAAA 43(9),16,78
orl-rat LD50:3730 mg/kg AIHAAP 23,95,62
ihl-rat LCLo:8000 ppm/4H AIHAAP 23,95,62
orl-mus LDLo:1 g/kg GISAAA 25(6),85,60
skn-rbt LD50:3540 mg/kg AIHAAP 23,95,62

SAFETY PROFILE: Moderately toxic by ingestion and skin contact. An experimental teratogen. A skin and severe eye irritant. When heated to decomposition it emits acrid smoke and fumes.

DVQ600 **CAS:16088-56-5** **HR: 3**
cis-2,3-p-DIOXANEDITHIOL-S,S-BIS(O,O-DIETHYL-PHOSPHORODITHIOATE)
mf: $C_{12}H_{26}O_6P_2S_4$ mw: 456.56

PROP: A liquid.

SYN: (E)-PHOSPHORODITHIOIC ACID-O,O-DIETHYL ESTER-S,S-DIESTER with p-DIOXANE-2,3-DIETHIOL

TOXICITY DATA WITH REFERENCE
scu-rat LD50:66 mg/kg TXAPA9 5,605,63

SAFETY PROFILE: Poison by subcutaneous route. When heated to decomposition it emits very toxic fumes of PO_x and SO_x. See also ESTERS.

DVQ709 **CAS:78-34-2** **HR: 3**
DIOXATHION
mf: $C_{12}H_{26}O_6P_2S_4$ mw: 456.56

PROP: Nonvolatile, stable solid or brown liquid (tech grade). D: 1.257 @ 26°/4°, mp: −20°, bp: 60–68° @ 0.5 mm. Nonflammable. Insol in water.

SYNS: BIS(DITHIOPHOSPHATE de O,O-DIETHYLE) de S,S′-(1,4-DIOXANNE-2,3-DIYLE) (FRENCH) ☐ DELNAV ☐ 1,4-DIOSSAN-2,3-DIYL-BIS(O,O-DIETIL-DITIOFOSFATO) (ITALIAN) ☐ 1,4-DIOXAAN-2,3-DIYL-BIS(O,O-DIETHYL-DITHIOFOSFAAT) (DUTCH) ☐ 2,3-p-DIOXANDITHIOL S,S-BIS(O,O-DIETHYL PHOSPHORODITHIOATE) ☐ 1,4-DIOXAN-2,3-DIYL-BIS(O,O-DIAETHYL-DITHIOPHOSPHAT) (GERMAN) ☐ 1,4-DIOXAN-2,3-DIYL-BIS(O,O-DIETHYLPHOSPHOROTHIOLOTHIONATE) ☐ 1,4-DIOXAN-2,3-DIYL-O,O,O′,O′-TETRAETHYL DI(PHOSPHORODITHIOATE) ☐ 2,3-p-DIOXANE-S,S-BIS(O,O-DIETHYLPHOSPHOROITHIOATE) ☐ p-DIOXANE-2,3-DITHIOL-S,S-DIESTER with O,O-DIETHYL PHOSPHORODITHIOATE ☐ p-DIOXANE-2,3-DIYL ETHYL PHOSPHORODITHIOATE ☐ ENT 22,897 ☐ NCI-C00395 ☐ PHOSPHORODITHIOIC ACID-S,S′-1,4-DIOXANE-2,3-DIYL O,O,O′,O′-TETRAETHYL ESTER

TOXICITY DATA WITH REFERENCE
mmo-sat 6667 μg/plate ENMUDM 8(Suppl 7),1,86
orl-hmn TDLo:9 mg/kg/60D:BLD 34ZIAG-,200,69
orl-rat LD50:20 mg/kg WRPCA2 9,119,70
ihl-rat LC50:1398 mg/m³/1H TXAPA9 5,605,63
skn-rat LD50:63 mg/kg TXAPA9 5,605,63
ipr-rat LD50:30 mg/kg TXAPA9 5,605,63
orl-mus LD50:176 mg/kg TXAPA9 5,605,63
ihl-mus LC50:340 mg/m³/1H TXAPA9 5,605,63
ipr-mus LD50:33 mg/kg PSEBAA 129,699,68
orl-dog LD50:10 mg/kg PCOC** -,427,66
skn-rbt LD50:85 mg/kg PCOC** -,427,66
orl-ckn LD50:170 mg/kg 32ZXAD 37,A10,75

CONSENSUS REPORTS: NCI Carcinogenesis Bioassay (feed); No Evidence: mouse, rat NCITR* NCI-CG-TR-125,78.

OSHA PEL: TWA 0.2 mg/m³ (skin)
ACGIH TLV: TWA 0.2 mg/m³ (skin)

SAFETY PROFILE: Poison by ingestion, inhalation, skin contact, and intraperitoneal routes. Mutation data reported. A cholinesterase inhibitor. When heated to decomposition it emits very toxic fumes of PO_x and SO_x. See also PARATHION.

D

DVQ759 **CAS:59261-17-5** **HR: 3**
cis-1,4-DIOXENEDIOXETANE
mf: $C_4H_6O_4$ mw: 118.09

SYN: 2,5,7,8-TETRAOXA[4.2.0]BICYCLOOCTANE

SAFETY PROFILE: Explodes at room temperature. When heated to decomposition it emits acrid smoke and fumes. See also PEROXIDES.

DVQ800 **CAS:17311-31-8** **HR: 2**
1,4-DI-N-OXIDE of DIHYDROXYMETHYLQUINOXALINE
mf: $C_{10}H_{12}N_2O_4$ mw: 222.22

SYNS: 2,3-BIS(HYDROXYMETHYL)QUINOXALINE DI-N-OXIDE ☐ 1,4-DI-N-OXIDE 2,3-BIS(OXYMETHYL)QUINOXLINE ☐ 1,4-DIOXIDE-2,3-QUINOXALINEDIMETHANOL ☐ DIOXIDIN ☐ DIOXIDINE ☐ DIOXYDINE

TOXICITY DATA WITH REFERENCE
mmo-sat 750 μg/L CYGEDX 14(1),57,80
mmo-esc 4 mg/L CYGEDX 14(1),57,80
dnr-esc 100 μg/L PCJOAU 15,721,82
dnd-esc 100 μg/L KHFZAN 16(10),11,82
pic-esc 30 mg/L TGANAK 16(6),38,82
scu-rat TDLo:300 mg/kg (11D preg):TER FATOAO 45(6),85,82
scu-rat TDLo:300 mg/kg (11D preg):REP FATOAO 45(6),85,82
ipr-mus LD50:750 mg/kg PCJOAU 14,440,80

CONSENSUS REPORTS: EPA Genetic Toxicology Program.

SAFETY PROFILE: Moderately toxic by intraperitoneal route. Experimental teratogenic and reproductive effects. Mutation data reported. When heated to decomposition it emits toxic fumes of NO_x.

DVR000 **CAS:107-61-9** **HR: 3**
4,4-DIOXIDE-1,4-OXATHIANE
mf: $C_4H_8O_3S$ mw: 136.18

SYNS: p-OXATHIANE-4,4-DIOXIDE □ USAF DO-38

TOXICITY DATA WITH REFERENCE
ipr-mus LD50:200 mg/kg NTIS** AD277-689

CONSENSUS REPORTS: Reported in EPA TSCA Inventory.

SAFETY PROFILE: Poison by intraperitoneal route. When heated to decomposition it emits toxic fumes of SO_x.

DVR200 **CAS:105-11-3** **HR: 2**
DIOXIME-p-BENZOQUINONE
mf: $C_6H_6N_2O_2$ mw: 138.14

PROP: Pale-yellow crystals. Mp: 240° (decomp).

SYNS: ACTOR Q □ 1,4-BENZOQUINONE DIOXINE □ 2,5-CYCLOHEXADIENE-1,4-DIONE DIOXIME □ DIBENZO PQD □ DIOXIME-1,4-CYCLOHEXADIENEDIONE □ DIOXIME-2,5-CYCLOHEXADIENE-1,4-DIONE □ G-M-F □ NCI-C03850 □ PQD □ QDO □ QUINONE DIOXIME □ p-QUINONE DIOXIME □ p-QUINONE OXIME

TOXICITY DATA WITH REFERENCE
mmo-sat 3300 ng/plate ENMUDM 7(Suppl 5),1,85
mma-sat 10 μg/plate ENMUDM 7(Suppl 5),1,85
dnr-bcs 1 mg/disc SAIGBL 26,147,84
orl-rat TDLo:14 g/kg/2Y-C:NEO NCITR* NCI-CG-TR-179,79
orl-mus TDLo:131 g/kg/104W-C:ETA NCITR* NCI-CG-TR-179,79
orl-rat LD50:464 mg/kg NCILB* NIH-NCI-E-C-72-3252,73
orl-mus LD50:1420 mg/kg GISAAA 29(10),15,64

CONSENSUS REPORTS: IARC Cancer Review: Group 3 IMEMDT 7,56,87; Animal Limited Evidence IMEMDT 29,185,82. NCI Carcinogenesis Bioassay (feed); Clear Evidence: rat NCITR* NCI-CG-TR-179,79; No Evidence: mouse NCITR* NCI-CG-TR-179,79. Reported in EPA TSCA Inventory.

SAFETY PROFILE: Moderately toxic by ingestion. Questionable carcinogen with experimental neoplastigenic and tumorigenic data. Mutation data reported. When heated to decomposition it emits toxic fumes of NO_x.

DVR600 **CAS:100-79-8** **HR: 1**
DIOXOLAN
mf: $C_6H_{12}O_3$ mw: 132.18

PROP: Water-white liquid. Mp: −26.4°, bp: 75°, flash p: 35°F (OC), d: 1.065, vap press: 70 mm @ 20°, vap d: 2.6.

SYNS: CYCLIC (HYDROXYMETHYL)ETHYLENE ACETAL ACETONE □ 2,2-DIMETHYL-1,3-DIOXOLANE-4-METHANOL □ 2,2-DIMETHYL-5-HYDROXYMETHYL-1,3-DIOXOLANE □ 2,2-DIMETHYL-4-OXYMETHYL-1,3-DIOXOLANE □ DIOXOLANE (DOT) □ GIE □ GLYCEROLACETONE □ GLYCEROL DIMETHYLKETAL □ 4-HYDROXYMETHYL-2,2-DIMETHYL-1,3-DIOXOLANE □ ISOPROPYLIDENE GLYCEROL □ 1,2-o-ISOPROPYLIDENE GLYCEROL □ SOLKETAL

TOXICITY DATA WITH REFERENCE
eye-rbt 100 mg TXAPA9 39,129,77
mnt-mus-ipr 1500 mg/kg TOLED5 21,349,84
orl-rat LD50:7 g/kg 85ESA3 11,820,89
ipr-rat LD50:3 g/kg 85ESA3 11,820,89
ivn-rat LDLo:3740 μg/kg DECRDP 9,895,83
ivn-rbt LDLo:8530 μg/kg DECRDP 9,895,83

CONSENSUS REPORTS: Reported in EPA TSCA Inventory.

SAFETY PROFILE: A poison by intravenous route. An eye irritant. Mutation data reported. A very dangerous fire hazard when exposed to heat or flame; can react vigorously with oxidizing materials. To fight fire, use alcohol foam, CO_2, dry chemical. When heated to decomposition it emits acrid smoke and fumes.

DVR800 **CAS:646-06-0** **HR: 2**
1,3-DIOXOLANE
DOT: UN 1166
mf: $C_3H_6O_2$ mw: 74.09

PROP: A liquid. D: 1.066 @ 15°/4°, fp: −95°, bp: 78° @ 750 mm, flash p: 35.6°F. Misc in water.

SYNS: 1,3-DIOXACYCLOPENTANE □ 1,3-DIOXOLAN □ ETHYLAENE GLYCOL FORMAL □ FORMAL GLYCOL □ GLYCOL FORMAL

TOXICITY DATA WITH REFERENCE
skn-rbt 530 mg open MLD UCDS** 12/17/71
eye-rbt 750 μg open SEV JIHTAB 31,60,49
orl-rat LD50:3000 mg/kg JIHTAB 31,60,49
ihl-rat LD50:20,650 mg/m³/4H 85GMAT -,70,82
dnd-rat-ipr 290 mg/kg STBIBN 107,205,85
ipr-rat LDLo:500 mg/kg JPPMAB 11,150,59
orl-mus LD50:3200 mg/kg 85GMAT -,70,82
ihl-mus LC50:104 g/m³ GTPZAB 19(8),45,75
ihl-rbt LCLo:32,000 ppm/4H UCDS** 12/17/71
skn-rbt LD50:8480 mg/kg UCDS** 12/17/71

CONSENSUS REPORTS: Reported in EPA TSCA Inventory.

DOT CLASSIFICATION: 3; *Label:* Flammable Liquid

SAFETY PROFILE: Moderately toxic by ingestion and intraperitoneal routes. Mildly toxic by skin contact and inhalation. A skin and severe eye irritant. Mutation data reported. A very dangerous fire hazard when exposed to heat or flame; can react with oxidizers. Used in lithium batteries. Potentially explosive reaction with lithium perchlorate. When heated to decomposition it emits acrid smoke and irritating fumes.

DVR909 **CAS:5464-28-8** **HR: 1**
1,3-DIOXOLANE-4-METHANOL
mf: $C_4H_8O_3$ mw: 104.12

SYNS: GF □ GLYCERINFORMALE □ GLYCEROL FORMAL □ SERICOSOL-N

TOXICITY DATA WITH REFERENCE
orl-rat TDLo:6600 mg/kg (female 7-17D post):REP TCMUD8 7,73,87
orl-rat TDLo:6 g/kg (6-15D preg):TER TXAPA9 56,93,80
orl-rat TDLo:6 g/kg (6-15D preg):REP TXAPA9 56,93,80
ipr-rat LD50:9500 mg/kg APFRAD 44,293,86
orl-mus LD50:8 g/kg APFRAD 44,293,86
ipr-mus LD50:7500 mg/kg APFRAD 44,293,86

SAFETY PROFILE: Mildly toxic by intraperitoneal

routes. An experimental teratogen. Other experimental reproductive effects. When heated to decomposition it emits acrid smoke and fumes.

DVS000 CAS:6988-21-2 HR: 3
o-(1,3-DIOXOLAN-2-YL)PHENYL METHYLCARBAMATE
mf: $C_{11}H_{13}NO_4$ mw: 223.25

PROP: Crystals. Mp: 114–115°.

SYNS: CIBA 8353 □ DIOXACARB □ 2-(1,3-DIOXOLANE-2-YL)PHENYL N-METHYLCARBAMATE □ 2-(1,3-DIOXOLAN-2-YL)PHENYL-N-METHYLCARBAMAT □ DU PONT INSECTICIDE 1519 □ ELOCRON □ ENT 27,389 □ FAMID □ NSC 190981

TOXICITY DATA WITH REFERENCE
mmo-smc 5 ppm RSTUDV 6,161,76
orl-rat LD50:25 mg/kg JTCEEM 6(3),175,86
ihl-rat LC50:160 mg/m³ 85JCAE-,951,86
skn-rat LD50:3 g/kg FMCHA2-,C122,91
ipr-rat LD50:8300 µg/kg PESTD5 17,351,76
orl-mus LD50:48 mg/kg PESTD5 17,351,76
skn-mus LD50:1660 mg/kg BESAAT 15,133,69
ipr-mus LD50:20 mg/kg PESTD5 17,351,76
skn-rbt LD50:1950 mg/kg BESAAT 15,133,69

SAFETY PROFILE: Poison by ingestion and intraperitoneal routes. Moderately toxic by skin contact. Mutation data reported. A toxic contact and systemic insecticide. When heated to decomposition it emits toxic fumes of NO_x.

DVS100 CAS:76059-11-5 HR: 3
3-(2-(1,3-DIOXO-2-METHYLINDANYL))GLUTARIMIDE
mf: $C_{15}H_{13}NO_4$ mw: 271.29

SYN: GLUTARIMIDE, 3-(1,3-DIOXO-2-METHYLINDAN-2-YL)-

TOXICITY DATA WITH REFERENCE
ipr-mus TDLo:10 mg/kg (female 9D post):REP ARPMAS 313,481,80
ipr-mus TDLo:10 mg/kg (female 9D post):TER ARPMAS 313,481,80
ipr-mus LD50:28 mg/kg ARPMAS 313,481,80

SAFETY PROFILE: Poison by intraperitoneal route. An experimental teratogen. Other experimental reproductive effects. When heated to decomposition it emits toxic fumes of NO_x.

DVS300 CAS:76059-13-7 HR: 3
3-(2-(1,3-DIOXO-2-PHENYLINDANYL))GLUTARIMIDE
mf: $C_{20}H_{15}NO_4$ mw: 333.36

SYN: GLUTARIMIDE, 3-(1,3-DIOXO-2-PHENYLINDAN-2-YL)-

TOXICITY DATA WITH REFERENCE
ipr-mus TDLo:2500 µg/kg (female 9D post):REP ARPMAS 313,481,80
ipr-mus TDLo:2500 µg/kg (female 9D post):TER ARPMAS 313,481,80
ipr-mus LD50:14 mg/kg ARPMAS 313,481,80

SAFETY PROFILE: Poison by ingestion. An experimental teratogen. Other experimental reproductive effects. When heated to decomposition it emits toxic fumes of NO_x.

DVS400 CAS:76059-14-8 HR: 3
3-(2-(1,3-DIOXO-2-PHENYL-4,5,6,7-TETRAHYDRO-4,7-DITHIAINDANYL))GLUTARIMIDE
mf: $C_{18}H_{15}NO_4S_2$ mw: 373.46

SYN: GLUTARIMIDE, 3-(5,7-DIOXO-6-PHENYL-2,3,6,7-TETRAHYDRO-5H-CYCLOPENTA-p-DITHIIN-6-YL)-

TOXICITY DATA WITH REFERENCE
ipr-mus TDLo:1200 µg/kg (female 9D post):REP ARPMAS 313,481,80
ipr-mus TDLo:2500 µg/kg (female 9D post):TER ARPMAS 313,481,80
ipr-mus LD50:7 mg/kg ARPMAS 313,481,80

SAFETY PROFILE: Poison by intraperitoneal route. An experimental teratogen. Other experimental reproductive effects. When heated to decomposition it emits toxic fumes of NO_x and SO_x.

DVS600 CAS:26581-81-7 HR: 2
2-(2,6-DIOXOPIPERIDEN-3-YL) PHTHALIMIDINE
mf: $C_{13}H_9N_2O_3$ mw: 241.24

SYNS: 3-(1,3-DIHYDRO-1-OXO-2H-ISOINDOL-2-YL)-2,6-DIOXOPIPERIDINE □ EM 12

TOXICITY DATA WITH REFERENCE
orl-rat TDLo:1 g/kg (9-12D preg):TER TJADAB 5,233,72
orl-rat TDLo:1 g/kg (9-12D preg):REP TJADAB 5,233,72
orl-rbt TDLo:400 mg/kg (female 8-15D post):TER ARZNAD 31,941,81
ivn-rat TDLo:15 mg/kg (female 9-11D post):TER TJADAB 5,233,72
ivn-rbt TDLo:40 mg/kg (female 9-12D post):TER TJADAB 5,233,72
orl-mus LD50:6000 mg/kg ARZNAD 31,941,81
ipr-mus LD50:1830 mg/kg ARZNAD 31,941,81

SAFETY PROFILE: Moderately toxic by intraperitoneal route. Mildly toxic by ingestion. Experimental teratogenic and reproductive effects. When heated to decomposition it emits toxic fumes of NO_x.

DVT400 CAS:13754-56-8 HR: 3
DIOXOPROMETHAZINE HYDROCHLORIDE
mf: $C_{17}H_{20}N_2O_2S{\cdot}ClH$ mw: 352.91

SYN: 5,5-DIOXO-10-(2-(DIMETHYLAMINO)PROPYL)PHENOTHIAZINE HYDROCHLORIDE

TOXICITY DATA WITH REFERENCE
orl-mus LD50:70 mg/kg PHARAT 25,91,70
ipr-mus LD50:49 mg/kg PHARAT 25,91,70
scu-mus LD50:96 mg/kg PHARAT 25,91,70
ivn-mus LD50:19 mg/kg PHARAT 25,91,70
orl-rbt LD50:165 mg/kg PHARAT 25,91,70
orl-gpg LD50:225 mg/kg PHARAT 25,91,70
scu-gpg LD50:33 mg/kg PHARAT 25,91,70
ivn-gpg LD50:8 mg/kg PHARAT 25,91,70

SAFETY PROFILE: Poison by ingestion, intraperitoneal, intravenous, and subcutaneous routes. When heated to decomposition it emits very toxic fumes of NO_x, SO_x, and HCl.

DVT459 **CAS:31083-55-3** **HR: 3**
1,3-DIOXO-2-(3-PYRIDYLMETHYLENE)INDAN
mf: $C_{15}H_9NO_2$ mw: 235.25

TOXICITY DATA WITH REFERENCE
ipr-mus TDLo:80 mg/kg (9D preg):REP ARPMAS 313,481,80
ipr-mus TDLo:40 mg/kg (9D preg):TER ARPMAS 313,481,80
ipr-mus LD50:200 mg/kg ARPMAS 313,481,80

SAFETY PROFILE: Poison by intraperitoneal route. An experimental teratogen. Other experimental reproductive effects. When heated to decomposition it emits toxic fumes of NO_x.

DVT500 **HR: 1**
DIOXYBIS(2,2′-DI-tert-BUTYLBUTANE
mf: $C_{24}H_{50}O_2$ mw: 370.74

SYN: 2,2-BIS-DI-tert-BUTYLPEROXYBUTANE

TOXICITY DATA WITH REFERENCE
skn-rbt 500 mg MLD SCCUR* -,2,61
orl-rat LD50:12,500 mg/kg FEPRA7 7,252,48
orl-mus LD50:17,500 mg/kg FEPRA7 7,252,48

SAFETY PROFILE: Mildly toxic by ingestion. A skin irritant. When heated to decomposition it emits acrid smoke and fumes. See also PEROXIDES.

DVT800 **CAS:12228-13-6** **HR: 3**
DIOXYGENYL TETRAFLUOROBORATE
mf: BFN_4O_2 mw: 118.81

SAFETY PROFILE: A very powerful oxidant. Explodes in methane or ethane. Ignites in benzene or 2-propanol. When heated to decomposition it emits toxic fumes of F^-. See also BORON COMPOUNDS and FLUORIDES.

DVU000 **CAS:580-74-5** **HR: 3**
1,4-DI-p-OXYPHENYL-2,3-DI-ISONITRILO-1,3-BUTADIENE
mf: $C_{18}H_{12}N_2O_2$ mw: 288.32

PROP: Yellow crystals.

SYNS: OPHTHOCILLIN □ XANTHOCILLIN □ XANTYRID

TOXICITY DATA WITH REFERENCE
orl-mus LD50:45 mg/kg ARZNAD 7,98,57
ipr-mus LD50:20 mg/kg ARZNAD 7,98,57

CONSENSUS REPORTS: Cyanide and its compounds are on the Community Right-To-Know List.

SAFETY PROFILE: Poison by ingestion and intraperitoneal routes. When heated to decomposition it emits toxic fumes of NO_x and CN^-. See also NITRILES.

DVU100 **CAS:519-65-3** **HR: 2**
DIOXYPYRAMIDON
mf: $C_{13}H_{17}N_3O_3$ mw: 263.33

PROP: Orthorhombic, translucent prisms from water. Somewhat bitter taste. Mp: 105.5° (softens at 96°), bp: (2) 194–201°. Solubility in water at 20° = 7.69 g/100 mL; at 37° = 48.2 g/100 mL. Also sol in alc.

SYNS: 1-ACETYL-2-PHENYL-1,5,5-TRIMETHYL-SEMIOXAMAZIDE □ (DIMETHYLAMINO)OXO-ACETIC ACID 2-ACETYL-2-METHYL-1-PHENYLHYDRAZIDE (9CI) □ DIOXOAMINOPYRINE □ DIOXYAMINOPYRINE

TOXICITY DATA WITH REFERENCE
orl-mus LD50:1631 mg/kg AEPPAE 213,501,51
scu-mus LD50:1066 mg/kg AEPPAE 213,501,51
ivn-mus LD50:698 mg/kg AEPPAE 213,501,51

SAFETY PROFILE: Moderately toxic by ingestion, subcutaneous, and intravenous routes. When heated to decomposition it emits toxic fumes of NO_x.

DVU200 **HR: 3**
DIPALLADIUM TRIOXIDE
mf: Pd_2O_3 mw: 260.80

SAFETY PROFILE: Hydrated oxide explodes when heated. See also PALLADIUM.

DVU300 **HR: 2**
DIPENICILLIN-G-ALUMINIUM-SULPHAMETHOXYPYRIDAZINE
mf: $C_{43}H_{45}AlN_8O_{11}S_3$ mw: 973.12

SYNS: AB 109 □ DIPENICILLINA-G-ALLUMINIO-SULFAMETOSSIPIRIDAZINA (ITALIAN)

TOXICITY DATA WITH REFERENCE
ipr-mus LD50:3580 mg/kg BCFAAI 98,453,59
scu-mus LD50:5720 mg/kg BCFAAI 98,453,59
ipr-gpg LD50:3000 mg/kg BCFAAI 98,453,59
scu-gpg LD50:4500 mg/kg BCFAAI 98,453,59

ACGIH TLV: TWA 2 $mg(Al)/m^3$

SAFETY PROFILE: Moderately toxic by intraperitoneal route. Mildly toxic by subcutaneous route. When heated to decomposition it emits toxic fumes of SO_x and NO_x. See also PENICILLIN and ORGANOMETALS.

DVU600 **CAS:18279-20-4** **HR: 3**
DIPENTYL LEAD DIACETATE
mf: $C_{14}H_{28}O_4Pb$ mw: 467.61

TOXICITY DATA WITH REFERENCE
orl-mus LD50:90 mg/kg CRSBAW 162,1456,68

CONSENSUS REPORTS: Lead and its compounds are on the Community Right-To-Know List.

SAFETY PROFILE: Poison by ingestion. When heated to decomposition it emits toxic fumes of Pb. See also LEAD COMPOUNDS.

DVV000 **CAS:2273-46-3** **HR: 3**
DIPENTYLOXOSTANNANE
mf: $C_{10}H_{22}OSn$ mw: 277.01

SYNS: DIPENTYLTIN OXIDE □ KYSLICNIK DI-n-AMYLCINICITY (CZECH)

TOXICITY DATA WITH REFERENCE
skn-rbt 500 mg/24H SEV 28ZPAK -,227,72
eye-rbt 100 mg/24H MOD 28ZPAK -,227,72
orl-rat LD50:55,200 µg/kg 28ZPAK -,227,72

OSHA PEL: TWA 0.1 mg(Sn)/m³ (skin)
ACGIH TLV: TWA 0.1 mg(Sn)/m³ (skin) (Proposed: TWA 0.1 mg(Sn)/m³; STEL 0.2 mg(Sn)/m³ (skin))
NIOSH REL: (Organotin Compounds) TWA 0.1 mg(Sn)/m³

SAFETY PROFILE: Poison by ingestion. An eye and severe skin irritant. When heated to decomposition it emits acrid smoke and fumes. See also TIN COMPOUNDS.

For occupational chemical analysis use NIOSH: Organotin Compounds 5504.

DVV109 **CAS:13403-01-5** **HR: 3**
2-(2,4-DI-tert-PENTYLPHENOXY)BUTYRIC ACID

TOXICITY DATA WITH REFERENCE
ipr-rat LD50:400 mg/kg KODAK* -,-,71

CONSENSUS REPORTS: Reported in EPA TSCA Inventory.

SAFETY PROFILE: Poison by intraperitoneal route. When heated to decomposition it emits acrid smoke and irritating fumes.

DVV200 **CAS:1118-42-9** **HR: 3**
DIPENTYLTIN DICHLORIDE
mf: $C_{10}H_{22}Cl_2Sn$ mw: 331.91

SYN: DICHLORODIPENTYLSTANNANE

TOXICITY DATA WITH REFERENCE
ivn-rat LDLo:10 mg/kg BJIMAG 15,15,58

OSHA PEL: TWA 0.1 mg(Sn)/m³ (skin)
ACGIH TLV: TWA 0.1 mg(Sn)/m³ (skin) (Proposed: TWA 0.1 mg(Sn)/m³; STEL 0.2 mg(Sn)/m³ (skin))
NIOSH REL: (Organotin Compounds) TWA 0.1 mg(Sn)/m³

SAFETY PROFILE: Poison by intravenous route. When heated to decomposition it emits toxic fumes of Cl^-. See also TIN COMPOUNDS and CHLORIDES.

For occupational chemical analysis use NIOSH: Organotin Compounds 5504.

DVV400 **HR: 3**
2,6-DIPERCHLORYL-4,4'-DIPHENOQUINONE
mf: $C_{12}H_6Cl_2O_8$ mw: 369.09

SAFETY PROFILE: A shock-sensitive explosive. When heated to decomposition it emits toxic fumes of Cl^-.

DVV500 **CAS:537-12-2** **HR: 3**
DIPERODON HYDROCHLORIDE
mf: $C_{22}H_{27}N_3O_4{\cdot}ClH$ mw: 433.98

PROP: Crystals; bitter taste followed by a sense of numbness. Decomp @ 195–200°. Sol in alc; sltly sol in water, acetone, and ethyl acetate; insol in benzene or ether.

SYNS: DIOTHANE HYDROCHLORIDE □ DIPERDON HYDROCHLORIDE □ 3-PIPERIDINO-1,2-PROPANEDIOL DICARBANILATE HYDROCHLORIDE □ 3-(1-PIPERIDYL)-1,2-PROPANEDIOL DICARBANILATE HYDROCHLORIDE □ PROCTODON

TOXICITY DATA WITH REFERENCE
scu-mus LD50:890 mg/kg JAPMA8 48,398,59
scu-rbt LDLo:300 mg/kg JPETAB 47,255,33
ivn-rbt LDLo:15 mg/kg JPETAB 47,255,33
scu-gpg LDLo:400 mg/kg JPETAB 47,255,33

SAFETY PROFILE: Poison by subcutaneous and intravenous routes. When heated to decomposition it emits toxic fumes of NO_x and HCl.

DVV550 **CAS:1711-42-8** **HR: 3**
DIPEROXYTEREPHTHALIC ACID
mf: $C_8H_6O_6$ mw: 198.13

SAFETY PROFILE: An impact- and heat-sensitive explosive. When heated to decomposition it emits acrid smoke and fumes. See also PEROXIDES.

DVV600 **CAS:82-66-6** **HR: 3**
DIPHENADIONE
mf: $C_{23}H_{16}O_3$ mw: 340.39

PROP: Pale-yellow crystals from alc. Mp: 147°. Sol in Me_2CO and AcOH; sltly sol in C_6H_6.

SYNS: DIDANDIN □ DIPAXIN □ DIPHACIN □ DIPHACINONE □ DIPHENACIN □ 2-DIPHENYLACETYL-1,3-DIKETOHYDRINDENE □ 2-(DIPHENYLACETYL)INDAN-1,3-DIONE □ 2-DIPHENYLACETYL-1,3-INDANDIONE □ 2-(DIPHENYLACETYL)-1H-INDENE-1,3(2H)-DIONE □ PID □ PROMAR □ RAMIK □ RATINDAN 1 □ U 1363

TOXICITY DATA WITH REFERENCE
orl-rat LD50:1500 µg/kg 85DPAN -,-,71/76
ihl-rat LC50:2 g/m³/4H PEMNDP 9,310,91
orl-mus LD50:28,300 µg/kg NNGADV 17,S319,92
orl-dog LD50:3 mg/kg 28ZEAL 5,84,76
orl-cat LD50:15 mg/kg PCOC** -,429,66
orl-rbt LD50:35 mg/kg 85DPAN -,-,71/76
orl-pig LD50:150 mg/kg 28ZEAL 5,84,76
orl-mam LD50:910 µg/kg SCIEAS 177,806,72

CONSENSUS REPORTS: EPA Extremely Hazardous Substances List.

SAFETY PROFILE: Poison by ingestion. Inhibits blood clotting, leading to hemorrhages. Action similar to coumadin (warfarin). A pesticide used in rodent control. When heated to decomposition it emits acrid smoke and irritating fumes.

DVV800 **CAS:1210-05-5** **HR: 2**
DIPHENALDEHYDE
mf: $C_{14}H_{10}O_2$ mw: 210.24

SYNS: 2,2′-BIPHENYLDICARBOXALDEHYDE □ 2,2′-DIFORMYLBIPHENYL

TOXICITY DATA WITH REFERENCE
skn-rbt 500 mg MOD IHFCAY 6,1,67
eye-rbt 500 mg SEV IHFCAY 6,1,67
orl-rat LD50:2830 mg/kg IHFCAY 6,1,67

SAFETY PROFILE: Moderately toxic by ingestion. A skin and severe eye irritant. When heated to decomposition it emits acrid smoke and irritating fumes. See also ALDEHYDES.

DVW000 **CAS:1798-49-8** **HR: 3**
DIPHENCHLOXAZINE HYDROCHLORIDE
mf: $C_{19}H_{22}ClNO_2{\cdot}ClH$ mw: 368.33

PROP: A solid. Mp: 140–141°.

SYN: PHENYL-p-CHLOROPHENYLTETRAHYDRO-1,4-OXAZINYLETHOXYMETHANE HYDROCHLORIDE

TOXICITY DATA WITH REFERENCE
orl-rat LD50:710 mg/kg 27ZQAG -,223,72
scu-mus LD50:368 mg/kg YKKZAJ 87,1109,67
ivn-mus LD50:140 mg/kg 27ZQAG -,223,72

SAFETY PROFILE: Poison by subcutaneous and intravenous routes. Moderately toxic by ingestion. When heated to decomposition it emits very toxic fumes of NO_x and Cl^-.

DVW100 **CAS:82-21-3** **HR: 1**
1,5-DIPHENOXYANTHRAQUINONE
mf: $C_{26}H_{16}O_4$ mw: 392.42

SYNS: 9,10-ANTHRACENEDIONE, 1,5-DIPHENOXY- □ 1,5-DIFENOXYANTHRACHINON

TOXICITY DATA WITH REFERENCE
eye-rbt 500 mg/24H MLD 85JCAE-,705,86
orl-mus LD:>12 g/kg GTPZAB 31(1),49,87

CONSENSUS REPORTS: Reported in EPA TSCA Inventory.

SAFETY PROFILE: An eye irritant. When heated to decomposition it emits acrid smoke and irritating fumes.

DVW600 **CAS:622-04-8** **HR: 3**
1,3-DIPHENOXY-2-PROPANOL
mf: $C_{15}H_{16}O_3$ mw: 244.31

SYN: GLYCERYL-α,γ-DIPHENYL ETHER

TOXICITY DATA WITH REFERENCE
skn-rbt 10 mg/24H open MLD AMIHBC 4,119,51
orl-rat LD50:1450 mg/kg AMIHBC 4,119,51
ivn-mus LD50:180 mg/kg CSLNX* NX#04474
skn-rbt LD50:15,800 mg/kg AMIHBC 4,119,51

CONSENSUS REPORTS: Reported in EPA TSCA Inventory.

SAFETY PROFILE: Poison by intravenous route. Moderately toxic by ingestion. Mildly toxic by skin contact. A skin irritant. When heated to decomposition it emits acrid smoke and irritating fumes.

DVW700 **CAS:132-18-3** **HR: 3**
DIPHENPYRALINE HYDROCHLORIDE
mf: $C_{19}H_{23}NO{\cdot}ClH$ mw: 317.89

PROP: Off-white crystals from 2-propanol/Et_2O. Mp: 206°. Sol in H_2O and EtOH; insol in C_6H_6 and Et_2O.

SYNS: DIAFEN □ DIAFEN (antihistamine) □ DIAFEN HYDROCHLORIDE □ DIAPHEN □ DIPHENYLPYRALINE HYDROCHLORIDE □ HISPRIL HYDROCHLORIDE □ SUMADIL

TOXICITY DATA WITH REFERENCE
orl-rat LD50:698 mg/kg NIIRDN 6,333,82
scu-rat LD50:278 mg/kg NIIRDN 6,333,82
ivn-rat LD50:28,800 µg/kg NIIRDN 6,333,82
ims-rat LD50:180 mg/kg NIIRDN 6,333,82
orl-mus LD50:202 mg/kg NIIRDN 6,333,82
scu-mus LD50:112 mg/kg NIIRDN 6,333,82
ivn-mus LD50:31,800 µg/kg NIIRDN 6,333,82
ims-mus LD50:64,300 µg/kg NIIRDN 6,333,82

CONSENSUS REPORTS: Reported in EPA TSCA Inventory.

SAFETY PROFILE: Poison by ingestion, subcutaneous, intramuscular, and intravenous routes. When heated to decomposition it emits toxic fumes of NO_x and HCi.

DVW750 **CAS:621-09-0** **HR: 3**
N,N′-DIPHENYLACETAMIDINE
mf: $C_{14}H_{14}N_2$ mw: 210.30

SYN: ACETAMIDINE, N,N′-DIPHENYL-

TOXICITY DATA WITH REFERENCE
orl-mus LD50:610 mg/kg THERAP 21,1327,66
scu-mus LD50:250 mg/kg APFRAD 40,231,82

CONSENSUS REPORTS: Reported in EPA TSCA Inventory.

SAFETY PROFILE: Poison by subcutaneous route. Moderately toxic by ingestion. When heated to decomposition it emits toxic vapors of NO_x.

DVW800 **CAS:117-34-0** **HR: 3**
DIPHENYLACETIC ACID
mf: $C_{14}H_{12}O_2$ mw: 212.26

PROP: White crystals or leaflets from alc. Mp: 149°, bp: sublimes, vap d: 7.3.

SYNS: α,α-DIPHENYLACETIC ACID □ α-PHENYLBENZENEACETIC ACID

TOXICITY DATA WITH REFERENCE
orl-rat LD50:5540 mg/kg GNAMAP 17,48,78
orl-mus LD50:3200 mg/kg GNAMAP 17,48,78
ipr-mus LD50:500 mg/kg FRPSAX 13,286,58
scu-mus LD50:400 mg/kg AIPTAK 116,154,58

CONSENSUS REPORTS: Reported in EPA TSCA Inventory.

SAFETY PROFILE: Poison by subcutaneous route. Moderately toxic by ingestion and intraperitoneal routes. When heated to decomposition it emits acrid smoke and irritating fumes.

DVW900 CAS:1984-87-8 HR: 2
DIPHENYLACETIC ACID 2-(BIS(2-HYDROXYETHYL) AMINO)ETHYL ESTER HYDROCHLORIDE
mf: $C_{20}H_{25}NO_4$•ClH mw: 379.92

SYN: ACETIC ACID, DIPHENYL-, 2-(BIS(2-HYDROXYETHYL)AMINO) ETHYL ESTER, HYDROCHLORIDE

TOXICITY DATA WITH REFERENCE
eye-cat 5% APPHAX 21,557,64
scu-mus LD50:1380 mg/kg APPHAX 21,557,64

SAFETY PROFILE: Moderately toxic by subcutaneous route. An eye irritant. When heated to decomposition it emits toxic fumes of NO_x and HCl.

DVX200 CAS:86-29-3 HR: 3
DIPHENYLACETONITRILE
mf: $C_{14}H_{11}N$ mw: 193.26

PROP: A solid. Mp: 75–76°, bp: 181° @ 12 mm.

SYNS: BENZYHYDRYLCYANIDE □ α-CYANODIPHENYLMETHANE □ DIPAN □ DIPHENATRILE □ DIPHENYL-α-CYANOMETHANE □ DIPHENYLMETHYLCYANIDE □ α-PHENYLBENZYLCYANIDE □ α-PHENYLPHENYLACETONITRILE □ USAF KF-13

TOXICITY DATA WITH REFERENCE
orl-mus TDLo:61 g/kg/78W-I:ETA NTIS** PB223-159
scu-mus TDLo:161 mg/kg:CAR NTIS** PB223-159
orl-rat LD50:3500 mg/kg RREVAH 10,97,65
ipr-mus LD50:200 mg/kg NTIS** AD691-490
ivn-mus LD50:100 mg/kg CSLNX* NX#04134

CONSENSUS REPORTS: Reported in EPA TSCA Inventory. Cyanide and its compounds are on the Community Right-To-Know List.

SAFETY PROFILE: Poison by ingestion, intraperitoneal, and intravenous routes. Moderately toxic by subcutaneous route. Questionable carcinogen with experimental carcinogenic and tumorigenic data. When heated to decomposition it emits toxic fumes of NO_x and CN^-. See also NITRILES.

DVX600 CAS:2510-95-4 HR: 3
2,3-DIPHENYLACRYLONITRILE
mf: $C_{15}H_{11}N$ mw: 205.27

SYNS: BENZAL-(BENZYL-CYANID) (GERMAN) □ BENZYLIDENEPHENYLACETONITRILE □ α-CYANOSTILBENE □ α,β-DIPHENYLACRYLONITRILE □ F 2387 □ α-PHENYLCINNAMONITRILE □ α-(PHENYLMETHYLENE)BENZENEACETONITRILE □ α-STILBENECARBONITRILE □ USAF A-9789

TOXICITY DATA WITH REFERENCE
ipr-mus LD50:100 mg/kg NTIS** AD277-689

CONSENSUS REPORTS: Reported in EPA TSCA Inventory. Cyanide and its compounds are on the Community Right-To-Know List.

SAFETY PROFILE: Poison by intraperitoneal route. When heated to decomposition it emits toxic fumes of NO_x and CN^-. See also NITRILES.

DVX800 CAS:122-39-4 HR: 3
DIPHENYLAMINE
mf: $C_{12}H_{11}N$ mw: 169.24

PROP: Crystals; floral odor. Mp: 52.9°, bp: 302.0°, flash p: 307°F (CC), d: 1.16, autoign temp: 1173°F, vap press: 1 mm @ 108.3°, vap d: 5.82. Sol in benzene, ether, and carbon disulfide; insol in water.

SYNS: ANILINOBENZENE □ BIG DIPPER □ C.I. 10355 □ DFA □ N,N-DIPHENYLAMINE □ DPA □ NO SCALD □ N-PHENYLANILINE □ N-PHENYLBENEZENAMINE □ SCALDIP

TOXICITY DATA WITH REFERENCE
orl-rat TDLo:7500 mg/kg (17-22D preg):TER PEREBL 4,448,70
orl-rat LD50:2 g/kg GISAAA 41(5),21,76
orl-mus LD50:1750 mg/kg GISAAA 41(5),21,76
orl-gpg LD50:300 mg/kg FMCHA2 -,C85,83

CONSENSUS REPORTS: Reported in EPA TSCA Inventory. EPA Genetic Toxicology Program.

OSHA PEL: TWA 10 mg/m³
ACGIH TLV: TWA 10 mg/m³

SAFETY PROFILE: Poison by ingestion. Experimental teratogenic effects. Action similar to aniline but less severe. Combustible when exposed to heat or flame. Can react violently with hexachloromelamine or trichloromelamine. Can react with oxidizing materials. To fight fire, use CO_2, dry chemical. When heated to decomposition it emits highly toxic fumes of NO_x. See also ANILINE, AMINES, and AROMATIC AMINES.

For occupational chemical analysis use OSHA: #22.

DVY000 CAS:587-84-8 HR: 3
DIPHENYLAMINE HYDROGEN SULFATE
mf: $C_{12}H_{11}N$•H_2O_4S mw: 267.32

PROP: White to yellowish powder. Mp: 123–125°. Insol in water; sol in alcohol and sulfuric acid.

SYNS: DIPHENYLAMINE SULFATE □ USAF EK-743

TOXICITY DATA WITH REFERENCE
ipr-mus LD50:200 mg/kg NTIS** AD277-689

CONSENSUS REPORTS: Reported in EPA TSCA Inventory.

SAFETY PROFILE: Poison by intraperitoneal route. When heated to decomposition it emits toxic fumes of NO_x. See also SULFATES.

DVY100 CAS:6217-24-9 HR: 3
DIPHENYLARSINOUS ACID
mf: $C_{12}H_{11}AsO$ mw: 246.15

SYNS: ARSINE, DIPHENYLHYDROXY- □ ARSINE, HYDROXYDIPHENYL- □ DIPHENYLHYDROXYARSINE

TOXICITY DATA WITH REFERENCE
ivn-mus LD50:10 mg/kg PHBUA9 2,19,54

OSHA PEL: TWA 0.5 mg(As)/m^3

SAFETY PROFILE: Poison by intravenous route. When heated to decomposition it emits toxic fumes of As.

DVY800 CAS:1103-05-5 HR: 3
DIPHENYLBIS(PHENYLTHIO)TIN
mf: $C_{24}H_{20}S_2Sn$ mw: 491.25

PROP: White solid from EtOH. Mp: 65–66°. Sol in org solvs.

SYN: DIPHENYLBIS(PHENYLTHIO)STANNANE

TOXICITY DATA WITH REFERENCE
ivn-mus LD50:18 mg/kg CSLNX* NX#01644

OSHA PEL: TWA 0.1 mg(Sn)/m^3 (skin)
ACGIH TLV: TWA 0.1 mg(Sn)/m^3 (skin) (Proposed: TWA 0.1 mg(Sn)/m^3; STEL 0.2 mg(Sn)/m^3 (skin))
NIOSH REL: (Organotin Compounds) TWA 0.1 mg(Sn)/m^3

SAFETY PROFILE: Poison by intravenous route. When heated to decomposition it emits toxic fumes of SO_x. See also TIN COMPOUNDS.

For occupational chemical analysis use NIOSH: Organotin Compounds 5504.

DVY900 CAS:20930-10-3 HR: 2
1,1-DIPHENYL-2-BUTYNYL-N-CYCLOHEXYLCARBAMATE
mf: $C_{23}H_{25}NO_2$ mw: 347.49

SYN: CYCLOHEXANECARBAMIC ACID, 1,1-DIPHENYL-2-BUTYNYL ESTER

TOXICITY DATA WITH REFERENCE
orl-rat TDLo:14,800 mg/kg/73W-C:ETA JJIND8 71,211,83
orl-rat TD:15,500 mg/kg/54W-C:ETA JJIND8 71,211,83

SAFETY PROFILE: Questionable carcinogen with experimental tumorigenic data. When heated to decomposition it emits toxic fumes of NO_x.

DVZ000 CAS:102-09-0 HR: 2
DIPHENYL CARBONATE
mf: $C_{13}H_{10}O_3$ mw: 214.23

PROP: Needles. Mp: 78°, bp: 306°.

SYNS: CARBONIC ACID, DIPHENYL ESTER □ PHENYL CARBONATE

TOXICITY DATA WITH REFERENCE
orl-mus TDLo:28 g/kg/78W-I:ETA NTIS** PB223-159
scu-mus TDLo:1000 mg/kg:NEO NTIS** PB223-159

CONSENSUS REPORTS: Reported in EPA TSCA Inventory.

SAFETY PROFILE: Questionable carcinogen with experimental neoplastigenic and tumorigenic data. When heated to decomposition it emits acrid smoke and irritating fumes.

DVZ100 CAS:883-40-9 HR: 3
1,1′-DIPHENYLDIAZOMETHANE
mf: $C_{13}H_{10}N_2$ mw: 194.25

SYNS: BENZENE, 1,1′-(DIAZOMETHYLENE)BIS- □ DIAZODIPHENYLMETHANE (DOT) □ METHANE, DIAZODIPHENYL-

DOT CLASSIFICATION: Forbidden

SAFETY PROFILE: An unstable substance forbidden for transport. When heated to decomposition it emits toxic vapors of NO_x.

DWA400 CAS:25868-47-7 HR: 3
DIPHENYLDICHLORO TIN DIPYRIDINE complex
mf: $C_{22}H_{20}Cl_2N_2Sn$ mw: 502.03

SYN: DICHLORODIPHENYLSTANNANE complex with PYRIDINE (1:2)

TOXICITY DATA WITH REFERENCE
ivn-mus LD50:14 mg/kg CSLNX* NX#02208

OSHA PEL: TWA 0.1 mg(Sn)/m^3 (skin)
ACGIH TLV: TWA 0.1 mg(Sn)/m^3 (skin) (Proposed: TWA 0.1 mg(Sn)/m^3; STEL 0.2 mg(Sn)/m^3 (skin))
NIOSH REL: (Organotin Compounds) TWA 0.1 mg(Sn)/m^3

SAFETY PROFILE: Poison by intravenous route. When heated to decomposition it emits very toxic fumes of Cl^- and NO_x. See also TIN COMPOUNDS.

DWA500 CAS:2652-77-9 HR: 3
DIPHENYLDIKETOPYRAZOLIDINE
mf: $C_{15}H_{15}N_2O_2$ mw: 252.29

PROP: Plates from EtOH. Mp: 178–179°.

SYNS: DA 339 □ DIFENILDICHETOPIRAZOLIDINA (ITALIAN) □ 1,2-DIPHENYL-3,5-DIOXOPYRAZOLIDIN (GERMAN) □ 1,2-DIPHENYL-3,5-PYRAZOLIDINEDIONE □ G 14744

TOXICITY DATA WITH REFERENCE
orl-rat LD50:1 g/kg MPHEAE 16,536,67
ipr-rat LD50:338 mg/kg AIPTAK 132,16,61
orl-mus LD50:574 mg/kg MPHEAE 16,536,67

SAFETY PROFILE: Poison by intraperitoneal route. Moderately toxic by ingestion. When heated to decomposition it emits toxic fumes of NO_x.

DWA600 CAS:14148-99-3 HR: 3
1,2-DIPHENYL-1-(DIMETHYLAMINO)ETHANE
mf: $C_{16}H_{19}N{\cdot}ClH$ mw: 261.82

PROP: A solid. Mp: 218–220°.

SYNS: (R) (−)-N,N-DIMETHYL-1,2-DIPHENYLETHYLAMINE HYDROCHLORIDE □ (R)-N,N-DIMETHYL-α-PHENYLBENZENEETHANAMINE, HYDROCHLORIDE □ SPA

TOXICITY DATA WITH REFERENCE
orl-rat LD50:300 mg/kg DRFUD4 2,39,77
scu-rat LD50:148 mg/kg AIPTAK 221,105,76
orl-mus LD50:176 mg/kg DRFUD4 2,39,77
scu-mus LD50:104 mg/kg DRFUD4 2,39,77
ivn-mus LD50:32,600 μg/kg DRFUD4 2,39,77

SAFETY PROFILE: Poison by ingestion, intravenous,

and subcutaneous routes. When heated to decomposition it emits very toxic fumes of HCl and NO_x.

DWA700 CAS:56767-15-8 HR: 2
3,3-DIPHENYL-3-DIMETHYLCARBAMOYL-1-PROPYNE
mf: $C_{18}H_{17}NO$ mw: 263.36

SYN: N,N-DIMETHYL-α-ETHYNYL-α-PHENYLBENZENEACETAMIDE

TOXICITY DATA WITH REFERENCE
orl-rat TDLo:780 mg/kg/21W-I:CAR CNREA8 35,2469,75

SAFETY PROFILE: Questionable carcinogen with experimental carcinogenic data. When heated to decomposition it emits toxic fumes of NO_x.

DWA800 HR: 3
1,3-DIPHENYL-1,3-EPIDIOXY-1,3-DIHYDROISOBENZOFURAN
mf: $C_{20}H_{14}O_3$ mw: 302.33

SAFETY PROFILE: Explosive. When heated to decomposition it emits acrid smoke and fumes.

DWB000 CAS:5959-42-2 HR: 3
DIPHENYLETHANOLAMINE HYDROCHLORIDE
mf: $C_{14}H_{15}NO \cdot ClH$ mw: 249.76

SYNS: α-(α-AMINOBENZYL)BENZYL ALCOHOL HYDROCHLORIDE ☐ 2-AMINO-1,2-DIPHENYLETHANOL HYDROCHLORIDE ☐ β-AMINO-α-PHENYLBENZENEETHANOL HYDROCHLORIDE

TOXICITY DATA WITH REFERENCE
scu-mus LDLo:336 mg/kg JAPMA8 39,354,50
ivn-rbt LDLo:60 mg/kg JACSAT 52,3317,30
scu-gpg LDLo:450 mg/kg JACSAT 52,3317,30

SAFETY PROFILE: Poison by intravenous and subcutaneous routes. When heated to decomposition it emits very toxic fumes of Cl^- and NO_x.

DWB300 CAS:24301-89-1 HR: 3
1,2-DIPHENYLETHYLAMINE HYDROCHLORIDE
mf: $C_{14}H_{15}N \cdot ClH$ mw: 233.76

TOXICITY DATA WITH REFERENCE
ivn-rat LD50:45 mg/kg JPETAB 77,317,43
ipr-mus LD50:175 mg/kg JPHYA7 98,424,40
scu-mus LDLo:222 mg/kg JAPMA8 39,354,50

SAFETY PROFILE: Poison by subcutaneous, intravenous, and intraperitoneal routes. When heated to decomposition it emits toxic fumes of NO_x and HCl. See also AMINES.

DWB400 CAS:150-61-8 HR: 2
N,N′-DIPHENYLETHYLENEDIAMINE
mf: $C_{14}H_{16}N_2$ mw: 212.32

SYNS: N,N′-DIFENYLETHYLENDIAMIN ☐ ETHYLENEDIAMINE, N,N′-DIPHENYL-

TOXICITY DATA WITH REFERENCE
orl-rat LDLo:500 mg/kg JPETAB 90,260,47

CONSENSUS REPORTS: Reported in EPA TSCA Inventory.

SAFETY PROFILE: Moderately toxic by ingestion. When heated to decomposition it emits toxic vapors of NO_x.

DWB800 CAS:1241-94-7 HR: 3
DIPHENYL-2-ETHYLHEXYL PHOSPHATE
mf: $C_{20}H_{27}O_4P$ mw: 362.44

SYNS: 2-ETHYL-1-HEXANOL ESTER with DIPHENYL PHOSPHATE ☐ 2-ETHYLHEXYL DIPHENYL ESTER PHOSPHORIC ACID ☐ 2-ETHYLHEXYL DIPHENYLPHOSPHATE ☐ SANTICIZER 141 (MONSANTO)

TOXICITY DATA WITH REFERENCE
orl-rat LDLo:10 g/kg TSCAT* OTS 206227
ipr-mus LD50:930 mg/kg TSCAT* OTS 206227
ivn rbt LDLo:272 mg/kg AMIHBC 8,170,53

CONSENSUS REPORTS: Reported in EPA TSCA Inventory.

SAFETY PROFILE: Poison by intravenous route. Moderately toxic by intraperitoneal route. When heated to decomposition it emits toxic fumes of PO_x.

DWB875 HR: 3
4,4-DIPHENYL-1-ETHYLPIPERIDINE MALEATE
mf: $C_{19}H_{23}N \cdot C_4H_4O_4$ mw: 381.51

TOXICITY DATA WITH REFERENCE
orl-mus LD50:150 mg/kg ARZNAD 34,233,84
ivn-mus LD50:45 mg/kg ARZNAD 34,233,84
scu-mus LD50:85 mg/kg ARZNAD 34,233,84

SAFETY PROFILE: Poison by ingestion, subcutaneous, and intravenous routes. When heated to decomposition it emits toxic fumes of NO_x.

DWC000 HR: 3
N-(1,2-DIPHENYLETHYL)-2-(PYRROLIDINYL)ACETAMIDE HYDROCHLORIDE
mf: $C_{20}H_{24}N_2O \cdot ClH$ mw: 344.92

SYN: C 3155

TOXICITY DATA WITH REFERENCE
eye-rbt 2% SEV ARZNAD 8,609,58
ipr-rat LD50:84 mg/kg ARZNAD 8,609,58
scu-mus LD50:247 mg/kg ARZNAD 8,609,58

SAFETY PROFILE: Poison by intraperitoneal and subcutaneous routes. A severe eye irritant. When heated to decomposition it emits very toxic fumes of NO_x and HCl.

DWC050 CAS:53067-74-6 HR: 3
3,3-DIPHENYL-2-ETHYL-1-PYRROLINE
mf: $C_{18}H_{19}N$ mw: 249.38

TOXICITY DATA WITH REFERENCE
ipr-rat LD50:335 mg/kg AIPTAK 115,332,58
ivn-rat LD50:42 mg/kg AIPTAK 115,332,58
orl-mus LD50:1350 mg/kg AIPTAK 115,332,58
ipr-mus LD50:118 mg/kg AIPTAK 115,332,58
ivn-mus LD50:57 mg/kg AIPTAK 115,332,58

SAFETY PROFILE: Poison by intravenous and intraperi-

D

toneal routes. Moderately toxic by ingestion. When heated to decomposition it emits toxic fumes of NO_x.

DWC100 **CAS:60662-79-5** **HR: 3**
3,3-DIPHENYL-3-(ETHYLSULFONYL)-N,N,1-TRIMETHYLPROPYLAMINE HYDROCHLORIDE
mf: $C_{20}H_{27}NO_2S \cdot ClH$ mw: 382.00

SYNS: γ-(ETHYLSULFONYL)-N,N,α-TRIMETHYL-γ-PHENYLBENZENEPROPANAMINE HYDROCHLORIDE □ I-C 26 □ WIN-1161-3

TOXICITY DATA WITH REFERENCE
ipr-mus LD50:36 mg/kg CPBTAL 6,109,58
scu-mus LD50:102 mg/kg CPBTAL 6,109,58
ivn-dog LD50:26 mg/kg CPBTAL 6,109,58

SAFETY PROFILE: Poison by subcutaneous, intravenous, and intraperitoneal routes. When heated to decomposition it emits toxic fumes of SO_x, NO_x, and HCl.

DWC600 **CAS:102-06-7** **HR: 3**
DIPHENYLGUANIDINE
mf: $C_{13}H_{13}N_3$ mw: 211.29

PROP: White powder or needles from alc. Mp: 150°, d: 1.115 @ 25°. Sol in Et_2O, $CHCl_3$, and dil acids; sltly sol in H_2O.

SYNS: N,N'-DIPHENYLGUANIDINE □ 1,3-DIPHENYLGUANIDINE □ DPG □ DPG ACCELERATOR □ DWUFENYLOGUANIDYNA (POLISH) □ MELANILINE □ NCI-C60924 □ USAF B-19 □ USAF EK-1270 □ VULCACID D □ VULKACIT D/C □ VULKAZIT

TOXICITY DATA WITH REFERENCE
mmo-sat 360 ng/plate JEPOEC 6,293,85
mma-sat 33 μg/plate ENMUDM 8(Suppl 7),1,86
bfa-mus/sat 36 ng/kg JEPOEC 6,293,85
orl-mus TDLo:56 mg/kg (male 7D pre):REP JTEHD6 11,869,83
orl-mus TDLo:190 mg/kg (female 1-19D post):REP JEPTDQ 4(1),451,80
orl-mus TDLo:196 mg/kg (male 49D pre):REP JTEHD6 11,869,83
orl-mus TDLo:140 mg/kg (male 35D pre):REP JTEHD6 11,869,83
orl-mus TDLo:76 mg/kg (1-19D preg):TER JEPTDQ 4(1),451,80
orl-rat LD50:375 mg/kg GISAAA 29(7),34,64
ipr-rat LD50:75 mg/kg MEPAAX 16,35,65
orl-mus LD50:150 mg/kg SCIEAS 36(1-4),10,89
ipr-mus LD50:25 mg/kg NTIS** AD277-689
orl-mam LDLo:250 mg/kg JIDHAN 13,87,31

CONSENSUS REPORTS: Reported in EPA TSCA Inventory.

SAFETY PROFILE: Poison by ingestion and intraperitoneal routes. Experimental teratogenic and reproductive effects. Mutation data reported. When heated to decomposition it emits toxic fumes of NO_x.

DWC625 **CAS:24245-27-0** **HR: 3**
sym-DIPHENYLGUANIDINE HYDROCHLORIDE
mf: $C_{13}H_{13}N_3 \cdot ClH$ mw: 247.75

SYN: GUANIDINE, 1,3-DIPHENYL-, MONOHYDROCHLORIDE

TOXICITY DATA WITH REFERENCE
scu-rat LDLo:59 mg/kg JPETAB 28,251,26
orl-uns LDLo:250 mg/kg JIDHAN 13,87,31

CONSENSUS REPORTS: Reported in EPA TSCA Inventory.

SAFETY PROFILE: A poison by ingestion and subcutaneous routes. When heated to decomposition it emits toxic vapors of NO_x and HCl.

DWD000 **HR: 3**
DIPHENYLHYDANTOIN and PHENOBARBITAL

SYNS: PHENOBARBITOL and DIPHENYLHDANTOIN □ PHENOBARBITONE and PHENOBARBITONE □ PHENYTOIN and PHENOBARBITONE

TOXICITY DATA WITH REFERENCE
orl-wmn TDLo:3203 mg/kg (1-44W preg):REP AJDCAI 127,758,74
unr-wmn TDLo:1884 mg/kg (1-39W preg):TER JOPDAB 84,254,74
unr-wmn TDLo:1624 mg/kg (1-40W preg):TER LANCAO 2,839,72
orl-wmn TDLo:621 mg/kg (1-39W preg):TER LANCAO 2,702,71
orl-wmn TDLo:2646 mg/kg (1-39W preg):TER JOPDAB 89,154,76
orl-wmn TDLo:3203 mg/kg (1-44W preg):TER AJDCAI 127,758,74

SAFETY PROFILE: A human teratogen. Human reproductive effects: developmental abnormalities of the eye, ear, central nervous system, craniofacial, musculoskeletal, urogenital and cardiovascular system. Effects on newborn including growth statistics and physical effects. See also BARBITURATES.

DWD200 **CAS:53421-38-8** **HR: 3**
3,3-DIPHENYL-3-HYDROXYPROPIONIC ACID DIETHYLAMINOETHYL ESTER HYDROCHLORIDE
mf: $C_{20}H_{27}NO_3 \cdot ClH$ mw: 365.94

SYN: DPE-HCl

TOXICITY DATA WITH REFERENCE
orl-rat LD50:2310 mg/kg NIIRDN 6,332,82
scu-rat LD50:330 mg/kg NIIRDN 6,332,82
ivn-rat LD50:27,500 μg/kg NIIRDN 6,332,82
orl-mus LD50:418 mg/kg NIIRDN 6,332,82
ipr-mus LD50:140 mg/kg JPETAB 74,274,42
scu-mus LD50:280 mg/kg NIIRDN 6,332,82
ivn-mus LD50:27,500 μg/kg NIIRDN 6,332,82

SAFETY PROFILE: Poison by subcutaneous, intraperitoneal, and intravenous routes. Moderately toxic by ingestion. When heated to decomposition it emits very toxic fumes of HCl and NO_x.

DWD400 **HR: 3**
3,4-DIPHENYL-1-ISOBUTYLPYRAZOLE-5-ACETIC ACID SODIUM SALT
mf: $C_{21}H_{21}N_2O_2{\cdot}Na$ mw: 356.43

SYNS: 1-ISOBUTYL-3,4-DIPHENYLPYRAZOLE-5-ACETIC ACID SODIUM SALT □ LM 22070

TOXICITY DATA WITH REFERENCE
orl-rat LD50:46 mg/kg AIPTAK 238,305,79
orl-mus LD50:320 mg/kg AIPTAK 238,305,79

SAFETY PROFILE: Poison by ingestion. When heated to decomposition it emits toxic fumes of NO_x and Na_2O.

DWD800 **CAS:587-85-9** **HR: 3**
DIPHENYLMERCURY
mf: $C_{12}H_{10}Hg$ mw: 354.81

PROP: White crystals or needles from alc. D: 2.318, mp: 124.5–125° (sublimes), bp: 204° @ 10.5 mm. Insol in water.

TOXICITY DATA WITH REFERENCE
orl-rat LDLo:500 mg/kg NCNSA6 5,30,53
ipr-rat LDLo:50 mg/kg NCNSA6 5,30,53
ipr-mus LDLo:250 mg/kg CBCCT* 4,231,52

CONSENSUS REPORTS: Reported in EPA TSCA Inventory. Mercury and its compounds are on the Community Right-To-Know List.

OSHA PEL: CL 0.1 mg(Hg)/m³ (skin)
ACGIH TLV: TWA 0.1 mg(Hg)/m³ (skin)
NIOSH REL: (Mercury, Aryl and Inorganic): CL 0.1 mg/m³ (skin)

SAFETY PROFILE: Poison by intraperitoneal route. Moderately toxic by ingestion. Incompatible with nonmetal oxides. When heated to decomposition it emits toxic fumes of Hg. See also MERCURY COMPOUNDS.

DWE200 **CAS:18656-25-2** **HR: 3**
2-(DIPHENYLMETHOXY)ACETAMIDOXIME HYDROGEN MALEATE
mf: $C_{15}H_{16}N_2O_2{\cdot}C_4H_4O_4$ mw: 372.41

TOXICITY DATA WITH REFERENCE
orl-mus LD50:450 mg/kg ARZNAD 17,1446,67
ivn-mus LD50:33 mg/kg ARZNAD 17,1446,67
ivn-cat LDLo:8 mg/kg ARZNAD 17,1446,67

SAFETY PROFILE: Poison by intravenous route. Moderately toxic by ingestion. When heated to decomposition it emits toxic fumes of NO_x.

DWE800 **CAS:524-83-4** **HR: 3**
3-(DIPHENYLMETHOXY)-8-ETHYLNORTROPANE
mf: $C_{22}H_{27}NO$ mw: 321.50

SYNS: 3-α-(DIPHENYLMETHOXY)-8-ETHYLNORTROPANE □ endo-3-(DIPHENYLMETHOXY)-8-ETHYL-8-AZABICYCLO(3.2.1)OCTANE □ ETHYBENZTROPINE □ ETHYLBENATROPINE □ ETHYLBENZTROPINE □ N-ETHYLNORTROPINE BENZHYDRYL ETHER □ ETYBENZATROPINE □ PANOLID □ PONALID □ PONALIDE □ UK-738 □ VK-738

TOXICITY DATA WITH REFERENCE
orl-rat LDLo:560 mg/kg 27ZQAG -,228,72
orl-mus LDLo:66 mg/kg 27ZQAG -,228,72
ivn-mus LD50:12 mg/kg EJPHAZ 9,304,70
orl-rbt LDLo:215 mg/kg 27ZQAG -,228,72
ivn-rbt LDLo:6 mg/kg 27ZQAG -,228,72

SAFETY PROFILE: Poison by ingestion and intravenous routes. When heated to decomposition it emits toxic fumes of NO_x. See also ETHERS.

DWF000 **CAS:606-90-6** **HR: 3**
4-(DIPHENYLMETHOXY)-1-METHYLPIPERIDINE CHLOROTHEOPHYLLINE
mf: $C_{19}H_{23}NO{\cdot}C_7H_7ClN_4O_2$ mw: 496.06

PROP: Minute crystals. Mp: 151°.

SYN: P 284

TOXICITY DATA WITH REFERENCE
orl-mus LD50:275 mg/kg ARZNAD 5,185,55
ivn-mus LD50:75 mg/kg ARZNAD 5,185,55

SAFETY PROFILE: Poison by ingestion and intravenous routes. When heated to decomposition it emits very toxic fumes of NO_x and Cl^-. See also THEOPHYLLINE.

DWF200 **CAS:1146-95-8** **HR: 3**
2-DIPHENYLMETHYLENEBUTYLAMINE HYDROCHLORIDE
mf: $C_{17}H_{19}N{\cdot}ClH$ mw: 273.83

PROP: A solid. Mp: 232°.

SYNS: EDPA HYDROCHLORIDE □ 2-ETHYL-3,3-DIPHENYL-2-PROPENYLAMINE HYDROCHLORIDE □ ETIFELMIN HYDROCHLORIDE □ GILUTENSIN □ Na III HYDROCHLORIDE □ TENSINASE D

TOXICITY DATA WITH REFERENCE
orl-rat LD50:139 mg/kg NIIRDN 6,112,82
ipr-rat LD50:23,400 μg/kg NIIRDN 6,112,82
scu-rat LD50:91,300 μg/kg NIIRDN 6,112,82
orl-mus LD50:115 mg/kg NIIRDN 6,112,82
ipr-mus LD50:45,500 μg/kg NIIRDN 6,112,82
scu-mus LD50:73,400 μg/kg NIIRDN 6,112,82
ivn-mus LD50:33,900 μg/kg NIIRDN 6,112,82

SAFETY PROFILE: Poison by ingestion, intravenous, intraperitoneal, and subcutaneous routes. When heated to decomposition it emits very toxic fumes of NO_x and HCl.

DWF700 **CAS:57726-65-5** **HR: 3**
2-(3,3-DIPHENYL-3-(5-METHYL-1,3,4-OXADIAZOL-2-YL)PROPYL)-2-AZABICYCLO(2.2.2)OCTANE
mf: $C_{25}H_{29}N_3O$ mw: 387.57

PROP: A solid. Mp: 121–123°.

SYNS: 2-(3-(5-METHYL-1,3,4-OXADIAZOL-2-YL)-3,3-DIPHENYLPROPYL)-2-AZABICYCLO(2.2.2)OCTANE □ NUFENOXOLE □ SC 27166 □ SEARLE 27166

TOXICITY DATA WITH REFERENCE
ivn-rat LD50:21 mg/kg JPETAB 210,327,79
orl-mus LD50:399 mg/kg NICHAS 28,1621,79
ipr-mus LD50:95 mg/kg NICHAS 38,1621,79

scu-mus LD50:258 mg/kg NICHAS 38,1621,79
ivn-mus LD50:12 mg/kg NICHAS 38,1621,79

SAFETY PROFILE: Poison by ingestion, intravenous, subcutaneous, and intraperitoneal routes. When heated to decomposition it emits toxic fumes of NO_x.

DWF865 CAS:19841-73-7 HR: 3
4-DIPHENYLMETHYLPIPERIDINE
mf: $C_{18}H_{21}N$ mw: 251.40

TOXICITY DATA WITH REFERENCE
ipr-mus LD50:30 mg/kg JMCMAR 15,690,72

CONSENSUS REPORTS: Reported in EPA TSCA Inventory.

SAFETY PROFILE: Poison by intraperitoneal route. When heated to decomposition it emits toxic fumes of NO_x.

DWF869 HR: 3
4,4-DIPHENYL-1-METHYLPIPERIDINE MALEATE
mf: $C_{18}H_{21}N \cdot C_4H_4O_4$ mw: 367.48

TOXICITY DATA WITH REFERENCE
orl-mus LD50:140 mg/kg ARZNAD 34,233,84
scu-mus LD50:80 mg/kg ARZNAD 34,233,84
ivn-mus LD50:45 mg/kg ARZNAD 34,233,84

SAFETY PROFILE: Poison by ingestion, subcutaneous, and intravenous routes. When heated to decomposition it emits toxic fumes of NO_x.

DWF875 CAS:101564-67-4 HR: 3
5,5-DIPHENYL-3-(3-(2-METHYLPIPERIDINO)PROPYL)-2-THIOHYDANTOIN HYDROCHLORIDE
mf: $C_{24}H_{29}N_3OS \cdot ClH$ mw: 444.08

SYNS: 5,5-DIPHENYL-3-(3-(2-METHYLPIPERIDINO)PROPYL)-2-THIOHYDANTOIN MONOHYDROCHLORIDE □ 3-γ-(α-METIL-1-PIPERIDINO)PROPIL-5,5-DIFENILTIOIDANTOINA CLORIDRATO (ITALIAN)

TOXICITY DATA WITH REFERENCE
ipr-mus LD50:25 mg/kg FRPSAX 15,809,60
scu-mus LD50:25 mg/kg FRPSAX 15,809,60
ivn-rbt LD50:3630 μg/kg FRPSAX 15,809,60

SAFETY PROFILE: Poison by subcutaneous, intravenous, and intraperitoneal routes. When heated to decomposition it emits toxic fumes of SO_x, NO_x, and HCl.

DWG600 HR: 3
α-2,2-DIPHENYL-4-(1-METHYL-2-PIPERIDYL)-1,3-DIOXOLANE HYDROCHLORIDE
mf: $C_{21}H_{25}NO_2 \cdot ClH$ mw: 359.93

TOXICITY DATA WITH REFERENCE
orl-mus LD50:400 mg/kg JMCMAR 9,127,66
ivn-mus LD50:25 mg/kg JMCMAR 9,127,66

SAFETY PROFILE: Poison by ingestion and intravenous routes. When heated to decomposition it emits very toxic fumes of NO_x and HCl.

DWH000 HR: 3
β-2,2-DIPHENYL-4-(1-METHYL-2-PIPERIDYL)-1,3-DIOXOLANE HYDROCHLORIDE
mf: $C_{21}H_{25}NO_2 \cdot ClH$ mw: 359.93

TOXICITY DATA WITH REFERENCE
orl-mus LD50:300 mg/kg JMCMAR 9,127,66
ivn-mus LD50:75 mg/kg JMCMAR 9,127,66

SAFETY PROFILE: Poison by ingestion and intravenous routes. When heated to decomposition it emits very toxic fumes of NO_x and HCl.

DWH200 HR: 3
α-2,2-DIPHENYL-4-(1-METHYL-2-PIPERIDYL)-1,3-DIOXOLANE METHYLIODIDE
mf: $C_{21}H_{25}NO_2 \cdot CH_3I$ mw: 465.41

TOXICITY DATA WITH REFERENCE
orl-mus LD50:800 mg/kg JMCMAR 9,127,66
ipr-mus LD50:75 mg/kg JMCMAR 9,127,66

SAFETY PROFILE: Poison by intraperitoneal route. Moderately toxic by ingestion. When heated to decomposition it emits very toxic fumes of NO_x and I^-. See also IODIDES.

DWH500 CAS:102280-81-9 HR: 3
3,3-DIPHENYL-2-METHYL-1-PYRROLINE
mf: $C_{17}H_{17}N$ mw: 235.35

TOXICITY DATA WITH REFERENCE
ipr-rat LD50:310 mg/kg AIPTAK 115,332,58
orl-mus LD50:560 mg/kg AIPTAK 115,332,58
ivn-mus LD50:47 mg/kg AIPTAK 115,332,58

SAFETY PROFILE: Poison by intravenous and intraperitoneal routes. Moderately toxic by ingestion. When heated to decomposition it emits toxic fumes of NO_x.

DWH600 CAS:545-91-5 HR: 3
4,4-DIPHENYL-6-MORPHOLINO-3-HEPTANONE HYDROCHLORIDE
mf: $C_{23}H_{29}NO_2 \cdot ClH$ mw: 387.99

PROP: A solid. Mp: 224–225° (decomp). Insol in C_6H_6.

TOXICITY DATA WITH REFERENCE
scu-rat LD50:132 mg/kg JPETAB 98,305,50
orl-mus LD50:208 mg/kg FSTEAI 5,185,50
ipr-mus LD50:131 mg/kg JACSAT 71,57,49
scu-mus LD50:110 mg/kg JPETAB 98,121,50
ivn-mus LD50:48 mg/kg JPETAB 98,305,50

SAFETY PROFILE: Poison by ingestion, intravenous, subcutaneous, and intraperitoneal routes. When heated to decomposition it emits very toxic fumes of Cl^- and NO_x.

DWH800 CAS:63765-88-8 HR: 3
4,4-DIPHENYL-6-MORPHOLINO-3-HEXANONE HYDROCHLORIDE
mf: $C_{22}H_{27}NO_2 \cdot ClH$ mw: 373.96

TOXICITY DATA WITH REFERENCE
orl-mus LD50:208 mg/kg BJPCAL 5,125,50

ipr-mus LD50:114 mg/kg JACSAT 71,57,49
scu-mus LD50:220 mg/kg JPPMAB 2,418,50
ivn-mus LD50:43 mg/kg BJPCAL 5,125,50

SAFETY PROFILE: Poison by ingestion, intraperitoneal, subcutaneous, and intravenous routes. When heated to decomposition it emits very toxic fumes of HCl and NO_x.

DWH875 **HR: 3**
2,3-DIPHENYL-3H-NAPHTHO(1,2-d)TRIAZOLIUM CHLORIDE
mf: $C_{22}H_{16}N_3 \cdot Cl$ mw: 357.86

SYN: α-β-NAPHTHO-2,3-DIPHENYL-TRIAZOLIUM CHLORID (GERMAN)

TOXICITY DATA WITH REFERENCE
scu-mus LDLo:2800 μg/kg SDMU** -,-,36
ivn-mus LDLo:2400 μg/kg SDMU** -,-,36
ivn-rbt LDLo:40 mg/kg SDMU** -,-,36
scu-gpg LDLo:12 mg/kg SDMU** -,-,36
scu-frg LDLo:45 mg/kg SDMU** -,-,36

SAFETY PROFILE: Poison by subcutaneous and intravenous routes. When heated to decomposition it emits toxic fumes of Cl^- and NO_x.

DWI000 **CAS:86-30-6** **HR: 3**
DIPHENYLNITROSAMINE
mf: $C_{12}H_{10}N_2O$ mw: 198.24

PROP: Yellow plates or green crystals. Mp: 66.5°.

SYNS: CURETARD A □ DELAC J □ DIPHENYLNITROSAMIN (GERMAN) □ DIPHENYL N-NITROSOAMINE □ N,N-DIPHENYLNITROSAMINE □ NAUGARD TJB □ NCI-C02880 □ NDPA □ NDPhA □ N-NITROSODIFENYLAMIN (CZECH) □ NITROSODIPHENYLAMINE □ N-NITROSODIPHENYLAMINE □ N-NITROSO-N-PHENYLANILINE □ NITROUS DIPHENYLAMIDE □ REDAX □ RETARDER J □ TJB □ VULCALENT A □ VULCATARD □ VULKALENT A (CZECH) □ VULTROL

TOXICITY DATA WITH REFERENCE
eye-rbt 500 mg/24H MLD 28ZPAK -,134,72
mma-sat 50 μg/plate CANCAR 49,1970,82
dnd-hmn:fbr 3 mmol/L ENMUDM 7,267,85
dns-rat:lvr 500 nmol/L CNREA8 42,3010,82
otr-ham:emb 6300 μg/L NCIMAV 58,243,81
orl-rat TDLo:140 g/kg/2Y-C:CAR NCITR* NCI-CG-TR-164,79
skn-mus TDLo:800 mg/kg/20W-I:ETA EJCAAH 16,695,80
orl-rat TD:170 g/kg/2Y-C:CAR NCITR* NCI-CG-TR-164,79
orl-rat TD:2800 g/kg/2Y-C:CAR IARC** 27,213,82
skn-mus TD:800 mg/kg/20W-I:ETA IARC** 27,213,82
ipr-rat TD:7700 mg/kg/77W-I:ETA IARC** 27,213,82
orl-mus TD:1750 g/kg/2Y-I:ETA IARC** 27,213,82
orl-rat TD:146 g/kg/2Y-C:CAR EESADV 3,29,79
orl-rat LD50:1825 mg/kg GISAAA 45(1),18,80
orl-mus LD50:1860 mg/kg GISAAA 45(1),18,80
ipr-mus LD50:1000 mg/kg PMRSDJ 1,682,81

CONSENSUS REPORTS: IARC Cancer Review: Group 3 IMEMDT 7,56,87; Animal Limited Evidence IMEMDT 27,213,82. NCI Carcinogenesis Bioassay (feed); Clear Evidence: rat NCITR* NCI-CG-TR-164,79; No Evidence: mouse NCITR* NCI-CG-TR-164,79. Reported in EPA TSCA Inventory. EPA Genetic Toxicology Program. Community Right-To-Know List.

SAFETY PROFILE: Moderately toxic by ingestion. An eye irritant. Questionable carcinogen with experimental carcinogenic and tumorigenic data. Human mutation data reported. Dangerous fire hazard when exposed to heat, flame, or oxidizing materials. Can react vigorously with oxidizing materials. When heated to decomposition it emits highly toxic fumes of NO_x. See also NITROSAMINES.

For occupational chemical analysis use OSHA: #23.

DWI200 **CAS:92-71-7** **HR: 2**
2,5-DIPHENYLOXAZOLE
mf: $C_{15}H_{11}NO$ mw: 221.27

PROP: Needles from pet ether. Mp: 74°. Sol in EtOH and Et_2O; prac insol in H_2O.

SYN: USAF EK-6775

TOXICITY DATA WITH REFERENCE
ipr-mus LD50:750 mg/kg NTIS** AD277-689

CONSENSUS REPORTS: Reported in EPA TSCA Inventory.

SAFETY PROFILE: Moderately toxic by intraperitoneal route. When heated to decomposition it emits toxic fumes of NO_x.

DWI400 **CAS:16230-71-0** **HR: 2**
3,3-DIPHENYL-2-OXETANONE
mf: $C_{15}H_{12}O_2$ mw: 224.27

SYNS: 2,2-DIPHENYL-3-HYDROXYPROPIONIC ACID LACTONE □ α,α-DIPHENYL-β-PROPIOLACTONE

TOXICITY DATA WITH REFERENCE
scu-rat TDLo:380 mg/kg/19W-I:ETA BJCAAI 15,85,61

SAFETY PROFILE: Questionable carcinogen with experimental tumorigenic data. When heated to decomposition it emits acrid smoke and irritating fumes.

DWI800 **HR: 3**
1,5-DIPHENYL-1,4-PENTAZDIENE
mf: $C_{12}H_{11}N_5$ mw: 225.26

SAFETY PROFILE: Explodes violently on warming, impact, or friction. When heated to decomposition it emits toxic fumes of NO_x.

DWJ300 **CAS:21413-28-5** **HR: 2**
5,5-DIPHENYL-1-PHENYLSULFONYLHYDANTOIN
mf: $C_{21}H_{16}N_2O_4S$ mw: 392.45

SYN: 5,5-DIPHENYL-1-(PHENYLSULFONYL)-2,4-IMIDAZOLIDINEDIONE (9CI)

TOXICITY DATA WITH REFERENCE
orl-rat LD50:1700 mg/kg ARZNAD 20,1579,70
ipr-rat LD50:695 mg/kg ARZNAD 20,1579,70
orl-mus LD50:1700 mg/kg ARZNAD 20,1579,70
ipr-mus LD50:1037 mg/kg ARZNAD 20,1579,70

SAFETY PROFILE: Moderately toxic by ingestion and

D

intraperitoneal routes. When heated to decomposition it emits toxic fumes of SO_x and NO_x.

DWJ400 **CAS:3736-92-3** **HR: 3**
1,2-DIPHENYL-4-PHENYLTHIOETHYL-3,5-PYRAZOLIDINEDIONE
mf: $C_{23}H_{20}N_2O_2S$ mw: 388.51

PROP: Crystals from EtOH. Mp: 110–113°.

SYN: 4-(PHENYLTHIOETHYL)-1,2-DIPHENYL-3,5-PYRAZOLIDINE-DIONE

TOXICITY DATA WITH REFERENCE
orl-rat LD50:450 mg/kg ANYAA9 86,263,60
ivn-rat LD50:190 mg/kg ANYAA9 86,263,60
orl-mus LD50:560 mg/kg ANYAA9 86,263,60
ivn-mus LD50:178 mg/kg ANYAA9 86,263,60
ivn-rbt LD50:100 mg/kg MDCHAG 5,391,65

SAFETY PROFILE: Poison by intravenous route. Moderately toxic by ingestion. When heated to decomposition it emits very toxic fumes of NO_x and SO_x.

DWK200 **CAS:972-02-1** **HR: 3**
α,α-DIPHENYL-1-PIPERIDINEBUTANOL
mf: $C_{21}H_{27}NO$ mw: 309.49

PROP: Needles from pet ether. Mp: 104–105°.

SYNS: DIFENIDOL □ DIPHENIDOL □ NOMETIC □ α-(3-PIPERIDINOPROPYL)BENZHYDROL □ SKF 478 □ SK&F No. 478-A □ VONTROL

TOXICITY DATA WITH REFERENCE
orl-rat LD50:815 mg/kg PHMCAA 5,265,63
scu-rat LD50:50 mg/kg TXAPA9 18,185,71
orl-mus LD50:450 mg/kg PHMCAA 5,265,63
ivn-mus LD50:32 mg/kg MPHEAE 13,325,65

SAFETY PROFILE: Poison by intravenous and subcutaneous routes. Moderately toxic by ingestion. When heated to decomposition it emits toxic fumes of NO_x.

DWK400 **CAS:467-60-7** **HR: 3**
α,α-DIPHENYL-2-PIPERIDINEMETHANOL
mf: $C_{18}H_{21}NO$ mw: 267.40

SYNS: DETARIL □ GERODYL □ MERATRAN □ MRD 108 □ α-(2-PIPERIDYL)BENZHYDROL □ PIPRADOL □ α-PIPRADOL □PIRIDROL □ PYRIDROL □ PYRIDROLE

TOXICITY DATA WITH REFERENCE
orl-rat LD50:180 mg/kg 27ZIAQ -,-,65
orl-mus LD50:74 mg/kg FRPSAX 12,853,57
ipr-mus LD50:74 mg/kg FRPSAX 12,853,57
ivn-cat LD50:15 mg/kg DAZEA2 122,283,82

SAFETY PROFILE: Poison by ingestion, intravenous, and intraperitoneal routes. When heated to decomposition it emits toxic fumes of NO_x.

DWK500 **CAS:101564-69-6** **HR: 3**
5,5-DIPHENYL-3-(3-PIPERIDINOPROPYL)-2-THIOHYDANTOIN HYDROCHLORIDE
mf: $C_{23}H_{27}N_3OS{\cdot}ClH$ mw: 430.05

SYN: 3-γ-(1-PIPERIDINO)PROPIL-5,5-DIFENILTIOIDANTOINA CLORIDRATO (ITALIAN)

TOXICITY DATA WITH REFERENCE
ipr-mus LD50:35 mg/kg FRPSAX 15,809,60
scu-mus LD50:30 mg/kg FRPSAX 15,809,60
ivn-rbt LD50:3350 μg/kg FRPSAX 15,809,60

SAFETY PROFILE: Poison by subcutaneous, intravenous, and intraperitoneal routes. When heated to decomposition it emits toxic fumes of SO_x, NO_x, and HCl.

DWK700 **CAS:24050-58-6** **HR: 3**
3-(3,3-DIPHENYLPROPYLAMINO)PROPYL-3′,4′,5′-TRIMETHOXYBENZOATE HYDROCHLORIDE
mf: $C_{28}H_{33}NO_5{\cdot}ClH$ mw: 500.08

SYNS: PF-26 □ 3,4,5-TRIMETHOXYBENZOIC ACID 3-((3,3-DIPHENYLPROPYL)AMINO)PROPYL ESTER HYDROCHLORIDE

TOXICITY DATA WITH REFERENCE
orl-rat TDLo:9 g/kg (30D male):REP GNRIDX 5,256,71
orl-rat TDLo:300 mg/kg (9-14D preg):TER GNRIDX 5,271,71
ipr-rat LD50:138 mg/kg ARZNAD 21,1628,71
ivn-rat LD50:31,500 μg/kg ARZNAD 21,1628,71
ivn-mus LD50:35 mg/kg ARZNAD 21,1628,71

SAFETY PROFILE: Poison by intravenous and intraperitoneal routes. Experimental teratogenic and reproductive effects. When heated to decomposition it emits toxic fumes of NO_x and HCl. See also ESTERS.

DWK900 **CAS:23903-11-9** **HR: 3**
3-(N-d,d-DIPHENYLPROPYL-N-METHYL)AMINOPROPAN-1-OL HYDROCHLORIDE
mf: $C_{19}H_{25}NO{\cdot}ClH$ mw: 319.91

SYN: PF-82

TOXICITY DATA WITH REFERENCE
orl-rat LD50:1400 mg/kg ARZNAD 24,166,74
ipr-rat LD50:160 mg/kg ARZNAD 24,166,74
orl-mus LD50:428 mg/kg ARZNAD 24,166,74
ipr-mus LD50:100 mg/kg ARZNAD 24,166,74
scu-mus LD50:364 mg/kg ARZNAD 24,166,74
ivn-mus LD50:51 mg/kg ARZNAD 24,166,74

SAFETY PROFILE: Poison by subcutaneous, intravenous, and intraperitoneal routes. Moderately toxic by ingestion. When heated to decomposition it emits toxic fumes of NO_x and HCl.

DWL200 **CAS:69-43-2** **HR: 3**
N-(3,3-DIPHENYLPROPYL)-α-METHYLPHENETHYLAMINE LACTATE
mf: $C_{24}H_{27}N{\cdot}C_3H_6O_3$ mw: 419.61

PROP: A solid. Mp: 140–142°.

SYN: PRENYLAMINE LACTATE

TOXICITY DATA WITH REFERENCE
orl-rat LD50:1000 mg/kg ARZNAD 26,2127,76
ipr-rat LD50:40 mg/kg ARZNAD 26,2127,76
orl-mus LD50:580 mg/kg ARZNAD 26,2127,76
ipr-mus LD50:40 mg/kg ARZNAD 26,2127,76

SAFETY PROFILE: Poison by intraperitoneal route. Moderately toxic by ingestion. When heated to decomposition it emits toxic fumes of NO_x.

DWL400 CAS:10087-89-5 HR: 3
1,1-DIPHENYL-2-PROPYNYL-N-CYCLOHEXYLCARBAMATE
mf: $C_{22}H_{23}NO_2$ mw: 333.46

PROP: A solid. Mp: 160–161°.

SYNS: 1,1-DIPHENYL-2-PROPYN-1-OL CYCLOHEXANECARBAMATE □ 1,1-DIPHENYL-2-PROPYNYL ESTER CYCLOHEXANECARBAMIC ACID □ ENPROMATE

TOXICITY DATA WITH REFERENCE
mma-sat 100 μg/plate PNASA6 72,5135,75
dns-rat:lvr 10 μmol/L ENMUDM 3,11,81
orl-rat TDLo:1690 mg/kg/27W-C:ETA PAACA3 10,35,69
skn-rat TDLo:31,500 mg/kg/67W-C:ETA JJIND8 71,211,83
scu-rat TDLo:1925 mg/kg/11W-C:ETA PAACA3 10,35,69
orl-rat TD:13,600 mg/kg/40W-C:ETA JJIND8 71,211,83
orl-mus LD50:1 g/kg GANMAX 2,261,66
ipr-mus LD50:374 mg/kg JMCMAR 11,1155,68

CONSENSUS REPORTS: EPA Genetic Toxicology Program.

SAFETY PROFILE: Poison by intraperitoneal route. Moderately toxic by ingestion. Questionable carcinogen with experimental tumorigenic data. Mutation data reported. When heated to decomposition it emits toxic fumes of NO_x. See also CARBAMATES.

DWL500 CAS:10473-64-0 HR: 2
1,1-DIPHENYL-2-PROPYNYL-N-ETHYLCARBAMATE
mf: $C_{18}H_{17}NO_2$ mw: 279.36

SYN: ETHYLCARBAMIC ACID 1,1-DIPHENYL-2-PROPYNYL ESTER

TOXICITY DATA WITH REFERENCE
orl-rat TDLo:8800 mg/kg/28W-C:ETA JJIND8 71,211,83

SAFETY PROFILE: Questionable carcinogen with experimental tumorigenic data. When heated to decomposition it emits toxic fumes of NO_x.

DWL525 CAS:10473-98-0 HR: 2
1,1-DIPHENYL-2-PROPYNYL 1-PYRROLIDINECARBOXYLATE
mf: $C_{20}H_{19}NO_2$ mw: 305.40

SYN: 3,3-DIPHENYL-3-(PYRROLIDINE-CARBONYLOXY)-1-PROPYNE

TOXICITY DATA WITH REFERENCE
orl-rat TDLo:2 g/kg/35W-C:ETA JJIND8 71,211,83

SAFETY PROFILE: Questionable carcinogen with experimental tumorigenic data. When heated to decomposition it emits toxic fumes of NO_x.

DWL600 CAS:3426-01-5 HR: 2
1,4-DIPHENYL-3,5-PYRAZOLIDINEDIONE
mf: $C_{15}H_{11}N_2O_2$ mw: 251.28

SYN: 1,4-DIPHENYL-3,5-DIOXO-PYRAZOLIDIN (GERMAN)

TOXICITY DATA WITH REFERENCE
scu-mus LD50:1003 mg/kg ARZNAD 4,249,54
ivn-mus LD50:740 mg/kg AEPPAE 233,365,58

SAFETY PROFILE: Moderately toxic by subcutaneous and intravenous routes. When heated to decomposition it emits toxic fumes of NO_x.

D

DWL800 CAS:4845-49-2 HR: 2
1,3-DIPHENYL-5-PYRAZOLONE
mf: $C_{15}H_{12}N_2O$ mw: 236.29

TOXICITY DATA WITH REFERENCE
ipr-mus LDLo:512 mg/kg CBCCT* 2,135,50

CONSENSUS REPORTS: Reported in EPA TSCA Inventory.

SAFETY PROFILE: Moderately toxic by intraperitoneal route. When heated to decomposition it emits toxic fumes of NO_x.

DWM000 CAS:57-96-5 HR: 3
DIPHENYLPYRAZONE
mf: $C_{23}H_{20}N_2O_3S$ mw: 404.51

SYNS: 1,2-DIPHENYL-4-(2′-PHENYLSULFINETHYL)-3,5-PYRAZOLIDINEDIONE □ 4-(PHENYLSULFOXYETHYL)-1,2-DIPHENYL-3,5-PYRAZOLIDINEDIONE □ SULFINPYRAZINE □ SULFOXYPHENYLPYRAZOLIDINE □ USAF GE-13

TOXICITY DATA WITH REFERENCE
sln-Mold-asn 1 g/L MUREAV 26,159,74
orl-hmn TDLo:29 mg/kg NEJMAG 303,702,80
orl-rat LD50:358 mg/kg DRUGAY 6,393,82
ivn-rat LD50:154 mg/kg ANYAA9 86,263,60
orl-mus LD50:298 mg/kg ANYAA9 86,263,60
ipr-mus LD50:100 mg/kg NTIS** AD414-344
ivn-mus LD50:240 mg/kg ANYAA9 86,263,60
ivn-rbt LD50:195 mg/kg MDCHAG 5,391,65

SAFETY PROFILE: Poison by ingestion, intravenous, and intraperitoneal routes. Mutation data reported. When heated to decomposition it emits very toxic fumes of NO_x and SO_x.

DWM400 CAS:10447-38-8 HR: 3
α,α-DIPHENYL-3-QUINUCLIDINEMETHANOL HYDROCHLORIDE
mf: $C_{20}H_{23}NO \cdot ClH$ mw: 329.90

PROP: A solid. Mp: 285–290°.

SYNS: FENCAROL □ FENKAROL □ PHENCAROL □ QUINUCLIDYL-3-DIPHENYLCARBINOL HYDROCHLORIDE

TOXICITY DATA WITH REFERENCE
orl-rat LD50:440 mg/kg RPTOAN 40,42,77
orl-mus LD50:370 mg/kg RPTOAN 40,42,77
ivn-mus LD50:62 mg/kg RPTOAN 40,42,77
orl-gpg LD50:860 mg/kg FATOAO 43,148,80

SAFETY PROFILE: Poison by ingestion and intravenous routes. When heated to decomposition it emits very toxic fumes of NO_x and HCl.

DWM600 **CAS:10504-99-1** **HR: 1**
DIPHENYLSELENONE
mf: $C_{12}H_{10}O_2Se$ mw: 265.17

CONSENSUS REPORTS: Selenium and its compounds are on the Community Right-To-Know List.

OSHA PEL: TWA 0.2 mg(Se)/m^3
ACGIH TLV: TWA 0.2 mg(Se)/m^3
DFG MAK: 0.1 mg(Se)/m^3

SAFETY PROFILE: Explodes weakly when heated. When heated to decomposition it emits toxic fumes of Se. See also SELENIUM COMPOUNDS.

DWM800 **CAS:1011-95-6** **HR: 3**
DIPHENYLSTANNANE
mf: $C_{12}H_{12}Sn$ mw: 274.93

PROP: Yellow powder or air and light sensitive crystals from pet ether/CH_2Cl_2. Mp: 226°. Insol in water.

SYNS: DIPHENYLTIN □ DIPHENYLTIN DIHYDRIDE

TOXICITY DATA WITH REFERENCE
ipr-rat LDLo:15 mg/kg BJIMAG 23,222,66

OSHA PEL: TWA 0.1 mg(Sn)/m^3 (skin)
ACGIH TLV: TWA 0.1 mg(Sn)/m^3 (skin) (Proposed: TWA 0.1 mg(Sn)/m^3; STEL 0.2 mg(Sn)/m^3 (skin))
NIOSH REL: (Organotin Compounds) TWA 0.1 mg(Sn)/m^3

SAFETY PROFILE: Poison by intraperitoneal route. Ignites on contact with fuming nitric acid. When heated to decomposition it emits acrid smoke and irritating fumes. See also TIN COMPOUNDS.

For occupational chemical analysis use NIOSH: Organotin Compounds 5504.

DWN000 **CAS:945-51-7** **HR: 2**
DIPHENYL SULFOXIDE
mf: $C_{12}H_{10}OS$ mw: 202.28

PROP: Crystals or prisms. Mp: 70.5°, bp: 210° @ 15 mm, vap d: 7.0.

TOXICITY DATA WITH REFERENCE
ipr-mus LD50:750 mg/kg IJRBA3 3,41,61

CONSENSUS REPORTS: Reported in EPA TSCA Inventory.

SAFETY PROFILE: Moderately toxic by intraperitoneal route. A fungicide. When heated to decomposition it emits toxic fumes of SO_x.

DWN200 **CAS:60-10-6** **HR: 3**
DIPHENYLTHIOCARBAZONE
mf: $C_{13}H_{12}N_4S$ mw: 256.35

PROP: Bluish-black crystalline powder from alc (aq). Mp: 165–169°. Sol in aq alkaline solns; sltly sol in EtOH, CCl_4, $CHCl_3$, and C_6H_6; insol in H_2O.

SYNS: CARBAZONE, DIPHENYLTHIO- □ DITHIZON □ DITHIZONE □ 3-FORMAZANTHIOL, 1,5-DIPHENYL- □ (PHENYLAZO)THIOFORMIC ACID, 2-PHENYLHYDRAZIDE □ SEMICARBAZIDE, 1-PHENYL-4-(PHENYLIMINO)-3-THIO- □ THIOFORMIC ACID, PHENYLAZO-, PHENYLHYDRAZIDE □ USAF EK-3092

TOXICITY DATA WITH REFERENCE
ipr-mus LD50:200 mg/kg NTIS** AD277-689
ivn-mus LD50:56 mg/kg CSLNX* NX#07955

CONSENSUS REPORTS: Reported in EPA TSCA Inventory.

SAFETY PROFILE: Poison by intravenous and intraperitoneal routes. Can cause eye injury and glycosuria. When heated to decomposition it emits highly toxic fumes of NO_x and SO_x.

DWN400 **CAS:622-03-7** **HR: 3**
1,5-DIPHENYL-3-THIOCARBOHYDRAZIDE
mf: $C_{13}H_{14}N_4S$ mw: 258.37

PROP: Crystals from EtOH. Mp: 156–158° (decomp). Sltly sol in EtOH and C_6H_6.

SYNS: DIPHENYL THIOCARBAZIDE □ USAF EK-3110

TOXICITY DATA WITH REFERENCE
orl-rat LD50:1500 mg/kg JPETAB 90,260,47
ipr-mus LD50:200 mg/kg NTIS** AD277-689

CONSENSUS REPORTS: Reported in EPA TSCA Inventory.

SAFETY PROFILE: Poison by intraperitoneal route. Moderately toxic by ingestion. When heated to decomposition it emits very toxic fumes of NO_x and SO_x.

DWN600 **CAS:21083-47-6** **HR: 2**
5,5-DIPHENYL-2-THIOHYDANTOIN
mf: $C_{15}H_{12}N_2OS$ mw: 268.35

TOXICITY DATA WITH REFERENCE
orl-rat TDLo:3500 mg/kg:ETA CNREA8 28,924,68

CONSENSUS REPORTS: Reported in EPA TSCA Inventory.

SAFETY PROFILE: Questionable carcinogen with experimental tumorigenic data. When heated to decomposition it emits very toxic fumes of NO_x and SO_x.

DWN800 **CAS:102-08-9** **HR: 2**
DIPHENYLTHIOUREA
mf: $C_{13}H_{12}N_2S$ mw: 228.33

PROP: White to faint gray powder or leaflets from alc. Mp: 154°, bp: decomp, d: 1.32 @ 25°.

SYNS: DFT □ N,N'-DIPHENYLTHIOCARBAMIDE □ sym-DIPHENYLTHIOCARBAMIDE □ N,N'-DIPHENYLTHIOUREA □ sym-DIPHENYLTHIOUREA □ 1,3-DIPHENYLTHIOUREA □ 1,3-DIPHENYL-2-THIOUREA □ 2-FENYLOTIOMOCZNIK (POLISH) □ RHENOCURE CA □ STABILISATOR C □ SULFOCARBANILIDE □ THIOCARBANILIDE □ USAF EK-245 □ VALKACIT CA

D

TOXICITY DATA WITH REFERENCE
orl-rat TDLo:1500 mg/kg (female 6-20D post):REP FAATDF 17,399,91
orl-rat TDLo:3 g/kg (female 6-20D post):REP FAATDF 17,399,91
orl-rat LD50:50 mg/kg ARTODN 54,275,83
ipr-rat LD50:1000 mg/kg MEPAAX 16,35,65
ipr-mus LD50:500 mg/kg NTIS** AD277-689
orl-cat LDLo:720 mg/kg JPETAB 17,349,21
orl-rbt LDLo:1500 mg/kg MEIEDD 10,487,83

CONSENSUS REPORTS: Reported in EPA TSCA Inventory.

SAFETY PROFILE: Moderately toxic by ingestion and intraperitoneal routes. Experimental reproductive effects. When heated to decomposition it emits highly toxic fumes of SO_x and NO_x.

DWO000 CAS:3898-08-6 HR: 3
1,1-DIPHENYL-2-THIOUREA
mf: $C_{13}H_{12}N_2S$ mw: 228.33

SYN: USAF EK-7087

TOXICITY DATA WITH REFERENCE
ipr-mus LD50:200 mg/kg NTIS** AD277-689

CONSENSUS REPORTS: Reported in EPA TSCA Inventory.

SAFETY PROFILE: Poison by intraperitoneal route. When heated to decomposition it emits very toxic fumes of NO_x and SO_x.

DWO400 CAS:1135-99-5 HR: 2
DIPHENYLTIN DICHLORIDE
mf: $C_{12}H_{10}Cl_2Sn$ mw: 343.81

PROP: Colorless crystals from pet ether. Decomp by water. Mp: 42°, bp: 333–337° (decomp).

SYN: DICHLORODIPHENYLSTANNANE

TOXICITY DATA WITH REFERENCE
sln-hmn:lyms 3 µmol/L MUREAV 246,109,91
orl-rat LDLo:410 mg/kg BJIMAG 23,222,66
orl-mus LDLo:470 mg/kg AECTCV 14,111,85
ipr-mus LD50:17,800 µg/kg JICSAH 67,740,90

OSHA PEL: TWA 0.1 mg(Sn)/m³ (skin)
ACGIH TLV: TWA 0.1 mg(Sn)/m³ (skin) (Proposed: TWA 0.1 mg(Sn)/m³; STEL 0.2 mg(Sn)/m³ (skin))
NIOSH REL: (Organotin Compounds) TWA 0.1 mg(Sn)/m³

SAFETY PROFILE: Moderately toxic by ingestion. Mutation data reported. When heated to decomposition it emits toxic fumes of Cl^-. See also TIN COMPOUNDS and CHLORIDES.

For occupational chemical analysis use NIOSH: Organotin Compounds 5504.

DWO600 CAS:31671-16-6 HR: 3
DIPHENYLTIN OXIDE POLYMER
mf: $(C_{12}H_{10}OSn)_x$

SYNS: DIPHENYLOXOSTANNANE, POLYMER □ OXODIPHENYLSTANNANE, POLYMER

TOXICITY DATA WITH REFERENCE
ivn-mus LD50:180 mg/kg CSLNX* NX#03497

OSHA PEL: TWA 0.1 mg(Sn)/m³ (skin)
ACGIH TLV: TWA 0.1 mg(Sn)/m³ (skin) (Proposed: TWA 0.1 mg(Sn)/m³; STEL 0.2 mg(Sn)/m³ (skin))
NIOSH REL: (Organotin Compounds) TWA 0.1 mg(Sn)/m³

SAFETY PROFILE: Poison by intravenous route. When heated to decomposition it emits acrid smoke and irritating fumes. See also TIN COMPOUNDS.

For occupational chemical analysis use NIOSH: Organotin Compounds 5504.

DWO800 CAS:136-35-6 HR: 2
1,3-DIPHENYLTRIAZENE
mf: $C_{12}H_{11}N_3$ mw: 197.26

PROP: Golden-yellow crystals from pet ether. Mp: 98–99°, bp: explodes, vap d: 6.8.

SYNS: CELLOFOR (CZECH) □ DAAB □ DIAZOAMINOBENZEN (CZECH) □ DIAZOAMINOBENZENE □ p-DIAZOAMINOBENZENE □ DIAZOAMINOBENZOL (GERMAN) □ N-(PHENYLAZO)ANILINE

TOXICITY DATA WITH REFERENCE
mmo-sat 300 ng/plate ENMUDM 9(Suppl 9),1,87
orl-mus TDLo:1480 mg/kg/59D-C:ETA GANNA2 29,209,35
skn-mus TDLo:30 g/kg/46W-I:ETA BJCAAI 2,290,48

CONSENSUS REPORTS: EPA Genetic Toxicology Program. Reported in EPA TSCA Inventory.

SAFETY PROFILE: Questionable carcinogen with experimental tumorigenic data. Mutation data reported. Strongly explosive when shocked or heated to 98°C. Mixture with acetic anhydride explodes when warmed. When heated to decomposition it emits toxic fumes of NO_x.

DWO875 CAS:34177-12-3 HR: 3
5,6-DIPHENYL-as-TRIAZIN-3-OL
mf: $C_{15}H_{11}N_3O$ mw: 249.29

TOXICITY DATA WITH REFERENCE
ipr-rat LD50:120 mg/kg AIPTAK 95,123,53
ivn-mus LD50:100 mg/kg CSLNX* NX#05146
ivn-rbt LDLo:175 mg/kg AIPTAK 95,123,53

SAFETY PROFILE: Poison by intravenous and intraperitoneal routes. When heated to decomposition it emits toxic fumes of NO_x.

DWO950 **CAS:2039-06-7** **HR: 3**
3,5-DIPHENYL-s-TRIAZOLE
mf: $C_{14}H_{11}N_3$ mw: 221.28

PROP: Prisms from EtOH (aq); needles from pet ether. Mp: 192°, bp: 280° (decomp). Sol in dil alkalis.

SYNS: 3,5-DIPHENYL-1H-1,2,4-TRIAZOLE □ 3,5-DIPHENYL-1,2,4-TRIAZOLE □ 3,5-DPT □ s-TRIAZOLE, 3,5-DIPHENYL- □ 1H-1,2,4-TRIAZOLE, 3,5-DIPHENYL-

TOXICITY DATA WITH REFERENCE
scu-ham TDLo:30 mg/kg (female 4-8D post):REP JMCMAR 26,1187,83
ipr-mus LD50:300 mg/kg JPMSAE 50,597,61
ivn-mus LD50:90 mg/kg JPMSAE 50,597,61

SAFETY PROFILE: Poison by intraperitoneal and intravenous routes. Experimental reproductive effects. When heated to decomposition it emits acrid smoke and irritating fumes.

DWP000 **CAS:2971-22-4** **HR: 2**
2,2-DIPHENYL-1,1,1-TRICHLOROETHANE
mf: $C_{14}H_{11}Cl_3$ mw: 285.60

SYN: DT

TOXICITY DATA WITH REFERENCE
orl-rat LDLo:1000 mg/kg JPETAB 88,359,46
orl-mus LD50:2000 mg/kg HCACAV 29,1317,46

CONSENSUS REPORTS: Reported in EPA TSCA Inventory.

SAFETY PROFILE: Moderately toxic by ingestion. When heated to decomposition it emits very toxic fumes of Cl^-.

DWP229 **CAS:13445-50-6** **HR: 3**
DIPHOSPHANE
mf: H_4P_2 mw: 65.98

PROP: Liquid which polymerizes when heated. Mp: −99°, bp: 51° @ 66.7 mm. Light sensitive.

SAFETY PROFILE: Ignites spontaneously in air. A concentration of 0.2% will cause flammable gases to ignite. When heated to decomposition it emits toxic fumes of PO_x and phosphine. See also PHOSPHINE.

DWP250 **CAS:5518-62-7** **HR: 3**
1,2-DIPHOSPHINOETHANE
mf: $C_2H_8P_2$ mw: 94.03

PROP: Foul-smelling liquid. Mp: −62.5°, bp: 114–117° @ 725 mm.

SAFETY PROFILE: Ignites spontaneously in air. When heated to decomposition it emits toxic fumes of PO_x. See also PHOSPHINE.

DWP300 **HR: 3**
DIPHTHERIA TOXIN

SYN: TOXIN, BACTERIUM CORYNE-BACTERIUM DIPHTHERIAE, DIPHTHERIA

TOXICITY DATA WITH REFERENCE
ipr-ham TDLo:11 μg/kg (1D pre):REP PSEBAA 162,170,79
par-cld LDLo:488 ng/kg/1W-I:PUL,KID,SKN ANYAA9 88,1093,60
ipr-mus LD50:300 ng/kg SCIEAS 144,1100,64
ipr-ham LD50:6500 ng/kg PSEBAA 162,170,79

SAFETY PROFILE: Poison by intraperitoneal route. Human systemic effects by parenteral routes: lungs consolidation, changes in kidney tubules, acute tubular necrosis, corrosive to skin. Experimental reproductive effects. A corrosive.

DWP500 **CAS:63665-41-8** **HR: 3**
2-β,16-β-DIPIPERIDINO-5-α-ANDROSTAN-3-α,17-β-DIOL DIPIVALATE HYDROCHLORIDE
mf: $C_{35}H_{62}N_2O_4 \cdot 2ClH$ mw: 647.91

TOXICITY DATA WITH REFERENCE
ipr-rat LD50:373 mg/kg CALEDQ 2,267,77
ipr-mus LD50:424 mg/kg CALEDQ 2,267,77
ipr-ham LD50:308 mg/kg CALEDQ 2,267,77

SAFETY PROFILE: Poison by intraperitoneal route. When heated to decomposition it emits toxic fumes of NO_x and HCl.

DWP559 **CAS:64019-93-8** **HR: 3**
DIPIVEFRIN HYDROCHLORIDE
mf: $C_{19}H_{29}NO_5 \cdot ClH$ mw: 387.95

PROP: A solid. Mp: 158–159°.

SYNS: (±)-2,2-DIMETHYL-PROPANOIC ACID-4-(1-HYDROXY-2-(METHYLAMINO)ETHYL)-1,2-PHENYLENE ESTER, HYDROCHLORIDE □ DIPIVEFRINE HYDROCHLORIDE

TOXICITY DATA WITH REFERENCE
scu-rat TDLo:13,500 μg/kg (female 17-22D post):REP OYYAA2 31,1083,86
scu-rat TDLo:75 mg/kg (4W pre):REP KSRNAM 20,25,86
scu-rat TDLo:18,900 μg/kg (4W male):REP KSRNAM 20,25,86
orl-rat LD50:183 mg/kg KSRNAM 20,25,86
ipr-rat LD50:8500 μg/kg KSRNAM 20,25,86
scu-rat LD50:21,200 μg/kg KSRNAM 20,25,86
orl-mus LD50:224 mg/kg KSRNAM 20,25,86
ipr-mus LD50:32,700 μg/kg KSRNAM 20,25,86
scu-mus LD50:35 mg/kg KSRNAM 20,25,86
ivn-mus LD50:4 mg/kg KSRNAM 20,25,86

SAFETY PROFILE: Poison by ingestion, subcutaneous, intravenous, and intraperitoneal routes. Experimental reproductive effects. When heated to decomposition it emits toxic fumes of NO_x and HCl.

DWP900 **CAS:78831-88-6** **HR: 3**
DIPOTASSIUM CYCLOOCTATETRAENE
mf: $C_8H_8K_2$ mw: 182.35

CH=CH(CH=CH)$_2$CH= CH•K_2 (ring)

SAFETY PROFILE: The dry solid explodes on contact with air or oxygen. Reacts with oxygen in THF solution to form a shock-sensitive explosive product. When heated to decomposition it emits toxic fumes of K_2O.

DWP950 CAS:76429-97-5 HR: 3
DIPOTASSIUM DIAZIRINE-3,3-DICARBOXYLATE
mf: $C_3K_2N_2O_4$ mw: 206.24

N=N C(CO•OK)$_2$

PROP: A solid. Mp: 195°.

SAFETY PROFILE: Explodes when triturated. Upon decomposition it emits toxic fumes of NO_x and K_2O.

DWQ000 CAS:7727-21-1 HR: 3
DIPOTASSIUM PERSULFATE
DOT: UN 1492
mf: $H_2O_8S_2$•2K mw: 272.34

PROP: White, odorless, colorless, triclinic crystals. Mp: decomp @ 100°, d: 2.477. Decomp on heating to $K_2S_2O_7$ with loss of O_2. Mod sol in H_2O.

SYNS: ANTHION □ PEROXYDISULFURIC ACID DIPOTASSIUM SALT □ POTASSIUM PEROXYDISULFATE □ POTASSIUM PEROXYDISULPHATE □ POTASSIUM PERSULFATE (DOT)

TOXICITY DATA WITH REFERENCE
orl-rat LD50:802 mg/kg 85INA8 5,468,86

CONSENSUS REPORTS: Reported in EPA TSCA Inventory.

ACGIH TLV: TWA 5 mg(S_2O_8)/m^3
DOT CLASSIFICATION: 5.1; *Label:* Oxidizer

SAFETY PROFILE: Moderately toxic by ingestion. An irritant and allergen. A powerful oxidizer. Flammable when exposed to heat or by chemical reaction. Can react with reducing materials. It liberates oxygen above 100° when dry or at about 50° when in solution. When heated to decomposition it emits highly toxic fumes of SO_x, S_2O_8, and K_2O.

DWQ800 CAS:3248-28-0 HR: 3
DIPROPIONYL PEROXIDE
mf: $C_6H_{10}O_4$ mw: 146.15

PROP: Crystals.
CH_3CH_2CO•OOCO•CH_2CH_3

SYNS: BIS(1-OXOPROPYL)PEROXIDE □ DIPROPIONYL PEROXIDE, >28% in solution (DOT) □ PEROXIDE, BIS(1-OXOPROPYL) □ PROPIONYL PEROXIDE (DOT)

TOXICITY DATA WITH REFERENCE
ihl-rat LCLo: 100 ppm BJIMAG 27,1,70

DOT CLASSIFICATION: Forbidden

SAFETY PROFILE: Moderately toxic by inhalation. The pure material explodes at room temperature. When heated to decomposition it emits acrid smoke and fumes. See also PEROXIDES.

DWQ850 CAS:90729-15-0 HR: 2
4,5-DIPROPOXY-2-IMIDAZOLIDINONE
mf: $C_9H_{18}N_2O_3$ mw: 202.29

SYN: SRC-7

TOXICITY DATA WITH REFERENCE
orl-mus LD50:1430 mg/kg CPBTAL 12,843,64
ipr-mus LD50:890 mg/kg CPBTAL 12,843,64
scu-mus LD50:980 mg/kg CPBTAL 12,843,64

SAFETY PROFILE: Moderately toxic by ingestion, subcutaneous, and intraperitoneal routes. When heated to decomposition it emits toxic fumes of NO_x.

DWQ875 CAS:106-19-4 HR: 2
DIPROPYL ADIPATE
mf: $C_{12}H_{22}O_4$ mw: 230.34

SYN: DI-n-PROPYL ADIPATE

TOXICITY DATA WITH REFERENCE
ipr-rat TDLo:757 mg/kg (5-15D preg):TER JPMSAE 62,1596,73
ipr-rat TDLo:1262 mg/kg (5-15D preg):REP JPMSAE 62,1596,73
ipr-rat TDLo:1262 mg/kg (5-15D preg):TER JPMSAE 62,1596,73
ipr rat LD50:3786 mg/kg JPMSAE 62,1596,73

CONSENSUS REPORTS: Reported in EPA TSCA Inventory.

SAFETY PROFILE: Moderately toxic by some routes. Experimental reproductive effects. An experimental teratogen. When heated to decomposition it emits acrid smoke and fumes.

DWR000 CAS:142-84-7 HR: 3
DIPROPYLAMINE
DOT: UN 2383
mf: $C_6H_{15}N$ mw: 101.22

PROP: Water-white liquid, amine odor. Mp: −63°, bp: 110°, flash p: 63°F (OC), d: 0.741 @ 20°, vap d: 3.5.

SYNS: DI-n-PROPYLAMINE □ n-DIPROPYLAMINE □ N-PROPYL-1-PROPANAMINE □ RCRA WASTE NUMBER U110

TOXICITY DATA WITH REFERENCE
skn-rbt 100 μg/24H open AIHAAP 23,95,62
orl-rat LD50:460 mg/kg 85GMAT-,63,82
ihl-rat LC50:4400 mg/m^3/4H 85GMAT -,63,82
ipr-rat LDLo:75 mg/kg FATOAO 31,238,68
ihl-mus LC50:3070 mg/m^3/2H 85GMAT -,63,82
skn-rbt LD50:1250 mg/kg AIHAAP 23,95,62
ihl-mam LC50:4400 mg/m^3 TPKVAL 14,80,75

CONSENSUS REPORTS: Reported in EPA TSCA Inventory.

DOT CLASSIFICATION: 3; *Label:* Flammable Liquid

SAFETY PROFILE: Poison by ingestion. Moderately toxic by skin contact and inhalation. A skin irritant. A very dangerous fire hazard when exposed to heat or flame. Can react with oxidizers. Explosion hazard is unknown. Keep away from heat and open flame. To fight fire, use foam, CO_2, dry chemical. When heated to decomposition it emits toxic fumes of NO_x. See also AMINES.

DWR200 **CAS:64140-51-8** **HR: 3**
1-DIPROPYLAMINOACETYLINDOLINE
mf: $C_{16}H_{24}N_2O$ mw: 260.42

SYN: 1-(N,N-DIPROPYLGLYCYL)INDOLINE

TOXICITY DATA WITH REFERENCE
eye-rbt 100 mg PCJOAU 11,785,77
ipr-mus LD50:240 mg/kg PCJOAU 11,785,77

SAFETY PROFILE: Poison by intraperitoneal route. An eye irritant. When heated to decomposition it emits toxic fumes of NO_x.

DWS200 **CAS:29091-21-2** **HR: 1**
N^3,N^3-DIPROPYL-2,4-DINITRO-6-TRIFLUOROMETHYL-m-PHENYLENEDIAMINE
mf: $C_{13}H_{17}F_3N_4O_4$ mw: 350.34

SYNS: BLOCKADE □ CN-11-2936 □ ENDURANCE □ MARATHON □ PRODIAMINE □ USB-3153

TOXICITY DATA WITH REFERENCE
orl-rat LD50:15,380 mg/kg 85ARAE 2,48,77

CONSENSUS REPORTS: Reported in EPA TSCA Inventory.

SAFETY PROFILE: Mildly toxic by ingestion. When heated to decomposition it emits very toxic fumes of F^- and NO_x.

DWS400 **CAS:94-91-7** **HR: 1**
α,α′-DIPROPYLENEDINITRILODI-o-CRESOL
mf: $C_{17}H_{18}N_2O_2$ mw: 282.37

SYNS: DISALICYLALPROPYLENEDIIMINE □ N,N′-DISALICYLIDENE-1,2-DIAMINOPROPANE □ N,N′-DISALICYCLIDENE-1,2-PROPANEDIAMINE

TOXICITY DATA WITH REFERENCE
orl-rat LD50:4560 mg/kg AEHLAU 6,324,63

CONSENSUS REPORTS: Reported in EPA TSCA Inventory.

SAFETY PROFILE: Mildly toxic by ingestion. When heated to decomposition it emits toxic fumes of NO_x.

DWS500 **CAS:25265-71-8** **HR: 1**
DIPROPYLENE GLYCOL
mf: $C_6H_{14}O_3$ mw: 134.20

SYN: PROPANOL, OXYBIS-

TOXICITY DATA WITH REFERENCE
orl-rat LD50:14,850 mg/kg 34ZIAG -,731,69
orl-uns LD50:15 g/kg GISAAA 39(4),86,74 FEREAC 54,7740,89

CONSENSUS REPORTS: Reported in EPA TSCA Inventory.

SAFETY PROFILE: Low toxicity by ingestion. When heated to decomposition it emits acrid smoke and irritating vapors.

DWS600 **CAS:29911-28-2** **HR: 2**
DIPROPYLENE GLYCOL BUTYL ETHER
mf: $C_{10}H_{22}O_3$ mw: 190.32

TOXICITY DATA WITH REFERENCE
eye-rbt 100 mg JACTDZ 1,172,92
orl-rat LDLo:2000 mg/kg 14CYAT 2,1576,63

CONSENSUS REPORTS: Reported in EPA TSCA Inventory. Glycol ethers are on the Community Right-To-Know List.

SAFETY PROFILE: Moderately toxic by ingestion. An eye irritant. When heated to decomposition it emits acrid smoke and irritating fumes. See also GLYCOL ETHERS.

DWS650 **CAS:57472-68-1** **HR: 2**
DIPROPYLENE GLYCOL DIACRYLATE
mf: $C_{12}H_{18}O_5$ mw: 242.30

SYN: 2-PROPENOIC ACID, OXYBIS(METHYL-2,1-ETHANEDIYL) ESTER

TOXICITY DATA WITH REFERENCE
skn-rbt 500 mg SEV EPASR* 8EHQ-0184-0459
eye-rbt 100 mg SEV EPASR* 8EHQ-0184-0459
msc-mus:lym 1300 μg/L EPASR* 8EHQ-1082-0460
orl-rat LD50:4600 mg/kg EPASR* 8EHQ-0184-0459

SAFETY PROFILE: Mildly toxic by ingestion. A severe skin and eye irritant. Mutation data reported. When heated to decomposition it emits acrid smoke and irritating fumes.

DWS800 **CAS:94-51-9** **HR: 1**
DIPROPYLENE GLYCOL DIBENZOATE
mf: $C_{20}H_{22}O_5$ mw: 342.42

SYNS: BENZOFLEX 9-88 □ BENZOFLEX 9-98 □ BENZOFLEX 9-88 SG □ BENZOIC ACID DIESTER with DIPROPYLENE GLYCOL □ BENZOIC ACID-n-DIPROPYLENE GLYCOL DIESTER □ DIBENZOYL DIPROPYLENE GLYCOL ESTER □ DIPROPANEDIOL DIBENZOATE □ K-FLEX DP □ 3,3′-OXYDI-1-PROPANOL DIBENZOATE

TOXICITY DATA WITH REFERENCE
orl-rat LD50:9800 mg/kg AIHAAP 23,95,62

SAFETY PROFILE: Mildly toxic by ingestion. When heated to decomposition it emits acrid smoke and fumes. See also ESTERS.

DWT000 **CAS:63716-17-6** **HR: 2**
DIPROPYLENE GLYCOL DIPELARGONATE

SYNS: EMERY X-88-R □ NONANOIC ACID OXYDI-3,1-PROPANEDIYL ESTER (9CI) □ NONANOIC ACID OXYDIPROPYLENE ESTER

TOXICITY DATA WITH REFERENCE
ivn-rat LD50:1060 mg/kg MRLR** No. 256,54
ivn-rbt LD50:1060 mg/kg MRLR** No. 256,54

SAFETY PROFILE: Moderately toxic by intravenous

route. When heated to decomposition it emits acrid smoke and irritating fumes.

DWT200 **CAS:34590-94-8** **HR: 2**
DIPROPYLENE GLYCOL METHYL ETHER
mf: $C_7H_{16}O_3$ mw: 148.23

PROP: Liquid. Bp: 190°, d: 0.951, vap d: 5.11, flash p: 185°F.

SYNS: ARCOSOLV □ DIPROPYLENE GLYCOL MONOMETHYL ETHER □ DOWANOL DPM □ DOWANOL-50B □ UCAR SOLVENT 2LM

TOXICITY DATA WITH REFERENCE
eye-hmn 8 mg MLD JTOTDO 2,229,83/84
skn-rbt 500 mg open MLD UCDS** 11/15/71
eye-rbt 238 mg MLD AMIHBC 9,509,54
orl-rat LD50:5660 mg/kg AIHAAP 23,95,62
orl-dog LD50:7500 mg/kg JPETAB 102,79,51
skn-rat LD50:9500 mg/kg DTLVS* 4,157,80

CONSENSUS REPORTS: Reported in EPA TSCA Inventory. Glycol ether compounds are on the Community Right-To-Know List.

OSHA PEL: TWA 100 ppm; STEL 150 ppm (skin)
ACGIH TLV: TWA 100 ppm; STEL 150 ppm (skin)
DFG MAK: 50 ppm (300 mg/m^3)

SAFETY PROFILE: Mildly toxic by ingestion and skin contact. An experimental skin and human eye irritant. A mild allergen. Combustible when exposed to heat or flame; can react with oxidizing materials. To fight fire, use dry chemical, CO_2, mist, foam. When heated to decomposition it emits acrid smoke and irritating fumes. See also GLYCOL ETHERS.

DWT400 **CAS:6976-50-7** **HR: 2**
N,N-DI-n-PROPYL ETHYL CARBAMATE
mf: $C_9H_{19}NO_2$ mw: 173.29

SYN: DIPROPYLCARBAMIC ACID ETHYL ESTER

TOXICITY DATA WITH REFERENCE
ipr-mus TDLo:6500 mg/kg/13W-I:ETA JNCIAM 9,35,48

SAFETY PROFILE: Questionable carcinogen with experimental tumorigenic data. When heated to decomposition it emits toxic fumes of NO_x. See also CARBAMATES.

DWT600 **CAS:123-19-3** **HR: 3**
DIPROPYL KETONE
DOT: UN 2710
mf: $C_7H_{14}O$ mw: 114.21

PROP: Colorless, refractive liquid. Bp: 144°, mp: −32.6°, vap press: 5.2 mm @ 20°, flash p: 120°F (CC), d: 0.815, vap d: 3.93.

SYNS: BUTYRONE (DOT) □ GBL □ HEPTAN-4-ONE □ 4-HEPTANONE □ PROPYL KETONE

TOXICITY DATA WITH REFERENCE
skn-rbt 500 mg/24H MLD 85JCAE-,286,86
eye-rbt 500 mg/24H MLD 85JCAE-,286,86
orl-rat LD50:3730 mg/kg TXAPA9 28,313,74
ihl-rat LCLo:4000 ppm/4H TXAPA9 28,313,74
skn-rbt LD50:5660 mg/kg TXAPA9 28,313,74

CONSENSUS REPORTS: Reported in EPA TSCA Inventory.

OSHA PEL: TWA 50 ppm
ACGIH TLV: TWA 50 ppm
DOT CLASSIFICATION: 3; *Label:* Flammable Liquid

SAFETY PROFILE: Moderately toxic by ingestion, inhalation, and skin contact. A skin and eye irritant. Flammable liquid when exposed to heat or flame; can react with oxidizing materials. To fight fire, use CO_2, dry chemical, alcohol foam, fog, and mist. When heated to decomposition it emits acrid smoke and fumes. See also KETONES.

DWU000 **CAS:628-85-3** **HR: 3**
DIPROPYL MERCURY
mf: $C_6H_{14}Hg$ mw: 286.79

PROP: Colorless liquid. Immiscible in water. D: 2.0208, bp: 190°. Sol in Et_2O; less sol in EtOH.

TOXICITY DATA WITH REFERENCE
ipr-mus LDLo:2 mg/kg CBCCT* 4,320,52

CONSENSUS REPORTS: Mercury and its compounds are on the Community Right To Know List.

OSHA PEL: TWA 0.01 mg(Hg)/m^3; STEL 0.03 mg/m^3 (skin)
ACGIH TLV: TWA 0.01 mg(Hg)/m^3; STEL 0.03 mg(Hg)/m^3
NIOSH REL: (Mercury, Organo): TWA 0.01 mg/m^3; STEL 0.03 mg/m^3 (skin)

SAFETY PROFILE: Poison by intraperitoneal route. Violent reaction with iodine. When heated to decomposition it emits toxic fumes of Hg. See also MERCURY COMPOUNDS, ORGANIC.

DWU200 **CAS:996-05-4** **HR: 3**
S,S-DIPROPYL METHYLPHOSPHONOTRITHIOATE
mf: $C_7H_{17}PS_3$ mw: 228.39

SYNS: ENT 25,979 □ V-C 3-670 □ VIRGINIA-CAROLINA 3-670

TOXICITY DATA WITH REFERENCE
orl-rat LD50:18 mg/kg ARSIM* 20,26,66
orl-ckn LD50:12 mg/kg TXAPA9 11,49,67

SAFETY PROFILE: Poison by ingestion. When heated to decomposition it emits very toxic fumes of PO_x and SO_x.

DWU400 **CAS:60580-30-5** **HR: 3**
O,O-DI-n-PROPYL-O-(4-METHYLTHIOPHENYL)PHOSPHATE
mf: $C_{13}H_{21}O_4PS$ mw: 304.37

SYNS: NK-1158 □ PHOSPHORIC ACID DIPROPYL-4-METHYLTHIOPHENYL ESTER

TOXICITY DATA WITH REFERENCE
orl-rat LD50:70 mg/kg HDIZAB 26,91,78
ipr-rat LD50:35 mg/kg HDIZAB 26,91,78

SAFETY PROFILE: Poison by ingestion and intraperito-

neal routes. When heated to decomposition it emits very toxic fumes of PO_x and SO_x. See also ESTERS.

DWU800 **CAS:53230-00-5** **HR: 2**
α-DIPROPYLNITROSAMINE METHYL ETHER

SYNS: 1-METHOXY-N-NITROSO-N-PROPYLPROPYLAMINE □ 1-METHOXYPROPYLPROPYLNITROSAMIN (GERMAN) □ 1-METHOXYPROPYLPROPYLNITROSAMINE □ 1-MPPN

TOXICITY DATA WITH REFERENCE
scu-ham TDLo:555 mg/kg/37W-I:NEO ZKKOBW 90,215,77
scu-ham LD50:458 mg/kg ZKKOBW 90,215,77

SAFETY PROFILE: Questionable carcinogen with experimental neoplastigenic data. Moderately toxic by subcutaneous route. Many nitrosamines are carcinogens. When heated to decomposition it emits toxic fumes of NO_x. See also NITROSAMINES and ETHERS.

DWV000 **CAS:7664-98-4** **HR: 3**
DIPROPYLOXOSTANNANE
mf: $C_6H_{14}OSn$ mw: 220.89

PROP: Polymeric powder.

SYNS: DIPROPYLTIN OXIDE □ KYSLICNIK DI-N-PROPYLCINICITY (CZECH)

TOXICITY DATA WITH REFERENCE
skn-rbt 500 mg/24H SEV 28ZPAK -,225,72
eye-rbt 100 mg/24H MOD 28ZPAK -,225,72
orl-rat LD50:36,800 μg/kg 28ZPAK -,225,72

OSHA PEL: TWA 0.1 mg(Sn)/m³ (skin)
ACGIH TLV: TWA 0.1 mg(Sn)/m³ (skin) (Proposed: TWA 0.1 mg(Sn)/m³; STEL 0.2 mg(Sn)/m³ (skin))
NIOSH REL: (Organotin Compounds) TWA 0.1 mg(Sn)/m³

SAFETY PROFILE: Poison by ingestion. An eye and severe skin irritant. When heated to decomposition it emits acrid smoke and irritating fumes. See also TIN COMPOUNDS.

For occupational chemical analysis use NIOSH: Organotin Compounds 5504.

DWV200 **CAS:29914-92-9** **HR: 3**
DIPROPYL PEROXIDE
mf: $C_6H_{14}O_2$ mw: 118.18

$CH_3CH_2CH_2OOCH_2CH_2CH_3$

SAFETY PROFILE: Potentially explosive. When heated to decomposition it emits acrid smoke and fumes. See also PEROXIDES.

DWV400 **CAS:16066-38-9** **HR: 2**
DI-n-PROPYL PEROXYDICARBONATE
mf: $C_8H_{14}O_6$ mw: 206.22

SYNS: PEROXYDICARBONIC ACID DIPROPYL ESTER □ n-PROPYL PERCARBONATE

TOXICITY DATA WITH REFERENCE
orl-rat LD50:3400 mg/kg BSPII* 1/75-19B
skn-rbt LD50:3500 mg/kg BSPII* 1/75-19B

CONSENSUS REPORTS: Reported in EPA TSCA Inventory.

SAFETY PROFILE: Moderately toxic by ingestion and skin contact. When heated to decomposition it emits acrid smoke and irritating fumes.

DWV500 **CAS:131-16-8** **HR: 2**
DIPROPYL PHTHALATE
mf: $C_{14}H_{18}O_4$ mw: 250.32

PROP: Bp: 317.5°, d: 1.078, flash p: >230°F.

SYNS: 1,2-BENZENEDICARBOXYLIC ACID, DIPROPYL ESTER □ DI-n-PROPYL PHTHALATE □ PHTHALIC ACID, DIPROPYL ESTER

TOXICITY DATA WITH REFERENCE
orl-mus TDLo:630 g/kg (male 15W pre): REP FAATDF 12,508,89
orl-mus TDLo:1260 g/kg (male 15W pre): REP FAATDF 12,508,89
ipr-mus LDLo:1251 mg/kg JPMSAE 56,1446,67

CONSENSUS REPORTS: Reported in EPA TSCA Inventory.

SAFETY PROFILE: Moderately toxic by intraperitoneal route. Experimental reproductive effects. An irritant. Combustible when exposed to heat and flame. When heated to decomposition it emits acrid smoke and irritating fumes.

DWV800 **CAS:925-15-5** **HR: 3**
DIPROPYL SUCCINATE
mf: $C_{10}H_{18}O_4$ mw: 202.28

SYNS: BUTANEDIOIC ACID DIPROPYL ESTER □ DI-N-PROPYL SUCCINATE □ SUCCINIC ACID DIPROPYL ESTER

TOXICITY DATA WITH REFERENCE
eye-rbt 500 mg AMIHBC 10,61,54
orl-rat LD50:6490 mg/kg AMIHBC 10,61,54
ipr-rat LD50:290 mg/kg NEPSBV 1,286,59

SAFETY PROFILE: Poison by intraperitoneal route. Mildly toxic by ingestion. An eye irritant. When heated to decomposition it emits acrid smoke and irritating fumes. See also ESTERS.

DWW000 **CAS:57-66-9** **HR: 3**
p-(DIPROPYLSULFAMOYL)BENZOIC ACID
mf: $C_{13}H_{19}NO_4S$ mw: 285.39

PROP: Crystals from EtOH (aq). Mp: 184–196°. Sol in $CHCl_3$.

SYNS: APURINA □ BENECID □ BENEMID □ BENURYL □ 4-((DIPROPYLAMINO)SULFONYL)BENZOIC ACID □ 4-(DIPROPYLSULFAMOYL)BENZOIC ACID □ p-(DIPROPYLSULFAMYL)BENZOIC ACID □ ETHAMIDE □ NCI-C56097 □ PROBECID □ PROBEN □ PROBENECID ACID □ PROBENEMID □ PROLONGINE □ SYNERGID R □ TUBOPHAN □ URICOSID

TOXICITY DATA WITH REFERENCE
hma-rat/smc 60 mg/kg MUREAV 28,57,75
orl-mus TDLo: 206 g/kg/2Y-C: NEO NTPTR* NTP-TR-395,91
orl-man TDLo: 50 mg/kg/1W-I: BLD JRHUA9 13,208,86
orl-rat LD50: 1600 mg/kg MEIEDD 10,1116,83
ipr-rat LDLo: 394 mg/kg CLDND*
scu-rat LDLo: 611 mg/kg CLDND*
orl-mus LDLo: 1666 mg/kg NIIRDN 6,735,82
ipr-mus LD50: 1000 mg/kg CPBTAL 16,1655,68
scu-mus LDLo: 1156 mg/kg CLDND*
ivn-mus LDLo: 458 mg/kg CLDND*
ivn-dog LDLo: 230 mg/kg CLDND*

CONSENSUS REPORTS: Reported in EPA TSCA Inventory. EPA Genetic Toxicology Program.

SAFETY PROFILE: Poison by intraperitoneal and intravenous routes. Moderately toxic by ingestion and subcutaneous routes. Human systemic effects: hemolysis with or without anemia. Questionable carcinogen with neoplastigenic data. Mutation data reported. A uricosuric that promotes the secretion of uric acid in the urine. When heated to decomposition it emits very toxic fumes of SO_x and NO_x.

DWW200 CAS:23795-03-1 HR: 3
p-(DIPROPYLSULFAMOYL)BENZOIC ACID SODIUM SALT
mf: $C_{13}H_{18}NO_4S \cdot Na$ mw: 307.37

SYNS: p-(DI-N-PROPYLSULFAMYL)BENZOIC ACID SODIUM SALT ◻ PROBENECID SODIUM SALT

TOXICITY DATA WITH REFERENCE
orl-man TDLo: 630 mg/kg/6W: KID ARPAAQ 94,241,72
orl-rat LD50: 1604 mg/kg JPETAB 102,208,51
ipr-rat LD50: 394 mg/kg JPETAB 102,208,51
scu-rat LD50: 611 mg/kg JPETAB 102,208,51
orl-mus LD50: 1666 mg/kg JPETAB 102,208,51
ipr-mus LD50: 500 mg/kg TXAPA9 24,37,73
scu-mus LD50: 1156 mg/kg JPETAB 102,208,51
ivn-mus LD50: 458 mg/kg JPETAB 102,208,51
ivn-dog LD50: 270 mg/kg JPETAB 102,208,51
ivn-rbt LD50: 304 mg/kg JPETAB 102,208,51

SAFETY PROFILE: Poison by intraperitoneal and intravenous routes. Moderately toxic by ingestion and subcutaneous routes. Human systemic effects by ingestion: proteinurea and damage to the kidney (glomeruli), ureter, and bladder. When heated to decomposition it emits very toxic fumes of NO_x, Na_2O, and SO_x.

DWW400 CAS:73927-87-4 HR: 3
DI-n-PROPYLTIN BISMETHANESULFONATE
mf: $C_8H_{20}O_6S_2Sn$ mw: 395.09

SYN: BIS((METHYLSULFONYL)OXY)DIPROPYLSTANNANE

TOXICITY DATA WITH REFERENCE
ivn-mus LD50: 17,800 μg/kg CSLNX* NX#02277

OSHA PEL: TWA 0.1 mg(Sn)/m³ (skin)
ACGIH TLV: TWA 0.1 mg(Sn)/m³ (skin) (Proposed: TWA 0.1 mg(Sn)/m³; STEL 0.2 mg(Sn)/m³ (skin))
NIOSH REL: (Organotin Compounds) TWA 0.1 mg(Sn)/m³

SAFETY PROFILE: Poison by intravenous route. When heated to decomposition it emits toxic fumes of SO_x. See also TIN COMPOUNDS and SULFONATES.

For occupational chemical analysis use NIOSH: Organotin Compounds 5504.

D

DWW500 CAS:628-91-1 HR: 3
DIPROPYL ZINC
mf: $C_6H_{14}Zn$ mw: 151.57

PROP: A liquid. D: 1.080 @ 20°/4°, bp: 157°.

CONSENSUS REPORTS: Zinc and its compounds are on the Community Right-To-Know List.

SAFETY PROFILE: Ignites in air if a large surface area is exposed. When heated to decomposition it emits toxic fumes of ZnO. See also ZINC COMPOUNDS.

DWW600 HR: 3
DIPYRIDINESODIUM
mf: $C_{10}H_{10}N_2Na$ mw: 181.19

SAFETY PROFILE: Ignites spontaneously in air. When heated to decomposition it emits toxic fumes of Na_2O.

DWW700 CAS:67730-10-3 HR: 3
DIPYRIDO(1,2-a:3′,2′-d)IMIDAZOL-2-AMINE
mf: $C_{10}H_8N_4$ mw: 184.22

SYNS: 2-AMINODIPYRIDO(1,2-a:3′,2′-d)-IMIDAZOLE ◻ GLU-P-2

TOXICITY DATA WITH REFERENCE
mma-sat 100 ng/plate MUREAV 136,23,84
pic-sat 10 μg/plate MUREAV 110,243,83
sce-hmn: lym 10 μmol/L MUREAV 116,137,83
dns-rat: lvr 1 μmol/L CALEDQ 20,283,83
dnd-mus: lvr 200 μmol/L JJCREP 76,835,85
orl-rat TDLo: 12,500 mg/kg/2Y-C: CAR,REP EVHPAZ 67,129,86
orl-mus TDLo: 23,100 mg/kg/83W-C: CAR EVHPAZ 67,129,86
orl-mus TD: 24,024 mg/kg/66W-C: CAR CRNGDP 5,815,84
orl-mus TD: 23,232 mg/kg/69W-C: CAR CRNGDP 5,815,84
orl-rat TD: 13 g/kg/2Y-C: CAR EVHPAZ 67,129,86

CONSENSUS REPORTS: IARC Cancer Review: Group 2B IMEMDT 7,56,87; Animal Sufficient Evidence IMEMDT 40,235,86.

SAFETY PROFILE: Confirmed carcinogen with experimental carcinogenic data. Experimental reproductive effects. Human mutation data reported. When heated to decomposition it emits toxic fumes of NO_x.

DWX000 CAS:21000-42-0 HR: 3
DIPYRIDYL HYDROGEN PHOSPHATE
mf: $C_{12}H_{14}N_2 \cdot 2C_2H_6O_4P$ mw: 436.2

SYN: DIPYRIDYL PHOSPHATE

TOXICITY DATA WITH REFERENCE
orl-rat LD50: 280 mg/kg GTPZAB 24(9),48,80

skn-rat LD50:460 mg/kg GTPZAB 24(9),48,80
orl-mus LD50:240 mg/kg GTPZAB 24(9),48,80
orl-rbt LD50:295 mg/kg GTPZAB 24(9),48,80
skn-rbt LD50:404 mg/kg GTPZAB 24(9),48,80

SAFETY PROFILE: Poison by ingestion. Moderately toxic by skin contact. When heated to decomposition it emits very toxic fumes of PO_x and NO_x.

DWX200 CAS:20738-78-7 HR: 3
DI-3-PYRIDYLMERCURY
mf: $C_{10}H_8HgN_2$ mw: 356.79

TOXICITY DATA WITH REFERENCE
ivn-mus LD50:180 mg/kg CSLNX* NX#05152

CONSENSUS REPORTS: Mercury and its compounds are on the Community Right-To-Know List.

OSHA PEL: CL 0.1 mg(Hg)/m³ (skin)
ACGIH TLV: TWA 0.1 mg(Hg)/m³ (skin)
NIOSH REL: (Mercury, Aryl and Inorganic): CL 0.1 mg/m³ (skin)

SAFETY PROFILE: Poison by intravenous route. When heated to decomposition it emits very toxic fumes of NO_x and Hg. See also MERCURY COMPOUNDS, ORGANIC.

DWX600 CAS:51-73-0 HR: 3
1,4-DIPYRROLIDINYL-2-BUTYNE
mf: $C_{12}H_{20}N_2$ mw: 192.34

PROP: A liquid. Bp: 116–116.5° @ 2.5 mm.

SYNS: BIOFERMIN □ 1,1′-(2-BUTYNYLENE)DIPYRROLIDINE □ 1,4-DIPYRROLIDINYL-2-BUTYNE □ TREMORINE

TOXICITY DATA WITH REFERENCE
ipr-mus LD50:25 mg/kg ARZNAD 21,1727,71
ivn-mus LD50:112 mg/kg CSLNX* NX12070
scu-cat LDLo:5 mg/kg AIPTAK 135,447,62
orl-bwd LD50:100 mg/kg TXAPA9 21,315,72

SAFETY PROFILE: Poison by ingestion, intraperitoneal, intravenous, and subcutaneous routes. When heated to decomposition it emits toxic fumes of NO_x. See also ACETYLENE COMPOUNDS.

DWX800 CAS:231-36-7 HR: 3
DIQUAT
mf: $C_{12}H_{12}N_2{\bullet}2Br$ mw: 344.08

PROP: Yellow crystals. Mp: 355°. Sol in water.

SYNS: AQUACIDE □ DEIQUAT □ DEXTRONE □ 9,10-DIHYDRO-8a,10,-DIAZONIAPHENANTHRENE DIBROMIDE □ 9,10-DIHYDRO-8a,10a-DIAZONIAPHENANTHRENE(1,1′-ETHYLENE-2,2′-BIPYRIDYLIUM) DIBROMIDE □ 5,6-DIHYDRO-DIPYRIDO(1,2a;2,1c)PYRAZINIUM DIBROMIDE □ 6,7-DIHYDROPYRIDO(1,2a;2′,1′-C)PYRAZINEDIUM DIBROMIDE □ DIQUAT DIBROMIDE □ 1,1′-ETHYLENE-2,2′-BIPYRIDYLIUM DIBROMIDE □ ETHYLENE DIPYRIDYLIUM DIBROMIDE □ 1,1-ETHYLENE 2,2-DIPYRIDYLIUM DIBROMIDE □ 1,1′-ETHYLENE-2,2′-DIPYRIDYLIUM DIBROMIDE □ FB/2 □ FEGLOX □ PREEGLONE □ REGLON □ REGLONE □ WEEDTRINE-D

TOXICITY DATA WITH REFERENCE
skn-rbt 400 mg/kg/20D MLD BJIMAG 27,51,70
eye-rbt 10 mg MLD BJIMAG 27,51,70
mmo-sat 100 nmol/plate TOLED5 3,169,79
dns-hmn:fbr 1 μmol/L MUREAV 42,161,77
ipr-rat TDLo:7 mg/kg (7D preg):TER 26UZAB 6,257,68
ivn-rat TDLo:15 mg/kg (female 17D post):REP TXAPA9 33,450,75
orl-rat LD50:120 mg/kg PRKHDK 1,31,75
skn-rat LD50:433 mg/kg FAATDF 7,299,86
ipr-rat LDLo:500 mg/kg PAREAQ 14,225,62
scu-rat LD50:20 mg/kg PAREAQ 14,225,62
orl-mus LD50:233 mg/kg BJIMAG 27,51,70
orl-dog LDLo:187 mg/kg BJIMAG 27,51,70
orl-rbt LD50:188 mg/kg BJIMAG 27,51,70

CONSENSUS REPORTS: EPA Genetic Toxicology Program.

OSHA PEL: TWA 0.5 mg/m³
ACGIH TLV: TWA 0.5 mg/m³; (Proposed: Total Dust: TWA 0.5 mg/m³; Respirable Dust: 0.1 mg/m³ (skin))

SAFETY PROFILE: Poison by ingestion, subcutaneous, intravenous, and intraperitoneal routes. Experimental teratogenic and reproductive effects. A skin and eye irritant. Human mutation data reported. When heated to decomposition it emits very toxic fumes of NO_x and Br^-. See also PARAQUAT.

DWY000 CAS:4032-26-2 HR: 3
DIQUAT DICHLORIDE
mf: $C_{12}H_{12}N_2{\bullet}2Cl$ mw: 255.16

SYN: 1,1′-ETHYLENE-2,2′-DIPYRIDINIUM DICHLORIDE

TOXICITY DATA WITH REFERENCE
scu-rat LD50:19 mg/kg BJIMAG 27,51,70
ivn-mus LD50:180 mg/kg CSLNX* NX#00223
orl-mky LDLo:100 mg/kg TXAPA9 51,277,79

SAFETY PROFILE: Poison by ingestion, intravenous, and subcutaneous routes. When heated to decomposition it emits very toxic fumes of NO_x and Cl^-.

DWY200 CAS:94-93-9 HR: 3
N,N′-DISALICYLIDENE ETHYLENEDIAMINE
mf: $C_{16}H_{16}N_2O_2$ mw: 268.34

PROP: Yellow crystals. Mp: 127–128°. Sol in C_6H_6, $CHCl_3$, Me_2CO, and alkalis; mod sol in EtOH; insol in Et_2O, CCl_4, and NH_3 (aq).

SYNS: N,N′-ETHYLENE DIIMINO DI(o-CRESOL) □ USAF DO-63

TOXICITY DATA WITH REFERENCE
orl-rat LDLo:500 mg/kg JPETAB 90,260,47
ipr-mus LD50:100 mg/kg NTIS** AD277-689

CONSENSUS REPORTS: Reported in EPA TSCA Inventory.

SAFETY PROFILE: Poison by intraperitoneal route. Moderately toxic by ingestion. When heated to decomposition it emits toxic fumes of NO_x.

DWY400 **CAS:63990-56-7** **HR: 3**
α,α′-DISELENOBIS-o-ACETOTOLUIDIDE

TOXICITY DATA WITH REFERENCE
orl-rat LDLo:25 mg/kg NCNSA6 5,10,53
ipr-rat LDLo:25 mg/kg NCNSA6 5,10,53

CONSENSUS REPORTS: Selenium and its compounds are on the Community Right-To-Know List.

OSHA PEL: TWA 0.2 mg(Se)/m^3
ACGIH TLV: TWA 0.2 mg(Se)/m^3
DFG MAK: 0.1 mg(Se)/m^3

SAFETY PROFILE: Poison by ingestion and intraperitoneal routes. When heated to decomposition it emits toxic fumes of Se. See also SELENIUM COMPOUNDS.

DWY600 **CAS:64046-56-6** **HR: 3**
2,2′-DISELENOBIS(N-PHENYLACETAMIDE)
mf: $C_{16}H_{16}N_2O_2Se_2$ mw: 426.26

TOXICITY DATA WITH REFERENCE
orl-rat LDLo:50 mg/kg NCNSA6 5,10,53
ipr-rat LDLo:25 mg/kg NCNSA6 5,10,53

CONSENSUS REPORTS: Selenium and its compounds are on the Community Right-To-Know List.

OSHA PEL: TWA 0.2 mg(Se)/m^3
ACGIH TLV: TWA 0.2 mg(Se)/m^3
DFG MAK: 0.1 mg(Se)/m^3

SAFETY PROFILE: Poison by ingestion and intraperitoneal routes. When heated to decomposition it emits very toxic fumes of NO_x and Se. See also SELENIUM COMPOUNDS.

DWY800 **CAS:1464-43-3** **HR: 3**
3,3′-DISELENODIALANINE
mf: $C_6H_{12}N_2O_4Se_2$ mw: 334.12

SYNS: SELENIUM CYSTINE □ SELENOCYSTINE

TOXICITY DATA WITH REFERENCE
slt-dmg-orl 10 μmol/L CNJGA8 17,55,75
sln-dmg-orl 2 μmol/L CNJGA8 11,677,69
ipr-rat LDLo:4 mg/kg CTOXAO 17,171,80

CONSENSUS REPORTS: Selenium and its compounds are on the Community Right-To-Know List.

OSHA PEL: TWA 0.2 mg(Se)/m^3
ACGIH TLV: TWA 0.2 mg(Se)/m^3
DFG MAK: 0.1 mg(Se)/m^3

SAFETY PROFILE: Poison by intraperitoneal route. Mutation data reported. When heated to decomposition it emits very toxic fumes of NO_x and Se. See also SELENIUM COMPOUNDS.

DWZ000 **CAS:35507-35-8** **HR: 3**
p,p′-DISELENODIANILINE
mf: $C_{12}H_{12}N_2Se_2$ mw: 342.18

TOXICITY DATA WITH REFERENCE
orl-rat LDLo:100 mg/kg NCNSA6 5,11,53
ipr-rat LDLo:25 mg/kg NCNSA6 5,11,53

CONSENSUS REPORTS: Selenium and its compounds are on the Community Right-To-Know List.

OSHA PEL: TWA 0.2 mg(Se)/m^3
ACGIH TLV: TWA 0.2 mg(Se)/m^3
DFG MAK: 0.1 mg(Se)/m^3

SAFETY PROFILE: Poison by ingestion and intraperitoneal routes. When heated to decomposition it emits very toxic fumes of NO_x and Se. See also SELENIUM COM POUNDS.

DWZ100 **CAS:70145-55-0** **HR: 3**
β,β′-DISELENODIPROPIONIC ACID, SODIUM SALT
mf: $C_6H_9O_4Se_2$•Na mw: 326.06

SYN: PROPIONIC ACID, 3,3′-DISELENODI-, SODIUM SALT

TOXICITY DATA WITH REFERENCE
ipr-rat LDLo:25 mg(Se)/kg JPETAB 63,357,38

OSHA PEL: TWA 0.2 mg(Se)/m^3
ACGIH TLV: TWA 0.2 mg(Se)/m^3

SAFETY PROFILE: Poison by intraperitoneal route. When heated to decomposition it emits toxic fumes of Se.

DXA000 **HR: 3**
DISILANE
mf: H_6Si_2 mw: 62.22

PROP: Gas, repulsive odor. Mp: −132.5°, bp: −14.5°, d: 0.686 @ −25°/4°.

SYN: SILICOETHANE

SAFETY PROFILE: Poison by inhalation. Dangerous when exposed to heat or flame or by chemical reaction; can react with oxidizing materials. Ignites spontaneously in air. Reacts violently with CCl_4, $CHCl_3$, O_2, and SF_6. See also HYDRIDES.

DXA500 **HR: 3**
DISILVER CYANAMIDE
mf: CAg_2N_2 mw: 255.76

CONSENSUS REPORTS: Silver and its compounds are on the Community Right-To-Know List.

SAFETY PROFILE: A heat- and light-sensitive explosive. When heated to decomposition it emits toxic fumes of NO_x. See also SILVER COMPOUNDS.

DXA600 **HR: 3**
DISILVER KETENIDE
mf: C_2Ag_2O mw: 255.76

$Ag_2C{=}C{=}O$

CONSENSUS REPORTS: Silver and its compounds are on the Community Right-To-Know List.

SAFETY PROFILE: The ketenide and its pyridine complex are heat- and impact-sensitive explosives. When heated to decomposition it emits acrid smoke and fumes. See also SILVER COMPOUNDS.

D

DXA800 **HR: 3**
DISILVER PENTATIN UNDECAOXIDE
mf: $Ag_2Sn_5O_{11}$ mw: 985.19

SYN: SILVER BETA-STANNATE

CONSENSUS REPORTS: Silver and its compounds are on the Community Right-To-Know List.

SAFETY PROFILE: Can explode on heating. See also SILVER COMPOUNDS and TIN COMPOUNDS.

DXB400 **CAS:25295-51-6** **HR: 3**
DISODIUM-4,4′-BIS((4-AMINO-6-(2-HYDROXYETHYL)AMINO-s-TRIAZIN-2-YL)AMINO)-2,2′-STILBENDISULFONIC ACID
mf: $C_{24}H_{26}N_{12}O_8S_2 \cdot 2Na$ mw: 720.72

TOXICITY DATA WITH REFERENCE
ipr-rat LD50:1000 mg/kg CTOXAO 13,171,78
scu-mus LD50:500 mg/kg CTOXAO 13,171,78
ivn-mus LD50:100 mg/kg CTOXAO 13,171,78

SAFETY PROFILE: Poison by intravenous route. Moderately toxic by intraperitoneal and subcutaneous routes. When heated to decomposition it emits very toxic fumes of SO_x, Na_2O, and NO_x. See also SULFONATES.

DXB450 **CAS:133-66-4** **HR: 2**
DISODIUM-4,4′-BIS((4,6-DIANILINO-1,3,5-TRIAZIN-2-YL)AMINO)STILBENE-2,2′-DISULFONATE
mf: $C_{44}H_{36}N_{12}O_6S_2 \cdot 2Na$ mw: 939.02

SYNS: BELOPHOR OD □ 4,4′-BIS((4,6-DIANILINO-s-TRIAZIN-2-YL) AMINO)-2,2′-STILBENEDISULFONIC ACID DISODIUM SALT □ BLANKOPHOR HZPA □ CALCOFLUOR WHITE MR □ CELLU-BRITE □ C.I. 40621 □ C.I. FLUORESCENT BRIGHTENER 9 □ COMPOUND 19-28 □ OZP 9 □ 2,2′-STILBENEDISULFONIC ACID, 4,4′-BIS((4,6-DIANILINO-s-TRIAZIN-2-YL)AMINO)-, DISODIUM SALT

TOXICITY DATA WITH REFERENCE
orl-rat LD50:>10 g/kg TXAPA9 5,176,63
ipr-rat LD50:1090 mg/kg GISAAA 49(1),85,84
orl-rbt LD50:>10 g/kg TXAPA9 5,176,63
orl-gpg LD50:>7 g/kg TXAPA9 5,176,63

CONSENSUS REPORTS: Reported in EPA TSCA Inventory.

SAFETY PROFILE: Moderately toxic by intraperitoneal route. Slightly toxic by ingestion. When heated to decomposition it emits toxic vapors of NO_x and SO_x.

DXC200 **CAS:7775-11-3** **HR: 3**
DISODIUM CHROMATE
mf: $CrO_4 \cdot 2Na$ mw: 161.98

PROP: Yellow crystals. Mp: 780°. Sol in H_2O; fairly insol in MeOH and EtOH.

SYNS: CHROMATE of SODA □ CHROMIUM DISODIUM OXIDE □ CHROMIUM SODIUM OXIDE □ NEUTRAL SODIUM CHROMATE □ SODIUM CHROMATE (DOT) □ SODIUM CHROMATE (VI)

TOXICITY DATA WITH REFERENCE
mmo-sat 33 μg/plate ENMUDM 7,185,85
dnr-sat 50 mmol/L TOLED5 7,439,81
sce-ham:lng 32 μg/L CRNGDP 4,605,83
ipr-rat TDLo:5 mg/kg (male 5D pre):REP TOLED5 51,269,90
ipr-rat LD50:57 mg/kg AIPTAK 154,243,65
ipr-mus LD50:32 mg/kg COREAF 257,791,63
ivn-dog LDLo:235 mg/kg EQSSDX 1,1,75
ivn-cat LD50:164 mg/kg AGSOA6 8,51,67
scu-rbt LDLo:243 mg/kg EQSFAP 1,1,75
ivn-rbt LDLo:32 mg/kg EQSSDX 1,1,75
idr-rbt LDLo:250 mg/kg JAPHAR 11,285,1877
skn-gpg LDLo:206 mg/kg AEHLAU 11,201,65
ipr-gpg LDLo:206 mg/kg AEHLAU 11,201,65
scu-gpg LDLo:30 mg/kg EQSSDX 1,1,75
idr-gpg LDLo:382 mg/kg JAPHAR 11,285,1877

CONSENSUS REPORTS: NTP 7th Annual Report On Carcinogens. IARC Cancer Review: Group 1 IMEMDT 49,49,90; Human Inadequate Evidence IMEMDT 23,205,80; Human Sufficient Evidence IMEMDT 49,49,90; Animal Inadequate Evidence IMEMDT 23,205,80. Reported in EPA TSCA Inventory. EPA Genetic Toxicology Program. Chromium and its compounds are on the Community Right-To-Know List.

OSHA PEL: Cl 0.1 $mg(CrO_3)/m^3$
ACGIH TLV: TWA 0.05 $mg(CrO_3)/m^3$
NIOSH REL: (Chromium(VI)) TWA 25 $\mu g(Cr(VI))/m^3$; CL 50 $\mu g/m^3$/15M

SAFETY PROFILE: Confirmed carcinogen. Poison by skin contact, intraperitoneal, intravenous, subcutaneous, and intradermal routes. Experimental reproductive effects. Mutation data reported. A powerful oxidizer. When heated to decomposition it emits toxic fumes of Na_2O. See also CHROMIUM COMPOUNDS.

For occupational chemical analysis use NIOSH: Chromium Hexavalent 7024.

DXC400 **CAS:144-33-2** **HR: 3**
DISODIUM CITRATE
mf: $C_6H_6O_7 \cdot 2Na$ mw: 236.10

PROP: White crystals or granular powder; odorless. Mp: loses water @ 150°, bp: decomp @ red heat. Sol in water; insol in alc.

SYNS: DISODIUM HYDROGEN CITRATE □ NATRIUM CITRICUM (GERMAN) □ SODIUM CITRATE (FCC)

TOXICITY DATA WITH REFERENCE
ipr-rat LD50:1724 mg/kg JPETAB 94,65,48
ipr-mus LD50:1771 mg/kg JPETAB 94,65,48
scu-mus LD50:2580 mg/kg ARZNAD 15,852,65
ivn-mus LD50:71 mg/kg JPETAB 94,65,48
ivn-rbt LD50:418 mg/kg JPETAB 94,65,48

CONSENSUS REPORTS: Reported in EPA TSCA Inventory.

SAFETY PROFILE: Poison by intravenous route. Moderately toxic by intraperitoneal and subcutaneous routes. When heated to decomposition it emits toxic fumes of Na_2O.

DXC600 **HR: 3**
DISODIUM-1,3-DIHYDROXY-1,3-BIS-(aci-NITROMETHYL)-2,2,4,4-TETRAMETHYLCYCLO BUTANE
mf: $C_{10}H_{16}N_2Na_2O_6$ mw: 306.23

SAFETY PROFILE: Explodes on contact with water. When heated to decomposition it emits toxic fumes of Na_2O.

DXC800 **HR: 3**
N,N′-DISODIUM N,N′-DIMETHOXYSULFONYLDIAMIDE
mf: $C_2H_6N_2Na_2O_2S$ mw: 168.13

SAFETY PROFILE: An unstable explosive. When heated to decomposition it emits toxic fumes of Na_2O.

DXD000 **CAS:129-67-9** **HR: 3**
DISODIUM-3,6-ENDOXOHEXAHYDROPHTHALATE
mf: $C_8H_8O_5 \cdot 2Na$ mw: 230.14

PROP: Water-sol solid. Mp: 144°.

SYNS: ACCELERATE □ AGUATHOL □ DES-I-CATE □ DINATRIUM-(3,6-EPOXY-CYCLOHEXAAN-1,2-DICARBOXYLAAT) (DUTCH) □ DINATRIUM-(3,6-EPOXY-CYCLOHEXAN-1,2-DICARBOXYLAT) (GERMAN) □ DISODIUM-3,6-EPOXYCYCLOHEXANE-1,2-DICARBOXYLATE □ DISODIUM-7-OXABICYCLO(2.2.1)HEPTANE-2,3-DICARBOXYLATE □ DISODIUM SALT of ENDOTHALL □ DISODIUM SALT of 7-OXABICYCLO (2.2.1)HEPTANE-2,3-DICARBOXYLIC ACID □ ENDOTAL □ ENDOTHAL □ ENDOTHAL-NATRIUM (DUTCH) □ ENDOTHAL-SODIUM □ ENDOTHAL WEED KILLER □ 3,6-ENDOXOHEXAHYDROPHTHALIC ACID DISODIUM SALT □ (3,6-EPOSSI-CICLOESAN-1,2-DICARBOSSILATO) DISODICO (ITALIAN) □ 3,6-EPOXY-CYCLOHEXANE 1,2-CARBOXYLATE DISODIQUE (FRENCH) □ HERBICIDE 273 □ HYDOUT □ HYDROTHOL □ NIAGARATHAL □ RCRA WASTE NUMBER P000 □ RIPENTHOL □ TRI-ENDOTHAL

TOXICITY DATA WITH REFERENCE
orl-rat LD50:51 mg/kg GUCHAZ 6,248,73
skn-rat LD50:750 mg/kg PHJOAV 185,361,60
ivn-dog LDLo:5 mg/kg FEPRA7 11,349,52
skn-rbt LD50:100 mg/kg AFDOAQ 16,3,52
ivn-rbt LDLo:5 mg/kg FEPRA7 11,349,52
orl-gpg LDLo:250 mg/kg HYSAAV 31,225,66

SAFETY PROFILE: Poison by ingestion, skin contact, and intravenous routes. Very irritating to eyes, skin, and mucous membranes. A defoliant and an herbicide. When heated to decomposition it emits toxic fumes of Na_2O.

DXD200 **CAS:142-59-6** **HR: 3**
DISODIUM ETHYLENE-1,2-BISDITHIOCARBAMATE
mf: $C_4H_6N_2S_4 \cdot 2Na$ mw: 256.34

PROP: Crystals. Sol in water.

SYNS: CARBON D □ CHEM BAM □ DINATRIUM-AETHYLENBISDITHIOCARBAMAT (GERMAN) □ DINATRIUM-(N,N′-AETHYLEN-BIS(DITHIOCARBAMAT)) (GERMAN) □ DINATRIUM-(N,N′-ETHYLEEN-BIS (DITHIOCARBAMAAT)) (DUTCH) □ DISODIUM ETHYLENEBIS(DITHIOCARBAMATE) □ DITHANE A-40 □ DITHANE D-14 □ DSE □ 1, 2-ETHANEDIYLBISCARBAMODITHIOIC ACID DISODIUM SALT □ N, N′-ETHYLENE BIS(DITHIOCARBAMATE de SODIUM) (FRENCH) □ ETHYLENEBIS(DITHIOCARBAMATE) DISODIUM SALT □ ETHYLENEBIS(DITHIOCARBAMIC ACID) DISODIUM SALT □ N,N′-ETILEN-BIS (DITIOCARBAMMATO) di SODIO (ITALIAN) □ NABAM □ NABAME (FRENCH) □ PARZATE □ SPRING-BAK

TOXICITY DATA WITH REFERENCE
mmo-omi 1000 ppm MMAPAP 50,233,73
scu-mus TDLo:194 mg/kg (6-14D preg):TER NTIS** Pb 223,-,160
scu-mus TDLo:418 mg/kg (6-14D preg):REP NTIS** PB223-160
orl-rat LD50:395 mg/kg FEPRA7 11,391,52
ipr-rat LD50:500 mg/kg 85DPAN -,-,71/76
orl-mus LD50:580 mg/kg PCOC** -,777,66

CONSENSUS REPORTS: Reported in EPA TSCA Inventory. EPA Genetic Toxicology Program.

SAFETY PROFILE: Poison by ingestion. Moderately toxic by intraperitoneal route. Experimental teratogenic and reproductive effects. Mutation data reported. When heated to decomposition it emits very toxic fumes of NO_x, Na_2O, and SO_x. See also CARBAMATES.

DXD400 **CAS:7414-83-7** **HR: 3**
DISODIUM ETIDRONATE
mf: $C_2H_6O_7P_2 \cdot 2Na$ mw: 249.99

SYNS: DIDRONEL R □ DISODIUM DIHYDROGEN-(1-HYDROXYETHYLIDENE)DIPHOSPHONATE □ DISODIUM ETHANOL-1,1-DIPHOSPHONATE □ DISODIUM ETHYDRONATE □ EITDRONATE DISODIUM □ ETHANE-1-HYDROXY-1,1-DIPHOSPHONIC ACID DISODIUM SALT □ (1-HYDROXYETHYLIDENE)DIPHOSPHONIC ACID DISODIUM SALT □ SODIUM ETHIDRONATE □ SODIUM ETIDRONATE □ SODIUM ETHYDRONATE

TOXICITY DATA WITH REFERENCE
orl-rat TDLo:2500 mg/kg (6-15D preg):REP TXAPA9 18,548,71
orl-rbt TDLo:1500 µg/kg (female 2-16D post):TER TXAPA9 18,548,71
orl-rbt TDLo:1500 µg/kg (female 2-16D post):REP TXAPA9 18,548,71
orl-rat TDLo:5500 mg/kg (female 1-22D post):REP TXAPA9 18,548,71
orl-rat LD50:1340 mg/kg TXAPA9 22,661,72
scu-rat LD50:372 mg/kg KSRNAM 23,1251,89
ivn-rat LD50:73 mg/kg KSRNAM 23,1251,89
orl-mus LD50:2050 mg/kg SIIPD4 10,447,83
ivn-mus LD50:49 mg/kg KSRNAM 23,1251,89
ivn-dog LDLo:32 mg/kg JPMSAE 73,1097,84

CONSENSUS REPORTS: Reported in EPA TSCA Inventory.

SAFETY PROFILE: Poison by intravenous and subcutaneous routes. Moderately toxic by ingestion. An experimental teratogen. Other experimental reproductive effects. When heated to decomposition it emits toxic fumes of PO_x and Na_2O.

DXD600 **CAS:10163-15-2** **HR: 3**
DISODIUM FLUOROPHOSPHATE
mf: $FO_3P \cdot 2Na$ mw: 143.95

PROP: A solid. Mp: 625°. Sol in H_2O. Insol in EtOH and Et_2O.

SYNS: DISODIUM MONOFLUOROPHOSPHATE □ DISODIUM

PHOSPHOROFLUORIDATE □ SODIUM FLUOROPHOSPHATE (Na_2PO_3F) □ SODIUM PHOSPHOROFLUORIDATE □ SODIUM PHOSPHOROFLURIDATE

TOXICITY DATA WITH REFERENCE
dlt-dmg-orl 20 μmol/L GENTAE 87,67,77
orl-rat LD50:570 mg/kg JDREAF 29,529,50
ipr-rat LD50:220 mg/kg JDREAF 29,529,50
orl-mus LD50:710 mg/kg CAREBK 12,177,78

CONSENSUS REPORTS: IARC Cancer Review: Group 3 IMEMDT 7,208,87; Animal Inadequate Evidence IMEMDT 27,237,82; Human Inadequate Evidence IMEMDT 27,237,82.

OSHA PEL: TWA 2.5 mg(F)/m^3
NIOSH REL: (Inorganic Fluorides) TWA 2.5 mg(F)/m^3

SAFETY PROFILE: Poison by intraperitoneal route. Moderately toxic by ingestion. Mutation data reported. Questionable carcinogen. An anticaries ingredient in dentifrices for children's teeth. When heated to decomposition it emits very toxic fumes of F^-, PO_x, and Na_2O.

DXD800 CAS:17013-01-3 HR: 2
DISODIUM FUMARATE
mf: $C_4H_2O_4$•2Na mw: 160.64

SYN: SODIUM FUMARATE

TOXICITY DATA WITH REFERENCE
orl-hmn TDLo:215 mg/kg:GIT JAPMA8 31,1,42
ipr-rat LDLo:2420 mg/kg JAPMA8 35,298,46
orl-mus LDLo:3680 mg/kg JAPMA8 31,12,42
ivn-rbt LDLo:500 mg/kg JAPMA8 31,1,42

SAFETY PROFILE: Moderately toxic by ingestion, intravenous, and intraperitoneal routes. Human systemic effects by ingestion: hypermotility, diarrhea, nausea or vomiting, and other gastrointestinal changes. When heated to decomposition it emits toxic fumes of Na_2O.

DXD875 CAS:71277-79-7 HR: 3
DISODIUM GLYCYRRHIZIN
mf: $C_{42}H_{62}O_{16}$•2Na mw: 869.02

SYNS: DISODIUM GLYCYRRHIZINATE □ GLYCYRRHIZINIC ACID DISODIUM SALT

TOXICITY DATA WITH REFERENCE
cyt-ham:fbr 4 g/L FCTOD7 22,623,84
cyt-ham:lng 1700 mg/L GMCRDC 27,95,81
ipr-mus LD50:144 mg/kg YKYUA6 32,1367,81

SAFETY PROFILE: Poison by intraperitoneal route. Mutation data reported. When heated to decomposition it emits toxic fumes of Na_2O.

DXE000 CAS:16893-85-9 HR: 3
DISODIUM HEXAFLUOROSILICATE
DOT: UN 2674
mf: F_6Si•2Na mw: 188.07

PROP: Colorless hexagonal crystals. Fluorescent when activated by Ti(IV.). Practically insol in H_2O; insol in EtOH.

SYNS: DESTRUXOL APPLEX □ (2-)-DISODIUM HEXAFLUOROSILICATE □ DISODIUM SILICOFLUORIDE □ ENS-ZEM WEEVIL BAIT □ ENT 1,501 □ FLUOSILICATE de SODIUM □ NATRIUMSILICOFLUORID (GERMAN) □ ORTHO EARWIG BAIT □ ORTHO WEEVIL BAIT □ PRODAN □ PSC CO-OP WEEVIL BAIT □ SAFSAN □ SALUFER □ SILICON SODIUM FLUORIDE □ SODIUM FLUOROSILICATE □ SODIUM FLUOSILICATE □ SODIUM HEXAFLUOROSILICATE □ SODIUM HEXAFLUOSILICATE □ SODIUM SILICOFLUORIDE (DOT) □ SUPER PRODAN

TOXICITY DATA WITH REFERENCE
skn-rbt 500 mg MLD FCTOD7 20,563,82
eye-rbt 100 mg SEV FCTOD7 20,573,82
eye-rbt 100 mg/4S rns SEV FCTOD7 20,573,82
orl-rat LD50:125 mg/kg ARSIM* 20,21,66
scu-rat LDLo:70 mg/kg JPETAB 39,246,30
orl-rbt LDLo:125 mg/kg JPETAB 39,246,30
scu-frg LDLo:448 mg/kg CRSBAW 124,133,37

CONSENSUS REPORTS: Reported in EPA TSCA Inventory.

OSHA PEL: TWA 2.5 mg(F)/m^3
NIOSH REL: (Inorganic Fluorides) TWA 2.5 mg(F)/m^3
DOT CLASSIFICATION: 6.1; *Label:* KEEP AWAY FROM FOOD

SAFETY PROFILE: Poison by ingestion and subcutaneous routes. A skin and severe eye irritant. An insecticide. When heated to decomposition it emits very toxic fumes of F^- and Na_2O.

DXE200 CAS:928-72-3 HR: 1
DISODIUM IMINODIACETATE
mf: $C_4H_5NO_4$•2Na mw: 177.08

SYN: IMINODIOCTAN SODNY (CZECH)

TOXICITY DATA WITH REFERENCE
eye-rbt 500 mg/24H MLD 28ZPAK -,128,72
orl-rat LD50:8070 mg/kg 28ZPAK -,128,72

CONSENSUS REPORTS: Reported in EPA TSCA Inventory.

SAFETY PROFILE: Mildly toxic by ingestion. An eye irritant. When heated to decomposition it emits toxic fumes of NO_x and Na_2O.

DXE500 CAS:4691-65-0 HR: 2
DISODIUM INOSINATE
mf: $C_{10}H_{11}N_4O_8P$•2Na mw: 392.20

PROP: Colorless to white crystals; characteristic taste. Sol in water; sltly sol in alc; insol in ether.

SYNS: DISODIUM-5′-INOSINATE □ DISODIUM IMP □ DISODIUM INOSINE-5′-MONOPHOSPHATE □ DISODIUM INOSINE-5′-PHOSPHATE □ IMP DISODIUM SALT □ 5′-IMP DISODIUM SALT □ IMP SODIUM SALT □ INOSINE-5′-MONOPHOSPHATE DISODIUM □ INOSIN-5′-MONOPHOSPHATE DISODIUM □ SODIUM INOSINATE □ SODIUM-5′-INOSINATE

TOXICITY DATA WITH REFERENCE
cyt-ham:fbr 1 g/L FCTOD7 22,623,84
ipr-mus TDLo:500 mg/kg (13D preg):TER JJPAAZ 22,201,72
ipr-mus TDLo:250 mg/kg (10D preg):TER JJPAAZ 22,201,72

orl-rat LD50:15,900 mg/kg AJINO* -,-,73
ipr-rat LD50:4850 mg/kg AJINO* -,-,73
scu-rat LD50:3900 mg/kg AJINO* -,-,73
ivn-rat LD50:2730 mg/kg AJINO* -,-,73
orl-mus LD50:12 g/kg TIDZAH 24,553,66
ipr-mus LD50:5400 mg/kg TIDZAH 24,553,66
scu-mus LD50:5480 mg/kg AJINO* -,-,73
ivn-mus LD50:3300 mg/kg TIDZAH 24,553,66

CONSENSUS REPORTS: Reported in EPA TSCA Inventory.

SAFETY PROFILE: Moderately toxic by several routes. An experimental teratogen. Mutation data reported. When heated to decomposition it emits toxic fumes of PO_x, NO_x, and Na_2O.

DXE600 CAS:144-21-8 HR: 3
DISODIUM METHANEARSENATE
mf: CH_3AsO_3•2Na mw: 183.94

PROP: Crystals or solid; water-sol hydrate. Mp: 132–139°, bp: 165°, fp: −6°, d: 1.15. Sol in H_2O and MeOH; pract insol in most org solvs.

SYNS: ANSAR 184 □ ANSAR DSMA LIQUID □ ARRHENAL □ ARSINYL □ ARSYNAL □ CACODYL NEW □ CHIPCO CRAB KLEEN □ CLOUT □ CRAB-E-RAD □ CRALO-E-RAD □ DAL-E-RAD 100 □ DIARSEN □ DIMET □ DINATE □ DISODIUM METHANEARSONATE □ DISODIUM METHYLARSENATE □ DISODIUM METHYLARSONATE □ DISODIUM MONOMETHYLARSONATE □ DISOMAR □ DI-TAC □ DMA □ DREXEL DSMA LIQUID □ DSMA LIQUID □ JON-TROL □ MAA SODIUM SALT □ METHAR □ METHARSINAT □ NAMATE □ NEOASYCODILE □ SODAR □ SODIUM METHANEARSONATE □ SODIUM METHARSONATE □ SODIUM METHYLARSONATE □ SOMAR □ STENOSINE □ TONARSEN □ VERSAR DSMA LQ □ WEED BROOM □ WEED-E-RAD □ WEED-HOE

TOXICITY DATA WITH REFERENCE
ipr-ham TDLo:500 mg/kg (12D preg):TER BECTA6 29,679,82
ipr-ham TDLo:600 mg/kg (9D preg):TER TJADAB 25(2),50A,82
orl-rat LD50:821 mg/kg FAATDF 7,299,86
orl-mus LD50:1150 mg/kg JPIFAN (3),5,70
skn-rbt LD50:10 g/kg FMCHA2 -,C114,89

CONSENSUS REPORTS: EPA Genetic Toxicology Program. Arsenic and its compounds are on the Community Right-To-Know List.

OSHA PEL: TWA 0.5 mg(As)/m^3
ACGIH TLV: TWA 0.2 mg(As)/m^3

SAFETY PROFILE: Confirmed human carcinogen. Moderately toxic by ingestion. Experimental teratogen. Dangerous fire hazard by spontaneous chemical reaction. Ignites spontaneously in dry air. Can react vigorously with oxidizing materials, e.g., air, Cl_2. An herbicide. When heated to decomposition it emits toxic fumes of As and Na_2O. See also ARSENIC COMPOUNDS.

DXE800 CAS:7631-95-0 HR: 3
DISODIUM MOLYBDATE
mf: MoO_4•2Na mw: 205.92

PROP: White solid. Mp: 686°. Sol in H_2O.

SYNS: MOLYBDIC ACID, DISODIUM SALT □ NATRIUMMOLYBDAT (GERMAN) □ SODIUM MOLYBDATE □ SODIUM MOLYBDATE(VI)

TOXICITY DATA WITH REFERENCE
pic-esc 16 mmol/L ENMUDM 6,59,84
sln-smc 80 mmol/L MUTAEX 1,21,86
itt-mus TDLo:16,474 µg/kg (1D male):REP JRPFA4 7,21,64
ipr-rat LD50:576 mg/kg EQSSDX 1,1,75
ipr-mus LD50:303 mg/kg EQSSDX 1,1,75
scu-mus LD50:570 mg/kg AEPPAE 244,17,62
ivn-cat LD50:917 mg/kg AGSOA6 8,51,67

CONSENSUS REPORTS: Reported in EPA TSCA Inventory.

OSHA PEL: TWA 5 mg(Mo)/m^3
ACGIH TLV: TWA 5 mg(Mo)/m^3

SAFETY PROFILE: Poison by intraperitoneal route. Moderately toxic by subcutaneous and intravenous routes. Experimental reproductive effects. Mutation data reported. When heated to decomposition it emits toxic fumes of Na_2O. See also MOLYBDENUM COMPOUNDS.

DXE875 CAS:10102-40-6 HR: 3
DISODIUM MOLYBDATE DIHYDRATE
mf: MoO_4•$_2$Na•$_2H_2O$ mw: 241.96

PROP: Orthorhombic crystals. Sol in H_2O.

SYN: SODIUM MOLYBDATE DIHYDRATE

TOXICITY DATA WITH REFERENCE
ivn-mus TDLo:968 g/kg (8D preg):TER ENVRAL 33,47,84
ipr-rat LD50:520 mg/kg AIPTAK 154,243,65
ipr-mus LD50:257 mg/kg AIPTAK 154,243,65

OSHA PEL: TWA Total Dust: 10 mg/m^3; Respirable Fraction: 5 mg/m^3
ACGIH TLV: TWA 10 mg(Mo)/m^3

SAFETY PROFILE: Poison by intraperitoneal route. An experimental teratogen. When heated to decomposition it emits toxic fumes of Na_2O. See also MOLYBDENUM COMPOUNDS.

DXF000 CAS:15467-20-6 HR: 3
DISODIUM NITRILOTRIACETATE
mf: $C_6H_7NO_6$•2Na mw: 235.12

SYNS: N,N-BIS(CARBOXYMETHYL)GLYCINE DISODIUM SALT □ DISODIUM HYDROGEN NITRILOTRIACETATE □ GLYCINE, N,N-BIS (CARBOXYMETHYL)-, DISODIUM SALT (9CI) □ KIRESUTO NTB □ NITRILOTRIACETIC ACID, DISODIUM SALT

TOXICITY DATA WITH REFERENCE
orl-rat LD50:1460 mg/kg TXAPA9 18,398,71

CONSENSUS REPORTS: IARC Cancer Review: Group 2B IMEMDT 48,181,90; Animal Sufficient Evidence IMEMDT 48,181,90. Reported in EPA TSCA Inventory.

SAFETY PROFILE: Confirmed carcinogen. Moderately toxic by ingestion. When heated to decomposition it emits toxic fumes of NO_x and Na_2O.

DXF200 CAS:12008-41-2 HR: 2
DISODIUM OCTABORATE, TETRAHYDRATE
mf: $B_8Na_2O_{13}\cdot4H_2O$ mw: 412.54

SYNS: POLYBOR □ POLYBOR 3

TOXICITY DATA WITH REFERENCE
orl-rat LD50:2000 mg/kg FMCHA2 -,C191,83
orl-gpg LD50:5300 mg/kg 28ZEAL 5,85,76

CONSENSUS REPORTS: Reported in EPA TSCA Inventory.

SAFETY PROFILE: Moderately toxic by ingestion. An insecticide. When heated to decomposition it emits toxic fumes of Na_2O. See also BORON COMPOUNDS.

DXF400 CAS:53778-51-1 HR: 2
DISODIUM-2-(p-(γ-PHENYLPROPYLAMINO)BENZENE-SULFONAMIDO) PYRIDINE
mf: $C_{20}H_{19}N_3O_8S_3\cdot2Na$ mw: 571.58

SYNS: DISODIUM CINNAMYLIDENE BISULFITE derivative of SULFAPYRIDINE □ 1-PHENYL-3-(p-2-PYRIDYLSULFAMOYLANILINO)-1,3-PROPANEDISULFONIC ACID DISODIUM SALT □ SOLUPYRIDINE □ SULFAPYRIDINE NEUTRAL SOLUBLE

TOXICITY DATA WITH REFERENCE
orl-mus LD50:7500 mg/kg JPETAB 84,203,45
scu-mus LD50:2680 mg/kg JPETAB 84,203,45
ivn-mus LD50:1280 mg/kg JPETAB 84,203,45

SAFETY PROFILE: Moderately toxic by ingestion, intravenous, and subcutaneous routes. When heated to decomposition it emits very toxic fumes of NO_x, Na_2O, and SO_x.

DXF600 CAS:26016-99-9 HR: 3
DISODIUM PHOSPHONOMYCIN
mf: $C_3H_7O_4P\cdot2Na$ mw: 184.05

PROP: A solid.

SYNS: DISODIUM FOSFOMYCIN □ (1R,2S)(−)-(1,2-EPOXYPROPYL)PHOSPHONIC ACID DISODIUM SALT □ FOM-Na □ FOSFOMYCIN DISODIUM □ FOSFOMYCIN DISODIUM SALT □ FOSFOMYCIN SODIUM SALT □ (2R-cis)-(3-METHYLOXIRANYL)PHOSPHONIC ACID DISODIUM SALT □ PHOSPHONOMYCIN DISODIUM SALT □ PHOSPHONOMYCIN SODIUM □ SODIUM FOSFOMYCIN

TOXICITY DATA WITH REFERENCE
mmo-klp 20 μmol/L MUREAV 89,269,81
ipr-rat TDLo:1375 mg/kg (female 7-17D post):REP JJANAX 32,155,79
ipr-rat TDLo:22,500 mg/kg (14-22D preg/21D post):REP JJANAX 32,171,79
ipr-rat TDLo:1375 mg/kg (female 7-17D post):TER JJANAX 32,155,79
ipr-rat TDLo:11 g/kg (63D male/14D pre-7D preg):TER JJANAX 32,164,79
orl-rat TD50:4550 mg/kg JJANAX 32,61,79
ipr-rat LD50:2000 mg/kg JJANAX 32,61,79
scu-rat LD50:4320 mg/kg JJANAX 32,61,79
ivn-rat LD50:1560 mg/kg JJANAX 32,61,79
ims-rat LD50:2460 mg/kg JJANAX 32,61,79
orl-mus LD50:7300 mg/kg JJANAX 32,61,79
ipr-mus LD50:2175 mg/kg IYKEDH 12,668,81
scu-mus LD50:5100 mg/kg IYKEDH 12,668,81
ivn-mus LD50:1225 mg/kg JJANAX 32,61,79
ims-mus LD50:2625 mg/kg JJANAX 32,61,79

SAFETY PROFILE: Poison by intraperitoneal and subcutaneous routes. Moderately toxic by intravenous and intramuscular routes. Mildly toxic by ingestion. Experimental teratogenic and reproductive effects. Mutation data reported. When heated to decomposition it emits toxic fumes of PO_x and Na_2O.

DXF700 CAS:50865-01-5 HR: 3
DISODIUM PROTOPORPHYRIN
mf: $C_{34}H_{32}H_4O_4\cdot2Na$ mw: 606.68

SYNS: 7,12-DIETHYENYL-3,8,13,17-TETRAMETHYL-21H,23H-PORPHINE-2,18-DIPROPANOIC ACID DISODIUM SALT □ PROTOPORPHYRIN DISODIUM □ PROTOPORPHYRIN SODIUM □ PROTOPORPHYRIN SODIUM SALT

TOXICITY DATA WITH REFERENCE
ivn-rat LD50:240 mg/kg NIIRDN 6,729,82
ipr-mus LD50:1029 mg/kg NIIRDN 6,729,82
scu-mus LD50:1147 mg/kg NIIRDN 6,729,82
ivn-mus LD50:484 mg/kg NIIRDN 6,729,82

SAFETY PROFILE: Poison by intravenous route. Moderately toxic by intraperitoneal and subcutaneous routes. When heated to decomposition it emits toxic fumes of NO_x and Na_2O.

DXF800 CAS:7758-16-9 HR: 3
DISODIUM PYROPHOSPHATE
mf: $H_2O_7P_2\cdot Na_2$ mw: 221.94

PROP: White, crystalline powder or monoclinic lattice. D: 1.862, mp: 220° (decomp). Sol in water.

SYNS: DINATRIUMPYROPHOSPHAT (GERMAN) □ DIPHOSPHORIC ACID, DISODIUM SALT □ DISODIUM DIHYDROGEN PYROPHOSPHATE □ DISODIUM DIPHOSPHATE □ SODIUM ACID PYROPHOSPHATE (FCC) □ SODIUM PYROPHOSPHATE

TOXICITY DATA WITH REFERENCE
orl-mus LD50:2650 mg/kg ARZNAD 7,445,57
scu-mus LD50:480 mg/kg ARZNAD 7,445,57
ivn-mus LD50:59 mg/kg ARZNAD 7,445,57
ivn-rbt LDLo:50 mg/kg AEPPAE 169,238,33

CONSENSUS REPORTS: Reported in EPA TSCA Inventory.

SAFETY PROFILE: Poison by intravenous route. Moderately toxic by ingestion and subcutaneous routes. An irritant to skin, eyes, and mucous membranes. When heated to decomposition it emits toxic fumes of PO_x and Na_2O.

DXG000 **CAS:13410-01-0** **HR: 3**
DISODIUM SELENATE
mf: $O_4Se{\cdot}2Na$ mw: 188.94

PROP: Colorless, rhombic crystals. D: 3.098. Very sol in H_2O.

SYNS: NATRIUMSELENIAT (GERMAN) ☐ P-40 ☐ SEL-TOX SSO2 and SS-20 ☐ SODIUM SELENATE

TOXICITY DATA WITH REFERENCE
mmo-sat 40 μmol/L ENVRAL 36,379,85
mma-sat 2 μmol/plate MUREAV 66,175,79
dnr-sat 10 μg/plate CALEDQ 10,75,80
mrc-bcs 50 μmol/plate MUREAV 66,175,79
dns-rat:lvr 100 μmol/L CALEDQ 10,75,80
orl-mus TDLo:60 mg/kg (female 8-12D post):REP TCMUD8 6,361,86
orl-mus TDLo:14 mg/kg (14D post):TER 32XPAD -,83,75
orl-rat TDLo:128 mg/kg/2Y-C:CAR JONUAI 101,1531,71
orl-wmn TDLo:53 mg/kg:CVS,GIT,LIV NZMJAX 87,354,78
orl-rat LD50:1600 μg/kg GISAAA 49(9),66,84
ipr-rat LDLo:8973 μg/kg JPETAB 58,454,36
scu-rat LDLo:11,336 μg/kg ARTODN 45,207,80
ivn-rat LDLo:4786 μg/kg JPETAB 60,449,37
scu-cat LDLo:20 mg/kg HBAMAK 4,1289,35
orl-rbt LD50:2250 μg/kg GISAAA 49(9),66,84
ivn-rbt LDLo:3600 μg/kg JPETAB 60,449,37

CONSENSUS REPORTS: IARC Cancer Review: Group 3 IMEMDT 7,56,87. Selenium and its compounds are on the Community Right-To-Know List. EPA Genetic Toxicology Program. Reported in EPA TSCA Inventory.

OSHA PEL: TWA 0.2 mg(Se)/m^3
ACGIH TLV: TWA 0.2 mg(Se)/m^3
DFG MAK: 0.1 mg(Se)/m^3
DOT CLASSIFICATION: 6.1; *Label:* Poison, Corrosive

SAFETY PROFILE: Poison by ingestion, intravenous, subcutaneous, and intraperitoneal routes. Questionable carcinogen with experimental carcinogenic and teratogenic data. Human systemic effects by ingestion: EKG changes, hypermotility, diarrhea, and liver impairment. Experimental reproductive effects. Effects similar to those of arsenic. Mutation data reported. A pesticide. When heated to decomposition it emits toxic fumes of Se and Na_2O. See also SELENIUM COMPOUNDS and ARSENIC COMPOUNDS.

DXG025 **CAS:2583-80-4** **HR: 1**
DISODIUM 2-(4-STYRYL-3-SULFOPHENYL)-7-SULFO-2H-NAPHTHO(1,2-d)TRIAZOLE
mf: $C_{24}H_{15}N_3O_6S_2{\cdot}2Na$ mw: 551.52

SYNS: 2H-NAPHTHO(1,2-d)TRIAZOLE, 2-(4-STYRYL-3-SULFOPHENYL)-7-SULFO-, DISODIUMSALT ☐ NAPHTHO(1,2-d)TRIAZOLE-7-SULFONIC ACID,2-(4-(2-PHENYLETHENYL)-3-SULFOPHENYL)-, DISODIUM ☐ 2-STILBENESULFONIC ACID, 4-(7-SULFO-2H-NAPHTHO(1,2-d)TRIAZOL-2-YL)-, DISODIUM SALT

TOXICITY DATA WITH REFERENCE
skn-rbt 500 mg/24H MLD MVCRB3 2,193,73
eye-rbt 100 mg MLD MVCRB3 2,193,73

CONSENSUS REPORTS: Reported in EPA TSCA Inventory.

SAFETY PROFILE: A skin and eye irritant. When heated to decomposition it emits toxic fumes of NO_x and SO_x

DXG035 **CAS:1330-43-4** **HR: 1**
DISODIUM TETRABORATE
mf: $B_4Na_2O_7$ mw: 201.22

SYNS: ANHYDROUS BORAX ☐ BORATES, TETRA, SODIUM SALT, anhydrous (OSHA) ☐ BORAX GLASS ☐ BORIC ACID, DISODIUM SALT ☐ FR 28 ☐ FUSED BORAX ☐ RASORITE 65 ☐ SODIUM BIBORATE ☐ SODIUM TETRABORATE ☐ SODIUM TETRABORATE ($Na_2B_4O_7$)

TOXICITY DATA WITH REFERENCE
orl-rat TDLo: 16,750 μg/kg (male 30D pre):REP EVHPAZ 13,59,76

OSHA PEL: TWA 10 mg/m^3
ACGIH TLV: TWA 1 mg/m^3

CONSENSUS REPORTS: Reported in EPA TSCA Inventory.

SAFETY PROFILE: A nuisance dust. Experimental reproductive effects. When heated to decomposition it emits toxic vapors of B.

DXG050 **CAS:68594-24-1** **HR: 3**
DISODIUM-5-TETRAZOLAZOCARBOXYLATE
mf: $C_2N_6Na_2O_2$ mw: 186.04

SAFETY PROFILE: An explosive. When heated to decomposition it emits toxic fumes of NO_x and Na_2O. See also EXPLOSIVES, HIGH.

DXG100 **CAS:2391-03-9** **HR: 3**
DISOMER MALEATE
mf: $C_{16}H_{19}BrN_2{\cdot}C_4H_4O_4$ mw: 435.36

PROP: Crystals. Mp: 103–113°.

SYNS: (+)-2-(p-BROMO-α-(2-(DIMETHYLAMINO)ETHYL)BENZYL) PYRIDINE MALEATE ☐ (S)-γ-(4-BROMOPHENYL)-N,N-DIMETHYL-2-PYRIDINEPROPANAMINE (Z)-2-BUTENEDIOATE (1:1) ☐ d-BROMPHENIRAMINE MALEATE ☐ DEXBROMPHENIRAMINE MALEATE

TOXICITY DATA WITH REFERENCE
orl-rat LD50:191 mg/kg CMTRAG 3,120,61
ipr-rat LD50:104 mg/kg CMTRAG 3,120,61
orl-mus LD50:176 mg/kg CMTRAG 3,120,61
ipr-mus LD50:106 mg/kg CMTRAG 3,120,61
ivn-mus LD50:25 mg/kg CMTRAG 3,120,61
orl-gpg LD50:259 mg/kg CMTRAG 3,120,61

SAFETY PROFILE: Poison by ingestion, intramuscular, and intraperitoneal routes. When heated to decomposition it emits toxic fumes of Br^- and NO_x.

DXG150 **HR: 3**
DISPHOLIDUS TYPHUS VENOM

SYN: VENOM, SNAKE, DISPHOLIDUS TYPHUS

TOXICITY DATA WITH REFERENCE
scu-mus LDLo:10 mg/kg SAMJAF 14,236,40
ivn-mus LD50:67 μg/kg 23EIAT 1,437,68
ims-mus LDLo:15 mg/kg SAMJAF 14,236,40

ivn-rbt LDLo: 5 μg/kg SAMJAF 14,236,40
ims-rbt LDLo: 500 μg/kg SAMJAF 14,236,40
ipr-pgn LDLo: 667 ng/kg SAMJAF 14,236,40
scu-pgn LDLo: 33 μg/kg SAMJAF 14,236,40
ivn-pgn LDLo: 667 ng/kg SAMJAF 14,236,40
ims-pgn LDLo: 33 μg/kg SAMJAF 14,236,40

SAFETY PROFILE: Poison by subcutaneous, intramuscular, intravenous, and intraperitoneal routes.

DXG200 **HR: 3**
3,6-DI(SPIROCYCLOHEXANE)TETRAOXANE
mf: $C_{12}H_{20}O_4$ mw: 228.29

SAFETY PROFILE: Explodes on impact. When heated to decomposition it emits acrid smoke and irritating fumes.

DXG600 **CAS:6576-51-8** **HR: 3**
DISTAMYCIN A HYDROCHLORIDE
mf: $C_{22}H_{27}N_9O_4\cdot ClH$ mw: 518.04

PROP: Crystals from dil HCl. Mp: 186–189°.

SYNS: N″-(2-AMIDINOETHYL)-4-FORMAMIDO-1,1′,1″-TRIMETHYL-(N,4′:N′,4″-TERPYRROLE)-2-CARBOXAMIDE HYDROCHLORIDE □ HEPERAL □ STALLIMYCIN HYDROCHLORIDE

TOXICITY DATA WITH REFERENCE
dni-mus:lym 1900 nmol/L CBINA8 8,183,74
oms-mus:lym 3200 nmol/L CBINA8 8,183,74
dnd-mam:lym 200 μmol/ EJBCAI 26,81,72L
ipr-rat LD50:169 mg/kg MDACAP 13,319,77
ipr-mus LD50:160 mg/kg MDACAP 13,319,77
ivn-mus LD50:75 mg/kg MEIEDD 11,1383,89

SAFETY PROFILE: Poison by intraperitoneal and intravenous routes. Mutation data reported. When heated to decomposition it emits very toxic fumes of NO_x and HCl.

DXG625 **CAS:107-64-2** **HR: 1**
DISTEARYL DIMETHYLAMMONIUM CHLORIDE
mf: $C_{38}H_{80}N\cdot Cl$ mw: 586.64

SYNS: ALIQUAT 207 □ AMMONIUM, DIMETHYLDIOCTADECYL-, CHLORIDE □ AROSURF TA 100 □ ARQUAD R 40 □ DIMETHYLDIOCTADECYLAMMONIUM CHLORIDE □ N,N-DIMETHYL-N-OCTADECYL-1-OCTADECANAMINIUM CHLORIDE □ GENAMIN DSAC □ KD 83 □ 1-OCTADECANAMINIUM, N,N-DIMETHYL-N-OCTADECYL-, CHLORIDE (9CI) □ Q-D 86P □ QUATERNIUM 5 □ TALOFLOC □ VARISOFT 100

TOXICITY DATA WITH REFERENCE
orl-rat LD50:11,300 mg/kg ESKHA5 (101),152,83

CONSENSUS REPORTS: Reported in EPA TSCA Inventory.

SAFETY PROFILE: Mildly toxic by ingestion. When heated to decomposition it emits toxic vapors of NO_x and Cl^-.

DXG700 **CAS:693-36-7** **HR: 2**
DISTEARYL 3,3′-THIODIPROPIONATE
mf: $C_{42}H_{82}O_4S$ mw: 683.30

SYNS: ADVASTAB 802 □ ADVASTAB PS 802 □ ANTIOK S □ ARBESTAB DSTDP □ CYANOX-STDP □ DIOCTADECYL THIODIPROPIONATE □ DIOCTADECYL 3,3′-THIODIPROPIONATE □ DISTEARYL THIOPROPIONATE □ DISTEARYL β-THIODIPROPIONATE □ DISTEARYL β,β′-THIODIPROPIONATE □ DSTDP □ DSTP □ HOSTANOX SE 2 □ HOSTANOX VP-SE 2 □ IRGANOX PS 802 □ LUSMIT SS □ NAUGARD DSTDP □ PLASTANOX STDP □ PLASTANOX STDP ANTIOXIDANT □ PROPANOIC ACID, 3,3′-THIOBIS-, DIOCTADECYL ESTER (9CI) □ PROPIONIC ACID, 3,3′-THIODI-, DIOCTADECYL ESTER □ PS 802 □ SEENOX DS □ SUMILIZER TPS □ YOSHINOX DSTDP

TOXICITY DATA WITH REFERENCE
orl-rat LD50:>2500 mg/kg AFREAW 3,197,51
orl-mus LD50:>2 g/kg AFREAW 3,197,51
ipr-mus LD50:>2 g/kg AFREAW 3,197,51

CONSENSUS REPORTS: Reported in EPA TSCA Inventory.

SAFETY PROFILE: Moderately toxic by ingestion and intraperitoneal routes. When heated to decomposition it emits toxic vapors of SO_x.

DXG800 **CAS:15876-67-2** **HR: 3**
DISTIGMINE BROMIDE
mf: $C_{22}H_{32}N_4O_4\cdot 2Br$ mw: 576.40

PROP: A solid. Mp: 149° (decomp).

SYNS: HEXAMARIUM □ 3,3′-(1,6-HEXANEDIYLBIS-((METHYLIMINO)CARBONYL)OXY)BIS(1-METHYLPYRIDINIUMDIBROMIDE) □ 3-HYDROXY-1-METHYLPYRIDINIUM BROMIDE HEXAMETHYLENEBIS (METHYLCARBAMATE) □ UBRETID □ UBRITIL

TOXICITY DATA WITH REFERENCE
orl-rat LD50:10 mg/kg OYYAA2 3,68,69
ipr-rat LD50:740 μg/kg DRUGAY 6,349,82
scu-rat LD50:1080 μg/kg DRUGAY 6,349,82
ivn-rat LD50:740 μg/kg OYYAA2 3,68,69
orl-mus LD50:10,500 μg/kg OYYAA2 3,68,69
ipr-mus LD50:310 μg/kg DRUGAY 6,349,82
ivn-mus LD50:300 μg/kg OYYAA2 3,68,69

SAFETY PROFILE: Poison by ingestion, intravenous, and intraperitoneal routes. When heated to decomposition it emits very toxic fumes of NO_x and Br^-.

DXG810 **CAS:64741-61-3** **HR: 3**
DISTILLATES (PETROLEUM), HEAVY CATALYTIC CRACKED

SYN: HEAVY CATALYTICALLY CRACKED DISTILLATE

CONSENSUS REPORTS: IARC Cancer Review: Animal Sufficient Evidence IMEMDT 45,39,89. Reported in EPA TSCA Inventory.

SAFETY PROFILE: Confirmed carcinogen. When heated to decomposition it emits acrid smoke and irritating vapors.

DXG820 **CAS:64742-80-9** **HR: 1**
DISTILLATES (PETROLEUM), HYDRODESULFURIZED MIDDLE

SYN: HYDRODESUFURIZED MIDDLE DISTILLATE

TOXICITY DATA WITH REFERENCE
skn-rbt 500 mg SEV JACTDZ 1,128,90
orl-rat LD:>5 g/kg JACTDZ 1,127,90

ihl-rat LC50:4600 mg/m³/4H JACTDZ 1,127,90
skn-rbt LD:>2 g/kg JACTDZ 1,127,90

CONSENSUS REPORTS: Reported in EPA TSCA Inventory.

SAFETY PROFILE: Low toxicity by ingestion, inhalation, and skin contact. A severe skin and eye irritant. When heated to decomposition it emits acrid smoke and irritating vapors.

DXG830 CAS:64742-46-7 HR: 2
DISTILLATES (PETROLEUM), HYDROTREATED MIDDLE

SYNS: AMOCO NT-45 PROCESS OIL ☐ KERMAC 600W (MINERAL SEAL OIL)

TOXICITY DATA WITH REFERENCE
mmo-sat 10 μL/plate EPASR* 8EHQ-0280-0333
skn-mus TDLo:416 g/kg/2Y-I:ETA EPASR* 8EHQ-1288-0775

CONSENSUS REPORTS: Reported in EPA TSCA Inventory.

SAFETY PROFILE: Questionable carcinogen with tumorigenic data reported. Mutation data reported. When heated to decomposition it emits acrid smoke and irritating vapors.

DXG840 CAS:64741-59-9 HR: 3
DISTILLATES (PETROLEUM), LIGHT CATALYTIC CRACKED

SYN: LIGHT CATALYTICALLY CRACKED DISTILLATE

TOXICITY DATA WITH REFERENCE
skn-rbt 500 mg SEV JACTDZ 1,129,90
orl-rat LD50:3200 mg/kg JACTDZ 1,130,90
ihl-rat LC50:3400 mg/m³/4H JACTDZ 1,130,90
skn-rbt LD:>2 g/kg JACTDZ 1,130,90

CONSENSUS REPORTS: IARC Cancer Review: Animal Sufficient Evidence IMEMDT 45,39,89. Reported in EPA TSCA Inventory.

SAFETY PROFILE: Confirmed carcinogen. Moderately toxic by ingestion. A severe skin and eye irritant. When heated to decomposition it emits acrid smoke and irritating vapors.

DXH200 CAS:150-60-7 HR: 1
DISULFIDE DIBENZYL
mf: $C_{14}H_{14}S_2$ mw: 246.40

PROP: A solid. Mp: 71–72°.

SYN: DIBENZYLDISULFID (CZECH)

TOXICITY DATA WITH REFERENCE
skn-rbt 500 mg/24H MLD 28ZPAK -,173,72
eye-rbt 500 mg/24H MLD 28ZPAK -,173,72

CONSENSUS REPORTS: Reported in EPA TSCA Inventory.

SAFETY PROFILE: A skin and eye irritant. When heated to decomposition it emits toxic fumes of SO_x.

DXH250 CAS:97-77-8 HR: 3
DISULFIRAM
mf: $C_{10}H_{20}N_2S_4$ mw: 296.56

PROP: Yellow-white crystals. Mp: 70°. Sol in CS_2, $CHCl_3$, C_6H_6, and EtOH.

SYNS: ABSTENSIL ☐ ABSTINYL ☐ ALCOPHOBIN ☐ ALK-AUBS ☐ ANTABUS ☐ ANTABUSE ☐ ANTADIX ☐ ANTAENYL ☐ ANTAETHAN ☐ ANTAETHYL ☐ ANTAETIL ☐ ANTALCOL ☐ ANTETAN ☐ ANTETHYL ☐ ANTETIL ☐ ANTEYL ☐ ANTIAETHAN ☐ ANTIETANOL ☐ ANTI-ETHYL ☐ ANTIETIL ☐ ANTIKOL ☐ ANTIVITIUM ☐ AVERSAN ☐ AVERZAN ☐ (BIS(DIETHYLAMINO)THIOXOMETHYL) DISULPHIDE ☐ BIS(N,N-DIETHYLTHIOCARBAMOYL) DISULFIDE ☐ BIS(DIETHYLTHIOCARBAMOYL) DISULFIDE ☐ BIS(N,N-DIETHYLTHIOCARBAMOYL) DISULPHIDE ☐ BONIBAL ☐ CONTRALIN ☐ CONTRAPOT ☐ CRONETAL ☐ DICUPRAL ☐ DISETIL ☐ DISULFAN ☐ DISULFURAM ☐ DISULPHURAM ☐ 1,1′-DITHIOBIS(N,N-DIETHYLTHIOFORMAMIDE) ☐ EKAGOM TEDS ☐ EPHORRAN ☐ ESPENAL ☐ ESPERAL ☐ ETABUS ☐ ETHYLDITHIOURAME ☐ ETHYLDITHIURAME ☐ ETHYL THIRAM ☐ ETHYL THIUDAD ☐ ETHYL THIURAD ☐ ETHYL TUADS ☐ ETHYL TUEX ☐ EXHORAN ☐ EXHORRAN ☐ HOCA ☐ KROTENAL ☐ NCI-C02959 ☐ NOCBIN ☐ NOXAL ☐ REFUSAL ☐ ROSULFIRAM ☐ STOPAETHYL ☐ STOPETHYL ☐ STOPETYL ☐ TATD ☐ TENURID ☐ TENUTEX ☐ TETD ☐ TETIDIS ☐ TETRADIN ☐ TETRADINE ☐ TETRAETHYLTHIOPEROXYDICARBONIC DIAMIDE ☐ TETRAETHYLTHIRAM DISULPHIDE ☐ TETRAETHYLTHIURAM ☐ TETRAETHYLTHIURAM DISULFIDE ☐ TETRAETHYLTHIURAM DISULPHIDE ☐ N,N,N′,N′-TETRAETHYLTHIURAM DISULPHIDE ☐ TETRAETIL ☐ TETURAM ☐ TETURAMIN ☐ THIOSAN ☐ THIOSCABIN ☐ THIRERANIDE ☐ THIURAM E ☐ THIURANIDE ☐ TILLRAM ☐ TIURAM ☐ TTD ☐ TTS ☐ USAF B-33

TOXICITY DATA WITH REFERENCE
eye-rbt 100 mg MLD FCTOD7 20,573,82
mmo-sat 25 μg/plate CBINA8 49,329,84
dni-ckn:emb 120 nmol/L BBACAQ 519,65,78
orl-mus TDLo:39,200 mg/kg (female 7-14D post):REP NTIS** PB86-197605
par-rat TDLo:400 mg/kg (female 4-11D post):TER BEXBAN 93,107,82
scu-mus TDLo:1278 mg/kg (female 6-14D post):TER NTIS** PB223-160
orl-rat TDLo:5 g/kg (female 3-12D post):TER JRPFA4 39,375,74
orl-rat TDLo:5 g/kg (female 3-12D post):REP JRPFA4 39,375,74
orl-mus TDLo:35 g/kg/78W-I:NEO NTIS** PB223-159
scu-mus TDLo:1000 mg/kg:NEO NTIS** PB223-159
orl-hmn LDLo:160 mg/kg BMJOAE 2,94,77
orl-wmn TDLo:90 mg/kg/18D-I:SYS JCLPDE 46,67,85
orl-man TDLo:150 mg/kg/6W-I:MUS ARHEAW 25,1494,82
orl-chd TDLo:150 mg/kg 34ZIAG -,230,69
orl-cld TDLo:150 mg/kg 34ZIAG -,230,69
orl-rat LD50:500 mg/kg ATXKA8 22,12,66
orl-mus LD50:1980 mg/kg AIPTAK 112,36,57
ipr-mus LD50:75 mg/kg NTIS** AD691-490

CONSENSUS REPORTS: IARC Cancer Review: Group 3 IMEMDT 7,56,87; Animal Inadequate Evidence IMEMDT 12,85,76. NCI Carcinogenesis Bioassay (feed); No Evidence: mouse, rat NCITR* NCI-CG-TR-16,79. Reported in EPA TSCA Inventory.

OSHA PEL: TWA 2 mg/m³
ACGIH TLV: TWA 2 mg/m³
DFG MAK: 2 mg/m³

SAFETY PROFILE: A human poison by ingestion. An experimental poison by intraperitoneal route. Toxic symptoms when accompanied by ingestion of alcohol. Human systemic effects by ingestion: jaundice, joint changes. An experimental teratogen. Other experimental reproductive effects. Questionable carcinogen with experimental neoplastigenic data. See also THIRAM.

DXH300 CAS:149-45-1 HR: 1
3,5-DISULFOCATECHOL DISODIUM SALT
mf: $C_6H_6O_8S_2$•2Na mw: 316.22

PROP: Crystals, nonhygroscopic. Produces water-sol, colored compounds with metal salts. Very freely sol in water; sltly sol in alc.

SYNS: 4,5-DIHYDROXY-1,3-BENZENEDISULFONIC ACID DISODIUM SALT □ SDD □ TIFERRON □ TIRON

TOXICITY DATA WITH REFERENCE
mmo-sat 100 μg/plate ABCHA6 45,327,81
mma-sat 100 μg/plate ABCHA6 45,327,81
ipr-mus TDLo:30 g/kg (female 6-15D post):REP RCOCB8 73,97,91
ipr-mus LD50:6103 mg/kg TOLED5 26,95,85

CONSENSUS REPORTS: Reported in EPA TSCA Inventory.

SAFETY PROFILE: Mildly toxic by intraperitoneal route. Experimental reproductive effects. Mutation data reported. Colorimetric reagent for iron, manganese, titanium, molybdenum. When heated to decomposition it emits toxic fumes of SO_x and Na_2O. See also SULFONATES.

DXH325 HR: 3
DISULFOTON
DOT: NA 2783
mf: $C_8H_{19}O_2PS_3$ mw: 274.42

SYNS: BAYER 19639 □ O,O-DIAETHYL-S-(2-AETHYLTHIO-AETHYL)-DITHIOPHOSPHAT (GERMAN) □ O,O-DIAETHYL-S-(3-THIA-PENTYL)-DITHIOPHOSPHAT (GERMAN) □ O,O-DIETHYL-S-(2-ETHYLTHIOETHYL) PHOSPHORODITHIOATE □ O,O-DIETHYL-S-(2-ETHYLTHIOETHYL) THIOTHIONOPHOSPHATE □ O,O-DIETHYL-S-(2-ETHYLMERCAPTOETHYL) DITHIOPHOSPHATE □ O,O-DIETHYL-S-(2-ETHYLTHIO-ETHYL)-DITHIOFOSFAAT (DUTCH) □ O,O-DIETHYL-2-ETHYLTHIOETHYL PHOSPHORODITHIOATE □ O,O-DIETHYL-S-2-(ETHYLTHIO)ETHYL PHOSPHORODITHIOATE □ O,O-DIETIL-S-(2-ETILTIO-ETIL)-DITIOFOSFATO (ITALIAN) □ DIMAZ □ DISULFATON □ DI-SYSTON □ DISYSTOX □ DITHIODEMETON □ DITHIOPHOSPHATE de O,O-DIETHYLE et de S-(2-ETHYLTHIO-ETHYLE) (FRENCH) □ DITHIOSYSTOX □ ENT 23,437 □ O,O-ETHYL-S-2(ETHYLTHIO) ETHYL PHOSPHORODITHIOATE □ S-2-(ETHYLTHIO)ETHYL O,O-DIETHYL ESTER of PHOSPHORODITHIOIC ACID □ FRUMIN AL □ M-74 □ RCRA WASTE NUMBER P039 □ S 276 □ SOLVIREX □ THIODEMETON □ THIODEMETRON

TOXICITY DATA WITH REFERENCE
mma-sat 5 mg/plate MUREAV 116,185,83
mmo-esc 5 μL/plate MUREAV 28,405,75
orl-rat LD50:2 mg/kg FMCHA2 -,C85,83
ihl-rat LC50:200 mg/m³ 85GYAZ -,25,71
skn-rat LD50:6 mg/kg TXAPA9 14,515,69
ipr-rat LD50:2 mg/kg AMIHAB 17,192,58
ivn-rat LD50:5500 mg/kg 13ZGAF -,206,62
orl-mus LD50:5500 μg/kg 85DPAN -,-,71/76
ipr-mus LD50:5500 μg/kg AMIHAB 17,192,58
orl-gpg LD50:10,800 μg/kg AMIHAB 17,192,58
ipr-gpg LD50:7 mg/kg AMIHAB 17,192,58
orl-qal LD50:12 mg/kg EESADV 8,551,84
orl-dck LD50:6500 μg/kg DOEAAH 35,25,79

CONSENSUS REPORTS: EPA Extremely Hazardous Substances List. EPA Genetic Toxicology Program.

OSHA PEL: TWA 0.1 mg/m³ (skin)
ACGIH TLV: TWA 0.1 mg/m³

SAFETY PROFILE: Poison by ingestion, inhalation, skin contact, intraperitoneal, and intravenous routes. Human mutation data reported. When heated to decomposition it emits very toxic SO_x and PO_x. See also various demeton entries and ESTERS.

DXH350 CAS:13172-31-1 HR: 3
DISULFUR DIBROMIDE
mf: Br_2S_2 mw: 223.93

PROP: Red solid or dark red liquid; dissociates on heating. Hydrol to HBr + SO_2 + S. Mp: −46°. Sol in CCl_4, C_6H_6, and CS_2.

SAFETY PROFILE: Violent reaction with potassium, sodium, aluminum, and antimony. Violent reaction with iron at 650°C. The moist bromide reacts violently with oxidants. When heated to decomposition it emits toxic fumes of Br^- and SO_x. See also BROMIDES.

DXH400 CAS:25474-92-4 HR: 3
DISULFUR DINITRIDE
mf: N_2S_2 mw: 92.14

SN=S=N (ring)

PROP: Colorless crystals. Polymerizes on standing. Subl *in vacuo.*

SAFETY PROFILE: An explosive sensitive to shock, friction, pressure, or heating above 30°C. Potentially explosive reaction with complexes with beryllium chloride or titanium tetrachloride. When heated to decomposition it emits toxic fumes of SO_x and NO_x. See also NITRIDES.

DXH600 HR: 1
DISULFUR HEPTAOXIDE
mf: O_7S_2 mw: 208.19

SYN: PEROXYDISULFIRIC ANHYDRIDE

SAFETY PROFILE: Explodes when exposed to moist air. When heated to decomposition it emits toxic fumes of SO_x. See also PEROXIDES.

DXH800 HR: 3
DISULFURYL DIAZIDE
mf: $N_6O_5S_2$ mw: 228.17

$N_3SO_2OSO_2N_3$

SAFETY PROFILE: An unstable explosive. Incompatible

with dilute alkali. When heated to decomposition it emits toxic fumes of SO_x and NO_x. See also AZIDES.

DXI000 **HR: 3**
1,2-DI(5-TETRAZOLYL)HYDRAZINE
mf: $C_2H_4N_{10}$ mw: 168.12

(HNN=NN=CNH—)$_2$

SAFETY PROFILE: Explodes on heating. When heated to decomposition it emits toxic fumes of NO_x. See also HYDRAZINE.

DXI200 **CAS:56929-36-3** **HR: 3**
1,3-DI(5-TETRAZOYL)TRIAZENE
mf: $C_2H_3N_{11}$ mw: 181.12

HNN=NN=CN=NNHC=NN=N NH

SAFETY PROFILE: The metal salts are heat- and friction-sensitive explosives (e.g., barium, copper, silver). The barium salt is a weak explosive. When heated to decomposition it emits toxic fumes of NO_x.

DXI400 **CAS:8018-01-7** **HR: 2**
DITHANE M-45
mf: $C_4H_6MnN_2S_4 \cdot C_4H_6N_2S_4Zn$ mw: 541.03

PROP: Contains 16% manganese, 2% zinc, and 62% ethylenebisdithiocarbamate ion/manganese ethylenebisdithiocarbamate plus zinc ion (85ARAE 4,52,76).

SYNS: CARMAZINE ☐ DITHANE S60 ☐ DITHANE SPC ☐ DITHANE ULTRA ☐ ETHYLENEBIS(DITHIOCARBAMIC ACID MANGANESE ZINC COMPLEX (8CI) ☐ F 2966 ☐ FORE ☐ GREEN-DAISEN M ☐ KARAMATE ☐ MANCOFOL ☐ MANCOZEB ☐ MANEB-ZINC ☐ MANEB-ZINEB-KOMPLEX (GERMAN) ☐ MANGAN-ZINK-AETHYLENDIAMIN-BIS-DITHIO-CARBAMAT (GERMAN) ☐ MANOSEB ☐ MANZATE 200 ☐ MANZEB ☐ MANZIN 80 ☐ MARZIN ☐ NEMISPOR ☐ POLICAR MZ ☐ POLICAR S ☐ TRIZIMAN ☐ TRIZIMAN D ☐ ZIMANAT ☐ZIMANEB ☐ ZIMMAN-DITHANE

TOXICITY DATA WITH REFERENCE
cyt-hmn:lym 4 mg/L MUREAV 116,341,83
cyt-rat-ipr 2500 μg/kg MUREAV 116,341,83
ihl-rat TCLo:500 mg/m³/6H (female 6-10D post):REP TXAPA9 84,355,86
orl-rat TDLo:1320 mg/kg (11D preg):TER TJADAB 14,171,76
orl-rat TDLo:1320 mg/kg (female 11D post):TER TJADAB 14,171,76
skn-mus TDLo:17,700 mg/kg/60W-I:NEO CALEDQ 53,191,90
orl-rat LD50:5 g/kg 85JFAN A251,86

CONSENSUS REPORTS: EPA Genetic Toxicology Program. Zinc and its compounds, as well as manganese and its compounds, are on the Community Right-To-Know List.

OSHA PEL: CL 5 mg(Mn)/m³
ACGIH TLV: TWA 5 mg(Mn)/m³

SAFETY PROFILE: An experimental teratogen. Other experimental reproductive effects. Questionable carcinogen with experimental tumorigenic data. Mutation data reported. When heated to decomposition it emits very toxic fumes of SO_x, ZnO, and NO_x. See also CARBAMATES, MANGANESE COMPOUNDS, and ZINC COMPOUNDS.

DXI500 **CAS:503-41-3** **HR: 3**
2,4-DITHIA-1,3-DIOXANE-2,2,4,4-TETRAOXIDE
mf: $C_2H_4O_6S_2$ mw: 188.17

PROP: Deliquescent crystals. Mp: 80°.

SYN: CARBYL SULFATE

SAFETY PROFILE: Potentially violent reaction with N-methyl-4-nitroaniline. When heated to decomposition it emits toxic fumes of SO_x.

DXI550 **CAS:505-29-3** **HR: 2**
p-DITHIANE
mf: $C_4H_8S_2$ mw: 120.24

SYNS: 1,4-DITHIACYCLOHEXANE ☐ 1,4-DITHIANE

TOXICITY DATA WITH REFERENCE
orl-rat LD50:2768 mg/kg NTIS** AD-A172-647

CONSENSUS REPORTS: Reported in EPA TSCA Inventory.

SAFETY PROFILE: Moderately toxic by ingestion. When heated to decomposition it emits toxic vapors of SO_x.

DXI600 **CAS:7187-55-5** **HR: 3**
DITHIAZANINE
mf: $C_{23}H_{23}N_2S_2$ mw: 391.60

TOXICITY DATA WITH REFERENCE
orl-mus LD50:2550 μg/kg RPOBAR 2,288,70
ipr-mus LD50:780 μg/kg RPOBAR 2,288,70
ivn-mus LD50:1800 μg/kg CSLNX* NX#01838

SAFETY PROFILE: Poison by ingestion, intravenous, and intraperitoneal routes. When heated to decomposition it emits very toxic fumes of NO_x and SO_x.

DXI800 **CAS:66788-41-8** **HR: 3**
(R,S)-α-(1-((3,3-DI-3-THIENYLALLYL)AMINO)ETHYL)) BENZYL ALCOHOL (+)-(α)-HYDROCHLORIDE
mf: $C_{20}H_{21}NOS_2 \cdot ClH$ mw: 392.00

SYNS: 1-3,3-BIS(3'-THIENYL)-2-PROPENYL-(3-HYDROXY-3-PHENYL-PROPYL-2)AMINE ☐ (+)-α-(1-((3,3-DI-3-THIENYLALLYL)AMINO)ETHYL)BENZYL ALCOHOL HYDROCHLORIDE ☐ d-8955 HYDROCHLORIDE ☐ NOVOCEBRIN HYDROCHLORIDE ☐ TINOFEDRINE HYDROCHLORIDE

TOXICITY DATA WITH REFERENCE
orl-rat LD50:6600 mg/kg DRFUD4 4,286,79
orl-mus LD50:1890 mg/kg DRFUD4 4,286,79
ipr-mus LD50:14 mg/kg DRFUD4 4,286,79
ivn-mus LD50:20,150 μg/kg DRFUD4 4,286,79
ivn-dog LD50:20 mg/kg DRFUD4 4,286,79

SAFETY PROFILE: Poison by intraperitoneal and intravenous routes. Moderately toxic by ingestion. A cerebral vasodilator. When heated to decomposition it emits very toxic fumes of SO_x and NO_x.

DXJ100 **HR: 3**
1-((3,3-DI-2-THIENYL-1-METHYL)ALLYL)PYRROLIDINE HYDROCHLORIDE
mf: $C_{16}H_{19}NS_2{\cdot}ClH$ mw: 325.94

TOXICITY DATA WITH REFERENCE
scu-rat LD50:92 mg/kg BJPCAL 8,2,53
scu-mus LD50:120 mg/kg BJPCAL 8,2,53
ivn-mus LD50:25 mg/kg BJPCAL 8,2,53

SAFETY PROFILE: Poison by subcutaneous and intravenous routes. When heated to decomposition it emits toxic fumes of NO_x, SO_x, and HCl.

DXJ400 **CAS:64059-02-5** **HR: 3**
2,2′-DITHIOBIS(N-(1-ADAMANTYLMETHYL)ACETAMIDINE) DIHYDROCHLORIDE HEMIHYDRATE
mf: $C_{26}H_{42}N_4S_2{\cdot}2ClH{\cdot}1/2H_2O$ mw: 556.77

TOXICITY DATA WITH REFERENCE
orl-mus LD50:325 mg/kg JMCMAR 15,1313,72
ipr-mus LD50:100 mg/kg JMCMAR 15,1313,72

SAFETY PROFILE: Poison by ingestion and intraperitoneal routes. When heated to decomposition it emits very toxic fumes of NO_x, SO_x, and HCl.

DXJ800 **CAS:1141-88-4** **HR: 3**
2,2′-DITHIOBISANILINE
mf: $C_{12}H_{12}N_2S_2$ mw: 248.38

PROP: Leaflets or needles from EtOH (aq). Mp: 93°. Sol in acids or EtOH; sltly sol in H_2O.

SYNS: BIS(o-AMINOPHENYL)DISULFIDE □ BIS(2-AMINOPHENYL) DISULFIDE □ 1,1′-BIS(2-AMINOPHENYL)DISULFIDE □ O,O′-DIAMINO DIPHENYL DISULFIDE □ O,O-DITHIO-BIS-ANILINE □ 2,2′-DITHIODIANILINE □ USAF AB-315

TOXICITY DATA WITH REFERENCE
eye-rbt 50 µg/24H SEV 28ZPAK -,172,72
ipr-mus LD50:50 mg/kg NTIS** AD277-689
ivn-mus LD50:178 mg/kg CSLNX* NX#00263

CONSENSUS REPORTS: Reported in EPA TSCA Inventory.

SAFETY PROFILE: Poison by intravenous and intraperitoneal routes. Moderately toxic by ingestion. A severe eye irritant. When heated to decomposition it emits very toxic fumes of NO_x and SO_x.

DXL800 **CAS:541-53-7** **HR: 3**
DITHIOBIURET
mf: $C_2H_5N_3S_2$ mw: 135.22

PROP: Crystals or needles from water. Mp: 181°, bp: decomp, d: 1.522 @ 30°. Sol in H_2O, EtOH, and Me_2CO.

SYNS: DTB □ RCRA WASTE NUMBER P049 □ 2-THIO-1-(THIOCARBAMOYL)UREA □ USAF B-44 □ USAF EK-P-6281

TOXICITY DATA WITH REFERENCE
orl-rat LD50:5 mg/kg JPETAB 90,260,47
ipr-rat LD50:29 mg/kg TXAPA9 57,63,81
ipr-mus LDLo:50 mg/kg NTIS** AD277-689

CONSENSUS REPORTS: Reported in EPA TSCA Inventory. EPA Extremely Hazardous Substances List.

SAFETY PROFILE: Poison by ingestion and intraperitoneal routes. When heated to decomposition it emits highly toxic fumes of SO_x and NO_x.

DXM000 **CAS:36551-21-0** **HR: 3**
DITHIOCARBONIC ACID-o-sec-BUTYL ESTER SODIUM SALT
mf: $C_5H_9OS_2{\cdot}Na$ mw: 172.25

SYN: sec-BUTYLXANTHIC ACID SODIUM SALT

TOXICITY DATA WITH REFERENCE
ipr-mus LDLo:500 mg/kg CBCCT* 6,230,54
par-mus LDLo:400 mg/kg CBCCT* 7,696,55

CONSENSUS REPORTS: Reported in EPA TSCA Inventory.

SAFETY PROFILE: Poison by parenteral route. Moderately toxic by intraperitoneal route. When heated to decomposition it emits toxic fumes of SO_x and Na_2O. See also ESTERS.

DXM100 **HR: 3**
DITHIOCARBOXYMETHYL-p-CARBAMIDOPHENYLARSENOUS OXIDE
mf: $C_9H_{11}AsN_2O_4S_2$ mw: 350.26

SYNS: ((p-ARSONOPHENYL)CARBAMOYL)DITHIOCARBAMIC ACID □ N-((p-ARSONPHENYL)CARBAMOYL)DITHIOGLYCINE

TOXICITY DATA WITH REFERENCE
orl-rat LD50:1 g/kg FEPRA7 6,306,47
ipr-rat LD50:75 mg/kg FEPRA7 6,306,47
ipr-mus LD50:100 mg/kg FEPRA7 6,306,47

CONSENSUS REPORTS: Arsenic and its compounds are on the Community Right-To-Know List.

SAFETY PROFILE: Poison by intraperitoneal route. Moderately toxic by ingestion. When heated to decomposition it emits toxic fumes of SO_x, NO_x, and As. See also ARSENIC COMPOUNDS.

DXM600 **CAS:1892-29-1** **HR: 3**
DITHIODIGLYCOL
mf: $C_4H_{10}O_2S_2$ mw: 154.26

PROP: Syrupy liquid. Mp: 17°, bp: 160–162° @ 0.1 mm.

SYNS: 2,2-DITHIODIETHANOL □ USAF TH-9

TOXICITY DATA WITH REFERENCE
ipr-mus LD50:100 mg/kg NTIS** AD277-689

CONSENSUS REPORTS: Reported in EPA TSCA Inventory.

SAFETY PROFILE: Poison by intraperitoneal route. When heated to decomposition it emits toxic fumes of SO_x.

DXN350 CAS:6892-68-8 HR: 3
1,4-DITHIOERYTHRITOL
mf: $C_4H_{10}O_2S_2$ mw: 154.26

SYNS: 2,3-BUTANEDIOL, 1,4-DIMERCAPTO-, (R*,S*)-(9CI) □ (R*,S*)-1,4-DIMERCAPTO-2,3-BUTANEDIOL □ DITHIOERYTHRITOL □ DTE □ ERYTHRITOL, 1,4-DITHIO-

TOXICITY DATA WITH REFERENCE
cyt-hmn:lyms 100 μmol/L HEREAY 91,105,79
ims-mus LD50:309 mg/kg JPPMAB 1,576,49

CONSENSUS REPORTS: Reported in EPA TSCA Inventory.

SAFETY PROFILE: A poison by intramuscular route. Mutation data reported. When heated to decomposition it emits acrid smoke and irritating vapors.

DXN400 CAS:14807-75-1 HR: 3
1,1′-DITHIOFORMAMIDINE DIHYDROCHLORIDE
mf: $C_2H_6N_4S_2$•2ClH mw: 223.16

SYNS: GUANYL DISULFIDE DIHYDROCHLORIDE □ USAF A-11074

TOXICITY DATA WITH REFERENCE
ipr-mus LD50:100 mg/kg NTIS** AD277-689

CONSENSUS REPORTS: Reported in EPA TSCA Inventory.

SAFETY PROFILE: Poison by intraperitoneal route. When heated to decomposition it emits very toxic fumes of NO_x and Cl^-.

DXN600 CAS:333-29-9 HR: 3
DITHIOLANE IMINOPHOSPHATE
mf: $C_7H_{14}NO_2PS_3$ mw: 271.37

SYNS: AC-43064 □ AMERICAN CYANAMID AC 43,064 □ CL-43,064 □ CYALANE □ CYCLIC ETHYLENE (DIETHOXYPHOSPHINOTHIOYL) DITHIOIMIDOCARBONATE □ CYCLIC ETHYLENE ESTER of (DIETHOXYPHOSPHINOTHIOYL)DITHIOIMIDOCARBONIC ACID □ CYLAN □ CYOLAN □ CYOLANE INSECTICIDE □ 2-(DIETHOXYPHOSPHINYLIMINO)-1,3-DITHIOLANE □ O,O-DIETHYL 1,3-DITHIOLAN-2-YLIDENEPHOSPHORAMIDOTHIOATE □ DIETHYL-N-1,3-DITHIOLANYL-2-IMINO PHOSPHATE □ DITHIOLANE □ 1,3-DITHIOLAN-2-YLIDENE-PHOSPHORAMIDOTHIOIC ACID DIETHYL ESTER □ 1,3-DITHIOLAN-2-YLIDENE-PHOSPHORAMIDOTHIOIC ACID-O,O-DIETHYL ESTER □ ENT 25,809 □ IMINOPHOSPHATE □ PHOSFOLAN

TOXICITY DATA WITH REFERENCE
orl-rat LD50:14 mg/kg IPCLBZ 22,40,80
orl-mus LD50:62 mg/kg AECTCV 14,111,85
skn-rbt LD50:23 mg/kg GUCHAZ 6,200,73
orl-ckn LD50:5200 μg/kg EXPEAM 30,63,74
orl-bwd LD50:1800 μg/kg TXAPA9 26,154,73
skn-bwd LD50:10 mg/kg TXAPA9 26,154,73

SAFETY PROFILE: Poison by ingestion and skin contact. An insecticide. When heated to decomposition it emits very toxic fumes of NO_x and SO_x.

DXN709 CAS:940-69-2 HR: 3
1,2-DITHIOLANE-3-VALERAMIDE
mf: $C_8H_{15}NOS_2$ mw: 205.36

SYNS: 1,2-DITHIOLANE-3-PENTANAMIDE (9CI) □ LIPAMIDE □ α-LIPAMIDE □ LIPOACIN □ LIPOAMID □ LIPOAMIDE □ α-LIPOAMIDE □ LIPOCTON □ α-LIPOIC ACID AMIDE □ LIPOICIN □ LIPOZYME □ LYPOARAN □ PATHOCLON □ THIOAMI □ THIOCTAMID □ THIOCTAMIDE □ THIOCTIC ACID AMIDE □ THIOTOMIN □ TICOLIN □ TIOCTAN

TOXICITY DATA WITH REFERENCE
orl-rat LD50:1980 mg/kg NIIRDN 6,454,82
ipr-rat LD50:250 mg/kg NIIRDN 6,454,82
scu-rat LD50:1700 mg/kg NIIRDN 6,454,82
orl-mus LD50:620 mg/kg NIIRDN 6,454,82
ipr-mus LD50:310 mg/kg NIIRDN 6,454,82
scu-mus LD50:3500 mg/kg NIIRDN 6,454,82

SAFETY PROFILE: Poison by intraperitoneal route. Moderately toxic by ingestion and subcutaneous routes. When heated to decomposition it emits toxic fumes of SO_x and NO_x.

DXN800 CAS:62-46-4 HR: 3
1,2-DITHIOLANE-3-VALERIC ACID
mf: $C_8H_{14}O_2S_2$ mw: 206.34

SYNS: ACETATE-REPLACING FACTOR □ BILETAN □ 5-(1,2-DITHIOLAN-3-YL)VALERIC ACID □ 6,8-DITHIOOCTANOIC ACID □ HEPARLIPON □ LIPOIC ACID □ α-LIPOIC ACID □ α-LIPONIC ACID □ α-LIPONSAEURE (GERMAN) □ LIPOSAN □ LIPOTHION □ THIOCTACID □ THIOCTIC ACID □ 6-THIOCTIC ACID □ 6,8-THIOCTIC ACID □ THIOCTIDASE □ THIOCTSAN □ THIOKTSAFURE (GERMAN) □ THIOOCTANOIC ACID □ 6-THIOTIC ACID □ 6,8-THIOTIC ACID □ TIOCTACID □ TIOCTIDASI □ TIOCTIDASI ACETATE REPLACING FACTOR

TOXICITY DATA WITH REFERENCE
scu-rat TDLo:480 mg/kg (30D male):REP NYKZAU 68,265,72
orl-rat LD50:1130 mg/kg NYKZAU 68,265,72
ipr-rat LD50:200 mg/kg NYKZAU 68,265,72
scu-rat LD50:230 mg/kg NYKZAU 68,265,72
ivn-rat LD50:180 mg/kg NYKZAU 68,265,72
orl-mus LD50:502 mg/kg ARTODN 41,79,78
ipr-mus LD50:160 mg/kg NIIRDN 6,454,82
scu-mus LD50:200 mg/kg NIIRDN 6,454,82
ivn-mus LD50:197 mg/kg NYKZAU 60,278,64

SAFETY PROFILE: Poison by intravenous, subcutaneous, and intraperitoneal routes. Moderately toxic by ingestion. Experimental reproductive effects. When heated to decomposition it emits toxic fumes of SO_x.

DXN850 CAS:3706-77-2 HR: 3
1,3-DITHIOLIUM PERCHLORATE
mf: $C_3H_3ClO_4S_2$ mw: 250.67

SAFETY PROFILE: A friction- and heat-sensitive explosive. Explodes at 250°C. When heated to decomposition it emits toxic fumes of Cl^- and SO_x. See also PERCHLORATES.

DXO000 CAS:572-48-5 HR: 3
DITHION
mf: $C_{17}H_{21}O_5PS$ mw: 368.41

PROP: Crystals nearly insol in water. Mp: 88°.

SYNS: O,O-DIETHYL-7-HYDROXY-3,4-TETRAMETHYLENE COUMAR-

INYL PHOSPHOROTHIOATE □ O,O-DIETHYL-O-(7,8,9,10-TETRAHYDRO-6-OXOBENZO(C)CHROMAN-3-YL)PHOSPHOROTHIOATE □ O,O-DIETHYL-O-(7,8,9,10-TETRAHYDRO-6-OXO-6H-DIBENZO(b,d)PYRAN-3-YL)PHOSPHOROTHIOATE □ O,O-DIETHYL-O-(3,4-TETRAMETHYLENECOUMARINYL-7) THIOPHOSPHATE □ DITHIONE □ ENT 24,986 □ 7-HYDROXY-3,4-TETRAMETHYLENECOUMARIN-O,O-DIETHYL THIOPHOSPHATE

TOXICITY DATA WITH REFERENCE
orl-rat LD50:67 mg/kg PHJOAV 185,361,60
orl-mus LD50:3800 mg/kg 28ZEAL 5,56,76
orl-dog LD50:400 mg/kg 28ZEAL 5,56,76
orl-rbt LD50:500 mg/kg 28ZEAL 5,56,76
orl-gpg LD50:200 mg/kg 28ZEAL 5,56,76

SAFETY PROFILE: Poison by ingestion. When heated to decomposition it emits very toxic fumes of PO_x and SO_x. See also PARATHION.

DXO200 CAS:79-40-3 HR: 3
DITHIOOXAMIDE
mf: $C_2H_4N_2S_2$ mw: 120.20

PROP: Red crystals.

SYNS: DITHIOOXALDIIMIDIC ACID □ DITHIOXAMIDE □ ETHANEDITHIOAMIDE □ HYDRORUBEANIC ACID □ RUBEANE □ RUBEANIC ACID □ RVK □ USAF EK-4394 □ USAF MK-6

TOXICITY DATA WITH REFERENCE
orl-rat LDLo:500 mg/kg NCNSA6 5,43,53
orl-mus LD50:350 mg/kg FATOAO 28,230,65
ipr-mus LD50:100 mg/kg NTIS** AD277-689
ivn-mus LD50:56 mg/kg CSLNX* NX#04475

CONSENSUS REPORTS: Reported in EPA TSCA Inventory.

SAFETY PROFILE: Poison by ingestion, intraperitoneal, and intravenous routes. When heated to decomposition it emits very toxic fumes of NO_x and SO_x.

DXO300 CAS:59-58-5 HR: 3
DITHIOPROPYLTHIAMINE
mf: $C_{15}H_{24}N_4O_2S_2$ mw: 356.55

PROP: Prisms from benzene. Mp: 128–129° (decomp). Sparingly sol in water. Soluble in organic solvents and lipids.

SYNS: ALINAMIN □ ALITON □ ANEURIMEC □ ARINAMINE □ AUSOVIT □ BETATRON □ BINOVA □ DITIOVIT □ LIPONEURINA □ MARINEURINA □ NERVILON □ NEURVITA □ NEVRITON □ NUVELBI V.C.A. □ OROBETINA □ PRONEURIN □ PROSULTHIAMINE □ PROSULTIAMINE □ SINTOTIAMINA □ THIAMINE PROPYL DISULFIDE □ THIAMIN PROPYL DISULFIDE □ TIOTIAMINE □ TIOVIT-B_1 □ TIPIDI □ TPD □ VITAMIN B_1 PROPYL DISULFIDE

TOXICITY DATA WITH REFERENCE
orl-mus LD50:2750 mg/kg NIIRDN 6,724,82
ipr-mus LD50:650 mg/kg NIIRDN 6,724,82
ivn-mus LD50:320 mg/kg NIIRDN 6,724,82

SAFETY PROFILE: Poison by intravenous route. Moderately toxic by ingestion and intraperitoneal routes. When heated to decomposition it emits toxic fumes of SO_x and NO_x.

DXO400 CAS:973-99-9 HR: 3
DITHIOPROPYLTHIAMINE HYDROCHLORIDE
mf: $C_{15}H_{24}N_4O_2S_2$•ClH mw: 393.01

PROP: A solid. Mp: 160–161° (decomp).

SYNS: DTP HYDROCHLORIDE □ THIAMINE PROPYL DISULFIDE HYDROCHLORIDE

TOXICITY DATA WITH REFERENCE
orl-mus LD50:2512 mg/kg IZVIAK 37,82,67
ivn-mus LD50:302 mg/kg IZVIAK 37,82,67

SAFETY PROFILE: Poison by intravenous route. Moderately toxic by ingestion. When heated to decomposition it emits very toxic fumes of HCl, SO_x, and NO_x.

DXO600 CAS:1076-98-8 HR: 3
DITHIOTEREPHTHALIC ACID
mf: $C_8H_6O_2S_2$ mw: 198.26

TOXICITY DATA WITH REFERENCE
ivn-mus LD50:320 mg/kg CSLNX* NX#01918

CONSENSUS REPORTS: Reported in EPA TSCA Inventory.

SAFETY PROFILE: Poison by intravenous route. When heated to decomposition it emits toxic fumes of SO_x.

DXO775 CAS:27565-41-9 HR: 3
DITHIOTHREITOL
mf: $C_4H_{10}O_2S_2$ mw: 154.26

SYNS: CLELAND'S REAGENT □ (R*,R*)-(±)-1,4-DIMERCAPTO-2,3-BUTANEDIOL (9CI) □ dl-threo-DIMERCAPTO-2,3-BUTANEDIOL □ dl-DITHIOTHREITOL □ dl-1,4-DITHIOTHREITOL □ rac-DITHIOTHREITOL □ DTT

TOXICITY DATA WITH REFERENCE
mmo-sat 100 µg/plate ABCHA6 45,327,81
mma-sat 100 µg/plate ABCHA6 45,327,81
dnd-omi 1 mmol/L BBRCA9 77,1150,77
mnt-nml:emb 2 mmol/L JCLBA3 60,497,74
cyt-nml:emb 2 mmol/L JCLBA3 60,497,74
dni-rat:lvr 1 mmol/L ABBIA4 166,400,75
ipr-mus LD50:169 mg/kg JMCMAR 15,600,72
scu-mus LD50:333 mg/kg ARZNAD 22,1434,72
ivn-mus LD50:94 mg/kg ARZNAD 22,1434,72

SAFETY PROFILE: Poison by subcutaneous, intravenous, and intraperitoneal routes. Mutation data reported. When heated to decomposition it emits toxic fumes of SO_x.

DXO800 CAS:3483-12-3 HR: 3
d-1,4-DITHIOTHREITOL
mf: $C_4H_{10}O_2S_2$ mw: 154.26

PROP: Bp: 115–116° @ 1 mm.

SYNS: CLELAND'S REAGENT □ D-DTT □ d-threo-1,4-DIMERCAPTO-2,3-BUTANEDIOL □ 1,4-DITHIOTHREITOL □ DTT □ SPUTOLYSIN

TOXICITY DATA WITH REFERENCE
ipr-mus LD50:154 mg/kg YKKZAJ 94,1419,74
ims-mus LD50:108 mg/kg JPPMAB 1,576,49

CONSENSUS REPORTS: Reported in EPA TSCA Inventory.

SAFETY PROFILE: Poison by intraperitoneal and intramuscular routes. When heated to decomposition it emits toxic fumes of SO_x.

DXP000 CAS:27755-15-3 HR: 2
DITOLYLETHANE
mf: $C_{16}H_{18}$ mw: 210.34

TOXICITY DATA WITH REFERENCE
orl-mus TDLo:9200 mg/kg/73W-C:ETA GTPZAB 15(5),49,71
orl-mus LD50:725 mg/kg GTPZAB 15(5),49,71
scu-mus LD50:10 g/kg GTPZAB 15(5),49,71

SAFETY PROFILE: Moderately toxic by ingestion. Questionable carcinogen with experimental tumorigenic data. When heated to decomposition it emits acrid smoke and irritating fumes.

DXP200 CAS:97-39-2 HR: 3
DI-o-TOLYLGUANIDINE
mf: $C_{15}H_{17}N_3$ mw: 239.35

PROP: White crystals from alc (aq). Mp: 179°, d: 1.10 @ 20°/4°, vap d: 8.24.

SYNS: DIORTHOTOLYLGUANIDINE □ 1,3-DI-o-TOLYLGUANIDINE □ DOTG ACCELERATOR □ USAF A-6598 □ VULKACIT DOTG/C

TOXICITY DATA WITH REFERENCE
orl-rat LD50:500 mg/kg JPETAB 90,260,47
ipr-mus LD50:25 mg/kg NTIS** AD277-689
orl-rbt LDLo:120 mg/kg JIDHAN 13,87,31
orl-gpg LDLo:120 mg/kg JIDHAN 13,87,31
orl-mam LDLo:120 mg/kg JIDHAN 13,87,31

CONSENSUS REPORTS: Reported in EPA TSCA Inventory.

SAFETY PROFILE: Poison by ingestion and intraperitoneal routes. When heated to decomposition it emits toxic fumes of NO_x.

DXP400 CAS:15017-02-4 HR: 3
N,N'-DI-o-TOLYL-p-PHENYLENE DIAMINE
mf: $C_{20}H_{20}N_2$ mw: 288.42

SYN: USAF GY-1

TOXICITY DATA WITH REFERENCE
ipr-mus LD50:300 mg/kg NTIS** AD277-689

CONSENSUS REPORTS: Reported in EPA TSCA Inventory.

SAFETY PROFILE: Poison by intraperitoneal route. When heated to decomposition it emits toxic fumes of NO_x.

DXP600 CAS:137-97-3 HR: 3
DI-o-TOLYLTHIOUREA
mf: $C_{15}H_{16}N_2S$ mw: 256.39

PROP: Crystals or needles from alc. Mp: 165–166°, vap d: 8.85. Sol in dichloroethane.

SYNS: N,N'-BIS(2-METHYLPHENYLTHIOUREA □ 1,3-BIS(o-TOLYL)-2-THIOUREA □ 2,2'-DIMETHYLTHIOCARBANILIDE □ DI-o-TOLUYLTHIOUREA □ USAF EK-1651

D

TOXICITY DATA WITH REFERENCE
ipr-mus LD50:100 mg/kg NTIS** AD277-689
orl-rbt LDLo:3000 mg/kg JPETAB 17,349,21

CONSENSUS REPORTS: Reported in EPA TSCA Inventory.

SAFETY PROFILE: Poison by intraperitoneal route. Moderately toxic by ingestion. When heated to decomposition it emits very toxic fumes of NO_x and SO_x.

DXP800 CAS:8015-54-1 HR: 3
DITRAN
mf: $C_{20}H_{29}NO_3 \cdot C_{20}H_{29}NO_3 \cdot 2ClH$ mw: 735.92

SYN: JB 329

TOXICITY DATA WITH REFERENCE
orl-hmn TDLo:40 µg/kg:EYE,CNS,GIT JNMDAN 131,428,60
ims-man TDLo:150 µg/kg:CNS FEPRA7 32,250,73
ipr-rat LD50:25 mg/kg 27ZQAG -,224,72
ivn-rat LD50:19 mg/kg 27ZQAG -,224,72
ims-rat LD50:1193 µg/kg BJPCBM 39,822,70
orl-mus LD50:300 mg/kg ARZNAD 21,1727,71
ipr-mus LD50:60 mg/kg 27ZQAG -,224,72
ivn-mus LD50:10 mg/kg 27ZQAG -,224,72
ims-mus LD50:1634 µg/kg BJPCBM 39,822,70
ivn-rbt LD50:12 mg/kg 27ZQAG -,224,72
ivn-gpg LD50:45 mg/kg 27ZQAG -,224,72
ims-gpg LD50:327 µg/kg BJPCBM 39,822,70

SAFETY PROFILE: Poison by intravenous, intraperitoneal, and intramuscular routes. Human systemic effects by ingestion and intramuscular routes: visual field changes, hallucinations, distorted perceptions, nausea and vomiting. When heated to decomposition it emits toxic fumes of NO_x and Cl^-.

DXQ000 CAS:5910-75-8 HR: 2
DITRIDECYLAMINE
mf: $C_{26}H_{55}N$ mw: 381.82

PROP: Crystals or powder. Sol in EtOH and Et_2O.

TOXICITY DATA WITH REFERENCE
skn-rbt 100 µg/24H open AIHAAP 23,95,62
orl-rat LD50:9850 mg/kg AIHAAP 23,95,62
skn-rbt LD50:3540 mg/kg AIHAAP 23,95,62

CONSENSUS REPORTS: Reported in EPA TSCA Inventory.

SAFETY PROFILE: Moderately toxic by skin contact. Mildly toxic by ingestion. A skin irritant. When heated to decomposition it emits toxic fumes of NO_x. See also AMINES.

DXQ200 **CAS:119-06-2** **HR: 2**
DITRIDECYL PHTHALATE
mf: $C_{34}H_{58}O_4$ mw: 530.92

PROP: D: 0.951 @ 20°/20°, bp: >285° @ 5 mm, flash p: 470°F (OC).

SYNS: 1,2-BENZENEDICARBOXYLIC ACID, DITRIDECYL ESTER □ DTDP □ JAYFLEX DTDP □ NUOPLAZ □ PHTHALIC ACID, DITRIDECYL ESTER □ POLYCIZER 962-BPA □ STALFEX DTDP □ 1-TRIDECANOL PHTHALATE □ TRUFLEX DTDP

TOXICITY DATA WITH REFERENCE
skn-rbt 10 mg/24H open MLD AIHAAP 23,95,62

CONSENSUS REPORTS: Reported in EPA TSCA Inventory.

SAFETY PROFILE: A skin irritant. Combustible when exposed to heat or flame. To fight fire, use dry chemical, CO_2. When heated to decomposition it emits acrid smoke and irritating fumes.

DXQ339 **HR: 3**
DI[TRIS-1,2-DIAMINOETHANECHROMIUM(III)]TRIPEROXODISULFATE
mf: $C_{12}H_{24}Cr_2N_{12}O_{24}S_6$ mw: 1016.78

CONSENSUS REPORTS: Chromium and its compounds are on the Community Right-To-Know List.

SAFETY PROFILE: Explodes. When heated to decomposition it emits toxic fumes of SO_x and NO_x. See also CHROMIUM COMPOUNDS, PEROXIDES, and SULFATES.

DXQ369 **HR: 3**
DI[TRIS-1,2-DIAMINOETHANECOBALT(III)]TRIPEROXODISULFATE
mf: $C_{12}H_{24}Co_2N_{12}O_{24}S_6$ mw: 1030.66

CONSENSUS REPORTS: Cobalt and its compounds are on the Community Right-To-Know List.

SAFETY PROFILE: Explodes. When heated to decomposition it emits toxic fumes of SO_x and NO_x. See also COBALT COMPOUNDS, PEROXIDES, and SULFATES.

DXQ400 **CAS:3648-20-2** **HR: 1**
DIUNDECYL PHTHALATE
mf: $C_{30}H_{50}O_2$ mw: 442.80

SYN: SANTICIZER 711

TOXICITY DATA WITH REFERENCE
eye-rbt 100 mg MLD MONS** 11/4/75

CONSENSUS REPORTS: Reported in EPA TSCA Inventory.

SAFETY PROFILE: An eye irritant. When heated to decomposition it emits acrid smoke and irritating fumes. See also ESTERS.

DXQ500 **CAS:330-54-1** **HR: 2**
DIURON
mf: $C_9H_{10}Cl_2N_2O$ mw: 233.11

PROP: Crystals. Mp: 153.5–155°. Sltly sol in water and hydrocarbon solvents.

SYNS: AF 101 □ CEKIURON □ CRISURON □ DAILON □ DCMU □ DIATER □ 3-(3,4-DICHLOOR-FENYL)-1,1-DIMETHYLUREUM (DUTCH) □ DICHLORFENIDIM □ 3-(3,4-DICHLOROPHENOL)-1,1-DIMETHYLUREA □ N'-(3,4-DICHLOROPHENYL)-N,N-DIMETHYLUREA □ 1-(3,4-DICHLOROPHENYL)-3,3-DIMETHYLUREE (FRENCH) □ 3-(3,4-DICHLOR-PHENYL)-1,1-DIMETHYL-HARNSTOFF (GERMAN) □ 3-(3,4-DICLORO-FENYL)-1,1-DIMETIL-UREA (ITALIAN) □ 1,1-DIMETHYL-3-(3,4-DICHLOROPHENYL)UREA □ DI-ON □ DIREX 4L □ DIUREX □ DIUROL □ DIURON 4L □ DMU □ DREXEL □ DREXEL DIURON 4L □ DURAN □ DYNEX □ FARMCO DIURON □ HERBATOX □ HW 920 □ KARMEX □ KARMEX DIURON HERBICIDE □ KARMEX DW □ MARMER □ SUP'R FLO □ TELVAR □ TELVAR DIURON WEED KILLER □ UNIDRON □ USAF P-7 □ USAF XR-42 □ VONDURON

TOXICITY DATA WITH REFERENCE
mma-sat 3 μg/plate MUREAV 58,353,78
dni-mus-orl 1 g/kg MUREAV 58,353,78
scu-mus TDLo:1935 mg/kg (female 6-14D post):TER NTIS** PB223-160
orl-rat TDLo:360 mg/kg (6-15D preg):TER GISAAA 49(8),83,84
orl-rat TDLo:1250 mg/kg (female 6-15D post):TER BECTA6 22,522,79
orl-mus TDLo:153 g/kg/78W-I:ETA NTIS** PB223-159
orl-rat LD50:1017 mg/kg JAFCAU 18,1104,70
ipr-mus LDLo:500 mg/kg NTIS** AD277-689

CONSENSUS REPORTS: Reported in EPA TSCA Inventory. EPA Genetic Toxicology Program. Chlorophenol compounds are on the Community Right-To-Know List.

OSHA PEL: TWA 10 mg/m³
ACGIH TLV: TWA 10 mg/m³
NIOSH REL: (Diuron) TWA 10 mg/m³

SAFETY PROFILE: Moderately toxic by ingestion and intraperitoneal routes. Questionable carcinogen with experimental tumorigenic and teratogenic data. Mutation data reported. When heated to decomposition it emits highly toxic fumes of Cl^- and NO_x. See also CHLOROPHENOLS.

DXQ740 **CAS:1321-74-0** **HR: 1**
DIVINYLBENZENE
mf: $C_{10}H_{10}$ mw: 130.20

SYNS: BENZENE, DIVINYL- □ VINYLSTYRENE

TOXICITY DATA WITH REFERENCE
orl-rat LDLo:4644 mg/kg AMIHAB 19,403,59

CONSENSUS REPORTS: Reported in EPA TSCA Inventory.

ACGIH TLV: TWA 10 ppm
NIOSH REL: (divinyl benzene) TWA 10 ppm

SAFETY PROFILE: Mildly toxic by ingestion. An eye irritant. Combustible. When heated to decomposition it emits acrid smoke and irritating fumes.

DXQ745 CAS:108-57-6 HR: 2
m-DIVINYLBENZENE
mf: $C_{10}H_{10}$ mw: 130.20

PROP: Pale straw-colored liquid. Bp: 195–200°, mp: −87°, d: 0.918, flash p: 165F°. Not misc in water; sol in ether and methanol.

SYNS: m-DIVINYLBENZEN □ m-DIVINYLBENZENE □ m-VINYLSTYRENE

TOXICITY DATA WITH REFERENCE
eye-rbt 500 mg/24H MLD 85JCAE-,38,86

CONSENSUS REPORTS: Reported in EPA TSCA Inventory.

OSHA PEL: 10 ppm
ACGIH TLV: 10 ppm

SAFETY PROFILE: An eye irritant. Combustible. When heated to decomposition it emits acrid smoke and irritating fumes.

DXQ850 CAS:6928-74-1 HR: 3
DIVINYL MAGNESIUM
mf: C_4H_6Mg mw: 78.40

$(H_2C{=}CH)_2Mg$

SAFETY PROFILE: May ignite spontaneously in air. See also MAGNESIUM COMPOUNDS.

DXR000 CAS:78-19-3 HR: 3
3,9-DIVINYLSPIROBI(m-DIOXANE)
mf: $C_{11}H_{16}O_4$ mw: 212.27

PROP: A solid. Mp: 40–45°, bp: 93–94° @ 1 mm.

SYN: 3,9-DIVINYL-2,4,8,10-TETRAOXASPIRO(5.5)UNDECANE

TOXICITY DATA WITH REFERENCE
orl-rat LD50:4066 mg/kg AIHAAP 30,470,69
ivn-mus LD50:320 mg/kg CSLNX* NX#01186
skn-rbt LD50:9908 mg/kg AIHAAP 30,470,69

SAFETY PROFILE: Poison by intravenous route. Mildly toxic by ingestion and skin contact. When heated to decomposition it emits acrid smoke and irritating fumes.

DXR200 CAS:77-77-0 HR: 3
DIVINYL SULFONE
mf: $C_4H_6O_2S$ mw: 118.16

PROP: Bp: 90–92°.

SYNS: TL 797 □ VINYL SULFONE

TOXICITY DATA WITH REFERENCE
skn-rbt 50 mg open MOD UCDS** 3/18/65
eye-rbt 5 mg/24H SEV 85JCAE-,1047,86
orl-rat LD50:32 mg/kg AIHAAP 23,95,62
ipr-rat LD50:3 mg/kg TXAPA9 31,222,75
scu-rat LD50:14 mg/kg JPETAB 93,1,48
ivn-rat LD50:12 mg/kg JPETAB 93,1,48
scu-mus LD50:16 mg/kg JPETAB 93,1,48
ihl-mus LCLo:990 mg/m³/10M NDRC** No.9-4-1-9,43
ivn-mus LD50:11 mg/kg JPETAB 93,1,48
skn-rbt LD50:22 mg/kg AIHAAP 23,95,62

SAFETY PROFILE: Poison by ingestion, skin contact, intravenous, subcutaneous, and intraperitoneal routes. A severe skin and eye irritant. See also SULFONATES.

DXR400 CAS:25724-33-8 HR: 2
2,5-DIVINYLTETRAHYDROPYRAN
mf: $C_9H_{14}O$ mw: 138.23

SYNS: 2,5-DIVINYLTETRAHYDRO-2H-PYRAN □ TETRAHYDRO-2,5-DIVINYL-2H-PYRAN □ TETRAHYDRO-2,5-DIVINYLPYRAN

TOXICITY DATA WITH REFERENCE
orl-rat LD50:2460 mg/kg AIHAAP 30,470,69
skn-rbt LD50:1410 mg/kg AIHAAP 30,470,69

SAFETY PROFILE: Moderately toxic by ingestion and skin contact. When heated to decomposition it emits acrid smoke and irritating fumes.

DXR500 CAS:1119-22-8 HR: 3
DIVINYL ZINC
mf: C_4H_6Zn mw: 119.48

$(H_2C{=}CH)_2Zn$

PROP: A liquid. Bp: 32° @ 22 mm.

CONSENSUS REPORTS: Zinc and its compounds are on the Community Right-To-Know List.

SAFETY PROFILE: Ignites spontaneously in air. When heated to decomposition it emits toxic fumes of ZnO. See also ZINC COMPOUNDS.

DXR800 CAS:60539-20-0 HR: 3
DIXYRAZINE DIHYDROCHLORIDE
mf: $C_{24}H_{33}N_3O_2S{\bullet}2ClH$ mw: 500.58

SYN: 2-(2-(4-(2-METHYL-3-PHENOTHIAZIN-10-YLPROPYL)-1-PIPERAZINYL)ETHOXY)ETHANOL DIHYDROCHLORIDE

TOXICITY DATA WITH REFERENCE
orl-mus LD50:420 mg/kg 27ZQAG -,22,72
ipr-mus LD50:102 mg/kg 27ZQAG -,22,72
ivn-mus LD50:48 mg/kg 27ZQAG -,22,72

SAFETY PROFILE: Poison by intravenous and intraperitoneal routes. Moderately toxic by ingestion. When heated to decomposition it emits very toxic fumes of Cl^-, SO_x, and NO_x.

DXS000 CAS:38001-34-2 HR: 2
(α-DIYLENE)POLY(p-AMINOBENZALDEHYDE-N)
mf: $C_{14}H_{12}N_2O{\bullet}(C_7H_5N)_n$

SYN: POLY-p-AMINOBENZALDEHYD (CZECH)

TOXICITY DATA WITH REFERENCE
eye-rbt 100 mg/24H SEV 28ZPAK -,254,72

SAFETY PROFILE: A severe eye irritant. When heated to decomposition it emits toxic fumes of NO_x. See also ALDEHYDES.

D

DXS200 **CAS:38222-35-4** **HR: 3**
DMA
mf: $C_8H_{13}NO_3$ mw: 171.22

SYNS: 2,5-DIHYDROXY-3-DIMETHYLAMINO-5-METHYL-2-CYCLOPENTEN-1-ONE ☐ DIMETHYLAMINO HEXOSE REDUCTIONE

TOXICITY DATA WITH REFERENCE
orl-mus LD50:400 mg/kg PSEBAA 106,656,61
ipr-mus LD50:300 mg/kg PSEBAA 106,656,61

SAFETY PROFILE: Poison by ingestion and intraperitoneal routes. When heated to decomposition it emits toxic fumes of NO_x.

DXS375 **CAS:52663-81-7** **HR: 3**
DOBUTAMINE HYDROCHLORIDE
mf: $C_{18}H_{23}NO_3$•ClH mw: 337.88

PROP: A solid. Mp: 184–186°.

SYNS: dl-3,4-DIHYDROXY-N-3-(4-HYDROXYPHENYL)-1-METHYL-n-PROPYL PHENETHYLAMINE HYDROCHLORIDE ☐ DOBUTREX ☐ (±)-4-(2-((3-(4-HYDROXYPHENYL)-1-METHYLPROPYL)AMINO)ETHYL)-1,2-BENZENEDIOL HYDROCHLORIDE ☐ 4-(2-((3-(p-HYDROXYPHENYL)-1-METHYLPROPYL)AMINO)ETHYL)PYROCATECHOL HYDROCHLORIDE ☐ INOTREX ☐ S-1000

TOXICITY DATA WITH REFERENCE
ivn-rat TDLo:300 mg/kg (female 17-22D post):REP YACHDS 7,1742,79
ivn-rat TDLo:22 mg/kg (7-17D preg):REP YACHDS 7,1707,79
ivn-rat TDLo:200 mg/kg (female 17-22D post-4D post):REP YACHDS 7,1742,79
ivn-rbt TDLo:390 mg/kg (female 6-18D post):TER YACHDS 7,1731,79
ivn-rat TDLo:770 mg/kg (7-17D preg):TER YACHDS 7,1707,70
ivn-rat TDLo:22 mg/kg (7-17D preg):TER YACHDS 7,1707,79
orl-rat LD50:2296 mg/kg IYKEDH 13,349,82
ipr-rat LD50:260 mg/kg IYKEDH 13,349,82
scu-rat LD50:368 mg/kg IYKEDH 13,349,82
ivn-rat LD50:59,600 μg/kg YACHDS 7,627,79
orl-mus LD50:1324 mg/kg IYKEDH 13,349,82
ipr-mus LD50:243 mg/kg YACHDS 7,627,79
scu-mus LD50:341 mg/kg YACHDS 7,627,79
ivn-mus LD50:34,300 μg/kg YACHDS 7,627,79

SAFETY PROFILE: Poison by subcutaneous, intravenous, and intraperitoneal routes. Moderately toxic by ingestion. Experimental reproductive effects. An experimental teratogen. When heated to decomposition it emits toxic fumes of NO_x and HCl.

DXS400 **HR: 3**
DODECACARBONYLDIVANADIUM
mf: $C_{12}O_{12}V_2$ mw: 438.02

SAFETY PROFILE: Ignites spontaneously in air. See also VANADIUM COMPOUNDS.

DXS600 **HR: 3**
DODECACARBONYLTRIIRON
mf: $C_{12}Fe_3O_{12}$ mw: 503.67

SAFETY PROFILE: On prolonged storage it will ignite spontaneously in air. When heated to decomposition it emits acrid smoke and fumes.

DXS700 **HR: 2**
Δ-DODECALACTONE
mf: $C_{12}H_{22}O_2$ mw: 198.31

PROP: Colorless to yellow liquid; coconut-fruity odor. Refr index: 1.458–1.461, flash p: 151°F. Very sol in alc, propylene glycol, veg oil; insol in water.

SYN: FEMA No. 2401

SAFETY PROFILE: Combustible liquid. When heated to decomposition it emits acrid smoke and irritating fumes.

DXS800 **CAS:141-63-9** **HR: 1**
DODECAMETHYLPENTASILOXANE
mf: $C_{12}H_{36}O_4Si_5$ mw: 384.93

SYN: PENTASILOXANE, DODECAMETHYL-

TOXICITY DATA WITH REFERENCE
orl-gpg LDLo:50 g/kg JIDHAN 30,332,48

CONSENSUS REPORTS: Reported in EPA TSCA Inventory.

SAFETY PROFILE: Slightly toxic by ingestion. When heated to decomposition it emits acrid smoke and irritating vapors.

DXT000 **CAS:112-54-9** **HR: 1**
1-DODECANAL
mf: $C_{12}H_{24}O$ mw: 184.36

PROP: Crystals. Mp: 44.5°, bp: 184–185° @ 100 mm. Reported in pine-needle, lime, sweet-orange, and a dozen other essential oils (FCTXAV 11,477,73). Colorless to light-yellow liquid; fatty odor. D: 0.826–0.836, refr index: 1.433–1.439, flash p: 180°F. Sol in alc, fixed oils, propylene glycol; insol in glycerin, water.

SYNS: C-12 ALDEHYDE, LAURIC ☐ 1-DODECYL ALDEHYDE ☐ DUODECYLIC ALDEHYDE ☐ FEMA No. 2615 ☐ LAURYL ALDEHYDE (FCC)

TOXICITY DATA WITH REFERENCE
skn-hmn 5 mg/48H MLD FCTXAV 11,1079,73
skn-rbt 500 mg/24H MOD FCTXAV 11,1079,73
orl-rat LD50:23 g/kg FCTXAV 11,483,73

CONSENSUS REPORTS: Reported in EPA TSCA Inventory.

SAFETY PROFILE: Mildly toxic by ingestion. A human and experimental skin irritant. Combustible liquid. When heated to decomposition it emits acrid smoke and irritating fumes. See also ALDEHYDES.

DXT200 CAS:112-40-3 HR: 2
DODECANE
mf: $C_{12}H_{26}$ mw: 170.38

PROP: A liquid. Fp: −12°, bp: 214.5°.

SYNS: ADAKANE 12 □ BIHEXYL □ DIHEXYL □ n-DODECAN (GERMAN) □ DUODECANE

TOXICITY DATA WITH REFERENCE
skn-mus TDLo:11 g/kg/22W-I:ETA TXAPA9 9,70,66

CONSENSUS REPORTS: Reported in EPA TSCA Inventory.

SAFETY PROFILE: Questionable carcinogen with experimental tumorigenic data. When heated to decomposition it emits acrid smoke and irritating fumes.

DXT400 CAS:2437-25-4 HR: 2
DODECANENITRILE
mf: $C_{12}H_{23}N$ mw: 181.36

SYN: LAURONITRILE

TOXICITY DATA WITH REFERENCE
orl-rat LDLo:500 mg/kg JPETAB 90,260,47

CONSENSUS REPORTS: Reported in EPA TSCA Inventory. Cyanide and its compounds are on the Community Right-To-Know List.

SAFETY PROFILE: Moderately toxic by ingestion. When heated to decomposition it emits toxic fumes of NO_x and CN−. See also NITRILES.

DXT800 CAS:25103-58-6 HR: 3
tert-DODECANETHIOL
mf: $C_{12}H_{26}S$ mw: 202.44

PROP: White to light-yellow liquid. Bp: 200–235°, flash p: 205°F (OC), d: 0.85 @ 25°/25°, vap d: 6.98.

SYNS: tert-DODECYLMERCAPTAN □ tert. DODECYLMERKAPTAN (CZECH) □ tert-DODECYLTHIOL □ 2,3,3,4,4,5-HEXAMETHYL-2-HEXANETHIOL

TOXICITY DATA WITH REFERENCE
skn-rbt 20 mg/24H MOD 85JCAE-,986,86
eye-rbt 500 mg/24H MLD 85JCAE-,986,86 28ZPAK -,169,72
ipr-rat LD50:1833 mg/kg 85GMAT -,64,82

CONSENSUS REPORTS: Reported in EPA TSCA Inventory.

SAFETY PROFILE: Poison by ingestion. Moderately toxic by intraperitoneal route. A skin and eye irritant. Combustible when exposed to heat or flame; can react vigorously with oxidizing materials. To fight fire, use foam, CO_2, dry chemical. When heated to decomposition it emits toxic fumes of SO_x. See also MERCAPTANS and SULFATES.

DXU200 CAS:18186-71-5 HR: 3
DODECATRIETHYLAMMONIUM BROMIDE
mf: $C_{18}H_{40}N{\cdot}Br$ mw: 350.50

TOXICITY DATA WITH REFERENCE
ipr-rat LD50:30 mg/kg JPMSAE 57,1431,68
orl-mus LD50:7 mg/kg JPMSAE 57,1431,68
ipr-mus LD50:8380 μg/kg JPMSAE 57,1431,68

SAFETY PROFILE: Poison by ingestion and intraperitoneal routes. When heated to decomposition it emits very toxic fumes of NO_x, NH_3, and Br^-. See also BROMIDES.

D

DXU250 CAS:76379-66-3 HR: 3
5,7,11-DODECATRIYN-1-OL
mf: $C_{12}H_{16}O$ mw: 176.28

TOXICITY DATA WITH REFERENCE
skn-rbt 500 mg/24H MLD JJATDK 8,35,88
orl-rat TDLo:510 mg/kg (female (2D pre)-15D post):REP JJATDK 8,35,88
orl-rat LD50:250 mg/kg JJATDK 8,35,88

SAFETY PROFILE: Poison by ingestion. Experimental reproductive effects. A skin irritant. When heated to decomposition it emits acrid smoke and irritating fumes.

DXU280 CAS:4826-62-4 HR: 2
2-DODECENAL
mf: $C_{12}H_{22}O$ mw: 182.34

SYN: β-OCTYL ACROLEIN

TOXICITY DATA WITH REFERENCE
skn-rbt 500 mg/24H SEV FCTOD7 21,849,83
skn-rbt LDLo:5 g/kg FCTOD7 21,849,83

CONSENSUS REPORTS: Reported in EPA TSCA Inventory.

SAFETY PROFILE: A severe skin irritant. When heated to decomposition it emits acrid smoke and irritating fumes.

DXU400 CAS:2855-19-8 HR: 2
DODECENE EPOXIDE
mf: $C_{12}H_{24}O$ mw: 184.36

SYN: 1,2-EPOXYDODECANE

TOXICITY DATA WITH REFERENCE
unr-mus TDLo:147 mg/kg:ETA RARSAM 3,193,63

CONSENSUS REPORTS: Reported in EPA TSCA Inventory.

SAFETY PROFILE: Questionable carcinogen with experimental tumorigenic data. When heated to decomposition it emits acrid smoke and irritating fumes.

DXU600 CAS:56530-48-4 HR: 2
9a-DODECENOATE
mf: $C_{32}H_{49}O_6$ mw: 528.80

SYN: 12-DEOXYPHORBOL-13-DODECENOATE

TOXICITY DATA WITH REFERENCE
skn-mus 42 ng open ARTODN 44,279,80
skn-mus 30 ng/4H APTOA6 37,250,75
skn-mus 100 ng/24H APTOA6 37,250,75

SAFETY PROFILE: A skin irritant requiring only very

small amounts for effect. When heated to decomposition it emits acrid smoke and irritating fumes.

DXU800 **CAS:20056-92-2** **HR: 1**
(Z)-7-DODECEN-1-OL
mf: $C_{12}H_{24}O$ mw: 184.36

PROP: A liquid.

SYN: LOOPLURE INHIBITOR

TOXICITY DATA WITH REFERENCE
skn-rbt 500 mg/24H TXAPA9 31,421,75

SAFETY PROFILE: A skin irritant. When heated to decomposition it emits acrid smoke and irritating fumes.

DXU830 **CAS:14959-86-5** **HR: 1**
cis-7-DODECENYL ACETATE
mf: $C_{14}H_{26}O_2$ mw: 226.40

SYNS: CABBLEMONE □ 7-DODECEN-1-OL, ACETATE, (Z)- □ (Z)-7-DODECENYL ACETATE □ (Z)-7-DODECEN-1-OLACETATE □ ENT 33266 □ LOOPLURE □ PHEROCON CL

TOXICITY DATA WITH REFERENCE
orl-rat LD50:13,430 mg/kg TXAPA9 31,421,75
ihl-rat LC50:>4500 mg/m³ TXAPA9 31,421,75
skn-rbt LD50:>2025 mg/kg TXAPA9 31,421,75

CONSENSUS REPORTS: Reported in EPA TSCA Inventory.

SAFETY PROFILE: Low toxicity by ingestion, inhalation, and skin contact. When heated to decomposition it emits acrid smoke and irritating vapors.

DXV000 **CAS:25377-73-5** **HR: 3**
DODECENYLSUCCINIC ANHYDRIDE
mf: $C_{16}H_{27}O_3$ mw: 266.38

PROP: Light-yellow, clear, visc oil. Bp: 180–182° @ 5 mm, flash p: 352°F (COC), d: 1.002 @ 25°/4°.

SYNS: DDS □ DDS A □ 2,5-FURANDIONE, 3-(DODECENYL)DIHYDRO-

TOXICITY DATA WITH REFERENCE
ihl-rat LCLo:1220 mg/m³/4H EPASR* 8EHQ-0282-0432
ipr-mus LD50:320 mg/kg NTIS** AD441-640

CONSENSUS REPORTS: Reported in EPA TSCA Inventory.

SAFETY PROFILE: Poison by inhalation and intraperitoneal route. An irritant and sensitizer. Combustible when exposed to heat or flame; can react with oxidizing materials. To fight fire, use foam, CO_2, dry chemical. When heated to decomposition it emits acrid smoke and irritating fumes.

DXV400 **CAS:112-66-3** **HR: 1**
n-DODECYL ACETATE
mf: $C_{14}H_{28}O_2$ mw: 228.42

SYNS: ACETATE C-12 □ ACETIC ACID, DODECYL ESTER □ DODECANOL ACETATE □ 1-DODECANOL ACETATE □ DODECAN-1-YL ACETATE □ DODECYL ACETATE □ DODECYL ALCOHOL ACETATE □ LAURYL ACETATE

TOXICITY DATA WITH REFERENCE
skn-rbt 500 mg/24H MLD FCTXAV 14,659,76

CONSENSUS REPORTS: Reported in EPA TSCA Inventory.

SAFETY PROFILE: A skin irritant. When heated to decomposition it emits acrid smoke and irritating fumes.

DXV600 **CAS:112-53-8** **HR: 2**
DODECYL ALCOHOL
mf: $C_{12}H_{26}O$ mw: 186.38

PROP: Crystals from EtOH (aq) or liquid above 24°; floral odor. Mp: 24°, bp: 145–148° @ 18 mm, d: 0.830–0.836, refr index: 1.440–1.444, flash p: 260°F, autoign temp: 527°F. Sol in 2 parts of 70% alc, fixed oils, propylene glycol; insol in water, glycerin.

SYNS: ALCOHOL C-12 □ ALFOL 12 □ CACHALOT L-50 □ CO 12 □ CO-1214 □ n-DODECANOL □ 1-DODECANOL □ n-DODECYL ALCOHOL □ DUODECYL ALCOHOL □ DYTOL J-68 □ EPAL 12 □ FEMA No. 2617 □ LAURIC ALCOHOL □ LAURINIC ALCOHOL □ LAURYL 24 □ LAURYL ALCOHOL (FCC) □ n-LAURYL ALCOHOL, PRIMARY □ LOROL □ MA-1214 □ SIPOL L12

TOXICITY DATA WITH REFERENCE
skn-hmn 75 mg/3D-I SEV 85DKA8 -,127,77
skn-mus TDLo:19 g/kg/39W-I:ETA TXAPA9 9,70,66
orl-rat LD50:12,800 mg/kg FCTXAV 11,95,73

CONSENSUS REPORTS: Reported in EPA TSCA Inventory.

SAFETY PROFILE: Moderately toxic by intraperitoneal route. Mildly toxic by ingestion. A severe human skin irritant. Questionable carcinogen with experimental tumorigenic data. Combustible when exposed to heat or flame; can react with oxidizing materials. To fight fire, use dry chemical, CO_2. When heated to decomposition it emits acrid smoke and irritating fumes.

DXW000 **CAS:124-22-1** **HR: 3**
DODECYLAMINE
mf: $C_{12}H_{27}N$ mw: 185.40

PROP: Oil, amine odor or crystals from C_6H_6. Fp: 28.3°, vap press: 64 mm @ 170°, mp: 27–28°, bp: 247–249°. Sol in EtOH, Et_2O, C_6H_6, and $CHCl_3$.

SYNS: ALAMINE 4 □ AMINE BB □ 1-AMINODODECANE □ ARMEEN 12D □ 1-DODECANAMINE (9CI) □ n-DODECYLAMINE □ 1-DODECYLAMINE □ KEMAMINE P690 □ LAURINAMINE □ LAURYLAMINE □ n-LAURYLAMINE □ MONODODECYLAMINE □ NISSAN AMINE BB

TOXICITY DATA WITH REFERENCE
skn-rbt 500 mg/24H SEV 28ZPAK -,62,72
eye-rbt 50 µg/24H SEV 28ZPAK -,52,72
eye-rbt 2 mg AEPPAE 219,119,53
orl-rat LD50:1020 mg/kg AEPPAE 219,119,53
orl-mus LD50:1160 mg/kg AEPPAE 219,119,53
ipr-mus LD50:50 mg/kg NTIS** AD691-490

CONSENSUS REPORTS: Reported in EPA TSCA Inventory.

SAFETY PROFILE: Poison by intraperitoneal route. Moderately toxic by ingestion. A severe skin and eye irritant. When heated to decomposition it emits toxic fumes of NO_x. See also AMINES.

DXW050 **CAS:2016-56-0** **HR: 2**
1-DODECYLAMINE ACETATE
mf: $C_{12}H_{27}N \cdot C_2H_4O_2$ mw: 245.46

SYNS: DODECANAMINE ACETATE ☐ DODECYLAMINE, ACETATE

TOXICITY DATA WITH REFERENCE
orl-mus LD50:1750 mg/kg CHTPBA 1,11,65

CONSENSUS REPORTS: Reported in EPA TSCA Inventory.

SAFETY PROFILE: Moderately toxic by ingestion. When heated to decomposition it emits toxic vapors of NO_x.

DXW200 **CAS:25155-30-0** **HR: 3**
DODECYL BENZENE SODIUM SULFONATE
mf: $C_{18}H_{29}O_3S \cdot Na$ mw: 348.52

PROP: White to light yellow flakes, granules, or powder.

SYNS: AA-9 ☐ ABESON NAM ☐ BIO-SOFT D-40 ☐ CALSOFT F-90 ☐ CONCO AAS-35 ☐ CONOCO C-50 ☐ DETERGENT HD-90 ☐ DODECYLBENZENESULFONIC ACID SODIUM SALT ☐ DODECYLBENZENESULPHONATE, SODIUM SALT ☐ DODECYLBENZENSULFONAN SODNY (CZECH) ☐ MERCOL 25 ☐ NACCANOL NR ☐ NECCANOL SW ☐ PILOT HD-90 ☐ PILOT SF-40 ☐ RICHONATE 1850 ☐ SANTOMERSE 3 ☐ SODIUM DODECYLBENZENESULFONATE (DOT) ☐ SODIUM DODECYLBENZENESULFONATE, dry ☐ SODIUM LAURYLBENZENESULFONATE ☐ SOLAR 40 ☐ SOL SODOWA KWASU LAURYLOBENZENOSULFONOWEGO (POLISH) ☐ SULFAPOL ☐ SULFAPOLU (POLISH) ☐ SULFRAMIN 85 ☐ SULFRAMIN 40 FLAKES ☐ SULFRAMIN 40 GRANULAR ☐ SULFRAMIN 1238 SLURRY ☐ p-1′,1′,4′,4′-TETRAMETHYLOKTYLBENZENSULFONAN SODNY (CZECH) ☐ ULTRAWET K

TOXICITY DATA WITH REFERENCE
skn-rbt 20 mg/24H MOD 85JCAE-,1063,86
eye-rbt 250 μg/24H SEV 28ZPAK -,195,72
eye-rbt 1% SEV JAPMA8 38,428,49
orl-rat LD50:438 mg/kg TRENAF 24,397,72
orl-mus LD50:1330 mg/kg TRENAF 24,397,72
ivn-mus LD50:105 mg/kg JAPMA8 38,428,49

CONSENSUS REPORTS: Reported in EPA TSCA Inventory.

SAFETY PROFILE: Poison by intravenous route. Moderately toxic by ingestion. A skin and severe eye irritant. When heated to decomposition it emits toxic fumes of Na_2O. See also SULFONATES.

DXW400 **CAS:1886-81-3** **HR: 2**
DODECYL BENZENESULFONATE
mf: $C_{18}H_{30}O_3S$ mw: 326.54

SYN: BENZENESULFONIC ACID, DODECYL ESTER

TOXICITY DATA WITH REFERENCE
orl-rat LD50:650 mg/kg ARTODN 32,245,74

CONSENSUS REPORTS: Reported in EPA TSCA Inventory.

SAFETY PROFILE: Moderately toxic by ingestion. When heated to decomposition it emits toxic fumes of SO_x.

DXW600 **CAS:28061-21-4** **HR: 1**
DODECYLBENZYL CHLORIDE
mf: $C_{19}H_{31}Cl$ mw: 294.95

SYN: CONOCO DBCL

TOXICITY DATA WITH REFERENCE
skn-rbt 500 mg MLD 34ZIAG -,233,69
skn-rbt LDLo:12,500 mg/kg 34ZIAG -,233,69

CONSENSUS REPORTS: Reported in EPA TSCA Inventory.

SAFETY PROFILE: Mildly toxic by skin contact. A skin and eye irritant. When heated to decomposition it emits toxic fumes of Cl^-.

DXW800 **CAS:2783-17-7** **HR: 2**
DODECYLDIAMINE
mf: $C_{12}H_{28}N_2$ mw: 200.42

PROP: A solid. Fp: 66–67°.

SYNS: 1,12′-DIAMINODECANE ☐ 1,12′-DODECAMETHYLENEDIAMINE ☐ 1,12-DODECANEDIAMINE ☐ 1,12′-DODECANEDIAMINE ☐ 1,12′-DODECYLENEDIAMINE

TOXICITY DATA WITH REFERENCE
eye-rbt 500 mg MLD JACTDZ 1,92,90
orl-rat LDLo:670 mg/kg JACTDZ 1,92,90
orl-mus LD50:1088 mg/kg GISAAA 43(5),18,78

CONSENSUS REPORTS: Reported in EPA TSCA Inventory.

SAFETY PROFILE: Moderately toxic by ingestion. An eye irritant. When heated to decomposition it emits toxic fumes of NO_x. See also AMINES.

DXX000 **CAS:538-71-6** **HR: 3**
DODECYLDIMETHYL(2-PHENOXYETHYL)AMMONIUM BROMIDE
mf: $C_{22}H_{40}NO \cdot Br$ mw: 414.54

PROP: A solid. Mp: 112–113°.

SYNS: PHENODODECINIUM BROMIDE ☐ β-PHENOXYETHYLDIMETHYLDODECYLAMMONIUM BROMIDE

TOXICITY DATA WITH REFERENCE
add-bac-esc 10 μmol/L MUREAV 89,95,81
ipr-rat LD50:40 mg/kg FEPRA7 6,307,47
ivn-rat LD50:18 mg/kg FEPRA7 6,307,47
ivn-mus LD50:31 mg/kg FEPRA7 6,307,47
ivn-rbt LDLo:11 mg/kg FEPRA7 6,307,47
ipr-gpg LDLo:10 mg/kg IIDTXAC 1,222,56

CONSENSUS REPORTS: Reported in EPA TSCA Inventory.

SAFETY PROFILE: Poison by intraperitoneal and intravenous routes. Mutation data reported. When heated to decomposition it emits very toxic fumes of NO_x, NH_3, and Br^-. See also BROMIDES.

DXX100 **CAS:56501-30-5** **HR: 3**
1-DODECYL-1-ETHYLPIPERIDINIUM BROMIDE
mf: $C_{19}H_{40}N \cdot Br$ mw: 362.51

TOXICITY DATA WITH REFERENCE
orl-mus LD50:198 mg/kg PSDTAP 15,331,74
ipr-mus LD50:19,980 µg/kg PSDTAP 15,331,74
ivn-mus LD50:4595 µg/kg PSDTAP 15,331,74

SAFETY PROFILE: Poison by ingestion, intravenous, and intraperitoneal routes. When heated to decomposition it emits toxic fumes of NO_x and Br^-.

DXX200 **CAS:1166-52-5** **HR: 3**
DODECYL GALLATE
mf: $C_{19}H_{30}O_5$ mw: 338.49

PROP: Sol in EtOH and Me_2CO.

SYNS: DODECYLESTER KYSELINY GALLOVE □ GALLIC ACID, DODECYL ESTER □ GALLIC ACID, LAURYL ESTER □ LAURYL GALLATE □ NIPAGALLIN LA □ PROGALLIN LA

TOXICITY DATA WITH REFERENCE
ipr-rat LD50:100 mg/kg FAONAU 38A,22,65

CONSENSUS REPORTS: Reported in EPA TSCA Inventory.

SAFETY PROFILE: A poison by intraperitoneal route. When heated to decomposition it emits acrid smoke and irritating fumes.

DXX400 **CAS:2439-10-3** **HR: 3**
N-DODECYLGUANIDINE ACETATE
mf: $C_{13}H_{29}N_3 \cdot C_2H_4O_2$ mw: 287.51

PROP: Crystals. Mp: 136°. Sol in hot water and alc; insol in nonpolar solvs.

SYNS: AC 5223 □ AMERICAN CYANAMID 5223 □ APADODINE □ CARPENE □ CURITAN □ CYPREX □ CYPREX 65W □ N-DODECYLGUANIDINACETAT (GERMAN) □ DODECYLGUANIDINE ACETATE □ DODGUADINE □ DODINE □ DODINE ACETATE □ DODINE, mixture with GLYODIN □ DOGQUADINE □ ENT 16,436 □ EXPERIMENTAL FUNGICIDE 5223 □ LAURYLGUANIDINE ACETATE □ MELPREX □ MILPREX □ SYLLIT □ TSITREX □ VENTUROL □ VONDODINE

TOXICITY DATA WITH REFERENCE
eye-rbt 100 mg SEV 34ZIAG -,234,69
scu-mus TDLo:1000 mg/kg:ETA NTIS** PB223-159
orl-rat LD50:566 mg/kg WRPCA2 7,135,68
orl-mus LD50:266 mg/kg 85GMAT-,64,82
ihl-mus LC50:129 mg/m³/2H 85GMAT-,64,82
orl-rbt LD50:535 mg/kg 85GMAT -,64,82
skn-rbt LD50:1500 mg/kg 28ZEAL 5,88,76
orl-gpg LD50:176 mg/kg 85GMAT -,64,82
skn-gpg LDLo:2 g/kg 85GMAT -,64,82

CONSENSUS REPORTS: Reported in EPA TSCA Inventory. EPA Genetic Toxicology Program.

SAFETY PROFILE: Poison by ingestion and inhalation. Moderately toxic by skin contact. A severe eye irritant. Questionable carcinogen with experimental tumorigenic data. A pesticide. When heated to decomposition it emits very toxic fumes of NO_x.

DXX600 **HR: 2**
DODECYLGUANIDINE ACETATE with SODIUM NITRITE (3:5)

SYNS: DODINE with SODIUM NITRITE (3:5) □ SODIUM NITRITE with DODECYL GUANIDINE ACETATE (5:3)

TOXICITY DATA WITH REFERENCE
orl-mus TDLo:112 mg/kg/6D-I:CAR CALEDQ 5,107,78
orl-mus TDLo:112 mg/kg (15-21D preg):CAR,TER CALEDQ 5,107,78

SAFETY PROFILE: Questionable carcinogen with experimental carcinogenic data. An experimental teratogen. When heated to decomposition it emits very toxic fumes of NO_x and Na_2O. See also SODIUM NITRITE.

DXX800 **CAS:13590-97-1** **HR: 3**
DODECYLGUANIDINE HYDROCHLORIDE
mf: $C_{13}H_{29}N_3 \cdot ClH$ mw: 263.91

TOXICITY DATA WITH REFERENCE
unr-mus LDLo:13 mg/kg ATMPA2 32,177,38

CONSENSUS REPORTS: Reported in EPA TSCA Inventory.

SAFETY PROFILE: Poison by unspecified route. When heated to decomposition it emits very toxic fumes of HCl and NO_x.

DXX875 **CAS:56501-36-1** **HR: 3**
1-DODECYLHEXAHYDRO-1H-AZEPINE-1-OXIDE
mf: $C_{18}H_{37}NO$ mw: 283.56

SYNS: 1-DODECYLHEXAMETHYLENIMINE-N-OXIDE □ HEXAHYDRO-1-DODECYL-1H-AZEPINE-1-OXIDE

TOXICITY DATA WITH REFERENCE
orl-mus LD50:94 mg/kg PESTD5 16,236,75
ipr-mus LD50:76 mg/kg PESTD5 16,236,75
ivn-mus LD50:33 mg/kg PESTD5 16,236,75

SAFETY PROFILE: Poison by ingestion, intravenous, and intraperitoneal routes. When heated to decomposition it emits toxic fumes of NO_x.

DXY000 **CAS:9002-92-0** **HR: 3**
α-DODECYL-ω-HYDROXY-POLYOXYETHYLENE
mf: $(C_2H_4O)_n \cdot C_{12}H_{26}O$

SYNS: ALDOSPERSE L 9 □ ATLAS G-2133 □ BASE LP 12 □ BL 9 □ BRIJ 30 □ CHIMIPAL AE 3 □ CIMAGEL □ DEHYDOL LS 4 □ DO 9 □ DODECANOL, ETHOXYLATE □ DODECANOL-ETHYLENE OXIDE (9.5 moles) CONDENSATE □ DODECANOL, POLYETHOXYLATED □ DODECYL ALCOHOL, ETHOXYLATED □ DODECYL-POLYAETHYLENOXYD-AETHER (GERMAN) □ DODECYL POLY(OXYETHYLENE) ETHER □ DU PONT WK □ EMULGEN 100 □ ETHAL LA-X □ ETHOSPERSE LA-4 □ ETHOXYLATED LAURYL ALCOHOL □ G 3707 □ G-2130A □ HYDROXYPOLYETHOXYDODECANE □ LA □ LAURETH □ LAUROMACROGOL 400 □ LAURYL ALCOHOL, ETHOXYLATED □ LAURYL POLYETHYLENE GLYCOL ETHER □ LIPAL 4LA □ LIPOCOL L-4 □ LUBROL 12A9 □ MARLIPAL 1217 □ MCI-C54875 □ NIKKOL BL □ NOIGEN 160 □ OXYETHYLENATED DODECYL ALCOHOL □ PLURA-

FAC RA 43 □ POLIDOCANOL □ POLYETHYLENE GLYCOL DODECYL ETHER □ POLY(ETHYLENE OXIDE) DODECYL ETHER □ POLYOXYETHYLENE LAURIC ALCOHOL □ POLYOXYETHYLENE LAURYL ETHER □ ROKANOL L □ SIMULOL 330 M □ SIPONIC L □ SURFACTANT WK □ TRYCOL LAL SERIES

TOXICITY DATA WITH REFERENCE
skn-hmn 6 mg/3D-I MOD 85DKA8 -,127,77
skn-rbt 500 mg/24H MLD 28ZPAK -,301,72
skn-rbt 75 mg/24H MLD TXAPA9 7,206,65
skn-rbt 500 mg/24H MOD TXAPA9 19,276,71
eye-rbt 100 mg TXAPA9 19,276,71
eye-rbt 750 µg/24H SEV 28ZPAK -,301,72
dnd-esc 50 mg/L MUREAV 89,95,81
orl-rat LD50:1 g/kg FCTXAV 8,125,70
ipr-rat LD50:125 mg/kg FCTXAV 8,125,70
orl-mus LD50:1170 mg/kg ARZNAD 7,162,57
ivn-mus LD50:100 mg/kg TXAPA9 7,206,65
ivn-rbt LD50:36 mg/kg ARZNAD 7,162,57
orl-gpg LD50:384 mg/kg ARZNAD 7,162,57
ivn-gpg LD50:38 mg/kg ARZNAD 7,162,57

CONSENSUS REPORTS: Reported in EPA TSCA Inventory. Glycol ethers are on the Community Right-To-Know List.

SAFETY PROFILE: Poison by ingestion, intraperitoneal, and intravenous routes. A human and experimental skin irritant. A severe eye irritant. Mutation data reported. When heated to decomposition it emits acrid smoke and irritating fumes. See also GLYCOL ETHERS.

DXY200 CAS:142-90-5 HR: 1
DODECYL METHACRYLATE
mf: $C_{16}H_{30}O_2$ mw: 254.46

SYNS: AGEFLEX FM 246 □ DODECYL-2-METHYL-2-PROPENOATE □ LAURYLESTER KYSELINYMETHAKRYLOVE (CZECH) □ LAURYL METHACRYLATE □ METHACRYLIC ACID DODECYL ESTER □ METHACRYLIC ACID LAURYL ESTER □ 2-METHYLACRYLIC ACID DODECYL ESTER

TOXICITY DATA WITH REFERENCE
skn-rbt 500 mg/24H MLD 28ZPAK -,44,72
eye-rbt 500 mg/24H MLD 28ZPAK -,44,72
ipr-rat LD50:12 g/kg AMPMAR 36,58,75
ipr-mus LD50:25 g/kg JPMSAE 62,778,73

CONSENSUS REPORTS: Reported in EPA TSCA Inventory.

SAFETY PROFILE: A skin and eye irritant. When heated to decomposition it emits acrid smoke and irritating fumes.

DXY300 CAS:2530-46-3 HR: 3
4-DODECYLMORPHOLINE-4-OXIDE
mf: $C_{16}H_{33}NO_2$ mw: 271.50

SYN: 4-DODECYLMORPHOLINE-N-OXIDE

TOXICITY DATA WITH REFERENCE
orl-mus LD50:1880 mg/kg PESTD5 16,236,75
ipr-mus LD50:110 mg/kg PESTD5 16,236,75
ivn-mus LD50:68 mg/kg PESTD5 16,236,75

SAFETY PROFILE: Poison by intravenous and intraperitoneal routes. When heated to decomposition it emits toxic fumes of NO_x.

DXY400 HR: 3
N-(2-DODECYLOXYETHYL)-N-METHYL-2-(PYRROLIDINYL)ACETAMIDE HYDROCHLORIDE
mf: $C_{21}H_{42}N_2O_2 \cdot ClH$ mw: 391.11

SYN: C 6606

TOXICITY DATA WITH REFERENCE
eye-rbt 2% SEV ARZNAD 9,113,59
scu-mus LD50:400 mg/kg ARZNAD 9,113,59

SAFETY PROFILE: Poison by subcutaneous route. A severe eye irritant. When heated to decomposition it emits very toxic fumes of NO_x and HCl.

DXY600 CAS:27193-86-8 HR: 2
DODECYLPHENOL (mixed isomers)
mf: $C_{18}H_{30}O$ mw: 262.48

PROP: Straw-colored liquid, phenolic odor. Bp: 154–168°, flash p: 325°F (OC), d. 0.93 @ 20°/20°, vap d: 9.04.

SYN: T-DET

TOXICITY DATA WITH REFERENCE
orl-rat LD50:2140 mg/kg AIHAAP 23,95,62
skn-rbt LD50:5000 mg/kg AIHAAP 23,95,62

CONSENSUS REPORTS: Reported in EPA TSCA Inventory.

SAFETY PROFILE: Moderately toxic by ingestion. Mildly toxic by skin contact. Combustible when exposed to heat or flame; can react with oxidizing materials. To fight fire, use CO_2, dry chemical. When heated to decomposition it emits acrid smoke and irritating fumes.

DXY700 CAS:56501-35-0 HR: 3
1-DODECYLPIPERIDINE-N-OXIDE
mf: $C_{17}H_{35}NO$ mw: 269.53

SYN: 1-DODECYLPIPERIDINE-1-OXIDE

TOXICITY DATA WITH REFERENCE
orl-mus LD50:619 mg/kg PSDTAP 15,331,74
ipr-mus LD50:67,920 µg/kg PSDTAP 15,331,74
ivn-mus LD50:23 mg/kg PESTD5 16,236,75

SAFETY PROFILE: Poison by intravenous and intraperitoneal routes. Moderately toxic by ingestion. When heated to decomposition it emits toxic fumes of NO_x.

DXY725 CAS:104-74-5 HR: 3
1-DODECYLPYRIDINIUM CHLORIDE
mf: $C_{17}H_{30}N \cdot Cl$ mw: 283.93

SYNS: C 2 □ DEHYQUART C □ DODECYLPYRIDINIUM CHLORIDE □ N-DODECYLPYRIDINIUM CHLORIDE □ DPC □ ELTREN □ LAURYLPYRIDINIUM CHLORIDE □ 1-LAURYLPYRIDINIUM CHLORIDE □ LPC □ QUATERNARIO LPC □ PYRIDINIUM, 1-DODECYL-, CHLORIDE

D

TOXICITY DATA WITH REFERENCE
unr-mus LD50:119 mg/kg PHARAT 40,273,85

CONSENSUS REPORTS: Reported in EPA TSCA Inventory.

SAFETY PROFILE: Poison by an unspecified route. When heated to decomposition it emits toxic vapors of NO_x and Cl^-.

DXY750 CAS:55257-88-0 HR: 2
1-DODECYL-2-PYRROLIDINONE
mf: $C_{16}H_{31}NO$ mw: 253.48

PROP: Bp: 202–205°/11 mm, d: 0.890, Flash p: >230°F.

SYNS: N-DODECYLPYRROLIDINONE □ 2-PYRROLIDINONE, 1-DODECYL-

TOXICITY DATA WITH REFERENCE
skn-rbt 500 mg SEV FCTOD7 26,475,88
eye-rbt 100 mg MOD FCTOD7 26,475,88
orl-rat LDLo:5 g/kg FCTOD7 26,475,88

SAFETY PROFILE: Mildly toxic by ingestion. A severe skin and moderate eye irritant. Combustible when exposed to heat and flame. When heated to decomposition it emits toxic fumes of NO_x.

DXZ000 CAS:7631-98-3 HR: 3
N-DODECYLSARCOSINE SODIUM SALT
mf: $C_{15}H_{30}NO_2$•Na mw: 279.45

SYN: SODIUM-N-LAURYL SARCOSINE

TOXICITY DATA WITH REFERENCE
ivn-mus LD50:180 mg/kg CSLNX* NX#00171

CONSENSUS REPORTS: Reported in EPA TSCA Inventory.

SAFETY PROFILE: Poison by intravenous route. When heated to decomposition it emits toxic fumes of NO_x and Na_2O.

DYA000 CAS:15826-16-1 HR: 2
DODECYL SODIUM ETHOXYSULFATE
mf: $C_{14}H_{30}O_5S$•Na mw: 333.49

SYNS: 2-(DODECYLOXY)ETHANOL HYDROGEN SULFATE SODIUM SALT □ SODIUM LAURYL ETHOXYSULPHATE

TOXICITY DATA WITH REFERENCE
orl-rat LD50:1995 mg/kg FCTXAV 5,763,67

CONSENSUS REPORTS: Reported in EPA TSCA Inventory.

SAFETY PROFILE: Moderately toxic by ingestion. When heated to decomposition it emits very toxic fumes of SO_x and Na_2O.

DYA200 CAS:765-15-1 HR: 2
n-DODECYL THIOCYANATE
mf: $C_{13}H_{25}NS$ mw: 227.45

SYNS: ENT 114 □ LAURYL RHODANATE □ LAURYL THIOCYANATE □ LORO □ LOROL THIOCYANATE □ 1-THIOCYANATODODECANE □ THIOCYANIC ACID, DODECYL ESTER

TOXICITY DATA WITH REFERENCE
skn-rat 350 mg/7D-I open MOD JPTLAS 49,363,39
orl-rat LD50:1250 mg/kg 28ZEAL 4,205,69
scu-mus LDLo:20 g/kg JIHTAB 18,310,36
orl-cat LDLo:2000 mg/kg JIHTAB 18,310,36

SAFETY PROFILE: Moderately toxic by ingestion. A skin irritant. When heated to decomposition it emits very toxic fumes of NO_x and SO_x. See also THIOCYANATES.

DYA400 CAS:5416-74-0 HR: 3
(DODECYLTHIO)PHENYLMERCURY
mf: $C_{18}H_{30}HgS$ mw: 479.13

SYN: DODECYL PHENYLMERCURI SULFIDE

CONSENSUS REPORTS: Mercury and its compounds are on the Community Right-To-Know List.

OSHA PEL: CL 0.1 mg(Hg)/m³ (skin)
ACGIH TLV: TWA 0.1 mg(Hg)/m³ (skin)
NIOSH REL: (Mercury, Aryl and Inorganic): CL 0.1 mg/m³ (skin)

SAFETY PROFILE: Probably a poison. When heated to decomposition it emits very toxic fumes of Hg and SO_x. See also MERCURY COMPOUNDS.

DYA600 CAS:1399-80-0 HR: 3
DODECYL-p-TOLYL TRIMETHYL AMMONIUM CHLORIDE

PROP: Aqueous preparation containing approximately 40% methyl dodecylbenzyl trimethyl ammonium chloride, and 10% methyl dodecylxylene bis(trimethyl ammonium chloride).

SYNS: ALKYL(C_{9-15})TOLYL METHYLTRIMETHYL AMMONIUM CHLORIDE □ HYAMINE 2389 □ METHYL DODECYL BENZYL AMMONIUM CHLORIDE

TOXICITY DATA WITH REFERENCE
skn-man 50 mg/2D MLD PSTGAW 20,16,53
eye-rbt 100 µg SEV PSTGAW 20,16,53
orl-rat LD50:389 mg/kg PCOC** -,592,66
ipr-rat LD50:10 mg/kg AEHA** 5176T6-66/67

SAFETY PROFILE: Poison by ingestion, intraperitoneal, and intravenous routes. A skin and severe eye irritant. When heated to decomposition it emits very toxic fumes of NO_x, NH_3, and Cl^-.

DYA800 CAS:4484-72-4 HR: 3
DODECYLTRICHLOROSILANE
DOT: UN 1771
mf: $C_{12}H_{25}Cl_3Si$ mw: 303.81

PROP: Colorless to yellow liquid. Bp: 152–153° @ 3 mm, d: 1.026 @ 25°/25°.

SYN: TRICHLORODODECYLSILANE

CONSENSUS REPORTS: Reported in EPA TSCA Inventory.

DOT CLASSIFICATION: 8; *Label:* Corrosive

SAFETY PROFILE: A poison. A corrosive irritant to the eyes, skin, and mucous membranes. Readily hydrolyzed by moisture with the production of hydrochloric acid. When heated to decomposition it emits toxic fumes of Cl^-. See also CHLOROSILANES.

DYA810 CAS:1119-94-4 HR: 3
DODECYLTRIMETHYLAMMONIUM BROMIDE
mf: $C_{15}H_{34}N \cdot Br$ mw: 308.41

SYN: AMMONIUM, DODECYLTRIMETHYL-, BROMIDE

TOXICITY DATA WITH REFERENCE
ivn-rat LD50:6800 µg/kg APTOA6 47,17,80
ivn-mus LD50:5200 µg/kg APTOA6 47,17,80

CONSENSUS REPORTS: Reported in EPA TSCA Inventory.

SAFETY PROFILE: Poison by intravenous route. When heated to decomposition it emits toxic vapors of NO_x and Br^-.

DYA850 HR: 2
DODICIN HYDROCHLORIDE
mf: $C_{18}H_{39}N_3O_2 \cdot ClH$ mw: 366.06

SYNS: DODECYLBIS(AMINOETHYL)GLYCINE HYDROCHLORIDE □ LEBON 15 HYDROCHLORIDE □ TEGO 51

TOXICITY DATA WITH REFERENCE
ipr-rat LD50:534 mg/kg KSRNAM 12,1821,78
scu-rat LD50:5262 mg/kg KSRNAM 12,1821,78
ipr-mus LD50:590 mg/kg KSRNAM 12,1821,78
scu-mus LD50:2145 mg/kg KSRNAM 12,1821,78

SAFETY PROFILE: Moderately toxic by intraperitoneal and subcutaneous routes. When heated to decomposition it emits toxic fumes of NO_x and HCl.

DYA875 HR: 3
DOG LAUREL

PROP: Deciduous or evergreen shrubs with alternating leaves and clusters of white or pink flowers. They grow wild in the coastal states from Virginia to Florida, Louisiana, Tennessee, and California.

SYNS: DOG HOBBLE □ FETTER BUSH □ LEUCOTHOE (VARIOUS SPECIES) □ PEPPER BUSH □ SWEET BELLS □ SWITCH IVY □ WHITE OSTER

SAFETY PROFILE: The leaves and nectar contain poisonous grayano toxins (andromedotoxins). Ingestion of these plant parts results in immediate pain in the mouth and may be followed several hours later by vomiting, diarrhea, headache, impaired vision, irregular heartbeat, severe low blood pressure, coma, convulsions, and death.

DYB000 CAS:482-49-5 HR: 2
DOISYNOLIC ACID
mf: $C_{18}H_{24}O_3$ mw: 288.42

SYNS: ACIDO DOISYNOLICO (SPANISH) □ 1-ETHYL-7-HYDROXY-2-METHYL-1,2,3,4,4a,9,10,10a-OCTAHYDROPHENANTHRENE-2-CARBOXYLIC ACID □ 1-ETHYL-1,2,3,4,4a,9,10,10a-OCTAHYDRO-7-HYDROXY-2-METHYL-2-PHENANTHRENECARBOXYLIC ACID □ 3-HYDROXY-16,7-SECOESTRA-1,3,5(10)-TRIEN-17-OIC ACID

TOXICITY DATA WITH REFERENCE
orl-gpg TDLo:72 mg/kg/12W-I:ETA,REP RSABAC 25,215,49

SAFETY PROFILE: Experimental reproductive effects. Questionable carcinogen with experimental tumorigenic data. When heated to decomposition it emits acrid smoke and irritating fumes.

DYB250 CAS:4956-15-4 HR: 3
DOLANTIN-N-OXIDE HYDROCHLORIDE
mf: $C_{15}H_{21}NO_3 \cdot ClH$ mw: 299.83

SYN: 1-METHYL-4-PHENYLISONIPECOTIC ACID ETHYL ESTER-1-OXIDE HYDROCHLORIDE

TOXICITY DATA WITH REFERENCE
orl-mus LD50:2 g/kg AITEAT 15,290,67
ipr-mus LD50:575 mg/kg AITEAT 15,290,67
scu-mus LD50:975 mg/kg AITEAT 15,290,67
ivn-mus LD50:300 mg/kg AITEAT 15,290,67

SAFETY PROFILE: Poison by intravenous. Moderately toxic by ingestion, subcutaneous, and intraperitoneal routes. When heated to decomposition it emits toxic fumes of NO_x and HCl. See also ESTERS.

DYB300 CAS:96081-07-1 HR: 2
DOLOCONCURASE

TOXICITY DATA WITH REFERENCE
orl-rat LD50:3681 mg/kg ACTTDZ 10,365,84
ivn-rat LD50:1301 mg/kg ACTTDZ 10,365,84
ims-rat LD50:3088 mg/kg ACTTDZ 10,365,84
orl-mus LD50:3273 mg/kg ACTTDZ 10,365,84
ivn-mus LD50:1213 mg/kg ACTTDZ 10,365,84
ims-mus LD50:2926 mg/kg ACTTDZ 10,365,84

SAFETY PROFILE: Moderately toxic by ingestion, intravenous, and intramuscular routes.

DYB400 CAS:17140-78-2 HR: 3
DOLOXENE
mf: $C_{10}H_8O_3S \cdot C_{22}H_{29}NO_2$ mw: 547.76

SYNS: DARVON-N □ 4-(DIMETHYLAMINO)-3-METHYL-1,2-DIPHENYL-2-BUTANOL PROPIONATE-2-NAPHTHALENESULFONATE □ PROPOXYPHENE N □ PROPOXYPHENE-2-NAPHTHALENESULFONATE □ PROPOXYPHENE NAPSYLATE

TOXICITY DATA WITH REFERENCE
orl-rat TDLo:3 g/kg (female 14-22D post):REP TXAPA9 19,471,71
orl-rat TDLo:12 g/kg (14-22D preg/21D post):REP TXAPA9 19,471,71
orl-rat TDLo:4 g/kg (female 6-15D post):REP TXAPA9 19,471,71
orl-man TDLo:129 mg/kg:BAH JTCTDW 23,347,85
orl-rat LD50:647 mg/kg TXAPA9 19,445,71
orl-mus LD50:915 mg/kg TXAPA9 19,445,71
orl-rbt LDLo:183 mg/kg TXAPA9 19,445,71

SAFETY PROFILE: Poison by ingestion. Human systemic effects by ingestion: somnolence, vascular changes. Experimental reproductive effects. When heated to decomposition it emits very toxic fumes of NO_x and SO_x. See also SULFONATES.

DYB600 **CAS:303-69-5** **HR: 3**
DOMINAL
mf: $C_{16}H_{19}N_3S$ mw: 285.44

PROP: A liquid. Bp: 217–219° @ 0.7 mm.

SYNS: D 206 ☐ DIMETHYLAMINO-N-PROPYL-THIOPHENYLPYRIDYLAMINE ☐ PROTHIPENDYL

TOXICITY DATA WITH REFERENCE
ivn-rat LD50:25 mg/kg THERAP 18,373,63
orl-mus LD50:415 mg/kg ARZNAD 8,489,58
ipr-mus LD50:155 mg/kg ARZNAD 8,489,58
scu-mus LD50:102 mg/kg AIPTAK 149,374,64

SAFETY PROFILE: Poison by intravenous, subcutaneous, and intraperitoneal routes. Moderately toxic by ingestion. When heated to decomposition it emits very toxic fumes of SO_x and NO_x.

DYB800 **CAS:1225-65-6** **HR: 3**
DOMINAL HYDROCHLORIDE
mf: $C_{16}H_{19}N_3S\cdot ClH$ mw: 321.90

SYNS: 10-(3-DIMETHYLAMINOPROPYL)-1-AZAPHENOTHIAZINE HYDROCHLORIDE ☐ ((4-DIMETHYLAMINOPROPYLPYRIDO (3,2b) BENZOTHIAZINE)) HYDROCHLORIDE ☐ PROTHIPENDYL HYDROCHLORIDE

TOXICITY DATA WITH REFERENCE
orl-rat LD50:610 mg/kg 27ZQAG -,292,72
ipr-rat LD50:115 mg/kg 27ZQAG -,292,72
scu-rat LD50:1323 mg/kg 27ZQAG -,292,72
ivn-rat LD50:110 mg/kg 27ZQAG -,292,72

SAFETY PROFILE: Poison by intravenous and intraperitoneal routes. Moderately toxic by ingestion and subcutaneous routes. When heated to decomposition it emits very toxic fumes of NO_x, SO_x, and HCl.

DYB875 **CAS:57808-66-9** **HR: 3**
DOMPERIDONE
mf: $C_{22}H_{24}ClN_5O_2$ mw: 425.96

PROP: Crystals from DMF/water. Mp: 242.5°.

SYNS: 5-CHLORO-1-(1-(3-(2-OXO-1-BENZIMIDAZOLINYL)PROPYL)-4-PIPERIDYL)-2-BENZIMIDAZOLINONE ☐ KW-5338 ☐ MOTILIUM

TOXICITY DATA WITH REFERENCE
orl-wmn TDLo:4800 μg/kg/3D-I:REP BMJOAE 286,1395,83
orl-rat TDLo:2200 mg/kg (female 7-17D post):REP YACHDS 8,4045,80
orl-mus TDLo:2 g/kg (female 6-15D post):REP YACHDS 8,4045,80
orl-mus TDLo:700 mg/kg (female 6-15D post):TER YACHDS 8,4045,80
orl-rat TDLo:770 mg/kg (female 7-17D post):TER YACHDS 8,4045,80
orl-mus TDLo:700 mg/kg (female 6-15D post):REP YACHDS 8,4045,80
orl-rat TDLo:81 mg/kg (male 60D pre):REP YACHDS 8,4045,80
ivn-rat TDLo:75 mg/kg (female 30D pre):REP YACHDS 8,4061,80
ivn-rat TDLo:600 mg/kg (male 30D pre):REP YACHDS 8,4061,80
ivn-rat TDLo:75 mg/kg (male 30D pre):REP YACHDS 8,4061,80
orl-rat TDLo:2200 mg/kg (female 7-17D post):TER YACHDS 8,4045,80
orl-rat TDLo:250 mg/kg (female 19-22D post):TER YACHDS 8,4045,80
orl-inf TDLo:1170 μg/kg/1D-I:EYE JOPDAB 108,630,86
ivn-hmn TDLo:714 μg/kg:CNS BMJOAE 288,1728,84
ivn-hmn LDLo:429 μg/kg:CNS BMJOAE 288,1728,84
ivn-wmn TDLo:1 mg/kg:CVS BMJOAE 289,1579,84
ivn-wmn TDLo:600 μg/kg/1D:CVS LANCAO 2,385,85
ims-man TDLo:200 μg/kg:CNS,EYE LANCAO 2,1259,80
rec-wmn TDLo:7200 μg/kg/30H-I:EYE JOPDAB 105,852,84
orl-rat LD50:5243 mg/kg YACHDS 8,3991,80
ipr-rat LD50:61,200 μg/kg IYKEDH 13,1128,82
ivn-rat LD50:41,700 μg/kg YACHDS 8,3991,80
ivn-mus LD50:46,500 μg/kg YACHDS 8,3991,80
ivn-dog LD50:42,700 μg/kg AIPTAK 244,130,80
ivn-gpg LD50:42,900 μg/kg AIPTAK 244,130,80

SAFETY PROFILE: Poison by intravenous and intraperitoneal routes. Human systemic effects by intravenous and intramuscular routes: eye effects, convulsions, muscle spasms, and cardiac arrhythmias. Experimental teratogenic and reproductive effects. When heated to decomposition it emits toxic fumes of Cl^- and NO_x.

DYC000 **CAS:87-52-5** **HR: 3**
DONAXINE
mf: $C_{11}H_{14}N_2$ mw: 174.27

PROP: A solid. Mp: 138–139°.

SYNS: β-DIMETHYLAMINOMETHYLINDOLE ☐ 3-(DIMETHYLAMINOMETHYL)INDOLE ☐ GRAMIN ☐ GRAMINE

TOXICITY DATA WITH REFERENCE
ipr-rat LDLo:250 mg/kg NCNSA6 5,11,53
ipr-mus LD50:122 mg/kg PSYPAG 16,385,70
ivn-mus LD50:46 mg/kg JPETAB 105,130,52

CONSENSUS REPORTS: Reported in EPA TSCA Inventory.

SAFETY PROFILE: Poison by intraperitoneal and intravenous routes. When heated to decomposition it emits toxic fumes of NO_x.

DYC200 **CAS:5796-14-5** **HR: 3**
l-DOPA HYDROCHLORIDE
mf: $C_9H_{11}NO_4\cdot ClH$ mw: 233.67

SYNS: l-3-(3,4-DIHYDROXYPHENYL)ALANINE ☐ 3-HYDROXY-l-TYROSINE HYDROCHLORIDE

TOXICITY DATA WITH REFERENCE
orl-mus LD50:1460 mg/kg TXAPA9 28,1,74

ivn-mus LD50:527 mg/kg TXAPA9 28,1,74
orl-rbt LDLo:950 mg/kg TXAPA9 28,1,74
ivn-rbt LD50:144 mg/kg TXAPA9 28,1,74

SAFETY PROFILE: Poison by intravenous route. Moderately toxic by ingestion. Used to treat Parkinson's disease. When heated to decomposition it emits very toxic fumes of NO_x and HCl.

DYC300 CAS:7101-51-1 HR: 2
l-DOPA METHYL ESTER
mf: $C_{10}H_{13}NO_4$ mw: 211.24

SYN: l-3-(3,4-DIHYDROXYPHENYL)ALANINE METHYL ESTER

TOXICITY DATA WITH REFERENCE
dni-hmn:oth 3300 nmol/L CNREA8 40,1414,80
dni-mus:oth 3300 nmol/L CNREA8 40,1414,80
ipr-mus LD50:1000 mg/kg CTRRDO 63,991,79

SAFETY PROFILE: Moderately toxic by intraperitoneal route. Human mutation data reported. When heated to decomposition it emits toxic fumes of NO_x.

DYC400 CAS:51-61-6 HR: 3
DOPAMINE
mf: $C_8H_{11}NO_2$ mw: 153.20

SYNS: 4-(2-AMINOETHYL)PYROCATECHOL □ 3-HYDROXYTYRAMINE

TOXICITY DATA WITH REFERENCE
dnd-esc 10 mg/L MUREAV 124,9,83
dnd-hmn:fbr 50 mg/L MUREAV 137,17,84
scu-rat TDLo:100 mg/kg (female 10-14D post):REP AJOGAH 104,578,69
ipr-rat TDLo:70 mg/kg (1-7D preg):REP TXAPA9 40,427,77
scu-rat TDLo:300 mg/kg (female 30D pre):REP AJOGAH 104,578,69
scu-rat TDLo:100 mg/kg (female 10-14D post):TER AJOGAH 104,578,69
ipr-rat LD50:163 mg/kg TXAPA9 88,433,87
ipr-mus LD50:950 mg/kg OYYAA2 8,835,74
ivn-mus LD50:59 mg/kg OYYAA2 8,835,74
icv-mus LD50:74 mg/kg TYKNAQ 27,131,80
ivn-dog LD50:79 mg/kg ARZNAD 28,2208,78

SAFETY PROFILE: Poison by intravenous, intracervical, and intraperitoneal routes. Experimental teratogenic and reproductive effects. Human mutation data reported. A neurotransmitter. An adrenergic agent. When heated to decomposition it emits toxic fumes of NO_x.

DYC600 CAS:62-31-7 HR: 3
DOPAMINE HYDROCHLORIDE
mf: $C_8H_{11}NO_2{\cdot}ClH$ mw: 189.66

PROP: A solid. Mp: 240–241° (decomp).

SYNS: 4-(2-AMINOETHYL)-1,2-BENZENEDIOL HYDROCHLORIDE □ 4-(2-AMINOETHYL)PYROCATECHOL HYDROCHLORIDE □ DOPAMINE CHLORIDE □ m-HYDROXYTYRAMINE HYDROCHLORIDE □ 3-HYDROXYTYRAMINE HYDROCHLORIDE □ INTROPIN

TOXICITY DATA WITH REFERENCE
dnd-mus:lvr 6 μmol/L JNCIAM 62,947,79
msc-mus:lym 93,800 μg/L MUREAV 124,9,83
ivn-rat TDLo:300 mg/kg (female 6-15D post):REP ARZNAD 28,2208,78
ipr-rat TDLo:120 mg/kg (30D pre):REP KSRNAM 8,2311,74
ipr-rat TDLo:6 mg/kg (male 30D pre):REP KSRNAM 8,2311,74
ivn-dom TDLo:361 mg/kg (female 15W post):TER AJOGAH 130,211,78
ipr-rat TDLo:2100 μg/kg (female 7-12D post):TER KSRNAM 8,2311,74
ipr-rat TDLo:7 mg/kg (1-7D preg):TER TXAPA9 38,251,76
orl-rat LD50:2859 mg/kg KSRNAM 8,2311,74
ipr-rat LD50:597 mg/kg KSRNAM 8,2311,74
scu-rat LD50:647 mg/kg KSRNAM 8,2311,74
ivn-rat LD50:4800 μg/kg KSRNAM 3,2311,74
orl-mus LD50:4361 mg/kg KSRNAM 8,2311,74
ipr-mus LD50:914 mg/kg KSRNAM 8,2311,74
scu-mus LD50:1366 mg/kg KSRNAM 8,2311,74
ivn-mus LD50:156 mg/kg KSRNAM 8,2311,74
ivn-rbt LD50:90 mg/kg KSRNAM 3,2311,74

SAFETY PROFILE: Poison by intravenous route. Moderately toxic by ingestion, intraperitoneal, and subcutaneous routes. Experimental teratogenic and reproductive effects. Mutation data reported. An antihypotensive. When heated to decomposition it emits very toxic fumes of NO_x and HCl. See also DOPAMINE.

DYC700 CAS:520-09-2 HR: 3
DOPAN
mf: $C_9H_{13}Cl_2N_3O_2$ mw: 266.15

PROP: Snow-white crystals. Decomp @ 178–179°. Practically insol in cold water, acetone, and benzene; sltly sol in alc.

SYNS: 5-(BIS(2-CHLOROETHYL)AMINO)-6-METHYLURACIL □ 5-(BIS(2-CLOROETIL)-AMINO)-6-METILURACILE (ITALIAN) □ CHLOROETHYLAMINOURACIL □ 2,6-DIHYDROXY-4-METHYL-5-BIS(2-CHLOROETHYL)AMINOPYRIMIDINE □ DOPANE □ ELDERFIELD PYRIMIDINE MUSTARD □ ENT 50,698 □ 4-METHYL-5-(BIS(β-CHLOROETHYL)AMINO)URACIL □ 6-METHYL-5-(BIS(2-CHLOROETHYL)AMINO)URACIL

TOXICITY DATA WITH REFERENCE
cyt-hmn:leu 2500 μg/L TSITAQ 15,1505,73
dni-mus:ast 2 mg/L NEOLA4 22,105,75
oms-mus:ast 2 mg/L NEOLA4 22,105,75
cyt-mky:kdy 100 mg/L TSITAQ 15,1505,73
orl-rat LD50:2400 μg/kg BCFAAI 101,630,62
ipr-mus LDLo:3 mg/kg FATOAO 34,83,71

SAFETY PROFILE: Poison by ingestion and intraperitoneal routes. Human mutation data reported. When heated to decomposition it emits toxic fumes of Cl^- and NO_x.

DYC800 CAS:77-21-4 HR: 3
DORIDEN
mf: $C_{13}H_{15}NO_2$ mw: 217.29

PROP: dl-Form: Crystals from ether or from ethyl acetate + pet ether. Mp: 84°. Freely sol in ethyl acetate, acetone, ether, chloroform; sol in ethanol, methanol.

D

Practically insol in water. d-Form: Crystals. Mp: 102.5–103°, refractive index: (α) (20/D) +176° (methanol). l-Form: Crystals. Mp: 102–103°, refr index: (α) (20/D) −181° (methanol).

SYNS: ALFIMID □ CC 11511 □ DORIDEN-SED □ ELRODORM □ 3-ETHYL-3-PHENYL-2,6-DIKETOPIPERIDINE □ 3-ETHYL-3-PHENYL-2,6-DIOXOPIPERIDINE □ α-ETHYL-α-PHENYLGLUTARIMIDE □ 2-ETHYL-2-PHENYLGLUTARIMIDE □ 3-ETHYL-3-PHENYL-2,6-PIPERIDINEDIONE □ GIMID □ GLIMID □ GLUTATHIMID □ GLUTETHIMID □ GLUTETHIMIDE □ GLUTETIMIDE □ NOXYRON □ 3-PHENYL-3-ETHYL-2,6-DIKETOPIPERIDINE □ 3-PHENYL-3-ETHYL-2,6-DIOXOPIPERIDINE □ α-PHENYL-α-ETHYLGLUTARIC ACID IMIDE □ 2-PHENYL-2-ETHYLGLUTARIC ACID IMIDE □ α-PHENYL-α-ETHYLGLUTARIMIDE □ SARODORMIN

TOXICITY DATA WITH REFERENCE
orl-rat TDLo:2250 mg/kg (6-14D preg):REP COREAF 256,1841,63
orl-rbt TDLo:1350 mg/kg (female 8-16D post):TER TXAPA9 10,244,67
orl-mus TDLo:2250 mg/kg (female 6-14D post):REP COREAF 256,1841,63
orl-rat TDLo:2250 mg/kg (6-14D preg):TER COREAF 256,1841,63
ipr-mus TDLo:300 mg/kg (female 10D post):TER CAJPBD 3,2,63
orl-chd TDLo:25 mg/kg:CNS AJDCAI 130,507,76
orl-hmn TDLo:171 mg/kg JAMAAP 214,1704,70
unr-man LDLo:147 mg/kg 85DCAI 2,73,70
orl-rat LD50:600 mg/kg 27ZQAG -,233,72
orl-mus LD50:360 mg/kg 27ZQAG -,233,72
ipr-mus LD50:208 mg/kg APPHAX 32,243,75
orl-dog LD50:500 mg/kg SMWOAS 85,305,55
orl-mky LDLo:300 mg/kg TXAPA9 29,75,74

CONSENSUS REPORTS: EPA Genetic Toxicology Program.

SAFETY PROFILE: Poison by ingestion and intraperitoneal routes. Human systemic effects by ingestion: pupillary dilation, ataxia, somnolence, coma, and blood pressure depression. An experimental teratogen. Other experimental reproductive effects. When heated to decomposition it emits toxic fumes of NO_x. Caution: May be habit forming. This is a controlled substance (depressant) listed in the U.S. Code of Federal Regulations, Title 21 Part 1308.13 (1985).

DYC875 CAS:113-53-1 HR: 3
DOSULEPIN
mf: $C_{19}H_{21}NS$ mw: 295.47

PROP: Crystals from EtOH/Et_2O. Mp: 226–227°, bp: 171–172° @ 0.05 mm.

SYNS: 3-DIBENZO(b,e)THIEPIN-11(6H)-YLIDENE-N,N-DIMETHYL-1-PROPAMINE □ 11-(3-DIMETHYLAMINOPROPYLIDENE)-6,11-DIHYDRODIBENZO(b,e)THIEPIN □ N,N-DIMETHYLDIBENZO(b,e)THIEPIN-$\Delta^{11(6H),\gamma}$PROPYLAMINE □ DOTHIEPIN □ IZ 914 □ PROTHIADEN □ PROTHIADEN SPOFA

TOXICITY DATA WITH REFERENCE
cyt-mus-ipr 100 mg/kg ACNSAX 15,114,73
dlt-mus-ipr 150 mg/kg ACNSAX 15,114,73
ipr-mus TDLo:150 mg/kg (6D male):REP ACNSAX 15,114,73
orl-inf TDLo:100 mg/kg:CNS,CVS BMJOAE 288,1800,84
orl-wmn TDLo:4500 μg/kg:CNS BMJOAE 2,97,77

CONSENSUS REPORTS: EPA Genetic Toxicology Program.

SAFETY PROFILE: Human systemic effects by ingestion: central nervous system including coma, and cardiac arrhythmia. Mutation data reported. Experimental reproductive effects. When heated to decomposition it emits toxic fumes of NO_x and SO_x.

DYC900 CAS:544-85-4 HR: 3
DOTRIACONTANE
mf: $C_{32}H_{66}$ mw: 450.98

TOXICITY DATA WITH REFERENCE
ivn-mus LD50:100 mg/kg CSLNX* NX#00741

CONSENSUS REPORTS: Reported in EPA TSCA Inventory.

SAFETY PROFILE: Poison by intravenous route. When heated to decomposition it emits acrid smoke and irritating vapors.

DYD200 CAS:35944-83-3 HR: 3
DOWCO 133
mf: $C_{10}H_{11}Cl_3NO_2PS$ mw: 346.60

SYN: N-ISOPROPYLPHOSPHORAMIDOTHIOIC ACID-O-(2,4,5-TRICHLOROPHENYL)ESTER

TOXICITY DATA WITH REFERENCE
orl-rat LDLo:31 mg/kg TXAPA9 21,315,72
orl-bwd LD50:7 mg/kg TXAPA9 21,315,72

SAFETY PROFILE: Poison by ingestion. A pesticide. When heated to decomposition it emits very toxic fumes of Cl^-, NO_x, PO_x, and SO_x.

DYD400 CAS:2213-84-5 HR: 3
DOWCO 159
mf: $C_9H_{11}Cl_3NO_3P$ mw: 318.53

SYN: ETHYLPHOSPHORAMIDIC ACID, METHYL-(2,4,5-TRICHLOROPHENYL) ESTER

TOXICITY DATA WITH REFERENCE
orl-bwd LD50:56 mg/kg TXAPA9 21,315,72

SAFETY PROFILE: Poison by ingestion. When heated to decomposition it emits very toxic fumes of Cl^-, NO_x, and PO_x. See also ESTERS.

DYD600 CAS:35944-82-2 HR: 3
DOWCO 160
mf: $C_8H_9Cl_3NO_3P$ mw: 304.50

SYN: ETHYLPHOSPHORAMIDIC ACID-(2,4,5-TRICHLOROPHENYL) ESTER

TOXICITY DATA WITH REFERENCE
orl-bwd LD50:5600 μg/kg TXAPA9 21,315,72

SAFETY PROFILE: Poison by ingestion. When heated to decomposition it emits very toxic fumes of Cl^-, NO_x, and PO_x. See also ESTERS.

DYD800 **CAS:35944-84-4** **HR: 3**
DOWCO 177
mf: $C_4H_{12}NO_2PS$ mw: 169.20

SYN: N-METHYL-PHOSPHORAMIDOTHIOIC ACID-O-ISOPROPYL ESTER

TOXICITY DATA WITH REFERENCE
orl-rat LD50:800 mg/kg TXAPA9 21,315,72
orl-bwd LD50:100 mg/kg TXAPA9 21,315,72

SAFETY PROFILE: Poison by ingestion. When heated to decomposition it emits very toxic fumes of NO_x, PO_x, and SO_x. See also ESTERS.

DYE200 **CAS:4492-96-0** **HR: 3**
DOWCO-183
mf: $C_{13}H_{20}ClNO_3P$ mw: 305.77

SYNS: 2-CHLORO-4-(1,1-DIMETHYLPROPYL)PHENYL METHYL METHYLPHOSPHORAMIDATE ☐ ENT 27,192 ☐ METHYLPHOSPHORAMIDIC-2-CHLORO-4-(1,1-DIMETHYLPROPYL)PHENYL METHYL ESTER

TOXICITY DATA WITH REFERENCE
orl-rat LDLo:500 mg/kg ARSIM* 20,9,66
scu-gpg LDLo:100 mg/kg JEENAI 61,1261,68

SAFETY PROFILE: Poison by subcutaneous route. Moderately toxic by ingestion. When heated to decomposition it emits very toxic fumes of Cl^- and NO_x.

DYE400 **CAS:53908-27-3** **HR: 3**
DOWFUME EB-5

PROP: Contains ethylene dichloride (30%), carbon tetrachloride (63%), and ethylene dibromide (7%) (AMIHAB 13,1,56).

TOXICITY DATA WITH REFERENCE
orl-rat LD50:660 mg/kg AMIHAB 13,1,56
ihl-rat LCLo:920 ppm/4H JAFCAU 2,1318,54
orl-rbt LD50:290 mg/kg JAFCAU 2,1318,54
skn-rbt LDLo:500 mg/kg JAFCAU 2,1318,54
orl-gpg LD50:280 mg/kg JAFCAU 2,1318,54
orl-ckn LD50:780 mg/kg JAFCAU 2,1318,54

SAFETY PROFILE: Poison by ingestion. Mildly toxic by skin contact and inhalation. When heated to decomposition it emits very toxic fumes of Cl^- and Br^-. See also components as listed.

DYE409 **CAS:1668-19-5** **HR: 3**
DOXEPIN
mf: $C_{19}H_{21}NO$ mw: 279.41

PROP: Oil. Bp: 154–157° @ 0.03 mm.

SYNS: 11-(3-DIMETHYLAMINOPROPYLIDENE)-6,11-DIHYDRODIBENZ(b,e)OXIPIN ☐ N,N-DIMETHYLDIBENZ(b,e)OXEPIN-$\Delta^{11(6H)}$-γ-PROPYLAMINE

TOXICITY DATA WITH REFERENCE
orl-hmn LDLo:60 mg/kg CTOXAO 10,327,77
orl-rat LD50:147 mg/kg ARZNAD 15,863,65
ipr-rat LD50:182 mg/kg ARZNAD 15,863,65
ivn-rat LD50:16 mg/kg ARZNAD 15,863,65
orl-mus LD50:135 mg/kg ARZNAD 15,863,65
ipr-mus LD50:79 mg/kg ARZNAD 15,863,65
ivn-mus LD50:26 mg/kg ARZNAD 15,863,65
ivn-rbt LD50:11 mg/kg ARZNAD 15,863,65

SAFETY PROFILE: Human poison by ingestion. Experimental poison by ingestion, intravenous, and intraperitoneal routes. A sedative and hypnotic used as an antianxiety agent. When heated to decomposition it emits toxic fumes of NO_x.

D

DYE415 **CAS:3094-09-5** **HR: 2**
DOXIFLURIDINE
mf: $C_9H_{11}FN_2O_5$ mw: 246.22

PROP: Needles. Mp: 192–193°.

SYNS: 5′-DEOXY-5-FLUOROURIDINE ☐ 5′-DFUR ☐ RO 21-9738

TOXICITY DATA WITH REFERENCE
orl-rat TDLo:1100 mg/kg (7-17D preg):REP YACHDS 13(Suppl 2),481,85
orl-rat TDLo:2200 mg/kg (7-17D preg):REP YACHDS 13(Suppl 2),481,85
orl-rat TDLo:550 mg/kg (7-17D preg):TER YACHDS 13(Suppl 2),481,85
orl-rat TDLo:5950 mg/kg (male 14W pre):REP YACHDS 13(Suppl 2),469,85
orl-rat LD50:3390 mg/kg YACHDS 13(Suppl 2),209,85
orl mus LDLo:5 g/kg YACHDS 13(Suppl 2),209,85

SAFETY PROFILE: Moderately toxic by ingestion. An experimental teratogen. Other experimental reproductive effects. When heated to decomposition it emits toxic fumes of F^- and NO_x.

DYE425 **CAS:564-25-0** **HR: 2**
DOXYCYCLINE
mf: $C_{22}H_{24}N_2O_8$ mw: 444.48

SYNS: α-6-DEOXY-5-HYDROXYTETRACYCLINE ☐ α-6-DEOXYOXYTETRACYCLINE ☐ 6-α-DEOXY-5-OXYTETRACYCLINE ☐ DOXICICLINA (ITALIAN) ☐ GS-3065 ☐ 5-HYDROXY-α-6-DEOXYTETRACYCLINE ☐ LIVIATIN ☐ VIBRAMYCIN

TOXICITY DATA WITH REFERENCE
dni-hmn:lym 3750 µg/L BCPHBM 16,127,83
orl-mus TDLo:750 mg/kg (female 2D post):TER ARZNAD 36,219,86
ipr-rat LD50:378 mg/kg THERAP 23,575,68
ivn-rat LD50:228 mg/kg PBPSDY 2,305,79
orl-mus LD50:1870 mg/kg GICTAL 17,276,70
ipr-mus LD50:410 mg/kg GICTAL 17,276,70

CONSENSUS REPORTS: EPA Genetic Toxicology Program.

SAFETY PROFILE: Moderately toxic by ingestion, and intraperitoneal routes. An experimental teratogen. Human mutation data reported. When heated to decomposition it emits toxic fumes of NO_x. See also TETRACYCLINE.

DYE500 **CAS:469-21-6** **HR: 3**
DOXYLAMINE
mf: $C_{17}H_{22}N_2O$ mw: 270.41

PROP: A liquid. Bp: 137–141° @ 0.5 mm. Sol in acids.

SYNS: 2-(α-(2-(DIMETHYLAMINO)ETHOXY)-α-METHYLBENZYL) PYRIDINE □ 2-DIMETHYLAMINOETHOXYPHENYLMETHYL-2-PICOLINE □ N,N-DIMETHYL-2-(1-PHENYL-1-(2-PYRIDINYL)ETHOXY)ETHANAMINE (9CI) □ NCI C60684 □ PHENYL-2-PYRIDYLMETHYL-β-N,N-DIMETHYLAMINOETHYL ETHER

TOXICITY DATA WITH REFERENCE
scu-rat LD50:440 mg/kg DPIRDU 2(5),17,82
orl-mus LD50:470 mg/kg DPIRDU 2(5),17,82
scu-mus LD50:460 mg/kg DPIRDU 2(5),17,82
ivn-mus LD50:62 mg/kg DPIRDU 2(5),17,82
orl-rbt LD50:250 mg/kg DPIRDU 2(5),17,82
ivn-rbt LD50:49 mg/kg DPIRDU 2(5),17,82

SAFETY PROFILE: Poison by ingestion and intravenous routes. Moderately toxic by subcutaneous route. When heated to decomposition it emits toxic fumes of NO_x.

DYE550 **CAS:54622-43-4** **HR: 2**
DPF
mf: $C_7H_{22}N_2O_{13}P_4$ mw: 466.19

SYNS: DPF-1 □ ((2-HYDROXY-1,3-PROPANEDIYL)BIS(NITRILOBIS (METHYLENE)))TETRAKISPHOSPHONIC ACID □ (2-HYDROXY-1,3-PROPYLENEDIAMINE)-N,N,N'',N''-TETRAMETHYLENEPHOSPHORIC ACID □ 2-OXY-1,3-PROPYLENEDIAMINE-N,N,N',N'-TETRAMETHYLENEPHOSPHONIC ACID

TOXICITY DATA WITH REFERENCE
orl-rat TDLo:21 g/kg (male 30D pre):REP GISAAA 53(6),77,88
orl-rat LD50:4300 mg/kg GTPZAB 29(4),45,85
orl-mus LD50:2800 mg/kg GISAAA 49(5),81,84
orl-gpg LD50:3800 mg/kg GTPZAB 29(4),45,85

CONSENSUS REPORTS: Cyanide and its compounds are on the Community Right-To-Know List.

SAFETY PROFILE: Moderately toxic by ingestion. Experimental reproductive effects. When heated to decomposition it emits toxic fumes of PO_x and NO_x. See also NITRILES.

DYE600 **CAS:523-87-5** **HR: 3**
DRAMAMINE
mf: $C_{17}H_{21}NO{\cdot}C_7H_7ClN_4O_2$ mw: 470.02

PROP: A solid. Mp: 102–107°.

SYNS: AMOSYT □ ANAUTINE □ ANDRAMINE □ AVIOMARIN □ o-BENZHYDRYLDIMETHYLAMINOETHANOL-8-CHLOROTHEOPHYLLINATE □ 2-(BENZHYDRYLOXY)-N,N-DIMETHYLETHYLAMINE with 8-CHLOROTHEOPHYLLINE □ CHLORANAUTINE □ DIAMARIN □ DIMENHYDRINATE □ DIPHENHYDRINATE □ DRAMAMIN □ DRAMARIN □ DRAMYL □ DROMYL □ ELDODRAM □ ETHYLAMINE-2-(DIPHENYLMETHOXY)-N,N-DIMETHYL, compound with 8-CHLOROTHEOPHYLLINE (1:1) □ GRAVINOL □ GRAVOL □ MENHYDRINATE □ NCI-C60639 □ NEO-NAVIGAN □ NOVAMIN □ NOVAMINE □ PERMITAL □ REISE-ENGLETTEN □ SUPREMAL □ TEODRAMIN □ TRAVELIN □ TRAVELMIN □ VOMEX A □ XAMAMINA

TOXICITY DATA WITH REFERENCE
mmo-sat 333 μg/plate ENMUDM 9(Suppl 9),1,87
orl-wmn LDLo:100 mg/kg:CNS,CVS,BAH AEMED3 22,1481,93
orl-man TDLo:11,400 μg/kg:CNS AJPSAO 128,1012,72
orl-rat LD50:1320 mg/kg NIIRDN 6,346,82
ivn-rat LD50:200 mg/kg CLDND* 6,346,82
ipr-mus LD50:149 mg/kg NIIRDN 6,346,82

CONSENSUS REPORTS: Reported in EPA TSCA Inventory.

SAFETY PROFILE: Poison by intraperitoneal and intravenous route. Moderately toxic by ingestion. A drug much used for motion sickness. Human systemic effects by ingestion: arrhythmias, convulsions, distorted perceptions, hallucinations, intracranial pressure increase. Mutation data reported. When heated to decomposition it emits very toxic fumes of NO_x and Cl^-. See also AMINES.

DYE700 **CAS:526-18-1** **HR: 3**
DRIOL
mf: $C_{13}H_{11}NO_3$ mw: 229.25

PROP: Crystals or light sensitive purplish-gray powder. Mp: 179°. Practically insol in cold water and acetic acid; sltly sol in warm water, benzene, and toluene; freely sol in methanol, ethanol, ether, and acetone.

SYNS: AUXOBIL □ BICHOL □ BILENE □ BILOCOL □ DRIBAZIL □ DRIOL-LABAZ □ ENIDRAN □ HYDROXYPHENYL SALICYLAMIDE □ N-(p-HYDROXYPHENYL)SALICYLAMIDE □ N-(4-HYDROXYPHENYL)SALICYLAMIDE □ p-HYDROXYPHENYLSALICYLAMIDE □ p'-HYDROXYSALICYLANILIDE □ 4'-HYDROXYSALICYLANILIDE □ KANOCHOL □ L 1718 □ OSALMID □ OSALMIDE □ OXAPHENAMID □ OXAPHENAMIDE □ PHPS □ SALMIDOCHOL □ SARYUURIN

TOXICITY DATA WITH REFERENCE
orl-rat LD50:6702 mg/kg NIIRDN 6,159,82
ipr-rat LD50:1621 mg/kg NIIRDN 6,159,82
orl-mus LD50:1050 mg/kg JJPAAZ 22,235,72
ipr-mus LD50:517 mg/kg NIIRDN 6,159,82
scu-mus LD50:1900 mg/kg AIPTAK 116,154,58
ivn-mus LD50:180 mg/kg CSLNX* NX#03234

SAFETY PROFILE: Poison by intravenous route. Moderately toxic by ingestion, intraperitoneal, and subcutaneous routes. When heated to decomposition it emits toxic fumes of NO_x.

DYF000 **CAS:125-72-4** **HR: 3**
l-DROMORAN TARTRATE
mf: $C_{17}H_{23}NO{\cdot}C_4H_6O_6$ mw: 407.51

SYNS: l-3-HYDROXY-N-METHYLMORPHINAN BITARTRATE □ l-3-HYDROXY-N-METHYLMORPHINAN TARTRATE □ LEMORAN □ LEVORPHAN TARTRATE □ LEVORPHANOL TARTRATE □ RO 1-5431/7

TOXICITY DATA WITH REFERENCE
oms-rat-ipr 2500 μg/kg RCOCB8 3,105,72
scu-mus TDLo:25 mg/kg (9D preg):TER DGDFA5 22,61,80
ivn-dog LD50:46 mg/kg AIPTAK 149,571,64
orl-rat LD50:150 mg/kg JPETAB 109,189,53
ipr-rat LD50:48 mg/kg TXCYAC 14,217,79
scu-rat LD50:110 mg/kg JPETAB 109,189,53
ivn-rat LD50:27 mg/kg AIPTAK 108,171,57
orl-mus LD50:285 mg/kg JPETAB 109,189,53
ipr-mus LD50:92 mg/kg JJPAAZ 11,101,62
scu-mus LD50:187 mg/kg JPETAB 109,189,53
ivn-mus LD50:32 mg/kg AIPTAK 105,221,56
ivn-rbt LD50:20 mg/kg AIPTAK 109,171,57

SAFETY PROFILE: Poison by ingestion, intravenous, intraperitoneal, and subcutaneous routes. Experimental teratogenic effects. Mutation data reported. When heated to decomposition it emits toxic fumes of NO_x.

DYF200 CAS:548-73-2 HR: 3
DROPERIDOL
mf: $C_{22}H_{22}FN_3O_2$ mw: 379.47

PROP: Crystals. Mp: 145–146.5°.

SYNS: DEHIDROBENZPERIDOL □ DEHYDROBENZPERIDOL □ DEIDROBENZPERIDOLO □ DHBP □ DIHIDROBENZPERIDOL □ DRIDOL □ DROLEPTAN □ 1-(1-(3-(p-FLUOROBENZOYL)PROPYL)-1,2,3,6-TETRAHYDRO-4-PYRIDYL)-2 -BENZIMIDAZOLINONE □ 1-(1-(4-(p-FLUOROPHENYL-4-OXOBUTYL))-1,2,3,6-TETRAHYDRO-4-PYRIDYL)-2-BENZIMIDAZOLINONE □ HALKAN □ INAPPIN □ INAPSIN □ INNOVAN □ INNOVAR □ INOPSIN □ INOVAL □ LEPTANAL □ LEPTOFEN □ MCN-JR-4749 □ PROPERIDOL □ R 4749 □ SINTOSIAN □ THALAMONAL □ VETKALM

TOXICITY DATA WITH REFERENCE
orl-hmn TDLo:223 μg/kg/12D-I:CNS ARZNAD 22,93,72
orl-rat LDLo:700 mg/kg TXAPA9 18,185,71
scu-rat LD50:20,520 mg/kg NIIRDN 6,440,82
ivn-rat LD50:30 mg/kg 27ZQAG -,187,72
ipr-mus LD50:80 mg/kg EJMCA5 13,387,78
scu-mus LD50:125 mg/kg TXAPA9 6,37,64
ivn-mus LD50:20 mg/kg 27ZQAG -,187,72

SAFETY PROFILE: Poison by intravenous, subcutaneous, and intraperitoneal routes. Moderately toxic by ingestion. Human systemic effects by ingestion: wakefulness, tremors, and muscle weakness. An antipsychotic agent. When heated to decomposition it emits very toxic fumes of F^- and NO_x.

DYF450 CAS:62-90-8 HR: 2
DURABOLIN
mf: $C_{27}H_{34}O_3$ mw: 406.61

PROP: Crystals. Mp: 95–96°.

SYNS: ACTIVIN □ DURABOL □ FENOBOLIN □ 17-β-HYDROXY-ESTR-4-ENE-3-ONE-3-PHENYLPROPIONATE □ 17-β-HYDROXY-ESTRA-4-EN-3-ONE-17-PHENYLPROPIONATE □ NANDROLIN □ NANDROLONE PHENPROPIONATE □ NANDROLONE PHENYLPROPIONATE □ NEROBIL □ NEROBIOLIL □ NORANDROLONE PHENYLPROPIONATE □ NORANDROSTENOLONE PHENYLPROPIONATE □ 19-NORTESTOSTERONE PHENYLPROPIONATE □ NPP □ NTPP □ 19NTPP □ (17-β)-17-(1-OXO-3-PHENYLPROPOXY)-ESTR-4-EN-3-ONE (9CI) □ PHENOBOLIN □ 17-β-PHENYLPROPIONYLOXY-4-ESTREN-3-ONE □ STRABOLENE □ SUPERANABOLON

TOXICITY DATA WITH REFERENCE
ims-rat TDLo:3750 μg/kg (female 9-14D post):REP OYYAA2 7,723,73
ims-rat TDLo:15 mg/kg (female 9-14D post):REP OYYAA2 7,723,73
ims-mus TDLo:240 mg/kg (female 7-12D post):REP OYYAA2 7,723,73
ims-rat TDLo:15 mg/kg (female 9-14D post):TER OYYAA2 7,723,73
scu-rat TDLo:60 mg/kg (female 7-12D post):REP STEDAM 18,731,71
scu-mus TDLo:1200 μg/kg (female 3D pre):REP ARZNAD 23,907,73
orl-rat TDLo:14,850 μg/kg (11D male):REP ARZNAD 26,1673,76
ims-rat TDLo:1400 μg/kg (male 14D pre):REP ARZNAD 23,907,73
ims-mus TDLo:60 mg/kg (female 7-12D post):TER OYYAA2 7,723,73
ims-rat TDLo:3750 μg/kg (female 9-14D post):TER OYYAA2 7,723,73
scu-rat TDLo:600 mg/kg (female 17-20D post):TER AVBIB9 13,71,74
ipr-rat LD50:595 mg/kg PCJOAU 20,143,86

SAFETY PROFILE: Moderately toxic by intraperitoneal routes. An experimental teratogen. Other experimental reproductive effects. When heated to decomposition it emits acrid smoke and fumes.

DYF500 CAS:68937-41-7 HR: 1
DURAD MP280[R] HYDRAULIC FLUID

SYNS: DURAN MP280[R] □ PHENOL, ISOPROPYLATED, PHOSPHATE (3:1)

TOXICITY DATA WITH REFERENCE
orl-rat LD:>5 g/kg JACTDZ 1,209,92
ihl-rat LC:>6350 mg/m³/4H JACTDZ 1,209,92
ipr-rat LD:>5 g/kg JACTDZ 1,209,92
skn-rbt LD:>2 g/kg JACTDZ 1,209,92

CONSENSUS REPORTS: Reported in EPA TSCA Inventory.

SAFETY PROFILE: Low toxicity by ingestion, inhalation, and skin contact. When heated to decomposition it emits acrid smoke and irritating vapors.

DYG000 HR: 3
DYNAMITE
DOT: UN 0081

PROP: Major constituent is nitroglycerin (85ESA3 8,739,68).

SYN: GELATINE DYNAMITE

DOT CLASSIFICATION: 1.1D; *Label:* EXPLOSIVE, BLASTING TYPE A

SAFETY PROFILE: A high explosive used industrially in construction and mining. The name generally refers to a mixture containing as its principal explosive ingredient either glyceryl trinitrate (nitroglycerin) or ammonium nitrate, suitably sensitized. It does not apply to black blasting powders, chlorate powders, and other deflagrating mixtures. While this material is a powerful explosive when detonated by shock or heat, it is only moderately hazardous. It can react vigorously with oxidizing materials. Dangerous; shock and heat will cause it to explode; when heated to decomposition it emits highly toxic fumes of NO_x and CO, etc. See also NITROGLYCERIN; EXPLOSIVES, HIGH; and NITRATES.

An ordinary blasting cap or an electric blasting cap is used for detonating a charge of dynamite. The various classes and grades of dynamite are made from mixtures composed of an explosive compound or a mixture of explosive compounds, a "dope," and an antiacid. If any of the explosive ingredients are in a liquid state they are referred to as the "explosive oil," which is usually composed of glyceryl trinitrate (nitroglycerin) and about 25–30% of ethylene glycol dinitrate. The latter compound depresses the freezing point of the nitroglycerin and renders the dynamite low-freezing. Other compounds may also be used as freezing point depressants. The explosive oil is absorbed by carbonaceous materials that have entirely replaced kieselguhr (diatomaceous earth), formerly used exclusively as the absorbent or "dope" in dynamites. This type of "dope" does not enter into the explosive reaction. Wood pulp is now most commonly used as the absorber, either alone or mixed in suitable proportions with flour, starch, etc.

The absorbents may be mixed with an oxidizer such as sodium nitrate, in which case an active "dope" is formed. For neutralizing any acid that may be present, about 1% of an antiacid (calcium carbonate or zinc oxide) is added to the mixture. The explosive oil is mixed into the "dope." The strength of a kieselguhr dynamite, when detonated, is derived only from the explosive oil, since kieselguhr is inert. A mixture of this kind is known as a straight dynamite. On the other hand, an active "dope" (an admixture of carbonaceous absorbents with an oxidizer) furnishes explosive strength in addition to that derived from the explosive ingredients.

By replacing a part of the explosive oil of a straight dynamite with ammonium nitrate, so that the latter becomes the principal explosive ingredient, a mixture known as an ammonia dynamite is obtained.

When the explosive oil is gelatinized the explosive is known as a gelatin or an ammonia gelatin dynamite.

Blasting gelatin is a gelatinized mass of an elastic nature obtained by incorporating nitrocotton with an explosive oil into which is mixed about 1% of antiacid.

Dynamites may be in bulk form (bag powder) or in cartridge form, the most common size being 1¼ inch in diameter and 8 inches long, although, for holes of small diameter, cartridges as small as ⅞ inch in diameter are also used. In large-diameter well-drill holes for quarry blasting, cartridge diameters up to 10 inches and lengths up to 30 inches may be used. These upper limits or 50 pounds in weight of each cartridge are imposed by the DOT Regulations, and the maximum length of 30 inches applies to all cartridge diameters between 4 and 10 inches.

An integral part of a stick of dynamite is the paraffined paper wrapper that not only holds the ingredients together but enters into the explosive reaction.

The wrapper also affords some measure of protection from moderate exposure to dampness. For blasting in wet operations, a gelatinized dynamite that resists the absorption of water is used.

The strength of straight dynamite is graded by its explosive oil content (% by weight), while for any other class of dynamite, the strength is determined experimentally in comparison with the various grades of the straight dynamites. For example, a 40% straight dynamite is one that contains 40% of explosive oil; a 40%-strength ammonia dynamite, as determined by tests, equals a 40% straight dynamite in strength. In other words, a 40%-strength ammonia dynamite will release the same energy as an equivalent weight of a 40% straight dynamite.

DYG400 **HR: 3**
DYSPROSIUM
af: Dy aw: 162.5

PROP: Bright, lustrous, silvery metal. Mp: 1409°, bp: 2335°, d: 8.540 @ 25°.

SAFETY PROFILE: It may exhibit an anticoagulant effect. Flammable; an active reducing agent. Reacts violently in air and with halogens. See also RARE EARTHS.

DYG600 **CAS:10025-74-8** **HR: 3**
DYSPROSIUM CHLORIDE
mf: Cl_3Dy mw: 268.85

PROP: Shiny, yellow crystals or hygroscopic white crystals. D: 3.67 @ 0°/4°, mp: 718°, bp: 1500°. A sol salt. Sol in water and alc.

TOXICITY DATA WITH REFERENCE
orl-mus LD50:5443 mg/kg EQSSDX 1,1,75
ipr-mus LD50:343 mg/kg AEHLAU 5,437,62
ipr-gpg LD50:196 mg/kg AEHLAU 5,437,62

CONSENSUS REPORTS: Reported in EPA TSCA Inventory.

SAFETY PROFILE: Poison by intraperitoneal route. Mildly toxic by ingestion. When heated to decomposition it emits toxic fumes of Cl^-. See also CHLORIDES and RARE EARTHS.

DYG800 **CAS:13074-91-4** **HR: 3**
DYSPROSIUM CITRATE
mf: $C_6H_5O_7{\cdot}Dy$ mw: 351.61

SYN: CITRIC ACID, DYSPROSIUM(3+) salt (1:1)

TOXICITY DATA WITH REFERENCE
ipr-mus LD50:113 mg/kg AEHLAU 5,437,62
ipr-gpg LD50:54 mg/kg AEHLAU 5,437,62

SAFETY PROFILE: Poison by intraperitoneal route. When heated to decomposition it emits acrid smoke and irritating fumes. See also RARE EARTHS.

DYH000 **CAS:35725-30-5** **HR: 3**
DYSPROSIUM(III) NITRATE HEXAHYDRATE (1:3:6)
mf: $N_3O_9{\cdot}Dy{\cdot}6H_2O$ mw: 456.65

SYN: NITRIC ACID DYSPROSIUM(3+) SALT HEXAHYDRATE

TOXICITY DATA WITH REFERENCE
orl-rat LD50:3100 mg/kg TXAPA9 5,750,63
ipr-rat LD50:295 mg/kg TXAPA9 5,750,63
ipr-mus LD50:310 mg/kg TXAPA9 5,750,63

SAFETY PROFILE: Poison by intraperitoneal route. Moderately toxic by ingestion. When heated to decomposition it emits very toxic fumes of NO_x. See also NITRATES and RARE EARTHS.

DYH100 **CAS:10143-38-1** **HR: 3**
DYSPROSIUM TRINITRATE
mf: $Dy(NO_3)_3$ mw: 348.60

PROP: Hygroscopic yellow solid. Sol in H_2O and EtOH.

SYN: DYSPROSIUM NITRATE

TOXICITY DATA WITH REFERENCE
orl-rat LD50:2386 mg/kg EQSSDX 1,1,75
ipr-rat LD50:227 mg/kg EQSSDX 1,1,75
ipr-mus LD50:238 mg/kg EQSSDX 1,1,75

SAFETY PROFILE: Poison by intraperitoneal route. Moderately toxic by ingestion. When heated to decomposition it emits toxic fumes of NO_x. See also NITRATES.

E

EAB200 **HR: 3**
EASTERN COTTONMOUTH VENOM

SYNS: AGKISTRODON PISCIVORUS PISCIVORUS VENOM □ VENOM, SNAKE, AGKISTRODON PISCIVORUS PISCIVORUS

TOXICITY DATA WITH REFERENCE
ipr-mus LD50:4844 µg/kg TOXIA6 9,131,71
scu-mus LD50:11 mg/kg TOXIA6 17,237,79
ivn-mus LD50:2044 µg/kg TOXIA6 9,131,71

SAFETY PROFILE: Poison by subcutaneous, intravenous, and intraperitoneal routes.

EAB225 **HR: 3**
EASTERN DIAMOND-BACK RATTLESNAKE VENOM

SYNS: C. ADAMANTEUS VENOM □ CROTALUS ADAMANTEUS VENOM □ VENOM, SNAKE, CROTALUS ADAMANTEUS

TOXICITY DATA WITH REFERENCE
ims-rat LDLo:25 mg/kg AJMSA9 236,204,58
scu-mus LD50:7700 µg/kg TOXIA6 17,661,79
ivn-mus LD50:1333 µg/kg TOXIA6 9,131,71
ims-mus LD50:28 mg/kg AJMSA9 239,1,60
scu-dog LDLo:5 mg/kg SURGAZ 47,975,60
ivn-dog LDLo:500 µg/kg 19DDA6 1,269,67
ims-dog LDLo:640 µg/kg AJMSA9 236,204,58
ims-mky LDLo:640 µg/kg AJMSA9 236,204,58
ivn-cat LDLo:80 µg/kg TOXIA6 1,99,63
ivn-rbt LDLo:125 µg/kg SCIEAS 117,47,53
ims-rbt LDLo:940 µg/kg AJMSA9 236,204,58
ims-pgn LDLo:1400 µg/kg AJMSA9 236,204,58
ims-ckn LDLo:1400 µg/kg AJMSA9 236,204,58
ipr-frg LDLo:190 mg/kg ANREAK 139,305,61

SAFETY PROFILE: A deadly poison by subcutaneous, intramuscular, intravenous, and intraperitoneal routes.

EAC500 **CAS:520-68-3** **HR: 3**
ECHIMIDINE
mf: $C_{20}H_{31}NO_7$ mw: 397.52

PROP: Glass or gum.

TOXICITY DATA WITH REFERENCE
sln-dmg-par 20 µmol/L ZEVBA5 91,74,60
ipr-rat LDLo:200 mg/kg CBINA8 12,299,76

SAFETY PROFILE: Poison by intraperitoneal route. Mutation data reported. When heated to decomposition it emits toxic fumes of NO_x.

EAD500 **CAS:512-64-1** **HR: 3**
ECHINOMYCIN
mf: $C_{51}H_{64}N_{12}O_{12}S_2$ mw: 1101.39

PROP: Crystals. Mp: 221–223°.

SYNS: ECHINOMYCIN A □ LEVOMYCIN □ NSC 526417 □ QUINOMYCIN A □ S-426-S (LEPETIT)

TOXICITY DATA WITH REFERENCE
pic-esc 600 ng/plate CNREA8 43,2819,83
dnd-mam:lym 400 nmol/L NATUAS 252,653,74
dni-mam:lym 27 µmol/L HXPHAU 38(Pt 2),623,75
ipr-mus LD50:280 µg/kg JANTAJ 31,465,68
scu-mus LD50:3800 µg/kg 85ERAY 1,311,78
ivn-mus LD50:629 µg/kg NTIS** PB83-114298
ivn-dog LDLo:89 µg/kg NTIS** PB82-114298

SAFETY PROFILE: A deadly poison by intraperitoneal, intravenous, and subcutaneous routes. Mutation data reported. When heated to decomposition it emits very toxic fumes of SO_x and NO_x.

EAD600 **HR: 3**
ECHIS CARINATUS VENOM

SYNS: E. CARINATUS VENOM □ VENOM, SNAKE, ECHIS CARINATUS

TOXICITY DATA WITH REFERENCE
scu-rat LD50:5130 µg/kg AIPTAK 137,299,62
ipr-mus LD50:196 µg/kg TOXIA6 18,384,80
ivn-mus LD50:120 µg/kg TOXIA6 22,373,84
ivn-dog LDLo:100 µg/kg TOXIA6 6,51,68
ivn-rbt LDLo:100 µg/kg TOXIA6 2,5,64
ivn-mam LD50:2300 µg/kg CLPTAT 8,849,67

SAFETY PROFILE: A deadly poison by subcutaneous, intravenous, and intraperitoneal routes.

EAD650 **HR: 3**
ECHIS COLORATA VENOM

SYNS: ECHIS COLORATUS VENOM □ VENOM, SNAKE, ECHIS COLORATUS

TOXICITY DATA WITH REFERENCE
ipr-mus LD50:1550 µg/kg AJTHAB 9,391,60
scu-mus LD50:5167 µg/kg AJTHAB 9,391,60
ivn-mus LD50:425 µg/kg TOXIA6 14,146,76

SAFETY PROFILE: A deadly poison by subcutaneous, intravenous, and intraperitoneal routes.

EAE000 **CAS:28558-28-3** **HR: 3**
ECONAZOLE NITRATE
mf: $C_{18}H_{15}Cl_3N_2O{\cdot}HNO_3$ mw: 444.72

SYNS: 1-(2-((4-CHLOROPHENYL)METHOXY)-2-(2,4-DICHLOROPHENYL)ETHYL)-1H-IMIDAZOLE NITRATE □ 1-(2,4-DICHLORO-β-((p-CHLOROBENZYL)OXY)-PHENETHYL)IMIDAZOLE NITRATE □ ECOSTATIN □ EPI-PEVARYL □ GYNO-PEVARYL □ IFENEC □ PALAVALE □ PEVARYL □ R 14,827 □ SPECTAZOLE

TOXICITY DATA WITH REFERENCE
eye-rbt 1% MLD IYKEDH 10,16,79
scu-rat TDLo: 22 mg/kg (female 7-17D post): REP IYKEDH 9,928,78
orl-rat TDLo: 2240 mg/kg (16-22D preg/21D post): REP ARZNAD 25,224,75
scu-rat TDLo: 2200 µg/kg (female 7-17D post): REP IYKEDH 9,928,78
scu-mus TDLo: 160 mg/kg (female 7-14D post): TER IYKEDH 9,955,78
scu-rat TDLo: 220 mg/kg (7-17D preg): TER IYKEDH 9,928,78
scu-mus TDLo: 160 mg/kg (female 7-14D post): REP IYKEDH 9,955,78
scu-rat TDLo: 750 mg/kg (30D male): REP IYKEDH 9,884,78
scu-rat TDLo: 110 mg/kg (female 7-17D post): TER IYKEDH 9,928,78
orl-rat LD50: 668 mg/kg ARZNAD 25,224,75
ipr-rat LD50: 240 mg/kg IYKEDH 9,643,78
scu-rat LD50: 1360 mg/kg IYKEDH 9,643,78
ivn-rat LD50: 50 mg/kg IYKEDH 9,643,78
orl-mus LD50: 463 mg/kg ARZNAD 25,224,75
ipr-mus LD50: 180 mg/kg IYKEDH 9,643,78
scu-mus LD50: 750 mg/kg IYKEDH 9,643,78
ivn-mus LD50: 38 mg/kg IYKEDH 9,643,78
orl-rbt LD50: 431 mg/kg YAKUD5 23,1253,81
scu-rbt LD50: 650 mg/kg IYKEDH 9,643,78
ivn-rbt LD50: 85 mg/kg IYKEDH 9,643,78
orl-gpg LD50: 272 mg/kg ARZNAD 25,224,75

SAFETY PROFILE: Poison by ingestion, intravenous, and intraperitoneal routes. Moderately toxic by subcutaneous route. Experimental teratogenic and reproductive effects. An eye irritant. An antimycotic drug. When heated to decomposition it emits very toxic fumes of Cl^- and NO_x. See also NITRATES.

EAE500 CAS:5036-03-3 HR: 2
EDROFURADENE
mf: $C_{10}H_{12}N_4O_5$ mw: 268.26

PROP: A solid. Mp: 199.5–201.5°.

SYNS: 1-(2-HYDROXYETHYL)-3-((5-NITROFURFURYLIDENE)AMINO)-2-IMIDAZOLIDINONE □ NF 1010 □ NIFURDAZIL

TOXICITY DATA WITH REFERENCE
mma-sat 100 ng/plate MUREAV 48,295,77
orl-rat TDLo: 27 g/kg/46W-C: CAR JNCIAM 51,403,73

CONSENSUS REPORTS: EPA Genetic Toxicology Program.

SAFETY PROFILE: Questionable carcinogen with experimental carcinogenic data. Mutation data reported. When heated to decomposition it emits toxic fumes of NO_x.

EAE600 CAS:116-38-1 HR: 3
EDROPHONIUM CHLORIDE
mf: $C_{10}H_{16}NO{\cdot}Cl$ mw: 201.72

PROP: A solid. Mp: 162–163°.

SYNS: ANTIREX □ DIMETHYLETHYL(m-HYDROXYPHENYL)AMMONIUM CHLORIDE □ N-ETHYL-3-HYDROXY-N,N-DIMETHYLBENZENAMINIUM CHLORIDE (9CI) □ ETHYL(m-HYDROXYPHENYL)DIMETHYLAMMONIUM CHLORIDE □ TENSILON □ TENSILON CHLORIDE

TOXICITY DATA WITH REFERENCE
ipr-mus LD50: 30 mg/kg PCJOAU 10,327,76
scu-mus LD50: 52 mg/kg JPETAB 123,121,58
ivn-mus LD50: 15 mg/kg NIIRDN 6,133,82

SAFETY PROFILE: Poison by subcutaneous, intravenous, and intraperitoneal routes. When heated to decomposition it emits toxic fumes of Cl^-, NH_3, and NO_x. See also CHLORIDES.

EAE675 CAS:55294-15-0 HR: 3
EDRUL
mf: $C_{11}H_{11}Cl_2N_3O$ mw: 272.15

PROP: Crystals from methanol. Mp: 127–129°.

SYNS: 3-AMINO-1-(3,4-DICHLORO-α-METHYLBENZYL)-2-PYRAZOLIN-5-ONE □ 5 AMINO-2-(1-(3,4-DICHLOROPHENYL)-ETHYL)-2,4-DIHYDRO-3H-PYRAZOL-3-ONE □ BAY G 2821 □ 2,4-DIHYDRO-5-AMINO-2-(1-(3,4-DICHLOROPHENYL)ETHYL)-3H-PYRAZOL-3-ONE □ LUTROL □ MUZOLIMINE

TOXICITY DATA WITH REFERENCE
orl-rat LD50: 1559 mg/kg CMROCX 4,716,77
ivn-rat LD50: 221 mg/kg CLNHBI 19(Suppl 1),20,83
orl-mus LD50: 1794 mg/kg CMROCX 4,716,77
ivn-mus LD50: 310 mg/kg CLNHBI 19(Suppl 1),20,83
orl-dog LD50: 2000 mg/kg CMROCX 4,716,77
orl-rbt LD50: 1250 mg/kg CMROCX 4,716,77

SAFETY PROFILE: Poison by intravenous route. Moderately toxic by ingestion. When heated to decomposition it emits toxic fumes of Cl^- and NO_x.

EAE775 CAS:68278-23-9 HR: 1
EFLORNITHINE HYDROCHLORIDE
mf: $C_6H_{12}F_2N_2O_2{\cdot}ClH$ mw: 218.66

PROP: Crystals from EtOH (aq). Mp: 183°.

SYNS: α-DIFLUOROMETHYLORNITHINE HYDROCHLORIDE □ 2-(DIFLUOROMETHYL)-dl-ORNITHINE HYDROCHLORIDE □ dl-ORNITHINE, 2-(DIFLUOROMETHYL)-, MONOHYDROCHLORIDE

TOXICITY DATA WITH REFERENCE
orl-rat TDLo: 11,250 mg/kg (female 10-18D post): TER TJADAB 35,55A,87
orl-rat TDLo: 1500 mg/kg (female 7-18D post): TER TJADAB 35,55A,87
orl-rbt TDLo: 900 mg/kg (female 11-19D post): REP TJADAB 35,55A,87
orl-rat LD50: >5 g/kg DRFUD4 6,142,81
ipr-rat LD50: >3 g/kg DRFUD4 6,142,81

SAFETY PROFILE: Low toxicity by ingestion and intraperitoneal routes. An experimental teratogen. Other experimental reproductive effects. When heated to decomposition it emits toxic fumes of NO_x and F^-.

EAE875 **CAS:1765-48-6** **HR: 3**
EICOSAFLUOROUNDECANOIC ACID
mf: $C_{11}H_2F_{20}O_2$ mw: 546.13

PROP: Crystals from toluene. Mp: 100–101°.

SYNS: 2,2,3,3,4,4,5,5,6,6,7,7,8,8,9,9,10,10,11,11-EICOSAFLUOROUNDECANOIC ACID □ 11-H-EICOSAFLUORUNDEKANSAEURE (GERMAN) □ ω-H-EICOSAFLUORUNDEKANSAEURE (GERMAN)

TOXICITY DATA WITH REFERENCE
orl-rat LD50:518 mg/kg ZHYGAM 26,9,80
ipr-mus LD50:181 mg/kg ZHYGAM 26,9,80
orl-rbt LD50:470 mg/kg ZHYGAM 26,9,80
orl-gpg LD50:470 mg/kg ZHYGAM 26,9,80

SAFETY PROFILE: Poison by intraperitoneal route. Moderately toxic by ingestion. When heated to decomposition it emits toxic fumes of F^-.

EAF000 **CAS:506-30-9** **HR: 2**
EICOSANOIC ACID
mf: $C_{20}H_{40}O_2$ mw: 312.60

PROP: White leaflets or crystals from alc. Mp: 77°, bp: 203–205° @ 1 mm°. Slt decomp in water; very sol in hot absolute alc and ether.

SYNS: ARACHIC ACID □ ARACHIDIC ACID

TOXICITY DATA WITH REFERENCE
imp-mus TDLo:1000 mg/kg:NEO CNREA8 26,105,66

CONSENSUS REPORTS: Reported in EPA TSCA Inventory.

SAFETY PROFILE: Questionable carcinogen with experimental neoplastigenic data by implant route. When heated to decomposition it emits acrid smoke and fumes.

EAG000 **CAS:23315-05-1** **HR: 3**
ELAIOMYCIN
mf: $C_{13}H_{26}N_2O_3$ mw: 258.41

PROP: Neutral yellow oil. Metabolite of *Streptomyces hepaticus* (NATUAS 221,765,69).

SYN: d-threo-METHOXY-3-(1-OCTENYL-O,N,N-AZOXY)-2-BUTANOL

TOXICITY DATA WITH REFERENCE
orl-rat TDLo:35 mg/kg:ETA NATUAS 221,765,69
orl-rat TD:35 mg/kg:ETA 29QKAZ 2,781,72
scu-mus LD50:63 mg/kg ANTCAO 4,338,54
ivn-mus LD50:44 mg/kg ANTCAO 4,338,54

SAFETY PROFILE: Poison by intravenous and subcutaneous routes. Questionable carcinogen with experimental tumorigenic data. Causes tumors of the brain. When heated to decomposition it emits toxic fumes of NO_x.

EAG500 **CAS:50814-62-5** **HR: 2**
ELASIOMYCIN

TOXICITY DATA WITH REFERENCE
ivn-rat TDLo:50 mg/kg (18D preg):ETA,TER IARCCD 4,100,73

SAFETY PROFILE: Questionable carcinogen with experimental tumorigenic and teratogenic data. When heated to decomposition it emits acrid smoke and fumes.

EAG875 **CAS:9004-06-2** **HR: 3**
ELASTASE

PROP: White, lyophilized powder with a sltly yellowish shade.

SYNS: E.C. 3.4.4.7 □ E.C. 3.4.2.1.11 □ ELASZYM □ PANCREATOPEPTIDASE E

TOXICITY DATA WITH REFERENCE
ipr-rat LD50:78 mg/kg IYKEDH 7,108,76
ivn-rat LD50:85 mg/kg IYKEDH 7,108,76
ipr-mus LD50:37 mg/kg IYKEDH 7,108,76
ivn-mus LD50:57 mg/kg IYKEDH 7,108,76

CONSENSUS REPORTS: Reported in EPA TSCA Inventory.

SAFETY PROFILE: Poison by intravenous and intraperitoneal routes. When heated to decomposition it emits toxic fumes of NO_x.

EAH100 **CAS:3877-86-9** **HR: 3**
ELATERICIN A
mf: $C_{30}H_{44}O_7$ mw: 516.74

PROP: Crystals from alc. Mp: 151–152°.

SYN: CUCURBITACINE (D)

TOXICITY DATA WITH REFERENCE
orl-rat LD50:8200 μg/kg AIPTAK 130,315,61
ipr-rat LD50:1300 μg/kg AIPTAK 130,315,61
scu-rat LD50:3400 μg/kg AIPTAK 130,315,61
orl-mus LD50:5 mg/kg CHTPBA 4,459,69
ipr-mus LD50:1750 μg/kg AIPTAK 130,315,61
scu-mus LD50:4600 μg/kg AIPTAK 130,315,61
ivn-mus LD50:960 μg/kg AIPTAK 130,315,61
ivn-dog LD50:1000 μg/kg AIPTAK 130,315,61
ivn-cat LD50:900 μg/kg AIPTAK 130,315,61

SAFETY PROFILE: Poison by ingestion, subcutaneous, intravenous, and intraperitoneal routes. When heated to decomposition it emits acrid smoke and fumes.

EAH500 **CAS:50-48-6** **HR: 3**
ELAVIL
mf: $C_{20}H_{23}N$ mw: 277.44

SYNS: AMITRIPTILINE □ AMITRIPTYLIN (GERMAN) □ AMITRIPTYLINE □ DAMILAN □ 3,10-DIHYDRO-5H-DIBENZO(a,d)CYCLOHEPTEN-5-YLIDENE-N,N-DIMETHYL-1-PROPANAMINE □ 10,11-DIHYDRO-5-(γ-DIMETHYLAMINOPROPYLIDENE)-5H-DIBENZO(a,d)CYCLOHEPTENE □ 10,11-DIHYDRO-N,N-DIMETHYL-5H-DIBENZO(a,d)HEPTALENE-Δ^5-γ-PROPYLAMINE □ 5-(3′-DIMETHYLAMINOPROPYLIDENE)-DIBENZO-(a,d)(1,4)-CYCLOHEPTADIENE □ 5-(γ-DIMETHYLAMINOPROPYLIDENE)-5H-DIBENZO(a,d)-10,11-DIHYDROCYCLOHEPTENE □ 5-(γ-DIMETHYLAMINOPROPYLIDENE)-10,11-DIHYDRO-5H-DIBENZO(A,D)CYCLOHEPTENE □ 5-(3-DIMETHYLAMINOPROPYLIDENE)-10,11-DIHYDRO-5H-DIBENZO(a,d)CYCLOHEPTENE □ 5-(3-DIMETHYLPROPYLIDENE)DIBENZO(a,d)(1,4)CYCLOHEPTADIENE □ ELANIL □ LAROXIL □ LAROXYL □ PROHEPTADIENE □ TRYPTIZOL

TOXICITY DATA WITH REFERENCE
orl-rbt TDLo:540 mg/kg (female 8-16D post):REP PSDTAP 10,235,69
orl-mus TDLo:252 mg/kg (6-14D preg):REP PSDTAP 10,235,69
orl-rat TDLo:75 mg/kg (4-18D preg):TER JNMDBO 2,271,71
orl-mus TDLo:504 mg/kg (6-14D preg):TER PSDTAP 10,235,69
orl-chd TDLo:4500 µg/kg:CNS BMJOAE 3,663,67
orl-wmn TDLo:16,800 µg/kg/2W-I:PNS,CNS LANCAO 1,426,68
orl-hmn TDLo:14 mg/kg:CVS PSDTAP 6,171,65
orl-inf TDLo:50 mg/kg:CNS,CVS AJDCAI 130,507,76
orl-wmn TDLo:13,500 µg/kg/3D-C:CNS LANCAO 1,390,68
ims-wmn TDLo:240 µg/kg/36H-C:CNS LANCAO 1,390,68
unr-man TDLo:45 mg/kg/21D:CVS,CNS LANCAO 2,1202,72
orl-rat LD50:320 mg/kg ARZNAD 15,863,65
ipr-rat LD50:72 mg/kg PJPPAA 27,503,75
orl-mus LD50:140 mg/kg THERAP 26,459,71
ipr-mus LD50:56 mg/kg ARZNAD 15,863,65
scu-mus LD50:140 mg/kg ARZNAD 19,458,69
ivn-mus LD50:16 mg/kg JMCMAR 17,65,74
scu-dog LDLo:50 mg/kg PHMCAA 11,283,69
ivn-rbt LD50:8600 µg/kg ARZNAD 15,863,65

SAFETY PROFILE: Human poison by an unspecified route. Poison experimentally by ingestion, intraperitoneal, subcutaneous, intramuscular, and intravenous routes. Human systemic effects by ingestion and intramuscular routes: changes in sleep, headache, paresthesia, convulsions, excitement, somnolence, muscle contractions, change in heart rate, and other cardiac changes. An experimental teratogen. Other experimental reproductive effects. An antidepressant. When heated to decomposition it emits toxic fumes of NO_x.

EAI000 CAS:549-18-8 HR: 3
ELAVIL HYDROCHLORIDE
mf: $C_{20}H_{23}N \cdot ClH$ mw: 313.90

PROP: Minute crystals. Mp: 196–197°. Very sol in H_2O.

SYNS: AMITID □ AMITRIL □ AMITRIPTYLINE CHLORIDE □ AMITRYPTYLINE HYDROCHLORIDE □ DAMILEN HYDROCHLORIDE □ DEPREX □ 10,11-DIHYDRO-N,N-DIMETHYL-5H-DIBENZO(a,d)-CYCLOHEPTENE-$\Delta^{5,\gamma}$-PROPYLAMINE HCL □ 3-(3-DIMETHYLAMINOPROPYLIDENE)-1:2-4:5-DIBENZOCYCLOHEPTA-1:4-DIENE □ 5-(3-DIMETHYLAMINOPROPYLIDENE)DIBENZO(a,d)(1,4)CYCLOHEPTADIENE HYDROCHLORIDE □ DOMICAL □ ELAVIL □ ENDEP □ LENTIZOL □ MIKETORIN □ PROHEPTADIEN MONOHYDROCHLORIDE □ SAROTEN □ SAROTENE □ SK-AMITRIPTYLINE □ TRYPTIZOL □ TRYPTIZOL HYDROCHLORIDE

TOXICITY DATA WITH REFERENCE
sln-dmg-orl 2 mg/2D SOGEBZ 11,718,75
scu-mus TDLo:60 mg/kg (female 9D post):TER DGDFA5 22,61,80
orl-rat TDLo:3300 mg/kg (female 33D pre):REP OYYAA2 6,899,72
orl-rat TDLo:330 mg/kg (33D male):REP OYYAA2 6,899,72
orl-mus TDLo:70 mg/kg (male 5D pre):REP IMSCE2 14,579,86
orl-rat TDLo:1650 mg/kg (male 33D pre):REP OYYAA2 6,899,72
ipr-ham TDLo:60 mg/kg (female 8D post):TER RCPBDC 5,275,80
orl-wmn TDLo:10 mg/kg CMAJAX 129,1203,83
orl-chd LDLo:62,500 µg/kg:CNS,CVS,PUL NEJMAG 267,1031,62
orl-chd TDLo:4167 µg/kg:CNS NEJMAG 267,1031,62
orl-hmn TDLo:200 mg/kg:CNS JAMAAP 230,1405,74
orl-chd LDLo:62,500 µg/kg:CNS,PUL AJDCAI 106,501,63
orl-wmn LDLo:19 mg/kg:CNS,PUL LANCAO 2,543,63
orl-man TDLo:14,286 µg/kg:CNS,PUL LANCAO 2,543,63
orl-rat LD50:240 mg/kg OYYAA2 6,889,72
ipr-rat LD50:72 mg/kg 27ZQAG -,61,72
scu-rat LD50:1290 mg/kg 27ZQAG -,61,72
ivn-rat LD50:14 mg/kg 27ZQAG -,61,72
orl-mus LD50:140 mg/kg WMWOA4 111,256,61
ipr-mus LD50:65 mg/kg JDGRAX 16,7,85
scu-mus LD50:80 mg/kg OYYAA2 6,889,72
ivn-mus LD50:21 mg/kg ARZNAD 21,808,71
ivn-mky LDLo:20,300 µg/kg IJEBA6 22,539,84
ivn-rbt LD50:9900 µg/kg 27ZQAG -,61,72

CONSENSUS REPORTS: Reported in EPA TSCA Inventory.

SAFETY PROFILE: Poison by ingestion, intraperitoneal, intravenous, and subcutaneous routes. Human systemic effects by ingestion: convulsions, respiratory depression, changes in sleep, hallucinations, muscle contractions, somnolence, blood pressure decrease, coma, cyanosis, dyspnea, and ataxia. An experimental teratogen. Other experimental reproductive effects. Mutation data reported. Used in the treatment of depression. When heated to decomposition it emits very toxic fumes of HCl and NO_x. See also ELAVIL.

EAI100 HR: 2
ELDERBERRY

PROP: A shrub that may grow to 12 feet. The compound leaves have about 7 serrated, oval leaflets. It bears clusters of small, white flowers and purple-black berries. Both flowers and berries may be on the plant simultaneously. Various species are found in every state in the United States, most of Canada, and the Greater Antilles.

SYNS: AMERICAN ELDER □ FLEUR SUREAU (CANADA, HAITI) □ SAMBUCUS CAERULEA □ SAMBUCUS CANADENSIS □ SAMBUCUS MELANOCARPA □ SAMBUCUS MEXICANA □ SAMBUCUS RACEMOSA □ SAUCO (MEXICO, PUERTO RICO) □ SAUCO BLANCO (CUBA) □ SUREAU (CANADA, HAITI)

SAFETY PROFILE: The whole plant contains poisonous cyanogenetic glycosides. Ingestion of small amounts of the raw or cooked fruit is harmless. Ingestion of the leaves, roots, bark or unripe berries may cause prolonged, severe diarrhea. Juice from the pressed berries may cause nausea, vomiting, abdominal cramps, dizziness, numbness and stupor. See also CYANIDE.

EAI500 CAS:8023-89-0 HR: 2
ELEMI OIL

PROP: Distilled from the crude resin of *Canarium*

commune or *Canarium luzonicum* (FCTXAV 14,659, 76).

SYN: ELEMI

TOXICITY DATA WITH REFERENCE
skn-rbt 500 mg/24H FCTXAV 14,755,76
orl-rat LD50:3370 mg/kg FCTXAV 14,755,76
skn-rbt LD50:5 g/kg FCTXAV 14,755,76

CONSENSUS REPORTS: Reported in EPA TSCA Inventory.

SAFETY PROFILE: Moderately toxic by ingestion. Mildly toxic by skin contact. A skin irritant. When heated to decomposition it emits acrid smoke and fumes.

EAI600 **HR: 1**
ELEPHANT'S EAR

PROP: Popular ornamental houseplant that has large, heart-shaped leaves and edible roots. They are grown indoors and outdoors in southern Florida, southern California, Hawaii, Guam, and the West Indies. One species, *A. macrorrhiza*, is grown for food.

SYNS: AHE POI (HAWAII) □ ALOCASIA (VARIOUS SPECIES) □ 'APE (HAWAII) □ CABEZA de BURRO □ CARAIBE (HAITI) □ CHINE APE □ COLOCASIA ESCULENTA □ COLOCASIA GIGANTEA □ DASHEEN □ EDDO □ KALO □ MALANGA CARA de CHIVO (CUBA) □ MALANGA de JARDIN (CUBA) □ MALANGA DEUX PALLES (HAITI) □ MALANGA ISLENA (CUBA) □ PAPAO-APAKA (GUAM) □ PAPAO-ATOLONG (GUAM) □ TARO □ TAYO BAMBOU (HAITI) □ YAUTIA MALANGA (PUERTO RICO)

SAFETY PROFILE: The leaves and stems contain calcium oxalate raphides. Chewing of these plant parts results in immediate burning pain in the lips, mouth and throat followed by inflammation and blistering. Systemic effects are usually not seen because the immediate pain prevents ingestion and the low solubility of calcium oxalate reduces its absorption. See also OXALATES and OXALIC ACID.

EAI800 **CAS:79818-59-0** **HR: 1**
ELEX 334

TOXICITY DATA WITH REFERENCE
orl-rat TDLo:1638 mg/kg (91D male):REP OYYAA2 11,901,76
orl-rat LD50:9150 mg/kg OYYAA2 11,901,76
orl-mus LD50:8910 mg/kg OYYAA2 11,901,76

SAFETY PROFILE: Mildly toxic by ingestion. Experimental reproductive effects.

EAI850 **CAS:519-23-3** **HR: 3**
ELLIPTISINE
mf: $C_{17}H_{14}N_2$ mw: 246.33

PROP: Bright yellow needles, orange rods or rosettes, from ethyl acetate. Mp: 311–315° (decomp).

SYNS: 5,11-DIMETHYL-6H-PYRIDO(4,3-b)CARBAZOLE □ ELLIPTICINE □ ICIG 770 □ NSC-71795

TOXICITY DATA WITH REFERENCE
mmo-sat 1 μg/plate BICMBE 60,1011,78
pic-esc 500 ng/plate CNREA8 43,2819,83
sce-ham:ovr 100 μg/L CNREA8 43,577,83
orl-mus LD50:178 mg/kg MEIEDD 10,512,83
ipr-mus LD50:150 mg/kg BIMDB3 21,101,74
ivn-mus LD50:19,500 μg/kg MEIEDD 10,512,83

SAFETY PROFILE: Poison by ingestion, intravenous, and intraperitoneal routes. Mutation data reported. When heated to decomposition it emits toxic fumes of NO_x.

EAI875 **CAS:3818-88-0** **HR: 3**
ELORINE CHLORIDE
mf: $C_{20}H_{32}NO•Cl$ mw: 337.98

PROP: Crystals from nitroethane. Mp: 159–164°.

SYNS: COMPOUND 14045 METHOCHLORIDE □ (±)-N-((3-CYCLOHEXYL-3-HYDROXY-3-PHENYL)PROPYL)-N-METHYLPYRROLIDINIUM CHLORIDE □ 1-(3-CYCLOHEXYL-3-HYDROXY-3-PHENYLPROPYL)-1-METHYL-PYRROLIDINIUM CHLORIDE □ 1-CYCLOHEXYL-1-PHENYL-3-PYRROLIDINO-1-PROPANOL METHYL CHLORIDE □ LERGINE CHLORIDE □ TRICOLOID CHLORIDE □ TRICYCLAMOL CHLORIDE □ TRICYCLAMOL METHOCHLORIDE

TOXICITY DATA WITH REFERENCE
orl-rat LD50:984 mg/kg 29ZVAB -,119,69
orl-mus LD50:395 mg/kg JAPMA8 43,408,54
ivn-mus LD50:11,700 μg/kg JAPMA8 43,408,54

SAFETY PROFILE: Poison by ingestion and intravenous routes. When heated to decomposition it emits toxic fumes of Cl^- and NO_x.

EAJ000 **CAS:548-43-6** **HR: 3**
ELYMOCLAVINE
mf: $C_{16}H_{18}N_2O$ mw: 254.36

PROP: A solid. Mp: 250–252° (decomp). A close chemical relative of LSD found in ergot fungi and bindweeds of the genus *Ipomoea*, Fam. convolvulaceae (JANSAG 41,1700,75).

SYNS: ELIMOCLAVIN □ ELYMOCLAVIN

TOXICITY DATA WITH REFERENCE
ipr-mus TDLo:60 mg/kg (10D preg):REP JANSAG 41,1700,75
ipr-mus TDLo:30 mg/kg (10D preg):TER JANSAG 41,1700,75
ipr-mus LD50:67 mg/kg JANSAG 41,1700,75
ivn-mus LD50:45 mg/kg NYKZAU 58,386,62

SAFETY PROFILE: A poison by intraperitoneal and intravenous routes. Experimental teratogenic and reproductive effects. When heated to decomposition it emits acrid smoke and fumes. See also N,N-DIETHYLLYSERGAMIDE (LSD).

EAJ100 **CAS:79458-80-3** **HR: 2**
EM 255
mf: $C_{13}H_{14}N_2O_2$ mw: 230.29

SYN: 3-(1,3-DIHYDRO-1-OXO-2H-ISOINDOL-2-YL)-2-OXOPIPERIDINE

TOXICITY DATA WITH REFERENCE
orl-rbt TDLo:400 mg/kg (8-15D preg):TER ARZNAD 31,941,81
orl-mus LD50:1200 mg/kg ARZNAD 31,941,81
ipr-mus LD50:494 mg/kg ARZNAD 31,941,81

SAFETY PROFILE: Moderately toxic by ingestion and other routes. An experimental teratogen. When heated to decomposition it emits toxic fumes of NO_x.

EAJ500 CAS:19526-81-9 HR: 2
EMAZOL RED B
mf: $C_{18}H_{16}N_2O_{10}S_3 \cdot 2Na$ mw: 562.52

TOXICITY DATA WITH REFERENCE
orl-rat TDLo:78 g/kg/86W-I:ETA TKORAS 3,53,67
scu-rat TDLo:10 g/kg/17W-I:ETA TKORAS 3,53,67
orl-mus TDLo:20 g/kg/17W-I:ETA TKORAS 3,53,67
scu-mus TDLo:135 g/kg/56W-I:ETA TKORAS 3,53,67

CONSENSUS REPORTS: Reported in EPA TSCA Inventory.

SAFETY PROFILE: Questionable carcinogen with experimental tumorigenic data. When heated to decomposition it emits very toxic fumes of NO_x, Na_2O, and SO_x.

EAJ600 CAS:550-24-3 HR: 3
EMBELIC ACID
mf: $C_{17}H_{26}O_4$ mw: 294.43

PROP: Glistening orange plates from alc, benzene, or acetic acid; orange crystals from MeOH or hexane/EtOH. Mp: 145–146°. Sol in the usual hot organic solvents or in alkali hydroxide solns; very sltly sol in petr ether; practically insol in water.

SYNS: 2,5-DIHYDROXY-3-UNDECYL-1,4-BENZOQUINONE □ 2,5-DIHYDROXY-3-UNDECYL-2,5-CYCLOHEXADIENE-1,4-DIONE (9CI) □ EMBELIN □ EMBERLINE

TOXICITY DATA WITH REFERENCE
orl-rbt TDLo:50 mg/kg (female 5D pre):REP JRIHDC 11(4),84,76
orl-rat TDLo:250 mg/kg (female 1-5D post):REP IJEBA6 18,1359,80
orl-rat TDLo:50 mg/kg (1-5D preg):REP JRIHDC 11(4),84,76
orl-rat TDLo:50 mg/kg (female 1-5D post):REP IJEBA6 16,1187,78
scu-rat TDLo:300 mg/kg (male 15D pre):REP CCPTAY 39,307,89
orl-dog TDLo:4 g/kg (male 14W pre):REP ANDRDQ 15,486,83
ipr-uns LD50:44 mg/kg IJPPAZ 21,31,77

SAFETY PROFILE: Poison by intraperitoneal route. Experimental reproductive effects. When heated to decomposition it emits acrid smoke and fumes.

EAK000 CAS:10429-82-0 HR: 3
EMBITOL
mf: $C_{11}H_{15}Cl_2N \cdot ClH$ mw: 268.63

SYNS: BENZYL-BIS(2-CHLOROETHYL)AMINE HYDROCHLORIDE □ N,N-BIS(2-CHLOROETHYL)BENZENEMETHANAMINE HYDROCHLORIDE □ N,N-BIS(2-CHLOROETHYL)BENZYLAMINE HYDROCHLORIDE

TOXICITY DATA WITH REFERENCE
mmo-asn 2500 µmol/L SOGEBZ 6,220,70
ipr-mus LD50:28 mg/kg CANCAR 2,1055,49

SAFETY PROFILE: Poison by intraperitoneal route. Mutation data reported. When heated to decomposition it emits very toxic fumes of Cl^- and NO_x.

EAK500 CAS:10433-59-7 HR: 3
EMBUTOX
mf: $C_{10}H_9Cl_2O_3 \cdot Na$ mw: 271.08

SYNS: BUTOXONE SB □ BUTYRAC □ 2,4-DB SODIUM SALT □ γ-(2,4-DICHLOROPHENOXY)BUTYRIC ACID, SODIUM SALT □ 2,4-DICHLOROPHENOXYBUTYRIC ACID, SODIUM SALT □ MB 2878

TOXICITY DATA WITH REFERENCE
orl-rat LD50:700 mg/kg GUCHAZ 6,153,73
orl-mus LD50:400 mg/kg PCOC** -,367,66

CONSENSUS REPORTS: EPA Genetic Toxicology Program.

SAFETY PROFILE: Poison by ingestion. An herbicide. When heated to decomposition it emits toxic fumes of Cl^- and Na_2O.

EAL100 CAS:1302-74-5 HR: 2
EMERY
mf: Al_2O_3 mw: 101.96

PROP: A varicolored mineral with transparent crystals that are very hard and resistant to attack by acids. D: 3.95–4.10, mp: 2050°, bp: 2977°.

SYNS: ALUMINUM OXIDE □ CORUNDUM □ ELECTROCORUNDUM □ EN 237 □ KER 710 □ KO 7 □ KORUND □ KU 5-3 □ MP 1 (refractory)

TOXICITY DATA WITH REFERENCE
ipr-rat TDLo:225 mg/kg/1W-I:CAR ZHPMAT 162,467,76

OSHA PEL: TWA Total Dust: 10 mg/m^3; Respirable Fraction: 5 mg/m^3
ACGIH TLV: TWA (nuisance particulate) 10 mg/m^3 of total dust (when toxic impurities are not present, e.g., quartz <1%)

SAFETY PROFILE: Questionable carcinogen with experimental carcinogenic data. May cause a pneumoconiosis. It is mainly a nuisance dust.

EAL500 CAS:483-18-1 HR: 3
EMETINE
mf: $C_{29}H_{40}N_2O_4$ mw: 480.71

PROP: White powder, lumps, or amorphous solid; bitter taste, darkens on exposure. Mp: 74° effervescent.

SYNS: CEPHAELINE METHYL ETHER □ (−)-EMETINE □ NSC-33669 □ 6',7',10,11-TETRAMETHOXYEMETAN

TOXICITY DATA WITH REFERENCE
skn-rbt 5 mg/24H rns TXCYAC 14,117,79
skn-rbt 1%/24H MOD NTIS** PB-274-082
eye-rbt 2000 ppm SEV AJOPAA 31,837,48

dnd-mam:lym 6600 μmol/L FOMIAZ 15,76,70
scu-man TDLo:10 mg/kg/10D:GIT,CNS,CVS CCROBU 57,423,73
unr-man LDLo:2941 μg/kg 85DCAI 2,73,70
orl-rat LD50:68 mg/kg ARZNAD 13,474,63
ipr-rat LD50:12 mg/kg JPETAB 104,421,52
scu-rat LD50:95 mg/kg ARZNAD 13,474,63
ipr-mus LD50:12 mg/kg JPETAB 104,421,52
scu-mus LD10:25 mg/kg EJCAAH 10,667,74
scu-cat LDLo:8 mg/kg HBAMAK 4,1289,35
ivn-cat LDLo:10 mg/kg HBAMAK 4,1289,35
scu-rbt LDLo:30 mg/kg HBAMAK 4,1289,35
ivn-rbt LDLo:2 mg/kg HBAMAK 4,1289,35
scu-gpg LDLo:70 mg/kg HBAMAK 4,1289,35
ivn-gpg LDLo:7 mg/kg HBAMAK 4,1289,35

CONSENSUS REPORTS: NCI Carcinogenesis Studies (ipr); Inadequate Studies: mouse, rat NCITR* NCI-CG-TR-43,77

SAFETY PROFILE: A human poison by an unspecified route. An experimental poison by ingestion, intraperitoneal, intravenous, and subcutaneous routes. Human systemic effects by subcutaneous route: muscle weakness, cardiac arrhythmias and gastrointestinal effects. Mutation data reported. A severe eye and moderate skin irritant. It is one of the two potent alkaloids obtained from the Brazilian plant ipecac. The therapeutic use of various ipecac preparations has caused many cases of poisoning, in some instances with fatal results. The toxic effects are particularly prominent if the drug is given intravenously. The symptoms of intoxication are gastrointestinal irritation and salivation, as well as general edema, which follows renal insufficiency, hemoptysis (blood-stained sputum), flaccid paralysis, peripheral neuritis (inflammation of the nerve endings), aphonia (loss of voice), difficulties in swallowing, delirium, coma and failure of the heart. The fatal dose is considered to be approximately 2 g, whether administered over a short or relatively long period. The drug seems to have a cumulative effect. When heated to decomposition it emits highly toxic fumes of NO_x.

EAM000 **HR: 3**
EMETINE ANTIMONY IODIDE

PROP: Percentage composition = 34% emetine and 14% antimony (AJTMAQ 10,249,30).

SYN: ANTIMONY EMETINE IODIDE

TOXICITY DATA WITH REFERENCE
orl-hmn TDLo:471 μg/kg:GIT AJTMAQ 10,249,30
orl-cat LDLo:15 mg/kg AJTMAQ 10,249,30
orl-rbt LDLo:10 mg/kg AJTMAQ 10,249,30

CONSENSUS REPORTS: Antimony and its compounds are on the Community Right-To-Know List.

OSHA PEL: TWA 0.5 mg(Sb)/m^3
ACGIH TLV: TWA 0.5 mg(Sb)/m^3
NIOSH REL: (Antimony) TWA 0.5 mg/m^3

SAFETY PROFILE: Poison by ingestion. Human systemic effects by ingestion: nausea, vomiting, and other gastrointestinal effects. When heated to decomposition it emits very toxic fumes of I^-, NO_x, and Sb. See also ANTIMONY COMPOUNDS, EMETINE, and IODIDES.

EAM500 **CAS:8001-15-8** **HR: 3**
EMETINE BISMUTH IODIDE
mf: $C_{29}H_{40}N_2O_4 \cdot BiI_3$ mw: 1070.39

PROP: Composition = 25% emetine and 17% bismuth (AJTMAQ 10,249,30).

SYNS: BISMUTH EMETINE IODIDE □ EMETINE with BISMUTH(III) TRIIODIDE □ EMETINE TRIIODOBISMUTH(III) □ NSC 44185 □ TRIIODO(6′,7′,10,11-TETRAMETHOXYEMETAN)BISMUTH

TOXICITY DATA WITH REFERENCE
orl-mus LD50:74 mg/kg JPPMAB 16,65,64
ipr-mus LD50:126 mg/kg NCISP* JAN86
orl-cat LDLo:20 mg/kg AJTMAQ 10,249,30
orl-rbt LDLo:40 mg/kg AJTMAQ 10,249,30

SAFETY PROFILE: A poison by ingestion and intraperitoneal routes. When heated to decomposition it emits very toxic fumes of I^- and NO_x. See also BISMUTH COMPOUNDS, EMETINE, and IODIDES.

EAN000 **CAS:316-42-7** **HR: 3**
1-EMETINE DIHYDROCHLORIDE
mf: $C_{29}H_{40}N_2O_4 \cdot 2ClH$ mw: 553.63

PROP: Needles. Mp: 255° (decomp); sinters at 2°.

SYNS: AMEBICIDE □ EMETINE, DIHYDROCHLORIDE □ (−)-EMETINE DIHYDROCHLORIDE □ EMETINE HYDROCHLORIDE □ NSC-33669

TOXICITY DATA WITH REFERENCE
eye-hmn 3250 μg BJIMAG 4,111,47
scu-wmn LDLo:2770 μg/kg/4D-I:BAH,GIT AIMDAP 17,420,16
scu-man LDLo:25 mg/kg/20D-I:GIT.PUL AIMDAP 17,420,16
orl-rat LD50:12 μg/kg ANTCAO 8,297,58
ipr-rat LD50:17 mg/kg JPETAB 94,431,48
orl-mus LD50:15 μg/kg ANTCAO 8,297,58
ipr-mus LD50:62 mg/kg JPETAB 94,431,48
scu-mus LD50:37 mg/kg JPPMAB 16,65,64
scu-mky LDLo:13,886 μg/kg IMGAAY 49,85,14
orl-cat LDLo:15 mg/kg AJTMAQ 10,249,30
orl-rbt LDLo:15 mg/kg AJTMAQ 10,249,30
scu-rbt LDLo:13,886 μg/kg IMGAAY 49,85,14
ivn-rbt LDLo:2500 μg/kg JAMAAP 62,501,14
orl-bwd LD50:56,200 μg/kg AECTCV 12,355,83

CONSENSUS REPORTS: EPA Extremely Hazardous Substances List. Reported in EPA TSCA Inventory.

SAFETY PROFILE: A poison by ingestion, intraperitoneal, subcutaneous, and intravenous routes. Human systemic effects: diarrhea, distorted perceptions, dyspnea, hallucinations, hypermotility, nausea or vomiting. A human eye irritant. When heated to decomposition it emits very toxic fumes of Cl^- and NO_x.

EAN500 CAS:73713-75-4 HR: 3
EMETINE DIHYDROCHLORIDE TETRAHYDRATE
mf: $C_{29}H_{40}N_2O_4 \cdot 2ClH \cdot 4H_2O$ mw: 625.71

SYNS: NCI-C01605 □ 6′,7′,10,11-TETRAMETHOXYEMETAN DIHYDROCHLORIDE TETRAHYDRATE

CONSENSUS REPORTS: NCI Carcinogenesis Bioassay (ipr); Inadequate Studies: mouse, rat NCITR* NCI-CG-TR-43,77.

SAFETY PROFILE: Probably a poison. When heated to decomposition it emits very toxic fumes of NO_x and HCl. See also EMETINE DIHYDROCHLORIDE.

EAN600 CAS:82-92-8 HR: 3
EMOQUIL
mf: $C_{18}H_{22}N_2$ mw: 266.42

PROP: Light-sensitive crystals from petr ether. Mp: 105.5–107.5°. Solubility in g/mL at 25°: chloroform 1.1; ether 0.17; ethanol 0.17; water <0.001.

SYNS: (N-BENZHYDRYL)(N′-METHYL)DIETHYLENEDIAMINE □ N-BENZHYDRYL-N-METHYL PIPERAZINE □ BW 47-83 □ CICLIZINA □ COMPOUND 47-83 □ CYCLIZINE □ 1-DIPHENYLMETHYL-4-METHYLPIPERAZINE □ MARAZINE □ MAREZINE □ MARZINE □ N-METHYL-N′-BENZYHYDRYLPIPERAZINE □ NAUTAZINE □ NEO-DEVOMIT □ VALOID □ WELLCOME PREPARATION 47-83

TOXICITY DATA WITH REFERENCE
orl-rat TDLo:450 mg/kg (10-15D preg):REP BCFAAI 109,323,70
orl-rat TDLo:450 mg/kg (10-15D preg):TER BCFAAI 109,323,70
orl-wmn TDLo:126 mg/kg/6W-I:BLD BMJOAE 292,174,86
orl-mus LD50:147 mg/kg 27ZQAG -,218,72

SAFETY PROFILE: Poison by ingestion. Human systemic effects: unspecified blood effects. An experimental teratogen. Other experimental reproductive effects. When heated to decomposition it emits toxic fumes of NO_x.

EAN700 CAS:38957-41-4 HR: 3
EMORFAZONE
mf: $C_{11}H_{17}N_3O_3$ mw: 239.31

PROP: Crystals from methanol/isopropyl ether. Mp: 89–91°.

SYNS: 4-ETHOXY-2-METHYL-5-MORPHOLINO-3(2H)-PYRIDAZINONE □ 4-ETHOXY-2-METHYL-5-(4-MORPHOLINYL)-3(2H)-PYRIDAZINONE □ M73101 □ NANDRON □ PENTOYL

TOXICITY DATA WITH REFERENCE
dlt-mus-orl 500 mg/kg IYKEDH 11,559,80
orl-rat TDLo:700 mg/kg (female 9-15D post):REP JTSCDR 3,69,78
orl-rat TDLo:4200 mg/kg (9-15D preg):REP JTSCDR 3,69,78
orl-rat TDLo:4200 mg/kg (9-15D preg):TER JTSCDR 3,69,78
orl-mus TDLo:500 mg/kg (male 1D pre):REP IYKEDH 11,559,80
orl-rat TDLo:6750 mg/kg (female 17-22D post):TER OYYAA2 18,429,79
orl-rat LD50:1900 mg/kg IYKEDH 14,838,83
ipr-rat LD50:600 mg/kg IYKEDH 14,838,83
scu-rat TDLo:720 mg/kg OYYAA2 16,1011,78
ivn-rat LD50:340 mg/kg IYKEDH 14,838,83
orl-mus LD50:960 mg/kg IYKEDH 14,838,83
ipr-mus LD50:620 mg/kg OYYAA2 16,1011,78
scu-mus LD50:750 mg/kg IYKEDH 14,838,83

SAFETY PROFILE: Poison by intravenous route. Moderately toxic by ingestion, subcutaneous, and intraperitoneal routes. Experimental teratogenic and reproductive effects. Mutation data reported. When heated to decomposition it emits toxic fumes of NO_x.

E

EAN800 HR: 2
EMULSOV O EXTRA P

SYN: RICINOLEIC ACID TRIESTER with GLYCEROL 12-ETHER with TRIDECAETHYLENE GLYCOL

TOXICITY DATA WITH REFERENCE
orl-rat LD50:37 g/kg APFRAD 18,53,60
ipr-rat LD50:7500 mg/kg APFRAD 18,53,60
ivn-mus LD50:3600 mg/kg APFRAD 18,53,60

CONSENSUS REPORTS: Glycol ether compounds are on the Community Right-To-Know List.

SAFETY PROFILE: Moderately toxic by intravenous route. Mildly toxic by ingestion and intraperitoneal routes. See also ESTERS and GLYCOL ETHERS.

EAO100 CAS:76095-16-4 HR: 2
ENALAPRIL MALEATE
mf: $C_{20}H_{28}N_2O_5 \cdot C_4H_4O_4$ mw: 492.58

PROP: Crystals from MeCN. Mp: 143–144.5°.

SYNS: N-((S)-1-ETHOXYCARBONYL-3-PHENYLPROPYL)-l-ALANYL-l-PROLINE MALEATE □ (S)-1-(N-(1-(ETHOXYCARBONYL)-3-PHENYLPROPYL)-l-ALANYL)-l-PROLINE MALEATE □ MK 421 □ MK 421 MALEATE

TOXICITY DATA WITH REFERENCE
orl-wmn TDLo:6800 μg/kg (female 32-34W post):TER AIMEAS 108,215,88
orl-rat TDLo:280 mg/kg (female 15-22D post):REP YACHDS 14,43,86
orl-rat TDLo:2520 mg/kg (female 15-22D post):REP YACHDS 14,43,86
orl-rat TDLo:14,400 mg/kg (female 6-17D post):TER YACHDS 13,519,85
orl-rat TDLo:10 g/kg (male 10W pre):REP JTEHD6 14,715,84
orl-rat TDLo:144 mg/kg (female 6-17D post):TER YACHDS 13,519,85
orl-wmn TDLo:400 μg/kg/2D-I:SKN BMJOAE 294,91,87
orl-man TDLo:143 μg/kg:SKN BMJOAE 294,91,87
orl-wmn TDLo:5600 μg/kg/4W-I:PUL LANCAO 2,1094,86
orl-man TDLo:71 μg/kg:CVS BMJOAE 291,1309,85
orl-wmn TDLo:5600 μg/kg/4W-I:SKN LANCAO 2,1395,86
orl-rat LD50:2973 mg/kg YACHDS 13,413,85
scu-rat LD50:1418 mg/kg YACHDS 13,413,85
ivn-rat LD50:849 mg/kg YACHDS 13,413,85
orl-mus LD50:3507 mg/kg YACHDS 13,413,85
scu-mus LD50:1160 mg/kg YACHDS 13,413,85
ivn-mus LD50:859 mg/kg YACHDS 13,413,85

SAFETY PROFILE: Moderately toxic by ingestion, subcutaneous, and intravenous routes. Human systemic effects by ingestion: blood pressure depression, dermatitis, cough. An experimental teratogen. Experimental reproductive effects. When heated to decomposition it emits toxic fumes of NO_x.

EAO500 **CAS:20311-78-8** **HR: 3**
ENANTHOTOXIN
mf: $C_{17}H_{22}O_2$ mw: 258.39

SYNS: 2,8,10-HEPTADECATRIENE-4,6-DIYNE-1,14-DIOL □ OENANTHOTOXIN

TOXICITY DATA WITH REFERENCE
ipr-rat LD50:2940 µg/kg AFTOD7 7,197,81
ipr-mus LD50:2940 µg/kg AFTOD7 7,197,81

SAFETY PROFILE: A deadly poison by intraperitoneal route. When heated to decomposition it emits acrid smoke and fumes. See also ACETYLENE COMPOUNDS.

EAP000 **CAS:8015-30-3** **HR: 3**
ENAVID
mf: $C_{21}H_{26}O_2 \cdot C_{20}H_{26}O_2$ mw: 608.93

PROP: Mixture of 98.5% (17-α)-19-norpregn-4-en-20-yn-3-one, 17-hydroxy- and 1.5% (17-α)-19-norpregna-1,3,5(10)-trien-20-yn-17-ol,3-methoxy- (IARC** 6,193,74).

SYNS: CONOVID □ CONOVID E □ ENIDREL □ ENOVID □ ENOVID-E □ ETHINYLESTRADIOL-3-METHYL ETHER and NORETHYNODRED (1:50) □ MESTRANOL mixed with NORETHYNODREL □ NORETHANDROL □ NORETHYNODREL and ETHINYLESTRADIOL-3-METHYL ETHER (50:1) □ NORETHYNODREL mixed with MESTRANOL

TOXICITY DATA WITH REFERENCE
oth-mus-par 3060 µg/kg/17D-C AJOGAH 120,390,74
orl-wmn TDLo:1040 µg/kg (female 20D pre):REP JRPFA4 4,229,62
orl-wmn TDLo:1014 µg/kg (female 20D pre):REP BMJOAE 2,1179,61
orl-wmn TDLo:4 mg/kg (female 20D pre):REP AJOGAH 85,427,63
orl-wmn TDLo:46,200 µg/kg (female 6-38W post):TER JCEMAZ 19,1369,59
scu-mus TDLo:102 µg/kg (female 10D post):REP INJFA3 15,182,70
orl-rat TDLo:4300 µg/kg (female 1-22D post):REP ENDOAO 80,447,67
orl-rat TDLo:75 mg/kg (female 60D pre):REP JGOBAC 1,141,72
orl-rat TDLo:1500 µg/kg (female 2-4D post):REP ENDOAO 77,873,65
orl-rat TDLo:14 mg/kg (female 1D pre):REP JRPFA4 12,473,66
orl-rat TDLo:75 mg/kg (female 60D pre):REP JGOBAC 1,141,72
orl-rat TDLo:2250 µg/kg (lactating female 9D post):REP JRPFA4 16,15,68
orl-rat TDLo:36 mg/kg (male 9D pre):REP PSEBAA 100,540,59
ipr-mus TDLo:120 mg/kg (male 30D pre):REP ANREAK 160,322,68
orl-rat TDLo:4300 µg/kg (female 1-22D post):TER ENDOAO 80,447,67
orl-wmn TDLo:20 mg/kg/2Y-I:NEO LANCAO 2,926,73
orl-mus TDLo:207 mg/kg/74W-C:CAR JNCIAM 43,671,69
scu-mus TDLo:728 mg/kg/91W-I:NEO BJCAAI 35,322,77
orl-mus TD:1250 mg/kg/89W-I:NEO SCIEAS 154,402,66
orl-wmn TD:50 mg/kg/5Y-I:NEO LANCAO 2,926,73
orl-wmn TD:71 mg/kg/7Y-I:NEO LANCAO 2,926,73
orl-mus TD:69 mg/kg/69W-I:ETA PEXTAR 11,440,69
orl-wmn TD:252 mg/kg/10Y-I:NEO LANCAO 1,1251,80

CONSENSUS REPORTS: IARC Cancer Review: Animal Sufficient Evidence IMEMDT 6,191,74. EPA Genetic Toxicology Program.

SAFETY PROFILE: Confirmed carcinogen producing liver tumors in women by ingestion. Experimental carcinogenic, neoplastigenic, and tumorigenic data. Human reproductive effects by ingestion: menstrual cycle changes or disorders; abnormalities of the uterus, cervix, and vagina; and changes in fertility. A human teratogen that causes developmental abnormalities of the urogenital system. Experimental reproductive effects. Mutation data reported. A steroid. When heated to decomposition it emits acrid smoke and fumes.

EAQ000 **HR: 2**
ENCEPHALARTOS HILDEBRANDTII

PROP: Flour made from the starchy kernels of *Ecephalartos hildebrandtii* (BJCAAI 22,563,68).

TOXICITY DATA WITH REFERENCE
orl-rat TDLo:450 g/kg/26W-C:CAR BJCAAI 22,563,68

SAFETY PROFILE: Questionable carcinogen with experimental carcinogenic data. When heated to decomposition it emits acrid smoke and fumes.

EAQ050 **CAS:1182-87-2** **HR: 3**
ENCORDIN
mf: $C_{30}H_{44}O_9$ mw: 548.74

PROP: Crystals from MeOH. Cardenolide (cardiac glycoside) found in seeds of the tropical garden plant *Thevetia peruviana (Pers.)* Spears from methanol + ether. Mp: 161–164°. One gram dissolves in about 2500 mL water. Freely sol in chloroform and acetone; sparingly sol in methanol and ethanol.

SYNS: CANNOGENIN-α-l-THEVETOSIDE □ 3-((6-DEOXY-3-O-METHYL-α-l-GLUCOPYRANOSYL)OXY)-14-HYDROXY-19-OXO-CARD-20(22)-ENOLIDE □ PERUVOSID □ PERUVOSIDE

TOXICITY DATA WITH REFERENCE
orl-hmn TDLo:24 µg/kg:CNS,GIT IJEBA6 5,31,67
orl-hmn TDLo:39 µg/kg/3D:CVS,GIT IJMRAQ 59,271,71
ivn-hmn TDLo:4285 ng/kg:CVS,GIT IJMRAQ 59,271,71
ivn-cat LD50:145 µg/kg AIPTAK 126,412,60
ivn-gpg LDLo:710 µg/kg AIPTAK 126,412,60
ivn-pgn LDLo:263 mg/kg AIPTAK 126,412,60

SAFETY PROFILE: Poison by intravenous route. Human systemic effects by ingestion and intravenous routes: anorexia, decreased pulse rate with fall in blood pressure, and nausea or vomiting. When heated to decomposition it emits acrid smoke and fumes.

EAQ100 **HR: 2**
ENDOD, EXTRACT

SYN: PHYTOLACCA DODECANDRA, extract

TOXICITY DATA WITH REFERENCE
iut-rat TDLo:500 μg/kg (female 1D post):REP CCPTAY 14,39,76
iut-rat TDLo:500 μg/kg (6D preg):REP CCPTAY 14,39,76
iut-rat TDLo:2500 μg/kg (4D preg):REP CCPTAY 14,39,76
orl-rat LD50:2200 mg/kg BWHOA6 42,597,70
orl-mus LD50:2600 mg/kg BWHOA6 42,597,70
orl-mky LDLo:3 g/kg BWHOA6 42,597,70
orl-ckn LDLo:2500 mg/kg BWHOA6 42,597,70

SAFETY PROFILE: Moderately toxic by ingestion. Experimental reproductive effects.

EAQ750 **CAS:115-29-7** **HR: 3**
ENDOSULFAN
mf: $C_9H_6Cl_6O_3S$ mw: 406.91

PROP: A mixture of 2 isomers, brown crystals, nearly insol in water; sol in most organic solvents. Mp (α): 106°, mp (β): 212°, d: 1.745 @ 20°/20°.

SYNS: BENZOEPIN ☐ BEOSIT ☐ BIO 5,462 ☐ CHLORTHIEPIN ☐ CRISULFAN ☐ CYCLODAN ☐ DEVISULPHAN ☐ ENDOCEL ☐ ENDOSOL ☐ ENDOSULPHAN ☐ ENSURE ☐ ENT 23,979 ☐ FMC 5462 ☐ 1,2,3,4,7,7-HEXACHLOROBICYCLO(2.2.1)HEPTEN-5,6-BIOXYMETHYLENESULFITE ☐ α,β-1,2,3,4,7,7-HEXACHLOROBICYCLO(2.2.1)-2-HEPTENE-5,6-BISOXYMETHYLENE SULFITE ☐ HEXACHLOROHEXAHYDROMETHANO 2,4,3-BENZODIOXATHIEPIN-3-OXIDE ☐ 6,7,8,9,10,10-HEXACHLORO-1,5,5a,6,9,9a-HEXAHYDRO-6,9-METHANO-2,4,3-BENZODIOXATHIEPIN-3-OXIDE ☐ 1,4,5,6,7,7-HEXACHLORO-5-NORBORNENE-2,3-DIMETHANOL CYCLIC SULFITE ☐ HILDAN ☐ HOE 2,671 ☐ INSECTOPHENE ☐ KOP-THIODAN ☐ MALIX ☐ NCI-C00566 ☐ NIA 5462 ☐ NIAGARA 5,462 ☐ OMS 570 ☐ RCRA WASTE NUMBER P050 ☐ SULFUROUS ACID, cyclic ester with 1,4,5,6,7,7-HEXACHLORO-5-NORBORNENE-2,3-DIMETHANOL ☐ THIFOR ☐ THIMUL ☐ THIODAN ☐ THIOFOR ☐ THIOMUL ☐ THIONEX ☐ THIOSULFAN ☐ THIOSULFAN TIONEL ☐ TIOVEL

TOXICITY DATA WITH REFERENCE
sln-dmg-orl 200 ppm/48H MUREAV 136,115,84
mmo-smc 100 mg/L TGANAK 18,455,84
sce-hmn:lym 1 μmol/L ARTODN 52,221,83
orl-rat TDLo:45 mg/kg (6-14D preg):REP APTOA6 42,150,78
orl-rat TDLo:600 mg/kg (60D male):REP TOLED5 18(Suppl 1),139,83
orl-rat TDLo:45 mg/kg (6-14D preg):TER APTOA6 42,150,78
orl-mus TDLo:330 mg/kg/78W-I:NEO NTIS** PB223-159
scu-mus TDLo:2 mg/kg:ETA NTIS** PB223-159
orl-man TDLo:86 mg/kg:BAH,PUL JTCTDW 26,265,88
orl-rat LD50:18 mg/kg ARSIM* 20,9,66
ihl-rat LC50:80 mg/m³/4H JOCMA7 9,35,67
skn-rat LD50:34 mg/kg 85GMAT -,72,82
ipr-rat LD50:8 mg/kg RREVAH 22,1,68
orl-mus LD50:7360 μg/kg 30ZDA9 -,267,71
ipr-mus LD50:7 mg/kg BECTA6 15,708,76
orl-dog LD50:76,700 μg/kg PEMNDP 8,335,87
orl-cat LD50:2 mg/kg 85DPAN -,-,71/76
orl-rbt LD50:28 mg/kg VETNAL 59(12),64,83
skn-rbt LD50:90 mg/kg JOCMA7 9,35,67
scu-rbt LD50:360 mg/kg RREVAH 22,1,68
orl-ham LD50:118 mg/kg EJTXAZ 7,159,74
ipr-ham LD50:80 mg/kg ARTODN 58,152,85

CONSENSUS REPORTS: EPA Extremely Hazardous Substances List. NCI Carcinogenesis Bioassay (feed); No Evidence: mouse, rat NCITR* NCI-CG-TR-62,77.

OSHA PEL: TWA 0.1 mg/m³ (skin)
ACGIH TLV: TWA 0.1 mg/m³ (skin)

SAFETY PROFILE: Poison by ingestion, inhalation, skin contact, intraperitoneal, and subcutaneous routes. Experimental teratogenic and reproductive effects. Questionable carcinogen with experimental tumorigenic and neoplastigenic data. Human systemic effects: convulsions, cyanosis. Human mutation data reported. A central nervous system stimulant producing convulsions. A highly toxic organochlorine pesticide that does not accumulate significantly in human tissue. Absorption is normally slow, but is increased by alcohols, oil, and emulsifiers. When heated to decomposition it emits toxic fumes of Cl^- and SO_x. See also CHLORIDES and SULFITES.

E

EAR000 **CAS:145-73-3** **HR: 3**
ENDOTHAL
mf: $C_8H_{10}O_5$ mw: 186.18

PROP: Solid. Mp: 144°. Sol in water.

SYNS: AQUATHOL ☐ 3,6-ENDOOXOHEXAHYDROPHTHALIC ACID ☐ ENDOTHALL ☐ ENDOTHAL TECHNICAL ☐ 3,6-ENDOXOHEXAHYDROPHTHALIC ACID ☐ 3,6-endo-EPOXY-1,2-CYCLOHEXANEDICARBOXYLIC ACID ☐ HEXAHYDRO-3,6-endo-OXYPHTHALIC ACID ☐ HYDOUT ☐ HYDROTHAL-47 ☐ 7-OXABICYCLO(2.2.1)HEPTANE-2,3-DICARBOXYLIC ACID ☐ RCRA WASTE NUMBER P088 ☐ TRI-ENDOTHAL

TOXICITY DATA WITH REFERENCE
orl-rat LD50:38 mg/kg PCOC** -,471,66

SAFETY PROFILE: Poison by ingestion. Very irritating to skin, eyes, and mucus membranes. Causes diarrhea. When heated to decomposition it emits acrid smoke and fumes.

EAS000 **CAS:2778-04-3** **HR: 3**
ENDOTHION
mf: $C_9H_{13}O_6PS$ mw: 280.25

SYNS: AC-18,737 ☐ O,O-DIMETHYL-S-(5-METHOXY-4-OXO-4H-PYRAN-2-YL)PHOSPHOROTHIOATE ☐ O,O-DIMETHYL-S-(5-METHOXYPYRON-2-YL)-METHYL)-THIOLPHOSPHAT (GERMAN) ☐ O,O-DIMETHYL-S-(5-METHOXYPYRONYL-2-METHYL) THIOPHOSPHATE ☐ ENDOCID ☐ ENDOCIDE ☐ ENT 24,653 ☐ EXOTHION ☐ 5-METHOXY-2-(DIMETHOXYPHOSPHINYLTHIOMETHYL)PYRONE-4 ☐ S-5-METHOXY-4-OXOPYRAN-2-YLMETHYL DIMETHYL PHOSPHOROTHIOATE ☐ S-((5-METHOXY-4H-PYRON-2-YL)-METHYL)-O,O-DIMETHYL-MONOTHIOFOSFAAT (DUTCH) ☐ S-((5-METHOXY-4H-PYRON-2-YL)-METHYL)-O,O-DIMETHYL-MONOTHIOPHOSPHAT (GERMAN) ☐ S-(5-METHOXY-4-PYRON-2-YLMETHYL) DIMETHYL PHOSPHOROTHIOLATE ☐ S-((5-METOSSI-4H-PIRON-2-IL)-METIL)-O,O-DIMETIL-MONOTIOFOSFATO (ITALIAN) ☐ NIA-5767 ☐ NIAGARA 5767 ☐ PHOSPHATE 100 ☐ PHOSPHOPYRON ☐ PHOSPHOPYRONE ☐ THIOPHOSPHATE de O,O-DIMETHYLE et de S-((5-METHOXY-4-PYRONYL)-METHYLE) (FRENCH)

TOXICITY DATA WITH REFERENCE
orl-rat LD50:23 mg/kg WRPCA2 4,36,64
skn-rat LD50:130 mg/kg PHJOAV 185,361,60
orl-mus LD50:17 mg/kg ARSIM* 20,10,66

CONSENSUS REPORTS: EPA Extremely Hazardous Substances List.

SAFETY PROFILE: A poison by ingestion and skin contact. A pesticide. When heated to decomposition it emits very toxic fumes of PO_x and SO_x.

EAS100 **HR: 3**
ENDOTOXIN

PROP: A purified lipopolysaccharide fraction isolated from *Salmonella typhosa* 0901 (TXAPA9 23,102,72).

SYNS: SALMONELLA TYPHI ENDOTOXIN □ SALMONELLA TYPHOSA ENDOTOXIN

TOXICITY DATA WITH REFERENCE
ipr-rat LD50:107 mg/kg BPJLAQ 7,7,8
ivn-rat LD50:4330 µg/kg CRSHAG 4,253,77
ipr-mus LD50:19 mg/kg TXAPA9 23,102,72
ivn-dog LD50:1 mg/kg BJPCBM 61,175,77

SAFETY PROFILE: Poison by intravenous and intraperitoneal routes.

EAS200 **HR: 3**
ENDOTOXIN, BACT. AERTRYCKE

PROP: The endotoxin was obtained by repeated freezing and thawing of the killed emulsions BJEPA5 16,454,35

TOXICITY DATA WITH REFERENCE
ipr-rat LDLo:750 mg/kg BJEPA5 16,454,35
scu-rat LDLo:15,000 mg/kg BJEPA5 16,454,35
ipr-rbt LDLo:400 mg/kg BJEPA5 16,454,35
ivn-rbt LDLo:75 mg/kg BJEPA5 16,454,35
ipr-pig LDLo:133 mg/kg BJEPA5 16,454,35
ipr-gpg LDLo:1500 mg/kg BJEPA5 16,454,35
scu-gpg LDLo:8 g/kg BJEPA5 16,454,35

SAFETY PROFILE: A poison by intraperitoneal route. When heated to decomposition it emits acrid smoke and irritating vapors.

EAS230 **HR: 1**
Δ-ENDOTOXIN, from BACILLUS THURINGIENSIS

TOXICITY DATA WITH REFERENCE
orl-rat LD50:>8 g/kg NNGADV 14,415,89
skn-rat LD50:>2 g/kg NNGADV 14,415,89
scu-rat LD50:>8 g/kg NNGADV 14,415,89
orl-mus LD50:>17,300 mg/kg NNGADV 14,415,89
ipr-mus LD50:1520 mg/kg NNGADV 14,415,89
scu-mus LD50:12,100 mg/kg NNGADV 14,415,89

SAFETY PROFILE: Low toxicity by ingestion and skin contact. When heated to decomposition it emits acrid smoke and irritating vapors.

EAS260 **HR: D**
ENDOTOXIN, klp

SYN: K. PNEUMONIAE ENDOTOXIN

TOXICITY DATA WITH REFERENCE
dns-hmn:lng 500 µg/L PSEBAA 171,109,82

SAFETY PROFILE: Mutation data reported. When heated to decomposition it emits acrid smoke and irritating vapors.

EAT500 **CAS:72-20-8** **HR: 3**
ENDRIN
mf: $C_{12}H_8Cl_6O$ mw: 380.90

PROP: White crystals. Mp: decomp @ 200°. Sol in Me_2CO, C_6H_6, xylene; sltly sol in CCl_4 and hexane.

SYNS: COMPOUND 269 □ ENDREX □ ENDRINE (FRENCH) □ ENT 17,251 □ HEXACHLOROEPOXYOCTAHYDRO-endo,endo-DIMETHANONAPHTHALENE □ 3,4,5,6,9,9-HEXACHLORO-1a,2,2a,3,6,6a,7,7a-OCTAHYDRO-2,7:3,6-DIMETHANONAPHTH(2,3-b)OXIRENE □ HEXADRIN □ MENDRIN □ NCI-C00157 □ NENDRIN □ RCRA WASTE NUMBER P051

TOXICITY DATA WITH REFERENCE
sce-ofs-mul 54 pmol/L MUREAV 118,61,83
cyt-rat-par 1 mg/kg BECTA6 9,65,73
orl-ham TDLo:15 mg/kg (female 5-14D post):REP TXCYAC 21,187,81
orl-mus TDLo:10 mg/kg (female 8-12D post):REP JTEHD6 10,541,82
orl-ham TDLo:7500 µg/kg (female 5-14D post):REP TXAPA9 48,A200,79
orl-ham TDLo:5 mg/kg (female 8D post):TER TJADAB 9,11,74
orl-mus TDLo:7 mg/kg (female 8D post):TER TCMUD8 5,3,85
orl-mus TDLo:2320 µg/kg (female 4D pre):REP OYYAA2 6,673,72
orl-ham TDLo:5 mg/kg (female 8D post):REP TJADAB 9,11,74
orl-rat TDLo:2320 µg/kg (4D pre):REP OYYAA2 6,673,72
orl-mus TDLo:10 mg/kg (8-12D preg):REP JTEHD6 10,541,82
orl-ham TDLo:5 mg/kg (female 8D post):TER TXAPA9 48,A201,79
orl-mus TDLo:2500 µg/kg (female 9D post):TER TJADAB 9,11,74
orl-mus TDLo:16,500 µg/kg (female 7-17D post):TER TXCYAC 21,141,81
orl-man LDLo:171 mg/kg HUTODJ 4,241,85
orl-rat LD50:3 mg/kg WRPCA2 9,119,70
skn-rat LD50:12 mg/kg SPEADM 78-1,13,78
orl-mus LD50:1370 µg/kg TXAPA9 25,42,73
ivn-mus LD50:2300 µg/kg TXAPA9 23,408,72
orl-mky LD50:3 mg/kg PCOC** -,475,66
orl-cat LDLo:5 mg/kg JAFCAU 3,842,55
orl-rbt LD50:7 mg/kg JAFCAU 3,842,55
skn-rbt LD50:60 mg/kg SPEADM 78-1,13,78
orl-gpg LD50:16 mg/kg PCOC** -,475,66

CONSENSUS REPORTS: IARC Cancer Review: Group 3 IMEMDT 7,56,87; Animal Inadequate Evidence IMEMDT 5,157,74; Human Inadequate Evidence IMEMDT 5,157,74. NCI Carcinogenesis Bioassay (feed);

No Evidence: mouse, rat NCITR* NCI-CG-TR-12,79. EPA Genetic Toxicology Program. EPA Extremely Hazardous Substances List.

OSHA PEL: TWA 0.1 mg/m³ (skin)
ACGIH TLV: TWA 0.1 mg/m³ (skin)
DFG MAK: 0.1 mg/m³

SAFETY PROFILE: Poison by ingestion, skin contact, and intravenous routes. Experimental teratogenic and reproductive effects. Questionable carcinogen. Mutation data reported. A central nervous system stimulant. Highly toxic to birds, fish, and humans. Many cases of fatal poisoning have been attributed to it. Does not accumulate in human tissue. In humans, ingestion of 1 mg/kg has caused symptoms. A dangerous fire hazard. Mixtures with parathion dissolve very exothermically in petroleum solvents and may cause an air-vapor explosion. See also ALDRIN.

EAT600 CAS:5836-29-3 HR: 3
ENDROCIDE
mf: $C_{19}H_{16}O_3$ mw: 292.35

SYNS: BAY 25634 □ BAY ENE 11183 B □ BAYER 25 634 □ COUMATETRALYL □ CUMATETRALYL (GERMAN, DUTCH) □ ENDOX □ ENDROCID □ ENE 11183 B □ 4-HYDROXY-3-(1,2,3,4-TETRAHYDRO-1-NAFTYL)-4-CUMARINE (DUTCH) □ 4-HYDROXY-3-(1,2,3,4-TETRAHYDRO-1-NAPHTHALENYL)-2H-1-BENZOPYRAN-2-ONE (9CI) □ 4-HYDROXY-3-(1,2,3,4-TETRAHYDRO-1-NAPHTHYL)CUMARIN □ 4-IDROSSI-3-(1,2,3,4-TETRAIDRO-1-NAFTIL)CUMARINA (ITALIAN) □ RACUMIN □ RAUCUMIN 57 □ RODENTIN □ 3-(1,2,3,4-TETRAHYDRO-1-NAPHTHYL)-4-HYDROXYCUMARIN (GERMAN) □ 3-(1,2,3,4-TETRAHYDRO-1-NAPHTYL)-4-HYDROXYCOUMARINE (FRENCH) □ 3-(α-TETRAL)-4-OXYCOUMARIN □ 3-(α-TETRALYL)-4-HYDROXYCOUMARIN □ 3-(d-TETRALYL)-4-HYDROXYCOUMARIN

TOXICITY DATA WITH REFERENCE
orl-rat LD50:16,500 μg/kg FMCHA2 -,C203,83
orl-cat LDLo:20 mg/kg 85GYAZ -,115,71
orl-gpg LDLo:250 mg/kg 85GYAZ -,115,71

CONSENSUS REPORTS: EPA Extremely Hazardous Substances List.

SAFETY PROFILE: Poison by ingestion. When heated to decomposition it emits acrid smoke and fumes. See also COUMARIN.

EAT800 CAS:11115-82-5 HR: 3
ENDURACIDIN

SYNS: ENRADIN □ ENRAMYCIN

TOXICITY DATA WITH REFERENCE
ipr-rat LD50:830 mg/kg TDKNAF 28,76,69
ivn-rat LD50:66,600 μg/kg TDKNAF 28,76,69
ipr-mus LD50:750 mg/kg TDKNAF 28,76,69
ivn-mus LD50:30 mg/kg TDKNAF 28,76,69
ivn-rbt LD50:92 mg/kg TDKNAF 28,76,69

SAFETY PROFILE: Poison by intravenous route. Moderately toxic by intraperitoneal route.

EAT810 CAS:34438-27-2 HR: 3
ENDURACIDIN A
mf: $C_{107}H_{138}Cl_2N_{26}O_{31}$ mw: 2355.61

PROP: Crystals.

SYNS: B-5477 □ ENRADINE □ ENRAMYCIN

TOXICITY DATA WITH REFERENCE
ipr-mus LD50:880 mg/kg 85GDA2 4(2),109,80
scu-mus LD50:3000 mg/kg 85GDA2 4(2),109,80
ivn-mus LD50:25 mg/kg 85GDA2 4(2),109,80

SAFETY PROFILE: Poison by intravenous route. Moderately toxic by intraperitoneal and subcutaneous routes. When heated to decomposition it emits toxic fumes of Cl^- and NO_x.

EAT900 CAS:13838-16-9 HR: 2
ENFLURANE
mf: $C_3H_2ClF_5O$ mw: 184.50

SYNS: ANESTHETIC COMPOUND No. 347 □ 2-CHLORO-1-(DIFLUOROMETHOXY)-1,1,2-TRIFLUOROETHANE □ 2-CHLORO-1,1,2-TRIFLUOROETHYL DIFLUOROMETHYL ETHER □ COMPOUND 347 □ ETHRANE □ METHYLFLURETHER □ NSC-115944 □ OHIO 347

TOXICITY DATA WITH REFERENCE
eye-rbt 100 mg MOD FEPRA7 35,729,76
cyt-hmn:lyms 1000 ppm ENVRAL 12,366,76
cyt-mus:fbr 1000 ppm ENVRAL 12,366,76
ihl-rat TCLo:1500 ppm/6H (1-20D post):REP AIPTAK 256,134,82
ihl-rat TCLo:16,500 ppm/6H (8-10D post):TER ANESAV 64,339,86
ihl-mus TCLo:10,000 ppm/4H (6-15D post):TER ANESAV 54,505,81
ihl-rat TCLo:16,500 ppm/6H (11-13D post):TER ANESAV 64,339,86
ihl-mus TCLo:3000 ppm/4H/78W-I:CAR ANESAV 56,9,82
ihl-hmn TCLo:1 pph/6H ANESAV 45,557,76
orl-rat LD50:5450 mg/kg DRUGAY 6,143,82
ihl-rat LC50:14,000 ppm/3H IYKEDH 12,668,81
ipr-rat LD50:6 g/kg DRUGAY 6,143,82
scu-rat LD50:19,500 mg/kg YKYUA6 32,491,81
orl-mus LD50:5 g/kg DRUGAY 6,143,82
ihl-mus LC50:8100 ppm/3H IYKEDH 12,668,81
ipr-mus LD50:3900 mg/kg DRUGAY 6,143,82
scu-mus LD50:38,800 mg/kg YKYUA6 32,491,81

CONSENSUS REPORTS: Reported in EPA TSCA Inventory.

ACGIH TLV: TWA 75 ppm
NIOSH REL: (Waste Anesthetic Gases and Vapors) CL 2 ppm/1H

SAFETY PROFILE: Mildly toxic by inhalation, ingestion, and subcutaneous routes. Experimental reproductive data by inhalation. Human systemic effects by inhalation: decreased urine volume or anuria. An experimental teratogen. Experimental reproductive effects. Human mutation data reported. An eye irritant. Questionable carcinogen with experimental carcinogenic data. An anesthetic. When heated to decomposition it emits very toxic fumes of F^- and Cl^-. See also ETHERS.

For occupational chemical analysis use OSHA: #29.

EAU000 **HR: 3**
ENHYDRINA SCHISTOSA VENOM

SYN: VENOM, SEA SNAKE, ENHYDRINA SCHISTOSA

TOXICITY DATA WITH REFERENCE
ivn-rat LD50:70 μg/kg TRSTAZ 54,50,60
ipr-mus LD50:110 μg/kg 85EGD4 -,341,78
scu-mus LD50:111 μg/kg TOXIA6 14,347,76
ivn-mus LD50:74 μg/kg TOXIA6 20,797,82
par-mus LDLo:125 μg/kg SCNEBK 110,355,76
ivn-rbt LD50:57 μg/kg TRSTAZ 54,40,60
ivn-gpg LD50:61 μg/kg TRSTAZ 54,50,60
par-frg LD50:20 μg/kg TRSTAZ 54,50,60
ivn-mam LD50:10 μg/kg CLPTAT 8,849,67

SAFETY PROFILE: A deadly poison by intraperitoneal, intravenous, parenteral, and subcutaneous routes.

EAU075 **CAS:55726-47-1** **HR: 3**
ENOCITABINE
mf: $C_{31}H_{55}N_3O_6$ mw: 565.89

PROP: Crystals from DMSO. Mp: 141–142°.

SYNS: N-(1-β-d-ARABINOFURANOSYL-1,2-DIHYDRO-2-OXO-4-PYRIMIDINYL)DOCOSANAMIDE □ N[4]-BEHENOYL-1-β-d-ARABINOFURANOSYLCYTOSINE □ BEHENOYLCYTOSINE ARABINOSIDE □ N[4]-BEHENOYLCYTOSINE ARABINOSIDE □ SUNRABIN

TOXICITY DATA WITH REFERENCE
ipr-rat LD50:1100 mg/kg GTKRDX 10,2077,83
ivn-rat LD50:380 mg/kg IYKEDH 14,484,83
ipr-mus LD50:525 mg/kg IYKEDH 14,484,83
scu-mus LD50:810 mg/kg GTKRDX 10,2077,83
ivn-mus LD50:500 mg/kg IYKEDH 14,484,83

SAFETY PROFILE: Poison by intravenous route. Moderately toxic by subcutaneous and intraperitoneal routes. When heated to decomposition it emits toxic fumes of NO_x.

EAU100 **CAS:74011-58-8** **HR: 3**
ENOXACIN
mf: $C_{15}H_{17}FN_4O_3$ mw: 320.36

PROP: Crystals from ethanol/methylene chloride. Mp: 246–248°.

SYNS: AT-2266 □ 1-ETHYL-6-FLUORO-1,4-DIHYDRO-4-OXO-7-(1-PIPERAZINYL)-1,8-NAPHTHYRIDINE-3-CARBOXYLIC ACID □ FLUMARK

TOXICITY DATA WITH REFERENCE
orl-rat TDLo:11 g/kg (female 7-17D post):TER NKRZAZ 32(Suppl 3),301,84
orl-rat TDLo:91 g/kg (male 13W pre):REP NKRZAZ 32(Suppl 3),293,84
orl-rat TDLo:275 g/kg (female 26W pre):REP NKRZAZ 32(Suppl 3),199,84
orl-rat TDLo:90 g/kg (male 30D pre):REP NKRZAZ 32(Suppl 3),199,84
ivn-rat LD50:236 mg/kg NKRZAZ 32(Suppl 3),192,84
scu-mus LD50:1100 mg/kg 43MKAT 1,456,80
ivn-mus LD50:327 mg/kg NKRZAZ 32(Suppl 3),192,84

SAFETY PROFILE: Poison by intravenous route. Moderately toxic by subcutaneous route. An experimental teratogen. Other experimental reproductive effects. When heated to decomposition it emits toxic fumes of F^- and NO_x.

EAU150 **HR: 3**
ENOXACIN HYDRATE (2:3)
mf: $C_{15}H_{17}FN_4O_3 \cdot 3/2H_2O$ mw: 347.39

TOXICITY DATA WITH REFERENCE
ivn-rat LD50:236 mg/kg IYKEDH 16,1461,85
scu-mus LD50:1237 mg/kg IYKEDH 16,1461,85
ivn-mus LD50:327 mg/kg IYKEDH 16,1461,85

SAFETY PROFILE: Poison by intravenous route. Moderately toxic by subcutaneous route. When heated to decomposition it emits toxic fumes of F^- and NO_x.

EAU500 **CAS:136-45-8** **HR: 3**
ENT 17591
mf: $C_{13}H_{17}NO_4$ mw: 251.31

SYNS: DIPROPYL ISOCINCHOMERONATE □ DI-N-PROPYL-ISOCINCHOMERONATE (GERMAN) □ DIPROPYL 2,5-PYRIDINEDICARBOXYLATE □ MGK 326 □ MGK REPELLENT-326 □ PYRIDIN-2,5-DICARBONSAEURE-DI-N-PROPYLESTER (GERMAN) □ 2,5-PYRIDINEDICARBOXYLIC ACID, DIPROPYL ESTER □ R-326 □ REPPER 333

TOXICITY DATA WITH REFERENCE
orl-rat LD50:5230 mg/kg 28ZEAL 5,94,76
skn-rat LD50:9400 mg/kg 28ZEAL 5,94,76
orl-mus LD50:1600 mg/kg YKYUA6 32,605,81
ipr-mus LD50:330 mg/kg YKYUA6 32,605,81
skn-rbt LD50:9500 mg/kg 85DPAN -,-,71/76

SAFETY PROFILE: Poison by intraperitoneal route. Moderately toxic by ingestion. Mildly toxic by skin contact. An insect repellent. When heated to decomposition it emits toxic fumes of NO_x.

EAV100 **CAS:23581-62-6** **HR: 3**
EO 122
mf: $C_{16}H_{22}N_2O \cdot ClH$ mw: 294.86

SYN: N-(2,6-DIMETHYLPHENYL)-2-AZABICYCLO(2.2.2)OCTANE-3-CARBOXAMIDE MONOHYDROCHLORIDE (9CI)

TOXICITY DATA WITH REFERENCE
ipr-mus LD50:66 mg/kg ANGIAB 31,410,80
ivn-mus LD50:22 mg/kg ANGIAB 31,410,80
ivn-rbt LD50:8500 μg/kg ANGIAB 31,410,80

SAFETY PROFILE: Poison by intravenous and intraperitoneal routes. When heated to decomposition it emits toxic fumes of NO_x and HCl.

EAV500 **CAS:33419-42-0** **HR: 3**
EPE
mf: $C_{29}H_{32}O_{13}$ mw: 588.61

PROP: Crystals from MeOH. Mp: 236–251°.

SYNS: DEMETHYL-EPIODOPHYLLOTOXIN ETHYLIDENE GLUCOSIDE □ 4-DEMETHYLEPIODOPHYLLOTOXIN-β,d-ETHYLIDENEGLUCOSIDE □ 4′-DEMETHYLEPIPODOPHYLLOTOXIN-9-(4,6-O-ETHYLIDENE-β-d-GLUCOPYRANOSIDE □ 4′-DEMETHYLEPIPODOPHYLLOTOXIN ETHYLIDENE-β,d-GLUCOSIDE □ 4-DEMETHYL-EPIPO-

DOPHYLLOTOXIN-β,d-ETHYLIDEN-GLUCOSIDE □ 4'-O-DEMETHYL-1-O-(4,6-O-ETHYLIDENE-β,d-GLUCOPYRANOSYL)EPIPODOPHYLLOTOXIN □ ETOPOSIDE □ NK 171 □ NSC 141540 □ VEPESID □ VP 16213

TOXICITY DATA WITH REFERENCE
mmo-sat 2 mg/plate TCMUD8 5,319,85
dnd-hmn:oth 2 μmol/L CNREA8 45,3106,85
dnd-ham:lng 40 μmol/L CNREA8 46,611,86
orl-rat TDLo:14,630 μg/kg (7-17D preg):REP KSRNAM 19,3897,85
orl-rat TDLo:312 mg/kg (17-21D preg/21D post):REP KSRNAM 19,3926,85
ipr-mus TDLo:1500 μg/kg (female 6D post):TER TJADAB 18,31,78
orl-rat TDLo:132 mg/kg (female 7-17D post):TER KSRNAM 19,3897,85
ipr-mus TDLo:1500 μg/kg (female 6D post):REP TJADAB 18,31,78
unr-rat TDLo:1400 mg/kg (male 28D pre):REP ARTODN 4,466,80
ipr-rat TDLo:14 mg/kg (male 5W pre):REP KSRNAM 19,3561,85
ipr-mus TDLo:1 mg/kg (female 8D post):TER TJADAB 18,31,78
ipr-mus TDLo:2 mg/kg (female 6D post):TER TJADAB 18,31,78
orl-rat TDLo:14,630 μg/kg (7-17D preg):TER KSRNAM 19,3897,85
orl-hmn TDLo:16 mg/kg/5D-I:BLD CANCAR 34,985,74
ivn-hmn TDLo:2630 μg/kg/10D-I:BLD CANCAR 34,985,74
ivn-cld TDLo:183 mg/kg/2H-C DICPBB 22,41,88
orl-rat LD50:1784 mg/kg KSRNAM 19,3473,85
ipr-rat LD50:39 mg/kg KSRNAM 19,3473,85
ivn-rat LD50:75 mg/kg KSRNAM 19,3473,85
orl-mus LD50:215 mg/kg NCISP* JAN86
ipr-mus LD50:64 mg/kg KSRNAM 19,3473,85
scu-mus LD50:223 mg/kg NCISP* JAN86
ivn-mus LD50:15,070 μg/kg NCISP* JAN86

SAFETY PROFILE: Poison by ingestion, intraperitoneal, intravenous, and subcutaneous routes. An experimental teratogen. Human systemic effects by ingestion and inhalation: agranulocytosis, aplastic anemia, and other changes in bone marrow. Experimental reproductive effects. Human mutation data reported. When heated to decomposition it emits acrid smoke and fumes.

EAV700 CAS:56839-43-1 HR: 3
EPERISONE HYDROCHLORIDE
mf: $C_{17}H_{25}NO \cdot ClH$ mw: 295.89

SYNS: E-646 □ EMPP □ 4'-ETHYL-2-METHYL-3-PIPERIDINOPROPIOPHENONE HYDROCHLORIDE □ MIONAL

TOXICITY DATA WITH REFERENCE
orl-rat LD50:1300 mg/kg OYYAA2 21,939,81
scu-rat LD50:490 mg/kg OYYAA2 21,939,81
ivn-rat LD50:51 mg/kg OYYAA2 21,939,81
ims-rat LD50:400 mg/kg OYYAA2 21,939,81
orl-mus LD50:940 mg/kg OYYAA2 21,939,81
scu-mus LD50:256 mg/kg OYYAA2 21,939,81
ivn-mus LD50:43 mg/kg OYYAA2 21,939,81
ims-mus LD50:280 mg/kg OYYAA2 21,939,81

SAFETY PROFILE: Poison by subcutaneous, intramuscular, and intravenous routes. Moderately toxic by ingestion. When heated to decomposition it emits toxic fumes of NO_x and HCl.

EAW000 CAS:299-42-3 HR: 3
EPHEDRINE
mf: $C_{10}H_{15}NO$ mw: 165.26

PROP: White granules. Mp: 79° (dl), mp: 40° (l), bp: 225° (decomp). Sol in ether and chloroform.

SYNS: BIOPHEDRIN □ ECIPHIN □ EFEDRIN □ EPHEDRAL □ EPHEDRATE □ EPHEDREMAL □ EPHEDRIN □ l-EPHEDRINE □ l(−)-EPHEDRINE □ EPHEDRITAL □ EPHEDROL □ EPHEDROSAN □ EPHEDROTAL □ EPHEDSOL □ EPHENDRONAL □ EPHOXAMIN □ FEDRIN □ α-HYDROXY-β-METHYL AMINE PROPYLBENZENE □ 1-HYDROXY-2-METHYLAMINO-1-PHENYLPROPANE □ l-SEDRIN □ ISOFEDROL □ KRATEDYN □ MANADRIN □ MANDRIN □ (−)-α-(1-METHYLAMINOETHYL)BENZYL ALCOHOL □ 1-α-(1-METHYLAMINOETHYL) BENZYL ALCOHOL □ 1-2-METHYLAMINO-1-PHENYLPROPANOL □ N-METHYLNOREPHEDRINE □ NASOL □ 1-PHENYL-2-METHYLAMINOPROPANOL □ SANEDRINE □ VENCIPON □ ZEPHROL

TOXICITY DATA WITH REFERENCE
ipr-rat TDLo:50 mg/kg (female 9D post):TER TJADAB 34,469,86
unr-man LDLo:9 mg/kg 85DCAI 2,73,70
orl-rat LD50:600 mg/kg 27ZQAG -,341,72
ipr-rat LDLo:150 mg/kg AEPPAE 195,647,40
scu-rat LD50:300 mg/kg AIPTAK 68,339,42
ivn-rat LDLo:130 mg/kg AEPPAE 120,189,27
orl-mus LD50:1250 mg/kg ARZNAD 23,1125,73
ipr-mus LD50:350 mg/kg PCJOAU 3,203,69
ivn-mus LD50:74 mg/kg APTOA6 38,474,76
par-mus LD50:170 mg/kg AEPPAE 195,647,40
ivn-dog LDLo:70 mg/kg AEPPAE 120,189,27
ims-rbt LDLo:340 mg/kg AEPPAE 120,189,27

CONSENSUS REPORTS: Reported in EPA TSCA Inventory.

SAFETY PROFILE: A human poison by an unspecified route. An experimental poison by intravenous, subcutaneous, intramuscular, and intraperitoneal routes. Moderately toxic by ingestion and parenteral routes. Causes rapid pulse, rise in blood pressure, and other actions similar to epinephrine. An experimental teratogen. Used in production of drugs of abuse. Has been known to cause allergic sensitization. When heated to decomposition it emits toxic fumes of NO_x.

EAW100 CAS:90-81-3 HR: 3
dl-EPHEDRINE
mf: $C_{10}H_{15}NO$ mw: 165.26

SYNS: BENZENEMETHANOL, α-(1-(METHYLAMINO)ETHYL)-, (R*,S*)-(±)-(9CI) □ (±)-EPHEDRINE □ EPHEDRINE, (±)- □ RACEPHEDRINE

TOXICITY DATA WITH REFERENCE
ipr-rat LDLo:170 mg/kg AEPPAE 195,647,40

CONSENSUS REPORTS: Reported in EPA TSCA Inventory.

SAFETY PROFILE: Poison by intraperitoneal route.

When heated to decomposition it emits toxic vapors of NO_x.

EAW200 **CAS:321-98-2** **HR: 3**
(+)-EPHEDRINE
mf: $C_{10}H_{15}NO$ mw: 165.26

SYNS: BENZENEMETHANOL, α-(1-(METHYLAMINO)ETHYL)-, (S-(R*, S*))-(9CI) □ d-EPHEDRINE □ EPHEDRINE, (+)- □ l-(+)-EPHEDRINE

TOXICITY DATA WITH REFERENCE
ipr-mus LD50:255 mg/kg JPMSAE 53,987,64
ivn-mus LD50:105 mg/kg JPETAB 148,158,65

CONSENSUS REPORTS: Reported in EPA TSCA Inventory.

SAFETY PROFILE: Poison by intravenous and intraperitoneal routes. When heated to decomposition it emits toxic vapors of NO_x.

EAW500 **CAS:50-98-6** **HR: 3**
EPHEDRINE HYDROCHLORIDE
mf: $C_{10}H_{15}NO \cdot ClH$ mw: 201.72

PROP: A solid. Mp: 218°.

SYNS: BENZENEMETHANOL, α-(1-(METHYLAMINO)ETHYL)-, HYDROCHLORIDE, (R-(R*,S*))- □ EPHEDRINE HYDROCHLORIDE □ l-EPHEDRINE, HYDROCHLORIDE □ N-METHYL-β-OXY-β-PHENYLISOPROPYLAMINHYDROCHLORID

TOXICITY DATA WITH REFERENCE
idr-hmn TDLo:5700 μg/kg:SKN JPETAB 76,295,42
ipr-rat LD50:165 mg/kg JPETAB 86,284,46
scu-rat LD50:1150 mg/kg JPETAB 95,336,49
ivn-rat LD50:69 mg/kg JAPMA8 33,80,44
orl-mus LD50:400 mg/kg JMCMAR 9,966,66
ipr-mus LD50:242 mg/kg JPETAB 158,135,67
scu-mus LD50:40,870 μg/kg JPETAB 87,214,46
ivn-mus LD50:95 mg/kg JPETAB 95,336,49
par-mus LDLo:500 mg/kg MAIZAB 3,1,25

CONSENSUS REPORTS: Reported in EPA TSCA Inventory.

SAFETY PROFILE: A poison by ingestion, intraperitoneal, subcutaneous, and intravenous routes. Human systemic effects by intradermal route: skin effects. When heated to decomposition it emits very toxic fumes of Cl^- and NO_x. See also EPHEDRINE.

EAW995 **CAS:24221-86-1** **HR: 3**
d-EPHEDRINE HYDROCHLORIDE
mf: $C_{10}H_{15}NO \cdot ClH$ mw: 201.72

PROP: A solid. Mp: 217–218°.

TOXICITY DATA WITH REFERENCE
scu-rat LDLo:160 mg/kg JPETAB 71,62,41
orl-mus LD50:785 mg/kg JMCMAR 9,966,66
ipr-mus LD50:248 mg/kg JPETAB 158,135,67
scu-mus LD50:425 mg/kg JMCMAR 9,966,66
ivn-mus LD50:175 mg/kg JMCMAR 9,966,66
ivn-rbt LDLo:80 mg/kg JPETAB 36,363,29

CONSENSUS REPORTS: Reported in EPA TSCA Inventory.

SAFETY PROFILE: A poison by intraperitoneal, subcutaneous, and intravenous routes. When heated to decomposition it emits very toxic fumes of HCl and NO_x. See also EPHEDRINE.

EAX000 **CAS:50-98-6** **HR: 3**
l-EPHEDRINE HYDROCHLORIDE
mf: $C_{10}H_{15}NO \cdot ClH$ mw: 201.72

SYNS: EPHEDRINE HYDROCHLORIDE □ (−)-EPHEDRINE HYDROCHLORIDE □ (R-(R*,S*))-α-(1-(METHYLAMINO)ETHYL)BENZENEMETHANOL HYDROCHLORIDE

TOXICITY DATA WITH REFERENCE
idr-hmn TDLo:6 μg/kg:CNS JPETAB 76,295,42
ipr-rat LD50:165 mg/kg JPETAB 86,284,46
ivn-rat LD50:69 mg/kg JAPMA8 33,80,44
orl-mus LD50:400 mg/kg JMCMAR 9,966,66
scu-mus LD50:600 mg/kg JPETAB 86,284,46
ivn-mus LD50:95 mg/kg JPETAB 95,336,49
par-mus LDLo:500 mg/kg MAIZAB 3,1,25
scu-rbt LD50:165 mg/kg JPETAB 86,284,46
ims-rbt LD50:175 mg/kg JPETAB 86,284,46

CONSENSUS REPORTS: Reported in EPA TSCA Inventory.

SAFETY PROFILE: Poison by ingestion, intraperitoneal, intravenous, subcutaneous, and intramuscular routes. Moderately toxic by parenteral route. Human systemic effects by intradermal route: local anesthetic. When heated to decomposition it emits very toxic fumes of HCl and NO_x. See also EPHEDRINE.

EAX500 **CAS:134-71-4** **HR: 3**
dl-EPHEDRINE HYDROCHLORIDE
mf: $C_{10}H_{15}NO \cdot ClH$ mw: 201.72

PROP: A solid. Mp: 188–189.5°.

SYNS: EPHETONIN □ EPHETONINE □ dl-α-(1-(METHYLAMINO) ETHYL) BENZYL ALCOHOL HYDROCHLORIDE □ 1-PHENYL-2-METHYLAMINOPROPANOL-1 □ RACEPHEDRINE HYDROCHLORIDE

TOXICITY DATA WITH REFERENCE
orl-wmn TDLo:1 mg/kg:CAR,GIT,SKN JPETAB 33,237,28
orl-man TDLo:1429 μg/kg JPETAB 33,237,28
scu-rat LD50:350 mg/kg AEPPAE 169,114,33
orl-mus LD50:700 mg/kg JMCMAR 9,966,66
ipr-mus LD50:254 mg/kg PHARAT 24,775,69
scu-mus LD50:520 mg/kg NYKZAU 55,653,59
ivn-mus LD50:135 mg/kg JMCMAR 9,966,66
ivn-rbt LDLo:60 mg/kg JPETAB 36,363,29

CONSENSUS REPORTS: Reported in EPA TSCA Inventory.

SAFETY PROFILE: Poison by subcutaneous, intravenous, and intraperitoneal routes. Moderately toxic by ingestion. Human systemic effects: cardiac changes, nausea or vomiting, sweating. When heated to decomposition it emits very toxic fumes of HCl and NO_x. See also EPHEDRINE.

EAY075 CAS:7234-07-3 HR: 2
l-EPHEDRINE PHOSPHATE (ESTER)
mf: $C_{10}H_{16}NO_4P$ mw: 245.24

SYN: l-α-(1-(METHYLAMINO)ETHYL)BENZYL PHOSPHATE

TOXICITY DATA WITH REFERENCE
orl-mus LD50:1065 mg/kg JMCMAR 9,966,66
ipr-mus LD50:2000 mg/kg JMCMAR 9,966,66
scu-mus LD50:1707 mg/kg JMCMAR 9,966,66
ivn-mus LD50:400 mg/kg JMCMAR 9,966,66

SAFETY PROFILE: Moderately toxic by ingestion, subcutaneous, intravenous, and intraperitoneal routes. When heated to decomposition it emits toxic fumes of PO_x and NO_x. See also EPHEDRINE.

EAY150 CAS:7234-08-4 HR: 2
d-EPHEDRINE PHOSPHATE (ESTER)
mf: $C_{10}H_{16}NO_4P$ mw: 245.24

SYN: d-α-(1-(METHYLAMINO)ETHYL)BENZYL PHOSPHATE

TOXICITY DATA WITH REFERENCE
orl-mus LD50:1800 mg/kg JMCMAR 9,966,66
ipr-mus LD50:815 mg/kg JMCMAR 9,966,66
scu-mus LD50:865 mg/kg JMCMAR 9,966,66
ivn-mus LD50:815 mg/kg JMCMAR 9,966,66

SAFETY PROFILE: Moderately toxic by ingestion, intraperitoneal, subcutaneous, and intravenous routes. When heated to decomposition it emits toxic fumes of PO_x and NO_x. See also EPHEDRINE and ESTERS.

EAY175 CAS:7234-09-5 HR: 2
dl-EPHEDRINE PHOSPHATE (ESTER)
mf: $C_{10}H_{16}NO_4P$ mw: 245.24

SYN: dl-α-(1-(METHYLAMINO)ETHYL)BENZYL PHOSPHATE

TOXICITY DATA WITH REFERENCE
orl-mus LD50:2000 mg/kg JMCMAR 9,966,66
ipr-mus LD50:790 mg/kg JMCMAR 9,966,66
scu-mus LD50:1150 mg/kg JMCMAR 9,966,66
ivn-mus LD50:840 mg/kg JMCMAR 9,966,66

SAFETY PROFILE: Moderately toxic by ingestion, intraperitoneal, subcutaneous, and intravenous routes. When heated to decomposition it emits toxic fumes of PO_x and NO_x. See also EPHEDRINE and ESTERS.

EAY500 CAS:134-72-5 HR: 3
1-EPHEDRINE SULFATE
mf: $C_{20}H_{30}N_2O_2 \cdot H_2O_4S$ mw: 428.60

SYNS: ISOFEDROL ☐ 1-α-(1-(METHYLAMINO)ETHYL)BENZYL ALCOHOL SULFATE ☐ NCI-C55652 ☐ 1-PHENYL-2-METHYLAMINE-PROPANOL-1-SULFATE

TOXICITY DATA WITH REFERENCE
ivn-hmn TDLo:1 mg/kg:CVS AEPPAE 120,189,27
orl-rat LD50:404 mg/kg JPETAB 94,150,48
scu-rat LD50:318 mg/kg JPETAB 94,150,48
ivn-rat LD50:102 mg/kg JPETAB 94,150,48
orl-mus LD50:812 mg/kg NTPTR* NTP-TR-307,86
ipr-mus LD50:400 mg/kg TXAPA9 28,227,74
orl-rbt LD50:825 mg/kg JPETAB 94,150,48
scu-rbt LD50:383 mg/kg JPETAB 94,150,48
ivn-rbt LD50:73 mg/kg JPETAB 94,159,48

CONSENSUS REPORTS: NTP Carcinogenesis Studies (feed); No Evidence: mouse, rat NTPTR* NTP-TR-307,86. Reported in EPA TSCA Inventory.

SAFETY PROFILE: Poison by intravenous, intraperitoneal, and subcutaneous routes. Moderately toxic by ingestion. Human systemic effects by intravenous route: increased pulse rate and blood pressure. When heated to decomposition it emits very toxic fumes of NO_x and SO_x. See also EPHEDRINE.

EAZ000 CAS:62-32-8 HR: 3
EPHININE HYDROCHLORIDE
mf: $C_9H_{13}NO_2 \cdot ClH$ mw: 203.69

PROP: Prisms from H_2O. Mp: 179–180°.

SYNS: 3,4-DIHYDROXYPHENYLETHYLMETHYLAMINE HYDROCHLORIDE ☐ 3,4-DIHYDROXYPHENYL-1-METHYLAMINO-2-ETHANE HYDROCHLORIDE ☐ 4-(2-METHYLAMINOETHYL)PYROCATECHOL HYDROCHLORIDE ☐ METHYL-(β-(3,4-DIHYDROXY PHENYL ETHYL) AMINE HYDROCHLORIDE ☐ N-METHYLDOPAMINE HYDROCHLORIDE

TOXICITY DATA WITH REFERENCE
scu-rat LDLo:160 mg/kg JPETAB 71,62,41
ipr-mus LD50:212 mg/kg JMCMAR 18,1194,75

SAFETY PROFILE: A poison by subcutaneous and intraperitoneal routes. When heated to decomposition it emits very toxic fumes of NO_x and HCl.

EAZ500 CAS:106-89-8 HR: 3
EPICHLOROHYDRIN
DOT: UN 2023
mf: C_3H_5ClO mw: 92.53

$ClCH_2CHO\,CH_2$ (epoxide ring bond between O and CH_2)

PROP: Colorless, mobile liquid; irritating chloroform-like odor. Bp: 117.9°, fp: −57.1°, flash p: 105.1°F (OC) (40°C), mp −25.6°C, d: 1.1761 @ 20°/20°, vap press: 10 mm @ 16.6°, vap d: 3.29.

SYNS: 1-CHLOOR-2,3-EPOXY-PROPAAN (DUTCH) ☐ 1-CHLOR-2,3-EPOXY-PROPAN (GERMAN) ☐ 1-CHLORO-2,3-EPOXYPROPANE ☐ 3-CHLORO-1,2-EPOXYPROPANE ☐ epi-CHLOROHYDRIN ☐ (CHLOROMETHYL)ETHYLENE OXIDE ☐ CHLOROMETHYLOXIRANE ☐ 2-(CHLOROMETHYL)OXIRANE ☐ CHLOROPROPYLENE OXIDE ☐ γ-CHLOROPROPYLENE OXIDE ☐ 3-CHLORO-1,2-PROPYLENE OXIDE ☐ 1-CLORO-2,3-EPOSSIPROPANO (ITALIAN) ☐ ECH ☐ EPICHLOOR-HYDRINE (DUTCH) ☐ EPICHLORHYDRIN (GERMAN) ☐ EPICHLORHYDRINE (FRENCH) ☐ α-EPICHLOROHYDRIN ☐ (dl)-α-EPICHLOROHYDRIN ☐ EPICHLOROHYDRYNA (POLISH) ☐ EPICHLOROPHYDRIN ☐ EPICLORIDRINA (ITALIAN) ☐ 1,2-EPOXY-3-CHLOROPROPANE ☐ 2,3-EPOXYPROPYL CHLORIDE ☐ GLYCEROL EPICHLORHYDRIN ☐ RCRA WASTE NUMBER U041 ☐ SKEKhG

TOXICITY DATA WITH REFERENCE
skn-rbt 10 mg/24H open JIHTAB 30,63,48
eye-rbt 100 mg/24H MOD 85JCAE-,769,86
dni-hmn:hla 2700 μmol/L MUREAV 92,427,82
sce-hmn:lym 10 nmol/L CARYAB 34,261,81
spm-mus-ihl 5 mg/m³ MUREAV 85,287,81

orl-mus TDLo:1200 mg/kg (female 6-15D post):TER JTEHD6 9,87,82
ihl-rat TCLo:50 ppm/6H (male 50D pre):REP TXAPA9 68,415,83
orl-rat TDLo:180 mg/kg (12D male):REP NATUAS 226,87,70
orl-rat TDLo:25 mg/kg (male 1D pre):REP ARTODN 53,71,83
scu-rat TDLo:75 mg/kg (male 1D pre):REP TXAPA9 70,67,83
orl-rat TDLo:60 g/kg/81W-I:CAR GANNA2 71,922,80
ihl-rat TCLo:100 ppm/6H/30D-C:CAR JJIND8 65,751,80
ipr-mus TDLo:2400 mg/kg/8W-I:NEO TXAPA9 82,19,86
scu-mus TDLo:720 mg/kg/18W-I:ETA JJIND8 48,1431,72
unr-mus TDLo:19 mg/kg:ETA RAREAE 3,193,63
scu-mus LD :2760 mg/kg/69W-I:NEO JJIND8 53,695,74
ihl-rat 11,100 ppm/6H/6W-I:ETA PAACA3 21,106,80
ihl-rat 11,100 ppm:ETA CHWKA9 123(24),25,78
ihl-rat TC :0 ppm/6H/57W-I:ETA JJIND8 65,751,80
orl-rat LD :36 g/kg/81W-I:ETA GANNA2 71,922,80
orl-rat LD :85,050 mg/kg/81W-C:NEO NAIZAM 32,270,81
orl-rat LD :42,525 mg/kg/81W-C:ETA NAIZAM 32,270,81
orl-rat LD :5150 mg/kg/2Y-I:ETA TXCYAC 36,325,85
ihl-hmn TCLo:40 ppm/2H:PUL 34ZIAG-,240,69
ihl-hmn TCLo:20 ppm:EYE 29ZWAE-,108,68
orl-rat LD50:90 mg/kg JIDHAN 30,63,48
ihl-rat LC50:250 ppm/8H NPIRI* 1,41,74
skn-rat LDLo:1 g/kg UCPHAQ 2,69,41
ipr-rat LD50:133 mg/kg TXAPA9 52,422,80
scu-rat LD50:150 mg/kg AMPMAR 28,505,67
ivn-rat LD50:154 mg/kg NPIRI* 1,41,74
orl-mus LD50:195 mg/kg GISAAA 33(1),46,68
skn-mus LD50:250 mg/kg 85GMAT-,64,82
orl-rbt LD50:345 mg/kg GISAAA 33(1),46,68
skn-rbt LD50:515 mg/kg WolMA# 18JAN77
ipr-rbt LD50:118 mg/kg JPMSAE 61,1712,72
orl-gpg LD50:280 mg/kg GISAAA 33(1),46,68
ipr-gpg LD50:118 mg/kg JPMSAE 61,1712,72

CONSENSUS REPORTS: NTP 7th Annual Report on Carcinogens. IARC Cancer Review: Group 2A IMEMDT 7,202,87; Animal Sufficient Evidence IMEMDT 11,131,76. EPA Genetic Toxicology Program. Community Right-To-Know List. EPA Extremely Hazardous Substances List. Reported in EPA TSCA Inventory.

OSHA PEL: TWA 2 ppm (skin)
ACGIH TLV: TWA 2 ppm (skin); (Proposed: TWA 0.5 ppm; Suspected Human Carcinogen)
DFG TRK: 3 ppm; Animal Carcinogen, Suspected Human Carcinogen
NIOSH REL: Minimize exposure
DOT CLASSIFICATION: 6.1; *Label:* Poison

SAFETY PROFILE: Confirmed carcinogen with experimental carcinogenic data. Poison by ingestion, skin contact, intravenous, and intraperitoneal routes. Moderately toxic by inhalation. An experimental teratogen. Other experimental reproductive effects. Human systemic effects by inhalation: respiratory, nose, and eyes. Human mutation data reported. A skin and eye irritant. A sensitizer. Flammable liquid when exposed to heat or flame. Explosive reaction with aniline. Reaction with trichloroethylene forms the explosive dichloroacetylene. Ignition on contact with potassium tert-butoxide. Violent reaction with sulfuric acid or isopropylamine. Exothermic polymerization on contact with strong acids, caustic alkalies, aluminum, aluminum chloride, iron(III) chloride, or zinc. When heated to decomposition it emits toxic fumes of Cl^-.

For occupational chemical analysis use NIOSH: Epichlorohydrin, 1010.

EBA100 **CAS:516-95-0** **HR: 2**
EPIDEHYDROCHOLESTERIN
mf: $C_{27}H_{48}O$ mw: 388.75

PROP: Needles from EtOH. Mp: 185–186°.

SYNS: CHOLESTAN-3-OL, (3-α-5α-)-(9CI) □ 5-α-CHOLESTAN-3-α-OL (8CI) □ α-CHOLESTANOL (7CI) □ epi-CHOLESTANOL □ EPICHOLESTANOL

TOXICITY DATA WITH REFERENCE
scu-mus TDLo:1 g/kg (female 6-15D post):TER KSRNAM 12,479,78
ipr-rat TDLo:700 mg/kg (female 35D pre):REP KSRNAM 12,456,78
orl-rat TDLo:5250 mg/kg (male 35D pre):REP KSRNAM 12,468,78
orl-rat LD50:>1500 mg/kg KSRNAM 13,1256,79

CONSENSUS REPORTS: Reported in EPA TSCA Inventory.

SAFETY PROFILE: Moderately toxic by ingestion. An experimental teratogen. Experimental reproductive effects. When heated to decomposition it emits acrid smoke and irritating fumes.

EBA600 **HR: 3**
1,4-EPIDIOXY-1,4-DIHYDRO-6,6-DIMETHYLFULVENE
mf: $C_8H_{10}O_2$ mw: 138.17

SAFETY PROFILE: Decomposes explosively above −10°C. When heated to decomposition it emits acrid smoke and fumes. See also PEROXIDES.

EBB100 **CAS:56420-45-2** **HR: 3**
4′-EPIDOXORUBICIN
mf: $C_{27}H_{29}NO_{11}$ mw: 543.57

SYNS: EPIRUBICIN □ EPI-DX □ FARMORUBICIN

TOXICITY DATA WITH REFERENCE
spm-mus-par 10 mg/kg MUREAV 160,39,86
dnd-hmn:leu 5 μmol/L CCPHDZ 30,51,92
ivn-rat LD50:14,270 μg/kg TXAPA9 79,412,85
ivn-mus LD50:16,070 μg/kg TXAPA9 79,412,85
ivn-dog LD50:2 mg/kg TXAPA9 79,412,85

SAFETY PROFILE: Poison by intravenous route. Human mutation data reported. When heated to decomposition it emits toxic fumes of NO_x.

EBB500 CAS:329-65-7 HR: 3
dl-EPINEPHRINE
mf: $C_9H_{13}NO_3$ mw: 183.23

PROP: Microscopic crystals. Sltly sol in H_2O.

SYNS: dl-ADRENALINE □ EPINEPHRINE racemic

TOXICITY DATA WITH REFERENCE
ivn-rat LD50:70 µg/kg JPETAB 95,502,49
ipr-mus LD50:7800 µg/kg PSEBAA 68,501,48
scu-mus LDLo:12 mg/kg 85IXA4-,57,48
ivn-mus LD50:3400 µg/kg JPETAB 95,502,49

SAFETY PROFILE: Very poisonous by intravenous and intraperitoneal routes. When heated to decomposition it emits toxic fumes of NO_x.

EBD500 CAS:2363-58-8 HR: 2
2-α,3-α-EPITHIO-5-α-ANDROSTAN-17-β-OL
mf: $C_{19}H_{30}OS$ mw: 306.55

PROP: Crystals from Me_2CO. Mp: 127–128°.

SYNS: 2,3-EPITHIOANDROSTAN-17-OL □ EPITIOSTANOL □ 10275-S □ THIODROL

TOXICITY DATA WITH REFERENCE
ims-rat TDLo:15 mg/kg (female 9-14D post):REP OYYAA2 7,723,73
ims-rat TDLo:60 mg/kg (female 9-14D post):REP OYYAA2 7,723,73
ims-mus TDLo:12,500 µg/kg (female 13-17D post):TER OYYAA2 7,723,73
scu-rat TDLo:100 mg/kg (female 40D pre):REP OYYAA2 7,805,73
scu-rat TDLo:800 mg/kg (female 40D pre):REP OYYAA2 7,805,73
ipr-rat TDLo:800 mg/kg (male 40D pre):REP OYYAA2 7,805,73
scu-rat TDLo:100 mg/kg (male 40D pre):REP OYYAA2 7,805,73
ims-rat TDLo:15 mg/kg (female 9-14D post):TER OYYAA2 7,723,73
ipr-rat LDLo:5 g/kg OYYAA2 7,805,73
ipr-mus LD50:1160 mg/kg OYYAA2 7,805,73

SAFETY PROFILE: Moderately toxic by intraperitoneal route. Experimental teratogenic and reproductive effects. A steroid. When heated to decomposition it emits very toxic fumes of SO_x.

EBD600 CAS:54096-45-6 HR: 3
4,5-EPITHIOVALERONITRILE
mf: C_5H_7NS mw: 113.19

SYN: 1-CYANO-3,4-EPITHIOBUTANE

TOXICITY DATA WITH REFERENCE
scu-rat TDLo:88 mg/kg (8-9D preg):TER FCTXAV 18,159,80
scu-rat TDLo:95 mg/kg (8D preg):REP FCTXAV 18,159,80
scu-rat LD50:109 mg/kg FCTXAV 18,159,80

CONSENSUS REPORTS: Cyanide and its compounds are on the Community Right-To-Know List.

SAFETY PROFILE: Poison by subcutaneous route. An experimental teratogen. Experimental reproductive effects. When heated to decomposition it emits toxic fumes of NO_x, CN^-, and SO_x. See also NITRILES.

EBD700 CAS:2104-64-5 HR: 3
EPN
mf: $C_{14}H_{14}NO_4PS$ mw: 323.32

PROP: Liquid or pale-yellow crystals with an aromatic odor. D: 1.268 @ 25°, mp: 36°. Nearly insol in water; sol in organic solvents.

SYNS: O-AETHYL-O-n(4-NITROPHENYL)-PHENYL-MONOTHIOPHOSPHONAT (GERMAN) □ ENT 17,798 □ O-ESTER-p-NITROPHENOL with O-ETHYL PHENYL PHOSPHONOTHIOATE □ ETHOXY-4-NITROPHENOXYPHENYLPHOSPHINE SULFIDE □ O-ETHYL-O-((4-NITROFENYL)-FENYL)-MONOTHIOFOSFONAAT (DUTCH) □ O-ETHYL O-(4-NITROPHENYL)BENZENETHIONOPHOSPHONATE □ ETHYL-p-NITROPHENYL BENZENETHIONOPHOSPHONATE □ ETHYL-p-NITROPHENYL BENZENETHIOPHOSPHATE □ ETHYL-p-NITROPHENYL BENZENETHIOPHOSPHONATE □ ETHYL-p-NITROPHENYL PHENYLPHOSPHONOTHIOATE □ O-ETHYL-O-p-NITROPHENYL PHENYLPHOSPHONOTHIOLATE □ O-ETHYL-O-(4-NITROPHENYL) PHENYLPHOSPHONOTHIOATE □ O-ETHYL-O-p-NITROPHENYL PHENYL PHOSPHOROTHIOATE □ ETHYL-p-NITROPHENYL THIONOBENZENEPHOSPHATE □ ETHYL-p-NITROPHENYL THIONOBENZENEPHOSPHONATE □ O-ETHYL PHENYL-p-NITROPHENYL THIOPHOSPHONATE □ O-ETIL-O-((4-NITRO-FENIL)-FENIL)-MONOTIOFOSFONATO (ITALIAN) □ PHENYLTHIOPHOSPHONATE de O-ETHYLE et O-4-NITROPHENYLE (FRENCH) □ PIN □ SANTOX □ THIONOBENZENEPHOSPHONIC ACID ETHYL-p-NITROPHENYL ESTER

TOXICITY DATA WITH REFERENCE
orl-mus TDLo:132 mg/kg (female 6-16D post):TER JESEDU 20,373,85
orl-rat LD50:7 mg/kg JPETAB 112,29,54
skn-rat LD50:25 mg/kg TXAPA9 2,88,60
ipr-rat LD50:7200 µg/kg APCRAW 4,117,61
orl-mus LD50:12,200 µg/kg ABCHA6 26,257,62
skn-mus LD50:348 mg/kg ABCHA6 26,257,62
ipr-mus LD50:8400 µg/kg IJBBBQ 15,336,78
orl-dog LD50:20 mg/kg PCOC** -,478,66
ipr-dog LDLo:35 mg/kg JPETAB 112,29,54
skn-cat LD50:45 mg/kg TOLED5 1000(Sp 1)141,80
skn-rbt LD50:30 mg/kg AFDOAQ 16,3,52
ipr-rbt LDLo:20 mg/kg JPETAB 112,29,54
ipr-gpg LDLo:20 mg/kg JPETAB 112,29,54

CONSENSUS REPORTS: EPA Farm Worker Field Reentry FEREAC 39, 16888,74. EPA Extremely Hazardous Substances List.

OSHA PEL: TWA 0.5 mg/m³ (skin)
ACGIH TLV: TWA 0.1 mg/m³ (skin)
DFG MAK: 0.5 mg/m³

SAFETY PROFILE: Poison by ingestion, skin contact, and intraperitoneal routes. An experimental teratogen. A cholinesterase inhibitor. This material is extremely hazardous on contact with skin, inhalation or ingestion. A highly toxic insecticide. When heated to decomposition it emits highly toxic fumes of SO_x, PO_x, NO_x, and phosphine. See also PARATHION, NITRO COMPOUNDS of AROMATIC HYDROCARBONS, PHOSPHINE, and SULFIDES.

E

For occupational chemical analysis use NIOSH: EPN, Malathion, and Parathion, 5012.

EBE100 **CAS:68401-82-1** **HR: 1**
EPOCELIN
mf: $C_{13}H_{12}N_5O_5S_2 \cdot Na$ mw: 405.41

SYNS: CEFTIZOXIME SODIUM □ CEFTIZOXIM-NATRIUM (GERMAN) □ FK 749 □ FR-13,479 □ SKF-88373

TOXICITY DATA WITH REFERENCE
scu-rat TDLo:10 g/kg (male 9W pre):REP ARZNAD 30,1669,80
scu-rat TDLo:27 g/kg (male 9W pre):REP JJANAX 34,466,81
scu-rat TDLo:1100 mg/kg (female 7-17D post):TER JJANAX 34,466,81
ipr-rat LD50:8130 mg/kg ARZNAD 30,1669,80
ivn-rat LD50:5570 mg/kg ARZNAD 30,1669,80
ipr-mus LD50:8930 mg/kg ARZNAD 30,1669,80
ivn-mus LD50:5150 mg/kg YAKUD5 24,905,82

SAFETY PROFILE: Mildly toxic by intravenous and intraperitoneal routes. Experimental teratogenic and reproductive effects. When heated to decomposition it emits toxic fumes of SO_x, NO_x, and Na_2O.

EBF000 **CAS:63089-76-9** **HR: 3**
EPON 562

TOXICITY DATA WITH REFERENCE
skn-rbt 500 mg/24H MOD AMIHAB 17,129,58
eye-rbt 100 mg SEV AMIHAB 17,129,58
eye-rbt 750 µg/24H SEV 85JCAE-,1402,86
orl-rat LD50:5000 mg/kg AMIHAB 17,129,58
ipr-rat LD50:380 mg/kg AMIHAB 17,129,58
orl-mus LD50:1870 mg/kg AMIHAB 17,129,58
ipr-mus LD50:303 mg/kg AMIHAB 17,129,58
orl-rbt LD50:4010 mg/kg AMIHAB 17,129,58
skn-rbt LD50:14,400 mg/kg 38MKAJ 2A,2239,81

SAFETY PROFILE: Poison by intraperitoneal route. Moderately toxic by ingestion. A severe skin and eye irritant. When heated to decomposition it emits acrid smoke and irritating fumes.

EBF500 **CAS:25068-38-6** **HR: 2**
EPON 820

TOXICITY DATA WITH REFERENCE
orl-rat LD50:13,600 mg/kg AMIHAB 17,129,58
ipr-rat LD50:1400 mg/kg AMIHAB 17,129,58
ipr-rat LD50:1780 mg/kg AMIHAB 17,129,58

CONSENSUS REPORTS: Reported in EPA TSCA Inventory.

SAFETY PROFILE: Moderately toxic by intraperitoneal route. See also EPOXY RESINS.

EBG000 **CAS:25068-38-6** **HR: 2**
EPON 1001
mf: $(C_{15}H_{16}O_2 \cdot C_3H_5ClO)_x$

TOXICITY DATA WITH REFERENCE
eye-rbt 100 mg MLD AMIHAB 17,129,58
orl-rat LD50:30 g/kg AMIHAB 17,129,58
ipr-rat LD50:2400 mg/kg AMIHAB 17,129,58
orl-mus LD50:20 g/kg AMIHAB 17,129,58

CONSENSUS REPORTS: Reported in EPA TSCA Inventory. EPA Genetic Toxicology Program.

SAFETY PROFILE: Moderately toxic by intraperitoneal route. Mildly toxic by ingestion. An eye irritant. See also EPOXY RESINS.

EBG500 **CAS:25068-38-6** **HR: 2**
EPON 1007
mf: $(C_{15}H_{16}O_2 \cdot C_3H_5ClO)_x$

TOXICITY DATA WITH REFERENCE
eye-rbt 100 mg MLD AMIHAB 17,129,58
orl-rat LD50:30 g/kg AMIHAB 17,129,58
ipr-rat LD50:2200 mg/kg AMIHAB 17,129,58
orl-mus LD50:20 g/kg AMIHAB 17,129,58

CONSENSUS REPORTS: Reported in EPA TSCA Inventory. EPA Genetic Toxicology Program.

SAFETY PROFILE: Moderately toxic by intraperitoneal route. Mildly toxic by ingestion. An eye irritant. See also EPOXY RESINS.

EBH400 **CAS:80471-63-2** **HR: 2**
EPOSTANE
mf: $C_{22}H_{31}NO_3$ mw: 357.54

PROP: Crystals from DMF (aq). Mp: 191–194°.

SYNS: ANDROST-2-ENE-2-CARBONITRILE, 3,17-DIHYDROXY-4,17-DIMETHYL-4,5-EPOXY-, (4-α-5-α- 17-β)- □ (4-α-5-α-17-β)-4,5-EPOXY-3,17-DIHYDROXY-4,17-DIMETHYLANDROST- 2-ENE-2-CARBONITRILE □ WIN 32729

TOXICITY DATA WITH REFERENCE
orl-wmn TDLo:96 mg/kg (female 7-8W post):REP CCPTAY 35,111,87
orl-rat TDLo:50 mg/kg (female 10D post):REP BIREBV 40,549,89
scu-dog TDLo:10 mg/kg (female 1D post):REP THGNBO 30,497,88
orl-mky TDLo:100 mg/kg (female 5D pre):REP CCPTAY 31,479,85
scu-dog TDLo:2500 µg/kg (female 1D post):REP THGNBO 30,497,88
orl-wmn TDLo:96 mg/kg CCPTAY 35,111,87

SAFETY PROFILE: Human toxic effects by ingestion. Experimental reproductive effects. When heated to decomposition it emits toxic fumes of NO_x.

EBH500 CAS:31305-88-1 HR: 2
EPOXIDE ERLA-0510
mf: $(C_{15}H_{19}NO_4)_x$

TOXICITY DATA WITH REFERENCE
skn-rbt 500 mg open MLD UCDS** 7/14/71
orl-rat LD50:1300 mg/kg UCDS** 7/14/71
skn-rbt LD50:640 mg/kg UCDS** 7/14/71

CONSENSUS REPORTS: Reported in EPA TSCA Inventory.

SAFETY PROFILE: Moderately toxic by ingestion and skin contact. A skin irritant.

EBH850 HR: 2
(E)-1-α-2-α-EPOXYBENZ(c)ACRIDINE-3-α-4-β-DIOL
mf: $C_{17}H_9NO_3$ mw: 275.27

SYN: BENZ(c)ACRIDINE 3,4-DIOL-1,2-EPOXIDE-2

TOXICITY DATA WITH REFERENCE
ipr-mus TDLo:5505 μg/kg/3D-I:NEO CNREA8 44,5161,84

SAFETY PROFILE: Questionable carcinogen with experimental neoplastigenic data. When heated to decomposition it emits toxic fumes of NO_x.

EBH875 HR: 2
(Z)-1-β,2-β-EPOXYBENZ(c)ACRIDINE-3-α-4-β-DIOL
mf: $C_{17}H_9NO_3$ mw: 275.27

SYN: BENZ(c)ACRIDINE 3,4-DIOL-1,2-EPOXIDE-1

TOXICITY DATA WITH REFERENCE
ipr-mus TDLo:11,561 μg/kg/3D-I:NEO CNREA8 44,5161,84

SAFETY PROFILE: Questionable carcinogen with experimental neoplastigenic data. When heated to decomposition it emits toxic fumes of NO_x.

EBJ500 CAS:930-22-3 HR: 3
3,4-EPOXY-1-BUTENE
mf: C_4H_6O mw: 70.10

PROP: Liquid. Mp: −135°, bp: 67°, flash p: <−58°F (CC), d: 0.869, autoign temp: 806°F, vap d: 2.41.

SYNS: BUTADIENE MONOEPOXIDE ☐ BUTADIENE MONOXIDE ☐ 1,2-EPOXYBUTENE-3 ☐ VINYLOXIRANE

TOXICITY DATA WITH REFERENCE
mmo-sat 1 μmol/plate BBRCA9 80,298,78
mma-sat 100 μmol/plate MUREAV 97,204,82
mmo-esc 20 μmol/L ARTODN 46,277,80
dnd-esc 1 μmol/L ARTODN 46,277,80
mmo-klp 1 μmol/L MUREAV 89,269,81
skn-mus TDLo:492 g/kg/41W-I:ETA JNCIAM 31,41,63
ipr-rat LD50:168 mg/kg TXAPA9 52,422,80

CONSENSUS REPORTS: Reported in EPA TSCA Inventory.

SAFETY PROFILE: A poison by intraperitoneal route. Questionable carcinogen with experimental tumorigenic data. Mutation data reported. A very dangerous fire hazard when exposed to heat or flame; can react with oxidizing materials. To fight fire, use CO_2, dry chemical, water spray. When heated to decomposition it emits acrid smoke and fumes.

EBK000 CAS:10138-34-8 HR: 2
2,3-EPOXYBUTYRIC ACID BUTYL ESTER
mf: $C_8H_{14}O_3$ mw: 158.22

TOXICITY DATA WITH REFERENCE
orl-rat LD50:500 mg/kg AIHAAP 23,95,62
skn-rbt LD50:2830 mg/kg AIHAAP 23,95,62

SAFETY PROFILE: Moderately toxic by ingestion and skin contact. When heated to decomposition it emits acrid smoke and fumes. See also EPOXY RESINS and ESTERS.

EBK500 CAS:6509-08-6 HR: 2
1,2-EPOXYBUTYRONITRILE
mf: C_4H_5NO mw: 83.10

SYN: 3,4-EPOXYBUTYRONITRILE

TOXICITY DATA WITH REFERENCE
scu-mus TDLo:324 mg/kg/81W-I:ETA JNCIAM 46,143,71

CONSENSUS REPORTS: Cyanide and its compounds are on the Community Right-To-Know List.

SAFETY PROFILE: Questionable carcinogen with experimental tumorigenic data. When heated to decomposition it emits toxic fumes of NO_x and CN^-. See also NITRILES.

EBL000 CAS:124-98-1 HR: 3
4,9-EPOXYCEVANE-3-α,4-β,12,14,16-β,17,20-HEPTOL
mf: $C_{27}H_{43}NO_8$ mw: 509.71

PROP: A solid. Mp: 172–176° (decomp) (anhyd), mp: 110° (hydrate). An alkamine isolated from *Veratrum album* (JPETAB 82,167,44).

SYNS: CEVIN ☐ CEVINE ☐ SABADININE

TOXICITY DATA WITH REFERENCE
ipr-rat LD50:67 mg/kg JPETAB 82,167,44
scu-mus LD50:160 mg/kg JPETAB 113,89,55
ivn-mus LD50:87 mg/kg JPETAB 82,167,44

SAFETY PROFILE: Poison by intraperitoneal, intravenous, and subcutaneous routes. When heated to decomposition it emits toxic fumes of NO_x.

EBL500 CAS:508-65-6 HR: 3
4,9-EPOXYCEVANE-3-β,4-β,7-α,14,15-α,16-β,20-HEPTOL
mf: $C_{27}H_{43}NO_8$ mw: 509.71

PROP: A solid. Mp: 219–220°. An alkamine isolated from *Veratrum album* (JPETAB 82,167,44).

SYNS: GERMIN ☐ GERMINE

TOXICITY DATA WITH REFERENCE
scu-rat LDLo:2000 mg/kg AEPPAE 189,397,38
ivn-mus LD50:139 mg/kg JPETAB 82,167,44
scu-frg LDLo:250 mg/kg AEPPAE 189,397,38

E

SAFETY PROFILE: Poison by subcutaneous and intravenous routes. When heated to decomposition it emits toxic fumes of NO_x.

EBM000 CAS:1250-95-9 HR: 2
EPOXYCHOLESTEROL
mf: $C_{27}H_{46}O_2$ mw: 402.73

SYNS: CHOLESTEROL-α-EPOXIDE □ CHOLESTEROL-5-α,6-α-EPOXIDE □ CHOLESTEROL OXIDE □ CHOLESTEROL-α-OXIDE □ 5-α,6-α-EPOXYCHOLESTANOL □ 5,6-α-EPOXY-5-α-CHOLESTAN-3-β-OL

TOXICITY DATA WITH REFERENCE
dns-hmn:fbr 10 mg/L/4H AJEBAK 56,287,78
cyt-hmn:fbr 500 μg/L/4H AJEBAK 56,287,78
otr-ham:emb 625 μg/L CALEDQ 6,143,79
scu-mus TDLo:750 mg/kg/9W-I:CAR JJIND8 19,977,57

CONSENSUS REPORTS: EPA Genetic Toxicology Program.

SAFETY PROFILE: Questionable carcinogen with experimental carcinogenic data. Human mutation data reported. When heated to decomposition it emits acrid smoke and fumes.

EBO000 CAS:3388-04-3 HR: 1
EPOXYCYCLOHEXYLETHYL TRIMETHOXY SILANE
mf: $C_{11}H_{22}O_4Si$ mw: 246.42

SYNS: β-(3,4-EPOXYCYCLOHEXYL)ETHYLTRIMETHOXYSILANE □ SILANE Y-4086 □ SILICONE A-186 □ UNION CARBIDE A-186

TOXICITY DATA WITH REFERENCE
skn-rbt 500 mg open MLD UCDS** 1/17/72
skn-mus TDLo:1587 g/kg/76W-I:CAR FAATDF 12,579,89
orl-rat LD50:12,300 mg/kg AIHAAP 30,470,69
skn-rbt LD50:6300 mg/kg UCDS** 2/11/64

CONSENSUS REPORTS: Reported in EPA TSCA Inventory.

SAFETY PROFILE: Mildly toxic by ingestion and skin contact. A skin irritant. When heated to decomposition it emits acrid smoke and fumes.

EBO050 CAS:2386-87-0 HR: 3
3,4-EPOXYCYCLOHEXYLMETHYL 3,4-EPOXYCYCLOHEXANE CARBOXYLATE
mf: $C_{14}H_{20}O_4$ mw: 252.34

SYNS: ERL-4221 □ 7-OXABICYCLO(4.1.0)HEPTANE-3-CARBOXYLIC ACID, 7-OXABICYCLO(4.1.0)HEPT-3-YLMETHYL ESTER

TOXICITY DATA WITH REFERENCE
orl-rat LD50:4490 mg/kg AIHAAP 24,305,63
skn-rbt LD50:20 mg/kg 38MKAJ 2A,2242,81

CONSENSUS REPORTS: Reported in EPA TSCA Inventory.

SAFETY PROFILE: Poison by skin contact. Mildly toxic by ingestion. When heated to decomposition it emits acrid smoke and irritating vapors.

EBP000 CAS:962-32-3 HR: 2
5,6-EPOXY-5,6-DIHYDROBENZ(a)ANTHRACENE
mf: $C_{18}H_{12}O$ mw: 244.30

SYNS: BENZ(a)ANTHRACENE-5,6-OXIDE □ BENZ(a)ANTHRA-5,6-OXIDE □ BENZO(a)ANTHRACENE-5,6-OXIDE □ 1a,11b-DIHYDROBENZ(3,4)ANTHRA(1,2-b)OXIRENE □ 3,4-DIHYDRO-3,4-EPOXY-1,2-BENZANTHRACENE

TOXICITY DATA WITH REFERENCE
mmo-sat 3 μg/plate IJCNAW 16,787,75
mma-sat 450 nmol/L BBRCA9 66,693,75
dni-omi 200 μg/L PNASA6 74,1378,77
sln-dmg-par 5 mmol/L CNREA8 33,2354,73
dns-hmn:fbr 10 μmol/L/3H IJCNAW 16,284,75
otr-mus:fbr 500 μg/L CNREA8 32,716,73
scu-mus TDLo:400 mg/kg/10W-I:CAR IJCNAW 2,500,67

CONSENSUS REPORTS: EPA Genetic Toxicology Program.

SAFETY PROFILE: Questionable carcinogen with experimental carcinogenic data. Human mutation data reported. When heated to decomposition it emits acrid smoke and fumes.

EBP500 CAS:1421-85-8 HR: 2
5,6-EPOXY-5,6-DIHYDRODIBENZ(a,h)ANTHRACENE
mf: $C_{22}H_{14}O$ mw: 294.36

SYNS: DBA-5,6-EPOXIDE □ DIBENZ(a,h)ANTHRACENE-5,6-OXIDE □ 5,6-DIHYDRO-5,6-EPOXYDIBENZ(a,h)ANTHRACENE

TOXICITY DATA WITH REFERENCE
mmo-sat 300 ng/plate IJCNAW 16,787,75
mma-sat 5 μg/plate PNASA6 72,5135,75
dns-hmn:hla 1 μmol/L CNREA8 38,2621,78
otr-mus:fbr 500 μg/L CNREA8 32,716,72
dnd-ham:lng 1 mg/L CBINA8 4,389,71/72
scu-mus TDLo:400 mg/kg/10W-I:CAR IJCNAW 2,500,67

CONSENSUS REPORTS: EPA Genetic Toxicology Program.

SAFETY PROFILE: Questionable carcinogen with experimental carcinogenic data. Human mutation data reported. When heated to decomposition it emits acrid smoke and fumes.

EBQ500 CAS:6921-35-3 HR: 2
1,3-EPOXY-2,2-DIMETHYLPROPANE
mf: $C_5H_{10}O$ mw: 86.15

PROP: Bp: 80–81°.

SYNS: 3,3-DIMETHYLOXETANE □ β,β-DIMETHYLTRIMETHYLENE OXIDE □ 3,3-DIMETHYLTRIMETHYLENE OXIDE

TOXICITY DATA WITH REFERENCE
scu-rat TDLo:1020 mg/kg/61W-I:ETA BJCAAI 17,100,63
ipr-mus LDLo:2000 mg/kg JMPCAS 1,355,59

SAFETY PROFILE: Moderately toxic by intraperitoneal route. Questionable carcinogen with experimental tumorigenic data. When heated to decomposition it emits acrid smoke and fumes.

EBQ550 **CAS:39597-90-5** **HR: 3**
endo-2,3-EPOXY-7,8-DIOXABICYCLO(2.2.2)OCT-5-ONE
mf: $C_6H_6O_3$ mw: 126.11

SYN: OXEPIN-3,6-ENDOPEROXIDE

SAFETY PROFILE: An unstable explosive. When heated to decomposition it emits acrid smoke and fumes. See also PEROXIDES and EXPLOSIVES.

EBR000 **CAS:96-09-3** **HR: 3**
1,2-EPOXYETHYLBENZENE
mf: C_8H_8O mw: 120.16

PROP: Colorless liquid. Bp: 194.2°, flash p: 165°F (OC), fp: −36.7°, d: 1.0469 @ 25°/4°, vap d: 4.14.

SYNS: EPOXYETHYLBENZENE (8CI) ☐ EPOXYSTYRENE ☐ α,β-EPOXYSTYRENE ☐ NCI-C54977 ☐ PHENETHYLENE OXIDE ☐ 1-PHENYL-1,2-EPOXYETHANE ☐ PHENYLETHYLENE OXIDE ☐ PHENYLOXIRANE ☐ 1-PHENYLOXIRANE ☐ 2-PHENYLOXIRANE ☐ STYRENE EPOXIDE ☐ STYRENE OXIDE ☐ STYRENE-7,8-OXIDE ☐ STYRYL OXIDE

TOXICITY DATA WITH REFERENCE
skn-rbt 10 mg/24H open MLD AMIHBC 10,61,54
skn-rbt 500 mg open MOD UCDS** 7/21/71
eye-rbt 500 mg open AMIHBC 10,61,54
dni-hmn:hla 4400 µmol/L MUREAV 93,447,82
oms-mus:fbr 1 µmol/L CRNGDP 6,1367,85
ihl-rat TCLo:100 ppm/7H (female 1-19D post):TER SWEHDO 9,94,83
ihl-rat TCLo:100 ppm/7H (female 1-19D post):REP SWEHDO 9,94,83
ihl-rbt TCLo:15 ppm/7H (female 1-24D post):REP NTIS** PB81-168510
orl-rat TDLo:65 g/kg/52-I:CAR tumors ANYAA9 534,203,88
orl-rat TDLo:10 g/kg/52W-I:CAR MELAAD 70,358,79
orl-mus TDLo:273 g/kg/2Y-C:CAR JJIND8 77,471,86
skn-mus TDLo:74 g/kg/62W-I:ETA JNCIAM 31,41,63
unr-mus TDLo:96 mg/kg:ETA RARSAM 3,193,63
orl-rat TD:52 g/kg/52W-I:CAR MELAAD 70,358,79
orl-rat TD:200 g/kg/2Y-C:CAR JJIND8 77,471,86
orl-rat LD50:2000 mg/kg NTIS** PB81-168510
ihl-rat LCLo:500 ppm/4H AMIHBC 10,61,54
ipr-rat LD50:460 mg/kg TXAPA9 18,321,71
orl-mus LD50:1500 mg/kg JJIND8 77,471,86
skn-rbt LD50:1060 mg/kg AMIHBC 10,61,54

CONSENSUS REPORTS: IARC Cancer Review: Group 2A IMEMDT 7,56,87; Animal Sufficient Evidence IMEMDT 36,245,85; Human No Adequate Data IMEMDT 36,245,85. Reported in EPA TSCA Inventory. EPA Genetic Toxicology Program.

SAFETY PROFILE: Confirmed carcinogen with experimental carcinogenic, tumorigenic, and teratogenic data. Moderately toxic by ingestion, inhalation, skin contact, and intraperitoneal routes. Experimental reproductive effects. Human mutation data reported. A skin and eye irritant. Flammable when exposed to heat, flame, or oxidizers; can react with oxidizing materials. To fight fire, use foam, CO_2, dry chemical. When heated to decomposition it emits acrid smoke and fumes.

EBR500 **CAS:13484-13-4** **HR: 2**
2-(α,β-EPOXYETHYL)-5,6-EPOXYBENZENE
mf: $C_8H_6O_2$ mw: 134.14

SYN: 2-(1,2-EPOXYETHYL)-5,6-EPOXYBENZENE

TOXICITY DATA WITH REFERENCE
skn-rbt 10 mg/24H open SEV AIHAAP 23,95,62
orl-rat LDLo:2830 mg/kg AIHAAP 23,95,62
skn-rbt LD50:620 mg/kg AIHAAP 23,95,62

SAFETY PROFILE: Moderately toxic by ingestion and skin contact. A severe skin irritant. When heated to decomposition it emits acrid smoke and fumes.

E

EBT500 **CAS:53940-49-1** **HR: D**
2′,3′-EPOXYEUGENOL
mf: $C_{10}H_{12}O_3$ mw: 180.22

SYNS: 4-(2,3-EPOXYPROPYL)-2-METHOXYPHENOL ☐ EUGENOL OXIDE ☐ EUGENO-2′,3′-OXIDE ☐ p-HYDROXY-m-METHOXYPHENYLPROPYLENE OXIDE ☐ PHENOL, 2-METHOXY-4-(OXIRANYLMETHYL)-

TOXICITY DATA WITH REFERENCE
mma-sat 20 nmol/plate CRSBAW 171,1041,77

CONSENSUS REPORTS: Reported in EPA TSCA Inventory.

SAFETY PROFILE: Mutation data reported. When heated to decomposition it emits acrid smoke and fumes.

EBU000 **CAS:503-09-3** **HR: 3**
1,2-EPOXY-3-FLUOROPROPANE
mf: C_3H_5FO mw: 76.08

SYN: EPIFLUOROHYDRIN

TOXICITY DATA WITH REFERENCE
mmo-esc 20 µmol/L ARTODN 46,277,80
mmo-klp 200 µmol/L MUREAV 89,269,81
mma-sat 500 µg/plate MUREAV 58,217,78
ihl-rat LCLo:111 mg/m^3 NDRC** -,12,43
ivn-mus LD50:178 mg/kg CSLNX* NX#00407

SAFETY PROFILE: Poison by inhalation and intravenous routes. Mutation data reported. When heated to decomposition it emits toxic fumes of F^-.

EBU100 **CAS:68071-23-8** **HR: 1**
EPOXYGUAIENE
mf: $C_{15}H_{24}O$ mw: 220.39

SYNS: AZULENE, 1,2,3,4,5,6,7,8-OCTAHYDRO-1,4-DIMETHYL-7-(1-METHYLETHYLIDENE)-, MONOEPOXIDE ☐ 1,2,3,4,5,6,7,8-OCTAHYDRO-1,4-DIMETHYL-7-(1-METHYLETHYLIDENE)AZULENEMONOEPOXIDE

TOXICITY DATA WITH REFERENCE
skn-rbt 500 mg/24H MOD FCTOD7 21,851,83
orl-rat LD50:>5 g/kg FCTOD7 21,851,83
skn-rbt LD50:>5 g/kg FCTOD7 21,851,83

CONSENSUS REPORTS: Reported in EPA TSCA Inventory.

SAFETY PROFILE: Low toxicity by ingestion and skin

contact. A skin irritant. When heated to decomposition it emits acrid smoke and irritating fumes.

EBU500 **HR: 2**
EPOXY HARDENER ZZL-0814

SYN: ZZL-0814

TOXICITY DATA WITH REFERENCE
skn-rbt 500 mg open MLD UCDS** 10/1/71
eye-rbt 15 mg SEV UCDS** 10/1/71
orl-rat LD50:3250 mg/kg UCDS** 10/1/71
skn-rbt LD50:2500 mg/kg UCDS** 10/1/71

SAFETY PROFILE: Moderately toxic by ingestion and skin contact. A skin and severe eye irritant.

EBV000 **HR: 2**
EPOXY HARDENER ZZL-0816

SYN: ZZL-0816

TOXICITY DATA WITH REFERENCE
skn-rbt 500 mg open MLD UCDS** 10/1/71
orl-rat LD50:4920 mg/kg UCDS** 10/1/71
skn-rbt LD50:2520 mg/kg UCDS** 10/1/71

SAFETY PROFILE: Moderately toxic by skin contact. Mildly toxic by ingestion. A skin irritant.

EBV100 **CAS:69136-21-6** **HR: 2**
EPOXY HARDENER ZZL-0822
mf: $C_{10}H_{24}N_2O_3$ mw: 220.36

SYNS: (OXYBIS(2,1-ETHANEDIYLOXY))BIS-PROPANAMINE □ POLYGLYCOLDIAMINE H 221 □ ZZL-0822

TOXICITY DATA WITH REFERENCE
skn-rbt 50 mg open MOD UCDS** 9/15/65
eye-rbt 5 mg SEV UCDS** 9/15/65
orl-rat LD50:4290 mg/kg UCDS** 9/15-65
skn-rbt LD50:2500 mg/kg UCDS** 9/15/65

SAFETY PROFILE: Moderately toxic by ingestion and skin contact. A skin and severe eye irritant. When heated to decomposition it emits toxic fumes of NO_x. See also AMINES.

EBV500 **HR: 2**
EPOXY HARDENER ZZL-0854

SYN: ZZL-0854

TOXICITY DATA WITH REFERENCE
skn-rbt 500 mg open MLD UCDS** 3/7/72
eye-rbt 15 mg SEV UCDS** 3/7/72
orl-rat LD50:2460 mg/kg UCDS** 3/7/72
skn-rbt LD50:5040 mg/kg UCDS** 3/7/72

SAFETY PROFILE: Moderately toxic by ingestion. Mildly toxic by skin contact. A skin and severe eye irritant.

EBW000 **CAS:42498-58-8** **HR: 2**
EPOXY HARDENER ZZLA-0334
mf: $C_9H_{10}O_3$ mw: 166.19

SYNS: 3a,4,7,7a-TETRAHYDRO-3a-METHYL-1,3-ISOBENZOFURANDIONE □ ZZLA-0334

TOXICITY DATA WITH REFERENCE
skn-rbt 500 mg open MLD UCDS** 6/22/66
eye-rbt 1 mg SEV UCDS** 6/22/66
orl-rat LD50:2460 mg/kg UCDS** 6/22/66
skn-rbt LD50:1590 mg/kg UCDS** 6/22/66

SAFETY PROFILE: Moderately toxic by ingestion and skin contact. A skin and severe eye irritant.

EBW500 **CAS:1024-57-3** **HR: 3**
EPOXYHEPTACHLOR
mf: $C_{10}H_5Cl_7O$ mw: 389.30

SYNS: ENT 25,584 □ HCE □ HEPTACHLOR EPOXIDE (USDA) □ 1,4,5,6,7,8,8-HEPTACHLORO-2,3-EPOXY-2,3,3a,4,7,7a-HEXAHYDRO-4,7-METHANOINDENE □ 1,4,5,6,7,8,8-HEPTACHLORO-2,3-EPOXY-3a,4,7,7a-TETRAHYDRO-4,7-METHANOINDAN □ 2,3,4,5,6,7,7-HEPTACHLORO-1a,1b,5,5a,6,6a-HEXAHYDRO-2,5-METHANO-2H-INDENO(1,2-b)OXIRENE □ VELSICOL 53-CS-17

TOXICITY DATA WITH REFERENCE
mma-hmn:fbr 10 µmol/L MUREAV 42,161,77
orl-mus TDLo:580 mg/kg/69W-C:CAR ARTODN 38,163,77
orl-mus TD:876 mg/kg/2Y-C:CAR ECEBDI 45,147,77
orl-rat LD50:15 mg/kg GISAAA 52(2),93,87
orl-mus LD50:39 mg/kg ARSIM* 20,27,66
ivn-mus LDLo:10 mg/kg JPETAB 107,266,53

CONSENSUS REPORTS: IARC Cancer Review: Human Inadequate Evidence IMEMDT 20,129,79; Animal Inadequate Evidence IMEMDT 5,173,74; Animal Limited Evidence IMEMDT 20,129,79. EPA Genetic Toxicology Program.

ACGIH TLV: 0.05 mg/m³ (skin); Animal Carcinogen

SAFETY PROFILE: Confirmed carcinogen with experimental carcinogenic data. Poison by ingestion and intravenous routes. Human mutation data reported. When heated to decomposition it emits toxic fumes of Cl^-. See also HEPTACHLOR.

EBX500 **CAS:7320-37-8** **HR: 2**
1,2-EPOXYHEXADECANE
mf: $C_{16}H_{32}O$ mw: 240.48

SYNS: HEXADECENE EPOXIDE □ NCI-C55538

TOXICITY DATA WITH REFERENCE
mic-mus:lym 24 mg/L EMMUEG 12,85,88
msc-mus:lym 18 mg/L EMMUEG 12,85,88
skn-mus TDLo:53 g/kg/44W-I:ETA JNCIAM 39,1217,67
unr-mus TDLo:240 mg/kg:ETA RARSAM 3,193,63

CONSENSUS REPORTS: Reported in EPA TSCA Inventory.

SAFETY PROFILE: Questionable carcinogen with experimental tumorigenic data. Mutation data reported.

When heated to decomposition it emits acrid smoke and fumes.

EBY500 CAS:37764-28-6 HR: 3
4,5-α-EPOXY-3-HYDROXY-17-METHYLMORPHINAN-6-ONE-N-OXIDE TARTRATE
mf: $C_{17}H_{19}NO_4$ mw: 301.37

SYNS: A/VI □ DIHYDROMORPHINON-N-OXYD-DITARTARAT (GERMAN)

TOXICITY DATA WITH REFERENCE
ivn-mus LD50:2000 mg/kg ARZNAD 7,594,57
ivn-rbt LD50:160 mg/kg ARZNAD 7,594,57

SAFETY PROFILE: Poison by intravenous route. When heated to decomposition it emits toxic fumes of NO_x.

EBY600 CAS:13647-35-3 HR: 3
4-α-5-EPOXY-17-β-HYDROXY-3-OXO-5-α-ANDROSTANE-2-α-CARBONITRILE
mf: $C_{20}H_{27}NO_3$ mw: 329.48

PROP: A solid. Mp: 258–270° (decomp).

SYNS: 5-α-ANDROSTANE-2-α-CARBONITRILE, 4-α-5-EPOXY-17-β-HYDROXY-3-OXO- □ ANDROSTANE-2-CARBONITRILE, 4,5 EPOXY 17 HYDROXY-3-OXO-, (2-α-4-α-5-α-17-β)- □ (2-α-4-α-5-α-17-β)-4,5-EPOXY-17-HYDROXY-3-OXOANDROSTANE-2-CARBONITRILE □ MODRENAL □ TRILOSTANE □ WIN 24450

TOXICITY DATA WITH REFERENCE
orl-man TDLo:48 mg/kg (male 14D pre):REP ARZNAD 33,754,83
orl-rat TDLo:1050 mg/kg (male 7D pre):REP ARZNAD 33,754,83
orl-mky TDLo:5 mg/kg (female 25-29D post):REP FESTAS 32,464,79
ipr-rat LD50:1275 mg/kg IYKEDH 17,365,86
scu-rat LD50:7050 mg/kg IYKEDH 17,365,86
ivn-rat LD50:102 mg/kg IYKEDH 17,365,86
ipr-mus LD50:1552 mg/kg IYKEDH 17,365,86
ivn-mus LD50:109 mg/kg IYKEDH 17,365,86

SAFETY PROFILE: Poison by intravenous route. Moderately toxic by intraperitoneal route. Experimental reproductive effects. When heated to decomposition it emits toxic fumes of NO_x.

ECA000 CAS:4247-30-7 HR: 2
4,5-EPOXY-3-HYDROXYVALERIC ACID-β-LACTONE
mf: $C_5H_6O_3$ mw: 114.11

TOXICITY DATA WITH REFERENCE
skn-mus TDLo:76 g/kg/63W-I:NEO JNCIAM 35,707,65
skn-mus TD:88 g/kg/73W-I:NEO 14JTAF -,275,64

SAFETY PROFILE: Questionable carcinogen with experimental neoplastigenic data. When heated to decomposition it emits acrid smoke and fumes.

ECA500 CAS:67195-51-1 HR: 2
11,12-EPOXY-3-METHYLCHOLANTHRENE
mf: $C_{21}H_{14}O$ mw: 282.35

SYNS: MCA-11,12-EPOXIDE □ MCA-11,12-OXIDE □ 3-METHYLCHOLANTHRENE-11,12-EPOXIDE □ 3-METHYL-11,12-EPOXYCHOLANTHRENE

TOXICITY DATA WITH REFERENCE
mmo-sat 50 μg/plate CNREA8 45,2600,85
pic-omi 23,400 nmol/L NNBYA7 234,186,71
otr-mus:fbr 750 μg/L CNREA8 32,716,72
msc-ham:lng 5 mmol/L PNASA6 68,3195,71
skn-mus TDLo:226 μg/kg:ETA JNCIAM 58,1051,77

SAFETY PROFILE: Questionable carcinogen with experimental tumorigenic data. Mutation data reported. When heated to decomposition it emits acrid smoke and fumes.

ECB000 CAS:141-37-7 HR: 2
3,4-EPOXY-6-METHYLCYCLOHEXYLMETHYL-3′,4′-EPOXY-6′-METHYLCYCLOHEXANE CARBOXYLATE
mf: $C_{16}H_{24}O_4$ mw: 280.40

SYNS: CHISSONOX 201 □ EP 201 □ EPOXIDE-201 □ 3,4-EPOXY-6-METHYLCYCLOHEXENECARBOXYLIC ACID (3,4-EPOXY-6-METHYLCYCLOHEXYLMETHYL) ESTER □ 3,4-EPOXY-6-METHYLCYCLOHEXYLMETHYL-3,4-EPOXY-6-METHYLCYCLOHEXANECARBOXYLATE □ 4,5-EPOXY-2-METHYLCYCLOHEXYLMETHYL-4,5-EPOXY-2-METHYLCYCLOHEXANECARBOXYLATE □ 6-METHYL-3,4-EPOXYCYCLOHEXYLMETHYL 6-METHYL-3,4-EPOXYCYCLOHEXANE CARBOXYLATE □ UNOX 201 □ UNOX EPOXIDE 201

TOXICITY DATA WITH REFERENCE
skn-rbt 10 mg/24H open MLD AIHAAP 23,95,62
skn-mus TDLo:67 g/kg/56W-I:ETA JNCIAM 39,1217,67
skn-mus TD:516 g/kg/43W-I:ETA AIHAAP 24,305,63
orl-rat LD50:4920 mg/kg UCDS** 6/8/65

CONSENSUS REPORTS: IARC Cancer Review: Group 3 IMEMDT 7,56,87; Animal Sufficient Evidence IMEMDT 11,147,76.

SAFETY PROFILE: Mildly toxic by ingestion. A skin irritant. Questionable carcinogen with experimental tumorigenic data. When heated to decomposition it emits acrid smoke and fumes.

ECB200 CAS:29804-22-6 HR: 2
cis-7,8-EPOXY-2-METHYLOCTADECANE
mf: $C_{19}H_{38}O$ mw: 282.57

SYNS: AI3-34886 □ ATRALYMON □ cis-2-DECYL-3-(5-METHYLHEXYL)OXIRANE □ DISPARLURE □ DISRUPT □ ENT 34886 □ OCTADECANE, 7,8-EPOXY-2-METHYL-, cis-(8CI) □ OXIRANE, 2-DECYL-3-(5-METHYLHEXYL)-, cis-

TOXICITY DATA WITH REFERENCE
orl-rat LD50:>34,600 mg/kg TXAPA9 31,421,75
ihl-rat LC50:>5000 mg/m³ TXAPA9 31,421,75
skn-rbt LD50:>2025 mg/kg TXAPA9 31,421,75 FEREAC 54,7740,89

CONSENSUS REPORTS: Reported in EPA TSCA Inventory.

SAFETY PROFILE: Moderately toxic by skin contact. Low toxicity by ingestion. When heated to decomposition it emits acrid smoke and irritating vapors.

E

ECC600 CAS:1192-22-9 HR: 2
2,3-EPOXY-2-METHYLPENTANE
mf: $C_6H_{12}O$ mw: 100.18

SYN: 2-METHYL-2,3-PENTYLENE OXIDE

TOXICITY DATA WITH REFERENCE
orl-rat LD50:3500 mg/kg GTPZAB 25(11),51,81
ihl-rat LC50:16,400 mg/m³ GTPZAB 25(11),51,81
ipr-rat LD50:1600 mg/kg GTPZAB 25(11),51,81
orl-mus LD50:3400 mg/kg GTPZAB 25(11),51,81
ihl-mus LC50:13,800 mg/m³ GTPZAB 25(11),51,81

SAFETY PROFILE: Moderately toxic by ingestion and intraperitoneal routes. When heated to decomposition it emits acrid smoke and fumes.

ECD500 CAS:2443-39-2 HR: 2
cis-9,10-EPOXYOCTADECANOIC ACID
mf: $C_{18}H_{34}O_3$ mw: 298.52

SYNS: cis-9,10-EPOXYOCTADECANOATE □ EPOXYOLEIC ACID □ 9,10-EPOXYSTEARIC ACID □ cis-9,10-EPOXYSTEARIC ACID □ cis-3-OCTYL-OXIRANEOCTANOIC ACID

TOXICITY DATA WITH REFERENCE
skn-mus TDLo:2880 mg/kg/24W-I:ETA JNCIAM 31,41,63
scu-mus TDLo:164 mg/kg/41W-I:ETA CNREA8 30,1037,70

CONSENSUS REPORTS: IARC Cancer Review: Group 3 IMEMDT 7,56,87; Animal Inadequate Evidence IMEMDT 11,153,76.

SAFETY PROFILE: Questionable carcinogen with experimental tumorigenic data. When heated to decomposition it emits acrid smoke and fumes.

ECE000 CAS:2984-50-1 HR: 2
1,2-EPOXYOCTANE
mf: $C_8H_{16}O$ mw: 128.24

SYN: OCTYLENE EPOXIDE

TOXICITY DATA WITH REFERENCE
unr-mus TDLo:103 mg/kg:ETA RARSAM 3,193,63

SAFETY PROFILE: Questionable carcinogen with experimental tumorigenic data. When heated to decomposition it emits acrid smoke and fumes.

ECE500 CAS:17397-89-6 HR: 3
2,3-EPOXY-4-OXO-7,10-DODECADIENAMIDE
mf: $C_{12}H_{17}NO_3$ mw: 223.30

PROP: Needles from CCl_4 or C_6H_6. Mp: 93°, bp: 120° @ 0.00001.

SYNS: CERULENIN □ (2R,3S)-2,3-EPOXY-4-OXO-7E,10E-DODECADIENAMIDE □ (2S)(3R)-2,3-EPOXY-4-OXO-7,10-DODECADIENOYLAMIDE □ HELICOCERIN □ (2R-(2-α,3-α(4E,7E)))-3-(1-OXO-4,7-NONADIENYL)OXIRANECARBOXAMIDE □ 3-(1-OXO-4,7-NONADIENYL) OXIRANECARBOXAMIDE

TOXICITY DATA WITH REFERENCE
mmo-omi 2 mg/L JOBAAY 133,472,78
orl-mus LD50:547 mg/kg JAJAAA 17,1,64
ipr-mus LD50:211 mg/kg JAJAAA 17,1,64
scu-mus LD50:245 mg/kg 85ERAY 3,1945,78
ivn-mus LD50:154 mg/kg JAJAAA 17,1,64

SAFETY PROFILE: Poison by intraperitoneal, intravenous, and subcutaneous routes. Moderately toxic by ingestion. Mutation data reported. When heated to decomposition it emits toxic fumes of NO_x.

ECE550 CAS:64011-46-7 HR: 3
4,5-EPOXY-2-PENTENAL
mf: $C_5H_6O_2$ mw: 98.11

SYN: 2-PENTENAL, 4,5-EPOXY-

TOXICITY DATA WITH REFERENCE
skn-rbt 2 mg/24H SEV 85JCAE-,771,86
eye-rbt 50 μg/24H SEV 85JCAE-,771,86
orl-rat LD50:62 mg/kg TXAPA9 28,313,74
ihl-rat LCLo:600 ppm/4H TXAPA9 28,313,74
skn-rbt LD50:40 mg/kg TXAPA9 28,313,74

SAFETY PROFILE: Poison by ingestion and skin contact. A severe skin and eye irritant. When heated to decomposition it emits acrid smoke and irritating fumes.

ECG100 CAS:67722-96-7 HR: 3
2,3-EPOXY PROPIONALDEHYDE OXIME
mf: $C_3H_5NO_2$ mw: 87.08

OCH_2 CHCH=NOH

SYN: OXIRANE CARBOXALDEHYDE OXIME

SAFETY PROFILE: Polymerization is violent or explosive. When heated to decomposition it emits toxic fumes of NO_x. See also ALDEHYDES.

ECG200 CAS:4711-95-9 HR: 3
4-(2,3-EPOXYPROPOXY)BUTANOL
mf: $C_7H_{14}O_3$ mw: 146.19

CH_2OCH OC_4H_8OH

SAFETY PROFILE: Potentially explosive reaction when heated with trichloroethylene. When heated to decomposition it emits acrid smoke and fumes. See also ALCOHOLS.

ECG500 CAS:63991-57-1 HR: 2
p-(2,3-EPOXYPROPOXY)-N-PHENYLBENZYLAMINE
mf: $C_{16}H_{17}NO_2$ mw: 255.34

SYN: N-(4-(2,3-EPOXYPROPOXY)PHENYL)BENZYLAMINE

TOXICITY DATA WITH REFERENCE
scu-rat TDLo:4750 mg/kg/18W-I:ETA ANYAA9 68,750,58

SAFETY PROFILE: Questionable carcinogen with experimental tumorigenic data. When heated to decomposition it emits toxic fumes of NO_x.

ECH000 CAS:2530-83-8 HR: 2
3-(2,3-EPOXYPROPOXY)PROPYLTRIMETHOXYSILANE
mf: $C_9H_{20}O_5Si$ mw: 236.38

SYNS: γ-GLYCIDOXYPROPYLTRIMETHOXYSILANE □ SILANE-Y-4087 □ SILICONE A-187 □ UNION CARBIDE A-187

TOXICITY DATA WITH REFERENCE
skn-rbt 500 mg open MLD UCDS** 1/19/72
eye-rbt 100 mg MLD UCDS** 1/19/72
orl-rat LD50:23 g/kg UCDS** 2/11/64
skn-rbt LD50:3970 mg/kg AIHAAP 30,470,69

CONSENSUS REPORTS: Reported in EPA TSCA Inventory.

SAFETY PROFILE: Moderately toxic by skin contact. Mildly toxic by ingestion. A skin and eye irritant. When heated to decomposition it emits acrid smoke and fumes.

ECH500 CAS:106-90-1 HR: 3
2,3-EPOXYPROPYL ACRYLATE
mf: $C_6H_8O_3$ mw: 128.14

PROP: Insol in water. Bp: 57.2° @ 2 mm, flash p: 141°F (OC), d: 1.1, vap d: 4.4.

SYNS: 2,3-EPOXY-1-PROPANOL ACRYLATE □ 2,3-EPOXYPROPYL ESTER ACRYLIC ACID □ GLYCIDYL ACRYLATE □ GLYCIDYL PROPENATE □ 2-PROPENOIC ACID OXIRANYLMETHYL ESTER

TOXICITY DATA WITH REFERENCE
skn-rbt 100 μg/24H open AIHAAP 23,95,62
skn-rbt 2 mg/24H SEV 85JCAE-,773,86
eye-rbt 1 mg SEV UCDS** 4/10/68
eye-rbt 50 μg/24H SEV 85JCAE-,773,86
cyt-rat-ipr 50 μg/kg BJPCAL 6,235,51
cyt-rat-ipr 5 mg/kg BJPCAL 6,235,51
orl-rat LD50:210 mg/kg AIHAAP 23,95,62
ihl-rat LCLo:125 ppm/4H AIHAAP 23,95,62
skn-rbt LD50:400 mg/kg UCDS** 4/10/68

CONSENSUS REPORTS: Reported in EPA TSCA Inventory.

SAFETY PROFILE: A poison by ingestion, inhalation, and skin contact. Mutation data reported. A skin and severe eye irritant. Flammable liquid when exposed to heat or flame. Can react vigorously with oxidizers. To fight fire, use foam, dry chemical, CO_2. When heated to decomposition it emits acrid smoke and fumes. See also ESTERS.

ECI000 CAS:106-91-2 HR: 3
2,3-EPOXYPROPYL METHACRYLATE
mf: $C_7H_{10}O_3$ mw: 142.17

SYNS: CP 105 □ 2,3-EPOXY-1-PROPANOL METHACRYLATE □ 2,3-EPOXYPROPYL ESTER METHACRYLIC ACID □ GLYCIDYL METHACRYLATE □ GLYCIDYL-α-METHYL ACRYLATE

TOXICITY DATA WITH REFERENCE
mmo-klp 200 μmol/L MUREAV 89,269,81
cyt-rat-ipr 3 mg/kg BJPCAL 6,235,51
orl-rat LD50:597 mg/kg GISAAA 50(2),67,85
orl-mus LD50:390 mg/kg GISAAA 50(2),67,85
skn-rbt LD50:469 mg/kg AIHAAP 30,470,69
orl-gpg LD50:697 mg/kg GISAAA 50(2),67,85

CONSENSUS REPORTS: Reported in EPA TSCA Inventory.

SAFETY PROFILE: Poison by ingestion. Moderately toxic by skin contact and intraperitoneal routes. Mutation data reported. When heated to decomposition it emits acrid smoke and fumes.

E

ECI600 CAS:6659-62-7 HR: 3
2,3-EPOXYPROPYL NITRATE
mf: $C_3H_5NO_4$ mw: 119.08

$OCH_2CHCH_2ONO_2$

SYN: OXIRANEMETHANOL NITRATE

TOXICITY DATA WITH REFERENCE
mmo-sat 500 μg/plate AEMIDF 43,144,82

SAFETY PROFILE: An explosive sensitive to shock and heating to 200°C. Mutation data reported. When heated to decomposition it emits toxic fumes of NO_x. See also NITRATES.

ECJ000 CAS:5431-33-4 HR: 2
2,3-EPOXYPROPYL OLEATE
mf: $C_{21}H_{38}O_3$ mw: 338.59

SYNS: 2,3-EPOXY-1-PROPANOL OLEATE □ 2,3-EPOXYPROPYL ESTER of OLEIC ACID □ GLYCIDOL OLEATE □ GLYCIDYL OCTADECENOATE □ GLYCIDYL OLEATE □ OLEIC ACID GLYCIDYL ESTER □ OXIRANYLMETHYL ESTER of 9-OCTADECENOIC ACID

TOXICITY DATA WITH REFERENCE
cyt-rat-ipr 10 mg/kg BJPCAL 6,235,51
scu-mus TDLo:1040 mg/kg/52W-I:ETA CNREA8 30,1037,70
orl-rat LD50:3520 mg/kg AIHAAP 24,305,63
skn-rbt LD50:8000 mg/kg AIHAAP 24,305,63

CONSENSUS REPORTS: IARC Cancer Review: Group 3 IMEMDT 7,56,87; Animal Inadequate Evidence IMEMDT 11,183,76.

SAFETY PROFILE: Moderately toxic by ingestion and subcutaneous routes. Mildly toxic by skin contact. Questionable carcinogen with experimental tumorigenic data. Mutation data reported. When heated to decomposition it emits acrid smoke and fumes. See also ESTERS.

ECJ100 CAS:76828-34-7 HR: 3
3-(2,3-EPOXYPROPYLOXY)-2,2-DINITROPROPYL AZIDE
mf: $C_6H_9N_5O_6$ mw: 247.17

$OCH_2CHCH_2OCH_2C(NO_2)_2CH_2N_3$

SAFETY PROFILE: An explosive sensitive to impact, shock, friction and heat. When heated to decomposition it emits toxic fumes of NO_x. See also AZIDES.

ECK000 CAS:5455-98-1 HR: 2
N-(2,3-EPOXYPROPYL)-PHTHALIMIDE
mf: $C_{11}H_9NO_3$ mw: 203.21

SYNS: N-GLYCIDYLPHTHALIMIDE □ 2-(OXIRANYLMETHYL)-1H-ISOINDOLE-1,3(2H)-DIONE

TOXICITY DATA WITH REFERENCE
skn-rbt 500 mg/24H MLD TXAPA9 51,197,79
eye-rbt 100 mg SEV TXAPA9 51,197,79
mmo-sat 500 ng/plate MUREAV 68,251,79
mma-sat 5 mg/plate MUREAV 68,251,79
mmo-klp 5 μmol/L MUREAV 89,296,81
orl-rat LD50:4700 mg/kg TXAPA9 51,197,79
ihl-rat LCLo:4400 mg/m³/4H JACTDZ 4(1),219,85

CONSENSUS REPORTS: Reported in EPA TSCA Inventory.

SAFETY PROFILE: Mildly toxic by ingestion and inhalation. A skin and severe eye irritant. Mutation data reported. When heated to decomposition it emits toxic fumes of NO_x.

ECL000 CAS:25068-38-6 HR: 2
EPOXY RESIN ERL-2795
mf: $(C_{15}H_{16}O_2 \cdot C_3H_5ClO)_x$

SYNS: ERL-2795 □ 4,4'-ISOPROPYLIDENE DIPHENOL POLYMER with 1-CHLORO-2,3-EPOXYPROPANE

TOXICITY DATA WITH REFERENCE
orl-rat LD50:1100 mg/kg UCDS** 6/6/69
skn-rbt LD50:>20 mL/kg UCDS** 6/6/69

SAFETY PROFILE: Moderately toxic by ingestion and skin contact. When heated to decomposition it emits acrid smoke and fumes. See also other epoxy resin entries.

ECL500 HR: D
EPOXY RESINS, CURED

SAFETY PROFILE: Most cured resins have little or no toxicity. If curing is incomplete there may be residues of highly toxic curing agents such as the organic amines: m-phenylene diamine, diethylene triamine, tetraethylene pentamine, and hexamethylene tetramine, as well as phthalic anhydride and related compounds. When heated to decomposition they emit highly toxic fumes. See also various epoxy hardeners and POLYMERS, INSOLUBLE.

ECM500 HR: 3
EPOXY RESINS, UNCURED

SYN: POLYMERS of EPICHLOROHYDRIN and 2,2-BIS(4-HYDROXYPHENYL)PIPERAZINE

SAFETY PROFILE: Animal experiments have shown disturbed blood formation. The degree of toxicity of uncured epoxy resins varies and is partly dependent on the extent of unreacted curing agents. See also other epoxy resin entries and POLYMERS, INSOLUBLE. When heated to decomposition they emit acrid smoke and fumes.

ECO500 CAS:123-36-4 HR: 2
9,10-EPOXYSTEARIC ACID ALLYL ESTER
mf: $C_{21}H_{38}O_3$ mw: 338.59

SYNS: ALLYL-9,10-EPOXYSTEARATE □ EP-145

TOXICITY DATA WITH REFERENCE
skn-rbt 500 mg open MLD UCDS** 3/5/59
orl-rat LD50:1198 mg/kg UCDS** 2,5,59
skn-rbt LD50:15,900 mg/kg AIHAAP 23,95,62

SAFETY PROFILE: Moderately toxic by ingestion. Mildly toxic by skin contact. A skin irritant. When heated to decomposition it emits acrid smoke and fumes. See also ALLYL COMPOUNDS and ESTERS.

ECP000 CAS:4509-11-9 HR: 2
3,4-EPOXYSULFOLANE
mf: $C_4H_6O_3S$ mw: 134.16

SYNS: 2,3-EPOXYTETRAMETHYLENE SULFONE □ 6-OXA-3-THIABICYCLO(3.1.0)HEXANE-3,3-DIOXIDE □ TETRAHYDRO-3,4-EPOXYTHIOPHENE-1,1-DIOXIDE

TOXICITY DATA WITH REFERENCE
scu-mus TDLo:88 mg/kg/22W-I:ETA JNCIAM 46,143,71
ipr-mus LD50:5500 mg/kg RPTOAN 41,257,78

SAFETY PROFILE: Mildly toxic by intraperitoneal route. Questionable carcinogen with experimental tumorigenic data. When heated to decomposition it emits toxic fumes of SO_x.

ECQ100 CAS:36504-68-4 HR: 2
9,10-EPOXY-7,8,9,10-TETRAHYDROBENZO(a)PYRENE
mf: $C_{20}H_{14}O$ mw: 270.34

SYNS: 7,8,9,10-TETRAHYDRO-BENZO(a)PYRENE-9,10-EPOXIDE □ 7,8,9,10-TETRAHYDRO-BENZO(a)PYRENE-9,10-EPOXYIDE □ 7,8,9,10-TETRAHYDRO-9,10-EPOXY-BENZO(a)PYRENE

TOXICITY DATA WITH REFERENCE
mmo-sat 100 pmol/plate CNREA8 40,642,80
mma-sat 300 pmol/plate CNREA8 36,3358,76
dni-omi 200 μg/L PNASA6 74,1378,77
msc-ham:lng 60 nmol/L CNREA8 36,3358,76
dnd-mam:lym 10 nmol/L CRNGDP 3,247,82
skn-mus TDLo:519 mg/kg/60W-I:ETA CNREA8 37,3356,77

CONSENSUS REPORTS: EPA Genetic Toxicology Program.

SAFETY PROFILE: Questionable carcinogen with experimental tumorigenic data by skin contact. Mutation data reported. When heated to decomposition it emits toxic fumes of NO_x.

ECQ150 CAS:66788-11-2 HR: 2
9,10-EPOXY-9,10,11,12-TETRAHYDROBENZO(e)PYRENE
mf: $C_{20}H_{14}O$ mw: 270.34

SYNS: B(e)P H4-9,10-EPOXIDE □ 9,10,11,12-TETRAHYDRO-9,10-EPOXY-BENZO(e)PYRENE

TOXICITY DATA WITH REFERENCE
mmo-sat 1 nmol/plate CNREA8 40,1985,80
mma-sat 13 pmol/plate JBCHA3 254,4408,79
msc-ham:lng 1100 nmol/L CNREA8 40,1985,80
dnd-mam:lym 10 µmol/L CRNGDP 3,247,82
scu-mus TDLo:10 mg/kg:ETA JJIND8 64,617,80

SAFETY PROFILE: Questionable carcinogen with experimental tumorigenic data. Mutation data reported. When heated to decomposition it emits acrid smoke and fumes.

ECQ200 CAS:67694-88-6 HR: 2
3,4-EPOXY-1,2,3,4-TETRAHYDROCHRYSENE
mf: $C_{18}H_{14}O$ mw: 246.32

TOXICITY DATA WITH REFERENCE
mma-sat 1 nmol/plate CNREA8 39,4069,79
msc-ham:lng 1 nmol/plate CNREA8 39,4069,79
ipr-mus TDLo:63 mg/kg/15D-I:NEO CNREA8 39,5063,79

SAFETY PROFILE: Questionable carcinogen with experimental neoplastigenic data. Mutation data reported. When heated to decomposition it emits acrid smoke and fumes. See also CHRYSENE.

ECR259 CAS:56179-80-7 HR: 2
1,2-EPOXY-1,2,3,4-TETRAHYDROPHENANTHRENE
mf: $C_{14}H_{12}O$ mw: 196.26

TOXICITY DATA WITH REFERENCE
mma-sat 1 nmol/plate CNREA8 39,4069,79
msc-ham:lng 1 nmol/plate CNREA8 39,4069,79
ipr-mus TDLo:50 mg/kg/15D-I:NEO CNREA8 39,5063,79

SAFETY PROFILE: Questionable carcinogen with experimental neoplastigenic data. Mutation data reported. When heated to decomposition it emits acrid smoke and fumes.

ECR500 CAS:66997-69-1 HR: 2
3,4-EPOXY-1,2,3,4-TETRAHYDROPHENANTHRENE
mf: $C_{14}H_{12}O$ mw: 196.26

SYN: PHENANTHRENETETRAHYDRO-3,4-EPOXIDE

TOXICITY DATA WITH REFERENCE
mma-sat 1 nmol/plate CNREA8 39,4069,79
msc-ham:lng 1 nmol/plate CNREA8 39,4069,79
skn-mus TDLo:16 mg/kg:NEO VOONAW 15(8),54,69

SAFETY PROFILE: Questionable carcinogen with experimental neoplastigenic data. Mutation data reported. When heated to decomposition it emits acrid smoke and fumes.

ECS000 CAS:72074-69-2 HR: 2
(±)-3-α,4-α-EPOXY-1,2,3,4-TETRAHYDRO-1-β,2-α-PHENANTHRENEDIOL
mf: $C_{14}H_{12}O_3$ mw: 228.26

SYN: (±)-1-β,2-α-DIHYDROXY-3-α,4-α-EPOXY-1,2,3,4-TETRAHYDROPHENANTHRENE

TOXICITY DATA WITH REFERENCE
ipr-mus TDLo:59 mg/kg/15D-I:ETA CNREA8 39,5063,79

SAFETY PROFILE: Questionable carcinogen with experimental tumorigenic data. When heated to decomposition it emits acrid smoke and fumes.

ECT500 CAS:3083-25-8 HR: 3
1,2-EPOXY-4,4,4-TRICHLOROBUTANE
mf: $C_4H_5Cl_3O$ mw: 175.44

SYNS: TRICHLOROBUTYLENE OXIDE □ 4,4,4-TRICHLORO-1,2-EPOXYBUTANE

TOXICITY DATA WITH REFERENCE
skn-rbt 720 mg/24H SEV OMCDS*-,-,76
eye-rbt 144 mg OMCDS* -,-,76
mmo-sat 666 µg/plate MUREAV 172,105,86
mmo-sat 500 µg/plate TSCAT* OTS 205875
orl-rat LD50:1500 mg/kg OMCDS*-,-,76
ihl-rat LCLo:200 g/m³/1H TSCAT* OTS 205875
ivn-mus LD50:56,200 µg/kg CSLNX* NX#02029

CONSENSUS REPORTS: Reported in EPA TSCA Inventory.

SAFETY PROFILE: Poison by intravenous route. An eye and severe skin irritant. Mutation data reported. When heated to decomposition it emits toxic fumes of Cl^-.

ECT600 CAS:16967-79-6 HR: 2
EPOXY-1,1,2-TRICHLOROETHANE
mf: C_2HCl_3O mw: 147.38

SYNS: TCEO □ 1,1,2-TRICHLOROEPOXYETHANE □ TRICHLOROETHYLENE EPOXIDE □ TRICHLOROETHYLENE OXIDE □ TRICHLORO-OXIRANE

TOXICITY DATA WITH REFERENCE
mmo-ssp 500 µmol/L 45OHAA -,333,80
otr-ham:emb 1100 µmol/L JJIND8 69,531,82
msc-ham:lng 50 µmol/L 45OHAA -,333,80
scu-mus TDLo:20 mg/kg/78W-I:CAR CNREA8 43,159,83

SAFETY PROFILE: Questionable carcinogen with experimental carcinogenic data. Mutation data reported. When heated to decomposition it emits toxic fumes of Cl^-.

For occupational chemical analysis use NIOSH: Hydrocarbons, Halogenated 1003.

ECU550 CAS:10402-53-6 HR: 3
EPRAZINONE DIHYDROCHLORIDE
mf: $C_{24}H_{32}N_2O_2 \cdot 2ClH$ mw: 453.50

PROP: A solid. Mp: 203–206° (anhyd).

SYNS: 1-(2-BENZOYLPROPYL)-2-(2-ETHOXY-2-PHENYLETHYL)PIPERAZINE DIHYDROCHLORIDE □ 746 CE □ EFTAPAN □ EPRAZINONE HYDROCHLORIDE □ 3-(4-(β-ETHOXYPHENETHYL)-1-PIPERAZINYL)-2-METHYL-1-PHENYL-1-PROPANONE DIHYDROCHLORIDE □ 3-(4-(2-ETHOXY-2-PHENYLETHYL)-1-PIPERAZINYL)-2-METHYL-1-PHENYL-1 PROPANONE DIHYDROCHLORIDE □ 2-(4-(β-ETHOXYPHENETHYL)-1-PIPERAZINYLMETHYL)PROPIOPHENONE DIHYDROCHLORIDE □ MUCITUX □ 1-(2-PHENYL-2-ETHOXY)ETHYL-4-(2 BENZYLOXY)PROPYLPIPERAZINE DIHYDROCHLORIDE

TOXICITY DATA WITH REFERENCE
orl-rat LD50:763 mg/kg IYKEDH 10,232,79

E

ipr-rat LD50:191 mg/kg IYKEDH 10,232,79
scu-rat LD50:238 mg/kg IYKEDH 10,232,79
orl-mus LD50:286 mg/kg IYKEDH 10,232,79
ipr-mus LD50:116 mg/kg PCIPDV 13(2),98,81
scu-mus LD50:300 mg/kg OYYAA2 2,314,68
ivn-mus LD50:20 mg/kg PCIPDV 13(2),98,81

SAFETY PROFILE: Poison by ingestion, subcutaneous, intravenous, and intraperitoneal routes. When heated to decomposition it emits toxic fumes of NO_x and HCl.

ECU600 CAS:27588-43-8 HR: 3
EPROZINOL DIHYDROCHLORIDE
mf: $C_{22}H_{30}N_2O_2{\cdot}2ClH$ mw: 427.46

PROP: A solid. Mp: 164°.

SYNS: ALECOR □ BROVEL □ EUPNERON □ 4-(β-METHOXYPHENETHYL)-α-PHENYL-1-PIPERAZINEPROPANOL DIHYDROCHLORIDE □ 1-(2-METHOXY-2-PHENYLETHYL)-4-(3-HYDROXY-3-PHENYLPROPYL) PIPERAZINE DIHYDROCHLORIDE □ 1-(2-PHENYL-2-METHOXY)ETHYL-4-(3-PHENYL-3-HYDROXY)PROPYL PIPERAZINE HYDROCHLORIDE □ 1-(2-PHENYL-2-METHYL)ETHYL-4-(3-PHENYL-3-HYDROXY)-PROPYLPIPERAZINE DIHYDROCHLORIDE

TOXICITY DATA WITH REFERENCE
orl-rat TDLo:750 mg/kg (female 30D pre):REP OYYAA2 11,463,76
orl-rat TDLo:2730 mg/kg (male 39W pre):REP OYYAA2 14,805,77
orl-rat TDLo:750 mg/kg (30D male):REP OYYAA2 11,463,76
orl-rat LD50:640 mg/kg OYYAA2 11,463,76
ipr-rat LD50:72 mg/kg OYYAA2 11,463,76
ims-rat LD50:140 mg/kg OYYAA2 11,463,76
orl-mus LD50:350 mg/kg OYYAA2 11,463,76
ipr-mus LD50:103 mg/kg OYYAA2 11,463,76
ims-mus LD50:122 mg/kg OYYAA2 11,463,76

SAFETY PROFILE: Poison by ingestion, intramuscular, and intraperitoneal routes. Experimental reproductive effects. When heated to decomposition it emits toxic fumes of NO_x and HCl.

ECU750 CAS:12126-59-9 HR: 3
EQUIGYNE

PROP: Conjugated forms of natural mixed estrogens, principally sodium estrone sulfate and sodium equilin sulfate, or synthetic estrogen piperazine estrone sulfate (IMEMDT** 21,147,79).

SYNS: AMNESTROGEN □ CES □ CLIMESTRONE □ CO-ESTRO □ CONEST □ CONESTRON □ CONJES □ CONJUGATED ESTROGENS □ CONJUTABS □ EQUIGYNE □ ESTRATAB □ ESTRIFOL □ ESTROATE □ ESTROCON □ ESTROMED □ ESTROPAN □ EVEX □ FEMACOID □ FEMEST □ FEM H □ FEMOGEN □ FORMATRIX □ GANEAKE □ GENISIS □ GLYESTRIN □ KESTRIN □ MENEST □ MENOGEN □ MENOTAB □ MENOTROL □ MILPREM □ MSMED □ NEO-ESTRONE □ NOVOCONESTRON □ OESTRILIN □ OESTROFEMINAL □ OESTROPAK MORNING □ OVEST □ PALOPAUSE □ PAR ESTRO □ PMB □ PREMARIN □ PRESOMEN □ PROMARIT □ SK-ESTROGENS □ SODESTRIN-H □ TAG-39 □ THEOGEN □ TRANSANNON □ TROCOSONE □ ZESTE

TOXICITY DATA WITH REFERENCE
unr-wmn TDLo:3 mg/kg (1-5D preg):REP JAMAAP 244,1336,80
orl-mky TDLo:4167 μg/kg (1-5D preg):REP AJOGAH 115,101,73
orl-wmn TDLo:108 mg/kg/4Y-C:CAR,CVS,LIV AIMDAP 137,357,77
orl-wmn TD:27 mg/kg/3Y-C:NEO,LIV NYSJAM 78,1933,78

CONSENSUS REPORTS: NTP 7th Annual Report on Carcinogens. IARC Cancer Review: Animal Limited Evidence IMEMDT 7,283,87.

SAFETY PROFILE: Confirmed human carcinogen producing tumors of the vascular system and liver. Human reproductive effects: changes in female fertility. When heated to decomposition it emits toxic fumes of Na_2O. See also individual components.

ECV000 CAS:517-09-9 HR: 2
EQUILENIN
mf: $C_{18}H_{18}O_2$ mw: 266.36

PROP: Leaflets from acetone and ethanol; very sltly sol in water.

SYNS: EQUILENINA (SPANISH) □ EQUILENINE □ 3-HYDROXYESTRA-1,3,5(10),6,8-PENATEN-17-ONE

TOXICITY DATA WITH REFERENCE
imp-gpg TDLo:1600 μg/kg:ETA,REP BSBSAS 8,142,51
imp-ham TDLo:640 mg/kg/38W-I:ETA CNREA8 43,5200,83
imp-gpg TD:17 mg/kg:ETA,REP RBBIAL 5,1,45

SAFETY PROFILE: Experimental reproductive effects. Questionable carcinogen with experimental tumorigenic data. When heated to decomposition it emits acrid smoke and irritating fumes.

ECV500 CAS:604-58-0 HR: 2
EQUILENIN BENZOATE
mf: $C_{25}H_{22}O_3$ mw: 370.47

PROP: Crystalline. Mp: 223° (in vacuo).

SYN: 3-HYDROXYESTRA-1,3,5,7,9-PENTAEN-17-ONE BENZOATE

TOXICITY DATA WITH REFERENCE
scu-mus TDLo:162 mg/kg/81W-I:NEO ZEKBAI 56,482,49
par-mus TDLo:86 mg/kg/43W-I:ETA CRSBAW 122,183,36
scu-mus TD:120 mg/kg/30W-I:ETA,REP YJBMAU 12,213,39

SAFETY PROFILE: Experimental reproductive effects. Questionable carcinogen with experimental neoplastigenic and tumorigenic data. When heated to decomposition it emits acrid smoke and irritating fumes. See also ESTERS.

ECW000 CAS:474-86-2 HR: 2
EQUILIN
mf: $C_{18}H_{20}O_2$ mw: 268.38

PROP: Crystals from EtOAc. Mp: 238–240°.

SYNS: 1,3,5,7-ESTRATETRAEN-3-OL-17-ONE □ 3-HYDROXYESTRA-1,3,5(10),7-TETRAEN-17-ONE

TOXICITY DATA WITH REFERENCE
scu-mus TDLo: 112 mg/kg/56W-I: NEO ZEKBAI 56,482,49
imp-ham TDLo: 640 mg/kg/38W-I: ETA CNREA8 43,5200,83
scu-mus TDLo: 168 mg/kg/42W-I: ETA YJBMAU 12,213,39

SAFETY PROFILE: Questionable carcinogen with experimental neoplastigenic and tumorigenic data. When heated to decomposition it emits acrid smoke and irritating fumes.

ECW500 CAS:6030-80-4 HR: 2
EQUILIN BENZOATE
mf: $C_{25}H_{24}O_3$ mw: 372.49

SYN: 3-HYDROXYESTRA-1,3,5(10),7-TETRAEN-17-ONE BENZOATE

TOXICITY DATA WITH REFERENCE
scu-mus TDLo: 96 mg/kg/48W-I: NEO ZEKBAI 56,482,49
par-mus TDLo: 42 mg/kg/21W-I: ETA CRSBAW 122,183,36
scu-mus TD: 148 mg/kg/37W-I: ETA,REP YJBMAU 12,213,39

SAFETY PROFILE: Experimental reproductive effects. Questionable carcinogen with experimental neoplastigenic and tumorigenic data. When heated to decomposition it emits acrid smoke and irritating fumes. See also ESTERS.

ECW520 CAS:16680-47-0 HR: 3
EQUILIN SODIUM SULFATE
mf: $C_{18}H_{20}O_5S{\cdot}Na$ mw: 371.43

SYNS: EQUILIN, SULFATE, SODIUM SALT (6CI) □ ESTRA-1,3,5(10),7-TETRAEN-17-ONE, 3-HYDROXY-, HYDROGEN SULFATE SODIUM SALT (8CI) □ ESTRA-1,3,5(10),7-TETRAEN-17-ONE, 3-(SULFOOXY)-, SODIUM SALT □ SODIUM EQUILIN 3-MONOSULFATE □ SODIUM EQUILIN SULFATE

CONSENSUS REPORTS: NTP 7th Annual Report On Carcinogens.

SAFETY PROFILE: Confirmed carcinogen. When heated to decomposition it emits toxic fumes of SO_x.

ECW600 CAS:153-87-7 HR: 3
EQUIPERTINE
mf: $C_{23}H_{29}N_3O_2$ mw: 379.49

SYNS: 5,6-DIMETHOXY-2-METHYL-3-(2-(4-PHENYL-1-PIPERAZINYL)ETHYL)-1H-INDOLE □ FORIT □ INTEGRIN □ OPERTIL □ OXYPERTIN □ OXYPERTINE □ WIN 18501-2

TOXICITY DATA WITH REFERENCE
orl-rat LD50: 1 g/kg NIIRDN 6,154,82
orl-mus LD50: 2300 mg/kg NIIRDN 6,154,82
ipr-mus LD50: 154 mg/kg BJPCAL 26,186,66

SAFETY PROFILE: Poison by intraperitoneal route. Moderately toxic by ingestion. When heated to decomposition it emits toxic fumes of NO_x.

ECX000 CAS:11094-61-4 HR: 3
ERABUTOXINA
mf: $C_{284}H_{442}N_{86}O_{95}S_8$ mw: 6838.60

PROP: Neurotoxic principles isolated from venom of seasnake, *Laticauda semifasciata* (19DDA6-,249,67).

TOXICITY DATA WITH REFERENCE
ims-rat LD50: 70 µg/kg 19DDA6-,249,67
ims-mus LD50: 150 µg/kg BIJOAK 99,624,66

SAFETY PROFILE: A deadly poison by intramuscular route.

ECX100 CAS:6673-35-4 HR: 3
ERALDIN
mf: $C_{14}H_{22}N_2O_3$ mw: 266.38

PROP: Mp: 134–136° (BuOAc). Sol in warm isopropanol.

SYNS: 1-(4-ACETAMIDOPHENOXY)-3-ISOPROPYLAMINO-2-PROPANOL □ AY 21011 □ DALZIC □ N-(4-(2-HYDROXY-3-(ISOPROPYLAMINO)PROPOXY)ACETAMIDE (9CI) □ 4'-(2-HYDROXY-3-(ISOPROPYLAMINO)PROPOXY)ACETANILIDE □ N-(4-(2-HYDROXY-3-((1-METHYLETHYL)AMINO)PROPOXY)PHENYL)ACETAMIDE □ ICI 50172 □ PRACTALOL □ PRACTOLOL □ PRAKTOLOLU (POLISH) □ TERANOL

TOXICITY DATA WITH REFERENCE
orl-rat TDLo: 300 mg/kg (9-14D preg): REP TOIZAG 21,567,74
orl-mus TDLo: 12 g/kg (7-12D preg): TER TOIZAG 21,567,74
orl-rat TDLo: 300 mg/kg (9-14D preg): TER TOIZAG 21,567,74
orl-wmn TDLo: 4 mg/kg: EYE BMJOAE 1,595,75
orl-man TDLo: 7550 mg/kg/2Y: EYE,SKN,GIT MJAUAJ 1,44,75
orl-rat LD50: 3458 mg/kg TOIZAG 20,126,73
ipr-rat LD50: 540 mg/kg TOIZAG 20,126,73
scu-rat LD50: 7500 mg/kg TOIZAG 20,126,73
ivn-rat LD50: 130 mg/kg TOIZAG 20,126,73
orl-mus LD50: 3661 mg/kg TOIZAG 20,126,73
ipr-mus LD50: 455 mg/kg PJPPAA 25,145,73
scu-mus LD50: 3357 mg/kg TOIZAG 20,126,73
ivn-mus LD50: 69 mg/kg ARZNAD 27,1022,77

CONSENSUS REPORTS: EPA Genetic Toxicology Program.

SAFETY PROFILE: Poison by intravenous route. Moderately toxic by ingestion, subcutaneous, and intraperitoneal routes. An experimental teratogen. Human systemic effects by ingestion: eye damage, peritonitis, and dermatitis. Experimental reproductive effects. When heated to decomposition it emits toxic fumes of NO_x.

ECX500 CAS:10138-41-7 HR: 3
ERBIUM CHLORIDE
mf: Cl_3Er mw: 273.61

PROP: Hygroscopic pink crystals. Mp: 776°. Sol in H_2O and EtOH.

SYN: ERBIUM TRICHLORIDE

TOXICITY DATA WITH REFERENCE
orl-mus LD50:4417 mg/kg EQSSDX 1,1,75
ipr-mus LD50:226 mg/kg AEHLAU 5,437,62
ipr-gpg LD50:128 mg/kg AEHLAU 5,437,62

CONSENSUS REPORTS: Reported in EPA TSCA Inventory.

SAFETY PROFILE: Poison by intraperitoneal and subcutaneous routes. Moderately toxic by ingestion. When heated to decomposition it emits toxic fumes of Cl^-. See also RARE EARTHS.

ECY000 CAS:3088-54-8 HR: 3
ERBIUM CITRATE
mf: $C_6H_8O_7$•Er mw: 359.40

SYN: 2-HYDROXY-1,2,3-PROPANETRICARBOXYLIC ACID EBRIUM (3+) salt (1:1)

TOXICITY DATA WITH REFERENCE
ipr-mus LD50:122 mg/kg AEHLAU 5,437,62
ipr-gpg LD50:63 mg/kg AEHLAU 5,437,62

SAFETY PROFILE: Poison by intraperitoneal route. When heated to decomposition it emits acrid smoke and irritating fumes. See also RARE EARTHS.

ECY500 CAS:10168-80-6 HR: 3
ERBIUM(III) NITRATE (1:3)
mf: N_3O_9•Er mw: 353.29

PROP: Reddish crystals or hygroscopic pink solid. Sol in H_2O and EtOH.

SYN: NITRIC ACID, ERBIUM (3+) SALT

TOXICITY DATA WITH REFERENCE
ivn-rat LD50:177 mg/kg EQSSDX 1,1,75
ivn-rat LD50:27,460 μg/kg EQSSDX 1,1,75

CONSENSUS REPORTS: Reported in EPA TSCA Inventory.

SAFETY PROFILE: Poison by intravenous and intraperitoneal routes. When heated to decomposition it emits toxic fumes of NO_x. See also RARE EARTHS and NITRATES.

ECZ000 CAS:13476-05-6 HR: 3
ERBIUM(III) NITRATE, HEXAHYDRATE (1:3:6)
mf: N_3O_9•Er•$6H_2O$ mw: 461.41

PROP: Mp: $-4H_2O$ @ 130°.

SYN: NITRIC ACID, ERBIUM (3^+) SALT, HEXAHYDRATE

TOXICITY DATA WITH REFERENCE
ipr-rat LD50:230 mg/kg TXAPA9 5,750,63
ivn-rat LDLo:30 mg/kg JPETAB 43,61,31
ipr-mus LD50:225 mg/kg TXAPA9 5,750,63

SAFETY PROFILE: Poison by intraperitoneal and intravenous routes. When heated to decomposition it emits toxic fumes of NO_x.

EDA500 CAS:35287-69-5 HR: 3
ERGOCHROME AA (2,2′)-5-β,6-α,10-β-5′,6′-α,10′-β
mf: $C_{32}H_{30}O_{14}$ mw: 638.62

PROP: Crystals from $CHCl_3$. Mp: 255–259° (hot stage), mp: 281–283° (sealed capillary).

SYNS: SAD □ SECALONIC ACID D

TOXICITY DATA WITH REFERENCE
mnt-mus-orl 30 mg/kg JEPTDQ 4(5-6),31,80
scu-rat TDLo:25 mg/kg (10D preg):TER TXCYAC 25,311,82
scu-rat TDLo:15 mg/kg (female 10D post):TER TXCYAC 25,311,82
ipr-mus TDLo:25 mg/kg (female 11D post):REP 55CXA9 -,549,85
scu-rat TDLo:25 mg/kg (10D preg):REP TXCYAC 25,311,82
scu-rat TDLo:25 mg/kg (female 9D post):TER TXCYAC 25,311,82
scu-rat TDLo:15 mg/kg (female 10D post):TER TXCYAC 25,311,82
scu-rat TDLo:25 mg/kg (10D preg):TER TXCYAC 25,311,82
orl-rat LD50:22 mg/kg TXAPA9 48,A14,79
orl-mus LD50:30 mg/kg TXAPA9 48,A14,79
ipr-mus LD50:26,500 μg/kg AEMIDF 39,285,80
ivn-mus LD50:25 mg/kg JTEHD6 5,1159,79

CONSENSUS REPORTS: EPA Genetic Toxicology Program.

SAFETY PROFILE: Poison by ingestion, intraperitoneal, and intravenous routes. An experimental teratogen. Other experimental reproductive effects. Mutation data reported. When heated to decomposition it emits acrid smoke and irritating fumes.

EDA600 CAS:564-36-3 HR: 3
ERGOCORNINE
mf: $C_{31}H_{39}N_5O_5$ mw: 561.75

PROP: A solid. Mp: 182–184° (decomp).

SYNS: ERGOCORNIN □ ERGOTAMAN-3′,6′,18-TRIONE, 12′-HYDROXY-2′,5′-BIS(1-METHYLETHYL)-, (5′-α)- □ 12′-HYDROXY-2′,5′-α-BIS(1-METHYLETHYL)ERGOTAMAN-3′,6′,18-TRIONE

TOXICITY DATA WITH REFERENCE
scu-rat TDLo:5 mg/kg (female 5D post):REP RCOCB8 7,701,74
unr-rat TDLo:10 mg/kg (lactating female 7D post):REP BIREBV 1,367,69
ivn-rbt LDLo:1170 μg/kg 85IXA4 -,304,48

SAFETY PROFILE: Poison by intravenous route. Experimental reproductive effects. When heated to decomposition it emits toxic fumes of NO_x.

EDB100 CAS:511-09-1 HR: 3
ERGOCRYPTINE
mf: $C_{32}H_{41}N_5O_5$ mw: 575.78

PROP: A solid. Mp: 211–212° (decomp) from MeOH. β-Ergocryptine: Rectangular plates from benzene.

SYNS: α-ERGOCRYPTINE □ α-ERGOKRYPTINE □ 12′-HYDROXY-2′-

(1-METHYLETHYL)-5'-α-(2-METHYLPROPYL)ERGOTAMAN-3',6',18-TRIONE

TOXICITY DATA WITH REFERENCE
scu-rat TDLo: 2 mg/kg (4D preg): REP JRPFA4 56,691,79
scu-rat TDLo: 1800 μg/kg (3D post): REP USXXAM #3752814
ivn-rbt LD50: 950 μg/kg USXXAM #3752814

SAFETY PROFILE: Poison by intravenous route. Experimental reproductive effects. When heated to decomposition it emits toxic fumes of NO_x.

EDB200 CAS:2624-03-5 HR: 3
ERGOSINE MONOMETHANESULFONATE
mf: $C_{30}H_{37}N_5O_5 \cdot CH_4O_3S$ mw: 643.83

SYNS: ERGOSINE METHANESULFONATE □ α ERGOSINE METHANESULFONATE □ ERGOTAMAN-3',6',18-TRIONE, 12'-HYDROXY-2'-METHYL-5'-(2-METHYLPROPYL)-, (5'-α)-, MONOMETHANESULFONATE (salt)

TOXICITY DATA WITH REFERENCE
scu-rat TDLo: 2525 μg/kg (female 4D post): REP JRPFA4 10,221,65
ipr-mus LD50: 206 mg/kg BAXXDU #2025960

SAFETY PROFILE: Poison by intraperitoneal route. Experimental reproductive effects. When heated to decomposition it emits toxic fumes of NO_x and SO_x.

EDB500 CAS:129-51-1 HR: 3
ERGOT
mf: $C_{19}H_{23}N_3O_2 \cdot C_4H_4O_4$ mw: 441.53

PROP: A solid. Mp: 167° (decomp). Composition: ergotamine, ergosine, ergocristine, ergocryptine, ergocornine, ergosinine, ergocristinine, ergocryptinine, ergotaminine, etc.

SYNS: CORNOCENTIN □ CRUDE ERGOT □ ERGOMETRINE ACID MALEATE □ ERGOMETRINE MALEATE □ ERGONOVINE, MALEATE (1:1) (SALT) □ ERGOTRATE □ ERGOTRATE MALEATE □ OXYTOCIC

TOXICITY DATA WITH REFERENCE
sce-ham:ovr 10 nmol/L TCMUD8 8,169,88
scu-rat TDLo: 100 mg/kg (5D post): REP ENDOAO 90,285,72
orl-rat TDLo: 669 g/kg/96W-C: ETA CNREA8 2,11,42
orl-rat TD: 773 g/kg/44W-C: ETA CNREA8 2,11,42
ims-inf TDLo: 40 μg/kg: GIT,PUL JAMAAP 250,729,83
unr-man LDLo: 15 mg/kg 85DCAI 2,73,70
ivn-mus LD50: 8260 μg/kg TXAPA9 23,537,72

SAFETY PROFILE: A deadly human poison. Experimental poison by intravenous route. Experimental reproductive effects. Human systemic effects: hypermotility, diarrhea, cyanosis, thirst, tachycardia, confusion, coma, central nervous system symptoms, gangrene; circulatory changes can follow ingestion. Questionable carcinogen with experimental tumorigenic data. Mutation data reported. When heated to decomposition it emits toxic fumes of NO_x. See also individual components.

EDC000 CAS:113-15-5 HR: 3
ERGOTAMINE
mf: $C_{33}H_{35}N_5O_5$ mw: 581.63

PROP: A solid. Mp: 213–214° (decomp). A specific alkaloid present in rye ergot (AIPTAK 27,459,23).

TOXICITY DATA WITH REFERENCE
cyt-hmn:lym 100 μg/L/24H MUREAV 48,205,77
ivn-rbt LD50: 100 mg/kg EXPEAM 33,1552,77
scu-frg LDLo: 33 mg/kg AIPTAK 27,459,23

SAFETY PROFILE: Poison by intravenous and subcutaneous routes. Human mutation data reported. When heated to decomposition it emits toxic fumes of NO_x.

EDC500 CAS:379-79-3 HR: 3
ERGOTAMINE TARTRATE
mf: $C_{66}H_{70}N_{10}O_{10} \cdot C_4H_6O_6$ mw: 1313.56

PROP: A solid. Mp: 203° (decomp). Analgesic specific for migraine.

SYNS: ERGAM □ ERGATE □ ERGOMAR □ ERGOSTAT □ ERGOTAMINE BITARTRATE □ ERGOTARTRATE □ ETIN □ EXMIGRA □ FEMERGIN □ GOTAMINE TARTRATE □ GYNERGEN □ LINGRAINE □ LINGRAN □ NEO-ERGOTIN □ RIGETAMIN □ SECAGYN □ SECUPAN

TOXICITY DATA WITH REFERENCE
cyt-hmn:lyms 100 μg/L MUREAV 48,205,77
sce-ham:ovr 10 nmol/L TCMUD8 8,169,88
scu-rat TDLo: 100 mg/kg (lactating female 5D post): REP ENDOAO 90,285,72
orl-rat TDLo: 100 mg/kg (6-15D preg): TER TJADAB 7,227,73
orl-rat TDLo: 300 mg/kg (female 5-16D post): REP RCOCB8 7,701,74
orl-rat TDLo: 100 mg/kg (6-15D preg): REP TJADAB 7,227,73
orl-mus TDLo: 300 mg/kg (female 6-15D post): REP TJADAB 7,227,73
orl-wmn TDLo: 11 mg/kg/13W-I: BPR PGMJAO 61,461,85
orl-hmn TDLo: 3700 μg/kg/26W-I: CNS,GIT HEADAE 18,95,78
orl-man TDLo: 214 μg/kg MMWOAU 75,736,28
orl-rat LDLo: 1 mg/kg TJADAB 7,227,73
ivn-rat LD50: 80 mg/kg BSAMA5 2,1,46
orl-mus LDLo: 300 mg/kg TJADAB 7,227,73
ipr-mus TDLo: 412 mg/kg BAXXDU #2025960
ivn-mus LD50: 62 mg/kg BSAMA5 2,1,46
scu-cat LD50: 11 mg/kg BSAMA5 2,1,46
orl-rbt LDLo: 1 mg/kg TJADAB 7,227,73

CONSENSUS REPORTS: Reported in EPA TSCA Inventory. EPA Genetic Toxicology Program. EPA Extremely Hazardous Substances List.

SAFETY PROFILE: Poison by ingestion, intravenous, and subcutaneous routes. An experimental teratogen. Human systemic effects by ingestion: hallucinations, distorted perceptions, convulsions, nausea or vomiting, blood pressure elevation. Experimental reproductive effects. Human mutation data reported. Used in production of drugs of abuse. When heated to decomposition it emits toxic fumes of NO_x. See also ERGOT.

EDC560 **CAS:28675-83-4** **HR: 2**
ERGOTERM TGO
mf: $C_{34}H_{52}O_4S_2Sn$ mw: 707.67

SYNS: ACETIC ACID, 2,2'-((BIS(PHENYLMETHYL)STANNYLENE)BIS (THIO))BIS-, DIISOOCTYL ESTER (9CI) □ ACETIC ACID, ((DIBENZYLSTANNYLENE)DITHIO)DI-, DIISOOCTYL ESTER □ D-BENZYL TG □ DIBENZYLTIN S,S'-BIS(ISOOCTYLMERCAPTOACETATE)

TOXICITY DATA WITH REFERENCE
orl-rat TDLo:378 mg/kg (female 1-21D post):TER TXAPA9 26,253,73
orl-rat LD50:1250 mg/kg RPZHAW 18,283,67

ACGIH TLV: TWA 0.1 mg(Sn)/m³; STEL 0.2 mg/m³ (skin)

OSHA PEL: TWA 0.1 mg(Sn)/m³
NIOSH REL: (Organotin compounds) TWA 0.1 mg(Sn)/m³

SAFETY PROFILE: Moderately toxic by ingestion. An experimental teratogen. When heated to decomposition it emits toxic fumes of SO_x and Sn.

For occupational chemical analysis use NIOSH: Organotin Compounds 5504.

EDC565 **CAS:8006-25-5** **HR: 3**
ERGOTOXINE

SYNS: ECBOLINE □ ERGOTOXIN

TOXICITY DATA WITH REFERENCE
ims-rat TDLo:20 mg/kg (female 5D post):REP PSEBAA 100,555,59
ivn-man TDLo:3570 ng/kg HXPHAU 6,107,38
scu-mus LDLo:107 mg/kg AEPPAE 176,171,34
ivn-mus LD50:33 mg/kg 85IXA4 -,285,48
ivn-rbt LDLo:1500 µg/kg HXPHAU 6,104,38

SAFETY PROFILE: Poison by intravenous and subcutaneous routes. Experimental reproductive effects. When heated to decomposition it emits acrid smoke and irritating fumes.

EDC575 **CAS:8047-28-7** **HR: 3**
ERGOTOXINE ETHANSULFONATE

SYNS: ECBOLINE ETHANESULFONATE □ ERGOTOXINE ETHANESULFONATE □ ERGOTOXINE ETHANESULPHONATE

TOXICITY DATA WITH REFERENCE
ipr-rat TDLo:10 mg/kg (female 12-21D post):REP AJPHAP 180,296,55
orl-mus TDLo:30 mg/kg (female 3-5D post):REP JRPFA4 18,81,69
scu-rat TDLo:5 mg/kg (female 1D pre):REP AJPHAP 180,47,55
ivn-mus LD50:40 mg/kg HXPHAU 6,103,38

SAFETY PROFILE: Poison by intravenous route. Experimental reproductive effects. When heated to decomposition it emits toxic fumes of SO_x. See also SULFONATES.

EDC600 **CAS:11052-01-0** **HR: 3**
ERIAMYCIN
mf: $C_{31}H_{23}NO_8$ mw: 537.55

TOXICITY DATA WITH REFERENCE
orl-mus LD50:50 mg/kg 85GDA2 3,372,80
ipr-mus LD50:500 µg/kg 85ERAY 1,164,78
scu-mus LD50:2 mg/kg 85GDA2 3,372,80

SAFETY PROFILE: Poison by ingestion, subcutaneous, and intraperitoneal routes. When heated to decomposition it emits toxic fumes of NO_x.

EDC650 **CAS:66733-21-9** **HR: 3**
ERIONITE
mf: $Al_2O_{18}Si_7$•1/2Ca•$7H_2O$•1/2Na mw: 715.68

CONSENSUS REPORTS: NTP 7th Annual Report on Carcinogens. IARC Cancer Review: Group 1 IMEMDT 7,203,87; Animal Sufficient Evidence, Human Sufficient Evidence IMEMDT 42,225,87.

SAFETY PROFILE: Confirmed carcinogen with experimental carcinogenic and tumorigenic data.

EDD500 **CAS:62796-23-0** **HR: 3**
EROCYANINE 540
mf: $C_{26}H_{32}N_3O_6S_2$•Na mw: 569.72

SYNS: 3(2H)-BENZOXAZOLEPROPANESULFONIC ACID, 2-(4-(1,3-DIBUTYLTETRAHYDRO-4,6-DIOXO-2-THIOXO-5(2H)-PYRIMIDINYLIDENE)-2-BUTENYLIDENE)-, SODIUM SALT □ MEROCYANINE 540 □ NK 2272

TOXICITY DATA WITH REFERENCE
ivn-mus LD50:92 mg/kg TXAPA9 44,225,78

CONSENSUS REPORTS: Reported in EPA TSCA Inventory.

SAFETY PROFILE: Poison by intravenous route. When heated to decomposition it emits very toxic fumes of Na_2O, SO_x, and NO_x.

EDE000 **CAS:63938-27-2** **HR: 3**
ERYSODINE HYDROCHLORIDE
mf: $C_{18}H_{21}NO_3$•ClH mw: 335.86

TOXICITY DATA WITH REFERENCE
orl-mus LD50:155 mg/kg JPETAB 80,53,44
scu-mus LD50:100 mg/kg JPETAB 80,53,44

SAFETY PROFILE: Poison by ingestion and subcutaneous routes. When heated to decomposition it emits very toxic fumes of HCl and NO_x.

EDE500 **CAS:63938-28-3** **HR: 3**
ERYSOPINE HYDROCHLORIDE
mf: $C_{17}H_{19}NO_3$•ClH mw: 321.83

SYN: (3-β)-1,2,6,7-TETRADEHYDRO-3-METHOXYERYTHRINAN-16-OL HYDROCHLORIDE

TOXICITY DATA WITH REFERENCE
orl-mus LD50:18 mg/kg JPETAB 80,53,44
scu-mus LD50:15 mg/kg JPETAB 80,53,44

SAFETY PROFILE: Poison by ingestion and subcutaneous routes. When heated to decomposition it emits very toxic fumes of Cl^-.

EDF000 CAS:31248-66-5 HR: 3
ERYTHRALINE HYDROBROMIDE
mf: $C_{18}H_{19}NO_3 \cdot BrH$ mw: 378.30

SYN: (3-β)-1,2,6,7-TETRADEHYDRO-3-METHOXY-15,16-(METHYLENEBIS(OXY))ERYTHRINAN, HYDROBROMIDE

TOXICITY DATA WITH REFERENCE
orl-mus LD50:80 mg/kg JPETAB 80,53,44
scu-mus LD50:72 mg/kg JPETAB 80,53,44

SAFETY PROFILE: Poison by ingestion and subcutaneous routes. When heated to decomposition it emits very toxic fumes of HBr and NO_x.

EDG500 CAS:466-81-9 HR: 3
β-ERYTHROIDINE
mf: $C_{16}H_{19}NO_3$ mw: 273.36

PROP: A solid. Mp: 99–100°.

SYN: 12,13-DIDEHYDRO-13,14-DIHYDRO-α-ERYTHROIDINE

TOXICITY DATA WITH REFERENCE
ipr-mus LD50:24 mg/kg JPETAB 93,362,48
ivn-rbt LD50:8600 μg/kg MEIEDD 7,419,60

SAFETY PROFILE: Poison by intraperitoneal and intravenous routes. When heated to decomposition it emits toxic fumes such as NO_x.

EDH000 CAS:596-11-2 HR: 3
β-ERYTHROIDINE HYDROCHLORIDE
mf: $C_{16}H_{19}NO_3 \cdot ClH$ mw: 309.82

SYN: 1,2,6,7-TETRADEHYDRO-14,17-DIHYDRO-3-METHOXY-16(15H)-OXAERYTHRINAN-15-ONE, HYDROCHLORIDE

TOXICITY DATA WITH REFERENCE
orl-rat LD50:510 mg/kg JPETAB 80,39,44
scu-rat LD50:1260 mg/kg JPETAB 80,39,44
ivn-rat LD50:39 mg/kg JPETAB 82,266,44
orl-mus LD50:75 mg/kg JPETAB 80,39,44
scu-mus LD50:48 mg/kg JPETAB 80,39,44
ivn-dog LD50:8800 μg/kg JPETAB 82,266,44
orl-cat LDLo:30 mg/kg JPETAB 80,39,44
scu-cat LDLo:20 mg/kg JPETAB 80,39,44
orl-rbt LDLo:200 mg/kg JPETAB 80,39,44
scu-rbt LDLo:50 mg/kg JPETAB 80,39,44
ivn-rbt LD50:8600 μg/kg JPETAB 82,266,44

SAFETY PROFILE: Poison by ingestion, intravenous, and subcutaneous routes. When heated to decomposition it emits very toxic fumes of HCl and NO_x.

EDH500 CAS:114-07-8 HR: 3
ERYTHROMYCIN
mf: $C_{37}H_{67}NO_{13}$ mw: 734.05

PROP: White or sltly yellow, crystalline powder; odorless. Crystals from Me_2CO (aq) or $CHCl_3$. Mp: 136–140°, mp: 190–193° (double mp). Freely sol in alc, chloroform, and ether; very sltly sol in water.

SYNS: DOTYCIN □ EM □ E-MYCIN □ ERYCIN □ ERYTHROCIN □ ERYTHROGRAN □ ERYTHROGUENT □ ERYTHROMYCIN A □ ILOTYCIN □ PANTOMICINA □ PROPIOCINE □ ROBIMYCIN

TOXICITY DATA WITH REFERENCE
dnr-esc 600 μg/disc MUREAV 97,1,82
scu-rat TDLo:50 mg/kg (female 6-10D post):TER OSDIAF 14,107,65
orl-rat TDLo:6 g/kg (10-15D preg):TER JJANAX 25,187,72
scu-rat TDLo:50 mg/kg (female 6-10D post):REP OSDIAF 14,107,65
orl-rat LD50:9272 mg/kg AMPMAR 39,259,78
scu-rat LDLo:427 mg/kg CLDND* 1,115,75
ipr-mus LD50:463 mg/kg ARZNAD 20,1751,70
scu-mus LD50:1800 mg/kg ANTCAO 2,281,52
ivn-mus LD50:426 mg/kg 85FZAT -,273,67
ims-mus LD50:394 mg/kg JAPMA8 44,199,55
ipr-gpg LD50:413 mg/kg JAPMA8 41,555,52
orl-ham LD50:3018 mg/kg JAPMA8 41,555,52

CONSENSUS REPORTS: EPA Genetic Toxicology Program.

SAFETY PROFILE: Poison by intravenous and intramuscular routes. Moderately toxic by ingestion, intraperitoneal, and subcutaneous routes. An experimental teratogen. Other experimental reproductive effects. Mutation data reported. When heated to decomposition it emits toxic fumes of NO_x.

EDI500 CAS:23067-13-2 HR: 3
ERYTHROMYCIN GLUCOHEPTONATE (1:1)
mf: $C_{37}H_{67}NO_{13} \cdot C_7H_{14}O_8$ mw: 960.26

PROP: A solid. Mp: 95–140°.

SYNS: ERYTHROHYCIN GLUCEPTATE □ GLUCOHEPTONIC ACID with ERYTHROMYCIN (1:1)

TOXICITY DATA WITH REFERENCE
orl-mus LD50:3112 mg/kg DRUGAY-,211,90
ivn-mus LD50:453 mg/kg JAPMA8 44,199,55

SAFETY PROFILE: Moderately toxic by ingestion and intravenous routes. When heated to decomposition it emits toxic fumes of NO_x. See also ERYTHROMYCIN.

EDJ000 CAS:14271-02-4 HR: 3
ERYTHROMYCIN HYDROCHLORIDE
mf: $C_{37}H_{67}NO_{13} \cdot ClH$ mw: 770.51

SYN: ILOTYCIN HYDROCHLORIDE

TOXICITY DATA WITH REFERENCE
scu-rat LD50:1442 mg/kg JAPMA8 41,555,52
ivn-rat LD50:209 mg/kg JAPMA8 41,555,52
orl-mus LD50:2927 mg/kg JAPMA8 41,555,52
ipr-mus LD50:490 mg/kg JAPMA8 41,555,52
scu-mus LD50:1849 mg/kg JAPMA8 41,555,52
ivn-mus LD50:377 mg/kg USXXAM #4393053
ivn-rbt LD50:183 mg/kg JAPMA8 41,555,52

SAFETY PROFILE: Poison by intravenous route. Moderately toxic by ingestion, intraperitoneal, and subcuta-

E

neous routes. When heated to decomposition it emits very toxic fumes of HCl and NO_x. See also ERYTHROMYCIN.

EDJ500 **CAS:643-22-1** **HR: 3**
ERYTHROMYCIN STEARATE
mf: $C_{37}H_{67}NO_{13}•C_{18}H_{36}O_2$ mw: 1018.59

PROP: A solid. Mp: 92–93°.

SYNS: BRISTAMYCIN □ DOWMYCIN E □ ERYPAR □ ERYTHROCIN STEARATE □ ERYTHROMYCIN OCTADECANOATE (salt) □ ERYTHROMYCIN STEARIC ACID SALT □ ETHRIL □ GALLIMYCIN □ NCI-C55674 □ OE 7 □ PFIZER-E □ QIDMYCIN □ SK-ERYTHROMYCIN

TOXICITY DATA WITH REFERENCE
orl-mus LD50:3112 mg/kg DRUGAY-,211,90
ipr-gpg LD50:413 mg/kg DRUGAY-,211,90

CONSENSUS REPORTS: NTP Carcinogenesis Studies (Feed); No Evidence; rat, mouse NCITR* NTP-TR-338,88

SAFETY PROFILE: Moderately toxic by ingestion and intraperitoneal routes. When heated to decomposition it emits toxic fumes of NO_x. See also ERYTHROMYCIN.

EDK600 **CAS:8057-51-0** **HR: 3**
ESBERICARD

SYN: CRATAEGUS, EXTRACT

TOXICITY DATA WITH REFERENCE
orl-rat LD50:33,800 mg/kg NIIRDN 6,212,82
ivn-rat LDLo:1000 mg/kg PLMEAA 43,105,81
orl-mus LD50:18,500 mg/kg NIIRDN 6,212,82
ivn-cat LDLo:1200 mg/kg PLMEAA 43,105,81
ivn-rbt LDLo:1000 mg/kg PLMEAA 43,105,81
ivn-gpg LDLo:1500 mg/kg PLMEAA 43,105,81
ivn-dck LDLo:300 mg/kg PLMEAA 43,105,81
par-frg LDLo:5 g/kg PLMEAA 43,105,81

SAFETY PROFILE: Poison by intravenous route. Mildly toxic by ingestion.

EDK650 **CAS:9003-98-9** **HR: D**
ESCHERICHIA COLI ENDONUCLEASE I

SYNS: ALKALINE DEOXYRIBONUCLEASE □ ALKALINE DNASE □ DEOXYRIBONUCLEASE □ DEOXYRIBONUCLEASE A □ DEOXYRIBONUCLEASE I □ DEOXYRIBONUCLEASE (pancreatic) □ DEOXYRIBONUCLEIC PHOSPHATASE □ DESOXYRIBONUCLEASE □ DNAASE □ DNA DEPOLYMERASE □ DNA ENDONUCLEASE □ DNA NUCLEASE □ DNASW □ DNASE I □ DORNASE □ DORNAVA □ DORNAVAC □ E.C. 3.1.4.5 □ ENDODEOXYRIBONUCLEASE I □ NUCLEASE, DEOXYRIBO- □ PANCREATIC DEOXYRIBONUCLEASE □ PANCREATIC DORNASE

TOXICITY DATA WITH REFERENCE
cyt-dmg:cells-uns 1 g/L NULSAK 2,85,59
cyt-grh:cells-uns 1 g/L NULSAK 2,85,59
dnd-ham:fbr 5000 ppm SFCRAO 23,346,70
sce-ham:ovr 2500 units MUREAV 307,315,94

CONSENSUS REPORTS: Reported in EPA TSCA Inventory.

SAFETY PROFILE: Mutation data reported. When heated to decomposition it emits acrid smoke and irritating vapors.

EDK700 **CAS:67924-63-4** **HR: 3**
ESCHERICHIA COLI ENDOTOXIN

SYNS: E. COLI ENDOTOXIN □ ENDOTOXIN, ESCHERICHIA COLI □ LPS

TOXICITY DATA WITH REFERENCE
dns-mus:oth 100 μg/L MMIYAO 168,201,80
dns-gpg:lng 1 mg/L PSEBAA 171,109,82
ivn-mky TDLo:5 mg/kg (female 24W post):TER AJOGAH 131,899,78
ipr-rat TDLo:1500 μg/kg (female 14-17D post):TER ANREAK 181,441,75
ivn-rat TDLo:10 μg/kg (female 12D post):TER SEIJBO 20,151,80
ivn-rat TDLo:1250 μg/kg (female 17D post):TER AJPAA4 42,357,63
ipr-rat TDLo:2250 μg/kg (female 11-17D post):REP ANREAK 181,441,75
ipr-rat TDLo:1500 μg/kg (female 14-17D post):REP ANREAK 181,441,75
ivn-rat TDLo:10 μg/kg (male 1D pre):REP JSTBBK 26,67,87
ipr-rat TDLo:2250 μg/kg (female 11-17D post):TER ANREAK 181,441,75
ivn-rbt TDLo:10 μg/kg (female 9D post):TER ARZNAD 29,1062,79
ivn-mus TDLo:100 mg/kg (female 8D post):TER APHGBP 9,141,59
ipr-rat LD50:7600 μg/kg CRSHAG 18,11,86
ivn-rat LD50:3700 μg/kg JJMCAQ 34,54,81
ipr-mus LD50:4550 μg/kg SCIEAS 229,869,85
scu-mus LD50:7700 mg/kg APJAAG 9,141,59
ivn-mus LD50:2 mg/kg JGMIAN 86,363,75
par-mus LD50:27,500 μg/kg EXPEAM 35,804,79
ivn-dog LD50:1 mg/kg CRSHAG 7,299,80
ivn-cat LD50:2200 μg/kg EJPHAZ 9,311,70
iat-cat LDLo:3 mg/kg MIVRA6 20,242,80

SAFETY PROFILE: Poison by intravenous, intraperitoneal, parenteral, and intraarterial routes. Moderately toxic by subcutaneous route. Experimental teratogenic and reproductive effects. Mutation data reported.

EDK750 **HR: 3**
ESCHERICHIA COLI LIPOPOLYSACCHARIDE

SYNS: E. COLI 0111: B4 LPS □ LIPOPOLYSACCHARIDE, ESCHERICHIA COLI

TOXICITY DATA WITH REFERENCE
ivn-rat TDLo:100 μg/kg (female 15D post):REP ESKHA5 (99),68,81
ipr-rat TDLo:500 μg/kg (female 15D post):REP PSEBAA 109,429,62
ivn-rat TDLo:1 μg/kg (female 12D post):REP ESKHA5 (99),68,81
ipr-rat TDLo:1 mg/kg (female 8D post):REP PSEBAA 109,429,62
ipr-rat LD50:10 mg/kg PSEBAA 109,429,62
ivn-mus LD50:7670 μg/kg MIIMDV 26,455,82

SAFETY PROFILE: A poison by intravenous and intraperitoneal routes. Experimental reproductive effects. Mutation data reported. When heated to decomposition it emits acrid smoke and irritating vapors.

EDK875 CAS:6805-41-0 HR: 3
ESCIN
mf: $C_{54}H_{84}O_{23}$ mw: 1101.38

PROP: α-Escin: Amorphous powder. Mp: 224–225°. Very sol in water. β-Escin: Leaflets from dil ethanol. Mp: 222–223°. Practically insol in water.

SYNS: A-4700 □ AESCIN (GERMAN) □ AESCUSAN □ AMORPHOUS AESCIN □ ESCINA (ITALIAN) □ REPARIL

TOXICITY DATA WITH REFERENCE
orl-rbt TDLo:325 mg/kg (female 6-18D post):TER KSRNAM 9,1227,75
orl-mus TDLo:36 mg/kg (female 7-16D post):TER BLLIAX 87,76,87
orl-rat LD50:833 mg/kg KSRNAM 8,118,74
ipr-rat LD50:10,150 μg/kg KSRNAM 8,118,74
scu-rat LD50:150 mg/kg KSRNAM 8,118,74
ivn-rat LD50:1600 μg/kg BCFAAI 115,272,76
orl-mus LD50:165 mg/kg KSRNAM 8,118,74
ipr-mus LD50:6700 μg/kg NIIRDN 6,APP 2,82
scu-mus LD50:38,590 μg/kg KSRNAM 8,118,74
ivn-mus LD50:2 mg/kg BCFAAI 115,272,76

SAFETY PROFILE: Poison by ingestion, subcutaneous, intravenous, and intraperitoneal routes. Experimental teratogenic effects. When heated to decomposition it emits acrid smoke and irritating fumes. See also other escin entries.

EDL000 CAS:66795-86-6 HR: 3
α-ESCIN
mf: $C_{54}H_{84}O_{23}$ mw: 1101.38

PROP: A form of triterpenglycosides isolated from *Aesculus hippocastanum L.* (ARZNAD 20,209,70).

SYNS: α-AESCIN □ α-AESCUSAN □ β-ESCINIC ACID

TOXICITY DATA WITH REFERENCE
orl-rat LD50:720 mg/kg ARZNAD 20,209,70
ivn-rat LD50:5 mg/kg ARZNAD 20,209,70
orl-mus LD50:320 mg/kg ARZNAD 20,209,70
ivn-mus LD50:3 mg/kg ARZNAD 20,209,70
ivn-dog LD50:2 mg/kg ARZNAD 20,209,70
ivn-rbt LD50:7800 μg/kg ARZNAD 20,209,70
orl-gpg LD50:475 mg/kg ARZNAD 20,209,70
ivn-gpg LD50:15,200 μg/kg ARZNAD 20,209,70

SAFETY PROFILE: Poison by ingestion and intravenous routes. When heated to decomposition it emits acrid smoke and irritating fumes. See also other escin entries.

EDL500 CAS:11072-93-8 HR: 3
β-ESCIN
mf: $C_{54}H_{84}O_{23}$ mw: 1101.38

PROP: A form of triterpenglycosides isolated from *Aesculus hippocastanum L.* (ARZNAD 20,209,70).

SYNS: β-AESCIN □ β-AESCUSAN □ β-REPARIL

TOXICITY DATA WITH REFERENCE
orl-mus TDLo:36 mg/kg (female 7-16D post):TER BLLIAX 87,76,87
orl-rat LD50:400 mg/kg ARZNAD 20,209,70
ivn-rat LD50:2 mg/kg ARZNAD 20,209,70
orl-mus LD50:134 mg/kg ARZNAD 20,209,70
ivn-mus LD50:1400 μg/kg ARZNAD 20,209,70
ivn-dog LD50:1 mg/kg ARZNAD 20,209,70
ivn-rbt LD50:3600 μg/kg ARZNAD 20,209,70
orl-gpg LD50:188 mg/kg ARZNAD 20,209,70
ivn-gpg LD50:7200 μg/kg ARZNAD 20,209,70

SAFETY PROFILE: Poison by ingestion and intravenous routes. An experimental teratogen. When heated to decomposition it emits acrid smoke and irritating fumes. See also other escin entries.

EDM000 CAS:20977-05-3 HR: 3
ESCIN, SODIUM SALT
mf: $C_{55}H_{85}O_{24}$•Na mw: 1153.39

PROP: A mixture of saponins occurring in the seed of the horse chestnut tree (ARZNAD 12,815,62).

SYNS: A-4760 □ AESCIN SODIUM SALT □ AESCUSAN SODIUM SALT □ Na-AESCINAT □ REPARIL SODIUM SALT □ SODIUM AESCINATE

TOXICITY DATA WITH REFERENCE
ivn-rbt TDLo:2250 μg/kg (8-16D preg):REP KSRNAM 8,269,74
ivn-rbt TDLo:4500 μg/kg (8-16D preg):REP KSRNAM 8,269,74
ivn-rbt TDLo:2250 μg/kg (8-16D preg):TER KSRNAM 8,269,74
orl-rat LD50:400 mg/kg YHTPAD 22,87,87
ipr-rat LD50:9180 μg/kg KSRNAM 8,114,74
scu-rat LD50:131 mg/kg KSRNAM 8,114,74
ivn-rat LD50:8131 μg/kg KSRNAM 8,114,74
orl-mus LD50:134 mg/kg YHTPAD 22,87,87
ipr-mus LD50:8299 μg/kg KSRNAM 8,114,74
scu-mus LD50:92,530 μg/kg KSRNAM 8,114,74
ivn-mus LD50:4730 mg/kg YHTPAD 22,87,87
ivn-rbt LDLo:5 mg/kg ARZNAD 10,263,60
ivn-gpg LD50:9130 μg/kg ARZNAD 10,263,60

SAFETY PROFILE: Poison by ingestion, intravenous, intraperitoneal, and subcutaneous routes. Experimental teratogenic and reproductive effects. When heated to decomposition it emits toxic fumes of Na_2O. See also SAPONIN and other escin entries.

EDM500 CAS:102505-08-8 HR: 3
ESCIN TRIETHANOLAMINE SALT
mf: $C_{54}H_{84}O_{23}$•$C_6H_{15}NO_3$ mw: 1250.60

SYN: AESCIN TRIETHANOLAMINE SALT

TOXICITY DATA WITH REFERENCE
ivn-rat LD50:30 mg/kg ARZNAD 10,263,60
ivn-mus LD50:50 mg/kg ARZNAD 10,263,60
ivn-rbt LD50:6 mg/kg ARZNAD 10,263,60

SAFETY PROFILE: Poison by intravenous route. When heated to decomposition it emits toxic fumes of NO_x. See also other escin entries.

EDN000 **HR: 3**
ESSENTIAL PHOSPHOLIPIDS

PROP: Chemically defined as diglyceride esters of choline phosphoric acid with a predominant content of unsaturated fatty acids (OYYAA2, 3,45,69).

SYN: EPL

TOXICITY DATA WITH REFERENCE
ipr-rat TDLo:70 g/kg (28D male):REP OYYAA2 3,45,69
orl-rat LD50:1840 mg/kg OYYAA2 3,45,69
ipr-rat LD50:600 mg/kg OYYAA2 3,45,69
scu-rat LD50:1480 mg/kg OYYAA2 3,45,69
orl-mus LD50:1450 mg/kg OYYAA2 3,45,69
ipr-mus LD50:305 mg/kg OYYAA2 3,45,69
scu-mus LD50:2300 mg/kg OYYAA2 3,45,69

SAFETY PROFILE: Poison by intraperitoneal route. Moderately toxic by ingestion and subcutaneous routes. Experimental reproductive effects. When heated to decomposition it emits very toxic fumes of PO_x.

EDN500 **HR: D**
ESTERS

PROP: A large group of organic compounds that correspond structurally to salts in inorganic chemistry. They are considered to be derived from acids by the replacement of hydrogen by an organic alkyl radical. Esters of acetic acid are called acetates and esters of carbonic acid are called carbonates. The esterification of a fatty acid RCOOH, by an alcohol R′OH, yields the fatty ester RCOOR′. The most common alcohol used is methanol, yielding the methyl ester $RCOOCH_3$. The methyl esters of fatty acids have higher vapor pressures than the corresponding acids.

SAFETY PROFILE: No general statement can be made as to the toxicity of esters. Many are highly volatile and hence can act as asphyxiants or narcotics. Skin contact, as well as inhalation, may be an important route of absorption for those esters that are volatile and have a high solvent action. The degree of toxicity ranges from mildly toxic to poison. Esters generally hydrolyze upon contact with moisture; hence, a rough guide to the toxicity of a given ester may be the sum of the toxicities of the products of hydrolysis. Incompatible with nitrates. When heated to decomposition they emit acrid smoke and fumes.

EDN600 **HR: 3**
ESTRACYT HYDRATE
mf: $C_{23}H_{30}Cl_2NO_6P•2Na•H_2O$ mw: 582.41

SYNS: ESTRAMUSTINE PHOSPHATE DISODIUM HYDRATE □ ESTRAMUSTINE PHOSPHATE SODIUM HYDRATE

TOXICITY DATA WITH REFERENCE
orl-rat LD50:5400 mg/kg IYKEDH 14,838,83
scu-rat LD50:9000 mg/kg IYKEDH 14,838,83
ivn-rat LD50:208 mg/kg IYKEDH 14,838,83
scu-mus LD50:5200 mg/kg YKYUA6 35,1347,84
ivn-mus LD50:380 mg/kg IYKEDH 14,838,83
orl-rbt LD50:5655 mg/kg YKYUA6 35,1347,84

SAFETY PROFILE: Poison by intravenous route. Mildly toxic by ingestion. When heated to decomposition it emits toxic fumes of PO_x, Cl^-, NO_x, and Na_2O.

EDO000 **CAS:50-28-2** **HR: 3**
ESTRADIOL
mf: $C_{18}H_{24}O_2$ mw: 272.42

PROP: Needles out of benzene, acetone; leaflets or needles from alc. Mp: 178°, bp: decomp. Sol in dioxone, alc, and ether.

SYNS: ALTRAD □ BARDIOL □ DIHYDROFOLLICULAR HORMONE □ DIHYDROFOLLICULIN □ DIHYDROMENFORMON □ DIHYDROTHEELIN □ 3,17-β-DIHYDROXYESTRA-1,3,5(10)-TRIENE □ 3,17-β-DIHYDROXY-1,3,5(10)-ESTRATRIENE □ DIHYDROXYESTRIN □ 3,17-β-DIHYDROXYOESTRA-1,3,5-TRIENE □ 3,17-β-DIHYDROXY-1,3,5(10)-OESTRATRIENE □ DIHYDROXYOESTRIN □ DIMENFORMON □ DIMENFORMON PROLONGATUM □ DIOGYN □ DIOGYNETS □ E[2] □ 3,17-EPIDIHYDROXYESTRATRIENE □ 3,17-EPIDIHYDROXYOESTRATRIENE □ ESTRADIOL-17-β □ α-ESTRADIOL □ β-ESTRADIOL □ cis-ESTRADIOL □ d-ESTRADIOL □ 3,17-β-ESTRADIOL □ 17-β-ESTRADIOL □ 17-β-OH-ESTRADIOL □ 17-β-OH-OESTRADIOL □ d-3,17-β-ESTRADIOL □ ESTRALDINE □ ESTRA-1,3,5(10)-TRIENE-3,17-β-DIOL □ 17-β-ESTRA-1,3,5(10)-TRIENE-3,17-DIOL □ 1,3,5-ESTRATRIENE-3,17-β-DIOL □ ESTROVITE □ FEMESTRAL □ FEMOGEN □ GYNERGON □ GYNESTREL □ GYNOESTRYL □ LAMDIOL □ MACRODIOL □ MACROL □ MICRODIOL □ NORDICOL □ NSC-9895 □ OESTERGON □ OESTRADIOL □ α-OESTRADIOL □ β-OESTRADIOL □ cis-OESTRADIOL □ d-OESTRADIOL □ 3,17-β-OESTRADIOL □ d-3,17-β-OESTRADIOL □ OESTRA-1,3,5(10)-TRIENE-3,17-β-DIOL □ OESTRADIOL R □ OESTRADIOL-17-β □ 17-β-OESTRA-1,3,5(10)-TRIENE-3,17-DIOL □ OESTROGLANDOL □ OESTROGYNAL □ OVAHORMON □ OVASTEROL □ OVASTEVOL □ OVOCICLINA □ OVOCYCLIN □OVOCYCLINE □ OVOCYLIN □ PRIMOFOL □ PROFOLIOL □ PROGYNON □ PROGYNON-DH □ SYNDIOL

TOXICITY DATA WITH REFERENCE
mnt-hmn:lym 5 μmol/L MUREAV 156,199,85
dns-hmn:mmr 10 nmol/L CNREA8 45,1644,85
dni-hmn:lym 10 μmol/L PSEBAA 146,401,74
dns-rat-par 10 μg/k ACHCBO 18,213,85
cyt-ham:ovr 50 μmol/L TOLED5 29,201,85
dns-rbt:oth 100 nmol/L CRNGDP 3,703,82
orl-wmn TDLo:4400 μg/kg (31W pre):REP ACEDAB 105,7,66
scu-mus TDLo:20 mg/kg (female 19D post):REP CNREA8 37,1099,77
ims-rbt TDLo:30 μg/kg (female 18-20D post):TER JRPFA4 53,237,78
scu-rat TDLo:14,400 ng/kg (female 5-16D post):TER RCOCB8 7,701,74
orl-rat TDLo:1 g/kg (4-8D preg):REP FESTAS 22,735,71
orl-mky TDLo:10 mg/kg (female 1-6D post):REP AJOGAH 115,101,73
ims-rbt TDLo:30 μg/kg (female 18-20D post):REP JRPFA4 53,237,78
scu-mus TDLo:10 mg/kg (male 5D pre):REP BIREBV 20,310,79
scu-mus TDLo:14,400 ng/kg (female 4-6D post):REP JRPFA4 5,239,63
scu-rat TDLo:500 μg/kg (female 1D pre):REP ARTODN 50,279,82
ims-rat TDLo:1800 mg/kg (female 15-20D post):REP JCEMAZ 29,30,69
imp-mus TDLo:1720 μg/kg (female 16-21D post):REP JRPFA4 22,509,70

orl-rat TDLo:750 μg/kg (female 3D pre):REP JRPFA4 13,101,67
ims-rat TDLo:6720 ng/kg (male 14D pre):REP IJEBA6 15,16,77
scu-rat TDLo:205 μg/kg (male 5D pre):REP ENDOAO 60,519,57
ims-rat TDLo:60 mg/kg (female 19-20D post):TER JCEMAZ 28,231,68
scu-rat TDLo:6250 μg/kg (female 16-20D post):TER JRPFA4 59,43,80
ipr-rat TDLo:1400 mg/kg/13W-I:ETA NIGHAE 52,655,83
imp-rat TDLo:100 mg/kg/52W-C:CAR JJCREP 78,134,87
orl-mus TDLo:84 mg/kg/20W-C:CAR JJCREP 82,1391,91
imp-mus TDLo:100 μg/kg:ETA BIMDB3 29,45,78
scu-gpg TDLo:7 mg/kg/12W-I:ETA CRSBAW 130,9,39
imp-gpg TDLo:1200 μg/kg:ETA RCBIAS 3,108,44
imp-ham TDLo:286 mg/kg:CAR CNREA8 12,274,52
imp-ham TD:360 mg/kg/15W-I:CAR MOPMA3 23,278,83
imp-gpg TD:2400 μg/kg:ETA BSBSAS 8,142,51
imp-ham TD:547 mg/kg/12W-I:CAR CNREA8 43,4638,83
imp-gpg TD:40 mg/kg:ETA LANCAO 1,1313,39
imp-gpg TD:100 mg/kg:ETA LANCAO 1,1313,39
orl-mus TD:44 mg/kg/52W-C:ETA JEPTDQ 1,1,77
imp-rat TD:62,500 μg/kg/36W-I:ETA JJIND8 73,123,84
imp-ham TD:160 mg/kg:ETA CANCAR 10,757,57
imp-mus TD:30 mg/kg:ETA REEBB3 16,425,71
imp mus TD:34 mg/kg:ETA EJCAAH 11,39,75
scu-rat TDLo:4200 μg/kg/12W-I YACHDS 20,3899,92

CONSENSUS REPORTS: NTP 7th Annual Report on Carcinogens. IARC Cancer Review: Human Limited Evidence IMEMDT 21,279,79; Animal Sufficient Evidence IMEMDT 21,279,79; IMEMDT 6,99,74. EPA Genetic Toxicology Program.

SAFETY PROFILE: Confirmed carcinogen with experimental carcinogenic, neoplastigenic, tumorigenic, and teratogenic data. A promoter. Human reproductive effects by ingestion: fertility effects. Experimental reproductive effects. Human mutation data reported. A steroid hormone much used in medicine. When heated to decomposition it emits acrid smoke and irritating fumes.

EDO500 **CAS:57-91-0** **HR: 2**
17-α-ESTRADIOL
mf: $C_{18}H_{24}O_2$ mw: 272.42

PROP: Needles from EtOH (aq). Mp: 223°.

SYNS: 3,17-DIHYDROXYESTRATRIENE □ 3,17-α-DIHYDROXYOESTRA-1,3,5(10)-TRIENE □ ESTRA-1,3,5(10)-TRIENE-3,17-α-DIOL □ 1,3,5-ESTRATRIENE-3,17-α-DIOL □ OESTRADIOL-17-α

TOXICITY DATA WITH REFERENCE
dni-hmn:oth 100 mg//L JTEHD6 10,143,82
imp-gpg TD:3 mg/kg:ETA,REP RBBIAL 5,1,45

SAFETY PROFILE: Experimental reproductive effects. Questionable carcinogen with experimental tumorigenic data. Human mutation data reported. When heated to decomposition it emits acrid smoke and irritating fumes. A steroid. See also ESTRADIOL.

EDP000 **CAS:50-50-0** **HR: 3**
ESTRADIOL-3-BENZOATE
mf: $C_{25}H_{28}O_3$ mw: 376.53

PROP: White or sltly yellow to brownish crystalline powder; odorless. Mp: 193°. Almost insol in water; sol in alc, acetone, and dioxane; sparingly sol in vegetable oils; sltly sol in ether.

SYNS: BENOVOCYLIN □ BENZHORMOVARINE □ BENZOATE d'OESTRADIOL (FRENCH) □ BENZOESTROFOL □ BENZOFOLINE □ BENZO-GYNOESTRYL □ BENZOIC ACID ESTRADIOL □ DIFFOLLISTEROL □ DIFOLLICULINE □ DIHYDROESTRIN BENZOATE □ DIHYDROFOLLICULIN BENZOATE □ DIMENFORMON BENZOATE □ DIMENFORMONE □ DIOGYN B □ EBZ □ ESTON-B □ ESTRADIOL BENZOATE □ β-ESTRADIOL BENZOATE □ ESTRADIOL-17-β-BENZOATE □ ESTRADIOL-17-β-3-BENZOATE □ β-ESTRADIOL-3-BENZOATE □ 17-β-ESTRADIOL BENZOATE □ 17-β-ESTRADIOL-3-BENZOATE □ ESTRADIOL MONOBENZOATE □ 17-β-ESTRADIOL MONOBENZOATE □ ESTRA-1,3,5(10)-TRIENE-3,17-DIOL (17-β)-3-BENZOATE □ ESTRA-1,3,5(10)-TRIENE-3,17-β-DIOL, 3-BENZOATE □ 1,3,5(10)-ESTRATRIENE-3,17-β-DIOL 3-BENZOATE □ FEMESTRONE □ FOLLICORMON □ FOLLIDRIN □ GRAAFINA □ de GRAAFINA □ GYNECORMONE □ GYNFORMONE □ HIDROESTRON □ HORMOGYNON □ HYDROXYESTRIN BENZOATE □ MEE □ ODB □ OESTRADIOL BENZOATE □ β-OESTRADIOL BENZOATE □ OESTRADIOL-3-BENZOATE □ β-OESTRADIOL-3-BENZOATE □ 17-β-OESTRADIOL-3-BENZOATE □ OESTRADIOL MONOBENZOATE □ OESTRAFORM (BDH) □ 1,3,5(10)-OESTRATRIENE-3,17-β-DIOL 3-BENZOATE □ OVAHORMON BENZOATE □ OVASTEROL-B □ OVEX □ OVOCYCLIN BENZOATE □ OVOCYCLIN M □ OVOCYCLIN-MB □ PRIMOGYN B □ PRIMOGYN BOLEOSUM □ PRIMOGYN I □ PROGYNON B □ PROGYNON BENZOATE □ RECTHORMONE OESTRADIOL □ SOLESTRO □ UNISTRADIOL

TOXICITY DATA WITH REFERENCE
dni-rat-scu 10 μg/kg JOENAK 65,45,75
ims-wmn TDLo:1 mg/kg (5D pre):REP FESTAS 26,405,75
ims-pig TDLo:3333 μg/kg (female 8-11W post):REP ENDKAC 69,347,77
ims-dog TDLo:60 μg/kg (female 42-62D post):REP ENDOAO 85,481,69
orl-rat TDLo:1063 μg/kg (42D pre-21D post):REP JAFCAU 22,969,74
ims-dom TDLo:364 μg/kg (female 20W post):REP THGNBO 5,289,76
ims-rat TDLo:2500 μg/kg (female 13D post):TER ACENA7 33,520,60
orl-mus TDLo:8 μg/kg (female 3-4D post):REP APTOA6 26(1),1,68
scu-rat TDLo:25 μg/kg (female 4D pre):REP BIREBV 7,260,72
ims-rat TDLo:3750 μg/kg (male 15D pre):REP INJFA3 29,98,84
scu-mus TDLo:2 μg/kg (female 8D post):REP JRPFA4 18,227,69
scu-rat TDLo:7 μg/kg (female 7D pre):REP BIREBV 23,733,80
ims-dom TDLo:136 μg/kg (female 20W post):REP AAAHAN 16,462,76
imp rat TDLo:10 mg/kg (female 24D pre):REP BIREBV 2,315,70
orl-rat TDLo:1500 ng/kg (female 3D pre):REP TXAPA9 21,68,72
par-dog TDLo:1120 μg/kg (male 16W pre):REP BIREBV 33,951,85

scu-rat TDLo:250 µg/kg (male 5D pre):REP JRFSAR 2,75,68
orl-rat TDLo:350 µg/kg (male 7D pre):REP TXAPA9 21,80,72
scu-mus TDLo:28 mg/kg (female 12-16D post):TER PSEBAA 97,809,58
ims-rat TDLo:5 mg/kg (female 19D post):TER ACENA7 33,520,60
imp-rat TDLo:25 mg/kg:NEO EJCAAH 13,1437,77
scu-mus TDLo:24 mg/kg/36W-I:CAR CNREA8 8,337,48
par-mus TDLo:38 mg/kg/57W-I:ETA CNREA8 1,359,41
imp-mus TDLo:60 mg/kg:ETA CNREA8 1,290,41
scu-gpg TDLo:240 µg/kg/8W-I:ETA,TER LANCAO 1,1313,39
imp-gpg TDLo:3000 µg/kg:ETA RCBIAS 3,108,44
imp-ham TDLo:80 mg/kg:ETA,TER JPBAA7 56,1,44
scu-mus TDLo:23 mg/kg/36W-I:ETA YJBMAU 17,75,44
scu-gpg TD:4 mg/kg/9W-I:ETA,TER CRSBAW 130,9,39
imp-rat TD:30 mg/kg:ETA NATUAS 175,724,55
imp-gpg TD:50 mg/kg:ETA CNREA8 1,367,41
imp-gpg TD:100 mg/kg:ETA CNREA8 1,367,41
imp-mus TD:80 mg/kg:ETA CNREA8 1,290,41
scu-mus TD:4790 µg/kg/36W-I:ETA YJBMAU 12,213,39
scu-mus TD:38 mg/kg/39W-I:CAR CNREA8 8,337,48
scu-mus TD:15 mg/kg/22W-I:ETA CNREA8 1,345,41
imp-rat TD:100 mg/kg:ETA JPTLAS 141,29,83

CONSENSUS REPORTS: IARC Cancer Review: Animal Sufficient Evidence IMEMDT 21,279,79.

SAFETY PROFILE: Confirmed carcinogen with experimental carcinogenic, tumorigenic, and teratogenic data. Human reproductive effects by intramuscular route: menstrual cycle changes and disorders. Experimental reproductive effects. Mutation data reported. A steroid. When heated to decomposition it emits acrid smoke and irritating fumes. See also ESTRADIOL.

EDP500 CAS:63042-19-3 HR: 2
ESTRADIOL-17-BENZOATE-3,n-BUTYRATE
mf: $C_{29}H_{34}O_4$ mw: 446.63

SYNS: 17-BENZOATE-3-n-BUTYRATE d'OESTRADIOL (FRENCH) □ ESTRA-1,3,5(10)-TRIENE-3,17-β-DIOL-17-BENZOATE-3-n-BUTYRATE

TOXICITY DATA WITH REFERENCE
scu-gpg TDLo:7 mg/kg/12W-I:ETA,REP CRSBAW 130,1466,39

SAFETY PROFILE: Experimental reproductive effects. Questionable carcinogen with experimental tumorigenic data. When heated to decomposition it emits acrid smoke and irritating fumes. See also ESTRADIOL.

EDQ000 CAS:8000-03-1 HR: 2
ESTRADIOL-3-BENZOATE mixed with PROGESTERONE (1:14 moles)

SYNS: ESTRADIOL BENZOATE mixed with PROGESTERONE (1:14 moles) □ ESTRA-1,3,5(10)-TRIETNE-3,17-β-DIOL-3-BENZOATE mixed with PROGESTERONE (1:14 moles) □ PROGESTERONE mixed with ESTRADIOL BENZOATE (14:1 moles) □ PROGESTERONE mixed with ESTRA-1,3,5(10)-TRIENE-3,17-β-DIOL-3-BENZOATE (14:1 moles)

TOXICITY DATA WITH REFERENCE
ims-mky TDLo:53 mg/kg (female 20-21D post):TER TJADAB 35,119,87
ims-mky TDLo:159 mg/kg (female 20-35D post):TER TJADAB 35,119,87
scu-mus TDLo:338 mg/kg/39W-I:ETA,REP YJBMAU 12,213,39

SAFETY PROFILE: An experimental teratogen. Other experimental reproductive effects. Questionable carcinogen with experimental tumorigenic data. When heated to decomposition it emits acrid smoke and irritating fumes. See also ESTRADIOL and PROGESTERONE.

EDQ500 CAS:63042-22-8 HR: 2
ESTRADIOL-17-CAPRYLATE
mf: $C_{26}H_{38}O_3$ mw: 398.64

SYNS: CAPRYLATE d'OESTRADIOL (FRENCH) □ ESTRA-1,3,5(10)-TRIENE-3,17,β-DIOL-17-OCTANOATE

TOXICITY DATA WITH REFERENCE
scu-gpg TDLo:800 µg/kg/13W-I:ETA,TER CRSBAW 130,1466,39
imp-gpg TDLo:2800 µg/kg:ETA RCBIAS 3,108,44

SAFETY PROFILE: Questionable carcinogen with experimental tumorigenic and teratogenic data. When heated to decomposition it emits acrid smoke and irritating fumes. See also ESTRADIOL.

EDR000 CAS:113-38-2 HR: 3
ESTRADIOL DIPROPIONATE
mf: $C_{24}H_{32}O_4$ mw: 384.56

PROP: Leaflets from MeOH (aq). Mp: 104–105°.

SYNS: AGOFOLLIN □ DIMENFORMON DIPROPIONATE □ DIOVOCYCLIN □ DIOVOCYLIN □ DIPROPIONATE d'OESTRADIOL (FRENCH) □ DIPROSTRON □ ENDOFOLLICOLINA D.P. □ β-ESTRADIOL DIPROPIONATE □ ESTRADIOL-3,17-DIPROPIONATE □ β-ESTRADIOL-3,17-DIPROPIONATE □ 3,17-β-ESTRADIOL DIPROPIONATE □ 17-β-ESTRADIOL DIPROPIONATE □ ESTRA-1,3,5(10)-TRIENE-3,17-DIOL (17-β)-DIPROPIONATE □ 1,3,5(10)-ESTRATRIENE-3,17-β-DIOL DIPROPIONATE □ ESTROICI □ ESTRONEX □ FOLLICYCLIN P □ NACYCLYL □ OESTRADIOL DIPROPIONATE □ β-OESTRADIOL DIPROPIONATE □ OESTRADIOL-3,17-DIPROPIONATE □ 3,17-β-OESTRADIOL DIPROPIONATE □ 17-β-OESTRADIOL DIPROPIONATE □ OVOCYCLIN DIPROPIONATE □ OVOCYCLIN-P □ PROGYNON-DP

TOXICITY DATA WITH REFERENCE
par-rat TDLo:5 mg/kg (female 14D post):REP CRSBAW 164,784,70
par-rat TDLo:1 mg/kg (female 14D post):TER CRSBAW 164,779,70
ims-rat TDLo:300 µg/kg (male 15D pre):REP CCPTAY 1,373,70
ivn-rbt TDLo:10 µg/kg (female 1D post):REP TJADAB 6,207,72
scu-rat TDLo:45,500 ng/kg (female 28D pre):REP IJEBA6 11,1,73
ims-rat TDLo:60 µg/kg (male 15D pre):REP CCPTAY 1,373,70
ims-rat TDLo:600 µg/kg (male 30D pre):REP CCPTAY 1,373,70

ims-rat TDLo: 12 μg/kg (male 30D pre): REP CCPTAY 1,373,70
scu-rat TDLo: 5 mg/kg (female 14D post): TER JOPHAN 55,237,63
orl-rat TDLo: 12,500 μg/kg (female 16-20D post): TER ARZNAD 36,1069,86
par-rat TDLo: 40 mg/kg/26W-I: ETA CNREA8 2,632,42
scu-mus TDLo: 36 mg/kg/36W-I: ETA YJBMAU 12,213,39
imp-gpg TDLo: 2200 μg/kg: ETA RCBIAS 3,108,44
scu-mus TD: 3000 mg/kg/25W-I: ETA JPBAA7 51,9,40
scu-mus TD: 54 mg/kg/27W-I: ETA YJBMAU 12,213,39

CONSENSUS REPORTS: IARC Cancer Review: Animal Sufficient Evidence IMEMDT 21,279,79. EPA Genetic Toxicology Program.

SAFETY PROFILE: Confirmed carcinogen with experimental carcinogenic, tumorigenic, and teratogenic data. A poison by intravenous and parenteral routes. Experimental reproductive effects. A drug for the treatment of menopause. When heated to decomposition it emits acrid smoke and irritating fumes. See also ESTRADIOL.

EDR500 CAS:22966-79-6 HR: 2
ESTRADIOL MUSTARD
mf: $C_{42}H_{50}Cl_4N_2O_4$ mw: 788.74

PROP: Mp: 40–65° (freeze dried).

SYNS: BIS((4-(BIS(2-CHLOROETHYL)AMINO)BENZENE)ACETATE) ESTRA-1,3,5(10)-TRIENE-3,17-DIOL(17-β) □ BIS((4-(BIS(2-CHLOROETHYL)AMINO)BENZENE)ACETATE)OESTRA-1,3,5(10)-TRIENE-3,17-DIOL(17-β) □ BIS((p-(BIS(2-CHLOROETHYL)AMINO)PHENYL)ACETATE)ESTRADIOL □ BIS((p-(BIS(2-CHLOROETHYL)AMINO)PHENYL) ACETATE)ESTRA-1,3,5(10)-TRIENE-3,17-β-DIOL □ BIS((p-(BIS(2-CHLOROETHYL)AMINO)PHENYL)ACETATE)OESTRADIOL □ BIS((p-BIS(2-CHLOROETHYL)AMINOPHENYL)ACETATE)OESTRA-1,3,5(10)-TRIENE-3,17-β-DIOL □ NCI-C01570 □ NSC 112259 □ OESTRADIOL MUSTARD

TOXICITY DATA WITH REFERENCE
orl-mus TDLo: 2340 mg/kg/52W-I: CAR NCITR* NCI-CG-TR-59,78
ipr-mus TDLo: 480 mg/kg/8W-I: NEO CNREA8 33,3069,73
orl-mus TD: 4680 mg/kg/1Y-I: CAR NCITR* NCI-CG-TR-59,78

CONSENSUS REPORTS: IARC Cancer Review: Group 3 IMEMDT 7,56,87; Animal Limited Evidence IMEMDT 9,217,75. NCI Carcinogenesis Bioassay (gavage); Clear Evidence: mouse NCITR* NCI-CG-TR-59,78; No Evidence: rat NCITR* NCI-CG-TR-59,78.

SAFETY PROFILE: Questionable carcinogen with experimental carcinogenic and neoplastigenic data. When heated to decomposition it emits very toxic fumes of Cl^- and NO_x. See also ESTRADIOL.

EDS000 CAS:28014-46-2 HR: 3
ESTRADIOL POLYESTER with PHOSPHORIC ACID
mf: $(C_{18}H_{24}O_2 \cdot H_3O_4P)_x$

SYNS: ESTRADIOL PHOSPHATE POLYMER □ ESTRADURIN □ (17-β)-ESTRA-1,3,5(10)-TRIENE-3,17-DIOL POLYMER with PHOSPHORIC ACID □ OESTRADIOL PHOSPHATE POLYMER □ OESTRADIOL POLYESTER with PHOSPHORIC ACID □ PEP □ POLY(ESTRADIOL PHOSPHATE) □ POLYOESTRADIOL PHOSPHATE

TOXICITY DATA WITH REFERENCE
ims-rat TDLo: 50 mg/kg (19D preg): TER ACENA7 33,520,60
ims-wmn TDLo: 173 mg/kg/9Y-I: CAR,LIV ACLRBL 7,287,75
scu-rat LD50: 5800 mg/kg FATOBP (15),56,80
scu-mus LD50: 6100 mg/kg FATOBP (15),56,80

CONSENSUS REPORTS: IARC Cancer Review: Animal Sufficient Evidence IMEMDT 21,279,79.

SAFETY PROFILE: Confirmed carcinogen producing liver tumors. An experimental teratogen. A drug used in cancer treatment. When heated to decomposition it emits toxic fumes of PO_x. See also ESTRADIOL, ESTERS, POLYMERS, and PHOSPHORIC ACID.

E

EDS100 CAS:979-32-8 HR: 3
ESTRADIOL-17-VALERATE
mf: $C_{24}H_{32}O_3$ mw: 368.56

PROP: A solid. Mp: 144–145°.

SYNS: ALTADIOL □ DELADIOL □ DELAHORMONE UNIMATIC □ DELESTROGEN □ DELESTROGEN 4X □ DURA-ESTRADIOL □ ESTRADIOL VALERATE □ ESTRADIOL 17-β-VALERATE □ ESTRADIOL VALERIANATE □ (17-β)-ESTRA-1,3,5(10)-TRIENE-3,17-DIOL-17-PENTANOATE (9CI) □ ESTRAVEL □ FEMOGEX □ NEOFOLLIN □ PHARLON □ PROGYNON □ PROGYNON-DEPOT □ PROGYNOVA

TOXICITY DATA WITH REFERENCE
scu-pig TDLo: 67 μg/kg (female 15-18D post): TER BEMTAM 88,385,75
ims-rat TDLo: 1 mg/kg (female 12-19D post): TER PTLGAX 5,183,73
scu-mus TDLo: 2 mg/kg (female 1D pre): REP BIREBV 30,556,84
ims-pig TDLo: 100 μg/kg (female 9-10D post): REP JRPFA4 80,133,87
scu-rat TDLo: 10 mg/kg (female 1D pre): REP BIREBV 31,587,84
ice-mus TDLo: 3200 μg/kg (female 1D pre): REP TJADAB 25,351,82
scu-rat TDLo: 104 mg/kg/2Y-I: CAR CRNGDP 5,1003,84
scu-mus TDLo: 400 μg/kg/W-I: CAR AVBIB9 22/23,359,79

SAFETY PROFILE: Suspected carcinogen with carcinogenic and teratogenic data. Experimental reproductive effects. When heated to decomposition it emits acrid smoke and irritating fumes. See also ESTRADIOL.

EDT100 CAS:52205-73-9 HR: 3
ESTRAMUSTINE PHOSPHATE SODIUM
mf: $C_{23}H_{32}Cl_2NO_6P \cdot 2Na$ mw: 566.41

SYNS: EMP □ ESTRACYT □ ESTRAMUSTINE PHOSPHATE DISODIUM □ ESTRA-1,3,5(10)-TRIENE-3,17-β-DIOL 3-(BIS(2-CHLOROETHYL) CARBAMATE)17-DISODIUM PHOSPHATE

TOXICITY DATA WITH REFERENCE
orl-rat TDLo: 5500 μg/kg (female 7-17D post): REP OYYAA2 20,1219,80
orl-rat TDLo: 22 mg/kg (female 7-17D post): REP OYYAA2 20,1219,80
orl-rat TDLo: 44 mg/kg (female 7-17D post): REP OYYAA2 20,1219,80

orl-rat TDLo:5500 μg/kg (female 7-17D post):REP OYYAA2 20,1219,80
orl-rat TDLo:44 mg/kg (female 7-17D post):TER OYYAA2 20,1219,80
orl-rat TDLo:350 mg/kg (35D pre):REP OYYAA2 21,211,81
orl-rat TDLo:30 mg/kg (60D male):REP OYYAA2 20,1211,80
orl-rat LD50:5400 mg/kg OYYAA2 20,1141,80
scu-rat LD50:9000 mg/kg OYYAA2 20,1141,80
ivn-rat LD50:208 mg/kg OYYAA2 20,1141,80
scu-mus LD50:5200 mg/kg OYYAA2 20,1141,80
ivn-mus LD50:380 mg/kg OYYAA2 20,1141,80
orl-rbt LD50:5655 mg/kg OYYAA2 20,1141,80

SAFETY PROFILE: Poison by intravenous route. Mildly toxic by ingestion. Experimental reproductive effects. When heated to decomposition it emits toxic fumes of Na_2O, NO_x, Cl^-, and PO_x.

EDT500 CAS:6639-99-2 HR: 2
α-ESTRA-1,3,5,7,9-PENTANE-3,17-DIOL
mf: $C_{18}H_{19}O_2$ mw: 267.37

PROP: A solid. Mp: 215–217°.

SYNS: α-DIHYDROEQUILENIN □ α-DIHYDROEQUILENINA (SPANISH)

TOXICITY DATA WITH REFERENCE
imp-gpg TDLo:20 mg/kg:ETA,REP RBBIAL 5,1,45

SAFETY PROFILE: Experimental reproductive effects. Questionable carcinogen with experimental tumorigenic data. When heated to decomposition it emits acrid smoke and irritating fumes.

EDU000 CAS:1423-97-8 HR: 2
β-ESTRA-1,3,5,7,9-PENTANE-3,17-DIOL
mf: $C_{18}H_{19}O_2$ mw: 267.37

SYNS: β-DIHYDROEQUILENIN □ β-DIHYDROEQUILENINA (SPANISH)

TOXICITY DATA WITH REFERENCE
imp-gpg TDLo:24 mg/kg:ETA,REP RBBIAL 5,1,45

SAFETY PROFILE: Experimental reproductive effects. Questionable carcinogen with experimental tumorigenic data. When heated to decomposition it emits acrid smoke and irritating fumes.

EDU500 CAS:50-27-1 HR: 3
ESTRIOL
mf: $C_{18}H_{24}O_3$ mw: 288.42

PROP: Small, white crystals. D: 0.965, mp: 214.6°, bp: 214.6°.

SYNS: AACIFEMINE □ COLPOVISTER □ DESTRIOL □ DEUSLON-A □ ESTRA-1,3,5(10)-TRIENE-3,16-α,17-β-TRIOL □ 1,3,5-ESTRATRIENE-3-β,16-α,17-β-TRIOL □ (16-α,17-β)-ESTRA-1,3,5(10)-TRIENE-3,16,17-TRIOL □ ESTRATRIOL □ 3,16-α,17-β-ESTRIOL □ 16-α,17-β-ESTRIOL □ ESTRIOLO (ITALIAN) □ FOLLICULAR HORMONE HYDRATE □ GYNAESAN □ HEMOSTYPTANON □ HOLIN □ HORMOMED □ HORMONIN □ 16-α-HYDROXYESTRADIOL □ 16-α-HYDROXYOESTRADIOL □ KLIMORAL □ NSC-12169 □ OE3 □ OESTRA-1,3,5(10)-TRIENE-3,16-α,17-β-TRIOL □ 1,3,5-OESTRATRIENE-3-β-3,16-α,17-β-TRIOL □ (16-α,17-β)-OESTRA-1,3,5(10)-TRIENE-3,16,17-TRIOL □ OESTRATRIOL □ OESTRIOL □ 3,16-α,17-β-OESTRIOL □ 16-α,17-β-OESTRIOL □ ORGASTYPTIN □ OVESTERIN □ OVESTIN □ OVESTINON □ OVESTRION □ STIPTANON □ SYNAPAUSE □ THEELOL □ THULOL □ TRIDESTRIN □ 3,16-α,17-β-TRIHYDROXY-Δ-1,3,5-ESTRATRIENE □ 3,16-α,17-β-TRIHYDROXYESTRA-1,3,5(10)-TRIENE □ 3,16-α,17-β-TRIHYDROXY-Δ-1,3,5-OESTRATRIENE □ TRIHYDROXYESTRIN □ 3,16-α,17-β-TRIHYDROXYOESTRA-1,3,5(10)-TRIENE □ TRIHYDROXYOESTRIN □ TRIODURIN □ TRIOVEX

TOXICITY DATA WITH REFERENCE
cyt-ham:ovr 50 μmol/L TOLED5 29,201,85
scu-rat TDLo:350 μg/kg (female 7D pre):REP FESTAS 7,301,56
scu-rat TDLo:250 μg/kg (female 4-8D post):TER ZDBEA9 (3),63,82
scu-rat TDLo:50 μg/kg (female 1D post):REP FESTAS 7,301,56
orl-rat TDLo:17,500 μg/kg (35D pre):REP INJFA3 10,327,65
scu-mus TDLo:100 mg/kg (male 10D pre):REP FESTAS 24,687,73
orl-rat TDLo:5 mg/kg (female 5D post):REP CCPTAY 14,487,76
orl-rat TDLo:3 mg/kg (female 1-4D post):REP JRPFA4 26,363,71
scu-rat TDLo:500 μg/kg (male 1D pre):REP ACENA7 49,193,65
orl-rat TDLo:8440 ng/kg (female 1D pre):REP JMCMAR 23,329,80
scu-rat TDLo:400 μg/kg (male 1D pre):REP ACENA7 49,193,65
orl-rat TDLo:577 μg/kg (female 16-19D post):TER AOGMAU 96,83,75
imp-mus TDLo:26 mg/kg:NEO EJCAAH 11,39,75
imp-gpg TDLo:20 mg/kg:ETA,TER RBBIAL 5,1,45
imp-ham TDLo:480 mg/kg/64W-C:ETA NCIMAV 1,1,59
imp-mus TD:34 mg/kg:NEO EJCAAH 11,39,75

CONSENSUS REPORTS: IARC Cancer Review: Animal Limited Evidence IMEMDT 21,327,79; Human Limited Evidence IMEMDT 21,327,79; Animal Inadequate Evidence IMEMDT 6,117,74.

SAFETY PROFILE: Suspected carcinogen with experimental carcinogenic, neoplastigenic, tumorigenic, and teratogenic data. Other experimental reproductive effects. Mutation data reported. A steroid drug for the treatment of menopause. When heated to decomposition it emits acrid smoke and irritating fumes.

EDV000 CAS:53-16-7 HR: 3
ESTRONE
mf: $C_{18}H_{22}O_2$ mw: 270.40

PROP: White crystals from EtOH trimorphic. Mp: 254°. Insol in water; sol in alc, benzene, ether, and chloroform.

SYNS: AQUACRINE □ CRINOvrL □ CRISTALLOVAR □CRYSTOGEN □ DESTRONE □ DISYNFORMON □ E' □ ENDOFOLLICULINA □ ESTERONE □ 1,3,5-ESTRATRIEN-3-OL-17-ONE □ 1,3,5(10)-ESTRATRIEN-3-OL-17-ONE □ Δ-1,3,5-ESTRATRIEN-3-β-OL-17-ONE □EST-RIN □ ESTROL □ ESTRON □ ESTRONA (SPANISH) □ ESTRONE-A □ ESTRUGENONE □ ESTRUSOL □ FEMESTRONE INJECTION □

FEMIDYN □ FOLIKRIN □ FOLIPEX □ FOLISAN □ FOLLESTRINE □ FOLLICULAR HORMONE □ FOLLICULIN □ FOLLICULINE BENZOATE □ FOLLICUNODIS □ FOLLIDRIN □ GLANDUBOLIN □ HIESTRONE □ HORMOFOLLIN □ HORMOVARINE □ 3-HYDROXYESTRA-1,3,5(10)-TRIEN-17-ONE □ 3-HYDROXY-17-KETO-ESTRA-1,3,5-TRIENE □ 3-HYDROXY-17-KETO-OESTRA-1,3,5-TRIENE □ 3-HYDROXY-OESTRA-1,3,5(10)-TRIEN-17-ONE □ 3-HYDROXY-1,3,5(10)-OESTRATRIEN-17-ONE □ KESTRONE □ KETODESTRIN □ KETOHYDROXY-ESTRATRIENE □ KETOHYDROXYESTRIN □ KETOHYDROXYOESTRIN □ KOLPON □ MENAGEN □ MENFORMON □ Δ-1,3,5-OESTRATRIEN-3-β-OL-17-ONE □ 1,3,5-OESTRATRIEN-3-OL-17-ONE □ 1,3,5(10)-OESTRATRIEN-3-OL-17-ONE □ OESTRIN □ OESTROFORM □ OESTRONE □ OESTROPEROS □ OVEX □ OVIFOLLIN □ PERLATAN □ SOLLICULIN □ THEELIN □ THELESTRIN □ THELYKININ □ THYNESTRON □ TOKOKIN □ UNDEN □ WNYESTRON

TOXICITY DATA WITH REFERENCE
dnd-rat-orl 870 nmol/kg CBINA8 23,13,78
cyt-rat-ipr 10 mg/kg CUSCAM 50,425,81
cyt-ham:ovr 50 μmol/L TOLED5 29,201,85
imp-man TDLo:1586 μg/kg (60D male):REP CCPTAY 25,591,82
scu-rat TDLo:50 μg/kg (female 9D post):REP BEXBAN 74,1255,72
scu-rat TDLo:10 μg/kg (female 2D post):TER INJFA3 14,56,69
orl-rbt TDLo:630 μg/kg (female 6-8D post):TER FESTAS 16,281,65
ims-rbt TDLo:13 iu/kg (female 9D post):TER AJOGAH 111,1083,71
scu-mus TDLo:28 mg/kg (female 12-16D post):REP PSEBAA 97,809,58
orl-ham TDLo:24 mg/kg (female 1-3D post):REP FESTAS 16,281,65
par-rat TDLo:25 μg/kg (female 2D post):REP FESTAS 12,178,61
orl-rbt TDLo:300 μg/kg (female 1-3D post):REP FESTAS 16,281,65
scu-rbt TDLo:5500 μg/kg (female 15-25D post):REP JPETAB 49,146,33
orl-rbt TDLo:630 μg/kg (female 6-8D post):REP FESTAS 16,281,65
orl-rat TDLo:100 μg/kg (1D pre):REP IJEBA6 15,1144,77
orl-rat TDLo:560 μg/kg (male 14D pre):REP ANYAA9 71,500,58
orl-rat TDLo:2240 μg/kg (male 14D pre):REP ANYAA9 71,500,58
scu-mus TDLo:28 mg/kg (female 12-16D post):TER PSEBAA 97,809,58
imp-rat TDLo:16 mg/kg:ETA CNREA8 13,147,53
orl-mus TDLo:11 mg/kg/68W-C:NEO BIMDB3 29,45,78
scu-mus TDLo:108 mg/kg/90W-I:NEO ZEKBAI 56,482,49
par-mus TDLo:1200 μg/kg/W-I:ETA COREAF 195,630,32
imp-mus TDLo:48 mg/kg:ETA YJBMAU 17,75,44
scu-gpg TDLo:40 mg/kg/18W-I:ETA,TER CRSBAW 130,9,39
imp-gpg TDLo:640 μg/kg:ETA,TER BSBSAS 8,142,51
imp-ham TDLo:320 mg/kg/59W-C:ETA NCIMAV 1,1,59
imp-gpg TD:2 mg/kg:ETA,TER RBBIAL 5,1,45
imp-gpg TD:1800 μg/kg:ETA RCBIAS 3,108,44
imp-rat TD:80 mg/kg:ETA PCCRA4 6,50,66
imp-ham TD:640 mg/kg/38W-I:ETA CNREA8 43,5200,83
scu-mus TD:48 mg/kg/24W-I:ETA JNCIAM 1,119,40

CONSENSUS REPORTS: NTP 7th Annual Report on Carcinogens. IARC Cancer Review: Human Limited Evidence IMEMDT 21,343,79; Animal Sufficient Evidence IMEMDT 6,123,74; IMEMDT 21,343,79. Reported in EPA TSCA Inventory.

SAFETY PROFILE: Confirmed carcinogen with experimental carcinogenic, neoplastigenic, tumorigenic, and teratogenic data. A poison by intraperitoneal and subcutaneous routes. Human reproductive effects by implantation: spermatogenesis and impotence. Mutation data reported. A steroid drug for the treatment of menopause and ovariectomy symptoms. When heated to decomposition it emits acrid smoke and irritating fumes.

E

EDV500 **CAS:2393-53-5** **HR: 3**
ESTRONE BENZOATE
mf: $C_{25}H_{26}O_3$ mw: 374.51

SYNS: BENZOATE d'OESTRONE (FRENCH) □ 3-(BENZOYLOXY)ESTRA-1,3,5(10)-TRIEN-17-ONE □ 3-HYDROXYESTRA-1,3,5(10)-TRIEN-17-ONE BENZOATE □ KETOHYDROXYESTRIN BENZOATE □ OESTRONBENZOAT (GERMAN)

TOXICITY DATA WITH REFERENCE
scu-rat TDLo:80 mg/kg/87W-I:ETA CRSBAW 137,325,43
scu-mus TDLo:760 μg/kg/19W-I:NEO ZEKBAI 56,482,49
par-mus TDLo:52 mg/kg/26W-I:ETA CRSBAW 122,183,36
scu-mus TD:6450 μg/kg/43W-I:ETA JPBAA7 42,169,36

CONSENSUS REPORTS: IARC Cancer Review: Animal Limited Evidence IMEMDT 21,343,79.

SAFETY PROFILE: Suspected carcinogen with experimental carcinogenic, neoplastigenic, and tumorigenic data. A steroid. When heated to decomposition it emits acrid smoke and irritating fumes.

EDV600 **CAS:438-67-5** **HR: 3**
ESTRONE SODIUM SULFATE
mf: $C_{18}H_{22}O_5S \cdot Na$ mw: 373.45

SYNS: CONESTORAL □ ESTRA-1,3,5(10)-TRIEN-17-ONE, 3-(SULFOXY)-, SODIUM SALT (9CI) □ ESTRONE, HYDROGEN SULFATE, SODIUM SALT □ ESTRONE SULFATE SODIUM □ ESTRONE SULFATE SODIUM SALT □ ESTRONE-3-SULFATE SODIUM SALT □ EVEX □ MORESTIN □ OESTRONE-3-SULPHATE SODIUM SALT □ SODIUM ESTRONE SULFATE □ SODIUM ESTRONE-3-SULFATE

TOXICITY DATA WITH REFERENCE
scu-ctl TDLo:504 μg/kg (female 9D pre):REP JRPFA4 55,191,79

CONSENSUS REPORTS: NTP 7th Annual Report on Carcinogens.

SAFETY PROFILE: Confirmed carcinogen. Experimental reproductive effects. When heated to decomposition it emits toxic fumes of SO_x.

EDV700 **CAS:15686-63-2** **HR: 3**
ETABENZARONE
mf: $C_{23}H_{27}NO_3$ mw: 365.51

SYNS: BENZOFURAN, 3-(p-(2-(DIETHYLAMINO)ETHOXY)BENZOYL)-2-ETHYL- □ p-(2-(DIETHYLAMINO)ETHOXY)PHENYL 2-ETHYL-3-BENZOFURANYL KETONE □ KETONE, p-(2-(DIETHYLAMINO)ETHOXY)PHENYL 2-ETHYL-3-BENZOFURANYL □ L 2642-LABAZ □ METH-

ANONE, (4-(2-(DIETHYLAMINO)ETHOXY)PHENYL)(2-ETHYL-3-BENZOFURANYL)-(9CI)

TOXICITY DATA WITH REFERENCE
ipr-mus LD50:225 mg/kg EJMCA5 14,517,79

DOT CLASSIFICATION: 3; *Label:* Flammable Liquid

SAFETY PROFILE: A poison by intraperitoneal route. A flammable liquid. When heated to decomposition it emits toxic vapors of NO_x.

EDW000 CAS:102534-95-2 HR: 3
ETABETACIN

PROP: An antibiotic extracted from cultures of a *Streptomyces* strain (AIPUAN 10,21,67).

TOXICITY DATA WITH REFERENCE
orl-mus LD50:500 μg/kg AIPUAN 10,21,67
ipr-mus LD50:250 μg/kg AIPUAN 10,21,67
ivn-mus LD50:250 μg/kg AIPUAN 10,21,67

SAFETY PROFILE: Poison by ingestion, intraperitoneal, and intravenous routes.

EDW300 CAS:34521-16-9 HR: 3
ETEAI
mf: $C_{11}H_{24}N_3S \cdot Br \cdot BrH$ mw: 391.27

SYNS: (2-((4,5-DIHYDRO-1H-IMIDAZOL-2-YL)THIO)ETHYL)TRIETHYLAMMONIUM BROMIDE HYDROBROMIDE □ 2-((4,5-DIHYDRO-1H-IMIDAZOL-2-YL)-THIO)-N,N,N-TRIETHYLETHANAMINIUM, BROMIDE HYDROBROMIDE □ 2-(2-TRIETHYLAMINOETHYLTHIO)-Δ^2-IMIDAZOLINE BROMIDE HYDROBROMIDE

TOXICITY DATA WITH REFERENCE
ipr-mus LD50:75 mg/kg CPBTAL 23,1639,75
scu-mus LD50:109 mg/kg CPBTAL 23,1639,75
ivn-mus LD50:53,900 μg/kg CPBTAL 23,1639,75

SAFETY PROFILE: Poison by subcutaneous, intravenous, and intraperitoneal routes. When heated to decomposition it emits toxic fumes of SO_x, NO_x, NH_3, and HBr.

EDW500 CAS:1837-57-6 HR: 3
ETHACRIDINE LACTATE
mf: $C_{15}H_{15}N_3O \cdot C_3H_6O_3$ mw: 343.42

PROP: Pale-yellow crystals from EtOH/Et_2O. Mp: 235°.

SYNS: ACRINOL □ ACROLACTINE □ 2-AETHOXY-6,9-DIAMINOACRIDINLACTAT (GERMAN) □ 2,5-DIAMINO-7-ETHOXYACRIDINE LACTATE □ 6,9-DIAMINO-2-ETHOXYACRIDINE LACTATE MONOHYDRATE □ ETHODIN □ 2-ETHOXY-6,9-DIAMINOACRIDINE LACTATE □ 2-ETHOXY-6,9-DIAMINOACRIDINE LACTATE HYDRATE □ 2-ETHOXY-6,9-DIAMINOACRIDINIUM LACTATE □ FLAVITROL □ METIFEX □ RIMAON □ RIVANOL □ RIVINOL □ VUCINE

TOXICITY DATA WITH REFERENCE
mma-sat 16 μg/plate MUREAV 144,9,85
dnd-esc 5 μmol/L MUREAV 89,95,81
mmo-omi 250 mg/L PNASA6 56,500,66
ipr-mus LD50:42 mg/kg NIIRDN 6,2,82
scu-mus LD50:120 mg/kg BJEPA5 28,1,47
ivn-rbt LDLo:30 mg/kg BSPHAV 40,582,33

CONSENSUS REPORTS: Reported in EPA TSCA Inventory.

SAFETY PROFILE: Poison by subcutaneous, intraperitoneal, and intravenous routes. Experimental reproductive effects. An antiseptic. When heated to decomposition it emits toxic fumes of NO_x.

EDW875 CAS:1070-11-7 HR: 3
ETHAMBUTOL DIHYDROCHLORIDE
mf: $C_{10}H_{24}N_2O_2 \cdot ClH$ mw: 240.82

PROP: A solid. Mp: 198.5–200.3°.

SYNS: CL 40881 □ DEXAMBUTOL □ EMB-FATOL □ ETAMBUTOL □ ETHAMBUTOL HYDROCHLORIDE □ ETHIBI □ ETHOPIAN □ MYAMBUTOL □ MYCOBUTOL

TOXICITY DATA WITH REFERENCE
mnt-mus-orl 50 mg/kg IRLCDZ 10,135,82
spm-mus-orl 12,500 μg/kg IRLCDZ 10,135,82
orl-man TDLo:45 mg/kg/3D-I:EYE BJDCAT 80,288,86
orl-man TDLo:46 mg/kg/3D-I:EYE PGMJAO 61,811,85
unr-wmn TDLo:720 mg/kg/48D:EYE ANOPB5 2,578,70
orl-rat LD50:6800 mg/kg OYYAA2 2,70,68
ipr-rat LD50:1200 mg/kg OYYAA2 2,70,68
ivn-rat LD50:300 mg/kg OYYAA2 2,70,68
orl-mus LD50:8900 mg/kg OYYAA2 2,70,68
scu-mus LD50:1800 mg/kg OYYAA2 2,70,68
ivn-mus LD50:230 mg/kg OYYAA2 2,70,68

SAFETY PROFILE: Poison by intravenous route. Moderately toxic by intraperitoneal and subcutaneous routes. Mildly toxic by ingestion. Human systemic effects: constriction of the pupil, visual field changes. Mutation data reported. When heated to decomposition it emits toxic fumes of NO_x and HCl. See also AMINES.

EDX000 CAS:51635-81-5 HR: 2
ETHAMON DS

PROP: A detergent (GISAAA 43(3),14,78).

TOXICITY DATA WITH REFERENCE
par-rat LD50:4425 mg/kg GISAAA 43(3),14,78
par-mus LD50:3500 mg/kg GISAAA 43(3),14,78

SAFETY PROFILE: Moderately toxic by parenteral route.

EDZ000 CAS:74-84-0 HR: 3
ETHANE
DOT: UN 1035/UN 1961
mf: C_2H_6 mw: 30.08

PROP: Colorless, odorless, flammable gas. Mp: −172°, bp: −88.6°, lel: 3.0%, uel: 12.5%, fp: −183.2°, d: 0.446 @ 0° (liquid), autoign temp: 959°F, vap d: 1.04, flash p: −202°F. Sol in EtOH, liquid O_2; sltly sol in H_2O.

SYNS: BIMETHYL □ DIMETHYL □ ETHANE, compressed (UN 1035) (DOT) □ ETHANE, refrigerated liquid (UN 1961) (DOT) □ ETHYL HYDRIDE □ METHYLMETHANE

CONSENSUS REPORTS: Reported in EPA TSCA Inventory.

DOT CLASSIFICATION: 2.1; *Label:* Flammable Gas

SAFETY PROFILE: A simple asphyxiant. See ARGON for properties of simple asphyxiants. A very dangerous fire hazard when exposed to heat or flame; can react vigorously with oxidizing materials. Moderate explosion hazard when exposed to flame. To fight fire, stop flow of gas. Incompatible with chlorine, dioxygenyl tetrafluoroborate, oxidizing materials, heat or flame. When heated to decomposition it emits acrid smoke and irritating fumes.

EEA000 CAS:557-30-2 HR: 3
ETHANEDIAL DIOXIME
mf: $C_2H_4N_2O_2$ mw: 88.08

PROP: Crystals. D: 1.547, mp: 178° (decomp). Very sol in H_2O, EtOH, and Et_2O.

SYNS: GLYOXAL, DIOXIME □ GLYOXIME □ PIK-OFF

TOXICITY DATA WITH REFERENCE
mmo-esc 20 µmol/plate MUREAV 164,263,86
orl-rat LD50:119 mg/kg 85ARAE 3,69,76/77
skn-rbt LD50:1580 mg/kg FMCHA2 -,C188,83

CONSENSUS REPORTS: Reported in EPA TSCA Inventory.

SAFETY PROFILE: Poison by ingestion. Moderately toxic by skin contact. Mutation data reported. When heated to decomposition it emits toxic fumes of NO_x.

EEA500 CAS:107-15-3 HR: 3
1,2-ETHANEDIAMINE
DOT: UN 1604
mf: $C_2H_8N_2$ mw: 60.12

PROP: Volatile, colorless, clear, thick, strongly alkaline, hygroscopic liquid; ammonia-like odor. Mp: 8.5°, bp: 117.2°, flash p: 110°F (CC), d: 0.8994 @ 20°/4°, vap press: 10.7 mm @ 20°, vap d: 2.07, autoign temp: 725°F. Sol in EtOH and H_2O (with hydration); insol in C_6H_6; sltly sol in Et_2O.

SYNS: AETHALDIAMIN (GERMAN) □ AETHYLENEDIAMIN (GERMAN) □ 1,2-DIAMINOAETHAN (GERMAN) □ 1,2-DIAMINO-ETHAAN (DUTCH) □ 1,2-DIAMINOETHANE □ 1,2-DIAMINO-ETHANO (ITALIAN) □ DIMETHYLENEDIAMINE □ ETHYLEENDIAMINE (DUTCH) □ ETHYLENEDIAMINE (OSHA) □ 1,2-ETHYLENEDIAMINE □ ETHYLENE-DIAMINE (FRENCH) □ NCI-C60402

TOXICITY DATA WITH REFERENCE
skn-rbt 450 mg open MOD UCDS** 12/15/71
skn-rbt 10 mg/24H open SEV AMIHBC 4,119,51
eye-rbt 675 µg SEV AJOPAA 29,1363,46
eye-rbt 750 µg/24H SEV 85JCAE -,440,86
mmo-sat 33 µg/plate ENMUDM 5(Suppl 1),3,83
mma-sat 1 mg/plate ENMUDM 5(Suppl 1),3,83
orl-mus TDLo:3200 mg/kg (female 6-13D post):REP TCMUD8 7,29,87
ihl-hmn TCLo:200 ppm:PNS AMIHBC 9,223,54
orl-rat LD50:500 mg/kg 85GMAT -,66,82
ihl-rat LCLo:4000 ppm/8H AMIHBC 4,119,51
ipr-rat LD50:76 mg/kg TXAPA9 21,454,72
scu-rat LD50:300 mg/kg 85GMAT -,66,82
ihl-mus LC50:300 mg/m³ 85GMAT -,66,82
ipr-mus LD50:200 mg/kg CHTPBA 4,136,69
scu-mus LD50:424 mg/kg ARZNAD 4,649,54
ivn-dog LDLo:100 mg/kg HBAMAK 4,1295,35
skn-rbt LD50:730 mg/kg AMIHBC 4,119,51
scu-rbt LDLo:500 mg/kg HBAMAK 4,1295,35
orl-gpg LD50:470 mg/kg JIHTAB 23,259,41

CONSENSUS REPORTS: Reported in EPA TSCA Inventory. EPA Extremely Hazardous Substances List.

OSHA PEL: TWA 10 ppm
ACGIH TLV: TWA 10 ppm
DFG MAK: 10 ppm (25 mg/m³)
DOT CLASSIFICATION: 8; *Label:* Corrosive, Flammable Liquid

SAFETY PROFILE: A poison in humans by inhalation. Experimental poison by inhalation, intraperitoneal, subcutaneous, and intravenous routes. Moderately toxic by ingestion and skin contact. Experimental reproductive effects. Corrosive. A severe skin and eye irritant. An allergen and sensitizer. Mutation data reported. Flammable liquid when exposed to heat, flame, or oxidizers. Can react violently with acetic acid, acetic anhydride, acrolein, acrylic acid, acrylonitrile, allyl chloride, CS_2, chlorosulfonic acid, epichlorohydrin, ethylene chlorohydrin, HCl, mesityl oxide, HNO_3, oleum, $AgClO_4$, H_2SO_4, β-propiolactone, or vinyl acetate. To fight fire, use CO_2, dry chemical, alcohol foam. When heated to decomposition it emits toxic fumes of NO_x and NH_3. See also AMINES.

For occupational chemical analysis use OSHA: #ID-60 or NIOSH: Ethylenediamine 2540.

EEA600 CAS:2224-15-9 HR: 2
1,2-ETHANEDIOL DIGLYCIDYL ETHER
mf: $C_8H_{14}O_4$ mw: 174.22

SYNS: 1,2-BIS(GLYCIDYLOXY)ETHANE □ DIGLYCIDYLETHYLENE GLYCOL □ 1,2-DIGLYCIDYLOXYETHANE □ ETHANE, 1,2-BIS(2,3-EPOXYPROPOXY)- □ 2,2'-(1,2-ETHANEDIYLBIS(OXYMETHYLENE))BISOXIRANE □ ETHYLENE DIGLYCIDYL ETHER □ ETHYLENE GLYCOL DIGLYCIDYL ETHER □ ETHYLENGLYKOLDIGLYCIDYLETHER □ GLYCOL DIGLYCIDYL ETHER □ OXIRANE, 2,2'-(1,2-ETHANEDIYLBIS (OXYMETHYLENE))BIS-(9CI)

TOXICITY DATA WITH REFERENCE
mmo-sat 300 nmol/plate MUREAV 231,205,90
oth-esc 1 mmol/L MUREAV 231,205,90
sce-ham:lng 6250 nmol/L MUREAV 249,55,91
orl-mus LD50:460 mg/kg 85JCAE -,775,86

CONSENSUS REPORTS: Reported in EPA TSCA Inventory.

SAFETY PROFILE: Moderately toxic by ingestion. Mutation data reported. When heated to decomposition it emits acrid smoke and irritating vapors.

EEB000 CAS:540-63-6 HR: 3
1,2-ETHANEDITHIOL
mf: $C_2H_6S_2$ mw: 94.20

PROP: A liquid. D: 1.124, bp: 146°.

SYNS: 1,2-DIMERCAPTOETHANE □ DITHIOETHYLENEGLYCOL □ DITHIOGLYCOL □ ETHYLENE DIMERCAPTAN □ α-ETHYLENE DIMERCAPTAN □ ETHYLENE DITHIOGLYCOL □ ETHYLENEDITHIOL □ ETHYL HYDROPERSULFIDE

TOXICITY DATA WITH REFERENCE
orl-mus LD50:342 mg/kg DCTODJ 3,249,80
ipr-mus LD50:50 mg/kg EJMCA5 17,235,82
ivn-mus LD50:56,200 μg/kg CSLNX* NX#02101

CONSENSUS REPORTS: Reported in EPA TSCA Inventory.

SAFETY PROFILE: Poison by ingestion, intraperitoneal, and intravenous routes. When heated to decomposition it emits very toxic fumes of SO_x. See also MERCAPTANS.

EEB100 CAS:2001-94-7 HR: 1
N,N′-1,2-ETHANEDIYLBIS(N-(CARBOXYMETHYL)GLYCINE), DIPOTASSIUM SALT
mf: $C_{10}H_{14}N_2O_8$•2K mw: 368.46

SYNS: ACETIC ACID, (ETHYLENEDINITRILO)TETRA-, DIPOTASSIUM SALT □ DIPOTASSIUM ETHYLENEDIAMINETETRAACETATE □ (ETHYLENEDINITRILO)TETRAACETATE DIPOTASSIUM SALT

TOXICITY DATA WITH REFERENCE
skn-rbt 500 mg MLD FCTOD7 20,563,82
eye-rbt 100 mg FCTOD7 20,573,82
eye-rbt 100 mg/30S RNS MLD FCTOD7 20,573,82

CONSENSUS REPORTS: Reported in EPA TSCA Inventory.

SAFETY PROFILE: A skin and eye irritant. When heated to decomposition it emits toxic vapors of NO_x.

EEB500 HR: 3
ETHANE HEXAMERCARBIDE
mf: $C_2Hg_6O_2(OH)_2$ mw: 1269.56

PROP: Yellowish-white powder. Vap d: 44.6, explodes @ 230°. Insol in water.

CONSENSUS REPORTS: Mercury and its compounds are on the Community Right-To-Know List.

SAFETY PROFILE: A poison. A dangerous explosion hazard. Explodes when shocked or heated to 230°C. Incompatible with oxidizing materials. When heated to decomposition or on contact with acid or acid fumes, below 230°C it emits highly toxic fumes of Hg. See also MERCURY COMPOUNDS, ORGANIC.

EEC000 CAS:594-44-5 HR: 3
ETHANESULFONYL CHLORIDE
mf: $C_2H_5ClO_2S$ mw: 128.58

PROP: Oil. D: 1.357, bp: 177.5°. Sltly sol (decomp) in water and alc; very sol in ether and CH_2Cl_2.

SYNS: ETHYLSULFOCHLORIDE □ TL 77

TOXICITY DATA WITH REFERENCE
ihl-mus LCLo:1220 mg/m³/10M NDRC** NDCrc-132,Aug,42

CONSENSUS REPORTS: Reported in EPA TSCA Inventory.

SAFETY PROFILE: Moderately toxic by inhalation. When heated to decomposition it emits very toxic fumes of Cl^- and SO_x.

EEC600 CAS:141-43-5 HR: 3
ETHANOLAMINE
DOT: UN 2491
mf: C_2H_7NO mw: 61.10

PROP: Colorless, viscous, hygroscopic liquid with ammonia-like odor. Bp: 170.5°, fp: 10.5°, flash p: 200°F (OC), d: 1.012 @ 25°/4°, vap press: 6 mm @ 60°, vap d: 2.11. Misc in water and alc; sltly sol in benzene; sol in chloroform.

SYNS: AETHANOLAMIN (GERMAN) □ 2-AMINOAETHANOL (GERMAN) □ 2-AMINOETANOLO (ITALIAN) □ 2-AMINOETHANOL (MAK) □ β-AMINOETHYL ALCOHOL □ COLAMINE □ ETANOLAMINA (ITALIAN) □ β-ETHANOLAMINE □ ETHANOLAMINE, solution (DOT) □ ETHYLOLAMINE □ GLYCINOL □ β-HYDROXYETHYLAMINE □ 2-HYDROXYETHYLAMINE □ MEA □ MONOAETHANOLAMIN (GERMAN) □ MONOETHANOLAMINE □ OLAMINE □ THIOFACO M-50 □ USAF EK-1597

TOXICITY DATA WITH REFERENCE
skn-rbt 505 mg open MOD UCDS** 1/13/73
eye-rbt 763 μg SEV AJOPAA 29,1363,46
cyt-hmn:lyms 100 μmol/L BMAOA3 39,422,86
sce-hmn:lyms 1 mmol/L CYGEDX 21(6),29,87
orl-rat TDLo:500 mg/kg (female 6-15D post):TER TCMUD8 6,403,86
orl-rat LD50:1720 mg/kg TXAPA9 42,417,77
ipr-rat LD50:67 mg/kg EVSSAV 2,289,68
scu-rat LD50:1500 mg/kg GTPZAB 23(9),55,79
ivn-rat LD50:225 mg/kg KBMEAL (4),44,68
ims-rat LD50:1750 mg/kg GTPZAB 23(9),55,79
orl-mus LD50:700 mg/kg TPKVAL 4,81,62
ipr-mus LD50:50 mg/kg NTIS** AD277-689
skn-rbt LD50:1000 mg/kg UCDS** 1/13/72

CONSENSUS REPORTS: Reported in EPA TSCA Inventory.

OSHA PEL: TWA 3 ppm; STEL 6 ppm
ACGIH TLV: TWA 3 ppm; STEL 6 ppm
DFG MAK: 3 ppm (8 mg/m³)
DOT CLASSIFICATION: 8; *Label:* Corrosive

SAFETY PROFILE: Poison by intraperitoneal route. Moderately toxic by ingestion, skin contact, subcutaneous, intravenous, and intramuscular routes. A corrosive irritant to skin, eyes, and mucous membranes. Human mutation data reported. Flammable when exposed to heat or flame. A powerful base. Reacts violently with acetic acid; acetic anhydride; acrolein; acrylic acid; acrylonitrile; cellulose; chlorosulfonic acid; epichlorohydrin; HCl; HF; mesityl oxide; HNO_3; oleum; H_2SO_4; β-propiolactone; vinyl acetate. To fight fire, use foam, alcohol foam, dry chemical. When heated to decomposition it emits toxic fumes of NO_x. See also AMINES.

For occupational chemical analysis use NIOSH: Aminoethanol Compounds 2007; Aminoethanol Compounds II 3509.

EEC700 CAS:2002-24-6 HR: 1
ETHANOLAMINE HYDROCHLORIDE
mf: C_2H_7NO•ClH mw: 97.56

SYNS: β-AMINOETHANOL HYDROCHLORIDE □ 2-AMINOETHANOL HYDROCHLORIDE □ COLAMINE HYDROCHLORIDE □ ETHANOL-

AMINE CHLORIDE □ ETHANOL, 2-AMINO-, HYDROCHLORIDE □ MEA HYDROCHLORIDE □ MONOETHANOLAMINE HYDROCHLORIDE

TOXICITY DATA WITH REFERENCE
scu-mus LD50:4053 mg/kg ARZNAD 4,649,54

CONSENSUS REPORTS: Reported in EPA TSCA Inventory.

SAFETY PROFILE: Mildly toxic by subcutaneous route. When heated to decomposition it emits toxic vapors of NO_x, HCl, and Cl^-.

EED000 CAS:1071-23-4 HR: 2
ETHANOLAMINE PHOSPHATE
mf: $C_2H_8NO_4P$ mw: 141.08

SYNS: o-PHOSPHOETHANOLAMINE □ PHOSPHORYLETHANOLAMINE

TOXICITY DATA WITH REFERENCE
ivn-mus LD50:639 mg/kg RPOBAR 2,316,70

CONSENSUS REPORTS: Reported in EPA TSCA Inventory.

SAFETY PROFILE: Moderately toxic by intravenous route. When heated to decomposition it emits very toxic fumes of PO_x and NO_x.

EED600 CAS:23471-13-8 HR: 3
ETHANOLMERCURY BROMIDE
mf: C_2H_5BrHgO mw: 325.57

SYNS: BROMO(2-HYDROXYETHYL)MERCURY □ 2-(BROMOMERCURI)ETHANOL

TOXICITY DATA WITH REFERENCE
orl-rat LD50:16 mg/kg PCJOAU 11,918,77
orl-mus LD50:16,500 μg/kg PCJOAU 11,918,77
ivn-mus LD50:56 mg/kg CSLNX* NX#05830

CONSENSUS REPORTS: Mercury and its compounds are on the Community Right-To-Know List.

OSHA PEL: TWA 0.01 mg(Hg)/m³; STEL 0.03 mg/m³ (skin)
ACGIH TLV: TWA 0.01 mg(Hg)/m³; STEL 0.03 mg(Hg)/m³
NIOSH REL: (Mercury, Organo): TWA 0.01 mg/m³; STEL 0.03 mg/m³ (skin)

SAFETY PROFILE: Poison by ingestion and intravenous routes. When heated to decomposition it emits toxic fumes of Br^- and Hg. See also MERCURY COMPOUNDS, ORGANIC and BROMIDES.

EEE000 CAS:20398-06-5 HR: 3
ETHANOL THALLIUM(1+) SALT
mf: $C_2H_6O{\cdot}Tl$ mw: 250.45

PROP: Sol in org solvs.

SYN: ETHYL ALCOHOL THALLIUM (I)

TOXICITY DATA WITH REFERENCE
orl-rat LDLo:50 mg/kg NCNSA6 5,43,53

CONSENSUS REPORTS: Reported in EPA TSCA Inventory.

OSHA PEL: TWA 0.1 mg(Tl)/m³ (skin)
ACGIH TLV: TWA 0.1 mg(Tl)/m³ (skin)

SAFETY PROFILE: Poison by ingestion. When heated to decomposition it emits toxic fumes of Tl. See also THALLIUM COMPOUNDS.

EEE025 CAS:52917-87-0 HR: 3
1-ETHANONE, 1-(3-BROMOADAMANTYL)-2-DIAZO-
mf: $C_{12}H_{15}BrN_2O$ mw: 283.20

SYNS: 3-BROMOADAMANTYL DIAZOMETHYL KETONE □ KETONE, 3-BROMO-1-ADAMANTYL DIAZOMETHYL

TOXICITY DATA WITH REFERENCE
ipr-mus LD50:1500 mg/kg PCJOAU 10,454,76

DOT CLASSIFICATION: 3; *Label:* Flammable Liquid

SAFETY PROFILE: Moderately toxic by intraperitoneal route. When heated to decomposition it emits toxic vapors of NO_x and Br^-.

EEE500 CAS:670-54-2 HR: 3
ETHENETETRACARBONITRILE
mf: C_6N_4 mw: 128.10

$(NC)_2C{=}C(CN)_2$

PROP: Colorless crystals from chlorobenzene. Subl at >120°, mp: 198–200°, bp: 223°.

SYNS: $\Delta^{1,1'}$-BIMALONONITRILE □ TETRACYANOETHENE □ 1,1,2,2-TETRACYANOETHENE □ TETRACYANOETHYLENE □ 1,1,2,2-TETRACYANOETHYLENE □ TETRAKYANETHYLEN

TOXICITY DATA WITH REFERENCE
orl-mus LD50:29 mg/kg KHZDAN 9,50,66
ivn-mus LD50:4500 μg/kg CSLNX* NX#01707

CONSENSUS REPORTS: Reported in EPA TSCA Inventory. Cyanide and its compounds are on the Community Right-To-Know List.

SAFETY PROFILE: Poison by ingestion and intravenous routes. Violent reaction with 1-methylsilacyclopenta-2,4-diene at 150°C. When heated to decomposition it emits toxic fumes of NO_x and CN^-. See also NITRILES.

EEE200 CAS:765-05-9 HR: 2
1-(ETHENYLOXY)DECANE
mf: $C_{12}H_{24}O$ mw: 184.36

SYNS: DECANE, 1-(ETHENYLOXY)- □ DECYL VINYL ETHER □ ETHER, DECYL VINYL (6CI,7CI,8CI)

TOXICITY DATA WITH REFERENCE
orl-rat LD50:3940 mg/kg FCTOD7 30,17S,92
skn-rbt LD50:>5 g/kg FCTOD7 30,17S,92

CONSENSUS REPORTS: Reported in EPA TSCA Inventory.

SAFETY PROFILE: Moderately toxic by ingestion. Low toxicity by skin contact. When heated to decomposition it emits acrid smoke and irritating vapors.

EEF000 **CAS:17088-21-0** **HR: 2**
1-ETHENYL PYRENE
mf: $C_{18}H_{12}$ mw: 228.2

SYNS: 1-VINYLPYRENE □ 3-VINYLPYRENE

TOXICITY DATA WITH REFERENCE
mma-sat 10 nmol/plate CNREA8 40,642,80
skn-mus TDLo:3651 μg/kg:ETA CNREA8 40,642,80

SAFETY PROFILE: Questionable carcinogen with experimental tumorigenic data. Mutation data reported. When heated to decomposition it emits acrid smoke and irritating fumes.

EEF500 **CAS:73529-25-6** **HR: 2**
4-ETHENYL PYRENE
mf: $C_{18}H_{12}$ mw: 228.2

SYN: 4-VINYLPYRENE

TOXICITY DATA WITH REFERENCE
skn-mus TDLo:3651 μg/kg:ETA CNREA8 40,642,80

SAFETY PROFILE: Questionable carcinogen with experimental tumorigenic data. When heated to decomposition it emits acrid smoke and irritating fumes.

EEG000 **CAS:88-12-0** **HR: 3**
1-ETHENYL-2-PYRROLIDINONE
mf: C_6H_9NO mw: 111.16

PROP: Colorless liquid, water-sol. Bp: 148° @ 100 mm, fp: 13.5°, flash p: 209°F (OC), d: 1.04 @ 25°, autoign temp: 687°F, vap d: 3.8, fire p: 213°F.

SYNS: VINYLBUTYROLACTAM □ N-VINYLPYRROLIDINONE □ N-VINYL-2-PYRROLIDINONE □ 1-VINYL-2-PYRROLIDINONE □ VINYLPYRROLIDONE □ N-VINYLPYRROLIDONE □ N-VINYL-2-PYRROLIDONE □ 1-VINYL-2-PYRROLIDONE □ V-PYROL

TOXICITY DATA WITH REFERENCE
eye-rbt 100 mg SEV BurLW# 02NOV78
orl-rat LD50:1470 mg/kg BurLW# 02NOV78
ihl-rat LC50:3200 mg/m³ BurLW# 02NOV78
skn-rbt LD50:560 mg/kg BurLW# 02NOV78

CONSENSUS REPORTS: IARC Cancer Review: Group 3 IMEMDT 7,56,87. Reported in EPA TSCA Inventory.

DFG MAK: Confirmed Animal Carcinogen, Suspected Human Carcinogen.

SAFETY PROFILE: Confirmed carcinogen. Moderately toxic by ingestion, inhalation, and skin contact. A severe eye irritant. Probably irritating and narcotic in high concentrations. Combustible when exposed to heat or flame; can react vigorously with oxidizing materials. To fight fire, use alcohol foam, CO_2, dry chemical. When heated to decomposition it emits highly toxic fumes of NO_x.

EEG500 **HR: 3**
ETHERS

PROP: Organic compounds in which an oxygen atom is interposed between two carbon atoms in the structure of the molecule.

SAFETY PROFILE: The simpler ethers such as ethyl ether, isopropyl ether, etc., are powerful narcotics that in large doses can cause death. The danger from ethers is usually acute and seldom chronic. Aftereffects to ether intoxication are uncommon although continued exposure to small concentrations (not enough to cause an overt symptom) has been known to cause loss of appetite, excessive thirst, and fatigue.

The most common ethers, such as ethyl, methyl, and diisopropyl, are particularly dangerous fire and explosion hazards when exposed to heat, flame, or sparks. They can react violently with strong oxidizers. Many plant and laboratory fires and explosions have resulted from their high flammability and tendency to form explosive peroxides. The common ethers are easily ignited and have low flash points. The diethyl, ethyl tert-butyl, ethyl tert-pentyl and diisopropyl ethers are very hazardous. Methyl tert-alkyl ethers are relatively safe. Besides the risk of explosion from air mixtures of ether vapors, ethers tend to form peroxides upon standing. For some ethers peroxide levels do not reach dangerous concentrations (e.g., diethyl ether; ethyl vinyl ether; tetrahydrofuran; p-dioxane; 1,1-diethoxyethane; and the dimethyl ethers of ethylene glycol). When ethers containing peroxides are heated they can detonate. It is necessary to control smoking, open flames, or even the use of hot plates in areas where low-molecular-weight ethers are apt to reach 1% concentration or more in air. Only electrical equipment of explosion-proof type (Group C classification) is permitted to be operated in ether areas. Ethers should not be stored near powerful oxidizers or in areas of high fire hazard. They should be kept cool and the containers electrically grounded to avoid sparks.

Dangerous; shock or heat can cause gaseous ethers to escape from their containers and create flammable or even explosive conditions. Incompatible with oxidizing materials, BI_3. See also ETHYL ETHER.

EEH000 **CAS:126-52-3** **HR: 3**
ETHINAMATE
mf: $C_9H_{13}NO_2$ mw: 167.23

PROP: Rods or needles from cyclohexane. Mp: 96–98°, bp: 118–122° @ 3 mm. Very sol in EtOH, sltly sol in hexane; very sltly sol in H_2O.

SYNS: CARBAMATE de l'ETHINYLCYCLOHEXANOL (FRENCH) □ 1-ETHINYLCYCLOHEXYL CARBAMATE □ 1-ETHINYLCYCLOHEXYL CARBONATE □ 1-ETHYNYLCYCLOHEXANOL CARBAMATE □ 1-ETHYNYLCYCLOHEXYL CARBAMATE □ 1-ETHYNYLCYCLOHEXYL ESTER CARBAMIC ACID □ ETINAMATE □ USAF EL-42 □ VALAMINA □ VALAMINETTEN □ VALMID □ VALMIDATE □ VOLAMIN

TOXICITY DATA WITH REFERENCE
ipr-mus TDLo:400 mg/kg (10D preg):TER CAJPBD 3,2,63
orl-hmn LDLo:57 mg/kg TOIZAG 7,513,60
orl-rat LD50:331 mg/kg JAPMA8 45,40,56
scu-rat LD50:390 mg/kg ARZNAD 4,477,54
ivn-rat LD50:157 mg/kg JAPMA8 45,40,56
orl-mus LD50:490 mg/kg JAPMA8 45,40,56
ipr-mus LD50:300 mg/kg NTIS** AD277-689
ivn-mus LD50:108 mg/kg JAPMA8 45,40,56
orl-dog LD50:190 mg/kg DRUGAY 6,111,82

CONSENSUS REPORTS: Reported in EPA TSCA Inventory.

SAFETY PROFILE: A deadly human poison. Experimental poison by ingestion, intravenous, subcutaneous, and intraperitoneal routes. An experimental teratogen. When heated to decomposition it emits toxic fumes of NO_x. See also CARBAMATES.

EEH500 CAS:57-63-6 HR: 3
ETHINYL ESTRADIOL
mf: $C_{29}H_{24}O_2$ mw: 296.44

PROP: Crystals from MeOH (aq). Mp: 145–146°.

SYNS: 3,17-β-DIHYDROXY-17-α-ETHYNYL-1,3,5(10)-ESTRATRIENE □ 3,17-β-DIHYDROXY-17 α ETHYNYL-1,3,5(10)-OESTRATRIENE □ ESTROGEN □ 17-α-ETHINYL-3,17-DIHYDROXY-$\Delta^{1,3,5}$-ESTRATRIENE □ 17-α-ETHINYL-3,17-DIHYDROXY-$\Delta^{1,3,5}$-OESTRATRIENE □ 17-ETHINYLESTRADIOL □ 17-ETHINYL-3,17-ESTRADIOL □ 17-α-ETHINYLESTRADIOL □ 17-α-ETHINYL-17-β-ESTRADIOL □ 17-α-ETHINYLESTRA-1,3,5(10)-TRIENE-3,17-β-DIOL □ ETHINYLESTRIOL □ ETHINYLOESTRADIOL □ 17-ETHINYL-3,17-OESTRADIOL □ ETHINYL-OESTRANOL □ 17-α-ETHINYLOESTRA-1,3,5(10)-TRIENE-3,17-β-DIOL □ 17-α-ETHINYL $\Delta^{1,3,5(10)}$OESTRATRIENE-3,17-β-DIOL □ ETHINYLOESTRIOL □ 17-ETHYNYL-3,17-DIHYDROXY-1,3,5-OESTRATRIENE □ ETHYNYLESTRADIOL □ 17-α-ETHYNYLESTRADIOL □ 17-α-ETHYNYLESTRADIOL-17-β □ 17-α-ETHYNYL-1,3,5(10)-ESTRATRIENE-3,17-β-DIOL □ 17-α-ETHYNYLESTRA-1,3,5(10)-TRIENE-3,17-β-DIOL □ ETHYNYLOESTRADIOL □ 17-ETHYNYLOESTRADIOL □ 17-α-ETHYNYLOESTRADIOL □ 17-α-ETHYNYL-17-β-OESTRADIOL □ 17-α-ETHYNYLOESTRADIOL-17-β □ 17-ETHYNYLOESTRA-1,3,5(10)-TRIENE-3,17-β-DIOL □ 17-α-ETHYNYL-1,3,5-OESTRATRIENE-3,17-β-DIOL □ 17-α-ETHYNYL-1,3,5(10)-OESTRATRIENE 3,17-β-DIOL □ 17-α-ETHYNYLOESTRA-1,3,5(10)-TRIENE-3,17-β-DIOL □ 19-NOR-17-α-PREGNA-1,3,5(10)-TRIEN-2-YNE-3,17-DIOL □ (17-α)-19-NORPREGNA-1,3,5(10)-TRIEN-20-YNE-3,17,DIOL

TOXICITY DATA WITH REFERENCE
mmo-ssp 50 μg/plate EGJBAY 20,29,79
dni-hmn:lym 50 μmol/L PSEBAA 146,101,74
orl-wmn TDLo:500 μg/kg (female 5D pre):REP CCPTAY 14,375,76
orl-rat TDLo:560 μg/kg (male 14D pre):REP ANYAA9 71,500,58
orl-mus TDLo:70 μg/kg (female 11-17D post):REP AJOGAH 127,832,77
orl-mus TDLo:200 μg/kg (female 8D post):TER TJADAB 23,233,81
unr-rbt TDLo:19 μg/kg (female 1-19D post):TER CCPTAY 2,85,70
orl-mus TDLo:40 μg/kg (female 3-4D post):REP APTOA6 26(1),1,68
orl-rat TDLo:55 μg/kg (female 11D pre):REP PSEBAA 116,343,64
orl-rat TDLo:250 μg/kg (female 1-4D post):REP ACENA7 41,351,62
orl-ham TDLo:20 mg/kg (female 4D pre):REP ACENA7 70,582,72
orl-rat TDLo:4500 ng/kg (female 1-9D post):REP CCPTAY 3,347,71
orl-ham TDLo:90 mg/kg (female 1-9D post):REP CCPTAY 3,347,71
scu-rat TDLo:5 μg/kg (female 1D pre):REP ACENA7 49,193,65
orl-rat TDLo:860 ng/kg (female 1D pre):REP JMCMAR 23,329,80
scu-rat TDLo:1 mg/kg (male 10D pre):REP INJFA3 8,589,63
orl-mus TDLo:200 μg/kg (female 8D post):TER TJADAB 23,233,81
orl-rat TDLo:750 μg/kg (female 15-17D post):TER IJEBA6 21,591,83
scu-rat TDLo:600 μg/kg (female 17-20D post):TER ACEDAB 240,104,81
orl-wmn TDLo:2738 μg/kg/10Y-I:CAR MJAUAJ 1,473,81
orl-rat TDLo:6 μg/kg/2Y-C:ETA JTEHD6 6,885,80
imp-rat TDLo:5 mg/kg:CAR JJIND8 67,455,81
imp-ham TDLo:621 mg/kg:NEO CNREA8 12,274,52
orl-rat TD:245 mg/kg/35W-C:ETA EMSUA8 22,28,64
imp-ham TD:800 mg/kg/34W-I:ETA CNREA8 43,5200,83
orl-wmn TD:102 mg/kg/5Y-I:CAR LANCAO 1,273,80
imp-rat TD:5 mg/kg:ETA CTRRDO 63,1180,79
orl-wmn TDLo:21 mg/kg/21D-I:GIN LANCAO 1,1479,73
orl-rat LD50:1200 mg/kg DRUGAY-,184,90
ipr-rat LD50:471 mg/kg YACHDS 18,2583,90
orl-mus LD50:1737 mg/kg TXAPA9 18,185,71
ipr-mus LD50:250 mg/kg YACHDS 18,2583,90
orl-mus LD50:1737 mg/kg TXAPA9 18,185,71

CONSENSUS REPORTS: NTP 7th Annual Report on Carcinogens. IARC Cancer Review: Human Limited Evidence IMEMDT 21,233,79; Animal Sufficient Evidence IMEMDT 6,77,74; IMEMDT 21,233,79. Reported in EPA TSCA Inventory.

SAFETY PROFILE: Confirmed carcinogen with experimental carcinogenic, tumorigenic, and neoplastigenic data. Poison by intraperitoneal route. Moderately toxic by ingestion. Human systemic effects by ingestion: glandular effects. An experimental teratogen. Experimental reproductive effects. Human mutation data reported. When heated to decomposition it emits acrid smoke and irritating fumes. See also ESTRADIOL.

EEH520 CAS:8015-12-1 HR: 3
ETHINYL ESTRADIOL and NORETHINDRONE ACETATE
mf: $C_{22}H_{28}O_3 \cdot C_{20}H_{24}O_2$ mw: 636.94

SYNS: ANOVLAR 21 □ CONTROVLAR □ ETHINYL OESTRADIOL mixed with NORETHISTERONE ACETATE □ GYN-ANOVLAR □ GYNONLAR 21 □ MINORLAR □ MINOVLAR □ NORETHINDRONE ACETATE and ETHINYLESTRADIOL □ NORETHISTERONE ACETATE mixed with ETHINYL OESTRADIOL □ NORLESTRIN □ PRIMODOS

TOXICITY DATA WITH REFERENCE
orl-wmn TDLo:1620 μg/kg (20D pre):REP JRPFA4 4,229,62
orl-wmn TDLo:214 μg/kg (20D pre):REP CCPTAY 25,463,82
orl-wmn TDLo:420 μg/kg (20D pre):REP CCPTAY 26,23,82
orl-wmn TDLo:1176 μg/kg (56D post):REP IJMRAQ 62,964,74
orl-wmn TDLo:51 mg/kg (2.7 Y pre):REP AJOGAH 107,717,70
orl-wmn TDLo:401 μg/kg (5 W post):TER NATUAS 216,83,67
orl-mky TDLo:24 mg/kg (23-46D post):REP NETOD7 5,301,83

orl-mky TDLo: 14,280 μg/kg (33-46D post): TER TJADAB 21,44A,80
orl-mky TDLo: 15,300 μg/kg (21-35D post): REP TJADAB 27,215,83
orl-mky TDLo: 1864 mg/kg (20-50D post): TER TJADAB 35,119,87
orl-wmn TDLo: 23,562 μg/kg/94W-I: CAR LANCAO 1,207,78
orl-rat TDLo: 11,087 mg/kg/1Y-C: ETA KNZOAU 23,57,82
orl-mus TDLo: 69 mg/kg/69W-I: ETA PEXTAR 11,440,69
orl-wmn TD: 41 mg/kg/2Y-I: NEO MJAUAJ 2,223,78
orl-rat TD: 22,174 mg/kg/1Y-C: ETA KNZOAU 23,57,82
orl-wmn TD: 81 mg/kg/8Y-I: NEO BMJOAE 3,7,74
orl-rat TD: 11,087 mg/kg/1Y-C: ETA GANNA2 71,576,80

SAFETY PROFILE: Suspected human carcinogen producing lung and liver tumors. Experimental neoplastigenic and tumorigenic data. Human and experimental teratogenic and reproductive effects. When heated to decomposition it emits acrid smoke and irritating fumes.

EEH550 CAS:68-23-5 HR: 3
17-α-ETHINYL-5,10-ESTRENOLONE

mf: $C_{20}H_{26}O_2$ mw: 298.46

PROP: Crystals from MeOH. Mp: 180–181.5°.

SYNS: 17-ETHINYL-5(10)-ESTRAENEOLONE ☐ 17-α-ETHINYL-ESTRA (5,10)ENEOLONE ☐ 17-α-ETHYNYL-5(10)-ESTREN-17-OL-3-ONE ☐ 17-α-ETHYNYLESTR-5(10)-EN-17-β-OL-3-ONE ☐ 17-α-ETHYNYL-ESTR-5 (10)-EN-3-ON-17-β-OL ☐ 17-α-ETHYNYL-17-HYDROXYESTR-5(10)-EN-3-ONE ☐ 17-α-ETHYNYL-17-HYDROXY-5(10)-ESTREN-3-ONE ☐ 17-α-ETHYNYL-17-β-HYDROXY-5(10)-ESTREN-3-ONE ☐ 17-α-ETHYNYL-17-β-HYDROXYESTR-5(10)-EN-3-ONE ☐ 17-α-ETHINYL-17-β-HYDROXY-$\Delta^{5(10)}$-ESTREN-3-ONE ☐ 17-α-ETHYNYL-17-β-HYDROXY-$\Delta^{5(10)}$-ESTREN-3-ONE ☐ 17-α-ETHYNYL-17-β-HYDROXY-3-OXO-$\Delta^{5(10)}$-ESTRENE ☐ 17-α-ETHYNYL-19-NOR-5(10)-ANDROSTEN-17-β-OL-3-ONE ☐ 17-α-ETHINYL-$\Delta^{5,10\ 19}$-NORTESTOSTERONE ☐ 17-β-HYDROXY-17-α-ETHINYL-5(10)-ESTREN-3-ONE ☐ 17-HYDROXY-19-NOR-17-α-PREGN-5(10)-EN-20-YN-3-ONE ☐ (17-α)-17-HYDROXY-19-NORPREGN-5(10)-EN-20-YN-3-ONE ☐ 17-HYDROXY(17-α)-19-NORPREGN-5(10)-EN-20-YN-3-ONE ☐ LYNESTROL ☐ NORETHINODREL ☐ 19-NOR-ETHINYL-5,10-TESTOSTERONE ☐ NORETHINYNODREL ☐ NORETHYNODRAL ☐ NORETHYNODREL ☐ 19-NORETHYNODREL ☐ NSC-15432 ☐ SC-4642

TOXICITY DATA WITH REFERENCE

dni-hmn: lyms 50 μmol/L PSEBAA 146,401,74
dns-rat: lvr 100 μmol/L CRNGDP 6,1201,85
oth-mus-par 16 μg/kg AJOGAH 120,390,74
cyt-mus: oth 10 mg/L AJOGAH 120,390,74
orl-wmn TDLo: 1 mg/kg (20D pre): REP FESTAS 16,158,65
orl-wmn TDLo: 8 mg/kg (16 W pre): REP AJOGAH 75,82,58
orl-wmn TDLo: 8 mg/kg (20D pre): REP PRSMA4 55,861,62
orl-rat TDLo: 150 mg/kg (10-12D post): REP NATUAS 197,308,63
orl-mus TDLo: 80 mg/kg (8-15D post): TER PSEBAA 121,455,66
orl-rat TDLo: 10 mg/kg (2-3D post): REP JOENAK 33,241,65
scu-rat TDLo: 66,400 ng/kg (10-17D post): REP IJEBA6 5,14,67
par-ham TDLo: 2400 μg/kg (3D pre): REP JRPFA4 37,269,74
orl-rbt TDLo: 50 μg/kg (1D pre): REP 13OPAL -,37,61
orl-rat TDLo: 150 mg/kg (10-12D post): REP NATUAS 197,308,63
orl-rat TDLo: 10 mg/kg (2-3D post): REP STEDAM 4,801,64
scu-rat TDLo: 60 mg/kg (17-20D post): REP AVBIB9 13,71,74
par-rat TDLo: 9600 μg/kg (48D pre): REP JOENAK 24,497,62
orl-rat TDLo: 25 mg/kg (1D pre): REP ENDOAO 75,359,64
scu-rat TDLo: 1400 μg/kg (14D pre): REP CCPTAY 5,57,72
orl-rat TDLo: 2100 μg/kg (14D pre): REP FESTAS 24,284,73
orl-rat TDLo: 12 mg/ (30D male pre): REP IJEBA6 9,132,71
orl-mus TDLo: 450 μg/kg (8-10D post): TER TJADAB 3,339,70
orl-rat TDLo: 100 mg/kg (17-20D post): TER ECJPAE 24,77,77
orl-rat TDLo: 60 mg/kg/17W-C: ETA PAACA3 21,76,80
scu-mus TDLo: 135 mg/kg/83W-C: ETA BJCAAI 21,153,67
par-mus TDLo: 120 mg/kg/78W-C: ETA PMDFA9 7,21,72
imp-mus TDLo: 119 mg/kg/77W-C: ETA NATUAS 212,686,66

CONSENSUS REPORTS: IARC Cancer Review: Animal Limited Evidence IMEMDT 21,461,79; Animal Sufficient Evidence IMEMDT 6,191,74.

SAFETY PROFILE: Suspected carcinogen with experimental tumorigenic data. Human and experimental reproductive effects. Mutation data reported. When heated to decomposition it emits acrid smoke and irritating fumes.

EEH575 CAS:8064-76-4 HR: 2
ETHINYLOESTRADIOL mixed with LYNOESTRENOL

mf: $C_{20}H_{28}O{\bullet}C_{20}H_{24}O_2$ mw: 580.92

SYNS: Ba 49249 ☐ C 49249Ba ☐ ETHYNYLESTRADIOL-LYNESTRENOL mixture ☐ FISIOQUENS ☐ FYSIOQUENS ☐ LYNOESTRENOL mixed with ETHINYLOESTRADIOL ☐ MINILYN ☐ OVOSTAT ☐ OVOSTAT 1375 ☐ OVOSTAT E ☐ YERMONIL

TOXICITY DATA WITH REFERENCE

orl-wmn TDLo: 4284 μg/kg (84D pre): REP BMJOAE 2,487,77
orl-wmn TDLo: 77,112 μg/kg/6Y-I: CAR LANCAO 1,273,80

SAFETY PROFILE: Questionable human carcinogen producing liver tumors. Human reproductive effects. When heated to decomposition it emits acrid smoke and irritating fumes.

EEH600 CAS:563-12-2 HR: 3
ETHION

mf: $C_9H_{22}O_4P_2S_4$ mw: 384.49

PROP: Oily liquid. Mp: −13°, d: 1.31 @ 20°/4°. Sol in Me_2CO, C_6H_6, $CHCl_3$, EtOH, and Et_2O; sltly sol in H_2O; mod sol in ligroin.

SYNS: AC 3422 ☐ BIS(S-(DIETHOXYPHOSPHINOTHIOYL)MERCAP-

TO)METHANE □ BLADAN □ DIETHION □ EMBATHION □ ENT 24, 105 □ ETHANOX □ ETHIOL □ ETHODAN □ ETHYL METHYLENE PHOSPHORODITHIOATE □ FMC-1240 □ FOSFONO 50 □HYLE-MOX □ ITOPAZ □ KWIT □ METHANEDITHIOL-S,S-DIESTER with O,O-DIETHYL ESTER PHOSPHORODITHIOIC ACID □ METHYLEEN-S,S'-BIS(O,O-DIETHYL-DITHIOFOSFAAT) (DUTCH) □ S,S'-METHYLEN-BIS(O,O-DIAETHYL-DITHIOPHOSPHAT) (GERMAN) □ METHYLENE-S,S'-BIS(O,O-DIAETHYL-DITHIOPHOSPHAT) (GERMAN) □ S,S'-METHYLENE O,O,O',O'-TETRAETHYL PHOSPHORODITHIOATE □ NIAGARA 1240 □ NIALATE □ PHOSPHOTOX E □ RHODIACIDE □ RHODOCIDE □ RODOCID □ RP 8167 □ SOPRATHION □ O,O,O',O'-TETRAAETHYL-BIS(DITHIOPHOSPHAT) (GERMAN) □ O,O,O',O'-TETRAETHYL S,S'-METHYLENEBISPHOSPHORDITHIOATE □ O,O,O',O'-TETRAETHYL-S,S'-METHYLENEBISPHOSPHORODITHIOATE □ TETRAETHYL S,S'-METHYLENE BIS(PHOSPHOROTHIOLOTHIONATE) □ O,O,O',O'-TETRAETHYL S,S'-METHYLENE DI(PHOSPHORODITHIOATE) □ VEGFRU FOSMITE

TOXICITY DATA WITH REFERENCE

orl-inf TDLo:15,700 µg/kg:CNS,PNS,MET TXMDAX 63,71,67
orl-hmn TDLo:100 µg/kg:BIO TXAPA9 22,286,72
orl-rat LD50:13 mg/kg PHJOAV 185,361,60
ihl-rat LC50:864 mg/m³ FMCHA2-,C128,91
skn-rat LD50:62 mg/kg TXAPA9 14,515,69
ipr-rat LD50:26 mg/kg PSEBAA 129,699,68
orl-mus LD50:40 mg/kg 85DPAN -,-,71/76
ipr-mus LD50:35 mg/kg PSEBAA 129,699,68
skn-rbt LD50:890 mg/kg TobJS# 9NOV73

CONSENSUS REPORTS: EPA: Farm Worker Field Reentry FEREAC 39,16888,74. EPA Genetic Toxicology Program. EPA Extremely Hazardous Substances List.

OSHA PEL: TWA 0.4 mg/m³ (skin)
ACGIH TLV: TWA 0.4 mg/m³ (skin)

SAFETY PROFILE: Poison by ingestion, skin contact, and intraperitoneal routes. Human systemic effects by ingestion: flaccid paralysis without anesthesia, motor activity changes, fever and inhibition of cholinesterase. When heated to decomposition it emits highly toxic fumes of SO_x and PO_x. See also PARATHION.

EEI000 CAS:67-21-0 HR: 3
dl-ETHIONINE
mf: $C_6H_{13}NO_2S$ mw: 163.26

PROP: Crystals from alc (aq). Mp: 267–268° decomp @ 273°.

SYNS: AETHIONIN □ 2-AMINO-4-(ETHYLTHIO)BUTYRIC ACID □ dl-2-AMINO-4-(ETHYLTHIO)BUTYRIC ACID □ CN 8676 □ ETH □ ETHIONIN □ ETHIONINE □ (±)-ETHIONINE □ S-ETHYL-HOMOCYSTEINE □ S-ETHYL-dl-HOMOCYSTEINE □ NSC 751 □ U-1434

TOXICITY DATA WITH REFERENCE

otr-rat:lvr 7500 µmol/L/12W-C ITCSAF 20,291,84
otr-mus:emb 9200 µmol/L MUREAV 152,113,85
scu-rat TDLo:765 mg/kg (female 12-20D post):REP NATUAS 210,1271,66
ipr-rat TDLo:1 g/kg (female 9-12D post):TER JEZOAO 150,135,62
unr-rat TDLo:800 mg/kg (female 7-9D post):TER PSEBAA 105,88,60
ipr-rat TDLo:2500 mg/kg (female 10D pre):REP ARPAAQ 81,569,66
ipr-rat TDLo:4750 mg/kg (male 19D pre):REP ARPAAQ 81,569,66
ipr-rat TDLo:800 mg/kg (male 4D pre):REP TOIZAG 7,551,60
orl-rat TDLo:8400 mg/kg (male 28D pre):REP AJPAA4 32,105,56
orl-rat TDLo:7200 mg/kg/34W-C:CAR CNREA8 42,4364,82
ipr-rat TDLo:5625 mg/kg/12W-I:ETA CNREA8 42,4364,82
orl-mus TDLo:44 g/kg/2Y-C:CAR PAACA3 25,78,84
orl-rat TD:28,800 mg/kg/69W-C:CAR CNREA8 42,4364,82
orl-rat TD:24 g/kg/23W-C:ETA CNREA8 28,1703,68
orl-mus TD:32 g/kg/30W-C:ETA PAACA3 25,252,84
orl-mus TD:57 g/kg/68W-C:CAR PAACA3 25,78,84
orl-rat LD:109 g/kg/2Y-C:CAR TOPADD 13,257,85
orl-rat LD:13,440 mg/kg/32W-C:CAR NAIZAM 24,27,73
orl-rat LDLo:5000 mg/kg CNREA8 26,619,66
ipr-mus LD50:4250 mg/kg NYKZAU 62,333,66

CONSENSUS REPORTS: EPA Genetic Toxicology Program. Reported in EPA TSCA Inventory.

SAFETY PROFILE: Mildly toxic by ingestion and intraperitoneal routes. Suspected carcinogen with experimental carcinogenic and tumorigenic data. An experimental teratogen. Experimental reproductive effects. Mutation data reported. When heated to decomposition it emits toxic fumes of SO_x and NO_x.

E

EEI025 CAS:29560-58-5 HR: 3
ETHMOSINE
mf: $C_{22}H_{25}N_3O_4S•ClH$ mw: 464.02

SYNS: EN 313 □ ETHMOZINE □ ETHYL ETHER of 10-(β-MORPHOLYLPROPIONYL)PHENTHIAZINECARBAMINO ACID HYDROCHLORIDE □ ETHYL 10-(3-MORPHOLINOPROPIONYL)PHENOTHIAZINE-2-CARBAMATE HYDROCHLORIDE □ ETMOZIN

TOXICITY DATA WITH REFERENCE

orl-rat LD50:1 g/kg FATOAO 53(3),30,90
ipr-rat LD50:105 mg/kg FATOAO 48(1),60,85
ivn-rat LD50:11 mg/kg FATOAO 53(3),30,90
ipr-mus LD50:131 mg/kg RPTOAN 35,74,72
ivn-mus LD50:36 mg/kg RPTOAN 35,74,72

CONSENSUS REPORTS: Reported in EPA TSCA Inventory.

SAFETY PROFILE: Poison by intravenous and intraperitoneal routes. An antiarrhythmic and tranquilizer. When heated to decomposition it emits toxic fumes of SO_x, NO_x and HCl. See also CARBAMATES.

EEI050 HR: 2
ETHISTERONE and DIETHYLSTILBESTROL
mf: $C_{21}H_{28}O_2•C_{18}H_{20}O_2$ mw: 580.87

SYNS: DIETHYLSTILBESTROL and ETHISTERONE □ DIETHYLSTILBESTROL and PRANONE □ PRANONE and DIETHYLSTILBESTROL □ PRANONE and STILBESTROL □ 17-α-PREGN-4-EN-20-YN-3-ONE, 17-HYDROXY-, and trans-α-α'-DIETHYL-4,4'-STILBENEDIOL □ STILBESTROL and PRANONE

TOXICITY DATA WITH REFERENCE

orl-wmn TDLo:1050 mg/kg (4-36 W post):TER JCEMAZ 19,1369,59
unr-wmn TDLo:164 mg/kg/9Y-I:CAR BJOGAS 82,421,75

SAFETY PROFILE: Questionable human carcinogen producing uterine tumors. An experimental teratogen. When heated to decomposition it emits acrid smoke and irritating fumes.

EEI060 CAS:53127-17-6 HR: 2
ETHODUOMEEN

TOXICITY DATA WITH REFERENCE
ipr-mus TDLo:600 mg/kg (female 9D post):REP DZZEA7 35,1070,80
ipr-mus LD50:614 mg/kg DZZEA7 35,1070,80

SAFETY PROFILE: Moderately toxic by intraperitoneal route. Experimental reproductive effects. When heated to decomposition it emits acrid smoke and irritating fumes.

EEI100 HR: 2
ETHODUOMEEN, HYDROFLUORIDE

TOXICITY DATA WITH REFERENCE
ipr-mus TDLo:400 mg/kg (female 9D post):REP DZZEA7 35,1070,80
ipr-mus LD50:470 mg/kg DZZEA7 35,1070,80

SAFETY PROFILE: Moderately toxic by intraperitoneal route. Experimental reproductive effects. When heated to decomposition it emits acrid smoke and irritating fumes.

EEJ000 CAS:61791-14-8 HR: 2
ETHOMEEN C/15

PROP: A polyoxyethylene (5%) cocoa amine in which alkyl bonds link C8–C18 carbons, which consists of dodecyl (47%), undecyl (18%), decyl (9%), octyl (8%), hexadecyl (10%), and octadecyl (5%) (FCTXAV 8,249,70).

TOXICITY DATA WITH REFERENCE
eye-rbt 100 mg MOD FCTXAV 8,249,70
orl-rat LD50:750 mg/kg FCTXAV 8,249,70

CONSENSUS REPORTS: Reported in EPA TSCA Inventory.

SAFETY PROFILE: Moderately toxic by ingestion. An eye irritant. When heated to decomposition it emits acrid smoke and irritating fumes.

EEJ500 CAS:61791-24-0 HR: 2
ETHOMEEN S/12

PROP: Polyoxyethylene (2%) soya amine alkyl links C14–C18 which consists of octadecadienyl (45%), octadecynyl (35%), octadecyl(10%) and hexadecyl (10%) (FCTXAV 8,249,70).

TOXICITY DATA WITH REFERENCE
eye-rbt 100 mg MOD FCTXAV 8,249,70
orl-rat LD50:1500 mg/kg FCTXAV 8,249,70

CONSENSUS REPORTS: Reported in EPA TSCA Inventory.

SAFETY PROFILE: Moderately toxic by ingestion. An eye irritant. When heated to decomposition it emits acrid smoke and irritating fumes.

EEK000 CAS:61791-24-0 HR: 2
ETHOMEEN S/15

PROP: Polyoxyethylene (5%) soya amine alkyl links C14-C18 which consists of octadecadienyl (45%), octadecenyl (35%), octadecyl(10%) and hexadecyl (10%) (FCTXAV 8,249,70).

TOXICITY DATA WITH REFERENCE
eye-rbt 100 mg MOD FCTXAV 8,249,70
orl-rat LD50:1000 mg/kg FCTXAV 8,249,70

CONSENSUS REPORTS: Reported in EPA TSCA Inventory.

SAFETY PROFILE: Moderately toxic by ingestion. An eye irritant. When heated to decomposition it emits acrid smoke and irritating fumes.

EEK050 CAS:61791-26-2 HR: 2
ETHOMEEN T/15

TOXICITY DATA WITH REFERENCE
eye-rbt 100 mg MOD FCTXAV 8,249,70
orl-rat LD50:500 mg/kg FCTXAV 8,249,70

CONSENSUS REPORTS: Reported in EPA TSCA Inventory.

SAFETY PROFILE: Moderately toxic by ingestion. An eye irritant. When heated to decomposition it emits acrid smoke and irritating vapors.

EEL000 CAS:927-80-0 HR: 3
ETHOXY ACETYLENE
mf: C_4H_6O mw: 70.09

$CH_3CH_2OC{\equiv}CH$

PROP: Lachrymatory liquid. Flash p: 19.4°F, d: 0.8, vap d: 2.4, bp: 50–52°. Insol in water.

SYN: ETHOXYETHYNE

SAFETY PROFILE: An eye irritant. Potentially explosive decomposition when heated above 100°C. Potentially explosive reaction with ethyl magnesium iodide. A very dangerous fire hazard when exposed to heat, flame, or oxidizers. To fight fire, use foam, dry chemical, CO_2. When heated to decomposition it emits acrid smoke and fumes. See also ACETYLENE COMPOUNDS.

EEL500 CAS:10031-82-0 HR: 2
p-ETHOXYBENZALDEHYDE
mf: $C_9H_{10}O_2$ mw: 150.18

PROP: Colorless. Mp: 13–14°, bp: 247–249°. Misc in alc and ether.

SYNS: ETHOXYBENZALDEHYDE □ 4-ETHOXYBENZALDEHYDE

TOXICITY DATA WITH REFERENCE
skn-rbt 500 mg/24H SEV FCTXAV 18,681,80
orl-rat LD50:2100 mg/kg FCTXAV 18,681,80

CONSENSUS REPORTS: Reported in EPA TSCA Inventory.

SAFETY PROFILE: Moderately toxic by ingestion. A severe skin irritant. When heated to decomposition it emits acrid smoke and irritating fumes. See also ALDEHYDES.

EEM000 CAS:938-73-8 HR: 3
2-ETHOXYBENZAMIDE
mf: $C_9H_{11}NO_2$ mw: 165.21

PROP: Crystals from EtOAc/hexane. Mp: 132–134°.

SYNS: ANOVIGAM □ ETAMIDE □ ETHBENZAMIDE □ ETHENZAMID □ ETHENZAMIDE □ ETHOSALICYL □ o-ETHOXYBENZAMIDE □ ETOCIL □ ETOSALICIL □ ETOSALICYL □ EUSAL □ H.P. 209 □ KATAGRIPPE □ LINDATOX □ LUCAMIDE □ PIROSOLVINA □ PROTOPYRIN □ TRANCALGYL

TOXICITY DATA WITH REFERENCE
cyt-ham:lng 500 mg/L/48H GMCRDC 27,95,81
cyt-ham:fbr 500 mg/L ESKHA5 96,55,78
orl-mus TDLo:544 g/kg/54W-C:CAR JJIND8 76,115,86
orl-rat LD50:2630 mg/kg ARZNAD 10,820,60
orl-mus LD50:700 mg/kg THERAP 7,27,52
ipr-mus LD50:400 mg/kg JPPMAB 4,872,52
orl-rbt LDLo:1500 mg/kg ARZNAD 10,820,60

CONSENSUS REPORTS: Reported in EPA TSCA Inventory.

SAFETY PROFILE: Poison by intraperitoneal route. Moderately toxic by ingestion. Questionable carcinogen with experimental carcinogenic data. Mutation data reported. When heated to decomposition it emits toxic fumes of NO_x. See also AMIDES.

EEN600 CAS:33077-69-9 HR: 3
4-(p-ETHOXYBENZOYL)PYRIDINE
mf: $C_{14}H_{13}NO_2$ mw: 227.28

SYNS: p-ETHOXYPHENYL 4-PYRIDYL KETONE □ KETONE, (p-ETHOXYPHENYL) 4-PYRIDYL □ PYRIDINE, 4-(p-ETHOXYBENZOYL)-

TOXICITY DATA WITH REFERENCE
ipr-mus LD50:490 mg/kg JMCMAR 14,551,71

DOT CLASSIFICATION: 3; *Label:* Flammable Liquid

SAFETY PROFILE: Moderately toxic by intraperitoneal route. A flammable liquid. When heated to decomposition it emits toxic vapors of NO_x.

EEO000 CAS:577-66-2 HR: 3
8-ETHOXYCAFFEINE
mf: $C_{10}H_{14}N_4O_3$ mw: 238.28

SYNS: EOC □ 1H-PURINE-2,6-DIONE, 8-ETHOXY-3,7-DIHYDRO-1,3,7-TRIMETHYL-(9CI)

TOXICITY DATA WITH REFERENCE
mma-sat 20 µg/plate MUREAV 60,349,79
sln-dmg-orl 15 mmol/L MUREAV 149,189,85
trn-dmg-orl 15 mmol/L MUREAV 149,189,85
cyt-ham:ovr 17 mmol/L MUREAV 60,349,79
cyt-ham:oth 12 mmol/L/2H-C MUREAV 12,463,71
ivn-mus LD50:56 mg/kg CSLNX* NX#04352

CONSENSUS REPORTS: EPA Genetic Toxicology Program.

SAFETY PROFILE: Poison by intravenous route. Mutation data reported. When heated to decomposition it emits toxic fumes of NO_x. See also CAFFEINE.

EEO500 CAS:13684-56-5 HR: 2
3-ETHOXYCARBONYLAMINOPHENYL-N-PHENYLCARBAMATE
mf: $C_{16}H_{16}N_2O_4$ mw: 300.34

PROP: Crystals. Mp: 120°.

SYNS: 3-(AETHOXYCARBONYLAMINOPHENYL)-N-PHENYL-CARBAMAT (GERMAN) □ BENTANEX □ BETANAL AM □ BETANEX □ m-CARBANILOYLOXYCARBANILIC ACID ETHYL ESTER □ DESMEDIPHAM □ EP-475 □ ETHYL-m-HYDROXYCARBANILATE CARBANILATE (ESTER) □ ETHYL PHENYLCARBAMOYLOXYPHENYLCARBAMATE □ SCHERING 38107 □ SN 38107

TOXICITY DATA WITH REFERENCE
orl-rat LD50:9600 mg/kg 85ARAE 2,92,77
skn-rbt LDLo:10 g/kg FMCHA2 -,C31,83
orl-qal LD50:2480 mg/kg 85DPAN -,-,71/76

SAFETY PROFILE: Moderately toxic by ingestion. Mildly toxic by skin contact. An herbicide. When heated to decomposition it emits toxic fumes of NO_x. See also CARBAMATES.

EEP000 CAS:73987-52-7 HR: 3
ETHOXY CARBONYL DIGOXIN

SYN: CARBAETHOXYDIGOXIN (GERMAN)

TOXICITY DATA WITH REFERENCE
orl-gpg LD50:3500 µg/kg ARZNAD 15,481,65
ivn-gpg LDLo:1400 µg/kg ARZNAD 15,481,65

SAFETY PROFILE: Poison by ingestion and intravenous routes. See also DIGOXIN.

EEQ000 CAS:63905-54-4 HR: 3
4-ETHOXYCARBONYL-1-(2-HYDROXY-3-PHENOXYPROPYL) 4-PHENYLPIPERIDINE HYDROCHLORIDE
mf: $C_{23}H_{29}NO_4{\cdot}ClH$ mw: 419.99

SYNS: B.D.H. 200 HYDROCHLORIDE □ ISONIPECOTIC ACID, 1-(2-HYDROXY-3-PHENOXYPROPYL)-4-PHENYL-, ETHYL ESTER, HYDROCHLORIDE

TOXICITY DATA WITH REFERENCE
orl-mus LD50:419 mg/kg JPPMAB 12,449,60
scu-mus LD50:145 mg/kg JPPMAB 12,449,60

SAFETY PROFILE: Poison by ingestion and subcutaneous routes. When heated to decomposition it emits very toxic fumes of HCl and NO_x.

EEQ100 CAS:60075-74-3 HR: 2
(2-(ETHOXYCARBONYL)-1-METHYL)ETHYL CARBONIC ACID-p-IODOBENZYL ESTER
mf: $C_{14}H_{17}IO_5$ mw: 392.21

TOXICITY DATA WITH REFERENCE
orl-rat LD50:1700 mg/kg GISAAA 41(6),95,76

orl-mus LD50:583 mg/kg GISAAA 41(6),95,76
ipr-mus LD50:3 g/kg JMCMAR 19,1362,76

SAFETY PROFILE: Moderately toxic by ingestion and intraperitoneal routes. When heated to decomposition it emits toxic fumes of I^-. See also ESTERS.

EEQ500 CAS:616-97-7 HR: 3
S-ETHOXYCARBONYLTHIAMINE HYDROCHLORIDE
mf: $C_{15}H_{22}N_4O_4S_2 \cdot ClH$ mw: 422.99

SYNS: S-AETHOXY-CARBONYLTHIAMIN HYDROCHLORID (GERMAN) □ S-CARBETHOXYTHIAMINE HYDROCHLORIDE □ SECT HYDROCHLORIDE

TOXICITY DATA WITH REFERENCE
orl-mus LD50:4250 mg/kg IZVIAK 37,82,67
ivn-mus LD50:339 mg/kg IZVIAK 37,82,67

SAFETY PROFILE: Poison by intravenous route. Mildly toxic by ingestion. When heated to decomposition it emits very toxic fumes of HCl, SO_x and NO_x.

EER000 CAS:1586-92-1 HR: 3
ETHOXY DIETHYL ALUMINUM
mf: $C_6H_{15}AlO$ mw: 129.16

$(CH_3CH_2)_2AlOCH_3CH_2$

SYN: DIETHYLETHOXYALUMINUM

SAFETY PROFILE: Ignites spontaneously in air in the pure form or in solutions with concentrations greater than 20%. When heated to decomposition it emits acrid smoke and fumes. See also ALUMINUM COMPOUNDS.

EER400 CAS:83053-57-0 HR: 2
11-ETHOXY-15,16-DIHYDRO-17-CYCLOPENTA(a)PHENANTHREN-17-ONE
mf: $C_{19}H_{16}O_2$ mw: 276.35

TOXICITY DATA WITH REFERENCE
mma-sat 20 µg/plate CRNGDP 3,677,82
skn-mus TDLo:1600 µg/kg:ETA CRNGDP 3,677,82

SAFETY PROFILE: Questionable carcinogen with experimental tumorigenic data. Mutation data reported. When heated to decomposition it emits acrid smoke and irritating fumes.

EER500 CAS:103-75-3 HR: 3
2-ETHOXY DIHYDROPYRAN
mf: $C_7H_{12}O_2$ mw: 128.19

PROP: D: 1.0, bp: 143°, flash p: 111°F (OC).

SYNS: 2-ETHOXY-2,3-DIHYDRO-γ-PYRAN □ 2-ETHOXY-3,4-DIHYDRO-1,2-PYRAN □ 2-ETHOXY-3,4-DIHYDRO-2H-PYRAN

TOXICITY DATA WITH REFERENCE
skn-rbt 10 mg/24H open MLD AIHAAP 23,95,62
eye-rbt 50 mg MOD UCDS** 7/15/71
orl-rat LD50:6160 mg/kg UCDS** 7/15/71
ihl-rat LCLo:8000 ppm/4H AIHAAP 23,95,62
skn-rbt LD50:3560 mg/kg UCDS** 7/15/71

CONSENSUS REPORTS: Reported in EPA TSCA Inventory.

SAFETY PROFILE: Moderately toxic by skin contact. Mildly toxic by ingestion and inhalation. A skin and eye irritant. Flammable liquid when exposed to flame, sparks, and oxidizers. To fight fire, use dry chemical, foam, fog. When heated to decomposition it emits acrid smoke and irritating fumes.

EES000 CAS:15769-72-9 HR: 3
ETHOXYDIISOBUTYLALUMINUM
mf: $C_{10}H_{23}AlO$ mw: 186.28

$CH_3CH_2OAl[CH_2CH(CH_3)_2]_2$

PROP: Air and moisture-sensitive colorless viscous liquid. Bp: 122–123° @ 2 mm.

SAFETY PROFILE: Ignites spontaneously in air. When heated to decomposition it emits acrid smoke and fumes. See also ALUMINUM COMPOUNDS.

EES100 CAS:69929-16-4 HR: 1
8-ETHOXY-2,6-DIMETHYLOCTENE-2
mf: $C_{12}H_{24}O$ mw: 184.36

SYNS: CITRONELLYL ETHYL ETHER □ 2-OCTENE, 8-ETHOXY-2,6-DIMETHYL-

TOXICITY DATA WITH REFERENCE
skn-rbt 500 mg/24H MLD FCTOD7 20,655,82
orl-rat LD50:>5 g/kg FCTOD7 20,655,82
skn-rbt LD50:>5 g/kg FCTOD7 20,655,82

CONSENSUS REPORTS: Reported in EPA TSCA Inventory.

SAFETY PROFILE: Low toxicity by ingestion and skin contact. A skin irritant. When heated to decomposition it emits acrid smoke and irritating fumes.

EES300 CAS:7495-45-6 HR: 2
ETHOXYDIPHENYLACETIC ACID
mf: $C_{16}H_{16}O_3$ mw: 256.32

PROP: Plates from Et_2O. Mp: 114–115°.

SYNS: ACIDE DIPHENYLETHOXYACETIQUE (FRENCH) □ α-ETHOXY-α-PHENYL-BENZENEACETIC ACID (9CI)

TOXICITY DATA WITH REFERENCE
orl-rat LD50:1 g/kg AIPTAK 116,154,58
ivn-rat LD50:600 mg/kg AIPTAK 116,154,58
orl-mus LD50:600 mg/kg AIPTAK 116,154,58
scu-mus LD50:580 mg/kg AIPTAK 116,154,58

SAFETY PROFILE: Moderately toxic by ingestion, intravenous, and subcutaneous routes. When heated to decomposition it emits acrid smoke and irritating fumes.

EES350 CAS:110-80-5 HR: 3
2-ETHOXYETHANOL
DOT: UN 1171
mf: $C_4H_{10}O_2$ mw: 90.14

$CH_3CH_2OCH_2CH_2OH$

PROP: Colorless liquid, practically odorless. Bp: 135.1°, lel: 1.8%, uel: 14%, fp: −70°, flash p: 202°F(CC), d: 0.9360 @ 15°/15°, autoign temp: 455°F, vap press: 3.8 mm @ 20°, vap d: 3.10. Misc in H_2O, EtOH, Et_2O, and Me_2CO.

SYNS: ATHYLENGLYKOL-MONOATHYLATHER (GERMAN) □ CELLOSOLVE (DOT) □ CELLOSOLVE SOLVENT □ DOWANOL EE □ EKTASOLVE EE □ ETHER MONOETHYLIQUE de l'ETHYLENE-GLYCOL (FRENCH) □ ETHYL CELLOSOLVE □ ETHYLENE GLYCOL ETHYL ETHER □ ETHYLENE GLYCOL MONOETHYL ETHER □ ETHYLENE GLYCOL MONOETHYL ETHER (DOT) □ ETOKSYETYLOWY ALKOHOL (POLISH) □ GLYCOL ETHER EE □ GLYCOL ETHYL ETHER □ GLYCOL MONOETHYL ETHER □ HYDROXY ETHER □ JEFFERSOL EE □ NCI-C54853 □ OXITOL □ POLY-SOLV EE

TOXICITY DATA WITH REFERENCE
eye-hmn 6000 ppm PHRPA6 45,1459,30
skn-rbt 500 mg open MLD UCDS** 5/20/66
eye-rbt 500 mg/24H MLD 85JCAE-,624,86
eye-rbt 50 mg MOD UCDS** 5/20/66
eye-gpg 10 µg MLD JPPMAB 11,150,59
ihl-rat TCLo:100 ppm/7H (female 14-20D post):REP TJADAB 21,58A,80
orl-mus TDLo:5600 mg/kg (female 8-14D post):REP TCMUD8 7,55,87
orl-mus TDLo:252 g/kg (female 18W pre):REP EVHPAZ 57,85,84
ihl-rat TCLo:600 ppm/7H (female 7-13D post):TER NRTXDN 2,231,81
orl-rat TDLo:1800 mg/kg (female 7-15D post):TER TOXID9 4,87,84
orl-rat TDLo:600 mg/kg (female 10-12D post):TER TOXID9 4,87,84
orl-rat TDLo:7820 mg/kg (female 1-21D post):REP ARZNAD 21,880,71
ihl-rat TCLo:10 ppm/6H (female 6-15D post):REP EVHPAZ 57,33,84
skn-rat TDLo:50 g/kg (female 7-16D post):REP DCTODJ 5,277,82
orl-rat TDLo:500 mg/kg (5D male):REP FAATDF 5,182,85
ihl-rat TCLo:100 ppm/7H (female 14-20D post):REP TJADAB 21,58A,80
orl-rat TDLo:4500 mg/kg (male 6W pre):REP FAATDF 7,348,86
orl-mus TDLo:25 g/kg (male 25D pre):REP SAIGBL 21,29,79
ihl-rbt TCLo:160 ppm/7H (female 1-18D post):TER SWEHDO 7(Suppl 4),66,81
skn-rat TDLo:50 g/kg (female 7-16D post):TER EVHPAZ 57,69,84
orl-mus TDLo:12,600 mg/kg (female 8-14D post):TER TJADAB 33,96C,86
orl-rat LD50:2125 mg/kg GTPZAB 32(3),48,88
ihl-rat LC50:2000 ppm/7H NPIRI* 1,54,74
skn-rat LD50:3900 mg/kg TOXID9 4,180,84
ipr-rat LD50:2800 mg/kg ARZNAD 21,880,71
ivn-rat LD50:2400 mg/kg NPIRI* 1,54,74
orl-mus LD50:2451 mg/kg KODAK* MSDS-10,170A,82
ihl-mus LC50:1820 ppm/7H JIHTAB 25,157,43
ipr-mus LD50:1707 mg/kg FEPRA7 6,342,47
orl-rbt LD50:1275 mg/kg GISAAA 53(10),78,88
skn-rbt LD50:3300 mg/kg NPIRI* 1,54,74
ihl-gpg LCLo:3000 ppm/24H PHRPA6 45,1459,30

CONSENSUS REPORTS: Reported in EPA TSCA Inventory. Glycol ether compounds are on the Community Right-To-Know List.

OSHA PEL: TWA 200 ppm (skin)
ACGIH TLV: TWA 5 ppm (skin)
DFG MAK: 20 ppm (75 mg/m³)
NIOSH REL: (Glycol Ethers): Reduce to lowest level
DOT CLASSIFICATION: 3; *Label:* Flammable Liquid

SAFETY PROFILE: Moderately toxic by ingestion, skin contact, intravenous, and intraperitoneal routes. Mildly toxic by inhalation and subcutaneous routes. An experimental teratogen. Other experimental reproductive effects. A mild eye and skin irritant. Combustible when exposed to heat or flame; can react with oxidizing materials. Moderate explosion hazard in the form of vapor when exposed to heat or flame. Mixture with hydrogen peroxide + polyacrylamide gel + toluene is explosive when dry. To fight fire, use alcohol foam, dry chemical. See also GLYCOL ETHERS.

For occupational chemical analysis use OSHA: #53 or NIOSH: Alcohols IV, 1403.

EES370 CAS:36306-87-3 HR: 1
4-(1-ETHOXYETHENYL)-3,3,5,5-TETRAMETHYLCYCLOHEXANONE
mf: $C_{14}H_{24}O_2$ mw: 224.38

SYNS: CYCLOHEXANONE, 4-(1-ETHOXYETHENYL)-3,3,5,5-TETRAMETHYL- □ KEPHALIS □ 3,3,5,5-TETRAMETHYL-4-ETHOXYVINYLCYCLOHEXANONE

TOXICITY DATA WITH REFERENCE
skn-rbt 500 mg/24H MOD FCTOD7 20,835,82
orl-rat LD50:>5 g/kg FCTOD7 20,835,82
skn-rbt LD50:>5 g/kg FCTOD7 20,835,82

CONSENSUS REPORTS: Reported in EPA TSCA Inventory.

SAFETY PROFILE: Low toxicity by ingestion and skin contact. A skin irritant. When heated to decomposition it emits acrid smoke and irritating vapors.

EES380 CAS:2556-10-7 HR: 1
(2-(1-ETHOXYETHOXY)ETHYL)BENZENE
mf: $C_{12}H_{18}O_2$ mw: 194.30

SYNS: ACETALDEHYDE ETHYL PHENETHYL ACETAL □ ACETALDEHYDE ETHYL 2-PHENYLETHYL ACETAL □ BENZENE, (2-(1-ETHOXYETHOXY)ETHYL)- □ HYACINTH BODY

TOXICITY DATA WITH REFERENCE
orl-rat LD50:>5 g/kg FCTOD7 30,1S,92
skn-rbt LD50:>5 g/kg FCTOD7 30,1S,92

CONSENSUS REPORTS: Reported in EPA TSCA Inventory.

E

SAFETY PROFILE: Low toxicity by ingestion and skin contact. When heated to decomposition it emits acrid smoke and irritating vapors.

EES400 CAS:111-15-9 HR: 3
2-ETHOXYETHYL ACETATE
DOT: UN 1172
mf: $C_6H_{12}O_3$ mw: 132.18

$CH_3CO{\cdot}OC_2H_4OCH_2CH_3$

PROP: Colorless liquid with a mild, pleasant, ester-like odor. Mp: −61°, bp: 156.4°, flash p: 117°F (COC), lel: 1.7%, fp: −61.7°, d: 0.9748 @ 20°/20°, autoign temp: 715°F, vap press: 1.2 mm @ 20°, vap d: 4.72.

SYNS: ACETATE de CELLOSOLVE (FRENCH) □ ACETATE de l'ETHER MONOETHYLIQUE DE L'ETHYLENE-GLYCOL (FRENCH) □ ACETATE d'ETHYLGLYCOL (FRENCH) □ ACETATO di CELLOSOLVE (ITALIAN) □ ACETIC ACID-2-ETHOXYETHYL ESTER □ 2-AETHOXY-AETHYLACETAT (GERMAN) □ AETHYLENGLYKOLAETHERACETAT (GERMAN) □ CELLOSOLVE ACETATE (DOT) □ CSAC □ EKTASOLVE EE ACETATE SOLVENT □ ETHOXY ACETATE □ 2-ETHOXYETHANOL ACETATE □ 2-ETHOXYETHANOL, ESTER with ACETIC ACID □ 2-ETH-OXY-ETHYLACETAAT (DUTCH) □ ETHOXYETHYL ACETATE □ β-E-THOXYETHYL ACETATE □ 2-ETHOXYETHYLE, ACETATE de (FRENCH) □ ETHYL CELLOSOLVE ACETAAT (DUTCH) □ ETHYLENE GLYCOL ETHYL ETHER ACETATE □ ETHYLENE GLYCOL MONOE-THYL ETHER ACETATE (MAK, DOT) □ ETHYLGLYKOLACETAT (GER-MAN) □ 2-ETOSSIETIL-ACETATO (ITALIAN) □ GLYCOL ETHER EE ACETATE □ GLYCOL MONOETHYL ETHER ACETATE □ OCTAN ETOKSYETYLU (POLISH) □ OXYTOL ACETATE □ POLY-SOLV EE AC-ETATE

TOXICITY DATA WITH REFERENCE
skn-rbt 490 mg open MLD UCDS** 1/13/67
eye-rbt 40 mg MOD UCDS** 1/13/67
ihl-rat TCLo:600 ppm/7H (female 7-15D post):TER EPASR* 8EHQ-0682-0450
ihl-rat TCLo:200 ppm/6H (female 6-15D post):TER FAATDF 10,20,88
ihl-rbt TCLo:200 ppm/6H (female 6-18D post):REP UCRR** 09OCT84
skn-rat TDLo:70 g/kg (female 7-16D post):REP EVHPAZ 57,69,84
orl-mus TDLo:100 g/kg (male 25D pre):REP SAIGBL 21,29,79
orl-mus TDLo:25 g/kg (male 25D pre):REP SAIGBL 21,29,79
ihl-rat TCLo:300 ppm/6H (female 6-15D post):TER UCRR** 09OCT84
ihl-rat TCLo:50 ppm/6H (female 6-15D post):TER UCRR** 09OCT84
orl-rat LD50:2900 mg/kg TXAPA9 51,117,79
ihl-rat LC50:12,100 mg/m³/8H AIHAAP 20,364,59
ipr-mus LD50:1420 mg/kg SCCUR* -,2,61
orl-rbt LD50:1950 mg/kg EPASR* 8EHQ-0682-0450
skn-rbt LD50:10,500 mg/kg UCDS** 1/13/67

CONSENSUS REPORTS: Reported in EPA TSCA Inventory. Glycol ether compounds are on the Community Right-To-Know List.

OSHA PEL: TWA 100 ppm (skin)
ACGIH TLV: TWA 5 ppm (skin)
DFG MAK: 20 ppm (110 mg/m³)
DOT CLASSIFICATION: 3; *Label:* Flammable Liquid

SAFETY PROFILE: Moderately toxic by ingestion and intraperitoneal routes. A skin and eye irritant. An experimental teratogen. Other experimental reproductive effects. Flammable liquid when exposed to heat or flame; can react with oxidizing materials. Moderate explosion hazard in the form of vapor when heated. Mild explosions have occurred at the end of distillations. To fight fire, use alcohol foam, CO_2, dry chemical. When heated to decomposition it emits acrid smoke and irritating fumes. See also GLYCOL ETHERS.

For occupational chemical analysis use OSHA: #53 or NIOSH: Esters I, 1450.

EES500 CAS:67262-62-8 HR: 3
2-(2-ETHOXYETHYLAMINO)-o-PROPIONOTOLUIDIDE
mf: $C_{14}H_{22}N_2O_2$ mw: 250.38

SYN: 2-(2-ETHOXYETHYLAMINO)-2'-METHYL-PROPIONANILIDE

TOXICITY DATA WITH REFERENCE
ipr-mus LD50:275 mg/kg JPMSAE 67,595,78
ivn-mus LD50:38 mg/kg JPMSAE 67,595,78

SAFETY PROFILE: Poison by intraperitoneal and intravenous routes. When heated to decomposition it emits toxic fumes of NO_x.

EET000 CAS:63716-10-9 HR: 1
ETHOXYETHYL ETHER of PROPYLENE GLYCOL
mf: $C_7H_{16}O_3$ mw: 148.23

SYN: 3-(2-ETHOXY)ETHOXY-2-PROPANOL

TOXICITY DATA WITH REFERENCE
skn-rbt 10 mg/24H MLD JIHTAB 31,60,49
eye-rbt 500 mg JIHTAB 31,60,49
orl-rat LD50:9330 mg/kg JIHTAB 31,60,49
skn-rbt LD50:12 g/kg JIHTAB 31,60,49

CONSENSUS REPORTS: Glycol ether compounds are on the Community Right-To-Know List.

SAFETY PROFILE: Mildly toxic by ingestion and skin contact. A skin and eye irritant. When heated to decomposition it emits acrid smoke and irritating fumes. See also GLYCOL ETHERS.

EET100 CAS:5417-82-3 HR: 3
(1-ETHOXYETHYLIDENE)MALONONITRILE
mf: $C_7H_8N_2O$ mw: 136.17

SYNS: 1-ETHOXYETHYLIDENEMALONONITRILE □ (1-ETHOXYE-THYLIDENE)PROPANEDINITRILE □ MALONONITRILE, (1-ETHOXYE-THYLIDENE)-(6CI,7CI,8CI) □ PROPANEDINITRILE, (1-ETHOXYETHY-LIDENE)-

TOXICITY DATA WITH REFERENCE
orl-rat LDLo:312 mg/kg EPASR* 8EHQ-0487-0663

CONSENSUS REPORTS: Reported in EPA TSCA Inventory.

SAFETY PROFILE: A poison by ingestion. When heated to decomposition it emits toxic vapors of NO_x.

EET500 **CAS:67465-42-3** **HR: 3**
N-(2-ETHOXY-3-HYDROXYMERCURIPROPYL)BARBITAL
mf: $C_{13}H_{22}HgN_2O_5$ mw: 486.96

SYN: HYDROXY(3-(5,5-DIETHYL-2,4,6-TRIOXO-(1H,3H,5H)-PYRIMIDINO)-2-ETHOXYPROPYL)MERCURY

TOXICITY DATA WITH REFERENCE
ivn-rat LDLo:18,600 µg/kg JAPMA8 37,333,48

CONSENSUS REPORTS: Mercury and its compounds are on the Community Right-To-Know List.

OSHA PEL: CL 0.1 mg(Hg)/m³ (skin)
ACGIH TLV: TWA 0.1 mg(Hg)/m³ (skin)
NIOSH REL: (Mercury, Aryl and Inorganic): CL 0.1 mg/m³ (skin)

SAFETY PROFILE: Poison by intravenous route. When heated to decomposition it emits very toxic fumes of Hg and NO_x. See also MERCURY COMPOUNDS, ORGANIC and BARBITURATES.

EET600 **CAS:13021-50-6** **HR: 2**
1-ETHOXY-3-ISOPROPOXYPROPAN-2-OL
mf: $C_8H_{18}O_3$ mw: 162.26

SYN: 1-ETHOXY-3-ISOPROPOXY-2-PROPANOL

TOXICITY DATA WITH REFERENCE
orl-rat LD50:5 g/kg AIPTAK 89,145,52
orl-mus LD50:5730 mg/kg AIPTAK 89,145,52
ipr-mus LD50:3180 mg/kg JPETAB 97,414,49
orl-rbt LDLo:6200 mg/kg AIPTAK 89,145,52

SAFETY PROFILE: Moderately toxic by intraperitoneal route. Mildly toxic by ingestion. When heated to decomposition it emits acrid smoke and irritating fumes.

EEV000 **CAS:63019-29-4** **HR: 2**
10-ETHOXYMETHYL-1:2-BENZANTHRACENE
mf: $C_{21}H_{18}O$ mw: 286.39

SYN: 7-(ETHOXYMETHYL)BENZ(a)ANTHRACENE

TOXICITY DATA WITH REFERENCE
skn-mus TDLo:1200 mg/kg/50W-I:ETA PRLBA4 129,439,40

SAFETY PROFILE: Questionable carcinogen with experimental tumorigenic data. When heated to decomposition it emits acrid smoke and irritating fumes.

EEV100 **CAS:2041-76-1** **HR: 3**
4-ETHOXY-2-METHYL-3-BUTYN-2-OL
mf: $C_7H_{12}O_2$ mw: 128.17

$CH_3CH_2OC{\equiv}CC(CH_3)_2OH$

SAFETY PROFILE: Potentially explosive decomposition above 115°C. Traces of acid increase the tendency to explode. When heated to decomposition it emits acrid smoke and irritating fumes. See also ACETYLENE COMPOUNDS.

EEV200 **CAS:87-13-8** **HR: 2**
ETHOXYMETHYLENEMALONIC ACID, ETHYL ESTER
mf: $C_{10}H_{16}O_5$ mw: 216.26

SYNS: DIETHYL EMME □ DIETHYL (ETHOXYMETHYLENE)MALONATE □ MALONIC ACID, (ETHOXYMETHYLENE)-, DIETHYL ESTER □ TL 1483

TOXICITY DATA WITH REFERENCE
skn-rbt 500 mg/24H MOD JACTDZ 1,37,90
orl-rat LD50:925 mg/kg JACTDZ 1,37,90
orl-mus LD50:2227 mg/kg JPMSAE 60,1810,71
ihl-mus LCLo:44 mg/m³ NDRC** 30101,11,45

CONSENSUS REPORTS: Reported in EPA TSCA Inventory.

SAFETY PROFILE: Moderately toxic by ingestion. A skin irritant. When heated to decomposition it emits acrid smoke and irritating fumes.

EEW000 **CAS:123-06-8** **HR: 3**
ETHOXYMETHYLENE MALONONITRILE
mf: $C_6H_6N_2O$ mw: 122.14

SYNS: USAF A-9230 □ USAF KF-10

TOXICITY DATA WITH REFERENCE
ipr-mus LD50:50 mg/kg NTIS** AD277-689

CONSENSUS REPORTS: Reported in EPA TSCA Inventory. Cyanide and its compounds are on the Community Right-To-Know List.

SAFETY PROFILE: Poison by intraperitoneal route. When heated to decomposition it emits very toxic fumes of NO_x and CN^-. See also NITRILES.

EEX000 **CAS:63020-27-9** **HR: 2**
7-ETHOXY METHYL-12-METHYL BENZ(a)ANTHRACENE
mf: $C_{22}H_{20}O$ mw: 300.42

SYN: 9-METHYL-10-ETHOXYMETHYL-1,2-BENZANTHRACENE

TOXICITY DATA WITH REFERENCE
scu-mus TDLo:80 mg/kg:ETA CNREA8 6,454,46

SAFETY PROFILE: Questionable carcinogen with experimental tumorigenic data. When heated to decomposition it emits acrid smoke and irritating fumes.

EEX500 **CAS:22960-71-0** **HR: 2**
N-ETHOXYMORPHOLINO DIAZENIUM FLUOROBORATE
mf: $C_6H_{13}N_2O_2{\cdot}BF_4$ mw: 232.02

TOXICITY DATA WITH REFERENCE
scu-rat TDLo:559 mg/kg:CAR ZKKOBW 80,17,73
scu-rat LD50:639 mg/kg ZKKOBW 80,17,73

SAFETY PROFILE: Moderately toxic by subcutaneous route. Questionable carcinogen with experimental carcinogenic data. When heated to decomposition it emits very toxic NO_x and F^-. See also BORON COMPOUNDS and FLUORIDES.

E

EEY000 **CAS:78109-88-3** **HR: 3**
O-ETHOXY-N-(3-MORPHOLINOPROPYL)BENZAMIDE
mf: $C_{16}H_{24}N_2O_3$ mw: 292.42

SYN: D-703

TOXICITY DATA WITH REFERENCE
scu-mus LD50:150 mg/kg ARZNAD 10,743,60
ivn-mus LD50:160 mg/kg ARZNAD 10,743,60

SAFETY PROFILE: Poison by subcutaneous and intravenous routes. When heated to decomposition it emits toxic fumes of NO_x.

EEY500 **CAS:93-18-5** **HR: 3**
2-ETHOXYNAPHTHALENE
mf: $C_{12}H_{12}O$ mw: 172.24

PROP: Plates. Mp: 37.5°, bp: 282°.

SYNS: BROMELIA ☐ ETHYL-β-NAPHTHOLATE ☐ ETHYL-β-NAPHTHYL ETHER ☐ ETHYL-2-NAPHTHYL ETHER ☐ β-NAPHTHOL ETHYL ETHER ☐ 2-NAPHTHOL ETHYL ETHER ☐ NEROLIN ☐ NEROLIN II ☐ NEROLINE ☐ NEROLIN NEW

TOXICITY DATA WITH REFERENCE
skn-rbt 500 mg/24H MOD FCTXAV 13,681,75
orl-rat LD50:3110 mg/kg FCTXAV 13,681,75
ipr-mus LD50:100 mg/kg NTIS** AD691-490
ivn-mus LD50:100 mg/kg CSLNX* NX#00191

CONSENSUS REPORTS: Reported in EPA TSCA Inventory.

SAFETY PROFILE: Poison by intraperitoneal and intravenous routes. Moderately toxic by ingestion. A skin irritant. When heated to decomposition it emits acrid smoke and irritating fumes. See also ETHERS.

EFA100 **CAS:622-62-8** **HR: 3**
4-ETHOXYPHENOL
mf: $C_8H_{10}O_2$ mw: 138.18

PROP: Leaflets. Mp: 66–67°, bp: 246–247°, d: 1.07 @ 25°/4°.

SYNS: ETHER MONOETHYLIQUE de l'HYDROQUINONE ☐ p-ETHOXYPHENOL ☐ 4-ETHYLOXYPHENOL ☐ HYDROQUINONE MONOETHYL ETHER ☐ p-HYDROXYPHENETOLE ☐ PHENOL, 4-ETHOXY-(9CI)

TOXICITY DATA WITH REFERENCE
orl-rat TDLo:667 mg/kg (female 11D post):REP TJADAB 41,43,90
ipr-mus LDLo:250 mg/kg RBPMAZ 22,1,52

CONSENSUS REPORTS: Reported in EPA TSCA Inventory.

SAFETY PROFILE: Poison by intraperitoneal route. Experimental reproductive effects. An irritant. When heated to decomposition it emits acrid smoke and irritating fumes.

EFC000 **CAS:21224-77-1** **HR: 3**
S-2-((4-(p-ETHOXYPHENYL)BUTYL)AMINO)ETHYL THIOSULFATE
mf: $C_{14}H_{23}NO_4S_2$ mw: 333.50

TOXICITY DATA WITH REFERENCE
orl-mus LD50:980 mg/kg JMCMAR 11,1190,68
ipr-mus LD50:22 mg/kg JMCMAR 11,1190,68

SAFETY PROFILE: Poison by intraperitoneal route. Moderately toxic by ingestion. When heated to decomposition it emits very toxic fumes of SO_x and NO_x. See also SULFATES.

EFC700 **CAS:32941-23-4** **HR: 3**
p-ETHOXYPHENYL 2-PYRIDYLKETONE
mf: $C_{14}H_{13}NO_2$ mw: 227.28

SYN: KETONE, p-ETHOXYPHENYL 2-PYRIDYL

TOXICITY DATA WITH REFERENCE
ipr-mus LD50:>500 mg/kg JMCMAR 14,551,71

DOT CLASSIFICATION: 3; *Label:* Flammable Liquid

SAFETY PROFILE: Moderately toxic by intraperitoneal route. A flammable liquid. When heated to decomposition it emits toxic vapors of NO_x.

EFC800 **CAS:32921-15-6** **HR: 3**
p-ETHOXYPHENYL 3-PYRIDYLKETONE
mf: $C_{14}H_{13}NO_2$ mw: 227.28

SYN: KETONE, p-ETHOXYPHENYL 3-PYRIDYL

TOXICITY DATA WITH REFERENCE
ipr-mus LD50:>1 g/kg JMCMAR 14,551,71

DOT CLASSIFICATION: 3; *Label:* Flammable Liquid

SAFETY PROFILE: Low toxicity by intraperitoneal route. A flammable liquid. When heated to decomposition it emits toxic vapors of NO_x.

EFE000 **CAS:150-69-6** **HR: 3**
4-ETHOXYPHENYLUREA
mf: $C_9H_{12}N_2O_2$ mw: 180.23

PROP: Needle-like crystals or plates from water. Mp: 174°.

SYNS: p-AETHOXYPHYLHARNSTOFF (GERMAN) ☐ DULCINE ☐ N-(4-ETHOXYPHENYL)UREA ☐ p-ETHOXYPHENYLUREA ☐ NCI-C02073 ☐ PHENETHYLCARBAMID (GERMAN) ☐ p-PHENETOLCARBAMID (GERMAN) ☐ p-PHENETOLCARBAMIDE ☐ p-PHENETOLECARBAMIDE ☐ p-PHENETYLUREA ☐ SUCROL ☐ SUESSTOFF ☐ VALZIN

TOXICITY DATA WITH REFERENCE
orl-rat TDLo:232 g/kg/59W-C:ETA JAPMA8 40,583,51
orl-wmn TDLo:600 mg/kg:CNS MEKLA7 43,105,48
orl-chd LDLo:400 mg/kg MEKLA7 43,105,48
orl-rat LD50:1900 mg/kg NCIMR* NIH-71-E-2144
orl-dog LDLo:1000 mg/kg HBAMAK 4,1345,35

CONSENSUS REPORTS: IARC Cancer Review: Group 3 IMEMDT 7,56,87; Animal Inadequate Evidence IMEMDT 12,97,76.

SAFETY PROFILE: Human poison by ingestion. Moderately toxic experimentally by ingestion. Human systemic effects by ingestion: somnolence, hallucinations, distorted perceptions, and changes in motor activity. In adults 20 to 40 g produces dizziness, nausea, methemoglobinemia, cyanosis, and hypotension. Questionable carcinogen with experimental tumorigenic data. When heated to decomposition it emits toxic fumes of NO_x.

EFE500 CAS:63815-42-9 HR: 3
4-ETHOXY-β-(1-PIPERIDYL)PROPIOPHENONE HYDROCHLORIDE
mf: $C_{16}H_{23}NO_2 \cdot ClH$ mw: 297.86

TOXICITY DATA WITH REFERENCE
ipr-mus LD50:60 mg/kg JPETAB 115,419,55
scu-mus LD50:73 mg/kg ARZNAD 5,559,55
ivn mus LD50:20 mg/kg JPETAB 115,419,55

SAFETY PROFILE: Poison by intraperitoneal, subcutaneous, and intravenous routes. When heated to decomposition it emits very toxic fumes of HCl and NO_x.

EFF000 CAS:1874-62-0 HR: 1
3-ETHOXY-1,2-PROPANEDIOL
mf: $C_5H_{12}O_3$ mw: 120.17

SYNS: α-ETHYL GLYCEROL ETHER □ GLYCEROL-α-ETHYL ETHER

TOXICITY DATA WITH REFERENCE
skn-rbt 530 mg/24H MLD AMIHBC 2,574,50
eye-rbt 106 mg AMIHBC 2,574,50
orl-mus LD50:9350 mg/kg FEPRA7 8,477,49

SAFETY PROFILE: Mildly toxic by ingestion. A skin and eye irritant. When heated to decomposition it emits acrid smoke and irritating fumes.

EFF500 CAS:1569-02-4 HR: 1
1-ETHOXY-2-PROPANOL
mf: $C_5H_{12}O_2$ mw: 104.17

SYN: PROPYLENE GLYCOL ETHYL ETHER

TOXICITY DATA WITH REFERENCE
orl-rat LD50:4400 mg/kg NPIRI* 1,104,74
skn-rbt LD50:8100 mg/kg NPIRI* 1,104,74

CONSENSUS REPORTS: Glycol ether compounds are on the Community Right-To-Know List.

SAFETY PROFILE: Mildly toxic by ingestion and skin contact. When heated to decomposition it emits acrid smoke and irritating fumes. See also GLYCOL ETHERS.

EFG000 CAS:111-35-3 HR: 2
3-ETHOXY-1-PROPANOL
mf: $C_5H_{12}O_2$ mw: 104.17

PROP: A solid. D: 0.917 @ 15°/4°, bp: 162.1°. Misc in water.

SYNS: β-PROPYLENE GLYCOL MONOETHYL ETHER □ PROPYLENE GLYCOL-β-MONOETHYL ETHER

TOXICITY DATA WITH REFERENCE
skn-rbt 452 mg open MLD UCDS** 7/28/66
skn-rbt 500 mg/24H MLD 85JCAE -,627,86
eye-rbt 20 mg/24H MOD 85JCAE -,627,86
orl-mus TDLo:24 g/kg (female 7-14D post):REP NTIS** PB86-197605
orl-rat LD50:7 g/kg JIDHAN 23,259,41
skn-rbt LD50:2558 mg/kg AIHAAP 30,470,69

CONSENSUS REPORTS: Glycol ether compounds are on the Community Right-To-Know List.

SAFETY PROFILE: Moderately toxic by skin contact. Slightly toxic by ingestion. Experimental reproductive effects. A skin irritant. When heated to decomposition it emits acrid smoke and irritating fumes. See also GLYCOL ETHERS.

E

EFG500 CAS:63918-98-9 HR: 3
ETHOXY PROPIONALDEHYDE
mf: $C_5H_{10}O_2$ mw: 102.15

PROP: Liquid. Mp: −69.4°, bp: 135.2°, flash p: 100°F (OC), d: 0.918 @ 20°/20°, vap d: 3.63, vap press: 5.5 mm @ 20°.

TOXICITY DATA WITH REFERENCE
skn-rbt 10 mg/24H open JIHTAB 30,63,48
eye-rbt 2 mg open SEV JIHTAB 30,63,48
orl-rat LD50:900 mg/kg JIHTAB 30,63,48
ihl-rat LC50:500 ppm/4H JIHTAB 30,63,48
orl-mus LD50:140 mg/kg ARZNAD 15,841,65
skn-rbt LD50:1000 mg/kg JIHTAB 30,63,48

SAFETY PROFILE: Poison by ingestion. Moderately toxic by inhalation and skin contact. A skin and severe eye irritant. A very dangerous fire hazard when exposed to heat or flame; can react with oxidizing materials. To fight fire, use alcohol foam, CO_2, dry chemical. When heated to decomposition it emits acrid smoke and irritating fumes. See also ALDEHYDES.

EFH000 CAS:1331-11-9 HR: 2
ETHOXYPROPIONIC ACID
mf: $C_5H_{10}O_3$ mw: 118.15

PROP: Liquid. Mp: −10.7°, bp: 219°, flash p: 225°F (OC), d: 1.0474, vap d: 4.08.

TOXICITY DATA WITH REFERENCE
skn-rbt 10 mg/24H open JIHTAB 30,63,48
eye-rbt 250 μg open SEV JIHTAB 30,63,48
orl-rat LD50:4800 mg/kg JIHTAB 30,63,48
skn-rbt LD50:750 mg/kg JIHTAB 30,63,48

SAFETY PROFILE: Moderately toxic by skin contact. Mildly toxic by ingestion. A skin and severe eye irritant. Combustible when exposed to heat or flame; can react with oxidizing materials. To fight fire, use alcohol foam, CO_2, dry chemical. When heated to decomposition it emits acrid smoke and irritating fumes.

EFH500 CAS:14631-45-9 HR: 2
β-ETHOXYPROPIONITRILE
mf: C_5H_9NO mw: 99.15

TOXICITY DATA WITH REFERENCE
orl-rat LD50:2860 mg/kg TNICS* 13,125,73
orl-mus LD50:2200 mg/kg TNICS* 13,125,73

CONSENSUS REPORTS: Cyanide and its compounds are on the Community Right-To-Know List.

SAFETY PROFILE: Moderately toxic by ingestion. When heated to decomposition it emits toxic fumes of NO_x and CN^-. See also NITRILES.

EFI000 **CAS:64050-15-3** **HR: 2**
ETHOXYPROPYLACRYLATE
mf: $C_8H_{14}O_3$ mw: 158.22

SYN: ETHOXYPROPYL ESTER ACRYLIC ACID

TOXICITY DATA WITH REFERENCE
skn-rbt 10 mg/24H open MLD AMIHBC 4,119,51
eye-rbt 100 mg open AMIHBC 4,119,51
orl-rat LD50:820 mg/kg AMIHBC 4,119,51
ihl-rat LC50:250 ppm/4H AMIHBC 4,119,51
skn-rbt LD50:1410 mg/kg AMIHBC 4,119,51

SAFETY PROFILE: Moderately toxic by ingestion, inhalation, and skin contact. A skin and eye irritant. When heated to decomposition it emits acrid smoke and irritating fumes.

EFJ000 **CAS:628-33-1** **HR: 3**
1-ETHOXY-2-PROPYNE
mf: C_5H_8O mw: 84.12

$CH_3CH_2OCH_2C{\equiv}CH$

PROP: Liquid with penetrating odor. Bp: 80°. Mod sol in H_2O; misc in EtOH.

SAFETY PROFILE: Peroxidizes in air to form a product which explodes when heated to 80°C. When heated to decomposition it emits acrid smoke and fumes. See also PEROXIDES and ACETYLENE COMPOUNDS.

EFJ500 **CAS:16357-59-8** **HR: 3**
2-ETHOXY-1(2H)-QUINOLINECARBOXYLIC ACID, ETHYL ESTER
mf: $C_{14}H_{17}NO_3$ mw: 247.32

PROP: A solid. Mp: 63.5–65°.

TOXICITY DATA WITH REFERENCE
ipr-mus LD50:32 mg/kg JMCMAR 14,49,71

CONSENSUS REPORTS: Reported in EPA TSCA Inventory.

SAFETY PROFILE: Poison by intraperitoneal route. When heated to decomposition it emits toxic fumes of NO_x. See also ESTERS.

EFJ600 **CAS:3463-21-6** **HR: 2**
ETHOXYSILATRANE
mf: $C_8H_{17}NO_4Si$ mw: 219.35

SYNS: AETHOXYSILATRAN □ 1-ETHOXYSILATRANE □ MIGUGEN □ SILICIC ACID, CYCLIC NITRILOTRIETHYLENE ETHYL ESTER □ 2,8,9-TRIOXA-5-AZA-1-SILABICYCLO(3.3.3)UNDECANE, 1-ETHOXY-

TOXICITY DATA WITH REFERENCE
orl-mus LD50:2800 mg/kg PHARAT 26,224,70
ipr-mus LD50:2300 mg/kg PHARAT 26,224,70

CONSENSUS REPORTS: Reported in EPA TSCA Inventory.

SAFETY PROFILE: Moderately toxic by ingestion and intraperitoneal routes. When heated to decomposition it emits toxic vapors of NO_x.

EFK000 **CAS:2593-15-9** **HR: 2**
5-ETHOXY-3-TRICHLOROMETHYL-1,2,4-THIADIAZOLE
mf: $C_5H_5Cl_3N_2OS$ mw: 247.53

PROP: Pale-yellow liquid. Mp: 20°.

SYNS: 5-AETHOXY-3-TRICHLORMETHYL-1,2,4-THIADIAZOL (GERMAN) □ DWELL □ ECHLOMEZOL □ ETHAZOLE (FUNGICIDE) □ ETMT □ ETRIDIAZOLE □ KOBAN □ MF-344 □ OLIN MATHIESON 2,424 □ OM 2424 □ PANSOIL □ TERRACHLOR-SUPER X □ TERRA-COAT □ TERRAFLO □ TERRAZOLE □ 3-(TRICHLOROMETHYL)-5-ETHOXY-1,2,4-THIADIAZOLE □ TRUBAN

TOXICITY DATA WITH REFERENCE
mmo-sat 400 μg/plate KHFKDF 8,551,80
mma-sat 400 μg/plate KHFKDF 8,551,80
orl-rat LD50:1077 mg/kg 28ZEAL 5,110,76
orl-mus LD50:2000 mg/kg GUCHAZ 6,486,73
orl-rbt LD50:779 mg/kg TXAPA9 56,164,80
skn-rbt LD50:1700 mg/kg FMCHA2 -,C137,83

SAFETY PROFILE: Moderately toxic by ingestion and skin contact. Mutation data reported. A fungicide. When heated to decomposition it emits very toxic fumes of Cl^-, SO_x, and NO_x.

EFL000 **CAS:112-50-5** **HR: 1**
ETHOXYTRIGLYCOL
mf: $C_8H_{18}O_4$ mw: 178.26

PROP: Bp: 255.4°, flash p: 275°F (OC), d: 1.0208 @ 20°/20°, vap press: 0.01 mm @ 20°.

SYNS: DOWANOL TE □ 2-(2-(2-ETHOXYETHOXY)ETHOXY)ETHANOL □ ETHOXYTRIETHYLENE GLYCOL □ POLY-SOLV TE □ TRIETHYLENE GLYCOL ETHYL ETHER □ TRIETHYLENE GLYCOL MONOETHYL ETHER □ TRIGLYCOL MONOETHYL ETHER

TOXICITY DATA WITH REFERENCE
eye-rbt 500 mg AJOPAA 29,1363,46
orl-rat LD50:10,610 mg/kg 34ZIAG -,730,69
skn-rbt LD50:8 g/kg AMIHBC 4,119,51

CONSENSUS REPORTS: Reported in EPA TSCA Inventory. Glycol ether compounds are on the Community Right-To-Know List.

SAFETY PROFILE: Mildly toxic by ingestion and skin contact. An eye irritant. Combustible when exposed to heat or flame; can react with oxidizing materials. To fight fire, use foam, alcohol foam, CO_2, dry chemical. When heated to decomposition it emits acrid smoke and irritating fumes. See also GLYCOL ETHERS.

EFL500 **CAS:1825-62-3** **HR: 3**
ETHOXYTRIMETHYLSILANE
mf: $C_5H_{14}OSi$ mw: 118.28

PROP: Bp: 75.7°, vap press: 100 mm @ 22.1°, vap d: 4.1.

TOXICITY DATA WITH REFERENCE
eye-rbt 79 mg JIHTAB 30,332,48
orl-rat LDLo:1400 mg/kg JIHTAB 30,332,48
ihl-rat LCLo:4000 ppm/8H JIHTAB 30,332,48

CONSENSUS REPORTS: Reported in EPA TSCA Inventory.

SAFETY PROFILE: Moderately toxic by ingestion and inhalation. An eye irritant. Flammable when exposed to heat or flame; can react with oxidizing materials. When heated to decomposition it emits acrid smoke and irritating fumes.

EFM000 CAS:625-50-3 HR: 2
ETHYLACETAMIDE
mf: C_4H_9NO mw: 87.14

PROP: Water white oily liquid. Mp: −32°, bp: 206–208°, flash p: 230°F, d: 0.920 @ 20°/20°, vap d: 3.0. Sol in H_2O, EtOH, and Et_2O.

SYNS: ACETAMIDOETHANE □ N-ETHYLACETAMIDE

TOXICITY DATA WITH REFERENCE
ipr-rat LD50:3130 mg/kg ARZNAD 20,1242,70
ipr-mus LD50:3560 mg/kg ARZNAD 20,1242,70
ivn-rbt LDLo:9410 mg/kg ARZNAD 20,1242,70
ivn-ckn LDLo:9600 mg/kg ARZNAD 20,1242,70

CONSENSUS REPORTS: Reported in EPA TSCA Inventory.

SAFETY PROFILE: Moderately toxic by intraperitoneal route. Combustible when exposed to heat or flame; can react with oxidizing materials. To fight fire, use alcohol foam. When heated to decomposition it emits toxic fumes of NO_x. See also AMIDES.

EFM500 CAS:102585-53-5 HR: 2
2-(2-(3-(N-ETHYLACETAMIDO)-2,4-6-TRIIODOPHENOXY)ETHOXY)ACETIC ACID SODIUM SALT
mf: $C_{14}H_{16}I_3NO_5$•Na mw: 682.00

TOXICITY DATA WITH REFERENCE
orl-mus LD50:1300 mg/kg FRPSAX 31,349,76
ivn-mus LD50:550 mg/kg FRPSAX 31,349,76

SAFETY PROFILE: Moderately toxic by ingestion and intravenous routes. When heated to decomposition it emits very toxic fumes of I^-, NO_x, and Na_2O. See also IODIDES.

EFN500 CAS:102585-52-4 HR: 2
2-(2-(3-(N-ETHYLACETAMIDO)-2,4,6-TRIIODOPHENOXY)ETHOXY)-2-PHENYL ACETIC ACID SODIUM SALT
mf: $C_{20}H_{20}I_3NO_5$•Na mw: 758.10

TOXICITY DATA WITH REFERENCE
orl-mus LD50:1600 mg/kg FRPSAX 31,349,76
ivn-mus DL50:405 mg/kg FRPSAX 31,349,76

SAFETY PROFILE: Moderately toxic by ingestion and intravenous routes. When heated to decomposition it emits toxic fumes of I^-, NO_x, and Na_2O. See also IODIDES.

EFO000 CAS:49642-61-7 HR: 2
2-(2-(3-(N-ETHYLACETAMIDO)-2,4,6-TRIIODOPHENOXY)ETHOXY)PROPIONIC ACID SODIUM SALT
mf: $C_{15}H_{18}I_3NO_5$•Na mw: 696.03

TOXICITY DATA WITH REFERENCE
orl-mus LD50:1600 mg/kg FRPSAX 31,349,76
ivn-mus LD50:425 mg/kg FRPSAX 31,349,76

SAFETY PROFILE: Moderately toxic by ingestion and intravenous routes. When heated to decomposition it emits very toxic fumes of I^-, NO_x, and Na_2O.

E

EFQ500 CAS:529-65-7 HR: 2
N-ETHYLACETANILIDE
mf: $C_{10}H_{13}NO$ mw: 163.24

PROP: White crystals, faint odor. Mp: 54°, bp: 258°, d: 0.994, vap d: 5.62.

SYNS: ACETETHYLANILIDE □ ETHYLACETANILIDE

TOXICITY DATA WITH REFERENCE
orl-mus LD50:409 mg/kg TXAPA9 19,20,71

CONSENSUS REPORTS: Reported in EPA TSCA Inventory.

SAFETY PROFILE: Moderately toxic by ingestion. Can react with oxidizing materials. To fight fire, use foam, CO_2, dry chemical. When heated to decomposition it emits toxic fumes of NO_x.

EFR000 CAS:141-78-6 HR: 3
ETHYL ACETATE
DOT: UN 1173
mf: $C_4H_8O_2$ mw: 88.12

$CH_3CH_2OCO{\cdot}CH_3$

PROP: A volatile, flammable, colorless liquid with fragrant fruity odor. Mp: −83.6°, bp: 77.15°, ULC: 85–90, lel: 2.2%, uel: 11%, flash p: 24°F, d: 0.8946 @ 25°, autoign temp: 800°F, vap press: 100 mm @ 27.0°, vap d: 3.04. Misc with alc, ether, glycerin, volatile oils, water @ 54° and most org solvs.

SYNS: ACETIC ETHER □ ACETIDIN □ ACETOXYETHANE □ AETHYLACETAT (GERMAN) □ ESSIGESTER (GERMAN) □ ETHYLACETAAT (DUTCH) □ ETHYL ACETIC ESTER □ ETHYLE (ACETATE d') (FRENCH) □ ETHYL ETHANOATE □ ETILE (ACETATO di) (ITALIAN) □ FEMA No. 2414 □ OCTAN ETYLU (POLISH) □ RCRA WASTE NUMBER U112 □ VINEGAR NAPHTHA

TOXICITY DATA WITH REFERENCE
eye-hmn 400 ppm JIHTAB 25,282,43
sln-smc 24,400 ppm MUREAV 149,339,85
cyt-ham:fbr 9 g/L FCTOD7 22,623,84
ihl-hmn TCLo:400 ppm:NOSE,EYE,PUL JIHTAB 25,282,43
orl-rat LD50:5620 mg/kg YKYUA6 32,1241,81
ihl-rat LC50:1600 ppm/8H 14CYAT 2,1879,63
scu-rat LDLo:5000 mg/kg BSIBAC 18,45,43
orl-mus LD50:4100 mg/kg GISAAA 48(4),66,83
ihl-mus LCLo:31 g/m³/2H AGGHAR 5,1,33
ipr-mus LD50:709 mg/kg SCCUR• -,5,61
ihl-cat LCLo:61 g/m³ HBTXAC 1,336,55
scu-cat LD50:3000 mg/kg AGGHAR 5,1,33

orl-rbt LD50:4935 mg/kg IMSUAI 41,31,72
orl-gpg LD50:5500 mg/kg GISAAA 48(4),66,83
ihl-gpg LCLo:77 mg/m^3/1H MELAAD 24,166,33
scu-gpg LD50:3000 mg/kg AGGHAR 5,1,33

CONSENSUS REPORTS: Reported in EPA TSCA Inventory. EPA Genetic Toxicology Program.

OSHA PEL: TWA 400 ppm
ACGIH TLV: TWA 400 ppm
DFG MAK: 400 ppm (1400 mg/m^3)
DOT CLASSIFICATION: 3; *Label:* Flammable Liquid

SAFETY PROFILE: Poison by inhalation. Moderately toxic by intraperitoneal and subcutaneous routes. Mildly toxic by ingestion. Human systemic effects by inhalation: olfactory changes, conjunctiva irritation, and pulmonary changes. Human eye irritant. Mutation data reported. Irritating to mucous surfaces, particularly the eyes, gums, and respiratory passages, and is also mildly narcotic. On repeated or prolonged exposures, it causes conjunctival irritation and corneal clouding. It can cause dermatitis. High concentrations have a narcotic effect and can cause congestion of the liver and kidneys. Chronic poisoning has been described as producing anemia, leucocytosis (transient increase in the white blood cell count), and cloudy swelling, and fatty degeneration of the viscera. A synthetic flavoring substance and adjuvant.

Highly flammable liquid. A very dangerous fire hazard when exposed to heat or flame; can react vigorously with oxidizing materials. Moderate explosion hazard when exposed to flame. Potentially explosive reaction with lithium tetrahydroaluminate. Ignites on contact with potassium tert-butoxide. Violent reaction with chlorosulfonic acid; ($LiAlH_2$ + 2-chloromethyl furan); oleum. To fight fire, use CO_2, dry chemical or alcohol foam. When heated to decomposition it emits acrid smoke and irritating fumes. See also ESTERS.

For occupational chemical analysis use NIOSH: Ethyl Acetate S49.

EFR100 CAS:6413-10-1 HR: 1
ETHYL ACETOACETATE ETHYLENE KETAL
mf: $C_8H_{14}O_4$ mw: 174.22

SYNS: 1,3-DIOXOLANE-2-ACETIC ACID, 2-METHYL-, ETHYL ESTER ☐ ETHYL 3-OXOBUTYRATE ETHYLENE KETAL ☐ FRUCTONE ☐ 2-METHYL-1,3-DIOXOLANE-2-ACETIC ACID ETHYL ESTER

TOXICITY DATA WITH REFERENCE
orl-rat LD50:>5 g/kg FCTOD7 26,315,88
skn-rbt LD50:>5 g/kg FCTOD7 26,315,88

CONSENSUS REPORTS: Reported in EPA TSCA Inventory.

SAFETY PROFILE: Low toxicity by ingestion and skin contact. When heated to decomposition it emits acrid smoke and irritating vapors.

EFS000 CAS:141-97-9 HR: 2
ETHYL ACETYL ACETATE
mf: $C_6H_{10}O_3$ mw: 130.16

$CH_3CO{\cdot}CH_2CO{\cdot}OCH_2CH_3$

PROP: Colorless liquid; fruity odor. Bp: 180.8°, fp: −45°, flash p: 185°F (COC), autoign temp: 563°F, d: 1.0282 @ 20°/20°, refr index: 1.418, vap press: 1 mm @ 28.5°, vap d: 4.48. Misc in most org solvents; sol in dil alkalis pptd with CO_2; sltly sol in H_2O.

SYNS: ACETOACETIC ACID, ETHYL ESTER ☐ ACETOACETIC ESTER ☐ ACTIVE ACETYL ACETATE ☐ DIACETIC ETHER ☐ EAA ☐ ETHYL ACETOACETATE (FCC) ☐ ETHYL ACETYLACETONATE ☐ ETHYL BENZYL ACETOACETATE ☐ ETHYL-3-OXOBUTANOATE ☐ ETHYL-3-OXOBUTYRATE ☐ FEMA No. 2415 ☐ 3-OXOBUTANOIC ACID ETHYL ESTER

TOXICITY DATA WITH REFERENCE
skn-rbt 510 mg open MLD UCDS** 3/12/69
eye-rbt 23 mg AJOPAA 29,1363,46
orl-rat LD50:3980 mg/kg JIHTAB 31,60,49
orl-mus LD50:5105 mg/kg JPMSAE 60,1810,71

CONSENSUS REPORTS: Reported in EPA TSCA Inventory.

SAFETY PROFILE: Moderately toxic by ingestion. A skin and eye irritant. Combustible liquid when exposed to heat or flame; can react with oxidizing materials. Explosive reaction when heated with Zn + trimbromoneopentyl alcohol or 2,2,2-tris(bromomethyl)ethanol. To fight fire, use alcohol foam, CO_2, dry chemical. When heated to decomposition it emits acrid smoke and irritating fumes. See also ESTERS.

EFS500 CAS:107-00-6 HR: 3
ETHYL ACETYLENE
mf: C_4H_6 mw: 54

PROP: A colorless, highly flammable gas. Bp: 8.3°, d: 0.669 @ 0°/0°, mp: −130°, flash p: <30°F (TOC), <7°C (Gas >8°C).

SYNS: 1-BUTYNE ☐ ETHYL ACETYLENE, INHIBITED ☐ ETHYLETHYNE

CONSENSUS REPORTS: Reported in EPA TSCA Inventory.

SAFETY PROFILE: Probably an asphyxiant. A very dangerous fire hazard when exposed to heat, open flame, or powerful oxidizers. A dangerous explosion hazard. To fight fire, stop flow of gas. See also ACETYLENE and ACETYLENE COMPOUNDS.

EFS600 CAS:539-88-8 HR: 1
ETHYL 3-ACETYLPROPIONATE
mf: $C_7H_{12}O_3$ mw: 144.19

PROP: Liquid. D: 1.012, bp: 205–206°. Very sol in water; miscible with alc.

SYNS: ETHYL KETOVALERATE ☐ ETHYL 4-KETOVALERATE ☐ ETHYL LAEVULINATE ☐ ETHYL LEVULATE ☐ ETHYL 4-OXOPENTANOATE ☐ ETHYL 4-OXOVALERATE ☐ LEVULINIC ACID, ETHYL ESTER ☐ PENTANOIC ACID, 4-OXO-, ETHYL ESTER (9CI)

TOXICITY DATA WITH REFERENCE
skn-rbt 500 mg/24H MLD FCTOD7 20,679,82

CONSENSUS REPORTS: Reported in EPA TSCA Inventory.

SAFETY PROFILE: A skin irritant. When heated to decomposition it emits acrid smoke and irritating fumes.

EFT000 CAS:140-88-5 HR: 3
ETHYL ACRYLATE
DOT: UN 1917
mf: $C_5H_8O_2$ mw: 100.13

PROP: Colorless liquid; acrid penetrating odor. Mp: −71.2°, bp: 99.8°, fp: <−72°, lel: 1.8%, flash p: 60°F (OC), 48.2°F, d: 0.916–0.919, vap press: 29.3 mm @ 20°, vap d: 3.45. Misc with alc, ether; sltly sol in water.

SYNS: ACRYLATE d'ETHYLE (FRENCH) □ ACRYLIC ACID ETHYL ESTER □ ACRYLSAEUREAETHYLESTER (GERMAN) □ AETHYLACRYLAT (GERMAN) □ ETHOXYCARBONYLETHYLENE □ ETHYLACRYLAAT (DUTCH) □ ETHYLAKRYLAT (CZECH) □ ETHYL PROPENOATE □ ETHYL-2-PROPENOATE □ ETIL ACRILATO (ITALIAN) □ ETILACRILATULUI (ROMANIAN) □ FEMA No. 2418 □ NCI-C50384 □ 2-PROPENOIC ACID, ETHYL ESTER (MAK) □ RCRA WASTE NUMBER U113

TOXICITY DATA WITH REFERENCE
eye-rat 1204 ppm/14H-I JIDHAN 31,317,49
eye-mky 1204 ppm/15H-I JIHTAB 31,317,49
skn-rbt 500 mg open MLD UCDS** 12/14/71
skn-rbt 10 mg/24H MLD JIHTAB 31,311,49
eye-rbt 45 mg MLD UCDS** 12/14/71
eye-rbt 1204 ppm/7H JIHTAB 31,317,49
mma-mus:lym 20 mg/L ENMUDM 8(Suppl 6),4,86
msc-mus:lym 20 mg/L ENMUDM 8(Suppl 6),4,86
orl-rat TDLo:51,500 mg/kg/2Y-I:CAR NTPTR* NTP-TR-259,86
orl-mus TDLo:103 g/kg/2Y-I:CAR NTPTR* NTP-TR-259,86
ihl-hmn TCLo:50 ppm:NOSE,EYE,PUL 34ZIAG -,75,69
orl-rat LD50:800 mg/kg BCTKAG 12,405,79
ihl-rat LC50:2180 ppm/4H:NOSE,EYE,PUL JTEHD6 16,811,85
skn-rat LDLo:1800 mg/kg PJPPAA 32,223,80
ipr-rat LD50:450 mg/kg AMPMAR 36,58,75
orl-mus LD50:1799 mg/kg TOLED5 11,125,82
ihl-mus LC50:16,200 mg/m³ GTPZAB 23(9),55,79
ipr-mus LD50:599 mg/kg JDREAF 51,526,72
ihl-rbt LCLo:1204 ppm/7H JIHTAB 31,317,49
ihl-gpg LCLo:1204 ppm/7H JIHTAB 31,317,49

CONSENSUS REPORTS: NTP 7th Annual Report on Carcinogens. IARC Cancer Review: Group 2B IMEMDT 7,56,87; Animal Sufficient Evidence IMEMDT 39,81,86; Animal Inadequate Evidence IMEMDT 19,47,79; Human Inadequate Evidence IMEMDT 19,47,79. NTP Carcinogenesis Studies (gavage); Clear Evidence: mouse, rat NTPTR* NTP-TR-259,86. Reported in EPA TSCA Inventory. Community Right-To-Know List.

OSHA PEL: TWA 5 ppm; STEL 25 ppm (skin)
ACGIH TLV: TWA 5 ppm; STEL 15 ppm; Suspected Human Carcinogen
DFG MAK: 5 ppm (20 mg/m³)
DOT CLASSIFICATION: 3; *Label:* Flammable Liquid

SAFETY PROFILE: Confirmed carcinogen with experimental carcinogenic data. Poison by ingestion and inhalation. Moderately toxic by skin contact and intraperitoneal routes. Human systemic effects by inhalation: eye, olfactory, and pulmonary changes. A skin and eye irritant. Characterized in its terminal stages by dyspnea, cyanosis, and convulsive movements. It caused severe local irritation of the gastroenteric tract; and toxic degenerative changes of cardiac, hepatic, renal, and splenic tissues were observed. It gave no evidence of cumulative effects. When applied to the intact skin of rabbits, the ethyl ester caused marked local irritation, erythema, edema, thickening, and vascular damage. Animals subjected to a fairly high concentration of these esters suffered irritation of the mucous membranes of the eyes, nose, and mouth as well as lethargy, dyspnea, and convulsive movements. A substance that migrates to food from packaging materials.

Flammable liquid. A very dangerous fire hazard when exposed to heat or flame; can react vigorously with oxidizing materials. Violent reaction with chlorosulfonic acid. To fight fire, use CO_2, dry chemical, or alcohol foam. When heated to decomposition it emits acrid smoke and irritating fumes. See also ESTERS.

For occupational chemical analysis use NIOSH: Esters I, 1450.

E

EFT500 CAS:462-95-3 HR: 3
ETHYLAL
DOT: UN 2373
mf: $C_5H_{12}O_2$ mw: 104.17

PROP: Bp: 89°, flash p: <69.8°F. Sol in H_2O.

SYN: DIETHOXYMETHANE (DOT)

TOXICITY DATA WITH REFERENCE
orl-rbt LD50:2604 mg/kg PSEBAA 29,730,32

CONSENSUS REPORTS: Reported in EPA TSCA Inventory.

DOT CLASSIFICATION: 3; *Label:* Flammable Liquid

SAFETY PROFILE: Moderately toxic by ingestion. Flammable when exposed to heat or flame; can react vigorously with oxidizers. When heated to decomposition it emits acrid smoke and irritating fumes.

EFU000 CAS:64-17-5 HR: 3
ETHYL ALCOHOL
DOT: UN 1170/UN 1986/UN 1987
mf: C_2H_6O mw: 46.08

PROP: Clear, colorless, very mobile liquid; fragrant odor and burning taste. Bp: 78.32°, ULC: 70, lel: 3.3%, uel: 19% @ 60°, fp: −117°, flash p: 55.6°F, d: 0.7893 @ 20°/4°, autoign temp: 793°F, vap press: 40 mm @ 19°, vap d: 1.59, refr index: 1.364. Misc in water, alc, chloroform, ether, and most org solvs.

SYNS: ABSOLUTE ETHANOL □ AETHANOL (GERMAN) □ AETHYLALKOHOL (GERMAN) □ ALCOHOL □ ALCOHOLS, n.o.s. (UN 1987) (DOT) □ ALCOHOL, anhydrous □ ALCOHOL, dehydrated □ ALCOOL ETHYLIQUE (FRENCH) □ ALCOOL ETILICO (ITALIAN) □ ALCOHOLS, toxic, n.o.s. (UN 1986) (DOT) □ ALGRAIN □ ALKOHOL (GER-

MAN) □ ALKOHOLU ETYLOWEGO (POLISH) □ ANHYDROL □ COLOGNE SPIRIT □ ETANOLO (ITALIAN) □ ETHANOL (MAK) □ ETHANOL 200 PROOF □ ETHANOL SOLUTIONS (UN 1170) (DOT) □ ETHYLALCOHOL (DUTCH) □ ETHYL ALCOHOL, anhydrous □ ETHYL ALCOHOL SOLUTIONS (UN 1170) (DOT) □ ETHYL HYDRATE □ ETHYL HYDROXIDE □ ETYLOWY ALKOHOL (POLISH) □ FERMENTATION ALCOHOL □ GRAIN ALCOHOL □ JAYSOL □ JAYSOL S □ METHYLCARBINOL □ MOLASSES ALCOHOL □ NCI-C03134 □ POTATO ALCOHOL □ SD ALCOHOL 23-HYDROGEN □ SPIRITS of WINE □ SPIRT □ TECSOL

TOXICITY DATA WITH REFERENCE
skn-rbt 20 mg/24H MOD 85JCAE-,189,86
skn-rbt 500 mg/24H SEV 28ZPAK -,34,72
eye-rbt 500 mg/24H MLD 85JCAE-,189,86
eye-rbt 100 mg/24H MOD 28ZPAK -,34,72
eye-rbt 100 mg/4S rns MOD FCTOD7 20,573,82
mmo-esc 140 g/L MUREAV 130,97,84
dni-hmn:lym 220 mmol/L PNASA6 79,1171,82
cyt-mus-orl 40 g/kg NATUAS 302,258,83
orl-wmn TDLo:41 g/kg (41W preg):REP AJDCAI 129,1075,75
iut-wmn TDLo:200 mg/kg (5D pre):REP INJFA3 14,280,69
ipr-rat TDLo:15 g/kg (female 8-13D post):REP TJADAB 36,31A,87
orl-gpg TDLo:90 g/kg (female 1-68D post):REP ANREAK 127,438,57
orl-rat TDLo:354 g/kg (lactating female 10D post):REP RCPBDC 2,119,77
orl-rat TDLo:132 g/kg (female 1-22D post):REP TJADAB 23,217,81
orl-rat TDLo:90 g/kg (female 1-15D post):REP TJADAB 24,13,81
orl-dog TDLo:21,600 mg/kg (female 1-60D post):REP FEPRA7 36,285,77
orl-dog TDLo:260 g/kg (female 1-62D post):REP ACRSDM 4,123,80
mul-rat TDLo:642 g/kg (female 1-21D post):REP DEPBA5 10,435,77
orl-rat TDLo:4 g/kg (13D preg):TER CYGEDX 15,23,81
ipr-mus TDLo:5622 μg/kg (female 10D post):TER AJOGAH 124,676,76
ipr-rat TDLo:2240 mg/kg (female 9-12D post):TER DABBBA 38,5835,78
orl-rat TDLo:12 g/kg (female 9-12D post):TER PLPSAX 7,311,79
ivn-rat TDLo:4 g/kg (female 6-7D post):TER RECYAR 8,105,71
orl-mky TDLo:78 g/kg (female 4-23W post):REP TJADAB 35,345,87
orl-rbt TDLo:3945 mg/kg (female 1D pre):REP JOENAK 34,275,66
itt-rat TDLo:400 mg/kg (male 1D pre):REP FESTAS 24,884,73
orl-rbt TDLo:3750 mg/kg (female 1D pre):REP IJMRAQ 58,501,70
ipr-mus TDLo:4300 mg/kg (female 10D post):REP NETOD7 2,227,80
orl-mky TDLo:206 g/kg (female 90D pre):REP SCIEAS 221,677,83
orl-rat TDLo:322 g/kg (male 35D pre):REP INJFA3 23,176,78
orl-rat TDLo:24 g/kg (female 14-16D post):TER TJADAB 23,41A,81
ipr-rat TDLo:600 mg/kg (female 8-15D post):TER IJEBA6 21,108,83
orl-rat TDLo:4 g/kg (female 6-15D post):TER JTEHD6 10,267,82
orl-rat TDLo:44 g/kg (female 7-17D post):TER ESKHA5 (103),10,85
ivn-dom TDLo:1 g/kg (female 18W post):TER AJOGAH 151,859,85
orl-mus TDLo:320 mg/kg/50W-I:ETA CALEDQ 13,345,81
rec-mus TDLo:120 g/kg/18W-I:ETA ZIETA2 59,203,28
orl-mus TD:400 g/kg/57W-I:ETA ZIETA2 59,203,28
orl-chd LDLo:2000 mg/kg ATXKA8 17,183,58
orl-cld TDLo:14,400 mg/kg/30M-I ACPAAN 74,977,85
orl-man TDLo:700 mg/kg NETOD7 8,77,86
orl-hmn LDLo:1400 mg/kg NPIRI* 1,44,74
orl-man TDLo:50 mg/kg:GIT JPETAB 56,117,36
orl-man TDLo:1430 μg/kg:CNS JPETAB 197,488,76
orl-wmn TDLo:256 g/kg/12W:CNS,END JAMAAP 238,2143,77
scu-inf LDLo:19,440 mg/kg:CNS,MET AJCPAI 5,466,35
orl-rat LD50:7060 mg/kg TXAPA9 16,718,70
ihl-rat LC50:20,000 ppm/10H NPIRI* 1,44,74
ipr-rat LD50:3750 mg/kg EVHPAZ 61,321,85
ivn-rat LD50:1440 mg/kg TXAPA9 18,60,71
orl-mus LD50:3450 mg/kg GISAAA 32(3),31,67
ihl-mus LC50:39 g/m³/4H GTPZAB 26(8),82
ipr-mus LD50:933 mg/kg SCCUR* -,5,61
scu-mus LD50:8285 mg/kg FAONAU 48A,99,70
ivn-mus LD50:1973 mg/kg HBTXAC 1,128,56
orl-dog LDLo:5500 mg/kg HBTXAC 1,130,56
ipr-dog LDLo:3000 mg/kg BJIMAG 1,207,44
scu-dog LDLo:6000 mg/kg HBTXAC 1,130,56

CONSENSUS REPORTS: IARC Cancer Review: Human Sufficient Evidence IMEMDT 44,259,88. Reported in EPA TSCA Inventory. EPA Genetic Toxicology Program.

OSHA PEL: TWA 1000 ppm
ACGIH TLV: TWA 1000 ppm
DFG MAK: 1,000 ppm (1,900 mg/m³)
DOT CLASSIFICATION: 3; *Label:* Flammable Liquid (UN 1987, UN 1170); DOT Class: 3; *Label:* Flammable Liquid, Poison (UN 1986)

SAFETY PROFILE: Confirmed human carcinogen for ingestion of beverage alcohol. Experimental tumorigenic and teratogenic data. Moderately toxic to humans by ingestion. Moderately toxic experimentally by intravenous and intraperitoneal routes. Mildly toxic by inhalation and skin contact. Human systemic effects by ingestion and subcutaneous route: sleep disorders, hallucinations, distorted perceptions, convulsions, motor activity changes, ataxia, coma, antipsychotic, headache, pulmonary changes, alteration in gastric secretion, nausea or vomiting, other gastrointestinal changes, menstrual cycle changes, and body temperature decrease. Can also cause glandular effects in humans. Human reproductive effects by ingestion, intravenous, and intrauterine routes: changes in female fertility index. Effects on newborn include: changes in apgar score, neonatal measures or effects, and drug dependence. Experimental reproductive effects. Human mutation data reported. An eye and skin irritant.

The systemic effect of ethanol differs from that of methanol. Ethanol is rapidly oxidized in the body to carbon dioxide and water, and, in contrast to methanol, no cumulative effect occurs. Though ethanol possesses narcotic properties, concentrations sufficient to produce this effect are not reached in industry. Concentrations below 1,000 ppm usually produce no signs of intoxication. Exposure to concentrations over 1,000 ppm may cause headache, irritation of the eyes, nose, and throat, and, if continued for an hour, drowsiness and lassitude, loss of appetite, and inability to concentrate. There is no concrete evidence that repeated exposure to ethanol vapor results in cirrhosis of the liver. Ingestion of large doses can cause alcohol poisoning. Repeated ingestions can lead to alcoholism. It is a central nervous system depressant.

Flammable liquid when exposed to heat or flame; can react vigorously with oxidizers. To fight fire, use alcohol foam, CO_2, dry chemical. Explosive reaction with the oxidized coating around potassium metal. Ignites and then explodes on contact with acetic anhydride + sodium hydrogen sulfate. Reacts violently with acetyl bromide (evolves hydrogen bromide), dichloromethane + sulfuric acid + nitrate or nitrite, disulfuryl difluoride, tetrachlorosilane + water, and strong oxidants. Ignites on contact with disulfuric acid + nitric acid, phosphorus(III) oxide, platinum, potassium-tert-butoxide + acids. Forms explosive products in reaction with ammonia + silver nitrate (forms silver nitride and silver fulminate), magnesium perchlorate (forms ethyl perchlorate), nitric acid + silver (forms silver fulminate), silver nitrate (forms ethyl nitrate), silver(I) oxide + ammonia or hydrazine (forms silver nitride and silver fulminate), sodium (evolves hydrogen gas). Incompatible with acetyl chloride, BrF_5, $Ca(OCl)_2$, ClO_3, CrO_3, $Cr(OCl)_2$, (cyanuric acid + H_2O), H_2O_2, HNO_3, (H_2O_2 + H_2SO_4), (I + CH_3OH + HgO), [$Mn(ClO_4)_2$ + 2,2-dimethoxy propane], $Hg(NO_3)_2$, $HClO_4$, perchlorates, (H_2SO_4 + permanganates), $HMnO_4$, KO_2, $KOC(CH_3)_3$, $AgClO_4$, NaH_3N_2, $UO_2(ClO_4)_2$.

For occupational chemical analysis use NIOSH: Alcohols I, 1400.

EFU050 CAS:563-43-9 HR: 1
ETHYL ALUMINUM DICHLORIDE
mf: $C_2H_5AlCl_2$ mw: 126.95

SYNS: ALUMINUM, DICHLOROETHYL- □ DICHLOROETHYLALUMINUM □ DICHLOROMONOETHYLALUMINUM □ ETHYLDICHLOROALUMINUM

CONSENSUS REPORTS: Reported in EPA TSCA Inventory.

ACGIH TLV: TWA 2 mg(Al)/m^3

SAFETY PROFILE: Mildly toxic by inhalation. When heated to decomposition it emits toxic vapors of Cl^-.

EFU100 CAS:2938-73-0 HR: 3
ETHYL ALUMINUM DIIODIDE
mf: $C_2H_5AlI_2$ mw: 309.85

SAFETY PROFILE: Ignites spontaneously in air. When heated to decomposition it emits toxic fumes of I^-. See also ALUMINUM and IODIDES.

EFU400 CAS:75-04-7 HR: 3
ETHYLAMINE
DOT: UN 1036/UN 2270
mf: C_2H_7N mw: 45.10

PROP: Colorless gas or liquid, strong ammonia-like odor. Bp: 16.6°, flammable, lel: 4.95%, uel: 20.75%, fp: −80.6°, flash p: −0.4°F, d: 0.662 @ 20°/4°, autoign temp: 725°F, vap d: 1.56. vap press: 400 mm @ 20°. Misc with water, alc, and ether; salted out by NaOH.

SYNS: AETHYLAMINE (GERMAN) □ AMINOETHANE □ 1-AMINOETHANE □ ETHANAMINE □ ETHYLAMINE (UN 1036) (DOT) □ ETHYLAMINE, aqueous solution with not <50% but not >70% ethylamine (UN 2270) (DOT) □ ETILAMINA (ITALIAN) □ ETYLOAMINA (POLISH) □ MONOETHYLAMINE (DOT) □ MONOETHYLAMINE, anhydrous (DOT)

TOXICITY DATA WITH REFERENCE
skn-rbt 500 mg/24H MLD 85JCAE -,429,86
eye-rbt 50 ppm/10D-I SEV AMIHBC 3,287,51
eye-rbt 250 μg/24H SEV 85JCAE-,429,86
orl-rat LD50:400 mg/kg AMIHBC 10,61,54
ihl-rat LCLo:3000 ppm/4H AEHLAU 1,343,60
skn-rbt LD50:390 mg/kg AEHLAU 1,343,60
ivn-rbt LDLo:350 mg/kg HBAMAK 4,1295,35
ihl-uns LC50:2300 mg/m^3 TPKVAL 14,80,75

CONSENSUS REPORTS: Reported in EPA TSCA Inventory.

OSHA PEL: TWA 10 ppm
ACGIH TLV: TWA 5 ppm; 15 ppm STEL (skin)
DFG MAK: 10 ppm (18 mg/m^3)
DOT CLASSIFICATION: 2.1; *Label:* Flammable Gas (UN 1036); DOT Class: 3; *Label:* Flammable Liquid (UN 2270)

SAFETY PROFILE: A poison by ingestion, skin contact, and intravenous routes. Moderately toxic by inhalation. A severe eye irritant. A very dangerous fire hazard when exposed to heat or flame. Moderate explosion hazard when exposed to spark or flame. Keep away from heat and open flame, can react vigorously with oxidizing materials. To fight fire, stop flow of gas, use alcohol foam, dry chemical. Incompatible with cellulose nitrate or oxidizers. When heated to decomposition it emits toxic fumes of NO_x. See also AMINES.

For occupational chemical analysis use OSHA: #36.

EFU500 CAS:75-23-0 HR: 2
ETHYLAMINE with BORON FLUORIDE (1:1)
mf: $C_2H_7BF_3N$ mw: 112.91

PROP: Crystals from Et_2O. Mp: 75–76°, bp: 147–148° @ 1 mm. Decomp rapidly when heated in air above mp. Sol in C_6H_6; sltly sol in Et_2O.

SYNS: BORONTRIFLUORIDE MONOETHYLAMINE □ BTFMEA

TOXICITY DATA WITH REFERENCE
ipr-mus LDLo:422 mg/kg NTIS** AD441-640

CONSENSUS REPORTS: Reported in EPA TSCA Inventory.

SAFETY PROFILE: Moderately toxic by intraperitoneal route. When heated to decomposition it emits very toxic fumes of NO_x and F^-. See also BORON TRIFLUORIDE and AMINES.

EFW000 **CAS:557-66-4** **HR: 2**
ETHYLAMINE HYDROCHLORIDE
mf: $C_2H_7N•ClH$ mw: 81.56

PROP: Crystals from alc (aq). Monoclinic. D: 1.227, mp: 108–109°. Sol in H_2O and EtOH; sltly sol in $CHCl_3$ and Me_2CO; prac insol in Et_2O.

SYNS: ETHYL AMMONIUM CHLORIDE □ MONOETHYLAMMONIUM CHLORIDE

TOXICITY DATA WITH REFERENCE
ipr-mus LD50:2610 mg/kg YKKZAJ 97,1117,77

CONSENSUS REPORTS: Reported in EPA TSCA Inventory.

SAFETY PROFILE: Moderately toxic by intraperitoneal route. When heated to decomposition it emits very toxic fumes of HCl, NH_3, and NO_x.

EFX000 **CAS:94-09-7** **HR: 3**
ETHYL-4-AMINOBENZOATE
mf: $C_9H_{11}NO_2$ mw: 165.21

PROP: Crystals or needles from alc. Mp: 92°, bp: 183–184° @ 14 mm.

SYNS: AMERICAINE □ p-AMINOBENZOIC ACID ETHYL ESTER □ 4-AMINOBENZOIC ACID ETHYL ESTER □ ANESTHESIN □ ANESTHONE □ BENZOCAINE □ ETHYL AMINOBENZOATE □ ETHYL-p-AMINOBENZOATE □ KELOFORM □ NORCAIN □ ORTHESIN □ PARATHESIN □ TOPCAINE

TOXICITY DATA WITH REFERENCE
skn-gpg 2%/48H MLD JSCCA5 28,357,77
rec-inf TDLo:12 mg/kg NEJMAG 263,454,60
ipr-mus LD50:216 mg/kg JMCMAR 17,900,74
orl-rbt LDLo:1150 mg/kg NIIRDN 6,33,82
orl-bwd LD50:56 mg/kg TXAPA9 21,315,72

CONSENSUS REPORTS: Reported in EPA TSCA Inventory.

SAFETY PROFILE: Poison by ingestion and intraperitoneal routes. Human systemic effects by rectal route: methemoglobinemia/carboxhemoglobinemia in infants. A skin irritant and a mild sensitizer. Local contact may cause contact dermatitis. Used as a topical anesthetic and as a sun-screening agent. When heated to decomposition it emits highly toxic fumes of NO_x. See also ETHYL ALCOHOL and ESTERS.

EFX500 **CAS:886-86-2** **HR: 3**
ETHYL-m-AMINOBENZOATE METHANE SULFONATE
mf: $C_{10}H_{15}NO_5S$ mw: 261.32

SYNS: 3-AMINOBENZOIC ACID ETHYL ESTER METHANESULFONATE □ ETHYL m-AMINOBENZOATE, METHANESULFONIC ACID SALT □ FINQUEL □ METACAINE □ MS-222 □ TRICAINE □ TRICAINE METHANE SULFONATE

TOXICITY DATA WITH REFERENCE
ivn-mus LD50:180 mg/kg CSLNX* NX#03233
ipr-frg LDLo:250 mg/kg PHMCAA 16,252,74

CONSENSUS REPORTS: Reported in EPA TSCA Inventory.

SAFETY PROFILE: Poison by intravenous and intraperitoneal routes. When heated to decomposition it emits very toxic fumes of SO_x and NO_x. See also ESTERS and SULFONATES.

EGA500 **CAS:110-73-6** **HR: 3**
2-ETHYLAMINOETHANOL
mf: $C_4H_{11}NO$ mw: 89.16

PROP: Oily liquid with faint odor, fumes in air. Bp: 169–170°. Flash p: 160°F (OC), d: 0.92, vap d: 3.06. Sol in H_2O.

SYNS: 2-(ETHYLAMINO)ETHANOL □ 2-N-MONOETHYLAMINOETHANOL

TOXICITY DATA WITH REFERENCE
skn-rbt 10 mg/24H open MLD AMIHBC 10,61,54
eye-rbt 250 µg open SEV AMIHBC 10,61,54
orl-rat LD50:1000 mg/kg FCTXAV 5,327,67
ipr-rat LD50:1170 mg/kg TXAPA9 12,486,68
skn-rbt LD50:360 mg/kg AMIHBC 10,61,54
orl-mam LD50:1200 mg/kg TXAPA9 8,344,66

CONSENSUS REPORTS: Reported in EPA TSCA Inventory.

SAFETY PROFILE: Poison by skin contact. Moderately toxic by ingestion and intraperitoneal routes. A skin and severe eye irritant. Flammable when exposed to heat or flame; can react vigorously with oxidizers. To fight fire, use alcohol foam, dry chemical, CO_2. When heated to decomposition it emits toxic fumes of NO_x.

EGC000 **CAS:52400-55-2** **HR: 3**
2-(2-(ETHYLAMINO)ETHYL)-2-METHYL-1,3-BENZODIOXOLE HYDROCHLORIDE
mf: $C_{12}H_{17}NO_2•ClH$ mw: 243.76

TOXICITY DATA WITH REFERENCE
ivn-rat LD50:40 mg/kg EJMCA5 12,413,77
ipr-mus LD50:100 mg/kg EJMCA5 12,413,77

SAFETY PROFILE: Poison by intravenous and intraperitoneal routes. When heated to decomposition it emits very toxic fumes of HCl and NO_x.

EGC500 **CAS:42145-91-5** **HR: 3**
dl-(±)-3-(2-ETHYLAMINO-1-HYDROXYETHYL)PHENYL PIVALATE HYDROCHLORIDE
mf: $C_{15}H_{23}NO_3•ClH$ mw: 301.85

PROP: Crystals from 2-propanol. Mp: 209.4°.

SYNS: 2,2-DIMETHYLPROPANOIC ACID-3-(2-(ETHYLAMINO)-1-HYDROXYETHYL)PHENYL ESTER HYDROCHLORIDE □ dl-(±)α-(ETHYLAMINOMETHYL)-3′-HYDROXYBENZYL ALCOHOL 3-(2,2-DIMETHYLPROPIONATE)HCl □ α-((ETHYLAMINO)METHYL)-m-HYDROXYBENZYL ALCOHOL 2,2-DIMETHYLPROPIONATE HYDROCHLORIDE □ ETILEFRINE PIVALATE HYDROCHLORIDE □ K-30052

TOXICITY DATA WITH REFERENCE
orl-rat LDLo:240 mg/kg DRFUD4 4,413,79
ivn-rat LDLo:20 mg/kg DRFUD4 4,413,79

orl-mus LDLo: 500 mg/kg DRFUD4 4,413,79
ivn-mus LDLo: 24 mg/kg DRFUD4 4,413,79
orl-dog LDLo: 140 mg/kg DRFUD4 4,413,79
ivn-dog LDLo: 8 mg/kg DRFUD4 4,413,79

SAFETY PROFILE: Poison by ingestion and intravenous routes. When heated to decomposition it emits very toxic fumes of Cl^- and NO_x.

EGD000 CAS:1610-17-9 HR: 2
2-ETHYLAMINO-4-ISOPROPYLAMINO-6-METHOXY-s-TRIAZINE
mf: $C_9H_{17}N_5O$ mw: 211.31

SYNS: ATRATON ☐ ATRATONE ☐ ATROTON ☐ 4-ETHYLAMINO-6-ISOPROPYLAMINO-2-METHOXY-s-TRIAZINE ☐ 6-ETHYLAMINO-4-ISOPROPYLAMINO-2-METHOXY-1,3,5-TRIAZINE ☐ N-ETHYL-6-METHOXY-N′-(1-METHYLETHYL)-1,3,5-TRIAZINE-2,4-DIAMINE ☐ GEIGY 32,293 ☐ GESATAMIN ☐ 2-METHOXY-4-ETHYLAMINO-6-ISOPROPYLAMINO-s-TRIAZINE ☐ 2-METHOXY-4-ISOPROPYLAMINO-6-ETHYLAMINO-s-TRIAZINE ☐ PRIMATOL

TOXICITY DATA WITH REFERENCE
orl-rat LD50: 1465 mg/kg FMCHA2 -,C19,83
orl-mus LD50: 905 mg/kg 85DPAN -,-,71/76

SAFETY PROFILE: Moderately toxic by ingestion. When heated to decomposition it emits toxic fumes of NO_x.

EGE500 CAS:709-55-7 HR: 3
α-((ETHYLAMINO)METHYL)-m-HYDROXYBENZYL ALCOHOL
mf: $C_{10}H_{15}NO_2$ mw: 181.26

PROP: A solid. Mp: 147–148°.

SYNS: EFFORTIL ☐ ETHYLADRIANOL ☐ ETHYL NORADRIANOL ☐ N-ETHYLNORPHENYLEPHRINE ☐ ETILEFRINE ☐ m-HYDROXYPHENYLETHANOLETHYLAMINE ☐ 1-(3′-HYDROXYPHENYL)-2-ETHYLAMINOETHANOL

TOXICITY DATA WITH REFERENCE
ims-gpg TDLo: 13 mg/kg (female 30-60D post): TER EOGRAL 30,173,89
ims-gpg TDLo: 13 mg/kg (female 30-60D post): REP EOGRAL 30,173,89
orl-rat LD50: 114 mg/kg AIPTAK 180,155,69
ipr-rat LD50: 820 mg/kg OYYAA2 3,27,69
scu-rat LD50: 244 mg/kg AIPTAK 180,155,69
idu-rat LD50: 200 mg/kg AIPTAK 180,155,69
orl-mus LD50: 770 mg/kg ARZNAD 22,869,72

SAFETY PROFILE: Poison by ingestion, subcutaneous, and intraduodenal routes. Moderately toxic by intraperitoneal route. An experimental teratogen. Other experimental reproductive effects. When heated to decomposition it emits toxic fumes of NO_x.

EGF000 CAS:943-17-9 HR: 3
α-((ETHYLAMINO)METHYL)-m-HYDROXYBENZYL ALCOHOL HYDROCHLORIDE
mf: $C_{10}H_{15}NO_2{\cdot}ClH$ mw: 217.72

PROP: A solid. Mp: 121°.

SYN: ETILEFRINE HYDROCHLORIDE

TOXICITY DATA WITH REFERENCE
orl-rat LDLo: 187 mg/kg DRFUD4 4,413,79
ivn-rat LDLo: 5 mg/kg DRFUD4 4,413,79
orl-mus LDLo: 345 mg/kg DRFUD4 4,413,79
ivn-mus LDLo: 11 mg/kg DRFUD4 4,413,79

SAFETY PROFILE: Poison by ingestion and intravenous routes. When heated to decomposition it emits very toxic fumes of HCl and NO_x. See also ALCOHOLS.

E

EGG000 CAS:69781-71-1 HR: 3
N-((3-(ETHYLAMINO)PROPOXY)METHYL)DIPHENYLAMINE
mf: $C_{18}H_{24}N_2O{\cdot}ClH$ mw: 320.90

SYN: ETHYLAMINOPROPYLDIPHENYLAMINOCARBINOL HYDROCHLORIDE

TOXICITY DATA WITH REFERENCE
ipr-rat LDLo: 400 mg/kg JPETAB 18,467,22
ipr-cat LDLo: 50 mg/kg JPETAB 18,467,22
scu-cat LDLo: 50 mg/kg JPETAB 18,467,22
ivn-cat LDLo: 10 mg/kg JPETAB 18,467,22
ivn-rbt LDLo: 17 mg/kg JPETAB 18,467,22

SAFETY PROFILE: Poison by intraperitoneal, subcutaneous, and intravenous routes. When heated to decomposition it emits very toxic fumes of NO_x and HCl. See also AMINES.

EGI000 CAS:13275-68-8 HR: 3
2-ETHYLAMINO-1,3,4-THIADIAZOLE
mf: $C_4H_7N_3S$ mw: 129.20

SYNS: CL 19217 4090L 7-5525 ☐ 2-ETHYLAMINOTHIADIAZOLE ☐ NSC 4730

TOXICITY DATA WITH REFERENCE
scu-rat TDLo: 150 mg/kg (10-11D preg): TER JHMJAX 130,95,72
ipr-rat LD50: 200 mg/kg CPCHAO 18,307,62
scu-rat LD50: 200 mg/kg JHMJAX 130,95,72
ipr-mus LD50: 795 mg/kg NCISP* JAN86

CONSENSUS REPORTS: Reported in EPA TSCA Inventory.

SAFETY PROFILE: Poison by intraperitoneal and subcutaneous routes. Experimental teratogenic effects. When heated to decomposition it emits very toxic fumes of NO_x and SO_x.

EGI500 CAS:457-87-4 HR: 3
ETHYLAMPHETAMINE
mf: $C_{11}H_{17}N$ mw: 163.29

PROP: Bp: 104.5–106 @ 14°.

SYNS: 1-PHENYL-2-AETHYLAMINO-PROPAN (GERMAN) ☐ α-PHENYL-β-ETHYLAMINOPROPANE ☐ 1-PHENYL-2-ETHYLAMINOPROPANE

TOXICITY DATA WITH REFERENCE
orl-man TDLo: 4286 μg/kg THERAP 34,205,79
orl-rat LD50: 250 mg/kg BBMS** -,-,48
ipr-rat LDLo: 80 mg/kg AEPPAE 195,647,40
ipr-mus LDLo: 80 mg/kg AEPPAE 195,647,40

SAFETY PROFILE: Poison by ingestion and intraperitoneal routes. When heated to decomposition it emits toxic fumes of NO_x. See also BENZEDRINE.

EGI750 CAS:541-85-5 HR: 3
ETHYL AMYL KETONE
mf: $C_8H_{16}O$ mw: 128.24

PROP: Liquid; mild, fruity odor. Bp: 157–162°, d: 0.822 @ 20°/20°, flash p: 138°F. Sol in many organic solvents.

SYNS: ETHYL sec-AMYL KETONE □ 3-HEPTANONE, 5-METHYL- □ 3-METHYL-5-HEPTANONE □ 5-METHYL-3-HEPTANONE

TOXICITY DATA WITH REFERENCE
skn-rbt 500 mg MLD SCCUR* -,6,61
orl-rat LD50:3500 mg/kg SCCUR* -,6,61
ihl-rat LCLo:3484 ppm/8H SCCUR* -,6,61
orl-mus LD50:3800 mg/kg SCCUR* -,6,61
ihl-mus LCLo:3484 ppm/4H SCCUR* -,6,61
orl-gpg LD50:2500 mg/kg BCPCA6 16,631,67

CONSENSUS REPORTS: Reported in EPA TSCA Inventory.

OSHA PEL: TWA 25 ppm
ACGIH TLV: TWA 25 ppm
DOT CLASSIFICATION: 3; *Label:* Flammable Liquid

SAFETY PROFILE: Moderately irritating to skin, eyes, and mucous membranes by inhalation and ingestion. Narcotic in high concentration. See also KETONES. Flammable liquid when exposed to heat, sparks, or flame. When heated to decomposition it emits acrid smoke. Combustible. To fight fire, use foam, CO_2, dry chemical. See also KETONES.

For occupational chemical analysis use NIOSH: Ketones II (Desorption in 99:1 CS_2:methanol) 1301.

EGJ000 CAS:10138-47-3 HR: 2
2-(1-ETHYLAMYLOXY)ETHANOL
mf: $C_9H_{20}O_2$ mw: 160.29

SYN: 2-((1-ETHYLPENTYL)OXY)ETHANOL

TOXICITY DATA WITH REFERENCE
skn-rbt 10 mg/24H open MLD AMIHBC 10,61,54
eye-rbt 250 µg open SEV AMIHBC 10,61,54
orl-rat LD50:2280 mg/kg AMIHBC 10,61,54

SAFETY PROFILE: Moderately toxic by ingestion. A skin and severe eye irritant. When heated to decomposition it emits acrid smoke and irritating fumes.

EGJ500 CAS:10138-87-1 HR: 2
2-(2-(1-ETHYLAMYLOXY)ETHOXY)ETHANOL
mf: $C_{11}H_{24}O_3$ mw: 204.35

SYN: 2-(((1-ETHYLPENTYL)OXY)ETHOXY)ETHANOL

TOXICITY DATA WITH REFERENCE
skn-rbt 10 mg/24H open MLD AMIHBC 10,61,54
eye-rbt 250 µg open SEV AMIHBC 10,61,54
orl-rat LD50:2940 mg/kg AMIHBC 10,61,54

SAFETY PROFILE: Moderately toxic by ingestion. A skin and severe eye irritant. When heated to decomposition it emits acrid smoke and irritating fumes.

EGK000 CAS:103-69-5 HR: 3
N-ETHYLANILINE
DOT: UN 2272
mf: $C_8H_{11}N$ mw: 121.20

PROP: Clear, yellow-brown oily liquid. Mp: −63.5°, bp: 204°, d: 0.963 @ 20°/4°, fp −80, vap press: 1 mm @ 38.5°, vap d: 4.18, flash p: 185°F (OC).

SYNS: AETHYLANILIN (GERMAN) □ ANILINOETHANE □ N-ETHYLAMINOBENZENE □ ETHYLANILINE □ N-ETHYLBENZENAMINE □ N-ETHYLBENZENAMINO □ ETHYLPHENYLAMINE

TOXICITY DATA WITH REFERENCE
orl-rat LD50:334 mg/kg MarJV# 29MAR77
skn-rat LD50:4700 mg/kg AGGHAR 15,447,57
ipr-rat LD50:180 mg/kg AGGHAR 15,447,57
ipr-mus LD50:242 mg/kg YKKZAJ 97,1117,77
unr-mam LD50:600 mg/kg GISAAA 48(6),22,83

CONSENSUS REPORTS: Reported in EPA TSCA Inventory.

DOT CLASSIFICATION: 6.1; *Label:* KEEP AWAY FROM FOOD

SAFETY PROFILE: Poison by ingestion and intraperitoneal routes. Moderately toxic by an unspecified route. Mildly toxic by skin contact. An allergen. Flammable when exposed to heat or flame; can react with oxidizing materials. To fight fire, use dry chemical, CO_2, foam. Hypergolic reaction with red fuming nitric acid. When heated to decomposition or on contact with acid or acid fumes it emits highly toxic fumes of aniline and NO_x.

EGK500 CAS:578-54-1 HR: 3
2-ETHYLANILINE
DOT: UN 2273
mf: $C_8H_{11}N$ mw: 121.20

PROP: Yellow liquid, darkens upon standing. Mp: −63.5°, bp: 215°, flash p: 185°F (OC), d: 0.98 @ 25°/25°, vap d: 4.17.

SYNS: o-AMINOETHYLBENZENE □ ANILINE, o-ETHYL-(8CI) □ BENZENAMINE, 2-ETHYL-(9CI) □ 2-ETHYL ANILINE □ 2-ETHYLANILINE (DOT) □ 2-ETHYLBENZENAMINE

TOXICITY DATA WITH REFERENCE
orl-rat LD50:1260 mg/kg TXAPA9 22,153,72
orl-bwd LD50:750 mg/kg AECTCV 12,355,83

CONSENSUS REPORTS: Reported in EPA TSCA Inventory.

DOT CLASSIFICATION: 6.1; *Label:* KEEP AWAY FROM FOOD

SAFETY PROFILE: A poison. Moderately toxic by ingestion. Flammable when exposed to heat or flame; can react with oxidizing materials. To fight fire, use foam, CO_2, dry chemical. When heated to decomposition it emits highly toxic fumes of aniline and NO_x. See also N-ETHYLANILINE.

EGL000 **CAS:589-16-2** **HR: 3**
4-ETHYLANILINE
mf: $C_8H_{11}N$ mw: 121.20

PROP: D: 0.963, mp: −5°, bp: 213–214°. Insol in water; misc in alc and ether.

SYNS: 1-AMINO-4-ETHYLBENZENE □ p-ETHYLANILINE

TOXICITY DATA WITH REFERENCE
dns-rat:lvr 100 μmol/L ENMUDM 5,803,83
orl-mus LD50:500 mg/kg JPETAB 89,153,47
ipr-mus LD50:133 mg/kg JMCMAR 17,900,74
ivn-mus LD50:56 mg/kg CSLNX* NX#02908
orl-qal LD50:422 mg/kg AECTCV 12,355,83
orl-bwd LD50:75 mg/kg AECTCV 12,355,83

CONSENSUS REPORTS: Reported in EPA TSCA Inventory.

SAFETY PROFILE: Poison by ingestion, intravenous, and intraperitoneal routes. Mutation data reported. When heated to decomposition it emits toxic fumes of NO_x. See also N-ETHYLANILINE.

EGL500 **CAS:84-51-5** **HR: 3**
2-ETHYL-9,10-ANTHRACENEDIONE
mf: $C_{16}H_{12}O_2$ mw: 236.28

SYNS: 2-ETHYLANTHRAQUINONE □ 2-ETHYL-9,10-ANTHRAQUINONE □ USAF SO-1

TOXICITY DATA WITH REFERENCE
ipr-mus LD50:200 mg/kg NTIS** AD277-689

CONSENSUS REPORTS: Reported in EPA TSCA Inventory.

SAFETY PROFILE: Poison by intraperitoneal route. When heated to decomposition it emits acrid smoke and irritating fumes.

EGM000 **CAS:87-25-2** **HR: 2**
ETHYL ANTHRANILATE
mf: $C_9H_{11}NO_2$ mw: 165.21

PROP: Colorless to amber liquid; floral, orange-blossom odor. D: 1.115–1.120, refr index: 1.563–1.566, flash p: 151°F. Sol in alc, fixed oils, propylene glycol.

SYNS: o-AMINOBENZOIC ACID, ETHYL ESTER □ ETHYL-o-AMINOBENZOATE □ FEMA No. 2421

TOXICITY DATA WITH REFERENCE
skn-rbt 500 mg/24H MOD FCTXAV 14,759,76
orl-rat LD50:3750 mg/kg FCTXAV 14,759,76

CONSENSUS REPORTS: Reported in EPA TSCA Inventory.

SAFETY PROFILE: Moderately toxic by ingestion. A skin irritant. Combustible liquid. When heated to decomposition it emits toxic fumes of NO_x.

EGM100 **CAS:42971-09-5** **HR: 3**
ETHYL APOVINCAMINATE
mf: $C_{22}H_{26}N_2O_2$ mw: 350.50

PROP: Crystals from benzene or alc. Mp: 148–152° (decomp).

SYNS: 3-α,16-α-APOVINCAMINIC ACID ETHYL ESTER □CAVIN-TON □ ETHYL APOVINCAMIN-22-OATE □ RGH-4405 □ TCV-3B □ VINPOCETINE

TOXICITY DATA WITH REFERENCE
orl-rat TDLo:100 mg/kg (female 14-21D post):REP ARZNAD 26,1938,76
orl-rat TDLo:1 g/kg (female 7-15D post):REP YACHDS 11,1125,83
orl-rat TDLo:1 g/kg (female 7-15D post):TER YACHDS 11,1125,83
orl-rat TDLo:450 mg/kg (female 7-15D post):TER ARZNAD 26,1938,76
orl-rbt TDLo:65 mg/kg (female 6-18D post):REP YACHDS 11,3585,83
orl-rbt TDLo:1750 mg/kg (female 14D pre):REP YACHDS 11,3585,83
orl-rat TDLo:4375 mg/kg (male 35D pre):REP YACHDS 10,1847,82
orl-rat LD50:503 mg/kg ARZNAD 26,1938,76
ipr-rat LD50:134 mg/kg ARZNAD 26,1938,76
ivn-rat LD50:42,600 mg/kg ARZNAD 26,1938,76
orl-mus LD50:534 mg/kg ARZNAD 26,1938,76
ipr-mus LD50:161 mg/kg ARZNAD 26,1938,76
ivn-mus LD50:58,700 μg/kg ARZNAD 26,1938,76

SAFETY PROFILE: Poison by intravenous and intraperitoneal routes. Moderately toxic by ingestion. Experimental teratogenic and reproductive effects. When heated to decomposition it emits toxic fumes of NO_x.

EGM500 **CAS:871-31-8** **HR: 3**
ETHYL AZIDE
mf: $C_2H_5N_3$ mw: 71.08

SAFETY PROFILE: A liquid. Bp: 50°. An explosive sensitive to rapid heating, shock, or impact. Has exploded when heated to room temperature. When heated to decomposition it emits toxic fumes of NO_x. See also AZIDES.

EGN000 **CAS:817-87-8** **HR: 3**
ETHYL AZIDOFORMATE
mf: $C_3H_5N_3O_2$ mw: 115.09

$CH_3CH_2OCO{\bullet}N_3$

PROP: Bp: 28° @ 20 mm.

SYN: ETHYL CARBONAZIDATE

SAFETY PROFILE: Explodes at its boiling point (114°C). When heated to decomposition it emits toxic fumes of NO_x. See also AZIDES.

E

EGN100 **CAS:81852-50-8** **HR: 3**
ETHYL-2-AZIDO-2-PROPENOATE
mf: $C_5H_7N_3O_2$ mw: 141.13

$CH_3CH_2OCO{\cdot}C(N_3)=CH_2$

SAFETY PROFILE: Potentially explosive when heated. When heated to decomposition it emits toxic fumes of NO_x. See also AZIDES.

EGO000 **CAS:56961-62-7** **HR: 2**
5-ETHYL-1,2-BENZANTHRACENE
mf: $C_{20}H_{16}$ mw: 256.36

SYN: 8-ETHYLBENZ(a)ANTHRACENE

TOXICITY DATA WITH REFERENCE
skn-mus TDLo:650 mg/kg/27W-I:ETA PRLBA4 123,343,37

SAFETY PROFILE: Questionable carcinogen with experimental tumorigenic data. When heated to decomposition it emits acrid smoke and irritating fumes.

EGO500 **CAS:3697-30-1** **HR: 2**
10-ETHYL-1,2-BENZANTHRACENE
mf: $C_{20}H_{16}$ mw: 256.36

PROP: Faintly green-yellow crystals from EtOH. Mp: 113.5–114°.

SYN: 7-ETHYLBENZ(a)ANTHRACENE

TOXICITY DATA WITH REFERENCE
mma-sat 50 µg/plate MUREAV 206,55,88
scu-mus TDLo:600 mg/kg:ETA JNCIAM 1,303,40

SAFETY PROFILE: Questionable carcinogen with experimental tumorigenic data. Mutation data reported. When heated to decomposition it emits acrid smoke and irritating fumes.

EGP000 **CAS:18868-66-1** **HR: 2**
12-ETHYLBENZ(a)ANTHRACENE
mf: $C_{20}H_{16}$ mw: 256.36

PROP: Crystals from EtOH. Mp: 107.4–108.4°.

TOXICITY DATA WITH REFERENCE
ims-rat TDLo:50 mg/kg:ETA CNREA8 29,506,69

SAFETY PROFILE: Questionable carcinogen with experimental tumorigenic data. When heated to decomposition it emits acrid smoke and irritating fumes.

EGP500 **CAS:100-41-4** **HR: 3**
ETHYL BENZENE
DOT: UN 1175
mf: C_8H_{10} mw: 106.18

PROP: Colorless liquid, aromatic odor. Bp: 136.2°, fp: −94.9°, flash p: 59°F, d: 0.8669 @ 20°/4°, autoign temp: 810°F, vap press: 10 mm @ 25.9°, vap d: 3.66, lel: 1.2%, uel: 6.8%. Misc in alc and ether; insol in NH_3; sol in SO_2.

SYNS: AETHYLBENZOL (GERMAN) □ EB □ ETHYLBENZEEN (DUTCH) □ ETHYLBENZOL □ ETILBENZENE (ITALIAN) □ ETYLOBENZEN (POLISH) □ NCI-C56393 □ PHENYLETHANE

TOXICITY DATA WITH REFERENCE
skn-rbt 15 mg/24H open MLD AIHAAP 23,95,62
eye-rbt 100 mg AJOPAA 29,1363,46
sce-hmn:lym 1 mmol/L MUREAV 116,379,83
ihl-rat TCLo:600 mg/m³/24H (female 7-15D post):TER ARTODN 8,425,85
ihl-rat TCLo:985 ppm/7H (1-19D preg):TER BATTL* JAN,81
ihl-rbt TCLo:1 g/m³/24H (female 7-20D post):REP ARTODN 8,425,85
ihl-rat TCLo:97 ppm/7H (15D preg):REP BATTL* JAN,81
ihl-rbt TCLo:99 ppm/7H (female 1-18D post):REP NTIS** PB83-208074
ihl-rat TCLo:600 mg/m³/24H (female 7-15D post):REP ARTODN 8,425,85
ihl-rat TCLo:96 ppm/7H (1-19D preg):TER BATTL* JAN,81
ihl-hmn TCLo:100 ppm/8H:EYE,CNS,PUL AIHAAP 31,206,70
orl-rat LD50:3500 mg/kg AMIHAB 14,387,56
ihl-rat LCLo:4000 ppm/4H AIHAAP 23,95,62
ihl-mus LCLo:50 g/m³/2H GTPZAB 5(5),3,61
ipr-mus LD50:2272 mg/kg ARTODN 58,106,85
skn-rbt LD50:17,800 mg/kg FCTXAV 13,803,75
ihl-gpg LCLo:10,000 ppm PHRPA6 45,1241,30

CONSENSUS REPORTS: Reported in EPA TSCA Inventory. EPA Genetic Toxicology Program. Community Right-To-Know List.

OSHA PEL: TWA 100 ppm; STEL 125 ppm
ACGIH TLV: TWA 100 ppm; STEL 125 ppm; BEI: 2 g(mandelic acid)/L in urine at end of shift; 2 ppm ethyl benzene in end-exhaled air prior to next shift.
DFG MAK: 100 ppm (440 mg/m³)
DOT CLASSIFICATION: 3; *Label:* Flammable Liquid

SAFETY PROFILE: Moderately toxic by ingestion and intraperitoneal routes. Mildly toxic by inhalation and skin contact. An experimental teratogen. Other experimental reproductive effects. Human systemic effects by inhalation: eye, sleep and pulmonary changes. An eye and skin irritant. Human mutation data reported. The liquid is an irritant to the skin and mucous membranes. A concentration of 0.1% of the vapor in air is an irritant to human eyes, and a concentration of 0.2% is extremely irritating at first, then causes dizziness, irritation of the nose and throat, and a sense of constriction in the chest. Exposure of guinea pigs to 1% concentration has been reported as causing ataxia, loss of consciousness, tremor of the extremities, and finally death through respiratory failure. The pathological findings were congestion of the brain and lungs with edema.

A very dangerous fire and explosion hazard when exposed to heat or flame; can react vigorously with oxidizing materials. To fight fire, use foam, CO_2, dry chemical. Emitted from modern building materials (CENEAR 69,22,91). When heated to decomposition it emits acrid smoke and irritating fumes.

For occupational chemical analysis use NIOSH: Hydrocarbons, Aromatic, 1501.

EGQ000 **CAS:93-54-9** **HR: 2**
α-ETHYLBENZENEMETHANOL
mf: $C_9H_{12}O$ mw: 136.21

SYNS: EJIBIL □ α-ETHYLBENZYL ALCOHOL □ ETHYL PHENYL CARBINOL □ FELICUR □ FELITROPE □ FENICOL □ α-HYDROXYPROPYLBENZENE □ LIVONAL □ PHENICOL □ PHENYCHOLON □ PHENYLAETHYLCARBINOL (GERMAN) □ 1-PHENYLPROPANOL □ 1-PHENYL-1-PROPANOL □ 1-PHENYLPROPYL ALCOHOL □ SH 261

TOXICITY DATA WITH REFERENCE
orl-rat LD50:1600 mg/kg ARZNAD 12,347,62
orl-mus LD50:500 mg/kg AIPTAK 116,154,58
scu-mus LD50:700 mg/kg AIPTAK 116,154,58

SAFETY PROFILE: Moderately toxic by ingestion and subcutaneous routes. When heated to decomposition it emits acrid smoke and irritating fumes.

EGR000 **CAS:93-89-0** **HR: 2**
ETHYL BENZOATE
mf: $C_9H_{10}O_2$ mw: 150.19

PROP: Colorless liquid; heavy fruity odor. Mp: −34.6°, bp: 213.4°, flash p: >204°F, d: 1.048 @ 20°/20°, fp: −34°, refr index: 1.502–1.506, vap press: 1 mm @ 44.0°, vap d: 5.17, autoign temp: 914°F. Sol in alc, fixed oils, and propylene glycol; insol in glycerin, water @ 212°; misc in petroleum, chloroform, and ether.

SYNS: BENZOIC ETHER □ ESSENCE of NIOBE □ FEMA No. 2422

TOXICITY DATA WITH REFERENCE
skn-rbt 10 mg/24H open MLD AMIHBC 10,61,54
eye-rbt 500 mg open AMIHBC 10,61,54
orl-rat LD50:2100 mg/kg JPETAB 84,358,45
skn-cat LDLo:10 g/kg JPETAB 84,358,45
orl-rbt LD50:2630 mg/kg JPETAB 84,358,45

CONSENSUS REPORTS: Reported in EPA TSCA Inventory.

SAFETY PROFILE: Moderately toxic by ingestion. Mildly toxic by skin contact. A skin and eye irritant. Combustible liquid when exposed to heat or flame; can react with oxidizing materials. To fight fire, use foam, CO_2, dry chemical. When heated to decomposition it emits acrid smoke and irritating fumes. See also ESTERS.

EGR500 **CAS:6797-13-3** **HR: 2**
2-ETHYLBENZOXAZOLE
mf: C_9H_9NO mw: 147.19

TOXICITY DATA WITH REFERENCE
orl-mus LD40:1 g/kg JACSAT 67,905,45

CONSENSUS REPORTS: Reported in EPA TSCA Inventory.

SAFETY PROFILE: Moderately toxic by ingestion. When heated to decomposition it emits toxic fumes of NO_x.

EGR600 **CAS:94-02-0** **HR: 1**
ETHYL BENZOYL ACETATE
mf: $C_{11}H_{12}O_3$ mw: 192.23

SYNS: ACETIC ACID, BENZOYL-, ETHYL ESTER □ BENZENEPROPANOIC ACID, β-OXO-, ETHYL ESTER (9CI) □ BENZOYLACETIC ACID ETHYL ESTER □ ETHYL β-OXOBENZENEPROPANOATE

TOXICITY DATA WITH REFERENCE
orl-mus LD50:6800 mg/kg 85GMAT -,65,82

CONSENSUS REPORTS: Reported in EPA TSCA Inventory.

SAFETY PROFILE: Mildly toxic by ingestion. When heated to decomposition it emits acrid smoke and irritating vapors.

EGS000 **CAS:33878-50-1** **HR: 2**
ETHYL-N-BENZOYL-N-(3,4-DICHLOROPHENYL)-2-AMINOPROPIONATE
mf: $C_{18}H_{17}Cl_2NO_3$ mw: 366.26

SYNS: N-BENZOYL-N-(3,4-DICHLOROPHENYL)-l-ALANINE ETHYL ESTER □ BENZOYLPROP ETHYL □ ENAVEN □ SD 30,053 □ SUFFIX □ SUFFIX 25 □ WL 17,731

TOXICITY DATA WITH REFERENCE
orl-mus LD50:716 mg/kg 28ZEAL 5,24,76

SAFETY PROFILE: Moderately toxic by ingestion. When heated to decomposition it emits very toxic fumes of NO_x and Cl^-.

EGS500 **CAS:59965-27-4** **HR: 2**
2-ETHYL-3,4-BENZPHENANTHRENE
mf: $C_{20}H_{16}$ mw: 256.36

SYN: 5-ETHYLBENZO(c)PHENANTHRENE

TOXICITY DATA WITH REFERENCE
skn-mus TDLo:620 mg/kg/26W-I:ETA PRLBA4 131,170,42

SAFETY PROFILE: Questionable carcinogen with experimental tumorigenic data. When heated to decomposition it emits acrid smoke and irritating fumes.

EGT000 **CAS:2521-01-9** **HR: 3**
ETHYL-N-BENZYLCYCLOPROPANECARBAMATE
mf: $C_{13}H_{17}NO_2$ mw: 219.31

PROP: Bp: 90–92° @ 0.08 mm.

SYNS: ENCYPRATE □ ETHYL-N-BENZYL-N-CYCLOPROPYLCARBAMATE

TOXICITY DATA WITH REFERENCE
orl-rat LD50:1380 mg/kg 27ZQAG -,391,72
ipr-rat LD50:370 mg/kg 27ZQAG -,391,72
orl-mus LD50:1100 mg/kg 27ZQAG -,391,72
ipr-mus LD50:510 mg/kg 27ZQAG -,391,72

SAFETY PROFILE: Poison by intraperitoneal route. Moderately toxic by ingestion. When heated to decomposition it emits toxic fumes of NO_x.

E

EGT500 **CAS:15718-71-5** **HR: 3**
1,2-ETHYL BIS-AMMONIUM PERCHLORATE
mf: $C_2H_{10}Cl_2N_2O_8$ mw: 261.02

$(—CH2N^+H3)2•2ClO4^-$

SYNS: ETHYLENE DIAMINE DIPERCHLORATE (DOT) □ ETHYLENEDIAMINE PERCHLORATE

DOT CLASSIFICATION: Forbidden

SAFETY PROFILE: Highly unstable. Much more powerful than TNT. When heated to decomposition it emits toxic fumes of NH_3. See also EXPLOSIVES, HIGH.

EGU000 **CAS:3590-07-6** **HR: 3**
ETHYLBIS(β-CHLOROETHYL)AMINE HYDROCHLORIDE
mf: $C_6H_{13}Cl_2N•ClH$ mw: 206.56

SYNS: HNl•HCl □ HN1 HYDROCHLORIDE □ TL 329 HYDROCHLORIDE

TOXICITY DATA WITH REFERENCE
scu-rat LD50:1 mg/kg JPETAB 91,224,47
ivn-rat LD50:500 µg/kg JPETAB 91,224,47
ipr-mus LDLo:1050 µg/kg NTIS** PB158-507
scu-mus LD50:1 mg/kg BJPCAL 1,247,46
orl-rbt LDLo:1500 µg/kg NTIS** PB158-507
ivn-rbt LD50:2 mg/kg JPETAB 91,224,47

CONSENSUS REPORTS: EPA Genetic Toxicology Program.

SAFETY PROFILE: Poison by ingestion, subcutaneous, intravenous, and intraperitoneal routes. When heated to decomposition it emits very toxic fumes of Cl^- and NO_x. See also AMINES.

EGV000 **CAS:105-36-2** **HR: 3**
ETHYL BROMACETATE
DOT: UN 1603
mf: $C_4H_7BrO_2$ mw: 167.02

PROP: Colorless to straw-colored liquid. Bp: 158.8°, fp: <−20°, flash p: 118°F, d: 1.514 @ 13°/4°, vap d: 5.8. Insol in water; misc in alc and ether.

SYNS: ANTOL □ BROMOACETIC ACID, ETHYL ESTER □ ETHOXYCARBONYLMETHYL BROMIDE □ ETHYL BROMOACETATE □ ETHYL-α-BROMOACETATE □ ETHYL MONOBROMOACETATE

TOXICITY DATA WITH REFERENCE
scu-mus TDLo:252 mg/kg/63W-I:NEO JNCIAM 53,695,74

CONSENSUS REPORTS: Reported in EPA TSCA Inventory.

DOT CLASSIFICATION: 6.1; *Label:* Poison

SAFETY PROFILE: A poison. An irritant to skin, eyes, and mucous membranes. Questionable carcinogen with experimental neoplastigenic data. Flammable liquid when exposed to heat, flame, and oxidizers. Will react with water or steam to produce toxic and corrosive fumes. To fight fire, use water as a fire blanket. When heated to decomposition or on contact with acid or acid fumes, it emits highly toxic fumes of Br^-. See also BROMIDES.

EGV400 **CAS:74-96-4** **HR: 3**
ETHYL BROMIDE
DOT: UN 1891
mf: C_2H_5Br mw: 108.98

PROP: Colorless, volatile liquid. Mp: −119°, bp: 38.4°, fp: −125.5°, lel: 6.7%, uel: 11.3%, flash p: <−4°F, d: 1.451 @ 20°/4°, autoign temp: 952°F, vap press: 400 mm @ 21°, vap d: 3.76.

SYNS: BROMIC ETHER □ BROMOETHANE □ BROMURE d'ETHYLE □ ETYLU BROMEK (POLISH) □ HALON 2001 □ HYDROBROMIC ETHER □ MONOBROMOETHANE □ NCI-C55481

TOXICITY DATA WITH REFERENCE
ipr-rat LD50:1750 mg/kg JPCEAO 320,133,78
ihl-mus LC50:16,230 ppm/1H AMRL** TR-72-62/72
ipr-mus LD50:2850 mg/kg JPCEAO 320,133,78
orl-rat LD50:1350 mg/kg 85GMAT-,65,82
ihl-rat LDLo:148,000 ppm/15M AMIHBC 6,435,52

CONSENSUS REPORTS: NTP Carcinogenesis Studies (inhalation); Clear Evidence: Mouse NTPTR* NTP-TR-363,89; (Inhalation); Some Evidence: Rat NTPTR* NTP-TR-363,89. EPA Genetic Toxicology Program. Reported in EPA TSCA Inventory.

OSHA PEL: TWA 200 ppm; STEL 250 ppm
ACGIH TLV: TWA 5 ppm (skin); Suspected Human Carcinogen
DFG MAK: Animal Carcinogen, Suspected Human Carcinogen.
DOT CLASSIFICATION: 6.1; *Label:* Poison

SAFETY PROFILE: Confirmed carcinogen. Moderately toxic by ingestion and intraperitoneal routes. Mildly toxic by inhalation. An eye and skin irritant. Physiologically, it is an anesthetic and narcotic. Its vapors are markedly irritating to the lungs on inhalation for even short periods. It can produce acute congestion and edema. Liver and kidney damage in humans has been reported. It is much less toxic than methyl bromide, but more toxic than ethyl chloride. It is a preparative hazard. Dangerously flammable by heat, open flame (sparks), oxidizers. Moderately explosive when exposed to flame. Reacts with water or steam to produce toxic and corrosive fumes. Vigorous reaction with oxidizing materials. To fight fire, use CO_2, dry chemical. Readily decomposes when heated to emit toxic fumes of Br^-. See also BROMIDES.

For occupational chemical analysis use NIOSH: Ethyl Bromide, 1011.

EGV500 **CAS:4824-78-6** **HR: 3**
ETHYL BROMOPHOS
mf: $C_{10}H_{12}BrCl_2O_3PS$ mw: 394.06

PROP: Pale-yellow liquid. D: 1.52–1.55 (tech) @ 20°, bp: 122–123° @ 0.004 mm.

SYNS: 4-BROMO-2,5-DICHLOROPHENOL-o-ESTER with O,O-DIETHYL PHOSPHOROTHIOATE □ O-(4-BROMO-2,5-DICHLOROPHENYL)-O,O-DIETHYL PHOSPHOROTHIOATE □ O-(4-BROMO-2,5-DICHLOROPHENYL)-O,O-DIETHYLPHOSPHOROTHIONATE □ BROMOFOS-ETHYL □ BROMOPHOSETHYL □ CELA S-2225 □ O,O-DIAETHYL-O-(4-BROM-2,5-DICHLOR)-PHENYL-MONOTHIOPHOSPHAT (GERMAN) □ O,O-DIAETHYL-O-(2,5-DICHLOR-4-BROMPHENYL)-

THIONOPHOSPHAT (GERMAN) □ O,O-DIETHYL-O-(4-BROOM-2,5-DICHLOOR-FENYL)-MONOTHIOFOSFAAT (DUTCH) □ O,O-DIETHYL O-2,5-DICHLORO-4-BROMOPHENYL-PHOSPHOROTHIOATE □ O,O-DIETHYL O-(2,5-DICHLORO-4-BROMOPHENYL) THIOPHOSPHATE □ O,O-DIETIL-O-(4-BROMO-2,5-DICLORO-FENIL)-MONOTIOFOSFATO (ITALIAN) □ ENT 27,258 □ FILARIOL □ NEXAGAN □ OMS-659 □ S 2225 □ THIOPHOSPHATE de O,O-DIETHYLE et de O-(2,5-DICHLORO-4-BROMO) PHENYLE (FRENCH)

TOXICITY DATA WITH REFERENCE
orl-rat LD50:52 mg/kg SPEADM 74-1,-,74
ihl-rat LC50:16,600 pph 85DPAN -,-,71/76
skn-rat LD50:1000 mg/kg WRPCA2 9,119,70
orl-mus LD50:210 mg/kg FMCHA2 -,C37,83
skn-rbt LD50:1366 mg/kg 28ZEAL 5,29,76

CONSENSUS REPORTS: Chlorophenol compounds are on the Community Right-To-Know List.

SAFETY PROFILE: Poison by ingestion. Moderately toxic by skin contact and inhalation. An insecticide. When heated to decomposition it emits very toxic fumes of Br^-, PO_x, SO_x, and Cl^-. See also CHLOROPHENOLS.

EGW000 CAS:97-95-0 HR: 3
2-ETHYLBUTANOL
DOT: UN 2275
mf: $C_6H_{14}O$ mw: 102.20

PROP: Clear liquid. Bp: 144–146°, flash p: 135°F (COC), d: 0.8328, vap press: 0.9 mm @ 20°, vap d: 3.4.

SYNS: 2-ETHYLBUTANOL-1 □ 2-ETHYL-1-BUTANOL □ 2-ETHYLBUTYL ALCOHOL □ sec-HEXANOL (DOT) □ sec-HEXYL ALCOHOL □ 3-METHYLOLPENTANE □ sec-PENTYLCARBINOL □ 3-PENTYLCARBINOL □ PSEUDOHEXYL ALCOHOL

TOXICITY DATA WITH REFERENCE
skn-rbt 415 mg open MLD UCDS** 12/14/71
skn-rbt 500 mg/24H MLD 85JCAE-,197,86
eye-rbt 250 μg open SEV AMIHBC 10,61,54
orl-rat LD50:1850 mg/kg AMIHBC 10,61,54
orl-rbt LD50:1200 mg/kg JPETAB 82,377,44
skn-rbt LD50:1260 mg/kg AMIHBC 10,61,54

CONSENSUS REPORTS: Reported in EPA TSCA Inventory.

DOT CLASSIFICATION: 3; *Label:* Flammable Liquid

SAFETY PROFILE: Moderately toxic by ingestion and skin contact. A skin and severe eye irritant. Flammable liquid when exposed to heat or flame; can react with oxidizing materials. To fight fire, use dry chemical, CO_2, foam, fog. When heated to decomposition it emits acrid smoke and irritating fumes. See also ALCOHOLS.

EGW500 CAS:760-21-4 HR: 3
2-ETHYL-1-BUTENE
mf: C_6H_{12} mw: 84.18

PROP: A liquid. Flash p: <−4°, autoign temp: 599°F, d: 0.69, vap d: 2.9, bp: 64.7°.

TOXICITY DATA WITH REFERENCE
eye-hmn 5 ppm JOCMA7 2,383,60

CONSENSUS REPORTS: Reported in EPA TSCA Inventory.

SAFETY PROFILE: A human eye irritant. A very dangerous fire hazard when exposed to heat, flames, or oxidizers. To fight fire, use dry chemical, CO_2, foam, spray. When heated to decomposition it emits acrid smoke and irritating fumes.

EGX000 CAS:4468-93-3 HR: 3
2-(2-ETHYLBUTOXY)ETHANOL
mf: $C_8H_{18}O_2$ mw: 146.26

PROP: Liquid. Bp: 197°, fp: −90°, flash p: 180°F (OC), d: 0.8954 @ 20°/20°, vap press: 0.17 mm @ 20°, vap d: 5.04.

TOXICITY DATA WITH REFERENCE
skn-rbt 10 mg/24H open MLD AMIHBC 10,61,54
eye-rbt 250 μg open SEV AMIHBC 10,61,54
orl-rat LD50:1910 mg/kg AMIHBC 10,61,54
skn-rbt LD50:320 mg/kg AMIHBC 10,61,54

SAFETY PROFILE: Poison by skin contact. Moderately toxic by ingestion. A skin and severe eye irritant. Flammable when exposed to heat or flame; can react with oxidizing materials. When heated to decomposition it emits acrid smoke and irritating fumes. To fight fire, use foam, CO_2, dry chemical.

EGY000 CAS:10213-74-8 HR: 2
3-(2-ETHYLBUTOXY)PROPIONIC ACID
mf: $C_9H_{18}O_3$ mw: 174.27

PROP: Water-white liquid, insol in water. D: 0.96 @ 20°/20°, bp: 200° @ 100 mm, vap press: <0.1 mm @ 20°, flash p: 280°F.

SYN: KYSELINA 3-(2-ETHYLBUTOXY)PROPIONOVA

TOXICITY DATA WITH REFERENCE
skn-rbt 10 mg/24H SEV AMIHBC 10,61,54
eye-rbt 750 μg SEV AMIHBC 10,61,54
orl-rat LD50:3730 mg/kg AMIHBC 10,61,54
skn-rbt LD50:530 mg/kg AMIHBC 10,61,54

SAFETY PROFILE: Moderately toxic by ingestion and skin contact. A severe skin and eye irritant. Combustible when exposed to heat or flame; can react vigorously with oxidizing materials. To fight fire, use dry chemical, CO_2. When heated to decomposition it emits acrid smoke and irritating fumes.

EGY500 CAS:10232-91-4 HR: 2
3-(2-ETHYLBUTOXY)PROPIONITRILE
mf: $C_9H_{17}NO$ mw: 155.27

TOXICITY DATA WITH REFERENCE
eye-rbt 500 mg AMIHBC 10,61,54
orl-rat LD50:2460 mg/kg AMIHBC 10,61,54
skn-rbt LD50:10 g/kg AMIHBC 10,61,54

CONSENSUS REPORTS: Cyanide and its compounds are on the Community Right-To-Know List.

SAFETY PROFILE: Moderately toxic by ingestion. Mildly toxic by skin contact. An eye irritant. When heated to decomposition it emits toxic fumes of NO_x and CN^-. See also NITRILES.

EGZ000 CAS:3953-10-4 HR: 3
2-ETHYLBUTYLACRYLATE
mf: $C_9H_{16}O_2$ mw: 156.25

PROP: Clear, colorless liquid. Bp: 82° @ 10 mm, fp: −70°, flash p: 125°F (OC), d: 0.8964 @ 20°/20°, vap press: 1.7 mm @ 20°.

SYNS: 2-ETHYLBUTYL ESTER, ACRYLIC ACID ☐ 2-ETHYLBUTYLESTER KYSELINY AKRYLOVE ☐ 2-PROPENOIC ACID-2-ETHYLBUTYL ESTER

TOXICITY DATA WITH REFERENCE
skn-rbt 10 mg/24H open SEV AMIHBC 4,119,51
eye-rbt 500 mg open AMIHBC 4,119,51
orl-rat LDLo:6490 mg/kg AMIHBC 4,119,51
skn-rbt LD50:5500 mg/kg AMIHBC 4,119,51

CONSENSUS REPORTS: Reported in EPA TSCA Inventory.

SAFETY PROFILE: Mildly toxic by ingestion and skin contact. An eye and severe skin irritant. Flammable liquid when exposed to heat or flame; can react with oxidizing materials. To fight fire, use foam, CO_2, dry chemical. When heated to decomposition it emits acrid smoke and irritating fumes. See also ESTERS.

EHA000 CAS:617-79-8 HR: 3
2-ETHYLBUTYLAMINE
mf: $C_6H_{15}N$ mw: 101.22

PROP: Water-white liquid. Bp: 125°, flash p: 64°F (OC), d: 0.739 @ 20°/20°, vap d: 3.5.

SYN: 2-ETHYL-1-BUTANAMINE

TOXICITY DATA WITH REFERENCE
skn-rbt 10 mg/24H open AMIHBC 10,61,54
skn-rbt 5 mg/24H SEV 85JCAE-,434,86
eye-rbt 250 µg open SEV AMIHBC 10,61,54
orl-rat LD50:310 mg/kg GISAAA 39(3),106,74
ihl-rat LC50:500 ppm/4H 85JCAE-,434,86
skn-rbt LD50:2000 mg/kg AMIHBC 10,61,54

SAFETY PROFILE: Poison by ingestion. Moderately toxic by skin contact. A skin and severe eye irritant. A very dangerous fire hazard when exposed to heat or flame; can react vigorously with oxidizing materials. Keep away from heat and open flame. To fight fire, use dry chemical, CO_2, foam. When heated to decomposition it emits toxic fumes of NO_x. See also AMINES.

EHA100 CAS:591-62-8 HR: 3
ETHYL BUTYLCARBAMATE
mf: $C_7H_{15}NO_2$ mw: 145.23

PROP: A liquid. Fp −22°, bp: 202–203° @ 765.5 mm.

SYNS: BUR ☐ 1-BUTYLURETHAN ☐ BUTYLURETHANE ☐ N-BUTYLURETHANE ☐ 1-BUTYLURETHANE ☐ ETHYL-N,N-BUTYLCARBAMATE

TOXICITY DATA WITH REFERENCE
mmo-bcs 5 g/L MUREAV 42,19,77
dnr-bcs 5 g/L MUREAV 42,19,77
scu-rat TDLo:100 mg/kg (15-21D preg):NEO,TER GANNA2 71,811,80
ipr-mus LD50:250 mg/kg NTIS** AD691-490

CONSENSUS REPORTS: EPA Genetic Toxicology Program.

SAFETY PROFILE: Poison by intraperitoneal route. An experimental teratogen. Questionable carcinogen with experimental tumorigenic data. Mutation data reported. When heated to decomposition it emits toxic fumes of NO_x. See also CARBAMATES and ESTERS.

EHA500 CAS:628-81-9 HR: 3
ETHYL BUTYL ETHER
DOT: UN 1179
mf: $C_6H_{14}O$ mw: 102.20

PROP: Colorless liquid. Bp: 92°, mp: −124°, flash p: 40°F, d: 0.7528 @ 20°/20°, vap d: 3.52. Insol in water; misc in alc and ether.

SYN: ETHER ETHYLBUTYLIQUE (FRENCH)

TOXICITY DATA WITH REFERENCE
skn-rbt 500 mg/24H MLD 85JCAE-,248,86
eye-rbt 500 mg/24H MLD 85JCAE-,248,86
skn-rbt 10 mg/24H open MLD AMIHBC 4,119,51
eye-rbt 500 mg open AMIHBC 4,119,51
orl-rat LD50:1870 mg/kg AMIHBC 4,119,51

CONSENSUS REPORTS: Reported in EPA TSCA Inventory.

DOT CLASSIFICATION: 3; *Label:* Flammable Liquid

SAFETY PROFILE: Moderately toxic by ingestion. A skin and eye irritant. A very dangerous fire hazard when exposed to heat or flame; can react vigorously with oxidizing materials. Keep away from heat and open flame. To fight fire, use alcohol foam, CO_2, dry chemical. When heated to decomposition it emits acrid smoke and irritating fumes. See also ETHERS.

EHA550 CAS:637-92-3 HR: 1
ETHYL tert-BUTYL ETHER
mf: $C_6H_{14}O$ mw: 102.20

SYNS: tert-BUTYL ETHYL ETHER ☐ 1,1-DIMETHYLETHYL ETHYL ETHER ☐ ETHER, tert-BUTYL ETHYL ☐ 2-ETHOXY-2-METHYLPROPANE ☐ ETHYL tert-BUTYL OXIDE ☐ ETHYL 1,1-DIMETHYLETHYL ETHER ☐ PROPANE, 2-ETHOXY-2-METHYL-(9CI)

TOXICITY DATA WITH REFERENCE
ihl-mus LC50:123 g/m³/15M ANESAV 11,455,50

CONSENSUS REPORTS: Reported in EPA TSCA Inventory.

SAFETY PROFILE: Low toxicity by inhalation. When heated to decomposition it emits acrid smoke and irritating vapors.

EHA600 CAS:106-35-4 HR: 3
ETHYL BUTYL KETONE
mf: $C_7H_{14}O$ mw: 114.21

PROP: Clear mobile liquid; fatty odor. Mp: −36.7°, bp: 149–152°, flash p: 115°F (OC), d: 0.8198 @ 20°/20°, vap d: 3.93. Misc with alc, ether, water @ 149°.

SYNS: AETHYLBUTYLKETON (GERMAN) ☐ n-BUTYL ETHYL KE-

TONE □ EPTAN-3-ONE (ITALIAN) □ ETHYLBUTYLCETONE (FRENCH) □ ETHYLBUTYLKETON (DUTCH) □ ETILBUTILCHETONE (ITALIAN) □ FEMA No. 2545 □ HEPTAN-3-ON (DUTCH, GERMAN) □ HEPTAN-3-ONE □ 3-HEPTANONE

TOXICITY DATA WITH REFERENCE
skn-rbt 500 mg open MLD UCDS** 3/12/69
skn-rbt 500 mg/24H MOD FCTXAV 16,731,78
eye-rbt 500 mg/24H MLD 85JCAE-,286,86
eye-rbt 100 mg MLD FCTXAV 16,731,78
orl-rat LD50:2760 mg/kg JIHTAB 31,60,49
ihl-rat LCLo:2000 ppm/4H JIHTAB 31,343,49

CONSENSUS REPORTS: Reported in EPA TSCA Inventory.

OSHA PEL: TWA 50 ppm
ACGIH TLV: TWA 50 ppm
DOT CLASSIFICATION: 3; *Label:* Flammable Liquid

SAFETY PROFILE: Moderately toxic by ingestion and inhalation. A skin and eye irritant. A flammable liquid. Can react with oxidizing materials. To fight fire, use foam, CO_2, dry chemical. See also KETONES.

For occupational chemical analysis use NIOSH: Ketones II (Desorption in 99:1 CS_2:methanol) 1301.

EHC000 CAS:4549-44-4 HR: 3
ETHYL-N-BUTYLNITROSAMINE
mf: $C_6H_{14}N_2O$ mw: 130.22

SYNS: AETHYL-N-BUTYL-NITROSOAMIN (GERMAN) □ N-ETHYL-N-NITROSOBUTYLAMINE □ N-NITROSO-N-BUTYLETHYLAMINE □ N-NITROSOETHYL-N-BUTYLAMINE

TOXICITY DATA WITH REFERENCE
mmo-sat 769 µmol/L ENMUDM 3,11,81
dns-rat:lvr 1 mmol/L ENMUDM 3,11,81
dni-mus-ipr 20 g/kg ARGEAR 51,605,81
orl-rat TDLo:1000 mg/kg/29W-C:ETA ARZNAD 19,1077,69
ivn-rat TDLo:1000 mg/kg/40W-I:ETA ARZNAD 19,1077,69
ipr-rat TDLo:1 g/kg/20W-I:ETA ZEKBAI 74,110,70
orl-mus TDLo:2360 mg/kg/34W-C:CAR NATWAY 50,717,63
orl-rat LD50:380 mg/kg NATWAY 50,100,63
ivn-rat LD50:380 mg/kg ZEKBAI 69,103,67

SAFETY PROFILE: Poison by ingestion and intravenous routes. Questionable carcinogen with experimental carcinogenic and tumorigenic data. Mutation data reported. When heated to decomposition it emits toxic fumes of NO_x. See also NITROSAMINES.

For occupational chemical analysis use OSHA: #38.

EHC800 HR: 2
ETHYL N-BUTYL-N-NITROSOSUCCINAMATE
mf: $C_{10}H_{18}N_2O_4$ mw: 230.30

SYNS: N-BUTYL-N-NITROSOSUCCINAMIC ACID ETHYL ESTER □ EBNS □ N-NITROSO-N-(3-CARBOETHOXYPROPIONYL)BUTYLAMINE

TOXICITY DATA WITH REFERENCE
scu-rat TDLo:70,500 µg/kg/10W-I:CAR GANNA2 73,687,82
scu-rat TD:87 mg/kg/10W-I:CAR IAPUDO 41,619,82

SAFETY PROFILE: Questionable carcinogen with experimental carcinogenic data. When heated to decomposition it emits toxic fumes of NO_x.

EHC900 CAS:68037-57-0 HR: 3
2-ETHYLBUTYL SILICATE

SYNS: POLYBIS(2-ETHYLBUTYL)SILOXANE □ SILICIC ACID, 2-ETHYLBUTYL ESTER

TOXICITY DATA WITH REFERENCE
skn-rbt 500 mg open MLD UCDS** 7/10/63
orl-rat LD50:19,700 mg/kg UCDS** 7/10/63

CONSENSUS REPORTS: Reported in EPA TSCA Inventory.

SAFETY PROFILE: Poison by ingestion route. A skin irritant. When heated to decomposition it emits acrid smoke and irritating fumes.

EHD000 CAS:63019-12-5 HR: 2
α-ETHYL-α′,sec-BUTYLSTILBENE
mf: $C_{20}H_{24}$ mw: 264.44

SYN: α-ETHYL-β-sec-BUTYLSTILBENE

TOXICITY DATA WITH REFERENCE
skn-mus TDLo:1250 mg/kg/52W-I:ETA NATUAS 148,142,41

SAFETY PROFILE: Questionable carcinogen with experimental tumorigenic data. When heated to decomposition it emits acrid smoke and irritating fumes.

EHE000 CAS:105-54-4 HR: 3
ETHYL n-BUTYRATE
DOT: UN 1180
mf: $C_6H_{12}O_2$ mw: 116.18

PROP: Colorless liquid; banana-pineapple odor. D: 0.900 @ 0°/4°, refr index: 1.391, mp: −100.8°, fp −93.3°, bp: 121.6°, flash p: 79°F. Sol in water, fixed oils, propylene glycol; misc in alc and ether; insol in glycerin @ 121°.

SYNS: BUTANOIC ACID ETHYL ESTER □ BUTYRIC ETHER □ ETHYL BUTANOATE □ ETHYL BUTYRATE (DOT,FCC) □ FEMA No. 2427

TOXICITY DATA WITH REFERENCE
skn-rbt 500 mg/24H MOD FCTXAV 12,703,74
orl-rat LD50:13 g/kg FCTXAV 2,327,64
orl-rbt LD50:5228 mg/kg IMSUAI 41,31,72

CONSENSUS REPORTS: Reported in EPA TSCA Inventory.

DOT CLASSIFICATION: 3; *Label:* Flammable Liquid

SAFETY PROFILE: Mildly toxic by ingestion. A skin irritant. Flammable liquid when exposed to heat or flame; can react vigorously with oxidizing materials. When heated to decomposition it emits acrid smoke and irritating fumes. See also ESTERS.

EHE500 **CAS:110-38-3** **HR: 1**
ETHYL CAPRATE
mf: $C_{12}H_{24}O_2$ mw: 200.36

PROP: Colorless liquid; oily, brandy odor. Bp: 243°, d: 0.863, refr index: 1.424, vap d: 6.9, flash p: 212°F. Sol in fixed oils; insol in glycerin, propylene glycol @ 243°.

SYNS: CAPRIC ACID ETHYL ESTER □ DECANOIC ACID, ETHYL ESTER □ ETHYL CAPRINATE □ ETHYL DECANOATE (FCC) □ ETHYL DECYLATE □ FEMA No. 2432

TOXICITY DATA WITH REFERENCE
skn-rbt 500 mg/24H MLD FCTXAV 16,733,78

CONSENSUS REPORTS: Reported in EPA TSCA Inventory.

SAFETY PROFILE: A skin irritant. Combustible liquid when exposed to heat or flame; can react with oxidizing materials. When heated to decomposition it emits acrid smoke and irritating fumes. See ESTERS and ETHERS.

EHF000 **CAS:123-66-0** **HR: 3**
ETHYL CAPROATE
mf: $C_8H_{16}O_2$ mw: 144.24

PROP: Colorless liquid; mild wine odor. Bp: 163°, flash p: 130°F (OC), d: 0.867–0.871, refr index: 1.406–1.409, vap d: 5.0. Sol in fixed oils; sltly sol in propylene glycol; insol in glycerin @ 166.

SYNS: ETHYL BUTYLACETATE (DOT) □ ETHYL HEXANOATE (FCC) □ FEMA No. 2439

TOXICITY DATA WITH REFERENCE
skn-rbt 500 mg/24H MOD FCTXAV 14,659,76

CONSENSUS REPORTS: Reported in EPA TSCA Inventory.

SAFETY PROFILE: A skin irritant. Flammable liquid when exposed to heat or flame; can react with oxidizing materials. When heated to decomposition it emits acrid smoke and irritating fumes. To fight fire, use CO_2, foam, dry chemical. See also ESTERS.

EHF500 **CAS:63833-90-9** **HR: 3**
2-(N-ETHYL CARBAMOYLHYDROXYMETHYL)FURAN
mf: $C_8H_{11}NO_3$ mw: 169.20

SYNS: CMF □ FURAN, 2-(N-ETHYLCARBAMOYLHYDROXYMETHYL)- □ 2-FURANMETHANOL, α-(N-ETHYLCARBAMOYL)-

TOXICITY DATA WITH REFERENCE
ipr-rat LD50:80 mg/kg BCPCA6 26,1909,77

SAFETY PROFILE: Poison by intraperitoneal route. When heated to decomposition it emits toxic fumes of NO_x.

EHG000 **CAS:4114-31-2** **HR: 3**
ETHYL CARBAZATE
mf: $C_3H_8N_2O_2$ mw: 104.13

SYNS: CARBAZIC ACID, ETHYL ESTER □ CARBETHOXYHYDRAZINE □ N-(CARBETHOXY)HYDRAZINE □ 1-CARBETHOXY HYDRAZINE □ CARBOETHOXYHYDRAZINE □ ETHOXYCARBONYL HYDRAZIDE □ (ETHOXYCARBONYL) HYDRAZINE □ N-(ETHOXYCARBONYL) HYDRAZINE □ 1-(ETHOXYCARBONYL) HYDRAZINE □ ETHYL CARBAZINATE □ HYDRAZINECARBOXYLIC ACID, ETHYL ESTER □ MONOCARBETHOXYHYDRAZINE

TOXICITY DATA WITH REFERENCE
scu-mus LD50:26 mg/kg ABMGAJ 21,635,68
ivn-mus LD50:56 mg/kg CSLNX* NX#02783
orl-qal LD50:100 mg/kg EESADV 6,149,82
orl-bwd LD50:23,700 μg/kg AECTCV 12,355,83

CONSENSUS REPORTS: Reported in EPA TSCA Inventory.

SAFETY PROFILE: Poison by ingestion, subcutaneous, and intravenous routes. When heated to decomposition it emits toxic fumes of NO_x. See also HYDRAZINE.

EHG050 **CAS:120-43-4** **HR: 3**
ETHYLCARBONYL PIPERAZINE
mf: $C_7H_{14}N_2O_2$ mw: 158.23

SYNS: 1-PIPERAZINECARBOXYLIC ACID, ETHYL ESTER □ PIPERAZINE ETHYLCARBOXYLATE

TOXICITY DATA WITH REFERENCE
ipr-mus LD50:138 mg/kg OYYAA2 33,825,87

CONSENSUS REPORTS: Reported in EPA TSCA Inventory.

SAFETY PROFILE: Poison by intraperitoneal route. When heated to decomposition it emits toxic vapors of NO_x.

EHG100 **CAS:9004-57-3** **HR: 1**
ETHYLCELLULOSE

PROP: Ethyl ether of cellulose. White to light tan powder. Sol in some organic solvents; insol in water, glycerin, and propylene glycol.

SYNS: AMPACET E/C □ CELLULOSE ETHYL □ CELLULOSE ETHYLATE □ ETs □ ETHOCEL □ ETHOCEL 150 □ ETHOCEL 890 □ ETHOCEL E7 □ ETHOCEL E50 □ ETHOCEL MED □ ETHOCEL N7 □ ETHOCEL N10 □ ETHOCEL N200 □ ETHOCEL STD □ ETs (POLYSACCHARIDE) □ G 50 □ G 200 □ G 50 (POLYSACCHARIDE) □ N 5 □ NIXON E/C □ SPT 50 CPS □ T 100 □ T 100 (POLYSACCHARIDE)

TOXICITY DATA WITH REFERENCE
skn-rbt 500 mg/24H MLD FCTXAV 19,97,81
orl-rat LD50:>5 g/kg FCTXAV 19,113,81
skn-rbt LD50:>5 g/kg FCTXAV 19,113,81

CONSENSUS REPORTS: Reported in EPA TSCA Inventory.

SAFETY PROFILE: Low toxicity by ingestion and skin contact. A skin irritant. When heated to decomposition it emits acrid smoke and irritating fumes.

EHG500 **CAS:105-39-5** **HR: 3**
ETHYL CHLORACETATE
DOT: UN 1181
mf: $C_4H_7ClO_2$ mw: 122.56

$CH_3CH_2OCO \cdot CH_2Cl$

PROP: Colorless liquid; fruity, pungent odor. Irritant to the eyes. Bp: 143.6°, fp: −26.6°, flash p: 100°F, d: 1.159

@ 20°/4°, vap press: 10 mm @ 37.5°, vap d: 4.3. Insol in water; misc in alc and ether.

SYNS: CHLOROACETIC ACID, ETHYL ESTER □ ETHYL CHLOROACETATE □ ETHYL-α-CHLOROACETATE □ ETHYL CHLOROETHANOATE □ ETHYL MONOCHLORACETATE □ ETHYL MONOCHLOROACETATE

TOXICITY DATA WITH REFERENCE
eye-rbt 250 μg/24H SEV 85JCAE-,587,86
ipr-mus TDLo:2940 mg/kg/8W-I:NEO CNREA8 39,391,79
scu-mus LD50:250 mg/kg JJPAAZ 3,99,54
skn-rbt LD50:230 mg/kg TXAPA9 42,417,77

CONSENSUS REPORTS: Reported in EPA TSCA Inventory.

DOT CLASSIFICATION: 6.1; *Label:* Poison

SAFETY PROFILE: Poison by skin contact and subcutaneous routes. A severe eye irritant. Questionable carcinogen with experimental neoplastigenic data. Flammable liquid; a dangerous fire hazard when exposed to heat or flame; can react vigorously with oxidizing materials. Will react with water or steam to produce toxic and corrosive fumes. Vigorous reaction with sodium cyanide. To fight fire, use water, foam, CO_2, dry chemical. When heated to decomposition it emits highly toxic fumes of Cl^-.

EHH000 **CAS:75-00-3** **HR: 3**
ETHYL CHLORIDE
DOT: UN 1037
mf: C_2H_5Cl mw: 64.52

PROP: Colorless liquid or gas which is volatile at room temp; ether-like odor, burning taste. Bp: 12.3°, lel: 3.8%, uel: 15.4%, fp: −142.5°, flash p: −58°F (CC), d: 0.917 @ 6°/6°, autoign temp: 966°F, vap press: 1000 mm @ 20°, vap d: 2.22.45; misc in alc and ether. Sltly sol in water.

SYNS: AETHYLCHLORID (GERMAN) □ AETHYLIS □ AETHYLIS CHLORIDUM □ ANODYNON □ CHELEN □ CHLOORETHAAN (DUTCH) □ CHLORETHYL □ CHLORIDUM □ CHLOROAETHAN (GERMAN) □ CHLOROETHANE □ CHLORURE d'ETHYLE (FRENCH) □ CHLORYL □ CHLORYL ANESTHETIC □ CLOROETANO (ITALIAN) □ CLORURO DI ETILE (ITALIAN) □ ETHER CHLORATUS □ ETHER HYDROCHLORIC □ ETHER MURIATIC □ ETYLU CHLOREK (POLISH) □ HYDROCHLORIC ETHER □ KELENE □ MONOCHLORETHANE □ MURIATIC ETHER □ NARCOTILE □ NCI-C06224

TOXICITY DATA WITH REFERENCE
ihl-rat TCLo:15,000 ppm/6H/2Y-I:ETA NTPTR* NTP-TR-346,89
ihl-mus TCLo:15,000 ppm/6H/2Y-I:CAR NTPTR* NTP-TR-346,89
ihl-rat LC50:160 g/m³/2H 85GMAT -,66,82
ihl-mus LC50:146 g/m³/2H 85GMAT -,66,82
ihl-gpg LCLo:40,000 ppm/45M XPHBAO 185,1,29

CONSENSUS REPORTS: Reported in EPA TSCA Inventory. Community Right-To-Know List.

OSHA PEL: TWA 1000 ppm
ACGIH TLV: TWA 1000 ppm
DFG MAK: Suspected carcinogen
NIOSH REL: (Chloroethane): Handle with caution
DOT CLASSIFICATION: 2.1; *Label:* Flammable Gas

SAFETY PROFILE: Suspected carcinogen with experimental carcinogenic and neoplastigenic data. Mildly toxic by inhalation. An irritant to skin, eyes, and mucous membranes. The liquid is harmful to the eyes and can cause some irritation. In the case of guinea pigs, the symptoms attending exposure are similar to those caused by methyl chloride, except that the signs of lung irritation are not as pronounced. It gives some warning of its presence because it is irritating, but it is possible to tolerate exposure to it until one becomes unconscious. It is the least toxic of all the chlorinated hydrocarbons. It can cause narcosis, although the effects are usually transient.

A very dangerous fire hazard when exposed to heat or flame; can react vigorously with oxidizing materials. Severe explosion hazard when exposed to flame. Reacts with water or steam to produce toxic and corrosive fumes. Incompatible with potassium. To fight fire, use carbon dioxide. When heated to decomposition it emits toxic fumes of phosgene and Cl^-. See also CHLORINATED HYDROCARBONS, ALIPHATIC.

EHH100 **CAS:638-07-3** **HR: 3**
ETHYL 4-CHLOROACETOACETATE
mf: $C_6H_9ClO_3$ mw: 164.60

SYNS: ACETOACETIC ACID, 4-CHLORO-, ETHYL ESTER □ BUTANOIC ACID, 4-CHLORO-3-OXO-, ETHYL ESTER (9CI) □ ETHYL γ-CHLOROACETOACETATE □ ETHYL 4-CHLORO-3-OXOBUTANOATE

TOXICITY DATA WITH REFERENCE
ipr-rat LD50:108 mg/kg OYYAA2 33,695,87
ipr-mus LD50:88 mg/kg OYYAA2 33,695,87

CONSENSUS REPORTS: Reported in EPA TSCA Inventory.

SAFETY PROFILE: Poison by intraperitoneal route. When heated to decomposition it emits toxic vapors of Cl^-.

EHH200 **CAS:687-46-7** **HR: 3**
ETHYL 2-CHLOROACRYLATE
mf: $C_5H_7ClO_2$ mw: 134.57

SYNS: ACRYLIC ACID, 2-CHLORO-, ETHYL ESTER □ 2-CHLOROACRYLIC ACID ETHYL ESTER □ ETHYL α-CHLOROACRYLATE □ 2-PROPENOIC ACID, 2-CHLORO-, ETHYL ESTER

TOXICITY DATA WITH REFERENCE
ivn-mus LD50:180 mg/kg CSLNX* NX#02176

CONSENSUS REPORTS: Reported in EPA TSCA Inventory.

SAFETY PROFILE: Poison by intravenous route. When heated to decomposition it emits toxic vapors of Cl^-.

EHH500 **CAS:1331-31-3** **HR: 3**
ETHYLCHLOROBENZENE
mf: C_8H_9Cl mw: 140.62

PROP: Clear, colorless liquid. Mp: −62.6°, bp: 184.3°, flash p: 147°F, d: 1.05 @ 25°/25°, vap press: 1 mm @ 19.2°, vap d: 4.86.

SYN: CHLOROETHYLBENZENE

TOXICITY DATA WITH REFERENCE
skn-rbt 10 mg/24H open JIHTAB 30,63,48
eye-rbt 500 mg AJOPAA 29,1363,46
orl-rat LD50:5000 mg/kg JIHTAB 30,63,48
skn-rbt LD50:18 g/kg JIHTAB 30,63,48

SAFETY PROFILE: Mildly toxic by ingestion and skin contact. A skin and eye irritant. Flammable liquid when exposed to heat or flame; can react vigorously with oxidizing materials. To fight fire, use foam, CO_2, dry chemical. When heated to decomposition it emits acrid smoke and irritating fumes. See also CHLORINATED HYDROCARBONS, AROMATIC; and CHLOROBENZENE.

EHI500 CAS:4310-69-4 HR: 3
7-(2-(ETHYL-2-CHLOROETHYL)AMINOETHYLAMINO) BENZ(c)ACRIDINE DIHYDROCHLORIDE
mf: $C_{23}H_{24}ClN_3 \cdot 2ClH$ mw: 450.87

SYNS: N′-BENZ(c)ACRIDIN-7-YL-N-(2-CHLOROETHYL)-N-ETHYL-1,2-ETHANEDIAMINE DIHYDROCHLORIDE □ 7-(2-(2-CHLOROETHYLETHYLAMINO)ETHYLAMINO)BENZ(c)ACRIDINE DIHYDROCHLORIDE □ ICR 311

TOXICITY DATA WITH REFERENCE
ivn-mus TDLo:4500 μg/kg:NEO CNREA8 36,2423,76
ivn-mus LDLo:4500 μg/kg CNREA8 36,2423,76

SAFETY PROFILE: Poison by intravenous route. Questionable carcinogen with experimental neoplastigenic data. When heated to decomposition it emits very toxic fumes of NO_x and Cl^-.

EHJ000 CAS:4251-89-2 HR: 3
7-(3-(ETHYL-2-(CHLOROETHYLAMINO)PROPYLAMINO))BENZ(c)ACRIDINE DIHYDROCHLORIDE
mf: $C_{24}H_{26}ClN_3 \cdot 2ClH$ mw: 464.90

SYNS: ICR 292 □ 1,3-PROPANEDIAMINE, N′-BENZ(c)ACRIDIN-7-YL-N-(2-CHLOROETHYL)-N-ETHYL-, DIHYDROCHLORIDE

TOXICITY DATA WITH REFERENCE
mmo-sat 500 ng/plate MUREAV 136,185,84
pic-esc 60 ng/plate CNREA8 43,2819,83
msc-ham:ovr 1 g/L CNREA8 39,4875,79
ivn-mus TDLo:4650 μg/kg:NEO CNREA8 36,2423,76
ivn-mus LDLo:4650 μg/kg CNREA8 36,2423,76

CONSENSUS REPORTS: EPA Genetic Toxicology Program.

SAFETY PROFILE: Poison by intravenous route. Questionable carcinogen with experimental neoplastigenic data. Mutation data reported. When heated to decomposition it emits very toxic fumes of NO_x and Cl^-.

EHJ500 CAS:38915-14-9 HR: 3
9-((3-ETHYL-2-CHLOROETHYL)AMINOPROPYLAMINO)-4-METHOXYACRIDINE DIHYDROCHLORIDE
mf: $C_{21}H_{26}ClN_3O \cdot 2ClH$ mw: 444.87

SYNS: ICR 377 □ 4-METHOXY-9-(3-(ETHYL-2-CHLOROETHYL)

TOXICITY DATA WITH REFERENCE
mmo-sat 500 ng/plate MUREAV 136,185,84
ivn-mus TDLo:2200 μg/kg:NEO CNREA8 36,2423,76
ivn-mus LDLo:2 mg/kg CNREA8 36,2423,76

SAFETY PROFILE: Poison by intravenous route. Questionable carcinogen with experimental neoplastigenic data. Mutation data reported. When heated to decomposition it emits very toxic fumes of Cl^- and NO_x.

EHJ600 CAS:92-49-9 HR: 3
ETHYL(CHLOROETHYL)ANILINE
mf: $C_{10}H_{14}ClN$ mw: 183.70

SYNS: ANILINE, N-(2-CHLOROETHYL)-N-ETHYL- □ BENZENAMINE, N-(2-CHLOROETHYL)-N-ETHYL-(9CI) □ N-(2-CHLOROETHYL)-N-ETHYLANILINE □ N-(2-CHLOROETHYL)-N-ETHYLBENZENAMINE □ EMERY 5770

TOXICITY DATA WITH REFERENCE
orl-rat LD50:616 mg/kg EPASR* 8EHQ-0578-0169S
ipr-rat LD50:200 mg/kg JMCMAR 8,167,65
ipr-mus LD50:325 mg/kg JMCMAR 8,167,65
skn-rbt LD50:200 mg/kg EPASR* 8EHQ-0578-0169S

CONSENSUS REPORTS: Reported in EPA TSCA Inventory.

SAFETY PROFILE: Poison by intraperitoneal and skin contact routes. Moderately toxic by ingestion. When heated to decomposition it emits toxic vapors of NO_x and Cl^-.

EHK000 CAS:63918-55-8 HR: 3
ETHYL-β-CHLOROETHYLETHYLENIMONIUM PICRYLSULFONATE
mf: $C_6H_{13}ClN \cdot C_6H_2N_3O_9S$ mw: 426.82

SYNS: ETHYL(2-CHLOROETHYL)ETHYLENIMONIUM PICRYLSULFONATE □ 1-ETHYL-1-(β-CHLOROETHYL)ETHYLENIMONIUM PICRYLSULFONATE

TOXICITY DATA WITH REFERENCE
ivn-rat LD50:500 μg/kg JPETAB 91,224,47
scu-mus LD50:2 mg/kg JPETAB 91,224,47
ivn-rbt LD50:3 mg/kg JPETAB 91,224,47

SAFETY PROFILE: Poison by intravenous and subcutaneous routes. When heated to decomposition it emits very toxic fumes of Cl^-, NO_x and SO_x.

EHK300 HR: 3
ETHYL-β-CHLOROETHYL-β-HYDROXYETHYLAMINE PICRYLSULFONATE
mf: $C_6H_{14}ClNO \cdot C_6H_3N_3O_9S$ mw: 444.84

SYNS: 2-((2-CHLOROETHYL)ETHYLAMINO)ETHANOL with 2,4,6-TRINITROBENZENESULFONIC ACID □ HN1 CHLOROHYDRIN

TOXICITY DATA WITH REFERENCE
ipr-mus LDLo:40 mg/kg NTIS** PB158-507
scu-mus LDLo:5 mg/kg NTIS** PB158-507
ivn-mus LDLo:8 mg/kg NTIS** PB158-507
ivn-rbt LD50:5 mg/kg NTIS** PB158-507

SAFETY PROFILE: Poison by subcutaneous, intravenous, and intraperitoneal routes. When heated to decomposition it emits toxic fumes of Cl^-, SO_x, and NO_x. See also SULFONATES.

E

EHK500 **CAS:541-41-3** **HR: 3**
ETHYL CHLOROFORMATE
DOT: UN 1182
mf: $C_3H_5ClO_2$ mw: 108.53

PROP: Colorless liquid. Mp: −80.6°, bp: 94°, flash p: 35.6°F, d: 1.1442 @ 15°/4°, vap d: 3.74, autoign temp: 932°F. Decomp in water; misc in alc, benzene, ether, and chloroform. Misc in EtOH, C_6H_6, $CHCl_3$, and Et_2O; prac insol in H_2O.

SYNS: CHLOROCARBONATE D'ETHYLE (FRENCH) □ CHLOROFORMIC ACID ETHYL ESTER □ CHLORAMEISENSAEUREAETHYLESTER (GERMAN) □ ECF □ ETHYLCHLOORFORMIAAT (DUTCH) □ ETHYL CHLOROCARBONATE (DOT) □ ETHYLE, CHLOROFORMIAT D' (FRENCH) □ ETIL CLOROCARBONATO (ITALIAN) □ ETIL CLOROFORMIATO (ITALIAN) □ TL 423

TOXICITY DATA WITH REFERENCE
orl-rat LD50:270 mg/kg TXAPA9 42,417,77
ihl-rat LD50:145 ppm/1H TXAPA9 42,417,77
ihl-mus LCLo:2260 mg/m³/10M NDRC** NDCrc-132,Nov,42
ipr-mus LDLo:15 mg/kg CBCCT* 6,220,54
skn-rbt LD50:7120 mg/kg TXAPA9 42,417,77

CONSENSUS REPORTS: Reported in EPA TSCA Inventory. Community Right-To-Know List.

DOT CLASSIFICATION: 6.1; *Label:* Poison, Flammable Liquid, Corrosive

SAFETY PROFILE: Poison by ingestion, inhalation, and intraperitoneal routes. Moderately toxic by skin contact. Corrosive. An eye, skin, and mucous membrane irritant. A very dangerous fire hazard when exposed to heat or flame; can react vigorously with oxidizing materials. Reacts with water or steam to produce toxic and corrosive fumes. To fight fire, use CO_2, dry chemical. When heated to decomposition it emits highly toxic fumes of Cl^-.

EHL000 **CAS:63019-53-4** **HR: 2**
7-ETHYL-10-CHLORO-11-METHYLBENZ(c)ACRIDINE
mf: $C_{20}H_{16}ClN$ mw: 305.82

SYN: 2-CHLORO-1-METHYL-10-ETHYL-7,8-BENZACRIDINE (FRENCH)

TOXICITY DATA WITH REFERENCE
scu-mus TDLo:10 mg/kg:ETA AICCA6 11,736,55
scu-mus TD:30 mg/kg/4W-I:ETA ACRSAJ 4,315,56

SAFETY PROFILE: Questionable carcinogen with experimental tumorigenic data. When heated to decomposition it emits very toxic fumes of Cl^- and NO_x.

EHL500 **CAS:10443-70-6** **HR: 2**
ETHYL-4-(4-CHLORO-2-METHYLPHENOXY)BUTYRATE
mf: $C_{13}H_{17}ClO_3$ mw: 256.75

SYNS: ETHYL 4-(4-CHLORO-2-METHYLPHENOXY)BUTYLATE □ MCPB-ETHYL

TOXICITY DATA WITH REFERENCE
orl-rat LD50:1570 mg/kg FMCHA2 -,C147,83
skn-rat LD50:4000 mg/kg FMCHA2 -,C147,83

SAFETY PROFILE: Moderately toxic by ingestion and skin contact. An herbicide. When heated to decomposition it emits toxic fumes of Cl^-. See also ESTERS.

EHM000 **CAS:7511-54-8** **HR: 2**
3-ETHYLCHOLANTHRENE
mf: $C_{22}H_{18}$ mw: 282.40

SYNS: 3-ETHYL-CHOLANTHRENE □ 20-ETHYLCHOLANTHRENE

TOXICITY DATA WITH REFERENCE
scu-mus TDLo:400 mg/kg:ETA AJCAA7 33,499,38

SAFETY PROFILE: Questionable carcinogen with experimental tumorigenic data. When heated to decomposition it emits acrid smoke and irritating fumes.

EHM100 **CAS:97-41-6** **HR: 2**
ETHYL CHRYSANTHEMUMATE
mf: $C_{12}H_{20}O_2$ mw: 196.32

SYNS: CYCLOPROPANECARBOXYLIC ACID, 2,2-DIMETHYL-3-(2-METHYLPROPENYL)-, ETHYL ESTER (8CI) □ CYCLOPROPANECARBOXYLIC ACID, 2,2-DIMETHYL-3-(2-METHYL-1-PROPENYL)-, ETHYLESTER □ CYCLOPROPANECARBOXYLIC ACID, 2,2-DIMETHYL-3-(2-METHYL-1-PROPENYL)-, ETHYLESTER (9CI) □ 2,2-DIMETHYL-3-(2-METHYLPROPENYL)CYCLOPROPANECARBOXYLIC ACID ETHYL ESTER □ ETHYL CHRYSANTHEMATE

TOXICITY DATA WITH REFERENCE
mrc-smc 1 pph NTIS** PB85-193761
orl-rat LD50:2600 mg/kg GISAAA 51(1),16,86 GTPZAB 31(7),53,87
orl-mus LD50:2600 mg/kg GISAAA 51(1),16,86
skn-mus LD50:>5 g/kg GISAAA 51(1),16,86
orl-gpg LD50:1900 mg/kg GISAAA 51(1),16,86

CONSENSUS REPORTS: Reported in EPA TSCA Inventory.

SAFETY PROFILE: Moderately toxic by ingestion. Mildly toxic by skin contact. Mutation data reported. When heated to decomposition it emits acrid smoke and irritating vapors.

EHN000 **CAS:103-36-6** **HR: 2**
ETHYL-trans-CINNAMATE
mf: $C_{11}H_{12}O_2$ mw: 176.23

PROP: Nearly colorless, oily liquid; faint cinnamon odor. D: 1.049 @ 20°/4°, refr index: 1.558–1.561, bp: 271°, mp: 9°, flash p: 212°F. Misc in alc, ether, fixed oils; insol in glycerin, water @ 272°.

SYNS: ETHYL CINNAMATE (FCC) □ ETHYL-β-PHENYLACRYLATE □ ETHYL-3-PHENYLPROPENOATE □ FEMA No. 2430

TOXICITY DATA WITH REFERENCE
orl-rat LD50:4000 mg/kg VPITAR 33(5),48,74
orl-mus LD50:4000 mg/kg VPITAR 33(5),48,74
orl-gpg LD50:4000 mg/kg VPITAR 33(5),48,74

CONSENSUS REPORTS: Reported in EPA TSCA Inventory.

SAFETY PROFILE: Moderately toxic by ingestion. Combustible liquid. When heated to decomposition it emits acrid smoke and irritating fumes. See also ESTERS.

EHN500 **CAS:41448-29-7** **HR: 2**
ETHYL CITRAL
mf: $C_{11}H_{18}O$ mw: 166.29

SYN: 3,7-DIMETHYL-2,6-NONADIEN-1-AL

TOXICITY DATA WITH REFERENCE
skn-rbt 500 mg/24H SEV FCTXAV 16,637,78

CONSENSUS REPORTS: Reported in EPA TSCA Inventory.

SAFETY PROFILE: A severe skin irritant. When heated to decomposition it emits acrid smoke and irritating fumes.

EHO000 **CAS:19780-25-7** **HR: 2**
2-ETHYLCROTONALDEHYDE
mf: $C_6H_{10}O$ mw: 98.16

SYN: 2-ETHYL-2-BUTENAL

TOXICITY DATA WITH REFERENCE
orl-rat LD50:2600 mg/kg TXAPA9 28,313,74
skn-rbt LD50:500 mg/kg TXAPA9 28,313,74

CONSENSUS REPORTS: Reported in EPA TSCA Inventory.

SAFETY PROFILE: Moderately toxic by ingestion and skin contact. When heated to decomposition it emits acrid smoke and irritating fumes. See also ALDEHYDES and ESTERS.

EHO200 **CAS:10544-63-5** **HR: 3**
ETHYL CROTONATE
DOT: UN 1862
mf: $C_6H_{10}O_2$ mw: 114.16

PROP: Bp: 142–143°, d: 0.918, flash p: 36°F.

SYNS: 2-BUTENOIC ACID, ETHYL ESTER □ CROTONIC ACID, ETHYL ESTER □ ETHYLESTER KYSELINY KROTONOVE

TOXICITY DATA WITH REFERENCE
eye-rbt 20 mg/24H MOD 85JCAE-,369,86
orl-rat LD50:3 g/kg NEZAAQ 34,183,79

CONSENSUS REPORTS: Reported in EPA TSCA Inventory.

DOT CLASSIFICATION: 3; *Label:* Flammable Liquid

SAFETY PROFILE: Slightly toxic by ingestion. Corrosive. An eye irritant and lachrymator. A flammable liquid. When heated to decomposition it emits acrid smoke and irritating fumes.

EHO500 **CAS:483-63-6** **HR: 2**
N-ETHYL-o-CROTONOTOLUIDINE
mf: $C_{13}H_{17}NO$ mw: 203.31

PROP: Yellow oil. Bp: 153–155°.

SYNS: CROTONYL-N-ETHYL-o-TOLUIDINE □ N-ETHYL-o-CROTONOTOLUIDIDE

TOXICITY DATA WITH REFERENCE
orl-rat LD50:1500 mg/kg 29ZVAB -,35,69
orl-mus LD50:1600 mg/kg 29ZVAB -,35,69

SAFETY PROFILE: Moderately toxic by ingestion. When heated to decomposition it emits toxic fumes of NO_x.

EHO700 **CAS:2884-67-5** **HR: 2**
2-ETHYL-trans-CROTONYLUREA
mf: $C_7H_{12}N_2O_2$ mw: 156.21

PROP: Crystals. Mp: 158°.

SYNS: ECTYLUREA □ trans-1-(2-ETHYLCROTONOYL)UREA □ MA-117

TOXICITY DATA WITH REFERENCE
orl-rat LD50:1130 mg/kg AIPTAK 114,418,58
ipr-rat LD50:575 mg/kg AIPTAK 114,418,58
orl-mus LD50:1280 mg/kg FRPSAX 14,418,58
ipr-mus LD50:990 mg/kg APSXAS 5,293,68
orl-dog LD50:1100 mg/kg AIPTAK 114,418,58

SAFETY PROFILE: Moderately toxic by ingestion and intraperitoneal routes. When heated to decomposition it emits toxic fumes of NO_x.

EHP000 **CAS:95-04-5** **HR: 2**
cis-(2-ETHYLCROTONYL) UREA
mf: $C_7H_{12}N_2O_2$ mw: 156.21

PROP: Needles from C_6H_6; crystals from 2-propanol. Mp: 198°. Sol in conc aq acids and alkalis; very sltly sol in org solvs.

SYNS: ACTINE □ (Z)-N-(AMINOCARBONYL)-2-ETHYL-2-BUTENAMIDE □ ASTYN □ CRONIL □ CROTURAL □ DISTESOL □DISTES-SOL □ ECTIDA □ ECTILUREA □ ECTON □ ECTYDA □ ECTYLCARBAMIDE □ ECTYLUREA □ ECTYN □ EKTYLCARBAMID □ (α-ETHYL-cis-CROTONYL)CARBAMIDE □ 2-ETHYL-cis-CROTONYLUREA □ 2-ETHYLCROTONYLUREA □ EUPLACID □ LEVANIL □ LEVIL □ MA-110 □ NASTYN □ NEOCROSEDIN □ NESTYN □ NEUROPROCIN □ NOSTAL □ NOSTIN □ PACETYN □ SEDAREX □ TRANZER □ U 8771

TOXICITY DATA WITH REFERENCE
orl-rat LD50:2500 mg/kg MEIEDD 10,507,83
ipr-rat LD50:900 mg/kg FEPRA7 12,357,53
orl-mus LD50:2500 mg/kg FRPSAX 14,845,59
ipr-mus LD50:1780 mg/kg 27ZQAG -,425,72
orl-dog LD50:2800 mg/kg AIPTAK 114,418,58
ipr-dog LD50:900 mg/kg FEPRA7 12,357,53
orl-rbt LD50:1200 mg/kg 27ZQAG -,425,72
ipr-rbt LD50:1400 mg/kg FEPRA7 12,357,53
ipr-gpg LD50:1100 mg/kg FEPRA7 12,357,53

SAFETY PROFILE: Moderately toxic by ingestion and intraperitoneal routes. A sedative. When heated to decomposition it emits toxic fumes of NO_x.

EHP500 **CAS:105-56-6** **HR: 3**
ETHYL CYANOACETATE
DOT: UN 2666
mf: $C_5H_7NO_2$ mw: 113.13

PROP: Colorless to pale straw-colored liquid. Mp: −22.5°, bp: 207°, flash p: 230°F, d: 1.06 @ 25°/25°, vap press: 1 mm @ 67.8°, vap d: 3.9. Insol in H_2O; sol in NH_3 aq.

SYNS: CYANACETATE ETHYLE (GERMAN) □ CYANOACETIC ACID ETHYL ESTER □ CYANOACETIC ESTER □ ESTERE CIANOACETICO

□ ETHYL CYANOACETATE □ ETHYL CYANOETHANOATE □ ETHYLESTER KYSELINY KYANOCTOVE □ MALONIC ACID ETHYL ESTER NITRILE □ USAF KF-25

TOXICITY DATA WITH REFERENCE
orl-rat LDLo:400 mg/kg 85JCAE-,921,86
ipr-mus LD50:500 mg/kg NTIS** AD277-689
scu-rbt LDLo:1500 mg/kg AIPTAK 5,161,1899
scu-frg LDLo:4000 mg/kg AIPTAK 5,161,1899

CONSENSUS REPORTS: Reported in EPA TSCA Inventory. Cyanide and its compounds are on the Community Right-To-Know List.

DOT CLASSIFICATION: 6.1; *Label:* KEEP AWAY FROM FOOD

SAFETY PROFILE: Poison by ingestion. Moderately toxic by intraperitoneal and subcutaneous routes. Combustible when exposed to heat or flame; can react with oxidizing materials. Will react with water or steam to produce toxic and flammable vapors. To fight fire, use CO_2, dry chemical. When heated to decomposition or on contact with acid or acid fumes it emits highly toxic fumes of CN^-. See also NITRILES.

EHQ500 CAS:148-87-8 HR: 2
N-ETHYL-N-(2-CYANOETHYL)ANILINE
mf: $C_{10}H_{14}N_2$ mw: 162.26

SYNS: ANILINE, N-(2-CYANOETHYL)-N-ETHYL- □ 3-(N-ETHYLANILINO)PROPIONITRILE □ N-ETHYL-N,β-CYANOETHYLANILINE □ N-ETHYL-N-2-KYANETHYLANILIN □ 3-(ETHYLPHENYLAMINO)PROPIONITRILE □ PROPANENITRILE, 3-(ETHYLPHENYLAMINO)-(9CI) □ PROPIONITRILE, 3-(N-ETHYLANILINO)-(6CI,7CI,8CI)

TOXICITY DATA WITH REFERENCE
orl-rat LD50:4840 mg/kg GTPZAB 32(9),32,88
orl-mus LD50:1510 mg/kg GTPZAB 30(1),50,86
orl-rbt LD50:1510 mg/kg GTPZAB 30(1),50,86

CONSENSUS REPORTS: Reported in EPA TSCA Inventory. Cyanide and its compounds are on the Community Right-To-Know List.

SAFETY PROFILE: Moderately toxic by ingestion. When heated to decomposition it emits very toxic fumes of NO_x and CN^-. See also CYANIDE.

EHR000 CAS:4806-61-5 HR: 3
ETHYL CYCLOBUTANE
mf: C_6H_{12} mw: 84.158

$CH_3CH_2CH(CH_2)_2CH_2$

PROP: A liquid. Flash p: <5°F, d: 0.7284 @ 20°/4°, autoign temp: 410°F, lel: 1.2%, uel: 7.7%, mp: −143.2°, bp: 71.5°. Insol in water.

SAFETY PROFILE: Probably an asphyxiant; see ARGON for a description of simple asphyxiants. A very dangerous fire and explosion hazard when exposed to heat or flame; can react vigorously with oxidizing materials. Easily forms explosive mixtures in air. To fight fire, use foam, spray, mist, dry chemical. When heated to decomposition it emits acrid smoke and fumes.

EHR500 CAS:1940-18-7 HR: 2
1-ETHYLCYCLOHEXANOL
mf: $C_8H_{16}O$ mw: 128.24

SYN: 1-AETHYL-CYCLOHEXANOL-(1) (GERMAN)

TOXICITY DATA WITH REFERENCE
orl-mus LD50:1300 mg/kg ARZNAD 4,477,54
scu-mus LD50:840 mg/kg ARZNAD 5,161,55

SAFETY PROFILE: Moderately toxic by ingestion and subcutaneous routes. When heated to decomposition it emits acrid smoke and irritating fumes.

EHS000 CAS:21722-83-8 HR: 2
ETHYLCYCLOHEXYL ACETATE
mf: $C_{10}H_{18}O_2$ mw: 170.28

SYNS: ACETIC ACID, CYCLOHEXYLETHYL ESTER □ CYCLOHEXANEETHANOL, ACETATE □ CYCLOHEXANE ETHYL ACETATE □ CYCLOHEXYLETHYL ACETATE □ HEXAHYDROPHENYL ETHYL ACETATE

TOXICITY DATA WITH REFERENCE
orl-rat LD50:3200 mg/kg FCTXAV 13,783,75
skn-rbt LD50:5000 mg/kg FCTXAV 13,783,75

CONSENSUS REPORTS: Reported in EPA TSCA Inventory.

SAFETY PROFILE: Moderately toxic by ingestion. Mildly toxic by skin contact. When heated to decomposition it emits acrid smoke and irritating fumes. See also ESTERS.

EHT000 CAS:5459-93-8 HR: 3
N-ETHYL(CYCLOHEXYL)AMINE
mf: $C_8H_{17}N$ mw: 127.26

PROP: A liquid with fishy odor. Flash p: 86°F (OC), bp: 164, d: 0.8, vap d: 4.4. Sltly sol in water

SYN: N-ETHYL-CYCLOHEXYLAMINE

TOXICITY DATA WITH REFERENCE
skn-rbt 100 µg/24H open AIHAAP 23,95,62
skn-rbt 5 mg/24H SEV 85JCAE-,461,86
eye-rbt 250 µg/24H SEV 85JCAE-,461,86
orl-rat LD50:590 mg/kg AIHAAP 23,95,62
ihl-rat LCLo:500 ppm/4H AIHAAP 23,95,62
skn-rbt LD50:750 mg/kg AIHAAP 23,95,62

CONSENSUS REPORTS: Reported in EPA TSCA Inventory.

SAFETY PROFILE: Moderately toxic by ingestion, inhalation, and skin contact. A severe skin and eye irritant. A very dangerous fire hazard when exposed to heat or flame; can react vigorously with oxidizing materials. To fight fire, use alcohol foam, mist, spray, dry chemical. See also AMINES.

EHT500 CAS:1134-23-2 HR: 2
S-ETHYL CYCLOHEXYLETHYLTHIOCARBAMATE
mf: $C_{11}H_{21}NOS$ mw: 215.39

PROP: A liquid with aromatic odor. D: 0.970, bp: 145° @ 10 mm. Sltly sol in H_2O. Misc in Me_2CO and C_6H_6.

SYNS: CYCLOATE □ CYCLOHEXYLETHYLCARBAMOTHIOIC ACID-S-ETHYL ESTER □ CYCLOHEXYLETHYLTHIOCARBAMIC ACID-S-ETHYL ESTER □ S-ETHYL-N-ETHYL-N-CYCLOHEXYLTHIOLCARBAMATE □ EUREX □ HEXYLTHIOCARBAM □ R 2063 □ RO-NEET □ RONIT

TOXICITY DATA WITH REFERENCE
cyt-mus-unr 200 mg/kg TGANAK 16(1),45,82
orl-rat LD50:1678 mg/kg FAATDF 7,299,86
ihl-rat LC50:90 g/m³/1H 85JFAN A106,83
skn-rat LD50:2467 mg/kg FAATDF 7,299,86
orl-mus LD50:1275 mg/kg GISAAA 51(10),77,86
skn-rbt LD50:3 g/kg GISAAA 50(3),26,85

SAFETY PROFILE: Moderately toxic by ingestion and skin contact. Mutation data reported. An herbicide. When heated to decomposition it emits very toxic fumes of SO_x and NO_x. See also CARBAMATES.

EHU000 CAS:1640-89-7 HR: 3
ETHYL CYCLOPENTANE
mf: C_7H_{14} mw: 98.16

$CH_3CH_2CH(CH_2)_3CH_2$ (ring)

PROP: A liquid. Autoign temp: 504°F, lel: 1.1%, uel: 6.7%, bp: 103–104°, mp: −137.9, d: 0.8, vap d: 3.4, flash p: <69.8°F.

SAFETY PROFILE: Probably an asphyxiant. See ARGON for a description of simple asphyxiants. A very dangerous fire and explosion hazard when exposed to heat or flame; can react vigorously with oxidizing materials. Easily forms explosive mixtures in air. To fight fire, use foam, dry chemical, mist. When heated to decomposition it emits acrid smoke and fumes.

EHU500 CAS:1191-96-4 HR: 3
ETHYL CYCLOPROPANE
mf: C_5H_{10} mw: 70.14

$CH_3CH_2CHCH_2CH_2$ (ring)

PROP: A volatile liquid. Fp: −149.6°, bp: 36.2°, flash p: <50°F.

SAFETY PROFILE: A very dangerous fire hazard when exposed to heat or flame; can react vigorously with oxidizers. When heated to decomposition it emits acrid smoke and fumes.

EHU600 CAS:868-59-7 HR: 2
ETHYL CYSTEINE HYDROCHLORIDE
mf: $C_5H_{11}NO_2S\cdot ClH$ mw: 185.69

SYNS: CYSTEINE ETHYL ESTER HYDROCHLORIDE □ ETHYL ESTER-1-CYSTEINE HYDROCHLORIDE (9CI)

TOXICITY DATA WITH REFERENCE
orl-rat LD50:5 g/kg NIIRDN 6,113,82
scu-rat LD50:2120 mg/kg NIIRDN 6,113,82
ivn-rat LD50:725 mg/kg NIIRDN 6,113,82
orl-mus LD50:3470 mg/kg NIIRDN 6,113,82
ipr-mus LD50:854 mg/kg CPBTAL 20,721,72
scu-mus LD50:1550 mg/kg NIIRDN 6,113,82
ivn-mus LD50:688 mg/kg NIIRDN 6,113,82

CONSENSUS REPORTS: Reported in EPA TSCA Inventory.

SAFETY PROFILE: Moderately toxic by ingestion, intraperitoneal, intravenous, and subcutaneous routes. When heated to decomposition it emits toxic fumes of SO_x, NO_x, and HCl. See also l-CYSTEINE and ESTERS.

EHV000 CAS:26747-87-5 HR: 3
ETHYL DECABORANE
mf: $C_2H_{18}B_{10}$ mw: 150.30

TOXICITY DATA WITH REFERENCE
ihl-rat LC50:23 ppm/4H NTIS** AD224-006
ihl-mus LC50:6.5 ppm/4H NTIS** AD224-006
ihl-dog LCLo:14 ppm/3H NTIS** AD224-006

SAFETY PROFILE: Poison by inhalation. See also BORON COMPOUNDS and BORANES.

EHV100 CAS:3025-30-7 HR: 1
ETHYL (2E,4Z)-DECADIENOATE
mf: $C_{12}H_{20}O_2$ mw: 196.32

SYNS: 2,4-DECADIENOIC ACID, ETHYL ESTER, (E,Z)- □ ETHYL (E,Z)-2,4-DECADIENOATE

TOXICITY DATA WITH REFERENCE
orl-rat LD50:>5 g/kg FCTOD7 26,317,88
skn-rbt LD50:>5 g/kg FCTOD7 26,317,88

CONSENSUS REPORTS: Reported in EPA TSCA Inventory.

SAFETY PROFILE: Low toxicity by ingestion and skin contact. When heated to decomposition it emits acrid smoke and irritating vapors.

EHV500 CAS:302-49-8 HR: 3
ETHYL(DI-(1-AZIRIDINYL)PHOSPHINYL)CARBAMATE
mf: $C_7H_{14}N_3O_3P$ mw: 219.21

PROP: Crystals from C_6H_6/cyclohexane. Mp: 88–90°.

SYNS: (BIS(1-AZIRIDINYL)PHOSPHINYL)CARBAMIC ACID, ETHYL ESTER □ BIS(ETHYLENIMIDO)PHOSPHORYLURETHAN □ ETHYL (BIS(1-AZIRIDINYL)PHOSPHINYL)CARBAMATE □ NSC 37095

TOXICITY DATA WITH REFERENCE
ipr-rat LD50:47 mg/kg JMCMAR 8,167,65
ipr-mus LD50:50 mg/kg JMCMAR 8,167,65

SAFETY PROFILE: Poison by intraperitoneal route. When heated to decomposition it emits very toxic fumes of PO_x and NO_x. See also CARBAMATES.

EHW000 CAS:63021-33-0 HR: 2
1-ETHYLDIBENZ(a,h)ACRIDINE
mf: $C_{23}H_{17}N$ mw: 307.41

SYN: 1'-ETHYL-1,2,5,6-DIBENZACRIDINE (FRENCH)

TOXICITY DATA WITH REFERENCE
skn-mus TDLo:780 mg/kg/26W-I:ETA BAFEAG 42,186,55

SAFETY PROFILE: Questionable carcinogen with experimental tumorigenic data. When heated to decomposition it emits toxic fumes of NO_x.

EHW500 CAS:63021-35-2 HR: 2
1-ETHYL-DIBENZ(a,j)ACRIDINE
mf: $C_{23}H_{17}N$ mw: 307.41

SYN: 1′-ETHYL-3,4,5,6-DIBENZACRIDINE (FRENCH)

TOXICITY DATA WITH REFERENCE
skn-mus TDLo:384 mg/kg/32W-I:ETA ACRSAJ 4,315,56
skn-mus TD:780 mg/kg/32W-I:ETA BAFEAG 42,186,55

SAFETY PROFILE: Questionable carcinogen with experimental tumorigenic data. When heated to decomposition it emits toxic fumes of NO_x.

EHX000 CAS:73927-60-3 HR: 2
8-ETHYL DIBENZ(a,h)ACRIDINE
mf: $C_{23}H_{17}N$ mw: 307.41

SYN: 1″-ETHYLDIBENZ(a,h)ACRIDINE

TOXICITY DATA WITH REFERENCE
skn-mus TDLo:312 mg/kg/26W-I:ETA ACRSAJ 4,315,56
scu-mus TDLo:27 mg/kg/26W-I:ETA ACRSAJ 4,315,56

SAFETY PROFILE: Questionable carcinogen with experimental tumorigenic data. When heated to decomposition it emits toxic fumes of NO_x.

EHY000 CAS:30812-87-4 HR: 2
ETHYL DIBROMOBENZENE
mf: $C_8H_8Br_2$ mw: 263.98

SYNS: ALKAZENE 42 □ DIAZENE 42 □ PRL-3191

TOXICITY DATA WITH REFERENCE
orl-rat LDLo:12,283 mg/kg WADTAA 55-251,23,55
orl-rbt LDLo:5536 mg/kg WADTAA 55-251,23,55
ihl-rbt LCLo:731 mg/m³/7H WADTAA 55-251,28,55

CONSENSUS REPORTS: Reported in EPA TSCA Inventory.

SAFETY PROFILE: Moderately toxic by ingestion and inhalation. When heated to decomposition it emits toxic fumes of Br^-.

EHY500 CAS:1331-29-9 HR: 2
ETHYLDICHLOROBENZENE
mf: $C_8H_8Cl_2$ mw: 175.06

PROP: Colorless liquid. Mp: <70°, bp: 220–224°, flash p: 205°F, d: 1.208 @ 25°/25°, vap d: 6.05.

SYN: DICHLOROETHYLBENZENE

TOXICITY DATA WITH REFERENCE
orl-rat LD50:5500 mg/kg JIHTAB 30,63,48
skn-rbt LD50:14 g/kg JIHTAB 30,63,48

SAFETY PROFILE: Moderately toxic by ingestion. Mildly toxic by skin contact. Combustible when exposed to heat or flame; can react vigorously with oxidizing materials. To fight fire, use water, foam, CO_2, dry chemical. When heated to decomposition it emits toxic fumes of Cl^-. See also CHLORINATED HYDROCARBONS, AROMATIC.

EIA000 CAS:78329-97-2 HR: 3
ETHYL-3-((DIETHYLAMINO)METHYL)-4-HYDROXYBENZOATE
mf: $C_{14}H_{21}NO_3$ mw: 251.36

SYN: 3-((DIETHYLAMINO)METHYL)-4-HYDROXYBENZOIC ACID ETHYL ESTER

TOXICITY DATA WITH REFERENCE
orl-mus LDLo:1000 mg/kg ARZNAD 11,85,61
scu-mus LDLo:375 mg/kg ARZNAD 11,85,61

SAFETY PROFILE: Poison by subcutaneous route. Moderately toxic by ingestion. When heated to decomposition it emits toxic fumes of NO_x. See also ESTERS.

EIB000 CAS:3553-80-8 HR: 2
ETHYL-N,N-DIETHYL CARBAMATE
mf: $C_7H_{15}NO_2$ mw: 145.23

TOXICITY DATA WITH REFERENCE
ipr-mus TDLo:4875 mg/kg/13W-I:ETA JNCIAM 9,35,48
scu-mus LD50:680 mg/kg AJEBAK 45,507,67

SAFETY PROFILE: Moderately toxic by subcutaneous route. Questionable carcinogen with experimental tumorigenic data. When heated to decomposition it emits toxic fumes of NO_x. See also CARBAMATES.

EIB500 CAS:4928-41-0 HR: 2
4′-ETHYL-N,N-DIETHYL-p-(PHENYLAZO)ANILINE
mf: $C_{18}H_{23}N_3$ mw: 281.44

SYN: 4′-ETHYL-N,N-DIMETHYL-4-AMINOAZOBENZENE

TOXICITY DATA WITH REFERENCE
otr-ham:kdy 2500 µg/L BJCAAI 38,34,78
orl-rat TDLo:7400 mg/kg/26W-C:ETA ARZNAD 12,270,62

SAFETY PROFILE: Questionable carcinogen with experimental tumorigenic data. Mutation data reported. When heated to decomposition it emits toxic fumes of NO_x.

EIC000 CAS:867-13-0 HR: 2
ETHYL (DIETHYLPHOSPHONO)ACETATE
mf: $C_8H_{17}O_5P$ mw: 224.22

PROP: A liquid. D: 1.135, bp: 152–153° @ 20 mm.

SYNS: ACETIC ACID, (DIETHOXYPHOSPHINYL)-, ETHYL ESTER (9CI) □ ACETIC ACID, PHOSPHONO-, TRIETHYL ESTER (9CI) □ ETHYL DIETHOXYPHOSPHORYL ACETATE □ ETHYL (DIETHYLPHOSPHONO)ACETATE □ TL 465 □ TRIETHYL PHOSPHONOACETATE

TOXICITY DATA WITH REFERENCE
orl-rat LD:>500 mg/kg NCNSA6 5,15,53
ihl-mus LCLo:1180 mg/m³/10M NDRC** NDCrc-132,Nov,42

CONSENSUS REPORTS: Reported in EPA TSCA Inventory.

SAFETY PROFILE: Moderately toxic by ingestion and inhalation. When heated to decomposition it emits toxic fumes of PO_x. See also ESTERS.

E

ECI200 CAS:15336-82-0 HR: 3
5-ETHYL-1,3-DIGLYCIDYL-5-METHYLHYDANTOIN
mf: $C_{12}H_{18}N_2O_4$ mw: 254.32

SYNS: BIS(2,3-EPOXYPROPYL)-5-ETHYL-5-METHYLHYDANTOIN □ HYDANTOIN, BIS(2,3-EPOXYPROPYL)-5-ETHYL-5-METHYL- □ 2,4-IMIDAZOLIDINEDIONE, 5-ETHYL-5-METHYL-1,3-BIS(OXIRANYLMETHYL)-

TOXICITY DATA WITH REFERENCE
eye-rbt 100 mg SEV EPASR* 8EHQ-0683-0418S
mma-sat 2 mg/plate EPASR* 8EHQ-0683-0418S
otr-mus:fbr 980 μg/L EPASR* 8EHQ-0683-0418S
msc-mus:lyms 1500 μg/L EPASR* 8EHQ-0683-0418S
cyt-ham-orl 5 g/kg/2D-C EPASR* 8EHQ-0683-0418S
orl-rat LD50:250 mg/kg EPASR* 8EHQ-0683-0418S
orl-mus LD50:479 mg/kg EPASR* 8EHQ-0683-0418S
orl-ham LD50:7488 mg/kg EPASR* 8EHQ-0683-0418S

CONSENSUS REPORTS: Reported in EPA TSCA Inventory.

SAFETY PROFILE: A poison by ingestion. A severe eye irritant. Mutation data reported. When heated to decomposition it emits toxic vapors of NO_x.

EID000 CAS:389-08-2 HR: 3
1-ETHYL-1,4-DIHYDRO-7-METHYL-4-OXO-1,8-NAPHTHYRIDINE-3-CARBOXYLIC ACID
mf: $C_{12}H_{12}N_2O_3$ mw: 232.26

PROP: Pale buff crystals. Mp: 229–230°. Sol in $CHCl_3$; mod sol in EtOH and MeOH.

SYNS: ACIDE 1-ETIL-7-METIL-1,8-NAFTIRIDIN-4-ONE-3-CARBOSSILICO (ITALIAN) □ ACIDE NALIDIXICO (ITALIAN) □ ACIDE NALIDIXIQUE (FRENCH) □ BETAXINA □ 3-CARBOXY-1-ETHYL-7-METHYL-1,8-NAPHTHIDIN-4-ONE □ CHINOIN □ CYBIS □ 1,4-DIHYDRO-1-ETHYL-7-METHYL-4-OXO-1,8-NAPHTHYRIDINE-3-CARBOXYLIC ACID □ DIXIBEN □ 1-ETHYL-7-METHYL-1,4-DIHYDRO-1,8-NAPHTHYRIDINE-4-ONE-3-CARBOXYLIC ACID □ 1-ETHYL-7-METHYL-1,4-DIHYDRO-1,8-NAPHTHYRIDIN-4-ONE-3-CARBOXYLIC ACID □ 1-ETHYL-7-METHYL-4-OXO-1,4-DIHYDRO-1,8-NAPHTHYRIDINE-3-CARBOXYLIC ACID □ EUCISTEN □ INNOXALON □ KUSNARIN □ NA □ NALIDIC ACID □ NALIDICRON □ NALIDIXIC ACID □ NALIDIXIN □ NALITUCSAN □ NARIGIX □ NCI-C56199 □ NEGRAM □ NEVIGRAMON □ NICELATE □ NOGRAM □ NSC-82174 □ POLEON □ SPECIFEN □ URALGIN □ URIBEN □ URODIXIN □ UROMAN □ URONEG □ WIN 18,320 □ WINTOMYLON

TOXICITY DATA WITH REFERENCE
mmo-bcs 100 mg/L MUREAV 95,191,82
dns-ssp 100 mg/L MGGEAE 187,96,82
sce-chd-unr 50 mg/kg/10D MUREAV 77,371,80
dnr-ham:oth 500 mg/L CRNGDP 5,187,84
orl-rat TDLo:3300 mg/kg (female 7-17D post):REP NKRZAZ 28,484,80
orl-rat TDLo:3300 mg/kg (female 7-17D post):TER NKRZAZ 28,484,80
orl-rat TDLo:1320 mg/kg (female 9-14D post):TER NKRZAZ 19,422,71
orl-rat TDLo:173 g/kg/2Y-C:CAR NTPTR* NTP-TR-368,89
orl-mus TDLo:173 g/kg/2Y-C:ETA NTPTR* NTP-TR-368,89
orl-wmn TDLo:160 mg/kg/2D:CNS,END,SKN BMJOAE 2,1518,77
unr-chd TDLo:1200 mg/kg/20D:BLD 34ZIAG -,414,69
orl-rat LD50:1160 mg/kg NIIRDN 6,540,82
ipr-rat LD50:319 mg/kg AACHAX -,117,70
scu-rat LD50:1584 mg/kg AACHAX -,117,70
ivn-rat LD50:88,400 μg/kg AACHAX -,117,70
orl-mus LD50:572 mg/kg TXAPA9 18,185,71
ipr-mus LD50:871 mg/kg AACHAX -,117,70
scu-mus LD50:500 mg/kg TXAPA9 18,185,71
ivn-mus LD50:101 mg/kg AACHAX -,117,70

CONSENSUS REPORTS: EPA Genetic Toxicology Program.

SAFETY PROFILE: Poison by intravenous and intraperitoneal routes. Moderately toxic by ingestion and subcutaneous routes. An experimental teratogen. Human systemic effects: convulsions, hyperglycemia, sweating, and blood changes in children. Experimental reproductive effects. Questionable carcinogen with experimental carcinogenic and tumorigenic data. Human mutation data reported. Used as an antibacterial agent and urinary tract antiseptic. When heated to decomposition it emits toxic fumes of NO_x.

EID100 HR: 3
ETHYL-3,4-DIHYDROXYBENZENE SULFONATE
mf: $C_8H_{10}O_5S$ mw: 218.22

$(OH)_2C_6H_3SO_2OCH_2CH_3$

SAFETY PROFILE: Reacts explosively when mixed with acetyl nitrate and oleum. When heated to decomposition it emits toxic fumes of SO_x. See also SULFONATES.

EID200 CAS:68-90-6 HR: 3
2-ETHYL-3-(3′,5′-DIIODO-4′-HYDROXYBENZOYL)-CUMARONE
mf: $C_{17}H_{12}I_2O_3$ mw: 518.09

SYNS: AETHYL-2-(3′,5′-DIJOD-4′-OXYBENZOYL)-3 CUMARON □ ALGOCOR □ AMPLIVIX □ BENZIODARON □ BENZIODARONE □ BENZOFURAN, 3-(3,5-DIIODO-4-HYDROXYBENZOYL)-2-ETHYL- □ CARDIVIX □ CAROFAM □ CORONAL-CRINOS □ 3,5-DIIODO-4-HYDROXYPHENYL 2-ETHYL-3-BENZOFURANYL KETONE □DILAFU-RANE □ DILA-VASAL □ KETONE, 3,5-DIIODO-4-HYDROXYPHENYL 2-ETHYL-3-BENZOFURANYL □ L 2329 □ 2329 LABAZ □ RETRANGOR

TOXICITY DATA WITH REFERENCE
orl-mus LD50:450 mg/kg YKYUA6 24,431,73
ipr-mus LD50:130 mg/kg YKYUA6 24,431,73
ivn-dog LDLo:10 mg/kg ARZNAD 15,1388,65

DOT CLASSIFICATION: 3; *Label:* Flammable Liquid

SAFETY PROFILE: A poison by intraperitoneal and intravenous routes. Moderately toxic by ingestion. A flammable liquid. When heated to decomposition it emits toxic vapors of I^-.

EID250 CAS:4568-83-6 HR: 3
ETHYL-2-(DIIODO-3,5 HYDROXY-4 BENZOYL)5-FURANNE
mf: $C_{13}H_{10}I_2O_3$ mw: 468.03

SYNS: DB 138 □ 3,5-DIIODO-4-HYDROXYPHENYL 5-ETHYL-2-FURYL KETONE □ KETONE, 3,5-DIIODO-4-HYDROXYPHENYL 5-ETHYL-2-FURYL

TOXICITY DATA WITH REFERENCE
ipr-mus LD50:275 mg/kg AIPTAK 147,497,64

ivn-gpg LDLo:161 mg/kg AIPTAK 147,497,64

DOT CLASSIFICATION: 3; *Label:* Flammable Liquid

SAFETY PROFILE: A poison by intraperitoneal and intravenous routes. When heated to decomposition it emits toxic vapors of NO_x and I^-.

EID500 CAS:2008-41-5 HR: 2
S-ETHYL N,N-DIISOBUTYLTHIOCARBAMATE
mf: $C_{11}H_{23}NOS$ mw: 217.41

PROP: A liquid. D: 0.942, bp: 138° @ 21 mm.

SYNS: BIS(2-METHYLPROPYL)CARBAMOTHIOIC ACID-S-ETHYL ESTER ☐ BUTILATE ☐ BUTYLATE ☐ DIISOBUTYLTHIOCARBAMIC ACID-S-ETHYL ESTER ☐ DIISOCARB ☐ S-ETHYL BIS(2-METHYLPROPYL)CARBAMOTHIOATE ☐ ETHYL-N,N-DIISOBUTYLTHIOCARBAMATE ☐ S-ETHYLDIISOBUTYL THIOCARBAMATE ☐ ETHYL-N,N-DIISOBUTYL THIOLCARBAMATE ☐ R-1910 ☐ STAUFFER R-1910 ☐ SUTAN

TOXICITY DATA WITH REFERENCE
cyt-mus-orl 1 g/kg CYGEDX 14(6),38,80
orl-rat LD50:4000 mg/kg WRPCA2 9,119,70
ipr-mus LD50:365 mg/kg JAFCAU 25,404,77

CONSENSUS REPORTS: Reported in EPA TSCA Inventory.

SAFETY PROFILE: Moderately toxic by ingestion and intraperitoneal route. Mutation data reported. Used as an herbicide. When heated to decomposition it emits very toxic fumes of NO_x and SO_x. See also CARBAMATES and ESTERS.

EIF000 CAS:77-81-6 HR: 3
ETHYL DIMETHYLAMIDOCYANOPHOSPHATE
mf: $C_5H_{11}N_2O_2P$ mw: 162.15

PROP: A colorless to brownish liquid. Bp: decomp @ 238°, fp: −49.4°, flash p: 172°F, d: 1.073 @ 25°, vap press: 0.07 mm @ 25°, vap d: 5.63.

SYNS: DIMETHYLAMIDOETHOXYPHOSPHORYL CYANIDE ☐ DIMETHYLAMINOCYANPHOSPHORSAEUREAETHYLESTER (GERMAN) ☐ DIMETHYLPHOSPHORAMIDOCYANIDIC ACID, ETHYL ESTER ☐ ETHYL N,N-DIMETHYLAMINO CYANOPHOSPHATE ☐ ETHYL DIMETHYLPHOSPHORAMIDOCYANIDATE ☐ ETHYL-N,N-DIMETHYLPHOSPHORAMIDOCYANIDATE ☐ GA ☐ GELAN I ☐ Le-100 ☐ MCE ☐ T-2104 ☐ TL 1578 ☐ TABOON A ☐ TABUN ☐ TRILON 83

TOXICITY DATA WITH REFERENCE
ihl-hmn LCLo:150 mg/m³ SCJUAD 4,33,67
skn-hmn LDLo:23 mg/kg SCJUAD 4,33,67
ivn-hmn LDLo:14 μg/kg SCJUAD 4,33,67
orl-rat LD50:3700 μg/kg NTIS** PB158-508
ihl-rat LC50:304 mg/m³/10M NTIS** PB158-508
skn-rat LDLo:18 mg/kg NTIS** PB158-508
scu-rat LD50:193 μg/kg AIPTAK 262,231,83
ivn-rbt LD50:66 μg/kg NTIS** PB158-508
ims-rat LD50:800 μg/kg JCINAO 37,350,58
ihl-mus LC50:15 mg/m³/30M DEGEA3 15,2179,60
skn-mus LD50:1 mg/kg NTIS** PB158-508
ipr-mus LD50:604 μg/kg 11FYAN 3,69,63
orl-dog LD50:200 μg/kg DEGEA3 15,2179,60
ihl-dog LC50:400 mg/m³/10M NTIS** PB158-508
skn-dog LDLo:30 mg/kg NTIS** PB158-508
ihl-mky LC50:250 mg/m³/10M NTIS** PB158-508
skn-mky LD50:9300 μg/kg NTIS** PB158-508

CONSENSUS REPORTS: EPA Extremely Hazardous Substances List. Cyanide and its compounds are on the Community Right-To-Know List. Reported in EPA TSCA Inventory.

SAFETY PROFILE: Human poison by inhalation, skin contact, and intravenous routes. Experimental poison by ingestion, inhalation, skin contact, subcutaneous, intravenous, intraperitoneal, and intramuscular routes. A nerve gas. Vapor does not penetrate skin; liquid does so rapidly. The primary physiological action is on the sympathetic nervous system, causing a vasoparesis (partial paralysis of the vasomotor nerves, which control the diameter of the blood vessels). Vapors when inhaled can cause nausea, vomiting, and diarrhea, which can be followed by muscular twitching and convulsions. Flammable when exposed to heat or flame; can react with oxidizing materials. When heated to decomposition it emits very toxic fumes of PO_x, CN^-, and NO_x. See also PARATHION and CYANIDE.

EIF450 CAS:93023-34-8 HR: 2
2′-ETHYL-4-DIMETHYLAMINOAZOBENZENE
mf: $C_{16}H_{19}N_3$ mw: 253.38

SYNS: ANILINE, p-((o-ETHYLPHENYL)AZO)-N,N-DIMETHYL- ☐ BENZENAMINE, N,N-DIMETHYL-2′-ETHYL-4-(PHENYLAZO)- ☐ BENZENAMINE, 4-((2-ETHYLPHENYL)AZO)-N,N-DIMETHYL- ☐ N,N-DIMETHYL-p-((o-ETHYLPHENYL)AZO)ANILINE

TOXICITY DATA WITH REFERENCE
orl-rat TDLo:13,541 mg/kg/52W C:CAR CBINA8 53,107,85

SAFETY PROFILE: Questionable carcinogen with experimental carcinogenic data. When heated to decomposition it emits toxic fumes of NO_x.

EIF500 CAS:20820-80-8 HR: 3
O-ETHYL-S-(2-DIMETHYL AMINO ETHYL)-METHYLPHOSPHONOTHIOATE
mf: $C_7H_{18}NO_2PS$ mw: 211.29

SYN: O-AETHYL-S-(2-DIMETHYLAMINOAETHYL)-METHYLPHOSPHONOTHIOATE (GERMAN)

TOXICITY DATA WITH REFERENCE
orl-rat LD50:122 μg/kg ABMGAJ 37,633,78
ipr-rat LD50:54 μg/kg ABMGAJ 37,633,78
ivn-rat LD50:17 μg/kg ABMGAJ 37,633,78
ims-rat LD50:24 μg/kg ABMGAJ 37,633,78

SAFETY PROFILE: Poison by ingestion, intraperitoneal, intravenous, and intramuscular routes. When heated to decomposition it emits very toxic fumes of NO_x, PO_x, and SO_x.

EIG000 CAS:50782-69-9 HR: 3
ETHYL-S-DIMETHYLAMINOETHYL METHYLPHOSPHONOTHIOLATE
mf: $C_{11}H_{26}NO_2PS$ mw: 267.41

SYNS: S-(2-DIISOPROPYLAMINOETHYL)-O-ETHYL METHYL PHOSPHONOTHIOLATE □ ETHYL-S-DIISOPROPYLAMINOETHYL METHYLTHIOPHOSPHONATE □ O-ETHYL-S-2-DIISOPROPYLAMINOETHYL METHYLPHOSPHONOTHIOTE □ METHYLPHOSPHONOTHIOIC ACID-S-(2-(BIS(METHYLETHYL)AMINO)ETHYL)o-ETHYL ESTER □ VX

TOXICITY DATA WITH REFERENCE
scu-rat TDLo:75 μg/kg (female 6-20D post):REP TJADAB 36,25A,87
scu-rat TDLo:75 μg/kg (female 6-20D post):TER TJADAB 36,25A,87
orl-man TDLo:4 μg/kg:GIT TXAPA9 27,241,74
skn-hmn LDLo:86 μg/kg WHOTAC -,24,70
scu-hmn TDLo:30 μg/kg 85IVAW 1,E1,82
ivn-man TDLo:1500 ng/kg:CNS,CVS,GIT TXAPA9 27,241,74
ims-hmn TDLo:3200 ng/kg:EYE,GIT 85IVAW 1,E1,82
scu-rat LD50:12 μg/kg AIPTAK 262,231,83
ipr-mus LD50:50 μg/kg WHOTAC -,39,70
scu-mus LD50:22 μg/kg APJUA8 30,151,80
scu-rbt LD50:14 μg/kg ABMHAM (Suppl),226,84
scu-gpg LD50:8400 ng/kg TXAPA9 40,109,77

CONSENSUS REPORTS: Reported in EPA TSCA Inventory. EPA Extremely Hazardous Substances List.

SAFETY PROFILE: Human poison by skin contact. Experimental poison by intraperitoneal and subcutaneous routes. An experimental teratogen. Other experimental reproductive effects. Human systemic effects by ingestion and intravenous route: hallucinations and distorted perceptions, blood pressure increase, hypermotility, diarrhea, nausea and vomiting, visual field changes, sleep disturbance. A chemical warfare agent. When heated to decomposition it emits very toxic fumes of SO_x and NO_x.

EIH000 CAS:27107-79-5 HR: 3
ETHYL-dl-trans-2-DIMETHYLAMINO-1-PHENYL-3-CYCLOHEXENE-1-CARBOXYLATE
mf: $C_{17}H_{23}NO_2•ClH$ mw: 309.87

PROP: Crystals from EtOAc/MeCOEt. Mp: 162°.

SYNS: dl-trans-2-DIMETHYLAMINO-1-PHENYL-CYCLOHEX-3-EN-trans-1-CARBONSAEUREAETHYLESTER HCl (GERMAN) □ dl-trans-2-DIMETHYLAMINO-1-PHENYL-CYCLOHEX-3-ENE-trans-CARBONIC ACID ETHYL ESTER HCl □ (±)-ETHYL-trans-2-2(DIMETHYLAMINO)-1-PHENYL-3-CYCLOHEXENE-1-CARBOXYLATE HYDROCHLORIDE □ GO-1261 □ TILIDATE □ TILIDINE HYDROCHLORIDE □ VALORON HYDROCHLORIDE □ W 5759A

TOXICITY DATA WITH REFERENCE
orl-rat TDLo:1450 mg/kg (15-22D preg/21D post):REP ARZNAD 20,983,70
orl-rat LD50:418 mg/kg ARZNAD 20,977,70
scu-rat LD50:400 mg/kg ARZNAD 20,977,70
ivn-rat LD50:74 mg/kg ARZNAD 20,977,70
orl-mus LD50:437 mg/kg ARZNAD 20,977,70
scu-mus LD50:490 mg/kg ARZNAD 20,977,70
ivn-mus LD50:52 mg/kg ARZNAD 20,977,70
orl-dog LD50:500 mg/kg ARZNAD 20,977,70

SAFETY PROFILE: Poison by ingestion, subcutaneous, and intravenous routes. Experimental reproductive effects. An analgesic. When heated to decomposition it emits very toxic fumes of HCl and NO_x.

EIH500 CAS:19622-19-6 HR: 2
S-ETHYL N-(3-DIMETHYLAMINOPROPYL)THIOL CARBAMATE HYDROCHLORIDE
mf: $C_8H_{18}N_2OS•ClH$ mw: 226.80

PROP: A solid. Mp: 120–121°. Very sol in H_2O.

SYNS: N-(3-DIMETHYLAMINOPROPYL)THIOCARBAMINSAEURE-S-AETHYLESTER-HYDROCHLORID (GERMAN) □ DYNONE □PREVI-CUR □ PROTHIOCARB □ SN-41703

TOXICITY DATA WITH REFERENCE
orl-rat LD50:1300 mg/kg 85ARAE 4,59,76/77
ihl-rat LC50:3300 mg/m³/4H 85JFAN A351,83
ipr-rat LD50:770 mg/kg 85DPAN -,-,71/76
orl-mus LD50:660 mg/kg 85DPAN -,-,71/76

SAFETY PROFILE: Moderately toxic by ingestion and intraperitoneal routes. When heated to decomposition it emits very toxic fumes of NO_x, SO_x, and HCl. See also CARBAMATES and ESTERS.

EII500 CAS:687-48-9 HR: 2
ETHYL-N,N-DIMETHYL CARBAMATE
mf: $C_5H_{11}NO_2$ mw: 117.17

PROP: A liquid. Bp: 140°.

SYNS: DIMETHYLCARBAMIC ACID ETHYL ESTER □ DI-N-METHYL ETHYL CARBAMATE

TOXICITY DATA WITH REFERENCE
mmo-sat 1 mg/plate MUREAV 245,227,90
mnt-mus-orl 114 mg/kg MUREAV 245,227,90
ipr-ham TDLo:504 mg/kg (8D preg):TER CNREA8 27,1696,67
ipr-mus TDLo:6500 mg/kg/13W-I:ETA JNCIAM 9,35,48
ipr-mus LD50:1110 mg/kg DPHFAK 23,25,71
scu-mus LD50:1050 mg/kg AJEBAK 45,507,67

SAFETY PROFILE: Moderately toxic by subcutaneous and intraperitoneal routes. Experimental teratogenic effects. Questionable carcinogen with experimental tumorigenic data. Mutation data reported. When heated to decomposition it emits toxic fumes of NO_x. See also CARBAMATES.

EII600 CAS:67465-27-4 HR: 3
3-ETHYL-2,2-DIMETHYLCYCLOBUTYL METHYL KETONE(E)-
mf: $C_{10}H_{18}O$ mw: 154.28

TOXICITY DATA WITH REFERENCE
ipr-mus LDLo:500 mg/kg CBCCT* 7,783,55

DOT CLASSIFICATION: 3; *Label:* Flammable Liquid

SAFETY PROFILE: Moderately toxic by intraperitoneal route. A flammable liquid. When heated to decomposition it emits acrid smoke and irritating vapors.

EII700 CAS:4951-97-7 HR: 3
3-ETHYL-2,2-DIMETHYLCYCLOBUTYL METHYL KETONE (Z)-
mf: $C_{10}H_{18}O$ mw: 154.28

SYN: CYCLOBUTANE, 1-ACETYL-2,2-DIMETHYL-3-ETHYL-, (cis)-

TOXICITY DATA WITH REFERENCE
ipr-mus LDLo:500 mg/kg CBCCT* 7,783,55

DOT CLASSIFICATION: 3; *Label:* Flammable Liquid

SAFETY PROFILE: Moderately toxic by intraperitoneal route. A flammable liquid. When heated to decomposition it emits acrid smoke and irritating vapors.

EIJ000 CAS:64037-50-9 HR: 3
N-ETHYL-N-1-DIMETHYL-3,3-DI-2-THIENYL-2-PROPENAMINE HYDROCHLORIDE
mf: $C_{15}H_{19}NS_2 \cdot ClH$ mw: 313.93

SYNS: 1C50 HYDROCHLORIDE □ N,1-DIMETHYL-3,3-DI-2-THIENYL-N-ETHYLALLYLAMINE HYDROCHLORIDE □ EMETHIBUTIN HYDROCHLORIDE □ N-ETHYL-N-1-DIMETHYL-3,3-DI-2-THIENYLALLYLAMINE HYDROCHLORIDE □ ETHYLMETHIAMBUTENE HYDROCHLORIDE □ 3-ETHYLMETHYLAMINO-1,1-DI(2'-THIENYL) BUT-1-ENE HYDROCHLORIDE □ ETHYLMETHYLTHIAMBUTENE HYDROCHLORIDE □ NIH-5145 HYDROCHLORIDE

TOXICITY DATA WITH REFERENCE
scu-rat LD50:63 mg/kg BJPCAL 8,2,53
orl-mus LD50:192 mg/kg MEIEDD 10,553,83
scu-mus LD50:88 mg/kg MEIEDD 10,553,83
ivn-mus LD50:17 mg/kg BJPCAL 8,2,53

SAFETY PROFILE: Poison by ingestion, intravenous, and subcutaneous routes. Used as a narcotic analgesic. When heated to decomposition it emits very toxic fumes of NO_x, SO_x, and HCl. See also ALLYL COMPOUNDS and AMINES.

EIJ500 CAS:617-38-9 HR: 3
ETHYL DIMETHYLDITHIOCARBAMATE
mf: $C_5H_{11}NS_2$ mw: 149.29

SYNS: EEDDKK □ ETHYL ESTER of DIMETHYLDITHIOCARBAMIC ACID

TOXICITY DATA WITH REFERENCE
orl-rat LD50:550 mg/kg FATOAO 28,230,65
orl-mus LD50:290 mg/kg FATOAO 28,230,65
ipr-mus LD50:190 mg/kg DPHFAK 23,25,71

SAFETY PROFILE: Poison by ingestion and intraperitoneal routes. When heated to decomposition it emits very toxic fumes of NO_x and SO_x. See also CARBAMATES.

EIJ600 CAS:67634-15-5 HR: 1
ETHYL DIMETHYLHYDROCINNAMALDEHYDE
mf: $C_{13}H_{18}O$ mw: 190.31

SYNS: BENZENEPROPANAL, α-α-DIMETHYL-4-ETHYL □ α α DIMETHYL-p-ETHYLPHENYLPROPANAL □ 2,2-DIMETHYL-3-(p-ETHYLPHENYL)PROPANAL □ FLORALOZONE

TOXICITY DATA WITH REFERENCE
orl-rat LDLo:5 g/kg FCTOD7 26,307,88
skn-rbt LD50:>5 g/kg FCTOD7 26,307,88

CONSENSUS REPORTS: Reported in EPA TSCA Inventory.

SAFETY PROFILE: Low toxicity by ingestion and skin contact. When heated to decomposition it emits acrid smoke and irritating vapors.

EIK000 CAS:78-78-4 HR: 3
ETHYLDIMETHYLMETHANE
DOT: UN 1265
mf: C_5H_{12} mw: 72.17

PROP: Colorless liquid with pleasant odor. Fp: −160.5°, bp: 30–30.2°, flash p: <−60°F (CC), D: 0.620 @ 20°/4°, vap press: 595 mm @ 21.1°, vap d: 2.48, lel: 1.4%, uel: 7.6%.

SYNS: ISOAMYLHYDRIDE □ ISOPENTANE (DOT) □ 2-METHYLBUTANE

TOXICITY DATA WITH REFERENCE
ihl-mus LCLo:419 g/m³/2H JPETAB 58,74,36

CONSENSUS REPORTS: Reported in EPA TSCA Inventory.

DFG MAK: 1000 ppm (2950 mg/m³)
NIOSH REL: (Alkanes) TWA 350 mg/m³
DOT CLASSIFICATION: 3; *Label:* Flammable Liquid

SAFETY PROFILE: Mildly toxic and narcotic by inhalation. See also PENTANE. Flammable Liquid. A very dangerous fire and explosion hazard when exposed to heat, flame, or oxidizers. Keep away from sparks, heat, or open flame; can react with oxidizing materials. To fight fire, use foam, CO_2, dry chemical. When heated to decomposition it emits acrid smoke and irritating fumes.

EIL000 CAS:1605-51-2 HR: 3
ETHYLDIMETHYLPHOSPHINE
mf: $C_4H_{11}P$ mw: 90.10

$CH_3CH_2P(CH_3)_2$

SAFETY PROFILE: A very dangerous fire hazard; ignites spontaneously in air. When heated to decomposition it emits toxic fumes of PO_x and phosphine. See also PHOSPHINE.

EIL100 HR: 2
2-ETHYL-3,5(6)-DIMETHYLPYRAZINE
mf: $C_8H_{12}N_2$ mw: 136.20

PROP: Colorless to sltly yellow liquid; roasted-cocoa odor. D: 0.950–0.980, refr index: 1.500, flash p: 154°F. Sol in water, organic solvents.

SYN: FEMA No. 3149

SAFETY PROFILE: Combustible liquid. When heated to decomposition it emits toxic fumes of NO_x.

E

EIL200 CAS:1500-91-0 HR: 3
4-ETHYL-3,5-DIMETHYLPYRROL-2-YL METHYL KETONE
mf: $C_{10}H_{15}NO$ mw: 165.26

SYNS: 2-ACETYL-3,5-DIMETHYL-4-ETHYL-PYRROLE □ KETONE, 4-ETHYL-3,5-DIMETHYLPYRROL-2-YL METHYL

TOXICITY DATA WITH REFERENCE
ipr-rat LD50:400 mg/kg JMCMAR 11,1251,68
ipr-mus LD50:767 mg/kg JMCMAR 11,1251,68

DOT CLASSIFICATION: 3; *Label:* Flammable Liquid

SAFETY PROFILE: Moderately toxic by intraperitoneal route. A flammable liquid. When heated to decomposition it emits toxic vapors of NO_x.

EIL500 CAS:19481-40-4 HR: 3
ETHYLDIMETHYL SULFONIUM IODIDE MERCURIC IODIDE ADDITION COMPOUND

SYN: DIMETHYLETHYL SULFONIUM IODIDE with MERCURY IODIDE (1:1)

TOXICITY DATA WITH REFERENCE
ivn-mus LD50:56 mg/kg CSLNX* NX#01852

CONSENSUS REPORTS: Mercury and its compounds are on the Community Right-To-Know List.

NIOSH REL: (Mercury, Aryl and Inorganic): CL 0.1 mg/m^3 (skin)

SAFETY PROFILE: Poison by intravenous route. When heated to decomposition it emits very toxic fumes of SO_x, I^-, and Hg. See also MERCURY COMPOUNDS, and IODIDES.

EIM000 CAS:17109-49-8 HR: 3
O-ETHYL-S,S-DIPHENYL DITHIOPHOSPHATE
mf: $C_{14}H_{15}O_2PS_2$ mw: 310.38

PROP: A clear yellow to light brown liquid. D: 1.23, bp: 154° @ 0.01 mm. Prac insol in water.

SYNS: O-AETHYL-S,S-DIPHENYL-DITHIOPHOSPHAT (GERMAN) □ BAYER 78418 □ DITHIOPHOSPHORSAEURE-O-AETHYL-S,S-DIPHENYLESTER (GERMAN) □ EDDP □ EDIFENPHOS □ EDIPHENPHOS □ O-ETHYL-S,S-DIPHENYL PHOSPHORODITHIOATE □ HINOSAN □ LUTROL □ SRA 7847

TOXICITY DATA WITH REFERENCE
cyt-rat-ipr 3 mg/kg MUREAV 321,103,94
cyt-mus-ipr 30 mg/kg CYTOAN 49,833,84
orl-rat LD50:100 mg/kg OYYAA2 2,76,68
ihl-rat LC50:650 mg/m^3/4H JPIFAN (17),25,73
ipr-rat LD50:26 mg/kg TXAPA9 23,519,72
orl-mus LD50:143 mg/kg TOIZAG 25,635,78
orl-rbt LD50:350 mg/kg 28ZEAL 5,91,76
orl-gpg LD50:350 mg/kg 28ZEAL 5,91,76

SAFETY PROFILE: Poison by ingestion and intraperitoneal routes. Mutation data reported. A cholinesterase inhibitor. When heated to decomposition it emits very toxic fumes of SO_x and PO_x. See also PARATHION.

EIN000 CAS:13194-48-4 HR: 3
O-ETHYL-S,S-DIPROPYLPHOSPHORODITHIOATE
mf: $C_8H_{19}O_2PS_2$ mw: 242.36

PROP: A yellow liquid. D: 1.094 @ 20°/4°, bp: 86–91° @ 0.2 mm. Sltly sol in H_2O; very sol in most org solvs.

SYNS: ENT 27,318 □ ETHOPROP □ ETHOPROPHOS □ O-ETHYL-S,S-DIPROPYL ESTER, PHOSPHORODITHIOIC ACID □ JOLT □ MOBIL V-C 9-104 □ MOCAP □ PROPHOS □ V-C CHEMICAL V-C 9-104 □ VIRGINIA CAROLINA VC 9-104

TOXICITY DATA WITH REFERENCE
orl-rat LD50:34 mg/kg ARSIM* 20,26,66
skn-rat LD50:60 mg/kg WRPCA2 9,119,70
orl-rbt LD50:55 mg/kg PEMNDP 9,349,91
skn-rbt LD50:26 mg/kg SPEADM 78-1,30,78
orl-pgn LD50:13,300 μg/kg ASTTA8 (680),157,79
orl-ckn LD50:5500 μg/kg TXAPA9 11,49,67
orl-qal LD50:7500 μg/kg ASTTA8 (680),157,79
orl-dck LD50:1260 μg/kg TXAPA9 47,451,79
skn-dck LD50:11 mg/kg TXAPA9 47,451,79
orl-bwd LD50:4210 μg/kg ASTTA8 (680),157,79

CONSENSUS REPORTS: EPA Extremely Hazardous Substances List.

SAFETY PROFILE: Poison by ingestion and skin contact. A cholinesterase inhibitor type of insecticide. When heated to decomposition it emits very toxic fumes of PO_x and SO_x. See also PARATHION.

EIN500 CAS:759-94-4 HR: 3
S-ETHYL-N,N-DI-N-PROPYLTHIOCARBAMATE
mf: $C_9H_{19}NOS$ mw: 189.35

PROP: A liquid. D: 0.955 @ 30°, bp: 127° @ 20 mm. Sltly sol in H_2O.

SYNS: S-AETHYL-N,N-DIPROPYLTHIOLCARBAMAT (GERMAN) □ DIPROPYLCARBAMOTHIOIC ACID-S-ETHYL ESTER □ N,N-DIPROPYLTHIOCARBAMIC ACID-S-ETHYL ESTER □ EPTAM □ EPTC □ S-ETHYL-N,N-DIPROPYLTHIOCARBAMATE □ ETHYL DI-N-PROPYLTHIOLCARBAMATE □ ETHYL-N,N-DIPROPYLTHIOLCARBAMATE □ ETHYL-N,N-DI-N-PROPYLTHIOLCARBAMATE □ FDA 1541 □ GENEP EPTC □ R-1608 □ STAUFFER R 1608 □ TORBIN

TOXICITY DATA WITH REFERENCE
mmo-esc 5 mg/L GNKAA5 22,2416,86
ihl-hmn TCLo:135 mg/m^3/90M HYSAAV 36(1-3)196,71
orl-rat LD50:916 mg/kg NTIS** PB82-913299
ihl-rat LCLo:200 mg/m^3/4H HYSAAV 36(1-3),196,71
skn-rat LD50:3200 mg/kg GISAAA 36(2),29,71
orl-mus LD50:750 mg/kg HYSAAV 36(1-3),196,71
ivn-mus LD50:320 mg/kg CSLNX* NX#03907
orl-cat LD50:112 mg/kg HYSAAV 36(1-3),196,71
ihl-cat LCLo:400 mg/m^3/4H HYSAAV 36(1-3),196,71
skn-rbt LD50:1460 mg/kg WRPCA2 9,119,70
orl-bwd LD50:100 mg/kg TXAPA9 21,315,72

CONSENSUS REPORTS: Reported in EPA TSCA Inventory. EPA Genetic Toxicology Program.

SAFETY PROFILE: Poison by ingestion, inhalation, and intravenous routes. Moderately toxic by skin contact route. Mutation data reported. An herbicide. When heated to decomposition it emits very toxic fumes of NO_x and SO_x. See also CARBAMATES.

EIO000 CAS:74-85-1 HR: 3
ETHYLENE
DOT: UN 1038/UN 1962
mf: C_2H_4 mw: 28.06

PROP: Colorless gas; odorless and tasteless. Bp: −103.9°, mp: −169.4°, lel: 2.7%, uel: 36%, d: 0.610 @ 0°, autoign temp: 914°F, vap d: 0.98, fp: −181°. Sltly sol in H_2O; very sol in EtOH, Et_2O; sol in Me_2CO and C_6H_6.

SYNS: ACETENE □ ATHYLEN (GERMAN) □ BICARBURETTED HYDROGEN □ ELAYL □ ETHENE □ ETHYLENE, compressed (DOT) □ ETHYLENE, refrigerated liquid (DOT) □ LIQUID ETHYENE □ OLEFIANT GAS

TOXICITY DATA WITH REFERENCE
ihl-mam LCLo:95,0000 ppm/5M AEPPAE 138,65,28

CONSENSUS REPORTS: Reported in EPA TSCA Inventory. Community Right-To-Know List.

DOT CLASSIFICATION: 2.1; *Label:* Flammable Gas

SAFETY PROFILE: A simple asphyxiant. High concentrations cause anesthesia. A common air contaminant. It is phytotoxic. A very dangerous fire hazard when exposed to heat or flame. Moderate explosion hazard when exposed to flame. A flammable gas. To fight fire, stop flow of gas, use CO_2, dry chemical or fine water spray. Mixtures with aluminum chloride explode in the presence of nickel catalysts, methyl chloride, or nitromethane. Explosive reaction with bromotrichloromethane (at 120°C/51 bar), carbon tetrachloride (25–100°C/30 bar). Explosive reaction with chlorine catalyzed by sunlight or UV light or in the presence of mercury(I) oxide, mercury(II) oxide, or silver oxide. Mixtures with chlorotrifluoroethylene polymerize explosively when exposed to 50 kv gamma rays at 308 krad/hr. Has been involved in industrial accidents. Violent polymerization is catalyzed by copper above 400°C/54 bar. Incompatible with $AlCl_3$, (CCl_4 + benzoyl peroxide), (bromtrichloromethane + $AlCl_3$), O_3, CCl_4, Cl_2, NO_x, tetrafluoroethylene trifluorohypofluorite. When heated to decomposition it emits acrid smoke and irritating fumes.

EIO500 CAS:126-39-6 HR: 2
ETHYLENEACETIC ACID
mf: $C_6H_{12}O_2$ mw: 116.18

PROP: Bp: 117°.

SYNS: CYCLIC ETHYLENE ACETAL-2-BUTANONE □ 2-ETHYL-2-METHYL-1,3-DIOXOLANE

TOXICITY DATA WITH REFERENCE
skn-rbt 500 mg/24H MLD 85JCAE-,797,86
eye-rbt 500 mg/24H MLD 85JCAE-,797,86
orl-rat LDLo:2830 mg/kg AIHAAP 23,95,62
ihl-rat LCLo:4000 ppm/4H AIHAAP 23,95,62

CONSENSUS REPORTS: Reported in EPA TSCA Inventory.

SAFETY PROFILE: Moderately toxic by ingestion and inhalation. A skin and eye irritant. When heated to decomposition it emits acrid smoke and irritating fumes.

EIP000 CAS:2274-11-5 HR: 3
ETHYLENE ACRYLATE
mf: $C_8H_{10}O_4$ mw: 170.18

SYNS: ACRYLIC ACID, ETHYLENE ESTER □ ACRYLIC ACID, ETHYLENE GLYCOL DIESTER □ ETHYLDIOL ACRILATE (RUSSIAN) □ ETHYLENE DIACRYLATE □ ETHYLENE GLYCOL DIACRYLATE □ 2-PROPENOIC ACID-1,2-ETHANEDIYL ESTER

TOXICITY DATA WITH REFERENCE
skn-rbt 500 mg/24H MLD 85JCAE-,372,86
eye-rbt 750 μg/24H SEV 85JCAE-,372,86
msc-mus:lym 16,500 nmol/L EMMUEG 17,264,91
orl-rat LD50:300 mg/kg GTPZAB 24(4),58,80
orl-mus LD50:300 mg/kg GTPZAB 24(4),58,80
ihl-mus LC50:350 mg/kg GTPZAB 24(4),58,80
skn-rbt LD50:570 mg/kg TXAPA9 28,313,74

CONSENSUS REPORTS: Reported in EPA TSCA Inventory.

SAFETY PROFILE: Poison by ingestion and inhalation. Moderately toxic by skin contact. A skin and severe eye irritant. Mutation data reported. When heated to decomposition it emits acrid smoke and irritating fumes. See also ESTERS.

EIQ000 CAS:124-05-0 HR: 2
ETHYLENE BIS(CHLOROFORMATE)
mf: $C_4H_4Cl_2O_4$ mw: 186.98

SYNS: 1,2-BIS((CHLOROCARBONYL)OXY)ETHANE □ CARBONOCHLORIDE ACID, 1,2-ETHANEDIYL ESTER □ ETHYLENE CHLOROFORMATE □ ETHYLENE GLYCOL DI(CHLOROFORMATE) □ ETHYLENE GLYCOL, BISCHLOROFORMATE

TOXICITY DATA WITH REFERENCE
orl-mus LD50:1100 mg/kg 37ASAA 4,758,78
ihl-mus LD50:5659 ppm/1H 37ASAA 4,758,78
skn-mus LD50:2000 mg/kg 37ASAA 4,758,78

CONSENSUS REPORTS: Reported in EPA TSCA Inventory.

DOT CLASSIFICATION: 6.1; *Label:* Poison, Corrosive

SAFETY PROFILE: Moderately toxic by ingestion and skin contact. Mildly toxic by inhalation. When heated to decomposition it emits toxic fumes of Cl^-.

EIQ200 CAS:4431-24-7 HR: 3
ETHYLENEBIS-(DIPHENYLARSINE)
mf: $C_{26}H_{24}As_2$ mw: 486.34

PROP: Colorless crystals from MeOH or EtOH. Mp: 100–103°.

SYN: ARSINE, ETHYLENEBIS(DIPHENYL)-

TOXICITY DATA WITH REFERENCE
ivn-mus LD50:180 mg/kg CSLNX* NX#04858

OSHA PEL: TWA 0.5 mg(As)/m^3

SAFETY PROFILE: Poison by intravenous route. An irritant. When heated to decomposition it emits toxic fumes of As.

E

EIR000 **CAS:12122-67-7** **HR: 2**
ETHYLENE BIS(DITHIOCARBAMATO)ZINC
mf: $C_4H_6N_2S_4{\cdot}Zn$ mw: 275.73

PROP: Moisture and light unstable pale yellow powder or crystals from $CHCl_3$/EtOH. Sol in CS_2 and Py; very sltly sol in H_2O; insol in water.

SYNS: ASPOR ☐ ASPORUM ☐ BERCEMA ☐ BLIGHTOX ☐BLITEX ☐ BLIZENE ☐ CARBADINE ☐ CHEM ZINEB ☐ CINEB ☐ CRITTOX ☐ CYNKOTOX ☐ DAISEN ☐ DIPHER ☐ DITHANE Z ☐ DITIAMINA ☐ ENT 14,874 ☐ ((1,2-ETHANEDIYLBIS(CARBAMODITHIOATO))(2-)ZINC ☐ 1,2-ETHANEDIYLBIS(CARBAMODITHIOATO)(2-)-S,S'-ZINC ☐ 1,2-ETHANEDIYLBISCARBAMODITHIOIC ACID, ZINC COMPLEX ☐ 1,2-ETHANEDIYLBISCARBAMOTHIOIC ACID, ZINC SALT ☐ ETHYLENEBIS(DITHIOCARBAMIC ACID), ZINC SALT ☐ ETHYL ZIMATE ☐ HEXATHANE ☐ KUPRATSIN ☐ KYPZIN ☐ LIROTAN ☐ LONACOL ☐ MICIDE ☐ MILTOX ☐ MILTOX SPECIAL ☐NOVIZIR ☐ NOVOSIR N ☐ PAMOSOL 2 FORTE ☐ PARZATE ☐ PEROSIN ☐ POLYRAM Z ☐ SPERLOX-Z ☐ THIODOW ☐ TIEZENE ☐ TRITOFTOROL ☐ TSINEB (RUSSIAN) ☐ Z-78 ☐ ZEBENIDE ☐ ZEBTOX ☐ ZIDAN ☐ ZIMATE ☐ ZINC ETHYLENEBISDITHIOCARBAMATE ☐ ZINC ETHYLENE-1,2-BISDITHIOCARBAMATE ☐ ZINEB ☐ ZINK-(N,N'-AETHYLEN-BIS(DITHIOCARBAMAT)) (GERMAN) ☐ ZINOSAN

TOXICITY DATA WITH REFERENCE
mmo-bcs 1 nmol/plate MSERDS 5,93,81
sce-hmn:lym 10 mg/L GESKAC 12,118,79
orl-rat TDLo:4 g/kg (13D preg):TER FCTXAV 11,239,73
orl-rat TDLo:4 g/kg (13D preg):REP FCTXAV 11,239,73
orl-rat TDLo:8 mg/kg (16D pre):REP TOERD9 3,279,81
orl-rat TDLo:8 mg/kg (female 16D pre):REP TOERD9 3,279,81
orl-rat TDLo:4 g/kg (female 13D post):TER EKMMA8 10,226,71
orl-rat TDLo:53,580 mg/kg/94W-I:CAR VPITAR 29,71,70
imp-rat TDLo:80 mg/kg:ETA VPITAR 29,71,70
orl-mus TDLo:7800 mg/kg/5W-I:ETA GISAAA 37(9),25,72
ipr-mus TDLo:2600 mg/kg/13W-I:ETA VRDEA5 (9),100,66
ipr-mus TDLo:1760 mg/kg (11-21D preg):ETA,TER VPITAR 30,49,71
orl-hmn LDLo:5 g/kg 85JFAN A420,84
orl-rat LD50:1850 mg/kg GISAAA 31(10),25,66
ihl-rat LCLo:800 mg/m³/4H 85GMAT -,121,82
orl-mus LD50:7600 mg/kg JTEHD6 4,93,78

CONSENSUS REPORTS: IARC Cancer Review: Group 3 IMEMDT 7,56,87; Animal Inadequate Evidence IMEMDT 12,245,76. Community Right-To-Know List. EPA Genetic Toxicology Program.

SAFETY PROFILE: Moderately toxic by ingestion. Experimental teratogenic and reproductive effects. Questionable carcinogen with experimental carcinogenic and tumorigenic data. Human mutation data reported. Used as a fungicide. When heated to decomposition it emits very toxic fumes of NO_x, ZnO, and SO_x. See also ZINC COMPOUNDS and CARBAMATES.

EIR500 **CAS:12275-13-7** **HR: 3**
ETHYLENEBIS(DITHIOCARBAMIC ACID) NICKEL(II) SALT
mf: $C_4H_6N_2S_4{\cdot}Ni$ mw: 269.07

TOXICITY DATA WITH REFERENCE
ipr-mus LDLo:250 mg/kg CBCCT* 4,377,52

CONSENSUS REPORTS: Nickel and its compounds are on the Community Right-To-Know List.

NIOSH REL: (Inorganic Nickel) TWA 0.015 mg(Ni)/m³

SAFETY PROFILE: Poison by intraperitoneal route. When heated to decomposition it emits very toxic fumes of NO_x and SO_x. See also NICKEL COMPOUNDS and CARBAMATES.

EIS000 **CAS:62207-76-5** **HR: 3**
N,N'-ETHYLENE BIS(3-FLUOROSALICYLIDENEIMINATO)COBALT(II)
mf: $C_{16}H_{12}CoF_2N_2O_2$ mw: 361.23

SYNS: BIS(3-FLUOROSALICYLALDEHYDE)-ETHYLENEDIIMINE-COBALT ☐ FLUOMINE ☐ FLUOMINE DUST

TOXICITY DATA WITH REFERENCE
skn-rbt 500 mg/24H MOD NTIS** AD-A083-929
eye-rbt 3 mg NTIS** AD-A083-929
orl-rat LD50:187 mg/kg AMRL** TR-74-78,74
ihl-rat LC50:112 mg/m³/6H AMRL** TR-74-78,74
orl-mus LD50:123 mg/kg AMRL** TR-74-78,74
ihl-mus LC50:416 mg/m³/6H NTIS** AD-A083-929
ihl-gpg LCLo:30 mg/m³/2H AMRL** TR-74-78,74

CONSENSUS REPORTS: Cobalt and its compounds are on the Community Right-To-Know List. EPA Extremely Hazardous Substances List. Reported in EPA TSCA Inventory.

SAFETY PROFILE: Poison by ingestion and inhalation. A skin and eye irritant. When heated to decomposition it emits very toxic fumes of F^- and NO_x. See also COBALT COMPOUNDS.

EIT000 **CAS:67-42-5** **HR: 3**
(ETHYLENEBIS(OXYETHYLENENITRILO))TETRAACETIC ACID
mf: $C_{14}H_{24}N_2O_{10}$ mw: 380.40

SYNS: 1,2-BIS(2-DICARBOXYMETHYLAMINOETHOXY)ETHANE ☐ 6,9-DIOXA-3,12-DIAZATETRADECANEDIOIC ACID, 3,12-BIS(CARBOXYMETHYL)-(9CI) ☐ EBONTA ☐ EGTA ☐ ETHYLENEDIOXYBIS(ETHYLENEAMINO)TETRAACETIC ACID ☐ ETHYLENE GLYCOL BIS(AMINOETHYL ETHER)TETRAACETATE ☐ ETHYLENE GLYCOL BIS(β-AMINOETHYL ETHER)TETRAACETATE ☐ ETHYLENE GLYCOL BIS(β-AMINOETHYL ETHER)-N,N'-TETRAACETIC ACID ☐ ETHYLENE GLYCOL BIS(2-AMINOETHYL ETHER)TETRAACETIC ACID ☐ ETHYLENE GLYCOL BIS(2-AMINOETHYL ETHER)-N,N,N',N'-TETRAACETIC ACID ☐ GLYCOL-ETHERDIAMINETETRAACETIC ACID

TOXICITY DATA WITH REFERENCE
orl-rat LD50:3587 mg/kg TXAPA9 16,807,70
ipr-mus LD50:150 mg/kg NTIS** AD691-490

CONSENSUS REPORTS: Glycol ether compounds are on the Community Right-To-Know List. Reported in EPA TSCA Inventory.

SAFETY PROFILE: Poison by intraperitoneal route. Moderately toxic by ingestion. When heated to decomposition it emits toxic fumes of NO_x. See also GLYCOL ETHERS.

EIT100 CAS:882-35-9 HR: 3
1,1′-ETHYLENEBIS(PYRIDINIUM)BROMIDE
mf: $C_{12}H_{14}N_2•2Br$ mw: 346.10

SYNS: 1,1′-ETHYLENEDIPYRIDINIUM DIBROMIDE □ G.L. 102 □ P.M. 346 □ PYRIDINIUM, 1,1′-ETHYLENEDI-, DIBROMIDE

TOXICITY DATA WITH REFERENCE
ivn-mus LD50:180 mg/kg CSLNX* NX#05031

CONSENSUS REPORTS: Reported in EPA TSCA Inventory.

SAFETY PROFILE: Poison by intravenous route. When heated to decomposition it emits toxic vapors of NO_x.

EIU000 CAS:10310-38-0 HR: 1
ETHYLENEBIS(TRIS(2-CYANOETHYL))PHOSPHONIUM BROMIDE)
mf: $C_{20}H_{28}N_6P_2•2Br$ mw: 574.30

SYN: 1,2-ETHANEDIYLBIS(TRIS(2-CYANOETHYL)PHOSPHONIUM DIBROMIDE

TOXICITY DATA WITH REFERENCE
orl-rat LD50:10 g/kg EPASR* 8EHQ-0780-0369

CONSENSUS REPORTS: Cyanide and its compounds are on the Community Right-To-Know List. Reported in EPA TSCA Inventory.

SAFETY PROFILE: Mildly toxic by ingestion. When heated to decomposition it emits very toxic fumes of PO_x, NO_x, and Br^-. See also CYANIDE and BROMIDES.

EIU800 CAS:107-07-3 HR: 3
ETHYLENE CHLOROHYDRIN
DOT: UN 1135
mf: C_2H_5ClO mw: 80.52

PROP: Colorless liquid; faint, ethereal odor. Mp: −69°, fp: −67.5°, bp: 128.8°, flash p: 140°F (OC), d: 1.197 @ 20°/4°, autoign temp: 797°F, vap press: 10 mm @ 30.3°, vap d: 2.78, lel: 4.9%, uel: 15.9%. Misc in water.

SYNS: AETHYLENECHLORHYDRIN (GERMAN) □ 2-CHLOORETHANOL (DUTCH) □ 2-CHLORAETHANOL (GERMAN) □ 2-CHLORETHANOL (GERMAN) □ Δ-CHLOROETHANOL □ 2-CHLOROETHANOL (MAK) □ β-CHLOROETHYL ALCOHOL □ 2-CHLOROETHYL ALCOHOL □ CHLOROETHYLOWY ALKOHOL (POLISH) □ 2-CLOROETANOLO (ITALIAN) □ ETHYLEEN-CHLOORHYDRINE (DUTCH) □ ETHYLENE GLYCOL, CHLOROHYDRIN □ GLICOL MONOCLORIDRINA (ITALIAN) □ GLYCOL CHLOROHYDRIN □ GLYCOLMONOCHLOORHYDRINE (DUTCH) □ GLYCOL MONOCHLOROHYDRIN □ GLYCOMONOCHLORHYDRIN □ MONOCHLORHYDRINE du GLYCOL (FRENCH) □ 2-MONOCHLOROETHANOL □ NCI-C50135

TOXICITY DATA WITH REFERENCE
skn-rbt 200 mg/2H MLD TXAPA9 16,382,70
eye-rbt 2 mg SEV AJOPAA 29,1363,46
eye-rbt 33 mg MOD TXAPA9 16,382,70
eye-rbt 9 mg/6H MOD BUYRAI 31,25,77
sln-asn 74,500 μmol/L MUREAV 138,33,84
oms-rat:lvr 12 g/L JACTDZ 1(3),37,82
orl-mus TDLo:1100 mg/kg (6-16D preg):TER JPFCD2 17,381,82
ihl-man LCLo:305 ppm/2H JIHTAB 26,277,44
orl-rat LD50:71 mg/kg HYSAAV 36,376,71
ihl-rat LC50:290 mg/m³ HYSAAV 26,376,71
ipr-rat LD50:58 mg/kg TXAPA9 21,454,72
scu-rat LD50:84 mg/kg HYSAAV 36(4-6),376,71
orl-mus LD50:81 mg/kg JPMSAE 60,568,71
ihl-mus LC50:385 mg/m³ HYSAAV 36,376,71
skn-mus LD50:18 mg/kg 85GMAT -,66,82
ipr-mus LD50:97 mg/kg JPMSAE 61,19,72
skn-rbt LD50:67 mg/kg JPMSAE 61,19,72
ivn-rbt LDLo:100 mg/kg BJIMAG 1,207,44

E

CONSENSUS REPORTS: NTP Carcinogenesis Studies (dermal); No Evidence: mouse, rat NTPTR* NTP-TR-275,85; Reported in EPA TSCA Inventory. EPA Genetic Toxicology Program. EPA Extremely Hazardous Substances List.

OSHA PEL: CL 1 ppm (skin)
ACGIH TLV: CL 1 ppm (skin)
DFG MAK: 1 ppm (3 mg/m³)
DOT CLASSIFICATION: 6.1; *Label:* Poison

SAFETY PROFILE: A poison by ingestion, inhalation, skin contact, intraperitoneal, intravenous, and subcutaneous routes. Moderately toxic to humans by inhalation. It can affect the nervous system, liver, spleen, and lungs. An experimental teratogen. Mutation data reported. A severe eye and mild skin irritant. Flammable liquid when exposed to heat, flame, or oxidizers. To fight fire, use alcohol foam, CO_2, dry chemical. Violent reaction with chlorosulfonic acid, ethylene diamine, sodium hydroxide. Reacts with water or steam to produce toxic and corrosive fumes. Potentially violent reaction with oxidizing materials. When heated to decomposition it emits highly toxic fumes of Cl^- and phosgene. See also CHLORINATED HYDROCARBONS, ALIPHATIC.

For occupational chemical analysis use NIOSH: Ethylene Chlorohydrin, 2513.

EIV000 CAS:64-02-8 HR: 3
N,N′-ETHYLENEDIAMINEDIACETIC ACID TETRASODIUM SALT
mf: $C_{10}H_{12}N_2O_8•4Na$ mw: 380.20

PROP: Amorphous powder.

SYNS: AQUAMOLLIN □ CALSOL □ CELON E □ CELON H □ CELON IS □ CHEELOX BF □ CHEELOX BR-33 □ CHELON 100 □ CHEMCOLOX 200 □ COMPLEXONE □ CONIGON BC □ DISTOL 8 □ EDATHANIL TETRASODIUM □ EDETATE SODIUM □ EDETIC ACID TETRASODIUM SALT □ EDTA, SODIUM SALT □ EDTA TETRASODIUM SALT □ ENDRATE TETRASODIUM □ N,N′-1,2-ETHANEDIYLBIS(N-(CARBOXYMETHYL))GLYCINE TETRASODIUM SALT □ ETHYLENEBIS (IMINODIACETIC ACID) TETRASODIUM SALT □ ETHYLENEDIAMINETETRAACETIC ACID, TETRASODIUM SALT □ HAMP-ENE 100 □ HAMP-ENE 215 □ HAMP-ENE 220 □ HAMP-ENE Na4 □ IRGALON □ KALEX □ KEPMPLEX 100 □ KOMPLXON □ METAQUEST C □ NERVANAID B LIQUID □ NERVANID B □ NULLAPON B □ NULLAPON BF-78 □ NULLAPON BFC CONC □ PERMA KLEER 50 CRYSTALS □ PERMA KLEER TETRA CP □ QUESTEX 4 □ SEQUESTRENE 30A □ SEQUESTRENE Na 4 □ SEQUESTRENE ST □ SODIUM EDETATE □ SODIUM EDTA □ SODIUM ETHYLENEDIAMINETETRAACETATE □ SODI-

UM ETHYLENEDIAMINETETRAACETIC ACID □ SODIUM SALT of ETHYLENEDIAMINETETRAACETIC ACID □ SYNTES 12A □ SYNTRON B □ TETRACEMIN □ TETRASODIUM EDTA □ TETRASODIUM ETHYLENEDIAMINETETRAACETATE □ TETRASODIUM ETHYLENEDIAMINETETRACETATE □ TETRASODIUM (ETHYLENEDINITRILO)TETRAACETATE □ TETRASODIUM SALT of EDTA □ TETRASODIUM SALT of ETHYLENEDIAMINETETRACETIC ACID □ TETRINE □ TRILON B □ TST □ TYCLAROSOL □ VERSENE 100 □ VERSENE POWDER □ WARKEELATE PS-43

TOXICITY DATA WITH REFERENCE
skn-rbt 500 mg/24H MOD 28ZPAK -,306,72
eye-rbt 1900 μg AAOPAF 48,681,52
eye-rbt 100 mg/24H MOD 28ZPAK -,306,72
ipr-mus LD50:330 mg/kg REPMBN 10,391,62

CONSENSUS REPORTS: Reported in EPA TSCA Inventory.

SAFETY PROFILE: Poison by intraperitoneal route. A skin and eye irritant. When heated to decomposition it emits toxic fumes of NO_x and Na_2O. See also ETHYLENEDIAMINETETRAACETIC ACID, DISODIUM SALT.

EIV100 CAS:1170-02-1 HR: 3
ETHYLENEDIAMINE-DI(o-HYDROXYPHENYL)ACETIC ACID
mf: $C_{18}H_{20}N_2O_6$ mw: 360.40

SYNS: CHEL 138 □ EDBPHA □ EDDHA □ EDHPA □ N,N'-ETHYLENEBIS(2-(o-HYDROXYPHENYL)GLYCINE) □ ETHYLENEDIAMINE-N,N'-BIS(2-HYDROXYPHENYLACETIC ACID) □ ETHYLENEDIAMINE-DI (2-HYDROXYPHENYL)ACETIC ACID □ GLYCINE, N,N'-ETHYLENEBIS (2-(o-HYDROXYPHENYL))-

TOXICITY DATA WITH REFERENCE
ipr-rat LD50:175 mg/kg JMCMAR 29,1231,86
ivn-rat LD50:47 mg/kg NTIS** PB82-163692
ipr-mus LD50:350 mg/kg NTIS** AD691-490
ivn-mus LD50:53 mg/kg FAATDF 6,292,86

CONSENSUS REPORTS: Reported in EPA TSCA Inventory.

SAFETY PROFILE: Poison by intravenous and intraperitoneal routes. When heated to decomposition it emits toxic vapors of NO_x.

EIV700 CAS:20829-66-7 HR: 3
ETHYLENEDIAMINEDINITRATE
mf: $C_2H_{10}N_4O_6$ mw: 186.12

SYN: 1,2-DIAMMONIOETHANE NITRATE

SAFETY PROFILE: Has been used as a military explosive. When heated to decomposition it emits toxic fumes of NO_x. See also EXPLOSIVES, HIGH; and NITRATES.

EIV750 CAS:27014-42-2 HR: 1
ETHYLENEDIAMINE ETHOXYLATE
mf: $(C_2H_4O)_n(C_2H_4O)_n(C_2H_4O)_n(C_2H_4O)_nC_{10}H_{24}N_2O_4$

SYNS: EDA 200 □ ETHYLENEDIAMINE ETHYLENE OXIDE ADDUCT □ POLY(OXY-1,2-ETHANEDIYL), α-α',α'',α'''-(1,2-ETHANEDIYLBIS(NITRILODI-2,1-ETHANEDIYL))TETRAKIS(ω-HYDROXY)- □ VERAMIN ED 4 □ VERAMIN ED 40

TOXICITY DATA WITH REFERENCE
skn-rbt 500 mg/24H MLD JACTDZ 1,19,90
eye-rbt 100 mg MLD JACTDZ 1,19,90
orl-rat LD50:9900 mg/kg JACTDZ 1,19,90
skn-rbt LD50:>8 g/kg JACTDZ 1,19,90

SAFETY PROFILE: Mildly toxic by ingestion. A skin and eye irritant. When heated to decomposition it emits toxic fumes of NO_x.

EIW000 CAS:333-18-6 HR: 3
ETHYLENEDIAMINE HYDROCHLORIDE
mf: $C_2H_8N_2{\bullet}2ClH$ mw: 133.04

PROP: Monoclinic prisms. Mp: subl. Sol in water; insol in alc and ether.

SYNS: CHLOR-ETHAMINE □ 1,2-DIAMINOETHANE DIHYDROCHLORIDE □ 1,2-ETHANEDIAMINE, DIHYDROCHLORIDE □ ETHYLENEDIAMMONIUM CHLORIDE

TOXICITY DATA WITH REFERENCE
orl-rat TDLo:10 g/kg (6-15D preg):TER TOXID9 4,165,84
orl-rat TDLo:10 g/kg (6-15D preg):REP TOXID9 4,165,84
orl-rat TDLo:91 g/kg (91D male):REP FAATDF 3,512,83
orl-rat LDLo:500 mg/kg JPETAB 90,260,47
ims-rat LD50:150 mg/kg EMSUA8 4,223,46
orl-mus LD50:1620 mg/kg FAATDF 3,512,83

CONSENSUS REPORTS: Reported in EPA TSCA Inventory.

SAFETY PROFILE: Poison by intramuscular route. Moderately toxic by ingestion. Experimental teratogenic and reproductive effects. When heated to decomposition it emits very toxic fumes of HCl and NO_x.

EIX000 CAS:60-00-4 HR: 3
ETHYLENEDIAMINETETRAACETIC ACID
mf: $C_{10}H_{16}N_2O_8$ mw: 292.28

PROP: Colorless crystals from water (dimorphic). Mp: 220° (decomp @ 240°). Sltly water-sol; insol in common organic solvents.

SYNS: ACIDE ETHYLENEDIAMINETETRACETIQUE (FRENCH) □ 3,6-BIS(CARBOXYMETHYL)-3,5-DIAZOOCTANEDIOIC ACID □ CELON A □ CELON ATH □ CHEELOX BF ACID □ CHEMCOLOX 340 □ COMPLEXON II □ EDATHAMIL □ EDETIC ACID □ EDTA (chelating agent) □ EDTA ACID □ ENDRATE □ N,N'-1,2-ETHANEDIYLBIS(N-(CARBOXYMETHYL))GLYCINE □ETHYLENEDIAMINETETRAACETATE □ ETHYLENEDIAMINE-N,N,N',N'-TETRAACETIC ACID □ ETHYLENEDINITRILOTETRAACETIC ACID □ HAMP-ENE ACID □ HAVIDOTE □ METAQUEST A □ NERVANAID B ACID □ NULLAPON BF ACID □ PERMA KLEER 50 ACID □ QUESTEX 4H □ SEQ 100 □ SEQUESTRENE AA □ SEQUESTRIC ACID □ SEQUESTROL □ TETRINE ACID □ TITRIPLEX □ TRICON BW □ TRILON BW □ VERSENE ACID □ VINKEIL 100 □ WARKEELATE ACID

TOXICITY DATA WITH REFERENCE
cyt-mus-ipr 50 mmol/L CISCB7 20,28,76
dni-rbt:kdy 250 nmol/L ECREAL 36,92,64
orl-rat TDLo:7632 mg/kg (female 7-14D post):TER TXAPA9 40,299,77

orl-rat TDLo:7632 mg/kg (female 7-14D post):REP TXAPA9 40,299,77
orl-rat TDLo:7632 mg/kg (7-14D preg):TER TXAPA9 40,299,77
ipr-rat LD50:397 mg/kg AHRTAN 13,295,62
ipr-mus LD50:250 mg/kg NTIS** AD691-490

CONSENSUS REPORTS: Reported in EPA TSCA Inventory. EPA Genetic Toxicology Program.

SAFETY PROFILE: Poison by intraperitoneal route. Experimental teratogenic and reproductive effects. Mutation data reported. A general-purpose chelating and complexing agent. When heated to decomposition it emits toxic fumes of NO_x.

EIX500 CAS:139-33-3 HR: 3
ETHYLENEDIAMINETETRAACETIC ACID, DISODIUM SALT
mf: $C_{10}H_{14}N_2O_8 \cdot 2Na$ mw: 336.24

PROP: White crystalline powder. Sol in water.

SYNS: CHELADRATE □ CHELAPLEX III □ CHELATON III □ COMPLEXON III □ d'E.D.T.A. DISODIQUE (FRENCH) □ DISODIUM DIACID ETHYLENEDIAMINETETRAACETATE □ DISODIUM DIHYDROGEN ETHYLENEDIAMINETETRAACETATE □ DISODIUM DIHYDROGEN(ETHYLENEDINITRILO)TETRAACETATE □ DISODIUM EDATHAMIL □ DISODIUM EDETATE □ DISODIUM EDTA (FCC) □ DISODIUM ETHYLENEDIAMINETETRAACETATE □ DISODIUM ETHYLENEDIAMINETETRAACETIC ACID □ DISODIUM (ETHYLENEDINITRILO)TETRAACETATE □ DISODIUM (ETHYLENEDINITRILO)TETRAACETIC ACID □ DISODIUM SALT of EDTA □ DISODIUM SEQUESTRENE □ DISODIUM TETRACEMATE □ DISODIUM VERSENATE □ DISODIUM VERSENE □ EDATHAMIL DISODIUM □ EDETATE DISODIUM □ EDTA, DISODIUM SALT □ ENDRATE DISODIUM □ N,N′-1,2-ETHANEDIYLBIS (N-(CARBOXYMETHYL)GLYCINE) DISODIUM SALT □ ETHYLENEBIS (IMINODIACETIC ACID) DISODIUM SALT □ ETHYLENEDIAMINETETRAACETATE DISODIUM SALT □ (ETHYLENEDINITRILO)-TETRAACETIC ACID DISODIUM SALT □ F 1 (complexon) □ KIRESUTO B □ METAQUEST B □ PERMA KLEER 50 CRYSTALS DISODIUM SALT □ SELEKTON B 2 □ SEQUESTRENE SODIUM 2 □ SODIUM VERSENATE □ TETRACEMATE DISODIUM □ TITRIPLEX III □ TRILON BD □ TRIPLEX III □ VERESENE DISODIUM SALT □ VERSENE SODIUM 2

TOXICITY DATA WITH REFERENCE
cyt-grh-par 1 mmol/L CISCB7 16,18,74
orl-rat TDLo:12,857 mg/kg (7-15D preg):TER SCIEAS 173,62,72
orl-rat TDLo:31,429 mg/kg (1-22D preg):REP SCIEAS 173,62,72
orl-rat TDLo:12,857 mg/kg (7-15D preg):REP SCIEAS 173,62,72
scu-rat TDLo:380 mg/kg (female 9D post):REP BCPCA6 22,407,73
orl-rat TDLo:12,857 mg/kg (female 7-15D post):TER SCIEAS 173,62,71
orl-rat LD50:2000 mg/kg FEPRA7 27,465,68
orl-mus LD50:2050 mg/kg NYKZAU 52,126S,56
ipr-mus LD50:260 mg/kg NYKZAU 52,126S,56
ivn-mus LD50:56 mg/kg CSLNX* NX#03781
orl-rbt LD50:2300 mg/kg NYKZAU 52,113,56
ivn-rbt LD50:47 mg/kg NYKZAU 52,113,56

CONSENSUS REPORTS: Reported in EPA TSCA Inventory. EPA Genetic Toxicology Program.

SAFETY PROFILE: Poison by intraperitoneal and intravenous route. Moderately toxic by ingestion. Experimental teratogenic and reproductive effects. Mutation data reported. The calcium disodium salt of EDTA is used as a chelating agent in treating lead poisoning. When heated to decomposition it emits toxic fumes of NO_x and Na_2O.

EIY500 CAS:106-93-4 HR: 3
1,2-ETHYLENE DIBROMIDE
DOT: UN 1605
mf: $C_2H_4Br_2$ mw: 187.88

PROP: Colorless, heavy liquid; sweet odor. Bp: 131.4°, fp: 9.3°, flash p: none, d: 2.178 @ 20°/4, mp: 10°, vap press: 17.4 mm @ 30°, vap d: 6.48.

SYNS: AETHYLENBROMID (GERMAN) □ BROMOFUME □ BROMURO di ETILE (ITALIAN) □ CELMIDE □ DBE □ 1,2-DIBROMAETHAN (GERMAN) □ 1,2-DIBROMOETANO (ITALIAN) □ α,β-DIBROMOETHANE □ sym-DIBROMOETHANE □ 1,2-DIBROMOETHANE (MAK) □ DIBROMURE d'ETHYLENE (FRENCH) □ 1,2-DIBROOMETHAAN (DUTCH) □ DOWFUME 40 □ DOWFUME EDB □ DOWFUME W-8 □ DWUBROMOETAN (POLISH) □ EDB □ EDB 85 □ E-D-BEE □ ENT 15,349 □ ETHYLENE BROMIDE □ FUMO-GAS □ GLYCOL BROMIDE □ GLYCOL DIBROMIDE □ ISCOBROME D □ KOPFUME □ NCI-C00522 □ NEPHIS □ PESTMASTER □ PESTMASTER EDB-85 □ RCRA WASTE NUMBER U067 □ SOILBROM-40 □ SOILBROM-85 □ SOILFUME □ UNIFUME

TOXICITY DATA WITH REFERENCE
skn-hmn 1538 mg/24H SEV AGGHAR 8,591,38
skn-rbt 1%/14D SEV AMIHBC 6,158,52
eye-rbt 1% AMIHBC 6,158,52
mmo-sat 1 μg/plate ENMUDM 7(Suppl 5),1,85
mma-esc 333 μg/plate ENMUDM 7(Suppl 5),1,85
ihl-man TCLo:88 ppb/8H (male 5 year(s) pre):REP BJIMAG 44,317,87
unr-rat TDLo:25 mg/kg (male 1D pre):REP TJADAB 27,27A,83
ihl-rat TCLo:66,670 ppb/4H (female 3-20D post):REP NETOD7 5,579,83
ihl-mus TCLo:38 ppm/23H (female 6-15D post):TER TXAPA9 45,347,78
orl-rat TDLo:50 mg/kg (male 5D pre):TER MUREAV 77,71,80
ihl-mus TCLo:20 ppm/23H (female 6-15D post):TER TXAPA9 46,173,78
ihl-rat TCLo:80 ppm/7H (female 3W pre):REP TXAPA9 49,97,79
ihl-rat TCLo:89 ppm/7H (male 10W pre):REP TXAPA9 49,97,79
ihl-rat TCLo:39 ppm/7H (female 3W pre):REP TXAPA9 49,97,79
orl-ctl TDLo:40 mg/kg (male 10D pre):REP JRPFA4 44,561,75
orl-rat TDLo:540 mg/kg /78W-C:CAR BANRDU 5,279,80
ihl-rat TCLo:10 ppm/6H/2Y-I:CAR NCITR* NCI-CG-TR-201,82
orl-mus TDLo:16 g/kg/53W-I:CAR NCITR* NCI-CG-TR-86,78
ihl-mus TDLo:40 ppm/6H/90W-I:CAR CALEDQ 12,121,81
skn-mus TDLo:190 g/kg/62W-I:NEO JJIND8 63,1433,79
ipr-mus TDLo:840 mg/kg/8W-I:NEO TXAPA9 82,19,86
orl-rat TD:16 g/kg/61W-I:CAR NCITR* NCI-CG-TR-86,78
orl-mus TD:23 g/kg/53W-I:CAR NCITR* NCI-CG-TR-86,78

orl-rat TD:24 g/kg/57W-I:CAR NCITR* NCI-CG-TR-86,78
orl-rat TD:6630 mg/kg/34W-I:CAR NCITR* NCI-CG-TR-86,78
orl-mus TD:28 g/kg/53W-I:CAR NCITR* NCI-CG-TR-86,78
ihl-rat TC:20 ppm/7H/78W-I:CAR TXAPA9 63,155,82
orl-rat TD:7000 mg/kg/47W-I:CAR NCITR* NCI-CG-TR-86,78
orl-rat TD:18 g/kg/44W-I:CAR NCITR* NCI-CG-TR-86,78
ihl-rat TCLo:10 ppm/6H/2Y-I:CAR NTPTR* NTP-TR-201,82
ihl-rat TC:40 ppm/6H/88W-I:CAR NCITR* NCI-CG-TR-201,82
orl-wmn LDLo:90 mg/kg:GIT,SYS 34ZIAG -,257,69
orl-rat LD50:108 mg/kg SPEADM 74-1,-,74
ihl-rat LC50:14,300 mg/m³/30M FATOBP 8,140,73
skn-rat LD50:300 mg/kg 85DPAN -,-,71/76
orl-mus LDLo:250 mg/kg TXAPA9 23,288,72
ipr-mus LD50:220 mg/kg JPCEAO 320(1),133,78
orl-rbt LD50:55 mg/kg AMIHBC 6,158,52
skn-rbt LD50:300 mg/kg AMIHBC 6,158,52
rec-rbt LDLo:2500 mg/kg JPETAB 34,223,28
ihl-gpg LCLo:400 ppm/3H AMIHBC 6,158,52

CONSENSUS REPORTS: NTP 7th Annual Report on Carcinogens. IARC Cancer Review: Group 2A IMEMDT 7,204,87; Animal Sufficient Evidence IMEMDT 15,195,77. NCI Carcinogenesis Bioassay (gavage); Clear Evidence: mouse, rat NCITR* NCI-CG-TR-86,78; NTP Carcinogenesis Bioassay (inhalation); Clear Evidence: mouse, rat NTPTR* NTP-TR-210,82. EPA Genetic Toxicology Program. Community Right-To-Know List. Reported in EPA TSCA Inventory.

OSHA PEL: TWA 20 ppm; CL 30 ppm; Pk 50 ppm/5M/8H
ACGIH TLV: Suspected Human Carcinogen
DFG TRK: 0.1 ppm; Animal Carcinogen, Suspected Human Carcinogen.
NIOSH REL: (EDB) 0.045 ppm; CL 1 mg/m³/15M
DOT CLASSIFICATION: 6.1; *Label:* Poison

SAFETY PROFILE: Confirmed carcinogen with experimental carcinogenic, neoplastigenic, and teratogenic data. Human poison by ingestion. Experimental poison by ingestion, skin contact, intraperitoneal, and possibly other routes. Moderately toxic by inhalation and rectal routes. Human systemic effects by ingestion: hypermotility, diarrhea, nausea or vomiting, decreased urine volume or anuria. Experimental reproductive effects. Human mutation data reported. A severe skin and eye irritant. An experimental eye irritant. Implicated in worker sterility. When heated to decomposition it emits toxic fumes of Br^-. See also ETHYLENE DICHLORIDE and BROMIDES.

For occupational chemical analysis use OSHA: #02 or NIOSH: Ethylene Dibromide, 1008.

EIY550 CAS:65313-36-2 HR: 3
ETHYLENEDICESIUM
mf: $C_2H_4Cs_2$ mw: 293.86

SAFETY PROFILE: Violent reaction on contact with water. When heated to decomposition it emits acrid smoke and irritating fumes. See also CESIUM.

EIY600 CAS:107-06-2 HR: 3
ETHYLENE DICHLORIDE
DOT: UN 1184
mf: $C_2H_4Cl_2$ mw: 98.96

PROP: Colorless, clear liquid; pleasant odor, sweet taste. Bp: 83.5°, ULC: 60–70, lel: 6.2%, uel: 15.9%, fp: −35.7°, flash p: 56°F, d: 1.257 @ 20°/4°, autoign temp: 775°F, vap press: 100 mm @ 29.4°, vap d: 3.35, refr index: 1.445 @ 20°. Sol in alc, ether, acetone, carbon tetrachloride; sltly sol in water.

SYNS: AETHYLENCHLORID (GERMAN) □ BICHLORURE d'ETHYLENE (FRENCH) □ BORER SOL □ BROCIDE □ CHLORURE d'ETHYLENE (FRENCH) □ CLORURO di ETHENE (ITALIAN) □ 1,2-DCE □ DESTRUXOL BORER-SOL □ 1,2-DICHLOORETHAAN (DUTCH) □ 1,2-DICHLOR-AETHAN (GERMAN) □ DICHLOREMULSION □ DI-CHLOR-MULSION □ α,β-DICHLOROETHANE □ sym-DICHLOROETHANE □ DICHLORO-1,2-ETHANE (FRENCH) □ 1,2-DICHLOROETHANE □ DICHLOROETHYLENE □ 1,2-DICLOROETANO (ITALIAN) □ DUTCH LIQUID □ DUTCH OIL □ EDC □ ENT 1,656 □ ETHANE DICHLORIDE □ ETHYLEENDICHLORIDE (DUTCH) □ ETHYLENE CHLORIDE □ 1,2-ETHYLENE DICHLORIDE □ GLYCOL DICHLORIDE □ NCI-C00511 □ RCRA WASTE NUMBER U077

TOXICITY DATA WITH REFERENCE
skn-rbt 500 mg/24H MLD 85JCAE-,93,86
eye-rbt 500 mg/24H MLD 85JCAE-,93,86
msc-hmn:lym 100 mg/L MUREAV 142,133,85
slt-mus-ipr 300 mg/kg MUREAV 117,201,83
ihl-rat TCLo:300 ppm/7H (6-15D preg):REP BANRDU 5,149,80
orl-rat TDLo:5286 mg/kg/69W-I:CAR BANRDU 5,35,80
ihl-rat TCLo:5 ppm/7H/78W-I:ETA BANRDU 5,3,80
orl-mus TDLo:3536 mg/kg/78W-I:CAR BANRDU 5,35,80
ihl-mus TCLo:5 ppm/7H/78W-I:ETA BANRDU 5,3,80
skn-mus TDLo:1120 g/kg/74W-I:NEO JJIND8 63,1433,79
orl-rat TD:38 g/kg/78W-I:CAR NCITR* NCI-CG-TR-55,78
orl-mus TD:76 g/kg/78W-I:CAR,TER NCITR* NCI-CG-TR-55,78
orl-rat TD:18 g/kg/78W-I:CAR NCITR* NCI-CG-TR-55,78
orl-mus TD:38 g/kg/78W-I:CAR,TER NCITR* NCI-CG-TR-55,78
ihl-hmn TCLo:4000 ppm/H:CNS,PNS,GIT PCOC** -,500,66
orl-hmn LDLo:286 mg/kg:GIT CLCEAL 86,203,47
orl-hmn TDLo:428 mg/kg:GIT,CNS,PUL SOMEAU 22,132,58
orl-man TDLo:892 mg/kg:GIT,LIV WILEAR 28,983,75
orl-man LDLo:714 mg/kg:CNS,CVS,PUL KLWOAZ 48,822,70
orl-rat LD50:670 mg/kg FMCHA2 -,C130,91
ihl-rat LC50:1000 ppm/7H AMIHBC 4,482,51
ipr-rat LD50:807 mg/kg GTPZAB 26(4),26,82
scu-rat LD50:1 g/kg FAONAU 48A,91,70
orl-mus LD50:489 mg/kg TOXID9 1,26,81
ihl-mus LCLo:5000 mg/m³/2H AEPPAE 141,19,29
scu-mus LDLo:380 mg/kg JPETAB 84,53,45
orl-dog LD50:5700 mg/kg FAONAU 48A,91,70
ivn-dog LDLo:175 mg/kg QJPPAL 7,205,34
orl-rbt LD50:860 mg/kg GUCHAZ 6,264,73
ihl-rbt LCLo:3000 ppm/7H JPETAB 84,53,45

CONSENSUS REPORTS: NTP 7th Annual Report on Carcinogens. IARC Cancer Review: Group 2B IMEMDT 7,56,87; Human Limited Evidence IMEMDT 20,429,79;

Animal Sufficient Evidence IMEMDT 20,429,79. NCI Carcinogenesis Bioassay (gavage); Clear Evidence: mouse-rat NCITR* NCI-CG-TR-55,78. EPA Genetic Toxicology Program. Reported in EPA TSCA Inventory.

OSHA PEL: TWA 1 ppm; STEL 2 ppm
ACGIH TLV: TWA 10 ppm
DFG MAK: Confirmed Animal Carcinogen, Suspected Human Carcinogen.
NIOSH REL: (Ethylene Dichloride) TWA 1 ppm; CL 2 ppm/15M
DOT CLASSIFICATION: 3; *Label:* Flammable Liquid, Poison

SAFETY PROFILE: Confirmed carcinogen with experimental carcinogenic, neoplastigenic and tumorigenic data. An experimental transplacental carcinogen. A human poison by ingestion. Poison experimentally by intravenous and subcutaneous routes. Moderately toxic by inhalation, skin contact, and intraperitoneal routes. Human systemic effects by ingestion and inhalation: flaccid paralysis without anesthesia (usually neuromuscular blockage), somnolence, cough, jaundice, nausea or vomiting, hypermotility, diarrhea, ulceration or bleeding from the stomach, fatty liver degeneration, change in cardiac rate, cyanosis, and coma. It may also cause dermatitis, edema of the lungs, toxic effects on the kidneys, and severe corneal effects. A strong narcotic. Experimental teratogenic and reproductive effects. A skin and severe eye irritant, and strong local irritant. Its smell and irritant effects warn of its presence at relatively safe concentrations. Human mutation data reported.
Flammable liquid. A dangerous fire hazard if exposed to heat, flame, or oxidizers. Moderately explosive in the form of vapor when exposed to flame. Violent reaction with Al, N_2O_4, NH_3, dimethylaminopropylamine. Can react vigorously with oxidizing materials and emit vinyl chloride and HCl. To fight fire, use water, foam, CO_2, dry chemicals. When heated to decomposition it emits highly toxic fumes of Cl^- and phosgene. See also CHLORINATED HYDROCARBONS, ALIPHATIC.

For occupational chemical analysis use OSHA: #03 or NIOSH: Hydrocarbons, Halogenated, 1003.

EJA000 CAS:6943-65-3 HR: 3
ETHYLENE DIISOTHIOURONIUM DIBROMIDE
mf: $C_4H_{10}N_4S_2{\cdot}2BrH$ mw: 340.14

SYNS: 1,2-ETHANEDIYL ESTER CARBAMIMIDOTHIOIC ACID DIHYDROBROMIDE □ 2,2′-ETHYLENE-BIS-(2-THIOPSEUDOUREA), DIHYDROBROMIDE □ ETHYLENE DIISOTHIOUREA DIHYDROBROMIDE □ 2,2-ETHYLENEDITHIODIPSEUDOUREA DIHYDROBROMIDE

TOXICITY DATA WITH REFERENCE
ipr-mus LD50:75 mg/kg NTIS** AD691-490
ivn-mus LD50:56 mg/kg CSLNX* NX#02909
par-mus LD50:160 mg/kg THERAP 8,929,53
unr-mus LDLo:200 mg/kg ATMPA2 32,177,38

SAFETY PROFILE: Poison by intraperitoneal, intravenous, parenteral, and possibly other routes. When heated to decomposition it emits very toxic fumes of NO_x, SO_x, and HBr. See also BROMIDES.

EJA100 CAS:3715-67-1 HR: 3
N,N-ETHYLENE-N′,N′-DIMETHYLUREA
mf: $C_5H_{10}N_2O$ mw: 114.17

SYNS: 1-AZIRIDINECARBOXAMIDE, N,N-DIMETHYL- □ N,N-DIMETHYL-1-AZIRIDINECARBOXAMIDE □ N-(DIMETHYLCARBAMOYL) AZIRIDINE □ 1-(DIMETHYLCARBAMYL)AZIRIDINE □ N,N-DIMETHYLETHYLENEUREA

TOXICITY DATA WITH REFERENCE
mmo-ssp 140 mmol/L ADWMAX -,193,62
ipr-rat TDLo:100 mg/kg (male 1D pre):REP 85GUAJ -,37,66
ipr-rat LD50:200 mg/kg BJPCAL 21,581,63

SAFETY PROFILE: Poison by intraperitoneal route. Mutation data reported. Experimental reproductive effects. When heated to decomposition it emits toxic fumes of NO_x.

EJA250 HR: 1
(ETHYLENEDINITRILO)TETRAACETATE DIPOTASSIUM SALT
mf: $C_{10}H_{16}N_2O_8{\cdot}2K$ mw: 370.48

SYN: DIPOTASSIUM ETHYLENEDIAMINETETRAACETATE

TOXICITY DATA WITH REFERENCE
skn-rbt 500 mg MLD FCTOD7 20,563,82
eye-rbt 100 mg MLD FCTOD7 20,573,82
eye-rbt 100 mg/30S rns MLD FCTOD7 20,573,82

SAFETY PROFILE: A skin and eye irritant. When heated to decomposition it emits toxic fumes of NO_x and K_2O.

EJA379 CAS:15708-41-5 HR: 1
((ETHYLENEDINITRILO)TETRAACETATO)-FERATE(1-), SODIUM
mf: $C_{10}H_{12}FeN_2O_8{\cdot}Na$ mw: 367.08

PROP: Crystals from EtOH (aq).

SYNS: CALMOSINE □ EDATHAMIL MONOSODIUM FERRIC SALT □ FERISAN □ FERRIC SODIUM EDETATE □ FERRIC SODIUM EDTA □ MONOSODIUM FERRIC EDTA □ REXENE □ SEQUESTRENE NaFe IRON CHELATE □ SODIUM FEREDETATE □ SODIUM FERRIC EDTA □ SODIUM IRON EDTA □ SYTRON

TOXICITY DATA WITH REFERENCE
orl-rat LD50:5 g/kg CIGET* -,-,77
orl-mus LD50:5000 mg/kg CIGET* -,-,77

CONSENSUS REPORTS: Reported in EPA TSCA Inventory.

SAFETY PROFILE: Mildly toxic by ingestion. When heated to decomposition it emits toxic fumes of NO_x and Na_2O.

EJA500 CAS:25481-21-4 HR: 3
(ETHYLENEDINITRILO)TETRA ACETIC ACID NICKEL (II) COMPLEX
mf: $C_{10}H_{12}N_2NiO_8{\cdot}2H$ mw: 348.96

SYNS: DIHYDROGEN (ETHYLENEDIAMINETETRAACETATO(4-)) NICKELATE (2-) □ NICKEL(II) EDTA COMPLEX

TOXICITY DATA WITH REFERENCE
ipr-mus LD50:88 mg(Ni)/kg PABIAQ 11,853,63

CONSENSUS REPORTS: Nickel and its compounds are on the Community Right-To-Know List. Reported in EPA TSCA Inventory.

SAFETY PROFILE: Poison by intraperitoneal route. When heated to decomposition it emits toxic fumes of NO_x. See also NICKEL COMPOUNDS.

EJB000 CAS:5766-67-6 HR: 2
(ETHYLENEDINITRILO)TETRAACETONITRILE
mf: $C_{10}H_{12}N_6$ mw: 216.28

PROP: Powder. Mp: 132°.

SYNS: ETHYLENEDIAMINETETRAACETONITRILE □ N,N,N′,N′-TETRACYANOMETHYLAETHYLENEDIAMIN (GERMAN)

TOXICITY DATA WITH REFERENCE
ipr-mus LD50:3030 mg/kg ARZNAD 16,734,66

CONSENSUS REPORTS: Cyanide and its compounds are on the Community Right-To-Know List. Reported in EPA TSCA Inventory.

SAFETY PROFILE: Moderately toxic by intraperitoneal route. When heated to decomposition it emits toxic fumes of NO_x and CN^-. See also NITRILES.

EJB100 HR: 3
6,6′-(ETHYLENEDIOXY)BIS(4-AMINOQUINALDINE) DIHYDROCHLORIDE
mf: $C_{22}H_{22}N_4O_2$•2ClH mw: 447.40

SYN: BIS-(2-METHYL-4-AMINO-6-QUINOLYLOXY)ETHANE DIHYDROCHLORIDE

TOXICITY DATA WITH REFERENCE
orl-mus LD50:1530 mg/kg ANTCAO 2,581,52
ipr-mus LD50:128 mg/kg ANTCAO 2,581,52
scu-mus LD50:130 mg/kg ANTCAO 2,581,52
ivn-mus LD50:12,100 μg/kg ANTCAO 2,581,52

SAFETY PROFILE: Poison by subcutaneous, intravenous, and intraperitoneal routes. Moderately toxic by ingestion. When heated to decomposition it emits toxic fumes of NO_x and HCl.

EJB500 CAS:111-21-7 HR: 1
2,2′-(ETHYLENEDIOXY)DI(ETHYL ACETATE)
mf: $C_{10}H_{18}O_6$ mw: 234.28

SYNS: ACETIC ACID, TRIETHYLENE GLYCOL DIESTER □ 2,2′-ETHYLENEDIOXYDIETHANOL DIACETATE □ TRIETHYLENE GLYCOL, DIACETATE □ TRIGLYCOL, DIACETATE

TOXICITY DATA WITH REFERENCE
skn-rbt 500 mg open MLD UCDS** 3/27/69
orl-mus TDLo:534 g/kg multi:REP FAATDF 18,602,92
orl-rat LD50:22,600 mg/kg AIHAAP 30,470,69
skn-rbt LD50:8 g/kg AIHAAP 30,470,69

CONSENSUS REPORTS: Reported in EPA TSCA Inventory.

SAFETY PROFILE: Mildly toxic by ingestion and skin contact. Experimental reproductive effects. A skin irritant. When heated to decomposition it emits acrid smoke and irritating fumes.

EJC000 CAS:52936-25-1 HR: 3
ETHYLENE DIPERCHLORATE
mf: $C_2H_4Cl_2O_8$ mw: 226.96

$(-CH_2OClO_3)_2$

SAFETY PROFILE: A highly sensitive, powerful explosive. Explodes on contact with small amounts of water. Upon decomposition it emits toxic fumes of Cl^-. See also PERCHLORATES.

EJC100 CAS:1852-14-8 HR: 1
1,1′-ETHYLENEDIUREA
mf: $C_4H_{10}N_4O_2$ mw: 146.18

SYNS: ETHANEDIUREA □ N,N″-1,2-ETHANEDIYLBISUREA □ 1,1′-ETHYLENEBISUREA □ ETHYLENEDIUREA □MONOETHYLENEDIUREA □ UREA, N,N″-1,2-ETHANEDIYLBIS-(9CI) □ UREA, 1,1′-ETHYLENEDI-

TOXICITY DATA WITH REFERENCE
ipr-mus LDLo:13,140 mg/kg JPETAB 54,188,35

CONSENSUS REPORTS: Reported in EPA TSCA Inventory.

SAFETY PROFILE: Slightly toxic by intraperitoneal route. When heated to decomposition it emits toxic vapors of NO_x.

EJC500 CAS:107-21-1 HR: 3
ETHYLENE GLYCOL
mf: $C_2H_6O_2$ mw: 62.08

PROP: Colorless, sweet-tasting, hygroscopic, viscid, poisonous liquid. Fp: −13°, mp: −15.6°, bp: 197.5°, lel: 3.2%, flash p: 232°F (CC), d: 1.113 @ 25°/25°, autoign temp: 752°F, vap d: 2.14, vap press: 0.05 mm @ 20°. Misc in H_2O, EtOH, MeOH, Me_2CO, AcOH, and Py. Immisc in $CHCl_3$, CCl_4, Et_2O, C_6H_6, CS_2 and ligroin.

SYNS: ATHYLENGLYKOL (GERMAN) □ 1,2-DIHYDROXYETHANE □ DOWTHERM SR 1 □ 1,2-ETHANEDIOL □ ETHYLENE ALCOHOL □ ETHYLENE DIHYDRATE □ GLYCOL □ GLYCOL ALCOHOL □ LUTROL-9 □ MACROGOL 400 BPC □ M.E.G. □ MONOETHYLENE GLYCOL □ NCI-C00920 □ NORKOOL □ TESCOL □ UCAR 17

TOXICITY DATA WITH REFERENCE
eye-rat 12 mg/m³/3D TXAPA9 16,646,70
skn-rbt 555 mg open MLD UCDS** 7/21/65
eye-rbt 500 mg/24H MLD 85JCAE-,205,86
eye-rbt 100 mg/1H MLD NTIS** LMF-69
eye-rbt 12 mg/m³/3D TXAPA9 16,646,70
eye-rbt 1440 mg/6H MOD BUYRAI 31,25,77
dni-hmn:lym 320 mmol/L PNASA6 79,1171,82
msc-mus:lym 100 mmol/L PAACA3 21,74,80
orl-mus TDLo:84 g/kg (female 1-21D post):REP TOXID9 4,136,84
orl-mus TDLo:88,720 mg/kg (female 7-14D post):REP EVHPAZ 57,141,84
orl-rat TDLo:8580 mg/kg (female 6-15D post):TER CHYCDW 20,289,86

orl-rat TDLo: 25 g/kg (female 6-15D post): REP NTIS** PB85-104594
orl-mus TDLo: 15 g/kg (female 6-15D post): REP NTIS** PB85-105385
orl-rat TDLo: 50 g/kg (female 6-15D post): REP NTIS** PB85-104594
orl-rat TDLo: 50 g/kg (6-15D preg): TER TXAPA9 81,113,85
orl-rat TDLo: 12,500 mg/kg (female 6-15D post): TER TJADAB 29(2),52A,84
orl-chd TDLo: 5500 mg/kg: CNS,PUL,KID PGMJAO 52,598,76
orl-hmn LDLo: 786 mg/kg EJTXAZ 9,373,76
orl-hmn LDLo: 398 mg/kg: CNS,GIT,LIV SMEZA5 26(2),48,83
ihl-hmn TCLo: 10,000 mg/m^3: EYE,PUL AGGHAR 5,1,33
unr-man LDLo: 1637 mg/kg 85DCAI 2,73,70
orl-rat LD50: 4700 mg/kg GTPZAB 26(6),28,82
ipr-rat LD50: 5010 mg/kg KRKRDT 9,36,81
scu-rat LD50: 2800 mg/kg NPIRI* 1,49,74
ivn-rat LD50: 3260 mg/kg KRKRDT 9,36,81
ims-rat LDLo: 3300 mg/kg JPETAB 41,387,31
orl-mus LD50: 7500 mg/kg JPETAB 65,89,39
ipr-mus LD50: 5614 mg/kg FEPRA7 6,342,47
scu-mus LDLo: 2700 mg/kg BJIMAG 1,207,44

CONSENSUS REPORTS: EPA Genetic Toxicology Program. Community Right-To-Know List. Reported in EPA TSCA Inventory.

OSHA PEL: CL 50 ppm
ACGIH TLV: CL 50 ppm (vapor)
DFG MAK: 10 ppm (26 mg/m^3)

SAFETY PROFILE: Human poison by ingestion. (Lethal dose for humans reported to be 100 mL.) Moderately toxic to humans by an unspecified route. Moderately toxic experimentally by ingestion, subcutaneous, intravenous, and intramuscular routes. Human systemic effects by ingestion and inhalation: eye lacrimation, general anesthesia, headache, cough, respiratory stimulation, nausea or vomiting, pulmonary, kidney, and liver changes. If ingested it causes initial central nervous system stimulation followed by depression. Later, it causes potentially lethal kidney damage. Very toxic in particulate form upon inhalation. An experimental teratogen. Other experimental reproductive effects. Human mutation data reported. A skin, eye, and mucous membrane irritant.

Combustible when exposed to heat or flame; can react vigorously with oxidants. Moderate explosion hazard when exposed to flame. Ignites on contact with chromium trioxide, potassium permanganate, and sodium peroxide. Mixtures with ammonium dichromate, silver chlorate, sodium chlorite, and uranyl nitrate ignite when heated to 100°C. Can react violently with chlorosulfonic acid, oleum, H_2SO_4, $HClO_4$, and P_2S_5. Aqueous solutions may ignite silvered copper wires that have an applied D.C. voltage. To fight fire, use alcohol foam, water, foam, CO_2, dry chemical. When heated to decomposition it emits acrid smoke and irritating fumes.

For occupational chemical analysis use NIOSH: Ethylene Glycol, 5500.

EJD000 CAS:3775-85-7 HR: 2
ETHYLENE GLYCOL BIS(2,3-EPOXY-2-METHYLPROPYL) ETHER
mf: $C_{10}H_{18}O_4$ mw: 202.28

SYNS: ETHYLENE GLYCOL DI(2,3-EPOXY-2-METHYLPROPYL) ETHER □ ETHYLENE GLYCOLIDE (2,3-EPOXY-2-METHYLPROPYL) ETHER

TOXICITY DATA WITH REFERENCE
skn-rbt 10 mg/24H open MLD AIHAAP 23,95,62
orl-rat LD50: 7460 mg/kg AIHAAP 24,305,63
skn-rbt LD50: 3150 mg/kg AIHAAP 23,95,62

CONSENSUS REPORTS: Glycol ether compounds are on the Community Right-To-Know List.

SAFETY PROFILE: Moderately toxic by skin contact. Mildly toxic by ingestion. A skin irritant. When heated to decomposition it emits acrid smoke and irritating fumes. See also GLYCOL ETHERS.

EJD759 CAS:111-55-7 HR: 2
ETHYLENE GLYCOL DIACETATE
mf: $C_6H_{10}O_4$ mw: 146.16

PROP: Colorless liquid or crystals. Mp: −31°, bp: 186–187°, flash p: 205°F (OC), fp: −31°, d: 1.128 @ 0°/4°, vap press: 1 mm @ 38.3°, vap d: 5.04. Misc in EtOH, Et_2O; sltly sol in H_2O.

SYNS: 1,2-ETHANEDIOL DIACETATE □ ETHYLENE ACETATE □ ETHYLENE GLYCOL ACETATE □ GLYCOL DIACETATE

TOXICITY DATA WITH REFERENCE
eye-rbt 555 mg AJOPAA 29,1363,46
orl-rat LD50: 6850 mg/kg UCDS** 6/6/69
ipr-mus LD50: 1190 mg/kg JPETAB 90,338,47
skn-rbt LD50: 8480 mg/kg UCDS** 6/6/69
orl-gpg LD50: 4940 mg/kg JIHTAB 23,259,41

CONSENSUS REPORTS: Reported in EPA TSCA Inventory.

SAFETY PROFILE: Moderately toxic by intraperitoneal route. Mildly toxic by ingestion and skin contact. An eye irritant. Combustible when exposed to heat or flame; can react with oxidizing materials. To fight fire, use alcohol foam, CO_2, dry chemical. When heated to decomposition it emits acrid smoke and irritating fumes.

EJE000 CAS:7529-27-3 HR: 2
ETHYLENE GLYCOL DIALLYL ETHER
mf: $C_8H_{14}O_2$ mw: 142.22

SYN: DIALLYLETHER ETHYLENGLYKOLU (CZECH)

TOXICITY DATA WITH REFERENCE
skn-rbt 500 mg/24H SEV 28ZPAK -,38,72
eye-rbt 250 µg/24H SEV 28ZPAK -,38,72

CONSENSUS REPORTS: Glycol ether compounds are on the Community Right-To-Know List. Reported in EPA TSCA Inventory.

SAFETY PROFILE: A severe skin and eye irritant. When heated to decomposition it emits acrid smoke and

E

irritating fumes. See also GLYCOL ETHERS and ALLYL COMPOUNDS.

EJE500 **CAS:629-14-1** **HR: 3**
ETHYLENE GLYCOL DIETHYL ETHER
DOT: UN 1153
mf: $C_6H_{14}O_2$ mw: 118.20

PROP: Colorless liquid, slight ethereal odor. Mp: −74°, bp: 123.5°, flash p: 95°F (OC), d: 0.8417 @ 20°/20°, autoign temp: 406°F, vap d: 6.56, vap press: 9.4 mm.

SYNS: 1,2-DIETHOXYETHANE □ DIETHYL CELLOSOLVE (DOT) □ ETHYL GLYME

TOXICITY DATA WITH REFERENCE
eye-rbt 17 mg AJOPAA 29,1363,46
orl-mus TDLo:23,640 mg/kg (female 6-13D post): REP TCMUD8 7,29,87
orl-mus TDLo:10 g/kg (female 6-15D post):TER NTIS** PB88-134093
orl-mus TDLo:23,640 mg/kg (7-14D preg):REP EVHPAZ 57,141,84
orl-mus TDLo:5000 mg/kg (female 6-15D post):TER NTIS** PB88-134093
orl-rat LD50:4390 mg/kg JIHTAB 23,259,41
ihl-rat LCLo:8000 ppm/4H JIHTAB 31,343,49
orl-gpg LD50:2440 mg/kg JIHTAB 23,259,41

CONSENSUS REPORTS: Reported in EPA TSCA Inventory. Glycol ether compounds are on the Community Right-To-Know List.

DOT CLASSIFICATION: 3; *Label:* Flammable Liquid

SAFETY PROFILE: Moderately toxic by ingestion. Mildly toxic by inhalation. An experimental teratogen. Experimental reproductive effects. An eye irritant. An aprotic solvent. A very dangerous fire hazard when exposed to heat or flame; can react with oxidizing materials. To fight fire, use CO_2, dry chemical. See also GLYCOL ETHERS and various cellosolve entries.

EJF000 **CAS:629-15-2** **HR: 3**
ETHYLENE GLYCOL DIFORMATE
mf: $C_4H_6O_4$ mw: 118.10

PROP: Liquid. Mp: −10°, bp: 177°, flash p: 200°F (OC), d: 1.2277 @ 20°/20°, vap d: 4.07.

SYNS: ETHYLENE FORMATE □ GLYCOL DIFORMATE

TOXICITY DATA WITH REFERENCE
eye-rbt 5 mg SEV AJOPAA 29,1363,46
orl-rat LD50:1510 mg/kg JIHTAB 23,259,41
orl-gpg LD50:390 mg/kg JIHTAB 23,259,41

CONSENSUS REPORTS: Reported in EPA TSCA Inventory.

SAFETY PROFILE: Poison by ingestion. A severe eye irritant. Flammable when exposed to heat or flame; can react with oxidizing materials. To fight fire, use CO_2, dry chemical. When heated to decomposition it emits acrid smoke and irritating fumes.

EJG000 **CAS:628-96-6** **HR: 3**
ETHYLENE GLYCOL DINITRATE
mf: $C_2H_4N_2O_6$ mw: 152.08

PROP: Yellow liquid. Mp: −22.3°, bp: 105.5° @ 19 mm, explodes @ 114°, d: 1.483 @ 8°, vap d: 5.25.

SYNS: DINITROGLICOL (ITALIAN) □ DINITROGLYCOL □ EGDN □ ETHANEDIOL DINITRATE □ ETHYLENE DINITRATE □ ETHYLENE NITRATE □ ETHYLENGLYKOLDINITRAT (CZECH) □ GLYCOLDINITRAAT (DUTCH) □ GLYCOL DINITRATE □ GLYCOL (DINITRATE DE) (FRENCH) □ GLYKOLDINITRAT (GERMAN) □ NITROGLYCOL □ NITROGLYKOL (CZECH)

TOXICITY DATA WITH REFERENCE
scu-cat LDLo:50 mg/kg AEPPAE 200,271,42
scu-rbt LDLo:300 mg/kg AEPPAE 200,271,42

CONSENSUS REPORTS: Reported in EPA TSCA Inventory.

OSHA PEL: STEL 0.1 mg/m^3 (skin)
ACGIH TLV: TWA 0.05 ppm (skin)
DFG MAK: 0.05 ppm (0.3 mg/m^3)
NIOSH REL: (Nitroglycerin) CL 0.1 mg/m^3/20M
DOT CLASSIFICATION: Forbidden

SAFETY PROFILE: Poison by subcutaneous route. Can cause lowered blood pressure leading to headache, dizziness, and weakness. Used as an explosive. When heated to decomposition it emits toxic fumes of NO_x. See also NITRATES.

For occupational chemical analysis use OSHA: #43 or NIOSH: Nitroglycerin and Ethylene Glycol Dinitrate, 2507.

EJG500 **CAS:26560-94-1** **HR: 2**
ETHYLENE GLYCOL MALEATE
mf: $C_6H_8O_5$ mw: 160.14

SYNS: ETHYLENE GLYCOL, MONO(HYDROGEN MALEATE) □ 2-HYDROXYETHYL ESTER MALEIC ACID □ MONO(HYDROXYETHYL) ESTER MALEIC ACID

TOXICITY DATA WITH REFERENCE
skn-rbt 500 mg/24H MLD 85JCAE-,657,86
eye-rbt 2 mg open SEV AMIHBC 10,61,54
orl-rat LD50:2460 mg/kg AMIHBC 10,61,54

CONSENSUS REPORTS: Reported in EPA TSCA Inventory.

SAFETY PROFILE: Moderately toxic by ingestion. A skin and severe eye irritant. When heated to decomposition it emits acrid smoke and irritating fumes. See also ESTERS.

EJH000 **CAS:868-77-9** **HR: 2**
ETHYLENE GLYCOL METHACRYLATE
mf: $C_6H_{10}O_3$ mw: 130.16

PROP: Bp: 71–73° @ 2 mm.

SYNS: ETHYLENE GLYCOL, MONOMETHACRYLATE □ GLYCOL METHACRYLATE □ GLYCOL MONOMETHACRYLATE □ 2-HYDROXYETHYL ESTER METHACRYLIC ACID □ HYDROXYETHYL METHACRYLATE □ β-HYDROXYETHYL METHACRYLATE □ 2-HYDROXYETHYL

METHACRYLATE □ MHOROMER □ MONOMER MG-1 □ MONOMETHACRYLIC ETHER of ETHYLENE GLYCOL

TOXICITY DATA WITH REFERENCE
orl-rat TDLo:3062 mg/kg (female 35W pre):REP GISAAA 52(11),81,87
orl-rat TDLo:3062 mg/kg (male 35W pre):REP GISAAA 52(11),81,87
orl-rat LD50:5050 mg/kg GISAAA 54(9),75,89
ipr-rat LD50:1250 mg/kg AMPMAR 36,58,75
orl-mus LD50:3275 mg/kg GISAAA 54(9),75,89
ipr-mus LD50:497 mg/kg JPMSAE 62,778,73

CONSENSUS REPORTS: Reported in EPA TSCA Inventory.

SAFETY PROFILE: Moderately toxic by intraperitoneal route. Mildly toxic by ingestion. Experimental reproductive effects. When heated to decomposition it emits acrid smoke and irritating fumes.

EJH500 CAS:109-86-4 HR: 3
ETHYLENE GLYCOL METHYL ETHER
DOT: UN 1188
mf: $C_3H_8O_2$ mw: 76.11

$CH_3OC_2H_4OH$

PROP: Colorless liquid; mild, agreeable odor. Misc in water, alc, ether, benzene. Bp: 124.5°, fp: −86.5°, flash p: 115°F (OC), lel: 2.5%, uel: 14%, d: 0.9660 @ 20°/4°, autoign temp: 545°F, vap press: 6.2 mm @ 20°, vap d: 2.62.

SYNS: AETHYLENGLYKOL-MONOMETHYLAETHER (GERMAN) □ DOWANOL EM □ EGM □ EGME □ ETHER MONOMETHYLIQUE de l'ETHYLENE-GLYCOL (FRENCH) □ ETHYLENE GLYCOL MONOMETHYL ETHER (MAK, DOT) □ GLYCOL ETHER EM □ GLYCOLMETHYL ETHER □ GLYCOL MONOMETHYL ETHER □ JEFFERSOL EM □ MECS □ 2-METHOXY-AETHANOL (GERMAN) □ 2-METHOXYETHANOL (ACGIH) □ METHOXYHYDROXYETHANE □ METHYL CELLOSOLVE (OSHA, DOT) □ METHYL ETHOXOL □ METHYL GLYCOL □ METHYLGLYKOL (GERMAN) □ METHYL OXITOL □ METIL CELLOSOLVE (ITALIAN) □ METOKSYETYLOWY ALKOHOL (POLISH) □ 2-METOSSIETANOLO (ITALIAN) □ MONOMETHYL ETHER of ETHYLENE GLYCOL □ POLY-SOLV EM □ PRIST

TOXICITY DATA WITH REFERENCE
skn-rbt 483 mg/24H MLD TXAPA9 19,276,71
eye-rbt 500 mg/24H MLD 85JCAE-,623,86
eye-gpg 10 µg MLD JPPMAB 11,150,59
dlt-rat-orl 500 mg/kg ENMUDM 6,390,84
spm-rat-orl 500 mg/kg ENMUDM 6,390,84
spm-mus-orl 500 mg/kg ENMUDM 6,390,84
ihl-rat TCLo:25 ppm/7H (female 7-13D post):REP TJADAB 27,65A,83
orl-rat TDLo:175 mg/kg (7-13D preg):REP TJADAB 32,33,85
ihl-rat TCLo:100 ppm/6H (female 6-17D post):REP TXAPA9 69,43,83
orl-mky TDLo:930 mg/kg (female 20-45D post):TER TJADAB 35,66A,87
orl-rat TDLo:350 mg/kg (female 9-15D post):TER TXAPA9 86,197,86
orl-mky TDLo:930 mg/kg (female 20-45D post):REP TJADAB 35,66A,87
ihl-mus TCLo:50 ppm/6H (female 6-15D post):REP TXAPA9 75,409,84
orl-mus TDLo:500 mg/kg (female 9D post):REP TXAPA9 80,108,85
orl-rat TDLo:250 mg/kg (male 1D pre):REP TOXID9 4,135,84
orl-rat TDLo:100 mg/kg (male 1D pre):REP TXAPA9 69,385,83
ihl-rbt TCLo:50 ppm/6H (female 6-18D post):TER TXAPA9 75,409,84
orl-rat TDLo:175 mg/kg (7-13D preg):TER TJADAB 32,33,85
ipr-rat TDLo:190 mg/kg (female 14D post):TER JJATDK 4,35,84
ipr-rat TDLo:330 mg/kg (female 12D post):TER TJADAB 29(2),54A,84
orl-hmn LDLo:3380 mg/kg JIHTAB 28,267,46
ihl-hmn TCLo:25 ppm:CNS JIHTAB 20,134,38
orl-rat LD50:2460 mg/kg JIHTAB 23,259,41
ihl-rat LC50:1500 ppm/7H NPIRI* 1,57,74
ipr-rat LD50:2500 mg/kg NPIRI* 1,57,74
ivn-rat LD50:2140 mg/kg AMIHAB 14,114,56
orl-mus LD50:2560 mg/kg GTPZAB 32(3),48,88
ihl-mus LC50:1480 ppm JIHTAB 25,157,43
ipr-mus LD50:2147 mg/kg FEPRA7 6,342,47
orl-rbt LD50:890 mg/kg AMIHAB 14,114,56
skn-rbt LD50:1280 mg/kg NPIRI* 1,57,74
orl-gpg LD50:950 mg/kg JIHTAB 23,259,41

CONSENSUS REPORTS: Reported in EPA TSCA Inventory. Community Right-To-Know List.

OSHA PEL: TWA 25 ppm (skin)
ACGIH TLV: TWA 5 ppm (skin)
DFG MAK: 5 ppm (15 mg/m³)
NIOSH REL: TWA (Glycol Ethers): Reduce to lowest level
DOT CLASSIFICATION: 3; *Label:* Flammable Liquid

SAFETY PROFILE: Moderately toxic to humans by ingestion. Moderately toxic experimentally by ingestion, inhalation, skin contact, intraperitoneal, and intravenous routes. Human systemic effects by inhalation: change in motor activity, tremors, and convulsions. Experimental teratogenic and reproductive effects. A skin and eye irritant. Mutation data reported. When used under conditions that do not require the application of heat, this material probably presents little hazard to health. However, in the manufacture of fused collars which require pressing with a hot iron, cases have been reported showing disturbance of the hemopoietic system with or without neurological signs and symptoms. The blood picture may resemble that produced by exposure to benzene. Two cases reported had severe aplastic anemia with tremors and marked mental dullness. The persons affected had been exposed to vapors of methyl "Cellosolve," ethanol, and methanol, ethyl acetate and petroleum naphtha.

Flammable liquid when exposed to heat or flame. A moderate explosion hazard. Can react with oxidizing materials to form explosive peroxides. To fight fire, use alcohol foam, CO_2, dry chemical. When heated to decomposition it emits acrid smoke and irritating fumes. See also GLYCOL ETHERS.

E

For occupational chemical analysis use OSHA: #53 or NIOSH: Alcohols IV, 1403.

EJI000 **CAS:542-59-6** **HR: 2**
ETHYLENE GLYCOL MONOACETATE
mf: $C_4H_8O_3$ mw: 104.12

PROP: Colorless, almost odorless liquid. Bp: 187–189°, flash p: 215°F (OC), d: 1.108 @ 15°, vap d: 3.59. Misc in EtOH and H_2O.

SYNS: 1,2-ETHANEDIOL, MONOACETATE ☐ ETHYLENE GLYCOL ACETATE ☐ GLYCOL MONOACETATE ☐ GLYCOL-MONOACETIN ☐ 2-HYDROXYETHYL ACETATE

TOXICITY DATA WITH REFERENCE
eye-rbt 100 mg SEV AJOPAA 29,1363,46
orl-rat LD50:8250 mg/kg JIHTAB 23,259,41
ipr-mus LD50:1310 mg/kg JPETAB 90,338,47
orl-gpg LD50:3800 mg/kg JIHTAB 23,259,41

CONSENSUS REPORTS: Reported in EPA TSCA Inventory.

SAFETY PROFILE: Moderately toxic by ingestion amd intraperitoneal routes. A severe eye irritant. Combustible when exposed to heat or flame; can react with oxidizing materials. To fight fire, use alcohol foam, foam, water, CO_2, dry chemical. When heated to decomposition it emits acrid smoke and irritating fumes.

EJI500 **CAS:622-08-2** **HR: 2**
ETHYLENE GLYCOL MONOBENZYL ETHER
mf: $C_9H_{12}O_2$ mw: 152.21

PROP: Water-white oily liquid; faint rose-like odor. Mp: −75°, bp: 137–138° @ 17 mm, flash p: 265°F (OC), d: 1.068, autoign temp: 665°F, vap d: 5.25.

SYNS: BENZYL "CELLOSOLVE" ☐ BENZYLCELOSOLV ☐ 2-BENZYLOXYETHANOL

TOXICITY DATA WITH REFERENCE
eye-rbt 2 mg SEV AJOPAA 29,1363,46
orl-rat LD50:1190 mg/kg JIHTAB 23,259,41

CONSENSUS REPORTS: Glycol ether compounds are on the Community Right-To-Know List. Reported in EPA TSCA Inventory.

SAFETY PROFILE: Moderately toxic by ingestion. A severe eye irritant. Combustible when exposed to heat or flame; can react with oxidizing materials. To fight fire, use CO_2, dry chemical. When heated to decomposition it emits acrid smoke and irritating fumes. See also GLYCOL ETHERS.

EJJ000 **CAS:7795-91-7** **HR: 2**
ETHYLENE GLYCOL MONO-sec-BUTYL ETHER
mf: $C_6H_{14}O_2$ mw: 118.20

SYN: 2-sec-BUTOXYETHANOL

TOXICITY DATA WITH REFERENCE
orl-rat LD50:790 mg/kg SCCUR* -,5,61
orl-mus LD50:1660 mg/kg SCCUR* -,5,61

CONSENSUS REPORTS: Glycol ether compounds are on the Community Right-To-Know List.

SAFETY PROFILE: Moderately toxic by ingestion. When heated to decomposition it emits acrid smoke and irritating fumes. See also GLYCOL ETHERS.

EJJ500 **CAS:110-49-6** **HR: 3**
ETHYLENE GLYCOL MONOMETHYL ETHER ACETATE
DOT: UN 1189
mf: $C_5H_{10}O_3$ mw: 118.15

PROP: Colorless liquid; pleasant, sweet, ether odor. Bp: 143°, fp: −70°, flash p: 111°F (CC), d: 1.005 @ 20°/20°, vap d: 4.07, lel: 1.7%, uel: 8.2%. Sol in water.

SYNS: ACETATE de L'ETHER MONOMETHYLIQUE de L'ETHYLENE-GLYCOL (FRENCH) ☐ ACETATE de METHYLE GLYCOL (FRENCH) ☐ ACETATO di METIL CELLOSOLVE (ITALIAN) ☐ AETHYLENGLYKOL-METHYLAETHERACETAT (GERMAN) ☐ ETHYLENE GLYCOL METHYL ETHER ACETATE ☐ GLYCOL ETHER EM ACETATE ☐ GLYCOL MONO-METHYL ETHER ACETATE ☐ MeCsAc ☐ 2-METHOXYAETHYLACETAT (GERMAN) ☐ 2-METHOXYETHANOL, ACETATE ☐ 2-METHOXY-ETHYL ACETAAT (DUTCH) ☐ 2-METHOXYETHYL ACETATE (ACGIH) ☐ 2-METHOXYETHYLE, ACETATE de (FRENCH) ☐ METHYL CELLOSOLYE ACETAAT (DUTCH) ☐ METHYL CELLOSOLVE ACETATE (OSHA, DOT) ☐ METHYL GLYCOL ACETATE ☐ METHYL GLYCOL MONOACETATE ☐ METHYLGLYKOLACETAT (GERMAN) ☐ 2-METOSSIETILACETATO (ITALIAN)

TOXICITY DATA WITH REFERENCE
eye-rbt 218 mg MLD UCDS** 3/20/73
sln-smc 56,600 ppm MUREAV 149,339,85
orl-mus TDLo:9800 mg/kg (female 6-13D post):REP TCMUD8 7,29,87
orl-mus TDLo:50 g/kg (25D male):REP SAIGBL 21,29,79
orl-mus TDLo:12,500 mg/kg (25D male):REP SAIGBL 21,29,79
ihl-hmn TCLo:1000 mg/m³:EYE,PUL AGGHAR 5,1,33
orl-rat LD50:3390 mg/kg JIHTAB 30,63,48
ihl-rat LCLo:7000 ppm/4H JIHTAB 30,63,48
ipr-rat LDLo:1200 mg/kg JPPMAB 11,150,59
ihl-cat LCLo:6 g/m³/7H AGGHAR 5,1,33
scu-cat LDLo:3000 mg/kg AGGHAR 5,1,33
skn-rbt LD50:5250 mg/kg UCDS** 3/20/73
orl-gpg LD50:1250 mg/kg JIHTAB 23,259,41
scu-gpg LDLo:5000 mg/kg AGGHAR 5,1,33

CONSENSUS REPORTS: Glycol ether compounds are on the Community Right-To-Know List. Reported in EPA TSCA Inventory.

OSHA PEL: TWA 25 ppm (skin)
ACGIH TLV: TWA 5 ppm (skin)
DFG MAK: 5 ppm (25 mg/m³)
DOT CLASSIFICATION: 3; *Label:* Flammable Liquid

SAFETY PROFILE: Moderately toxic by ingestion, intraperitoneal, and subcutaneous routes. Mildly toxic by inhalation and skin contact. Human systemic effects by inhalation: eye lacrimation, cough, and pulmonary changes. Experimental reproductive effects. Mutation data reported. An inhalation irritant in humans. An eye irritant. Flammable liquid when exposed to heat or flame; can react with oxidizing materials. A moderate explosion hazard. To fight fire, use CO_2, dry chemical.

When heated to decomposition it emits acrid smoke and irritating fumes. See also GLYCOL ETHERS.

For occupational chemical analysis use OSHA: #53 or NIOSH: Methyl Cellosolve Acetate S39.

EJK000 **CAS:10137-96-9** **HR: 2**
ETHYLENE GLYCOL MONO-2-METHYLPENTYL ETHER
mf: $C_8H_{18}O_2$ mw: 146.26

SYNS: ETHYLENE GLYCOL MONOMETHYLPENTYL ETHER ☐ 2-METHYLPENTYL CELLOSOLVE ☐ 2-((2-METHYLPENTYL)OXY)ETHANOL

TOXICITY DATA WITH REFERENCE
skn-rbt 10 mg/24H open MLD AIHAAP 23,95,62
eye-rbt 100 mg SEV 34ZIAG -,729,69
orl-rat LD50:3730 mg/kg AIHAAP 23,95,62
skn-rbt LD50:440 mg/kg AIHAAP 23,95,62

CONSENSUS REPORTS: Glycol ether compounds are on the Community Right-To-Know List.

SAFETY PROFILE: Moderately toxic by ingestion and skin contact. A skin and severe eye irritant. When heated to decomposition it emits acrid smoke and irritating fumes. See also GLYCOL ETHERS.

EJK500 **CAS:23495-12-7** **HR: 1**
ETHYLENEGLYCOL MONOPHENYL ETHER PROPIONATE
mf: $C_{11}H_{14}O_3$ mw: 194.25

SYNS: 2-PHENOXYETHANOL PROPIONATE ☐ PHENOXYETHYL PROPIONATE ☐ PROPIONIC ACID-2-PHENOXYETHYL ESTER

TOXICITY DATA WITH REFERENCE
orl-rat LD50:4400 mg/kg FCTXAV 14,659,76

CONSENSUS REPORTS: Glycol ether compounds are on the Community Right-To-Know List. Reported in EPA TSCA Inventory.

SAFETY PROFILE: Mildly toxic by ingestion. When heated to decomposition it emits acrid smoke and irritating fumes. See also GLYCOL ETHERS.

EJL000 **CAS:10137-98-1** **HR: 2**
ETHYLENE GLYCOL MONO-2,6,8-TRIMETHYL-4-NONYL ETHER
mf: $C_{14}H_{30}O_2$ mw: 230.44

SYN: 2-((1-ISOBUTYL-3,5-DIMETHYLHEXYL)OXY)ETHANOL

TOXICITY DATA WITH REFERENCE
skn-rbt 10 mg/24H open MLD AIHAAP 23,95,62
orl-rat LD50:5360 mg/kg AIHAAP 23,95,62
skn-rbt LD50:3150 mg/kg AIHAAP 23,95,62

CONSENSUS REPORTS: Glycol ether compounds are on the Community Right-To-Know List.

SAFETY PROFILE: Moderately toxic by skin contact. Mildly toxic by ingestion. A skin irritant. When heated to decomposition it emits acrid smoke and irritating fumes. See also GLYCOL ETHERS.

EJL500 **CAS:764-48-7** **HR: 2**
ETHYLENE GLYCOL MONOVINYL ETHER
mf: $C_4H_8O_2$ mw: 88.12

SYNS: 2-(ETHENYLOXY)ETHANOL ☐ ETHYLENE GLYCOL VINYL ETHER ☐ ETHYLENGLYCOL MONOVINYL ESTER (RUSSIAN) ☐ 2-HYDROXYETHYL VINYL ETHER ☐ MVEEG (RUSSIAN) ☐ VINYLOXYETHANOL ☐ 2-(VINYLOXY)ETHANOL

TOXICITY DATA WITH REFERENCE
orl-rat LD50:3910 mg/kg GISAAA (3),12,77
orl-mus LD50:2900 mg/kg GISAAA (3),12,77
ihl-mus LC50:29 g/m^3 GISAAA 39(11),94,74

CONSENSUS REPORTS: Glycol ether compounds are on the Community Right-To-Know List.

SAFETY PROFILE: Moderately toxic by ingestion. Mildly toxic by inhalation. When heated to decomposition it emits acrid smoke and irritating fumes. See also GLYCOL ETHERS.

EJM500 **CAS:111-60-4** **HR: 3**
ETHYLENE GLYCOL STEARATE
mf: $C_{20}H_{40}O_3$ mw: 328.60

SYNS: CLINDROL SEG ☐ EMEREST 2350 ☐ EMPILAN 2848 ☐ ETHYLENE GLYCOL MONOSTEARATE ☐ GLYCOL MONOSTEARATE ☐ GLYCOL STEARATE ☐ 2-HYDROXYETHYL ESTER STEARIC ACID ☐ IVORIT ☐ LIPO EGMS ☐ MONTHYBASE ☐ MONTHYLE ☐ PARASTARIN ☐ PRODHYBASE ETHYL ☐ S 151 ☐ SEDETOL ☐ STEARIC ACID, MONOESTER with ETHYLENE GLYCOL ☐ TEGO-STEARATE ☐ USAF KE-11

TOXICITY DATA WITH REFERENCE
skn-rbt 500 mg/24H MLD JACTDZ 1(2),1,82
ipr-mus LD50:200 mg/kg NTIS** AD277-689

CONSENSUS REPORTS: Reported in EPA TSCA Inventory.

SAFETY PROFILE: Poison by intraperitoneal route. A skin irritant. Used in cosmetics. When heated to decomposition it emits acrid smoke and irritating fumes. See also ESTERS.

EJM900 **CAS:151-56-4** **HR: 3**
ETHYLENEIMINE
DOT: UN 1185
mf: C_2H_5N mw: 43.08

PROP: Oily, water-white liquid. Pungent ammonia-like odor. Bp: 55–56°, fp: −71.5°, flash p: 12°F, d: 0.832 @ 20°/4°, autoign temp: 608°F, vap press: 160 mm @ 20°, vap d: 1.48, lel: 3.6%, uel: 46%. Misc in water.

SYNS: AETHYLENIMIN (GERMAN) ☐ AMINOETHYLENE ☐ AZACYCLOPROPANE ☐ AZIRANE ☐ AZIRIDIN (GERMAN) ☐ AZIRIDINE ☐ DIHYDROAZIRENE ☐ DIHYDRO-1H-AZIRINE ☐ DIMETHYLENEIMINE ☐ DIMETHYLENIMINE ☐ EI ☐ ENT 50,324 ☐ ETHYLEENIMINE (DUTCH) ☐ ETHYLENE IMINE, INHIBITED (DOT) ☐ ETHYLENIMINE ☐ ETHYLIMINE ☐ ETILENIMINA (ITALIAN) ☐ RCRA WASTE NUMBER P054 ☐ TL 337

TOXICITY DATA WITH REFERENCE
eye-mus 2 ppm/1M JIHTAB 30,7,48
skn-rbt 10 mg/24H open JIHTAB 30,63,48
eye-rbt 2 mg SEV AJOPAA 29,1363,46

sln-dmg-mul 1 pph/8H-C GNKAA5 21,958,85
mnt-nml-mul 5000 ppm/8D-C MUREAV 125,275,84
cyt-hmn:lng 1 µmol/L EVSRBT 24,433,81
dns-mus-par 5 mg/kg EVSRBT 24,943,81
dlt-mus-ipr 5 mg/kg EVSRBT 24,943,81
msc-ham:ovr 2 mg/L MUREAV 94,449,82
ihl-rat TCLo:800 µg/m³/24H (1D male):REP TPKVAL 14,16,75
scu-mus TDLo:41,760 µg/kg (6-14D preg):TER NTIS** PB223-160
scu-rat TDLo:20 mg/kg/67D-I:NEO BJPCAL 9,306,54
orl-mus TDLo:6500 µg/kg/5D-I:CAR VOONAW 27(5),88,81
orl-mus TDLo:235 mg/kg/76W-C:CAR JNCIAM 42,1101,69
scu-rat TD:10 mg/kg/8W-I:ETA BJPCAL 9,306,54
orl-rat LD50:15 mg/kg JIHTAB 23,259,41
ihl-rat LC50:100 mg/m³/2H 85GMAT -67,82
ipr-rat LD50:3500 µg/kg BJPCAL 21,581,63
ihl-mus LC50:400 mg/m³/2H 85GMAT -,67,82
ihl-rbt LCLo:100 mg/m³/2H 85GMAT -,67,82
skn-rbt LDLo:10 mg/kg 85GMAT -,67,82
ihl-gpg LCLo:25 ppm/8H JIHTAB 30,2,48
skn-gpg LD50:14 mg/kg JIHTAB 30,63,48

CONSENSUS REPORTS: IARC Cancer Review: Group 3 IMEMDT 7,56,87; Animal Sufficient Evidence IMEMDT 9,37,75. Community Right-To-Know List. EPA Extremely Hazardous Substances List. Reported in EPA TSCA Inventory. EPA Genetic Toxicology Program.

OSHA PEL: TWA 1 mg/m³ (skin); Cancer Suspect Agent
ACGIH TLV: TWA 0.5 ppm (skin)
DFG TRK: Animal Carcinogen, Suspected Human Carcinogen.
NIOSH REL: (Ethyleneimine): TWA use 29 CFR 1910.1012
DOT CLASSIFICATION: 6.1; *Label:* Poison, Flammable Liquid

SAFETY PROFILE: Confirmed carcinogen with experimental carcinogenic, neoplastigenic, tumorigenic, and teratogenic data. Other experimental reproductive effects. Poison by ingestion, skin contact, inhalation, and intraperitoneal routes. Human mutation data reported. A skin, mucous membrane, and severe eye irritant. An allergic sensitizer of skin. Causes opaque cornea, keratoconus, and necrosis of cornea (experimentally). Has been known to cause severe human eye injury. Drinking of carbonated beverages is recommended as an antidote to this material in stomach.

A very dangerous fire and explosion hazard when exposed to heat, flame, or oxidizers. Reacts violently with acids, aluminum chloride + substituted anilines, acetic acid, acetic anhydride, acrolein, acrylic acid, allyl chloride, CS_2, Cl_2, chlorosulfonic acid, epichlorohydrin, glyoxal, HCl, HF, HNO_3, oleum, β-propiolactone, Ag, NaOCl, H_2SO_4, vinyl acetate. Reacts with chlorinating agents (e.g., sodium hypochlorite solution) to form the explosive 1-chloroaziridine. Reacts with silver or its alloys to form explosive silver derivatives. Dangerous; heat and/or the presence of catalytically active metals or chloride ions can cause a violent exothermic reaction. To fight fire, use alcohol foam, CO_2, dry chemical. When heated to decomposition it emits acrid smoke and irritating fumes.

For occupational chemical analysis use NIOSH: Ethylenimine P&CAM 300.

EJN400 CAS:13279-24-8 HR: 3
N,N-ETHYLENE-N'-METHYLUREA
mf: $C_4H_8N_2O$ mw: 100.14

SYNS: 1-AZIRIDINECARBOXAMIDE, N-METHYL- □ N-METHYL-1-AZIRIDINECARBOXAMIDE □ N-METHYLETHYLENEUREA

TOXICITY DATA WITH REFERENCE
ipr-rat TDLo:50 mg/kg (male 1D pre):REP 85GUAJ -,37,66
ipr-rat LD50:90 mg/kg BJPCAL 21,581,63

SAFETY PROFILE: Poison by intraperitoneal route. Experimental reproductive effects. When heated to decomposition it emits toxic fumes of NO_x.

EJN500 CAS:75-21-8 HR: 3
ETHYLENE OXIDE
DOT: UN 1040
mf: C_2H_4O mw: 44.06

PROP: Colorless gas at room temperature. Mp: −111.3°, bp: 10.7°, ULC: 100, lel: 3.0%, uel: 100%, flash p: −4°F, d: 0.8711 @ 20°/20°, autoign temp: 804°F, vap press: 1095 mm @ 20°, vap d: 1.52. Misc in water and alc; very sol in ether.

SYNS: AETHYLENOXID (GERMAN) □ AMPROLENE □ANPROLENE □ ANPROLINE □ DIHYDROOXIRENE □ DIMETHYLENE OXIDE □ ENT 26,263 □ E.O. □ 1,2-EPOXYAETHAN (GERMAN) □ EPOXYETHANE □ 1,2-EPOXYETHANE □ ETHENE OXIDE □ ETHYLEENOXIDE (DUTCH) □ ETHYLENE (OXYDE d') (FRENCH) □ ETILENE (OSSIDO di) (ITALIAN) □ ETO □ ETYLENU TLENEK (POLISH) □ FEMA No. 2433 □ MERPOL □ NCI-C50088 □OXACYCLOPROPANE □ OXANE □ OXIDOETHANE □ α,β-OXIDOETHANE □ OXIRAAN (DUTCH) □ OXIRANE □ OXYFUME □ OXYFUME 12 □ RCRA WASTE NUMBER U115 □ STERILIZING GAS ETHYLENE OXIDE 100% □ T-GAS

TOXICITY DATA WITH REFERENCE
skn-hmn 1%/7S AMIHBC 2,549,50
eye-rbt 18 mg/6H MOD BUYRAI 31,25,77
mmo-omi 540 mg/L 47YKAF 8,273,84
dns-hmn:leu 4 mmol/L CBINA8 47,265,83
sce-hmn:lym 4 pph TCMUD8 6,15,86
sce-hmn:lym 10 mg/L PHMGBN 25,214,82
dnd-mus-ipr 100 mg/kg ENMUDM 8(Suppl 6),74,86
dlt-mus-ihl 500 ppm/6H/4D-C ENMUDM 8,1,86
ipr-mus TDLo:750 mg/kg (male 25D pre):REP MUREAV 73,133,80
ihl-rat TCLo:100 ppm/6H (female pre):REP TXAPA9 48,A84,79
ihl-mus TCLo:1200 ppm/90M (female 1D post):TER MUREAV 176,269,87
ihl-rat TCLo:100 ppm/6H (6-15D preg):TER TXAPA9 48,A84,79
ivn-mus TDLo:450 mg/kg (female 10-12D post):REP TXAPA9 56,16,80
ihl-mus TCLo:1200 ppm/90M (female 1D post):REP MUREAV 176,269,87
ihl-rat TCLo:3600 µg/m³/24H (male 60D pre):REP TPKVAL 14,11,75

ihl-mky TCLo:50 ppm/7H (male 96W pre):REP ATAREK 8-84,7,84
ihl-rat TCLo:150 ppm/7H (female 7-16D post):TER NTIS** PB83-258038
orl-rat TDLo:1186 mg/kg/2Y-I:CAR BJCAAI 46,924,82
ihl-rat TCLo:33 ppm/6H/2Y-I:CAR TXAPA9 75,105,84
ihl-mus TDLo:50 ppm/6H/2Y:CAR tumors NTPTR* NTP-TR-326,87
scu-mus TDLo:292 mg/kg/95W-I:CAR ZHPMAT 174,383,81
scu-mus TD:1090 mg/kg/91W-I:NEO BJCAAI 39,588,79
scu-mus TD:908 mg/kg/95W-I:CAR ZHPMAT 174,383,81
scu-mus TD:2576 mg/kg/95W-I:CAR ZHPMAT 174,383,81
orl-rat TD:5112 mg/kg/2Y-I:CAR BJCAAI 46,924,82
ihl-rat TC:50 ppm/7H/2Y-I:CAR TXAPA9 76,69,84
ihl-rat TC:33 ppm/6H/2Y-I:ETA NRTXDN 6,117,85
ihl-rat TC:33 ppm/6H/2Y-I:CAR FCTOD7 24,145,86
ihl-hmn TCLo:12,500 ppm/10S:NOSE JOHYAY 32,409,32
ihl-wmn TCLo:500 ppm/2M:CNS,GIT,PUL DICPBB 15,384,81
orl-rat LD50:72 mg/kg SPEADM 78-1,17,78
ihl-rat LC50:800 ppm/4H 34ZIAG -,258,69
scu-rat LD50:187 mg/kg GISAAA 48(1),23,83
ihl-mus LC50:836 ppm/4H NTIS** PB214-270
ipr-mus LD50:175 mg/kg GISAAA 48(1),23,83
ivn-mus LD50:290 mg/kg APTOA6 43,69,78
ihl-dog LC50:960 ppm/4H AMIHAB 13,237,56
scu-cat LDLo:100 mg/kg HDWU** -,-,33
ivn-rbt LDLo:175 mg/kg JOHYAY 32,409,32
ihl-gpg LC50:1500 mg/m³/4H 85GMAT -,67,82

CONSENSUS REPORTS: NTP 7th Annual Report On Carcinogens. IARC Cancer Review: Group 2A IMEMDT 7,205,87; Animal Inadequate Evidence IMEMDT 11,157,76; Human Inadequate Evidence IMEMDT 36,189,85; Animal Sufficient Evidence IMEMDT 36,189,85. Community Right-To-Know List. EPA Extremely Hazardous Substances List. Reported in EPA TSCA Inventory. EPA Genetic Toxicology Program.

OSHA PEL: TWA 1 ppm; Cancer Hazard
ACGIH TLV: TWA 1 ppm; Suspected Human Carcinogen.
DFG TRK: 3 ppm; Animal Carcinogen, Suspected Human Carcinogen.
NIOSH REL: (Ethylene Oxide) TWA 0.1 ppm; CL 5 ppm/10M/D
DOT CLASSIFICATION: 2.3; *Label:* Poison Gas, Flammable Gas

SAFETY PROFILE: Confirmed human carcinogen with experimental carcinogenic, tumorigenic, neoplastigenic, and teratogenic data. Poison by ingestion, intraperitoneal, subcutaneous, and intravenous routes. Moderately toxic by inhalation. Human systemic effects by inhalation: convulsions, nausea, vomiting, olfactory and pulmonary changes. Experimental reproductive effects. Mutation data reported. A skin and eye irritant. An irritant to mucous membranes of respiratory tract. High concentrations can cause pulmonary edema.

Highly flammable liquid or gas. Severe explosion hazard when exposed to flame. To fight fire, use alcohol foam, CO_2, dry chemical. Violent polymerization occurs on contact with ammonia; alkali hydroxides; amines; metallic potassium; acids; covalent halides (e.g., aluminum chloride; iron(III) chloride; tin(IV) chloride; aluminum oxide; iron oxide; rust). Explosive reaction with glycerol at 200°. Rapid compression of the vapor with air causes explosions. Incompatible with bases; alcohols; air; m-nitroaniline; trimethyl amine; copper; iron chlorides; iron oxides; magnesium perchlorate; mercaptans; potassium; tin chlorides; contaminants; alkane thiols; bromoethane. When heated to decomposition it emits acrid smoke and irritating fumes.

For occupational chemical analysis use OSHA: #30, SUPERSEDED BY #50 or NIOSH: Ethylene Oxide 1614 or (Portable GC) 3702.

EJO000 CAS:8070-50-6 HR: 3
ETHYLENE OXIDE, mixed with CARBON DIOXIDE
DOT: UN 1041/UN 1952

PROP: Contains less than 10% carbon dioxide (NTIS** PB225–283).

SYNS: ANHYDRIDE CARBONIQUE et OXYDE d'ETHYLENE MELANGES (FRENCH) □ ETHYLENE OXIDE and CARBON DIOXIDE MIXTURES (DOT) □ OXYFUME 20 □ OXYFUME 30

TOXICITY DATA WITH REFERENCE
ihl-mus LC50:836 ppm/4H NTIS** PB214-270
ihl dog LC50:973 ppm/4H NTIS** PB214 270
ihl-gpg LCLo:7000 ppm/2.5H PHRPA6 45,1832,30

DOT CLASSIFICATION: 2.1; *Label:* Flammable Gas (UN 1041)

SAFETY PROFILE: A poison. Mildly toxic by inhalation. Used for the sterilization of vacuum chambers. See also ETHYLENE OXIDE.

EJO500 CAS:289-14-5 HR: 3
ETHYLENE OZONIDE
mf: $C_2H_4O_3$ mw: 76.05

$OCH_2OO\ CH_2$ (ring)

SYN: 1,2,4-TRIOXOLANE

SAFETY PROFILE: Explodes violently on heating, friction, or shock. Explodes at room temperature. Stable at 0°C. May explode when poured. When heated to decomposition it emits acrid smoke and irritating fumes. See also OZONE.

EJP000 CAS:1072-53-3 HR: 2
ETHYLENE SULFATE
mf: $C_2H_4O_4S$ mw: 124.12

PROP: Crystals from $CHCl_3$. Mp: 96.5–97.5°, bp: 110–142° @ 3 mm.

SYNS: 2,2-DIOXIDE-1,3,2-DIOXATHIOLANE □ ETHYLENE GLYCOL, CYCLIC SULFATE □ GLYCOL SULFATE □ SULFURIC ACID, CYCLIC ETHYLENE ESTER

TOXICITY DATA WITH REFERENCE
mmo-sat 1 µmol/plate CBINA8 19,241,77
hma-mus/sat 5 mmol/kg CBINA8 19,241,77
ipr-mus TDLo:128 mg/kg/64W-I:ETA JNCIAM 53,695,74
scu-mus TDLo:840 mg/kg/42W-I:NEO JNCIAM 53,695,74

CONSENSUS REPORTS: EPA Genetic Toxicology Program.

SAFETY PROFILE: Questionable carcinogen with experimental tumorigenic and neoplastigenic data. Mutation data reported. When heated to decomposition it emits toxic fumes of SO_x. See also SULFATES and ESTERS.

EJP500 **CAS:420-12-2** **HR: 3**
ETHYLENE SULFIDE
mf: C_2H_4S mw: 60.12

PROP: Colorless liquid. Bp: 55–56° decomp, d: 1.0368 @ 0°/4°, vap d: 2.07.

SYNS: AETHYLENSULFID (GERMAN) ☐ 2,3-DIHYDROTHIIRENE ☐ ETHYLENE EPISULFIDE ☐ ETHYLENE EPISULPHIDE ☐ ETHYLENE SULPHIDE ☐ THIACYCLOPROPANE ☐ THIIRANE

TOXICITY DATA WITH REFERENCE
eye-rbt 100 mg SEV AJOPAA 29,1363,46
scu-rat TDLo:400 mg/kg/50W-I:ETA ZEKBAI 74,241,70
orl-rat LD50:178 mg/kg AIHAAP 25,560,64
ihl-rat LC50:690 ppm/6H BECTA6 6,509,71
ipr-rat LD50:42 mg/kg AIHAAP 25,560,64
scu-rat LD50:90 mg/kg ZEKBAI 74,241,70

CONSENSUS REPORTS: IARC Cancer Review: Group 3 IMEMDT 7,56,87; Animal Limited Evidence IMEMDT 11,257,76. Reported in EPA TSCA Inventory.

SAFETY PROFILE: Poison by ingestion, intraperitoneal, and subcutaneous routes. Mildly toxic by inhalation. A skin, eye and mucous membrane irritant. Questionable carcinogen with experimental tumorigenic data. Can react with oxidizing materials. When heated to decomposition, or on contact with acid or acid fumes, it emits highly toxic fumes of SO_x. See also SULFIDES.

EJQ000 **CAS:33813-20-6** **HR: 3**
ETHYLENE THIURAM MONOSULFIDE
mf: $C_4H_4N_2S_3$ mw: 176.28

PROP: Yellow crystals or powder from $CHCl_3$. Mp: 121–124°.

SYNS: 5,6-DIHYDRO-3H-IMIDAZO(2,1-c)-1,2,4-DITHIAZOLE-3-THIONE ☐ ENDODAN ☐ ETEM ☐ ETHYLENE BISTHIURAM MONOSULFIDE ☐ ETHYLENETHIOCARBAMYL SULFIDE ☐ ETHYLENE THIURAM MONOSULPHIDE ☐ ETM ☐ ETM (heterocycle) ☐ HORTOCRITT

TOXICITY DATA WITH REFERENCE
mma-sat 7500 ng/plate MUREAV 157,13,85
mrc-bcs 2 μg/disc/24H MUREAV 40,19,76
sce-hmn:lym 20 mg/L MUREAV 157,13,85
cyt-ham-orl 93 mg/kg MUREAV 157,13,85
orl-rat TDLo:304 mg/kg (16W pre):REP JHEMA2 24,295,80
orl-rat LD50:380 mg/kg ATSUDG 4,459,80

CONSENSUS REPORTS: EPA Genetic Toxicology Program.

SAFETY PROFILE: Poison by ingestion. Experimental reproductive effects. Human mutation data reported. Used as a pesticide. When heated to decomposition it emits very toxic fumes of SO_x and NO_x. See also SULFIDES.

EJQ100 **CAS:822-38-8** **HR: 2**
ETHYLENE TRITHIOCARBONATE
mf: $C_3H_4S_3$ mw: 136.25

SYNS: CARBONIC ACID, TRITHIO-, CYCLIC ETHYLENE ESTER ☐ CYCLIC ETHYLENE TRITHIOCARBONATE ☐ 1,3-DITHIOLANE-2-THIONE ☐ TRITHIOCARBONIC ACID, CYCLIC ETHYLENE ESTER

TOXICITY DATA WITH REFERENCE
ipr-mus LD50:500 mg/kg EJMCA5 17,235,82

CONSENSUS REPORTS: Reported in EPA TSCA Inventory.

SAFETY PROFILE: Moderately toxic by intraperitoneal route. When heated to decomposition it emits toxic vapors of SO_x.

EJQ500 **CAS:105-95-3** **HR: 1**
ETHYLENE UNDECANE DICARBOXYLATE
mf: $C_{15}H_{26}O_4$ mw: 270.41

SYNS: ASTRATONE ☐ EMERESSENCE 1150 ☐ ETHYLENE BRASSYLATE ☐ MUSK-T ☐ TRIDECANEDIOIC ACID, CYCLIC ETHYLENE ESTER ☐ 1,1′-UNDECANEDICARBOXYLIC ACID ESTER with ETHYLENE GLYCOL

TOXICITY DATA WITH REFERENCE
skn-rbt 500 mg/24H MOD FCTXAV 13,91,75

CONSENSUS REPORTS: Reported in EPA TSCA Inventory.

SAFETY PROFILE: A skin irritant. Used as a fragrance. When heated to decomposition it emits acrid smoke and irritating fumes. See also ESTERS.

EJR500 **CAS:530-35-8** **HR: 3**
1-N-ETHYLEPHEDRINE HYDROCHLORIDE
mf: $C_{12}H_{19}NO{\cdot}ClH$ mw: 229.78

PROP: Crystals. Mp: 183–184°.

SYNS: α-(1-(ETHYLMETHYLAMINO)ETHYL)BENZYL ALCOHOL HYDROCHLORIDE (−) ☐ MENETYL ☐ 2-METHYLETHYLAMINO-1-PHENYL-1-PROPANOL HYDROCHLORIDE ☐ NETHAMINE HYDROCHLORIDE ☐ NOVEDRIN HYDROCHLORIDE ☐ 1-1-PHENYL-2-METHYLETHYLAMINOPROPAN-1-OL HYDROCHLORIDE

TOXICITY DATA WITH REFERENCE
orl-rbt LDLo:550 mg/kg JPETAB 75,289,42
ivn-rbt LD50:70 mg/kg JAPMA8 39,382,50

SAFETY PROFILE: Poison by intravenous route. Moderately toxic by ingestion. When heated to decomposition it emits very toxic fumes such as Cl^- and NO_x.

EJS000 **CAS:19780-35-9** **HR: 2**
ETHYL-2,3-EPOXYBUTYRATE
mf: $C_6H_{10}O_3$ mw: 130.16

PROP: A liquid. Bp: 106.5° @ 73 mm.

SYN: 2,3-EPOXYBUTYRIC ACID, ETHYL ESTER

TOXICITY DATA WITH REFERENCE
mmo-sat 5 g/L MUREAV 89,269,81
scu-mus TDLo:272 mg/kg/68W-I:ETA JNCIAM 46,143,71
orl-rat LD50:500 mg/kg AIHAAP 24,305,63
skn-rbt LDLo:2830 mg/kg AIHAAP 24,305,63

SAFETY PROFILE: Moderately toxic by ingestion and skin contact. Questionable carcinogen with experimental tumorigenic data. Mutation data reported. When heated to decomposition it emits acrid smoke and irritating fumes. See also ESTERS.

EJS100 CAS:67658-42-8 HR: 3
(8-β)-6-ETHYLERGOLINE-8-ACETAMIDE TARTRATE
mf: $C_{18}H_{23}N_3O•C_4H_6O_6$ mw: 447.54

SYN: ERGOLINE-8-ACETAMIDE, 6-ETHYL-, (8-β)-, (R-(R*,R*))-2,3-DIHYDROXYBUTANEDIOATE (1:1)

TOXICITY DATA WITH REFERENCE
orl-rat TDLo:30 μg/kg (female 5D post):REP CCCCAK 44,3385,79
orl-rat LD50:38,500 μg/kg CCCCAK 44,3385,79

SAFETY PROFILE: Poison by ingestion. Experimental reproductive effects. When heated to decomposition it emits toxic fumes of NO_x.

EJT000 CAS:67466-28-8 HR: 2
ETHYL ESTER of 1,2,5,6-DIBENZANTHRACENE-endo-α,β-SUCCINO GLYCINE
mf: $C_{30}H_{21}NO_4$ mw: 459.52

TOXICITY DATA WITH REFERENCE
scu-mus TDLo:200 mg/kg:ETA CNREA8 1,685,41

SAFETY PROFILE: Questionable carcinogen with experimental tumorigenic data. When heated to decomposition it emits toxic fumes of NO_x. See also ESTERS.

EJT500 HR: 2
ETHYL ESTER of 3-METHYLCHOLANTHRENE-endo-α,β-SUCCINOGLYCINE
mf: $C_{29}H_{23}NO_4$ mw: 449.53

TOXICITY DATA WITH REFERENCE
scu-mus TDLo:200 mg/kg:ETA CNREA8 1,685,41

SAFETY PROFILE: Questionable carcinogen with experimental tumorigenic data. When heated to decomposition it emits toxic fumes of NO_x. See also ESTERS.

EJT575 HR: 2
11-β-ETHYLESTRADIOL
mf: $C_{20}H_{28}O_2$ mw: 300.48

SYNS: ESTRA-1,3,5(10)-TRIENE-3,17-β-DIOL, 11-β-ETHYL- □ 11-β-ETHYLESTRA-1,3,5(10)-TRIENE-3,17-β-DIOL

TOXICITY DATA WITH REFERENCE
otr-mus:fbr 30 μmol/L CNREA8 47,2583,87
imp-ham TDLo:400 mg/kg/30W C:NEO CNREA8 47,2583,87

SAFETY PROFILE: Questionable carcinogen with experimental neoplastigenic data. Mutation data reported. When heated to decomposition it emits acrid smoke and irritating fumes.

EJU000 CAS:60-29-7 HR: 3
ETHYL ETHER
DOT: UN 1155
mf: $C_4H_{10}O$ mw: 74.14

$CH_3CH_2OCH_2CH_3$

PROP: A clear, volatile liquid; sweet, pungent odor. Mp: −116.2°, bp: 34.6°, ULC: 100, lel: 1.85%, uel: 36%, flash p: −49°F, d: 0.7135 @ 20°/4°, autoign temp: 320°F, vap press: 442 mm @ 20°, vap d: 2.56. Sol in H_2SO_4; sltly sol in H_2O; misc in most org solvs.

SYNS: AETHER □ ANAESTHETIC ETHER □ ANESTHESIA ETHER □ ANESTHETIC ETHER □ DIAETHYLAETHER (GERMAN) □ DIETHYL ETHER (DOT) □ DIETHYL OXIDE □ DWUETYLOWY ETER (POLISH) □ ETERE ETILICO (ITALIAN) □ ETHER □ ETHER ETHYLIQUE (FRENCH) □ ETHOXYETHANE □ 1,1′-OXYBISETHANE □ OXYDE d'ETHYLE (FRENCH) □ RCRA WASTE NUMBER U117 □ SOLVENT ETHER

TOXICITY DATA WITH REFERENCE
eye-hmn 100 ppm JIHTAB 25,282,43
skn-rbt 360 mg open MLD UCDS** 4/5/73
eye-rbt 100 mg MOD FEPRA7 35,729,76
skn-gpg 50 mg/24H SEV HIFUAG 22,373,80
dnr-esc 50 μL/well/16H CBINA8 15,219,76
dyt-smc 100 mmol/tube HEREAY 33,457,47
oms-ham:fbr 1 pph ANESAV 43,21,75
orl-man LDLo:260 mg/kg 85DCAI 2,73,70
orl-hmn LDLo:420 mg/kg 32ZWAA 8,275,74
ihl-hmn TCLo:200 ppm:NOSE JIHTAB 25,282,43
orl-rat LD50:1215 mg/kg TXAPA9 19,699,71
ihl-rat LC50:73,000 ppm/2H TXAPA9 17,275,70
ihl-mus LC50:6500 ppm/99M TXAPA9 17,275,70
ipr-mus LD50:2420 mg/kg PWPSA8 27,511,84
scu-mus LDLo:8 mg/kg HDAMAK 1,1295,35
ivn-mus LD50:996 mg/kg JPMSAE 67,566,78
ihl-dog LCLo:76,000 ppm HBAMAK 4,1294,35
ihl-rbt LCLo:10,6000 ppm HBAMAK 4,1294,35
ipr-gpg LDLo:2000 mg/kg AIHAAP 35,21,74
scu-frg LDLo:24 g/kg HBAMAK 4,1295,35

CONSENSUS REPORTS: IARC Cancer Review: Animal No Adequate Data IMEMDT 7,93,87. Reported in EPA TSCA Inventory. EPA Genetic Toxicology Program.

OSHA PEL: TWA 400 ppm; STEL 500 ppm
ACGIH TLV: TWA 400 ppm; STEL 500 ppm
DFG MAK: 400 ppm (1200 mg/m³)
DOT CLASSIFICATION: 3; *Label:* Flammable Liquid

SAFETY PROFILE: Moderately toxic to humans by ingestion. Poison experimentally by subcutaneous route. Moderately toxic by intraperitoneal and intravenous routes. Mildly toxic by inhalation. Human systemic effects by inhalation: olfactory changes. Mutation data reported. A severe eye and moderate skin irritant. Ethyl ether is not corrosive or dangerously reactive. It must not be considered safe for individuals to inhale or ingest. It is a depressant of the central nervous system and is capable of producing intoxication, drowsiness, stupor, and unconsciousness. Death due to respiratory failure may result from severe and continued exposure.

A very dangerous fire and explosion hazard when exposed to heat or flame. A storage hazard. It autooxidizes to form explosive polymeric 1-oxy-peroxides. Explosive reaction with boron triazide, bromine trifluoride, bromine pentafluoride, perchloric acid, uranyl nitrate + light, wood pulp extracts + heat. Violent reaction or ignition on contact with halogens (e.g., bromine, chlorine), interhalogens (e.g., iodine heptafluoride), oxidants (e.g., silver perchlorate, nitrosyl perchlorate, nitryl perchlorate, chromyl chloride, fluorine nitrate, permanganic acid, nitric acid, hydrogen peroxide, peroxodisulfuric acid, iodine(VII) oxide, sodium peroxide, ozone, and liquid air), sulfur and sulfur compounds (e.g., sulfur when dried with peroxidized ether, sulfuryl chloride). Can react vigorously with acetyl peroxide, air, bromoazide, ClF_3, CrO_3, $Cr(OCl)_2$, $LiAlH_2$, $NOClO_4$, O_2, $NClO_2$, (H_2SO_4 + permanganates), K_2O_2, [$(C_2H_5)_3Al$ + air], [$(CH_3)_3Al$ + air]. To fight fire, use alcohol foam, CO_2, dry chemical. Used in production of drugs of abuse. When heated to decomposition it emits acrid smoke and irritating fumes. See also ETHERS.

For occupational chemical analysis use NIOSH: Ethyl Ether, 1610.

EJV000 CAS:52125-53-8 HR: 2
ETHYL ETHER of PROPYLENE GLYCOL
mf: $C_5H_{12}O_2$ mw: 104.17

PROP: Vap d: 3.59.

SYNS: ETHOXYPROPANOL □ PROPANOL, ETHOXY-(9CI) □ PROPYLENE GLYCOL MONOETHYL ETHER

TOXICITY DATA WITH REFERENCE
eye-rbt 20 mg AJOPAA 29,1363,46

CONSENSUS REPORTS: Glycol ether compounds are on the Community Right-To-Know List.

SAFETY PROFILE: An eye irritant. Combustible when exposed to heat or flame; can react with oxidizers. When heated to decomposition it emits acrid smoke and irritating fumes. See also GLYCOL ETHERS.

EJV400 HR: 2
11-β-ETHYL-17-α-ETHINYLESTRADIOL
mf: $C_{22}H_{28}O_2$ mw: 324.50

SYN: 11-β-ETHYL-19-NOR-17-α-PREGNA-1,3,5(10)-TRIEN-20-YNE-3,17-DIOL

TOXICITY DATA WITH REFERENCE
imp-ham TDLo:400 mg/kg/39W C:CAR CNREA8 47,2583,87

SAFETY PROFILE: Questionable carcinogen with experimental carcinogenic data. When heated to decomposition it emits acrid smoke and irritating fumes.

EJV500 CAS:763-69-9 HR: 1
ETHYL-β-ETHOXYPROPIONATE
mf: $C_7H_{14}O_3$ mw: 146.21

PROP: Liquid. Mp: −100°, bp: 170.1°, flash p: 180°F (OC), d: 0.9496 @ 20°/20°, vap d: 5.03.

SYNS: ETHOXYPROPIONIC ACID, ETHYL ESTER □ 3-ETHOXYPROPIONIC ACID, ETHYL ESTER

TOXICITY DATA WITH REFERENCE
skn-rbt 10 mg/24H open MLD AMIHBC 4,119,51
eye-rbt 500 mg open AMIHBC 4,119,51
orl-rat LD50:5000 mg/kg AMIHBC 4,119,51
skn-rbt LD50:10 g/kg AMIHBC 4,119,51

CONSENSUS REPORTS: Reported in EPA TSCA Inventory.

SAFETY PROFILE: Mildly toxic by ingestion and skin contact. A skin and eye irritant. Flammable when exposed to heat or flame; can react with oxidizing materials. To fight fire, use foam, CO_2, dry chemical. When heated to decomposition it emits acrid smoke and irritating fumes. See also ESTERS.

EJW500 CAS:623-78-9 HR: 2
ETHYL-N-ETHYL CARBAMATE
mf: $C_5H_{11}NO_2$ mw: 117.17

PROP: A liquid. Bp: 170°.

SYN: ETHYLCARBAMIC ACID, ETHYL ESTER

TOXICITY DATA WITH REFERENCE
ipr-mus TDLo:6500 mg/kg/13W-I:ETA JNCIAM 9,35,48
scu-mus LD50:860 mg/kg AJEBAK 45,507,67

CONSENSUS REPORTS: Reported in EPA TSCA Inventory.

SAFETY PROFILE: Moderately toxic by subcutaneous routes. Questionable carcinogen with experimental tumorigenic data. When heated to decomposition it emits toxic fumes of NO_x. See also CARBAMATES and ESTERS.

EJY000 CAS:17013-37-5 HR: 3
5-ETHYL-5-(1-ETHYLPROPYL)BARBITURIC ACID
mf: $C_{11}H_{18}N_2O_3$ mw: 240.32

SYNS: 5-ETHYL-5-(1-ETHYLPROPYL)2,4,6(1H,3H,5H)-PYRIMIDINETRIONE □ ISOMEBUMAL

TOXICITY DATA WITH REFERENCE
ipr-rat LDLo:110 mg/kg JPHAA3 26,317,37
ipr-rbt LDLo:75 mg/kg JACSAT 56,1139,34

SAFETY PROFILE: Poison by intraperitoneal route. When heated to decomposition it emits toxic fumes of NO_x. See also BARBITURATES.

EJZ000 CAS:102584-01-0 HR: 3
1-ETHYL-4-(p-(p-((p-((1-ETHYLPYRIDINIUM-4-YL)AMINO)-2-AMINOPHENYL)CARBAMOYL)CINNAMAMIDO)ANILINO)PYRIDINIUM, DIBROMIDE
mf: $C_{36}H_{37}N_7O_2$•Br mw: 759.62

TOXICITY DATA WITH REFERENCE
dnd-mus:lym 1500 nmol/L JMCMAR 22,134,79
ipr-mus LD10:7500 μg/kg JMCMAR 22,134,79

SAFETY PROFILE: Poison by intraperitoneal route. Mutation data reported. When heated to decomposition it emits very toxic fumes of Br^- and NO_x. See also BROMIDES.

EKA000 CAS:68772-29-2 HR: 3
1-ETHYL-4-(p-(p-((1-ETHYLPYRIDINIUM-4-YL)AMINO) BENZAMIDO)ANILINO)QUINOLINIUM DIBROMIDE
mf: $C_{31}H_{31}N_5O \cdot 2Br$ mw: 649.49

TOXICITY DATA WITH REFERENCE
dnd-mus:lym 1600 nmol/L JMCMAR 22,134,79
ipr-mus LD10:10 mg/kg JMCMAR 22,134,79

SAFETY PROFILE: Poison by intraperitoneal route. Mutation data reported. When heated to decomposition it emits very toxic fumes of NO_x and Br^-. See also BROMIDES.

EKA500 CAS:68772-10-1 HR: 3
1-ETHYL-4-(p-((p-((1-ETHYLPYRIDINIUM-4-YL)AMINO)PHENYL)CARBAMOYL)ANILINO)QUINOLINIUM, DIBROMIDE
mf: $C_{31}H_{31}N_5O \cdot 2Br$ mw: 649.49

TOXICITY DATA WITH REFERENCE
dnd-mus:lym 2 µmol/L JMCMAR 22,134,79
ipr-mus LD10:15 mg/kg JMCMAR 22,134,79

SAFETY PROFILE: Poison by intraperitoneal route. Mutation data reported. When heated to decomposition it emits very toxic fumes of NO_x and Br^-. See also BROMIDES.

EKB000 CAS:20719-23-7 HR: 3
1-ETHYL-4-(p-(p-((p-((1-ETHYLPYRIDINIUM-4-YL)AMINO)PHENYL)CARBAMOYL)CINNAMAMIDO)ANILINO)PYRIDINIUM, DI-p-TOLUENE SULFONATE
mf: $C_{36}H_{36}N_6O_2 \cdot 2C_7H_7O_3S$ mw: 927.18

TOXICITY DATA WITH REFERENCE
dnd-mus:lym 430 nmol/L JMCMAR 22,134,79
ipr-mus LD10:8 mg/kg JMCMAR 22,134,79

SAFETY PROFILE: Poison by intraperitoneal route. Mutation data reported. When heated to decomposition it emits very toxic fumes of SO_x and NO_x. See also SULFONATES.

EKC500 CAS:68772-21-4 HR: 3
1-ETHYL-4-(p-((p-(1-ETHYLPYRIDINIUM-4-YL)PHENYL)CARBAMOYL)ANILINO)QUINOLINIUM), DI-p-TOLUENE SULFONATE
mf: $C_{31}H_{30}N_4O \cdot 2C_7H_7O_3S$ mw: 817.05

TOXICITY DATA WITH REFERENCE
dnd-mus:lym 890 nmol/L JMCMAR 22,134,79
ipr-mus LD10:10 mg/kg JMCMAR 22,134,79

SAFETY PROFILE: Poison by intraperitoneal route. Mutation data reported. When heated to decomposition it emits very toxic fumes of NO_x and SO_x. See also SULFONATES.

EKD000 CAS:18355-54-9 HR: 3
1-ETHYL-7-((p-(p-((1-ETHYLQUINOLINIUM-7-YL)CARBAMOYL)BENZAMIDO)BENZAMIDO)QUINOLINIUM), DI-p-TOLUENE SULFONATE
mf: $C_{37}H_{33}N_5O_3 \cdot 2C_7H_7O_3S$ mw: 938.15

TOXICITY DATA WITH REFERENCE
dnd-mus:lym 160 nmol/L JMCMAR 22,134,79
ipr-mus LD10:30 mg/kg JMCMAR 22,134,79

SAFETY PROFILE: Poison by intraperitoneal route. Mutation data reported. When heated to decomposition it emits very toxic fumes of NO_x and SO_x. See also SULFONATES.

EKD500 CAS:16758-28-4 HR: 3
1-ETHYL-6-(p-(p-((1-ETHYLQUINOLINIUM-6-YL)CARBAMOYL)BENZAMIDO)BENZAMIDO)QUINOLINIUM, DI-p-TOLUENE SULFONATE
mf: $C_{37}H_{33}N_5O_3 \cdot 2C_7H_7O_3S$ mw: 938.15

TOXICITY DATA WITH REFERENCE
dnd-mus:lym 370 nmol/L JMCMAR 22,134,79
ipr-mus LD10:45 mg/kg JMCMAR 22,134,79

SAFETY PROFILE: Poison by intraperitoneal route. Mutation data reported. When heated to decomposition it emits very toxic fumes of SO_x and NO_x. See also SULFONATES.

EKF600 CAS:29177-84-2 HR: 2
ETHYL FLUCLOZEPATE
mf: $C_{18}H_{14}ClFN_2O_3$ mw: 360.79

PROP: Crystals from ether. Mp: 193–194° (decomp).

SYNS: CM 6912 ☐ ETHYL-7-CHLORO-5-(o-FLUOROPHENYL)-2,3-DIHYDRO-2-OXO-1H-1,4-BENZODIAZEPINE-3-CARBOXYLATE ☐ ETHYL LOFLAZEPATE ☐ VICTAN

TOXICITY DATA WITH REFERENCE
orl-rat TDLo:2640 mg/kg (17-22D preg/21D post):REP KSRNAM 20,1683,86
orl-rbt TDLo:1300 mg/kg (female 6-18D post):TER KSRNAM 20,1655,86
orl-rbt TDLo:1300 mg/kg (female 6-18D post):REP KSRNAM 20,1655,86
orl-rat TDLo:146 g/kg (female 26W pre):REP KSRNAM 20,1463,86
orl-rat TDLo:24 g/kg (male 30D pre):REP KSRNAM 20,1418,86
orl-rat TDLo:2200 mg/kg (7-17D preg):TER KSRNAM 20,1655,86
ipr-rat LD50:645 mg/kg IJCPB5 19,453,81
scu-rat LD50:2 g/kg KSRNAM 20,1411,86
orl-mus LD50:5506 mg/kg KSRNAM 20,1411,86
ipr-mus LD50:712 mg/kg IJCPB5 19,453,81
scu-mus LD50:1795 mg/kg KSRNAM 20,1411,86

SAFETY PROFILE: Moderately toxic by intraperitoneal and subcutaneous routes. Mildly toxic by ingestion. Experimental teratogenic and reproductive effects. Note: This is a controlled substance (depressant) listed in the U.S. Code of Federal Regulations, Title 21 Part 1308.14 (1985). When heated to decomposition it emits toxic fumes of F^-, Cl^-, and NO_x.

E

EKG500 **CAS:459-72-3** **HR: 3**
ETHYL FLUOROACETATE
mf: $C_4H_7FO_2$ mw: 106.11

SYN: ETHYLESTER KYSELINY FLUOROCTOVE

TOXICITY DATA WITH REFERENCE
ipr-mus LD50:19 mg/kg 11FYAN 3,59,63

CONSENSUS REPORTS: Reported in EPA TSCA Inventory.

SAFETY PROFILE: Poison by intraperitoneal route. When heated to decomposition it emits toxic fumes of F^-. See also FLUORIDES.

EKI000 **CAS:353-03-7** **HR: 3**
ETHYL-10-FLUORODECANOATE
mf: $C_{12}H_{23}FO_2$ mw: 218.35

SYNS: ETHYL-ω-FLUORODECANOATE ☐ ETHYL-9-FLUORONONANECARBOXYLATE

TOXICITY DATA WITH REFERENCE
ipr-mus LD50:1650 μg/kg JOCEAH 21,883,56
par-mus LD50:10 mg/kg JCSOA9 -,1471,49
par-rbt LD50:200 μg/kg JCSOA9 -,1471,49

SAFETY PROFILE: Poison by intraperitoneal and parenteral routes. When heated to decomposition it emits toxic fumes of F^-. See also FLUORIDES.

EKK500 **CAS:332-97-8** **HR: 3**
ETHYL-8-FLUORO OCTANOATE
mf: $C_{10}H_{19}FO_2$ mw: 190.29

SYNS: ETHYL-ω-FLUOROOCTANOATE ☐ 8-FLUOROOCTANOIC ACID, ETHYL ESTER

TOXICITY DATA WITH REFERENCE
ipr-mus LD50:1750 μg/kg JOCEAH 21,883,56
scu-mus LD50:9 mg/kg NATUAS 172,1139,53
par-mus LD50:9 mg/kg JCSOA9 -,1471,49

SAFETY PROFILE: Poison by intraperitoneal, subcutaneous, and parenteral routes. When heated to decomposition it emits toxic fumes of F^-. See also ESTERS and FLUORIDES.

EKK550 **CAS:371-69-7** **HR: 2**
ETHYL FLUOROSULFATE
mf: $C_2H_5FO_3S$ mw: 128.12

PROP: Bp: 112–113°.

SAFETY PROFILE: Explodes violently at room temperature. When heated to decomposition it emits toxic fumes of F^- and SO_x. See also FLUORIDES and SULFATES.

EKK600 **CAS:627-45-2** **HR: 2**
N-ETHYLFORMAMIDE
mf: C_3H_7NO mw: 73.11

SYNS: N-AETHYLFORMAMID ☐ ETHYLFORMAMIDE ☐ FORMAMIDE, N-ETHYL- ☐ N-FORMYLETHYLAMINE

TOXICITY DATA WITH REFERENCE
unr-rat LD50:3000 mg/kg ARZNAD 18,645,68
ipr-mus LD:>1200 mg/kg THERAP 27,873,72

CONSENSUS REPORTS: Reported in EPA TSCA Inventory.

SAFETY PROFILE: Moderately toxic by intraperitoneal and possibly other routes. When heated to decomposition it emits toxic vapors of NO_x.

EKL000 **CAS:109-94-4** **HR: 3**
ETHYL FORMATE
DOT: UN 1190
mf: $C_3H_6O_2$ mw: 74.09

PROP: Colorless, mobile flammable liquid; sharp, pleasant, rum-like odor. Mp: −79°, bp: 54.3°, lel: 2.7%, uel: 13.5%, flash p: −4°F (CC), d: 0.9236 @ 20°/20°, refr index: 1.359, autoign temp: 851°F, vap press: 100 mm @ 5.4°, vap d: 2.55. Misc in EtOH, Et_2O, C_6H_6; sltly sol in and gradually hyd by H_2O.

SYNS: AETHYLFORMIAT (GERMAN) ☐ AREGINAL ☐ ETHYLE (FORMIATE d') (FRENCH) ☐ ETHYLFORMIAAT (DUTCH) ☐ ETHYL FORMIC ESTER ☐ ETHYL METHANOATE ☐ ETILE (FORMIATO di) (ITALIAN) ☐ FEMA No. 2434 ☐ FORMIC ACID, ETHYL ESTER ☐ FORMIC ETHER ☐ MROWCZAN ETYLU (POLISH)

TOXICITY DATA WITH REFERENCE
skn-rbt 460 mg open MLD UCDS** 4/10/68
eye-rbt 100 mg/24H MOD 85JCAE-,350,86
skn-mus TDLo:110 g/kg/9W-I:ETA BJCAAI 9,177,55
orl-rat LD50:1850 mg/kg FCTXAV 2,327,64
ihl-rat LCLo:8000 ppm/4H AMIHBC 10,61,54
orl-rbt LD50:2075 mg/kg IMSUAI 41,31,72
skn-rbt LD50:20 g/kg FCTXAV 16,637,78
scu-rbt LDLo:1000 mg/kg FCTXAV 16,737,78
orl-gpg LD50:1110 mg/kg FCTXAV 2,327,64

CONSENSUS REPORTS: Reported in EPA TSCA Inventory.

OSHA PEL: TWA 100 ppm
ACGIH TLV: TWA 100 ppm
DFG MAK: 100 ppm (300 mg/m³)
DOT CLASSIFICATION: 3; *Label:* Flammable Liquid

SAFETY PROFILE: Moderately toxic by ingestion and subcutaneous routes. Mildly toxic by skin contact and inhalation. A powerful inhalation irritant in humans. A skin and eye irritant. Questionable carcinogen with experimental tumorigenic data. Highly flammable liquid. A very dangerous fire and explosion hazard when exposed to heat, flame, or oxidizers. To fight fire, use alcohol foam, spray, mist, dry chemical. When heated to decomposition it emits acrid smoke and irritating fumes. See also ESTERS.

For occupational chemical analysis use NIOSH: Ethyl Formate S36.

EKL250 **CAS:74920-78-8** **HR: 2**
1-ETHYL-1-FORMYLHYDRAZINE
mf: $C_3H_8N_2O$ mw: 88.13

SYNS: EFH ☐ N-ETHYL-N-FORMYLHYDRAZINE ☐ FORMIC ACID, 1-ETHYLHYDRAZIDE

TOXICITY DATA WITH REFERENCE
orl-mus TDLo:18,345 mg/kg/52W C:CAR CRNGDP 1,61,80

SAFETY PROFILE: Questionable carcinogen with experimental carcinogenic data. When heated to decomposition it emits toxic fumes of NO_x.

EKL500 **CAS:2407-43-4** **HR: 2**
5-ETHYL-2(5H)-FURANONE
mf: $C_6H_8O_2$ mw: 112.14

SYN: 4-HYDROXYHEX-2-ENOIC ACID LACTONE

TOXICITY DATA WITH REFERENCE
scu-rat TDLo:2560 mg/kg/64W-I:ETA BJCAAI 15,85,61

SAFETY PROFILE: Questionable carcinogen with experimental tumorigenic data. When heated to decomposition it emits acrid smoke and irritating fumes.

EKM000 **CAS:614-99-3** **HR: 3**
ETHYL FUROATE
mf: $C_7H_8O_3$ mw: 140.15

PROP: (A) Leaflets. D: 1.117, mp: 34°, bp: 195°, Insol in water; misc in alc and ether. (B) Liquid. D: 1.38, bp: 65–67°.

TOXICITY DATA WITH REFERENCE
ipr-rat LDLo:75 mg/kg JPETAB 58,174,36
ivn-mus LD50:180 mg/kg CSLNX* NX#00372

CONSENSUS REPORTS: Reported in EPA TSCA Inventory.

SAFETY PROFILE: Poison by intraperitoneal and intravenous routes. When heated to decomposition it emits acrid smoke and irritating fumes.

EKM100 **CAS:831-61-8** **HR: 1**
ETHYL GALLATE
mf: $C_9H_{10}O_5$ mw: 198.19

SYNS: BENZOIC ACID, 3,4,5-TRIHYDROXY-, ETHYL ESTER (9CI) ☐ ETHYLESTER KYSELINY GALLOVE ☐ ETHYL 3,4,5-TRIHYDROXYBENZOATE ☐ GALLIC ACID, ETHYL ESTER ☐ NIPAGALLIN A ☐ NIPA NO. 48 ☐ PHYLLEMBLIN ☐ PROGALLIN A

TOXICITY DATA WITH REFERENCE
orl-mus LD50:5810 mg/kg 85JCAE -,668,86

CONSENSUS REPORTS: Reported in EPA TSCA Inventory.

SAFETY PROFILE: Mildly toxic by ingestion. When heated to decomposition it emits acrid smoke and irritating vapors.

EKM200 **CAS:4016-11-9** **HR: 3**
ETHYL GLYCIDYL ETHER
DOT: UN 2752
mf: $C_5H_{10}O_2$ mw: 102.15

SYNS: 1,2-EPOXY-3-ETHOXYPROPANE ☐ 1,2-EPOXY-3-ETHOXYPROPANE (DOT) ☐ (ETHOXYMETHYL)OXIRANE ☐ PROPANE, 1,2-EPOXY-3-ETHOXY- ☐ OXIRANE, (ETHOXYMETHYL)-(9CI)

TOXICITY DATA WITH REFERENCE
mmo-sat 8 mmol/L CBINA8 45,153,83
mmo-klp 500 μmol/L MUREAV 89,269,81

CONSENSUS REPORTS: Reported in EPA TSCA Inventory.

DOT CLASSIFICATION: 3; *Label:* Flammable Liquid

SAFETY PROFILE: Mutation data reported. A flammable liquid. When heated to decomposition it emits acrid smoke and irritating vapors.

EKM500 **CAS:623-50-7** **HR: 2**
ETHYL GLYCOLATE
mf: $C_4H_8O_3$ mw: 104.12

PROP: Colorless liquid. D: 1.087, bp: 160°, very sol in alc and ether.

SYN: HYDROXYACETIC ACID ETHYL ESTER

TOXICITY DATA WITH REFERENCE
eye-gpg 10 μg MLD JPPMAB 11,150,59
ipr rat LDLo:1500 mg/kg JPPMAB 11,150,59

CONSENSUS REPORTS: Reported in EPA TSCA Inventory.

SAFETY PROFILE: Moderately toxic by intraperitoneal route. An eye irritant. When heated to decomposition it emits acrid smoke and irritating fumes.

EKN000 **CAS:2642-71-9** **HR: 3**
ETHYL GUTHION
mf: $C_{12}H_{16}N_3O_3PS_2$ mw: 345.40

PROP: Needles. D 1.284 @ 20°/4°, mp: 53°, bp: 111° @ 0.001 mm.

SYNS: ATHYL-GUSATHION ☐ AZINFOS-ETHYL (DUTCH) ☐ AZINOS ☐ AZINPHOS-AETHYL (GERMAN) ☐ AZINPHOS ETHYL ☐ AZINPHOS-ETILE (ITALIAN) ☐ BAY 16225 ☐ BAYER 16259 ☐ BENZOTRIAZINE derivative of an ETHYL DITHIOPHOSPHATE ☐ COTNION-ETHYL ☐ CYRSTHION ☐ O,O-DIAETHYL-S-(4-OXOBENZOTRIAZIN-3-METHYL)-DITHIOPHOSPHAT (GERMAN) ☐ O,O-DIAETHYL-S-((4-OXO-3H-1,2,3-BENZOTRIAZIN-3-YL)-METHYL)-DITHIOPHOSPHAT (GERMAN) ☐ O,O-DIETHYL-S-(4-OXO-3H-1,2,3-BENZOTRIAZINE-3-YL)-METHYL-DITHIOPHOSPHATE ☐ O,O-DIETHYL-S-(4-OXOBENZOTRIAZINO-3-METHYL)PHOSPHORODITHIOATE ☐ O,O-DIETHYL PHOSPHORODITHIOATE S-ester with 3-(MERCAPTOMETHYL)-1,2,3-BENZOTRIAZIN-4(3H)-ONE ☐ O,O-DIETHYL-S-((4-OXO-3H-1,2,3-BENZOTRIAZIN-3-YL)-METHYL)-DITHIOFOSFAAT (DUTCH) ☐ O,O-DIETIL-S-((4-OXO-3H-1,2,3-BENZOTRIAZIN-3-IL)-METIL)-DITIOFOSFATO (ITALIAN) ☐ 3,4-DIHYDRO-4-OXO-3-BENZOTRIAZINYLMETHYL O,O-DIETHYL PHOSPHORODITHIOATE ☐ S-(3,4-DIHYDRO-4-OXO-1,2,3-BENZOTRIAZIN-3-YLMETHYL) O,O-DIETHYL PHOSPHORODITHIOATE ☐ ENT 22,014 ☐ ETHYL GUSATHION ☐ GUSATHION A ☐ GUTHION (ETHYL) ☐ R 1513 ☐ TRIAZOTION (RUSSIAN)

E

TOXICITY DATA WITH REFERENCE
orl-rat LD50:7 mg/kg ARSIM* 20,2,66
ihl-rat LC50:390 mg/m^3 85GYAZ -,16,71
skn-rat LD50:250 mg/kg GUCHAZ 6,24,73
ipr-rat LD50:7500 μg/kg GUCHAZ 6,24,73
orl-ckn LD50:34 mg/kg TXAPA9 11,49,67

CONSENSUS REPORTS: EPA Extremely Hazardous Substances List.

SAFETY PROFILE: Poison by ingestion, inhalation, skin contact, and intraperitoneal route. A cholinesterase inhibitor type of insecticide. When heated to decomposition it emits toxic fumes of SO_x, PO_x, and NO_x. See also PARATHION.

EKN050 CAS:106-30-9 HR: 3
ETHYL HEPTANOATE
mf: $C_9H_{18}O_2$ mw: 158.24

PROP: Colorless liquid; wine-brandy odor. D: 0.867–0.872, refr index: 1.411, fp: −66.1°, bp: 188.6°, flash p: 149°F. Misc in alc, chloroform, fixed oils; sltly sol in propylene glycol.

SYNS: COGNAC OIL □ ENANTHYLIC ETHER □ ETHYL ENANTHATE □ ETHYL HEPTANOATE □ ETHYL n-HEPTANOATE □ ETHYL HEPTOATE □ ETHYL HEPTYLATE □ ETHYL OENANTHATE □ ETHYL OENANTHYLATE □ FEMA No. 2437 □ OENANTHIC ETHER

TOXICITY DATA WITH REFERENCE
orl-rat LD50:>34,640 mg/kg FCTXAV 2,327,64
skn-rbt LD50:>5 g/kg FCTXAV 19,247,81

CONSENSUS REPORTS: Reported in EPA TSCA Inventory.

SAFETY PROFILE: Low toxicity by ingestion and skin contact. Flammable liquid when exposed to heat, sparks, or flame. When heated to decomposition it emits acrid smoke and irritating fumes.

EKN100 CAS:23489-00-1 HR: 3
1-ETHYL-1-HEPTYLPIPERIDINIUM BROMIDE
mf: $C_{14}H_{30}N\cdot Br$ mw: 292.36

TOXICITY DATA WITH REFERENCE
orl-mus LD50:288 mg/kg PSDTAP 15,331,74
ipr-mus LD50:67,934 μg/kg PSDTAP 15,331,74
ivn-mus LD50:4541 μg/kg PSDTAP 15,331,74

SAFETY PROFILE: Poison by ingestion, intravenous, and intraperitoneal routes. When heated to decomposition it emits toxic fumes of NO_x and Br^-. See also BROMIDES.

EKN500 CAS:124-03-8 HR: 3
ETHYL HEXADECYL DIMETHYL AMMONIUM BROMIDE
mf: $C_{20}H_{44}N\cdot Br$ mw: 378.56

PROP: Crystals. Sol in H_2O.

SYNS: AMMONYX DME □ BRETOL □ CDA: CETYLCIDE □ CETYL DIMETHYL ETHYL AMMONIUM BROMIDE □ CETYL ETHYL DIMETHYLAMMONIUM BROMIDE □ DIMETHYL ETHYL HEXADECYL AMMONIUM BROMIDE □ ETHYL CETAB □ RADIOL GERMICIDAL SOLUTION

TOXICITY DATA WITH REFERENCE
orl-rat LD50:500 mg/kg PCOC** -,206,66
orl-gpg LD50:158 mg/kg PCOC** -,206,66

SAFETY PROFILE: Poison by ingestion. When heated to decomposition it emits toxic fumes of NO_x, NH_3, and Br^-. See also BROMIDES.

EKN600 CAS:56501-33-8 HR: 3
1-ETHYL-1-HEXADECYLPIPERIDINIUM BROMIDE
mf: $C_{23}H_{48}N\cdot Br$ mw: 418.63

TOXICITY DATA WITH REFERENCE
orl-mus LD50:235 mg/kg PSDTAP 15,331,74
ipr-mus LD50:3993 μg/kg PSDTAP 15,331,74
ivn-mus LD50:2296 μg/kg PSDTAP 15,331,74

SAFETY PROFILE: Poison by ingestion, intravenous, and intraperitoneal routes. When heated to decomposition it emits toxic fumes of NO_x and Br^-. See also BROMIDES.

EKO000 CAS:63867-09-4 HR: 3
ETHYL HEXAFLUORO-2-BROMOBUTYRATE
mf: $C_5H_3BrF_6O_2$ mw: 288.99

SYN: 2-BROMO-2,3,3,4,4,4-HEXAFLUOROBUTYRIC ACID METHYL ESTER

TOXICITY DATA WITH REFERENCE
orl-mus LD50:980 mg/kg TXAPA9 12,486,68
ipr-mus LD50:33 mg/kg TXAPA9 14,114,69
orl-mus LD50:980 mg/kg TXAPA9 14,114,69

SAFETY PROFILE: Poison by intraperitoneal route. Moderately toxic by ingestion. When heated to decomposition it emits very toxic fumes of Br^- and F^-.

EKO500 CAS:2212-67-1 HR: 3
sec-ETHYL HEXAHYDRO-1H-AZEPINE-1-CARBOTHIOATE
mf: $C_9H_{17}NOS$ mw: 187.33

PROP: A liquid. D: 1.065 @ 20°/20°, bp: 202° @ 10 mm. Sol in Me_2CO, C_6H_6, and MeOH; sltly sol in H_2O.

SYNS: S-AETHYL-N-HEXAHYDRO-1H-AZEPINTHIOLCARBAMAT (GERMAN) □ ETHYL-1-HEXAMETHYLENEIMINECARBOTHIOLATE □ S-ETHYL-1-HEXAMETHYLENEIMINOTHIOCARBAMATE □ S-ETHYL-N-HEXAMETHYLENETHIOCARBAMATE □ FELAN □ HYDRAM □ JALAN □ MOLINATE □ MOLMATE □ ORDRAM □ R-4572 □ STAUFFER R-4,572 □ YALAN □ YULAN

TOXICITY DATA WITH REFERENCE
cyt-hmn:lym 20 μg/L CYGEDX 11(4),62,77
ihl-rat TCLo:600 μg/m^3/6H (male 20D pre):REP TOXID9 4,80,84
ihl-rat TClo:640 μg/m^3/6H (20D male):REP TOXID9 4,80,84
orl-rat LD50:369 mg/kg PEMNDP 8,578,87
ihl-rat LC50:2100 mg/m^3/1H 85JFAN A282,83
scu-rat LD50:1167 mg/kg VRDEA5 (1),119,69
orl-mus LD50:530 mg/kg VRDEA5 (1),119,69
ihl-cat LCLo:200 mg/m^3 GISAAA 35(8),35,70

skn-rbt LD50:3536 mg/kg FMCHA2 -,C173,83

SAFETY PROFILE: Poison by inhalation. Moderately toxic by ingestion, skin contact, and subcutaneous routes. Experimental reproductive effects. Human mutation data reported. An herbicide. When heated to decomposition it emits very toxic fumes of NO_x and SO_x.

EKO600 CAS:760-67-8 HR: 2
2-ETHYLHEXANOIC ACID CHLORIDE
mf: $C_8H_{15}ClO$ mw: 162.68

SYNS: 2-ETHYLCAPROYL CHLORIDE □ 2-ETHYLHEXANOYL CHLORIDE □ HEXANOYL CHLORIDE, 2-ETHYL-

TOXICITY DATA WITH REFERENCE
ihl-rat LC50:1260 mg/m³ EPASR* 8EHQ-0387-0656
orl-uns LD50:1500 mg/kg EPASR* 8EHQ-0387-0656
skn-uns LD50:>2 g/kg EPASR* 8EHQ-0387-0656

CONSENSUS REPORTS: Reported in EPA TSCA Inventory.

SAFETY PROFILE: Moderately toxic by ingestion and inhalation. Slightly toxic by skin contact. When heated to decomposition it emits toxic vapors of Cl^-.

EKQ000 CAS:104-76-7 HR: 2
2-ETHYLHEXANOL
mf: $C_8H_{18}O$ mw: 130.26

PROP: Clear liquid. Bp: 184–185°, mp: <−76°, flash p: 178°F (81°C), n (20/D) 1.4300, d: 0.834 @ 20°/20°, vap press: 0.2 mm @ 20°, vap d: 4.49. Sol in about 720 parts water and in many organic solvents.

SYNS: 1-AETHYLHEXANOL (GERMAN) □ 2-ETHYL-1-HEXANOL □ 2-ETHYLHEXYL ALCOHOL

TOXICITY DATA WITH REFERENCE
skn-rbt 415 mg open MLD UCDS** 11/11/69
skn-rbt 500 mg/24H MOD FCTXAV 17,775,79
eye-rbt 4165 µg SEV AJOPAA 29,1363,46
eye-rbt 20 mg/24H MOD 28ZPAK -,36,72
mmo-sat 500 µmol/L EVHPAZ 45,111,82
orl-mus TDLo:12 g/kg (female 7-14D post):REP NTIS** PB85-220143
orl-mus TDLo:12,200 mg/kg (female 6-13D post):REP TCMUD8 7,29,87
orl-rat TDLo:1628 mg/kg (female 12D post):TER TJADAB 35,41,87
orl-rat LD50:2049 mg/kg AIHAAP 30,470,69
ipr-rat LD50:500 mg/kg HYDRDA 3,201,78
scu-rat LD50:650 mg/kg NPIRI* 1,61,74
par-rat LD50:4600 mg/kg FCTXAV 17,775,79
orl-mus LD50:2500 mg/kg ZHYGAM 20,575,74
ipr-mus LD50:759 mg/kg ZHYGAM 20,575,74
par-mus LD50:1670 mg/kg FCTXAV 17,775,79
orl-rbt LD50:1180 mg/kg ZHYGAM 20,575,74
skn-rbt LD50:1970 mg/kg NPIRI* 1,61,74
orl-gpg LD50:1860 mg/kg ZHYGAM 20,575,74

CONSENSUS REPORTS: Reported in EPA TSCA Inventory.

SAFETY PROFILE: Moderately toxic by ingestion, skin contact, intraperitoneal, subcutaneous, and parenteral routes. An experimental teratogen. Other experimental reproductive effects. A severe eye and moderate skin irritant. Mutation data reported. A dangerous fire hazard when exposed to heat or flame; can react vigorously with oxidizing materials. To fight fire, use foam, CO_2, dry chemical. When heated to decomposition it emits acrid smoke and fumes. See also ALCOHOLS.

EKQ500 CAS:115-82-2 HR: 2
2-ETHYL-1-HEXANOL SILICATE
mf: $C_{32}H_{68}O_4Si$ mw: 545.09

PROP: Insol in water. Bp: 350–370°, fp: −90°, flash p: 390°F (OC).

SYNS: TETRA(2-ETHYLHEXOXY)SILANE □ TETRA(2-ETHYLHEXYL) ORTHOSILICATE □ TETRA(2-ETHYLHEXYL)SILICATE

TOXICITY DATA WITH REFERENCE
skn-rbt 500 mg open MLD UCDS** 7/14/65
eye-rbt 500 mg/24H MLD 85JCAE-,1239,86

CONSENSUS REPORTS: Reported in EPA TSCA Inventory.

SAFETY PROFILE: A skin and eye irritant. Combustible when exposed to heat or flame. To fight fire, use spray, dry chemical, CO_2. When heated to decomposition it emits acrid smoke and irritating fumes.

EKR000 CAS:645-62-5 HR: 2
2-ETHYL-2-HEXENAL
mf: $C_8H_{14}O$ mw: 126.22

PROP: Colorless liquid, powerful odor. Bp: 175°, flash p: 155°F (OC), d: 0.848 @ 20°/4°, vap d: 4.35, vap press: 1.0 mm @ 20°.

SYNS: 2-ETHYLHEXENAL □ α-ETHYL-β-n-PROPYLACROLEIN □ 2-ETHYL-3-PROPYL ACROLEIN

TOXICITY DATA WITH REFERENCE
skn-rbt 10 mg/24H open JIHTAB 26,269,44
eye-rbt 85 mg AJOPAA 29,1363,46
orl-rat LD50:3000 mg/kg JIHTAB 26,269,44

CONSENSUS REPORTS: Reported in EPA TSCA Inventory.

SAFETY PROFILE: Moderately toxic by ingestion. A skin, eye, and mucous membrane irritant. Combustible when exposed to heat or flame; can react with oxidizing materials. To fight fire, use alcohol foam, CO_2, dry chemical. When heated to decomposition it emits acrid smoke and irritating fumes.

EKR500 CAS:1632-16-2 HR: 3
2-ETHYL-1-HEXENE
mf: C_8H_{16} mw: 112.24

PROP: Colorless liquid. Bp: 120°, d: 0.7270 @ 20°/20°, vap d: 3.87.

SYNS: 2-ETHYL HEXENE-1 □ USAF DO-21

TOXICITY DATA WITH REFERENCE
skn-rbt 100 mg/24H MOD 85JCAE-,14,86
eye-rbt 500 mg/24H MLD 85JCAE-,14,86

ihl-rat LCLo:4000 ppm/4H JIHTAB 31,343,49
ipr-mus LD50:100 mg/kg NTIS** AD277-689

CONSENSUS REPORTS: Reported in EPA TSCA Inventory.

SAFETY PROFILE: Poison by intraperitoneal route. Mildly toxic by inhalation. A skin and eye irritant. Combustible when exposed to heat or flame; can react with oxidizing materials. When heated to decomposition it emits acrid smoke and irritating fumes.

EKS000 **CAS:5309-52-4** **HR: 2**
2-ETHYL-2-HEXENOIC ACID
mf: $C_8H_{14}O_2$ mw: 142.22

TOXICITY DATA WITH REFERENCE
skn-rbt 10 mg/24H open MLD AMIHBC 4,119,51
eye-rbt 150 μg open SEV AMIHBC 4,119,51
orl-rat LDLo:5660 mg/kg AMIHBC 4,119,51
skn-rbt LD50:2750 mg/kg AMIHBC 4,119,51

SAFETY PROFILE: Moderately toxic by skin contact. Mildly toxic by ingestion. A skin and severe eye irritant. When heated to decomposition it emits acrid smoke and irritating fumes.

EKS100 **CAS:28069-74-1** **HR: 1**
ETHYL cis-3-HEXENYL ACETAL
mf: $C_{10}H_{20}O_2$ mw: 172.30

SYNS: ACETALDEHYDE ETHYL cis-3-HEXENYL ACETAL □ (Z)-1-ETHOXY-1-(3-HEXENYLOXY)ETHANE □ 3-HEXENE, 1-(1-ETHOXYETHOXY)-, (Z)- □ cis-3-HEXENYL ETHYL ACETAL

TOXICITY DATA WITH REFERENCE
orl-rat LDLo:5 g/kg FCTOD7 26,275,88
skn-rbt LD50:>5 g/kg FCTOD7 26,275,88

CONSENSUS REPORTS: Reported in EPA TSCA Inventory.

SAFETY PROFILE: Low toxicity by ingestion and skin contact. When heated to decomposition it emits acrid smoke and irritating vapors.

EKS120 **CAS:54484-73-0** **HR: 1**
ETHYL HEXYL ACETAL
mf: $C_{10}H_{22}O_2$ mw: 174.32

SYNS: ACETALDEHYDE ETHYL HEXYL ACETAL □ 1-ETHOXY-1-HEXYLOXYETHANE □ HEXANE, 1-(1-ETHOXYETHOXY)- □ LIVERT

TOXICITY DATA WITH REFERENCE
orl-rat LDLo:13,312 mg/kg FCTOD7 26,277,88
skn-rbt LDLo:5 g/kg FCTOD7 26,277,88

CONSENSUS REPORTS: Reported in EPA TSCA Inventory.

SAFETY PROFILE: Low toxicity by ingestion and skin contact. When heated to decomposition it emits acrid smoke and irritating vapors.

EKS150 **CAS:29214-60-6** **HR: 1**
ETHYL-2-HEXYL ACETOACETATE
mf: $C_{12}H_{22}O_3$ mw: 214.34

SYNS: ETHYL 2-ACETYLCAPRYLATE □ ETHYL 2-ACETYLOCTANOATE □ ETHYL α-HEXYLACETOACETATE □ OCTANOIC ACID, 2-ACETYL-, ETHYL ESTER

TOXICITY DATA WITH REFERENCE
orl-rat LD50:>5 g/kg FCTXAV 13,454,75
skn-rbt LD50:>5 g/kg FCTXAV 13,454,75

CONSENSUS REPORTS: Reported in EPA TSCA Inventory.

SAFETY PROFILE: Low toxicity by ingestion and skin contact. When heated to decomposition it emits acrid smoke and irritating vapors.

EKS500 **CAS:104-75-6** **HR: 3**
2-ETHYL HEXYLAMINE
DOT: UN 2276
mf: $C_8H_{19}N$ mw: 129.28

PROP: A clear, miscible liquid. Bp: 169.2°, flash p: 140°F (OC), d: 0.7894 @ 20°/20°, vap press: 1.2 mm @ 20°, vap d: 4.45.

SYN: 1-AMINO-2-ETHYLHEXAN (CZECH)

TOXICITY DATA WITH REFERENCE
skn-rbt 750 μg/24H SEV 85JCAE-,436,86
eye-rbt 50 μg/24H SEV 85JCAE-,436,86
skn-rbt 500 mg/24H SEV 28ZPAK -,62,72
eye-rbt 50 μg/24H SEV 28ZPAK -,62,72
orl-rat LD50:450 mg/kg UCDS** 2/20/63
ihl-rat LCLo:250 ppm/4H AEHLAU 1,343,60
ipr-mus LDLo:4 mg/kg CBCCT* 2,189,50
skn-rbt LD50:600 mg/kg UCDS** 2/20/63

CONSENSUS REPORTS: Reported in EPA TSCA Inventory.

DOT CLASSIFICATION: 8; *Label:* Corrosive

SAFETY PROFILE: Poison by intraperitoneal route. Moderately toxic by ingestion, inhalation, and skin contact. Corrosive. A severe skin and eye irritant. Flammable liquid when exposed to heat or flame; can react with oxidizing materials. To fight fire, use alcohol foam, CO_2, dry chemical. When heated to decomposition it emits toxic fumes of NO_x. See also AMINES.

EKT500 **CAS:144-00-3** **HR: 3**
5-ETHYL-5-HEXYLBARBITURIC ACID SODIUM SALT
mf: $C_{12}H_{19}N_2O_3{\cdot}Na$ mw: 262.32

SYNS: 5-ETHYL-5-HEXYL-2,4,6-(1H,3H,5H)-PYRIMIDINETRIONE MONOSODIUM SALT □ HEBARAL □ HEXETHAL SODIUM □ ORTAL SODIUM □ SODIUM-5-ETHYL-5-HEXYLBARBITURATE □ SODIUM HEXETHAL □ SODIUM-N-HEXYLETHYL BARBITURATE

TOXICITY DATA WITH REFERENCE
orl-rat LDLo:2000 mg/kg JPETAB 60,125,37
ipr-rat LD50:250 mg/kg MEIEDD 10,680,83
ivn-rat LDLo:210 mg/kg JPETAB 60,125,37
ipr-mus LDLo:230 mg/kg JPETAB 31,455,27
scu-dog LDLo:150 mg/kg JPETAB 60,439,37

ivn-dog LDLo: 40 mg/kg JPETAB 60,439,37
ivn-cat LDLo: 50 mg/kg JPETAB 60,439,37
ipr-rbt LDLo: 75 mg/kg JPETAB 60,439,37
ivn-rbt LDLo: 50 mg/kg JPETAB 60,439,37

SAFETY PROFILE: Poison by intraperitoneal, intravenous, and subcutaneous routes. Moderately toxic by ingestion. A sedative and hypnotic agent. When heated to decomposition it emits toxic fumes of NO_x and Na_2O. See also BARBITURATES.

EKU000 CAS:123-04-6 HR: 3
2-ETHYLHEXYL-1-CHLORIDE
mf: $C_8H_{17}Cl$ mw: 148.70

PROP: Colorless liquid. Bp: 172.9°, fp: −135°, flash p: 140°F (OC), d: 0.8833, @ 20°, vap d: 5.14.

SYNS: 1-CHLORO-2-ETHYLHEXANE □ 3-CHLOROMETHYLHEPTANE

TOXICITY DATA WITH REFERENCE
skn-rbt 500 mg MLD 34ZIAG -,745,69
skn-rbt 100 mg/24H MOD 85JCAE-,102,86
eye-rbt 100 mg/24H MLD 85JCAE-,102,86
orl-rat LD50: 7340 mg/kg AMIHBC 4,119,51
ihl-rat LCLo: 2000 ppm/4H AMIHBC 4,119,51
skn-rbt LD50: 15,800 mg/kg 34ZIAG -,745,69

CONSENSUS REPORTS: Reported in EPA TSCA Inventory.

SAFETY PROFILE: Mildly toxic by ingestion, inhalation, and skin contact. A skin and eye irritant. Flammable liquid when exposed to heat or flame; can react vigorously with oxidizing materials. To fight fire, use foam, CO_2, dry chemical. When heated to decomposition it emits highly toxic fumes of phosgene and Cl^-. See also CHLORINATED HYDROCARBONS, ALIPHATIC.

EKU100 CAS:2350-24-5 HR: 1
2-ETHYLHEXYL-6-CHLORIDE
mf: $C_8H_{17}Cl$ mw: 148.70

SYN: HEPTANE, 1-CHLORO-5-METHYL-

TOXICITY DATA WITH REFERENCE
skn-rbt 10 mg/24H open MLD AMIHBC 4,119,51
eye-rbt 500 mg open AMIHBC 4,119,51
orl-rat LD50: 7340 mg/kg AMIHBC 4,119,51
ihl-rat LCLo: 4000 ppm/4H JIDHAN 31,343,49

CONSENSUS REPORTS: Reported in EPA TSCA Inventory.

SAFETY PROFILE: Mildly toxic by ingestion and inhalation routes. A skin and eye irritant. When heated to decomposition it emits toxic vapors of Cl^-.

EKV000 CAS:94-96-2 HR: 2
ETHYL HEXYLENE GLYCOL
mf: $C_8H_{18}O_2$ mw: 146.26

PROP: Practically colorless, somewhat viscous, odorless liquid. Bp: 243.1°, flash p: 260°F (OC), fp: −40°, d: 0.9422 @ 20°/20°, vap press: <0.01 mm @ 20°, vap d: 5.03. Sltly sol in H_2O; misc in $CHCl_3$, EtOH, and Et_2O.

SYNS: CARBIDE 6-12 □ COMPOUND 6-12 INSECT REPELLENT □ ENT 375 □ ETHOHEXADIOL □ ETHYL HEXANEDIOL □ 2-ETHYL-HEXANEDIOL-1,3 □ 2-ETHYLHEXANE-1,3-DIOL □ 2-ETHYL-1,3-HEXANEDIOL □ 2-ETHYL-3-PROPYL-1,3-PROPANEDIOL □ 3-HYDROXYMETHYL-n-HEPTAN-4-OL □ 6-12-INSECT REPELLENT □ OCTYLENE GLYCOL □ REPELLENT 612 □ RUTGERS 612

TOXICITY DATA WITH REFERENCE
skn-rbt 500 mg MLD 34ZIAG -,731,69
eye-rbt 5 mg SEV AJOPAA 29,1363,46
orl-rat TDLo: 20 g/kg (female 6-15D post): TER EPASR* 8EHQ-1288-0778
orl-rat TDLo: 20 g/kg (female 6-15D post): REP EPASR* 8EHQ-1288-0778
orl-rat LD50: 1400 mg/kg SPEADM 78-1,52,78
orl-mus LD50: 1900 mg/kg SPEADM 78-1,52,78
orl-rbt LD50: 2600 mg/kg PCOC** -,508,66
skn-rbt LD50: 2000 mg/kg 31ZOAD 1,208,68
orl-gpg LD50: 1900 mg/kg JPETAB 93,26,48
orl-ckn LD50: 1400 mg/kg JPETAB 93,26,48

CONSENSUS REPORTS: Reported in EPA TSCA Inventory.

SAFETY PROFILE: Moderately toxic by ingestion and skin contact. Experimental teratogenic and reproductive effects. A skin and severe eye irritant. Used as an insecticide, insect repellent, and in hair care preparations. Combustible when exposed to heat or flame; can react with oxidizing materials. To fight fire, use alcohol foam, foam, dry chemical. When heated to decomposition it emits acrid smoke and irritating fumes.

EKV500 CAS:141-38-8 HR: 1
2-ETHYLHEXYL EPOXYSTEARATE
mf: $C_{26}H_{50}O_3$ mw: 410.76

SYNS: 9,10-EPOXYSTEARIC ACID-2-ETHYLHEXYL ESTER □ 2 ETHYLHEXYL-9,10-EPOXYOCTADECANOATE

TOXICITY DATA WITH REFERENCE
skn-rbt 10 mg/24H open MLD AIHAAP 23,95,62
orl-rat LD50: 30,800 mg/kg AIHAAP 23,95,62

CONSENSUS REPORTS: Reported in EPA TSCA Inventory.

SAFETY PROFILE: Mildly toxic by ingestion. A skin irritant. When heated to decomposition it emits acrid smoke and irritating fumes.

EKW000 CAS:7425-14-1 HR: 1
2-ETHYLHEXYL-2-ETHYLHEXANOATE
mf: $C_{16}H_{32}O_2$ mw: 256.48

PROP: Liquid. Vap d: 8.85.

SYNS: 2-ETHYLHEXANOIC ACID, 2-ETHYLHEXYL ESTER □ 2-ETHYLHEXYLESTER KYSELINY 2-ETHYLKAPRONOVE

TOXICITY DATA WITH REFERENCE
skn-rbt 10 mg/24H open MLD AIHAAP 23,95,62
eye-rbt 500 mg AJOPAA 29,1363,46
orl-rat LD50: 27 g/kg AIHAAP 23,95,62

CONSENSUS REPORTS: Reported in EPA TSCA Inventory.

SAFETY PROFILE: Mildly toxic by ingestion. A skin and eye irritant. Combustible when exposed to heat or flame; can react with oxidizing materials. When heated to decomposition it emits acrid smoke and irritating fumes.

EKW300 CAS:7659-86-1 HR: 3
2-ETHYLHEXYL MERCAPTOACETATE
mf: $C_{10}H_{20}O_2S$ mw: 204.36

SYNS: 2-ETHYLHEXYL THIOGLYCOLATE □ MERCAPTOACETIC ACID-2-ETHYLHEXYL ESTER □ THIOGLYCOLIC ACID-2-ETHYLHEXYL ESTER □ THIOGLYKOLSAEURE-2-AETHYLHEXYL ESTER (GERMAN)

TOXICITY DATA WITH REFERENCE
orl-rat LD50:303 mg/kg ZHYGAM 20,575,74
ipr-rat LD50:265 mg/kg ZHYGAM 20,575,74
orl-mus LD50:1430 mg/kg ZHYGAM 20,575,74
ipr-mus LD50:865 mg/kg ZHYGAM 20,575,74
orl-rbt LD50:534 mg/kg ZHYGAM 20,575,74
orl-gpg LD50:955 mg/kg ZHYGAM 20,575,74

SAFETY PROFILE: Poison by ingestion and intraperitoneal routes. When heated to decomposition it emits toxic fumes of SO_x. See also MERCAPTANS and ESTERS.

EKX500 CAS:1559-35-9 HR: 2
2-(2-ETHYLHEXYLOXY)ETHANOL
mf: $C_{10}H_{22}O_2$ mw: 174.32

SYN: 2-((2-ETHYLHEXYL)OXY)ETHANOL

TOXICITY DATA WITH REFERENCE
skn-rbt 10 mg/24H open SEV AMIHBC 10,61,54
eye-rbt 250 μg open SEV AMIHBC 10,61,54
orl-rat LD50:3080 mg/kg AMIHBC 10,61,54
skn-rbt LD50:2120 mg/kg AMIHBC 10,61,54

CONSENSUS REPORTS: Reported in EPA TSCA Inventory.

SAFETY PROFILE: Moderately toxic by ingestion and skin contact. A severe skin and eye irritant. When heated to decomposition it emits acrid smoke and irritating fumes.

EKY000 CAS:2549-90-8 HR: 1
4-(2-ETHYLHEXYLOXY)-2-HYDROXYBENZOPHENONE
mf: $C_{21}H_{26}O_3$ mw: 326.47

SYN: 2-HYDROXY-4-(2′-ETHYLHEXOXY)BENZOFENON (CZECH)

TOXICITY DATA WITH REFERENCE
skn-rbt 500 mg/24H MLD 28ZPAK -,100,72
eye-rbt 500 mg/24H MLD 28ZPAK -,100,72

SAFETY PROFILE: A skin and eye irritant. When heated to decomposition it emits acrid smoke and irritating fumes.

EKZ000 CAS:10213-75-9 HR: 1
3-(2-ETHYLHEXYLOXY)PROPIONITRILE
mf: $C_{11}H_{21}NO$ mw: 183.33

SYN: 3-((2-ETHYLHEXYL)OXY)PROPANENITRILE

TOXICITY DATA WITH REFERENCE
eye-rbt 500 mg open AMIHBC 10,61,54
orl-rat LD50:4920 mg/kg AMIHBC 10,61,54
skn-rbt LD50:5990 mg/kg AMIHBC 10,61,54

CONSENSUS REPORTS: Cyanide and its compounds are on the Community Right-To-Know List. Reported in EPA TSCA Inventory.

SAFETY PROFILE: Mildly toxic by ingestion and skin contact. An eye irritant. When heated to decomposition it emits toxic fumes of NO_x and CN^-. See also NITRILES.

ELA000 CAS:5397-31-9 HR: 3
2-ETHYLHEXYLOXYPROPYLAMINE
mf: $C_{11}H_{25}NO$ mw: 187.37

PROP: Bp: 239°, mp: −90°, flash p: 210°F (OC), d: 0.8483, vap d: 6.47.

SYNS: 2-ETHYLHEXYL-3-AMINOPROPYL ETHER □ 3-((2-ETHYLHEXYL)OXY)PROPYLAMINE

TOXICITY DATA WITH REFERENCE
skn-rbt 10 mg/24H open SEV AMIHBC 4,119,51
eye-rbt 50 μg open SEV AMIHBC 4,119,51
orl-rat LD50:320 mg/kg AMIHBC 4,119,51
skn-rbt LD50:360 mg/kg AMIHBC 4,119,51

CONSENSUS REPORTS: Reported in EPA TSCA Inventory.

SAFETY PROFILE: Poison by ingestion and skin contact. A severe skin and eye irritant. Combustible when exposed to heat or flame; can react vigorously with oxidizing materials. When heated to decomposition it emits toxic fumes of NO_x. See also ETHERS and AMINES.

ELA600 CAS:24234-06-8 HR: 3
1-ETHYL-1-HEXYLPIPERIDINIUM BROMIDE
mf: $C_{13}H_{28}N{\cdot}Br$ mw: 278.33

TOXICITY DATA WITH REFERENCE
orl-mus LD50:408 mg/kg PSDTAP 15,331,74
ivn-mus LD50:4496 μg/kg PSDTAP 15,331,74
ipr-mus LD50:42,958 μg/kg PSDTAP 15,331,74

SAFETY PROFILE: Poison by intravenous and intraperitoneal routes. Moderately toxic by ingestion. When heated to decomposition it emits toxic fumes of NO_x and Br^-.

ELB000 CAS:118-60-5 HR: 3
2-ETHYLHEXYL SALICYLATE
mf: $C_{15}H_{22}O_3$ mw: 250.37

SYNS: SALICYLIC ACID-2-ETHYLHEXYL ESTER □ USAF DO-11 □ WMO

TOXICITY DATA WITH REFERENCE
skn-rbt 500 mg/24H MLD FCTXAV 14,659,76
ipr-mus LD50:200 mg/kg NTIS** AD277-689

CONSENSUS REPORTS: Reported in EPA TSCA Inventory.

SAFETY PROFILE: Poison by intraperitoneal route. A

skin irritant. When heated to decomposition it emits acrid smoke and irritating fumes. See also ESTERS.

ELB400 CAS:72214-01-8 HR: 2
2-ETHYLHEXYL SULFATE
mf: $C_8H_{18}O_4S$ mw: 210.32

SYN: SULFURIC ACID, MONO(2-ETHYLHEXYL)ESTER

TOXICITY DATA WITH REFERENCE
orl-mus TDLo:1747 g/kg/2Y C:ETA EVHPAZ 65,271,86
orl-rat LD50:4125 mg/kg 34ZIAG-,690,69

SAFETY PROFILE: Moderately toxic by ingestion. Questionable carcinogen with experimental tumorigenic data. When heated to decomposition it emits toxic fumes of SO_x.

ELB500 CAS:103-44-6 HR: 3
2-ETHYLHEXYL VINYL ETHER
mf: $C_{10}H_{20}O$ mw: 156.30

PROP: Liquid. Mp: −100°, bp: 177.5°, flash p: 135°F (OC), d: 0.810, autoign temp: 395°F, vap d: 5.4.

SYNS: 1-ETHENOXY-2-ETHYLHEXANE ☐ VINYL-2-ETHYLHEXYL ETHER

TOXICITY DATA WITH REFERENCE
skn-rbt 10 mg/24H open MLD AMIHBC 10,61,54
skn-rbt 500 mg open MLD UCDS** 3/23/73
eye-rbt 500 mg open AMIHBC 10,61,54
orl-rat LD50:1350 mg/kg AMIHBC 10,61,54
ihl-rat LCLo:1005 ppm/8H UCDS** 3/23/73
skn-rbt LD50:3560 mg/kg AMIHBC 10,61,54

SAFETY PROFILE: Moderately toxic by ingestion and skin contact. A skin and eye irritant. Flammable liquid when exposed to heat or flame; can react with oxidizing materials. Potentially explosive. To fight fire, use alcohol foam, foam, CO_2, dry chemical. When heated to decomposition it emits acrid smoke and irritating fumes. See also ETHERS.

ELC000 CAS:18413-14-4 HR: 2
ETHYLHYDRAZINE HYDROCHLORIDE
mf: $C_2H_8N_2 \cdot ClH$ mw: 96.58

TOXICITY DATA WITH REFERENCE
orl-mus TDLo:11 g/kg/64W-C:CAR IJCNAW 13,500,74

SAFETY PROFILE: Questionable carcinogen with experimental carcinogenic data. When heated to decomposition it emits very toxic fumes of HCl and NO_x. See also HYDRAZINE.

ELC500 CAS:3413-58-9 HR: 3
ETHYLHYDROCUPREINE HYDROCHLORIDE
mf: $C_{21}H_{28}N_2O_2 \cdot ClH$ mw: 376.97

PROP: Rhombic crystals from Me_2CO/Et_2O. Mp: 252–254°.

SYNS: HYDROCUPREINE ETHYL ESTER HYDROCHLORIDE ☐ NEUMOLISINA ☐ NUMOQUIN HYDROCHLORIDE ☐ OPTOCHIN HYDROCHLORIDE ☐ OPTOQUINHYDROCHLORIDE ☐ RHOMBIC

TOXICITY DATA WITH REFERENCE
ivn-man TDLo:31 mg/kg/3D:EYE,CVS JPETAB 8,53,16
scu-mus LDLo:500 mg/kg JPETAB 8,53,16
unr-frg LDLo:300 mg/kg JPETAB 8,53,16

SAFETY PROFILE: Poison by unspecified route. Moderately toxic by subcutaneous route. Human systemic effects by intravenous route: visual field changes and arteriolar constriction. An antiseptic. When heated to decomposition it emits very toxic fumes such as Cl^- and NO_x.

E

ELC600 CAS:626-86-8 HR: 1
ETHYL HYDROGEN ADIPATE
mf: $C_8H_{14}O_4$ mw: 174.22

SYNS: ADIPIC ACID, MONOETHYL ESTER ☐ HEXANOIC ACID, MONOETHYL ESTER (9CI) ☐ MONOETHYL ADIPATE ☐ MONOETHYLADIPIC ACID ESTER ☐ MONOETHYL HEXANEDIOATE

TOXICITY DATA WITH REFERENCE
orl-uns LD50:4100 mg/kg GTPZAB 21(10),39,77

CONSENSUS REPORTS: Reported in EPA TSCA Inventory.

SAFETY PROFILE: Mildly toxic by ingestion. When heated to decomposition it emits acrid smoke and irritating vapors.

ELD000 CAS:3031-74-1 HR: 3
ETHYL HYDROPEROXIDE
mf: $C_2H_6O_2$ mw: 62.07

CH_3CH_2OOH

SYNS: ETHYL HYDROGEN PEROXIDE ☐ HYDROPEROXIDE, ETHYL

DOT CLASSIFICATION: Forbidden

SAFETY PROFILE: Explodes violently when superheated. The barium salt is heat- and impact-sensitive. Explosive reaction with hydroiodic acid or finely divided silver. When heated to decomposition it emits acrid smoke and irritating fumes. See also PEROXIDES.

ELD100 CAS:38692-98-7 HR: 3
1-ETHYL-3-(HYDROXYACETYL)INDOLE
mf: $C_{12}H_{13}NO_2$ mw: 203.26

SYN: KETONE, 1-ETHYL-3-INDOLYL HYDROXYMETHYL

TOXICITY DATA WITH REFERENCE
ipr-mus LDLo:600 mg/kg PCJOAU 6,33,72

DOT CLASSIFICATION: 3; *Label:* Flammable Liquid

SAFETY PROFILE: Moderately toxic by intraperitoneal route. A flammable liquid. When heated to decomposition it emits toxic vapors of NO_x.

ELD500 CAS:2497-34-9 HR: 2
4'-ETHYL-4-HYDROXYAZOBENZENE
mf: $C_{14}H_{14}N_2O$ mw: 226.30

TOXICITY DATA WITH REFERENCE
orl-rat TDLo:9780 mg/kg/43W-C:ETA ARZNAD 12,270,62

SAFETY PROFILE: Questionable carcinogen with experimental tumorigenic data. When heated to decomposition it emits toxic fumes of NO_x.

ELE500 CAS:54897-62-0 HR: 2
N-ETHYL-N-(4-HYDROXYBUTYL)NITROSOAMINE
mf: $C_6H_{14}N_2O_2$ mw: 146.22

SYNS: EHBN □ 4-(ETHYLNITROSOAMINO)-1-BUTANOL

TOXICITY DATA WITH REFERENCE
mma-sat 35 µmol/plate CNREA8 37,399,77
orl-rat TDLo:4375 mg/kg/20W-C:ETA GANNA2 67,175,76
orl-mus TDLo:4900 mg/kg/14W-C:CAR GANNA2 72,647,81
orl-ham TDLo:24 g/kg/20W-C:ETA GANNA2 67,175,76
orl-rat TD:59 g/kg/20W-C:ETA GANNA2 67,825,76
orl-rat TD:5300 mg/kg/20W-I:ETA GANNA2 65,565,74
orl-mus TD:9800 mg/kg/14W-C:CAR GANNA2 72,647,81
orl-mus TD:7 g/kg/120W-C:ETA GANNA2 67,175,76
orl-mus TD:7000 mg/kg/20W-C:CAR GANNA2 72,647,81
orl-rat TD:4704 mg/kg/20W-C:ETA GANNA2 72,539,81

CONSENSUS REPORTS: EPA Genetic Toxicology Program.

SAFETY PROFILE: Questionable carcinogen with experimental carcinogenic and tumorigenic data. Mutation data reported. When heated to decomposition it emits toxic fumes of NO_x. See also NITROSAMINES.

ELE600 CAS:7159-96-8 HR: 3
ETHYL 3-HYDROXYCARBANILATE
mf: $C_9H_{11}NO_3$ mw: 181.21

SYNS: CARBAMIC ACID, (3-HYDROXYPHENYL)-, ETHYL ESTER (9CI) □ CARBANILIC ACID, m-HYDROXY-, ETHYL ESTER □ 3-CARBETHOXYAMINOPHENOL □ m-(ETHOXYCARBONYLAMINO)PHENOL □ 3-(ETHOXYCARBONYLAMINO)PHENOL □ ETHYL m-HYDROXYCARBANILATE □ ETHYL (3-HYDROXYPHENYL)CARBAMATE □ ETHYL N-(3-HYDROXYPHENYL)CARBAMATE □ m-HYDROXYCARBANILIC ACID ETHYL ESTER

TOXICITY DATA WITH REFERENCE
ivn-mus LD50:56 mg/kg CSLNX* NX#06547

CONSENSUS REPORTS: Reported in EPA TSCA Inventory.

SAFETY PROFILE: A poison by intravenous route. When heated to decomposition it emits toxic vapors of NO_x.

ELF100 CAS:33765-68-3 HR: 2
16-ETHYL-17-HYDROXYESTER-4-EN-3-ONE
mf: $C_{20}H_{30}O_2$ mw: 302.50

PROP: Crystals from ether. Mp: 152–153°.

SYNS: 16-β-ETHYL-17-β-HYDROXYESTR-4-EN-3-ONE □ 16-β-ETHYL-17-β-HYDROXY-4-ESTREN-3-ONE □ 16-β-ETHYL-19-NORTESTOSTERONE □ OXENDOLONE □ PROSTETIN □ TSAA 291

TOXICITY DATA WITH REFERENCE
ims-rat TDLo:150 mg/kg (female 30D pre):REP YACHDS 7,943,79
ims-rat TDLo:15 mg/kg (female 30D pre):REP YACHDS 7,943,79
ims-rat TDLo:750 mg/kg (male 30D pre):REP INJFA3 34,235,89
ims-rat TDLo:150 mg/kg (30D pre):REP YACHDS 7,943,79
orl-rat LD50:12,105 mg/kg YACHDS 7,937,79
ipr-rat LD50:5879 mg/kg YACHDS 7,937,79
orl-mus LD50:5480 mg/kg IYKEDH 12,1204,81
ipr-mus LD50:2925 mg/kg YACHDS 7,937,79

SAFETY PROFILE: Moderately toxic by intraperitoneal route. Mildly toxic by ingestion. Experimental reproductive effects. A steroid. When heated to decomposition it emits acrid smoke and irritating fumes. See also ESTERS.

ELF110 CAS:33765-80-9 HR: 1
16-β-ETHYL-17-β-HYDROXYESTER-4-EN-3-ONE ACETATE
mf: $C_{22}H_{32}O_3$ mw: 344.54

SYNS: 17-β-ACETOXY-16-β-ETHYLESTR-4-EN-3-ONE □ (16-β,17-β)-17-(ACETYLOXY)-16-ETHYL-ESTR-4-EN-3-ONE (9CI) □ TSAA-328

TOXICITY DATA WITH REFERENCE
ims-rat TDLo:2500 mg/kg (5D male):REP TAKHAA 32,330,73
ipr-rat LD50:6100 mg/kg TAKHAA 32,330,73
ipr-mus LD50:7500 mg/kg TAKHAA 32,330,73
ims-mus LDLo:10 g/kg TAKHAA 32,330,73

SAFETY PROFILE: Mildly toxic by intraperitoneal and intramuscular routes. Experimental reproductive effects. When heated to decomposition it emits acrid smoke and irritating fumes.

ELF500 CAS:1164-38-1 HR: 3
ETHYL (2-HYDROXYETHYL)DIMETHYLAMMONIUM BENZILATE CHLORIDE
mf: $C_{20}H_{26}NO_3 \cdot Cl$ mw: 363.92

PROP: Crystals from Me_2CO/EtOH. Mp: 213°.

SYNS: BENZILIC ACID, ester with ETHYL (2-HYDROXYETHYL)DIMETHYLAMMONIUM CHLORIDE □ BENZILYLOXYETHYLDIMETHYLETHYLAMMONIUM CHLORIDE □ E-3 □ N-ETHYL-2-((HYDROXYDIPHENYLACETYL)OXY)-N,N-DIMETHYLETHANAMINIUM CHLORIDE □ ETHYL(2-HYDROXYETHYL)DIMETHYLAMMONIUM CHLORIDE BENZILATE □ LACHESIN □ LACHESINE CHLORIDE □ LAXESIN

TOXICITY DATA WITH REFERENCE
orl-mus LD50:1 g/kg MEIEDD 10,767,83
ipr-mus LD50:40 mg/kg MEIEDD 10,767,83
scu-mus LD50:160 mg/kg MEIEDD 10,767,83

SAFETY PROFILE: Poison by intraperitoneal and subcutaneous routes. Moderately toxic by ingestion. When heated to decomposition it emits very toxic fumes of Cl^-, NH_3, and NO_x.

ELG000 CAS:63918-38-7 HR: 3
ETHYL-β-HYDROXYETHYLETHYLENIMONIUM PICRYLSULFONATE
mf: $C_6H_{14}NO \cdot C_6H_2N_3O_9S$ mw: 408.38

SYNS: 1-ETHYL-1-(2-HYDROXYETHYL)AZIRIDINIUM-2,4,6-TRINI-

TROBENZENESULFONATE □ 1-ETHYL-1-(2-HYDROXYETHYL)AZIRIDINIUM SALT with 2,4,6-TRINITROBENZENESULFONIC ACID □ ETHYL (2-HYDROXYETHYL)ETHYLENIMONIUM PICRYLSULFONATE □ 1-ETHYL-1-(β-HYDROXYETHYL)ETHYLENIMONIUM PICRYLSULFONATE

TOXICITY DATA WITH REFERENCE
scu-mus LD50:5500 μg/kg JPETAB 91,224,47
ivn-mus LD50:5 mg/kg JPETAB 91,224,47
ivn-rbt LDLo:5 mg/kg JPETAB 91,224,47

SAFETY PROFILE: Poison by subcutaneous and intravenous routes. When heated to decomposition it emits very toxic fumes of NO_x and SO_x. See also SULFONATES, PICRIC ACID, and NITRO COMPOUNDS of AROMATIC HYDROCARBONS.

ELG500 CAS:13147-25-6 HR: 3
ETHYL-2-HYDROXYETHYLNITROSAMINE
mf: $C_4H_{10}N_2O_2$ mw: 118.16

$CH_3CH_2N(N:O)C_2H_4OH$

SYNS: AETHYL-AETHANOL-NITROSOAMIN (GERMAN) □ EENA □ EHEN □ N-ETHYL-N-HYDROXYETHYLNITROSAMINE □ 2-(ETHYLNITROSAMINO)ETHANOL □ N-NITROSOAETHYLAETHANOLAMIN (GERMAN) □ N-NITROSOETHYLETHANOLAMINE □ N-NITROSOETHYL-2-HYDROXYETHYLAMINE □ N-NITROSO-N-ETHYL-N-(2-HYDROXYETHYL)AMINE

TOXICITY DATA WITH REFERENCE
mma-sat 20 μg/plate MUREAV 66,1,79
orl-rat TDLo:1680 mg/kg/14D-C:CAR GANNA2 70,817,79
ivn-rat TDLo:50 mg/kg/10W-I:CAR CRNGDP 7,1313,86
orl-rat TD:840 mg/kg/2W-C:NEO JJIND8 72,483,84
orl-rat TD:700 mg/kg/1W-C:ETA JJIND8 70,477,83
orl-rat TD:700 mg/kg/2W-C:CAR NAIZAM 31,361,80
orl-rat TD:1400 mg/kg/2W-C:CAR NAIZAM 31,361,80
orl-rat TD:840 mg/kg/2W-C:CAR SAIGAK 37,1771,82
orl-rat TD:5146 mg/kg/23W-C:CAR CRNGDP 8,719,87
orl-rat TD:1400 mg/kg/2W-C:ETA CRNGDP 4,523,83
orl-rat TD:1400 mg/kg/2W-C:NEO GANNA2 74,607,83

CONSENSUS REPORTS: IARC Cancer Review: Animal Limited Evidence IMEMDT 17,83,78. EPA Genetic Toxicology Program.

SAFETY PROFILE: Suspected carcinogen with experimental carcinogenic, neoplastigenic, and tumorigenic data. Mutation data reported. Explodes when heated to 170°C. When heated to decomposition it emits toxic fumes of NO_x. See also NITROSAMINES.

ELH600 CAS:119-41-5 HR: 2
ETHYL-7-HYDROXYFLAVONE
mf: $C_{19}H_{16}O_5$ mw: 324.35

PROP: Crystals from 50% ethanol. Mp: 123–124°. Sol in the usual organic solvents; sltly sol in water.

SYNS: 7-α-(ACETOXYETHANE)OXYFLAVONE □ ANGORLISIN □ CORIL □ CORLIN □ COROSANIN □ DILATAN KORE □ DOMUCOR □ EFLOXATE □ ETHYL FLAVONE-7-OXYACETATE □ ETHYL-7-FLAVONOXYACETATE □ ETHYL FLAVON-7-YLOXYACETATE □ ETHYL FLAVONYL-7-OXYACETATE □ FLACETHYLE □ 7-FLAVONE ETHYL HYDROXYACETATE □ FLAVONE-7-ETHYLOXYACETATE □ 7-FLAVONOXYACETIC ACID ETHYL ESTER □ OXIFLAVIL □ ((4-OXO-2-PHENYL-4H-1-BENZOPYRAN-7-YL)OXY)ACETIC ACID ETHYL ESTER □ OXYFLAVIL □ RE 1-0185 □ REC 1-0185 □ RECORDIL

TOXICITY DATA WITH REFERENCE
ipr-rat LD50:3200 mg/kg MEIEDD 10,509,83
ipr-mus LD50:3200 mg/kg FRPSAX 13,561,58
ipr-gpg LD50:2 g/kg NIIRDN 6,125,82

SAFETY PROFILE: Moderately toxic by intraperitoneal route. When heated to decomposition it emits acrid smoke and irritating fumes.

ELH700 CAS:80-55-7 HR: 2
ETHYL 2-HYDROXYISOBUTYRATE
mf: $C_6H_{12}O_3$ mw: 132.18

SYNS: ETHYL α-HYDROXYISOBUTYRATE □ ETHYL 2-HYDROXY-2-METHYLPROPANOATE □ ETHYL 2-METHYLLACTATE □ LACTIC ACID, 2-METHYL-, ETHYL ESTER □ 2-METHYLLACTIC ACID ETHYL ESTER □ PROPANOIC ACID, 2-HYDROXY-2-METHYL-, ETHYL ESTER (9CI)

TOXICITY DATA WITH REFERENCE
ims-gpg LDLo:2200 mg/kg JPETAB 76,189,42

SAFETY PROFILE: Moderately toxic by intramuscular route. When heated to decomposition it emits acrid smoke and irritating vapors.

ELI500 CAS:10029-04-6 HR: 3
ETHYL-2-(HYDROXYMETHYL)ACRYLATE
mf: $C_6H_{10}O_3$ mw: 130.16

PROP: A liquid. Bp: 47° @ 0.8 mm.

SYNS: ETHYL-α-(HYDROXYMETHYL)ACRYLATE □ 2-(HYDROXYMETHYL)ACRYLIC ACID, ETHYL ESTER

TOXICITY DATA WITH REFERENCE
skn-rbt 500 mg MOD IHFCAY 6,1,67
eye-rbt 500 mg SEV IHFCAY 6,1,67
orl-rat LD50:620 mg/kg IHFCAY 6,1,67
skn-rbt LD50:360 mg/kg IHFCAY 6,1,67

SAFETY PROFILE: Poison by skin contact. Moderately toxic by ingestion. A skin and severe eye irritant. When heated to decomposition it emits acrid smoke and irritating fumes. See also ESTERS.

ELI600 CAS:39544-02-0 HR: 3
1-(2-ETHYL-7-(2-HYDROXY-3-((1-METHYLETHYL)AMINO)PROPOXY)-4-BENZOFURANYL) ETHANONE
mf: $C_{18}H_{25}NO_4$ mw: 319.44

SYNS: ETHANONE, 1-(2-ETHYL-7-(2-HYDROXY-3-((1-METHYLETHYL)AMINO)PROPOXY)-4-BENZOFURANYL)- □ 2-ETHYL-4-ACETYL-7-(2-HYDROXY-3-ISOPROPYLAMINOPROPOXY)BENZOFURAN □ KETONE, 2-ETHYL-7-(2-HYDROXY-3-(ISOPROPYLAMINO)PROPOXY)-4-BENZOFURANYL METHYL

TOXICITY DATA WITH REFERENCE
ivn-mus LD50:120 mg/kg GWXXBX #2223184

DOT CLASSIFICATION: 3; *Label:* Flammable Liquid

SAFETY PROFILE: A poison by intravenous route. A flammable liquid. When heated to decomposition it emits toxic vapors of NO_x.

ELJ500 CAS:1005-93-2 HR: 3
2-ETHYL-2-(HYDROXYMETHYL)-1,3-PROPANEDIOL, CYCLIC PHOSPHATE (1:1)
mf: $C_6H_{11}O_4P$ mw: 178.14

PROP: Crystals from H_2O or Me_2CO. Mp: 207–208°.

SYNS: 4-AETHYL-1-PHOSPHA-2,6,7-TRIOXABICYCLO(2.2.2)OCTAN-1-OXID (GERMAN) ☐ 4-ETHYL-1-PHOSPHA-2,6,7-TRIOXABICYCLO (2.2.2)OCTANE-1-OXIDE ☐ 4-ETHYL-2,6,7-TRIOXA-1-PHOSPHABICYCLO(2.2.2)OCTANE-1-OXIDE

TOXICITY DATA WITH REFERENCE
orl-rat LD50:3080 μg/kg ARTODN 35,149,76
ihl-rat LC50:30 mg/m³/1H ARTODN 35,149,76
skn-rat LD50:50 mg/kg ARTODN 35,149,76
ipr-rat LD50:960 μg/kg ARTODN 35,149,76
orl-mus LD50:3550 μg/kg ARTODN 35,149,76
ipr-mus LD50:1 mg/kg SCIEAS 182,1135,73
ivn-mus LDLo:500 μg/kg EJMCA5 13,207,78
orl-dog LD50:1 mg/kg ARTODN 35,149,76
orl-rbt LD50:5 mg/kg ARTODN 35,149,76

SAFETY PROFILE: Poison by ingestion, inhalation, skin contact, intraperitoneal, and intravenous routes. When heated to decomposition it emits toxic fumes of PO_x.

ELK000 CAS:78330-02-6 HR: 2
ETHYL-4-HYDROXY-3-MORPHOLINOMETHYLBENZOATE
mf: $C_{13}H_{19}NO_4$ mw: 253.33

SYN: 4-HYDROXY-3-MORPHOLINOMETHYLBENZOIC ACID ETHYL ESTER

TOXICITY DATA WITH REFERENCE
orl-mus LDLo:3000 mg/kg ARZNAD 11,85,61
scu-mus LDLo:460 mg/kg ARZNAD 11,85,61

SAFETY PROFILE: Moderately toxic by ingestion and subcutaneous routes. When heated to decomposition it emits toxic fumes of NO_x.

ELL500 CAS:70-70-2 HR: 3
ETHYL-p-HYDROXYPHENYL KETONE
mf: $C_9H_{10}O_2$ mw: 150.19

PROP: Needles or prisms from H_2O. Mp: 149°.

SYNS: FRENANTOL ☐ FRENOHYPON ☐ H-365 ☐ p-HYDROXYPHENYL-1-PROPANONE ☐ 1-(4-HYDROXYPHENYL)-1-PROPANONE ☐ HYDROXYPROPIOPHENONE ☐ p-HYDROXYPROPIOPHENONE ☐ 4-HYDROXYPROPIOPHENONE ☐ HYPOPHENON ☐ p-OXYPROPIOPHENONE ☐ PAROXON ☐ PAROXYPROPIONE ☐ PHP ☐ POP ☐ PROFENONE ☐ p-PROPIONYLPHENOL ☐ USAF EK-3302

TOXICITY DATA WITH REFERENCE
ims-rbt TDLo:1600 mg/kg (female 15-30D post):REP AOGNAX 66,286,61
orl-mus LD50:3 mg/kg 85JDAH -,197,74
ipr-mus LD50:200 mg/kg NTIS** AD277-689
scu-mus LD50:1130 μg/kg AIPTAK 124,212,60
par-frg LD50:91 mg/kg AIPTAK 124,212,60

CONSENSUS REPORTS: Reported in EPA TSCA Inventory.

DOT CLASSIFICATION: 3; *Label:* Flammable Liquid

SAFETY PROFILE: Poison by intraperitoneal, subcutaneous, and parenteral routes. An experimental teratogen. Other experimental reproductive effects. A flammable liquid. When heated to decomposition it emits acrid smoke and irritating fumes. See also KETONES.

ELM500 CAS:624-85-1 HR: 3
ETHYL HYPOCHLORITE
mf: C_2H_5ClO mw: 80.52

CH_3CH_2OCl

PROP: Mobile, yellow liquid with irritating odor. Bp: 36° @ 752 mm. Misc in Et_2O, $CHCl_3$, and C_6H_6.

SAFETY PROFILE: Very unstable. The vapor explodes on contact with flame, spark, or upon rapid heating. The cold liquid explodes on contact with copper. Incompatible with light. When heated to decomposition it emits toxic fumes of Cl^-. See also HYPOCHLORITES.

ELN500 CAS:75-37-6 HR: 3
ETHYLIDENE DIFLUORIDE
mf: $C_2H_4F_2$ mw: 66.06

PROP: Colorless gas. Mp: −117.0°, bp: −26.5°, d: 1.004 @ 25°, vap d: 2.28.

SYNS: ALGOFRENE TYPE 67 ☐ DIFLUOROETHANE ☐ 1,1-DIFLUOROETHANE ☐ ETHYLENE FLUORIDE ☐ ETHYLIDENE FLUORIDE ☐ FC 152a ☐ FREON 152 ☐ GENETRON 100 ☐ HALOCARBON 152A

TOXICITY DATA WITH REFERENCE
sln-dmg-ihl 98 pph/10M ENVRAL 7,275,74
ihl-rat LCLo:64,000 ppm/4H JIDHAN 31,343,49
ihl-mus LC50:977 g/m³/2H 85GMAT -,54,82

CONSENSUS REPORTS: Reported in EPA TSCA Inventory. EPA Genetic Toxicology Program.

SAFETY PROFILE: Mildly toxic by inhalation. Mutation data reported. Narcotic in high concentration. A very dangerous fire hazard, when exposed to heat or flame; can react vigorously with oxidizing materials. See also FLUORIDES.

ELN600 CAS:55044-04-7 HR: 3
ETHYLIDENE DINITRATE
mf: $C_2H_4N_2O_6$ mw: 152.06

SAFETY PROFILE: A heat-sensitive explosive. When heated to decomposition it emits toxic fumes of NO_x. See also NITRATES.

ELO000 CAS:539-71-9 HR: 2
ETHYLIDENE DIURETHAN
mf: $C_8H_{16}N_2O_4$ mw: 204.26

PROP: Needles from H_2O. Sltly sol in H_2O. Mp: 125–126°, bp: 170–178° @ 20 mm.

SYNS: N,N'-ETHYLIDENE-BIS(ETHYL CARBAMATE) ☐ ETHYLIDENEDICARBAMIC ACID, DIETHYL ESTER

TOXICITY DATA WITH REFERENCE
ipr-mus TDLo:6500 mg/kg/13W-I:ETA JNCIAM 9,35,48

SAFETY PROFILE: Questionable carcinogen with experimental tumorigenic data. When heated to decomposition it emits toxic fumes of NO_x.

ELO500 **CAS:16219-75-3** **HR: 2**
ETHYLIDENE NORBORNENE
mf: C_9H_{12} mw: 120.21

PROP: Bp: 70.2–70.4° @ 58 mm.

SYNS: 5-ETHYLIDENEBICYCLO(2.2.1)HEPT-2-ENE □ 5-ETHYLIDENE-2-NORBORNENE

TOXICITY DATA WITH REFERENCE
skn-rbt 445 mg open MLD UCDS** 11/28/67
ihl-hmn TCLo:6 ppm/30M:NOSE,EYE,TONG TXAPA9 20,250,71
orl-rat LD50:2527 mg/kg AIHAAP 30,470,69
ihl-rat LC50:1246 ppm/4H TXAPA9 20,250,71
orl-mus LD50:3250 mg/kg GTPZAB 18(10),52,74
ihl-mus LC50:732 ppm/4H TXAPA9 20,250,71
ihl-rbt LC50:3104 ppm/4H TXAPA9 20,250,71
skn-rbt LD50:8189 mg/kg AIHAAP 30,470,69
ihl-gpg LC50:2896 ppm/4H TXAPA9 20,250,71

CONSENSUS REPORTS: Reported in EPA TSCA Inventory.

OSHA PEL: CL 5 ppm
ACGIH TLV: CL 5 ppm

SAFETY PROFILE: Moderately toxic by ingestion. Mildly toxic by inhalation and skin contact. Human systemic effects by inhalation: conjuctiva, olfactory, and taste changes. A skin irritant. When heated to decomposition it emits acrid smoke and irritating fumes.

ELP000 **CAS:139-87-7** **HR: 2**
N-ETHYL-2,2′-IMINODIETHANOL
mf: $C_6H_{15}NO_2$ mw: 133.22

PROP: A liquid. Mp: −50°, bp: 246–252°.

SYNS: DIETHANOLETHYLAMINE □ ETHANOL, 2,2′-(ETHYLIMINO)BIS-(9CI) □ ETHYLBIS(2-HYDROXYETHYL)AMINE □ ETHYLDIETHANOLAMINE □ N-ETHYLDIETHANOLAMINE □ 2-(N-ETHYL-N-2-HYDROXYETHYLAMINO)ETHANOL □ 2,2′-(ETHYLIMINO)BISETHANOL □ 2,2′-(ETHYLIMINO)DIETHANOL

TOXICITY DATA WITH REFERENCE
skn-rbt 10 mg/24H open MLD AMIHBC 10,61,54
eye-rbt 750 µg open SEV AMIHBC 10,61,54
orl-rat LD50:4570 mg/kg AMIHBC 10,61,54

CONSENSUS REPORTS: Reported in EPA TSCA Inventory.

SAFETY PROFILE: Mildly toxic by ingestion. A skin and severe eye irritant. When heated to decomposition it emits toxic fumes of NO_x.

ELP500 **CAS:75-03-6** **HR: 3**
ETHYL IODIDE
mf: C_2H_5I mw: 155.97

PROP: Clear, refractive, heavy, colorless liquid with ethereal odor; turns brownish-red on exposure to light. Mp: −108°, bp: 72.4°, d: 1.948 @ 15°, vap press: 100 mm @ 18.0°, vap d: 5.38. Misc in org solvs; very sltly sol in H_2O (gradual decomp).

SYNS: ETHYL IODIDE □ ETHYLJODID □ HYDRIODIC ETHER □ IODOETHANE □ JODETHAN

TOXICITY DATA WITH REFERENCE
mmo-esc 20 µmol/L ARTODN 46,277,80
dnd-esc 1 µmol/L ARTODN 46,277,80
ihl-rat LC50:65,000 mg/m³/30M FAVUAI 7,35,75
ipr-rat LD50:330 mg/kg 85GMAT -,68,82
ipr-mus LD50:560 mg/kg 85GMAT -,68,82
scu-mus LD50:1000 mg/kg JJPAAZ 3,99,54
ipr-gpg LD50:322 mg/kg 85GMAT -,68,82

CONSENSUS REPORTS: Reported in EPA TSCA Inventory.

SAFETY PROFILE: Poison by intraperitoneal route. Moderately toxic by subcutaneous route. Mildly toxic by inhalation. Mutation data reported. A skin, eye, and mucous membrane irritant. Narcotic in high concentration. Flammable when exposed to heat or flame, a preparative hazard. Will react with water or steam to produce toxic and corrosive fumes; can react vigorously with oxidizing materials. Incompatible with silver chlorite. To fight fire, use water, CO_2, dry chemical. When heated to decomposition it emits highly toxic fumes of I^-. See also IODIDES.

E

ELQ000 **CAS:623-48-3** **HR: 3**
ETHYL IODOACETATE
mf: $C_4H_7IO_2$ mw: 214.01

PROP: Dense, colorless liquid. Bp: 179°, d: 1.80, vap press: 0.54 mm @ 20°, vap d: 7.4.

SYNS: ETHYL MONOIODOACETATE □ IODOACETIC ACID ETHYL ESTER

TOXICITY DATA WITH REFERENCE
ipr-mus LD50:45 mg/kg JNCIAM 31,297,63

CONSENSUS REPORTS: Reported in EPA TSCA Inventory.

SAFETY PROFILE: Poison by intraperitoneal route. A skin, eye, and mucous membrane irritant. Will react with water or steam to produce toxic and corrosive fumes. When heated to decomposition or on contact with acid or acid fumes it emits highly toxic fumes of I^-. See also IODIDES and ESTERS.

ELQ100 **HR: 3**
ETHYLIODOMETHYLARSINE
mf: C_3H_8AsI mw: 245.92

$CH_3CH_2As(I)CH_3$

CONSENSUS REPORTS: Arsenic and its compounds are on the Community Right-To-Know List.

SAFETY PROFILE: Arsenic compounds are poisons. Ignites spontaneously in air. When heated to decomposition it emits toxic fumes of I^- and As. See also ARSENIC COMPOUNDS and IODIDES.

ELQ500 **CAS:99-79-6** **HR: 2**
ETHYL-10-(p-IODOPHENYL)UNDECYLATE
mf: $C_{19}H_{29}IO_2$ mw: 416.38

PROP: Viscous liquid. D: 1.240–1.263 @ 20°/20°, bp: 196–198° @ 1 mm. Sltly sol in H_2O; sol in EtOH, C_6H_6, and $CHCl_3$.

SYNS: ETHIODAN ☐ ETHYL 10-(p-IODOPHENYL)UNDECANOATE ☐ IOFENDYLATE ☐ IOPHENDYLATE ☐ MULSOPAQUE ☐ MYODIL ☐ MYODYL ☐ NEUROTRAST ☐ PANTOPAQUE

TOXICITY DATA WITH REFERENCE
orl-rat LD50:2100 mg/kg DRUGAY 6,65,82
ipr-rat LD50:1130 mg/kg RPTOAN 34,38,71
ipr-mus LD50:5640 mg/kg RPTOAN 34,38,71
ivn-rbt LD50:1 g/kg RPTOAN 34,38,71

SAFETY PROFILE: Moderately toxic by ingestion, intraperitoneal, and intravenous routes. When heated to decomposition it emits toxic fumes of I^-.

ELS000 **CAS:97-62-1** **HR: 3**
ETHYL ISOBUTYRATE
DOT: UN 2385
mf: $C_6H_{12}O_2$ mw: 116.18

PROP: Colorless, volatile liquid; fruity, aromatic odor. Mp: −88°, bp: 110–111°, d: 0.869, vap press: 40 mm @ 33.8°, vap d: 4.01, refr index: 1.385, flash p: <64.4°F.

SYNS: ETHYL ISOBUTANOATE ☐ ETHYLISOBUTYRATE (DOT) ☐ ETHYL-2-METHYLPROPANOATE ☐ ETHYL-2-METHYLPROPIONATE ☐ FEMA No. 2428 ☐ ISOBUTYRIC ACID, ETHYL ESTER ☐ 2-METHYLPROPIONIC ACID, ETHYL ESTER

TOXICITY DATA WITH REFERENCE
skn-rbt 500 mg/24H MOD FCTXAV 16,741,78
ipr-mus LD50:800 mg/kg FCTXAV 16,741,78

CONSENSUS REPORTS: Reported in EPA TSCA Inventory.

DOT CLASSIFICATION: 3; *Label:* Flammable Liquid

SAFETY PROFILE: Moderately toxic by intraperitoneal route. A skin irritant. Flammable liquid. A very dangerous fire hazard when exposed to heat or flame; can react vigorously with oxidizing materials. To fight fire, use foam, CO_2, dry chemical. When heated to decomposition it emits acrid smoke and irritating fumes. See also ESTERS.

ELS500 **CAS:109-90-0** **HR: 3**
ETHYL ISOCYANATE
DOT: UN 2481
mf: C_3H_5NO mw: 71.09

PROP: Pungent smelling liquid. Bp: 60°, d: 0.90 @ 20°/4°, vap d: 2.45.

SYNS: ETHYL ISOCYANATE (DOT) ☐ ISOCYANATOETHANE ☐ ISOCYANIC ACID, ETHYL ESTER

TOXICITY DATA WITH REFERENCE
mmo-sat 50 µg/plate ABCHA6 44,3017,80
ivn-mus LD50:56 mg/kg CSLNX* NX#02910

CONSENSUS REPORTS: Reported in EPA TSCA Inventory.

DOT CLASSIFICATION: 3; *Label:* Flammable Liquid, Poison; DOT Class: 6.1; *Label:* Poison; DOT Class: 6.1; *Label:* Poison, Flammable Liquid; DOT Class: 3; *Label:* Flammable Liquid, Poison

SAFETY PROFILE: Poison by intravenous route. Mutation data reported. A flammable liquid. When heated to decomposition it emits toxic fumes of NO_x. See also CYANATES.

ELT000 **CAS:624-79-3** **HR: 3**
ETHYL ISOCYANIDE
mf: C_3H_5N mw: 55.08

$CH_3CH_2N{=}C:$

PROP: Colorless liquid. D: 0.7402 @ 20°/4°, mp: <−66°, bp: 79°. Sltly sol in water and organic solvents.

SYNS: ETHYL ISONITRILE ☐ ISOCYANOETHANE

CONSENSUS REPORTS: Cyanide and its compounds are on the Community Right-To-Know List.

SAFETY PROFILE: Probably very toxic. Can explode upon heating. When heated to decomposition it emits toxic fumes of CN^-. See also NITRILES.

ELT500 **CAS:106-67-2** **HR: 2**
2-ETHYLISOHEXANOL
mf: $C_8H_{18}O$ mw: 130.26

SYNS: 2-ETHYL-4-METHYLPENTANOL ☐ 2-ETHYL-4-METHYL-1-PENTANOL

TOXICITY DATA WITH REFERENCE
skn-rbt 10 mg/24H open MLD AIHAAP 23,95,62
orl-rat LD50:4290 mg/kg AIHAAP 23,95,62
orl-mus LDLo:1600 mg/kg KODAK* -,-,71

CONSENSUS REPORTS: Reported in EPA TSCA Inventory.

SAFETY PROFILE: Moderately toxic by ingestion. A skin irritant. When heated to decomposition it emits acrid smoke and irritating fumes. See also ALCOHOLS.

ELU000 **CAS:1570-45-2** **HR: 3**
ETHYL ISONICOTINATE
mf: $C_8H_9NO_2$ mw: 151.18

PROP: Bp: 219–220°.

SYNS: ISONICOTINIC ACID, ETHYL ESTER ☐ 4-PYRIDINECARBOXYLIC ACID, ETHYL ESTER

TOXICITY DATA WITH REFERENCE
ivn-mus LD50:56 mg/kg CSLNX* NX#03434

CONSENSUS REPORTS: Reported in EPA TSCA Inventory.

SAFETY PROFILE: Poison by intravenous route. When heated to decomposition it emits toxic fumes of NO_x.

ELX000 CAS:76-76-6 HR: 3
ETHYL ISOPROPYLBARBITURIC ACID
mf: $C_9H_{14}N_2O_3$ mw: 198.25

PROP: Needles. Mp: 197–198°.

SYNS: 5-ETHYL-5-ISOPROPYLBARBITURIC ACID □ 5-ETHYL-5-(1-METHYLETHYL)-2,4,6(1H,3H,5H)-PYRIMIDINETRIONE □ IPRAL □ IRENAL □ PROBARBITAL □ PROBARBITONE □ VASALGIN

TOXICITY DATA WITH REFERENCE
ipr-rat LDLo: 110 mg/kg JPETAB 44,325,32
scu-rat LDLo: 110 mg/kg JPETAB 26,371,25
ipr-mus LDLo: 250 mg/kg 27ZWAY -,-,36
orl-cat LDLo: 140 mg/kg 27ZWAY -,-,36
orl-rbt LDLo: 150 mg/kg JPETAB 44,337,32
ipr-rbt LDLo: 110 mg/kg JPETAB 44,325,32
scu-rbt LDLo: 200 mg/kg JACSAT 45,243,23
ivn-rbt LDLo: 140 mg/kg JPPGAR 30,364,32
orl-bwd LD50: 24 mg/kg TXAPA9 21,315,72

SAFETY PROFILE: Poison by ingestion, intraperitoneal, subcutaneous, and intravenous routes. When heated to decomposition it emits toxic fumes of NO_x. See also BARBITURATES.

ELX100 HR: 3
ETHYL ISOPROPYL FLUOROPHOSPHONATE
mf: $C_5H_{12}FOP$ mw: 138.14

SYN: ISOPROPYL ETHANE FLUOROPHOSPHONATE

TOXICITY DATA WITH REFERENCE
ihl-rat LC50: 260 mg/m³/10M NTIS** PB158-508
ihl-mus LC50: 245 mg/m³/5M NTIS** PB158-508
skn-mus LD50: 1700 µg/kg NTIS** PB158-508
scu-mus LD50: 400 µg/kg NTIS** PB158-508
ihl-dog LC50: 230 mg/m³/10M NTIS** PB158-508
ihl-mky LC50: 210 mg/m³/10M NTIS** PB158-508
ihl-cat LC50: 170 mg/m³/10M NTIS** PB158-508
ihl-rbt LC50: 230 mg/m³/10M NTIS** PB158-508
ihl-gpg LC50: 350 mg/m³/10M NTIS** PB158-508

SAFETY PROFILE: Poison by inhalation, skin contact, and subcutaneous routes. When heated to decomposition it emits toxic fumes of F^- and PO_x. See also ESTERS.

ELX500 CAS:16339-04-1 HR: 2
ETHYL ISOPROPYLNITROSAMINE
mf: $C_5H_{12}N_2O$ mw: 116.19

PROP: Bp: 70° @ 11 mm.

SYNS: AETHYL-ISOPROPYL-NITROSOAMIN (GERMAN) □ 1-METHYL-N-NITROSODIETHYLAMINE □ N-NITROSOETHYLISOPROPYLAMINE

TOXICITY DATA WITH REFERENCE
orl-rat TDLo: 3920 mg/kg/56W-C:CAR NATWAY 50,100,63
orl-rat TD: 3700 mg/kg/53W-C:ETA ARZNAD 19,1077,69
orl-rat LD50: 1100 mg/kg NATWAY 50,100,63

SAFETY PROFILE: Moderately toxic by ingestion. Questionable carcinogen with experimental carcinogenic and tumorigenic data. When heated to decomposition it emits toxic fumes of NO_x. See also NITROSAMINES.

ELX530 CAS:24066-82-8 HR: 3
ETHYL ISOTHIOCYANOACETATE
mf: $C_5H_7NO_2S$ mw: 145.19

SYNS: ACETIC ACID, ISOTHIOCYANATO-, ETHYL ESTER □ CARBETHOXYMETHYL ISOTHIOCYANATE □ CARBOETHOXYMETHYL ISOTHIOCYANATE □ ETHYL ISOTHIOCYANATOACETATE

TOXICITY DATA WITH REFERENCE
scu-rat LDLo: 350 mg/kg AIPTAK 35,314,29

CONSENSUS REPORTS: Reported in EPA TSCA Inventory.

SAFETY PROFILE: A poison by subcutaneous route. When heated to decomposition it emits toxic vapors of NO_x and SO_x.

ELY550 CAS:22722-03-8 HR: 3
S-ETHYLISOTHIOURONIUM HYDROGEN SULFATE
mf: $C_3H_{10}N_2O_4S_2$ mw: 202.24

$CH_3CH_2SC(:N^+H_2)NH_2HSO_4^-$

SYN: ETHYL CARBAMIMONIOTHIOATE HYDROGEN SULFATE

SAFETY PROFILE: Reacts with chlorine to produce the dangerously explosive nitrogen trichloride. When heated to decomposition it emits toxic fumes of SO_x and NO_x. See also SULFATES.

ELY575 CAS:21704-44-9 HR: 3
S-ETHYLISOTHIURONIUM METAPHOSPHATE
mf: $C_3H_8N_2S{\cdot}HO_3P$ mw: 184.17

SYNS: CARBAMIMIDOTHIOIC ACID, ETHYL ESTER with METAPHOSPHORIC ACID (1:1) □ 2-ETHYL-2-THIO-PSEUDOUREA with METAPHOSPHORIC ACID (1:1)

TOXICITY DATA WITH REFERENCE
ipr-rat LD50: 338 mg/kg FATOAO 43,212,80
orl-mus LD50: 2490 mg/kg FATOAO 43,212,80
ipr-mus LD50: 455 mg/kg FATOAO 43,212,80
scu-mus LD50: 472 mg/kg FATOAO 43,212,80
ims-mus LD50: 450 mg/kg FATOAO 43,212,80

SAFETY PROFILE: Poison by intraperitoneal route. Moderately toxic by ingestion, subcutaneous, and intramuscular routes. When heated to decomposition it emits toxic fumes of NO_x, SO_x, and PO_x.

ELY700 CAS:106-33-2 HR: 2
ETHYL LAURATE
mf: $C_{14}H_{28}O_2$ mw: 228.37

PROP: Colorless, oily liquid; fruity-floral odor. D: 0.858, refr index: 1.430, bp: 163° @ 25 mm, flash p: 212°F. Misc in alc, chloroform, ether; insol in water @ 269°.

SYNS: ETHYL DODECANOATE □ FEMA No. 2441

SAFETY PROFILE: Combustible liquid. When heated to decomposition it emits acrid smoke and irritating fumes.

ELZ000 **CAS:10339-55-6** **HR: 1**
ETHYL LINALOOL
mf: $C_{11}H_{20}O$ mw: 168.31

SYN: 3,7-DIMETHYL-16-NONADIEN-3-OL

TOXICITY DATA WITH REFERENCE
skn-rbt 500 mg/24H MOD FCTXAV 14,767,76

CONSENSUS REPORTS: Reported in EPA TSCA Inventory.

SAFETY PROFILE: A skin irritant. When heated to decomposition it emits acrid smoke and irritating fumes.

ELZ050 **CAS:40910-49-4** **HR: 1**
ETHYLLINALYL ACETAL
mf: $C_{14}H_{26}O_2$ mw: 226.40

SYNS: ACETALDEHYDE ETHYL LINALYL ACETAL □ 3,7-DIMETHYL-3-(1-ETHOXYETHOXY)-1,6-OCTADIENE □ ELINTAAL □ 1-ETHOXY-1-LINALYLOXETHANE □ 1,6-OCTADIENE, 3,7-DIMETHYL-3-(1-ETHOXYETHOXY)-

TOXICITY DATA WITH REFERENCE
orl-rat LD50:>5 g/kg FCTOD7 26,281,88
skn-rbt LD50:>5 g/kg FCTOD7 26,281,88

CONSENSUS REPORTS: Reported in EPA TSCA Inventory.

SAFETY PROFILE: Low toxicity by ingestion and skin contact. When heated to decomposition it emits acrid smoke and irritating vapors.

ELZ100 **HR: 3**
ETHYLLITHIUM
mf: C_2H_5Li mw: 36.00

SAFETY PROFILE: Ignites spontaneously in air. See also LITHIUM COMPOUNDS. When heated to decomposition it emits acrid smoke and irritating fumes.

EMA000 **CAS:10467-10-4** **HR: 3**
ETHYL MAGNESIUM IODIDE
mf: C_2H_5IMg mw: 180.27

CH_3CH_2MGI

SAFETY PROFILE: Mixtures with ethoxyacetylene in ether are explosive. When heated to decomposition it emits toxic fumes of I^-. See also IODIDES and MAGNESIUM COMPOUNDS.

EMA500 **CAS:105-53-3** **HR: 2**
ETHYL MALONATE
mf: $C_7H_{12}O_4$ mw: 160.19

PROP: Clear, colorless liquid; fruit-like odor. Bp: 198.9°, fp: −49.8°, flash p: 200°F (OC), d: 1.055 @ 20°/4°, refr index: 1.413–1.416, vap press: 1 mm @ 40.0°, vap d: 5.52. Sol in fixed oils, propylene glycol; sltly sol in alc, water; insol in glycerin, mineral oil @ 200°.

SYNS: CARBETHOXYACETIC ESTER □ DICARBETHOXYMETHANE □ DIETHYL MALONATE (FCC) □ DIETHYL PROPANEDIOATE □ FEMA No. 2375 □ MALONIC ACID, DIETHYL ESTER □ MALONIC ESTER □ METHANEDICARBOXYLIC ACID, DIETHYL ESTER □ PROPANEDIOIC ACID, DIETHYL ESTER

TOXICITY DATA WITH REFERENCE
skn-rbt 500 mg/24H MLD FCTXAV 14,745,76
orl-rat LD50:15 g/kg AIHAAP 30,470,69
orl-mus LD50:6400 mg/kg BIJOAK 34,1196,40

CONSENSUS REPORTS: Reported in EPA TSCA Inventory.

SAFETY PROFILE: Mildly toxic by ingestion. A skin irritant. Combustible liquid when exposed to heat or flame; can react with oxidizing materials. To fight fire, use water to blanket fire, foam, CO_2, dry chemical. When heated to decomposition it emits acrid smoke and irritating fumes. See also ESTERS.

EMA600 **CAS:4940-11-8** **HR: 2**
ETHYL MALTOL
mf: $C_7H_8O_3$ mw: 140.15

PROP: White crystalline powder; sweet fruity taste. Mp: 90°. Sol in water, alc, propylene glycol, chloroform.

SYNS: 2-ETHYL-3-HYDROXY-4H-PYRAN-4-ONE □ 2-ETHYL PYROMECONIC ACID □ 3-HYDROXY-2-ETHYL-4-PYRONE

TOXICITY DATA WITH REFERENCE
mmo-sat 1 mg/plate MUREAV 67,367,79
orl-rat LD50:1150 mg/kg TXAPA9 15,604,69
orl-mus LD50:780 mg/kg TXAPA9 15,604,69
scu-mus LD50:910 mg/kg CPBTAL 22,1008,74
orl-ckn LD50:1270 mg/kg TXAPA9 15,604,69

CONSENSUS REPORTS: Reported in EPA TSCA Inventory.

SAFETY PROFILE: Moderately toxic by ingestion and subcutaneous routes. Mutation data reported. When heated to decomposition it emits acrid smoke and irritating fumes.

EMB000 **CAS:774-40-3** **HR: 2**
ETHYL MANDELATE
mf: $C_{10}H_{11}O_2$ mw: 163.21

PROP: Crystals.

SYNS: MANDELIC ACID, ETHYL ESTER □ MANDELSAEUREAETHYLESTER (GERMAN)

TOXICITY DATA WITH REFERENCE
eye-rbt 5 mg AJOPAA 29,1363,46
orl-rat LD50:3750 mg/kg ARZNAD 12,347,62

SAFETY PROFILE: Moderately toxic by ingestion. An eye irritant. Combustible when exposed to heat or flame; can react with oxidizing materials. When heated to decomposition it emits acrid smoke and irritating fumes. See also ESTERS.

EMB100 **CAS:75-08-1** **HR: 3**
ETHYL MERCAPTAN
DOT: UN 2363
mf: C_2H_6S mw: 62.14

PROP: Colorless, volatile, liquid with penetrating garlic-like odor. Fp: −148°, mp: −121°, bp: 36.1°, lel: 2.8%, uel: 18.2%, d: 0.83907 @ 20°/4°, autoign temp: 570°F, vap d: 2.14, flash p: <−0.4°F. Sol in EtOH, Et_2O, alkalis; very sltly sol in H_2O.

SYNS: AETHANETHIOL (GERMAN) □ AETHYLMERCAPTAN (GERMAN) □ ETANTIOLO (ITALIAN) □ ETHAANTHIOL (DUTCH) □ ETHANETHIOL □ ETHYL HYDROSULFIDE □ ETHYLMERCAPTAAN (DUTCH) □ ETHYLMERKAPTAN (CZECH) □ ETHYL SULFHYDRATE □ ETHYL THIOALCOHOL □ ETILMERCAPTANO (ITALIAN) □ LPG ETHYL MERCAPTAN 1010 □ THIOETHANOL □ THIOETHYL ALCOHOL

TOXICITY DATA WITH REFERENCE
skn-rbt 500 mg/24H MLD 85JCAE-,982,86
eye-rbt 84 mg AIHAAP 19,171,58
eye-rbt 100 mg/24H MOD 85JCAE-,982,86
orl-rat LD50:1960 mg/kg 28ZPAK -,166,72
ihl-rat LC50:4420 ppm/4H AIHAAP 19,171,58
ipr-rat LD50:450 mg/kg AIHAAP 19,171,58
ihl-mus LD50:2770 mg/kg AIHAAP 19,171,58

CONSENSUS REPORTS: Reported in EPA TSCA Inventory.

OSHA PEL: TWA 0.5 ppm
ACGIH TLV: TWA 0.5 ppm
DFG MAK: 0.5 ppm (1 mg/m³)
NIOSH REL: (n-Alkane Mono Thiols) CL 0.5 ppm/15M
DOT CLASSIFICATION: 3; *Label:* Flammable Liquid; DOT Class: 6.1; *Label:* Poison, Flammable Liquid

SAFETY PROFILE: Moderately toxic by ingestion, inhalation, and intraperitoneal routes. A skin and eye irritant. Inhalation causes central nervous system effects in humans. A very dangerous fire hazard when exposed to heat or flame; can react vigorously with oxidizing materials. A moderate explosion hazard when exposed to spark or flame. Violent reaction with $Ca(OCl)_2$. Will react with water or steam to produce toxic and flammable vapors. To fight fire, use CO_2, dry chemical. When heated to decomposition or on contact with acid or acid fumes it emits highly toxic fumes of SO_x. See also MERCAPTANS.

EMB200 **CAS:623-51-8** **HR: 3**
ETHYL-2-MERCAPTOACETATE
mf: $C_4H_8O_2S$ mw: 120.18

PROP: Bp: 156–158°.

SYNS: ETHYL MERCAPTOACETATE □ ETHYL-α-MERCAPTOACETATE □ ETHYL MERCAPTOACETIC ACID □ ETHYL THIOGLYCOLATE □ MERCAPTOACETIC ACID ETHYL ESTER □ THIOGLYCOLIC ACID ETHYL ESTER □ THIOGLYKOLSAEURE-AETHYLESTER (GERMAN) □ USAF EK-2070

TOXICITY DATA WITH REFERENCE
orl-rat LD50:178 mg/kg ZHYGAM 20,575,74
ipr-rat LD50:176 mg/kg ZHYGAM 20,575,74
ipr-mus LD50:100 mg/kg NTIS** AD277-689

SAFETY PROFILE: Poison by ingestion and intraperitoneal routes. When heated to decomposition it emits toxic fumes of SO_x. See also MERCAPTANS.

EMC000 **CAS:5840-95-9** **HR: 3**
ETHYL (2-MERCAPTOETHYL) CARBAMATE S-ESTER with O,O-DIMETHYL PHOSPHORODITHIOATE
mf: $C_7H_{16}NO_4PS_2$ mw: 273.33

SYNS: (2-((DIETHOXYPHOSPHINOTHIOYL)THIO)ETHYL)CARBAMIC ACID, ETHYL ESTER □ S-(O,O-DIMETHYLPHOSPHORODITHIOATE) of N-(2-MERCAPTOETHYL)ETHYLCARBAMATE □ ENT 25,801 □ ETHYL (2-((DIETHOXYPHOSPHINOTHIOYL)THIO)ETHYL)CARBAMATE □ ETHYL-N-(2-(O,O-DIMETHYLPHOSPHORODITHIOYL)ETHYL)CARBAMATE □ (2-MERCAPTOETHYL)CARBAMIC ACID, ETHYL ESTER, S-ESTER with O,O-DIMETHYL PHOSPHORODITHIOATE □ R-3422-S □ STAUFFER R-3442-S

TOXICITY DATA WITH REFERENCE
orl-rat LD50:108 mg/kg 28ZEAL 4,220,69
orl-mus LDLo:470 mg/kg AECTCV 14,111,85
orl-ckn LD50:255 mg/kg TXAPA9 7,606,65

SAFETY PROFILE: Poison by ingestion. When heated to decomposition it emits very toxic fumes of NO_x, PO_x, and SO_x. See also CARBAMATES.

EMC500 **CAS:5427-20-3** **HR: 3**
9-ETHYL-6-MERCAPTOPURINE
mf: $C_7H_8N_4S$ mw: 180.25

SYNS: 9-ETHYL-1,9-DIHYDRO-6H-PURINE-6-THIONE □ 9-ETHYL-6-MP □ 9-ETHYL-9H-PURINE-6-THIOL □ 9-ETHYL-9H-PURINE-6(1H)-THIONE □ NSC 14575

TOXICITY DATA WITH REFERENCE
ipr-rat LD50:400 mg/kg ADTEAS 3,181,68
ipr-mus LD50:289 mg/kg NCISP* JAN86

SAFETY PROFILE: Poison by intraperitoneal route. When heated to decomposition it emits very toxic fumes of NO_x and SO_x. See also MERCAPTANS.

EMD000 **CAS:109-62-6** **HR: 3**
ETHYLMERCURIC ACETATE
mf: $C_4H_8HgO_2$ mw: 288.71

PROP: Crystals from CCl_4. Mp: 69–69.8°.

SYNS: (ACETATO-O)ETHYLMERCURY □ ETHYLMERKURIACETAT

TOXICITY DATA WITH REFERENCE
orl-ckn LDLo:29 mg/kg TXAPA9 3,459,61

CONSENSUS REPORTS: Mercury and its compounds are on the Community Right-To-Know List.

OSHA PEL: TWA 0.01 mg(Hg)/m³; STEL 0.03 mg/m³ (skin)
ACGIH TLV: TWA 0.01 mg(Hg)/m³; STEL 0.03 mg(Hg)/m³
NIOSH REL: (Mercury, Organo): TWA 0.01 mg/m³; STEL 0.03 mg/m³ (skin)

SAFETY PROFILE: Poison by ingestion. When heated to decomposition it emits toxic fumes of Hg. See also MERCURY COMPOUNDS.

EME000 **CAS:21082-50-8** **HR: 3**
ETHYLMERCURIC CYSTEINE
mf: $C_5H_{11}NO_2S{\cdot}Hg$ mw: 349.82

SYN: ETHYL(HYDROGEN CYSTEINATO)MERCURY

TOXICITY DATA WITH REFERENCE
cyt-hmn:hla 1 mg/L JJEMAG 39,47,69
scu-rat LD50:131 mg/kg JJEMAG 39,47,69

CONSENSUS REPORTS: Mercury and its compounds are on the Community Right-To-Know List.

OSHA PEL: TWA 0.01 mg(Hg)/m³; STEL 0.03 mg/m³ (skin)
ACGIH TLV: TWA 0.01 mg(Hg)/m³; STEL 0.03 mg(Hg)/m³

SAFETY PROFILE: Poison by subcutaneous route. Human mutation data reported. When heated to decomposition it emits very toxic fumes of NO_x, SO_x, and Hg. See also MERCURY COMPOUNDS.

EME050 **CAS:2597-93-5** **HR: 3**
ETHYLMERCURICHLORENDIMIDE
mf: $C_{11}H_7Cl_6HgNO_2$ mw: 598.48

SYNS: 50-CS-46 □ EMMI □ N-(ETHYLMERCURI)-1,4,5,6,7,7-HEXACHLOROBICYCLO(2.2.1)HEPT-5-ENE-2,3-DICARBOXIMIDE □ N-ETHYLMERCURI-3,4,5,6,7,7-HEXACHLORO-3,6-ENDOMETHYLENE-1,2,3,6- TETRAHYDROPHTHALIMIDE □ N-ETHYLMERCURI-1,2,3,6-TETRAHYDRO-3,6-ENDOMETHANO-3,4,5,6,7,7-HEXACHLOROPHTHALIMIDE □ 1,4,5,6,77-HEXACHLORO-N-(ETHYLMERCURI)-5-NORBORNENE-2,3-DICARBOXIMIDE

TOXICITY DATA WITH REFERENCE
orl-rat LD50:150 mg/kg PCOC** -,465,66

CONSENSUS REPORTS: Mercury and its compounds are on the Community Right-To-Know List.

OSHA PEL: CL 0.1 mg(Hg)/m³ (skin)
ACGIH TLV: TWA 0.1 mg(Hg)/m³ (skin)
NIOSH REL: (Mercury, Aryl and Inorganic): CL 0.1 mg/m³ (skin)

SAFETY PROFILE: Poison by ingestion. When heated to decomposition it emits very toxic fumes of Cl^-, Hg, and NO_x. See also MERCURY COMPOUNDS.

EME100 **CAS:2235-25-8** **HR: 3**
ETHYLMERCURIC PHOSPHATE
mf: $C_2H_7HgO_4P$ mw: 326.65

SYNS: EMP □ ETHYLMERCURY PHOSPHATE □ GRANOSAN M □ LIGNASAN □ N. I. CERESAN □ RUBERON □ SOILSIN

TOXICITY DATA WITH REFERENCE
scu-mus TDLo:40 mg/kg (10D preg):TER CAJPBD 7,53,67
orl-hmn LDLo:8614 µg/kg/13W TXCYAC 6,155,76
orl-rat LD50:48 mg/kg NYKZAU 59,452,63
orl-mus LD50:48 mg/kg NYKZAU 59,452,63
scu-mus LD50:76 mg/kg CAJPBD 7,53,67

CONSENSUS REPORTS: Mercury and its compounds are on the Community Right-To-Know List.

OSHA PEL: TWA 0.01 mg(Hg)/m³; STEL 0.03 mg/m³ (skin)
ACGIH TLV: TWA 0.01 mg(Hg)/m³; STEL 0.03 mg(Hg)/m³
NIOSH REL: (Mercury, Organo): TWA 0.01 mg/m³; STEL 0.03 mg/m³ (skin)

SAFETY PROFILE: Human poison by ingestion. Experimental poison by subcutaneous route. An experimental teratogen. When heated to decomposition it emits toxic fumes of PO_x and Hg. See also MERCURY COMPOUNDS and PHOSPHATES.

EME500 **CAS:517-16-8** **HR: 3**
ETHYLMERCURY-p-TOLUENE SULFONAMIDE
mf: $C_{15}H_{17}HgNO_2S$ mw: 475.98

PROP: Crystals from EtOH; pungent, garlic-like odor. Mp: 156°. Practically insol in water.

SYNS: CERESAN M □ COMPOUND-1452-F □ EMTS □ N-ETHYLMERCURI-N-PHENYL-p-TOLUENESULFONAMIDE □ N-(ETHYLMERCURI)-p-TOLUENESULFONANILIDE □ N-(ETHYLMERCURI)-p-TOLUENESULPHONANILIDE □ ETHYLMERCURY p-TOLUENESULFANILIDE □ ETHYLMERCURY-p-TOLUENESULFONANILIDE □ ETHYL(N-PHENYL-p-TOLUENESULFONAMIDATO)MERCURY □ ETHYL(N-PHENYL-p-TOLUENESULFONAMIDO)MERCURY □ ETHYL(p-TOLUENESULFONANILIDATO)MERCURY □ GRANOSAN M □ (N-PHENYL-p-TOLUENESULFONAMIDO)ETHYLMERCURY

TOXICITY DATA WITH REFERENCE
sln-dmg-orl 40 mg/kg MUREAV 40,31,76
orl-rat LD50:100 mg/kg PCOC** -,204,66
ipr-mus LD50:128 mg/kg TOIZAG 7,71,60

CONSENSUS REPORTS: Mercury and its compounds are on the Community Right-To-Know List. EPA Genetic Toxicology Program.

OSHA PEL: TWA 0.01 mg(Hg)/m³; STEL 0.03 mg/m³ (skin)
ACGIH TLV: TWA 0.01 mg(Hg)/m³; STEL 0.03 mg(Hg)/m³
NIOSH REL: (Mercury, Organo): TWA 0.01 mg/m³; STEL 0.03 mg/m³ (skin)

SAFETY PROFILE: Poison by ingestion and intraperitoneal route. Mutation data reported. A fungicide. When heated to decomposition it emits very toxic fumes of Hg, NO_x, and SO_x. See also MERCURY COMPOUNDS.

EMF000 **CAS:97-63-2** **HR: 3**
ETHYL METHACRYLATE
DOT: UN 2277
mf: $C_6H_{10}O_2$ mw: 114.16

PROP: A liquid. Mp: <−75°, bp: 119°, lel: 1.8%, uel: saturation, flash p: 68°F (OC), d: 0.911 @ 25°/25°, vap d: 3.94.

SYNS: ETHYL METHACRYLATE, INHIBITED (DOT) □ ETHYL-α-METHYL ACRYLATE □ ETHYL-2-METHYLACRYLATE □ ETHYL-2-METHYL-2-PROPENOATE □ 2-METHYL-2-PROPENOIC ACID, ETHYL ESTER □ RCRA WASTE NUMBER U118 □ RHOPLEX AC-33 (ROHM and HAAS)

TOXICITY DATA WITH REFERENCE
skn-rbt 10 g/kg open JIHTAB 23,343,41

ipr-rat TDLo:735 mg/kg (5-15D preg):TER JDREAF 51,1632,72
ipr-rat TDLo:366 mg/kg (5-15D preg):REP JDREAF 51,1632,72
ipr-rat TDLo:366 mg/kg (5-15D preg):TER JDREAF 51,1632,72
orl-rat LD50:14,800 mg/kg JIHTAB 23,343,41
ihl-rat LC50:8300 ppm/4H JTEHD6 16,811,85
ipr-rat LD50:1223 mg/kg JDREAF 51,1632,72
scu-rat LDLo:25 g/kg JIHTAB 23,343,41
orl-mus LD50:7836 mg/kg TOLED5 11,125,82
ipr-mus LD50:1369 mg/kg JPMSAE 62,778,73
orl-rbt LDLo:3630 mg/kg JIHTAB 23,343,41

CONSENSUS REPORTS: Reported in EPA TSCA Inventory.

DOT CLASSIFICATION: 3; *Label:* Flammable Liquid

SAFETY PROFILE: Moderately toxic by ingestion and intraperitoneal routes. Mildly toxic by inhalation. Experimental teratogenic and reproductive effects. A skin irritant. A very dangerous fire and explosion hazard when exposed to heat, sparks, or flame; can react with oxidizing materials. To fight fire, use CO_2, dry chemical. When heated to decomposition it emits acrid smoke and irritating fumes.

EMF500 CAS:62-50-0 HR: 3
ETHYL METHANESULFONATE
mf: $C_3H_8O_3S$ mw: 124.17

SYNS: EMS □ ENT 26,396 □ ETHYL ESTER of METHANESULFONIC ACID □ ETHYL ESTER of METHYLSULFONIC ACID □ ETHYL ESTER of METHYLSULPHONIC ACID □ ETHYL METHANESULPHONATE □ ETHYL METHANSULFONATE □ ETHYL METHANSULPHONATE □ HALF-MYLERAN □ METHANESULPHONIC ACID ETHYL ESTER □ METHYLSULFONIC ACID, ETHYL ESTER □ NSC 26805 □ RCRA WASTE NUMBER U119

TOXICITY DATA WITH REFERENCE
sln-dmg-orl 500 ppm ENMUDM 7(Suppl 3),76,85
oms-hmn:lym 400 µmol/L MUREAV 155,75,85
ipr-rat TDLo:200 mg/kg (female 15D post):REP JRPFA4 18,15,69
ipr-rat TDLo:200 mg/kg (female 15D post):TER JRPFA4 18,15,69
orl-mus TDLo:1750 mg/kg (male 7D pre):REP JACTDZ 2(2),209,83
orl-mus TDLo:1 g/kg (male 5D pre):TER EVHPAZ 21,71,77
ipr-mus TDLo:175 mg/kg (female 11D post):TER ARTODN 51,1,82
ipr-mus TDLo:200 mg/kg (female 1D pre):REP MUREAV 13,171,71
orl-mus TDLo:1 g/kg (male 5D pre):REP EVHPAZ 21,71,77
ipr-rat TDLo:50 mg/kg (male 1D pre):REP TCMUD8 7,497,87
orl-rat TDLo:500 mg/kg (5D male):REP TXAPA9 70,303,83
ipr-ham TDLo:25 mg/kg (male 5D pre):REP JTEHD6 8,929,81
ipr-mus TDLo:195 mg/kg (female 11D post):TER ARTODN 51,1,82
ipr-rat TDLo:200 mg/kg (female 13D post):TER JRPFA4 17,325,68
par-mus TDLo:100 mg/kg (female 10D post):TER EVHPAZ 24,113,78
orl-rat TDLo:1050 g/kg/12W-C:CAR CALEDQ 7,79,79
ipr-rat TDLo:300 mg/kg:CAR BJCAAI 29,50,74
ivn-rat TDLo:100 mg/kg (21D post):ETA EXPTAX 16,157,78
ivn-rat TDLo:1650 mg/kg/2W-I:ETA 43XWAI -,15,78
par-rat TDLo:800 mg/kg/3W-I:ETA RCOCB8 7,25,74
ipr-mus TDLo:373 mg/kg:NEO CBINA8 3,117,71
ipr-rat TD:825 mg/kg/10D-I:NEO NATUAS 223,947,69
ipr-rat TD:413 mg/kg/10D-I:ETA NATUAS 223,947,69
orl-rat TD:3353 mg/kg/13W-C:CAR CRNGDP 2,1223,81
ipr-mus TD:600 mg/kg/6W-I:ETA BECCAN 39,77,61
ipr-mus TD:400 mg/kg:ETA BIJOAK 174,1031,78
ivn-rat TD:1650 mg/kg/30W-I:ETA EXPTAX 16,157,78
orl-rat LD :1050 g/kg/12W-C:CAR CALEDQ 7,79,79
ipr-rat LD50:350 mg/kg CPBTAL 8,807,60
orl-mus LD50:470 mg/kg MUREAV 223,373,89
ipr-mus LD50:435 mg/kg CBINA8 3,117,71

CONSENSUS REPORTS: NTP 7th Annual Report on Carcinogens. IARC Cancer Review: Group 2B IMEMDT 7,56,87; Animal Sufficient Evidence IMEMDT 7,245,74. Reported in EPA TSCA Inventory. EPA Genetic Toxicology Program.

SAFETY PROFILE: Confirmed carcinogen with experimental carcinogenic, neoplastigenic, tumorigenic, and teratogenic data. Poison by ingestion and intraperitoneal routes. Experimental reproductive effects. Human mutation data reported. When heated to decomposition it emits toxic fumes of SO_x. See also SULFONATES and ESTERS.

EMF600 CAS:64988-06-3 HR: 1
ETHYL o-METHOXYBENZYL ETHER
mf: $C_{10}H_{14}O_2$ mw: 166.24

SYNS: BENZENE, 1-(ETHOXYMETHYL)-2-METHOXY- □ 1-(ETHOXYMETHYL)-2-METHOXYBENZENE □ ETHYL 2-METHOXYBENZYL ETHER □ o-METHOXYBENZYL ETHYL ETHER □ 2-METHOXYBENZYL ETHYL ETHER

TOXICITY DATA WITH REFERENCE
orl-rat LDLo:5 g/kg FCTOD7 30,33S,92
skn-rbt LD50:>5 g/kg FCTOD7 30,33S,92

CONSENSUS REPORTS: Reported in EPA TSCA Inventory.

SAFETY PROFILE: Low toxicity by ingestion and skin contact. When heated to decomposition it emits acrid smoke and irritating vapors.

EMG000 CAS:102504-44-9 HR: 3
2′-ETHYL-2-(2-METHOXY BUTYLAMINO) PROPIONANILIDE
mf: $C_{16}H_{26}N_2O_2$ mw: 278.44

TOXICITY DATA WITH REFERENCE
ipr-mus LD50:150 mg/kg JPMSAE 67,595,78
ivn-mus LD50:26 mg/kg JPMSAE 67,595,78

SAFETY PROFILE: Poison by intraperitoneal and intravenous routes. When heated to decomposition it emits toxic fumes of NO_x.

EMG500 CAS:67262-69-5 HR: 3
2′-ETHYL-3-(2-METHOXYETHYL)AMINOBUTYRANILIDE HYDROCHLORIDE
mf: $C_{15}H_{24}N_2O_2 \cdot ClH$ mw: 300.87

TOXICITY DATA WITH REFERENCE
ipr-mus LD50:225 mg/kg JPMSAE 67,595,78
ivn-mus LD50:40 mg/kg JPMSAE 67,595,78

SAFETY PROFILE: Poison by intraperitoneal and intravenous routes. When heated to decomposition it emits very toxic fumes of HCl and NO_x.

EMH000 CAS:67262-72-0 HR: 3
2′-ETHYL-4-(2-METHOXYETHYL)AMINOBUTYRANILIDE HYDROCHLORIDE
mf: $C_{15}H_{24}N_2O_2 \cdot ClH$ mw: 300.87

TOXICITY DATA WITH REFERENCE
ipr-mus LD50:400 mg/kg JPMSAE 67,595,78
ivn-mus LD50:80 mg/kg JPMSAE 67,595,78

SAFETY PROFILE: Poison by intraperitoneal and intravenous routes. When heated to decomposition it emits very toxic fumes of HCl and NO_x.

EMH500 CAS:67262-71-9 HR: 3
2′-ETHYL-3-(2-METHOXYETHYL)AMINO-3-METHYLBUTYRANILIDE CYCLAMATE
mf: $C_{16}H_{26}N_2O_2 \cdot C_6H_{13}NO_3S$ mw: 457.70

SYN: 2′-ETHYL-3-(2-METHOXYETHYL)AMINO-3-METHYLBUTYRANILIDE CYCLOHEXANE SULFAMATE

TOXICITY DATA WITH REFERENCE
ipr-mus LD50:125 mg/kg JPMSAE 67,595,78
ivn-mus LD50:22 mg/kg JPMSAE 67,595,78

SAFETY PROFILE: Poison by intraperitoneal and intravenous routes. When heated to decomposition it emits very toxic fumes of NO_x and SO_x.

EMI000 CAS:67262-64-0 HR: 3
2′-ETHYL-2-(2-METHOXYETHYLAMINO)PROPIONANILIDE
mf: $C_{14}H_{22}N_2O_2$ mw: 250.38

TOXICITY DATA WITH REFERENCE
ipr-mus LD50:250 mg/kg JPMSAE 67,595,78
ivn-mus LD50:38 mg/kg JPMSAE 67,595,78

SAFETY PROFILE: Poison by intraperitoneal and intravenous routes. When heated to decomposition it emits toxic fumes of NO_x.

EMI500 CAS:102504-45-0 HR: 3
2′-ETHYL-2-(2-METHOXYETHYLAMINO)-PROPIONANILIDE HYDROCHLORIDE
mf: $C_{14}H_{22}N_2O_2 \cdot ClH$ mw: 286.84

TOXICITY DATA WITH REFERENCE
ipr-mus LD50:250 mg/kg JPMSAE 67,595,78
ivn-mus LD50:65 mg/kg JPMSAE 67,595,78

SAFETY PROFILE: Poison by intraperitoneal and intravenous routes. When heated to decomposition it emits very toxic fumes of HCl and NO_x.

EMJ500 CAS:54350-48-0 HR: 1
ETHYL all-trans-9-(4-METHOXY-2,3,6-TRIMETHYLPHENYL)-3,7-DIMETHYL-2,4,6,8-NONATETRAENOATE
mf: $C_{23}H_{30}O_3$ mw: 354.53

PROP: A solid. Mp: 104–105°.

SYNS: 3,7-DIMETHYL-9-(4-METHOXY-2,3,6-TRIMETHYLPHENYL)-2,4,6,8-NONANETETRAENOIC ACID ETHYL ESTER □ ETHYL ETRINOATE □ ETRETINATE □ Ro 10-9359 □ TIGASON

TOXICITY DATA WITH REFERENCE
sce-hmn:fbr 5 mg/L MUREAV 58,317,78
orl-mus TDLo:40 mg/kg (female 7-16D post):REP 46IZA7 -,49,81
orl-mus TDLo:80 mg/kg (female 7-16D post):TER 46IZA7 -,49,81
orl-rat TDLo:40 mg/kg (female 1D pre):TER YACHDS 10,5117,82
orl-rat TDLo:40 mg/kg (female 1D pre):REP YACHDS 10,5117,82
orl-rat TDLo:44 mg/kg (female 7-17D post):TER YACHDS 10,5095,82
orl-rat TDLo:40 mg/kg (female 1D pre):TER YACHDS 10,5117,82
orl-rat TDLo:11 mg/kg (7-17D preg):TER YACHDS 10,5095,82
orl-mus TDLo:100 mg/kg (female 11D post):TER TJADAB 34,412,86
orl-hmn TDLo:78 mg/kg/26W-I:SKN ARZNAD 32,842,82
orl-wmn TDLo:60 mg/kg/17W-I:SKN CUTIBC 35,466,85
orl-wmn TDLo:80 µg/kg/8D-I:SKN,SYS BJDEAZ 112,373,85

CONSENSUS REPORTS: EPA Genetic Toxicology Program.

SAFETY PROFILE: Human systemic effects: dermatitis, nail and hair changes, increased body temperature. Experimental teratogenic and reproductive effects. Human mutation data reported. When heated to decomposition it emits acrid smoke and irritating fumes.

EMK600 CAS:689-93-0 HR: 3
ETHYL METHYL ARSINE
mf: C_3H_9As mw: 120.03

$CH_3CH_2AsHCH_3$

CONSENSUS REPORTS: Arsenic and its compounds are on the Community Right-To-Know List.

SAFETY PROFILE: Arsenic compounds are poisons. Ignites spontaneously in air. When heated to decomposition it emits toxic fumes of As. See also ARSENIC COMPOUNDS and ARSINE.

EMM000 CAS:63039-89-4 HR: 2
7-ETHYL-9-METHYLBENZ(c)ACRIDINE
mf: $C_{20}H_{17}N$ mw: 271.38

SYNS: 3-METHYL-10-ETHYLBENZ(c)ACRIDINE □ 3-METHYL-10-ETHYL-7,8-BENZACRIDINE (FRENCH)

TOXICITY DATA WITH REFERENCE
skn-mus TDLo:348 mg/kg/29W-I:ETA ACRSAJ 4,315,56
scu-mus TDLo:72 mg/kg/9W-I:NEO MUREAV 66,307,79
scu-mus TD:250 mg/kg/10D-I:ETA BAFEAG 34,22,47

SAFETY PROFILE: Questionable carcinogen with experimental neoplastigenic and tumorigenic data. When heated to decomposition it emits toxic fumes of NO_x.

EMM500 CAS:16354-50-0 HR: 2
7-ETHYL-12-METHYLBENZ(a)ANTHRACENE
mf: $C_{21}H_{18}$ mw: 270.39

TOXICITY DATA WITH REFERENCE
cyt-rat-ivn 50 mg/kg GANNA2 64,637,73
ims-rat TDLo:50 mg/kg:NEO PNASA6 58,2253,67
ims-rat TD:50 mg/kg:ETA BCPCA6 16,607,67

SAFETY PROFILE: Questionable carcinogen with experimental neoplastigenic and tumorigenic data. Mutation data reported. When heated to decomposition it emits acrid smoke and irritating fumes.

EMN000 CAS:16354-55-5 HR: 2
12-ETHYL-7-METHYLBENZ(a)ANTHRACENE
mf: $C_{21}H_{18}$ mw: 270.39

TOXICITY DATA WITH REFERENCE
cyt-rat-ivn 50 mg/kg GANNA2 64,637,73
ims-rat TDLo:50 mg/kg:NEO PNASA6 58,2253,67

SAFETY PROFILE: Questionable carcinogen with experimental neoplastigenic data. Mutation data reported. When heated to decomposition it emits acrid smoke and irritating fumes.

EMO500 CAS:125-42-8 HR: 3
5-ETHYL-5-(1-METHYL-1-BUTENYL)BARBITURATE
mf: $C_{11}H_{16}N_2O_3$ mw: 224.29

PROP: A solid. Mp: 162–163°.

SYNS: 5-ETHYL-5-(1-METHYL-1-BUTENYL)BARBITURIC ACID □ 5-ETHYL-5-(1-METHYL-1-BUTENYL)-2,4,6(1H,3H,5H)-PYRIMIDINETRIONE

TOXICITY DATA WITH REFERENCE
orl-mus LD50:190 mg/kg JACSAT 61,776,39
ipr-mus LD50:180 mg/kg JACSAT 61,776,39

SAFETY PROFILE: Poison by ingestion and intraperitoneal routes. When heated to decomposition it emits toxic fumes of NO_x. See also BARBITURATES.

EMO875 CAS:17013-35-3 HR: 3
5-ETHYL-5-(1-METHYL-2-BUTENYL)BARBITURIC ACID
mf: $C_{11}H_{16}N_2O_3$ mw: 224.29

TOXICITY DATA WITH REFERENCE
orl-rat LD50:20 mg/kg JMPCAS 1,31,37
ipr-rat LD50:3 mg/kg JMPCAS 1,31,59
ipr-mus LD50:21 mg/kg JMPCAS 1,31,59

SAFETY PROFILE: Poison by ingestion and intraperitoneal routes. When heated to decomposition it emits toxic fumes of NO_x. See also BARBITURATES.

E

EMP550 CAS:22457-23-4 HR: 2
ETHYL 2-METHYLBUTYL KETOXINE
mf: $C_8H_{17}NO$ mw: 143.26

SYNS: 3-HEPTANONE, 5-METHYL-, OXIME □ 5-METHYL-3-HEPTANONE OXIME □ STEMONE

TOXICITY DATA WITH REFERENCE
orl-rat LD50:3800 mg/kg FCTOD7 30,87S,92
skn-rbt LDLo:5 g/kg FCTOD7 30,87S,92

CONSENSUS REPORTS: Reported in EPA TSCA Inventory.

SAFETY PROFILE: Moderately toxic by ingestion. Low toxicity by skin contact. When heated to decomposition it emits toxic vapors of NO_x.

EMP600 CAS:7452-79-1 HR: 2
ETHYL 2-METHYLBUTYRATE
mf: $C_7H_{14}O_2$ mw: 130.19

PROP: Colorless liquid; strong, apple-like odor. D: 0.861–0.866, refr index: 1.396, bp: 133.5°, flash p: 153°F. Sol in alc, propylene glycol; misc in fixed oils; very sltly sol in water.

SYN: FEMA No. 2443

SAFETY PROFILE: Combustible liquid. When heated to decomposition it emits acrid smoke and irritating fumes.

EMQ500 CAS:105-40-8 HR: 2
ETHYL-N-METHYLCARBAMATE
mf: $C_4H_9NO_2$ mw: 103.14

PROP: Needles. Mp: 54°, bp: 170°.

SYNS: ETHYLESTER KYSELINY METHYLKARBAMINOVE □ ETHYL METHYLCARBAMATE □ METHYLCARBAMIC ACID, ETHYL ESTER □ N-METHYL URETHAN □ METHYLURETHANE □ N-METHYLURETHANE

TOXICITY DATA WITH REFERENCE
mic-mus:lym 5 mmol/L MUREAV 174,285,86
ipr-ham TDLo:495 mg/kg (8D preg):TER CNREA8 27,1696,67
skn-mus TDLo:10 g/kg/1W-I:ETA BJCAAI 9,177,55
ipr-mus TDLo:6500 mg/kg/13W-I:ETA JNCIAM 9,35,48
scu-mus LD50:1360 mg/kg AJEBAK 45,507,67

CONSENSUS REPORTS: Reported in EPA TSCA Inventory.

SAFETY PROFILE: Moderately toxic by subcutaneous route. Experimental teratogenic effects. Questionable carcinogen with experimental tumorigenic data. Mutation data reported. When heated to decomposition it emits toxic fumes of NO_x. See also CARBAMATES.

EMR000 **CAS:2698-38-6** **HR: 2**
ETHYL-2-METHYL-4-CHLOROPHENOXYACETATE
mf: $C_{11}H_{13}ClO_3$ mw: 228.69

PROP: A liquid. Bp: 115–117° @ 1 mm.

SYNS: 4-CHLORO-2-METHYLPHENOXYACETIC ACID, ETHYL ESTER □ ((4-CHLORO-o-TOLYL)OXY)ACETIC ACID, ETHYL ESTER □ MCPA-ETHYL □ MCPEE

TOXICITY DATA WITH REFERENCE
orl-rat TDLo:800 mg/kg (8-15D preg):REP TXAPA9 23,326,72
orl-rat TDLo:400 mg/kg (8-15D preg):TER TXAPA9 23,326,72
orl-rat TDLo:800 mg/kg (8-15D preg):TER TXAPA9 23,326,72
orl-mus LD50:1290 mg/kg YKYUA6 30,985,79

SAFETY PROFILE: Moderately toxic by ingestion. Experimental teratogenic and reproductive effects. When heated to decomposition it emits toxic fumes of Cl^-. See also ESTERS.

EMR600 **CAS:39562-70-4** **HR: 3**
ETHYL METHYL 1,4-DIHYDRO-2,6-DIMETHYL-4-(m-NITROPHENYL)-3,5-PYRIDINEDICARBOXYLATE
mf: $C_{18}H_{20}N_2O_6$ mw: 360.40

PROP: A solid. Mp: 158–159°.

SYNS: BAY e 5009 □ 1,4-DIHYDRO-2,6-DIMETHYL-4-(3-NITROPHENYL)-3,5-PYRIDINEDICARBOXYLIC ACID ETHYL METHYL ESTER □ NITRENDIPINE □ 3,5-PYRIDINEDICARBOXYLIC ACID, 1,4-DIHYDRO-2,6-DIMETHYL-4-(3-NITROPHENYL)-, ETHYL METHYL ESTER

TOXICITY DATA WITH REFERENCE
orl-rat TDLo:1720 mg/kg (female 17-22D post):REP OYYAA2 36,133,88
orl-rat TDLo:330 mg/kg (female 7-17D post):REP OYYAA2 36,145,88
orl-rat TDLo:1720 mg/kg (female 17-22D post):TER OYYAA2 36,133,88
orl-rat TDLo:11,000 mg/kg (female 7-17D post):REP OYYAA2 36,145,88
orl-rat TDLo:1100 mg/kg (female 7-17D post):TER OYYAA2 36,145,88
orl-rat LD50:15,370 mg/kg OYYAA2 36,121,88
ipr-rat LD50:205 mg/kg OYYAA2 36,121,88
scu-rat LD50:5166 mg/kg OYYAA2 36,121,88
ivn-rat LD50:12,600 μg/kg 52EDA6 -,25,84
orl-mus LD50:2540 mg/kg 52EDA6 -,25,84
ipr-mus LD50:303 mg/kg OYYAA2 36,121,88
scu-mus LD50:7613 mg/kg OYYAA2 36,121,88
ivn-mus LD50:34,500 μg/kg FRPSAX 42,697,87
orl-dog LD50:250 mg/kg 52EDA6 -,25,84

SAFETY PROFILE: Poison by ingestion and intraperitoneal routes. An experimental teratogen. Experimental reproductive effects. When heated to decomposition it emits toxic fumes of NO_x.

EMS000 **CAS:6030-03-1** **HR: 2**
4′-ETHYL-2-METHYL-4-DIMETHYLAMINOAZOBENZENE
mf: $C_{17}H_{21}N_3$ mw: 267.41

SYN: 4′-ETHYL-N,N-DIMETHYL-4-(PHENYLAZO)-m-TOLUIDINE

TOXICITY DATA WITH REFERENCE
orl-rat TDLo:4040 mg/kg/15W-C:ETA ARZNAD 12,270,62

SAFETY PROFILE: Questionable carcinogen with experimental tumorigenic data. When heated to decomposition it emits toxic fumes of NO_x.

EMT000 **CAS:540-67-0** **HR: 3**
ETHYL METHYL ETHER
DOT: UN 1039
mf: C_3H_8O mw: 60.11

PROP: Colorless liquid or gas at room temp. Bp: 10.8°, lel: 2.0%, uel: 10.1%, flash p: −35°F (CC), d: 0.7260 @ 0°/4°, autoign temp: 374°F, vap d: 2.07.

SYNS: ETHOXYMETHANE □ ETHYL METHYL ETHER (DOT) □ METHOXYETHANE □ METHYL ETHYL ETHER (DOT)

DOT CLASSIFICATION: 2.1; *Label:* Flammable Gas

SAFETY PROFILE: Has anesthetic properties. A very dangerous fire and moderate explosion hazard when exposed to heat or flame; can react vigorously with oxidizing materials (e.g., air, O_2). To fight fire, use alcohol foam, CO_2, dry chemical. See also ETHERS.

EMT500 **CAS:139-88-8** **HR: 2**
7-ETHYL-2-METHYL-4-HENDECANOL SULFATE SODIUM SALT
mf: $C_{14}H_{29}O_4S{\cdot}Na$ mw: 316.48

PROP: A waxy solid. Sltly water-sol.

SYNS: 7-ETHYL-2-METHYL-4-UNDECANOL SULFATE SODIUM SALT □ OBLITEROL □ SODIUM-7-ETHYL-2-METHYL-4-UNDECANOL SULFATE □ SODIUM-7-ETHYL-2-METHYLUNDECYL-4-SULFATE □ SODIUM-2-METHYL-7-ETHYLUNDECANOL-4-SULFATE □ SODIUM-2-METHYL-7-ETHYLUNDECYL SULFATE-4 □ SODIUM SOTRADECOL □ SOTRADECOL □ STS □ TERGITOL □ TERGITOL ANIONIC 4 □ TERGITOL PENETRANT 4 □ TROMBOVAR □ VARICOL

TOXICITY DATA WITH REFERENCE
skn-rbt 500 mg open SEV UCDS** 12/13/63
eye-rbt 250 μg MLD AROPAW 34,99,45
orl-rat LD50:1250 mg/kg JIHTAB 23,478,41
orl-gpg LD50:650 mg/kg JIHTAB 23,478,41
skn-gpg LD50:650 mg/kg JIHTAB 23,478,41

CONSENSUS REPORTS: Reported in EPA TSCA Inventory.

SAFETY PROFILE: Moderately toxic by ingestion and skin contact. An eye and severe skin irritant. When heated to decomposition it emits toxic fumes of SO_x and Na_2O. See also SULFATES.

EMT600 CAS:5921-54-0 **HR: 3**
ETHYL METHYL KETONE AZINE
mf: $C_8H_{16}N_2$ mw: 140.26

SYNS: 2-BUTANONE, AZINE □ 2-BUTANONE, (1-METHYLPROPYLIDENE)HYDRAZONE □ METHYLETHYL KETAZINE □ METHYL ETHYL KETONE AZINE □ METHYL ETHYL KETONE KETAZINE

TOXICITY DATA WITH REFERENCE
ihl-mus LC:>2500 mg/kg NDRC** NDCrc-132,FEB42

DOT CLASSIFICATION: 3; *Label:* Flammable Liquid

SAFETY PROFILE: Low toxicity by inhalation. A flammable liquid. When heated to decomposition it emits toxic vapors of NO_x.

EMU500 CAS:96-29-7 **HR: 3**
ETHYL METHYL KETOXIME
mf: C_4H_9NO mw: 87.14

$CH_3C(:NOH)CH_2CH_3$

PROP: A liquid. D: 0.9232 @ 20°/4°, mp: −29.5°, bp: 152°.

SYNS: ETHYL METHYL KETONE OXIME □ ETHYL-METHYLKETONOXIM □ 2-BUTANONE, OXIME □ MEK-OXIME □ METHYL ETHYL KETOXIME □ SKINO #2 □ TROYKYD ANTI-SKIN B □ USAF AM-3 □ USAF DO-44 □ USAF EK-906

TOXICITY DATA WITH REFERENCE
scu-rat LD50:2702 mg/kg NJMSAG 29,393,67
ipr-mus LD50:200 mg/kg NTIS** AD277-689

CONSENSUS REPORTS: Reported in EPA TSCA Inventory.

DOT CLASSIFICATION: 3; *Label:* Flammable Liquid

SAFETY PROFILE: Poison by intraperitoneal route. Moderately toxic by subcutaneous route. May explode if heated. Reacts with sulfuric acid to form an explosive product. When heated to decomposition it emits toxic fumes of NO_x.

EMW100 **HR: 3**
N-ETHYL-6-METHYL-α-(METHYLSULFONYL)ERGOLINE-8-β-PROPIONAMIDE
mf: $C_{21}H_{29}N_3O_3S$ mw: 403.59

SYN: ERGOLINE-8-β-PROPIONAMIDE, N-ETHYL-6-METHYL-α-(METHYLSULFONYL)-

TOXICITY DATA WITH REFERENCE
orl-rat TDLo:3 mg/kg (female 5D post):REP ARZNAD 33,1094,83
orl-mus LD50:400 mg/kg ARZNAD 33,1094,83

SAFETY PROFILE: Poison by ingestion. Experimental reproductive effects. When heated to decomposition it emits toxic fumes of NO_x and SO_x

EMY000 CAS:10024-78-9 **HR: 3**
4-ETHYL-1-METHYLOCTYLAMINE
mf: $C_{11}H_{25}N$ mw: 171.37

SYNS: 2-AMINO-5-ETHYLNONANE □ 1-METHYL-4-ETHYLOCTYLAMINE

TOXICITY DATA WITH REFERENCE
skn-rbt 10 mg/24H open AMIHBC 10,61,54
skn-rbt 5 mg/24H SEV 85JCAE-,437,86
eye-rbt 50 µg open SEV AMIHBC 10,61,54
orl-rat LD50:730 mg/kg AMIHBC 10,61,54
skn-rbt LD50:380 mg/kg AMIHBC 10,61,54

SAFETY PROFILE: Poison by skin contact. Moderately toxic by ingestion. A skin and severe eye irritant. When heated to decomposition it emits toxic fumes of NO_x. See also AMINES.

E

EMY100 CAS:95524-59-7 **HR: 1**
2-ETHYL-1-(3-METHYL-1-OXO-2-BUTENYL)PIPERIDINE
mf: $C_{12}H_{21}NO$ mw: 195.34

SYNS: AI3-36175-Ga □ PIPERIDINE, 2-ETHYL-1-(3-METHYL-1-OXO-2-BUTENYL)-

TOXICITY DATA WITH REFERENCE
skn-rbt 500 mg MLD NTIS** AD-A002-053

SAFETY PROFILE: A skin irritant. When heated to decomposition it emits toxic fumes of NO_x.

EMZ000 CAS:106-67-2 **HR: 2**
2-ETHYL-4-METHYL-1-PENTANOL
mf: $C_8H_{18}O$ mw: 130.26

SYNS: 2-ETHYLISOHEXANOL □ 2-ETHYL-4-METHYLPENTANOL

TOXICITY DATA WITH REFERENCE
skn-rbt 10 mg/24H open MLD AIHAAP 23,95,62
orl-rat LD50:4290 mg/kg AIHAAP 23,95,62
orl-mus LDLo:1600 mg/kg KODAK* 21MAY71

CONSENSUS REPORTS: Reported in EPA TSCA Inventory.

SAFETY PROFILE: Moderately toxic by ingestion. A skin irritant. When heated to decomposition it emits acrid smoke and irritating fumes.

ENA000 CAS:67526-05-0 **HR: 3**
5-ETHYL-5-(1-METHYL-1-PENTENYL)BARBITURIC ACID
mf: $C_{12}H_{18}N_2O_3$ mw: 238.32

TOXICITY DATA WITH REFERENCE
orl-mus LD50:210 mg/kg JACSAT 61,776,39
ipr-mus LD50:160 mg/kg JACSAT 61,776,39

SAFETY PROFILE: Poison by ingestion and intraperitoneal routes. When heated to decomposition it emits toxic fumes of NO_x. See also BARBITURATES.

ENA500 CAS:70299-48-8 **HR: 3**
ETHYL METHYL PEROXIDE
mf: $C_3H_8O_2$ mw: 76.10

$CH_3CH_2OOCH_3$

PROP: Volatile liquid with ethereal odor. Bp: 40° @ 740 mm.

SAFETY PROFILE: Both the liquid and the vapor are

shock-sensitive explosives. Also explodes on superheating. When heated to decomposition it emits acrid smoke and irritating fumes. See also PEROXIDES.

ENB000 CAS:2058-66-4 HR: 2
N-ETHYL-N-METHYL-p-(PHENYLAZO)ANILINE
mf: $C_{15}H_{17}N_3$ mw: 239.35

SYNS: p-ETHYLMETHYLAMINOAZOBENZENE ☐ N-ETHYL-N-METHYL-p-AMINOAZOBENZENE ☐ 4-ETHYLMETHYLAMINOAZOBENZENE ☐ N-METHYL-N-ETHYL-p-AMINOAZOBENZENE ☐ 4-(METHYLETHYL) AMINOAZOBENZENE

TOXICITY DATA WITH REFERENCE
dni-mus-orl 20 g/kg ARGEAR 51,605,81
orl-rat TDLo:4684 mg/kg/17W-C:CAR JEMEAV 87,139,48
orl-rat TD:2100 mg/kg/16W-C:ETA CNREA8 8,141,48

SAFETY PROFILE: Questionable carcinogen with experimental carcinogenic and tumorigenic data. Mutation data reported. When heated to decomposition it emits toxic fumes of NO_x.

ENB500 CAS:115-38-8 HR: 3
5-ETHYL-N-METHYL-5-PHENYLBARBITURIC ACID
mf: $C_{13}H_{14}N_2O_3$ mw: 246.29

PROP: A solid. Mp: 176°.

SYNS: ENFENEMAL ☐ ENPHENEMAL ☐ N-ETHYLMETHYLPHENYLBARBITURIC ACID ☐ 5-ETHYL-1-METHYL-5-PHENYLBARBITURIC ACID ☐ 5-ETHYL-1-METHYL-5-PHENYL-2,4,6(1H,3H,5H)-PYRIMIDINETRIONE ☐ 5-ETHYL-5-PHENYL-N-METHYLBARBITURIC ACID ☐ ISONAL ☐ ISONAL (ROUSSEL) ☐ MEBARAL ☐ MEBEREL ☐ MENTA-BAL ☐ MEPHOBARBITAL ☐ MEPHOBARBITONE ☐ MEPHYTAL ☐ METHYL-CALMINAL ☐ 1-METHYL-5-ETHYL-5-PHENYLBARBITURIC ACID ☐ METHYLPHENOBARBITAL ☐ N-METHYLPHENOBARBITAL ☐ 1-METHYLPHENOBARBITAL ☐ METHYLPHENOBARBITONE ☐ N-METHYLPHENOBARBITOL ☐ METHYLPHENYLBARBITURIC ACID ☐ N-METHYL-5-PHENYL-5-ETHYLBARBITAL ☐ 1-METHYL-5-PHENYL-5-ETHYLBARBITURIC ACID ☐ METYLFENEMAL ☐ METYNA ☐ MORBUSAN ☐ PHEMETONE ☐ PHEMITON ☐ PHEMITONE ☐ 5-PHENYL-5-ETHYL-3-METHYLBARBITURIC ACID ☐ PROMINAL

TOXICITY DATA WITH REFERENCE
unr-wmn TDLo:3360 mg/kg (1-40W preg):TER LANCAO 2,839,72
ipr-rat LD50:130 mg/kg PHMCAA 5,237,63
orl-mus LD50:300 mg/kg HPPAAL 19,241,61
orl-cat LD50:230 mg/kg NIIRDN 6,851,82
orl-rbt LD50:400 mg/kg NIIRDN 6,851,82

SAFETY PROFILE: Poison by ingestion and intraperitoneal routes. A human teratogen by an unspecified route with developmental abnormalities of the cardiovascular (circulatory) system. When heated to decomposition it emits toxic NO_x. See also BARBITURATES.

ENC000 CAS:77-83-8 HR: 2
ETHYL METHYLPHENYLGLYCIDATE
mf: $C_{12}H_{14}O_3$ mw: 206.26

PROP: Colorless to yellowish liquid; strawberry-like odor. D: 1.086–1.112, refr index: 1.504–1.513, flash p: 273°F. Sol in fixed oils, propylene glycol; insol in glycerin.

SYNS: C-16 ALDEHYDE ☐ EMPG ☐ α-β-EPOXY-β-METHYLHYDROCINNAMIC ACID, ETHYL ESTER ☐ ETHYL α,β-EPOXY-β-METHYLHYDROCINNAMATE ☐ ETHYL 2,3-EPOXY-3-METHYL-3-PHENYLPROPIONATE ☐ ETHYL ESTER of 2,3-EPOXY-3-PHENYLBUTANOIC ACID ☐ FEMA No. 2444 ☐ FRAESEOL ☐ 3-METHYL-3-PHENYLGLYCIDIC ACID ETHYL ESTER ☐ STRAWBERRY ALDEHYDE

TOXICITY DATA WITH REFERENCE
cyt-ham:ovr 50 mg/L EMMUEG 10(Suppl 10),1,87
sce-ham:ovr 16 mg/L EMMUEG 10(Suppl 10),1,87
orl-rat LD50:5470 mg/kg FCTXAV 2,327,64
orl-gpg LD50:4050 mg/kg FCTXAV 2,327,64

CONSENSUS REPORTS: Reported in EPA TSCA Inventory.

SAFETY PROFILE: Mildly toxic by ingestion. Mutation data reported. Combustible liquid. When heated to decomposition it emits acrid smoke and irritating fumes. See also ALDEHYDES.

ENC500 CAS:5696-06-0 HR: 2
5-ETHYL-1-METHYL-5-PHENYLHYDANTOIN
mf: $C_{12}H_{14}N_2O_2$ mw: 218.28

PROP: A solid. Mp: 210°.

SYNS: DELTOIN ☐ METHETOIN

TOXICITY DATA WITH REFERENCE
orl-mus LD50:500 mg/kg 29ZVAB -,75,69
orl-rbt LD50:1400 mg/kg 29ZVAB -,75,69

SAFETY PROFILE: Moderately toxic by ingestion. When heated to decomposition it emits toxic fumes such as NO_x.

ENC600 CAS:18886-42-5 HR: 3
α-ETHYL-1-METHYL-α-PHENYL-3-PYRROLIDINEMETHANOL PROPIONATE FUMARATE
mf: $C_{17}H_{25}NO_2 \cdot C_4H_4O_4$ mw: 391.51

SYNS: AHR-1767 ☐ α-1-(1-METHYL-3-PYRROLIDINYL)-1-PHENYLPROPYL PROPIONATE FUMARATE

TOXICITY DATA WITH REFERENCE
orl-rat LD50:460 mg/kg AIPTAK 178,446,69
ipr-rat LD50:120 mg/kg AIPTAK 178,446,69
orl-mus LD50:230 mg/kg AIPTAK 178,446,69
ipr-mus LD50:144 mg/kg AIPTAK 178,446,69

SAFETY PROFILE: Poison by ingestion and intraperitoneal routes. When heated to decomposition it emits toxic fumes of NO_x.

END000 CAS:104-89-2 HR: 3
5-ETHYL-2-METHYLPIPERIDINE
mf: $C_8H_{17}N$ mw: 127.26

SYNS: COPELLIDIN ☐ 3-ETHYL-6-METHYLPIPERIDINE

TOXICITY DATA WITH REFERENCE
skn-rbt 10 mg/24H open AMIHBC 10,61,54
skn-rbt 5 mg/24H SEV 85JCAE-,846,86
eye-rbt 250 μg SEV AMIHBC 10,61,54
orl-rat LD50:540 mg/kg AMIHBC 10,61,54
ihl-rat LCLo:250 ppm/4H AMIHBC 10,61,54
skn-rbt LD50:630 mg/kg AMIHBC 10,61,54

scu-rbt LDLo: 100 mg/kg BDCGAS 34,2408,01

CONSENSUS REPORTS: Reported in EPA TSCA Inventory.

SAFETY PROFILE: Poison by subcutaneous route. Moderately toxic by ingestion, skin contact, and inhalation. A severe skin and eye irritant. When heated to decomposition it emits toxic fumes of NO_x.

END500 CAS:77791-37-8 HR: 3
N-ETHYL-2-(2-METHYLPIPERIDINO)-N-(1-(2,4-XYLYLOXY)-2-PROPYL) ACETAMIDE HYDROCHLORIDE
mf: $C_{21}H_{34}N_2O_2{\cdot}ClH$ mw: 383.03

SYN: C 2103

TOXICITY DATA WITH REFERENCE
eye-rbt 2% MLD ARZNAD 9,70,59
scu-mus LD50: 317 mg/kg ARZNAD 9,70,59

SAFETY PROFILE: Poison by subcutaneous route. An eye irritant. When heated to decomposition it emits very toxic fumes of NO_x and HCl.

ENE500 CAS:55283-68-6 HR: 1
N-ETHYL-N-(2-METHYL-2-PROPENYL)-2,6-DINITRO-4-(TRIFLUOROMETHYL)BENZENAMINE
mf: $C_{13}H_{14}F_3N_3O_4$ mw: 333.30

PROP: Yellow-orange crystals. Mp: 55–56°. Very sltly sol in H_2O; sol in most org solvs.

SYNS: EL-161 □ ETHALFLURLIN □ ETHALFLURALIN □ SOMILAN □ SONALAN □ SONALEN

TOXICITY DATA WITH REFERENCE
orl-rat LD50: 10,000 mg/kg 85ARAE 2,52,77

CONSENSUS REPORTS: Reported in EPA TSCA Inventory.

SAFETY PROFILE: Mildly toxic by ingestion. When heated to decomposition it emits very toxic fumes of F^- and NO_x.

ENF000 CAS:78-28-4 HR: 2
1-ETHYL-1-METHYLPROPYL CARBAMATE
mf: $C_7H_{15}NO_2$ mw: 145.23

PROP: Needles from EtOH (aq) with camphoraceous odor. Mp: 56–58°, bp: 35° @ 1 mm. Sol in EtOH, Et_2O, C_6H_6; sltly sol in H_2O.

SYNS: DIETHYL METHYL CARBINOLURETHAN □ tert-HEXANOL CARBAMATE □ METHYL DIETHYL CARBINOLURETHAN □ 3-METHYL-3-PENTANOL CARBAMATE

TOXICITY DATA WITH REFERENCE
ipr-rat LD50: 430 mg/kg 27ZQAG -,413,72
orl-mus LD50: 760 mg/kg 27ZQAG -,413,72
ipr-mus LD50: 550 mg/kg QJSAAP 21,223,60

SAFETY PROFILE: Moderately toxic by ingestion and intraperitoneal routes. When heated to decomposition it emits toxic fumes of NO_x. See also CARBAMATES.

ENF200 CAS:15707-23-0 HR: 2
2-ETHYL-3-METHYLPYRAZINE
mf: $C_7H_{10}N_2$ mw: 122.17

PROP: Colorless to sltly yellow liquid; strong, raw potato odor. D: 0.980–0.999 @ 20°, refr index: 1.502. Sol in water, organic solvents.

SYN: FEMA No. 3155

TOXICITY DATA WITH REFERENCE
orl-rat LD50: 600 mg/kg DCTODJ 3,249,80

CONSENSUS REPORTS: Reported in EPA TSCA Inventory.

SAFETY PROFILE: Moderately toxic by ingestion. When heated to decomposition it emits acrid smoke and irritating fumes.

ENG500 CAS:77-67-8 HR: 2
2-ETHYL-2-METHYLSUCCINIMIDE
mf: $C_7H_{11}NO_2$ mw: 141.19

SYNS: AETHOSUXIMIDE (GERMAN) □ ASAMID □ ATYSMAL □ CAPITUS □ CI 366 □ EMESIDE □ EPILEO PETIT MAL □ ETHOSUCCIMIDE □ ETHOSUCCINIMIDE □ ETHOSUXIDE □ETHOSUXIMIDE □ 3-ETHYL-3-METHYLPYRROLIDINE-2,5-DIONE □ 3-ETHYL-3-METHYL-2,5-PYRROLIDINE-DIONE □ α-ETHYL-α-METHYLSUCCINIMIDE □ ETHYMAL □ ETOMAL □ ETOSUXIMIDA □ H-490 □ H 940 □ MESENTOL □ 3-METHYL-3 ETHYLPYRROLIDINE-2,5-DIONE □ γ-METHYL-γ-ETHYL-SUCCINIMIDE □ PEMAL □ PEMALIN □ PENTINIMID □ PETINIMID □ PETNIDAN □ PM 671 □ PYKNOLEPSINUM □ RONTON □ SIMATIN(E) □ SUCCIMAL □ SUCCIMITIN □ SUXILEP □ SUXIMAL □ SUXIN □ SUXINUTIN □ THETAMID □ THILOPEMAL □ ZARAONDAN □ ZARODAN □ ZARONDAN-SAFT □ ZARONTIN □ ZARTALIN

TOXICITY DATA WITH REFERENCE
cyt-hmn:lyms 30 mg/L MUREAV 204,623,88
orl-rat TDLo: 2250 mg/kg (female 9-17D post): TER EAMJAV 60,407,83
orl-rat TDLo: 33,300 μg/kg (6-14D preg): TER 40YJAX -,59,79
orl-rat TDLo: 66,600 μg/kg (6-14D preg): REP 40YJAX -,59,70
unr-mus TDLo: 2626 mg/kg (female 8-10D post): TER FEPRA7 38,438,79
orl-mus TDLo: 3960 mg/kg (female 6-16D post): TER TXAPA9 40,365,77
orl-rat TDLo: 33,300 μg/kg (6-14D preg): TER 40YJAX -,59,79
orl-rat TDLo: 166 mg/kg (female 6-14D post): TER 40YJAX -,59,79
orl-mus LD50: 1530 mg/kg EPILAK 4(4),66,63
ipr-mus LD50: 1752 mg/kg EPILAK 29,198,88
scu-mus LD50: 1810 mg/kg NIIRDN 6,117,82
ivn-mus LD50: 780 mg/kg NEPHBW 24,427,85

SAFETY PROFILE: Moderately toxic by ingestion, intravenous, subcutaneous, and intraperitoneal routes. Experimental teratogenic and reproductive effects. Mutation data reported. An anticonvulsant. When heated to decomposition it emits toxic fumes of NO_x.

ENH000 CAS:73696-65-8 HR: 3
N-ETHYL-N'-(3-METHYL-2-THIAZOLIDINYLIDENE) UREA
mf: $C_7H_{13}N_3OS$ mw: 187.29

TOXICITY DATA WITH REFERENCE
orl-mus LD50:203 mg/kg JMCMAR 23,773,80
ivn-mus LD50:1110 mg/kg JMCMAR 23,773,80

SAFETY PROFILE: Poison by ingestion. Moderately toxic by intravenous route. When heated to decomposition it emits very toxic fumes of SO_x and NO_x.

ENI175 CAS:2591-57-3 HR: 3
ETHYLMETHYLTHIOPHOS
mf: $C_9H_{12}NO_3PS$ mw: 245.25

SYNS: O-ETHYL-O-METHYL-O-p-NITROFENYLESTER KYSELINY THIOFOSFORECNE □ O-METHYL O-ETHYL O-p-NITROPHENYL THIOPHOSPHATE □ METHYLETHYLTHIOFOS □ METHYLETHYLTHIOPHOS □ PHOSPHOROTHIOIC ACID, O-ETHYL O-METHYL O-(4-NITROPHENYL) ESTER (9CI)

TOXICITY DATA WITH REFERENCE
orl-rat LD50:2800 μg/kg 85GMAT -,83,82
orl-mus LD50:4200 μg/kg 85GMAT -,83,82
orl-cat LD50:5600 μg/kg 85GMAT -,83,82
skn-rbt LDLo:200 mg/kg 85GMAT -,83,82

SAFETY PROFILE: Poison by ingestion and skin contact. When heated to decomposition it emits toxic fumes of NO_x, PO_x, and SO_x.

ENI500 CAS:3568-56-7 HR: 3
O-ETHYL-O-(4-METHYLTHIO-m-TOLYL) METHYLPHOSPHORAMIDOTHIOATE
mf: $C_{11}H_{18}NO_2PS_2$ mw: 291.39

SYNS: BAY 34042 □ ENT 25,610 □ 4-(METHYLTHIO)-m-CRESOL-O-ESTER with O-ETHYL METHYLPHOSPHORAMIDOTHIOATE

TOXICITY DATA WITH REFERENCE
orl-rat LD50:5 mg/kg TXAPA9 21,315,72
orl-bwd LD50:1800 μg/kg TXAPA9 21,315,72

SAFETY PROFILE: Poison by ingestion. When heated to decomposition it emits very toxic fumes of NO_x, PO_x, and SO_x.

ENJ000 CAS:458-24-2 HR: 3
N-ETHYL-α-METHYL-m-(TRIFLUOROMETHYL)PHENETHYLAMINE
mf: $C_{12}H_{16}F_3N$ mw: 231.29

SYNS: FENFLURAMINE □ 3-(TRIFLUOROMETHYL)-N-ETHYL-α-METHYL PHENETHYL AMINE

TOXICITY DATA WITH REFERENCE
orl-rat TDLo:280 mg/kg (7-20D preg):REP TJADAB 17,39A,78
orl-hmn LDLo:50 mg/kg LANCAO 2,1306,69
orl-man TDLo:4286 μg/kg:ANS,CNS THERAP 34,205,79
orl-rat LD50:130 mg/kg JMCMAR 18,177,75
orl-mus LD50:170 mg/kg ISYAM* -,75,70
ipr-mus LD50:53 mg/kg ISYAM* -,75,70
orl-dog LD50:100 mg/kg ISYAM* -,75,70

SAFETY PROFILE: A human poison by ingestion. An experimental poison by ingestion and intraperitoneal routes. Human systemic effects by ingestion: hallucinations, distorted perceptions, and autonomic nervous system effects (an adrenergic stimulant). Experimental reproductive effects. When heated to decomposition it emits very toxic fumes of F^- and NO_x. See also AMINES.

ENJ500 CAS:4171-13-5 HR: 2
2-ETHYL-3-METHYLVALERAMIDE
mf: $C_8H_{17}NO$ mw: 143.26

PROP: A solid. Mp: 114–115°.

SYNS: ETHYLMETHYL VALERAMIDE □ VALMETHAMIDE

TOXICITY DATA WITH REFERENCE
orl-mus TDLo:700 mg/kg (female 6-12D post):REP CRSBAW 159,6,65
orl-rbt TDLo:600 mg/kg (female 6-8D post):REP CRSBAW 159,6,65
orl-rat LD50:760 mg/kg 27ZQAG -,429,72
ipr-rat LD50:580 mg/kg 27ZQAG -,429,72
orl-mus LD50:999 mg/kg NYKZAU 62,404,66
ipr-mus LD50:540 mg/kg 27ZQAG -,429,72

SAFETY PROFILE: Moderately toxic by ingestion and intraperitoneal routes. Experimental reproductive effects. When heated to decomposition it emits toxic fumes of NO_x.

ENK000 CAS:76-58-4 HR: 3
ETHYLMORPHINE
mf: $C_{19}H_{23}NO_3$ mw: 313.43

PROP: A solid. Mp: 199–201°.

SYNS: CODETHYLINE □ (5-α,6-α)-7,8-DIDEHYDRO-4,5-EPOXY-3-ETHOXY-17-METHYLMORPHINAN-6-OL □ DIONIN □ DIONINE □ 3-o-ETHYLMORPHINE

TOXICITY DATA WITH REFERENCE
orl-rat LD50:810 mg/kg JPPMAB 25,929,73
ipr-rat LD50:110 mg/kg PHARAT 31,655,76
scu-rat LD50:200 mg/kg JPPMAB 25,929,73
ivn-rat LD50:62 mg/kg JPPMAB 25,929,73
ipr-mus LD50:120 mg/kg APFRAD 8,261,50
scu-mus LD50:136 mg/kg 28ZNAE 138,8,38

SAFETY PROFILE: Poison by intraperitoneal, intravenous, and subcutaneous routes. Moderately toxic by ingestion. When heated to decomposition it emits toxic fumes of NO_x. See also MORPHINE.

ENK500 CAS:6746-59-4 HR: 3
ETHYL MORPHINE HYDROCHLORIDE DIHYDRATE
mf: $C_{19}H_{23}NO_3 \cdot ClH \cdot 2H_2O$ mw: 385.93

PROP: White, microscopic, crystalline powder. Mp: 125° (decomp), vap d: 13.3.

SYNS: 7,8-DIDEHYDRO-4,5-α-EPOXY-3-ETHOXY-17-METHYLMORPHINAN-6-α-OL HYDROCHLORIDE DIHYDRATE □ DIONIN □ ETHYLMORPHINE HYDROCHLORIDE

TOXICITY DATA WITH REFERENCE
scu-ham TDLo:182 mg/kg (female 8D post):TER AJOGAH 123,705,75
scu-mus LD50:200 mg/kg 29ZVAB -,53,69
scu-rbt LDLo:550 mg/kg HBAMAK 4,1289,35
scu-gpg LDLo:300 mg/kg HBAMAK 4,1289,35

SAFETY PROFILE: Poison by subcutaneous route. Can be habit forming. An experimental teratogen. When heated to decomposition it emits very toxic fumes of NO_x and HCl. See also CODEINE and MORPHINE.

ENL000 CAS:100-74-3 HR: 3
N-ETHYLMORPHOLINE
mf: $C_6H_{13}NO$ mw: 115.20

PROP: Colorless liquid. Bp: 138°, flash p: 89.6°F (OC), d: 0.916 @ 20°/20°, vap d: 4.00.

SYNS: 4-ETHYLMORPHOLINE □ NEM

TOXICITY DATA WITH REFERENCE
skn-rbt 453 mg open MLD UCDS** 11/3/71
eye-rbt 2 mg open SEV AMIHBC 10,61,54
orl-rat LD50:1780 mg/kg DTLWS* 4,190,82
orl-mus LD50:1200 mg/kg TPKVAL 15,116,79
ihl-mus LC50:18,000 mg/m³/2H TPKVAL 15,116,79
ivn-mus LD50:180 mg/kg CSLNX* NX#04778

CONSENSUS REPORTS: Reported in EPA TSCA Inventory.

OSHA PEL: TWA 5 ppm (skin)
ACGIH TLV: TWA 5 ppm (skin)

SAFETY PROFILE: Poison by intravenous route. Moderately toxic by ingestion. Mildly toxic by inhalation. A skin and severe eye irritant. A very dangerous fire hazard when exposed to heat or flame; can react vigorously with oxidizing materials. To fight fire, use alcohol foam, foam, CO_2, dry chemical. When heated to decomposition it emits toxic fumes of NO_x.

ENL100 CAS:309-29-5 HR: 3
1-ETHYL-4-(2-MORPHOLINOETHYL)-3,3-DIPHENYL-2-PYRROLIDINONE
mf: $C_{24}H_{30}N_2O_2$ mw: 378.56

SYNS: AHR-619 □ DOPRAM □ DOXAPRAM □ 2-PYRROLIDINONE, 1-ETHYL-4-(2-MORPHOLINOETHYL)-3,3-DIPHENYL- □ 2-PYRROLIDINONE, 1-ETHYL-4-(2-(4-MORPHOLINYL)ETHYL)-3,3-DIPHENYL-(9CI)

TOXICITY DATA WITH REFERENCE
ivn-dom TDLo:1800 µg/kg (female 16W post):TER BJOGAS 84,48,77
ipr-mus LD50:268 mg/kg JPPMAB 30,522,78

SAFETY PROFILE: Poison by intraperitoneal route. An experimental teratogen. When heated to decomposition it emits toxic fumes of NO_x.

ENL850 CAS:124-06-1 HR: 2
ETHYL MYRISTATE
mf: $C_{16}H_{23}O_2$ mw: 256.42

PROP: Colorless to pale-yellow liquid; waxy odor or crystals from (Me_2CO). D: 0.857, mp: 12.3°, bp: 295°, refr index: 1.434, flash p: 212°F.

SYN: FEMA No. 2445

SAFETY PROFILE: Combustible liquid. When heated to decomposition it emits acrid smoke and irritating fumes.

ENL900 CAS:2122-70-5 HR: 2
ETHYL 1-NAPHTHYLACETATE
mf: $C_{14}H_{14}O_2$ mw: 214.28

SYNS: ETHYL 1-NAPHTHALENEACETATE □ 1-NAPHTHALENEACETIC ACID, ETHYL ESTER

TOXICITY DATA WITH REFERENCE
orl-rat LD50:3580 mg/kg PESTC* 9,10,80

CONSENSUS REPORTS: Reported in EPA TSCA Inventory.

SAFETY PROFILE: Moderately toxic by ingestion. When heated to decomposition it emits acrid smoke and irritating vapors.

ENM500 CAS:625-58-1 HR: 3
ETHYL NITRATE
mf: $C_2H_5NO_3$ mw: 91.08

PROP: Colorless liquid, pleasant odor, sweet taste. Mp: −112°, bp: 87.7°, lel: 3.8%, flash p: 50°F (CC), d: 1.004 @ 20°/4°, vap d: 3.14. Sol in water.

SYNS: NITRIC ACID, ETHYL ESTER □ NITRIC ETHER

TOXICITY DATA WITH REFERENCE
mmo-sat 10 µmol/plate FCTOD7 21,707,83
mma-sat 10 µmol/plate FCTOD7 21,707,83
sln-dmg-ihl 1200 ppm/3D FCTOD7 21,707,83
ipr-cat LD50:300 mg/kg 85JCAE-,393,86

CONSENSUS REPORTS: Reported in EPA TSCA Inventory.

SAFETY PROFILE: A poison by intraperitoneal route. Mutation data reported. A very dangerous fire hazard when exposed to heat or flame; can react vigorously with oxidizing materials. A moderate explosion hazard when exposed to heat (explodes @ 185°F). To fight fire, use foam, CO_2, dry chemical, water to blanket fire. Incompatible with Lewis acids. When heated to decomposition it emits toxic fumes of NO_x. See also NITRATES and ESTERS.

ENN000 CAS:109-95-5 HR: 3
ETHYL NITRITE
DOT: UN 1194
mf: $C_2H_5NO_2$ mw: 75.08

PROP: Colorless or yellowish liquid or gas; highly aromatic, ethereal odor. Decomp on standing. Very sltly sol in water; misc in alc and ether. Bp: 17°, lel: 3.0%, uel: 50%, explodes at 194°F, flash p: −31°F (CC), d: 0.900 @ 15.5°, autoign temp: 194°F, vap d: 2.59. Can explode >90°C.

SYNS: ETHYLESTER KYSELINY DUSITE □ ETHYL NITRITE □ ETH-

YL NITRITE SOLUTIONS (DOT) □ NITROSYL ETHOXIDE □ NITROUS ETHER □ NITROUS ETHYL ETHER

TOXICITY DATA WITH REFERENCE
ihl-rat LC50:160 ppm/4H FAATDF 8,101,87

CONSENSUS REPORTS: Reported in EPA TSCA Inventory.

DOT CLASSIFICATION: 3; *Label:* Flammable Liquid, Poison

SAFETY PROFILE: Poison by inhalation and ingestion. Narcotic in high concentrations. Lowers blood pressure. Methemoglobinemia has been reported. A very dangerous fire and severe explosion hazard when exposed to heat or flame. A powerful oxidizer. May explode when heated above 90°C. Highly dangerous when heated to decomposition or on contact with acid or acid fumes. To fight fire, use foam, CO_2, dry chemical, or water spray. When heated to decomposition it emits toxic fumes of NO_x. See also NITRITES and ETHERS.

ENN100 CAS:626-35-7 HR: 2
ETHYL NITROACETATE
mf: $C_4H_7NO_4$ mw: 133.12

SYN: ACETIC ACID, NITRO-, ETHYL ESTER

TOXICITY DATA WITH REFERENCE
ipr-rat LD50:2275 mg/kg VINIT* #6802-83

CONSENSUS REPORTS: Reported in EPA TSCA Inventory.

SAFETY PROFILE: Moderately toxic by intraperitoneal route. When heated to decomposition it emits toxic vapors of NO_x.

ENO000 CAS:99-77-4 HR: 3
ETHYL-p-NITROBENZOATE
mf: $C_9H_9NO_4$ mw: 195.19

PROP: Triclinic crystals or leaflets from ethanol. Mp: 57°. Insol in water; sol in alc and ether.

SYN: ETHYL NITROBENZOATE, PARA ESTER

TOXICITY DATA WITH REFERENCE
ipr-mus LD50:250 mg/kg NTIS** AD691-490

CONSENSUS REPORTS: Reported in EPA TSCA Inventory.

SAFETY PROFILE: Poison by intraperitoneal route. An insecticide. See also NITRO COMPOUNDS of AROMATIC HYDROCARBONS.

ENO100 CAS:838-57-3 HR: 1
ETHYL (4-NITROBENZOYL)ACETATE
mf: $C_{11}H_{11}NO_5$ mw: 237.23

SYNS: ACETIC ACID, (p-NITROBENZOYL)-, ETHYL ESTER □ BENZENEPROPANOIC ACID, 4-NITRO-β-OXO-, ETHYL ESTER (9CI) □ ETHYL (p-NITROBENZOYL)ACETATE □ ETHYL 4-NITRO-β-OXOBENZENEPROPANOATE

TOXICITY DATA WITH REFERENCE
orl-mus LD50:8620 mg/kg GISAAA 51(1),79,86

CONSENSUS REPORTS: Reported in EPA TSCA Inventory.

SAFETY PROFILE: Mildly toxic by ingestion. When heated to decomposition it emits toxic vapors of NO_x.

ENQ000 CAS:546-71-4 HR: 3
ETHYL-4-NITROPHENYL ETHYLPHOSPHONATE
mf: $C_{10}H_{14}NO_5P$ mw: 259.22

SYNS: ARMINE □ ETHOXY-4-NITROPHENYLOXYETHYLPHOSPHINEOXIDE □ ETHYLPHOSPHONIC ACID ETHYL-p-NITROPHENYL ESTER

TOXICITY DATA WITH REFERENCE
scu-rat LD50:330 μg/kg FATOAO 42(3),299,79
ivn-rat LD50:250 μg/kg FATOAO 42(3),299,79
ipr-mus LD50:1862 μg/kg PHARAT 35,806,80
scu-mus LD50:330 μg/kg RPTOAN 42,106,79
ivn-mus LD50:250 μg/kg RPTOAN 42,106,79

SAFETY PROFILE: Poison by subcutaneous, intravenous, and intraperitoneal routes. When heated to decomposition it emits very toxic fumes of NO_x and PO_x. See also ESTERS.

ENQ500 CAS:3015-75-6 HR: 3
ETHYL-p-NITROPHENYLPENTYLPHOSPHONATE
mf: $C_{13}H_{20}NO_5P$ mw: 301.31

SYN: p-NITROPHENYL ETHYL PENTYLPHOSPHONATE

TOXICITY DATA WITH REFERENCE
ivn-rbt LD50:118 μg/kg BCPCA6 13,1229,64
ivn-gpg LD50:299 μg/kg BCPCA6 13,1229,64

SAFETY PROFILE: Poison by intravenous route. When heated to decomposition it emits very toxic fumes of PO_x and NO_x.

ENR000 CAS:35363-12-3 HR: 2
3-ETHYL-4-NITROPYRIDINE-1-OXIDE
mf: $C_7H_8N_2O_3$ mw: 168.17

TOXICITY DATA WITH REFERENCE
dnd-mus:fbr 500 μmol/L CNREA8 35,521,75
scu-mus TDLo:1840 mg/kg/15W-I:ETA GANNA2 70,799,79

SAFETY PROFILE: Questionable carcinogen with experimental tumorigenic data. Mutation data reported. When heated to decomposition it emits toxic fumes of NO_x.

ENR500 CAS:65986-80-3 HR: 2
(ETHYLNITROSAMINO)METHYL ACETATE
mf: $C_5H_{10}N_2O_3$ mw: 146.17

SYNS: ACETOXYMETHYLETHYLNITROSAMINE □ N-(ACETOXY)METHYL-N-ETHYLNITROSAMINE □ N-ACETOXYMETHYL-N-NITROSOETHYLAMINE □ N-(1-ACETOXYMETHYL)-N-NITROSOETHYL AMINE □ AETHYL ACETOXYMETHYLNITROSAMIN (GERMAN) □ EAMN □ ETHYL ACETOXYMETHYLNITROSAMINE □ N-ETHYL-N-(ACETOXYMETHYL)NITROSAMINE

TOXICITY DATA WITH REFERENCE
mmo-esc 1 μmol/plate GANNA2 71,124,80

dnr-bcs 5 μmol/plate GANNA2 70,663,79
dns-rat:oth 10 μmol/L CBINA8 53,99,85
dnd-mus:fbr 70 μmol/L GANNA2 73,565,82
msc-ham:lng 100 μmol/L GANNA2 72,531,81
orl-rat TDLo:380 mg/kg/90D-I:ETA ZKKOBW 91,317,78
scu-rat TDLo:50 mg/kg/10W-I:CAR JCROD7 104,13,82
scu-rat TD:55 mg/kg/10W-I:CAR IAPUDO 41,619,82
orl-rat LD50:810 mg/kg ZKKOBW 91,317,78

SAFETY PROFILE: Moderately toxic by ingestion. Questionable carcinogen with experimental carcinogenic and tumorigenic data. Mutation data reported. When heated to decomposition it emits toxic fumes of NO_x. See also NITROSAMINES.

ENS000 CAS:41735-29-9 HR: 3
5-(N-ETHYL-N-NITROSO)AMINO-3-(5-NITRO-2-FURYL)-s-TRIAZOLE
mf: $C_8H_8N_6O_4$ mw: 252.22

SYN: N-(3-(5-NITRO-2-FURYL)-s-TRIAZOL-5-YL)-N-NITROSOETHYLAMINE

TOXICITY DATA WITH REFERENCE
ipr-rat LD50:235 mg/kg JMCMAR 9,42,66
orl-mus LD50:440 mg/kg JMCMAR 16,312,73
ipr-mus LD50:150 mg/kg JMCMAR 16,312,73

SAFETY PROFILE: Poison by intraperitoneal route. Moderately toxic by ingestion. Many N-nitroso compounds are carcinogens. When heated to decomposition it emits toxic fumes of NO_x. See also N-NITROSO COMPOUNDS.

ENS500 CAS:20689-96-7 HR: 3
N-ETHYL-N-NITROSOBENZYLAMINE
mf: $C_9H_{12}N_2O$ mw: 164.23

SYNS: N-NITROSO-N-ETHYLBENZYLAMIN (GERMAN) □ N-NITROSO-N-ETHYLBENZYLAMINE

TOXICITY DATA WITH REFERENCE
orl-rat TDLo:250 mg/kg/36W-C:ETA ZKKOBW 92,235,78
orl-rat LD50:250 mg/kg ZKKOBW 92,235,78

SAFETY PROFILE: Poison by ingestion. Questionable carcinogen with experimental tumorigenic data. Many N-nitroso compounds are carcinogens. When heated to decomposition it emits toxic fumes of NO_x. See also AMINES and N-NITROSO COMPOUNDS.

ENT000 CAS:32976-88-8 HR: 2
N-ETHYL-N-NITROSOBIURET
mf: $C_4H_8N_4O_3$ mw: 160.16

SYNS: ENBU □ ETHYLNITROSOBIURET □ N-NITROSO-N-ETHYL BIURET

TOXICITY DATA WITH REFERENCE
orl-rat TDLo:300 mg/kg (14D preg):TER IARCCD 4,45,73
orl-rat TDLo:400 mg/kg:ETA ZKKOBW 76,45,71
orl-rat TDLo:100 mg/kg (22D preg):CAR,TER IARCCD 4,45,73
orl-rat TD:100 mg/kg (15D preg):ETA,TER ZKKOBW 76,45,71
orl-rat LD50:1050 mg/kg IARCCD 4,45,73

SAFETY PROFILE: Moderately toxic by ingestion. Questionable carcinogen with experimental carcinogenic, and tumorigenic data. Experimental teratogenic effects. When heated to decomposition it emits toxic fumes of NO_x. See also N-NITROSO COMPOUNDS.

ENT500 CAS:38434-77-4 HR: 3
ETHYLNITROSOCYANAMIDE
mf: $C_3H_5N_3O$ mw: 99.11

SYNS: N-CYANO-N-NITROSOETHYLAMINE □ ENC □ NITROSOETHANECARBAMONITRILE

TOXICITY DATA WITH REFERENCE
mmo-sat 1 μmol/L MUREAV 48,131,77
orl-rat TDLo:1800 mg/kg/52W-I:NEO JJIND8 62,1523,79
ipr-rat LD50:15 mg/kg JOCEAH 38,1325,73

CONSENSUS REPORTS: EPA Genetic Toxicology Program. Cyanide and its compounds are on the Community Right-To-Know List.

SAFETY PROFILE: Poison by intraperitoneal route. Questionable carcinogen with experimental neoplastigenic data. Many N-nitroso compounds are carcinogens. Mutation data reported. When heated to decomposition it emits toxic fumes of NO_x and CN^-. See also N-NITROSO COMPOUNDS and NITRILES.

ENU000 CAS:4245-77-6 HR: 2
N-ETHYL-N-NITROSO-N′-NITROGUANIDINE
mf: $C_3H_7N_5O_3$ mw: 161.15

SYNS: N-AETHYL-N′-NITRO-N-NITROSOGUANIDIN (GERMAN) □ ENNG □ N-ETHYL-N′-NITRO-N-NITROSOGUANIDINE □ NSC 38191

TOXICITY DATA WITH REFERENCE
mma-sat 5 μg/plate TCMUE9 1,13,84
mmo-esc 2 μg/plate KSRNAM 19,4465,85
pic-esc 200 μg/L TCMUE9 1,91,84
sce-hmn:lym 1 μmol/L NGCJAK 15,1085,80
dnd-ham:ovr 20 μmol/L MUREAV 132,41,84
sce-ham-orl 10 mg/kg MUREAV 113,33,83
sce-ham:lng 10 mg/L CNREA8 44,3270,84
orl-rat TDLo:504 mg/kg/12W-C:ETA JJCREP 78,126,87
scu-rat TDLo:55 mg/kg/5W-I:NEO GANNA2 69,277,78
orl-mus TDLo:3 g/kg/43W-C:CAR JNCIAM 52,519,74
skn-mus TDLo:630 mg/kg/21W-I:NEO JNCIAM 46,973,71
ipr-mus TDLo:160 mg/kg:ETA BJCAAI 23,757,69
orl-dog TDLo:195 mg/kg/37W-C:ETA RKGEDW 1,571,81
rec-dog TDLo:1275 mg/kg/23W-I:ETA IJCNAW 32,255,83
orl-mky TDLo:2600 mg/kg/48W-C:ETA JJIND8 77,179,86
orl-ham TDLo:1900 mg/kg/33W-C:ETA ZEKBAI 81,29,74
orl-dog TD:5320 mg/kg/73W-I:ETA GANNA2 66,683,75
orl-dog TD:191 g/kg/34W-I:ETA ZEKBAI 90,241,77
orl-dog TD:3000 mg/kg/34W-C:ETA GANNA2 65,163,74
orl-mus TD:6300 mg/kg/63D-C:NEO GANMAX 25,89,80
orl-dog TD:1200 mg/kg/34W-C:ETA GANNA2 71,349,80
rec-dog TD:1440 mg/kg/26W-I:ETA IJCNAW 32,255,83
orl-dog TD:2400 mg/kg/34W-C:ETA NIPAA4 79,765,82
orl-mus TD:780 mg/kg/26W-I:ETA SACAB7 24,413,80
orl-dog TD:560 mg/kg/12W-C:NEO JCREA8 110,87,85
orl-dog TD:1500 mg/kg/17W-C:ETA GANNA2 72,880,81

CONSENSUS REPORTS: EPA Genetic Toxicology Program.

SAFETY PROFILE: Questionable carcinogen with experimental carcinogenic, neoplastigenic, and tumorigenic data. Human mutation data reported. When heated to decomposition it emits toxic fumes of NO_x. See also N-NITROSO COMPOUNDS.

ENV000 CAS:759-73-9 HR: 3
1-ETHYL-1-NITROSOUREA
mf: $C_3H_7N_3O_2$ mw: 117.13

PROP: Pale yellow crystals. Mp: 103° (decomp) 1% water soln @ 20°.

SYNS: AENH (GERMAN) □ AETHYLNITROSO-HARNSTOFF (GERMAN) □ ENU □ N-ETHYL-N-NITROSOCARBAMIDE □ ETHYLNITROSOUREA □ N-ETHYL-N-NITROSO-UREA □ NEU □ NITROSOETHYLUREA □ NSC 45403 □ RCRA WASTE NUMBER U176

TOXICITY DATA WITH REFERENCE
dni-hmn:hla 100 μmol/L MUREAV 92,427,82
cyt-ham-orl 100 mg/kg TRENAF 36,396,85
ipr-rat TDLo:15 mg/kg (female 18D post):REP NETOD7 2,297,80
ivn-rat TDLo:50 mg/kg (female 20D post):REP JONRA9 29,771,77
ivn-mus TDLo:80 mg/kg (female 15D post):REP NUPOBT 14,115,76
scu-mus TDLo:351 mg/kg (female 6-16D post):TER TCMUD8 3,219,83
ipr-rat TDLo:60 mg/kg (female 18D post):REP CNREA8 38,1263,78
ipr-rbt TDLo:100 mg/kg (female 18-27D post):REP JJIND8 65,607,80
ipr-rat TDLo:15 mg/kg (female 18D post):REP CNREA8 38,1263,78
ipr-rat TDLo:40 mg/kg (female 15D post):TER EXPEAM 33,521,77
orl-rat TDLo:125 mg/kg (female 7-16D post):TER RCOCB8 41,265,83
orl-rat TDLo:75 mg/kg (female 7-16D post):TER RCOCB8 41,265,83
orl-rat TDLo:440 mg/kg (female 7-17D post):TER OYYAA2 23,981,82
orl-rat TDLo:150 mg/kg (female 7-16D post):REP RCOCB8 41,265,83
orl-rat TDLo:125 mg/kg (female 7-16D post):REP RCOCB8 41,265,83
orl-rat TDLo:440 mg/kg (female 7-17D post):REP OYYAA2 23,981,82
ipr-mus TDLo:50 mg/kg (male 1D pre):REP TOLED5 17,77,83
ipr-mus TDLo:58,500 μg/kg (female 8D post):TER CNREA8 34,151,74
unr-rat TDLo:40 mg/kg (female 9D post):TER NAGZAC 53,45,78
orl-rat TDLo:100 mg/kg (female 7-16D post):TER RCOCB8 41,265,83
ivn-rat TDLo:80 mg/kg (female 15D post):TER NATUAS 210,1378,66
orl-rat TDLo:10 mg/kg (female 19D post):NEO,TER CALEDQ 2,93,76
orl-rat TDLo:374 mg/kg/2W-C:CAR JJIND8 75,743,85
skn-rat TDLo:350 mg/kg (female 16-22D post):NEO,TER ZEKBAI 81,169,74
ipr-rat TDLo:10 mg/kg (female 15D post):NEO,TER SCIEAS 218,682,82
ipr-rat TDLo:45 mg/kg:CAR ANTRD4 4,5,84
scu-rat TDLo:5 mg/kg:CAR CRNGDP 6,769,85
ivn-rat TDLo:17 mg/kg:ETA EXPTAX 17,394,79
ivn-rat TDLo:50 mg/kg (15D preg):CAR,TER ZEKBAI 73,371,70
unr-rat TDLo:5 mg/kg (13D preg):ETA,TER XENOBH 3,271,73
orl-mus TDLo:20 mg/kg/24W-I:NEO TXAPA9 72,313,84
skn-mus TDLo:375 mg/kg/40W-I:CAR JCROD7 102,13,81
ipr-mus TDLo:20 mg/kg (19D preg):CAR,TER VOONAW 23(3),41,77
ipr-mus TDLo:234 mg/kg:NEO CBINA8 3,117,71
par-mus TDLo:100 mg/kg/(18D preg):ETA,TER PAACA3 17,122,76
par-mus TDLo:58,565 μg/kg (18D preg):CAR,TER JNCIAM 51,1965,73
ipr-dog TDLo:100 mg/kg (53D preg):ETA,TER CALEDQ 12,161,81
ivn-dog TDLo:30 mg/kg (59D preg):ETA,TER ZAPPAN 121,54,77
ivn-mky TDLo:610 mg/kg/2Y-I:ETA PAACA3 18,53,77
ipr-rbt TDLo:100 mg/kg (15-24D preg):NEO,TER JJIND8 65,607,80
ivn-rbt TDLo:80 mg/kg (25D preg):NEO,TER BEXBAN 85,369,78
par-rbt TDLo:160 mg/kg (25D preg):ETA,TER ZAPPAN 121,54,77
ivn-pig TDLo:120 mg/kg (20-31D preg):ETA,TER ARGEAR 34,25,69
orl-ham TDLo:60 mg/kg (11D preg):ETA,TER ZEKBAI 71,320,68
skn-ham TDLo:107 mg/kg/20W-C:NEO ZKKOBW 90,233,77
ipr-ham TDLo:800 mg/kg (16D preg):NEO,TER ZKKOBW 90,233,77
ipr-ham TDLo:274 mg/kg:NEO CNREA8 43,829,83
ivn-ham TDLo:80 mg/kg (15D preg):ETA,TER ZAPPAN 121,54,77
ipr-rat TD:10 mg/kg (21D preg):NEO,REP JSONDX 5,396,84
ipr-mus TD:30 mg/kg:CAR CALEDQ 18,131,83
orl-rat TD:180 mg/kg:ETA ANYAA9 381,250,82
orl-rat TD:520 mg/kg/1Y-I:ETA PPTCBY 2,73,72
ivn-rat TD:20 mg/kg/(12D preg):ETA,TER NAGZAC 54,130,79
ipr-mus TD:240 mg/kg:NEO BIJOAK 174,1031,78
ivn-rbt TD:150 mg/kg (8-10D preg):ETA,TER ZKKOBW 89,331,77
ivn-rat TD:10 mg/kg (16D preg):ETA,TER ANPTAL 34,21,76
ivn-rat TD:20 mg/kg (17D preg):ETA,TER CALEDQ 1,345,76
ivn-rbt TD:70 mg/kg (25D preg):ETA,TER NEOLA4 25,453,78
orl-rat LD50:300 mg/kg PPTCBY 2,73,72
scu-rat LD50:240 mg/kg ZEKBAI 74,141,70
ivn-rat LD50:240 mg/kg NATUAS 222,1064,69
orl-mus LD50:960 mg/kg MUREAV 223,377,89
ipr-mus LD50:490 mg/kg MUREAV 223,377,89

CONSENSUS REPORTS: NTP 7th Annual Report On Carcinogens. IARC Cancer Review: Group 2A IMEMDT 7,56,87; Human Limited Evidence IMEMDT 17,191,78; Animal Sufficient Evidence IMEMDT 17,191,78; IMEMDT 1,135,72. EPA Genetic Toxicology Program. Community Right-To-Know List. Reported in EPA TSCA Inventory.

SAFETY PROFILE: Confirmed carcinogen with experimental carcinogenic, neoplastigenic, tumorigenic, and teratogenic data. Poison by ingestion, subcutaneous, intraperitoneal, and intravenous routes. Human mutation data reported. When heated to decomposition it emits toxic fumes of NO_x. See also N-NITROSO COMPOUNDS.

ENV500 **CAS:139-94-6** **HR: 2**
1-ETHYL-3-(5-NITRO-2-THIAZOLYL) UREA
mf: $C_6H_8N_4O_3S$ mw: 216.24

PROP: A solid. Mp: 228° (decomp).

SYNS: N-ETHYL-N'-(5-NITRO-2-THIAZOLYL)UREA □ HEPZIDE □ NCI-C03792 □ NITHIAZID □ NITHIAZIDE

TOXICITY DATA WITH REFERENCE
mmo-sat 75 µg/plate ENMUDM 5(Suppl 1),3,83
mma-sat 75 µg/plate ENMUDM 5(Suppl 1),3,83
orl-rat TDLo:41 g/kg/94W-I:CAR NCITR* NCI-CG-TR-146,79
orl-mus TDLo:395 g/kg/94W-I:CAR NCITR* NCI-CG-TR-146,79
orl-mus LD50:2150 mg/kg NCIUS* NIH-NCI-E-C-72-3252,73

CONSENSUS REPORTS: IARC Cancer Review: Group 3 IMEMDT 7,56,87; Animal Limited Evidence IMEMDT 31,179,83. NCI Carcinogenesis Bioassay (feed); Clear Evidence: mouse, rat NCITR* NCI-CG-TR-146,79.

SAFETY PROFILE: Suspected carcinogen with experimental carcinogenic data. Moderately toxic by ingestion. Mutation data reported. When heated to decomposition it emits very toxic fumes of NO_x and PO_x.

ENW000 **CAS:123-29-5** **HR: 2**
ETHYL NONANOATE
mf: $C_{11}H_{22}O_2$ mw: 186.33

PROP: Colorless liquid; fruity, cognac odor. D: 0.863–0867, refr index: 1.420, flash p: 185°F. Misc with alc, propylene glycol; insol in water.

SYNS: ETHYL NONYLATE □ ETHYL PELARGONATE □ FEMA No. 2447 □ NONANOIC ACID, ETHYL ESTER □ WINE ETHER

TOXICITY DATA WITH REFERENCE
skn-rbt 500 mg/24H MOD FCTXAV 16,747,78
orl-gpg LD50:24 g/kg FCTXAV 2,327,64

CONSENSUS REPORTS: Reported in EPA TSCA Inventory.

SAFETY PROFILE: Mildly toxic by ingestion. A skin irritant. Combustible liquid. When heated to decomposition it emits acrid smoke and irritating fumes.

ENW500 **CAS:103-08-2** **HR: 1**
5-ETHYL-2-NONANOL
mf: $C_{11}H_{24}O$ mw: 172.35

SYN: (3-ETHYL-N-HEPTYL)METHYLCARBINOL

TOXICITY DATA WITH REFERENCE
skn-rbt 415 mg open MLD UCDS** 2/21/58
eye-rbt 42 mg MOD UCDS** 2/21/58
skn-rbt LD50:4760 mg/kg UCDS** 2/21/58

SAFETY PROFILE: Mildly toxic by skin contact. A skin and eye irritant. When heated to decomposition it emits acrid smoke and irritating fumes.

ENX000 **CAS:10137-90-3** **HR: 1**
5-ETHYL-3-NONEN-2-ONE
mf: $C_{11}H_{20}O$ mw: 168.31

TOXICITY DATA WITH REFERENCE
skn-rbt 10 mg/24H open MLD AMIHBC 10,61,54
eye-rbt 500 mg open AMIHBC 10,61,54
orl-rat LD50:8120 mg/kg AMIHBC 10,61,54
skn-rbt LD50:8480 mg/kg AMIHBC 10,61,54

SAFETY PROFILE: Mildly toxic by ingestion and skin contact. A skin and eye irritant. When heated to decomposition it emits acrid smoke and irritating fumes.

ENX100 **CAS:23489-02-3** **HR: 3**
1-ETHYL-1-NONYLPIPERIDINIUM BROMIDE
mf: $C_{16}H_{34}N{\cdot}Br$ mw: 320.42

TOXICITY DATA WITH REFERENCE
orl-mus LD50:162 mg/kg PSDTAP 15,331,74
ipr-mus LD50:42,955 µg/kg PSDTAP 15,331,74
ivn-mus LD50:3895 µg/kg PSDTAP 15,331,74

SAFETY PROFILE: Poison by ingestion, intravenous, and intraperitoneal routes. When heated to decomposition it emits toxic fumes of NO_x and Br^-.

ENX500 **CAS:3198-07-0** **HR: 3**
ETHYLNORADRENALINE HYDROCHLORIDE
mf: $C_{10}H_{15}NO_3{\cdot}ClH$ mw: 233.72

SYNS: α-(1-AMINOPROPYL)PROTOCATECHUYL ALCOHOL HYDROCHLORIDE □ BRONKEPHRINE HYDROCHLORIDE □ BUTANEFRINE HYDROCHLORIDE □ 1-(3,4-DIHYDROXYPHENYL)-2-AMINO-1-BUTANOL HYDROCHLORIDE □ 1-(3,4-DIHYDROXYPHENYL)-1-HYDROXY-2-AMINOBUTANE HYDROCHLORIDE □ E.N.E. □ E.N.S. □ ETHYL NOREPINEPHRINE HYDROCHLORIDE □ α-ETHYLNOREPINEPHRINE HYDROCHLORIDE □ ETHYLNORSUPRARENIN HYDROCHLORIDE

TOXICITY DATA WITH REFERENCE
scu-hmn TDLo:36 µg/kg:CVS JPETAB 81,269,44
ivn-hmn TDLo:9 µg/kg:CVS JPETAB 81,269,44
ims-hmn TDLo:18 µg/kg:CVS JPETAB 81,269,44
scu-rat LDLo:160 mg/kg JPETAB 81,269,44
ivn-mus LD50:117 mg/kg JPETAB 81,269,44

SAFETY PROFILE: Poison by intravenous and subcutaneous routes. Human systemic effects by subcutaneous, intravenous, and intramuscular routes: heart rate change, blood pressure decrease, and pulse pressure increase. When heated to decomposition it emits very toxic fumes of Cl^- and NO_x. See also NOREPINEPHRINE.

ENX875 **CAS:56501-34-9** **HR: 3**
1-ETHYL-1-OCTADECYLPIPERIDINIUM BROMIDE
mf: $C_{25}H_{50}N \cdot Br$ mw: 444.67

TOXICITY DATA WITH REFERENCE
orl-mus LD50:238 mg/kg PSDTAP 15,331,74
ipr-mus LD50:1847 μg/kg PSDTAP 15,331,74
ivn-mus LD50:4493 μg/kg PSDTAP 15,331,74

SAFETY PROFILE: Poison by ingestion, intravenous, and intraperitoneal routes. When heated to decomposition it emits toxic fumes of NO_x and Br^-. See also BROMIDES.

ENY000 **CAS:106-32-1** **HR: 2**
ETHYL OCTANOATE
mf: $C_{10}H_{20}O_2$ mw: 172.30

PROP: Colorless liquid; wine-brandy fruit odor. D: 0.865–0.869, refr index: 1.417, flash p: 185°F. Sol in fixed oils; sltly sol in propylene glycol; insol in glycerin, water @ 209°.

SYNS: ETHYL CAPRYLATE □ ETHYL OCTYLATE □ FEMA No. 2449 □ OCTANOIC ACID, ETHYL ESTER

TOXICITY DATA WITH REFERENCE
skn-rbt 500 mg/24H MOD FCTXAV 14,763,76
orl-rat LD50:25,960 mg/kg FCTXAV 14,763,76

CONSENSUS REPORTS: Reported in EPA TSCA Inventory.

SAFETY PROFILE: Mildly toxic by ingestion. A skin irritant. Combustible liquid. When heated to decomposition it emits acrid smoke and irritating fumes. See also ESTERS.

ENY100 **CAS:23489-01-2** **HR: 3**
1-ETHYL-1-OCTYLPIPERIDINIUM BROMIDE
mf: $C_{15}H_{32}N \cdot Br$ mw: 306.328

TOXICITY DATA WITH REFERENCE
orl-mus LD50:210 mg/kg PSDTAP 15,331,74
ivn-mus LD50:4793 μg/kg PSDTAP 15,331,74
ipr-mus LD50:48,453 μg/kg PSDTAP 15,331,74

SAFETY PROFILE: Poison by ingestion, intravenous, and intraperitoneal routes. When heated to decomposition it emits toxic fumes of NO_x and Br^-.

ENY500 **CAS:122-51-0** **HR: 3**
ETHYL ORTHOFORMATE
DOT: UN 2524
mf: $C_7H_{16}O_3$ mw: 148.23

PROP: Clear liquid, pungent sweet odor. Fp: 30°, bp: 145.9°, flash p: 86°F (CC), d: 0.895 @ 20°/20°, vap press: 10 mm @ 40.5°, vap d: 5.11.

SYNS: AETHON □ ETHONE □ ETHYLESTER KYSELINY ORTHOMRAVENCI (CZECH) □ 1,1′,1′-(METHYLIDYNETRIS(OXY))TRIS(ETHANE) □ ORTHOFORMIC ACID, ETHYL ESTER □ ORTHOFORMIC ACID, TRIETHYL ESTER □ ORTHOMRAVENCAN ETHYLNATY (CZECH) □ TRIETHOXYMETHANE □ TRIETHYL ORTHOFORMATE

TOXICITY DATA WITH REFERENCE
skn-rbt 10 mg/24H open MLD AMIHBC 4,119,51
skn-rbt 500 mg/24H MLD 85JCAE-,352,86
eye-rbt 100 mg/24H MOD 85JCAE-,352,86
eye-rbt 100 mg/24H MOD 28ZPAK -,44,72
orl-rat LD50:2920 mg/kg 28ZPAK -,44,72
ihl-rat LCLo:4000 ppm/8H AMIHBC 4,119,51
skn-rbt LD50:20 g/kg AMIHBC 4,119,51
scu-rbt LD50:20 g/kg FCTXAV 17,917,79

CONSENSUS REPORTS: Reported in EPA TSCA Inventory.

DOT CLASSIFICATION: 3; *Label:* Flammable Liquid

SAFETY PROFILE: Moderately toxic by ingestion. Mildly toxic by inhalation, skin contact, and subcutaneous routes. A skin and eye irritant. A very dangerous fire hazard when exposed to heat or flame; can react vigorously with oxidizing materials. To fight fire, use foam, CO_2, dry chemical. When heated to decomposition it emits acrid smoke and irritating fumes. See also ESTERS.

ENZ000 **CAS:97-81-4** **HR: 2**
ETHYL-3-OXATRICYCLO-(3.2.1.0^{2,4})OCTANE-6-CARBOXYLATE
mf: $C_{10}H_{14}O_3$ mw: 182.24

SYNS: 3,4-EPOXY-2,5-ENDOMETHYLENECYCLOHEXANECARBOXYLIC ACID, ETHYL ESTER □ 5,6-EPOXY-2-NORBORNANECARBOXYLIC ACID, ETHYL ESTER □ 3-OXATRICYCLO(3.2.1.0^{2,4})OCTANE-6-CARBOXYLIC ACID, ETHYL ESTER

TOXICITY DATA WITH REFERENCE
skn-rbt 10 mg/24H open MLD AIHAAP 23,95,62
orl-rat LD50:4760 mg/kg AIHAAP 24,305,63
skn-rbt LD50:3540 mg/kg AIHAAP 24,305,63

SAFETY PROFILE: Moderately toxic by skin contact. Mildly toxic by ingestion. A skin irritant. When heated to decomposition it emits acrid smoke and irritating fumes. See also ESTERS.

EOA000 **CAS:35629-44-8** **HR: 3**
1-ETHYL-3-(2-OXAZOLYL)UREA
mf: $C_6H_9N_3O_2$ mw: 155.18

TOXICITY DATA WITH REFERENCE
orl-mus LD50:400 mg/kg JMCMAR 14,1075,71
ipr-mus LD50:300 mg/kg JMCMAR 14,1075,71

SAFETY PROFILE: Poison by ingestion and intraperitoneal routes. When heated to decomposition it emits toxic fumes of NO_x.

EOA500 **CAS:17243-64-0** **HR: 2**
cis-2-(3-ETHYL-4-OXO-5-PIPERIDINO-2-THIAZOLIDINYLIDENE)ACETIC ACID
mf: $C_{12}H_{18}N_2O_3S$ mw: 270.38

PROP: Crystals. Mp: 86–87°.

SYNS: Z-(3-AETHYL-4-OXO-5-PIPERIDINO-THIAZOLIDIN-2-YLI-

DEN)-ESSIGSAURE-AETHYLESTER (GERMAN) □ ETHYL 3-ETHYL-4-OXO-5-PIPERIDINO-$\Delta^{2,\alpha}$-THIAZOLIDINEACETATE □ ETHYL (Z)-(3-ETHYL-4-OXO-5-PIPERIDINOTHIAZOLIDIN-2-YLIDENE)ACETATE □ (Z)-2-(3-ETHYL-4-OXO-5-PIPERIDINO-2-THIAZOLIDINYLIDENE) ACETIC ACID □ (3-ETHYL-4-OXO-5-(1-PIPERIDINYL)-2-THIAZOLIDINYLIDENE)ACETIC ACID ETHYL ESTER □ Go 919 □ PIPROZOLIN □ PIPROZOLINE □ PROBILIN □ W 3699

TOXICITY DATA WITH REFERENCE
orl-rat LD50:3256 mg/kg ARZNAD 27,467,77
orl-mus LD50:1070 mg/kg ARZNAD 27,467,77

SAFETY PROFILE: Moderately toxic by ingestion. A choleretic agent. When heated to decomposition it emits very toxic fumes of NO_x and SO_x.

EOB000 CAS:17243-64-0 HR: 2
2-(3-ETHYL-4-OXO-5-PIPERIDINO-2-THIAZOLIDINYLIDENE)ACETIC ACID ETHYL ESTER
mf: $C_{14}H_{22}N_2O_3S$ mw: 298.44

TOXICITY DATA WITH REFERENCE
orl-rat LD50:3256 mg/kg ARZNAD 27,463,77
orl-mus LD50:1310 mg/kg ARZNAD 27,463,77

SAFETY PROFILE: Moderately toxic by ingestion. When heated to decomposition it emits very toxic fumes of NO_x and SO_x. See also ESTERS.

EOB100 CAS:28853-06-7 HR: 3
ETHYL PENTABORANE (9)
mf: $C_2H_{14}B_5$ mw: 91.17

SAFETY PROFILE: Ignites spontaneously in air. See also BORANES and BORON COMPOUNDS.

EOB200 CAS:56501-32-7 HR: 3
1-ETHYL-1-PENTADECYLPIPERIDINIUM BROMIDE
mf: $C_{22}H_{46}N{\cdot}Br$ mw: 404.60

TOXICITY DATA WITH REFERENCE
orl-mus LD50:195 mg/kg PSDTAP 15,331,74
ipr-mus LD50:3096 μg/kg PSDTAP 15,331,74
ivn-mus LD50:4692 μg/kg PSDTAP 15,331,74

SAFETY PROFILE: Poison by ingestion, intravenous, and intraperitoneal routes. When heated to decomposition it emits toxic fumes of NO_x and Br^-. See also BROMIDES.

EOB300 CAS:609-27-8 HR: 3
3-ETHYL-2-PENTANOL
mf: $C_7H_{16}O$ mw: 116.23

SYN: 2-PENTANOL, 3-ETHYL-

TOXICITY DATA WITH REFERENCE
ivn-mus LD50:180 mg/kg CSLNX* NX#03027

CONSENSUS REPORTS: Reported in EPA TSCA Inventory.

SAFETY PROFILE: Poison by intravenous route. When heated to decomposition it emits acrid smoke and irritating vapors.

EOD000 CAS:22750-93-2 HR: 3
ETHYL PERCHLORATE
mf: $C_2H_5ClO_4$ mw: 128.52

$CH_3CH_2OClO_3$

PROP: Oil. Bp: 74°.

SYN: PERCHLORIC ACID, ETHYL ESTER

DOT CLASSIFICATION: Forbidden

SAFETY PROFILE: Possibly the most explosive chemical known. Very sensitive to impact, friction, and heat. Upon decomposition it emits toxic fumes of Cl^-. See also PERCHLORATES.

EOD500 CAS:66922-67-6 HR: 2
ETHYLPHENACETIN
mf: $C_{12}H_{17}NO_2$ mw: 207.30

SYN: N-ETHYL-p-ACETOPHENETIDIDE

TOXICITY DATA WITH REFERENCE
orl-rbt LDLo:2500 mg/kg AEXPBL 33,216,1894
orl-gpg LDLo:2700 mg/kg AEXPBL 33,216,1894
orl-frg LDLo:4545 mg/kg AEXPBL 33,216,1894

SAFETY PROFILE: Moderately toxic by ingestion. When heated to decomposition it emits toxic fumes of NO_x.

EOE100 CAS:123-07-9 HR: 3
4-ETHYLPHENOL
mf: $C_8H_{10}O$ mw: 122.18

SYN: PHENOL, p-ETHYL-

TOXICITY DATA WITH REFERENCE
ipr-mus LD50:138 mg/kg JMCMAR 10,060,75

CONSENSUS REPORTS: Reported in EPA TSCA Inventory.

DOT CLASSIFICATION: 6.1; *Label:* KEEP AWAY FROM FOOD

SAFETY PROFILE: Poison by intravenous route. When heated to decomposition it emits acrid smoke and irritating vapors.

EOF500 CAS:101491-82-1 HR: 3
N-ETHYL-N-(1-PHENOXY-2-PROPYL)CARBAMIC ACID-2-(DIETHYLAMINO)ETHYL ESTER HYDROCHLORIDE
mf: $C_{18}H_{30}N_2O_3{\cdot}ClH$ mw: 358.96

SYN: C 2137

TOXICITY DATA WITH REFERENCE
eye-rbt 2% MOD ARZNAD 9,113,59
scu-mus LD50:210 mg/kg ARZNAD 9,113,59

SAFETY PROFILE: Poison by subcutaneous route. An eye irritant. When heated to decomposition it emits very toxic fumes of HCl and NO_x. See also ESTERS.

EOG000 CAS:101491-83-2 HR: 3
N-ETHYL-N-(1-PHENOXY-2-PROPYL)CARBAMIC ACID-2-(2-METHYLPIPERIDINO)ETHYL ESTER HYDROCHLORIDE
mf: $C_{20}H_{32}N_2O_3 \cdot ClH$ mw: 385.00

SYN: C 2126

TOXICITY DATA WITH REFERENCE
eye-rbt 2% MLD ARZNAD 9,113,59
scu-mus LD50:300 mg/kg ARZNAD 9,113,59

SAFETY PROFILE: Poison by subcutaneous route. An eye irritant. When heated to decomposition it emits very toxic fumes of NO_x and HCl. See also CARBAMATES.

EOG500 CAS:101651-79-0 HR: 3
N-ETHYL-N-(1-PHENOXY-2-PROPYL)-2-(2-METHYLPIPERIDINO) ACETAMIDE HYDROCHLORIDE
mf: $C_{19}H_{30}N_2O_2 \cdot ClH$ mw: 354.97

SYNS: C 2054 □ N-ETHYL-2-(METHYLPIPERIDINO)-N-(1-PHENOXY-2-PROPYL)ACETAMIDE HYDROCHLORIDE

TOXICITY DATA WITH REFERENCE
eye-rbt 2% MLD ARZNAD 9,70,59
scu-mus LD50:165 mg/kg ARZNAD 9,70,59

SAFETY PROFILE: Poison by subcutaneous route. An eye irritant. When heated to decomposition it emits very toxic fumes of NO_x and HCl.

EOH000 CAS:101-97-3 HR: 2
ETHYL PHENYLACETATE
mf: $C_{10}H_{12}O_2$ mw: 164.22

PROP: Colorless liquid; sweet, honey odor. Bp: 227°, d: 1.033 @ 20°, refr index: 1.496–1.500, vap d: 5.67, flash p: 100° C. Sol in fixed oils; insol in glycerin, propylene glycol, water.

SYNS: BENZENEACETIC ACID, ETHYL ESTER (9CI) □ ETHYL BENZENEACETATE □ ETHYL PHENACETATE □ ETHYL-2-PHENYLETHANOATE □ ETHYL-α-TOLUATE □ FEMA No. 2452 □ PHENYLACETIC ACID, ETHYL ESTER □ α-TOLUIC ACID, ETHYL ESTER

TOXICITY DATA WITH REFERENCE
dnr-bcs 21 mg/disc OIGZSE 34,267,85
orl-rat LD50:3300 mg/kg FCTXAV 13,99,75

CONSENSUS REPORTS: Reported in EPA TSCA Inventory.

SAFETY PROFILE: Moderately toxic by ingestion. Combustible liquid. Mutation data reported. When heated to decomposition it emits acrid smoke and irritating fumes. See also ESTERS.

EOH500 CAS:2058-67-5 HR: 2
N-ETHYL-p-(PHENYLAZO)ANILINE
mf: $C_{14}H_{15}N_3$ mw: 225.32

SYNS: EAB □ 4-(ETHYLAMINOAZOBENZENE) □ N-ETHYL-4-AMINOAZOBENZENE □ N-ETHYL-4-(PHENYLAZO)BENZENAMINE

TOXICITY DATA WITH REFERENCE
ipr-mus TDLo:400 mg/kg (10-11D preg):TER KAIZAN 37,179,62
ipr-mus TDLo:400 mg/kg (10-11D preg):REP KAIZAN 37,179,62
orl-rat TDLo:376 mg/kg/5W-I:ETA CNREA8 39,3411,79

SAFETY PROFILE: Experimental teratogenic and reproductive effects. Questionable carcinogen with experimental tumorigenic data. When heated to decomposition it emits toxic fumes of NO_x.

EOI000 CAS:17010-65-0 HR: 2
p-((m-ETHYLPHENYL)AZO)-N,N-DIMETHYLANILINE
mf: $C_{16}H_{19}N_3$ mw: 253.38

SYNS: N,N-DIMETHYL-p((m-ETHYLPHENYL)AZO)ANILINE □ N,N-DIMETHYL-p-(3′-ETHYLPHENYLAZO)ANILINE □ N,N-DIMETHYL-3′-ETHYL-4-(PHENYLAZO)BENZENAMINE □ 3′-ETHYL-DAB □ 3′-ETHYL-4-DIMETHYLAMINOAZOBENZENE

TOXICITY DATA WITH REFERENCE
orl-rat TDLo:3175 mg/kg/12W-C:CAR CBINA8 53,107,85
orl-rat TD:2261 mg/kg/17W-I:ETA CNREA8 30,1520,70
orl-rat TD:4858 mg/kg/17W-C:CAR CRNGDP 1,419,80

SAFETY PROFILE: Questionable carcinogen with experimental carcinogenic and tumorigenic data. When heated to decomposition it emits toxic fumes of NO_x.

EOI500 CAS:5302-41-0 HR: 2
p-((p-ETHYLPHENYL)AZO)-N,N-DIMETHYLANILINE
mf: $C_{16}H_{19}N_3$ mw: 253.38

SYNS: N,N-DIMETHYL-4′-ETHYL-4-AMINOAZOBENZENE □ N,N-DIMETHYL-p-((4-ETHYLPHENYL)AZO)ANILINE □ 4′-ETHYL-DAB □ 4′-ETHYL-4-DIMETHYLAMINOAZOBENZENE

TOXICITY DATA WITH REFERENCE
orl-rat TDLo:4858 mg/kg/17W-C:CAR CRNGDP 1,419,80
orl-rat TD:2142 mg/kg/17W-C:ETA JNCIAM 27,663,61

SAFETY PROFILE: Questionable carcinogen with experimental carcinogenic and tumorigenic data. When heated to decomposition it emits toxic fumes of NO_x.

EOJ000 CAS:55398-27-1 HR: 2
p-(4-ETHYLPHENYLAZO)-N-METHYLANILINE
mf: $C_{15}H_{17}N_3$ mw: 239.35

SYNS: 4′-ETHYL-N-METHYL-4-AMINOAZOBENZENE □ N-METHYL-4′-ETHYL-p-AMINOAZOBENZENE

TOXICITY DATA WITH REFERENCE
orl-rat TDLo:1890 mg/kg/15W-C:ETA JNCIAM 15,67,54
scu-rat TDLo:181 mg/kg/8W-I:ETA CNREA8 35,880,75

SAFETY PROFILE: Questionable carcinogen with experimental tumorigenic data. When heated to decomposition it emits toxic fumes of NO_x.

EOJ500 CAS:6368-72-5 HR: 2
N-ETHYL-1-((p-(PHENYLAZO)PHENYL)AZO)-2-NAPHTHYLAMINE
mf: $C_{24}H_{21}N_5$ mw: 379.50

SYNS: CERES RED 7B □ C.I. 26050 □ C.I. SOLVENT RED 19 □ N-ETHYL-1-((p-(PHENYLAZO)PHENYL)AZO)-2-NAPHTHALENAMINE □ N-ETHYL-1-((4-(PHENYLAZO)PHENYL)AZO)-2-NAPHTHALENAMINE □ N-ETHYL-1-((4-(PHENYLAZO)PHENYL)AZO)-2-NAPHTHYLAMINE □

FAT RED 7B □ HEXATYPE CARMINE B □ LACQUER RED V3B □ OIL VIOLET □ ORGANOL BORDEAUX B □ (PHENYLAZO-4-PHENYLAZO)-1-ETHYLAMINO-2-NAPHTHALENE □ 1-(4-PHENYLAZO-PHENYLAZO)-2-ETHYLAMINONAPHTHALENE □ SOLVENT RED 19 □ SPECIAL BLUE X 2137 □ SUDAN RED 7B □ SUDANROT 7B □ TYPOGEN CARMINE

TOXICITY DATA WITH REFERENCE
mma-sat 500 μg/plate EPASR* 8EHQ-0982-0455
mma-mus:lym 300 mg/L EPASR* 8EHQ-0982-0455
orl-rat TDLo:15 g/kg/50W-C:ETA XPHPAW 149,403,69

CONSENSUS REPORTS: IARC Cancer Review: Group 3 IMEMDT 7,56,87; Animal Inadequate Evidence IMEMDT 8,253,75. Reported in EPA TSCA Inventory.

SAFETY PROFILE: Questionable carcinogen with experimental tumorigenic data. Mutation data reported. When heated to decomposition it emits toxic fumes of NO_x.

EOK000 CAS:50-06-6 HR: 3
5-ETHYL-5-PHENYLBARBITURIC ACID
mf: $C_{12}H_{12}N_2O_3$ mw: 232.26

PROP: Crystals in three modifications, one stable and two unstable. Mp: 156–157°. Sol in EtOH; mod sol in Et_2O; sltly sol in $CHCl_3$; prac insol in C_6H_6.

SYNS: ACIDO-5-FENIL-5-ETILBARBITURICO (ITALIAN) □ADON-AL □ AGRYPNAL □ AMYLOFENE □ APHENYLBARBIT □ AUSTROMINAL □ BARBAPIL □ BARBENYL □ BARBILEHAE (BARBILETTAE) □ BARBIPHENYL □ BARBITA □ BARBONAL □ BARBOPHEN □ BARTOL □ BIALMINAL □ CABRONAL □ CALMINAL □ CODIBARBITA □ CRATECIL □ DEZIBARBITUR □ DORMIRAL □ DUNERYL □ENSO-DORM □ EPIDORM □ EPISEDAL □ ESKABARB □ 5-ETHYL-5-PHENYL-2,4,6-(1H,3H,5H)PYRIMIDINETRIONE □ ETILFEN □ FENBITAL □ FENOBARBITAL □ FENYLETTAE □ GARDEPANYL □ GLYSOLETTEN □ HAPLOS □ HENNOLETTEN □ HYPNOGEN □ HYPNO-TABLINETTEN □ LEFEBAR □ LEPHEBAR □ LEPINAL □ LINASEN □ LIQUITAL □ LUBERGAL □ LUMEN □ LUMESETTES □ LUMINAL □ LUMOFRIDETTEN □ LURAMIN □ NEUROBARB □ NOPTIL □ NOVA-PHENO □ PARKOTAL □ PHENAEMAL □ PHENOBAL □ PHENOBARBITAL □ PHENOBARBITONE □ PHENOBARBITURIC ACID □ PHENOLURIC □ PHENOMET □ PHENONYL □ PHENOTURIC □ PHENYLETHYLBARBITURATE □ PHENYL-ETHYL-BARBITURIC ACID □ 5-PHENYL-5-ETHYLBARBITURIC ACID □ PHENYLETHYLMALONYLUREA □ PHENYLETTEN □ PHENYRAL □ PHOB □ POLCOMINAL □ PROMPTONAL □ SEDA-TABLINEN □ SEDICAT □ SEDIZORIN □ SEDOFEN □ SEDONAL □ SEDOPHEN □ SEVENAL □ SK-PHENOBARBITAL □ SOMBUTOL □ SOMNOLETTEN □ SOMNOSAN □ SOMONAL □STARIFEN □ STENTAL EXTENTABS □ TALPHENO □ THENOBARBITAL □ THEOMINAL □ TRIABARB □ TRIDEZIBARBITUR □ VERSOMNAL □ ZADONAL

TOXICITY DATA WITH REFERENCE
mmo-sat 100 μg/plate MUREAV 147,255,85
cyt-hmn:lym 388 mg/L AGMGAK 21,305,72
otr-ham:emb 100 mg/L PMRSDJ 5,665,85
orl-wmn TDLo:3600 μg/kg (female 26W post):REP AJOGAH 159,1491,88
orl-wmn TDLo:151 mg/kg (27-39W preg):REP JOPDAB 80,190,72
unr-wmn TDLo:907 mg/kg (female 1-36W post):REP LANCAO 2,839,72
unr-wmn TDLo:491 mg/kg (female 1-39W post):TER LANCAO 2,839,72
unr-wmn TDLo:907 mg/kg (female 1-36W post):TER LANCAO 2,839,72
unr-wmn TDLo:529 mg/kg (female 1-42W post):TER LANCAO 2,839,72
orl-wmn TDLo:3600 μg/kg (female 26W post):TER AJOGAH 159,1491,88
orl-wmn TDLo:151 mg/kg (27-39W preg):TER JOPDAB 80,190,72
orl-rat TDLo:320 mg/kg (female 7-10D post):REP JPETAB 227,274,83
orl-rat TDLo:280 mg/kg (lactating female 1-7D post):REP SCIEAS 211,719,81
scu-rat TDLo:320 mg/kg (female 12-19D post):REP EXMDA4 (551),250,81
orl-mus TDLo:3600 mg/kg (female 9-18D post):REP PSCHDL 68,301,80
unr-rat TDLo:40 mg/kg (female 17D post):REP SCIEAS 216,640,82
orl-mus TDLo:33 mg/kg (female 9-19D post):REP IJMDAI 14,929,78
orl-rat TDLo:60 mg/kg (female 7-18D post):REP JPETAB 227,274,83
orl-rat TDLo:320 mg/kg (female 7-10D post):REP JPETAB 227,274,83
ims-rat TDLo:200 mg/kg (female 18-20D post):TER TJADAB 26,21,82
orl-rat TDLo:800 mg/kg (female 6-13D post):TER TJADAB 24,1,81
ipr-rat TDLo:1 g/kg (female 1-20D post):TER AJOGAH 107,1250,70
orl-rat TDLo:880 mg/kg (female 7-17D post):TER SEIJBO 19,178,79
ipr-rat TDLo:180 mg/kg (female 19-21D post):TER TXAPA9 73,457,84
orl-rat TDLo:300 mg/kg (female 4-8D post):REP FESTAS 22,735,71
orl-rbt TDLo:450 mg/kg (female 8-16D post):REP TXAPA9 10,244,67
scu-rat TDLo:320 mg/kg (multi):REP PPHAD4 1,55,80
ipr-ham TDLo:140 mg/kg (female 1D pre):REP PSEBAA 162,170,79
orl-mus TDLo:210 mg/kg (male 5D pre):REP FCTOD7 21,335,83
par-mus TDLo:60 μg/kg (female 1D pre):REP PPHAD4 5,157,85
orl-rbt TDLo:450 mg/kg (female 8-16D post):TER TXAPA9 10,244,67
orl-mus TDLo:400 mg/kg (female 6-15D post):TER TXCYAC 6,323,76
orl-rat TDLo:440 mg/kg (female 7-17D post):TER TJADAB 20,157,79
scu-rat TDLo:320 mg/kg (female 12-19D post):TER EXMDA4 (551),250,81
orl-rat TDLo:7560 mg/kg/36W-C:ETA GANNA2 69,679,78
orl-mus TDLo:22 g/kg/1Y-C:NEO JNCIAM 51,1349,73
orl-mus TD:38 g/kg/90W-C:NEO FCTXAV 11,433,73
orl-rat TD:4200 mg/kg/20W-C:ETA CALEDQ 5,139,78
orl-rat TD:30 mg/kg/78W-C:ETA CRNGDP 10,183,89
orl-rat TD:3990 mg/kg/19W-C:CAR CRNGDP 4,935,83
orl-rat TD:2520 mg/kg/12W-C:NEO CRNGDP 4,935,83
orl-rat TD:2100 mg/kg/12W-C:CAR IGAYAY 123,1069,82
orl-mus LD:5200 mg/kg/52W-C:ETA 53HJAC -,145,84
orl-wmn TDLo:46 mg/kg CPEDAM 24,678,85
orl-wmn LDLo:25,272 μg/kg/13D-I CMAJAX 33,635,35
orl-chd TDLo:10 mg/kg:CNS AJDCAI 130,507,76

orl-hmn TDLo: 18 mg/kg: SKN,PUL,MET JAMAAP 116,700,41
orl-man LDLo: 6485 μg/kg/2W-I: SKN,MET AIMEAS 13,1243,40
ims-inf TDLo: 50 mg/kg PEDIAU 77,848,86
orl-rat LD50: 162 mg/kg TXAPA9 18,185,71
ipr-rat LD50: 151 mg/kg ARZNAD 30,477,80
scu-rat LD50: 200 mg/kg AEPPAE 152,341,30
ivn-rat LD50: 209 mg/kg ARTODN 40,211,78
rec-rat LD50: 284 mg/kg AACRAT 46,395,67
orl-mus LD50: 137 mg/kg PCJOAU 10,41,76
ipr-mus LD50: 128 mg/kg FRPSAX 14,269,59
scu-mus LD50: 228 mg/kg ARZNAD 30,477,80
ivn-mus LD50: 218 mg/kg AIPTAK 191,405,71
orl-dog LD50: 150 mg/kg SMWOAS 85,305,55

CONSENSUS REPORTS: EPA Genetic Toxicology Program. IARC Cancer Review: Group 2B IMEMDT 7,313,87; Animal Sufficient Evidence IMEMDT 7,313,87; Human Inadequate Evidence IMEMDT 13,157,77.

SAFETY PROFILE: Confirmed carcinogen with experimental carcinogenic, neoplastigenic, tumorigenic, and teratogenic data. A human poison by ingestion. An experimental poison by ingestion, intraperitoneal, subcutaneous, intravenous, and rectal routes. Human systemic effects by ingestion: somnolence, motor activity changes, pulmonary changes, allergic dermatitis, and fever. Human reproductive effects by ingestion: drug dependence and other postnatal measures or effects. Human teratogenic effects include developmental abnormalities of the central nervous system, body wall, musculoskeletal, respiratory, gastrointestinal, and urogenital systems. Experimental reproductive effects. Human mutation data reported. Used as a drug in the treatment of epilepsy, and as a hypnotic and sedative. When heated to decomposition it emits toxic fumes of NO_x. See also BARBITURATES.

EOK500 CAS:21224-57-7 HR: 3
S-2-((4-(p-ETHYLPHENYL)BUTYL)AMINO)ETHYL THIOSULFATE

mf: $C_{14}H_{23}NO_3S_2$ mw: 317.50

TOXICITY DATA WITH REFERENCE
orl-mus LD50: 1000 mg/kg JMCMAR 11,1190,68
ipr-mus LD50: 15 mg/kg JMCMAR 11,1190,68

SAFETY PROFILE: Poison by intraperitoneal route. Moderately toxic by ingestion. When heated to decomposition it emits very toxic fumes of NO_x and SO_x. See also THIOSULFATES.

EOK550 CAS:13037-20-2 HR: 3
ETHYL PHENYLDITHIOCARBAMATE

mf: $C_9H_{11}NS_2$ mw: 197.33

PROP: Plates from EtOH. Mp: 60–61°. Very sol in EtOH; insol in H_2O.

SYNS: S-AETHYL-N-PHENYL-DITHIOCARBAMAT □ CARBAMODITHIOIC ACID, PHENYL-, ETHYL ESTER □ CARBANILIC ACID, DITHIO-, ETHYL ESTER □ DITHIOCARBANILIC ACID ETHYL ESTER □ ETHYL N-PHENYLDITHIOCARBAMATE □ ZEPC

TOXICITY DATA WITH REFERENCE
orl-rat TDLo: 1125 mg/kg (female 7-15D post): REP JTSCDR 8,337,83
ipr-rat LDLo: 300 mg/kg ARZNAD 16,1092,66

SAFETY PROFILE: Poison by intraperitoneal route. Experimental reproductive effects. When heated to decomposition it emits toxic fumes of SO_x.

EOK600 CAS:121-39-1 HR: 2
ETHYL PHENYLGLYCIDATE

mf: $C_{11}H_{12}O_3$ mw: 192.23

PROP: Colorless liquid; strong strawberry odor. D: 1.120, refr index: 1.516–1.521. Sol in alc, chloroform, ether; insol in water.

SYNS: ETHYL-α,β-EPOXYHYDROCINNAMATE □ ETHYL-α,β-EPOXY-α-PHENYLPROPIONATE □ ETHYL-3-PHENYLGLYCIDATE □ FEMA No. 2454

TOXICITY DATA WITH REFERENCE
otr-mus: fbr 103 mg/L MUREAV 138,1,84
msc-ham: ovr 103 mg/L MUREAV 138,1,84
orl-rat LD50: 2300 mg/kg FCTXAV 13,101,75

CONSENSUS REPORTS: Reported in EPA TSCA Inventory.

SAFETY PROFILE: Moderately toxic by ingestion. Mutation data reported. When heated to decomposition it emits acrid smoke and irritating fumes. See also ESTERS.

EOL000 CAS:631-07-2 HR: 3
ETHYLPHENYLHYDANTOIN

mf: $C_{11}H_{12}N_2O_2$ mw: 204.25

PROP: Colorless, odorless, crystalline powder. Mp: 199–200°.

SYNS: 5-ETHYL-5-PHENYLHYDANTOIN □ 5-ETHYL-5-PHENYL-2,4-IMIDAZOLIDINEDIONE □ NIRVANOL

TOXICITY DATA WITH REFERENCE
orl-mus LDLo: 500 mg/kg LDTU** -,-,31
scu-mus LDLo: 500 mg/kg HDTU** -,-,33
scu-dog LDLo: 200 mg/kg HBAMAK 4,1289,35
scu-frg LDLo: 100 mg/kg HBAMAK 4,1289,35

SAFETY PROFILE: Poison by subcutaneous route. Moderately toxic by ingestion. A skin, eye, and mucous membrane irritant. Combustible when exposed to heat or flame; can react with oxidizing materials. When heated to decomposition it emits toxic fumes such as NO_x.

EOL050 CAS:65567-32-0 HR: 3
l-5-ETHYL-5-PHENYLHYDANTOIN

mf: $C_{11}H_{12}N_2O_2$ mw: 204.25

PROP: Crystals from EtOH (aq). Mp: 235–237°.

SYNS: (−)-5-ETHYL-5-PHENYLHYDANTOIN □ HYDANTOIN, 5-ETHYL-5-PHENYL-, (−)- □ 2,4-IMIDAZOLIDINEDIONE, 5-ETHYL-5-PHENYL-, (R)- (9CI) □ (−)-NIRVANOL □ l-NIRVANOL □ (R)-NIRVANOL

TOXICITY DATA WITH REFERENCE
ipr-mus TDLo:97 mg/kg (female 10D post):TER JPETAB 221,228,82
ipr-mus TDLo:97 mg/kg (female 10D post):REP JPETAB 221,228,82
orl-cld TDLo:105 mg/kg:SKN,SYS JPETAB 47,209,33
ipr-mus LD50:500 mg/kg BBIADT 46,623,87
scu-rbt LDLo:125 mg/kg JPETAB 47,209,33
scu-gpg LDLo:400 mg/kg JPETAB 47,209,33

SAFETY PROFILE: Poison by subcutaneous route. Moderately toxic by intraperitoneal route. Human systemic effects by ingestion: allergic dermatitis, increased body temperature. An experimental teratogen. Experimental reproductive effects. When heated to decomposition it emits toxic fumes of NO_x.

EOL100 CAS:86-35-1 HR: 2
3-ETHYL-5-PHENYLHYDANTOIN
mf: $C_{11}H_{12}N_2O_2$ mw: 204.25

PROP: Stout prisms from water. Mp: 94°. Sparingly sol in cold water, more sol in hot water. Freely sol in alc, ether, benzene, dil aq solns of alkali hydroxides.

SYNS: ETHOTOIN □ 1-ETHYL-2,5-DIOXO-4-PHENYLIMIDAZOLIDINE □ 3 ETHYL 5 PHENYLIMIDAZOLIDIN-2,4-DIONE □PEGANONE □ PEGOANONE

TOXICITY DATA WITH REFERENCE
ipr-mus TDLo:858 mg/kg (8-10D preg):TER TXAPA9 64,271,82
unr-mus TDLo:1195 mg/kg (female 8-10D post): TER FEPRA7 38,438,79
ipr-mus TDLo:1194 mg/kg (8-10D preg):TER TXAPA9 64,271,82
orl-rat LD50:1500 mg/kg NIIRDN 6,117,82
ipr-rat LD50:625 mg/kg NIIRDN 6,117,82
scu rat LD50:1 g/kg NIIRDN 6,117,82
orl-mus LD50:1750 mg/kg NIIRDN 6,117,82
ipr-mus LD50:923 mg/kg NIIRDN 6,117,82
scu-mus LD50:1060 mg/kg NIIRDN 6,117,82

SAFETY PROFILE: Moderately toxic by ingestion, subcutaneous, and intraperitoneal routes. Experimental teratogenic effects. When heated to decomposition it emits toxic fumes of NO_x.

EOL500 CAS:93-55-0 HR: 3
ETHYL PHENYL KETONE
mf: $C_9H_{10}O$ mw: 134.19

PROP: Water-white to light amber liquid or crystals. Mp: 19–20°, bp: 218.0°, d: 1.009 @ 25°/25°, vap press: 1 mm @ 50.0°. Insol in water; misc in alc, ether.

SYNS: PHENYL ETHYL KETONE □ 1-PHENYL-1-PROPANONE □ PROPIONYLBENZENE □ PROPIOPHENONE □ USAF EK-1235

TOXICITY DATA WITH REFERENCE
skn-rbt 500 mg/24H MLD 85JCAE-,291,86
eye-rbt 500 mg/24H MLD 85JCAE-,291,86
orl-rat LD50:4490 mg/kg TXAPA9 28,313,74
ipr-mus LD50:100 mg/kg NTIS** AD277-689
scu-mus LD50:2250 mg/kg ARZNAD 5,559,55
skn-rbt LD50:4490 mg/kg TXAPA9 28,313,74

CONSENSUS REPORTS: Reported in EPA TSCA Inventory.

DOT CLASSIFICATION: 3; *Label:* Flammable Liquid

SAFETY PROFILE: Poison by intraperitoneal route. Moderately toxic by subcutaneous route. Mildly toxic by ingestion and skin contact. A skin and eye irritant. A flammable liquid when exposed to heat or flame; can react with oxidizing materials. To fight fire use foam, CO_2, dry chemical. When heated to decomposition it emits acrid smoke and irritating fumes. See also KETONES.

EOL600 CAS:69095-83-6 HR: 2
5-(2-ETHYLPHENYL)-3-(3-METHOXYPHENYL)-s-TRIAZOLE
mf: $C_{17}H_{17}N_3O$ mw: 279.37

SYNS: DL 111 □ DL-111-IT □ (3-ETHYLPHENYL)-5-(3-METHOXYPHENYL)-1,2,4-TRIAZOLE □ 3-(o-ETHYLPHENYL)-5-(m-METHOXYPHENYL)-s-TRIAZOLE □ 3-(2-ETHYLPHENYL)-5-(3-METHOXYPHENYL)-1H-1,2,4-TRIAZOLE □ 5-(2-ETHYLPHENYL)-3-(3-METHOXYPHENYL) 1H 1,2,4-TRIAZOLE

TOXICITY DATA WITH REFERENCE
scu-ham TDLo:1500 μg/kg (female 5D post):TER CCPTAY 33,263,86
scu-rat TDLo:12,500 μg/kg (female 6-10D post):TER CCPTAY 23,163,81
ims-mky TDLo:10 mg/kg (female 53-56D post):REP 45JVAR -,344,80
scu-ham TDLo:250 μg/kg (female 4D post):REP 45JVAR -,344,80
orl-rat TDLo:125 mg/kg (6-10D preg):REP CCPTAY 23,163,81
orl-rat TDLo:125 mg/kg (female 6-10D post):REP JMCMAR 26,1187,83
scu-rat TDLo:35 mg/kg (male 7D pre):REP CCPTAY 33,89,86
scu-rat TDLo:37,500 μg/kg (male 15D pre):REP CCPTAY 33,79,86
ipr-rat LD50:1180 mg/kg 45JVAR -,344,80
scu-rat LD50:3190 mg/kg DRFUD4 7,875,82
ims-rat LD50:2000 mg/kg DRFUD4 7,875,82
ipr-mus LD50:650 mg/kg 45JVAR -,344,80
scu-mus LD50:3910 mg/kg DRFUD4 7,875,82
scu-ham LD50:2000 mg/kg DRFUD4 7,875,82

SAFETY PROFILE: Moderately toxic by intraperitoneal, subcutaneous and intramuscular routes. An experimental teratogen. Experimental reproductive effects. Used as a non-hormonal post implantation antifertility agent. When heated to decomposition it emits toxic fumes of NO_x.

EOM000 CAS:2240-14-4 HR: 3
N-ETHYL-3-PHENYL-2-NORBORNANAMINE HYDROCHLORIDE
mf: $C_{15}H_{21}N \cdot ClH$ mw: 251.83

PROP: Crystals from Me_2CO. Mp: 192°.

SYNS: 2-AETHYLAMINO-3-PHENYL-NOR-CAMPHAN (GERMAN) □ 2-ETHYLAMINO-3-PHENYL-NORCAMPHANE HYDROCHLORIDE □ FENCAMFAMINE HYDROCHLORIDE □ H 610 □ NORCAMPHANE

TOXICITY DATA WITH REFERENCE
orl-rat LD50:83 mg/kg JOPDAB 69,663,66
scu-rat LD50:69 mg/kg ARZNAD 11,20,61
ivn-rat LD50:24 mg/kg ARZNAD 11,20,61
orl-mus LD50:135 mg/kg ARZNAD 11,20,61
scu-mus LD50:86 mg/kg ARZNAD 11,20,61
ivn-mus LD50:16 mg/kg ARZNAD 11,20,61
orl-dog LD50:30 mg/kg ARZNAD 11,20,61
ivn-dog LD50:15 mg/kg ARZNAD 11,20,61
orl-cat LD50:34 mg/kg ARZNAD 11,20,61

SAFETY PROFILE: Poison by ingestion, subcutaneous, and intravenous routes. A central nervous system stimulant. When heated to decomposition it emits very toxic fumes of HCl and NO_x.

EOM100 CAS:5075-13-8 HR: 3
ETHYL P-PHENYLPHOSPHONOCHLORIDOTHIOATE
mf: $C_8H_{10}ClOPS$ mw: 220.66

SYNS: PHENYLPHOSPHONOCHLORIDOTHIOIC ACID O-ETHYL ESTER □ PHOSPHONOCHLORIDOTHIOIC ACID, PHENYL-, O-ETHYL ESTER

TOXICITY DATA WITH REFERENCE
ivn-mus LD50:56 mg/kg CSLNX* NX#06040

CONSENSUS REPORTS: Reported in EPA TSCA Inventory.

SAFETY PROFILE: A poison by intravenous route. When heated to decomposition it emits toxic vapors of NO_x, SO_x, PO_x, and Cl^-.

EOM700 CAS:5556-57-0 HR: 2
(±)-5-ETHYL-5-PHENYLPYRROLID-2-ONE
mf: $C_{12}H_{15}NO$ mw: 189.28

SYN: (±)-5-ETHYL-5-PHENYL-2-PYRROLIDINONE

TOXICITY DATA WITH REFERENCE
orl-rat LD50:600 mg/kg BJPCAL 25,790,65
orl-mus LD54:812 mg/kg JMCMAR 9,52,66
ipr-mus LD50:750 mg/kg BJPCAL 25,790,65

SAFETY PROFILE: Moderately toxic by ingestion and intraperitoneal routes. When heated to decomposition it emits toxic fumes of NO_x.

EON000 CAS:593-68-0 HR: 3
ETHYL PHOSPHINE
mf: C_2H_7P mw: 62.05

$CH_3CH_2PH_2$

PROP: Colorless volatile liquid with unpleasant odor. Bp: 25°, d: 1, vap d: 2.14.

SYN: PHOSPHINOETHANE

SAFETY PROFILE: Probably a poison. A very dangerous fire hazard when exposed to heat or flame; ignites spontaneously in air. Ignites on contact with concentrated acids. To fight fire, use foam, CO_2, dry chemical. Explodes on contact with chlorine; bromine; or fuming nitric acid. Can react vigorously with oxidizing materials. When heated to decomposition it emits highly toxic fumes of PO_x and phosphine. See also PHOSPHINE.

EON500 CAS:35335-60-5 HR: 3
ETHYL PHOSPHONAMIDOTHIONIC ACID-4-(METHYLTHIO)-m-TOLYL ESTER

SYN: BAY HOL 0574

TOXICITY DATA WITH REFERENCE
orl-bwd LD50:13,300 µg/kg AECTCV 12,355,83
orl-bwd LD50:3200 µg/kg TXAPA9 26,154,73
skn-bwd LD50:75 mg/kg TXAPA9 26,154,73

SAFETY PROFILE: Poison by ingestion and skin contact. When heated to decomposition it emits very toxic fumes of PO_x, SO_x, and NO_x. See also ESTERS.

EOO000 CAS:2984-65-8 HR: 3
ETHYL PHOSPHONODITHIOIC ACID-O-METHYL-S-(p-TOLYL) ESTER
mf: $C_{10}H_{15}OPS_2$ mw: 246.34

SYNS: BAY 42903 □ BAYER 42903 □ ENT 27,250 □ O-METHYL-S-(4-METHYLPHENYL) ETHYLPHOSPHONODITHIOATE □ N 4328 □ STAUFFER N-4328

TOXICITY DATA WITH REFERENCE
orl-rat LD50:93 mg/kg ARSIM* 20,22,66
orl-mus LDLo:710 mg/kg AECTCV 14,111,85
orl-gpg LDLo:25 mg/kg JEENAI 61,1261,68
scu-gpg LDLo:25 mg/kg JEENAI 61,1261,68
orl-bwd LD50:5600 µg/kg TXAPA9 21,315,72

SAFETY PROFILE: Poison by ingestion and subcutaneous routes. When heated to decomposition it emits very toxic fumes of PO_x and SO_x. See also ESTERS.

EOP500 CAS:50335-09-6 HR: 3
ETHYL PHOSPHONOTHIOIC ACID-(2-DIETHYLAMINOMETHYL)-4,6-DICHLOROPHENYL ESTER

SYN: BAY 50519

TOXICITY DATA WITH REFERENCE
orl-bwd LD50:4200 µg/kg TXAPA9 26,154,73
skn-bwd LD50:5600 µg/kg TXAPA9 26,154,73

SAFETY PROFILE: Poison by ingestion and skin contact. When heated to decomposition it emits very toxic fumes of PO_x, SO_x, and Cl^-. See also ESTERS.

EOP600 CAS:993-43-1 HR: 2
ETHYL PHOSPHONOTHIOIC DICHLORIDE
DOT: NA 2927
mf: $C_2H_5Cl_2PS$ mw: 163.00

SYNS: DICHLOROETHYLPHOSPHINE SULFIDE □ ETHYL PHOSPHONOTHIOIC DICHLORIDE, anhydrous (DOT) □ ETHYLPHOSPHONOTHIONIC DICHLORIDE □ ETHYL PHOSPHONOTHIOYL DICHLORIDE □ ETHYLTHIONOPHOSPHONYL DICHLORIDE □ ETHYLTHIOPHOSPHONIC DICHLORIDE □ PHOSPHONOTHIOIC DICHLORIDE, ETHYL-

CONSENSUS REPORTS: Reported in EPA TSCA Inventory.

DOT CLASSIFICATION: 6.1; *Label:* Poison, Corrosive

SAFETY PROFILE: A corrosive. When heated to decomposition it emits toxic vapors of SO_x, PO_x, and Cl^-.

EOQ000 CAS:1498-40-4 HR: 2
ETHYL PHOSPHONOUS DICHLORIDE
DOT: UN 2845
mf: $C_2H_5Cl_2P$ mw: 130.94

PROP: A liquid with pungent, foul, odor. D: 1.26 @ 20°/4°, bp: 111–112° @ 722 mm.

SYNS: DICHLOROETHYLPHOSPHINE □ DICHLOROMETHYL PHOSPHINE □ ETHYL PHOSPHONOUS DICHLORIDE, anhydrous (DOT) □ TL 373

TOXICITY DATA WITH REFERENCE
ihl-mus LCLo:1990 mg/m³/10M NDRC** NDCrc-132,Sept,42

DOT CLASSIFICATION: 6.1; *Label:* Poison, Spontaneously Combustible

SAFETY PROFILE: Moderately toxic by inhalation. Corrosive. A severe irritant to skin, eyes, and mucous membranes. When heated to decomposition it emits very toxic fumes of PO_x, Cl^-, and phosphine. See also PHOSPHINE.

EOQ500 CAS:36031-66-0 HR: 3
ETHYL PHOSPHORAMIDIC ACID-2,4-DICHLOROPHENYL ESTER
mf: $C_8H_{10}Cl_2NO_3$ mw: 239.09

SYN: DOWCO 161

TOXICITY DATA WITH REFERENCE
orl-pgn LD50:75 mg/kg TXAPA9 21,315,72
orl-dck LD50:13 mg/kg TXAPA9 21,315,72
orl-bwd LD50:7500 μg/kg TXAPA9 21,315,72

SAFETY PROFILE: Poison by ingestion. When heated to decomposition it emits very toxic fumes of Cl^- and NO_x. See also ESTERS.

EOR000 CAS:1498-51-7 HR: 2
ETHYL PHOSPHORODICHLORIDATE
DOT: NA 2927
mf: $C_2H_5Cl_2O_2P$ mw: 162.94

PROP: Colorless pungent liquid. D: 1.38, bp: 167°.

SYNS: DICHLOROPHOSPHORIC ACID, ETHYL ESTER □ PHOSPHORODICHLORIDIC ACID, ETHYL ESTER

CONSENSUS REPORTS: Reported in EPA TSCA Inventory.

DOT CLASSIFICATION: 6.1; *Label:* Poison, Corrosive

SAFETY PROFILE: A corrosive material that is very toxic to tissue. A severe eye, skin, and mucous membrane irritant. When heated to decomposition it emits very toxic fumes of Cl^- and PO_x.

EOR500 CAS:5022-29-7 HR: 3
N-ETHYLPHTHALIMIDE
mf: $C_{10}H_9NO_2$ mw: 175.20

PROP: Needles from EtOH. Mp: 78–79°, bp: 285° @ 758 mm.

TOXICITY DATA WITH REFERENCE
orl-rat LD:>500 mg/kg NCNSA6 5,23,53
ivn-mus LD50:320 mg/kg CSLNX* NX#04571

CONSENSUS REPORTS: Reported in EPA TSCA Inventory.

SAFETY PROFILE: Poison by intravenous route. Moderately toxic by ingestion. When heated to decomposition it emits toxic fumes of NO_x.

EOS000 CAS:104-90-5 HR: 3
5-ETHYL-α-PICOLINE
DOT: UN 2300
mf: $C_8H_{11}N$ mw: 121.20

$N{=}C(CH_3)CH{=}CHC(CH_2CH_3){-}CH$ (ring)

PROP: Liquid. Bp: 174°, d: 0.9184 @ 23°/4°, flash p: 165° (OC).

SYNS: ALDEHYDECOLLIDINE □ ALDEHYDINE □ COLLIDINE, ALDEHYDECOLLIDINE □ 3-ETHYL-6-METHYLPYRIDINE □ 5-ETHYL-2-METHYLPYRIDINE □ 5-ETHYL-2-PICOLINE □ MEP □ 2-METHYL-5-ETHYLPYRIDINE □ 6-METHYL-3-ETHYLPYRIDINE □ METHYL ETHYL PYRIDINE (DOT) □ 2-METHYL-5-ETHYLPYRIDINE (DOT)

TOXICITY DATA WITH REFERENCE
skn-rbt 10 mg/24H open SEV AMIHBC 4,119,51
skn-rbt 500 mg open SEV UCDS** 6/29/66
eye-rbt 250 μg open SEV AMIHBC 4,119,51
orl-rat LD50:368 mg/kg 85GMAT -,83,82
ihl-rat LCLo:1000 ppm/4H AMIHBC 4,119,51
scu-rat LD50:826 mg/kg 85GMAT -,83,82
orl-mus LD50:282 mg/kg 85GMAT -,83,82
scu-mus LD50:294 mg/kg 85GMAT -,83,82
skn-rbt LD50:1000 mg/kg UCDS** 6/29/66
skn-gpg LD50:2500 mg/kg 85JCAE ,846,86

CONSENSUS REPORTS: Reported in EPA TSCA Inventory. EPA Genetic Toxicology Program.

DOT CLASSIFICATION: 6.1; *Label:* KEEP AWAY FROM FOOD

SAFETY PROFILE: Poison by ingestion and subcutaneous routes. Moderately toxic by skin contact. Mildly toxic by inhalation. Corrosive. A severe skin and eye irritant. Flammable when exposed to heat or flame; can react vigorously with oxidizers. Potentially explosive reaction with nitric acid at 145°C/14.5 bar. To fight fire, use alcohol foam. When heated to decomposition it emits acrid smoke and irritating fumes. See also ALDEHYDES.

EOS100 CAS:72320-60-6 HR: 3
2-(4-ETHYL-1-PIPERAZINYL)-4-PHENYLQUINOLINE DIHYDROCHLORIDE
mf: $C_{21}H_{23}N_3{\cdot}2ClH$ mw: 390.39

SYN: D 1308

TOXICITY DATA WITH REFERENCE
orl-mus LD50:568 mg/kg AIPTAK 245,283,80
ipr-mus LD50:119 mg/kg AIPTAK 245,283,80
ivn-mus LD50:38 mg/kg AIPTAK 245,283,80

SAFETY PROFILE: Poison by intravenous and intraperitoneal routes. Moderately toxic by ingestion. When heated to decomposition it emits toxic fumes of NO_x and HCl.

EOS500 CAS:766-09-6 HR: 3
1-ETHYLPIPERIDINE
DOT: UN 2386
mf: $C_7H_{15}N$ mw: 113.23

PROP: A liquid. D: 0.824 @ 20°/4°, bp: 128°, flash p: 66.2°F.

SYN: N-AETHYLPIPERIDIN (GERMAN)

TOXICITY DATA WITH REFERENCE
eye-rbt 50 mg/5M BJIMAG 23,153,66
ivn-mus LD50:56 mg/kg CSLNX* NX#00374
scu-rbt LDLo:100 mg/kg BDCGAS 34,2408,01

CONSENSUS REPORTS: Reported in EPA TSCA Inventory.

DOT CLASSIFICATION: 3; *Label:* Flammable Liquid

SAFETY PROFILE: Poison by intravenous and subcutaneous routes. An eye irritant. A very dangerous fire hazard when exposed to heat or flame; can react vigorously with oxidizing materials. When heated to decomposition it emits toxic fumes of NO_x.

EOU500 CAS:77791-38-9 HR: 3
N-ETHYL-2-(PIPERIDINO)-N-(1-(2,4-XYLYLOXY)-2-PROPYL)ACETAMIDE HYDROCHLORIDE
mf: $C_{20}H_{32}N_2O_2 \cdot ClH$ mw: 369.00

SYN: C 2102

TOXICITY DATA WITH REFERENCE
eye-rbt 2% MLD ARZNAD 9,70,59
scu-mus LD50:157 mg/kg ARZNAD 9,70,59

SAFETY PROFILE: Poison by subcutaneous route. An eye irritant. When heated to decomposition it emits very toxic fumes of NO_x and HCl.

EOV000 CAS:78219-57-5 HR: 3
1-ETHYL-4-PIPERIDYL-p-AMINOBENZOATE HYDROCHLORIDE
mf: $C_{14}H_{20}N_2O \cdot ClH$ mw: 268.82

SYN: p-AMINO-BENZOIC ACID 1-ETHYL-4-PIPERIDYL ESTER, HYDROCHLORIDE

TOXICITY DATA WITH REFERENCE
ivn-rat LDLo:35 mg/kg JACSAT 51,922,29
scu-mus LDLo:25 mg/kg JACSAT 51,922,29

SAFETY PROFILE: Poison by intravenous and subcutaneous routes. When heated to decomposition it emits very toxic fumes of NO_x and HCl. See also ESTERS.

EOW000 CAS:78219-58-6 HR: 3
1-ETHYL-4-PIPERIDYL BENZOATE HYDROCHLORIDE
mf: $C_{14}H_{19}NO_2 \cdot ClH$ mw: 269.80

SYN: BENZOIC ACID-1-ETHYL-4-PIPERIDYL ESTER, HYDROCHLORIDE

TOXICITY DATA WITH REFERENCE
ivn-rat LDLo:25 mg/kg JACSAT 51,922,29
scu-mus LDLo:300 mg/kg JACSAT 51,922,29

SAFETY PROFILE: Poison by intravenous and subcutaneous routes. When heated to decomposition it emits very toxic fumes of NO_x and HCl. See also ESTERS.

EOY000 CAS:129-77-1 HR: 3
1-ETHYL-3-PIPERIDYL DIPHENYLACETATE HYDROCHLORIDE
mf: $C_{21}H_{25}NO_2 \cdot ClH$ mw: 359.93

PROP: A solid. Mp: 195–196°.

SYNS: AN 1087 □ DACTIL □ DACTIL HYDROCHLORIDE □ DIPHENYLACETIC ACID-1-ETHYL-3-PIPERIDYL ESTER HYDROCHLORIDE □ 1-ETHYL-3-PIPERIDINOL DIPHENYLACETATE HYDROCHLORIDE □ 1-ETHYL-3-PIPERIDINYL ESTER-α-PHENYLBENZENEACETIC ACID HYDROCHLORIDE □ N-ETHYL-3-PIPERIDYL DIPHENYLACETATE HYDROCHLORIDE □ 1-ETHYL-3-PIPERIDYL ESTER-α-PHENYLBENZENEACETIC ACID HYDROCHLORIDE □ J.B. 305 □ PIPERIDILATE HYDROCHLORIDE □ PIPERIDOLATE HYDROCHLORIDE

TOXICITY DATA WITH REFERENCE
ivn-rat LD50:19 mg/kg 29ZVAB -,97,69
orl-mus LD50:1040 mg/kg CLDND* 104,269,52
ivn-dog LD50:35 mg/kg 29ZVAB -,97,69

SAFETY PROFILE: Poison by intravenous route. Moderately toxic by ingestion. When heated to decomposition it emits very toxic fumes of NO_x and HCl.

EOY100 CAS:78219-33-7 HR: 3
β-2-(N-ETHYLPIPERIDYL)ETHYLBENZOATE HYDROCHLORIDE
mf: $C_{16}H_{23}NO_2 \cdot ClH$ mw: 297.86

SYNS: BENZOIC ACID-2-(1-ETHYL-2-PIPERIDYL)ETHYL ESTER HYDROCHLORIDE □ β-(1-ETHYL-2-PIPERIDYL)ETHYL BENZOATE HYDROCHLORIDE

TOXICITY DATA WITH REFERENCE
scu-mus LDLo:1700 mg/kg ANESAV 1,305,40
ivn-mus LDLo:46 mg/kg JACSAT 61,961,39
ivn-rbt LDLo:20 mg/kg ANESAV 1,305,40
isp-rbt LDLo:29,980 μg/kg ANESAV 1,305,40
scu-gpg LDLo:439 mg/kg ANESAV 1,305,40

SAFETY PROFILE: Poison by intravenous and intraspinal routes. Moderately toxic by subcutaneous route. When heated to decomposition it emits toxic fumes of NO_x and HCl.

EOY500 CAS:928-55-2 HR: 3
ETHYL-1-PROPENYL ETHER
mf: $C_5H_{10}O$ mw: 86.15

PROP: Flash p: >19°F (OC), d: 0.8, bp: 158°F.

SYN: ETHYL PROPENYL ETHER

TOXICITY DATA WITH REFERENCE
skn-rbt 8276 μg/24H open MLD AIHAAP 23,95,62
eye-rbt 500 mg/24H MLD 85JCAE-,248,86
orl-rat LD50:4660 mg/kg TXAPA9 28,313,74
ihl-rat LCLo:8000 ppm/4H TXAPA9 28,313,74
skn-rbt LD50:3970 mg/kg TXAPA9 28,313,74

SAFETY PROFILE: Moderately toxic by skin contact. Mildly toxic by ingestion and inhalation. A skin and eye irritant. A very dangerous fire hazard when exposed to heat, flame, or oxidizers. To fight fire, use water spray or mist, dry chemical, foam, CO_2. When heated to decomposition it emits acrid smoke and irritating fumes. See also ETHERS.

EPB500 CAS:105-37-3 HR: 3
ETHYL PROPIONATE
DOT: UN 1195
mf: $C_5H_{10}O_2$ mw: 102.15

PROP: Colorless liquid; fruity, rum odor. Mp: −72.6°, bp: 99°, flash p: 54°F (CC), d: 0.891 @ 20°/4°, fp: −73°, refr index: 1.383, autoign temp: 824°F, vap press: 40 mm @ 27.2°, vap d: 3.52, lel: 1.9%, uel: 11%. Misc with alc, ether, propylene glycol; sol in water and fixed oils.

SYNS: FEMA No. 2456 □ PROPIONATE d'ETHYLE (FRENCH) □ PROPIONIC ACID, ETHYL ESTER □ PROPIONIC ETHER

TOXICITY DATA WITH REFERENCE
skn-rbt 500 mg/24H MOD FCTXAV 16,749,78
eye-rbt 100 mg MOD JACTDZ 1,161,92
orl-rat LD50:8732 mg/kg JACTDZ 1,174,92
skn-rbt LDLo:14,256 mg/kg JACTDZ 1,174,92
ipr-rat LD50:1200 mg/kg FCTXAV 16,749,78
ipr-mus LD50:1300 mg/kg 14CYAT 2,1879,63
orl-rbt LD50:3500 mg/kg 14CYAT 2,1879,63

CONSENSUS REPORTS: Reported in EPA TSCA Inventory.

DOT CLASSIFICATION: 3; *Label:* Flammable Liquid

SAFETY PROFILE: Moderately toxic by ingestion and intraperitoneal routes. A skin and eye irritant. A flammable liquid. A very dangerous fire and explosion hazard when exposed to heat or flame; can react vigorously with oxidizing materials. To fight fire, use foam, CO_2, dry chemical. When heated to decomposition it emits acrid smoke and irritating fumes. See also ETHERS.

EPC000 CAS:67050-97-9 HR: 3
5-ETHYL-5-(1-PROPYL-1-BUTENYL)BARBITURIC ACID
mf: $C_{13}H_{20}N_2O_3$ mw: 252.35

TOXICITY DATA WITH REFERENCE
orl-mus LD50:220 mg/kg JACSAT 61,776,39
ipr-mus LD50:130 mg/kg JACSAT 61,776,39

SAFETY PROFILE: Poison by ingestion and intraperitoneal routes. When heated to decomposition it emits toxic fumes of NO_x. See also BARBITURATES.

EPC050 CAS:623-85-8 HR: 2
ETHYL-N,N-PROPYLCARBAMATE
mf: $C_6H_13NO_2$ mw: 131.20

PROP: Bp: 191.5–192.5°.

SYN: N,N-PROPYL ETHYL CARBAMATE

TOXICITY DATA WITH REFERENCE
ipr-mus TDLo:6500 mg/kg/13W-I:ETA JNCIAM 9,35,48
scu-mus LD50:540 mg/kg AJEBAK 45,507,67

SAFETY PROFILE: Moderately toxic by subcutaneous route. Questionable carcinogen with experimental tumorigenic data. When heated to decomposition it emits toxic fumes of NO_x. See also CARBAMATES.

EPC100 HR: 2
N-ETHYL-N-PROPYLCARBAMOYL CHLORIDE
mf: $C_6H_{12}ClNO$ mw: 293.61

SAFETY PROFILE: Probably an eye, skin, and mucous membrane irritant. Vigorous reaction with water evolves large amounts of gases. Releases HCl in contact with water. When heated to decomposition it emits toxic fumes of Cl^- and NO_x.

EPC115 HR: 3
ETHYL N-((5R,8S,10R)-6-PROPYL-8-ERGOLINYL) CARBAMATE
mf: $C_{20}H_{27}N_3O_2$ mw: 341.50

SYNS: ERGOLINE-8-CARBAMIC ACID, 6-PROPYL-, ETHYL ESTER, (8S)- □ (8S)-6-PROPYLERGOLINE-8-CARBAMIC ACID ETHYL ESTER

TOXICITY DATA WITH REFERENCE
mmo-sat 400 μg/L CCCCAK 52,2983,87
orl-rat TDLo:20 μg/kg (female 5D post):REP CCCCAK 52,2983,87
orl-rat TDLo:60 μg/kg (lactating female 4D post): REP CCCCAK 52,2983,87
ivn-mus LD50:11,400 μg/kg CCCCAK 52,2983,87

SAFETY PROFILE: Poison by intraperitoneal route. Experimental reproductive effects. Mutation data reported. When heated to decomposition it emits toxic fumes of NO_x.

EPC125 CAS:628-32-0 HR: 3
ETHYL PROPYL ETHER
DOT: UN 2615
mf: $C_5H_{12}O$ mw: 88.15

PROP: A liquid. D: 0.8, bp: 63.6°, flash p: <−4°F; lel: 1.7%, uel: 9%.

SYNS: 1-ETHOXYPROPANE □ ETHYL n-PROPYL ETHER □ PROPANE, 1-ETHOXY-(9CI) □ PROPYL ETHYL ETHER

TOXICITY DATA WITH REFERENCE
ihl-mus LC50:220 g/m³/15M ANESAV 11,455,50

DOT CLASSIFICATION: 3; *Label:* Flammable Liquid

SAFETY PROFILE: A slight inhalation hazard. Very dangerous fire and explosion hazard when exposed to heat or open flame. To fight fire, use alcohol foam.

When heated to decomposition it emits acrid smoke and fumes. See also ETHERS.

EPC135 **CAS:66859-63-0** **HR: 2**
2-ETHYL-2-PROPYLTHIOBUTYRAMIDE
mf: $C_9H_{19}NOS$ mw: 189.35

TOXICITY DATA WITH REFERENCE
orl-mus LD50:1576 mg/kg JMCMAR 6,351,63
ipr-mus LD50:419 mg/kg JMCMAR 6,351,63

SAFETY PROFILE: Moderately toxic by intraperitoneal route. Mildly toxic by ingestion. When heated to decomposition it emits toxic fumes of NO_x and SO_x.

EPC150 **CAS:55365-87-2** **HR: 3**
O-ETHYL PROPYLTHIOCARBAMATE
mf: $C_6H_{13}NOS$ mw: 147.26

SYN: EPTAM

TOXICITY DATA WITH REFERENCE
orl-rat LD50:1660 mg/kg 85GMAT -,68,82
skn-rat LD50:3200 mg/kg 85GMAT -,69,82
orl-mus LD50:750 mg/kg 85GMAT -,68,82
orl-cat LD50:112 mg/kg 85GMAT -,68,82

SAFETY PROFILE: Poison by ingestion. Moderately toxic by skin contact. When heated to decomposition it emits toxic fumes of SO_x and NO_x. See also CARBAMATES and ESTERS.

EPC175 **CAS:38524-82-2** **HR: 3**
O-ETHYL-S-PROPYL-O-(2,4,6-TRICHLOROPHENYL) PHOSPHOROTHIOATE
mf: $C_{11}H_{14}Cl_3O_3PS$ mw: 363.63

SYNS: ENT 29,118 □ RH-8218 □ ROHM & HAAS RH-218 □ TH-218 □ TRIFENOFOS

TOXICITY DATA WITH REFERENCE
orl-rat LD50:200 mg/kg SPEADM 78-1,35,78
skn-rat LD50:250 mg/kg SPEADM 78-1,35,78
orl-mus LD50:200 mg/kg PCBPBS 6,85,76
skn-rbt LD50:108 mg/kg SPEADM 78-1,35,78

SAFETY PROFILE: Poison by ingestion and skin contact. When heated to decomposition it emits toxic fumes of PO_x, Cl^-, and SO_x.

EPC500 **CAS:297-97-2** **HR: 3**
ETHYL PYRAZINYL PHOSPHOROTHIOATE
mf: $C_8H_{13}N_2O_3PS$ mw: 248.26

PROP: Amber liquid or oil. Mp: −1.7°, bp: 80°, n (25/D) 1.5131, vap press @ 30°: 0.003 mm Hg. Sltly sol in water; misc with most organic solvents.

SYNS: AC 18133 □ AMERCIAN CYANAMID 18133 □ CL 18133 □ CYNEM □ O,O-DIAETHYL-O-(PYRAZIN-2YL)-MONOTHIOPHOSPHAT (GERMAN) □ O,O-DIAETHYL-O-(2-PYRAZINYL)-THIONOPHOSPHAT (GERMAN) □ O,O-DIETHYL-O,2-PYRAZINYL PHOSPHOROTHIOATE □ DIETHYL-O-2-PYRAZINYL PHOSPHOROTHIONATE □ O,O-DIETHYL-O-2-PYRAZINYL PHOSPHOTHIONATE □ O,O-DIETHYL-O-PYRAZINYL THIOPHOSPHATE □ EN 18133 □ ENT 25,580 □ EXPERIMENTAL NEMATOCIDE 18,133 □ NEMAFOS □ NEMAPHOS □ NEMATOCIDE □ PHOSPHOROTHIOIC ACID-O,O-DIETHYL-O-2-PYRAZINYL ESTER □ PYRAZINOL-O-ESTER with O,O-DIETHYL PHOSPHOROTHIOATE □ RCRA WASTE NUMBER P404 □ THIONAZIN □ ZINOPHOS

TOXICITY DATA WITH REFERENCE
orl-rat LD50:3500 μg/kg TXAPA9 14,515,69
skn-rat LD50:8 mg/kg SPEADM 78-1,46,78
ocu-rbt LDLo:50 mg/kg 85GYAZ -,36,71
skn-gpg LD50:10 mg/kg GUCHAZ 6,498,73
orl-pgn LD50:2370 μg/kg ASTTA8 (680),157,79
orl-qal LD50:3160 μg/kg ASTTA8 (680),157,79
orl-dck LD50:1680 μg/kg TXAPA9 47,451,79
skn-dck LD50:7 mg/kg TXAPA9 47,451,79
orl-bwd LD50:2420 μg/kg ASTTA8 (680),157,79

CONSENSUS REPORTS: EPA Extremely Hazardous Substances List. Reported in EPA TSCA Inventory. Community Right-To-Know List.

SAFETY PROFILE: Poison by ingestion, skin contact, and ocular routes. A cholinesterase inhibitor type of insecticide. When heated to decomposition it emits highly toxic fumes of NO_x, PO_x, and SO_x. See also PARATHION.

EPC700 **CAS:2687-91-4** **HR: 2**
1-ETHYL-2-PYRROLIDINONE
mf: $C_6H_{11}NO$ mw: 113.18

PROP: A liquid. Bp: 97°/20 mm, d: 0.992, flash p: 169°F.

SYNS: N-ETHYLPYRROLIDINONE □ N-ETHYLPYRROLIDONE □ 2-PYRROLIDINONE, 1-ETHYL-

TOXICITY DATA WITH REFERENCE
eye-rbt 100 mg MOD FCTOD7 26,475,88
orl-rat LD50:1350 mg/kg FCTOD7 26,475,88

CONSENSUS REPORTS: Reported in EPA TSCA Inventory.

SAFETY PROFILE: Moderately toxic by ingestion. An eye irritant. Combustible. When heated to decomposition it emits toxic fumes of NO_x.

EPC875 **HR: 3**
N-ETHYL-2-(PYRROLIDINYL)-o-ACETOTOLUIDIDE HYDROCHLORIDE
mf: $C_{15}H_{22}N_2O•ClH$ mw: 282.85

SYN: C 5124

TOXICITY DATA WITH REFERENCE
eye-rbt 2% MLD ARZNAD 8,609,58
ipr-rat LD50:63 mg/kg ARZNAD 8,609,58
scu-mus LD50:87 mg/kg ARZNAD 8,609,58

SAFETY PROFILE: Poison by subcutaneous and intraperitoneal routes. An eye irritant. When heated to decomposition it emits toxic fumes of NO_x and HCl.

EPC950 **CAS:75236-19-0** **HR: 2**
4′-ETHYL-4-N-PYRROLIDINYLAZOBENZENE
mf: $C_{18}H_{21}N_3$ mw: 279.42

SYN: PYRROLIDINE, 1-(p-((p-ETHYLPHENYL)AZO)PHENYL)-

TOXICITY DATA WITH REFERENCE
mma-sat 10 µg/plate CRNGDP 3,559,82
orl-rat TDLo: 11 g/kg/34W C: CAR CRNGDP 1,419,80

SAFETY PROFILE: Questionable carcinogen with experimental carcinogenic data. Mutation data reported. When heated to decomposition it emits toxic fumes of NO_x.

EPC999 CAS:19137-90-7 HR: 3
3(1-ETHYL-2-PYRROLIDINYL)INDOLE HYDROCHLORIDE
mf: $C_{14}H_{18}N_2 \cdot ClH$ mw: 250.80

TOXICITY DATA WITH REFERENCE
ipr-rat LD50: 71 mg/kg JMCMAR 7,415,64
ipr-mus LD50: 56 mg/kg JMCMAR 7,415,64

SAFETY PROFILE: Poison by intraperitoneal route. When heated to decomposition it emits toxic fumes of NO_x and Cl^-.

EPD100 CAS:53583-79-2 HR: 3
N-((1-ETHYL-2-PYRROLIDINYL)METHYL)-5-(ETHYLSULFONYL)-o-ANISAMIDE
mf: $C_{17}H_{26}N_2O_4S$ mw: 354.51

SYNS: o-ANISAMIDE, N-((1-ETHYL-2-PYRROLIDINYL)METHYL)-5-(ETHYLSULFONYL)- □ BARNETIL □ BENZAMIDE, N-((1-ETHYL-2-PYRROLIDINYL)METHYL)-5-(ETHYLSULFONYL)-2-METHOXY- □ LIN 1418 □ SULTOPRIDE

TOXICITY DATA WITH REFERENCE
scu-rat TDLo: 51,300 µg/kg (female 19D pre): REP THERAP 30,231,75
scu-rat TDLo: 793 mg/kg (female 25D pre): REP THERAP 30,231,75
orl-mus LD50: 665 mg/kg DRFUD4 10,758,85
ipr-mus LD50: 300 mg/kg EJMCA5 17,437,82

SAFETY PROFILE: Poison by intraperitoneal route. Moderately toxic by ingestion. Experimental reproductive effects. When heated to decomposition it emits toxic fumes of NO_x and SO_x.

EPD500 CAS:15676-16-1 HR: 3
N-((1-ETHYL-2-PYRROLIDINYL)METHYL)-5-SULFAMOYL-o-ANISAMIDE
mf: $C_{15}H_{23}N_3O_4S$ mw: 341.47

PROP: A solid. Mp: 175–182° (decomp). Insol in H_2O, Et_2O, $CHCl_3$, and C_6H_6.

SYNS: ABILIT □ AIGLONYL □ 5-(AMINOSULFONYL)-N-((1-ETHYL-2-PYRROLIDINYL)METHYL)-2-METHOXYBENZAMIDE □ COOLSPAN □ DOBREN □ DOGMATIL □ DOGMATYL □ EGLONYL □ N-((1-ETHYL-2-PYRROLIDINYL)METHYL)-2-METHOXY-5-SULFAMOYLBENZAMIDE □ GUASTIL □ MIRADOL □ MIRBANIL □ MISULVAN □ OMPERAN □ PYRKAPPL □ R.D. 1403 □ SERNEVIN □ SPLOTIN □ SULPIRID □ SULPIRIDE □ SULPYRID □ SURSUMID □ TRILAN

TOXICITY DATA WITH REFERENCE
scu-rat TDLo: 181 mg/kg (female 1D pre): REP THERAP 30,231,75
scu-rat TDLo: 3 g/kg (female 30D pre): REP OYYAA2 7,1379,73
scu-rat TDLo: 1015 mg/kg (female 29D pre): REP THERAP 30,231,75
orl-rat TDLo: 4550 mg/kg (91D pre): REP OYYAA2 25,803,83
orl-rat TDLo: 45,500 mg/kg (male 91D pre): REP OYYAA2 25,803,83
orl-rat TDLo: 11,375 mg/kg (male 91D pre): REP OYYAA2 25,803,83
orl-hmn TDLo: 2143 µg/kg/D: CNS,GIT,MET JMGZAI 10(7),11,73
ims-hmn TDLo: 1429 µg/kg/D: CNS,GIT,MET JMGZAI 10(7),11,73
orl-rat LD50: 9800 mg/kg JMGZAI 10(7),11,73
ipr-rat LD50: 210 mg/kg NIIRDN 6,384,82
scu-rat LD50: 360 mg/kg NIIRDN 6,384,82
ivn-rat LD50: 40 mg/kg IYKEDH 4,193,73
orl-mus LD50: 1700 mg/kg NIIRDN 6,384,82
ipr-mus LD50: 231 mg/kg ARZNAD 27,404,77
scu-mus LD50: 290 mg/kg NIIRDN 6,384,82
ivn-mus LD50: 48 mg/kg IYKEDH 4,193,73
orl-dog LD50: 2 g/kg JMGZAI 10(7),11,73
scu-dog LD50: 350 mg/kg JMGZAI 10(7),11,73
ivn-dog LD50: 137 mg/kg JMGZAI 10(7),11,73
orl-rbt LD50: 4 g/kg JMGZAI 10(7),11,73
scu-rbt LD50: 2 g/kg JMGZAI 10(7),11,73
ivn-rbt LD50: 63 mg/kg JMGZAI 10(7),11,73

SAFETY PROFILE: Poison by intraperitoneal, subcutaneous and intravenous routes. Moderately toxic by ingestion. Human systemic effects by ingestion and intramuscular route: tremors, hypermotility, diarrhea and fever. Experimental reproductive effects. Used as a digestive aid and psychotropic drug. When heated to decomposition it emits very toxic fumes of NO_x and SO_x.

EPD600 CAS:606-55-3 HR: 3
1-ETHYLQUINALDINIUM IODIDE
mf: $C_{12}H_{14}N \cdot I$ mw: 299.17

SYN: QUINALDINIUM, 1-ETHYL-, IODIDE

TOXICITY DATA WITH REFERENCE
ivn-mus LD50: 18 mg/kg CSLNX* NX#01580

CONSENSUS REPORTS: Reported in EPA TSCA Inventory.

SAFETY PROFILE: Poison by intravenous route. When heated to decomposition it emits toxic vapors of NO_x and I^-.

EPF500 CAS:7648-01-3 HR: 3
3-ETHYLRHODANINE
mf: $C_5H_7NOS_2$ mw: 161.25

PROP: A solid. Mp: 36–36.5°.

SYN: 3-ETHYLRODANIN

TOXICITY DATA WITH REFERENCE
ipr-rat LDLo: 150 mg/kg ARZNAD 19,558,69
orl-mus LD50: 940 mg/kg FRZKAP 17(1),36,62
ivn-mus LD50: 180 mg/kg CSLNX* NX#03762

CONSENSUS REPORTS: Reported in EPA TSCA Inventory.

SAFETY PROFILE: Poison by intravenous and intraperitoneal routes. Moderately toxic by ingestion. When heated to decomposition it emits very toxic fumes of NO_x and SO_x.

EPF550 **CAS:78-10-4** **HR: 3**
ETHYL SILICATE
DOT: UN 1292
mf: $C_8H_{20}O_4Si$ mw: 208.37

PROP: Colorless liquid. Mp: −77°, bp: 165–166°. flash p: 125°F (52°C), d: 0.933 @ 20°/4°, n (25/D) 1.3818. Viscosity 0.6 cps. Practically insol in water with slow decomp. Miscible with alc.

SYNS: ETHYL ORTHOSILICATE ☐ ETYLU KRZEMIAN (POLISH) ☐ EXTREMA ☐ SILICATE D'ETHYLE (FRENCH) ☐ SILICIC ACID TETRAETHYL ESTER ☐ TEOS ☐ TETRAETHOXYSILANE ☐ TETRAETHYL ORTHOSILICATE ☐ TETRAETHYL ORTHOSILICATE (DOT) ☐ TETRAETHYL SILICATE ☐ TETRAETHYL SILICATE (DOT)

TOXICITY DATA WITH REFERENCE
eye-hmn 3000 ppm JIHTAB 22,288,40
skn-rbt 500 mg/24H MOD UCDS** 7/23/70
eye-rbt 500 mg/24H MLD 85JCAE-,1231,86
eye-gpg 2500 ppm/2H SEV JIHTAB 22,288,40
orl-rat LD50:6270 mg/kg JIHTAB 31,60,49
ihl-rat LCLo:1000 ppm/4H JIHTAB 31,343,49
skn-rbt LD50:5878 mg/kg UCDS** 7/23/70
ivn-rbt LDLo:200 mg/kg INMEAF 6,660,37
ihl-gpg LCLo:700 ppm/6H JIHTAB 22,288,40

CONSENSUS REPORTS: Reported in EPA TSCA Inventory.

OSHA PEL: TWA 10 ppm
ACGIH TLV: TWA 10 ppm
DFG MAK: 20 ppm (170 mg/m³)
DOT CLASSIFICATION: 3; *Label:* Flammable Liquid

SAFETY PROFILE: Poison by intravenous route. Moderately toxic by other routes. A skin, mucous membrane and severe eye irritant. Narcotic in high concentrations. Flammable liquid when exposed to heat or flame; can react vigorously with oxidizing materials. When heated to decomposition it emits acrid smoke and fumes. See also ESTERS.

EPF600 **CAS:676-54-0** **HR: 3**
ETHYL SODIUM
mf: C_2H_5Na mw: 52.05

PROP: Colorless crystals. Decomp below mp. Insol.

SAFETY PROFILE: The powder ignites spontaneously in air. When heated to decomposition it emits toxic fumes of Na_2O.

EPF700 **CAS:111-61-5** **HR: 1**
ETHYL STEARATE
mf: $C_{20}H_{40}O_2$ mw: 312.60

SYNS: ETHYL OCTADECANOATE ☐ ETHYL n-OCTADECANOATE ☐ STEARIC ACID, ETHYL ESTER

TOXICITY DATA WITH REFERENCE
skn-rbt 500 mg/24H MOD FCTXAV 17,781,79

CONSENSUS REPORTS: Reported in EPA TSCA Inventory.

SAFETY PROFILE: A skin irritant. When heated to decomposition it emits acrid smoke and irritating vapors.

EPG000 **CAS:7525-62-4** **HR: 3**
m-ETHYLSTYRENE
mf: $C_{10}H_{12}$ mw: 132.22

PROP: Water-white liquid. Bp: 191.5°, fp: −127°, d: 0.8955 @ 20°, vap press: 2.17 mm @ 40°, vap d: 4.56.

SYNS: m-ETHYL VINYLBENZEN (CZECH) ☐ m-ETHYLVINYLBENZENE

TOXICITY DATA WITH REFERENCE
eye-rbt 500 mg/24H MLD 28ZPAK -,24,72

SAFETY PROFILE: An eye irritant. Flammable when exposed to heat or flame; can react with oxidizing materials. To fight fire, use foam, CO_2, dry chemical. When heated to decomposition it emits acrid smoke and irritating fumes.

EPH000 **CAS:352-93-2** **HR: 3**
ETHYL SULFIDE
DOT: UN 2375
mf: $C_4H_{10}S$ mw: 90.20

PROP: Liquid, garlic-like odor. Mp: −102°, fp: −102.05°, bp: 92–93°, d: 0.837 @ 20°/4°, vap d: 3.11. Flash p: 14°F. Insol in water.

SYNS: DIETHYLSULFID (CZECH) ☐ DIETHYL SULFIDE (DOT) ☐ DIETHYLTHIOETHER ☐ ETHYL MONOSULFIDE ☐ ETHYLTHIOETHANE ☐ ETHYL THIOETHER ☐ SULFODOR (CZECH) ☐ 3-THIAPENTANE ☐ 1,1′-THIOBISETHANE ☐ THIOETHYL ETHER

TOXICITY DATA WITH REFERENCE
skn-rbt 20 mg/24H MOD 85JCAE-,991,86
eye-rbt 500 mg/24H MLD 85JCAE-,991,86
orl-rat LD50:5930 mg/kg 28ZPAK -,170,72

CONSENSUS REPORTS: Reported in EPA TSCA Inventory.

DOT CLASSIFICATION: 3; *Label:* Flammable Liquid

SAFETY PROFILE: Mildly toxic by ingestion. A skin and eye irritant. A very dangerous fire hazard when exposed to heat, flame, or sparks; can react vigorously with oxidizers. Reacts with water, steam, acids, or acid fumes to produce toxic and flammable vapors. To fight fire, use water spray or mist, dry chemical, CO_2, foam. When heated to decomposition it yields highly toxic fumes of SO_x. See also SULFIDES.

EPI300 **CAS:842-00-2** **HR: 2**
4-(ETHYLSULFONYL)-1-NAPHTHALENE SULFONAMIDE
mf: $C_{12}H_{13}NO_4S_2$ mw: 299.38

SYNS: ENS ☐ 4-ETHYLSULPHONYLNAPHTHALENE-1-SULFONAMIDE ☐ 4-ETHYLSULPHONYLNAPHTHALENE-1-SULPHONAMIDE ☐ HPA

TOXICITY DATA WITH REFERENCE
dns-mus-orl 20 mg/kg BIJOAK 111,12P,69
oms-mus-orl 20 mg/kg BIJOAK 111,12P,69
cyt-mus-orl 40 mg/kg CTKIAR 2,249,69
orl-mus TDLo:4200 mg/kg/50W-C:CAR BJCAAI 23,772,69
imp-mus TDLo:77 mg/kg:CAR BJCAAI 19,311,65
orl-mus TD:12 g/kg/86W-C:CAR BJCAAI 28,227,73
orl-mus TD:440 mg/kg/52W-C:NEO JNCIAM 51,2007,73
orl-mus TD:5460 mg/kg/65W-C:ETA BCPCA6 16,619,67
orl-mus TD:4368 mg/kg/39W-C:ETA BJURAN 39,26,64

SAFETY PROFILE: Questionable carcinogen with experimental carcinogenic, neoplastigenic, and tumorigenic data. Mutation data reported. When heated to decomposition it emits very toxic fumes of NO_x and SO_x.

EPI400 CAS:67465-28-5 HR: 3
3-(3-ETHYLSULFONYL)PENTYL PIPERIDINO KETONE
mf: $C_{13}H_{25}NO_3S$ mw: 275.45

SYN: KETONE, 3-(3-ETHYLSULFONYL)PENTYL PIPERIDINO

TOXICITY DATA WITH REFERENCE
ipr-mus LD50:387 mg/kg JMCMAR 6,351,63

DOT CLASSIFICATION: 3; *Label:* Flammable Liquid

SAFETY PROFILE: A poison by intraperitoneal route. A flammable liquid. When heated to decomposition it emits toxic vapors of NO_x and SO_x.

EPI500 CAS:70-29-1 HR: 2
ETHYL SULFOXIDE
mf: $C_4H_{10}OS$ mw: 106.20

PROP: A liquid. Mp: 4–6°, bp: 88–89° @ 15 mm.

SYNS: DESO □ DIETHYL SULPHOXIDE □ ETHANE-1,1'-SULFINYL BIS

TOXICITY DATA WITH REFERENCE
orl-rat LD50:5650 mg/kg COREAF 260,327,65
ipr-rat LD50:4370 mg/kg COREAF 260,327,65
ivn-rat LD50:4990 mg/kg COREAF 260,327,65
orl-mus LD50:3610 mg/kg COREAF 260,327,65
ipr-mus LD50:2500 mg/kg IJRBA3 3,41,61
ivn-mus LD50:4370 mg/kg COREAF 260,327,65

SAFETY PROFILE: Moderately toxic by ingestion and intraperitoneal routes. Mildly toxic by intravenous route. When heated to decomposition it emits toxic fumes of SO_x.

EPJ000 CAS:20941-65-5 HR: 2
ETHYL TELLURAC
mf: $C_{20}H_{40}N_4S_8 \cdot Te$ mw: 720.72

PROP: Orange-yellow powder. D: 1.44, mp: 108–118°.

SYNS: DIETHYLDITHIO CARBAMIC ACID TELLURIUM SALT □ NCI-C02857 □ TELLURIUM DIETHYLDITHIOCARBAMATE □ TETRAKIS (DIETHYLCARBAMODITHIOATO-S,S')TELLURIUM □ TETRAKIS(DIETHYLDITHIOCARBAMATO)TELLURIUM

TOXICITY DATA WITH REFERENCE
orl-mus TDLo:113 g/kg/107W-C:ETA NCITR* NCI-CG-TR-152,79

CONSENSUS REPORTS: IARC Cancer Review: Group 3 IMEMDT 7,56,87; Animal Inadequate Evidence IMEMDT 12,115,76. NCI Carcinogenesis Bioassay (feed); No Evidence: mouse, rat NCITR* NCI-CG-TR-152,79; Results Indefinite: Mouse, Rat NCITR* NCI-CG-TR-152,79. Reported in EPA TSCA Inventory.

OSHA PEL: TWA 0.1 mg(Te)/m³
ACGIH TLV: TWA 0.1 mg(Te)/m³

SAFETY PROFILE: Questionable carcinogen with experimental tumorigenic data. When heated to decomposition it emits very toxic fumes of NO_x, SO_x, and Te. See also TELLURIUM COMPOUNDS and CARBAMATES.

EPJ600 CAS:56501-31-6 HR: 3
1-ETHYL-1-TETRADECYLPIPERIDINIUM BROMIDE
mf: $C_{21}H_{44}N \cdot Br$ mw: 309.57

TOXICITY DATA WITH REFERENCE
orl-mus LD50:140 mg/kg PSDTAP 15,331,74
ipr-mus LD50:11,984 μg/kg PSDTAP 15,331,74
ivn-mus LD50:5091 μg/kg PSDTAP 15,331,74

SAFETY PROFILE: Poison by ingestion, intraperitoneal, and intravenous routes. When heated to decomposition it emits toxic fumes of NO_x and Br^-.

EPK000 CAS:15547-17-8 HR: 3
2-ETHYL-5,6,7,8-TETRAHYDROANTHRAQUINONE
mf: $C_{16}H_{16}O_2$ mw: 240.32

SYNS: 6-ETHYL-1,2,3,4-TETRAHYDRO-9,10-ANTHRACENEDIONE □ USAF SO-2

TOXICITY DATA WITH REFERENCE
ipr-mus LD50:200 mg/kg NTIS** AD277-689

CONSENSUS REPORTS: Reported in EPA TSCA Inventory.

SAFETY PROFILE: Poison by intraperitoneal route. When heated to decomposition it emits acrid smoke and irritating fumes.

EPL000 CAS:101651-73-4 HR: 3
1-ETHYL-2,2,6,6-TETRAMETHYL-4-(N-ACETYL-N-PHENYL)PIPERIDINE
mf: $C_{19}H_{30}N_2O$ mw: 302.51

SYNS: 4-(N-ACETYL)PHENYLAMINO-1-ETHYL-2,2,6,6-TETRAMETHYLPIPERIDINE □ N-(1-ETHYL-2,2,6,6-TETRAMETHYLPIPERIDIN-4-YL)ACETANILIDE

TOXICITY DATA WITH REFERENCE
scu-mus LD50:590 mg/kg PCJOAU 8,82,74
ivn-mus LD50:86 mg/kg PCJOAU 8,82,74

SAFETY PROFILE: Poison by intravenous route. Moderately toxic by subcutaneous route. When heated to decomposition it emits toxic fumes of NO_x.

EPL600 **HR: 3**
1-ETHYL-2,2,6,6-TETRAMETHYLPIPERIDINE HYDROCHLORIDE
mf: $C_{11}H_{23}N \cdot ClH$ mw: 205.81

SYN: COMPOUND 26539 HYDROCHLORIDE

TOXICITY DATA WITH REFERENCE
orl-mus LD50:180 mg/kg BJPCAL 13,501,58
ipr-mus LD50:58 mg/kg BJPCAL 13,501,58
ivn-mus LD50:52 mg/kg BJPCAL 13,501,58

SAFETY PROFILE: Poison by ingestion, intravenous and intraperitoneal routes. When heated to decomposition it emits toxic fumes of NO_x and HCl.

EPL700 **HR: 3**
1-ETHYL-2,2,6,6-TETRAMETHYLPIPERIDINE HYDROGEN TARTRATE
mf: $C_{11}H_{23}N \cdot C_4H_6O_4$ mw: 287.45

SYN: 1-ETHYL-2,2,6,6-TETRAMETHYL-PIPERIDINE TARTRATE

TOXICITY DATA WITH REFERENCE
orl-mus LD50:236 mg/kg BJPCAL 13,501,58
ipr-mus LD50:73 mg/kg BJPCAL 13,501,58
ivn-mus LD50:63 mg/kg BJPCAL 13,501,58

SAFETY PROFILE: Poison by ingestion, intravenous, and intraperitoneal routes. When heated to decomposition it emits toxic fumes of NO_x.

EPM000 **CAS:52098-56-3** **HR: 3**
1-ETHYL-2,2,6,6-TETRAMETHYL-4-(N-PROPIONYL-N-BENZYLAMINO)PIPERIDINE
mf: $C_{21}H_{34}N_2O$ mw: 330.57

SYN: 4-(N-PROPIONYL)BENZYLIMINO-1-ETHYL-2,2,6,6-TETRAMETHYLPIPERIDINE

TOXICITY DATA WITH REFERENCE
scu-mus LD50:520 mg/kg PCJOAU 8,82,74
ivn-mus LD50:100 mg/kg PCJOAU 8,82,74

SAFETY PROFILE: Poison by intravenous route. Moderately toxic by subcutaneous route. When heated to decomposition it emits toxic fumes of NO_x.

EPM550 **HR: 3**
1-ETHYL-1,1,3,3-TETRAMETHYLTETRAZENIUM
mf: $C_6H_{17}BF_4N_4$ mw: 232.03

$(CH_3)_2NN{=}NN^+CH_2CH_2(CH_3)_2BF_4^-$

SAFETY PROFILE: An explosive salt. When heated to decomposition it emits toxic fumes of F^- and NO_x. See also BORON COMPOUNDS.

EPM590 **HR: 3**
2-ETHYLTETRAZOLE
mf: $C_3H_6N_4$ mw: 98.11

$CH{=}NNCH_2CH_2N{=}N$ (ring)

SAFETY PROFILE: Explodes on contact with aluminum hydride. When heated to decomposition it emits toxic fumes of NO_x.

EPM600 **CAS:50764-78-8** **HR: 3**
5-ETHYLTETRAZOLE
mf: $C_3H_6N_4$ mw: 98.11

$HNN{=}NN{=}CCH_2CH_3$ (ring)

SAFETY PROFILE: Explodes on contact with aluminum hydride. When heated to decomposition it emits toxic fumes of NO_x.

EPN500 **CAS:4147-51-7** **HR: 1**
2-ETHYLTHIO-4,6-BIS(ISOPROPYLAMINO)-s-TRIAZINE
mf: $C_{11}H_{21}N_5S$ mw: 255.43

PROP: A solid. Mp: 104–106°.

SYNS: 2,4-BIS(ISOPROPYLAMINO)-6-ETHYLTHIO-s-TRIAZINE ☐ COTOFOR ☐ DIPROPETRYN ☐ DIPROPETRYNE ☐ 6-(ETHYLTHIO) N,N'-BIS(1-METHYLETHYL)-1,3,5-TRIAZINE-2,4-DIAMINE ☐ GS-16068 ☐ SANCAP

TOXICITY DATA WITH REFERENCE
orl-rat LD50:7144 mg/kg FMCHA2 -,C210,83
skn-rbt LD50:10 g/kg CIGET* -,-,77

SAFETY PROFILE: Mildly toxic by ingestion and skin contact. An herbicide. When heated to decomposition it emits very toxic fumes of NO_x and SO_x.

EPP000 **CAS:542-90-5** **HR: 3**
ETHYL THIOCYANATE
mf: C_3H_5NS mw: 87.15

PROP: A liquid. D: 1.020 @ 16°, mp: −85.5, bp: 145°. Insol in water; misc in alc and ether.

SYNS: AETHYLRHODANID (GERMAN) ☐ ETHYL RHODANATE ☐ ETHYL SULFOCYANATE ☐ THIOCYANATOETHANE ☐ THIOCYANIC ACID, ETHYL ESTER

TOXICITY DATA WITH REFERENCE
orl-rat LDLo:200 mg/kg JIHTAB 18,310,36
scu-rat LDLo:40 mg/kg JIHTAB 18,310,36
ipr-mus LD50:10 mg/kg JJPAAZ 3,99,54
scu-mus LDLo:70 mg/kg JJPAAZ 3,99,54
ivn-mus LD50:18 mg/kg CSLNX* NX#02865
orl-cat LDLo:10 mg/kg JIHTAB 18,310,36
scu-rbt LDLo:15 mg/kg AEPPAE 150,257,30

CONSENSUS REPORTS: Community Right-To-Know List. Reported in EPA TSCA Inventory.

SAFETY PROFILE: Poison by ingestion, subcutaneous, intraperitoneal, and intravenous routes. When heated to decomposition it emits very toxic fumes of NO_x and SO_x. See also THIOCYANATES.

EPP500 **CAS:110-77-0** **HR: 2**
2-(ETHYLTHIO)ETHANOL
mf: $C_4H_{10}OS$ mw: 106.20

PROP: Bp: 182–184°.

SYNS: ETHYL-2-HYDROXYETHYL SULFIDE ☐ ETHYL-2-HYDROXYETHYL THIOETHER ☐ β-ETHYLMERKAPTOETHANOL (CZECH) ☐ β-HYDROXYDIETHYL SULFIDE

TOXICITY DATA WITH REFERENCE
skn-rbt 2 mg/24H SEV 85JCAE-,998,86
eye-rbt 750 μg/24H SEV 28ZPAK -,171,72

CONSENSUS REPORTS: Reported in EPA TSCA Inventory.

SAFETY PROFILE: A severe skin and eye irritant. When heated to decomposition it emits toxic fumes of SO_x. See also SULFIDES.

EPQ000 CAS:536-33-4 HR: 3
2-ETHYLTHIOISONICOTINAMIDE
mf: $C_8H_{10}N_2S$ mw: 166.26

PROP: Yellow crystals from EtOH. Mp: 164–166° (decomp). Very sltly sol in Et_2O; sltly sol in MeOH and EtOH; sol in Py.

SYNS: AETINA □ AETIVA □ AMIDAZIN □ BAYER 5312 □ ETH □ ETHIMIDE □ ETHINA □ ETHINAMIDE □ ETHIONIAMIDE □ α-ETHYLISONICOTINIC ACID THIOAMIDE □ 2-ETHYLISONICOTINIC ACID THIOAMIDE □ 2 ETHYLISONICOTINIC THIOAMIDE □ α-ETHYLISONICOTINOYLTHIOAMIDE □ ETHYLISOTHIAMIDE □ α-ETHYLISOTHIONICOTINAMIDE □ 2-ETHYLISOTHIONICOTINAMIDE □ 2-ETHYL-4-PYRIDINECARBOTHIOAMIDE □ 2-ETHYL-4-THIOAMIDYLPYRIDINE □ 2-ETHYL-4-THIOCARBAMOYLPYRIDINE □ α-ETHYLTHIOISONICOTINAMIDE □ ETHYONOMIDE □ ETIMID □ ETIOCIDAN □ ETIONAMID □ ETIONIZINA □ ETP □ FATOLIAMID □ F.I. 58-30 □ IRIDOCIN □ IRIDOZIN □ ISOTHIN □ ISOTIAMIDA □ ITIOCIDE □ NCI-C01694 □ NICOTION □ NISOTIN □ NIZOTIN □ RIGENICID □ SERTINON □ TEBERUS □ TH 1314 □ THIANIDE □ THIOAMIDE □ THIONIDEN □ TRECATOR □ TRESCATYL □ TRESCAZIDE □ TUBERMIN □ TUBEROID □ TUBEROSON

TOXICITY DATA WITH REFERENCE
cyt-ham:fbr 400 mg/L ESKHA5 96,55,78
cyt-ham:lng 540 mg/L GMCRDC 27,95,81
unr-wmn TDLo:5400 mg/kg (12W pre/1-38W preg):REP POMJAC 5,1152,66
unr-wmn TDLo:600 mg/kg (36-39W preg):TER POMJAC 5,1152,66
orl-rbt TDLo:900 mg/kg (female 8-16D post):REP PSDTAP 10,235,69
orl-rat TDLo:1800 mg/kg (female 6-14D post):TER SEIJBO 5,236,65
orl-rat TDLo:900 mg/kg (female 6-14D post):TER OSDIAF 16,313,67
orl-mus TDLo:450 mg/kg (female 6-14D post):REP PSDTAP 10,235,69
orl-rbt TDLo:108 mg/kg (female 7-14D post):REP DIPHAH 23,383,71
orl-rbt TDLo:450 mg/kg (female 8-16D post):REP PSDTAP 10,235,69
orl-rat TDLo:1800 mg/kg (6-14D preg):TER OSDIAF 16,313,67
scu-rat TDLo:486 mg/kg (female 6-14D post):TER DIPHAH 23,383,71
orl-mus TDLo:24 g/kg/50W-I:CAR LAPPA5 24,145,64
orl-hmn TDLo:856 mg/kg:LIV ARDSBL 84,890,61
orl-hmn TDLo:482 mg/kg/5W:LIV ARDSBL 87,896,63
orl-rat LD50:1320 mg/kg NIIRDN 6,111,82
ipr-rat LD50:490 mg/kg NIIRDN 6,111,82
scu-rat LD50:1350 mg/kg NIIRDN 6,111,82
orl-mus LD50:1000 mg/kg FEPRA7 21,452,62
ipr-mus LD50:1350 mg/kg DPHFAK 23,383,71
scu-mus LD50:1580 mg/kg NIIRDN 6,111,82
scu-gpg LD50:550 mg/kg UBZHD4 52,593,80

CONSENSUS REPORTS: IARC Cancer Review: Group 3 IMEMDT 7,56,87; Animal Limited Evidence IMEMDT 13,83,77. NCI Carcinogenesis Bioassay (feed); No Evidence: mouse, rat NCITR* NCI-CG-TR-46,78.

SAFETY PROFILE: A human systemic poison. Moderately toxic by ingestion, intraperitoneal, and subcutaneous routes. Human systemic effects by ingestion: jaundice and liver function impairment. It affects the human peripheral nervous system. Experimental teratogenic and reproductive effects. Questionable carcinogen with experimental carcinogenic data. Mutation data reported. Used to treat tuberculosis. When heated to decomposition it emits very toxic fumes of SO_x and NO_x.

EPR000 CAS:29973-13-5 HR: 3
(2-ETHYLTHIOMETHYLPHENYL)-N-METHYLCARBAMATE
mf: $C_{11}H_{15}NO_2S$ mw: 225.33

PROP: Yellow oil or solid. D: 1.147 @ 20°/4°, mp: 33.4°. Sltly sol in H_2O.

SYNS: BAY-HOX-1901 □ CRONETON □ ETHIOFENCARB □ ETHIOPHENCARP □ 2-ETHYLMERCAPTOMETHYLPHENYL-N-METHYLCARBAMATE □ 2-((ETHYLTHIO)METHYL)PHENOL METHYLCARBAMATE □ 2-((ETHYLTHIO)METHYL)PHENYL METHYLCARBAMATE □ HOX 1901

TOXICITY DATA WITH REFERENCE
orl-rat LD50:200 mg/kg FMCHA2 -,C64,83
ihl-rat LCLo:97 mg/m³ GISAAA 49(10),23,84
skn-rat LD50:1000 mg/kg FMCHA2 -,C64,83
orl-mus LD50:71 mg/kg 85ALAU -,106,76
ihl-cat LCLo:97 mg/m³ GISAAA 49(10),23,84
skn-rbt LD50:2500 mg/kg GISAAA 49(10),23,84

SAFETY PROFILE: Poison by ingestion and inhalation routes. Moderately toxic by skin contact. An insecticide. When heated to decomposition it emits very toxic fumes of NO_x and SO_x. See also CARBAMATES.

EPR100 CAS:99422-01-2 HR: 3
5-((2-ETHYLTHIO)PROPYL)-2-(1-OXOPROPYL)-1,3-CYCLOHEXANEDIONE
mf: $C_{14}H_{22}O_3S$ mw: 270.42

SYNS: 1,3-CYCLOHEXANEDIONE, 5-(2-(ETHYLTHIO)PROPYL)-2-(1-OXOPROPYL)- □ RE-45550

TOXICITY DATA WITH REFERENCE
orl-rat LD50:170 mg/kg EPASR* 8EHQ-0186-0584

CONSENSUS REPORTS: Reported in EPA TSCA Inventory.

SAFETY PROFILE: A poison by ingestion. When heated to decomposition it emits toxic vapors of SO_x.

E

EPR200 **CAS:70303-46-7** **HR: 2**
(ETHYLTHIO)TRIOCTYLSTANNANE
mf: $C_{26}H_{56}SSn$ mw: 519.57

SYNS: STANNANE, (ETHYLTHIO)TRIOCTYL- □ TRIOCTYL(ETHYLTHIO)STANNANE

TOXICITY DATA WITH REFERENCE
ipr-mus LD50:2945 mg/kg RPTOAN 42,73,79

OSHA PEL: TWA 0.1 mg(Sn)/m^3
ACGIH TLV: TWA 0.1 mg(Sn)/m^3; STEL 0.2 mg/m^3 (skin)
NIOSH REL: (Organotin Compounds): 10H TWA 0.1 mg(Sn)/m^3

SAFETY PROFILE: Moderately toxic by intraperitoneal route. When heated to decomposition it emits toxic fumes of SO_x and Sn.

For occupational chemical analysis use NIOSH: Organotin Compounds 5504.

EPR600 **CAS:625-53-6** **HR: 3**
ETHYL THIOUREA
mf: $C_3H_8N_2S$ mw: 104.19

SYN: 1-ETHYLTHIOUREA

TOXICITY DATA WITH REFERENCE
mmo-sat 150 µg/plate ABCHA6 44,3017,80
dnr-omi 10 mg/plate BIZNAT 95,463,76
orl-rat TDLo:500 mg/kg (12D preg):TER TJADAB 23,335,81
orl-rat TDLo:1 g/kg (12D preg):TER TJADAB 23,335,81
orl-rat LD50:100 mg/kg JPETAB 90,260,47
orl-mus LDLo:500 mg/kg TJADAB 23,335,81

CONSENSUS REPORTS: Reported in EPA TSCA Inventory.

SAFETY PROFILE: Poison by ingestion. Experimental teratogenic effects. Mutation data reported. When heated to decomposition it emits toxic fumes of SO_x and NO_x.

EPS000 **CAS:1066-57-5** **HR: 3**
ETHYLTIN TRICHLORIDE
mf: $C_2H_5Cl_3Sn$ mw: 254.11

PROP: Colorless liquid. Mp: −10°, bp: 196–198°.

SYNS: AETHYLZINNTRICHLORID (GERMAN) □ ETHYLTRICHLOROSTANNANE □ ETHYLTRICHLOROTIN □ MONOETHYLTIN TRICHLORIDE □ TRICHLOROETHYLSTANNANE □ TRICHLOROETHYLTIN

TOXICITY DATA WITH REFERENCE
orl-rat LDLo:200 mg/kg TRIPA7-,73
ipr-rat LDLo:200 mg/kg BJPCAL 10,16,55
ipr-mus LD50:79 mg/kg TXAPA9 8,337,66

OSHA PEL: TWA 0.1 mg(Sn)/m^3 (skin)
ACGIH TLV: TWA 0.1 mg(Sn)/m^3 (skin) (Proposed: TWA 0.1 mg(Sn)/m^3; STEL 0.2 mg(Sn)/m^3 (skin))
NIOSH REL: (Organotin Compounds) TWA 0.1 mg(Sn)/m^3

SAFETY PROFILE: Poison by ingestion and intraperitoneal routes. When heated to decomposition it emits toxic fumes of Cl^-. See also TIN COMPOUNDS and CHLORIDES.

For occupational chemical analysis use NIOSH: Organotin Compounds 5504.

EPS500 **CAS:611-14-3** **HR: 3**
2-ETHYLTOLUENE
mf: C_9H_{12} mw: 120.21

PROP: Autoign temp: 824°F, d: 0.88, vap d: 4.15, bp: 164.1°.

SYNS: 1-ETHYL-2-METHYLBENZENE □ o-ETHYL METHYLBENZENE □ o-ETHYLTOLUENE □ o-METHYLETHYLBENZENE

TOXICITY DATA WITH REFERENCE
orl-rat LDLo:5000 mg/kg 28ZRAQ -,57,60
ihl-mus LC50:54 g/m^3/4H 85GMAT -,69,82
ihl-cat LC50:50 g/kg/2H 85GMAT -,69,82

CONSENSUS REPORTS: Reported in EPA TSCA Inventory.

SAFETY PROFILE: Mildly toxic by ingestion and inhalation. Flammable when exposed to heat, flame or oxidizers. To fight fire, use water spray or mist, dry chemical, CO_2. When heated to decomposition it emits acrid smoke and irritating fumes.

EPT000 **CAS:622-96-8** **HR: 1**
p-ETHYLTOLUENE
mf: C_9H_{12} mw: 120.21

PROP: A liquid. Mp: −62.4°, bp: 162.2°, autoign temp: 887°F.

SYNS: 1-ETHYL-4-METHYLBENZENE □ p-ETHYLMETHYLBENZENE □ 4-ETHYLTOLUENE □ p-METHYLETHYLBENZENE □ 4-METHYLETHYLBENZENE

TOXICITY DATA WITH REFERENCE
orl-rat LDLo:5000 mg/kg 28ZRAQ -,57,60

CONSENSUS REPORTS: Reported in EPA TSCA Inventory.

SAFETY PROFILE: Mildly toxic by ingestion. Flammable when exposed to heat or flame; can react vigorously with oxidizing materials. To fight fire use water spray or mist, dry chemical, CO_2. When heated to decomposition it emits acrid smoke and irritating fumes.

EPT500 **CAS:17010-63-8** **HR: 2**
p-((3-ETHYL-p-TOLYL)AZO)-N,N-DIMETHYLANILINE
mf: $C_{17}H_{21}N_3$ mw: 267.41

SYNS: N,N-DIMETHYL-p-(3′-ETHYL-4′-METHYLPHENYLAZO)ANILINE □ N,N-DIMETHYL-p-((3-ETHYL-p-TOLYL)AZO)ANILINE

TOXICITY DATA WITH REFERENCE
orl-rat TDLo:6497 mg/kg/17W-C:CAR CBINA8 53,107,85
orl-rat TDLo:2005 mg/kg/50D-I:ETA CNREA8 30,1520.70

SAFETY PROFILE: Questionable carcinogen with experimental carcinogenic and tumorigenic data. When heated to decomposition it emits toxic fumes of NO_x.

EPU000 **CAS:17010-62-7** **HR: 2**
p-((4-ETHYL-m-TOLYL)AZO)-N,N-DIMETHYLANILINE
mf: $C_{17}H_{21}N_3$ mw: 266.40

SYNS: N,N-DIMETHYL-p-(4′-ETHYL-3′-METHYLPHENYLAZO)ANILINE □ N,N-DIMETHYL-p-((4-ETHYL-m-TOLYL)AZO) ANILINE

TOXICITY DATA WITH REFERENCE
orl-rat 1109 mg/kg/8W-C:CAR CBINA8 53,107,85
orl-rat TDLo:500 mg/kg/50D-I:ETA CNREA8 30,1520,70

SAFETY PROFILE: Questionable carcinogen with experimental carcinogenic and tumorigenic data. When heated to decomposition it emits toxic fumes of NO_x.

EPV000 **CAS:101491-84-3** **HR: 3**
N-ETHYL-N-(1-(o-TOLYLOXY-2-PROPYL)CARBAMIC ACID)-2-(2-METHYLPIPERIDINO)ETHYL ESTER HYDROCHLORIDE
mf: $C_{21}H_{34}N_2O_3 \cdot ClH$ mw: 399.03

SYN: C 2138

TOXICITY DATA WITH REFERENCE
eye-rbt 2% MOD ARZNAD 9,113,59
scu-mus LD50:250 mg/kg ARZNAD 9,113,59

SAFETY PROFILE: Poison by subcutaneous route. An eye irritant. When heated to decomposition it emits very toxic fumes of HCl and NO_x. See also CARBAMATES and ESTERS.

EPW000 **CAS:50707-40-0** **HR: 3**
1-ETHYL-3-p-TOLYLTRIAZENE
mf: $C_9H_{13}N_3$ mw: 163.25

TOXICITY DATA WITH REFERENCE
ipr-mus TDLo:20 mg/kg/I:NEO JNCIAM 54,495,75
ipr-mus LDLo:200 mg/kg StoGD# 27May75

SAFETY PROFILE: Poison by intraperitoneal route. Questionable carcinogen with experimental neoplastigenic data. When heated to decomposition it emits toxic fumes of NO_x.

EPW500 **CAS:80-40-0** **HR: 2**
ETHYL TOSYLATE
mf: $C_9H_{12}O_3S$ mw: 200.27

PROP: Crystals from alc. Mp: 33°, bp: 173° @ 15 mm, flash p: 316°F (CC), d: 1.17, vap d: 6.98.

SYNS: ETHYL-p-METHYL BENZENESULFONATE □ ETHYL PTS □ ETHYL-p-TOLUENESULFONATE □ ETHYL-p-TOSYLATE □ p-TOLUOLSULFONSAEUREAETHYL ESTER (GERMAN)

TOXICITY DATA WITH REFERENCE
mmo-sat 2000 μg/plate JNCIAM 62,893,79
mma-sat 1700 μg/plate PNASA6 72,5135,75
dnr-esc 50 mg/L JNCIAM 62,873,79
mrc-smc 1 pph JNCIAM 62,901,79
otr-ham:emb 500 μg/L CRNGDP 1,323,80
scu-rat TDLo:50 mg/kg:ETA FCTXAV 6,576,68
scu-rat TD:3250 mg/kg/65W-I:ETA ZEKBAI 74,241,70
scu-rat LD50:500 mg/kg ZEKBAI 74,241,70
ipr-mus LD50:1000 mg/kg JJIND8 62,911,79

CONSENSUS REPORTS: Reported in EPA TSCA Inventory. EPA Genetic Toxicology Program.

SAFETY PROFILE: Moderately toxic by subcutaneous and intraperitoneal routes. Questionable carcinogen with experimental tumorigenic data. Mutation data reported. Combustible when exposed to heat or flame; can react with oxidizing materials. To fight fire, use CO_2, dry chemical. When heated to decomposition it emits highly toxic fumes of SO_x. See also SULFONATES and ESTERS.

EPX000 **CAS:63449-87-6** **HR: 3**
ETHYL TRI-n-BUTYLAMMONIUM HYDROXIDE
mf: $C_{14}H_{32}N \cdot HO$ mw: 231.48

TOXICITY DATA WITH REFERENCE
scu-mus LDLo:24 mg/kg JPETAB 28,367,26

CONSENSUS REPORTS: Reported in EPA TSCA Inventory.

SAFETY PROFILE: Poison by subcutaneous route. When heated to decomposition it emits toxic fumes of NO_x and NH_3.

EPY000 **CAS:327-98-0** **HR: 3**
ETHYL TRICHLOROPHENYLETHYLPHOSPHONOTHIOATE
mf: $C_{10}H_{12}Cl_3O_2PS$ mw: 333.60

PROP: Amber liquid. D: 1.365 @ 20°/4°, bp: 108° @ 0.01 mm.

SYNS: O-AETHYL-O-(2,4,5-TRICHLORPHENYL)-AETHYLTHIONOPHOSPHONAT (GERMAN) □ AGRISIL □ AGRITOX □ BAYER 5081 □ BAYER 37289 □ BAYER S 4400 □ CHEMAGRO 37289 □ ENT 25,712 □ O-ETHYL-O-2,4,5-TRICHLOROPHENYL ETHYLPHOSPHONOTHIOATE □ FENOPHOSPHON □ PHYTOSOL □ STAUFFER N 3049 □ TRICHLORONAT □ 2,4,5-TRICHLOROPHENOL-O-ESTER with O-ETHYL ETHYLPHOSPHONOTHIOATE □ WIRKSTOFF 37289

TOXICITY DATA WITH REFERENCE
orl-rat LD50:15 mg/kg ARSIM* 20,4,66
ihl-rat LCLo:700 mg/m³/4H 85GYAZ -,35,71
skn-rat LD50:64 mg/kg WRPCA2 9,119,70
orl-mus LD50:40 mg/kg 85DPAN -,-,71/76
orl-cat LD50:10 mg/kg 85DPAN -,-,71/76
orl-rbt LD50:25 mg/kg 85GYAZ -,35,71

CONSENSUS REPORTS: Chlorophenol compounds are on the Community Right-To-Know List. EPA Extremely Hazardous Substances List.

SAFETY PROFILE: Poison by ingestion and skin contact. Moderately toxic by inhalation. An insecticide. When heated to decomposition it emits very toxic fumes of Cl^-, PO_x, and SO_x. See also CHLOROPHENOLS.

EPY500 **CAS:115-21-9** **HR: 3**
ETHYL TRICHLOROSILANE
DOT: UN 1196
mf: $C_2H_5Cl_3Si$ mw: 163.51

PROP: Liquid. Mp: −105.6°, bp: 99.5°, flash p: 72°F (OC), d: 1.24 @ 25°/25°, vap d: 5.6.

E

SYNS: ETHYL SILICON TRICHLORIDE □ ETHYLTRICHLOROSILANE (DOT) □ SILANE, TRICHLOROETHYL- □ SILICANE, TRICHLOROETHYL- □ TRICHLOROETHYLSILANE □ TRICHLOROETHYLSILICANE

TOXICITY DATA WITH REFERENCE
skn-rbt 500 mg/24H MLD 85JCAE-,1220,86
eye-rbt 250 µg/24H SEV 85JCAE-,1220,86
ihl-rat LCLo:500 ppm/4H JIHTAB 31,343,49
orl-rat LD50:1330 mg/kg JIHTAB 31,60,49
ipr-rat LDLo:30 mg/kg JIHTAB 30,332,48
ihl-mus LC50:300 µg/m³/2H TPKVAL (3),23,61

CONSENSUS REPORTS: EPA Extremely Hazardous Substances List. Reported in EPA TSCA Inventory.

DOT CLASSIFICATION: 3; *Label:* Flammable Liquid, Corrosive

SAFETY PROFILE: Poison by inhalation and intraperitoneal routes. Moderately toxic by ingestion. A skin and severe eye irritant. A very dangerous fire hazard when exposed to heat, flame, or oxidizers; will react with water or steam to produce heat and toxic and corrosive fumes; can react vigorously with oxidizing materials. To fight fire, use CO_2, dry chemical. When heated to decomposition it emits highly toxic fumes of Cl^- and phosgene. See also CHLOROSILANES.

EPY600 CAS:993-86-2 HR: 3
ETHYLTRICHLORPHON
mf: $C_6H_{12}Cl_3O_4P$ mw: 285.50

PROP: A solid. Mp: 57–58°.

SYNS: AETHYLTRICHLORPHON (GERMAN) □ ETC □ 1-HYDROXY-2,2,2-TRICHLOROETHYL DIETHYL PHOSPHONITE

TOXICITY DATA WITH REFERENCE
ipr-mus TDLo:960 mg/kg (9D preg):REP ZHYGAM 22,565,76
ipr-mus TDLo:240 mg/kg (9D preg):TER ZHYGAM 22,565,76
ipr-mus LD50:400 mg/kg ZHYGAM 22,565,76

SAFETY PROFILE: Poison by intraperitoneal route. An experimental teratogen. Other experimental reproductive effects. When heated to decomposition it emits toxic fumes of Cl^- and PO_x. See also ESTERS.

EPZ000 CAS:313-93-9 HR: 2
3-ETHYLTRICYCLOQUINAZOLINE
mf: $C_{23}H_{16}N_4$ mw: 348.43

TOXICITY DATA WITH REFERENCE
skn-mus TDLo:1200 mg/kg/50W-I:CAR BCPCA6 14,323,65

SAFETY PROFILE: Questionable carcinogen with experimental carcinogenic data. When heated to decomposition it emits toxic fumes of NO_x.

EQA000 CAS:78-07-9 HR: 1
ETHYLTRIETHOXYSILANE
mf: $C_8H_{20}O_3Si$ mw: 192.37

PROP: A liquid. D: 0.915 @ 20°/4°, bp: 158°.

SYNS: NSC 5230 □ SILANE, ETHYL TRIETHOXY A-15 □ SILANE, TRIETHOXYETHYL- □ TRIETHOXY-ETHYLSILANE □ UNION CARBIDE A-15

TOXICITY DATA WITH REFERENCE
skn-rbt 500 mg open MLD UCDS** 4/27/64
eye-rbt 500 mg/24H MLD 85JCAE-,1231,86
orl-rat LD50:14 g/kg JIHTAB 31,60,49
skn-rbt LD50:16 g/kg JIHTAB 31,60,49

CONSENSUS REPORTS: Reported in EPA TSCA Inventory.

SAFETY PROFILE: Mildly toxic by ingestion and skin contact. A skin and eye irritant. When heated to decomposition it emits acrid smoke and irritating fumes.

EQB500 CAS:28781-86-4 HR: 3
ETHYL-2,2,3-TRIFLUORO PROPIONATE
mf: $C_5H_7F_3O_2$ mw: 188.22

$CH_3CH_2OCO{\bullet}CF_2CH_2F$

SAFETY PROFILE: Decomposes violently on contact with sodium hydride. When heated to decomposition it emits toxic fumes of F^-. See also FLUORIDES.

EQC000 CAS:1923-76-8 HR: 2
α-ETHYL-β-(2,4,6-TRIIODO-3-BUTYRAMIDOPHENYL) ACRYLIC ACID SODIUM SALT
mf: $C_{15}H_{15}I_3NO_3{\bullet}Na$ mw: 661.00

SYNS: BILI-ORAL □ BUNAIOD □ BUNAMIODYL □ BUNIODYL □ 3-BUTYRAMIDO-α-ETHYL-2,4,6-TRIIODOCINNAMIC ACID SODIUM SALT □ 2-(3-BUTYRAMIDO-2,4,6-TRIIODOPHENYLMETHYLENE)BUTYRIC ACID SODIUM SALT □ 3-(3-BUTYRYLAMINO-2,4,6-TRIIODOPHENYL)-2-ETHYLACRYLIC ACID SODIUM SALT □ INTRABILIX □ ORABILEX □ SODIUM-3-BUTYRAMIDO-α-ETHYL-2,4,6-TRIIODOCINNAMATE □ α-(2,4,6-TRIIODO-3-BUTYRYLAMINOBENZYLIDENE)BUTYRIC ACID SODIUM SALT

TOXICITY DATA WITH REFERENCE
orl-rat LD50:2800 mg/kg TXAPA9 14,232,69
ivn-rat LD50:480 mg/kg TXAPA9 14,232,69
orl-mus LD50:2780 mg/kg JMCMAR 13,997,70
ivn-mus LD50:570 mg/kg MEIEDD 10,206,83

SAFETY PROFILE: Moderately toxic by ingestion and intravenous routes. When heated to decomposition it emits very toxic fumes of I^-, Na_2O, and NO_x.

EQC600 CAS:51-93-4 HR: 3
ETHYLTRIMETHYLAMMONIUM IODIDE
mf: $C_5H_{14}N{\bullet}I$ mw: 215.10

SYNS: AMMONIUM, ETHYLTRIMETHYL-, IODIDE □ ETHANAMINIUM, N,N,N-TRIMETHYL-, IODIDE (9CI) □ IODURE d'ETHYL-TRIMETHYL-AMMONIUM □ N,N,N-TRIMETHYLETHANAMINIUM IODIDE

TOXICITY DATA WITH REFERENCE
ipr-mus LD50:40 mg/kg 85IXA4 -,675,48

CONSENSUS REPORTS: Reported in EPA TSCA Inventory.

SAFETY PROFILE: Poison by intraperitoneal route. When heated to decomposition it emits toxic vapors of NO_x and I^-.

EQD000 CAS:41096-46-2 HR: 1
ETHYL-3,7,11-TRIMETHYLDODECA-2,4-DIENOATE
mf: $C_{17}H_{30}O_2$ mw: 266.47

SYNS: ALTOZAR □ ENT 70,459 □ ETHYL(2E,4E)-3,7,11-TRIMETHYL-2,4-DODECADIENOATE □ ETHYL (2E,4E)-3,7,11-TRIMETHYL-DODECA-2-4-DIENOATE □ HYDROPRENE □ ZR 512

TOXICITY DATA WITH REFERENCE
ipr-mus TDLo:1 g/kg (9D preg):REP LIFSAK 15,1649,74
ipr-mus TDLo:1 g/kg (9D preg):TER LIFSAK 15,1649,74
orl-rat LD50:34,000 mg/kg 85ARAE 1,98,77
skn-rbt LD50:4550 mg/kg FMCHA2 -,C12,83

SAFETY PROFILE: Mildly toxic by ingestion and skin contact. Experimental teratogenic and reproductive effects. When heated to decomposition it emits acrid smoke and irritating fumes.

EQD100 CAS:67590-56-1 HR: 3
4-ETHYL-2,6,7-TRIOXA-1-ARSABICYCLO(2.2.2)OCTANE
mf: $C_6H_{11}AsO_3$ mw: 206.09

SYN: 2,6,7-TRIOXA-1-ARSABICYCLO(2.2.2)OCTANE, 4-ETHYL-

TOXICITY DATA WITH REFERENCE
ivn-mus LD50:31 mg/kg EJMCA5 13,207,78

OSHA PEL: TWA 0.5 mg(As)/m^3

SAFETY PROFILE: Poison by intravenous route. When heated to decomposition it emits toxic fumes of As.

EQD200 CAS:692-86-4 HR: 1
ETHYL 10-UNDECENOATE
mf: $C_{13}H_{24}O_2$ mw: 212.37

SYNS: ETHYL 10-HENDECENOATE □ ETHYL UNDECENOATE □ ETHYL UNDECYLENATE □ 10-UNDECENOIC ACID, ETHYL ESTER

TOXICITY DATA WITH REFERENCE
skn-rbt 500 mg/24H MLD FCTOD7 20,687,82
orl-rat LD50:>5 g/kg FCTOD7 20,687,82
skn-rbt LD50:>5 g/kg FCTOD7 20,687,82

CONSENSUS REPORTS: Reported in EPA TSCA Inventory.

SAFETY PROFILE: Slightly toxic by ingestion. A skin irritant. When heated to decomposition it emits acrid smoke and irritating vapors.

EQD600 CAS:57420-66-3 HR: 3
1-ETHYL-1-UNDECYLPIPERIDINIUM BROMIDE
mf: $C_{18}H_{38}N \cdot Br$ mw: 348.48

TOXICITY DATA WITH REFERENCE
orl-mus LD50:188 mg/kg PSDTAP 15,331,74
ipr-mus LD50:25,970 μg/kg PSDTAP 15,331,74
ivn-mus LD50:4417 μg/kg PSDTAP 15,331,74

SAFETY PROFILE: Poison by ingestion, intravenous, and intraperitoneal routes. When heated to decomposition it emits toxic fumes of NO_x and Br^-. See also BROMIDES.

EQD875 CAS:2884-67-5 HR: 1
ETHYLUREA
mf: $C_3H_8N_2O$ mw: 88.13

SYNS: N-ETHYLUREA □ 1-ETHYLUREA

TOXICITY DATA WITH REFERENCE
mmo-clr 400 mmol/L FOMIAZ 20,452,75
orl-mus TDLo:2 g/kg (10D preg):REP TJADAB 23,335,81
orl-rat LD:>250 mg/kg NCNSA6 5,47,53
par-mus LDLo:6610 mg/kg JPETAB 52,216,34

CONSENSUS REPORTS: EPA Genetic Toxicology Program. Reported in EPA TSCA Inventory.

SAFETY PROFILE: Mildly toxic by parenteral routes. Experimental reproductive effects. Mutation data reported. When heated to decomposition it emits toxic fumes of NO_x.

EQE000 HR: 3
ETHYLUREA and SODIUM NITRITE (2:1)

SYNS: AETHYLHARNSTOFF und NATRIUMNITRIT (GERMAN) □ AETHYLHARNSTOFF und NITRIT (GERMAN) □ SODIUM NITRITE and ETHYLUREA (1:2)

TOXICITY DATA WITH REFERENCE
orl-rat TDLo:22,274 mg/kg (1-22D preg/21D post):REP FCTXAV 15,283,77
orl-rat TDLo:100 mg/kg (9-10D preg):TER IARCCD 4,112,73
orl-rat TDLo:1650 mg/kg (13-23D preg):CAR,TER IARCCD 4,92,73
orl-rat TDLo:750 mg/kg (15D preg):NEO,TER EAFFAN 121,61,77
orl-rbt TDLo:450 mg/kg/(17-19D preg):ETA,TER JNCIAM 59,427,77
orl-ham TDLo:300 mg/kg (15D preg):CAR,TER JNCIAM 55,1389,75
orl-rat TD:1650 mg/kg (13-23D preg):ETA,TER NATWAY 57,460,70
orl-rat TD:25 g/kg/35D-C:ETA ARZNAD 21,1707,71
orl-ham TDLo:600 mg/kg (12-15D preg):NEO,TER ZKKOBW 85,201,76

SAFETY PROFILE: Suspected carcinogen with experimental carcinogenic, neoplastigenic, tumorigenic, and teratogenic data. Experimental reproductive effects. When heated to decomposition it emits toxic fumes of NO_x and Na_2O. See also SODIUM NITRITE.

EQE500 CAS:617-05-0 HR: 3
ETHYL VANILLATE
mf: $C_{10}H_{12}O_4$ mw: 196.22

PROP: Mp: 44°, bp: 291–293°, Insol in water; sol in alk; very sol in alc and ether.

SYNS: ETHYL-4-HYDROXY-3-METHOXYBENZOATE □ 3-METHOXY-4-HYDROXYBENZOIC ACID, ETHYL ESTER

TOXICITY DATA WITH REFERENCE
scu-rat LD50:2000 mg/kg HBTXAC 1,138,56
ivn-mus LD50:56 mg/kg CSLNX* NX#06395
orl-rbt LDLo:3000 mg/kg HBTXAC 1,138,56

SAFETY PROFILE: Poison by intravenous route. Moder-

ately toxic by ingestion and subcutaneous routes. When heated to decomposition it emits acrid smoke and irritating fumes. See also ESTERS.

EQF000 **CAS:121-32-4** **HR: 2**
ETHYL VANILLIN
mf: $C_9H_{10}O_3$ mw: 166.19

PROP: Fine, crystalline needles; vanilla odor. Mp: 76.5°, flash p: 212°F. Sol in alc, chloroform, ether, propylene glycol; sltly sol in water.

SYNS: BOURBONAL □ ETHAVAN □ ETHOVAN □ 3-ETHOXY-4-HYDROXYBENZALDEHYDE □ ETHYLPROTAL □ FEMA No. 2464 □ 4-HYDROXY-3-ETHOXYBENZALDEHYDE □ PROTOCATECHUIC ALDEHYDE ETHYL ETHER □ QUANTROVANIL □ VANILLAL □ VANIROM

TOXICITY DATA WITH REFERENCE
skn-hmn 10 mg/48H MLD FCTXAV 13,103,75
cyt-ham:fbr 250 mg/L FCTOD7 22,623,84
orl-rat LD50:1590 mg/kg FAONAU 44A,39,67
scu-rat LDLo:1800 mg/kg JAPMA8 29,425,40
ipr-mus LD50:750 mg/kg CTOXAO 10,61,77
ivn-dog LDLo:760 mg/kg COREAF 238,2576,54
orl-rbt LDLo:3000 mg/kg JAPMA8 29,425,40
ipr-gpg LD50:1140 mg/kg COREAF 238,2576,54

CONSENSUS REPORTS: Reported in EPA TSCA Inventory.

SAFETY PROFILE: Moderately toxic by ingestion, intraperitoneal, subcutaneous, and intravenous routes. A human skin irritant. Mutation data reported. When heated to decomposition it emits acrid smoke and irritating fumes. See also ALDEHYDES and ETHERS.

EQF100 **CAS:72207-94-4** **HR: 1**
ETHYL VANILLIN ACETATE
mf: $C_{11}H_{12}O_4$ mw: 208.23

SYNS: 4-ACETOXY-3-ETHOXYBENZALDEHYDE □ BENZALDEHYDE, 4-(ACETYLOXY)-3-ETHOXY-

TOXICITY DATA WITH REFERENCE
orl-rat LDLo:5 g/kg FCTOD7 26,283,88
skn-rbt LD50:>5 g/kg FCTOD7 26,283,88

CONSENSUS REPORTS: Reported in EPA TSCA Inventory.

SAFETY PROFILE: Low toxicity by ingestion and skin contact. When heated to decomposition it emits acrid smoke and irritating vapors.

EQF200 **CAS:3195-79-7** **HR: 2**
N-ETHYL-N-VINYLACETAMIDE
mf: $C_6H_{11}NO$ mw: 113.18

TOXICITY DATA WITH REFERENCE
orl-rat LD50:2460 mg/kg AIHAAP 30,470,69
ihl-rat LCLo:2000 ppm/4H AIHAAP 30,470,69
skn-rbt LD50:1250 mg/kg AIHAAP 30,470,69

SAFETY PROFILE: Moderately toxic by ingestion and skin contact. Mildly toxic by inhalation. When heated to decomposition it emits toxic fumes of NO_x.

EQF500 **CAS:109-92-2** **HR: 3**
ETHYL VINYL ETHER
DOT: UN 1302
mf: C_4H_8O mw: 72.12

PROP: Colorless, volatile liquid. Fp: −115°, bp: 35.6°, flash p: <−50°F, d: 0.754, autoign temp: 395°F, vap press: 428 mm @ 20°, lel: 1.7%, uel: 28%, vap d: 2.5. Sltly sol in water.

SYNS: ETHOXY ETHENE □ EVE □ VINAMAR □ VINYL ETHYL ETHER □ VINYL ETHYL ETHER, inhibited (DOT)

TOXICITY DATA WITH REFERENCE
skn-rbt 500 mg open MLD UCDS** 11/15/71
sce-ham:ovr 17,900 ppm ANESAV 50,426,79
orl-rat LD50:6153 mg/kg AIHAAP 30,470,69

CONSENSUS REPORTS: Reported in EPA TSCA Inventory.

DOT CLASSIFICATION: 3; *Label:* Flammable Liquid

SAFETY PROFILE: Mildly toxic by ingestion. Mutation data reported. A skin irritant. A very dangerous fire and explosion hazard when exposed to heat or flame; can react vigorously with oxidizing materials. To fight fire, use alcohol foam, foam, CO_2, dry chemical. Explosive polymerization is catalyzed by methane sulfonic acid. When heated to decomposition it emits acrid smoke and irritating fumes. See also ETHERS.

EQG500 **CAS:5408-74-2** **HR: 2**
5-ETHYL-2-VINYLPYRIDINE
mf: $C_9H_{11}N$ mw: 133.21

SYN: 3-ETHYL-6-VINYLPYRIDINE

TOXICITY DATA WITH REFERENCE
skn-rbt 10 mg/24H open AMIHBC 10,61,54
skn-rbt 5 mg/24H SEV 85JCAE-,847,86
eye-rbt 20 mg open AMIHBC 10,61,54
eye-rbt 100 mg/24H MOD 85JCAE-,847,86
orl-rat LD50:1230 mg/kg AMIHBC 10,61,54
ihl-rat LCLo:8000 ppm/4H AMIHBC 10,61,54
skn-rbt LD50:890 mg/kg AMIHBC 10,61,54

SAFETY PROFILE: Moderately toxic by ingestion and skin contact. Mildly toxic by inhalation. An eye and severe skin irritant. When heated to decomposition it emits toxic fumes of NO_x.

EQI000 **CAS:101491-85-4** **HR: 3**
N-ETHYL-N-(1-(3,5-XYLYLOXY)-2-PROPYLCARBAMIC ACID-2-(DIETHYLAMINO)ETHYL ESTER HYDROCHLORIDE
mf: $C_{20}H_{34}N_2O_3{\bullet}ClH$ mw: 387.02

SYN: C 2142

TOXICITY DATA WITH REFERENCE
eye-rbt 2% SEV ARZNAD 9,113,59
scu-mus LD50:202 mg/kg ARZNAD 9,113,59

SAFETY PROFILE: Poison by subcutaneous route. A severe eye irritant. When heated to decomposition it emits very toxic fumes of HCl and NO_x. See also CARBAMATES and ESTERS.

EQI500 CAS:101491-86-5 **HR: 3**
N-ETHYL-N-(1-(3,5-XYLYLOXY)-2-PROPYL)CARBAMIC ACID-2-(2-METHYLPIPERIDINO)ETHYL ESTER HYDROCHLORIDE
mf: $C_{22}H_{36}N_2O_3 \cdot ClH$ mw: 413.06

SYN: C 2127

TOXICITY DATA WITH REFERENCE
eye-rbt 2% SEV ARZNAD 9,113,59
scu-mus LD50:352 mg/kg ARZNAD 9,113,59

SAFETY PROFILE: Poison by subcutaneous route. A severe eye irritant. When heated to decomposition it emits very toxic fumes of HCl and NO_x. See also CARBAMATES and ESTERS.

EQI600 CAS:2482-80-6 **HR: 3**
ETHYMIDINE
mf: $C_8H_9ClN_4$ mw: 196.66

SYNS: 2,4-BIS(1-AZIRIDINYL)-6-CHLOROPYRIMIDINE □ 2,6-BIS(1-AZIRIDINYL)-4-CHLOROPYRIMIDINE □ 4-CHLORO-2,6-BIS-ETHYLENEIMINOPYRIMIDINE □ 2,6-DIETHYLENEIMINO-4-CHLOROPYRIMIDINE □ ETIMIDIN

TOXICITY DATA WITH REFERENCE
mmo-asn 2500 μmol/L MUREAV 14,115,72
oms-rat-ipr 2500 μg/kg BJPCAL 6,357,51
cyt-rat-ipr 2500 μg/kg BJPCAL 6,357,51
ivn-rat LDLo:15 mg/kg PMDCAY 8,61,71
ivn-mus LDLo:15 mg/kg PMDCAY 8,61,71

CONSENSUS REPORTS: EPA Genetic Toxicology Program.

SAFETY PROFILE: Poison by intravenous route. Mutation data reported. When heated to decomposition it emits toxic fumes of NO_x and Cl^-.

EQJ000 **HR: 2**
ETHYNERONE mixed with MESTRANOL (20:1)
mf: $C_{20}H_{23}ClO_2 \cdot C_{21}H_{26}O_2$ (20:1) mw: 641.0

SYN: MESTRANOL mixed with ETHYNERONE (1:20)

TOXICITY DATA WITH REFERENCE
orl-dog TDLo:243 mg/kg/47W-I:ETA JJIND8 65,137,80

SAFETY PROFILE: Questionable carcinogen with experimental tumorigenic data. When heated to decomposition it emits toxic fumes of Cl^-.

EQJ100 CAS:1231-93-2 **HR: 3**
ETHYNODIOL
mf: $C_{20}H_{28}O_2$ mw: 300.48

PROP: Crystals from Me_2CO/hexane. Mp: 211–214°.

SYNS: ED □ ETHINODIOL □ 17-α-ETHYNYL-19-NORANDROST-4-ENE-3-β,17-β-DIOL

TOXICITY DATA WITH REFERENCE
unr-wmn TDLo:12 mg/kg (6-20W preg):REP SCIEAS 211,1171,81
scu-rbt TDLo:200 μg/kg (female 1D pre):REP 85GRAA -,57,65
orl-rat TDLo:100 mg/kg (17-20D preg):TER ECJPAE 24,77,77

SAFETY PROFILE: Human reproductive effects by an unspecified route: biochemical and metabolic disorders in newborn. An experimental teratogen. Experimental reproductive effects. When heated to decomposition it emits acrid smoke and irritating fumes.

E

EQJ500 CAS:297-76-7 **HR: 3**
ETHYNODIOL ACETATE
mf: $C_{24}H_{32}O_4$ mw: 384.56

PROP: Crystals from MeOH (aq) or crystals from Me_2CO/hexane. Mp: 129–132°.

SYNS: CERVICUNDIN □ 3 β,17 β DIACETOXY 17 α ETHYNYL-4-OESTRENE □ 3-β,17-β-DIACETOXY-19-NOR-17-α-PREGN-4-EN-20-YNE □ ETHINODIOL DIACETATE □ ETHYNODIOL DIACETATE □ β-ETHYNODIOL DIACETATE □ 17-α-ETHYNYL-3,17-DIHYDROXY-4-ESTRENE DIACETATE □ 17-α-ETHYNYLESTR-4-ENE-3-β,17-β-DIOL ACETATE □ 17-α-ETHYNYL-4-ESTRENE-3-β,17-DIOL DIACETATE □ 17-α-ETHYNYL-4-ESTRENE-3-β,17-β-DIOL DIACETATE □ 17-α-ETHYNYL-19-NORANDROST-4-ENE-3-β,17-β-DIOL DIACETATE □ FEMULEN □ LUTO-METRODIOL □ METRODIOL □ METRODIOL DIACETATE □ (3-β,17-α)-19-NORPREGN 4 EN 20 YNE 3,17 DIOL DIACETATE □ OVULEN 50

TOXICITY DATA WITH REFERENCE
oth-mus-par 16 μg/kg AJOGAH 120,390,74
cyt-ctl:oth 10 mg/L AJOGAH 120,390,74
orl-wmn TDLo:840 μg/kg (21D pre):REP SCIEAS 138,439,62
scu-mus TDLo:1560 μg/kg (female 1-3D post):REP JRPFA4 9,277,65
scu-rat TDLo:1400 μg/kg (female 14D pre):REP CCPTAY 5,57,72
orl-rat TDLo:700 μg/kg (14D pre):REP FESTAS 24,284,73
orl-rat TDLo:6 mg/ (30D male pre):REP IJEBA6 9,132,71
scu-rat TDLo:48 mg/kg (male 12D pre):REP INJFA3 13,95,68
orl-rat TDLo:6 mg/ (30D male pre):REP IJEBA6 9,132,71

CONSENSUS REPORTS: IARC Cancer Review: Animal Limited Evidence IMEMDT 21,387,79; Animal Sufficient Evidence IMEMDT 6,173,74.

SAFETY PROFILE: Suspected carcinogen. Human reproductive effects by ingestion: menstrual cycle changes. Experimental reproductive effects. Mutation data reported. A steroid. When heated to decomposition it emits acrid smoke and irritating fumes.

EQK010 **HR: 2**
ETHYNODIOL DIACETATE mixed with MESTRANOL

SYNS: MESTRANOL mixed with ETHYNODIOL DIACETATE □ 19-NOR-17-α-PREGN-4-EN-20-YNE-3-β,17-DIOL DIACETATE mixed with 3-METHYOXY-19-NOR-17-α-PREGNA-1,3,5(10)-TRIEN-20-YN-17-OL

TOXICITY DATA WITH REFERENCE
orl-wmn TDLo:15 mg/kg/78W-I:NEO,LIV JAMAAP 235,730,76
ipr-mus LD50:3900 mg/kg 27ZTAP 3,69,69

SAFETY PROFILE: Questionable carcinogen producing liver tumors. Moderately toxic by intraperitoneal route. A steroid.

EQK100 **CAS:8056-92-6** **HR: 2**
ETHYNODIOL mixed with MESTRANOL
mf: $C_{24}H_{32}O_4 \cdot C_{21}H_{26}O_2$ mw: 695.03

SYNS: MESTRANOL mixed with ETHYNODIOL □ 10-NOR-17-α-PREGN-4-EN-20-YNE-3-β,17-DIOL mixed with 3-METHOXY-17-α-19-NOR-PREGNA-1-3-5(10)-TRIEN-20-YN-17-OL □ OVULEN

TOXICITY DATA WITH REFERENCE
orl-wmn TDLo:63,504 μg/kg/12Y-I:CAR,LIV,MET LANCAO 1,273,80
orl-wmn TDLo:220 μg/kg/2W:SKN ARDEAC 113,333,76
orl-mus TDLo:186 mg/kg/93W-C:NEO GMCRDC 17,205,75

SAFETY PROFILE: Human systemic effects by ingestion: skin dermatitis, weight loss or decreased weight gain. Questionable carcinogen producing liver tumors. A steroid. When heated to decomposition it emits acrid smoke and fumes.

EQL000 **CAS:77-75-8** **HR: 3**
2-ETHYNYL-2-BUTANOL
mf: $C_6H_{10}O$ mw: 98.16

PROP: Colorless, mobile liquid; acrid odor; burning taste. Sol in water, ether, etc. Mp: −30.6°, bp: 122°, flash p: 101°F (OC), d: 0.8688 @ 20°, vap d: 3.38.

SYNS: 2-AETHINYLBUTANOL □ ALLOTROPAL □ ANTI-STRESS □ APRIDOL □ ATEMPOL □ (BDH) □ COMESA □ DALGOL □ DORMIDIN □ DORMIPHEN □ DORMOSAN □ 2-ETHINYLBUTANOL-2 □ ETHINYLMETHYLETHYLCARBINOL □ 3-ETHYLBUTINOL □ 3-ETHYLBUTYNOL □ HESOFEN □ INSOMNOL □ MECAROL □ MEPENTAMATO □ MEPENTIL □ METHYLETHYLACETYLENYLCARBINOL □ METHYLETHYLETHYNYLCARBINOL □ 3-METHYLPENTIN-3-OL □ 3-METHYLPENT-1-YN-3-OL □ 3-METHYL-1-PENTYN-3-OL □ MIRAMEL □ NOXOKRATIN □ OBLEVIL □ m-OBLIVON □ PENTADORM □ PENTYDORM □ m-PENTYNOL □ PENTYREST □ PLACIDAL □ RIPOSON □ SOMNESIN □ TRUSONO □ UTIL

TOXICITY DATA WITH REFERENCE
orl-mus LD50:525 mg/kg PSCBAY 2,17,63
ipr-mus LD50:525 mg/kg AEPPAE 218,427,53
scu-mus LD50:284 mg/kg YKKZAJ 76,181,56

CONSENSUS REPORTS: Reported in EPA TSCA Inventory.

SAFETY PROFILE: Moderately toxic by intraperitoneal and subcutaneous routes. Used as a soporific. Average doses may produce dermatitis, eructations (belching), psychoses and central nervous system abnormalities. Overdoses can produce coma and death. Flammable liquid when exposed to heat or flame; can react with oxidizing materials. To fight fire use alcohol foam, mist, fog. When heated to decomposition it emits acrid smoke and fumes.

EQL500 **CAS:78-27-3** **HR: 3**
1-ETHYNYL-1-CYCLOHEXANOL
mf: $C_8H_{12}O$ mw: 124.20

PROP: Crystals. Mp: 30–31°, bp: 180°, vap d: 3.73.

SYNS: 1-ETHYNYLCYCLOHEXANOL □ 1-ETHYNYLCYCLOHEXAN-1-OL □ NSC 8194

TOXICITY DATA WITH REFERENCE
skn-rbt 10 mg/24H open MLD AIHAAP 23,95,62
orl-rat LD50:600 mg/kg AIHAAP 23,95,62
ipr-mus LDLo:500 mg/kg CBCCT* 4,228,52
skn-rbt LD50:1000 mg/kg AIHAAP 23,95,62

CONSENSUS REPORTS: Reported in EPA TSCA Inventory.

SAFETY PROFILE: Moderately toxic by ingestion, skin contact, and intraperitoneal routes. A skin irritant. Flammable when exposed to heat or flame; can react with oxidizing materials. To fight fire, use foam, CO_2, dry chemical. When heated to decomposition it emits acrid smoke and irritating fumes.

EQM500 **CAS:37270-71-6** **HR: 3**
ETHYNYLESTRADIOL mixed with NORETHINDRONE
mf: $C_{20}H_{26}O_2 \cdot C_{20}H_{24}O_2$ mw: 594.90

SYNS: GYNOVLAR □ NORETHINDRONE mixed with ETHYNYLESTRADIOL □ NORETHISTERONE mixed with ETHINYL OESTRADIOL (60:1) □ 19-NOR-17-α-PREGN-4-EN-20-YN-3-ONE, 17-HYDROXY-, mixed with 19-NOR-17-α-PREGNA-1,3,5(10)-TRIEN-2-YNE-3,17-DIOL (60:1)

TOXICITY DATA WITH REFERENCE
oms-ctl:oth 17,410 μg/L AJOGAH 120,390,74
oms-dom:oth 17,410 μg/L AJOGAH 120,390,74
oms-mam:oth 10,100 μg/L AJOGAH 120,390,74
orl-wmn TDLo:420 μg/kg (20D pre):REP CCPTAY 26,23,82
unr-wmn TDLo:63,298 μg/kg (11-34W preg):TER LANCAO 2,438,60
orl-wmn TDLo:138 mg/kg/9Y-I:CAR,LIV LANCAO 1,273,80
orl-wmn TD:138 mg/kg/9Y-I:CAR,LIV BMJOAE 4,496,75
orl-mus TDLo:120 mg/kg/84W-I:ETA SCIEAS 154,402,66
orl-wmn TDLo:55 mg/kg/5Y-I:PUL,GIT,MET LANCAO 1,1479,73

SAFETY PROFILE: Human teratogenic effects by an unspecified route: developmental abnormalities of the urogenital system. Human systemic effects by ingestion: dyspnea, nausea or vomiting, and fever. Experimental reproductive effects. Questionable human carcinogen producing liver tumors. Mutation data reported. A steroid. When heated to decomposition it emits acrid smoke and irritating fumes.

EQM600 **CAS:2028-63-9** **HR: 3**
1-ETHYNYLETHANOL
mf: C_4H_6O mw: 70.10

SYN: 1-BUTYN-3-OL

TOXICITY DATA WITH REFERENCE
orl-mus LD50:30 mg/kg ARZNAD 7,85,57

CONSENSUS REPORTS: Reported in EPA TSCA Inventory.

SAFETY PROFILE: Poison by ingestion. When heated to decomposition it emits acrid smoke and irritating vapors.

EQN000 **CAS:18649-64-4** **HR: 3**
2-ETHYNYLFURAN
mf: C_6H_4O mw: 92.10

OCH=CHCH= CC≡CH

PROP: A liquid. Bp: 105–106°.

SAFETY PROFILE: Explodes on heating or on contact with concentrated nitric acid. When heated to decomposition it emits acrid smoke and irritating fumes. See also ACETYLENE COMPOUNDS.

EQN225 **HR: 2**
α-ETHYNYL-p-METHOXYBENZYL ALCOHOL ACETATE
mf: $C_{12}H_{12}O_3$ mw: 204.24

SYN: 1′-ACETOXY-2′,3′-DEHYDROESTRAGOLE

TOXICITY DATA WITH REFERENCE
mmo-sat 150 nmol/plate CRNGDP 7,2089,86
ipr-mus TDLo:10,212 μg/kg:NEO CNREA8 47,2275,87

SAFETY PROFILE: Questionable carcinogen with experimental neoplastigenic data. Mutation data reported. When heated to decomposition it emits acrid smoke and irritating fumes.

EQN230 **CAS:127-66-2** **HR: 2**
α-ETHYNYL-α-METHYLBENZYL ALCOHOL
mf: $C_{10}H_{10}O$ mw: 146.20

SYNS: 3-BUTYN-2-OL, 2-PHENYL- □ 3-PHENYL-BUTIN-1-OL-(3)

TOXICITY DATA WITH REFERENCE
orl-mus LD50:620 mg/kg ARZNAD 4,477,54

CONSENSUS REPORTS: Reported in EPA TSCA Inventory.

SAFETY PROFILE: Moderately toxic by ingestion. When heated to decomposition it emits acrid smoke and irritating vapors.

EQN300 **CAS:37172-05-7** **HR: 2**
1-ETHYNYL-2-(1-METHYLPROPYL)CYCLOHEXYL ACETATE
mf: $C_{14}H_{22}O_2$ mw: 222.36

SYNS: AMBRATE □ CYCLOHEXANOL, 1-ETHYNYL-2-(1-METHYLPROPYL)-, ACETATE □ 1-ETHYNYL-2-(1-METHYLPROPYL)CYCLOHEXANOL ACETATE

TOXICITY DATA WITH REFERENCE
orl-rat LD50:680 mg/kg FCTOD7 30,3S,92
skn rbt LD50:2 g/kg FCTOD7 30,3S,92

CONSENSUS REPORTS: Reported in EPA TSCA Inventory.

SAFETY PROFILE: Moderately toxic by ingestion. Low toxicity by skin contact. When heated to decomposition it emits acrid smoke and irritating vapors.

EQN500 **HR: 3**
ETHYNYL VINYL SELENIDE
mf: C_4H_4Se mw: 131.04

HC≡CSeCH=CH_2

CONSENSUS REPORTS: Selenium and its compounds are on the Community Right-To-Know List.

OSHA PEL: TWA 0.2 mg(Se)/m³
ACGIH TLV: TWA 0.2 mg(Se)/m³
DFG MAK: 0.1 mg(Se)/m³

SAFETY PROFILE: Decomposes violently when heated. Upon decomposition it emits toxic fumes of Se. See also SELENIUM COMPOUNDS and ACETYLENE COMPOUNDS.

EQN600 **CAS:40054-69-1** **HR: 2**
ETIZOLAM
mf: $C_{17}H_{15}ClN_4S$ mw: 342.87

PROP: Crystals from toluene. Mp: 144–146°.

SYNS: AHR 3219 □ 6-(o-CHLOROPHENYL)-8-ETHYL-1-METHYL-4H-sec-TRIAZOLO(3,4-c)THIENO(2,3-e)(1,4)-DIAZEPINE □ 6-(o-CHLORPHENYL)-8-AETYL-1-METHYL-4H sec-TRIAZOLO(3,4-c)THIENO(2,3-e)(1,4)DIAZEPIN (GERMAN) □ 4-(2-CHLORPHENYL)-2-ETHYL-9-METHYL-6H-THIENO(3,2-f)(1,2,4)TRIAZOLO(4,3-a)(1,4)DIAZEPINE □ DEPAS □ Y-7131

TOXICITY DATA WITH REFERENCE
orl-rat TDLo:135 mg/kg (female 17-22D post):REP OYYAA2 17,787,79
orl-rat TDLo:55 mg/kg (female 7-17D post):REP OYYAA2 17,763,79
orl-rat TDLo:1100 mg/kg (7-17D preg):TER OYYAA2 17,763,79
orl-rat TDLo:9100 mg/kg (26W pre):REP OYYAA2 16,1021,78
orl-mus TDLo:3 g/kg (female 7-12D post):TER OYYAA2 17,763,79
orl-rat LD50:3509 mg/kg ARZNAD 28,1158,78
ipr-rat LD50:825 mg/kg OYYAA2 16,1021,78
orl-mus LD50:4258 mg/kg OYYAA2 16,1021,78
ipr-mus LD50:783 mg/kg OYYAA2 16,1021,78

SAFETY PROFILE: Moderately toxic by ingestion and intraperitoneal routes. Experimental teratogenic and reproductive effects. When heated to decomposition it emits toxic fumes of Cl^-, SO_x, and NO_x.

EQN700 **CAS:32458-57-4** **HR: 3**
ETMA
mf: $C_8H_{20}N_3S \cdot Br \cdot BrH$ mw: 351.20

SYNS: (2-(((ETHYLAMINO)IMINOMETHYL)THIO)ETHYL)TRIMETHYL AMMONIUM BROMIDE HYDROBROMIDE □ 2-(((ETHYLAMINO)IMINOMETHYL)THIO)-N,N,N-TRIMETHYLETHANAMINIUM BROMIDE, MONOHYDROBROMIDE □ S-(2-TRIMETHYLAMINOETHYL)-1′-ETHYLISOTHIURONIUM BROMIDE HYDROBROMIDE

TOXICITY DATA WITH REFERENCE
ipr-mus LD50:95 mg/kg CPBTAL 23,1639,75
scu-mus LD50:102 mg/kg CPBTAL 23,1639,75
ivn-mus LD50:81,800 μg/kg CPBTAL 23,1639,75

SAFETY PROFILE: Poison by subcutaneous, intrave-

E

nous, and intraperitoneal routes. When heated to decomposition it emits toxic fumes of SO_x, NO_x, NH_3, and HBr.

EQN750 **CAS:2053-25-0** **HR: 3**
ETONITAZENE HYDROCHLORIDE
mf: $C_{22}H_{28}N_4O_3$•ClH mw: 433.00

SYNS: BENZIMIDAZOLE, 1-(2-DIETHYLAMINOETHYL)-2-(p-ETHOXYBENZYL)-5-NITRO-, HYDROCHLORIDE □ C 20684 □ 1-(2-DIETHYLAMINOETHYL)-2-(p-ETHOXYBENZYL)-5-NITROBENZIMIDAZOLE HYDROCHLORIDE

TOXICITY DATA WITH REFERENCE
scu-rat TDLo:6700 ng/kg (male 1D pre):REP JPETAB 198,340,76
par-mus LD50:126 mg/kg JMCMAR 11,889,68

SAFETY PROFILE: Poison by parenteral route. Experimental reproductive effects. When heated to decomposition it emits toxic fumes of NO_x and HCl.

EQO000 **CAS:57775-22-1** **HR: 3**
ETOPERIDONE
mf: $C_{19}H_{28}ClN_5O$•ClH mw: 414.43

SYNS: CLOPRADONE □ ST-1191 □ TRAZOLINONE

TOXICITY DATA WITH REFERENCE
orl-rat TDLo:3 g/kg (6-15D preg):TER TXCYAC 8,87,77
orl-rat TDLo:3 g/kg (6-15D preg):REP TXCYAC 8,87,77
orl-rat LD50:397 mg/kg ARZNAD 29,294,79
ipr-rat LD50:120 mg/kg ARZNAD 28,417,78
scu-rat LD50:543 mg/kg ARZNAD 29,294,79
ivn-rat LD50:62 mg/kg ARZNAD 28,417,78
orl-mus LD50:518 mg/kg ARZNAD 29,294,79
ipr-mus LD50:135 mg/kg ARZNAD 28,417,78
ivn-mus LD50:68 mg/kg ARZNAD 29,294,79

SAFETY PROFILE: Poison by ingestion, intravenous, and intraperitoneal routes. An experimental teratogen. Experimental reproductive effects. When heated to decomposition it emits very toxic fumes of Cl^- and NO_x.

EQO450 **CAS:14521-96-1** **HR: 3**
7-α-ETORPHINE
mf: $C_{25}H_{33}NO_4$ mw: 411.59

PROP: Off-white prisms from 2-ethoxyethanol. Mp: 215–216°.

SYNS: 7,8-DIHYDRO-7-α-(1-(R)-HYDROXY-1-METHYLBUTYL)-O⁶-METHYL-6,14-endo-ETHENOMORPHINE □ 6,14-endo-ETHENOTETRAHYDROORIPAVINE, 7-α-(1-HYDROXY-1-METHYLBUTYL)- □ ETORPHINE □ (−)-ETORPHINE □ 7-α-(1-(R)-HYDROXY-1-METHYLBUTYL)-6,14-endo-ETHENOTETRAHYDROORIPAVINE □ ORIPAVINE, 6,14-endo-ETHYLENETETRAHYDRO-7-(1-HYDROXY-1-METHYLBUTYL) -(7CI) □ 19-PROPYLORVINOL □ TETRAHYDRO-7-α-(1-HYDROXY-1-METHYLBUTYL)-6,14-endo-ETHENOORIPAVINE □ TETRAHYDRO-7-α-(2-HYDROXY-2-PENTYL)-6,14-endo-ETHENOORIPAVINE

TOXICITY DATA WITH REFERENCE
scu-rat TDLo:3200 ng/kg (male 1D pre):REP JPETAB 198,340,76
ipr-mus LD50:200 mg/kg AGACBH 6,755,76

SAFETY PROFILE: Poison by intraperitoneal route. Experimental reproductive effects. When heated to decomposition it emits toxic fumes of NO_x.

EQO500 **CAS:13764-49-3** **HR: 3**
ETORPHINE HYDROCHLORIDE
mf: $C_{25}H_{33}NO_4$•ClH mw: 448.05

SYNS: 6,14-ENDOETHENO-7-(2-HYDROXY-2-PENTYL)-TETRAHYDRO-ORIPAVINE HYDROCHLORIDE □ 7-α-(1-(R)-HYDROXY-1-METHYLBUTYL)-6,14-ENDOETHENOTETRAHYDRO-ORIPAVINE HYDROCHLORIDE □ PROPYLORVINOL HYDROCHLORIDE

TOXICITY DATA WITH REFERENCE
orl-rat LD50:72 mg/kg BJPCAL 30,11,67
scu-rat LD50:53 mg/kg BJPCAL 30,11,67
ivn-rat LD50:5300 μg/kg BJPCAL 30,11,67
orl-mus LD50:1856 mg/kg BJPCAL 30,11,67
scu-mus LD50:425 mg/kg BJPCAL 30,11,67
ivn-mus LD50:80 mg/kg BJPCAL 30,11,67

SAFETY PROFILE: Poison by ingestion, subcutaneous, and intravenous routes. When heated to decomposition it emits very toxic fumes of NO_x and HCl.

EQP000 **CAS:29767-20-2** **HR: 3**
ETP
mf: $C_{32}H_{32}O_{13}S$ mw: 656.70

PROP: Crystals from EtOH. Mp: 242–246°.

SYNS: 4′-DEMETHYLEPIPODOPHYLLOTOXIN-9-(4,6-O-2-THENYLIDENE-β-d-GLUCOPYRANOSIDE □ 4′-DEMETHYL-EPIPODOPHYLLOTOXIN-β-d-THENYLIDENE-GLUCOSIDE □ 4′-DEMETHYL 1-O-(4,6-O, O-(2-THENYLIDENE)-β-d-GLUCOPYRANOSYL)EPIPODOPHYLLOTOXIN □ EPT □ NSC-122819 □ PTG □ TENIPOSIDE □ VEHAM-SANDOZ □ VEHEM □ VM-26 □ VUMON

TOXICITY DATA WITH REFERENCE
mmo-sat 100 μg/plate TCMUD8 5,319,85
dnd-hmn:oth 500 nmol/L CNREA8 45,3106,85
cyt-hmn:lym 5 mg/L CRSBAW 179,331,85
dnd-mus/leu 50 μg/kg CNREA8 40,4225,80
dnd-mus:leu 2 μmol/L CNREA8 44,3360,84
cyt-mus:lym 500 ng/L ENMUDM 8(Suppl 6),24,86
sce-ham:ovr 10 μg/L CNREA8 43,577,83
ipr-mus TDLo:1 mg/kg (7D preg):TER TJADAB 18,31,78
ipr-mus TDLo:1 mg/kg (7D preg):REP TJADAB 18,31,78
orl-hmn TDLo:9579 mg/kg:CNS,GIT,SKN EJCAAH 14,1395,78
ivn-hmn TDLo:26 mg/kg/10D-I:BLD CANCAR 34,985,74
ivn-hmn TDLo:132 mg/kg/7W-I:GIT,BLD CTRRDO 63,7,79
ipr-mus LD17:50 mg/kg CTRRDO 60,1127,76
ipr-mus LD50:29,570 μg/kg NCISP* JAN86
scu-mus LD50:31,560 μg/kg NCISP* JAN86

SAFETY PROFILE: Poison by intraperitoneal and subcutaneous routes. An experimental teratogen. Human systemic effects by ingestion and intravenous route: anorexia, nausea or vomiting, leukopenia, agranulocytosis and aplastic anemia of the blood, bone marrow changes, and hair changes. Experimental reproductive effects. Human mutation data reported. When heated to decomposition it emits very toxic fumes of SO_x.

EQP100 **HR: 3**
E. TYPHOSA LIPOPOLYSACCHARIDE

TOXICITY DATA WITH REFERENCE
ipr-rat TDLo: 1 mg/kg (12D preg): TER PSEBAA 109,429,62
ipr-rat TDLo: 1 mg/kg (12D preg): REP PSEBAA 109,429,62
ipr-rat LD50: 6 mg/kg PSEBAA 109,429,62

SAFETY PROFILE: Poison by intraperitoneal route. Experimental teratogenic and reproductive effects.

EQP500 **CAS:500-34-5** **HR: 3**
β-EUCAINE
mf: $C_{15}H_{21}NO_2$ mw: 247.37

PROP: Plates from pet ether. Mp: 70–71°.

SYNS: EUCAINE B □ EUKAIN B □ TRIMETHYLBENZOLOXYPIPERIDINE □ 2,2,6-TRIMETHYL-4-PIPERIDINOL BENZOATE (ESTER)

TOXICITY DATA WITH REFERENCE
ivn-rat LDLo: 15 mg/kg PHREA7 12,190,32
ivn-mus LDLo: 71 mg/kg WDMU** -,-,36
ivn-cat LDLo: 10 mg/kg PHREA7 12,190,32
scu rbt LDLo: 400 mg/kg PHREA7 12,190,32
ipr-gpg LDLo: 18 mg/kg PHREA7 12,190,32
scu-gpg LDLo: 310 mg/kg PHREA7 12,190,32
ivn-gpg LDLo: 30 mg/kg PHREA7 12,190,32

SAFETY PROFILE: Poison by intravenous, subcutaneous, and intraperitoneal routes. When heated to decomposition it emits toxic fumes of NO_x.

EQQ000 **CAS:8000-48-4** **HR: 3**
EUCALYPTUS OIL

PROP: From steam distillation of leaves of *Eucalyptus globulus* Labillardiere. Chief constituent is eucalyptol (FCTXAV 13,19,75). Colorless to pale-yellow liquid; spicy odor and taste. Composition: eucalyptol, aldehydes, d-pinene. Mp: −15.4° (approx) d: 0.905–0.925 @ 25°/25°.

SYNS: DINKUM OIL □ EUKALYPTUS OEL (GERMAN) □ OIL of EUCALYPTUS

TOXICITY DATA WITH REFERENCE
skn-rbt 500 mg/24H MOD FCTXAV 13,91,75
orl-man LDLo: 375 mg/kg: BAH,GIT,SKN ADCHAK 28,475,53
orl-chd TDLo: 218 mg/kg: EYE,CNS,PUL ADCHAK 28,475,53
orl-rat LD50: 2480 mg/kg FCTXAV 13,91,75
skn-rbt LD50: 2480 mg/kg FCTOD7 26,323,88

CONSENSUS REPORTS: Reported in EPA TSCA Inventory.

SAFETY PROFILE: A human poison by ingestion. Moderately toxic by skin contact. Human systemic effects by ingestion: ciliary eye spasms, nausea or vomiting, respiratory depression, somnolence, sweating. A skin irritant. When heated to decomposition it emits acrid smoke and irritating fumes. See also ALDEHYDES.

EQQ100 **CAS:1641-74-3** **HR: 3**
EUCAST
mf: $C_{12}H_{18}N_2O_2 \cdot C_6H_8O_7$ mw: 414.46

SYNS: 53-11 C □ EUCLIDAN □ NICAMETATE CITRATE □ NICAMETATE DIHYDROGEN CITRATE □ PROVASAN □ SOCLIDAN

TOXICITY DATA WITH REFERENCE
orl-rat LD50: 8400 mg/kg NIIRDN 6,542,82
scu-rat LD50: 3600 mg/kg NIIRDN 6,542,82
ivn-rat LD50: 327 mg/kg NIIRDN 6,542,82
orl-mus LD50: 11,900 mg/kg NIIRDN 6,542,82
scu-mus LD50: 2600 mg/kg NIIRDN 6,542,82
ivn-mus LD50: 316 mg/kg NIIRDN 6,542,82

SAFETY PROFILE: Poison by intravenous route. Moderately toxic by subcutaneous route. Mildly toxic by ingestion. When heated to decomposition it emits toxic fumes of NO_x. See also ESTERS.

EQQ500 **CAS:18984-80-0** **HR: 3**
EUCUPIN DIHYRDOCHLORIDE
mf: $C_{24}H_{32}N_2O_2 \cdot 2ClH$ mw: 453.50

SYNS: EUKUPIN DIHYDROCHLORIDE □ ISOAMYL HYDROCUPREINE DIHYDROCHLORIDE □ 4-((1-METHOXYBUTOXY)(5-VINYL-2-QUINUCLIDINYL)METHYL)-6-QUINOLINOL DIHYDROCHLORIDE □ OTODYNE

TOXICITY DATA WITH REFERENCE
scu-rat LDLo: 800 mg/kg CLDND* 11,257,20
orl-cat LDLo: 25 mg/kg ZGEMAZ 11,257,20
scu-cat LDLo: 50 mg/kg ZGEMAZ 11,257,20
ivn-cat LDLo: 6820 μg/kg ZGEMAZ 11,257,20

SAFETY PROFILE: Poison by ingestion and intravenous routes. Moderately toxic by subcutaneous route. When heated to decomposition it emits very toxic fumes of HCl and NO_x.

EQR000 **CAS:28399-17-9** **HR: 3**
EUDESMA-3,11(13)-DIEN-12-OIC ACID
mf: $C_{15}H_{22}O_2$ mw: 234.37

SYN: 12-CARBOXYEUDESMA-3,11(13)-DIENE

TOXICITY DATA WITH REFERENCE
orl-mus LD50: 1000 mg/kg JMCMAR 13,1221,70
ipr-mus LD50: 200 mg/kg JMCMAR 13,1221,70

SAFETY PROFILE: Poison by intraperitoneal route. Moderately toxic by ingestion. When heated to decomposition it emits acrid smoke and irritating fumes.

EQR500 **CAS:97-53-0** **HR: 2**
EUGENOL
mf: $C_{10}H_{12}O_2$ mw: 164.22

PROP: Colorless or yellowish liquid or oil; pungent, clove odor. D: 1.064–1.070, refr index: 1.540, fp: −9°, bp: 248, flash p: 219°F. Sol in alc, chloroform, ether, volatile oils; very sltly sol in water.

SYNS: 4-ALLYLGUAIACOL □ 4-ALLYL-1-HYDROXY-2-METHOXYBENZENE □ 4-ALLYL-2-METHOXYPHENOL □ CARYOPHYLLIC ACID □ EUGENIC ACID □ Fa 100 □ FEMA No. 2467 □ 1-HYDROXY-2-METHOXY-4-ALLYLBENZENE □ 4-HYDROXY-3-METHOXYALLYLBEN-

E

ZENE □ 1-HYDROXY-2-METHOXY-4-PROP-2-ENYLBENZENE □ 2-METHOXY-4-ALLYLPHENOL □ 2-METHOXY-4-PROP-2-ENYLPHENOL □ 2-METHOXY-4-(2-PROPENYL)PHENOL □ 2-METOKSY-4-ALLILOFENOL (POLISH) □ NCI-C50453 □ SYNTHETIC EUGENOL

TOXICITY DATA WITH REFERENCE
skn-hmn 40 mg/48H MLD FCTXAV 13,545,75
skn-man 16 mg/48H MOD CTOIDG 94(8),41,79
skn-rbt 100 mg/24H SEV CTOIDG 94(8),41,79
skn-pig 50 mg/48H MLD CTOIDG 94(8),41,79
mma-sat 50 μg/plate NTIS** AD-A116-715
oms-ham:ovr 400 mg/L CALEDQ 14,251,81
cyt-ham:fbr 125 mg/L FCTOD7 22,623,84
cyt-ham:ovr 400 mg/L CALEDQ 14,251,81
orl-mus TDLo:37,080 mg/kg/2Y-I:ETA NTPTR* NTP-TR-223,82
orl-rat LD50:1930 mg/kg PSEBAA 73,148,50
ipr-rat LDLo:800 mg/kg RMSRA6 16,449,1896
scu-rat LDLo:5000 mg/kg RMSRA6 16,449,1896
orl-mus LD50:3000 mg/kg FCTXAV 2,327,64
ipr-mus LD50:500 mg/kg COREAF 250,1148,60
orl-dog LDLo:500 mg/kg GASTAB 15,481,50
orl-gpg LD50:2130 mg/kg FCTXAV 2,327,64

CONSENSUS REPORTS: IARC Cancer Review: Group 3 IMEMDT 7,56,87; Animal Limited Evidence IMEMDT 36,75,85. NTP Carcinogenesis Studies (feed); Equivocal Evidence: mouse NTPTR* NTP-TR-223,83; No Evidence: rat NTPTR* NTP-TR-223,83. Reported in EPA TSCA Inventory. EPA Genetic Toxicology Program.

SAFETY PROFILE: Moderately toxic by ingestion, intraperitoneal, and subcutaneous routes. Human mutation data reported. A human skin irritant. Questionable carcinogen with experimental carcinogenic and tumorigenic data. Combustible liquid. When heated to decomposition it emits acrid smoke and irritating fumes. See also ALLYL COMPOUNDS.

EQS000 CAS:93-28-7 HR: 2
EUGENOL ACETATE
mf: $C_{12}H_{14}O_3$ mw: 206.26

PROP: Solid or pale-yellow liquid or plates from alc; mild clove odor. D: 1.87, mp: 30–31°, bp: 281.2°, flash p: 151°F. Insol in water; sol in alc and ether.

SYNS: ACETEUGENOL □ 1-ACETOXY-2-METHOXY-4-ALLYLBENZENE □ ACETYLEUGENOL □ 4-ALLYL-2-METHOXYPHENOL ACETATE □ 1,3,4-EUGENOL ACETATE □ EUGENYL ACETATE □ FEMA No. 2469

TOXICITY DATA WITH REFERENCE
skn-rbt 500 mg/24H MOD FCTXAV 12,807,74
orl-rat LD50:1670 mg/kg FCTXAV 2,327,64

CONSENSUS REPORTS: Reported in EPA TSCA Inventory.

SAFETY PROFILE: Moderately toxic by ingestion. A skin irritant. Combustible liquid. When heated to decomposition it emits acrid smoke and irritating fumes. See also EUGENOL, ALLYL COMPOUNDS, and ESTERS.

EQS100 CAS:10031-96-6 HR: 2
EUGENOL FORMATE
mf: $C_{11}H_{12}O_3$ mw: 192.23

SYNS: 4-ALLYL-2-METHOXYPHENOL FORMATE □ EUGENYL FORMATE □ PHENOL, 4-ALLYL-2-METHOXY-, FORMATE (ester) □ 4-(2-PROPENYL)-2-METHOXYPHENYL FORMATE

TOXICITY DATA WITH REFERENCE
skn-rbt 500 mg/24H MLD FCTOD7 20,689,82
orl-rat LD50:3400 mg/kg FCTOD7 20,689,82

SAFETY PROFILE: Moderately toxic by ingestion. A skin irritant. When heated to decomposition it emits acrid smoke and irritating fumes.

EQS500 CAS:534-76-9 HR: 3
EULICIN
mf: $C_{24}H_{52}N_8O_2$ mw: 484.84

SYNS: 9-((AMINOIMINOMETHYL)AMINO)-N-(10-((AMINOIMINOMETHYL)AMINO)-1-(3-AMINOPROPYL)-20-HYDROXYDECYL)NONANAMIDE □ N-(1-(3-AMINOPROPYL)-10-GUANIDINO-2-HYDROXYDECYL)-9-GUANIDINONONANAMIDE □ N-METHOXY-N-METHYLNONANAMIDE

TOXICITY DATA WITH REFERENCE
ipr-mus LD50:17 mg/kg 85ERAY 2,1142,78
scu-mus LD50:46 mg/kg 85ERAY 2,1142,78
ivn-mus LD50:3 mg/kg 85ERAY 2,1142,78
ims-mus LD50:12 mg/kg 85ERAY 2,1142,78

SAFETY PROFILE: Poison by intraperitoneal, subcutaneous, intravenous, and intramuscular routes. A fungicide. When heated to decomposition it emits toxic fumes of NO_x.

EQT000 CAS:1403-51-6 HR: 3
EUMYCETIN

PROP: An antifungal antibiotic produced by the strain *Streptomyces sp.* 108 (85RAY 2,1126,78).

TOXICITY DATA WITH REFERENCE
ipr-mus LD50:1200 μg/kg 85ERAY 2,1126,78
scu-mus LD50:3 mg/kg JAJAAA 7,165,54

SAFETY PROFILE: Poison by intraperitoneal and subcutaneous routes. When heated to decomposition it emits acrid smoke and irritating fumes.

EQT500 HR: 2
EUPHORBIA ABYSSINICA LATEX

PROP: Acetone soluble portion of *Euphorbia abyssinica* latex.

TOXICITY DATA WITH REFERENCE
skn-mus TDLo:2600 mg/kg/26W-I:ETA CNREA8 21,338,61

SAFETY PROFILE: Questionable carcinogen with experimental tumorigenic data. When heated to decomposition it emits acrid smoke and irritating fumes.

EQU000 **HR: 2**
EUPHORBIA CANARIENSIS LATEX

PROP: Acetone soluble fraction of latex from *Euphorbia canariensis.*

TOXICITY DATA WITH REFERENCE
skn-mus TDLo:2600 mg/kg/26W-I:ETA CNREA8 21,338,61

SAFETY PROFILE: Questionable carcinogen with experimental tumorigenic data. When heated to decomposition it emits acrid smoke and irritating fumes.

EQU500 **HR: 2**
EUPHORBIA CANDELABRIUM LATEX

PROP: Acetone soluble portion of latex from *Euphorbia candelabrum.*

TOXICITY DATA WITH REFERENCE
skn-mus TDLo:2600 mg/kg/26W-I:ETA CNREA8 21,338,61

SAFETY PROFILE: Questionable carcinogen with experimental tumorigenic data. When heated to decomposition it emits acrid smoke and irritating fumes.

EQV000 **HR: 2**
EUPHORBIA ESULA LATEX

PROP: Acetone soluble portion of latex from *Euphorbia esula.*

TOXICITY DATA WITH REFERENCE
skn-mus TDLo:3500 mg/kg/20W-I:ETA TUMOAB 64,99,78

SAFETY PROFILE: Questionable carcinogen with experimental tumorigenic data. When heated to decomposition it emits acrid smoke and irritating fumes.

EQV500 **HR: 2**
EUPHORBIA GRANDIDENS LATEX

PROP: Acetone soluble fraction of latex from *Euphorbia grandides.*

TOXICITY DATA WITH REFERENCE
skn-mus TDLo:2600 mg/kg/26W-I:ETA CNREA8 21,338,61

SAFETY PROFILE: Questionable carcinogen with experimental tumorigenic data. When heated to decomposition it emits acrid smoke and irritating fumes.

EQW000 **HR: 2**
EUPHORBIA LATHYRIS LATEX

PROP: Oil from the seeds of *Euphorbia lathyris.*

SYN: CAPER SPURGE

TOXICITY DATA WITH REFERENCE
skn-mus 13 μg MLD CNREA8 28,2338,68
skn-mus TDLo:5680 mg/kg/W-I:ETA CNREA8 28,2338,68

SAFETY PROFILE: Questionable carcinogen with experimental tumorigenic data. A skin irritant. When heated to decomposition it emits acrid smoke and irritating fumes.

EQW500 **HR: 2**
EUPHORBIA OBOVALIFOLIA LATEX

PROP: Acetone soluble portion of latex from *Euphorbia obovalifolia.*

TOXICITY DATA WITH REFERENCE
skn-mus TDLo:2600 mg/kg/26W-I:ETA CNREA8 21,338,61

SAFETY PROFILE: Questionable carcinogen with experimental tumorigenic data. When heated to decomposition it emits acrid smoke and irritating fumes.

EQX000 **HR: 1**
EUPHORBIA PULCHERRIMA WILLD.

PROP: The plant contains sterols and latex that have 7–15% caoutchouc.

SYNS: ANNUAL POINSETTIA ☐ CHRISTMAS FLOWER ☐ EASTER FLOWER ☐ EUPHORBIA POINSETTIS BUIST ☐ MEXICA FLAME LEAF ☐ MEXICAN FLAME TREE ☐ MEXICAN FLOWER PLANT ☐ POINSETTIA ☐ POINSETTIA PULCHERRIMA GRAH

TOXICITY DATA WITH REFERENCE
skn-rbt 11,200 mg/14D MOD CTOXAO 13,27,78

SAFETY PROFILE: A skin irritant. When heated to decomposition it emits acrid smoke and irritating fumes.

EQX500 **HR: 2**
EUPHORBIA SERRATA LATEX

PROP: Acetone soluble fraction of latex from *Euphorbia serrata.*

TOXICITY DATA WITH REFERENCE
skn-mus TDLo:6400 mg/kg/16W-I:ETA TUMOAB 64,99,78

SAFETY PROFILE: Questionable carcinogen with experimental tumorigenic data. When heated to decomposition it emits acrid smoke and irritating fumes.

EQY000 **HR: 2**
EUPHORBIA TIRUCALLI LATEX

PROP: Acetone soluble portion of *Euphorbia tirucalli* latex. Contains all biological activity.

SYN: BLEISTIFTBAUMS (GERMAN)

TOXICITY DATA WITH REFERENCE
skn-mus 26 ng MLD PLMEAA 22,241,72
skn-mus TDLo:2400 mg/kg/12W-I:ETA PLMEAA 22,241,72
skn-mus TD:2800 mg/kg/28W-I:ETA CNREA8 21,338,61

SAFETY PROFILE: Questionable carcinogen with experimental tumorigenic data. A skin irritant. When heated to decomposition it emits acrid smoke and irritating fumes.

E

EQY500 **HR: 2**
EUPHORIA WULFENII LATEX

PROP: Acetone soluble portion of latex from *Euphoria wulfenii.*

TOXICITY DATA WITH REFERENCE
skn-mus TDLo:2600 mg/kg/26W-I:ETA CNREA8 21,338,61

SAFETY PROFILE: Questionable carcinogen with experimental tumorigenic data. When heated to decomposition it emits acrid smoke and irritating fumes.

EQY600 **CAS:32665-36-4** **HR: 3**
EUPNERON
mf: $C_{22}H_{30}N_2O_2$ mw: 354.54

SYNS: EPROZINOL □ 4-(2-METHOXY-2-PHENYLETHYL)-α-PHENYL-1-PIPERAZINEPROPANOL (9CI)

TOXICITY DATA WITH REFERENCE
orl-rat LD50:640 mg/kg OYYAA2 19,503,80
ipr-rat LD50:72 mg/kg OYYAA2 19,503,80
ims-rat LD50:140 mg/kg OYYAA2 19,503,80
orl-mus LD50:350 mg/kg OYYAA2 19,503,80
ipr-mus LD50:103 mg/kg OYYAA2 19,503,80
ims-mus LD50:122 mg/kg OYYAA2 19,503,80

SAFETY PROFILE: Poison by ingestion, intramuscular and intraperitoneal routes. When heated to decomposition it emits toxic fumes of NO_x.

EQZ000 **CAS:53198-87-1** **HR: 3**
EURAZYL
mf: $C_{19}H_{28}N_2 \cdot ClH$ mw: 320.95

PROP: Crystals from $EtOH/Et_2O$. Mp: 200–203°.

SYNS: α-sec-BUTYL-α-PHENYL-1-PIPERIDINEBUTYRONITRILE HYDROCHLORIDE □ α-PHENYL-α-(2-PIPERIDINOETHYL)-β-ETHYLBUTYRIC ACID NITRILE HYDROCHLORIDE □ α-PHENYL-α-(2-PIPERIDINOETHYL)-β-ETHYLBUTYRONITRILE HYDROCHLORIDE

TOXICITY DATA WITH REFERENCE
orl-mus LD50:700 mg/kg JAPMA8 49,298,60
ivn-dog LDLo:64 mg/kg JAPMA8 49,298,60

CONSENSUS REPORTS: Cyanide and its compounds are on the Community Right-To-Know List.

SAFETY PROFILE: Poison by intravenous route. Moderately toxic by ingestion. When heated to decomposition it emits very toxic fumes of HCl, NO_x, and CN^-. See also NITRILES.

ERA100 **HR: 3**
EUROPEAN MISTLETOE

PROP: A parasitic plant that grows mostly on the trunks of deciduous trees such as the apple. The thick leaves are usually a pale yellow-green. The white berry is sticky. It is native to Europe and now grows in Sonoma County, California.

SYN: VISCUM ALBUM

SAFETY PROFILE: The leaves and stems contain the poisonous viscumin and viscotoxins, toxalbumins that inhibit protein synthesis. The berries may be edible. Ingestion of the leaves causes after a few hours: abdominal pain, diarrhea and possibly necrotic lesions throughout the gastroenteric tract. See also VISCOTOXIN and ABRIN.

ERA309 **HR: 3**
EUROPEAN SPINDLE TREE

PROP: A small tree native to Europe and now cultivated in the northern United States. It has tooth-edged leaves, yellow-green flowers and a pink seed capsule.

SYN: EUONYMUS EUROPAEUS

SAFETY PROFILE: The plant contains the poison evomonoside, a digitalis-like cardiac glycoside, several alkaloids including evonine, and a protein that inhibits protein synthesis. Ingestion of the fruit has caused (within 12 hours) diarrhea, vomiting, fever, hallucinations, sleepiness, coma, and convulsions. Cardiac glycosides may cause death by their effect on heart function. See also DIGITALIS.

ERA500 **CAS:10025-76-0** **HR: 3**
EUROPIUM CHLORIDE
mf: Cl_3Eu mw: 258.31

PROP: Hygroscopic yellow needles. Mp: 623°.

SYN: EUROPIC CHLORIDE

TOXICITY DATA WITH REFERENCE
orl-mus LD50:3527 mg/kg EQSSDX 1,1,75
ipr-mus LD50:387 mg/kg AEHLAU 5,437,62
ipr-gpg LD50:156 mg/kg AEHLAU 5,437,62

CONSENSUS REPORTS: Reported in EPA TSCA Inventory.

SAFETY PROFILE: Poison by intraperitoneal route. Moderately toxic by ingestion. When heated to decomposition it emits very toxic fumes of Cl^-. See also RARE EARTHS.

ERB000 **CAS:13240-06-7** **HR: 3**
EUROPIUM CITRATE

TOXICITY DATA WITH REFERENCE
ipr-mus LD50:187 mg/kg AEHLAU 5,437,62
ipr-gpg LD50:72 mg/kg AEHLAU 5,437,62

SAFETY PROFILE: Poison by intraperitoneal route. When heated to decomposition it emits acrid smoke and irritating fumes. See also RARE EARTHS.

ERB500 **CAS:15158-64-2** **HR: 3**
EUROPIUM EDETATE

TOXICITY DATA WITH REFERENCE
ipr-mus LD50:240 mg/kg AEHLAU 5,437,62
ipr-gpg LD50:118 mg/kg AEHLAU 5,437,62

SAFETY PROFILE: Poison by intraperitoneal route. When heated to decomposition it emits acrid smoke and irritating fumes. See also RARE EARTHS.

ERC000 **CAS:10031-53-5** **HR: 3**
EUROPIUM(III) NITRATE, HEXAHYDRATE (1:3:6)
mf: $N_3O_9 \cdot Eu \cdot 6H_2O$ mw: 446.11

SYN: NITRIC ACID, EUROPIUM(3+) SALT, HEXAHYDRATE

TOXICITY DATA WITH REFERENCE
ipr-rat LD50:210 mg/kg TXAPA9 5,750,63
ipr-mus LD50:320 mg/kg TXAPA9 5,750,63

SAFETY PROFILE: Poison by intraperitoneal route. When heated to decomposition it emits very toxic fumes of NO_x. See also NITRATES and RARE EARTHS.

ERC500 **CAS:12020-65-4** **HR: 3**
EUROPIUM(II) SULFIDE
mf: EuS mw: 184.02

PROP: Black powder (golden hue by reflected light).

SAFETY PROFILE: Ignites or explodes spontaneously in air. When heated to decomposition it emits toxic fumes of SO_x. See also RARE EARTHS and SULFIDES.

ERC550 **CAS:10138-01-9** **HR: 3**
EUROPIUM TRINITRATE
mf: $Eu(NO_3)_3$ mw: 421.98

SYN: EUROPIUM NITRATE

TOXICITY DATA WITH REFERENCE
orl-rat LD50:3828 mg/kg EQSSDX 1,1,75
ipr-rat LD50:162 mg/kg EQSSDX 1,1,75
ipr-mus LD50:215 mg/kg EQSSDX 1,1,75

SAFETY PROFILE: Poison by intraperitoneal route. Moderately toxic by ingestion. When heated to decomposition it emits toxic fumes of NO_x. See also NITRATES and RARE EARTHS.

ERC600 **CAS:62851-59-6** **HR: 3**
EUROTIN (A)

SYN: YUROTIN A

TOXICITY DATA WITH REFERENCE
ipr-mus LD50:3 mg/kg 85FZAT -,281,67
scu-mus LD50:6 mg/kg 85FZAT -,281,67
ivn-mus LD50:3 mg/kg 85FZAT -,281,67

SAFETY PROFILE: Poison by subcutaneous, intravenous, and intraperitoneal routes.

ERC800 **CAS:39340-46-0** **HR: 3**
EVERNINOMICIN D
mf: $C_{66}H_{99}Cl_2NO_{35}$ mw: 1537.56

PROP: Amorphous solid.

TOXICITY DATA WITH REFERENCE
ims-rat LD50:400 mg/kg 85ERAY 1,208,78
orl-mus LD50:3750 mg/kg 85ERAY 1,208,78
ipr-mus LD50:3750 mg/kg 85ERAY 1,208,78
scu-mus LD50:3750 mg/kg 85ERAY 1,208,78
ivn-mus LD50:125 mg/kg 85ERAY 1,208,78
ims-dog LD50:140 mg/kg 85ERAY 1,208,78
ims-rbt LD50:300 mg/kg 85ERAY 1,208,78

SAFETY PROFILE: Poison by intramuscular and intravenous routes. Moderately toxic by ingestion, intraperitoneal, and subcutaneous routes. When heated to decomposition it emits very toxic fumes of Cl^- and NO_x.

ERD000 **CAS:53296-30-3** **HR: 2**
EVERNINOMYCIN-B
mf: $C_{66}H_{99}Cl_2NO_{36}$ mw: 1553.56

PROP: A solid. Mp: 184–185°.

SYN: R-451-B

TOXICITY DATA WITH REFERENCE
ipr-mus LD50:880 mg/kg 85GDA2 1,354,80
scu-mus LD50:1700 mg/kg 85GDA2 1,354,80
ivn-mus LD50:875 mg/kg 85GDA2 1,354,80

SAFETY PROFILE: Moderately toxic by intraperitoneal, subcutaneous and intravenous routes. When heated to decomposition it emits toxic fumes of $Cl^-$and NO_x.

ERD500 **CAS:56-29-1** **HR: 3**
EVIPAL
mf: $C_{12}H_{16}N_2O_3$ mw: 236.30

PROP: Prisms. Mp: 145–147°.

SYNS: BARBIDORM □ CITODON □ CITOPAN □ 5-(1-CYCLOHEXEN-1-YL)-1,5-DIMETHYLBARBITURIC ACID □ 5-(1-CYCLOHEXEN-1-YL)-1,5-DIMETHYL-2,4,6(1H,3H,5H)-PYRIMIDINETRIONE □ 5-(1-CYCLOHEXENYL-1)-1-METHYL-5-METHYLBARBITURIC ACID □ 5-(Δ-1,2-CYCLOHEXENYL)-5-METHYL-N-METHYL-BARBITURSAEURE (GERMAN) □ CYCLONAL □ CYCLOPAN □ 1,5-DIMETHYL-5-(1-CYCLOHEXENYL) BARBITURIC ACID □ DORICO □ ENHEXYMAL □ ESOBARBITALE (ITALIAN) □ EVIPAN □ HEXABARBITAL □ HEXANASTAB ORAL □ HEXENAL □ HEXENAL (barbiturate) □ HEXOBARBITAL □ HEXOBARBITONE □ METHEXENYL □ N-METHYL-5-CYCLOHEXENYL-5-METHYLBARBITURIC ACID □ METHYLHEXABARBITAL □ METHYLHEXABITAL □ NARCOSAN □ NOCTOVANE □ SOMBUCAPS □ SOMBULEX □ SOMNALERT

TOXICITY DATA WITH REFERENCE
cyt-hmn:leu 1500 mg/L TGANAK 18(1),13,84
ipr-rat LD50:330 mg/kg AIPTAK 184,5,70
scu-rat LDLo:400 mg/kg AEPPAE 182,348,36
orl-mus LD50:468 mg/kg JPETAB 106,444,52
ipr-mus LD50:270 mg/kg TXAPA9 5,790,63
scu-mus LD50:250 mg/kg ARZNAD 15,688,65
ivn-mus LD50:133 mg/kg AIPTAK 163,11,66
ipl-mus LDLo:340 mg/kg JPETAB 134,95,61
orl-rbt LDLo:1200 mg/kg JPETAB 60,189,37
ivn-rbt LDLo:80 mg/kg JPETAB 60,189,37
rec-rbt LDLo:175 mg/kg JPETAB 60,189,37
ipr-frg LDLo:30 mg/kg PHREA7 19,472,39

SAFETY PROFILE: Poison by subcutaneous, intraperitoneal, intravenous, intrapleural, and rectal routes. Moderately toxic by ingestion. Human mutation data reported. When heated to decomposition it emits toxic fumes of NO_x. See also BARBITURATES.

ERE000 **CAS:50-09-9** **HR: 3**
EVIPAL SODIUM
mf: $C_{12}H_{15}N_2O_3 \cdot Na$ mw: 258.28

PROP: Hygroscopic powder. Very sol in H_2O.

SYNS: trans-2-BUTENOIC ACID □ 5-(1-CYCLOHEXEN-1-YL)-1,5-DIMETHYLBARBITURIC ACID SODIUM SALT □ 5-(1-CYCLOHEXEN-1-YL)-1,5-DIMETHYL-2,4,6(1H,3H,5H,)-PYRIMIDINETRIONE MONOSODIUM SALT □ 5-(1-CYCLOHEXEN-1-YL)-1,5-DIMETHYL-2,4,6(1H,3H,5H)-PYRIMIDINETRIONE SODIUM SALT (9CI) □ CYCLONAL SODIUM □ 1,5-DIMETHYL-5-CYCLOHEXENYL-1'-BARBITURIC ACID, SODIUM SALT □ DORICO SOLUBLE □ ENHEXYMAL □ ENHEXYMAL NFN □ EVIPAN SODIUM □ HEXANAL □ HEXANASTAB □ HEXENAL SODIUM □ HEXOBARBITAL Na □ HEXOBARBITAL SODIUM □ HEXOBARBITONE Na □ HEXOBARBITONE SODIUM □ METHEXENYL SODIUM □ NARCOSAN SOLUBLE □ NOCTIVANE SODIUM □ PRIVENAL □ SODIUM-5-(1-CYCLOHEXEN-1-YL)-1,5-DIMETHYLBARBITURATE □ SODIUM EVIPAL □ SODIUM EVIPAN □ SODIUM HEXABARBITAL □ SODIUM HEXOBARBITONE □ SODIUM-N-METHYL CYCLOHEXENYLMETHYBARBITURATE □ SODIUM METHYLHEXABITAL □ SODIUM METHYLHEXABITOL

TOXICITY DATA WITH REFERENCE
orl-rat LD50:468 mg/kg MEIEDD 10,680,83
ipr-rat LDLo:160 mg/kg JPETAB 50,347,34
orl-mus LD50:1325 mg/kg TXAPA9 27,70,74
ipr-mus LD50:270 mg/kg JPETAB 87,265,46
scu-mus LD50:410 mg/kg TXAPA9 27,70,74
ivn-mus LD50:145 mg/kg TXAPA9 23,537,72
ivn-dog LDLo:85 mg/kg JPETAB 60,125,37
par-frg LDLo:30 mg/kg JPETAB 50,347,34

SAFETY PROFILE: Poison by intraperitoneal, parenteral, and intravenous routes. Moderately toxic by ingestion and subcutaneous routes. Used intravenously as a general anesthetic. When heated to decomposition it emits toxic fumes of NO_x and Na_2O. See also EVIPAL and BARBITURATES.

ERE100 CAS:59333-90-3 HR: 3
EXAPROLOL HYDROCHLORIDE
mf: $C_{18}H_{29}NO_2{\cdot}ClH$ mw: 327.94

PROP: Crystals from EtOH or $MeOH/Et_2O$. Mp: 178–179°.

SYNS: 1-(2-(1-CYCLOHEXEN-1-YL)PHENOXY)-3-((1-METHYLETHYL) AMINO)-2-PROPANOL HYDROCHLORIDE □ 1-ISOPROPYLAMINO-3-(2-CYCLOHEXYLPHENOXY)-2-PROPANOL HYDROCHLORIDE □ M.G. 8823

TOXICITY DATA WITH REFERENCE
orl-rat LD50:1850 mg/kg ARZNAD 26,506,76
ipr-rat LD50:54 mg/kg ARZNAD 26,506,76
orl-mus LD50:860 mg/kg ARZNAD 26,506,76
ipr-mus LD50:78 mg/kg ARZNAD 26,506,76

SAFETY PROFILE: Poison by intraperitoneal route. Moderately toxic by ingestion. When heated to decomposition it emits toxic fumes of NO_x and HCl.

ERE200 CAS:3478-44-2 HR: 2
EXIPROBEN SODIUM
mf: $C_{16}H_{23}O_5{\cdot}Na$ mw: 318.38

PROP: A solid. Mp: 147°.

SYNS: DCH 21 □ DROCTIL □ 3-HEXOXY-1-(2'-CARBOXYPHENOXY)-PROPANOL-(2) SODIUM SALT □ o-(3-(HEXYLOXY)-2-HYDROXYPROPOXY)BENZOIC ACID MONOSODIUM SALT

TOXICITY DATA WITH REFERENCE
orl-rat LD50:2150 mg/kg ARZNAD 24,111,74
ipr-rat LD50:645 mg/kg ARZNAD 24,111,74
orl-mus LD50:1660 mg/kg ARZNAD 24,111,74
ipr-mus LD50:650 mg/kg ARZNAD 24,111,74

SAFETY PROFILE: Moderately toxic by ingestion and intraperitoneal routes. When heated to decomposition it emits toxic fumes of Na_2O.

ERF000 HR: 3
EXPLOSIVES, HIGH

SAFETY PROFILE: High explosives (HE) are those that decompose by detonation. This is a very rapid (nearly instantaneous), and hence violent, process. An explosion may be initiated by sudden shock, high temperatures, or a combination of the two. The conditions under which many explosives will explode are well known.

An explosion may be initiated by elevated temperature alone, as in the following cases.

(1) Mercury fulminate by 15-second exposure to 200°C or 1-second exposure to 340°C will be set off.

(2) Trinitrotoluene will be set off by exposure to 500°C for 1 second.

(3) Tetryl will detonate in 1000 seconds at 160°C or in 0.1 second at 500°C.

(4) Picric acid will detonate in 9 seconds at 300°C or 1 second at 355°C.

An explosion of HE may also be initiated by severe shock. Sensitivity of explosives to shock may be measured in several ways, such as the impact pendulum method and the drop test. In the impact pendulum test, a heavy pendulum swings down over a sample of explosive in a dished, inclined container so arranged that there is very little clearance between the pendulum and the sample. Thus, the effect of contact between the sample and the pendulum bob is one of a combination of shock and rubbing. The height from which the pendulum is allowed to swing to explode the sample is a measure of the sensitivity of the sample to this test. The drop test consists of placing a sample upon an anvil and allowing a 5-pound weight to drop on it. The height from which the weight must drop to explode the sample is a measure of the sample's sensitivity to shock.

The table below shows the results of a drop test upon several samples. These results must be considered as relative and not by any means absolute. A solid explosive in a tightly fitting container is much more sensitive to shock.

(1) mercury fulminate = 2 in. at 5 lbs.
(2) nitroglycerin = 4 in. at 5 lbs.
(3) tetryl = 8 in. at 5 lbs.
(4) picric acid = 14 in. at 5 lbs.
(5) trinitrotoluene = 20 in. at 5 lbs.
(6) black powder (a low explosive) = 30 in. at 5 lbs.*

*From *Explosions, Their Anatomy and Destructiveness*, by C. S. Robinson (McGraw-Hill).

Another test for explosives is the speed at which a detonation travels. This speed is usually in the range of thousands of m/sec. Speed of detonation is found to be a function of the kind of explosive and state of compaction. There is an optimum state of compaction beyond which the explosive tends to become "deadpressed," in which state it is difficult to make the whole sample explode. Below the point of optimum compaction the rate of detonation is found to be directly proportional to the density of the sample. Some maximum detonation rates are listed below in m/sec for some common explosives:

(1) nitroglycerin, 8500
(2) PETN, 8100
(3) tetryl, 7700
(4) picric acid, 7400
(5) trinitrotoluene, 7400
(6) lead azide, 4900
(7) mercury fulminate, 4800
(8) ammonlium nitrate, 1100
(9) low explosives, 1000

It has been found that upon detonation, an explosive can cause a nearby sample of explosive to detonate "sympathetically." The distance over which one charge can detonate another is a function of the amount of energy produced by the first explosion and the medium through which the shock wave is propagated to the second charge of explosive. For instance, the relationship for air (very approximately) would be expected to be: weight of explosive in lbs/(distance in ft)^3 = 4. Thus, to calculate the maximum distance for a possible sympathetic detonation of 40,000 lbs of explosive, the calculation is:

$D^3 = (40{,}000)/4$
$D^3 = 10{,}000$
$D = 22$ ft (approximately).

According to C. S. Robinson, the formula is more nearly:

weight of explosive $= 4 \times (\text{distance})^{2.25}$

The power of the shock wave is much more rapidly attenuated in water, wood, etc., than in air, which means that if a shield of water or wood is interposed between piles of explosive the distance between them may be lessened.

Liquid Oxygen: Though not itself explosive, liquid oxygen can be dangerous when blended with highly flammable or carbonaceous materials. In this combination it is used in coal mining, quarrying, strip mining, open-cut ore mining, and in rocket fuels. Its use underground or in confined places is not recommended by the U.S. Bureau of Mines because it evolves a great deal of carbon monoxide. This type of explosive has many safety advantages. For instance, it is not itself an explosive until mixed with a flammable absorbent a process that can be done at the last moment before firing. However, once the explosive has been made up, it is very flammable and when it catches fire it will usually detonate. Liquid oxygen explosives are not stored, as they deteriorate rapidly and lose a great deal of their explosive power in a short time.

A very dangerous fire hazard when exposed to heat or by chemical reaction with powerful oxidizing or reducing agents.

A moderate to dangerous explosion hazard when severely shocked or heated, depending upon the kind of explosive, state of compaction, degree of confinement, etc. Practically all high explosives used commercially require a detonator or cap to set them off, as compared to an igniter needed to set off black blasting powder.

Detonating Devices: To develop the desired disruptive effect of an explosive, some means must be adopted to "set off," "fire," or "detonate" it without killing or maiming the persons doing the blasting. Several devices or methods are being utilized, all with a view to having this work done as safely and efficiently as possible. There are two general types of devices or methods of getting explosives into action, namely, igniters and detonators. The former merely conveys a flame to the explosive mass and ignites it, while the latter transmits (originally through ignition of a small quantity of highly explosive substance by an arc, a flame, or a spark) a sharp blow that causes the explosive to disassociate, or detonate, or burn with very great rapidity. Igniters include squibs (plain and electric), fuse, and delay igniters; detonators include blasting caps (plain and electric), delay electric blasting caps, delay electric igniters with caps, and Cordeau-Bickford detonating fuse.

The squib is a small-diameter tube of straw or paper filled with quick-burning powder and having a relatively slow-burning "match head" attached to one end; the latter is ignited or lighted by an ordinary match or other flame, and its relatively slow burning allows the person handling the ignition to retire before the fire is communicated to the quick burning material in the tube. Squibs are by no means either safe or efficient, even though still used to a considerable extent, especially in coal mining. Electric squibs are somewhat similar to ordinary squibs, except that the ignition is accomplished by means of an electric arc; electric squibs are much more satisfactory from a safety viewpoint than ordinary squibs.

A fuse (or, as it is sometimes called, "safety fuse") consists of a fine-grained black powder core covered with cotton hemp or jute to form a ropelike material about 3/16 inch in diameter, one end of the fuse is brought in contact with the powder charge or with a detonating "cap," and the other end (usually several feet away from the explosive) is lighted by a flame from a match or open light. The fine-grained black powder burns gradually and somewhat slowly (about 30 to 40 seconds to the linear foot of fuse) until it reaches the explosive (black powder) or the detonating cap (if some form of dynamite is used), giving the blaster time to get in the clear before the main explosion takes place. Fuses are much safer than squibs, but have their own hazards and must be used with care.

Delay electric igniters usually are a combination of electric igniters and fuses, the latter being ignited by the igniters within the blasting hole, the fuse transmitting the ignition to the explosive. Delay igniters usually are much safer than fuses, particularly for coal-mine use; but they, too, have their hazards. Delay blasting is by no means a safe procedure in coal mining, though it is a standard and relatively safe practice in metal mining and tunneling, if sensible precautions are taken.

Blasting caps or detonators are metallic cylinders (usually copper) closed at one end, about 3/16 inch in diameter, and usually less than 2 inches in length, partly filled with a small amount of relatively easily fired or "detonated" compound, the resultant shock or blow when fired being sufficient, when embedded in dynamite, to fire or detonate the dynamite mass. Ordinary blasting caps usually are fired or detonated by the flame of the fuse, the end of the latter being inserted into the open end of the detonator or cap and placed in contact with the highly explosive material in the interior of the metallic capsule or cap. Caps are extremely hazardous to handle, as they are likely to be detonated by heat, friction, or a relatively moderate blow; however, they are relatively safe if handled carefully. Partial proof of this is the fact that they are manufactured and shipped by the thousands daily and accidents are decidedly rare, primarily because the caps are at all times handled with utmost care.

Electric blasting caps are somewhat similar to ordinary caps or detonators, but the cap is fired by electricity. The electric wires are so placed in the capsule or cap that when attached to an electric current an arc is formed within the cap, which detonates the sensitive explosive material in the cap. A hazard in the use of electric blasting caps is unexpected explosions due to radio or radar-induced electric currents that may activate the cap.

Delay electric blasting caps or detonators are somewhat similar to ordinary electric blasting caps, except that several time intervals in blasting are obtained by having the electric arc ignite a short piece of fuse or some slow-burning substance before it reaches the highly sensitive detonating material in the capsule or cap. Numerous time-interval delays are obtained; in general, delay electric blasting caps are relatively safe and effective even in wet holes, though they ought not be used in coal mining if explosions of gas or dust are to be avoided. Delay electric igniters with caps or detonators are a combination of electric igniter and blasting cap, usually with suitable lengths of fuse between to give the desired delay; they have some advantages but are relatively unsafe and should not be used in coal mining.

The Cordeau-Bickford denotating fuse is a combined fuse and detonator in the form of a lead tube about 1/4 inch in diameter filled throughout its length with a high explosive, trinitrotoluene (TNT). It is fired by a fuse and an ordinary detonator or cap or by an electric cap; when fired, it detonates throughout its length (which may be up to or over 100 feet) almost instantaneously, the explosion wave traveling at a rate of about 17,500 ft/sec. Although somewhat expensive, it is relatively safe to handle and is particularly effective in deep-well drill holes in quarry and similar work, as it detonates simultaneously throughout its length, adding effectiveness to the main body of explosive that it detonates. It fires black powder as well as high explosives (dynamite, etc.), and is obtainable in lengths of approximately 500 feet wound on spools.

See also EXPLOSIVES, PERMITTED; DYNAMITE, NITROGLYCERIN, AMMONIUM NITRATE, and NITRATES.

ERF500 **HR: 3**

EXPLOSIVES, LOW

DOT: UN 0027/UN 0028

PROP: Black powder is composed of saltpeter, charcoal and sulfur in the approximate proportions of 6:1:1. ("A" blasting powder uses KNO_3 and "B" blasting powder uses $NaNO_3$.)

SYNS: "A" BLASTING POWDER □ "B" BLASTING POWDER □ BLACK POWDER, compressed (DOT) □ BLACK POWDER, granular or as a meal (UN 0027) (DOT) □ BLACK POWDER, in pellets (UN 0028) (DOT) □ BLASTING POWDER □ GUNPOWDER □ GUNPOWDER, compressed (UN 0028) (DOT) □ GUNPOWDER, granular or as a meal (UN 0027) (DOT) □ GUNPOWDER, in pellets (UN 0028) (DOT) □ RIFLE POWDER □ BLACK BLASTING POWDER

DOT CLASSIFICATION: EXPLOSIVE 1.1D; *Label:* EXPLOSIVE 1.1D

SAFETY PROFILE: Low explosives are explosives that deflagrate; this differentiates them both in composition and properties from high explosives, which detonate. A deflagrating explosive is one that burns progressively over a relatively sustained period of time, in contrast with a detonating explosive, which decomposes almost instantaneously. A dangerous fire hazard when exposed to heat or flame or by chemical reaction.

Black powder is the most treacherous explosive material used today and it is regarded as one of the worst explosive hazards known. When ignited unconfined it burns with explosive violence and will explode if ignited under even slight confinement. It can be ignited easily by very small sparks, heat, and friction. It is the slowest acting of all explosives. It has a shearing and heaving action tending to blast materials into large, firm fragments. The action derives from a relatively slow development of gas pressure so that it must be carefully loaded and closely confined. It is subject to rapid deterioration in the presence of moisture, but if kept dry it retains its explosive properties for many years. It is used to ignite smokeless powder, propelling charges, airplane flares, and bursting charges of hand grenades, as a bursting charge in shrapnel, practice bombs, practice trench-mortar shells, in saluting charges, smoke-puff charges, time and percussion fuses, pellets, primers and primer detonators, and in expelling charges of pyrotechnic signals.

Although most safety experts now look upon black blasting powder with disfavor, it is one of the oldest and most generally used explosives in commercial work. It burns with extreme rapidity instead of detonating as high explosives do. It is highly sensitive to flame, sparks, or friction and gives off much flame, which is hot and of great length of duration. These properties make it extremely hazardous for use in mines (especially coal mines) and quarries. The gases given off in detonation are not only very hot but frequently contain harmful constituents. Notwithstanding its numerous deficiencies, from a safety standpoint it has action characteristics that make it valuable in both coal mining and quarrying, though it has relatively little utility in metal mining. It is difficult to use effectively in wet places and this is its main disadvantage from an efficiency standpoint.

Most black powder fires start from sparks. Ignition results in an explosion so quickly that no attempt can be made to fight the fire. Every effort should be made to prevent fires from reaching stores of black powder; but if this fails, fire-fighting forces should be withdrawn to a distance of at least 800 feet from the fire and should protect themselves against an explosion by seeking any cover available or by lying flat on the ground. If an explosion does occur, every effort should be made to prevent flames from spreading to neighboring magazines. Fire-fighting forces should be cautioned against approaching a fire that may involve black powder to avoid being trapped or injured by an explosion.

The following safety rules should be strictly enforced and obeyed. Open no containers in a magazine in which explosives or ammunitions are stored. This should be done only in a building free from all other explosives or ammunitions, or in suitable weather in the open, at least 100 feet from the nearest magazine. The quantity at or near such an operation should be limited to 100 pounds. Only safety tools should be used in opening or closing containers or in other operations involving black powder. Processes should be so laid out as to bring about frequent grounding of all operators handling this material. Safety shoes (non-insulating) should be worn in all rooms where black powder is handled and by all persons engaged in handling black powder. The wearing of all nonconductive shoes, such as rubber, is prohibited. If black powder is handled on over a concrete floor, the floor should be covered with a tarpaulin or other suitable material. Loose black powder is extremely dangerous. Whenever it is necessary to handle loose black powder, not over 50 pounds should be permitted at or near such operations. If black powder is spilled on benches or floors, all work should be stopped until it has been removed and the explosive hazard of any remaining dust or particles has been neutralized with water. Rooms or buildings in which black powder is handled should be inspected frequently for dust, and all such dust should be immediately removed with water. The empty powder containers should be washed out, as explosions are said to have occurred from "empty" containers.

If dry and in good condition, black powder burns rapidly, especially in small grain size, with a yellow or pinkish-blue flame and dense smoke.

Pellet powder is black blasting powder in consolidated (pellet or stick) form rather than in grains or granules, and it has few if any real advantages over black blasting powder, notwithstanding the fairly prevalent idea that it is a "safe" explosive.

Smokeless powders have a composition somewhat different from that of black blasting powder and are used chiefly for sportsmen's ammunition and, more widely, for military purposes. They are decidedly sensitive to flame and impact but ordinarily are so packaged that if reasonable judgment is used they are relatively harmless.

ERG000 HR: 3
EXPLOSIVES, PERMITTED

SAFETY PROFILE: "Permissible" explosives are essentially high explosives (dynamite) modified by the introduction of "dopes." The function of the dopes in general is to decrease flame temperature, and to a smaller extent, the length and duration of flame, when the explosive is converted from a solid into a gas, i.e., when it is fired or detonated. The designation "permissible" is given to an explosive of modified dynamite type after it has passed certain tests designated by the Federal Bureau of Mines. The permissible character of such explosives depends not only upon the ingredients in the explosive, but also on certain well-defined specifications as to handling and use. As with the dynamites, there are several different types and grades: "permissibles," hydrated "permissibles," organic nitrate "permissibles," nitroglycerin "permissibles," ammonium perchlorate "permissibles," and gelatin "permissibles." Essentially all of those now used to any extent are in either the ammonium nitrate or the gelatin classes. See also DYNAMITE.

The ammonium nitrate "permissible" explosives contain relatively little nitroglycerin and relatively large proportions of ammonium nitrate. The latter is an explosive but one less sensitive to impact, sparks, and flames than nitroglycerin. This type of permissible explosive is now used extensively, as it has a rather wide range of strength, rate of detonation, density, size of cartridge, etc., and can be utilized not only in dry but also to some extent in fairly wet holes if charged carefully and fired promptly.

Gelatin "permissible," explosives are more suitable than ammonium nitrate "permissible" ones for wet holes, and in general are stronger and more violent than the ammonium nitrate types.

All "permissible" explosives are strong, and must be used in relatively small quantities (less than 1.5 pounds) per hole to retain their permissibility. They give off considerable quantities of toxic gases on detonation, and, while much safer than black blasting powder or dynamite, must be stored, handled, and used with care.

Classification upon Basis of Toxic Gases: All "permissible" explosives, when detonated, emit some toxic gases and a much larger volume of nontoxic gases. In order that the toxic products may not become a menace to the life or health of miners, no explosive is now or can become "permissible" if upon detonation it evolves more than 158 liters (5.5 cu ft) of toxic gases per 1.5 pound charge as determined by tests in the Bichel pressure gauge. Classification upon the basis of the volume of toxic gases produced by 580 g (1.5-pound) of explosive is as follows: *Class A,* not more than 53 liters; *Class B,* between 53 and 106 liters; and *Class C,* between 106 and 158 liters. (These classifications are not to be confused with the I.C.C. Classification of explosives).

Field tests were made with a 1.5 pounds charge of a "permissible" explosive that produced, in the Bichel gauge, the maximum allowable quantity of poisonous gases (158 liters per 1.5 pounds); these tests indicated that in a narrow entry, without artificial ventilation, 1800 ppm of carbon monoxide (the only poisonous gas present) was produced, as shown by analysis of an air sample taken 2 minutes after the shot. Another sample of the air taken 2 minutes later contained 800 ppm of carbon monoxide. Under no conditions should miners or shot firers return to the place until the poisonous gases have been removed by adequate ventilation.

It is provided further that, in accordance with the provisions and conditions, explosives enumerated on the "permissible" lists of the Bureau of Mines are "permissible" in use only when they satisfy the following requirements:

1. The explosive must be in all respects similar to the sample submitted by the manufacturer for test, and that the diameters of the cartridges used must be those that have been approved.

2. Electric detonators (not fuse and detonators) must be used of not less efficiency than No. 6, the detonation charge of which shall consist of a 1 g mixture of 80 parts of mercury fulminate and 20 parts of potassium chlorate

(or their equivalents), and the required electric firing must be done by means of a "permissible" type blasting unit.

3. The explosive must be stored in surface magazines under proper conditions, so that it will not undergo change in character, and after being taken underground it must be used in less than 36 hours.

4. The coal to be blasted must be undercut or equivalently relieved; to prevent blow-through, all portions of the borehole must be at least 18 inches from relief in any direction; to prevent blowouts, the charge must be properly confined with not less than 2 feet of clay (if the length of the hole will not permit the charge desired and 2 feet of stemming, at least half the length of the hole shall be filled with stemming) or other incombustible stemming and not be on the solid; that, to prevent the hole being on the solid; it shall be at least 6 inches shorter than the depth of the undercut or equivalent relief, and, when placed adjacent to the roofs, ribs, or floor, all but 12 inches at the rear of the hole must be at least 6 inches from the adjacent surface as projected into the coal to be blasted, and all parts of the hole shall be free from the adjacent surface as projected into the coal to be blasted; the shot must be not be a dependent shot; and the shot hole must be cleaned before charging.

5. The quantity used for a shot (1) must not be in excess of 680 g (1.5 pounds) when fired in accordance with these requirements and (2) when used under certain additional requirements or restrictions must not be in excess of 1,361 g (3 pounds). For charges of over 1.5 pounds, the following additional requirements must be observed: (a) shot holes must be 6 feet or more in length; (b) explosives must be charged in a continuous train, with no cartridges deliberately deformed or crushed, with all cartridges in contact with each other, and with the end cartridges touching the rear of the hole and the stemming, respectively; (c) examination for gas must be made in the blasting area before and after a shot is fired; (d) the "permissible" explosive must be one showing toxic gas emission that will place it either in Class A or Class B.

6. The region in which the blasting is done must be kept well protected by rock dust or otherwise be in accordance with Bureau of Mines inspection standards.

7. The shot must not be fired in the presence of a dangerous percentage of firedamp. Examination for firedamp is to be made at the blasting area before shooting in a gassy mine.

See also AMMONIUM NITRATE, AZIDES, DYNAMITE, FULMINATES. NITRATES, NITROGLYCERIN, PENTAERYTHRITOL TETRANITRATE, and PICRIC ACID.

ERG100 CAS:3486-30-4 HR: 2
EXT D and C BLUE No. 3
mf: $C_{37}H_{36}N_2O_6S_2 \cdot Na$ mw: 691.86

SYNS: ACID BLUE O □ ACID SKY BLUE O □ ACID TURQUOISE BLUE A □ AIZEN BRILLIANT ACID BLUE AFH □ ALPHAZURINE A (6CI) □ BENZENEMETHANAMINIUM, N-(4-((2,4-DISULFOPHENYL)(4-(ETHYL(PHENYLMETHYL)AMINO)PHENYL)METHYLENE)-2, 5-CYCLOHEXADIEN-1-YLIDENE)-N-ETHYL-, HYDROXIDE, inner salt, SODIUM SALT □ BLEKIT TURKUSOWY A □ BRILLIANT ACID BLUE AS □ BRILLIANT ACID BLUE N EXTRA □ BUCACID PATENT BLUE AF □ CALCOCID BLUE AX □ CARMINE BLUE AF □ C.I. 42080 □ C.I. ACID BLUE 7 (7CI) □ C.I. ACID BLUE 7, SODIUM SALT (8CI) □ D and C BLUE No. 3 □ DISULPHINE BLUE AN □ DISULPHINE LAKE BLUE AN □ ERIO GLAUCINE X □ ERIO GLAUCINE XS □ FENAZO BLUEXG □ HIDACID BLUE A □ HIDACID BLUE AF □ KAYACYL PURE BLUE FGA □ KITON BLUE A □ LAKE BLUE AFX □ MERANTINE BLUE AF □ PATENT BLUE A □ PATENT BLUE AF □ PONTACYL BRILLIANT BLUE A □ SANDOLAN TURQUOISE E-AS □ SHIMAZAKI PATENT BLUE AFX □ TERTRACID CARMINE BLUE A □ VULCOL BLUE BZ □ XYLENE BLUE AS

TOXICITY DATA WITH REFERENCE
mma-sat 100 μg/plate MUREAV 147,285,85
mnt-mus-ipr 38 mg/kg BCTKAG 18,280,85
dlt-mus-ipr 220 mg/kg MUREAV 147,285,85
ipr-mus LD50:437 mg/kg BCTKAG 18,280,85

CONSENSUS REPORTS: Reported in EPA TSCA Inventory.

SAFETY PROFILE: Moderately toxic by intraperitoneal route. Mutation data reported. When heated to decomposition it emits toxic vapors of NO_x and SO_x.

F

FAB000 **CAS:88-27-7** **HR: 2**
F 1 (antioxidant)
mf: $C_{17}H_{29}NO$ mw: 263.47

SYNS: AGIDOL 3 □ p-CRESOL, 2,6-DI-tert-BUTYL-α-(DIMETHYLAMINO)- □ 2,6-DI-tert-BUTYL-α-(DIMETHYLAMINO)-p-CRESOL □ ETHYL 703 □ ETHYL ANTIOXIDANT 703 □ F 1 □ OMI □ PHENOL, 4-((DIMETHYLAMINO)METHYL)-2,6-BIS(1,1-DIMETHYLETHYL)-(9CI)

TOXICITY DATA WITH REFERENCE
orl-rat LD50:1030 mg/kg IPSTB3 3,93,76

CONSENSUS REPORTS: Reported in EPA TSCA Inventory.

SAFETY PROFILE: Moderately toxic by ingestion. When heated to decomposition it emits toxic vapors of NO_x.

FAB010 **HR: 3**
F III (sugar fraction)

PROP: Sugar fraction of the extract obtained from *Sasa albamarginata* with a molecular weight between 5000 and 10,000 (YKKZAJ 99,663,79).

TOXICITY DATA WITH REFERENCE
orl-mus LD50:10 g/kg YKKZAJ 99,663,77
ipr-mus LD50:2750 mg/kg YKKZAJ 99,663,77
scu-mus TD50:6400 mg/kg YKKZAJ 99,663,77
ivn-mus LD50:360 mg/kg YKKZAJ 99,663,77

SAFETY PROFILE: Poison by intravenous route. Moderately toxic by intraperitoneal route. Mildly toxic by ingestion and subcutaneous routes. When heated to decomposition it emits acrid smoke and irritating fumes.

FAB100 **HR: 3**
FALSE HELLEBORE

PROP: Top (plant) material from *Veratrum californicum (Durand)* (CJBIAE 44,829,66). Tall, perennial herbs with pleated leaves and white flowers with green marks. The winged seeds are held in a small pod. The various species grow wild in the region bounded by the Aleutian Islands, most of Canada, New Mexico, and Georgia.

SYNS: AMERICAN WHITE HELLEBORE □ CORN LILY □ EARTH GALL □ GREEN HELLEBORE □ INDIAN POKE □ ITCH WEED □ PEPPER-ROOT □ RATTLESNAKE WEED □ SKUNK CABBAGE □ SWAMP HELLEBORE □ TICKLE WEED □ VERATRUM ALBUM □ VERATRUM CALIFORNICUM □ VERATRUM VIRIDE □ WHITE HELLEBORE

TOXICITY DATA WITH REFERENCE
orl-dom TDLo:236 g/kg(1-20D preg):TER AJVRAH 24,1164,63
orl-mus LD50:900 mg/kg AFECAT 76(815),65,83
orl-ham LDLo:15 mg/kg AMZOAF 9,1134,69

SAFETY PROFILE: Poison by ingestion. Experimental teratogenic effects. All parts of the plant contain poisonous veratrum alkaloids. Ingestion of any part of the plant causes an immediate pain in the abdomen followed by nausea and vomiting, blurred vision, confusion, slowed heartbeat, and low blood pressure. The subject may feel as if he is having a heart attack. Effects usually last only 24 hours.

FAB400 **CAS:13171-21-6** **HR: 2**
FAMFOS
mf: $C_{10}H_{19}ClNO_5P$ mw: 299.72

PROP: Pale yellow liquid. D: 1.21 @ 25°/4°, bp: 162° @ 1.5 mm. Sol in H_2O; most org solvs.

SYNS: APAMIDON □ C 570 □ (2-CHLOOR-3-DIETHYLAMINO-1-METHYL-3-OXO-PROP-1-EN-YL)-DIMETHYL-FOSFAAT □ (2-CHLOR-3-DIAETHYLAMINO-1-METHYL-3-OXO-PROP-1-EN-YL)-DIMETHYL-PHOSPHAT □ 2-CHLORO-2-DIETHYLCARBAMOYL-1-METHYLVINYL DIMETHYLPHOSPHATE □ 1-CHLORO-DIETHYLCARBAMOYL-1-PROPEN-2-YL DIMETHYL PHOSPHATE □ (2-CLORO-3-DIETILAMINO-1-METIL-3-OXO-PROP-1-EN-IL)-DIMETIL-FOSFATO □ CIBA 570 □ CROTONAMIDE, 2-CHLORO-N,N-DIETHYL-3-HYDROXY-, DIMETHYL PHOSPHATE □ DIMECRON □ DIMECRON 100 □ DIMETHYL 2-CHLORO-2-DIETHYLCARBAMOYL-1-METHYLVINYL PHOSPHATE □ O,O-DIMETHYL O-(2-CHLORO-2-(N,N-DIETHYLCARBAMOYL)-1-METHYLVINYL) PHOSPHATE □ DIMETHYL DIETHYLAMIDO-1-CHLOROCROTONYL (2) PHOSPHATE □ O,O-DIMETHYL-O-(1-METHYL-2-CHLOR-2-N,N-DIAETHYL-CARBAMOYL)-VINYL-PHOSPHAT □ (O,O-DIMETHYL-O-(1-METHYL-2-CHLORO-2-DIETHYLCARBAMOYL-VINYL) PHOSPHATE) □ DIMETHYL PHOSPHATE of 2-CHLORO-N,N-DIETHYL-3-HYDROXYCROTONAMIDE □ DIXON □ ENT 25,515 □ FOSFAMIDON □ FOSFAMIDONE □ FOSZFAMIDON □ ML 97 □ NCI-C00588 □ OMS 1325 □ OR 1191 □ PHOSPHAMIDON □ PHOSPHATE de DIMETHYLE et de (2-CHLORO-2-DIETHYLCARBAMOYL-1-METHYL-VINYLE)

TOXICITY DATA WITH REFERENCE
mmo-sat 5 μg/plate IJEBA6 24,305,86
mma-sat 9 mg/plate ENMUDM 5(Suppl 1),3,83
cyt-hmn:lyms 1900 μg/L MUREAV 31,103,75
cyt-hmn:leu 5 μmol/L IJEBA6 18,1145,80
bfa-rat:sat 5800 μg/kg IJEBA6 24,305,86
mnt-mus-orl 10 mg/kg/24H BECTA6 25,277,80
orl-mus TDLo:5 mg/kg (7D post):TER TOLED5 42,101,88
orl-rat TDLo:5400 mg/kg/80W-C:ETA NCITR* NCI-TR-16,79
orl-rat LD50:8 mg/kg 85JCAE-,1138,86
ihl-rat LC50:135 mg/m³/4H EGESAQ 24,173,80
skn-rat LD50:125 mg/kg PHJOAV 185,361,60
ipr-rat LD50:8700 μg/kg PESTD5 17,351,76
scu-rat LD50:15 mg/kg IPPABX 4,253,68
orl-mus LD50:6 mg/kg SPEADM 78-1,28,78
ihl-mus LC50:30 mg/m³/1H PSDTAP 15,239,74
ipr-mus LD50:5800 μg/kg TXAPA9 13,37,68
scu-mus LD50:13,200 μg/kg AIPTAK 132,180,61
ivn-mus LD50:6 mg/kg ATXKA8 18,316,60

orl-rbt LD50:70 mg/kg 85GYAZ-,33,71
skn-rbt LD50:80 mg/kg 85JCAE-,1138,86
ihl-gpg LC50:1300 mg/m³/4H PSDTAP 15,239,74
orl-dck LD50:3100 µg/kg DOEAAH 35,25,79
skn-dck LD50:26 mg/kg TXAPA9 47,451,79
orl-brd LD50:1800 µg/kg TXAPA9 21,315,72

CONSENSUS REPORTS: NTP Carcinogenesis Bioassay (feed): No Evidence: mouse NCITR* NCI-TR-16,79; Inadequate Studies: rat NCITR* NCI-TR-16,79

SAFETY PROFILE: Poison by ingestion and most other routes. An experimental teratogen. Experimental reproductive data. Questionable carcinogen with experimental tumorigenic data. Mutation data reported. When heated to decomposition it emits toxic fumes of NO_x and PO_x.

FAB500 CAS:76824-35-6 HR: 2
FAMOTIDINE
mf: $C_8H_{15}N_7O_2S_3$ mw: 337.48

PROP: Mp: 163–164°. Solubility at 20° (%, w/v): 80 in DMF; 50 in acetic acid; 0.3 in methanol; 0.1 in water; <0.01 in ethanol, ethyl acetate, chloroform.

SYNS: 3-(((2-((AMINOIMINOMETHYL)AMINO)-4-THIAZOLYL) METHYL)THIO)-N-(AMINOSULFONYL)PROPANIMIDAMIDE ☐ 3-(((2-((DIAMINOMETHYLENE)AMINO)-4-THIAZOLYL)METHYL)THIO)-N[2]-SULFAMOYLPROPIONAMIDINE ☐ FAMODIL ☐ GASTER ☐ GASTRIDIN ☐ MOTIAX ☐ PEPCID ☐ PEPCIDINE ☐ PEPDUL ☐ YM-11170

TOXICITY DATA WITH REFERENCE
orl-rat TDLo:2800 mg/kg (15-22D preg/21D post):REP OYYAA2 26,543,83
orl-rat TDLo:22 g/kg (female 7-17D post):REP OYYAA2 26,489,83
orl-rat TDLo:248 g/kg (male 12W pre):REP OYYAA2 26,551,83
orl-rat TDLo:1100 mg/kg (multi) :TER OYYAA2 26,489,83
orl-rat TDLo:5500 mg/kg (female 7-17D post):TER OYYAA2 26,489,83
ivn-rbt TDLo:130 mg/kg (female 6-18D post):REP OYYAA2 26,573,83
orl-rbt TDLo:6500 mg/kg (female 6-18D post):TER OYYAA2 26,565,83
orl-rat LD50:4907 mg/kg IYKEDH 16,590,85
ipr-rat LD50:800 mg/kg DIGEBW 32(Suppl 1),7,85
scu-rat LD50:800 mg/kg DIGEBW 32(Suppl 1),7,85
ivn-rat LD50:440 mg/kg YKYUA6 36,653,85
orl-mus LD50:4686 mg/kg IYKEDH 16,590,85
ipr-mus LD50:778 mg/kg OYYAA2 26,147,83
ivn-mus LD50:410 mg/kg YKYUA6 36,653,85

SAFETY PROFILE: Moderately toxic by intraperitoneal, subcutaneous, and intravenous routes. An experimental teratogen. Experimental reproductive effects. When heated to decomposition it emits toxic fumes of NO_x and SO_x.

FAB600 CAS:52-85-7 HR: 3
FAMPHUR
mf: $C_{10}H_{16}NO_5PS_2$ mw: 325.36

PROP: Crystalline powder from toluene/cyclohexane. Mp: 52.5–53.5°. Very sol in chloroform and carbon tetrachloride; sltly sol in water.

SYNS: AC 38023 ☐ AMERICAN CYANAMID-38023 ☐ AMERICAN CYANAMID CL-38,023 ☐ BO-ANA ☐ CL-38023 ☐ CYFLEE ☐ O-(4-((DIMETHYLAMINO)SULFONYL)PHENYL) O,O-DIMETHYL PHOSPHOROTHIOATE ☐ O,O-DIMETHYL-O-(p-(N,N-DIMETHYLSULFAMOYL) PHENYL)PHOSPHOROTHIOATE ☐ DOVIP ☐ ENT 25,644 ☐ FAMFOS ☐ FAMOPHOS ☐ FAMOPHOS WARBEX ☐ FAMPHOS ☐ FANFOS ☐ RCRA WASTE NUMBER P097 ☐ WARBEX ☐ WARBEXOL

TOXICITY DATA WITH REFERENCE
orl-rat LD50:28 mg/kg FAATDF 7,299,86
skn-rat LD50:400 mg/kg FAATDF 7,299,86
orl-mus LD50:9500 µg/kg ABCHA6 34,1697,70
skn-rbt LD50:1460 mg/kg WRPCA2 9,119,70
ims-dom LD50:59 mg/kg VETNAL 54(12),94,78
ims-mam LD50:64 mg/kg VETNAL 54(12),94,78
orl-bwd LD50:1800 µg/kg TXAPA9 21,315,72

CONSENSUS REPORTS: Reported in EPA TSCA Inventory.

SAFETY PROFILE: Poison by ingestion and intramuscular routes. Moderately toxic by skin contact. A cholinesterase inhibitor. When heated to decomposition it emits toxic fumes of NO_x and SO_x. See also PARATHION.

FAB800 CAS:4602-84-0 HR: 2
FARNESOL
mf: $C_{15}H_{26}O$ mw: 222.41

PROP: trans-Farnesol: Light-yellow liquid; mild oily odor. Bp: 111°, n (25/D) 1.4872. Commercial farnesol: Bp: 110–113°, d: (20/4) 0.8871, n (20/D) 1.4870., refr index: 1.487–1.492. Insol in water @ 263°.

SYNS: FARNESYL ALCOHOL ☐ FEMA No. 2478 ☐ 3,7,11-TRIMETHYL-2,6,10-DODECATRIEN-1-OL

TOXICITY DATA WITH REFERENCE
dni-oin:ovr 100 µmol/L ABCHA6 43,1285,79
orl-rat LD50:6000 mg/kg THERAP 27,893,72
orl-mus LD50:7400 mg/kg THERAP 27,893,72
ipr-mus LD50:443 mg/kg THERAP 27,893,72

CONSENSUS REPORTS: Reported in EPA TSCA Inventory.

SAFETY PROFILE: Moderately toxic by intraperitoneal route. Mildly toxic by ingestion. Mutation data reported. When heated to decomposition it emits acrid smoke and irritating fumes.

FAB830 CAS:1064-48-8 HR: D
FAST SULON BLACK BN
mf: $C_{22}H_{16}N_6O_9S_2 \cdot 2Na$ mw: 618.54

SYNS: ACIDAL BLACK 10B ☐ ACIDAL NAVY BLUE 3BR ☐ ACID BLACK 10A ☐ ACID BLACK 10B ☐ ACID BLACK 12B ☐ ACID BLACK 10BA ☐ ACID BLACK BASE M ☐ ACID BLACK 4BN ☐ ACID BLACK 4BNU ☐ ACID BLACK 10BN ☐ ACID BLACK BRX ☐ ACID BLACK BX ☐ ACID BLACK H ☐ ACID BLACK 1 ☐ ACID BLACK JVS ☐ ACID BLUE BLACK B ☐ ACID BLUE BLACK 10B ☐ ACID BLUE BLACK BG ☐ ACID BLUE BLACK DOUBLE 600 ☐ ACID LEATHER BLUE IGW ☐ ACID LEATHER DARK BLUE G ☐ ACID LEATHER FAST BLUE BLACK G ☐ AIREDALE BLACK 2BG ☐ AMACID BLACK 10BR ☐ ATUL ACID BLACK 10BX ☐ ATUL ACID BLACK BX ☐ AZANOL FAST ACID BLACK

10B □ AZO DARK BLUE C 2B □ AZO DARK BLUE HR □ AZO DARK BLUE S □ AZO DARK BLUE SH □ BLUE BLACK 12B □ BLUE BLACK SX □ BORUTA BLACK A □ BRASILAN BLACK BS □ BUCACID BLUE BLACK □ CALCOCID BLUE BLACK □ CALCOCID BLUE BLACK 2R □ CERN KYSELA 1 □ C.I. 20470 □ C.I. ACID BLACK 1 (7CI) □ C.I. ACID BLACK 1, DISODIUM SALT (8CI) □ COLACID BLACK 10A □ COMACID BLUE BLACK B □ DIACID BLUE BLACK 10B □ ENIACID BLACK IVS □ ENIACID BLACK SH □ ERIOSIN BLUE BLACK B □ FENAZO BLUE BLACK □ 2,7-NAPHTHALENEDISULFONIC ACID, 4-AMINO-5-HYDROXY-3-((4-NITROPHENYL)AZO)-6-(PHENYLAZO)-, DISODIUM SALT

TOXICITY DATA WITH REFERENCE
mmo-sat 50 µg/plate EMMUEG 22,188,93
mma-sat 50 µg/plate EMMUEG 22,188,93

CONSENSUS REPORTS: Reported in EPA TSCA Inventory.

SAFETY PROFILE: Mutation data reported. When heated to decomposition it emits toxic vapors of NO_x and SO_x.

FAB920 CAS:61788-72-5 HR: 1
FATTY ACID, TALL OIL, EPOXIDIZED, OCTYL ESTER

SYNS: ADMEX 741 □ ADMEX 746 □ DRAPEX 4.4 □ OCTYL EPOXYTALLATE □ PLASTOLEIN 9214 □ PX-806

TOXICITY DATA WITH REFERENCE
orl-rat LD50:32 g/kg NPIRI* 2,80,75

CONSENSUS REPORTS: Reported in EPA TSCA Inventory.

SAFETY PROFILE: Mildly toxic by ingestion. When heated to decomposition it emits acrid smoke and irritating fumes.

FAC050 CAS:2014-22-4 HR: 3
FC 402
mf: $C_{11}H_{21}NO_2 \cdot ClH$ mw: 271.82

SYN: N,N-DIETHYLGLYCINE-2,6 XYLYL ESTER HYDROCHLORIDE

TOXICITY DATA WITH REFERENCE
skn-rbt 200 mg MOD BCFAAI 107,310,68
eye-rbt 1 g MOD BCFAAI 107,310,68
scu-mus LD50:1350 mg/kg BCFAAI 107,310,68
ivn-mus LD50:35 mg/kg BCFAAI 107,310,68

SAFETY PROFILE: Poison by intravenous route. Moderately toxic by subcutaneous route. An eye and skin irritant. When heated to decomposition it emits toxic fumes of NO_x and HCl. See also ESTERS.

FAC060 CAS:2173-44-6 HR: 3
FC-410
mf: $C_{15}H_{22}N_2O_2 \cdot 2ClH$ mw: 335.31

SYN: 4-METHYL-1-PIPERAZINEACETIC ACID 2,6-XYLYL ESTER DIHYDROCHLORIDE

TOXICITY DATA WITH REFERENCE
skn-rbt 200 mg MOD BCFAAI 107,310,68
eye-rbt 1 g MOD BCFAAI 107,310,68
scu-mus LD50:500 mg/kg BCFAAI 107,310,68
ivn-mus LD50:22 mg/kg BCFAAI 107,310,68

SAFETY PROFILE: Poison by intravenous route. Moderately toxic by subcutaneous route. A skin and eye irritant. When heated to decomposition it emits toxic fumes of NO_x and HCl. See also ESTERS.

FAC100 CAS:2014-27-9 HR: 2
FC 457
mf: $C_{15}H_{23}NO_2 \cdot ClH$ mw: 285.85

SYN: 1-N,N-DIETHYLALANINE-2,6-XYLYL ESTER HYDROCHLORIDE

TOXICITY DATA WITH REFERENCE
skn-rbt 200 mg MLD BCFAAI 107,310,68
eye-rbt 1 g MLD BCFAAI 107,310,68
scu-mus LD50:1275 mg/kg BCFAAI 107,310,68

SAFETY PROFILE: Moderately toxic by subcutaneous route. A skin and eye irritant. When heated to decomposition it emits toxic fumes of NO_x and HCl. See also ESTERS.

FAC130 CAS:20684-29-1 HR: 3
FC 480
mf: $C_{17}H_{26}N_2O_2 \cdot ClH$ mw: 326.91

SYNS: 2,6-DIMETILFENILICO DELL'ACIDO α-N-METILPIPERAZINOBUTIRRICO IDOCLORIDRAT (ITALIAN) □ α-ETHYL-4-METHYL-1-PIPERAZINEACETIC ACID-2,6-XYLYL ESTER HYDROCHLORIDE

TOXICITY DATA WITH REFERENCE
skn-rbt 200 mg MOD BCFAAI 107,310,68
eye-rbt 1 g MOD BCFAAI 107,310,68
orl-mus LD50:230 mg/kg BCFAAI 107,310,68
ipr-mus LD50:82 mg/kg BCFAAI 107,310,68
scu-mus LD50:410 mg/kg BCFAAI 107,310,68
ivn-mus LD50:20 mg/kg BCFAAI 107,310,68

SAFETY PROFILE: Poison by ingestion, intravenous, and intraperitoneal routes. Moderately toxic by subcutaneous route. An eye and skin irritant. When heated to decomposition it emits toxic fumes of NO_x and HCl. See also ESTERS.

FAC150 CAS:2085-83-8 HR: 2
FC 590
mf: $C_{20}H_{25}NO_2 \cdot ClH$ mw: 347.92

SYN: N,N-DIETHYL-2-PHENYL-GLYCINE-2,6-XYLYL ESTER HYDROCHLORIDE

TOXICITY DATA WITH REFERENCE
skn-rbt 200 mg MLD BCFAAI 107,310,68
eye-rbt 1 g MLD BCFAAI 107,310,68
scu-mus LD50:1200 mg/kg BCFAAI 107,310,68

SAFETY PROFILE: Moderately toxic by subcutaneous route. A skin and eye irritant. When heated to decomposition it emits toxic fumes of NO_x and HCl. See also ESTERS.

FAC155 CAS:19245-07-9 HR: 3
FC 591
mf: $C_{22}H_{28}N_2O_3 \cdot ClH$ mw: 404.98

SYN: 4-(2-HYDROXYETHYL)-α-PHENYL-1-PIPERAZINEACETIC ACID-2,6-XYLYL ESTER HYDROCHLORIDE

TOXICITY DATA WITH REFERENCE
skn-rbt 200 mg SEV BCFAAI 107,310,68
eye-rbt 1 g SEV BCFAAI 107,310,68
scu-mus LD50:65 mg/kg BCFAAI 107,310,68

SAFETY PROFILE: Poison by subcutaneous route. A severe eye and skin irritant. When heated to decomposition it emits toxic fumes of NO_x and HCl. See also ESTERS.

FAC157 CAS:2282-89-5 HR: 2
FC 642
mf: $C_{17}H_{27}NO_2•ClH$ mw: 313.91

SYN: l-N,N-DIETHYLALANINE-2,6-DIETHYLPHENYL ESTER HYDROCHLORIDE

TOXICITY DATA WITH REFERENCE
skn-rbt 200 mg MOD BCFAAI 107,310,68
eye-rbt 1 g MOD BCFAAI 107,310,68
scu-mus LD50:1750 mg/kg BCFAAI 107,310,68

SAFETY PROFILE: Moderately toxic by subcutaneous route. An eye and skin irritant. When heated to decomposition it emits toxic fumes of NO_x and HCl. See also ESTERS.

FAC160 CAS:1877-53-8 HR: 3
FC 646
mf: $C_{20}H_{32}N_2O_2•2ClH$ mw: 405.46

SYN: α,4-DIMETHYL-1-PIPERAZINEACETIC ACID-2,6-DIISOPROPYLPHENYL ESTER DIHYDROCHLORIDE

TOXICITY DATA WITH REFERENCE
skn-rbt 200 mg SEV BCFAAI 107,310,68
eye-rbt 1 g SEV BCFAAI 107,310,68
scu-mus LD50:600 mg/kg BCFAAI 107,310,68
ivn-mus LD50:42 mg/kg BCFAAI 107,310,68

SAFETY PROFILE: Poison by intravenous route. Moderately toxic by subcutaneous route. A severe eye and skin irritant. When heated to decomposition it emits toxic fumes of NO_x and HCl. See also ESTERS.

FAC163 CAS:2409-35-0 HR: 2
FC 650
mf: $C_{15}H_{23}NO_4•ClH$ mw: 317.85

SYN: l-N,N-DIETHYLALANINE-2,6-DIMETHOXYPHENYL ESTER HYDROCHLORIDE

TOXICITY DATA WITH REFERENCE
skn-rbt 200 mg MOD BCFAAI 107,310,68
eye-rbt 1 g MOD BCFAAI 107,310,68
scu-mus LD50:1425 mg/kg BCFAAI 107,310,68

SAFETY PROFILE: Moderately toxic by subcutaneous route. An eye and skin irritant. When heated to decomposition it emits toxic fumes of NO_x and HCl. See also ESTERS.

FAC165 CAS:1877-52-7 HR: 3
FC 651
mf: $C_{19}H_{30}N_2O_2•2ClH$ mw: 391.43

SYN: α-ETHYL-4-METHYL-1-PIPERAZINEACETIC ACID-2,6-DIETHYLPHENYL ESTER DIHYDROCHLORIDE

TOXICITY DATA WITH REFERENCE
skn-rbt 200 mg MOD BCFAAI 107,310,68
eye-rbt 1 g MOD BCFAAI 107,310,68
scu-mus LD50:540 mg/kg BCFAAI 107,310,68
ivn-mus LD50:34 mg/kg BCFAAI 107,310,68

SAFETY PROFILE: Poison by intravenous route. Moderately toxic by subcutaneous route. An eye and skin irritant. When heated to decomposition it emits toxic fumes of NO_x and HCl. See also ESTERS.

FAC166 CAS:2085-85-0 HR: 2
FC 652
mf: $C_{14}H_{21}NO_4•ClH$ mw: 303.82

SYN: N,N-DIETHYLGLYCINE-2,6-DIMETHOXYPHENYL ESTER HYDROCHLORIDE

TOXICITY DATA WITH REFERENCE
skn-rbt 200 mg MLD BCFAAI 107,310,68
eye-rbt 1 g MLD BCFAAI 107,310,68
scu-mus LD50:550 mg/kg BCFAAI 107,310,68

SAFETY PROFILE: Moderately toxic by subcutaneous route. A skin and eye irritant. When heated to decomposition it emits toxic fumes of NO_x and HCl. See also ESTERS.

FAC170 CAS:2014-32-6 HR: 2
FC 657
mf: $C_{15}H_{23}NO_2•ClH$ mw: 285.85

SYN: N,N-DIETHYLGLYCINE MESITYL ESTER HYDROCHLORIDE

TOXICITY DATA WITH REFERENCE
skn-rbt 200 mg MOD BCFAAI 107,310,68
eye-rbt 1 g MOD BCFAAI 107,310,68
scu-mus LD50:2000 mg/kg BCFAAI 107,310,68

SAFETY PROFILE: Moderately toxic by subcutaneous route. An eye and skin irritant. When heated to decomposition it emits toxic fumes of NO_x and HCl. See also ESTERS.

FAC175 CAS:2014-33-7 HR: 2
FC 659
mf: $C_{16}H_{23}NO_2•ClH$ mw: 297.86

SYN: MESITYL ESTER of 1-PIPERIDINEACETIC ACID HYDROCHLORIDE

TOXICITY DATA WITH REFERENCE
skn-rbt 200 mg MOD BCFAAI 107,310,68
eye-rbt 1 g MOD BCFAAI 107,310,68
scu-mus LD50:2000 mg/kg BCFAAI 107,310,68

SAFETY PROFILE: Moderately toxic by subcutaneous route. An eye and skin irritant. When heated to decomposition it emits toxic fumes of NO_x and HCl. See also ESTERS.

FAC179 CAS:2173-47-9 HR: 2
FC 660
mf: $C_{16}H_{25}NO_2{\cdot}ClH$ mw: 299.88

SYN: 1-N,N-DIETHYLALANINE MESITYL ESTER HYDROCHLORIDE

TOXICITY DATA WITH REFERENCE
skn-rbt 200 mg MLD BCFAAI 107,310,68
eye-rbt 1 g MLD BCFAAI 107,310,68
scu-mus LD50:2000 mg/kg BCFAAI 107,310,68

SAFETY PROFILE: Moderately toxic by subcutaneous route. A skin and eye irritant. When heated to decomposition it emits toxic fumes of NO_x and HCl. See also ESTERS.

FAC185 CAS:2014-31-5 HR: 3
FC 668
mf: $C_{15}H_{21}ClN_2O_2{\cdot}2ClH$ mw: 369.75

SYN: α,4-DIMETHYL-1-PIPERAZINEACETIC ACID-6-CHLORO-o-TOLYL ESTER DIHYDROCHLORIDE

TOXICITY DATA WITH REFERENCE
skn-rbt 200 mg MOD BCFAAI 107,310,68
eye-rbt 1 g MOD BCFAAI 107,310,68
scu-mus LD50:180 mg/kg BCFAAI 107,310,68

SAFETY PROFILE: Poison by subcutaneous route. An eye and skin irritant. When heated to decomposition it emits toxic fumes of NO_x and HCl. See also ESTERS.

FAC195 CAS:1940-89-2 HR: 3
FC 681
mf: $C_{18}H_{28}N_2O_2{\cdot}2ClH$ mw: 377.40

SYN: α-ETHYL-4-METHYL-1-PIPERAZINEACETIC ACID MESITYL ESTER DIHYDROCHLORIDE

TOXICITY DATA WITH REFERENCE
skn-rbt 200 mg MOD BCFAAI 107,310,68
eye-rbt 1 g MOD BCFAAI 107,310,68
scu-mus LD50:250 mg/kg BCFAAI 107,310,68

SAFETY PROFILE: Poison by subcutaneous route. An eye and skin irritant. When heated to decomposition it emits toxic fumes of NO_x and HCl. See also ESTERS.

FAE000 CAS:3844-45-9 HR: 2
FD&C BLUE No. 1
mf: $C_{37}H_{36}N_2O_9S_3{\cdot}2Na$ mw: 794.91

PROP: Dark-purple to bronze powder with metallic luster. Sol in water, ether, conc sulfuric acid.

SYNS: ACID SKY BLUE A ☐ AIZEN FOOD BLUE No. 2 ☐ 1206 BLUE ☐ BRILLIANT BLUE FCD No. 1 ☐ BRILLIANT BLUE FCF ☐ CANACERT BRILLIANT BLUE FCF ☐ C.I. 42090 ☐ C.I. ACID BLUE 9, DISODIUM SALT ☐ C.I. FOOD BLUE 2 ☐ COGILOR BLUE 512.12 ☐ COSMETIC BLUE LAKE ☐ D&C BLUE No. 4 ☐ DISPERSED BLUE 12195 ☐ DOLKWAL BRILLIANT BLUE ☐ EDICOL BLUE CL 2 ☐ ERIOGLAUCINE G ☐ FENAZO BLUE XI ☐ FOOD BLUE 2 ☐ FOOD BLUE DYE No. 1 ☐ HEXACOL BRILLIANT BLUE A ☐ INTRACID PURE BLUE L ☐ MERANTINE BLUE EG ☐ USACERT BLUE No. 1

TOXICITY DATA WITH REFERENCE
cyt-ham:lng 4400 mg/L GANMAX 27,95,81
scu-rat TDLo:5500 mg/kg/97W-I:NEO ZEKBAI 64,287,61
par-rat TDLo:4580 mg/kg/62W-I:ETA FAONAU 38B,27,66
scu-rat TD:2 g/kg/94W-I:ETA FAONAU 38B,27,66
scu-mus LD50:4600 mg/kg ZEKBAI 64,287,61

CONSENSUS REPORTS: IARC Cancer Review: Group 3 IMEMDT 7,56,87; Animal Sufficient Evidence IMEMDT 16,171,78. Reported in EPA TSCA Inventory.

SAFETY PROFILE: Questionable carcinogen with experimental neoplastigenic and tumorigenic data. Mutation data reported. When heated to decomposition it emits very toxic fumes of NO_x, Na_2O and SO_x.

FAE100 CAS:860-22-0 HR: 3
FD&C BLUE No. 2
mf: $C_{16}H_{10}N_2O_8S_2{\cdot}2Na$ mw: 468.38

PROP: Blue-brown to red-brown powder or dark blue solid. Sol in H_2O; sltly sol in EtOH; insol in C_6H_6, $CHCl_3$.

SYNS: ACID BLUE W ☐ ACID LEATHER BLUE IC ☐ A.F. BLUE No. 2 ☐ AIRDALE BLUE IN ☐ AMACID BRILLIANT BLUE ☐ ANILINE CARMINE POWDER ☐ ATUL INDIGO CARMINE ☐ 1311 BLUE ☐ 12070 BLUE ☐ BUCACID INDIGOTINE B ☐ CANACERT INDIGO CARMINE ☐ CARMINE BLUE (BIOLOGICAL STAIN) ☐ C.I. 73015 ☐ C.I. 7581 ☐ C.I. ACID BLUE 74 ☐ C.I. FOOD BLUE 1 ☐ DISODIUM INDIGO-5,5-DISULFONATE ☐ DISODIUM SALT of 1-INDIGOTIN-S,S'-DISULPHONIC ACID ☐ DOLKWAL INDIGO CARMINE ☐ E 132 ☐ GRAPE BLUE A GEIGY ☐ INDIGO CARMINE ☐ INDIGO CARMINE (BIOLOGICAL STAIN) ☐ INDIGO CARMINE DISODIUM SALT ☐ INDIGO EXTRACT ☐ INDIGO-KARMIN (GERMAN) ☐ 5,5'-INDIGOTIN DISULFONIC ACID ☐ INDIGOTINE ☐ INDIGOTINE DISODIUM SALT ☐ INTENSE BLUE ☐ L-BLAU 2 (GERMAN) ☐ MAPLE INDIGO CARMINE ☐ SACHSISCHBLAU ☐ SCHULTZ Nr. 1309 (GERMAN) ☐ SODIUM-5,5'-INDIGOTIDISULFONATE ☐ SOLUBLE INDIGO ☐ USACERT BLUE No. 2

TOXICITY DATA WITH REFERENCE
mmo-sat 320 μg/plate MUREAV 89,21,81
ihl-rat TCLo:100 mg/m³/13W-I:NEO,REP GTPZAB 7(2),54,63
orl-rat TDLo:660 g/kg/43W-C:ETA GASTAB 23,1,53
orl-rat LD:>3 g/kg GTPZAB 7(2),54,63
ivn-rat LD50:93 mg/kg NIIRDN 6,83,82

CONSENSUS REPORTS: Reported in EPA TSCA Inventory. EPA Genetic Toxicology Program.

SAFETY PROFILE: Poison by intravenous route. Moderately toxic by ingestion and subcutaneous routes. Questionable carcinogen with experimental neoplastigenic data. Mutation data reported. When heated to decomposition it emits very toxic fumes of SO_x, NO_x, and Na_2O.

FAE950 CAS:4680-78-8 HR: 2
FD&C GREEN No. 1
mf: $C_{37}H_{36}N_2O_6S_2{\cdot}Na$ mw: 691.86

SYNS: ACIDAL GREEN G ☐ ACID GREEN 3 ☐ ACID GREEN ☐ ACID GREEN B ☐ ACID GREEN 2G ☐ ACID GREEN G ☐ ACID GREEN L ☐ ACID GREEN S ☐ ACID LEATHER GREEN F ☐ ACID LEATHER GREEN 3G ☐ ACILAN GREEN B ☐ A.F. GREEN NO. 1 ☐ AMACID GREEN B ☐ BRILLIANT GREEN 3EMBL ☐ BUCACID GUINEA GREEN BA ☐ CALCOCID GREEN G ☐ C.I. 42085 ☐ C.I. ACID GREEN 3 ☐ C.I. ACID GREEN 3, MONOSODIUM SALT ☐ C.I. ACID GREEN 3, SODIUM SALT ☐ C.I. FOOD GREEN 1 ☐ FDC GREEN 1 ☐ FENAZO GREEN L ☐ FOOD GREEN 1 ☐ GUINEA GREEN ☐ GUINEA GREEN B ☐ GUINEA GREEN BA ☐ HIDACID EMERALD GREEN ☐ HISPACID GREEN GB ☐ INTRACID GREEN F ☐ KITON GREEN F ☐ KITON

F

GREEN FC □ LEATHER GREEN B □ LISSAMINE GREEN G □ MERANTINE GREEN G □ NAPHTHALENE GREEN G □ NAPHTHALENE LAKE GREEN G □ NAPHTHALENE LEATHER GREEN G □ NERAN BRILLIANT GREEN G □ PONTACYL GREEN BL □ SULFACID BRILLIANT GREEN 1B □ SULPHO GREEN 2B □ VONDACID GREEN L □ ZELEN KYSELA 3 □ ZELEN POTRAVINARSKA 1

TOXICITY DATA WITH REFERENCE
mma-sat 320 μg/plate MUREAV 89,21,81
ihl-rat TCLo: 100 mg/m³/13W-I:NEO,REP GTPZAB 7(2),54,63
orl-rat TDLo: 660 g/kg/43W-C:ETA GASTAB 23,1,53
orl-rat LD:>3 g/kg GTPZAB 7(2),54,63

CONSENSUS REPORTS: IARC Cancer Review: Group 3 IMEMDT 7,56,87; Animal Sufficient Evidence IMEMDT 16,199,78. Reported in EPA TSCA Inventory. Community Right-To-Know List.

SAFETY PROFILE: Questionable carcinogen with experimental tumorigenic data. Experimental reproductive effects. Mutation data reported. When heated to decomposition it emits very toxic fumes of SO_x, Na_2O, and NO_x.

FAF000 CAS:5141-20-8 HR: 2
FD&C GREEN No. 2
mf: $C_{37}H_{36}N_2O_9S_3 \cdot 2Na$ mw: 794.91

SYNS: ACIDAL LIGHT GREEN SF □ ACID BRILLIANT GREEN SF □ ACID GREEN A □ ACID GREEN N □ ACILAN GREEN SFG □ A.F. GREEN No. 2 □ AMACID GREEN G □ C.I. 42095 □ C.I. ACID GREEN 5 □ C.I. ACID GREEN 5, DISODIUM SALT □ C.I. FOOD GREEN 2 □ D&C GREEN No. 4 □ FAS □ ACID GREEN N □ FD&C GREEN No. 2-ALUMINUM LAKE □ FENAZO GREEN 7G □ FOOD GREEN 2 □ GREEN No. 203 □ LEATHER GREEN SF □ LICHTGRUEN (GERMAN) □ LIGHT GREEN FCF YELLOWISH □ LIGHT GREEN LAKE □ LIGHT SF YELLOWISH (BIOLOGICAL STAIN) □ LISSAMINE LAKE GREEN SF □ MERANTINE GREEN SF □ MY/68 □ PENCIL GREEN SF □ SULFO GREEN J □ SUMITOMO LIGHT GREEN SF YELLOWISH □ WOOL BRILLIANT GREEN SF

TOXICITY DATA WITH REFERENCE
mma-sat 320 μg/plate MUREAV 89,21,81
orl-rat TDLo: 819 g/kg (male 65W pre): REP JPPMAB 8,417,56
orl-rat TDLo: 1300 g/kg/86W-C:CAR GASTAB 23,1,53
scu-rat TDLo: 6825 mg/kg/46W-I:NEO JNCIAM 24,769,60
orl-rat LD50:>2 g/kg MEIEDD 11,864,89
scu-mus LD50: 525 mg/kg ZEKBAI 64,287,61
ivn-mus LD50: 700 mg/kg TXAPA9 44,225,78

CONSENSUS REPORTS: IARC Cancer Review: Group 3 IMEMDT 7,56,87; Animal Sufficient Evidence IMEMDT 16,209,78. Reported in EPA TSCA Inventory. EPA Genetic Toxicology Program.

SAFETY PROFILE: Moderately toxic by subcutaneous and intravenous routes. Questionable carcinogen with experimental carcinogenic and neoplastigenic data. Experimental reproductive effects. Mutation data reported. When heated to decomposition it emits very toxic fumes of NO_x, Na_2O, and SO_x.

FAG000 CAS:2353-45-9 HR: 2
FD&C GREEN No. 3
mf: $C_{37}H_{36}N_2O_{10}S_3 \cdot 2Na$ mw: 810.91

PROP: Red to brown-violet powder or dark green crystals. Mp: 290° (decomp). Sol in water, conc sulfuric acid.

SYNS: AIZEN FOOD GREEN No. 3 □ C.I. 42053 □ C.I. FOOD GREEN 3 □ FAST GREEN FCF □ 1724 GREEN □ SOLID GREEN FCF

TOXICITY DATA WITH REFERENCE
mma-sat 10 mg/plate FCTOD7 22,623,84
cyt-ham: fbr 4 g/L FCTOD7 22,623,84
cyt-ham: ovr 20 μmol/L/5H-C ENMUDM 1,27,79
cyt-ham: lng 2 g/L GMCRDC 27,95,81
scu-rat TDLo: 5925 mg/kg/48W-I:NEO JNCIAM 24,769,60

CONSENSUS REPORTS: IARC Cancer Review: Group 3 IMEMDT 7,56,87; Animal Sufficient Evidence IMEMDT 16,187,78. Reported in EPA TSCA Inventory. EPA Genetic Toxicology Program.

SAFETY PROFILE: Questionable carcinogen with experimental neoplastigenic data. Mutation data reported. When heated to decomposition it emits very toxic fumes of NO_x and SO_x.

FAG010 CAS:523-44-4 HR: 2
FD&C ORANGE No. 1
mf: $C_{16}H_{11}N_2O_4S \cdot Na$ mw: 350.34

PROP: Reddish-brown powder giving orange soln. Sol in H_2O; sltly sol in EtOH; insol in common org solvs.

SYNS: ACID LEATHER ORANGE I □ ACID PHOSPHINE CL □ A.F. ORANGE No. 1 □ AIZEN FOOD ORANGE No. 1 □ AIZEN NAPHTHOL ORANGE I □ AIZEN ORANGE I □ CERTIQUAL ORANGE I □ C.I. 14600 □ C.I. ACID ORANGE 20 □ C.I. ACID ORANGE 20, MONOSODIUM SALT □ D&C ORANGE No. 3 □ DYE ORANGE No. 1 □ EGACID ORANGE GG □ ELGACID ORANGE 2G □ ENIACID ORANGE I □ EXT. D&C ORANGE No. 3 □ FDC ORANGE I □ HISPACID ORANGE 1 □ 4-((4-HYDROXY-1-NAPHTHALENYL)AZO)BENZENESULPHONIC ACID, MONOSODIUM SALT □ p-((4-HYDROXY-1-NAPHTHYL)AZO) BENZENESULFONIC ACID, MONOSODIUM SALT □ p-((4-HYDROXY-1-NAPHTHYL)AZO)BENZENESULPHONIC ACID, MONOSODIUM SALT □ p-((4-HYDROXY-1-NAPHTHYL)AZO)BENZENESULPHONIC ACID, SODIUM SALT □ JAVA ORANGE I □ NANKAI ACID ORANGE I □ NAPHTHALENE ORANGE I □ NAPHTHOL ORANGE □ α-NAPHTHOL ORANGE □ NEKLACID ORANGE 1 □ 1333 ORANGE □ ORANGE I □ SODIUM AZO-α-NAPHTHOLSULFANILATE □ SODIUM AZO-α-NAPHTHOLSULPHANILATE □ 4-p-SULFOPHENYLAZO-1-NAPHTHOL MONOSODIUM SALT □ TERTRACID ORANGE I □ TROPAEOLIN 1

TOXICITY DATA WITH REFERENCE
mma-sat 100 μg/plate MUREAV 56,249,78
scu-rat TDLo: 9360 mg/kg/2Y-I:ETA FEPRA7 16,367,57
ipr-rat LD50: 1 g/kg FAONAU 38B,67,66

CONSENSUS REPORTS: IARC Cancer Review: Group 3 IMEMDT 7,56,87; Animal Limited Evidence IMEMDT 8,173,75. Reported in EPA TSCA Inventory. EPA Genetic Toxicology Program.

SAFETY PROFILE: Low toxicity by ingestion. Questionable carcinogen with experimental tumorigenic data. Mutation data reported. When heated to decomposition it emits very toxic fumes of NO_x, SO_x, and Na_2O.

FAG018 **CAS:3564-09-8** **HR: 3**
FD&C RED No. 1
mf: $C_{19}H_{16}N_2O_7S_2 \cdot 2Na$ mw: 494.47

SYNS: A.F. RED No. 1 ☐ CERVEN KUMIDINOVA ☐ C.I. 16155 ☐ C.I. FOOD RED 6 ☐ C.I. FOOD RED 6, DISODIUM SALT ☐ DISODIUM-3-HYDROXY-4-((2,4,5-TRIMETHYLPHENYL)AZO)-2,7-NAPHTHALENEDISULFONATE ☐ DISODIUM-3-HYDROXY-4-((2,4,5-TRIMETHYLPHENYL)AZO)-2,7-NAPHTHALENEDISULFONIC ACID ☐ DISODIUM-3-HYDROXY-4-((2,4,5-TRIMETHYLPHENYL)AZO)-2,7-NAPHTHALENEDISULPHONATE ☐ DISODIUM-3-HYDROXY-4-((2,4,5-TRIMETHYLPHENYL)AZO)-2,7-NAPHTHALENEDISULPHONIC ACID ☐ DOLKWAL PONCEAU 3R ☐ EXT. D&C RED No. 15 ☐ 3-HYDROXY-4-((2,4,5-TRIMETHYLPHENYL)AZO)-2,7-NAPHTHALENEDISULFONIC ACID, DISODIUM SALT ☐ 3-HYDROXY-4-((2,4,5-TRIMETHYLPHENYL)AZO)-2,7-NAPHTHALENEDISULPHONIC ACID, DISODIUM SALT ☐ MAPLE PONCEAU 3R ☐ PONCEAU 3R ☐ SODIUM CUMENEAZO-β-NAPHTHOL DISULPHONATE ☐ USACERT RED No. 1

TOXICITY DATA WITH REFERENCE
mma-sat 660 nmol/plate INFIBR 23,686,79
oth-esc 300 μmol/L SKEZAP 12,298,71
orl-rat TDLo:730 g/kg/2Y-C:CAR TXAPA9 5,105,63
orl-mus TDLo:1640 g/kg/65W-C:CAR TXAPA9 3,509,61
imp-mus TDLo:80 mg/kg:CAR BJCAAI 17,127,63
orl-mus TD:1140 g/kg/68W-C:ETA TXAPA9 5,105,63
orl-rat TD:200 g/kg/24W-C:NEO TXAPA9 5,105,63

CONSENSUS REPORTS: IARC Cancer Review: Group 2B IMEMDT 7,56,87; Animal Sufficient Evidence IMEMDT 8,199,75.

SAFETY PROFILE: Confirmed carcinogen with experimental carcinogenic and tumorigenic data. Mutation data reported. When heated to decomposition it emits toxic fumes of NO_x and SO_x.

FAG020 **CAS:915-67-3** **HR: 2**
FD&C RED No. 2
mf: $C_{20}H_{11}N_2O_{10}S_3 \cdot 3Na$ mw: 604.48

PROP: Dark-red-brown powder. Sol in H_2O.

SYNS: ACETACID RED 2BR ☐ ACID AMARANTH ☐ ACILAN RED SE ☐ AIZEN AMARANTH ☐ AMACID AMARANTH ☐ AMARANT ☐ AMARANTH ☐ AMARANTHE USP (biological stain) ☐ AZO RED R ☐ BORDEAUX ☐ CALCOCID AMARANTH ☐ CANACERT AMARANTH ☐ CERVEN KYSELA 27 ☐ CERVEN POTRAVINARSKA 9 ☐ C.I. 184 ☐ C.I. 16185 ☐ C.I. ACID RED 27 ☐ C.I. FOOD RED 9 ☐ CILEFA RUBINE 2B ☐ DAISHIKI AMARANTH ☐ D&C RED 2 ☐ DOLKWAL AMARANTH ☐ DYE FDC RED 2 ☐ DYE RED RASPBERRY ☐ EDICOL AMARANTH ☐ EUROCERT AMARANTH ☐ FD&C RED No. 2-ALUMINIUM LAKE ☐ FOOD RED 2 ☐ FOOD RED 9 ☐ FRUIT RED A GEIGY ☐ HD AMARANTH B ☐ HEXACERT RED No. 2 ☐ HIDACID AMARANTH ☐ 2-HYDROXY-1,1′-AZONAPHTHALENE-3,6,4′-TRISULFONIC ACID TRISODIUM SALT ☐ 3-HYDROXY-4-((4-SULFO-1-NAPHTHALENYL)AZO)-2,7-NAPHTHLENEDISULFONIC ACID, TRISODIUM SALT ☐ 3-HYDROXY-4-((4-SULFO-1-NAPHTHYL)AZO)-2,7-NAPHTHALENEDISULFONIC ACID, TRISODIUM SALT ☐ 3-HYDROXY-4-((4-SULPHO-1-NAPHTHALENYL)AZO)-2,7-NAPHTHALENEDISULPHONIC ACID, TRISODIUM SALT ☐ 3-HYDROXY-4-((4-SULPHO-1-NAPHTHYL)AZO)-2,7-NAPHTHALENEDISULPHONIC ACID, TRISODIUM SALT ☐ JAVA AMARANTH ☐ KAYAKU AMARANTH ☐ KCA FOODCOL AMARANTH A ☐ KITON RUBINE S ☐ LISSAMINE AMARANTH AC ☐ MAPLE AMARANTH ☐ NAPHTHOL RED B ☐ NAPTHOLROT S ☐ NEKLACID RED A ☐ RAKUTO AMARANTH ☐ RASPBERRY RED for JELLIES ☐ RED DYE No. 2 ☐ RED No. 2 ☐ SHIKISO AMARANTH ☐ 1-(4-SULPHO-1-NAPHTHYLAZO)-2-NAPHTHOL-3,6-DISULPHONIC ACID, TRISODIUM SALT ☐ TAKAOKA AMARANTH ☐ TERTRACID RED A ☐ TOYO AMARANTH ☐ TRISODIUM SALT of 1-(4-SULFO-1-NAPHTHYLAZO)-2-NAPHTHOL-3,6-DISULFONIC ACID ☐ VICTORIA RUBINE O ☐ WHORTLEBERRY RED ☐ WOOL BORDEAUX 6RK ☐ WOOL RED

TOXICITY DATA WITH REFERENCE
mmo-sat 20 μg/L ENVIDV 9,145,83
mmo-esc 20 μg/L ENVIDV 9,145,83
mrc-smc 20 mg/L ENVIDV 9,145,83
cyt-ham:fbr 1 g/L/48H MUREAV 48,337,77
orl-rat TDLo:296 mg/kg (multi):REP TXCYAC 3,115,75
orl-rat TDLo:547 mg/kg (multi):REP VPITAR 29,66,70
orl-rat TDLo:65 mg/kg (multi):TER TXCYAC 3,129,75
orl-rat TDLo:33 mg/kg (female 1-22D post):REP VPITAR 31(5),28,72
ipr-mus TDLo:250 mg/kg (male 1D pre):REP FCTXAV 14,163,76
orl-rat TDLo:548 mg/kg (female 52W pre):REP VPITAR 31(5),28,72
orl-rat TDLo:548 mg/kg (male 52W pre):REP VPITAR 31(5),28,72
orl-rat TDLo:65 mg/kg (multi):TER TXCYAC 3,129,75
orl-rat TDLo:680 g/kg/14W-I:ETA VPITAR 29,61,70
orl-rat TD:1200 g/kg/78W-C:CAR GASTAB 23,1,53
orl-rat TD:1643 g/kg/78W-C:ETA NEZAAQ 37,714,82
ipr-rat LD50:1 g/kg SCPHA4 47,39,79
ivn-rat LD50:1 g/kg SCPHA4 47,39,79
ipr-mus LD50:1000 mg/kg FCTXAV 14,163,76

CONSENSUS REPORTS: IARC Cancer Review: Group 3 IMEMDT 7,56,87, Animal Inadequate Evidence IMEMDT 8,41,75. Reported in EPA TSCA Inventory.

SAFETY PROFILE: Moderately toxic by intraperitoneal route. Questionable carcinogen with experimental carcinogenic and tumorigenic data. An experimental teratogen. Other experimental reproductive effects. Mutation data reported. When heated to decomposition it emits toxic fumes of NO_x and SO_x.

FAG040 **CAS:16423-68-0** **HR: 3**
FD&C RED No. 3
mf: $C_{20}H_6I_4O_5 \cdot 2Na$ mw: 879.84

PROP: Brown powder. Sol in water giving cherry red solution or in conc sulfuric acid.

SYNS: AIZEN ERYTHROSINE ☐ CALCOCID ERYTHROSINE N ☐ CANACERT ERYTHROSINE BS ☐ 9-(o-CARBOXYPHENYL)-6-HYDROXY-2,4,5,7-TETRAIODO-3-ISOXANTHONE ☐ C.I. 45430 ☐ C.I. ACID RED 51 ☐ CILEFA PINK B ☐ D&C RED No. 3 ☐ DOLKWAL ERYTHROSINE ☐ DYE FD&C RED No. 3 ☐ E 127 ☐ EBS ☐ EDICOL SUPRA ERYTHROSINE A ☐ ERYTHROSIN ☐ ERYTHROSINE B-FO (BIOLOGICAL STAIN) ☐ FOOD RED 14 ☐ HEXACERT RED No. 3 ☐ HEXACOL ERYTHROSINE BS ☐ LB-ROT 1 ☐ MAPLE ERYTHROSINE ☐ NEW PINK BLUISH GEIGY ☐ 1427 RED ☐ 1671 RED ☐ 2′,4′,5′,7′-TETRAIODOFLUORESCEIN, DISODIUM SALT ☐ TETRAIODOFLUORESCEIN SODIUM SALT ☐ USACERT RED No. 3

TOXICITY DATA WITH REFERENCE
mrc-smc 100 mg/L MUREAV 138,153,84
dni-hmn:leu 500 mg/L NEZAAQ 30,574,75
cyt-ham:fbr 600 mg/L FCTOD7 22,623,84
orl-rat TDLo:482 g/kg multi:REP FCTOD7 28,813,90
orl-rat TDLo:1798 g/kg/2Y-C:NEO FCTOD7 25,723,87
orl-rat LD50:1840 mg/kg SCPHA4 47,39,79
ivn-rat LD50:200 mg/kg SCPHA4 47,39,79

F

orl-mus LD50:1264 mg/kg EAPHA6 24,125,81
ivn-mus LD50:370 mg/kg APPHAX 9,127,52

CONSENSUS REPORTS: Reported in EPA TSCA Inventory. EPA Genetic Toxicology Program.

SAFETY PROFILE: Poison by intravenous route. Moderately toxic by ingestion. Questionable carcinogen with experimental tumorigenic data. Experimental reproductive effects. Human mutation data reported. When heated to decomposition it emits very toxic fumes of Na_2O and I^-.

FAG050 CAS:4548-53-2 HR: 2
FD&C RED No. 4
mf: $C_{18}H_{14}N_2O_7S_2$•2Na mw: 480.44

SYNS: CERTICOL PONCEAU SXS ☐ CERVEN POTRAVINARSKA 1 ☐ C.I. 14700 ☐ C.I. FOOD RED 1 ☐ C.I. FOOD RED 1, DISODIUM SALT ☐ 3-((2,4-DIMETHYL-5-SULFOPHENYL)AZO)-4-HYDROXY-1-NAPHTHALENESULFONIC ACID, DISODIUM SALT ☐ 3-((2,4-DIMETHYL-5-SULPHOPHENYL)AZO)-4-HYDROXY-1-NAPHTHALENESULPHONIC ACID, DISODIUM SALT ☐ DYE FD & C RED No. 4 ☐ DYE FD AND C RED No. 4 ☐ EDICOL SUPRA PONCEAU SX ☐ FD & C RED No. 4-ALUMINIUM LAKE ☐ FOOD RED 4 ☐ HEXACOL PONCEAU SX ☐ 4-HYDROXY-3-((5-SULFO-2,4-XYLYL)AZO)-1-NAPHTHALENESULFONIC ACID, DISODIUM SALT ☐ 4-HYDROXY-3-((5-SULPHO-2,4-XYLYL) AZO)-1-NAPHTHALENESULPHONIC ACID, DISODIUM SALT ☐ PONCEAU SX ☐ PURPLE 4R ☐ 1306 RED ☐ 12101 RED ☐ RED No. 1 ☐ RED No. 4 ☐ 2-(6-SULFO-2,4-XYLYLAZO)-1-NAPHTHOL-4-SULFONIC ACID, DISODIUM SALT ☐ USACERT FD & C RED No. 4

TOXICITY DATA WITH REFERENCE
orl-rat TDLo:1200 g/kg/78W-C:CAR GASTAB 23,1,53
orl-dog TDLo:182 g/kg/1Y-C:REP TXAPA9 8,306,66
orl-rat TD:524 g/kg/73W-C:ETA VPITAR 29,61,70
orl-rat LD50:>2 g/kg MEIEDD 11,1209,89

CONSENSUS REPORTS: IARC Cancer Review: Group 3 IMEMDT 7,56,87; Animal Sufficient Evidence IMEMDT 8,207,75. Reported in EPA TSCA Inventory.

SAFETY PROFILE: Low toxicity by ingestion. Questionable carcinogen with experimental carcinogenic and tumorigenic data. When heated to decomposition it emits toxic fumes of NO_x and SO_x.

FAG070 CAS:81-88-9 HR: 2
FD&C RED No. 19
mf: $C_{28}H_{31}N_2O_3$•Cl mw: 479.06

PROP: Violet crystals or powder. Sol in H_2O, EtOH; sltly sol in Me_2CO.

SYNS: ACID BRILLIANT PINK B ☐ ADC RHODAMINE B ☐ AIZEN RHODAMINE BH ☐ AKIRIKU RHODAMINE B ☐ BASIC VIOLET 10 ☐ BRILLIANT PINK B ☐ CALCOZINE RED BX ☐ CALCOZINE RHODAMINE BX ☐ 9-o-CARBOXYPHENYL-6-DIETHYLAMINO-3-ETHYLIMINO-3-ISOXANTHENE, 3-ETHOCHLORIDE ☐ (9-(o-CARBOXYPHENYL)-6-(DIETHYLAMINO)-3H-XANTHEN-3-YLIDENE) DIETHYLAMMONIUM CHLORIDE ☐ CERISE TONER X1127 ☐ CERTIQUAL RHODAMIEN ☐ C.I. 749 ☐ C.I. BASIC VIOLET 10 ☐ C.I. FOOD RED 15 ☐ COGILOR RED 321.10 ☐ COSMETIC BRILLIANT PINK BLUISH D CONC ☐ D&C RED No. 19 ☐ DIABASIC RHODAMINE B ☐ DIETHYL-m-AMINO-PHENOLPHTHALEIN HYDROCHLORIDE ☐ EDICOL SUPRA ROSE B ☐ ELCOZINE RHODAMINE B ☐ ERIOSIN RHODAMINE B ☐ FOOD RED 15 ☐ GERANIUM LAKE N ☐ HEXACOL RHODAMINE B EXTRA ☐ IKADA RHODAMINE B ☐ IRAGEN RED L-U ☐ MITSUI RHODAMINE BX ☐ 11411 RED ☐ RED NO 213 ☐ RHEONINE B ☐ RHODAMINE ☐ RHODAMINE S (RUSSIAN) ☐ SICILIAN CERISE TONER A-7127 ☐ SYMULEX MAGENTA F ☐ SYMULEX PINK F ☐ TAKAOKA RHODAMINE B ☐ TETRAETHYLDIAMINO-o-CARBOXY-PHENYL-XANTHENYL CHLORIDE ☐ TETRAETHYLRHODAMINE

TOXICITY DATA WITH REFERENCE
sln-dmg-orl 1000 ppm AMNTA4 87,295,53
cyt-mam:fbr 2 mg/L MUREAV 88,211,81
ipr-mus TDLo:60 mg/kg (female 7-10D post):REP TJADAB 40,143,89
scu-rat TDLo:3600 mg/kg/68W-I:ETA GANNA2 47,51,56
scu-rat TD:3870 mg/kg/68W-I:ETA GANNA2 46,369,55

CONSENSUS REPORTS: IARC Cancer Review: Group 3 IMEMDT 7,56,87. Reported in EPA TSCA Inventory. EPA Genetic Toxicology Program.

SAFETY PROFILE: Questionable carcinogen with experimental carcinogenic and tumorigenic data. Experimental reproductive effects. Mutation data reported. When heated to decomposition it emits very toxic fumes of NO_x, NH_3, and Cl^-.

FAG080 CAS:85-82-5 HR: 2
FD&C RED No. 32
mf: $C_{18}H_{16}N_2O$ mw: 276.36

PROP: Needles with green highlights from EtOH. Mp: 150–151°.

SYNS: FD&C ACID RED 32 ☐ OIL RED XO ☐ 1-XYLYLAZO-2-NAPHTHOL ☐ 1-(2,5-XYLYLAZO)-2-NAPHTHOL

TOXICITY DATA WITH REFERENCE
orl-mus TDLo:29 g/kg/52W:ETA BJCAAI 10,653,56

SAFETY PROFILE: Questionable carcinogen with experimental tumorigenic data. When heated to decomposition it emits toxic fumes of NO_x.

FAG120 CAS:1694-09-3 HR: 3
FD&C VIOLET No. 1
mf: $C_{39}H_{41}N_3O_6S_2$•Na mw: 734.94

SYNS: ACID VIOLET ☐ A.F. VIOLET No 1 ☐ AIZEN FOOD VIOLET No 1 ☐ BENZYL VIOLET ☐ BENZYL VIOLET 3B ☐ CALCOCID VIOLET 4BNS ☐ C.I. 42640 ☐ C.I. FOOD VIOLET 2 ☐ COOMASSIE VIOLET ☐ DISPERSED VIOLET 12197 ☐ FORMYL VIOLET S4BN ☐ PERGACID VIOLET 2B ☐ SOLAR VIOLET 5BN ☐ WOOL VIOLET

TOXICITY DATA WITH REFERENCE
mma-sat 320 μg/plate MUREAV 89,21,81
orl-rat TDLo:498 g/kg/28W-C:CAR JNCIAM 51,1337,73
scu-rat TDLo:9360 mg/kg/2Y-I:ETA FEPRA7 16,367,57

CONSENSUS REPORTS: IARC Cancer Review: Group 2B IMEMDT 7,56,87; Animal Sufficient Evidence IMEMDT 16,153,78. EPA Genetic Toxicology Program. Reported in EPA TSCA Inventory.

SAFETY PROFILE: Confirmed carcinogen with experimental carcinogenic and tumorigenic data. Mutation data reported. When heated to decomposition it emits very toxic fumes of NO_x, NH_3, Na_2O, and SO_x.

FAG130 **CAS:85-84-7** **HR: 2**
FD&C YELLOW No. 3
mf: $C_{16}H_{13}N_3$ mw: 247.32

PROP: Red plates from EtOH. Mp: 102–104°. Sol in EtOH and AcOH.

SYNS: A.F YELLOW No. 2 ☐ 1-BENZENE-AZO-β-NAPHTHYLAMINE ☐ 1-BENZENEAZO-2-NAPHTHYLAMINE ☐ CERISOL YELLOW AB ☐ C.I. 11380 ☐ C.I. FOOD YELLOW 10 ☐ C.I. SOLVENT YELLOW 5 ☐ DOLKWAL YELLOW AB ☐ EXT. D&C YELLOW No. 9 ☐ GRASAL YELLOW ☐ JAUNE AB ☐ OIL YELLOW A ☐ 1-(PHENYLAZO)-2-NAPHTHALENAMINE ☐ 1-(PHENYLAZO)-2-NAPHTHYLAMINE ☐ YELLOW AB ☐ YELLOW No. 2

TOXICITY DATA WITH REFERENCE
dns-rat-orl 500 mg/kg ENMUDM 7,101,85
orl-dog TDLo:45,625 mg/kg/1Y-C:REP TXAPA9 5,16,63
orl-rat TDLo:8190 mg/kg/65W-C:ETA JPPMAB 7,591,55
orl-rbt LDLo:1000 mg/kg JBCHA3 27,403,16
scu-rbt LDLo:1000 mg/kg JBCHA3 27,403,16

CONSENSUS REPORTS: IARC Cancer Review: Group 3 IMEMDT 7,56,87; Animal No Evidence IMEMDT 8,279,75. Reported in EPA TSCA Inventory. EPA Genetic Toxicology Program.

SAFETY PROFILE: Moderately toxic by ingestion and subcutaneous routes. Questionable carcinogen with experimental tumorigenic data. Mutation data reported. When heated to decomposition it emits toxic fumes of NO_x.

FAG135 **CAS:131-79-3** **HR: 3**
FD&C YELLOW No. 4
mf: $C_{17}H_{15}N_3$ mw: 261.35

SYNS: A.F. YELLOW No. 3 ☐ CERISOL YELLOW TB ☐ C.I. 11390 ☐ C.I. FOOD YELLOW 11 ☐ DOLKWAL YELLOW OB ☐ EXT. D&C YELLOW No. 10 ☐ JAUNE OB ☐ 1-(2-METHYLPHENYL)AZO-2-NAPHTHALENAMINE ☐ 1-((2-METHYLPHENYL)AZO)-2-NAPHTHALENAMINE ☐ 1-(2-METHYLPHENYL)AZO-2-NAPHTHYLAMINE ☐ OIL YELLOW OB ☐ o-TOLUENE-1-AZO-2-NAPHTHYLAMINE ☐ 1-(o-TOLYLAZO)-2-NAPHTHYLAMINE ☐ YELLOW OB ☐ ZLUT MASELNA OB ☐ ZLUT ROZPOUSTEDLOVA 6

TOXICITY DATA WITH REFERENCE
mma-sat 50 μg/plate CANCAR 49,1970,82
orl-rat TDLo:36,400 mg/kg/2Y-C:REP TXAPA9 5,16,63
orl-dog TDLo:45,625 mg/kg/1Y-C:REP TXAPA9 5,16,63
scu-rat TDLo:700 mg/kg/2Y-I:ETA FEPRA7 16,367,57
orl-rat LD50:120 mg/kg 85JCAE-,1315,86
orl-mus LDLo:1 g/kg 85JCAE-,1315,86
orl-rbt LDLo:1 g/kg JBCHA3 27,403,16
ipr-rbt LDLo:1 g/kg JBCHA3 27,403,16
scu-rbt LDLo:1 g/kg JBCHA3 27,403,16

CONSENSUS REPORTS: IARC Cancer Review: Group 3 IMEMDT 7,56,87; Animal Sufficient Evidence IMEMDT 8,287,75. EPA Genetic Toxicology Program.

SAFETY PROFILE: A poison by ingestion. Moderately toxic by intraperitoneal and subcutaneous routes. Questionable carcinogen with experimental tumorigenic data. May be contaminated with the carcinogen β-naphthylamine. Mutation data reported. When heated to decomposition it emits toxic fumes of NO_x. See also AROMATIC AMINES.

FAG140 **CAS:1934-21-0** **HR: 1**
FD&C YELLOW No. 5
mf: $C_{16}H_9N_4O_9S_2 \cdot 3Na$ mw: 534.38

PROP: Yellow-orange powder. Sol in water, conc sulfuric acid.

SYNS: ACID LEATHER YELLOW T ☐ ACID YELLOW 23 ☐ ACID YELLOW T ☐ ACILAN YELLOW GG ☐ A.F. YELLOW NO. 4 ☐ AIREDALE YELLOW T ☐ AIZEN TARTRAZINE ☐ AMACID YELLOW T ☐ ATUL TARTRAZINE ☐ BUCACID TARTRAZINE ☐ CALCOCID YELLOW MCG ☐ CALCOCID YELLOW XX ☐ CANACERT TARTRAZINE ☐ 3-CARBOXY-5-HYDROXY-1-p-SULFOPHENYL-4-p-SULFOPHENYLAZOPYRAZOLE TRISODIUM SALT ☐ CERTICOL TARTRAZOL YELLOW S ☐ CILEFA YELLOW T ☐ C.I. 640 ☐ C.I. 19140 ☐ C.I. ACID YELLOW 23 ☐ C.I. ACID YELLOW 23, TRISODIUM SALT ☐ C.I. FOOD YELLOW 4 ☐ CURON FAST YELLOW 5G ☐ D and C YELLOW NO. 5 ☐ DOLKWAL TARTRAZINE ☐ DYE FD and C YELLOW NO. 5 ☐ E 102 ☐ EDICOL SUPRA TARTRAZINE N ☐ EGG YELLOW A ☐ ERIO TARTRAZINE ☐ EUROCERT TARTRAZINE ☐ FD & C YELLOW NO. 5 TARTRAZINE ☐ FENAZO YELLOW T ☐ FOOD YELLOW 4 ☐ FOOD YELLOW 5 ☐ FOOD YELLOW NO. 4 ☐ HD TARTRAZINE ☐ HD TARTRAZINE SUPRA ☐ HEXACERT YELLOW NO. 5 ☐ HEXACOL TARTRAZINE ☐ HIDAZID TARTRAZINE ☐ HISPACID FAST YELLOW T ☐ HYDRAZINE YELLOW ☐ HYDROXINE YELLOW L ☐ KAKO TARTRAZINE ☐ KAYAKU FOOD COLOUR YELLOW NO. 4 ☐ KAYAKU TARTRAZINE ☐ KCA FOODCOL TARTRAZINE PF ☐ KCA TARTRAZINE PF ☐ KITON YELLOW T ☐ LAKE YELLOW ☐ LEMON YELLOW A ☐ LEMON YELLOW A GEIGY ☐ L-GELB 2 ☐ MAPLE TARTRAZOL YELLOW ☐ MITSUI TARTRAZINE ☐ NAPHTOCARD YELLOW O ☐ NEKLACID YELLOW T ☐ OXANAL YELLOW T ☐ SAN-EI TARTRAZINE ☐ SCHULTZ NO. 737 ☐ SUGAI TARTRAZINE ☐ TARTAR YELLOW N ☐ TARTAR YELLOW S ☐ TARTAR YELLOW FS ☐ TARTAR YELLOW PF ☐ TARTRAN YELLOW ☐ TARTRAPHENINE ☐ TARTRAZINE ☐ TARTRAZINE A EXPO T ☐ TARTRAZINE B ☐ TARTRAZINE B.P.C. ☐ TARTRAZINE EXTRA PURE A ☐ TARTRAZINE FD & C YELLOW #5 ☐ TARTRAZINE FQ ☐ TARTRAZINE G ☐ TARTRAZINE LAKE ☐ TARTRAZINE LAKE YELLOW N ☐ TARTRAZINE M ☐ TARTRAZINE MCGL ☐ TARTRAZINE N ☐ TARTRAZINE NS ☐ TARTRAZINE O ☐ TARTRAZINE T ☐ TARTRAZINE XX ☐ TARTRAZINE XXX ☐ TARTRAZINE YELLOW ☐ TARTRAZOL BPC ☐ TARTRAZOL YELLOW ☐ TARTRINE YELLOW O ☐ TRISODIUM 3-CARBOXY-5-HYDROXY-1-p-SULFOPHENYL-4-p-SULFOPHENYLAZOPYRAZOLE ☐ TRISODIUM SALT of 3-CARBOXY-5-HYDROXY-1-SULFOPHENYLAZOPYRAZOLE ☐ UNITERTRACID YELLOW TE ☐ USACERT YELLOW NO. 5 ☐ VONDACID TARTRAZINE ☐ WOOL YELLOW ☐ XYLENE FAST YELLOW GT ☐ Y-4 ☐ 1310 YELLOW ☐ 1409 YELLOW ☐ YELLOW LAKE 69 ☐ YELLOW NO. 5 ☐ YELLOW NO. 5 FDC ☐ ZLUT KYSELA 23 ☐ ZLUT PIGMENT 100 ☐ ZLUT POTRAVINARSKA 4

TOXICITY DATA WITH REFERENCE
cyt-hmn:lym 100 mg/L SOGEBZ 11,528,75
mic-mus:lym 1 g/L SCIEAS 236,933,87
cyt-ham:lng 2 g/L GANMAX 27,95,81
orl-rat LD50:>10 g/kg FCTXAV 5,747,67
ipr-rat LD50:3800 mg/kg FCTXAV 5,747,67
orl-mus LD50:>6 g/kg FCTXAV 5,747,67
ipr-mus LD50:4600 mg/kg FCTXAV 5,747,67

CONSENSUS REPORTS: Reported in EPA TSCA Inventory. EPA Genetic Toxicology Program.

SAFETY PROFILE: Low toxicity by ingestion. Mutation data reported. When heated to decomposition it emits very toxic fumes of NO_x, SO_x, and Na_2O.

F

FAG150 **CAS:2783-94-0** **HR: 1**
FD&C YELLOW No. 6
mf: $C_{16}H_{10}N_2O_7S_2Na_2$ mw: 452.36

PROP: Orange powder. Sol in water, conc sulfuric acid; sltly sol in abs alc.

SYNS: ACID YELLOW TRA □ A.F. YELLOW NO. 5 □ AIZEN FOOD YELLOW NO. 5 □ ALABASTER NO. 3 □ ATUL SUNSET YELLOW FCF □ CANACERT SUNSET YELLOW FCF □ CERTICOL SUNSET YELLOW CFS □ CERTOLAKE SUNSET YELLOW □ C.I. 15985 □ C.I. FOOD YELLOW 3 □ C.I. FOOD YELLOW 3, DISODIUM SALT □ CILEFA ORANGE S □ DISODIUM SALT of 1-p-SULPHOPHENYLAZO-2-NAPHTHOL-6-SULPHONIC ACID □ DISPERSED ORANGE 11348 □ DISPERSED YELLOW 12116 □ DOLKWAL SUNSET YELLOW □ DYE FDC YELLOW LAKE 6 □ DYE FD & C YELLOW LAKE 6 □ DYE FDC YELLOW NO. 6 □ DYE FD & C YELLOW NO. 6 □ DYE SUNSET YELLOW □ E 110 □ EDICOL SUPRA YELLOW FC □ ENIACID SUNSET YELLOW □ EUROCERT ORANGE FCF □ FD & C NO. 6 □ FD and C NO. 6 □ FD and C YELLOW 6 □ FD and C YELLOW LAKE NO. 6 □ FD and C YELLOW NO. 6 □ FD & C YELLOW NO. 6 ALUMINIUM LAKE □ FDC YELLOW NO. 6 □ FOODCOL SUNSET YELLOW FCF □ FOOD YELLOW 3 □ FOOD YELLOW 6 □ GELBORANGE-S □ HD SUNSET YELLOW FCF □ HD SUNSET YELLOW FCF SUPRA □ HEXACOL SUNSET YELLOW F & F SUPRA □ HEXACOL SUNSET YELLOW FCF □ HEXACOL SUNSET YELLOW FCF SUPRA □ HEXACOL SUNSET YELLOW FCP □ 6-HYDROXY-5-((p-SULFOPHENYL)AZO)-2-NAPHTHALENESULFONIC ACID, DISODIUM SALT □ 6-HYDROXY-5-((4-SULFOPHENYL)AZO)-2-NAPHTHALENESULFONIC ACID, DISODIUM SALT □ 6-HYDROXY-5-((p-SULPHOPHENYL)AZO)-2-NAPHTHALENESULPHONIC ACID, DISODIUM SALT □ 6-HYDROXY-5-((4-SULPHOPHENYL)AZO)-2-NAPHTHALENESULPHONIC ACID, DISODIUM SALT □ KCA FOODCOL SUNSET YELLOW FCF □ JAUNE ORANGE S □ JAUNE SOLEIL □ L-ORANGE 2 □ L. ORANGE Z2010 □ MAPLE SUNSET YELLOW FCF □ NCI-C53907 □ ORANGE PAL □ ORANGE II R □ ORANGE RGL CONC. SPECIALLY PURE □ ORANGE YELLOW S □ ORANGE YELLOW S.AF □ ORANGE YELLOW S.FQ □ PARA ORANGE □ STANDACOL SUNSET YELLOW FCF □ 1-p-SULFOPHENYLAZO-2-HYDROXYNAPHTHALENE-6-SULFONATE, DISODIUM SALT □ 1-p-SULFOPHENYLAZO-2-NAPHTHOL-6-SULFONIC ACID, DISODIUM SALT □ 1-p-SULPHOPHENYLAZO-2-NAPHTHOL-6-SULPHONIC ACID, DISODIUM SALT □ SUN ORANGE A GEIGY □ SUNSET YELLOW □ SUNSET YELLOW BSS □ SUNSET YELLOW FCF □ SUNSET YELLOW FCF SUPRA □ SUNSET YELLOW FU □ SUNSET YELLOW FU SUPRA □ SUNSET YELLOW LAKE □ SUN YELLOW □ SUN YELLOW A-CE □ SUN YELLOW A-FDC □ SUN YELLOW EXTRA CONC. A EXPORT □ SUN YELLOW EXTRA PURE A □ SUN YELLOW FCF □ USACERT YELLOW NO. 6 □ USACERT FD & C YELLOW NO. 6 □ USALAKE FD & C YELLOW NO. 6 LAKE □ 1351 YELLOW □ 1899 YELLOW □ YELLOW NO. 6 □ YELLOW ORANGE S □ YELLOW ORANGE S SPECIALLY PURE □ YELLOW ORANGE SPECIALLY PURE 85 □ YELLOW SF FOR FOOD □ YELLOW SUN □ YELLOW SY FOR FOOD □ ZLUT POTRAVINARSKA 3

TOXICITY DATA WITH REFERENCE
mic-mus:lym 1 g/L SCIEAS 236,933,87
cyt-ham:lng 2 g/L GANMAX 27,95,81
orl-rat LD50:>10 g/kg FCTXAV 5,747,67
ipr-rat LD50:3800 mg/kg FCTXAV 5,747,67
orl-mus LD50:>6 g/kg FCTXAV 5,747,67
ipr-mus LD50:4600 mg/kg FCTXAV 5,747,67

CONSENSUS REPORTS: IARC Cancer Review: Group 3 IMEMDT 7,56,87; Animal Inadequate Evidence IMEMDT 8,257,75. Reported in EPA TSCA Inventory.

SAFETY PROFILE: Low toxicity by ingestion and intraperitoneal routes. Mutation data reported. When heated to decomposition it emits very toxic fumes of NO_x and SO_x.

FAJ100 **CAS:4551-59-1** **HR: 3**
FELDENE
mf: $C_{15}H_{13}N_3O_4S$ mw: 331.37

PROP: Oil or crystals from methanol. Mp: 198–200°, bp: 182–188° @ 3 mm.

SYNS: CP 16171 □ 4-HYDROXY-2-METHYL-N-(2-PYRIDYL)-2H-1,2-BENZOTHIAZIN-3-CARBOXYAMID-1,1-DIOXID (GERMAN) □ 4-HYDROXY-2-METHYL-N-(2-PYRIDYL)-2H-1,2-BENZOTHIAZINE-3-CARBOXAMIDE-1,1-DIOXIDE □ PIROXICAM □ ROXICAM □ SOLOCALM

TOXICITY DATA WITH REFERENCE
orl-rat TDLo:110 mg/kg (female 7-17D post):REP YACHDS 8,4655,80
orl-rat TDLo:55 mg/kg (female 7-17D post):REP YACHDS 8,4655,80
orl-rat TDLo:110 mg/kg (7-17D preg):REP YACHDS 8,4655,80
orl-rat TDLo:35 mg/kg (15-21D preg):REP TXCYAC 30,59,84
orl-man TDLo:7636 mg/kg/6W-I:KID AJNED9 5,142,85
orl-cld TDLo:7143 µg/kg SAMJAF 66,31,84
orl-man TDLo:52 mg/kg/26W-I:SYS AIMEAS 99,282,83
orl-wmn TDLo:1200 µg/kg/3D-I:SKN JRHUA9 11,554,84
orl-man LDLo:3714 mg/kg/13D-I:BLD,SKN NEJMAG 309,795,83
orl-wmn LDLo:2800 µg/kg/1W-I:KID AIMDAP 144,63,84
orl-wmn TDLo:28 mg/kg AMSVAZ 216,335,84
orl-rat LD50:216 mg/kg ARZNAD 28,1714,78
ipr-rat LD50:335 mg/kg YACHDS 8,4639,80
scu-rat LD50:148 mg/kg YACHDS 8,4639,80
rec-rat LD50:400 mg/kg YACHDS 8,4639,80
orl-mus LD50:350 mg/kg YACHDS 8,4639,80
ipr-mus LD50:350 mg/kg YACHDS 8,4639,80
scu-mus LD50:300 mg/kg YACHDS 8,4639,80
orl-dog LD50:108 mg/kg YACHDS 8,4639,80

SAFETY PROFILE: A human poison by ingestion. Poison experimentally by ingestion, subcutaneous, intraperitoneal, and rectal routes. Human systemic effects by ingestion: interstitial nephritis, metabolic acidosis, dermatitis, agranulocytosis. An experimental teratogen. Other experimental reproductive effects. When heated to decomposition it emits toxic fumes of SO_x and NO_x.

FAJ150 **CAS:4551-59-1** **HR: 3**
FEMAMIDE
mf: $C_{19}H_{30}N_2O_3$ mw: 334.51

PROP: Bp: 182–188°. Sol in acetone, methanol, ethanol, ethyl acetate, benzene, chloroform, ether, mineral acids, practically insol in water, alkalies.

SYNS: N-(2-(DIETHYLAMINO)ETHYL)-2-ETHYL-2-PHENYLMALONAMIC ACID ETHYL ESTER □ FENALAMIDE □ FENAMIDE □ PHENAMIDE □ PHENYLAETHYLMALONSAEURE-AETHYLESTER-DIAETHYLAMINOAETHYL-AMID (GERMAN) □ Sch 5706 □ SH 30858 □ SPASMAMIDE

TOXICITY DATA WITH REFERENCE
orl-rat LD50:610 mg/kg AEPPAE 237,264,59
ipr-rat LD50:250 mg/kg AEPPAE 237,264,59
orl-mus LD50:400 mg/kg AEPPAE 237,264,59
ipr-mus LD50:210 mg/kg BCFAAI 111,293,72

SAFETY PROFILE: Poison by ingestion and intraperito-

neal routes. When heated to decomposition it emits toxic fumes of NO_x. See also ESTERS.

FAJ200 **CAS:56222-04-9** **HR: 2**
FEMOXETINE HYDROCHLORIDE
mf: $C_{20}H_{25}NO_2 \cdot ClH$ mw: 347.92

PROP: A solid. Mp: 191–193°.

SYNS: FG 4963 □ (3R-trans)-3-((4-METHOXYPHENOXY)METHYL)-1-METHYL-4-PHENYLPIPERIDINE HYDROCHLORIDE □ PIPERIDINE, 3-((4-METHOXYPHENOXY)METHYL)-1-METHYL-4-PHENYL-, HYDROCHLORIDE, (3R-trans)-

TOXICITY DATA WITH REFERENCE
orl-ham TDLo:21,600 mg/kg (female 6-15D post):REP LIFSAK 32,1193,83
orl-wmn TDLo:2576 mg/kg/23W-I APTOA6 58,253,86

SAFETY PROFILE: Human toxic effects by ingestion. Experimental reproductive effects. When heated to decomposition it emits toxic fumes of NO_x and HCl.

FAK000 **CAS:22224-92-6** **HR: 3**
FENAMIPHOS
mf: $C_{13}H_{22}NO_3PS$ mw: 303.39

PROP: A solid. Mp: 49°. Sltly sol in H_2O.

SYNS: O-AETHYL-O-(3-METHYL-4-METHYLTHIOPHENYL)-ISOPROPYLAMIDO-PHOSPHORSAEURE ESTER (GERMAN) □ BAY 68138 □ ENT 27,572 □ ETHYL-3-METHYL-4-(METHYLTHIO)PHENYL(1-METHYLETHYL)PHOSPHORAMIDATE □ ETHYL 4-(METHYLTHIO)-m-TOLYL ISOPROPYL PHOSPHOR AMIDATE □ ISOPROPYLAMINO-O-ETHYL-(4-METHYLMERCAPTO-3-METHYLPHENYL)PHOSPHATE □ 1-(METHYLETHYL)-ETHYL 3-METHYL-4-(METHYLTHIO)PHENYL PHOSPHORAMIDATE □ NEMACUR □ NSC 195106 □ PHANAMIPHOS

TOXICITY DATA WITH REFERENCE
orl-rat LD50:8 mg/kg BESAAT 15,116,69
ihl-rat LC50:91 mg/m³/4H 85DPAN -,-,71/76
skn-rat LD50:80 mg/kg FMCHA2-,C216,91
orl-mus LD50:22,700 μg/kg 85DPAN -,-,71/76
orl-dog LD50:10 mg/kg 28ZEAL 5,112,76
orl-cat LD50:10 mg/kg 85DPAN -,-,71/76
skn-rbt LD50:178 mg/kg DTLWS* 4,191,82
orl-gpg LD50:75 mg/kg 28ZEAL 5,112,76
orl-qal LD50:1 mg/kg EESADV 8,551,84
orl-dck LD50:1680 μg/kg TXAPA9 47,451,79
skn-dck LD50:24 mg/kg TXAPA9 47,451,79

CONSENSUS REPORTS: EPA Extremely Hazardous Substances List.

OSHA PEL: TWA 0.1 mg/m³ (skin)
ACGIH TLV: TWA 0.1 mg/m³ (skin)

SAFETY PROFILE: Poison by ingestion, inhalation, and skin contact. When heated to decomposition it emits very toxic fumes of NO_x, PO_x, and SO_x.

FAK100 **CAS:60168-88-9** **HR: 2**
FENARIMOL
mf: $C_{17}H_{12}Cl_2N_2O$ mw: 331.21

PROP: White, odorless crystals. Mp: 117–119°. Practically insol in water; sol in most organic solvents.

SYNS: BLOC □ (2-CHLOROPHENYL)-α-(4-CHLOROPHENYL)-5-PYRIMIDINEMETHANOL □ α-(2-CHLOROPHENYL)-α-(4-CHLOROPHENYL)-5-PYRIMIDINEMETHANOL □ EL 222 □ RIMIDIN □ RUBIGAN

TOXICITY DATA WITH REFERENCE
sln-asn 6 mg/L MUREAV 79,169,80
cyt-mus-orl 450 mg/kg DBANAD 36,1351,83
orl-rat TDLo:980 mg/kg (male 4D pre):REP TXAPA9 86,391,86
orl-rat LD50:2500 mg/kg FMCHA2 -,C254,89
orl-mus LD50:4500 mg/kg FMCHA2 -,C254,89

CONSENSUS REPORTS: Reported in EPA TSCA Inventory.

SAFETY PROFILE: Moderately toxic by ingestion. Experimental reproductive effects. Mutation data reported. When heated to decomposition it emits toxic fumes of Cl^- and NO_x.

FAL000 **CAS:13669-70-0** **HR: 3**
FENAZOXINE
mf: $C_{17}H_{19}NO$ mw: 253.37

SYNS: NEFOPAM □ 3,4,5,6,7 TETRAHYDRO-5-METHYL-1-PHENYL-1H-2,5-BENZOXAZOCINE

TOXICITY DATA WITH REFERENCE
orl-wmn TDLo:15 mg/kg:CNS,CVS BMJOAE 283,1508,81
orl-man TDLo:17 mg/kg:CNS,CVS BMJOAE 283,1508,81
orl-rat LD50:178 mg/kg DRUGAY 19,249,80
ivn-rat LD50:28 mg/kg DRUGAY 19,249,80
ims-rat LD50:57 mg/kg DRUGAY 19,249,80
orl-mus LD50:119 mg/kg DRUGAY 19,249,80
ipr-mus LD50:50 mg/kg DDEVD6 3,10,79
ivn-mus LD50:45 mg/kg DRUGAY 19,249,80
ims-mus LD50:53 mg/kg DRUGAY 19,249,80
orl-dog LD50:100 mg/kg DRUGAY 19,249,80

SAFETY PROFILE: Poison by ingestion, intravenous, intraperitoneal, and intramuscular routes. Human systemic effects by ingestion: hallucinations, distorted perceptions, excitement, motor activity changes, increased pulse rate without blood pressure fall and heart rate changes. When heated to decomposition it emits toxic fumes of NO_x.

FAM000 **CAS:63918-50-3** **HR: 3**
FENCAMINE HYDROCHLORIDE
mf: $C_{20}H_{28}N_6O_2 \cdot ClH$ mw: 421.00

SYNS: SICOCLOR □ 1-(1,3,7-TRIMETHYL-2,6-DIOXOPURIN-8-YL)-4-(2-PHENYL-1-METHYL)ETHYL-4-METHYL-ETHYLENEDIAMINE, HYDROCHLORIDE

TOXICITY DATA WITH REFERENCE
orl-rat LD50:508 mg/kg 27ZQAG -,229,72
ipr-rat LD50:93 mg/kg 27ZQAG -,229,72
orl-mus LD50:418 mg/kg 27ZQAG -,229,72
ipr-mus LD50:82 mg/kg 27ZQAG -,229,72

SAFETY PROFILE: Poison by intraperitoneal route. Moderately toxic by ingestion. A central nervous system stimulant. When heated to decomposition it emits very toxic fumes of Cl^- and NO_x.

F

FAM300 CAS:4695-62-9 HR: 1
d-FENCHONE
mf: $C_{10}H_{16}O$ mw: 152.26

SYNS: BICYCLO(2.2.1)HEPTAN-2-ONE, 1,3,3-TRIMETHYL-, (1R)-(9CI) □ (+)-FENCHONE □ d(+)-FENCHONE □ 2-NORBORNANONE, 1,3,3-TRIMETHYL-, (1R,4S)-(+)- □ 1,3,3-TRIMETHYLBICYCLO(2.2.1)HEPTAN-2-ONE □ (1R)-1,3,3-TRIMETHYLBICYCLO(2.2.1)HEPTAN-2-ONE □ d-1,3,3-TRIMETHYL-2-NORBORNANONE □ d-1,3,3-TRIMETHYL-2-NORCAMPHANONE

TOXICITY DATA WITH REFERENCE
orl-rat LD50:6160 mg/kg FCTXAV 2,327,64

CONSENSUS REPORTS: Reported in EPA TSCA Inventory.

SAFETY PROFILE: Low toxicity by ingestion. When heated to decomposition it emits acrid smoke and irritating vapors.

FAO000 CAS:13851-11-1 HR: 1
FENCHYL ACETATE
mf: $C_{12}H_{19}O_2$ mw: 195.31

SYNS: 1,3,3-TRIMETHYL-2-NORBORNANOL ACETATE □ 1,3,3-TRIMETHYL-2-NORBORNANYL ACETATE

TOXICITY DATA WITH REFERENCE
skn-rbt 500 mg/24H MOD FCTXAV 14,773,76

CONSENSUS REPORTS: Reported in EPA TSCA Inventory.

SAFETY PROFILE: A skin irritant. When heated to decomposition it emits acrid smoke and irritating fumes.

FAO100 CAS:53597-27-6 HR: 2
FENDOSAL
mf: $C_{25}H_{19}NO_3$ mw: 381.45

PROP: Crystals from acetic acid. Mp: 223–225° (decomp).

SYNS: ALNOVIN □ 3-(3-CARBOXY-4-HYDROXYPHENYL)-2-PHENYL-4,5-DIHYDRO-3H-BENZ(e)INDOLE □ 5-(4,5-DIHYDRO-2-PHENYL-3H-BENZ(e)INDOL-3-YL)-2-HYDROXYBENZOIC ACID □ 5-(4,5-DIHYDRO-2-PHENYL-3H-BENZ(e)INDOL-3-YL)SALICYLIC ACID □ FENDOZAL □ HP 129 □ P 71-0129

TOXICITY DATA WITH REFERENCE
orl-rat LD50:450 mg/kg AGACBH 8,209,78
orl-mus LD50:740 mg/kg AGACBH 8,209,78
ipr-mus LD50:510 mg/kg AGACBH 8,209,78
orl-rbt LD50:560 mg/kg AGACBH 8,209,78

SAFETY PROFILE: Moderately toxic by ingestion and intraperitoneal routes. When heated to decomposition it emits toxic fumes of NO_x.

FAO200 CAS:7698-97-7 HR: 3
FENESTREL
mf: $C_{16}H_{20}O_2$ mw: 244.36

PROP: A solid. Mp: 158–163°.

SYNS: 2-METHYL-3-ETHYL-4-PHENYL-4-CYCLOHEXENE CARBOXYLIC ACID □ 2-METHYL-3-ETHYL-4-PHENYL-Δ^4-CYCLOHEXENECARBOXYLIC ACID □ ORF 3858

TOXICITY DATA WITH REFERENCE
unr-rbt TDLo:290 µg/kg (female 1-29D post):TER CCPTAY 2,85,70
orl-mky TDLo:10 mg/kg (female 1-6D post):REP AJOGAH 115,101,73
orl-rat TDLo:22,500 ng/kg (1-9D preg):REP CCPTAY 3,347,71
orl-rat TDLo:100 ng/kg (female 7D pre):REP CCPTAY 3,347,71
orl-rat TDLo:27 mg/kg/52W-C:ETA TXAPA9 19,412,71
orl-rat LD50:2350 mg/kg TXAPA9 19,412,71
orl-mus LD50:680 mg/kg TXAPA9 19,412,71
ipr-mus LD50:150 mg/kg TXAPA9 19,412,71
ivn-dog LDLo:20 mg/kg TXAPA9 19,412,71

SAFETY PROFILE: Poison by intravenous and intraperitoneal routes. Moderately toxic by ingestion. An experimental teratogen. Other experimental reproductive effects. Questionable carcinogen with experimental tumorigenic data. When heated to decomposition it emits acrid smoke and irritating fumes.

FAP000 CAS:8006-84-6 HR: 2
FENNEL OIL

PROP: From steam distillation of *Foeniculum vulgare* Miller (Fam. Umbelliferae) (FCTXAV 12,807,74). Colorless to pale-yellow liquid; odor and taste of fennel.

SYNS: BITTER FENNEL OIL □ FENCHEL OEL (GERMAN) □ OIL of FENNEL

TOXICITY DATA WITH REFERENCE
skn-mus 100% SEV FCTXAV 12,879,74
skn-rbt 500 mg/24H MOD FCTXAV 12,879,74
skn-rbt 500 mg/24H FCTXAV 14,309,76
mma-sat 2500 µg/plate JAFCAU 30,563,82
orl-rat LD50:3120 mg/kg PHARAT 14,435,59
orl-mus LD50:3100 mg/kg TOFOD5 8,91,85

CONSENSUS REPORTS: Reported in EPA TSCA Inventory.

SAFETY PROFILE: Moderately toxic by ingestion. Mutation data reported. A severe skin irritant. When heated to decomposition it emits acrid smoke and irritating fumes.

FAP100 CAS:53746-45-5 HR: 3
FENOPROFEN CALCIUM DIHYDRATE
mf: $C_{30}H_{26}O_6\cdot Ca\cdot 2H_2O$ mw: 558.68

SYNS: CALCIUM-2-(m-PHENOXYPHENYL)PROPIONATE DIHYDRATE □ FENOPROFEN CALCIUM SALT DIHYDRATE □ FENOPRON □ FEPRONA □ LILLY 69323 □ (±)-α-METHYL-3-PHENOXYBENZENEACETIC ACID CALCIUM SALT DIHYDRATE □ NALFON □ NALGESIC □ dl-2-(3-PHENOXYPHENYL)-PROPIONIC ACID CALCIUM SALT, DIHYDRATE □ PROGESIC □ YM-09229

TOXICITY DATA WITH REFERENCE
orl-rat TDLo:330 mg/kg (7-17D preg):REP KSRNAM 14,4435,80
orl-rat TDLo:60 mg/kg (female 17-22D post):REP KSRNAM 14,4452,80

orl-rat TDLo: 90 mg/kg (female 17-22D post): REP KSRNAM 14,4452,80
orl-rbt TDLo: 975 mg/kg (female 6-18D post): TER KSRNAM 14,2950,80
orl-rat TDLo: 330 mg/kg (7-17D preg): TER KSRNAM 14,4435,80
orl-rbt TDLo: 50 mg/kg (female 1D pre): REP FESTAS 38,238,82
orl-rat TDLo: 18 mg/kg (female 17-22D post): REP KSRNAM 14,4452,80
orl-wmn TDLo: 5832 mg/kg/35W-I: GIT,KID AJMEAZ 72,81,82
orl-wmn TDLo: 720 mg/kg/30D-I JAMAAP 242,1896,79
orl-man TDLo: 521 mg/kg/17W-I: KID AIMEAS 93,508,80
orl-wmn LDLo: 8748 mg/kg/35W-I: KID,PUL AJMEAZ 72,81,82
orl-wmn TDLo: 1898 mg/kg/12W-I: KID AIMEAS 92,72,80
orl-rat LD50: 380 mg/kg IYKEDH 13,1128,82
ipr-rat LD50: 234 mg/kg YKYUA6 34,363,83
scu-rat LD50: 366 mg/kg IYKEDH 13,1128,82
ivn-rat LD50: 526 mg/kg YKYUA6 34,363,83
orl-mus LD50: 439 mg/kg YHTPAD 21,36,86
ipr-mus LD50: 286 mg/kg YKYUA6 34,363,83
scu-mus LD50: 463 mg/kg YKYUA6 34,363,83
ivn-mus LD50: 471 mg/kg YKYUA6 34,363,83

SAFETY PROFILE: Poison by ingestion, subcutaneous, and intraperitoneal routes. Moderately toxic by intravenous route. Human systemic effects by ingestion: dyspnea, nausea or vomiting, acute renal failure, acute tubular necrosis, interstitial nephritis, and other kidney damage. Experimental reproductive effects. An experimental teratogen. When heated to decomposition it emits acrid smoke and irritating fumes.

FAQ000 CAS:34691-31-1 HR: 3
FENOPROFEN SODIUM
mf: $C_{15}H_{13}O_3 \cdot Na$ mw: 264.27

SYN: d,l-2-(3-PHENOXYPHENYL)PROPIONIC ACID SODIUM SALT

TOXICITY DATA WITH REFERENCE
orl-rat LD50: 800 mg/kg TXAPA9 25,444,73
ipr-rat LD50: 234 mg/kg KSRNAM 14,4385,80
scu-rat LD50: 500 mg/kg TXAPA9 25,444,73
ivn-rat LD50: 500 mg/kg TXAPA9 25,444,73
ipr-mus LD50: 286 mg/kg KSRNAM 14,4385,80
scu-mus LD50: 463 mg/kg KSRNAM 14,4385,80
ivn-mus LD50: 471 mg/kg KSRNAM 14,4385,80

SAFETY PROFILE: Poison by intraperitoneal route. Moderately toxic by ingestion, subcutaneous, and intravenous routes. When heated to decomposition it emits toxic fumes of Na_2O. See also FENOPROFEN CALCIUM DIHYDRATE.

FAQ100 CAS:1944-12-3 HR: 3
FENOTEROL HYDROBROMIDE
mf: $C_{17}H_{21}NO_4 \cdot BrH$ mw: 384.31

PROP: Crystals from MeOH/Et_2O. Mp: 222–223°.

SYNS: BEROTEC □ BEROTEC HYDROBROMIDE □ 1-(3,5-DIHYDROXY-PHENYL-2-((1-(4-HYDROXYBENZYL)ETHYL)AMINO)-ETHANOL) HYDROBROMIDE □ FENOTEROL BROMIDE □ PARTUSISTEN □ PHENOTEROL HYDROBROMIDE □ Th 1165a □ TH 1165a HYDROBROMIDE

TOXICITY DATA WITH REFERENCE
orl-wmn TDLo: 200 μg/kg (1D pre): REP RDCNBM 5,31,81
orl-rat TDLo: 275 mg/kg (female 7-17D post): REP IYKEDH 12,742,81
orl-rat TDLo: 650 mg/kg (female 17-21D post): REP IYKEDH 12,742,81
orl-rat TDLo: 2600 mg/kg (female 17-21D post): REP IYKEDH 12,742,81
orl-rbt TDLo: 32,500 μg/kg (female 6-18D post): TER IYKEDH 12,742,81
orl-rat TDLo: 11 g/kg (female 7-17D post): TER ARZNAD 32,1518,82
orl-rat TDLo: 150 mg/kg (male 60D pre): REP IYKEDH 12,742,81
orl-rat TDLo: 11 g/kg (female 7-17D post): TER ARZNAD 32,1518,82
orl-rat TDLo: 5 g/kg (female 13-17D post): TER ARZNAD 32,1518,82
orl-rat LD50: 1600 mg/kg IYKEDH 11,542,80
scu-rat LD50: 1080 mg/kg IYKEDH 15,1140,84
ivn-rat LD50: 65 mg/kg IYKEDH 11,542,80
orl-mus LD50: 1990 mg/kg TXAPA9 18,185,71
scu-mus LD50: 1100 mg/kg TXAPA9 18,185,71

SAFETY PROFILE: Poison by intravenous route. Moderately toxic by ingestion and subcutaneous routes. An experimental teratogen. Human reproductive effects by ingestion: changes in the uterus, cervix or vagina. Experimental reproductive effects. When heated to decomposition it emits toxic fumes of Br^- and NO_x.

FAQ800 CAS:115-90-2 HR: 3
FENSULFOTHION
mf: $C_{11}H_{17}O_4PS_2$ mw: 308.37

PROP: Yellow oil. D: 1.202 @ 20°/4°, bp: 138–141° @ 0.01 mm. Sol in most org solvs; sltly sol in H_2O.

SYNS: BAY 25141 □ BAYER S767 □ CHEMAGRO 25141 □ DASANIT □ O,O-DIAETHYL-O-4-METHYLSULFINYL-PHENYL-MONOTHIOPHOSPHAT (GERMAN) □ O,O-DIETHYL-O-(p-(METHYLSULFINYL) PHENYL) PHOSPHOROTHIOATE □ O,O-DIETHYL-O-p-(METHYLSULFINYL)PHENYL THIOPHOSPHATE □ DMSP □ ENT 24,945 □ S 767 □ TERRACUR P

TOXICITY DATA WITH REFERENCE
ipr-rat TDLo: 30 mg/kg/60D-C: REP,ETA TXAPA9 6,78,64
orl-rat LD50: 2 mg/kg FMCHA2 -,C70,83
ihl-rat LC50: 113 mg/m³/1H 85JFAN A202,83
skn-rat LD50: 3 mg/kg WRPCA2 9,119,70
orl-gpg LDLo: 9 mg/kg TXAPA9 6,78,64
orl-qal LD50: 1200 μg/kg EESADV 8,551,84
orl-dck LD50: 747 μg/kg TXAPA9 22,556,72
skn-dck LD50: 3 mg/kg TXAPA9 47,451,79

CONSENSUS REPORTS: EPA Genetic Toxicology Program. EPA Extremely Hazardous Substances List.

OSHA PEL: TWA 0.1 mg/m³
ACGIH TLV: TWA 0.1 mg/m³

SAFETY PROFILE: A poison by ingestion, inhalation, and skin contact. Experimental reproductive effects. A

F

pesticide. When heated to decomposition it emits very toxic fumes of SO_x and PO_x.

FAQ900 **CAS:55-38-9** **HR: 3**
FENTHION
mf: $C_{10}H_{15}O_3PS_2$ mw: 278.34

PROP: A liquid with slight garlic odor. D: 1.25 @ 20°/4°, bp: 87° @ 0.01 mm. Very sltly sol in H_2O.

SYNS: BAY 29493 □ BAYCID □ BAYER 9007 □ BAYTEX □ O,O-DIMETHYL-O-4-(METHYLMERCAPTO)-3-METHYLPHENYL PHOSPHOROTHIOATE □ O,O-DIMETHYL-p-4-(METHYLMERCAPTO)-3-METHYLPHENYL THIOPHOSPHATE □ O,O-DIMETHYL-O-(3-METHYL-4-METHYLMERCAPTOPHENYL)PHOSPHOROTHIOATE □ O,O-DIMETHYL-O-(3-METHYL-4-METHYLTHIO-FENYL)-MONOTHIOFOSFAAT (DUTCH) □ O,O-DIMETHYL-O-(3-METHYL-4-METHYLTHIOPHENYL)-MONOTHIOPHOSPHAT (GERMAN) □ O,O-DIMETHYL-O-3-METHYL-4-METHYLTHIOPHENYL PHOSPHOROTHIOATE □ O,O-DIMETHYL-O-(3-METHYL-4-METHYLTHIO-PHENYL)-THIONOPHOSPHAT (GERMAN) □ O,O-DIMETHYL-O-(4-METHYLTHIO-3-METHYLPHENYL) PHOSPHOROTHIOATE □ O,O-DIMETHYL-O-(4-(METHYLTHIO)-m-TOLYL) PHOSPHOROTHIOATE □ O,O-DIMETIL-O-(3-METIL-4-METILTIO-FENIL)-MONOTIOFOSFATO (ITALIAN) □ DMTP □ ENT 25,540 □ENTEX □ LEBAYCID □ MERCAPTOPHOS □ 4-METHYLMERCAPTO-3-METHYLPHENYL DIMETHYL THIOPHOSPHATE □ MPP □ NCI-C08651 □ OMS 2 □ PHOSPHOROTHIOIC ACID-O,O-DIMETHYL-O-(3-METHYL-4-METHYLTHIOPHENYLE) (FRENCH) □ QUELETOX □ S 1752 □ SPOTTON □ TALODEX □ THIOPHOSPHATE de O,O-DIMETHYLE et de O-(3-METHYL-4-METHYLTHIOPHENYLE) (FRENCH) □ TIGUVON

TOXICITY DATA WITH REFERENCE
mma-sat 333 μg/plate ENMUDM 8(Suppl 7),1,86
sce-ham:lng 40 mg/L ENMUDM 4,621,82
orl-mus TDLo:1050 mg/kg (MGN):REP TXAPA9 26,29,73
ipr-mus TDLo:40 mg/kg (11D preg):TER TXAPA9 24,324,73
orl-mus TDLo:1050 mg/kg (MGN):REP TXAPA9 26,29,73
ipr-mus TDLo:40 mg/kg (11D preg):TER TXAPA9 24,324,73
orl-mus TDLo:1730 mg/kg/103W-C:ETA NCITR* NCI-CG-TR-103,79
orl-man TDLo:257 mg/kg:CVS,GIT JTCTDW 19,425,82
unr-hmn LD50:50 mg/kg DTLVS* 4,191,80
orl-rat LD50:180 mg/kg KSKZAN 16(2),59,78
ihl-rat LC50:800 mg/m³/4H 85INA8 6,646,91
skn-rat LD50:330 mg/kg TXAPA9 2,88,60
ipr-rat LDLo:260 mg/kg TXAPA9 6,86,64
orl-mus LD50:88,100 μg/kg HOEKAN 30,53,80
ihl-mus LCLo:1 g/m³/2H 85GYAZ -,27,71
skn-mus LD50:500 mg/kg OYYAA2 1,74,67
ipr-mus LD50:125 mg/kg TXAPA9 6,86,64
ivn-mus LD50:320 mg/kg CSLNX* NX#00142
ihl-rbt LCLo:1 g/m³/2H 85GYAZ -,27,71

CONSENSUS REPORTS: NCI Carcinogenesis Bioassay Completed; Results Negative: rat NCITR* NCI-CG-TR-103,79; Indefinite: mouse NCITR* NCI-CG-TR-103,79. EPA Genetic Toxicology Program.

OSHA PEL: TWA 0.2 mg/m³ (skin)
ACGIH TLV: TWA 0.2 mg/m³ (skin)
DFG MAK: 0.2 mg/m³

SAFETY PROFILE: A human poison by an unspecified route. Poison experimentally by ingestion, skin contact, intraperitoneal, intravenous, and intramuscular routes. Moderately toxic by inhalation. Human systemic effects by ingestion: pulse rate increase, hypermotility, diarrhea, nausea or vomiting. Experimental teratogenic and reproductive effects. Questionable carcinogen with experimental tumorigenic data. Mutation data reported. When heated to decomposition it emits very toxic fumes of PO_x and SO_x. See also MERCAPTANS and ESTERS.

FAQ930 **CAS:8066-27-1** **HR: 2**
FENTHIURAM
mf: $C_6H_{12}N_2S_4 \cdot C_6H_6Cl_6 \cdot C_6H_3Cl_3O \cdot 1/2Cu$ mw: 793.43

SYNS: FENTIURAM □ PHENTHIURAM □ THIOPEROXYDICARBONIC DIAMIDE, TETRAMETHYL-, mixture with (1-α-2-α-3-β,4-α- 5-α-6-β)-1,2,3,4,5,6-HEXACHLOROCYCLOHEXANE and TRICHLOROPHENOL COPPER(2+) SALT □ UNYSH A

TOXICITY DATA WITH REFERENCE
mnt-mus-ipr 50 mg/kg CYGEDX 21(1),57,87
mnt-ham:lng 15 mg/L CYGEDX 21(1),57,87
orl-rat TDLo:220 mg/kg (female 1-10D post):TER VETNAL 53(12),67,76
ipr-mus LD50:650 mg/kg CYGEDX 21(1),57,87

SAFETY PROFILE: Moderately toxic by intraperitoneal route. An experimental teratogen. Mutation data reported. When heated to decomposition it emits toxic fumes of SO_x and Cl^-.

FAQ950 **CAS:80830-42-8** **HR: 3**
FENTIAPRIL
mf: $C_{13}H_{15}NO_4S_2$ mw: 313.41

PROP: Crystals from EtOAc. Mp: 146–148° (decomp).

SYNS: (2R,4R)-2-(o-HYDROXYPHENYL)-3-(3-MERCAPTOPROPIONYL)-4-THIAZOLIDINECARBOX YLIC ACID □ PRESCRIN □ RENTIAPRIL □ SA 446 □ 4-THIAZOLIDINECARBOXYLIC ACID, 2-(2-HYDROXYPHENYL)-3-(3-MERCAPTO-1-OXOPROPYL)-, (2R-cis)-

TOXICITY DATA WITH REFERENCE
orl-rat TDLo:2700 mg/kg (female 17-22D post):REP ARZNAD 37,164,87
orl-rat TDLo:156 g/kg (female 13W pre):REP OYYAA2 31,245,86
orl-rat TDLo:39 g/kg (male 13W pre):REP OYYAA2 31,245,86
orl-rat LD50:8290 mg/kg OYYAA2 31,235,86
ipr-rat LD50:921 mg/kg OYYAA2 31,235,86
scu-rat LD50:1369 mg/kg OYYAA2 31,235,86
ivn-rat LD50:884 mg/kg OYYAA2 31,235,86
orl-mus LD50:8050 mg/kg OYYAA2 31,235,86
ivn-mus LD50:946 mg/kg OYYAA2 31,235,86
ivn-dog LD50:200 mg/kg OYYAA2 31,235,86

SAFETY PROFILE: Poison by intravenous route. Moderately toxic by subcutaneous and intraperitoneal routes. Experimental reproductive effects. When heated to decomposition it emits toxic fumes of NO_x and SO_x.

FAR000 **CAS:68990-15-8** **HR: 1**
FENUGREEK ABSOLUTE

PROP: Found in the seed of the plant *Trigonella foenum graecum L.* (FCTXAV 16,637,78).

TOXICITY DATA WITH REFERENCE
skn-rbt 500 mg/24H MOD FCTXAV 16,755,78

CONSENSUS REPORTS: Reported in EPA TSCA Inventory.

SAFETY PROFILE: A skin irritant. When heated to decomposition it emits acrid smoke and irritating fumes.

FAR100 CAS:51630-58-1 HR: 3
FENVALERATE
mf: $C_{25}H_{22}ClNO_3$ mw: 419.93

PROP: Clear, yellow, viscous liquid at 23°. D: 1.17, n (20/D) 1.5533, bp: 300° @ 37 mm. Solubility at 20° (g/L): acetone, >450; chloroform, >450; methanol, >450; hexane, 77. Insol in water. Decomp gradually between 150 and 300°.

SYNS: BELMARK □ α-CYANO-3-PHENOXYBENZYL-2-(4-CHLOROPHENYL)ISOVALERATE PYDRIN □ α-CYANO-3-PHENOXYBENZYL-2-(4-CHLOROPHENYL)-3-METHYLBUTYRATE □ CYANO(3-PHENOXYPHENYL)METHYL 4-CHLORO-α-(1-METHYLETHYL)BENZENEACETATE □ ECTRIN □ PHENVALERATE □ PYDRIN □ S 5602 □ SANMARTON □ SD 43775 □ SUMICIDIN □ SUMIFLY □ SUMIPOWER □ WL 43775

TOXICITY DATA WITH REFERENCE
cyt-hmn:lym 2 mg/L MUTAEX 4,72,89
sce-hmn:lym 10 mg/L MUTAEX 4,72,89
orl-dog TDLo:1138 mg/kg (26W pre):REP FAATDF 4,577,84
par-mus TDLo:100 mg/kg (male 5D pre):REP MUREAV 222,149,89
orl-rat LD50:70,200 μg/kg CHYCDW 21,215,87
ivn-rat LDLo:50 mg/kg ARTODN 45,325,80
orl-mus LD50:185 mg/kg FAATDF 5,278,85
ice-mus LDLo:200 μg/kg TXAPA9 66,290,82
skn-rbt LD50:2500 mg/kg FMCHA2 -,C104,83

CONSENSUS REPORTS: Cyanide and its compounds are on the Community Right-To-Know List.

SAFETY PROFILE: Poison by ingestion, intravenous, and intracerebral routes. Moderately toxic by skin contact. Experimental reproductive effects. Mutation data reported. Highly toxic to fish and bees. Corrosive, causes eye damage. A skin irritant. When heated to decomposition it emits toxic fumes of Cl^-, NO_x, and CN^-. See also CYANIDE.

FAR200 CAS:2441-88-5 HR: 3
FENYRIPOL HYDROCHLORIDE
mf: $C_{12}H_{13}N_3O{\cdot}ClH$ mw: 251.74

PROP: Crystals from EtOH. Mp: 167–168°.

SYNS: 2-(β-HYDROXY-β-PHENETHYLAMINO)-PYRIMIDINE HYDROCHLORIDE □ 2-(β-HYDROXY-β-PHENYL-ETHYL-AMINO)-PYRIMIDINE CHLORHYDRATE (FRENCH) □ IN 836 □ α-((2-PYRIMIDINYLAMINO)METHYL)BENZYL ALCOHOL HYDROCHLORIDE

TOXICITY DATA WITH REFERENCE
ipr-rat LD50:235 mg/kg AIPTAK 141,83,63
orl-mus LD50:850 mg/kg AIPTAK 141,83,63
ipr-mus LD50:455 mg/kg AIPTAK 141,83,63
ivn-mus LD50:137 mg/kg AIPTAK 141,83,63
ivn-dog LD50:125 mg/kg AIPTAK 141,83,63

SAFETY PROFILE: Poison by intravenous and intraperitoneal routes. Moderately toxic by ingestion. When heated to decomposition it emits toxic fumes of NO_x and HCl.

FAS000 CAS:14484-64-1 HR: 3
FERBAM
mf: $C_9H_{18}N_3S_6{\cdot}Fe$ mw: 416.51

PROP: Black solid or powder. Mp: 180° (decomp). Decomposes upon prolonged storage or in contact with moisture. Sltly sol in H_2O; sol in Me_2CO, $CHCl_3$, Py, and MeCN.

SYNS: AAFERTIS □ BERCEMA FERTAM 50 □ CARBAMATE □ DIMETHYLCARBAMODITHIOIC ACID, IRON COMPLEX □ DIMETHYLCARBAMODITHIOIC ACID, IRON(3+) SALT □ DIMETHYLDITHIOCARBAMIC ACID, IRON SALT □ DIMETHYLDITHIOCARBAMIC ACID, IRON(3+) SALT □ EISENDIMETHYLDITHIOCARBAMAT (GERMAN) □ EISEN(III)-TRIS(N,N-DIMETHYLDITHIOCARBAMAT) (GERMAN) □ ENT 14,689 □ FERBAM 50 □ FERBAM, IRON SALT □ FERBECK □ FERMATE FERBAM FUNGICIDE □ FERMOCIDE □ FERRADOW □ FERRIC DIMETHYLDITHIOCARBAMATE □ FUKLASIN ULTRA □ HEXAFERB □ HOKMATE □ IRON DIMETHYLDITHIOCARBAMATE □ KARBAM BLACK □ KNOCKMATE □ NIACIDE □ SUP'R FLO FERBAM FLOWABLE □ TRIFUNGOL □ TRIS(DIMETHYLCARBAMODITHIOATO-S,S')IRON □ TRIS(DIMETHYLDITHIOCARBAMATO)IRON □ TRIS (N,N-DIMETHYLDITHIOCARBAMATO) IRON(III) □ VANCIDE FE95

TOXICITY DATA WITH REFERENCE
mmo-sat 50 μg/plate CSHCAL 4,267,77
mmo-omi 1000 ppm MMAPAP 50,233,73
orl-rat LDLo:536 mg/kg (16-22D preg/21D post):REP TXAPA9 35,83,76
orl-rat TDLo:3479 mg/kg (16-22D preg/21D post):REP TXAPA9 35,83,76
scu-mus TDLo:41,760 μg/kg (female 6-14D post):TER NTIS** PB223-160
orl-rat TDLo:1140 mg/kg (6-15D preg):TER TXAPA9 35,83,76
orl-rat TDLo:1140 mg/kg (6-15D preg):REP TXAPA9 35,83,76
orl-mus TDLo:3500 mg/kg/78W-I:ETA NTIS** PB223-159
scu-mus TDLo:100 mg/kg:CAR NTIS** PB223-159
orl-rat LD50:1130 mg/kg FATOAO 32,356,69
ipr-rat LD50:2700 mg/kg JAPMA8 41,662,52
orl-mus LD50:3400 mg/kg JTEHD6 4,93,78
ipr-mus LDLo:63 mg/kg CBCCT* 4,228,52
orl-rbt LDLo:3 g/kg JAPMA8 41,662,52
ipr-rbt LDLo:1500 mg/kg JAPMA8 41,662,52

CONSENSUS REPORTS: IARC Cancer Review: Group 3 IMEMDT 7,56,87; Animal Inadequate Evidence IMEMDT 12,121,76. Reported in EPA TSCA Inventory. EPA Genetic Toxicology Program.

OSHA PEL: TWA Total Dust: 10 mg/m³; Respirable Fraction: 5 mg/m³
ACGIH TLV: TWA 10 mg/m³
DFG MAK: 15 mg/m³

SAFETY PROFILE: Poison by intraperitoneal route. Moderately toxic by ingestion. Experimental teratogenic and reproductive effects. Questionable carcinogen with experimental carcinogenic and tumorigenic data. Mutation data reported. A fungicide. When heated to decomposition it emits very toxic fumes of NO_x and SO_x. See also CARBAMATES.

FAS100 **CAS:50767-79-8** **HR: 1**
FERODIN SL
mf: $C_{16}H_{28}O_2$ mw: 252.44

SYNS: LITLURE A □ PHERODIN SL □ PRODLURE □ PT-2 □ 9,11-TETRADECADIEN-1-YL, ACETATE, (E,Z)-

TOXICITY DATA WITH REFERENCE
orl-rat LD50:>1 g/kg JPIFAN (40),32,82
orl-mus LD50:>1 g/kg JPIFAN (40),32,82

CONSENSUS REPORTS: Reported in EPA TSCA Inventory.

SAFETY PROFILE: Low toxicity by ingestion. When heated to decomposition it emits acrid smoke and irritating vapors.

FAS700 **CAS:1185-57-5** **HR: 1**
FERRIC AMMONIUM CITRATE
mf: $C_6H_8O_7 \cdot xFe \cdot xH_4N$

PROP: A complex salt of undetermined structure. Transparent green scales, granules, powder, or crystals; ammonia-like odor, mild iron-metallic taste. Sol in water; insol in alc. Deliquescent.

SYNS: CITRIC ACID, AMMONIUM IRON(3+) SALT □ FAC □ FERRIC AMMONIUM CITRATE, GREEN □ IRON(III) AMMONIUM CITRATE

CONSENSUS REPORTS: Reported in EPA TSCA Inventory.

ACGIH TLV: TWA 1 mg(Fe)/m^3

CONSENSUS REPORTS: Reported in EPA TSCA Inventory.

SAFETY PROFILE: When heated to decomposition it emits acrid smoke and irritating fumes.

FAU000 **CAS:7705-08-0** **HR: 3**
FERRIC CHLORIDE
DOT: UN 1773/UN 2582
mf: Cl_3Fe mw: 162.20

PROP: Black-brown solid or hygroscopic dark green or black crystals. Mp: 303°, bp: 315°, d: 2.90 @ 25°, vap press: 1 mm @ 194.0°. Aq solns are strongly acidic. Sol in H_2O to give hydrates; sol in MeOH and Et_2O.

SYNS: CHLORURE PERRIQUE □ FERRIC CHLORIDE (UN 1733) (DOT) □ FERRIC CHLORIDE, solution (UN 2582) (DOT) □ FLORES MARTIS □ IRON CHLORIDE □ IRON(III) CHLORIDE □ IRON TRICHLORIDE □ PERCHLORURE de FER

TOXICITY DATA WITH REFERENCE
oth-esc 500 nmol/tube LAMEDS 6,252,86
ivg-rat TDLo:29 mg/kg (1D pre):REP CCPTAY 4,91,71
itt-rat TDLo:12,976 μg/kg (1D male):REP JRPFA4 7,21,64
orl-rat LD50:450 mg/kg GISAAA 39(5),16,74
orl-mus LD50:895 mg/kg TRENAF 27,159,76
ivn-mus LD50:58 mg/kg YKKZAJ 87,677,67

CONSENSUS REPORTS: Reported in EPA TSCA Inventory. EPA Genetic Toxicology Program.

OSHA PEL: TWA 1 mg(Fe)/m^3
ACGIH TLV: TWA 1 mg(Fe)/m^3
DOT CLASSIFICATION: 8; *Label:* Corrosive

SAFETY PROFILE: Poison by ingestion and intravenous routes. Experimental reproductive effects. Corrosive. Probably an eye, skin, and mucous membrane irritant. Mutation data reported. Reacts with water to produce toxic and corrosive fumes. Catalyzes potentially explosive polymerization of ethylene oxide; chlorine + monomers (e.g., styrene). Forms shock-sensitive explosive mixtures with some metals (e.g., potassium, sodium). Violent reaction with allyl chloride. When heated to decomposition it emits highly toxic fumes of HCl.

FAW000 **CAS:10025-77-1** **HR: 3**
FERRIC CHLORIDE HEXAHYDRATE
mf: $Cl_3Fe \cdot 6H_2O$ mw: 270.32

PROP: Orange-brown very hygroscopic crystals. Mp: 37°. Sol in H_2O, EtOH, Me_2CO, and Et_2O.

SYNS: FERRIC TRICHLORIDE HEXAHYDRATE □ IRON(3+) CHLORIDE HEXAHYDRATE □ IRON(III), CHLORIDE HEXAHYDRATE □ IRON TRICHLORIDE HEXAHYDRATE

TOXICITY DATA WITH REFERENCE
dni-hmn:lym 4800 μmol/L IAAAAM 79,83,86
cyt-rat/ast 500 mg/kg GANNA2 54,155,63
orl-rat LDLo:900 mg/kg EQSSDX 1,1,75
ipr-mus LD50:260 mg/kg JAFCAU 14,512,66
ivn-rbt LDLo:7200 μg/kg EQSSDX 1,1,75

OSHA PEL: TWA 1 mg(Fe)/m^3
ACGIH TLV: TWA 1 mg(Fe)/m^3

SAFETY PROFILE: Poison by intraperitoneal and intravenous routes. Moderately toxic by ingestion. Human mutation data reported. Used as an astringent. When heated to decomposition it emits toxic fumes of Cl^-. See also CHLORIDES.

FAX000 **CAS:7783-50-8** **HR: 3**
FERRIC FLUORIDE
mf: F_3Fe mw: 112.85

PROP: White or green crystals. D: 3.87, mp: 1000°. Sltly sol in H_2O, HF, EtOH, and Et_2O.

SYNS: IRON FLUORIDE □ IRON TRIFLUORIDE

TOXICITY DATA WITH REFERENCE
oth-esc 250 nmol/tube LAMEDS 6,252,86
ipr-mus LD50:601 mg/kg COREAF 256,1043,63

CONSENSUS REPORTS: Reported in EPA TSCA Inventory.

OSHA PEL: TWA 2.5 mg(F)/m^3; TWA 1 mg(Fe)/m^3
ACGIH TLV: TWA 2.5 mg(F)/m^3; 1 mg(Fe)/m^3
NIOSH REL: TWA (Inorganic Fluorides) 2.5 mg(F)/m^3

SAFETY PROFILE: Moderately toxic by intraperitoneal route. Mutation data reported. When heated to decomposition it emits toxic fumes of F^-. See also FLUORIDES.

FAY000 **HR: 2**
FERRIC HYDROXIDE NITRILOTRIPROPIONIC ACID COMPLEX

PROP: Complex with 20.0% Fe (ONCOBS 19,239,65).

SYN: IRON(+3) HYDROXIDE COMPLEX with NITRILO-TRI-PROPIONIC ACID

TOXICITY DATA WITH REFERENCE
ims-mus TDLo:45 mg(Fe)/kg/3W-I:ETA ONCOAR 19,239,65

SAFETY PROFILE: Questionable carcinogen with experimental tumorigenic data. When heated to decomposition it emits toxic fumes of NO_x.

FAY200 **CAS:10421-48-4** **HR: 2**
FERRIC NITRATE
DOT: UN 1466
mf: N_3O_9•Fe mw: 241.88

SYNS: FERRIC NITRATE (DOT) □ IRON NITRATE □ IRON (III) NITRATE, ANHYDROUS □ IRON TRINITRATE □ NITRIC ACID, IRON (3+) SALT

TOXICITY DATA WITH REFERENCE
oth-esc 1250 mmol/tube LAMEDS 6,252,86

ACGIH TLV: TWA 1 mg/(Fe)/m^3
DOT CLASSIFICATION: 5.1; *Label:* Oxidizer

CONSENSUS REPORTS: Reported in EPA TSCA Inventory.

SAFETY PROFILE: Mutation data reported. A reactive oxidizer.

FAZ000 **HR: 2**
FERRIC NITROSODIMETHYL DITHIOCARBAMATE and TETRAMETHYL THIURAM DISULFIDE

PROP: 58.5% main component, 6.5% secondary component (NTIS** PB223-159).

SYNS: BIS(DIMETHYLTHIOCARBAMOYL)DISULFIDE and NITROSOTRIS(DIMETHYLDITHIOCARBAMATO)IRON □ TETRAMETHYLTHIURAM DISULFIDE mixed with FERRIC NITROSODIMETHYLDITHIOCARBAMATE □ VANGUARD GF

TOXICITY DATA WITH REFERENCE
orl-mus TDLo:27 g/kg/78W-I:CAR NTIS** PB223-159
scu-mus TDLo:46 mg/kg:ETA NTIS** PB223-159

SAFETY PROFILE: Questionable carcinogen with experimental carcinogenic and tumorigenic data. When heated to decomposition it emits very toxic fumes of NO_x and SO_x. See also N-NITROSO COMPOUNDS, CARBAMATES, and SULFIDES.

FBB000 **CAS:9007-73-2** **HR: 2**
FERRITIN

PROP: Prepared from rat liver protein by precipitation with a cadmium salt (BECCAN 39,74,61).

TOXICITY DATA WITH REFERENCE
cyt-ham:ovr 27 mg/L CNREA8 41,1628,81
scu-rat TDLo:224 mg/kg/15W-I:NEO BJCAAI 18,667,64

CONSENSUS REPORTS: Reported in EPA TSCA Inventory.

SAFETY PROFILE: Questionable carcinogen with experimental neoplastigenic data. Mutation data reported. When heated to decomposition it emits acrid smoke and irritating fumes.

FBB100 **HR: 2**
FERRLECIT

TOXICITY DATA WITH REFERENCE
ipl-rat TD:80 mg/kg:CAR CNREA8 2,157,86
ipr-mus LD50:229 mg(Fe)/kg NNAPBA 270(Suppl),R50,71

SAFETY PROFILE: Poison by intraperitoneal route. Questionable carcinogen with experimental carcinogenic data. When heated to decomposition it emits acrid smoke and irritating fumes.

FBC000 **CAS:102-54-5** **HR: 3**
FERROCENE
mf: $C_{10}H_{10}Fe$ mw: 186.05

PROP: Orange crystals from alc (aq); camphor odor. Mp: 172.5–173°, bp: 249°, subl @ >100°, volatile in steam. Insol in water; sol in alcohol and ether.

SYNS: BISCYCLOPENTADIENYLIRON □ DI-2,4-CYCLOPENTADIEN-1-YL IRON □ DICYCLOPENTADIENYL IRON (OSHA, ACGIH) □ IRON BIS(CYCLOPENTADIENE) □ IRON DICYCLOPENTADIENYL

TOXICITY DATA WITH REFERENCE
sln-dmg-par 100 ppm ENMUDM 7,87,85
trn-dmg-par 100 ppm ENMUDM 7,87,85
sce-ham:ovr 130 μg/L ENMUDM 7,1,85
ims-rat TDLo:5175 mg/kg/2Y-I:ETA NCIUS* PH 43-64-886,AUG,69
orl-rat LD50:1320 mg/kg SCCUR* -,5,61
ipr-rat LD50:500 mg/kg NCIUS* PH 43-64-886,JAN,65
orl-mus LD50:832 mg/kg BJPCAL 24,352,65
ipr-mus LD50:335 mg/kg NCIUS* PH 43-64-886,JAN,65
ivn-mus LD50:178 mg/kg CSLNX* NX#02382

CONSENSUS REPORTS: Reported in EPA TSCA Inventory.

OSHA PEL: TWA Total Dust: 10 mg/m^3; Respirable Fraction: 5 mg/m^3
ACGIH TLV: TWA 10 mg/m^3

SAFETY PROFILE: Poison by intraperitoneal and intravenous routes. Moderately toxic by ingestion. Questionable carcinogen with experimental tumorigenic data. Mutation data reported. Flammable; reacts violently with NH_4ClO_4. When heated to decomposition it emits acrid smoke and irritating fumes.

FBC100 **CAS:1336-80-7** **HR: 3**
FERROCHOLINATE
mf: $C_6H_{10}FeO_{10}$•$C_5H_{14}NO$ mw: 402.21

PROP: Greenish-brown, reddish-brown, or brown amorph solid or powder with glistening surface upon fracture. Sol in water, acids, and alkalies.

F

SYNS: CHELAFER ☐ CHEL-IRON ☐ FERRIC CHOLINE CITRATE ☐ FERROLIP ☐ IRON CHOLINE CITRATE COMPLEX

TOXICITY DATA WITH REFERENCE
orl-mus LD50:5500 mg/kg AJMSA9 241,296,61
ipr-mus LD50:151 mg/kg AJMSA9 241,296,61
ivn-mus LD50:210 mg/kg AJMSA9 241,296,61
ivn-dog LD50:140 mg/kg AJMSA9 241,296,61

SAFETY PROFILE: Poison by intravenous and intraperitoneal routes. Mildly toxic by ingestion. When heated to decomposition it emits toxic fumes of NO_x. See also CHOLINE.

FBD000 **CAS:11114-46-8** **HR: 3**
FERROCHROME (exothermic)

SYNS: CARBON FERROCHROMIUM ☐ CHROME FERROALLOY ☐ CHROMIUM ALLOY, Cr,C,Fe,N,Si ☐ CHROMIUM ALLOY, BASE, Cr,C,Fe, N,Si (FERROCHROMIUM) ☐ exothermic FERROCHROME ☐ FERROCHROME ☐ FERROCHROMIUM

CONSENSUS REPORTS: IARC Cancer Review: Animal Inadequate Evidence IMEMDT 23,205,80. Reported in EPA TSCA Inventory. Chromium and its compounds are on the Community Right-To-Know List.

OSHA PEL: TWA 1 mg(Cr)/m^3
ACGIH TLV: TWA 0.5 mg(Cr)/m^3; Not Classifiable as a Carcinogen.

SAFETY PROFILE: Poison by inhalation. Questionable carcinogen. See also CHROMIUM COMPOUNDS.

FBD100 **HR: 1**
FERROCYANIDES

CONSENSUS REPORTS: Cyanide and its compounds are on the Community Right-To-Know List.

SAFETY PROFILE: Ferrocyanides in general have low toxicity, but highly toxic decomposition products can form upon mixing them with hot concentrated acids. Acid, basic or neutral solutions of ferrocyanides liberate hydrocyanic acid upon strong irradiation. Fusion of mixtures of metal cyanides with metal chlorates, perchlorates, nitrates, or nitrites may cause violent explosions. When heated to decomposition or on contact with acids or acid fumes they emit toxic fumes of CN^-. See also CYANIDE and specific compounds.

FBD500 **CAS:17169-60-7** **HR: 3**
FERROGLYCINE SULFATE
mf: $C_2H_4FeNO_6S{\cdot}H$ mw: 226.99

SYNS: FERRATE(1-), (GLYCINATO-N,O)(SULFATO(2-)-O',O')-, HYDROGEN, (T-4)-(9CI) ☐ FERROGLYCINE SULFATE ☐ FERROGLYCINE SULFATE COMPLEX ☐ FERRONORD ☐ FERROSANOL ☐ GLYFERRO

TOXICITY DATA WITH REFERENCE
orl-rat LD50:5590 mg/kg AJMSA9 241,296,61
orl-mus LD50:1940 mg/kg AJMSA9 241,296,61
ipr-mus LD50:365 mg/kg AJMSA9 241,296,61

ACGIH TLV: TWA 1 mg(Fe)/m^3

SAFETY PROFILE: Poison by intraperitoneal route. Moderately toxic by ingestion. When heated to decomposition it emits toxic fumes of SO_x and NO_x.

FBE000 **CAS:12604-53-4** **HR: 3**
FERROMANGANESE (exothermic)

SYN: exothermic FERROMANGANESE (DOT)

CONSENSUS REPORTS: Reported in EPA TSCA Inventory. Manganese and its compounds are on the Community Right-To-Know List.

SAFETY PROFILE: The dust will burn violently and give off toxic fumes of MnO_2. See also MANGANESE COMPOUNDS.

FBG000 **CAS:8049-17-0** **HR: 3**
FERROSILICON
DOT: UN 1408
mf: FeSi mw: 83.90

PROP: Crystalline, metallic solid. Fe + Si, d: 5.4. Containing 30% or more but not more than 70% silicon (FEREAC 41,15972,76).

SYN: FERROSILICON, containing more than 30% but less than 90% SILICON (DOT)

TOXICITY DATA WITH REFERENCE
skn-rbt LD50:>20 g/kg NTIS** AD-A062-138
ihl-rat TCLo:194 mg/m^3/6H/26W-I GTPZAB 22(6),28,78

CONSENSUS REPORTS: Reported in EPA TSCA Inventory.

DOT CLASSIFICATION: 4.3; *Label:* Dangerous When Wet

SAFETY PROFILE: Moderate inhalation hazard. Low skin toxicity. Reaction with moisture releases hydrogen and acetylene gases, which then ignite; impurities in the alloy may liberate such poisonous and reactive gases as phosphine and arsine. Dry mixtures with sodium hydroxide react incandescently when water is added. Reaction with acid, acid fumes, or oxidizing materials can emit toxic fumes. Reaction hazards increase with decreasing particle size.

FBG200 **CAS:15669-07-5** **HR: 2**
FERROTREMOLITE
mf: $Ca_2Fe_5H_2O_{24}Si_8$ mw: 970.15

SYN: FERROACTINOLITE

TOXICITY DATA WITH REFERENCE
ipl-rat TDLo:80 mg/kg:CAR TOLED5 13,143,82
itr-rat TDLo:24 mg/kg/12W-I:CAR TOLED5 13,143,82

SAFETY PROFILE: Questionable carcinogen with experimental carcinogenic data.

FBH000 **CAS:3094-87-9** **HR: 2**
FERROUS ACETATE
mf: $C_4H_6O_4{\cdot}Fe$ mw: 173.95

PROP: White or colorless crystals.

SYNS: ACETIC ACID, IRON(2+) SALT ☐ IRON(2+) ACETATE ☐ IRON(II) ACETATE ☐ IRON DIACETATE

TOXICITY DATA WITH REFERENCE
scu-rat LD50:492 mg/kg 27ZWAY 3.2,1268,-

CONSENSUS REPORTS: Reported in EPA TSCA Inventory.

OSHA PEL: TWA 1 mg(Fe)/m^3
ACGIH TLV: TWA 1 mg(Fe)/m^3

SAFETY PROFILE: Moderately toxic by subcutaneous route. When heated to decomposition it emits acrid smoke and irritating fumes.

FBH050 CAS:14536-17-5 HR: 1
FERROUS ASCORBATE

PROP: Blue-violet solid.

SYN: IRON(II) ASCORBATE

SAFETY PROFILE: A nuisance dust.

FBH100 CAS:563-71-3 HR: 1
FERROUS CARBONATE
mf: $CFeO_3$ mw: 115.86

PROP: White solid; odorless; gray solid. Decomposes yielding CO_2 + FeO at 2°. Insol in H_2O; sol acids to give CO_2; sol in H_2O saturated with CO_2 to give $Fe(HCO_3)_2$ which then oxidizes.

SYN: IRON(II) CARBONATE

SAFETY PROFILE: A nuisance dust.

FBI000 CAS:7758-94-3 HR: 3
FERROUS CHLORIDE
DOT: UN 1759/UN 1760
mf: Cl_2Fe mw: 126.75

PROP: White crystals when pure; hygroscopic. Green to yellow, deliquescent crystals. Mp: 676°, bp: 1012°, d: 3.16, vap press: 10 mm @ 700°. Sol in H_2O; insol in Et_2O; sltly sol in C_6H_6.

SYNS: IRON(II) CHLORIDE (1:2) ☐ FERROUS CHLORIDE, solution (NA 1760) (DOT) ☐ FERROUS CHLORIDE, solid (NA 1759) (DOT) ☐ IRON DICHLORIDE ☐ IRON PROTOCHLORIDE

TOXICITY DATA WITH REFERENCE
otr-ham:emb 2500 μmol/L CNREA8 39,193,79
orl-rat LD50:450 mg/kg GISAAA 39(5),16,74
ipr-mus LD50:59 mg/kg AEPPAE 244,17,62

CONSENSUS REPORTS: Reported in EPA TSCA Inventory. EPA Genetic Toxicology Program.

OSHA PEL: TWA 1 mg(Fe)/m^3
ACGIH TLV: TWA 1 mg(Fe)/m^3
DOT CLASSIFICATION: 8; *Label:* Corrosive

SAFETY PROFILE: Poison by ingestion and intraperitoneal route. Mutation data reported. Corrosive. Probably an irritant to the eyes, skin, and mucous membranes. Can react violently with ethylene oxide, K, Na. When heated to decomposition it emits toxic fumes of Cl^-. See also CHLORIDES and IRON.

FBJ000 CAS:13478-10-9 HR: 3
FERROUS CHLORIDE TETRAHYDRATE
mf: $Cl_2Fe{\cdot}4H_2O$ mw: 198.83

PROP: Blue green crystals; readily oxidized in soln. Sol in H_2O, EtOH.

SYNS: IRON CHLORIDE TETRAHYDRATE ☐ IRON (II) CHLORIDE TETRAHYDRATE ☐ IRON(2+) CHLORIDE TETRAHYDRATE ☐ IRON DICHLORIDE TETRAHYDRATE

TOXICITY DATA WITH REFERENCE
rec-rat LDLo:498 mg/kg EQSSDX 1,1,75
ipr-mus LD50:93 mg/kg AEPPAE 244,17,62
orl-rbt LDLo:890 mg/kg EQSSDX 1,1,75
scu-rbt LDLo:189 mg/kg EQSSDX 1,1,75
rec-rbt LDLo:984 mg/kg EQSSDX 1,1,75

OSHA PEL: TWA 1 mg(Fe)/m^3
ACGIH TLV: TWA 1 mg(Fe)/m^3

SAFETY PROFILE: Poison by intraperitoneal and subcutaneous routes. Moderately toxic by ingestion and rectal routes. When heated to decomposition it emits toxic fumes of Cl^-.

FBJ100 CAS:141-01-5 HR: 3
FERROUS FUMARATE
mf: $C_4H_2O_4{\cdot}Fe$ mw: 169.91

PROP: Reddish-orange to reddish-brown granular powder or solid; odorless, almost tasteless. D: 2.435. Solubility at 25° in water: 0.14 g/100 mL; in alc <0.01 g/100 mL. Solubility in acid is limited by liberation of fumaric acid. Insol in EtOH; sltly sol in H_2O.

SYNS: CPIRON ☐ ERCO-FER ☐ ERCOFERRO ☐ FEOSTAT ☐ FEROTON ☐ FERROFUME ☐ FERRONAT ☐ FERRONE ☐ FERROTEMP ☐ FERRUM ☐ FERSAMAL ☐ FIRON ☐ FUMAFER ☐ FUMAR-F ☐ FUMIRON ☐ GALFER ☐ HEMOTON ☐ IRCON ☐ IRON FUMARATE ☐ METERFER ☐ METERFOLIC ☐ ONE-IRON ☐ PALAFER ☐ TOLERON ☐ TOLFERAIN ☐ TOLIFER

TOXICITY DATA WITH REFERENCE
orl-wmn LDLo:400 mg/kg:BAH,PUL HUTODJ 7,281,88
orl-rat LD50:3850 mg/kg NIIRDN 6,683,82
ipr-rat LD50:185 mg/kg NIIRDN 6,683,82
scu-rat LD50:500 mg/kg NIIRDN 6,683,82
orl-mus LD50:1570 mg/kg AJMSA9 241,296,61
ipr-mus LD50:480 mg/kg AJMSA9 241,296,61

CONSENSUS REPORTS: Reported in EPA TSCA Inventory.

OSHA PEL: TWA 1 mg(Fe)/m^3
ACGIH TLV: TWA 1 mg(Fe)/m^3

SAFETY PROFILE: Poison by intraperitoneal route. Moderately toxic by ingestion and subcutaneous routes. Human systemic effects by ingestion: dyspnea, nausea or vomiting, somnolence. When heated to decomposition it emits acrid smoke and irritating fumes. See also FUMARIC ACID.

FBK000 **CAS:299-29-6** **HR: 3**
FERROUS GLUCONATE
mf: $C_{12}H_{22}O_{14}$•Fe mw: 446.19

PROP: Yellowish-gray or pale greenish-yellow, fine powder or granules with slt odor of burned sugar. Sol in water and glycerin; insol in alc.

SYNS: FERGON □ FERGON PREPARATIONS □ FERLUCON □ FERRONICUM □ GLUCO-FERRUM □ IROMIN □ IRON GLUCONATE □ IROX (GADOR) □ NIONATE □ RAY-GLUCIRON

TOXICITY DATA WITH REFERENCE
scu-mus TDLo: 2600 mg/kg/13W-I: ETA,REP JNCIAM 24,109,60
orl-chd TDLo: 162 mg/kg: GIT JAMAAP 218,1179,71
orl-rat LD50: 2237 mg/kg NTIS** UR-3490-168
orl-mus LD50: 3950 mg/kg AJMSA9 241,296,61
ipr-mus LD50: 160 mg/kg AJMSA9 241,296,61
ivn-mus LD50: 199 mg/kg AJMSA9 241,296,61

CONSENSUS REPORTS: Reported in EPA TSCA Inventory.

OSHA PEL: TWA 1 mg(Fe)/m^3
ACGIH TLV: TWA 1 mg(Fe)/m^3

SAFETY PROFILE: Poison by intraperitoneal and intravenous routes. Moderately toxic by ingestion. Human systemic effects by ingestion: hypermotility, diarrhea, nausea, and vomiting. Experimental reproductive effects. Questionable carcinogen with experimental tumorigenic data. When heated to decomposition it emits acrid smoke and irritating fumes.

FBL000 **CAS:22830-45-1** **HR: 3**
FERROUS GLUCONATE DIHYDRATE
mf: $C_{12}H_{22}O_{14}$•Fe•$2H_2O$ mw: 483.23

PROP: Yellow-gray powder. Sol in H_2O; insol in EtOH.

TOXICITY DATA WITH REFERENCE
orl-rat LD50: 4500 mg/kg AJMSA9 230,491,55
orl-mus LD50: 3700 mg/kg AJMSA9 230,491,55
ivn-mus LD50: 98 mg/kg AJMSA9 230,491,55

OSHA PEL: TWA 1 mg(Fe)/m^3
ACGIH TLV: TWA 1 mg(Fe)/m^3

SAFETY PROFILE: Poison by intravenous route. Moderately toxic by ingestion. When heated to decomposition it emits acrid smoke and irritating fumes.

FBM000 **CAS:2896-87-9** **HR: 2**
FERROUS GLUTAMATE
mf: $C_5H_9FeNO_4$ mw: 203.00

SYN: GLUTAMIC ACID, IRON (2+) SALT (1:1)

TOXICITY DATA WITH REFERENCE
scu-mus TDLo: 2600 mg/kg/I: ETA BECCAN 40,30,62

SAFETY PROFILE: Questionable carcinogen with experimental tumorigenic data. When heated to decomposition it emits toxic fumes of NO_x.

FBN000 **CAS:15438-31-0** **HR: D**
FERROUS ION
mf: Fe mw: 55.85

SYNS: FERROUS □ IRON(2+) □ IRON (Fe 2+) □ IRON(II) ION

TOXICITY DATA WITH REFERENCE
mmo-nsc 200 μmol/L MAGDA3 10,249,79

SAFETY PROFILE: Mutation data reported.

FBN100 **CAS:7720-78-7** **HR: 3**
FERROUS SULFATE
mf: O_4S•Fe mw: 151.91

PROP: Grayish white to buff powder. Slowly sol in water; insol in alc.

SYNS: COPPERAS □ DURETTER □ DUROFERON □ EXSICCATED FERROUS SULFATE □ EXSICCATED FERROUS SULPHATE □ FEOSOL □ FEOSPAN □ FER-IN-SOL □ FERO-GRADUMET □ FERRALYN □ FERRO-GRADUMET □ FERROSULFAT (GERMAN) □ FERROSULFATE □ FERRO-THERON □ FERSOLATE □ GREEN VITRIOL □ IRON MONOSULFATE □ IRON PROTOSULFATE □ IRON(II) SULFATE (1:1) □ IRON VITRIOL □ IROSPAN □ IROSUL □ SLOW-FE □ SULFERROUS □ SULFURIC ACID, IRON(2^+) SALT (1:1)

TOXICITY DATA WITH REFERENCE
mmo-smc 100 mmol/L MUREAV 117,149,83
cyt-ham: fbr 1250 mg/L FCTOD7 22,623,84
orl-rat TDLo: 7200 mg/kg: TER OYYAA2 17,483,79
itt-rat TDLo: 12,153 mg/kg (1D male): REP JRPFA4 7,21,64
scu-mus TDLo: 1600 mg/kg/16W-I: ETA JNCIAM 24,109,60
orl-chd LDLo: 390 mg/kg JOPDAB 64,218,64
orl-wmn TDLo: 60 mg/kg: CNS,GIT JAMAAP 229,1333,74
orl-wmn TDLo: 10,560 μg/kg: GIT JAMAAP 236,2320,76
orl-chd TDLo: 20 mg/kg: BRN,CNS JOPDAB 94,147,79
orl-chd TDLo: 150 mg/kg: CNS,GIT NEJMAG 273,1124,65
orl-wmn TDLo: 600 mg/kg JAMAAP 229,1333,74
unr-man LDLo: 441 mg/kg 85DCAI 2,73,70
orl-rat LD50: 319 mg/kg JOPDAB 69,663,66
scu-rat LD50: 155 mg/kg NIIRDN 6,888,82
orl-mus LD50: 680 mg/kg BJPCAL 24,352,65
ipr-mus LD50: 289 mg/kg COREAF 256,1043,63
scu-mus LD50: 60,300 μg/kg NIIRDN 6,6888,82
ivn-mus LD50: 112 mg/kg AJMSA9 241,296,61
ivn-dog LD50: 79 mg/kg AJMSA9 241,296,61
idu-rbt LDLo: 200 mg/kg TXAPA9 66,329,82

CONSENSUS REPORTS: Reported in EPA TSCA Inventory. EPA Genetic Toxicology Program.

OSHA PEL: TWA 1 mg(Fe)/m^3
ACGIH TLV: TWA 1 mg(Fe)/m^3

SAFETY PROFILE: A human poison by ingestion. Moderately toxic to humans by an unspecified route. An experimental poison by ingestion, intraduodenal, intraperitoneal, intravenous, and subcutaneous routes. Human systemic effects by ingestion: aggression, somnolence, brain recording changes, diarrhea, nausea or vomiting, bleeding from the stomach, coma. Questionable carcinogen with experimental tumorigenic data. Experimental teratogenic and reproductive effects. Mutation data reported. Potentially explosive reaction with methyl isocyanoacetate at 25°. May ignite on contact with arsenic trioxide + sodium nitrate. When heated to decomposition it emits toxic fumes of SO_x. See also IRON COMPOUNDS.

FBO000 CAS:7782-63-0 HR: 3
FERROUS SULFATE HEPTAHYDRATE
mf: $O_4S{\cdot}Fe{\cdot}7H_2O$ mw: 278.05

PROP: Pale blue-green monoclinic, hygroscopic crystals or granules; odorless with a salt taste. D: 2.99–3.08, mp: 64°. Sol in H_2O to give $[Fe(H_2O)_6]_2$; acidic soln insol in EtOH.

SYNS: COPPERAS □ FEOSOL □ FER-IN-SOL □ FERO-GRADUMET □ FERROUS SULFATE (FCC) □ FESOFOR □ FESOTYME □ GREEN VITROL □ HAEMOFORT □ IRONATE □ IRON(II) SULFATE (1:1), HEPTAHYDRATE □ IRON VITROL □ IROSUL □ MOL-IRON □ PRESFERSUL □ SULFERROUS

TOXICITY DATA WITH REFERENCE
mmo-esc 30 μmol/L CIWYAO 49,144,50
orl-rat LDLo:1389 mg/kg EQSSDX 1,1,75
rec-rat LDLo:697 mg/kg EQSSDX 1,1,75
orl-mus LD50:1520 mg/kg AJMSA9 230,491,55
ipr-mus LD50:245 mg/kg ARZNAD 17,748,67
ivn-mus LD50:51 mg/kg AJMSA9 230,491,55
orl-rbt LDLo:2778 mg/kg EQSSDX 1,1,75
scu-rbt LDLo:279 mg/kg EQSSDX 1,1,75
ivn-rbt LDLo:99 mg/kg EQSSDX 1,1,75

OSHA PEL: TWA 1 mg(Fe)/m^3
ACGIH TLV: TWA 1 mg(Fe)/m^3

SAFETY PROFILE: Poison by intravenous, intraperitoneal, and subcutaneous routes. Moderately toxic by ingestion and rectal routes. Mutation data reported. When heated to decomposition it emits toxic fumes of SO_x.

FBP000 CAS:12604-58-9 HR: 2
FERROVANADIUM DUST

PROP: A gray to black dust.

CONSENSUS REPORTS: Reported in EPA TSCA Inventory.

OSHA PEL: TWA 1 mg/m^3; STEL 3 mg/m^3
ACGIH TLV: TWA 1 mg/m^3; STEL 3 mg/m^3
DFG MAK: 1 mg/m^3
NIOSH REL: (Vanadium) TWA 1.0 mg(V)/m^3

SAFETY PROFILE: Can cause pulmonary damage. Combustible when exposed to heat or flame. See also VANADIUM and IRON.

FBP050 CAS:53858-86-9 HR: 2
FERRUM

TOXICITY DATA WITH REFERENCE
ivn-rat LD50:1140 mg/kg KSRNAM 5,226,71
ims-rat LD50:1820 mg/kg KSRNAM 5,226,71
ivn-mus LD50:540 mg/kg KSRNAM 5,226,71
ims-mus LD50:4060 mg/kg KSRNAM 5,226,71

SAFETY PROFILE: Moderately toxic by intravenous and intramuscular routes.

FBP100 CAS:2624-43-3 HR: 2
FERTODUR
mf: $C_{23}H_{24}O_4$ mw: 364.47

PROP: Crystals from ethanol. Mp: 135–136°.

SYNS: 4-((4-(ACETYLOXY)PHENYL)CYCLOHEXYLIDENEMETHYL) PHENOL ACETATE □ BIS-(p-ACETOXYPHENYL)-CYCLOHEXYLIDENEMETHANE □ BIS-(p-HYDROXYPHENYL)-CYCLOHEXYLIDENEMETHANE DIACETATE □ CYCLOFENIL □ CYCLOFENYL □ CYCLOPENIL □ CYCLOPHENYL □ F 6066 □ H 3452 □ ICI 48213 □ NEOCLYM □ OGINEX □ ONDOGYNE □ ONDONID □ REHIBIN □ SANOCRISIN □ SEXADIENO □ SEXOVAR □ SEXOVID

TOXICITY DATA WITH REFERENCE
orl-man TDLo:1577 mg/kg (male 46D pre):REP ASUPAZ 73,199,68
orl-rat TDLo:48 mg/kg (female 9-14D post):REP OYYAA2 4,635,70
orl-mus TDLo:48 mg/kg (female 1-6D post):TER OYYAA2 4,645,70
orl-rat TDLo:24 mg/kg (female 1-6D post):TER OYYAA2 4,635,70
orl-mus TDLo:20 mg/kg (female 3-4D post):REP APTOA6 26(1),1,68
orl-mus TDLo:32 mg/kg (female 1-2D post):REP APTOA6 25,65,67
scu-rat TDLo:300 mg/kg (male 30D pre):REP APTOA6 23,365,66
orl-ham TDLo:900 mg/kg (female 1-9D post):REP CCPTAY 3,347,71
orl-ham TDLo:1800 mg/kg (female 1-9D post):REP CCPTAY 3,347,71
orl-rat TDLo:375 mg/kg (female 75D pre):REP APTOA6 23,365,66
orl-mus TDLo:6400 μg/kg (female 5D pre):REP APTOA6 23,365,66
orl-rat TDLo:175 mg/kg (male 7D pre):REP OYYAA2 4,835,70
orl-rat TDLo:24 mg/kg (female 1-6D post):TER OYYAA2 4,635,70
orl-rat TDLo:96 mg/kg (female 9-14D post):TER OYYAA2 4,635,70
ipr-rat LD50:1700 mg/kg OYYAA2 4,821,70
ipr-mus LD50:1080 mg/kg OYYAA2 4,821,70

SAFETY PROFILE: Moderately toxic by intraperitoneal route. Experimental teratogenic and reproductive effects. When heated to decomposition it emits acrid smoke and irritating fumes.

FBP200 CAS:537-98-4 HR: 2
trans-FERULIC ACID
mf: $C_{10}H_{10}O_4$ mw: 194.20

PROP: Orthorhombic needles (H_2O). Mp: 174°.

SYNS: FERULIC ACID □ 4-HYDROXY-3-METHOXYCINNAMIC ACID □ 3-(4-HYDROXY-3-METHOXYPHENYL)PROPENOIC ACID

TOXICITY DATA WITH REFERENCE
cyt-ham:ovr 25 g/L CALEDQ 14,251,81
orl-mam TDLo:9600 mg/kg (12D pre):REP SCIEAS 195,575,77
par-mus LDLo:1200 mg/kg CBCCT* 7,688,55

SAFETY PROFILE: Moderately toxic by parenteral route.

F

Experimental reproductive effects. Mutation data reported. When heated to decomposition it emits acrid smoke and irritating fumes.

FBP300 **CAS:483-57-8** **HR: 3**
FERVENULIN
mf: $C_7H_7N_5O_2$ mw: 193.19

PROP: Yellow, orthorhombic crystals. Mp: 178–179°. Sol in most common organic solvents; sol in cold water to about 2 mg/mL; in hot water to about 40 mg/mL. Practically insol in hydrocarbons.

SYNS: 10204-BII □ COMPOUND 7215 □ 6,8-DIMETHYLPYRIMIDO (5,4-e)-as-TRIAZINE-5,7(6H,8H)-DIONE □ 6,8-DIMETHYL-PYRIMIDO (5,4-e)-1,2,4-TRIAZINE-5,7(6H,8H)-DIONE □ FERVENULINE □ PLANOMYCIN □ PULANOMYCIN □ U-7118

TOXICITY DATA WITH REFERENCE
ipr-mus LD50:11,200 µg/kg 85GDA2 5,198,81
ivn-mus LD50:15 g/kg 85GDA2 5,199,81

SAFETY PROFILE: Poison by intraperitoneal route. When heated to decomposition it emits toxic fumes of NO_x.

FBP520 **HR: 3**
FETTERBUSH

PROP: Evergreen shrubs or small trees with simple, leathery leaves. They produce clusters of white flowers. *P. floribunda* is a bush that grows wild in the Atlantic coastal states of the United States from Virginia to Georgia. *P. japonica* is a small tree native to Japan that is cultivated in temperate areas of the United States, particularly along the Pacific coast.

SYNS: LILY-OF-THE-VALLEY BUSH □ PIERIS FLORIBUNDA □ PIERIS JAPONICA

SAFETY PROFILE: The leaves and nectar contain poisonous grayanotoxins (andromedotoxins). Ingestion of the leaves or honey made from the nectar causes an immediate burning pain that may be followed after a delay period of several hours by vomiting, diarrhea, headache, muscle weakness, vision impairment, slowed heartbeat, severe low blood pressure, convulsions, and coma. Fatalities have been reported among children.

FBP850 **CAS:33237-74-0** **HR: 3**
FIBORAN
mf: $C_{22}H_{30}N_2{\cdot}ClH$ mw: 359.00

SYNS: APRINDINE HYDROCHLORIDE □ N,N-DIETHYL-N′-2-INDANYL-N′-PHENYL-1,3-PROPANEDIAMINE HYDROCHLORIDE □ N-(2,3-DIHYDRO-1H-INDEN-2-YL)-N′,N′-DIETHYL-N-PHENYL-2,3-PROPANEDIAMINE MONOHYDROCHLORIDE □ MS-5075

TOXICITY DATA WITH REFERENCE
orl-rat TDLo:405 mg/kg (female 17-22D post):REP YACHDS 13,2687,85
orl-rat TDLo:810 mg/kg (female 17-22D post):REP YACHDS 13,2687,85
orl-rat TDLo:405 mg/kg (female 17-22D post):TER YACHDS 13,2687,85
orl-rat TDLo:440 mg/kg (7-17D preg):TER YACHDS 13,2661,85
ivn-rbt TDLo:78 mg/kg (female 6-18D post):REP YACHDS 15,1133,87
orl-rat TDLo:175 mg/kg (male 35D pre):REP YACHDS 13,2555,85
orl-rat TDLo:440 mg/kg (7-17D preg):TER YACHDS 13,2661,85
unr-wmn TDLo:82 mg/kg/31D-I:BLD AIMDAP 143,241,83
mul-wmn TDLo:468 mg/kg/8W-I:BLD AIMDAP 143,229,83
orl-rat LD50:525 mg/kg OYYAA2 27,353,84
ipr-rat LD50:68,200 µg/kg OYYAA2 27,353,84
scu-rat LD50:198 mg/kg OYYAA2 27,353,84
ivn-rat LD50:16,600 µg/kg OYYAA2 27,353,84
orl-mus LD50:262 mg/kg OYYAA2 27,353,84
ipr-mus LD50:56,600 µg/kg OYYAA2 27,353,84
scu-mus LD50:80,200 µg/kg OYYAA2 27,353,84
ivn-mus LD50:17,100 µg/kg OYYAA2 27,353,84
orl-dog LDLo:30 mg/kg OYYAA2 27,353,84

SAFETY PROFILE: Poison by ingestion, subcutaneous, intravenous, and intraperitoneal routes. Human systemic effects: agranulocytosis. An experimental teratogen. Experimental reproductive effects. When heated to decomposition it emits toxic fumes of NO_x and HCl.

FBQ000 **HR: 2**
FIBROUS GLASS

PROP: Is of a borosilicate variety, of low alkalinity, and consists of calcia-alumina-silicate (85INA8 5,270,86).

SYNS: FIBERGLASS □ FIBROUS GLASS DUST (ACGIH) □ GLASS □ GLASS FIBERS

TOXICITY DATA WITH REFERENCE
oms-hmn:fbr 10 mg/L MUREAV 116,369,83
oms-ham:ovr 10 mg/L MUREAV 116,369,83
ihl-rat TCLo:5 mg/m³/7H/90W-I:CAR NTIS** PB83-258111
ipr-rat TDLo:50 mg/kg:ETA IAPUDO 30,337,80
imp-rat TDLo:200 mg/kg:NEO JJIND8 67,965,81
ipr-rbt TDLo:25 mg/kg:ETA IAPUDO 30,337,80
ipr-ham TDLo:400 mg/kg:ETA IAPUDO 30,337,80
imp-rat TD:200 mg/kg:ETA IARCCD 8,289,73
imp-mus TD:1600 mg/kg:ETA BJURAN 36,225,64
ipl-rat TD:100 mg/kg:ETA IAPUDO 30,311,80

OSHA PEL: TWA 15 mg/m³ (total dust); 5 mg/m³ (nuisance dust)
ACGIH TLV: TWA 10 mg/m³ (dust)
NIOSH REL: TWA 5 mg/m³ (total fibrous glass)

SAFETY PROFILE: Suspected carcinogen with experimental carcinogenic, neoplastigenic, and tumorigenic data by inhalation and other routes. Human mutation data reported. Used as thermal and acoustic insulation.

The possibility of lung problems due to inhalation of fine particles or flakes or fibers of fiberglass has often been raised. The extensive medical research so far reported has shown no consistent evidence of chronic health effects in workers who are exposed to manmade vitreous fibers. In some studies where massive doses of fine-diameter fibers were implanted into mice, cancer development in the pleura was noted. Also some animal studies involving injection of fibers into the trachae resulted in a minimal fibrosis.

Exposure to glass fibers sometimes causes irritation of the skin and, less frequently, irritation of the eyes, nose, or throat. This is not an allergic reaction, but simply a mechanical irritation. Skin irritation typically is experienced by individuals who are newly exposed to fibrous glass and it usually diminishes after several days of exposure. Good personal and industrial hygiene practices minimize the amount of discomfort experienced.

FBS000 CAS:9001-33-6 HR: 3
FICIN

PROP: A proteolytic enzyme in the crude latex of the fig tree *Ficus* (JPETAB 71,20,41). White powder. Very sol in water.

SYNS: DEBRICIN □ FICUS PROTEASE □ FICUS PROTEINASE □ HIGUEROXYL DELABARRE □ TL 367

TOXICITY DATA WITH REFERENCE
orl-rat LD50:10 g/kg MEIEDD 10,585,83
orl-mus LD50:10 g/kg MEIEDD 10,585,83
ihl-mus LCLo:290 mg/m^3/10M NDRC** NDCrc-132,Sept,42
ihl-cat LCLo:290 mg/m^3/10M NDRC** NDCrc-132,Sept,42
orl-rbt LD50:5000 mg/kg JPETAB 71,20,41
ihl-rbt LCLo:290 mg/m^3/10M NDRC** NDCrc-132,Sept,42
orl-gpg LD50:5000 mg/kg JPETAB 71,20,41
ihl-gpg LCLo:290 mg/m^3/10M NDRC** NDCrc-132,Sept,42
ivn-mam LDLo:50 mg/kg JPETAB 71,20,41

CONSENSUS REPORTS: Reported in EPA TSCA Inventory.

SAFETY PROFILE: Poison by inhalation and intravenous routes. Mildly toxic by ingestion. When heated to decomposition it emits toxic fumes.

FBS100 HR: 3
FIGWORT

PROP: Annual or perennial herbs that grow to 3 feet and produce yellow, white, or occasionally red flowers. They grow in swampy areas throughout North America and Hawaii.

SYNS: BASSINET (CANADA) □ BLISTER FLOWER □ BLISTER WORT □ BOUTON d'OR (CANADA) □ BUTTER CRESS □ BUTTERCUP □ BUTTER DAISY □ CRAIN □ CROWFOOD □ DEVIL'S CLAWS □ GOLDBALLS □ GOLDWEED □ HORSE GOLD □ HUNGER WEED □ LESSER CELANDINE □ PILEWORT □ RAM'S CLAWS □ RANUNCULUS ACRIS □ RANUNCULUS BULBOSUS □ RANUNCULUS SCELERATUS □ RENONCULE (CANADA) □ SITFAST □ SPEARWORT □ ST. ANTHONY'S TURNIP □ STARVE-ACRE □ WATER CROWFOOT □ YELLOW GOWAN

SAFETY PROFILE: The roots contain the poison protoanemonin, which is a direct irritant and drying agent on the skin and mucous membranes. Ingestion causes pain and blistering of the mouth possibly followed by vomiting, diarrhea, cramps, bloody and painful urination, dizziness, fainting, confusion, and convulsions.

FBS200 CAS:8007-47-4 HR: 1
FIR BALSAM ABSOLUTE

SYNS: BALSAM OF FIR □ BALSAMS, CANADA □ CANADIAN BALSAM

TOXICITY DATA WITH REFERENCE
orl-rat LD50:>5 g/kg FCTXAV 13,449,75
skn-rbt LD50:>5 g/kg FCTXAV 13,449,75 FEREAC 54,7740,89

CONSENSUS REPORTS: Reported in EPA TSCA Inventory.

SAFETY PROFILE: Low toxicity by ingestion and skin contact. When heated to decomposition it emits acrid smoke and irritating vapors.

FBU000 CAS:59536-65-1 HR: 3
FIREMASTER BP-6

PROP: Consists mainly of penta-, hexa-, and heptabromobiphenyl, with lesser amounts of tetra- and other brominated biphenyls (ENVRAL 10,390,75).

SYNS: HEXABROMOBIPHENYL (technical grade) □ PBB □ POLYBROMINATED BIPHENYLS

TOXICITY DATA WITH REFERENCE
pic-esc 3100 ng/well MUREAV 260,349,91
dnd-sal:oth 8 μmol/L EVHPAZ 23,51,78
orl-rat TDLo:24,600 μg/kg (female 6-22D post):REP TOXID9 4,86,84
orl-pig TDLo:4 mg/kg (female 8-16W post):REP AJVRAH 42,183,81
orl-rat TDLo:215 mg/kg (8-22D preg/28D post):REP TXAPA9 59,300,81
orl-pig TDLo:41 mg/kg (female 8-16W post):REP AJVRAH 42,183,81
orl-mus TDLo:312 mg/kg (female 4-16D post):TER TXAPA9 37,171,76
orl-rat TDLo:11,250 μg/kg (female 7-15D post):TER EVHPAZ 23,295,78
orl-mus TDLo:312 mg/kg (female 4-16D post):REP TXAPA9 37,171,76
orl-rat TDLo:400 mg/kg (female 11D post):TER ASBDD9 2,211,79
orl-mus TDLo:144 mg/kg (female 7-18D post):TER ENVRAL 10,390,75
orl-rat TDLo:21,500 μg/kg (female 8-22D post):TER TXAPA9 59,300,81
orl-rat TD:1000 mg/kg:ETA EVHPAZ 23,265,78
orl-rat TD:1200 mg/kg/18W-I:CAR JJIND8 66,535,81
orl-qal LD50:15 μg/kg BECTA6 33,308,84

CONSENSUS REPORTS: IARC Cancer Review: Animal Inadequate Evidence IMEMDT 18,107,78. Polybrominated biphenyl compounds are on the Community Right-To-Know List.

SAFETY PROFILE: Poison by ingestion. Experimental teratogenic and reproductive effects. Questionable carcinogen with experimental carcinogenic and tumorigenic data. Mutation data reported. When heated to decomposition it emits very toxic Br^-. See also POLYBROMINATED BIPHENYLS.

FBU509 **CAS:67774-32-7** **HR: 3**
FIREMASTER FF-1

PROP: 2,4,5,2′,4′,5′-hexabromobiphenyl is the predominant isomer (LANCAO 2,602,77).

SYNS: 2,4,5,2′,4′,5′-HEXABROMOBIPHENYL □ PBB □ POLYBROMINATED BIPHENYL □ POLYBROMINATED BIPHENYL (FF-1)

TOXICITY DATA WITH REFERENCE
orl-rat TDLo: 10 mg/kg (female 20D pre): REP TOLED5 24,229,85
orl-rat TDLo: 1600 mg/kg (7-14D preg): REP JTEHD6 14,695,84
orl-mus TDLo: 900 mg/kg (female 6-15D post): REP TOXID9 4,87,84
orl-mus TDLo: 360 mg/kg (female 6-15D post): TER TOXID9 4,87,84
orl-rat TDLo: 375 mg/kg/26W-I: CAR NTPTR* NTP-TR-244,83
orl-mus TDLo: 1250 mg/kg/26W-I: CAR NTPTR* NTP-TR-244,83
orl-rat TD: 1000 mg/kg: NEO LANCAO 2,602,77
orl-rat TD: 1250 mg/kg/26W-I: CAR NTPTR* NTP-TR-244,83

CONSENSUS REPORTS: NTP 7th Annual Report on Carcinogens. IARC Cancer Review: Group 2B IMEMDT 7,321,87; Human Inadequate Evidence IMEMDT 41,261,86; Animal Sufficient Evidence IMEMDT 41,261,86. NTP Carcinogenesis Studies (gavage); Clear Evidence: mouse, rat NTPTR* NTP-TR-244,83. Polybrominated biphenyl compounds are on the Community Right-To-Know List.

SAFETY PROFILE: Confirmed carcinogen with experimental carcinogenic, neoplastigenic, and teratogenic data. Experimental reproductive effects. When heated to decomposition it emits very toxic fumes of Br^-. See also POLYBROMINATED BIPHENYLS.

FBV000 **CAS:8021-29-2** **HR: 1**
FIR NEEDLE OIL, SIBERIAN

PROP: Found in the needles and twigs of *Abies sibirica* Ledeb. (Fam. Pinaceae) (FCTXAV 13,449,75). Colorless to faintly yellow liquid. Sol in fixed oils, mineral oil; insol in glycerin, propylene glycol.

SYN: PINE NEEDLE OIL

TOXICITY DATA WITH REFERENCE
skn-hmn 12,500 µg/48H FCTXAV 13,450,75
skn-rbt 500 mg/24H MOD FCTXAV 13,450,75
orl-rat LD50: 10,200 mg/kg FCTXAV 13,450,75

CONSENSUS REPORTS: Reported in EPA TSCA Inventory.

SAFETY PROFILE: Mildly toxic by ingestion. A human and experimental skin irritant. When heated to decomposition it emits acrid smoke and irritating fumes.

FBV100 **HR: 1**
FISCHER'S SOLUTION

PROP: An amino acid solution for hepatic failure (IYKEDH 13,11,82).

SYN: GO-80

TOXICITY DATA WITH REFERENCE
ivn-rbt TDLo: 750 g/kg (male 30D pre): REP IYKEDH 13,24,82
ivn-rbt LDLo: 200 g/kg IYKEDH 13,11,82

SAFETY PROFILE: Very slight toxicity by intravenous route. Experimental reproductive effects. When heated to decomposition it emits acrid smoke and irritating fumes.

FBW000 **CAS:528-48-3** **HR: 3**
FISETIN
mf: $C_{15}H_{10}O_6$ mw: 286.25

PROP: Coloring matter isolated from *Rhus cotinus L. Anacardiaceae*. Yellow needles from dil alc. Decomp @ 330°. Sol in alc, acetone, acetic acid, solns of fixed alkali hydroxides; practically insol in water, ether, benzene, chloroform, and petr ether.

SYNS: BOIS BLEUDE HONQRIE □ C.I. 75620 □ C.I. NATURAL BROWN 1 □ COTININ □ 5-DESOXYQUERCETIN □ FIETIN □ FISETHOLZ □ FUSTEL □ FUSTET □ JUNGER FUSTIK □ SUPERFUSTEL □ SUPERFUSTEL K □ 3,3′,4′,7-TETRAHYDROXYFLAVONE □ UNGARISCHES GELBHOLZ □ VENTIN SUMACH □ VISET □ YOUNG FUSTIC □ YOUNG FUSTIC CRYSTALS □ ZANTE FUSTIC

TOXICITY DATA WITH REFERENCE
mmo-sat 100 µg/plate BCSTB5 5,1489,77
mma-sat 100 µg/plate MUREAV 58,231,78
dni-hmn:fbr 50 mg/L BCPCA6 33,3823,84
ivn-mus LD50: 180 mg/kg CSLNX* NX#01998

CONSENSUS REPORTS: Reported in EPA TSCA Inventory.

SAFETY PROFILE: Poison by intravenous route. Human mutation data reported. When heated to decomposition it emits acrid smoke and fumes.

FBW100 **HR: 2**
FISHTAIL PALM

PROP: A palm with long (up to 20 feet) leaves formed of small fishtail-shaped leaflets. It produces red or black berries clustered on long strings. They are native to Asia and grow wild and under cultivation in Florida, Hawaii, the West Indies, and Guam.

SYNS: CARIOTA (CUBA) □ CARYOTA (VARIOUS SPECIES) □ C. MITIS □ C. URENS □ WINE PALM

SAFETY PROFILE: The berries contain toxic calcium oxalate raphides. Chewing the berries results in burning pain in the lips, mouth and throat, possibly followed by inflammation and blistering. Systemic effects are usually not seen because of the insolubility of calcium oxalate. Skin contact may result in pain and itching. See also OXALATES.

FBW150 **CAS:487-26-3** **HR: 3**
FLAVANONE
mf: $C_{15}H_{12}O_2$ mw: 224.27

SYNS: 4H-1-BENZOPYRAN-4-ONE, 2,3-DIHYDRO-2-PHENYL-(9CI)

□ 2,3-DIHYDRO-2-PHENYL-4H-1-BENZOPYRAN-4-ONE □ 4-FLAVANONE

TOXICITY DATA WITH REFERENCE
orl-brd LD50:75 mg/kg AECTCV 12,355,83

CONSENSUS REPORTS: Reported in EPA TSCA Inventory.

SAFETY PROFILE: Poison by ingestion. When heated to decomposition it emits acrid smoke and irritating vapors.

FBY000 CAS:114-42-1 HR: 3
FLAVASPIDIC ACID
mf: $C_{24}H_{30}O_8$ mw: 446.54

PROP: Yellow or colorless crystals. Mp: 92°, mp: 156° (double mp).

SYNS: FLAVASPIDSAEURE (GERMAN) □ FLAVISPIDIC ACID BB □ POLYSTICHOCITRIN □ TOXIFREN

TOXICITY DATA WITH REFERENCE
orl-mus LD50:690 mg/kg APTOA6 25,33,67
ivn-mus LD50:94 mg/kg JPMSAE 54,1362,65

SAFETY PROFILE: Poison by intravenous route. Moderately toxic by ingestion. When heated to decomposition it emits acrid smoke and irritating fumes.

FBZ000 CAS:29382-82-9 HR: 3
FLAVENSOMYCIN
mf: $C_{47}H_{65}O_{14}$ mw: 854.12

PROP: Pale-yellow crystals. Mp: 152°, mp: 250–300° (decomp). An antibiotic produced by a strain of *Streptomyces sp* (85ERAY 2,1141,78).

TOXICITY DATA WITH REFERENCE
orl-mus LD50:5 mg/kg 85GDA2 6,412,81
ipr-mus LD50:1 mg/kg 85GDA2 6,412,81
scu-mus LD50:2 mg/kg 85FZAT -,1291,67

SAFETY PROFILE: Poison by ingestion, subcutaneous, and intraperitoneal routes. When heated to decomposition it emits acrid smoke and irritating fumes.

FCA000 CAS:11006-22-7 HR: 3
FLAVOFUNGIN (1:10)

PROP: Crystals. Mp: 210° (decomp).

TOXICITY DATA WITH REFERENCE
orl-mus LD50:1200 mg/kg 85ERAY 2,996,78
ipr-mus LD50:25 mg/kg 85ERAY 2,996,78

SAFETY PROFILE: Poison by intraperitoneal route. Moderately toxic by ingestion. When heated to decomposition it emits acrid smoke and irritating fumes.

FCA100 CAS:81-11-8 HR: 1
FLAVONIC ACID
mf: $C_{14}H_{14}N_2O_6S_2$ mw: 370.42

SYNS: AMSONIC ACID □ BENZENESULFONIC ACID, 2,2'-(1,2-ETHYLENEDIYL)BIS(5-AMINO)-(9CI) □ DASD □ 4,4'-DIAMINO-2,2'-STILBENEDISULFONIC ACID □ 2,2'-(1,2-ETHYLENEDIYL)BIS(5-AMINOBENZENESULFONIC ACID) □ NCI-C60162 □ 2,2'-STILBENEDISULFONIC ACID, 4,4'-DIAMINO- □ TINOPAL BHS

TOXICITY DATA WITH REFERENCE
ipr-rat TDLo:300 mg/kg (female 1D pre):REP JTEHD6 36,13,92
orl-gpg LD50:47 g/kg GISAAA 45(3),73,80

CONSENSUS REPORTS: Reported in EPA TSCA Inventory.

SAFETY PROFILE: Slightly toxic by ingestion. Experimental reproductive effects. When heated to decomposition it emits toxic vapors of NO_x and SO_x.

F

FCA200 CAS:7336-20-1 HR: 2
FLAVONIC ACID DISODIUM SALT
mf: $C_{14}H_{12}N_2O_6S_2 \cdot 2Na$ mw: 414.38

SYNS: AMSONIC ACID DISODIUM SALT □ BENZENESULFONIC ACID, 2,2'-(1,2-ETHANEDIYL)BIS(5-AMINO)-, DISODIUM SALT □ p,p'-DIAMINOSTILBENE-o,o'-DISULFONIC ACID DISODIUM SALT □ DIAMINOSTILBENE DISULPHONATE DISODIUM SALT □ 2,2'-DISULFO-4,4'-STILBENEDIAMINE DISODIUM SALT □ 2,3'-(1,2-ETHENEDIYL)BIS(5-AMINOBENZENESULFONIC ACID) DISODIUM SALT □ 2,2'-STILBENEDISULFONIC ACID, 4,4'-DIAMINO-, DISODIUM SALT

TOXICITY DATA WITH REFERENCE
orl-rat TDLo:42 g/kg/14D-C NTPTR* NTP-TR-412,92
orl-mus TDLo:168 g/kg/14D-C NTPTR* NTP-TR-412,92

CONSENSUS REPORTS: Reported in NTP Carcinogenesis Studies (feed); No Evidence: rat, mouse NTPTR* NTP-TR-412,92. Reported in EPA TSCA Inventory.

SAFETY PROFILE: Questionable carcinogen with negative NTP carcinogenesis bioassay. When heated to decomposition it emits toxic vapors of NO_x and SO_x.

FCB100 CAS:3717-88-2 HR: 3
FLAVOXATE HYDROCHLORIDE
mf: $C_{24}H_{25}NO_4 \cdot ClH$ mw: 427.96

SYNS: BLADDERON □ DW-61 □ FLAROXATE HYDROCHLORIDE □ 3-METHYL-4-OXO-2-PHENYL-4H-1-BENZOPYRAN-8-CARBOXYLATE 1-PIPERIDINEETHANOL HYDROCHLORIDE □ NSC-114649 □ PIPERIDINOETHYL-3-METHYLFLAVONE-8-CARBOXYLATE HYDROCHLORIDE □ 2-PIPERIDINOETHYL-3-METHYL-4-OXO-2-PHENYL-4H-1-BENZOPYRAN-8-CARBOXYLATE HYDROCHLORIDE □ REC 7-0040 □SPASUR-ET □ URISPAS

TOXICITY DATA WITH REFERENCE
mmo-sat 79 µg/plate YACHDS 19,773,91
orl-mus TDLo:6 g/kg (female 7-12D post):REP YACHDS 3,1214,75
orl-rat TDLo:4500 mg/kg (female 9-14D post):TER YACHDS 3,1214,75
orl-mus TDLo:1920 mg/kg (female 7-12D post):REP YACHDS 3,1214,75
orl-mus TDLo:9 g/kg (female 7-12D post):REP YACHDS 3,1214,75
orl-rat TDLo:9990 mg/kg (female 30D pre):REP KSRNAM 9,2695,75
orl-rat TDLo:30 g/kg (30D male):REP KSRNAM 9,2695,75
orl-rat TDLo:4500 mg/kg (female 9-14D post):TER YACHDS 3,1214,75

orl-rat LD50:1040 mg/kg YACHDS 15,5271,87
ipr-rat LD50:170 mg/kg YACHDS 15,5271,87
scu-rat LD50:1010 mg/kg IYKEDH 10,232,79
ivn-rat LD50:25 mg/kg IYKEDH 10,232,79
orl-mus LD50:740 mg/kg JPETAB 130,356,60
ipr-mus LD50:350 mg/kg AIPTAK 131,187,61
scu-mus LD50:610 mg/kg IYKEDH 10,232,79
ivn-mus LD50:28 mg/kg IYKEDH 10,232,79
ivn-rbt LD50:25 mg/kg YACHDS 3,1214,75

SAFETY PROFILE: Poison by intravenous and intraperitoneal routes. Moderately toxic by ingestion and subcutaneous routes. Experimental teratogenic and reproductive effects. Mutation data reported. A smooth muscle relaxant. When heated to decomposition it emits toxic fumes of NO_x and HCl. See also ESTERS.

FCC000 CAS:57608-59-0 HR: 3
FLAVUMYCIN B

PROP: A solid. Mp: 232–234°. Produced by *A. flavus var. geptinicus* 4 LIA-073 (JANTAJ 29,76–91,76).

TOXICITY DATA WITH REFERENCE
ipr-mus LD50:6500 μg/kg JANTAJ 29,76-91,76
ivn-mus LD50:2630 μg/kg JANTAJ 29,76-91,76

SAFETY PROFILE: Poison by intraperitoneal and intravenous routes. When heated to decomposition it emits acrid smoke and irritating fumes.

FCC100 CAS:8013-07-8 HR: 1
FLEXOL EPO

TOXICITY DATA WITH REFERENCE
skn-rbt 500 mg open MLD UCDS** 4/17/70
orl-rat LD50:23 g/kg UCDS** 4/17/70
skn-rbt LD50:>20 g/kg UCDS** 4,17,20

CONSENSUS REPORTS: Reported in EPA TSCA Inventory.

SAFETY PROFILE: Low toxicity by ingestion and skin contact. A skin irritant. When heated to decomposition it emits acrid smoke and irritating vapors.

FCD512 CAS:18268-70-7 HR: 1
FLEXOL 4GO
mf: $C_{24}H_{46}O_7$ mw: 446.70

SYNS: POLYETHYLENE GLYCOL 200 DI(2-ETHYLHEXOATE) □ TETRAETHYLENE GLYCOL DI(2-ETHYLHEXOATE)

TOXICITY DATA WITH REFERENCE
skn-rbt 500 mg open MLD UCDS** 7/21/71
orl-rat LD50:18,000 mg/kg NPIRI* 2,85,75

CONSENSUS REPORTS: Reported in EPA TSCA Inventory.

SAFETY PROFILE: Mildly toxic by ingestion. A skin irritant. When heated to decomposition it emits acrid smoke and irritating fumes.

FCD560 CAS:94-28-0 HR: 1
FLEXOL PLASTICIZER 3GO
mf: $C_{22}H_{42}O_6$ mw: 402.64

SYNS: BIS(ETHYLHEXANOATE)TRIETHYLENE GLYCOL □ TRIETHYLENE GLYCOL DI(2-ETHYLHEXOATE)

TOXICITY DATA WITH REFERENCE
skn-rbt 500 mg open MLD UCDS** 7/21/71
orl-rat LD50:31 g/kg JIHTAB 23,259,41
orl-gpg LD50:21 g/kg JIHTAB 23,259,41

CONSENSUS REPORTS: Reported in EPA TSCA Inventory.

SAFETY PROFILE: A mild skin irritant. When heated to decomposition it emits acrid smoke and fumes. See also TRIETHYLENE GLYCOL and ESTERS.

FCN000 CAS:101564-02-7 HR: 1
FLEXOL PLASTICIZER CC-55

PROP: Liquid. Mp: −53°, bp: 216° @ 5 mm, flash p: 425°F (OC), d: 0.9586 @ 20°/20°, vap d: 13.7.

TOXICITY DATA WITH REFERENCE
skn-rbt 500 mg open MLD UCDS** 1/12/72

SAFETY PROFILE: A skin irritant. Combustible when exposed to heat or flame; can react with oxidizing materials. To fight fire, use foam, CO_2, dry chemical. When heated to decomposition it emits acrid smoke and irritating fumes. See also ESTERS.

FCN050 CAS:8036-63-3 HR: 1
FLEXOL PLASTICIZER 810

TOXICITY DATA WITH REFERENCE
skn-rbt 500 mg open MLD UCDS** 7/23/71
skn-rbt LD50:20 mg/kg UCDS** 7/23/71

SAFETY PROFILE: A skin irritant. When heated to decomposition it emits acrid smoke and irritating fumes.

FCN100 HR: 2
FLOMOXEF SODIUM
mf: $C_{15}H_{17}F_2N_6O_7S_2\cdot Na$ mw: 518.49

SYNS: FLOMOXEF □ 5-OXA-1-AZABICYCLO(4.2.0)OCT-2-ENE-2-CARBOXYLIC ACID, 7-(2-((DIFLUOROMETHYL)THIO)ACETAMIDO)-3-(((1-(2-HYDROXYETHYL)-1H-TETRAZOL-5-YL)THIO)METHYL)-7-METHOXY-8-OXO-, SODIUM SALT, (6R,7R)- □ 6315-S

TOXICITY DATA WITH REFERENCE
ivn-rat TDLo:41,600 mg/kg (female 17-22D post):REP NKRZAZ 35(Suppl 1),404,87
ivn-rat TDLo:8800 mg/kg (female 7-17D post):REP NKRZAZ 35(Suppl 1),370,87
ivn-rbt TDLo:81 mg/kg (female 6-18D post):TER OYYAA2 34,475,87
ivn-rat TDLo:8800 mg/kg (female 7-17D post):TER NKRZAZ 35(Suppl 1),370,87
scu-mus TDLo:8 g/kg (female 6-15D post):TER OYYAA2 34,335,87
ipr-rat TDLo:72 g/kg (female 26W pre):REP NKRZAZ 35(Suppl 1),270,87

ivn-rat TDLo:21 g/kg (female 17-22D post):REP NKRZAZ 35(Suppl 1),404,87
scu-mus TDLo:2 g/kg (female 6-15D post):TER OYYAA2 34,335,87
ivn-rat LD50:7360 mg/kg NKRZAZ 35(Suppl 1),207,87
scu-mus LD50:13,500 mg/kg IYKEDH 19,735,88
ivn-mus LD50:7086 mg/kg IYKEDH 19,735,88

SAFETY PROFILE: Moderately toxic by intravenous route. An experimental teratogen. Experimental reproductive effects. When heated to decomposition it emits toxic fumes of NO_x, SO_x, F^-, and Cl^-.

FDA000 CAS:68916-09-6 HR: 1
FLOUVE OIL

PROP: Found in the upper part of the dried grass *Flouve odorante* (FCTXAV 16,637,78).

TOXICITY DATA WITH REFERENCE
skn-rbt 500 mg/24H MOD FCTXAV 16,757,78
orl-rat LD50:4100 mg/kg FCTXAV 16,757,78
orl-rbt LD50:5000 mg/kg FCTXAV 16,757,78

CONSENSUS REPORTS: Reported in EPA TSCA Inventory.

SAFETY PROFILE: Mildly toxic by ingestion. A skin irritant. When heated to decomposition it emits acrid smoke and irritating fumes.

FDA100 CAS:1847-24-1 HR: 2
FLOXAPEN SODIUM
mf: $C_{19}H_{16}ClFN_3O_5S{\cdot}Na$ mw: 475.88

SYNS: FLOXACILLIN SODIUM □ FLUCLOXACILLIN SODIUM □ FLUCLOXACILLIN SODIUM SALT □ 5-METHYL-3-(2-CHLORO-6-FLUOROPHENYL)-4-ISOXAZOLYLPENICILLIN SODIUM □ MFI-PC □ MONOSODIUM FLUCLOXACILLIN □ SODIUM FLUCLOXACILLIN

TOXICITY DATA WITH REFERENCE
orl-rat TDLo:6 g/kg (9-14D preg):TER NKRZAZ 17,1523,69
orl-wmn TDLo:120 mg/kg/3D-I:SYS AJGAAR 82,483,87
orl-rat LD50:11 g/kg NIIRDN 6,697,82
ipr-rat LD50:1100 mg/kg NIIRDN 6,697,82
scu-rat LD50:2800 mg/kg NIIRDN 6,697,82
ivn-rat LD50:680 mg/kg NIIRDN 6,697,82
orl-mus LD50:7600 mg/kg NIIRDN 6,697,82
ipr-mus LD50:1350 mg/kg NIIRDN 6,697,82
scu-mus LD50:1450 mg/kg NKRZAZ 17,1523,69
ivn-mus LD50:1360 mg/kg NIIRDN 6,697,82
ivn-dog LD50:760 mg/kg NKRZAZ 17,1523,69

SAFETY PROFILE: Moderately toxic by intravenous, intraperitoneal, and subcutaneous routes. Mildly toxic by ingestion. Human systemic effects: cholestatic jaundice. An experimental teratogen. When heated to decomposition it emits toxic fumes of F^-, Cl^-, SO_x, NO_x, and Na_2O.

FDA875 CAS:5107-49-3 HR: 3
FLUALAMIDE
mf: $C_{17}H_{23}F_3N_2O_2$ mw: 344.42

SYNS: 2-(ALLYLOXY)-N-(2-(DIETHYLAMINO)ETHYL)-α,α,α-TRIFLUORO-p-TOLUAMIDE □ 305CE □ 4-TRIFLUOROMETHYL ALLYLOXY-2 N-(β-DIETHYLAMINO-ETHYL)BENZAMIDE (FRENCH)

TOXICITY DATA WITH REFERENCE
orl-mus LD50:1375 mg/kg MPHEAE 15,241,66
scu-mus LD50:249 mg/kg JJPAAZ 20,1,70
ivn-mus LD50:86 mg/kg MPHEAE 15,241,66
ivn-dog LDLo:40 mg/kg JJPAAZ 20,1,70

SAFETY PROFILE: Poison by subcutaneous and intravenous routes. Moderately toxic by ingestion. When heated to decomposition it emits toxic fumes of F^- and NO_x.

F

FDA880 CAS:17160-71-3 HR: 3
FLUANISONE HYDROCHLORIDE
mf: $C_{21}H_{25}FN_2O_2{\cdot}ClH$ mw: 392.94

SYNS: ANTI-PICA □ 4'-FLUORO-4-(4-(o-METHOXYPHENYL)-1-PIPERAZINYL)BUTYROPHENONE HYDROCHLORIDE □ HALOANISONE COMPOSITUM

TOXICITY DATA WITH REFERENCE
scu-rat LD50:420 mg/kg 27ZQAG -,189,72
ivn-rat LD50:20 mg/kg 27ZQAG -,189,72
ipr-mus LD50:200 mg/kg 27ZQAG -,189,72
scu-mus LD50:150 mg/kg 27ZQAG -,189,72
ivn-mus LD50:25 mg/kg 27ZQAG -,189,72

SAFETY PROFILE: Poison by subcutaneous, intraperitoneal, and intravenous routes. When heated to decomposition it emits very toxic fumes of HCl, NO_x, and F^-.

FDA885 CAS:69806-50-4 HR: 2
FLUAZIFOP-BUTYL
mf: $C_{19}H_{20}F_3NO_4$ mw: 383.38

SYNS: BUTYL 2-(4-(5-TRIFLUOROMETHYL-2-PYRIDINYLOXY)PHENOXY)PROPANOATE □ FUSILADE □ HACHE UNO SUPER □ IH 773B □ ONECIDE EC □ ONECIDE □ PP 009 □ PROPIONIC ACID, 2 (p-((5-(TRIFLUOROMETHYL)-2-PYRIDYL)OXY)PHENOXY)-, BUTYL ESTER □ SL-236 □ TF 1169 □ TS-7236

TOXICITY DATA WITH REFERENCE
orl-rat LD50:2910 mg/kg NNGADV 15,305,90
ipr-rat LD50:1620 mg/kg NNGADV 15,305,90
orl-mus LD50:1490 mg/kg PEMNDP 9,400,91
ipr-mus LD50:1240 mg/kg NNGADV 15,305,90
orl-rbt LD50:621 mg/kg 85JFAN A535,84
orl-gpg LD50:2659 mg/kg 85JFAN A535,84

SAFETY PROFILE: Moderately toxic by ingestion and intraperitoneal routes. When heated to decomposition it emits toxic fumes of F^- and NO_x.

FDA887 CAS:31430-15-6 HR: 2
FLUBENDAZOLE
mf: $C_{16}H_{12}FN_3O_3$ mw: 313.31

PROP: A solid. Mp: 260°.

SYNS: 2-BENZIMIDAZOLECARBAMIC ACID, 5-(p-FLUOROBENZOYL)-, METHYL ESTER □ CARBAMIC ACID, (5-(4-FLUOROBENZOYL)-1H-BENZIMIDAZOL-2-YL)-, METHYL ESTER (9CI) □ (5-(4-FLUOROBENZOYL)-1H-BENZIMIDAZOLE-2-YL)CARBAMIC ACID METHYL ESTER □ METHYL N-(5-(p-FLUOROBENZOYL)-2-BENZIMIDAZOLYL)CARBAMATE □ R 17899

TOXICITY DATA WITH REFERENCE
mmo-sat 500 nmol/L PCBRD2 340E,225,90
sce-mus-ipr 100 mg/kg PCBRD2 340E,225,90
orl-rat TDLo:320 mg/kg (female 8-15D post):TER TXCYAC 43,133,87
orl-rat TDLo:1280 mg/kg (female 8-15D post):REP TXCYAC 43,133,87
orl-rat LD50:2560 mg/kg YRTMA6 9,11,78
orl-rbt LD50:2560 mg/kg YRTMA6 9,11,78

SAFETY PROFILE: Moderately toxic by ingestion. An experimental teratogen. Experimental reproductive effects. Mutation data reported. When heated to decomposition it emits toxic fumes of NO_x and F^-.

FDA900 **CAS:33245-39-5** **HR: 2**
FLUCHLORALIN
mf: $C_{12}H_{13}ClF_3N_3O_4$ mw: 355.73

PROP: Orange-yellow, crystalline solid. Mp: 42–43°. Solubility in water: 10 ppm.

SYNS: BAS 392-H □ BASALIN □ N-(2-CHLOROETHYL)-2,6-DINITRO-N-PROPYL-4-(TRIFLUOROMETHYL)ANILINE □ N-(2-CHLOROETHYL)-2,6-DINITRO-N-PROPYL-4-(TRIFLUOROMETHYL)BENZENAMIDE □ N-(2-CHLOROETHYL)-α,α,α-TRIFLUORO-2,6-DINITRO-N-PROPYL-p-TOLUIDINE □ N-PROPYL-N-(2-CHLOROETHYL)-2,6-DINITRO-4-TRIFLUOROMETHYLANILINE □ N-PROPYL-N-(2-CHLOROETHYL)-α,α,α-TRIFLUORO-2,6-DINITRO-p-TOLUIDINE

TOXICITY DATA WITH REFERENCE
orl-rat LD50:2940 mg/kg FMCHA2-,C37,91
ihl-rat LC50:8400 mg/m³/4H PEMNDP 9,403,91
orl-mus LD50:730 mg/kg PEMNDP 9,403,91

SAFETY PROFILE: Moderately toxic by ingestion. Low toxicity by inhalation. When heated to decomposition it emits toxic fumes of F^-, Cl^-, and NO_x. See also NITRO COMPOUNDS of AROMATIC HYDROCARBONS.

FDB000 **CAS:2825-60-7** **HR: 2**
FLUDERMA
mf: $C_{29}H_{38}ClFO_8$ mw: 569.12

PROP: Crystals from Et_2O/petr ether. Mp: 180–182°.

SYNS: CORTOCIN-F □ CUTISTEROL □ DEFLAMENE □ FI 6341 □ FLUOROFORMYLON □ FORMOCORTAL

TOXICITY DATA WITH REFERENCE
ipr-mus LD50:537 mg/kg MEIEDD 10,605,83
scu-mus LD50:490 mg/kg MEIEDD 10,605,83

SAFETY PROFILE: Moderately toxic by intraperitoneal and subcutaneous routes. Used as a glucocorticoid (a corticosteroid which affects carbohydrate, protein and fat metabolism with many other side effects). When heated to decomposition it emits very toxic fumes of F^- and Cl^-.

FDB100 **CAS:3900-31-0** **HR: 3**
FLUDIAZEPAM
mf: $C_{16}H_{12}ClFN_2O$ mw: 302.75

PROP: Colorless prisms from n-hexane/isopropanol. Mp: 88–92°.

SYNS: 7-CHLORO-5-(o-FLUOROPHENYL)-1,3-DIHDYRO-1-METHYL-2H-1,4-BENZODIAZEPIN-2-ONE □ 7-CHLORO-5-(2-FLUOROPHENYL)-1-METHYL-1H-1,4-BENZODIAZEPIN-2(3H)-ONE □ ERISPAN □ ID 540 □ RO 5-3438

TOXICITY DATA WITH REFERENCE
orl-rat TDLo:150 mg/kg (30D pre):REP KSRNAM 11,529,77
orl-rat TDLo:13,650 mg/kg (91D male):REP KSRNAM 11,529,77
orl-rat LD50:1740 mg/kg YAKUD5 22,811,80
ipr-rat LD50:315 mg/kg IYKEDH 11,811,80
scu-rat LD50:7600 mg/kg IYKEDH 11,811,80
orl-mus LD50:830 mg/kg IYKEDH 11,811,80
ipr-mus LD50:360 mg/kg NIIRDN 6,698,82
scu-mus LD50:1150 mg/kg NIIRDN 6,698,82

SAFETY PROFILE: Poison by intraperitoneal route. Moderately toxic by ingestion and subcutaneous routes. Experimental reproductive effects. Used as a tranquilizer. Note: This is a controlled substance (depressant) listed in the U.S. Code of Federal Regulations, Title 21 Part 1308.14 (1985). When heated to decomposition it emits toxic fumes of F^-, Cl^-, and NO_x. See also DIAZEPAM.

FDB200 **CAS:4301-50-2** **HR: 3**
FLUENETIL
mf: $C_{16}H_{15}FO_2$ mw: 258.31

SYNS: 4-BIPHENYLACETIC ACID, 2-FLUOROETHYL ESTER □ (1,1'-BIPHENYL)-4-ACETIC ACID, 2-FLUOROETHYL ESTER □ FLUENYL □ LAMBROL

TOXICITY DATA WITH REFERENCE
ipr-mus LD50:42 mg/kg EVHPAZ 14,93,76
orl-rat LD50:6 mg/kg FMCHA2 -,C138,83

CONSENSUS REPORTS: EPA Extremely Hazardous Substances List.

SAFETY PROFILE: Poison by ingestion and intraperitoneal routes. When heated to decomposition it emits toxic fumes of F^-. See also ESTERS.

FDD080 **CAS:30484-77-6** **HR: 3**
FLUNARIZINE DIHYDROCHLORIDE
mf: $C_{26}H_{26}F_2N_2 \cdot 2ClH$ mw: 477.46

PROP: Crystals from EtOH/2-propanol. Mp: 251.5°.

SYNS: (3)-1-(BIS(p-FLUOROPHENYL)METHYL)-4-CINNAMYLPIPERAZINE DIHYDROCHLORIDE □ (E)-1-(BIS(4-FLUOROPHENYL)METHYL)-4-(3-PHENYL-2-PROPENYL)PIPERAZINE DIHYDROCHLORIDE □ FLUNARIZINE HYDROCHLORIDE □ KW-3149 □ R 14950

TOXICITY DATA WITH REFERENCE
orl-rat TDLo:33 mg/kg (female 7-17D post):REP KSRNAM 16,1840,82
orl-rat TDLo:120 mg/kg (female 18-21D post):REP KSRNAM 16,1860,82
orl-rat TDLo:330 mg/kg (7-17D preg):TER KSRNAM 16,1840,82
orl-rat LD50:503 mg/kg ARZNAD 25,1408,75
ipr-rat LD50:305 mg/kg YKYUA6 36,825,85
scu-rat LD50:1793 mg/kg YKYUA6 36,825,85
ivn-rat LD50:35 mg/kg YKYUA6 36,825,85

orl-mus LD50:1203 mg/kg YKYUA6 36,825,85
ipr-mus LD50:379 mg/kg YKYUA6 36,825,85
ivn-mus LD50:27 mg/kg YKYUA6 36,825,85

SAFETY PROFILE: Poison by intravenous and intraperitoneal routes. Moderately toxic by ingestion and subcutaneous routes. An experimental teratogen. Experimental reproductive effects. When heated to decomposition it emits toxic fumes of Cl^-, NO_x, and F^-.

FDD085 CAS:3385-03-3 HR: 1
FLUNISOLIDE
mf: $C_{24}H_{31}FO_6$ mw: 434.55

SYNS: 6-α-FLUORO-11-β,16-α,17,21-TETRAHYDROXYPREGNA-1,4-DIENE,-3,20-DIONE, CYCLIC 16,17-ACETAL with ACETONE □ LUNIS □ NASALIDE □ RHINALAR □ SYNTARIS

TOXICITY DATA WITH REFERENCE
orl-rat TDLo:26 µg/kg (female 17-21D post):REP OYYAA2 24,661,82
orl-rat TDLo:550 µg/kg (female 7-17D post):REP OYYAA2 24,643,82
orl-rat TDLo:1100 µg/kg (female 7-17D post):TER OYYAA2 24,643,82
scu-mus TDLo:400 µg/kg (female 6-15D post):TER OYYAA2 24,741,82
orl-rat TDLo:1050 µg/kg (female 21D pre):REP OYYAA2 24,631,82
orl-rat TDLo:630 µg/kg (63D male):REP OYYAA2 24,631,82
scu-mus TDLo:80 mg/kg (female 6-15D post):TER OYYAA2 24,741,82
ihl-cld TCLo:21 mg/kg/30W-I:NOSE JOPDAB 105,840,84

SAFETY PROFILE: Human systemic effects by inhalation: deviated nasal septum, ulcerated nasal septum. An experimental teratogen. Experimental reproductive effects. A steroid. When heated to decomposition it emits toxic fumes of F^-.

FDD100 CAS:1622-62-4 HR: 2
FLUNITRAZEPAM
mf: $C_{16}H_{12}FN_3O_3$ mw: 313.31

PROP: Pale yellow needles from methylene chloride-hexane. Mp: 166–167°.

SYNS: 1,3-DIHYDRO-5-(o-FLUOROPHENYL)-1-METHYL-7-NITRO-2H-1,4-BENZODIAZEPIN-2-ONE □ 1-METHYL-7-NITRO-5-(2-FLUOROPHENYL)-3H-1,4-BENZODIAZEPIN-2(1H)-ONE □ NARCOZEP □ PRIMUN □ RO 5-4200 □ ROHYPNOL □ ROIPNOL

TOXICITY DATA WITH REFERENCE
mmo-sat 3 mg/plate BSIBAC 60,2247,84
mma-sat 2 mg/plate BSIBAC 60,2247,84
ims-wmn TDLo:560 µg/kg (39W preg):REP THERAP 36,305,81
orl-wmn TDLo:310 µg/kg (16-39W preg):REP THERAP 36,305,81
orl-rat TDLo:288 mg/kg (male 9W pre):REP TXAPA9 53,92,80
orl-rat TDLo:275 mg/kg (female 7-17D post):REP KSRNAM 17,2585,83
orl-rat TDLo:350 mg/kg (female 1-7D post):TER KSRNAM 17,2585,83
orl-rat TDLo:275 mg/kg (female 7-17D post):TER KSRNAM 17,2585,83
orl-rbt TDLo:260 mg/kg (female 6-18D post):TER KSRNAM 17,2585,83
orl-rat TDLo:225 mg/kg (female 7-15D post):TER TXAPA9 53,92,80
orl-wmn TDLo:127 µg/kg:BAH AIMDAP 145,663,85
orl-rat LD50:415 mg/kg KSRNAM 19,1277,85
ipr-rat LD50:1060 mg/kg KSRNAM 19,1277,85
orl-mus LD50:1200 mg/kg KSRNAM 19,1277,85
ipr-mus LD50:1050 mg/kg KSRNAM 19,1277,85

SAFETY PROFILE: Moderately toxic by ingestion and intraperitoneal routes. An experimental teratogen. Human reproductive effects by ingestion and intramuscular routes: apgar score of newborn and other neonatal and postnatal effects. Experimental reproductive effects. Mutation data reported. Human systemic effects by ingestion: coma. Used as a hypnotic agent. Note: This is a controlled substance (depressant) listed in the U.S. Code of Federal Regulations, Title 21 Part 1308.14 (1985). When heated to decomposition it emits toxic fumes of F^- and NO_x. See also DIAZEPAM.

FDD125 CAS:16872-11-0 HR: 2
FLUOBORIC ACID
DOT: UN 1775
mf: $BF_4 \cdot H$ mw: 87.82

SYNS: BORATE(1-), TETRAFLUORO-, HYDROGEN □ BOROFLUORIC ACID □ FLUOBORIC ACID (DOT) □ HYDROFLUOBORIC ACID □ HYDROGEN TETRAFLUOROBORATE □ TETRAFLUOROBORIC ACID

CONSENSUS REPORTS: Reported in EPA TSCA Inventory.

OSHA PEL: TWA 2.5 mg(F)/m³
NIOSH REL: (Fluorides, inorganic): TWA 2.5 mg(F)/m³
DOT CLASSIFICATION: 8; *Label:* Corrosive

SAFETY PROFILE: A corrosive acid. When heated to decomposition it emits toxic vapors of B and F^-.

FDD150 CAS:356-12-7 HR: 3
FLUOCINOLIDE
mf: $C_{26}H_{32}F_2O_7$ mw: 494.58

PROP: Crystals from methanol. Mp: 308–311°.

SYNS: FLUOCINOLONE ACETONIDE ACETATE □ FLUOCINOLONE ACETONIDE 21-ACETATE □ FLUOCINONIDE □ METOSYN

TOXICITY DATA WITH REFERENCE
dni-mus-skn 4 g/kg BJDEAZ 94,1,76
scu-rat TDLo:3 mg/kg (3D pre):REP NYKZAU 84,553,84
orl-rat LD50:14 mg/kg NIIRDN 6,694,82
ipr-rat LD50:300 µg/kg NIIRDN 6,694,82
scu-rat LD50:720 µg/kg IYKEDH 6,386,75
ipr-mus LD50:160 mg/kg YAKUD5 17,849,75
scu-mus LD50:165 mg/kg YKYUA6 26,741,75

SAFETY PROFILE: Poison by ingestion, subcutaneous, and intraperitoneal routes. Mutation data reported. Experimental reproductive effects. A steroid. When heated to decomposition it emits toxic fumes of F^-.

FDF000 **CAS:206-44-0** **HR: 3**
FLUORANTHENE
mf: $C_{16}H_{10}$ mw: 202.26

PROP: A polycyclic hydrocarbon. Colorless solid. Needles or plates from alc. Mp: 110°, bp: 250–251° @ 60 mm, vap press: 0.01 mm @ 20°.

SYNS: 1,2-BENZACENAPHTHENE ☐ BENZO(jk)FLUORENE ☐ IDRYL ☐ 1,2-(1,8-NAPHTHALENEDIYL)BENZENE ☐ 1,2-(1,8-NAPHTHYLENE)BENZENE ☐ RCRA WASTE NUMBER U120

TOXICITY DATA WITH REFERENCE
mma-sat 5 μg/plate MUREAV 156,61,85
msc-hmn:lym 2 μmol/L DTESD7 10,277,82
msc-ham:ovr 20 mg/L ENMUDM 6,539,84
skn-mus TDLo:280 mg/kg/58W-I:ETA JNCIAM 56,1237,76
orl-rat LD50:2000 mg/kg AIHAAP 23,95,62
ivn-mus LD50:100 mg/kg CSLNX* NX#00205
skn-rbt LD50:3180 mg/kg AIHAAP 23,95,62

CONSENSUS REPORTS: IARC Cancer Review: Group 3 IMEMDT 7,56,87; Animal No Evidence IMEMDT 32,355,83. Reported in EPA TSCA Inventory. EPA Genetic Toxicology Program.

SAFETY PROFILE: Poison by intravenous route. Moderately toxic by ingestion and skin contact. Questionable carcinogen with experimental tumorigenic data. Human mutation data reported. Combustible when exposed to heat or flame. When heated to decomposition it emits acrid smoke and irritating fumes.

For occupational chemical analysis use NIOSH: Polynuclear Aromatic Hydrocarbons (HPLC), 5506; (GC), 5515.

FDI000 **CAS:153-78-6** **HR: 3**
FLUOREN-2-AMINE
mf: $C_{13}H_{11}N$ mw: 181.25

PROP: Crystals from EtOH (aq). Mp: 129°.

SYNS: AMINOFLUOREN (GERMAN) ☐ 2-AMINOFLUORENE ☐ 2-FLUORENAMINE ☐ 2-FLUORENEAMINE

TOXICITY DATA WITH REFERENCE
mma-sat 150 ng/plate CBINA8 54,71,85
msc-rat:lvr 50 μmol/L MUREAV 130,53,84
orl-rat TDLo:3600 mg/kg/32W-C:CAR CNREA8 15,188,55
skn-rat TDLo:240 mg/kg/73W-I:CAR JNCIAM 10,1201,50
scu-rat TDLo:400 mg/kg/26W-I:ETA CNREA8 7,453,47
orl-mus TDLo:100 mg/kg/47W-C:ETA CNREA8 7,453,47
skn-mus TDLo:11 g/kg/34W-C:NEO BJCAAI 14,195,60
imp-mus TDLo:50 mg/kg:CAR BJCAAI 12,222,58
orl-rat TD:4000 mg/kg/23W-C:ETA CNREA8 7,730,47
orl-rat TD:3200 mg/kg/58W-C:ETA CNREA8 7,453,47
orl-rat TD:2420 mg/kg/23W-C:NEO JNCIAM 10,1201,50
imp-mus TD:100 mg/kg:ETA BMBUAQ 14,147,58
skn-rat TD:18 g/kg/30W-I:ETA BJEPA5 25,1,44
skn-rat TD:1080 mg/kg/30W-I:ETA ENDOAO 76,1027,65
ipr-mus LD50:132 mg/kg JJIND8 62,911,79

CONSENSUS REPORTS: EPA Genetic Toxicology Program.

SAFETY PROFILE: Suspected carcinogen with experimental carcinogenic, neoplastigenic, and tumorigenic data. Poison by intraperitoneal route. Mutation data reported. When heated to decomposition it emits toxic fumes of NO_x. See also AMINES.

FDI100 **CAS:86-73-7** **HR: 1**
9H-FLUORENE
mf: $C_{13}H_{10}$ mw: 166.23

SYNS: o-BIPHENYLENEMETHANE ☐ o-BIPHENYLMETHANE ☐ DIPHENYLENEMETHANE ☐ FLUORENE ☐ 2,2′-METHYLENEBIPHENYL

TOXICITY DATA WITH REFERENCE
mma-mus:lyms 19,500 nmol/L MUTAEX 3,193,88
otr-mus:mmr 1 μg/L CNREA8 39,1784,79
dnd-mus:lyms 150 μmol/L MUREAV 203,155,88
msc-mus:lyms 584 μmol/L MUTAEX 3,193,88
cyt-ham:lng 25 mg/L MUREAV 259,103,91
ipr-mus LD50:2 g/kg RPTOAN 48,143,85
par-mus LD50:>2 g/kg RPTOAN 52,112,89

CONSENSUS REPORTS: IARC Cancer Review: Group 3 IMEMDT 7,56,87; Animal Inadequate Evidence IMEMDT 32,365,83; Human No Adequate Data IMEMDT 32,365,83

CONSENSUS REPORTS: Reported in EPA TSCA Inventory.

SAFETY PROFILE: Slightly toxic by intraperitoneal and parenteral routes. Mutation data reported. When heated to decomposition it emits acrid smoke and irritating vapors.

For occupational chemical analysis use NIOSH: Polynuclear Aromatic Hydrocarbons (HPLC), 5506; (GC), 5515.

FDK000 **HR: 3**
9-FLUORENECARBOXYLATE-3-QUINUCLIDINOL HYDROCHLORIDE
mf: $C_{21}H_{21}NO_2 \cdot ClH$ mw: 355.89

SYNS: FLUORENE-9-CARBOXYLIC ACID-3-QUINUCLIDINYL ESTER ☐ PAVATRINE ☐ RO 2-3208

TOXICITY DATA WITH REFERENCE
ipr-mus LD50:165 mg/kg JPETAB 104,284,52
ivn-mus LD50:26 mg/kg JPETAB 104,284,52

SAFETY PROFILE: Poison by intraperitoneal and intravenous routes. When heated to decomposition it emits very toxic fumes of NO_x and HCl. See also ESTERS.

FDM000 **CAS:525-64-4** **HR: 2**
FLUORENE-2,7-DIAMINE
mf: $C_{13}H_{11}N_2$ mw: 195.26

PROP: Plates from C_6H_6 or Et_2O. Mp: 164–165°. Sol in acids.

SYNS: 2,7-DIAMINOFLUORENE ☐ 2,7-FLUORENEDIAMINE ☐ 2,7-FLUOROENEDIAMINE

TOXICITY DATA WITH REFERENCE
mmo-sat 10 μg/plate MUREAV 143,213,85
mma-sat 5 μg/plate MUREAV 143,213,85
dns-hmn:lym 10 μmol/L SHYCD4 (40),47,83
dns-rat:lvr 500 nmol/L ENMUDM 3,11,81

dnd-mus:ast 5 μmol/L MUREAV 89,95,81
orl-rat TDLo:1000 mg/kg:CAR CNREA8 26,619,66
orl-rat TD:3600 mg/kg/27D-I:CAR CNREA8 28,924,68
orl-rat LDLo:1000 mg/kg CNREA8 26,619,66

CONSENSUS REPORTS: EPA Genetic Toxicology Program.

SAFETY PROFILE: Moderately toxic by ingestion. Questionable carcinogen with experimental carcinogenic data. Human mutation data reported. When heated to decomposition it emits toxic fumes of NO_x. See also AMINES.

FDN000 CAS:66686-30-4 HR: 2
1,1′-(9H-FLUORENE-2,7-DIYL)BIS(2-(DIETHYLAMINO) ETHANONE) DIHYDROCHLORIDE TRIHYDRATE
mf: $C_{25}H_{32}N_2O_2$•2ClH•$3H_2O$ mw: 519.57

SYNS: RMI 11002 □ RMI 11002 DA

TOXICITY DATA WITH REFERENCE
orl-mus LD50:5000 mg/kg ALACBI 12,77,79
scu-mus LD50:930 mg/kg ALACBI 12,77,79

SAFETY PROFILE: Moderately toxic by subcutaneous route. Mildly toxic by ingestion. When heated to decomposition it emits very toxic fumes of NO_x and HCl.

FDO000 CAS:486-25-9 HR: 2
9H-FLUOREN-9-ONE
mf: $C_{13}H_8O$ mw: 180.21

PROP: Yellow, rhombic crystals. Mp: 83–84°, bp: 341.5°. Insol in water; very sol in alc and ether.

SYNS: FLUOREN-9-ONE □ 9-FLUORENONE

TOXICITY DATA WITH REFERENCE
scu-rat TDLo:300 mg/kg/26W-I:ETA CNREA8 7,453,47
ipr-mus LD50:2 g/kg RPTOAN 48,143,85

CONSENSUS REPORTS: Reported in EPA TSCA Inventory. EPA Genetic Toxicology Program.

SAFETY PROFILE: Questionable carcinogen with experimental tumorigenic data. When heated to decomposition it emits acrid smoke and irritating fumes. See also KETONES.

FDP000 CAS:206-00-8 HR: 2
FLUORENO(9,1-gh)QUINOLINE
mf: $C_{19}H_{11}N$ mw: 253.31

SYN: PYRIDO(3′,2′:3,4)FLUORANTHENE

TOXICITY DATA WITH REFERENCE
scu-mus TDLo:72 mg/kg/9W-I:ETA COREAF 259,3387,64

SAFETY PROFILE: Questionable carcinogen with experimental tumorigenic data. When heated to decomposition it emits toxic fumes of NO_x.

FDQ000 CAS:28314-03-6 HR: 2
N-FLUOREN-1-YL ACETAMIDE
mf: $C_{15}H_{13}NO$ mw: 223.29

SYNS: 1-FLUORENYLACETAMIDE □ N-1-FLUORENYLACETAMIDE

TOXICITY DATA WITH REFERENCE
orl-rat TDLo:4400 mg/kg/44W-C:ETA,REP JNCIAM 24,149,60
ipr-rat TDLo:627 mg/kg/4W-I:ETA CNREA8 30,1485,70

SAFETY PROFILE: Experimental reproductive effects. Questionable carcinogen with experimental tumorigenic data. When heated to decomposition it emits toxic fumes of NO_x.

FDR000 CAS:53-96-3 HR: 3
N-FLUOREN-2-YL ACETAMIDE
mf: $C_{15}H_{13}NO$ mw: 223.29

SYNS: AAF □ 2-AAF □ 2-ACETAMIDOFLUORENE □ 2-ACETAMINOFLUORENE □ ACETOAMINOFLUORENE □ 2-ACETYLAMINOFLUOREN (GERMAN) □ N-ACETYL-2-AMINOFLUORENE □ 2-ACETYLAMINOFLUORENE (OSHA) □ AZETYLAMINOFLUOREN (GERMAN) □ FAA □ 2-FAA □ 2-FLUORENYLACETAMIDE □ N-2-FLUORENYLACETAMIDE □ RCRA WASTE NUMBER U005

TOXICITY DATA WITH REFERENCE
mmo-sat 100 μg/plate CBINA8 22,297,78
dnr-esc 5 μmol/L MUREAV 89,95,81
dnd-hmn:fbr 1 mmol/L ENMUDM 7,267,85
sce-hmn:lym 4400 μg/L CNREA8 40,4775,80
otr-rat:emb 100 μg/L JJIND8 67,1303,81
otr-rat:lvr 100 μmol/L CNREA8 43,5087,83
sce-mus-ipr 20 mg/kg MUREAV 157,181,85
ipr-mus TDLo:100 mg/kg (10D preg):TER KAIZAN 37,239,62
ipr-mus TDLo:100 mg/kg (female 8D post):REP KAIZAN 37,239,62
ipr-mus TDLo:625 mg/kg (male 5D pre):REP MUREAV 124,235,83
ipr-mus TDLo:2500 mg/kg (male 5D pre):REP MUREAV 124,235,83
orl-rat TDLo:300 mg/kg (9D preg):TER IARCCD 4,112,73
ipr-mus TDLo:100 mg/kg (10D preg):TER KAIZAN 37,239,62
orl-rat TDLo:672 mg/kg/8W-C:CAR TXAPA9 74,63,84
skn-rat TDLo:260 mg/kg/71W-I:CAR JNCIAM 10,1201,50
ipr-rat TDLo:192 mg/kg/4W-I:CAR CNREA8 32,1554,72
par-rat TDLo:1700 mg/kg/17W-I:ETA CNREA8 6,617,46
imp-rat TDLo:22 mg/kg:ETA CNREA8 33,2489,73
orl-mus TDLo:7840 mg/kg/49W-C:CAR EJCAAH 5,41,69
ipr-mus TDLo:60 mg/kg/2W-I:NEO TXAPA9 82,19,86
scu-mus TDLo:400 mg/kg (15D preg):ETA,TER IJCNAW 20,293,77
imp-mus TDLo:96 mg/kg:ETA GMCRDC 17,383,75
orl-dog TDLo:2625 mg/kg/25W-C:NEO CNREA8 10,266,50
orl-cat TDLo:11 g/kg/69W-C:ETA CNREA8 11,280,51
orl-rbt TDLo:6500 mg/kg/65W-I:CAR PAMIAD 32,177,68
ipr-rbt TDLo:3600 mg/kg/40W-I:ETA CNREA8 27,838,67
orl-ham TDLo:9240 mg/kg/44W-C:CAR GANNA2 59,239,68

CONSENSUS REPORTS: NTP 7th Annual Report On Carcinogens. Community Right-To-Know List. EPA Genetic Toxicology Program. Reported in EPA TSCA Inventory.

OSHA PEL: Cancer Suspect Agent
NIOSH REL: (2-Acetylaminofluorene): TWA use 29 CFR 1910.1014

F

SAFETY PROFILE: Confirmed human carcinogen with experimental carcinogenic, neoplastigenic, tumorigenic, and teratogenic data. Moderately toxic by ingestion and intraperitoneal routes. Experimental reproductive effects. Human mutation data reported. When heated to decomposition it emits toxic fumes of NO_x.

FDS000 **CAS:6292-55-3** **HR: 2**
3-FLUORENYL ACETAMIDE
mf: $C_{15}H_{13}NO$ mw: 223.29

SYNS: N-3-FLUORENYL ACETAMIDE □ N-FLUOREN-3-YL ACETAMIDE □ N-9H-FLUOREN-3-YL ACETAMIDE

TOXICITY DATA WITH REFERENCE
orl-rat TDLo:6100 mg/kg/66W-C:ETA JNCIAM 24,149,60
ipr-rat TDLo:654 mg/kg/4W-I:ETA CNREA8 30,1485,70
imp-rat TDLo:22 mg/kg:ETA CNREA8 37,111,77

SAFETY PROFILE: Questionable carcinogen with experimental tumorigenic data. When heated to decomposition it emits toxic fumes of NO_x.

FDT000 **CAS:22251-01-0** **HR: 2**
1-FLUORENYL ACETHYDROXAMIC ACID
mf: $C_{15}H_{13}NO_2$ mw: 239.29

PROP: Flakes from EtOH (aq). Mp: 78–80°.

SYNS: N-FLUOREN-1-YL ACETOHYDROXAMIC ACID □ N-HYDROXY-1-FLUORENYL ACETAMIDE

TOXICITY DATA WITH REFERENCE
ipr-rat TDLo:603 mg/kg/4W-I:NEO CNREA8 30,1485,70

SAFETY PROFILE: Questionable carcinogen with experimental neoplastigenic data. When heated to decomposition it emits toxic fumes of NO_x.

FDU000 **CAS:22225-32-7** **HR: 2**
3-FLUORENYL ACETHYDROXAMIC ACID
mf: $C_{15}H_{13}NO_2$ mw: 239.29

PROP: Prisms from EtOH (aq). Mp: 132–134°.

SYNS: N-FLUOREN-3-YL ACETOHYDROXAMIC ACID □ N-HYDROXY-3-FAA □ N-HYDROXY-3-FLUORENYL ACETAMIDE

TOXICITY DATA WITH REFERENCE
orl-rat TDLo:3000 mg/kg/21W-C:CAR CNREA8 35,447,75
ipr-rat TDLo:280 mg/kg/4W-I:CAR JNCIAM 60,433,78
imp-rat TDLo:24 mg/kg:NEO CNREA8 37,111,77
ipr-rat TD:270 mg/kg/4W-I:NEO CNREA8 30,1485,70
imp-rat TD:24 mg/kg:ETA CNREA8 33,2489,73
imp-rat TD:40 mg/kg:ETA JNCIAM 60,433,78

SAFETY PROFILE: Questionable carcinogen with experimental carcinogenic, neoplastigenic, and tumorigenic data. When heated to decomposition it emits toxic fumes of NO_x.

FDU875 **CAS:14751-87-2** **HR: 2**
N-FLUOREN-2-YLACETOHYDROXAMIC ACID, COBALT(2+) COMPLEX
mf: $C_{30}H_{24}N_2O_4 \cdot Co$ mw: 535.49

SYNS: COBALT, BIS(N-9H-FLUOREN-2-YL-N-HYDROXYACETAMIDATO-O,O')-(9CI) □ COBALT N-FLUOREN-2-YLACETOHYDROXAMATE □ COBALT SALTS □ N-HYDROXY-2-ACETYLAMINOFLUORENE, COBALTOUS CHELATE

TOXICITY DATA WITH REFERENCE
scu-rat TDLo:160 mg/kg/4W-I:NEO CNREA8 25,527,65

SAFETY PROFILE: Questionable carcinogen with experimental neoplastigenic data. When heated to decomposition it emits toxic fumes of NO_x and Co.

FDV000 **CAS:16808-85-8** **HR: 2**
N-FLUOREN-2-YL ACETOHYDROXAMIC ACID SULFATE
mf: $C_{15}H_{13}NO_5S$ mw: 319.35

SYNS: SULFATE ESTER of N-HYDROXY-N-2-FLUORENYL ACETAMIDE □ N-SULFONOXY-AAF □ N-SULFONOXY-N-ACETYL-2-AMINOFLUORENE

TOXICITY DATA WITH REFERENCE
mmo-bcs 6600 mmol MOPMA3 4,411,68
dnd-bcs 6600 mmol MOPMA3 4,411,68
scu-rat TDLo:255 mg/kg/15W-I:ETA CNREA8 37,1461,77

CONSENSUS REPORTS: EPA Genetic Toxicology Program.

SAFETY PROFILE: Questionable carcinogen with experimental tumorigenic data. Mutation data reported. When heated to decomposition it emits very toxic fumes of NO_x and SO_x. See also ESTERS and SULFATES.

FDX000 **CAS:3671-78-1** **HR: 2**
N-(2-FLUORENYL)BENZAMIDE
mf: $C_{20}H_{15}NO$ mw: 285.36

SYNS: 2-BENZOYLAMINOFLUORENE □ N-FLUOREN-2-YL BENZAMIDE □ N-9H-FLUOREN-2-YL-BENZAMIDE (9CI)

TOXICITY DATA WITH REFERENCE
mma-sat 100 ng/plate BBRCA9 71,1201,76
ipr-rat TDLo:540 mg/kg/4W-I:ETA CNREA8 27,1443,67
ipr-rat TD:540 mg/kg/4W-I:ETA NATUAS 209,202,66

SAFETY PROFILE: Questionable carcinogen with experimental tumorigenic data. Mutation data reported. When heated to decomposition it emits toxic fumes of NO_x.

FDY000 **CAS:29968-64-7** **HR: 2**
N-FLUOREN-1-YL BENZOHYDROXAMIC ACID
mf: $C_{20}H_{15}NO_2$ mw: 301.36

PROP: Red-orange flakes from EtOH (aq). Mp: 147° (decomp).

SYN: N-HYDROXY-1-FLUORENYL BENZAMIDE

TOXICITY DATA WITH REFERENCE
ipr-rat TDLo:823 mg/kg/4W-I:NEO CNREA8 30,1485,70

SAFETY PROFILE: Questionable carcinogen with experimental neoplastigenic data. When heated to decomposition it emits toxic fumes of NO_x.

FDZ000 CAS:3671-71-4 HR: 2
N-FLUOREN-2-YL BENZOHYDROXAMIC ACID
mf: $C_{20}H_{15}NO_2$ mw: 301.36

PROP: A solid. Mp: 187–190°.

SYNS: N-BENZOYLOXY-ACETYLAMINOFLUORENE □ N-(2-FLUORENYL)BENZOHYDROXAMIC ACID □ N-9H-FLUOREN-2-YL-N-HYDROXYBENZAMIDE □ N-HYDROXY-2-BENZOYLAMINOFLUORENE □ N-HYDROXY-N-2-FLUORENYLBENZAMIDE

TOXICITY DATA WITH REFERENCE
mma-sat 1500 ng/plate CBINA8 54,71,85
mmo-bcs 10 mol MOPMA3 4,411,68
oms-bcs 10 mol MOPMA3 4,411,68
orl-rat TDLo:5200 mg/kg/23W-I:CAR CNREA8 30,1485,70
ipr-rat TDLo:540 mg/kg/4W-I:CAR CNREA8 27,1443,67
ipr-rat TD:540 mg/kg/4W-I:CAR NATUAS 209,202,66

SAFETY PROFILE: Questionable carcinogen with experimental carcinogenic data. Mutation data reported. When heated to decomposition it emits toxic fumes of NO_x.

FEE000 CAS:391-57-1 HR: 2
N,N′-FLUOREN-2,7-YLENE BIS(TRIFLUOROACETAMIDE)
mf: $C_{17}H_{10}F_6N_2O_2$ mw: 388.29

TOXICITY DATA WITH REFERENCE
orl-rat TDLo:4850 mg/kg/37W-C:CAR JNCIAM 30,143,63
orl-rat TD:5 g/kg/48W-C:ETA JJIND8 71,211,83

SAFETY PROFILE: Questionable carcinogen with experimental carcinogenic and tumorigenic data. When heated to decomposition it emits toxic fumes of $F^-$and NO_x. See also FLUORIDES.

FEE100 CAS:13929-01-6 HR: 2
N-(9-FLUORENYL)-N-ETHYL-β-CHLOROETHYLAMINE HYDROCHLORIDE
mf: $C_{17}H_{18}ClN{\cdot}ClH$ mw: 308.27

SYNS: N-(2-CHLOROETHYL)-N-ETHYLFLUOREN-9-AMINE HYDROCHLORIDE □ N-(2-CHLOROETHYL)-N-ETHYL-9-FLUORENAMINE HYDROCHLORIDE □ 2-CHLORO-N-(9-FLUORENYL)DIETHYLAMINE HYDROCHLORIDE □ DIETHYLAMINE, 2-CHLORO-N-(9-FLUORENYL)-, HYDROCHLORIDE □ FLUOREN-9-AMINE, N-(2-CHLOROETHYL)-N-ETHYL-, HYDROCHLORIDE □ 9H-FLUOREN-9-AMINE, N-(2-CHLOROETHYL)-N-ETHYL-, HYDROCHLORIDE (9CI) □ SKF-501 HYDROCHLORIDE

TOXICITY DATA WITH REFERENCE
scu-rat TDLo:20 mg/kg (female 1D pre):REP ENDOAO 74,309,64
scu-mus LD50:800 mg/kg JPETAB 97,25,49

SAFETY PROFILE: Moderately toxic by subcutaneous route. Experimental reproductive effects. When heated to decomposition it emits toxic fumes of NO_x, HCl, and Cl^-.

FEF000 CAS:6957-71-7 HR: 2
N-FLUOREN-2-YL FORMAMIDE
mf: $C_{14}H_{11}NO$ mw: 209.26

SYNS: N,2-FLUORENYL FORMAMIDE □ 2-FORMYLAMINOFLUORENE

TOXICITY DATA WITH REFERENCE
mma-sat 100 ng/plate BBRCA9 71,1201,76
orl-rat TDLo:4400 mg/kg/35W-C:CAR CNREA8 22,1002,62

SAFETY PROFILE: Questionable carcinogen with experimental carcinogenic data. Mutation data reported. When heated to decomposition it emits toxic fumes of NO_x.

F

FEG000 CAS:67176-33-4 HR: 2
N-(2-FLUORENYL)FORMOHYDROXAMIC ACID
mf: $C_{14}H_{11}NO_2$ mw: 225.26

SYNS: N-FORMYL-N-2-FLUORENYLHYDROXYLAMINE □ N-HYDROXY-N-2-FORMYLAMINOFLUORENE

TOXICITY DATA WITH REFERENCE
mmo-sat 10 μg/plate CRNGDP 3,233,82
mma-sat 5 μg/plate CNREA8 40,1204,80
par-rat TDLo:2253 μg/kg:CAR NTIS** PB86-115920
par-rat TDLo:2815 μg/kg:CAR CRNGDP 3,233,82

SAFETY PROFILE: Questionable carcinogen with experimental carcinogenic data. Mutation data reported. When heated to decomposition it emits toxic fumes of NO_x.

FEH000 CAS:51029-30-2 HR: 2
3-FLUORENYLHYDROXYLAMINE
mf: $C_{13}H_{11}NO$ mw: 197.25

SYNS: N-FLUOREN-3-YL HYDROXYLAMINE □ N-HYDROXY-3-AMINOFLUORENE

TOXICITY DATA WITH REFERENCE
imp-rat TDLo:20 mg/kg:ETA CNREA8 33,2489,73

SAFETY PROFILE: Questionable carcinogen with experimental tumorigenic data. When heated to decomposition it emits toxic fumes of NO_x.

FEI000 CAS:34461-49-9 HR: 2
N-FLUOREN-2-YL-HYDROXYLAMINE-o-GLUCURONIDE
mf: $C_{19}H_{19}NO_8$ mw: 389.39

SYN: N-2-FLUORENYLHYDROXYLAMINE-o-GLUCURONIDE

TOXICITY DATA WITH REFERENCE
mmo-bcs 750 μg/L MUREAV 21,63,73
uns-bac-bcs 750 μg/L MUREAV 21,63,73
scu-rat TDLo:750 mg/kg/9W-I:ETA CNREA8 31,1645,71

SAFETY PROFILE: Questionable carcinogen with experimental tumorigenic data. Mutation data reported. When heated to decomposition it emits toxic fumes of NO_x.

FEI100 CAS:28920-43-6 HR: 3
9-FLUORENYLMETHYL CHLOROFORMATE
mf: $C_{15}H_{11}ClO_2$ mw: 258.71

SYNS: CARBONOCHLORIDIC ACID, 9H-FLUOREN-9-YLMETHYL ESTER □ FORMIC ACID, CHLORO-, FLUOREN-9-YLMETHYL ESTER

TOXICITY DATA WITH REFERENCE
mma-sat 100 μg/plate MUREAV 170,23,86

DOT CLASSIFICATION: 6.1; *Label:* Poison, Corrosive

SAFETY PROFILE: A poison. Mutation data reported. A corrosive. When heated to decomposition it emits toxic vapors of Cl^-.

FEI200 CAS:781-73-7 HR: 3
2-FLUORENYL METHYL KETONE
mf: $C_{15}H_{12}O$ mw: 208.27

SYNS: 2-ACETOFLUORENE □ 2-ACETYLFLUORENE □ ETHANONE, 1-(9H-FLUOREN-2-YL) □ KETONE, FLUOREN-2-YL METHYL

TOXICITY DATA WITH REFERENCE
dnr-esc 80 mg/L MUREAV 119,135,83

DOT CLASSIFICATION: 3; *Label:* Flammable Liquid

SAFETY PROFILE: Mutation data reported. A flammable liquid. When heated to decomposition it emits acrid smoke and irritating vapors.

FEI500 CAS:63019-68-1 HR: 2
2-FLUORENYLMONOMETHYLAMINE
mf: $C_{14}H_{13}N$ mw: 195.28

SYNS: 2-METHYLAMINOFLUORENE □ N-MONOMETHYL-2-AMINOFLUORENE □ 2-MONOMETHYLAMINOFLUORENE

TOXICITY DATA WITH REFERENCE
orl-rat TDLo:1260 mg/kg/24W-I:ETA BJCAAI 6,89,52
orl-rat TD:4300 mg/kg/47W-C:ETA JBCHA3 221,845,56

SAFETY PROFILE: Questionable carcinogen with experimental tumorigenic data. When heated to decomposition it emits toxic fumes of NO_x.

FEM000 CAS:63224-45-3 HR: 2
N-(2-FLUORENYL)MYRISTOHYDROXAMIC ACID ACETATE
mf: $C_{29}H_{39}NO_3$ mw: 449.69

SYNS: N-ACETOXY-2-MYRISTOYL-AMINOFLUORENE □ N-(ACETYLOXY)-N-9H-FLUOREN-2-YL-TETRADECANAMIDE

TOXICITY DATA WITH REFERENCE
mmo-sat 1 nmol/plate CNREA8 37,1461,77
dns-hmn:fbr 100 μmol/L/5H IJCNAW 16,284,75
scu-rat TDLo:114 mg/kg/6W-I:CAR CNREA8 37,1461,77

CONSENSUS REPORTS: EPA Genetic Toxicology Program.

SAFETY PROFILE: Questionable carcinogen with experimental carcinogenic data. Human mutation data reported. When heated to decomposition it emits toxic fumes of NO_x.

FEM100 CAS:15860-31-8 HR: 3
1-FLUORENYL PHENYL KETONE
mf: $C_{20}H_{14}O$ mw: 270.34

SYNS: 2-BENZOYLFLUORENE □ 9H-FLUOREN-2-YLPHENYLMETHANONE □ KETONE, FLUOREN-2-YL PHENYL □ METHANONE, 9H-FLUOREN-2-YLPHENYL-

TOXICITY DATA WITH REFERENCE
ipr-mus LD50:>2 g/kg PCJOAU 23,325,89

DOT CLASSIFICATION: 3; *Label:* Flammable Liquid

SAFETY PROFILE: Low toxicity by intraperitoneal route. A flammable liquid. When heated to decomposition it emits acrid smoke and irritating vapors.

FEN000 CAS:2485-10-1 HR: 2
N-FLUORENYL-2-PHTHALIMIC ACID
mf: $C_{21}H_{15}NO_3$ mw: 329.37

SYNS: 2-BENZOYLAMIDOFLUORENE-2′-CARBOXYLATE □ 2-BENZOYLAMINOFLUORENE-2′-CARBOXYLATE □ N-(2-FLUORENYL) PHTHALAMIC ACID

TOXICITY DATA WITH REFERENCE
orl-rat TDLo:5000 mg/kg/42W-C:ETA AICCA6 20,1364,64
orl-rat TD:6000 mg/kg/42W-C:ETA CNREA8 20,1252,60

SAFETY PROFILE: Questionable carcinogen with experimental tumorigenic data. When heated to decomposition it emits toxic fumes of NO_x.

FEO000 CAS:52663-84-0 HR: 2
N-(2-FLUORENYL)PROPIONOHYDROXAMIC ACID
mf: $C_{16}H_{15}NO_2$ mw: 253.32

SYNS: N-HYDROXY-N-2-PROPIONYLAMINO FLUORENE □ N-PROPIONYL-N-2-FLUORENYLHYDROXYLAMINE

TOXICITY DATA WITH REFERENCE
mmo-sat 10 μg/plate CRNGDP 3,233,82
mma-sat 5 μg/plate CNREA8 40,1204,80
par-rat TDLo:2533 μg/kg:CAR NTIS** PB86-115920
par-rat TDLo:3166 μg/kg:ETA CRNGDP 3,233,82

SAFETY PROFILE: Questionable carcinogen with experimental carcinogenic and tumorigenic data. Mutation data reported. When heated to decomposition it emits toxic fumes of NO_x.

FEP000 CAS:59935-47-6 HR: 2
N-2-FLUORENYL SUCCINAMIC ACID
mf: $C_{17}H_{15}NO_3$ mw: 281.33

TOXICITY DATA WITH REFERENCE
orl-rat TDLo:6200 mg/kg/72W-C:NEO JNCIAM 24,149,60

SAFETY PROFILE: Questionable carcinogen with experimental neoplastigenic data. When heated to decomposition it emits toxic fumes of NO_x.

FER000 CAS:363-17-7 HR: 2
N-FLUOREN-2-YL-2,2,2-TRIFLUOROACETAMIDE
mf: $C_{15}H_{10}F_3NO$ mw: 277.26

SYNS: N-(2-FLUORENYL)-2,2,2-TRIFLUOROACETAMIDE □ 2-TRI-

FLUOROACETYLAMINOFLUORENE □ 2,2,2-TRIFLUORO-N-(FLUOREN-2-YL)ACETAMIDE

TOXICITY DATA WITH REFERENCE
orl-rat TDLo:3750 mg/kg/44W-C:CAR JNCIAM 24,149,60
orl-rat TD:4500 mg/kg/26W-C:CAR CNREA8 22,1002,62

SAFETY PROFILE: Questionable carcinogen with experimental carcinogenic data. When heated to decomposition it emits very toxic F^- and NO_x. See also FLUORIDES.

FEV000 CAS:2321-07-5 HR: 3
FLUORESCEIN
mf: $C_{20}H_{12}O_5$ mw: 332.32

PROP: Orange-red, crystalline powder. Mp: 314–316° with decomp.

SYNS: 9-(o-CARBOXYPHENYL)-6-HYDROXY-3-ISOXANTHENONE □ 9-(o-CARBOXYPHENYL)-6-HYDROXY-3H-XANTHEN-3-ONE □ C.I. 45330 □ C.I. 45350 (FREE ACID) □ C.I. SOLVENT YELLOW 94 □ D&C YELLOW No. 7 □ 3',6'-DIHYDROXYFLUORAN □ DIHYDROXYFLUORANE □ 3',6'-DIHYDROXYSPIRO(ISOBENZOFURAN-1(3H),9'(9H)-XANTHEN)-3-ONE □ 3,6-FLUORANDIOL □ 3',6'-FLUORANDIOL □ FLUORESCEINE □ HIDACID FLUORESCEIN □ RESORCINOLPHTHALEIN □ SOAP YELLOW F □ 11712 YELLOW

TOXICITY DATA WITH REFERENCE
dnd-esc 15 µmol/L MUREAV 89,95,81
ipr-rat LDLo:600 mg/kg IJLEAG 2,257,34
ivn-rbt LDLo:300 mg/kg IJLEAG 2,257,34

CONSENSUS REPORTS: Reported in EPA TSCA Inventory. EPA Genetic Toxicology Program.

SAFETY PROFILE: Poison by intravenous route. Moderately toxic by intraperitoneal route. Mutation data reported. When heated to decomposition it emits acrid smoke and irritating fumes. See also FLUORESCEIN SODIUM.

FEV100 CAS:3570-80-7 HR: 3
FLUORESCEIN MERCURIC ACETATE
mf: $C_{24}H_{16}Hg_2O_9$ mw: 849.58

SYNS: FLUORESCEIN MERCURIACETATE □ FLUORESCEIN MERCURY ACETATE □ FMA □ FMA (analytical reagent) □ MERCURY, BIS (ACETATO)(mu-(3',6'-DIHYDROXY-2',7'-FLUORANDIYL))DI-

TOXICITY DATA WITH REFERENCE
dnd-mmo-omi 200 nmol/L BBACAQ 454,309,76
dnd-uns:lyms 100 nmol/L BBACAQ 454,309,76

CONSENSUS REPORTS: EPA TSCA Chemical Inventory.

ACGIH TLV: TWA 0.1 mg(Hg)/m³ (skin)
NIOSH REL: (Mercury, Aryl and Inorganic): CL 0.1 mg/m³ (skin)

SAFETY PROFILE: Highly toxic. Mutation data reported. When heated to decomposition it emits toxic fumes of Hg.

FEW000 CAS:518-47-8 HR: 2
FLUORESCEIN SODIUM
mf: $C_{20}H_{10}O_5{\cdot}2Na$ mw: 376.28

PROP: Orange-red, hygroscopic powder. Sol in water; sltly sol in alc.

SYNS: AIZEN URANINE □ CALCOCID URANINE B4315 □ 9-o-CARBOXYPHENYL-6-HYDROXY-3-ISOXANTHONE, DISODIUM SALT □ CERTIQUAL FLUORESCEINE □ C.I. 766 □ C.I. ACID YELLOW 73 □ C.I. 45350 DISODIUM SALT □ D&C YELLOW No. 8 □ DISODIUM-6-HYDROXY-3-OXO-9-XANTHENE-o-BENZOATE □ FLUORESCEIN SODIUM B.P □ FLUORESCEIN, soluble □ FLUOR-I-STRIP A.T. □ FUL-GLO □ FUNDUSCEIN □ FURANIUM □ HIDACID URANINE □ NCI-C54706 □ RESORCINOL PHTHALEIN SODIUM □ SODIUM FLUORESCEIN □ SODIUM FLUORESCEINATE □ SODIUM SALT of HYDROXY-o-CARBOXY-PHENYL-FLUORONE □ SOLUBLE FLUORESCEIN □ SPIRO (ISOBENZOFURAN)-1(3H),9'-(9H)XANTHENE-3-ONE, 3',6'-DIHYDROXY-DISODIUM SALT □ URANIN □ URANINE A EXTRA □ URANINE USP XII □ URANINE YELLOW □ 11824 YELLOW □ 12417 YELLOW

TOXICITY DATA WITH REFERENCE
dnd-esc 15 µmol/L MUREAV 89,95,81
ivn-rbt TDLo:933 mg/kg (5-8D preg):REP AJOPAA 84,847,77
scu-rat TDLo:19 g/kg/79W-I:ETA GANNA2 45,446,54
scu-rat TD:16 g/kg/1Y-I:ETA GANNA2 47,51,56
ivn-man TDLo:14 mg/kg/10M-I:CVS AACRAT 66,283,87
ivn-man TDLo:7142 µg/kg:EYE,GIT AJOPAA 103,111,87
orl-rat LD50:6721 mg/kg JOPRAJ 48,228,77
orl-mus LD50:4738 mg/kg JOPRAJ 48,228,77
ipr-mus LD50:1800 mg/kg TXAPA9 24,37,73

CONSENSUS REPORTS: Reported in EPA TSCA Inventory.

SAFETY PROFILE: Moderately toxic by intraperitoneal route. Mildly toxic by ingestion. Human systemic effects by intravenous route: arrhythmias, eye hemorrhage, nausea or vomiting. Experimental reproductive effects. Questionable carcinogen with experimental tumorigenic data. Mutation data reported. When heated to decomposition it emits toxic fumes of Na_2O.

FEX875 CAS:16984-48-8 HR: 3
FLUORIDE
af: F aw: 18.9984

SYNS: FLUORIDE(1-) □ FLUORIDE ION □ FLUORIDE ION(1-) □ PERFLUORIDE

TOXICITY DATA WITH REFERENCE
dni-mus:fbr 1300 µmol/L APTOA6 45,96,79
orl-hmn TDLo:3 mg/kg:LIV,PUL,CNS IMEMDT** 27,237,82
orl-hmn LDLo:50 mg/kg:PUL,LIV IMEMDT** 27,237,82
ivn-mus LD50:22,800 µg/kg MIVRA6 8,320,74

CONSENSUS REPORTS: EPA Genetic Toxicology Program.

OSHA PEL: TWA 2.5 mg(F)/m³
ACGIH TLV: TWA 2.5 mg(F)/m³
DFG MAK: 2.5 mg/m³
NIOSH REL: (Fluorides, inorganic) TWA 2.5 mg(F)/m³

SAFETY PROFILE: Human poison by ingestion. Human systemic effects by ingestion: convulsions, changes in

the respiratory system, liver, and kidneys. Mutation data reported. When heated to decomposition it emits toxic fumes of F^-.

For occupational chemical analysis use NIOSH: Fluorides 7902; Fluoride in Urine 8308.

FEY000 **HR: 2**
FLUORIDES

OSHA PEL: TWA 2.5 mg(F)/m³
ACGIH TLV: TWA 2.5 mg(F)/m³
DFG MAK: 2.5 mg/m³; BAT: 7 mg/kg creatinine in urine at end of exposure; 4 mg/kg creatinine in urine about 16 hours after end of exposure
NIOSH REL: TWA 2.5 mg(F)/m³

SAFETY PROFILE: Inorganic fluorides are generally highly irritating and toxic. Acute effects resulting from exposure to fluorine compounds are due to HF. Chronic fluorine poisoning, or "fluorosis," occurs among miners of cryolite, and consists of a sclerosis of the bones, caused by fixation of the calcium by the fluorine. There may also be some calcification of the ligaments. The teeth are mottled, and there is osteosclerosis and ostemalacia. The bony and ligamentous changes are demonstrable by x-ray. The estimated human lethal dose is 2.5 to 5.9 g of F^-. Large doses can cause very severe nausea, vomiting, diarrhea, abdominal burning, and cramp-like pains. It is not taken up by the thyroid and does not interfere with iodine uptake. Can cause or aggravate attacks of asthma and severe bone changes, making normal movements painful. Some signs of pulmonary fibrosis are noted. Some enzyme systems effects are reported. Irritants to the eyes, skin, and mucous membranes. Loss of weight, anorexia, anemia, wasting and cachexia, and dental defects are among the common findings in chronic fluorine poisoning. There may be an eosinophilia and impairment of growth in young workers. Symptoms of intoxication include gastric, intestinal, circulatory, respiratory and nervous complaints, and skin rashes. When heated to decomposition, or on contact with acid or acid fumes, they emit highly toxic fumes of F^-.

Organic fluorides are generally less toxic than other halogenated hydrocarbons. Fluorocarbons are chemically inert to most materials but can react violently with barium, sodium, and potassium. Fluoroamides react violently with lithium tetrahydroaluminate and with sodium at very high temperatures. Some fluorinated cyclopropenyl methyl ethers react violently with water or methanol. Some fluorodinitro compounds of methane and ethane are sensitive explosives. When heated to decomposition they emit toxic fumes of F^-. Common air contaminants.

FEZ000 **CAS:7782-41-4** **HR: 3**
FLUORINE
DOT: UN 1045
mf: F_2 mw: 38.00

PROP: Pale-yellow gas (turning white at −2°) which reacts with most organic and inorganic materials. Powerful oxidant. Mp: −218°, bp: −187°, d: 1.14 @ −200°, 1.108 @ −188°, vap d: 1.695.

SYNS: BIFLUORIDEN (DUTCH) □ FLUOR (DUTCH, FRENCH, GERMAN, POLISH) □ FLUORINE, compressed (DOT) □ FLUORO (ITALIAN) □ FLUORURES ACIDE (FRENCH) □ FLUORURI ACIDI (ITALIAN) □ RCRA WASTE NUMBER P056 □ SAEURE FLUORIDE (GERMAN)

TOXICITY DATA WITH REFERENCE
eye-hmn 25 ppm/5M MLD AIHAAP 29,11,68
eye-rat 140 ppm/30M AIHAAP 29,11,68
eye-mus 467 ppm/5M AIHAAP 29,11,68
eye-dog 68 ppm/1H AIHAAP 29,11,68
mmo-sat 1 mg/plate CYGEDX 16(6),41,82
ihl-rat LC50:185 ppm/1H AIHAAP 29,11,68
ihl-mus LC50:150 ppm/1H AIHAAP 29,11,68
ihl-rbt LC50:270 ppm/30M AIHAAP 29,11,68
ihl-gpg LC50:170 ppm/1H AIHAAP 29,11,68

CONSENSUS REPORTS: Reported in EPA TSCA Inventory.

OSHA PEL: TWA 0.1 ppm
ACGIH TLV: TWA 1 ppm; STEL 2 ppm
DFG MAK: 0.1 ppm (0.2 mg/m³)
DOT CLASSIFICATION: 2.3; *Label:* Poison Gas, Oxidizer

SAFETY PROFILE: A poison gas. A skin, eye, and mucous membrane irritant. A most powerful caustic irritant to tissue. Mutation data reported. A very dangerous fire and explosion hazard. A powerful oxidizer. Reacts violently with many materials.

Explosive or potentially explosive reaction with ammonia, cesium fluoride + fluorocarboxylic acids, cesium heptafluoropropoxide, 1- or 2-fluoriminoperfluoropropane, graphite, halocarbons (e.g., carbon tetrachloride, chloroform, perfluorocyclobutane, iodoform, 1,2-dichlorotetrafluoroethane), liquid hydrocarbons (e.g., anthracene, turpentine), hydrogen, hydrogen + oxygen, hydrogen fluoride + seleninyl fluoride + heat, nitric acid, silver cyanide, sulfur dioxide, carbon monoxide, sodium acetate, sodium bromate, stainless steel, water.

Reacts to form explosive products with alkanes + oxygen (forms peroxides), cyanoguanidine, perchloric acid (forms fluorine perchlorate gas), potassium chlorate (forms fluorine perchlorate gas), potassium hydroxide (forms potassium trioxide). Forms explosive mixtures with acetonitrile + chlorine fluoride, ice.

Ignition or violent reaction on contact with acetylene, ceramic materials, covalent halides (e.g., chromyl chloride, phosphorus pentachloride, phosphorus trichloride, phosphorus trifluoride, boron trichloride, silicon tetrachloride), halogens (e.g., bromine, iodine, chlorine + spark or heating to 100°C), dicyanogen, gaseous hydrocarbons (e.g., town gas, methane, benzene), hydrogen halide gases or concentrated solutions (e.g., hydrogen bromide, hydrogen chloride, hydrogen iodide, hydrogen fluoride), metal acetylides and carbides (e.g., monocesium acetylide, cesium acetylide, lithium acetylide, rubidium acetylide, tungsten carbide, ditungsten carbide, zirconium dicarbide, uranium dicarbide), metal cyano complexes [e.g., potassium hexacyanoferrate(II), lead hexacyanoferrate(III), potassium hexacyanoferrate(III)], metal hydrides (e.g., copper hydride, potassium hydride, sodium hydride), metal iodides (e.g., lead iodide, calcium iodide, mercury iodide, potassium iodide), metals, metal salts, metal silicides (e.g., calcium disilicide, lithium hexasilicide), nickel(IV) oxide, nonmetals (e.g., boron, yellow or red phosphorus, selenium, tellurium, silicon, carbon, charcoal, sulfur), oxygenated organic compounds (e.g., methanol, ethanol, 3-methyl butanol, acetaldehyde, trichloroacetaldehyde, acetone, lactic acid, benzoic acid, salicylic acid, ethyl acetate, methyl borate), nonmetal oxides (e.g., arsenic trioxide, nitrogen oxide, dinitrogen tetroxide), oxygen + polymers [e.g., phenol-formaldehyde resins (bakelite), polyacrylonitrile-butadiene (Buna N), polyamides (nylons), polychloropene (neoprene), polyethylene, polytrifluoropropylmethylsiloxane, polyvinylchloride-vinyl acetate (Tygon), polyvinylidene fluoride-hexafluoropropylene (Viton), polyurethane foam, polymethyl methacrylate (Perspex), polytetrafluooethylene (Teflon)], sulfides (e.g., antimony trisulfide, carbon disulfide vapor, chromium (II) sulfide, hydrogen sulfide, barium sulfide, potassium sulfide, zinc sulfide, molybdenum sulfide), xenon + catalysts (e.g., nickel fluoride, silver difluoride, nickel(III) oxide, silver (I) oxide).

Incandescent reaction with boron nitride, hexalithium disilicide + heat, metal borides, metal oxides (e.g., nickel(II) oxide, alkali metal oxides, alkaline earth oxides), nitrogenous bases (e.g., aniline, dimethylamine, pyridine), gallic acid.

Incompatible with cesium heptafluoro propoxide, cyanoguanidine, halocarbons, hexalithium disilicide, seleninyl fluoride, hydrogen sulfide, oxygen, sodium acetate, sodium bromate, sodium dicyanamides, most organic matter, H-containing molecules, oxides of S, N, P, alkali metals,and alkaline earths. It reacts violently with halogen acids, hydrazine, ClO_2, coke, cyanamide, cyanides, KNO_3, (PbO + glycerol), CCl_4, silicides, silicates, trinitromethane, alkenes, alkyl benzenes, CS_2, $Cr(OCl)_2$, Al, Tl, Sn, Sb, As, natural gas, liquid air, perfluoropropionyl fluoride, polyvinyl chloride acetate. Many reactions go on even at <−160°. Reacts with water or steam to produce heat and toxic and corrosive fumes. Used as one component of liquid rocket fuel and in chemical lasers. See also FLUORIDES.

FFA000 CAS:14986-60-8 HR: 3
FLUORINE AZIDE
mf: FN_3 mw: 61.02

PROP: A green gas. Mp: −154°, bp: −82°.

SYN: AZIDO FLUORINE

SAFETY PROFILE: A poison. An irritant to the eyes, skin, and mucous membranes. An extremely unstable explosive sensitive to light and heat. Usually explodes on vaporization at −82°C. When heated to decomposition it emits toxic fumes of F^- and NO_x. See also FLUORINE and AZIDES.

FFB000 CAS:13536-85-1 HR: 3
FLUORINE FLUORO SULFATE
mf: F_2O_3S mw: 118.06

FSO_2OF

PROP: Colorless gas or vapor. Mp: −158.5°, bp: −31.3°.

SYN: FLUOROSULFURYL HYPOFLUORITE

SAFETY PROFILE: Unstable it may explode at room temperature. When heated to decomposition it emits toxic fumes of F^- and SO_x. See also FLUORINE and SULFATES.

FFD000 CAS:10049-03-3 HR: 3
FLUORINE PERCHLORATE
mf: $ClFO_4$ mw: 118.45

O_3ClOF

PROP: Colorless gas, pungent, acrid odor. Mp: −167.3°, bp: −15.9°.

SYNS: CHLORINE TETROXYFLUORIDE □ PERCHLORYL HYPOFLUORITE

SAFETY PROFILE: A poison by ingestion and inhalation. Corrosive to the skin, eye, and mucous membranes. Very unstable. A powerful oxidizer that can react violently with oxidizable materials. A very dangerous explosion hazard; it explodes on slightest provocation. The liquid explodes on freezing at −167°C. The gas explodes when exposed to sparks, flame, or on contact with grease, dust, rubber, or aqueous potassium iodide. Ignites in contact with hydrogen gas. When heated to decomposition it emits highly toxic fumes of F^- and Cl^-. See also FLUORINE and PERCHLORATES.

FFE000 CAS:1544-46-3 HR: 3
FLUOROACETALDEHYDE
mf: C_2H_3FO mw: 62.05

PROP: Viscous liquid; rapidly polymerizing to a white solid. Bp: 19–23°.

TOXICITY DATA WITH REFERENCE
ipr-mus LD50:6 mg/kg JACSAT 78,4996,56
scu-mus LD50:4800 μg/kg JCSOA9 ,773,49

SAFETY PROFILE: Poison by intraperitoneal and subcutaneous routes. When heated to decomposition it emits toxic fumes of F^-. See also FLUORIDES and ALDEHYDES.

FFF000 CAS:640-19-7 HR: 3
FLUOROACETAMIDE
mf: C_2H_4FNO mw: 77.07

PROP: Needles from $CHCl_3$. Mp: 107–109°. Sol in H_2O and Me_2CO. Sltly sol in $CHCl_3$.

SYNS: AFL 1081 □ COMPOUND 1081 □ FAA □ FLUORAKIL 100 □ 2-FLUOROACETAMIDE □ FLUOROACETIC ACID AMIDE □FUSSOL □ MEGATOX □ MONOFLUOROACETAMIDE □ NAVRON □ RCRA WASTE NUMBER P057 □ RODEX □ YANOCK

TOXICITY DATA WITH REFERENCE
spm-rat-ipr 4 mg/kg VAAZA2 1,346,68
cyt-mam:lng 300 μmol/L HKXUDL 3,94,83
orl-rat TDLo:90 mg/kg (30D male):REP EXPEAM 20,492,64
orl-hmn LDLo:2 mg/kg SKEZAP 9,1,68
unr-hmn LDLo:5 mg/kg:CNS,GIT 34ZIAG -,274,69
orl-rat LD50:5750 μg/kg PCOC** -,538,66
skn-rat LD50:80 mg/kg WRPCA2 9,119,70
ipr-rat LD50:12 mg/kg JSFAAE 8,400,57

orl-mus LD50:25 mg/kg ABCHA6 31,1294,67
ihl-mus LC50:550 mg/m³ ABCHA6 31,1294,67
skn-mus LD50:34 mg/kg SKEZAP 9,1,68
ipr-mus LD50:85 mg/kg BCPCA6 12,1201,63
scu-mus LD50:34 mg/kg NHOZAX 26,203,72
ivn-rbt LD50:250 μg/kg JCSOA9 -,912,49

CONSENSUS REPORTS: EPA Extremely Hazardous Substances List. Reported in EPA TSCA Inventory.

SAFETY PROFILE: A human poison by an unspecified route. Poison experimentally by ingestion, skin contact, intraperitoneal, subcutaneous, and intravenous routes. Human systemic effects by unspecified route: convulsions, coma, nausea and vomiting. Experimental reproductive effects. Mutation data reported. Used as an insecticide and rodenticide. When heated to decomposition it emits very toxic fumes of F^- and NO_x. See also FLUORIDES.

FFG000 CAS:343-89-5 HR: 2
7-FLUORO-2-ACETAMIDO-FLUORENE
mf: $C_{15}H_{12}FNO$ mw: 241.28

SYNS: 7-FLUORO-2-ACETYLAMINOFLUORENE □ N-(7-FLUOROFLUORENE-2-YL)ACETAMIDE

TOXICITY DATA WITH REFERENCE
dnd-rat-par 27 μmol/kg CBINA8 40,27,82
orl-rat TDLo:450 mg/kg/11W-C:CAR CNREA8 26,2239,66
orl-rat TD:1200 mg/kg/15W-C:ETA CNREA8 18,469,58
orl-rat TD:1200 mg/kg/14W-C:CAR CNREA8 15,188,55

SAFETY PROFILE: Questionable carcinogen with experimental carcinogenic and tumorigenic data. Mutation data reported. When heated to decomposition it emits toxic fumes of F^- and NO_x. See also FLUORIDES.

FFH000 CAS:330-68-7 HR: 3
FLUOROACETANILIDE
mf: C_8H_8FNO mw: 153.17

SYNS: AFL 1082 □ 2-FLUOROACETANILIDE □ 2-FLUORO-N-PHENYLACETAMIDE □ TL 1312

TOXICITY DATA WITH REFERENCE
orl-rat LDLo:3 mg/kg NCNSA6 5,10,53
ipr-rat LD50:6 mg/kg JSFAAE 8,400,57
orl-mus LD50:25 mg/kg YKKZAJ 88,1620,68
ihl-mus LCLo:480 mg/m³/10M NDRC** 30101,4,45

SAFETY PROFILE: Poison by ingestion and intraperitoneal routes. Moderately toxic by inhalation. When heated to decomposition it emits very toxic fumes of F^- and NO_x. See also FLUORIDES.

FFJ000 CAS:503-20-8 HR: 3
FLUOROACETONITRILE
mf: C_2H_2FN mw: 59.05

SYN: FLUOROMETHYL CYANIDE

TOXICITY DATA WITH REFERENCE
ipr-mus LD50:25 mg/kg JACSAT 78,3484,56
scu-mus LD50:25 mg/kg CLDND*

CONSENSUS REPORTS: Cyanide and its compounds are on the Community Right-To-Know List.

SAFETY PROFILE: Poison by intraperitoneal and subcutaneous routes. When heated to decomposition it emits very toxic fumes of F^-, CN^-, and NO_x. See also NITRILES and FLUORIDES.

FFL000 CAS:2824-10-4 HR: 2
1-FLUORO-2-ACETYLAMINOFLUORENE
mf: $C_{15}H_{12}FNO$ mw: 241.28

SYNS: 1-FLUORO-2-FAA □ N-(1-FLUOROFLUOREN-2-YL)ACETAMIDE

TOXICITY DATA WITH REFERENCE
orl-rat TDLo:1400 mg/kg/19W-C:CAR CNREA8 22,1002,62

SAFETY PROFILE: Questionable carcinogen with experimental carcinogenic data. When heated to decomposition it emits very toxic fumes of F^- and NO_x. See also AMINES and FLUORIDES.

FFM000 CAS:2823-93-0 HR: 2
3-FLUORO-2-ACETYLAMINOFLUORENE
mf: $C_{15}H_{12}FNO$ mw: 241.28

SYNS: 3-FLUORO-2-FAA □ N-(3-FLUOROFLUOREN-2-YL)ACETAMIDE

TOXICITY DATA WITH REFERENCE
orl-rat TDLo:3900 mg/kg/26W-C:CAR CNREA8 22,1002,62

SAFETY PROFILE: Questionable carcinogen with experimental carcinogenic data. When heated to decomposition it emits very toxic fumes of F^- and NO_x. See also AMINES and FLUORIDES.

FFN000 CAS:2823-91-8 HR: 2
4-FLUORO-2-ACETYLAMINOFLUORENE
mf: $C_{15}H_{12}FNO$ mw: 241.28

SYNS: 4-FLUORO-2-FAA □ N-(4-FLUOROFLUOREN-2-YL)ACETAMIDE

TOXICITY DATA WITH REFERENCE
orl-rat TDLo:2560 mg/kg/17W-C:CAR CNREA8 22,1002,62

SAFETY PROFILE: Questionable carcinogen with experimental carcinogenic data. When heated to decomposition it emits very toxic fumes of F^- and NO_x. See also AMINES and FLUORIDES.

FFO000 CAS:2823-90-7 HR: 2
5-FLUORO-2-ACETYLAMINOFLUORENE
mf: $C_{15}H_{12}FNO$ mw: 241.28

SYNS: 5-FLUORO-2-FAA □ N-(5-FLUOROFLUOREN-2-YL)ACETAMIDE

TOXICITY DATA WITH REFERENCE
orl-rat TDLo:3900 mg/kg/26W-C:CAR CNREA8 22,1002,62

SAFETY PROFILE: Questionable carcinogen with experimental carcinogenic data. When heated to decomposition it emits very toxic fumes of F^- and NO_x. See also AMINES and FLUORIDES.

FFP000 CAS:2823-94-1 HR: 2
6-FLUORO-2-ACETYLAMINOFLUORENE
mf: $C_{15}H_{12}FNO$ mw: 241.28

SYNS: 6-FLUORO-2-FAA □ N-(6-FLUOROFLUOREN-2-YL)ACETAMIDE

TOXICITY DATA WITH REFERENCE
orl-rat TDLo:3900 mg/kg/26W-C:CAR CNREA8 22,1002,62

SAFETY PROFILE: Questionable carcinogen with experimental carcinogenic data. When heated to decomposition it emits very toxic fumes of F^- and NO_x. See also AMINES and FLUORIDES.

FFQ000 CAS:2823-95-2 HR: 2
8-FLUORO-2-ACETYLAMINOFLUORENE
mf: $C_{15}H_{12}FNO$ mw: 241.28

SYNS: 8-FLUORO-2-FAA □ N-(8-FLUOROFLUOREN-2-YL)ACETAMIDE

TOXICITY DATA WITH REFERENCE
orl-rat TDLo:3900 mg/kg/26W-C:CAR CNREA8 22,1002,62

SAFETY PROFILE: Questionable carcinogen with experimental carcinogenic data. When heated to decomposition it emits very toxic fumes of F^- and NO_x. See also FLUORIDES and AMINES.

FFR000 CAS:359-06-8 HR: 3
FLUOROACETYL CHLORIDE
mf: C_2H_2ClFO mw: 96.49

TOXICITY DATA WITH REFERENCE
ihl-mus LCLo:200 mg/m³/10M NDRC** No. 9-4-1-9,43
ihl-gpg LCLo:100 mg/m³/10M NDRC** No. 9-4-1-9,43

CONSENSUS REPORTS: Reported in EPA TSCA Inventory.

SAFETY PROFILE: Poison by inhalation. When heated to decomposition it emits very toxic fumes of Cl^- and F^-. See also FLUORIDES and CHLORIDES.

FFS000 CAS:2713-09-9 HR: 3
FLUORO ACETYLENE
mf: C_2HF mw: 44.03

FC≡CH

PROP: A gas. Fp: −196°, bp: −105°.

SAFETY PROFILE: Can explode violently. The silver and mercury salts are heat-sensitive explosives. May explode spontaneously close to its boiling point −80°C. The gas is stable. Ignites on contact with solutions of bromine in carbon tetrachloride. When heated to decomposition it emits toxic fumes of F^-. See also ACETYLENE COMPOUNDS and FLUORIDES.

FFT000 CAS:364-71-6 HR: 3
o-(FLUOROACETYL)SALICYLIC ACID
mf: $C_9H_7FO_4$ mw: 198.16

SYN: SALICYLIC ACID, FLUOROACETATE

TOXICITY DATA WITH REFERENCE
orl-rat LDLo:10 mg/kg NCNSA6 5,8,53
scu-mus LD50:15 mg/kg JCSOA9 -,1773,48

SAFETY PROFILE: Poison by ingestion and subcutaneous routes. When heated to decomposition it emits toxic fumes of F^-. See also FLUORIDES.

FFT100 HR: 3
α-FLUORO-β-ALANINE HYDROCHLORIDE
mf: $C_3H_6FNO_2 \cdot ClH$ mw: 143.56

SYNS: 3-AMINO-2-FLUORO-PROPANOIC ACID HYDROCHLORIDE □ FLUORO-β-ALANINE HYDROCHLORIDE

TOXICITY DATA WITH REFERENCE
ipr-rat LD50:218 mg/kg JPMSAE 73,212,84
ipr-mus LD50:167 mg/kg JPMSAE 73,212,84
ipr-dog LD50:24,500 μg/kg JPMSAE 73,212,84
ipr-cat LD50:9400 μg/kg JPMSAE 73,212,84
ipr-rbt LD50:27 mg/kg JPMSAE 73,212,84

SAFETY PROFILE: Poison by intraperitoneal route. When heated to decomposition it emits toxic fumes of F^-, NO_x, and HCl.

FFV000 CAS:592-79-0 HR: 3
5-FLUORO AMYLAMINE
mf: $C_5H_{12}FN$ mw: 105.18

SYN: 5-FLUOROPENTYLAMINE

TOXICITY DATA WITH REFERENCE
ipr-mus LD50:50 mg/kg JACSAT 78,3487,56
scu-mus LD50:50 mg/kg CLDND*

SAFETY PROFILE: Poison by intraperitoneal and subcutaneous routes. When heated to decomposition it emits very toxic fumes of F^- and NO_x.

FFW000 CAS:407-98-7 HR: 3
5-FLUOROAMYL CHLORIDE
mf: $C_5H_{10}ClF$ mw: 124.60

SYN: 1-CHLORO-5-FLUOROPENTANE

TOXICITY DATA WITH REFERENCE
ipr-mus LD50:30,400 μg/kg CJBPAZ 36,339,58
scu-mus LD50:32 mg/kg CLDND*

SAFETY PROFILE: Poison by intraperitoneal and subcutaneous routes. When heated to decomposition it emits very toxic fumes of Cl^- and F^-.

FFX000 CAS:661-18-7 HR: 3
5-FLUOROAMYL THIOCYANATE
mf: $C_6H_{10}FNS$ mw: 147.23

SYN: 5-FLUOROPENTYL THIOCYANATE

TOXICITY DATA WITH REFERENCE
ipr-mus LD50:30 mg/kg JACSAT 78,3843,56
scu-mus LD50:30 mg/kg CLDND*

SAFETY PROFILE: Poison by intraperitoneal and subcutaneous routes. When heated to decomposition it emits very toxic F^-, NO_x, and SO_x. See also THIOCYANATES and FLUORIDES.

FFY000 **CAS:371-40-4** **HR: 3**
4-FLUOROANILINE
mf: C_6H_6FN mw: 111.13

PROP: D: 1.1724, mp: −0.82°, bp: 184–186°.

SYNS: BENZENAMINE, 4-FLUORO-(9CI) □ 4-FLUORANILIN □ p-FLUOROANILINE □ 4-FLUOROBENZENAMINE □ p-FLUOROPHENYLAMINE

TOXICITY DATA WITH REFERENCE
skn-rbt 2 mg/24H SEV 85JCAE-,611,86
eye-rbt 250 µg/24H SEV 28ZPAK -,95,72
mma-sat 1 µmol/plate MUREAV 77,317,80
orl-rat LD50:417 mg/kg CEHYAN 23,168,78
orl-bwd LD50:100 mg/kg TXAPA9 21,315,72

CONSENSUS REPORTS: Reported in EPA TSCA Inventory. EPA Genetic Toxicology Program.

SAFETY PROFILE: Poison by ingestion. Mutation data reported. A severe skin and eye irritant. When heated to decomposition it emits very toxic fumes of NO_x and F^-.

FFZ000 **CAS:388-72-7** **HR: 2**
4-FLUOROBENZANTHRACENE
mf: $C_{18}H_{11}F$ mw: 246.29

SYNS: 4-FLUOROBENZ(a)ANTHRACENE □ 4′-FLUORO-1,2-BENZANTHRACENE

TOXICITY DATA WITH REFERENCE
scu-rat TDLo:8 mg/kg:NEO CNREA8 23,229,63
scu-mus TDLo:41 mg/kg:ETA CNREA8 23,229,63

SAFETY PROFILE: Questionable carcinogen with experimental neoplastigenic and tumorigenic data. When heated to decomposition it emits toxic fumes of F^-.

FGA000 **CAS:462-06-6** **HR: 3**
FLUOROBENZENE
DOT: UN 2387
mf: C_6H_5F mw: 96.11

PROP: Colorless liquid. D: 1.024, mp: −40°, bp: 85.2°, flash p: 5°F, d: 1.024, vap d: 3.31. Insol in water; misc in alc and ether.

SYN: PHENYL FLUORIDE

TOXICITY DATA WITH REFERENCE
orl-rat LD50:4399 mg/kg GTPZAB 19(9),36,75
ihl-rat LC50:26,908 mg/m³ GTPZAB 19(9),36,75
ihl-mus LC50:45 g/m³/2H IZSBAI 3,91,65

CONSENSUS REPORTS: Reported in EPA TSCA Inventory.

DOT CLASSIFICATION: 3; *Label:* Flammable Liquid

SAFETY PROFILE: Mildly toxic by ingestion and inhalation. A very dangerous fire hazard when exposed to heat, flame, or oxidizers. To fight fire, use water spray, mist, foam, dry chemical, CO_2. When heated to decomposition it emits toxic fumes of F^-.

FGA100 **CAS:5430-13-7** **HR: 3**
4-FLUOROBENZENEARSONIC ACID
mf: $C_6H_6AsFO_3$ mw: 220.04

PROP: A solid. Mp: 240° (decomp).

SYN: BENZENEARSONIC ACID, p-FLUORO-

TOXICITY DATA WITH REFERENCE
ivn-mus LD50:180 mg/kg CSLNX* NX#05114

OSHA PEL: TWA 0.5 mg(As)/m³

SAFETY PROFILE: Poison by intravenous route. When heated to decomposition it emits toxic fumes of As and F^-.

FGB000 **CAS:52831-45-5** **HR: 2**
2-FLUORO-BENZO(e)(1)BENZOTHIOPYRANO(4,3-b) INDOLE
mf: $C_{19}H_{10}FNS$ mw: 303.36

TOXICITY DATA WITH REFERENCE
mma-sat 30 µg/plate MUREAV 66,307,79
scu-mus TDLo:72 mg/kg/9W-I:NEO MUREAV 66,307,79

SAFETY PROFILE: Questionable carcinogen with experimental neoplastigenic data. Mutation data reported. When heated to decomposition it emits very toxic fumes of SO_x, NO_x, and F^-.

FGD000 **CAS:52831-56-8** **HR: 2**
3-FLUORO-BENZO(e)(1)BENZOTHIOPYRANO(4,3-b) INDOLE
mf: $C_{19}H_{10}FNS$ mw: 303.36

TOXICITY DATA WITH REFERENCE
mma-sat 30 µg/plate MUREAV 66,307,79
scu-mus TDLo:72 mg/kg/9W-I:NEO MUREAV 66,307,79

SAFETY PROFILE: Questionable carcinogen with experimental neoplastigenic data. Mutation data reported. When heated to decomposition it emits very toxic fumes of NO_x, SO_x, and F^-.

FGF000 **CAS:52831-68-2** **HR: 2**
4-FLUORO-BENZO(e)(1)BENZOTHIOPYRANO(4,3-b) INDOLE
mf: $C_{19}H_{10}FNS$ mw: 303.36

TOXICITY DATA WITH REFERENCE
mma-sat 30 µg/plate MUREAV 66,307,79
scu-mus TDLo:72 mg/kg/9W-I:NEO MUREAV 66,307,79

SAFETY PROFILE: Questionable carcinogen with experimental neoplastigenic data. Mutation data reported. When heated to decomposition it emits very toxic fumes of F^-, NO_x, and SO_x.

FGG000 **CAS:52831-65-9** **HR: 2**
4-FLUORO-BENZO(g)(1)BENZOTHIOPYRANO(4,3-b) INDOLE
mf: $C_{19}H_{10}FNS$ mw: 303.36

TOXICITY DATA WITH REFERENCE
mma-sat 30 µg/plate MUREAV 66,307,79
scu-mus TDLo:72 mg/kg/9W-I:NEO MUREAV 66,307,79

SAFETY PROFILE: Questionable carcinogen with experimental neoplastigenic data. Mutation data reported. When heated to decomposition it emits very toxic fumes of F^-, NO_x, and SO_x.

FGH000 CAS:445-29-4 HR: 3
o-FLUOROBENZOIC ACID
mf: $C_7H_5FO_2$ mw: 140.12

PROP: Needles from water. Mp: 126.5°, d: 1.460 @ 25°/4°. Sltly sol in hot water; sol in alc and ether.

SYN: o-FLUORBENZOESAEURE (GERMAN)

TOXICITY DATA WITH REFERENCE
scu-mus LDLo: 700 mg/kg AEPPAE 183,427,36
ivn-mus LD50: 180 mg/kg CSLNX* NX#01726

CONSENSUS REPORTS: Reported in EPA TSCA Inventory.

SAFETY PROFILE: Poison by intravenous route. Moderately toxic by subcutaneous route. When heated to decomposition it emits toxic fumes of F^-.

FGH100 CAS:1194-02-1 HR: 2
4-FLUOROBENZONITRILE

SYNS: BENZONITRILE, p-FLUORO- □ p-CYANOFLUOROBENZENE □ p-FLUOROBENZONITRILE

TOXICITY DATA WITH REFERENCE
orl-mus LD50:>300 mg/kg JMCMAR 21,906,78
ipr-mus LD50: 1 g/kg FRPSAX 41,41,86

CONSENSUS REPORTS: On Community Right-To-Know List. Reported in EPA TSCA Inventory.

SAFETY PROFILE: Moderately toxic by ingestion. Mildly toxic by intraperitoneal route. When heated to decomposition it emits toxic vapors of CN^- and F^-.

FGI000 CAS:61735-77-1 HR: 2
3-FLUOROBENZO(rst)PENTAPHENE
mf: $C_{24}H_{13}F$ mw: 320.37

TOXICITY DATA WITH REFERENCE
scu-mus TDLo: 20 mg/kg: NEO JFLCAR 8,513,76

SAFETY PROFILE: Questionable carcinogen with experimental neoplastigenic data. When heated to decomposition it emits toxic fumes of F^-.

FGI100 HR: 3
6-FLUOROBENZO(a)PYRENE
mf: $C_{20}H_{11}F$ mw: 270.31

TOXICITY DATA WITH REFERENCE
scu-rat TDLo: 35 mg/kg/1W-I: CAR JJIND8 71,309,83
skn-mus TDLo: 49 mg/kg/60W-I: NEO JJIND8 71,309,85
scu-mus TDLo: 9 mg/kg: CAR JJIND8 71,309,83
skn-mus TD: 16 mg/kg/60W-I: ETA JJIND8 71,309,83

SAFETY PROFILE: Suspected carcinogen with experimental carcinogenic and neoplastigenic data. When heated to decomposition it emits toxic fumes of F^-.

FGJ000 CAS:52831-39-7 HR: 2
2-FLUORO-(1)BENZOTHIOPYRANO(4,3-b)INDOLE
mf: $C_{15}H_8FNS$ mw: 253.30

SYN: 2-FLUOROTHIOPYRANO(4,3-b)BENZ(e)INDOLE

TOXICITY DATA WITH REFERENCE
mma-sat 30 µg/plate MUREAV 66,307,79
scu-mus TDLo: 72 mg/kg/9W-I: NEO MUREAV 66,307,79

SAFETY PROFILE: Questionable carcinogen with experimental neoplastigenic data. Mutation data reported. When heated to decomposition it emits very toxic fumes such as F^-, SO_x, and NO_x.

FGL000 CAS:52831-62-6 HR: 2
4-FLUORO-(1)BENZOTHIOPYRANO(4,3-b)INDOLE
mf: $C_{15}H_8FNS$ mw: 253.30

SYN: 4-FLUOROTHIOPYRANO(4,3-b)BENZ(e)INDOLE

TOXICITY DATA WITH REFERENCE
mma-sat 30 µg/plate MUREAV 66,307,79
scu-mus TDLo: 72 mg/kg/9W-I: NEO MUREAV 66,307,79

SAFETY PROFILE: Questionable carcinogen with experimental neoplastigenic data. Mutation data reported. When heated to decomposition it emits very toxic fumes such as F^-, SO_x, and NO_x.

FGO000 CAS:52831-58-0 HR: 2
4-FLUORO-6H-(1)BENZOTHIOPYRANO(4,3-b)QUINOLINE
mf: $C_{16}H_{10}FNS$ mw: 267.33

TOXICITY DATA WITH REFERENCE
mma-sat 90 µg/plate MUREAV 66,307,79
scu-mus TDLo: 72 mg/kg/9W-I: NEO MUREAV 66,307,79

SAFETY PROFILE: Questionable carcinogen with experimental neoplastigenic data. Mutation data reported. When heated to decomposition it emits very toxic fumes such as F^-, NO_x, and SO_x.

FGP000 CAS:393-52-2 HR: 3
o-FLUOROBENZOYL CHLORIDE
mf: C_7H_4ClFO mw: 158.56

PROP: A liquid. Mp: 4°, bp: 206°.

SYN: 2-FLUOROBENZOYL CHLORIDE

TOXICITY DATA WITH REFERENCE
mma-sat 100 µg/plate ENMUDM 5(Suppl 1),3,83
ivn-mus LD50: 100 mg/kg CSLNX* NX#04779

CONSENSUS REPORTS: Reported in EPA TSCA Inventory.

SAFETY PROFILE: Poison by intravenous route. Mutation data reported. When heated to decomposition it emits very toxic fumes of Cl^- and F^-.

FGQ000 CAS:56390-16-0 HR: 3
2-(3-(p-FLUOROBENZOYL)-1-PROPYL)-5-α,9-α-DIMETHYL-2′-HYDROXY-6,7-BENZOMORPHAN
mf: $C_{24}H_{28}FNO_2$ mw: 381.53

SYN: ID-1229

TOXICITY DATA WITH REFERENCE
orl-rat LD50:1000 mg/kg DRFUD4 3,298,78
ipr-rat LD50:82 mg/kg ARZNAD 25,795,75
scu-rat LD50:520 mg/kg ARZNAD 25,795,75
ivn-rat LD50:24 mg/kg ARZNAD 25,795,75
orl-mus LD50:500 mg/kg DRFUD4 3,298,78
ipr-mus LD50:112 mg/kg ARZNAD 25,795,75
scu-mus LD50:270 mg/kg ARZNAD 25,795,75
ivn-mus LD50:32 mg/kg DRFUD4 3,298,78
scu-rbt LD50:75 mg/kg ARZNAD 25,795,75
ivn-rbt LD50:7850 μg/kg ARZNAD 25,795,75

SAFETY PROFILE: Poison by intravenous, intraperitoneal, and subcutaneous routes. Moderately toxic by ingestion. When heated to decomposition it emits very toxic fumes of F^- and NO_x.

FGU000 HR: 3
1-(1-(3-(p-FLUOROBENZOYL)PROPYL)-4-PIPERIDYL)-2-BENZIMIDAZOLINONE, HYDROCHLORIDE MONOHYDRATE
mf: $C_{22}H_{24}FN_3O_2•ClH•H_2O$ mw: 435.97

SYNS: BENPERIDOL □ BENZOPERIDOL □ BENZPERIDOL □ 8089 CB □ CONCILIUM □ FRENACTIL □ FRENACTYL □ GLIANIMON □ MCN-JR-4584 □ R 4584

TOXICITY DATA WITH REFERENCE
dnd-mus-skn 192 μmol/kg CRNGDP 5,231,84
scu-rat LD50:218 mg/kg 27ZQAG -,185,72
ivn-rat LD50:21 mg/kg 27ZQAG -,185,72
orl-mus LD50:1000 mg/kg 27ZQAG -,185,72
scu-mus LD50:210 mg/kg 27ZQAG -,185,72
ivn-mus LD50:27 mg/kg 27ZQAG -,185,72

SAFETY PROFILE: Poison by subcutaneous and intravenous routes. Moderately toxic by ingestion. Mutation data reported. Used as an antipsychotic agent. When heated to decomposition it emits very toxic fumes of F^-, NO_x, and HCl.

FGV000 CAS:20977-50-8 HR: 3
1-(3-(4-FLUOROBENZOYL)PROPYL)-4-PIPERIDYL-N-ISOPROPYL CARBAMATE
mf: $C_{19}H_{27}FN_2O_3$ mw: 350.48

PROP: A solid. Mp: 105–105.3°.

SYN: AL-1021

TOXICITY DATA WITH REFERENCE
orl-hmn TDLo:2140 μg/kg:PSY CTCEA9 11,779,69
orl-rat LD50:310 mg/kg 27ZQAG -,184,72
orl-mus LD50:265 mg/kg 27ZQAG -,184,72
ivn-mus LD50:44 mg/kg 27ZQAG -,184,72
orl-dog LD50:25 mg/kg 27ZQAG -,184,72
orl-rbt LD50:200 mg/kg 27ZQAG -,184,72

SAFETY PROFILE: Poison by ingestion and intravenous routes. Human systemic effects by ingestion: psychotropic effects. When heated to decomposition it emits very toxic F^- and NO_x. See also CARBAMATES.

FGW000 CAS:59921-81-2 HR: 3
1-(4′-FLUOROBENZOYL)-3-PYRROLIDINYLPROPANE MALEATE
mf: $C_{14}H_{18}FNO•C_4H_4O_4$ mw: 351.41

SYNS: 1-(4′-FLUOROBENZOIL)-3-PIRROLIDINOPROPANO MALEATO (ITALIAN) □ 4′-FLUORO-4-(1-PYRROLIDINYL)BUTYROPHENONE MALEATE

TOXICITY DATA WITH REFERENCE
orl-rat LD50:335 mg/kg FRPSAX 31,442,76
orl-mus LD50:235 mg/kg FRPSAX 31,442,76
ivn-mus LD50:47 mg/kg FRPSAX 31,442,76

SAFETY PROFILE: Poison by ingestion and intravenous routes. When heated to decomposition it emits very toxic fumes of F^- and NO_x.

FGW100 CAS:1426-40-0 HR: 3
FLUOROBIS(TRIFLUOROMETHYL)PHOSPHINE
mf: C_2F_7P mw: 187.98

PROP: A gas. Bp: −11°.

SAFETY PROFILE: Ignites spontaneously in air. When heated to decomposition it emits toxic fumes of F^-, phosphine, and NO_x. See also PHOSPHINE.

FGX000 CAS:1072-85-1 HR: 1
1-FLUORO-2-BROMOBENZENE
mf: C_6H_4BrF mw: 175.01

PROP: Colorless liquid. D: 1.597, mp: −8°, bp: 152°. Insol in water; very sol in alc and ether.

SYN: 2-BROMFLUORBENZEN (CZECH)

TOXICITY DATA WITH REFERENCE
skn-rbt 500 mg/24H MLD 28ZPAK -,31,72
eye-rbt 100 mg/24H MOD 28ZPAK -,31,72

CONSENSUS REPORTS: Reported in EPA TSCA Inventory.

SAFETY PROFILE: A skin and eye irritant. When heated to decomposition it emits very toxic fumes of Br^- and F^-.

FGY000 CAS:1073-06-9 HR: 2
1-FLUORO-3-BROMOBENZENE
mf: C_6H_4BrF mw: 175.01

PROP: Colorless liquid. D: 1.597, mp: −8°, bp: 152°. Insol in water, very sol in alcohol and ether.

SYN: 3-BROMFLUORBENZEN (CZECH)

TOXICITY DATA WITH REFERENCE
skn-rbt 500 mg/24H MOD 28ZPAK -,32,72
eye-rbt 20 mg/24H MOD 28ZPAK -,32,72

CONSENSUS REPORTS: Reported in EPA TSCA Inventory.

SAFETY PROFILE: A skin and eye irritant. When heated to decomposition it emits very toxic fumes of F^- and Br^-.

FHA000 **CAS:462-72-6** **HR: 3**
4-FLUOROBUTYL BROMIDE
mf: C_4H_8BrF mw: 155.03

TOXICITY DATA WITH REFERENCE
ipr-mus LD50:8200 µg/kg JOCEAH 21,748,56
scu-mus LD50:8 mg/kg CLDND*

SAFETY PROFILE: Poison by intraperitoneal and subcutaneous routes. When heated to decomposition it emits very toxic fumes of Br^- and F^-.

FHB000 **CAS:462-73-7** **HR: 3**
4-FLUOROBUTYL CHLORIDE
mf: C_4H_8ClF mw: 110.57

TOXICITY DATA WITH REFERENCE
ipr-mus LD50:1250 µg/kg JOCEAH 21,748,56
scu-mus LD50:1250 mg/kg CLDND*

SAFETY PROFILE: Poison by intraperitoneal route. Moderately toxic by subcutaneous route. When heated to decomposition it emits very toxic fumes of Cl^- and F^-.

FHC000 **CAS:372-91-8** **HR: 3**
4-FLUOROBUTYL IODIDE
mf: C_4H_8FI mw: 202.02

TOXICITY DATA WITH REFERENCE
ipr-mus LD50:5200 µg/kg JOCEAH 21,748,56
scu-mus LD50:5 mg/kg CLDND*

SAFETY PROFILE: Poison by intraperitoneal and subcutaneous routes. When heated to decomposition it emits very toxic fumes of F^- and I^-.

FHC200 **CAS:353-16-2** **HR: 3**
3-FLUOROBUTYL ISOCYANATE
DOT: UN 2206/UN 2207/UN 2478/UN 3080
mf: C_5H_8FNO mw: 117.14

SYN: ISOCYANIC ACID, 4-FLUOROBUTYL ESTER

TOXICITY DATA WITH REFERENCE
ipr-mus LD50:4700 µg/kg JACSAT 79,1956,57

DOT CLASSIFICATION: 6.1; *Label:* KEEP AWAY FROM FOOD (UN2207); 6.1; *Label:* Poison (UN2206); 6.1; *Label:* Poison, Flammable Liquid (UN3080); 3; *Label:* Flammable Liquid, Poison (UN2478)

SAFETY PROFILE: A poison. A flammable liquid. When heated to decomposition it emits toxic vapors of NO_x and CN^-.

FHD000 **CAS:353-17-3** **HR: 3**
4-FLUOROBUTYL THIOCYANATE
mf: C_5H_8FNS mw: 133.20

TOXICITY DATA WITH REFERENCE
ipr-mus LD50:2600 µg/kg JACSAT 78,3843,56
scu-mus LD50:2600 µg/kg CLDND*

SAFETY PROFILE: Poison by intraperitoneal and subcutaneous routes. When heated to decomposition it emits very toxic fumes of F^-, NO_x, and SO_x. See also THIOCYANATES and FLUORIDES.

FHF000 **CAS:407-83-0** **HR: 3**
4-FLUOROBUTYRONITRILE
mf: C_4H_6FN mw: 87.11

SYN: γ-FLUOROBUTYRONITRILE

TOXICITY DATA WITH REFERENCE
ipr-mus LD50:16 mg/kg JACSAT 78,3484,56
scu-mus LD50:16 mg/kg CLDND*

CONSENSUS REPORTS: Cyanide and its compounds are on the Community Right-To-Know List.

SAFETY PROFILE: Poison by intraperitoneal and subcutaneous routes. When heated to decomposition it emits very toxic fumes of F^-, CN^-, and NO_x. See also NITRILES.

F

FHG000 **CAS:1893-33-0** **HR: 3**
FLUOROBUTYROPHENONE
mf: $C_{21}H_{30}FN_3O_2$ mw: 375.54

SYNS: DIPIPERAL □ DIPIPERON □ DIPIPERONE □ FLOROPIPAMIDE □ 1'-(3-(p-FLUOROBENZOYL)PROPYL)(1,4'-BIPIPERIDINE)-4'-CARBOXAMIDE □ 1-(3-(p-FLUOROBENZOYL)PROPYL)-4-PIPERIDINOISONIPACOTAMIDE □ 4'-FLUORO-4-(4-N-PIPERIDINO-4-CARBAMIDOPIPERIDINO)BUTYROPHENONE □ p-FLUORO-γ-(4-PIPERIDINO-4-CARBAMOYLPIPERIDINO)BUTYROPHENONE □ FPA □ MCN-JR 3345 □ PIPAMPERONE □ PIPANEPERONE □ PIPERONYL □ PROPITAN □ R 3345

TOXICITY DATA WITH REFERENCE
orl-mus TDLo:1500 mg/kg (10-12D preg):TER TOIZAU 20,621,81
orl-rat LD50:160 mg/kg ANPBAZ 61,611,61
scu-rat LD50:160 mg/kg ANPBAZ 61,611,61
ivn-rat LD50:48 mg/kg 27ZQAG -,188,72
orl-mus LD50:490 mg/kg PHMGBN 15,485,77
scu-mus LD50:160 mg/kg 27ZQAG -,188,72
ivn-mus LD50:66 mg/kg 27ZQAG -,188,72

SAFETY PROFILE: Poison by ingestion, subcutaneous, and intravenous routes. An experimental teratogen. When heated to decomposition it emits very toxic fumes of F^- and NO_x.

FHH000 **HR: 2**
FLUOROCHLOROCARBON LIQUID

TOXICITY DATA WITH REFERENCE
orl-mus LD50:7600 mg/kg GISAAA 39(4),114,74
ihl-mus LC50:930 mg/m³/2H GISAAA 39(4),114,74
ivn-mus LD50:490 mg/kg GISAAA 39(4),114,74

SAFETY PROFILE: Moderately toxic by intravenous route. Mildly toxic by ingestion and inhalation. When heated to decomposition it emits very toxic fumes of F^-.

FHH025 **CAS:67639-45-6** **HR: 2**
5-FLUORO-7-CHLOROMETHYL-12-METHYLBENZ(a) ANTHRACENE
mf: $C_{20}H_{14}ClF$ mw: 308.79

SYN: 7-(CHLOROMETHYL)-5-FLUORO-12-METHYLBENZ(a)ANTHRACENE

TOXICITY DATA WITH REFERENCE
ivn-mus TDLo:1239 µg/kg:NEO CNREA8 40,782,80

SAFETY PROFILE: Questionable carcinogen with experimental neoplastigenic data. When heated to decomposition it emits toxic fumes of Cl^- and F^-.

FHH100 CAS:127-31-1 HR: 3
FLUOROCORTISONE
mf: $C_{21}H_{29}FO_5$ mw: 380.50

PROP: Crystals from alc. Mp: 260–262° (decomp). Sol in water.

SYNS: ALFLORONE ☐ F-COL ☐ F-CORTEF ☐ FLORINEF ☐ FLUDROCORTISONE ☐ FLUDROCORTONE ☐ FLUODROCORTISONE ☐ FLUOHYDRISONE ☐ FLUOHYDROCORTISONE ☐ 9-α-FLUOROHYDROCORTISONE ☐ 9-α-FLUORO-17-HYDROXYCORTICOSTERONE ☐ 9-FLUORO-11-β,17,21-TRIHYDROXYPREGN-4-ENE-3,20-DIONE ☐ 9-α-FLUORO-11-β,17-α,21-TRIHYDROXY-4-PREGNENE-3,20-DIONE ☐ U 5963

TOXICITY DATA WITH REFERENCE
orl-rbt TDLo: 1 mg/kg (female 13-16D post): TER 28QFAD -,85,73
par-rat TDLo: 8750 μg/kg (female 7D pre): REP JEMEAV 102,347,55
orl-mus TDLo: 100 mg/kg (female 6-15D post): TER 28QFAD -,85,73
ipr-mus LD50: 170 mg/kg CLDND*

SAFETY PROFILE: Poison by intraperitoneal route. Experimental reproductive effects. When heated to decomposition it emits toxic fumes of F^-. See also PREDNISONE.

FHI000 CAS:2022-85-7 HR: 2
5-FLUOROCYTOSINE
mf: $C_4H_4FN_3O$ mw: 129.11

PROP: A solid. Mp: 295–297° (decomp).

SYNS: ALCOBON ☐ 4-AMINO-5-FLUORO-2(1H)-PYRIMIDINONE ☐ ANCOBON ☐ ANCOTIL ☐ 5-FC ☐ FLUCYTOSINE ☐ 5-FLUOROCYSTOSINE ☐ 2-HYDROXY-4-AMINO-5-FLUOROPYRIMIDINE ☐ RO 2-9915

TOXICITY DATA WITH REFERENCE
dni-omi 100 mg/L JMMIAV 12,83,79
oms-omi 100 mg/L JMMIAV 12,83,79
mmo-smc 8 mg/L MGGEAE 146,253,76
ipr-rat LD50: 3811 mg/kg IYKEDH 10,710,79
scu-rat LD50: 3600 mg/kg NIIRDN 6,699,82
ipr-mus LD50: 1190 mg/kg AACHAX -,566,63
scu-mus LD50: 1000 mg/kg CHTHBK 25,54,79
ivn-mus LD50: 500 mg/kg AACHAX -,566,63

SAFETY PROFILE: Moderately toxic by intraperitoneal, subcutaneous, and intravenous routes. Mutation data reported. An antifungal agent. When heated to decomposition it emits very toxic fumes of F^- and NO_x.

FHN000 CAS:334-62-3 HR: 3
10-FLUORODECYL CHLORIDE
mf: $C_{10}H_{20}ClF$ mw: 194.75

SYN: 1-CHLORO-10-FLUORO-DECANE

TOXICITY DATA WITH REFERENCE
ipr-mus LD50: 5 mg/kg JOCEAH 21,748,56
scu-mus LD50: 5 mg/kg CLDND*

SAFETY PROFILE: Poison by intraperitoneal and subcutaneous routes. When heated to decomposition it emits very toxic fumes of Cl^- and F^-.

FHO000 CAS:10356-76-0 HR: 2
5-FLUORO-2′-DEOXYCYTIDINE
mf: $C_9H_{12}FN_3O_4$ mw: 245.24

PROP: Crystals from EtOH. Mp: 195–196.5°.

SYNS: FCdR ☐ FCDR ☐ 5-FLUOR-DESOXYCYTIDIN (GERMAN) ☐ 5-FLUORODEOXYCYTIDINE

TOXICITY DATA WITH REFERENCE
dni-mus: leu 1 mg/L CPBTAL 30,1018,82
sce-ham: ovr 10 μmol/L CHROAU 85,603,82
ipr-mus TDLo: 20 mg/kg (15-16D preg): REP TJADAB 24(2),50A,81
ipr-mus TDLo: 10 mg/kg (female 9D post): TER AUCMBJ 61,207,73
ipr-mus TDLo: 6200 μg/kg (female 9D post): REP DZZEA7 32,861,77
ipr-ham TDLo: 2500 μg/kg (female 9D post): TER SEIJBO 11,41,71
ipr-mus TDLo: 1500 μg/kg (female 10D post): TER TJADAB 8,191,73
ipr-ham TDLo: 1250 μg/kg (female 9D post): TER SEIJBO 11,41,71
unr-mus TDLo: 10 mg/kg (female 9D post): TER TJADAB 14,374,76
ipr-mus TDLo: 156 μg/kg (8D preg): TER TJADAB 1,311,68
ipr-rat LD50: 2000 mg/kg ADTEAS 3,181,68

SAFETY PROFILE: Moderately toxic by intraperitoneal route. Experimental teratogenic and reproductive effects. Mutation data reported. When heated to decomposition it emits very toxic fumes of F^- and NO_x.

FHP000 CAS:1764-39-2 HR: 2
6-FLUORODIBENZ(a,h)ANTHRACENE
mf: $C_{22}H_{13}F$ mw: 296.35

SYN: 4-FLUORO-1,2:5,6-DIBENZANTHRACENE

TOXICITY DATA WITH REFERENCE
scu-mus TDLo: 200 mg/kg/I: ETA BECCAN 40,30,62

SAFETY PROFILE: Questionable carcinogen with experimental tumorigenic data. When heated to decomposition it emits toxic fumes of F^-.

FHQ000 CAS:321-25-5 HR: 2
2-FLUORO-4-DIMETHYLAMINOAZOBENZENE
mf: $C_{14}H_{14}FN_3$ mw: 243.31

SYN: N,N-DIMETHYL-2-FLUORO-4-PHENYLAZOANILINE

TOXICITY DATA WITH REFERENCE
orl-rat TDLo: 3200 mg/kg/12W-C: ETA CNREA8 13,93,53

SAFETY PROFILE: Questionable carcinogen with experimental tumorigenic data. When heated to decomposition it emits toxic fumes of NO_x and F^-.

FHQ010 CAS:331-91-9 HR: 2
2′-FLUORO-4-DIMETHYLAMINOAZOBENZENE
mf: $C_{14}H_{14}FN_3$ mw: 243.31

SYN: N,N-DIMETHYL-p-(2-FLUOROPHENYLAZO)ANILINE

TOXICITY DATA WITH REFERENCE
orl-rat TDLo:3150 mg/kg/13W-C:ETA CNREA8 9,652,49
orl-rat TD:3200 mg/kg/12W-C:ETA CNREA8 13,93,53

SAFETY PROFILE: Questionable carcinogen with experimental tumorigenic data. When heated to decomposition it emits very toxic fumes of F^- and NO_x. See also FLUORIDES.

FHQ100 CAS:332-54-7 HR: 2
3′-FLUORO-4-DIMETHYLAMINOAZOBENZENE
mf: $C_{14}H_{14}FN_3$ mw: 243.31

SYN: m-FLUORODIMETHYLAMINOAZOBENZENE

TOXICITY DATA WITH REFERENCE
scu-mus TDLo:1050 mg/kg (female 9D post):TER OFAJAE 36,195,60
ipr-mus TDLo:600 mg/kg (10-11D preg):TER KAIZAN 37,179,62
ipr-mus TDLo:600 mg/kg (10-11D preg):REP KAIZAN 37,179,62
ipr-mus TDLo:600 mg/kg (female 8-9D post):TER KAIZAN 37,193,62
orl-rat TDLo:5300 mg/kg/21W-C:NEO CNREA8 17,387,57
orl-rat TD:3200 mg/kg/12W-C:ETA CNREA8 13,93,53
orl-rat TD:3150 mg/kg/13W-C:ETA CNREA8 9,652,49
scu-mus LDLo:1200 mg/kg OFAJAE 36,195,60

SAFETY PROFILE: Moderately toxic by subcutaneous route. An experimental teratogen. Experimental reproductive effects. Questionable carcinogen with experimental neoplastigenic and tumorigenic data. When heated to decomposition it emits toxic fumes of F^- and NO_x.

FHR000 CAS:64038-39-7 HR: 2
10-FLUORO-9,12-DIMETHYLBENZ(a)ACRIDINE
mf: $C_{19}H_{14}FN$ mw: 275.34

SYNS: 2,10-DIMETHYL-3-FLUORO-5,6-BENZACRIDINE □ 9,12-DIMETHYL-10-FLUOROBENZ(a)ACRIDINE □ 3-FLUORO-2,10-DIMETHYL-5,6-BENZACRIDINE

TOXICITY DATA WITH REFERENCE
skn-mus TDLo:324 mg/kg/27W-I:ETA AICCA6 11,736,55
skn-mus TD:370 mg/kg/31W-I:ETA ACRSAJ 4,315,56

SAFETY PROFILE: Questionable carcinogen with experimental tumorigenic data. When heated to decomposition it emits toxic F^- and NO_x.

FHS000 CAS:68141-57-1 HR: 2
1-FLUORO-7,12-DIMETHYLBENZ(a)ANTHRACENE
mf: $C_{20}H_{15}F$ mw: 274.35

SYN: 7,12-DIMETHYL-1-FLUOROBENZ(a)ANTHRACENE

TOXICITY DATA WITH REFERENCE
msc-ham:lng 1 mg/L MUREAV 136,65,84
skn-mus TDLo:110 µg/kg:ETA CNREA8 39,411,79

CONSENSUS REPORTS: EPA Genetic Toxicology Program.

SAFETY PROFILE: Questionable carcinogen with experimental tumorigenic data. Mutation data reported. When heated to decomposition it emits toxic fumes of F^-.

FHU000 CAS:959-73-9 HR: 2
2′-FLUORO-N,N-DIMETHYL-4-STILBENAMINE
mf: $C_{16}H_{16}FN$ mw: 241.33

SYNS: 2′-FLUORO-4-DIMETHYLAMINOSTILBENE □ 2-FLUORO-4-STILBENYL-N,N-DIMETHYLAMINE

TOXICITY DATA WITH REFERENCE
orl-rat TDLo:490 mg/kg/46W-C:ETA ABMGAJ 9,87,62

SAFETY PROFILE: Questionable carcinogen with experimental tumorigenic data. When heated to decomposition it emits very toxic fumes of F^- and NO_x. See also FLUORIDES.

FHV000 CAS:405-86-7 HR: 2
4′-FLUORO-N,N-DIMETHYL-4-STILBENAMINE
mf: $C_{16}H_{16}FN$ mw: 241.33

SYNS: 4′-FLUORO-4-DIMETHYLAMINOSTILBENE □ 4′-FLUORO-4-STILBENYL-N,N-DIMETHYLAMINE

TOXICITY DATA WITH REFERENCE
orl-rat TDLo:440 mg/kg/42W-C:ETA ABMGAJ 9,87,62

SAFETY PROFILE: Questionable carcinogen with experimental tumorigenic data. When heated to decomposition it emits very toxic fumes of F^- and NO_x. See also FLUORIDES.

FHV300 HR: 3
1-FLUORO-1,1-DINITRO-2-BUTENE
mf: $C_4H_5FN_2O_4$ mw: 164.09

$FC(NO_2)_2CH{=}CHCH_3$

SAFETY PROFILE: An explosive. When heated to decomposition it emits toxic fumes of F^- and NO_x.

FHV800 CAS:68795-10-8 HR: 3
2-FLUORO-1,1-DINITROETHANE
mf: $C_2H_3FN_2O_4$ mw: 138.06

SAFETY PROFILE: Explosive reaction with air at 75°C. When heated to decomposition it emits toxic fumes of F^- and NO_x.

FHW000 CAS:17003-75-7 HR: 3
2-FLUORO-2,2-DINITROETHANOL
mf: $C_2H_3FN_2O_5$ mw: 154.06

$FC(NO_2)_2CH_2OH$

PROP: Bp: 55–57° @ 1.5 mm.

SYNS: 2,2-DINITRO-2-FLUOROETHANOL □ 2-FLUORO-2,2-DINITROETHANOL

TOXICITY DATA WITH REFERENCE
ipr-mus LD50:54 mg/kg KHFZAN 11(1),73,77

CONSENSUS REPORTS: Reported in EPA TSCA Inventory.

SAFETY PROFILE: A poison by intraperitoneal route. A vesicant. Potentially explosive. When heated to decomposition it emits toxic fumes of F^- and NO_x.

FHX000 CAS:18139-02-1 HR: 3
2-FLUORO-2,2-DINITROETHYLAMINE
mf: $C_2H_4FN_3O_4$ mw: 153.08

$FC(NO_2)_2CH_2NH_2$

SAFETY PROFILE: The pure material may explode at room temperature. Concentrated solutions in dichloromethane may decompose violently in storage. When heated to decomposition it emits toxic fumes of F^- and NO_x.

FHY000 CAS:7182-87-8 HR: 3
FLUORO DINITROMETHANE
mf: $CHFN_2O_4$ mw: 124.03

$FCH(NO_2)_2$

SAFETY PROFILE: Potentially explosive. When heated to decomposition it emits toxic fumes of F^- and NO_x.

FHZ000 CAS:17003-82-6 HR: 3
FLUORO DINITROMETHYL AZIDE
mf: CFN_5O_4 mw: 165.04

$F(NO_2)_2CN_3$

PROP: Bp: 45° @ 60 mm.

SAFETY PROFILE: Unstable at room temperature. When heated to decomposition it emits toxic fumes of F^- and NO_x. See also AZIDES.

FHZ200 CAS:22692-30-4 HR: 3
1-FLUORO-1,1-DINITRO-2-PHENYLETHANE
mf: $C_8H_7FN_2O_4$ mw: 214.15

$FC(NO_2)_2CH_2C_6H_6$

SAFETY PROFILE: Explosive. When heated to decomposition it emits toxic fumes of F^- and NO_x. See also NITRO COMPOUNDS of AROMATIC HYDROCARBONS.

FIA000 CAS:334-71-4 HR: 3
12-FLUORO DODECANO NITRILE
mf: $C_{12}H_{22}FN$ mw: 199.35

SYN: NC11F

TOXICITY DATA WITH REFERENCE
ipr-mus LD50:80 mg/kg JACSAT 78,3484,56
scu-mus LD50:80 mg/kg CLDND*

CONSENSUS REPORTS: Cyanide and its compounds are on the Community Right-To-Know List.

SAFETY PROFILE: Poison by intraperitoneal and subcutaneous routes. When heated to decomposition it emits very toxic fumes of F^-, NO_x, and CN^-. See also FLUORIDES and NITRILES.

FIA500 CAS:1881-37-4 HR: 2
4-FLUOROESTRADIOL
mf: $C_{18}H_{23}FO_2$ mw: 290.41

PROP: Needles from C_6H_6. Mp: 190–190.5°.

SYNS: ESTRA-1,3,5(10)-TRIENE-3,17-DIOL, 4-FLUORO-, (17-β)-(9CI) □ 4-FLUOROESTRA-1,3,5-(10)-TRIENE-3,17-β-DIOL

TOXICITY DATA WITH REFERENCE
imp-ham TDLo:360 mg/kg/15W-I:CAR MOPMA3 23,278,83

SAFETY PROFILE: Questionable carcinogen with experimental carcinogenic data. When heated to decomposition it emits toxic fumes of F^-.

FIB000 CAS:353-36-6 HR: 3
FLUOROETHANE
DOT: UN 2453
mf: C_2H_5F mw: 48.06

PROP: A gas with ethereal odor. Mp: −143.2°, bp: −37.7°, d: 0.8158 @ −37.7°, vap d: 1.66. Sol in EtOH, and Et_2O; sltly sol in H_2O.

SYNS: ETHYL FLUORIDE (DOT) □ MONOFLUOROETHANE □ R161

TOXICITY DATA WITH REFERENCE
ihl-rat LCLo: 26 pph/4H JIDHAN 31,343,49

CONSENSUS REPORTS: Reported in EPA TSCA Inventory.

DOT CLASSIFICATION: 2.1; *Label:* Flammable Gas

SAFETY PROFILE: A poison by inhalation. A very dangerous fire hazard when exposed to heat, flames, or oxidizers. To fight fire, stop flow of gas. When heated to decomposition it emits toxic fumes of F^-.

FIC000 CAS:144-49-0 HR: 3
FLUOROETHANOIC ACID
DOT: UN 2642
mf: $C_2H_3FO_2$ mw: 78.05

PROP: Colorless solid or crystals. Mp: 35.3°, d: 1.393 @ 36 mm, bp: 167–168°. Sol in water and alc.

SYNS: ACIDE-MONOFLUORACETIQUE (FRENCH) □ ACIDO MONOFLUOROACETIO (ITALIAN) □ CYMONIC ACID □ FAA □ FLUOROACETATE □ FLUOROACETIC ACID □ 2-FLUOROACETIC ACID □ FLUOROACETIC ACID (DOT) □ GIFBLAAR POISON □ HFA □ MFA □ MONOFLUORAZIJNZUUR (DUTCH) □ MONOFLUORESSIGSAEURE (GERMAN) □ MONOFLUOROACETATE □ MONOFLUOROACETIC ACID

TOXICITY DATA WITH REFERENCE
orl-rat LD50:4680 μg/kg TXAPA9 13,189,68
orl-mus LD50:7 mg/kg TXAPA9 13,189,68
ipr-mus LD50:6600 μg/kg JOCEAH 21,883,56
scu-mus LD50:281 mg/kg JPETAB 95,62,49

ivn-mus LD50:13 mg/kg CSLNX* NX#03014
ivn-rbt LD50:250 μg/kg NATUAS 158,382,46
orl-gpg LD50:468 μg/kg TXAPA9 13,189,68

CONSENSUS REPORTS: EPA Extremely Hazardous Substances List. Reported in EPA TSCA Inventory.

DOT CLASSIFICATION: 6.1; *Label:* Poison

SAFETY PROFILE: Poison by ingestion, subcutaneous, intraperitoneal, and intravenous routes. Affects the human central nervous system, causing convulsions and ventricular fibrillation. When heated to decomposition it emits toxic fumes of F^- and Na_2O. See also SODIUM FLUOROACETATE.

FID000 CAS:63919-01-7 HR: 3
FLUOROETHANOL
mf: C_2H_5FO mw: 64.07

SYNS: MONOFLUORETHANOL □ MONOFLUOROETHANOL

TOXICITY DATA WITH REFERENCE
ipr-rat LD50:5 mg/kg JPETAB 106,464,52
ihl-mus LC50:1100 mg/m³/10M 11FYAN 3,61,63
scu-mus LD50:19 mg/kg JPETAB 95,62,49

SAFETY PROFILE: Poison by intraperitoneal and subcutaneous routes. Mildly toxic by inhalation. When heated to decomposition it emits very toxic fumes of F^-.

FIE000 CAS:371-62-0 HR: 3
2-FLUOROETHANOL
mf: C_2H_5FO mw: 64.07

PROP: A liquid. Fp: −26.45°, bp: 103.35° @ 757 mm. Sol in H_2O.

SYNS: β-FLUOROETHANOL □ TL 741

TOXICITY DATA WITH REFERENCE
orl-rat LD50:5 mg/kg 85JCAE-,515,86
ihl-rat LC50:200 mg/m³/10M NTIS** PB158-508
ihl-mus LC50:1100 mg/m³/10M JCSOA9 -,773,49
ipr-mus LD50:10 mg/kg JOCEAH 21,739,56
scu-mus LD50:15 mg/kg NATUAS 172,1139,53
ihl-dog LC50:7 mg/m³/10M NTIS** PB158-508
ihl-mky LC50:1500 mg/m³/10M NTIS** PB158-508
ihl-cat LC50:35 mg/m³/10M NTIS** PB158-508
ihl-rbt LC50:25 mg/m³/10M NTIS** PB158-508
ivn-rbt LDLo:250 μg/kg JCSOA9 -,1279,49
ihl-gpg LC50:150 mg/m³/10M NTIS** PB158-508

CONSENSUS REPORTS: EPA Extremely Hazardous Substances List. Reported in EPA TSCA Inventory.

SAFETY PROFILE: Poison by inhalation, intraperitoneal, subcutaneous, and intravenous routes. When heated to decomposition it emits very toxic fumes of F^-.

FIH000 CAS:63884-92-4 HR: 3
β-FLUOROETHYL-N-(β-CHLOROETHYL)-N-NITROSOCARBAMATE
mf: $C_4H_8ClFN_2O_3$ mw: 186.59

SYNS: (2-CHLOROETHYL)NITROSOCARBAMIC ACID-2-FLUOROETHYL ESTER □ TL 821

TOXICITY DATA WITH REFERENCE
ihl-gpg LCLo:300 mg/m³/10M NDRC** No. 9-4-1-9,43

SAFETY PROFILE: Poison by inhalation. Many N-nitroso compounds are carcinogens. When heated to decomposition it emits very toxic fumes of Cl^-, F^-, and NO_x. See also N-NITROSO COMPOUNDS and CARBAMATES.

FIH100 CAS:462-27-1 HR: 3
2-FLUOROETHYL CHLOROFORMATE
mf: $C_3H_4ClFO_2$ mw: 126.52

SYNS: CHLOROFORMIC ACID 2-FLUOROETHYL ESTER □ 2-FLUORETHYLESTER KYSELINY CHLORMRAVENCI □ FORMIC ACID, CHLORO-, 2-FLUOROETHYL ESTER □ TL 751

TOXICITY DATA WITH REFERENCE
ihl-mus LDLo:200 mg/kg/10M NDRC** No. 9-4-1-9,43
ihl-gpg LDLo:100 mg/kg/10M NDRC** No. 9-4-1-9,43

DOT CLASSIFICATION: 6.1; *Label:* Poison, Corrosive

SAFETY PROFILE: Poison by inhalation. A corrosive. When heated to decomposition it emits toxic vapors of F^- and Cl^-.

FIK875 CAS:60553-18-6 HR: 3
FLUOROETHYLENE OZONIDE
mf: $C_2H_3FO_3$ mw: 94.04

SYN: 3-FLUORO 1,2,4-TRIOXOLANE

SAFETY PROFILE: Spontaneously explosive at room temperature. When heated to decomposition it emits toxic fumes of F^-. See also OZONE.

FIM000 CAS:459-99-4 HR: 3
β-FLUOROETHYL FLUOROACETATE
mf: $C_4H_6F_2O_2$ mw: 124.10

SYNS: 2-FLUOROETHYL FLUOROACETATE □ TL 855

TOXICITY DATA WITH REFERENCE
ihl-rat LC50:200 mg/m³/10M NTIS** PB158-508
ihl-mus LC50:450 μg/m³ NATUAS 160,179,47
scu-mus LD50:8500 μg/kg NATUAS 160,179,47
par-mus LD50:8500 μg/kg JCSOA9 -,1471,49
ihl-rbt LC50:50 mg/m³ JCSOA9 -,916,49
ihl-gpg LC50:70 mg/m³/10M NTIS** PB158-508

SAFETY PROFILE: Poison by inhalation, subcutaneous, and parenteral routes. When heated to decomposition it emits toxic fumes of F^-. See also ESTERS and FLUORIDES.

FIN000 CAS:371-29-9 HR: 3
2-FLUORO ETHYL-γ-FLUORO BUTYRATE
mf: $C_6H_{10}F_2O_2$ mw: 152.16

SYN: β-FLUOROETHYL-γ-FLUOROBUTYRATE

TOXICITY DATA WITH REFERENCE
ihl-rat LC50:200 mg/m³/10M NTIS** PB158-508
ihl-mus LC50:73 mg/m³/10M NDRC** No. 9-4-1-19,44
ihl-dog LC50:25 mg/m³/10M NTIS** PB158-508
ihl-mky LC50:500 mg/m³/10M NTIS** PB158-508
ihl-cat LC50:25 mg/m³/10M NTIS** PB158-508

ihl-gpg LC50:35 mg/m³/10M NTIS** PB158-508

SAFETY PROFILE: Poison by inhalation. When heated to decomposition it emits toxic fumes of F^-.

FIO000 CAS:63765-78-6 HR: 3
2-FLUOROETHYL-5-FLUOROHEXOATE
mf: $C_8H_{14}F_2O_2$ mw: 180.22

TOXICITY DATA WITH REFERENCE
ihl-rat LD50:200 mg/m³ NATUAS 160,179,47
ims-rat LD50:1800 μg/kg NATUAS 160,179,47
ihl-mus LC50:150 μg/m³ NATUAS 160,179,47
scu-mus LD50:2500 μg/kg NATUAS 160,179,47
ihl-rbt LC50:20 μg/m³ NATUAS 160,179,47
ivn-rbt LD50:100 μg/kg NATUAS 160,179,47

SAFETY PROFILE: Poison by inhalation, intramuscular, subcutaneous, and intravenous routes. When heated to decomposition it emits very toxic F^-.

FIP999 CAS:4242-33-5 HR: 3
β-FLUOROETHYLIC ESTER of XENYLACETIC ACID
mf: $C_{16}H_{15}FO_2$ mw: 258.31

PROP: A brown solid, sol in organic solvents. Mp: 60.6°.

SYNS: 2-FLUOROETHYL ESTER DIPHENYLACETIC ACID □ LAMBROL □ M 2060

TOXICITY DATA WITH REFERENCE
skn-rat LD50:4 mg/kg BESAAT 15,102,69
orl-mus LD50:35 mg/kg BESAAT 15,102,69
orl-dog LD50:2 mg/kg BESAAT 15,102,69
orl-rbt LD50:600 μg/kg SPEADM 74-1,-,74
skn-rbt LD50:7 mg/kg BESAAT 15,102,69
orl-gpg LD50:700 μg/kg 85DPAN -,-,71/76

SAFETY PROFILE: Poison by ingestion and skin contact. An insecticide. When heated to decomposition it emits toxic fumes of F^-. See also FLUORIDES.

FIQ000 CAS:762-51-6 HR: 3
2-FLUOROETHYL IODIDE
mf: C_2H_4FI mw: 173.96

PROP: A liquid. D: 2.14 @ 25°/4°, bp: 98–102°.

SYNS: 1-FLUOR-2-JODETHAN □ 1-FLUORO-2-IODOETHANE

TOXICITY DATA WITH REFERENCE
ipr-mus LD50:28 mg/kg JOCEAH 21,748,56
scu-mus LD50:28 mg/kg CLDND*

SAFETY PROFILE: Poison by intraperitoneal and subcutaneous routes. When heated to decomposition it emits very toxic fumes of F^- and I^-. See also IODIDES.

FIS000 CAS:63982-15-0 HR: 3
2-FLUOROETHYL-N-METHYL-N-NITROSOCARBAMATE
mf: $C_4H_7FN_2O_3$ mw: 150.13

PROP: Crystals from Me_2CO/EtOH or Me_3CN. Mp: 198–199° (decomp).

SYN: TL 790

TOXICITY DATA WITH REFERENCE
ihl-mus LCLo:500 mg/m³/10M NDRC** No. 9-4-1-9,43
ihl-gpg LCLo:100 mg/m³/10M NDRC** No. 9-4-1-9,43

SAFETY PROFILE: Poison by inhalation. Many N-nitroso compounds are carcinogens. When heated to decomposition it emits very toxic fumes of F^- and NO_x. See also N-NITROSO COMPOUNDS and CARBAMATES.

FIT200 CAS:2508-18-1 HR: 2
7-FLUORO-2-N-(FLUORENYL)ACETHYDROXAMIC ACID
mf: $C_{15}H_{12}FNO_2$ mw: 257.28

SYNS: 7-FLUORO-2-FAA □ 7-FLUORO-N-(FLUOREN-2-YL)ACETOHYDROXAMIC ACID □ N-(7-FLUORO-2-FLUORENYL)ACETOHYDROXAMIC ACID □ 7-FLUORO-N-HYDROXY-N-2-ACETYLAMINOFLUORENE □ N-HYDROXY-7-FLUORO-2-ACETYLAMINOFLUORENE

TOXICITY DATA WITH REFERENCE
mmo-sat 15 nmol/plate MUREAV 67,85,79
orl-rat TDLo:675 mg/kg/10W-C:CAR CNREA8 26,2239,66

SAFETY PROFILE: Questionable carcinogen with experimental carcinogenic data. Mutation data reported. When heated to decomposition it emits toxic fumes of NO_x and F^-.

FIW000 CAS:31540-62-2 HR: 3
4′-FLUORO-4-(8-FLUORO-2,3,4,5-TETRAHYDRO-1H-PYRIDO(4,3-b)INDOL-2-YL)BUTYROPHENONE HYDROCHLORIDE
mf: $C_{21}H_{20}F_2N_2O•ClH$ mw: 390.89

SYN: ABBOTT-30360

TOXICITY DATA WITH REFERENCE
ims-rat LD50:76 mg/kg AIPTAK 190,124,71
orl-mus LD50:335 mg/kg AIPTAK 190,124,71
ivn-mus LD50:47 mg/kg AIPTAK 190,124,71
ims-mus LD50:112 mg/kg AIPTAK 190,124,71

SAFETY PROFILE: Poison by intramuscular, ingestion, and intravenous routes. When heated to decomposition it emits very toxic fumes of HCl, NO_x, and F^-.

FIX000 CAS:661-11-0 HR: 3
1-FLUOROHEPTANE
mf: $C_7H_{15}F$ mw: 118.22

PROP: A liquid. Fp: −73°, bp: 119° @ 755 mm.

TOXICITY DATA WITH REFERENCE
ipr-mus LD50:35 mg/kg JACSAT 79,2311,57

CONSENSUS REPORTS: Reported in EPA TSCA Inventory.

SAFETY PROFILE: Poison by intraperitoneal route. When heated to decomposition it emits very toxic fumes of F^-.

FIY000 CAS:334-44-1 HR: 3
7-FLUOROHEPTANONITRILE
mf: $C_7H_{12}FN$ mw: 129.20

TOXICITY DATA WITH REFERENCE
ipr-mus LD50:2700 µg/kg JACSAT 78,3484,56
scu-mus LD50:2700 µg/kg CLDND*

CONSENSUS REPORTS: Cyanide and its compounds are on the Community Right-To-Know List.

SAFETY PROFILE: Poison by intraperitoneal and subcutaneous routes. When heated to decomposition it emits very toxic fumes of F^-, NO_x, and CN^-. See also NITRILES.

FIZ000 CAS:353-21-9 HR: 3
7-FLUOROHEPTYLAMINE
mf: $C_7H_{16}FN$ mw: 133.24

TOXICITY DATA WITH REFERENCE
ipr-mus LD50:47 mg/kg CJBPAZ 35,407,57
scu-mus LD50:50 mg/kg CLDND*

SAFETY PROFILE: Poison by intraperitoneal and subcutaneous routes. When heated to decomposition it emits very toxic fumes of NO_x and F^-. See also AMINES.

FJA000 CAS:373-14-8 HR: 3
1-FLUOROHEXANE
mf: $C_6H_{13}F$ mw: 104.19

PROP: A liquid. Bp: 91–92° @ 760 mm.

SYN: FLUOROHEXANE

TOXICITY DATA WITH REFERENCE
ipr-mus LD50:1700 µg/kg JACSAT 79,2311,57

CONSENSUS REPORTS: Reported in EPA TSCA Inventory.

SAFETY PROFILE: Poison by intraperitoneal route. When heated to decomposition it emits toxic fumes of F^-. See also FLUORIDES.

FJI500 CAS:78343-32-5 HR: 3
1-FLUOROIMINOHEXAFLUOROPROPANE
mf: C_3F_7N mw: 183.03

SAFETY PROFILE: Explosive reaction on contact with fluorine. When heated to decomposition it emits toxic fumes of F^- and NO_x. See also FLUORIDES.

FJI510 CAS:2802-70-2 HR: 3
2-FLUOROIMINOHEXAFLUOROPROPANE
mf: C_3F_7N mw: 183.03

SAFETY PROFILE: Explosive reaction on contact with fluorine. When heated to decomposition it emits toxic fumes of F^- and NO_x. See also FLUORIDES.

FJK000 CAS:593-53-3 HR: 3
FLUOROMETHANE
mf: CH_3F mw: 34.03

PROP: Colorless gas; agreeable, ether-like odor. D: (liquid) 0.8774 @ −78°, (gas) 1.1951 (air = 1), (gas) 1.0813 (oxygen = 1), mp: −141.8°, bp: −75.7° @ 872 mm, −78.2° @ 760 mm. Mod sol in H_2O; sol in EtOH and Et_2O.

SYNS: FREON 41 □ METHYL FLUORIDE

CONSENSUS REPORTS: Reported in EPA TSCA Inventory.

SAFETY PROFILE: Narcotic in high concentrations. Acts as a simple asphyxiant. Burns with evolution of hydrogen fluoride. The flame is about as colorless as that of alcohol. When heated to decomposition it emits toxic fumes of F^-.

F

FJN000 CAS:1994-57-6 HR: 2
2-FLUORO-7-METHYLBENZ(a)ANTHRACENE
mf: $C_{19}H_{13}F$ mw: 260.32

SYN: 7-METHYL-2-FLUOROBENZ(a)ANTHRACENE

TOXICITY DATA WITH REFERENCE
ims-rat TDLo:10 mg/kg:NEO NATUAS 273,566,78

SAFETY PROFILE: Questionable carcinogen with experimental neoplastigenic data. When heated to decomposition it emits toxic fumes of F^-.

FJO000 CAS:2606-87-3 HR: 2
3-FLUORO-7-METHYLBENZ(a)ANTHRACENE
mf: $C_{19}H_{13}F$ mw: 260.32

SYN: 3′-FLUORO-10-METHYL-1,2-BENZANTHRACENE

TOXICITY DATA WITH REFERENCE
scu-rat TDLo:9 mg/kg:NEO CNREA8 23,229,63
skn-mus TDLo:120 mg/kg/20W-I:ETA CNREA8 23,229,63
scu-mus TDLo:43 mg/kg:NEO CNREA8 23,229,63

SAFETY PROFILE: Questionable carcinogen with experimental neoplastigenic and tumorigenic data. When heated to decomposition it emits toxic fumes of F^-.

FJP000 CAS:2541-68-6 HR: 2
6-FLUORO-7-METHYLBENZ(a)ANTHRACENE
mf: $C_{19}H_{13}F$ mw: 260.32

TOXICITY DATA WITH REFERENCE
scu-rat TDLo:9 mg/kg:NEO CNREA8 23,229,63
ims-rat TDLo:10 mg/kg:NEO NATUAS 273,566,78
skn-mus TDLo:120 mg/kg/20W-I:NEO CNREA8 23,229,63
scu-mus TDLo:43 mg/kg:NEO CNREA8 23,229,63

SAFETY PROFILE: Questionable carcinogen with experimental neoplastigenic data. When heated to decomposition it emits toxic fumes of F^-.

FJQ000 CAS:1881-75-0 HR: 2
9-FLUORO-7-METHYLBENZ(a)ANTHRACENE
mf: $C_{19}H_{13}F$ mw: 260.32

SYN: 6-FLUORO-10-METHYL-1,2-BENZANTHRACENE

TOXICITY DATA WITH REFERENCE
ims-rat TDLo:10 mg/kg:NEO NATUAS 273,566,78
skn-mus TDLo:120 mg/kg/20W-I:NEO CNREA8 23,229,63

SAFETY PROFILE: Questionable carcinogen with experimental neoplastigenic data. When heated to decomposition it emits toxic fumes of F^-.

FJR000 **CAS:1881-76-1** **HR: 2**
7-FLUORO-10-METHYL-1,2-BENZANTHRACENE
mf: $C_{19}H_{13}F$ mw: 260.32

SYN: 10-FLUORO-7-METHYLBENZ(a)ANTHRACENE

TOXICITY DATA WITH REFERENCE
scu-rat TDLo: 9 mg/kg: NEO CNREA8 23,229,63
skn-mus TDLo: 120 mg/kg/20W-I: NEO CNREA8 23,229,63
scu-mus TDLo: 43 mg/kg: NEO CNREA8 23,229,63

SAFETY PROFILE: Questionable carcinogen with experimental neoplastigenic data. When heated to decomposition it emits toxic fumes of F^-.

FJT000 **CAS:52831-60-4** **HR: 2**
4-FLUORO-7-METHYL-6H-(1)BENZOTHIOPYRANO(4,3-b)QUINOLINE
mf: $C_{17}H_{12}FNS$ mw: 281.36

TOXICITY DATA WITH REFERENCE
mma-sat 100 µg/plate MUREAV 66,307,79
scu-mus TDLo: 72 mg/kg/9W-I: NEO MUREAV 66,307,79

SAFETY PROFILE: Questionable carcinogen with experimental neoplastigenic data. Mutation data reported. When heated to decomposition it emits very toxic fumes of F^-, SO_x, and NO_x.

FJT100 **CAS:91503-79-6** **HR: 3**
2-FLUORO-α-METHYL-(1,1′-BIPHENYL)-4-ACETIC ACID 1-(ACETYLOXY)ETHYL ESTER
mf: $C_{19}H_{19}FO_4$ mw: 330.38

SYNS: 1-ACETOXYETHYL 2-(2-FLUORO-4-BIPHENYLYL)PROPIONATE □ (1,1′-BIPHENYL)-4-ACETIC ACID, 2-FLUORO-α-METHYL-, 1-(ACETYLOXY)ETHYL ESTER □ 4-BIPHENYLACETIC ACID, 2-FLUORO-α-METHYL-, 1-ACETOXYETHYL ESTER □ FLURBIPROFEN AXETIL □ FP-83 □ LFP 83 □ LIPFEN □ LIPO-FLURBIPROFEN AXETIL □ LIPOSOMAL FLURBIPROFEN AXETIL □ ROPION

TOXICITY DATA WITH REFERENCE
ivn-rat TDLo: 1 mg/kg: TER IYKEDH 20,42,89
orl-rat LD50: 66 mg/kg KSRNAM 22,3949,88
scu-rat LD50: 66 mg/kg KSRNAM 22,3949,88
ivn-rat LD50: 88 mg/kg KSRNAM 22,3949,88
orl-mus LD50: 69 mg/kg KSRNAM 22,3949,88
scu-mus LD50: 183 mg/kg KSRNAM 22,3949,88
ivn-mus LD50: 194 mg/kg KSRNAM 22,3949,88

SAFETY PROFILE: Poison by ingestion, subcutaneous, and intravenous routes. An experimental teratogen. When heated to decomposition it emits toxic fumes of F^-.

FJU000 **CAS:73771-72-9** **HR: 2**
2-FLUORO-3-METHYLCHOLANTHRENE
mf: $C_{21}H_{15}F$ mw: 286.36

TOXICITY DATA WITH REFERENCE
ims-rat TDLo: 50 mg/kg: NEO NATUAS 273,566,78

SAFETY PROFILE: Questionable carcinogen with experimental neoplastigenic data. When heated to decomposition it emits toxic fumes of F^-.

FJV000 **CAS:73771-73-0** **HR: 2**
6-FLUORO-3-METHYLCHOLANTHRENE
mf: $C_{21}H_{15}F$ mw: 286.36

TOXICITY DATA WITH REFERENCE
ims-rat TDLo: 50 mg/kg: NEO NATUAS 273,566,78

SAFETY PROFILE: Questionable carcinogen with experimental neoplastigenic data. When heated to decomposition it emits toxic fumes of F^-.

FJW000 **CAS:73771-74-1** **HR: 2**
9-FLUORO-3-METHYLCHOLANTHRENE
mf: $C_{21}H_{15}F$ mw: 286.36

TOXICITY DATA WITH REFERENCE
ims-rat TDLo: 50 mg/kg: NEO NATUAS 273,566,78

SAFETY PROFILE: Questionable carcinogen with experimental neoplastigenic data. When heated to decomposition it emits toxic fumes of F^-.

FJY000 **CAS:64977-44-2** **HR: 2**
1-FLUORO-5-METHYLCHRYSENE
mf: $C_{19}H_{13}F$ mw: 260.32

TOXICITY DATA WITH REFERENCE
mma-sat 20 µg/plate JMCMAR 21,38,78
skn-mus TDLo: 1200 µg/kg/20D-I: ETA CNREA8 38,1694,78
skn-mus TD: 40 mg/kg/I: ETA CALEDQ 1,147,76

CONSENSUS REPORTS: EPA Genetic Toxicology Program.

SAFETY PROFILE: Questionable carcinogen with experimental tumorigenic data. Mutation data reported. When heated to decomposition it emits toxic fumes of F^-. See also CHRYSENE.

FKA000 **CAS:64977-46-4** **HR: 2**
6-FLUORO-5-METHYLCHRYSENE
mf: $C_{19}H_{13}F$ mw: 260.32

TOXICITY DATA WITH REFERENCE
mma-sat 20 µg/plate JMCMAR 21,38,78
skn-mus TDLo: 1200 µg/kg/20D-I: ETA CNREA8 38,1694,78

CONSENSUS REPORTS: EPA Genetic Toxicology Program.

SAFETY PROFILE: Questionable carcinogen with experimental tumorigenic data. Mutation data reported. When heated to decomposition it emits toxic fumes of F^-. See also CHRYSENE.

FKB000 **CAS:64977-47-5** **HR: 2**
7-FLUORO-5-METHYLCHRYSENE
mf: $C_{19}H_{13}F$ mw: 260.32

TOXICITY DATA WITH REFERENCE
mma-sat 20 μg/plate JJIND8 63,855,79
skn-mus TDLo:4000 μg/kg/20D-I:ETA CNREA8 38,1694,78

SAFETY PROFILE: Questionable carcinogen with experimental tumorigenic data. Mutation data reported. When heated to decomposition it emits toxic fumes of F^-. See also CHRYSENE.

FKC000 **CAS:64977-48-6** **HR: 2**
9-FLUORO-5-METHYLCHRYSENE
mf: $C_{19}H_{13}F$ mw: 260.32

TOXICITY DATA WITH REFERENCE
mma-sat 20 μg/plate JMCMAR 21,38,78
skn-mus TDLo:1200 μg/kg/20D-I:ETA CNREA8 38,1694,78

CONSENSUS REPORTS: EPA Genetic Toxicology Program.

SAFETY PROFILE: Questionable carcinogen with experimental tumorigenic data. Mutation data reported. When heated to decomposition it emits toxic fumes of F^-. See also CHRYSENE.

FKD000 **CAS:64977-49-7** **HR: 2**
11-FLUORO-5-METHYLCHRYSENE
mf: $C_{19}H_{13}F$ mw: 260.32

TOXICITY DATA WITH REFERENCE
mma-sat 20 μg/plate JMCMAR 21,38,78
skn-mus TDLo:1200 μg/kg/20D-I:ETA CNREA8 38,1694,78

CONSENSUS REPORTS: EPA Genetic Toxicology Program.

SAFETY PROFILE: Questionable carcinogen with experimental tumorigenic data. Mutation data reported. When heated to decomposition it emits toxic fumes of F^-. See also CHRYSENE.

FKE000 **CAS:61413-38-5** **HR: 2**
12-FLUORO-5-METHYLCHRYSENE
mf: $C_{19}H_{13}F$ mw: 260.32

TOXICITY DATA WITH REFERENCE
mma-sat 20 μg/plate JMCMAR 21,38,78
skn-mus TDLo:4 mg/kg/I:NEO CALEDQ 1,147,76
skn-mus TD:40 mg/kg/3W-I:ETA CCSUDL 1,325,76

CONSENSUS REPORTS: EPA Genetic Toxicology Program.

SAFETY PROFILE: Questionable carcinogen with experimental neoplastigenic and tumorigenic data. Mutation data reported. When heated to decomposition it emits toxic fumes of F^-. See also CHRYSENE.

FKF800 **CAS:937-25-7** **HR: 2**
p-FLUORO-N-METHYL-N-NITROSOANILINE
mf: $C_7H_7FN_2O$ mw: 154.16

SYN: N-NITROSO-N-METHYL-4-FLUOROANILINE

TOXICITY DATA WITH REFERENCE
mma-sat 500 μg/plate MUREAV 89,255,81
orl-rat TDLo:1140 mg/kg/50W-I:ETA CRNGDP 4,157,83

SAFETY PROFILE: Questionable carcinogen with experimental tumorigenic data. Mutation data reported. When heated to decomposition it emits toxic fumes of NO_x and F^-.

FKI000 **CAS:1622-79-3** **HR: 3**
4′-FLUORO-4-(4-METHYLPIPERIDINO)BUTYROPHENONE HYDROCHLORIDE
mf: $C_{16}H_{22}FNO•ClH$ mw: 299.85

PROP: A solid. Mp: 209–211°.

SYNS: BURONIL □ EUNERPAN □ FG 5111 □ METHYLPERONE HYDROCHLORIDE □ γ-(4-METHYLPIPERIDINE)-p-FLUOROBUTYROPHENONE HYDROCHLORIDE

TOXICITY DATA WITH REFERENCE
orl-rbt TDLo:195 mg/kg (6-18D preg):TER FRPPAO 29,586,74
orl-rat LD50:330 mg/kg APTOA6 23,109,65
scu-rat LD50:220 mg/kg APTOA6 23,109,65
ivn-rat LD50:40 mg/kg APTOA6 23,109,65
orl-mus LD50:230 mg/kg APTOA6 23,109,65
scu-mus LD50:230 mg/kg 27ZQAG ,191,72
ivn-mus LD50:35 mg/kg APTOA6 23,109,65

SAFETY PROFILE: Poison by ingestion, subcutaneous, and intravenous routes. Experimental teratogenic effects. A neuroleptic drug used to treat anxiety and confusion. When heated to decomposition it emits very toxic fumes of F^-, NO_x, and HCl.

FKK100 **CAS:364-76-1** **HR: 2**
4-FLUORO-3-NITROANILINE
mf: $C_6H_5FN_2O_2$ mw: 156.13

SYNS: ANILINE, 4-FLUORO-3-NITRO- □ BENZENAMINE, 4-FLUORO-3-NITRO- □ 4-F-3NA

TOXICITY DATA WITH REFERENCE
orl-rat LD50:1100 mg/kg TXAPA9 72,400,84

CONSENSUS REPORTS: Reported in EPA TSCA Inventory.

SAFETY PROFILE: Moderately toxic by ingestion. When heated to decomposition it emits toxic vapors of NO_x and F^-.

FKL000 **CAS:350-46-9** **HR: 3**
1-FLUORO-4-NITROBENZENE
mf: $C_6H_4FNO_2$ mw: 141.11

PROP: Pale yellow crystals. D: 1.34 @ 17.2°, mp: 27°, bp: 214–216°.

SYNS: p-FLUORONITROBENZENE □ 4-FLUORONITROBENZENE □ p-NITROFLUOROBENZENE □ 4-NITROFLUOROBENZENE

F

TOXICITY DATA WITH REFERENCE
mmo-sat 160 nL/plate MUREAV 116,217,83
orl-rat LDLo: 250 mg/kg NCNSA6 5,19,53

CONSENSUS REPORTS: Reported in EPA TSCA Inventory.

SAFETY PROFILE: Poison by ingestion. Mutation data reported. When heated to decomposition it emits very toxic fumes of F^- and NO_x. See also NITRO COMPOUNDS of AROMATIC HYDROCARBONS.

FKO000 CAS:17576-63-5 HR: 2
3-FLUORO-4-NITROQUINOLINE-1-OXIDE
mf: $C_9H_5FN_2O_3$ mw: 208.16

TOXICITY DATA WITH REFERENCE
cyt-omi 10 μmol/L GANNA2 60,155,69
cyt-rat:ast 100 nmol/L GMCRDC 17,31,75
dns-ham:oth 2 μmol/L NATUAS 229,416,71
scu-mus TDLo: 347 mg/kg/I:ETA CPBTAL 17,544,69

CONSENSUS REPORTS: EPA Genetic Toxicology Program.

SAFETY PROFILE: Questionable carcinogen with experimental tumorigenic data. Mutation data reported. When heated to decomposition it emits very toxic fumes of F^- and NO_x.

FKP000 CAS:19789-69-6 HR: 2
8-FLUORO-4-NITROQUINOLINE-1-OXIDE
mf: $C_9H_5FN_2O_3$ mw: 208.16

TOXICITY DATA WITH REFERENCE
scu-mus TDLo: 60 mg/kg/I:ETA CPBTAL 17,544,69

SAFETY PROFILE: Questionable carcinogen with experimental tumorigenic data. When heated to decomposition it emits very toxic fumes of F^- and NO_x.

FKQ000 HR: 3
FLUORONIUM PERCHLORATE
mf: $ClFH_2O_4$ mw: 120.5

$H_2F^+ClO_4^-$

SAFETY PROFILE: Complex explodes on contact with water. When heated to decomposition it emits very toxic fumes of F^- and Cl^-. See also FLUORIDES and PERCHLORATES.

FKQ100 CAS:399-24-6 HR: 3
9-FLUORONONYL PHENYL KETONE
mf: $C_{16}H_{23}FO$ mw: 250.39

SYN: DECANOPHENONE, 10-FLUORO-

TOXICITY DATA WITH REFERENCE
ipr-mus LD50: 90 mg/kg JACSAT 79,1959,57

DOT CLASSIFICATION: 3; *Label:* Flammable Liquid

SAFETY PROFILE: A poison by intraperitoneal route. A flammable liquid. When heated to decomposition it emits toxic vapors of F^- and Cl^-.

FKS000 CAS:593-12-4 HR: 3
8-FLUOROOCTYL BROMIDE
mf: $C_8H_{16}BrF$ mw: 211.15

SYN: 1-BROMO-8-FLUOROOCTANE

TOXICITY DATA WITH REFERENCE
ipr-mus LD50: 20 mg/kg JOCEAH 21,748,56
scu-mus LD50: 20 mg/kg CLDND*

SAFETY PROFILE: Poison by intraperitoneal and subcutaneous routes. When heated to decomposition it emits very toxic fumes of Br^- and F^-. See also FLUORIDES and BROMIDES.

FKT000 CAS:593-14-6 HR: 3
8-FLUOROOCTYL CHLORIDE
mf: $C_8H_{16}ClF$ mw: 166.69

SYN: 1-CHLORO-8-FLUOROOCTANE

TOXICITY DATA WITH REFERENCE
ipr-mus LD50: 2300 μg/kg JOCEAH 21,748,56
scu-mus LD50: 2300 μg/kg CLDND*

SAFETY PROFILE: Poison by intraperitoneal and subcutaneous routes. When heated to decomposition it emits very toxic fumes of Cl^- and F^-. See also FLUORIDES and CHLORIDES.

FKT050 CAS:326-52-3 HR: 3
8-FLUOROOCTYL PHENYL KETONE
mf: $C_{15}H_{21}FO$ mw: 236.36

SYN: NONANOPHENONE, 9-FLUORO-

TOXICITY DATA WITH REFERENCE
ipr-mus LD50: 100 mg/kg JACSAT 79,1959,57

DOT CLASSIFICATION: 3; *Label:* Flammable Liquid

SAFETY PROFILE: A poison by intraperitoneal route. A flammable liquid. When heated to decomposition it emits toxic vapors of F^-.

FKT100 CAS:367-12-4 HR: 2
2-FLUOROPHENOL
mf: C_6H_5FO mw: 112.11

SYNS: o-FLUOROPHENOL □ PHENOL, o-FLUORO-

TOXICITY DATA WITH REFERENCE
ipr-mus LD50: 537 mg/kg JMCMAR 18,868,75

CONSENSUS REPORTS: Reported in EPA TSCA Inventory.

SAFETY PROFILE: Moderately toxic by intraperitoneal route. When heated to decomposition it emits toxic vapors of F^-.

FKV000 CAS:371-41-5 HR: 3
4-FLUOROPHENOL
mf: C_6H_5FO mw: 112.11

PROP: A solid. Mp: 48°, bp: 189°.

TOXICITY DATA WITH REFERENCE
orl-rat TDLo:1 g/kg (female 11D post):REP TJADAB 41,43,90
skn-mus TDLo:10 g/kg/25W-I:CAR CNREA8 19,413,59
ipr-mus LD50:312 mg/kg JMCMAR 18,868,75

CONSENSUS REPORTS: Reported in EPA TSCA Inventory.

SAFETY PROFILE: A poison by intraperitoneal route. Experimental reproductive effects. Questionable carcinogen with experimental carcinogenic data. When heated to decomposition it emits toxic fumes of F^-. See also PHENOL.

FKX000 CAS:725-04-2 HR: 2
2′-FLUORO-4′-PHENYLACETANILIDE
mf: $C_{14}H_{12}FNO$ mw: 229.27

SYN: 3-FLUORO-4-ACETYLAMINOBIPHENYL

TOXICITY DATA WITH REFERENCE
orl-rat TDLo:3700 mg/kg/26W-C:CAR CNREA8 22,1002,62

SAFETY PROFILE: Questionable carcinogen with experimental carcinogenic data. When heated to decomposition it emits very toxic fumes of NO_x and F^-. See also FLUORIDES.

FKY000 CAS:725-06-4 HR: 2
4′-(m-FLUOROPHENYL)ACETANILIDE
mf: $C_{14}H_{12}FNO$ mw: 229.27

SYN: 3′-FLUORO-4-ACETYLAMINOBIPHENYL

TOXICITY DATA WITH REFERENCE
orl-rat TDLo:2580 mg/kg/21W-C:CAR CNREA8 22,1002,62

SAFETY PROFILE: Questionable carcinogen with experimental carcinogenic data. When heated to decomposition it emits very toxic fumes of F^- and NO_x. See also FLUORIDES.

FKZ000 CAS:398-32-3 HR: 2
4′-(p-FLUOROPHENYL)ACETANILIDE
mf: $C_{14}H_{12}FNO$ mw: 229.27

PROP: A solid. Mp: 205–205.5°.

SYNS: 4′-FLUORO-4-ACETYLAMINOBIPHENYL □ N-4-(4′-FLUORO)BIPHENYLACETAMIDE □ N-(4′-FLUORO-4-BIPHENYLYL)ACETAMIDE

TOXICITY DATA WITH REFERENCE
orl-rat TDLo:4032 mg/kg/24W-C:ETA AJPAA4 100,317,80
orl-rat TD:8064 mg/kg/48W-C:ETA JJIND8 64,1537,80
orl-rat TD:6048 mg/kg/36W-C:ETA EXPEAM 32,217,76
orl-rat TD:5 g/kg/36W-I:ETA JNCIAM 54,427,75
orl-rat TD:6048 mg/kg/36W-C:ETA BECTA6 23,464,79
orl-rat TD:5040 mg/kg/36W-C:ETA JNCIAM 54,1223,75

SAFETY PROFILE: Questionable carcinogen with experimental tumorigenic data. When heated to decomposition it emits very toxic fumes of NO_x and F^-. See also FLUORIDES.

FLC000 CAS:459-22-3 HR: 3
p-FLUOROPHENYLACETONITRILE
mf: C_8H_6FN mw: 135.15

PROP: Bp: 228–230°.

SYNS: 4-FLUOROBENZENEACETONITRILE □ p-FLUOROBENZYL CYANIDE □ 4-FLUOROBENZYLCYANIDE □ 4-FLUOROPHENYLACETONITRILE

TOXICITY DATA WITH REFERENCE
ivn-mus LD50:42 mg/kg CSLNX* NX#07881

CONSENSUS REPORTS: Cyanide and its compounds are on the Community Right-To-Know List. Reported in EPA TSCA Inventory.

SAFETY PROFILE: Poison by intravenous route. When heated to decomposition it emits very toxic fumes of F^-, NO_x, and CN^-. See also FLUORIDES and NITRILES.

FLG000 CAS:2924-67-6 HR: 3
p-FLUOROPHENYL ETHYL SULFONE
mf: $C_8H_9FO_2S$ mw: 188.23

PROP: Crystals. Mp: 41°.

SYNS: BRIPADON □ CADUCID □ ETHYL-p-FLUOROPHENYL SULFONE □ 1-(ETHYLSULFONYL)-4-FLUOROBENZENE □ FLUORESONE

TOXICITY DATA WITH REFERENCE
orl-mus LD50:542 mg/kg NEPSBV 3,326,64
ipr-mus LD50:343 mg/kg NEPSBV 3,326,64
ivn-mus LD50:320 mg/kg NEPSBV 3,317,64
ivn-rbt LD50:110 mg/kg NEPSBV 3,317,64

SAFETY PROFILE: Poison by intraperitoneal and intravenous routes. Moderately toxic by ingestion. Used as an anti-epileptic agent and as a tranquilizer. When heated to decomposition it emits very toxic fumes of F^- and SO_x. See also FLUORIDES.

FLG100 CAS:5104-49-4 HR: 3
3-FLUORO-4-PHENYLHYDRATROPIC ACID
mf: $C_{15}H_{13}FO_2$ mw: 244.28

SYNS: 4-BIPHENYLACETIC ACID, 2-FLUORO-α-METHYL- □ (1,1′-BIPHENYL)-4-ACETIC ACID, 2-FLUORO-α-METHYL- (9CI) □ BTS 18322 □ 2-(2-FLUORO-4-BIPHENYLYL)PROPIONIC ACID □ 2-FLUORO-α-METHYL-4-BIPHENYLACETIC ACID □ 2-FLUORO-α-METHYL(1,1′-BIPHENYL)-4-ACETIC ACID □ FLURBIPROFEN □ FP 70 □ FROBEN

TOXICITY DATA WITH REFERENCE
orl-rat TDLo:10 mg/kg (female 21D post):TER PEREBL 22,567,87
orl-man TDLo:30 mg/kg/2W-I:BAH BMJOAE 4,496,74
orl-wmn TDLo:14 mg/kg/7D-I:SKN AIMEAS 112,550,90
orl-rat LD50:117 mg/kg KSRNAM 9,2641,75
ipr-rat LD50:108 mg/kg KSRNAM 9,2641,75
scu-rat LD50:100 mg/kg KSRNAM 9,2641,75
orl-mus LD50:640 mg/kg KSRNAM 9,2641,75

SAFETY PROFILE: Poison by ingestion, subcutaneous, and intraperitoneal routes. Human systemic effects: dermatitis, distorted perceptions, hallucinations, headache. An experimental teratogen. When heated to decomposition it emits toxic fumes of F^-.

FLH100 CAS:1493-23-8 HR: 3
4-FLUOROPHENYLLITHIUM
mf: C_6H_4FLi mw: 102.04

PROP: A solid. Very sol in Et_2O.

SAFETY PROFILE: The pure material is explosive. When heated to decomposition it emits toxic fumes of F^-. See also LITHIUM COMPOUNDS.

FLJ000 CAS:2804-00-4 HR: 3
8-(4-p-FLUORO PHENYL-4-OXOBUTYL)-2-METHYL-2, 8-DIAZASPIRO(4.5)DECANE-1,3-DIONE
mf: $C_{19}H_{23}FN_2O_3$ mw: 346.44

SYNS: F-33 □ FR-33 □ R 7158

TOXICITY DATA WITH REFERENCE
ivn-rat LD50:92 mg/kg 27ZQAG -,189,72
orl-mus LD50:193 mg/kg 27ZQAG -,189,72
ivn-mus LD50:90 mg/kg 27ZQAG -,189,72
ivn-dog LD50:56 mg/kg 27ZQAG -,189,72

SAFETY PROFILE: Poison by ingestion and intravenous routes. When heated to decomposition it emits very toxic fumes of F^- and NO_x.

FLK100 CAS:2062-84-2 HR: 3
1-(1-(4-(4-FLUOROPHENYL)-4-OXOBUTYL)-4-PIPERIDINYL)-1,3-DIHYDRO-2H-BENZIMIDAZOL-2-ONE
mf: $C_{22}H_{24}FN_3O_2$ mw: 381.49

PROP: Solid. Mp: 170–171.8°.

SYNS: ANQUIL □ BENPERIDOL □ BENZOPERIDOL □ BENZPERIDOL □ 8089 C.B. □ 1-(1-(3-(p-FLUOROBENZOYL)PROPYL)-4-PIPERIDYL)-2-BENZIMIDAZOLINONE □ FRENACTIL □ FRENACTYL □ GLIANIMON □ GLIANIMON MITE □ MCM-JR-4584 □ R 4584

TOXICITY DATA WITH REFERENCE
scu-rat LD50:220 mg/kg MDCHAG 4(2),199,67
ipr-mus LD50:250 mg/kg FRPSAX 35,605,80
ivn-mus LD50:20 mg/kg CSLNX* NX#11942

SAFETY PROFILE: Poison by subcutaneous, intravenous, and intraperitoneal routes. When heated to decomposition it emits toxic fumes of F^- and NO_x.

FLL000 CAS:2354-61-2 HR: 3
4′-FLUORO-4-(1-(4-PHENYL)PIPERAZINO)BUTYROPHENONE
mf: $C_{20}H_{23}FN_2O$ mw: 326.45

SYNS: BPZ □ BUTROPIPAZON □ BUTROPIPAZONE □ 1-(4-FLUOROPHENYL)-4-(4-PHENYL-1-PIPERAZINYL)-1-BUTANONE □ GALVANISONE □ R 1892

TOXICITY DATA WITH REFERENCE
orl-rat LD50:550 mg/kg PHMGBN 15,485,77
scu-rat LD50:550 mg/kg MDCHAG 4(2),199,67
ivn-rat LD50:32 mg/kg 27ZQAG -,186,72
orl-mus LD50:650 mg/kg PHMGBN 15,485,77
ipr-mus LD50:76 mg/kg 27ZQAG -,186,72
ivn-mus LD50:24 mg/kg 27ZQAG -,186,72

SAFETY PROFILE: Poison by intravenous and intraperitoneal routes. Moderately toxic by ingestion, and subcutaneous routes. When heated to decomposition it emits very toxic fumes of F^- and NO_x.

FLN000 CAS:3781-28-0 HR: 3
4′-FLUORO-4-(4-PIPERIDINO-4-PROPIONYLPIPERIDINO)BUTYROPHENONE
mf: $C_{23}H_{33}FN_2O_2$ mw: 388.58

SYNS: FLOROPIPETON □ PROPYPERONE

TOXICITY DATA WITH REFERENCE
orl-rat LD50:780 mg/kg 27ZQAG -,192,72
scu-rat LD50:350 mg/kg 27ZQAG -,192,72
orl-mus LD50:275 mg/kg 27ZQAG -,192,72
scu-mus LD50:420 mg/kg 27ZQAG -,192,72

SAFETY PROFILE: Poison by ingestion and subcutaneous routes. When heated to decomposition it emits very toxic fumes of F^- and NO_x.

FLQ000 CAS:5675-31-0 HR: 3
2-FLUORO-2-PROPEN-1-OL
mf: C_3H_5FO mw: 76.08

TOXICITY DATA WITH REFERENCE
orl-rat LD50:130 mg/kg AIHAAP 23,95,62
ihl-rat LCLo:1000 ppm/1H AIHAAP 23,95,62
skn-rbt LD50:3 mg/kg AIHAAP 23,95,62

SAFETY PROFILE: Poison by ingestion and skin contact. Mildly toxic by inhalation. When heated to decomposition it emits toxic fumes of F^-.

FLR000 CAS:461-56-3 HR: 3
3-FLUOROPROPIONIC ACID
mf: $C_3H_5FO_2$ mw: 92.08

PROP: Oily liquid with sharp, characteristic odor. D: 1.2406 @ 20°/4°, bp: 97° @ 29 mm. Very sol in H_2O and org solvents.

SYN: ω-FLUOROPROPIONIC ACID

TOXICITY DATA WITH REFERENCE
ipr-mus LD50:60 mg/kg JOCEAH 21,739,56

CONSENSUS REPORTS: Reported in EPA TSCA Inventory.

SAFETY PROFILE: Poison by intraperitoneal route. When heated to decomposition it emits toxic fumes of F^-.

FLR100 CAS:407-99-8 HR: 3
3-FLUOROPROPYL ISOCYANATE
DOT: UN 2206/UN 2207/UN 2478/UN 3080
mf: C_4H_6FNO mw: 103.11

SYN: ISOCYANIC ACID, 3-FLUOROPROPYL ESTER

TOXICITY DATA WITH REFERENCE
ipr-mus LD50:10 mg/kg JACSAT 79,1956,57

DOT CLASSIFICATION: 6.1; *Label:* KEEP AWAY FROM FOOD (UN2207); 6.1; *Label:* Poison (UN2206); 6.1; *Label:* Poison, Flammable Liquid (UN3080); 3; *Label:* Flammable Liquid, Poison (UN2478)

SAFETY PROFILE: A poison by intraperitoneal route. A flammable liquid. When heated to decomposition it emits toxic vapors of NO_x and CN^-.

FLT100 **CAS:372-48-5** **HR: 3**
2-FLUOROPYRIDINE
mf: C_5H_4FN mw: 97.09

CH=CHCH=CHCF=N

PROP: A liquid. Bp: 125°.

TOXICITY DATA WITH REFERENCE
mmo-sat 5 mg/plate MUREAV 176,185,87

SAFETY PROFILE: Mutation data reported. Reaction with bromine trifluoride forms a spontaneously ignitable product. When heated to decomposition it emits toxic fumes of F^- and NO_x.

FLU000 **CAS:1649-18-9** **HR: 3**
4′-FLUORO-4-(4-(2-PYRIDYL)-1-PIPERAZINYL)BUTYROPHENONE
mf: $C_{19}H_{22}FN_3O$ mw: 327.44

PROP: A solid. Mp: 73-75°.

SYNS: AZAPERONE (USDA) □ AZEPERONE □ EUCALMYL □ FLUOPERIDOL □ 1-(3-(4-FLUOROBENZOYL)PROPYL)-4-(2-PYRIDYL) PIPERAZINE □ 1-(4-FLUOROPHENYL)-4-(4-(2-PYRIDINYL)-1-PIPERAZINYL)-1-BUTANONE □ R 1929 □ STRESNIL □ SUICALM

TOXICITY DATA WITH REFERENCE
orl-rat LD50:245 mg/kg ARZNAD 24,1798,74
scu-rat LD50:450 mg/kg ARZNAD 24,1798,74
ivn-rat LD50:25 mg/kg 27ZQAG -,185,72
orl-mus LD50:385 mg/kg ARZNAD 24,1798,74
ipr-mus LD50:63 mg/kg FRPSAX 35,605,80
scu-mus LD50:179 mg/kg 27ZQAG -,185,72
ivn-mus LD50:42 mg/kg ARZNAD 24,1798,74

SAFETY PROFILE: Poison by ingestion, intravenous, intraperitoneal, and subcutaneous routes. When heated to decomposition it emits very toxic fumes of F^- and NO_x.

FLV000 **CAS:2266-22-0** **HR: 3**
4′-FLUORO-4-(n-(4-PYRROLIDINAMIDO)-4-m-TOLYPIPERIDINO)BUTYROPHENONE
mf: $C_{27}H_{33}FN_2O_2$ mw: 436.62

SYNS: MEPERIDIDE □ METHYLPERIDIDE

TOXICITY DATA WITH REFERENCE
scu-rat LD50:275 mg/kg 27ZQAG -,191,72
ivn-rat LD50:9300 μg/kg 27ZQAG -,191,72
scu-mus LD50:140 mg/kg 27ZQAG -,191,72
ivn-mus LD50:14 mg/kg 27ZQAG -,191,72

SAFETY PROFILE: Poison by subcutaneous and intravenous routes. When heated to decomposition it emits very toxic fumes of F^- and NO_x.

FLY000 **CAS:10010-36-3** **HR: 2**
4′-FLUORO-4-STILBENAMINE
mf: $C_{14}H_{12}FN$ mw: 213.27

TOXICITY DATA WITH REFERENCE
scu-rat TDLo:200 mg/kg/W-I:ETA BMBUAQ 14,141,58

SAFETY PROFILE: Questionable carcinogen with experimental tumorigenic data. When heated to decomposition it emits very toxic fumes of F^- and NO_x. See also FLUORIDES.

FLY100 **HR: 3**
FLUOROSULFONATES

SAFETY PROFILE: Probably highly toxic, as they are salts of a strong acid. Probably strong irritants to the eyes, skin, and mucous membranes. They react with water or steam to produce toxic and corrosive fumes. When heated to decomposition they emit toxic fumes of F^- and SO_x. See also specific compounds.

FLY200 **CAS:2489-52-3** **HR: 3**
m-FLUOROSULFONYLBENZENESULFONYL CHLORIDE
mf: $C_6H_4ClFO_4S_2$ mw: 258.67

SYNS: BENZENE-1,3-DISULFONYL CHLORIDE FLUORIDE □ BENZENESULFONYL CHLORIDE, m-(FLUOROSULFONYL)-

TOXICITY DATA WITH REFERENCE
ivn-mus LD50:56 mg/kg CSLNX* NX#04699

CONSENSUS REPORTS: Reported in EPA TSCA Inventory.

SAFETY PROFILE: Poison by intravenous route. When heated to decomposition it emits toxic vapors of NO_x, SO_x, F^-, and Cl^-.

FLZ000 **CAS:7789-21-1** **HR: 3**
FLUOROSULFURIC ACID
DOT: UN 1777
mf: FHO_3S mw: 100.07

PROP: Colorless, fumes in moist air; highly corrosive liquid. Mp: −89°, bp: 163°, d: 1.726 @ 20°. Sol in HOAc, $PhNO_2$, Et_2O; insol in CCl_4, CS_2.

SYNS: FLUOROSUFONIC ACID (DOT) □ FLUOSULFONIC ACID (DOT)

CONSENSUS REPORTS: Reported in EPA TSCA Inventory.

OSHA PEL: TWA 2.5 mg(F)/m³
NIOSH REL: (Inorganic Fluorides) TWA 2.5 mg(F)/m³
DOT CLASSIFICATION: 8; *Label:* Corrosive

SAFETY PROFILE: Probably a poison by inhalation. A corrosive irritant to the skin, eyes, and mucous membranes. See also FLUORIDES, SULFURIC ACID, and FLUOROSULFONATES.

FLZ050 CAS:17902-23-7 HR: 3
5-FLUORO-1-(TETRAHYDROFURAN-2-YL)URACIL
mf: $C_8H_9FN_2O_3$ mw: 200.19

SYNS: CARZONAL □ CITOFUR □ COPAROGIN □ EXONAL □ FENTAL □ F-5-FU □ FLUOROFUR □ 5-FLUORO-1-(TETRAHYDRO-2-FURANYL)-2,4-PYRIMIDINEDIONE □ 5-FLUORO-1-(TETRAHYDRO-2-FURANYL)-2,4(1H,3H)-PYRIMIDINEDIONE □ 5-FLUORO-1-(TETRAHYDRO-3-FURYL)URACIL □ FRANROZE □ FTORAFUR □ FULAID □ FULFEEL □ FURAFLUOR □ FUROFUTRAN □ FUTRAFUL □ LAMAR □ LIFRIL □ MJF-12264 □ NEBERK □ NITOBANIL □ NSC-148958 □ PYRIMIDINE-DEOXYRIBOSE N1-2'-FURANIDYL-5-FLUOROURACIL □ RIOL □ SINOFLUROL □ SUNFRAL □ TEFSIEL C □ TEGAFUR □ 1-(TETRAHYDROFURAN-2-YL)-5-FLUOROURACIL □ N'-(2-TETRAHYDROFURYL)-5-FLUOROURACIL □ THFU

TOXICITY DATA WITH REFERENCE
slt-dmg-mul 10 mg/L TAKHAA 44,96,85
dns-hmn:fbr 1 g/L STBIBN 78,165,80
orl-rat TDLo:810 mg/kg (17-22D preg/21D post): REP OYYAA2 22,109,81
ivn-mus TDLo:210 mg/kg (female 6-12D post):TER OYYAA2 5,555,71
orl-rat TDLo:550 mg/kg (female 7-17D post):TER KSRNAM 14,1373,80
ivn-rat TDLo:420 mg/kg (female 7-13D post):TER OYYAA2 5,555,71
ivn-rat TDLo:630 mg/kg (female 7-13D post):REP OYYAA2 5,555,71
orl-rat TDLo:3500 mg/kg (male 35D pre):REP YACHDS 13(Suppl 2),293,85
orl-rat TDLo:1260 mg/kg (18W male):REP OYYAA2 17,169,79
orl-man TDLo:11,046 mg/kg/4.7Y-C:CAR,GIT ARSUAX 118,1454,83
ivn-hmn LDLo:640 mg/kg/8D CANCAR 36,103,75
ivn-hmn TDLo:23 mg/kg:CNS CANCAR 36,103,75
orl-rat LD50:930 mg/kg YACHDS 6,2911,78
ipr-rat LD50:700 mg/kg YACHDS 6,2911,78
scu-rat LD50:600 mg/kg YACHDS 6,2911,78
ivn-rat LD50:685 mg/kg GTKRDX 10,1987,83
orl-mus LD50:775 mg/kg YHTPAD 22,27,87
ipr-mus LD50:493 mg/kg NCISP* JAN86
scu-mus LD50:760 mg/kg OYYAA2 5,569,71
orl-dog LD50:34 mg/kg OYYAA2 20,1009,80

CONSENSUS REPORTS: Reported in EPA TSCA Inventory.

SAFETY PROFILE: Poison by ingestion. Moderately toxic to humans by intravenous route. Moderately toxic experimentally by intraperitoneal, intravenous, and subcutaneous routes. Experimental teratogenic data. Human systemic effects: nausea and vomiting. Experimental reproductive effects. Questionable human carcinogen producing gastrointestinal tumors. Human mutation data reported. Used as an anti-cancer agent. When heated to decomposition it emits very toxic fumes of F^- and NO_x.

FLZ100 CAS:95-52-3 HR: 3
o-FLUOROTOLUENE
mf: C_7H_7F mw: 110.14

SYNS: BENZENE, 1-FLUORO-2-METHYL-(9CI) □ 1-FLUORO-2-METHYLBENZENE □ 2-FLUOROTOLUENE □ TOLUENE, o-FLUORO-

TOXICITY DATA WITH REFERENCE
orl-brd LD50:100 mg/kg AECTCV 12,355,83

CONSENSUS REPORTS: Reported in EPA TSCA Inventory.

SAFETY PROFILE: Poison by ingestion. When heated to decomposition it emits toxic vapors of F^-.

FMC000 CAS:352-32-9 HR: 3
p-FLUOROTOLUENE
mf: C_7H_7F mw: 110.14

PROP: Colorless liquid. D: 1.001, mp: −56°, bp: 116–117°, flash p: 50°F. Insol in water; sol in alc and ether.

TOXICITY DATA WITH REFERENCE
par-mus LDLo:500 mg/kg 11FYAN 3,91,63

CONSENSUS REPORTS: Reported in EPA TSCA Inventory.

SAFETY PROFILE: Moderately toxic by parenteral route. A very dangerous fire hazard when exposed to heat or flame; can react vigorously with oxidizing materials. When heated to decomposition it emits toxic fumes of F^-. See also FLUORIDES.

FME000 CAS:1983-10-4 HR: 3
FLUOROTRIBUTYLSTANNANE
mf: $C_{12}H_{27}FSn$ mw: 309.08

SYNS: TRIBUTYLSTANNANE FLUORIDE □ TRIBUTYLTIN FLUORIDE

TOXICITY DATA WITH REFERENCE
oth-ham:ovr 60 μg/L MUREAV 300,5,93
orl-mus LDLo:320 mg/kg AECTCV 14,111,85
orl-rbt LDLo:50 mg/kg SAIGBL 15,3,73

CONSENSUS REPORTS: Reported in EPA TSCA Inventory.

OSHA PEL: TWA 0.1 mg(Sn)/m³ (skin)
ACGIH TLV: TWA 0.1 mg(Sn)/m³ (skin) (Proposed: TWA 0.1 mg(Sn)/m³; STEL 0.2 mg(Sn)/m³ (skin))
DFG MAK: 0.002 ppm (0.05 mg/m³)
NIOSH REL: (Organotin Compounds) TWA 0.1 mg(Sn)/m³

SAFETY PROFILE: Poison by ingestion. Many tributyl tin compounds are highly toxic to marine life. Mutation data reported. When heated to decomposition it emits toxic fumes of F^-. See also TIN COMPOUNDS and FLUORIDES.

For occupational chemical analysis use NIOSH: Organotin Compounds 5504.

FMF000 CAS:313-95-1 HR: 2
2-FLUOROTRICYCLOQUINAZOLINE
mf: $C_{21}H_{11}FN_4$ mw: 338.36

TOXICITY DATA WITH REFERENCE
skn-mus TDLo:48 mg/kg:ETA BECCAN 41,420,63

SAFETY PROFILE: Questionable carcinogen with ex-

perimental tumorigenic data. When heated to decomposition it emits very toxic fumes of NO_x and F^-. See also FLUORIDES.

FMG000 **CAS:803-57-6** **HR: 2**
3-FLUORO-TRICYCLOQUINAZOLINE
mf: $C_{21}H_{11}FN_4$ mw: 338.36

TOXICITY DATA WITH REFERENCE
skn-mus TDLo:920 mg/kg/1Y-I:NEO BJCAAI 16,275,62

SAFETY PROFILE: Questionable carcinogen with experimental neoplastigenic data. When heated to decomposition it emits very toxic fumes of F^- and NO_x. See also FLUORIDES.

FMH000 **CAS:19982-87-7** **HR: 3**
4-FLUORO-4′-TRIFLUOROMETHYLBENZOPHENONE GUANYLHYDRAZONE HYDROCHLORIDE
mf: $C_{15}H_{12}F_4N_4{\cdot}ClH$ mw: 360.77

SYNS: FTBG ☐ WR 09792

TOXICITY DATA WITH REFERENCE
orl-rat LD50:199 mg/kg TXAPA9 15,614,69
ipr-rat LD50:23 mg/kg TXAPA9 15,614,69
orl-mus LD50:114 mg/kg TXAPA9 15,614,69
ipr-mus LD50:12 mg/kg TXAPA9 15,614,69
orl-gpg LD50:132 mg/kg TXAPA9 15,614,69
ipr-gpg LD50:19 mg/kg TXAPA9 15,614,69

SAFETY PROFILE: Poison by ingestion and intraperitoneal routes. When heated to decomposition it emits very toxic fumes of HCl, F^- and NO_x.

FMI000 **CAS:1840-42-2** **HR: 3**
FLUOROTRINITROMETHANE
mf: CFN_3O_6 mw: 169.04

PROP: D: 1.59 @ 20°, bp: 83–84°.

TOXICITY DATA WITH REFERENCE
ipr-mus LD50:57,300 µg/kg KHFZAN 10(6),53,76

CONSENSUS REPORTS: Reported in EPA TSCA Inventory.

SAFETY PROFILE: Poison by intraperitoneal route. Mixtures with nitrobenzene are liquid explosives. When heated to decomposition it emits very toxic fumes of F^- and NO_x. See also FLUORIDES and NITRO COMPOUNDS.

FMJ000 **CAS:139-26-4** **HR: 3**
3-FLUOROTYROSIN
mf: $C_9H_{10}FNO_3$ mw: 199.20

SYNS: m-FLUOROTYROSINE ☐ 3-FLUOROTYROSINE ☐ FLUORTHYRIN ☐ 3-FLUORTYROSIN (GERMAN)

TOXICITY DATA WITH REFERENCE
orl-man LDLo:143 mg/kg AEPPAE 183,427,36
scu-rat LDLo:12 mg/kg JPETAB 73,176,41
orl-mus LDLo:14 mg/kg AEPPAE 183,427,36
skn-mus LDLo:300 mg/kg AEPPAE 183,427,36
scu-mus LDLo:11 mg/kg AEPPAE 183,427,36
scu-gpg LDLo:8 mg/kg AEPPAE 183,427,36

SAFETY PROFILE: Human poison by ingestion. Experimental poison by ingestion, skin contact, and subcutaneous routes. When heated to decomposition it emits very toxic fumes of F^- and NO_x. See also FLUORIDES.

FMM000 **CAS:51-21-8** **HR: 3**
FLUOROURACIL
mf: $C_4H_3FN_2O_2$ mw: 130.09

PROP: Crystals from H_2O. Mp: 282–283° (decomp).

SYNS: ADRUCIL ☐ ARUMEL ☐ CARZONAL ☐ EFFLUDERM (free base) ☐ EFUDEX ☐ EFUDIX ☐ 5-FLUORACIL (GERMAN) ☐ FLUOROBLASTIN ☐ FLUOROPLEX ☐ 5-FLUORO-2,4-PYRIMIDINEDIONE ☐ 5-FLUORO-2,4(1H,3H)-PYRIMIDINEDIONE ☐ 5-FLUOROURACIL ☐ 5-FLUORPROPYRIMIDINE-2,4-DIONE ☐ 5-FLUORURACIL (GERMAN) ☐ FLURACIL ☐ FLURI ☐ FLURIL ☐ 5-FU ☐ NSC-19893 ☐ RO 2-9757 ☐ TIMAZIN ☐ U-8953 ☐ ULUP

TOXICITY DATA WITH REFERENCE
skn-hmn 84 mg/3W CTRRDO 63,619,79
dnr-bcs 300 µg/plate TAKHAA 44,96,85
dns-hmn:oth 1 mmol/L CNREA8 44,3414,84
ivn-wmn TDLo:150 mg/kg (20-31W preg):REP JAMAAP 217,214,71
ivn-wmn TDLo:240 mg/kg (11-14W preg):TER AJOGAH 137,747,80
ipr-mus TDLo:30 mg/kg (female 12D post):REP TCMUD8 7,7,87
ipr-mus TDLo:50 mg/kg (female 13D post):TER TJADAB 28,42A,83
orl-mus TDLo:245 mg/kg (female 7-13D post):TER KSRNAM 8,2603,74
orl-rat TDLo:175 mg/kg (female 7-13D post):TER KSRNAM 8,2603,74
ipr-mus TDLo:30 mg/kg (female 9D post):REP DZZEA7 32,861,77
orl-mus TDLo:245 mg/kg (female 7-13D post):REP KSRNAM 8,2603,74
ims-ham TDLo:28 mg/kg (female 9D post):REP JEEMAF 43,47,78
orl-rat TDLo:700 mg/kg (female 35D pre):REP YACHDS 13(Suppl 2),293,85
rec-rat TDLo:5400 mg/kg (male 26W pre):REP KSRNAM 12,1309,78
ivn-mus TDLo:67 mg/kg (male 1D pre):REP CNREA8 42,122,82
orl-rat TDLo:420 mg/kg (84D male):REP OYYAA2 17,153,79
ipr-mus TDLo:10 mg/kg (female 10D post):TER TJADAB 6,5,72
ipr-rat TDLo:30 mg/kg (female 12D post):TER SEIJBO 21,105,81
orl-rat TDLo:35 mg/kg (7-13D preg):TER KSRNAM 8,2603,74
ipr-mus TDLo:1500 mg/kg/50W-I:CAR TUMOAB 76,179,90
orl-hmn TDLo:450 mg/kg/30D:GIT,BLD CANCAR 39,1936,77
ivn-hmn TDLo:6 mg/kg/3D:CVS,PUL BMJOAE 1,547,78
ivn-wmn TDLo:150 mg/kg/17W-I:BLD BJHEAL 65,357,87

F

ivn-man TDLo: 39 mg/kg/1D-I: CVS AHJOA2 114,433,87
ivn-wmn TDLo: 27 mg/kg/4D-C: CVS BMJOAE 294,125,87
orl-rat LD50: 230 mg/kg IYKEDH 4,90,73
ipr-rat LD50: 70 mg/kg OYYAA2 5,569,71
scu-rat LD50: 217 mg/kg IYKEDH 4,90,73
ivn-rat LD50: 245 mg/kg OYYAA2 5,569,71
par-rat LD50: 500 mg/kg RRCRBU 52,76,75
rec-rat LD50: 884 mg/kg KSRNAM 12,1309,78
orl-mus LD50: 115 mg/kg JMCMAR 21,738,78
ipr-mus LD50: 100 mg/kg EKFMA7 7,100,77
scu-mus LD50: 169 mg/kg IYKEDH 4,90,73
ivn-mus LD50: 81 mg/kg IYKEDH 4,90,73
orl-dog LD50: 30 mg/kg OYYAA2 16,303,78

CONSENSUS REPORTS: IARC Cancer Review: Group 3 IMEMDT 7,210,87; Human Inadequate Evidence IMEMDT 26,217,81; Animal Inadequate Evidence IMEMDT 26,217,81. Reported in EPA TSCA Inventory. EPA Genetic Toxicology Program. EPA Extremely Hazardous Substances List.

SAFETY PROFILE: Poison by ingestion, intraperitoneal, subcutaneous, and intravenous routes. Moderately toxic by parenteral and rectal routes. Experimental teratogenic and reproductive effects. Human systemic effects: EKG changes, bone marrow changes, cardiac, pulmonary, and gastrointestinal effects. Human mutation data reported. A human skin irritant. Questionable carcinogen. When heated to decomposition it emits very toxic fumes of F^- and NO_x.

FMN000 CAS:316-46-1 HR: 3
5-FLUOROURIDINE
mf: $C_9H_{11}FN_2O_6$ mw: 262.22

SYNS: FUR □ 5-FUR

TOXICITY DATA WITH REFERENCE
dns-hmn: hla 10 μmol/L BCPCA6 14,205,65
dni-mus-ipr 500 μmol/kg CNREA8 39,2406,79
ipr-rat LD50: 400 mg/kg ADTEAS 3,181,68
ipr-mus LD50: 160 mg/kg CNREA8 39,2406,79
scu-mus LD50: 384 mg/kg CNCRA6 6,94,60

SAFETY PROFILE: Poison by intraperitoneal and subcutaneous routes. Human mutation data reported. When heated to decomposition it emits very toxic fumes of F^- and NO_x.

FMO000 CAS:353-13-9 HR: 3
5-FLUOROVALERONITRILE
mf: C_5H_8FN mw: 101.14

TOXICITY DATA WITH REFERENCE
ipr-mus LD50: 1 mg/kg JACSAT 78,3484,56
scu-mus LD50: 1 mg/kg CLDND*

CONSENSUS REPORTS: Cyanide and its compounds are on the Community Right-To-Know List.

SAFETY PROFILE: Poison by intraperitoneal and subcutaneous routes. When heated to decomposition it emits very toxic fumes of F^-, NO_x, and CN^-. See also FLUORIDES and NITRILES.

FMO050 HR: 1
FLUOSOL-DA 20%

TOXICITY DATA WITH REFERENCE
ivn-rat TDLo: 97,500 mg/kg (male 13W pre): REP KSRNAM 16,3934,82
ivn-rat LD50: 128 g/kg KSRNAM 16,3899,82
ivn-mus LD50: 128 g/kg KSRNAM 16,3899,82

SAFETY PROFILE: Slightly toxic by intravenous route. Experimental reproductive effects. When heated to decomposition it emits acrid smoke and irritating fumes.

FMO100 CAS:671-35-2 HR: 3
FLUOXYDINE
mf: $C_4H_3FN_2O$ mw: 114.09

SYNS: 5-FLUORO-4(1H)-PYRIMIDINONE □ FLUOXIDINE

TOXICITY DATA WITH REFERENCE
orl-mus LD50: 118 mg/kg YHTPAD 15(6),7,80
ipr-mus LD50: 72 mg/kg YHTPAD 15(6),7,80
ivn-mus LD50: 1435 mg/kg YHTPAD 15(6),7,80

SAFETY PROFILE: Poison by ingestion and intraperitoneal routes. Moderately toxic by intravenous route. When heated to decomposition it emits toxic fumes of F^- and NO_x.

FMO129 CAS:2709-56-0 HR: 3
cis-(Z)-FLUPENTHIXOL
mf: $C_{23}H_{25}F_3N_2OS$ mw: 434.56

SYNS: EMERGIL □ FLUANXOL □ FLUPENTHIXOL □ (α,β)-FLUPENTHIXOL □ FLUPENTHIXOLE □ FLUPENTIXOL □FLURENTIX-OL □ FLUXANXOL □ LC 44 □ N 7009 □ SIPLARIL □ SIPLAROL □ 2-TRIFLUOROMETHYL-9-(3-(4-(β-HYDROXYETHYL)-1-PIPERAZINYL) PROPYLIDENE)THIOXANTHENE □ 4-(3-(2-(TRIFLUOROMETHYL)THIOXANTHEN-9-YLIDENE)PROPYL)-1-PIPERAZINEETHANOL □ 4-(3-(2-(TRIFLUOROMETHYL)-9H-THIOXANTHEN-9-YLIDENE)PROPYL)-1-PIPERAZINEETHANOL

TOXICITY DATA WITH REFERENCE
orl-chd TDLo: 900 μg/kg: CNS DICPBB 15,388,81
orl-hmn TDLo: 367 μg/kg/12D-I: CNS ARZNAD 22,93,72
orl-mus LD50: 300 mg/kg JMCMAR 23,878,80
ipr-mus LD50: 150 mg/kg ARZNAD 27,2121,77
ivn-mus LD50: 87 mg/kg APTOA6 41,369,77

SAFETY PROFILE: Poison by ingestion, intraperitoneal, and intravenous routes. Human systemic effects by ingestion: central nervous system effects. When heated to decomposition it emits very toxic fumes of F^-, NO_x, and SO_x.

FMO150 CAS:2413-38-9 HR: 3
FLUPENTIXOL HYDROCHLORIDE
mf: $C_{23}H_{25}F_3N_2OS \cdot 2ClH$ mw: 507.48

SYNS: FLUPENTIXOL DIHYDROCHLORIDE □ FX 703 □ N 7009 □ 4-(3-(2-(TRIFLUOROMETHYL)THIOXANTHEN-9-YLIDENE)PROPYL)-1-PIPERAZINEETHANOL DIHYDROCHLORIDE □ 4-(3-(2-TRIFLUOROMETHYLTHIOXANTH-9-YLIDENE)PROPYL)-1-PIPERAZINEETHANOL DIHYDROCHLORIDE

TOXICITY DATA WITH REFERENCE
orl-rat LD50:791 mg/kg NIIRDN 6,704,82
scu-rat LD50:258 mg/kg IYKEDH 4,90,73
ivn-rat LD50:37 mg/kg IYKEDH 4,90,73
orl-mus LD50:423 mg/kg IYKEDH 4,90,73
scu-mus LD50:425 mg/kg IYKEDH 4,90,73
ivn-mus LD50:94 mg/kg APTOA6 19,87,62

SAFETY PROFILE: Poison by subcutaneous and intravenous routes. Moderately toxic by ingestion. When heated to decomposition it emits toxic fumes of F^-, SO_x, NO_x, and HCl.

FMP000 CAS:146-56-5 HR: 3
FLUPHENAZINE DIHYDROCHLORIDE
mf: $C_{22}H_{26}F_3N_3OS{\cdot}2ClH$ mw: 510.49

PROP: Crystals from EtOH. Mp: 235–237°.

SYNS: A 4077 □ FLUPHENAZINE HYDROCHLORIDE □ 10-(3-(4-(2-HYDROXYETHYL)PIPERAZINYL)PROPYL)-2-TRIFLUOROMETHYLPHENOTHIAZINE DIHYDROCHLORIDE □ MODITEN □ PERMITIL HYDROCHLORIDE □ PROLIXIN □ SQUIBB 4918 □ 4-(3-(2-(TRIFLUOROMETHYL)PHENOTHIAZIN-10-YL)PROPYL)-1-PIPERAZINEETHANOL, DIHYDROCHLORIDE

TOXICITY DATA WITH REFERENCE
mnt-mus orl 55 mg/kg MUREAV 89,237,81
cyt-mus-ipr 21,250 μg/kg IRLCD7 9,701,81
orl-man TDLo:429 μg/kg (male 2D pre):REP AEMED3 14,600,85
orl-hmn TDLo:2800 μg/kg/2W-C:MSK BMJOAE 2,1071,66
orl-mus LD50:220 mg/kg 27ZQAG -,24,72
ipr-mus LD50:89 mg/kg 27ZQAG -,24,72
ivn-mus LD50:56 mg/kg CSLNX* NX#00179
ipr-gpg LD50:299 mg/kg PHARAT 38,749,83

SAFETY PROFILE: Poison by ingestion, intraperitoneal, and intravenous routes. Moderately toxic by subcutaneous route. Human systemic effects by ingestion: musculo-skeletal changes. Mutation data reported. When heated to decomposition it emits very toxic fumes of HCl, SO_x, NO_x, and F^-.

FMP100 HR: 2
FLUPIRTINE MALEATE
mf: $C_{15}H_{17}FN_4O_2{\cdot}C_4H_4O_4$ mw: 420.44

SYNS: (2-AMINO-6-(((4-FLUOROPHENYL)METHYL)AMINO)-3-PYRIDINYL)CARBAMIC ACID ETHYL ESTER MALEATE □ D 9998 MALEATE □ ETHYL-N-(2-AMINO-6-(4-FLUOROPHENYLMETHYLAMINO)PYRIDIN-3-YL)CARBAMATE MALEATE □ ETHYL-N-(2-AMINO-6-(4-FLUOR-PHENYLMETHYLAMINO)PYRIDIN-3-YL)CARBAMAT MALEAT (GERMAN) □ FLUPIRTIN-MALEAT (GERMAN)

TOXICITY DATA WITH REFERENCE
orl-rat LD50:1660 mg/kg ARZNAD 35,30,85
orl-mus LD50:603 mg/kg ARZNAD 35,30,85
scu-mus LD50:432 mg/kg ARZNAD 35,30,85

SAFETY PROFILE: Moderately toxic by ingestion and subcutaneous routes. When heated to decomposition it emits toxic fumes of F^- and NO_x. See also ESTERS and CARBAMATES.

FMQ000 CAS:17617-23-1 HR: 3
FLURAZEPAM
mf: $C_{21}H_{23}ClFN_3O$ mw: 387.92

PROP: White rods from ether-petr ether. Mp: 77–82°.

SYNS: 7-CHLORO-1-(2-(DIETHYLAMINO)ETHYL)-5-(2-FLUOROPHENYL)-1H-1,4-BENZODIAZEPIN-2(3H)-ONE □ FELMANE □ NOCTOSOM □ Ro-5-6901/3 □ STAURODERM

TOXICITY DATA WITH REFERENCE
orl-mus TDLo:504 mg/kg (1-21D preg):REP PLRCAT 9,325,77
orl-rat LD50:980 mg/kg OYYAA2 14,637,77
ipr-rat LD50:600 mg/kg OYYAA2 14,637,77
ivn-rat LD50:38,700 μg/kg IYKEDH 6,530,75
orl-mus LD50:500 mg/kg THERAP 26,439,71
ipr-mus LD50:540 mg/kg OYYAA2 14,637,77
scu-mus LD50:3844 mg/kg OYYAA2 14,637,77
ivn-mus LD50:59,100 μg/kg IYKEDH 6,530,75

SAFETY PROFILE: Poison by intravenous route. Moderately toxic by ingestion and subcutaneous routes. Experimental reproductive effects. Caution. May be habit-forming. This is a controlled substance (depressant) listed in the U.S. Code of Federal Regulations, Title 21 Part 1308.14. When heated to decomposition it emits very toxic fumes of Cl^-, F^-, and NO_x. See also DIAZEPAM.

FMQ100 CAS:36105-20-1 HR: 3
FLURAZEPAM MONOHYDROCHLORIDE
mf: $C_{21}H_{23}ClFN_3O{\cdot}ClH$ mw: 424.38

TOXICITY DATA WITH REFERENCE
orl-rat TDLo:330 mg/kg (30D pre):REP KSRNAM 7,2820,73
orl-rat TDLo:32,400 mg/kg (90D male):REP KSRNAM 7,2820,73
orl-rat LD50:978 mg/kg KSRNAM 7,2820,73
ipr-rat LD50:179 mg/kg KSRNAM 7,2820,73
scu-rat LD50:859 mg/kg KSRNAM 7,2820,73
ivn-rat LD50:40 mg/kg KSRNAM 7,2820,73
orl-mus LD50:684 mg/kg KSRNAM 7,2820,73
ipr-mus LD50:205 mg/kg OYYAA2 7,381,73
scu-mus LD50:598 mg/kg KSRNAM 7,2820,73
ivn-mus LD50:58 mg/kg OYYAA2 7,381,73
orl-rbt LD50:460 mg/kg KSRNAM 7,2820,73

SAFETY PROFILE: Poison by intravenous and intraperitoneal routes. Moderately toxic by ingestion and subcutaneous routes. Experimental reproductive effects. When heated to decomposition it emits toxic fumes of F^-, NO_x, and HCl. See also DIAZEPAM.

FMQ200 CAS:59756-60-4 HR: 1
FLURIDONE
mf: $C_{19}H_{14}F_3NO$ mw: 329.34

SYNS: 1-METHYL-3-PHENYL-5-(3-(TRIFLUOROMETHYL)PHENYL) 4(1H)-PYRIDINONE □ PRIDE □ 4(1H)-PYRIDINONE, 1-METHYL-3-PHENYL-5-(3-(TRIFLUOROMETHYL)PHENYL)- □ SONAR

TOXICITY DATA WITH REFERENCE
orl-rat LD50:>10 g/kg PEMNDP 9,418,91
orl-mus LD50:>10 g/kg PEMNDP 9,418,91

F

orl-dog LD50:>500 mg/kg PEMNDP 9,418,91
orl-cat LD50:>250 mg/kg PEMNDP 9,418,91
skn-rbt LD50:>500 mg/kg PEMNDP 9,418,91
orl-qal LD50:>2 g/kg PEMNDP 9,418,91

CONSENSUS REPORTS: Reported in EPA TSCA Inventory.

SAFETY PROFILE: Low toxicity by ingestion. When heated to decomposition it emits toxic vapors of NO_x, F^-, and Cl^-.

FMR050 **CAS:13311-84-7** **HR: 3**
FLUTAMIDE
mf: $C_{11}H_{11}F_3N_2O_3$ mw: 276.24

PROP: Crystals from benzene. Mp: 111.5–112.5°.

SYNS: DROGENIL ☐ FLUGERIL ☐ FUGEREL ☐ 2-METHYL-N-(4-NITRO-3-(TRIFLUOROMETHYL)PHENYL)PROPANAMIDE (9CI) ☐ NIFTHOLIDE ☐ NIFTOLIDE ☐ 4'-NITRO-3'-TRIFLUOROMETHYLISOBUTYRANILIDE ☐ SCH ☐ SCH 13521 ☐ SEBATROL ☐ α,α,α-TRIFLUORO-2-METHYL-4'-NITRO-m-PROPIONOTOLUIDIDE

TOXICITY DATA WITH REFERENCE
dni-rat:lvr 50 µmol/L CRNGDP 13,373,92
ims-rat TDLo:65 mg/kg (female 10-22D post):REP ENDOAO 103,1702,78
scu-rat TDLo:70 mg/kg (male 7D pre):REP IJEBA6 15,788,77
orl-rat TDLo:150 mg/kg (male 30D pre):REP IJEBA6 25,81,87
scu-rat TDLo:525 mg/kg (female 14-20D post):TER BIREBV 20(Suppl 1),116A,79
orl-rat TDLo:1050 mg/kg (14-20D preg):TER AAMMAU 64,27,75
orl-rat TDLo:2730 mg/kg/1Y-I:ETA OYYAA2 45,141,93
orl-man TDLo:310 mg/kg/31D-I AJPSAO 143,1498,86
orl-rat LD50:787 mg/kg OYYAA2 45,135,93
ipr-rat LD50:289 mg/kg OYYAA2 45,135,93

CONSENSUS REPORTS: EPA Genetic Toxicology Program.

SAFETY PROFILE: A poison by intraperitoneal route. Moderately toxic by ingestion. An experimental teratogen. Experimental reproductive effects. Questional carcinogen with tumorigenic data. Mutation data reported. When heated to decomposition it emits toxic fumes of F^- and NO_x.

FMR075 **CAS:27060-91-9** **HR: 2**
FLUTAZOLAM
mf: $C_{19}H_{18}ClFN_2O_3$ mw: 376.84

PROP: Crystals from CH_2Cl_2/hexane: mp: 183–184°. White prisms from toluene: mp: 142–147°. Freely sol in chloroform, ethanol; moderately sol in acetone, benzene, methanol, practically insol in water.

SYN: MS-4101

TOXICITY DATA WITH REFERENCE
orl-rat TDLo:270 mg/kg (female 17-22D post):REP OYYAA2 16,395,78
orl-rat TDLo:1050 mg/kg (14D pre/1-7D preg):REP OYYAA2 16,395,78
orl-rbt TDLo:13 g/kg (female 6-18D post):REP OYYAA2 16,709,78
orl-dog TDLo:1820 mg/kg (female 26W pre):REP OYYAA2 15,1081,78
orl-rat TDLo:1500 mg/kg (female 17-22D post):REP OYYAA2 16,395,78
orl-rbt TDLo:875 mg/kg (male 35D pre):REP OYYAA2 16,693,78
orl-rat TDLo:80 mg/kg (female 8-15D post):TER YACHDS 6,1692,78
ipr-rat LD50:4050 mg/kg YACHDS 6,1997,78
orl-mus LD50:1910 mg/kg YACHDS 6,2335,78
ipr-mus LD50:1680 mg/kg YACHDS 6,1997,78

SAFETY PROFILE: Moderately toxic by ingestion and intraperitoneal routes. An experimental teratogen. Experimental reproductive effects. When heated to decomposition it emits toxic fumes of F^-, Cl^-, and NO_x.

FMR100 **CAS:25967-29-7** **HR: 2**
FLUTOPRAZEPAM
mf: $C_{19}H_{16}ClFN_2O$ mw: 342.82

PROP: Crystals. Mp: 86–88°.

SYNS: 7-CHLORO-1-CYCLOPROPYLMETHYL-1,3-DIHYDRO-5-(2-FLUOROPHENYL)-2H-1,4-BENZODIAZEPIN-2-ONE ☐ KB-509 ☐ RESTAS

TOXICITY DATA WITH REFERENCE
orl-rat TDLo:110 mg/kg (female 7-17D post):REP OYYAA2 21,13,81
orl-rat TDLo:2700 mg/kg (female 17-22D post):REP OYYAA2 21,27,81
orl-rat TDLo:9800 mg/kg (female 14D pre):REP OYYAA2 21,1,81
orl-rat TDLo:3640 mg/kg (female 26W pre):REP OYYAA2 20,1061,80
orl-dog TDLo:2275 mg/kg (female 91D pre):REP OYYAA2 20,1105,80
orl-rat TDLo:10,920 mg/kg (male 91D pre):REP OYYAA2 20,1061,80
orl-dog TDLo:91 mg/kg (male 91D pre):REP OYYAA2 20,1105,80
orl-rat TDLo:11 mg/kg (female 7-17D post):TER OYYAA2 21,13,81
orl-rat LD50:10,060 mg/kg OYYAA2 20,1055,80
ipr-rat LD50:2230 mg/kg OYYAA2 20,1055,80
orl-mus LD50:2430 mg/kg OYYAA2 20,1055,80
ipr-mus LD50:2110 mg/kg OYYAA2 20,1055,80
orl-rbt LD50:1000 mg/kg OYYAA2 20,1055,80

SAFETY PROFILE: Moderately toxic by ingestion and intraperitoneal routes. Experimental teratogenic and reproductive effects. When heated to decomposition it emits toxic fumes of F^-, Cl^-, and NO_x. See also DIAZEPAM.

FMR300 **CAS:63516-07-4** **HR: 3**
FLUTROPIUM BROMIDE HYDRATE
mf: $C_{24}H_{29}FNO_3{\cdot}Br{\cdot}H_2O$ mw: 496.47

SYNS: 8-AZONIABICYCLO(3.2.1)OCTANE, 8-(2-FLUOROETHYL)-3-((HYDROXYDIPHENYLACETYL)OXY)-8-METHYL-, BROMIDE, (endo, syn)-, MONOHYDRATE □ Ba 598 BROMIDE HYDRATE □ (8R)-8-(2-FLUOROETHYL)-3-α-HYDROXY-1-α-H,5-α-H-TROPANIUM BROMIDE BENZILATE H2O

TOXICITY DATA WITH REFERENCE
orl-rat TDLo: 13,500 mg/kg (female 17-22D post): REP KSRNAM 20,8174,86
orl-rat TDLo: 5500 mg/kg (female 7-17D post): TER KSRNAM 20,8143,86
orl-rat TDLo: 11 g/kg (female 2W pre): REP KSRNAM 20,8132,86
orl-rat TDLo: 31,500 mg/kg (male 9W pre): REP KSRNAM 20,8132,86
orl-rat LD50: 2900 mg/kg KSRNAM 20,8123,86
ipr-rat LD50: 77 mg/kg KSRNAM 20,8123,86
scu-rat LD50: 615 mg/kg KSRNAM 20,8123,86
ivn-rat LD50: 12,500 μg/kg KSRNAM 20,8123,86
orl-mus LD50: 930 mg/kg KSRNAM 20,8123,86
ipr-mus LD50: 53 mg/kg KSRNAM 20,8123,86
scu-mus LD50: 228 mg/kg KSRNAM 20,8123,86
ivn-mus LD50: 12,500 μg/kg KSRNAM 20,8123,86

SAFETY PROFILE: Poison by intravenous and intraperitoneal routes. Moderately toxic by ingestion and subcutaneous routes. An experimental teratogen. Experimental reproductive effects. When heated to decomposition it emits toxic fumes of NO_x, Br^-, and Cl^-.

FMR500 CAS:1391-62-4 HR: 2
FLUVOMYCIN

SYNS: EFSIOMYCIN □ RIOMYCIN □ VIVICIL

TOXICITY DATA WITH REFERENCE
scu-mus LD50: 1250 mg/kg ANTCAO 3,765,53
ivn-mus LD50: 1300 mg/kg ANTCAO 3,765,53
ims-mus LD50: 750 mg/kg ANTCAO 3,765,53

SAFETY PROFILE: Moderately toxic by subcutaneous, intravenous, and intramuscular routes.

FMR700 CAS:76050-49-2 HR: 2
FOGARD

SYN: FOGARD S

TOXICITY DATA WITH REFERENCE
ipr-mus TDLo: 2600 μg/kg/39D-I: ETA PATHAB 73,707,81
scu-mus TDLo: 2600 μg/kg/39D-I: ETA PATHAB 73,707,81

SAFETY PROFILE: Questionable carcinogen with experimental tumorigenic data. When heated to decomposition it emits acrid smoke and irritating fumes.

FMS000 CAS:8031-00-3 HR: 1
FOIN ABSOLUTE

PROP: Found in several common fodder grasses, contains coumarin (FCTXAV 14,659,76).

SYNS: FOIN COUPE □ HAY ABSOLUTE

TOXICITY DATA WITH REFERENCE
skn-rbt 500 mg/24H MLD FCTXAV 14,777,76

CONSENSUS REPORTS: Reported in EPA TSCA Inventory.

SAFETY PROFILE: A skin irritant. When heated to decomposition it emits acrid smoke and irritating fumes. See also COUMARIN.

FMS875 CAS:370-14-9 HR: 3
FOLEDRIN
mf: $C_{10}H_{15}NO$ mw: 165.26

PROP: Crystals from methanol; acrid, burning taste. Mp: 162–163°. Sltly sol in water. Sol in alc, ether; readily sol in dil acids.

SYNS: EPIFEN □ p-HYDROXY-N,α-DIMETHYLPHENETHYLAMINE □ p-HYDROXYMETHAMPHETAMINE □ p-HYDROXY-N-METHYLBENZEDRINE □ β-(p-HYDROXYPHENYL)ISOPROPYLMETHYLAMINE □ α-(p-HYDROXYPHENYL)-β-METHYLAMINOPROPANE □ 1-(p-HYDROXYPHENYL)-2-METHYLAMINOPROPANE □ ISODRINE □ ISODRINUM □ KNOLL H75 □ p-(2-METHYLAMINOPROPYL)PHENOL □ N-METHYLPAREDRINE □ 1-(p-OXYPHENYL)-2-METHYLAMINOPROPAN (GERMAN) □ α-(p-OXYPHENYL)-β-METHYL AMINOPROPANE □ PAREDRINOL □ PHOLEDRINE □ PHOLETONE □ PRESSITAN □ PROMETHIN □ PROMETIN □ PULSOTYL □ STIMATONE □ SYMPROPAMIN □ SYNCORDAN □ TERAPINYL □ VARITOL □ VERITAIN □ VERITOL

TOXICITY DATA WITH REFERENCE
ipr-rat LD50: 100 mg/kg AEPPAE 195,647,40
scu-rat LD50: 400 mg/kg AIPTAK 159,442,66
ipr-mus LDLo: 100 mg/kg AEPPAE 195,647,40

SAFETY PROFILE: Poison by subcutaneous and intraperitoneal routes. When heated to decomposition it emits toxic fumes of NO_x. See also AMINES.

FMT000 CAS:59-30-3 HR: 3
FOLIC ACID
mf: $C_{19}H_{19}N_7O_6$ mw: 441.45

PROP: A member of the vitamin B complex. Odorless orange-yellow needles or platelets from H_2O at pH 3. Sol in dilute alkali hydroxide and carbonate solns; sltly sol in water; insol in lipid solvents, acetone, alc, chloroform, ether.

SYNS: 1-N-(p-(((-2-AMINO-4-HYDROXY-6-PTERIDINYL)METHYL)AMINO)BENZOYL)GLUTAMIC ACID □ FOLACIN □ FOLATE □ FOLCYSTEINE □ NSC 3073 □ PTEGLU □ PTEROYLGLUTAMIC ACID □ PTEROYL-l-GLUTAMIC ACID □ PTEROYLMONOGLUTAMIC ACID □ PTEROYL-l-MONOGLUTAMIC ACID □ USAF CB-13 □ VITAMIN Bc □ VITAMIN M

TOXICITY DATA WITH REFERENCE
dns-rat-ipr 150 mg/kg CNREA8 29,136,69
dns-mus-ipr 250 mg/kg CNREA8 29,136,69
par-rat TDLo: 150 mg/kg (10D preg): TER FEPRA7 23,292,64
orl-mus LD50: 10 g/kg TXAPA9 23,537,72
ipr-mus LD50: 85 mg/kg EXPEAM 41,72,85
scu-mus LDLo: 200 mg/kg PSEBAA 71,544,49
ivn-mus LD50: 282 mg/kg TXAPA9 23,537,72

CONSENSUS REPORTS: Reported in EPA TSCA Inventory.

SAFETY PROFILE: Poison by intraperitoneal and intravenous routes. Experimental teratogenic effects. Mutation data reported. When heated to decomposition it emits toxic fumes of NO_x.

F

FMU000 CAS:24600-36-0 HR: 3
FOMINOBEN HYDROCHLORIDE
mf: $C_{21}H_{24}ClN_3O_3 \cdot ClH$ mw: 438.39

PROP: A solid. Mp: 206–208° (decomp).

SYNS: 3′-CHLORO-α-(METHYL((MORPHOLINOCARBONYL)METHYL)AMINO)-o-BENZOTOLUIDIDE HYDROCHLORIDE ☐ 3′-CHLORO-α-(N-METHYL-N-((MORPHOLINOCARBONYL)METHYL)AMINOMETHYL) BENZANILIDE HYDROCHLORIDE ☐ 3′-CHLORO-2′-(N-METHYL-N-((MORPHOLINOCARBONYL)METHYL)AMINOMETHYL)BENZANILIDE HYDROCHLORIDE ☐ NOLEPTAN ☐ OLEPTAN ☐ PB 89 CHLORIDE ☐ PB-89 HYDROCHLORIDE

TOXICITY DATA WITH REFERENCE
orl-rat TDLo:16,500 mg/kg (7-17D preg):REP IYKEDH 9,724,78
orl-rat TDLo:90 g/kg (9D pre):REP IYKEDH 9,736,78
ipr-rat LD50:1000 mg/kg IYKEDH 14,297,83
scu-rat LD50:11 g/kg IYKEDH 14,297,83
ivn-rat LD50:57 mg/kg IYKEDH 14,297,83
orl-mus LD50:2200 mg/kg ARZNAD 23,296,73
ipr-mus LD50:630 mg/kg ARZNAD 23,296,73
scu-mus LD50:1770 mg/kg OYYAA2 8,1491,74
ivn-mus LD50:130 mg/kg IYKEDH 14,297,83
orl-dog LD50:3300 mg/kg IYKEDH 14,297,83

SAFETY PROFILE: Poison by intravenous route. Moderately toxic by ingestion, subcutaneous, and intraperitoneal routes. Experimental teratogenic and reproductive effects. An antitussive. When heated to decomposition it emits very toxic fumes of HCl and NO_x.

FMU039 CAS:13115-40-7 HR: 3
FONAZINE MESYLATE
mf: $C_{19}H_{23}N_3O_2S_2 \cdot CH_4O_3S$ mw: 487.70

SYNS: DIMETHIOTAZINE ☐ DIMETHOTHIAZINE MESYLATE ☐ DIMETHOTHIAZINE METHANESULFONATE ☐ IL-6302 MESYLATE ☐ MIGRISTENE ☐ PROMAQUID ☐ RP 8599 ☐ RP 8599 MESYLATE

TOXICITY DATA WITH REFERENCE
orl-rbt TDLo:135 mg/kg (8-16D preg):REP OYYAA2 4,373,70
orl-rbt TDLo:540 mg/kg (8-16D preg):TER OYYAA2 4,373,70
orl-mus LD50:740 mg/kg YAKUD5 22,1953,80
ipr-mus LD50:190 mg/kg YAKUD5 22,1953,80
scu-mus LD50:473 mg/kg YAKUD5 22,1953,80
ivn-mus LD50:61,500 μg/kg KSRNAM 5,325,71

SAFETY PROFILE: Poison by intravenous route. Moderately toxic by ingestion. An experimental teratogen. Experimental reproductive effects. When heated to decomposition it emits toxic fumes of SO_x and NO_x. See also SULFONATES.

FMU045 CAS:944-22-9 HR: 3
FONOFOS
mf: $C_{10}H_{15}OPS_2$ mw: 246.34

SYNS: O-AETHYL-S-PHENYL-AETHYL-DITHIOPHOSPHONAT (GERMAN) ☐ DIFONATE ☐ DYFONATE ☐ DYPHONATE ☐ ENT 25,796 ☐ O-ETHYL-S-PHENYL ETHYLDITHIOPHOSPHONATE ☐ O-ETHYL-S-PHENYL ETHYLPHOSPHONODITHIOATE ☐ N 2790 ☐ STAUFFER N 2790

TOXICITY DATA WITH REFERENCE
orl-rat LD50:3 mg/kg WRPCA2 9,119,70
ihl-rat LC50:1900 mg/m³/1H 85JFAN A214,83
skn-rat LD50:147 mg/kg WRPCA2 9,119,70
skn-rbt LD50:25 mg/kg FMCHA2 -,D121,80
skn-gpg LD50:278 mg/kg 28ZEAL 5,118,76
orl-pgn LD50:13,300 μg/kg ASTTA8 (680),157,79
orl-qal LD50:12 mg/kg EESADV 8,551,84

CONSENSUS REPORTS: EPA Genetic Toxicology Program. EPA Extremely Hazardous Substances List.

OSHA PEL: TWA 0.1 mg/m³ (skin)
ACGIH TLV: TWA 0.1 mg/m³ (skin)

SAFETY PROFILE: Poison by ingestion and skin contact. An insecticide. When heated to decomposition it emits very toxic fumes of PO_x and SO_x.

FMU059 CAS:2650-18-2 HR: 3
FOOD BLUE 1
mf: $C_{37}H_{36}N_2O_9S_3 \cdot 2H_3N$ mw: 783.01

PROP: Dark blue crystals or powder. Sol in H_2O.

SYNS: ACID BLUE 9 ☐ ACILAN TURQUOISE BLUE AE ☐ A.F. BLUE No. 1 ☐ AIZEN BRILLIANT BLUE FCF ☐ ALPHAZURINE ☐ AMACID BLUE FG CONC ☐ BLEU BRILLIANT FCF ☐ 11388 BLUE ☐ BRILLIANT BLUE ☐ BUCACID AZURE BLUE ☐ CALCOCID BLUE EG ☐ C.I. 671 ☐ C.I. 42090 ☐ C.I. ACID BLUE 9, DIAMMONIUM SALT ☐ C.I. DIRECT BROWN 78, DIAMMONIUM SALT ☐ C.I. FOOD BLUE 2 ☐ D&C BLUE No. 4 ☐ DISULPHINE LAKE BLUE EG ☐ EDICOL SUPRA BLUE E6 ☐ ERIOGLAUCINE ☐ ERIOSKY BLUE ☐ FENAZO BLUE XR ☐ HIDACID AZURE BLUE ☐ H.K. FORMULA No. K. 7117 ☐ KITON PURE BLUE L ☐ MAPLE BRILLIANT BLUE FCF ☐ NEPTUNE BLUE BRA CONCENTRATION ☐ PATENT BLUE AE ☐ PEACOCK BLUE X-1756 ☐ SCHULTZ No. 770 ☐ TRIANTINE LIGHT BROWN 3RN ☐ XYLENE BLUE VSG

TOXICITY DATA WITH REFERENCE
dns-rat:lvr 200 μmol/L ENMUDM 7,101,85
dns-rat-orl 500 mg/kg ENMUDM 7,101,85
cyt-ham:fbr 5 g/L FCTOD7 22,623,84
scu-rat TDLo:10 g/kg/77W-I:NEO TXAPA9 8,29,66
ivn-hmn LDLo:33 μg/kg:CNS,PUL 34ZIAG -,87,69

CONSENSUS REPORTS: Community Right-To-Know List. Reported in EPA TSCA Inventory. EPA Genetic Toxicology Program.

SAFETY PROFILE: Human poison by intravenous route. Human systemic effects by intravenous route: muscle contractions or spasticity and dyspnea. Questionable carcinogen with experimental neoplastigenic and tumorigenic data. Mutation data reported. When heated to decomposition it emits very toxic fumes of NH_3, NO_x, and SO_x.

FMU070 CAS:3761-53-3 HR: 2
FOOD RED No. 101
mf: $C_{18}H_{14}N_2O_7S_2 \cdot 2Na$ mw: 480.44

PROP: Dark red crystals or powder. Sol in H_2O; slty sol in EtOH and Me_2CO; insol in Et_2O and C_6H_6.

SYNS: ACETACID RED J ☐ ACIDAL PONCEAU G ☐ ACID LEATHER RED KPR ☐ ACID PONCEAU R ☐ ACID RED 26 ☐ ACID SCARLET ☐ ACILAN PONCEAU RRL ☐ AHCOCID FAST SCARLET R ☐ AIZEN PON-

CEAU RH □ AMACID LAKE SCARLET 2R □ BRILLIANT PONCEAU G □ CALCOCID 2RIL □ CALCOLAKE SCARLET 2R □ CERTICOL PONCEAU MXS □ CERVEN KYSELA 26 □ C.I. 79 □ C.I. 16150 □ C.I. ACID RED 26 □ C.I. ACID RED 26, DISODIUM SALT □ C.I. FOOD RED 5 □ COLACID PONCEAU SPECIAL □ D&C RED No. 5 □ 4-((2,4-DIMETHYLPHENYL)AZO)-3-HYDROXY-2,7-NAPHTHALENEDISULFONIC ACID, DISODIUM SALT □ 4-((2,4-DIMETHYLPHENYL)AZO)-3-HYDROXY-2,7-NAPHTHALENEDISULPHONIC ACID, DISODIUM SALT □ DISODIUM (2,4-DIMETHYLPHENYLAZO)-2-HYDROXYNAPHTHALENE-3,6-DISULFONATE □ DISODIUM (2,4-DIMETHYLPHENYLAZO)-2-HYDROXYNAPHTHALENE-3,6-DISULPHONATE □ DISODIUM SALT of 1-(2,4-XYLYLAZO)-2-NAPHTHOL-3,6-DISULFONIC ACID □ DISODIUM SALT of 1-(2,4-XYLYLAZO)-2-NAPHTHOL-3,6-DISULPHONIC ACID □ EDICOL PONCEAU RS □ FENAZO SCARLET 2R □ HEXACOL PONCEAU MX □ HIDACID SCARLET 2R □ 3-HYDROXY-4-(2,4-XYLYLAZO)-3,7-NAPHTHALENEDISULFONIC ACID, DISODIUM SALT □ 3-HYDROXY-4-(2,4-XYLYLAZO)-3,7-NAPHTHALENEDISULPHONIC ACID, DISODIUM SALT □ JAVA PONCEAU 2R □ KITON PONCEAU R □ LAKE PONCEAU □ NAPHTHALENE LAKE SCARLET R □ NEKLACID RED RR □ NEW PONCEAU 4R □ PAPER RED HRR □ PIGMENT PONCEAU R □ PONCEAU BNA □ PONCEAU R (BIOLOGICAL STAIN) □ PONCEAU XYLIDINE (BIOLOGICAL STAIN) □ 1695 RED □ RED R □ SCARLET R □ SCHULTZ No. 95 □ TERTRACID PONCEAU 2R □ XYLIDINE PONCEAU □ 1-XYLYLAZO-2-NAPHTHOL-3,6-DISULFONIC ACID, DISODIUM SALT □ 1-XYLYLAZO-2-NAPHTHOL-3,6-DISULPHONIC ACID, DISODIUM SALT □ 1-(2,4-XYLYLAZO)-2-NAPHTHOL-3,6-DISULPHONIC ACID, DISODIUM SALT

TOXICITY DATA WITH REFERENCE
mma-sat 100 µg/plate CANCAR 49,1970,82
oth-esc 300 µmol/L SKEZAP 12,298,71
orl-mus TDLo:136 g/kg/81W-C:CAR FCTXAV 6,591,68
orl-mus TD:35 g/kg/52W:ETA BJCAAI 10,653,56
ipr-mus LD50:2000 mg/kg FCTXAV 4,375,66

CONSENSUS REPORTS: IARC Cancer Review: Group 3 IMEMDT 7,56,87; Animal Sufficient Evidence IMEMDT 8,189,75. Reported in EPA TSCA Inventory.

SAFETY PROFILE: Moderately toxic by intraperitoneal route. Questionable carcinogen with experimental carcinogenic and tumorigenic data. Mutation data reported. When heated to decomposition it emits toxic fumes of NO_x and SO_x.

FMU080 CAS:2611-82-7 HR: 2
FOOD RED No. 102
mf: $C_{20}H_{14}N_2O_{10}S_3{\bullet}3Na$ mw: 607.51

SYNS: ACIDAL BRIGHT PONCEAU 3R □ ACID BRILLIANT SCARLET 3R □ ACID PONCEAU 4R □ ACID RED 18 □ ACID SCARLET 3R □ ACILAN SCARLET V3R □ AIZEN BRILLIANT SCARLET 3RH □ ATUL ACID SCARLET 3R □ BRILLIANT PONCEAU 3R □ BRILLIANT SCARLET □ BUCACID BRILLIANT SCARLET 3R □ CALCOCID BRILLIANT SCARLET 3RN □ CERTICOL PONCEAU 4RS □ CERVEN KOSENILOVA A □ CILEFA PONCEAU 4R □ COCCINE □ COCHENILLEROT A □ COCHINEAL RED A □ COLACID PONCEAU 4R □ C.I. 185 □ C.I. 16255 □ C.I. ACID RED 18 □ C.I. FOOD RED 7 □ CRIMSON SX □ CUROL BRIGHT RED 4R □ DAISHIKI BRILLIANT SCARLET 3R □ EDICOL SUPRA PONCEAU 4R □ EUROCERT COCHINEAL RED A □ FENAZO SCARLET 3R □ FOOD RED 6 □ FOOD RED 7 □ HD PONCEAU 4R □ HEXACOL PONCEAU 4R □ HIDACID FAST SCARLET 3R □ HISPACID BRILLIANT SCARLET 3RF □ JAVA SCARLET 3R □ KAYAKU ACID BRILLIANT SCARLET 3R □ KITON SCARLET 4R □ KOCHINEAL RED A FOR FOOD □ 1,3-NAPHTHALENEDISULFONIC ACID, 7-HYDROXY-8-((4-SULFO-1-NAPHTHYL)AZO)-, TRISODIUM SALT □ NAPHTHALENE INK SCARLET 4R □ NEKLACID RED 3R □ NEUCOCCIN □ NEW COCCIN □ PONCEAU 4R □ PONCEAU 4R ALUMINUM LAKE □ PONTACYL SCARLET RR □ PURPLE RED □ ROUGE de COCHENILLE A □ SAN-EI BRILLIANT SCARLET 3R □ STRAWBERRY RED A GEIGY □ SYMULON ACID BRILLIANT SCARLET 3R □ TAKAOKA BRILLIANT SCARLET 3R □ VICTORIA SCARLET 3R

TOXICITY DATA WITH REFERENCE
cyt-ham:fbr 1 g/L FCTOD7 22,623,84
orl-rat TDLo:428 g/kg/64W-I:ETA VPITAR 29,61,70
ipr-rat LD50:600 mg/kg FCTXAV 5,187,67
ivn-rat LD50:1 g/kg SCPHA4 47,39,79
unr-rat LD50:2 g/kg NAIZAM 30,179,79
orl-mus LDLo:8 g/kg SCPHA4 47,39,79
ipr-mus LD50:1600 mg/kg FCTXAV 5,187,67

CONSENSUS REPORTS: Reported in EPA TSCA Inventory.

SAFETY PROFILE: Moderately toxic by intraperitoneal route. Questionable carcinogen with experimental tumorigenic data. Mutation data reported. When heated to decomposition it emits toxic fumes of NO_x and SO_x.

FMU200 HR: 1
FOOL'S PARSLEY

PROP: A carrot-like plant 8 to 24 inches tall with leaves like parsley but with a garlic smell. It is native to Europe but grows wild in the northern United States and Canada.

SYNS: AETHUSA CYNAPIUM □ DOG PARSLEY □ DOG POISON □ FALSE PARSLEY □ FOOL'S CICELY □ LESSER HEMLOCK

SAFETY PROFILE: The whole plant contains a toxic mixture of unsaturated aliphatic alcohols, similar to cicutoxin, and traces of coniine. Human systemic effects by ingestion: nausea, vomiting, sweating, and headache. See also CICUTOXIN and 2-PROPYLPIPERIDINE.

FMU225 CAS:22514-23-4 HR: 3
FOPIRTOLINE HYDROCHLORIDE
mf: $C_{11}H_{15}ClN_2OS{\bullet}ClH$ mw: 295.25

SYNS: 4-(2-((6-CHLORO-2-PYRIDYL)THIO)ETHYL)MORPHOLINE MONOHYDROCHLORIDE □ D-1126 □ FOPIRTOLINA (SPANISH)

TOXICITY DATA WITH REFERENCE
orl-rat LD50:768 mg/kg DRFUD4 7,806,82
orl-mus LD50:1513 mg/kg DRFUD4 7,806,82
ivn-mus LD50:140 mg/kg DRFUD4 7,806,82
orl-dog LD50:100 mg/kg DRFUD4 7,806,82

SAFETY PROFILE: Poison by ingestion and intravenous routes. When heated to decomposition it emits toxic fumes of SO_x, NO_x, and HCl.

FMU409 CAS:3614-69-5 HR: 3
FORHISTAL MALEATE
mf: $C_{20}H_{24}N_2{\bullet}C_4H_4O_4$ mw: 408.54

PROP: A solid. Mp: 159–161°.

SYNS: DIMETHINDENE MALEATE □ DIMETHINDEN MALEATE □ DIMETHPYRINDENE MALEATE □ N,N-DIMETHYL-3-(1-(2-PYRIDINYL) ETHYL)-1H-INDENE-2-ETHANAMINE (Z)-2-BUTENEDIOATE (1:1) □ DIMETINDENE MALEATE □ FENISTIL □ FENISTIL-RETARD □ FENOSTIL

TOXICITY DATA WITH REFERENCE
orl-rat LD50:618 mg/kg NIIRDN 6,344,82
ivn-rat LD50:27 mg/kg MEIEDD 10,468,83
orl-gpg LD50:880 mg/kg NIIRDN 6,344,82

SAFETY PROFILE: Poison by intravenous route. Moderately toxic by ingestion. When heated to decomposition it emits toxic fumes of NO_x.

FMV000 CAS:50-00-0 HR: 3
FORMALDEHYDE
DOT: UN 1198/UN 2209
mf: CH_2O mw: 30.03

PROP: Clear, water-white, very sltly acid gas or liquid; pungent odor. Pure formaldehyde is not available commercially because of its tendency to polymerize. It is sold as aqueous solns containing from 37 to 50% formaldehyde by weight and varying amounts of methanol. Some alcoholic solns are used industrially and the physical properties and hazards may be greatly influenced by the solvent. Lel: 7.0%, uel: 73.0%, autoign temp: 806°F, mp: −92, d: 1.083, bp: −21°, flash p: (37%, methanol-free): 185°F, flash p: (15%, methanol-free): 122°F. Sol in H_2O and most org solvs except pet ether.

SYNS: ALDEHYDE FORMIQUE (FRENCH) □ ALDEIDE FORMICA (ITALIAN) □ BFV □ FA □ FANNOFORM □ FORMALDEHYD (CZECH, POLISH) □ FORMALDEHYDE, solution (DOT) □ FORMA-LIN □ FORMALIN 40 □ FORMALIN (DOT) □ FORMALINA (ITALIAN) □ FORMALINE (GERMAN) □ FORMALIN-LOESUNGEN (GERMAN) □ FORMALITH □ FORMIC ALDEHYDE □ FORMOL □ FYDE □ HOCH □ IVALON □ KARSAN □ LYSOFORM □ METHANAL □ METHYL ALDEHYDE □ METHYLENE GLYCOL □ METHYLENE OXIDE □ MORBOCID □ NCI-C02799 □ OPLOSSINGEN (DUTCH) □ OXOMETHANE □ OXYMETHYLENE □ PARAFORM □ POLYOXYMETHYLENE GLYCOLS □ RCRA WASTE NUMBER U122 □ SUPERLYSOFORM

TOXICITY DATA WITH REFERENCE
skn-hmn 150 μg/3D-I MLD 85DKA8 -,127,77
eye-hmn 4 ppm/5M IAPWAR 4,79,61
eye-hmn 1 ppm/6M nse MLD AIHAAP 44,463,83
skn-rbt 2 mg/24H SEV 85JCAE-,264,86
skn-rbt 540 mg open MLD UCDS** 4/21/67
skn-rbt 50 mg/24H MOD TXAPA9 21,369,72
eye-rbt 750 μg/24H SEV 85JCAE-,264,86
eye-rbt 10 mg SEV TXAPA9 55,501,80
mma-sat 5 μL/plate BIMADU 6,129,85
dni-esc 5 mmol/L MUREAV 156,153,85
dnd-hmn:fbr 100 μmol/L ENMUDM 7,267,85
ihl-rat TCLo:50 μg/m³/4H (female 1–19D post):REP TPKVAL 12,78,71
ihl-rat TCLo:12 μg/m³/24h (female 1–22D post):REP HYSAAV 33(7-9),112,68
ihl-rat TCLo:12 μg/m³/24H (female 15D pre):REP HYSAAV 33(1-3),327,68
ihl-rat TCLo:1 mg/m³/24H (1–22D preg):TER HYSAAV 34(5),266,69
ipr-mus TDLo:240 mg/kg (female 7–14D post):TER TJADAB 30(1),34A,84
itt-rat TDLo:400 mg/kg (male 1D pre):REP FESTAS 24,884,73
ims-mus TDLo:259 mg/kg (female 11D post):REP ANREAK 142,479,62
ihl-rat TCLo:35 μg/m³/8H (male 60D pre):REP PRKHDK 4,101,79
orl-rat TDLo:200 mg/kg (1D male):REP TJADAB 26(3),14A,82
ipr-rat TDLo:80 mg/kg (male 10D pre):REP JRBED2 7,42,87
ipr-mus TDLo:240 mg/kg (female 7–14D post):TER TJADAB 28,37A,83
ipr-mus TDLo:160 mg/kg (female 7–14D post):TER TJADAB 30(1),34A,84
ihl-rat TCLo:14,300 ppb/6H/2Y-I:CAR CNREA8 43,4382,83
scu-rat TDLo:1170 mg/kg/65W-I:ETA GANNA2 45,451,54
ihl-mus TCLo:14,300 ppm/6H/2Y-I:ETA CNREA8 43,4382,83
ihl-rat TC:15 ppm/6H/78W-I:CAR CNREA8 49,3398,80
scu-rat TD:350 mg/kg/78W-I:ETA FAONAU 50A,77,72
ihl-rat TC:6 ppm/6H/2Y-I:ETA EVSRBT 25,353,82
ihl-rat TC:15 ppm/6H/86W-I:CAR TXAPA9 81,401,85
ihl-rat TC:14 ppm/6H/84W-I:CAR JJIND8 68,597,82
ihl-rat TC:18,750 μg/m³/2Y-I:ETA GISAAA 48(4),60,83
ihl-mus TC:15 ppm/6H/104W-I:ETA EVSRBT 25,353,82
ihl-rat TC:15 ppm/6H/2Y-I:CAR CIIT** DOCKET #10992,82
ihl-rat TC:5600 ppb/6H/2Y-I:ETA CNREA8 43,4382,83
ihl-rat TC:14,300 ppb/6H/2Y-I:ETA 50EXAK -,111,83
orl-wmn LDLo:108 mg/kg 29ZWAE -,328,68
ihl-hmn TCLo:17 mg/m³/30M:EYE,PUL JAMAAP 165,1908,57
ihl-man TCLo:300 μg/m³:NOSE,CNS GTPZAB 12(7),20,68
unr-man LDLo:477 mg/kg 85DCAI 2,73,70
orl-rat LD50:100 mg/kg FCTOD7 26,447,88
ihl-rat LC50:590 mg/m³ GISAAA 41(6),103,76
scu-rat LD50:420 mg/kg APTOA6 6,299,50
ivn-rat LD50:87 mg/kg AEPPAE 221,166,54
orl-mus LD50:42 mg/kg NTIS** AD-A125-539
ihl-mus LC50:400 mg/m³/2H 85GMAT -,69,82
ipr-mus LDLo:16 mg/kg TXAPA9 23,288,72
scu-mus LD50:300 mg/kg APTOA6 6,299,50
scu-dog LDLo:350 mg/kg IPSTB3 3,93,76
ihl-cat LCLo:400 mg/m³/2H 85GMAT -,69,82
skn-rbt LD50:270 mg/kg UCDS** 4/21/67
scu-rbt LDLo:240 mg/kg JAMAAP 62,984,14
orl-gpg LD50:260 mg/kg JIHTAB 23,259,41

CONSENSUS REPORTS: NTP 7th Annual Report on Carcinogens. IARC Cancer Review: Group 2A IMEMDT 7,211,87; Human Inadequate Evidence IMEMDT 29,345,82; Animal Sufficient Evidence IMEMDT 29,345,82. EPA Genetic Toxicology Program. Reported in EPA TSCA Inventory.

OSHA PEL: TWA 0.75 ppm; STEL 2 ppm
ACGIH TLV: TWA 1 ppm; Suspected Human Carcinogen (Proposed: CL 0.3 ppm; Suspected Human Carcinogen)
DFG MAK: 0.5 ppm (0.6 mg/m³); Suspected Carcinogen
NIOSH REL: (Formaldehyde) Limit to lowest feasible level
DOT CLASSIFICATION: 9; *Label:* None (UN 2209); DOT Class: 3; *Label:* Flammable Liquid (UN 1198)

SAFETY PROFILE: Confirmed carcinogen with experimental carcinogenic, tumorigenic, and teratogenic data. Human poison by ingestion. Experimental poison by ingestion, skin contact, inhalation, intravenous, intraperitoneal, and subcutaneous routes. Human systemic effects by inhalation: lacrimation, olfactory changes, aggression, and pulmonary changes. Experimental reproductive effects. Human mutation data reported. A human skin and eye irritant. If swallowed it causes violent vomiting and diarrhea that can lead to collapse. Frequent or prolonged exposure can cause hypersensitivity leading to contact dermatitis, possibly of an eczematoid nature. An air concentration of 20 ppm is quickly irritating to eyes. A common air contaminant.

Flammable liquid when exposed to heat or flame; can react vigorously with oxidizers. A moderate explosion hazard when exposed to heat or flame. The gas is a more dangerous fire hazard than the vapor. Should formaldehyde be involved in a fire, irritating gaseous formaldehyde may be evolved. When aqueous formaldehyde solutions are heated above their flash points, a potential for an explosion hazard exists. High formaldehyde concentration or methanol content lowers the flash point. Reacts with sodium hydroxide to yield formic acid and hydrogen. Reacts with NO_x at about 180°; the reaction becomes explosive. Also reacts violently with perchloric acid + aniline; performic acid; nitromethane; magnesium carbonate; H_2O_2. Moderately dangerous because of irritating vapor that may exist in toxic concentrations locally if storage tank is ruptured. To fight fire, stop flow of gas (for pure form); alcohol foam for 37% methanol-free form. When heated to decomposition it emits acrid smoke and fumes. See also ALDEHYDES.

For occupational chemical analysis use OSHA: #ID-102 or NIOSH: Formaldehyde (Oxazolidine), 2502; (Chromotropic Acid), 3500.

FMV100 **CAS:58567-11-6** **HR: 1**
FORMALDEHYDE CYCLODODECYL ETHYL ACETAL
mf: $C_{15}H_{30}O_2$ mw: 242.45

SYNS: BOISAMBRENE FORTE □ CYCLODODECANE, (ETHOXYMETHOXY)- □ (ETHOXYMETHOXY)CYCLODODECANE

TOXICITY DATA WITH REFERENCE
orl-rat LD50:>5 g/kg FCTOD7 26,325,88
skn-rbt LD50:>5 g/kg FCTOD7 26,325,88

CONSENSUS REPORTS: Reported in EPA TSCA Inventory.

SAFETY PROFILE: Low toxicity by ingestion and skin contact. When heated to decomposition it emits acrid smoke and irritating vapors.

FMV200 **CAS:42604-12-6** **HR: 1**
FORMALDEHYDE CYCLODODECYL METHYL ACETAL
mf: $C_{14}H_{28}O_2$ mw: 228.42

SYNS: BOISAMBRENE □ CYCLODODECANE, (METHOXYMETHOXY)- □ (METHOXYMETHOXY)CYCLODODECANE

TOXICITY DATA WITH REFERENCE
orl-rat LD50:>5 g/kg FCTOD7 26,327,88
skn-rbt LD50:>5 g/kg FCTOD7 26,327,88

CONSENSUS REPORTS: Reported in EPA TSCA Inventory.

SAFETY PROFILE: Low toxicity by ingestion and skin contact. When heated to decomposition it emits acrid smoke and irritating vapors.

FMW000 **CAS:149-44-0** **HR: 2**
FORMALDEHYDE HYDROSULFITE
mf: $CH_4O_3S{\cdot}Na$ mw: 119.10

PROP: Crystals.

SYNS: ALDANIL □ DISCOLITE □ FORMALDEHYDE SODIUM BISULFITE ADDUCT □ FORMALDEHYDE SODIUM SULFOXYLATE □ FORMALDEHYDESULFOXYLIC ACID SODIUM SALT □ FORMOPAN □ HYDROLIT □ HYDROSULFITE AWC □ HYDROXYMETHANESULFINIC ACID SODIUM SALT □ RONGALIT □ RONGALITE C □ SODIUM FORMALDEHYDE SULFOXYLATE □ SODIUM HYDROXYMETHANESULFINATE □ SODIUM METHANALSULFOXYLATE □ SODIUM SULFOXYLATE FORMALDEHYDE

TOXICITY DATA WITH REFERENCE
orl-mus LD50:4 g/kg YKYUA6 31,959,80

CONSENSUS REPORTS: Reported in EPA TSCA Inventory. EPA Genetic Toxicology Program.

SAFETY PROFILE: Moderately toxic by ingestion. When heated to decomposition it emits toxic fumes of SO_x and Na_2O. See also FORMALDEHYDE and SULFITES.

FMW300 **HR: 3**
FORMALDEHYDE OXIDE POLYMER
mf: $(CH_2O_2)_n$

SAFETY PROFILE: A shock-sensitive explosive. When heated to decomposition it emits acrid smoke and irritating fumes. See also PEROXIDES.

FMW330 **CAS:25214-70-4** **HR: 1**
FORMALDEHYDE, POLYMER with BENZENAMINE
mf: $(C_6H_7N{\cdot}CH_2O)_x$

SYNS: AF 10 □ ANILINE-FORMALDEHYDE CONDENSATE □ ANILINE-FORMALDEHYDE POLYMER □ ANILINE, POLYMER with FORMALDEHYDE (8CI) □ FORMALDEHYDE-ANILINE COPOLYMER □ JEFFAMINE AP22 □ JEFFAMINE AP27 □ MDA 150 □ MDA 220 □ POLYAMINE T

TOXICITY DATA WITH REFERENCE
skn-rbt 500 mg open MLD UCDS** 7/16/65
orl-rat LD50:7460 mg/kg UCDS** 7/16/65
skn-rbt LD50:20 mg/kg UCDS** 7/16/65

CONSENSUS REPORTS: Reported in EPA TSCA Inventory.

SAFETY PROFILE: Low toxicity by ingestion and skin contact. A skin irritant. When heated to decomposition it emits toxic vapors of NO_x.

FMX000 **CAS:541-66-2** **HR: 3**
FORMAL-γ-TRIMETHYLAMMONIUM PROPANEDIOL
mf: $C_7H_{16}NO_2{\cdot}I$ mw: 273.14

PROP: Crystals. Mp: 158–160°.

SYNS: DILVASENE □ ((1,3-DIOXOLAN-4-YL)METHYL)TRIMETHYLAMMONIUM IODIDE □ 2249F □ OXAPROPANIUM IODIDE □ N,N,N-

TRIMETHYL-1,3-DIOXOLANE-4-METHANAMINIUM IODIDE □ VASO-DILATATEUR 2249F

TOXICITY DATA WITH REFERENCE
scu-rat LDLo:25 mg/kg BSCIA3 26,516,44
orl-mus LDLo:300 mg/kg BSCIA3 26,516,44
scu-mus LDLo:35 mg/kg BSCIA3 26,516,44
ivn-mus LDLo:2500 μg/kg BSCIA3 26,516,44
scu-gpg LDLo:5 mg/kg BSCIA3 26,516,44

SAFETY PROFILE: Poison by ingestion, subcutaneous, and intravenous routes. When heated to decomposition it emits very toxic fumes of NO_x, NH_3, and I^-.

FMY000 CAS:75-12-7 HR: 3
FORMAMIDE
mf: CH_3NO mw: 45.05

PROP: Colorless, odorless, hygroscopic, sltly viscous, oily liquid. Mp: 2.5°, fp: 2.6°, vap press: 29.7 mm @ 129.4°, flash p: 310°F (COC), bp: 70.5° @ 1 mm, d: 1.134 @ 20°/40°, 1.1292 @ 25°/4°. Misc in H_2O, MeOH; very sltly sol in Et_2O, C_6H_6; insol in $CHCl_3$, hexane.

SYNS: CARBAMALDEHYDE □ METHANAMIDE

TOXICITY DATA WITH REFERENCE
eye-rbt 23 mg AJOPAA 29,1363,46
oms-nml:oth 500 mmol/L CAANAT 56,712,72
cyt-nml:oth 500 mmol/L CAANAT 56,712,72
skn-rat TDLo:1200 mg/kg (female 10-11D post):TER TXAPA9 41,35,77
orl-rat TDLo:2 g/kg (7D preg):TER 85DJA5 -,95,71
orl-rat TDLo:2 g/kg (7D preg):REP 85DJA5 -,95,71
orl-rat TDLo:7980 mg/kg (female 7-12D post):TER 85DJA5 -,95,71
orl-rat LD50:5570 mg/kg 85GMAT -,70,82
ipr-rat LD50:5700 mg/kg TXAPA9 26,596,73
orl-mus LD50:3150 mg/kg 85GMAT -,70,82
ipr-mus LD50:2450 mg/kg AEPPAE 230,559,57
ivn-dog LDLo:1500 mg/kg HBAMAK 4,1289,35
skn-rbt LDLo:6 mg/kg PJPPAA 32,223,80
ipr-gpg LD50:1250 mg/kg 85GMAT -,70,82
scu-frg LDLo:30 mg/kg HBAMAK 4,1289,35

CONSENSUS REPORTS: Reported in EPA TSCA Inventory. EPA Genetic Toxicology Program.

OSHA PEL: TWA 20 ppm; STEL 30 ppm
ACGIH TLV: TWA 10 ppm (skin)

SAFETY PROFILE: Poison by skin contact and subcutaneous routes. Moderately toxic by ingestion, intraperitoneal, and intramuscular routes. An irritant to skin, eyes, and mucous membranes. Experimental teratogenic and reproductive effects. An eye irritant. Mutation data reported. Combustible when exposed to heat or flame; can react vigorously with oxidizing materials. Incompatible with I_2, pyridine, SO_3. When heated to decomposition it emits toxic fumes of NO_x. Has exploded while in storage.

FMZ000 CAS:4312-87-2 HR: 2
FORMHYDROXAMIC ACID
mf: CH_3NO_2 mw: 61.05

PROP: Waxy leaflets. Sol in H_2O, EtOH; sltly sol in Et_2O; insol in $CHCl_3$, ligroin, and C_6H_6.

SYNS: FORMHYDROXAMSAEURE (GERMAN) □ N-FORMYLHYDROXYLAMINE

TOXICITY DATA WITH REFERENCE
par-rat TDLo:680 mg/kg (female 13D post):TER TJADAB 33,69C,86
ipr-rat TDLo:500 mg/kg (13D preg):TER TJADAB 26(3),9A,82
scu-rat TDLo:500 mg/kg (10D preg):TER APEPA2 257,296,67
scu-rat LD50:570 mg/kg APEPA2 257,296,67

SAFETY PROFILE: Moderately toxic by subcutaneous route. Experimental teratogenic effects. When heated to decomposition it emits toxic fumes of NO_x.

FNA000 CAS:64-18-6 HR: 3
FORMIC ACID
DOT: UN 1779
mf: CH_2O_2 mw: 46.03

PROP: Colorless, fuming liquid; pungent, penetrating odor. Bp: 100.8°, fp: 8.2°, flash p: 156°F (OC), d: 1.220 @ 20°/4°, 1.220 @ 20°/4°, mp: 8.4°, autoign temp: 1114°F, vap press: 40 mm @ 24.0°, vap d: 1.59, flash p: (90% soln): 122°F, autoign temp (90% soln) 813°F, lel (90% soln) 18%, uel (90% soln) 57%. Misc in H_2O, EtOH, Et_2O; mod sol in C_6H_6.

SYNS: ACIDE FORMIQUE (FRENCH) □ ACIDO FORMICO (ITALIAN) □ AMEISENSAEURE (GERMAN) □ AMINIC ACID □ FORMYLIC ACID □ HYDROGEN CARBOXYLIC ACID □ KWAS METANIOWY (POLISH) □ METHANOIC ACID □ MIERENZUUR (DUTCH) □ RCRA WASTE NUMBER U123

TOXICITY DATA WITH REFERENCE
skn-rbt 610 mg open MLD UCDS** 5/8/68
eye-rbt 122 mg SEV UCDS** 5/8/68
mmo-esc 70 ppm/3H AMNTA4 85,119,51
pic-esc 100 mmol/L MDMIAZ 31,11,79
sln-dmg-ihl 1000 ppm/24H THAGA6 39,330,69
oms-nml:oth 100 mmol/L CAANAT 56,712,72
ihl-man TCLo:7300 μg/m³/8H ARTODN 66,522,92
orl-wmn LDLo:2440 μg/kg AJEMEN 7,286,89
orl-rat LD50:1100 mg/kg GTPZAB 23(12),49,79
ihl-rat LC50:15 g/m³/15M GTPZAB 23(12),49,79
orl-mus LD50:700 mg/kg GTPZAB 23(12),49,79
ihl-mus LC50:6200 mg/m³/15M GTPZAB 23(12),49,79
ipr-mus LD50:940 mg/kg IGIBA5 11,507,62
ivn-mus LD50:145 mg/kg ZERNAL 9,332,69
orl-dog LD50:4000 mg/kg AMIHAB 20,517,59
ivn-dog LDLo:3000 mg/kg HBTXAC 1,146,56

CONSENSUS REPORTS: Reported in EPA TSCA Inventory.

OSHA PEL: TWA 5 ppm
ACGIH TLV: 5 ppm; STEL 10 ppm
DFG MAK: 5 ppm (9 mg/m³)
DOT CLASSIFICATION: 8; *Label:* Corrosive

SAFETY PROFILE: Poison by inhalation, intravenous, and intraperitoneal routes. Moderately toxic by ingestion. Mutation data reported. Corrosive. A skin and severe eye irritant. A substance migrating to food from packaging materials. Combustible liquid when exposed to heat or flame; can react vigorously with oxidizing materials. Explosive reaction with furfuryl alcohol; H_2O_2; $Tl(NO_3)_3{\cdot}3H_2O$; nitromethane; P_2O_5. To fight fire, use CO_2, dry chemical, alcohol foam. When heated to decomposition it emits acrid smoke and irritating fumes.

For occupational chemical analysis use OSHA: #ID-112 or NIOSH: Formic Acid S173.

FNB000 CAS:32852-21-4 HR: 2
FORMIC ACID (2-(4-METHYL-2-THIAZOLYL))HYDRAZIDE
mf: $C_5H_7N_3OS$ mw: 157.21

TOXICITY DATA WITH REFERENCE
pic-esc 10 mg/L MUREAV 26,3,74
orl-rat TDLo: 11 g/kg/46W-C:NEO JNCIAM 47,437,71

CONSENSUS REPORTS: EPA Genetic Toxicology Program.

SAFETY PROFILE: Questionable carcinogen with experimental neoplastigenic data. Mutation data reported. When heated to decomposition it emits very toxic fumes of SO_x and NO_x.

FNC000 CAS:2142-94-1 HR: 1
FORMIC ACID, NERYL ESTER
mf: $C_{11}H_{18}O_2$ mw: 182.29

SYNS: 3,7-DIMETHYL-2,6-OCTADIEN-1-OL, FORMATE (cis) □ NERYL FORMATE

TOXICITY DATA WITH REFERENCE
skn-rbt 500 mg/24H MOD FCTXAV 14,627,76
orl-rat LD50: >5 g/kg FCTXAV 14,627,76
skn-rbt LD50: >5 g/kg FCTXAV 14,627,76

CONSENSUS REPORTS: Reported in EPA TSCA Inventory.

SAFETY PROFILE: Low toxicity by ingestion and skin contact. A skin irritant. When heated to decomposition it emits acrid smoke and irritating fumes. See also ESTERS and FORMIC ACID.

FND100 CAS:10176-39-3 HR: 3
FORMILOXINE
mf: $C_{46}H_{64}O_{19}$ mw: 921.10

SYNS: AC 2770 □ FORMILOXIN □ GITOFORMATE □ PENTAFORMYLGITOXIN

TOXICITY DATA WITH REFERENCE
orl-mus LD50: 2390 μg/kg AIPTAK 153,436,65
scu-mus LD50: 2270 μg/kg AIPTAK 153,436,65
orl-gpg LDLo: 1835 μg/kg AIPTAK 153,436,65
ivn-gpg LDLo: 734 μg/kg AIPTAK 153,436,65

SAFETY PROFILE: Poison by ingestion, subcutaneous, and intravenous routes. When heated to decomposition it emits acrid smoke and irritating fumes.

FNE100 CAS:43229-80-7 HR: 3
FORMOTEROL FUMARATE DIHYDRATE
mf: $C_{19}H_{24}N_2O_4{\cdot}1/2C_4H_4O_4{\cdot}2H_2O$ mw: 422.48

PROP: Crystals from 2-propanol (aq). Mp: 140°.

SYN: BD 40A

TOXICITY DATA WITH REFERENCE
orl-rat TDLo: 1200 μg/kg (female 17-22D post): REP OYYAA2 27,375,84
orl-rat TDLo: 162 mg/kg (female 17-22D post): REP OYYAA2 27,375,84
orl-rat TDLo: 36 mg/kg (female 17-22D post): REP OYYAA2 27,375,84
orl-rat TDLo: 66 mg/kg (female 7-17D post): TER OYYAA2 27,239,84
orl-rat TDLo: 810 mg/kg (female 17-22D post): REP OYYAA2 27,375,84
orl-rat TDLo: 2200 μg/kg (7-17D preg): TER OYYAA2 27,239,84
orl-rat LD50: 3130 mg/kg OYYAA2 26,811,83
ipr-rat LD50: 170 mg/kg OYYAA2 26,811,83
scu-rat LD50: 1000 mg/kg OYYAA2 26,811,83
ivn-rat LD50: 98 mg/kg OYYAA2 26,811,83
orl-mus LD50: 6700 mg/kg OYYAA2 26,811,83
ipr-mus LD50: 210 mg/kg OYYAA2 26,811,83
scu-mus LD50: 640 mg/kg OYYAA2 26,811,83
ivn-mus LD50: 71 mg/kg OYYAA2 26,811,83

SAFETY PROFILE: Poison by intraperitoneal and intravenous routes. Moderately toxic by ingestion and subcutaneous routes. Experimental teratogenic and reproductive effects. Used as a bronchodilator. When heated to decomposition it emits toxic fumes of NO_x.

FNE500 CAS:17702-57-7 HR: 3
FORMPARANATE
mf: $C_{12}H_{17}N_3O_2$ mw: 235.32

SYNS: N,N-DIMETHYL-N′-(2-METHYL-4-(((METHYLAMINO)CARBONYL)OXY)PHENYL)METHANIMIDAMIDE □ ENT 27,305 □ SCHERING 36103 □ UC-25074

TOXICITY DATA WITH REFERENCE
orl-mus LD50: 16,600 μg/kg EJTXAZ 7,152,74
orl-rat LD50: 7200 μg/kg ARSIM* 20,26,66

CONSENSUS REPORTS: EPA Extremely Hazardous Substances List.

SAFETY PROFILE: Poison by ingestion. When heated to decomposition it emits toxic fumes of NO_x.

FNF000 CAS:104-06-3 HR: 2
4′-FORMYLACETANILIDE THIOSEMICARBAZONE
mf: $C_{10}H_{12}N_4OS$ mw: 236.32

SYNS: p-ACETAMIDOBENZALDEHYDE THIOSEMICARBAZONE □ p-ACETAMINOBENZYLIDENETHIOSEMICARBAZONE □ p-ACETYLAMINOBENZALDEHYDE THIOSEMICARBAZONE □ AKTIVAN □ AMBATHIZON □ N-(4-(((AMINOTHIOXOMETHYL)HYDRAZONO)METHYL)PHENYL)ACETAMIDE □ AMITHIOZONE □ AMITIOZON □ ANTIB

F

□ BENTHIOZONE □ BENZOTHIOZANE □ BENZOTHIOZON □ BERCULON A □ BERKAZON □ CONTEBEN □ DIASAN □ DIAZAN □ DOMAGK'S T.B.1 CONTEBEN □ DOMAKOL □ ILBION □ LIVAZONE □ MAGK'S T.B.1 CONTEBEN □ MIVIZON □ MYVIZONE □NEOTIBIL □ NEUSTAB □ NOVAKOL □ NUCLON ARGENTINIAN □ PANRONE □ PARAZONE □ RP 4207 □ SDT 1041 □ SERODEN □ SIOCARBAZONE □ SQ 2321 □ TB 1 (BAYER) □ TEBALON □ TEBECURE □ TEBEMAR □ TEBESONE I □ TEBETHIONE □ TEBEZON □ THIACETAZONE □ THIBONE □ THIOACETAZONE □ THIOCARBAZIL □ THIOMICID □ THIONICID □ THIOPARAMIZONE □ THIOSEMICARBARZONE □ THIOSEMICARBAZONE (PHARMACEUTICAL) □ THIOTEBESIN □ THIZONE □ TIACETAZON □ TIBICUR □ TIBIONE □ TIBIZAN □ TIOACETAZON □ TIOCARONE □TUBI-GAL □ TUBIN

TOXICITY DATA WITH REFERENCE
cyt-mus-orl 3300 μg/kg NULSAK 22,96,79
orl-mus LD50:950 mg/kg CRSBAW 144,1310,50
scu-mus LD50:1 g/kg JPPMAB 2,764,50

SAFETY PROFILE: Moderately toxic by ingestion and subcutaneous routes. Mutation data reported. A tuberculostatic antibacterial agent. When heated to decomposition it emits very toxic fumes of NO_x and SO_x.

FNJ000 CAS:103-70-8 HR: 3
FORMYLANILINE
mf: C_7H_7NO mw: 121.15

PROP: Crystals. Mp: 50°, bp: 216° @ 120 mm.

SYNS: CARBANILALDEHYDE □ FORMAMIDOBENZENE □ FORMANILIDE □ N-FORMYLANILINE □ PHENYL FORMAMIDE □ N-PHENYLFORMAMIDE

TOXICITY DATA WITH REFERENCE
orl-dog LDLo:400 mg/kg XPHBAO 271,19,41
ivn-dog LDLo:400 mg/kg XPHBAO 271,19,41
orl-frg LDLo:800 μg/kg XPHBAO 271,19,41

CONSENSUS REPORTS: Reported in EPA TSCA Inventory.

SAFETY PROFILE: Poison by ingestion and intravenous routes. When heated to decomposition it emits toxic fumes of NO_x.

FNK000 CAS:63040-55-1 HR: 2
6-FORMYLANTHANTHRENE
mf: $C_{25}H_{14}O$ mw: 330.39

SYN: DIBENZO(def,mno)CHRYSENE-12-CARBOXALDEHYDE

TOXICITY DATA WITH REFERENCE
scu-mus TDLo:72 mg/kg 9W-I:ETA COREAF 252,1711,61

SAFETY PROFILE: Questionable carcinogen with experimental tumorigenic data. When heated to decomposition it emits acrid smoke and irritating fumes.

FNK010 CAS:119-67-5 HR: 2
2-FORMYLBENZOIC ACID
mf: $C_8H_6O_3$ mw: 150.14

SYNS: BENZOIC ACID, 2-FORMYL-(9CI) □ o-CARBOXYBENZALDEHYDE □ 2-CARBOXYBENZALDEHYDE □ o-FORMYLBENZOIC ACID □ PHTHALALDEHYDIC ACID

TOXICITY DATA WITH REFERENCE
orl-rat LD50:7500 mg/kg JANTAJ 27,665,74
scu-rat LD50:2430 mg/kg JANTAJ 27,665,74
orl-mus LD50:4480 mg/kg JANTAJ 27,665,74
scu-mus LD50:1860 mg/kg JANTAJ 27,665,74

CONSENSUS REPORTS: Reported in EPA TSCA Inventory.

SAFETY PROFILE: Moderately toxic by subcutaneous route. Mildly toxic by ingestion. When heated to decomposition it emits acrid smoke and irritating vapors.

FNK025 CAS:100-50-5 HR: 3
4-FORMYLCYCLOHEXENE
DOT: UN 2498
mf: $C_7H_{10}O$ mw: 110.17

PROP: Oil. Mp: −96.1°, bp: 163.5–164.5°.

SYNS: 3-CYCLOHEXENE-1-CARBOXALDEHYDE □ 1,2,3,6-TETRAHYDROBENZALDEHYDE (DOT) □ 1,2,5,6-TETRAHYDROBENZALDEHYDE

TOXICITY DATA WITH REFERENCE
skn-rbt 10 mg/24H open MLD AIHAAP 23,95,62
orl-rat LD50:2460 mg/kg AIHAAP 23,95,62
skn-rbt LD50:1770 mg/kg AIHAAP 23,95,62
ihl-rat LC50:2000 ppm/4H 85JCAE-,276,86

CONSENSUS REPORTS: Reported in EPA TSCA Inventory.

DOT CLASSIFICATION: 3; *Label:* Flammable Liquid

SAFETY PROFILE: Moderately toxic by ingestion, inhalation, and skin contact. Corrosive. An eye, skin, and mucous membrane irritant. Flammable liquid. When heated to decomposition it emits acrid smoke and irritating fumes. See also ALDEHYDES.

FNK040 CAS:2454-11-7 HR: 3
FORMYLDIENOLONE
mf: $C_{21}H_{28}O_4$ mw: 344.49

PROP: Crystals from ethyl acetate. Mp: 209–212°. Sol in H_2O; insol in pet ether and C_6H_6.

SYNS: 11-α,17-β-DIHYDROXY-17-METHYL-3-OXOANDROSTA-1,4-DIENE-2-CARBOXALDEHYDE □ ESICLENE □ FORMEBOLONE □ 2-FORMYL-11-α-HYDROXY-Δ^1-METHYLTESTOSTERONE □ 2-FORMYL-17-α-METHYLANDROSTA-1,4-DIENE-11-α,17-β-DIOL-3-ONE

TOXICITY DATA WITH REFERENCE
ipr-rat LD50:104 mg/kg MEIEDD 10,606,83
scu-rat LD50:270 mg/kg MEIEDD 10,606,83
ipr-mus LD50:187 mg/kg MEIEDD 10,606,83
scu-mus LD50:293 mg/kg MEIEDD 10,606,83

SAFETY PROFILE: Poison by subcutaneous and intraperitoneal routes. Used as an anabolic steroid. When heated to decomposition it emits acrid smoke and irritating fumes.

FNK050 HR: 3
3-FORMYL-DIGITOXIGENIN
mf: $C_{24}H_{34}O_5$ mw: 402.58

SYN: 3-β,14-DIHYDROXY-5-β-CARD-20(22)-ENOLIDE-3-FORMATE

TOXICITY DATA WITH REFERENCE
orl-mus LD50:33,630 µg/kg AIPTAK 153,436,65
scu-mus LD50:16,600 µg/kg AIPTAK 153,436,65
ivn-gpg LDLo:7711 µg/kg AIPTAK 153,436,65

SAFETY PROFILE: A deadly poison by ingestion, subcutaneous, and intravenous routes. When heated to decomposition it emits acrid smoke and irritating fumes.

FNK075 **HR: 3**
3-12-FORMYL-DIGOXIGENIN
mf: $C_{25}H_{34}O_7$ mw: 446.59

SYN: 3-β,12-β,14-TRIHYDROXY-5-β-CARD-20(22)-ENOLIDE-3,12-DIFORMATE

TOXICITY DATA WITH REFERENCE
orl-mus LD50:33,680 µg/kg AIPTAK 153,436,65
scu-mus LD50:19,420 µg/kg AIPTAK 153,436,65
ivn-gpg LDLo:4750 µg/kg AIPTAK 153,436,65

SAFETY PROFILE: Poison by ingestion, subcutaneous, and intravenous routes. When heated to decomposition it emits acrid smoke and irritating fumes.

FNK150 **CAS:564-94-3** **HR: 3**
2-FORMYL-6,6-DIMETHYLBICYCLO(3.1.1)HEPT-2-ENE
mf: $C_{10}H_{14}O$ mw: 150.24

SYNS: BENIHINAL □ BICYCLO(3.1.1)HEPT-2-ENE-2-CARBOXALDEHYDE, 6,6-DIMETHYL- □ 6,6-DIMETHYLBICYCLO(3.1.1)HEPT-2-ENE-2-CARBOXALDEHYDE □ MYRTENAL □ 2-NORPINENE-2-CARBOXALDEHYDE, 6,6-DIMETHYL-

TOXICITY DATA WITH REFERENCE
orl-rat LD50:2300 mg/kg FCTOD7 26,329,88
ivn-mus LD50:170 mg/kg FCTOD7 26,329,88
skn-rbt LD50:>5 g/kg FCTOD7 26,329,88

CONSENSUS REPORTS: Reported in EPA TSCA Inventory.

SAFETY PROFILE: Poison by intravenous route. Moderately toxic by ingestion. Slightly toxic by skin contact. When heated to decomposition it emits acrid smoke and irritating vapors.

FNK200 **CAS:58243-85-9** **HR: 2**
FORMYLETHYLTETRAMETHYLTETRALIN
mf: $C_{17}H_{24}O$ mw: 244.41

SYNS: 6-ETHYL-7-FORMYL-1,1,4,4-TETRAMETHYL-1,2,3,4-TETRAHYDRONAPHTHALENE □ FETT □ 5,6,7,8-TETRAHYDRO-3-ETHYL-5,5,8,8-TETRAMETHYL-2-NAPHTHALENECARBOXALDEHYDE □ TETRAMETHYLETHYLFORMYLTETRALIN

TOXICITY DATA WITH REFERENCE
skn-rbt 500 mg/24H MLD FCTOD7 21,853,83
eye-rbt 100 mg MLD FCTOD7 21,853,83
orl-rat LD50:3200 mg/kg FCTOD7 21,853,83

SAFETY PROFILE: Moderately toxic by ingestion. A skin and eye irritant. When heated to decomposition it emits acrid smoke and irritating fumes.

FNM000 **CAS:621-59-0** **HR: 2**
5-FORMYLGUAIACOL
mf: $C_8H_8O_3$ mw: 152.16

SYNS: 3-HYDROXY-p-ANISALDEHYDE □ 3-HYDROXY-4-METHOXYBENZALDEHYDE □ ISOVANILLIN □ ISOVANILLINE □ OXY-3-METHOXY-4 BENZALDEHYDE (FRENCH)

TOXICITY DATA WITH REFERENCE
sce-hmn:lyms 1500 µmol/L MUREAV 206,17,88
ipr-rat LD50:1276 mg/kg COREAF 243,609,56
ivn-dog LDLo:1470 mg/kg APFRAD 14,456,56

CONSENSUS REPORTS: Reported in EPA TSCA Inventory.

SAFETY PROFILE: Moderately toxic by intraperitoneal and intravenous routes. Mutation data reported. When heated to decomposition it emits acrid smoke and irritating fumes.

FNN000 **CAS:624-84-0** **HR: 3**
FORMYLHYDRAZINE
mf: CH_4N_2O mw: 60.07

PROP: Crystals from EtOH. Mp: 54°. Very sol in alc and ether; sol in benzene.

SYNS: CARBAZALDEHYDE □ FORMAL HYDRAZINE □ FORMHYDRAZID (GERMAN) □ FORMHYDRAZIDE □ FORMIC ACID, HYDRAZIDE □ FORMIC HYDRAZIDE □ FORMOHYDRAZIDE □ FORMYLHYDRAZIDE □ N-FORMYLHYDRAZINE □ HYDRAZINECARBOXALDEHYDE

TOXICITY DATA WITH REFERENCE
orl-mus TDLo:73 g/kg/43W-C:NEO BJCAAI 37,960,78
scu-rat LD50:120 mg/kg ARZNAD 18,645,68
ipr-mus LD50:65 mg/kg JMPCAS 4,259,61

CONSENSUS REPORTS: Reported in EPA TSCA Inventory.

SAFETY PROFILE: Poison by intraperitoneal and subcutaneous routes. Questionable carcinogen with experimental neoplastigenic data. When heated to decomposition it emits toxic fumes of NO_x. See also HYDRAZINE.

FNO000 **CAS:689-13-4** **HR: 1**
N-FORMYL-N-HYDROXYGLYCINE
mf: $C_3H_5NO_4$ mw: 119.09

PROP: Unstable crystals from Me_2CO/pet ether. Mp: 119–120°.

SYNS: ASYMMETRIN □ N-FORMYL HYDROXYAMINOACETIC ACID □ HADACIDIN □ HADACIDINE □ HADACIN □ NFHAA □ NSC 521778

TOXICITY DATA WITH REFERENCE
ipr-mus TDLo:2 g/kg (female 10D post):TER PCBRD2 110,365,83
ipr-mus TDLo:3 g/kg (female 8-11D post):TER CRSBAW 164,2171,70
par-rat TDLo:5 g/kg (female 10D post):TER TJADAB 2,267,69
ipr-rat TDLo:1 g/kg (female 12D post):TER FAATDF 4,352,84

ipr-ham TDLo: 2400 mg/kg (female 6-13D post): TER CRSBAW 164,2171,70
ipr-rat TDLo: 4 g/kg (female 11D post): TER JEEMAF 15,193,66
ipr-rat TDLo: 6750 mg/kg (female 7-15D post): TER CRSBAW 164,1919,70
ipr-ham TDLo: 2400 mg/kg (female 6-13D post): TER CRSBAW 164,2171,70
ipr-rat TDLo: 2250 mg/kg (12D preg): TER ABILAE 80,167,69
ipr-rat LDLo: 5000 mg/kg ADTEAS 3,181,68

SAFETY PROFILE: Mildly toxic by intraperitoneal route. Experimental teratogenic effects. When heated to decomposition it emits toxic fumes of NO_x.

FNO100 CAS:487-89-8 HR: 2
3-FORMYLINDOLE
mf: C_9H_7NO mw: 145.17

SYNS: INDOLE-3-ALDEHYDE □ INDOLE-3-CARBALDEHYDE □ INDOLE-3-CARBOXALDEHYDE □ 1H-INDOLE-3-CARBOXALDEHYDE (9CI) □ β-INDOLYLALDEHYDE

TOXICITY DATA WITH REFERENCE
ipr-mus LDLo: 600 mg/kg PCJOAU 6,33,72

CONSENSUS REPORTS: Reported in EPA TSCA Inventory.

SAFETY PROFILE: Moderately toxic by intraperitoneal route. When heated to decomposition it emits toxic vapors of NO_x.

FNP000 CAS:66409-98-1 HR: 3
N-FORMYLJERVINE
mf: $C_{28}H_{39}NO_4$ mw: 453.68

TOXICITY DATA WITH REFERENCE
orl-ham TDLo: 85 mg/kg (7D preg): REP 41CIAR -,409,78
orl-ham TDLo: 85 mg/kg (7D preg): TER JAFCAU 26,564,78
orl-ham LDLo: 85 mg/kg JAFCAU 26,564,78

SAFETY PROFILE: Poison by ingestion. Experimental teratogenic and reproductive effects.

FNQ000 CAS:63040-58-4 HR: 2
6-FORMYL-12-METHYLANTHANTHRENE
mf: $C_{26}H_{16}O$ mw: 344.42

SYN: 6-METHYLDIBENZO(def,mno)CHRYSENE-12-CARBOXALDEHYDE

TOXICITY DATA WITH REFERENCE
scu-mus TDLo: 72 mg/kg/9W-I: ETA COREAF 252,1711,61

SAFETY PROFILE: Questionable carcinogen with experimental tumorigenic data. When heated to decomposition it emits acrid smoke and irritating fumes.

FNR000 CAS:2732-09-4 HR: 2
7-FORMYL-9-METHYLBENZ(c)ACRIDINE
mf: $C_{19}H_{13}NO$ mw: 271.33

SYN: 9-METHYLBENZ(c)ACRIDINE-7-CARBOXALDEHYDE

TOXICITY DATA WITH REFERENCE
scu-mus TDLo: 120 mg/kg/9W-I: ETA CHDDAT 267,981,68

SAFETY PROFILE: Questionable carcinogen with experimental tumorigenic data. When heated to decomposition it emits toxic fumes of NO_x.

FNS000 CAS:18936-78-2 HR: 2
7-FORMYL-11-METHYLBENZ(c)ACRIDINE
mf: $C_{19}H_{13}NO$ mw: 271.33

SYN: 11-METHYLBENZ(c)ACRIDINE-7-CARBOXALDEHYDE

TOXICITY DATA WITH REFERENCE
scu-mus TDLo: 120 mg/kg/9W-I: ETA CHDDAT 267,981,68

SAFETY PROFILE: Questionable carcinogen with experimental tumorigenic data. When heated to decomposition it emits toxic fumes of NO_x.

FNT000 CAS:13345-61-4 HR: 2
7-FORMYL-12-METHYLBENZ(a)ANTHRACENE
mf: $C_{20}H_{14}O$ mw: 270.34

SYN: 12-METHYLBENZ(a)ANTHRACENE-7-CARBOXALDEHYDE

TOXICITY DATA WITH REFERENCE
dni-omi 200 µg/L PNASA6 74,1378,77
scu-rat TDLo: 20 mg/kg/39D-I: NEO CNREA8 31,1951,71
skn-mus TDLo: 8000 µg/kg: NEO JJIND8 61,135,78
scu-mus TDLo: 120 mg/kg/6W-I: CAR IJCNAW 2,500,67

SAFETY PROFILE: Questionable carcinogen with experimental carcinogenic and neoplastigenic data. Mutation data reported. When heated to decomposition it emits acrid smoke and irritating fumes.

FNU000 CAS:63040-56-2 HR: 2
5-FORMYL-8-METHYL-3,4:9,10-DIBENZOPYRENE
mf: $C_{26}H_{16}O$ mw: 344.42

SYN: 8-METHYLBENZO(rst)PENTAPHENE-5-CARBOXALDEHYDE

TOXICITY DATA WITH REFERENCE
scu-mus TDLo: 72 mg/kg/9W-I: ETA COREAF 252,1236,61

SAFETY PROFILE: Questionable carcinogen with experimental tumorigenic data. When heated to decomposition it emits acrid smoke and irritating fumes.

FNV000 CAS:63040-57-3 HR: 2
5-FORMYL-10-METHYL-3,4:8,9-DIBENZOPYRENE
mf: $C_{26}H_{16}O$ mw: 344.42

SYN: 14-METHYLDIBENZO(b,def)CHRYSENE-7-CARBOXALDEHYDE

TOXICITY DATA WITH REFERENCE
scu-mus TDLo: 72 mg/kg/9W-I: ETA COREAF 252,1711,61

SAFETY PROFILE: Questionable carcinogen with experimental tumorigenic data. When heated to decomposition it emits acrid smoke and irritating fumes.

FNW000 **CAS:758-17-8** **HR: 3**
N-FORMYL-N-METHYLHYDRAZINE
mf: $C_2H_6N_2O$ mw: 74.10

SYNS: FORMIC ACID, METHYLHYDRAZIDE □ 1-FORMYL-1-METHYLHYDRAZINE □ N-METHYL-N-FORMYLHYDRAZINE □ MFH

TOXICITY DATA WITH REFERENCE
mmo-sat 100 μmol/plate TXCYAC 26,155,83
mma-sat 100 μmol/plate TXCYAC 26,155,83
orl-mus TDLo:14 mg/kg/62W-C:CAR JNCIAM 60,201,78
scu-mus TDLo:400 mg/kg/40W-I:CAR NEOLA4 39,437,83
orl-ham TDLo:6100 mg/kg/80W-C:CAR JCROD7 93,109,79
orl-mus TD:7840 mg/kg/70W-C:CAR MYCPAH 68,121,79
orl-mus TD:3360 mg/kg/2Y-C:CAR NEOLA4 27,25,80
orl-mus TD:3158 mg/kg/94W-C:CAR NEOLA4 27,25,80
orl-mus TD:3920 mg/kg/2Y-C:CAR NEOLA4 27,25,80
scu-mus TD:180 mg/kg:CAR JTEHD6 6,577,80
orl-rat TD:64 mg/kg/80W-C:CAR MYCPAH 78,11,82
orl-mus TD:84 mg/kg/70W-C:CAR MYCPAH 78,11,82
orl-mus TD:187 mg/kg/78W-C:CAR MYCPAH 78,11,82
orl-mus TD:158 mg/kg/79W-C:CAR MYCPAH 78,11,82
orl-mus LD50:118 mg/kg TXAPA9 45,429,78

SAFETY PROFILE: Suspected carcinogen with experimental carcinogenic data. Poison by ingestion route. Mutation data reported. When heated to decomposition it emits toxic fumes of NO_x. See also HYDRAZINE.

FNX000 **CAS:4845-14-1** **HR: 2**
N-FORMYL-N-METHYL-p-(PHENYLAZO)ANILINE
mf: $C_{14}H_{13}N_3O$ mw: 239.30

SYN: 4-FORMYLMONOMETHYLAMINOAZOBENZENE

TOXICITY DATA WITH REFERENCE
orl-rat TDLo:9100 mg/kg/34W-C:ETA CNREA8 9,652,49

SAFETY PROFILE: Questionable carcinogen with experimental tumorigenic data. When heated to decomposition it emits toxic fumes of NO_x.

FNZ000 **CAS:51-15-0** **HR: 3**
2-FORMYL-1-METHYLPYRIDINIUM CHLORIDE OXIME
mf: $C_7H_9N_2O{\cdot}Cl$ mw: 172.63

PROP: Crystals from $EtOH/Et_2O$. Mp: 235–238° (decomp). Sol in H_2O.

SYNS: 2-FORMYL-N-METHYLPYRIDINIUM OXIME CHLORIDE □ 2-(HYDROXYIMINOMETHYL)-1-METHYLPYRIDINIUM CHLORIDE □ 1-METHYL-2-ALDOXIMINOPYRIDINIUM CHLORIDE □ 1-METHYL-2-FORMYLPYRIDINIUM CHLORIDE OXIME □ 1-METHYL-2-PYRIDINIUM ALDOXIME CHLORIDE □ N-METHYLPYRIDINIUM CHLORIDE-2-ALDOXIME □ 2-PAM CHLORIDE □ PRALIDOXIME CHLORIDE □ PROTOPAM CHLORIDE □ 2-PYRIDINEALDOXIME CHLORIDE □ PYRIDINE-2-ALDOXIME METHOCHLORIDE □ 2-PYRIDINE ALDOXIME METHYL CHLORIDE □ PYRIDINIUM ALDOXIME METHOCHLORIDE

TOXICITY DATA WITH REFERENCE
ivn-man TDLo:14 mg/kg:CNS,PUL 34ZIAG -,449,69
ivn-hmn TDLo:15 mg/kg:CVS AEHLAU 15,599,67
ivn-rat LD50:96 mg/kg TXAPA9 16,40,70
ims-rat LD50:150 mg/kg TXAPA9 16,40,70
orl-mus LD50:4100 mg/kg JPMSAE 53,1143,64
ipr-mus LD50:155 mg/kg ARZNAD 14,5,64
ivn-mus LD50:90 mg/kg ARZNAD 14,5,64
ims-mus LD50:100 mg/kg DCTODJ 8,431,85
ims-dog LD50:75 mg/kg FAATDF 4(2, Pt 2),S106,84
ivn-rbt LD50:95 mg/kg JPETAB 132,50,61
ims-gpg LD50:168 mg/kg TXAPA9 16,40,70

SAFETY PROFILE: Poison by intravenous, intramuscular, and intraperitoneal routes. Moderately toxic by ingestion. Human systemic effects by intravenous route: coma, blood pressure increase, bronchiolar constriction and cyanosis. Used as a cholinesterase reactivator. When heated to decomposition it emits very toxic fumes of Cl^- and NO_x.

F

FOA100 **CAS:4394-85-8** **HR: 1**
4-FORMYLMORPHOLINE
mf: $C_5H_9NO_2$ mw: 115.15

SYNS: N-FORMYLMORFOLIN □ N-FORMYLMORPHOLINE □ 4-MORPHOLINECARBOXALDEHYDE

TOXICITY DATA WITH REFERENCE
skn-rbt 500 mg/24H MLD 85JCAE-,888,86
eye-rbt 500 mg/24H MLD 85JCAE-,888,86

CONSENSUS REPORTS: Reported in EPA TSCA Inventory.

SAFETY PROFILE: A skin and eye irritant. When heated to decomposition it emits toxic vapors of NO_x.

FOD000 **CAS:42540-40-9** **HR: 2**
(6R-(6-α,7-β(R)))-7-(((FORMYLOXY)PHENYLACETYL)AMINO)-3-(((1-METHYL-1H-TETRAZOL-5-YL)THIO)METHYL)-8-OXO-5-THIA-1-AZABICYCLO(4.2.0)OCT-2-ENE-2-CARBOXYLIC ACID MONOSODIUM SALT
mf: $C_{19}H_{18}N_6O_6S_2{\cdot}Na$ mw: 513.54

PROP: Needles. Mp: 190° (decomp).

SYNS: CEFAMANDOLE NAFATE □ CEFAMANDOL NAFATO □ CEPHAMANDOLE NAFATE □ O-FORMYLCEFAMANDOLE SODIUM □ MANDOL

TOXICITY DATA WITH REFERENCE
ivn-wmn TDLo:240 mg/kg/2D-I SMJOAV 78,1268,85
ivn-man TDLo:400 μg/kg/1W-I:SYS DICPBB 19,553,85
unr-wmn TDLo:408 mg/kg/4D-I:BLD AIMDAP 146,1125,86
unr-man TDLo:771 mg/kg/9D-I:BLD AIMDAP 146,1125,86
ivn-rat LD50:2562 mg/kg JIDIAQ 137,S51,78
scu-mus LD50:7 g/kg JIDIAQ 137,S51,78
ivn-mus LD50:3915 mg/kg JIDIAQ 137,S51,78

SAFETY PROFILE: Moderately toxic by intravenous route. Human systemic effects: clotting factor change, jaundice, joints. When heated to decomposition it emits very toxic fumes of NO_x, Na_2O, and SO_x.

FOE000 **CAS:17977-68-3** **HR: 3**
FORMYLOXYTRIBENZYLSTANNANE
mf: $C_{23}H_{24}O_2Sn$ mw: 451.16

SYNS: (FORMYLOXY)TRIS(PHENYLMETHYL)STANNANE □ MRAVENCAN TRIBENZYLCINICITY (CZECH) □ TRIBENZYLTIN FORMATE

TOXICITY DATA WITH REFERENCE
skn-rbt 500 mg/24H MLD 28ZPAK -,232,72
eye-rbt 100 mg/24H MOD 28ZPAK -,232,72
orl-rat LD50:312 mg/kg 28ZPAK -,232,72

OSHA PEL: TWA 0.1 mg(Sn)/m^3 (skin)
ACGIH TLV: TWA 0.1 mg(Sn)/m^3 (skin) (Proposed: TWA 0.1 mg(Sn)/m^3; STEL 0.2 mg(Sn)/m^3 (skin))
NIOSH REL: (Organotin Compounds) TWA 0.1 mg(Sn)/m^3

SAFETY PROFILE: Poison by ingestion. A skin and eye irritant. When heated to decomposition it emits acrid smoke and irritating fumes. See also TIN COMPOUNDS.

For occupational chemical analysis use NIOSH: Organotin Compounds 5504.

FOF000 CAS:123-08-0 HR: 2
p-FORMYLPHENOL
mf: $C_7H_6O_2$ mw: 122.13

PROP: Needles from water. D: 1.129, mp: 115–116°, bp: sublimes.

SYNS: 4-FORMYLPHENOL □ p-HYDROXYBENZALDEHYDE □ 4-HYDROXYBENZALDEHYDE □ p-OXYBENZALDEHYDE □ PARAHYDROXYBENZALDEHYDE □ USAF M-6

TOXICITY DATA WITH REFERENCE
sce-hmn:lym 1 mmol/L MUREAV 206,17,88
ipr-mus LD50:500 mg/kg NTIS** AD277-689

CONSENSUS REPORTS: Reported in EPA TSCA Inventory.

SAFETY PROFILE: Moderately toxic by intraperitoneal route. Mutation data reported. When heated to decomposition it emits acrid smoke and irritating fumes. See also ALDEHYDES.

FOH000 CAS:2591-86-8 HR: 3
1-FORMYLPIPERIDINE
mf: $C_6H_{11}NO$ mw: 113.18

PROP: A liquid. Bp: 222°.

SYN: N-FORMYLPIPERIDIN (GERMAN)

TOXICITY DATA WITH REFERENCE
orl-rat TDLo:6600 mg/kg (female 6-20D post):REP FAATDF 18,96,92
scu-rbt LDLo:300 mg/kg BDCGAS 34,2408,01

CONSENSUS REPORTS: Reported in EPA TSCA Inventory.

SAFETY PROFILE: Poison by subcutaneous route. Experimental reproductive effects. When heated to decomposition it emits toxic fumes of NO_x.

FOI000 CAS:6804-07-5 HR: 2
2-FORMYLQUINOXALINE-1,4-DIOXIDE CARBOMETHOXYHYDRAZONE
mf: $C_{11}H_{10}N_4O_4$ mw: 262.25

PROP: Minute yellow crystals. Mp: 239.5–240°. Insol in H_2O.

SYNS: CARBADOX (USDA) □ FORTIGRO □ GS 6244 □ MECADOX □ (2-QUINOXALINYLMETHYLENE)-HYDRAZINECARBOXYLIC ACID METHYL ESTER-N,N'-DIOXIDE

TOXICITY DATA WITH REFERENCE
dns-hmn:oth 50 mg/L JTEHD6 10,143,82
mnt-mus-orl 100 mg/kg MUREAV 144,81,85
orl-rat TDLo:5475 mg/kg/1Y-C:CAR IGSBAL 127,283,93
orl-rat LD50:850 mg/kg CKFRAY 30,26,81
orl-mus LD50:2810 mg/kg CKFRAY 30,26,81

SAFETY PROFILE: Moderately toxic by ingestion. Questionable carcinogen with experimental carcinogenic effects. Human mutation data reported. When heated to decomposition it emits toxic fumes of NO_x.

FOJ000 CAS:2302-84-3 HR: 2
1-FORMYL-3-THIOSEMICARBAZIDE
mf: $C_2H_5N_3OS$ mw: 119.16

TOXICITY DATA WITH REFERENCE
orl-rat TDLo:43 g/kg/46W-C:ETA JNCIAM 47,437,71

CONSENSUS REPORTS: Reported in EPA TSCA Inventory.

SAFETY PROFILE: Questionable carcinogen with experimental tumorigenic data. When heated to decomposition it emits very toxic fumes of NO_x and SO_x.

FOJ100 CAS:71522-58-2 HR: 1
FORPHENICINOL
mf: $C_9H_{11}NO_4$ mw: 197.21

SYNS: BENZENEACETIC ACID, α-AMINO-3-HYDROXY-4-(HYDROXYMETHYL)-, (S)- □ BF 121

TOXICITY DATA WITH REFERENCE
orl-rbt TDLo:5200 mg/kg (female 6-18D post):TER IYKEDH 17,693,86
ipr-rat LD50:5500 mg/kg JANTAJ 35,1049,82
ipr-mus LD50:5000 mg/kg JANTAJ 35,1049,82

SAFETY PROFILE: Mildly toxic by intraperitoneal route. An experimental teratogen. When heated to decomposition it emits toxic fumes of NO_x.

FOK000 CAS:55779-06-1 HR: 3
FORTIMICIN A
mf: $C_{17}H_{35}N_5O_6$ mw: 405.57

PROP: Amorphous powder. Produced by *Micromonospora olivoasterospora* MK-70 (JANTAJ 30,77–7,77).

SYNS: ANTIBIOTIC KW-1070 □ KW-1070 □ XK-70-1

TOXICITY DATA WITH REFERENCE
ims-rat TDLo:1210 mg/kg (7-17D preg):REP NKRZAZ 29(Suppl 2),167,81

ims-rat TDLo:5170 mg/kg (22D pre):TER NKRZAZ 29(Suppl 2),167,81
ims-rat TDLo:21 g/kg (male 8W pre):REP NKRZAZ 29(Suppl 2),167,81
ims-rat TDLo:5170 mg/kg (22D pre):REP NKRZAZ 29(Suppl 2),167,81
ims-rat TDLo:275 mg/kg (7-17D preg):TER NKRZAZ 29(Suppl 2),167,81
ipr-rat LD50:913 mg/kg NKRZAZ 29(Suppl 2),167,81
ivn-rat LD50:209 mg/kg NKRZAZ 29(Suppl 2),167,81
orl-mus LD50:13,600 mg/kg NKRZAZ 29(Suppl 2),167,81
ipr-mus LD50:533 mg/kg NKRZAZ 29(Suppl 2),167,81
scu-mus LD50:400 mg/kg JANTAJ 30,77-7,77
ims-mus LD50:427 mg/kg NKRZAZ 29(Suppl 2),167,81
ivn-dog LD50:214 mg/kg NKRZAZ 29(Suppl 2),167,81

SAFETY PROFILE: Poison by subcutaneous and intravenous routes. Moderately toxic by intraperitoneal and intramuscular routes. Mildly toxic by ingestion. Experimental teratogenic and reproductive effects. When heated to decomposition it emits toxic fumes of NO_x.

FOL000 CAS:72275-67-3 HR: 3
FORTIMICIN A SULFATE
mf: $C_{17}H_{35}N_5O_6{\cdot}2H_2O_4S$ mw: 601.73

SYN: ABBOTT-44747

TOXICITY DATA WITH REFERENCE
ivn-rat TDLo:1100 mg/kg (female 7-17D post):TER KSRNAM 20,4274,86
ivn-rat TDLo:550 mg/kg (female 7-17D post):TER KSRNAM 20,4274,86
scu-rat LD50:1365 mg/kg IYKEDH 16,866,65
ivn-rat LD50:86 mg/kg TXAPA9 53,399,80
orl-mus LD50:13,600 mg/kg JJANAX 35,1402,82
ipr-mus LD50:533 mg/kg IYKEDH 16,866,85
scu-mus LD50:653 mg/kg IYKEDH 16,866,85
ivn-mus LD50:94 mg/kg TXAPA9 53,399,80
ims-mus LD50:436 mg/kg TXAPA9 53,399,80
ivn-dog LD50:214 mg/kg IYKEDH 16,866,85
ims-dog LD50:750 mg/kg IYKEDH 16,866,85

SAFETY PROFILE: Poison by intravenous route. Moderately toxic by intraperitoneal, subcutaneous, and intramuscular routes. Mildly toxic by ingestion. An experimental teratogen. When heated to decomposition it emits very toxic fumes of SO_x and NO_x.

FOL100 CAS:35322-07-7 HR: 3
FOSAZEPAM
mf: $C_{18}H_{18}ClN_2O_2P$ mw: 360.80

PROP: Crystals from cyclohexane. Mp: 174–175°.

SYNS: 7-CHLORO-1-((DIMETHYLPHOSPHINYL)METHYL)-1,3-DIHYDRO-5-PHENYL-2H-1,4-BENZODIAZEPINE-2-ONE ☐ HR 930

TOXICITY DATA WITH REFERENCE
orl-rat LD50:3110 mg/kg OYYAA2 10,265,75
ipr-rat LD50:300 mg/kg OYYAA2 10,265,75
scu-rat LD50:1740 mg/kg OYYAA2 10,265,75
ivn-rat LD50:160 mg/kg OYYAA2 10,265,75
orl-mus LD50:2 g/kg OYYAA2 10,265,75
ipr-mus LD50:350 mg/kg OYYAA2 10,265,75
scu-mus LD50:550 mg/kg OYYAA2 10,265,75
ivn-mus LD50:285 mg/kg OYYAA2 10,265,75

SAFETY PROFILE: Poison by intravenous and intraperitoneal routes. Moderately toxic by ingestion and subcutaneous routes. When heated to decomposition it emits toxic fumes of Cl^-, PO_x, and NO_x. See also DIAZEPAM.

FOL200 HR: 2
FOSFOMYCIN DISODIUM HYDRATE
mf: $C_3H_5O_4P{\cdot}2Na{\cdot}H_2O$ mw: 200.05

SYNS: DISODIUM (−)-(1R,2S)-(1,2-EPOXYPROPYL)PHOSPHONATE HYDRATE ☐ DISODIUM FOSFOMYCIN HYDRATE ☐ DISODIUM PHOSPHONOMYCIN HYDRATE ☐ FOSFOMYCIN SODIUM HYDRATE ☐ SODIUM FOSFOMYCIN HYDRATE

TOXICITY DATA WITH REFERENCE
orl-rat LD50:4700 mg/kg IYKEDH 12,668,81
ipr-rat LD50:2060 mg/kg IYKEDH 12,668,81
scu-rat LD50:5100 mg/kg IYKEDH 12,668,81
ivn-rat LD50:1650 mg/kg IYKEDH 12,668,81
ims-rat LD50:2630 mg/kg IYKEDH 12,668,81
orl-mus LD50:8020 mg/kg IYKEDH 12,668,81
ipr-mus LD50:2175 mg/kg IYKEDH 12,668,81

SAFETY PROFILE: Moderately toxic by intraperitoneal, intravenous, and intramuscular routes. Mildly toxic by ingestion. When heated to decomposition it emits toxic fumes of PO_x and Na_2O.

FOM050 CAS:1332-10-1 HR: 3
FOWLER'S SOLUTION

SYNS: ARSENICAL solution ☐ POTASSIUM ARSENITE solution

CONSENSUS REPORTS: NTP 7th Annual Report On Carcinogens. IARC Cancer Review: Group 1 IMEMDT 7, 100,87.

SAFETY PROFILE: Confirmed carcinogen.

FOM100 HR: 3
FOXGLOVE

PROP: These hardy plants grow wild throughout the northern United States, Canada, and Alaska. They are cultivated for the production of digitalis. The flowers are purple or pink and hang down from a central stalk.

SYNS: DIGITALIS ☐ DIGITALIS PURPUREA ☐ FAIRY BELLS ☐ FAIRY CAP ☐ FAIRY GLOVE ☐ FAIRY THIMBLES ☐ FOLKS GLOVE ☐ LADIE'S THIMBLES ☐ LION'S MOUTH ☐ POP-DOCK ☐ RABBIT FLOWER ☐ THIMBLES ☐ THROATWORT ☐ WITCHES' THIMBLES

SAFETY PROFILE: The whole plant contains poisonous digitalis glycosides and irritant saponins. Human systemic effects by ingestion include: mouth pain, nausea, vomiting, abdominal pain, cramps, and diarrhea. Cardiac glycosides may cause death by their effect on heart function. See also DIGITALIS and SAPONIN.

FOM200 CAS:28808-62-0 HR: 3
FRAXINELLONE
mf: $C_{14}H_{16}O_3$ mw: 232.30

PROP: Crystals from EtOH/Et_2O. Mp: 116°.

F

SYNS: 1(3H)-ISOBENZOFURANONE, 3-(3-FURANYL)-3a,4,5,6-TETRAHYDRO-3a,7-DIMETHYL-, (3R-cis)-(9CI) □ PHTHALIDE, 3-(3-FURYL)-3a,4,5,6-TETRAHYDRO-3a,7-DIMETHYL-

TOXICITY DATA WITH REFERENCE
orl-rat TDLo:424 mg/kg (female 5-8D post):REP PLMEAA 53,399,87
orl-rat LD50:274 mg/kg PLMEAA 53,399,87
ipr-rat LD50:116 mg/kg PLMEAA 53,399,87
orl-mus LD50:430 mg/kg PLMEAA 53,399,87
ipr-mus LD50:355 mg/kg PLMEAA 53,399,87

SAFETY PROFILE: Poison by ingestion and intraperitoneal routes. Experimental reproductive effects. When heated to decomposition it emits acrid smoke and irritating fumes.

FON100 **HR: 2**
FRAXINUS JAPONICA Blume, bark extract

TOXICITY DATA WITH REFERENCE
orl-rat LDLo:20 g/kg KSRNAM 4,253,70
scu-rat LD50:5400 mg/kg KSRNAM 4,253,70
ivn-rat LD50:1620 mg/kg KSRNAM 4,253,70
orl-mus LD50:16,300 mg/kg KSRNAM 4,253,70
scu-mus LD50:6300 mg/kg KSRNAM 4,253,70
ivn-mus LD50:3280 mg/kg KSRNAM 4,253,70

SAFETY PROFILE: Moderately toxic by intravenous route. Mildly toxic by ingestion and subcutaneous routes.

FON200 **CAS:13254-34-7** **HR: 2**
FREESIOL
mf: $C_9H_{20}O$ mw: 144.29

SYNS: DIMETOL □ 2,6-DIMETHYL-2-HEPTANOL □ 2-HEPTANOL, 2,6-DIMETHYL- □ LOLITOL

TOXICITY DATA WITH REFERENCE
skn-rbt 500 mg SEV FCTOD7 30,23S,92
eye-rbt 100 mg SEV FCTOD7 30,23S,92
orl-rat LD50:6800 mg/kg FCTOD7 30,23S,92
skn-rbt LD50:>5 g/kg FCTOD7 30,23S,92

CONSENSUS REPORTS: Reported in EPA TSCA Inventory.

SAFETY PROFILE: Low toxicity by ingestion and skin contact. A severe skin and eye irritant. When heated to decomposition it emits acrid smoke and irritating vapors.

FOO000 **CAS:76-13-1** **HR: 1**
FREON 113
mf: $C_2Cl_3F_3$ mw: 187.37

PROP: Colorless gas. Mp: −36.4°, bp: 45.8°, d: 1.5702, autoign temp: 1256°F.

SYNS: ARCTON 63 □ ARKLONE P □ DAIFLON S 3 □ FLUOROCARBON 113 □ FREON 113TR-T □ FRIGEN 113a □ GENETRON 113 □ HALOCARBON 113 □ ISCEON 113 □ KAISER CHEMICALS 11 □ KHLADON 113 □ R 113 □ REFRIGERANT 113 □ TRICHLOROTRIFLUOROETHANE □ 1,1,2-TRICHLORO-1,2,2-TRIFLUOROETHANE (OSHA, ACGIH, MAK) □ UCON 113 □ UCON FLUOROCARBON 113 □ UCON 113/HALOCARBON 113

TOXICITY DATA WITH REFERENCE
skn-rbt 500 mg open MLD UCDS** 7/10/70
orl-rat LD50:43 g/kg JMCMAR 7,378,64
ihl-rat LCLo:87,000 ppm/6H JOCMA7 4,262,62
ihl-mus LCLo:25 pph/90S ANASAB 16,3,61

CONSENSUS REPORTS: Reported in EPA TSCA Inventory.

OSHA PEL: TWA 1000 ppm; STEL 1250 ppm
ACGIH TLV: TWA 1000 ppm; STEL 1250 ppm
DFG MAK: 500 ppm (3800 mg/m³)

SAFETY PROFILE: Mildly toxic by ingestion and inhalation. Affects the central nervous system in humans. A skin irritant. Combustible when exposed to heat or flame. Incompatible with Al; Ba; Li; Sm; NaK alloy; Ti. See also CHLORINATED HYDROCARBONS, ALIPHATIC; and FLUORIDES.

For occupational chemical analysis use NIOSH: see 1,1,2-Trichloro-1,2,2-trifluoroethane, 1020.

FOO509 **CAS:76-14-2** **HR: 1**
FREON 114
mf: $C_2Cl_2F_4$ mw: 170.92

PROP: Colorless, practically odorless, noncorrosive, nonirritating, nonflammable gas. Faint, ether-like odor in high concentrations. D: 1.5312, mp: −94°, bp: 4.1°, n (0/D) 1.3092. Insol in water; sol in alc and ether.

SYNS: ARCTON 33 □ ARCTON 114 □ CRYOFLUORAN □ CRYOFLUORANE □ sym-DICHLOROTETRAFLUOROETHANE □ 1,2-DICHLORO-1,1,2,2-TETRAFLUOROETHANE (MAK) □ DICHLOROTETRAFLUOROETHANE (OSHA, ACGIH) □ F 114 □ FC 114 □ FLUORANE 114 □ FLUOROCARBON 114 □ FRIGEN 114 □ FRIGIDERM □ GENETRON 114 □ GENETRON 316 □ HALOCARBON 114 □ LEDON 114 □ PROPELLANT 114 □ R 114 □ 1,1,2,2-TETRAFLUORO-1,2-DICHLOROETHANE □ UCON 114

TOXICITY DATA WITH REFERENCE
ihl-rat LC50:72 pph/30M EJTXAZ 9,385,76
ihl-mus LC50:70 pph/30M EJTXAZ 9,385,76
ihl-rbt LC50:75 pph/30M EJTXAZ 9,385,76

CONSENSUS REPORTS: Reported in EPA TSCA Inventory.

OSHA PEL: TWA 1000 ppm
ACGIH TLV: TWA 1000 ppm
DFG MAK: 1000 ppm (7000 mg/m³)

SAFETY PROFILE: An asphyxiant. See also DICHLOROTETRAFLUOROETHANE.

For occupational chemical analysis use NIOSH: see Dichlorodifluoromethane and 1,2-Dichlorotetrafluoroethane 1018.

FOO510 **CAS:39432-81-0** **HR: 1**
FREON 502
DOT: UN 1973
mf: $C_2ClF_5 \cdot CHClF_2$ mw: 240.94

SYNS: CHLORODIFLUOROMETHANE and CHLOROPENTAFLUOROETHANE MIXTURE (DOT) □ ETHANE, CHLOROPENTAFLUORO-, mixt. with CHLORODIFLUOROMETHANE □ R502 (DOT) □ REFRIGERANT 502

DOT CLASSIFICATION: 2.2; *Label:* Nonflammable Gas

SAFETY PROFILE: A simple asphyxiant. When heated to decomposition it emits toxic vapors of NO_x and Cl^-.

FOO515 CAS:50815-73-1 HR: 1
FREON 503
DOT: UN 2599

SYNS: CHLOROTRIFLUOROMETHANE mixed with TRIFLUOROMETHANE ☐ CHLOROTRIFLUOROMETHANE and TRIFLUOROMETHANE AZEOTROPIC MIXTURE (DOT) ☐ METHANE, CHLOROTRIFLUORO-, mixt. with TRIFLUOROMETHANE (9CI) ☐ METHANE, TRIFLUORO-, mixt. with CHLOROTRIFLUOROMETHANE ☐ R503 (DOT)

DOT CLASSIFICATION: 2.2; *Label:* Nonflammable Gas

SAFETY PROFILE: A simple asphyxiant. When heated to decomposition it emits toxic vapors of NO_x and Cl^-.

FOO525 CAS:124-73-2 HR: 1
FREON 114B2
mf: $C_2Br_2F_4$ mw: 259.84

SYNS: 1,2-DIBROMOPERFLUOROETHANE ☐ sym-DIBROMOTETRAFLUOROETHANE ☐ 1,2-DIBROMOTETRAFLUOROETHANE ☐ 1,2-DIBROMO-1,1,2,2-TETRAFLUOROETHANE ☐ ETHANE, 1,2-DIBROMO-TETRAFLUORO- ☐ ETHANE, 1,2-DIBROMO-1,1,2,2-TETRAFLUORO-(9CI) ☐ F-114B2 ☐ FC 114B2 ☐ FLUOBRENE ☐ HALON 2402 ☐ KHLADON 114B2 ☐ R 114B2

TOXICITY DATA WITH REFERENCE
ihl-rat LC50:869 g/m³/2H GISAAA 55(2),17,90
ihl-mus LC50:300 g/m³/2H 85JCAE -,136,86

CONSENSUS REPORTS: Reported in EPA TSCA Inventory.

SAFETY PROFILE: Slightly toxic by inhalation. When heated to decomposition it emits toxic vapors of Br^- and F^-.

FOO550 CAS:1717-00-6 HR: 1
FREON 141
mf: $C_2H_3Cl_2F$ mw: 116.95

SYNS: 1,1-DICHLORO-1-FLUOROETHANE ☐ ETHANE, 1,1-DICHLORO-1-FLUORO-

TOXICITY DATA WITH REFERENCE
ihl-rat LD50:240 g/m³/2H 85GMAT-,46,82
ihl-mus LC50:151 g/m³/2H 85JCAE-,134,86

SAFETY PROFILE: Slightly toxic by inhalation. When heated to decomposition it emits toxic vapors of F^- and Cl^-.

FOO600 CAS:9007-81-2 HR: D
FREUND'S ADJUVANT

TOXICITY DATA WITH REFERENCE
par-rat TDLo:1 g/kg (female 1D pre):REP JRPFA4 14,147,67

SAFETY PROFILE: Experimental reproductive effects. When heated to decomposition it emits acrid smoke and irritating fumes.

FOO875 CAS:23191-75-5 HR: 3
FTORIN
mf: $C_{20}H_{22}F_3NO_4$ mw: 397.43

SYNS: 1,4-DIHYDRO-2,6-DIMETHYL-4-(α,α,α-TRIFLUORO-o-TOLYL)-3,5-PYRIDINEDICARBOXYLIC ACID DIETHYL ESTER ☐ FTORIN (PHARMACEUTICAL) ☐ SKF 24260 ☐ 4-(2-TRIFLUOROMETHYLPHENYL)-3,5-DICARBETHOXY-2,6-DIMETHYL-1,4-DIHYDROPYRIDINE

TOXICITY DATA WITH REFERENCE
orl-rat LD50:1225 mg/kg JMCMAR 17,956,74
ivn-rat LDLo:5 mg/kg JMCMAR 17,956,74
orl-mus LD50:1480 mg/kg JMCMAR 17,956,74
ipr-mus LD50:38 mg/kg PCJOAU 16,817,82
ivn-dog LDLo:700 µg/kg JMCMAR 17,956,74

SAFETY PROFILE: Poison by intravenous and intraperitoneal routes. Moderately toxic by ingestion. When heated to decomposition it emits toxic fumes of F^- and NO_x.

FOP000 HR: 3
FUEL OIL
DOT: NA 1993

PROP: A petroleum fraction consisting of a complex mixture of aromatic, paraffinic, olefinic, and naphthenic hydrocarbons. Brown, sltly viscous liquid. Flash p: 100°F, d: <1, autoign temp: 494°F.

SYNS: AUTOMOTIVE DIESEL OIL ☐ DIESEL FUEL (DOT) ☐ DIESEL OIL (PETROLEUM) ☐ DIESEL OILS ☐ DIESEL TEST FUEL ☐ FUELS, DIESEL ☐ OLEJ NAPEDOWY III

TOXICITY DATA WITH REFERENCE
skn-rbt 500 mg MOD NTIS** AD-A172-198
orl-rat LD50:9 g/kg 52MLA2 1,1,83

CONSENSUS REPORTS: IARC Cancer Review: Group 3 IMEMDT 45,219,89; Human Inadequate Evidence IMEMDT 45,219,89

DOT CLASSIFICATION: 3; *Label:* None

SAFETY PROFILE: Mildly toxic by ingestion. A moderate skin irritant. Flammable when exposed to heat or flame; can react vigorously with oxidizing materials. To fight fire, use CO_2, dry chemical. When heated to decomposition it emits acrid smoke and irritating fumes. See also DIESEL EXHAUST, DIESEL EXHAUST EXTRACT, DIESEL EXHAUST PARTICLES, DIESEL FUEL MARINE.

FOP050 CAS:68476-33-5 HR: 3
FUEL OIL, RESIDUAL

SYN: RESIDUAL(HEAVY) FUEL OIL

CONSENSUS REPORTS: IARC Cancer Review: Group 2B IMEMDT 45,239,89; Animal Sufficient Evidence IMEMDT 45,239,89. Reported in EPA TSCA Inventory.

SAFETY PROFILE: Confirmed carcinogen. When heated to decomposition it emits acrid smoke and irritating vapors.

F

FOP100 **HR: 2**
FUEL OIL, pyrolyzate

SYN: WATER QUENCH PYROLYSIS FUEL OIL

TOXICITY DATA WITH REFERENCE
skn-mus TDLo:977 g/kg/8W-I:CAR AIHAAP 38,730,77

SAFETY PROFILE: Questionable carcinogen with experimental carcinogenic data. When heated to decomposition it emits acrid smoke and irritating fumes.

FOQ000 **CAS:4368-28-9** **HR: 3**
FUGU POISON
mf: $C_{11}H_{17}N_3O_8$ mw: 319.31

PROP: Crystals.

SYNS: MACULOTOXIN □ SPHEROIDINE □ TARICHATOXIN □ TETRODONTOXIN □ TETRODOTOXIN □ TETRODOXIN □ TTX

TOXICITY DATA WITH REFERENCE
orl-mus LD50:435 µg/kg JJPAAZ 17,267,67
ipr-mus LD50:8 µg/kg SCIEAS 144,1100,64
scu-mus LD50:8 µg/kg CTOXAO 4,331,71
ivn-mus LD50:9 µg/kg JJPAAZ 17,267,67

SAFETY PROFILE: Poison by ingestion, intraperitoneal, subcutaneous, and intravenous routes. When heated to decomposition it emits toxic fumes of NO_x.

FOR000 **CAS:3309-87-3** **HR: 3**
FUJITHION
mf: $C_8H_{10}ClO_3PS$ mw: 252.66

SYNS: S-(p-CHLOROPHENYL)-O,O-DIMETHYL PHOSPHOTHIOATE □ O,O-DIMETHYL-S-p-CHLOROPHENYL PHOSPHOROTHIOATE

TOXICITY DATA WITH REFERENCE
orl-mus LD50:94 mg/kg BESAAT 15,118,69
skn-mus LD50:920 mg/kg BESAAT 15,118,69

SAFETY PROFILE: Poison by ingestion. Moderately toxic by skin contact. When heated to decomposition it emits very toxic fumes of Cl^-, PO_x, and SO_x.

FOS000 **HR: 3**
FULMINATES

SAFETY PROFILE: Variable toxicity. A very dangerous fire hazard when exposed to heat or flame. Severe explosion hazard when shocked or exposed to heat or flame. See also various fulminates and EXPLOSIVES, HIGH.

The fulminates are a group of explosives that are very sensitive to heat, impact, and friction when dry. They should be kept moist until ready for use. If compressed beyond 25,000 psi they become what is known as "dead pressed," i.e., not capable of being exploded by flame. Fulminates are subject to deterioration when stored in hot climates. They decompose completely and violently when detonated. They can be ignited with a flame or "spit," with a fuse, or with an electrically heated wire. They are widely used as initiators or primers for detonation of high explosives or the ignition of powder. They are commonly used in combination with substances that provide a more prolonged blow and a bigger flame than fulminates alone. In the reinforced type of detonator, fulminates are made more effective by the addition of a more sensitive and powerful high explosive such as tetryl. This material is generally used in the manufacture of caps and detonators for initiating explosions for military, industrial, and sporting purposes.

All precautions required for protection of magazines apply to storage of these materials. They should not be handled when frozen. Wet fulminate of mercury or wet floor coverings containing small quantities of fulminates may be burned on windrows of flammable material. Nonexplosive products are formed by neutralizing fulminates with cold sodium thiosulfate. All floors, tables, and walls where the dry fulminates have been used should be washed with this solution. In the manufacture of mercury fulminate, the fumes given off are toxic and flammable. Care is required to prevent fulminate dust from being carried off in the exhaust system: deposits thus made have caused explosions. Careful attention should be given to cleanliness as foreign or gritty materials in the product may cause an unexpected explosion. The floors on which fulminates are used should be covered with 1/16-inch cloth-inserted rubber packing or its equal. All cracks and crevices should be covered. The walls of these rooms should be covered with glazed, waterproof material. Frequent washing with neutralizing solution is necessary. In manufacture, the fulminate is dried on muslin squares on a drying table. Drying tables may be heated with hot water or the dry house may be heated with an air blower system to between 50 and 60°. Primer caps and detonators loaded with fulminate of mercury are less sensitive than the dry bulk material but must be handled with great care. Fires involving these assemblies should be treated the same as for the bulk material. They will explode as soon as fire reaches them. Stocks in an assembly or loading room should be kept as small as possible. Examples of fulminates commonly used in the explosive industry are mercury fulminate, copper fulminate, and silver fulminate.

FOS050 **CAS:506-85-4** **HR: 3**
FULMINIC ACID
mf: CHNO mw: 43.02

CONSENSUS REPORTS: Cyanide and its compounds are on the Community Right-To-Know List.

DOT CLASSIFICATION: Forbidden

SAFETY PROFILE: An unstable explosive sensitive to heat, shock, or friction. When heated to decomposition it emits toxic fumes of NO_x.

FOS100 **CAS:14976-57-9** **HR: 3**
FULUMINOL
mf: $C_{21}H_{26}ClNO \cdot C_4H_4O_4$ mw: 460.01

PROP: Crystals from MeOH. Mp: 177–178° (decomp).

SYNS: AGASTEN □ ALOGINAN □ ALPHAMIN □ ANHISTAN □ (+)-2-(2-((p-CHLORO-α-METHYL-α-PHENYLBENZYL)OXY)ETHYL)-1-METHYL PYRROLIDINE FUMARATE □ CLEMANIL □ CLEMASTINE FUMARATE □ CLEMASTINE HYDROGEN FUMARATE □ HS 592 □ IN-

BESTAN □ KINOTOMIN □ LACRETIN □ LECASOL □ MAIKOHIS □ MALLERMIN-F □ MARSTHINE □ MASLETINE □ MECLASTINE HYDROGEN FUMARATE □ PILORAL □ RECONIN □ TAVEGIL □ TAVEGYL □ TELGIN-G □ TIVIST □ TRABEST □ XOLAMIN

TOXICITY DATA WITH REFERENCE
orl-rat LD50:3550 mg/kg BCFAAI 106,467,67
ivn-rat LD50:82 mg/kg BCFAAI 106,467,67
orl-mus LD50:730 mg/kg BCFAAI 106,467,67
ivn-mus LD50:43 mg/kg BCFAAI 106,467,67
orl-dog LD50:175 mg/kg BCFAAI 106,467,67
orl-rbt LD50:1 g/kg NIIRDN 6,230,82
ivn-rbt LD50:19 mg/kg BCFAAI 106,467,67

SAFETY PROFILE: Poison by ingestion and intravenous routes. When heated to decomposition it emits toxic fumes of Cl^- and NO_x.

FOS300 CAS:72443-10-8 HR: 3
6-FULVENOSELONE
mf: C_6H_4Se mw: 155.06

CONSENSUS REPORTS: Selenium and its compounds are on the Community Right-To-Know List.

OSHA PEL: TWA 0.2 mg(Se)/m^3
ACGIH TLV: TWA 0.2 mg(Se)/m^3
DFG MAK: 0.1 mg(Se)/m^3

SAFETY PROFILE: Polymerizes explosively at −196°C. When heated to decomposition it emits toxic fumes of Se. See also SELENIUM COMPOUNDS.

FOT000 CAS:6029-87-4 HR: 3
FULVINE
mf: $C_{16}H_{23}NO_5$ mw: 309.40

PROP: Prisms from Me_2CO. Mp: 212–213°.

SYN: CRISPATINE

TOXICITY DATA WITH REFERENCE
sln-dmg-par 10 mmol/L JOGNAU 59,273,66
trn-dmg-par 10 mmol/L JOGNAU 59,273,66
ipr-rat TDLo:20 mg/kg (female 9-12D post):TER EXPTAX 9,59,74
ipr-rat TDLo:40 mg/kg (9-12D preg):REP EXPTAX 9,59,74
ipr-rat DL50:40 mg/kg PAREAQ 22,429,70

SAFETY PROFILE: Poison by intraperitoneal route. Experimental teratogenic and reproductive effects. Mutation data reported. When heated to decomposition it emits toxic fumes of NO_x.

FOU000 CAS:110-17-8 HR: 3
FUMARIC ACID
mf: $C_4H_4O_4$ mw: 116.08

PROP: White, monoclinic, prismatic, crystals, needles or leaflets; odorless. Mp: 300–302° (sealed tube), d: 1.635 @ 20°/4°, bp: 290°. Sol in EtOH; sltly sol in H_2O, Et_2O, and Me_2CO; prac insol in C_6H_6.

SYNS: ALLOMALEIC ACID □ BOLETIC ACID □ trans-BUTENEDIOIC ACID □ (E)-BUTENEDIOIC ACID □ trans-1,2-ETHYLENEDICARBOXYLIC ACID □ (E)1,2-ETHYLENEDICARBOXYLIC ACID □ KYSELINA FUMAROVA (CZECH) □ LICHENIC ACID □ NSC-2752 □ U-1149 □ USAF EK-P-583

TOXICITY DATA WITH REFERENCE
skn-rbt 500 mg/24H MLD 28ZPAK -,51,72
eye-rbt 100 mg/24H MOD 28ZPAK -,51,72
dni-rat-ivn 40 mg/kg JJCREP 77,750,86
orl-rat LD50:9300 mg/kg TXAPA9 42,417,77
ipr-rat LDLo:587 mg/kg JAPMA8 35,298,46
ipr-mus LD50:100 mg/kg NTIS** AD277-689
orl-rbt LDLo:5000 mg/kg IECHAD 15,628,23
skn-rbt LD50:20 g/kg TXAPA9 42,417,77

CONSENSUS REPORTS: Reported in EPA TSCA Inventory.

SAFETY PROFILE: Poison by intraperitoneal route. Mildly toxic by ingestion and skin contact. A skin and eye irritant. Mutation data reported. Combustible when exposed to heat or flame; can react vigorously with oxidizing materials. When heated to decomposition it emits acrid smoke and irritating fumes.

FOW000 CAS:130-86-9 HR: 3
FUMARINE
mf: $C_{20}H_{19}NO_5$ mw: 353.40

PROP: Crystals from $CHCl_3$/EtOH. Mp: 207°.

SYNS: BIFLORINE □ CORYDININE □ MACLEYINE □ 7-METHYL-2,3:9,10-BIS(METHYLENEDIOXY)-7,13a-SECOBERBIN-13a-ONE □ PROTOPINE □ 4,6,7,14-TETRAHYDRO-5-METHYL-BIS(1,3)BENZODIOXOLO(4,5-c:5′,6′-g)AZECIN-13(5H)-ONE

TOXICITY DATA WITH REFERENCE
ipr-rat LDLo:100 mg/kg NCNSA6 5,24,53
ipr-mus LD50:482 mg/kg YHTPAD 16(6),7,81
orl-gpg LD50:237 mg/kg THERAP 28,767,73
ipr-gpg LD50:116 mg/kg THERAP 28,767,73

SAFETY PROFILE: Poison by ingestion and intraperitoneal routes. When heated to decomposition it emits toxic fumes of NO_x.

FOX000 CAS:764-42-1 HR: 3
FUMARONITRILE
mf: $C_4H_2N_2$ mw: 78.08

PROP: Needles. Mp: 96°, bp: 186°. Sol in EtOH, Et_2O, and C_6H_6.

SYNS: 2-BUTENEDINITRILE, (E)- □ trans-1,2-DICYANOETHENE □ (E)-1,2-DICYANOETHYLENE □ FUMARIC ACID DINITRILE □ FUMARONITRILE □ FUMARSAEUREDINITRIL

TOXICITY DATA WITH REFERENCE
sln-smc 17 mg/L MUREAV 241,255,90
orl-rat LD50:132 mg/kg TOLED5 12,157,82
ivn-mus LD50:56 mg/kg CSLNX* NX#05718

CONSENSUS REPORTS: Reported in EPA TSCA Inventory. Cyanide and its compounds are on the Community Right-To-Know List.

SAFETY PROFILE: Poison by ingestion and intravenous routes. Mutation data reported.When heated to decomposition it emits toxic fumes of NO_x and CN^-. See also NITRILES.

F

FOY000 **CAS:627-63-4** **HR: 2**
FUMARYL CHLORIDE
DOT: UN 1780
mf: $C_4H_2Cl_2O_2$ mw: 152.96

PROP: Clear, straw-colored liquid. Mp: 158–160°, d: 1.408 @ 20°/4°.

SYNS: CHLORURE de FUMARYLE (FRENCH) □ DICHLORID KYSELINY FUMAROVE (CZECH) □ FUMAROYL CHLORIDE □ FUMARYLCHLORID (CZECH) □ TL 189

TOXICITY DATA WITH REFERENCE
skn-rbt 750 μg/24H SEV 85JCAE-,327,86
eye-rbt 5 mg/24H SEV 85JCAE-,327,86
orl-rat LD50:810 mg/kg AIHAAP 30,470,69
ihl-rat LCLo:500 ppm/4H AIHAAP 30,470,69
ihl-mus LCLo:1000 mg/m³/10M NDRC** NDCrc-132,May,42
skn-rbt LD50:1410 mg/kg AIHAAP 30,470,69

CONSENSUS REPORTS: Reported in EPA TSCA Inventory.

DOT CLASSIFICATION: 8; *Label:* Corrosive

SAFETY PROFILE: Moderately toxic by ingestion, inhalation, and skin contact. A skin, eye and mucous membrane irritant. A corrosive agent. Will react with water or steam to produce toxic and corrosive fumes. When heated to decomposition it emits highly toxic fumes of phosgene and HCl.

FOZ000 **CAS:23110-15-8** **HR: 2**
FUMIDIL
mf: $C_{26}H_{34}O_7$ mw: 458.60

PROP: Yellow needles (MeOH aq). Mp: 189–194°. Sol in dil alkalies; insol in hydrocarbons and H_2O. Isolated from *A. fumigatus* (ANTCAO 1,54,51).

SYNS: AMEBACILIN □ FUGILIN □ FUMADIL B □ FUMAGILLIN

TOXICITY DATA WITH REFERENCE
unk-rat TDLo:25 mg/kg (11D preg):TER 85DJA5 -,95,71
unk-rat TDLo:25 mg/kg (11D preg):REP 85DJA5 -,95,71
orl-mus LD50:2000 mg/kg 85ERAY 3,1834,78
scu-mus LD50:800 mg/kg ANTCAO 1,54,51

SAFETY PROFILE: Moderately toxic by ingestion and subcutaneous routes. An experimental teratogen. Experimental reproductive effects. When heated to decomposition it emits acrid smoke and fumes.

FPB875 **CAS:35554-44-0** **HR: 3**
FUNGAFLOR
mf: $C_{14}H_{14}Cl_2N_2O$ mw: 297.20

PROP: Solidified oil. Sltly sol in organic solvents; poorly sol in water.

SYNS: (±)-1-(β-(ALLYLOXY)-2,4-DICHLOROPHENETHYL)IMIDAZOLE □ 1-(2-(2,4-DICHLOROPHENYL)-2-(2-PROPENYLOXY)ETHYL)-1H-IMIDAZOLE □ 1-(2-(2,4-DICHLORPHENYL)-2-(PROPENYLOXY)AETHYL)-1H-IMIDAZOLE □ ENILOCONAZOL (SP) □ IMAVEROL □ IMAZALIL □ R 23979

TOXICITY DATA WITH REFERENCE
eye-rbt 49 mg MOD ARZNAD 31,309,81
orl-rat TDLo:2240 mg/kg (16-22D preg/21D post):REP ARZNAD 31,309,81
orl-rbt TDLo:32,500 μg/kg (female 6-18D post):REP ARZNAD 31,309,81
orl-rat TDLo:560 mg/kg (16-22D preg/21D post):REP ARZNAD 31,309,81
orl-rbt TDLo:8190 μg/kg (female 6-18D post):REP ARZNAD 31,309,81
orl-rat LD50:227 mg/kg ARZNAD 31,309,81
ihl-rat LC50:16 g/m³/4H PEMNDP 9,482,91
skn-rat LD50:4200 mg/kg PEMNDP 8,471,87
ipr-rat LD50:155 mg/kg ARZNAD 31,309,81

SAFETY PROFILE: Poison by ingestion and intraperitoneal routes. Experimental reproductive effects. A skin and eye irritant. When heated to decomposition it emits toxic fumes of Cl^- and NO_x.

FPC000 **CAS:6834-98-6** **HR: 3**
FUNGICHROMIN
mf: $C_{35}H_{58}O_{12}•H_2O$ mw: 688.95

PROP: Needles from MeOH; hydrate pale-yellow crystals. Mp: 235° (decomp).

SYNS: ANTIBIOTIC A-246 □ COGOMYCIN □ FUNGICHROMIN, HYDRATE □ 14-HYDROXYFILIPIN □ LAGOSIN □ MOLDCIDIN B □ NSC-105388 □ PENTAMYCIN

TOXICITY DATA WITH REFERENCE
orl-mus LD50:1624 mg/kg 85GDA2 2,212,80
ipr-mus LDLo:16,400 μg/kg ABANAE 2,716,54/55

SAFETY PROFILE: Poison by intraperitoneal route. Moderately toxic by ingestion. When heated to decomposition it emits acrid smoke and irritating fumes.

FPC100 **HR: 3**
FUNGINON
mf: $C_6H_8O_4$ mw: 144.14

TOXICITY DATA WITH REFERENCE
orl-mus LD50:109 mg/kg 85GDA2 5,371,81
ipr-mus LD50:20 mg/kg 85GDA2 5,371,81
ivn-mus LD50:24 mg/kg 85GDA2 5,371,81

SAFETY PROFILE: Poison by ingestion, intravenous, and intraperitoneal routes. When heated to decomposition it emits acrid smoke and irritating fumes.

FPD000 **CAS:11055-06-4** **HR: 3**
FUNICOLOSIN
mf: $C_{27}H_{41}NO$ mw: 395.69

PROP: Fine needles. Mp: 165–166°.

TOXICITY DATA WITH REFERENCE
orl-rat LD50:5 mg/kg JANTAJ 31,533,78
ipr-rat LD50:5 mg/kg JANTAJ 31,533,78
orl-mus LD50:5 mg/kg JANTAJ 31,533,78
ipr-mus LD50:4 mg/kg 85ERAY 3,1803,78

SAFETY PROFILE: Poison by ingestion and intraperitoneal routes. When heated to decomposition it emits toxic fumes of NO_x.

FPD100 **HR: 2**
FUQUA

PROP: Creeping vines with tendrils, deeply cut leaves, and tubular yellow flowers. The yellow-orange fruit is pear-shaped or oval, has warts and a bright red pulp. The fruit splits open when ripe. They are common weeds in the Gulf coast states, Hawaii, Guam, and the West Indies.

SYNS: BALSAM APPLE ☐ BALSAM PEAR ☐ BITTER CUCUMBER ☐ BITTER GOURD ☐ CUNDEAMOR (CUBA, PUERTO RICO) ☐ MOMORDICA BALSAMINA ☐ MOMORDICA CHARANTIA ☐ MOMORDIQUE A FUEILLES de VIGNE ☐ SORCI ☐ SORROSIE (HAITI) ☐ WILD BALSAM APPLE ☐ YESQUIN (HAITI)

SAFETY PROFILE: The seeds and skin of the fruit contain the poisonous momordin, a toxalbumin that inhibits protein synthesis in the intestinal wall. The red pulp surrounding the seeds and the boiled leaves are edible. Ingestion of the seeds or skin of the fruit causes after a delay period: nausea, vomiting, diarrhea, and low blood sugar. See also ABRIN as an example toxalbumin.

FPE100 **CAS:405-22-1** **HR: 3**
FURADROXYL
mf: $C_8H_{10}N_4O_5$ mw: 242.22

PROP: Bright orange plates from alc. Mp: 214–216° (decomp). Solubility in water 1:2000.

SYNS: NIDROXYZONE ☐ 5-NITRO-2-FURALDEHYDE-2-(2-HYDROXYETHYL)SEMICARBAZONE ☐ USAF EA-3

TOXICITY DATA WITH REFERENCE
orl-mus TDLo:5520 mg/kg (female 23D pre):REP ENDOAO 69,331,61
orl-mus TDLo:3360 mg/kg (female 14D pre):REP ENDOAO 69,331,61
orl-rat TDLo:1800 mg/kg (30D male):REP ENDOAO 56,727,55
orl-mus TDLo:2520 mg/kg (male 21D pre):REP ENDOAO 69,331,61
orl-mus LD50:600 mg/kg JAPMA8 39,313,50
ipr-mus LD50:200 mg/kg NTIS** AD277-689

SAFETY PROFILE: Poison by intraperitoneal route. Moderately toxic by ingestion. Experimental reproductive effects. When heated to decomposition it emits toxic fumes of NO_x. See also ALDEHYDES.

FPF000 **CAS:556-12-7** **HR: 3**
FURALAZIN
mf: $C_9H_7N_5O_3$ mw: 233.2

PROP: Red powder from Me_2CO. Mp: 270° (decomp).

SYNS: 3-AMINO-6-(2-(5-NITRO-2-FURYL)VINYL)-as-TRIAZINE ☐ 3-AMINO-6-(2-(5-NITRO-2-FURYL)VINYL)-1,2,4-TRIAZINE ☐ NFT ☐ PANFURAN ☐ 1,2,4-TRIAZIN-3-AMINE, 6-(2-(5-NITRO-2-FURANYL)ETHENYL)-(9CI)

TOXICITY DATA WITH REFERENCE
plc-esc 100 µg/L MUREAV 26,3,74
mmo-esc 300 µg/L CJMIAZ 11,185,65
mmo-eug 2 mg/L JPROAR 17,129,70
orl-rat LD50:250 mg/kg YKKZAJ 83,778,63
orl-mus LD50:500 mg/kg AACHAX-,275,62

SAFETY PROFILE: A poison by ingestion. Mutation data reported. When heated to decomposition it emits toxic fumes of NO_x.

FPH000 **CAS:5428-37-5** **HR: 3**
2-FURALDEHYDE AZINE
mf: $C_{10}H_8N_2O_2$ mw: 188.20

TOXICITY DATA WITH REFERENCE

SYNS: 2-FURANCARBOXALDEHYDE, (2-FURANYLMETHYLENE)HYDRAZONE (9CI) ☐ FURFURALDAZINE ☐ FURFURALDEHYDE AZINE ☐ FURFURYL AZINE

TOXICITY DATA WITH REFERENCE
ipr-mus LDLo:125 mg/kg CBCCT* 6,220,54
ivn-mus LD50:180 mg/kg CSLNX* NX#00398

CONSENSUS REPORTS: Reported in EPA TSCA Inventory.

SAFETY PROFILE: Poison by intravenous and intraperitoneal routes. When heated to decomposition it emits toxic fumes of NO_x. See also ALDEHYDES.

FPI000 **CAS:139-91-3** **HR: 3**
FURALTADONE
mf: $C_{13}H_{16}N_4O_6$ mw: 324.33

PROP: Yellow crystals from 95% ethanol. Decomp 206°. Sparingly sol in water: about 75 mg/100 mL at 25°.

SYNS: ALTABACTINA ☐ ALTAFUR ☐ F-150 ☐ FURAZOLIN ☐ FURAZOLINE ☐ FURMETHANOL ☐ FURMETHONOL ☐ FURMETONOL ☐ IBIFUR ☐ MEDIFURAN ☐ 5-MORPHOLINOMETHYL-3-(5-NITRO-2-FURFURYLIDINE-AMINO)-2-OXAZOLIDINONE ☐ NF 260 ☐ NITRALDONE ☐ NITROFURMETHONE ☐ NITROFURMETON ☐ OTIFURIL ☐ SEPSINOL ☐ ULTRAFUR ☐ UNIFUR ☐ VALSYN

TOXICITY DATA WITH REFERENCE
mmo-sat 2500 ng/plate MUREAV 136,1,84
mmo-esc 10 µg/plate MUREAV 26,3,74
pic-esc 5 mg/L CJMIAZ 10,932,64
mmo-omi 5 mg/L CUMIDD 10,19,84
orl-mus LD50:600 mg/kg FRPSAX 19,269,64
ipr-mus LD50:1000 mg/kg JPPMAB 16,663,64
ivn-mus LD50:400 mg/kg FRPSAX 19,269,64

CONSENSUS REPORTS: EPA Genetic Toxicology Program.

SAFETY PROFILE: A poison by intravenous route. Moderately toxic by ingestion and intraperitoneal route. Mutation data reported. When heated to decomposition it emits toxic fumes of NO_x.

FPI100 **CAS:3759-92-0** **HR: 3**
FURALTADONE HYDROCHLORIDE
mf: $C_{13}H_{16}N_4O_6$•ClH mw: 360.79

SYNS: 5-MORPHOLINOMETHYL-3-(5-NITROFURFURYLIDINE)AMINO-2-OXAZOLIDINONE HYDROCHLORIDE ☐ NF-269 ☐ NF-902 HYDROCHLORIDE

TOXICITY DATA WITH REFERENCE
mmo-esc 100 µmol/L RAREAE 75,424,78
orl-mus LD50:1000 mg/kg JPPMAB 16,663,64
ipr-mus LD50:190 mg/kg JPPMAB 16,663,64

CONSENSUS REPORTS: EPA Genetic Toxicology Program.

SAFETY PROFILE: Poison by intraperitoneal route. Moderately toxic by ingestion. Mutation data reported. When heated to decomposition it emits toxic fumes of NO_x and HCl.

FPI150 **CAS:3031-51-4** **HR: 3**
I-FURALTADONE HYDROCHLORIDE
mf: $C_{13}H_{16}N_4O_6{\cdot}ClH$ mw: 360.79

PROP: Yellow crystals. Decomposes @ 206°.

SYNS: FURMETHONOL □ I-5-(MORPHOLINOMETHYL)-3-((5-NITROFURFURYLIDENE)AMINO)-2-OXAZOLIDINONEHYDROCHLORIDE □ NF-260

TOXICITY DATA WITH REFERENCE
mmo-eug 10 mg/L JPROAR 17,129,70
orl-rat TDLo:25 g/kg/46W-C:CAR JNCIAM 51,403,73
orl-mus LD50:600 mg/kg FRPSAX 19,269,64
ivn-mus LD50:400 mg/kg FRPSAX 19,269,64

CONSENSUS REPORTS: IARC Cancer Review: Group 2B IMEMDT 7,56,87; Animal Limited Evidence IMEMDT 7,161,74.

SAFETY PROFILE: Suspected carcinogen with experimental carcinogenic data. Poison by intravenous route. Moderately toxic by ingestion. Mutation data reported. When heated to decomposition it emits very toxic fumes of HCl and NO_x.

FPK000 **CAS:110-00-9** **HR: 3**
FURAN
DOT: UN 2389
mf: C_4H_4O mw: 68.08

PROP: Water white volatile liquid. Mp: −85.65°, bp: 32°, lel: 2.3%, uel: 14.3%, flash p: −32°F, d: 0.964 @ 0°, vap d: 2.35. Sol in EtOH, Et_2O; insol in H_2O.

SYNS: DIVINYLENE OXIDE □ FURFURAN □ NCI-C56202 □ OXACYCLOPENTADIENE □ OXOLE □ RCRA WASTE NUMBER U124 □ TETROLE

TOXICITY DATA WITH REFERENCE
cyt-ham:ovr 184 mmol/L CALEDQ 13,89,81
orl-rat TDLo:240 mg/kg/16D-I NTPTR* NTP-TR-402,93
ipr-rat LD50:5200 µg/kg AIHAAP 40,310,79
orl-mus TDLo:1950 mg/kg/13W-I NTPTR* NTP-TR-402,93
ihl-mus LC50:120 mg/m³/1H AIHAAP 40,310,79
ipr-mus LD50:7 mg/kg AIHAAP 40,310,79

CONSENSUS REPORTS: EPA Extremely Hazardous Substances List. Reported in EPA TSCA Inventory.

DOT CLASSIFICATION: 3; *Label:* Flammable Liquid

SAFETY PROFILE: Poison by inhalation and intraperitoneal routes. Moderately toxic by ingestion and skin contact. A narcotic. Mutation data reported. The exposure concentration limit of 10 ppm together with its low boiling point requires that adequate ventilation be provided in areas where this chemical is handled. Contact with liquid must be avoided since this chemical can be absorbed through the skin. Washing thoroughly with soap and water followed by prolonged rinsing should be done immediately after accidental contact.

A very dangerous fire hazard when exposed to heat or flame; can react with oxidizing materials. Unstabilized, it may form unstable peroxides on exposure to air and should always be tested before distillation. Washing with an aqueous solution of ferrous sulfate slightly acidified with sodium bisulfate will remove these peroxides. Confirm by test. Contact with acids can initiate a violent exothermic reaction. Moderate explosion hazard when exposed to flame. Furan's low boiling point makes it easy to obtain explosive concentrations of the vapor in inadequately ventilated areas. Highly dangerous upon exposure to heat or flame; can react vigorously with oxidizing materials. To fight fire, use CO_2, dry chemical. When heated to decomposition it emits acrid smoke and irritating fumes. See also PEROXIDES.

FPK050 **CAS:539-47-9** **HR: 3**
2-FURANACRYLIC ACID
mf: $C_7H_6O_3$ mw: 138.13

TOXICITY DATA WITH REFERENCE
mmo-sat 10 µg/plate JOPHDQ 1,15,78
ipr-mus LD50:276 mg/kg YKKZAJ 104,793,84

CONSENSUS REPORTS: Reported in EPA TSCA Inventory.

SAFETY PROFILE: Poison by intravenous route. Mutation data reported. When heated to decomposition it emits acrid smoke and irritating vapors.

FPK100 **CAS:50892-99-4** **HR: 3**
FURAN-2-AMIDOXIME
mf: $C_5H_6N_2O_2$ mw: 126.11

OCH=CHCH=CC(NH_2):NOH

SAFETY PROFILE: Explodes when heated above 100°C. Exothermic reaction if heated to 65°C. When heated to decomposition it emits toxic fumes of NO_x.

FPK200 **CAS:3658-77-3** **HR: 3**
FURANEOL
mf: $C_6H_8O_3$ mw: 128.14

SYNS: ALLETONE □ COE 536 □ 2,5-DIMETHYL-4-HYDROXY-3(2H)-FURANONE □ FEMA 3174 □ 3(2H)-FURANONE, 2,5-DIMETHYL-4-HYDROXY- □ 4-HYDROXY-2,5-DIMETHYL-3(2H)FURANONE □ PINEAPPLE KETONE

TOXICITY DATA WITH REFERENCE
mmo-sat 2 mg/plate CHYCDW 22,85,88
mma-sat 2 mg/plate CHYCDW 22,85,88
spm-mus-ipr 232 mg/kg CHYCDW 22,85,88
orl-mus LD50:1608 mg/kg CHYCDW 22,85,88

CONSENSUS REPORTS: Reported in EPA TSCA Inventory.

DOT CLASSIFICATION: 3; *Label:* Flammable Liquid

SAFETY PROFILE: Mutation data reported. A flammable liquid. When heated to decomposition it emits acrid smoke and irritating vapors.

FPM000 **CAS:98-02-2** **HR: 3**
2-FURANMETHANETHIOL
DOT: UN 1228/ UN 3071
mf: C_5H_6OS mw: 114.17

PROP: A liquid with very disagreeable odor. Bp: 160°.

SYNS: FURFURYL MERCAPTAN ☐ USAF B-58

TOXICITY DATA WITH REFERENCE
orl-rat TDLo:1260 mg/kg (42D male):REP FCTXAV 15,383,77
ipr-mus LD50:100 mg/kg NTIS** AD277-689

CONSENSUS REPORTS: Reported in EPA TSCA Inventory.

DOT CLASSIFICATION: 3; *Label:* Flammable Liquid, Poison (UN 1228); DOT Class: 6.1; *Label:* Poison, Flammable Liquid (UN 3071)

SAFETY PROFILE: Poison by intraperitoneal route. Experimental reproductive effects. Used as a flavoring in chocolate, fruit, nuts, and coffee. When heated to decomposition it emits toxic fumes of SO_x. See also MERCAPTANS.

FPO100 **CAS:75888-03-8** **HR: 3**
FURAPYRIMIDONE
mf: $C_9H_{10}N_4O_4$ mw: 238.23

SYNS: 1-((5-NITROFURANYL-2)METHYLENEAMINO)TETRAHYDROPYRIMIDONE-2-ONE ☐ TETRAHYDRO-1-((5-NITROFURFURYLIDENE) AMINO)-2(1H)-PYRIMIDINONE ☐ TETRAHYDRO-1-(5-NITROFURFURYLIDENEAMINO)-2-PYRIMIDONE

TOXICITY DATA WITH REFERENCE
mmo-sat 100 ng/plate CYLPDN 4,201,83
mma-sat 100 ng/plate CYLPDN 4,201,83
orl-rat TDLo:1050 mg/kg (7-11D preg):TER CYLPDN 4,201,83
orl-mus LD50:243 mg/kg CYLPDN 1,56,80
ipr-mus LD50:720 mg/kg JPPMAB 16,663,64

SAFETY PROFILE: Poison by ingestion. Moderately toxic by intraperitoneal route. Experimental teratogenic effects. Mutation data reported. When heated to decomposition it emits toxic fumes of NO_x.

FPP100 **CAS:19237-84-4** **HR: 3**
FURAZOSIN HYDROCHLORIDE
mf: $C_{19}H_{21}N_5O_4$•ClH mw: 419.91

SYNS: ABBOTT-45975 ☐ 1-(4-AMINO-6,7-DIMETHOXY-2-QUINAZOLINYL)-4-(2-FURANYLCARBONYL)PIPERAZINE HYDROCHLORIDE ☐ 1-(4-AMINO-6,7-DIMETHOXY-2-QUINAZOLINYL)-4-(2-FUROYL)PIPERAZINE MONOHYDROCHLORIDE ☐ 2-(4-(2-FUROYL)PIPERAZIN-1-YL)-4-AMINO-6,7-DIMETHOXYQUINAZOLINE HYDROCHLORIDE ☐ PRAZOSIN HYDROCHLORIDE ☐ TERAZOSIN

TOXICITY DATA WITH REFERENCE
orl-rat TDLo:8700 mg/kg (female 14-22D post):REP OYYAA2 17,57,79
orl-rat TDLo:1450 mg/kg (female 14-22D post):REP OYYAA2 17,57,79
orl-rat TDLo:300 mg/kg (female 9-14D post):TER OYYAA2 17,57,79
orl-rat TDLo:4150 mg/kg (male 8W pre):TER OYYAA2 17,57,79
imp-rat TDLo:30 mg/kg (male 3D pre):REP JRPFA4 70,643,84
scu-rat TDLo:12 mg/kg (female 4-5D post):REP JRPFA4 81,51,87 (TER-neg)
scu-rat TDLo:4 mg/kg (female 22D post):REP JRPFA4 76,415,86
orl-rat TDLo:14,560 mg/kg (26W male):REP OYYAA2 17,39,79
orl-man TDLo:1714 µg/kg HUTODJ 4,53,85
orl-hmn TDLo:285 µg/kg:CVS JAMAAP 238,157,77
orl-wmn TDLo:13 mg/kg/6W-I:CNS,PSY BMJOAE 293,1347,86
orl-wmn TDLo:10 µg/kg DICPBB 21,723,87
orl-rat LD50:1950 mg/kg IYKEDH 12,933,81
ipr-rat LD50:102 mg/kg NIIRDN 6,688,82
ivn-rat LD50:277 mg/kg NIIRDN 6,688,82
ipr-mus LD50:60 mg/kg NIIRDN 6,688,82
scu-mus LD50:3100 mg/kg NIIRDN 6,688,82

SAFETY PROFILE: Poison by intravenous and intraperitoneal routes. Moderately toxic by ingestion and subcutaneous routes. Human systemic effects by ingestion: blood pressure depression, encephalitis, somnolence, toxic psychosis. An experimental teratogen. Experimental reproductive effects. When heated to decomposition it emits toxic fumes of NO_x and HCl.

FPQ000 **CAS:9000-21-9** **HR: 2**
FURCELLERAN GUM

PROP: Vegetable gum from *Furcellaria fastigiata* (Fam. Rhodophyceae) available as an odorless white powder. Sol in warm water.

SYN: BURTONITE 44

TOXICITY DATA WITH REFERENCE
orl-rat LD50:5000 mg/kg FDRLI* 124,-,76
orl-mus LD50:6000 mg/kg FDRLI* 124,-,76
orl-rbt LD50:3400 mg/kg FDRLI* 124,-,76
orl-ham LD50:4800 mg/kg FDRLI* 124,-,76

CONSENSUS REPORTS: Reported in EPA TSCA Inventory.

SAFETY PROFILE: Moderately toxic by ingestion. When heated to decomposition it emits acrid smoke and fumes.

FPQ100 **CAS:2385-81-1** **HR: 3**
FURETHIDINE
mf: $C_{21}H_{31}NO_4$ mw: 361.47

PROP: Bp: 175–183° @ 0.3 mm, mp: 28°, n (20/D) 1.5219.

SYN: ETHYL 4-PHENYL-1-(2-TETRAHYDROFURFURYLOXYETHYL)PIPERIDINE-4-CARBOXYLATE

TOXICITY DATA WITH REFERENCE
orl-rat LD50:135 mg/kg BJPCAL 15,254,60
scu-rat LD50:26 mg/kg BJPCAL 15,254,60
ivn-mus LD50:15,500 µg/kg BJPCAL 15,254,60

SAFETY PROFILE: Poison by ingestion, subcutaneous,

F

and intravenous routes. When heated to decomposition it emits toxic fumes of NO_x. Can be habit forming. This is a controlled substance (opiate) listed in the U.S. Code of Federal Regulations, Title 21 Part 1308.11 (1985). When heated to decomposition it emits toxic fumes of NO_x. See also ESTERS.

FPQ875 **CAS:98-01-1** **HR: 3**
FURFURAL
DOT: UN 1199
mf: $C_5H_4O_2$ mw: 96.09

OCH=CHCH=CCO•H

PROP: Colorless to yellowish liquid; almond-like odor. Bp: 161.7° @ 764 mm, lel: 2.1%, uel: 19.3%, flash p: 140°F (CC), d: 1.154–1.158, refr index: 1.522–1.528, autoign temp: 600°F, vap d: 3.31. Sol in water; misc with alc.

SYNS: ARTIFICIAL ANT OIL ☐ FEMA No. 2489 ☐ FURAL ☐ 2-FURALDEHYDE ☐ FURALE ☐ 2-FURANALDEHYDE ☐ 2-FURANCARBONAL ☐ 2-FURANCARBOXALDEHYDE ☐ 2-FURFURAL ☐ FURFURALDEHYDE ☐ FURFURALE (ITALIAN) ☐ FURFUROL ☐ FURFUROLE ☐ 2-FURIL-METANALE (ITALIAN) ☐ FUROLE ☐ α-FUROLE ☐ 2-FURYL-METHANAL ☐ NCI-C56177 ☐ PYROMUCIC ALDEHYDE ☐ RCRA WASTE NUMBER U125

TOXICITY DATA WITH REFERENCE
skn-rbt 20 mg/24H MOD 85JCAE-,788,86
eye-rbt 100 mg/24H MOD 85JCAE-,788,86
skn-rbt 500 mg/24H MLD FCTXAV 16,759,78
eye-rbt 20 mg/24H MOD 28ZPAK -,139,72
eye-rbt 50 mg MLD 34ZIAG -,279,69
sln-dmg-par 100 ppm ENMUDM 7,677,85
sce-hmn:lym 70 μmol/L MUREAV 156,233,85
ihl-hmn TCLo:310 μg/m³ GISAAA 26(6),3,61
orl-rat LD50:65 mg/kg BCTKAG 13,371,80
ihl-rat LCLo:153 ppm/4H 28ZPAK -,139,72
ipr-rat LD50:20 mg/kg FCTXAV 16,759,78
scu-rat LD50:148 mg/kg 34ZIAG -,279,69
orl-mus LD50:400 mg/kg BIJOAK 34,1196,40
ihl-mus LCLo:370 ppm/6H 34ZIAG -,279,69
ipr-mus LD50:102 mg/kg FCTXAV 16,759,78
scu-mus LD50:119 mg/kg FCTXAV 16,759,78
ivn-mus LD50:152 mg/kg FCTXAV 16,759,78
orl-dog LD50:950 mg/kg 34ZIAG -,279,69
ihl-dog LC50:370 ppm/6H 34ZIAG -,279,69
skn-rbt LDLo:620 mg/kg FCTXAV 16,759,78

CONSENSUS REPORTS: NTP Carcinogenesis Studies (gavage); Clear Evidence: Mouse NCITR* NTP-TR-382,90, Some Evidence: Rat NCITR* NTP-TR-382,90. EPA Genetic Toxicology Program. Reported in EPA TSCA Inventory.

OSHA PEL: TWA 2 ppm (skin)
ACGIH TLV: TWA 2 ppm (skin)
DFG MAK: 5 ppm (20 mg/m³)
DOT CLASSIFICATION: 3; *Label:* Flammable Liquid

SAFETY PROFILE: Poison by ingestion, intraperitoneal, subcutaneous, intravenous, and intramuscular routes. Moderately toxic by inhalation and skin contact. Human mutation data reported. A skin and eye irritant. Mutation data reported. The liquid is dangerous to the eyes. The vapor is irritating to mucous membranes and is a central nervous system poison. However, its low volatility reduces its toxicity effect. Ingestion of furfural has produced cirrhosis of the liver in rats. In industry there is a tendency to minimize the danger of acute effects resulting from exposure to it. This is particularly true because of its low volatility.

Flammable liquid when exposed to heat or flame; can react with oxidizing materials. Moderate explosion hazard when exposed to heat or flame or by chemical reaction. An exothermic polymerization of almost explosive violence can occur upon contact with strong mineral acids or alkalies. Keep away from heat and open flames. Mixture with sodium hydrogen carbonate ignites spontaneously. To fight fire, use alcohol foam, CO_2, dry chemical. When heated to decomposition it emits acrid smoke and irritating fumes.

For occupational chemical analysis use OSHA: #ID-72 or NIOSH: Furfural, 2529.

FPQ900 **CAS:28438-99-5** **HR: 3**
FURFURAL ACETONE MONOMER FA

SYNS: FA ☐ FA MONOMER ☐ FURFURAL-ACETONE ADDUCT ☐ FURFURAL-ACETONE MONOMER ☐ 1:1 FURFURAL-ACETONE MONOMER ☐ FURFUROLATSETONOVYI MONOMER FA ☐ MONOMER FA

TOXICITY DATA WITH REFERENCE
skn-rbt 500 mg/24H MLD 28ZPAK-,275,72
eye-rbt 500 mg/24H MLD 28ZPAK-,275,72
orl-rat LD50:592 mg/kg 28ZPAK-,275,72
skn-rat LD50:2600 mg/kg GTPZAB 18(5),53,74
orl-mus LD50:980 mg/kg GTPZAB 18(5),53,74
orl-rbt LD50:285 mg/kg GTPZAB 18(5),53,74
skn-rbt LD50:900 mg/kg GTPZAB 18(5),53,74

SAFETY PROFILE: Poison by ingestion. Moderately toxic by skin contact. A skin and eye irritant. When heated to decomposition it emits acrid smoke and irritating fumes.

FPR000 **CAS:1121-47-7** **HR: 3**
FURFURAL OXIME
mf: $C_5H_5NO_2$ mw: 111.11

TOXICITY DATA WITH REFERENCE
ipr-mus LD50:500 mg/kg JPETAB 119,522,57
ivn-mus LD50:180 mg/kg CSLNX* NX#03758

CONSENSUS REPORTS: Reported in EPA TSCA Inventory.

SAFETY PROFILE: Poison by intravenous route. Moderately toxic by intraperitoneal route. When heated to decomposition it emits toxic fumes such as NO_x.

FPS000 **CAS:494-47-3** **HR: 3**
FURFURAMIDE
mf: $C_{15}H_{12}N_2O_3$ mw: 268.29

PROP: Needles from alc. Mp: 117–21°, bp: 250° decomp. Insol in water; decomp in acid; sol in alc and ether.

SYNS: 2-(BIS(FURFURYLIDENAMINO))METHYLFURAN □ HYDROFURAMIDE

TOXICITY DATA WITH REFERENCE
orl-rat LD50:400 mg/kg HYSAAV 29,37,64
orl-mus LD50:1200 mg/kg BIJOAK 34,1196,40
orl-rbt LDLo:50 mg/kg HYSAAV 29,37,64

SAFETY PROFILE: Poison by ingestion. A skin, eye, and mucous membrane irritant. Causes intense pulmonary irritation and reported to cause liver and kidney damage. When heated to decomposition it emits toxic fumes of NO_x. A component of fungicides. See also AMINES and AMIDES.

FPT000 CAS:699-17-2 HR: 3
FURFURYLACETONE
mf: $C_8H_{10}O_2$ mw: 138.18

PROP: Oil with fruity odor. D: 1.036 @ 19°/4°, bp: 203°.

SYN: 4-(2-FURYL)-2-BUTANONE

TOXICITY DATA WITH REFERENCE
ipr-mus LDLo:62,500 µg/kg CBCCT* 6,217,54

CONSENSUS REPORTS: Reported in EPA TSCA Inventory.

SAFETY PROFILE: Poison by intraperitoneal route. When heated to decomposition it emits acrid smoke and irritating fumes.

FPT100 CAS:525-79-1 HR: 2
N-FURFURYLADENINE
mf: $C_{10}H_9N_5O$ mw: 215.24

SYNS: ADENINE, N-FURFURYL- □ FAP □ N^6-FURFURYLADENINE □ 6-FURFURYLADENINE □ N^6-(FURFURYLAMINO)PURINE □ 6-(FURFURYLAMINO)PURINE □ KINETIN □ KINETIN (PLANT HORMONE)

TOXICITY DATA WITH REFERENCE
dns-hmn:leu 1 µmol/L EXPEAM 32,29,76
oth-hmn:leu 1 µmol/L EXPEAM 32,29,76
dni-hmn:leu 100 µmol/L EXPEAM 32,29,76
ipr-mus LD50:450 mg/kg NYKZAU 61(2),43S,65

CONSENSUS REPORTS: Reported in EPA TSCA Inventory.

SAFETY PROFILE: Moderately toxic by intraperitoneal route. Human mutation data reported. When heated to decomposition it emits toxic vapors of NO_x.

FPU000 CAS:98-00-0 HR: 3
FURFURYL ALCOHOL
DOT: UN 2874
mf: $C_5H_6O_2$ mw: 98.11

OCH=CHCH=CCH$_2$OH (ring: O to C)

PROP: Clear, colorless, mobile liquid. Mp: −31°, lel: 1.8%, uel: 16.3% (between 72 and 122°), bp: 171° @ 750 mm, flash p: 167°F (OC), d: 1.129 @ 20°/4°, autoign temp: 915°F, vap press: 1 mm @ 31.8°, vap d: 3.37. Misc in H_2O; very sol in EtOH and Et_2O.

SYNS: 2-FURANCARBINOL □ 2-FURANMETHANOL □ FURFURAL ALCOHOL □ 2-FURFURYLALKOHOL (CZECH) □ FURYL ALCOHOL □ α-FURYLCARBINOL □ 2-FURYLCARBINOL □ (2-FURYL)METHANOL □ 2-HYDROXYMETHYLFURAN □ NCI-C56224

TOXICITY DATA WITH REFERENCE
eye-rbt 100 mg/24H MOD 85JCAE-,787,86
cyt-ham:ovr 2500 µmol/L CALEDQ 13,89,81
orl-rat LD50:177 mg/kg GTPZAB 25(9),52,81
orl-rat LD50:88,300 µg/kg 28ZPAK -,139,72
ihl-rat LC50:233 ppm/4H AIHAAP 19,91,58
ipr-rat LD50:650 mg/kg NPIRI* 1,64,74
scu-rat LD50:85 mg/kg 34ZIAG -,280,69
orl-mus LD50:160 mg/kg BIJOAK 34,1196,40
ihl-mus LCLo:597 ppm/6H 34ZIAG -,280,69
skn-rbt LD50:400 mg/kg 34ZIAG -,280,69
ivn-rbt LD50:650 mg/kg FEPRA7 8,294,49

CONSENSUS REPORTS: Reported in EPA TSCA Inventory.

OSHA PEL: TWA 10 ppm; STEL 15 ppm (skin)
ACGIH TLV: TWA 10 ppm; STEL 15 ppm (skin)
DFG MAK: 10 ppm (40 mg/m^3)
NIOSH REL: (Furfuryl Alcohol) TWA 200 mg/m^3
DOT CLASSIFICATION: 6.1; *Label:* KEEP AWAY FROM FOOD

SAFETY PROFILE: Poison by ingestion, skin contact, and subcutaneous routes. Moderately toxic by inhalation and intraperitoneal routes. Mutation data reported. An eye irritant. Flammable when exposed to heat or flame; can react with oxidizing materials. Moderate explosion hazard when exposed to heat or flame. Reacts violently with acids (e.g., formic acid; cyanoacetic acid + heat). Ignites on contact with 85% hydrogen peroxide. To fight fire, use alcohol foam, CO_2, dry chemical. When heated to decomposition it emits acrid smoke and fumes.

For occupational chemical analysis use NIOSH: Furfuryl Alcohol 2505.

FPV000 CAS:10427-00-6 HR: 3
FURFURYL ALCOHOL PHOSPHATE (3:1)
mf: $C_{15}H_{27}O_7P$ mw: 350.39

TOXICITY DATA WITH REFERENCE
ivn-mus LD50:320 mg/kg CSLNX* NX#03980

CONSENSUS REPORTS: Reported in EPA TSCA Inventory.

SAFETY PROFILE: Poison by intravenous route. When heated to decomposition it emits toxic fumes of PO_x.

FPW000 CAS:617-89-0 HR: 3
FURFURYLAMINE
DOT: UN 2526
mf: C_5H_7NO mw: 97.13

PROP: Light straw-colored liquid or oil. Bp: 146°, flash p: 99°F (OC), fp: −70°, d: 1.0502 @ 25°, vap d: 3.35. Misc in water.

SYNS: 2-FURANMETHYLAMINE □ 1-(2-FURYL)METHYLAMINE □ USAF Q-1

TOXICITY DATA WITH REFERENCE
ipr-mus LD50:200 mg/kg NTIS** AD277-689

CONSENSUS REPORTS: Reported in EPA TSCA Inventory.

DOT CLASSIFICATION: 3; *Label:* Flammable Liquid

SAFETY PROFILE: Poison by intraperitoneal route. A skin, eye, and mucous membrane irritant. A dangerous fire hazard when exposed to heat or flame; can react with oxidizing materials. To fight fire, use foam, CO_2, dry chemical. When heated to decomposition it emits toxic fumes of NO_x. See also AMINES.

FPX000 **CAS:67227-30-9** **HR: 3**
FURFURYL-BIS(2-CHLOROETHYL)AMINE HYDROCHLORIDE
mf: $C_9H_{13}Cl_2NO{\cdot}ClH$ mw: 258.59

SYNS: N,N-BIS(2-CHLOROETHYL)FURFURYLAMINE HYDROCHLORIDE □ 2,2′-DICHLORO-N-FURFURYLDIETHYLAMINE HYDROCHLORIDE □ FURFURYL-BIS(β-CHLOROETHYL)AMINE HYDROCHLORIDE □ TL 1055

TOXICITY DATA WITH REFERENCE
ihl-mus LCLo:500 mg/m³/10M NDRC** 30101,5,45
ipr-mus LD50:12 mg/kg CANCAR 2,1055,49
scu-mus LDLo:25 mg/kg NTIS** PB158-507

SAFETY PROFILE: Poison by intraperitoneal and subcutaneous routes. Moderately toxic by inhalation. When heated to decomposition it emits very toxic fumes of Cl^- and NO_x.

FPX025 **CAS:4437-22-3** **HR: 3**
FURFURYL ETHER
mf: $C_{10}H_{10}O_3$ mw: 178.20

SYNS: DIFURFURYL ETHER (7CI) □ 2,2′-DIFURFURYL ETHER □ FURAN, 2,2′-(OXYBIS(METHYLENE))BIS- □ FURAN, 2,2′-(OXYDIMETHYLENE)DI-(6CI,8CI) □ 2,2′-(OXYBIS(METHYLENE))BISFURAN

TOXICITY DATA WITH REFERENCE
orl-rat LD50:210 mg/kg JACTDZ 1,93,90

CONSENSUS REPORTS: Reported in EPA TSCA Inventory.

SAFETY PROFILE: A poison by ingestion. When heated to decomposition it emits acrid smoke and irritating vapors.

FPX050 **CAS:874-66-8** **HR: 2**
FURFURYLIDINE-2-PROPANAL
mf: $C_8H_8O_2$ mw: 136.16

SYNS: 3-FURANACROLEIN, 2-METHYL- □ α-METHYLFURYLACROLEIN □ α-METHYL-β-FURYLACROLEIN □ 2-METHYL-3-FURYLACROLEIN □ 2-METHYL-3-(2-FURYL)ACROLEIN □ 2-METHYL-3-(2-FURYL) PROPENAL □ 2-PROPENAL, 3-(2-FURANYL)-2-METHYL-

TOXICITY DATA WITH REFERENCE
orl-rat LD50:1400 mg/kg JACTDZ 1,3,90

CONSENSUS REPORTS: Reported in EPA TSCA Inventory.

SAFETY PROFILE: Moderately toxic by ingestion. When heated to decomposition it emits acrid smoke and irritating vapors.

FPX100 **CAS:1438-94-4** **HR: 3**
N-FURFURYL PYRROLE
mf: C_9H_9NO mw: 147.19

SYNS: 1-FURFURYLPYRROLE □ N-(2-FURFURYL)PYRROLE □ PYRROLE, 1-FURFURYL-

TOXICITY DATA WITH REFERENCE
orl-mus LD50:380 mg/kg DCTODJ 3,249,80

CONSENSUS REPORTS: Reported in EPA TSCA Inventory.

SAFETY PROFILE: Poison by ingestion. When heated to decomposition it emits toxic vapors of NO_x.

FPY000 **CAS:541-64-0** **HR: 3**
FURFURYLTRIMETHYLAMMONIUM IODIDE
mf: $C_8H_{14}NO{\cdot}I$ mw: 267.13

PROP: A solid. Mp: 116–117°.

SYNS: FT □ FURAMON □ FURAMON IODIDE □ FURANOL □ FURMETHIDE □ FURMITHIDE IODIDE □ FURTHRETHONIUM IODIDE □ FURTRETHONIUM IODIDE □ FURTRIMETHONIUM IODIDE □ N,N,N-TRIMETHYL-2-FURANMETHANAMINIUM IODIDE □ TRIMETHYLFURFURYLAMMONIUM IODIDE

TOXICITY DATA WITH REFERENCE
scu-rat LD50:90 mg/kg JPETAB 68,231,40
ipr-mus LD50:50 mg/kg PCJOAU 6,501,72
ivn-mus LD50:9750 μg/kg TXAPA9 37,184,76
scu-dog LDLo:5 mg/kg JPETAB 68,231,40

SAFETY PROFILE: Poison by intraperitoneal, subcutaneous, and intravenous routes. When heated to decomposition it emits very toxic fumes of NO_x, NH_3, and I^-. See also IODIDES.

FPZ000 **CAS:492-94-4** **HR: 3**
FURIL
mf: $C_{10}H_6O_4$ mw: 190.16

PROP: Yellow needles from C_6H_6 or EtOH. Mp: 165–166°.

SYNS: BIPYROMUCYL □ DI-2-FURANYLETHANEDIONE □ DI-2-FURYLGLYOXAL □ α-FURIL □ 2,2′-FURIL

TOXICITY DATA WITH REFERENCE
ivn-mus LD50:56 mg/kg CSLNX* NX#03858

CONSENSUS REPORTS: Reported in EPA TSCA Inventory.

SAFETY PROFILE: Poison by intravenous route. When heated to decomposition it emits acrid smoke and irritating fumes.

FQB000 **CAS:4339-69-9** **HR: 3**
α-FURILMONOXIME
mf: $C_{10}H_7NO_4$ mw: 205.18

PROP: Yellow crystals. Mp: 97–98°.

SYNS: DI-2-FURYLGLYOXAL MONOXIME □ 2,2′-OXALYDIFURAN OXIME

TOXICITY DATA WITH REFERENCE
eye-rbt 100 mg/24H MOD 28ZPAK -,141,72

orl-rat LD50:2580 mg/kg 28ZPAK -,141,72
ivn-mus LD50:56 mg/kg CSLNX* NX#04786

SAFETY PROFILE: Poison by intravenous route. Moderately toxic by ingestion. An eye irritant. When heated to decomposition it emits toxic fumes of NO_x.

FQC000 **CAS:523-50-2** **HR: 3**
2H-FURO(2,3-h)(1)BENZOPYRAN-2-ONE
mf: $C_{11}H_6O_3$ mw: 186.17

PROP: A solid. Mp: 138–139.5°.

SYNS: ANGECIN □ ANGELICIN (coumarin derivative) □ FURO(5′,4′,7,8)COUMARIN □ ISOPSORALIN

TOXICITY DATA WITH REFERENCE
mmo-esc 40 mg/L MUREAV 58,23,78
mmo-smc 50 μmol/L BUCABS 67,245,80
orl-rat LD50:322 mg/kg IJMRAQ 63,833,75
ipr-rat LD50:165 mg/kg IJMRAQ 63,833,75
ipr-mus LD50:254 mg/kg IJMRAQ 63,833,75

CONSENSUS REPORTS: IARC Cancer Review: Group 3 IMEMDT 7,56,87; Animal Inadequate Evidence IMEMDT 40,291,86. EPA Genetic Toxicology Program.

SAFETY PROFILE: Poison by ingestion and intraperitoneal routes. Questionable carcinogen. Mutation data reported. A tranquilizer, sedative, and anticonvulsant. When heated to decomposition it emits acrid smoke and irritating fumes.

FQE000 **CAS:26447-28-9** **HR: 2**
FUROIC ACID
mf: $C_5H_4O_3$ mw: 112.09

PROP: White solid. Mp: 133°, bp: 230–232°.

TOXICITY DATA WITH REFERENCE
orl-mus LD50:1000 mg/kg BIJOAK 34,1196,40

CONSENSUS REPORTS: Reported in EPA TSCA Inventory.

SAFETY PROFILE: Moderately toxic by ingestion. A bactericide. When heated to decomposition it emits acrid smoke and irritating fumes.

FQF000 **CAS:88-14-2** **HR: 3**
2-FUROIC ACID
mf: $C_5H_4O_3$ mw: 112.09

PROP: Monoclinic prisms or leaflets. Mp: 133–134°; bp: 230–232°. Mod sol in cold water, alc and ether; very sol in hot water.

SYNS: 2-CARBOXYFURAN □ α-FURANCARBOXYLIC ACID □ α-FUROIC ACID □ PYROMUCIC ACID

TOXICITY DATA WITH REFERENCE
mmo-sat 10 μg/plate JOPHDQ 1,15,78
ipr-mus LD50:100 mg/kg PHREEB 2,233,85

CONSENSUS REPORTS: A poison by intraperitoneal route. Reported in EPA TSCA Inventory.

SAFETY PROFILE: Mutation data reported. When heated to decomposition it emits acrid smoke and irritating fumes.

FQI000 **CAS:552-86-3** **HR: 3**
FUROIN
mf: $C_{10}H_8O_4$ mw: 192.18

PROP: Pale-brown needles of methyl alcohol. Mp: 138–139°, bp: decomp. Very sltly sol in hot water; sltly sol in hot alc and in hot toluene; sltly sol in ether.

SYN: 2-FURYL α-HYDROXYFURFURYL KETONE

TOXICITY DATA WITH REFERENCE
ipr-mus LDLo:64 mg/kg CBCCT* 2,301,50

CONSENSUS REPORTS: Reported in EPA TSCA Inventory.

SAFETY PROFILE: Poison by intraperitoneal route. When heated to decomposition it emits acrid smoke and irritating fumes.

FQJ000 **CAS:2578-75-8** **HR: 2**
FUROTHIAZOLE
mf: $C_8H_6N_4O_4S$ mw: 254.24

SYN: N-(5-(5-NITRO-2-FURYL)-1,3,4-THIADIAZOL-2-YL)ACETAMIDE

TOXICITY DATA WITH REFERENCE
mma-sat 100 ng/plate MUREAV 48,295,77
mmo-esc 300 nmol/L CNREA8 34,2266,74
orl-rat TDLo:33 g/kg/40W-C:CAR JNCIAM 54,841,75
orl-mus TDLo:80 g/kg/46W-C:NEO CNREA8 33,1593,73

SAFETY PROFILE: Questionable carcinogen with experimental carcinogenic and neoplastigenic data. Mutation data reported. When heated to decomposition it emits very toxic fumes of SO_x and NO_x.

FQJ025 **CAS:20762-98-5** **HR: 3**
2-FUROYL AZIDE
mf: $C_5H_3N_3O_2$ mw: 137.10

OCH=CHCH=CCO•N_3 (ring closure from O to C)

SAFETY PROFILE: Explodes violently when heated. When heated to decomposition it emits toxic fumes of NO_x. See also AZIDES.

FQJ050 **CAS:1300-32-9** **HR: 3**
FUROYL CHLORIDE
mf: $C_5H_3ClO_2$ mw: 130.53

OCH=CHCH=CO•Cl (ring closure from O to C)

SAFETY PROFILE: Can explode spontaneously in storage. When heated to decomposition it emits toxic fumes of Cl^-. See also CHLORIDES.

FQJ100 **CAS:804-30-8** **HR: 2**
FURSULTIAMIN
mf: $C_{17}H_{26}N_4O_3S_2$ mw: 398.59

PROP: Crystalline powder with coffee flavor. Mp:

148–150°. Decomp @ 132°. Sparingly sol in water; sol in organic solvents, dil mineral acids.

SYNS: ADVENTAN □ ALINAMIN F □ DITEFTIN □ FURSULTIAMINE □ JUDOLOR □ LINAMIN □ RETAR-B_1 □ THIAMINE TETRAHYDROFURFURYL DISULFIDE □ TTFD

TOXICITY DATA WITH REFERENCE
orl-mus LD50:2200 mg/kg NIIRDN 6,699,82
ipr-mus LD50:540 mg/kg TAKHAA 30,242,71
ivn-mus LD50:430 mg/kg NIIRDN 6,699,82

SAFETY PROFILE: Moderately toxic by ingestion, intraperitoneal, and intravenous routes. When heated to decomposition it emits toxic fumes of SO_x and NO_x.

FQK000 CAS:3878-19-1 HR: 3
2-(2-FURYL)BENZIMIDAZOLE
mf: $C_{11}H_8N_2O$ mw: 184.21

PROP: Crystals or powder. Mp: 286° (decomp). Very sltly sol in H_2O; sltly sol in org solvs.

SYNS: BAYER 33172 □ FUBERIDATOL □ FUBERIDAZOLE □ FUBERISAZOL □ FUBRIDAZOLE □ 2-(2-FURANYL)-1H-BENZIMIDAZOLE □ FURIDAZOL □ FURIDAZOLE □ 2-(2'-FURYL)-BENZIMIDAZOLE □ VORONITE □ W VII/117

TOXICITY DATA WITH REFERENCE
mmo-sat 100 μg/plate MUREAV 15,273,72
orl-rat LD50:500 mg/kg FMCHA2-,C148,91
ihl-rat LC50:330 mg/m^3/4H 85DPAN -,-,71/76
skn-rat LD50:500 mg/kg 85DPAN -,-,71/76
ipr-rat LD50:100 mg/kg GUCHAZ 6,290,73
orl-mus LD50:825 mg/kg 85DPAN -,-,71/76

CONSENSUS REPORTS: EPA Extremely Hazardous Substances List.

SAFETY PROFILE: Poison by inhalation and intraperitoneal routes. Moderately toxic by ingestion and skin contact. Mutation data reported. A fungicide. When heated to decomposition it emits toxic fumes of NO_x.

FQL100 CAS:32954-58-8 HR: 3
1-(3-FURYL)-4-HYDROXYPENTANONE
mf: $C_9H_{12}O_3$ mw: 168.21

SYNS: 1-(β-FURYL)-4-HYDROXYPENTANONE □ IPOMEANOL □ 4-IPOMEANOL

TOXICITY DATA WITH REFERENCE
ipr-rat LD50:17 mg/kg TXCYAC 19,85,81
orl-mus LD50:38 mg/kg BBACAQ 337,184,74
ipr-mus LD50:11 mg/kg TXAPA9 66,193,82
ivn-mus LD50:21 mg/kg BBACAQ 337,184,74
ipr-rbt LD50:30 mg/kg TXCYAC 19,85,81

SAFETY PROFILE: Poison by ingestion, intravenous, and intraperitoneal routes. When heated to decomposition it emits acrid smoke and irritating fumes. See also KETONES.

FQL200 CAS:4682-94-4 HR: 3
2-FURYL p-HYDROXYPHENYL KETONE
mf: $C_{11}H_8O_3$ mw: 188.19

SYNS: DB 133 □ HYDROXY-4 BENZOYL-2-FURANNE □ KETONE, 2-FURYL p-HYDROXYPHENYL

TOXICITY DATA WITH REFERENCE
ipr-mus LD50:300 mg/kg AIPTAK 147,497,64

DOT CLASSIFICATION: 3; *Label:* Flammable Liquid

SAFETY PROFILE: A poison by intraperitoneal route. A flammable liquid. When heated to decomposition it emits acrid smoke and irritating vapors.

FQM000 CAS:63905-60-2 HR: 3
2-FURYLISOPROPYLAMINE SULFATE
mf: $C_7H_{11}NO•H_2O_4S$ mw: 223.27

SYN: β-(2-FURYL)ISOPROPYLAMINE SULFATE

TOXICITY DATA WITH REFERENCE
orl-man TDLo:290 μg/kg:CNS;CVS JPETAB 72,265,41
ipr-mus LD50:348 mg/kg JPETAB 72,265,41

SAFETY PROFILE: Poison by intraperitoneal route. Human systemic effects by ingestion of very small amounts: changes in EEG, excitement, and unspecified vascular effects. When heated to decomposition it emits very toxic fumes of SO_x and NO_x. See also SULFATES and AMINES.

FQN000 CAS:3688-53-7 HR: 3
2-(2-FURYL)-3-(5-NITRO-2-FURYL)ACRYLAMIDE
mf: $C_{11}H_8N_2O_5$ mw: 248.21

SYNS: AF-2 (preservative) □ FF □ FURYLAMIDE □ FURYLFURAMIDE □ α-2-FURYL-5-NITRO-2-FURANACYRLAMIDE □ 2-(2-FURYL)-3-(5-NITRO-2-FURYL)ACRYLIC ACID AMIDE □ α-(FURYL)-β-(5-NITRO-2-FURYL)ACRYLIC AMIDE □ TOFURON

TOXICITY DATA WITH REFERENCE
mmo-sat 4 μg/L MUREAV 147,219,85
sce-hmn:lym 500 μg/L CNREA8 40,4775,80
orl-mus TDLo:5750 mg/kg (male 5W pre):REP SKEZAP 18,455,77
scu-mus TDLo:150 mg/kg (female 13-17D post):REP NATUAS 258,610,75
orl-rat TDLo:600 mg/kg (1-20D preg):REP JTSCDR 2,149,77
par-ham TDLo:50 mg/kg (female 12D post):TER IDZAAW 53,59,78
orl-rat TDLo:6 g/kg (1-20D preg):TER JTSCDR 2,149,77
orl-mus TDLo:28,750 mg/kg (multi) :REP SKEZAP 18,455,77
orl-mus TDLo:150 mg/kg (female 8D post):REP SEIJBO 15,41,75
orl-rat TDLo:6600 mg/kg (female 1-21D post):REP SEIJBO 16,249,76
orl-mus TDLo:28,750 mg/kg (multi) :REP SKEZAP 18,455,77
orl-rat TDLo:6 g/kg (1-20D preg):TER JTSCDR 2,149,77
orl-rat TDLo:52 g/kg/40W-C:CAR CALEDQ 3,115,77
orl-mus TDLo:156 g/kg/44W-C:CAR ZKKOBW 89,61,77
scu-mus TDLo:150 mg/kg (13-17D preg):NEO,TER NATUAS 258,610,75

orl-ham TDLo:63 g/kg/94W-C:CAR FCTXAV 17,339,79
orl-rat TD:25 g/kg/78W-C:NEO TOLED5 1,11,77
orl-mus TD:211 g/kg/63W-C:CAR GANNA2 68,825,77
orl-mus TD:158 g/kg/47W-C:CAR CALEDQ 3,115,77
orl-mus TD:42 g/kg/63W-C:NEO GANNA2 68,825,77
orl-ham TD:127 g/kg/94W-C:CAR FCTXAV 17,339,79
orl-ham TD:116 g/kg/80W-C:CAR ZKKOBW 89,61,77
orl-mus TD:32,400 mg/kg/72W-C:CAR ESKHA5 (100),80,82
orl-rat LD50:1554 mg/kg TJEMAO 103,331,71
orl-mus LD50:221 mg/kg NEZAAQ 28,463,73

CONSENSUS REPORTS: IARC Cancer Review: Group 2B IMEMDT 7,56,87; Human Inadequate Evidence IMEMDT 31,47,83; Animal Sufficient Evidence IMEMDT 31,47,83. EPA Genetic Toxicology Program.

SAFETY PROFILE: Confirmed carcinogen with experimental carcinogenic, neoplastigenic, and teratogenic data. Poison by ingestion. Experimental reproductive effects. Human mutation data reported. When heated to decomposition it emits toxic fumes of NO_x.

FQO050 CAS:6453-98-1 HR: 3
3-FURYL PHENYL KETONE
mf: $C_{11}H_8O_2$ mw: 172.19

SYNS: 3-BENZOYLFURAN □ 3-FURANYLPHENYLMETHANONE □ KETONE, 3-FURYL PHENYL □ METHANONE, 3-FURFANYLPHENYL- (9CI)

TOXICITY DATA WITH REFERENCE
ipr-mus LD50:29,788 µg/kg JANSAG 68,1072,90

DOT CLASSIFICATION: 3; *Label:* Flammable Liquid

SAFETY PROFILE: A poison by intraperitoneal route. A flammable liquid. When heated to decomposition it emits acrid smoke and irritating vapors.

FQQ100 CAS:75884-37-6 HR: 2
N-(4-(2-FURYL)-2-THIAZOLYL)ACETAMIDE
mf: $C_9H_8N_2O_2S$ mw: 208.25

SYN: FTA

TOXICITY DATA WITH REFERENCE
orl-mus TDLo:1008 mg/kg/12W-C:ETA CNREA8 41,1397,81

SAFETY PROFILE: Questionable carcinogen with experimental tumorigenic data. When heated to decomposition it emits toxic fumes of NO_x and SO_x.

FQQ400 CAS:77503-17-4 HR: 2
N-(4-(2-FURYL)-2-THIAZOLYL)FORMAMIDE
mf: $C_8H_6N_2O_2S$ mw: 194.22

SYN: FAFT

TOXICITY DATA WITH REFERENCE
orl-mus TDLo:1008 mg/kg/12W-C:ETA CNREA8 41,1397,81

SAFETY PROFILE: Questionable carcinogen with experimental tumorigenic data. When heated to decomposition it emits toxic fumes of NO_x and SO_x.

FQR000 CAS:23255-69-8 HR: 3
FUSARENONE X
mf: $C_{17}H_{22}O_8$ mw: 354.39

PROP: Crystals. Mp: 181–184°. Isolated from the culture filtrate of *Fusarium nivale* (34ZHAD -,163,71).

SYNS: 4-ACETYLOXY-12,13-EPOXY-3,7,15-TRIHYDROXY-(3-α,4-β,7-β)-TRICHOTHEC-9-EN-8-ONE □ NIVALENOL-4-O-ACETATE □ 3,7,15-TRIHYDROXY-4-ACETOXY-8-OXO-12,13-EPOXY-Δ^9-TRICHOTHECENE □ 3,7,15-TRIHYDROXYSCIRP-4-ACETOXY-9-EN-8-ONE

TOXICITY DATA WITH REFERENCE
dnd-hmn:hla 32 mg/L/1H JJEMAG 42,527,72
orl-mus TDLo:2100 µg/kg (1-7D preg):TER JJEMAG 50,167,80
orl-mus TDLo:2400 µg/kg (female 7-14D post):TER JJEMAG 50,167,80
orl-rat TDLo:146 mg/kg/2Y-C:ETA JJEMAG 50,293,80
orl-mus TDLo:22,750 µg/kg/1Y-I:ETA,REP MAIKD3 (21),38,85
orl-rat TDLo:146 mg/kg/2Y-C:ETA JJEMAG 50,293,80
orl-rat LD50:4 mg/kg JJEMAG 41,521,71
orl-mus LD50:4500 µg/kg JJEMAG 41,521,71
ipr-mus LD50:3300 µg/kg PACHAS 49,1737,77
scu-mus LD50:4200 µg/kg JJEMAG 41,521,71
ivn-mus LD50:3400 µg/kg JJEMAG 41,521,71
scu-cat LDLo:5 mg/kg JJEMAG 41,521,71
ipr-gpg LDLo:500 µg/kg 34ZHAD -,163,71
orl-dck LDLo:5 mg/kg JJEMAG 41,521,71

CONSENSUS REPORTS: IARC Cancer Review: Group 3 IMEMDT 7,56,87; Animal Inadequate Evidence IMEMDT 11,169,76; Animal Inadequate Evidence IMEMDT 31,153,83.

SAFETY PROFILE: Poison by ingestion, subcutaneous, intravenous, and intraperitoneal routes. An experimental teratogen. Experimental reproductive effects. Questionable carcinogen with experimental tumorigenic data. Human mutation data reported. When heated to decomposition it emits acrid smoke and irritating fumes.

FQR100 CAS:21813-99-0 HR: 3
FUSARIC ACID CALCIUM SALT
mf: $C_{20}H_{24}N_2O_4 \cdot Ca \cdot H_2O$ mw: 414.56

SYNS: 5-BUTYLPICOLINIC ACID CALCIUM SALT HYDRATE □ 5-BUTYL-2-PYRIDINECARBOXYLIC ACID CALCIUM SALT HYDRATE □ CALCIUM 5-BUTYLPICOLINATE HYDRATE □ CALCIUM FUSARATE □ FA-Ca □ FUSARIC ACID-Ca □ PICOLINIC ACID, 5-BUTYL-, CALCIUM SALT, HYDRATE

TOXICITY DATA WITH REFERENCE
orl-rat TDLo:875 mg/kg (female 8-14D post):REP JJANAX 29,801,76
orl-rat TDLo:140 mg/kg (female 8-14D post):TER JJANAX 29,543,76
orl-mus TDLo:140 mg/kg (female 7-13D post):TER JJANAX 29,543,76
orl-mus TDLo:875 mg/kg (female 7-13D post):REP JJANAX 29,543,76
orl-rat LD50:930 mg/kg JJANAX 29,439,76
orl-mus LD50:235 mg/kg JJANAX 29,439,76
ipr-mus LD50:125 mg/kg JANTAJ 22,228,69
scu-mus LD50:1440 mg/kg JJANAX 29,439,76

ims-mus LD50:125 mg/kg JANTAJ 22,228,69
orl-dog LD50:570 mg/kg JJANAX 29,439,76

SAFETY PROFILE: Poison by ingestion, intramuscular, and intraperitoneal routes. Moderately toxic by subcutaneous route. An experimental teratogen. Experimental reproductive effects. When heated to decomposition it emits toxic fumes of NO_x.

FQS000 CAS:21259-20-1 HR: 2
FUSARIOTOXIN T 2
mf: $C_{24}H_{34}O_9$ mw: 466.58

PROP: Needles. Mp: 151–152°. A strain of *F. tricinctum* isolated from infected corn (AJVRAH 32,1843,71).

SYNS: 4,15-DIACETOXY-8-(3-METHYLBUTYRYLOXY)-12,13-EPOXY-Δ-9-TRICHOTHECEN-3-OL ☐ 4-β,15-DIACETOXY-8-α-(3-METHYLBUTYRYLOXY)-3-α-HYDROXY-12,13-EPOXYTRICHOTHEC-9-ENE ☐ 3-HYDROXY-4,15-DIACETOXY-8-(3-METHYLBUTYRYLOXY)-12,13-EPOXY-Δ^9-TRICHOTHECENE ☐ INSARIOTOXIN ☐ 8-ISOVALERATE ☐ 8-(3-METHYLBUTYRYLOXY)-DIACETOXYSCIRPENOL ☐ NSC 138780 ☐ T-2 MYCOTOXIN ☐ TOXIN T2 ☐ T^2-TRICHOTHECENE

TOXICITY DATA WITH REFERENCE
skn-rat 20 ng/24H MLD JANCA2 57,1121,74
skn-rbt 20 ng/24H MOD JANCA2 57,1121,74
skn-gpg 4667 pg MLD FAATDF 4(2, Pt 2),S124,84
dnd-mus-ipr 3 mg/kg MUREAV 88,115,81
dnd-mus:lym 5 μg/L MUREAV 88,115,81
orl-mus TDLo:1500 μg/kg (female 11D post):REP TJADAB 33,75C,86
orl-mus TDLo:3 mg/kg (female 7D post):TER JJATDK 7,281,87
orl-mus TDLo:500 μg/kg (female 9D post):TER JJATDK 7,281,87
orl-mus TDLo:1500 μg/kg (female 11D post):REP CJVRE9 51,399,87
orl-mus TDLo:500 μg/kg (female 9D post):REP JJATDK 7,281,87
ipr-mus TDLo:1 mg/kg (female 10D post):TER RCOCB8 10,743,75
orl-mus TDLo:179 mg/kg/71W-C:NEO FCTOD7 25,593,87
orl-rat LD50:2700 μg/kg DFSCDX 4,135,83
skn-rat LD50:2560 μg/kg NTIS** AD-A158-874
ipr-rat LD50:900 μg/kg TOXIA6 24,933,86
scu-rat LD50:560 μg/kg FEPRA7 41,924,82
ivn-rat LD50:740 μg/kg JPETAB 232,786,85
ims-rat LD50:470 μg/kg FEPRA7 41,924,82
ice-rat LD50:60 μg/kg ARTODN 58,40,85
orl-mus LD50:3800 μg/kg BIBIAU 10,445,68
ihl-mus LCLo:140 ppb/30M FAATDF 4,S124,84
ims-mky LD20: 650 μg/kg TXAPA9 82,532,86
iat-pig LDLo:1200 μg/kg TOXIA6 24,13,86

CONSENSUS REPORTS: IARC Cancer Review: Group 3 IMEMDT 7,56,87; Animal Inadequate Evidence IMEMDT 31,265,83. EPA Genetic Toxicology Program.

SAFETY PROFILE: Poison by ingestion, intramuscular, subcutaneous, intraperitoneal, intracerebral, and intravenous routes. Moderately toxic by inhalation. Experimental teratogenic and reproductive effects. A skin irritant. Questionable carcinogen with experimental neoplastigenic data. Mutation data reported. When heated to decomposition it emits acrid smoke and irritating fumes.

FQT000 CAS:8013-75-0 HR: 3
FUSEL OIL
DOT: UN 1201

PROP: Colorless to pale-yellow liquid; odorless. D: 0.807–0.813, refr index: 1.405–1.410. Composition of grain fusel oil is methanol, ethanol, acetaldehyde, and other alcohols (ARGEAR 33,49,69).

SYNS: FEMA No. 2497 ☐ FUSELOEL (GERMAN) ☐ FUSEL OIL, REFINED (FCC) ☐ HUILE de FUSEL (FRENCH)

TOXICITY DATA WITH REFERENCE
mmo-esc 7000 ppm ARGEAR 33,49,69
dlt-mus-scu 12,500 mg/kg/5D-C ARGEAR 33,49,69
scu-mus TDLo:12,500 mg/kg (male 5D pre):REP ARGEAR 33,49,69

CONSENSUS REPORTS: Reported in EPA TSCA Inventory.

DOT CLASSIFICATION: 3; *Label:* Flammable Liquid

SAFETY PROFILE: May contain carcinogens. Experimental reproductive effects. Mutation data reported. Flammable liquid when exposed to heat or flame; can react vigorously with oxidizing materials. When heated to decomposition it emits acrid smoke and fumes. See also individual components.

FQU000 CAS:6990-06-3 HR: 3
FUSIDINE
mf: $C_{31}H_{48}O_6$ mw: 516.79

PROP: Crystals from Et_2O. Mp: 192–193°.

SYNS: FUSIDIC ACID ☐ FUZIDIN ☐ RAMYCIN ☐ SQ 16603

TOXICITY DATA WITH REFERENCE
cyt-dmg:oth 100 mg/L CLDFAT 2,97,73
unr-rat TDLo:15 g/kg (9-14D preg):TER FATOAO 45,83,82
unr-rat TDLo:15 g/kg (9-14D preg):REP FATOAO 45,83,82
orl-mus LD50:1500 mg/kg LANCAO 1,928,62
ipr-mus LD50:165 mg/kg 85GDA2 6,157,81
scu-mus LD50:1200 mg/kg 85ERAY 3,1959,78
ivn-mus LD50:180 mg/kg 85GDA2 6,156,81

SAFETY PROFILE: Poison by intraperitoneal and intravenous routes. Moderately toxic by ingestion and subcutaneous routes. Experimental teratogenic and reproductive effects. Mutation data reported. When heated to decomposition it emits acrid smoke and irritating fumes.

FQU875 CAS:13674-87-8 HR: 2
FYROL FR 2
mf: $C_9H_{15}Cl_6O_4P$ mw: 430.91

PROP: Viscous liquid. Bp: (5) 236–237°, n (20/D) 1.5022. Solubility in water: 100 ppm.

SYNS: 1,3-DICHLORO-2-PROPANOL PHOSPHATE (3:1) ☐ EMULSION 212 ☐ FOSFORAN TROJ-(1,3-DWUCHLOROIZOPROPYLOWY)

(POLISH) □ PF 38 □ PHOSPHORIC ACID TRIS(1,3-DICHLORO-2-PROPYL)ESTER □ TCPP □ TDCPP □ TRIS(1-CHLOROMETHYL-2-CHLOROETHYL)PHOSPHATE □ TRIS(1,3-DICHLOROISOPROPYL) PHOSPHATE □ TRIS-(1,3-DICHLORO-2-PROPYL)-PHOSPHATE

TOXICITY DATA WITH REFERENCE
mmo-sat 100 μmol/plate MUREAV 66,373,79
mma-sat 100 μg/plate SCIEAS 200,785,78
otr-ham:emb 20 μmol/L APTOA6 56,20,85
orl-rat TDLo:3600 mg/kg (7-15D preg):TER JTSCDR 8,339,83
orl-rat TDLo:3600 mg/kg (7-15D preg):REP JTSCDR 8,339,83
orl-rat TDLo:58,400 mg/kg/2Y-C:CAR EPASR* 8EHQ-1280-0401S
orl-rat LD50:1850 mg/kg BCTKAG 9,141,76
orl-mus LD50:2250 mg/kg ESKHA5 (107),36,89

CONSENSUS REPORTS: EPA Genetic Toxicology Program. Reported in EPA TSCA Inventory.

SAFETY PROFILE: Moderately toxic by ingestion. An experimental teratogen. Experimental reproductive effects. Questionable carcinogen with experimental carcinogenic data. Mutation data reported. When heated to decomposition it emits toxic fumes of Cl^- and PO_x.

G

GAA100 **CAS:51909-61-6** **HR: 3**
G-52
mf: $C_{20}H_{39}N_5O_7$ mw: 461.64

SYNS: o-2-AMINO-2,3,4,6-TETRADEOXY-6-(METHYLAMINO)-α-d-glycero-HEX-4-ENOPYRANOSYL-(1-4)-o-(3-DEOXY-4-C-METHYL-3-(METHYLAMINO)-β-l-ARABINOPYRANOSYL-(1-6))-2-DEOXY-d-STREPTAMINE □ ANTIBIOTIC G-52 □ SCH-17726

TOXICITY DATA WITH REFERENCE
ipr-mus LD50:200 mg/kg 85GDA2 1,180,80
scu-mus LD50:400 mg/kg 85GDA2 1,180,80
ivn-mus LD50:50 mg/kg 85GDA2 1,180,80

SAFETY PROFILE: Poison by subcutaneous, intravenous, and intraperitoneal routes. When heated to decomposition it emits toxic fumes of NO_x.

GAA120 **CAS:51922-16-8** **HR: 3**
G 52 SULFATE
mf: $C_{20}H_{39}N_5O_7{\bullet}(H_2O_4S)_7$ mw: 1148.07

SYNS: 4-o-(3-AMINO-3,4-DIHYDRO-6-((METHYLAMINO)METHYL)-2H-PYRAN-2-YL)-2-DEOXY-6-o-(3-DEOXY-4-C-METHYL-3-(METHYLAMINO)-β-l-ARABINOPYRANOSYL)-d-STREPTAMINE (2S-cis)-, SULFATE (salt) □ ANTIBIOTIC G-52 SULFATE

TOXICITY DATA WITH REFERENCE
ipr-mus LD50:200 mg/kg JANTAJ 29,483,76
scu-mus LD50:400 mg/kg JANTAJ 29,483,76
ivn-mus LD50:50 mg/kg JANTAJ 29,483,76

SAFETY PROFILE: Poison by subcutaneous, intravenous, and intraperitoneal routes. When heated to decomposition it emits toxic fumes of NO_x and SO_x.

GAC000 **CAS:1952-11-0** **HR: 3**
G 3063 HYDROCHLORIDE
mf: $C_{18}H_{25}NO_2{\bullet}ClH$ mw: 323.90

SYNS: G 3063 □ 1-PHENYLCYCLOPENTANECARBOXYLIC ACID-1-METHYL-4-PIPERIDINYL ESTER HYDROCHLORIDE □ 1-PHENYLCYCLOPENTANECARBOXYLIC ACID-1-METHYL-4-PIPERIDYL ESTER HYDROCHLORIDE

TOXICITY DATA WITH REFERENCE
ims-rat LD50:2483 μg/kg BJPCBM 39,822,70
ims-mus LD50:730 μg/kg BJPCBM 39,822,70
ims-gpg LD50:157 μg/kg BJPCBM 39,822,70

SAFETY PROFILE: A deadly poison by intramuscular route. When heated to decomposition it emits toxic fumes of NO_x and HCl. See also ESTERS.

GAD000 **CAS:3060-41-1** **HR: 2**
p-GABA HYDROCHLORIDE
mf: $C_{10}H_{13}NO_2{\bullet}ClH$ mw: 215.70

SYNS: β-(AMINOMETHYL)-BENZENEPROPANOIC ACID HYDROCHLORIDE □ β-(AMINOMETHYL)-HYDROCINNAMIC ACID HYDROCHLORIDE □ FENIBUT HYDROCHLORIDE □ FENIGAM HYDROCHLORIDE □ PHENIBUT HYDROCHLORIDE □ PHENIGAM HYDROCHLORIDE □ PHENIGAMA HYDROCHLORIDE □ PHENYBUT HYDROCHLORIDE □ PHENYGAM HYDROCHLORIDE □ PHENYL-GAMMA HYDROCHLORIDE □ PHGABA HYDROCHLORIDE

TOXICITY DATA WITH REFERENCE
ipr-rat LD50:900 mg/kg PCJOAU 10,1703,76
ipr-mus LD50:1000 mg/kg PCJOAU 10,1703,76

SAFETY PROFILE: Moderately toxic by intraperitoneal route. When heated to decomposition it emits very toxic fumes of Cl^- and NO_x.

GAD400 **CAS:56974-61-9** **HR: 3**
GABEXATE MESYLATE
mf: $C_{16}H_{23}N_3O_4{\bullet}CH_4O_3S$ mw: 417.53

PROP: Crystals.

SYNS: ETHYL-p-(6-GUANIDINOHEXANOYLOXY) BENZOATE METHANESULFONATE □ FOY □ GABEXATE MESILATE

TOXICITY DATA WITH REFERENCE
ivn-mus TDLo:600 mg/kg (7-12D preg):TER OYYAA2 9,743,75
ivn-rat TDLo:560 mg/kg (14D pre):REP OYYAA2 9,743,75
ivn-rbt TDLo:1680 mg/kg (male 56D pre):REP OYYAA2 9,743,75
ivn-rat TDLo:24 mg/kg (9-14D preg):TER OYYAA2 9,743,75
orl-rat LD50:6480 mg/kg OYYAA2 9,743,75
scu-rat LD50:4020 mg/kg OYYAA2 9,743,75
ivn-rat LD50:79 mg/kg OYYAA2 9,743,75
orl-mus LD50:8 g/kg IYKEDH 8,680,77
scu-mus LD50:4550 mg/kg OYYAA2 9,743,75
ivn-mus LD50:248 mg/kg OYYAA2 9,743,75

SAFETY PROFILE: Poison by intravenous route. Mildly toxic by ingestion. Experimental teratogenic and reproductive effects. When heated to decomposition it emits toxic fumes of SO_x and NO_x.

GAF000 **CAS:7440-54-2** **HR: 2**
GADOLINIUM
af: Gd aw: 157.25

PROP: A yellow-white, malleable, lustrous, and ductile metallic element. Tarnishes in moist air. A rare earth, stable in dry air; reacts slowly with H_2O and O_2. Dissolves in acids. Mp: 1312°, bp: 3273°, d: 7.898 @ 25°.

TOXICITY DATA WITH REFERENCE
imp-mus TDLo:25 g/kg:ETA PSEBAA 135,426,70

CONSENSUS REPORTS: Reported in EPA TSCA Inventory.

SAFETY PROFILE: Questionable carcinogen with experimental tumorigenic data. It may act as an anticoagulant. It can react violently with air and halogens. See also RARE EARTHS.

GAH000 CAS:10138-52-0 HR: 3
GADOLINIUM CHLORIDE
mf: Cl_3Gd mw: 263.60

PROP: White, monoclinic, colorless crystals. D: 4.52 @ 0°, bp: 1580°, mp: approx 609°. Sol in water and alc.

SYN: GADOLINIUM TRICHLORIDE

TOXICITY DATA WITH REFERENCE
skn-rbt 500 mg MOD BJPCAL 17,526,61
eye-rbt 1 mg/1H MLD BJPCAL 17,526,61
ipr-mus LD50:378 mg/kg AEHLAU 5,437,62

CONSENSUS REPORTS: Reported in EPA TSCA Inventory.

SAFETY PROFILE: Poison by intraperitoneal route. A skin and eye irritant. When heated to decomposition it emits very toxic fumes of Cl^-. See also CHLORIDES, GADOLINIUM, and RARE EARTHS.

GAJ000 CAS:3088-53-7 HR: 3
GADOLINIUM CITRATE

TOXICITY DATA WITH REFERENCE
ipr-mus LD50:153 mg/kg AEHLAU 5,437,62
ipr-gpg LD50:60 mg/kg AEHLAU 5,437,62

SAFETY PROFILE: Poison by intraperitoneal route. When heated to decomposition it emits acrid smoke and irritating fumes. See also GADOLINIUM and RARE EARTHS.

GAL000 CAS:10168-81-7 HR: 3
GADOLINIUM(III) NITRATE (1:3)
mf: $N_3O_9{\cdot}Gd$ mw: 343.28

PROP: Hygroscopic white solid. Sol in EtOH and H_2O.

SYN: NITRIC ACID, GADOLINIUM(3+) SALT

TOXICITY DATA WITH REFERENCE
orl-rat LD50:3805 mg/kg EQSSDX 1,1,75
ipr-rat LD50:175 mg/kg EQSSDX 1,1,75

CONSENSUS REPORTS: Reported in EPA TSCA Inventory.

SAFETY PROFILE: Poison by intraperitoneal route. Moderately toxic by ingestion. When heated to decomposition it emits toxic fumes of NO_x. See also GADOLINIUM, RARE EARTHS, and NITRATES.

GAN000 CAS:19598-90-4 HR: 3
GADOLINIUM(III) NITRATE, HEXAHYDRATE (1:3:6)
mf: $N_3O_9{\cdot}Gd{\cdot}6H_2O$ mw: 451.40

PROP: Deliqescent, colorless, triclinic crystals. Mp: 91–92°, d: 2.406 @ 15°, d: 2.332. Very sol in H_2O; sol in EtOH and Me_2CO.

TOXICITY DATA WITH REFERENCE
ipr-rat LD50:230 mg/kg TXAPA9 5,750,63
ipr-mus LD50:300 mg/kg TXAPA9 5,750,63

SAFETY PROFILE: Poison by intraperitoneal route. When heated to decomposition it emits very toxic fumes of NO_x. See also RARE EARTHS, NITRATES, and GADOLINIUM.

G

GAP000 CAS:12064-62-9 HR: 1
GADOLINIUM OXIDE
mf: Gd_2O_3 mw: 362.50

PROP: White to cream colored, hygroscopic powder. D: 7.407 @ 15°, mp: 2339°, bp: 3900°. Insol in water; sol in acids.

SYNS: DIGADOLINIUM TRIOXIDE □ GADOLINIA □ GADOLINIUM(III) OXIDE □ GADOLINIUM(3+) OXIDE □ GADOLINIUM SESQUIOXIDE □ GADOLINIUM TRIOXIDE

TOXICITY DATA WITH REFERENCE
eye-rbt 100 mg MLD JACTDZ 12,620,93
orl-rat LD50:>5 g/kg JACTDZ 12,620,93

CONSENSUS REPORTS: Reported in EPA TSCA Inventory.

SAFETY PROFILE: Low toxicity by ingestion. An eye irritant. See also RARE EARTHS.

GAR000 CAS:526-99-8 HR: 1
GALACTARIC ACID
mf: $C_6H_{10}O_8$ mw: 210.16

PROP: Crystalline powder. Mp: 213°. Decomp @ approx 230° when rapidly heated. Sol in alkalies, water. Practically insol in alc and ether.

SYNS: GALACTOSACCHARIC ACID □ MUCIC ACID □ SACCHAROLACTIC ACID □ SCHLEIMSAURE □ TETRAHYDROXYADIPIC ACID

TOXICITY DATA WITH REFERENCE
orl-mus LD50:8000 mg/kg BIJOAK 34,1196,40

CONSENSUS REPORTS: Reported in EPA TSCA Inventory.

SAFETY PROFILE: Mildly toxic by ingestion. When heated to decomposition it emits acrid smoke and irritating fumes.

GAT000 CAS:1772-03-8 HR: 2
d-GALACTOSAMINE HYDROCHLORIDE
mf: $C_6H_{13}NO_5{\cdot}ClH$ mw: 215.66

PROP: A solid. Mp: 178–190° (decomp).

SYN: 2-AMINO-2-DEOXY-d-GALACTOSE HYDROCHLORIDE

TOXICITY DATA WITH REFERENCE
oth-rat:lvr 100 mmol/L CBINA8 51,63,84
oth-rat-icb 21,500 µg/kg ABMGAJ 39,5,80
ipr-rat TDLo:135 g/kg/77W-I:ETA VAAZA2 12,285,73
ipr-mus LD50:2660 mg/kg CTYAD8 11,262,80

CONSENSUS REPORTS: Reported in EPA TSCA Inventory.

SAFETY PROFILE: Moderately toxic by intraperitoneal route. Questionable carcinogen with experimental tumorigenic data. Mutation data reported. When heated to decomposition it emits very toxic fumes of NO_x and Cl^-.

GAV000 **CAS:59-23-4** **HR: D**
GALACTOSE
mf: $C_6H_{12}O_6$ mw: 180.18

PROP: A solid. Mp: 118–120° (monohydrate). (α form): Prisms from water or ethanol. Mp: 167°. Freely sol in hot water; sol in pyridine; sltly sol in alc. (β form): Crystals. Mp: 167°.

SYN: d-GALACTOSE

TOXICITY DATA WITH REFERENCE
orl-rat TDLo:1000 g/kg (female 3-22D post):TER SCIEAS 214,1145,81
orl-rat TDLo:440 g/kg (female 1-22D post):REP JOPDAB 67,438,65
orl-mus TDLo:16,500 mg/kg (female 8-12D post):REP TCMUD8 7,7,87
orl-rat TDLo:475 g/kg (female 3-22D post):REP SCIEAS 214,1145,81
orl-mus TDLo:1260 g/kg (female 1-21D post):REP TJADAB 31,29A,85
orl-rat TDLo:240 g/kg (female 11-22D post):TER JOPDAB 70,676,67
orl-mus TDLo:840 g/kg (female 14D pre):REP FESTAS 49,522,88
orl-rat TDLo:188 g/kg (female 8-10D post):REP ANENAG 20,203,59

CONSENSUS REPORTS: Reported in EPA TSCA Inventory.

SAFETY PROFILE: Experimental teratogenic and reproductive effects. When heated to decomposition it emits acrid smoke and irritating fumes.

GAV100 **CAS:9031-11-2** **HR: 2**
β-GALACTOSIDASE

SYNS: E.C. 3.2.1.23 □ LACTASE

TOXICITY DATA WITH REFERENCE
ipr-rat LD50:660 mg/kg KSRNAM 4,725,70
scu-rat LD50:4090 mg/kg KSRNAM 4,725,70
ipr-mus LD50:630 mg/kg KSRNAM 4,729,70
scu-mus LD50:1470 mg/kg KSRNAM 4,729,70

CONSENSUS REPORTS: Reported in EPA TSCA Inventory.

SAFETY PROFILE: Moderately toxic by subcutaneous and intraperitoneal routes.

GAX000 **CAS:7681-28-9** **HR: 2**
GALACTURONIC ACID with α-(6-METHOXY-4-QUINOLYL)-5-VINYL-2-QUINUCLIDINEMETHANOL
mf: $C_{20}H_{24}N_2O_2{\cdot}xC_6H_{10}O_7$ mw: 1683.58

SYNS: CARDIOQUIN □ GALACTOQUIN □ α-(6-METHOXY-4-QUINOLYL)-5-VINYL-2-QUINUCLIDINEMETHANOL GALACTURONATE (salt) □ NATICARDINA □ QUINIDINE POLYGALACTURONATE □ SINEFLUTTER

TOXICITY DATA WITH REFERENCE
orl-rat LD50:3200 mg/kg ANTCAO 9,97,59
orl-mus LD50:2680 mg/kg ANTCAO 9,97,59

SAFETY PROFILE: Moderately toxic by ingestion. Used as a cardiac depressant. When heated to decomposition it emits toxic fumes of NO_x.

GBA000 **CAS:69353-21-5** **HR: 3**
GALANTHAMINE HYDROBROMIDE
mf: $C_{17}H_{21}NO_3{\cdot}BrH$ mw: 368.31

PROP: Crystals from water. Decomp @ 246–247°.

SYNS: JILKON HYDROBROMIDE □ LYCOREMINE HYDROBROMIDE □ NIVALIN □ 1,2,3,4,6,7,7a,11c-OCTAHYDRO-9-METHOXY-2-METHYL-BENZOFURO(4,3,2-efg)(2)BENZAZOCIN-6-OL HBr

TOXICITY DATA WITH REFERENCE
orl-mus LD50:18,700 µg/kg MEIEDD 10,620,83
ipr-mus LD50:14,900 µg/kg AAREAV 22,285,65
ivn-mus LD50:5200 µg/kg AAREAV 22,285,65

SAFETY PROFILE: Poison by ingestion, intravenous, and intraperitoneal routes. Used as a cholinesterase inhibitor. When heated to decomposition it emits very toxic fumes of HBr and NO_x.

GBB500 **CAS:3691-74-5** **HR: 3**
GALATONE
mf: $C_{12}H_{13}N_3O_6$ mw: 295.28

PROP: Plates and rods from methanol, needles from abs ethanol. Decomp at 150–160°. Freely sol in water; practically insol in cold alc.

SYNS: GATALONE □ GLUCAZIDE □ GLURONAZID □ GLURONAZIDE □ GLYCONIAZIDE □ GUIDAZIDE □ INH-G □ HYDRONSAN □ N-ISONICOTINOYL-N'-GLUCURONSAEURE-γ-LACTON-HYDRAZON (GERMAN) □ MYCOBACTYL

TOXICITY DATA WITH REFERENCE
orl-rat LD50:2990 mg/kg ARZNAD 26,409,76
ivn-rat LD50:820 mg/kg ARZNAD 26,409,76
orl-mus LD50:348 mg/kg ARZNAD 26,409,76
ivn-mus LD50:298 mg/kg ARZNAD 26,409,76
orl-gpg LD50:530 mg/kg ARZNAD 26,409,76

SAFETY PROFILE: Poison by ingestion and intravenous routes. When heated to decomposition it emits toxic fumes of NO_x.

GBC000 **CAS:8023-91-4** **HR: 1**
GALBANUM OIL

PROP: Found in the dried resinous exudate of *Ferula galbaniflua boiss & buhse* and other *Ferula* species (FCTXAV 16,637,78).

TOXICITY DATA WITH REFERENCE
skn-rbt 500 mg/24H MLD FCTXAV 16,765,78

CONSENSUS REPORTS: Reported in EPA TSCA Inventory.

SAFETY PROFILE: A skin irritant. When heated to decomposition it emits acrid smoke and irritating fumes.

GBE000 CAS:149-91-7 HR: 3
GALLIC ACID
mf: $C_7H_6O_5$ mw: 170.13

PROP: Needles from MeOH or $CHCl_3$. White to pale, fawn-colored, odorless crystals from water; somewhat water-sol. D: 1.694, mp: 253° (decomp). Sol in Me_2CO, EtOH; sltly sol in H_2O; insol in C_6H_6, $CHCl_3$.

SYN: 3,4,5-TRIHYDROXYBENZOIC ACID

TOXICITY DATA WITH REFERENCE
mmo-sat 100 μg/plate ABCHA6 45,327,81
mrc-smc 100 mg/L MUREAV 135,109,84
scu-rat TDLo:5 mg/kg (1D pre):REP ENDOAO 57,466,55
scu-rat LDLo:5 g/kg JAFCAU 17,497,69
ipr-mus LD50:4300 mg/kg PHTXA6 64,247,89
ivn-mus LD50:320 mg/kg CSLNX* NX#02597
orl-rbt LD50:5 g/kg AJVRAH 23,1264,62

CONSENSUS REPORTS: Reported in EPA TSCA Inventory. EPA Genetic Toxicology Program.

SAFETY PROFILE: Poison by intravenous route. Moderately toxic by intraperitoneal route. Mildly toxic by ingestion. Experimental reproductive effects. Mutation data reported. When heated to decomposition it emits acrid smoke and irritating fumes.

GBG000 CAS:7440-55-3 HR: 3
GALLIUM
DOT: UN 2803
af: Ga aw: 69.72

PROP: Soft sltly blue-white metal. Stable in air and not attacked by H_2O. Contracts on melting. Longest liquid range of all elements. Wets glass, porcelain, and most other surfaces except graphite, quartz, and Teflon. Mp: 29.8°, bp: 2403°. Sol in acids and alkalies.

TOXICITY DATA WITH REFERENCE
dni-hmn:lyms 480 μmol/L CNREA8 48,3014,88

PROP: A beautiful, lustrous, silvery liquid or a gray solid. Mp: 29.78°, bp: 2403°, d (solid): 5.904 @ 29.6°, d (liquid): 6.095 @ 29.8°.

CONSENSUS REPORTS: Reported in EPA TSCA Inventory.

DOT CLASSIFICATION: 8; *Label:* Corrosive

SAFETY PROFILE: Poison by subcutaneous and intravenous routes. Mutation data reported. Corrosive; probably an eye, skin, and mucous membrane irritant. It has a metallic taste, causes dermatitis and depression of bone marrow function. Potentially explosive reaction with hydrogen peroxide + hydrochloric acid. Violent or vigorous reaction with halogens. Forms an amalgam with aluminum alloys. See also GALLIUM COMPOUNDS.

GBK000 CAS:1303-00-0 HR: 3
GALLIUM ARSENIDE
mf: AsGa mw: 144.64

PROP: Black solid or cubic crystals with dark-gray metallic sheen. Mp: 1238°, d: 5.31.

SYN: GALLIUM MONOARSENIDE

TOXICITY DATA WITH REFERENCE
ipr-rat LD30:10 g/kg GTPZAB 24(3),45,80
ipr-mus LD50:4700 mg/kg GISAAA 45(10),13,80

CONSENSUS REPORTS: Reported in EPA TSCA Inventory. Arsenic and its compounds are on the Community Right-To-Know List.

OSHA PEL: Cancer Hazard
NIOSH REL: (Gallium Arsenide) CL 0.002 $mg(As)/m^3/15M$

SAFETY PROFILE: Confirmed carcinogen. Mildly toxic by intraperitoneal route. Most arsenic compounds are poisons. Can react with steam, acids, and acid fumes to evolve the deadly poisonous arsine. Molten gallium arsenide attacks quartz. When heated to decomposition it emits very toxic fumes of As. See also ARSENIC COMPOUNDS and GALLIUM COMPOUNDS.

GBM000 CAS:13450-90-3 HR: 3
GALLIUM(3+) CHLORIDE
mf: Cl_3Ga mw: 176.07

PROP: Colorless needles. Mp: 78°.

SYN: GALLIUM CHLORIDE

TOXICITY DATA WITH REFERENCE
dnr-bcs 2350 μg/disc MUREAV 264,163,91
ihl-rat LCLo:316 $mg/m^3/3H$ JPETAB 95,487,49
scu-rat LD50:306 mg/kg EQSSDX 1,1,75
ivn-rat LD50:47 mg/kg EQSSDX 1,1,75
ipr-mus LD50:93,400 μg/kg COREAF 256,1043,63
ivn-dog LD50:41 mg/kg EQSSDX 1,1,75
scu-rbt LD50:245 mg/kg EQSSDX 1,1,75
ivn-rbt LD50:43 mg/kg EQSSDX 1,1,75

CONSENSUS REPORTS: Reported in EPA TSCA Inventory. EPA Extremely Hazardous Substances List.

SAFETY PROFILE: Poison by inhalation, subcutaneous, intravenous, and intraperitoneal routes. Mutation data reported. When heated to decomposition it emits very toxic fumes of Cl^-. See also GALLIUM COMPOUNDS.

GBO000 CAS:27905-02-8 HR: 3
GALLIUM CITRATE
mf: $C_6H_5O_7{\cdot}Ga$ mw: 258.83

SYN: CITRIC ACID, GALLIUM SALT (1:1)

TOXICITY DATA WITH REFERENCE
scu-rat LD50:220 mg/kg 14CYAT 2,1040,63
scu-mus LD50:2250 mg/kg EQSSDX 1,1,75

scu-dog LD50:37,500 μg/kg EQSSDX 1,1,75
scu-dom LD50:37,500 μg/kg EQSSDX 1,1,75

SAFETY PROFILE: Poison by subcutaneous route. When heated to decomposition it emits acrid smoke and irritating fumes. See also GALLIUM COMPOUNDS.

GBO500 **HR: 3**
GALLIUM COMPOUNDS

SAFETY PROFILE: Preliminary investigations were done with the oxide, tartrate, benzoate, and anthranilate, which were used by some investigators in the experimental treatment of syphilis. Amounts up to 15 mg/kg of body weight were injected intravenously and were tolerated without harm by laboratory animals. Larger doses produced hemorrhagic nephritis. In the case of gallium lactate, work done at the Naval Medical Research Institute showed that intravenous injections of about 40 mg of gallium per kg of body weight in rats or rabbits was lethal. Metallic gallium as well as the nitrate produced no skin injury, and subcutaneous injections of relatively large amounts could be tolerated both by rabbits and rats without evidence of injury. It has, however, been demonstrated that gallium remains in the tissues for long periods of time following intramuscular injections of soluble gallium salts. Tissue distribution experiments indicate that it behaves like bismuth and mercury in that one respect.

GBS000 **CAS:13494-90-1** **HR: 3**
GALLIUM(III) NITRATE (1:3)
mf: N_3O_9•Ga mw: 255.75

PROP: White, deliquescent crystals. Mp: decomp @ 110°, bp: releases Ga_2O_3 @ 200°.

SYNS: GALLIUM NITRATE □ NITRIC ACID, GALLIUM(3+) SALT

TOXICITY DATA WITH REFERENCE
skn-mam 500 mg SEV GISAAA 45(10),13,80
dnd-mam:lym 40 μmol/L JCHODP 7,411,76
ivn-hmn TDLo:7 mg/kg:GIT,KID CTRRDO 62,1449,78
ivn-hmn TDLo:144 mg/kg:BLD CTRRDO 62,1449,78
ipr-rat LD50:67,500 μg/kg EQSSDX 1,1,75
scu-rat LDLo:72 mg/kg INMEAF 12,7,43
orl-mus LD50:4360 mg/kg GISAAA 45(10),13,80
scu-mus LD50:600 mg/kg GISAAA 45(10),13,80
ivn-mus LD50:55 mg/kg CTRRDO 70,1311,86

CONSENSUS REPORTS: Reported in EPA TSCA Inventory.

SAFETY PROFILE: Poison by intraperitoneal, intravenous, and subcutaneous routes. Moderately toxic by ingestion. Human systemic effects by intravenous route: nausea or vomiting, renal function changes, proteinuria, normocytic anemia, and thrombocytopenia. A severe skin irritant. Mutation data reported. When heated to decomposition it emits toxic fumes of NO_x. See also GALLIUM COMPOUNDS and NITRATES.

GBS050 **CAS:12024-21-4** **HR: 1**
GALLIUM OXIDE
mf: Ga_2O_3 mw: 187.44

SYNS: DIGALLIUM TRIOXIDE □ GALLIA □ GALLIUM SESQUIOXIDE □ GALLIUM TRIOXIDE

TOXICITY DATA WITH REFERENCE
orl-mus LD50:10 g/kg GISAAA 45(10),13,80
ipr-uns LDLo:10 g/kg GTPZAB 31(6),58,87

CONSENSUS REPORTS: Reported in EPA TSCA Inventory.

SAFETY PROFILE: Low toxicity by ingestion and intraperitoneal routes. When heated to decomposition it emits toxic vapors of Ga.

GBS100 **CAS:13494-91-2** **HR: 2**
GALLIUM SULFATE
mf: $O_{12}S_3$•2Ga mw: 427.62

SYNS: DIGALLIUM TRISULFATE □ SULFURIC ACID, GALLIUM SALT (3:2)

TOXICITY DATA WITH REFERENCE
ivn-ham TDLo:40 mg/kg (female 8D post):REP TXAPA9 16,166,70
ipr-rat LD50:2 g/kg JTCEEM 7,411,87
ipr-mus LD50:2330 mg/kg JTCEEM 7,411,87

CONSENSUS REPORTS: Reported in EPA TSCA Inventory.

SAFETY PROFILE: Moderately toxic by intraperitoneal route. Experimental reproductive effects. When heated to decomposition it emits acrid smoke and irritating fumes.

GBU000 **CAS:1222-05-5** **HR: 1**
GALOXOLIDE
mf: $C_{18}H_{26}O$ mw: 258.44

SYN: 1,3,4,6,7,8-HEXAHYDRO-4,6,6,7,8,8-HEXAMETHYL-CYCLOPENTA-γ-2-BENZOPYRAN

TOXICITY DATA WITH REFERENCE
skn-rbt 500 mg/24H MOD FCTXAV 14,793,76

CONSENSUS REPORTS: Reported in EPA TSCA Inventory.

SAFETY PROFILE: A skin irritant. When heated to decomposition it emits acrid smoke and irritating fumes.

GBU600 **HR: 2**
GANCIDIN (unpurified)

PROP: Produced by the strain *Streptomyces sp.* AAK-84 (85ERAY 2,1354,78)

TOXICITY DATA WITH REFERENCE
ipr-mus LD50:1500 mg/kg 85ERAY 2,1354,78
scu-mus LD50:2800 mg/kg 85ERAY 2,1354,78
ivn-mus LD50:700 mg/kg 85ERAY 2,1354,78

SAFETY PROFILE: Moderately toxic by intravenous, intraperitoneal, and subcutaneous routes.

GBU800 **HR: 1**
GARLIC OIL

PROP: From steam distillation of *Allium sativum* L. (Fam. Liliaceae). Clear to yellow liquid; strong odor and taste of garlic. Sol in fixed oils, mineral oil; insol in glycerin, alc, propylene glycol.

SAFETY PROFILE: An eye irritant. When heated to decomposition it emits acrid smoke and irritating fumes.

GBU850 **HR: 2**
GARLIC POWDER

SYN: ALLIUM SATIVUM Linn., powder

TOXICITY DATA WITH REFERENCE
orl-rat TDLo:9 g/kg (male 45D pre):REP IJEBA6 20,534,82
orl-rat TDLo:14 g/kg (male 70D pre):REP IJEBA6 20,534,82
ivn-mus LD50:1650 mg/kg PJPHEO 5(1),21,88

SAFETY PROFILE: Moderately toxic by intravenous route. Experimental reproductive effects. When heated to decomposition it emits acrid smoke and irritating fumes.

GBW000 **CAS:64741-44-2** **HR: 2**
GAS OIL

PROP: Yellow liquid. Flash p: 150°F, d: 1, lel: 6.0%, uel: 13.5%, autoign temp: 640°F, boiling range: 230–250°.

SYNS: DISTILLATES (PETROLEUM), STRAIGHT-RUN MIDDLE □ STRAIGHT RUN MIDDLE DISTILLATE

TOXICITY DATA WITH REFERENCE
skn-rbt 500 mg MOD JACTDZ 1,126,90
skn-mus TDLo:114 g/kg/38W-I:CAR FAATDF3 7,228,86
skn-mus TD:114 g/kg/38W-I:CAR FAATDF 7,228,86
orl-rat LD:>5 g/kg JACTDZ 1,126,90
ihl-rat LC50:1700 mg/m^3/4H JACTDZ 1,126,90
skn-rbt LD:>2 g/kg JACTDZ 1,126,90

CONSENSUS REPORTS: Reported in EPA TSCA Inventory.

SAFETY PROFILE: Low toxicity by ingestion and skin contact. Questionable carcinogen with experimental carcinogenic data. A skin irritant. Pulmonary aspiration can cause severe pneumonitis. Combustible when exposed to heat or flame; can react vigorously with oxidizing materials. A moderate explosion hazard when exposed to heat or flame. To fight fire, use foam, CO_2, dry chemical. See also KEROSENE.

GBW005 **CAS:64741-57-7** **HR: 3**
GAS OILS (PETROLEUM), heavy vacuum

SYNS: HEAVY VACUUM DISTILLATE □ HEAVY VACUUM GAS OIL (PETROLEUM)

CONSENSUS REPORTS: IARC Cancer Review: Animal Sufficient Evidence IMEMDT 45,39,89. Reported in EPA TSCA Inventory.

SAFETY PROFILE: Confirmed carcinogen. When heated to decomposition it emits acrid smoke and irritating vapors.

GBW010 **CAS:64742-86-5** **HR: 2**
GAS OILS (petroleum), hydrodesulfurized heavy vacuum

SYN: HYDRODESULFURIZED HEAVY VACUUM GAS OIL

TOXICITY DATA WITH REFERENCE
skn-mus TDLo:104 g/kg/26W-I:NEO EPASR* 8EHQ-0887-0687
skn-mus TDLo:172 g/kg/43W-I:CAR EPASR* 8EHQ-1287-0710
skn-mus TD:104 g/kg/26W-I:NEO EPASR* 8EHQ-0887-0687

SAFETY PROFILE: Questionable carcinogen with experimental carcinogenic and neoplastigenic data. When heated to decomposition it emits acrid smoke and irritating fumes.

GBW025 **CAS:64741-58-8** **HR: 3**
GAS OILS (petroleum), light vacuum

SYN: LIGHT GAS OIL

TOXICITY DATA WITH REFERENCE
skn-mus TDLo:306 g/kg/2Y-I:CAR FAATDF 7,228,86
skn-mus LD :306 g/kg/2Y-I:CAR FAATDF 7,228,86

CONSENSUS REPORTS. IARC Cancer Review: Animal Sufficient Evidence IMEMDT 45,39,89. Reported in EPA TSCA Inventory.

SAFETY PROFILE: Confirmed carcinogen with experimental carcinogenic data. Pulmonary aspiration can cause severe pneumonitis. A flammable liquid. When heated to decomposition it emits acrid smoke and irritating fumes.

GBY000 **CAS:8006-61-9** **HR: 3**
GASOLINE
DOT: UN 1203/UN 1257

PROP: Clear, aromatic, volatile liquid; a mixture of aliphatic hydrocarbons. Flash p: −50°F, d: <1.0, vap d: 3.0–4.0, ULC: 95–100, lel: 1.3%, uel: 6.0%, autoign temp: 536–853°F, bp: initially 39°, after 10% distilled = 60°, after 50% = 110°, after 90% = 170°, final bp: = 204°. Insol in water; freely sol in abs alc, ether, chloroform, and benzene.

SYNS: MOTOR SPIRIT (DOT) □ NATURAL GASOLINE (DOT) □ PETROL (DOT)

TOXICITY DATA WITH REFERENCE
eye-man 500 ppm/1H MOD AEHLAU 1,548,60
eye-hmn 140 ppm/8H MLD JIHTAB 25,225,43
ihl-man TCLo:900 ppm/1H:EYE,CNS,PUL JIHTAB 25,225,43
par-man TDLo:53 mg/kg JTCTDW 21,409,83/84
ihl-rat LC50:300 g/m^3/5M NTIS** PB158-508
ihl-mus LC50:300 g/m^3/5M NTIS** PB158-508
ihl-gpg LC50:300 g/m^3/5M NTIS** PB158-508
ihl-mam LCLo:30,000 ppm/5M AEPPAE 138,65,28

G

CONSENSUS REPORTS: Reported in EPA TSCA Inventory.

OSHA PEL: TWA 300 ppm; STEL 500 ppm
ACGIH TLV: TWA 300 ppm; STEL 500 ppm
DOT CLASSIFICATION: 3; *Label:* Flammable Liquid

SAFETY PROFILE: Mildly toxic by inhalation. Human systemic effects by inhalation: cough, conjunctiva irritation, hallucinations or distorted perceptions. Repeated or prolonged dermal exposure causes dermatitis. Can cause blistering of skin. Questionable carcinogen. Inhalation or ingestion can cause central nervous system depression. Pulmonary aspiration can cause severe pneumonitis. Some addiction has been reported from inhalation of fumes. Even brief inhalations of high concentrations can cause a fatal pulmonary edema. The vapors are considered to be moderately poisonous. If its concentration in air is sufficiently high to reduce the oxygen content below that needed to maintain life, it acts as a simple asphyxiant. A human eye irritant. Gasoline is a common air contaminant. A very dangerous fire and explosion hazard when exposed to heat or flame; can react vigorously with oxidizing materials. To fight fire, use foam, CO_2, dry chemical.

GCA000 **HR: 3**
GASOLINE (100–130 octane)

PROP: Flash p: −50°F, autoign temp: 824°F, lel: 1.3%, uel: 7.1%.

SAFETY PROFILE: Moderately toxic by inhalation. Pulmonary aspiration can cause severe pneumonitis. A very dangerous fire and explosion hazard when exposed to heat, flame or oxidizers. To fight fire, use water spray or mist, CO_2, dry chemical. See also GASOLINE.

GCC000 **HR: 3**
GASOLINE (115–145 octane)

PROP: Flash p: −50°F, autoign temp: 880°F, lel: 1.2%, uel: 7.1%.

SAFETY PROFILE: Moderately toxic by inhalation. Pulmonary aspiration can cause severe pneumonitis. A very dangerous fire and explosion hazard when exposed to heat or flame. To fight fire, use water spray or mist, CO_2, dry chemical. See also GASOLINE.

GCE000 **HR: 2**
GASOLINE ENGINE EXHAUST "TAR"

SYNS: AUTOMOBILE EXHAUST CONDENSATE □ GASOLINE ENGINE EXHAUST CONDENSATE

TOXICITY DATA WITH REFERENCE
skn-mus TDLo: 110 g/kg/69W-I:CAR CANCAR 15,103,62
itr-ham TDLo: 469 mg/kg/60W-I:NEO CANCAR 40,203,77
itr-ham TD: 860 mg/kg/60W-I:NEO CANCAR 40,203,77
itr-ham TD: 420 mg/kg/78W-I:ETA JCROD7 105,24,83
itr-ham TD: 643 mg/kg/78W-I:ETA JCROD7 105,24,83

SAFETY PROFILE: Questionable carcinogen with experimental carcinogenic, neoplastigenic, and tumorigenic data.

GCE100 **HR: 3**
GASOLINE, UNLEADED

SYNS: UNLEADED GASOLINE □ UNLEADED MOTOR GASOLINE

TOXICITY DATA WITH REFERENCE
skn-rbt 500 mg/24H MLD 52MLA2 1,1,83
ihl-rat TCLo: 1501 ppm/78W-C:CAR AETODY 7,65,84
ihl-mus TCLo: 2056 ppm/6H/78W-I:CAR NTIS** PB86-209152
ihl-rat TC: 2056 ppm/6H/78W-I:CAR NTIS** PB86-209152

SAFETY PROFILE: Suspected carcinogen with experimental carcinogenic data. Moderately toxic by inhalation. Pulmonary aspiration can cause severe pneumonitis. Skin irritant. Flammable liquid. When heated to decomposition it emits acrid smoke and irritating fumes.

GCE500 **CAS:29868-97-1** **HR: 3**
GASTROZEPIN
mf: $C_{19}H_{21}N_5O_2 \cdot 2ClH$ mw: 424.37

PROP: A solid. Mp: 240–243°.

SYNS: L-S 519 □ LS 519 DIHYDROCHLORIDE □ 5,11-DIHYDRO-11-((4-METHYL-1-PIPERAZINYL)ACETYL)-6H-PYRIDO(2,3-b)(1,4)BENZODIAZEPIN-6-ONE DIHYDROCHLORIDE □ PIRENZEPINE HYDROCHLORIDE

TOXICITY DATA WITH REFERENCE
ivn-rat TDLo: 810 mg/kg (female 17-22D post):REP IYKEDH 17,859,86
ivn-rat TDLo: 1350 mg/kg (female 17-22D post):REP IYKEDH 17,859,86
orl-rat TDLo: 6750 mg/kg (female 17-22D post):REP IYKEDH 11,424,80
orl-rbt TDLo: 81,250 µg/kg (female 6-18D post):REP IYKEDH 11,424,80
orl-rat TDLo: 675 mg/kg (female 17-22D post):REP IYKEDH 11,424,80
orl-rbt TDLo: 1125 mg/kg (female 8-16D post):TER OYYAA2 9,377,75
orl-rbt TDLo: 1125 mg/kg (female 8-16D post):REP OYYAA2 9,377,75
orl-rat TDLo: 91 g/kg (female 26W pre):REP IYKEDH 11,328,80
orl-rat TDLo: 875 mg/kg (male 35D pre):REP IYKEDH 11,328,80
orl-rat LD50: 5 g/kg IYKEDH 11,328,80
ipr-rat LD50: 440 mg/kg YAKUD5 23,1999,81
scu-rat LD50: 3 g/kg IYKEDH 19,735,88
ivn-rat LD50: 92 mg/kg IYKEDH 19,735,88
orl-mus LD50: 2600 mg/kg IYKEDH 11,328,80
ipr-mus LD50: 407 mg/kg DRUGAY 6,645,82
scu-mus LD50: 2100 mg/kg IYKEDH 19,735,88
ivn-mus LD50: 96 mg/kg IYKEDH 19,735,88
orl-rbt LD50: 3000 mg/kg OYYAA2 9,377,75

SAFETY PROFILE: Poison by intravenous route. Moderately toxic by ingestion and intraperitoneal routes. An experimental teratogen. Experimental reproductive effects. When heated to decomposition it emits toxic fumes of NO_x and HCl. See also DIAZEPAM.

GCE600 **CAS:83-81-8** **HR: 3**
GEASTIGMOL
mf: $C_{16}H_{24}N_2O_2$ mw: 276.42

SYNS: ANALETIL □ 1,2-BENZENEDICARBOXAMIDE, N,N,N',N'-TETRAETHYL-(9CI) □ BIS-DIETHYLAMID KYSELINY FTALOVE □ neo-CARDIAMINE □ CARDIOVITAL □ CORETONIN □ GEASTIMOL □ NEO-CARDIAMINE □ NEOSPIRAN □ PHTHALAMIDE, N,N,N',N'-TETRAETHYL- □ PHTHALETHAMIDE □ o-PHTHALIC ACID BIS(DIETHYLAMIDE) □ o-PHTHALYLBIS(DIETHYLAMIDE) □ N,N,N',N'-TETRAETHYL-1,2-BENZENEDICARBOXAMIDE □ N,N,N',N'-TETRAETHYLPHTHALAMIDE □ UNISPIRAN

TOXICITY DATA WITH REFERENCE
scu-rbt LD50:30 mg/kg 85JCAE -,344,86

CONSENSUS REPORTS: Reported in EPA TSCA Inventory.

SAFETY PROFILE: Poison by subcutaneous route. When heated to decomposition it emits toxic vapors of NO_x.

GCG300 **CAS:427-01-0** **HR: 3**
GEISSOSPERMINE
mf: $C_{40}H_{48}N_4O_3$ mw: 632.92

PROP: Anhydrous form, crystals from absolute acetone. Mp: 213–214° (decomp) sinters at 1°. Anhydrous 217–212° decomp. Sltly sol in water, ether; sol in alc.

TOXICITY DATA WITH REFERENCE
orl-mus LD50:450 mg/kg APFRAD 19,104,61
ipr-mus LD50:125 mg/kg APFRAD 19,104,61
scu-mus LD50:150 mg/kg APFRAD 19,104,61
ivn-mus LD50:20 mg/kg APFRAD 19,104,61

SAFETY PROFILE: Poison by subcutaneous, intravenous, and intraperitoneal routes. Moderately toxic by ingestion. When heated to decomposition it emits toxic fumes of NO_x.

GCI000 **CAS:30562-34-6** **HR: 3**
GELDANAMYCIN
mf: $C_{29}H_{40}N_2O_9$ mw: 560.71

PROP: Yellow needles. Mp: 252–255°.

SYN: U-29135

TOXICITY DATA WITH REFERENCE
dni-mus:lym 1 mg/L JANTAJ 35,886,82
orl-rat LD50:2500 mg/kg 85ERAY 1,820,78
ipr-mus LD50:1 mg/kg UPJOH* 2(6),-,71

CONSENSUS REPORTS: EPA Genetic Toxicology Program.

SAFETY PROFILE: Poison by intraperitoneal route. Moderately toxic by ingestion. Mutation data reported. When heated to decomposition it emits toxic fumes of NO_x.

GCK000 **CAS:509-15-9** **HR: 3**
GELSEMINE
mf: $C_{20}H_{22}N_2O_2$ mw: 322.44

PROP: An alkaloid. Mp: 178°. Sltly sol in water; sol in alc, benzene, chloroform, ether, acetone, and dilute acids.

SYN: GELSEMIN

TOXICITY DATA WITH REFERENCE
ipr-mus LD50:49 mg/kg ARZNAD 7,349,57
scu-rbt LDLo:100 μg/kg MEIEDD 10,625,83

SAFETY PROFILE: A deadly poison by subcutaneous and intraperitoneal routes. A poisonous alkaloid. Can cause muscular weakness and respiratory arrest. Used as a central nervous system stimulant. When heated to decomposition it emits toxic fumes of NO_x.

GCK300 **CAS:25812-30-0** **HR: 2**
GEMFIBROZIL
mf: $C_{15}H_{22}O_3$ mw: 250.37

PROP: Crystals from hexane. Mp: 61–63°, bp: 158–159°.

SYNS: CI-719 □ 5-(2,5-DIMETHYLPHENOXY)-2,2-DIMETHYLPENTANOIC ACID (9CI) □ 2,2-DIMETHYL-5-(2,5-XYLYLOXY)VALERIC ACID □ GEVILON □ LIPUR □ LOPID

TOXICITY DATA WITH REFERENCE
orl-rat TDLo:2648 mg/kg (female 15-22D post):REP FAATDF 8,454,87
orl-rat TDLo:18,910 mg/kg (male 61D pre):REP FAATDF 8,454,87
orl-rat TDLo:218 g/kg/2Y-C:CAR JJIND8 67,1105,81
orl-mus TDLo:16 g/kg/78W-C:ETA JJIND8 67,1105,81
orl-rat TD:22 g/kg/2Y-C:NEO JJIND8 67,1105,81
orl-rat LD50:4786 mg/kg PRSMA4 69(Suppl 2),15,76
orl-mus LD50:3162 mg/kg PRSMA4 69(Suppl 2),15,76

SAFETY PROFILE: Moderately toxic by ingestion. Experimental reproductive effects. Questionable carcinogen with experimental carcinogenic, neoplastigenic, and tumorigenic data.

GCM000 **CAS:8023-80-1** **HR: 1**
GENET ABSOLUTE

PROP: Extracted from the flowers of *Spartium junceum* (FCTXAV 14,659,76).

SYNS: BROOM ABSOLUTE □ SPANISH BROOM

TOXICITY DATA WITH REFERENCE
skn-rbt 500 mg/24H MOD FCTXAV 14,779,76

CONSENSUS REPORTS: Reported in EPA TSCA Inventory.

SAFETY PROFILE: A skin irritant. When heated to decomposition it emits acrid smoke and irritating fumes.

GCM300 **CAS:6902-77-8** **HR: 3**
GENIPIN
mf: $C_{11}H_{14}O_5$ mw: 226.25

PROP: Crystals from MeOH. Mp: 120–121°.

TOXICITY DATA WITH REFERENCE
orl-rat TDLo:1750 mg/kg (35D male):REP OYYAA2 19,259,80

G

orl-mus LD50:237 mg/kg YKKZAJ 94,157,74
ipr-mus LD50:190 mg/kg YKKZAJ 94,157,74
ivn-mus LD50:153 mg/kg YKKZAJ 94,157,74

SAFETY PROFILE: Poison by ingestion, intravenous, and intraperitoneal routes. Experimental reproductive effects. When heated to decomposition it emits acrid smoke and irritating fumes. See also ESTERS.

GCO000 **CAS:1403-66-3** **HR: 3**
GENTAMYCIN

SYNS: GARAMYCIN □ GENTAMYCIN □ GENTAMYCIN-CREME (GERMAN) □ UROMYCINE

TOXICITY DATA WITH REFERENCE
mmo-esc 250 μg/L MUREAV 140,13,84
dnd-rat-ipr 70 mg/kg/7D AMACCQ 24,586,83
scu-rat TDLo:660 mg/kg (female 10-15D post):REP TJADAB 33,13A,86
ims-rat TDLo:900 mg/kg (female 10-21D post):REP TXCYAC 43,301,87
ipr-rat TDLo:750 mg/kg (7-18D preg):REP DPTHDL 7(Suppl 1),89,84
unr-rat TDLo:90 mg/kg (female 9-14D post):TER ANTBAL 27,926,82
ims-dog TDLo:1456 mg/kg (female 91D pre):REP JZKEDZ 4,73,78
ims-rat TDLo:900 mg/kg (female 10-21D post):TER TXCYAC 43,301,87
ivn-wmn TDLo:8200 μg/kg/12D:EAR,CNS APMIBM 81(Suppl 241),73,73
ivn-wmn TDLo:8200 μg/kg/12D:EAR,BAH AMBPBZ 81(Suppl 241),73,73
ivn-hmn TDLo:2 mg/kg:KID NEJMAG 303,1002,80
ivn-hmn TDLo:23 mg/kg/1Y-I:PNS AIMDAP 138,1621,78
ims-man TDLo:8 mg/kg/2W-I:EYE,SKN JAMAAP 211,123,70
ims-inf TDLo:20 mg/kg:KID PEDIAU 77,848,86
orl-rat LD50:6600 mg/kg AMPMAR 39,259,78
ivn-rat LD50:70 mg/kg JIMRBV 2,100,74
ims-rat LD50:463 mg/kg JJANAX 35,461,82
orl-mus LD50:10 g/kg JJANAX 30,386,77
ipr-mus LD50:235 mg/kg JIMRBV 2,100,74
scu-mus LD50:274 mg/kg TXAPA9 25,398,73
ivn-mus LD50:51 mg/kg JJANAX 30,386,77
ims-mus LD50:167 mg/kg JJANAX 35,461,82

CONSENSUS REPORTS: EPA Genetic Toxicology Program.

SAFETY PROFILE: Poison by intravenous, intraperitoneal, intramuscular, and subcutaneous routes. Mildly toxic by ingestion. Experimental teratogenic and reproductive effects. Mutation data reported. Human systemic effects: change in motor activity, changes in vestibular functions, distorted perceptions, eye hemorrhage, hallucinations, kidney changes, motor activity changes, trigeminal nerve sensory changes, vestibular function changes, visual field changes. Affects the peripheral nervous system by intravenous route. An antibiotic. When heated to decomposition it emits acrid smoke and irritating fumes. See also other GENTAMYCIN entries.

GCO200 **CAS:11097-82-8** **HR: 3**
GENTAMICIN C COMPLEX

TOXICITY DATA WITH REFERENCE
ipr-mus LD50:430 mg/kg AMACCQ 2,464,72
scu-mus LD50:485 mg/kg AMACCQ 2,464,72
ivn-mus LD50:55,900 μg/kg JANTAJ 35,94,82

SAFETY PROFILE: Poison by intravenous route. Moderately toxic by intraperitoneal and subcutaneous routes. See also other GENTAMYCIN entries.

GCS000 **CAS:1405-41-0** **HR: 3**
GENTAMYCIN SULFATE

SYNS: GARAMYCIN □ GENOPTIC □ GENOPTIC S.O.P. □ GM SULFATE □ NSC-82261 □ SCH 9724

TOXICITY DATA WITH REFERENCE
dnd-esc 5 mg/L MUREAV 89,95,81
spm-rat-unr 9600 μg/kg/8D JOURAA 112,348,74
scu-rat TDLo:660 mg/kg (female 15-20D post):REP ARTODN 62,274,88
ipr-rat TDLo:375 mg/kg (female 14-18D post):REP PPHAD4 5,229,86
scu-rat TDLo:660 mg/kg (female 10-15D post):TER ARTODN 62,274,88
ivn-wmn TDLo:45 mg/kg/1W-I:SYS DICPBB 18,596,84
ivn-man TDLo:21 mg/kg/6D-I:SYS DICPBB 18,596,84
ipr-rat LD50:630 mg/kg JJANAX 26,221,73
ivn-rat LD50:96 mg/kg JZKEDZ 8,219,82
ims-rat LD50:384 mg/kg TXAPA9 18,185,71
ipr-mus LDLo:245 mg/kg JZKEDZ 8,219,82
scu-mus LD50:478 mg/kg JANTAJ 23,551,70
ivn-mus LD50:47 mg/kg JJANAX 39,3164,86
ims-mus LD50:250 mg/kg JZKEDZ 8,219,82

SAFETY PROFILE: Poison by intravenous, intraperitoneal, and intramuscular routes. Moderately toxic by subcutaneous route. Human systemic effects: level changes for metals other than Na/K/Fe/Ca/P/Cl. An experimental teratogen. Other experimental reproductive effects. Mutation data reported. When heated to decomposition it emits very toxic fumes of SO_x. See also other GENTAMYCIN entries.

GCU000 **CAS:490-79-9** **HR: 3**
GENTISIC ACID
mf: $C_7H_6O_4$ mw: 154.13

PROP: Needles, monoclinic prisms from water. Mp: 204.5–205°. Sol in water (more so in hot water), alc, ether; practically insol in carbon disulfide, chloroform, ether.

SYNS: 2,5-DHBA □ 2,5-DIHYDROXYBENZOIC ACID □GENTISATE □ HYDROQUINONECARBOXYLIC ACID □ 5-HYDROXYSALICYLIC ACID

TOXICITY DATA WITH REFERENCE
dni-hmn:lyms 1 mmol/L BCPCA6 29,1275,80
scu-rat TDLo:380 mg/kg (9D preg):REP BCPCA6 22,407,73
scu-rat TDLo:642 mg/kg (11D preg):TER RCOCB8 38,209,82
ipr-rat LD50:3000 mg/kg BCFAAI 112,53,73

orl-mus LD50:4500 mg/kg BJPCAL 8,30,53
ivn-mus LD50:374 mg/kg JAPMA8 42,254,53

CONSENSUS REPORTS: Reported in EPA TSCA Inventory.

SAFETY PROFILE: Poison by intravenous route. Moderately toxic by ingestion and intraperitoneal routes. Experimental teratogenic and reproductive effects. Mutation data reported. When heated to decomposition it emits acrid smoke and irritating fumes.

GCU050 **CAS:4955-90-2** **HR: 2**
GENTISIC ACID, MONOSODIUM SALT
mf: $C_7H_5O_4 \cdot Na$ mw: 176.11

SYNS: BENZOIC ACID, 2,5-DIHYDROXY-, MONOSODIUM SALT (9CI) ☐ CASATE ☐ CASATE SODIUM ☐ GABAIL ☐ GENSALATE SODIUM ☐ GENTALPIN ☐ GENTASOL ☐ GENTIDOL ☐ GENTINATRE ☐ GENTISAN ☐ GENTISATE SODIUM ☐ GENTISOD ☐ 5-HYDROXYSALICYLATE SODIUM ☐ LEGENTIAL ☐ MONOSODIUM 2,5-DIHYDROXYBENZOATE ☐ NAGENT ☐ NAGENTIS ☐ SODIUM 2,5-DIHYDROXYBENZOATE ☐ SODIUM-GENT ☐ SODIUM GENTISATE

TOXICITY DATA WITH REFERENCE
ipr-mus LD50:3735 mg/kg JAPMA8 43,334,54
scu-mus LD50:3900 mg/kg CRSBAW 146,466,52

CONSENSUS REPORTS: Reported in EPA TSCA Inventory.

SAFETY PROFILE: Moderately toxic by intraperitoneal and subcutaneous routes. When heated to decomposition it emits acrid smoke and irritating vapors.

GCU100 **CAS:141-27-5** **HR: 2**
GERANALDEHYDE
mf: $C_{10}H_{16}O$ mw: 152.26

SYNS: CITRAL α ☐ α-CITRAL ☐ (E)-CITRAL ☐ trans-CITRAL ☐ trans-3,7-DIMETHYL-2,6-OCTADIENAL ☐ GERANIAL ☐ 2,6-OCTADIENAL, 3,7-DIMETHYL-, (E)-

TOXICITY DATA WITH REFERENCE
orl-rat LD50:500 mg/kg FRXXBL #2448856

CONSENSUS REPORTS: Reported in EPA TSCA Inventory.

SAFETY PROFILE: Moderately toxic by ingestion. When heated to decomposition it emits acrid smoke and irritating vapors.

GCW000 **CAS:459-80-3** **HR: 2**
GERANIC ACID
mf: $C_{10}H_{16}O_2$ mw: 168.26

SYNS: 3,7-DIMETHYL-2,6-OCTADIENOIC ACID ☐ 3,7-DIMETHYL-2,7-OCTADIENOIC ACID

TOXICITY DATA WITH REFERENCE
skn-rbt 500 mg/24H MOD FCTXAV 17,785,79
orl-mus LD50:4000 mg/kg BIJOAK 34,1196,40
skn-rbt LD50:1750 mg/kg FCTXAV 17,785,79

CONSENSUS REPORTS: Reported in EPA TSCA Inventory.

SAFETY PROFILE: Moderately toxic by ingestion and skin contact. A skin irritant. When heated to decomposition it emits acrid smoke and irritating fumes.

GCY000 **CAS:105-86-2** **HR: 1**
GERANIOL FORMATE
mf: $C_{11}H_{18}O_2$ mw: 182.29

PROP: Colorless to pale-yellow liquid or oil; rose odor. D: 0.906–0.920, refr index: 1.457–1.466, flash p: 205°F, bp: 113–114° @ 15 mm. Sol in alc, fixed oils; insol in glycerin, propylene glycol, water @ 216°.

SYNS: trans-3,7-DIMETHYL-2,6-OCTADIEN-1-OL FORMATE ☐ 3,7-DIMETHYL-2,6-OCTADIENYL ESTER FORMIC ACID (E) ☐ trans-3,7-DIMETHYL-2,6 OCTADIEN-1-YL FORMATE ☐ FEMA No. 2514 ☐ FORMIC ACID, GERANIOL ESTER ☐ GERANYL FORMATE (FCC)

TOXICITY DATA WITH REFERENCE
skn-hmn 10 mg/48H MLD FCTXAV 12,893,74
eye-rbt 100 mg/24H MLD FCTXAV 12,893,74
orl-rat LD50:>6 g/kg FCTXAV 12,893,74
skn-rbt LD50:>5 g/kg FCTXAV 12,893,74

CONSENSUS REPORTS: Reported in EPA TSCA Inventory.

SAFETY PROFILE: Low toxicity by ingestion and skin contact. A human skin irritant and an experimental eye irritant. Combustible liquid. When heated to decomposition it emits acrid smoke and irritating fumes. See also ESTERS.

GDA000 **CAS:8000-46-2** **HR: 1**
GERANIUM OIL ALGERIAN TYPE

PROP: From steam distillation of leaves from *Pelargonium graveolens* l'Her (Fam. Geraniaceae). Contains geraniol and geranyl tiglate (FCTXAV 14,659,76). Yellow liquid; odor of rose and geraniol. D: 0.886–0.898, refr index: 1.454–1.472 @ 20°. Sol in fixed oils, mineral oil; insol in glycerin.

SYNS: GERANIUM OIL ☐ OIL of GERANIUM ☐ OIL of PELARGONIUM ☐ OIL of ROSE GERANIUM ☐ OIL ROSE GERANIUM ALGERIAN ☐ PELARGONIUM OIL ☐ ROSE GERANIUM OIL ALGERIAN

TOXICITY DATA WITH REFERENCE
skn-rbt 500 mg/24H MLD FCTXAV 14,781,76
skn-gpg 100% MLD FCTXAV 14,781,76

CONSENSUS REPORTS: Reported in EPA TSCA Inventory.

SAFETY PROFILE: A skin irritant. When heated to decomposition it emits acrid smoke and irritating fumes.

GDC000 **HR: 2**
GERANIUM OIL BOURBON

PROP: Chief constituents are geraniol, citronellol (FCTXAV 12,807,74).

SYNS: OIL GERANIUM REUNION ☐ OIL ROSE GERANIUM

TOXICITY DATA WITH REFERENCE
skn-rbt 500 mg/24H MOD FCTXAV 12,883,74
skn-rbt LD50:2500 mg/kg FCTXAV 12,883,74

SAFETY PROFILE: Moderately toxic by skin contact. A skin irritant. When heated to decomposition it emits acrid smoke and irritating fumes.

GDE000 **HR: 1**
GERANIUM OIL MOROCCAN

PROP: Found in the leaves and stems of *Pelargonium roseum* (FCTXAV 13,449,75).

TOXICITY DATA WITH REFERENCE
skn-rbt 500 mg/24H MLD FCTXAV 13,451,75
skn-gpg 100% MLD FCTXAV 13,451,75

SAFETY PROFILE: A skin irritant. When heated to decomposition it emits acrid smoke and irritating fumes.

GDE300 **HR: 2**
GERANIUM THUNBERGII Sieb. et Zucc., extract

TOXICITY DATA WITH REFERENCE
scu-rat LD50:3700 mg/kg KSRNAM 11,458,77
ivn-rat LD50:461 mg/kg KSRNAM 11,458,77
scu-mus LD50:524 mg/kg KSRNAM 11,458,77
ivn-mus LD50:841 mg/kg KSRNAM 11,458,77

SAFETY PROFILE: Moderately toxic by intravenous and subcutaneous routes.

GDE400 **CAS:689-67-8** **HR: 1**
GERANYL ACETONE
mf: $C_{13}H_{22}O$ mw: 194.35

SYNS: DIHYDROPSEUDOIONONE □ α-β-DIHYDROPSEUDOIONONE □ 6,10-DIMETHYL-UNDECA-5,9-DIEN-2-ONE □ 5,9-UNDECADIEN-2-ONE, 6,10-DIMETHYL-

TOXICITY DATA WITH REFERENCE
skn-rbt 500 mg/24H MOD FCTXAV 17,787,79

CONSENSUS REPORTS: Reported in EPA TSCA Inventory.

SAFETY PROFILE: A skin irritant. When heated to decomposition it emits acrid smoke and irritating fumes.

GDE800 **HR: 2**
GERANYL BENZOATE
mf: $C_{17}H_{22}O_2$ mw: 258.36

PROP: Sltly yellow liquid; floral odor resembling that of ylang ylang oil. D: 0.978–0.984, refr index: 1.513–1.518, flash p: 212°F. Misc in alc, chloroform; insol in water @ 305°.

SYNS: 3,7-DIMETHYL-2,6-OCTADIEN-1-YL BENZOATE □ FEMA No. 2511

SAFETY PROFILE: Combustible liquid. When heated to decomposition it emits acrid smoke and irritating fumes.

GDE810 **CAS:106-29-6** **HR: 1**
GERANYL BUTANOATE
mf: $C_{14}H_{24}O_2$ mw: 224.38

SYNS: BUTANOIC ACID, 3,7-DIMETHYL-2,6-OCTADIENYL ESTER, (E)-(9CI) □ BUTYRIC ACID, 3,7-DIMETHYL-2,6-OCTADIENYL ESTER, (E)- □ trans-3,7-DIMETHYL-2,6-OCTADIEN-1-YL BUTYRATE □ GERANIOL BUTYRATE □ GERANYL BUTYRATE □ GERANYL n-BUTYRATE

TOXICITY DATA WITH REFERENCE
orl-rat LD50:10,660 mg/kg FCTXAV 2,327,64
skn-rbt LD50:5 g/kg FCTXAV 12,889,74

CONSENSUS REPORTS: Reported in EPA TSCA Inventory.

SAFETY PROFILE: Mildly toxic by ingestion and skin contact. When heated to decomposition it emits acrid smoke and irritating vapors.

GDE825 **HR: 1**
GERANYL BUTYRATE
mf: $C_{14}H_{24}O_2$ mw: 224.34

PROP: Colorless to pale-yellow liquid; fruity, roselike odor. D: 0.889–0.904, refr index: 1.455–1.462, flash p: 199°F. Sol in alc, fixed oils; insol in glycerin, propylene glycol, water @ 253°.

SYNS: 3,7-DIMETHYL-2,6-OCTADIEN-1-YL BUTYRATE □ FEMA No. 2512

SAFETY PROFILE: Combustible liquid. When heated to decomposition it emits acrid smoke and irritating fumes.

GDG000 **CAS:10032-02-7** **HR: 1**
GERANYL CAPROATE
mf: $C_{16}H_{28}O_2$ mw: 252.44

SYNS: (E)-3,7-DIMETHYLOCTA-2,6-DIEN-1-YL ESTER, HEXANOIC ACID □ (E)-3,7-DIMETHYLOCTA-2,6-DIEN-1-YL-n-HEXANOATE □ GERANYL HEXANOATE

TOXICITY DATA WITH REFERENCE
skn-rbt 500 mg/24H MLD FCTXAV 14,783,76

CONSENSUS REPORTS: Reported in EPA TSCA Inventory.

SAFETY PROFILE: A skin irritant. When heated to decomposition it emits acrid smoke and irritating fumes.

GDG100 **CAS:40267-72-9** **HR: 1**
GERANYL ETHYL ETHER
mf: $C_{12}H_{22}O$ mw: 182.34

SYNS: 1-ETHOXY-3,7-DIMETHYL-2,6-OCTADIENE □ ETHYL GERANYL ETHER □ 2,6-OCTADIENE, 1-ETHOXY-3,7-DIMETHYL-

TOXICITY DATA WITH REFERENCE
skn-rbt 500 mg/24H MLD FCTOD7 20,693,82
orl-rat LD50:>5 g/kg FCTOD7 20,693,82
skn-rbt LD50:>5 g/kg FCTOD7 20,693,82

CONSENSUS REPORTS: Reported in EPA TSCA Inventory.

SAFETY PROFILE: Low toxicity by ingestion and skin contact. A skin irritant. When heated to decomposition it emits acrid smoke and irritating fumes.

GDG200 CAS:51-77-4 **HR: 2**
GERANYL FARNESYL ACETATE
mf: $C_{27}H_{44}O_2$ mw: 400.71

SYNS: DA-688 □ GEFARNATE □ GEFARNIL □ GEFARNYL □ 4,8,12-TETRADECATRIENOIC ACID, 5,9,13-TRIMETHYL-, 3,7-DIMETHYL-2,6-OCTADIENYLESTER, (E,E,E)- □ (E,E,E)-5,9,13-TRIMETHYL-4,8,12-TETRADECATRIENOIC ACID 3,7 DIMETHYL-2,6-OCTADIENYL ESTER

TOXICITY DATA WITH REFERENCE
orl-rat LD50:>9 g/kg DRUGAY 6,266,82
ivn-rat LD50:2040 mg/kg DRUGAY 6,266,82
ims-rat LD50:>13,500 mg/kg DRUGAY 6,266,82
orl-mus LD50:>8 g/kg JMCMAR 6,457,63
ipr-mus LD50:>4 g/kg JMCMAR 6,457,63
ivn-mus LD50:2821 mg/kg JMCMAR 6,457,63
ims-mus LD50:>13,500 mg/kg DRUGAY 6,266,82

CONSENSUS REPORTS: Reported in EPA TSCA Inventory.

SAFETY PROFILE: Moderately toxic by intravenous route. Mildly toxic by other routes. When heated to decomposition it emits acrid smoke and irritating vapors.

GDG300 CAS:1113-21-9 **HR: 1**
GERANYLINALOOL
mf: $C_{20}H_{34}O$ mw: 290.54

SYNS: 1,6,10,14-HEXADECATETRAEN-3-OL, 3,7,11,15-TETRAMETHYL-, (E,E)- □ E,E-3,7,11,15-TETRAMETHYL-1,6,10,14-HEXADECATETRAEN-3-OL

TOXICITY DATA WITH REFERENCE
skn-rbt 500 mg FCTOD7 30,41S,92
eye-rbt 100 mg FCTOD7 30,41S,92
orl-rat LD50:>5 g/kg FCTOD7 30,41S,92
ipr-mus LD50:>2 g/kg FCTOD7 30,41S,92
skn-rbt LD50:>5 g/kg FCTOD7 30,41S,92

CONSENSUS REPORTS: Reported in EPA TSCA Inventory.

SAFETY PROFILE: Low toxicity by ingestion and skin contact. A skin irritant. When heated to decomposition it emits acrid smoke and irritating vapors.

GDI000 CAS:2345-26-8 **HR: 1**
GERANYL ISOBUTYRATE
mf: $C_{14}H_{24}O_2$ mw: 224.38

SYNS: trans-3,7-DIMETHYL-2,6-OCTADIEN-1-OL ISOBUTYRATE □ trans-3,7-DIMETHYL-2,6-OCTADIENYL ESTER ISOBUTYRIC ACID □ trans-3-7-DIMETHYL-2,6-OCTADIENYL ISOBUTYRATE □ (E)2-METHYL-3,7-DIMETHYL-2,6-OCTADIENYL ESTER PROPANOIC ACID

TOXICITY DATA WITH REFERENCE
skn-rbt 500 mg/24H MLD FCTXAV 13,45175
orl-rat LD50:>5 g/kg FCTXAV 13,451,75
skn-rbt LD50:>5 g/kg FCTXAV 13,451,75

CONSENSUS REPORTS: Reported in EPA TSCA Inventory.

SAFETY PROFILE: Low toxicity by ingestion and skin contact. A skin irritant. When heated to decomposition it emits acrid smoke and irritating fumes.

GDK000 CAS:109-20-6 **HR: 1**
GERANYL ISOVALERATE
mf: $C_{15}H_{26}O_2$ mw: 238.41

SYNS: trans-3,7-DIMETHYL-2,6-OCTADIENYL ISOPENTANOATE □ (E)-ISOVALERIC ACID-3,7-DIMETHYL-2,6-OCTADIENYL ESTER □ (E)-3-METHYLBUTYRIC ACID-3,7-DIMETHYL-2,6-OCTADIENYL ESTER

TOXICITY DATA WITH REFERENCE
skn-rbt 500 mg/24H MLD FCTXAV 14,785,76

CONSENSUS REPORTS: Reported in EPA TSCA Inventory.

SAFETY PROFILE: A skin irritant. When heated to decomposition it emits acrid smoke and irritating fumes. See also ESTERS.

GDM000 CAS:5585-39-7 **HR: 2**
GERANYL NITRILE
mf: $C_{10}H_{15}N$ mw: 149.26

SYNS: (E)-3,7-DIMETHYL-2,6-OCTADIENENITRILE □ GERANONITRILE

TOXICITY DATA WITH REFERENCE
orl-rat LD50:3100 mg/kg FCTXAV 14,787,76
skn-rbt LD50:4300 mg/kg FCTXAV 14,787,76

CONSENSUS REPORTS: Reported in EPA TSCA Inventory. Cyanide and its compounds are on the Community Right To-Know List.

SAFETY PROFILE: Moderately toxic by ingestion. Mildly toxic by skin contact. When heated to decomposition it emits toxic fumes of NO_x and CN^-. See also NITRILES.

GDM100 CAS:65405-73-4 **HR: 1**
GERANYL OXYACETALDEHYDE
mf: $C_{12}H_{20}O_2$ mw: 196.32

SYNS: ACETALDEHYDE, ((3,7-DIMETHYL-2,6-OCTADIENYL)OXY)-, (E)- □ GERANOXY ACETALDEHYDE

TOXICITY DATA WITH REFERENCE
skn-rbt 500 mg/24H MOD FCTXAV 17,789,79

CONSENSUS REPORTS: Reported in EPA TSCA Inventory.

SAFETY PROFILE: A skin irritant. When heated to decomposition it emits acrid smoke and irritating fumes.

GDM400 CAS:102-22-7 **HR: 1**
GERANYL PHENYLACETATE
mf: $C_{18}H_{24}O_2$ mw: 272.39

PROP: Yellow liquid; honey-rose odor. D: 0.971–0.978, refr index: 1.507–1.511, flash p: 212°F. Misc in alc, chloroform, ether; insol in water.

G

SYNS: ACETIC ACID, PHENYL-, 3,7-DIMETHYL-2,6-OCTADIENYL ESTER, (E)-(8CI) □ BENZENEACETIC ACID, 3,7-DIMETHYL-2,6-OCTADIENYL ESTER, (E)- □ trans-3,7-DIMETHYL-2,6-OCTADIEN-1-YL PHENYLACETATE □ 3,7-DIMETHYL-2,6-OCTADIEN-1-YL PHENYLACETATE □ GERANYL α-TOLUATE □ FEMA No. 2516

TOXICITY DATA WITH REFERENCE
orl-rat LD50:>5 g/kg FCTXAV 12,895,74
skn-rbt LD50:>5 g/kg FCTXAV 12,895,74

CONSENSUS REPORTS: Reported in EPA TSCA Inventory.

SAFETY PROFILE: Low toxicity by ingestion and skin contact. Combustible liquid. When heated to decomposition it emits acrid smoke and irritating fumes.

GDM450 CAS:105-90-8 HR: 2
GERANYL PROPIONATE
mf: $C_{13}H_{22}O_2$ mw: 210.32

PROP: Colorless liquid; rosy, fruity odor. D: 0.896–0.913, refr index: 1.456–1.464, flash p: 212°F. Sol in alc, fixed oils; insol in glycerin, propylene glycol, water @ 253°.

SYNS: (E)-3,7-DIMETHYL-2,6-OCTADIEN-1-OL PROPIONATE □ 3,7-DIMETHYL-2,6-OCTADADIEN-1-YL PROPIONATE □ trans-3,7-DIMETHYL-2,6-OCTADIEN-1-YL PROPIONATE □ 2,6-OCTADIEN-1-OL, 3,7-DIMETHYL-, PROPIONATE, (E)-(8CI) □ FEMA No. 2517

TOXICITY DATA WITH REFERENCE
orl-rat LD50:>5 g/kg FCTXAV 12,897,74
skn-rbt LD50:>5 g/kg FCTXAV 12,897,74

CONSENSUS REPORTS: Reported in EPA TSCA Inventory.

SAFETY PROFILE: Low toxicity by ingestion and skin contact. Combustible liquid. When heated to decomposition it emits acrid smoke and irritating fumes.

GDO000 CAS:7785-33-3 HR: 1
GERANYL TIGLATE
mf: $C_{15}H_{24}O_2$ mw: 236.39

SYNS: cis-α,β-DIMETHYL ACRYLIC ACID, GERANIOL ESTER □ trans-3,7-DIMETHYL-2,6-OCTADIEN-1-YL cis-α,β-DIMETHYL ACRYLATE □ (E,E)-3,7-DIMETHYL-2,6-OCTADIENYL ESTER-2-METHYL-2-BUTENOIC ACID □ TIGLIC ACID, GERANIOL ESTER

TOXICITY DATA WITH REFERENCE
skn-rbt 500 mg/24H MLD FCTXAV 12,899,74
orl-rat LD50:>5 g/kg FCTXAV 12,899,74
skn-rbt LD50:>5 g/kg FCTXAV 12,899,74

CONSENSUS REPORTS: Reported in EPA TSCA Inventory.

SAFETY PROFILE: Low toxicity by ingestion and skin contact. A skin irritant. When heated to decomposition it emits acrid smoke and irritating fumes. See also ESTERS.

GDO800 CAS:39236-46-9 HR: 2
GERMALL 115
mf: $C_{11}H_{16}N_8O_8$ mw: 388.35

SYNS: METHANEBIS(N,N'-(5-UREIDO-2,4-DIKETOTETRAHYDROIMIDAZOLE)-N,N-DIMETHYLOL) □ N,N''-(METHYLENEBIS(N'-(1-(HYDROXYMETHYL)-2,5-DIOXO-4-IMIDAZOLIDINYL)UREA)

TOXICITY DATA WITH REFERENCE
orl-rat LD50:11,300 mg/kg JEPTDQ 4(4),133,80
ipr-rat LDLo:4000 mg/kg JEPTDQ 4(4),133,80
orl-mus LD50:7200 mg/kg JEPTDQ 4(4),133,80

CONSENSUS REPORTS: Reported in EPA TSCA Inventory.

SAFETY PROFILE: Moderately toxic by intraperitoneal route. Mildly toxic by ingestion. When heated to decomposition it emits toxic fumes of NO_x.

GDS000 CAS:1310-53-8 HR: 3
GERMANIC OXIDE (crystalline)

TOXICITY DATA WITH REFERENCE
ipr-gpg LDLo:300 mg/kg JPETAB 42,277,31

CONSENSUS REPORTS: Reported in EPA TSCA Inventory.

SAFETY PROFILE: Poison by intraperitoneal route. When heated to decomposition it emits acrid smoke and irritating fumes.

GDU000 CAS:7440-56-4 HR: 3
GERMANIUM
mf: Ge mw: 72.59

PROP: A gray-white, lustrous metalloid; crystalline and brittle, stable @ room temp. Mp: 945° (937.2° best value), bp: 2850° @ 28°, d: 5.323 @ 25°. Insol in water, hydrochloric acid, dilute alc, hydroxides. Relatively stable.

SYNS: GERMANIUM ELEMENT □ GERMANIUM, metal powder

TOXICITY DATA WITH REFERENCE
ihl-rat LCLo:3860 mg/m³/4H FCTOD7 28,571,90

CONSENSUS REPORTS: Reported in EPA TSCA Inventory.

SAFETY PROFILE: Inhaled dust is rapidly absorbed. Explosive reaction when heated with potassium chlorate or potassium nitrate. Violent reaction with nitric acid. Ignites on contact with chlorine or bromine + heat. Incandescent reaction when heated with oxygen or potassium hydroxide. Incompatible with aqua regia; concentrated sulfuric acid; fused alkalies; nitrates or carbonates; halogens; oxidants. See also GERMANIUM COMPOUNDS.

GDW000 CAS:13450-92-5 HR: 3
GERMANIUM BROMIDE
mf: Br_4Ge mw: 392.23

PROP: Gray-white, octahedral crystals. Mp: 26.1°, bp: 186.5°, d: 3.232 @ 29°/29°. Decomp in H_2O; sol in EtOH, Et_2O, and C_6H_6.

SYN: GERMANIUM TETRABROMIDE

TOXICITY DATA WITH REFERENCE
ivn-mus LD50:56 mg/kg CSLNX* NX#02624

CONSENSUS REPORTS: Reported in EPA TSCA Inventory.

SAFETY PROFILE: Poison by intravenous route. When heated to decomposition it emits very toxic fumes of Br^-. See also GERMANIUM COMPOUNDS.

GDY000 CAS:10038-98-9 HR: 3
GERMANIUM CHLORIDE
mf: Cl_4Ge mw: 214.39

PROP: Colorless, mobile liquid. Fumes in air. Peculiar acidic odor but can be distinguished from that of concentrated HCl. Mp: −49.5°, bp: 83.1°, d: 1.879 @ 20°/20°. Volatile @ room temp. Decomp in H_2O; sol in EtOH, Et_2O. Very sol in dil HCl; insol in conc HCl, and H_2SO_4.

SYN: EXTREMA □ GERMANIUM TETRACHLORIDE

TOXICITY DATA WITH REFERENCE
eye-rbt 50 mg SEV GTPZAB 8(4),57,64
ihl-mus LC50:44 g/m³/2H GTPZAB 12(5),51,68
ivn-mus LD50:56 mg/kg CSLNX* NX#02625

CONSENSUS REPORTS: Reported in EPA TSCA Inventory. EPA Genetic Toxicology Program.

SAFETY PROFILE: Poison by intravenous route. Mildly toxic by inhalation. A skin, severe eye, and mucous membrane irritant. Will react violently with water or steam to produce toxic and corrosive fumes. When heated to decomposition it emits toxic fumes of Cl^-. See also GERMANIUM COMPOUNDS.

GEA000 HR: 2
GERMANIUM COMPOUNDS

SAFETY PROFILE: Germanium compounds are considered to be of a low order of toxicity, but rare instances of poisoning have been reported in the literature. Experimental LD50 values are typically about 100–1000 mg/kg for parenteral route and 500–5000 mg/kg for ingestion. The experimental animals suffer from hypothermia, diarrhea, and respiratory and cardiac failure. Inhalation of large amounts of $GeCl_4$ by experimental animals causes necrosis of the tracheal epithelium, bronchitis, and interstitial pneumonia. These effects were not apparent with chronic inhalation of 7 mg/m³. The tetrachloride and tetrafluoride are eye, skin, and mucous membrane irritants. Alkyl germanium compounds are much less toxic than the corresponding tin or lead compounds. Tributyl germanium and germanium tetrachloride are mutagens. Dimethyl germanium is a teratogen. Chronic ingestion of 1000 ppm or 100 ppm of germanium dioxide in water has been shown to inhibit growth in chickens. No effect was seen at 5 ppm. It has been found that the dioxide stimulates the generation of red blood cells, but it is believed to be relatively nontoxic. Buffered germanium dioxides in solution have been found to be nonirritating to the skin. Germanium hydride is a hemolytic gas and has been shown to have toxic properties at a concentration of 100 ppm. It can cause death at a concentration of 150 ppm. Otherwise, little is known about the toxicity of organic germanium compounds except that they may resemble other organometals in having higher toxicity than inorganic forms. When germanium is given in sublethal amounts, it causes a pronounced tolerance. Interest is high in this material because of its close chemical relationship to arsenic.

GEC000 CAS:1310-53-8 HR: 3
GERMANIUM DIOXIDE
mf: GeO_2 mw: 104.59

PROP: Soluble form: Hexagonal, colorless crystals. Mp: 1115.0°, d: 4.703 @ 18°. Insoluble form: Tetragonal crystals, mp: 1086 ± 5°, d: 6.239.

SYNS: GERMANIA □ GERMANIC ACID □ GERMANIUM OXIDE □ GERMANIUM OXIDE (GeO_2)

TOXICITY DATA WITH REFERENCE
unr-rat TDLo:133 mg/kg (7-13D preg):REP OYYAA2 21,797,81
scu-rat TDLo:5250 mg/kg (35D male):REP OYYAA2 21,773,81
unr-rat TDLo:266 mg/kg (7-13D preg):TER OYYAA2 21,773,81
orl-rat LD50:1250 mg/kg 31ZUAV -,95,64
ipr-rat LD50:750 mg/kg AMIHBC 8,466,53
scu-rat LD50:1910 mg/kg OYYAA2 21,773,81
orl-mus LD50:1250 mg/kg 85GMAT-,71,82
ipr-mus LD50:1550 mg/kg OYYAA2 16,671,78
scu-mus LD50:2550 mg/kg OYYAA2 21,773,81
scu-rbt LD50:845 mg/kg EQSSDX 1,1,75
ipr-gpg LDLo:400 mg/kg EQSSDX 1,1,75

CONSENSUS REPORTS: Reported in EPA TSCA Inventory.

SAFETY PROFILE: Poison by intraperitoneal route. Moderately toxic by ingestion and subcutaneous routes. Experimental teratogenic and reproductive effects. When heated to decomposition it emits acrid smoke and irritating fumes. See also GERMANIUM COMPOUNDS.

GEI000 CAS:12025-32-0 HR: 3
GERMANIUM(II) SULFIDE
mf: GeS mw: 104.64

SAFETY PROFILE: Yellow-red amorphous or black rhombic pyramids. Mp: 530°, sublimes at 4°.

SAFETY PROFILE: Explosive reaction when heated with potassium nitrate. When heated to decomposition it emits toxic fumes of SO_x. See also GERMANIUM COMPOUNDS.

GEI100 CAS:7782-65-2 HR: 3
GERMANIUM TETRAHYDRIDE
DOT: UN 2192
mf: GeH_4 mw: 76.63

PROP: Colorless gas. Mp: −165°, bp: −90°, d: 1.523 @ −142°/4°. Insol in H_2O; sol in aq NH_3, aq NaOCl; sltly sol in hot HCl.

G

SYNS: GERMANE (DOT) □ GERMANIUM HYDRIDE □ MONOGERMANE

TOXICITY DATA WITH REFERENCE
orl-mus LD50:1250 mg/kg GTPZAB 18(2),56,74
ihl-mus LC50:1380 mg/m³ GTPZAB 18(2),56,74

CONSENSUS REPORTS: Reported in EPA TSCA Inventory.

ACGIH TLV: TWA 0.2 ppm
DOT CLASSIFICATION: 2.3; *Label:* Poison Gas, Flammable Gas

SAFETY PROFILE: Poison by inhalation. Moderately toxic by ingestion. A hemolytic gas. Ignites spontaneously in air. Incompatible with Br_2. See also HYDRIDES, GERMANIUM COMPOUNDS, and GERMANIUM.

GEK000 CAS:306-67-2 HR: 3
GERONTINE TETRAHYDROCHLORIDE
mf: $C_{10}H_{26}N_4{\bullet}4ClH$ mw: 348.24

PROP: Crystals from EtOH. Mp: 312–314.5°.

SYNS: 1,4-BIS(AMINOPROPYL)BUTANEDIAMINE TETRAHYDROCHLORIDE □ MUSCULAMINE TETRAHYDROCHLORIDE □ NEURIDINE TETRAHYDROCHLORIDE □ SPERMINE TETRAHYDROCHLORIDE

TOXICITY DATA WITH REFERENCE
dnd-esc 8900 nmol/L BIPMAA 5,227,67
dnd-srm 8900 nmol/L BIPMAA 5,227,67
dnd-mam:lym 16,700 nmol/L BIPMAA 5,227,67
ipr-mus LD50:370 mg/kg FEPRA7 41,1575,82

CONSENSUS REPORTS: Reported in EPA TSCA Inventory.

SAFETY PROFILE: A poison by intraperitoneal route. Mutation data reported. When heated to decomposition it emits very toxic fumes of HCl and NO_x.

GEK200 CAS:137-86-0 HR: 3
GEROSTOP
mf: $C_{23}H_{36}N_4O_5S_3$ mw: 544.81

PROP: Crystals or solid with slt bitter taste. Mp: 106–109°.

SYNS: NEUVITAN □ OCTOTIAMINE □ TATD

TOXICITY DATA WITH REFERENCE
orl-mus LD50:2590 mg/kg NIIRDN 6,159,82
ipr-mus LD50:522 mg/kg NIIRDN 6,159,82
scu-mus LD50:1410 mg/kg NIIRDN 6,159,82
ivn-mus LD50:399 mg/kg NIIRDN 6,159,82

SAFETY PROFILE: Poison by intravenous route. Moderately toxic by ingestion, subcutaneous, and intraperitoneal routes. When heated to decomposition it emits toxic fumes of SO_x and NO_x. See also ESTERS.

GEK500 CAS:434-03-7 HR: 3
GESTORAL
mf: $C_{21}H_{28}O_2$ mw: 312.49

PROP: Crystals from $CHCl_3/Me_2CO$ or EtOAc; needles from EtOH. Mp: 270–272°. Practically insol in water; sltly sol in alc, acetone, ether, chloroform, vegetable oils.

SYNS: AETHISTERON □ ANHYDROHYDROXYPROGESTERONE □ ANHYDROXYPROGESTERONE □ COLUTOID □ ETHINONE □ ETHINYLTESTOSTERONE □ 17-ETHINYLTESTOSTERONE □ ETHISTERONE □ ETHYNYLTESTOSTERONE □ 17-α-ETHYNYLTESTOSTERONE □ 17-β-HYDROXY-17-α-ETHYNYL-4-ANDROSTER-3-ONE □ 17-HYDROXY-17-α-PREGN-4-EN-20-YN-3-ONE □ LUCORTEUM ORAL □ LUTIDON ORAL □ LUTOCYLOL □ NALUTORAL □ NUGESTORAL □ ORA-LUTIN □ PRANONE □ PREGNENINOLONE □ PREGNIN □ PRIMOLUT C □ PRODOXAN □ PRODROXAN □ PRODUXAN □ PROGESTAB □ PROGESTIN P □ PROGESTOLETS □ PROGESTORAL □ PROLUTOL □ PROLUTON C □ PRONE □ SYNGESTROTABS □ TROSINONE

TOXICITY DATA WITH REFERENCE
orl-wmn TDLo:168 mg/kg (3W pre):REP AJOGAH 76,626,58
orl-wmn TDLo:386 mg/kg (6-28W preg):TER AJOGAH 84,962,62
orl-rat TDLo:200 mg/kg (female 17-20D post):TER ECJPAE 24,77,77
orl-rat TDLo:10 mg/kg (female 1D pre):REP FESTAS 5,282,54
scu-rat TDLo:280 mg/kg (female 14D pre):REP CCPTAY 5,57,72
scu-rat TDLo:140 mg/kg (female 14D pre):REP CCPTAY 5,57,72
orl-rat TDLo:288 mg/kg (male 9D pre):REP PSEBAA 100,540,59

SAFETY PROFILE: Human female reproductive and teratogenic effects: menstrual cycle changes or disorders and developmental abnormalities of the fetal urogenital system. Other experimental reproductive effects. A steroid. When heated to decomposition it emits acrid smoke and irritating fumes.

GEK510 CAS:1253-28-7 HR: 3
GESTRONOL CAPROATE
mf: $C_{26}H_{38}O_4$ mw: 414.64

PROP: Crystals. Mp: 123–124°.

SYNS: 17-β-ACETYL-17-HYDROXYESTR-4-ENE-3-ONE HEXANOATE □ DEPOSTAT □ GESTONORONE CAPROATE □ GESTONORONE CAPRONATE □ GESTRONOL HEXANOATE □ 17-HYDROXY-19-NORPREGN-4-ENE-3,20-DIONE HEXANOATE □ 17-HYDROXY-19-NOR-4-PREGNENE-3,20-DIONE HEXANOATE □ 17-α-HEXANOYLOXY-19-NOR-4-PREGNENE-3,20-DIONE □ 17-α-HYDROXY-19-NORPROGESTERONE CAPROATE □ PRIMOSTAT □ SH 582 □ SH 80582

TOXICITY DATA WITH REFERENCE
scu-rat TDLo:700 mg/kg (28W male):REP IYKEDH 9,307,78
scu-rat TDLo:140 mg/kg (female 28W pre):REP IYKEDH 9,307,78
ipr-rat LD50:6800 mg/kg IYKEDH 13,349,82
ims-rat LD50:400 mg/kg IYKEDH 8,296,77
scu-mus LD50:10 g/kg IYKEDH 13,349,82

SAFETY PROFILE: Poison by intramuscular route. Experimental reproductive effects. A steroid. When heated to decomposition it emits acrid smoke and irritating fumes.

GEK600 **HR: 3**
GEUM ELATUM (Royle) Hook. f., extract

PROP: Indian plant belonging to the family Rosaceae (IJEBA6 18,594,80).

TOXICITY DATA WITH REFERENCE
orl-rat TDLo:150 mg/kg (12-14D preg):REP IJEBA6 18,594,80
orl-ham TDLo:500 mg/kg (1-5D preg):REP IJEBA6 18,594,80
ipr-mus LD50:375 mg/kg IJEBA6 18,594,80

SAFETY PROFILE: Poison by intraperitoneal route. Experimental reproductive effects.

GEK875 **HR: 1**
GFX-E

PROP: Consists of glucose, fructose, and xylitol (4:2:1) in a 23.3% solution with electrolytes (IYKEDH 16,328,85).

TOXICITY DATA WITH REFERENCE
ipr-rat LD50:42,600 mg/kg IYKEDH 16,328,85
ivn-rat LD50:40,600 mg/kg IYKEDH 16,328,85
ivn-rbt LD50:40 g/kg IYKEDH 16,328,85

SAFETY PROFILE: Mildly toxic by intravenous and intraperitoneal routes. When heated to decomposition it emits acrid smoke and irritating fumes. See also individual components.

GEK880 **HR: 1**
GFX-ES

PROP: Consists of glucose, fructose, and xylitol (4:2:1) in a 23.3% solution with electrolytes (IYKEDH 16,320,85).

TOXICITY DATA WITH REFERENCE
ipr-rat LD50:57,600 mg/kg IYKEDH 16,320,85
ivn-rat LD50:61 g/kg IYKEDH 16,320,85
ivn-rbt LD50:59,300 mg/kg IYKEDH 16,320,85

SAFETY PROFILE: Mildly toxic by intravenous and intraperitoneal routes. When heated to decomposition it emits acrid smoke and irritating fumes. See also individual components.

GEM000 **CAS:77-06-5** **HR: 2**
GIBBERELLIC ACID
mf: $C_{19}H_{22}O_6$ mw: 346.41

PROP: A plant-growth-promoting hormone. White crystals or crystalline powder from alc/pet ether. Mp: 233–235° (decomp). Sltly sol in water, ether; sol in methanol, ethanol, acetone, aq solns of sodium bicarbonate and sodium acetate; moderately sol in ethyl acetate.

SYNS: BERELEX ☐ BRELLIN ☐ CEKUGIB ☐ FLORALTONE ☐ GA ☐ GIBBERELLIN ☐ GIBBREL ☐ GIB-SOL ☐ GIB-TABS ☐ GROCEL ☐ NCI-C55823 ☐ PRO-GIBB ☐ 2,4a,7-TRIHYDROXY-1-METHYL-8-METHYLENEGIBB-3-ENE-1,10-CARBOXYLIC ACID 1-4-LACTONE

TOXICITY DATA WITH REFERENCE
dnd-sal:spr 1 mmol/L PYTCAS 11,3135,72
dnd-mam:lym 1 mmol/L PYTCAS 11,3135,72
orl-mus TDLo:142 g/kg/78W-I:ETA NTIS** PB223-159
orl-rat LD50:6300 mg/kg 85ARAE 3,43,76/77

CONSENSUS REPORTS: EPA Genetic Toxicology Program. Reported in EPA TSCA Inventory.

SAFETY PROFILE: Mildly toxic by ingestion. Questionable carcinogen with experimental tumorigenic data. Mutation data reported. When heated to decomposition it emits acrid smoke and irritating fumes.

GEO000 **CAS:12002-43-6** **HR: 1**
GILSONITE

PROP: A black solid hydrocarbon mineral formed from petroleum millions of years ago by geologic processes.

SYN: NCI-C55185

SAFETY PROFILE: A skin, eye, and mucous membrane irritant. An allergen. Has been known to cause photosensitization of skin. Flammable when exposed to heat or open flame. To fight fire, use water, foam, dry chemical, and CO_2. When heated to decomposition it emits acrid smoke and irritating fumes.

GEO200 **CAS:77879-90-4** **HR: 3**
GILVOCARCIN V
mf: $C_{27}H_{26}O_9$ mw: 494.53

PROP: Yellow needles. Mp: 264–267°.

SYNS: ANTIBIOTIC B 21085 ☐ B-21085 ☐ DC-38-A ☐ DC-38-V ☐ TOROMYCIN

TOXICITY DATA WITH REFERENCE
dni-bcs 500 μg/L JANTAJ 35,545,82
oms-bcs 500 μg/L JANTAJ 35,545,82
ipr-mus LD50:1500 mg/kg 85GDA2 3,369,80
ivn-mus LD50:300 mg/kg JANTAJ 34,266,81

SAFETY PROFILE: Poison by intravenous route. Mutation data reported. When heated to decomposition it emits acrid smoke and fumes.

GEO600 **CAS:4880-82-4** **HR: 3**
GINDARINE HYDROCHLORIDE
mf: $C_{21}H_{25}NO_4 \cdot ClH$ mw: 391.93

SYNS: GINDARIN HYDROCHLORIDE ☐ (S)-5,8,13,13a-TETRAHYDRO-2,3,9,10-TETRAMETHOXY-6H-DIZENZO(a,g)QUINOLIZINE HYDROCHLORIDE ☐ 2,3,9,10-TETRAMETHOXY-13a,α-BERBINE HYDROCHLORIDE

TOXICITY DATA WITH REFERENCE
orl-rat TDLo:20 mg/kg (female 1-20D post):REP FATOAO 46(4),107,83
orl-rat TDLo:100 mg/kg (1-20D preg):REP FATOAO 46(4),107,83
orl-rat TDLo:1 g/kg (female 1-20D post):TER FATOAO 46(4),107,83
orl-rat LD50:580 mg/kg FATOAO 46(4),107,83
ipr-rat LD50:330 mg/kg FATOAO 46(4),107,83
orl-mus LD50:1190 mg/kg FATOAO 46(4),107,83

G

ipr-mus LD50:550 mg/kg IJMRAQ 60,472,72
ipr-gpg LD50:460 mg/kg FATOAO 46(4),107,83

SAFETY PROFILE: Poison by intraperitoneal route. Moderately toxic by ingestion. An experimental teratogen. Experimental reproductive effects. When heated to decomposition it emits toxic fumes of NO_x and HCl.

GEQ000 CAS:8007-08-7 HR: 2
GINGER OIL

PROP: From steam distillation of ground rhizomes of *Zingiber officinale* Roscoe (Fam. Zingiberaceae) (FCTXAV 12,807,74). Yellow liquid; odor of ginger. D: 0.870–0.882, refr index: 1.488 @ 20°. Sol in fixed oils, mineral oil, alc; insol in glycerin, propylene glycol.

TOXICITY DATA WITH REFERENCE
skn-rbt 500 mg/24H MOD FCTXAV 12,901,74
dnr-bcs 5 μL/disc TOFOD5 8,91,85
orl-mus LD50:3450 mg/kg CTYAD8 19,407,88
ipr-mus LD50:1230 mg/kg CTYAD8 19,407,88

CONSENSUS REPORTS: Reported in EPA TSCA Inventory.

SAFETY PROFILE: Moderately toxic by ingestion and intraperitoneal routes. A skin irritant. Mutation data reported. When heated to decomposition it emits acrid smoke and irritating fumes.

GEQ400 HR: 3
GINSENG

SYN: PANAX

TOXICITY DATA WITH REFERENCE
orl-rat LD50:750 mg/kg FRMTAL 31(6),45,82
orl-mus LD50:200 mg/kg FRMTAL 31(6),45,82
ipr-mus LD50:54 mg/kg FRMTAL 31(6),45,82

SAFETY PROFILE: Poison by ingestion and intraperitoneal routes.

GEQ425 HR: 3
GINSENG ROOT-NEUTRAL SAPONINS

PROP: A mixture of neutral saponins composed of ginsenoside-rb$_1$, -rb$_2$, and -rc (JJPAAZ 23,29,73).

SYNS: GINSENG, ROOT EXTRACT ☐ GINSENGWURZEL, EXTRACT (GERMAN) ☐ GNS ☐ NEUTRAL SAPONINS of PANAX GINSENG ROOT ☐ PANAX GINSENG, ROOT EXTRACT ☐ SONG-SAM, ROOT EXTRACT

TOXICITY DATA WITH REFERENCE
oms-rat-ipr 50 mg/kg CPBTAL 25,1665,77
ipr-rat LDLo:200 mg/kg JJPAAZ 23,29,73
ipr-mus LD50:545 mg/kg JJPAAZ 23,29,73
ivn-mus LD50:367 mg/kg JJPAAZ 23,29,73
par-mus LDLo:32,500 mg/kg AEPPAE 170,443,33
par-frg LDLo:8500 mg/kg AEPPAE 170,443,33

SAFETY PROFILE: Poison by intravenous and intraperitoneal routes. Mutation data reported. See also SAPONIN.

GEQ500 CAS:123-46-6 HR: 2
GIRARD T REAGENT
mf: $C_5H_{14}N_3O{\bullet}Cl$ mw: 167.67

SYNS: AMMONIUM, (CARBOXYMETHYL)TRIMETHYL-, CHLORIDE, HYDRAZIDE ☐ BETAINE HYDRAZIDE HYDROCHLORIDE ☐ (CARBOXYMETHYL)TRIMETHYLAMMONIUM CHLORIDE HYDRAZIDE ☐ ETHANAMINIUM, 2-HYDRAZINO-N,N,N-TRIMETHYL-2-OXO-, CHLORIDE (9CI) ☐ GIRARD REAGENT T ☐ GIRARD'S REAGENT T ☐ TRIMETHYLACETHYDRAZIDE AMMONIUM CHLORIDE ☐ TRIMETHYLAMINOACETOHYDRAZIDE CHLORIDE ☐ TRIMETHYLAMMONIUM ACETYL HYDRAZIDE CHLORIDE

TOXICITY DATA WITH REFERENCE
orl-rat LD:>500 mg/kg NCNSA6 5,25,53

CONSENSUS REPORTS: Reported in EPA TSCA Inventory.

SAFETY PROFILE: Moderately toxic by ingestion. When heated to decomposition it emits toxic vapors of NO_x and Cl^-.

GES000 CAS:1391-75-9 HR: 3
GITALIN
mf: $C_{35}H_{56}O_{12}$ mw: 668.91

PROP: Amorph powder. Readily sol in alc; slt to slowly sol in water.

SYN: VERODIGEN

TOXICITY DATA WITH REFERENCE
scu-mus LDLo:29 mg/kg 27ZWAY E.1,78,-
ivn-mus LDLo:2400 μg/kg ARZNAD 6,182,56
orl-cat LD50:1230 μg/kg JAPMA8 44,607,55
ivn-cat LDLo:1040 μg/kg JAPMA8 44,607,55
ivn-pgn LDLo:1 mg/kg JPHAA3 25,611,36

SAFETY PROFILE: Poison by ingestion, subcutaneous, and intravenous routes. When heated to decomposition it emits acrid smoke and irritating fumes.

GES100 CAS:3261-53-8 HR: 3
GITALOXIN
mf: $C_{42}H_{64}O_{15}$ mw: 809.06

PROP: Crystals from Me_2CO/pet ether. Mp: 250–253°.

SYNS: 16-FORMYL-GITOXIN ☐ GITALOXIGENIN-TRIDIGITOXOSID (GERMAN) ☐ GITALOXIN-16-FORMATE ☐ GITOXIGENIN TRIDIGITOXOXIDE-16-FORMATE

TOXICITY DATA WITH REFERENCE
orl-mus LD50:2870 μg/kg AIPTAK 153,436,65
scu-mus LD50:2830 μg/kg AIPTAK 153,436,65
ivn-cat LDLo:900 μg/kg JMPCAS 5,988,62
orl-gpg LDLo:2455 μg/kg AIPTAK 153,436,65
ivn-gpg LDLo:982 μg/kg AIPTAK 153,436,65

SAFETY PROFILE: A deadly poison by ingestion, subcutaneous, and intravenous routes. When heated to decomposition it emits acrid smoke and irritating fumes.

GEU000 **CAS:4562-36-1** **HR: 3**
GITOXIN
mf: $C_{41}H_{64}O_{14}$ mw: 781.05

PROP: Crystals or stout prisms from chloroform and methanol. Decomp @ 285° (rapid heating). Almost insol in chloroform, ethyl acetate, and acetone. Dissolves in a mixture of chloroform and alc or pyridine or dil alc.

SYNS: ANHYDROGITALIN ☐ BIGITALIN ☐ GITOXIGENIN-TRIDIGITOXOSID (GERMAN) ☐ PSEUDODIGITOXIN

TOXICITY DATA WITH REFERENCE
orl-cat LDLo:880 μg/kg 27ZWAY E.1,78,-
par-cat LDLo:465 μg/kg ARZNAD 6,182,56
ivn-gpg LD50:68 mg/kg ARZNAD 6,182,56
par-pgn LDLo:1210 μg/kg CPBTAL 11,613,63

SAFETY PROFILE: Poison by ingestion, intravenous, and parenteral routes. When heated to decomposition it emits acrid smoke and irritating fumes.

GEW000 **CAS:7242-04-8** **HR: 3**
GITOXIN PENTAACETATE
mf: $C_{51}H_{74}O_{19}$ mw: 991.25

PROP: Rhombic crystals. Mp: 151–155°.

SYNS: CARNACID-COR ☐ CORDOVAL ☐ PENGITOXIN ☐ PENTAACETYLGITOXIN ☐ PENTA-o-ACETYLGITOXIN ☐ PENTAGIT

TOXICITY DATA WITH REFERENCE
ivn-rat LD50:21 mg/kg AIPTAK 155,165,65
ipr-mus LD50:6400 μg/kg AIPTAK 155,165,65
orl-cat LD50:200 μg/kg AIPTAK 159,1,66
ipr-cat LD50:230 μg/kg AIPTAK 159,1,66
orl-gpg LD50:1 mg/kg AIPTAK 159,1,66
ivn-gpg LD50:450 μg/kg ARZNAD 6,182,56

SAFETY PROFILE: A deadly poison by ingestion, intraperitoneal, and intravenous routes. Used as a cardiotonic agent. When heated to decomposition it emits acrid smoke and irritating fumes.

GEW700 **CAS:1448-23-3** **HR: 3**
GLARUBIN
mf: $C_{25}H_{36}O_{10}$ mw: 496.61

PROP: Crystals. Mp: 250–255° (decomp).

SYNS: GLAUCARUBIN ☐ GLAUMEBA ☐ α-KIRONDRIN ☐ β-KIRONDRIN ☐ MK-33 ☐ SIMARUBACEAE

TOXICITY DATA WITH REFERENCE
orl-rat LD50:800 mg/kg 85GDA2 8(2),172,82
orl-mus LD50:1200 mg/kg AIPTAK 114,307,58
scu-mus LD50:28 mg/kg 85GDA2 8(2),172,82

SAFETY PROFILE: Poison by subcutaneous route. Moderately toxic by ingestion. When heated to decomposition it emits acrid smoke and fumes. See also ESTERS.

GEW800 **HR: 2**
GLORY LILY

PROP: Climbing lilies with tuberous roots. The leaves have tendrils on the tips. The flowers are bright red and yellow with well separated petals. They are cultivated in Hawaii, the southernmost parts of the continental United States, and the West Indies.

SYNS: CLIMBING LILY ☐ GLORIOSA LILY ☐ GLORIOSA ROTHSCHILDIANA ☐ GLORIOSA SUPERBA ☐ PIPA de TURCO (CUBA)

SAFETY PROFILE: The whole plant and especially the tubers contain the poison colchicine. Ingestion of any part of the plant causes a burning pain in the mouth, intense thirst, nausea, vomiting, abdominal cramps, severe diarrhea, and sometimes kidney damage. There may be extensive fluid and electrolyte loss. Colchicine is excreted slowly so the effects may persist for some time. See also COLCHICINE.

G

GEW875 **CAS:9007-92-5** **HR: 1**
GLUCAGON
mf: $C_{151}H_{224}N_{42}O_{50}S$ mw: 3460.23

PROP: A polypeptide hormone produced in the alpha cells of the islets of Langerhans in the pancreas. Rhombic dodecahedra. Stable. Practically insol in water; sol in acidic, basic media, in the range below pH 3 and above pH 9.5.

TOXICITY DATA WITH REFERENCE
dns-rat:lvr 75 mg/L EJBCAI 34,474,73
scu-rat TDLo:3600 μg/kg (16-21D preg):TER AJOGAH 106,656,70
ims-man TDLo:28 μg/kg AIPTAK 218,312,75

SAFETY PROFILE: Human systemic effects by intramuscular route: leukopenia (reduced white blood cell count). An experimental teratogen. Mutation data reported. When heated to decomposition it emits toxic fumes of SO_x and NO_x.

GFA000 **CAS:15879-93-3** **HR: 3**
α-d-GLUCOCHLORALOSE
mf: $C_8H_{11}Cl_3O_6$ mw: 309.54

PROP: Needles from EtOH or Et_2O. Mp: 187°.

SYNS: AGC ☐ ALFAMAT ☐ ANHYDROGLUCOCHLORAL ☐ APHOSAL ☐ CHLORALOSANE ☐ α-CHLORALOSE ☐ CHLOROALOSANE ☐ DULCIDOR ☐ GLUCOCHLORAL ☐ GLUCOCHLORALOSE ☐ KALMETTUMSOMNIFERUM ☐ MONOTRICHLOR-AETHYLIDEN-α-GLUCOSE (GERMAN) ☐ MUREX ☐ SOMIO ☐ 1,2-o-(2,2,2-TRICHLOROETHYLIDENE)-α-d-GLUCOFURANOSE

TOXICITY DATA WITH REFERENCE
scu-mus TDLo:215 mg/kg:ETA NTIS** PB223-159
orl-rat LDLo:400 mg/kg 85ESA3 9,260,76
scu-rat LDLo:200 mg/kg 27ZIAQ -,70,73
orl-mus LD50:200 mg/kg FMCHA2-,C68,91
ipr-mus LD50:175 mg/kg ARZNAD 21,1727,71
orl-dog LD50:250 mg/kg RMVEAG 154,137,78
ipr-dog LDLo:400 mg/kg AIPTAK 3,191,1897
orl-cat LD50:250 mg/kg RMVEAG 154,137,78
ipr-cat LDLo:150 mg/kg AIPTAK 3,191,1897

CONSENSUS REPORTS: Reported in EPA TSCA Inventory.

SAFETY PROFILE: Poison by ingestion, subcutaneous, and intraperitoneal routes. Questionable carcinogen with experimental tumorigenic data. When heated to decomposition it emits toxic fumes of Cl^-.

GFC000 CAS:124-99-2 HR: 3
GLUCOPROSCILLARIDIN A
mf: $C_{36}H_{52}O_{13}$ mw: 692.88

PROP: Crystals from EtOH. Mp: 270°.

SYNS: 14-β-HYDROXY-3-β-SCILLOBIOSIDOBUFA-4,20,22-TRIENOLIDE □ SCILLAGLYKOSID A (GERMAN) □ SCILLAREN A □ SCILLARENIN-3,6-DEOXY-4-o-β-d-GLUCOPYRANOSYL-α-l-MANNOPYRANOSIDE □ SCILLAREN & RHAMNOSE & GLUCOSE (GERMAN) □ 3-β-SCILLOBIOSIDO-14-β-HYDROXY-Δ-4,20,22-BUFATRIENOLID (GERMAN) □ TRANSVAALIN

TOXICITY DATA WITH REFERENCE
ivn-rat LD50:15 mg/kg ARZNAD 11,848,61
ivn-cat LDLo:143 mg/kg MEIEDD 10,1208,83
ivn-gpg LDLo:353 μg/kg AEPPAE 252,314,66

SAFETY PROFILE: Poison by intravenous route. When heated to decomposition it emits acrid smoke and irritating fumes.

GFC100 CAS:554-35-8 HR: 2
2-(β-d-GLUCOPYRANOSYLOXY)ISOBUTYRONITRILE
mf: $C_{10}H_{17}NO_6$ mw: 247.28

PROP: Needles. Mp: 143–144°.

SYNS: 2-(β-d-GLUCOPYRANOSYLOXY)-2-METHYLPROPANENITRILE □ LINAMARIN □ PHASEOLUNATIN □ PROPANENITRILE, 2-(β-d-GLUCOPYRANOSYLOXY)-2-METHYL-

TOXICITY DATA WITH REFERENCE
orl-ham TDLo:120 mg/kg (female 8D post):TER TJADAB 31,241,85
orl-rat LDLo:500 mg/kg TXAPA9 42,539,77

SAFETY PROFILE: Moderately toxic by ingestion. An experimental teratogen. When heated to decomposition it emits toxic fumes of NO_x.

GFG000 CAS:50-99-7 HR: 2
d-GLUCOSE
mf: $C_6H_{12}O_6$ mw: 180.18

$CHO(CHOH)_4CH_2OH$

PROP: Colorless crystals or white crystalline or granular powder; odorless with sweet taste. D: 1.544, mp: 146°. Sol in water; sltly sol in alc. α Form: (monohydrate) crystals from water. Mp: 83°. α Form: (anhydrous) crystals from hot ethanol or water. Mp: 146°. Very sparingly sol in abs alc, ether, acetone; sol in hot glacial acetic acid, pyridine, aniline. β Form: crystals from hot H_2O + ethanol, from dil acetic acid, or from pyridine. Mp: 148–155°.

SYNS: CARTOSE □ CERELOSE □ CORN SUGAR □ DEXTROPUR □ DEXTROSE (FCC) □ DEXTROSE, anhydrous □ DEXTROSOL □ GLUCOLIN □ GLUCOSE □ d-GLUCOSE, anhydrous □ GLUCOSE LIQUID □ GRAPE SUGAR □ SIRUP

TOXICITY DATA WITH REFERENCE
mmo-sat 25 mg/plate NARHAD 12,2127,84
oms-omi 1 mol/L ARMKA7 91,305,73
ivn-wmn TDLo:2 g/kg (female 39W post):TER BJOGAS 89,27,82
ivn-wmn TDLo:1057 μg/kg (female 39W post):TER BJOGAS 91,1014,84
ipr-rat TDLo:300 g/kg (female 30D pre):REP OYYAA2 6,251,72
ipr-ham TDLo:20 g/kg (female 6-8D post):TER TJADAB 33,74C,86
scu-rat TDLo:15,400 g/kg/22W-C:ETA GANNA2 30,419,36
orl-rat LD50:25,800 mg/kg 85AIAL -,39,73
ipr-mus LDLo:18 g/kg PSEBAA 35,98,36
ivn-mus LD50:9 g/kg ARZNAD 18,666,68
orl-dog LDLo:8000 mg/kg HBTXAC 1,150,55
orl-rbt LDLo:20,000 mg/kg HBTXAC 1,150,55
ivn-rbt LDLo:12,000 mg/kg HBTXAC 1,150,55

CONSENSUS REPORTS: Reported in EPA TSCA Inventory. EPA Genetic Toxicology Program.

SAFETY PROFILE: Mildly toxic by ingestion. An experimental teratogen. Experimental reproductive effects. Questionable carcinogen with experimental tumorigenic data. Mutation data reported. Potentially explosive reaction with potassium nitrate + sodium peroxide when heated in a sealed container. Mixtures with alkali release carbon monoxide when heated. When heated to decomposition it emits acrid smoke and irritating fumes.

GFG100 CAS:9001-37-0 HR: 3
GLUCOSE OXIDASE

PROP: Amorph powder or crystals. Freely sol in water giving yellowish-green solns.

SYNS: CORYLOPHYLINE □ DEOXIN-1 □ E.C. 1.1.3.4 □ GLUCOSE AERODEHYDROGENASE □ β-d-GLUCOSE OXIDASE □ MICROCID □ NOTATIN □ OXIDASE GLUCOSE □ PENATIN

TOXICITY DATA WITH REFERENCE
ipr-mus LD50:3 mg/kg 85GDA2 4(2),302,80
scu-mus LD50:4500 μg/kg 85GDA2 4(2),302,80
ivn-mus LD50:13 mg/kg BJPCAL 1,225,46
scu-rbt LD50:7500 μg/kg BJPCAL 1,225,46

CONSENSUS REPORTS: Reported in EPA TSCA Inventory.

SAFETY PROFILE: Poison by subcutaneous, intravenous, and intraperitoneal routes.

GFG200 HR: 1
GLUCOSE-RINGER'S SOLUTION (23.3%)

PROP: Consists of Ringer's solution containing 93.2 g/400 mL glucose (IYKEDH 16,320,85).

SYNS: GR-23 □ RINGER'S GLUCOSE SOLUTION

TOXICITY DATA WITH REFERENCE
ipr-rat LD50:53,300 mg/kg IYKEDH 16,320,85
ivn-rat LD50:57,200 mg/kg IYKEDH 16,320,85
ivn-rbt LDLo:50 g/kg IYKEDH 16,320,85

SAFETY PROFILE: Very mildly toxic by intravenous and intraperitoneal routes.

GFG205 **HR: 1**
GLUCOSE-RINGER'S SOLUTION (29.2%)

PROP: Consists of Ringer's solution containing 116.8 g/400 mL glucose (IYKEDH 16,328,85).

SYN: GR-29

TOXICITY DATA WITH REFERENCE
ipr-rat LD50:41,500 mg/kg IYKEDH 16,328,85
ivn-rat LD50:38,700 mg/kg IYKEDH 16,328,85
ivn-rbt LDLo:45 g/kg IYKEDH 16,328,85

SAFETY PROFILE: Very mildly toxic by intravenous and intraperitoneal routes.

GFK000 **CAS:20408-97-3** **HR: 1**
α-d-GLUCOTHIOPYRANOSE
mf: $C_6H_{12}O_5S$ mw: 196.24

SYNS: 5-THIO-α-d-GLUCOPYRANOSE □ 5-THIO-d-GLUCOSE

TOXICITY DATA WITH REFERENCE
spm-mus-orl 1400 mg/kg/35D JOHEA8 70,325,79
spm-mus-ipr 840 mg/kg/21D-C JOHEA8 72,347,81
orl-mus TDLo:2450 mg/kg (male 49D pre):REP CCPTAY 17,123,78
orl-mus TDLo:693 mg/kg (male 21D pre):REP IJEBA6 22,401,84
orl-mus TDLo:840 mg/kg (male 21D pre):REP JRPFA4 45,69,75
orl-rat TDLo:2800 mg/kg (56D male):REP BIREBV 17,697,77
ipr-mus LDLo:5500 mg/kg IOBPD3 8,589,82

CONSENSUS REPORTS: EPA Genetic Toxicology Program.

SAFETY PROFILE: Mildly toxic by intraperitoneal route. Mutation data reported. Experimental reproductive effects. When heated to decomposition it emits very toxic fumes of SO_x.

GFM000 **CAS:32449-92-6** **HR: 2**
GLUCURONIC ACID LACTONE
mf: $C_6H_8O_6$ mw: 176.14

PROP: Crystals from ethanol. Mp: 176–178° (commercial grades, mp: 172°), d: 1.76. Sol in water; sltly sol in ethanol; very sltly sol in abs ethanol in glacial acetic acid.

SYNS: DICURONE □ GLUCOXY □ GLUCURON □ GLUCURONE □ GLUCURONIC ACID-γ-LACTONE □ d-GLUCURONIC ACID LACTONE □ d-GLUCURONIC ACID-γ-LACTONE □ GLUCURONOLACTONE □ d-GLUCURONOLACTONE □ GLUCURONOSAN □ GLYCURONE □ GURONSAN □ REULATT S.S.

TOXICITY DATA WITH REFERENCE
orl-rat LD50:10,700 mg/kg NIIRDN 6,225,82
scu-rat LD50:4700 mg/kg NIIRDN 6,225,82
ivn-rat LD50:3200 mg/kg NIIRDN 6,225,82
ipr-mus LD50:5797 mg/kg NIIRDN 6,225,82
scu-dog LD50:4700 mg/kg NIIRDN 6,225,82
ivn-dog LD50:940 mg/kg NIIRDN 6,225,82
scu-rbt LD50:4700 mg/kg NIIRDN 6,225,82
ivn-rbt LD50:940 mg/kg NIIRDN 6,225,82

CONSENSUS REPORTS: Reported in EPA TSCA Inventory.

SAFETY PROFILE: Moderately toxic by intravenous route. Mildly toxic by ingestion and subcutaneous routes. When heated to decomposition it emits acrid smoke and irritating fumes.

GFM200 **CAS:1492-02-0** **HR: 3**
GLUDIASE
mf: $C_{12}H_{15}N_3O_2S_2$ mw: 297.42

PROP: Needles. Mp: 163°.

SYNS: AN 1324 □ 2-BENZENESULFONAMIDO-5-tert-BUTYL-1,3,4-THIADIAZOLE □ 2-BENZENESULFONAMIDO-5-TERTIOBUTYL-1-THIA-3,4-DIAZOLE □ N-(5-tert-BUTYL-1,3,4-THIADIAZOL-2-YL)BENZENESULFONAMIDE □ DESAGLYBUZOLE □ GLYBUZOLE □ RP 7891 □ TH-1395

TOXICITY DATA WITH REFERENCE
orl-mus TDLo:2100 mg/kg (female 7-13D post):TER YIKUAO 18,21,69
orl-mus TDLo:1050 mg/kg (female 7-13D post):TER YIKUAO 18,21,69
orl-rat LD50:500 mg/kg YIKUAO 18,21,69
ipr-rat LD50:219 mg/kg YIKUAO 18,21,69
scu-rat LD50:310 mg/kg YIKUAO 18,21,69
orl-mus LD50:730 mg/kg YIKUAO 18,21,69
ipr-mus LD50:235 mg/kg YIKUAO 18,21,69
scu-mus LD50:248 mg/kg YIKUAO 18,21,69
ivn-mus LD50:193 mg/kg YIKUAO 18,21,69
orl-rbt LD50:967 mg/kg YIKUAO 18,21,69
ipr-rbt LD50:300 mg/kg YIKUAO 18,21,69
ivn-rbt LD50:118 mg/kg YIKUAO 18,21,69

SAFETY PROFILE: Poison by subcutaneous, intravenous, and intraperitoneal routes. Moderately toxic by ingestion. An experimental teratogen. When heated to decomposition it emits toxic fumes of SO_x and NO_x.

GFM300 **CAS:617-65-2** **HR: 1**
dl-GLUTAMIC ACID (9CI)
mf: $C_5H_9NO_4$ mw: 147.15

SYNS: GLUTAMIC ACID, dl- □ (±)-GLUTAMIC ACID

TOXICITY DATA WITH REFERENCE
orl-hmn TDLo:71 mg/kg SCIEAS 163,826,69

CONSENSUS REPORTS: Reported in EPA TSCA Inventory.

SAFETY PROFILE: Human systemic effects by ingestion: headache. When heated to decomposition it emits toxic vapors of NO_x.

GFO000 **CAS:56-86-0** **HR: 1**
l-GLUTAMIC ACID
mf: $C_5H_9NO_4$ mw: 147.15

PROP: A nonessential amino acid present in all complete proteins. White rhombic crystals from alc (aq), or crystalline powder. Mp (dl form): 194°, d (dl form): 1.4601 @ 20°/4°, mp (l form): 224–225°, d (l form): 1.538 @ 20°/4°. Sltly sol in water.

SYNS: α-AMINOGLUTARIC ACID □ l-2-AMINOGLUTARIC ACID □ 2-AMINOPENTANEDIOIC ACID □ 1-AMINOPROPANE-1,3-DICARBOXYLIC ACID □ GLUSATE □ GLUTACID □ GLUTAMIC ACID □ α-GLUTAMIC ACID □ d-GLUTAMIENSUUR □ GLUTAMINIC ACID □ l-GLUTAMINIC ACID □ GLUTAMINOL □ GLUTATON

TOXICITY DATA WITH REFERENCE
orl-hmn TDLo:71 mg/kg:CNS SCIEAS 163,826,69
ivn-hmn TDLo:117 mg/kg:GIT AJMSA9 214,281,47

CONSENSUS REPORTS: Reported in EPA TSCA Inventory.

SAFETY PROFILE: Human systemic effects by ingestion and intravenous routes: headache and nausea or vomiting. When heated to decomposition it emits toxic fumes of NO_x.

GFO050 **CAS:56-85-9** **HR: 1**
GLUTAMINE
mf: $C_5H_{10}N_2O_3$ mw: 146.17

PROP: l-Form (natural): Fine opaque needles from water or dil ethanol. Decomp at 185–186°. Sol in water; practically insol in methanol, ethanol, ether, benzene, acetone, ethyl acetate, chloroform. dl-Form: prisms from dil acetone. Mp: 185–186°.

SYNS: 2-AMINOGLUTARAMIC ACID □ l-2-AMINOGLUTARAMIDIC ACID □ CEBROGEN □ GLUMIN □ GLUTAMIC ACID AMIDE □ GLUTAMIC ACID-5-AMIDE □ γ-GLUTAMINE □ l-GLUTAMINE (9CI, FCC) □ LEVOGLUTAMID □ LEVOGLUTAMIDE □ STIMULINA

TOXICITY DATA WITH REFERENCE
orl-man TDLo:27 mg/kg/1W-I:BAH AJPSAO 141,1302,84
orl-rat LD50:7500 mg/kg NIIRDN 6,228,82
orl-mus LD50:21,700 mg/kg NIIRDN 6,228,82

CONSENSUS REPORTS: Reported in EPA TSCA Inventory.

SAFETY PROFILE: Mildly toxic by ingestion. Human systemic effects: euphoria. When heated to decomposition it emits toxic fumes of NO_x.

GFO100 **HR: 2**
N^2-(γ-l-(+)-GLUTAMYL)-4-CARBOXYPHENYLHYDRAZINE
mf: $C_{12}H_{15}N_3O_5$ mw: 281.30

SYNS: ANTHGLUTIN □ l-GLUTAMIC ACID, 5-(2-(4-CARBOXYPHENYL)HYDRAZIDE)

TOXICITY DATA WITH REFERENCE
orl-mus TDLo:72,800 mg/kg/52W-I:CAR ANTRD4 6,917,86

SAFETY PROFILE: Questionable carcinogen with experimental carcinogenic data. When heated to decomposition it emits toxic fumes of NO_x.

GFO200 **CAS:60762-50-7** **HR: 2**
1-(l-α-GLUTAMYL)-2-ISOPROPYLHYDRAZINE
mf: $C_8H_{17}N_3O_3$ mw: 203.28

SYN: RO 4-1385

TOXICITY DATA WITH REFERENCE
orl-mus LD50:500 mg/kg 27ZQAG -,429,72
scu-mus LD50:1400 mg/kg 27ZQAG -,429,72
ivn-mus LD50:1000 mg/kg 27ZQAG -,429,72

SAFETY PROFILE: Moderately toxic by ingestion, subcutaneous, and intravenous routes. When heated to decomposition it emits toxic fumes of NO_x. See also HYDRAZINE.

GFQ000 **CAS:111-30-8** **HR: 3**
GLUTARALDEHYDE
mf: $C_5H_8O_2$ mw: 100.13

PROP: Oil. Bp: 71–72° @ 10 mm.

SYNS: CIDEX □ GLUTARAL □ GLUTARALDEHYD (CZECH) □ GLUTARDIALDEHYDE □ GLUTARIC DIALDEHYDE □ NCI-C55425 □ 1,5-PENTANEDIAL □ 1,5-PENTANEDIONE □ POTENTIATED ACID GLUTARALDEHYDE □ SONACIDE

TOXICITY DATA WITH REFERENCE
skn-hmn 6 mg/3D-I SEV 85DKA8 -,127,77
skn-rbt 13 mg open MLD UCDS** 1/30/70
skn-rbt 500 mg/24H SEV 28ZPAK -,42,72
eye-rbt 1 mg SEV UCDS** 1/30/70
eye-rbt 250 μg/24H SEV 28ZPAK -,42,72
oms-nml:oth 50 mmol/L MUREAV 148,25,85
sce-ham:ovr 110 μg/L ENMUDM 7,1,85
orl-mus TDLo:8 g/kg (female 6-15D post):TER TJADAB 22,51,80
orl-rat TDLo:4370 mg/kg (female 35D pre):REP OYYAA2 12,11,76
orl-rat TDLo:875 mg/kg (35D male):REP OYYAA2 12,11,76
orl-mus TDLo:50 g/kg (female 6-15D post):TER TJADAB 22,51,80
orl-rat LD50:134 mg/kg OYYAA2 19,503,80
ipr-rat LD50:17,900 μg/kg IYKEDH 10,232,79
scu-rat LD50:2390 mg/kg OYYAA2 19,503,80
ivn-rat LD50:9800 μg/kg EPASR* 8EHQ-1290-1008
orl-mus LD50:100 mg/kg OYYAA2 19,503,80
ipr-mus LD50:13,900 μg/kg IYKEDH 10,232,79
scu-mus LD50:1430 mg/kg OYYAA2 19,503,80
ivn-mus LD50:15,400 μg/kg OYYAA2 19,503,80
skn-rbt LD50:2560 mg/kg AIHAAP 23,95,62

CONSENSUS REPORTS: Reported in EPA TSCA Inventory.

OSHA PEL: CL 0.2 ppm
ACGIH TLV: CL 0.2 ppm (Proposed: CL 0.05 ppm)
DFG MAK: 0.2 ppm (0.8 mg/m³)

SAFETY PROFILE: Poison by ingestion, intravenous, and intraperitoneal routes. Moderately toxic by inhalation, skin contact, and subcutaneous routes. Experimental teratogenic and reproductive effects. A severe eye and human skin irritant. Mutation data reported. When heated to decomposition it emits acrid smoke and irritating fumes. See also ALDEHYDES.

For occupational chemical analysis use OSHA: #ID-64 or NIOSH: Glutaraldehyde 2531.

GFS000 **CAS:110-94-1** **HR: 1**
GLUTARIC ACID
mf: $C_5H_8O_4$ mw: 132.13

PROP: Colorless crystals/needles or prisms from C_6H_6 or $CHCl_3$. D: 1.429 @ 15°/4°, mp: 97.5°, bp: 200°. Very sol in abs alc and in ether; sol in benzene chloroform, alc, and ether. Large monoclinic prisms.

SYNS: PENTANDIOIC ACID □ PENTANEDIOIC ACID □ 1,5-PENTANEDIOIC ACID □ 1,3-PROPANEDICARBOXYLIC ACID

TOXICITY DATA WITH REFERENCE
orl-mus LD50:6000 mg/kg BIJOAK 34,1196,40

CONSENSUS REPORTS: Reported in EPA TSCA Inventory.

SAFETY PROFILE: Mildly toxic by ingestion. When heated to decomposition it emits acrid smoke and irritating fumes.

GFU000 **CAS:108-55-4** **HR: 2**
GLUTARIC ANHYDRIDE
mf: $C_5H_6O_3$ mw: 114.11

PROP: Crystals from Et_2O. Mp: 56°, bp: 144–146° @ 13 mm, d: 0.989. Sol in benzene and toluene; highly sol in water on complete hydrolysis.

TOXICITY DATA WITH REFERENCE
orl-rat LDLo:4460 mg/kg AIHAAP 23,95,62
skn-rbt LDLo:1780 mg/kg AIHAAP 23,95,62

CONSENSUS REPORTS: Reported in EPA TSCA Inventory.

SAFETY PROFILE: Moderately toxic by skin contact. Mildly toxic by ingestion. When heated to decomposition it emits acrid smoke and irritating fumes. See also ANHYDRIDES.

GFU200 **CAS:64624-44-8** **HR: 3**
GLUTARYL DIAZIDE
mf: $C_5H_6N_6O_2$ mw: 182.14

$N_3CO{\cdot}(CH_2)_3CO{\cdot}N_3$

SAFETY PROFILE: Explodes when heated. Upon decomposition it emits toxic fumes of NO_x. See aiso AZIDES.

GFW000 **CAS:70-18-8** **HR: 2**
GLUTATHIONE
mf: $C_{10}H_{17}N_3O_6S$ mw: 307.36

PROP: Colorless prisms out of alc. Mp: 195° decomp in hot water; insol in abs alc, ether, and acid. Freely sol in H_2O, dil alc, liquid ammonia, and dimethylformamide.

SYNS: COPREN □ DELTATHIONE □ GLUTATHIONE (reduced) □ GLUTATIOL □ GLUTATIONE □ GLUTIDE □ GLUTINAL □ GSH □ ISETHION □ NEUTHION □ TATHIONE □ TRIPTIDE

TOXICITY DATA WITH REFERENCE
mma-sat 6 mmol/L SCIEAS 220,961,83
dns-hmn:fbr 1 mmol/L CALEDQ 5,199,78
cyt-ham:ovr 1 mmol/L CALEDQ 5,199,78
sce-ham:ovr 100 μmol/L MUREAV 68,351,79
dni-mam:lym 10 mmol/L CBINA8 31,265,80
orl-rat TDLo:1250 mg/kg (1-22D preg):REP AJANA2 110,29,62
orl-mus LD50:5 g/kg 85IPAE -,93,72
ipr-mus LD50:4020 mg/kg 85IPAE -,93,72
scu-mus LD50:5 g/kg 85IPAE -,93,72
ivn-mus LD50:2238 mg/kg JJANAX 38,137,85

CONSENSUS REPORTS: EPA Genetic Toxicology Program. Reported in EPA TSCA Inventory.

SAFETY PROFILE: Moderately toxic by intravenous route. Experimental reproductive effects. Human mutation data reported. When heated to decomposition it emits very toxic fumes of SO_x and NO_x.

G

GFY100 **CAS:26944-48-9** **HR: 2**
GLUTRIL
mf: $C_{10}H_{26}N_2O_4S$ mw: 366.52

PROP: Crystals. Mp: 192–195° (ethanol loses water).

SYNS: GLIBORNURIDE □ GLUBORID □ 1-((1R)-2-endo-HYDROXY-3-endo BORNYL)-3-(p-TOLYLSULFONYL)UREA □ RO 6-4563 □ RO-6-4563/8

TOXICITY DATA WITH REFERENCE
orl-rat TDLo:70 mg/kg (7-13D preg):REP KSRNAM 6,1968,72
orl-rbt TDLo:2250 mg/kg (female 8-16D post):TER KSRNAM 6,1983,72
orl-rat LD50:18 g/kg KSRNAM 6,1925,72
ipr-rat LD50:1360 mg/kg KSRNAM 6,1925,72
scu-rat LD50:10,800 mg/kg KSRNAM 6,1925,72
ipr-mus LD50:1530 mg/kg KSRNAM 6,1925,72
scu-mus LD50:20 g/kg KSRNAM 6,1925,72

SAFETY PROFILE: Moderately toxic by intraperitoneal route. Mildly toxic by ingestion. Experimental teratogenic and reproductive effects. When heated to decomposition it emits toxic fumes of SO_x and NO_x.

GFY200 **CAS:56-82-6** **HR: 1**
dl-GLYCERALDEHYDE
mf: $C_3H_6O_3$ mw: 90.09

SYNS: GLYCERALDEHYDE, (±)- □ dl-GLYCERIC ALDEHYDE

TOXICITY DATA WITH REFERENCE
mmo-sat 100 μg/plate ABCHA6 47,2461,83
ipr-rat LD50:2 g/kg JPPMAB 17,814,65

CONSENSUS REPORTS: Reported in EPA TSCA Inventory.

SAFETY PROFILE: Mildly toxic by intraperitoneal route. Mutation data reported.

GGA000 **CAS:56-81-5** **HR: 3**
GLYCERIN
mf: $C_3H_8O_3$ mw: 92.11

$HOCH_2CHOHCH_2OH$

PROP: Colorless or pale-yellow liquid; odorless, syrupy, sweet and warm taste. Mp: 17.9 (solidifies at a much lower temp), bp: 290° (part decomp), ULC: 10–20, flash p: 320°F, d: 1.260 @ 20°/4°, autoign temp: 698°F, vap press: 0.0025 mm @ 50°, vap d: 3.17. Misc in H_2O, EtOH; insol in C_6H_6, $CHCl_3$, and CCl_4.

SYNS: GLYCERIN, anhydrous ☐ GLYCERINE ☐ GLYCERIN, synthetic ☐ GLYCERITOL ☐ GLYCEROL ☐ GLYCYL ALCOHOL ☐ GROCOLENE ☐ MOON ☐ 1,2,3-PROPANETRIOL ☐ STAR ☐ SUPEROL ☐ SYNTHETIC GLYCERIN ☐ 90 TECHNICAL GLYCERINE ☐ TRIHYDROXYPROPANE ☐ 1,2,3-TRIHYDROXYPROPANE

TOXICITY DATA WITH REFERENCE
skn-rbt 500 mg/24H MLD 28ZPAK -,37,72
eye-rbt 126 mg MLD BIOFX* 9-4/70
eye-rbt 500 mg/24H MLD 28ZPAK -,37,72
dni-hmn:lym 200 mmol/L PNASA6 79,1171,82
itt-rat TDLo:1600 mg/kg (male 1D pre):REP CCPTAY 29,291,84
orl-rat TDLo:100 mg/kg (male 1D pre):REP TGANAK 19,436,85
itt-rat TDLo:280 mg/kg (male 2D pre):REP CCPTAY 29,291,84
orl-hmn TDLo:1428 mg/kg:CNS,GIT 34ZIAG -,288,69
orl-rat LD50:12,600 mg/kg FEPRA7 4,142,45
ipr-rat LD50:4420 mg/kg RCOCB8 56,125,87
scu-rat LD50:100 mg/kg NIIRDN 6,215,82
orl-mus LD50:4090 mg/kg FRZKAP (6),56,77
ipr-mus LD50:8982 mg/kg ARZNAD 26,1581,76
scu-mus LD50:91 mg/kg NIIRDN 6,215,82
ivn-mus LD50:4250 mg/kg JAPMA8 39,583,50
ivn-rbt LD50:53 g/kg NIIRDN 6,215,82
orl-gpg LD50:7750 mg/kg JIHTAB 23,259,41

CONSENSUS REPORTS: Reported in EPA TSCA Inventory.

OSHA PEL: TWA Total Mist: 10 mg/m³; Respirable Fraction: 5 mg/m³
ACGIH TLV: TWA 10 mg/m³ (vapor)

SAFETY PROFILE: Poison by subcutaneous route. Mildly toxic by ingestion. Human systemic effects by ingestion: headache and nausea or vomiting. Experimental reproductive effects. Human mutation data reported. A skin and eye irritant. In the form of mist it is a nuisance particulate and inhalation irritant.

Combustible liquid when exposed to heat, flame, or powerful oxidizers. Mixtures with hydrogen peroxide are highly explosive. Ignites on contact with potassium permanganate, calcium hypochlorite. Mixture with nitric acid + sulfuric acid forms the explosive glyceryl nitrate. Mixture with perchloric acid + lead oxide forms explosive perchlorate esters. Confined mixture with chlorine explodes if heated to 70–80°. Can react violently with acetic anhydride; aniline + nitrobenzene; $Ca(OCl)_2$; CrO_3; Cr_2O_3; F_2 + PbO; phosphorus triiodide; ethylene oxide + heat; $KMnO_4$; K_2O_2; $AgClO_4$; Na_2O_2; NaH. Energetic reaction with sodium hydride. Mixture with nitric acid + hydrofluoric acid is a storage hazard due to gas evolution. To fight fire, use alcohol foam, CO_2, dry chemical. When heated to decomposition it emits acrid smoke and fumes.

For occupational chemical analysis use NIOSH: Nuisance Dust, Total 0500; Nuisance Dust, Respirable, 0600.

GGA100 CAS:25395-31-7 HR: 1
GLYCEROL DIACETATE
mf: $C_7H_{12}O_5$ mw: 176.19

SYNS: ACETIN, DI- ☐ DIACETIN ☐ DIACETYLGLYCEROL ☐ GLYCERIN DIACETATE ☐ GLYCERINE DIACETATE ☐ GLYCERYL DIACETATE ☐ 1,2,3-PROPANETRIOL, DIACETATE (9CI)

TOXICITY DATA WITH REFERENCE
scu-rat LD50:4 g/kg PSEBAA 46,26,41
orl-mus LD50:8500 mg/kg BCFAAI 125,401,86
ipr-mus LD50:2300 mg/kg BCFAAI 125,401,86
scu-mus LD50:2500 mg/kg PSEBAA 46,26,41
ivn-dog LD50:3 g/kg 85JCAE-,667,86

CONSENSUS REPORTS: Reported in EPA TSCA Inventory.

SAFETY PROFILE: Low toxicity by ingestion and other routes. When heated to decomposition it emits acrid smoke and irritating vapors.

GGA200 CAS:623-87-0 HR: 3
GLYCEROL-1,3-DINITRATE
mf: $C_3H_6N_2O_7$ mw: 182.11

SYNS: 1,3-DINITROGLYCERIN ☐ 1,3-DNG ☐ GLYCEROL, 1,3-DINITRATE ☐ 1,3-GLYCERYL DINITRATE ☐ 1,2,3-PROPANETRIOL, 1,3-DINITRATE (9CI)

TOXICITY DATA WITH REFERENCE
orl-rat LD50:1065 mg/kg NTIS** AD-B011-150
orl-mus LD50:676 mg/kg NTIS** AD-B011-150

DOT CLASSIFICATION: Forbidden

SAFETY PROFILE: Moderately toxic by ingestion. An explosive forbidden for transport. When heated to decomposition it emits toxic vapors of NO_x.

GGA800 CAS:9001-62-1 HR: 3
GLYCEROL ESTER HYDROLASE

SYNS: AMANO N-AP ☐ BUTYRINASE ☐ E.C.3.1.1.3. ☐ GA 56 (enzyme) ☐ LIPASE ☐ LIPAZIN ☐ MEITO MY 30 ☐ REMZYME PL 600 ☐ STEAPSIN ☐ TAKEDO 1969-4-9 ☐ TRIACETINASE ☐ TRIACYLGLYCEROL HYDROLASE ☐ TRIACYLGLYCEROL LIPASE ☐ TRIBUTYRASE ☐ TRIBUTYRINASE ☐ TRIBUTYRIN ESTERASE ☐ TRIGLYCERIDE HYDROLASE ☐ TRIGLYCERIDE LIPASE ☐ TRIOLEIN HYDROLASE ☐ TWEENASE ☐ TWEEN ESTERASE ☐ TWEEN HYDROLASE

TOXICITY DATA WITH REFERENCE
orl-mus TDLo:56 g/kg (female 7-13D post):REP NYKZAU 70,107,74
orl-rat TDLo:56 g/kg (8-14D preg):REP NYKZAU 70,107,74
orl-rat TDLo:4550 g/kg (male 26W pre):REP NYKZAU 70,89,74
ipr-rat LD50:634 mg/kg NYKZAU 69,191,73
ipr-mus LD50:395 mg/kg NYKZAU 69,191,73
scu-mus LD50:2050 mg/kg NYKZAU 69,191,73

SAFETY PROFILE: Poison by intraperitoneal route. Moderately toxic by subcutaneous route. Experimental reproductive effects.

GGA915 **CAS:544-62-7** **HR: 2**
GLYCEROL MONOOCTADECYL ETHER
mf: $C_{21}H_{44}O_3$ mw: 344.65

SYNS: BATILOL □ BATYL ALCOHOL □ MONOOCTADECYL ETHER of GLYCEROL □ α-OCTADECYLETHER of GLYCEROL □ 1-O-OCTADECYLGLYCEROL □ 3-(OCTADECYLOXY)-1,2-PROPANEDIOL □ 1,2-PROPANEDIOL, 3-(OCTADECYLOXY)-

TOXICITY DATA WITH REFERENCE
ipr-mus LD50:750 mg/kg NTIS** AD691-490

CONSENSUS REPORTS: Reported in EPA TSCA Inventory.

SAFETY PROFILE: Moderately toxic by intraperitoneal route. When heated to decomposition it emits acrid smoke and irritating vapors.

GGG000 **CAS:96-11-7** **HR: 2**
GLYCEROL TRIBROMOHYDRIN
mf: $C_3H_5Br_3$ mw: 280.81

PROP: A liquid. Mp: 16.5°, bp: 220°, d: 2.43 @ 23°. Insol in H_2O; sol in EtOH and Et_2O.

SYNS: GLYCERYL TRIBROMOHYDRIN □ sym-TRIBROMOPROPANE □ 1,2,3-TRIBROMOPROPANE

TOXICITY DATA WITH REFERENCE
mmo-sat 1 μmol/plate ENMUDM 2,59,80
mma-sat 500 ng/plate ENMUDM 7(Suppl 3),15,85
orl-rat TDLo:250 mg/kg (5D male):REP MUREAV 101,321,82
orl-rat LDLo:500 mg/kg MUREAV 101,321,82

CONSENSUS REPORTS: EPA Genetic Toxicology Program. Reported in EPA TSCA Inventory.

SAFETY PROFILE: Moderately toxic by ingestion. Experimental reproductive effects. Mutation data reported. When heated to decomposition it emits toxic fumes of Br^-. See also BROMIDES.

GGI000 **CAS:38571-73-2** **HR: 3**
GLYCEROL (TRI(CHLOROMETHYL))ETHER
mf: $C_6H_{11}Cl_3O_3$ mw: 237.52

SYN: TRIS-1,2,3-(CHLOROMETHOXY)PROPANE

TOXICITY DATA WITH REFERENCE
skn-mus TDLo:8640 mg/kg/72W-I:ETA CNREA8 35,2553,75
ipr-mus TDLo:910 mg/kg/76W-I:NEO CNREA8 35,2553,75
scu-mus TDLo:970 mg/kg/81W-I:NEO CNREA8 35,2553,75

CONSENSUS REPORTS: IARC Cancer Review: Group 2A IMEMDT 7,56,87; Animal Sufficient Evidence IMEMDT 15,301,77.

SAFETY PROFILE: Confirmed carcinogen with experimental neoplastigenic and tumorigenic data. When heated to decomposition it emits toxic fumes of Cl^-. See also ETHERS.

GGK000 **CAS:621-70-5** **HR: 3**
GLYCEROL TRIHEXANOATE
mf: $C_{21}H_{38}O_6$ mw: 386.59

PROP: A liquid. Mp: −25°.

SYNS: CAPROIC TRIGLYCERIDE □ GLYCEROL TRICAPROATE □ GLYCERYL TRICAPROATE □ HEXANOIC ACID, 1,2,3-PROPANETRIYL ESTER (9CI) □ HEXANOIN, TRI-(6CI,7CI,8CI) □ TRICAPROIN □ TRICAPRONIN □ TRICAPROYLGLYCEROL □ TRIHEXANOIN □ TRIHEXANOYLGLYCEROL

TOXICITY DATA WITH REFERENCE
ivn-mus LD50:122 mg/kg APSCAX 40,338,57

CONSENSUS REPORTS: Reported in EPA TSCA Inventory.

SAFETY PROFILE: Poison by intravenous route. When heated to decomposition it emits acrid smoke and irritating fumes.

GGO000 **CAS:106-61-6** **HR: 2**
GLYCERYL ACETATE
mf: $C_5H_{10}O_4$ mw: 134.15

PROP: Colorless, very hygroscopic liquid; characteristic odor. D: 1.206 @ 20°/4°, bp: 158° @ 17 mm. Very sol in water and alc; sltly sol in ether; insol in benzene.

SYNS: 1-ACETATE-1,2,3-PROPANETRIOL □ ACETIC ACID, MONOGLYCERIDE □ ACETIN □ 2,3-DIHYDROXYPROPYL ACETATE □ GLYCEROL-1-ACETATE □ GLYCEROL MONOACETATE □ GLYCEROL-α-MONOACETATE □ GLYCEROL-1-MONOACETATE □ GLYCERYL MONOACETATE □ MONOACETIN □ α-MONOACETIN □ 1-MONOACETIN □ MONOACETYL GLYCERINE □ 1,2,3-PROPANETRIOL MONOACETATE

TOXICITY DATA WITH REFERENCE
mmo-sat 3333 μg/plate NTPTB* JAN 82
mma-sat 100 μg/plate NTPTB* JAN 82
scu-rat LD50:5500 mg/kg PSEBAA 46,26,41
scu-mus LD50:3500 mg/kg PSEBAA 46,26,41

SAFETY PROFILE: Moderately toxic by subcutaneous route. Mutation data reported. When heated to decomposition it emits acrid smoke and irritating fumes.

GGQ000 **CAS:136-44-7** **HR: 1**
GLYCERYL-p-AMINOBENZOATE
mf: $C_{10}H_{13}NO_4$ mw: 211.24

PROP: Crystals from alc/pet ether. Mp: 116–117°. Semisolid, waxy mass or syrup. Faint aromatic odor. Liquifies and congeals very slowly. Sol in methanol, ethanol, isopropanol, glycerol, propylene glycol; insol in water, oils, fats.

SYNS: p-AMINOBENZOIC ACID MONOGLYCERYL ESTER □ GLYCEROL-1-p-AMINOBENZOATE □ MONOGLYCEROL-p-AMINOBENZOATE

TOXICITY DATA WITH REFERENCE
skn-hmn 15 mg/3D-I MLD 85DKA8 -,127,77

CONSENSUS REPORTS: Reported in EPA TSCA Inventory.

SAFETY PROFILE: A human skin irritant. When heated to decomposition it emits toxic fumes of NO_x.

G

GGR200 **CAS:25496-72-4** **HR: 1**
GLYCERYL MONOOLEATE
mf: $C_{21}H_{40}O_4$ mw: 356.61

SYNS: ADCHEM GMO □ AJAX GMO □ ALDO 40 □ ALDO MO-FG □ DUR-EM 204 □ EMCOL O □ EMERY OLEIC ACID ESTER 2221 □ EMRITE 6009 □ GLYCERINE MONOOLEATE □ GLYCERIN MONOOLEATE □ GLYCEROL MONOOLEATE □ GLYCEROL OLEATE □ GLYCERYL OLEATE □ GMO 8903 □ HAROWAX L 9 □ LOXIOL G 10 □ MONOGLYCERYL OLEATE □ MONOOLEIN □ MONOOLEOYLGLYCEROL □ 9-OCTADECENOIC ACID (Z)-, MONOESTER with 1,2,3-PROPANETRIOL (9CI) □ OLEIC ACID GLYCEROL MONOESTER □ OLEIC ACID MONOGLYCERIDE □ OLEOYLGLYCEROL □ OLEYLMONOGLYCERIDE □ OLICINE □ RIKEMAL O 71D □ RIKEMAL OL 100 □ S 1096 □ S 1097 □ SINNOESTER OGC □ S 1096R □ SUNSOFT O 30B □ SUPEOL

TOXICITY DATA WITH REFERENCE
skn-rbt 500 mg MLD JACTDZ 5(5),391,86
eye-rbt 100 mg MLD JACTDZ 5(5),391,86

CONSENSUS REPORTS: EPA TSCA Chemical Inventory, JUNE 1993

CONSENSUS REPORTS: Reported in EPA TSCA Inventory.

SAFETY PROFILE: A skin and eye irritant. When heated to decomposition it emits acrid smoke and irritating fumes.

GGS000 **CAS:59-47-2** **HR: 3**
GLYCERYL-o-TOLYL ETHER
mf: $C_{10}H_{14}O_3$ mw: 182.24

SYNS: A 1141 □ AGEFLEX CGE □ ANATENSIN □ ANXINE □ ATENSIN □ AVESYL □ AVOXYL □ BDH 312 □ CRESODIOL □ o-CRESOL GLYCERYL ETHER □ CRESOSSIDIOLO □ CRESOSSIPROPANDIOLO □ CRESOXYDIOL □ CRESOXYPROPANEDIOL □ o-CRESYL-α-GLYCERYL ETHER □ CURARIL □ CURARYTHAN □ DASERD □ DASEROL □ DECONTRACTIL □ α,β-DIHYDROXY-γ-(2-METHYLPHENOXY)PROPANE □ 1,2-DIHYDROXY-3-(2-METHYLPHENOXY)PROPANE □ DILOXOL □ FINDOLAR □ GLUKRESIN □ GLYOTOL □ KINAVOSYL □ o-KRESOL-GLYCERINAETHER (GERMAN) □ KRESOXYPROPANDIOL □ LISSENPHAN □ MC 2303 □ MEFENSINA □ MEPHATE □ MEPHEDAN □ MEPHELOR □ MEPHENSIN □ MEPHOSAL □ MEPHSON □ 3-(2-METHYLPHENOXY)-1,2-PROPANEDIOL □ MIANESINA □ MOCTYNOL □ MYANIL □ MYCOCURAN □ MYODETENSINE □ MYOLAX □ MYOPAN □ MYOSEROL □ MYOXANE □ NEMBUSEN □ NEPHELOR □ ORANIXON □ ORTOL □ PROLAX □ PROLOXIN □ RELAXANT □ RELAXAR □ RENARCOL □ REX REGULANS □ RP 3602 □ SANSDOLOR □ SECONESINZ □ SINAN □ SPARTOLOXYN □ SQ 1156 □ STILALGIN □ THIOXIDIL □ TOLANSIN □ TOLCIL □ TOLOFREN □ 3-o-TOLOXY-1,2-PROPANEDIOL □ TOLSEROL □ TOLULOX □ TOLYDRIN □ 1-o-TOLYLGLYCEROL ETHER □ α-(o-TOLYL)GLYCERYL ETHER □ 3-(o-TOLYLOXY)PROPANE-1,2-DIOL □ TOLYNOL □ TOLYSPAZ □ TORULOX □ WALKO-NESIN □ XERAL

TOXICITY DATA WITH REFERENCE
orl-rat LD50:625 mg/kg AIPTAK 130,280,61
ipr-rat LD50:283 mg/kg JPETAB 129,75,60
ivn-rat LD50:133 mg/kg PSEBAA 85,323,54
orl-mus LD50:720 mg/kg ARZNAD 17,242,67
ipr-mus LD50:320 mg/kg NYKZAU 55,1272,59
scu-mus LD50:285 mg/kg APTOA6 19,247,62
ivn-mus LD50:175 mg/kg COREAF 248,3642,59
orl-rbt LDLo:2300 mg/kg AIPTAK 89,145,52
ivn-rbt LD50:125 mg/kg IJNEAQ 5,305,66
orl-ham LD50:821 mg/kg JPETAB 129,75,60
ipr-ham LD50:322 mg/kg JPETAB 129,75,60

SAFETY PROFILE: Poison by intraperitoneal, intravenous, and subcutaneous routes. Moderately toxic by ingestion. When heated to decomposition it emits acrid smoke and irritating fumes. See also ETHERS.

GGW000 **CAS:765-34-4** **HR: 3**
GLYCIDALDEHYDE
DOT: UN 2622
mf: $C_3H_4O_2$ mw: 72.07

PROP: Colorless liquid. Bp: 113°, d: 1.1403 @ 20°/4°.

SYNS: EPIHYDRINALDEHYDE □ EPIHYDRINE ALDEHYDE □ 2,3-EPOXYPROPANAL □ 2,3-EPOXY-1-PROPANAL □ 2,3-EPOXYPROPIONALDEHYDE □ GLYCIDAL □ GLYCIDYLALDEHDYE □ OXIRANE-CARBOXALDEHYDE □ RCRA WASTE NUMBER U126

TOXICITY DATA WITH REFERENCE
eye-hmn 1 ppm/5M MOD AEHLAU 2,23,61
mmo-esc 10 μg/plate ENMUDM 6(Suppl 2),1,84
mma-esc 33,300 ng/plate ENMUDM 6(Suppl 2),1,84
skn-rbt 100 mg/24H MOD 85JCAE-,770,86
scu-rat TDLo:13 g/kg/77W-I:CAR JNCIAM 39,1213,67
skn-mus TDLo:17 g/kg/48W-I:CAR JNCIAM 35,707,65
scu-mus TDLo:8844 mg/kg/67W-I:NEO JNCIAM 37,825,66
skn-mus TD:26 g/kg/71W-I:CAR 14JTAF -,275,65
skn-mus TD:26 g/kg/22W-I:ETA JNCIAM 39,1217,67
scu-rat TDLo:390 mg/kg/78W-I:ETA JNCIAM 37,825,66
ihl-hmn TCLo:5 ppm:BRN,EYE,CNS 34ZIAG -,289,69
orl-rat LDLo:50 mg/kg AJHYA2 76,209,62
ihl-rat LCLo:251 ppm/4H 14CYAT 2,1636,63
ipr-mus LD50:200 mg/kg JJIND8 62,911,79
skn-rbt LD50:249 mg/kg AEHLAU 2,23,61
ivn-rbt LDLo:20 mg/kg AEHLAU 2,23,61

CONSENSUS REPORTS: IARC Cancer Review: Group 2B IMEMDT 7,56,87; Animal Sufficient Evidence IMEMDT 11,175,76. EPA Genetic Toxicology Program.

DOT CLASSIFICATION: 3; *Label:* Flammable Liquid, Poison

SAFETY PROFILE: Confirmed carcinogen with experimental carcinogenic, neoplastigenic, and tumorigenic data. Poison by ingestion, skin contact, intraperitoneal, and intravenous routes. Moderately toxic by inhalation. Human systemic effects by inhalation: changes in central nervous system electrical activity, olfactory changes, and excitement. Mutation data reported. A human eye irritant. Powerful skin sensitizer and mucous membrane irritant. Flammable when exposed to heat, flame, or oxidizing materials. When heated to decomposition it emits acrid smoke and irritating fumes. See also ALDEHYDES.

GGW500 **CAS:556-52-5** **HR: 3**
GLYCIDOL
mf: $C_3H_6O_2$ mw: 74.09

OCH_2CHCH_2OH

PROP: Colorless liquid. D: 1.165 @ 0°/4°, bp: 167° (decomp). Entirely sol in water, alc, and ether.

SYNS: EPIHYDRIN ALCOHOL ☐ 2,3-EPOXYPROPANOL ☐ 2,3-EPOXY-1-PROPANOL ☐ 2,3-EPOXY-1-PROPANOL (OSHA) ☐ GLYCIDE ☐ GLYCIDYL ALCOHOL ☐ 3-HYDROXY-1,2-EPOXYPROPANE ☐ METHANOL, OXIRANYL- ☐ NCI-C55549

TOXICITY DATA WITH REFERENCE
skn-rbt 558 mg/3D MOD AMIHAB 14,250,56
mmo-klp 200 µmol/L MUREAV 89,269,81
cyt-hmn:lym 400 µmol/L MUREAV 91,243,81
orl-mus TDLo:2 g/kg (6-15D preg):TER JTEHD6 9,87,82
orl-rat TDLo:180 mg/kg (12D male):REP NATUAS 226,87,70
orl-rat TDLo:9600 mg/kg/16D-I:REP NTPTR* NTP-TR-374,90
orl-rat TDLo:38,625 mg/kg/2Y-C:CAR NTPTR* NTP-TR-374,90
orl-mus TDLo:25,750 mg/kg/2Y-C:CAR NTPTR* NTP-TR-374,90
orl-rat LD50:420 mg/kg FCTXAV 19,347,81
ihl-rat LC50:580 ppm/8H AMIHAB 14,250,56
ipr-rat LD50:200 mg/kg FCTXAV 19,347,81
orl-mus LD50:431 mg/kg GTPZAB 24(3),42,80
ihl-mus LC50:450 ppm/4H AMIHAB 14,250,56
ipr-mus LDLo:500 mg/kg PSEBAA 35,98,36
skn-rbt LD50:1980 mg/kg AMIHAB 14,250,56

CONSENSUS REPORTS: NTP 7th Annual Report On Carcinogens. Reported in EPA TSCA Inventory. EPA Genetic Toxicology Program.

OSHA PEL: TWA 25 ppm
ACGIH TLV: TWA 25 ppm (Proposed: 2 ppm; Animal Carcinogen)
DFG MAK: 50 ppm (150 mg/m³)

SAFETY PROFILE: Confirmed carcinogen with carcinogenic data reported. Poison by intraperitoneal route. Moderately toxic by ingestion, inhalation, and skin contact. Experimental teratogenic and reproductive effects. A skin irritant. Human mutation data reported. Animal experiments suggest somewhat lower toxicity than for related epoxy compounds. Readily absorbed through the skin. Causes nervous excitation followed by depression. Explodes when heated or in the presence of strong acids, bases, metals (e.g., copper, zinc), and metal salts (e.g., aluminum chloride, iron(III) chloride, tin(IV) chloride). When heated to decomposition it emits acrid smoke and fumes. See also DIGLYCIDYL ETHER.

For occupational chemical analysis use NIOSH: Glycidol, 1608.

GGW800 **CAS:2917-91-1** **HR: 2**
N-GLYCIDYL DIETHYL AMINE
mf: $C_7H_{15}NO$ mw: 129.23

SYNS: 3-DIETHYLAMINO-1,2-EPOXYPROPANE ☐ EPIHYDRINAMINE, N,N-DIETHYL- ☐ N-(2,3-EPOXYPROPYL)DIETHYLAMINE ☐ GLYCIDYLDIETHYLAMINE ☐ PROPYLAMINE, 2,3-EPOXY-N,N-DIETHYL-

TOXICITY DATA WITH REFERENCE
skn-rbt 100 µg/24H open AIHAAP 23,95,62
skn-rbt 5 mg/24H SEV 85JCAE-,774,86
eye-rbt 1 mg MLD UCDS** 5/13/59
eye-rbt 250 µg/24H SEV 85JCAE-,774,86
orl-rat LD50:420 mg/kg UCDS** 5/13/59
ihl-rat LCLo:2000 ppm/4H AIHAAP 23,95,62
skn-rbt LD50:790 mg/kg AIHAAP 23,95,62

SAFETY PROFILE: Moderately toxic by ingestion and skin contact. A severe skin and eye irritant. When heated to decomposition it emits toxic fumes of NO_x.

G

GGY000 **CAS:17526-74-8** **HR: 2**
GLYCIDYL ESTER of HEXANOIC ACID
mf: $C_9H_{16}O_3$ mw: 172.25

TOXICITY DATA WITH REFERENCE
scu-rat TDLo:2000 mg/kg/7W-I:ETA ANYAA9 68,750,58

SAFETY PROFILE: Questionable carcinogen with experimental tumorigenic data. When heated to decomposition it emits acrid smoke and irritating fumes. See also ESTERS.

GGY100 **CAS:2461-15-6** **HR: 1**
GLYCIDYL 2-ETHYLHEXYL ETHER
mf: $C_{11}H_{22}O_2$ mw: 186.33

SYNS: 2-ETHYLHEXYL GLYCIDYL ETHER ☐ (((2-ETHYLHEXYL)OXY)METHYL)OXIRANE ☐ OXIRANE, (((2-ETHYLHEXYL)OXY)METHYL)-(9CI) ☐ PROPANE, 1,2-EPOXY-3-((2-ETHYLHEXYL)OXY)-

TOXICITY DATA WITH REFERENCE
mma-sat 100 µg/plate MUREAV 172,105,86
orl-rat LD50:7800 mg/kg 38MKAJ 2A,2210,81

CONSENSUS REPORTS: Reported in EPA TSCA Inventory.

SAFETY PROFILE: Slightly toxic by ingestion. Mutation data reported. When heated to decomposition it emits acrid smoke and irritating vapors.

GGY200 **CAS:3033-77-0** **HR: 3**
GLYCIDYL-TRIMETHYL-AMMONIUM CHLORIDE
mf: $C_6H_{14}NO{\cdot}Cl$ mw: 151.66

SYNS: (2,3-EPOXYPROPYL)TRIMETHYLAMMONIUM CHLORIDE ☐ GLYTAC ☐ GLYTAC A 100 ☐ G-MAC ☐ OXIRANEMETHANAMINIUM, N,N,N-TRIMETHYL-, CHLORIDE (9CI) ☐ TRIMETHYLGLYCIDYLAMMONIUM CHLORIDE ☐ N,N,N-TRIMETHYLOXIRANEMETHANAMINIUM CHLORIDE

TOXICITY DATA WITH REFERENCE
mmo-klp 2 mmol/L MUREAV 89,269,81
cyt-rat:lvr 10 mg/L MUREAV 153,57,85
scu-mus LD50:90 mg/kg JCSOA9 -,176,47

CONSENSUS REPORTS: Reported in EPA TSCA Inventory.

DFG MAK: Confirmed Animal Carcinogen, Suspected Human Carcinogen

SAFETY PROFILE: Suspected carcinogen. Poison by subcutaneous route. Mutation data reported. When heated to decomposition it emits toxic fumes of Cl^-, NH_3, and NO_x. See also AMMONIUM CHLORIDE.

GHA000 **CAS:56-40-6** **HR: 2**
GLYCINE
mf: $C_2H_5NO_2$ mw: 75.08

PROP: White crystals from alc (aq); odorless, sweet taste. The simplest amino acid and the principal amino acid in sugar cane. Mp: 262° (decomp), d: 1.1607. Sol in water; insol in alc and ether.

SYNS: AMINOACETIC ACID ☐ GLYCOLIXIR ☐ HAMPSHIRE GLYCINE

TOXICITY DATA WITH REFERENCE
sce-hmn:lym 100 mg/L MUREAV 280,279,92
orl-rat LD50:7930 mg/kg YACHDS 5,1502,77
scu-rat LD50:5200 mg/kg YACHDS 5,1502,77
ivn-rat LD50:2600 mg/kg YACHDS 5,1502,77
orl-mus LD50:4920 mg/kg YACHDS 5,1502,77
ipr-mus LD50:4450 mg/kg YACHDS 5,1502,77
scu-mus LD50:5060 mg/kg YACHDS 5,1502,77
ivn-mus LD50:2370 mg/kg YACHDS 5,1502,77
ivn-cat LDLo:3000 mg/kg JAPMA8 31,306,42

CONSENSUS REPORTS: Reported in EPA TSCA Inventory.

SAFETY PROFILE: Moderately toxic by intravenous route. Mildly toxic by ingestion. Mutation data reported. When heated to decomposition it emits toxic fumes of NO_x.

GHA050 **CAS:107-43-7** **HR: 2**
GLYCINE BETAINE
mf: $C_5H_{11}NO_2$ mw: 117.17

SYNS: ABROMINE ☐ BETAINE ☐ (CARBOXYMETHYL)TRIMETHYLAMMONIUM HYDROXIDE, inner salt ☐ α-EARLEINE ☐ GLYCOCOLL BETAINE ☐ GLYCYLBETAINE ☐ GLYKOKOLLBETAIN ☐ JORTAINE ☐ LORAMINE AMB 13 ☐ LYCINE ☐ OXYNEURINE ☐ RUBRINE C ☐ TRIMETHYLGLYCINE ☐ TRIMETHYLGLYCOCOLL

TOXICITY DATA WITH REFERENCE
scu-mus LD50:10,800 mg/kg ABMGAJ 3,28,59
ivn-mus LD50:830 mg/kg MPHEAE 16,529,67

CONSENSUS REPORTS: Reported in EPA TSCA Inventory.

SAFETY PROFILE: Moderately toxic by intravenous route. Mildly toxic by subcutaneous route. When heated to decomposition it emits toxic vapors of NO_x.

GHA100 **CAS:623-33-6** **HR: 2**
GLYCINE, ETHYL ESTER, HYDROCHLORIDE
mf: $C_4H_9NO_2•ClH$ mw: 139.60

SYN: USAF DO-10

TOXICITY DATA WITH REFERENCE
ipr-mus LD50:750 mg/kg NTIS** AD277-689

CONSENSUS REPORTS: Reported in EPA TSCA Inventory.

SAFETY PROFILE: Moderately toxic by intraperitoneal route. When heated to decomposition it emits toxic vapors of NO_x, HCl, and Cl^-.

GHE000 **CAS:2619-97-8** **HR: 3**
GLYCINE NITROGEN MUSTARD
mf: $C_6H_{11}Cl_2NO_2•ClH$ mw: 236.54

SYNS: N,N-BIS(β-CHLOROETHYL)GLYCINE HYDROCHLORIDE ☐ N,N-BIS(2-CHLOROETHYL)GLYCINE HYDROCHLORIDE ☐ GLYCINE MUSTARD ☐ NSC 17661

TOXICITY DATA WITH REFERENCE
ice-rat LD50:113 μg/kg JPPMAB 18,760,66
ipr-rat LD50:15 mg/kg PHBUA9 2,275,54
ipr-mus LD50:9700 μg/kg NCISA* PH-43-63-1132
ivn-dog LDLo:1 mg/kg CCSUBJ 2,201,65
ivn-mky LDLo:1 mg/kg CCSUBJ 2,201,65

SAFETY PROFILE: A deadly poison by intraperitoneal, intravenous, and intracerebral routes. When heated to decomposition it emits very toxic fumes of Cl^- and NO_x.

GHG000 **CAS:6000-44-8** **HR: 2**
GLYCINE, SODIUM SALT
mf: $C_2H_4NO_2•Na$ mw: 97.06

TOXICITY DATA WITH REFERENCE
ivn-mus LD50:564 mg/kg RPOBAR 2,292,70

CONSENSUS REPORTS: Reported in EPA TSCA Inventory.

SAFETY PROFILE: Moderately toxic by intravenous route. When heated to decomposition it emits toxic fumes of NO_x and Na_2O. See also GLYCINE.

GHK000 **CAS:6000-43-7** **HR: 2**
GLYCOCOIL HYDROCHLORIDE
mf: $C_2H_5NO_2•ClH$ mw: 111.54

SYN: GLYCOHYDROCHLORIDE

TOXICITY DATA WITH REFERENCE
orl-rat LD50:3340 mg/kg JPMSAE 62,49,73

CONSENSUS REPORTS: Reported in EPA TSCA Inventory.

SAFETY PROFILE: Moderately toxic by ingestion. When heated to decomposition it emits very toxic fumes of Cl^- and NO_x.

GHK200 **CAS:3459-20-9** **HR: 2**
GLYCODIAZINE SODIUM SALT
mf: $C_{13}H_{14}N_3O_4S•Na$ mw: 331.35

PROP: A solid. Mp: 221–222° (sesquihydrate mp: 86°).

SYNS: 2-BENZENESULFONAMIDO-5-(β-METHOXYETHOXY)PYRIMIDINE SODIUM SALT ☐ GLYCONORMAL ☐ GLYMIDINE SODIUM SALT ☐ GONDAFON ☐ LYCANOL ☐ N-(5-(2-METHOXYETHOXY)-2-PYRIMIDINYL)BENZENESULFONAMIDE SODIUM SALT ☐ REDUL ☐ SH 717

TOXICITY DATA WITH REFERENCE
orl-rat LD50:2850 mg/kg ARZNAD 14,377,64

ipr-rat LD50:3120 mg/kg NIIRDN 6,221,82
scu-rat LD50:2800 mg/kg NIIRDN 6,221,82
ivn-rat LD50:2000 mg/kg ARZNAD 14,377,64
orl-mus LD50:5300 mg/kg ARZNAD 14,377,64
ipr-mus LD50:3210 mg/kg NIIRDN 6,221,82
scu-mus LD50:3340 mg/kg NIIRDN 6,221,82
ivn-mus LD50:1480 mg/kg ARZNAD 14,377,64

SAFETY PROFILE: Moderately toxic by ingestion, intraperitoneal, subcutaneous, and intravenous routes. When heated to decomposition it emits toxic fumes of SO_x, NO_x, and Na_2O.

GHK300 **CAS:9005-79-2** **HR: D**
GLYCOGEN

SYNS: ANIMAL STARCH □ LIVER STARCH □ LYOGLYCOGEN □ PHYTOGLYCOGEN

TOXICITY DATA WITH REFERENCE
par-mus TDLo:160 μg/kg (female 2-9D post):REP SCYYDZ 11(4),7,91

CONSENSUS REPORTS: Reported in EPA TSCA Inventory.

SAFETY PROFILE: Experimental reproductive effects. When heated to decomposition it emits acrid smoke and irritating vapors.

GHK500 **CAS:4746-61-6** **HR: 2**
GLYCOLANILIDE
mf: $C_8H_9NO_2$ mw: 151.18

SYN: 2-HYDROXY-N-PHENYLACETAMIDE

TOXICITY DATA WITH REFERENCE
orl-hmn TDLo:14,286 μg/kg:PUL JAPMA8 35,50,46
orl-rat LD50:1700 mg/kg JAPMA8 35,50,46
orl-mus LD50:2300 mg/kg JAPMA8 35,50,46

SAFETY PROFILE: Moderately toxic by ingestion. Human systemic effects by ingestion: cyanosis. When heated to decomposition it emits toxic fumes of NO_x.

GHM000 **CAS:96-49-1** **HR: 2**
GLYCOL CARBONATE
mf: $C_3H_4O_3$ mw: 88.07

PROP: Colorless liquid or crystalline solid. Needles from Et_2O. Mp: 38.5–39°, fp: 35.7°, bp: 244° @ 740 mm, flash p: 290°F (OC), d: 1.322 @ 40°/20°, vap press: 0.01 mm @ 20°, vap d: 3.04.

SYNS: CARBONIC ACID, CYCLIC ETHYLENE ESTER □ CYCLIC ETHYLENE CARBONATE □ 1,3-DIOXOLAN-2-ONE □ DIOXOLONE-2 □ ETHYLENE CARBONATE □ ETHYLENE CARBONIC ACID □ ETHYLENE GLYCOL CARBONATE □ ETHYLENE GLYCOL, CYCLIC CARBONATE

TOXICITY DATA WITH REFERENCE
skn-rbt 660 mg open MLD UCDS** 7/21/71
eye-rbt 20 mg open AMIHBC 10,61,54
eye-rbt 100 mg MOD 34ZIAG -,255,69
orl-rat LD50:10 g/kg UCDS** 7/21/71
ipr-mus LDLo:500 mg/kg CBCCT* 5,338,53

CONSENSUS REPORTS: Reported in EPA TSCA Inventory.

SAFETY PROFILE: Moderately toxic by intraperitoneal route. Mildly toxic by ingestion. A skin and eye irritant. Combustible when exposed to heat or flame; can react with oxidizing materials. To fight fire, use alcohol foam, CO_2, dry chemical. When heated to decomposition it emits acrid smoke and irritating fumes.

GHN000 **HR: 3**
GLYCOL ETHERS

CONSENSUS REPORTS: Glycol ether compounds are on the Community Right-To-Know List.

SAFETY PROFILE: The acute toxic effects of ethylene glycol monomethyl ether (2-methoxyethanol or 2ME) in humans are irritation of the eyes, nose, and throat; drowsiness; weakness; and shaking. Ingestion of 2ME may be fatal. Prolonged or repeated exposures may cause headache, drowsiness, weakness, fatigue, staggering, personality change, and decreased mental ability. Exposed workers have suffered encephalopathy (degenerative brain disease), bone marrow depression, and pancytopenia (reduced levels of all blood cells). 2ME and cellosolve (2-ethoxyethanol or 2EE) have the potential to cause adverse reproductive effects in male and female workers. They have been shown to cause embryotoxicity and other reproductive effects in several species of animals exposed by different routes of administration. The exposure of pregnant animals to concentrations of 2ME or 2EE at or below their OSHA permissible exposure limits led to increased incidences of embryonic death, teratogenesis, or growth retardation. Exposure of male animals resulted in testicular atrophy and sterility. They can be absorbed through the skin. Structurally related glycol ethers are 2-methoxyethyl acetate; 2-ethoxyethyl acetate; 2-butoxyethanol; 2-phenoxyethanol; ethylene glycol dimethyl ether; bis(2-methoxyethyl)ether; 2-(2-ethoxyethoxy)ethanol; 1-methoxy-2-propanol; propylene glycol monomethyl ether. Although there is limited experimental information on the reproductive effects of these individual compounds, much of the information that is available is consistent with the reproductive effects caused by 2ME and 2EE. The acetate esters of 2ME and 2EE (2MEA and 2EEA) have caused male reproductive toxicity equivalent to that of 2ME and 2EEK in male mice. 2EEA appears to have fetotoxicity and teratogenicity equivalent to that of 2EE. Flammable or combustible when exposed to heat or flame; can react vigorously with oxidizing materials. When heated to decomposition they emit acrid smoke and fumes.

GHO000 **CAS:79-14-1** **HR: 2**
GLYCOLIC ACID
mf: $C_2H_4O_3$ mw: 76.06

PROP: Rhombic leaflets from ether; needles from water. Odorless. Bp: decomp, mp (α): 63°, mp (β): 79°. Sol in H_2O, methanol, alc, acetone, acetic acid, ether.

SYNS: HYDROXYACETIC ACID □ HYDROXYETHANOIC ACID

TOXICITY DATA WITH REFERENCE
eye-rbt 2 mg SEV AJOPAA 29,1363,46
orl-rat LD50:1950 mg/kg JIHTAB 23,259,41
orl-gpg LD50:1920 mg/kg JIHTAB 23,259,41

CONSENSUS REPORTS: Reported in EPA TSCA Inventory.

SAFETY PROFILE: Moderately toxic by ingestion. A severe eye irritant. A skin and mucous membrane irritant. When heated to decomposition it emits acrid smoke and irritating fumes.

GHQ000 CAS:5847-48-3 HR: 3
(GLYCOLOYLOXY)TRIBUTYLSTANNANE
mf: $C_{14}H_{30}O_3Sn$ mw: 365.13

SYNS: TRIBUTYL(GLYCOLOYLOXY)STANNANE □ TRIBUTYL(GLYCOLOYLOXY)TIN

TOXICITY DATA WITH REFERENCE
orl-mus LDLo:470 mg/kg AECTCV 14,111,85
ivn-mus LD50:18 mg/kg CSLNX* NX#03601

OSHA PEL: TWA 0.1 mg(Sn)/m³ (skin)
ACGIH TLV: TWA 0.1 mg(Sn)/m³ (skin) (Proposed: TWA 0.1 mg(Sn)/m³; STEL 0.2 mg(Sn)/m³ (skin))
NIOSH REL: (Organotin Compounds) TWA 0.1 mg(Sn)/m³

SAFETY PROFILE: Poison by intravenous route. Moderately toxic by ingestion. Tributyl tin compounds are extremely toxic to marine life. When heated to decomposition it emits acrid smoke and irritating fumes. See also TIN COMPOUNDS.

For occupational chemical analysis use NIOSH: Organotin Compounds 5504.

GHR609 CAS:631-27-6 HR: 2
GLYCOLPYRAMIDE
mf: $C_{11}H_{14}ClN_3O_3S$ mw: 303.79

PROP: Plates from EtOH (aq). Mp: 199–201°.

SYNS: 4-CHLORO-N-((1-PYRROLIDINYLAMINO)CARBONYL)BENZENESULFONAMIDE (9CI) □ CPBU 7 □ DEAMELIN S

TOXICITY DATA WITH REFERENCE
orl-mus TDLo:560 mg/kg (female 7-13D post):REP KSRNAM 11,1620,77
orl-rat TDLo:900 mg/kg (female 30D pre):REP KSRNAM 11,1605,77
orl-rat TDLo:4500 mg/kg (male 30D pre):REP KSRNAM 11,1605,77
orl-rat LD50:4100 mg/kg KSRNAM 11,1605,77
ipr-rat LD50:1620 mg/kg KSRNAM 11,1605,77
orl-mus LD50:8910 mg/kg KSRNAM 11,1605,77
ipr-mus LD50:1600 mg/kg KSRNAM 11,1605,77
scu-mus LD50:5500 mg/kg KSRNAM 11,1605,77

SAFETY PROFILE: Moderately toxic by intraperitoneal route. Mildly toxic by ingestion. Experimental reproductive effects. When heated to decomposition it emits toxic fumes of Cl^-, NO_x, and SO_x.

GHS000 CAS:9036-19-5 HR: 3
GLYCOLS, POLYETHYLENE, MONO((1,1,3,3-TETRAMETHYLBUTYL)PHENYL) ETHER
mf: $(C_2H_4O)_n$ $C_{14}H_{22}O$

SYNS: CHARGER E □ ETHOXYLATED OCTYL PHENOL □ ETHYLAN CP □ IGEPAL CA □ NEUTRONYX 622 □ NONIDET P40 □ NONION HS 206 □ OCTYLPHENOXYPOLY(ETHOXYETHANOL) □ tert-OCTYLPHENOXYPOLY(ETHOXYETHANOL) □ OCTYLPHENOXYPOLY(ETHYLENEOXY)ETHANOL □ tert-OCTYLPHENOXY POLY(OXYETHYLENE)ETHANOL □ OP 1062 □ POLYETHYLENE GLYCOL MONO(OCTYLPHENYL) ETHER □ POLYETHYLENE GLYCOL OCTYLPHENYL ETHER □ POLY(ETHYLENE OXIDE)OCTYLPHENYL ETHER □ POLYOXYETHYLENE MONOOCTYLPHENYL ETHER □ POLY(OXYETHYLENE)OCTYLPHENOL ETHER □ SECOPAL OP 20 □ SYNPERONIC OP □ T 45 (POLYGLYCOL) □ α-((1,1,3,3-TETRAMETHYLBUTYL)PHENYL)-ω-HYDROXY-POLY(OXY-1,2-ETHANEDIYL) □ TRITON X 15

TOXICITY DATA WITH REFERENCE
eye-rbt 1% SEV JAPMA8 38,428,49
dni-hmn:lym 5 ppm ENPBBC 5,84,75
dni-mus:oth 10 ppm ENPBBC 5,84,75
orl-rat LD50:4190 mg/kg FCTOD7 22,665,84
ipr-rat LD50:770 mg/kg FCTOD7 22,665,84
orl-mus LD50:3500 mg/kg JAPMA8 38,428,49
ivn-mus LD50:70 mg/kg JAPMA8 38,428,49

CONSENSUS REPORTS: Reported in EPA TSCA Inventory. Glycol ether compounds are on the Community Right-To-Know List.

SAFETY PROFILE: Poison by intravenous route. Moderately toxic by ingestion and intraperitoneal routes. Human mutation data reported. A severe eye irritant. When heated to decomposition it emits acrid smoke and irritating fumes. See also GLYCOL ETHERS.

GHY000 CAS:9038-95-3 HR: 1
GLYCOLS, POLYETHYLENE POLYPROPYLENE, MONOBUTYL ETHER (nonionic)

SYNS: TERGITOL NONIONIC XD □ TERGITOL XD (nonionic)

TOXICITY DATA WITH REFERENCE
skn-rbt 500 mg open MLD UCDS** 9/22/64
orl-rat LD50:12 g/kg UCDS** 9/22/64

CONSENSUS REPORTS: Reported in EPA TSCA Inventory. Glycol ether compounds are on the Community Right-To-Know List.

SAFETY PROFILE: Mildly toxic by ingestion. A skin irritant. When heated to decomposition it emits acrid smoke and irritating fumes. See also GLYCOL ETHERS.

GIA000 CAS:36734-19-7 HR: 2
GLYCOPHEN
mf: $C_{13}H_{13}Cl_2N_3O_3$ mw: 330.19

PROP: Crystals. Mp: 136°.

SYNS: CHIPCO 26019 □ 3-(3,5-DICHLOROPHENYL)-N-(1-METHYLETHYL)-2,4-DIOXO-1-IMIDAZOLIDINECARBOXAMIDE □ GLYCOPHENE □ IPRODIONE □ 1-ISOPROPYL CARBAMOYL-3-(3,5-DICHLOROPHENYL)-HYDANTOIN □ LFA 2043 □ MRC 910 □ PROMIDIONE □ ROP 500 F □ ROVRAL □ RP 26019

TOXICITY DATA WITH REFERENCE
orl-rat LD50:4400 mg/kg FMCHA2 -,C132,83
orl-mus LD50:4000 mg/kg FMCHA2 -,C132,83

SAFETY PROFILE: Moderately toxic by ingestion. When heated to decomposition it emits very toxic fumes of NO_x and Cl^-.

GIC000 **CAS:596-51-0** **HR: 3**
GLYCOPYRRONIUM BROMIDE
mf: $C_{19}H_{28}NO_3$•Br mw: 398.39

PROP: Crystals from butanone. Mp: 193.2–194.5°.

SYNS: ASECRYL ☐ 1,1-DIMETHYL-3-HYDROXYPYRROLIDINIUM BROMIDE-α-CYCLOPENTYLMANDELATE ☐ GASTRODYN ☐ GLYCOPYRROLATE ☐ GLYCOPYRROLATE BROMIDE ☐ NODAPTON ☐ ROBANUL ☐ ROBINUL ☐ TARODYL ☐ TARODYN

TOXICITY DATA WITH REFERENCE
orl-rat TDLo:4004 mg/kg (26W pre):REP OYYAA2 7,627,73
orl-mus TDLo:10 mg/kg (1D post):REP OYYAA2 15,721,78
ipr-rat LD50:196 mg/kg TXAPA9 17,361,70
scu-rat LD50:833 mg/kg NIIRDN 6,349,82
orl-mus LD50:570 mg/kg JMPCAS 2,523,60
ipr-mus LD50:90 mg/kg JMPCAS 2,523,60
scu-mus LD50:122 mg/kg YKYUA6 26,741,75
ivn-mus LD50:15 mg/kg 29ZVAB -,55,69
orl-rbt LD50:2360 mg/kg OYYAA2 7,627,73
ivn-rbt LD50:29,100 μg/kg OYYAA2 7,627,73

SAFETY PROFILE: Poison by intravenous and intraperitoneal routes. Moderately toxic by ingestion and subcutaneous routes. Experimental reproductive effects. When heated to decomposition it emits very toxic fumes of NO_x and Br^-. See also BROMIDES.

GIE000 **CAS:471-53-4** **HR: 3**
α-GLYCYRRHETINIC ACID
mf: $C_{30}H_{46}O_4$ mw: 470.76

PROP: Crystals from MeOH. Mp: 300–304°.

SYNS: BIOSONE ☐ ENOXOLONE ☐ 3-β-HYDROXY-11-OXOOLEAN-12-EN-30-OIC ACID ☐ GLYCYRRHETIC ACID ☐ 18-β-GLYCYRRHETIC ACID ☐ GLYCYRRHETIN ☐ GLYCYRRHETINIC ACID ☐ 18-β-GLYCYRRHETINIC ACID ☐ URALENIC ACID

TOXICITY DATA WITH REFERENCE
ipr-mus LD50:308 mg/kg DRUGAY-,319,90
ivn-mus LD50:56 mg/kg CSLNX* NX#02067

CONSENSUS REPORTS: Reported in EPA TSCA Inventory.

SAFETY PROFILE: Poison by intravenous and intraperitoneal routes. When heated to decomposition it emits acrid smoke and irritating fumes.

GIE050 **CAS:1449-05-4** **HR: 3**
β-GLYCYRRHETINIC ACID
mf: $C_{30}H_{46}O_4$ mw: 470.76

PROP: Crystals from $CHCl_3$/MeOH. Mp: 330–335°.

SYN: 3-β-HYDROXY-11-OXO-18-α-OLEAN-12-EN-30-OIC ACID

TOXICITY DATA WITH REFERENCE
orl-mus LD50:560 mg/kg CPBTAL 28,3449,80
ipr-mus LD50:455 mg/kg CPBTAL 28,3449,80
ivn-mus LD50:100 mg/kg CSLNX* NX#02068

SAFETY PROFILE: Poison by intravenous route. Moderately toxic by ingestion and intraperitoneal routes. When heated to decomposition it emits acrid smoke and irritating fumes.

GIE100 **CAS:53956-04-0** **HR: 2**
GLYCYRRHIZIC ACID, AMMONIUM SALT
mf: $C_{42}H_{63}O_{16}$•xH_3N mw: 943.33

PROP: Needles from alc (aq).

SYNS: AMMONIATED GLYCYRRHIZIN ☐ AMMONIUM GLYCYRRHIZINATE ☐ α-D-GLUCOPYRANOSIDURONIC ACID, (3-β,20-β)-20-CARBOXY-11-OXO-30-NOROLEAN-12-EN-3-YL 2-O-β-D-GLUCOPYRANURONOSYL-, AMMONIATE ☐ MONOAMMONIUM GLYCYRRHIZINATE

TOXICITY DATA WITH REFERENCE
dlt-rat-orl 54,600 mg/kg/10W-C ENMUDM 8,357,86
dlt-mus-orl 350 g/kg/10W-C NTIS** PB279-650
orl-rat TDLo:350 g/kg (male 10W pre):REP ENMUDM 8,357,86
orl-rat TDLo:256 mg/kg (female 7-17D post):TER FCTOD7 26,435,88
ipr-mus LDLo:1 g/kg YHTPAD 22,449,87
ivn-mus LD50:540 mg/kg YHTPAD 22,449,87

CONSENSUS REPORTS: Reported in EPA TSCA Inventory.

SAFETY PROFILE: Moderately toxic by intravenous route. An experimental teratogen. Experimental reproductive effects. Mutation data reported. When heated to decomposition it emits toxic fumes of NH_3.

GIG000 **CAS:1405-86-3** **HR: 3**
GLYCYRRHIZINIC ACID
mf: $C_{42}H_{62}O_{16}$ mw: 823.04

PROP: Crystals from glacial acetic acid or hygroscopic powder. Mp: 220° (approx). Intensely sweet taste. Freely sol in hot water, alc; practically insol in ether. The active component of licorice (BMJOAE 1,488,77).

SYNS: GLYCYRON ☐ GLYCYRRHETINIC ACID GLYCOSIDE ☐ GLYCYRRHIZIC ACID ☐ GLYCYRRHIZIC ACID (8CI) ☐ 18-β-GLYCYRRHIZIC ACID ☐ GLYCYRRHIZIN ☐ β-GLYCYRRHIZIN ☐ GLYCYRON ☐ GLYCYRRHETINIC ACID GLYCOSIDE ☐ LIQUORICE

TOXICITY DATA WITH REFERENCE
orl-hmn TDLo:280 mg/kg/4W:CNS,MET BMJOAE 1,488,77
orl-rat LDLo:3 g/kg YACHDS 5,2041,77
ipr-rat LDLo:2 g/kg YACHDS 5,2041,77
orl-mus LDLo:4 g/kg YACHDS 5,2041,77
ipr-mus LDLo:1 g/kg YACHDS 5,2041,77
ivn-mus LDLo:300 mg/kg YACHDS 5,2041,77

CONSENSUS REPORTS: Reported in EPA TSCA Inventory.

SAFETY PROFILE: Poison by intravenous route. Moderately toxic by ingestion and intraperitoneal routes. Human systemic effects by ingestion: somnolence and

G

changes in the metabolism of phosphorus. When heated to decomposition it emits acrid smoke and irritating fumes.

GII000 **CAS:556-22-9** **HR: 2**
GLYODIN
mf: $C_{20}H_{40}N_2{\cdot}C_2H_4O_2$ mw: 368.68

PROP: Light orange crystals. Mp: 62–68°, d: 1.035 @ 20°. The base is a soft greasy wax (mp: 94°). Insol in water, acetone, toluene; sol in isopropanol.

SYNS: CRAG 341 □ CRAG FRUIT FUNGICIDE 341 □ EXPERIMENTAL FUNGICIDE 341 □ GLYODIN ACETATE □ GLYOXIDE □ GLYOXIDE DRY □ 2-HEPTADECYL-4,5-DIHYDRO-1H-IMIDAZOLYL MONOACETATE □ 2-HEPTADECYL GLYOXALIDINE ACETATE □ 2-HEPTADECYL-2-IMIDAZOLINE ACETATE

TOXICITY DATA WITH REFERENCE
orl-rat LD50:4600 mg/kg FMCHA2 -,C116,83
unr-mam LD50:1000 mg/kg 30ZDA9 -,416,71

CONSENSUS REPORTS: Reported in EPA TSCA Inventory.

SAFETY PROFILE: Moderately toxic by unspecified route. Mildly toxic by ingestion. A skin, eye, and mucous membrane irritant. A fungicide that can damage the cornea. When heated to decomposition it emits toxic fumes of NO_x.

GIK000 **CAS:107-22-2** **HR: 3**
GLYOXAL
mf: $C_2H_2O_2$ mw: 58.04

PROP: Yellow prisms or irregular pieces turning white on cooling. D: 1.29 @ 20°/4°. Opaque @ 10°, mp: 15°, bp: (776) 51°. The vapors are green and burn with a purple flame, n (20.5/D) 1.3826. Sol in anhyd solvents, pH of a 40% aq soln: 2.1–2.7; d: (20/4) 1.27.

SYN: AEROTEX GLYOXAL 40

TOXICITY DATA WITH REFERENCE
skn-rbt 258 mg open MLD UCDS** 11/12/65
orl-rat LD50:7070 mg/kg UCDS** 11/12/65
skn-rbt LD50:10 g/kg UCDS** 11/12/65

CONSENSUS REPORTS: Reported in EPA TSCA Inventory.

SAFETY PROFILE: Low toxicity by ingestion and skin contact. A skin irritant. A powerful reducing agent. May explode on contact with air. Polymerizes violently on contact with water. During storage it may spontaneously polymerize and ignite. Reacts violently with chlorosulfonic acid; ethylene imine; HNO_3; oleum; NaOH; can cause violent reactions. Can explode during manufacture. When heated to decomposition it emits acrid smoke and irritating fumes. See also ALDEHYDES.

GIO000 **CAS:105-28-2** **HR: 2**
GLYOXIDE
mf: $C_{20}H_{40}N_2$ mw: 308.62

SYNS: CRAG FRUIT FUNGICIDE 34 □ GLIODIN □ GLYODIN □ GLYOXALIDIN □ 2-HEPTADECYL GLYOXALIDINE □ 2-HEPTADECYL-2-IMIDAZOLINE

TOXICITY DATA WITH REFERENCE
eye-rbt 250 µg SEV AJOPAA 29,1363,46
orl-rat LD50:3170 mg/kg WRPCA2 9,119,70

CONSENSUS REPORTS: Reported in EPA TSCA Inventory.

SAFETY PROFILE: Moderately toxic by ingestion. A severe eye irritant. A fruit fungicide. When heated to decomposition it emits toxic fumes of NO_x.

GIQ000 **CAS:298-12-4** **HR: 3**
GLYOXYLIC ACID
mf: $C_2H_2O_3$ mw: 74.04

PROP: In anhydrous form as monoclinic crystals from water. Very deliquescent prisms giving yellow aq soln. Mp: 104–107°. D: 1.42 @ 20°/4°, mp: 73°. Deliquesces quickly and forms a syrup on short exposure to air. Freely sol in water; insol in ether and hydrocarbons.

SYN: KYSELINA GLYOXYLOVA

TOXICITY DATA WITH REFERENCE
mmo-sat 200 µg/plate ABCHA6 47,2461,83
ims-rat LDLo:25 mg/kg BSIBAC 36,1937,60

CONSENSUS REPORTS: Reported in EPA TSCA Inventory.

SAFETY PROFILE: Poison by intramuscular route. Mutation data reported. A skin, eye, and mucous membrane irritant. When heated to decomposition it emits acrid smoke and fumes.

GIS000 **CAS:7440-57-5** **HR: 1**
GOLD
af: Au aw: 196.97

PROP: Cubic, yellow, ductile, metallic crystals. Forms red, blue, or violet colloidal suspensions. Physical properties depend on mechanical treatment. Mp: 1064.76°, bp: 2700°, d: 19.3 (liquid) 17.0 @ 1063°, vap press: 1 mm @ 1869°, hardness: (Mohs') 2.5–3.0, (Brinell's) 18.5.

SYNS: BURNISH GOLD □ C.I. 77480 □ C.I. PIGMENT METAL 3 □ COLLOIDAL GOLD □ GOLD FLAKE □ GOLD LEAF □ GOLD POWDER □ MAGNESIUM GOLD PURPLE □ SHELL GOLD

TOXICITY DATA WITH REFERENCE
imp-rat TDLo:200 mg/kg:ETA NATWAY 42,75,55
imp-mus TDLo:21 g/kg:ETA NATWAY 42,75,55
imp-rat TD:4730 mg/kg:ETA NATWAY 42,75,55
ivn-rat LDLo:58 mg/kg ZEKBAI 63,586,60

CONSENSUS REPORTS: Reported in EPA TSCA Inventory.

SAFETY PROFILE: Poison by intravenous route. Questionable carcinogen with experimental tumorigenic data by implantation. Can form explosive compounds with NH_3, NH_4OH + aqua regia, H_2O_2. Incompatible with mixtures containing chlorides, bromides, or iodides (if they can generate nascent halogens), some oxidizing materials (especially those containing halogens), alkali cyanides, thiocyanate solutions, and double cyanides. See also GOLD COMPOUNDS.

GIT000 **CAS:70950-00-4** **HR: 3**
GOLD(I) ACETYLIDE
mf: C_2Au_2 mw: 417.96

AuC≡CAu

PROP: Yellow powder. Insol in common org solvs.

SAFETY PROFILE: An unstable explosive with high shattering power. It is easily detonated by light, impact, friction, or rapid heating to 83°C. See also GOLD COMPOUNDS and EXPLOSIVES.

GIW176 **CAS:13453-07-1** **HR: 2**
GOLD CHLORIDE
mf: $AuCl_3$ mw: 303.33

PROP: Claret-red crystals; mp: 254° (decomp), bp: subl at 265°, d: 3.9.

SYNS: AURIC CHLORIDE □ GOLD(III) CHLORIDE □ GOLD TRICHLORIDE

TOXICITY DATA WITH REFERENCE
dns-hmn:lym 6 mg/L IAAAAM 77,459,85
scu-mus TDLo:22,106 μg/kg (30D male):REP JRPFA4 7,21,64

CONSENSUS REPORTS: Reported in EPA TSCA Inventory.

SAFETY PROFILE: Experimental reproductive effects. Human mutation data reported. Reaction with ammonia or ammonium salts yields fulminating gold, a heat-, friction-, and impact-sensitive explosive similar to mercury and silver fulminates. See also GOLD COMPOUNDS and CHLORIDES. When heated to decomposition it emits toxic fumes of Cl^-.

GIW179 **HR: 1**
GOLD COMPOUNDS

SAFETY PROFILE: Gold poisoning is rare. The few recorded cases of fatalities are the result of therapeutic overdose. Human systemic effects are similar to those of arsenic exposure and include: violent diarrhea, gastritis, colitis, dermatitis, blood dyscrasias, leukopenia, agranulocytosis, and aplastic anemia. The therapeutic use of gold compounds has been associated with serious effects of the kidney, liver, and other vital organs. Gold sodium thiomalate, aurothioglucose, gold thioglycoanilid, and gold sodium thiosulfate are used to treat rheumatoid arthritis and lupus erythematosus. Generally, gold compounds are poorly absorbed when ingested. Effects are usually greater by intramuscular and intravenous routes.

GIW189 **CAS:506-65-0** **HR: 3**
GOLD(I) CYANIDE
mf: CAuN mw: 222.98

CONSENSUS REPORTS: Cyanide and its compounds are on the Community Right-To-Know List.

PROP: Lemon-yellow crystalline powder. Stable in air. Unstable to light when moist. Sltly sol in H_2O; sol in alkali cyanide solns.

SAFETY PROFILE: Explosive reaction when heated with magnesium. When heated to decomposition it emits toxic fumes of CN^- and cyanogen. See also CYANIDE and GOLD COMPOUNDS.

GIW195 **HR: 2**
GOLDEN CHAIN

PROP: A tree that may grow to 30 feet. It blooms in masses of golden flowers, and produces bean-type seedpods.

SYNS: BEAN TREE □ LABURNUM □ LABURNUM ANAGYROIDES

SAFETY PROFILE: All parts of the plant and particularly the seeds contain the poison cytisine (related to nicotine). Ingestion of any part of the plant rapidly causes vomiting, sleepiness, poor muscular control, headache, sweating, dilated pupils, and increased heart rate. See also CYTISINE, various cytisine compounds, and NICOTINE.

GIW200 **HR: 3**
GOLDEN DEWDROP

PROP: A large shrub commonly cultivated as a hedge. It produces small light blue or white flowers and masses of orange berries. It grows wild in southern Florida and southern Texas of the United States, and the West Indics.

SYNS: AZOTA CABALLO □ BOIS JAMBETTE (HAITI) □ CUENTAS de ORO (PUERTO RICO) □ DURANTA REPENS □ GARBANCILLO (CUBA) □ MAIS BOUILLI (HAITI) □ PIGEON BERRY □ SKY FLOWER □ VELO de NOVIA (MEXICO)

SAFETY PROFILE: The berries contain poisonous saponins. Ingestion of the berries can cause sleepiness, fever, increased heart rate, and convulsions. See also SAPONIN.

GIW300 **HR: 2**
GOLDEN SHOWER

PROP: A tropical tree (up to 30 feet tall) with long leaves composed of 4 to 8 pairs of 2- to 6-inch long leaflets. It produces large numbers of gold flowers on drooping racemes. The fruit is in the form of a long, thin pod holding up to 100 flat seeds embedded in a sticky matrix. The tree is cultivated in the United States in south Florida, the southern coast of California, and Hawaii, and the West Indies. Related species are found in the West Indies, Hawaii, and Guam.

SYNS: CANAFISTOLA (CUBA) □ CASSE (HAITI) □ CASSIA FISTULA □ GOLDEN RAIN □ INDIAN LABURNUM □ PUDDING-PIPE TREE □ PURGING FISTULA

SAFETY PROFILE: The sticky pulp of the seeds, and to a lesser extent the leaves and bark, contains toxic emodin glycosides. Human systemic effects by ingestion include nausea, vomiting, abdominal cramps, dizziness, and diarrhea. Emodin also causes a harmless discoloration of the urine.

GIX300 **HR: 3**
GOLD(III) HYDROXIDE-AMMONIA
mf: $2AuH_3O_3 \cdot 3H_3N$ mw: 547.07

SAFETY PROFILE: A sensitive explosive. Heating in water forms the more explosive $Au_2O_3 \cdot 2NH_3$. Dry heating first forms the explosives $Au_2O_3 \cdot 3NH_3$ and $Au_2O \cdot 4NH_3$. When heated to decomposition it emits toxic fumes of NH_3. See also GOLD COMPOUNDS and EXPLOSIVES.

GIY300 **HR: 3**
GOLD(I) NITRIDE-AMMONIA
mf: $Au_2N_3 \cdot H_3N$ mw: 452.98

SAFETY PROFILE: An explosive. Upon decomposition it emits toxic fumes of NO_x and NH_3. See also NITRIDES, EXPLOSIVES, and other gold compounds.

GIZ000 **HR: 3**
GOLD(III) NITRIDE TRIHYDRATE
mf: $Au_3N_2 \cdot 3H_2O$ mw: 672.96

SAFETY PROFILE: Very explosive when dry. When heated to decomposition it emits toxic fumes of NO_x. See also NITRIDES, EXPLOSIVES, and GOLD COMPOUNDS.

GIZ100 **CAS:15189-51-2** **HR: 3**
GOLD SODIUM CHLORIDE
mf: $AuCl_4 \cdot Na$ mw: 361.76

SYNS: AURATE(1-), TETRACHLORO-, SODIUM □ AURATE(1-), TETRACHLORO-, SODIUM, (SP-4-1)-(9CI) □ GOLD CHLORIDE SODIUM □ HYDROCHLOROAURIC ACID, SODIUM SALT □ SODIUM CHLOROAURATE □ SODIUM GOLD CHLORIDE □ SODIUM TETRACHLOROAURATE □ SODIUM TETRACHLOROAURATE(1-) □ SODIUM TETRACHLOROAURATE(3+) □ TETRACHLOROAURIC(3+) ACID, SODIUM SALT

TOXICITY DATA WITH REFERENCE
ipr-mus LD50:72 mg/kg TXAPA9 63,461,82

CONSENSUS REPORTS: Reported in EPA TSCA Inventory.

SAFETY PROFILE: A poison by intraperitoneal route. When heated to decomposition it emits toxic vapors of NaO_2 and Cl^-.

GJC000 **CAS:12244-57-4** **HR: 3**
GOLD SODIUM THIOMALATE
mf: $C_4H_3AuO_4S \cdot 2Na$ mw: 390.08

PROP: Yellowish-white powder. Sol in H_2O; insol in EtOH and Et_2O.

SYNS: AuTM □ ((1,2-DICARBOXYETHYL)THIO)GOLD DISODIUM SALT □ (DIHYDROGEN MERCAPTOSUCCINATO)GOLD DISODIUM SALT □ DISODIUM AUROTHIOMALATE □ (MERCAPTOBUTANEDIOATO(1-))GOLD DISODIUM SALT □ MERCAPTOSUCCINIC ACID, GOLD SODIUM SALT □ MYOCHRYSINE □ MYOCRISIN □ SODIUM AUROTHIOMALATE □ TAURE(o)DON

TOXICITY DATA WITH REFERENCE
ipr-rat TDLo:75 mg/kg (4-6D preg):REP JPETAB 214,250,80
scu-rat TDLo:900 mg/kg (female 6-15D post):REP VTPHAK 15(Suppl 5),89,78
scu-rbt TDLo:257 mg/kg (female 6-13D post):REP VTPHAK 15(Suppl 5),97,78
scu-rat TDLo:250 mg/kg (6-15D preg):TER VTPHAK 15(Suppl 5),89,78
orl-man TDLo:5500 μg/kg:PNS,KID,BLD ARHEAW 19,936,76
ims-wmn TDLo:182 μg/kg:SKN BMJOAE 2,1294,76
ivn-wmn TDLo:2700 μg/kg/4W-I:CNS JRHUA9 11,235,84
ims-wmn TDLo:20 mg/kg/1Y:CNS,KID ARPAAQ 96,133,73
ims-man TDLo:5 mg/kg/13W:BLD,MSK JAMAAP 133,754,47
ims-wmn LDLo:600 μg/kg/1W-I:PUL,SYS JRSMD9 77,960,84
par-wmn TDLo:3366 μg/kg/2W:CNS,PNS NEURAI 20,455,70
par-man TDLo:12,500 μg/kg/7W-I:CNS,PNS JRHUA9 11,233,84
par-man TDLo:7857 μg/kg/11W-I:SKN BMJOAE 286,1547,83
unr-cld TDLo:11,900 μg/kg/18W-I: JRHUA9 13,224,86
unr-hmn TDLo:9750 μg/kg/23W-I:BLD BMJOAE 1,1266,76
unr-man TDLo:12 mg/kg:BLD JRHUA9 13,225,86
unr-wmn TDLo:1600 μg/kg/3W-I:SYS,BLD ANZJB8 16,72,86
unr-wmn TDLo:7 mg/kg/7W-I:BLD JRHUA9 12,180,85
unr-man TDLo:7857 μg/kg/11W-I:ALR BJRHDF 24,367,85
scu-rat LD50:303 mg/kg NIIRDN 6,208,82
ivn-rat LD50:440 mg/kg NIIRDN 6,208,82
ims-rat LDLo:185 mg/kg PSEBAA 49,121,42
scu-mus LD50:930 mg/kg NIIRDN 6,208,82
ivn-mus LD50:855 mg/kg NIIRDN 6,208,82
ims-mus LD50:800 mg/kg VTPHAK 15(Suppl 5),1,78

SAFETY PROFILE: Poison by subcutaneous and intramuscular routes. Moderately toxic by intravenous route. Human systemic effects: aggression, agranulocytosis, aplastic anemia, cell count changes, changes in circulation, cholestatic jaundice, dermatitis, encephalitis, fasciculations, flaccid paralysis without anesthesia, hemorrhage, hepatitis (hepatocellular necrosis), increased body temperature, interstitial fibrosis, muscle weakness, proteinuria, recording from peripheral motor nerve, depressed renal function tests, somnolence, structural changes in nerve sheath, thrombocytopenia, uncharacterized allergic reaction, changes in blood, teeth, and supporting structures. Experimental teratogenic and reproductive effects. When heated to decomposition it emits very toxic Na_2O and SO_x.

GJE000 **CAS:10233-88-2** **HR: 3**
GOLD SODIUM THIOSULFATE
mf: $O_6S_4 \cdot Au \cdot 3Na$ mw: 490.18

PROP: White crystals, odorless.

SYN: THIOSULFURIC ACID, GOLD(1+) SODIUM SALT (2:1:3)

TOXICITY DATA WITH REFERENCE
ipr-rat LD50:78 mg/kg JAPMA8 41,105,52
orl-mus LD50:35 mg/kg JINCAO 40,2081,78
ipr-mus LD50:245 mg/kg TXAPA9 49,41,79
ivn-mus LDLo:140 mg/kg JAPMA8 41,105,52

CONSENSUS REPORTS: Reported in EPA TSCA Inventory.

SAFETY PROFILE: Poison by ingestion, intraperitoneal, and intravenous routes. When heated to decomposition it emits very toxic fumes of SO_x and Na_2O. See also GOLD SODIUM THIOSULFATE DIHYDRATE.

GJG000 CAS:10210-36-3 HR: 3
GOLD SODIUM THIOSULFATE DIHYDRATE
mf: $O_6S_4 \cdot Au \cdot 3Na \cdot 2H_2O$ mw: 526.22

SYNS: AURICIDINE □ AUROCIDIN □ AUROLIN □ AUROPEX □ AUROPIN □ AUROSAN □ AUROTHION □ CRISALBINE □ NOVACRYSIN □ SANOCHRYSINE □ SODIUM AUROTHIOSULPHATE DIHYDRATE □ SOLFOCRISOL □ THIOCHRYSINE □ THIOSULFURIC ACID, GOLD(1+) SODIUM SALT(2:1:3), DIHYDRATE

TOXICITY DATA WITH REFERENCE
ims-hmn TDLo:400 μg/kg/1W:SKN,BLD AIMEAS 37,323,52
scu-rat LDLo:30 mg/kg 27ZWAY 3.3,2134,-
ivn-rat LDLo:80 mg/kg 27ZWAY 3.3,2134,-
ims-rat LDLo:35 mg/kg PSEBAA 49,121,42
ipr-mus LD50:110 mg/kg JRHUA9 12,274,85
ivn-mus LDLo:100 mg/kg EMSUA8 3,146,45
ivn-rbt LDLo:59 mg/kg 27ZWAY 3.3,2134,-
scu-gpg LDLo:60 mg/kg ZGEMAZ 57,77,27

SAFETY PROFILE: Poison by subcutaneous, intraperitoneal, intravenous, and intramuscular routes. Human systemic effects by intramuscular route: dermatitis, granulocytopenia, and thrombocytopenia. Used as an antirheumatic agent. When heated to decomposition it emits very toxic fumes of SO_x and Na_2O.

GJI250 CAS:500-64-1 HR: 3
GONOSAN
mf: $C_{14}H_{14}O_3$ mw: 230.28

PROP: Prisms from Et_2O/MeOH. Mp: 106°. (+)- Form: Rods from methanol + ether. Bp: 195–197°. Practically insol in water; sol in acetone, ether, methanol; sltly sol in hexane. (+/-)- Form: Needles from methanol. Mp: 146–147°.

SYNS: (R)-5,6-DIHYDRO-4-METHOXY-6-STYRYL-2H-PYRAN-2-ONE □ 5-HYDROXY-3-METHOXY-7-PHENYL-2,6-HEPTADIENOIC ACID γ-LACTONE □ KAVAIN □ (+)-KAVAIN □ KAWAIN □ 4-METHOXY-6-(β-PHENYLVINYL)-5,6-DIHYDRO-α-PYRONE

TOXICITY DATA WITH REFERENCE
orl-mus LD50:1130 mg/kg AIPTAK 177,261,69
ipr-mus LD50:420 mg/kg AIPTAK 177,261,69
ivn-mus LD50:69 mg/kg AIPTAK 177,261,69

SAFETY PROFILE: Poison by intravenous route. Moderately toxic by ingestion and intraperitoneal routes. When heated to decomposition it emits acrid smoke and irritating fumes.

GJI400 CAS:477-73-6 HR: 3
GOSSYPIMINE
mf: $C_{20}H_{19}N_4 \cdot Cl$ mw: 350.88

SYNS: BASIC RED 2 □ BRILLIANT SAFRANINE BR □ BRILLIANT SAFRANINE G □ BRILLIANT SAFRANINE GR □ CALCOZINE RED Y □ CERVEN ZASADITA 2 □ C.I. 50240 □ C.I. BASIC RED 2 □ 2,8-DIMETHYLPHENOSAFRANINE □ HIDACO SAFRANINE □ LEATHER RED HT □ MITSUI SAFRANINE □ NIPPON KAGAKU SAFRANINE GK □ NIPPON KAGAKU SAFRANINE T □ PHENAZINIUM, 3,7-DIAMINO-2,8-DIMETHYL-5-PHENYL-, CHLORIDE □ SAFRANIN □ SAFRANINE □ SAFRANINE A □ SAFRANINE B □ SAFRANINE G □ SAFRANINE GF □ SAFRANINE J □ SAFRANINE O □ SAFRANINE OK □ SAFRANINE SUPERFINE G □ SAFRANINE T □ SAFRANINE TH □ SAFRANINE TN □ SAFRANINE Y □ SAFRANINE YN □ SAFRANINE ZH □ SAFRANIN T □ TOLUSAFRANINE

TOXICITY DATA WITH REFERENCE
mmo-sat 16 μg/plate TRENAF 27,153,76
mma-sat 16 μg/plate TRENAF 27,153,76
dnr-esc 4 μg/well ENMUDM 3,429,81
dnr-bcs 2 mg/disc TRENAF 27,153,76
ivn-rat LD50:28,740 μg/kg SMBUA9 9,96,51
orl-mus LDLo:1600 mg/kg JPMSAE 69,327,80
ivn-mus LD50:24,020 μg/kg SMBUA9 9,96,51
ivn-rbt LD50:26,940 μg/kg SMBUA9 9,96,51

CONSENSUS REPORTS: Reported in EPA TSCA Inventory.

SAFETY PROFILE: Poison by intravenous route. Moderately toxic by ingestion. Mutation data reported. When heated to decomposition it emits toxic vapors of NO_x and Cl^-.

GJK000 CAS:50933-33-0 HR: 1
GOSSYPLURE
mf: $C_{18}H_{32}O_2$ mw: 280.50

SYNS: GOSSYPLURE H.F. □ 7,11-HEXADECADIEN-1-OL, ACETATE □ NOMATE PBW

TOXICITY DATA WITH REFERENCE
orl-rat LD50:15 g/kg FMCHA2 -,D219,80

CONSENSUS REPORTS: Reported in EPA TSCA Inventory.

SAFETY PROFILE: Mildly toxic by ingestion. When heated to decomposition it emits acrid smoke and irritating fumes.

GJM000 CAS:303-45-7 HR: 2
GOSSYPOL
mf: $C_{30}H_{30}O_8$ mw: 518.60

PROP: A polyphenolic yellow pigment isolated from cottonseed pigment glands (JAOCA7 40,571,63). Mp: 180° from ether, mp: 199° from chloroform, mp: 214° from ligroin. Very sltly sol in methanol, ethanol, ether, chloroform, DMF; freely sol (with slow decomp) in dilute solns of ammonia, sodium carbonate; insol in water.

SYNS: 2,2′-BIS(1,6,7-TRIHYDROXY-3-METHYL-5-ISOPROPYL)-8-ALDEHYDONAPHTHALENE □ 8-FORMYL-1,6,7-TRIHYDROXY-5-ISOPROPYL-3-METHYL-2,2′-BISNAPHTHALENE

TOXICITY DATA WITH REFERENCE
dni-hmn:hla 10 mg/L CNREA8 44,35,84
cyt-man:lym 9 μg/plate CPHPA5 12,293,81
sce-man:lym 1 μg/plate CPHPA5 12,293,81
orl-man TDLo:17 mg/kg (60D male):REP CMJODS 4,417,78
orl-rat TDLo:1050 mg/kg (male 35D pre):REP NFGZAD 26,393,81
orl-ham TDLo:700 mg/kg (male 70D pre):REP FESTAS 37,311,82
orl-rat TDLo:700 mg/kg (35D male):REP CDESDK 3,#278,82
orl-rat LD50:2315 mg/kg JAOCA7 37,40,60
orl-pig LD50:550 mg/kg JAOCA7 40,571,63

CONSENSUS REPORTS: EPA Genetic Toxicology Program.

SAFETY PROFILE: Moderately toxic by ingestion. Human reproductive effects by ingestion: spermatogenesis and male fertility index changes. Experimental reproductive effects. Human mutation data reported. Can be irritating to the gastrointestinal tract. In experimental animals, large doses cause edema of lungs, shortness of breath, paralysis. When heated to decomposition it emits acrid smoke and irritating fumes.

GJM025 CAS:20300-26-9 HR: 3
(+)-GOSSYPOL
mf: $C_{30}H_{30}O_8$ mw: 518.60

PROP: Pale-yellow needles from pet ether; deep-yellow prisms from Me_2CO; large elongated plates from Me_2CO (aq). Mp: 181–183°.

TOXICITY DATA WITH REFERENCE
orl-rat TDLo:50 mg/kg (1-5D preg):REP CUSCAM 50,64,81
ipr-rat TDLo:10 mg/kg (female 1D post):REP IJEBA6 22,367,84
ipr-mus LD50:35 mg/kg CUSCAM 50,64,81

SAFETY PROFILE: Poison by intraperitoneal route. Experimental reproductive effects. When heated to decomposition it emits acrid smoke and irritating fumes. See also GOSSYPOL.

GJM259 CAS:12542-36-8 HR: 3
GOSSYPOL ACETIC ACID

SYN: GOSSYPOL ACETATE

TOXICITY DATA WITH REFERENCE
dnd-hmn:leu 15 mg/L/1H MUREAV 164,71,86
sce-hmn:lym 1 mg/L ENMUDM 7(Suppl 3),66,85
sce-mus-ipr 20 mg/kg TCMUD8 6,83,86
orl-man TDLo:12,800 μg/kg (male 12W pre):REP CCPTAY 37,153,88
orl-rat TDLo:1200 mg/kg (female 60D pre):REP DCTODJ 8,469,85
orl-rat TDLo:10 mg/kg (female 10D post):TER TJADAB 32,251,85
ivg-rbt TDLo:1 mg/kg (female 1D pre):REP CCPTAY 22,659,80
orl-ham TDLo:420 mg/kg (male 42D pre):REP CCPTAY 24,97,81
orl-rat TDLo:215 mg/kg (female 1-5D post):REP YHTPAD 20,464,85
orl-rat TDLo:1080 mg/kg (male 36D pre):REP INJFA3 27,213,82
orl-rat TDLo:1240 mg/kg (male 62D pre):REP CCPTAY 37,269,88
orl-rat TDLo:1080 mg/kg (male 36D pre):REP INJFA3 27,213,82
orl-man TDLo:282 mg/kg/63W-I FESTAS 48,459,87
orl-mus LDLo:3 g/kg BIMADU 12,1,84
ivn-cat LDLo:75 mg/kg JAFCAU 17,497,69

SAFETY PROFILE: Poison by intravenous route. Human reproductive effects by ingestion: spermatogenesis. An experimental teratogen. Experimental reproductive effects. Human mutation data reported. See also GOSSYPOL.

GJO000 CAS:1405-97-6 HR: 3
GRAMICIDIN
mf: $C_{148}H_{210}N_{30}O_{26}$ mw: 2825.88

PROP: Spear-shaped or lenticular platelets. Mp: 229–230°. Almost insol in water; sol in lower alc, acetic acid, and pyridine; practically insol in ether or hydrocarbons.

TOXICITY DATA WITH REFERENCE
add-bac-esc 2200 nmol/L MUREAV 89,95,81
ivn-mus LD50:1500 μg/kg SCIEAS 103,419,46
par-mus LD50:17 μg/kg 85ERAY 3,1542,78

SAFETY PROFILE: Poison by intravenous, intraperitoneal, and parenteral routes. An antibiotic. Mutation data reported. When heated to decomposition it emits very toxic fumes of NO_x.

GJO025 CAS:11029-61-1 HR: 3
GRAMICIDIN A
mf: $C_{148}H_{210}N_{30}O_{26}$ mw: 2825.88

SYN: VALYL GRAMICIDIN A

TOXICITY DATA WITH REFERENCE
orl-mus LD50:1000 mg/kg 85GDA2 4(1),240,80
ipr-mus LD50:60 mg/kg 85GDA2 4(1),240,80
ivn-mus LD50:5 mg/kg 85GDA2 4(1),240,80

SAFETY PROFILE: Poison by intravenous and intraperitoneal routes. Moderately toxic by ingestion. An antibiotic. When heated to decomposition it emits toxic fumes of NO_x.

GJQ100 CAS:1481-70-5 HR: 3
GRAMINIC ACID
mf: $C_{66}H_{87}N_{13}O_{13}\cdot ClH$ mw: 1307.12

PROP: Needles or rods.

SYN: TYROCIDINE A, HYDROCHLORIDE

TOXICITY DATA WITH REFERENCE
orl-mus LD50:1000 mg/kg 85GDA2 4(1),269,80
ipr-mus LD50:40 mg/kg 85GDA2 4(1),269,80
ivn-mus LD50:15 mg/kg 85GDA2 4(1),269,80

SAFETY PROFILE: Poison by intravenous and intraperi-

toneal routes. Moderately toxic by ingestion. When heated to decomposition it emits toxic fumes of NO_x and HCl.

GJS000 **CAS:19879-06-2** **HR: 3**
GRANATICIN
mf: $C_{22}H_{20}O_{10}$ mw: 444.42

PROP: Deep-red, garnet-like crystals from Me_2CO or C_6H_6. Mp: 223–225°.

SYNS: ANTIBIOTIC WR 141 ☐ GRANATICIN A ☐ LITMOMYCIN ☐ WR 141

TOXICITY DATA WITH REFERENCE
dni-bcs 600 nmol/L ZBPHA6 212,259,69
oms-bcs 600 nmol/L ZBPHA6 212,259,69
ipr-mus LD50:25 mg/kg 85ERAY 1,135,78
scu-mus LD50:250 mg/kg MEIEDD 10,652,83

SAFETY PROFILE: Poison by subcutaneous and intraperitoneal routes. Mutation data reported. An antibiotic. When heated to decomposition it emits acrid smoke and irritating fumes.

GJS200 **CAS:22345-47-7** **HR: 3**
GRANDAXIN
mf: $C_{22}H_{26}N_2O_4$ mw: 382.50

PROP: Colorless to light cream crystalline powder from isopropyl alc. Mp: 156–157°.

SYNS: 1-(3,4-DIMETHOXYPHENYL)-5-ETHYL-7,8-DIMETHOXY-4-METHYL-5H-2,3-BENZODIAZEPINE ☐ EGYT 341 ☐ SERIEL ☐ TF ☐ TOFISOPAM

TOXICITY DATA WITH REFERENCE
orl-rat TDLo:270 mg/kg (female 17-22D post):REP IYKEDH 12,565,81
orl-rat TDLo:110 mg/kg (7-17D preg):REP IYKEDH 12,565,81
orl-rat TDLo:12 g/kg (male 30D pre):REP IYKEDH 12,547,81
orl-rat LD50:825 mg/kg IYKEDH 12,547,81
ipr-rat LD50:1270 mg/kg IYKEDH 12 547,81
ivn-rat LD50:103 mg/kg IYKEDH 12,547,81
orl-mus LD50:3800 mg/kg IYKEDH 12,547,81
ipr-mus LD50:1950 mg/kg IYKEDH 12,547,81
ivn-mus LD50:415 mg/kg IYKEDH 12,547,81

SAFETY PROFILE: Poison by intravenous route. Moderately toxic by ingestion and intraperitoneal route. Experimental reproductive effects. When heated to decomposition it emits toxic fumes of NO_x. See also DIAZEPAM.

GJU000 **CAS:8016-20-4** **HR: 2**
GRAPEFRUIT OIL

PROP: From the fresh peel of *Citrus paradisi* Macfayden (*Citrus decumana* L.). Yellow liquid. Sol in fixed oils, mineral oil; sltly sol in propylene glycol; insol in glycerin.

SYNS: GRAPEFRUIT OIL, coldpressed ☐ GRAPEFRUIT OIL, expressed ☐ OIL of GRAPEFRUIT ☐ OIL of SHADDOCK

TOXICITY DATA WITH REFERENCE
skn-rbt 500 mg/24H MLD FCTXAV 12,743,74
dnr-bcs 20 mg/disc TOFOD5 8,91,85
skn-mus TDLo:280 g/kg/33W-I:ETA JNCIAM 24,1389,60

CONSENSUS REPORTS: Reported in EPA TSCA Inventory.

SAFETY PROFILE: A skin irritant. Questionable carcinogen with experimental tumorigenic data. Mutation data reported. When heated to decomposition it emits acrid smoke and irritating fumes.

GJU050 **CAS:16974-11-1** **HR: 1**
GRAPEMONE
mf: $C_{14}H_{26}O_2$ mw: 226.40

SYNS: (Z)-ACETATE-9-DODECEN-1-OL ☐ BOCEP VITI ☐ Z-9-DDA ☐ 9-DODECEN-1-OL, ACETATE, (Z)- ☐ cis-9-DODECENYL ACETATE ☐ (Z)-9-DODECENYL ACETATE ☐ RAK 1

TOXICITY DATA WITH REFERENCE
orl-rat LD50:>5 g/kg FMCHA2-,C261,91

CONSENSUS REPORTS: Reported in EPA TSCA Inventory.

SAFETY PROFILE: Low toxicity by ingestion. When heated to decomposition it emits acrid smoke and irritating vapors.

GJU460 **HR: 3**
GREEN LILY

PROP: Perennial bulb-producing herbs. The grass-like leaves grow directly from the bulb. Pale-green to yellow-white flowers grow in heavy clusters from a leafless stalk. The seed pod has 3 lobes and holds 4 or more seeds. The various species grow in the area from southern New Mexico to Peru.

SYNS: CEBOLLEJA (MEXICO) ☐ SCHOENOCAULON DRUMMONDII ☐ SCHOENOCAULON OFFICIANLIS ☐ SCHOENOCAULON TEXANUM

SAFETY PROFILE: The whole plant and especially the seeds are thought to contain poisonous veratridine alkaloids. Ingestion of the seeds may cause severe vomiting, catharsis, slowed heartbeat, low blood pressure and, in some cases, death. See also VERATRIDINE.

GJU475 **HR: 2**
GREEN LOCUST

PROP: A large tree that may grow to 80 feet. The leaves are compound with many 1-inch leaflets. There are 2 thorns on the branch where each leaf stem is attached. It produces clusters of white, fragrant flowers and a flat, red-brown, 4-inch long seed pod that stays on the tree through the winter. It is native to the Smoky Mountains and the Ozarks of the United States. It is commonly cultivated in the temperate regions of the United States and southern Canada.

SYNS: BASTARD ACACIA ☐ BLACK ACACIA ☐ BLACK LOCUST ☐ FALSE ACACIA ☐ PEA FLOWER LOCUST ☐ POST LOCUST ☐ ROBI-

NIA PSEUDOACACIA □ SILVER CHAIN □ TREESAIL □ WHITE HONEY FLOWER □ WHITE LOCUST □ WHYO TREE □ YELLOW LOCUST

SAFETY PROFILE: The bark, seeds, and leaves contain the poison robin, a plant lectin (toxalbumin) that inhibits protein synthesis in the intestinal wall. Ingestion of these plant parts may cause after a delay period of several hours: nausea, vomiting, diarrhea, and intestinal dysfunction. There is a potential for massive fluid and electrolyte loss. See also ABRIN as an example of toxalbumin.

GJU500 **HR: 3**
GREEN MAMBA VENOM

SYNS: DENDROASPIS VIRIDIS VENOM □ VENOM, SNAKE, DENDROASPIS VIRIDIS

TOXICITY DATA WITH REFERENCE
scu-mus LD50:790 µg/kg 19DDA6 1,223,67
ivn-mus LD50:667 µg/kg 23EIAT 1,437,68
ivn-rbt LDLo:700 µg/kg TOXIA6 2,5,64

SAFETY PROFILE: A deadly poison by subcutaneous and intravenous routes.

GJU600 **CAS:68085-85-8** **HR: 3**
GRENADE
DOT: NA 0349/UN 0110/UN 0284/UN 0285/UN 0292/UN 0293/UN 0318/UN 0372/UN 0452
mf: $C_{23}H_{19}ClF_3NO_3$ mw: 449.88

SYNS: CYCLOPROPANECARBOXYLIC ACID, 3-(2-CHLORO-3,3,3-TRIFLUORO-1-PROPENYL)-2,2-DIMETHYL-, CYANO(3-PHENOXYPHENYL) METHYL ESTER □ CYHALOTHRIN □ CYHALOTHRINE □ GRENADES, empty primed (NA0349) (DOT) □ GRENADES, hand or rifle, with bursting charge (UN0284, UN0285, UN0292, UN0293) (DOT) □ GRENADES, practice, hand or rifle (UN0452, UN0110, UN0318, UN0372) (DOT) □ ICI 146814 □ ICI-PP 563 □ PP 563

TOXICITY DATA WITH REFERENCE
orl-rat LD50:144 mg/kg PEMNDP 9,201,91
ihl-rat LC50:83 mg/m^3/4H PEMNDP 9,201,91
orl-rbt LD50:>1 g/kg PEMNDP 9,201,91
skn-rbt LD50:>2500 mg/kg PEMNDP 9,201,91

DOT CLASSIFICATION: Explosive 1.4S; *Label:* None (NA 0349, UN 0110); Explosive 1.1D; *Label:* Explosive 1.1D (UN 0284); Explosive 1.1F; *Label:* Explosive 1.1F (UN 0292); Explosive 1.2F; *Label:* Explosive 1.2F (UN 0293); Explosive 1.4G; *Label:* Explosive 1.4G (UN 0452); Explosive 1.3G; *Label:* Explosive 1.3G (UN 0318); Explosive 1.2G; *Label:* Explosive 1.2G (UN0372); Explosive 1.2D; *Label:* Explosive 1.2D (UN0285)

SAFETY PROFILE: A poison by ingestion and inhalation. An explosive. When heated to decomposition it emits toxic vapors of NO_x, F^-, and Cl^-.

GJU800 **CAS:1391-82-8** **HR: 3**
GRISEIN
mf: $C_{40}H_{61}FeN_{10}O_{20}S$ mw: 1090.02

PROP: Antibiotic substance produced by strains of *Streptomyces griseus.* Amorph, red powder. Sol in water; sltly sol in 95% alc; insol in abs alc, ether, acetone, chloroform, benzene.

SYNS: ALBOMYCIN A1 □ CORMOGRIZIN □ GRISIN □ KORMOGRIZEIN

TOXICITY DATA WITH REFERENCE
orl-mus LD50:600 mg/kg MEIEDD 10,653,83
ipr-mus LD50:600 mg/kg 85GDA2 4(1),442,80
scu-mus LD50:34 mg/kg MEIEDD 10,653,83

SAFETY PROFILE: Poison by subcutaneous route. Moderately toxic by ingestion and intraperitoneal routes. An antibiotic. When heated to decomposition it emits toxic fumes of SO_x and NO_x.

GJW000 **CAS:2072-68-6** **HR: 3**
GRISEOLUTEIN B
mf: $C_{17}H_{16}N_2O_6$ mw: 344.35

TOXICITY DATA WITH REFERENCE
orl-mus LD50:800 mg/kg 85ERAY 1,763,78
scu-mus LD50:400 mg/kg 85GDA2 5,144,81
ivn-mus LD50:200 mg/kg 85GDA2 5,144,81

SAFETY PROFILE: Poison by intravenous and subcutaneous routes. Moderately toxic by ingestion. When heated to decomposition it emits toxic fumes of NO_x.

GJY000 **CAS:1393-89-1** **HR: 3**
GRISEOMYCIN
mf: $C_{25}H_{46}ClNO_8$ mw: 524.17

PROP: Produced by *Streptomyces sp.* (ANTCAO 3,1243,53). Crystalline base, bitter taste, platelets, decomp @ 76–80°, alkaline reaction; freely sol in chloroform, ethanol, butanol, benzene, acetone, ethyl acetate.

SYN: LOMYCIN

TOXICITY DATA WITH REFERENCE
ipr-mus LD50:210 mg/kg 85ERAY 1,98,78
scu-mus LD50:1330 mg/kg 85ERAY 1,98,78

SAFETY PROFILE: Poison by intraperitoneal route. Moderately toxic by subcutaneous route. When heated to decomposition it emits very toxic fumes of NO_x and Cl^-.

GJY100 **CAS:73297-70-8** **HR: 3**
GRISEORUBIN COMPLEX

TOXICITY DATA WITH REFERENCE
ipr-mus LD50:50 mg/kg 85GDA2 3,202,80
scu-mus LD50:100 mg/kg 85GDA2 3,202,80
ivn-mus LD50:10 mg/kg 85GDA2 3,202,80

SAFETY PROFILE: Poison by subcutaneous, intravenous, and intraperitoneal routes.

GKA000 **CAS:101563-93-3** **HR: 3**
GRISEORUBIN I HYDROCHLORIDE

PROP: Produced by *Streptomyces griseus* (JANTAJ 31,78–111,78).

TOXICITY DATA WITH REFERENCE
ipr-mus LD50:50 mg/kg JANTAJ 31,78,78
scu-mus LD50:100 mg/kg JANTAJ 31,78-111,78
ivn-mus LD50:10 mg/kg JANTAJ 31,78,78

SAFETY PROFILE: Poison by intraperitoneal, subcutaneous, and intravenous routes. When heated to decomposition it emits very toxic fumes of NO_x and HCl.

GKC000 **CAS:53216-90-3** **HR: 3**
GRISEOVIRIDIN
mf: $C_{22}H_{27}N_3O_7S$ mw: 477.58

PROP: Polymorphic crystals from MeOH. Mp: 161–163° decomp depending on the crystal modification. Sol in pyridine; mod sol in lower alcs; very sltly sol in water and nonpolar solvents.

TOXICITY DATA WITH REFERENCE
ipr-mus LD50:100 mg/kg 85ERAY 1,322,78
scu-mus LD50:100 mg/kg 85ERAY 1,322,78
ivn-mus LD50:75 mg/kg ABANAE 2,790,54/55

SAFETY PROFILE: Poison by intraperitoneal, subcutaneous, and intravenous routes. When heated to decomposition it emits very toxic fumes of SO_x and NO_x.

GKE000 **CAS:126-07-8** **HR: 3**
GRISOFULVIN
mf: $C_{17}H_{17}ClO_6$ mw: 352.79

PROP: Octahedra crystals from C_6H_6. Mp: 225–226°.

SYNS: AMUDANE □ BIOGRISIN-FP □ 7-CHLORO-4,6,2'-TRIMETHOXY-6'-METHYLGRIS-2'-EN-3,4'-DIONE □ DELMOFULVINA □ FULCIN □ FULCINE □ FULVICAN GRISACTIN □ FULVICIN □ FULVINA □ FULVISTATIN □ FUNGIVIN □ GREOSIN □ GRESFEED □GRI-CIN □ GRIFULVIN □ GRISACTIN □ GRISCOFULVIN □ GRISEFULINE □ GRISEO □ (+)-GRISEOFULVIN □ GRISEOFULVIN-FORTE □ GRISEOFULVINUM □ GRISETIN □ GRISOVIN □ GRIS-PEG □ GRYSIO □ GUSERVIN □ LAMORYL □ LIKUDEN □ MURFULVIN □ NEO-FULCIN □ NSC 34533 □ PONCYL □ SPIROFULVIN □ SPOROSTATIN □ USAF SC-2

TOXICITY DATA WITH REFERENCE
dnr-bcs 100 µL/plate MUREAV 97,1,82
dni-hmn:fbr 20 mg/L/3D-C KAMJDW 2,127,76
dni-hmn:lym 20 mg/L/3D-C KAMJDW 2,127,76
cyt-hmn:lym 40 mg/L/3D MUREAV 25,123,74
cyt-ham:fbr 10 mg/L CRNGDP 3,499,82
orl-rat TDLo:1250 mg/kg (6-15D preg):REP SCIEAS 175,1483,72
orl-rat TDLo:500 mg/kg (female 9D post):TER ANTBAL 14,44,69
orl-rat TDLo:1250 mg/kg (6-15D preg):TER SCIEAS 175,1483,72
orl-rat TDLo:12,500 mg/kg (female 6-15D post):REP SCIEAS 175,1483,72
orl-rat TDLo:2 g/kg (11-14D preg):TER ANTBAL 14,44,69
orl-rat TDLo:12,500 mg/kg (female 6-15D post): TER SCIEAS 175,1483,72
orl-rat TDLo:462 g/kg/2Y-I:NEO BJCAAI 38,237,78
orl-mus TDLo:440 g/kg/52W-C:NEO CNREA8 26,721,66
scu-mus TDLo:120 mg/kg/49W-I:NEO CNREA8 27,1900,67
orl-mus TD:730 g/kg/52W-C:NEO CNREA8 26,721,66
ivn-rat LD50:400 mg/kg NATUAS 182,1320,58
ipr-mus LD50:200 mg/kg NTIS** AD277-689
scu-mus LD50:1200 mg/kg 85GDA2 6,290,81
ivn-mus LD50:280 mg/kg 85ERAY 3,1766,78

CONSENSUS REPORTS: IARC Cancer Review: Group 2B IMEMDT 7,56,87; Animal Sufficient Evidence IMEMDT 10,153,76. EPA Genetic Toxicology Program.

SAFETY PROFILE: Confirmed carcinogen with experimental neoplastigenic and teratogenic data. Poison by intravenous and intraperitoneal routes. Moderately toxic by subcutaneous route. Human mutation data reported. Experimental reproductive effects. Used as an antibiotic, pharmaceutical, and veterinary drug. When heated to decomposition it emits toxic fumes of Cl^-.

G

GKE900 **CAS:14567-61-4** **HR: 2**
GRUNERITE

PROP: Fibrous or non-fibrous gray, dark green or brown crystals.

TOXICITY DATA WITH REFERENCE
ipl-rat TDLo:80 mg/kg:ETA CNREA8 2,157,86

SAFETY PROFILE: Questionable carcinogen with experimental tumorigenic data.

GKG300 **CAS:17471-82-8** **HR: 3**
GUABENXANE
mf: $C_{10}H_{13}N_3O_2{\cdot}1/2H_2O_4S$ mw: 256.26

PROP: Crystals from H_2O. Mp: 205°.

SYNS: GUANIDINO-6-METHYL-1,4-BENZODIOXANE □ 6-GUANIDINOMETHYL 1,4-BENZODIOXANE SULFATE □ GUANIDINO METHYL-6-BENZODIOXANNE-1,4 (FRENCH) □ L'HEMISULFATE de GUANIDINO METHYL-6-BENZODIOXANNE-1,4 (FRENCH) □ LM 433

TOXICITY DATA WITH REFERENCE
ipr-rat LD50:243 mg/kg EJMCA5 12,241,77
orl-mus LD50:535 mg/kg EJMCA5 12,241,77
ipr-mus LD50:179 mg/kg EJMCA5 12,241,77
ivn-mus LD50:46 mg/kg EJMCA5 12,241,77

SAFETY PROFILE: Poison by intravenous and intraperitoneal routes. Moderately toxic by ingestion. When heated to decomposition it emits toxic fumes of NO_x and SO_x. See also SULFATES.

GKG400 **CAS:61789-17-1** **HR: 1**
GUAIAC ACETATE

SYNS: GUAIACWOOD ACETATE □ GUAIOL ACETATE □ OILS, GUAIACWOOD, ACETATES

TOXICITY DATA WITH REFERENCE
orl-rat LD50:>5 g/kg FCTXAV 12,903,74
skn-rbt LD50:>5 g/kg FCTXAV 12,903,74

CONSENSUS REPORTS: Reported in EPA TSCA Inventory.

SAFETY PROFILE: Low toxicity by ingestion and skin

contact. When heated to decomposition it emits acrid smoke and irritating vapors.

GKI000 **CAS:90-05-1** **HR: 3**
GUAIACOL
mf: $C_7H_8O_2$ mw: 124.15

PROP: Clear, pale-yellow liquid, solid, prisms, or needles. Characteristic odor, darkens on exposure to air and light. D (crystals): 1.129, d (liquid): about 1.112, (needles), mp: 32°, (prisms), bp: 202–209°, fp: −3.2° flash p: 180°F (OC), d: 1.097 @ 25°/25°. Misc with alc, chloroform, ether, oils, glacial acetic acid; sltly sol in petr ether; sol in NaOH soln.

SYNS: GUAICOL □ o-HYDROXYANISOLE □ 2-HYDROXYANISOLE □ 1-HYDROXY-2-METHOXYBENZENE □ o-METHOXYPHENOL □ 2-METHOXYPHENOL □ METHYLCATECHOL □ PYROGUAIAC ACID

TOXICITY DATA WITH REFERENCE
skn-rbt 500 mg/24H SEV FCTOD7 20(Suppl),697,82
eye-rbt 5 mg MLD FCTOD7 20(Suppl),697,82
sce-hmn:lym 500 µmol/L MUREAV 169,129,86
orl-hmn LDLo:43 mg/kg:CNS,GIT 34ZIAG -,295,69
orl-rat LD50:520 mg/kg FOMDAK 32,309,91
orl-mus LD50:621 mg/kg DRFUD4 5,539,80
ihl-mus LC50:7570 mg/m³ FCTOD7 20(Suppl),697,82
ivn-mus LD50:170 mg/kg FCTOD7 20(Suppl),697,82
skn-rbt LD50:4600 mg/kg FCTOD7 20(Suppl),697,82
scu-gpg LDLo:900 mg/kg FCTOD7 20(Suppl),697,82

CONSENSUS REPORTS: Reported in EPA TSCA Inventory. EPA Genetic Toxicology Program.

SAFETY PROFILE: Human poison by ingestion. Experimental poison by intravenous route. Mildly toxic by skin contact and inhalation. Human systemic effects by ingestion: tremors and gastrointestinal changes. Human mutation data reported. An eye and severe skin irritant. Ingestion produces burning in the mouth and throat. Flammable when exposed to heat or flame; can react with oxidizing materials. To fight fire, use foam, CO_2, dry chemical. Protect from light. Used as an expectorant. When heated to decomposition it emits acrid smoke and irritating fumes. See also PHENOL.

GKK000 **CAS:532-03-6** **HR: 2**
GUAIACOL GLYCERYL ETHER CARBAMATE
mf: $C_{11}H_{15}NO_5$ mw: 241.27

PROP: Crystals from C_6H_6. Mp: 92–94°.

SYNS: AHR 85 □ CARBAMIC ACID, 2-HYDROXY-3-(o-METHOXYPHENOXY)PROPYL ESTER □ ETROFLEX □ GLYCERYLGUAIACOLATE CARBAMATE □ GLYCERYLGUAJACOL-CARBAMAT □ GUIACOL-GLICERILETERE MONOCARBAMMATO □ 2-HYDROXY-3-(o-METHOXYPHENOXY)PROPYL 1-CARBAMATE □ LUMIRELAX □ METHOCARBAMOL □ 3-(2-METHOXYPHENOXY)-1-GLYCERYL CARBAMATE □ 3-(o-METHOXYPHENOXY)-2-HYDROXYPROPYL CARBAMATE □ 3-(o-METHOXYPHENOXY)-1,2-PROPANEDIOL 1-CARBAMATE □ METOCARBAMOL □ METOCARBAMOLO □ METOFENINA □ MIOLAXENE □ MIORILAS □ MIOWAS □ MYOLAXENE □ NEURAXIN □ PERILAX □ REFLEXYN □ RELAX □ RELESTRID □ ROBAXAN □ ROBAXIN □ ROBAXINE □ ROBAXON □ ROBINAX □ TRESORTIL

TOXICITY DATA WITH REFERENCE
orl-rat LD50:1320 mg/kg JPETAB 129,75,60
ipr-rat LD50:815 mg/kg ARZNAD 17,242,67
orl-mus LD50:812 mg/kg DRUGAY 6,836,82
scu-mus LD50:780 mg/kg APTOA6 19,247,62
ivn-mus LD50:774 mg/kg DRUGAY 6,836,82

CONSENSUS REPORTS: Reported in EPA TSCA Inventory.

SAFETY PROFILE: Moderately toxic by ingestion, intraperitoneal, subcutaneous, and intravenous routes. When heated to decomposition it emits toxic fumes of NO_x. See also ETHERS and CARBAMATES.

GKM000 **CAS:8016-23-7** **HR: 1**
GUAIAC WOOD OIL

PROP: From steam distillation of *Bulnesia sarmienti lor.* wood chips or sawdust (FCTXAV 12,807,74).

TOXICITY DATA WITH REFERENCE
skn-rbt 500 mg/24H MOD FCTXAV 12,905,74
orl-rat LD50:>5 g/kg FCTXAV 12,905,74
skn-rbt LD50:>5 g/kg FCTXAV 12,905,74

CONSENSUS REPORTS: Reported in EPA TSCA Inventory.

SAFETY PROFILE: Low toxicity by ingestion and skin contact. A skin irritant. When heated to decomposition it emits acrid smoke and irritating fumes.

GKO000 **CAS:88-84-6** **HR: 2**
GUAIA-1(5),7(11)-DIENE
mf: $C_{15}H_{24}$ mw: 204.39

PROP: Oil. Bp: 137–139° @ 15.5 mm.

SYNS: GUAIENE □ β-GUAIENE □ (1S,cis)1,2,3,4,5,6,7,8-OCTAHYDRO-1,4-DIMETHYL-7-(1-METHYLETHYLIDENE)-AZULENE

TOXICITY DATA WITH REFERENCE
skn-rbt 500 mg/24H SEV FCTXAV 17,371,79
orl-rat LD50:>5 g/kg FCTXAV 17,371,79
skn-rbt LD50:>5 g/kg FCTXAV 17,371,79

CONSENSUS REPORTS: Reported in EPA TSCA Inventory.

SAFETY PROFILE: Low toxicity by ingestion and skin contact. A severe skin irritant. When heated to decomposition it emits acrid smoke and irritating fumes.

GKO500 **HR: 3**
GUAIMERCOL
mf: $C_{11}H_{11}Hg_2NO_8$ mw: 1102.18

SYN: 6-ACETOXYMERCURI-5-NITROGUAIACOL and 4,6-DIACETOXYMERCURI-5-NITROGUAIACOL

TOXICITY DATA WITH REFERENCE
orl-rat LD50:125 mg/kg JAPMA8 38,270,49
ipr-rat LD50:6 mg/kg JAPMA8 38,270,49
ivn-rbt LD50:2600 µg/kg JAPMA8 38,270,49

CONSENSUS REPORTS: Mercury and its compounds are on the Community Right-To-Know List.

OSHA PEL: CL 0.1 mg(Hg)/m³ (skin)
ACGIH TLV: TWA 0.1 mg(Hg)/m³ (skin)

NIOSH REL: TWA 0.05 mg(Hg)/m³

SAFETY PROFILE: Poison by ingestion, intravenous, and intraperitoneal routes. When heated to decomposition it emits toxic fumes of NO_x and Hg. See also MERCURY COMPOUNDS, ORGANIC.

GKO750 CAS:23256-50-0 HR: 3
GUANABENZ ACETATE
mf: $C_8H_8Cl_2N_4 \cdot C_2H_4O_2$ mw: 291.16

PROP: A solid. Mp: 192.5° (decomp).

SYNS: BR 750 □ ((2,6-DICHLOROBENZYLIDENE)AMINO)GUANIDINE ACETATE □ 2-((2,6-DICHLOROPHENYL)METHYLENE)HYDRAZINECARBOXIMIDAMIDE MONOACETATE (9CI) □ GUANABENZ □ WY-8678 □ WY 8678 ACETATE

TOXICITY DATA WITH REFERENCE
orl-rat TDLo:170 mg/kg (female 6-22D post):REP JTSCDR 7(Suppl 2),123,82
orl-rat TDLo:272 mg/kg (female 6-22D post):REP JTSCDR 7(Suppl 2),123,82
orl-rat TDLo:330 mg/kg (female 7-17D post):REP JTSCDR 7(Suppl 2),107,82
orl-rat TDLo:165 mg/kg (7-17D preg):TER JTSCDR 7(Suppl 2),107,82
orl-rat TDLo:330 mg/kg (female 7-17D post):REP JTSCDR 7(Suppl 2),107,82
orl-chd TDLo:1 mg/kg:CNS,CVS AIMEAS 102,787,85
orl-wmn TDLo:1 mg/kg:CNS,CVS AIMEAS 102,787,85
orl-rat LD50:238 mg/kg YACHDS 10,4571,82
ipr-rat LD50:62 mg/kg YACHDS 10,4571,82
scu-rat LD50:84 mg/kg YACHDS 10,4571,82
orl-mus LD50:260 mg/kg YACHDS 10,4571,82
ipr-mus LD50:75 mg/kg YACHDS 10,4571,82
scu-mus LD50:89 mg/kg YACHDS 10,4571,82

SAFETY PROFILE. Poison by ingestion, subcutaneous, and intraperitoneal routes. Human systemic effects by ingestion: sleep, pulse rate, and blood pressure changes. An experimental teratogen. Experimental reproductive effects. When heated to decomposition it emits toxic fumes of Cl^- and NO_x. A hypotensive drug.

GKO800 CAS:32059-15-7 HR: 3
GUANAZODINE
mf: $C_9H_{20}N_4$ mw: 184.33

SYNS: 1-AZACYCLOOCT-2-YL METHYL GUANIDINE □ ((OCTAHYDRO-2-AZOCINYL)METHYL)GUANIDINE

TOXICITY DATA WITH REFERENCE
orl-mus TDLo:150 mg/kg (7-12D preg):REP OYYAA2 15,333,78
orl-mus TDLo:2400 mg/kg (female 7-12D post):TER OYYAA2 15,333,78
orl-mus TDLo:600 mg/kg (female 7-12D post):TER OYYAA2 15,333,78
ipr-mus LD50:136 mg/kg DRFUD4 2,592,77

SAFETY PROFILE: Poison by intraperitoneal route. An experimental teratogen. Experimental reproductive effects. When heated to decomposition it emits toxic fumes of NO_x.

GKQ000 CAS:55-65-2 HR: 3
GUANETHIDINE
mf: $C_{10}H_{22}N_4$ mw: 198.36

SYNS: 1-(2-GUANIDINOETHYL)HEPTAMETHYLENIMINE □ 1-(2-GUANIDINOETHYL)OCTAHYDROAZOCINE □ ISMELIN □ OCTATENSIN □ OCTATENZINE □ SU 5864

TOXICITY DATA WITH REFERENCE
orl-mus LD50:845 mg/kg USXXAM #3856778
ipr-mus LD50:100 mg/kg DRFUD4 4,185,79
scu-mus LD50:224 mg/kg USXXAM #3856778
ivn-mus LD50:28 mg/kg PCJOAU 15,349,81
ivn-cat LDLo:50 mg/kg 27ZIAQ -,112,73
ivn-rbt LDLo:50 mg/kg 27ZIAQ -,112,73

SAFETY PROFILE: Poison by intraperitoneal, intravenous, and subcutaneous routes. Moderately toxic by ingestion. When heated to decomposition it emits toxic fumes of NO_x.

GKS000 CAS:60-02-6 HR: 3
GUANETHIDINE BISULFATE
mf: $C_{20}H_{44}N_8 \cdot H_2O_4S$ mw: 494.80

PROP: Crystals. Mp: 276–281° (decomp).

SYNS: β-1-AZACYCLOOCTYLETHYLGUANIDINE SULFATE □ GUANETHIDINE SULFATE □ 1-(2-GUANIDINOETHYL)OCTAHYDROAZOCINE SULFATE (2:1) □ (2-(HEXAHYDRO-1(2H)-AZOCINYL)ETHYL)GUANIDINE SULFATE □ ISMELIN □ ISMELIN SULFATE □ISOME-LIN □ NSC-29863 □ (2-(OCTAHYDRO-1-AZOCINYL)ETHYL)GUANIDINE SULFATE □ SU-5864

TOXICITY DATA WITH REFERENCE
unr-man TDLo:391 mg/kg (78W male):REP JSXRAJ 3,69,67
par-rat TDLo:900 mg/kg (14D pre/1-22D preg):REP JRPFA4 56,715,79
ims-rat TDLo:1400 mg/kg (56D male):REP BJPCBM 48,30,73
orl-rat LD50:1000 mg/kg JPETAB 128,22,60
ivn-rat LD50:23 mg/kg JPETAB 128,22,60
ipr-mus LD50:137 mg/kg RPTOAN 32,11,69
scu-mus LD50:750 mg/kg YKKZAJ 83,629,63
ivn-mus LD50:18 mg/kg CSLNX* NX#08547

SAFETY PROFILE: Poison by intravenous and intraperitoneal routes. Moderately toxic by ingestion and subcutaneous routes. Human reproductive effects by unspecified route. Experimental reproductive effects. Used to treat hypertension. When heated to decomposition it emits very toxic fumes of SO_x and NO_x.

GKU000 CAS:645-43-2 HR: 3
GUANETHIDINE MONOSULFATE
mf: $C_{10}H_{22}N_4 \cdot H_2O_4S$ mw: 296.44

SYNS: N-(2-GUANIDINO ETHYL)HEPTAMETHYLENIMINE SULFATE □ (2-(HEXAHYDRO-1(2H)-AZOCINYL)ETHYL) GUANIDINE HYDROGEN SULFATE □ 2-(OCTAHYDRO-1-AZOCINYL)ETHYL GUANIDINE SULPHATE

G

TOXICITY DATA WITH REFERENCE
eye-hmn 420 mg/6W-I MLD BMJOAE 4,592,67
ipr-rat LD50:290 mg/kg TXAPA9 24,37,73
orl-mus LD50:1450 mg/kg BCFAAI 111,353,72
ipr-mus LD50:180 mg/kg BCFAAI 111,353,72
ivn-mus LD50:18,500 µg/kg BCFAAI 111,353,72

SAFETY PROFILE: Poison by intraperitoneal and intravenous routes. Moderately toxic by ingestion. A human eye irritant. When heated to decomposition it emits very toxic fumes of SO_x and NO_x.

GKU300 **CAS:29110-48-3** **HR: 3**
GUANFACINE HYDROCHLORIDE
mf: $C_9H_9Cl_2N_3O \cdot ClH$ mw: 282.57

SYNS: N-AMIDINO-2-(2,6-DICHLOROPHENYL)ACETAMIDE HYDROCHLORIDE ☐ N-(AMINOIMINOMETHYL)-2,6-DICHLOROBENZENEACETAMIDE HYDROCHLORIDE ☐ BS100-141 ☐ ESTULIC ☐ LON 798

TOXICITY DATA WITH REFERENCE
orl-mus TDLo:10 mg/kg (female 6-15D post):REP JZKEDZ 6,105,80
orl-mus TDLo:25 mg/kg (multi) :REP JZKEDZ 6,117,80
orl-mus TDLo:20 mg/kg (female 6-15D post):TER JZKEDZ 6,105,80
orl-mus TDLo:10 mg/kg (female 6-15D post):TER JZKEDZ 6,105,80
orl-mus TDLo:20 mg/kg (female 6-15D post):REP JZKEDZ 6,105,80
orl-mus TDLo:5 mg/kg (female 6-15D post):REP JZKEDZ 6,105,80
orl-rat TDLo:105 mg/kg (female 35D pre):REP KSRNAM 14,4511,80
orl-rat TDLo:54,600 µg/kg (26W male):REP KSRNAM 14,4531,80
orl-rat LD50:145 mg/kg IYKEDH 16,357,85
scu-rat LD50:114 mg/kg KSRNAM 14,4511,80
ivn-rat LD50:5800 µg/kg KSRNAM 14,4511,80
orl-mus LD50:15,300 µg/kg IYKEDH 16,357,85
scu-mus LD50:46 mg/kg KSRNAM 14,4511,80
ivn-mus LD50:25 mg/kg KSRNAM 14,4511,80

SAFETY PROFILE: Poison by ingestion, subcutaneous and intravenous routes. Experimental teratogenic and reproductive effects. When heated to decomposition it emits toxic fumes of NO_x and HCl.

GKW000 **CAS:113-00-8** **HR: 3**
GUANIDINE
mf: CH_5N_3 mw: 59.09

PROP: Hygroscopic, deliquescent, colorless crystals. Mp: 50°. Very sol in water, alc, and acids (aq soln).

SYNS: AMINOFORMAMIDINE ☐ AMINOMETHANAMIDINE ☐ CARBAMAMIDINE ☐ CARBAMIDINE ☐ IMINOUREA

TOXICITY DATA WITH REFERENCE
ipr-rat LDLo:22,500 µg/kg AEXPBL 90,129,21
scu-rat LDLo:150 mg/kg HBAMAK 4,1352,35
scu-mus LDLo:266 mg/kg HBAMAK 4,1352,35
ipr-cat LDLo:10 mg/kg AEXPBL 90,129,21
orl-rbt LDLo:500 mg/kg MEIEDD 10,657,83
scu-frg LDLo:3000 mg/kg HBAMAK 4,1352,35

SAFETY PROFILE: Poison by intraperitoneal and subcutaneous routes. Moderately toxic by ingestion. On heating to 160° it converts to melamine and NH_3. Keep well closed. When heated to decomposition it emits toxic fumes of NO_x.

GKW100 **CAS:593-85-1** **HR: 3**
GUANIDINE CARBONATE
mf: $CH_5N_3 \cdot 1/2CH_2O_3$ mw: 493.30

SYNS: AI3-14631 ☐ BISGUANIDINIUM CARBONATE ☐ CARBONIC ACID, compd. with GUANIDINE (1:2) ☐ DIGUANIDINIUM CARBONATE ☐ GUANIDINIUM CARBONATE

TOXICITY DATA WITH REFERENCE
orl-mus LD50:350 mg/kg CKFRAY 1,434,52
scu-rbt LDLo:500 mg/kg HBAMAK 4,1352,35

CONSENSUS REPORTS: Reported in EPA TSCA Inventory.

SAFETY PROFILE: Poison by ingestion. Moderately toxic by subcutaneous route. When heated to decomposition it emits toxic vapors of NO_x.

GKY000 **CAS:50-01-1** **HR: 3**
GUANIDINE, MONOHYDROCHLORIDE
mf: $CH_5N_3 \cdot HCl$ mw: 72.11

PROP: White powder. Mp: 183°. Freely sol in water, alc.

SYNS: AMINOFORMAMIDINE HYDROCHLORIDE ☐ AMINOMETHANAMIDINE HYDROCHLORIDE ☐ CARBAMIDINE HYDROCHLORIDE ☐ GUANIDINE CHLORIDE ☐ GUANIDINE HYDROCHLORIDE ☐ GUANIDINIUM CHLORIDE ☐ GUANIDINIUM HYDROCHLORIDE ☐ IMINOUREA HYDROCHLORIDE ☐ USAF EK-749

TOXICITY DATA WITH REFERENCE
skn-rbt 500 mg/24H SEV NTIS** AD-A166-306
eye-rbt 81,400 µg MOD NTIS** AD-A184-068
mmo-smc 4 mmol/L BBRCA9 53,531,73
orl-rat LD50:475 mg/kg NTIS** AD-A165-747
scu-rat LDLo:404 mg/kg JPETAB 28,251,26
orl-mus LD50:571 mg/kg JACTDZ 12,598,93
ipr-mus LD50:500 mg/kg NTIS** AD277-689
scu-dog LDLo:200 mg/kg HBAMAK 4,1352,35
scu-gpg LDLo:100 mg/kg HBAMAK 4,1352,35

CONSENSUS REPORTS: Reported in EPA TSCA Inventory.

SAFETY PROFILE: Poison by subcutaneous route. Moderately toxic by ingestion and intraperitoneal routes. Mutation data reported. An eye and severe skin irritant. Can cause nausea, diarrhea, and neurological disturbances. When heated to decomposition it emits highly toxic fumes of HCl and NO_x.

GLA000 **CAS:506-93-4** **HR: 3**
GUANIDINE MONONITRATE
DOT: UN 1467
mf: $CH_5N_3 \cdot HNO_3$ mw: 122.11

PROP: White granules. Mp: 214°.

SYN: GUANIDINE NITRATE (DOT)

TOXICITY DATA WITH REFERENCE
skn-rbt 500 mg SEV JACTDZ 12,592,93
eye-rbt 92 mg MLD NTIS** AD-A166-449
orl-rat LD50:730 mg/kg JACTDZ 12,594,93
orl-mus LD50:1028 mg/kg NTIS** AD-A197-827
skn-rbt LD-: >2 g/kg JACTDZ 12,592,93

CONSENSUS REPORTS: Reported in EPA TSCA Inventory.

DOT CLASSIFICATION: 5.1; *Label:* Oxidizer

SAFETY PROFILE: Moderately toxic by ingestion. A severe skin and an eye irritant. A powerful oxidizer. Flammable when shocked or exposed to heat or flame. A stable, flashless, non-hygroscopic high explosive used as a blasting explosive in combination with charcoal and inorganic nitrates. Keep away from heat and open flame. When heated to decomposition it emits very toxic fumes of HNO_3 and NO_x. See also NITRATES, GUANIDINE MONOHYDROCHLORIDE, and EXPLOSIVES, HIGH.

GLB100 **CAS:27698-99-3** **HR: 3**
GUANIDINIUM DICHROMATE
mf: $C_2H_{12}Cr_2N_6O_7$ mw: 336.14

CONSENSUS REPORTS: Chromium and its compounds are on the Community Right-To-Know List.

SAFETY PROFILE: Explodes violently when heated in a sealed container. When heated to decomposition it emits toxic fumes of NO_x. See also CHROMIUM COMPOUNDS.

GLB300 **CAS:52470-25-4** **HR: 3**
GUANIDINIUM NITRATE
mf: $CH_6N_4O_3$ mw: 122.08

PROP: A solid. Mp: 214°.

SAFETY PROFILE: Decomposes explosively when heated. When heated to decomposition it emits toxic fumes of NO_x. See also NITRATES.

GLC000 **CAS:10308-84-6** **HR: 3**
GUANIDINIUM PERCHLORATE
mf: $CH_6ClN_3O_4$ mw: 159.53

SAFETY PROFILE: A very sensitive, powerful, and unstable explosive. Decomposes violently at 350°C. Mixtures with 10% iron are more thermally sensitive. Upon decomposition it emits toxic fumes of Cl^- and NO_x. See also PERCHLORATES.

GLC100 **HR: 2**
p-GUANIDINOBENZOIC ACID 4-METHYL-2-OXO-2H-1-BENZOPYRAN-7-YL ESTER
mf: $C_{18}H_{15}N_3O_4$ mw: 337.36

SYNS: BENZOIC ACID, p-GUANIDINO-, 4-METHYL-2-OXO-2H-1-BENZOPYRAN-7-YL ESTER ☐ COUMARIN, 4-METHYL-7-HYDROXY-, p-GUANIDINOBENZOATE ☐ 4-METHYLUMBELLIFERONE 4-GUANIDINOBENZOATE ☐ MUGB

TOXICITY DATA WITH REFERENCE
imp-mus TDLo:2 mg/kg (female 2D pre):REP CCPTAY 26,137,82
ipr-mus LD50:600 mg/kg CCPTAY 26,137,82

SAFETY PROFILE: Moderately toxic by intraperitoneal route. Experimental reproductive effects. When heated to decomposition it emits toxic fumes of NO_x.

GLI000 **CAS:73-40-5** **HR: 2**
GUANINE
mf: $C_5H_5N_5O$ mw: 151.15

PROP: Crystals. Usually amorph. Decomp: >360° with partial subl. Very sol in ammonia water, aq KOH solns, dil acids; very sltly sol in alc, ether; insol in water.

SYNS: 2-AMINOHYPOXANTHINE ☐ MEARLMAID

TOXICITY DATA WITH REFERENCE
sln-hmn:lym 30 μmol/L MUTAEX 1,99,86
cyt-mus-ipr 15 ng/kg NULSAK 19,40,76
scu-rat TDLo:1300 mg/kg/26W-I:ETA CNREA8 27,925,67

CONSENSUS REPORTS: Reported in EPA TSCA Inventory. EPA Genetic Toxicology Program.

SAFETY PROFILE: Questionable carcinogen with experimental tumorigenic data. Human mutation data reported. When heated to decomposition it emits toxic fumes of NO_x.

GLK100 **CAS:18905-29-8** **HR: 3**
GUANINE-3-N-OXIDE
mf: $C_5H_5N_5O_2$ mw: 167.15

TOXICITY DATA WITH REFERENCE
ipr-rat TDLo:75 mg/kg (female 11D post):TER ARPAAQ 86,395,68
ipr-rat LD50:75 mg/kg ARPAAQ 86,395,68

SAFETY PROFILE: Poison by intraperitoneal route. An experimental teratogen. When heated to decomposition it emits toxic fumes of NO_x.

GLM000 **CAS:5227-68-9** **HR: 3**
GUANINE-7-N-OXIDE
mf: $C_5H_5N_5O_2$ mw: 167.15

PROP: Brownish crystals from $3H_2O$.

SYN: 7-HYDROXYGUANINE

TOXICITY DATA WITH REFERENCE
scu-rat TDLo:390 mg/kg/26W-I:NEO CNREA8 27,925,67
ipr-rat LD50:95 mg/kg ADTEAS 3,181,68
ipr-mus LD50:40 mg/kg JANTAJ 38,972,85

SAFETY PROFILE: Poison by intraperitoneal route. Questionable carcinogen with experimental neoplastigenic data. When heated to decomposition it emits very toxic fumes of NO_x.

G

GLO000 **CAS:19039-44-2** **HR: 2**
GUANINE-3-N-OXIDE HEMIHYDROCHLORIDE
mf: $C_5H_5O_2 \cdot 1/2ClH$ mw: 115.33

TOXICITY DATA WITH REFERENCE
scu-rat TDLo: 1040 mg/kg/24W-I:NEO CNREA8 30,184,70

SAFETY PROFILE: Questionable carcinogen with experimental neoplastigenic data. When heated to decomposition it emits toxic fumes of Cl^-.

GLS000 **CAS:118-00-3** **HR: 3**
GUANOSINE
mf: $C_{10}H_{13}N_5O_5$ mw: 283.28

PROP: A component of nucleic acids. Mp: 239° (decomp).

SYNS: GR ☐ GUANINE, 9-β-D-RIBOFURANOSYL- ☐ GUANINE RIBOSIDE ☐ 2(3H)-IMINO-9-β-D-RIBOFURANOSYL-9H-PURIN-6(1H)-ONE ☐ INOSINE, 2-AMINO- ☐ RIBOFURANOSIDE, GUANINE-9, β-D- ☐ 9-β-D-RIBOFURANOSYLGUANINE ☐ VERNINE ☐ USAF CB-11

TOXICITY DATA WITH REFERENCE
pic-esc 1 g/L ZAPOAK 12,583,72
oms-hmn:oth 1 mmol/L JIDEAE 65,52,75
ipr-mus TDLo: 750 mg/kg (13D preg):TER JJPAAZ 22,201,72
ipr-mus TDLo: 1 g/kg (13D preg):TER JJPAAZ 22,201,72
ipr-mus LD50: 500 mg/kg NTIS** AD277-689
ivn-mus LD50: 180 mg/kg CSLNX* NX#03206

CONSENSUS REPORTS: Reported in EPA TSCA Inventory.

SAFETY PROFILE: Poison by intravenous route. Moderately toxic by intraperitoneal route. Experimental teratogenic effects. Human mutation data reported. When heated to decomposition it emits toxic fumes of NO_x.

GLS700 **CAS:1021-11-0** **HR: 3**
GUANOXYFEN SULFATE
mf: $C_{20}H_{30}N_6O_2 \cdot H_2O_4S$ mw: 484.61

PROP: A solid. Mp: 145–146°.

SYNS: C.I. 515 ☐ EA 166 ☐ (3-PHENOXYPROPYL)GUANIDINE SULFATE

TOXICITY DATA WITH REFERENCE
orl-mus LD50: 402 mg/kg JPETAB 143,374,64
ipr-mus LD50: 123 mg/kg JPETAB 143,374,64
ivn-mus LD50: 17 mg/kg JPETAB 143,374,64
ims-mus LD50: 122 mg/kg JPETAB 143,374,64

SAFETY PROFILE: Poison by intravenous, intramuscular, and intraperitoneal routes. Moderately toxic by ingestion. See also SULFATES.

GLS750 **CAS:85-32-5** **HR: 2**
5′-GUANYLIC ACID
mf: $C_{10}H_{14}N_5O_8P$ mw: 363.26

SYNS: GMP ☐ 5′-GMP ☐ GUANIDINE MONOPHOSPHATE ☐ GUANOSINE MONOPHOSPHATE ☐ GUANOSINE 5′-MONOPHOSPHATE ☐ GUANOSINE 5′-MONOPHOSPHORIC ACID ☐ GUANOSINE 5′-PHOSPHATE ☐ GUANYLIC ACID

TOXICITY DATA WITH REFERENCE
oth-hmn:oth 1 mmol/L JIDEAE 65,52,75
ipr-mus LDLo: 1500 mg/kg ANYAA9 60,251,54

CONSENSUS REPORTS: Reported in EPA TSCA Inventory.

SAFETY PROFILE: Moderately toxic by intraperitoneal route. Human mutation data reported. When heated to decomposition it emits toxic vapors of NO_x and PO_x.

GLS800 **CAS:5550-12-9** **HR: 2**
GUANYLIC ACID SODIUM SALT
mf: $C_{10}H_{14}N_5O_8P \cdot 2Na$ mw: 409.24

PROP: Colorless to white crystals; characteristic taste. Sol in water; sltly sol in alc; insol in ether.

SYNS: DISODIUM GMP ☐ DISODIUM-5′-GMP ☐ DISODIUM GUANYLATE (FCC) ☐ DISODIUM-5′-GUANYLATE ☐ GMP DISODIUM SALT ☐ 5′-GMP DISODIUM SALT ☐ GMP SODIUM SALT ☐ SODIUM GMP ☐ SODIUM GUANOSINE-5′-MONOPHOSPHATE ☐ SODIUM GUANYLATE ☐ SODIUM-5′-GUANYLATE

TOXICITY DATA WITH REFERENCE
cyt-ham:fbr 1 g/L FCTOD7 22,623,84
ipr-rat LD50: 3880 mg/kg AJINO* -,-,73
scu-rat LD50: 3400 mg/kg AJINO* -,-,73
ivn-rat LD50: 2720 mg/kg AJINO* -,-,73
orl-mus LD50: 15 g/kg AJINO* -,-,73
ipr-mus LD50: 5010 mg/kg AJINO* -,-,73
scu-mus LD50: 5050 mg/kg AJINO* -,-,73
ivn-mus LD50: 3580 mg/kg AJINO* -,-,73

CONSENSUS REPORTS: Reported in EPA TSCA Inventory.

SAFETY PROFILE: Moderately toxic by intraperitoneal, subcutaneous, and intravenous routes. Mildly toxic by ingestion. Mutation data reported. When heated to decomposition it emits toxic fumes of PO_x, NO_x, and Na_2O.

GLU000 **CAS:9000-30-0** **HR: 1**
GUAR GUM

PROP: Yellowish-white powder, dispersible in hot or cold water, obtained from the ground endosperms of *Cyanopsis tetragonoloan* L. Taub (Fam. Leguminosae). White powder; odorless. Sol in water; insol in oils, grease, hydrocarbons, ketones, esters.

SYNS: 1212A ☐ A-20D ☐ BURTONITE V-7-E ☐ CYAMOPSIS GUM ☐ DEALCA TP1 ☐ DEALCA TP2 ☐ DECORPA ☐ GALACTASOL ☐ GENDRIV 162 ☐ GUAR ☐ GUARAN ☐ GUAR FLOUR ☐ GUM CYAMOPSIS ☐ GUM GUAR ☐ INDALCA AG ☐ INDALCA AG-BV ☐ INDALCA AG-HV ☐ JAGUAR ☐ JAGUAR 6000 ☐ JAGUAR A 20 B ☐ JAGUAR A 20D ☐ JAGUAR A 40F ☐ JAGUAR GUM A-20-D ☐ JAGUAR No. 124 ☐ JAGUAR PLUS ☐ J 2Fp ☐ LYCOID DR ☐ NCI-C50395 ☐REGONOL ☐ REIN GUARIN ☐ SUPERCOL G.F. ☐ SUPERCOL U POWDER ☐ SYNGUM D 46D ☐ UNI-GUAR

TOXICITY DATA WITH REFERENCE
orl-rat TDLo: 228 g/kg/13W-C:REP NTPTR* NTP-TR-229,82
orl-rat LD50: 6770 mg/kg FCTXAV 19,287,81
orl-mus LD50: 8100 mg/kg FDRLI* 124,-,76
orl-rbt LD50: 7000 mg/kg FDRLI* 124,-,76
orl-ham LD50: 6000 mg/kg FDRLI* 124,-,76

CONSENSUS REPORTS: NTP Carcinogenesis Bioassay (feed); No Evidence: mouse, rat NTPTR* NTP-TR-229,82. Reported in EPA TSCA Inventory. EPA Genetic Toxicology Program.

SAFETY PROFILE: Mildly toxic by ingestion. Experimental reproductive effects. When heated to decomposition it emits acrid smoke and irritating fumes.

GLW000 **HR: 2**
GUAVA

PROP: Material extracted with hot water from the unripe fruits of *P. guajava* (JNCIAM 60,683,78).

SYN: PSIDIUM GUAJAVA

TOXICITY DATA WITH REFERENCE
scu-rat TDLo: 10 g/kg/72W-I: ETA JNCIAM 60,683,78

SAFETY PROFILE: Questionable carcinogen with experimental tumorigenic data. When heated to decomposition it emits acrid smoke and irritating fumes.

GLY000 **CAS:9000-28-6** **HR: 1**
GUM GHATTI

PROP: The gummy exudation from the stem of *Anogeissus latifolia*. Colorless to pale-yellow tears; almost odorless. Sltly sol in water.

SYN: INDIAN GUM

TOXICITY DATA WITH REFERENCE
orl-rat LD50: 17 g/kg FDRLI* 124,-,76
orl-rbt LD50: 7000 mg/kg FDRLI* 124,-,76

CONSENSUS REPORTS: Reported in EPA TSCA Inventory.

SAFETY PROFILE: Mildly toxic by ingestion. When heated to decomposition it emits acrid smoke and irritating fumes.

GLY100 **CAS:9000-29-7** **HR: 2**
GUM GUAIAC

PROP: From wood of *guajacum officinale* L. or *Guajacum sanctum* L. (Fam. Zygophyllaceae). Brown solid; balsamic odor, sltly acrid taste. Sol in alc, ether, chloroform, solns of alkalies; sltly sol in carbon disulfide, benzene.

SYN: GUAIAC GUM

TOXICITY DATA WITH REFERENCE
orl-gpg LD50: 1120 mg/kg AFREAW 3,197,51

SAFETY PROFILE: Moderately toxic by ingestion. When heated to decomposition it emits acrid smoke and irritating fumes.

GMC000 **CAS:73341-70-5** **HR: 3**
GUNACIN
mf: $C_{17}H_{16}O_8$ mw: 348.33

PROP: Orange powder.

TOXICITY DATA WITH REFERENCE
dni-omi 200 μg/L JANTAJ 32,1104,79
ipr-mus LD50: 16 mg/kg JANTAJ 32,1104,79
ivn-mus LD50: 12 mg/kg JANTAJ 32,1104,79

SAFETY PROFILE: Poison by intraperitoneal and intravenous routes. Mutation data reported. When heated to decomposition it emits acrid smoke and irritating fumes.

GME000 **CAS:8030-55-5** **HR: 1**
GURJUN BALSAM

PROP: Oleoresin from various species of *Dipterocarpus* (FCTXAV 14,789,76).

SYNS: BALSAM GURJUN □ EAST INDIAN COPAIBA □ WOOD OIL

TOXICITY DATA WITH REFERENCE
skn-rbt 500 mg/24H MOD FCTXAV 14,789,76

CONSENSUS REPORTS: Reported in EPA TSCA Inventory.

SAFETY PROFILE: A skin irritant. When heated to decomposition it emits acrid smoke and irritating fumes.

GME300 **CAS:11048-92-3** **HR: 3**
α-2-GUTTIFERIN
mf: $C_{33}H_{38}O_8$ mw: 562.71

PROP: A solid. Mp: 113–115°.

SYNS: α-GUTTIFERIN (9CI) □ A"2-GUTTIFERIN □ Y-GUTTIFERIN

TOXICITY DATA WITH REFERENCE
ipr-rat LD50: 91 mg/kg IJEBA6 5,96,67
scu-rat LD50: 279 mg/kg IJEBA6 5,96,67
ivn-rat LD50: 105 mg/kg IJEBA6 5,96,67
ipr-mus LD50: 83 mg/kg IJEBA6 5,96,67
scu-mus LD50: 400 mg/kg 85GDA2 8(1),331,82
ivn-mus LD50: 97 mg/kg IJEBA6 5,96,67

SAFETY PROFILE: Poison by subcutaneous, intravenous, and intraperitoneal routes. When heated to decomposition it emits acrid smoke and fumes.

GMG000 **CAS:639-14-5** **HR: 3**
GYPSOGENIN
mf: $C_{30}H_{46}O_4$ mw: 470.76

PROP: Needles or leaflets from methanol; crystals from MeOH. Mp: 274–276°.

SYNS: ALBSAPOGENIN □ ASTRANTIAGENIN D □ GITHAGENIN □ GYPSOPHILASAPOGENIN □ GYPSOPHILASAPONIN □ 3-β-HYDROXY-23-OXO-OLEAN-12-EN-28-OIC ACID □ SAPONIN-GYPSOPHILA

TOXICITY DATA WITH REFERENCE
orl-mus LDLo: 2 g/kg HBAMAK 4,1289,35
scu-mus LDLo: 100 mg/kg HBAMAK 4,1289,35
ivn-mus LDLo: 15 mg/kg HBAMAK 4,1289,35

SAFETY PROFILE: Poison by subcutaneous and intravenous routes. Moderately toxic by ingestion. When heated to decomposition it emits acrid smoke and irritating fumes.

GMG100 **CAS:24237-00-1** **HR: 1**
GYRANE
mf: $C_{11}H_{20}O$ mw: 168.31

SYNS: 6-BUTYL-2,4-DIMETHYLDIHYDROPYRANE □ 2H-PYRAN, 3,6-DIHYDRO-6-BUTYL-2,4-DIMETHYL-

TOXICITY DATA WITH REFERENCE
orl-rat LD50:>5 g/kg FCTOD7 26,289,88
skn-rbt LD50:>5 g/kg FCTOD7 26,289,88

CONSENSUS REPORTS: Reported in EPA TSCA Inventory.

SAFETY PROFILE: Low toxicity by ingestion and skin contact. When heated to decomposition it emits acrid smoke and irritating vapors.

A Guide to Using This Book

Entry Code: – Entries are indexed in order by this alphanumeric code.
See Introduction: paragraph l, p. xiii.

Entry Name: – A complete entry name and synonym cross-index is located in Section 3.
See Introduction: paragraph 2, p. xiii.

DOT: – The four-digit hazard code assigned by the U.S. DOT. A complete DOT cross-index is located in Section 1.
See Introduction: paragraph 5, p. xiii.

mf: – The molecular formula
mw: – The molecular weight
See Introduction: paragraphs 6 and 7, p. xiv.

PROP: – Physical properties, including solubility and flammability data. May contain a definition of the entry.
See Introduction: paragraph 9, p. xiv.

SYNS: – Synonyms for the entry. A complete synonym cross-index is located in Section 3.
See Introduction: paragraph 10, p. xiv.

Toxicity Data: – Data for skin and eye irritation, mutation, teratogenic, reproductive, carcinogenic, human, and acute lethal effects.
See Introduction: paragraphs 11 - 15 pp. xiv-xxii.

Standards and Recommendations: – OSHA PEL, ACGIH TLV, DFG MAK, NIOSH REL workplace air levels, and U.S. DOT Classification and labels are listed here.
See Introduction: paragraph 18, p. xxv.